The Proceedings of the Seventh (1997) International OFFSHORE AND POLAR ENGINEERING CONFERENCE

VOLUME I, 1997

Offshore Resource & Technology, Sakhalin Oil & Gas, Resources & Energy (Wave Energy, Gas Hydrates and Deep-Ocean Mining), TLP, Very Large Floating Structure (VLFS) and FPSO, Offshore Systems, Environment (Deep-Sea Impact, Waste and Transport, Pollutant, Contamination, Ecology), Geotechnical Engineering

edited by:

Jin S. Chung, Colorado School of Mines, Golden, Colorado, USA
Braja M. Das, California State University, Sacramento, California, USA
Tamotsu Matsui, Osaka University, Osaka, Japan
H. Thiel, Alfred Wegener Institute for Polar and Marine Research, Bremerhaven, Germany

presented at:
The Seventh (1997) International Offshore and Polar Engineering Conference
held in Honolulu, USA, May 25-30, 1997

organized by:
International Society of Offshore and Polar Engineers

sponsored by:
International Society of Offshore and Polar Engineers (ISOPE)
with cooperating societies and associations

International Society of Offshore and Polar Engineers (ISOPE)
P.O. Box 1107 Golden, Colorado 80402-1107 U.S.A.

International Standard Book Number: ISBN 1-880653-28-1 (Set)
ISBN 1-880653-29-X (Vol. I)

Library of Congress Catalog Card Number: 97-70292

Cooperating Organizations:

Canadian Association of Petroleum Producers (CAPP)
American Society of Civil Engineers (ASCE) - Engineering Mechanics Division
Korea Committee for Ocean Resources and Engineering (KCORE)
Canadian Society of Civil Engineers (CSCE) - Engineering Mechanics Division
Chinese Society of Ocean Engineers (CSOE)
Chinese Society of Naval Architects and Marine Engineers (CSNAME)
Chinese Society of Theoretical and Applied Mechanics (CSTAM)
Russian Academy of Sciences (RAS)
Singapore Structural Steel Society (SSSS)
Norwegian Petroleum Society (NPF)
The Institution of Engineers Australia (IE Australia)
Kansai Society of Naval Architects, Japan (KSNAJ)
IRO (The Netherlands)
Technical Research Centre of Finland (VTT)
The Society of Materials Science, Japan (JSMS)
The Offshore Engineering Society (OES), United Kingdom
Ukraine Society of Mechanical Engineers (USME)
IFREMER, France
Scott Polar Research Institute (SPRI), UK
The Institute of Engineers Indonesia (PII)

Printed and bound in USA

FOREWORD

We greatly appreciate the excellent responses and help we have received from colleagues around the world in the successful organization of the 7th International Offshore and Polar Engineering Conference (ISOPE-97), Honolulu, May 25 -30, 1997. The Conference features **91** sessions of **517** refereed papers and **10** plenary sessions from 40 countries.

The purposes of the ISOPE conference are to:

- promote technological progress and activities, international technological transfer and cooperation, and opportunities for engineers to maintain and improve technical competence; and
- provide a timely international forum for technical activities, cooperation, opportunity and fellowship among researchers and engineers,

by developing focused session topics with papers of high quality in both originality and significance of the results which were accepted after rigorous review. The conference promotes interdisciplinary interaction between academia and industry.

The International Society of Offshore and Polar Engineers (ISOPE) has already held 12 successful international meetings:

- the 1st (1990) European Offshore Mechanics Symposium (ISOPE EUROMS-90) in Trondheim, 1990;
- the 1st (1990) Pacific/Asia Offshore Mechanics Symposium (ISOPE-PACOMS-90) Seoul; PACOMS-94 (SOSC-94) Beijing; and PACOMS-96 Pusan.
- the Annual ISOPE conferences, starting in Edinburgh, 1991 were held in San Francisco, Singapore, Osaka, The Hague and Los Angeles. Since 1992, the annual ISOPE Conference has been held with the world's largest technical program of its kind with refereed papers.
- the 1st (1995) ISOPE Ocean Mining Symposium (ISOPE OMS-95), Tsukuba; the 2nd ISOPE OMS-97 will be held in Seoul, November 24-25, 1997.
- the 1st (1996) International Deep-Ocean Technology (IDOT-96) Symposium and Workshop, Los Angeles.

This year the ISOPE-97 TPC expanded greatly the number of papers on recent worldwide developments in environment, advanced marine technology, coastal engineering, Sakhalin oil and gas. We shall continue to add other emerging topics to the future conferences.

On behalf of all the individual TPC members (listed in this volume), we would like to thank the reviewers (listed in this volume) for their help in maintaining the technical accuracy and quality of papers. We greatly appreciate the help from colleagues around the world, particularly from the authors and reviewers in meeting the deadlines. The efficient handling of the conference organization by the ISOPE staff is greatly appreciated. TPC received 517 mats, examined individual papers, page by page, paginated, and retyped all individual titles according to the ISOPE format, all in 3 weeks, ready for the printer. Only about 5% of the papers were redone by the authors, following the ISOPE mat format and font size. It was possible with the help of fax, international express mail and wordprocessor.

It is our pleasure to welcome participants from all over the world to the ISOPE-97 Conference in Honolulu. We hope that the participants will also join us at the ISOPE-98 Conference to be held in Montréal, Canada, May 24-29, 1998.

Co-chairmen of the ISOPE-97 Conference:

Jin Chung, Colorado School of Mines, USA
Bob Frederking, National Research Council, Canada
Ron Knapp, University of Hawaii, USA
Michel Olagnon, Ifremer, France
Yukio Ueda, Kinki University, Japan

Acknowledgement

The success of the 7th International Offshore and Polar Engineering Conference (ISOPE-97), USA, May 25-30, 1997, is credited to the teamwork of the members of the Technical Program Committee (TPC), the conference cosponsors, the cooperating organizations and the reviewers (see the list).

The TPC would like to express its sincere appreciation to the TPC members' organizations for their support of the TPC members, and University of Hawaii TPC colleagues for their support toward the successful conduct of the conference events, and to the ISOPE staff for the efficient handling of the conference organization and proceedings.

ISOPE-97 Technical Program Committee (TPC)

Prof. T. Yao, Hiroshima Univ., Higashi-Hiroshima City, Japan
Prof. R.W. Yeung, Univ. of California, Berkeley, CA, USA
Prof. S.C.S. Yim, Oregon State Univ., Corvallis, OR, USA

ISOPE-97 Executive Committee

Prof. Jin S. Chung (Chairman), Colorado School of Mines, USA
Dr. Robert M.W. Frederking, National Research Council, Canada
Prof. Ronald H. Knapp, University of Hawaii, USA
Prof.-Dr. Ram S. Puthli, Karlsruhe University, Germany

Conference Co-chairmen

Jin S. Chung, Colorado School of Mines, USA
Robert M.W. Frederking, National Research Council, Canada
Ron H. Knapp, University of Hawaii, Honolulu, USA
Michel Olagnon, Ifremer, France
Yukio Ueda, Kinki University, Wakayama, Japan

Hawaii Committee

Prof. R.H. Knapp (Chairman), University of Hawaii at Manoa, Honolulu, HI, USA
Prof. H.-J. Krock, University of Hawaii at Manoa, Honolulu, HI, USA
Prof. H.R. Riggs, University of Hawaii at Manoa, Honolulu, HI, USA

Partial List of Reviewers

Abe, N
Abel, A
Akyarli, A
Albert, D G
Alessandrini, B
Ali, P
Allyn, N
Andronicou, L
Aoki, S
Aoki, S I
Arai, M
Asai, K
Aso, K
Athanassios, T
Athanassoulis, G A
Bahlavouni, A
Bai, Y
Baier, V
Balasubramanian, S
Bangaru, N V
Barthelemy, J
Bea, R G
Bekker, A T
Berhault, C
Bertsimas, D J
Bitner-Gregersen, E
Bjørnsen, T
Blackmore, R
Boo, S Y
Borekci, O
Boswell, L F
Bouchard, G
Bowen, A
Brandes, H G
Bretz, G
Brown, A J
Bryden, I G
Burcharth, H F
Burgess, J J
Busch, W
Capanoglu, C
Caridis, P A
Carter, D J
Cavanié, A
Chan, D B
Chan, E S
Chandler, B
Chaplin, J R
Che, X L
Chen, H C
Cheung, K F
Chiang, E
Chien, L K
Chin, Y P
Cho, S R
Choi, H S
Choo, Y S
Chow, K W
Christensen, J
Chun, H H
Chung, J S
Chwang, A T
Clément, A
Cole, I S
Contento, G
Cornett, A M
Cox, D T
Croasdale, K R
Crookshank, N E
Dalane, J I
Damy, G
Das, B M
Deguchi, I
Demirbilek, Z
Den Haag, T
Dier, A
Dijkstra, G J
Dijkstra, O D
Dillon, W P
Doi, Y
Dos Santos, J F
Dursthoff, W
Dutta, P K
Duvat, G
Easson, W J
Engebretsen, T
Ertekin, R C
Fairchild, D P
Falzarano, J M
Farzaneh, M
Ferrant, P
Ferrante, A J
Fish, A M
Fithen, R
Fithen, R M
Ford, S F
Forristall, G Z
Francescutto, A
Frederking, R
Fujikubo, M
Fujino, M
Fujisaku, H
Fukumoto, M
Fuwa, T
Fylling, I
Garza-Rios, L
Goda, Y
Goel, A P
Gokce, T
Gokkus, U
Gouleilmos, A M
Grass, A J
Gresnigt, A M
Grilli, S
Grundy, P
Guerin, G
Gunbak, A R
Guo, T
Gutfraind, R
Handa, K
Hardy, C
Haritos, N
Harrison, G E
Haver, S
Herion, S
Hessler, R R
Higo, Y
Hiraishi, T
Hobbs, R E
Holder, G D
Hollebone, B
Hong, S W
Hong, S Y
Honji, H
Hopkins, M
Hosoyamada, T
Hover, F
Huang, T
Huang, T L
Huang, X L
Hyodo, M
Hyun, B S
Ifuku, M
Ikeda, Y
Isaacson, M
Iwata, K
Jacobsen, V
Jakobsen, J B
Jang, Y S
Jiang Q
Jin, D
Jo, C H
Johnson, B H
Jones, S J
Ju, F
Juhl, J
Kagawa, K
Kajiwara, H
Kaneko, K
Kao, C C
Kapdasli, S
Karahan, E
Karr, L A
Kashiwagi, M
Kawamura, Y

Kawano, K
Kershenbaum, N Y
Kielland, P
Kim, C H
Kim, D H
Kim, M H
Kim, S B
Kitamura, M
Kitamura, R
Kite-Powell, H L
Kiyokawa, T
Knapp, R H
Kioka, W
Kobayashi, A
Kobayashi, T
Kokawa, H
Kolle, J J
Koo, J Y
Koterayama, W
Krock, H J
Krogstad, H E
Kubo, M
Kujala, P
Kuriyama, Y
Kuroiwa, T
Kvamme, B
Kvenvolden, K A
Kyozuka, Y
Labeyrie, J
Laflamme, J
Laforte, J L
Lampela, K
Landrini, M
Langen, I
Le Boulluec, M
Le Goff, J P
Lee, S J
Lee, S K
Lee, Y G
Leira, B
Leonard, J
Li, J C
Li, Y C
Lian, D
Liu, Y H
Lin, G
Lin, M C
Lin, S S
Ling, S
Lipp, S
Liu, Y L
Liukkonen, S
Lozowski, E P
Luton, M J
Ma, Z L
Maeda, H
Magda, W
Makkonen, L
Manabe, Y
Mansard, E P D
Maple, J A
Marcus, H S
Martin, A J
Mase, H
Masson, D
Masuda, K
Masuda, Y
Matsui, T
McClimans, T
McClintock, F
McClure, G
McComber, P
Mebarkia, S A
Meon, G
Meylan, M H
Miao, G P
Mimura, N
Mitsuyasu, H
Miura, K
Mizutani, N
Modi, V J
Moe, G
Mogridge, G R
Mohr, W C
Molin, B
Moo-Young, H
Mori, K
Moshagen, H
Muggeridge, D B
Murakawa, H
Nabeshima, Y
Nadaoka, K
Naito, S
Nakai, T
Nakamura, M
Nam, D
Narayanan, R
Natvig, B J
Nerzic, R
Nestegard, A
Niedzwecki, J M
Niho, O
Nomura, S
Nwogu, N
O'Donoghue, T
Ogawa, K
Ogawa, Y
Oh, S K
Ohtaka, K
Olagnon, M
Onitsuka, K
Onoufriou, T
Oppenheim, B
Orisamolu, I R
Oshiwa, K
Otsuka, K
Oumeraci, H
Paik, J K
Pamukcu, S
Papanikolaou, A
Park, I K
Parsey, M R
Paté-Cornell, M E
Patrikalakis, N M
Pedersen, P T
Penna, R
Perkins, N C
Pervizpour, M
Pesce, C P
Powers, E J
Pradhan, B S
Prakash, S
Prasad, S
Prevosto, M
Puri, V K
Puthli, R S
Randall, R E
Raschke, S
Rigaud, V
Riggs, H R
Rijken, O
Riska, K
Roby, K
Rodd, J L
Roesset, J M
Ronold, K
Ryerson, C C
Saeki, H
Sakai, T
Sakakiyama, T
Sakamoto, Y
San, K
Sanni, S
Sarmento, A
Savenije, R Ph A C
Sayed, M
Sayer, P G
Scarsi, G
Schjølberg, I
Schriever, G
Senjanovic, I
Sergy, G
Setoguchi, T
Shen, F
Shen, H T
Shibayama, T
Shigematsu, T
Shin, J G
Shin, M S
Shinoda, T
Shogaki, T
Sibetheros, I
Sigbjörnsson, R
Silver, A L
Skjong, R
Sklenicka, V
Skop, R A
Skourup, J
Snijder, H H
Sodhi, D S
Sørensen, J D
Spencer, P
Squire, V A
Stansberg, C T
Stear, J
Strømsem, K
Su, M Y
Suga, Y
Sumer, B M
Sun, Z C
Suzuki, H
Suzuki, K
Szilder, K
Tabeta, S
Takagi, K
Takahashi, H
Takaki, M
Talley, W K
Tanaka, H
Tanaka, Y
Tanizawa, K
Tedeschi, R
Temperville, A
Teng, B
Thayamballi, A
Thiel, H
Timco, G W
Tomita, H
Tomita, Y
Toyosada, M
Triantafyllou, M S
Tsai, C P
Tsai, T K
Tsuboi, H
Tsunekawa, Y
Tsurusaki, K
Tuhkuri, J
Tulin, M
Uchida, K
Uchida, M
Ueda, Y
Utsunomiya, T
van der Vegte, G J
Vassalos, D
Venkataramana, K
Verley, R
Wadhams, P
Wang, B P
Wang, J
Wang, K H
Wang, Y S
Wang, Y X
Wang, Z Z
Watabe, T
Weave, T J
White, H B
Williams, A N
Williams, F M
Winterstein, S
Wright, B D
Wu, K Q
Xu, T
Yajima, H
Yamaguchi, S
Yamazaki, T
Yao, T
Yashima, A
Yasuzawa, Y
Yeung, R W
Yim, J Z
Yoneyama, H
Yoon, H K
Yoon, J H
Zdravkovich, M M
Zhang, D H
Zhang, H X
Zhang, J
Zhao, Y N
Zhou, L D

ISOPE Publication Order Information

International Journal of Offshore and Polar Engineers - Quarterly (March, June, September, December) (ISSN 1053-5381).

Editors: Jin S. Chung, J. Wardenier, R.M.W. Frederking, W. Koterayama. **Topics:** *offshore engineering, ocean engineering, environment, offshore mechanics, polar (Arctic and Antarctic) engineering, advanced marine technology, materials and technology and energy*. **Yearly subscription rates:** US$120. (Member, complimentary) and US$100 per a set of annual back issues.

Proceedings of the 1st (1990) European Offshore Mechanics Symposium (EUROMS-90), Trondheim, Norway, 20-22 August 1990. A single volume: (ISBN 0-9626104-4-5; LCCN 90-084691); approx. 590 pp., US$80. (Member, US$60.)

Proceedings of the 1st (1990) Pacific/Asia Offshore Mechanics Symposium (PACOMS '90), Seoul, Korea, 24-28 June 1990. For a set of 3 volumes, (ISBN 0-9626104-0- 2; LCCN 90-082039): approx. 1260 pp., US$150 per set and US$70 (Member, US$50) per volume.

Proceedings of Special Offshore Symposium China (SOSC-94) / 3rd (1994) Pacific/Asia Offshore Mechanics Symposium (PACOMS-94), Beijing, China, 16-18 April 1994. A single volume (ISBN 1-880653-15-X); 797 pp., US$100.

Proceedings of 4th (1996) Pacific/Asia Offshore Mechanics Symposium (PACOMS-96), Pusan, Korea, October 31-November 1, 1996. A single volume (ISBN 1-880653-27-3); 325 pp., US$60 (Member, US$50.)

Proceedings of 1st (1995) SOPE Ocean Mining Symposium, Tsukuba, Japan, November 21-22, 1995. A single volume (ISBN 1-880653-21-4, LCCN 95-77708): 238 pp., US$100 (Member, US$80.)

Proceedings of 2nd (1997) ISOPE Ocean Mining Symposium, Seoul, Korea, November 24-25, 1997. A single volume (ISBN 1-880653-33-8): 250 pp. est., US$100 (Member, US$80.)

Proceedings of the 1st (1991) International Offshore and Polar Engineering Conference (ISOPE-91), Edinburgh, U.K., 11-16 August 1991. For a set of 4 volumes, (ISBN 0-9626104-5-3; LCCN 91-071635): approx. 2190 pp., $320 (US$240; member) and $100 ($80; member) for a single volume.

Proceedings of the 2nd (1992) International Offshore and Polar Engineering Conference (ISOPE-92), San Francisco, USA, June 14-19, 1992. For a set of 4 volumes, (ISBN 1-880653-00-1; LCCN 91-78280): 2,899 pp. est., $340 (Member, $260) and $100 ($80; member) for a single volume. See below for postage and handling.

Proceedings of the 3rd (1993) International Offshore and Polar Engineering Conference (ISOPE-93), Singapore, 6-11 June 1993. For a set of 4 volumes, (ISBN 1-880653-05-2; LCCN 92-76219): 2,965 pp., $340 (ISOPE Member, $260) and $100 ($80; member) for a single volume. See below for postage and handling.

Proceedings of the 4th (1994) International Offshore and Polar Engineering Conference (ISOPE-94), Osaka, Japan, April 10-15, 1994. For a set of 4 volumes, (ISBN 1-880653-10-9; LCCN 93-80555): 2,700 pp. est., $340 (ISOPE Member, $260) and $100 ($80; member) for a single volume. See below for postage and handling.

Proceedings of the 5th (1995) International Offshore and Polar Engineering Conference (ISOPE-95), The Hague, June 11-16, 1995. For a set of 4 volumes, (ISBN 1-880653-16-8; LCCN 94-73796): 2,543 pp., $340 (ISOPE Member, $260). 2,589 pp., $340 (ISOPE Member, $260) and $100 ($80; member) for a single volume. See below for postage and handling.

Proceedings of the 6th (1996) International Offshore and Polar Engineering Conference (ISOPE-96), Los Angeles, May 26-31, 1996. For a set of 4 volumes, (ISBN 1-880653-22-2): 2,326 pp., $360 (ISOPE Member, $280). For a single volume, $110 (ISOPE Member, $80). See below for postage and handling.

Proceedings of the 7th (1997) International Offshore and Polar Engineering Conference (ISOPE-97), Honolulu, May 25-30, 1997. For a set of 4 volumes, (ISBN 1-880653-28-1): 3,548 pp., $380 (ISOPE Member, $300). For a single volume, $120 (ISOPE Member, $80). See below for postage and handling.

Postage (surface mail) and handling: add for each of ISOPE-92 through ISOPE-97 Conference Proceedings orders

A set	USA $15	Canada $20	Other countries $35
A volume	USA $8	Canada $10	Other countries $15

How To Order

To order, check or tick ☑ the appropriate boxes, fill in the information below, enclose the correct amount of payment or purchase order, and mail to ISOPE Order Department, P.O. Box 1107, Golden, Colorado 80402-1107 USA

❑ Member rate ISOPE Member No. _______ ❑ Non-member rate

❑ **International Journal of Offshore and Polar Engineering**
❑ **Proceedings of the 1st (1990) Pacific/Asia Offshore Mechanics Symposium**
❑ Vol. I ❑ Vol. II ❑ Vol. III ❑ A Set
❑ **Proceedings of the 3rd (1994) Pacific/Asia Offshore Mechanics Symposium (Special Offshore Symposium China)**
❑ **Proceedings of the 4th (1996) Pacific/Asia Offshore Mechanics Symposium**
❑ **Proceedings of the 1st (1990) European Offshore Mechanics Symposium**
❑ **Proceedings of the 1st (1995) ISOPE Ocean Mining Symposium**
❑ **Proceedings of the 2nd (1997) ISOPE Ocean Mining Symposium**
❑ **Proceedings of the 1st (1991) International Offshore and Polar Engineering Conference**
❑ **Proceedings of the 2nd (1992) International Offshore and Polar Engineering Conference**
❑ **Proceedings of the 3rd (1993) International Offshore and Polar Engineering Conference**
❑ **Proceedings of the 4th (1994) International Offshore and Polar Engineering Conference**
❑ **Proceedings of the 5th (1995) International Offshore and Polar Engineering Conference**
❑ **Proceedings of the 6th (1996) International Offshore and Polar Engineering Conference**
❑ **Proceedings of the 7th (1997) International Offshore and Polar Engineering Conference**

Proceedings of ISOPE-91~97 (*check*):
❑ Vol. I ❑ Vol. II ❑ Vol. III ❑ Vol. IV ❑ A Set

Method of Payment: Advance payment must accompany order (including the postage and handling charge above for 1992-97 Proceedings.) International money order in U.S. dollars, bank checks in U.S. dollars drawn on a U.S. bank, or credit card (Visa or MasterCard only) (*payable to ISOPE*) will be accepted. Remit with the mailing address including the postal (or zip) code for shipping.

❑ international money order in U.S. dollars is enclosed.
❑ bank check in U.S. dollars through a U.S. bank is enclosed.

❑ credit card no. ______________________ *Circle one*: Visa MasterCard American Express

Exp. Date ________ Cardholder's name ______________________

Signature ______________________ Date ______________

Ship To Name ______________________

Company ______________________ Dept. ______________________

Address ______________________

City ______________ State (Province) ______________

Country ______________ Postal (or Zip) Code ______________

Prospective authors are invited to submit abstract to

ISOPE-98 Montréal, Canada, May 24-29

The 8th (1998) International

Offshore and Polar Engineering Conference

Montréal, Canada, May 24-29, 1998

First

Call For Papers

Abstract Deadline	**August 1, 1997**
Manuscript (for review) Deadline	**November 30, 1997**

The International Society of Offshore and Polar Engineers (ISOPE) is organizing about 60 sessions of refereed papers in cooperation with international technical program committee (TPC) members and cooperating organizations. Delegates from more than 35 countries are expected. All manuscripts will be peer-reviewed prior to the final acceptance for the conference proceedings. Papers of archival value may be further reviewed for the *International Journal of Offshore and Polar Engineering*.

Main fields of interest are:

Offshore Technology
Offshore Mechanics
Ocean Engineering
Polar Engineering
Deep-Ocean Tech
Underwater Systems
Measurements
Environment
Geotechnics
Energy, Resources
TLP, SPAR, FPS
Adv Marine Tech
Hydrodynamics
Underwater Eng
Materials & Welding
Tubular Structures
Risers, Pipelines, Cables
Comp Methods/CAD
Safety, Risk , Reliability
Coastal Engineering
Emerging Topics

The abstract should stress the originality and the significance of the results, and include the complete names, addresses and fax numbers of all authors.

Send 2 copies of your abstract in 300-400 words to: (1) One of the TPC members (session organizers) or (2) **ISOPE-97 TPC**, ISOPE, P.O. Box 1107, Golden, Colorado 80402-1107 USA: FAX 1-303-420-3760.

Sponsors: *ISOPE*. **Cooperating Organizations and Societies**: *ASCE (USA), CAPP (Canada), CSCE (Canada), CSOE (China), CSNAME (China), CSTAM (China), IE Australia, IFREMER (France), IRO (The Netherlands), JSMS (Japan), KCORE (Korea), KSNAJ (Japan), NPF (Norway), OES (UK), PII (Indonesia), RAS (Russia), Scott Polar Res Inst (UK), SSSS (Singapore), and VTT (Finland).*

The Proceedings of
The Seventh (1997) International
OFFSHORE AND POLAR ENGINEERING CONFERENCE

Honolulu, Hawaii, USA, May 25-30, 1997

VOLUME I

OFFSHORE RESOURCE & TECHNOLOGY
SAKHALIN OIL & GAS
RESOURCES & ENERGY (WAVE ENERGY, GAS HYDRATES, DEEP-OCEAN MINING)
TLP, VERY LARGE FLOATING PLATFORMS (VLFS) AND FPSO
OFFSHORE SYSTEMS
ENVIRONMENT (DEEP-SEA IMPACT, WASTE AND TRANSPORT, POLLUTANT, CONTAMINATION, ECOLOGY)
GEOTECHNICAL ENGINEERING

VOLUME II

OFFSHORE AND ARCTIC PIPELINES
MECHANICS OF RISERS, CABLES AND MOORING
UNDERWATER VEHICLES, CONTROL AND SYSTEMS
POLAR AND ICE ENGINEERING
JOIA PROJECTS
ARCTIC AND ANTARCTIC SYSTEMS
ARCTIC OIL POLLUTION
ICE REMOTE SENSING

VOLUME III

NUMERICAL WAVES
NONLINEAR WAVES
WAVE STATISTICS AND MEASUREMENTS
INTERNAL WAVES
HYDRODYNAMIC FORCES
DYNAMIC RESPONSES
AIRGAP AND METOCEAN
VORTEX SHEDDING AND FLOW-INDUCED VIBRATIONS
COASTAL HYDRODYNAMICS

VOLUME IV

TUBULAR STRUCTURES
WELDING AND ROBOTICS
FATIGUE AND FRACTURE
STEEL, CORROSION AND NDT
COMPOSITE MATERIALS
MECHANICS AND STRUCTURES
RELIABILITY, RISK AND SAFETY
COLLISION/IMPACT/DAMAGE
MARINE & UNDERWATER SYSTEMS
EARTHQUAKE ENGINEERING

CONTENT

GEOTECHNICAL ENGINEERING

ADDITIONAL PAPERS

Proceedings of the Seventh (1997) International Offshore and Polar Engineering Conference
Honolulu, USA, May 25-30, 1997

ISBN 1-880653-28-1 (Set); ISBN 1-880653-29-X (Vol. I)

Environmental Protection of the Deep Sea: A Challenge for Engineers, Scientists and Politicians

Hjalmar Thiel
Alfred Wegener Institute for Polar and Marine Research
Bremerhaven, Germany

ABSTRACT

The actual and potential use of the deep sea for mining of metalliferous resources and for disposal of waste materials are described, potential environmental problems are elucidated, and necessary ocean research activities are presented for each of these deep-sea intrusions. The large scale of all contemplated actions and disturbances gives rise to the most severe concerns. Large-scale experimental impacts at great depths and careful monitoring of individual pilot mining and waste disposal operations are proposed for precautionary reasons. International and multi-disciplinary cooperation of engineers, oceanographers, politicians, and economists is emphasized to channel all efforts into mutually beneficial synergistic effects. It is a challenging task for all those involved to collaborate in order to further a sustainable and environmentally friendly development of deep-sea utilization. Specific responsibility and regulatory functions could be assumed by the International Seabed Authority of the United Nations.

INTRODUCTION

Controversy ranging between the extremes of total disapproval and enthusiastic consent typically governs most discussions on uses of the deep sea. This "inner space" represents the largest niche on the Earth, yet is the least understood, the hardest to imagine, and the most difficult to investigate.

Since mankind uses the terrestrial and shallow marine 40% of the planet, there is little ethical argument to support restraint in use of the remaining 60% that constitutes the deep sea. These abyssal (3000 - 6000 m) and hadal (6000-11,000 m) realms are far removed from most human settlements. Such regions are characterized by total darkness - except for some faint flashes of light resulting from animal and bacterial bioluminescence - and temperatures below 2° C, in some places even below 0° C. Food resources, and consequently life itself, are rare. Using the deep sea through exploitation of the available metalliferous resources and deposition of all types of waste materials for final storage are magnificent ideas!

Life, however, exists everywhere! Even the deepest portions of the oceans contain delicate organisms - bacteria and fungi, unicellular life forms, numerous invertebrates, and primitive to highly advanced vertebrates - associated with each other in a complex and extremely sensitive ecosystem. Density of individuals is usually low, but species diversity is typically very high. Until recently, much of this remained unrevealed because we had not yet developed the capability of conceiving of these tiny lifeforms in their micro-habitats.

These conditions demand a careful approach to any penetration of the deep sea and call for mutual cooperation between engineers, economists, politicians, and scientists. If our advance into the deep ocean results in generation of severe environmental impacts, sustainability of this fragile ecosystem must be the overriding aim and must be balanced against any beneficial effects of exploitation.

Human wastes have entered the ocean as long as populations lived close to the sea and ships were used for transportation. Since the deep sea and the shallow waters are contiguous and are undergoing regular, often centuries long in duration exchange processes, contaminants in ocean surface waters may eventually reach the deep sea. During the steamship era, tremendous amounts of clinker and coal were shoveled over board, and some of this material is occassionally recovered by biologists using deep-sea trawls. Contamination by oil is a more severe and recent problem. Depending on the type of oil, some components may sink into the deep sea along the shipping routes. Other massive sources of wastes include sunken ships from the commercial and naval fleets. These amount to more than a million tons per year, although most of these ships are lost in shelf regions (Hurd 1929, Institute of London Underwriters 1995).

However, severe impacts on the deep sea can also be expected from regular and repetetive uses such as dumping of wastes and mining of metalliferous resources (Table 1). A general overview is presented by Thiel et al. (1997).

Table 1: Actual and potential uses of the deep sea

Disposal	Mining
Radioactive wastes	Polymetallic nodules
Sewage sludge	Polymetallic crusts
Dredge spoils	Metalliferous muds
Munitions	Massive sulfides
Offshore platforms	Phosphorite nodules
Carbon dioxide	

1. Disposal

All storage of waste in the deep ocean is characterized by two common attributes:

- The amounts transported and disposed of must be immense in order to render these undertakings economically feasible.

- The storage must be considered permanent without any chance to retrieve the material in case new information suggests such action.

1.1 Radioactive wastes

Low level radioactive wastes have been dumped in the NE Atlantic Ocean between 1949 and 1982 at depths ranging from about 4500 m to 5300 m. More than 100,000 barrels, filled with concrete, but leaving a small chamber for some waste material, have been released at several prescribed sites. Other dump sites exist in even shallower depths.

By international regulation through the Intermaritime Organization (IMO), further dumping is currently prohibited, but Britain and France have voted for reconsideration of the disposal option, i.e. ocean storage of radioactive wastes has not been totally abandoned (for reviews see NEA 1982, Saunders 1992).

Some general research has been done in the disposal localities, but not enough, in my opinion, to allow a thorough re-evaluation of this permanent waste storage concept. It is necessary to closely inspect the barrels on the seafloor, and this can be carried out only by submersible or remotely controlled vehicles.

1.2 Sewage sludge and dredge spoils

Contaminated sedimentary materials which are not suitable for land fills exist in large quantities, but again, disposal is prohibited by IMO regulation. However, available disposal areas on land are decreasing and other options must be discussed. Technical and economic evaluations are still or are again underway. This makes clear that ocean disposal remains a future option. Environmental studies should be conducted well in advance and before acting on any further consideration. One recently completed broad study (Valent and Young 1995) covering all aspects of waste isolation is a magnificent example of collaborative economic and impact assessment, although the final evaluation should include the long-time monitoring of dump sites following large-scale disposal tests.

1.3 Carbon dioxide

This greenhouse gas in the atmosphere stands in equilibrium with its concentration in the oceans, where it becomes part of biological production (anabolic) and destruction (catabolic) processes. Large amounts are naturally stored in the deep sea. To reduce carbon dioxide (CO_2) in the atmosphere discussions are underway to short cut the natural processes and deliver the CO_2 directly into the deep sea in a fluid or compact state. Although energy would be spent for liquid CO_2 or dry ice production, under certain circumstances this approach to greenhouse effect reduction seems to be economically feasible.

However, we have little knowledge on the reaction of plants and animals to an increase in CO_2 concentration in the upper or deep ocean. Numerical modelling may elucidate the transport and dilution of the CO_2 in the water masses, and laboratory experiments will help to explain the reactions of organisms to increased CO_2 levels, but the true reaction of the larger oceanic systems will be understood only with the use of large-scale experiments involving the disposal of vast amounts of CO_2 into a fjord or into the deep ocean, and by careful monitoring of such an artificial impact (for review see Haugan & Drange 1996, Ormerod & Angel 1996).

1.4 Munitions

Not much is known about the dumping of munitions by the world's navies and probably other branches of military forces. What worries me is the fact that:

- these components of human society have for a long time been exempted from IMO regulations, and

- nothing or probably very little is known about the general localities, the actual sites, and the reaction of the environment in those places where dumping was conducted.

What we need is an inventory of all such dump sites and funding of monitoring studies sponsored by the responsible parties.

Whether individual grenades in large numbers are dumped or a ship's hull loaded with munition is scuttled, a severe and large scale impact occures on the seafloor in both instances. Some of these impacts should be selected for research. Through such large-scale "experiments", science could learn much about deep-sea communities, recolonization of disturbed habitats, and time scales of geochemical and biological processes in the deep sea. But even today, every munitions dumping event is kept confidential. Is there a reason for such secrecy?

1.5 Offshore platforms

Offshore platforms such as drilling or oil storage rigs exist in great numbers. Several years ago it was suggested that their decommissioning might end with disposal in the deep sea. Discussions over deep-sea dumping of such large structures escalated emotionally during the Brent Spar affair in 1995. It is now known that most of the offshore platforms will be rebuild and recycled, and that minor parts may be deposited permanently on land. A few of these large structures may be dumped in deep waters, but decisions on this should be discussed by a multidisciplinary decommissioning expert group. Little is known about potential impact of such disposal on the sea floor, but investigating the impacts of sunken ships should allow some prediction (NERC 1996).

2. Mining

Manganese nodules are metal-rich concretions discovered some 120 years ago in the deep Atlantic and later in the Pacific Oceans. Although they were analyzed for their contents, they had to be rediscovered in the 1960s as a potential resource for several important metals. Since then, other potential metal-rich resources have been discovered in the Exclusive Economic Zones of several countries and in high sea regions.

Since crusts, massive sulfides and phosphorite nodules have less frequently or rarely been considered for commercial mining, this discussion is restricted to manganese or polymetallic nodules and metalliferous sediments. The environmental problems are similar to some extent, but since phosphorite nodules occur in a few hundred meter of water depth, the impact from their mining could severely disturb life in the euphotic zone, specifically where the sediment plume is transported in an upwelling system.

2.1 Polymetallic nodules

In addition to manganese and iron, these nodules contain up to a total of 3 % of other important metals including nickel, copper, and cobalt, and nodule mining may become economically feasible one day. The technology has been developed, but may be reevaluated and redesigned when deep-sea mining can be conducted economically in two or more decades. Pre-pilot mining tests (PPMT) were successfully achieved in 1978/1979 but further, more extensive mining tests such as Pilot Mining Operations (PMO) are currently postponed for economic reasons (for review see Thiel et al. 1991).

Environmental studies have been conducted together with PPMTs, but it soon became apparent that great difficulties exist in trying to extrapolate normal-scale oceanographic research to assessment of large-scale impacts. Therefore, a new approach was developed, the Large-scale Experimental Impact Study (LEXIS), in which a large-scale test disturbance elucidates the reactions of the ecosystem. Two types of LEXIS have been conducted: plowing of the deep seafloor (Thiel & Schriever 1990, Thiel 1991, Thiel 1995, Schriever et al. 1997) and sediment plume creation and resettlement observations (Fukushima et al. 1995, Kaneko et al. 1997, Trueblood & Ozturgut 1997).

The main concern is based on the massive transport of sediment into the water column. About 1 km2 will be mined and totally disturbed per day. Between 5000 an 15,000 metric tons (mt) of nodules will be collected, and 25,000 mt of sediment, conservatively calculated, will be processed by the collector system. Most of the sediment will create a near-bottom plume, a smaller part will be transported to the mining ship together will the nodules. This part of the mud and nodule fines must be released as tailings into the ocean.

Investigations on the impacts are underway, and it is important that they are conducted before any mining is started. It is the precautionary attitude which tries to assess impacts early and help to minimize them. In this respect, LEXIS is of utmost importance.

However, the most important step will be the monitoring of a PMO, because this will be the first industrial activity to move large masses of nodules and sediment. Monitoring a PMO will allow extrapolation of impacts to full scale mining and verification of numerical models. Monitoring a POM for environmental assessment will be very expensive and therefore should be conducted in international cooperation (Thiel et al. 1993, Thiel & Schriever, 1993).

2.2 Metalliferous muds

The warm metalliferous muds of the Red Sea, specifically at a depth of 2200m in the Atlantis II Deep, have been known for about 50 years. They were intensively explored in the seventies and a PPMT was sucsessfully conducted in 1979. Some material was pumped to the surface and processed on board the mining ship to concentrate the metalliferous components.

For economic mining, approximately 100,000 mt must be recovered per day. Only 2 - 3 % of this material will constitute the concentrate to be shipped to a smelter on shore, while 97 % or more needs to be discharged back into the ocean. If mining were to start, the depression containing the sulfidic muds (60 km2 in area) would be mined out in about 15 years. Whereas the impact of this anoxic and warm (about 64^{0} C) mud will be severe, the environmental impact on the Red Sea will originate predominantly from the discharge of the partly oxidized muds and dissolution of metallic and other components. Whereas industry tended to allow the tailings to be washed into surface waters, ecologists convinced them that a deep discharge at about 1000 m depth had to be accepted. No large scale industrial test such as a PMO has yet been conducted, and the true behavior of the tailings is not known (Thiel et al. 1986, Karbe 1987).

3. Precaution, responsibility, challenge

There is some uniqueness in the development of deep ocean uses. The conscious penetration of society into the deep sea has involved ocean scientists in the mid-sixties to asses potential impacts before the disposal of radioactive wastes and later to monitor effects on the environment (NEA 1980, Feldt et al. 1985, Roe et al. 1987, Sibuet & Coic 1989). Early mining tests in 1972 were accompanied by environmental studies (Amos et al. 1973, Roels et al. 1973, Amos & Roels 1977) and broad oceanographic research was conducted together with the PPMTs in DOMES (Deep Ocean Mining Environmental Study) (Ozturgut et al. 1981) and MESEDA (Metalliferous Sediments Atlantis II Deep) (Thiel et al. 1986, Karbe 1987). These investigations have delivered important oceanographic results on Atlantic, Pacific and Red Sea regions, but it became evident that the extrapolation of results, collected with the normal oceanographic methods, would not allow extrapolation of environmental disturbance to full-scale commercial disposal of wastes or mining of metallliferous resources. Governed by this general issue, two large-scale environmental impact experiments, DISCOL (Disturbance and Recolonization Experiment in a Manganese Nodule Area of the deep Southeastern Pacific Ocean) (Thiel 1991) and BIE (Benthic Impact Experiment) (Fukushima 1995, Trueblood 1997), were stimulated.

Additionally, it became apparent that early uses of the deep sea must be carefully monitored. The next essential step in the development of deep-sea mining will be the execution of a smaller-scale, but several month in duratio, PMO. Such industrial activity must be accompanied by intensive oceanographic envirommental assessments. A PMO will perhaps provide the only chance to verify predictive impact models, e.g. on the creation and the fate of sediment plumes, on differential particle settlement, far-reaching transport with currents, and the true effects on the seafloor.

For impact evaluation of deep-sea disposal, equivalent large-scale experiments are essential. Dump site 106 off New Jersey in about 2000 m depth, where municipal sewage sludge was released, was studied intensively (Van Dover et al. 1992, Takizawa et al. 1993, Bothner et al. 1994). But other disposal or accident sites were not monitored. Radioactive wastes in the deep sea in the form of dumped waste drums and nuclear submersibles should be investigated by close-up inspections. Munition dumps could be utilized to teach us about the large impacts to be expected from their arrival and resting upon the deep sea floor. Ecologists would evaluate such events as first steps in large-scale experiments. They would inquire into changes of the current system, the sedimentary regime, the distribution of organisms, and the modification of processes. The series of causes and effects and the reactions of the organisms would teach them about basic functioning of the deep-sea system and would guide the precautionary environmental impact assessments.

All this sums up to two essential implications:

- international cooperation, since the costs and efforts for deep-sea research and for deep-sea monitoring are high, and
- multi-disciplinary cooperation, since the best results are to be expected when the efforts of oceanographers, engineers, politicians and economists are channeled into synergistic action.

It should be a challenging task for all those involved to collaborate in order to further a sustainable and environmentally friendly development of deep-sea uses. Specific responsibility and regulatory functions could be assumed by the Seabed Authority of the United Nations.

REFERENCES

Amos, AF, Garside, C, Gerard, RD, Levitus, S, Malone, TC, Paul, AZ, and Roels, OA (1973). "Study of the impact of manganese nodule mining on the seabed and water column," in: *Phase I Report, Inter-University Program of Rsearch on Ferromanganese Deposits of the Ocean Floor*, NSF/IDOE, Washington, DC, pp 221 - 265.

Amos, AF, and Roels, OA (1977). "Environmental aspects of manganese nodule mining," *Mar. Policy*, 1,2, pp.156-163.

Bothner, MH, Takada, H, Knight, IT, Hill, RT, Butmann, B, Farrington, JW, Colwell, RR and Grassle, JF (1994). "Sewage Contamination in Sediments beneath a deep-ocean dump site off New York," *Mar. Environm. Res.*, 38, pp 43 - 59.

Feldt, W, Kanisch, G, Kanisch, M, and Vobach, M (1985). "Radioecological studies of sites in the Northeast Atlantic used for dumping of low-level radioactive wastes - results of the research cruises of FRV "Walther Herwig" 1980 - 1984," *Arch. FischWiss. 35*, pp 91 - 195.

Fukushima, T (1995). "Overview "Japan Deep-sea Impact Experiment = JET," *Proceed. First ISOPE Ocean Mining Symposium*, Tsukuba, Japan, pp 47- 53.

Haugan, PM, and Drange, H (1996). "Effects of CO_2 on the ocean environment," *Energy Conversion Management 37*, pp 1019-1022.

Hurd, A (1929). "The Marchant Navy," *John Murray*, Vol 3, London.

Institute of London Underwriters (1995). *Annual Report for 1995.*

Kaneko, T, Maejima, Y, and Teishima, H (1997). "The Abundance and vertical Distribution of abyssal benthic fauna in the deep-sea Japan Environmental Impact Experiment," *Proceed, Seventh Intern. Offshore Polar Engineering Conference*, Honolulu, Hawaii, in print.

Karbe, L (1987). "Hot brines and the deep sea environment," in: A.J. Edwards and S.M. Head (Eds) Red Sea (Key Environment Series), *Pergamon Press*, Oxford, pp 70 - 89.

NEA (1982). "Disposal of radioactive waste: an overview of principles involved," Paris: OECD, pp 22.

NEA (1980). "Review of the continued suitability of the dumping site for radioactive waste in the North-East Atlantic," Paris: OECD Nuclear Energy Agency, pp 100.

NERC (1996). "Scientific group on decommissioning offshore structures," *First Report*, April 1996, pp 76.

Ormerod, B, and Angel, MV (Eds) (1996). "Ocean storage of carbon dioxide," *Workshop 2 - Environmental impact. IEA Greenhouse Gas R & D Programme*, Chaltenham, pp 131.

Ozturgut, E, Lavelle, J, and Burns, RE (1981). "Impacts of manganese nodule mining on the environment: Results from pilot-scale mining tests in the north equatorial Pacific", in: R.A. Geyer, (Ed.), Marine environmental pollution. 2. Dumping and mining, Amsterdam: Elsevier Oceanography Series, 27B, *Elsevier*, pp 437-474.

Roe, HSJ and Great Meteor East Research Group (1987). "Great Meteor East: a biological characterisation. (Ocean disposal of high level radioactive waste. A research report prepared for the Department of the Environment)," *Institute of Oceanographic Sciences Deacon Labortory Report*, No. 248, pp 332 and microfiche.

Roels, OA, AMOS, AF, Anderson, OR, Garside, C, Haines, KC, Malone TC, Paul, AZ, and Rice, GE (1973). "The environmental impact of deep-sea mining: progress report." *NOAA Tech. Memo.*, ERL 29-OD-11, pp 185.

Saunders, P (1992). "Managing radioactive waste in the European community," Brussels: *CEC*, pp 40.

Schriever, G, Ahnert, A, Bluhm, H, Borowski, C, and Thiel, H (1997). "Results of the large-scale deep-sea environmental impact study DISCOL during eight years of investigation," *Proceed. Seventh Intern. Offshore Polar Engineering Conference*, Honolulu, Hawaii, in print.

Sibuet, M, and Coic, D (1989). "Photographic prospection of the NEA dumpsite in the northeast Atlantic: quantitative distribution of epibenthic megafauna," Pp. 167 - 177 in *Interim oceanographic description of the North-east Atlantic site for the disposal of low-level radioactive waste*, Volume 3, Paris: OECD Nuclear Energy Agency, pp 374.

Takizawa, M, Straube, WL, Hill, RT, and Colwell, RR (1993). "Near bottom pelagic bacteria at a deep-water sewage sludge disposal site", *Appl. Environm. Microbiol.*, 59, 10, pp 3406 -3410.

Thiel, H, Weikert, H, and Karbe, L (1986). "Risk asessment for mining metalliferous muds in the deep Red Sea," *Ambio*, 15, 1, pp 34-41.

Thiel, H, and Schriever, G (1990). "Environmental protection of the deep sea and the DISCOL project", *Ambio* 19, 5, pp 245-250.

Thiel, H, Foell, EJ, and Schriever, G (1991). "Potential environmental effects of deep seabed mining," *Ber. Zentrum Meeres. u. Klimaforsch.*, UniversitSt Hamburg, 26, pp 243.

Thiel, H (1991). "From MESEDA to DISCOL: a new approach to deep-sea mining risk assessments." *Mar. Mining,* 10, pp 369-386.

Thiel, H, Foell, EJ, and Schriever, G (1993). "Economics and the environment: cooperative aspects of future deep-sea mining activities," *Proceedings, 25th Annual Offshore Technology Conference*, Houston, Texas, OTC Paper 7102, pp 203-212.

Thiel, H, and Schriever, G (1993). "Environmental consequences of deep-sea mining," *International Challenges* , 13, 1, pp 54 - 70.

Thiel, H and Forschungsverbund Tiefsee-Umweltschutz (1995). "The German environmental impact research for manganese nodule mining in the SE Pacific Ocean", *Proceed. First ISOPE Ocean Mining Symposium,* Tsukuba, Japan, pp 39 - 45.

Thiel, H, Angel, MV, Foell, EJ, Rice, AL, and Schriever, G (1997). "Environmental risks from large-scale ecological research in the deep sea. A desk study." *Report to the EU; MAST II, MAS2 CT94 0086*, (in prep).

Trueblood, DD, and Ozturgut, E (1997). "The Benthic Impact Experiment: A study of the ecological impacts of deep seabed mining on abyssal benthic communities," *Proceed. Seventh Intern. Offshore Polar Engineering Conference,* Honolulu, Hawaii, in print.

Valent, PJ, and Young, DK (1995). "Abyssal seafloor waste isolation: Environmental report," *Naval Research Labortory,* Stennis Space Center, NRL/MR/7401-95-7576, pp 479.

Van Dover, CL, Grassle, JF, Fry, B, Garritt, RH, and Starczak, VR (1992). "Stable isotopic evidence for entry of sewage-derived organic material into a deep-sea food web," *Nature*, 360, pp 153-156.

Proceedings of the Seventh (1997) International Offshore and Polar Engineering Conference
Honolulu, USA, May 25-30, 1997

ISBN 1-880653-28-1 (Set); ISBN 1-880653-29-X (Vol. I)

The Tokyo Bay Tunnel of the Trans-Tokyo Bay Highway

T. Funasaki, Y. Izumi and T. Masuda
Trans-Tokyo Bay Highway Corp.
Tokyo, Japan

M. Kanai
Tokyo Bay Tunnel Kisarazu South J.V., Obayashi Corp.
Tokyo, Japan

ABSTRACT

The Trans-Tokyo Bay Highway project is the first phase of an ambitious plan to connect Kawasaki and Chiba by a 15-km crossing of Tokyo Bay. The crossing involves a 5-km bridge, a 10-km undersea tunnel, and two manmade islands in the middle of the bay. The construction method for the undersea tunnel must take into account the large external diameter of the primary lining (approx. 14 m); the extremely soft ground under the sea; the extremely high water pressure to which the tunnel will be subjected; and active seismic conditions in the Tokyo Bay area. This paper discusses the state-of-the-art in planning, engineering and construction of the tunnel.

KEY WORDS: tunnel, slurry shield, watertight and durable lining, aseismatic design, automation in tunneling

INTRODUCTION

The Tokyo metropolitan region, inclusive of the three neighboring prefectures (Kanagawa, Chiba and Saitama), has the highest concentration of population and industry in Japan; with an area of 13,500 km^2 or only 3.6 percent of the national land, but a population of 31.8 million or 26 percent of the national total, and a Gross Regional Product (GRP) of 146 trillion yen (US$1.33 trillion) or 36 percent of the overall GDP of Japan as of 1990. Tokyo Bay, embraced by the metropolitan region as shown in Fig. 1 (a), (b), is open to the Pacific Ocean as a marine gateway and is ringed by large cities and coastal industrial zones together with port facilities, all constituting a concentration of prime importance for socio-economic activities.

The Trans-Tokyo Bay Highway (TTB highway), which crosses Tokyo Bay and provides a direct transportation link between the west and east sides of the bay, has long been a sort of "dream-come-true" project in Japan. As shown in Fig. 2, linked to the Tokyo Bay Shore Highway, Metropolitan Inter-city Expressway, Tokyo Outer Ring Road, Tateyama Expressway, and others, it will play an important role as an integral part of the regional highway network to help promote spatial redistribution and multipolarization of urban functions, thus strengthening cooperation between cities and restructuring the metropolitan region.

The TTB Highway is a project of 15.1 km in length designed as a toll highway to cross the bay by tunnel and bridge, and will require ten years to complete at a total cost of more than 1,482.3 billion yen (US$12.353 billion).

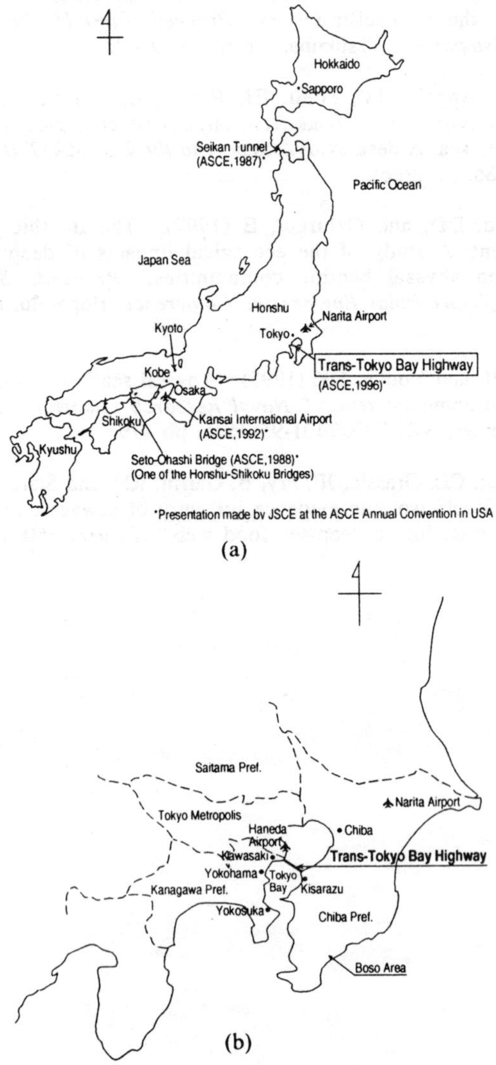

Fig. 1 Location of Trans-Tokyo Bay Highway

Following more than 20 years of feasibility and environmental assessment studies, construction of this giant offshore highway project commenced in May 1989 as the first large-scale highway project with private-sector participation under the prevailing Road Law in Japan, and is now steadily proceeding toward the goal of completion in fiscal 1997.

This report outlines the entire TTB Highway project, including the waterproofing measures against high water pressure and the rapid construction method applied in the construction of shield tunnel sections.

OUTLINE OF THE HIGHWAY STRUCTURES

During the preliminary study of the highway, various structural options for tunnel and bridge were evaluated before commencement of the project. The final design was determined on the basis of the following reasons:

1) To ensure the safety of aircraft taking off from and landing at the Tokyo International Airport at Haneda, airspace restrictions were imposed in the area between the Ukishima Access and the Kawasaki Man-made Island (Kawasaki Island). This posed problems for building a long-span bridge.

2) Shipping traffic surveys indicated that of the some 1,400 vessels crossing the planned line of the highway, about 90 percent were concentrated in a 10-km section on the Kawasaki side. This meant that a bridge spanning the bay could produce adverse effects in terms of maritime safety and vessels trying to seek refuge in stormy weather.

3) Soft subsoil conditions extending to considerable depths posed difficulties in selecting a suitably engineered and cost effective foundation.

The general plan of the highway is illustrated in Fig. 3, and having a highway length of 15.1 km in total, of which the marine section is 14.3 km. At the Kawasaki end, the land section of the highway joins the Tokyo Bay Shore Highway at the Ukishima Interchange. At its Kisarazu end, it is linked to the Tateyama Expressway by the TTB Highway Connector Road (7.1 km). The land sections at the Kawasaki and Kisarazu ends are approximately 0.3 and 0.6 km respectively. An outline of the project under the terms of the project license is given in Table 1 below.

The highway is being constructed as a four-lane facility: two lanes in each direction. For this, two parallel tunnels are being driven in the undersea section. However, in order to accommodate the expected future increase of traffic, the dimensions of major structures such as the man-made islands and the bridge have been determined to make future widening feasible.

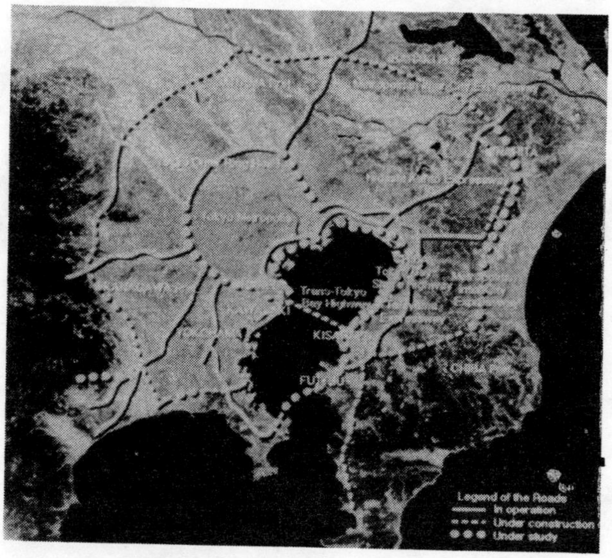

Fig. 2 Metropolitan Highway Network

Table 1 Outline of the Project under the Terms of the Project License

Name of highway	Trans-Tokyo Bay Highway
Official route designation	National Route 409
Origin and destination	From Ukishima, Kawasaki City To Nakajima, Kisarazu City
Length of highway	15.1 km (Marine section 14.3km)
Number of lanes	Dual 2-lane (6 lanes at final phase)
Design speed	80 kilometers per hour
Design load	TL–20t and TT–43t
Forecast traffic volume	33,000 vehicles per day in the first year of operation (64,000 vehicles per day after 20 years of operation)
Construction period	About 10 years from fiscal 1986 (Completion expected in fiscal 1997)
Total project cost	¥1,482.3 billion (US$12.353 billion)

Exchange rate : US$1 = JPN¥120

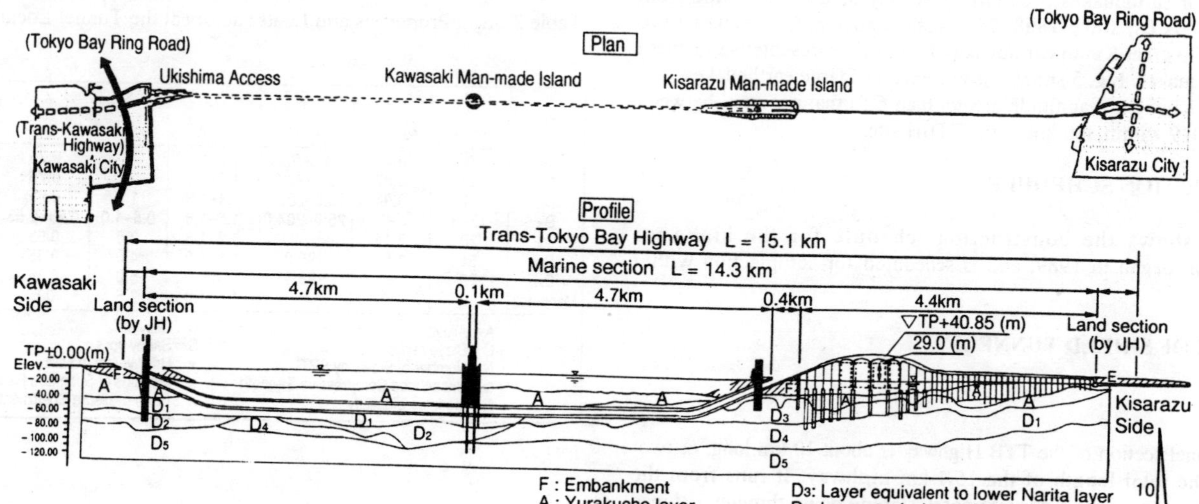

Fig. 3 Project Plan and Geological Profile

Marine Section

The structural components of the marine section of the highway will consist of a shield-driven undersea tunnel, approximately 10 km in length, for the western two-thirds of the highway where shipping traffic is heavy; and a bridge about 4.4 km in length for the remaining section. Near the middle of the tunnels, a ventilation facility has been provided in the form of the Kawasaki Island, and where the tunnel and bridge meet, the Kisarazu Man-made Island (Kisarazu Island) is being constructed. The Kisarazu Island will serve as a rest area for users of the highway. Both islands have also been used as launching bases for the shield machines driving the tunnel. Fig. 4 shows the highway structures of the marine section.

Land Section

The land section on the Kawasaki side serves as an approach section to the tunnel from the interchange, and has structural components of a large U-shaped retaining wall and a box culvert. Whereas the section on the Kisarazu side is an embankment structure, where a tollgate is being constructed. Muck excavated during the shield tunnel work was effectively used to build an approximately 11-m high embankment.

NATURAL RESTRICTIONS AND PLANNING CONDITIONS

Topography and Geology (Fig. 3)

The topography of the seabed along the planned highway link is extremely gentle, generally conforming to the shape of a ship's hull, with a maximum depth of approximately 28 m. Geologically, from Ukishima on the Kawasaki side to the center of the Bay, there is a very soft 20-to 30-m-deep layer of alluvial soil, known as the Yurakucho layer; on the Kisarazu side, a relatively dense sand layer has accumulated from the surface. The upper stratum of the Kazura formation, a sandy layer with an N-value greater than 70 at depths below TP-80 to-90 m, is considered to be a suitable bearing stratum for engineering designs.

Along the tunnel route, the geology mainly consists of alluvial and diluvial clay soil layers on the Kawasaki side, with a diluvial sandy layer sandwiched between the two. The Ukishima and Kisarazu ramp sections are on reclaimed land.

Earthquake Activity

Earthquake activity is common in the Tokyo Bay area. It is believed that 32 major earthquakes occurred in the Bay area in the period from 818 to 1867 A.D.; since 1868, 23 damage-causing earthquakes have occurred. The great Kanto earthquake of 1923 is representative of these major earthquakes. Fig. 5 shows the epicenters of large earthquakes (i.e., those having a magnitude greater than 6.5) that occurred between 1885 and 1979 within 300 km of the TTBH site.

CONSTRUCTION SCHEDULE

Fig. 6 shows the construction schedule for the highway. Construction began in 1989, and is scheduled for complection within fiscal 1997.

OUTLINE OF SHIELD TUNNELS

Outline

The tunnel section of the TTB Highway is about 10 km long, or two-thirds of the total length of the 15.1-km highway. It runs from the Ukishima Access on the Kawasaki coast, sloping down through a sloped section into the seabed, then passing through a level section and finally up

(a) Ukishima Access

(b) Shield Tunnels

Fig. 4 Highway Structures of the Marine Section

Table 2 Soil Properties and Load Factors at the Tunnel Location

Soils	N Blow	q kg / cm	E kg / cm	γ_r ton /m^3	κ	λ	Water in Soils
A_{c1}	0	0.44	4.8–9.6	1.3–1.5	0	0.75	Combined
A_{c2}	0	0.87	5.7–22.9	1.6–1.7	0	0.75	Do.
D_{ic}	12	0.96	35.9–289.0	1.4–1.8	1.5	0.65	Do.
D_{is}	15–5	–	25.0–204.0	1.6–1.8	0.5–4.0	0.35–0.55	Separated
D_{3c}	20	2.11	–	1.7–1.8	2.0	0.65	Combined
D_{3s}	54	1.57	296.2	1.8–1.9	4.0	0.35	Separated
D_{3g}	78	–	195.5	1.7	5.0	0.35	Do.
Manmade	–	7.5–21.3	–	1.6–1.8	3.5	0.6	Combined

A : Alluvial
C : Clayey soils
N : Number of Blows by SPT
qu: Unconfined Compressive Strength
E : Modulus Elasticity
D : Diluvial
S : Sandy soils
γ_r : Unit Weight of soils
κ : Modulus of Subgrade Reaction
λ : Earth-pressure Coefficient

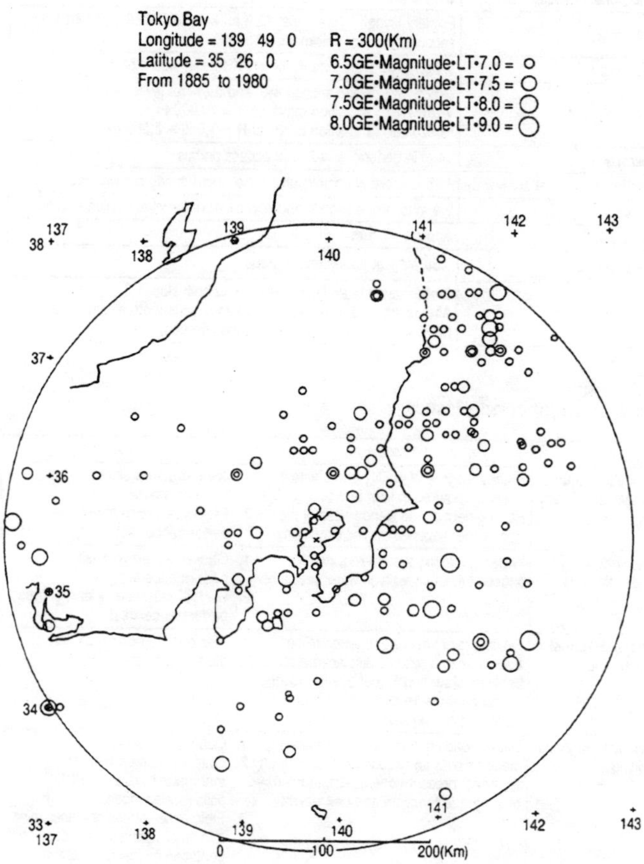

Fig. 5 Earthquakes of Magnitude ≥ 6.5 that Have Occurred within a 300-km Radius of the Center of Tokyo Bay from 1885 to 1990

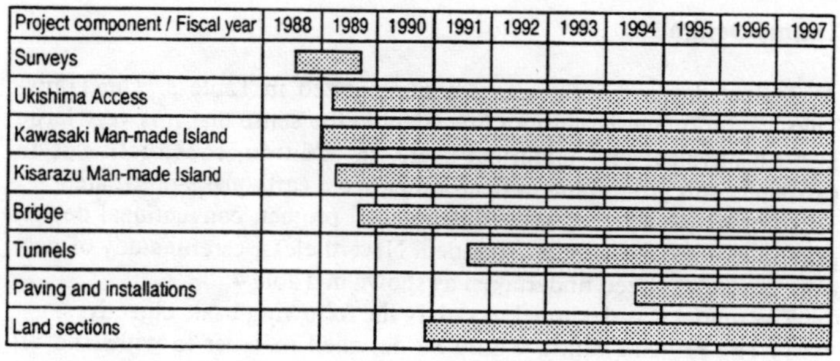

Fig. 6 Construction Schedule

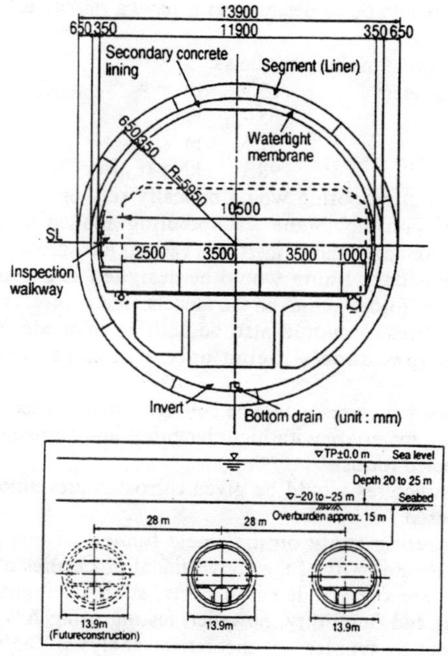

Fig. 7 Tunnel Cross Section and Traffic Lanes

through a second sloped section, to the Kisarazu Island. Constructed midway along the level section between the Ukishima Access and the Kisarazu Island (at a point about 5 km offshore from the Ukishima Access) is the Kawasaki Island, where ventilation facilities are located. The horizontal alignment of the tunnel is basically straight. The sloped sections have been designed with a 4 percent grade in order to shorten the sections as much as possible. The level seabed section has a 1.0 D-thick overburden (where D is the tunnel outer diameter), so that the combined weight of the overburden and tunnel structure will counteract buoyancy forces.

Fig. 7 shows the basic tunnel cross section. The upper section of each tunnel is 10.5 m wide and accommodates the roadway, comprising the left-hand shoulder (2.5 m wide), carriageway (dual 3.5-m traffic lanes) and the right-hand shoulder (1.0 m wide). The lower half of the tunnel consists of a central service and evacuation passage and two flanking sections containing telecommunications, electric power lines and other facilities. Ventilation is provided by a longitudinal system with electrostatic precipitator installed on the ceiling of the tunnel.

The wider left-hand shoulder for the entire section of the tunnel (2.5 m in width) will serve a continuous lay-by and enable drivers to evacuate from the carriageways when an emergency arises. In addition, the evacuation passage prepared under the carriageway slab does not necessitate construction of evacuation cross passages connecting two tunnel tubes, thereby saving construction cost and time.

Basic Plan

For the basic structure of the tunnel, a comparative study was conducted between a submerged tunnel and a shield tunnel and the shield tunnel was finally adopted for the highway based on the following reasons:

1) A submerged tunnel requires dredging a trench in the seabed, backfilling, and towing and sinking the tunnel tube sections, all of which would necessitate placing prolonged restrictions on shipping traffic. This was not a good choice for a congested shipping environment.

2) With technology for shield tunneling established in the late 1970s and the method since proven in the construction of long, large diameter tunnels under high water pressure, shield tunneling was also determined to be applicable to the execution of the TTB Highway project.

3) Although the tunnels will initially carry four traffic lanes, plans are for future expansion of the roadway to six lanes in order to cope with future traffic demand increase. Shield tunneling solution can easily meet the requirements for such an extension by construction of an additional tunnel.

Tunnel Design

Tunnel design parameters are summarized in Table 3. The Trans-Tokyo Bay Tunnel is without precedent in the sense that it is very large in both physical and technical scale. In addition, construction is in extremely soft ground under deep water in an earthquake-prone area.

In order to succeed with this colossal project, conventional design principles were not simply extended. Nevertheless, careful study of past experiences has been undertaken as shown in Table 4.

The tunnel was designed to satisfy the following basic objectives:

1) The primary lining would be designed in order to withstand all loads, and for the purpose of earthquake resistance in soft ground, it would be built as a highly rigid ring structure. Each liner ring would be divided into 11 equal-sized segments. Fig. 8 shows a segment configuration. Segment joints would be so designed as to have a rigidity equal to that of the segment body.

Segment dimensions are as follows:

a. Outer diameter: 13.9 m
b. Thickness: 65 cm
c. Width: 1.5 m
d. Weight: 98.1 kN per piece

2) Tunnel waterproofing would basically rely on hydro-expansive segment seals, with membrane waterproofing applied in between the primary and secondary tunnel liners as a backup measure.

3) The secondary lining would be designed as a counterweight against buoyancy forces acting on the tunnels, and to protect the primary lining against fires. It would also be built to maintain longitudinal flexibility of the tunnels, being only allowed to crack during earthquakes.

4) Since the structures would be built for a planned service life of at least 100 years, materials with high durability and corrosion resistance should be selected for use.

5) Steel structures would be given corrosion-prevention treatment wherever exposed.

The engineering study on the shield tunnels of this project was implemented on the basis of the conventional procedures of design and construction. For certain design aspects, such as rigidity, strength, watertightness and durability, however, testing using full size models was carried out to supplement theoretical analysis. Table 4 lists the engineering issues which were studied and the various tests and experiments performed.

Since the tunnels would be driven under the severe conditions outlined below, careful and extensive deliberation went into the design and execution work, including meetings of a committee that included advisory experts.

1) The roadway configuration would necessitate construction of a shield tunnel having a large cross section (13.9 m in diameter), and requiring long-distance tunneling operations.

2) The tunnels would be driven through soft ground, and in a seismically active area.

3) Since the water depth at the site is high (the maximum head would reach 50 to 60 m), and because there is no impermeable layer, the tunnels would be subject to extremely high water pressure (0.49 to 0.59 MPa).

4) The tunnels would experience the corrosive effects of salts contained in the sea water.

5) The structure being a twin tunnel spaced 0.5 to 1.0 D along the sloping sections and 1.0 D along the level seabed section could cause construction work on one tunnel very easily affecting the other. Shield tunnel cross effects had to be considered in the tunnel design.

6) The tunnels being long, several shield machines would have to advance concurrently from the vertical shafts at the Ukishima Access, Kawasaki Island and Kisarazu Island in order to shorten the construction period. Underground connections would therefore be made below the seabed and under high water pressure.

7) To enhance both safety and work precision, there would be a need to automate various operations such as excavation and assembly of segmental ring liners.

Table 3 Tunnel Design Parameters

Finished inner diameter	ϕ 11.9 m
Tunnel lining	Primary lining (outer dia. 13.9 m, inner dia. 12.6 m, width 1.5 m); secondary lining (thickness 0.35 m)
Tunnel length	9.1km (Kawasaki side 4,600 m; Kisaraza side 4,510 m)
Tunnel alignment	Grade : slope section 4%, and 0.2% for level section Vertical : transition curve of R = 10,000 m Horizontal : transition curve of R = 1,650 – 2,500 m
Tunnel spacing	14 m in general, and 7 m at access portals
Safety factor against buoyancy	1.25 or more at completion ; 1.1 or more during construction
Overburden	9.4 m for slope section and approx. 15m for level section
Water depth	Approx. 27.5 m
Water pressure	0.59 MPa at tunnel invert grade
Excavation soils	Embankment fill and improved soil for both slopes Alternating layers of clayey and sandy soils (both alluvial and pleistocene origin) [along level seabed section]

Table 4 Engineering Issues

Item	Issue	Test / Experiment
Rigidity & ultimate strength of primary lining	Highest degree of rigidity and ultimate strength required to joint and body of the RC segment due to slenderness of the lining under severest loading conditions	1. Joint structure selection test (full-size model) 2. Prototype segment ring performance test
Watertight lining	Watertight lining system requried for subsea tunnel under high water pressure	1. Gasket watertightness performance test 2. Backfill materials watertightness performance trest
Shaft-and-Tunnel joint structure	Flexible joint structure required to be able to absorb relative displacements between shaft and tuunel due to motions during earthquake and settlements of tuunel supporting soils	1. Flexibility performance test of flexible segmental ring
Estimation of loads on lining	Design loads determination for man-made fill soils and subsea soft soils such as lateral pressure coefficient, modulus of subgrade reaction, and unit weight	1. Geological survey 2. Load evaluation test for manmade soils and soft soils (scaled model) 3. Field observations of tunnel lining and surrounding aoils under similar condition by instrumentation
Durability of liner elements	Corrosion control of lining elements for subsea tunnel under highly corrosive environment	1. Corrosion performance test of corrosion controlled bolts 2. Durebility performance test of primary lining concrete with slag cement
Aseismatic design of tunnel lining	1. Seismic wave propagation in man-made fill ground with limited width, and in boundary area of man-made and natural ground with abrupt rigidity change 2. Optimum design of secondary lining not to damage primary lining, and not to deteriorate longitudinal flexibility of primary lining	1. Tunnel behaviour test in boundary of man-made and natural ground under seismic motion 2. Interaction evaluation tests of primary and secondary lining system under seismic motion 3. Test to evaluate isolation effect of membrane between primary and secondary lining

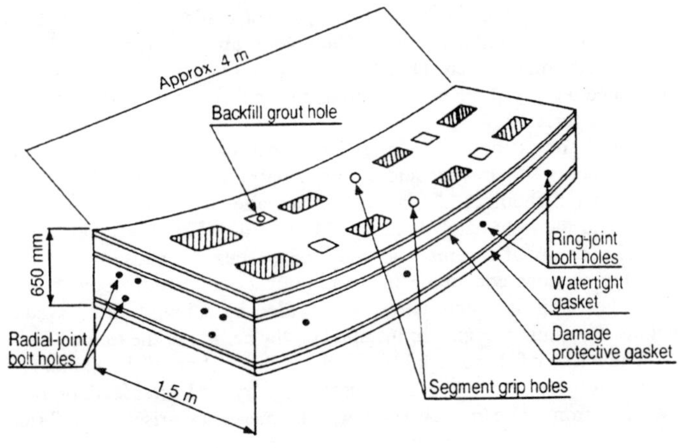

Fig. 8 Segment Configuration

SHIELD TUNNELING AND CONSTRUCTION

Measures to Ensure High Cross-sectional Rigidity and Strength

Measures to increase cross-sectional rigidity and strength included the following:

1) To increase rigidity at segment joints, six high-strength bolts (36 mm in diameter) would be used for radial joints, that is between segments within a liner ring. The bolts would be prestressed to give them an axial tension of 80 percent of the allowable stress or approximately 192.2 kN. The ring joints, between adjoining segmental rings, would be made using a total of 44 high-strength bolts, with an axial tension of 60 percent of the allowable stress or approximately 144.2 kN.

2) The strength of joints would be obtained within the range from the axial tension given to the bolts (235 MPa) to the yield point (883 MPa).

3) Stresses occurring at joints would be transmitted directly from the bolts to joint reinforcing metal plates and then to anchor bars and to the segment body. This would increase joint rigidity and thereby prevent cracking around the joints. It would also, in the event of joints having opened up or cracking having occurred around them, prevent a sudden drop in rigidity, causing joints and even connected segments themselves to fail.

4) The design strength of joints would be made at least 60 percent that of the segment body, with the bolts being the weakest spots in the cross section of the liner ring.

5) The liner ring would be divided into equal-sized segments, with the key segment (K segment) given the same dimensions as the others, and a minimum number of joints (eleven) would be used. This would allow a maximum spliced-joint effect when segments in adjoining rings were placed in a staggered configuration, to yield an increase in the rigidity of the rings as a whole. The K segment would, in addition, be designed to be inserted axially to prevent them from becoming dislodged in limit states as shown in Fig. 9.

6) The watertight gaskets and damage protective gaskets should fill 85 to 100 percent by volume of the space between joint surfaces previous to swelling, and remain trim within spaces when the number of joint openings is zero.

Measures to Achieve Longitudinal Flexibility

Measures to provide longitudinal flexibility for the tunnels were as follows:

1) Long bolts (effective length - 500 mm) would be used for connections between liner rings to produce a structure that would easily absorb longitudinal deformations. In order to maintain a sufficient margin of safety for allowable elastic limits, high-strength bolts would be given a low axial tension (60 percent of the allowable stress).

2) Where ground stiffness or tunnel rigidity changed abruptly, so that long bolts alone might not be capable of absorbing the required longitudinal strains, elastic washers would be installed between the segmental rings to absorb the additional strains.

3) At shaft/tunnel connections, where the structural flexibility offered by long bolts and elastic washers would not be sufficient to absorb strains, a flexible segmental ring would be used to absorb the relative displacements expected to arise in between shaft and tunnel body. Fig. 10 shows the structure of the flexible segmental ring.

4) The steel reinforcement in the secondary lining should be minimum enough in volume and continuous longitudinally to disperse any cracking of the secondary lining over a wider area in the longitudinal direction, and to prevent longitudinal strains from concentrating in any particular joints of the primary lining. This was determined experimentally.

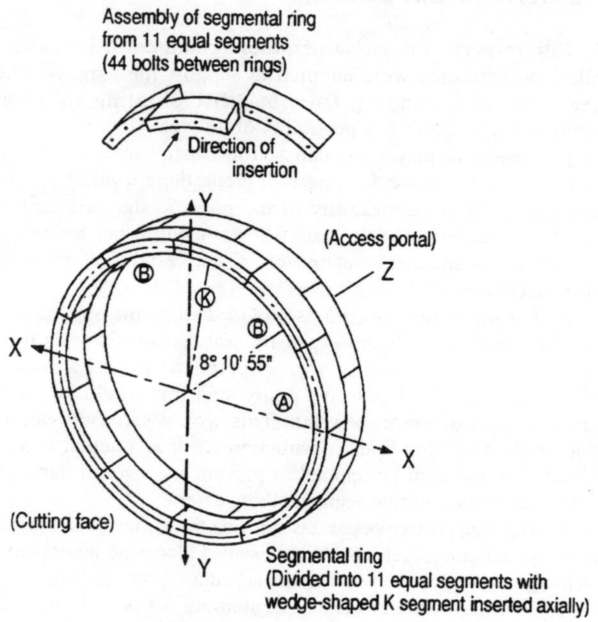

Fig. 9 Segmental Ring Structure

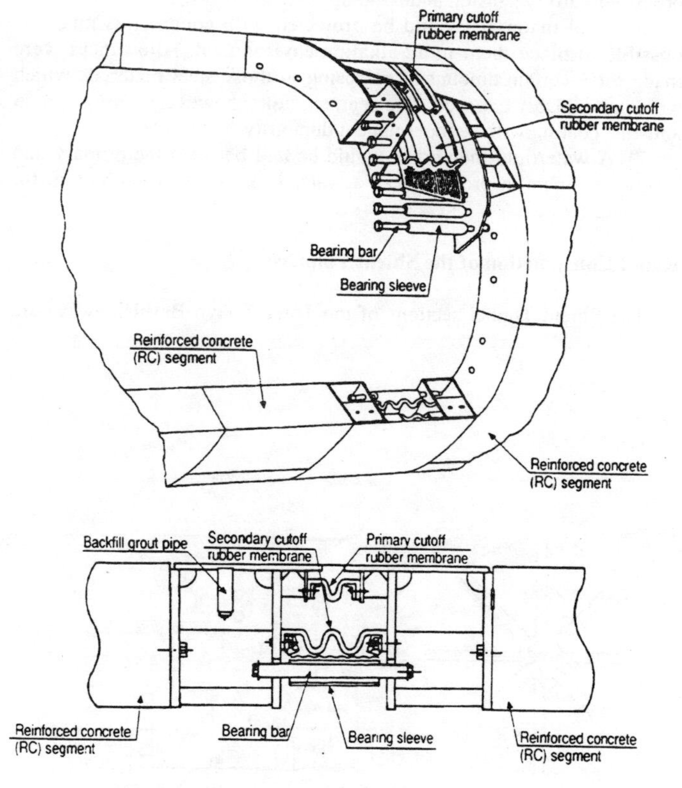

Fig. 10 Flexible Segmental Ring

Measures to Enhance Durability

With respect to tunnel waterproofing and corrosion control, the following measures were adopted to support the various watertight measures used (ranging from backfill grouting to membrane waterproofing) against any possibility of leakage:

1) Even with plastic (as opposed to elastic) materials, considered the most reliable choice for a backfill grout, there would be a change in the coefficient of permeability of the grout in the long term. It was therefore decided, that the backfill grout could not be expected to provide semi-permanent waterproofing, but could only act as a short-term water stop.

2) For the watertight gaskets provided along the outer perimeter of each liner segment, a hydro-expansive seal that swells when it becomes wet was chosen. A sealer was selected whose hydro-expansive component would not dissolve easily, and this would be installed in strips 35 mm or more in width. This would satisfy waterproofing require-ments during both construction work and actual operation. A lubricant would also be applied to prevent mechanical damage to the watertight gaskets during segmental ring assembly

3) Damage protective gaskets carrying the same amount of resilience as the watertight gaskets would be installed along the inner perimeter of each segment to prevent segment damage during installation.

4) Normal portland cement containing 50 percent by weight of blast-furnace slag would be used to produce the reinforced concrete segments. This would give the concrete a coefficient of permeability of about one-tenth that of normal portland cement concrete, thereby increasing its resistance against saltwater corrosion.

5) Each segmental ring was provided with four backfill grout holes. Around the perimeter of each hole, a butyl-based rubber seal would be set. When concrete was placed, this would react with the concrete to produce hydro-expansion and bonding.

6) Steel materials would be protected with concrete as much as possible to place them in an alkaline environment. Allowances were made for a certain amount of corrosion in those steel materials which were unavoidably exposed. The segment bolts, however, were given an anti-rust coating with semipermanent durability.

7) A watertight membrane would be laid between the primary and secondary liners in order to lead a water leak down to the bottom for drainage.

Rapid Construction of the Shield Tunnels

The Shield Tunnel sections of the Trans-Tokyo Bay Highway are being constructed by means of the Pressurized Slurry Shield method. This employs a tightly sealed shield in which pressurized slurry maintains a stabilizing force against the working face while the cutters excavate the soil and the shield machine moves forward. The particular characteristic of this method is that while the slurry is forced against the working face, excavated material is pumped up in slurry form to a treatment plant above ground. In the treatment plant, the constituent sand is separated out from the slurry, which is the returned by pump to the working face, after having its density and specific gravity adjusted.

With the object of opening the tunnel for use as soon as possible, a number of measures were adopted for the excavation and the lining operations. Those adopted at the planning stage, those related to the installations, and those related to the execution of the work are here outlined separately.

Measures adopted at the planning stage

1) The tunnel was divided into eight sections, each one 3,000 m or less in length. This would ensure that if by any chance one of the tunneling machines should suffer a breakdown and be unable to continue boring, the machine coming from the opposite direction to meet it would be able to complete that section.

2) Temporary facilities were designed to permit excavation and lining to be carried out simultaneously, and a process introduced by which the lining could be completed without interruption when the excavation operations came to an end.

3) To reduce the amount of time taken to install the lining, the liner rings were divided into eleven equal segments and installed automatically.

Measures Taken Regarding the Installations

SHIELD MACHINES

(1) Standard specifications

Twin parallel tunnels are being driven between the Ukishima Access and the Kawasaki Island, and between the Kawasaki and Kisarazu Islands, making a total of eight work sections. Tunneling involves eight shield machines, each starting from one of the vertical access shafts at the Ukishima Access, Kawasaki Island and Kisarazu Island, to eventually connect below the seabed.

Fig. 11 shows the front view and cross section of the shield machine, and Fig. 12 shows a shield machine assembled at the fabricating plant.

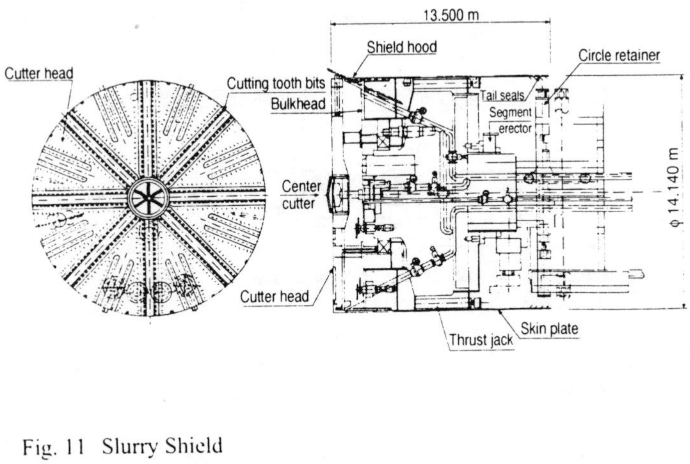

Fig. 11 Slurry Shield

Fig. 12 Shield Machine

① Basic configuration
a. Bore diameter: 14.14 m
b. Shield machine length: 13.5 m
c. Total weight: 29.4 to 31.4 MN

② Driving machinery
a. Total thrust: 235 MN (48 jacks with 4.9 MN capacity)
b. Jack stroke: 2,550 mm

③ Cutting head
a. Cutting teeth: Pin-jointed cutting tooth bits are being used to facilitate any necessary replacement. The cutter head is mounted also with leading bits, peripheral bit-abrasion protection bits, peripheral lateral-protection bits, and abrasion-sensor dummy bits.
b. Cutter head: The cutter head is operated by electric motors with a speed of rotation ranging from 0.39 to 0.45 min^{-1}.

④ Tail seal: The mechanical shield is equipped with four lows of wire-brush tail seals to improve abrasion resistance, durability and water pressure resistance of the tail seals.

⑤ Auxiliaries: The shield machine is also equipped with a circle retainer, a man lock, backfill grout injectors, and equipment for tunnel connection.

(2) Special specifications for the shield machine

(A) Special specifications for underground tunnel connections

There was a choice between using one shield machine to excavate each section between vertical access shafts, and using two, one from either end of each section. As the result of an overall evaluation that considered, for example, the work schedule, durability of the shield machine, reliability of execution and safety, the latter option was chosen. This meant that the TTB Highway, with its two tunnels for eastbound and westbound traffic, required a total of four underground tunnel connections to be made.

After comparing three alternative methods; ①freezing method combined with center cutter retraction;
② shield hood extension method; ③connection ring method, it was decided to adopt a method which had already been used with considerable success, one combining ground freezing with retraction of the shield's center cutter (①). In this method, retracting the center cutter of each shield machine allows two shields driving from opposite directions to be brought very close to one another. As a result, it is possible to minimize the size of the freezing zone.

It was determined that for connecting two shield machines midway in the tunnel section, the machines would be brought opposite each other, and then the surrounding natural ground frozen by means of ground freezing pipes extended from the first-arrived shield machine.

To perform tunnel connections in this way, the shield machines were equipped with the following instruments:

① Center cutter retractor
In order to expedite the ground freezing process, the shields were equipped with a center cutter retractor to allow the cutting heads of two opposing shields to come within about 300 mm of each other.

② Instrumentation to confirm relative shield positions
When two shield machines were brought within 50 m of each other, a horizontal bore of about 300 mm diameter was driven between the two machines to measure the relative position and direction (attitude angle).

③ Freezing pipes and pipe insertion jacks
The shield machines were equipped with freezing pipes along the inner surface of the shield hood skin plate and the cutter head perimeter plate. In addition, the shield hood skin plate of the body of the shield machine was given 48 holes to permit radial insertion of freezing pipes.

(B) Special specifications for automatic segment assembly

Segment erection work was automated for greater work safety, labor economy, assembly precision and work efficiency. Reducing primary lining erection time was a particularly important issue in constructing the shield tunnels. Automation permitted the time required to assemble one segmental ring to be reduced to less than 110 minutes.

Features of liner segments and the shield machine's segment erector necessary for robotic assembly are listed below, many of which were developed specifically for the TTB Highway project:

① Segments
a. Bolt dislodgement prevention devices
b. Division of liner ring into equal segments to shorten erection time
c. Furnishing of sensing grooves for accurate segment position detection
d. Special metal parts for mechanical gripping of segments

② Segment erector
a. Use of a mechanism and control system to enable high positioning accuracy
b. Adoption of automatic bolt fastening devices
c. Installation of bolt delivery mechanism inside segments
d. Nut and washer feed mechanism.

Measures Taken Regarding Execution of the Work

1) Excavation was carried out on six days of the week, with Sundays given to maintenance of the machines.
2) Each day's boring was done in two shifts, both of ten hours' duration.
3) Excavated soil was used for the toll gate embankments or as covering material for industrial waste disposal sites. It was removed in such a way as to avoid interference with the rate of shield progress.

CONCLUSION

Cross-sectional Rigidity and Strength

Highest degree of primary lining rigidity was verified by the shape of assembled segmental ring. Fig. 13 is an example of measured inner circle of the assembled ring, which shows the difference of vertical and horizontal diameter is within 5 mm. The difference was maintained all along the tunnel irrespective of manmade soils or natural soils, sandy soils or clayey soils, depth of tunnel cover, and soft soils or hard soils.

Longitudinal Flexibility

Designed tunnel flexibility by lengthy ring bolts, elastic washer and flexible segmental ring is yet to be verified because the tunnel has not experienced an earthquake of expected magnitude.

Watertight Lining

Designed watertightness of the primary lining was achieved all along the tunnel, which is also shown in Fig. 14 in which a completely dry tunnel lining is observed.

Durability

Expected durability of 100 years or over is yet to be verified, however the achieved dry tunnel lining implies that the durability would be almost ensured due to no water leakage into the lining.

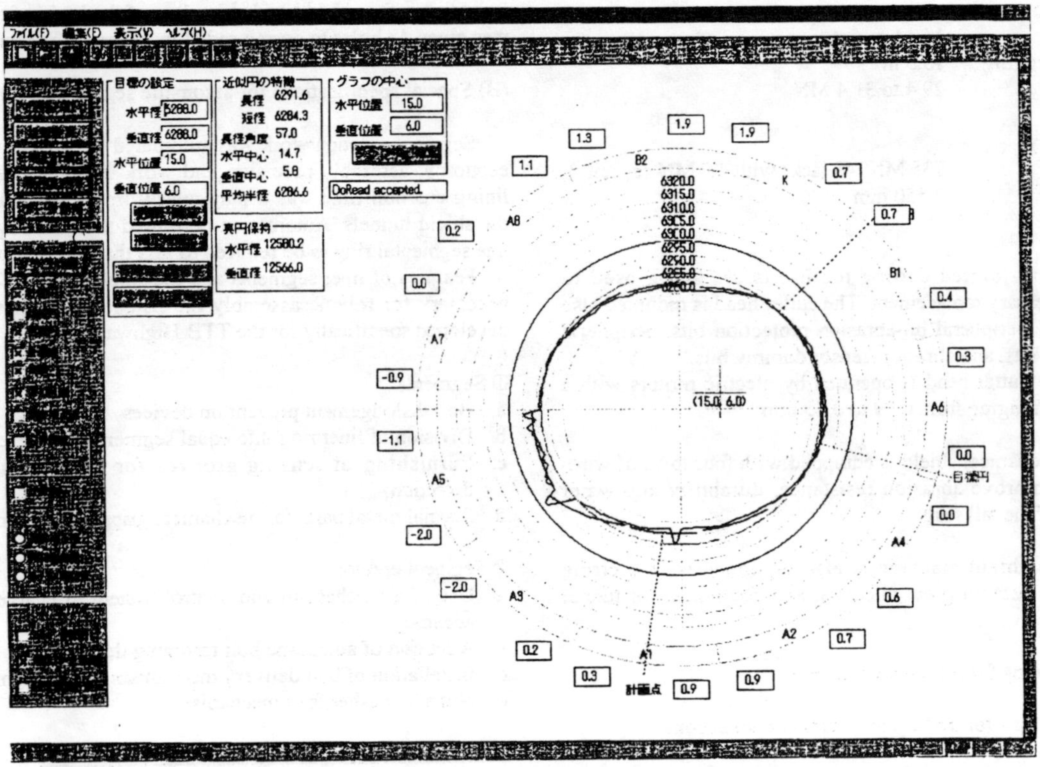

Fig. 13 Measured Inner Circle of the Assembled Ring

Fig. 14 Assembled Primary Lining

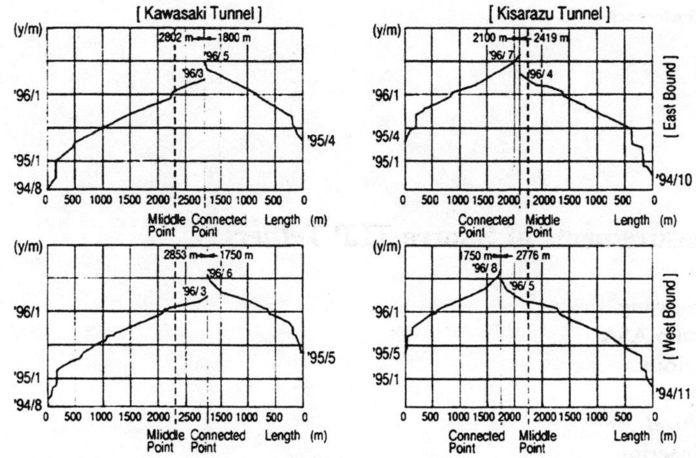

Fig. 15 Tunneling Construction Process

Rapid Tunneling

Fig. 15 shows the tunneling progress record of those 8 shields, which bored tunneling length of 1,750 m to 2,853 m. The shields advanced average 150 m or more monthly, which was beyond expected schedule. The best monthly progress is 310.5 m.

Automation

For labor saving, control accuracy and stability, rapid operation, and occupational safety, automation technologies were introduced in various field of the tunneling such as slurry treatment, slurry transportation, tunnel face pressure control, tunnel logistic control, segment assembly, and tunnel alignment control.

Assembled segmental ring accuracy and rapid operation implies that the newly developed segment assembly robot fulfilled its original purpose. However to increase the robot reliability by making the system more simple, it is essential to develop robot conforming segment joint structure which at the same time fulfill design requirements of highest ring rigidity and tunnel longitudinal flexibility.

Measured alignment error is within 10 cm of the plan, and especially the error at four underwater connections is within 5 cm, which verifies the completeness of the automated tunnel alignment control system.

AFTERWORD

The Trans- Tokyo Bay Highway will be a main road in the metropolitan highway network to help promote a balanced growth of the region. It also is expected to achieve a rearrangement of the transport function in the metropolitan area. Twenty years after the preliminary feasibility study, the project finally started with a newly organized system taking advantage of vital capacity of the private firms, and it has now achieved over 90 % of completion. It is expected that the highway will be open at earliest time in fiscal year of 1997.

In the project implementation, adequate considerations have been given to the natural and living environment, and navigational safety in the area, and the construction safety.

The authors wish to express their indebtedness and thanks to each and every personnel of various organizations involved in the project.

Proceedings of the Seventh (1997) International Offshore and Polar Engineering Conference
Honolulu, USA, May 25-30, 1997

ISBN 1-880653-28-1 (Set); ISBN 1-880653-29-X (Vol. I)

Lessons to Learn from Full Scale Measurements of Snorre TLP Tethers

Arnljot Skogvang
Saga Petroleum ASA
Sandvika, Norway

Horst Vogel
Aker Engineering
Oslo, Norway

1. ABSTRACT

Measured data from the Snorre Platform are compared with analytical results. The available measurements are time series of wave and wind data, platform motion and draught, tether and riser top tension. In addition, local strain measurements are available for two of the 16 tethers.

The data has been gathered during a period of one year and includes 2201 seastates with storm conditions up to 10 year return period.

The large number of seastates considered provides the statistical basis for the comparison with analytical results. Emphasis has been put on the verification of analytical procedures and results, applied for the design of the Snorre platform. Extreme and fatigue response of tethers was extracted from the measurements in order to be compared with the original design values. The quality, design reserves and possible conservatism of design methods have been evaluated.

Particular attention has been given to the non-linear effects like springing and ringing response of the tethers. The measurements have been post-processed to identify such effects and to compare with available analytical results.

Tether extreme response is obtained from linear frequency domain analysis including frequency dependent hydrodynamic coefficients. Spectral and rainflow analysis of the measured response were used to calculate tether fatigue damage which were compared with the original design values.

2. INTRODUCTION

The Snorre tension leg platform is located about 15 km northeast of the Statfjord A platform in the Norwegian sector of the North Sea. The waterdepth is 307 meters, and the platform is anchored with 4 tethers in each corner. The platform is designed to remain in safe operation during horisontal platform excursions up to around 10 % of the waterdepth, and the design of the tethers was a great challenge.
An Offshore Verification Program (OVP) was undertaken to verify platform motion characteristics, together with the structural response in
selected areas of hull, deck, tethers and risers, as well as the behaviour of the 4 concrete suction anchors for the tethers.

The measurement program lasted for about 4 years, and one key issue was to collect simultaneous data for environmental conditions, platform motion and structural response. Collecting only structural response data is limited value. Without information about the environmental conditions and the platform motion, structural response data can only be used as a monitoring tool to see if the response is lower than the design limits. In order to analyse why design limits are not exceeded, also simultaneous data for environmental conditions, platform motion and structural response must be recorded.

The comparison of measured and theoretical tether data are focused on the following parameters describing the dynamic tether tension:

- Max-, min- and standard deviation
- Rayleigh factors
- Trends for max tension versus significant wave heigth.

From this comparison the design margins are extracted, and the conservatism inherent in the design methods is estimated.

3. DATA ACQUISITION SYSTEM

The key task of the data recording system is to store timeseries of 35 minutes duration for all the sensors in the measurement system. The sampling rate varies between 1, 2, 4 and 5 hz, and data are collected

every 3. hour.The raw data are stored on a magnetic tape (video8), which is sent from the platform once a week. Verification of data quality is achieved by comparing trends for waves, platform motion and structural response.
These trends are strongly correlated, and errors in the measurement system are readily identified.
To verify the behaviour of the tethers, data have been collected by the following devices:

- Wave (directional) buoy for wave data
- Cup anemometer and wind vane for wind data
- Inertial Navigation System (INS) for platform motion
- Differential GPS for platform position
- Loadcells for tether top tension
- LVDT sensors for local tether wall strain

Most of this equipment is off the shelf items, but part of the software as well as the setup for measurement of local tether wall strain had to be specially made.

The wave buoy is located about 400 m west of the platform, and the reference station for the DGPS is located onboard the Statfjord C platform. The wind sensors are located in the top of the drilling derrick, about 90 m above the sealevel.

A central unit (workstation) onboard is collecting the data using a local network (Ethernet). Most of the data are transferred to the central workstation as files after the logging period of 35 minutes.

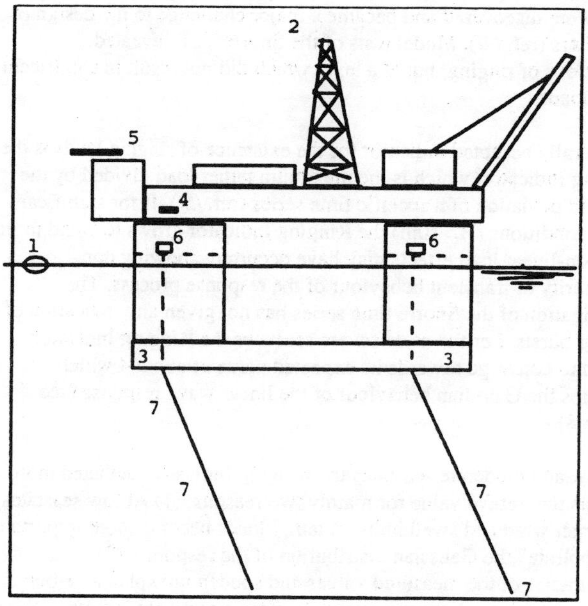

Legend:

1. Wave buoy
2. Wind sensor
3. Draught sensor
4. INS
5. GPS antenna
6. Tether load cells
7. Tether wall strain

Figure 1. Location of Measurement Devices

4. VERIFICATION OF TETHER EXTREME AND FATIGUE RESPONSE

4.1 TETHER EXTREME RESPONSE

Method

The Snorre tethers are designed for extreme load conditions described by storms of 100 year return period. These conditions are defined by H_s/T_p - contour curves for which the pertaining tether design loads were established, using analytical methods which were developed and applied for the Snorre Field Development Project(ref. /1/. The basic features of this methods are

- Non-linear static offset analysis due to static environmental loads.
- Linear diffraction/radiation analysis of a panellised platform model to obtain frequency dependent wave excitation forces and moments, added mass, potential damping and wave drift force coefficients.
- Linearized frequency domain analysis using frequency dependent hydrodynamic coefficients and linear spring tether models to obtain statistical properties of tether load and TLP motion response.

The same methods and analysis tools were used to analyse the tether response for an as-built platform model and for selected measured environmental conditions (40 seastates measured during 1995 and 1996). The comparison with the measured results validates the analytical approach and its ability to predict response for non-measured extreme environmental conditions. Based on this comparison in can be concluded that calculated tether loads for extreme conditions (100-year storms) will be predicted with the same accuracy as obtained for the measured environmental conditions.

Hence, the comparison of measured tether response with design values is performed in two steps:

1. To compare actually measured response with corresponding analytical results by applying the environmental load models as close as possible to the measured conditions. This step is needed to verify the analytical tools and procedures (see Figures 5, 6, 7).

2. To establish trends of measured response which are compared with 100-year design values in order to identify the safety margins of measured response versus design values.

Comparison of Measured Data with Analysis Results

A total of 40 storm conditions have been investigated which cover moderate seastates of H_s=3m to 10-year storm conditions of H_s=12 m as shown by Fig. 2 and 3. For each condition all 16 tethers have been analysed and been compared with the measurements. It meant that wave direction was important for the outcome of the comparison. However, the directional measurements of the wave environment were influenced of the direct wind driven sea and low frequency swell components which sometimes gave an unreliable total direction. To overcome this problem the individual tether response was calculated for various wave headings for each considered seastate to obtain the best fit of individual tether response. A typical comparison of wave frequency standard deviation load of measured versus analysed results is shown in Figure 4. The maximum difference of any one tether is used as relevant parameter to compare measured and analysed results for a specific environmental condition.

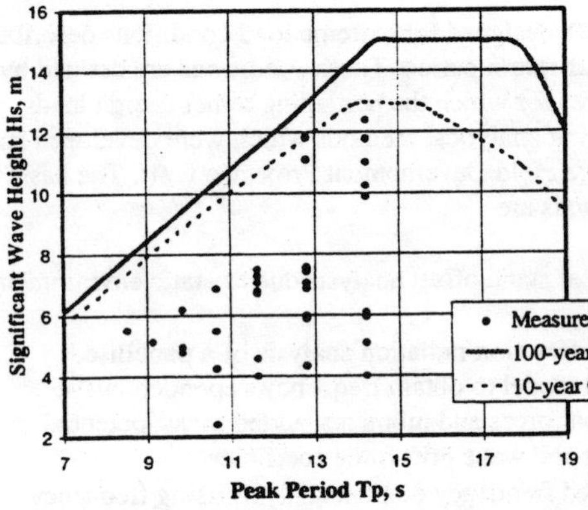

Figure 2. Seastate Distribution of selected Measured Conditions vs. 100- and 10 years Design Curves

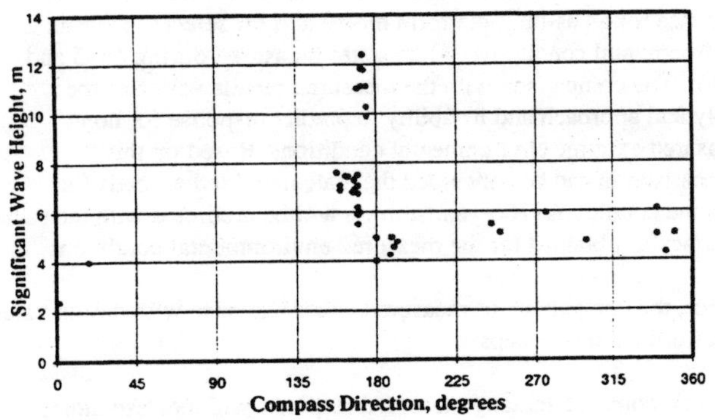

Figure 3. Wave Direction Distribution of selected measured conditions versus H_s

The analysis results were provided both for the measured numerical wave spectrum and a corresponding JONSWAP spectrum based on the H_s and T_p parameters. Although the measured wave spectra were quite different to the standard JONSWAP shape, the statistical spread of the tether response based on the measured - and the JONSWAP spectrum was rather small. It showed that for the purpose of tether response analysis the JONSWAP spectrum is sufficient to model the wave environment. For steeper seastates than the ones considered, one may have to use more general spectral models to obtain accurate results.

The results of standard deviation, maximum and minimum tether loads are shown in Figures 5, 6 and 7. One will observe that the difference for the standard deviations is for some cases more than 10%, but mainly for the moderate seastates. For the more extreme seastates the comparison is better and well within the 10% limit. For the extreme tether loads which include pretension and all environmental effects

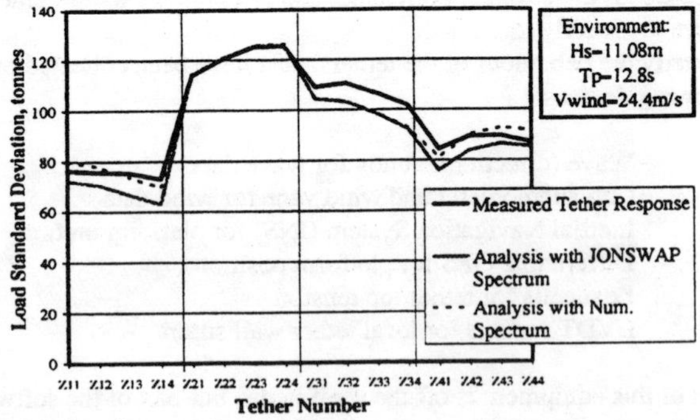

Figure 4 : Tether Tension - Measured versus Analysis

(e.g. wind and wave drift) the differences are only a few per cent. Also here one may observe that the comparison improves for the more severe seastates.

During the design of the Heidrun concrete TLP significant ringing loads were discovered and became a major challenge to the design of the tethers (ref. /4/). Model tests of the Snorre TLP revealed indications of ringing, but at a level which did not result in significant tether loads.

A generally accepted indicator for the existence of ringing loads is the Ringing Indicator, which is the maximum tether load divided by the standard deviation of a specific time series (ref. /5/). If for significant storm conditions (H_s>10m) the Ringing Indicator grows to 8 and more then non-linear load effects may have occurred, showing non-stationarity or transient behaviour of the response process. The investigation of the Snorre time series has not given any indication of ringing bursts. For the more severe seastates the Ringing Indicator tended to converge towards its expected value of about 4 which indicates the Gaussian behaviour of the linear wave response (see Figure 8).

For low and moderate seastates the Ringing Indicator deviated more from its theoretical value for mainly two reasons: (1) At low seastates the direct wind and swell induced tether loads become more important and "pollute" the Gaussian distribution of the response. (2) Inaccuracies of the measured values and sudden unexplainable bursts of the load cell readings are relatively more significant in low seastates.

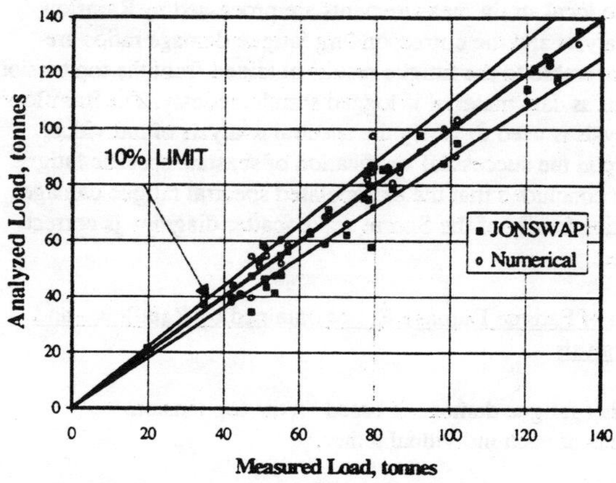

Figure 5. Comparison of Wave-Frequency Response - Standard Deviation of Tether Load

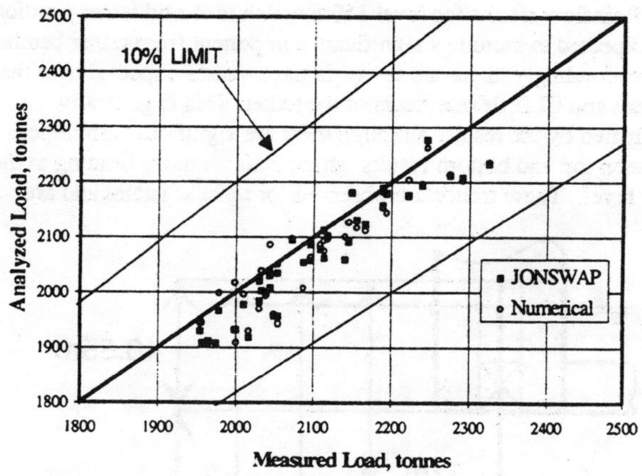

Figure 6. Comparison of Maximum Tether Load

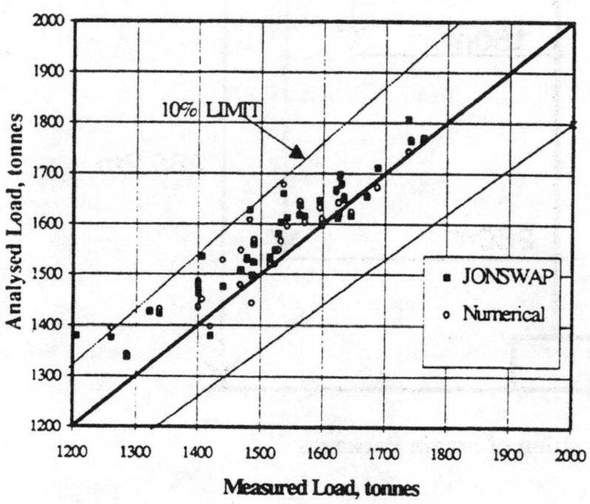

Figure 7. Comparison of Minimum Tether Load

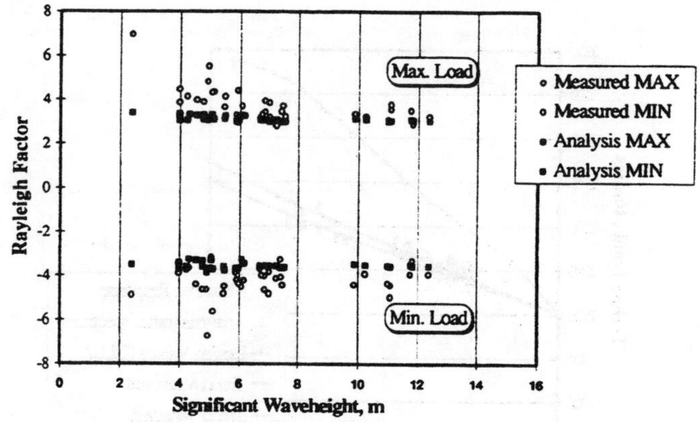

Figure 8. Ringing Indicator versus Significant Wave Height (H_s)

Comparison of Measured Data with Design Values

In order to compare measured tether loads with design values the measured values are adjusted for the constant load components which are required by the design procedure and the design code format. Such contributions are for example, the operational pretension range, the accuracy allowance of measurements and extreme tidal variations. However, safety factors prescribed by the design code are not included as such factors are part of the safety margins identified between the measured results and design values.

The measurements and the corresponding analysis results are trended to predict results for higher seastates (Figure 9 and Figure 10). These trends indicate that both maxima and minima are slightly underestimated by analysis. However, compared to the design limits also shown there are sufficient margins and reserves both for maximum and minimum tension which means that the inherent safety format of the design procedure covers for the uncertainties of the analytical models. The Design Limits are based on Design Seastates and appropriate loadfactors. One may observe that for minimum tension the reserves are smaller than for maximum tension. The reason is that the design code prescribes no particular safety factor on minimum tension because slack tethers are not considered to represent a severe failure mode. The design criteria for minimum tension imply that slack tethers shall not occur in 100-year storm conditions which means that the minimum design value shall be larger than zero. Figure 10 hows that the minimum design value is more than 800 tonnes per tether which represents a considerable reserve and a potential for pretension reduction and corresponding payload increase.

Until the 100-year design storms have actually occurred and have been measured, any verification of extreme design values leaves doubts. However, in the author's opinion the amount of results investigated justify the conclusion that it is very unlikely that any 100-year storm, whenever it may occur, could cause tether loads to exceed the design limits.

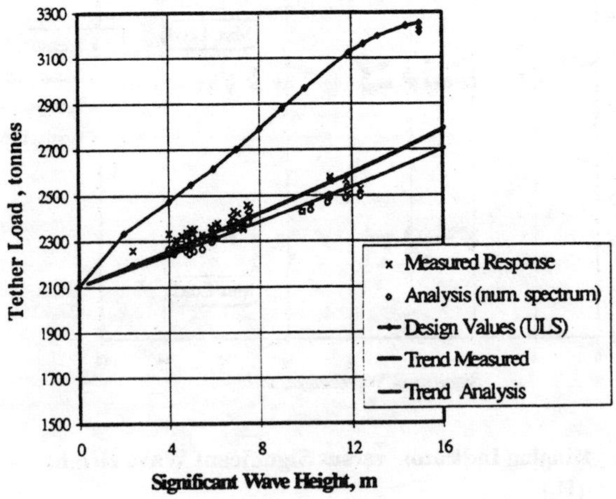

Figure 9. Extreme Design Verification - Maximum Tension

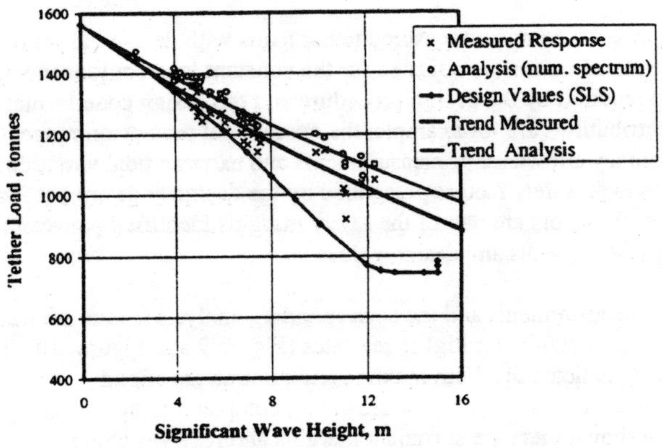

Figure 10. Extreme Design Verification - Minimum tension

4.2 TETHER FATIGUE RESPONSE

Method

The basis of tether fatigue verification is the comparison of accumulated fatigue damage resulting from the design load history and corresponding results from the full scale measurements. Accumulated fatigue damage is obtained by running both types of load input through a spectral analysis and using the Miner-Palgren rule to calculate fatigue damage ratios. This method allows to split the tether load history into its different frequency content which typically consists of high frequency (springing) and wave frequency response. Fatigue damage calculations are performed both for the individual high- and wave-frequency loads as well as for the combined total response.

The Snorre tethers are also equipped with strain gauges which provide stress results from three cross-sections along the tether pipe. This measurements include the local effect of tether bending which are excluded in the top tension measurements used for the spectral fatigue analysis. The local strain measurements are processed by Rainflow counting analysis and the corresponding fatigue damage ratios are directly comparable to the fatigue results obtained from the top tension measurement as data material is logged simultaneously. The Rainflow fatigue analysis is used to verify the spectral analysis of individual seastates. Upon the successful verification of seastate specific fatigue damage it is concluded that the extrapolated spectral fatigue damage for the seastate family of the Snorre wave scatter diagram is correct and validated.

Comparison of Fatigue Damage Ratios obtained by Rainflow - and Spectral Analysis

Spectral tether fatigue damage is based on the top tension measurements of each individual tether.

Rainflow fatigue damage is calculated at two levels of each of the two instrumented tethers, i.e. 51m and 280m at Z14 and 51m and 150m at Z44. Figure 11 indicates the position of the strain packages which consist of four strain gauges placed on the circumference of the tether pipe. The measurements of each package (four channels) are averaged prior to be compared with the spectral results.

The Rainflow results for level 150m which is at mid tether position, are expected to include a significant component from tether bending. Correspondingly, there are lower damage values expected near the bottom and CLB flexelements of the tether. This is generally confirmed by the results although there are significant differences between top and bottom results which indicate more bending at the 51m level. These trends are observed for most seastates and are

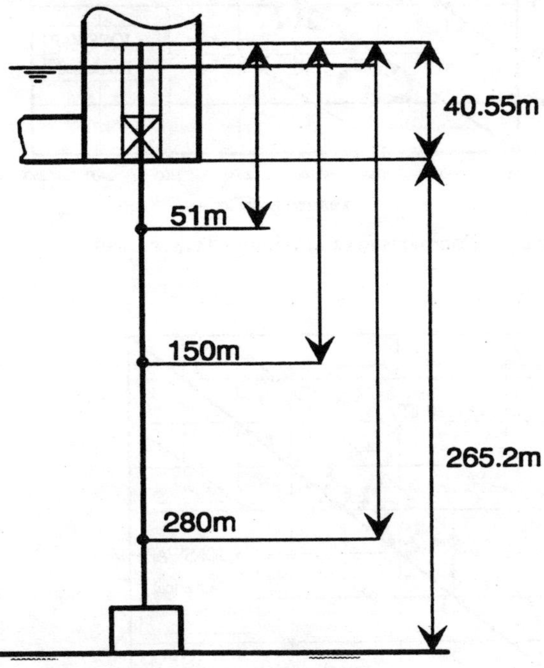

Figure 11. Position of Strain Packages

caused by the platform pendulum motion which introduces increasing drag forces towards the top of the tethers. Similar differences have been predicted by design analysis, but somewhat less than for the measured response(see Table 1)

At level 150m the Rainflow calculated damage has its maximum due to the presence of significant bending. The spectral results still overpredict, although there are three cases which are slightly underpredicted. This shows that the spectral approach is conservative even if bending effects are neglected. The results confirm that the spectral approach is suitable to assess total accumulated fatigue damage of the measured response which can be compared with design analysis predictions.

Fatigue Spot Level (see 11)	Rainflow Fatigue Damage as % of Spectral Damage		Design Fatigue Damage as % of Mid Tether Fatigue
	Tether Z14	Tether Z44	
51m	73%	68%	83%
150m	-	97%	100%
280m	50%	-	70%

Table 1. **Relative Difference of Rainflow -, Spectral - and Design Fatigue Damage Along the Tether Pipe**

Fatigue Damage/Life Predictions based on Measured and Design Response Data

As the spectral analysis method has been shown to give reasonable and conservative fatigue results, the method may now be used to calculate the total fatigue damage using the Snorre scatter diagram for 30 years service life. Figure 12 shows the fatigue damage ratio related to the sum-frequency, wave-frequency and combined loading. The corresponding fatigue life due to the measured response is 708 years compared to 190 years which was the design value.

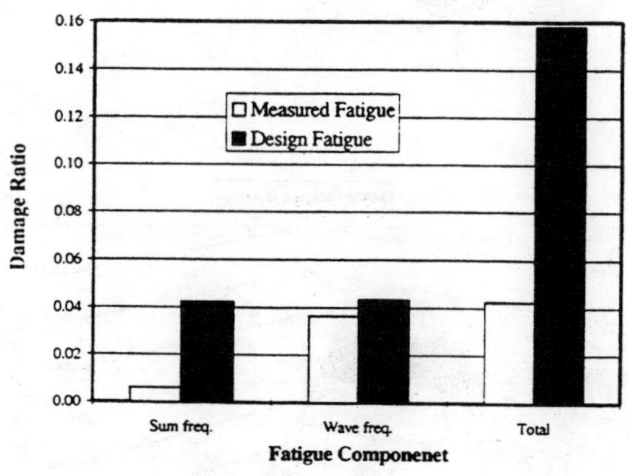

Figure 12. **Fatigue Damage of Measured Response vs. Design Results**

The considerable higher fatigue life from the measured response is mainly a consequence of the much lower sum frequency influence than predicted by design analysis. The sum frequency damage was predicted about 7 times higher than what can be seen from the measured response. The relation of the measured fatigue damage versus design values is also shown by Figure 13. One may observe that the wave frequency fatigue is predicted very close to the measured results. The sum frequency peak as predicted for seastates Hs=2.5m and wave peak periods 4.5 and 5.5 s was not all confirmed by measurements. However, the general much lower fatigue damage for the majority of seastates is the main reason for the low sum frequency contribution.

The Snorre project was first to use sum frequency analysis for the fatigue design of tethers, hence little experience or other results for calibration were available. An overprediction of this effect was almost inevitable in order to cope with the uncertainties and to document a required safety level for the tethers. Even so, the conservatism, as it may seen in retrospective, did not have serious consequences for the design or cost-effectiveness of the tether system, since other parameters were as much controlling during the sizing and selection process.

The results of this comparison are a valuable contribution for the assessment of second order effects and may help to increase the competitiveness of TLP concepts by reducing the sum frequency "penalty allowance" for future TLP studies.

Figure 13. **Fatigue Damage vs. H_s and T_p - Measured Response vs. Design Values**

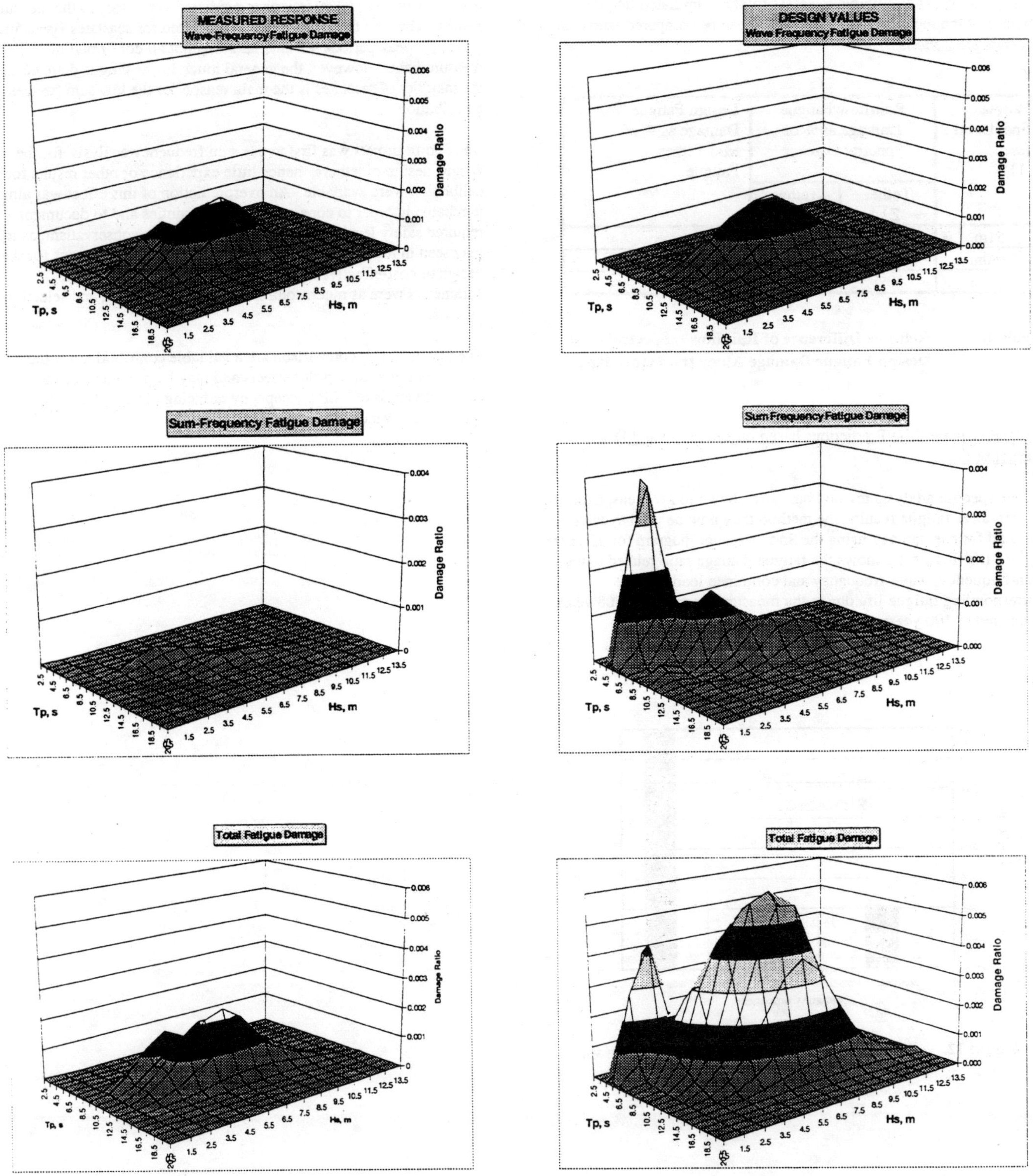

5. CONCLUSIONS

Extreme tether tension and tether fatigue life are crucial parameters of a TLP and therefore, are studied closely during design and in-service monitoring. The design procedure to predict these parameters are based on theoretical models with simplifications and deficiencies compared to the real physical processes they describe. This will give rise to questions how accurate and reliable the results are and whether the design achieved is both safe and cost-effective. Full scale measurements are as far as one can come the truth and the answer to such questions. They provide valuable information to verify and improve design tools for future TLP design.

5.1.1 Tether Extreme Response Verification

The prediction of extreme tether response during design is based on extreme environmental loads with a statistical occurrence once in 100 years for Ultimate Limit State (ULS) or once in 10,000 years for Progressive Limit State (PLS) conditions respectively. Such storm conditions have a low statistical probability and cannot be expected to occur during the limited period of an Offshore Verification Program. In order to verify the design load it is necessary to extrapolate the analytical results for existing measured conditions into the range of statistical extreme conditions. This implies that the analysis models are calibrated as best as possible, to reproduce the measured results and that there is a representative number of seastate conditions investigated. Despite the inevitable simplifications of the analysis model it has been shown that the dynamic force has been predicted within a 10% error range. The safety format implied by the design procedure does account for such errors and leaves sufficient reserves for other uncertainties.

For the Snorre TLP, for which significant ringing loads are not expected, the extreme loads calculated by linear frequency domain analysis can be combined by the SRSS method assuming wave frequency, slow drift and wind gust loads with Rayleigh distributed extremes without any further correction. The slow drift load has a non-Gaussian character and therefore, is not correctly represented by the SRSS method. This non-conservative method is however compensated by the overall conservatism within the design procedure. Whether or not the SRSS method is generally acceptable in TLP tether design cannot be confirmed on the basis of the limited number of measurements, but should be validated by third part investigations, preferably using full scale data of a different TLP.

To summarise the conclusion from the extreme load verification the following items seem important:

- Frequency domain tether response analysis based on linear diffraction analysis is an adequate tool to predict dynamic tether response of the Snorre TLP.

- The total dynamic load (wave frequency, slow drift and wind gust) is adequately predicted by the square root sum of the individual components squared (SRSS) where individual load components can be assumed Rayleigh distributed.(Note! The SRSS method may turn out to be too inaccurate for design purposes if slow drift components are significant compared to wave frequency effects, e.g. TLP motion response.)

- Extrapolations of measured tether loads into the range of 100-year conditions indicate that design predictions of minimum and maximum tether loads will not be exceeded.

- The safety format implied by the limit state requirements (ULS, SLS, PLS) is appropriate and leads to design values which are unlikely to be exceeded.

5.1.2 Tether Fatigue Response Verification

Fatigue assessments are always difficult due to the complex nature of the fatiguing process itself. Any analytical approach must account for the numerous uncertainties by being conservative which may easily result in over-estimating effects and fatigue damages. The objective of this full scale comparison was to verify the loads and load components which mainly contribute to tether fatigue damage. The measured tension time series have been divided into their high and wave frequency components and are described in terms of standard deviation and zero upcrossing period. By calculating the individual fatigue damage ratios and compare these with design results, it is possible to verify the design load analysis.

<u>Wave Frequency Fatigue</u>

The wave frequency induced fatigue damage has been shown to agree well for measured and analysed response. The linear frequency response of tether axial dynamics is sufficient to describe the fatigue loading from direct wave impact. Tether dynamics (bending) seem to contribute significantly (60 to 70%) to total damage, but the inherent conservatism of a spectral analysis almost covers this additional effect which has been shown by comparing with Rainflow fatigue damage obtained from strain measurements. Nevertheless, any tether fatigue analysis based on the spectral method should include the fatigue component due to tether dynamics. Design fatigue analysis must be conservative and certainly include the well documented first order effects in the total fatigue assessment.

<u>Sum Frequency Fatigue</u>

The sum-frequency loading has been shown to be much lower than expected. The quadratic transfer functions (QTF) for a one cylinder model which were used for the design analysis are known to be conservative which also was confirmed by comparison with QTF results from other analyses after the Snorre tether design was completed. However, the magnitude of over-prediction which in terms of fatigue damage reaches a factor of 7, was a surprise. Whilst analysis values identified sum-frequency damage as big as wave frequency, it showed to be only 16% for the measured response. In addition, the measured high frequency resonant response may as well be caused by first order loads so that the real second order effect is even lower. The total fatigue damage due to the measured response was only one fourth of the design fatigue damage.

These results are encouraging as they confirm that the uncertainties and assumptions of the Snorre fatigue load analysis in general, and the sum frequency analysis in particular, were on the conservative side.

There are some interesting aspects of measured sum frequency response which although not entirely clear, instigate these reflections:

- The directional dependence of sum frequency response is insignificant, i.e. the variation of sum-frequency loading between columns is small. This has also been confirmed by the Heidrun model tests, but is in strong contradiction to QTF analysis results.

- TLP resonant response at the natural heave, roll and pitch period is not entirely caused by second order forcing. At short period seastates for which second order response is expected to have its maximum also first order resonant response occurs. Seastates with low H_s are often multi-peaked (have significant swell contributions) so that normalisation by H_s^2 of the measured high frequency tether response is inaccurate and not well suited to extrapolate response for arbitrary H_s conditions.

- Model tests are most suited to predict sum frequency response. For long wave periods extremely good agreement is achieved, for short periods some over-estimation is seen, probably because of too low damping. Analysis results known to the author are not yet reliable. Analytical results tend to strongly over-predict sum frequency effects (see Figure 14).

- Damping is important for the correct analysis of sum frequency response, but cannot explain the overprediction by analytical models. Random decrement analysis of Heidrun motion data have shown only 20 to 30% higher full scale damping compared to analytical results (ref./4/).

- Heidrun and Snorre high frequency loads are compared in Figure 14. The Heidrun results are in fact close to the Snorre measurements if one would adjust (i.e. offset) for the different natural periods (Snorre=2.2s, Heidrun=3.3s). In terms of fatigue damage the high frequency effect is almost identical if one also adjusts for the 30% larger tether cross-section of Heidrun tethers (i.e. using stresses instead of loads). This may be a mere coincidence, but it is amazing that two completely different platforms experience high frequency tether stress of the same order of magnitude. One is tempted to believe that sum frequency effects are much more constant and affected by fewer parameters than assumed in the past. If this trend can be confirmed by further comparisons of measurements, it would much facilitate and improve the assessment of sum-frequency induced tether fatigue for future TLP concepts.

- The comparison between Heidrun and Snorre high frequency tether response shows that the influence of higher TLP natural periods (>3s). is probably less dramatic. This may lead to the conclusion that even higher natural periods of a TLP (e.g. deep water concepts) may not cause high frequency tether response to grow out of manageable proportions.

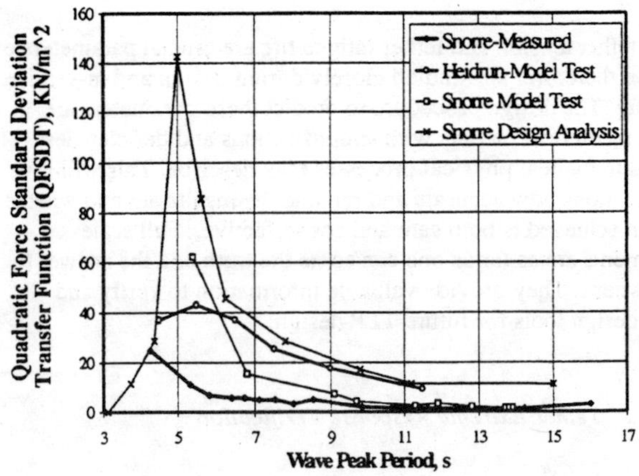

Figure 14. High Frequency Response per Tether - Snorre and Heidrun

ACKNOWLEDGEMENT

We like to express our gratitude to Saga Petroleum and the other Snorre partners Statoil, Esso, Deminex, Idemitsu, Norsk Hydro, Elf Acquitaine, Amerada Hess and Enterprise Oil for support and approval to present this paper.

REFERENCES

/1/ Olav T. Lappegård,Dag Juell Pettersen, Terry A. Gaul, Inge Bertin Almeland, Horst Vogel: Snorre TLP Configuration and Analysis Technology, OTC 1991

/2/ Marthinsen T. and Skogvang A.: Offshore Verification Programme for the Snorre TLP, Thermie Workshop and Structural Requalification and Diagnostics, Aberdeen Oct. 1992

/3/ Marthinsen T and Muren J.: Snorre TLP - Analysis of Measured Response, ISOPE-94, Osaka Japan

/4/ Matten R B, Strømsem K C, Lassabliere J, Vogel H: The Heidrun Field: TLP global Performance and Hydrodynamics, OTC 1996

/5/ Haver, S, Dalane J I, Sjursen A, Teigen P: Heidrun TLP - Experience during the First Winter in Operation, Offshore Platform Design and Construction, London March 1996

Proceedings of the Seventh (1997) International Offshore and Polar Engineering Conference
Honolulu, USA, May 25-30, 1997

ISBN 1-880653-28-1 (Set); ISBN 1-880653-29-X (Vol. I)

Norway's Remaining Petroleum Resources: Perspectives and Challenges

Finn Roar Aamodt
Norwegian Petroleum Directorate
Stavanger, Norway

Introduction

The Norwegian continental shelf, which covers an area of more than 1 mill km^3, is subdivided into 3 separate petroleum provinces (fig. 1):

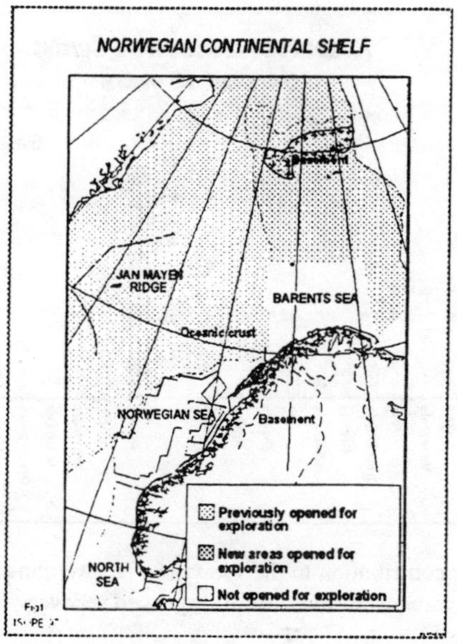

- The North Sea
- The Norwegian Sea including the Jan Mayen Ridge
- The Barents Sea including the islands of Spitsbergen

The geological characteristics, petroleum potential and technological challenges vary within these areas.

Exploration for hydrocarbons offshore Norway started in the North Sea 30 years ago and moved into areas in the Norwegian Sea and the Barents Sea in 1980. New large areas were opened for exploration in deep waters (1000-2000m) in the Norwegian Sea in 1994. Exploration in these areas will start in 1997 and industry interest is high.

A total of 520 wildcat wells have been drilled and almost 200 discoveries have been made on the Norwegian continental shelf. This gives a very high success rate (40%), which is a direct consequence of efficient exploration strategies employed over time. Important elements here are:

- Sound balance between high and low risk exploration (no discoveries made where probability of discovery is lower than 10%)
- Sequenced approach in the drilling of new prospects
- Liberal data exchange policies
- Extensive technological and operational cooperation between oil companies

Resources

The total recoverable discovered and undiscovered Norwegian petroleum potential amounts to approximately 12.5 billion Sm^3 o.e. with an upside potential of 16 billion Sm^3 (fig. 2). Only 15% (2 billion Sm^3 o.e.) of these resources are so far produced.

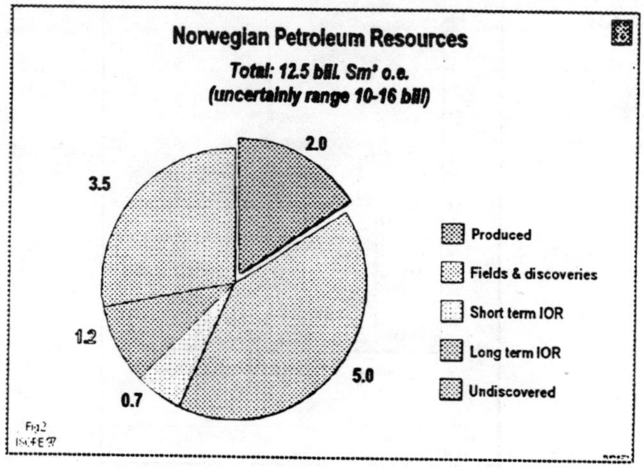

Discovered resources amount to approximately 9 billion Sm^3 oil equivalents, including a potential for the improved recovery of oil and gas of almost 2 billion Sm^3.

The majority of the fields and discoveries and most of the resources are located in the mature North Sea Basin. However, significant resources have also been discovered in the Norwegian Sea and in the Barents Sea (fig. 3).

DISCOVERED RESOURCES

	Exploration wells	Discoveries	Resources (10^6 Sm^3 o.e)	
			Oil/NGL	Gas
North Sea	392	159	3600	2830
Norwegian Sea	81	21	640	430
Barents Sea	47	16	30	230
Total	*520*	*196*	*7760* (incl. Short term IOR)	

Fig.3
ISOFE 97

39 fields are in production and 12 other fields are currently being developed. The majority of these fields are large and approximately 80% of the discovered recoverable resources are contained in these fields. There is more oil (60%) than gas in these fields.

Most of the remaining discoveries (142) are smaller and contain more gas (57%) than oil (43%). There are firm plans to develop between 30-40 of these discoveries within the next 2-3 years.

The undiscovered recoverable resources are calculated by the NPD to be in the order of 3.5 bill Sm^3 o.e. with a level of uncertainty ranging from 2-6 bill. Sm^3. 40% of the undiscovered resources are expected to be oil (fig. 4).

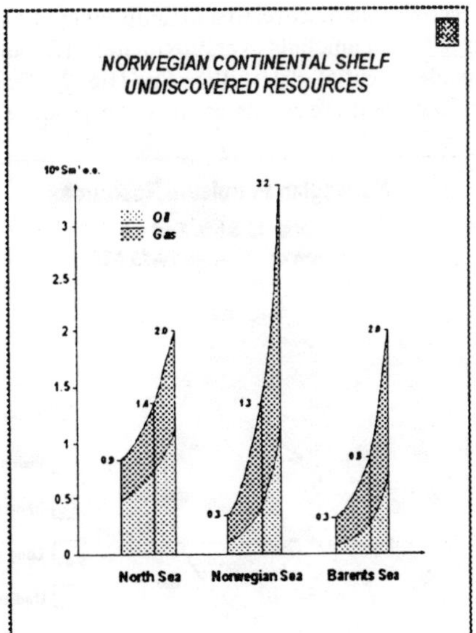

Fig.4
ISOFE 97

The growth in Norwegian petroleum resources has been very encouraging in recent years.

A significant part of this growth is the result of exploration activity (fig. 5).

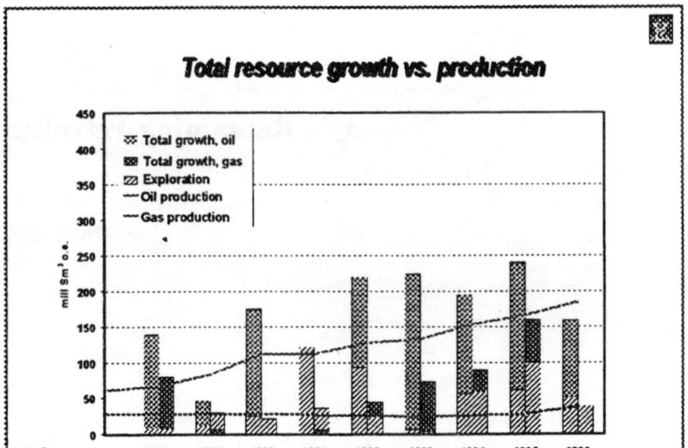

Fig.5
ISOFE 97

If we look at global discovery statistics for the last 10 years, Norway is sixth on the list of discovered oil resources and fifth on the list of discovered gas resources (fig. 6). This gives Norway an interesting competitive position.

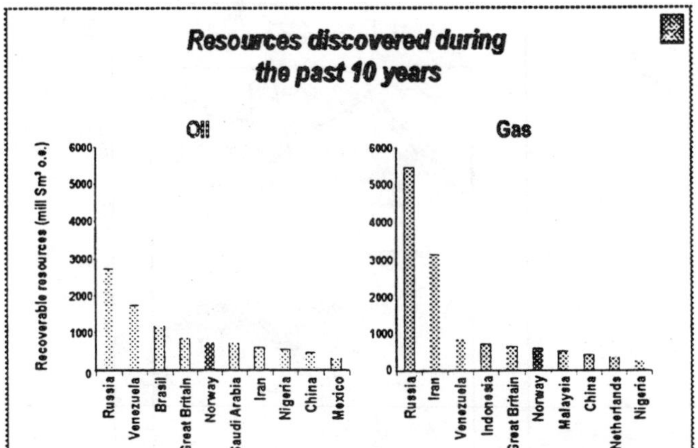

Fig.6
ISOFE 97

The major contribution to the growth in Norwegian petroleum resources comes, however, from improved recovery from producing fields (fig. 5).

The average recovery factor for Norwegian oil fields has increased from 34% in 1991 to 41% in 1996. This is due to the development and application of several new technologies in the management of production.

The most important technological improvements are:

- Enhanced reservoir descriptions through the use of 3D seismic
- Advanced wells (horizontal and long reach wells)
- Gas injection
- Water injection

- Alternate water and gas injection (WAG)
- Foam and gel treatment

As a consequence of encouraging exploration results, more cost effective field development schemes and improved production technologies, Norwegian oil production has increased considerably in recent years.

The current level of production is approximately 3.2 mill. bbls/day (NGL not included). This puts Norway at sixth among the oil producing countries in the world and second, only behind Saudi Arabia, as far as net export is concerned.

Norwegian oil production accounts for approximately 4.5% of global consumption.

The production level for Norwegian gas has, however, been moderate and stable (25-30 bill. Sm^3/year) during the past 15 years.

Future perspectives

According to the latest prognoses, Norwegian oil production will increase to a level of approximately 3.7 mill. bbls/day (excluding NGL) within the next 3-4 years (fig. 7).

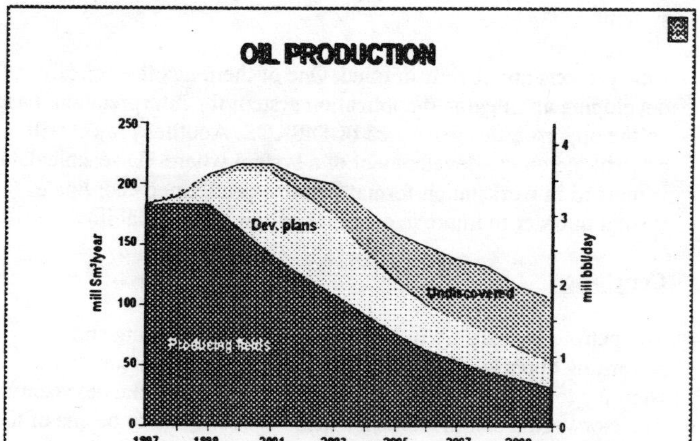

Fig 7
ISOPE 97

Shortly after the turn of the century oil production is expected to decline.

The implementation of current and new technologies in existing and future field operations is, however, expected significantly to limit the decline in Norwegian petroleum production.

Based on these assumptions, Norway will continue to be a major oil producer several decades into the next century.

Norwegian gas exports are expected to increase from today's level of 40 bill. Sm^3 pr. year to about 70 bill. Sm^3/year within the next 3-4 years. Further growth in the Norwegian gas exports is quite likely (fig. 8).

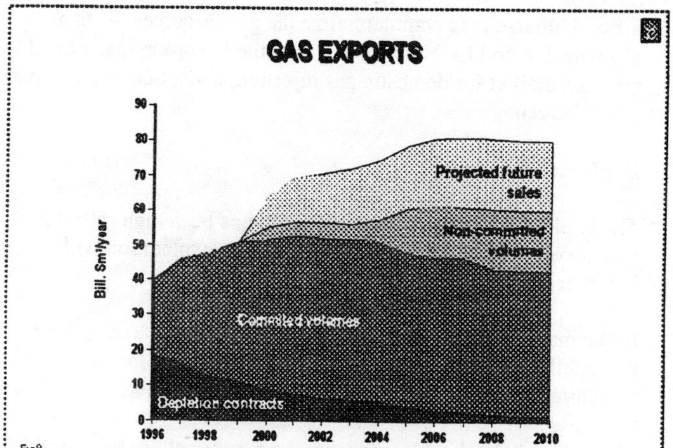

Fig 8
ISOPE 97

To put Norway's remaining gas resources into perspective, the following scenario may be considered. Based on a future average production of 80 bill. Sm^3 gas pr. year, Norwegian gas resources will last for approximately 70 years.

Challenges

The Norwegian petroleum industry is facing interesting and rewarding opportunities.

The challenges related to future activities are, however, very demanding and will require further technological and operational improvements.

Improved recovery

The average recovery factors for Norwegian oil fields have increased from 34-41% during the past 5 years. This is encouraging and the NPD believes that it is possible to increase the average recovery factor to 50% within the next 10 years. The resulting potential resource growth is almost 2 bill. Sm^3 o.e. Managing this growth will require the following:

- 4D seismic monitoring of all fields in combination with long reach, horizontal and multipath wells and further improvements in completion technology
- Use of seabottom seismic
- Further improvements in the use of gas, water and alternate water/gas injection
- Air injection
- More extensive use of foam, gel, soap and other types of chemical treatment
- Microbiological methods

New field developments

All the largest discovered fields are already in production or have their development approved, but more than 140 smaller fields are still awaiting a development solution. Most of these fields contain both oil and gas. The main problems in developing these fields have been the high unit cost and the utilization of the gas resources.

Recent field developments show that the cost issue is being taken care of and quite small oil fields are now considered economic. The great challenge is to commercialize the gas resources. With a growing demand for Norwegian gas in the European market and the growing market for domestic gas injection, a solution to this issue is also foreseen.

Exploration

The exploration success rate in Norway has been high (50%) in recent years and this is encouraging. Future exploration will, however, be more demanding than ever before.

In the mature areas of the North Sea the remaining traps are small and subtle. The profit margins are correspondingly low. It is therefore extremely important to limit the exploration costs.

This can only be done by reducing the number of dry wells and by making better assessments of the volumes and type of hydrocarbons in each prospect.

Statistics from Norway show that the industry is very poor in making such estimates. Another interesting observation is that only a minority of exploration wells in Norway are based on Direct Hydrocarbon Indicators (DHI).

Obviously, there is a significant potential for improvement here and the NPD is actively persuing these issues.

In the more immature areas of the Norwegian_and Barents Seas the challenges are different.

The undiscovered potential in the Norwegian Sea is uncertain but gigantic structures have been identified in the deep water areas in the west.

The first wells will be drilled in 1997.

Information from these wells will be very important for improving our understanding of the geology and petroleum potential of these areas, which are considered to be among the most promising exploration areas in the world.

The prospectivity of our deep water areas will have a great impact on future Norwegian oil and gas production and on the Norwegian economy.

The main technological challenges are related to improved mapping, managing the seabottom conditions (soft sediments on high seabottom gradients) and to deep water development and production.

Exploration activity in the Barents Sea has been very low in recent years, as a result of discouraging drilling results since the mid-eighties.

The problems in the Barents Sea are related primarily to severe uplift and erosion. The area is, however, poorly explored and the seismic data quality is limited in many areas.

New licences will be awarded in the area in the spring of 1997. The main challenge in the next 2-3 years will be to acquire high quality seismic data to mature prospects with a high probability of finding oil. Gas accumulations are less attractive in the Barents Sea due to the distance from markets.

Data management

Modern exploration techniques generate large quantities of information, but the logistics of traditional data storage are inefficient and costly. People are spending too much time searching for and collecting data, which is often of uncertain quality.

The efficient management of data is, therefore, more critical than ever before. The Norwegian Petroleum Directorate, in cooperation with active oil companies in Norway and IBM, is developing a data management system called DISKOS.

By means of this repository, data will be transferred directly to workstations at a high speed and low cost. Data can be traded by simply changing ownership rights within the database.

DISKOS contains cultural data, seismic data and well data. This is only the beginning and represents the basic building block in the development of more efficient and integrated data management systems.

New projects are already defined. One of them has the objective of developing an integrated application system for interpretation, based on the opportunities generated by DISKOS. Another project will concentrate on the development of a system where 3D seismic data is merged in workstation format within a predefined coordinate system in order to improve data integration and availability.

Conclusion

The petroleum industry in Norway will face interesting and rewarding opportunities in the years to come and Norway will continue to be a major oil and gas producer far into the next century. The Norwegian continental shelf will also continue to be one of the most important areas in the world for the development of new technology and of more efficient ways of managing exploration and exploitation of hydrocarbons. This will benefit the oil industry at large, the Norwegian society and the consumers of Norwegian petroleum products.

Proceedings of the Seventh (1997) International Offshore and Polar Engineering Conference
Honolulu, USA, May 25-30, 1997

ISBN 1-880653-28-1 (Set); ISBN 1-880653-29-X (Vol. I)

Emergency Management for Coastal Earthquake Disaster from the Viewpoint of the Sea

Ichiro Tanaka
Osaka University and Former President of KSNAJ
Osaka, Japan

Summary

The Southern Hyogo-Prefecture Earthquake occurred in Japan on January 17, 1995 with the magnitude 7.2 in the Richter scale and brought a great deal of disastrous damages over Kobe-Osaka area, which was thus named "Hanshin-Awaji Earthquake Disaster". The Kansai Society of Naval Architects, Japan, KSNAJ, organized a committee to investigate various aspects of the disasters and to propose methods of crisis and risk management for such a coastal earthquake disaster, especially from the viewpoint from the sea. The items to be investigated are the rescue activity by means of ships, marine transportation, harbor damages, activities by various organizations, risk management planning in coastal city, etc. The present article is an overview of the discussions and proposals on these items.

1. Introduction

As is well known, Jan. at 5:46 am on 17, 1995, an earthquake suddenly attacked the area around Kobe, Akashi, Awaji-island, and several cities between Kobe and Osaka, Japan,. The quake was named "Southern Hyogo Prefecture Earthquake". The quake was characterized by the magnitude 7.2 in the Richter scale, the highest quake level 7 in Japanese Intensity Scale, the very shallow epicenter of about 10 Km depth beneath the Akashi strait, together with several earth cracks near there. The maximum accelerations were reported to be 600 through 800 Gal in horizontal direction and 330 Gal in vertical one, and the shake continued about 20 seconds in Kobe area.

The quake and the subsequent fires brought a tremendous amount of damages over the areas around Kobe, so it was named "Hanshin-Awaji Earthquake Disaster". The summarized disaster just after one month reported by Hyogo Prefecture Government were as follows; 5373 people were dead, 2 people unfound, 34,568 people injured, 320,298 people took refuge in 1,247 rescue camps, 82,105 homes lost, 98,892 houses damaged, 531 fires, 100 ha area burned out, 21 bridges lost, 15 trains derailments, 5 tunnels lost, 40 railway stations damaged, express way damaged at 10 locations, 24 harbor piers damaged, electricity stopped for 1,000,000 homes, gas stopped for 857,400 homes, water supply ceased for 1,219,000 homes, 473,500 telephone lines disconnected, etc. These damages were accounted to be about 9,963 Billion Japanese Yen. (Final numerals increased further.)

Just after the quake occurrence, various facilities, organizations and voluntary peoples had begun to rescue the damaged cities and peoples, as were reported in media. Since the most members of the Kansai Society of Naval Architects, Japan lived in the Kobe-Osaka areas, KSNAJ immediately took the action of collecting life information on the society members and to organize voluntary group to help the restoration of the damaged member organizations. Then the society organized a research committee to investigate how ships were used in the rescue activity of the disaster, and the damages of marine transportation around Kobe Port due to the earthquake.

The committee also discussed the future problems of the risk management of disasters in the coastal city areas, e.g., desirable rescue system using ships, conceptual viewpoints on the coastal rescue system, establishment of temporary marine transportation route, importance of sea and port in the coastal city disaster, etc. The committee finally proposed a couple of examples of rescue ships and fleets. Some of these items will be briefly described in this paper.

2. Activities of Ships in the Disaster

2.1 Rescue Activity Using Ships

Various ships were used for the rescue and supplemental activity in the quake disaster, i.e., transportation of patients and rescued people from the disastrous area, transportation of materials, fire fightings, temporary hospitals, hotels for rescued people, substitution for land transportation systems, etc. The followings are these facts, analyses and problems.

(1) Emergency Escapes

On the day of the quake, operations of most public transportation stopped, so most people wanted to escape from their places by using their own cars, but this caused tremendous traffic jams, making the disaster much worse.

Almost nobody thought to use ships to escape. Although the public ship transportation also ceased to operates and the most ports suffered from heavy damages, on the 2nd day after the quake a couple of temporary ports were set in Kobe area and fast small passenger ships were used for people to escape from the devastated area to neighboring safe cities.

Problems may lie on the facts that there was no idea in people's mind to use boats to escape in such an emergency, and that there was no available port nor passenger ship operated on that day, although no ships were damaged by the quake.

(2) Transportation of Rescue Parties and Materials

Various small ships, such as fire boats, fishing boats, and pleasure boats were used for bringing the rescue parties and materials onto the disaster area from the other side of Osaka Bay in the evening of the 1st day. Then on the 2nd day other various small vessels, such as tug boats, fishing vessels, coast guard boats were used to convey the materials and peoples.

These were quite helpful means of transportation since almost no route was available on land at that time. Up to about 30,000 ton materials were brought to Kobe Port by the end of February (about 40 days). However, lack of the port facility for the use of small vessels can be pointed out as one of the problems.

(3) Fire Fighting by Ship

Road traffic jam, lack of water, so many fire places prevented the fire fightings and there was almost no effective way to fight the widely spread fires. However, a couple of fire boats contributed by pumping up the sea water and supplying it to the fire engines, and also by putting out directly the seaside fire. The difference of the junction units between fire boats and fire engines was pointed out as a problem to be fixed.

(4) Medical Care and Sanitary Facility

Hospitals in the quake area became almost panic due to lack of information, lack of staffs, lack of medical materials, lack of transportation, and full of patients. Navy ships was never used as a temporary hospital although some of them were nearby the ports and had some medical facilities. Such a medical facility of ordinary ship should have been used more at emergency.

(5) Hotel Facility

Three passenger ships were provided as emergency camps for the people in the disaster areas, staying at the seashore. Only 800 people utilized these ships. The reason why many people did not want to get on ships might be such fears that they are separated from their own houses, though damaged, where their properties are still left, and also that the ship might move anytime. It would be necessary to remove such a psychological uncomfortable feeling.

However, many passenger ships and navy ships were used as the hotel facility for the rescue staffs and parties.

2.2 Marine Transportation during the Disaster Period

(1) Human Transportation

Within 4 days after the quake many temporary passenger ship routes were established by the Ministry of Transportation in which high-speed small passenger boats were used. This measure, which worked well, was due to the facts that there had been well-established small craft routes in Osaka Bay, and that there were so many available boats for that purpose.

(2) Domestic Marine Transportation

In domestic cargo transportation, car ferry and ro/ro ships substituted for the track convoy system skipping the land routes around the destroyed area. This sudden change brought, however, various troubles at harbor and ports. An example was that Osaka port, being a substitute of Kobe port, had to accept too much tracks and ferries to handle so congestion was terrible. It was noticed that a modal shift from tracks to ships should have been made much earlier.

(3) International Marine Transportation

About 35 container piers in Kobe port were completely broken, so the container cargoes bound to Kobe were distributed to other terminals like Osaka, Nagoya, Yokohama, and even Pusan.

2.3 Activities of Governmental Organizations

Ministry of Transportation made a quick response to the disaster by setting a temporary passenger route from Kobe Port to other cities and established a Kobe-Osaka route at the 2nd day after the quake. Japanese Coast Guard put many patrol boats to transport people and materials with no accidents in Osaka Bay area during the disastrous period. Self-Defense Navy also made their full capability to help the recovery of the disaster in transporting the rescue materials and water, under its legal limitations.

2.4 Damages of Ports

Almost all ports in greater Kobe port were completely collapsed by the quake. Within a week, however, small piers left available were facilitated to a temporary ports to accept peoples and materials by using small boats, and then other ports were rapidly recovered except for the specially facilitated container yards, which took longer period to restore to the ordinary level.

The experience of the Kobe Port damages teaches us the necessity of building strong piers and safe access-roads to them.

2.5 Utilization of Ships at the Disaster

As stated in the above, various ships were utilized in the Hanshin-Awaji Earthquake Disaster. Table 1 shows a results of how and what kind of ships were utilized in the disaster.

Table 1 Utilization of ships in the disaster

Activity	Kinds of ship	Level
Transport of Patients	Fast Boat	Rare
Medical Care	Hospital Ship	Rare
Fire Fighting	Fire Fighting Ship	Little
Hotel	Ferry, Passenger Ship	Fully Used
Life Support Center	Ferry, Passenger Ship	Fully Used
Transport of Supply Goods	Cargo Ship, Water Supply Ship	Fully Used
Transport of Passenger	Passenger Ship	Fully Used

It is clear from Table 1 that ships were used for transport of various rescue people and materials, and used as a center of the rescue activity. However, as for the emergency medical care, ships were not so utilized even in this coastal area, and fire boat was not fully utilized either. The information facility which every ship naturally has was not utilized in the emergency information systems. These points should be considered in the future risk management planning.

3. Maritime Support to Urgent Coastal Crises

3.1 Conceptual View of Maritime Support

When a crisis such as quakes takes place in a coastal city area there should arise various needs for the rescue and recovery of the area. Such needs can be divided into the following 6 categories according to the time scale they require. Utilization of various type ships will be discussed on each item.

(1) Rescue and Fire Fighting

This should be needed immediately or within a week after the crisis takes place. For this purpose, small boats especially traffic boats and fire boats should be provided by governmental organizations.

(2) Medical Care

An emergency medical care would be necessary probably within month after the crisis. Small boat will be useful especially in the very early stage of the medical care. For this purpose as well as the cooperation between the boats and the ambulances an inland canal route would be quite efficient. Any kind of passenger ships would be used as a hospital ship which can be anchored along any coastal line.

(3) Life Support

For the support of the people's living, the transportation of necessaries of life like water, energy, food, and many rescue materials, would be inevitable. This support may continue about 6 months. Ordinary cargo ships, ferries, passenger ships, restaurant ships, and small boats would be utilized for this aim. The time interval of these ship operation should be adjusted properly.

(4) Traffic and Transportation

About 2 months or sometimes much longer depending on disasters, there would be a need of ships as substitutes of land traffic and transportation, since the roads in the crisis area would be damaged and useless in some periods. Fast ships, passenger ships and ferries would serve as transportation means in a coastal area.

(5) Space Utilization

There should be needs of space for various purposes such as emergency medical care, living places, rescue center, stores, etc.. These needs would require a half or a year long period. Relatively large container ships, ferries, passenger ships should be utilized for these purposes.

(6) Recovery and Restoration

It would take several years at least for a damaged city to recover to its ordinary state from the crisis. Various types of work vessels would be needed according to the missions. Cargo ships would also be needed to accept various materials for the recovery of the city life.

Utilization of these various ships in coastal city crisis should only be effective when these ships are to be used in daily life in many ways. For the emergency use of these vessels it is required the following conditions are met as the city functions:

1) Citizen's recognition of emergency use of vessels,
2) Functional variety of ports,
3) Flexibility of land-sea interface,
4) Arrangement of crisis management,
5) Establishment of principle of sea utilization,
6) Introduction of the viewpoint from the sea in the city planning.

Fig. 1 illustrates a conceptual view of the maritime rescue support from the sea. Inland canal system would be quite effective for the rescue from the sea as well as for the city functions stated above. This concept should be taken into account in the planning of the coastal city risk management system.

3.2 Function of Ships and Its Utilization

We have learned in the early rescue stage of the Hanshin-Awaji Disaster that the troubles in the information networks and in the road traffic jam prevented effective rescue activities. Kobe area is an elongated region between northern mountains and southern coastal line. Instead of thinking from land point of view, we should have thought of the rescue activity from the view point of the sea.

Ships normally have several functions as their own facilities, such as 1) self-completed living capability, 2) marine transportation facility, and 3) information communication ability. Those functions should be used in the crisis management in a coastal area.

The first function, self-completed living capability, offers an opportunity for people to live in the ship by using the hospital function on board. This function should be applied to the rescue and medical care at the crisis occasion. The 2nd function, i.e., the marine transportation, can of course be used as the transportation of the people and material to and from the disastrous area.

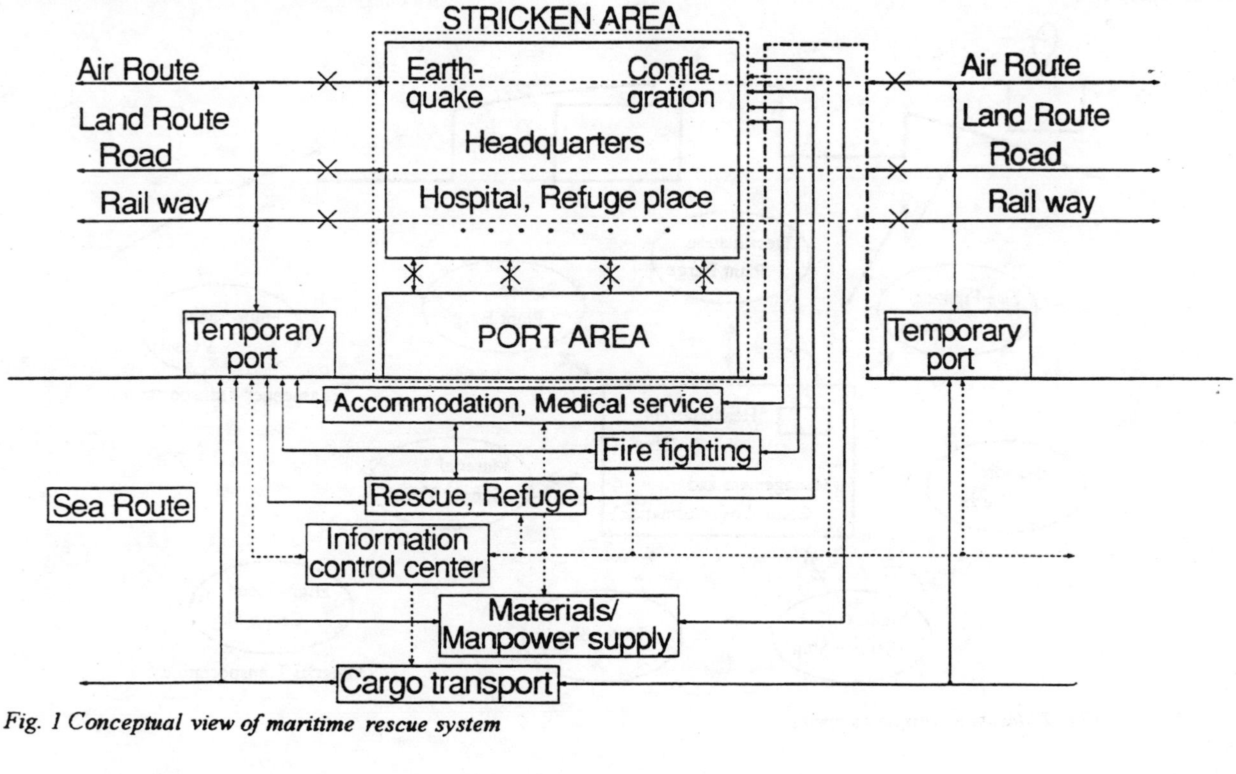

Fig. 1 Conceptual view of maritime rescue system

High speed small craft could also be used for the emergency transportation for the medical staffs and patients. The 3rd function, i.e., communication ability should fully be utilized in all the crisis stages. Ships usually have a satellite communication, telephone, facsimile, TV, wide band radio, and VHF type wireless system. These facilities are cooperated with the land communication system, which would make the ship to become a communication center for the crisis management.

Fig. 2 shows an image view of a marine rescue system, in which a headquarters-ship acts as an information center handling all information, accompanied by the various vessels with their own special missions, such as drinking water producing plant barge, power plant barge, various materials supply vessel, hotel ship, medical service ship, etc. In normal time these ships can be used according to their ordinary missions and could be called up immediately when a crisis occurs.

3.3 Utilization of Small Boat

Various small boats made a great deal of contribution to the rescue activity in the Hanshin-Awaji Disaster as stated in the preceding chapter. The features of the small boats can be characterized as free from the damaged ports, substitute to the land routes, high speed, easiness of decision to action, big force in massive use of a variety of missions, etc., which can be summarized as high mobility.

This property should more be used in the early stage of the crisis management since the most big cities in Japan are located within several kilometers from the coastal line. Small boats should especially be operated complementary to land rescue system.

In order to use small boats for the transportation of people and rescue materials in the early stage of a disaster, it is first of all necessary to secure the accessibility to the land route from the sea or inland river. For this purpose seashore should be reconsidered as the interface between the land route and the sea region, where the easy accessibility of both small boats and people gathering there should be developed.

Fire fighting with small boats is effective in the vicinity of coastal lines, and cooperation with fire engines will also be necessary in the inland canal area. Joint practice training as well as the facility standards matching should be helpful for the emergency.

Medical care activity would also be effective when small boats are available for the transportation of patients, medical staffs and materials in early stage of the crisis of a coastal area. Here again road-access to the seashore should be secured.

The coast guard should act to secure the safety in the sea using small boat with high mobility. Its role implies to watch the crisis states of the sea, to remove the floating obstacles, to secure sea routes, to maintain the traffic control, etc.

In order for these functions of small boats to be effective in the early stage of the crisis management, it is plausible the coastal line is used by citizens in their daily life in normal time. Seashore should be an attractive place for citizens in such a way that there are many attractive facilities like marine parks, pleasure boats, restaurants, swimming pools, yacht harbors, etc.

3.4 Emergency Management of Marine Transportation Routes

In a coastal crisis it is highly important to secure immediately substitutes of the traffic and transportation routes utilizing maritime potentialities. Marine traffic routes can be established providing quick investigation of available ports and berths, setting of a new route for ships, request of opening the route to shipping company, making of temporary time table and fare, and setting the method of compensating the deficit, etc. These would only be possible under good cooperation between the Ministry of Transportation and shipping companies. In the usual time various risk management should be done such as listing up of the available ships, standardization of the rampways and port facilities, maintenance of berths for emergency use, etc.

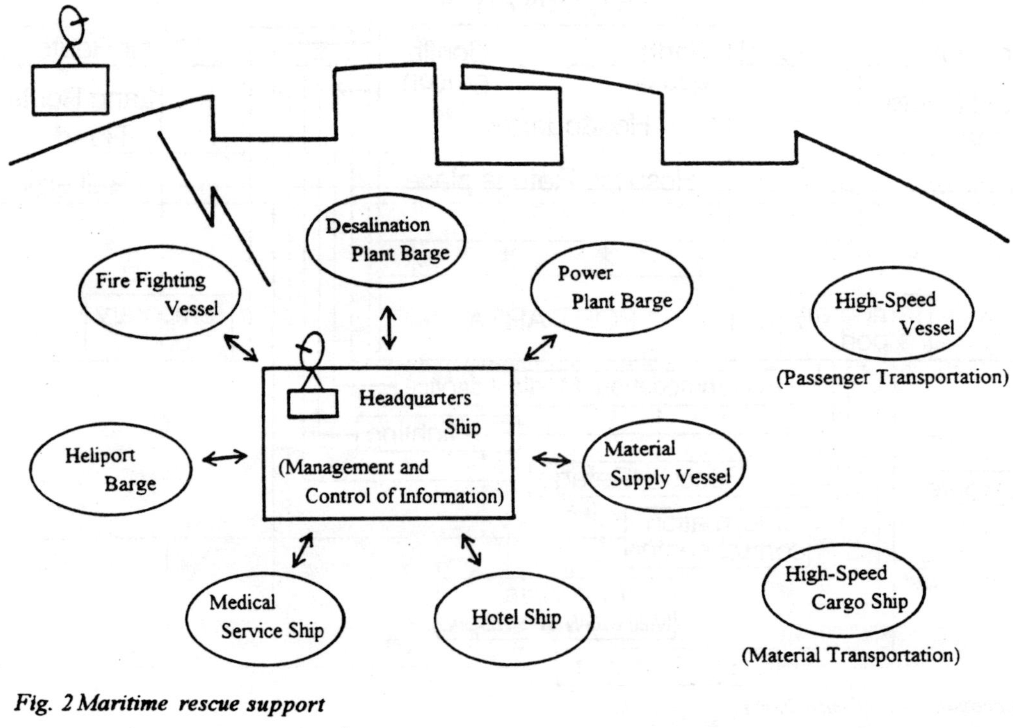

Fig. 2 Maritime rescue support

Marine transportation routes should also be activated in case of coastal disaster, in which mid- and long-range car ferries should effectively be utilized. For this purpose, it should be noticed that the modal shift from track convoy to car ferry system should be made in normal time. Standardization of ro/ro ship cargo handling system are desirable for possible utility at emergency.

4. Proposals for Emergency Management of Coastal Crises

According to the preceding discussions, some proposals for the emergency management of the coastal crises due to, say, quakes, can be made from the viewpoint of the sea.

4.1 Utilization of ship functions and a new notion of sea borne axis

While there are several core transportation routes on land, which are called here "land axes", it would be worthwhile to recognize another route in the sea, i.e., the transportation route along the coastal line. This new axis is named here "the sea borne axis". The function of the sea borne axis consists of three factors from the standpoint of the system theory, i.e., the route itself, the intersection between the sea and the lands, and the transportation means, which is ship itself. In view of crisis management, it should be important to locate geographically the sea borne axes along the coastal line parallel to the land axes. Some structures, floating or fixed, or ships are positioned at the gates as well as at nodes to set the axis. The new sea borne axis thus set, including gates, nodes and ships can perform the following roles in an emergency:

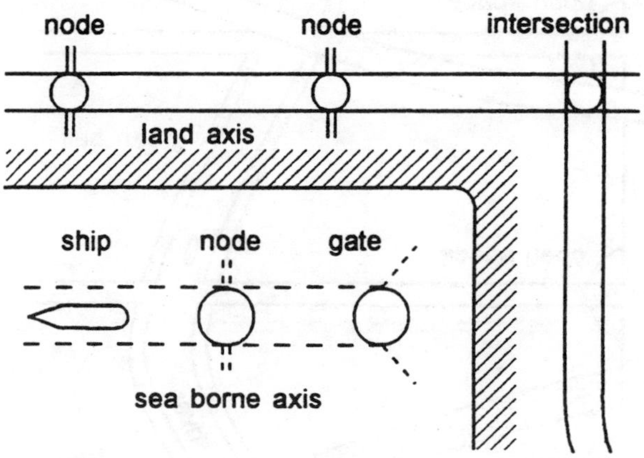

Fig. 3 Sea borne axis

(1) The transportation of emergency supply as water, food or rescue materials.
(2) The transportation of the rescue workers
(3) Traffic support for the resident in the disaster region.
(4) Accommodations as a hotel or as a shelter for the evacuees.
(5) A place for water and sanitation station
(6) A base of the operation vehicles.
(7) An emergency medical service center.
(8) A headquarters as information and instruction center as proposed in Fig. 2.
(9) Transportation of people, foods and materials needed for restoration.

These ship functions and abilities to contribute the rescue activities, should be reconsidered in the crisis management in the coastal area with the combination of the new seaborne axis. Ships can operate themselves not only in the sea borne route but also at the intersection gates and nodes between the land and the sea under the control of the traffic center.

4.2 City Planning for Crisis Prevention

The Hanshin-Awaji Disaster has taught us new aspects for a city to prevent big damages against disaster. The transportation systems were destroyed and all the necessities of life were disrupted instantly. It was very hard or impossible to get out of the disaster area for evacuees to safer places. It is also hard for rescue parties to get to the damaged area from outside of the area on land. Generally cities in Japan or in the world have been built up along river. In a crisis the river could be used as a transportation route when the regular roads were destroyed. Along the main river, sub-rivers are recommended to be constructed for the emergency transportation routes. River fronts serve, of course, as places for comfort and relaxation places in our daily life, too. In conjunction with this proposal, we need to build shallow bottom and low height boats to go through the rivers. We would like to suggest to the civil engineers, the bridges are to be constructed so that boats can pass under them. We hope when the civil engineers plan to build or reconstruct a town, they pay attention to this proposal from the view point of the sea and river. This concept is shown in Fig. 4.

4.3 Enlightenment of attractiveness and importance of seashore

As one of the first report of the earthquake, announcement is normally made about Tsunami. It is obvious that this is quite important especially to those people living close to the seashore. After the alarm of Tsunami has gone, the seashore can become a safe and convenient place to be a transit gate to supply materials and to evacuate to safer places. We have to notice that when the land traffic and transportation routes are disrupted, the route to and from the seashore is the only and best way to enter and get out to the disaster area..

Therefore, it is very important for us to remember this and behave ourselves in that way instinctively at a disaster. For realizing this, It is necessary and important to make people to recognize that the seashore is relievable and relaxing place and to habituate to the seashore in normal days. This would be a natural training of preparing for the coastal disaster, and this would make the crisis operations more smooth.

5. Conclusions

A brief overview on the Hanshin-Awaji Earthquake Disaster is given especially in focus to ship activity. Discussions and some proposals are also made for the emergency managements of coastal crisis from the viewpoint of the sea. We can summarize the conclusions as follows:

(1) Ships played a remarkable active part in the rescue operation in collapsed area at the Hanshin-Awaji Earthquake Disaster and have shown their importance for emergency management in coastal region.

(2) The seaborne routes must be recognized as a substitute for the ordinary transportation routes on land,. Both routes should be

encouraged for use in the normal and urgent situations.

(3) For the land or countries with long coastal line, a ship for the emergency operation, namely an emergency headquarters ship, is needed, where the information is collected and the orders are given.

(4) Importance of seashore at emergency in rescue work, evacuation and restoration must be enlightened and stressed from the viewpoint of emergency management of coastal cities.

The research was conducted by the following committee members;

Prof. Kiyoshi HARA,	University of Kobe Mercantile Marine
Prof. Yoji HIMENO,	Osaka Prefecture University
Prof. Ryusuke HOSODA,	Osaka Prefecture University
Prof. Yoshiho IKEDA,	Osaka Prefecture University
Prof. Kinzo INOUE,	University of Kobe Mercantile Marine
Prof. Shigeru NAITO,	Osaka University
Dr. Koji OTSUKA,	Osaka Prefecture University
Mr. Saburo SHIBAHARA,	Mitsui Engineering & Shipbuilding Co., Ltd.
Prof. Em. Ichiro TANAKA (Chairman)	Osaka University

The author would like to thank to many contributors in governmental, naval architecture and maritime worlds who have given invaluable data and comments to this project. The author also thanks to many governmental and maritime organizations for their cooperation with the present research. The names are so many and not shown in this English text

The committee work was supported by 80th anniversary fund of the Kansai Society of Naval Architects, Japan.

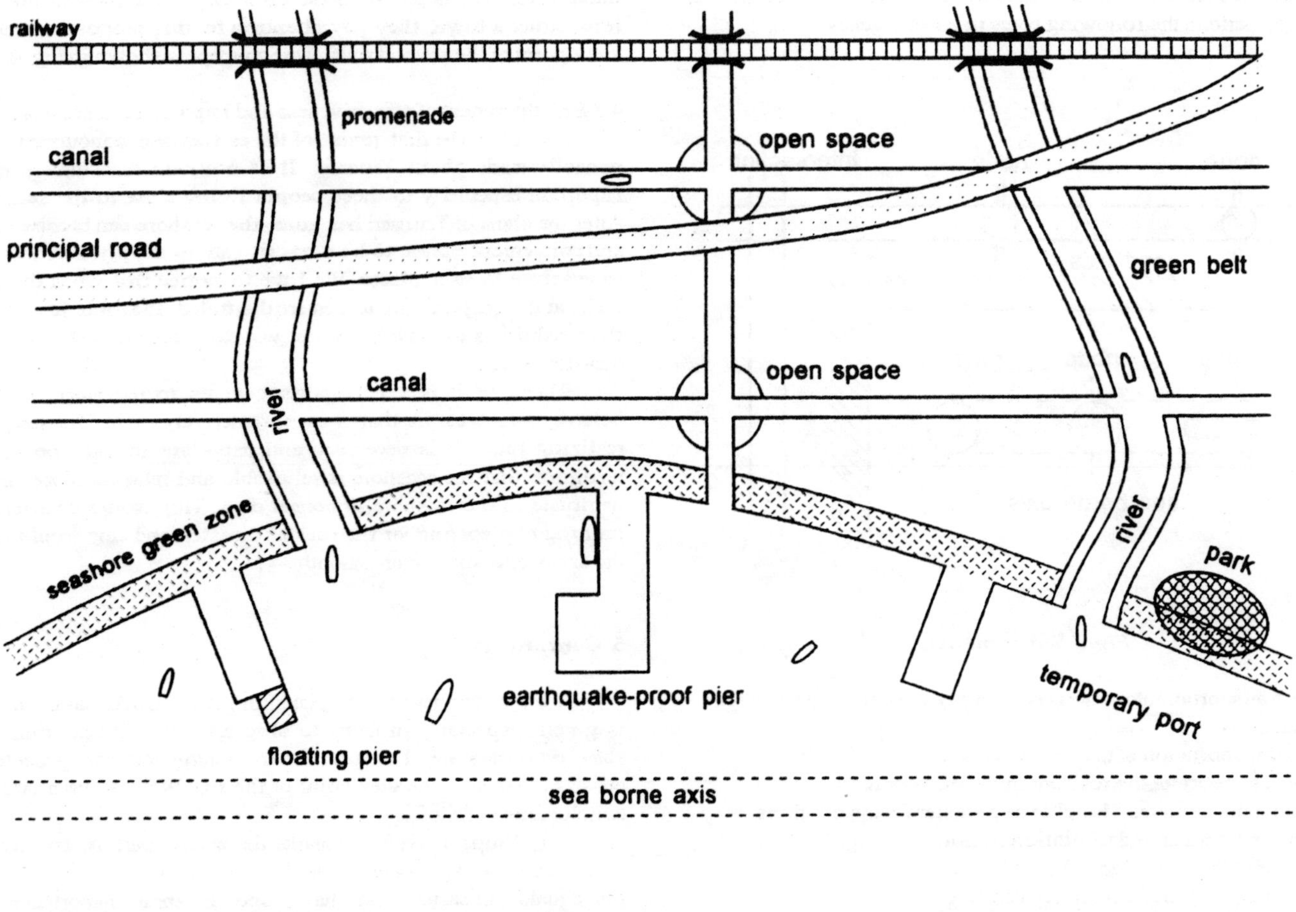

Fig. 4 Planning concept of the city from the viewpoint of the sea

Proceedings of the Seventh (1997) International Offshore and Polar Engineering Conference
Honolulu, USA, May 25-30, 1997

ISBN 1-880653-28-1 (Set); ISBN 1-880653-29-X (Vol. I)

Neftegorsk Earthquake, 28 May 1995, Sakhalin Island

V.V. Kharakhinov, S.G. Litvinov and S.P. Beketsky
Sakhalin Oil and Gas Institute
Okha, Russia

ABSTRACT

On May 28, 1995. at 01:04 a.m. local time on the Sakhalin Island has occurred the strong earthquake with magnitude of 7,6 degrees (on Richter scale).

The earthquake has occurred on a large part of territory of Sakhalin and on the territory of Khabarovsk Region, adjacent to the Tatar Strait. It has caused the terrific consequences in Neftegorsk, being closest to the epicentre, where a significant number of buildings and facilities were completely destroyed and numerous human victims happened. The intensity of shocks in this village was 9-9,5 degrees as per 12 degrees scale of MSK-64. The intensity of shocks of earthquake was felt to be 7 degrees in Sabo and Tungor, 6-6,5 degrees in Okha, 5,5 degrees in Nogliki and Nyvrovo, 4,5 degrees in Nikolaevsk-on Amur, Lazarev (Khabarovsk Region), 4 degrees in Nysh, 3,5 degrees in Tymovskoe, 3 degrees in Alexandrovsk-Sakhalinsky, 2,5 degrees in Uglegorsk and Shakhtersk.

There have had place damages (breaks) of oil and gas pipelines, power lines, as well as numerous serious damages of the narrow-gauge railway Okha-Nogliki on a section from Mukhto to Garomay.

The earthquake has caused visible changes of landscape, resulting, in particular, in occurrence, southward of Neftegorsk, of surface fracture - right side shift up to 30 km in length with amplitude of relative displacement of its margins (wings) up to 3-7 м, occurrence of numerous small sandy islands in Piltun Bay, formation of numerous cracks in ground and large discharges of sands from them on Piltun Bay bar to the south of Vrangel Island.

Neftegorsk earthquake on May 28, 1995 had significantly higher magnitude than all previous earthquakes, registered till now on Sakhalin. Its energy more than 100 time exceeds the energy of the earthquake in Nogliki on October 2, 1964 magnitude 5,8 degrees, being till now the strongest seismic event on the north of Sakhalin Island.

The study of Neftegorsk earthquake proceeds.

NEFTEGORSK EARTHQUAKE

A destructive earthquake of M 7.5 with focus of H=15-20 km occurred in the northern Sakhalin on 28 May 1995, 1:04. Neftegorsk was the nearest point to the epicentre and 17 apartments were collapsed. The clear surface fault about 3-4 km south-west of Neftegorsk showed up and the length of the major fault reached more than 30 km, the north-western flank was uplifted from 0.4 to 1.2 meters with the displacement to north-east of up to 0.5-1.5 m. Such dislocations are accompanied with earthquakes of M 9-12 on a 12-point scale. According to scale of seismic activities MSK shakings were registed of M 8-9 with destruction of living apartments. The population of Neftegorsk by that time was in excess of 3000 and 1989 lives were lost mostly in the collapsed 17 five-storeys apartments as a result of the strong vertical underground shock. Two-storeys self-contained housing apartments were undamaged. During that destructive earthquake people from those houses were out-doors, but shocks were so strong that people fell to the ground. The furniture, refrigerators, televisions fell down. All people out-doors felt the earth's vibration. Trough fractures appeared in some houses, balconies came off the houses. The cause of destruction is the difference in constructions: five-storeys houses were constructed in 60-70s and the antiseismicity was not take into consideration. At that time the North Sakhalin was considered to be asseismic and two-storeys living department were built later with consideration of seismicity (by that time the region was considered to be the seismic with the probability of earthquakes up to M 7. The earthquake was felt over the most part of Sakhalin - to north of Uglegorsk-Vakhrushev, on the territory of Khabarovsk Krai and adjacent Tatar Strait.

The eyewitnesses being on the Sea of Okhotsk Coast marked the ominous stillness before the first shock, but during the earthquake they heard a droning and saw a bright fluorescence.

Tectonically the region was enough active. The epicentral area of the earthquake is within the North-Sakhalinskiy Cenozoic trough which is considered to be the northern link of the

Hokkaido-Sakhalinskaja folded system. The trough is filled with later Cretaceous and Cenozoic deposits with thickness from 3.5 to 12 km. Within this territory a net of rupture dislocations is developed; regional, zonal and local faults are identified. Regional faults, with the length of hundreds km, cut the earth crust over the entire thickness and determine many features of the geological period during the Cenozoic. Zonal faults up to tens km length are developed during one or two geological epochs. Local faults up to several km length are observed within one or two stratigraphic complexes at one stage of the faulting being rapid relative to the geological time.

Rupture dislocations on the North Sakhalin are combined into a set of different-trend systems. The system combines faults of different ages and ranges.

The East-Sakhalin fault system is located within the north-eastern offshore Sakhalin with the extension from the east coast of the Shmidt Peninsula to the south-east up to 500-600 km. This systems combines a set of large shears, reversed faults, thrusts which form the overophiolite structural zone, characterized by the intensive dislocation.

The Hokkaido-Sakhalin fault system is of the submeridional trend along the Sakhalin eastern coast as an arrow zone of intensive rupture dislocations and folded deformations. Its northern link is within the region. Extensional structures (near-shear grabens of Daekhuriinskiy and early Okobykaiskiy-Oligocene-Miocene) and compressional structures (near-shear folds of the late Daghinskiy [Miocen] and Pliocene period, large reverse faults with the displacement amplitude up to 1.5 km) were formed at the main stages of the activity within the fault zone. Destructive fields of the Hokkaido-Sakhalin fault system are characterized by the concentration of disjunctives in the narrow (up to 5-10 km width) zone.

The West-Sakhalin fault system is composed of submeridional disjunctives striking along the Sakhalin western coast. Within the region it is represented by the West-Baikalskiy pre-middle Miocene fault, which is reflected into the upper Cenozoic deposits as the high-tilted flexural zone up to 5-10 km width. In water areas of the Sakhalinskiy Bay the fault is splitted into a set of large normal faults - shears, which were active in the late Cenozoic time.

The West-Okhotsk fault system transacts the western Sea of Okhotsk and Sakhalin Island north to south and is considered to be the main boundary between the Amur and Okhotsk Sea lithospheric plates.

The Verkhne-Piltun fault is the constituent part of this system on the discussed territory and adjoins the Hokkaido-Sakhalin fault south to west in the vicinity of Neftegorsk. The continually rotating counter-clock wise the Amur and Sea of Okhotsk plates relative to each other caused the movement - detachment within a zone of Verkhne-Piltun fault, which lead to the outburst of the elastic energy, that is the earthquake, being accumulated in the earth's crust for a long time.

The epicentre of Neftegorsk earthquake was in the conjunction zone of West-Okhotsk and Hokkaido-Sakhalin fault systems. In this geological situation the earthquake has caused the activity on the most part of the territory with widely-developed seismogenic structures.

The analysis of the seismodislocations as the result of Neftegorsk earthquake reflects the definite relationships in their development and position. One might distinguish three large blocks, which are involved in the seismotectonic process and are characterized by different levels and types of seismodislocations.

In the north-western (uplift) block restricted in the south-east and east by disjunctives of the Verkhne-Piltun and Hokkaido-Sakhalin (northern link) of the fault systems, a set of seismogenic N-E trend fractures (micrograbens 3-5 m length and grabens base 0.5-1.0 m width) are well-developed. These seismodislocations are small right-lateral faults having the amplitude of the vertical offset 0.3-0.7 m. Their distribution areal is controlled by a set of normal faults-shears of the South-Baikalskaja fault system.

In the south-eastern (downdip) block restricted by faults of the Verkhne-Piltun (north to west) and the Hokkaido-Sakhalin (from the east) systems the seismogenic fracturing has expressed as the dextral shears with the amplitude of the vertical offset and horizontal displacement up to 0.5 m and 1.0 m, respectively. The south-eastern flanks are uplifted in this block unlike the north-western block.

A sharp change in the pattern of seismodislocations is observed in the eastern (uplift) block located east off faults of the Hokkaido-Sakhalin system. The block was under the seismic action within the band of Mukhto-Goromai interfluve restricted by closen transverse faults with the crystalline foundation. Large landslides and collapses are confined to creeks and rivers, creek valleys, controlled by neotectonic fractures. Thick sand outbursts (30-100 m in diameter) and fresh water gryphons on the eastern coast of Piltun Bay are also confined to these neotectonic fractures.

A sharp predominance of seismo-gravitational dislocations probable is due to strong influence of the north-western block, shifted along the Verkhne-Piltun fault about 3-7 meters and collided with the block, which is located east off the Hokkaido-Sakhalin fault. The shearing motion ceased within a zone of their collision: its energy is resulted in formation of large gravitational seismotectonic dislocations of the eastern block and fracturing in the northern link of the Hokkaido-Sakhalin fault. The eastern directions of slumps sharp deviation of the railroad in the conjunction area of two large faults and the western trend in the Goromai River confirm this nature of seismodislocations. The analysis of the geologic-tectonic situation within the region of Neftegorsk Earthquake leads to the conclusion, that the activity of asthenospheric processes on the boundary between the Amur and Sea of Okhotsk plates is considered to be the major geodynamic tendency of the region; deep geology of the region should be investigated in future. The force of Neftegorsk earthquake has exceeded all the preceding and registered earthquakes on the North Sakhalin.

Its energy is in excess of 100 times than the energy of Nogliky earthquake, registered on October 2, 1964 (M 5.8). The relief the elastic energy like aftershocks may continue for a long time decreasing the probability of the new strong earthquake. The comprehensive study of Neftegorsk earthquake will permit seismologysts to obtain the new real materials which will be used for investigations, forecast and prediction of underground perturbations.

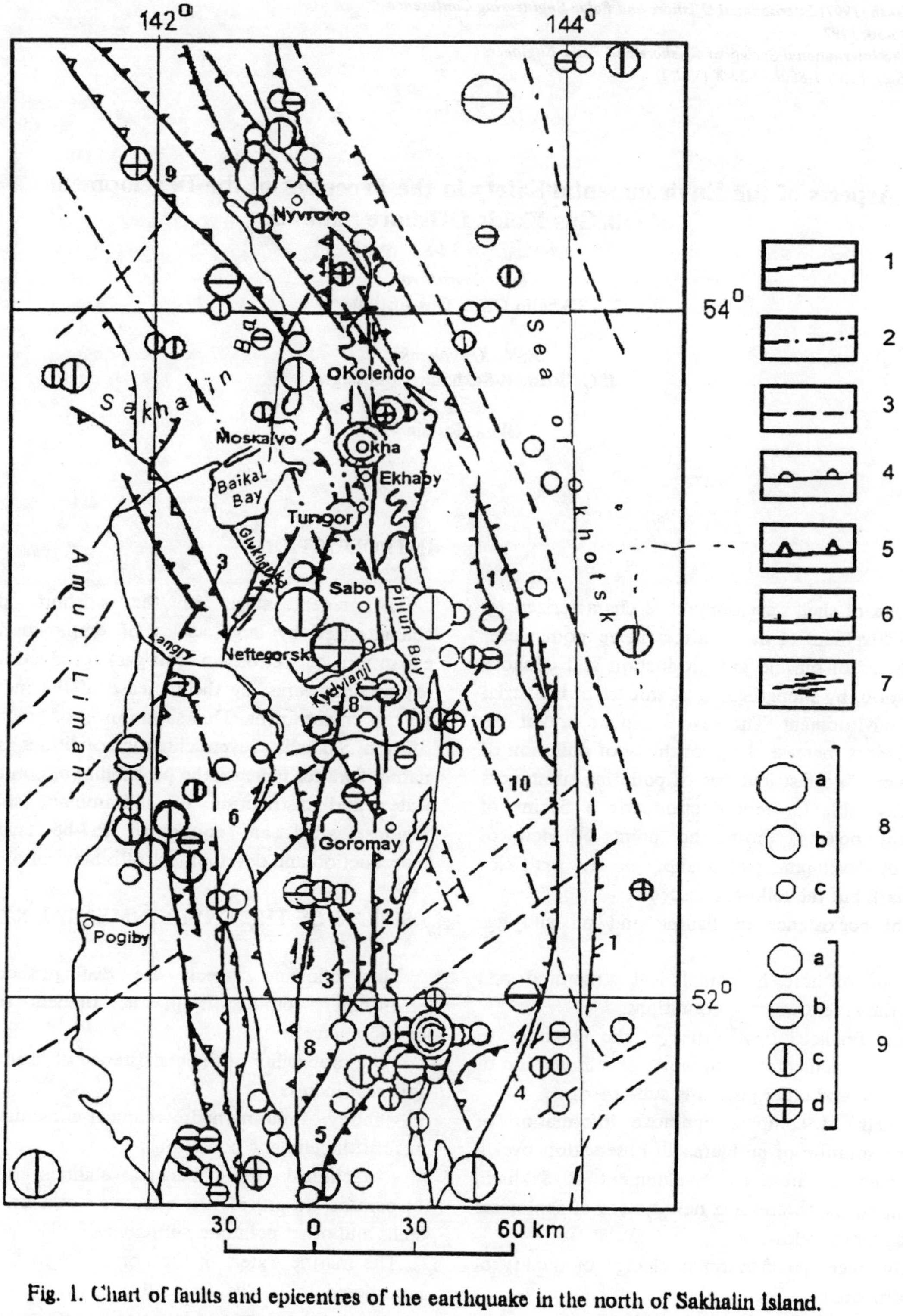

Fig. 1. Chart of faults and epicentres of the earthquake in the north of Sakhalin Island.

1-3 - faults developed: 1 - in the entire strata of the earth crust, 2 - in the lower part of the sedimentary strata and deeper parts of the earth crust, 3 - in the consolidated part of the earth crust;
4-7 - kinematic type of faults: 4 - thrusts, 5 - reverse faults, 6 - normal faults, 7-shears;
8 - earthquake classification by the magnitude: a) 6.4 ≥ M ≥ 5.5; b) 5.4 ≥ M ≥ 4.5; c) 4.4 = M = 3.5;
9 - earthquake classification by the focus depths, km: a) non-determined, b) 0-10, c) 11-20, d) 21-30.
Faults: 1 - East-Sakhalin, 2 - Hokkaido-Sakhalin, 3 - West-Baikalskiy, 4 - Aukun-Lunskiy, 5 - Tymskiy, 6 - Gyrgylanjinskiy, 7 - South Baikalskiy, 8 - Verkhne-Piltunskiy, 9 - West-Shmidtovskiy, 10 - West-Odoptinskiy.

Proceedings of the Seventh (1997) International Offshore and Polar Engineering Conference
Honolulu, USA, May 25-30, 1997

ISBN 1-880653-28-1 (Set); ISBN 1-880653-29-X (Vol. I)

Aspects of the Environmental Safety in the Processes of the Development of Oil/Gas Fields Offshore Sakhalin

T.V. Gnezdova
Sakhalin Oil & Gas Institute

A.N. Kravchenko
JSC "Rosneft-Sakhalinmorneftegas

Okha, Russia

ABSTRACT

The present stage of shelf development is characterized by the increase of involvement of its resources in economic use, by expansion of areas of oil and gas production and offshore geological exploration, by strengthening of impact of industrial activity onto the environment. The severe climate conditions and some other factors increase the probability of pollution of offshore environment and distributions of polluting substances to the regions, favorable for reproduction and a fishing of valuable biological objects. From the point of view of ecological safety by development of offshore oil and gas fields it is possible to mark out the following aspects:

- Possibility of coexistence of fishing and oil and gas industries;
- The ability of offshore hydrotechnical structured and facilities to resist the severe weather conditions;
- Technical and financial possibilities of the Operator to ensure the effective implementation of works on elimination of accidental spills of oil and other polluting substances

The establishment of complex operative information of control during the solution of problems of observation over a state of environment in areas of development of Sakhalin offshore fields will enable to create a new system of ecological monitoring, which would allow:

- to constantly trace the data on a change of quality of environmental components;
- to create the ecological data bank within a short period of time;
- to make various kinds of forecasts with a high degree of reliability;
- quickly to respond in case of appearance of negative changes of environment or abnormal parameters of nature impacts.

INTRODUCTION

The recent stage in the offshore development is characterized by large scales of employment of resources, expansion of petroleum (oil/gas) production and offshore exploration works, by the increase of the industrial influence on the environment. This situation is of great importance for offshore Sakhalin. Severe climatic conditions and a lot of other nature features increase the probability of contamination of sea water and distribution of contaminant substances within regions, which are considered to be favourable for the reproduction and development of biological objects.

ASPECTS OF THE ENVIRONMENTAL SAFETY

The following aspects are distinguished according to environment protection in the process of oil/gas fields development:

- the probability of the coexistence of fishing and petroleum (oil/gas) industry;
- ability of marine hydrotechnical constructions to hold out against the climatic conditions;
- technical and financial abilities of the company "Operator" to provide the effective works on liquidation of oil spills and other pollutant substances.

The marine waters of the norst-eastern Sakhalin Island are considered to be the favourable spawning area for industrial types fish ("Ecology and trofic relationships of industrial fishes of the Far Eastern Seas", 1994). The total volume of fish catch is 200-250 thousand tonnes per year. The salmon type of industrial fish is the most important and their spawning is usually during June-November and this period coincides with exploration works offshore Sakhalin. Near the shores of the north-eastern Sakhalin there is one of 5 large spawning places of mintae fish of the norther part of the Sea of Okhotsk. Here one might observe places of industrial gastropodsa (Gastropoda) mollusk - "trumpeters" with the biomass of 63 thousand tonnes ("Discharges of drilling muds and Environment protection", 1994).

Among the mammels in the stidied region there is a lot seals, about 100-200 thousand specimen, but rarely one might meet the Sea of Okhotsk whales, the banded family: blue whale, grey whale and their herds are restored.

The marine mammals are very significat due to their high population in the sea ecosystem as one of the main consumer at the highest trofic level (TINRO - 70, 1995).

The petroleum production experience abroad gives positive results of influence on the biological sea resources ("The compehensive investigations of marine hydrobiontos and their living conditions", 1994 and "Discharges of drilling muds and Environment protection", 1994). For example, in Mexico Bay the main fish catch is carried out near the artificial offshore constructions, that is explained by the formation of the new living medium, occupied by the other marine organisms ("Discharges of drilling muds and Environment protection", 1994). The analysis of the influences on the elements of the nature environment has shown a low interaction of marine oil/gas constructions with the processes of the exploitation of the sea biological resources and with the conditional changes in the nature environment.

Bioproduction of marine echosystems and ecology are under low influences of the engineering methods of marine petroleum (oil/gas) constructions. Taking into consideration the variety of natural conditions and specific features in each individual region it is difficult to confirm the existance of such positive effect in the process of exploitation of fields offshore Sakhalin. At that time in the Russian practice there is a lack of investigations on the interaction of exploration works and biological resources under conditions of the northers seas, which might prove the negative moments. Only the exploration experience on such fields, including Sakhalin fields, may give real results.

The level of the influence of petroleum objects on the ecology of the offshore zone is determined by the accepted constructive decisions and projected methods of construction engineering.

The project standards and demands ensure the minimization of the probable negative influence (Areshev E.G., Kuznetsov A.M. and others, 1984).

The negative influences on quality changes in the nature environment are not expected due to the planning and development of oil/gas fields offshore Sakhalin on the level of the recent scientific and technical achievements. The fulfilment of the Russian demands on the environment protection, enlisting of companies - leaders of the world oil/gas production and employment of the best world technologies in exploration and production processes are considered to be the guarantee of the environment protection.

The level of probable quality alterations in different natural components under the normal exploitation regime of the marine oil/gas producing complex, estimated by the Technical Economic Engineering "Sakhalin-2" and Technical Economic Estimations "Sakhalin-1" is turned to be lower than the limiting permissible level according to Russian standards and documents.

The evaluation of effects of all probable sources of blowouts on the quality of the atmospheric air was made according to the program "Ecologyst" (company "Integral", S.-Petersburg), allowed to determine nearsurface concentrations of pollutant substances. It was suggested, that the estimated concentrations of pollutant substances reflected the level of the probable environment pollution effect on water areas and onshore part.

The estimation was made for the purpose of combining of the maximum probable quantity of blowouts and the most non-favourable meteorological conditions under the normal exploitation regime of the oil/gas producing complex. The astimated analysis has shown that for platforms of floating rigs and marine ice-fixed drilling rigs, the maximum limiting concentrations won't exceed the standards within the working zone. The full dispersion of the pollutant substances is at the distance of 1,5-2,0 km from the blowout sources and it does not reach the shoreline.

The area of the influence of pollutant substances on the atmospheric air will be small and a will depend upon the weather conditions, such as speed and trend of air streams. The discovered oil offshore Sakhakin contains practically no sulphur, so the mot toxin components such as sulphur dioxide and hydrocarbon sulphide are absent.

In estimations of the influence on water because of the industrial fish catch the legislated lemiting permissible concentration of pollutant substances was considered be the criterion. The dispersion of pollutant substances with the concentration not in the excess of the criterion is considered to be the quarantee of the environment protection and the corresponding degree of the safety for marine organism or community. The estimations have shown that the real discharge of industrial drilling flowing waters is less than the limiting permissible discharges.

The technological and environmental protection decisions, the control for their fulfilment should minimize the probable negative changes in the quality of components of the environment setting it permits to distinguist them as the small, local and short-term ones.

It should be noted that under the normal exploration and exploitation regime in water areas, the departure from the norm is easily liquidated by stopping works and repairing the equipment. The monitoring of the echological setting is considered to be main task in order to prevent catastrophic consequences.

The second aspect of the environmental safety in the process of the offshore drilling is the reliability of drilling constructions. The reliability of technological equipments and constructions is determined by the possibility to resist the climatic conditions (Astafiev V.N., Kuznetsov A.M., 1984). The high level of the echological risk in the process of field development offshore Sakhalin is caused by the seismicity, active neotectonics, tsunami, ice, wind and wave loads, bottom gas shows, shoreline migration, lithodynamic conditions.

By the materials of the seismic zonation the North-East Sakhalin is located within MSK-64 8-9 number zone. In 1995 the strongest earthquake of 9.5 (epicenter in the sea) was registered near Neftegorsk settlement, having caused mass destruction living houses with a lot of victims, damage of pipelines and bridges on the north, forming of fractures, slumps, changes in the configuration of the coastal zone. The earthquake of M 7.5 in the Sea of Okhotsk may cause the tsunami. According to prelimenaty estimations, most of offshore fields are located within a zone with the extreme parameters (Kulikov N.V., Mansurov M.N., 1992). The tsunami may be expected in bays of the north-eastern coasts as a wave jump across the spits and refraction waves while tsunami penetration across the strait. The average duration of the ice period on the North Sakhalin is 200 days and decreases up to 160-180 days to the central part of the island.

The drift ice and grounded hummocks are dangerous for submarine pipe-lines and basements of hydrotechnical constructions (Kulikov N.V., Mansurov M.N., 1991).

Tidal currents and winds are considered to be the main factor causing the ice drifting in the Sea of Okhotsk. The ice from the north-eastern, northern and north-western points is drifting along the western coast with the average speed of 6-7 miles per day. The velocity of the drifting ice in the coastal sea part is of 25 miles per day, and at the eastern shore is of 108 miles per day (Truskov P.A., 1992).

The hummocking under the influence of tidal phenomenon is typical for regions of the Sakhalinskiy Bay and north-east Sakhalin. The maximum hummocking is observed under the influence of currents, wind and tidal currents. Such conjunction of factors is characteristic of the north-eastern offshore Sakhalin. Zones with the hummocking of 3 points predominate here in January. The height of hummocks in the open sea is not in excess of 1 m and in areas of drift ice edges and at the fast-ice edge it may reach 1.5-3.0 m. At depths up to 20 m grounded hummocks are observed.

The north-eastern Sakhalin is characterized by the shoreline migration and movements of sea-bottom deposits near the coast. The intensive deformations cause the washout and lost the resistance of the pipe-line. The investigations have shown that the average migrative velocity is in range of the first meters per year. Lithodynamic deformations of the floor on the Sakhalin north-eastern coast are characterized by the maximum value of the rupture up to 1-2 m and are usually up to the isobath of 7-10 m (Truskov P.A., 1987).

The active shows of the gaseous hydrocarbons, relief areas of gas-saturated underground waters (submarine sources) and associated gas-water gryphons, hills and cones have been identified on some fields of the north-eastern Sakhalin. The existance of migration zones is due to the decompaction of the sedimentary rocks in domes of structures.

The second zone of gas shows is associated mainly with paleovalleys. These features are also dangerous for marine constructions and are investigated. Due to this variety of nature phenomenon of the echological risk for exploration works offshore Sakhalin their estimation is desired from the point of view of the maximum influence on the resistance of constructions and their exploitation safety. The exploration experience in the Norther Seas has shown the significant influences of earthquakes, ice conditions, washouts and accretion of marine deposits. While design exploration and exploitation works all these loads are included in the main estimations. So a very high coefficient of the reserve strength of constructions, probabilities of the exception of destructive effects on the oil/gas producing objects are considered (Areshev E.G., Kuznetsov A.M., 1984). One of the major aspects of the environment safety is the long-term monitoring for the state of constructions, measuring of parameters of external factors and the degree of their influence on marine constructions.

CONCLUSIONS

Unfortunately, the technical level can't exclude the possibility of the emergency, but only the probability and the scale of the influence are minimized. The oil spills cause the very negative effect and one of the requirements specifying to companies is the elimenation of oil spills in a short time. So companies should have the corresponding experience, equipment, monitoring and finances. The elimination of oil spills will be delayed due to the absence of one of these factors for so much time; and scales and consequences will be under the influence of these factors.

The enviroument safety in the process of the development offshore Sakhalin is connected with the monitoring of the environment setting.

The joint-stock-company "Sakhalinmorneftegas" will develop new offshore fields including projects "Sakhalin-1" and "Sakhalin-2" which came into force in 1996, so the aspect of the monitoring is of great importance.

The decision on formation of the special monitoring service in the process of exploration and exploitation of offshore fields was accepted. For this purpose "Sakhalinmorneftegas" has enough funds, technical and laboratory facilities, high proficiency specialists and scientists and the institute "SakhalinNIPImorneft" and fleet.

The main task of the echological monitoring is the constant control during the year and 24-hour control in exploration areas of the joint-stock-company "Sakhalinmorneftegas".

The observations of changes in the environment setting under exploitation of drilling rigs and platforms including drilling mud discharge should be conducted by independent means due to economical efficiency and advantage. The botton self-sufficient stations may be used as the technical means for measuring and record of parameters of the sea water.

The period of data gathering may be from several months up to half a year. The information will be transmitted by the hydroacoustic channel on the ship board or by the buoy with satellites to the data gathering and processing centers.

The self-sufficient stations will permit to recieve the full-scale quality data and to make monitoring cheapen.

Two problems should be solved as the initial stage of the observation system and control of the environment setting. The first one is the choice of the most indicative hydrodynamic parameters, which determine the living existence of the biological resources. For this purpose the correlation of changes in the hydrochemical and biological components by nature and laboratory investigations is reguired. The second one is the design of new informative systems and mathemetical models.

These problems should be solved by the corporation of the Joint Stock Company "SMNG" with the special Design Bureau of the Automation of the Marine Research of the Far Eastern Department of the Academy Sciences.

This comprehensive operative informative maintenance of observation for the environment setting within drilling areas offshore Sakhalin will lead to building of a new system of the echologic monitoring allowing:

1) to record data on guality changes in components of nature setting;

2) to develop the echologic data bank in a short period;

3) to make forecasts with a high reliability level;

4) to eliminate negative changes in the environment setting or abnormal parameters of the nature influences.

REFERENCES

Berljand, ME (1985). "Forecast and monitoring of the atmosphere pollution," Leningrad: Cidrometeoizdat (in Russian).

Guseinov, TI, Alekperov, PA (1989). "Environment protection under offshore oil/gas fields development," - Moscow, Nedra (in Russian).

"The compehensive investigations of marine hydrobiontos and their living conditions," Vladivostok, TINRO. (1994) (in Russian).

Nelson-Smit, A (1977). "Oil and Sea Ecology,"Moscow, Progress (in Russian).

Panov, GE, Petrjashin, LF, Lysjany, GN (1986). "Environment Protection on Enterprises of the oil and gas industry," Moscow, Nedra (in Russian).

"Discharges of drilling muds and Environment protection". *Reports to the international congress Mobil and SakhTINRO.* Yuzhno-Sakhalinsk, (1994) (in Russian).

TINRO - 70. - Vladivostok, 1995. (in Russian).

Tronov V.P. (1983). "Field oil treatment abroad," Moscow, Nedra (in Russian)..

"Ecology and trofic relationships of industrial fishes of the Far Eastern Seas," TINRO, Vol. 116. Vladivostok, (1994) (in Russian).

Proceedings of the Seventh (1997) International Offshore and Polar Engineering Conference
Honolulu, USA, May 25-30, 1997

ISBN 1-880653-28-1 (Set); ISBN 1-880653-29-X (Vol. I)

The Role of Oil and Gas Projects of Sakhalin Shelf in Supplement of Far-Eastern Region of Russia with Fuel-Energy Resources

L.S. Brodski
Sakhalinmorneftegas-Shelf Co. Ltd.
I.Ph. Vasyanovich
Sakhalinnipimorneft Institute
E.K. Terpugov
JSC "Rosneft-Sakhalinmorneftegas

Okha, Russia

ABSTRACT

The fuel-energy complex (FEC) is a basis of economical development of Far-Eastern areas. The specific weight of FEC branches in industrial production of some territories of Far-East is near 50%. The Far-Eastern economical region during continuous time (1986-1995) need additional fuel resources. The development of coal industry and atomic energetic is stressed on solution of fuel supplement problem.

KEYWORDS: Gas, fuel, Far-East, Sakhalin, project, energy.

INTRODUCTION

The coal exploration on Far-East have increased up to 1988 when the peak mining (57 mln.t) was achieved. In spite of constant increasement of coal mining the fuel deficiency was constantly observed and coal deliver increased continuously. If up to 1985 the coal deliver was not over 9 mln.t than in 1985 it increased 3 times. At period 1988-1990 the coal mining decreased on 7.2 mln.t and delivery was 7.5 mln.t.

At period 1976-1986 a problem of atomic power plants in Primorje and Khabarovsk Regions was actively considered. But after Chernobyl catastrophethe the project development was finished.

At the same time the coal specific weight in using fuel in Far-East at period 1990-1995 have increased from 55% up to 59% and of black mineral oil have decreased from 25% to 21%. Owing to large part of coal and black mineral oil the concentration of dust substances exceed the several times the permissible norms from fuel burning in all large cities.

At present time the consumption of boiler-stove fuel in Far-Eastern area (without Amour Region and Republic of Sakha) is 27 mln.t of reference fuel (1994) and at period 1990-1994 have decreased to 17% and stove black mineral oil - 34%, natural gas - 12%.

CONSUMPTION STRUCTURE OF FAR-EAST

The most great consumers of fuel are Khabarovsk and Primorje Regions. The consumer part of stove-boiler black mineral oil decreased and part of Sakhalin Region is 14%, of other regions (Kamchatka and Magadan Regions, Republic of Sakha) - only 18%. Table 1 show the consumption structure of main kinds of stove-boiler oil (coal, gas, black mineral oil, other kinds) corresponding to regions.

The Far-East and, first of all, Khabarovsk and Primorje Regions constantly have a fuel resource deficiency. At period 1976-1986 the problem of atomic power plant construction was discussed. The development of atomic energetic is supposed on the basis of new reactors which are more reliable but more expensive and naturally move the available construction time of atomic plant on 15-20 years. The adopted on Government level decisions concerning FEC on Far-East was not done in practice.

The objects of FEC (mining plants, mining areas, power plants, oil treating plants) have high physical and moral wear and tear and not response to modern requirements of ecology and efficiency of fuel burning. One of the main reasons of such situation in FEC is insufficient investments on reconstruction and new construction of energetic objects and also the non-rational structure of stove-boiler fuel balance. So. the average in Russia gas part in consumption structure of stove-boiler fuel is 54% in 1990 and on Far-East is not over 7%.

The own coal resources provide for not over 15% of total need of Khabarovsk Region with this fuel kind and other amount (8-9 mln.t) is delivered from other regions of Russia Federation, mainly, from Zabaikalje and Republic of Sakha (37-40%) and also from Eastern Siberia by railway.

It can be noted that the black mineral oil and other oil products, which is delivered from Khabarovsk oil treating plant, even accounting for transport expenses have more high cost in comparison with plants of Eastern Siberia.

The main part of stove-boiler fuel is used on thermal power stations and boilers (71-71%). The municipal consumer sector of city and village use 10-12% and industry 6-9%. Table 2 show the difference of consumption structure of stove-boiler fuel Regions.

The regional power plants today also use the non-designed coal kinds, as a rule, which increase the equipment wear and tear and greatly decrease the technical and economical indexes of it work. The great increasement of prices on coal and other kinds of stove-boiler fuel in 1995 (3,000-5,000 times in comparison with 1991) and also the more greater increasement of transport tariff (12,000-15,000 times) cause the decreasement of solvent possibilities of consumers, which made more hard the problems of fuel and energy supplement.

SOLUTION OF PROBLEMS

The following measurements are need:

- overcome of crisis events and stabilisation of economics up to period 2000 with following grow of industrial production and strengthening of speliazation branches;
- expand of natural-row basis for economic development and possibility of it structural build, supplement of conversion profile;
- more effective use of capital investments with supplement of state guaranty for investor;
- development of social and industrial infrastructure, increasement of living level;
- improvement of export basis and export potential;
- improvement of ecological situation by reduce of dust into atmosphere and water;
- overcome of payment crisis, state regulation of prices on production and service influencing essentially on economic state and social sphere;

The potential consumers of hydrocarbons of Sakhalin shelf on internal market are Sakhalin, Khabarovsk and Primorje Regions. The following factors are considered under definition of rational gas volumes to consumers of these Regions:

- occurrence or absence of own fuel resources, it cost, expenses of delivery;
- level and concentration of thermal loads on Regions planned for gas supplement;
- fuel economy as a result of replacement by gas of expensive black mineral oil and delivering coal;
- increasement of reliability of consumer fuel supplement;
- improvement of air basin of large industrial centres - main consumers of fuel.

The design gas need of Sakhalin, Khabarovsk and Primorje Regions even today is 12-13 mlrd. m^3 per year; to 2010 it will be probably 15-16 mlrd. m^3 per year, and to 2020 - 19-20 mlrd. m^3 per year. The design gas need of Sakhalin Region (without gas and chemical complex of western coast) is evaluated on level near 4 mlrd. m^3 in 2010 and 4.5 mlrd. m^3 in 2020, Khabarovsk - 6.0-6.5 mlrd. m^3 and 7.5-8.0 mlrd. m^3 and Primorje Region - 5.0-5.5 mlrd. m^3 and 7.0-7.5 mlrd. m^3 correspondingly in 2010 and 2020.

The gas capacity of Primorje and Khabarovsk consumers is defined on the base of gas extraction volumes on land and shelf of Sakhalin island including requirements of Sakhalin Region, export for credit payment and supplement of investor profit.

The following way of gas supplement of region is suggested:

- gas delivery to consumers of central and southern areas of Sakhalin Region;
- gas supplement of consumers of Komsomolsk industrial centre of Khabarovsk Region;
- gas delivery to consumers of Khabarovsk industrial centre;
- gas delivery to northern industrial zone of Primorje Region (city Dalnerechensk territory;
- gas supplement of other areas of Primorje Region near gas pipeline Khabarovsk-Vladivostok.

Gas delivery on south of Sakhalin Region and volume increasement of gas to consumers of Komsomolsk industrial centre can be done from first exploration year of fields according to projects "Sakhalin-1" and "Sakhalin-2". The gasification of Khabarovsk industrial centre will be possible after 5-6 years of extraction beginning on island shelf.

Only after fulfilment at nearest 15-20 years of these projects the gas delivery to Primorje Region is possible only with extraction beginning for purchase on Arkutun-Dagin oilfield or after 13-14 years from work beginning.

The gas requirements of Sakhalin, Khabarovsk and Primorje Regions according to projects "Sakhalin-1" and "Sakhalin-2" can be supplied approximately on 70% in 2010, and 75% in 2020. The full requirements will be fulfilled after exploration of other oilfields of shelf.

Table 3 show the forecast structure of fuel balance for Sakhalin, Khabarovsk and Primorje Regions up to 2000 accounting for shelf exploration (projects "Sakhalin-1", "Sakhalin-2", "Sakhalin-3" and others).

CONCLUSIONS

The fulfilment in time of projects "Sakhalin-1" and "Sakhalin-2", and then of project "Sakhalin-3" and "Sakhalin-4" allow to create the conditions for reliable fuel supplement of Sakhalin, Khabarovsk and Primorje Regions. Also on first stage of gasification it is supposed to supply by gas over 70% of total requirement of Sakhalin Region. High level of gasification is explained by route of gas pipeline along railway and roads with high density of population and will achieve 86%.

There is no need to coal delivery from other regions according to analysis also for reduction of the coal extraction to 1.5 mln.t per year in Region. The black mineral oil requirement will reduce over 100,000 t per year after own production on small block devices of oil treatment with unit capacity for initial raw material 250,000-300,000 t per year.

The gas delivery to Khabarovsk Region is designed for consumers of Khabarovsk and Komsomolsk industrial centres, where over 70% of regional industrial potential is concentrated. In future the own fuel resources of Region (coal, black mineral oil, other fuel kinds) are evaluated on level 5-6 mln.t of reference fuel when the need is 13-14 mln.t (38-42%). Using of Sakhalin gas in Khabarovsk Region in design volumes allow to decrease the need in stove black mineral oil on 1.4-1.5 mln.t in 2010 and on 2.0-2.1 mln.t in 2020, and coal delivery from other regions of Russia on 7 and 8 mln.t correspondingly. The Sakhalin gas can supply 55% of total need of Region in 2010 and near 67% in 2020.

The fulfilment of gasification program in Primorje Region allow to decrease the need in delivering black mineral oil on 1 mln.t in 2010 and on 1.6-1.7 in 2020 and provide for Region with fuel in practice even during decreasement of coals in Primorje on 35-40% from level of 1995.

Table 1

Consumption structure of main kinds of stove-boiler fuel in Far-Eastern area in percents

Fuel kind	Far-East*		Khabarovsk Region		Primorje Region		Sakhalin Region		Kamchatka and Magadan Regions, Republic of Sakha	
	1990	1995	1990	1995	1990	1995	1990	1995	1990	1995
Coal	55.3	58.7	48.7	55.0	58.1	61.8	61.7	61.5	58.7	58.4
Gas	6.0	6.6	8.9	9.7	-	-	20.8	22.6	-	-
Black mineral oil	24.8	21.2	29.0	23.0	31.4	26.6	5.1	5.4	17.1	18.6
Other	13.9	13.5	13.4	12.3	10.5	11.6	12.4	10.5	24.2	23.0
TOTAL	100.0	100.0	100.0	100.0	100.0	100.0	100.0	100.0	100.0	100.0

* Without Amour Region and Republic of Sakha

Table 2

Consumption structure of stove-boiler fuel for different use in Far-Eastern area in percents

Consumers	Far-East*		Khabarovsk Region		Primorje Region		Sakhalin Region		Kamchatka and Magadan Regions, Republic of Sakha	
	1990	1995	1990	1995	1990	1995	1990	1995	1990	1995
Power plants and boilers	71.1	73.0	68.3	71.4	67.7	69.2	74.3	76.6	82.1	82.5
Municipal	10.1	12.1	9.3	11.0	10.4	12.2	12.6	13.7	8.0	8.4
Industry (for techn.aims)	9.0	5.9	11.6	7.1	10.3	7.2	3.7	2.7	4.4	3.2
Other	9.8	9.0	10.8	10.5	11.7	14.0	9.4	7.0	5.5	5.9
TOTAL	100.0	100.0	100.0	100.0	100.0	100.0	100.0	100.0	100.0	100.0

Table 3

Forecast consumption structure in mln.t of reference fuel

	2000	2010	2020
Far-East*			
Coal	58,8	33,6	26,0
Gas	7,0	46,2	58,0
Black mineral oil	21,4	10,2	7,4
Other	12,8	10.0	8,6
Total	100,0	100,0	100,0
Khabarovsk Region			
Coal	54,0	24,4	18,2
Gas	9,7	55,3	66,5
Black mineral oil	24,6	12,5	8,8
Other	11.7	7.8	6.5
Total	100,0	100,0	100,0
Primorje Region			
Coal	62,8	36,8	26,2
Gas	-	40,4	57,2
Black mineral oil	25,8	12,2	7,4
Other	11.4	10.6	9.2
Total	100,0	100,0	100,0
Sakhalin Region			
Coal	61,3	22,5	10,4
Gas	24,2	72,0	85,8
Black mineral oil	5,5	2,3	1,8
Other	9,0	3,2	2,0
Total	100,0	100,0	100,0

* Gasification of southern areas of Kamchatka Region is accounted after 2000 with use of gas of Kshuk and Nizhne-Kvakchik gasfields in volume 500 mln. m^3 per year

Proceedings of the Seventh (1997) International Offshore and Polar Engineering Conference
Honolulu, USA, May 25-30, 1997

ISBN 1-880653-28-1 (Set); ISBN 1-880653-29-X (Vol. I)

Oil and Gas Exploration on the North Sakhalin and Its Shelf

V.M. Zakalsky
JSC "Rosneft-Sakhalinmorneftegas"

V.S. Kovalchuk, A.V. Kuchin and Yu. S. Mavrinsky
Sakhalin Oil & Gas Institute

Okha, Russia

ABSTRACT

On Northern Sakhalin a large volume of gravimetric, magnitometric, electroprospecting and seismic works is carried out. Seismic surveys have revealed 195 anticlinal structures, on 138 of them exploration drilling has been carried out. 53 oil and gas deposits connected with terrigenous sediments of Miocene age were discovered.

Oil, gas and condensate deposits are discovered at depths up to 5000 m. There are totally 2150 wildcats and exploration wells drilled on Northern Sakhalin.

Offshore of Northern Sakhalin seismic surveys have revealed 39 anticline folds, the exploration drilling (51 wells totally) was carried out on 9 structures. 7 oil, gas and condensate deposits, also dated to terrigenous rocks of Miocene age, were discovered. Offshore of Northern Sakhalin there are discovered the larger deposits, than those onshore. The total explored reserves of hydrocarbons for 7 offshore deposits about 4 times exceed the initial reserves of all deposits onshore.

There exist the objective geological data for the growth of exploration works and discovery of new oil and gas fields, including the larger ones.

OIL AND GAS EXPLORATION

The North Sakhalin and adjacent shelf are considered to be the single region on the Far East with high potential resources, commercial hydrocarbon resources, oil and gas/condensate production. A large volume of geologic-geophysical works and deep drilling was conducted during the 70-s years. Geological surveys have been carried out on the North Sakhalin territory. The structural drilling was carried out on some anticlinal structures. Since 1948 the North Sakhalin territory was under the survey by geophysical methods: seismic exploration, gravimetric and magnitometric survey and electric prospection (electroprofiling). The seismic prospection (reflection method) has played the most important role in discovery and preparation of anticlinal structures for drilling works. The total volume of seismic prospecting (regional, wildcating, detailed) during 50 years was in excess of 50 thousand km. The level of investigation on the North Sakhalin territory is 1.6 p.km/km^2. The largest net of seismic profiles covered the North Eastern Sakhalin (up to 3 p.km/km^2). Onshore island 195 structures have been revealed by seismic survey (reflection method), and 138 of them have been under exploration drilling.

Geophysical investigations with the purpose of the regional study, search and preparation of anticlinal structures have begun since 1954. A large volume of geophysical works (seismic survey: reflection method, CDP-method 24-multiple and 48-multiple, gravimetric, magnitometric and geochemical prospecting). The prospecting and detailed survey were the main aspects in the discovery and preparation of structures for drilling works. The prospecting seismic surveys (reflection method, CDP-method) were conducted by the net of profiles (4x6 km) with the density of 0.4-0.5 p.km/km^2. 18 structures were revealed offshore North Sakhalin by seismic exploration. Detailed seismic works were carried out on anticlinal structures by the net 1-1.5x2 km with the density of 1.5-2 p.km/km^2 by multiple profiling (24-48).

In general offshore North East Sakhalin 39 anticlinal structures were discovered, 22 structures have been prepared for drilling works, 10 structures were under drilling and into 7 anticlinal folds oil and gas fields were discovered.

The total volume of seismic profiles within Sakhalin oil/gas bearing region is about 213 thousand p. km, and more than 1/2 of them is offshore North Sakhalin.

On perspective areas of the North Sakhalin 2151 wells were drilled during 70s years (stratigraphic, wildcat and exploration) with the total footage of 3955410 - thousand p. meters. The level of investigation of the North Sakhalin territory is about 164.8 p.m/km^2 or 13 km^2 per for one well. The very high level of investigation by deep drilling is noted within north-eastern part (354 p.m/km^2 and 2.6 km^2/well). The Paleogene and upper Cretaceous deposits are poor investigated, they have been encountered only by single wells. On 138 objects, prepared for drilling works (anticlinal structures) 53 oil and gas fields were discovered.

The first offshore well was drilled in 1977 on Odoptinskaja meganticline, and the first offshore Sakhalin oil/gas/condensate field was discovered in the same year. The next drilling period (1977-1992) offshore North Sakhalin was marked by drilling of 51 wells with the total footage of 124667 m (about 71% of the total drilling volume offshore Far Eastern Seas). Due to categories 20 wildcat wells (volume of 52486 m), 29 exploration (volume of 67311 m) and 2 appraisal wells (volume of 4870 m) were drilled.

Depths of wildcat wells were in range of 1811 to 3482 m, with average depth of 2614 m. Since 1977-1992 offshore North-Eastern Sakhalin 9 perspective structures were under wildcat drilling, 7 fields were discovered and two structures (Khanguzinskaja and Borisovskaja) were non-producing.

Among 20 drilled wildcat wells in 16 wells (80%) hydrocarbon pools have been identified. After testing of the first wildcat well 6 offshore fields were discovered (Odoptinskoe, Chaivinskoe, Piltun-Astokhskoe, Lunskoe, Kirinskoe, Arkutun-Daghinskoe).

Ultimate potential oil resources (recoverable) on the North Sakhalin (E.G.Koblov) are estimated of 267 million tonnes, development of oil resources is 60%. Ultimate potential free gas resources are 325 billion m^3, development of gas resources is 37%.

Potential oil resources (recoverable) offshore North Sakhalin are 608 million tonnes, 2872 billion m^3 gas. Development of oil resources and free gas is 44.5% and 28.5%, respectively.

The distribution analysis of commercial and potential hydrocarbon resources on the North Sakhalin has shown, that a high share of reserves and resources is associated with depths up to 5 km with deposits of Nizhenutovskiy-Okobykaiskiy and Daghinskiy horizons in the north-eastern part of the island. Large thicknesses of Neogene deposits (up to 6-7 km), composed of marine clayey-sandy rocks are characteristic of the commercial region. Deposits are dislocated into anticlinal folds, grouped into anticlinal zones of the meridional and submeridional strike. Three main groups of oil and gas traps are identified: anticlinal, disjunctive and lithological.

Anticlinal traps are controlled by the main oil and gas fields on the Sakhalin Island. According to sizes of the area all anticlinal folds are subdivided into three groups: small (up to 20 km^2) - 52%, average (20-50 km^2) - 31%, large (in excess of 50 km^2) - 17%. A lot of anticlinal folds offshore North Sakhalin have areas within limits of 50-150 km^2. By hight the anticlinal folds are subdivided into low-amplitude (up to 50 m) - 37%, average-amplitude (up to 100 m) - 28% and high-amplitude (more than 100 m) - 35%. Many anticlinal traps are faulted (with amplitudes in range of some meters up to hundred meters), separating producing beds into a set of oil and gas pools.

Disjunctive oil and gas traps within Sakhalin Island are developed in zones of regional and zonal faults and subdivided into two groups: underthrust and traps of monoclinal blocks. Underthrust traps are confined to large upthrow-thrusts of the submeridional strike with a complicated and durable formation history. The underthrust traps are associated with oil pools on East-Ekhabinskoe, Mukhtinskoe and Paromaiskoe fields.

Traps of the monoclinal blocks, sealed by large growth faults (Monginskiy type), have been formed in the middle-upper Miocene. On the background of the total extension the subsidences developed along regional upthrows of narrow grabens with the submeridional strike, composed of the middle Miocene deep water clays. The extension of the growth fault of the monginskiy type reaches 130 km, amplitude up to 1500 m.

Oil and gas traps are composed of faults with gemianticlines and monoclinal blocks, subdivided by a set of growth faults. Traps of the monginskiy type control the oil and gas pools on Monginskoe, Nizhne-Daghinskoe, R.S.Mirzoev, Ust-Evai fields.

Lithological traps are formed in pinching out zones or faces replacement of separate reservoir-beds. Within the North Sakhalin oil and gas pools are characterized by a small distribution and are opened on flanks or periclinal ends on the Odoptinskoe and Kolendinskoe fields.

The offshore North Sakhalin, according to the development of potential hydrocarbon resources is characterized by high discovery possibilities of large fields. Commercial oil reserves (recoverable) of the C_1 - category are discovered in 6 offshore fields (Odoptinskoe, Chaivinskoe, Arkutun-Daghinskoe, Piltun-Astokhskoe, Lunskoe and Kirinskoe) and are estimated be about 130 million tonnes, 587 billion m^3 gas. Aijashskaja, South-Lunskaja, South-Kirinskaja, Mynginskaja and Astrakhanovskaja folds belong to perspective anticlinal structures. Hydrocarbon pools are predicted at depths up to 3-4 km into deposits of Nizhenutovskiy, Okobykaiskiy, Daghinskiy and Daekhuriinskiy horizons.

During the development period, 1928-1995 (including) about 100 million tonnes oil and 32 billion m^3 free gas are discovered on the North Sakhalin. The hydrocarbon production from fields offshore Sakhalin was not carried out. We have the objective assumptions for exploration and discovery of new oil and gas fields taking into consideration the resources basis of the North Sakhalin and its shelf.

The wildcat drilling on perspective structures (Vostochno-Ossoiskaja, Vostochno-Pribrezhnaja, Astrakhanovskaja (sea) structures) and inclined drilling of the wildcat well № 2 on the northern dome of the structure of the oil/gas/condensate field Odoptu-sea are suggested in 1996-1998 (Fig. 1).

The discovery of new fields will permit to stabilize the onshore oil/gas production and shipment to the mainland.

The development of offshore fields (Piltun-Astokhskoe, Lunskoe) will be conducted by the concerssium "Sakhalin Energy" the project "Sakhalin 2".

Fields (Odoptinskoe, Chaivinskoe, Arkutun-Daghinskoe) after the additional survey and technico-economic estimation will be developed by the Russian side and concerssium (company "Exxon" and "Sodeko") by the program "Sakhalin-1". The exploration on Kirinskiy sector (Yuzhno-Kirinskaja, Mynginskaja structures), sine 1996 will be carried out by foreign companies "Mobil" and "Texaco" by the project "Sakhalin-3". The participation of foreign investors in the resources development offshore North Sakhalin will permit to produce up to 30 million tonnes oil and to 28 billion m^3 free gas per year in the nearest decade.

REFERENCES

Varnavskiy, VG, Koblov, EG, Butsenko, PL et al (1990). "Lithologic - petrophysical criteria of the petroleum content (oil/gas)," *Nauka*, Moscow, 270 p. (in Russian)

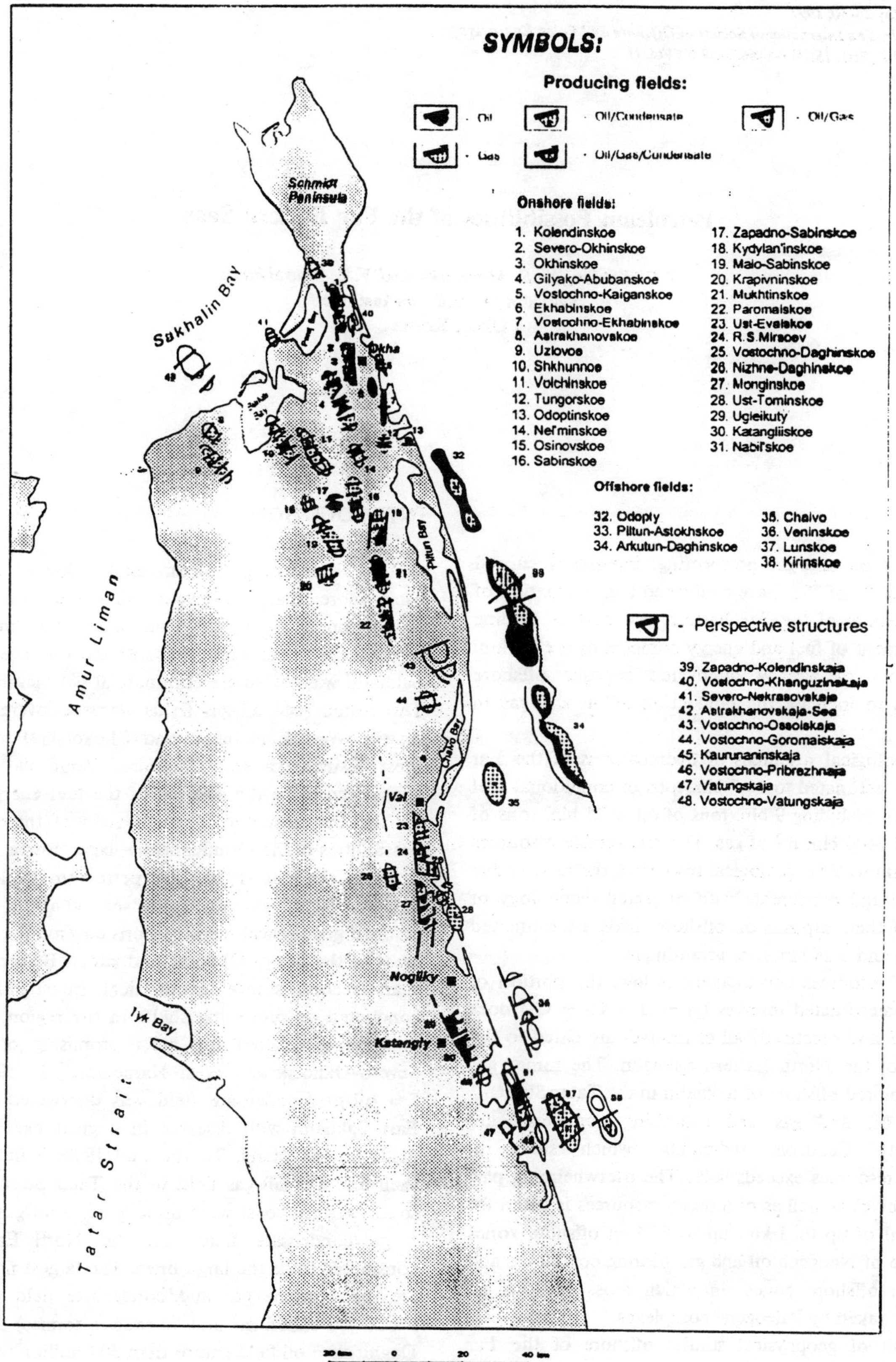

Fig. 1. North Sakhalin. General map.

Proceedings of the Seventh (1997) International Offshore and Polar Engineering Conference
Honolulu, USA, May 25-30, 1997

ISBN 1-880653-28-1 (Set); ISBN 1-880653-29-X (Vol. I)

Petroleum Possibilities of the Far Eastern Seas

E.G. Koblov, Yu.S. Mavrinsky and V.V. Kharakhinov
Sakhalin Oil and Gas Institute
Okha, Russia

ABSTRACT

The results of oil and gas prospecting, intensively carried out since the middle of 70's, have confirmed high evaluation of oil and gas potential of the Far Eastern seas and its leading role in development of fuel and energy complex of the region. The development of already discovered deposits offshore Sakhalin allows to increase the production of oil and gas in 15-20 times.

The total geological resources of hydrocarbons of the Far Eastern seas are estimated to 16,7 bln. tons of conditional fuel (in terms of oil), including 9 bln. tons of oil, 0,35 bln. tons of condensate and 7400 bln. m^3 of gas. The recoverable resources of gas correspond to their geological resources, the recoverable resources of oil and condensate with projected technology of development of their deposits on offshore fields are estimated to 2,4 bln. tons and 0,24 bln. tons accordingly.

The level of resources development is low, the portion of explored and preestimated reserves ($A + B + C_1 + C_2$) does not exceed 13% and practically all of reserves are dated to the offshore zone of the North Eastern Sakhalin. The minor gas reserves are explored offshore of Sakhalin in the Tatar Strait.

The main oil and gas and promising complexes are connected with Cenozoic sediments, which share in distribution of resources exceeds 90%. The overwhelming part of explored reserves as well as of forecast resources is dated for a depth interval of up to 3 km: up to 87% in offshore zones with prevalence of Neogene oil and gas bearing complexes and up to 95% in offshore zones, in which cross sections an essential part is taken by Paleogene complexes.

In the result of geophysical studies offshore of the Far Eastern seas there are revealed about 60 anticlinal zones, which unite the local structures (more than 300), being promising for exploration of large oil and gas deposits.

INTRODUCTION

The history of petroleum exploration of the region goes back more than 100 years and the commercial oil/gas production is about 70 years. Up to the middle of 1970-s exploration works were concentrated onshore and the North Sakhalin was the single commercial petroleum bearing region. The commercial oil/gas fields were discovered only in 1980s on the West Kamchatka and Chukotski Peninsula. But the West Kamchatka and Chukotski Peninsula regions are not considered to be the sources of the fuel-energy raw materials for the Far East due to small sizes of these fields and low possibilities of the significant raw-base growth.

Since the mid 1970s the petroleum exploration offshore Sakhalin Island was very intensive; and at the end of 1970s the planned geophysical investigations covered the remaining water areas of the Sea of Okhotsk and later - Bering Sea. The recent years reconnaissance geophysical investigations have been conducted on promising shelfs in the region, but the seismic prospecting covered the most promising offshore Sakhalin, Severo-Okhotsk and West Kamchatka. In 1977 the Odoptu-Sea oil/gas/condensate field was discovered offshore North-East Sakhalin with reserves in a great excess of the largest fields on the island. To the mid 1990s 7 fields on the noted shelf and a small gas field in the Tatar Strait at the Sakhalin south-western coast were discovered. 5 oil/gas/condensate and 1 gas/condensate fields on the North-Eastern shelf are considered to be the large ones. The largest fields in the region are the Lunskoe gas and condensate field (balance reserves about 400 billion m^3 and 50 million tonnes) and the Arkutun-Daghinskoe oil field (more than 500 million tonnes).

These discoveries have proved high petroleum (oil/gas) potential estimations of the Far Eastern Seas and its role as the development basis of the fuel-energy complex of the region. The development of already discovered fields offshore Sakhalin permits to increase oil and gas production 15-20 times and to make requirements in fuel-energy resources available for the region.

PETROLEUM RESOURCES AND THEIR STRUCTURE

The total geological hydrocarbon resources of the Far Eastern Seas are estimated about 16.7 billion tonnes of the tentative fuel (in terms of oil), including 9 billion tonnes oil, 0.35 billion tonnes condensate and 7.4 trillion m^3 gas. The recoverable gas resources correspond to the geological ones; recoverable oil and condensate resources are estimated of 2.4 billion tonnes and 0.24 billion tonnes, respectively (tab.1) due to the projected technology of pools development on offshore fields.

The resources development is low, a share of explored and pre-estimated reserves ($A+B+C_1+C_2$) is not in excess of 13% and practically all these reserves are confined to offshore North-Eastern Sakhalin. The minor gas reserves have been explored on sakhalinski shelf of the Tatar Strait.

The C_3+D_1 categories resources are 1/4 off the ultimate potential resources of the region and are identified in offshore areas being under exploration drilling with discovery of oil/gas fields (offshore North Sakhalin and Tatar Strait) or in areas with the identification of the commercial oil/gas potential of sedimentary complexes on the extensions of the tectonic structures from offshore to onshore (offshore Sakhalin Bay, South Sakhalin, West Kamchatka, Anadyr and Khatyrskiy sectors of the Chukot shelf). The main share of the ultimate potential resources (62%) is the D_2 - category resources that proves the low level of the offshore development by the stratigraphic and wildcat drilling.

The resources structure under occurrence terms (tab.2) reflects insights into conditions of oil/gas formation and accumulation within different stratigraphic complexes in accordance with the burial depths, derived from the analysis of the geological-geochemical information and spacing features of the explored hydrocarbon reserves.

The main petroleum-bearing and perspective complexes are associated with the Cenozoic deposits with the share in the resources distribution in excess of 90%.

On the Sea of Okhotsk water areas the Neogene complexes predominate with the share in the resources distribution from 63% offshore Kamchatka up to 90% offshore North Sakhalin. The Neogene and Paleogene complexes on the Bering Sea water areas have approximately the equal resources distribution.

PROMISING WATER ARCAS

The distribution of hydrocarbon resources according to burial depths is determined by the observable patterns of the changable properties into sealed and reservoir rocks in accord with the degree of their catagenetic transformation, age and depositional conditions. The substantial differences in the critical distribution depths of the porous reservoir rocks are observed in the Neogene and Paleogene deposits of the region. For the Paleogene these depths are not in excess of 3-3.5 km, but in the Neogene deposits they may reach 5-5.5 km under favourable conditions, but the optimum oil/gas accumulation zone with maximum resources concentrations is limited up to 3-3.5 km (Lithologic-petrophysic criteria ..., 1990). This fact explains the coincidence of the main part of explored reserves and forecast resources with the depths interval up to 3 km: up to 87% on shelfs with the predominant Neogene petroleum bearing complexes and up to 95% on shelfs with the predominant Paleogene complexes within their sections.

The shallow-water offshore parts of the Far Eastern Seas (sea depths to 50 m) have a small share in their area and resources. Among the promising offshore areas the shoaliness is clear observed only offshore North Sakhalin with the resources share on this shelf of 60%. The predominant resources quota (more than 90%) offshore Khabarovsk and Magadan is associated with sea depths in excess of 100 m and sea depths offshore Kamchatka are of 50-200 m (65%).

The resources distribution across the offshore area reflects the pattern being characteristic of other petroleum bearing regions that is the resources concentration within regions and zones being distinguished by the optimum conjunction of favourable conditions for oil/gas accumulation and formation. The main share (about 80%) of ultimate potential resources in the region is associated with high-promising offshore areas of the Sea of Okhotsk: North Sakhalin, Khabarovsk, Magadan and Kamchatka. They are characterized by a large thickness (max. 10-14 km) of the sedimentary cover mainly of the Cenozoic age.

Thick strata of marine argillaceous and argillaceous-siliceous source rocks with the high degree of the organic matter and high oil/gas generative potential are present within this section. The amount of migrated hydrocarbons, being added together across the entire section reaches 15-25 million tonnes (in terms of oil hydrocarbons per km^2).

The reservoir strata are present (offshore Sakhalin) or predicted (offshore Khabarovsk, Magadan and Kamchatka) within the section of these water areas containing the bedded reservoirs with the optimum ratio of reservoir rocks and cap rocks (1:4 - 2:3) or the massive bedded reservoirs. The degree of the catagenetic transformation of deposits within the main oil/gas bearing and perspective horizons corresponds to catagenesis levels PK_3-MK_1. Large sizes and synsedimentary formation are characteristic of traps in deposits of these horizons.

The North-Sakhalin Shelf, the promising area about 74 thousand km^2 (to sea depth of 500 meters) in common with the North Sakhalin territory is included in the North- Sakhalin commercial oil/gas bearing basin and it covers the adjacent part of the Deriugin basin. Large oil and gas/condensate fields (Odoptu-sea, Chaivo, Piltun-Astokhskoe, Arkutun-Daghinskoe, Lunskoe) and Kirinskoe gas/condensate field are discovered offshore basin.

The Miocene deposits of the Daghinskiy horizon and the lower Notovskiy subhorizon are considered to be the main producing and promising complexes. The evaluated possibilities are associated with pore-fractured and fractured reservoirs into siliceous deposits of the Daekhuriinskiy horizon (Upper Oligocene) and to basement rocks, and with pore reservoirs of the Machigarskiy horizon and upper Nutovskiy subhorison.

The offshore oil/gas bearing zones: Odoptinskaja, Veninskaja, Mynginskaja, East-Odoptinskaja, Espenbergskaja, Astrakhanovskaja, offshore extension of Volchinsko-Sabinskaja (Nekrasovskaja zone) zones characterized by optimum conditions for oil/gas generation and accumulation and high resources concentrations are distinguished and considered to be perspective for discovery of large fields.

These zones are responsible for 80% shelf resources. Offshore Khabarovsk with the promising area of 152 thousand km^2 corresponds to Shantar, Kukhtuiski and north-western Deriugin basins, which are probable oil/gas bearing. Three promising complexes: Paleocene-Eocene, Oligocene - middle Miocene and middle - upper Miocene are distinguished within the section. The Mesozoic deposits are considered to be perspective for the exploration of gas/condensate pools in fractured reservoir rocks within petroleum accumulation zones of Shantar sector.

The most perspective oil/gas accumulation zones are associated with Kukhtuiskiy basins: Urakskaja, Kukhtuiskaja, Gavantsevskaja, Ermolaiskaja, Ushakovskaja zones.

The Magadan shelf, with the promising area of 163 thousand km^2 corresponds to Magadan, Gizhigin, partly West Kamchatka basins, including Tinro basin which is probable oil/gas bearing one. Perspective complexes identical with ones offshore Khabarovsk are distinguished, but the main possibilities are associated with Oligocene-middle Miocene and middle - upper Miocene deposits.The Paleocene-Eocene complex is included into the main perspective complexes in the eastern offshore. By results of the zonal forecast a set of large oil/gas accumulation zones is identified with a high share of shelf resources (more than 60% oil and 80% gas) (Zavjalovskaja, Khmitjevskaja, Nagaevskaja, Olninskaja, Umarskaja, Pjaginskaja and Evenskaja zones).

Offshore Kamchatka has the promising area of 126 thousand km^2 and covers West Kamchatka commercial oil/gas bearing basin and probably oil/gas bearing Penzhin and Golygin basins. Three main perspective complexes: Paleocene-Eocene, low-middle Miocene and middle-upper Miocene are distinguished within the section. The commercial gas and gas/condensate pools are discovered on West Kamchatka.

The upper Cretaceous, Oligocene and Pliocene deposits are perspective in local areas. The pore reservoirs predominate, the pore-fractured type is present in the upper Cretaceous and Paleocene-Eocene deposits. The fractured-pore or fractured reservoir is probable in siliceous mudstones and opoka of the upper Oligocene.

The largest oil/gas accumulation zones (Togatonskaja, Sukhanovskaja, Inushskaja and Kunzhinskaja set of oil/gas accumulation zones) being promising for discovery of large and middle fields are confined to West Kamchatka and Shelikhov troughs.

The promising water areas cover Khabarovsk and Sakhalin shelfs of the Tatar strait, South Sakhalin shelf of the Sea of Okhotsk, Chukot and Kamchatka offshore areas of the Bering Sea, Kamchatka and South Kurile shelfs of the Pacific Ocean with the total perspective area of 292 thousand km^2. The share of total and forecast resources in this offshore group is about 20% of the resources of the Far Eastern Seas. The lower oil/gas generative potential is due to small thicknesses of the sedimentary cover (to 7-8 km). The gas of serervoir rocks in their sections is considered to be the negative feature for offshore Tatar strait, South Sakhalin shelf and Navarin sector of the Chukot shelf. The main perspective complexes offshore Tatar strait and South-Sakhalin of the Sea of Okhotsk are associated with Miocene-Oligocene deposits. But the stratigraphic range of the probable petroleum content covers the interval from the upper Cretaceous up to the lower Miocene deposits (nizhnemarujamskiy subhorizon) offshore Sakhalin of the Tatar Strait.

The non-commercial oil influx up to 60 litres per day (Uglovskaja area) was produced from the same deposits in Krasnogorsk region, Sakhalin Island. Non-commercial oil influxes with a high paraffin content were produced from Paleogene deposits (North-Nevelskaja, Nevelskaja and Shebuninskaja areas). The short-term gas flow was produced from deposits of the Krasnojarskaja suite (upper Cretaceous) by drilling works on Bolshekholmskaja area).

The Miocene and Eocene-Oligocene deposits are considered to be perspective complexes offshore Bering Sea with the commercial petroleum content being proved by the onshore deep drilling within Anadyr and Khatyrskiy troughs. The oil/gas content is suggested in the Paleocene and upper Cretaceous deposits.

MAJOR TRENDA OF EXPLORATION WORKS

The marked resources distribution determines the major exploration trend, that is the discovery of the forecast oil/gas resources offshore North Sakhalin, Khabarovsk, Magadan and Kamchatka of the Sea of Okhotsk. The amount of resources, their concentration and discoverable probability of large fields is the great advantage of the North Sakhalin shelf being the single water area with proved oil/gas content, and more than 1/2 of forecast resources belong to the D_1 - category. The oil/gas potential is predicted on the basis of geological parameters, including a set of parameters for offshore Far Eastern Seas except of Sakhalin and partly West Kamchatka shelfs. The same feature is characteristic of several oil/gas accumulation zones offshore Sakhalin. A new information, including the stratigraphic drilling may allow to put corrections into trends and paramount significance of objects. The analysis of forecast estimations of resources offshore Sakhalin has proved that the input of geological information about the structure and composition of the section is the main factor of their sharp variation. The stratigraphic drilling is considered to be the major aspect of exploration works offshore Far Eastern Seas. The drilling results in conjunction with geophysical investigation data on the prediction of the geological section may stabilize forecast resources estimations and increase the exploration efficiency.

PROMISING OFFSHORE AREAS - THE URGENT LICENSED EXPLORATION OBJECTS

The quantitative forecast of probable fields, their maximum sizes and distribution due to evaluation of recoverable reserves is made on the basis of estimations of perspective and forecast resources and their density. Results are in tab. 3.

Estimations of localized resources in perspective traps have been used for Sakhalin, Khabarovsk, Magadan and Kamchatka offshore areas of the Sea of Okhotsk. These materials are considered to be the basis for identification of the most perspective water areas of the Far Eastrn Seas.

The East-Shmidtovskiy, Astrakhanovskiy and Veninskiy sectors are of great interest besides the licensed prospects Sakhalin-1, Sakhalin-2 and Sakhalin-3 on the North Sakhalin shelf.

The East-Shmidtovskiy sector (it includes Khanguzinskaja and East Shmidtovskaja zones) (Fig. 1) is of great interest for the exploration of small fields (reserves: 10-30 million tonnes, billion m^3, 1 oil and 10 gas fields with average reserves of 17 million tonnes, billion m^3) into Pliocene and Miocene deposits at 1.5-2.5 km depths.

In the south-eastern part of the section gas pools may be associated with the Miocene-Oligocene fractured reservoirs at depths of 4-5.5 km. Sea depths on the forecast fields are in range of 30 m to 120 m, at a shore distance of 15-75 km.

The Astrakhanovskiy sector includes Astrakhanovskaja, south-western Espenbergskaja and Nekrasovskaja oil/gas accumulation zones. The Miocene deposits with porous reservoir rocks are considered to be perspective for discovery of large gas/condensate fields (with reserves of 50-100 billion m^3, Astrakhanovskaja zone - 2) and of average-size oil and gas fields (Espenbergskaja and Nekrasovskaja zones - 2 oil fields with average reserves of 15 million tonnes, 2 gas fields with average reserves of 17 billion m^3). These fields have total recoverable reserves about 30 million tonnes oil and 190 billion m^3 gas. Sea depths are 10 m - 30 m, at a shore distance of 3-44 km.

The Veninskiy sector includes the large Veninskaja structure with the area of 306 km^2. The Miocene deposits with pore reservoir rocks into tectonically screened traps on the southern and northern periclines and underthrust western flank of the structure are perspective. The oil pools are predicted due to estimations of localized resources at depths of 1-3 km with total recoverable reserves of 36 million tonnes; gas/condensate pools are probable at depths of 0.8-4.3 km with total gas reserves about 265 billion m^3 and condensate of 16 million tonnes. Sea depth is 25 m; at a shore distance of 7-11 km.

The major portion of perspective and forecast resources are not yet developed within the licensed prospects Sakhalin-1, Sakhalin-2, Sakhalin-3 including Odoptinskaja, East-Odoptinskaja, Lunskaja and Mynginskaja oil/gas accumulation zones. Besides the explored fields (Odoptu-sea, Piltun-Astokhskoe, Arkutun-Daghinskoe, Lunskoe and Kirinskoe) more than 30 perspective traps have been identified with the forecast discovery of 1 large and 5 middle-size oil fields, 10 large gas fields (reserves about 30-300 billion m^3) and 10 middle-size gas fields. The total recoverable reserves from the forecast fields are evaluated about 120 million tonnes oil and 1030 billion m^3 gas.

The deposits of the Nutovskiy horizon with the bedded reservoir and porous reservoir rock (Odoptinskaja, East-Odoptinskaja zones), Daghinskiy horison with the massive bedded reservoir and porous reservoir rock (Lunskaja and Mynginskaja zones), siliceous Daghinsko-Daekhuriinskiy complexes with fractured reservoir and massive reservoir - rock (Odoptinskaja and East- Odoptinskaja zones), the upper Cretaceous with the fractured reservoir rock and massive reservoir (Mynginskaja, East- Odoptinskaja zones) are considered to be the promising ones.

Occurrence depths for oil pools are of 1.5-2.5 km, gas pools - 1.3-5.4 km.

The Pogranichny sector of the offshore South Sakhalin covers three oil and gas accumulation zones: Shatuno-Khunguzinskaja (Okruzhnaja), Bellinsgauzena (Borisovskaja), Polevogo. Traps within there zones are perspective for fields discovery with average reserves (2 oil and 2 gas fields with reserves of 12 and 20 million tonnes and 13 and 18 billion m^3, respectively) from the Miocene and Eocene porous reservoir rocks at depths of 1.5-2.5 km. Sea depths are in range of 20 to 127 m, at a shore distance - 4-35 km.

According to exploration possibilities of large fields offshore Magadan and Khabarovsk (Fig. 2), potential resources and spacing features of forecast fields one may identify four large sectors: Priokhotskiy, Khmitevskoy, Magadanskiy and Pjaginskiy.

The Priokhotskiy sector includes oil/gas accumulation zones being promising for fields exploration of large and middle sizes: Urakskaja, Magadan, Kukhtuiskaja, Gavantsevskaja, Ushakovskaja, Ermolaiskaja, South-Shilkinskaja, Central zones. The Miocene-Eocene deposits (seismic complexes B.C.D) with predominant porous reservoir rocks are considered to be perspective.

5 oil fields with average reserves of 18 million tonnes, 4 large gas fieds with average reserves of 50 billion m^3 and 8 middle gas fields with average reserves of 17 billion m^3 are predicted.

The total recoverable reserves from these fields are 90 million tonnes oil, 340 billion m^3 gas. The occurrence interval for oil and gas pools is about 1-2.5 km and 0.5-3.5 km, respectively. Sea depths maily are of 100-150 km, at a shore distance of 45-145 km.

The Khmitevskoj sector includes Spafarjevskaja, Khmitevskaja, Zavjalovskaja and Nagaevskaja oil/gas accumulation zones and it consideres to be perspective mainly for middle fields predominantly in porous reservoir rocks.

The probable 3 oil fields (average reserves of 12-14 million tonnes) and 13 gas fields (reserves od 11-25 billion m^3) are predicted.

Some traps in Spafarjevskaja and Khmitevskaja zones are perspective for the exploration of large oil/gas fields with reserves of 30-100 million tonnes or billion m^3. The total reserves from middle-size and large fields are 90 million tonnes oil, 280 billion m^3 gas. Sea depth on perspective traps are of 75-140 m at a shore distance 25-140 km.

Occurrence depths for oil/gas pools are 1-2.5 km and 0.5-3 km, respectively.

The Magadan sector covers Olninskaja, Siglanovskaja, Umarskaja oil/gas accumulation zones where middle and large oil/gas fields (1 oil field, reserves of 30-100 million tonnes, 5 gas fields, reserves 30-300 billion m^3) and 4 oil and 4 gas fields with reserves in limits of 10-30 million tonnes or billion m^3 are predicted.

The total recoverable reserves of these fields are 130 million tonnes oil, 430 billion m^3 gas. The Miocene deposits (seismic complex B and C) are promising, but the Eocene-Paleocene deposits (seismic complex D) are less promising with porous reservoir rocks. Occurrence depths for oil and gas pools are 0.5-2.5 km and 0.5-3 km, respectively. Sea depths are 105-155 m, at a shore distance 25-85 km.

The Pogranichny sector covers oil/gas accumulation zones, associated structurally with the Pjagin uplift: Kekurnaja, Evenskaja, Pjaginskaja zones. The Evenskaja and Pjaginskaja zones are related genetically with thick generative hydrocarbon sources within Shelikhov trough; these zones will be perspective for the exploration of middle and large fields in case of reservoir strata presence in their sections.

The discoverable probability of a large oil field (30-100 million tonnes) and 3 gas fields with reserves of 30 to 300 billion m^3, 2 oil fields with average reserves and 2 gas fields are predicted.

The total recoverable reserves from these fields are 90 million tonnes oil, 280 billion m^3 gas. Deposits mainly of the middle Miocene - Oligocene and Eocene - Paleocene complexes (seismic complexes D, C) are perspective. Sea depths are 75-400 m, at a shore distance 15-120 km.

Offshore Kamchatka the main potential hydrocarbon resources are concetrated within Ichinsko-Kolpakovskiy oil/gas bearing region with their large concentration share.

Two sectors are identified here: Utkholokskiy, related genetically with Utkholovskoe uplift and oil/gas generative kitchens within Shelikhov trough and Kolpakovskiy sector including oil/gas accumulation zones of the West Kamchatka trough.

The Utkholokskiy sector covers Sukhanovskaja and Inushskaja oil/gas accumulation zones. The Miocene and Oligocene-Eocene deposits are perspective and on the subsided structures of the Shelikhov trough low-Pliocene deposits are also perspective. The probability of descovery of two large oil and one gas fields (30-100 million tonnes, billion m^3), two oil and 6 gas fields with average reserves (10-30 million tonnes, billion m^3) is predicted. The total recoverable reserves on these fields are of 100 million tonnes oil and 140 billion m^3 gas. Occurrence depths for oil pools are 1.5-2.5 km, for gas pools - 1-3.5 km. Sea depths are in range of 56 to 345 m, at shore distance 30-94 km.

The Kolpakovskiy sector includes Togatonskaja, Kalavajamskaja and Kunzhinskaja oil/gas accumulation zones. The Miocene-Oligocene and Eocene deposits are considered to be the promising. The probability of discovery of 1 oil field (30-100 million tonnes) and 3 gas (30-300 billion m^3) fields, and also 3 oil and 5 gas fields with average reserves (10-30 mill. tonnes, billion m^3) is predicted. The total recoverable reserves of fields are oil - 85 mill. tonnes, gas - 315 billion m^3.

The occurrence depth for oil and gas pools is 0.5-2.5 km, and 1-4 km, respectively. Sea depths on the perspective traps are of 30-170 m, at a shore distance of 15-75 km.

The remaining offshore of the Far East Seas are the perspective for fields exploration, mainly of the middle - sizes (10-30 mill. tonnes, billion m^3). These remaining offshore areas cover the Tatar Strait shelf, Shantarskiy sector of offshore Khabarovsk of the Sea of Okhotsk, Khatyrskiy, Navarinskiy and Anadyrskiy water areas of the Chukutsk shelf of the Bering Sea (Fig. 3). The first three areas are characterized by the high discoverable probability of the single large gas fields (30-100 billion m^3).

THE GROWTH POSSIBILITIES OF THE RESOURCE BASE

The total recoverable resources of the identified perspective areas are estimated of 830 mill. tonnes and 35 billion m^3 gas. They are considered to be the main resource basis for reserves growth as the result of the exploration drilling. In near-term outlook the resources of North-Sakhalinskiy shelf and Pogranichny area of the South-Sakhalinskiy shelf (oil - 235 million tonnes, gas - 1.7 billion m^3) are considered to be the source for fulfilment of the resources basis. In more far-term outlook the resources basis way be extended due to resources development of Khabarovskiy, Magadanskiy and Kamchatskiy shelf the Sea of Okhotsk: 595 million tonnes oil and 18 billion m^3 gas.

REFERENCES

Varnavskiy, VG, Koblov, EG, Butsenko, PL, and oth. (1990). "Lithologic - petrophysical criteria of the petroleum content (oil/gas)," Nauka, Moscow, 270 p.

Table 1

Ultimate potential hydrocarbon resources in administrative regionsoffshore Far Eastern Seas

	Oil, million tonnes				Condensate, million tonnes				Free gas			
	A+B+C_1	C_2	C_3+D_1	D_2	A+B+C_1	C_2	C_3+D_1	D_2	A+B+C_1	C_2	C_3+D_1	D_2
Nort Sakhalin Shelf, Sea of Okhotsk	547.3 143.9	657.7 129.1	962 228	785 177	65.9 45.8	25.4 17.9	119.2 86.2	35.1 24.5	597.3	230.4	1552	559
South Sakhalin Shelf, Sea of Okhotsk			200 30	230 62			0.4 0.3	0.2 0.1			27	123
Sakhalin Shelf, Tatar Strait			109 24	159 35			1.1 0.7	1.0 0.7	3.8	0.8	63	92
Khabarovsk Shelf, Tatar Strait			73 16	132 30			0.6 0.4	1.7 1.2			60	113
Khabarovsk Shelf, Sea of Okhotsk				886 266				14 8.8				804
Magadan Shelf, Sea of Okhotsk				1815 545				42.2 26.5				1450
Kamchatskiy Shelf, Sea of Okhotsk			540 162	685 205			8 5	12.4 8.3			350	538
Chukotskiy Shelf, Bering Sea			35 10	890 265			0.7 0.5	16.3 12.5			25	590
Kamchatka Shelf, Bering Sea				145 45				2.9 2.0				100
Kamchatka Shelf, Pacific Ocean				70 25				0.4 0.3				30
South Kurile Shelf, Pacific Ocean				29 7				0.2 0.1				51

Table 2

Distribution of ultimate potential hydrocarbon resources of the Far Eastern Seas by stratigraphic complexes, occurence depths and sea depths (% of the total in offshore areas)

Offshore areas	By stratigraphic complexes			By occurence depths, km			By sea depths, m			
	Crctaceous	Paleogen	Neogen	up to 3	3-5	5-7	up to 50	50-100	100-200	200-500
Nort Sakhalin Shelf, Sea of Okhotsk	3	7	90	87	12	1	59	22	15	4
South SakhalinShelf, Sea of Okhotsk		42	58	95	5		35	20	42	3
Sakhalin Shelf, Tatar Strait	7	9	84	84	16		21	28	28	23
Khabarovsk Shelf, Tatar Strait	1	18	81	95	5		3	12	51	34
Khabarovsk Shelf, Sea of Okhotsk	12	15	73	87	13		1	8	56	35
Magadan Shelf, Sea of Okhotsk	1	16	83	87	13		1	6	53	40
Kamchatskiy Shelf, Sea of Okhotsk	5	32	63	95	5		19	34	31	16
Chukotskiy Shelf, Bering Sea	5	42	53	94	6		25	45	26	4
Kamchatka Shelf, Bering Sea		66	34	95	5		67	24	7	2
Kamchatka Shelf, Pacific Ocean		50	50	100			30	19	51	
South Kurile Shelf, Pacific Ocean		10	90	95	5		28	49	11	12

Table 3

Distribution of hydrocarbon accumulation with large and everage reserves within offshore areas

Offshore areas	Classes of fields due to recoverable resources				
	oil, mln. tonnes		gas, billion m^3		
	30-100	10-30	100-300	30-100	10-30
Nort Sakhalin Shelf, Sea of Okhotsk	2 / 74	9 / 147	3 / 510	10 / 550	27 / 460
South SakhalinShelf, Sea of Okhotsk		2 / 32			2 / 31
Sakhalin Shelf, Tatar Strait		1 / 45		1 / 45	4 / 60
Khabarovsk Shelf, Tatar Strait		1 / 15		1 / 45	2 / 30
Khabarovsk Shelf, Sea of Okhotsk		5 / 90		5 / 250	13 / 221
Magadan Shelf, Sea of Okhotsk	3 / 150	9 / 154	2 / 300	7 / 350	21 / 348
Kamchatskiy Shelf, Sea of Okhotsk	3 / 120	7 / 105	1 / 140	4 / 200	13 / 195
Chukotskiy Shelf, Bering Sea		7 / 110		1 / 37	13 / 200
Kamchatka Shelf, Bering Sea					2 / 30
South-Kurile Shelf, Pacific Ocean					1 / 13

Numerator - quantity of forecast accumulation into a class.

Dominator - their resources.

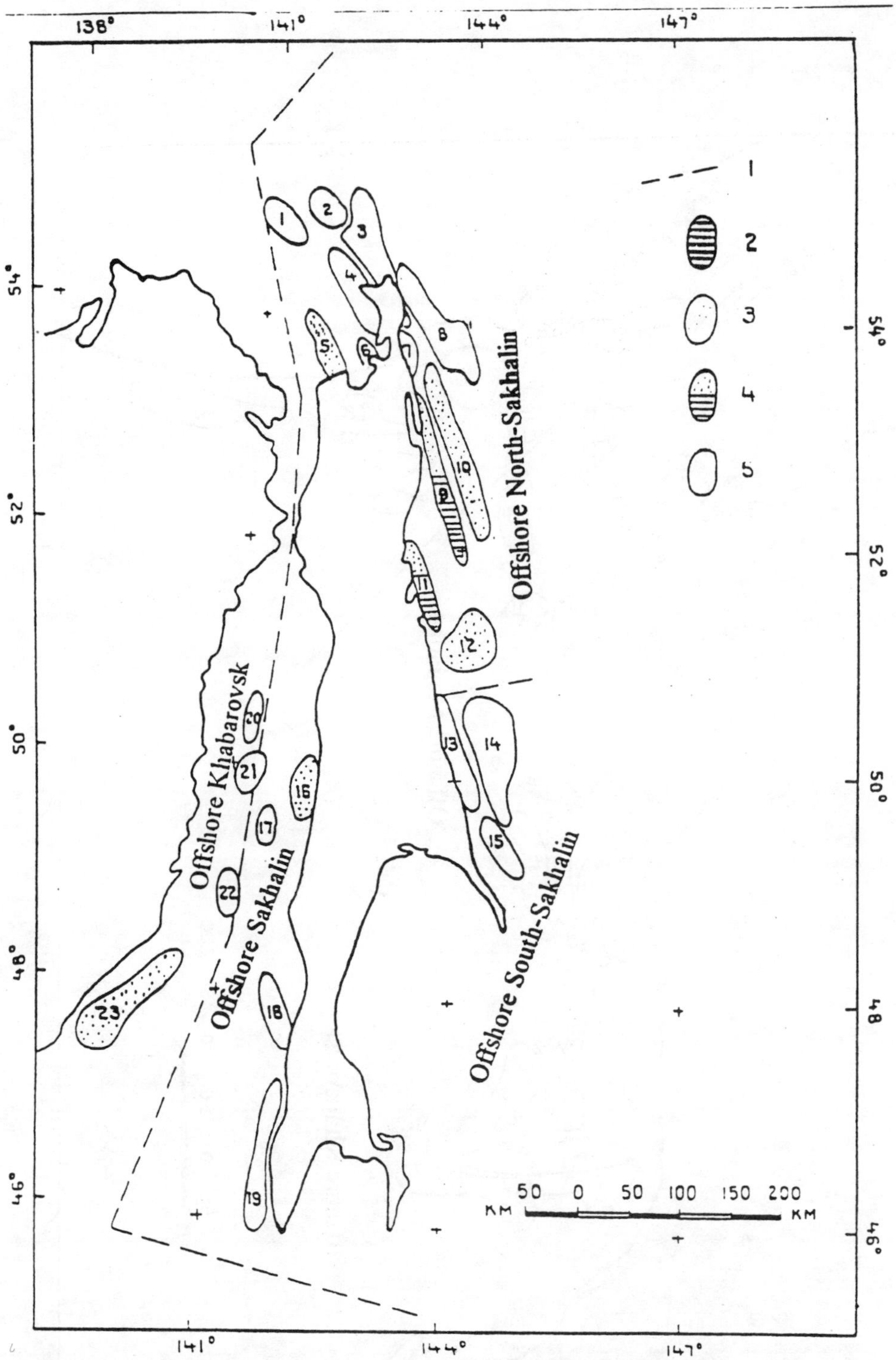

Fig. 1. Offshore Sakhalin and Tatar Strait. A chart of spacing of oil/gas accumulation zones being perspective for exploration of large and middle fields.

1 - boundaries of administrative offshore areas;
2-4 - oil/gas accumulation zones being perspective for fields exploration with large oil reserves (2), gas (3), oil and gas (4);
5 - oil/gas accumulation zones being perspective for fields exploration with average reserves.

Oil/gas accumulation zones: 1 - Sakhalinskaja, 2 - North-Espenbergskaja, 3 - Trjekh-bratkaja, 4 - Espenbergskaja, 5 - Astrakhanovskaja, 6 - Nekrasovskaja, 7 - Khanguzinskaja, 8 - East-Shmidtovskaja, 9 - Odoptinskaja, 10 - East- Odoptinskaja, 11 - Veninskaja, 12 - Mynginskaja, 13 - Shatuno-Khuzinskaja (Okruzhnaja), 14 - Polevogo, 15 - Bellinsgau-zena (Borisovskaja), 16 - Boshnjakovskaja, 17 - Yuzhnaja, 18 - Staromajachninskaja, 19 - Kuznetsovskaja, 20 - Chamganskaja, 21 - Sjurkumskaja, 22 - Innokentjevskaja, 23 - Primorskaja.

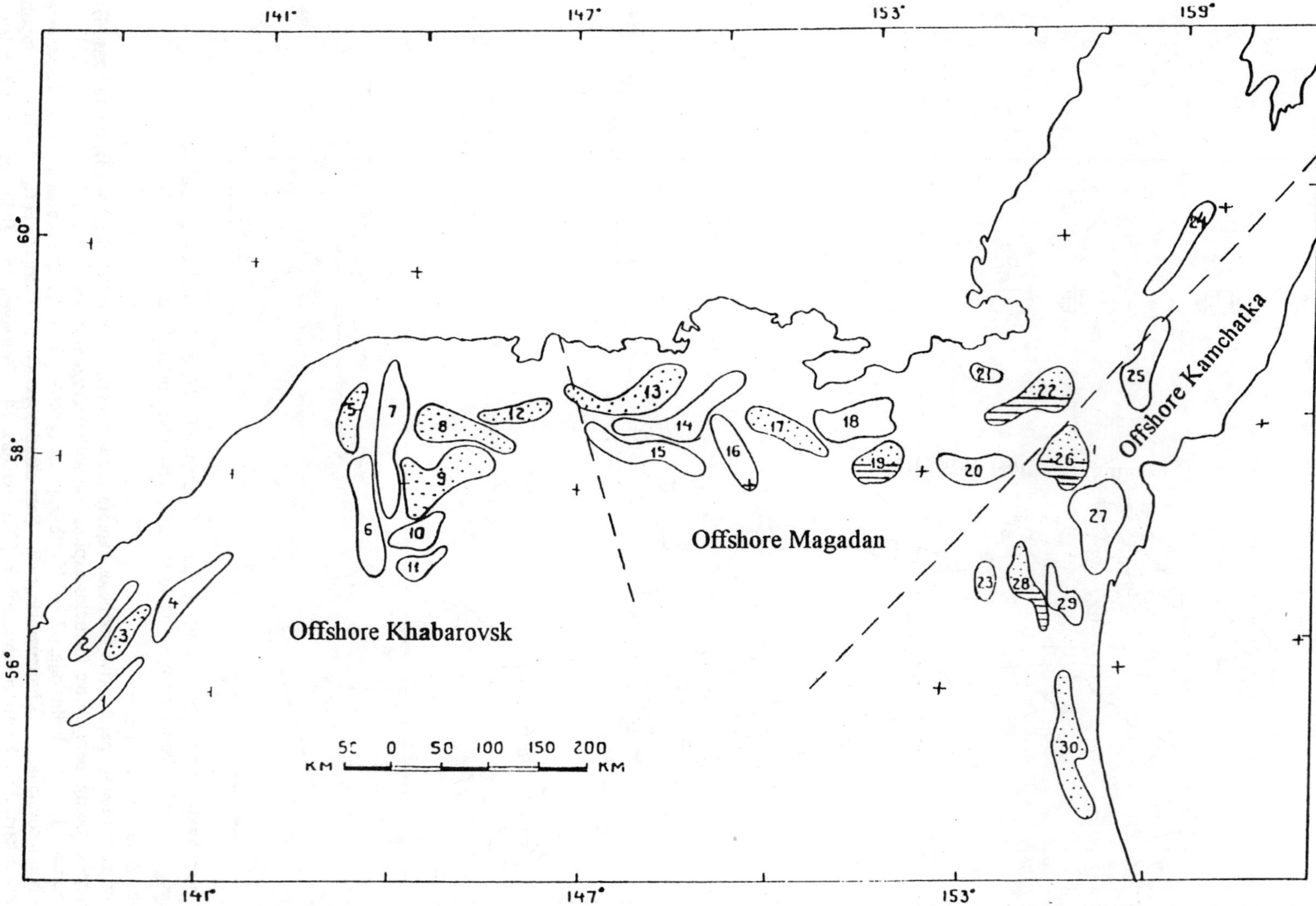

Fig. 2. **North-Okhotskiy shelf. A chart of spacing of oil/gas accumulation zones being perspective for exploration of large and middle fields**

Oil/gas accumulation zones: 1 - **Ulbinskaja**, 2 - **Ajanskaja**, 3 - **Shantarskaja**, 4 - **North-Ulbinskaja**, 5 - **Urakskaja**, 6 - **Nongdanskaja**, 7 - **Kukhtuiskaja**, 8 - **Gavantsevskaja**, 9 - **Ushakovskaja**, 10 - **South-Shilkinskaja**, 11 - **Central**, 12 - **Ermolaiskaja**, 13 - **Spafarjevskaja**, 14 - **Zavjalovskaja**, 15 - **Khmitevskaja**, 16 - **Nagaevskaja**, 17 - **Olninskaja**, 18 - **Siglanovskaja**, 19 - **Umarskaja**, 20 - **Pjaginskaja**, 21 - **Kekurnaja**, 22 - **Evenskaja**, 23 - **Siglanskaja**, 24 - **Gorkinskaja**, 25 - **Najakhanskaja**, 26 - **Sukhanovskaja**, 27 - **Inushskaja**, 28 - **Togatonskaja**, 29 - **Kalavajamskaja**, 30 - **Kunzhikskaja**.

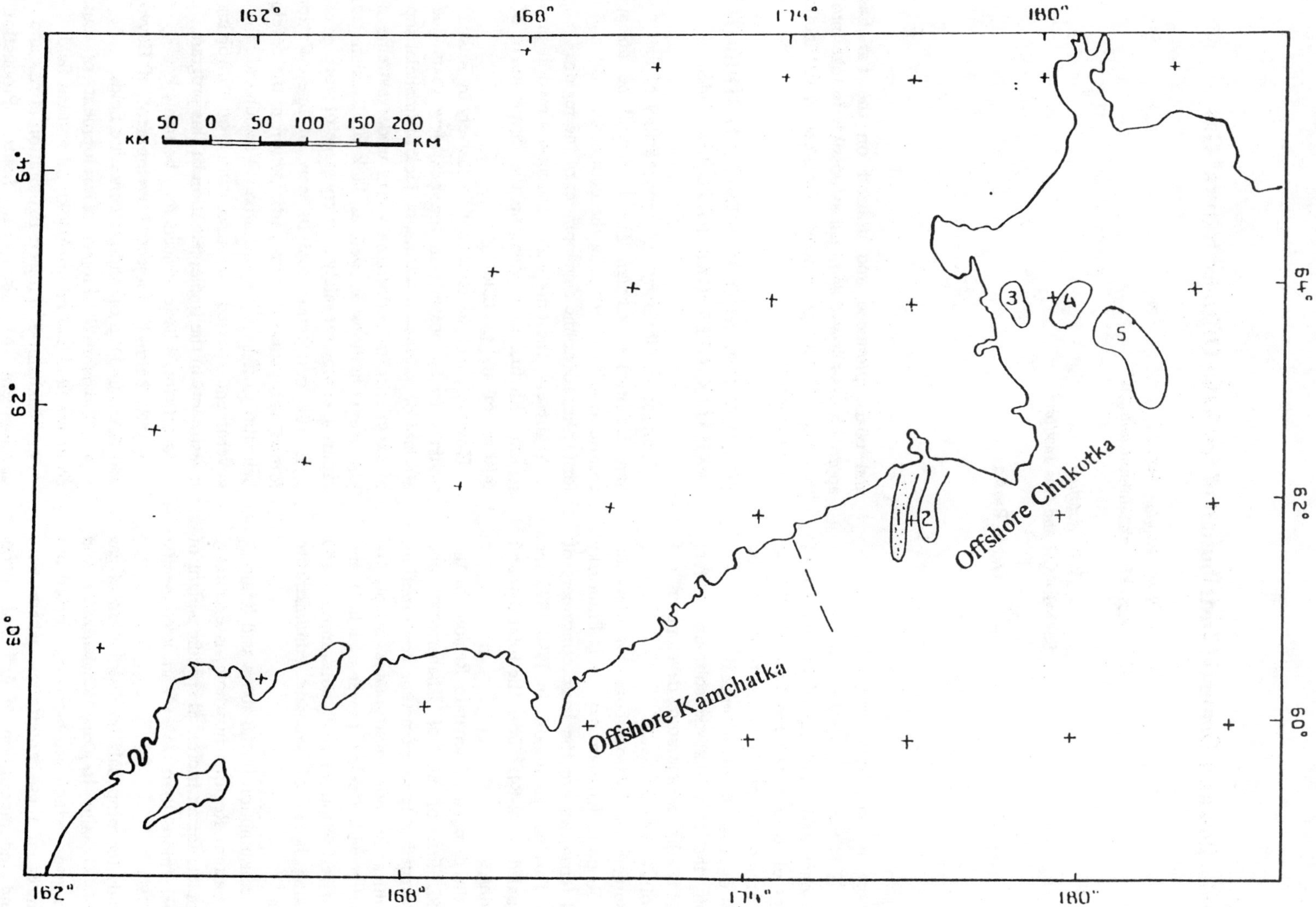

Fig. 3. Offshore Bering Sea. A chart of spacing of oil/gas accumulation zones being perspective for exploration of large and middle fields. Oil/gas accumulation zones: 1 - Mainopylginskaja, 2 - East-Khanguzinskaja, 3 - Tumanskaja, 4 - Vasilevskaja, 5 - East-Anadyrskaja.

Proceedings of the Seventh (1997) International Offshore and Polar Engineering Conference
Honolulu, USA, May 25-30, 1997

ISBN 1-880653-28-1 (Set); ISBN 1-880653-29-X (Vol. I)

Production and Domestic Distribution of Sakhalin Offshore Natural Gas

S.M. Bogdanchikov
JSC "Rosneft-Sakhalinmorneftegas"

V.N. Astafiev
Sakhalin Oil and Gas Institute

Okha, Russia

ABSTRACT

The increase of efficiency and reliability of functioning of fuel and energy complex of the Russian Far East will be the essential factor to overcome economic crisis and to provide the subsequent development of economy of the region.

In present conditions the only realistic solution of fuel supply to the region is the accelerated development of hydrocarbon resources offshore of Sakhalin.

The potential consumers of hydrocarbons on domestic market are Sakhalin Region, Khabarovsky and Primorsky Regions. The specified territories are the basic consumers of boiler fuel (BF) in the Far Eastern region. In 1990-1995 the consumption of BF equaled to 82-84% from the total demand of the Far Eastern regions.

In the recent years the demand of separate Regions of the Russian Far East and, first of all, of Khabarovsky and Primorsky Regions exceeded a level of actual consumption, which created difficulties in non-interruptable supply of enterprises and population with electric power and heat and resulted in necessity of restriction in their consumption and even change of working hours of separate administrative districts and large cities.

The structure of BF consumption in the above said Regions, as well as in the Far Eastern Regions in whole, is extremely non-effective and is notable for an increased specific weight of coal and liquefied fuel (more than 75%) and low specific weight of natural gas (7%).

Therefore in the study the proposals on distribution of gas for the enterprises of Sakhalin Region, Khabarovsky and Primorsky Regions are given, which are based on worked out fuel and energy balances of these enterprises, where were considered the demand and production of fuel and energy resources and power streams. There are developed not only the balance of Sakhalin gas, but also the balances of oil products and coals, produced and utilized on the Far East. Such approach has allowed as much as possible to take into account the factor of shortage of power resources by distribution of gas for enterprises.

PRODUCTION AND DOMESTIC DISTRIBUTION OF SAKHALIN OFFSHORE NATURAL GAS

Increase in efficiency and serviceability of fuel and energy complex in the Russian Far East will be an appreciable contribution to overcoming the economic crisis and providing for further economic development of the region.

At present, the only viable decision aimed at providing the region with fuel is to speed-up the development of Sakhalin offshore oil and gas fields.

However, such development exclusively by Russian efforts is unlikely to be possible in the next few years due to large amount of required investments, lack of financial support, lack of experience in performing works under extreme meteocean and ice conditions as well as lack of Russian technologies ensuring ecological safety at every stage of work.

In this connection and in view of interest expressed by foreign companies towards participation in the development of Sakhalin offshore hydrocarbon resources, the Russian government passed a resolution on involving foreign investments in the implementation of this program.

At present, 4 basic projects may be named:

- "Sakhalin-1" Project: Development of Cayvo, Odoptu and Arkutun-Dagi oil and gas condense fields.
- "Sakhalin-2" Project: Development of Lunskoe gas condense field and Piltun-Astokh oil and gas field.
- "Sakhalin-3" Project: Exploration at risk and subsequent development of discovered fields. Production Sharing Agreement(PSA) has been worked out for this project and

submitted to the government of the Russian Federation for approval.

- "Sakhalin-4" Project: Exploration and subsequent development of fields in Sakhalin Bay, Schmidt peninsular and Veni.

By the end of 1995 the President of the Russian federation signed the "PSA" law which established a legal basis for the relations arising out of Russian and foreign investments to prospecting, exploration and production of mineral resources in the territory of the Russian Federation and on Russian continental shelf.

Agreements on "Sakhalin-1" and "Sakhalin-2" Projects which are unique and unprecedented agreements in the Russian oil market, entered their implementation stage in June, 1996. They ensure Russian and foreign investments to the economy of the region in the amount of 25 billion USD and completely meet demands for natural gas and oil products in the region.

The following facilities are to be constructed under "Sakhalin-1" and "Sakhalin-2" Projects:

- offshore ice-resistant fixed platforms.
- subsea wellhead systems
- offshore pipelines
- onshore facilities for conditioning oil, gas and gas condensate
- main and gathering oil and gas pipelines.
- liquefied natural gas (LNG) plants
- shipment terminals for oil, condensate and LNG

Basic design decisions will take into account initial data and special conditions of design, construction and operation of oil and gas production and transport facilities for offshore fields located in freezing water areas with prolonged periods of self-containment.

Fig.1 gives integrated production and distribution levels for gas from Sakhalin offshore ("Sakhalin-1" and "Sakhalin-2" Projects) and onshore fields in 2000-2025.

Table gives expected hydrocarbon production levels under "Sakhalin-1" and "Sakhalin-2" Projects (including onshore fields) and hydrocarbon distribution in domestic market.

years	oil +condensate, mln t		gas, billion cu m	
	production	domestic market	production	domestic market
2000	1.5	1.5	6.3	5.4
2005	17.6	5.4	19.6	10.5
2010	30.6	5.2	20.0	10.9
2015	23.5	4.9	28.1	14.5
2020	12.2	4.6	28.2	14.8

Expected hydrocarbon production levels under "Sakhalin-1" and "Sakhalin-2" Projects (ncluding onshore fields) and hydrocarbon distribution in domestic market.

Sakhalin region, Khabarovsky Krai and Primorsky Krai are potential domestic consumers of hydrocarbons. These areas are the main consumers of boiler-furnace fuel. Consumption of boiler-furnace fuel in 1990-1995 covered 82-84% of the total boiler-furnace fuel demand in the Far East.

For the last years, boiler-furnace fuel demands in some areas of the Far East and primarily, in Khabarovsky Krai and Primorsky Krai have exceeded the actual consumption levels, which resulted in problems related to uninterrupted supply of citizens with electric power and heat during winter seasons and necessitated the introduction of reduced consumption levels and even changes to working schedules for separate districts and cities.

The structure for boiler-furnace fuel consumption in the above-mentioned entities of the Russian Federation as well as in the Far East as a whole, is extremely ineffective and includes large amount of coal (over 55%) and liquid fuel (over 20%) and small amount of natural gas(7%).

Fig.2 gives natural gas distribution among consumers of Sakhalin region, Khabarovsky Krai and Primorsky Krai with account for economic efficiency and power efficiency of natural gas consumption by various categories of consumers. Future gas demands are determined for each enterprise based on current fuel consumption, new conditions and requirements for economic development and influence of conversion process on growth rates of fuel and energy demands.

Proposals on gas distribution among enterprises in Sakhalin region, Khabarovsky Krai and Primorsky Krai are based on estimated energy balances for these enterprises which take into account required consumption levels as well as energy fluxes. Balances were estimated not only for Sakhalin gas but also for oil products and coal produced and consumed in the Far East. Such approach takes maximum account of resources deficiency factor when distributing gas among the enterprises.

Energy efficiency factor should also be taken into account for optimum gas distribution.

Based on these considerations, amounts of gas to be supplied to consumers in Sakhalin region, Khabarovsky Krai and Primorsky Krai were estimated.

Khabarovsky Krai is one of the most developed industrial area in the Far East. Industry occupies the larger portion in the economic structure of the area, its share covering two-thirds of the integrated gross product.

Basically, industrial production concentrates in two areas: Komsomolsk-on-Amur and Khabarovsk.

Fig.3 gives estimated demands for boiler-furnace fuel in Sakhalin region, Khabarovsky Krai and Primorsky Krai till 2020.

Provision of gas supply to Komsomolsk industrial complex will amount to 82% by 2020.

Provision of gas supply to Khabarovsk industrial complex by 2020 will be as much as 77%, which includes 53% in industry, 80% in power industry and 80% in household system.

Gas distribution in domestic market will ensure primary coverage of gas demands in Sakhalin region with the exception of western Sakhalin coast where local coal consumption is planned to be continued based on study of Macroeconomics Institute, with the provision for liquefied propane-butane supply to the household system.

Proposals of Sakhalin Administration on promotion of gas chemistry and cement production were taken into consideration when estimating demands for boiler-furnace fuel in Sakhalin region.

It should be noted that the development program for Sakhalin economics planned by the Administration for 2005-2010 is very strenuous and requires large investments. In this connection, the second option for gas supply to the south of Sakhalin has been proposed.

Under this option, the priority in gas supply is given to Sakhalin consumers. However, plans on the construction of gas processing facilities in the south of Sakhalin were proposed to be postponed in order to increase gas supply to the continent.

Under this option, boiler-furnace balance for Khabarovsk industrial complex and Primorsky Krai will improve considerably.

It should be remembered that gas production may be initiated in 2010-2015 in gas fields under "Sakhalin-3" and "Sakhalin-4" Projects and other potential projects.

Gas supply to consumers in Primorsky Krai at a rate of 1.0-1.7 billion cubic meters per year is expected to start in 2015 upon the beginning of the development of Arkutun-Dagi gas deposits.

Basically, gas is planned to be supplied to Dalnerechensk in the north of Primorsky Krai to convert the State Power Plant in Primorsky Krai to a gas-based plant.

Thermal power plants which are major fuel consumers noted for marked seasonal variations in fuel consumption are considered as terminal gas suppliers.

In view of potential deficiency of power plant coals, gas application in boiler furnaces of thermal power plants will result in appreciable economic efficiency, which is rather attributed to reducing cost of air protection and hard fuel production, transport and conditioning than to decreasing unit rates of fuel consumption.

Distribution of Russian share of gas in domestic market requires active marketing and timely construction of the main gas pipeline system and distribution networks in cities and villages.

Consumers in southern and central areas of Sakhalin region will be supplied with gas by the main gas pipeline 1067 mm in diameter from onshore facilities to Prigorodnoye. Construction of this pipeline will be financed under "Sakhalin-1" and "Sakhalin-2" Projects.

The main gas pipeline from onshore gathering and conditioning facilities to Prigorodnoye will be laid along the central valley of the island. The full length of the pipeline will be fitted with equipment for distributing gas among the areas of Sakhalin region.

Construction of gas pipeline branches over 100 km in length, incoming and outcoming gas pipe networks about 380 km in length as well as other gas supply facilities are required for arranging gas supply to consumers in Sakhalin region.

Gas supply to consumers of Sakhalin region is proposed to be implemented in two stages:

1 stage - gas supply to large populated areas and major gas consumers to cover over 70% of total gas demands

2 stage - gas supply to the remaining of the populated areas and minor gas consumers.

Investments to the construction of gas pipeline branches, gas distribution stations, in-village and intervillage gas pipe networks and other gas supply facilities for Sakhalin region are evaluated in the amount of 110 billion USD.

The first stage of gas supply to consumers of Komsomolsk industrial complex will require upgrading the existing "Okha-Komsomolsk" gas pipeline: construction of gas-compressor stations, additional pipeline crossing in Nevelskoy Straight, gas pipeline from onshore facilities to "Okha-Komsomolsk" gas pipeline, gas pipeline branches 110 km in total length to Amursk, Solnechny, Elban and other populated areas and local in-village gas pipe networks 60 km in total length. Total cost of updating "Okha-Komsomolsk" pipeline and constructing pipeline branches and local in-town and in-village gas pipe networks is evaluated as 200 billion USD.

"Okha-Komsomolsk-Khabarovsk" gas pipeline 1020 mm in diameter and 1000 km in length has been designed to supply gas to consumers in Khabarovsk industrial complex.

This gas pipeline will be routed to Komsomolsk in parallel to the existing gas pipeline. Smaller diameter gas pipeline from Khabarovsk to the State Power Plant in Primorsky Krai is planned for construction.

The decision on the construction of "Sakhalin - Khabarovsk - Primorsky Krai " gas pipeline may be taken after 2000 upon the decision on initiating development of other Sakhalin offshore fields ("Sakhalin-3", "Sakhalin-4" and other projects).

In compliance with the resolutions passed by the governors of Sakhalin region and Khabarovsky Krai, JSC "Sakhainmorneftegas" will act as a general customer to place orders for feasibility study of gas supply to industrial complexes and household systems, development of design plans and specifications, construction of gas transport facilities capable of gas gathering (JSC "SMNG" will be the operator of these facilities).

Currently, the federal program on providing gas supply to Sakhalin region, Khabarovsky Krai and Primorsky Krai is being developed by JSC"Sakhalinmorneftegas" and Oil Company "Rosneft". The development of the program is expected to be completed within the second half of 1996. In the near future, this program will be the only decision to provide the Russian Far East with fuel based on Sakhalin offshore natural gas resources.

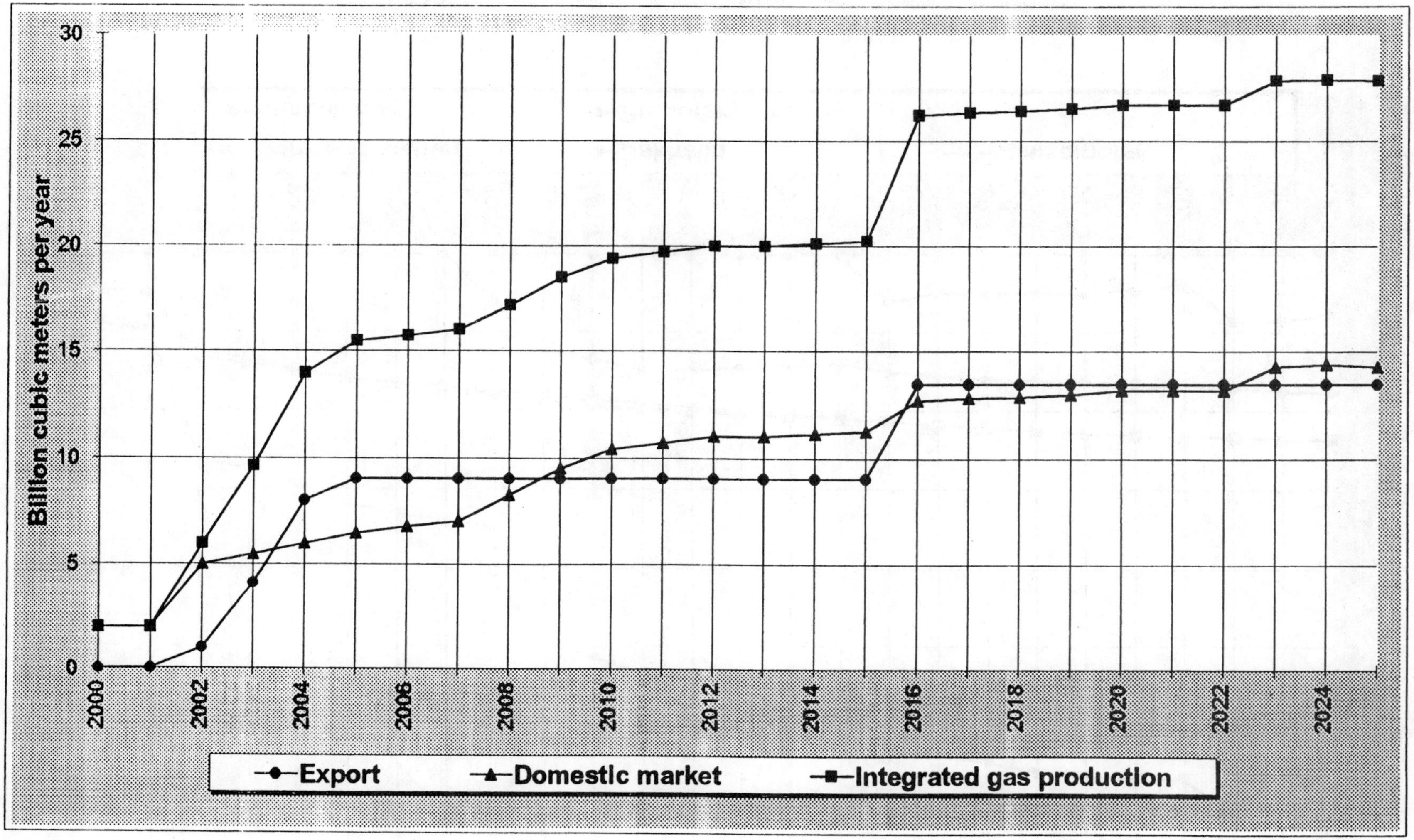

Fig.1. Integrated gas production and distribution levels for Sakhalin offshore ("Sakhalin-1" and "Sakhalin-2" Projects) and onshore fields in 2000-2025

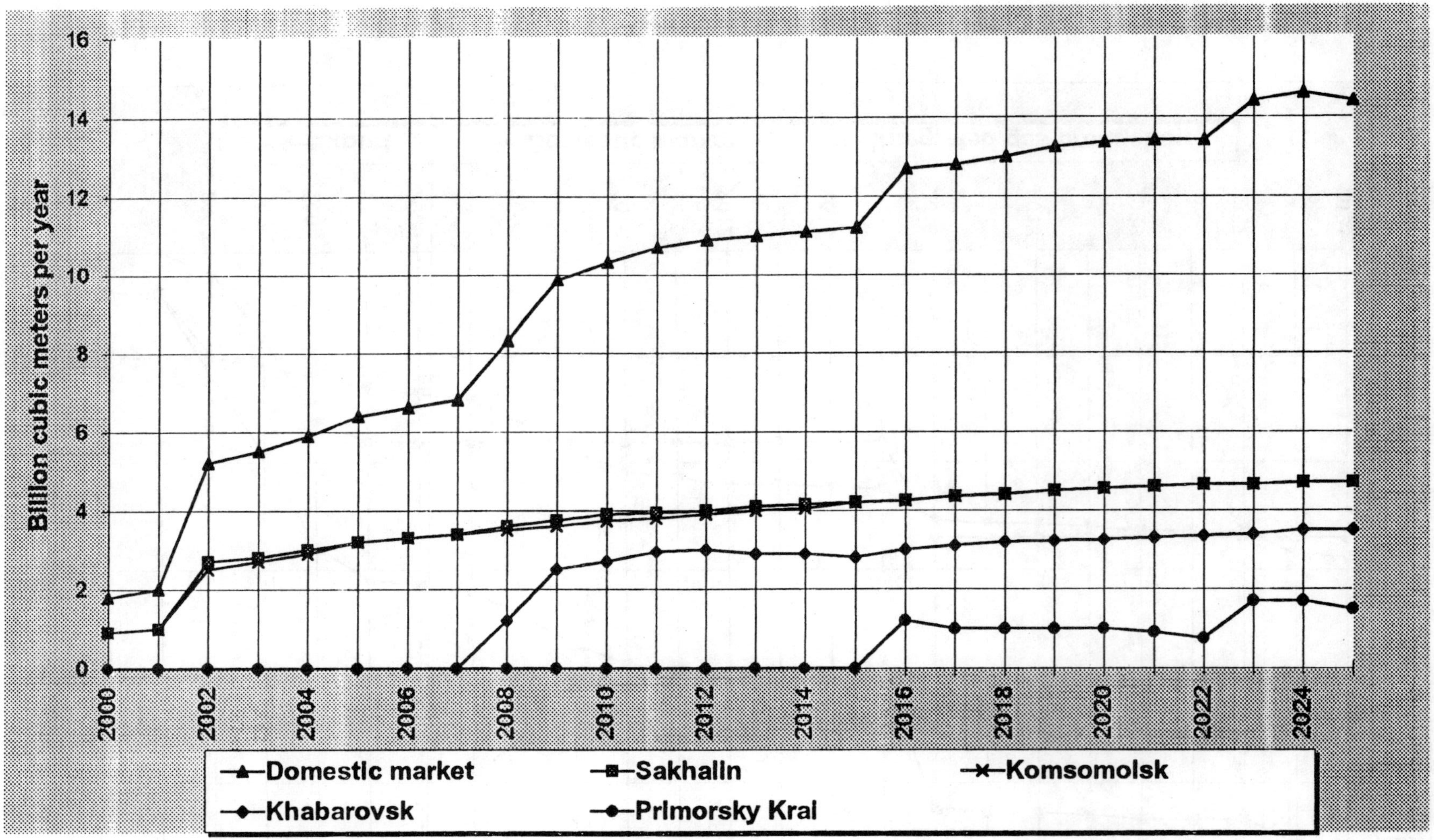

Fig.2. Distribution of gas from Sakhalin offshore ("Sakhalin-1" and "Sakhalin-2" Projects) and onshore fields among the Far East areas in 2000-2025

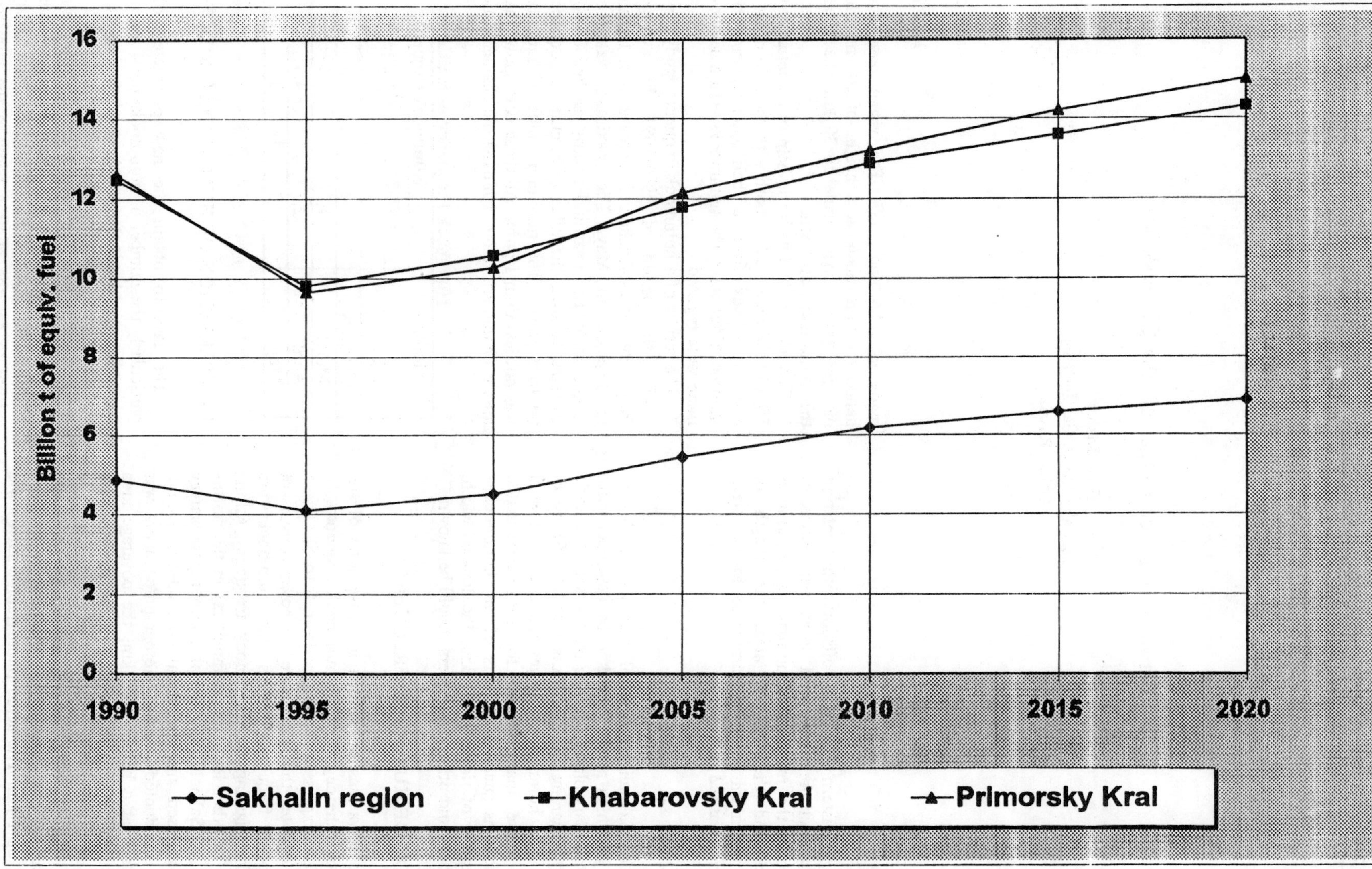

Fig.3.Demands for boiler-furnace fuel in Sakhalin region, Khabarovsky Krai and Primorsky Krai till 2020

Proceedings of the Seventh (1997) International Offshore and Polar Engineering Conference
Honolulu, USA, May 25-30, 1997

ISBN 1-880653-28-1 (Set); ISBN 1-880653-29-X (Vol. I)

Parameters of Ice Ridges of the Okhotsk Sea

G.A. Surkov
Sakhalin Oil & Gas Institute
Okha, Russia

ABSTRACT

For the north-eastern Sakhalin offshore the relation between the hummock frequency and some parameters of hummocks is found: the numbers of hummocks per unit of observation trace and the unit of area, the length of hummocks. The minimum height of hummocks for the areas under investigations was 0.3 m.

INTRODUCTION

In offshore of Sakhalin many oil and gas fields are discovered most of which are supposed to be developed with the help of the stationary platforms. The ice loads are the basic parameters when designing the platforms for the offshore of the Arctic seas. The hummocked ridges make the essential contribution to these loads. Thus, for the substantiated designing of platforms such parameters of the hummocked ridges as the number of ridges per 1 km, the length, width, height of hummocks and other parameters should be known.

INVESTIGATION OF HUMMOCK FREQUENCY

During the observations of the sea ice cover one of the basic registed data is the degree of the hummock frequency. In Russia this index is expressed by points (from 0 to 5). The 5 point frequency denotes that the entire ice cover consists of hummocks, 0 point frequency means that hummocks are absent. When investigating the hummock frequency (along with researching of other ice cover parameters) in the north-eastern offshore of Sakhalin the aerial photographic survey from a plane was used and the visual observations from a helicopter were conducted. Aerial photographic survey was made from a plane at a height of 3 km, the coverage area across the course made up 1 km. Photographic surveys were made in a scale of 1 : 7500. The routes of aerial photographic survey with the length of 10-20 km began from the coast and passed perpendicularly the coastal line. The distance between the routes made up 1 km. The aerial photographic survey was made in 1986 and 1987. The visual observations from a helicopter were made at a height of 200 m. The maximum distance from a coast made up 30 km, the distance between the routes made up 5 km.

The area studied with the help of an aerial survey made up 5% of total area studied both with the help of an aerial survey and the survey from a helicopter. During the study of hummock frequency the hummocks with a height of 0.3 m and more were registed.

The results of a hummock frequency study are in Table 1.

For the period of 1986-1991 the average hummock frequency was established to increase from 1 point in January to 3 points in May. The maximum value of hummock frequency for the areas under study reached 4.5 points in April. In separate areas the value of hummock frequency may reach up to 5 points. The minimum value of hummock frequency was not less than 1 point. All these data shows that the present area is characterized by high dynamic characteristics.

Table 1

Hummock Frequency of Ice Cover

Month	Hummock frequency, points		
	Average	Maximum	Minimum
January	1	2,5	1
February	2	3	1
March	1,7	4	1.5
April	2,7	4.5	2.5
May	3	3.5	2

INVESTIGATION OF HUMMOCKED RIDGES

The task to obtain the new parameters of hummocks particularly the number of hummocks per 1 km of observation trace is related to the need of designing of ice-resistant stationary platforms. To obtain the reliable data the series of observations as many as possible should be taken (according to Russian standards not less than 5 years). Thus, the aerial photographic surveys of 1986, 1987 were analyzed by the author to get the relation between the hummock frequency and other parameters of hummocked ridges. For this purpose the areas with the appropriate values of hummock frequency (from 1 to 4 points) were distinguished in the aerial photographic surveys referring to different regions of the north-eastern offshore of Sakhalin. The average number of hummocks per 100×100 m² was counted on these areas. The average, minimum and maximum length was also defined. Furthermore, on these distinguished areas the lines were drawn to different directions and the number of the ridge-line intersections was counted.

Table 1 represents the results of these investigations. Figure 1 represents the relation between the average number of hummocks per 1 km of observation trace and the average number of hummocks per area of 100×100 m². Figure 2 shows the length (average, minimum, maximum) - hummock frequency relation.

According to the obtained data the average number of hummocks per trace of 1 km varieties from 9 points in January to 25 points in May.

CONCLUSION

The relation obtained allow to increase substantially the series of observations of hummocked ridges parameters for the north-eastern offshore of Sakhalin. Furthermore, the similar approach may be realized in other regions that may substantially enrich the scientific ice cover information.

Table 2

Relation of hummock parameters with hummock frequency

Hummock frequency, points (1 point = 20% of hummocks)	Average hummock number per km	Hummock length, m			Hummock number per 100×100 м²
		Average	Maximum	Minimum	
1	9	130	500	27	1
2	18	109	380	25	4.2
3	25	80	266	20	7.9
4	31	49	144	12	11.6

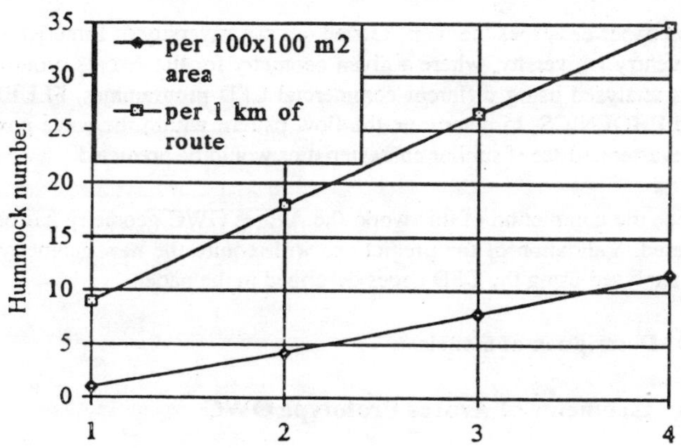

Fig.1. Average hummock number vs. 1 km of route and 100×100 m² area.

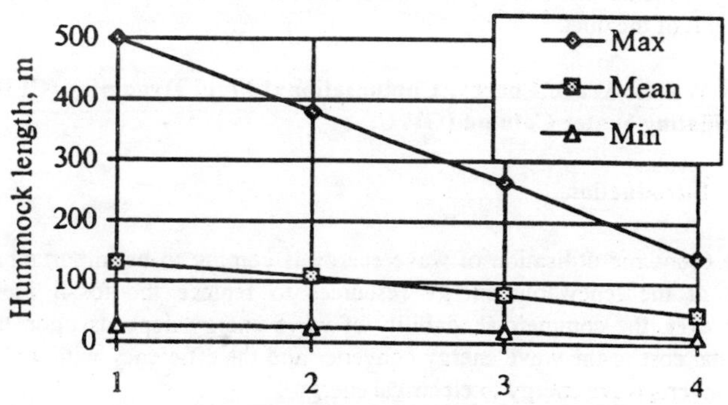

Fig.2. Hummock length vs. hummock frequency.

Proceedings of the Seventh (1997) International Offshore and Polar Engineering Conference
Honolulu, USA, May 25-30, 1997

ISBN 1-880653-28-1 (Set); ISBN 1-880653-29-X (Vol. I)

Utilisation of CFD to Study an Oscillating Water Column

P. White and M. Webster
Coventry University
Coventry, United Kingdom

A. Thakker and M. O'Dowd
University of Limerick, Ireland

Abstract

This paper describes how computational fluid dynamics (CFD) can be applied to the study and optimisation of flow through an oscillating water column (OWC). A simplified version of the Azores prototype OWC was studied, using two different commercial CFD programmes, to see if identical characteristics would be predicted. Although it will be necessary to compare the CFD results with experimental results when they become available, the predicted results were very encouraging. The two codes predicted flow with the same general characteristics, the most distinctive feature being a large vortex motion , present in the wall region of the upper section of the oscillation chamber, which persists on entry to the turbine duct, with two contra rotating vortices continuing to the exit of the duct.

Key Words: Wave Energy, Computational Fluid Dynamics (CFD), Oscillating Water Column (OWC)

1.0 Introduction

The economic utilisation of wave energy is coming to be important as one of the renewable energy resources to replace the fossil fuels. However, the commercial viability of wave energy depends upon the capital cost of the wave energy converter and the efficiency with which it converts wave energy to electrical energy.

In order to maximise the turbine output, the turbine characteristics have to be matched to the pneumatic output of the wave energy converter (White, P.R.S., 1996). The optimisation of this process can be accomplished using model testing, full scale testing or prediction. One of the predictive techniques involves using commercially available Computational Fluid Dynamics (CFD) codes based on the solution of the Navier Stokes equations in three dimensions, with the addition of turbulence modelling.

Analysis of the geometry of one of the European Wave Energy Pilot Plants currently under construction, the Azores prototype Oscillating Water Column (OWC), has recently been carried out using CFD. One problem with these plants that has been observed, in both experimental and theoretical models, is that if the flow is maldistributed it reduces turbine efficiency (Curran, *et al.*, 1995; White, 1995). It is hoped that CFD will become a useful tool to predict the airflow through the wave power converter to the turbine in order to optimise the turbine output by changing the structure of the wave energy converter and hence the airflow distribution whilst still at the design stage.

This paper describes the work carried out at University of Limerick and Coventry University, where a given geometry for the Azores prototype was analysed using different commercial CFD programmes, FLUENT and PHOENICS, to determine the flow pattern within the wave power converter and see if similar characteristics would be predicted

Since the completion of this work, the Azores OWC geometry has been altered. Validation of the predictions will require the new geometry to be analysed using the CFD codes described in the paper.

2.0 Description of models

2.1 Geometry of Azores Prototype OWC

The initially proposed geometry of the Azores prototype is shown in Figure 1. The air enters through the bottom face marked 'ABCD' due to the action of the ocean waves. This air then flows up from the square oscillation chamber, into the narrow rectangle duct and then makes a right angle turn into the horizontal circular pipe section via a smooth bellmouth entry. The turbine is mounted in the ellipsoid, around which the air has to flow to create the flow annulus, before exhausting to atmosphere at the exit from the circular duct.

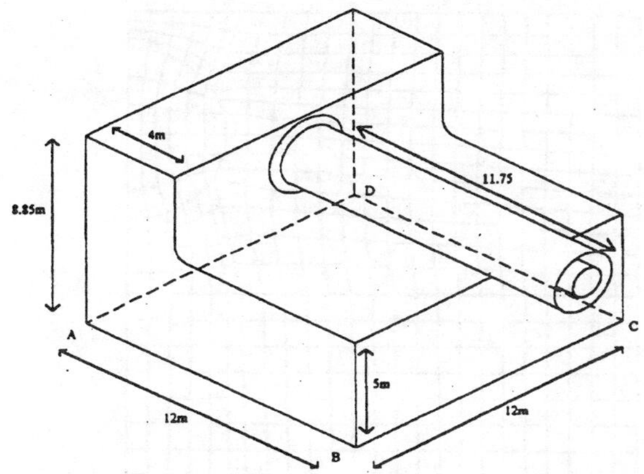

Initially Proposed Geometry For The Azores Prototype OWC

Figure 1

2.2 Geometry Modelled Using CFD Codes

The models to be compared were of a simplified version of the geometry (see Figures 2 and 3), with a relatively small grid size. It was thought that once a simple model had been proved to be realistic, the predictions of more complicated models could be analysed with confidence.

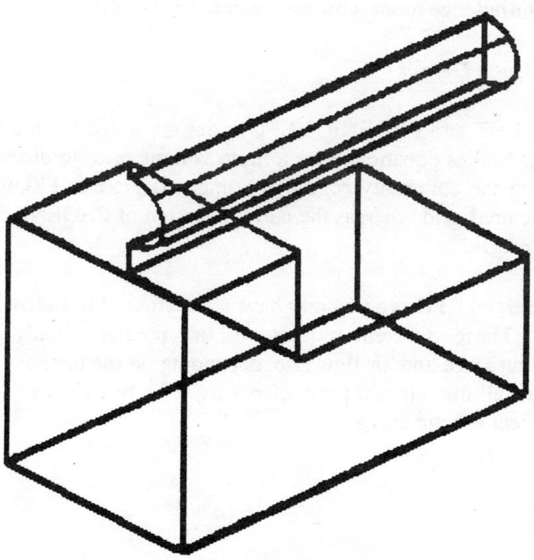

A Simplified Version Of The Azores Prototype OWC - Half Model

Figure 2

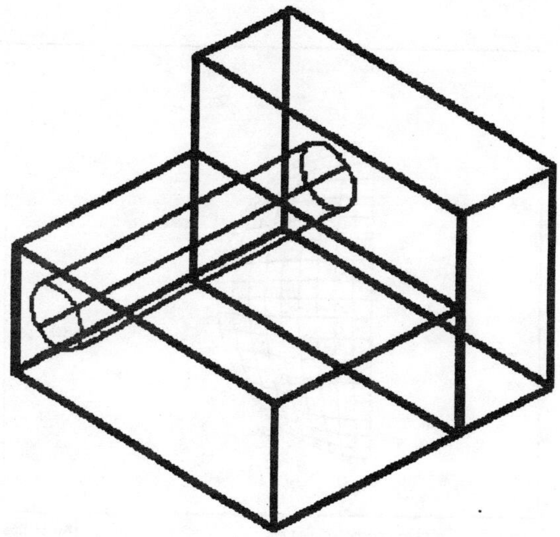

A Simplified Version Of The Azores Prototype OWC - Whole Model

Figure 3

A simplifying assumption can be made which reduces the difficulty of the computational analysis. This is that the flow through the geometry can be considered to be quasi steady, i.e. unidirectional and steady despite its oscillating nature. This assumption is made widely in respect of Wells turbine testing and research into its effects have been investigated (Raghunathan and Ombaka, 1984).

As can be seen, the geometry has a vertical plane of symmetry through the circular section centreline. This means that a model of just half the geometry can be analysed, reducing the overall grid size and computational resources necessary. The computational grids used are shown in Figures 4 to 7.

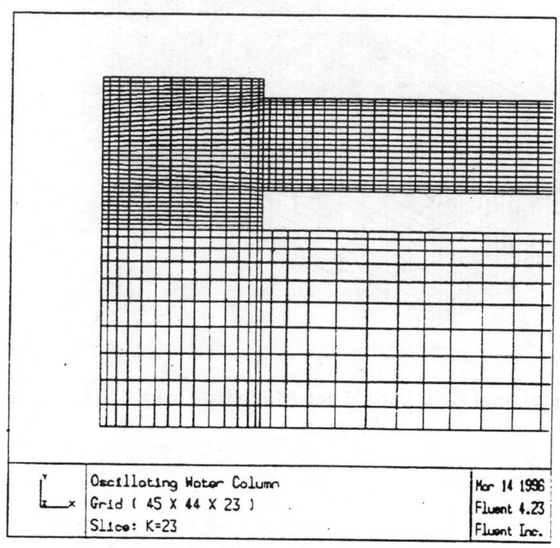

The Computational Grid Used When Modelled Using FLUENT

Figure 4

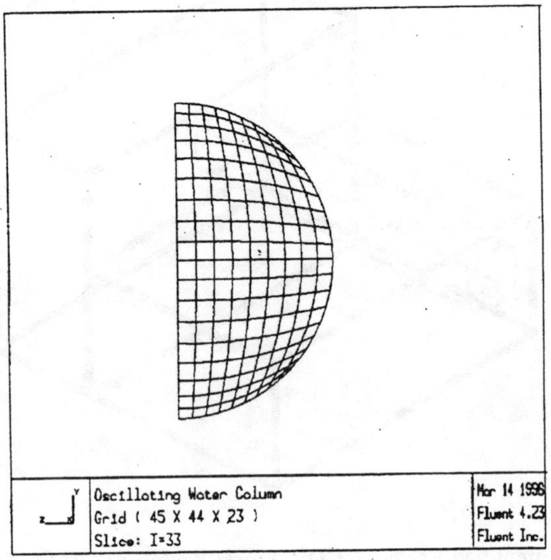

The Computational Grid Used When Modelled Using FLUENT

Figure 5

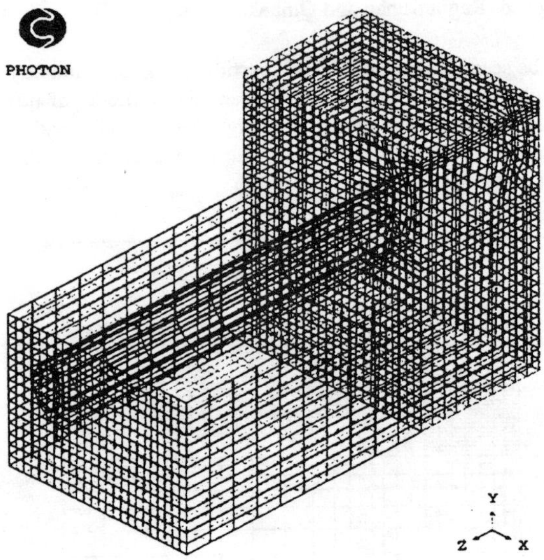

The Computational Grid Used When Modelled Using PHOENICS
Figure 6

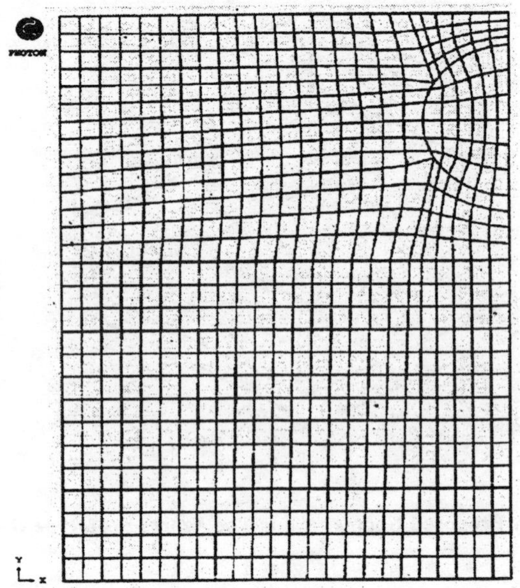

The Computational Grid Used When Modelled Using PHOENICS

Figure 7

The problem under consideration is a steady, incompressible, adiabatic study of the air flow in the oscillating water column geometry. The only variables which are of interest are the pressure and velocities throughout the flow field. It was found that the flow could be considered to be fully turbulent, (Thakker and O'Dowd, 1994) so a turbulence model had to be used with each programme.

The basic model described above was modified and extended by the researchers at the University of Limerick and Coventry University. In particular, it was thought necessary to evaluate the effect of using different turbulence models on the predicted flow field.

2.3 CFD Code Details

Both FLUENT and PHOENICS CFD codes are based upon integrating the Navier Stokes equations over a finite volume prior to discretisation, so ensuring the continuity condition is met. At present, FLUENT uses an unstructured grid whereas the current version of PHOENICS utilises a structured grid

All commercial CFD programmes have a selection of turbulence models available. These different models will all predict slightly different values of pressure and air flow rate, depending on the methods they use to calculate all the relevant parameters - e.g. effective viscosity, velocity and turbulent kinetic energy.

The work at Coventry University was conducted using two different turbulence models, a LVEL model and a KE model. The KE turbulence model is the one that would normally be applied to this type of problem. It is a standard high Reynolds number K-Epsilon model. However the user support group at CHAM, the suppliers of PHOENICS, advised that the LVEL model might prove to be more suitable, considering how coarse the grid was, especially in the near wall region. It was necessary to use a fairly coarse grid due the necessary compromise between workstation capability, program requirements and convergence time. The LVEL model is so named because all one needs for calculating the effective viscosity are L, the distance from the nearest wall, VEL, the local velocity and the laminar viscosity. It is a zero equation length scale based model. It can be an advantage to use this model when there is a lack of resolution in the near wall region.

At the University of Limerick, the work was conducted using three different turbulence models, a KE (standard) model - using one turbulence and one dissipation equation, a KE-RNG model, which uses renormalisation group theory and a Reynolds Stress Model, which is the standard high Reynolds number version - using six stress equations and one dissipation equation.

Additional work was undertaken to assess the effect of grid density, grid dependency (modelling the whole model), plus including the fillet radius and bellmouth entry to the turbine duct. The assessment of the outcome of all these modifications are discussed in section 4.0.

3.0 Results

Comparisons of the models produced by the two codes, FLUENT and PHOENICS, show that the flows predicted have the same general characteristics.

The velocity of the flow at entry to the domain (the sea water/air boundary, bottom plane ABCD in Figure 1) is small (2 m/s) due to the large cross-sectional area and the sea state under consideration. The fluid velocity increases as it converges toward the narrower upper section of the oscillation chamber (see Figures 8 and 9). As the flow enters the upper section of the OWC the vertical and horizontal components of velocity combine to induce a horizontal vortex centred near the front face. On entry to the turbine duct, the vortex motion persists.

Although no other geometries have been examined, it is thought that the strength of this vortex is dependant upon the relative areas of the top and bottom sections of the OWC. If these sections were made more similar in size. it is believed the strength of the vortex would be reduced. Now that such a strong vortex motion has been discovered within this particular design, it is a feature which designers of future OWC's should consider. This demonstrates the value of conducting a CFD exercise prior to construction.

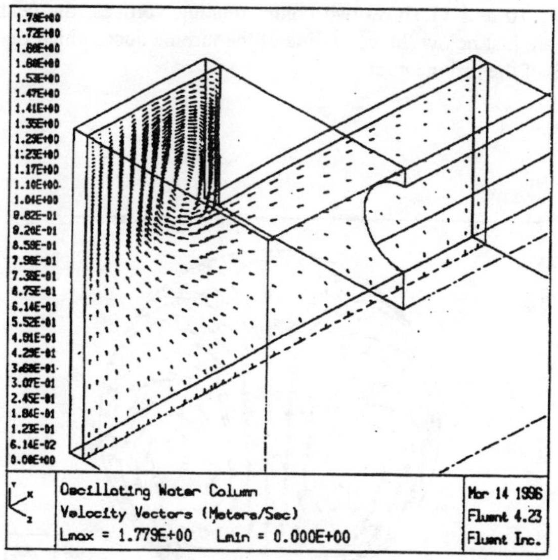

Velocity Vectors In A Vertical Plane Near The Wall Of The OWC

Figure 8

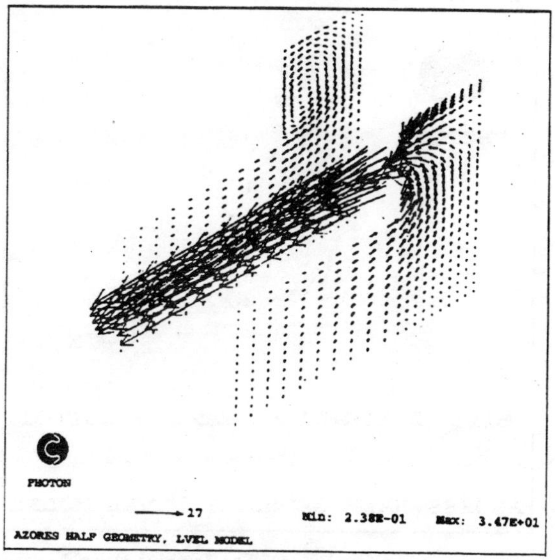

Velocity Vectors In Vertical Planes At The Outer Wall And The Plane Of Symmetry Of The OWC

Figure 9

Figures 10 and 11 show two contra rotating vortices, the centres of which are just below the centreline of the turbine duct, which persist to the exit of the turbine duct.

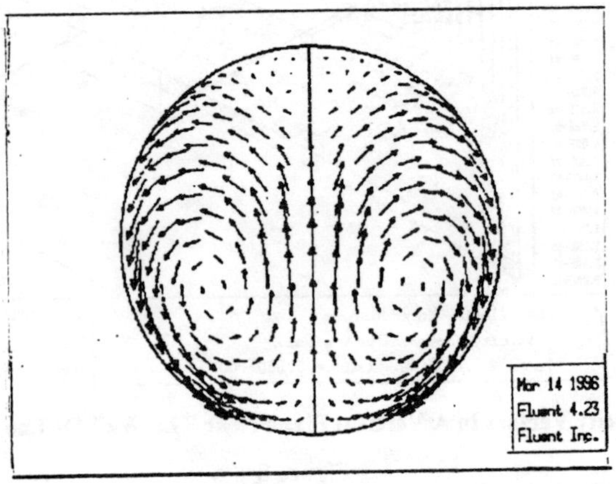

Velocity Vectors In The Turbine Duct, At The Turbine Location, Normal To Mean Flow Direction

Figure 10

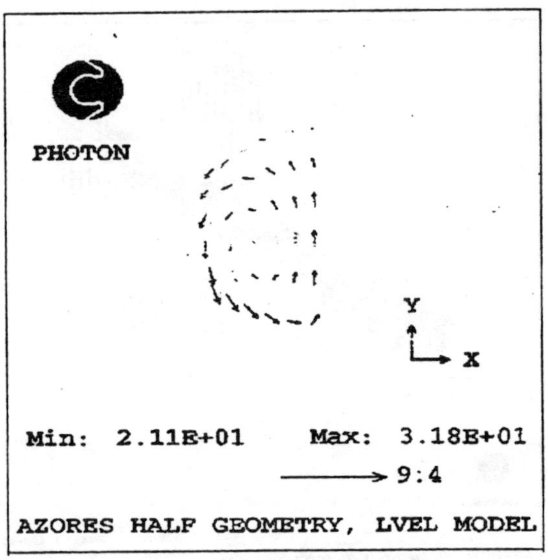

Velocity Vectors At The Exit Of The Turbine Duct, Normal To Mean Flow Direction

Figure 11

This swirl motion is superimposed on a mean axial flow which is biased towards the upper section of the turbine duct. (Figures 12 and 13). The flow is therefore non-axisymmetric and its effect on the turbine output performance should be determined.

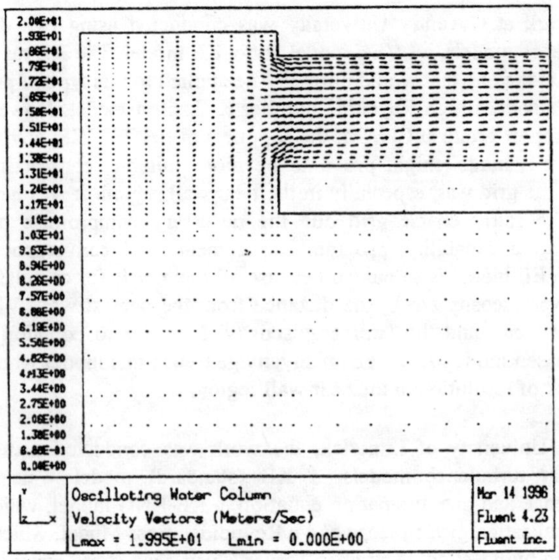

Velocity Vectors Along The Plane Of Symmetry Of The OWC

Figure 12

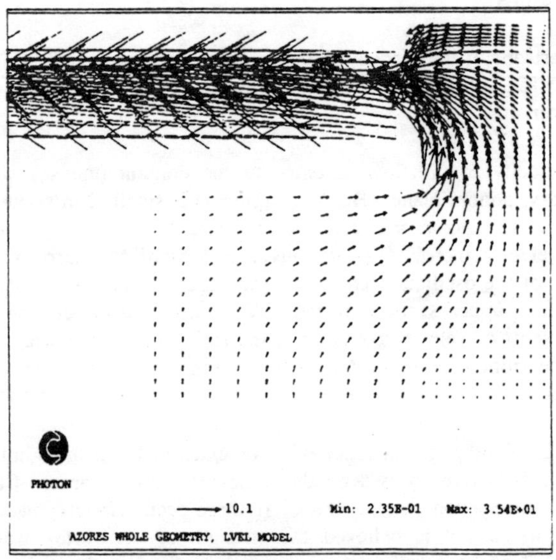

Velocity Vectors Along The Plane Of Symmetry Of The OWC

Figure 13

It can be seen (referring to Figures 8 and 9) that in the main chamber and entrance to the turbine duct there is low velocity and therefore high pressure. Further investigation reveals (Figure 14) that as the flow accelerates into the duct, there is a decrease in static pressure followed by a slight recovery as the flow progresses along the duct.

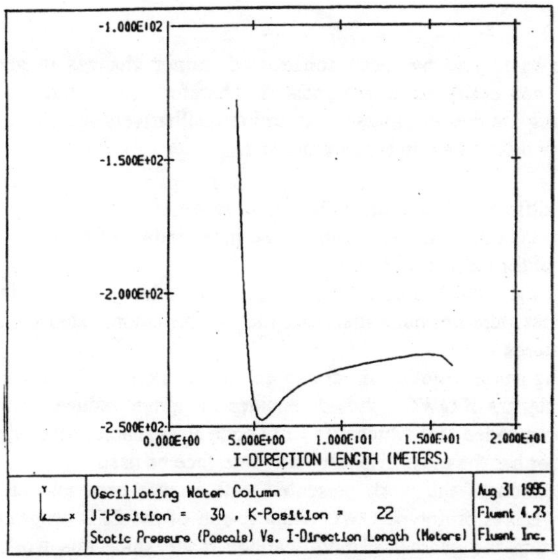

Values Of Pressure Through The Centre Of The Model (The Same Cross Section As Shown In Figure 3.5), From The Chamber Through To The End Of The Turbine Duct

Figure 14

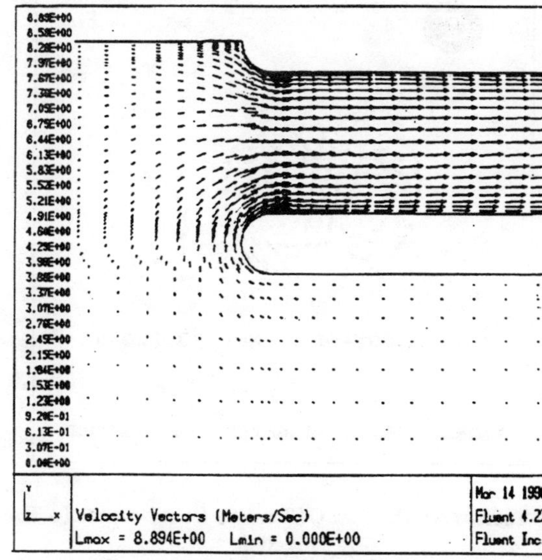

Velocity Vectors Along The Plane Of Symmetry Of The OWC - Model With Bellmouth Entry Added

Figure 15

4.0 Extensions To The Basic Model

4.1 University Of Limerick

Three different turbulence models were used to compare the same case. It was concluded that the velocity and pressure profiles at the turbine location were of the same shape with the magnitudes differing. The profiles were not symmetric with the fluid in the upper half of the duct travelling marginally faster than the fluid in the lower half. There was also some swirl component to the velocity in the duct. Variation in turbulence quantities at inlet had no effect on the flow quality at the turbine location.

However, whilst the KE-RNG and Reynold Stress Models gave similar results, those of the KE (standard) model gave significantly different predictions of turbulence kinetic energy and also affected predictions of axial velocity profiles at the turbine location. When a bellmouth entry was added to the model, the asymmetry of the velocity profiles at the turbine location was reduced, plus the separation of the flow at entry to the circular duct was removed. (see figure 15)

4.2 Coventry University

In addition to modelling half the geometry, (the other half being symmetrical) the whole of the geometry was modelled in order to test the CFD programme for grid dependency. These results proved similar enough to those obtained previously for it to be concluded that the programme is grid independent. (comparing Figure 16 with Figure 9 and Figure 17 with Figure 11)

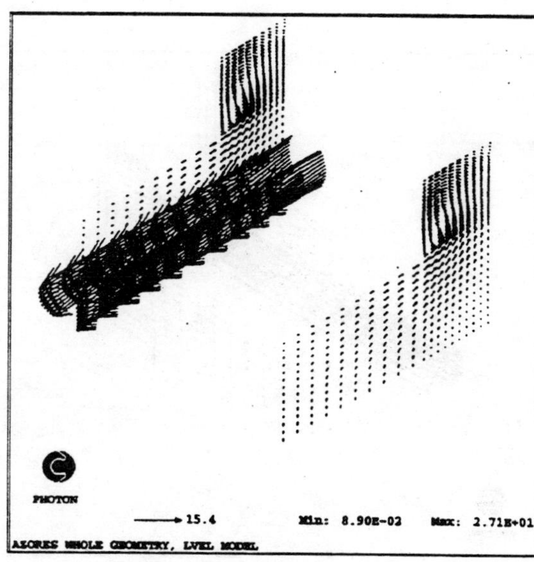

Velocity Vectors In Vertical Planes At The Two Outer Walls Of The OWC.

Figure 16

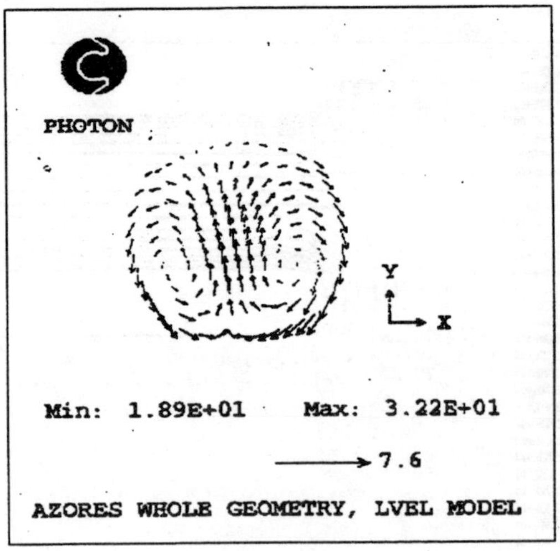

Velocity Vectors At The Exit Of Turbine Duct Normal To The Flow Direction

Figure 17

Another model was tried which had twice as many cells along the y-axis of the turbine duct, to assess the effect of grid density. Again the results proved to be similar, although, the more cells there are in a model, the more detailed a picture it is possible to get of the flow and pressure patterns. However this should be balanced against the increase in the computational time required to obtain a converged solution. These various cases were all evaluated using two slightly different turbulence models, a LVEL model and a KE model. Both models proved to give almost identical results. (comparing Figure 18 with Figure 9)

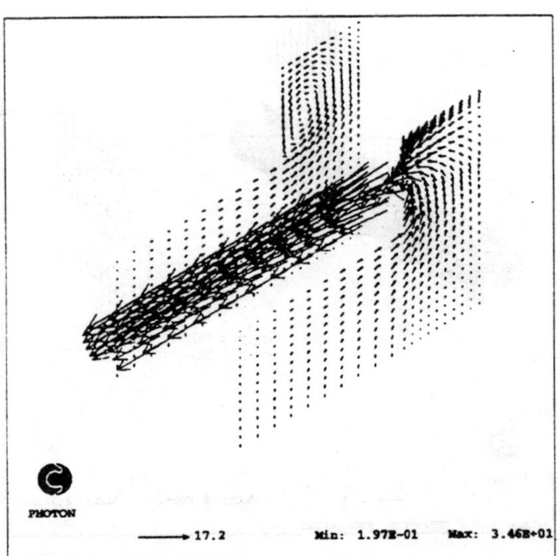

Velocity Vectors In Vertical Planes At The Outer Wall And The Plane Of Symmetry Of The OWC

Figure 18

5.0 Conclusions

Once the basic grid has been constructed, minor changes in generic geometry can easily be accommodated. Therefore CFD codes can be used during the design process to examine qualitatively the changes to the airflow distribution in the turbine duct.

- The two different CFD codes gave similar results.
- The solutions obtained were found to be independant of grid size, grid type and of the turbulence model used
- Although the actual predicted pressure and velocity values in the turbine duct were not quite identical, they both demonstrated the same major features.
- The strong vortex motion described in this work is a feature which future designers of OWC's should consider and aim to reduce.
- It is recommended that whichever CFD code is available, best suits the problem, or has the most 'friendly' user interface be used.
- The validation of the work presented will require experimental data from the Azores prototype OWC and a re-run of the CFD programmes incorporating the geometric modifications made since this work was undertaken.

Acknowledgements

The work here was financially supported by the EC under contract number JOU2-CT93-0333. Additional facilities were provided by the University of Limerick and Coventry University.

References

Curran, R., Raghunathan, S. and Whittaker, T.J.T. (1995) 'Importance Of Flow Profile Uniformity To The Islay Wells Turbine,' *Second European Wave Power Conference, Lisbon, Portugal.*

Raghunathan, S. and Ombaka, O.O. (1984) 'The Wells Turbine In An Oscillating Airflow' *Proceedings of the 19th Intersociety Energy Conversion Conference.*

Thakker, A. and O'Dowd, M. (1994) 'Wells Turbine Research and Computational Dynamics Study Of Air Flow In An Oscillating Column Device' *Air Turbine Development and Assessment for Wave Power Plants - Progress Report for EC contract JOU2-CT93-0333.*

White, P.R.S. (1995) 'The Effect Of Non-Axisymmetric Flow On Wells Turbine Performance,' *Second European Wave Power Conference, Lisbon, Portugal.*

White, P.R.S. (1996) 'Air Turbine Development And Assessment For Wave Power Plants,' *Final report for EC contract JOU2-CT93-0333.*

Proceedings of the Seventh (1997) International Offshore and Polar Engineering Conference
Honolulu, USA, May 25-30, 1997

ISBN 1-880653-28-1 (Set); ISBN 1-880653-29-X (Vol. I)

Air Turbine with Cambered Blades for Wave Energy Conversion: Performances in Steady Flow Condition

Kenji Kaneko, Manabu Takao and Toshiaki Setoguchi
Saga University
Saga, Japan

ABSTRACT

Experimental investigations directed towards improvement of the performance of the Wells turbine for wave energy conversion are reported. The Wells turbine has inherent disadvantages: lower efficiency, poorer starting characteristics and higher axial thrust in comparison with conventional turbines. In this report, a new idea of a shape-controlled airfoil to be applied to turbine blade is proposed. With this flexible airfoil, blade camber can be controlled so as to optimize to inlet flow condition. In order to get a basic design data and to estimate the characteristics of the air turbine with cambered blades, the effects of blade shape and solidity on the turbine characteristics are investigated. The results are compared with those of the Wells turbine. As a result, it is found that the starting characteristics is extremely improved by use of cambered blades. Furthermore, a suitable choice of the design factors is suggested.

Key words : Fluid Machinery, Axial Air Turbine, Cambered Blade, Wave Energy, Conversion, Shape-controlled Blade, Shape Memory Alloys

INTRODUCTION

Today, natural energy resources are again in focus mainly from the viewpoint of global environmental problem. Once it had been a world wide topic after the oil crisis attacked all over the world in 1973. Since then, various research programs have been conducted among several countries to save fossil fuel resources.

Wave energy has relatively higher energy density than other natural energy resources. Many researchers have proposed various wave energy generation systems in these twenty years. Among them, an air turbine called Wells turbine combine with OWC(oscillating water column) is recognized as a promising wave energy generation system. In this system, wave energy is primarily converted into air pressure energy and then Wells turbine absorbs pressure energy to drive the generator.

Wells turbine is a self-rectifying air turbine invented specifically for wave energy generator by Dr.A.A.Wells in 1976 in UK, Fig.1. It is a kind of axial turbine with non-staggered symmetrical airfoil.

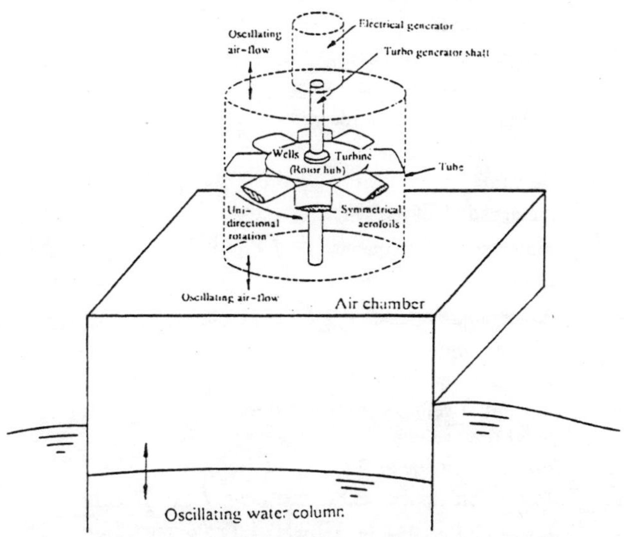

Fig. 1 Wells turbine

For this rotor configuration it is possible to rotate in a same direction under the alternating air flow induced by OWC. This wave generation system has very simple construction, that is a big merit for easy fabrication and maintenance. There are several reports on the performance of the Wells turbine (Raghnathan, et al.,1985,1986, Inoue, et al.,1986a,b, Kaneko, et al.,1986, Setoguchi, et al.,1986). According to these results, the Wells turbine has inherent weak points: lower turbine efficiency and poorer starting characteristics in comparison with the conventional turbine. In order to improve these weak points, newly devised self-rectifying turbines were proposed until now (Hamakawa, et al., 1988, Inoue, et al., 1986c, Kaneko, et al., 1991, Setoguchi, et al., 1993, 1994, 1996 and Maeda, et al., 1995) But each turbine has both its own advantage and disadvantage.

If the symmetrical airfoil shape of the Wells turbine rotor can be controlled to be optimized to alternating flow condition in phase with wave motion, it is reasonably expected that the running or starting characteristics may be improved. One of the possible ways to realize this idea is to adopt the shape memory alloy to the rotor blade. As shown in Fig.2, the camber of the airfoil is to be controlled to vary positive and negative according to the flow direction. For high angle of attack, cambered airfoil would produce a higher lift force than the symmetrical airfoil and it must leads to a higher torque.

The objective of this study is to develop the air turbine with variable geometry blade using shape memory alloy. Specifically a variable camber airfoil is firstly considered because camber has important role in the aerodynamic characteristic of an airfoil. In order to obtain the basic design data, the performance of the Wells turbine with solid cambered blade is necessary. For this purpose an experimental investigation was carried out for three classical cambered airfoils, i.e. NACA5520, Göttingen642, and NACA63$_3$-618, which indicate relatively high aerodynamic performance in single airfoil. The test results obtained under one directional steady flow condition were compared with those of the Wells Turbine with blade profile of NACA0020. Furthermore, the effect of airfoil shape on the starting characteristics was clarified by the quasi-steady analysis.

NOMENCLATURE

AR : aspect ratio of rotor blade
b : span of blade
C_T : torque coefficient defined by Eq.(3)
f : frequency of sinusoidal wave
I : moment of inertia of rotor
l : blade chord length
Q : flow rate
r_R : mean radius of rotor
S : dimensionless frequency $= f r_R / V_a$
t : time
t^* : dimensionless time $= t f$
T : output torque
T_L : loading torque
U_R : tangential velocity at mean radius
v_a : axial flow velocity
V_a : maximum value of v_a
X_I : dimensionless moment of inertia $= I / (\pi \rho r_R^5)$
X_L : dimensionless loading torque $= T_L / (\pi \rho r_R^3 V_a^2)$
z : number of blades
Δp_0 : total pressure drop across turbine
ν : hub-to-tip ratio
ρ : density of air
σ_R : rotor solidity at mean radius
ϕ : flow coefficient $= v_a / U_R$
Φ : flow coefficient $= V_a / U_R$
ω : angular velocity of rotor
ω^* : dimensionless angular velocity $= \omega / f$

OUTLINE OF SHAPE-CONTROLLED BLADE

The Wells turbine rotor consists of unstaggered blade row with symmetrical profile, as shown in Fig.1. Adopting a symmetrical profile is a necessary condition in the design of the Wells turbine in order to get a symmetrical operating condition in a reciprocating flow direction. On the other hand, it is true that the performance of the Wells turbine is limited by relatively small stall angle of attack of the symmetrical airfoil. This limited characteristic is possibly improved by introducing a cambered airfoil because the angle of attack for cambered airfoil is generally greater than that for symmetrical airfoil and also greater lift to drag ratio is expected. But in this case it is necessary to change the sign of the camber angle periodically for the turbine to suitably operate under reciprocating flow direction, as shown in Fig.2. The airfoil shape drawn by solid line shows the condition when air flows upward. For opposite flow direction, the airfoil shape changes as shown by dashed line.

One possible way to build up this variable camber airfoil is realized by use of a shape memory alloys. For example, a sheet spring and shape memory alloy are combined to shape a camber line of a airfoil. By controlling the temperature of the shape memory alloy, it is easy to control camber from positive to negative for a flexible airfoil. In the Wells turbine periodical temperature control is to be related to the air flow direction from OWC via pressure change of the air chamber.

TEST APPARATUS AND PROCEDURE

The performance test of the turbine with cambered blades was carried out by use of a specifically designed wind tunnel which simulates OWC. It also enables to produce both steady and oscillating air flow by controlling the motion of a 1.4m-diameter piston in a cylinder as instructed by a microcomputer. In the present experiment, steady unidirectional flow was supplied to turbine. The turbine was set up at the exit of the wind tunnel as shown in Fig.3. In testing the turbine characteristics, the variations of turbine performance with flow coefficient were examined by changing the rotational speed of the rotor step by step from a low speed to high speed, so as to cover the effective operating range of the turbine. During the test, the turbine output torque, the air flow rate and the total pressure drop across the rotor were also measured, and these signals were logged and analyzed with an on-line microcomputer.

The four tested blade profiles in this experiment are shown in Fig.4, they are NACA0020, NACA5520, Göttingen 624 and NACA63$_3$ -618. Here, NACA0020 airfoil (non-cambered blade) was adopted to Wells turbine for comparison of the characteristics. The selected cambered airfoils have relatively low drag to lift ratio according to the available test data of two-dimensional isolated airfoils (Abbot, et al., 1956 and Wieselsberger, et al., 1923).

The specifications of all test rotor are summarized in Table 1. They are arranged in order to clarify the effects of airfoil profile and solidity at mean radius on the performance. Tip clearance between rotor and outer casing is 1mm. The test Reynolds number is about Re= 2.0×10^5 based on the tip chord and tip velocity.

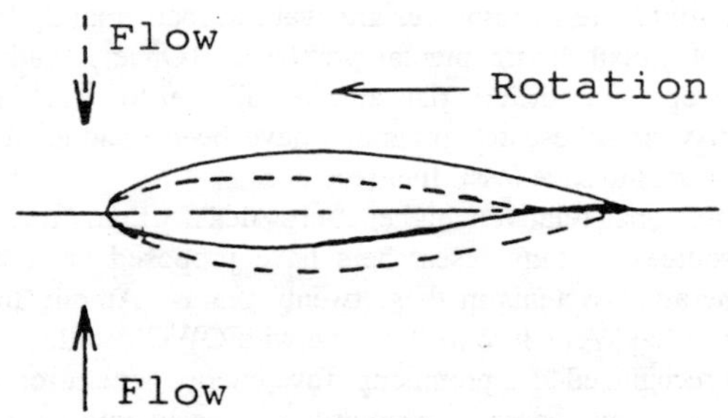

Fig.2 Shape-controlled airfoil

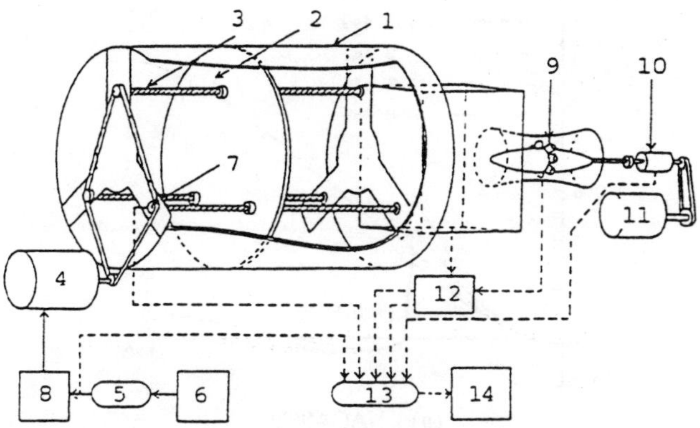

Fig.3 Wind tunnel for turbine test

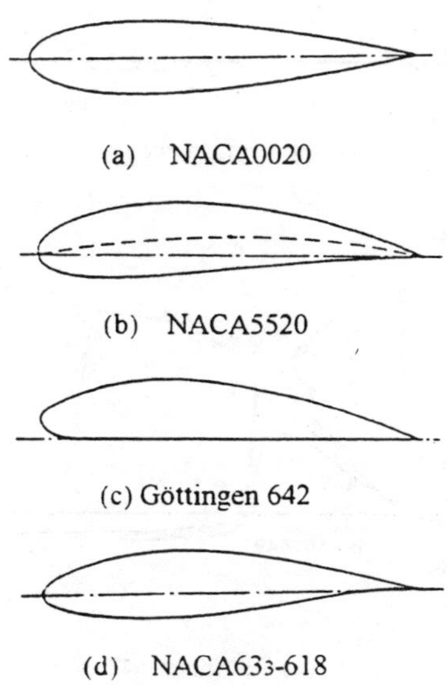

Fig 4 Blade profiles for rotor

Table 1 Specifications of rotor

Rotor		
Chord length	l	: 90mm
Solidity at mean radius	σ_R	: 0.449, 0.562, 0.674, 0.786
Blade profile		: NACA0020, NACA5520
		Göttingen624, NACA63_3 – 618
Aspect ratio	AR	: 0.5
Hub to tip ratio	ν	: 0.71
Tip diameter	D_t	: 298mm

RESULTS AND DISCUSSIONS

Running Characteristics

When the turbine operates at a constant rotating speed in an unidirectional steady air flow, the running characteristics of the turbine are evaluate by efficiency η, which is defined as:

$$\eta = T\omega / \Delta p_o Q \qquad (1)$$

Figure 5 shows the effects of the blade profile and solidity on the efficiency, where the experiments were carried out under the unidirectional flow condition. The peak efficiency decreases with σ_R for any blade. For NACA0020 the maximum peak efficiency is obtained with σ_R=0.449. The efficiency of cambered blade is lower than that of non-cambered blade. However, the optimum blade profile should be determined taking account of both the running and starting characteristics. As will be shown in next section, the turbines with σ_R=0.449 and 0.562 (NACA0020) do not start by themselves. Among the turbines with self-starting, the highest efficiency is obtained in the case of σ_R=0.674 for NACA0020. Comparably high efficiency is also obtained with cambered blades in the case of σ_R=0.562 for NACA5520 and NACA63 3 -618. The comparison of these three favorable turbines shows that the recommendable cambered blade is NACA5520 with σ_R=0.562. This is because turbine efficiency keeps comparably high value over wide operating range of flow coefficient Φ and the value of Φ at the peak efficiency point is the highest among them. The latter means that the design rotational speed of the turbine is the lowest. This is advantageous from the standpoint of noise reduction.

Starting Characteristics

The advantage of using cambered blades is discussed from the standpoint of the starting characteristics. In general, the starting characteristics of the turbine in a sinusoidally reciprocating flow can be evaluated with the equation of motion for the rotating system. The equation (Inoue,et al.,1986a) is given in the dimensionless form as :

$$S^2 X_I (d\omega^*/dt^*) + X_L = 2C_T\sigma_R \frac{1-\nu}{1+\nu}\{\sin^2(2\pi t^*) + \frac{1}{\Phi^2}\} \qquad (2)$$

where C_T is the torque coefficient defined as:

$$C_T = 2T / \{\rho(v_a^2 + U_R^2) b l r_R z\} \qquad (3)$$

Equation (2) can be solved numerically as an initial value problem when C_T and wave motion are given. In the calculation, the flow condition is assumed to be quasi-steady (strictly speaking, this is not correct because of hysteresis in $C_T - \phi$ characteristics, which will be shown in other paper near future). Figure 6 shows the $C_T - \phi$ characteristics obtained under steady flow conditions which were used in the calculation. Furthermore, it is assumed, for simplicity, that there is no delay in the behavior of changing blade profile, namely, the blade shape changes instantaneously itself to the required one corresponding to the flow direction.
Figure 7 shows the starting characteristics calculated. The ordinate and abscissa represent nondimensional angular velocity ω^* and

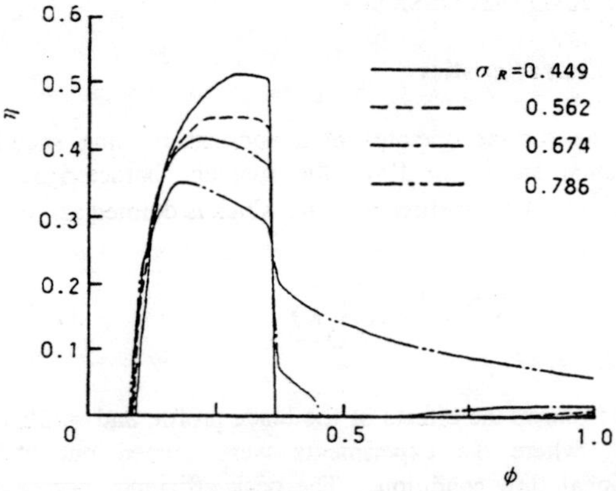

(a) NACA0020

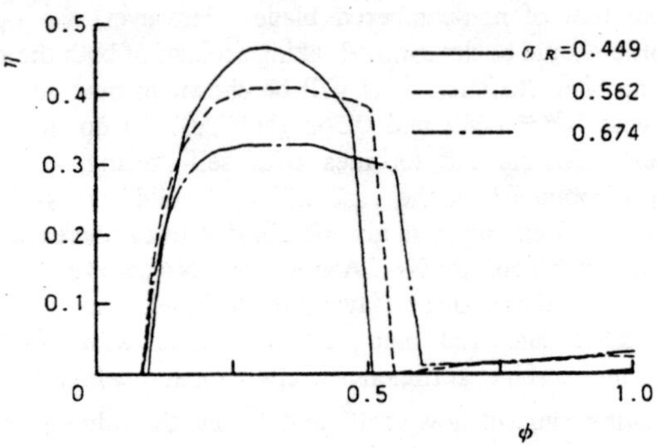

(b) NACA5520

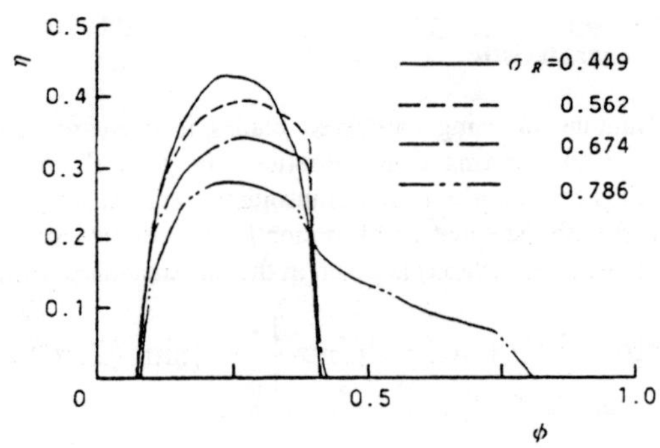

(c) Göttingen 642

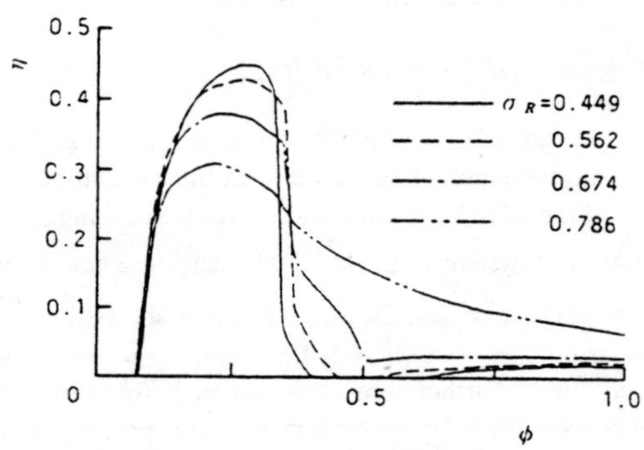

(d) NACA633-618

Fig.5 Running characteristics

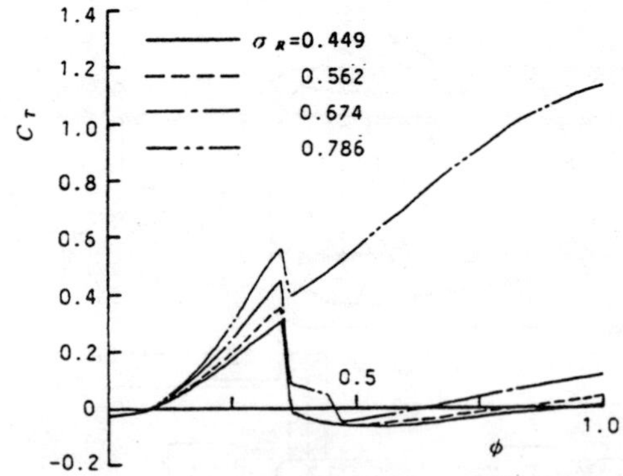

(a) NACA0020

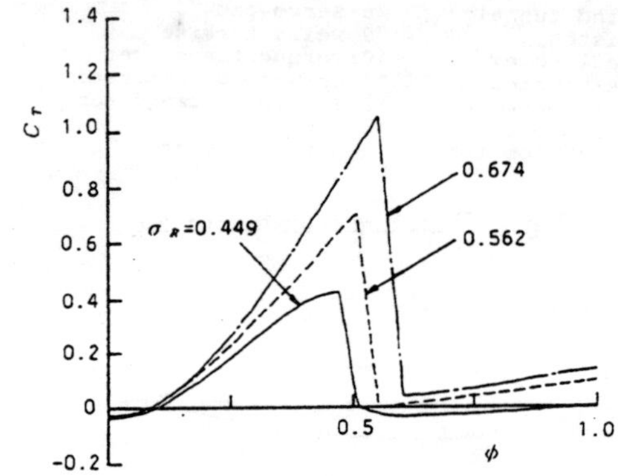

(b) NACA5520

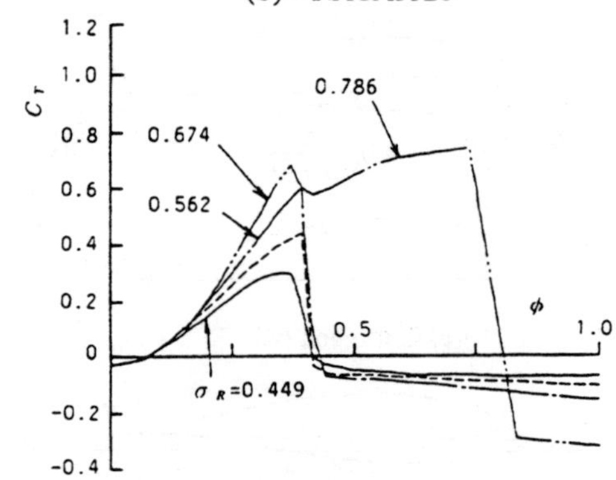

(c) Göttingen 642

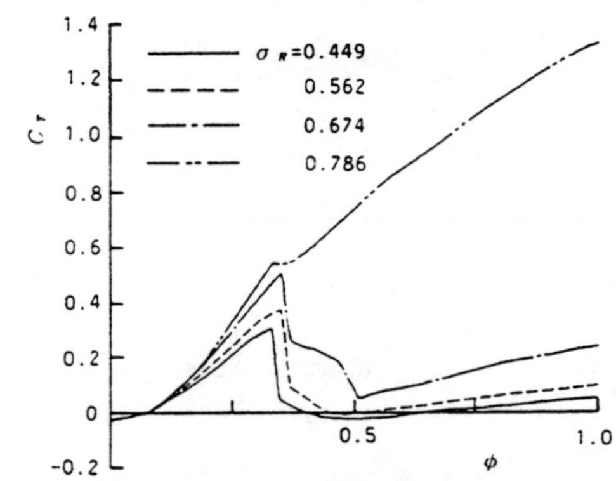

(d) NACA633-618

Fig.6 Torque coefficient characteristics

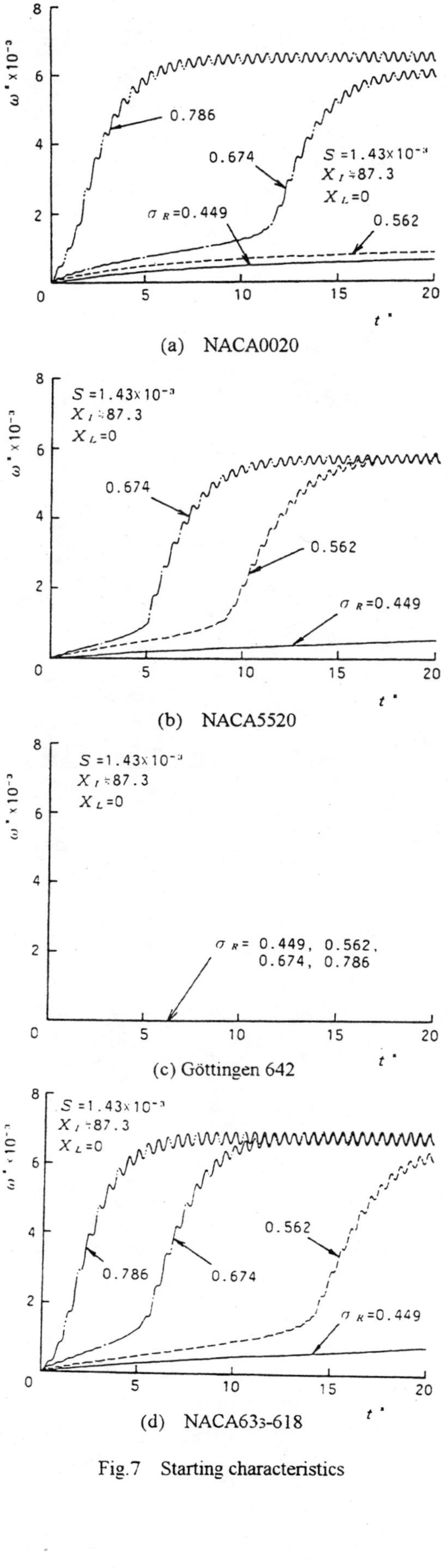

Fig.7 Starting characteristics

nondimensional time t^*, respectively. The turbine using non-cambered blades cannot start for $\sigma_R \leq 0.562$, Fig.7(a). However, it is clear that the rotor using NACA5520 with $\sigma_R = 0.562$ can start, and the asymptotic angular velocity is lower than that for NACA0020. Therefore, it can be said that the starting characteristics, in general, are improved by using cambered blades except for Göttingen624 blades. The rotor using Göttingen624 blades is not suitable for this type of turbine for wave energy conversion because it can not start for any σ_R. The starting characteristics depend on tangential component (parallel to the chord) of lift force (this corresponds to C_T). In the case of NACA5520 and NACA63₃ -618, the values of C_T are higher over wide range of ϕ than those of non-cambered blade, as shown in Fig.6. It is noted that the turbine with Gö ttingen624 blades might rotate in an opposite direction when starting from rest. This is inferred from the fact that the value of C_T keeps negative in the region of high ϕ-value as shown in Fig.6(c).

CONCLUSIONS

The performance of the wave-power air turbine using cambered blades has been investigated by model testing in steady unidirectional flows. The results have been compared with that of the turbine using non-cambered blades(Wells turbine). The conclusions are summarized as follows:

(1) The turbine using cambered blades is, in general, advantageous from a viewpoint of starting characteristics in comparison with the turbine using non-cambered blades (Wells turbine). Furthermore, it is superior in running characteristics to the Wells turbine from a viewpoint of low operation speed.

(2) The recommendable geometrical parameters of the turbine using cambered blades are blade profile: NACA5520 with a solidity at mid span: $\sigma_R = 0.562$.

REFERENCES

Abbott, I. H., et al (1956). "Theory of Wing Section," *Dover Pub.*

Hamakawa, H., Kaneko, K., Setoguchi, T., and Inoue, M (1988). "Aerodynamic Performance of Air Turbine with Self-Pitch-Controlled Blades for Wave Power Generator," *Proc. of 1st KSME-JSME Thermal and Fluid Eng. Conf.*, pp.127-132.

Inoue, M., Kaneko, K., Setoguchi, T. and Raghunathan, S (1986a). "Simulation of Starting Characteristics of the Wells Turbine," *AIAA/ASME 4th Fluid Mech., Plasma Dynamics and Lasers Conf.*, Paper No.88-1122.

Inoue, M., Kaneko, K. and Setoguchi, T (1986b). "Determination of Optimum Geometry of Wells Turbine Rotor for Wave Power Generator (Part 2)," *Proc. of Current Practice & New Tech. in Ocean Eng.*, pp.441-446.

Inoue, M., Kaneko, K., Setoguchi, T. and Raghunathan, S (1986c). " Starting and Running Characteristics of Biplane Wells Turbine," *Proc. of 5th Int. Offshore and Arctic Eng. Symp.*, pp.574-579.

Kaneko, K., Setoguchi, T. and Inoue, M (1986). "Performance of Wells Turbine in Oscillating Flow," *Proc. of Current Practice & New Tech. in Ocean Eng.*, pp.447-452.

Kaneko, K., Setoguchi, T., Hamakawa, H. and Inoue, M (1991). "Biplane Axial Turbine for Wave Power Generator," *Int. J. of Offshore and Polar Eng.*, Vol.1, No.2, pp.122-128.

Maeda, H., Setoguchi, T., Kaneko, K., Kim, T. W. and Inoue, M (1995). "Effect of Turbine Geometry on the Performance of Impulse Turbine with Self-Pitch-Controlled Guide Vanes for Wave Power Conversion," *Int. J. of Offshore and Polar Eng.*, Vol.5, No.1, pp.72-74.

Raghunathan, S., Tan, C. P. and Ombaka, O. O (1985). "Performance of the Wells Self-Rectifying Air Turbine, *Aeronautical J.*, Vol.89, No.890, pp.369-379.

Raghunathan, S (1986). "The Aerodynamic Aspects of Wells Turbine," *The 3rd Int. Symp. on Wave, Tidal, OTEL and Small Scale Hydroenergy*, Paper No.20, Brighton.

Setoguchi, T., Kaneko, K. and Inoue, M (1986). "Determination of Optimum Geometry of Wells Turbine Rotor for Wave Power Generator (Part 1)," *Proc. of Current Practice & New Tech. in Ocean Eng.*, pp.435-440.

Setoguchi, T., Kaneko, K., Maeda, H., Kim, T. W. and Inoue, M (1993). "Impulse turbine with Self-Pitch-Controlled Guide Vanes for Wave Power Conversion: Performance of Mono-Vane Type," *Int. J. of Offshore and Polar Eng.*, Vol.3, No.1, pp.73-78.

Setoguchi, T., Kaneko, K., Maeda, H., Kim, T. W. and Inoue, M (1994). "Impulse Turbine with Self-Pitch-Controlled Tandem Guide Vanes for Wave Power Conversion," *Int. J. of Offshore and Polar Eng.*, Vol.4, No.1,pp.76-80.

Setoguchi, T., Kaneko, K., Taniyama, H., Maeda, H. and Inoue, M (1996). "Impulse Turbine with Self-Pitch-Controlled Guide Vanes for Wave Power Conversion: Guide Vanes Connected by Link," *Int. J. of Offshore and Polar Eng.*, Vol.6, No.1, pp.76-80.

Wieselsberger, C., et al (1923). "Ergebnisse der Aerodynamisehen Versuchsanstalt zu Göttingen," *Druck und Verlag*.

Proceedings of the Seventh (1997) International Offshore and Polar Engineering Conference
Honolulu, USA, May 25-30, 1997

ISBN 1-880653-28-1 (Set); ISBN 1-880653-29-X (Vol. I)

Effect of Guide Vanes on the Performance of a Wells Turbine for Wave Energy Conversion

T. Setoguchi, M. Takao and K. Kaneko
Saga University
Saga, Japan

M. Inoue
Kyushu University
Fukuoka, Japan

ABSTRACT

In order to improve the performance of a Wells turbine, effects of the gap between rotor and guide vane and solidity of the guide vane have been investigated by model testing. The results have been compared with those of the case without guide vane. It is found that the overall characteristics of the turbine are considerably improved by the guide vanes. A suitable choice of design parameters such as the gap and the solidity of the guide vane has been suggested.

KEY WORDS: Fluid machinery, guide vane, natural energy, Wells turbine, wave energy conversion, ocean energy

NOMENCLATURE

AR : aspect ratio of rotor blade
b : blade height
C_A : input coefficient
C_T : torque coefficient
D : drag
D_t : tip diameter of rotor
f : frequency of wave
G : gap between rotor and guide vane
I : moment of inertia of rotor
l_g : chord length of guide vane
l_r : chord length of rotor blade
L : lift
Q : flow rate
r_R : mean radius of rotor
S : dimensionless frequency $= f r_R / V_a$
SPL : sound pressure level
t : time
t^* : dimensionless time $= tf$
T : output torque
T_L : loading torque
U_R : blade speed at mean radius
v_a : mean axial flow velocity
v_1 : absolute velocity at inlet
v_2 : absolute velocity at exit
V_a : maximum value of v_a
w_1 : relative velocity at inlet
w_2 : relative velocity at exit
X_I : dimensionless moment of inertia $= I / (\pi \rho r_R^5)$
X_L : dimensionless loading torque $= T_L / (\pi \rho r_R^3 V_a^2)$
z : number of rotor blades
α : angle of attack
β_1 : absolute inlet flow angle
γ : stagger angle
Δp : total pressure drop between settling chamber and atmosphere
$\bar{\eta}$: mean turbine efficiency
ν : hub-to-tip ratio
ρ : density of air
σ_{gR} : solidity of guide vane at mean radius
σ_{rR} : solidity of rotor blade at mean radius
ϕ : flow coefficient $= v_a / U_R$
Φ : flow coefficient $= V_a / U_R$
ω : angular velocity
ω^* : dimensionless angular velocity $= \omega / f$

INTRODUCTION

Several of the wave energy devices currently studied in the United Kingdom, Japan, Portugal, India and other countries make use of the principle of oscillating water-air column for converting wave energy to low pressure pneumatic energy which in turn can be converted into mechanical energy by a Wells turbine. This turbine rotates in a single direction in an oscillating airflow and therefore does not require a system of non-return valves. There are several reports which describe the performance of the Walls turbine and factors which influence the performance (Inoue et al., 1986; Kaneko et al., 1986; Setoguchi et al.,

1986; Raghunathan et al., 1987, 1994). According to these results, the turbine efficiency is lower in comparison with the usual turbines. In this case, the guide vanes before and after rotor may be one of the most effective equipment to improve the turbine performance.

The performances of the Wells turbine with guide vanes were studied theoretically (Sturge, 1977; Gato et al., 1990) and experimentally (Inoue et al., 1985; Arakawa et al., 1987) under steady operating conditions so far. But there are many unknown aspects about the favorable configurations of guide vanes.

In this paper, in order to clarify the performance of the Wells turbine with guide vanes, the effects of the gap between rotor and guide vane and solidity of the guide vane have been investigated by model testing.

EXPERIMENTAL METHOD AND PROCEDURE

The test rig consists of a large piston cylinder, a settling chamber and a 300-mm-dia test section with a bell-mouth entry and diffuser exit (Fig.1). The details are shown in Kaneko et al. (1986). The turbine rotor with hub-to-tip ratio of 0.71 was placed at the center of the test section and tested at a constant rotational speed both under steady flow condition and sinusoidal flow condition (frequency of wave : $f = 0.1$Hz). The flow rate was evaluated by integrating the flow velocity distribution over a cross-sectional area which was obtained by a Pitot tube survey. The overall performance was evaluated by torque T, the flow rate Q, the rotor angular velocity ω and the total pressure drop between the settling chamber and atmosphere Δp. Furthermore, in order to clarify the noise characteristics under operating condition, the sound pressure level was measured at 1m distance from the turbine inlet with azimuth angle of 45° in a semi-reverberant chamber by a sound noise meter (RION NA-61). Tests were performed in the range of Δp up to 1800N/m^2, ω up to 523rad/s and Q up to 0.476 m^3/s. The Reynolds number based on the blade chord was approximately equal to 2×10^5. The uncertainty in turbine efficiency is about ±2%.

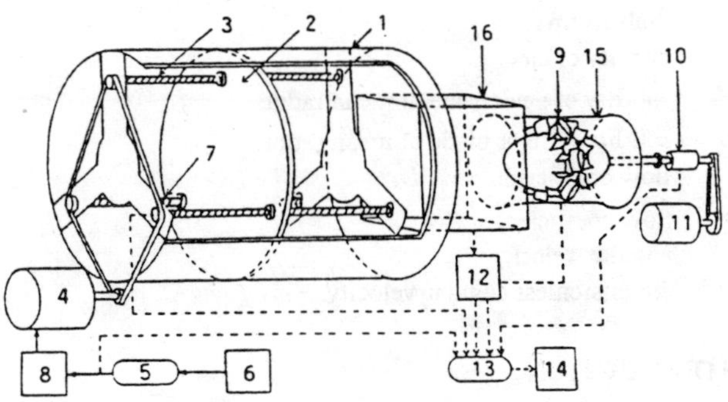

Fig.1 Outline of test apparatus

The guide vanes with the chord length of $l_g = 91$mm are symmetrically installed at the distance G downstream and upstream of the rotor as shown in Fig.2. The detailed information of guide vane used in the experiments is as follows; circular-arc profile with constant thickness ratio of 0.011, solidity at mean radius of $0.454 \le \sigma_{gR} \le 2.5$, camber angle of 60°, aspect ratio of 0.5, gap to chord ratio of $0.122 \le G/l_r \le 0.589$, stagger angle of $8.3° \le \gamma \le 20.3°$ (the design stagger angle: 16.3°), and leading and trailing edges are both rounded to a semi-circle with radius of 0.5mm (see Table 1). These geometries were determined by using the two-dimensional cascade data (Ikui et al., 1971) such that the torque coefficient at mean radius, which will be defined later, is 0.2 (this corresponds to the condition showing the peak efficiency point for the turbine without guide vanes).

The turbine rotor adopted in the experiments is shown in Table 1. That is, the blade profile: NACA0020 with chord length of 90mm, solidity at mean radius: 0.674, aspect ratio: 0.5, tip diameter: 298mm, tip clearance: 1mm. Note that this rotor was considered as the most promising one in the previous studies (Kaneko et al., 1986; Setoguchi et al., 1986).

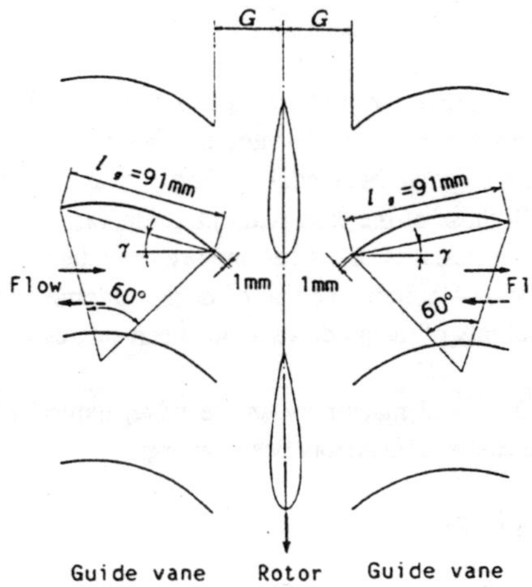

Fig.2 Configuration and dimensions of guide vanes

Table 1 Specifications of turbine

Rotor		
Chord length	l_r	: 90mm
Solidity at mean radius	σ_{rR}	: 0.674
Blade profile		: NACA0020
Aspect ratio	AR	: 0.5
Hub to tip ratio	ν	: 0.71
Tip diameter	D_t	: 298mm
Guide vane		
Chord length	l_g	: 91mm
Solidity at mean radius	σ_{gR}	: 0.454, 0.908, 1.25, 1.81, 2.5
Gap to chord ratio	G/l_r	: 0.122, 0.2, 0.278, 0.433, 0.589
Stagger angle	γ	: 8.3°, 12.3°, 16.3°, 20.3°

EXPERIMENTAL RESULTS AND DISCUSSIONS

Running Characteristics

The effect of guide vanes on the running characteristics is evaluated by the relationship between mean efficiency $\bar{\eta}$ and flow coefficient Φ defined as:

$$\bar{\eta} = \left(f\int_0^{1/f} T\omega \, dt\right) \Big/ \left(f\int_0^{1/f} \Delta p Q \, dt\right) \quad (1)$$

$$\Phi = V_a / U_R \quad (2)$$

where V_a and U_R are the maximum axial velocity and the blade speed at mean radius (see Fig.3).

Fig.4 shows the effect of stagger angle γ on the mean efficiency $\bar{\eta}$. The highest efficiency in the running condition is obtained in the case of $\gamma = 16.3°$ that is the design stagger angle. This fact suggests the validity of the design method of the guide vanes of Wells turbine, in which the torque coefficient is set at mean radius and suitable circular-arc blades are selected from the two-dimensional cascade data.

From the above result, the following investigations are carried out for $\gamma = 16.3°$.

The effect of the gap between rotor and guide vane G on $\bar{\eta}$ is shown in Fig.5, where the case without guide vanes is denoted by the solid line for comparison. The mean efficiency in the case with guide vanes is larger than that of the guide-vaneless turbine except for $G/l_r = 0.122$. Furthermore, in the range of $\Phi < 0.4$, $\bar{\eta}$ increases with G/l_r, and the maximum peak efficiency is obtained for $G/l_r = 0.433$. On the other hand, $\bar{\eta}$ increases with decreasing G/l_r for $\Phi > 0.4$. Now let us consider the above reasons on the basis of experimental results obtained under steady flow conditions.

The torque coefficient C_T and input coefficient C_A of the turbine rotor are defined using the relative inlet velocity $w_1 = \{(U_R - v_a \tan\beta_1)^2 + v_a^2\}^{1/2}$ as:

$$C_T = T / (\rho b l_r r_R w_1^2 z / 2) \quad (3)$$

$$C_A = \Delta p Q / (\rho b l_r v_a w_1^2 z / 2) \quad (4)$$

Here, the values of w_1 and β_1 were not measured, but were determined using v_a, U_R and γ by velocity diagrams shown in Fig.3.

Firstly, let us show the effect of guide vanes before and after rotor on the performance. The variations of C_T and C_A with ϕ are shown in Fig.6, where the dashed and solid lines represent the cases with and without guide vanes, respectively. As is evident from Fig.6(a), the variations of C_T with ϕ are qualitatively the same in both cases. However, the torque coefficient of the turbine with guide vanes keeps high values over a wide range of ϕ. This is due to the fact that the tangential force of ($L\sin\alpha - D\cos\alpha$) (this corresponds to C_T, see Fig.3) of the turbine increases by equipping the guide vanes, because of increase in angle of attack. Furthermore, in the case with guide vanes, the decrease in C_T due to stalling is low in comparison with guide-vaneless turbine. This fact means that the starting and transitional characteristics to running conditions (Inoue et al., 1986) are improved by use of guide vanes. The starting characteristics will

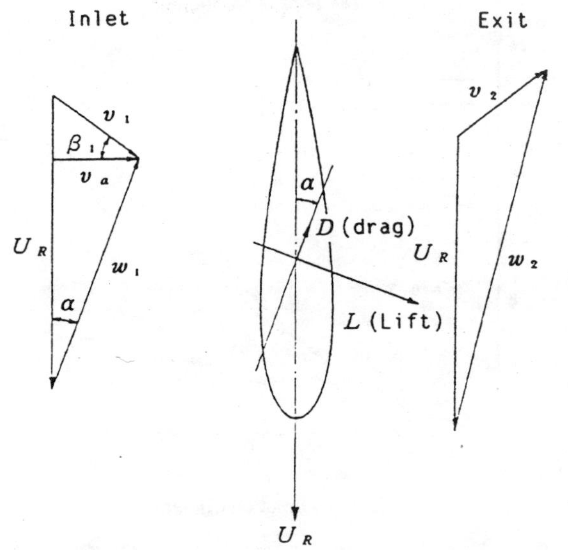

Fig.3 Velocity diagrams before and after rotor (v_1, v_2 : absolute velocities, w_1, w_2 : relative velocities)

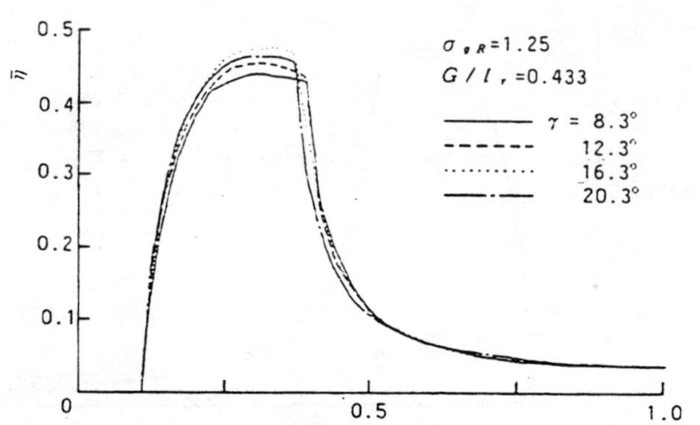

Fig.4 Effect of stagger angle of guide vane on mean efficiency

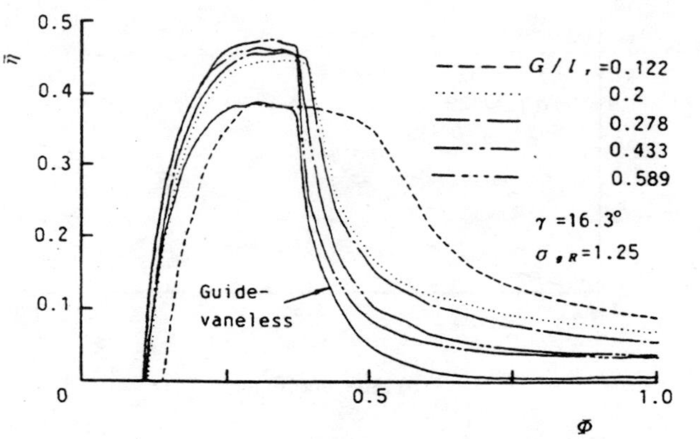

Fig.5 Effect of gap between rotor and guide vane on mean efficiency

be shown more clearly later. The relationship between C_A and ϕ (Fig.6(b)) has the similar tendency to that of $C_T-\phi$ characteristics. This fact means that axial force of $(L\cos\alpha + D\sin\alpha)$ (this corresponds to C_A, see Fig.3) of the turbine increases also by the guide vanes. From the above results, the effect of guide vanes shown in Fig.5 is due to the fact that the increase in C_T is larger than that in C_A because the efficiency under steady flow condition is estimated by

$$\eta = T\omega / (\Delta p Q) = C_T / (C_A \phi) \quad (5)$$

The effects of upstream and downstream guide vanes on turbine characteristics are also shown in Fig.6, where the dotted and chain lines denote the cases with upstream and downstream guide vanes, respectively. In the running condition shown in Fig.6, C_T and C_A with upstream and downstream guide vanes are almost the same as those with upstream guide vanes, and are much larger than that with downstream ones. This means that the improvement of running characteristics is mainly attributed to the effect of the upstream guide should be noted that the value of C_T just after stalling in the case

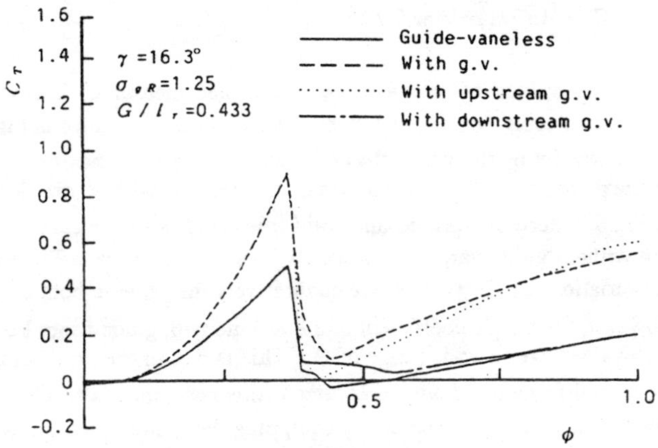

(a) Torque coefficient

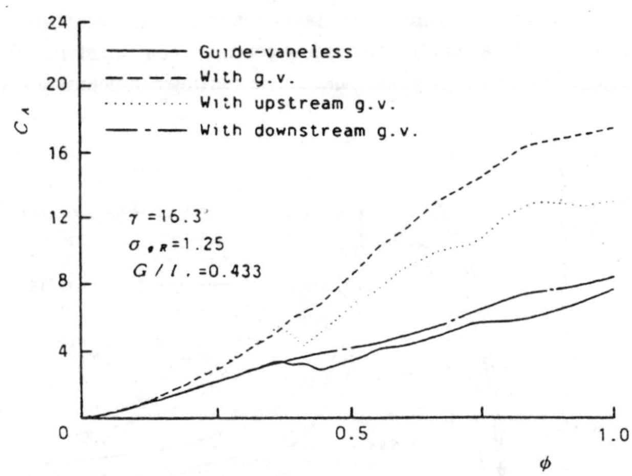

(b) Input coefficient

Fig.6 Turbine characteristics under steady flow conditions (Effect of guide vanes before and after rotor)

vanes, and the diffuser effect on it is small. On the other hand, it with downstream guide vanes is larger than that with upstream ones. This fact means that the starting characteristics of the turbine is improved by the downstream guide vanes.

The turbine characteristics under steady flow conditions are shown

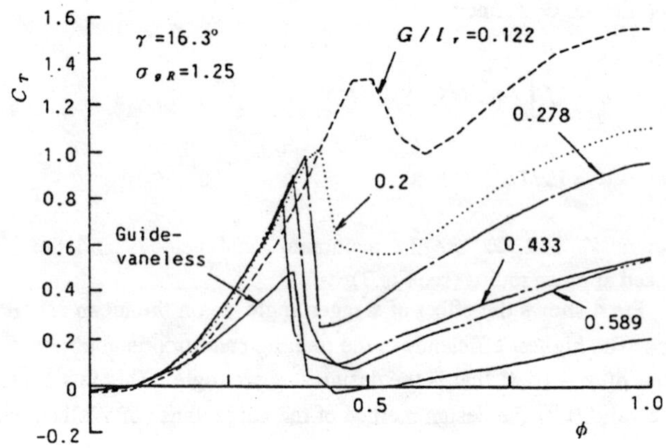

(a) Torque coefficient

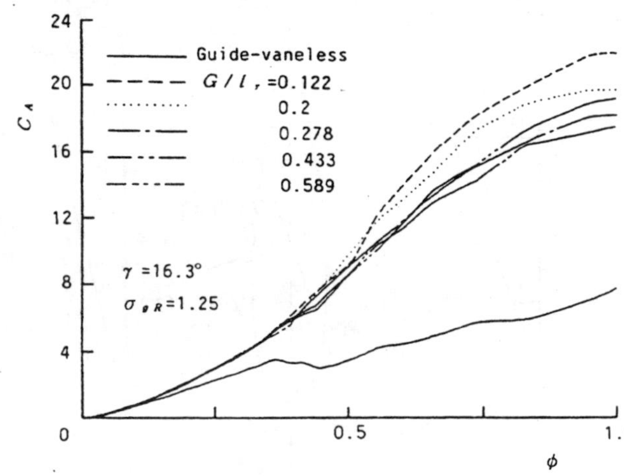

(b) Input coefficient

Fig.7 Turbine characteristics under steady flow conditions (Effect of gap between rotor and guide vane)

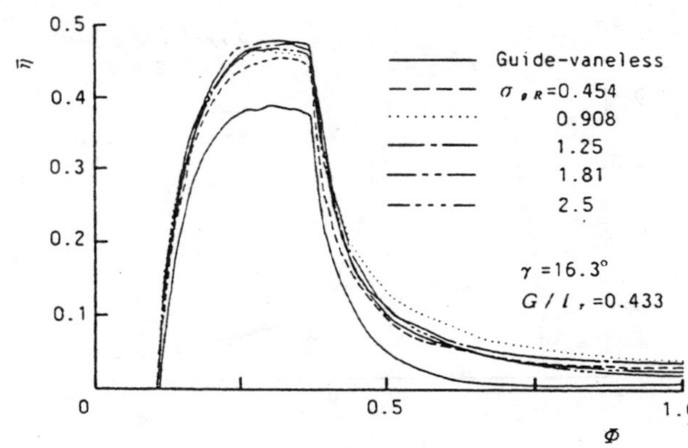

Fig.8 Effect of solidity of guide vane on mean efficiency

in Fig.7, where the parameter is G/l_r. The figure shows that the C_T increases with G/l_r in the running condition although the C_A does not depend on it. From these results, we can easily explain the behavior of $\bar{\eta}-\Phi$ characteristics shown in Fig.5.

In general, the stall angle increases with decreasing G/l_r. The smaller the gap is, the larger the turbulence level becomes (Raghunathan et al., 1987). This may be said that the increase in stall angle is due to high turbulence level of incoming axial flow.

The effect of the solidity of guide vane σ_{gR} on mean efficiency $\bar{\eta}$ is shown in Fig.8. The peak efficiency increases with σ_{gR}, and then it becomes almost the same value for $\sigma_{gR} \geq 0.908$. In this case, the optimum solidity for the running characteristics from a viewpoint of economical reason can be said that $\sigma_{gR}=0.908$ because smaller solidity results in less number of guide vanes. The relationship between $\bar{\eta}$ and σ_{gR} is also discussed by the turbine characteristics obtained under the steady flow conditions (Fig.9).

Starting Characteristics

The starting characteristics of the turbine are evaluated by the variation of the rotational speed from the rest point. Fig.10(a) shows the effect of G/l_r on the starting characteristics under the sinusoidal flow condition, where the abscissa and the ordinate represent the dimensionless time t^* and the dimensionless angular velocity ω^*, respectively. X_I, X_L and S in the figure denote the nondimensional moment of inertia, the nondimensional loading torque and the nondimensional frequency, respectively (Inoue et al., 1986). As all turbines with guide vanes can start in a very short time, it is clear that the starting characteristics can be improved by the guide vanes. This is because the value of C_T at just after stalling position is positive in the cases with guide vanes although it is negative for the guide-vaneless turbine as shown in Fig.7(a). The smaller the G/l_r is, the more improved the starting characteristics are, and the lower the rotational speed in the running condition is. These are because the value of C_T at stalling point increases with decreasing G/l_r for the former case, and the flow coefficient at loading-free condition (i.e. $\bar{\eta}=0$) is the largest in the case of $G/l_r=0.122$ (see Fig.5) for the latter case. Fig.10(b) shows the effect of solidity σ_{gR} on the starting characteristic, where the optimum gap to chord ratio for the running

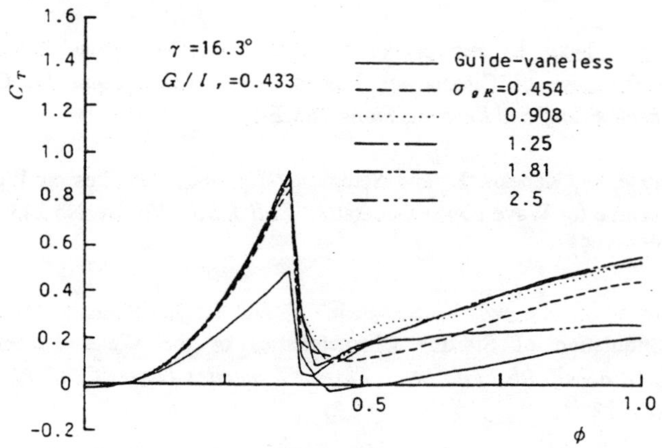

(a) Torque coefficient

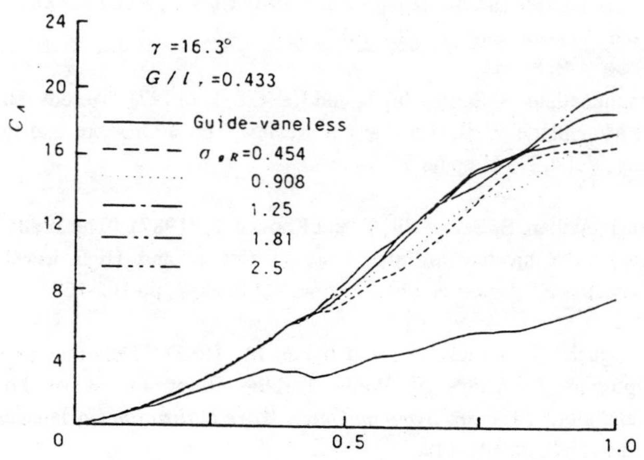

(b) Input coefficient

Fig.9 Turbine characteristics under steady flow conditions (Effect of solidity of guide vane)

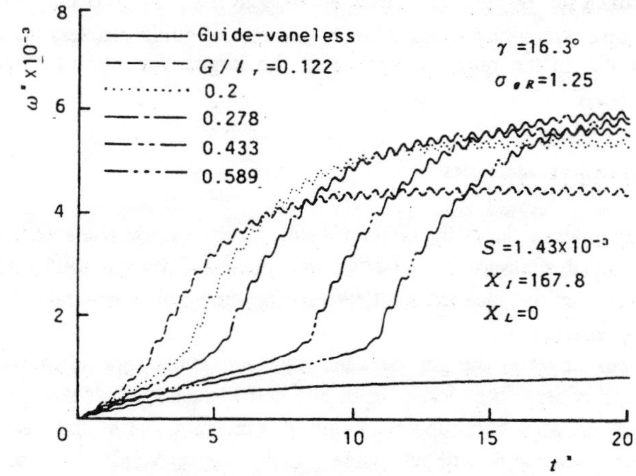

(a) Effect of gap between rotor and guide vane

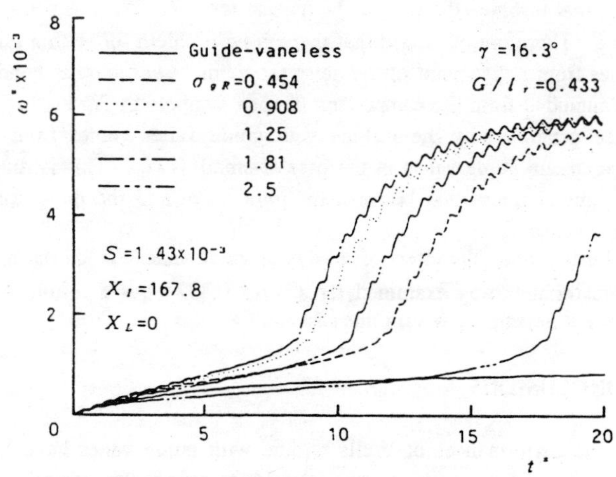

(b) Effect of solidity of guide vane

Fig.10 Starting characteristics

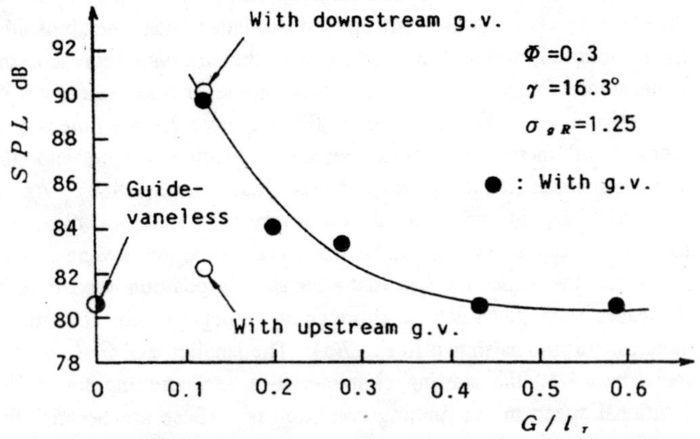

Fig.11 Effect of gap between rotor and guide vane on sound pressure level

characteristics; $G/l_r = 0.433$ is used. The figure shows that all turbines with guide vanes can start, and the optimum solidity is obtained for $\sigma_{gR} = 1.25$. These are evident from the facts that C_T in the case with guide vanes is larger than that of guide-vaneless turbine and C_T in the range of stalling is the largest for $\sigma_{gR} = 1.25$ (see Fig.9(a)).

Noise Characteristics

In order to clarify the effect of guide vanes on noise characteristics, the sound pressure level (*SPL*) was measured by the sound noise meter. In the measurement, the background noise was taken into consideration.

The effect of the gap between rotor and guide vane is shown in Fig.11, where the closed, open and half closed symbols denote the cases of with both upstream and downstream, with upstream or downstream and without guide vanes, respectively. The sound pressure level decrease with increasing G/l_r. However, *SPL* becomes the same to that of the guide-vaneless turbine for $G/l_r > 0.4$. Note that turbine efficiency is the highest for $G/l_r \fallingdotseq 0.4$ as shown in Fig.5. Therefore, it is said that there is no problem for setting guide vanes from a viewpoint of the noise problems. On the other hand, it is concluded from the comparison of open symbols in Fig.11 that the noise generated by the turbine with guide vanes comes from the downstream guide vanes in the case of small G/l_r. This is due to the mutual interaction between the rotor wake and the downstream guide vanes.

Furthermore, the effect of solidity of guide vane σ_{gR} on the noise characteristics was examined for $G/l_r = 0.433$. As a result, *SPL* does not depend on σ_{gR} in this case.

CONCLUSIONS

The performances of Wells turbine with guide vanes have been investigated by a model testing in order to clarify the effects of the gap between rotor and guide vane and solidity of guide vane. The conclusions obtained are summarized as follows.

(1) There are the optimum conditions of gap to chord ratio (G/l_r) and solidity (σ_{gR}) for the running characteristics, that is, $G/l_r \fallingdotseq 0.433$ and $\sigma_{gR} \fallingdotseq 0.908$.

(2) The smaller gap gives more favorable effect on the starting characteristics. However, there is the optimum solidity that is $\sigma_{gR} \fallingdotseq 1.25$.

(3) The sound pressure level decreases with increasing G/l_r (<0.433), and keeps almost the same value for $G/l_r \geq 0.433$.

(4) Considering of the running, starting and noise characteristics, the recommendable geometrical parameters of G/l_r and σ_{gR} are : $G/l_r \fallingdotseq 0.433$ and $\sigma_{gR} \fallingdotseq 1.25$.

REFERENCES

Arakawa, C, Suzuki, M, and Tagori, T (1987). "Improvement of Performances of Wells Turbine," *Proc 6th Conf on Energy and Resources* (in Japanese), pp 39-44.

Gato, L, and Falcao, A, F (1990). "Performance of Wells Turbine with Double Row of Guide Vanes," *JSME Int J Ser II* Vol 33, No 2, pp 262-271.

Ikui, T, Inoue, M, and Kaneko, K (1971). "Two-Dimensional Cascade Performance of Circular-arc Blades," *Proc Tokyo Joint Int Gas Turbine Conf and Products Show*, JSME-9.

Inoue, M, Kaneko, K, and Setoguchi, T (1985). "Studies on Wells Turbine for Wave Power Generator." *Bull JSME*, Vol 28, No 243, pp 1986-1991.

Inoue, M, Kaneko, K, Setoguchi, T, and Raghunathan, S (1986). "Simulation of Starting Characteristics of the Wells Turbine," *AIAA/ASME 4th Fluid Mech Plasma Dynamics Lasers Conf*, AIAA-86-1122.

Kaneko, K, Setoguchi, T, and Inoue, M (1986). "Performance of Wells Turbine in Oscillating Flow," *Proc Current Practices and New Tech in Ocean Eng*, ASME, Vol 2, pp 477-452.

Raghunathan, S, Setoguchi, T, and Kaneko, K (1997). "Aerodynamics of Monoplane Wells Turbine - A Review," Int J Offshore and Polar Eng, Vol 4, No 1, pp 68-75.

Raghunathan, S, Setoguchi, T, and Kaneko, K (1987). "The Wells Air Turbine Subjected to Inlet Flow Distortion and High Level of Turbulence," *J Heat and Fluid Flow*, Vol 8, No 2, pp 165-167.

Setoguchi, T, Kaneko, K, and Inoue, M (1986). "Determination of Optimum Geometry of Wells Turbine Rotor for Wave Power Generator," *Proc 3rd Symp on Ocean Wave Utilization* (in Japanese), JAMSTEC, pp 141-149.

Sturge, D, P (1977). "Turbine for an Oscillating Water Column Wave Power System," *CEGB Report*, No.MM/MECH/TA41.

Proceedings of the Seventh (1997) International Offshore and Polar Engineering Conference
Honolulu, USA, May 25-30, 1997

ISBN 1-880653-28-1 (Set); ISBN 1-880653-29-X (Vol. I)

Some Considerations to Ocean Wave Power Converter

Tomiji Watabe
T-Wave Consulting Volunteer
Noboribetsu, Japan

ABSTRACT

Several wave power converters have been proposed though no converters have come into a practical stage except small scale ones. It would be a warning that we should go back to fundamental studies again to find a way to the goal. This study is one of them to look for it.

Considering properties of ocean wave power, wave-resonant principle is basic to get a good converter. "the spring and mass coupled with damping (SMD) system" is one of the best styles. Differential-phase drive of SMD system can generate high quality electricity. Quasi-optimal control is possible by the differential-phase drive system not provided with wave sensors. A concrete design was also introduced.

KEY WORDS: Ocean wave, power, converter, system synthesis, optimal control, sensor.

INTRODUCTION

Several wave power converters have been proposed though no converters have come into a practical stage except small scales devices. The reasons considered are: ① the proposed converters have not proved the cost effectiveness for generation of commercial electricity and ② a really appreciable converter would be still behind a curtain. So, we must go back to a basic investigation i.e. "how the ocean wave energy systems should be".

Typical wave power converters are classified in a forced vibration system constructed with ① mass, ② spring and ③ damping parts.

The converter's behavior is the results of the response of the wave-driven system. Therefore, the system behavior can be known if both properties of the converter and the waves are given. Meanwhile, wave data could be derived inversely from the behavior of the converter. Putting this concept into application, a converter controlled by wave information will become much simpler because the converter does not need conventional wave sensors, however it is completed with the converter itself. Feasibility of this idea was proved by showing an actual system design applied it.

In order to make a converter really practical, electricity generated by it should be free from the pulsation of waves. However, the converter should be simple and should perform power conversion as well. These fundamental problems were also studied.

HOW THE NATURAL ENERGY SYSTEMS SHOULD BE

Generally, utilization of natural energy requires a big investment compared with the energy utilized. For instance, building a wave power plant requires great resources such as a certain space of sea or land, physical substances, energies, labors and so on—

Therefore, it is essential that the plant must generate more electricity than the resources lost to build the plant. It means that the parallel operation with a conventional power plant and a natural energy plant can save resources by mutual use of the facilities. Considering those described above, the basic notes for utilization of natural energy are summarized in Table. 1.

Table. 1 Basic notes for utilization of natural energy

considered from basic importance	1. Never waste the other resources more than natural energy created. 2. The system must be gentle to the environment. 3. It must make careful use of resources and be financially sound. 4. Never be too inconvenient to utilize it. 5. The system can work with conventional systems.
considered from practical importance	1. The electricity can be supplied through conventional grids. 2. Quality of the electricity must clear an authorized level. 3. Never make other plants troublesome when the system operates with them in parallel.

The amount of an investment for a wave power plant could easily become big due to severe sea conditions. However, it is necessary to develop an excellent wave power system which can be cheap and can operate very effectively too. Of course, the system should satisfy the notes of Table. 1.

NATURE OF OCEAN WAVE POWER

Ocean waves can be considered to be composed of numerous linear waves of various heights and periods. Ocean wave power consists of power by all elementary waves. Mean wave power $\overline{W}$ is written next for a determined crest width B.

$$\overline{W} = \int_0^\infty d\overline{W} = \frac{\rho g B}{2\pi}\int_0^\infty C_g S(\omega) d\omega \qquad (1)$$

here, $\overline{W}$: mean wave power [W]

ρ : density of sea water [kg/m³]

g : acceleration of gravity [m/s²]

B : crest width [m]

C_g : group velocity [m/s]

$S(\omega)$: power spectral density [m²·s]

ω : circular frequency of a component of waves [s⁻¹], $\omega = 2\pi/T$

T : period [s]

$$C_g = \frac{1}{2}\left(1+\frac{2k_o h}{\sinh 2k_o h}\right)\frac{\omega}{k_o} \qquad (2)$$

$$S(\omega) = \lim_{t_M \to \infty} \frac{1}{t_M}\left|X_t(\omega)\right|^2 \qquad (3)$$

here, $X_t(\omega)$ is the spectrum of an elementary wave amplitude a_t [m] which was observed in period t_M.

$$X_t(\omega) = \int_{-\infty}^{\infty} a_t(t) e^{-i\omega t} dt \qquad [\text{m·s}] \quad (4)$$

There is a relationship between ω and wave number $k_o[\text{m}^{-1}]=2\pi/\lambda$. (λ : wave length [m])

$$\frac{\omega^2}{g} = k_o \tanh k_o h \qquad (5)$$

h : water depth [m]

Fig. 1 is an example of power spectrum measured at Muroran (Japan). In Fig. 1, the spectrum takes a peak at ω_p (core frequency) so that the wave power is concentrated around an elementary wave of which frequency is ω_p (Watabe et al, 1996).

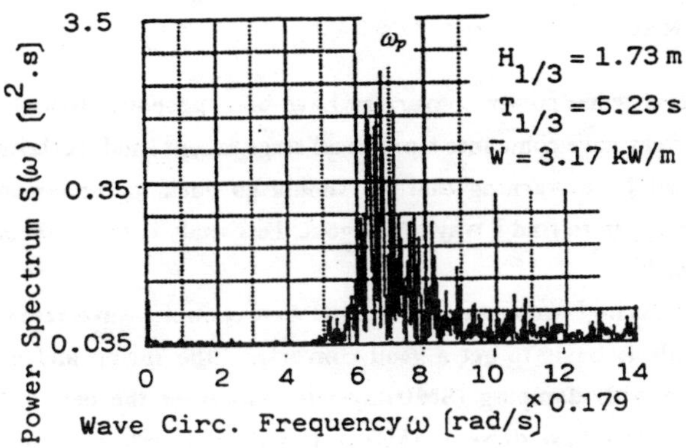

Fig. 1 Power Spectrum (Muroran)

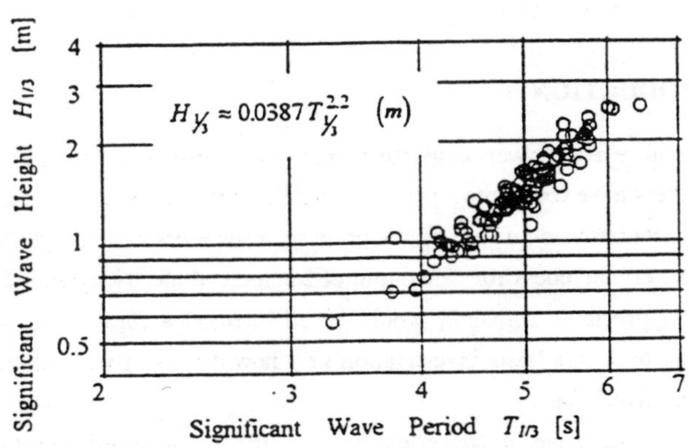

Fig. 2 Wave Height (Muroran)

Fig. 2 shows significant wave height $H_{1/3}$ observed statistically at Muroran. As it is seen, there is a relationship between $H_{1/3}$ and significant wave period $T_{1/3}$ that can be expressed approximately in the next form.

$$H_{1/3} \approx 0.0387\, T_{1/3}^{2.2} \quad \text{[m]} \quad (6)$$

Wave power $\overline{W}$ calculated by equations (1) ~ (5) is plotted in Fig. 3. The power $\overline{W}$ is written by equation (7) when wave width B is 2.3m.

$$\overline{W} \approx 2.82\, H_{1/3}^{1.9} \quad \text{[kW]} \quad (7)$$

(B= 2.3 [m])

Substituting equation (6), equation (7) becomes equation (8).

$$\overline{W} \approx 5.85 \times 10^{-3}\, T_{1/3}^{4.2} \quad \text{[kW]} \quad (8)$$

Considering the ocean waves statistically, an elemental regular wave which has the core frequency ω_p can be a representative of random sea. The frequency ω_p is almost same to $2\pi/T_{1/3}$ so that $T_{1/3}$ can be used instead of ω_p.

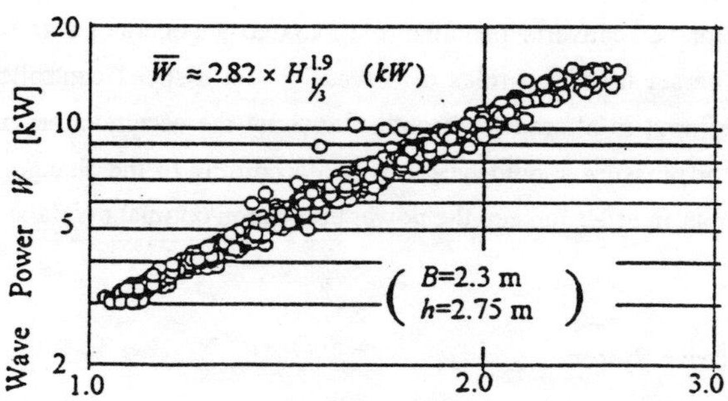

Fig. 3 Wave Power (Muroran)

INDISPENSABLE PROPERTIES FOR A GOOD WAVE POWER CONVERTER

It is very reasonable that the wave-resonant principle has been applied to the typical wave power converters because of the periodic wave properties. Therefore, the natural frequency of the system should correspond to the core frequency ω_p to operate the system in resonance (Ando et al, 1984). For actual use, the system also should satisfy some aspects described as follows.

Better Survivability with Less-Investment

Wave power $\overline{W}$ of equation (8) was derived from the data measured at twenty minutes each. (However, the greatest wave power can be instantaneously 2 ~ 4 times larger than equation (8).) It is possible that the power could grow so great (encountering a stormy condition) that the peak is over ten times of normal. For this reason, better survivability of the system with less investment is an important subject.

So far in wind energy utilization, wind turbines are protected by cutting the input whenever the wind velocity is greater than a critical speed. The variable pitch propeller mechanism is familiar in the application (Mitsubishi, 1990). This idea seems the best way to keep the natural energy converter safe against such huge input. Although it fits for wind, the idea can hardly be applied in a wave energy converter. The main reasons not to expect reliable operation in wave systems is that stubborn problems constantly arise in such a wild and unpredictable environment.

Therefore it is not easy to get a converter which is satisfied with a good survivability and economy. In order to solve it, "active damping of the wave input by the wave motion" was studied (Watabe et al. 1986). The damping works like this ① part of the water of incident waves comes into the backward chamber when the wave height becomes bigger than a critical level, ② then the water in the chamber pushes the pendulum against the incident waves by using damping pressure. This damping system does not need any moving parts such as the variable propeller.

Smoothing of the Output

In order to drive a converter, standing waves are suitable for the converter to extract the wave power efficiently, because the whole incident wave power is concentrated (in oscillating water motion) at a certain place. On the contrary, smoothing of the output is an important subject concerned with power oscillatory problems by the standing waves.

As long as the forced vibration system is used here, the converter performs the best absorption of power when ① the system is on a resonance and ② the load is so adjusted to get the resistance match (Asano, 1980). However, the output should be kept smooth even when the system is working in the best absorption. There are three typical systems which will be explained later.

For long-term fluctuation of the power $\overline{W}$, synchronized operation with a conventional power plant is cost efficient to solve the change (Watabe et al, 1992).

Optimal Operation

Behavior of a wave power converter, a mass and spring system coupled with load—such as OWC, is shown by the dynamic equation (9). Here, wave force drives the converter.

$$\Sigma M\ddot{x} + \left(N_o + N\right)\dot{x} + Kx = F \sin \omega t \quad (9)$$

here, ΣM : moving mass including adding water [kg]

N_o : damping factor by wave generation (wave) [N·s/m]

N : damping factor by electricity generation (load) [N·s/m]

K : spring constant of the system [N/m]

F : exciting amplitude by the wave [N]

x : displacement of the moving body (measured from neutral position) [m]

ω : circular frequency of the wave [*red* /s]

The system resonates with the wave when the natural frequency ω_o (of the system) coincides with the wave frequency ω. At the same time, if damping (load) N is so adjusted as it is in a same intensity of damping (wave) N_o, the power flow from sea to a generator will become maximum. This is called "resistance match" to make the system most efficient. So, the followings are for the optimal condition (Ando et al, 1984).

$$\omega_o = \sqrt{\frac{K}{\Sigma M}} = \omega \quad (10) \qquad N = N_o \quad (11)$$

However, equation (9) is for a linear case which is somewhat unfamiliar for a real system being non-linear. Referring the results of the linear case, the optimal condition of the real system also can be found. It is explained in Table. 2 (Watabe & Kondo, 1989).

Table. 2 Optimization of a wave power converter (Non-linear system)

1. Utilization of whole wave power (Standing waves are preferable as the object of power source.)
2. Semi-resonant operation: $\omega_o = \omega_p$ (Adjusting ω_o : natural frequency of the system to correspond with ω_p : core frequency of the random waves)
3. Resistance match: $\int_0^{T_o} N_o \dot{x}^2 dt = \int_0^{T_o} N(t)\dot{x}^2 dt$ (13) here, T_o : certain period [s] (Control of the extracted energy for electricity to correspond with an amount of energy used to wave generation)

COMPARISON WITH TYPICAL WAVE POWER SYSTEMS

In order to produce high quality power of electricity, the following steps are necessary for the forced vibration system.

(1) In order to make one way rotation from reciprocating water flow, the system should work as a rectifier.

(2) For the generator, constant speed rotation is preferable so that a variable speed power transmission should be involved in the system.

(3) Accumulators are connected in the power circuits of the system to absorb the oscillation caused by wave pulse.

(4) All the steps explained in (1) ~ (3) should never conflict with the resistance match of the system.

The forced vibration system will be investigated further by comparing it with typical wave power systems.

Direct Drive System

As it is seen in Fig. 4, the direct drive system has a primitive style in which the components are connected in series to transmit power.

Fig. 4(1) shows an OWC converter which is very simple and that accepts special turbines: the Wells turbine or the McCormic turbine. The turbines can take a role of rectifier too. Furthermore, the turbine speed can be controlled by adjusting pitch angle of the turbine blades. On the other hand, smoothing output is not done well because the OWC converter does not have effective accumulators. In the case of Fig. 4(1), excessive power is cut away.

Fig. 4(2) shows an early stage of the Pendulor in which an oil pump and a motor set converts pendular motion into a generator's rotary motion. The set takes the roles of a rectifier and a speed controller. Pulse like input is altered to smooth output by the accumulator, but accumulator pressure should be controlled according to the change of average input in order to keep the power absorption optimal (Watabe & Kondo, 1989).

Parallel Drive System

Prof. Salter and his staff invented a system shown in Fig. 5 in order to improve the direct drive system. In Fig. 5, power is transmitted to a generator through two lines: a mechanical one and a hydraulic one. The system uses a variable pitch Wells turbine so that turbine speed is controlled constantly. The mechanical line transmits the motion through a shaft, therefore the line is unable to modify the output's smoothness.

The hydraulic line consists of an oil pump/motor and an accumulator. The pump/motor is basically an individual component which can be a pump or a motor according with its valve timing. When the power is greater than a certain level, the pump/motor delivers oil to the accumulator working as a pump. And when the power is smaller than the level, the pump/motor drives the generator being as a motor. Since the displacement of the pump/motor is variable from zero to maximum, the resultant output by the two-lines-connection can be kept fairly smooth (Rampen, Salter et al, 1995).

The control depends on a computer. The power flow of the hydraulic

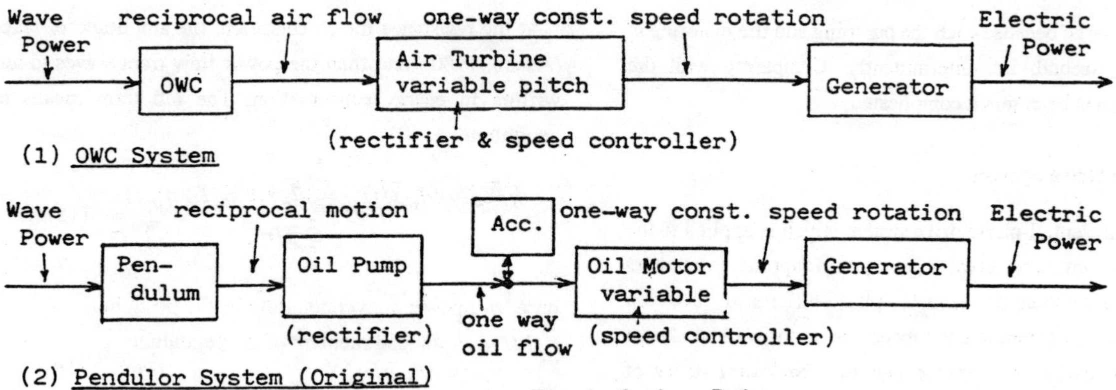

Fig. 4 Series Drive

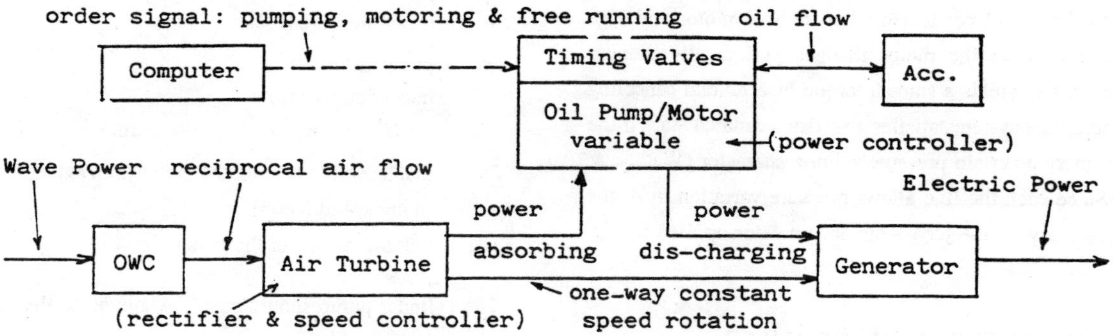

Fig. 5 Parallel Drive (Edinburgh University)

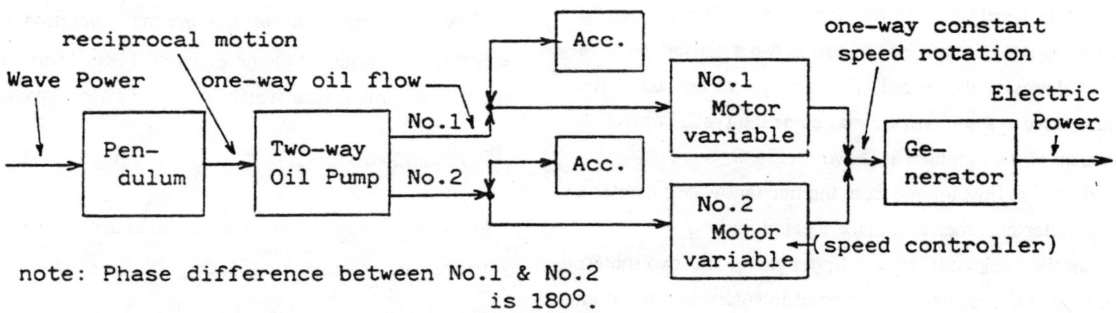

Fig. 6 Differential-Phase Drive (Muroran Institute of Technology)

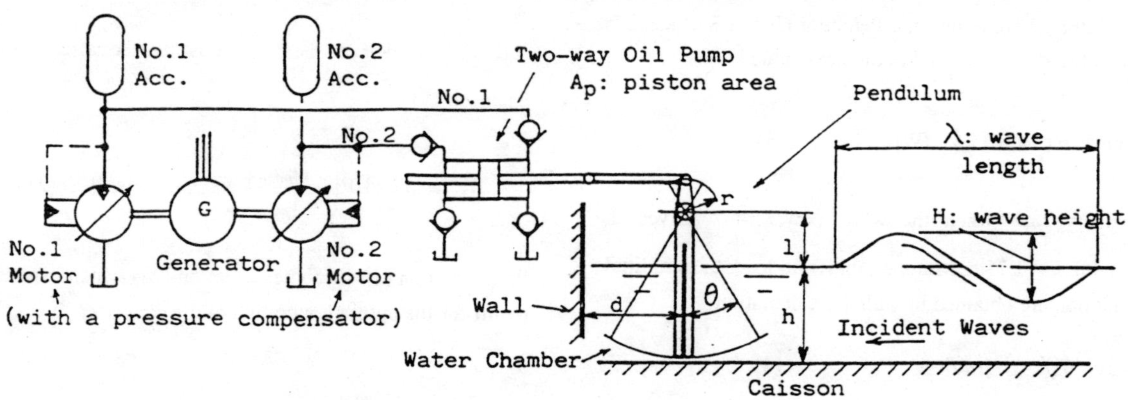

Fig. 7 Pendulor (Muroran Institute of Technology)

line likely becomes large because each the pumping and the motoring is rushed into a few second, i.e. intermittently. Comparing with the original, the profile has been much complicated.

Differential-Phase Drive System

Fig. 6 shows a differential-phase drive system which is applied to the Pendulor. As it is shown there, a two-way oil pump supplies two motors pressurized-oil by turns, then, there exists a phase-difference between the motors because No.1 motor corresponds to a forward stroke of standing waves, otherwise No.2 motor does to a backward stroke of them. The two motors together drive a generator.

Since each line has an accumulator, No.1 motor can continue running while the pump delivers oil not to No.1 but to No.2 motor and vice versa. The torque of each the motor changes periodically though, superposition of the two yields a smooth torque by a mutual canceling effect. Furthermore, this system satisfies a resistance match state if the accumulator has mere a certain pressure/volume character (Watabe & Kondo, 1989). Since each the line allows pressure variation in it, the system is not necessary to use such a big accumulator applied in Fig. 4(2).

QUASI-OPTIMAL CONTROL OF WAVE POWER CONVERTER WITHOUT A WAVE SENSOR

Generally, installation of a wave sensor requires much investment. In order to save it, it is worthwhile to develop a cost effective method which can do the optimal control not to use a wave sensor. Here, a Pendulor which belongs to the forced vibration system and takes the differential-phase drive system, will be studied as the typical object. A concrete mechanism of the Pendulor is shown in Fig. 7.

In Fig. 7, reciprocating flow altered from the incident waves excites a pendulum in the water chamber at a place located apart $d = \lambda/4$ (λ : wave length) from the back wall. By a cylinder pump and two motors combination, the pendular motion is converted to rotary motion of an electric generator.

Supposing a state that the incident waves of equation (6) (derived by observation of Muroran) come into the Pendulor (B = 2.3m) steadily, the mean input by the waves $\overline{W}_{1/3}$ takes the next value.

$$\overline{W}_{1/3} = \frac{1}{8}\rho g B C_g H_{1/3}^2 \approx 16.1 T_{1/3}^{4.4} \text{ [W]} \qquad (13)$$

For an imaginary case when the wave power $\overline{W}_{1/3}$ drives the pendulum (l = 4.4m, h = 2.75m, Muroran) under no-load, the amplitude θ_a of pendular motion are obtained by author's calculation.

$$\theta_a \approx 2.86 \times 10^{-3} T_{1/3}^3 \quad [rad] \qquad (14)$$

At the resistance match condition, the amplitude θ_o takes $\theta_o = \theta_a/2$ (Asano, 1980), and then the power flow from waves to motors can be written in energy conservation. The left term means input to the pendulum.

$$\eta_1 \overline{W}_{1/3} = 4\theta_o \overline{M}/T_{1/3} = 2\theta_a r A_p \overline{p}/T_{1/3}\eta_p = \overline{Q}\,\overline{p}/\eta_p \qquad (15)$$

here, η_1 : power conversion ratio by the pendulum

$\overline{M}$: mean load moment of the pendulum $\approx r A_p \overline{p}/\eta_p$ [N·m] (16)

r : arm radius (Fig. 7) [m]

A_p : piston area of the cylinder [m²]

$\overline{p}$: mean pressure [Pa]

η_p : pump efficiency

$\overline{Q}$: mean delivery flow $= 2D_m n_m/\eta_{vm}$ [m³/s] (17)

D_m : displacement of each the motor [m³/*rev*]

n_m : motor speed [*rev*/s]

η_{vm} : volumetric eff. of the motor

The cylinder pump should supply whole oil to the motors so that the next equation is written.

$$\overline{Q} = 2r\theta_a A_p/T_{1/3} \quad \text{[m}^3\text{/s]} \qquad (18)$$

Flow $\overline{Q}$ controlled for the optimal operation can be found by substitute equation (14) for equation (18). Therefore, equation (19) shows the optimal flow which fits to the given wave period.

$$\overline{Q} \approx 5.72 \times 10^{-3} r A_p T_{1/3}^2 \quad \text{[m}^3\text{/s]} \qquad (19)$$

Pressure $\overline{p}$ controlled for the optimal operation can also be found by substitute equations (13), (14) and (19) for equation (15).

$$\overline{p} \approx 2.81 \times 10^3 \frac{\eta_1 \eta_p}{r A_p} T_{1/3}^{2.4} \quad \text{[Pa]} \qquad (20)$$

and $T_{1/3}^{2.4}$ is derived next by substitute equations (14) and (17) for equation (15).

$$T_{1/3}^{2.4} \approx 1.13 \times 10^3 \left(\frac{D_m n_m}{r A_p \eta_{vm}} \right)^{1.2} \quad \text{[s}^{2.4}\text{]} \qquad (21)$$

then equation (20) becomes equation (22) which explains an important result for the optimal operation of the converter.

$$\bar{p} \approx 3.18 \times 10^6 \frac{\eta_1 \eta_p (D_m n_m)^{1.2}}{\eta_{vn}^{1.2} (rA_p)^{2.2}} \quad [P_a] \quad (22)$$

However, equation (22) shows that pressure $\bar{p}$ is proportional to $D_m^{1.2}$ when the converter is operating in the optimal condition. From this fact, it is understood that the converter can keep optimal operation when displacement of the motor D_m is so adjusted by pressure $\bar{p}$ as equation (22) be satisfied. Since $\bar{p}$ is of the main passage, $\bar{p}$ can be used not only as a control signal but also as a power medium when it flows even a fraction of the oil. In Fig. 7, the two motors are provided with the pressure compensators which adjust D_m by signal $\bar{p}$. Thin tubes transmit both control signal and control power. The actual feature can be very simple and the system is completed without wave sensors.

CONCLUSIONS

In order to get a practical wave power converter, some fundamental researches were studied. The conclusions are summarized into the following articles.

(1) Through static observation, some regularity is found in the ocean waves even they seem random. Wave power is concentrated around a core wave component of which period can be representative of the ocean wave properties.

(2) For this reason, the forced vibration system is suitable as a wave power converter to be fitted with the ocean waves.

(3) Differential-phase drive system can achieve efficient power conversion and nice smoothing of the output. The feature looks rather simple.

(4) It was clarified; a quasi-optimal control for power absorption is possible even when the system does not provide wave sensors. A concrete system of its example was introduced in case of power line pressure was used as the control signal.

The wave data referred in this paper were observed by Cold Region Port & Harbor Eng. Research Center, Muroran Inst. of Tech., Narasaki Co. Ltd. and T-Wave Consulting Volunteer JAPAN.

REFERENCES

Ando, A. et al (1984). "On a Flap-type Wave Energy Converter", Proc. of ECOR Int. conference.

Asano, S. (1980). "Power Absorption Ratio by a Flap-type Converter installed in a Break Water", Report of Japan ship builder eng. society, SK60-14. (in Japanese)

Mitsubishi (1990). "MITSUBISHI, MWT-250, Mitsubishi Wind Turbine", Catalogue.

Rampen, Salter, Almond, Taylor & Ehsan (1995). "Development status of the High Efficiency Computer Controlled Digital Hyd. Pump/Motor for Wave Energy Power Conversion Use", Proc. 2nd European Wave Power Conference 1995 Lisbon. (in printing)

Watabe, T. et al (1996). "Studies on a Large Displacement Pump for an Ocean Wave Power Converter", Proc. of JHPS Int. Symp. on Fluid Power Yokohama.

Watabe, T. et al (1986). "US Patent 4,580,400".

Watabe, T. et al (1992). "A Hybrid Power Supply Concept of the Pendulor Device", Proc. of Renewable Energy, Vol. 5, Technology & Environment U. K..

Watabe, T. & Kondo, H. (1987). "Hydraulic Technology and Utilization of Ocean Wave Power", Proc. JHPS Int. Symp. on Fluid Power Tokyo.

Watabe, T. et al (1984). "US Patent 4,490,621".

Proceedings of the Seventh (1997) International Offshore and Polar Engineering Conference
Honolulu, USA, May 25-30, 1997

ISBN 1-880653-28-1 (Set); ISBN 1-880653-29-X (Vol. I)

Gas Hydrates in the Messoyakha Gas Field of the West Siberian Basin - A Re-Examination of the Geologic Evidence

Timothy S. Collett
U.S. Geological Survey
Denver, CO, USA

Gabriel D. Ginsburg
VNIIOkeangeologia
St. Petersburg, Russia

ABSTRACT

The amount of natural gas within the gas hydrate accumulations of the world is believed to greatly exceed the volume of known conventional natural gas reserves. The hydrocarbon production history of the Russian Messoyakha field, located in the West Siberian Basin, has been used as evidence that gas hydrates are an immediate source of natural gas that can be produced by conventional means. Re-examination of available geologic, geochemical, and hydrocarbon production data suggests, however, that gas hydrates may not have contributed to gas production in the Messoyakha field. More field and laboratory studies are needed to assess the historical contribution of gas hydrate production in the Messoyakha field.

KEY WORDS: Gas hydrates, natural gas, West Siberia, Russia, resources

INTRODUCTION

The Messoyakha gas field in the northern part of the West Siberian Basin is often used as an example of a hydrocarbon accumulation from which gas has been produced from in-situ natural gas hydrates. Production data and other pertinent geologic information have been used to document the presence of gas hydrates in the Messoyakha field (Makogon and others, 1972; Makogon, 1981, 1988; Cherskiy and others, 1985; Krason and Ciesnik, 1985; Krason and Finley, 1993). It has also been suggested that the production history of the Messoyakha field demonstrates that gas hydrates are an immediate producible source of natural gas and that production can be started and maintained by conventional methods. Recently, however, several studies suggest that gas hydrates may not be contributing to gas production in the Messoyakha field and that the potential resource significance of gas hydrates may have been overestimated (Verkhovsky and others, 1988; Ginsburg and others, 1990; Ginsburg, 1993).

In this paper we re-examine the evidence for gas hydrates in the Messoyakha field and critically review the available geologic data in order to determine if gas hydrates have contributed gas production in the Messoyakha field. This paper begins with a technical review of gas hydrate physical properties and description of known gas hydrate occurrences, which is followed by a description of the geologic setting of the Messoyakha field. The main body of the paper includes the review of the geologic evidence for and against the occurrence of gas hydrates in the Messoyakha field. This paper ends with a summary discussion in which the available geologic data are compared and contrasted.

GAS HYDRATE TECHNICAL REVIEW

Gas hydrates are crystalline substances composed of water and gas in which a solid water-lattice accommodates gas molecules in a cage-like structure, or clathrate. Gas hydrates are widespread in permafrost regions and beneath the sea in sediment of outer continental margins. Although methane, propane, and other gases can be included in the clathrate structure, methane hydrates appear to be the most common in nature (Kvenvolden, 1988). Under appropriate conditions of temperature and pressure, gas hydrates usually form one of two basic crystal structures known as Structure I and Structure II. Each unit cell of Structure I gas hydrate consists of 46 water molecules that form two small dodecahedral voids and six large tetradecahedral voids. Structure I gas hydrates can only hold small gas molecules, such as methane and ethane, that have molecular diameters not exceeding 5.2 angstroms. The unit cell of Structure II gas hydrate consists of 16 small dodecahedral and 8 large hexakaidecahedral voids formed by 136 water molecules. Structure II gas hydrates can contain gases that have molecular dimensions ranging from 5.9 to 6.9 angstroms, such as propane and isobutane. At standard conditions (STP), one volume of saturated methane hydrate can contain as many as 164 volumes of methane gas; because of this large gas storage capacity, gas hydrates are thought to represent an important source of natural gas. For a complete description of the structure and properties of hydrates see the summary by Sloan (1990).

Onshore gas hydrates are "known" to be present in the West Siberian Basin (Makogon and others, 1972) and are believed to be present in other permafrost areas of northern Russia, including the Timan-Pechora province, the eastern Siberian Craton, and the northeastern Siberian and Kamchatka areas (Cherskiy and others, 1985). Permafrost-associated gas hydrates are also present in the North American Arctic. Direct evidence for gas hydrates on the North Slope of Alaska comes from a core-test, and indirect evidence comes from drilling and open-hole industry well logs that suggest the presence of numerous gas hydrate layers in the area of the Prudhoe Bay and Kuparuk River oil fields (Collett, 1993). Well-log responses attributed to the presence of gas hydrates have been obtained in about one-fifth of the wells drilled in the Mackenzie Delta, and in the Arctic Islands more than half of the wells are inferred to contain gas hydrates (Judge, 1988). The combined information from Arctic gas-hydrate studies shows that, in permafrost regions, gas hydrates may be present at subsurface depths ranging from about 130 to 2,000 m (reviewed by Kvenvolden, 1988).

The presence of gas hydrates in offshore continental margins has been inferred mainly from anomalous seismic reflectors that coincide with the predicted phase boundary at the base of the gas-hydrate stability zone. This reflector is commonly called a bottom-simulating reflector, or BSR. Bottom-simulating reflectors have been mapped at depths below the sea floor ranging from about 100 to 1,100 m. (Kvenvolden, 1988). Gas hydrates have been recovered during gravity core surveys (reviewed by Kvenvolden, 1988) within 10 m of the sea floor in sediment of the Gulf of Mexico, the offshore portion of the Eel River Basin near California, the Black Sea, the Caspian Sea, and the Sea of Okhotsk. Also, gas hydrates have been recovered at greater sub-bottom depths (reviewed by Kvenvolden, 1988) during research coring along the southeastern coast of the United States on the Blake Outer Ridge, in the Gulf of Mexico, in the Cascadia basin near Oregon, in the Middle America Trench, offshore Peru, and on both the eastern and western margins of Japan.

Because large quantities of gas hydrates are present in permafrost regions and in offshore marine sediments, they may be a potential energy resource. World estimates for the amount of natural gas in gas hydrate deposits range from $1.4x10^{13}$ to $3.4x10^{16}$ cubic meters ($5.0x10^2$ - $1.2x10^6$ trillion cubic feet) for permafrost areas and from 3.1×10^{15} to $7.6x10^{18}$ cubic meters ($1.1x10^5$ - $2.7x10^8$ trillion cubic feet) for oceanic sediments (reviewed by Kvenvolden, 1993). Current estimates of the amount of methane in the world gas hydrate accumulations are in rough accord at about $2x10^{16}$ cubic meters ($7x10^5$ trillion cubic feet) (reviewed by Kvenvolden, 1993). If these estimates are valid, then the amount of methane in gas hydrates is almost two orders of magnitude larger than the estimated total remaining recoverable conventional methane resource, which is estimated to be about $2.5x10^{14}$ cubic meters ($9x10^3$ trillion cubic feet) (Masters and others, 1991).

GEOLOGIC SETTING OF THE MESSOYAKHA FIELD

The Messoyakha gas field (figure 1), discovered in 1968, was the first producing field in the northern part of the West Siberian Basin. By the mid-1980's, more than 66 gas fields had been discovered in the West Siberian Basin, containing an estimated total gas reserve of 22 trillion cubic meters (777 trillion cubic feet), approximately one-third of the world's reserves (Grace and Hart, 1986). Between 1969 and 1987 about 14.4 billion cubic meters (508 billion cubic feet) of gas were produced from the Messoyakha field and delivered by pipeline to the city of Norilsk (Makogon, 1988). The geology, stratigraphy (figure 2), and petroleum geochemistry of the West Siberian Basin have been described in detail in many English language publications (Kortsenshteyn, 1970; Makogon and others, 1972; Alekseyev, 1974; Trofimuk and others, 1977; Kruglikov and others, 1983; Krason and Ciesnik, 1985; Grace and Hart, 1986; Makogon, 1988; Peterson and Clarke, 1989). Production in the northern part of the West Siberian Basin is principally from Neocomian reservoirs of the Megion "suite" (average depth 2,800 m) and Cenomanian reservoirs of the Pokur "suite" (average depth 1,100 m); about two-thirds of the region's gas production is from the Cenomanian reservoirs (Grace and Hart, 1986). The Pokur "suite" is a 700- to 800-m-thick complex of interbedded marine and nonmarine sandstone and shale deposited during an Aptian to Cenomanian marine regression. Coal and plant debris are prevalent within the nonmarine parts of the Pokur "suite." The Pokur "suite" is overlain by the shale sequence of the Turonian Kuznetsov "suite," which forms a regional seal for most of the gas in the underlying sandstone reservoirs.

EVIDENCE FOR GAS HYDRATES IN THE MESSOYAKHA FIELD

In this section we review the available geologic data that support the hypothesis that gas hydrates have contributed to gas production in the Messoyakha field. This data synthesis relies mainly on the work of Makogon and others (1972), Sapir and others (1973), Makogon (1981, 1988), Cherskiy and others (1985), Krason and Ciesnik (1985), and Krason and Finley (1993).

Regional analysis of subsurface temperatures, formation-pressures, and pore-water salinities suggests that the methane-hydrate stability zone in the northern part of the West Siberian Basin may extend to depths of approximately 1,000 m (figure 1; modified from Makogon and others, 1972). The Messoyakha gas accumulation is confined to the Dolgan Formation of the Pokur "suite," and production has been established from the depth interval between 720 and 820 m. The upper part (about 40 m) of the Messoyakha field lies within the zone of predicted methane-hydrate stability (figure 3). By assuming a reservoir pressure of 78 kg/cm^2, Makogon and others (1972) determined that the 10°C isotherm defines the lower limit of in-situ gas hydrates and separates the Messoyakha field into an upper gas-hydrate accumulation and a lower free-gas accumulation (figure 4).

Unusually low gas yields from production tests in the upper part of the Messoyakha reservoir were the first physical evidence of possible in-situ gas-hydrate occurrences (production data discussed in more detail later in this paper). Analysis of spontaneous potential and caliper well logs from 62 wells drilled in the Messoyakha field reveal the presence of apparently "frozen" rock intervals within the Dolgan Formation (Makogon and others, 1972). Because these "frozen" layers are more than 250 to 350 m below the zone of permafrost and are at formation temperatures near 10°C, they have been interpreted to contain in-situ gas hydrates rather then ice. Further analysis of electrical resistivity well logs also indicated the presence of gas hydrates in the upper part of the Messoyakha gas accumulation (Makogon, 1981). The gas-hydrate and free-gas parts of the Messoyakha field are depicted in the generalized cross section in figure 4. The detailed cross sections in figure 5 (Makogon and others, 1972) reveal that the well-log inferred gas hydrates are within a series of discrete, laterally continuous sandstone sequences separated by interbedded shale and siltstone. The Messoyakha structure, as described by Makogon and others (1972), is a dome having dimensions of 12.5 by 19 km (figure 6). Studies of cores, well cuttings, and well logs show that the free-gas part of the Messoyakha reservoir is characterized by porosities ranging from 16 to 38 percent; permeabilities average about 125 millidarcies. Water saturations in the Messoyakha field average about 40 percent (Makogon and others, 1972). Prior to production, calculated total gas reserves within the gas-hydrate and free-gas parts of the Messoyakha accumulation were estimated to be about 80×10^9 cubic meters (2.8 trillion cubic feet), with about one-third of the reserves within the gas hydrates (Krason and Ciesnik, 1985).

Several authors (Kortsenshteyn, 1970; Trofimuk and others, 1977; Du Rochet, 1980) have suggested that conditions conducive to the formation of permafrost and gas hydrates have persisted in the northern part of the West Siberian Basin since the early Pleistocene (about 1.65 Ma). Russian geologists generally conclude that pre-existing free-gas accumulations were converted to gas hydrates during Pleistocene glacial advances and that the gas-hydrate pools were transformed back into free-gas accumulations during subsequent interglacial periods. The present-day Messoyakha field is in

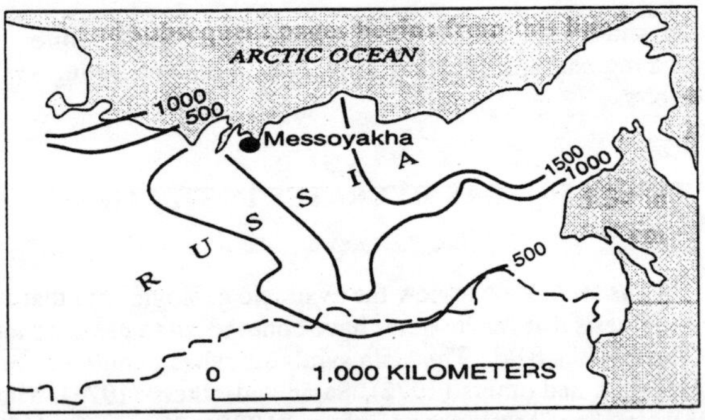

Figure 1. Depth (in meters) to the base of the methane-hydrate stability field in continental areas of Russian (modified from Makogon and others, 1972). Also shown is the location of the Messoyakha gas field. Dashed line is southern boundary of the former U.S.S.R.

System	Series	Stage	Grace and Hart (1986)
Overlying beds			Tibeysalin "suite" (Tertiary)
CRETACEOUS	Upper	Maastrichtian	Gankin "suite"
		Campanian	Berezov "suite"
		Santonian	
		Coniacian	
		Turonian	Kuznetsov "suite"
		Cenomanian	Pokur "suite"
	Lower	Albian	
		Aptian	Vartov "suite"
		Barremian	
	Neocomian	Hauterivian	Megion "suite"
		Valanginian	
		Berriasian	
Underlying beds			Bazhenov "suite" (Jurassic)

Figure 2. Stratigraphic column for northern West Siberian Basin (modified from Grace and Hart, 1986).

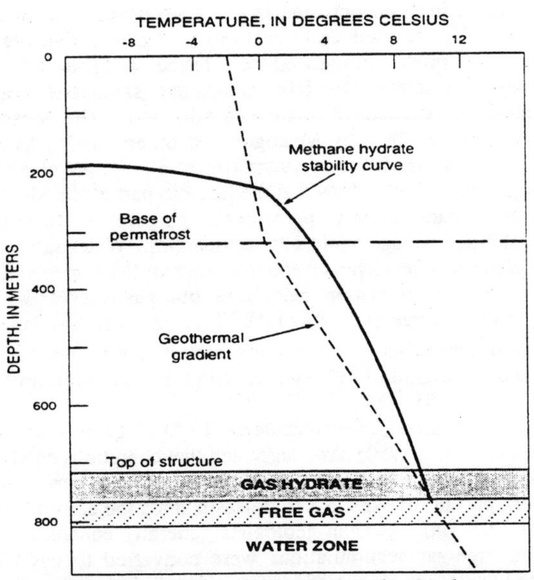

Figure 3. Phase diagram illustrating the depth-temperature zone in which methane hydrates are stable in the area of the Messoyakha field of the West Siberian Basin (modified from Makogon, 1988).

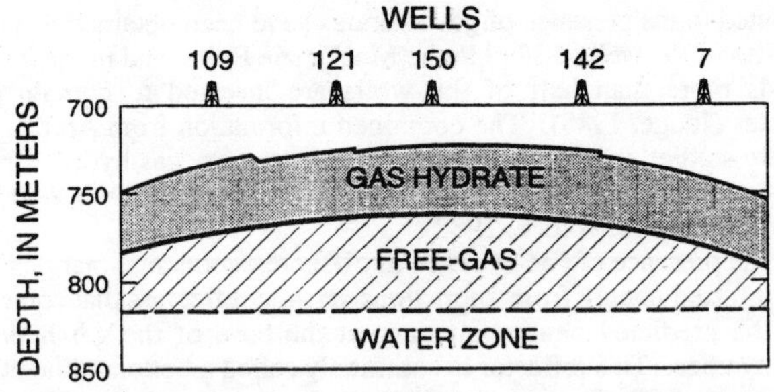

Figure 4. Generalized cross section illustrating the distribution of the gas-hydrate and free-gas parts of the Messoyakha field (modified from Makogon, 1988).

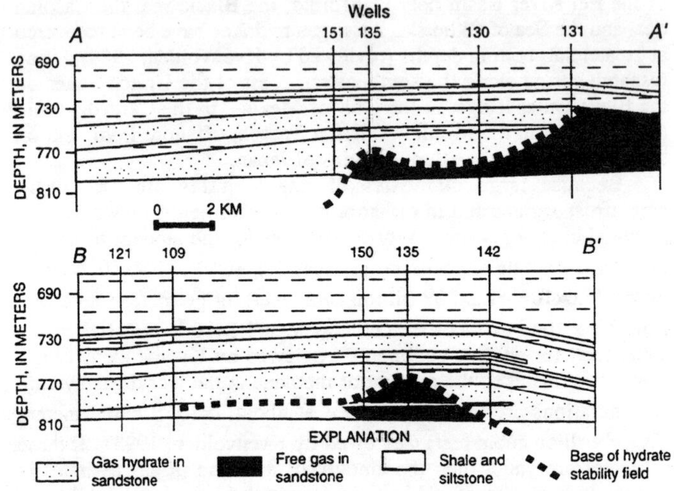

Figure 5. Multi-well cross-sections showing the lateral and vertical extent of gas hydrates and underlying free-gas in the Messoyakha field (modified from Makogon and others, 1972). Location of cross sections are shown in figure 6.

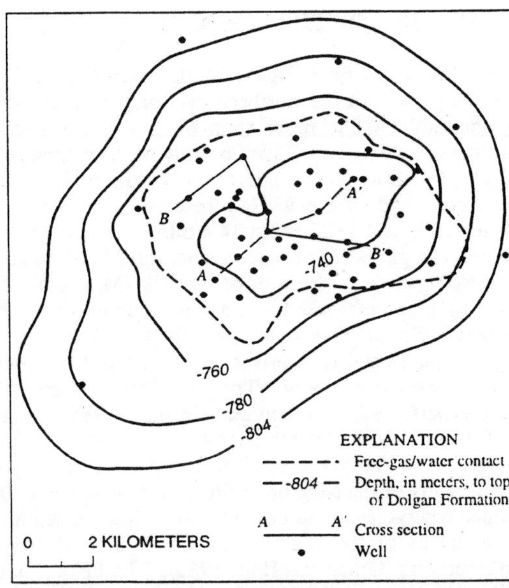

Figure 6. Structure map of the top of the gas-hydrate-bearing Dolgan Formation in the Messoyakha field (modified from Krason and Ciesnik, 1985). Dots show well locations. Location of cross-sections in figure 5 are also shown.

transition from a gas-hydrate to a free-gas accumulation and it has not completely converted back to its pre-glacial free-gas state.

Production tests in the early development history of the Messoyakha field yielded highly variable results. Measured flow rates from the free-gas part of the reservoir were substantially greater than those from the gas-hydrate part. To confirm the presence of gas hydrates within the upper part of the Messoyakha field, a series of hydrate-inhibitor injection tests was conducted. During these tests, substances, such as methanol and calcium chloride, that destabilize and prevent the formation of gas hydrates were injected into the suspected gas-hydrate-bearing rock units. Most of these tests (summarized by Sumetz, 1974, and Makogon, 1981) resulted in dramatic increases in production rates, that were attributed to the dissociation of the in-situ gas hydrates.

Long-term production from the gas-hydrate part of the Messoyakha field has been achieved by a simple depressurization scheme. The reservoir-pressure and production history of the Messoyakha field are depicted in figure 7. When production began in 1969, the reservoir pressure decline curve followed the predicted path; however, in 1971 measured reservoir-pressures began to deviate from predicted values. This deviation has been attributed to the liberation of free gas from dissociating gas hydrates. The seventeen-year production history of the Messoyakha field has been divided into five stages (figure 7):

Stage I. From 1969 to 1971, reservoir-pressure did not fall below gas-hydrate stability conditions, and gas production was only from the deeper free-gas part of the field.

Stage II. From 1971 to 1975, actual reservoir-pressures exceeded predicted reservoir pressures. This departure marks the start of gas-hydrate dissociation and gas production from the gas-hydrate part of the field.

Stage III. From 1976 to 1977, the volume of gas withdrawn from the reservoir was equal to the amount of gas liberated from dissociating gas hydrates.

Stage IV. From 1978 to 1981, production from the Messoyakha field was reduced and eventually shut-in. Reservoir-pressures began to rise as gas hydrates continued to dissociate.

Stage V. Since 1982 there has only been modest production from the Messoyakha field. During this period the amount of gas liberated from the hydrates is equal to the amount of gas produced.

Throughout the production history of the Messoyakha field it is estimated that about 36 percent (about 5.17×10^9 cubic meters, or about 183 billion cubic feet) of the gas withdrawn from the field has come from gas hydrates (Makogon, 1988).

EVIDENCE AGAINST GAS HYDRATES IN THE MESSOYAKHA FIELD

The Messoyakha field is believed to contain gas hydrates mainly because initial reservoir pressures and temperatures fall within conditions suitable for gas hydrate formation (Makogon and others, 1972). According to Ginsburg and others (1990), however, it is impossible to determine with any degree of certainty if the conditions necessary for gas hydrate formation exist within the Messoyakha field, primarily because of limited reservoir pressure and temperature data and to the highly variable compositional nature of the sampled reservoir gases and waters.

Ginsburg (1993) also suggested that other gas hydrate indicators in the Messoyakha field, such as from logging records and reservoir production data, may not be due to the presence of gas hydrates. Low gas yields from production tests and the presence of well-log-inferred "frozen" rock intervals in the upper part of the Messoyakha reservoir were used by Makogon and others (1972) to divide the Messoyakha field into two parts, an upper gas-hydrate accumulation and lower free-gas accumulation. However, Ginsburg and others (1990) documented significant reservoir heterogeneities (figure 8) that could account for the observed reservoir production anomalies and well log responses. Core samples from the Messoyakha field confirm that the upper part of the reservoir has relatively low permeabilities ranging from 10 to 100 millidarcies (Bro and Sorokov, 1970), whereas the lower part of the reservoir is characterized by permeabilities ranging from 100 to 1,000 millidarcies. Thus, variations in permeabilities could account for the lower production rates in the upper part of the Messoyakha reservoir. In addition, the apparent well-log increase in electrical resistivity and reduction in the spontaneous potential observed within the upper part of the Messoyakha field (figure 9) may be due to lower reservoir porosities and higher interstitial water contents, not the presence of "frozen" gas hydrates.

As previously discussed, the production history of the Messoyakha field is characterized by periods of reservoir pressure fluctuations believed to be related to gas hydrate dissociation (Makogon, 1988). Ginsburg and others (1990) showed, however, that the observed Messoyakha production and reservoir pressure history can be attributed to gas cross-flow from the upper to the lower part of the Messoyakha reservoir. In most Messoyakha wells, both the upper and lower parts of the reservoir have been perforated and are open to production. Due to reservoir permeability heterogeneities, long-term gas production has caused a differential pressure drop between the upper and lower part of the Messoyakha reservoir. This production-induced differential pressure reduction has allowed gas to flow from the upper part of the Messoyakha reservoir into the under-pressured lower part of the Messoyakha reservoir. Thus, the reservoir pressure increases observed during shut-in periods may be due to "recharge" from other parts of the reservoir. Evidence for reservoir cross-flow includes reduced formation temperatures (figure 10) in the upper part of the Messoyakha reservoir, which can be attributed to gas flow and expansion. Geochemical changes in the composition of the produced gases (discussed later in this paper) also suggest that cross-flow has contributed to "recharging" the lower part of the Messoyakha field. Ginsburg and others (1990) indicated that gas cross-flow and associated temperature declines have contributed to the formation of "technogenic" gas hydrates within the Messoyakha reservoir. Ginsburg and others (1990) believed that increases in gas production rates observed during hydrate-inhibitor injection tests in the Messoyakha field (summarized by Sumetz, 1974, and Makogon, 1981) are due to the dissociation of "technogenic" gas hydrates rather than "natural" gas hydrates.

A great deal of water has also been produced from the Messoyakha field (Makogon, 1988). This water can be attributed to the dissociation of gas hydrates; however, Ginsburg and others (1990) showed that, if all the water produced (4 L of water per cubic meter of gas produced) was attributed to gas hydrate dissociation, it would account for no more than 0.1 percent of all the gas recovered, this conclusion suggests that gas hydrates have not contributed significantly to gas production in the Messoyakha field.

In 1983 VNIIOkeangeologia and Norilskgasprom initiated a study in which the composition of gas produced from the Messoyakha field was analyzed on a regular basis (Ginsburg and others, 1990). This study was undertaken because of variations in gas composition and the appearance of anomalous gases. Geochemical analyses of the gas produced from the Messoyakha field reveal that the concentration of methane (about 99 percent) has remained relatively unchanged throughout the production history of the field; however, other hydrate forming gases, including ethane, propane, and carbon dioxide, have exhibited small concentration changes. On initial start-up (figure 7, Stage I, 1969-1971), the produced gas contained about 0.05 percent ethane, 0.022 percent propane, and 0.47 percent carbon dioxide. By 1982 (figure 7, Stage V), the concentration of ethane and propane had decreased to 0.03 and 0.005 percent, whereas the concentration of carbon dioxide had increased to

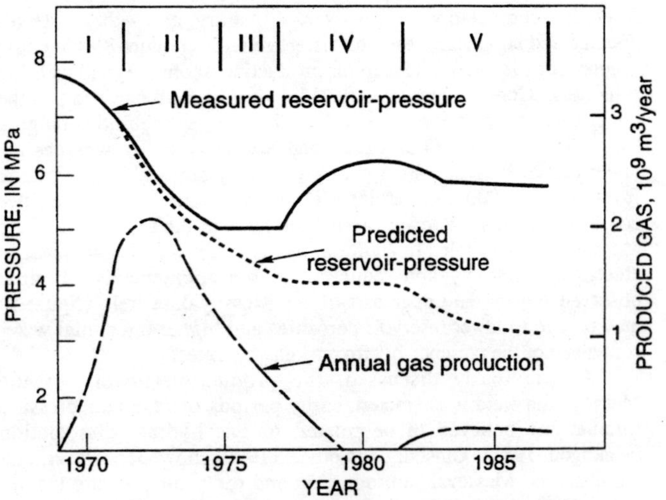

Figure 7. A plot of reservoir-pressures and production history as a function of time for the Messoyakha field (modified from Makogon, 1988). The roman numerals across the top of the figure denote the stages of production discussed in the text. MPa, megapascal.

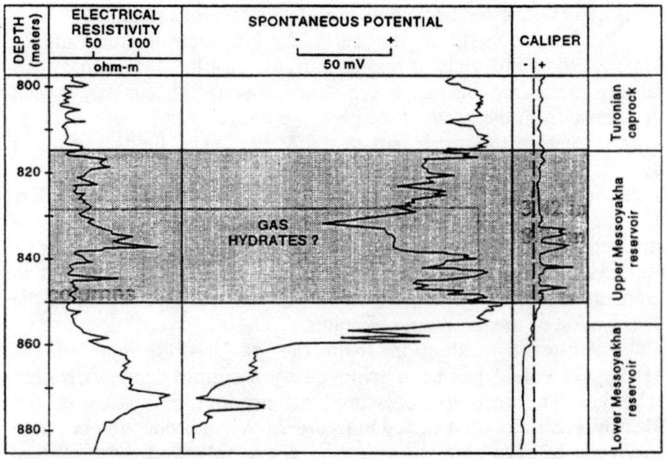

Figure 9. Well logs from Well-131 in the Messoyakha field showing a possible gas-hydrate-bearing interval.

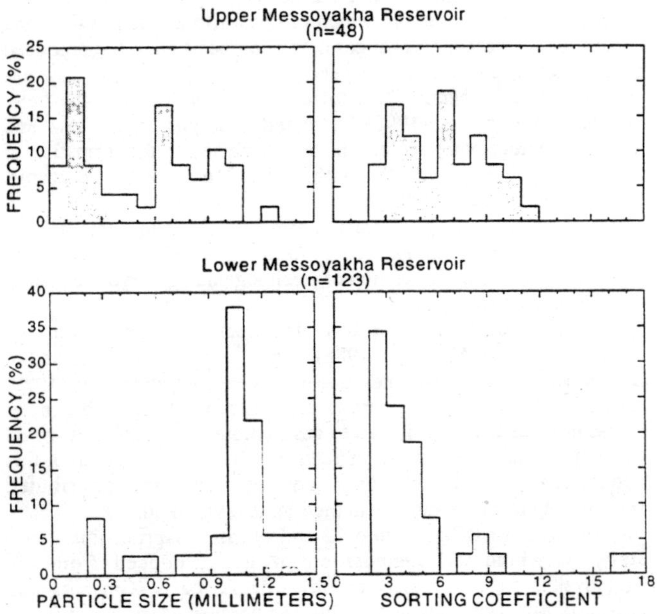

Figure 8. Particle size (medium diameter) and sorting coefficient of sandy and silty sediments of the upper and lower part of the Messoyakha reservoir. n, number of samples.

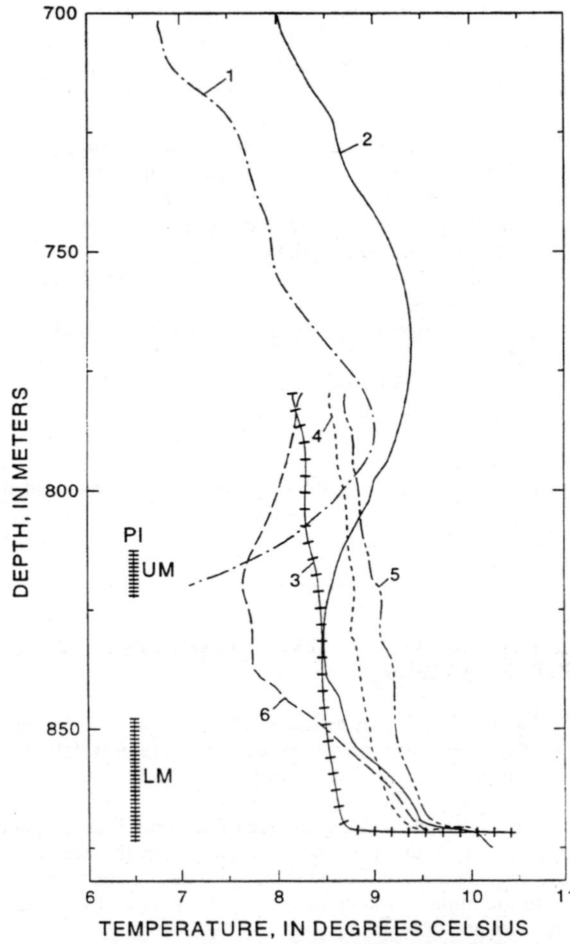

Figure 10. Temperature well logs (1976, 1985) from Well-142 in the Messoyakha field. Log surveys 1, 2, and 6 were taken when the well was temporarily shut-in (note temperature drop in the upper part of the Messoyakha reservoir). Log surveys 3, 4, and 5 were ran while the well was under production. UM, upper Messoyakha reservoir. LM, lower Messoyakha reservoir. PI, perforated zones open to production.

0.61 percent. If gas hydrates are in chemical equilibrium with the free-gas in the Messoyakha field, the decrease in ethane and propane concentrations cannot be attributed to the dissociation of Structure I or Structure II gas hydrates because we would expect to see a relative increase in ethane and propane if gas hydrates were contributing to gas production (figure 11). Even though carbon dioxide can be included in a Structure I hydrate, it does not appear that hydrates are the source of the carbon dioxide; rather, it is believed that the production-induced reservoir pressure drop has permitted the carbon dioxide dissolved in the formation waters to come out of solution and enter the vapor phase.

Ginsburg and others (1990) believed that helium concentrations in produced reservoir gases can be used to assess the presence of gas hydrates because helium does not enter into a hydrate structure. On the initial production start-up of the Messoyakha field, the helium concentration in the produced gas ranged from about 0.005 to 0.041 percent, which is within the normal range for other Cenomanian gas deposits of the West Siberian Basin. By the late 1970's (figure 7, Stage IV), however, the helium concentration in the gases had changed, and concentrations ranged from low (0.0002 percent) to high (0.6 percent) values. Additional geochemical sampling of shut-in wells reveals that the helium-enriched gases are limited to the crest of the Messoyakha structure within the upper, potentially gas-hydrate-bearing, part of the reservoir (figures 12, 13A, and 13B). Ginsburg and others (1990) interpreted the elevated helium concentrations as evidence of cross-flow from the upper to lower part of the Messoyakha field. Detailed analysis of gas sampled from a shut-in well (figure 14, Well Number 142) reveals that within a 80-day period the helium concentration increased from about 0.02 percent to more than 0.12 percent, this increase also indicates that cross-flow from the upper part of the Messoyakha reservoir is "recharging" the lower part of the reservoir. Since helium does not enter into a hydrate structure, Ginsburg and others (1990) interpreted that the upper part of the Messoyakha reservoir does not contain gas hydrates because the crest of the structure is enriched in helium (figures 12, 13A, and 13B). It is possible, however, that the low helium concentrations observed around the margin of the Messoyakha structure (figures 12, 13A, and 13B) are a product of decomposing gas hydrates.

SUMMARY DISCUSSION

The evidence for and against the occurrence of gas hydrates in the Messoyakha field centers around five primary questions: (1) Does the Messoyakha reservoir lie within the zone of predicted gas hydrate stability? (2) Do well logs indicate the presence of gas hydrates in the Messoyakha field? (3) Has the production history of the Messoyakha field been affected by the dissociation of "natural" gas hydrates? (4) What does the volume of water produced from the Messoyakha field tell us about the potential for in-situ gas hydrates? (5) Can the observed historical changes in gas chemistry within the Messoyakha field be attributed to the dissociation of "natural" gas hydrates?

Most published studies indicate that some portion of the Messoyakha field lies within the zone of predicted gas hydrate stability; however, some researchers question the quality of available Messoyakha reservoir temperature and pressure data, which is required to assess the potential occurrence of gas hydrates. Additional data, including formation temperatures and pressures and gas chemistries, are needed to conclusively determine if the Messoyakha field lies within the zone of gas hydrate stability.

Using available Messoyakha well-log data, it is impossible to distinguish between the physical properties unique to either gas hydrates or the reservoir rocks. Closer examination of the available well log data and the addition of acoustic logs and other more sophisticated devices may, however, yield pertinent gas hydrate and reservoir data.

The production history of the Messoyakha field and results of associated production tests have long been used as evidence for the presence of gas hydrates. Ginsburg and others (1990) have provided a viable non-hydrate alternative to explain the Messoyakha production history. This alternative explanation shows that reservoir production histories can be misleading and cannot be used as conclusive evidence for gas hydrates.

The volume of water produced from the Messoyakha field suggests that only a limited amount of the produced gas can be attributed to gas hydrate production. It is possible, however, that some of the water associated with the produced gas hydrates remains in the Messoyakha reservoir.

Although changes in ethane, propane, and carbon dioxide concentrations in the produced gases suggest that gas hydrates have not contributed to gas production in the Messoyakha field, the observed gas compositional changes are relatively small and may not be statistically significant. The observed changes in helium concentrations are relatively dramatic. The authors of this paper do not agree between themselves that helium concentrations in produced gases can be used to assess the presence of gas hydrates. Our point of contention deals with the theory that, because helium does not enter into a hydrate structure, the occurrence of helium indicates that gas hydrates are not present. It is possible, however, that gas hydrates may not completely fill all of the pore-space within a reservoir and the non-hydrate-bearing portion of the reservoir may contain either gas enriched in helium or water that has a high dissolved helium content. More field and laboratory studies are needed to evaluate the utility of helium as a gas hydrate assessment tool.

In conclusion, all of the researchers who have studied the Messoyakha field believe that some part of the field contains gas hydrates. It remains unclear, however, how much gas hydrate is present within the Messoyakha field or if gas hydrates have contributed significantly to the historical gas production.

ACKNOWLEDGMENT

This work was funded by the Russian State Program "Global Natural and Climatic Change" and the U.S. Geological Survey Global Change and Climate History Program.

REFERENCES

Alekseyev, F.A., (1974). Zonality in oil and gas formation in the earth's crust based on isotopic studies: Geologiya Nefti i Gaza, April, p. 62-67; Journal of Petroleum Geology, v. 12, p. 191-192 (English translation).

Bro, E.G., and Sorokov, D.S., (1970). Znacheniye nekotorykh osobennostej produktivnoj tolschi Messoyakhskogo gazovogo mestoroxhdeniya dlya ratsional'noj razrabotki [Some peculiarities of the pay section of the Messoyakhskoye gas filed - significance for development]: Uchonyie zapiski NIIGA, Regional'naya Geologiya, v. 17, Leningrad, p. 25-30.

Cherskiy, N.V., Tsarev, V.P., and Nikitin, S.P., (1985). Investigation and prediction of conditions of accumulation of gas resources in gas-hydrate pools: Petroleum Geology, v. 21, p. 65-89.

Collett, T.S., (1993). Natural gas hydrates of the Prudhoe Bay and Kuparuk River area, North Slope, Alaska: American Association of Petroleum Geologists Bulletin, v. 77, no. 5, p. 793-812.

Du Rochet, J., (1980). Une explication paleoclimatique des grands gisments superficiels de gaz sec du norde de la Siberie Occidentale [A paleoclimatic explanation for the shallow giant gas reserves of northern West Siberia]: Bulletin des Centres de Recherches Exploration-Production Elf-Aquitains, v. 4, p. 119-142.

Ginsburg, G.D., (1993). Challenging the presence of natural gas hydrate in the Messoyakha pool: American Association of Petroleum Geologists Bulletin, v. 77, no. 9, p. 1625.

Ginsburg, G.D., Borisov, V.V., and Novozhilov, A.A., (1990). On hydrate-bearing of the Messoyakha gas deposit, *in* Sources of non-traditional gas resources and problems of their development: USSR Ministry of Geology, All-Union Research Institute for the Exploration, Labour Red Banner Publishers, p. 211-223.

Grace, J.D., and Hart, G.F., (1986). Giant gas fields of northern West Siberia: American Association of Petroleum Geologists Bulletin, v. 70, no. 7, p. 830-852.

Judge, A.S., (1988). Mapping the distribution and properties of natural gas hydrates in Canada: American Chemical Society Third Chemical Congress of the North American Continent, June 6-7, Toronto, Ontario, Proceedings, Abstract Number 29.

Kortsenshteyn, V.N., (1970). Effect of periodic glaciations on the formation of the enormously large gas fields in the northern part of the Tyumen Oblast: Akademiya Nauk SSSR Doklady, v. 191, p. 1366-1369.

Krason, J., and Ciesnik, M., (1985). Geological evolution and analysis of confirmed or suspected gas hydrate localities, Volume 5--Gas hydrates in the Russian literature: Report for U.S. Department of Energy, Office of Fossil Energy, Morgantown Energy Technology Center, Morgantown, West Virginia, Contract DE-AC21-84MC21181, 164 p.

Krason, J., and Finley, P.D., (1993). Messoyakha gas field -- Russia, West Siberian Basin, *in* Beaumont, E.A., and Foster, N.H., eds., Structural traps II -- Treatise of petroleum geology atlas of oil and gas fields: American Association of Petroleum Geologists, Publisher, Tulsa, Oklahoma, p. 197-220.

Kruglikov, N.M., Prasolov, E.M., Yakovlev, O.N., (1983). Gidrogeologicheskaya sostavlyayuchaya prognoza neftegazonosnosti: All Union Scientific-Research Institute for Geologic Exploration, p. 4-27; Hydrogeological aspect of petroleum resource assessment: Journal of Petroleum Geology, v. 20, p. 481-485 (English translation).

Kvenvolden, K.A., (1988). Methane hydrate--A major reservoir of carbon in the shallow geosphere?: Chemical Geology, v. 71, p. 41-51.

Kvenvolden, K.A., (1993). Gas hydrates as a potential energy resource--A review of their methane content, *in* Howell, D.G., ed., The future of energy gases: U.S. Geological Survey Professional Paper 1570, p. 555-561.

Makogon, Y.F., (1981). Hydrates of natural gas: Tulsa, PennWell Publishing Company, 237 p.

Makogon, Y.F., (1988). Natural gas hydrates--the state of study in the USSR and perspectives for its use: Third Chemical Congress of North America, Toronto, Canada, June 5-10, 18 p.

Makogon, Y.F., Trebin, F.A., Trofimuk, A.A., Tsarev, V.P., and Cherskiy, N.V., (1972). Detection of a pool of natural gas in a solid (hydrate gas) state: Doklady Academy of Sciences U.S.S.R., Earth Science Section, v. 196, p. 197-200.

Masters, C.D., Root, D.H., and Attanasi, E.D., (1991). Resource constraints in petroleum production potential: Science, v. 253, p. 146-152.

Peterson, J.A., and Clarke, J.W., (1989). West Siberian oil-gas province: U.S. Geological Survey Open-File Report 89-192, 142 p.

Sapir, M.H., Khramenkov, E.N., Yefremov, I.D., Ginsburg, G.D., Beniaminovich, A.E., Lenda, S.M., and Kislova, V.L., (1973). Geologicheskie i promislovo -- geofizicheskie osobennosti gazogidratnoi zalezhi Messoiakhskogo gazovogo mestorozhdenia [Geologic and geophysical features of the gas hydrate deposits in the Messoykha field]: Geologiia Nefti i Gaza, v. 6, p. 26-34.

Sloan, E.D., (1990). Clathrate hydrates of natural gases: Marcel Dekker, Inc., Publisher, New York, New York, 641 p.

Sumetz, V.I., (1974). Prevention of hydrate formation in gas wells zone: Gazovaya Promishlennost, v. 2, p. 24-26.

Trofimuk, A.A., Chersky, N.V., and Tsaryov, V.P., (1977). The role of continental glaciation and hydrate formation on petroleum occurrence, *in* Meyer, R.F., ed., The future supply of nature-made petroleum and gas: Pergamon Press, New York, p. 919-926.

Verkhovsky, A.B., Ginsburg, G.D., and Prasolov, E.M., (1988). On the diagnostics of the hydrate-forming process and gas-hydrates at depth with inert components: Doklady Academy Science USSR, v. 301, no. 4, p. 959-962.

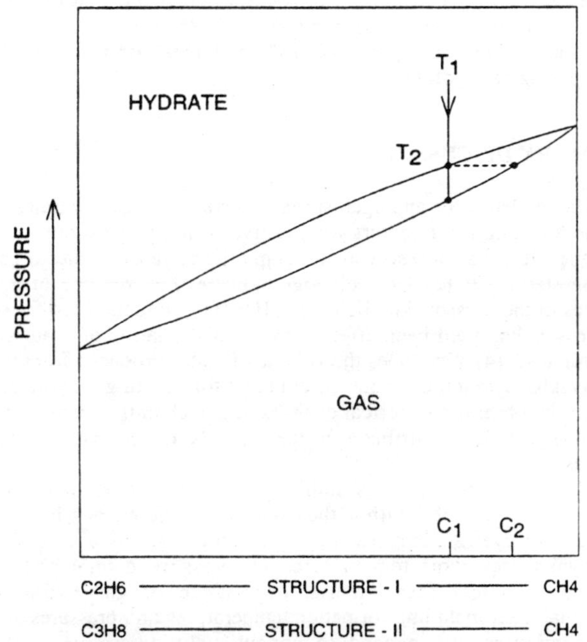

Figure 11. Hypothetical gas hydrate pressure-composition phase diagram. At initial production start-up (T1), reservoir pressures decline until gas hydrates begin to dissociate (T2). At T2, the composition (C1) of the dissociating gas hydrates is enriched in ethane (C2H6) and propane (C3H8) relative to the composition (C2) of the free-gas portion of the field.

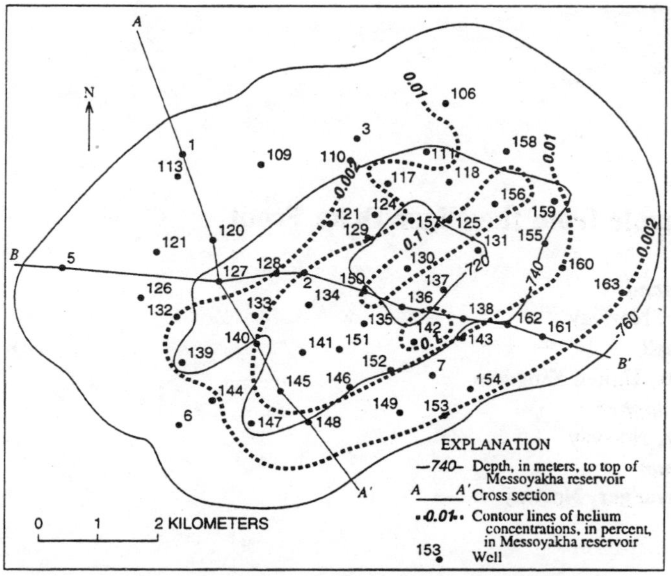

Figure 12. Structure map of the top of the Dolgan Formation in the Messoyakha field area depicting the helium concentration in gases from shut-in wells (modified from Ginsburg and others, 1990). Location of cross-sections in figure 13A are also shown.

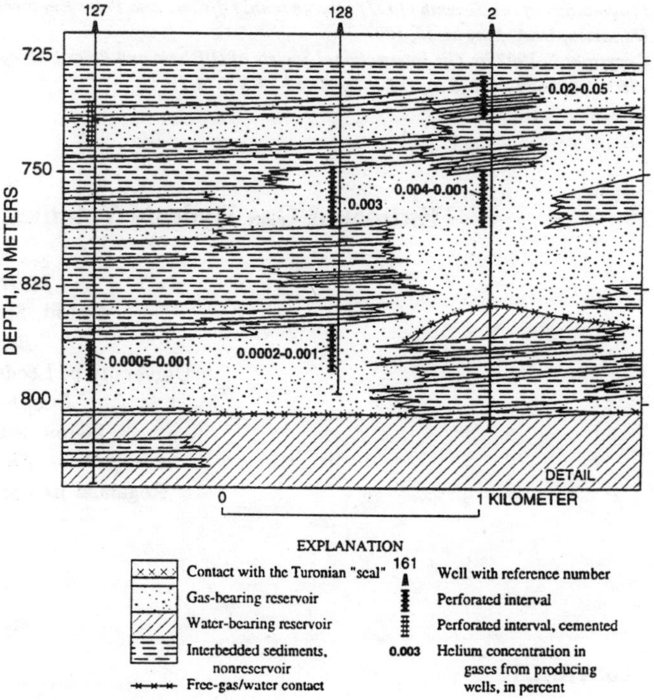

Figure 13B. Detailed cross-section insert from figure 13A, shows helium concentrations in produced reservoir gases. Location of cross sections are shown in figure 12.

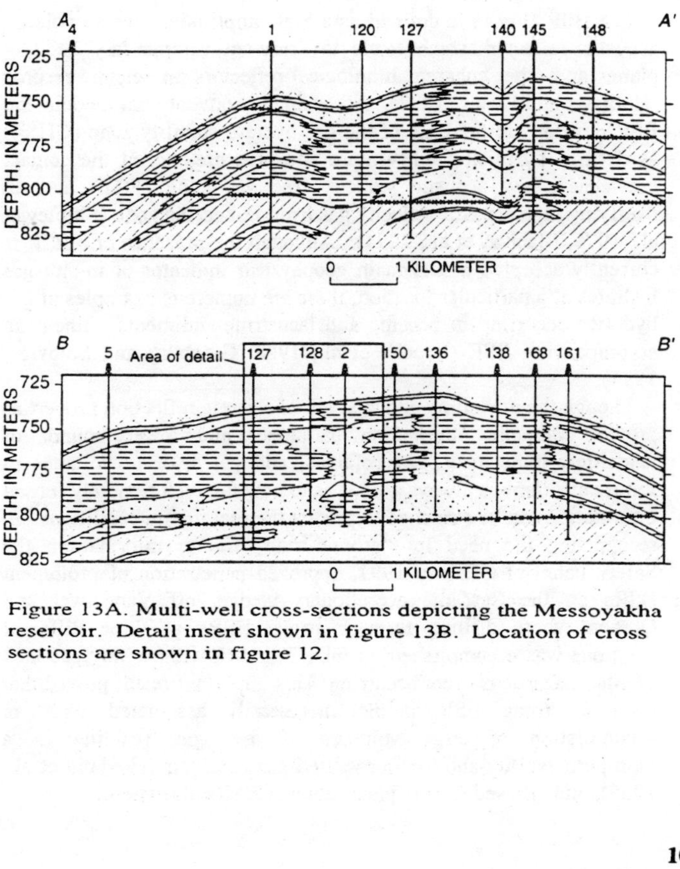

Figure 13A. Multi-well cross-sections depicting the Messoyakha reservoir. Detail insert shown in figure 13B. Location of cross sections are shown in figure 12.

Figure 14. Helium concentration in gases sampled from a shut-in Messoyakha well (Well-142).

Proceedings of the Seventh (1997) International Offshore and Polar Engineering Conference
Honolulu, USA, May 25-30, 1997

ISBN 1-880653-28-1 (Set); ISBN 1-880653-29-X (Vol. I)

Estimated Gas Hydrate Volumes: Example from the Niger Delta Front

Martin Hovland
Statoil, Stavanger, Norway
Ben Clennell
University of Leeds, Leeds, United Kingdom
Joseph W. Gallagher
Statoil, Stavanger, Norway
Knut Lekvam
Rogaland Research, Stavanger, Norway

ABSTRACT

We present a model of natural gas hydrate occurrence that incorporates capillary effects arising from the co-existence of liquid water, gaseous hydrocarbons and solid hydrate phases in porous media. Based on a review of recent gas hydrate observations and on our model, the maximum amount of gas hydrates residing in sediments above a bottom simulating reflector (BSR) mapped on the Niger Delta front is suspected to be only 4% by volume. The accompanying free gas volume in sediments below the BSR is suspected to be only about 6% by volume. Because these resources are most probably occurring inhomogeneously and finely dispersed in the sediments, there is currently no commercial potential for recovering and exploiting this trapped energy source.

KEY WORDS: gas hydrates, BSR, Niger Delta, capillary effects, water depletion.

INTRODUCTION

A vast reservoir of methane is undoubtedly trapped in the form of natural gas hydrates residing in submarine sediments of the continental margins (Kvenvolden, 1993). Gas hydrates recovered from beneath the sea floor display a wide range of growth habits and tend to be patchily distributed within the host sediment (Booth et al., 1996a). Even though there is a very extensive zone of potential stability in the subsurface, the occurrence is evidently limited either by the availability of sufficient quantities of guest gases or the in-situ availability of liquid water (Clennell et al., 1995). Little attention has so far been paid to the distribution of hydrate on a small scale, or to the role played by the host sediment physical properties on the stability and habitat of these minerals.

We have developed a model incorporating capillary effects arising from the co-existence of liquid water, gaseous hydrocarbons, and solid hydrate phases in porous media. In this paper we report the results of applying the model for gas hydrate and free gas volume prediction in a natural gas hydrate prone environment on the Niger Delta front, where BSRs (bottom simulating reflectors) are common.

GAS HYDRATES AND THE BSR

The BSR (Fig.1) is defined as a high-amplitude, reverse polarity acoustic reflector which tends to cross-cut sedimentary bedding planes and other coherent lithological reflectors on seismic records (Hollister and Ewing, 1972). According to conventional models, the BSR develops at the base of the gas hydrate stability zone (GHSZ) as defined by in-situ temperature and pressure, i.e. at the contact between a low acoustic velocity free gas zone (FGZ) and an overlying high acoustic velocity gas hydrate-bearing zone (Shipley et al., 1979; MacKay et al., 1994). Although the presence of a BSR is currently accepted as the main geophysical indicator of in-situ gas hydrates at a particular location, there are numerous examples of gas hydrates occurring in oceanic and lacustrine sediments without an accompanying BSR (Brooks et al., 1988; Ginsburg and Soloviev, 1994).

Though the modelling and analysis of seismic reflection properties provide valuable information on upper and lower bounds of velocities, impedance, gas volumes, and gas hydrate volumes in sediments, there is always a limit as to how far such remotely sensed information can be stretched to predict the real in-situ conditions. In recognising the need for "ground-truth" in-situ information, the Safety Panel of ODP, in 1991, approved penetration of prominent BSRs on the Cascadia accretionary wedge, off Vancouver and Oregon where drilling through strong BSRs at three different locations was accomplished on ODP Leg 146 without any gas kick or other hazardous event occurring. This important result proved that even a strong BSR is not necessarily associated with an accumulation of large volumes of free gas residing in a high-permeability and high-pressured gas reservoir (Hovland et al., 1995), and allowed further penetration of BSRs elsewhere.

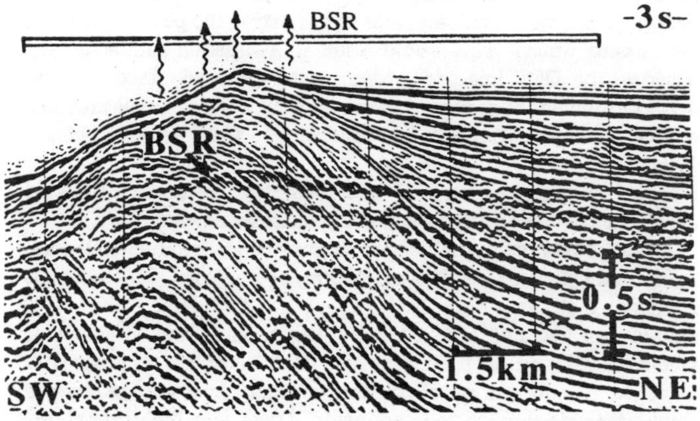

Fig. 1 A classical BSR from the study area on the Niger Delta front. Wavy arrows in the water column indicate locations of suspected fluid escape. Horizontal bar at top indicates the extent of BSR.

ODP Leg 164 was therefore targeted on the archetype-BSR of the Blake Ridge, off eastern USA (Dillon et al., 1991). Here, three strengths of BSR were penetrated: faint, strong, and very strong BSR. The finding was that they all had a similar lithology and there were no noticeable effects on drilling when penetrating the BSRs (Paull et al., 1996). Only the downhole acoustic velocity differed, with lowest velocity below the strongest BSR (Paull et al., 1996).

However, results from ODP Leg 146, on the Cascadia accretionary wedge, off Oregon, revealed that there was a significant difference in sediment water-content above and below the BSR. The sediments above the BSR had up to two metres-thick discrete water-saturated sediment layers (called "soupy layers"), caused by dissociation of dispersed gas hydrates (Westbrook et al., 1994). However, some of the sampled sediments below the BSR were anomalously water-depleted (Hovland et al., 1995). Furthermore, sediment sampling became much more difficult after penetration of the BSR, most probably because the sediments were densily fractured, over-consolidated and brittle. A drill-string packer test integrated over the depth interval 94-146 metres below seabed, well below the BSR, was performed at Site 892, Leg 146. The results showed the in-situ pressure to be at least 0.2 Mpa above hydrostatic pressure (Screaton and Carson, 1995), demonstrating an overpressured situation below the BSR, which may possibly account for the fractured nature of the sediments.

Another important finding from these two ODP Legs was the relatively low content of bulk gas hydrates residing within the sediments above the GHSZ. ODP Leg 146 only found visible amounts of gas hydrates, in the form of centimetric nodules and smaller "hailstone"- type pellets within the upper 20 m of soft shale at one of the BSR locations (Hovland et al., 1995). Otherwise on that leg, there was only indirect in-situ gas hydrate evidence in the form of "soupy layers", layers with chloride depleted pore-waters, and layers of anomalously low core temperatures (Westbrook et al., 1994). Similar results were found on Leg 164, where, on the basis of chloride depletion results and direct measurements with a pressure core sampler (PCS), a maximum volume of only 9% of the bulk pore space was estimated to be filled with gas hydratres (Paull et al., 1996; Dickens et al., 1997). Based on results from ODP Legs 146 and 164, and on modern seismic analyses, we therefore conclude that the total amount of bulk gas hydrate to be expected within the GHSZ is only 1 - 4% by volume, at these drillsites, combined with a small amount, i.e., 1 - 6% of sediment bulk volume free gas below the BSR (Holbrook et al., 1996; Dickens et al., 1997). The thickness of this sub-BSR "gas reservoir" is estimated to be at least up to 50 m (MacKay et al., 1994).

GAS HYDRATES AND FLUID ESCAPE

Numerous occurrences of gas hydrates exist in oceanic sediments without an associated BSR (Brooks et al., 1988; Brooks et al., 1994; Ginsburg and Soloviev, 1994; Ginsburg et al., 1993). Booth et al. (1996a) made a thorough literature survey of all documented occurrences of natural gas hydrates and constructed a data base including general and specific geological conditions. They found that in nearly three quarters of the occurrences hydrates had been sampled proximal to faults or vent systems. Furthermore, they note that authigenic carbonates, another indicator of seepage or vertical gas migration, are very often associated with gas hydrate zones.

However, not all occurrences of oceanic sediment gas hydrates are seep-related. In a recent study (Booth et al., 1996b), gas hydrates residing in sediments well above the phase boundary (GHSZ) were studied and it was found that some of these so-called "internal hydrates" are products of site-specific enrichments of autochthonous microbial methane production. The volume of these bacterially induced hydrate accumulations are limited by the organic content of the local sediments. Because of the limited resource potential for such deposits, we will omit further discussion of this hydrate type, and rather focus attention on the potentially more voluminous types, i.e., those formed by migrating allochthonous light hydrocarbons (methane, ethane, propane, and butane) of either bacterial, thermogenic, or a mixture of both gas sources.

The clearest evidence of a strong link between natural gas hydrates and fluid escape comes from the observations of submarine mud diapirs and mud volcanoes. In the Gulf of Mexico, Brooks et al. (1988) first described BSR-free gas hydrates of which the Bush Hill occurrence is an example: "The site is named after large chemosynthetic tube worm bushes that occur on the seafloor. These tube worm communities derive their energy from hydrogen resulting from biodegradation of seeping oil at the site (MacDonald et al., 1989). The site is only a few kilometres away from Jolliet Field, one of the first deep water oil discoveries and deep water production facilities in the Gulf of Mexico. The topography of the site results from migration of oil and gas up an antithetic fault associated with deep faulting." (Brooks et al., 1994, p. 387).

Similar evidence also exists in the Caspian and Black seas: Ginsburg and Soloviev (1994) discovered gas hydrates in pockmark fields on the summit of two large mud volcanoes, Buzdag and Elm, at between 475 and 660 m water depth, in the Caspian Sea. Here, hydrates were observed in 19 out of 20 gravity cores, each of 1.2 m length, from the Buzdag, and in 5 out of 7 cores from the Elm mud volcanoes: "In both cases hydrates were observed in clay breccia as ice-like semi-transparent mat inclusions of various shapes and sizes. Irregular plates up to several mm thick and up to 12 cm long, sub-isometric angular pieces up to 5 cm and very small scales and needles were found." (Ginsburg and Soloviev, 1994, p. 96).

Based on these and numerous other observations and documentations, we tend to agree with the conclusions of Brooks et al. (1994): "Of importance is the fact that gas hydrates in the shallow subsurface may not act as major barriers to the venting of gas into the water column since gas seeps appear to be common, although sometimes localized, in areas contatining shallow gas hydrates." (Brooks et al., 1994, p. 386).

Therefore, we conclude this review stating that although gas hydrates may form where there is local abundance of autochthonous biogenic methane production, significant (large) amounts of gas hydrates in marine sediments with or without BSRs only form as a concequence of gas migration and fluid escape through the seafloor.

MODEL FOR GAS HYDRATE OCCURRENCES

We now link recent natural gas hydrate observations with a theoretical examination of the role of host sediment properties and sediment water content on the stability and growth of gas hydrates. The small size of pores found in typical deep marine sediments inhibits the nucleation and growth of hydrates with respect to the bulk conditions of the laboratory, so that their temperature of formation becomes lower at a given pressure (Clennell et al., 1995). These considerations incorporate capillary effects arising from the coexistence of liquid water, gaseous hydrocarbon, and solid hydrate phases in porous media. In fine-grained sediments, it is unlikely that nucleation of hydrate will occur in isolated small pores, but preferentially in larger pores or fractures. When the space in all voids larger than a critical size is filled, the hydrate will tend to grow as a displacive inclusion or segregated layer rather than penetrating into smaller pores (Clennell et al., 1995). The requirement to minimise surface energy also explains the growth forms commonly observed in hydrate samples, i.e., nodules, pellets, sheets, and lenses (Ginsburg and Soloviev, 1996; Hovland et al., 1995).

Because growth of gas hydrates will sequester water and gas from surrounding sediments in a way that is analogous to ice lensing in frozen soil, this will lead to the formation of layers and zones of anomalously low porewater content in the sediments, which further leads to the conclusion that the water-content and water migration caused by gas hydrate formation may represent a key element in the understanding of the seismic response of the sediments above and below the BSR (Clennell et al., 1996). The analogy between gas hydrate formation and water movement in soils exposed to freezing (permafrost) conditions have been investigated and proposed previously by Trofimuk et al. (1972) and more recently by Ginsburg and Soloviev (1996) and Clennell et al. (1995). Porewater in freezing soil, adjacent to an ice lense is subjected to suction forces caused by the freezing front (Konrad and Seto, 1994). Similarly it is likely that suction forces act in porewater of soils within the GHSZ. When free gas molecules are available, gas hydrate will start forming and water will be drawn from the surrounding sediment matrix towards the gas hydrate forming location in the sediments. The resulting water depletion in the sediments adjacent to the gas hydrate "front" will also cause consolidation to start (Clennell et al., 1995). This water depletion model agrees with the finding of "soupy layers" enriched in water, by the formation of minute crystals of gas hydrate, and also the finding of anomalously dry sediment zones, especially below the BSR (Hovland et al., 1995). Furthermore, below the BSR it is likely that water may have been depleted to such an extent that gas hydrates no longer form, partly because of lack of "free" water, even though the region is within the theoretical GHSZ.

Following these lines of thought further, the BSR and sediments immediately below, will be depleted in water and they will be over-consolidated, however, with an important modification, on the condition that there is more or less continuous replenishment of gases from below: the sediments will tend to fracture, in a similar manner to hydrofracturing, but because of the observed over-pressure, with small amounts of gas taking the place of some of the missing water. This would result in a situation similar to that observed (on ODP Leg 146), where water-depleted, fractured, and over-consolidated sediments below the BSR contain small amounts of free and interstitial gas (Hovland et al., 1995; MacKay et al., 1994).

Based on the observational and theoretical evidence discussed above, we conclude with the following main statements:

Although relatively small accumulations of gas hydrates will form as internal hydrates (Booth et al., 1996b) by biogenic, autochthonous methane, there is a very high probability that large volumes of natural gas hydrates and extensive BSRs only form at locations where relative vertical fluid migration takes place, i.e., in areas with diffusive percolation of allochthonous light hydrocarbon-charged fluids upwards through the sediments. There is a high probability that extensive and strong BSRs only occur in areas where there is percolation or diffusive transport of fluid movement upwards through the sediments.

Because most hydrates in marine sediments will tend to occur where there is an upward component in the fluid migration, there is also an over-representation of BSRs and gas hydrate occurrences in areas with dipping strata, along faultplanes conducive to fluid migration, and along other vertical migration pathways, such as mud volcanic feeder pipes and vertical high-permeability inhomogeneities. This model explains why BSRs mostly seem to occur in areas where there is some dip to the sediment bedding planes: the gas is transported to the GHSZ via the dipping sediment structure - and therefore the BSR is seen as a reflector crossing the sediment bedding planes.

VOLUME ESTIMATES OFF NIGERIA

The outer shelf and slope of the Niger Delta are recognised as a zone of active, gravity driven shale tectonics characterised by linked extensional, compressional and minor diapiric structures (Cohen and McClay, 1996; Townsend et al., 1996). Within our study area (Fig.2) there are shale cored compressional structures, extensional structures related to collapse of deeper folds, and mud volcanoes which pierce through all younger sedimentary strata and which may be several km long and create up to 100 m sea-floor relief by extruded material (Cohen and McClay, 1996). A region of elongated patches of BSRs and mud diapirs have been mapped on the westward slope of the Niger Delta (Fig. 2). The summary map (Fig.2) shows water depth contours every 500 m, the location of anticlinal crests, piercement shale diapirs associated with mud volcanoes, and the location of BSRs. The map is based on a 2D reflection seismic dataset which has a grid of less than 5x5 km. This SW section of the delta front reaches a depth of over 2500 m. Mud volcanism, fluid escape, and sea-floor channel development has rendered parts of the delta front topographically complex. The BSRs tend to follow the updipping strata formed by the anticlinal fold structures, whereas the mud volcanoes tend to form at the very summit of these folds.

About 75% of the BSR patches straddle or occur very close to the anticlinal crests and several of the patches include mud volcanoes (Fig. 2). The size of the patches vary from small circular "blobs", about 2 km wide, to large 70 km long and 5 km wide patches. The total coverage of all the mapped patches is 568 square km, which represents 8.5% of the total study area shown in Fig. 2.

Fig. 1 shows a typical profile from this region with a relatively strong BSR with a depth below sea-floor varying between 510 ms

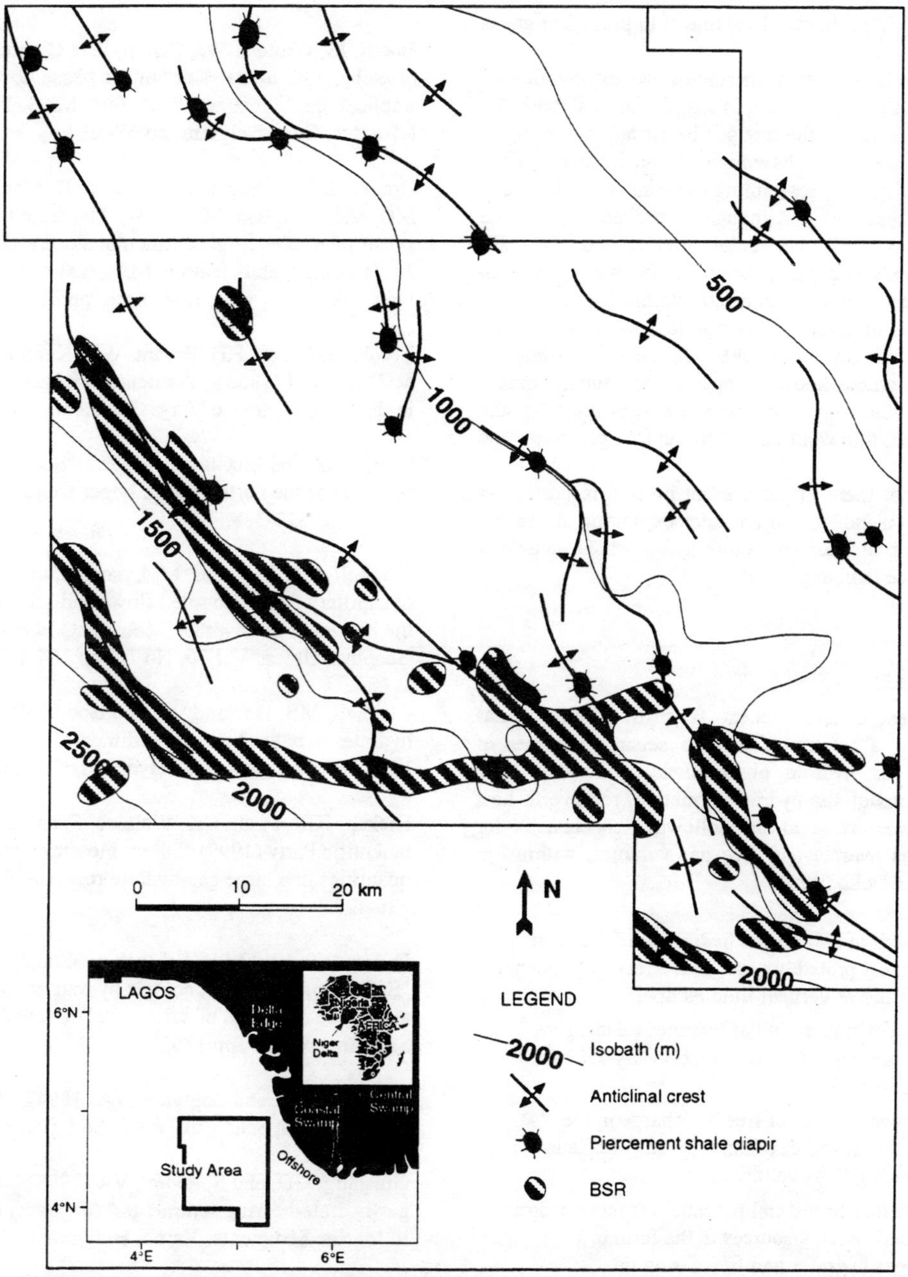

Fig. 2 Map over the study area on the western Niger Delta front, showing locations of BSRs, water depths, crests of intrusive shale anticlines, and piercement shale diapirs. BSRs occur at water depths >1000 m and cover a total of 8.5% of the study area.

two-way-time (twt) in the NE, to about 390 ms twt in the SW. It starts in the NE, where the bedding becomes inclined, runs straight through the strongly dipping strata associated with the fold, and terminates below the deepest part of the sea-floor slope. Based on our gas hydrate model discussed earlier, the following interpretation of this section is given: hydrocarbon gases migrate up-dip from the NE and form gas hydrates within the GHSZ with excess gas traversing through the BSR and emitting diffusively to the ocean through the crestal area of the sea-floor anticline (Fig. 1). According to our assumptions, which are mainly based on ODP drilling through strong BSRs, the sediments above the BSR contain a total of 1 - 4% gas hydrates by volume, and a normal or higher than normal water content, for similar deep ocean sediments. Based on similar assumptions, the sediments below the BSR are suspected to contain 1 - 6% free gas, by volume, have less than normal water content, and are suspected to be over-consolidated and fractured. We have used the same interpretation for the whole of the study area in order to arrive at the following maximum resource estimate:

The mean sediment thickness above the BSR is 340 m, using a mean twt of 400 ms, at a mean velocity of 1700 m/s. With a mean maximum in-situ gas hydrate concentration of 4%, the total volume in place gas hydrate is 7.7 km^3, which represents a considerable amount of gas when converted to standard atmospheric pressure. For the calculation of the total free gas volumes beneath the BSRs, a mean maximum gas charge of 6% has been used. With a mean "reservoir" thickness of 50 m, the total volume of in place free gas is calulated to be 1.7 km^3.

Even though these volumes seem enormous, the exploitation of them may not be very attractive. This is mainly because both the gas hydrates above the BSR and the free gas below are, according to our model, contained in a finely disseminated and inhomogeneous fashion, in the sediment matrix, resembling more an ore-body than a common hydrocarbon reservoir. Although it probably would be technically feasible to sink horizontal wells into the gas hydrate-bearing sediments and pump brine or hot water, or even glycol through the deposits in order to dissociate the hydrates (Max and Lowrie, 1996), it is probably not energetically economic to do so. The most probable method of mining the deposits, is assumed to be normal excavation methods known to open pit and tunnel mining on land. Because of the very poor reservoir conditions in the underlying free gas zone, this would apply to the free gas below the BSR as well.

We thus conclude that there are at present no real resources, at least for the hydrocarbon industry to consider exploiting at present on the Niger Delta front, at least not before hydrocarbons achieve a considerably higher price than at present.

CONCLUSIONS

On the basis of results from recent scientific drillings that penetrate strong BSRs, a review of modern seismic analyses of BSRs, a review of gas hydrate observations, and some new theoretical considerations of gas hydrate formation, relative to host sediment properties, we arrive at the following conclusions for estimating in place gas hydrate and free gas volumes, within our study area, on the Niger Delta front:

1. The largest deposits of natural gas hydrates and the most extensive BSRs most probably occur as a direct consequence of focused and diffusive vertical fluid escape.
2. The mean maximum amount of gas hydrates residing in sediments above the BSR is most probably only 4% by volume.
3. The mean maximum amount of free gas charge in the sediments below even well developed (strong) BSRs is suspected to be only 6% by volume.
4. Currently there is no commercial potential for recovering and exploiting "trapped" energy sources in the form of BSR-associated gas hydrates and free gas on the Niger Delta front, mainly because they are suspected to occur as dispersed, rather than concentrated deposits.

ACKNOWLEDGEMENTS

Seismic data courtesy of TGS-CALIBRE Geophysical Company and Mabon Limited, (TGSI/Mabon). Statoil and the BP/Statoil alliance are greatly thanked for the release of these data. Phil Townsend is thanked for constructive comments.

REFERENCES

Booth, JS, Rowe, MM, and Fischer, KM (1996a). "Offshore gas hydrate sample database," USGS Open file report, No 96-272.

Booth, JS, Winters, WJ, Dillon, WP, Clennell, MB, and Rowe, MM (1996b). "Gas hydrates not of the phase boundary: their origins and implications," (abstract) Proc. First MASTER workshop, Gas Hydrates, Gent, Belgium, pp 58-59.

Brooks, JM, Anderson, AL, Sassen, R, MacDonald, IR, Kennicutt, MC, and Guinasso, ML (1994). "Hydrate occurrences in shallow subsurface cores from continental slope sediments." In Sloan, ED Jr., Happel, J, and Hnatow, MA (eds) "Natural gas hydrates" Annals New York Acad. Sci., New York, pp 381-391.

Brooks, JM, Cox, HB, Bryant, WR, Kennicutt, MC, Mann, RG, and McDonald TJ (1988). "Association of gas hydrates and oil seepage in the Gulf of Mexico" Org. Geochem., Vol 10, pp 221-234.

Cohen, HA and McClay, K (1996). "Sedimentation and shale tectonics of the northwestern Niger Delta front," Mar. Petrol. Geol., Vol 13, pp 313-328.

Clennell, MB, Hovland, M, Lysne, D, and Booth, JS (1995). "Role of capillary forces, coupled flows and sediment-water depletion in the habitat of gas hydrate," (abstract) EOS Transactions of Am. Geophys. Union, Vol 76, No 16, pp 164-165.

Clennell, MB, Hovland, M, and Booth, JS (1996). "Natural gas hydrates in natural marine sediments," (abstract) Proc. First MASTER workshop, Gas Hydrates, Gent, Belgium, pp 40.

Dickens, GR, Paull, CK, Wallace, P, and the ODP Leg 164 Scientific Party (1997) "Direct measurement of *in situ* methane quantities in a large gas-hydrate reservoir," Nature, Vol 385, pp 426-428.

Dillon, WP, Lee, MW, Fehlhaber, K, and Hutchinson, DR (1991). "Estimation of amounts of gas hydrate in marine sediments using amplitude reduction of seismic reflections," (abstract) J. Acoustic Soc. Am., Vol 89, pp 1853.

Ginsburg, GD, and Soloviev, VA (1994). "Mud volcano gas hydrates in the Caspian Sea" Bull. Geol. Soc. Denmark, Vol 41, pp 95-100.

Ginsburg, GD, and Soloviev, VA (1996). "Structures of gas-hydrate-bearing sediments," (abstract) Proc. 4th Int. Conf. Gas in Marine Sediments, Varna, Bulgaria.

Ginsburg, GD, Soloviev, VA, Cranston, RE, Lorenson, TD, and Kvenvolden, KA (1993). "Gas hydrates from the continental slope offshore from Sakhalin Island, Okhotsk Sea," Geo-Marine Letters, Vol 13, pp 41-48.

Holbrook, WS, Hoskins, H, Wood, WT, Stephen, RA, Lizarralde, D, and Leg 164 Scientific Party (1996). "Methane hydrate and free gas on the Blake Ridge from vertical seismic profiling," Science, Vol 273, pp 1,840-1,843.

Hollister, CD, and Ewing, JI (1972). "Sites 102, 103, 104, Blake-Bahama Outer Ridge (Northern end)," Initial report of the Deep Sea Drilling Project, Vol 11, pp 135-218.

Hovland, M, Lysne, D, and Whitaker, MJ (1995). "Gas hydrate and sediment gas composition, Hole 892A," In: Carson, B, Westbrook,

GK, Musgrave, RJ, and Suess, E (eds) Proceedings of the Ocean Drilling Program, Scientific results, Vol 146, pp 151-161.

Konrad, JM, and Seto, JTC (1994). "Frost heave characteristics of undisturbed sensitive Champlain Sea clay," Can. Geotechnical J., Vol 31, pp 285-298.

Kvenvolden, KA (1993). "Gas hydrates - geological perspective and global change," Reviews of Geophysics, Vol 31, No 2, pp 173-187.

MacDonald, IR, Boland, GS, Baker, JS, Brooks, JM, Kennicutt II, MC, and Bidigare, RR (1989). "Gulf of Mexico hydrocarbon seep communities II. Spatial distribution of seep organisms and hydrocarbons at Bush Hill," Mar. Biol., Vol 101, pp 235-247.

MacKay, ME, Jarrard, RD, Westbrook, GK, Hyndman, RD, and the Shipboard Scientific Party of ODP Leg 146 (1994). "Origin of bottom-simulating reflectors: Geophysical evidence from the Cascadia accretionary prism," Geology, Vol 22, pp 459-462.

Max, MD and Lowrie, A (1996). "Oceanic methane hydrates: a "frontier" gas resource," J. Petrol. Geol., Vol 19, pp 41-56.

Paull, CK, and the ODP Leg 164 Scientific Party (1996). "Assessing the marine gas hydrate inventory: Preliminary results of ODP Leg 164," (abstract) EOS Transactions Am. Geophys. Union supplement, April 23, Spring Meeting, Am. Geophys. Union, S163.

Screaton, EJ, and Carson, B (1995). "Hydrogeologic properties of a thrust fault within the Oregon Accretionary Prism," J. Geophys. Res., Vol 100, No B10, pp 20,025- 20,035.

Shipley, TH, Houston, RT, Buffler, RT, Shaub, FJ, McMillen, KJ, Ladd, JW, and Worzel, JL (1979). "Seismic evidence for widespread possible gas hydrate horizons on continental slopes and rises," American Association of Petroleum Geologists Bull., Vol 63, pp 2,204-2,213.

Stoll, RD, Ewing, J, and Bryan, GM (1971). "Anomalous wave velocities in sediments containing gas hydrates," Journ. Geophys. Res., Vol 76, No 8, pp 2,090-2,094.

Townsend, P, Philip, N, Owusu, K, Hovland, M, Dutta, N, and Brumfield, D (1996). "Risking topseal failure in a frontier basin, an example from deepwater Nigeria," (abstract) Conf. Proceedings: Hydrocarbon seals - importance for exploration and production Trondheim, Norwegian Petroleum Soc., Norway.

Trofimuk, AA, Cherskiy, NV, Makogon, YF, and Tsarev, VP (1972). "Possible mechanism of the accumulation of natural gas," Int. Geol. Rev. 1968-1981, AGI Reprint Series, Vol VI, pp 72-76.

Westbrook, GK, Carson, B, Musgrave, RJ, and the ODP Leg 146 Scientific Party (1994). Proceedings of the Ocean Drilling Program, Initial reports, College Station, Texas, Vol 146, Pt 1, pp 611.

Proceedings of the Seventh (1997) International Offshore and Polar Engineering Conference
Honolulu, USA, May 25-30, 1997

ISBN 1-880653-28-1 (Set); ISBN 1-880653-29-X (Vol. I)

A Study and Hypothesis of Hydrate Growth Using a Low Dosage Hydrate Inhibitor (During a Shut-In Period in a Multiphase Pipeline)

Are Lund, SINTEF, Applied Chemistry, Trondheim, Norway
Olav Urdahl, Statoil Research Centre, Trondheim, Norway
Olav Lier, SINTEF, Applied Chemistry, Trondheim, Norway
Lars Henrik Gjertsen, Statoil Research Centre, Trondheim, Norway
Thorvald Jakobsen, Norsk Hydro a.s., Bergen, Norway
John Andreas Støvneng, Norwegian University of Science and Technology, Trondheim, Norway

ABSTRACT

If a multiphase subsea pipeline has to shut-in at high pressure, conditions may after some time be favorable for hydrate formation. A thin hydrate layer is then formed on the oil/water interphase. However, addition of a low dosage hydrate inhibitor to avoid problems may result in a plugged pipeline due to a possible increase in the hydrate growth on the oil/water interface. In this work this phenomenon is studied with low dosage of sorbitans in a given condensate/gas/water system.

KEYWORDS: Gas hydrates, kinetic, flow simulator, sorbitans, multiphase, hydrate plug, stagnant conditions

INTRODUCTION

The development of offshore mature basins such as the North Sea is increasingly characterized by marginal reservoirs. Feasible economic development of these reservoirs requires a shift towards total subsea production systems without fixed or floating production platforms. Unprocessed or minimum processed reservoir fluids will be transported to a central processing facility or ultimately to shore. One of the key issues of total subsea production systems is multiphase flow technology with particular emphasis on hydrate control.

Gas hydrates are representatives of a larger class of compounds, known as clathrates or inclusion compounds (Sloan, 1990). Common for all hydrates is that water molecules, due to hydrogen bonding, form a three dimensional shell whose voids are occupied by non-polar molecules with low molecular weight and small rotational diameter. Whereas the lattice is usually not stable itself, the encaged gas molecules will contribute to stabilize it. At high pressure and/or at low temperatures, thermodynamically stable hydrates will form. If the pressure-temperature conditions in the subsea-pipeline favor the formation of these hydrates, extensive formation may take place and this may result in plugging of the pipeline (Nygård, 1989; Austvik, 1992).

Subsea transportation of unprocessed or minimum processed well fluids over long distances today requires the use of large amounts of methanol or glycols for hydrate inhibition. In the recent years new chemical additives have been identified/developed which are claimed to be effective at low concentrations. These additives may act as hydrate kinetic inhibitors (Lederhos et al, 1996, Duncum et al, 1995) and/or as dispersants (Klomp et al, 1995). A dispersant works by converting free water in a pipeline to dry powder hydrates without any hydrate deposits on the pipe wall, or highly viscous bulk intermediates.

In a multiphase subsea pipeline these new hydrate inhibitors have to be effective both at flowing conditions and during shut-in situations. Experimental testing of some of the new chemical additives has shown that even though they might work well in a turbulent multiphase flow, they may lead to a complete plugging of the pipeline during, or after, a shut-in period. In this work the problem is illustrated through experiments with low dosages of sorbitans to a given condensate/gas/water system.

EXPERIMENTAL SETUP

The experiments were carried out in a high pressure loop formed as a wheel at Petreco A/S, Stjørdal, Norway. The test wheel was filled with the desired fluid at a specified temperature and pressure, and then set in rotation. A relative velocity is created between the pipe wall and the fluid thus simulating transport through a pipeline.

The high pressure wheel is made from stainless steel with an internal diameter of 52.2 mm and a wheel diameter of 2.0 m. The volume is 13.4 liters. The wheel includes two high pressure windows for visual inspection. One of these is equipped with a Sony V800E Hi-8 video camera. The flow simulator is placed in a temperature controlled chamber. The temperature is controlled by use of a programmable regulator, a heating fan and a refrigeration system. The temperature development in the chamber as function of time is preset on the regulator.

The wheel is attached to a motor/gearbox system enabling a variation of the peripheral velocity of the wheel between 0.3 m/s and 5.0 m/s. A torque sensor is installed as a part of the rotational shaft enabling torque measurements to be performed during rotation. Pressure and temperature sensors on the wheel have ranges of 0-250 bara and -10 to +150 °C respectively. All signals are transferred through cables and slip rings to a real time PC-based data acquisition system.

The accuracy of the measurements is estimated to be ±0.2 Nm for torque, ±0.5 bar for pressure, ±0.1 °C for fluid temperature in the wheel and ±1.0 °C for temperature in the chamber. The reproducibility of an experiment has been discussed in en earlier paper for a system without chemicals (Lund et al, 1996). The 2" ID flow simulator used is shown in Figure 1.

EXPERIMENTAL WORK

The objective of the experiments was to study the influence on hydrate formation and the behavior of a given hydrocarbon gas/water/model condensate system when different sorbitans were added. In the experiments the basic system (65 bara at 10 °C) consisted of approximately 31.3 mol methane, 6.4 mol ethane, 3.7 mol propane, 1.0 liter (55.5 mol) distilled water, and 5.0 liters (24.5 mol) Exxsol D60 (Table 1).

Two essentially different kinds of experiments were performed for each system tested:

> The wheel was rotated with a constant peripheral velocity of 1.0 m/s as the temperature was decreased from approximately 60 °C to 4 °C. In the following, these experiments are referred to as continuous flow experiments. The experiments simulate flow in a long pipeline where the fluid is gradually cooled to sea bed temperature.
>
> The system was first heated while the wheel was rotating (peripheral velocity of 1 m/s) to about 60 °C. The wheel was then stopped and cooled down to a temperature of 4 °C where it was kept constant for a period of a minimum of 12 hours. The wheel was then restarted and rotated at a constant peripheral velocity of 1.0 m/s for the rest of the experiment. These experiments are referred to as restart experiments. This simulates a restart after a shutdown period.

In order to minimize the number of adjustable parameters for the experiments, only one given cooling rate (temperature in chamber from 60 °C to 4 °C in 1.5 hours) was used for all experiments.

The sorbitans evaluated were: Sorbitan monolaurate (Span 20), sorbitan monopalmitate (Span 40), sorbitan monostearate (Span 60), sorbitan tristearate (Span 65), sorbitan monooleate (Span 80) and sorbitan trioleate (Span 85). The sorbitans were from Fluka Chemika or Atlas-Goldschmidt GMBH and tested at 0.5 wt% to the water phase.

Table 1 Composition and physical properties, Exxsol D60.

Weight %	C_9	C_{10}	C_{11}	C_{12}	C_{13}	Total C
Normal + iso parafin hydrocarbons	0	5	23	23	6	57
Naphtenic hydrocarbons	1	13	18	8	3	43
Minimum 99.5 % aliphatic hydrocarbons Maximum 0.5 % aromatic hydrocarbons						
Density 15°C Viscosity 25 °C Average molecular weight				0.79 kg/l 1.28 mPa 161		

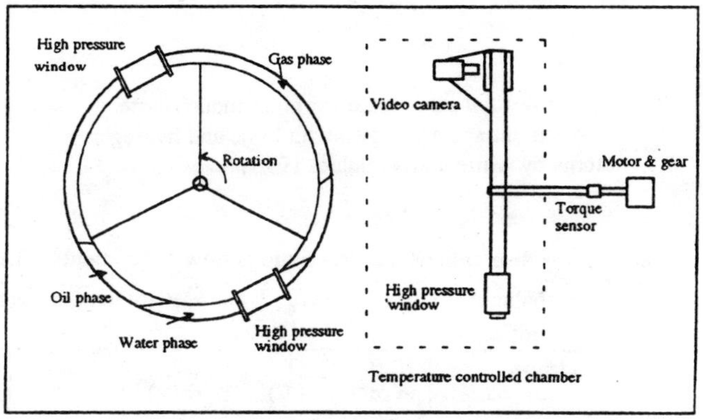

Fig.1 2" ID flow simulator

RESULTS AND DISCUSSION

Experimental observations are given in Table 2 for the continuous flow experiments while the restart experiments are given in Table 3.

Continuous flow experiments

Compared to the experiment without any chemicals, the sorbitans tested did not delay hydrate formation in the continuous flow experiments. However, the sorbitans seem to work as dispersants with properties dependent on the hydrophilic/lipophilic balance (HLB). The best result was obtained by sorbitan monostearate (Figure 2) having a HLB value of 4.7. This value makes it preferable for location at the interface with most of its hydrophobic alkane tail (C18) into the oil phase, and the hydrophilic sorbitan group into the water phase. This position should be favorable for a chemical to be adsorbed on a growing hydrate crystal, if the hydrate formation process is accepted to take place on the interface area near the oil phase (Kvamme, 1996). Video recordings from the experiment with sorbitan monostearate showed the water bulk phase to convert to dry hydrate grains in the oil phase without intermediate macroscopic structures. This might be due to the sorbitan being adsorbed to the growing hydrate crystals at the oil/water interface. The result is that the crystals become hydrophobic before they reach a given size. If the sorbitan is too oil soluble (low HLB value), the inhibitor seems to be laying on top of the hydrate formation zone at the interface. Similar, if it is to water soluble (high HLB value), it seems to be laying below the hydrate formation zone at the interface.

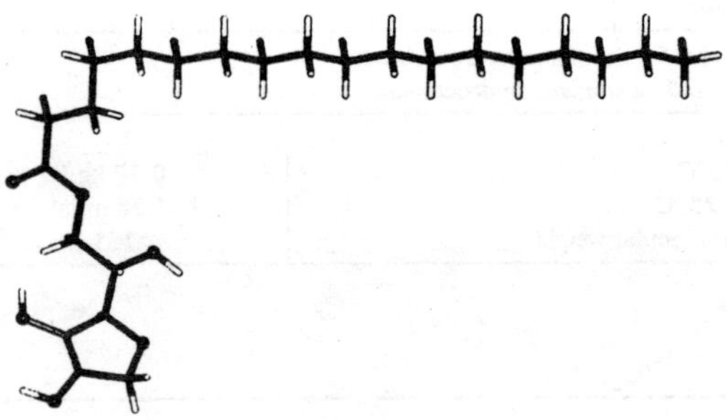

Fig. 2 The molecular structure of sorbitan monostearte (oxygen atoms are shown with balls and hydrogen atoms by white ends) (Nohre, 1996).

The mechanism for a sorbitan group to adsorb on a growing hydrate lattice is today not well known. However, it is reasonable to assume that adsorption takes place via hydrogen bonds between one or more functional groups in sorbitan and water molecules in the hydrate lattice. Theoretical studies (Nohre, 1996), based on density- functional quantum-chemical calculations, have been done to illustrate how the rich structure of sorbitan enables effective hydrogen bonding. An example is given in Figure 3. A cluster with 18 water molecules is chosen as a model of the neighborhood of an adsorption site on the lattice surface of a type II hydrate, and the hydrophobic tail on sorbitan is represented by a methyl group to speed up the calculations. The sorbitan may interact with the lattice through three hydrogen bonds. Between oxygens in the lattice surface and the two hydroxyl hydrogens in sorbitan, and between a hydrogen in the surface and the carbonyl oxygen in sorbitan. The hydrophobic tail (here: the methyl group) points away from the surface and may enter the oil phase of the system. Among the configurations modeled, this is one of the most stable, with a binding energy of about -41.9 kJ/mol.

Table 2 Observations from continuous flow experiments. The results are ranged after increasing HLB (hydrophilic/lipophilic balance) values. T_0, ΔT_0, T and ΔP are temperature, degree of subcooling, pressure and overpressurization at hydrate initiation.

Additive (to the water phase)	HLB value	T_0 (°C)	P_0 (bara)	ΔT (°C)	ΔP (bar)	Observations
No		11.2	66.4	5.7	38.5	Before hydrate formation separate oil and water were observed. After hydrate initiation slushlike hydrate/water lumps were formed in the liquid flow. At the same time hydrates started to deposit on the pipe wall. The pipe was plugged 45 min. after hydrate initiation.
0.5 wt% sorbitan trioleate	1.8	12.4	66.8	4.9	35.5	Before hydrate formation a transparent oil phase and a coarse water phase were observed. After hydrate initiation slushlike hydrate/water lumps were formed in the liquid flow. At the same time hydrates deposited on the pipe wall. Later on the lumps converted to dry hydrate powder and small hydrate balls.
0.5 wt% sorbitan tristearate	2.1	10.5	65.5	6.7	41.8	A white water phase and a transparent oil phase were observed prior to hydrate formation. After hydrate initiation most hydrates formed deposited on the pipe wall.
0.5 wt% sorbitan monooleate	4.3	10.6	66.5	7.6	45.5	The oil and water phases were mixed before hydrate formation. After hydrate initiation the water phase was converted through loose slushlike lumps to powderlike hydrates in the liquid flow. During the conversion process hydrates deposited on the pipe wall.
0.5 wt% sorbitan monostearate	4.7	10.3	66.3	6.5	41.3	A mixture of oil and water phases was observed prior to hydrate formation. After hydrate initiation the water phase was converted to powderlike hydrates in the liquid flow. During the conversion process only some hydrate grains deposited on the pipe wall.
0.5 wt% sorbitan monopalmitate	6.7	10.9	66.4	6.2	40.3	The oil and water phases were mixed before hydrate formation. After hydrate initiation the water phase was converted through loose slushlike lumps to powderlike hydrates in the liquid flow. During the conversion process a thin layer of hydrates deposited on the pipe wall.
0.5 wt% sorbitan monolaurate	8.6	9.8	66.0	7.2	43.6	The oil and water phases were mixed before hydrate formation. After hydrate initiation the water phase was converted through loose slushlike lumps to powderlike hydrates in the liquid flow. During the conversion process hydrates deposited on the pipe wall.

Table 3 Observations from restart experiments. The results are ranged after increasing HLB (hydrophilic/lipophilic balance) values. At startup the degree of subcooling was about 13 °C.

Additive (to the water phase)	HLB value	Time at subcooled conditions before restart (hours)	Observations during the stagnant period	Observations after restart
No	-	16	A thin layer of hydrates on the oil/water interface.	After restart, slushlike hydrate/water lumps were formed in the liquid flow. At the same time hydrates started to deposit on the pipe wall. 10 min. after restart the pipe was plugged.
0.5 wt% sorbitan trioleate	1.8	16	A white layer of hydrates on the oil/water interface.	After restart, slushlike hydrate/water lumps were formed in the liquid flow. At the same time hydrates deposited on the pipe wall. 30 min. after restart, the lumps were converted to dry hydrate powder and small hydrate balls.
0.5 wt% sorbitan tristearate	2.1	16	A solid hydrate lump grew from the oil/water interface and into the oil phase after 4 hours and 28 min. at subcooled conditions.	The pipe was plugged by hydrates prior to restart.
0.5 wt% sorbitan monooleate	4.3	15	A thick hydrate layer was formed on the oil/water interface after 1 hour and 34 min. at subcooled conditions.	After restart the water phase was converted through loose slushlike lumps (nearly plugging the pipe 12 min. after restart) to powderlike hydrates in the liquid flow. During the conversion process a thin layer of hydrates deposited on the pipe wall.
0.5 wt% sorbitan monostearate	4.7	15.5	A thick hydrate layer was formed on the oil/water interface after 3 hours and 22 min. at subcooled conditions.	After restart the water phase was converted through loose slushlike lumps to powderlike hydrates in the liquid flow. During the conversion process a layer of hydrates deposited on the pipe wall.
0.5 wt% sorbitan monopalmitate	6.7	15	A thick hydrate layer (lumps) was formed on the oil/water interface after 3 hours and 10 min. at subcooled conditions.	After restart, slushlike hydrate/water lumps were formed in the liquid flow. At the same time hydrates deposited on the pipe wall. The pipe was plugged 2 min. after restart.
0.5 wt% sorbitan monolaurate	8.6	18	A solid hydrate plugg was formed on the oil/water interface after 5 hours and 56 min. at subcooled conditions.	The pipe was plugged by hydrates prior to restart.

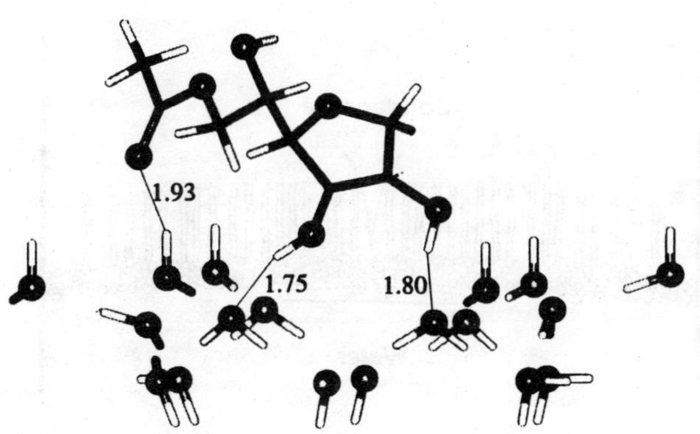

Fig. 3 Possible configuration for adsorption of sorbitan on the lattice surface of hydrate structure II. Numbers are hydrogen bond lengths in Å (Nohre, 1996).

Restart experiments

In the experiments a hydrate layer was formed at the oil/water interface during the stagnant period at subcooled conditions in all restart experiments with the sorbitans. Except for the experiment with sorbitan trioleate (the most oil soluble sorbitan), all hydrate layers had a thickness causing a large resistance, or complete plugging of the pipe, after restart.

From analysis of video recordings buildup of a hydrate layer on the oil/water interface appears to occur in the following way:

1. The hydrate surface is assumed normally to be hydrophilic. In hydrocarbon-water systems without surfactants, at proper non flowing conditions, a hydrate film forms at the interface between the two phases. This film is probably formed by nucleation and growth of a few hydrate crystals growing together in a "single-crystal". The solid film is almost diffusion tight, stopping further hydrate formation. As a result of the volume expansion by hydrate formation, the film is stressed and cracks are formed. Because the hydrate crystal surface is hydrophilic, water is sucked

up by the cracks, and wet the crystal surface at the hydrocarbon side (Figure 4). This water is quickly transformed to hydrates which blocks the cracks. After some time the high stress areas are released through cracking and a thin hydrate layer is formed, practically preventing further hydrate formation.

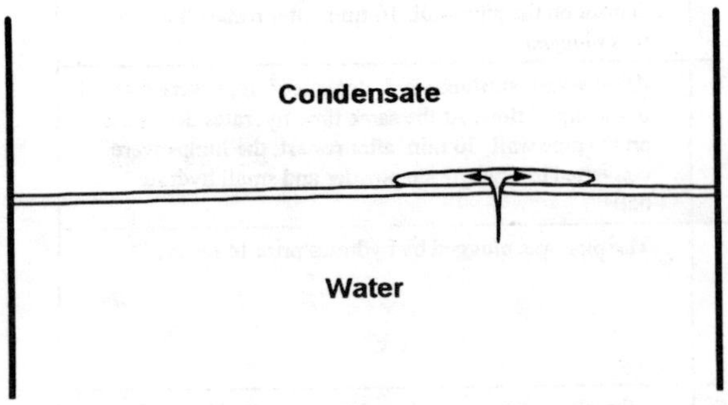

Fig. 4 Build-up of a thin hydrate layer on a condensate-water interface by cracking of the hydrate film. Water is sucked on top of the hydrophilic crystal surface.

2. The sorbitans (surfactants) most likely influence on the hydrate crystal growth, not the hydrate crystal nucleation. In systems containing sorbitans, the first hydrate film on the oil/water interface is probably made up of a high number of small hydrate crystals, more or less agglomerated together. This "polycrystalline" film is porous and thicker than in the pure water system (Figure 5). At the hydrocarbon side, the surfactants adsorb at the surface of the hydrate crystals, making the surface hydrophobic. Inside the "polycrystalline" film the crystal growth has probably consumed most of the surfactant, making the internal crystal surfaces hydrophilic. Water from the water phase is sucked into the hydrophilic pores in the hydrate film, but do not wet the hydrophobic surface against the hydrocarbon phase. Instead water forms small bubbles into the hydrocarbon phase.

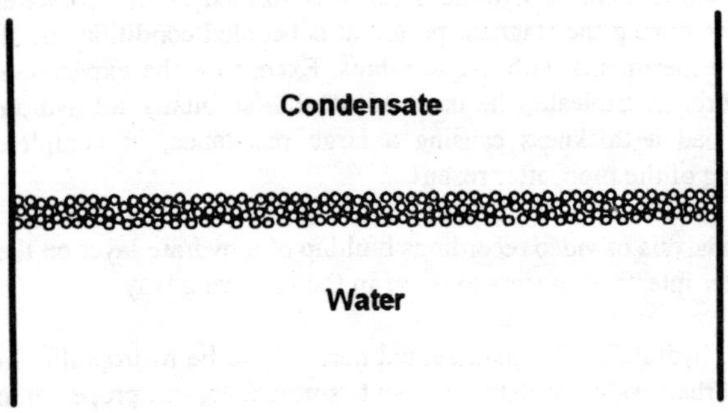

Fig. 5 "Polycrystalline" hydrate film at condensate/water interface.

3. A special phenomenon observed was the formation of tiny hollow hydrate tubes, growing into the hydrocarbon phase (Figure 6). The tubes are probably formed on the water bubble-hydrocarbon interface by hydrate crystal growth, starting at the hydrate film. This growth, making a hydrate tube hydrophilic inside and hydrophobic outside, lifts the water droplets and results in further tube growth. Large amounts of these tubes result in the hydrocarbon side of the film being shaggy, and this water transfer creates a thick hydrate film.

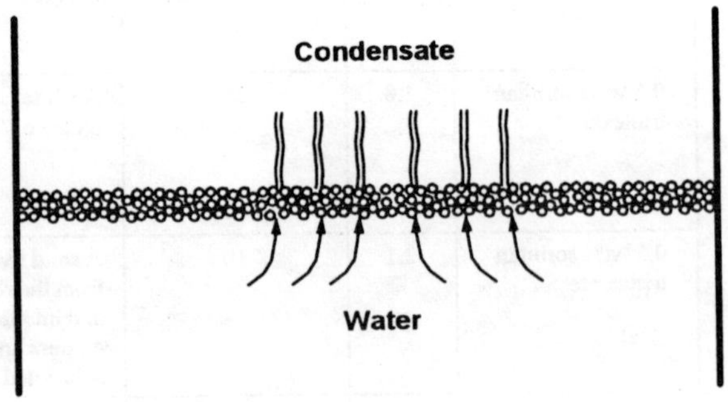

Fig. 6 Growth of hydrate tubes into the condensate phase from the "polycrystalline" film at condensate/water interface.

4. Due to the formation of small hydrate crystals, and the porous structure, only some bending of the hydrate film towards the hydrocarbon phase was observed until the film got really thick. At a given thickness of the film, stress start to initiate cracking. Because the film bends towards the hydrocarbon side, the cracks are initiated at this side. The surfactant (dissolved in the condensate phase) adsorbs on the crack surface and makes it hydrophobic. When the cracks reach the water side, hydrocarbons flow into the water phase (Figure 7), initiating an almost explosive hydrate formation. This rapid hydrate formation cracks the hydrate film further, making more hydrocarbons flow into the water phase and so on until the process is apparently terminated, or until all water is consumed.

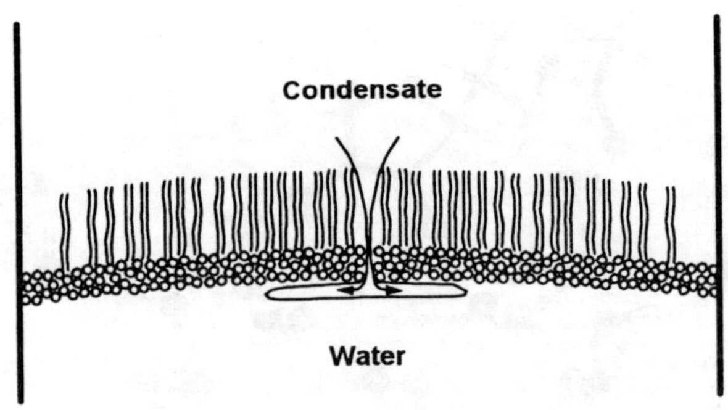

Fig. 7 Hydrocarbon flow into the water phase at cracking of the hydrate film at the condensate/water interface.

The behavior described here for stagnant conditions is not special for sorbitans. We have also observed the same behavior using other low concentration inhibitors effective at continuous flow conditions at high subcooling. As far as we can see this is a phenomenon that is common for a lot of the low-dosage hydrate inhibitors available on the market today.

In the North Sea, low concentration inhibitor use will mainly be to prevent hydrate formation at shut-in situations. It is hence important to address these topics in the further development.

CONCLUSION

In this paper we have investigated the influence of different sorbitans on the hydrate formation and growth in a condensate/gas/water system. We have focused on what is happening in the system during stagnant conditions at a high subcooling (13 ^{0}C).

Based on the experimental results we have postulated a hypothesis in order to explain the observed physical appearance of the hydrate formation when the sorbitans are present. This is based on a process taking place in 4 subsequent steps leading to an explosive hydrate growth and, in most cases, plugging of the pipe.

However, the main issue here is that this is a phenomenon that can occur when low-dosage hydrate inhibitors are applied at "high" subcoolings if a shut-in situation takes place. In development and design of hydrate inhibitor packages this problem has to be addressed.

ACKNOWLEDGMENTS

The Statoil, Saga and Hydro collaboration within multiphase transportation is acknowledged for giving us permission to publish the work.

REFERENCES

Austvik, T (1992). "Hydrate Formation and Behaviour in Pipes", *Dr.ing. Thesis*, The Norwegian Institute of Technology, Trondheim, Norway.

Duncum, SN, Edwards, AR and Osborne CG (1995). "Hydrate Inhibition", *Patent* WO 95/19408.

Kvamme, B (1996). "A New Theory for the Kinetics of Hydrate Formation", *Proc 2nd Int Conf Nat Gas Hydrates*, Toulouse, France, pp 139-146.

Klomp, UC, Kruka, VR, Reijnhart, R and Weisenborn, AJ (1995). "A Method for Inhibiting the Plugging of Conduits by Gas Hydrates", *Patent* WO 95/17579.

Lederhos, JP, Long, JP, Sum, A, Christiansen, RL and Sloan, EDJr (1996). "Effective Kinetic Inhibitors for Natural Gas Hydrates", *Chem Engng Sci*, Vol 51, No 8, pp 1221-1229.

Lund, A, Urdahl, O and Kirkhorn, SS (1996). " Inhibition of Gas Hydrate Formation by Means of Chemical Additives-II. An Evaluation of the Screening Method", *Chem Engng Sci*, Vol 51, No 13, pp 3449-3458.

Nohre A (1996). *Molecular Modeling of Inhibitors Preventing the Formation of Clathrate Hydrates,* Norwegian University of Science and Technology, Dep of Ind Chem, Trondheim, Norway.

Nygård, HF (1989). "Transportability of Hydrates in Multiphase Systems", *Proc 4th Intern Conf on Multiphase Flow*, Nice, France.

Sloan. EDJr. (1990). *Chlathrate Hydrates of Natural Gas*, Marcel Dekker, New York.

Proceedings of the Seventh (1997) International Offshore and Polar Engineering Conference
Honolulu, USA, May 25-30, 1997

ISBN 1-880653-28-1 (Set); ISBN 1-880653-29-X (Vol. I)

Sea Bottom Radioactivity Measurement Systems in Japan

Mutsuo Hattori
Japan Marine Science and Technology Center, Yokosuka, Japan
Yoshii Kobayashi
Japan Marine Science Foundation, Mutsu, Japan
Masaharu Okano
Japan Marine Science and Technology Center, Yokosuka, Japan

ABSTRACT

Four sea bottom *in-situ* real time radioactivity measurement systems are operational in Japan.

The main objectives of these systems are to measure real time marine environment natural and artificial radioactivities around Japan to know radioactive levels around nuclear facilities and various geological conditions such as hot vent areas, cold seep areas, active fault zones and relatively common areas .

Examples of total intencities (total cps) of gamma ray around Japan are:

1. Vent sites such as Iheya knolls and Izena hole in Okinawa Trough are ranging from 200 to 9000 cps and rich in U-series radionuclides especially Rn.

2. Cold water seep areas along active fault zones such as Sagami Bay and Suruga Bay are ranging from 18 to 120 cps , wherese total counts of bottom water is ranging from 10 to 15 cps.

3. Active fault or thrust zones of eastern Nankai Trough accretional belts are characterized rather high total gamma ray intencities which are ranging 50 to 235 cps.

KEY WORDS: Sea bottom radioactivity, gamma ray, ROV, manned submersible

INTRODUCTION

Deep sea *in-situ* radioactiovity measurements are initiated in 1964 at sunken USN atomic power submarine Thresher by using Geiger counters (Wakelin, 1964). *In-situ* radioactivity monitorings utilizing Tow fish and remote sea bottom stations are carried out in various countries. But they are mainly used for surveys of artificial radioactivities around atomic power plants, radioactive waste disposal sites and sunken nuclear objects (Jones *et. al.*, 1995). Deep sea gamma ray surveys for scientific use are intensively carried out by ODP logging for determination of rock types.

One of the purpose of ROV MERIS is to measure artificial radiactivities around atomic power facilities. Main purpose of gamma ray measurement in JAMSTEC is to establish a data base of background natural sea bottom radioactivities around Japan and to carry out precise surveys at the sites that are tectonically active areas such as off Tokai and off Kanto districts to know the present status and the change of gamma ray total intensities and contents of K, U-, Th-series radionuclides. Because the intensity anomaly of Rn is strictly correspondant with occurrence of eathquakes (Igarashi *et. al.*, 1995) and it is supposed that total intensity changes of gamma ray are also correspondant with occurrence of earthquakes.

SPECIFICATIONS OF RADIOACTIVITY MEASUREMENT SYSTEMS

General Descriotions of Systems

Four sea bottom *in-situ* real time radioactivity measurement systems are operational in Japan. They are MERIS (Marine Environment Radioaitivity Investigation System) of JMSF (Japan Marine Science Foundation) and three systems of JAMSTEC (Japan Marine science and Technology Center).

MERIS is completed in 1993 by JAERI (Japan Atomic Energy Research search Institute) under the auspices of the Science and Technology Agency of Japan and assinged to JMSF in 1996. MERIS was constructed as 200m depth ROV for measure gamma ray of sea bottom.

JAMSTEC has three sea bottom gamma ray measurement systems, they are:

- O The system equipped on the manned submersible "Shinkai 2000", 2000m depth capability .
- O The system eauipped on the manned submersible "Shinkai 6500", 6500m depth capability.
- O The system eauipped on the "JAMSTEC deep tow system" on JAERI's consignment, 5000m depth capability. But the system is not operational now. A new system for the 3300m deep ROV DOLPHIN-3K is under construction. The gamma ray detectors of the "JAMSTEC deep tow system" is going to carnivarize for construction of DOLPHIN-3K gamma ray measurement system.

Four systems are utilizing NaI(TI) sensors and measured data are recorded in voice channels of 8 mm video tapes or removable hard disks (in case of MERIS). Data files are compiled as spectra of 1024 channels. Data of energy spectra are analysed as total counts and counts of K,U-, and Th- etc. Finally, abundance of nuclear species are calculated.

MERIS RADIOACTIVITY MEASUREMENT SYSTEM

MERIS is completed and successfully tested in 1993 as a ROV of 200 m depth capability (Kobayashi et al., 1994).

Total system of ROV is composed of vehicle, underwater cable, control/navigation van, launch and recovery units and the acoustic navigation unit. The system utilizes 3 inch spherical NaI(TI) scintillator as a radioactivity sensor for measure gamma ray of marine environment.
The sensor, electronics and power supply are installed in a pressure bottle of 170 mm diameter and 1.7 m length. The bottle is attached at the center front -bottom of the vehicle frame work. The measured pulse height signal from scintillator is converted into digital signals by the digital processing unit in the bottle. The signals from the digital processing unit are processed to display as pulse amplitude distribution and bar graph of the radiactive data. These data and under water TV images are displayed on the TV monitor in the control/navigation van. The main specifications of the ROV is shown in Table 1.

Table 1 Main specifications of the MERIS

Dimension	190 x 130 x 120 cm
Weight	760 kg
Operational depth	200 m
Speed	2.5 kt forward
Propulsion:	
forward-reverse	3 x 3HP (Oil fille brushless DC-motor)
vertical	1 x 3HP (Oil fille brushless DC-motor)
Instrumentaion	Radioactivity sensor, Color TV camera with pan & tilt unit, low light b/w TVcamera with tilt unit, still camera with strobe, CTD, current meter, water sampler, tube core sampler
Cable	Fiber-optic-electro-mechanichal cable, diameter: 22 mm, length: 800 m
Navigation	SSBL acoustic navigation system, flux gate compass, angular velocity sensor, depth sensor, altitude sensor
Shipboard component	Control/navigation van, launch and recovery system, winch for cable
Support ship	Tugboat "Mizuho Maru", 191.53 tons 31.72(L) x 8.60(W) x 3.50(D)

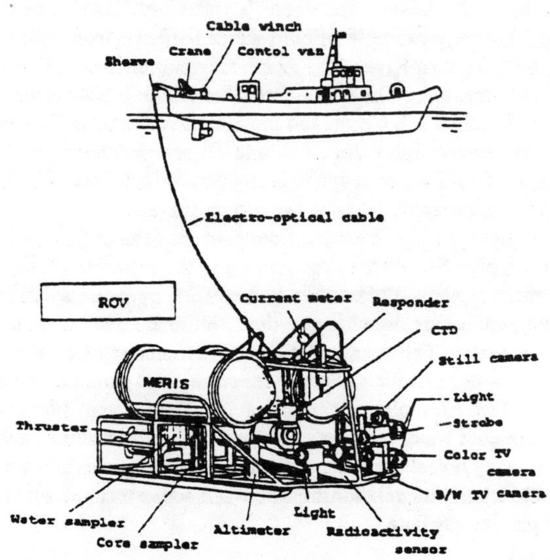

Fig. 1 The total system of the ROV MERIS.

The total system of the ROV MERIS is shown in Fig. 1. Fig. 2 shows the radioactivity measurement system of the MERIS. The ROV utilizes fiber-optic-electro-mechanical cable of 800 m length. Signals of two TV camera, sensors and control signals from ship board to vehicle are transmitted via optical fiber cable. The MERIS will be used for radiactivity and oceanographic surveys around Japan.

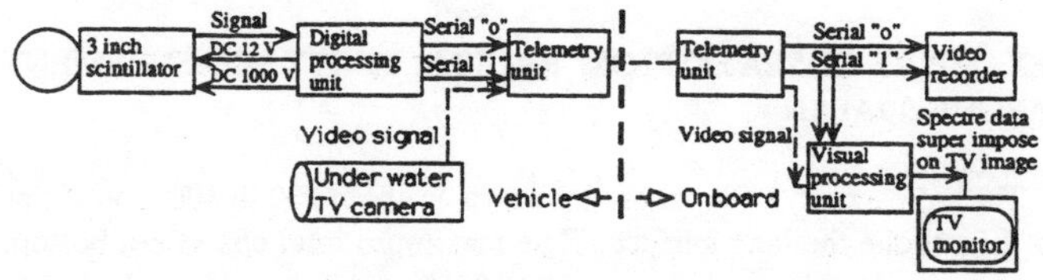

Fig. 2 The radioactivity measurement system of the MERIS.

"Shinkai 2000" Radioactivity Measurement System

The manned submersible "Shinkai 2000" has been operational since 1983. The radioactivity measurement system for "Shinkai 2000" has been installed in 1994. The radioactivity measurement system also utilizes 3x3 inch NaI(TI) scintillator as radioactivity sensor. The sensor and the signal processing unit are installed in a pressure bottle. The pressure bottle is fixed in the landing sled of the "Shinkai 2000". The schematic of the system is shown in Fig. 3. The digital signal from the pressure bottle is recorded on the 2 channel sound track of a 8 mm video tape and images of under water TV camera that is equipped "Shinkai 2000" is recorded in the same video tape. The image records of energy spectre are also recorded as the superimposed images on the under water TV images.

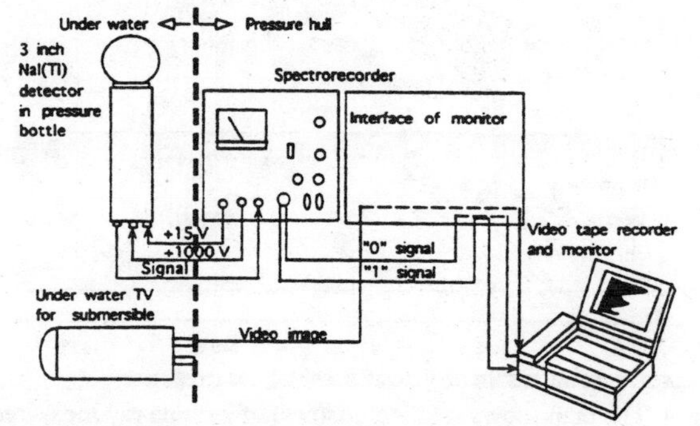

Fig. 3 The radioactivity measurement system of "Shinkai 2000".

"SHINKAI 6500" Radioactivity Measurement System

The manned submersible "Shinkai 6500" has been operational since 1991. The radioactivity measurement system for "Shinkai 6500" has been tested since 1994. The system utilizing 2 inch NaI(TI) scintillator as radioactivity sensor, because there are not so much room to install 3 inch sensor in the sled of "Shinkai 6500". The other components are nearly the same as the system of "Shinkai 2000".

The radioactivity data collected by "Shinkai 6500" are now under reanalysis and re-processing, because data colledted by "Shinkai 6500" are not so reliable and stable as these of "Shinkai 2000" , the reasons of the unstability are now under investigation.

"JAMSTEC Deep Tow " Radioactivity Measurement System

The first radioactivity measurement system in JAMSTEC has been developed for "JAMSTEC deep tow". The "JAMSTEC deep tow" was used for pre-site surveys of "Shinkai 2000" dives. Radioactivities were measured some times and high radioactive areas such as Izena hole and Iheya knolls are discovered by the system. Detectors of the measurement system for "JAMSTEC deep tow" is now carnivarized by measurement system for ROV DOLPHIN-3K and the system will be operational in this April.

EXAMPLES OF SEA BOTTOM RADIOACTIVITY MEASUREMENT AROUND JAPAN

The radioactivity of Japanese land area is about 100 to 400 in total cps at 1 m abobe the land surface. The maximum total cps of sea bottom measurements are ranging from 18 to 9000 cps. Fig. 4 shows the maximum total counts of sea bottom radioactivity that are measured in 1996 by "Shinkai 2000". The data colled by "Shinkai 6500" are not shown in Fig. 4, because the data are now under analysis and processing. Cold water seep area of Sagami Bay, where the Phillippine Sea Plate is subsiding under the Eurasia Plate, shows relatively low value of under 100 cps(Hattori and Okano, 1995). Vent sites such as Izena hole, Iheya Knolls and Kagoshima Bays are high in maximum total counts, the highest value is up to 9000 cps in Izena hole of Okinawa Trough.Some examples of analysis are shown in Figs. 5 to 8, dive numbers of these examples are shown in Fig. 4.

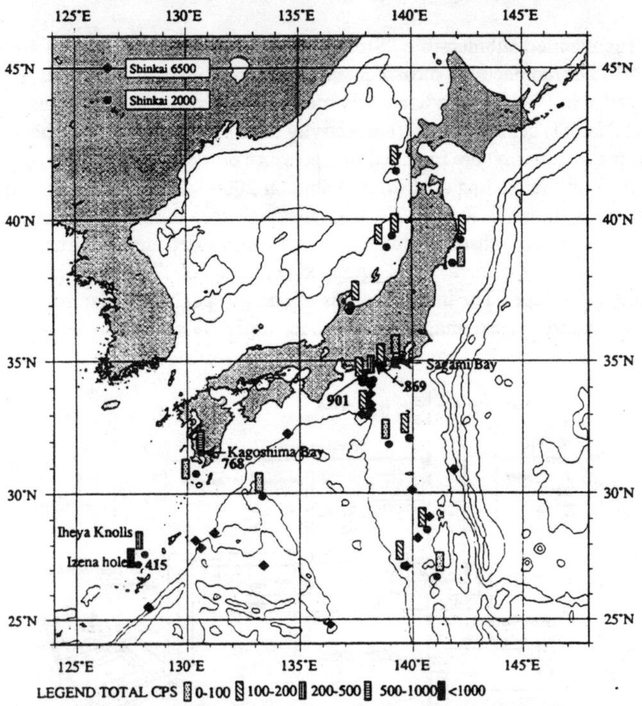

Fig. 4 The map shows total counts(cps) of gamma ray measured by "Shinkai 2000" and dive sites of sea bottom gamma ray measurement by "Shinkai 2000" and "Shinkai 6500". Block letters, 415, 768, 869, 901 are dive numbers and related to graphs of Figs. 5 to 8.

Fig. 5 shows an example of total cps in a dive (Dive No.415 of "Shinkai2000") in Izena hole of Okinawa Trough. The graph shows 100 to 3000 cps of total counts.The value of 3000 cps is measured side of a vent chimney. The maximum total counts from the same site exceeds 9000cps in another dive.

Fig. 6 shows a graph of total cps in Kagoshima Bay vent site. The graph shows a maximum value of near 700 cps at vent site. Two vent sites

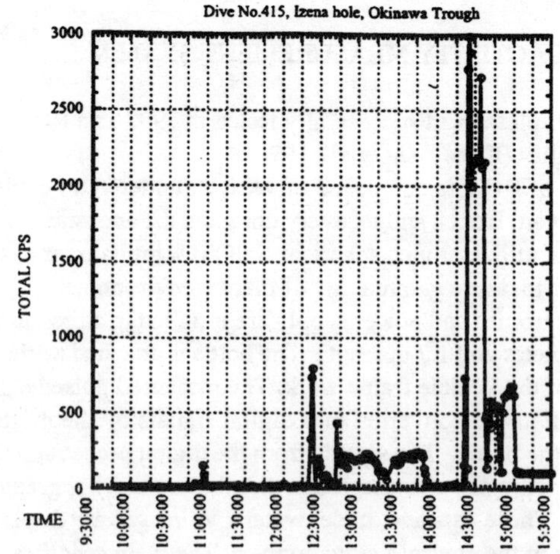

Fig. 5 Graph shows an example of total cps in a dive (Dive No.415 of "Shinkai2000") in Izena hole of Okinawa Trough.

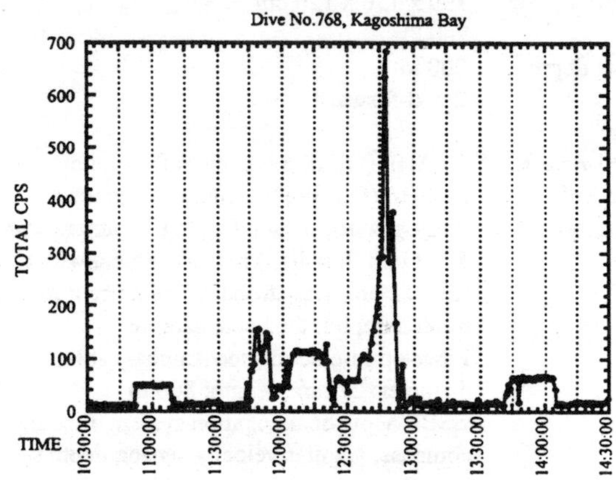

Fig. 6 Graph shows total cps in Kagoshima Bay vent site. Dive No. is 768.

show large value of total cps when "Shinkai 2000" landed near vent.

Fig. 7 shows a example of cold water seep site from eastern Nankai Trough off Tenryu River. The graph shows contents of K, U- and Th-series radionuclides. The maximum value of K is 0.024 %, the maximum value of U-series is 8.6 ppm and the maximum value of Th-series is 6.2 ppm. Relatively high value of U- and Th-series radionuclides indicates presence of cold water seapage along active fault zone which is ascertained by topographic features and airgun surveys.

Fig. 8 shows another example from cold seep site of Suruga Bay, where the Phillippine Sea Plate is converging to the Eurasian Plate. Th-series radionuclides show a maximum value of 2.7 ppm and average value of 1.2 ppm, values are not change either "Shinkai 2000" is landed on bottom or in water. This means that Th-series radionuclides are contained in cold seep water that fuels chemosynthetic animal community observed at the site. Two examples of gamma ray analysis of vent sites and two examples of cold water seep sites show characteristic features, more precise analysis may reveal more detail features of diving sites and we hope that we could clalify the relationships between active tectonic zones and measured gamma ray data.

CONCLUDING REMARKS

Radioactivity measuement surveys in the sea bottom are very important to evaluate and monitor active tectonics of tectonically active areas such as Sagami Trough, Nankai Trough and areas of active vent sites. We are going to continue precise surveys and spot surveys around Japan using manned submersibles, ROVs and long term sea bottom lavoratories that will be installed at sea bottom around Japan in near feature.

REFERENCES

Hattori, M. and Okano, S., (1996). "A preliminary reports of *in-situ* radioactivity survey in a cold seep environment off Hatsushima Isl. Sagami Bay (in Japanese),"*KAIKO-TOKAI News letter*, No.8.

Igarashi, G. , Saeki, S., Takahata, N., Sumikawa, K., Tasaka, S., Sasaki, Y., Takahashi, M. and Sano, Y (1995). "Ground-water radon anomaly before the Kobe earthquake in Japan," *Science*, 269.

Jones, D.G., Roberts, P.D. and Miller, J.M.,(1988). "The distributuion of Gamma-emitting Radionuclides in Surface Subtidal Sediments near the Sellafield Plant," *Estuarine, Coastal and Shelf Science*, 27.

Kobayashi, Y., Nakai, S., Hattori, M., Okano, S., Inada, T., Watanabe, N. and Sakakibara, Y., (1994). "Ouitline of Marine Environmental Radioactivity Investigation System," *Proc., UJNR MFP., 19th Joint meeting*, No.24.

Wakelin, J. H. Jr.,(1964). "Thresher: Lesson and Challenge," *National Georaphic*, Vol 25, No 6.

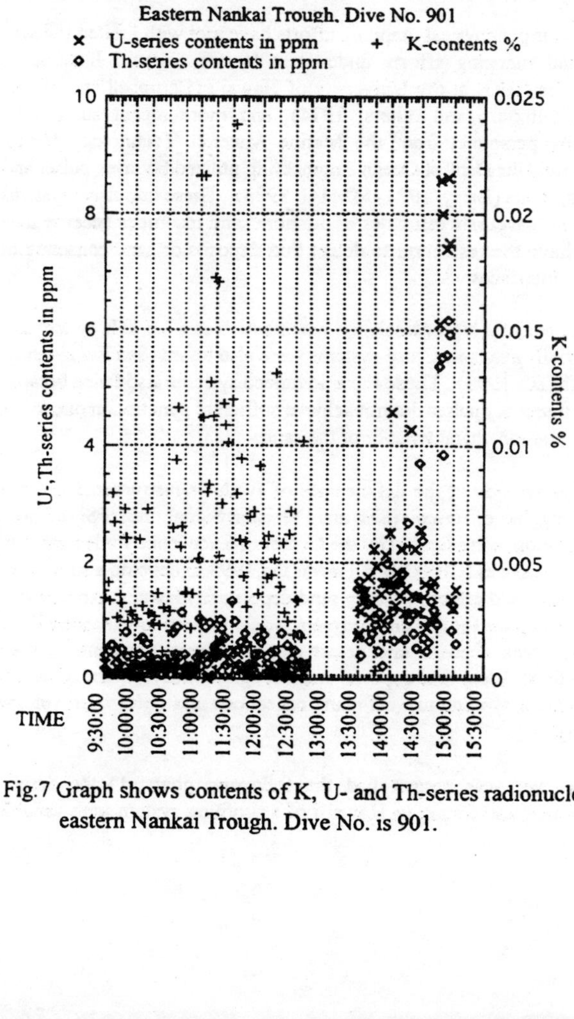

Fig.7 Graph shows contents of K, U- and Th-series radionucleids at eastern Nankai Trough. Dive No. is 901.

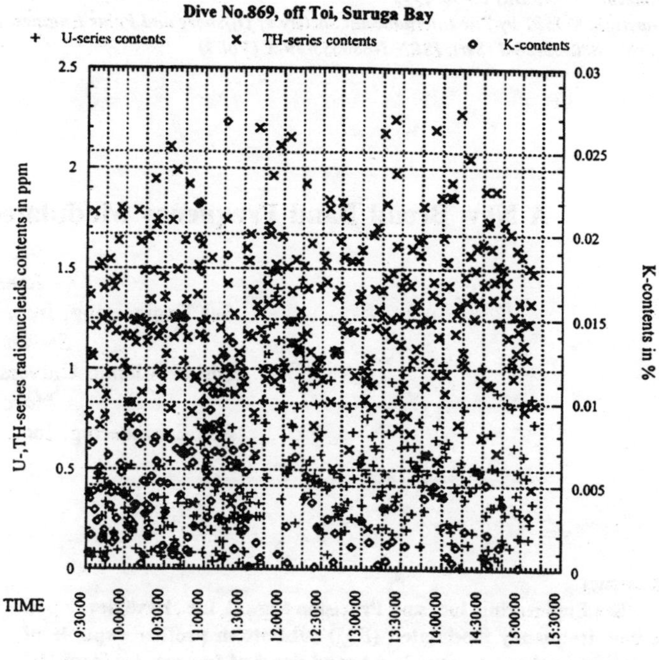

Fig. 8 Graph shows contents of K, U- and Th-series radionucleids at Suruga Bay. Dive No. is 869.

Proceedings of the Seventh (1997) International Offshore and Polar Engineering Conference
Honolulu, USA, May 25-30, 1997

ISBN 1-880653-28-1 (Set); ISBN 1-880653-29-X (Vol. I)

A New Broad Band Frequency Modulated Subbottom Profiler Developed for Hawaii

James Barry
Sea Engineering, Inc., Waimanalo, Hawaii, USA
Steven G. Schock
Florida Atlantic University, Boca Raton, FL, USA
Marc Ericksen
Sea Engineering, Inc., Waimanalo, Hawaii, USA

Abstract

Sea Engineering, Inc., and Precision Signal, Inc., have developed a new frequency modulated (FM) subbottom profiler capable of transmitting shaped pulses in a broad band of frequencies from 300 to 10,000 Hz. This project was funded by the Center of Excellence for Research in Ocean Sciences (CEROS) because off-the-shelf commercial systems were not able to generate images of Hawaiian shelf sediments with sufficient penetration or resolution for identifying sand resources. A new tow vehicle design contains four parallel 1.5 meter line arrays providing about 12 dB greater directivity than a streamer hydrophone array of the same length. The 12 dB reduction in scattering provides a large improvement in image quality. A range of pulses can be designed and individually selected to target specific survey requirements. During sea trials of the sonar, tests showed that the best images of the top 40 meters of sediments were collected using an FM pulse with a center frequency of 1.4 to 2.0 kHz and a bandwidth of 2 to 3 kHz. Pulses with higher frequencies produced degraded images because attenuation and rough subsurface boundaries caused weak reflections from the subsurface structure. Preliminary surveys on O`ahu have produced the first clear images of sand and other calcareous sedimentary features, and have also delineated the underlying reef structure. The system is calibrated, and allows real-time quantitative analysis of the sonar data. Robust sediment classification software has been developed that uses the reflection data to provide surficial and subsurface measurements of compressional wave attenuation, acoustic impedance, and volume scattering. These three acoustic parameters are used to predict physical properties of the sediment such as grain size, bulk density, and porosity. Predictions are based upon empirical relationships derived from correlation studies of the acoustic parameters and sediment properties. Extensive ground-truth probing, and sediment sampling and analysis have established the accuracy of the system. The ability to select from an assortment of pulses with a wide range of spectral characteristics enables the profiler to be adapted to any geologic environment with ease, and to be used for a multitude of profiling purposes.

Introduction

Studies by various researchers in recent years have shown that offshore sand resource assessment in Hawai`i is a demanding task for geophysical instrumentation. For several reasons, the geologic character of offshore sand deposits on Hawai`i's shelf presents difficult problems for subbottom profiling:

- The sand deposits vary in thickness from less than one meter to more than 40 meters;

- The sands are predominantly calcareous, clean (with little or no silt and clay), and highly attenuating;

- The deposits may be underlain by acoustically similar material such as cemented sands, or by coral rubble or irregular paleo-reef surfaces that cause weak non-coherent acoustic reflections that are difficult to detect.

Previous geophysical mapping efforts have met with limited success. The initial mapping efforts undertaken by the Hawai`i Institute of Geophysics (HIG) at the University of Hawai`i (Campbell and others, 1970; Coulbourn and others, 1974), and more recent surveys by University personnel from the Marine Minerals Technology Center (MMTC) and the Dept. of Ocean Engineering utilized boomer, pulser and sparker systems (Barry, 1991; MMTC, 1993). These impulsive systems resulted in inaccurate estimates of offshore sand resources because they did not have the resolution to detect thin deposits or sand/consolidated sediment interfaces.

More recently, higher resolution systems such as a 3.5 kHz system and commercially available Chirp systems were also tested (Sea Engineering, 1993; MMTC, 1993). These more advanced systems also failed because the high frequency pulses did not achieve sufficient penetration due to the high attenuation characteristics of the sand.

In recognition of the deficiencies of available instrumentation, Sea Engineering, Inc. of Hawai`i (SEI) and Precision Signal, Inc. (PSI) of Boca Raton, Florida, were funded by the Center of Excellence for Research in the Ocean Sciences (CEROS) in Hawai`i in 1993 to develop a subbottom profiler that could effectively map sand deposits in Hawai`i. An effective subbottom profiler for this environment would require low frequencies to penetrate thick deposits and detect gradual density gradients, a wide bandwidth to detect thin deposits and stratigraphic layers, and a narrow beam pattern to minimize off-plane reflections and noise from volume scattering.

The two year project had the following goals: 1) the design, development and testing in Hawai`i of a profiling system with variable array geometries and pulse frequencies; 2) experimentation to establish the appropriate frequency bands, and the receiving and transmitting aperture sizes necessary for imaging and classifying sands overlying a coral basement; 3) implementation of state-of-the-art sediment classification techniques previously developed by Drs. Steven Schock and Lester LeBlanc at Florida Atlantic University under an Office of Naval Research grant (Dr. J. Kravitz, program manager); and 4) final field testing with the profiler and sediment classification techniques. The project was completed in April 1996 (SEI and PSI, 1996).

Sonar System Design

The system developed during this project is a modification of the X-Star Full Spectrum Sonar developed by PSI and manufactured by Edgetech. The X-Star is a wide-band frequency modulated, or chirp, system that has evolved from previous work of PSI personnel (Schock and LeBlanc, 1986; Schock, LeBlanc, and Mayer, 1989; Schock and LeBlanc, 1990). In contrast to impulsive sonar systems, where the frequency content of the source is released almost instantaneously, the source signal of chirp systems is generated by sweeping the pulse frequencies over the frequency range of the system. The overlapping chirp waveforms reflected from the sea bottom and substrate are compressed using a matched filter that correlates the chirp return signal with a replica of the outgoing pulse, producing an interpretable seismic record. The compressed pulse resulting from the signal processing has a time duration approximately equal to the inverse of the bandwidth of the chirp pulse. This determines the system resolution. In addition to pulse compression, correlation processing achieves a signal processing gain over background noise that is approximately ten times the log of the time-bandwidth product.

The chirp pulses can be shaped to achieve the desired frequency content, and signal power can be increased by lengthening the duration of the pulse. Inherent noise problems such as source ringing and pulse coloring due to system electronics and transducer characteristics are solved by compensation of the output signal with the system transfer function. The chirp wavelet is weighted in the frequency domain to have a Gaussian-like shape. Although energy is lost as the Gaussian-shaped spectrum is attenuated by the sediment, the bandwidth is nearly preserved, and the resolution of the system is not significantly reduced at the bottom of the subbottom image.

The subbottom system developed for imaging and classifying Hawaiian shelf sediments is schematically illustrated in Figure (1). It includes the following key features:

1) The system operating band is 400 Hz to 10 kHz, which is generated by two piston sources driven with one power amplifier through a crossover network. The wide frequency range provides sufficient penetration into thick deposits to detect basement, while at the same time allows resolution of thin stratigraphic units.

2) A large 1.5 x 1.5 meter planar receiving aperture and acoustic baffle greatly reduces scattering noise from the sediments and seasurface, but is still deployable from a relatively small vessel. We routinely survey using our 32 foot workboat with a mast and boom. The receiving aperture consists of four line arrays, each containing two 0.75 meter array segments, mounted inside the wings of the towed vehicle. This design allows the operator to connect any combination of the arrays to the receiving preamplifiers.

3) An innovative V-wing configuration in the tow vehicle minimizes the target strength of the vehicle and reduces multiples. This design results in the two outer line arrays being located above the two inner line arrays. The vertical offset necessitates two receiving channels: one for the upper arrays and one for the lower arrays. After digitization, the signal from the upper arrays is shifted in time before summing so the signals from the line arrays add coherently for normal incidence subbottom data.

4) The X-Star Full Spectrum Sonar top side processor was modified by Precision Signal, Inc. The software modifications allow selection of transmitting pulse bands and pulse lengths for the V-wing sonar vehicle and provide sediment property prediction capability.

Field Tests

The Halekūlani sand channel off of Waikīkī, O`ahu was chosen as the primary field test site for the project; it is a large, well defined and well studied sand deposit conveniently located directly offshore of Waikīkī Beach, a managed shoreline requiring routine nourishment. We collected sediment samples and jet probed sand depths to establish accurate ground truth data for the field tests. To determine optimal pulse frequencies for imaging the sand deposits, we completed 58 profiles across the test lines, experimenting with 14 different operating bands within the range of 400 Hz to 10 kHz. These tests illustrated the dramatic impact that bandwidth and center frequency have on the quality of the subbottom image. As the bandwidth of the source pulse increases, the resolution of the image and the resolved strata greatly increase. However, as the center frequency increases, penetration into the deposit decreases. Pulses with frequency ranges typical of off-the-shelf subbottom profilers, in the range of 1 to 10 kHz, were rapidly attenuated and could not detect most subbottom structures. We found that a pulse center frequency in the range of 1400 to 2000 Hz and a bandwidth of 2000 to 3000 Hz were necessary to adequately image the sand deposits. Pulses with frequencies of 400-2400 Hz, 800-3200 Hz, 400-2900 Hz and 400-3300 Hz produced the best images during our tests; pulses with bandwidths of less than 2000 Hz did not resolve the sedimentary strata, while pulses with center frequencies higher than 2000 Hz did not adequately penetrate the sand or generated reflections that were too weak to be detected.

Additional tests were conducted to determine the effects of receiving array geometry on the subbottom image. Four array configurations were tested: 1.5 meters x 1.5 meters, 1.5 meters long x 0.75 meters wide, 0.75 meters long x 1.5 meters wide, and 0.75 meters x 0.75 meters. Increasing the array size increased directivity of the signal and thereby reduced scattering noise. Our tests indicated that reducing the receiving array size significantly reduced the reflected energy to scattering noise ratio, causing noisy images. It was originally hoped that one of the smaller array configurations, and therefore a smaller tow vehicle size, would be sufficient to image Hawai`i's sand deposits. Our experiments show, however, that the full array configuration (four parallel 1.5 meter line arrays) is necessary to generate high quality images of Hawai`i's subbottom structure.

Comparison with Ground Truth Data and Other Profilers

Comparison of images of the test area generated by a 400 Hz pulser type source, a 3.5 kHz system, a 2-10 kHz Chirp system and our system illustrate the problems that were previously encountered in Hawai`i, and the improvement in imaging technology resulting from this project. Source ringing of the 400 Hz pulser inhibits resolution of the sedimentary structure, while the high frequency signals of 3.5 kHz and 2-10 kHz Chirp systems are rapidly attenuated, preventing any imaging of the sand deposit. On the other hand, our low frequency Chirp system achieves penetration of 45 meters or more into the sand channel while at the same time resolving sedimentary features less than 30 cm thick. The ability to quickly and easily select from a wide range of pulses within the band of 400 Hz to 10 kHz provides important flexibility that allows the collection of the best possible subbottom imagery. Using a pulse with a wider bandwidth (400-3300 Hz pulse) allows higher resolution, while a lower central frequency (400-2400 Hz) allows greater penetration of the substrate.

The subbottom images were also compared with water jet probes of sand depths conducted by divers across the test line in the Halekūlani sand channel (Figure 2). The measured probe penetrations closely match the uppermost layer imaged by the profiler, and indicate that the subbottom system can effectively detect and map the interface between sand and underlying consolidated sediments.

The system has already proven useful to the scientific community. Images generated during our field tests clearly reveal thick wedges of sand burying notches and terraces formed in paleoreef complexes during ancient lower sea level stands. Profiles in Kailua Bay, collected during ongoing research by the U.S. Geological Survey (Dr. M. Hampton, principal investigator), show sediments over 20 meters thick burying numerous coral terraces at depths between 60 and 110 meters (Figure 3). Survey lines running parallel to the leeward coast of O`ahu reveal the existence of classic channels cut into the reef and filled with sand. These images are the clearest ever produced of Hawai`i's subbottom structure, and have revealed important new information on offshore sedimentary features.

Sediment Classification

Generating a calibrated, repeatable sonar pulse with the X-Star allows quantitative analysis of bottom and subbottom returns. Sediment classification software has been coded, tested, and installed in the X-Star subbottom system as part of this project. The sediment classification algorithms are based on quantitative measurement of the amplitude and phase of the bottom and subbottom reflections. Sediment physical properties are predicted using three acoustic properties of the sediments to provide a robust methodology for sediment classification. Sediment acoustic impedance, compressional wave attenuation and sediment volume scattering are computed from the reflected sonar signal, and then empirically related with a system database to sediment physical properties such as grain size, porosity, rigidity, and bulk density. As more information becomes available with time, the system database can be updated and calibrated to specific depositional environments.

Surficial sediment acoustic impedance is calculated by measuring the amplitude of the acoustic return and computing signal losses. In linear terms, the reflection coefficient *(R)* can be calculated by,

$$R = \frac{Pe(2r)}{Ps} \qquad (1)$$

where *Pe* is the measured pressure amplitude of the echo, *Ps* is the measured far field pressure amplitude of the source, referenced to 1 meter from the source, and 2*r* is the spherical spreading correction.

The X-Star system measures the amplitude of the seabed reflection, measures the sonar height above the seafloor, *r*, and uses equation (1) to calculate the pressure reflection coefficient of the sediment water interface. Surficial impedance is calculated using the pressure reflection coefficient measurement by,

$$Z_2 = Z_1 \frac{1 + R}{1 - R} \qquad (2)$$

where Z_2 is the specific acoustic impedance of the surficial sediments, and Z_1 is the specific acoustic impedance of seawater. However, for coarser sediments, such as sand, the seabed acts as a porous solid and equation (2) is an approximation that can cause errors in calculating the surficial impedance of up to 20%. According to Biot theory (Biot, 1962), for a sandy seabed fast and slow compressional waves are generated at the fluid-porous solid boundary; this energy conversion reduces the measured reflection coefficient resulting in an underestimation of the sediment impedance. Tables generated by Biot theory are therefore used to convert the reflection coefficient to impedance.

To calculate the subbottom acoustic impedance as a function of depth, the interlayer reflection coefficient for the nth subbottom interface, R_n is related ABS-ALL.WPDto the measured reflection amplitudes (corrected for attenuation and spherical spreading) by,

$$R_n = \frac{A_n R_{n-1}}{A_{n-1}(1 - R_{n-1}^2)} \qquad (3)$$

where A_n is the reflection amplitude of the nth interface (A_1 being the reflection amplitude of the sediment-water interface).

The impedances of the subbottom layers are calculated after the interlayer reflection coefficients are determined. The impedance of the nth layer is related to Z_{sw}, the impedance of seawater, and the interlayer reflection coefficient, R_n by

$$Z_N = Z_{sw} \prod_{n=1}^{N} \frac{1 + R_n}{1 - R_n} \qquad (4)$$

In practice, vertical profiles of impedance are computationally intensive and cannot be performed in real time. The steps involved for calculating a vertical impedance profile include: 1) selection of 50 acoustic returns from the image for processing, 2) determination of layer interfaces using a fuzzy logic tracking algorithm, 3) discarding reflections not associated with stratigraphy, 4) determination of the amplitude and phase of each layer interface, and 5) calculation of the impedance profile.

Compressional wave attenuation is calculated by measuring the phase of the signal returns. Attenuation is a measure of the rate at which acoustic energy is converted to thermal energy. This conversion is caused by viscous losses and intergranular friction, and is a function of the physical properties of the sediment. Attenuation of the Gaussian-shaped chirp pulse as it propagates through sediment not only causes a loss of total energy, but also causes the center frequency of the pulse spectrum to shift to a lower frequency. Measurement of the shift of the center frequency per unit travel time of the reflected signal therefore gives an estimate of signal attenuation. LeBlanc, Panda, and Schock (1992) present a robust method for calculating signal attenuation where the rate of change of the signal phase can be used to calculate the center frequency of the signal.

In complex form, the chirp signal can be represented by,

$$z(t) = E(t)e^{i\phi(t)} \qquad (5)$$

where E(t) is the envelope function of the chirp pulse and $\phi(t)$ is the phase. The phase is therefore given by,

$$\phi(t) = -i \log[\frac{z(t)}{E(t)}] \qquad (6)$$

The instantaneous frequency is given by,

$$f_i = \frac{1}{2\pi} \frac{\partial\phi}{\partial t} \qquad (7)$$

The instantaneous frequency will of course oscillate over the frequency range of the pulse spectrum. Furthermore, the oscillation of the instantaneous frequency is stochastic due to random phase and amplitude of the many reflected pulses arriving from volume scatterers within the marine sediment. However, the local mean value of the instantaneous frequency will represent the center frequency of the spectrum, f_c, and will steadily decrease as the pulse attenuates.

If attenuation is proportional to frequency, and if the transmitted spectrum is Gaussian, the change in center frequency per unit travel time is given by,

$$\frac{\Delta f_c}{\Delta t} = \frac{kcW^2}{4} \tag{8}$$

where k is the attenuation constant (nepers/m/Hz), W is the bandwith in Hertz, and c is the compressional wave velocity (m/s). The attenuation constant is therefore calculated using,

$$k = \frac{4}{cW^2} \frac{\Delta f_c}{\Delta t} \tag{9}$$

The volume scattering coefficient of sediment can be used to generate an along-track plot of sediment inhomogeneities. Scattering strength is an indication of sediment type and can be used to estimate physical properties. The volume scattering coefficient is defined as,

$$S_v = 10 \log(\frac{I_r}{I_i}) \tag{10}$$

which is the acoustic intensity of scattering originating from a cubic meter of sediment, divided by the intensity of the acoustic wave illumination, converted to a dB scale.

The volume scattering strength can be extracted from the measured acoustic return by using the sonar equation for the volume reverberation level (RL),

$$SL - 40 \log(r) + S_v + 10 \text{ l} \tag{11}$$

The sediment volume, V, illuminated at an instant in time at range r is given by

$$V = \frac{c\tau}{2} \psi r^2 \tag{12}$$

where c is the sound speed, τ is the pulse length, r is the range from the sonar to the scattering field, and ψ is the effective two-way beamwidth (solid angle in steradians) of the sonar transmitting and receiving arrays. For arrays with a two-way beam pattern that is approximately conical, the effective solid angle is related to the effective beam width in radians by

$$\psi = 2\pi(1 - \cos(2\theta)) \tag{13}$$

where 2θ is the two-way effective beam width of a conical beam pattern measured in degrees.

Equation (13) can be used to derive the following expression relating the square amplitudes of the reflected signal to the scattering coefficient:

$$\frac{r^2 \sum_{n=1}^{N} (x_n r_n)^2}{2N (x_0 r_0)^2 (1 - R)(1 + R)} \tag{14}$$

where R is the pressure reflection coefficient and $x_n r_n$ is the correlated signal corrected for spherical spreading. Equation (14) is used to deterimine the average scattering coefficient for a subbottom layer selected by the user. N samples of the envelope x_n are squared and corrected for two-way spherical spreading, and then summed before applying corrections for the two-way boundary losses and for the volume of sediments illuminated by the beam at each instant in time. The scattering coefficient (s_v) of the selected subbottom depth interval is displayed in real time using an along-track plot.

These measurements provide three independent acoustic parameters: 1) the pressure reflection coefficient (from which impedance is derived), 2) pulse attenuation, and 3) sediment volume scattering coefficient, which can be empirically related to sediment physical properties of grain size, bulk density and shear strength. The ability to measure three acoustic properties greatly improves classification capabilities by eliminating ambiguities associated with any one parameter. For example, sediments with different mean grain sizes can have the same impedance (cemented clays versus sands, for example) but will usually have different attenuation coefficients. Also, different sediments may have identical attenuation coefficients, but have significantly different scattering strengths (gassy muds versus sand).

LeBlanc *et al.* (1992) derived theoretical relationships between the porosity and rigidity properties of a given marine sediment sample, from which bulk density and grain size can be derived, to acoustic impedance and sound velocity. They found good correlation with empirical relationships derived by Faas (1969) and Hamilton (1970). LeBlanc, Panda, and Schock (1992) were successful in predicting sediment characteristics based upon attenuation modelling. There is currently little data relating volume scattering to sediment properties.

Prior to this project, empirical relationships between sediment properties and acoustic properties had been derived for continental environments far different from the carbonate reef environments typical of Hawai`i. Field testing of the sediment classification software enabled a comparison to be made with previous work. Field tests were conducted by collecting subbottom data with the profiler at twenty-one sites where we had previously collected and analyzed surficial sediment samples. The sampling sites were selected to span the range of sedimentary environments in the islands and included silts from Honolulu harbor, and fine to coarse sands in deposits along the south and west coasts of O`ahu. The sediment classification software was used to measure the reflection coefficient and calculate the acoustic impedance at each sample site. The surficial impedance data were then plotted against (a) measured sediment bulk density and (b) mean grain size, and a linear regression used to generate a set of empirical relationships for the Hawai`i sedimentary environments (Figure 4). These data provide the empirical relationships between the physical properties of Hawai`i nearshore sediments, and impedance and reflection coefficient.

The plots illustrate two important points. First, the Honolulu Harbor silts are easily distinguishable from sand: the subbottom profiler and sediment classification software can accurately distinguish between the harbor silts and the offshore sands. The impedance variations within the various sand samples, however, appear to be too small to distinguish between the sand samples using the existing data.

Second, the plots demonstrate that the local acoustic-physical property relationships can vary significantly from the published database (LeBlanc *et al.*, 1992; Hamilton, 1970). The grain size relationship shows a large differnce, while the bulk density curve diverges primarily for the smaller values of bulk density. This underscores the importance of collecting data locally to enter into the system database.

Subsurface Sediment Classification

Since the sediment sampling survey was limited to surficial samples, the subbottom classification capability of the software was tested by using a sand channel above a hard reef surface with a known impedance. The Maili sand channel off the west coast of O`ahu was selected for this test because both the exposed reef surrounding the sand channel, and the buried reef underlying the sand, were relatively flat and provided coherent, repeatable acoustic reflection measurements, and met the coherent layer criteria of the fuzzy logic tracking algorithm.

An impedance profile was generated with the inversion techniques of the sediment classification software; the impedance of the buried reef was calculated to be 4.5 Pa-sec/m, and compares well with the 4.0 to 5.5 Pa-sec/m values calculated using the surface reflection coefficient of the surrounding exposed reef. The impedance of the sand was calculated to be 2.8 Pa-sec/m. This was then used to estimate sediment physical properties using the Hawai`i empirical relationships added to the system database. The software predicts sand with a bulk density of 1.76 gm/ml and a mean grain size of 2.2 phi (.22 mm). These values compare well with measured values of sand samples collected in this area: mean grain sizes of Mā`ili sediment samples range from 1.92 to 2.24 phi, while bulk density values range from 1.81 to 1.84 gm/ml.

The Mā`ili sand channel was also used to test the volume scattering and compressional wave attenuation algorithms of the sediment classification software. Figure (5) is an example of along-track sediment classification parameters displayed by the X-Star software. Note that the second layer shown is that selected for attenuation and scattering measurements. The exposed reef surface is characterized by a high reflection coefficient and a high scattering coefficient (about 10 times that of sand), however it shows no significant compressional wave attenuation.

Attenuation is much higher in the sand deposit. The compressional wave attenuation is calculated using the difference between the center frequency of the sediment-water interface reflection, and the center frequency of the tracking layer. Volume scattering drops significantly in the sand because it is relatively homogenous, generating small internal reflections, and has high attenuation.

Conclusions

An improved frequency-modulated subbottom reflection profiler has been developed that can successfully image and classify the shelf sediments of Hawai`i. The development was funded by CEROS because off-the-shelf commercial systems such as sparkers, boomers, a 400 Hz pulser, 3.5 kHz profilers and other chirp sonars were not able to generate images of Hawai`i shelf sediments with sufficient penetration or resolution to accurately indentify sand resources. An X-Star Full Spectrum sonar was modified to process two receiving channels and generate imagery using the band of 400 Hz to 10 kHz. A new tow vehicle design contains four parallel 1.5 meter long line arrays providing 12 dB greater directivity than a streamer hydrophone array of the same length. The 12 dB reduction in scattering provides a large improvement in image quality. During the sea trials of the sonar, the spectral content of the chirp pulse was varied within the range of 400 Hz to 10 kHz. The tests showed that the best images of the top 40 meters of sediment were collected using an FM pulse with a center frequency of 1.4 to 2.0 kHz and a bandwidth of 2 to 3 kHz. Pulses with higher frequencies provided degraded images because attenuation and rough subsurface boundaries caused weak reflections from the subsurface structure. Tests also showed that reducing the number of line arrays or the length of the line arrays increased the scattering noise in the images. Sediment classification software modules were coded and tested to provide surficial and subsurface measurements of impedance, attenuation and volume scattering from the reflection data, and predictions of physical properties such as grain size and density using empirical relationships.

The capabilities of the sediment classification software and the ability to select from an assortment of pulses with a wide range of spectral characteristics are not available elsewhere and represent an important advance in the state-of-the-art of subbottom profiling. The profiler can be adapted to any geologic environment with ease, and can be used for a multitude of profiling purposes. The system's high acoustic directivity and wide range of pulse bands make it uniquely effective for: 1) tropical coralline environments characterized by calcareous sands with coral rubble and coralline limestone basements; 2) areas with thick sands and gravel that rapidly attenuate the signals of higher frequency systems; 3) applications requiring deep penetration and high resolution of the substrate. We have used the system in Hawai`i for sand mapping, sea level history investigations, and assessment of outfall pipeline route stability. Other applications of interest include dredging surveys, foundation investigations for offshore structures, directional drilling planning, geohazard surveys, and pipeline and cable location.

Acknowledgements

Work on this project has been funded by grants from the National Defense Center of Excellence for Research in Ocean Sciences (CEROS). Their support and encouragement are gratefully acknowledged.

References

Barry, J. H. 1991; "Characterization of deep sand deposits in the tropical reef environment of Hawaii." MTS'91 An Ocean Cooperative: Industry, Government and Academia, Proceedings.

Biot, M.A. 1962; " Generalized theory of acoustic propagation in porous dissipative media," J. Acoustic Soc. Am. 34:1254-1264.

Campbell, J.F., W.T. Coulbourn, R.M. Moberly, and B.R. Rosendahl, 1970; "Reconnaissance sand inventory: off Leeward O`ahu. HIG-70-16. Hawai`i Institute of Geophysics, University of Hawai`i.

Coulbourn, W.T., J.F. Campbell, R.M. Moberly, 1974; "Hawaiian submarine terraces, canyons, and Quaternary history evaluated by seismic profiling. Marine Geology 17:215-234.

Faas, R.W. 1969; "Analysis of the relationship between acoustic reflectivity and sediment porosity," Geophysics 34: 546-553.

Hamilton, E.L. 1970; "Reflection coefficients and bottom losses at normal incidence computed from Pacific sediment properties," Geophysics 35:995-1004.

LeBlanc, L. R., L. Mayer, M. Rufino, S. Schock and J. King, 1992; "Marine sediment classification using the chirp sonar" JASA 91(1) pp. 107-115.

LeBlanc, L. R., S. Panda and S. Schock, 1992; "Sonar attenuation modeling for classification of marine sediments" JASA 91(1) pp. 116-126.

Marine Minerals Technology Center, 1993; "Characterization of Hawaiian sand deposits on the insular shelf of Southern Oahu and Penguin Bank." MMTC Project 1511, University of Hawai`i, 1991.

Schock, S. G. and L. R. LeBlanc, 1990; "Chirp sonar: new technology for subbottom profiling," Sea Technology, Sept. 1990, pp. 35-43.
Schock, S.G., L.R. LeBlanc, L. Mayer, 1989; "Chirp subbottom profiler for quantitative sediment analysis," Geophysics, V. 54 no. 4, April 1989.

Schock, S.G., L.R. LeBlanc, 1986; "Sediment classification using a wideband, frequency-modulated sonar system;" 18th Offshore Technology Conference Proceedings, 1986; OTC 5125.

Sea Engineering,Inc. and Precision Signal, Inc., 1996; "Development of a broadband FM subbottom profiler for seafloor imaging and sediment classification - phase 2." Report prepared for the National Defense Center of Excellence for Research in Ocean Sciences (CEROS).

Sea Engineering,Inc. and Precision Signal, Inc., 1995; "Development of a broadband FM subbottom profiler for seafloor imaging and sediment classification." Report prepared for the National Defense Center of Excellence for Research in Ocean Sciences (CEROS).

Sea Engineering, 1993; "Beach nourishment viability study". Prepared for the Office of State Planning, Coastal Zone Management Program.

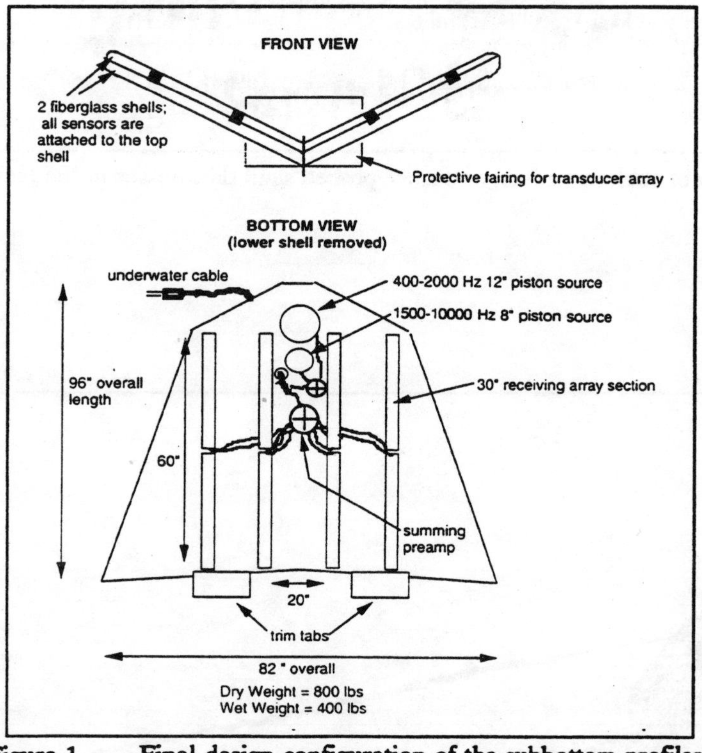

Figure 1. Final design configuration of the subbottom profiler.

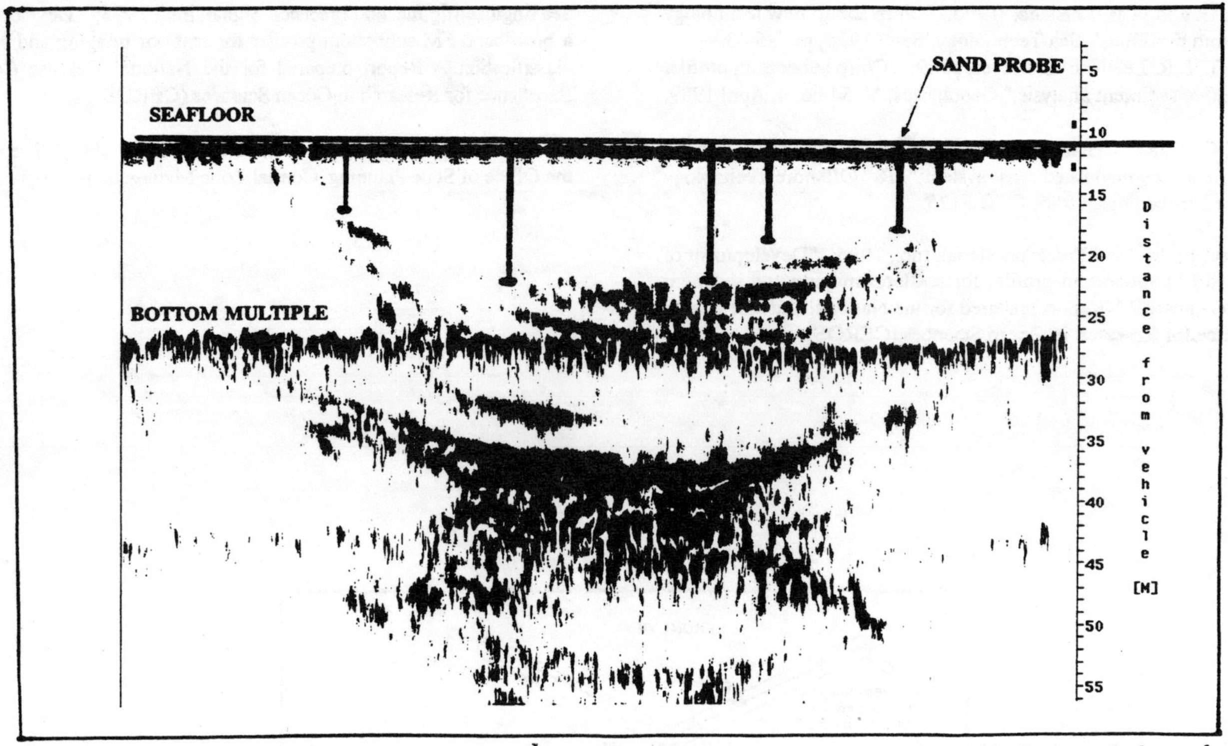

Figure 2. **Comparison of subbottom image with jet-probed sand thicknesses in the Halekūlani sand channel.**

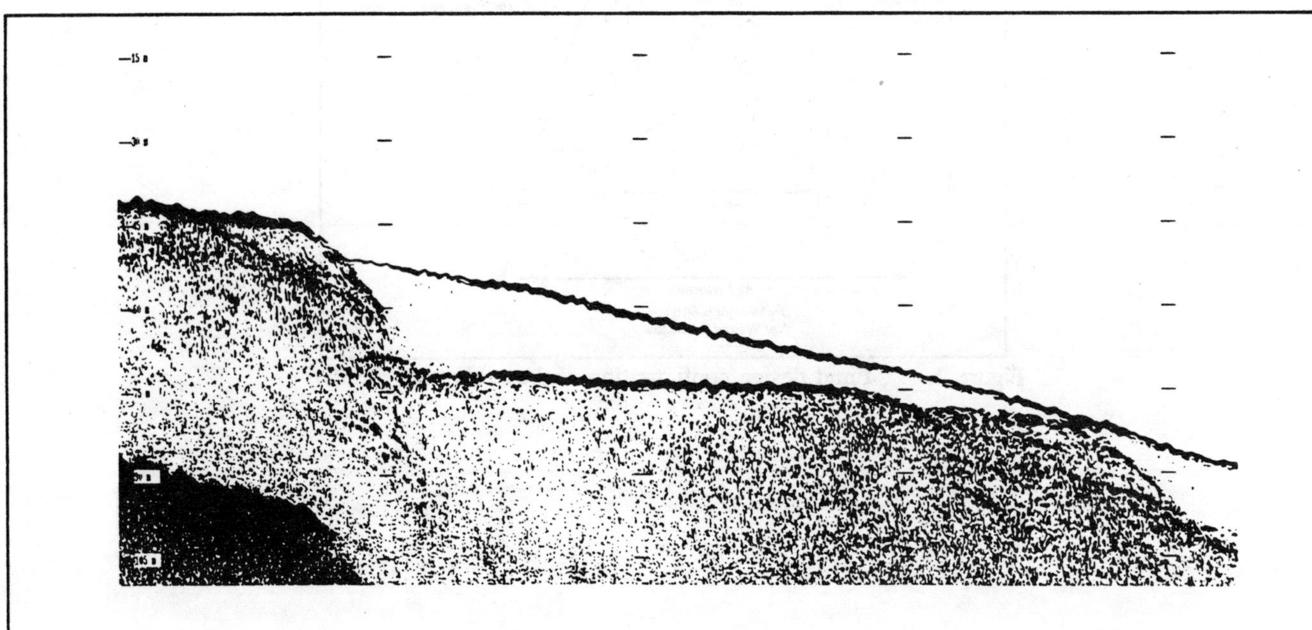

Figure 3. **Subbottom profile of Kailua Bay, O'ahu.**

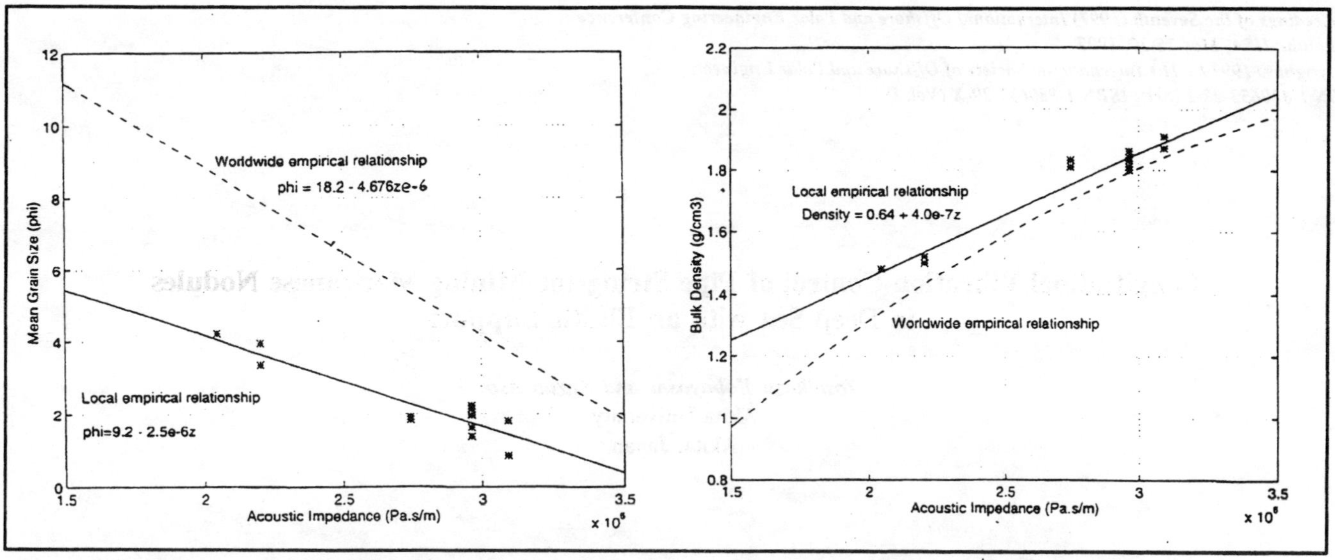

Figure 4. Acoustic impedance, bulk density and grain size data collected at Hawai'i test sites.

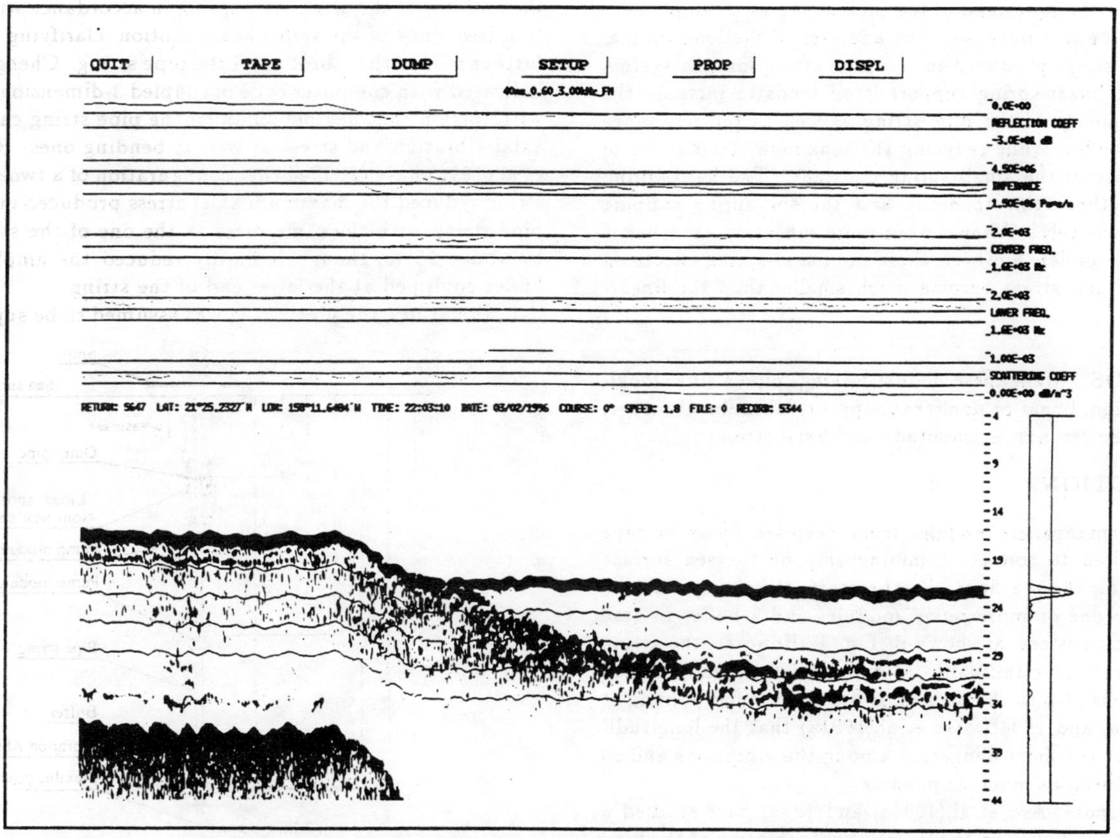

Figure 5. Along track graphs of acoustic properties. Layer 2 is being used to generate the layer frequency and volume scattering graphs.

Proceedings of the Seventh (1997) International Offshore and Polar Engineering Conference
Honolulu, USA, May 25-30, 1997

ISBN 1-880653-28-1 (Set); ISBN 1-880653-29-X (Vol. I)

Longitudinal Vibration-Control of Pipe String for Mining Manganese Nodules in Deep Sea with an Elastic Support

Yoshikazu Kobayashi and Kazuo Aso
Akita University
Akita, Japan

ABSTRACT

In order to reduce the longitudinal vibration of a pipe string for mining manganese nodules from deep sea floors, the pipe string was assumed to be supported by a linear or nonlinear spring at its upper part. Furthermore, a vibration absorber was attached to the buffer at the lower end of the pipe string as a complementary measure for this purpose. The analyses of the longitudinal vibration and stress produced in the pipe string for this system show that the linear-spring support itself tends to increase the resonance frequency of the pipe string as well as the resonance amplitude of buffer, while reducing the maximum axial stress of the pipe in case of the small spring constant. The hard-spring support strengthens this tendency and the soft-spring support tends to act reversely. When an optimum vibration absorber is attached to the buffer, however, those nonlinear-spring effects on the amplitude and stress become much smaller than the linear-spring ones.

KEY WORDS : Manganese nodule mining, pipe string, longitudinal vibration, linear or nonlinear support, vibration absorber, reduction of the resonance amplitude and axial stress.

INTRODUCTION

For mining manganese nodules from deep-sea floors, a pipe string is required to connect a mining ship on the sea surface with a miner on the sea floor. Furthermore, the string must be equipped with one or more pump modules and a buffer in case of the pump-lift system, as shown in Fig. 1. However, the string is easy to vibrate longitudinally, laterally and torsionally due to the ship motions. Up to the present, it has been found (Chung et al.,1981, 1994a, and 1994b; Aso et al.,1991a) that the longitudinal vibration is the most dangerous among the vibrations and so it must be reduced as much as possible.

For this purpose, Aso et al.(1991a, and 1993) have studied a shape-effect of the buffer on the longitudinal vibration of the pipe string, identifying that the buffer whose shape causes a higher drag force more reduces the vibration. They (1992,1994a, and 1995) have also found the effect of the passive control by a vibration absorber on the vibration, indicating that the absorber could reduce the resonance amplitude of the pipe string by about 40 % and its maximum axial stress by about 17 % as compared with those of the string without any vibration absorbers. Furthermore, they (1994b) suggested a semi-active control system by changing the stiffness of the absorber's spring in accordance with an excitation frequency of the ship's heave motion, clarifying that it could further reduce the vibration of the pipe string. Cheng et al.(1995) indicated with computer code of coupled 3-dimensional vibrations with torsion that flexible joints on the pipe string can reduce the axial vibration and stress as well as bending ones. Recently, Aso et al.(1996) have clarified that configuration of a two-stepped pipe string reduced the maximum axial stress produced in the uniform pipe string with the same mass as the one of the stepped string by about 25 %, though it hardly reduced the amplitude of the buffer equipped at the lower end of the string.

In this study, the pipe string was assumed to be supported with

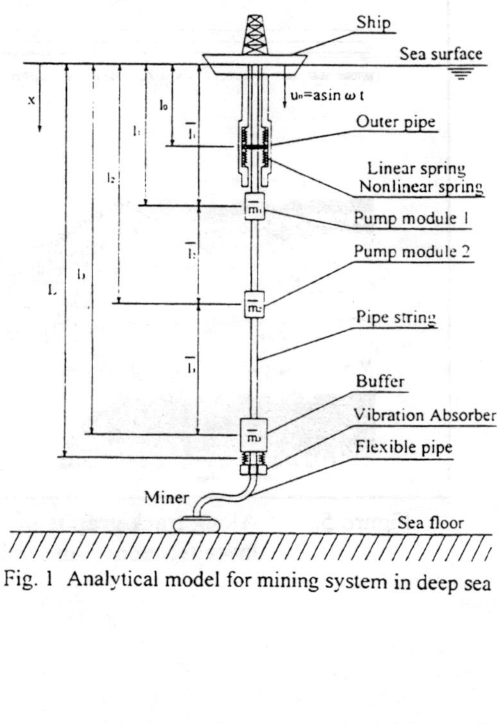

Fig. 1 Analytical model for mining system in deep sea

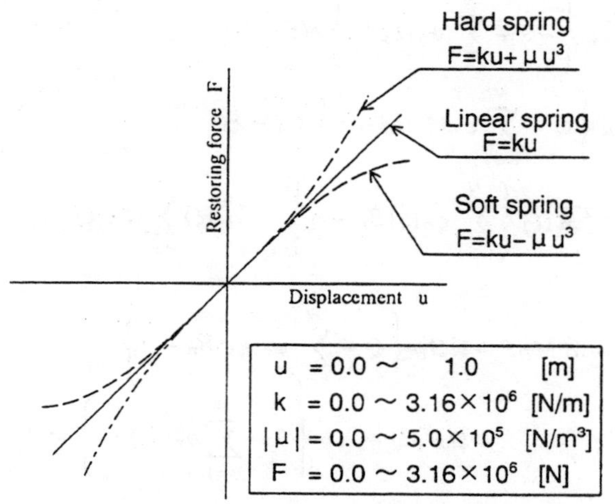

Fig. 2 Characteristics of linear and nonlinear springs

a linear or nonlinear spring at the upper part, as shown in Fig. 1, for reducing the resonance amplitude of the pipe string and its maximum axial stress induced. The nonlinear spring used for the support was the hard or soft spring characterized by Fig.2. The hard spring is composed of a linear part and a positively acting nonlinear part, while the soft one is composed of a linear part and a negatively acting nonlinear part. Furthermore, a vibration absorber was attached to the buffer at the lower end of the pipe string as a complementary measure to reduce its longitudinal vibration and axial stress. Then, the longitudinal vibration as well as the axial stress caused by the ship's vertical-motion were theoretically analyzed for the above-mentioned pipe string in the pump-lift system. The results obtained were compared with those for the pipe string without any supports and absorbers to clarify the advantage of using the support with a linear or nonlinear spring at the upper part of the string.

ANALYSIS

In order to analyze the longitudinal vibration of the pipe string with one of the supports and a vibration absorber attached only to the buffer in this system shown in Fig. 1, it is assumed that both the pump modules and buffer equipped on the pipe string are considered as concentrated masses, the fluid inside the pipe string vibrates together with the string, the flexible pipe does not affect the vibration of the pipe string because of its much smaller mass and rigidity than those of the pipe string (e.g. Chung and Whitney, 1983), and that the top of the pipe string is forced to vibrate vertically in the harmonic function of $u_0(t) = a\sin\omega t$ by the vertical motion of the ship. Furthermore, the fluid forces acting on the pump modules and the buffer are calculated by the Morison's equation with linearized drag term (Morison et al.,1950; Aso et al., 1991a,1991b, and 1993).

Then, the equations of motion for the longitudinal vibration of the pipe string and the absorbers' mass m_B in this problem can be expressed as follows:

$$EA\frac{\partial^2 u}{\partial x^2} - \left\{k(u-u_0) + \mu(u-u_0)^3\right\}\delta(x-l_0) + g_B(\dot{x}_B - \dot{u})\delta(x-l_3) + k_B(x_B - u)\delta(x-l_3) = M(x)\ddot{u} + C(x)\dot{u} \quad (1)$$

$$m_B\ddot{x}_B + g_B\{\dot{x}_B - \dot{u}]_{x=l_3}\} + c_B\dot{x}_B + k_B\{x_B - u]_{x=l_3}\} = 0 \quad (2)$$

where

$$M(x) = m_0 + m_1\delta(x-l_1) + m_2\delta(x-l_2) + m_3\delta(x-l_3)$$
$$C(x) = c_0 + c_1\delta(x-l_1) + c_2\delta(x-l_2) + c_3\delta(x-l_3)$$

and where m_0, c_0, E and A are the mass per unit length of the pipe string, the coefficient of viscous damping of the string, Young's modulus of the pipe material and the cross-sectional area of the pipe string, respectively. m_j ($j = 1 \sim 3$) are the masses of the pump-module 1, pump-module 2 and buffer, respectively. These masses include their added masses due to the ambient water, which have been determined by the previous study (Aso et al.,1991a,1991b, and 1993). c_j($j = 1 \sim 3$) are the equivalent linear damping coefficients of the pump-modules 1,2 and buffer (Aso et al.,1991a) due to the drag forces of the ambient water. m_B, g_B, c_B and k_B are the mass of the vibration absorber including the added mass due to the ambient water, the viscous damping coefficient of the absorber's damper, the equivalent linear damping coefficient of the absorber's mass due to the ambient water, and the spring constant of the absorber's spring, respectively. k and μ are the stiffness coefficients of the linear and nonlinear parts of the support at the upper part of the string. In addition, the dot attached on the variables represents the differentiation with respect to time, $\delta(\cdot)$ is the Dirac delta function, u and x_B are the displacements of the pipe string and absorber's mass in the x-direction, and t is the time.

The boundary conditions for this problem are expressed in the following equations:

$$u(0,t) = u_0(t) = a\sin\omega t \quad , \quad EA\frac{\partial u(L,t)}{\partial x} = 0 \quad (3)$$

where $u_0(t)$ is the forced displacement applied at the top of the pipe string, as mentioned previously.

Now, the solution of Eq. 1 can be assumed to take the following form:

$$u(x,t) = u_v(x,t) + u_0(t) \quad (4)$$

Then, substituting Eq. 4 into Eqs. 1 and 2, and defining the following nondimensional quantities,

$$\left.\begin{aligned}
&\xi = \frac{x}{L},\ \xi_i = \frac{x_i}{L},\ X = \frac{x_B}{L},\ U_v = \frac{u_v}{L}\\
&U_0 = \frac{u_0}{L},\ \gamma = \frac{a}{L},\ K = \frac{kL}{EA},\ K_B = \frac{k_B L}{EA}\\
&\tau = \frac{t}{L}\sqrt{\frac{EA}{m_0}},\ \Omega = \omega L\sqrt{\frac{m_0}{EA}},\ D_0 = \frac{c_0 L}{\sqrt{EAm_0}}\\
&D_i = \frac{c_i}{\sqrt{EAm_0}},\ D_B = \frac{c_B}{\sqrt{EAm_0}},\ G_B = \frac{g_B}{\sqrt{EAm_0}}\\
&M_i = \frac{m_i}{m_0 L},\ M_B = \frac{m_B}{m_0 L},\ \varepsilon = \frac{\mu L^3}{EA} \quad (i = 1 \sim 3)
\end{aligned}\right\} \quad (5)$$

the nondimensional equations of motion for U_v , X are obtained as follows:

$$\frac{\partial^2 U_v}{\partial \xi^2} - (KU_v + \varepsilon U_v^{\,3})\delta(\xi-\xi_0) + G_B\frac{\partial}{\partial\tau}(X - U_v - U_0)\delta(\xi-\xi_3) + K_B(X - U_v - U_0)\delta(\xi-\xi_3) = \overline{M}(\xi)\frac{\partial^2}{\partial\tau^2}(U_v + U_0) + \overline{D}(\xi)\frac{\partial}{\partial\tau}(U_v + U_0) \quad (6)$$

$$M_B\frac{\partial^2 X}{\partial\tau^2}+G_B\frac{\partial}{\partial\tau}\{X-U_v(\xi_3,\tau)-U_0(\tau)\}$$
$$+D_B\frac{\partial X}{\partial\tau}+K_B\{X-U_v(\xi_3,\tau)-U_0(\tau)\}=0 \quad (7)$$

where

$$\overline{M}(\xi)=\{1+M_1\delta(\xi-\xi_1)+M_2\delta(\xi-\xi_2)+M_3\delta(\xi-\xi_3)\}$$
$$\overline{D}(\xi)=\{D_0+D_1\delta(\xi-\xi_1)+D_2\delta(\xi-\xi_2)+D_3\delta(\xi-\xi_3)\}$$

Furthermore, the non-dimensional boundary conditions are obtained as follows:

$$U_v(0,\tau)=0 \quad , \quad \frac{\partial}{\partial\xi}U_v(1,\tau)=0 \quad (8)$$

Therefore, the approximate solutions for the equations can be assumed to take the following forms:

$$\left.\begin{aligned} U_0(\tau)&=\gamma\sin\Omega\tau\\ U_v(\xi,\tau)&=\sum_{n=1}^{N}\phi_n(\xi)(A_n\cos\Omega\tau+B_n\sin\Omega\tau)\\ X(\tau)&=S\cos\Omega\tau+Q\sin\Omega\tau\end{aligned}\right\} \quad (9)$$

where A_n, B_n, S, Q are constants to be determined, and $\phi_n(\xi)$ could be assumed as the following function which satisfies the nondimensional boundary conditions:

$$\phi_n(\xi)=\sin\frac{(2n-1)\pi}{2}\xi \quad (n=1,2,\ldots,N) \quad (10)$$

Then, the approximate solutions of Eq. 9 are substituted into the nondimensional equations of motion, Eqs. 6 and 7, and the resulting equations are further reduced by the following approximate relations because only the first resonance of vibration is practically interested and the contribution of higher harmonics is considered to be much smaller than that of the first harmonic.

$$\begin{aligned}\cos^3\Omega\tau&=\frac{3}{4}\cos\Omega\tau+\frac{1}{4}\cos 3\Omega\tau\\ \sin^3\Omega\tau&=\frac{3}{4}\sin\Omega\tau-\frac{1}{4}\sin 3\Omega\tau\end{aligned} \quad (11)$$

Thus, the following four equations are obtained by equating the cosine and sine terms, respectively.

$$\sum_{n=1}^{N}\frac{\partial^2\phi_n(\xi)}{\partial\xi^2}A_n-K\sum_{n=1}^{N}\phi_n(\xi)A_n\delta(\xi-\xi_0)$$
$$-\frac{3}{4}\varepsilon\sum_{n=1}^{N}\phi_n(\xi)A_n\left\{(\sum_{n=1}^{N}\phi_n(\xi)A_n)^2+(\sum_{n=1}^{N}\phi_n(\xi)B_n)^2\right\}\delta(\xi-\xi_0)$$
$$+\Omega G_B\left\{Q-\sum_{n=1}^{N}\phi_n(\xi)B_n-\gamma\right\}\delta(\xi-\xi_3)$$
$$+K_B\left\{S-\sum_{n=1}^{N}\phi_n(\xi)A_n\right\}\delta(\xi-\xi_3)$$
$$=-\Omega^2\overline{M}(\xi)\sum_{n=1}^{N}\phi_n(\xi)A_n+\Omega\overline{D}(\xi)\left\{\sum_{n=1}^{N}\phi_n(\xi)B_n+\gamma\right\} \quad (12)$$

$$\sum_{n=1}^{N}\frac{\partial^2\phi_n(\xi)}{\partial\xi^2}B_n-K\sum_{n=1}^{N}\phi_n(\xi)B_n\delta(\xi-\xi_0)$$
$$-\frac{3}{4}\varepsilon\sum_{n=1}^{N}\phi_n(\xi)B_n\left\{(\sum_{n=1}^{N}\phi_n(\xi)A_n)^2+(\sum_{n=1}^{N}\phi_n(\xi)B_n)^2\right\}\delta(\xi-\xi_0)$$
$$+\Omega G_B\left\{-S+\sum_{n=1}^{N}\phi_n(\xi)A_n\right\}\delta(\xi-\xi_3)$$
$$+K_B\left\{Q-\sum_{n=1}^{N}\phi_n(\xi)B_n-\gamma\right\}\delta(\xi-\xi_3)$$
$$=-\Omega^2\overline{M}(\xi)\left\{\sum_{n=1}^{N}\phi_n(\xi)B_n+\gamma\right\}-\Omega\overline{D}(\xi)\sum_{n=1}^{N}\phi_n(\xi)A_n \quad (13)$$

$$-\Omega^2 M_B S+\Omega G_B\left\{Q-\sum_{n=1}^{N}\phi_n(\xi_3)B_n-\gamma\right\}$$
$$+\Omega D_B Q+K_B\left\{S-\sum_{n=1}^{N}\phi_n(\xi_3)A_n\right\}=0 \quad (14)$$

$$-\Omega^2 M_B Q+\Omega G_B\left\{-S+\sum_{n=1}^{N}\phi_n(\xi_3)A_n\right\}$$
$$-\Omega D_B S+K_B\left\{Q-\sum_{n=1}^{N}\phi_n(\xi_3)B_n-\gamma\right\}=0$$
$$(15)$$

From Eqs. 14 and 15, S and Q can be found in the following forms:

$$S=S_1Z_1(\xi_3)+S_2Z_2(\xi_3) \quad (16)$$
$$Q=Q_1Z_1(\xi_3)+Q_2Z_2(\xi_3)=-S_2Z_1(\xi_3)+S_1Z_2(\xi_3) \quad (17)$$

where

$$Z_1(\xi_3)=\sum_{n=1}^{N}\phi_n(\xi_3)A_n \quad , \quad Z_2(\xi_3)=\sum_{n=1}^{N}\phi_n(\xi_3)B_n+\gamma$$

$$S_1=\left\{K_B(K_B-\Omega^2M_B)+\Omega^2G_B(G_B+D_B)\right\}/D_T=Q_2$$
$$S_2=\left\{\Omega G_B(K_B-\Omega^2M_B)-K_B\Omega(G_B+D_B)\right\}/D_T=-Q_1$$
$$D_T=(K_B-\Omega^2M_B)^2+\Omega^2(G_B+D_B)^2$$

Now, substituting Eqs. 16 and 17 into Eqs. 12 and 13, then applying the Galerkin's method to the resulting equations. and defining the following quantities,

$$\Delta_1=\sum_{n=1}^{N}\phi_n(\xi_0)A_n \quad (18)$$

$$\Delta_2=\sum_{n=1}^{N}\phi_n(\xi_0)B_n \quad (19)$$

the following simultaneous equations with respect to A_n and B_n are obtained.

$$\left\{\Omega^2-\frac{(2k-1)^2\pi^2}{4}\right\}\frac{A_k}{2}+\Omega^2\sum_{i=1}^{3}M_i\sum_{n=1}^{N}\phi_n(\xi_i)\phi_k(\xi_i)A_n$$
$$+\{-S_2\Omega G_B+K_B(S_1-1)\}\sum_{n=1}^{N}\phi_n(\xi_3)A_n\phi_k(\xi_3)$$
$$+\{\Omega G_B(S_1-1)+K_BS_2\}\sum_{n=1}^{N}\phi_n(\xi_3)\phi_k(\xi_3)B_n$$

$$-\Omega\sum_{i=1}^{3}D_i\sum_{n=1}^{N}\phi_n(\xi_i)\phi_k(\xi_i)B_n-\Omega D_0\frac{B_k}{2}$$
$$=\frac{2\Omega D_0\gamma}{(2k-1)\pi}+\Omega\gamma\sum_{i=1}^{3}D_i\phi_k(\xi_i)$$
$$+\left\{K\Delta_1+\frac{3}{4}\varepsilon\Delta_1(\Delta_1^2+\Delta_2^2)\right\}\phi_k(\xi_0)$$
$$-\{\Omega G_B(S_1-1)+K_BS_2\}\gamma\phi_k(\xi_3)\quad(k=1,2,\ldots,N)\qquad(20)$$

$$\frac{D_0\Omega}{2}A_k+\{\Omega G_B(1-Q_2)+K_BQ_1\}\sum_{n=1}^{N}\phi_n(\xi_3)\phi_k(\xi_3)A_n$$
$$+\Omega\sum_{i=1}^{3}D_i\sum_{n=1}^{N}\phi_n(\xi_i)\phi_k(\xi_i)A_n+\left\{\Omega^2-\frac{(2k-1)^2\pi^2}{4}\right\}\frac{B_k}{2}$$
$$+\{\Omega G_BQ_1+K_B(Q_2-1)\}\sum_{n=1}^{N}\phi_n(\xi_3)\phi_k(\xi_3)B_n$$
$$+\Omega^2\sum_{i=1}^{3}M_i\sum_{n=1}^{N}\phi_n(\xi_i)\phi_k(\xi_i)B_n$$
$$=-\frac{2\Omega^2\gamma}{(2k-1)\pi}+\left\{K\Delta_2+\frac{3}{4}\varepsilon\Delta_2({\Delta_1}^2+{\Delta_2}^2)\right\}\phi_k(\xi_0)$$
$$-\{\Omega G_BQ_1+K_B(Q_2-1)\}\gamma\phi_k(\xi_3)$$
$$-\gamma\Omega^2\sum_{i=1}^{3}M_i\phi_k(\xi_i)\quad(k=1,2,\ldots,N)\qquad(21)$$

Finally, the constants A_n and B_n can be determined from Eqs. 20 and 21 by the following procedure:
(a) For a fixed Ω, an initial value of Δ_2 is assumed.
(b) Further assumption of Δ_1 in addition to the above Ω and Δ_2 gives the values of A_n and B_n from Eqs. 20 and 21.
(c) The iteration for Δ_1 is carried on until the assumed Δ_1 coincides with the calculated Δ_1 by Eq. 18 with the above-obtained A_n within an allowable error.
(d) Another iteration for Δ_1 is done until the assumed Δ_2 coincides with the calculated Δ_2 by Eq. 19 with the above-obtained B_n within an allowable error.
(e) Similarly, the values of Δ_1 for satisfying Eqs. 18 and 19 are obtained for different Δ_2's, and then, two curves for the two equations are depicted in Δ_1-Δ_2 diagram. The intersection of these two curves gives the true values of Δ_1 and Δ_2 for the initially fixed Ω.
(f) With the obtained true-values of them, the true values of A_n and B_n can be calculated by solving the simultaneous equations of Eqs. 20 and 21.

Furthermore, the true values of S and Q can be determined by substituting those values of A_n and B_n into Eqs. 16 and 17. Then, the longitudinal displacement of the pipe string, $u(x,t)$, and the displacement of the absorber's mass, $x_B(t)$, for a given ω can be obtained by substituting the above values of the constants into Eq. 9 and using the relations of Eqs. 4 and 5.

The axial stress, $\sigma(x,t)$, produced in the pipe string can be obtained by the following equation:

$$\sigma(x,t)=E\frac{\partial u(x,t)}{\partial x}+\frac{W_b+w_0(L-x)}{A}\qquad(22)$$

where W_b stands for the sum of the submerged weights of the pump modules, buffer and vibration absorber equipped below the position considered along the pipe string, and w_0 represents the submerged weight per unit length of the pipe string. Here, it must be noted that the submerged weight, W_b, for the pump modules or buffer was calculated by the gravity force subtracted by the buoyancy caused by the pressure difference between the upper and lower ends of the pump modules or buffer.

Table 1 Physical properties of the pipe string,punp module, buffer and absorber

Quantity	Value	Unit
Pipe-string:		
Outer diameter, d_0	0.2445	m
Inner diameter, d_i	0.2005	m
Density of the pipe material, ρ_p	7.850×10^3	Kg/m^3
Young's modulus of the pipe material, E	0.206×10^{12}	N/m^2
Coefficient of structural damping, c_0	0.049×10^3	Ns/m^2
Density of the inner fluid, ρ_f	1.200×10^3	Kg/m^3
Density of the ambient fluid, ρ	1.030×10^3	Kg/m^3
Pump-modules:		
Shapes;Cylinders		
Outer diameters, D	1.500	m
Lengths, L_m	10.50	m
Ratio of L_m/D	7	
Masses, $\overline{m}_1$, $\overline{m}_2$	50.0	t
Buffer:		
Shape;Cylinder		
Outer diameter, D	3.000	m
Length, L_m	9.000	m
Ratio of L_m/D	3	
Mass, $\overline{m}_3$	300.0	t
Vibration absorber:		
Shape;Hollow cylinder		
Outer diameter	3.000	m
Inner diameter	0.750	m
Mass, $\overline{m}_B$	60.0	t
Added-mass coefficient	1.00	
Drag coefficient	1.16	

Table 2 Locations of the support and layouts of the pump modules and buffer

	l_0	l_1	l_2	l_3	$\bar{l}_1$	$\bar{l}_2$	$\bar{l}_3$
Case 1	1000m	1000m	2000m	5000m	1000m	1000m	3000m
Case 2	1500m	1500m	2500m	5000m	1500m	1000m	2500m
Case 3	2000m	2000m	3000m	5000m	2000m	1000m	2000m
Total length of the pipe-string, L : 5000m							

RESULTS AND DISCUSSIONS

According to the analytical procedure mentioned above, the longitudinal displacement, $u(x,t)$, and the axial stress, $\sigma(x,t)$, of the pipe string supported with a linear or nonlinear spring and equipped with a vibration absorber as shown in Fig. 1 were analyzed for the practical case shown in Table 1. In this analysis, the vibration absorber whose mass is 60 t was attached to the buffer only, because of much smaller effect of the absorber attached to the pump module on the vibration of the pipe string (Aso et al., 1992, and 1994a). The masses of the pump-modules 1,2 and buffer were fixed as 50 t, 50 t,and 300 t, respectively. Furthermore, three kinds of locations for the support, which correspond to the three layouts of the pump-modules and buffer, as shown in Table 2, were used in this study. The amplitude of the ship's vertical-motion was fixed as 1 m throughout this study, and the frequency range to be examined was within 0.10 ~ 0.20 Hz from the practical point of view.

The results obtained are shown in Figs. 3~11. Figs. 3 and 4 indicate the amplitude of the buffer and the maximum axial stress

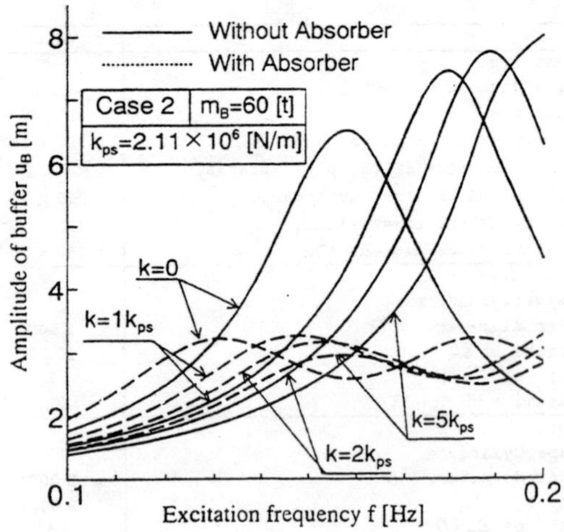

Fig. 3 Frequency response of the buffer attached to the pipe string with the linear-spring support in Case 2

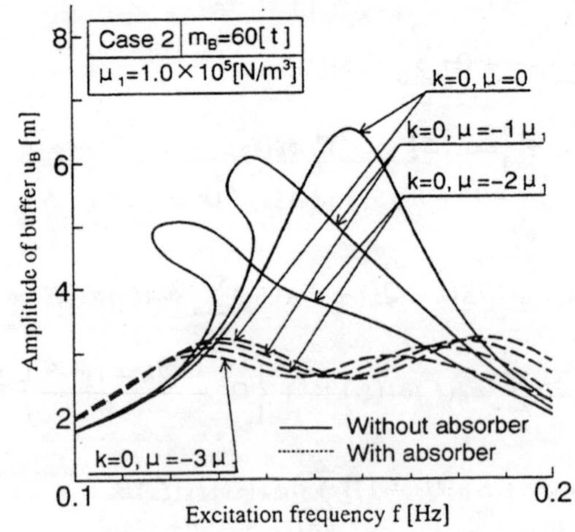

Fig. 5 Frequency response of the buffer attached to the pipe string supported with the soft spring having the nonlinear part only in Case 2

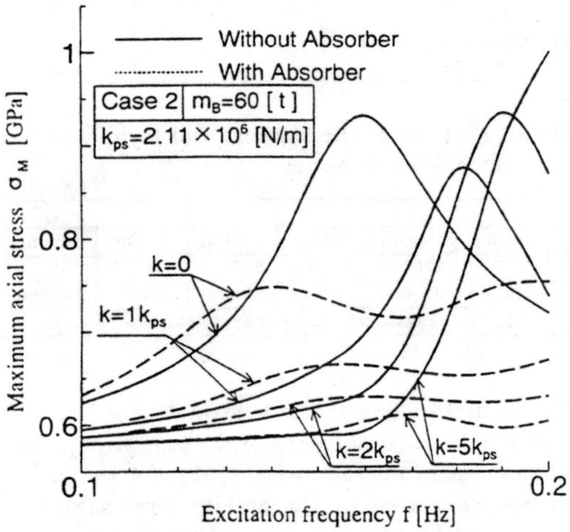

Fig. 4 Maximum axial stress in the pipe string with the linear-spring support in Case 2

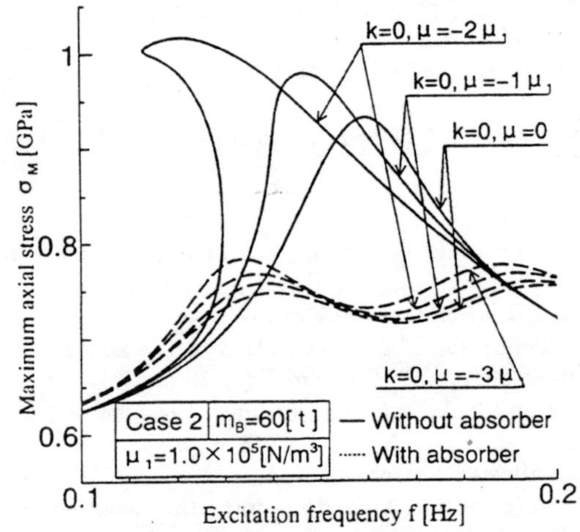

Fig. 6 Maximum axial stress in the pipe string supported with the soft spring having the nonlinear part only in Case 2

of the pipe string supported with the linear spring at the 1.500 m depth of water (Case 2). In these figures, the solid lines show the results of the buffer without any vibration absorbers and the dotted lines do those of the buffer with the optimum absorber which minimizes the resonance amplitude of buffer. Moreover, k_{ps} stands for the spring constant equivalent to the stiffness of the pipe string up to the supporting point from the sea surface. From Fig. 3, it can be seen that the support with the linear spring shifts the resonance frequency of the pipe string to the higher one. and that the support with larger spring constant causes greater amplitude of the buffer without the absorber than that of the string without any supports. This support also causes a little smaller amplitude of the buffer with the optimum absorber, as compared with that of the pipe string without any supports ($k = 0$). The corresponding maximum axial stresses in these cases are shown in Fig. 4, which indicates that the resonance maximum-axial stress produced in the pipe string having the buffer without the absorber first decreases and then increases as the spring constant of the support increases. This phenomenon is attributable to the change of the position where the maximum axial stress occurs. The stress occurs at the top of the pipe string in case of the linear-spring support whose spring constant is up to about $1k_{ps}(k_{ps} = 2.11 \times 10^6$ N/m), while the stress occurs slightly below the position of the support when its spring constant is greater than the value. If the optimum absorber of 60 t is attached to the buffer of the non-supported pipe string, the maximum axial stress can be reduced by about 17 % and further reduced by supporting the pipe string with the linear spring. The maximum stress in the pipe string supported with the linear spring and equipped with the buffer having the optimum absorber decreases as the spring constant k of the support increases. Thus, more than 28 % reduction of the maximum axial stress of the pipe string can be expected if both the buffer with the absorber and the linear-spring support are applied to the string. Therefore, the pipe string supported with a linear spring at the upper part and equipped with the buffer having an absorber could be considered as a quite useful measure for reducing the maximum axial stress in the pipe string.

Next, the effect of the above-mentioned support with the soft spring was examined, and the results are shown in Figs. 5 and 6. Fig. 5 shows the amplitudes of buffers with and without the optimum absorber in cases of various nonlinear supports with soft springs, whose spring constants are zero for the linear part and $-1\mu_1$ ($\mu_1 = 1.0 \times 10^5$ N/m^3), $-2\mu_1$, $-3\mu_1$ for the nonlinear part, respectively. These cases were examined for clarifying the net effect of the soft spring. The cases seem to be fictitious ones, but they may be realized by using a nonlinear elastic joint built in the pipe string. In Fig. 5, the solid lines show the amplitudes of the buffer without the absorber, indicating that the resonance amplitude remarkably decreases as the absolute value,$|\mu|$, of the soft spring increases, as compared with the amplitude of the buffer attached to the string without any supports. Furthermore, the dotted lines show the amplitudes of buffer with the optimum absorber, clarifying that the peak amplitude of buffer a little decreases as $|\mu|$ increases. However, the maximum axial stress in the pipe string with the buffer having no absorber remarkably increases for this case as $|\mu|$ increases, as shown by the solid lines in Fig. 6. Moreover, the dotted lines pertaining to the buffer with the optimum absorber indicate that the peak maximum-axial stress of buffer a little increases as $|\mu|$ increases. Hence, the nonlinear support with a soft spring at the upper part of the pipe string is not advantageous from the viewpoint of reducing the axial stress, though it significantly reduces the resonance amplitude of buffer.

The effect of the nonlinear support with a hard spring at the upper part of the pipe string in Case 2 was also examined, and the results are shown in Figs. 7 and 8. Fig. 7 shows that the nonlinear support with the hard spring at 1,500 m depth of water considerably shifts the resonance frequency toward the higher one and increases the resonance amplitude of the buffer without the absorber. However, the effect on the amplitude of buffer is hardly recognized when the optimum absorber is attached to the buffer. The corresponding maximum axial stresses in the pipe string are shown in Fig. 8, which indicates that the resonance maximum-axial stress pertaining to the buffer without the absorber increases if the positive nonlinear part is added to the linear part of the supporting spring whose spring constant is k_{ps}. However, this effect is very small when the absorber is attached to the buffer.

Furthermore, the effects of the afore-mentioned nonlinear supports at other locations in Cases 1 and 3 on the amplitude of buffer as well as on the maximum axial stress in the pipe string were also examined, and the results obtained indicated the same kinds of phenomena as the above-mentioned ones in Case 2. Consequently, it has been found that the effect of the nonlinear part of the supporting spring on the maximum stress is very small in case of using the buffer with the absorber. Thus, considering the simplicity in the mechanism and handling as well as the low initial and running costs, the linear-spring support applied at the upper part of the pipe string is considered as the most promising measure to reduce the maximum axial stress produced in the pipe string among the above-mentioned three types of supports.

Then, the effects of location of the linear-spring support on the amplitude of buffer as well as on the maximum axial stress in the pipe string were analyzed, and the results are shown in Figs. 9~11. Fig. 9 indicates the amplitudes of buffers with and without the absorber, attached to the pipe string with the support in Case 3. From this figure, it can be seen that the tendencies of variation for the amplitudes of buffers are the same as those for the amplitudes in Case 2 shown in Fig. 3, though the peak

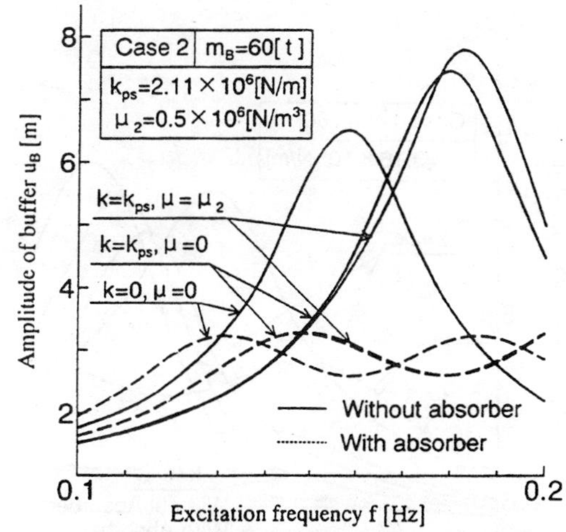

Fig. 7 Frequency response of the buffer attached to the pipe string with the hard-spring support in Case 2

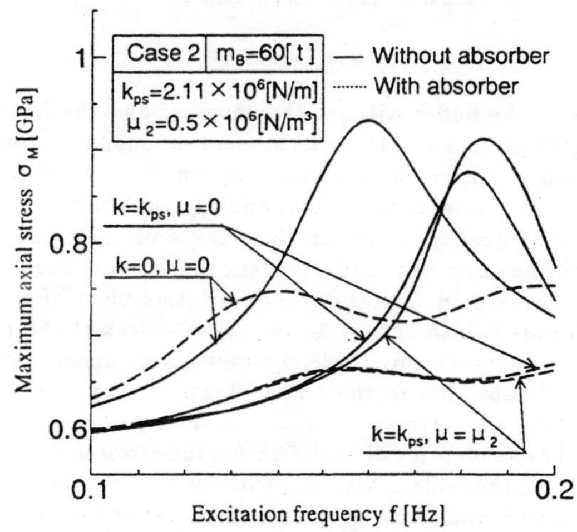

Fig. 8 Maximum axial stress in the pipe string with the hard-spring support in Case 2

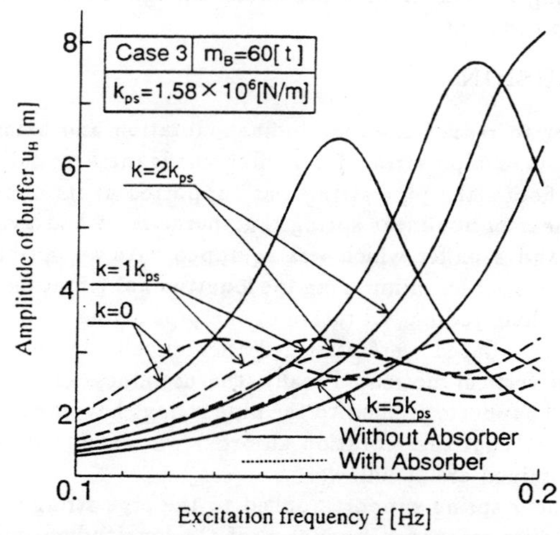

Fig. 9 Frequency response of the buffer attached to the pipe string with the linear-spring support in Case 3

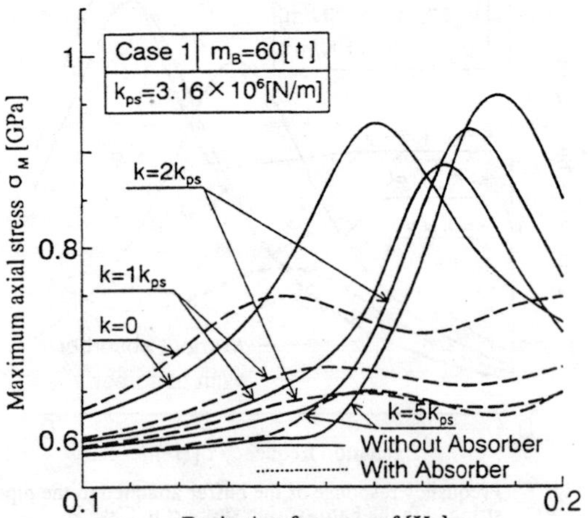

Fig. 10 Maximum axial stress in the pipe string with the linear-spring support in Case 1

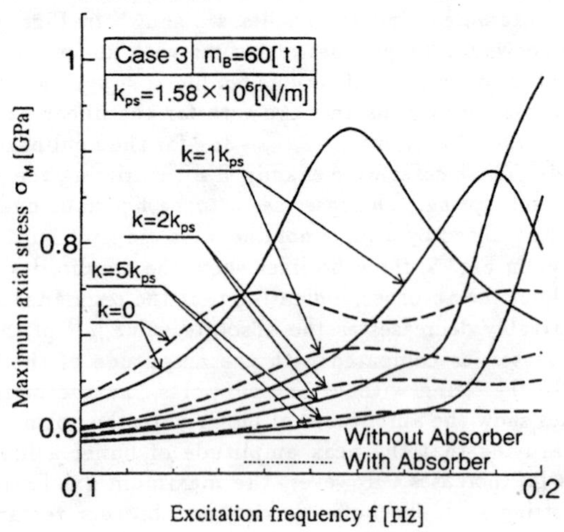

Fig. 11 Maximum axial stress in the pipe string with the linear-spring support in Case 3

amplitude of the buffer without the absorber and the frequency shift are greater in case of lower location of the support. However, the peak amplitude of buffer with the optimum absorber in Case 3 is almost the same as the corresponding amplitude in Case 2, though the frequency, at which the peak amplitude occurs, is different in the cases. The same tendency for the peak amplitude was also observed in Case 1. Therefore, the effect of location of the support, which relates to the layout-effect of the pump-modules, on the peak amplitude of buffer is very small in case of attaching the absorber to the buffer. Figs. 10 and 11 represent the maximum axial stresses produced in the pipe string in Cases 1 and 3. The tendencies of variation for the stresses in the pipe strings having the buffers with and without the absorber in these cases are also found to be the same as those for the stresses in Case 2. However, the peak maximum-axial stress pertaining to the buffer with the optimum absorber is a little smaller in Case 2 than the corresponding stresses in Cases 1 and 3, suggesting that there is an optimum location of the support along the pipe string for reducing the maximum axial stress, though it is a function of the spring constant k.

CONCLUSION

In order to reduce the longitudinal vibration and axial stress produced in a pipe string for mining manganese nodules from deep sea floors, the pipe string was supported at its upper part with a linear or nonlinear spring. Furthermore, it had two pump modules and a buffer which was equipped with an optimum vibration absorber for minimizing the longitudinal vibration. Then, the forced longitudinal vibration of the pipe string as well as its axial stress induced were analyzed theoretically by introducing the ship's vertical-motion. Finally, the usefulness of the above-mentioned supports applied to the pipe strings having the buffers with and without the vibration absorber was examined, and the results obtained are as follows:

1. The linear-spring support applied to the pipe string considerably shifts its resonance frequency of the longitudinal vibration toward the higher value, increasing the resonance amplitude of buffer without any absorbers. However, this effect on the amplitude of buffer becomes very small, when the optimum absorber is attached to the buffer.
2. The resonance maximum-axial stress produced in the pipe string, supported with a linear spring and equipped with the buffer having no absorbers, decreases until the spring constant increases up to an certain value, and then increases as the constant increases beyond the value. On the other hand, the effect of this support on the stress is quite large in case of the pipe string with the buffer having the optimum absorber, reducing the peak maximum-axial stress by more than 28 %, as compared with that of the string without any supports and vibration absorbers.
3. The soft-spring support applied to the pipe string shifts its resonance frequency of the vibration toward the lower one, decreasing the resonance amplitude of buffer significantly with increasing the absolute value of nonlinear-stiffness of the spring in case of the string having the buffer without any absorbers. However, the effect of this support on the resonance amplitude of buffer becomes quite small if the optimum absorber is attached to the buffer.
4. The resonance maximum-axial stress in the string supported with a soft spring remarkably increases when the buffer has no absorbers, and a little increases when the buffer has the optimum absorber as the absolute value of the nonlinear-stiffness of the soft spring increases. Therefore, the soft-spring support is not advantageous for the purpose of this study.
5. The hard-spring support applied to the pipe string shifts the resonance frequency toward the higher one, increasing the resonance amplitude of the buffer without absorbers. However, this effect on the amplitude of buffer is very small as compared with that of the linear-spring support, when the optimum absorber is attached to the buffer.
6. The maximum axial stress of the pipe string supported with a hard spring in case of the buffer without absorbers increases more than the corresponding stress pertaining to the linear-spring support, though this increase is hardly recognized in case of the buffer with the absorber.
7. Consequently, the linear-spring support can be considered as the most promising measure for the purpose of this study among the three types of the supports considered. Furthermore, there

is the optimum location of the linear-spring support along the pipe string for reducing the maximum axial stress, though it is a function of the spring constant.

ACKNOWLEDGMENT

The authors wish to express their deep appreciation and profound gratitude to Mr. Katsushige Kan, Associate Professor of Akita University and to Kohji Narita, Technician of Akita University for their great assistance during this study and the preparation of this paper.

REFERENCES

Aso,K, Kan,K, Doki,H, and Mori,M (1991a). "The Shape-Effect of Buffer on the Longitudinal Vibration of a Pipe String in the Deep Sea," *Int J Offshore and Polar Eng*, Vol 1, No 2, pp 154-160.

Aso,K, Kan,K Doki,H, and Mori,M (1991b). "The Drag and Added-Mass Coefficients of Buffers Vibrating Axially in Water, " *J Energy Resources Tech, ASME*, Vol 113, pp 80-86.

Aso,K, Kan,K, Doki,H, and Iwato,K (1992). "The Effects of Vibration Absorbers on the Longitudinal Vibration of a Pipe String in the Deep Sea–Part 1: In Case of Mining Cobalt Crusts, " *Int J Offshore and Polar Eng*,Vol 2,No 4,pp 309-317.

Aso,K, Kan,K, Doki,H, Ohkoshi,T, and Natori,T (1993). "Vibration-Control of a Pipe String for Mining Manganese Nodules by the Shape of Buffer," Proc 3rd *Int Offshore and Polar Eng Conf*, Singapore, Vol I, pp 143-148.

Aso,K, Kan,K, Doki,H, and Ohkoshi,T (1994a). "The Effects of Vibration Absorbers on the Longitudinal Vibration of a Pipe String in the Deep Sea–Part 2: A Case for Mining Manganese Nodules, " *Int J Offshore and Polar Eng*, Vol 4, No 1, pp 62-67.

Aso,K, Kan,K, Doki,H, and Natori,T (1994b). "A Method for Semi-Active Control of the Longitudinal Vibration of a Pipe String for Mining Manganese Nodules," *Proc 4th Int Offshore and Polar Eng Conf*, Osaka, Japan, Vol I, pp 70-75.

Aso,K, Kan,K, Doki,H, and Natori,T (1995). "The Effects of Vibration Absorbers with Nonlinear Springs on the Longitudinal Vibration of a Pipe String for Mining Manganese Nodules, " *Proc 5th Int Offshore and Polar Eng Conf*, The Hague, The Netherlands, Vol I, pp 303-309.

Aso,K, Kan,K, Doki,H, and Maeda,T (1996). "The Optimum Design of a Stepped Pipe String for Mining Manganese Nodules, with Consideration for Reducing Its Maximum Axial Stress," *Proc 6th Int Offshore and Polar Eng Conf*, Los Angeles, USA, Vol I, pp 29-36.

Cheng,B-R, Chung,J S, and Zheng,Z-C (1995). "Effects of Flexible Joints on the 3-D Nonlinear Coupled Responses of a Long Vertical Pipe, " *Proc 5th Int Offshore a Polar Eng Conf*, The Hague, The Netherlands, Vol II, pp 236-243.

Chung,J S, and Whitney,A K (1981,1983)."Axial Stretching Oscillation of an 18,000-Ft Vertical Pipe in the Ocean," *Proc. Offshore Tech. Conf, Houston*, and *J Energy Resources Tech*, ASME, Vol 105, pp 195-200.

Chung,J S, Cheng,B, and Huttelmaier,H -P (1994a). "Three-Dimensional Coupled Responses of a Vertical Deep-Ocean Pipes: Part I. Excitation at Pipe Ends and External Torsion, " *Int J Offshore and Polar Eng*, Vol 4, No 4, pp 320-330.

Chung,J S, Cheng,B, and Huttelmaier,H -P (1994b). "Three-Dimensional Coupled Responses of a Vertical Deep-Ocean Pipes: Part II. Excitation at Pipe Top and External Torsion, " *Int J Offshore and Polar Eng*, Vol 4, No 4, pp 331-339.

Morison,J R, O'brien, M P, Johnson, J W, and Schaaf, S A (1950). "The Force Exerted by Surface Waves on Piles," *Petro Trans AIME*, Vol 189, pp 149-154.

Proceedings of the Seventh (1997) International Offshore and Polar Engineering Conference
Honolulu, USA, May 25-30, 1997

ISBN 1-880653-28-1 (Set); ISBN 1-880653-29-X (Vol. I)

To the Problem of Application of Ecologically Good Methods of Building and Titanium-Magnetite Sands Mining in the Shelf

Y.N. Niskovskiy and E.V. Niskovskaya
Far-Eastern State Technical University

A.M. Vasianovitch
Corporation "Primorskugol"

Vladivostok, Russia

ABSTRACT

The brief description of titanium-magnetite placers in the shelf of the Far- Eastern Seas is adduced. The appraisal of world and home experience of mining marine deposits is made. A new technological scheme of titanium-magnetite sands mining excluding disturbances of the sea ground is offered.

KEY WORDS: Shelf, placer, titanium-magnetite, sea-floor, technologies.

INTRODUCTION

The Far-Eastern seas shelf is rich in alluvial deposits. Titanium-magnetite iron, vanadium, gold, tin ores and other rare and precious metals are their components.

The strict observance of ecological demands excluding negative influence of mining works on fauna and flora of water area is the main technological and technical and economic principal in working out of technological approach in the design of mining-marine enterprises carrying out mining of marine deposits.

CONDITIONS OF TITANIUM-MAGNETITE SANDS OCCURENCE

Examination of geological and economic study and comparative estimation of 1 metal content in placers of the Far-Eastern and Arctic shelf zones shows that those areas were weakly explored. Metal content of the deposits was determined in the process of search works in the beach and under-water shelf slopes of the Far-Eastern seas. The most prospective titanium-magnetite deposits are situated in Kamchatka, Kuril Islands, in western shelf of the Tartar strait, in Primorskiy territory. Productivity of under-water slope along that ore-bearing beach area of those deposits is different. Deposits of under-water slope propagate from the shore to the depth of 50m. Thickness of deposits varies from 4 to 16m. They are covered with overburden of 2-20m thick.

New large deposit of marine vanadium titanium magnetite placers in the Tartar strait is particularly prospective. It was determined in the process of magmatic formations exploration in the middle part of the Eastern Sikhote-Alin volcanic belt by officials of the Acaemy of Sciences. Beach titanium-magnetite placers in the shelf of the Tartar strait are concentrated not in the beach, usually extremely narrow, but in the floor. Total deposits of the placers retracting along 700km from the cape Tumaniy to the cape Lazarev may be rather rich. In the destruction of 1km 3 of volcanic rocks, containing 2% of titanium-magnetite dissemination 50mln t of that mineral are released. In addition, titanium-magnetite contains high content of vanadium (0,4-0,6%), that considerably increases value of placers. Ore concentrate of placers is easily separated into magnetic and electromagnetic fraction being 50-80% of concentrate mass, electro-magnetic is 20-50% of concentrate. Magnetic fraction of ore concentrates consists of 50-60% of magnetic grains without aggregation of ilmenite and 2-9% of free ilmenite grains. The high content of vanadium penta-oxide (0,4-0,6%) is an important feature of magnetic fraction. Mass portion of gross iron in magnetite fraction differs from 7 to 13%. Thus, magnetic fraction of the Tartar strait placers is rich complex vanadium, iron and titanium ore. Titanium pigment dioxide is the main product for production of expensive titanium paint, which are extremely necessary in aircraft industry, ship-building industry, space industry and other branches of the economy. In addition, according to the investigations of specialists, titanium-magnetic sands may be applied as weighting material for clay solution in drilling for oil, in oil-well cement as a filling agent, in coal mining for heavy suspension in coal dressing, in tin and polymetalic ore concentration, for powder metallurgy it is used as components, additions into the resistant powder materials to manufacture different types of resistant parts of engines and for manufacturing of electrodes for manufacturing for alloying of steel products etc.

It is impossible to carry out development of titanium-magnetic, magnetic placers and placers of other minerals without reliable, effective and ecologically good technologies and technical means, However, one must not think that traditional methods and technical means of under-water mining of mineral deposits of the sea floor such as suction-tube dredge, clamshell dredge etc. can be used for this purpose nowadays without serious damage of the environment.

Production development of those placers demands working out of new technological principles, technical, economic and ecological estimations of titanium-magnetite deposits development efficiency.

NEW TECHNOLOGY OF TITANIUM-MAGNETITE SANDS MINING

World experience of placers mining examination shows that application of different types of dredge and suction dredges results in destruction of the sea floor and fauna and flora, that excludes the reproduction of fishes for many years.

In connection with the above, method of borehole hydraulic mining excluding disturbances of the floor in working out of marine placers should be widely practiced. (Fig. 1).

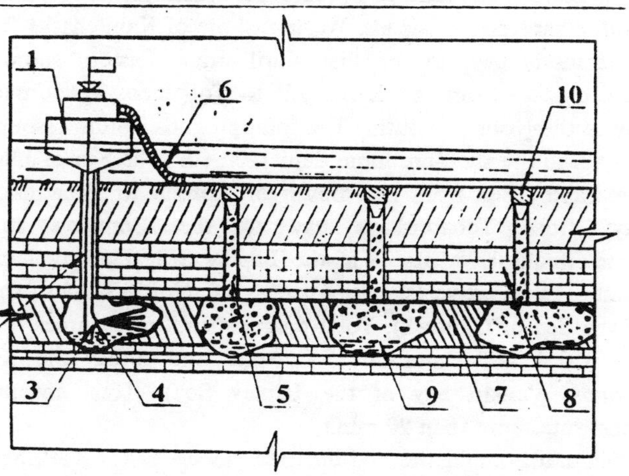

Fig.1. 1 - SGD module; 2- production well; 3 - air lift; 4 - monitor; 5 - stripped well; 6 - pulp pipe; 7 - useful minerals; 8 - roofing; 9 - hydraulic holling; 10 - cement cork

However improvement of chamber roof support reliability if its rocks are not steady enough and water flow in holes is rather high, is still a problem.

Officials of FESTU have offered method of roof stability improvement in mining coal and placers confirmed by author's certificate for the invention. The method is as follows: opening up of productive seam by borehole, placing hydraulic giant and washing out of cavity near the upper limit of the seam is followed by formation of supporting layer from filler in the cavity. The density of the filler is less than that of the water and as a consequence of the above, Archimedian force appears in the filler. The supporting layer is formed from separate pieces of filler not jointing with each other, supplying from the hydraulic giant column. This method provides regulation of the rate of supporting force of the roof by floating layer which gives possibility to increase the supporting area of the roof at any moment of a chamber development, makes the pollution of the extracted mineral to be lower, reduces losses of mineral, excludes disturbances of the sea floor.

The above method of mining with leaving natural ceiling and pillars in borehole hydraulic mining may be applied in working out of not valuable ores and placers to preserve sea fauna for example in extraction of 5 billion tons of titanium-magnetite sands under sea floor in the shelf of the Sea of Okhotsk and the Sea of Japan. In this case drilling of holes will be carried out from floating platform. Pulp, transported along the pipeline to the shore will go through magnetic dressing on the shore. The sand being got after dressing may be used in construction or as a filler or may be transported to the chamber, assisting in production without waste.

Application of the above methods of extraction provides the improvement of ecological conditions of the environment in mining useful minerals under the sea floor.

Thus, in accordance with the «Program of working out of ecologically good technologies for mining, dressing and complex processing of titanium-magnetic sands of the shelf of the Far-Eastern seas for 1995-2000» it is necessary to underline three main tasks:

- carrying out of full cycle of geological prospecting and exploration works, beginning with searching and finishing with estimation of the deposits;
- carrying out of the necessary value of scientific-research, and experimental works in the field of working out of new progressive ecologically safe technologies and technical means of under-water mining and processing of titanium magnetic sands resulting in production of concentrates;
- continuation of research works in complex metallurgical processing of titanium-magnetic raw material (concentrates).

All that will provide to estimate more objectively and precisely not only necessity to attract titanium magnetic deposits of the sea shelf to the economy but to substantiate ae number of technical, economic and social aspects of shelf mineral resources development in the Far East.

REFERENCES

Zhukov, AV, Papulov, VI, Menshikov, AI, Makhmutova, RM (1990). «The main technological, technical and economic principles of titanium-magnetite sands mining in the shelf of the Far Eastern seas,» *Collection of articles.* Complex application of mineral resources of the Far East. Vladivostok, DVO AN USSR. pp 131-134(in Russian).

Niskovskiy, YN, Vasianovitch, AM (1996). «Investigation of possibility to apply untraditional and ecologically good methods of coal mining under sea floor,» *Int.J.Offshore and Polar Engrg.*,Vol.1, pp 51-53(in Russian).

Nikolaichuk, N, Niskovskiy, YN, Zvonarev, M, (1991). «Method of borehole extraction of materials from thick underground formations,» A.C. № 1694903, *bulletin №44.* (in Russian).

Niskovskiy, YN (1990). «Application of modelling methods in mineral mining under reservoirs,» Far Eastern Branch of the Academy of Sciences of the USSR (in Russian).

Proceedings of the Seventh (1997) International Offshore and Polar Engineering Conference
Honolulu, USA, May 25-30, 1997

ISBN 1-880653-28-1 (Set); ISBN 1-880653-29-X (Vol. I)

The Project of Marine Dressing Complex for Mining the Deposits of the Far Eastern and North-Eastern Seas Shelf

Anatoliy V. Zhukov, Vladimir T. Lutsenko and Mikhail I. Zvonarev
Far-Eastern State Technical University
Vladivostok, Russia

ABSTRACT

The brief analysis of the mineral raw material base of continental shelf of the Far-Eastern and North-Eastern seas is given in this paper. The main technological, technical and economic principles of the extraction and complex use of titanium-magnetite and ilmenite mineral raw material are explained, the strict observance of the above principles is necessary in working out of new technologies, regulations, technical conditions and projects of sea mines.

The principle propositions of the project of the marine dressing complex of module type and creation of specialized modules for extraction, dressing and other processing operations of solid useful minerals on the base of application of existing and untraditional methods, used in home and foreign science and practice are stated.

A number of calculated values of the main parameters of the floating base of the marine dressing complex are explained.

KEY WORDS: Mineral resources, shelf, principles of the extraction and processing, marine dressing complex, specialized modules and methods.

SHORT DESCRIPTION OF MINERAL RAW MATERIAL BASE, PROBLEMS AND OBJECTS OF INVESTIGATION

Deposits of different useful minerals from which considerable amount of iron ore, coal, sulfur, rutile, titanium-magnetite, gold, platinum, diamonds, and building materials are produced nowadays have been selected in bedrock and sedimentary rocks forming the ocean bed and sea bed of the Far-Eastern and North-Eastern seas.

The existence of the following ore deposits in the shelf of the Far-Eastern and North-Eastern seas of Russia was determined in prognostication estimation of the deposits of solid useful minerals: gold (bays of Primorye, the Sakhalin Gulf- Shantarskiye Islands, Western shore of Kamtchatka from Penzhinskiy bay, to the first Kuril strait; Eastern shore of Kamtchatka; from the Krest gulf to Toigunen, Tchaunskaya bay to the Kolima mouth.) The total prognostication resources of gold in the shelf are more than 120t; tin (The Sea of Japan, the Laptev Sea from the mouth of Kolima to Tchaunskaya bay; from Tchaunskaya bay to the Beringov strait, Tchekurdakhskaya placer in the Laptev Sea; Dmitriy Laptev strait - Likhovskiye Islands; Anadir gulf; North-Western shore of the Sea of Okhotsk). Total prognostication resources of tin are 350 - 400 thousand t. Ilmenite and rutile (the Kuril Islands, Vankin bay of the Laptev Sea). Total resources constitute more than 20 mln.t

Titanium-magnetite and ilmenite placers of the shelf are of the most interest as far as production is concerned and they are the most profoundly geologically explored. The prospective provinces of titanium-magnetite are in Kamtchatka, the Kuril Islands, western cost of the Tartar strait, in Primorskiy region. According to the available data of JSC «Dalmorgeologiya» the prospected and prognostication reserves of titanium-magnetite placers in the shore and shelf of the Far-Eastern seas are more than 800mln t. And if coastal placers are taken into account, they are more than 1 billion t.

Under strict ecological requirements of the nature preservation bodies in Russia to marine dressing complexes it is impossible to apply foreign mining and dressing equipment for mining of shelf deposits. Ecologically good technologies and technical means applied to certain mine and geological and technical conditions of the placers being mined worked out by us permit from systematic point of view to model different variants of the scheme of extraction and dressing of mineral, to substantiate technological and technical parameters of separate modules and technical equipment of the marine mining and dressing complex.

THE MAIN TECHNOLOGICAL, TECHNICAL AND ECONOMIC PRINCIPLES OF TITANIUM-MAGNETITE AND ILMENITE DEPOSITS MINING

Variety of natural factors, characterizing sea placers in shelf are as follows: origin and geological construction, morphology, physical and mechanical characteristic of enclosing rock, hydrology of the areas of occurrence and lithodynamics of sea bed sedimentation, meteorological conditions and geographical location (how far they are from large industrial centers) and level of ecological condition of the area of water where it is supposed to carry out works of underwater extraction of solid useful minerals - all that result in necessity to create technologies and technical means where all the above and other not less important conditions such as preservation of resources and production without waste are to be taken into consideration.

Industrial development of the shelf deposits demands working out of new technological principles, technical and economic and ecological estimation of the efficiency of the development of titanium-magnetite and ilmenite placers. The main principles in working out of technological schemes, regulations, technical conditions and projects of the marine mines are as follows:

strict observance of ecological demands, excluding negative influence of sea mining on fauna and flora of the areas of water;

complex application of valuable components of titanium-magnetite concentrates including utilization of processing products (working out of technologies of production without waste);

working out of technologies and equipment for extraction and dressing with maximum extraction of useful components from productive sands of ilmenite and titanium-magnetite placers;

working out of highly efficient technology of concentrate dressing to the conditions, stated by requirements of metallurgical processing;

working out and technical and practical basing of technologies and technical means for mining and technical and biological treatment of the sea bed area, disturbed by sea mining;

formation of rock dumps and tailings for wastes of mining providing maximum technical and biological safety on the bed of the shelf and above-water area.

Applying the above principles for the hydrological and meteorological conditions of the most active coastal zone of the shelf of the northern and freezing seas with heavy ice conditions, it is necessary without doubt to create floating means with strengthened hull and mining equipment of high resistance to ice.

Considering problem of creation of sea mining complexes taking into consideration creation of dressing and processing equipment, it should be noted that for dressing of metal-bearing sands of ore placers (tin, gold, ilmenite, titanium magnetite etc.) it is necessary to mount on board of the ship special dressing machines and apparatus, ensuring highly effective extraction of ore and production of concentrate only of certain metal. But if there appears possibility to unify within some limits technical means of extraction of productive sands of different mineral structure, dressing of those sands using the same equipment is practically impossible.

TECHNOLOGICAL STRUCTURE OF MARINE DRESSING COMPLEX

Depending on the type and physical and mechanic characteristic of the extracted mineral raw material , values and level of dressing, structure of the extraction of useful components structure of mining and dressing complex, technical and technological parameters of separate modules may be different, Mining equipment includes cutter and spoil moving element used together with transportation unit of auger conveyor type mounted in pipe housing excluding propagation of suspended solid particles in water environment. Dressing plant depending on certain conditions contains disintegrated and dressing modules. Disintegrated module represents sizing screen - disintegrator, ensuring crumbling of mass to the determined level with simultaneous division of it to classes according to size (e.g. dredge cask). Dressing process is based on the application of gravitational and magnetic separators, that excludes necessity to use chemical substances (flocculants, acids, salts etc.) thus excludes danger of ejection of such substances to the environment together with tailings. Modules of gravitational dressing (e.g. spiral or cone separators) and modules of magnetic dressing - disk magnetic separators with many drums, which, when mounting of the disks from the materials of different intensity of magnetic field permit to separate a wide range of placer material) are located in succession one after another and any of such dressing line works with its class of size of mining mass. Number of dressing lines is determined by definite characteristics of the useful mineral taking into consideration ensuring of optimum level (from the economic point of view) of extraction of the useful component. «Concentrate» output of dressing lines are connected with collecting bin for the concentrate (or collecting bin if output in bulk is accepted), which, in their turn are connected with transportation module by re-loaders. It is seemed to be expedient especially in mining of placers with many components to store finished concentrate to the containers, collected on the container site, that permits to produce concentrate as a wide range of products. Ejection of tailing is ensured by units, the construction of which includes the elements, excluding propagation of the suspended solid particles in the water environment, they are manufactured as pipe housing connected by hinges with corresponding collecting bins for tailing. The length of housing exceeds the depth of the water area of the mining area, and housings are supplied with hoisting equipment.

Technological schemes of mining underwater deposits include complex of measures of localization of the most small and most mobile solid particles at the site of mining, contact of the moving material with water environment while lifting of material of placer on the board and ejection of tailing is excluded for that purpose. The tailings are ejected directly to the bottom of the trench left after extraction of the material of placer in the order, ensuring initial fill of tailing of the smallest size of fractions with its subsequent «burial» under the tailings of more and less mobile fraction.

Depending of the rate of protection of the water area, cross dredging of the floating base is carried out using its own power unit or by pile and rope method. Extraction of the material of the placer is carried out in the following order: spoil moving element is lowered to the bottom and engaged into the operation at the same time as far as the output of placer or the achievement of the level predetermined before according to technical and economic data.

Then spoil moving element is moved along the determined trajectory with simultaneous moving of floating base along the axis of the stope. Then floating base 1 is moved to the next stope, turned and extraction is made to the location of the start of the first stope and it is extracted till the full extraction of the area. Rotating spoil moving element cutters by its cutter 2 the mining mass which by auger conveyor 3 is lifted along the pipe housing 4 on to the board of the floating base 1. Then mining mass is moved to the dredge cask 5, where it is disintegrated and separated to a number of classes according to their size, and planting waste and pebbles are moved to tailings at once but fraction of determined size is dressed. If placer material requires additional dewatering and slime removal the necessary module is included to the mine and dressing complex, e.g. spiral classifier which is located in front of dressing module 6.

Then mining mass is moved to the cone separators. The produced concentrate, containing titanium-magnetite, rutile, ilmenite and other useful components with large specific gravity is moved to the concentrate collecting bins, and tailing of that stage of dressing are moved to the magnetic separators. To pick out weakly magnetic minerals electro-magnetic separation is used. Applying separators with different magnetic field intensity (H), definite minerals with definite characteristics are extracted. For example, when H = 200 кА/м chromate will be extracted to the magnetic fraction of the separation process, with H=380 кА/м - ilmenite, with H=1280кА/м - rutile, etc. If it is necessary to pick out zirkon or sphenium electrical separation is to be used.

Concentrate, after magnetic dressing is moved to the concentrate collecting bins 7. Depending on the determined scheme of the mass of rough concentrate formation, units of gravitational and magnetic dressing are unloaded to the same or different collecting bins. In the first case magnetic and non-magnetic fractions of useful components are mixed; in the second case magnetic and non-magnetic fractions are not mixed.

Tailings of the units of magnetic dressing, collecting in tailings collecting bins are thrown to the bottom of the trench formed by the operation of the spoil moving element 2. For that loose ends (tags) of unloading housings 8 are lowered to the trench 9. Thus, they appeared to be located lower than the level of the reservoir bottom 10. In the operation zone of spoil moving element 2 the trench 9 is left on the bottom of the reservoir 10 , the width of the trench on the top is determined by the angle of natural moving of material, forming its stopes (sides).

Floating base 1 is to be moved so, as housings 8 do not «go out» of the trench 9 in the process of moving of the mining equipment along the stope. Filling of the tailings to the trench is carried out with outlet of the housing 8 lowered directly to the trench to pour small fraction directly to the bottom of the trench 9, and to pour larger fraction to the area with small fraction, larger fraction will form coating, excluding considerable wash-out of small fraction and, consequently, turbidity of the water by current and roughness near the bottom.

Resulting from the mine works on the bottom of the reservoir trenches are left, their depth is considerably less than initial depth of the trenches 9, and there is a layer of particles of gangue rock, its size excludes their transfer to the suspended stage and turbidity is excluded in the zone subjected to the influence of the flow (currents) and roughness.

CHOICE OF THE MAIN PARAMETERS OF FLOATING MEANS FOR THE MARINE MINING AND DRESSING COMPLEX

In calculation of ship parameters the method based on the taking into consideration influence of hydrometeorological conditions on the efficiency of technology of mining was used. Such ship is the a foundation (base) for mounting of the equipment for rock processing, that is why efficiency of its work will depend on the stability, i.e. on the influence of wave disturbance on the process of concentration, gravity and magnetic separation.

Parameter of oscillating motion of the ship and so the value of disturbances to parameters of the processing at equal hydrometeorological conditions depends on the main dimensions of the ship. At the known hydrometeorological conditions of the operation area a ship should ensure maximum quantity of concentrate at minimum cost. Otherwise, the output of concentrate may be increased but its quality will be lower.

The efficiency function of scarce deposits of useful mineral mining will have the form:

$$\sum_{i=0}^{m} S_i - \sum_{i=0}^{m} Z_i \rightarrow max \quad (1)$$

where S_i and Z_i - are profitable and expense parts of the exploitation of the deposit.

The expenses on the ship exploitation depending on its main dimensions are included to the expense part.

The investigations carried out before have shown that dependence of extraction and concentrate output on the angle of the apparatus stope and the angle of the ship heeling will have the form:

$$\varphi_k = \sqrt{0{,}62 + 0{,}04\eta_{из}\left(1 - \gamma_п / \gamma_{max}\right)} - 0{,}79 \quad (2)$$

where $\gamma_п$ - value of production minimum; γ_{max} - value of the content of useful component at the mapping area.

The calculations have shown that operation of apparatus with dressing in thin layer is possible at the angle of oscillating motions up to 5 grades with seas up to 5.0 number.

The ships with two hulls - catamarans having hightened sea-going qualities with sufficient capacity, good stability and floodability and redundant width satisfy most fully the requirements of marine mining and dressing complex.

The authors have carried out searching work to design mining and dressing complex of catamaran type on the base of two hulls of one of the widely practiced projects of fishing vessels.

Architecturally the vessel is built with upper volume joint structure combined with forecastle, with bow position of the superstructure and is separated to functional units, practically isolated from each other: dwelling unit in the bow with full set of rooms for dwelling, hose-hold and for every-day necessities of life, tanks with drinking water and water for washing; dressing equipment with one sector of open deck, enough for mounting of the dressing plant and hold for sumps of pulp and product (located nearer to the midsection for less influence of oscillating motions and on the area of the most full completeness); dressing equipment and engine and boiler-room equipment (mining is carried out from the stern).

Basing on the calculations the ship with the width of the deck of 23.8m and total displacement 2035 t and of KM+Л2 IA2 class - catamaran was designed. Displacement and weight load, principal dimensions in two approximations, characteristics of rolling motion at the windward side and heaving motion for the ship turned broadside on to the wave, propulsive quality of the ship (resistance of water to the ship movement and tow-rope horse power, coefficient of ship's hull and screw interaction, calculation of (screw) propeller), curves of the elements of theoretical drawing, for-and-aft trimming at various loading, boarding and initial stability at different loads, minimum freeboard and jointing structures were calculated for that ship, necessary ship units were selected and the problems of further examination were planned.

CONCLUSIONS

1. Complex application of placers and produced concentrates - rich complex iron ore, titanium ore, vanadium, scandium and rare-earth elements becomes a topical problem in consideration of the development of solid mineral resources of the Far-Eastern and North-Eastern Sea Shelf and production development of ilmenite and titanium-magnetic deposits. In the areas where utilization of waste after dressing of metal-bearing sands as building materials is possible, the efficiency of mining sea placers may considerably increase.
2. Modifying various modules in accordance with technological problems, it is possible to arrange dressing complexes efficiently from them taking into consideration definite mining and geological, hydrologic and exploitation situations and ecological requirements as well.

REFERENCES

Edited by Rzhevskiy, VV (1978). "Technology of mining useful minerals from the bed of the lakes, seas and oceans," Nedra, Moscow. (in Russian)

Zhukov, AV, Papulov, VI, Menschikov, AI (1991). "Creation of technological complexes of module type for sub-water useful minerals extraction from the placers of the Far-Eastern shelf. Intensification of the mining and dressing," DVO AN USSR, Vladivostok, Russia. (in Russian)

Zhukov, AV, Lutsenko, VT, Zvonarev, MI (1995). "Principles of designig of mining and dressing complexes of module type". XXXY Scientific and technical conference of FETU. *Theses of the reports. Economy and management*, Part 2, Vladivostok, Russia. (in Russian)

Lobanov, VA (1983). "Reference book on the technique of the shelf developmen". Shipbuilding, Leningrad. (in Russian)

Bronnikov, AV (1991). "Ship designing". Shipbuilding, Leningrad. (in Russian)

Dubrovskiy, VA (1978). "Multi-hull Ships" Shipbuilding Leningrad. (in Russian)

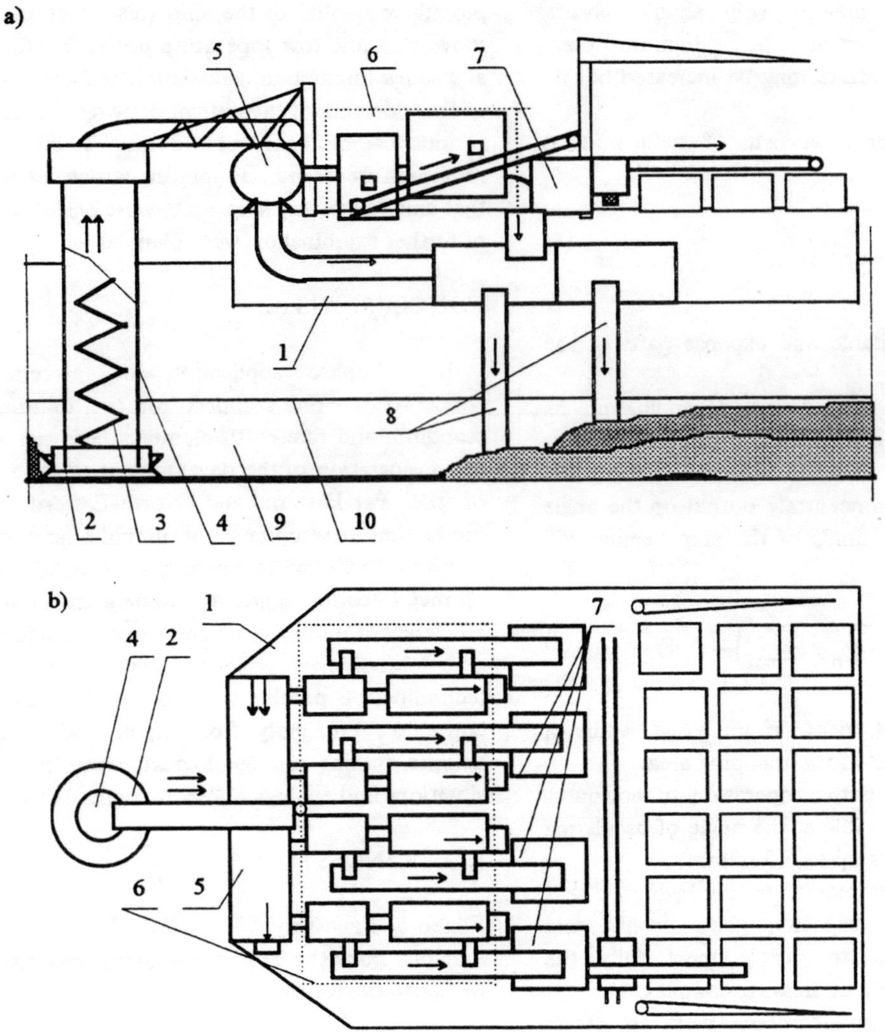

Fig. 1. Diagram of marine dressing complex

a) longitudinal section; b) view from above

1 - floating base; 2- spoil moving element; 3- (auger) screw conveyor; 4 - pipe housing; 5- dredge cask; 6 - dressing module; 7 - collecting bin for concentrate; 8 - unloading housing; 9 - trench; 10 - reservoir (water area)

Proceedings of the Seventh (1997) International Offshore and Polar Engineering Conference
Honolulu, USA, May 25-30, 1997

ISBN 1-880653-28-1 (Set); ISBN 1-880653-29-X (Vol. I)

Mathematical Simulation of Run Operations with Cable-Container System for Deep Ocean Mining of Hard Minerals

Ivan F. Glumov, Russian Committee of Geology, Moscow, Russia
Vadim S. Tikhonov, Central Research Institute of Geological Prospecting for Base and Precious Metals, Moscow, Russia
Alexander I. Safronov, NTC "Energocosmos", Moscow, Russia
Yuri P. Semyonov and Valery E. Lyubinsky, Rocket-Space Corporation "Energy", Kaliningrad, Moscow Region, Russia

ABSTRACT

Development of efficient technologies for deep ocean mining of hard minerals is still an acute problem, though an interest to exploration of ocean fields is not as strong as it was before. This work studied the dynamics of an absolutely new cable-container system (CCS) of hard minerals lifting from the ocean floor so as to validate its capacity and serviceability.

We have obtained 3-D nonlinear dynamic motion equations for the CCS basic elements namely two cables, a container and a platform. These equations form a system of three interconnected nonlinear boundary-value problems. A transformation of independent coordinates allowing to perform finite difference discretization of equations on a grid with a constant step on length was used for numerical approximation of equations of cables as variable length bodies.

The CCS dynamics was studied in a time domain, with the Newmark method used for integration of transient processes. The analysis of container run operations shows that all worries about possible entanglement of cables, which can influence badly a normal use of a system are groundless.

KEY WORDS: Deep ocean mining of hard minerals, cable-container system, run operations, dynamics, cable of a variable length, numerical methods.

INTRODUCTION

In 1990, comparative studies with various deep ocean mining system designs were conducted by the USSR Ministry of Geology in cooperation with the Rocket-Space Corporation "Energy". In the course of the studies the CCS deep ocean mining set was designed and recommended for further constructing (Fig.1). The operation of the above set is described below (Lyubinsky, Semyonov et al, 1993).

The platform 3 being the basic unit of the set is suspended from the mining ship 4 at altitude of 10-15 m from the seafloor by means of an umbilical 1, reinforced by a synthetic fiber. The umbilical, while affording the linkage between the mining ship and the platform (as a simple cable), also provides for power supply, control and transmission of telemetry and TV information. The upper end of the umbilical is connected to the winch

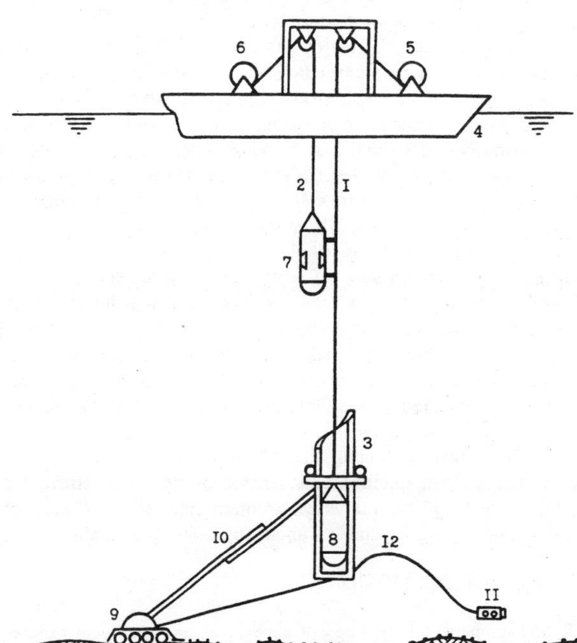

cable length, L_1= 6000 m;
diameter: D_1= 0.16 m, D_2= 0.11 m;

5 which is used for running up/down operations. Water-jet propellers mounted on the platform ensure its manoeuvrability and dynamic positioning. At the same moment, mining ship moves simultaneously along the line of ore collection.

Tracked or wheeled self-propulsion collector 9 collects and washes the ore . All the collected and washed ore is transported by a transportation line 10 to the platform, where it is charged to the transportation container 8 installed on the platform.

Small ROV 11 fastened with the tether 12 and furnished with TV camera and floodlight moves fore of the platform and helps to study seafloor and to identify any obstacles.

At the same time, another transportation container 7 runs down from the mining ship under its own weight suspended by synthetic fiber cable 2. The upper end of cable 2 is fastened on the winch 6, which is used to run down the empty container 7 and run up the container 8 charged with ore. This container 7 slides along the umbilical 1 tightened by a platform weight.

Controlled rudders are installed on container to prevent its spinning about umbilical 1, and water-jet propellers are used for its stabilization at low speed.

Entering in the platform guides, the container 7 detaches from the cable and is placed for charging. Cable 2 is switched from the empty container 7 to a charged one by means of a special rigid link on the lower end of the cable (Lyubinsky, Semenov et al, 1993). The cable is therewith automatically detached from the empty container immediately upon the aforementioned link becomes securely attached to the charged container as the empty one enters the platform guides.

It takes 10-15 min to run empty container 6.0 km down and about 30 min to run charged one up. Onboard the ship the container, which has arrived, is discharged, and another empty container hooked to the cable runs down. Unlike the traditional pipe mining unit, the above system does not break ore, which mitigates the environment pollution and any loss of valuable ore fractions.

In case of storms, lifting of undersea units onboard is not necessary, it is just enough to lift it a bit over seafloor.

All designing carried out in the Rocket-Space Corporation "Energy" proves that construction of a mining set on the basis of the offered CCS system is quite possible in the shortest time.

Run operations take the major part of the CCS operation and are connected with a complex interrelation of two cables, one of them being pulled along the other one with a very high speed, which incurs a risk of cable cross-whip. Numerical simulation, of run operations in ordinary working conditions is carried out. It helps gaining better insight into mechanical specifications of the CCS and modeling the cables interaction.

The result of our study was a 3-D model of the CCS dynamics with due account of geometrical nonlinearity of cables, one of them cable, being considered as a body of variable length, and of a nonlinearity of hydrodynamical characteristics of container. Dynamics study was performed in a time domain and reduced to the solution of three boundary-value problems, equations of the ship's motion, winches, platform and container with taken as boundary conditions. Finite differences were used for numerical approximation of the cable motion equations. Time integration was performed by the Newmark method with the use of iteration techniques.

It is shown that at running speed higher than 8.5 m/s, container looses its longitudinal stability, and tension of the cable relieves up to zero, thus "breaking" the numerical solution (Howell, 1992). In other cases all run operations are stable without any cable cross-whip.

EQUATIONS OF CCS MOTION

By the moment, a large body of experience has been accumulated in the field of numerical simulation of dynamic behavior of long cables and pipes being in common use in ocean technologies. Noteworthy are the reviews concerning the simulation of cable (Choo and Casarella, 1973) and riser (Chakrabarti and Frampton, 1982) dynamics. A great contribution to the studies on the dynamics of an ocean mining pipe was made by Chung et al, 1980-81, 1993-95. In the majority of works in this field, the length of a cable or a pipe is taken to be constant. However, in CCS cable length may be changed by thousands of metres with a velocity of 8-9 m/s; this fact hinders the use of standard methods.

The basic equation of CCS motion are equations of the cable and the umbilical. Equations of ship's motion, a platform, a container and other elements of the CCS determine boundary conditions in equations of cables, connecting all CCS elements together. An influence of cables on the ship's dynamics and of the tether on the platform is omitted. In this case the motion of suspension points of both cables can be considered as predetermined by ship's path and oscillating motions.

Equation of cables motion

We will consider the cables as perfectly flexible, ignoring the case of possible umbilical tension slackening down to zero. Also neglected is the twisting of cables, since these exhibit low torsional rigidity, and there is a possibility to counteract roll moments acting on the platform and the container from the cables with the help of water-jet propellers.

To describe cables 1 and 2 motion, a fixed coordinate system OXYZ with basis (**i, j, k**) is to be introduced. OX axis is directed vertically down. A natural trihedron ($\boldsymbol{\tau}_i, \mathbf{v}_i, \boldsymbol{\beta}_i$), i = 1, 2, is to be connected to every point of the center line of every cable. Hereinafter, index i = 1 is related to umbilical 1, while index i = 2 is related to cable 2. Unit vector $\boldsymbol{\tau}_i$ is to be directed along the tangential to the axis line of cable, while unit vector $\mathbf{v}_i$ is placed within OXY plane.

Arc coordinate s_i that determines center line points position of the cable in deformed state is counted from the lower end. It is assumed that both cables can be stretched according to the Hooke's law:

$$\partial s_i/\partial\sigma_i = 1+T_i^*/(EF)_i \qquad (1)$$

where σ_i is the coordinate of the cable center line points position in undeformed state; T_i^* = effective tension of the cable; $(EF)_i$ = the cable's rigidity in tension.

Then, the equation of forces acting on a cable elements ds_i may be presented as:

$$m_i\partial^2\mathbf{r}_i/\partial t^2 = \partial(T_i^*\boldsymbol{\tau}_i)/\partial\sigma_i\,[1+T_i^*/(EF)_i]+\mathbf{f}_i$$

$$\sigma_1 \in (0,L_c)\cup(L_c+L, L_1),\ \sigma_1 \in (0, L_2) \qquad (2)$$

where $\mathbf{r}_i$ = radius-vector of a center line point of the cable; L_i = cable length in unstreched state; $L_c = L_c(t)$ = distance along σ_1 from point of the platform suspension to the lower support of container 7; L = distance between lower and upper supports of the container (Fig. 2); $\mathbf{f}_i$ = vector of external forces acting per unit of cable's length; m_i = mass per unit of cable's length; t = time.

The below geometric relationships are to be added to Eqs. (2):

$$\partial x_i/\partial s_i = \cos\vartheta_i\cos\varphi_i \qquad (3)$$

$$\partial y_i/\partial s_i = \cos\vartheta_i\sin\varphi_i \qquad (4)$$

$$\partial z_i/\partial s_i = -\sin\vartheta_i \qquad (5)$$

where x_i, y_i, z_i are projections of raduis-vector $\mathbf{r}_i$ on axis OX, OY, OZ; φ_i = angle between unit vector **j** and the normal $\mathbf{v}_i$; ϑ_i = angle between unit vector **k** and the binormal $\boldsymbol{\beta}_i$.

Taken into account as components of forces $\mathbf{f}_i$ are weight of cable in seawater, tangential flow friction and hydrodynamic drag of viscous nature, as well as inertia hydrodynamic force of potential nature. Inertia force due to variations of pressure gradient within seawater moving with acceleration as well as effects due to vortex shedding from the cable are to be neglected.

Taking into account the above assumption, we can write down the expression for force $\mathbf{f}_i$ as:

$$\mathbf{f}_i = w_i\,\mathbf{i}-0.5\rho\pi D_iC_{ti}|\mathbf{v}_i|(\mathbf{v}_i\boldsymbol{\tau}_i)-0.5\rho D_i\{C_{ni}|\mathbf{v}_i-(\mathbf{v}_i\boldsymbol{\tau}_i)\boldsymbol{\tau}_i|+\pi C_{di}|\mathbf{v}_i|[\boldsymbol{\tau}_i\times(\mathbf{v}_i\boldsymbol{\tau}_i)]\}$$
$$-0.25\rho\pi D_i^2\{\boldsymbol{\tau}_i\times\{[\partial/\partial t-(\partial\mathbf{r}_i/\partial t\,\boldsymbol{\tau}_i)\,\partial/\partial s_i]^2\,\mathbf{r}_i\times\boldsymbol{\tau}_i\}\} \qquad (6)$$

where $\mathbf{v}_i = \partial\mathbf{r}_i/\partial t - \mathbf{v}_f$ = vector of overtaking flow, taken with reverse sign; $\mathbf{v}_f$ = water particles velocity vector; w_i = weight per unit of the cable length in water; ρ = sea water density; D_i = cable diameter; C_{ti} = coefficient of tangential flow friction; C_{ni} and C_{di} are coefficients of hydrodynamic drag [when angles between vectors $\mathbf{v}_i$ and $\boldsymbol{\tau}_i$ are smaller than 3° drag component proportional to C_{di} dominates, when the larger angles - a component proportional to C_{ni} is dominant (Kennedy, 1987)].

Components of drag and inertia force in Eq. (6) are written in assumption that longitudinal velocity of cable in water is high enough (Saltanov, 1974; Dowling, 1988). It is evident while pulling the cable 2, but as for umbilical 1 of constant length, convecting term in inertia force expression (6) can be neglected, and coefficient C_{di} can be equal to zero. In this case, the expression for the hydrodynamic force takes on the ordinary form (Chung et al, 1981).

Boundary conditions for umbilical and cable in points $\sigma_1 = L_1$, $L_1 =$ const, and $\sigma_2 = L_2(t)$ could be presented as:

$$\mathbf{r}_1(L_1,t) = \mathbf{r}_{s1} \tag{7}$$

$$\mathbf{r}_2(L_2,t) = \mathbf{r}_{s1} + \Delta\mathbf{r} \tag{8}$$

where $\mathbf{r}_{s1}$ = raduis-vector of umbilical suspension point; $\Delta\mathbf{r}$ = radius-vector of cable suspension point in regard to umbilical suspension point. When examining longitudinal oscillations of umbilical suspended via compensator of heaving, the equation describing a displacement of the upper end of cables along the tangential in (7) and (8) should be to changed (Bugaenko, 1979). Change in cable length with respect to time can be given directly [the first problem of hydrodynamics of a variable length body (Goroshko and Savin, 1971)], and could be derived from equation of the winch motion (the second problem of dynamics of a variable length body):

$$J_w d^2L_2/dt^2 = a[M_w - aT_2^*(L_2,t)] \tag{9}$$

where J_w = reduced inertia moment of rotating of the winch; a = radius of the winch drum; M_w = control moment at the winch shaft.

Boundary conditions at points $s_1 = L_c \cup L_c + L$ and $s_2 = 0$ are derived from equation of the container 7 motion, and at the point $s_1 = 0$ from equation of the platform motion.

Equations of the container motion

Transport container intended for lifting ore onboard the ship is a vessel in the form of a torpedo-shaped body of revolution, consisting of a semisphere facing down, a cylinder and a cone (Fig. 2). In running up, container is stuffed with ore, and when running down - with water.

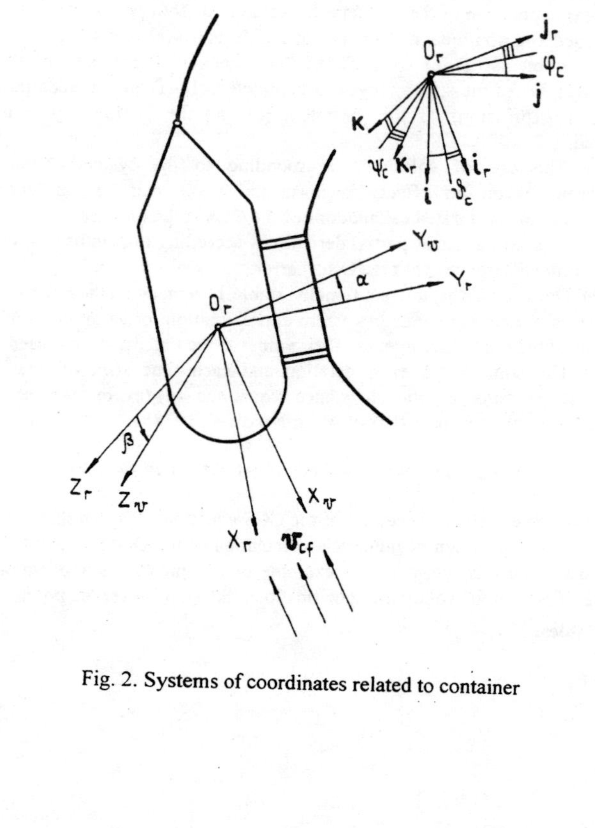

Fig. 2. Systems of coordinates related to container

Container is connected with the umbilical by two supports, and it can move along the umbilical using it as a guide line. Friction at the points of supports contact with umbilical may be neglected. Cable is connected to the top of the cone. Container has small hydrodynamic edges - hydrorudders spaced crosswise at 45^0 with regard to supports' surface, and is provided with water-jet propellers for stabilization of angle positioning of the container with respect to its the longitudinal axis.

Let as assume that the containers center of mass coinsides with the center of buoyancy and is located at the longitudinal axis of container.

To describe container's motion, two coordinate systems $O_rX_rY_rZ_r$ and $O_rX_vY_vZ_v$ with bases ($\mathbf{i}_r$, $\mathbf{j}_r$, $\mathbf{k}_r$) and ($\mathbf{i}_v$, $\mathbf{j}_v$, $\mathbf{k}_v$) correspondingly are to be introduced. Reference point O_r is placed at the center of mass of the container. Axis O_rX_r will be connected with longitudinal axis of container towards to the top of semisphere, and axis O_rY_r will be oriented colinearly to the supports. Velocity axis O_rX_v will have direction opposite to velocity vector of the overtaking flow, and position of axis O_rY_v and axis O_rZ_v with regard to related axes O_rY_r and O_rZ_r will be determined by angles of attack α and slip β.

Equations of forces and of moments are as follows (Pantov et al, 1973):

$$M_c dv_c/dt = \mathbf{F}_{ca} + \mathbf{F}_{ch} + \mathbf{F}_{c1} + \mathbf{F}_{c2} + \mathbf{F}_{cg} \tag{10}$$

$$J_c d\omega_c/dt = \mathbf{M}_{ca} + \mathbf{M}_{ch} + \mathbf{M}_{c1} + \mathbf{M}_{c2} + \mathbf{M}_{cs} \tag{11}$$

where $\mathbf{v}_c$ and ω_c are vectors of linear and angular velocity of the container's center of mass; $\mathbf{M}_c$ = container mass; $\mathbf{J}_c = \mathrm{diag}(J_x, J_y, J_y)$; J_x and J_y are container inertia moment about axes O_rX_r and O_rY_r (centrifugal inertia moment J_{xy} is to be neglected); $\mathbf{F}_{ca}$ = vector of inertia hydrodynamic force ($\mathbf{M}_{ca}$ = moment) of potential nature; $\mathbf{F}_{ch}$ = vector of hydrodynamic force ($\mathbf{M}_{ch}$ = moment) of viscous nature; $\mathbf{F}_{ci}$, i = 1, 2, are vectors of forces ($\mathbf{M}_{ch}$ = moments) acting on the container from the umbilical 1 and the cable 2; $\mathbf{F}_{cg} = G_c\mathbf{i}$ = vector of the container weight in water G_c; $\mathbf{M}_{cs} = M_{cs}\mathbf{i}$ = vector of stabilizing moment of water-jet propellers M_{cs} installed on container.

Vectors of force $\mathbf{F}_{ca}$ and of moment $\mathbf{M}_{ca}$ are derived from added masses of container λ_{jk} within $O_rX_rY_rZ_r$ coordinate system using known formulae [Korotkin, 1986, formulae (1.43)-(1.48)].

Dependences force $\mathbf{F}_{ch}$ and moment $\mathbf{M}_{ch}$ on positioning and damping components within a wide range of pitch angles $0 \le \alpha \le 180^0$ and slip $0 \le \beta \le 90^0$ are rather complicated. To simplify it we'll restrict us to just those dependences having prevailing influence on the dynamics of container (Fedyaevski, 1974). Let the positioning components be presented as 2-3 terms of Fourier-series expansion, and damping components - as a linear function of angular velocities (Ageev et al, 1981).

At the points of contact umbilical/container supports a couple of forces acts on the container from the umbilical. To simplify it, let's assume that umbilical tensions in these points are equal each other:

$$T_1^*(L_c,t) = T_1^*(L_c+L,t)$$

Then vectors of force $\mathbf{F}_{c1}$ and of moment $\mathbf{M}_{c1}$ can be presented as:

$$\mathbf{F}_{c1} = T_1^*(L_c,t)[\boldsymbol{\tau}_1(L_c+L,t) - \boldsymbol{\tau}_1(L_c,t)]$$

$$\mathbf{M}_{c1} = T_1^*(L_c,t)\{\mathbf{r}_u \times [\boldsymbol{\tau}_1(L_c+L,t)+\mathbf{i}_r] - \mathbf{r}_d \times [\boldsymbol{\tau}_1(L_c,t)+\mathbf{i}_r]\}$$

where $\mathbf{r}_u$ and $\mathbf{r}_d$ are radius-vectors of the upper and the lower points of contact of umbilical with respect to center of mass of the container.

Vectors of force $\mathbf{F}_{c2}$ and of moment $\mathbf{M}_{c2}$ acting on the container from the cable:

$$\mathbf{F}_{c2} = T_2^*(0,t)\boldsymbol{\tau}_2(0,t)$$

$$\mathbf{M}_{c2} = T_2^*(0,t)[\mathbf{r}_t \times \boldsymbol{\tau}_2(0,t)]$$

where $\mathbf{r}_t$ = radius-vector of the point of fastening of the cable with respect to center of mass of the container.

Geometrical relationships determining position of the contact points of cables with respect to center of mass of the container are to be added to equations of forces and of moments (10), (11):

$$\mathbf{r}_1(L_c,t) = \mathbf{r}_c(t)+\mathbf{r}_d \quad (12)$$

$$\mathbf{r}_1(L_c+L,t) = \mathbf{r}_c(t)+\mathbf{r}_u \quad (13)$$

$$\mathbf{r}_2(0,t) = \mathbf{r}_c(t)+\mathbf{r}_t \quad (14)$$

Equations of the platform motion

As a heavy platform we'll consider a circular cylinder-shaped body. By spacing the point of suspension and the center of mass of the platform at enough distance, as well as by the platform weight its axis remains practically vertical, any time. Then, excluding equations of angular motion out of consideration, we can write equation of force acting on the platform as:

$$d[\mathbf{M}_p{}^*\partial\mathbf{r}_1(0,t)/\partial t]/dt = \mathbf{F}_{ph}+\mathbf{F}_{p1}+\mathbf{F}_{pg}+\mathbf{F}_{pt}+\mathbf{F}_{ps} \quad (15)$$

where $\mathbf{M}_p{}^*$ = diag ($M_{px}{}^*$, $M_{py}{}^*$, $M_{py}{}^*$); $M_{px}{}^*$ and $M_{py}{}^*$ are effective masses of the platform in direction of longitudinal and transverse axes; $\mathbf{F}_{ph}$ = vector of hydrodynamic force of viscous nature; $\mathbf{F}_{p1} = -T_1{}^*(0,t)\boldsymbol{\tau}_1(0,t)$ = vector of force acting on the platform from the umbilical; $\mathbf{F}_{pg} = \mathbf{G}_p\mathbf{i}$ = vector of platform weight in water Gp; $\mathbf{F}_{pt}$ = vector of force acting on the platform from the transporter; $\mathbf{F}_{ps}$ = vector of control moment of the water-jet propellers mounted on the platform.

Thus, a system of three boundary-value problems determined within segments $[0, L_1]$, $[0, L_c]$ and $[L_c+L, L_2]$ is obtained that describes dynamics of the 3-D motion of CCS.

METHOD OF CCS EQUATIONS SOLUTION

A derived system of equations of CCS was solved numerically. Let's underline some key points of numerical scheme construction.

(1) Iteration-by-iteration splitting of the each cable motion equations into equations of longitudinal and transverse modes, describing a motion of points of cables' center line along the vertical and horizontal plane respectively (Tikhonov et al, 1995). The above splitting is justified, as the orientation of center line of both cables is close to the vertical and it is hardly possible that the slope of a tangent line relative to the vertical is larger than 45^0.

At each iteration step a system of equations of the CCS was split into 6 quasi-independent systems, which were solved in the following order:

(a) Boundary-value problem of the second order (2)-(3) with respect to x_2 with boundary conditions $x_2(0,t)$ (10), $x_2(L_2,t)$ (8) together with equation of winch motion (9). As a result of solution, the $x_2(\sigma_2,t)$, $T_2{}^*(\sigma_2,t)$ and the $L_2(t)$ variables and

$$x_1(L_c,t) = x_2(0,t) - x_f + x_d$$

are determined.

(b) Boundary-value problem of the second order (2)-(3) with respect to x_1, with boundary conditions $x_1(0,t)$ (15), $x_1(L_1,t)$ (7). As a result of the solution, the $x_1(\sigma_1,t)$ and the $T_1{}^*(\sigma_1,t)$ variables are determined. Then the equation,

$$\partial\sigma_1/\partial x_1 = [1-T_1{}^*/(EF)_1]/\cos\varphi_1\cos\vartheta_1$$

obtained by a transformation of (1) and (3), is integrated on the segment $[x_1(0,t), x_1(L_c,t)]$ and $L_c = \sigma_1[x_1(L_c,t)]$ is determined.

(c) Boundary-value problem of the fourth order (2)-(3) with respect to variables y_2, z_2, $\sigma_2 \in [0, L_2]$, with boundary conditions $y_2(L_2, t)$, $z_2(L_2, t)$ (8), the solution of which is recorded as a function of the unknowns $y_2(0, t)$, $z_2(0, t)$.

(d) Boundary-value problem of the fourth order (2)-(3) with respect to variables y_1, z_1, $\sigma_1 \in [L_c+L, L_1]$, with boundary conditions $y_1(L_1, t)$, $z_1(L_1, t)$ (7), the solution of which is recorded as a function of the unknowns $y_1(L_c+L, t)$, $z_1(L_c+L, t)$.

(e) Boundary-value problem of the fourth order (2)-(3) with respect to variables y_1, z_1, $\sigma_1 \in [0, L_c]$, with boundary conditions $y_1(0, t)$, $z_1(0, t)$ (15), the solution of which is recorded as a function of the unknowns $y_1(L_c, t)$, $z_1(L_c, t)$.

(f) The system of eleven Eqs. (10)-(14), connected the variables: $y_1(L_c, t)$, $z_1(L_c, t)$, $y_1(L_c + L, t)$, $z_1(L_c + L, t)$, $y_2(0, t)$, $z_2(0, t)$, y_c, z_c, angles of container pitch φ_c, yaw ϑ_c and roll ψ_c.

(2) A use of new variables (Tikhonov et al, 1995):

$$\xi_1 = \sigma_1/L_c(t)$$

for interval $\sigma_1 \in [0, L_c(t)]$,

$$\xi_2 = -(\sigma_1 - L_1 + L)/[L_1 - L - L_c(t)]$$

for interval $\sigma_1 \in [L_c(t) + L, L_1]$ and

$$\xi_3 = \sigma_2/L_2(t)$$

for interval $\sigma_2 \in [0, L_2(t)]$.

It allows for integration of all boudary-value problems described in item (1) in the ξ_j intervals, j = 1-3, with the constant step on length h_j.

(3) The quasi-linearization of nonlinear equations. The expressions which can profoundly affect the convergence of iteration process were linearized with use of the Newton iterations. Such nonlinearities are: the expressions which determine a dependence of hydrodynamic forces and moments of viscous nature acting on cables, container and platform, trigonometric functions of container roll ψ_c and the expressions, including variable $L_2(t)$ in the boundary-value problem (a). Simple iterations were used for linearization of other nonlinearities.

The solution of the quasi-linearized problem (a) is included into nonlinear expression of the problem (b), solution of both problems (a), (b) is included into nonlinear expressions of problems (c)-(e), and solutions of all five problems (a)-(e) are included into nonlinear expressions of the system (f). When the solution cycle is completed, (b)-(f) are included into nonlinear expressions of the problem (a) and the iterative cycle is repeated.

(4) The use of solution corresponding to the system's static equilibrium when no effects in horizontal plane exist, as an zero approximation for iterated calculations of the CCS at the moment t = 0.

(5) Approximation of partial derivatives according to lengths ξ_j with use of finite differences with the $O(h_j^2)$ error.

(6) The use of the unconditionally stable Newmark method for the integration of grid equations. As a zero approximation for variables at the moment of $t+\Delta t$, Δt -time interval, their values at the t moment were used.

(7) The solution, at each iteration and each time step, of quasi-linearized systems of the difference equations (a)-(e) by the non-monotonic method of pass (Samarsky & Nikolaev, 1978).

CALCULATION AND ANALYSIS OF TRANSIENT PROCESSES

The above described method of the CCS equatious solution provided a basis for the program of numerical calculation of the DOMC3 transient processes. You can see below an example of this program's application for the CCS with the following specifications (the ship is over the point).

Cables:

mass per unit length: $m_1 = 27.3$ kg/m, $m_2 = 11.6$ kg/m;
rigidity in tension: $(EF)_1 = 10.11 \cdot 10^8$ N, $(EF)_2 = 4.77 \cdot 10^8$ N;
flow friction coefficient: $C_{t1} = 0$, $C_{t2} = 0.003$;
drag coefficients: $C_{n1} = 1.8$, $C_{d1} = 0$, $C_{n2} = 1.3$, $C_{d2} = 0.003$;
radius-vector of the umbilical suspension point; $\mathbf{r}_{s1} = 1.32\ \mathbf{k}$ m;
radius-vector of the cable suspension point, $\Delta\mathbf{r} = -2.64\ \mathbf{k}$ m.

Cable 2 winch:
reduced inertia moment, $J_w = 6.37$ t·m²;
drum radius, a = 2.4 m.

Container:
mass at run down/up operations, $M_c = 94/129$ t;
moment of inertia at run down/up operations: $J_x = 117/161$ t·m², $J_y = 564.6/771$ t·m²;
coefficients of added masses: $\lambda_{11} = 8.2$ t, $\lambda_{22} = 56$ t, $\lambda_{44} = 13.8$ t·m², $\lambda_{66} = 126.9$ t·m², $\lambda_{26} = 0$;
expansion coefficients of drag characteristic in Fourier-series expansion with respect to cosine of angle of attack or slip $[p_i \cos(\alpha \cap \beta)^i]$: $p_0 = p_2 = 0$, $p_1 = 0.2415/0.2634$;
expansion coefficients of pitching moment characteristic in Fourier-series expansion with respect to angle of attack or slip $[q_i \sin(\alpha \cap \beta)^i]$: $q_0 = q_1 = 0$, $q_2 = 0.208/0.192$;
derivatives of coefficient of lift force with respect to angle of attack and pitching velocity: $C_y^{\alpha} = 1.119/1.119$ and $C_y^{\varphi_z} = 0.416/0.383$;
derivative of coefficient of rolling moment with respect to rolling velocity, $m_x^{\varphi_x} = 0.091$;
derivative of coefficient of pitching moment with respect to pitching velocity, $m_z^{\varphi_z} = 0.431/0.425$;
radius-vectors of the upper and the lower contact points between supports and umbilical: $\mathbf{r}_u = -2.5\mathbf{i}_r + 2.64\mathbf{j}_r$ m, $\mathbf{r}_d = 2.45\mathbf{i}_r + 2.64\mathbf{j}_r$ m;
radius-vector of the cable fastening point, $\mathbf{r}_t = -6.42\mathbf{i}_r$ m;
stabilizing moment of water-jets, $M_{cs} = 0$.

Platform:
diameter, $D_p = 8.5$ m;
height, $L_p = 18.5$ m;
effective masses at container run down/up operations: $M^*_{px} = 250/215$ t, $M^*_{py} = 1025/990$ t;
drag coefficients in direction of longitudinal and transverse axes: $C_{px} = 0.1$, $C_{py} = 1.7$;
vector of force, acting from transporter: $\mathbf{F}_{pt} = -23$ kN $\mathbf{i}$.

Distribution in depth of current velocity vector components in the horizontal plane V_{fy} and V_{fz} are shown in Fig.3 ($V_{fx} = 0$).

Fig.4 shows transient processes of container run down velocity $V_{cx} = dr_c/dt \cdot \mathbf{i}$, pitch angle φ_c and tension $T_2^*(0,t)$ at different values of controlling moment in the winch shaft $\Delta M_w = M_w - M_{w0}$, where $\mathbf{r}_c$ = radius-vector of center of mass of the container; M_{w0} = winch moment in statics.

At $\Delta M_w = 100$ kN·m, firstly the container run down velocity increases up to 6.8 m/s, and then it slowly decreases with an increase in the cable length and of hydrodynamic frictional force acting on the cable. At ship's heaving of 1.5 m amplitude and of 0.75 1/s frequency, the transient process $V_{cx}(t)$ is superimposed by oscillations at heaving amplitude of no more than 0.35 m/s.

At increase of ΔM_w up to 150 kN·m, the character of transient processes changes drastically. Firstly, a container quickly accelerates to 8.6 m/s velocity, however, then a change of the $\varphi_c(t)$ pitch takes a form of oscillations with sharply increasing amplitude. As a result, the angle of attack increases up to 55-60°, which leads to an increase in container drag. The run down velocity sharply falls and cable tension is becoming a zero one. To avoid any instability of numerical solution at cable zero tension, it is necessary either to account for its flexural rigidity (Howell, 1992) or to exclude its acting on container from consideration having defined $T_2^*(0,t) = 0$ (Tikhonov and Safronov, 1995). However, in any case, cable slackening will be followed by a sharp jerk, fraught with possible cable rupture.

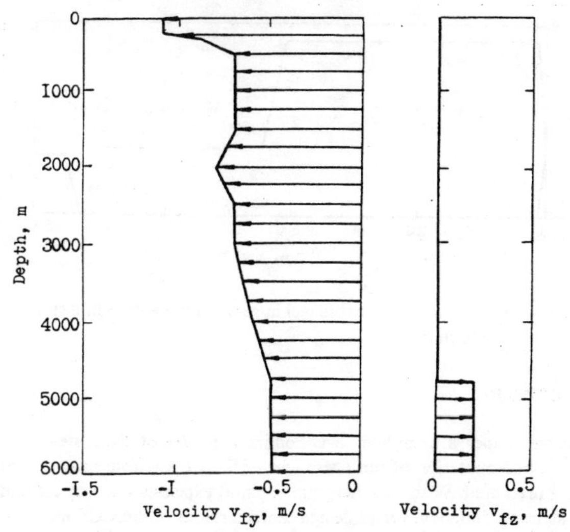

Fig. 3. Distribution of components of a current velocity vector in depth

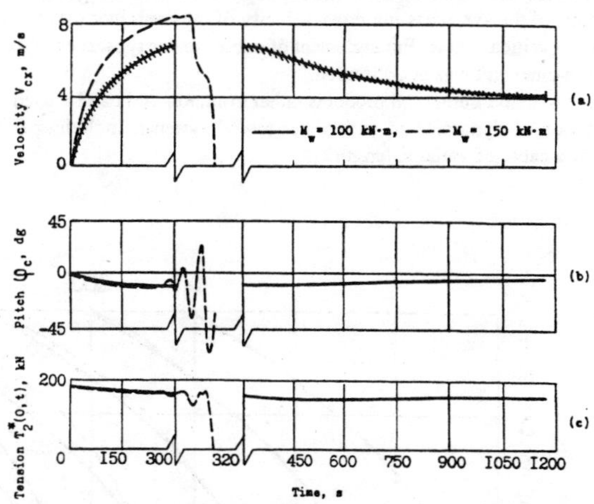

Fig. 4. Variations of container run down velocity (a), pitch angle (b) and tension of cable T_2 (0,t) (c) with reference to time by different values of the winch control moment (area of run down velocity variations in case of the ship's heaving is shaded)

When analyzing a character of pitch changes (Fig. 4b), we can suppose that at rather high run down velocity a loss of longitudinal static stability of container occurs. The V_{cx}^{cr}, critical velocity, can be assessed by the following formula (Tikhonov and Safronov, 1995):

$$V_{cx}^{cr} = \{[T_1^*(L_c,t)L + T_2^*(0,t)x_f]/[0.5\rho V(q_1 + 2q_2) + \lambda_{22} - \lambda_{11}]\}^{1/2}$$

To eliminate exceeding the velocity V_{cx}^{cr}, it is necessary to control an angular velocity of winch drum rather than the moment of the ΔM_w winch. Fig. 5 shows laws of change of the cable reel in/reel out velocity dL_2/dt, which were used for study of container run processes.

Cables configuration for different time moments in projections of the OXY and the OXZ planes, when running container down and up, are shown in Figs. 6 and 7. All transient processes connected with container pitch, yaw and roll are shown in Figs. 8 and 9. It is well seen, that no anomalies at container run are observed. Most of the time, cables are separated and they can intercross only once at the ψ_c roll change by 180° (Figs. 8, 9). In this case, the cross-whip of cables 1 and 2 is impossible.

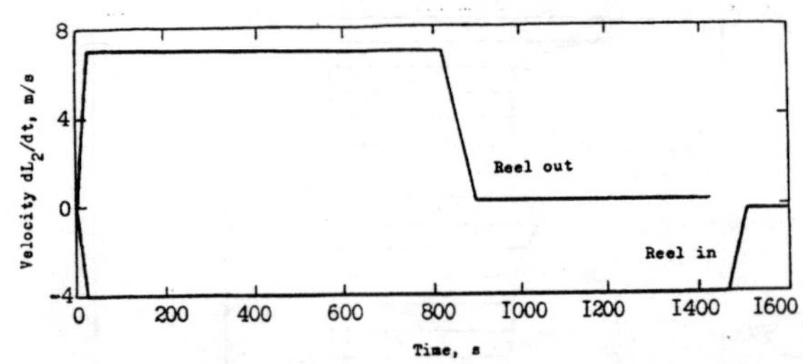

Fig. 5. Variations of the cable reel in/reel out velocity while run containers

CONCLUSIONS

In this paper a complete 3-D nonlinear model of dynamics and a method of computation of run operations of the CCS within time domain area are presented. With their help and minimal expenses one can estimate the system's efficiency, verify design characteristics of the CCS elements, predict probable entanglement of cables, model evaluation of transient process under an extreme situation, etc.

Numerical examples discussed in the paper show that one of the main parameters of the system is run down velocity of the container that has not to exceed a critical value. Entanglement of cables under typical conditions of the container run was not observed.

The discussed numerical model is rather common and can be used for computation of dynamics of underwater towed systems, including those towed by a cable of variable length.

This paper doesn't try to do a comprehensive study of the CCS dynamics: operations of deployment of underwater equipment, collection of ore from the seabed, etc. were not discussed. In models of cables, no flexural and torsion rigidity were taken into consideration, and the model of hydrodynamic forces and moments acting on the container could be more complete. Nevertheless, in any case, the model and numerical method of estimation described in this paper can be a basis for solving more complicated problems of the CCS dynamics.

REFERENCES

Ageev, MD, Kasatkin,BA, Kiselyov, LV, et al (1981). *Unmanned Free Sumbersibles* (in Russian),"Sudostroenie", Leningrad, pp 88 - 96.

Bugaenko, BA, (1979). *Ship Launching/Hoisting Equipment: Constructing and Designing* (in Russian), P1, NKI, Nikolaev.

Chacrabarti, SK, and Frampton, RE (1982). "Review of Riser Analysis Techniques", *Appl Ocean Res,* Vol 4, No 2, pp 73-90.

Choo, Y, and Casarella, MJ (1973). "Survey of Analytical Methods for Dynamic Simulation of Cable-Body Systems", *J Hydronautics,* Vol 7, No 4, pp 137-144.

Chung, JS, Whitney, AK, and Loden, WA (1981). "Nonlinear Transient Motion of Deep Ocean Mining Pipe, *J Energy Resources Tech,* Vol 103, 1981, pp 2-10.

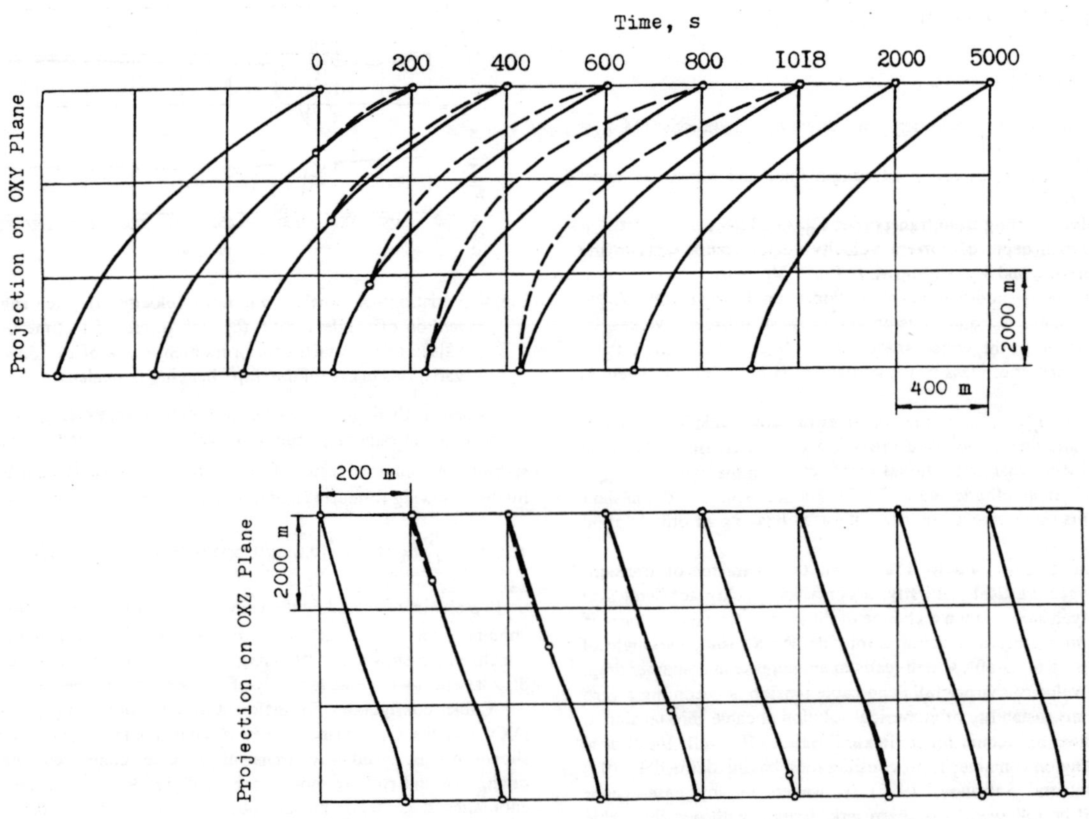

Fig. 6. Cable configurations when container is running down for different time moments

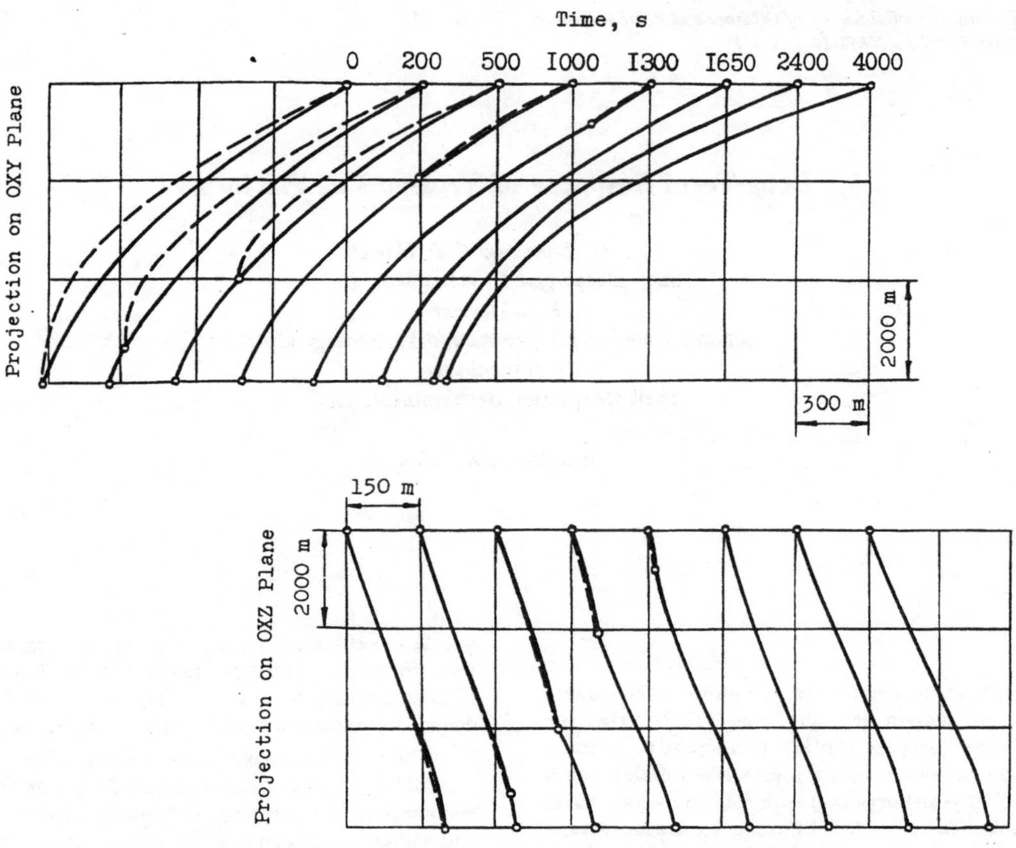

Fig. 7. Cable configurations when container is running up for different time moments

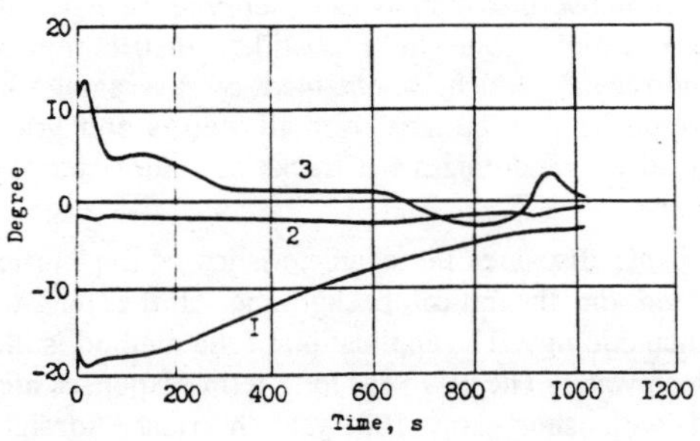

Fig. 8. Variations angles of pitch (1), yaw (2) and roll (3) when container is running down

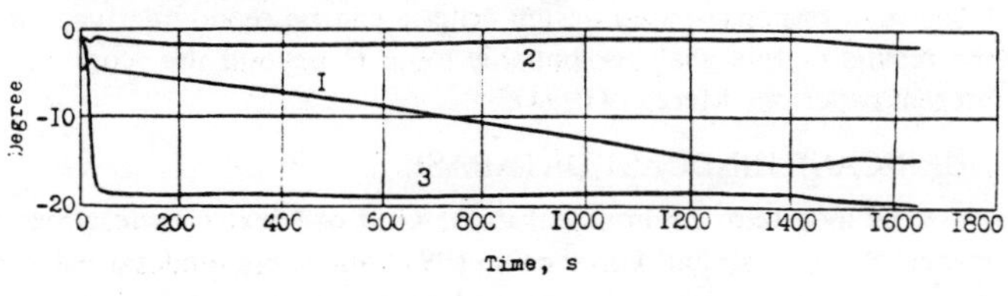

Fig. 9. Variations angles of pitch (1), yaw (2) and roll (3) when container is running up

Dowling, AP (1988). "The Dynamics of Towed Flexible Cylinders: Part1. Neutrally Buoyant Element", *J Fluid Mech,* Vol 187, pp 507-532.

Fedyaevski,KK (1974). *Selected Transacions* (in Russian), "Sudostroenie", Leningrad.

Goroshko, OA, and Savin, GN (1971). *Introduction in 1-D Bodies with Variable Length* (in Russian), "Naukova Dumka", Kiev, 224p.

Howell, CT (1992). "Investigation of Large Amplitude Nonlinear Dynamics of Hangins Chains", *Proc 2nd Int Offshore and Polar Eng Conf,* ISOPE, San-Francisco, Vol 2, pp 316-320.

Kennedy, RM (1987). "Forced Vibration of a Thin Flexible Cylinder in Viscous Flow", *J Sound Vib,* Vol 115, No 2, pp 189-201.

Korotkin, AI (1986). *Added Masses of Ship: Handbook* (in Russian), "Sudostroenie", Leningdrad, pp 8-17.

Lyubinsky, VE, Semyonov, YuP, Gorshkov, LA, Glumov, IF, et al (1993). *Underwater Mining System,* Russian Patent No 20011276C1.

Pantov,EN, Makhin, NN, and Sheremetov, BB (1973). *Theory Principles of Underwater Vehicle Motion* (in Russian), "Sudostroenie", Leningdrad, 216 p.

Saltanov, NV (1974). *Flexible Strings in Flows* (in Russian), "Naukova Dumka", Kiev, pp 75-78.

Samarsky, AA, and Nikolaev, ES (1978). *Methods of Solution of Difference Equations* (in Russian), "Nauka", Moscow, pp 84-103.

Tikhonov, VS, and Safronov, AI (1995). Dynamics of Descent-and-Hoist Well-streamlined Body on String with Variable Length in Fluid (in Russian), *Hydromekhanika,* "Naukova Dumka", Kiev (in Print).

Tikhonov, VS, Safronov, AI, Kamyshev, MA, and Figarov, NG (1995). "Numerical Analysis of Pipeline Dynamics in Seabed Laying", *Proc 5th Int Offshore and Polar Eng Conf,* ISOPE, The Hague, Vol 2, pp 53-62.

Proceedings of the Seventh (1997) International Offshore and Polar Engineering Conference
Honolulu, USA, May 25-30, 1997

ISBN 1-880653-28-1 (Set); ISBN 1-880653-29-X (Vol. I)

Long-Term Response of Tension Leg Platforms

J.J.M. Baar and C.T. Howell
Shell Deepwater Development, Inc.
R.S. Mercier
Shell Exploration & Production Technology Co.
G. Rodenbusch
Shell Deepwater Development, Inc.

Houston, TX, USA

ABSTRACT

The current state-of-the-practice in the global performance design verification of tension leg platforms (TLP) relies on conventional short-term response analysis techniques to estimate the extreme platform responses during a prescribed design storm environment (e.g. a 100 year hurricane) with collinear wind, waves and current conditions. However, the TLP responses are sensitive to non-aligned wind, wave and current, which may result in significant variations in extreme airgap (minimum under deck wave clearance) and extreme tendon tensions. In the Gulf of Mexico such conditions can occur during the passage of the eye of a hurricane. Their effect on the global performance of TLP's can be evaluated by investigating the long-term response characteristics.

The paper discusses a simple and efficient long-term response analysis method, which can be used to derive consistent design criteria and can be readily incorporated in the design process. The method is applied to TLP in deep water. The 100 year long-term extreme responses are compared with the short-term 100 year hurricane design estimates. Moreover, for a given storm at a particular site, the temporal variations of extreme airgap and tendon tension are explored to provide enhanced insight into the sensitivity of TLP responses to varying wind, waves and current.

KEY WORDS

Tension leg platform, Long-Term Response, Gulf of Mexico, Design Criteria

INTRODUCTION

Some extreme responses (e.g. minimum airgap and tendon tension) of tension leg platforms are sensitive to non-aligned wind, waves and current conditions as discussed, for example, by Schott *et al* (1994). The global response design verification often relies on conventional short-term response analysis techniques to estimate the extreme platform responses in an extreme design storm event (e.g. 100 year hurricane) with *collinear* wind, waves and current, see API (1987). In the Gulf of Mexico such non-collinear conditions are often associated with the passage of a hurricane. Their combined effect on global platform responses may be assessed by means of a proper long-term response analysis. The objective of this paper is to present a simple and efficient long-term response analysis method, which can be used to derive consistent design criteria. Moreover, the proposed method can also be readily incorporated into the design process, a common drawback of most long-term analysis methods.

The basis for the analysis is a hindcast database of storms in the Gulf of Mexico. The database covers 35 storms over an 85 year period and contains 11,322 records with hourly hindcast values of wind, wave and current parameters at 98 grid points in the Gulf of Mexico. Forristall *et al* (1991) showed that the arrival of hurricanes in the Gulf of Mexico, and hence platform responses, is Poisson distributed. The long-term response distribution can therefore be calculated based on Poisson's equation from the probability distribution of a "single random hurricane" which is obtained by averaging the short-term extreme response distributions over all storms and grid points. This process in effect randomizes the tracks of future storms relative to a particular site.

The paper discusses the characteristics of the hurricane hindcast database and the theoretical background of the long-term response analysis methodology. The application of the method is illustrated for a TLP in deep water. The 100 year long-term responses are derived and compared with short-term 100 year hurricane design values. In addition, some typical storm samples are examined for the temporal variation of minimum airgap and extreme tendon tension in order to assess their sensitivity to varying wind, wave and current conditions. Consistent response-based design criteria can be readily derived using the results of this analysis, but this topic is beyond the scope of the present paper, see Mercier (1994).

HURRICANE HINDCAST DATABASE

The long-term environment in the Gulf of Mexico is described by means of Forristall and Larrabee's (1991) hurricane hindcast database. This database provides a hindcast of 35 storms with significant wave heights in excess of 25 ft during an 85 year period since 1900. Table 1 lists the 35 storms and their respective maximum significant wave heights and wind speeds. The original database contains in excess of 90,000 records with the hourly hindcast values of wind, wave and current parameters. It was derived from a hindcast on 784 grid points in the Gulf of Mexico. For the present purpose, the database was culled to 11,322 records by reducing the number of grid points to 98.

Each record in the database contains hourly hindcast values for the following eight parameters:

1. significant wave height
2. spectral peak wave period
3. spectral peakedness parameter
4. one-hour mean wind speed at 33 ft
5. surface current speed
6. wave heading w.r.t. true north
7. wind heading w.r.t. true north
8. current heading w.r.t. true north

The effect of wave spreading is not considered here. Table 2 provides some basic statistics for the wind, wave and current parameters. Table 3 lists database queries for:

- the 5 highest waves,
- the 5 steepest waves,
- the 5 highest wind conditions, and
- the 5 highest current conditions.

Figure 1 shows histograms of wave height, wave period, wave steepness, wind speed, current speed, wave direction, wind direction and current direction. On average, the waves, wind and current are fairly collinear but on an individual basis large variations can occur. The change over time in the relative wind, wave and current directions associated with the passage of the eye of a hurricane is of particular relevance since such conditions may induce significant variations in extreme TLP responses.

LONG-TERM RESPONSE ANALYSIS

Consider the distribution of wave height. The basic assumption is that the future will be similar to the past or, in other words, that the sample distribution calculated from the hindcast will approximate the true distribution of wave heights. The wave height distribution can thus be averaged over sites to produce a smooth distribution for an arbitrary site.

The same assumptions can also be applied to any response Y that is dependent on the environment. Let $P_{ijk}(y)$ be the "*short-term*" cumulative probability that the response is less than y during hour k (k = 1..N_{ij}) of storm i (i = 1..85) at grid point j (j = 1..98). For the response to be less than y during the N_{ij} hours that storm i hits grid point j, it must be less than y for each hour in the storm. Thus the "*medium term*" cumulative probability distribution for the maximum response during storm i at grid point j is given by a simple multiplication of the hourly distributions:

$$P_{ij}(y) = \prod_k P_{ijk}(y)$$

This equation also accounts for the likelihood that the maximum response may not occur when the environment is most severe.

Even small differences in the positions of the sites with respect to the tracks of the strongest storms can cause the upper tails of the medium term distributions to differ from each other. Since the historical storm tracks are not expected to be exactly duplicated in the future, the averaging of distribution functions from many sites in effect randomizes the tracks of future storms relative to a particular site. The distribution function for the entire 85-year history of the hindcast database can be calculated by taking averages over both storms and sites to get a "*single random storm*" (SRS) distribution:

$$P_{SRS}(y) = \frac{1}{35}\sum_i \frac{1}{98}\sum_j P_{ij}(y)$$

In order to derive the distribution function of maximum response for exposures other than 85 years, it is assumed that the frequency of arrival of hurricanes, and hence responses, is Poisson distributed. Hence, the "*long-term*" extreme response distribution for an exposure of T years at a random Gulf of Mexico site is given by:

$$P_T(y) = \exp[-\nu T(1 - P_{SRS}(y))]$$

where $\nu = N_S/N_G = 35/85$ is the arrival rate of hurricanes (i.e. the expected number of storms per year).

It can be shown that the most probable (MPM), median and expected (mean) maximum responses correspond to risk levels (probability of exceedance) of 63.2%, 50% and 43% respectively, see, for example, Baar (1992).

Figure 2 shows a schematic flow diagram of the long-term response methodology. The process can also be inverted to derive a response-based design storm, i.e. a particular combination of wind, wave and current parameters which will most likely yield a certain prescribed response level, see Mercier (1994) for further details.

SHORT-TERM RESPONSE ANALYSIS

Three characteristic TLP responses will be considered here:

1. minimum airgap, i.e. the clearance between the deck and the top of the disturbed wave crest
2. minimum tendon bottom (effective) tension which usually occurs in a downwave tendon
3. maximum tendon top tension which usually occurs in a upwave tendon

In view of nonlinear effects such as TLP springing and ringing, the short-term probability distributions for these responses can only be derived from model tests.

Minimum Airgap

Short-term airgap is computed from:

$$A = \Gamma - S - \gamma H_S$$

where

Γ = still water deck clearance
S = mean platform setdown
A = airgap
H_S = significant wave height

The term γH_S represents the disturbed wave crest elevation and incorporates the combined effects of wave diffraction around the TLP hull columns and skewness of the undisturbed extreme wave profile (crests are higher than troughs). Model tests suggest that the 3-hourly expected value of γ is around 1.6, see Schott et al (19.

Consider the normalized disturbed wave crest

$$y = (\Gamma - S - A) / H_S$$

The measurement of the disturbed wave crest within a TLP is very difficult which leads to uncertainty in estimating the short-term distribution of y. Therefore a simplified ad-hoc approach is adopted here to demonstrate the effect of the underlying short-term variability on the long-term airgap. Three alternative short-term probability distribution models will be *assumed*:

1. "*deterministic*" model: the short-term variability of airgap is ignored and the distribution of y degenerates to a simple unit step function, y = 1.6 (constant). This model is likely to yield an unconservative estimate of the long-term airgap.
2. "*Gaussian*" model: the distribution of y is assumed to be non-skewed and Gaussian with mean 1.6 and an assumed standard deviation 0.1 (COV = 6.25%). This model provides a first-cut insight into the effect of short-term variability.
3. "*Weibull*" model: in this most-conservative model, the distribution of y is a skewed 3-parameter Weibull with a 3-hour MPM of 1.6.

Extreme Tension

The short-term extreme tendon tension is computed from:

$$T = T_{MEAN} \pm T_{DYN}$$

where the + or - sign applies to maximum or minimum tension respectively. Furthermore,

1. The mean tension T_{MEAN} includes the initial pretension as well as the combined effects of static platform offset and setdown due to mean wind, wave and current forces.
2. The dynamic tension T_{DYN} is estimated from, see Schott *et al* (1995)

$$T_{DYN} = K\sqrt{f\ T_{RMS}^2 + T_{RRA}^2}$$

where K is a 3-hour MPM peak factor, f is a calibration factor on the computed wave plus low frequency RMS tension response T_{RMS}, and T_{RRA} is a measured resonant response allowance for the nonlinear high frequency ringing and springing response.

In a long-term analysis each tendon needs to be processed separately since there is no uniquely defined upwave or downwave tendon. The emphasis in the interpretation of the model tests is therefore not so much on deriving conservative algorithms for short-term design checks but more on capturing the "average" results of all model tests for extreme sea states.

Consider the normalized maximum tendon tension

$$y_{MAX} = \frac{T_{MAX} - T_{MEA}}{\sqrt{f^2 T_{RMS}^2 + T_{RRA}^2}}$$

Likewise, the normalized minimum tendon tension is

$$y_{MIN} = \frac{T_{MEAN} - T_{MIN}}{\sqrt{f^2 T_{RMS}^2 + T_{RRA}^2}}$$

From analysis of extensive model test experiments (with collinear wind, waves and current) the 3-hour short-term distributions of y_{MAX} and y_{MIN} have been expressed as skewed 3-parameter Weibull distributions.

DYNAMIC ANALYSIS

Referring to figure 2 and the preceding discussion. some efficient and accurate method is required to estimate the mean and RMS response for arbitrary wind. wave and current conditions. Based on extensive Shell experience with the global performance analysis of TLP's an efficient method has been developed with the following main features:

- A coupled model is used where the TLP is represented as a rigid body with six degrees of freedom and with mass. damping and stiffness properties. while the tendons and risers are represented by means of slender rod finite elements with six degrees of freedom. Efficiency is further improved by lumping the tendons on each hull column into one corner tendon and using only one finite element per lumped corner tendon.
- A linearized analysis is performed in the frequency domain. The hydrodynamic load model incorporates first order wave diffraction loads on the TLP. mean wave drift. wind and current forces. and the slow wave and wind drift forces including linearized contributions from viscous wave-current interaction effects.

The accuracy of the simplified dynamic analysis model has been carefully checked against results from both model tests and more detailed analytical models. For one set of wind, wave and current parameters the required CPU-time is on the order of 10 seconds on a modern workstation. Analyzing all 11,322 records in the hurricane hindcast database therefore requires less than 24 hours CPU-time. The main advantage of this procedure is in that it avoids the cumbersome and time-consuming task of fitting potentially inaccurate response functions to results derived from a much more limited number of dynamic response analyses, see Forristall *et al* (1991).

RESULTS

Long-Term Response

The calculated long-term response distributions are shown in figure 3. The 100 year long-term MPM response design extremes are tabulated in table 4 and compared with the short-term MPM design extremes for the collinear 100 year hurricane design condition. The following observations can be made:

1. The "deterministic" long-term airgap estimate of 18.7 ft agrees well with the 100 year hurricane short-term design value of 25.0 ft, as should be the case since the underlying short-term variability is ignored in the long-term analysis.
2. However, the long-term airgap is seen to be sensitive to the modeling of short-term variability of the disturbed wave crest. If the "Gaussian" and "Weibull" short-term variability models were adopted, the 100 year airgap would be reduced to 15.8 ft and 7.1 ft respectively.
3. The 100 year maximum tendon tension is 5,221 kip for the SE tendon, while the 100 year minimum tension is 330 kip for the NW tendon. These particular tendon locations make sense in view of the fact that most hurricanes approach the platform from the southeast, see figure 1.
4. The 100 year long-term *maximum* tension of 5,221 kip compares well with the 100 year hurricane short-term estimate of 5,183 kip suggesting that the short-term design analysis is only marginally unconservative.
5. The 100 year long-term *minimum* tension of 330 kip is 80 kip lower than the 100 year hurricane short-term estimate of 410 kip suggesting that the short-term design analysis is somewhat unconservative for minimum tension. The reason for the larger discrepancy between long and short-term estimates of minimum tension is the increased sensitivity of minimum tension to non-aligned wind, wave and current conditions as will be discussed in the following section.

Short-Term Response

As mentioned in the introduction, the derivation of response-based storms is beyond the scope of the present paper. However, some interesting observations can be inferred from a consideration of the short and medium term response variations. Table 5 summarizes the basic statistics of the 3-hour MPM short-term responses for all 11,322 records in the hindcast database. Figure 4 shows the correlation between significant wave height and the worst-case extremes that occurred during each of the 35 storms in the database at any of the 98 grid points. It is seen that:

1. Minimum airgap is highly correlated with significant wave height (apparent since the airgap is in effect a deterministic function of wave height). The small variations are all due to the effect of platform setdown. The worst-case short-term minimum airgap is 0.3 ft and occurs during storm 9 (October 1916).
2. Maximum tension is also fairly well correlated with significant wave height. The worst-case short-term maximum tension is 5991 kip and occurs during storm 30 (Camille 1969).
3. There is a very poor correlation between minimum tension and significant wave height. The worst-case short-term minimum tension is 392 kip and occurs during storm 26 (Carla 1961). This value is higher than the long-term 100 year extreme of 330 kip!

With regards to the last item, it should be noticed that the minimum tendon tension is highly sensitive to the underlying short-term variability and therefore the long-term 100-year estimate of 330 kip stretches well beyond what one might expect from a deterministic approach. For example, if the short-term variability were eliminated, a long-term response would be recovered that is well within the range of short-term extremes predicted for individual records in the database. However, as the short-term variability becomes more and more important, the long-term response distribution becomes more skewed and the resulting 100-year estimate will eventually fall outside the range of short-term extremes predicted for any individual record in the database. If the size of the database were increased, one might expect to encounter individual storm events with short-term response extremes falling beyond the long-term estimates.

Medium Term Response

The medium term response is understood to describe the temporal variation of the 3-hour MPM short-term extreme for a given storm at a particular site. In this section the three storm sequences that resulted in worst-case short-term extreme responses will be considered in more detail (see table 5 for the definition of wind, wave and current parameters during these events).

Worst-case minimum airgap storm

Figure 5 shows the medium term variation of the environmental parameters and responses during the October 1916 storm. It should be noticed that:

- This storm has the highest significant wave height on record of 50.1 ft.
- The current speed continues to increase with wave height and wind speed, but then maintains a constant and high value of around 7 ft/s, even as wave height and wind speed decrease.
- Both wind and current are reasonably well aligned with the waves.
- The minimum airgap decreases from an initial value of 46 ft to 0 ft before it increases again to 43 ft, all of this happening in fewer than 11 hours.

Worst-case maximum tension storm

Figure 6 shows the medium term variation of the environmental parameters and responses during hurricane Camille (1969). It should be noticed that:

- This storm has the highest wind speed on record of 151 ft/s.
- Again, the current speed continues to increase with wave height and wind speed, but again maintains a constant and high value of around 8 ft/s, even as wave height and wind speed decrease.
- Both wind and current are reasonably well aligned with the waves.
- The maximum tension increases from an initial value of around 4,000 kip to almost 6,000 kip and back to around 4,900 kip in less than 10 hours.

Worst-case minimum tension storm

Figure 7 shows the medium term variation of the environmental parameters and responses during hurricane Carla (1961). It should be noticed that:

- This storm has rather different characteristics from the previous ones. It hits this particular grid point for over 40 hours.
- Wave height increases from 26 ft to a maximum of 44 ft in around 15 hours before it starts to decrease. Wind speed shows a similar trend. However, the wave steepness ($2\pi H_S/gT_P^2$) increases during the first 30 hours of the storm before it decreases, while the current speed increases from 2.6 ft/s to 8 ft/s in just over 20 hours and then drops to a negligible value of less than 1 ft/s.
- Wind is well aligned with the waves, but the current veers all the way around the compass after the wave height and wind speed start to decrease ("passage of the eye"). The maximum wave steepness occurs when the current is about 90 degrees to the right of the waves which is explained by considering the Doppler shift effect on wave period.
- The lowest minimum tension of 392 kip also occurs when the waves are steepest and current is 90-135 degrees to the right of the waves. At this point both wave height and wind speed are fairly moderate.

CONCLUSION

1. The presented method for performing a long-term response analysis is simple, efficient and accurate. The results can be incorporated into the design in the form of a long-term response allowance or be used to derive consistent response-based design criteria.
2. Since the platform setdown is mostly a deterministic function of wave height, the 100 year hurricane short-term design estimate of minimum airgap agrees well with the 100 year long-term estimate, provided that the underlying short-term variability of the disturbed wave crest elevation is ignored. However, this assumption is non-conservative. Inclusion of short-term variability in the analysis reduces the 100 year long-term minimum airgap. A better understanding of the short-term variability of airgap would reduce design uncertainties and allowances.
3. The 100 year hurricane short-term design estimate of maximum tension is only slightly conservative when compared with the 100 year long-term extreme. Maximum tension is fairly well correlated with significant wave height or wind speed and is rather insensitive to non-collinear wind, waves and current conditions.
4. The 100 year hurricane short-term design estimate of minimum tension is somewhat under-conservative when compared to the 100 year long-term extreme. Minimum tension correlates poorly with wave height, but is very sensitive to wave steepness and the relative wave-current direction.
5. Three representative storm samples have been extracted from the hurricane hindcast database which are representative for worst-case scenarios for minimum airgap, maximum tension and minimum tension. These samples can be used in a limited scope analysis for *qualitative* assessment of the sensitivity of airgap and extreme tension to the environment.

ACKNOWLEDGMENT

Shell Offshore, Inc., sponsored this investigation and granted permission to publish the results.

REFERENCES

API (1987): "API RP 2T Recommended Practice for Planning, Designing and Constructing Tension Leg Platforms", American Petroleum Institute, Washington, 1st Edition, April 1, 1987 (Supplement 1, April 1, 1992).

Baar, J.J.M. (1992): "Extreme Values of Morison-Type Processes", *Applied Ocean Research*, Vol. 14, pp. 65-58.

Forristall, G.Z., Larrabee, R.D. and Mercier, R.S. (1991): "Combined Oceanographic Criteria for Deepwater Structures in the Gulf of Mexico", *Proc. 23rd Annual Offshore Technology Conference*, Houston, Paper No. OTC6541.

Mercier, R.S. (1994): "Generation of Hurricane Design Criteria for Deepwater Platforms", to be published.

Schott, W.E., Mercier, R.S. and Howell, C.T. (1995): "Resonant Tendon Tension Response Observed in the Scale Model Testing of the Mars Tension Leg Platform", *Tension Leg Platform Technology Symposium*, SNAME Texas Section, Houston.

Schott, W.E., Rodenbusch, G., Mercier, R.S. and Webb, C.M. (1994): "Global Design and Analysis of the Auger Tension Leg Platform", *Proc. 26th Annual Offshore Technology Conference*, Houston, Paper No. OTC7621.

storm	month/name	year	$H_{s,max}$ (ft)	$V_{w,max}$ (ft/s)
1	September	1900	39.4	136
2	August	1901	31.7	104
3	September	1906	41.7	123
4	July	1909	40.0	130
5	August	1915	42.3	126
6	September	1915	47.4	137
7	July	1916	35.9	110
8	August	1916	36.1	126
9	October	1916	**50.1**	144
10	September	1917	39.1	114
11	August	1918	29.0	109
12	September	1919	40.1	131
13	September	1920	26.6	90
14	June	1921	38.8	133
15	August	1926	36.6	120
16	August	1932	29.5	115
17	September	1933	42.3	129
18	June	1934	33.8	105
19	September	1941	35.8	117
20	August	1942	39.0	128
21	September	1947	39.5	123
22	October	1949	39.6	126
23	Flossy	1956	30.0	97
24	Audrey	1957	31.2	108
25	September	1960	33.1	120
26	Carla	1961	45.1	130
27	Hilda	1964	37.4	126
28	Betsy	1965	47.0	133
29	Beulah	1967	31.8	95
30	Camille	1969	43.7	**154**
31	Edith	1971	33.0	110
32	Carmen	1974	34.2	127
33	Eloise	1975	38.9	123
34	Frederic	1979	38.8	124
35	Allen	1980	39.5	122
1-35		1900-1985	50.1	154

Table 1: 35 Storms in hurricane hindcast database

parameter	unit	long term	short term
"deterministic" airgap	ft	18.7	
"Gaussian" airgap	ft	15.8	
"Weibull" airgap	ft	7.1	
minimum airgap	ft		19.0
NE maximum tension	kip	5116	
SE maximum tension	kip	5221	
SW maximum tension	kip	4952	
NW maximum tension	kip	5147	
maximum tension	kip		5183
NE minimum tension	kip	748	
SE minimum tension	kip	729	
SW minimum tension	kip	471	
NW minimum tension	kip	330	
minimum tension	kip		410
Note: Short term estimates for collinear 100 year hurricane w/ H_S = 40 ft, T_P = 14.2 s, V_W = 120 ft/s, V_C = 5.7 ft/s			

Table 4: Long and short term MPM responses

parameter	symbol	unit	mean	rms	minimum	maximum
significant wave height	H_s	ft	29.7	4.1	25.0	50.1
peak wave period	T_p	s	12.6	1.0	9.8	17.3
wave steepness	$2\pi H_s/gT_p^2$		0.0366	0.0045	0.0195	0.0516
spectral peakedness	Q_p		2.69	0.23	1.89	3.36
wind speed	V_w	ft/s	91.4	14.7	4.3	154.4
current speed	V_c	ft/s	4.38	1.61	0.09	9.07
wave direction	ϑ_{wave}	degrees	-63.2	47.4	-179.9	180.0
wind direction	ϑ_{wind}	degrees	-59.6	73.8	-180.0	179.9
current direction	ϑ_{curr}	degrees	-39.7	65.7	-179.9	179.9
wind-wave direction	$\vartheta_{wind}-\vartheta_{wave}$	degrees	-16.7	25.0	-176.4	172.1
current-wave direction	$\vartheta_{curr}-\vartheta_{wave}$	degrees	22.9	37.2	-179.4	178.4

Table 2: Basic statistics of 11,322 wind, wave and current parameters

record	storm	grid	hour	H_s	T_p	$2\pi H_s/gT_p^2$	Q_p	V_w	V_c	ϑ_{wave}	ϑ_{wind}	ϑ_{curr}
5 highest wave conditions:												
2493	6	334	8	**47.4**	15.7	0.037	2.73	137	7.0	-84	-102	-87
3459	9	726	4	**47.4**	16.1	0.036	2.16	127	6.2	-86	-98	-105
3498	9	750	5	**49.7**	16.1	0.037	2.09	133	6.9	-85	-87	-108
3535	9	774	4	**48.6**	16.6	0.035	1.99	144	6.0	-104	-129	-134
3536	9	774	5	**50.1**	16.2	0.037	1.99	134	7.1	-84	-68	-110
5 steepest wave conditions:												
5908	15	438	13	25.2	9.9	**0.051**	2.47	96	7.0	-9	39	55
5909	15	438	14	25.1	9.8	**0.051**	2.47	97	7.1	-3	39	67
5958	15	454	18	25.6	9.8	**0.052**	2.78	99	6.3	5	51	69
5959	15	454	19	25.7	9.9	**0.052**	2.78	99	6.5	10	49	80
10657	30	286	8	29.4	10.6	**0.051**	2.74	110	7.3	9	43	-19
5 highest wind conditions:												
10517	30	166	5	43.1	13.9	0.043	2.89	**153**	7.6	-85	-99	-93
10567	30	206	5	43.7	14.1	0.043	2.92	**154**	7.5	-84	-96	-96
10615	30	246	5	42.5	14.1	0.042	2.84	**151**	7.0	-80	-96	-93
10616	30	246	6	42.7	13.5	0.045	2.84	**151**	8.4	-58	-52	-76
10654	30	286	5	43.1	14.5	0.040	2.74	**154**	7.0	-81	-102	-95
5 highest current conditions:												
2259	6	238	12	35.4	12.3	0.046	2.86	109	**9.1**	-17	-5	7
2260	6	238	13	33.6	12.1	0.045	2.86	103	**9.0**	-12	-3	19
2362	6	278	13	36.4	12.7	0.044	2.82	111	**9.1**	-19	-8	6
2363	6	278	14	34.6	12.2	0.045	2.82	105	**9.0**	-14	-6	18
2464	6	318	14	37.4	13.1	0.042	2.78	113	**9.1**	-18	-7	5

Table 3: Queries from hurricane hindcast database

parameter	unit	mean	rms	minimum	maximum
airgap	ft	38.6	14.3	0.3	48.2
NE maximum tension	kip	4159	307	3575	5660
SE maximum tension	kip	4246	373	3508	5991
SW maximum tension	kip	4130	268	3635	5304
NW maximum tension	kip	4291	299	3603	5528
maximum tension	kip	4371	306	3794	5991
NE minimum tension	kip	1240	130	596	1595
SE minimum tension	kip	1218	132	584	1632
SW minimum tension	kip	1065	211	419	1633
NW minimum tension	kip	903	204	392	1453
minimum tension	kip	801	101	392	1185

Table 5: Basic statistics of 11,322 short term MPM responses

hour	H_s	T_p	Q_p	V_w	V_c	ϑ_{wave}	ϑ_{wind}	ϑ_{curr}
worst case minimum airgap storm (October 1916)								
1	26.1	14.8	1.99	89.4	2.5	-114	-164	-141
2	31.4	14.6	1.99	104.1	3.4	-112	-163	-142
3	38.5	17.3	1.99	124.1	4.5	-111	-157	-140
4	48.6	16.6	1.99	143.9	6.0	-104	-129	-134
5	50.1	16.2	1.99	134.3	7.1	-84	-68	-110
6	41.0	15.9	1.99	114.5	7.2	-73	-38	-81
7	33.9	15.2	1.99	96.6	7.0	-68	-22	-61
8	30.5	14.8	1.99	86.5	6.9	-63	-19	-44
9	28.7	14.8	1.99	80.7	7.0	-59	-20	-29
10	27.4	14.6	1.99	75.3	6.9	-57	-23	-16
11	26.1	13.2	1.99	70.5	6.9	-56	-26	-4
worst case maximum tension storm (Camille 1969)								
1	26.4	13.1	2.84	93.9	3.2	-90	-138	-99
2	29.4	12.8	2.84	105.1	3.8	-92	-136	-101
3	33.0	13.3	2.84	117.9	4.6	-92	-130	-101
4	37.5	13.7	2.84	134.5	5.6	-90	-120	-99
5	42.5	14.1	2.84	150.8	7.0	-80	-96	-93
6	42.7	13.5	2.84	151.1	8.4	-58	-52	-76
7	35.5	11.9	2.84	132.1	8.6	-30	-1	-49
8	30.9	11.0	2.84	114.1	8.3	-6	24	-25
9	28.6	10.8	2.84	104.0	8.1	11	28	-6
10	26.8	10.6	2.84	95.0	8.2	18	29	9
worst case minimum tension storm (Carla 1961)								
1	25.9	13.0	2.85	78.2	2.6	-107	-150	-95
2	27.0	13.1	2.85	81.4	2.7	-108	-150	-96
3	28.2	13.3	2.85	85.1	2.8	-108	-150	-97
4	29.5	13.6	2.85	88.5	3.0	-108	-149	-98
5	30.8	13.7	2.85	92.3	3.2	-108	-148	-99
6	32.1	14.0	2.85	95.8	3.4	-108	-147	-99
7	33.5	14.1	2.85	99.9	3.7	-107	-145	-98
8	35.0	14.4	2.85	104.3	4.0	-106	-143	-98
9	36.7	14.6	2.85	108.2	4.3	-105	-140	-96
10	38.5	14.7	2.85	113.5	4.6	-104	-136	-94
11	39.9	14.8	2.85	116.9	5.0	-102	-132	-91
12	41.3	14.8	2.85	120.4	5.4	-100	-128	-88
13	42.4	14.9	2.85	123.4	5.8	-97	-123	-83
14	43.4	14.9	2.85	126.4	6.2	-94	-116	-78
15	44.1	14.9	2.85	128.3	6.5	-91	-110	-72
16	44.4	14.8	2.85	129.2	6.9	-86	-101	-66
17	44.4	14.7	2.85	129.4	7.2	-82	-93	-58
18	44.1	14.7	2.85	129.3	7.5	-77	-86	-49
19	43.7	14.6	2.85	128.5	7.8	-72	-78	-41
20	43.2	14.6	2.85	127.6	7.9	-67	-72	-32
21	42.5	14.4	2.85	125.7	8.0	-63	-66	-23
22	41.6	13.9	2.85	123.6	8.0	-58	-62	-15
23	40.8	13.6	2.85	122.2	7.9	-54	-56	-5
24	40.1	13.4	2.85	121.0	7.8	-50	-51	3
25	39.3	13.4	2.85	119.3	7.7	-46	-47	12
26	38.8	13.3	2.85	118.2	7.5	-43	-44	21
27	38.2	13.2	2.85	116.2	7.1	-39	-41	30
28	37.5	13.2	2.85	114.3	6.8	-37	-38	38
29	37.0	13.0	2.85	112.7	6.4	-33	-36	47
30	36.3	12.6	2.85	110.4	5.9	-31	-34	57
31	35.7	12.3	2.85	108.9	5.5	-28	-32	73
32	35.2	12.2	2.85	107.1	4.9	-26	-31	91
33	34.8	12.1	2.85	105.4	4.3	-24	-29	106
34	34.3	12.1	2.85	103.2	3.7	-22	-27	123
35	33.6	12.1	2.85	99.3	3.1	-20	-25	140
36	32.8	12.0	2.85	95.7	2.6	-18	-24	157
37	32.1	12.0	2.85	91.9	2.0	-17	-22	173
38	31.1	12.0	2.85	88.3	1.6	-16	-21	185
39	30.1	11.9	2.85	85.2	1.2	-14	-20	204
40	28.9	11.9	2.85	82.0	1.0	-13	-18	217
41	27.4	11.7	2.85	77.9	1.0	-12	-16	230
42	26.0	11.2	2.85	74.1	1.2	-11	-14	245

Table 6: Storms with worst-case short term responses

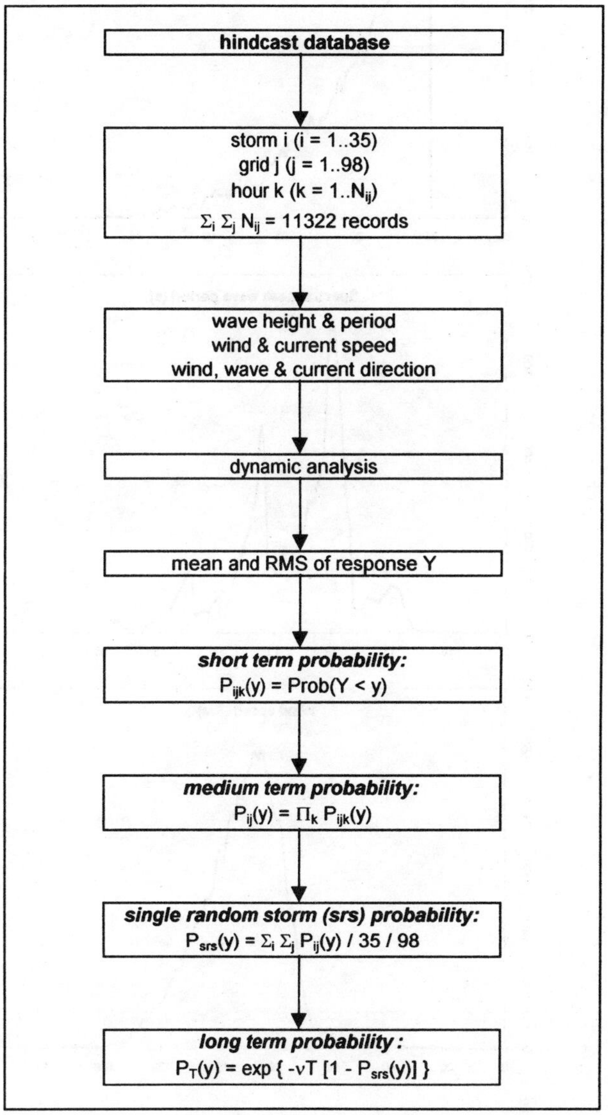

Figure 2: Long Term Response Analysis Methodology

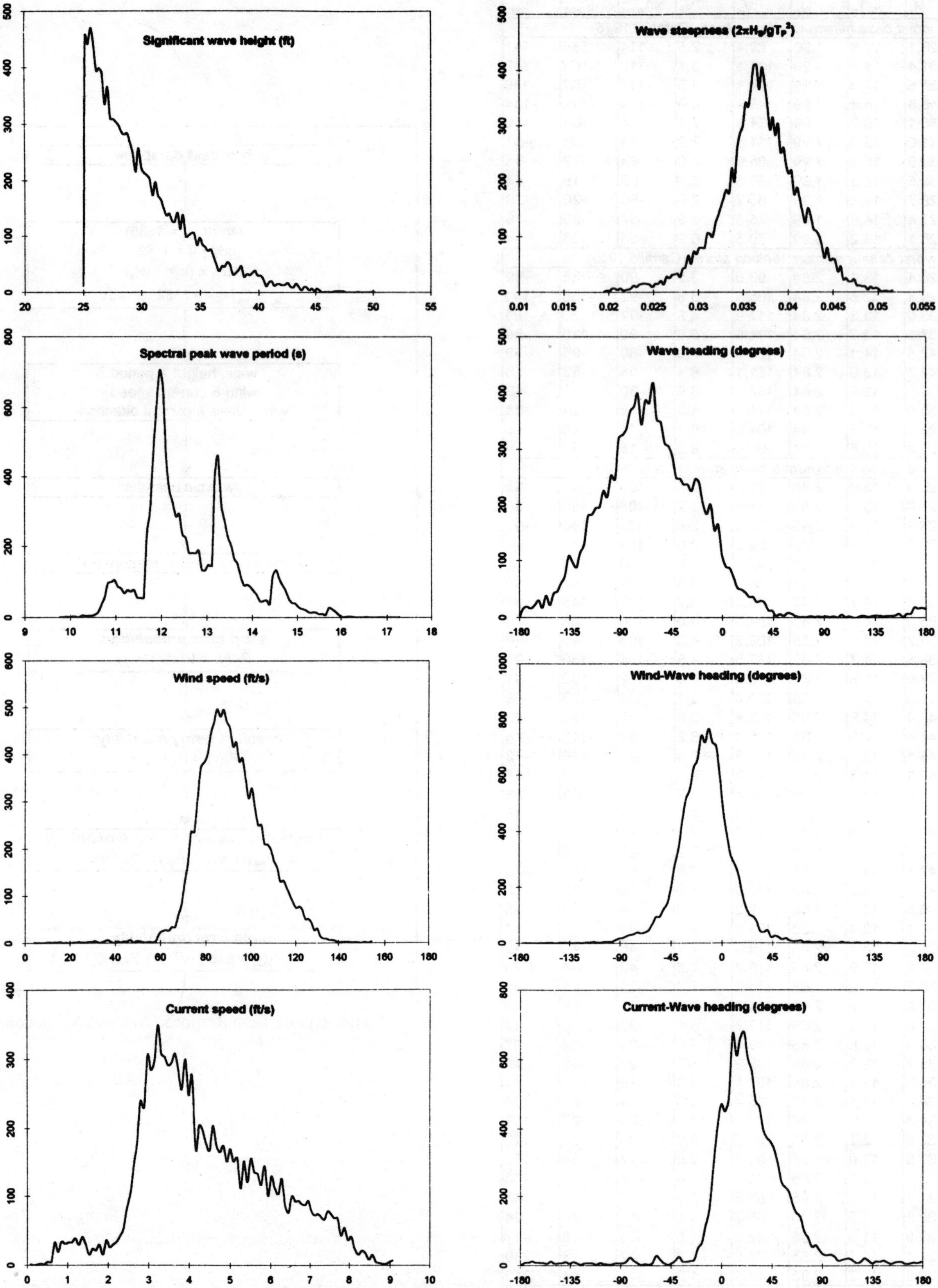

Figure 1: Histograms of wind, wave and current parameters in hurricane hindcast database.

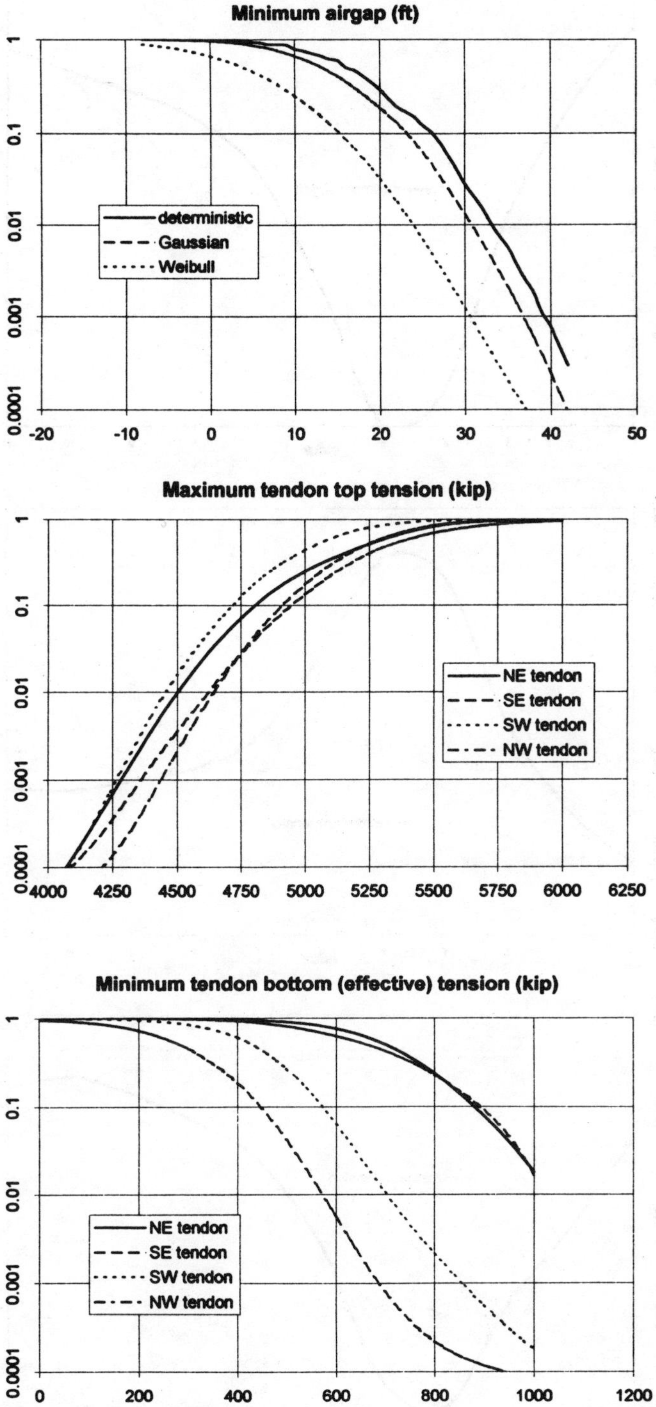

Figure 3: 100 year long-term cumulative distribution functions

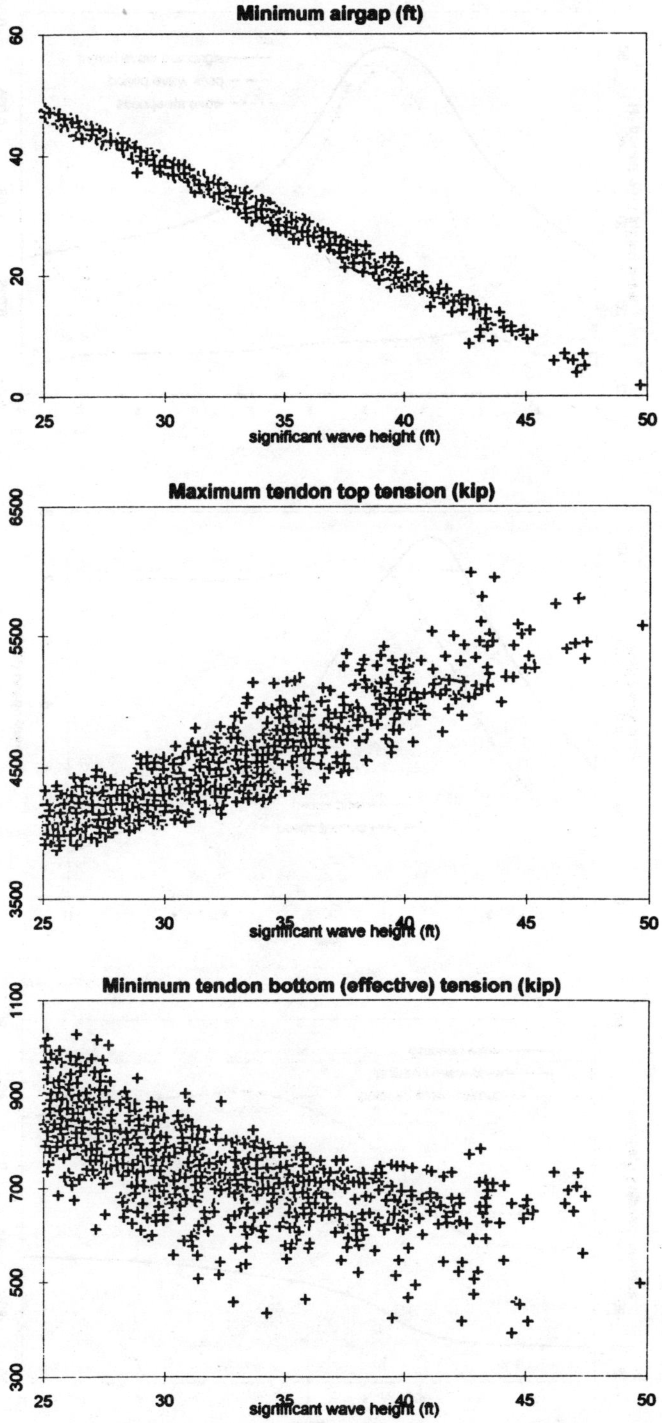

Figure 4: Correlation between significant wave height and short-term extreme responses

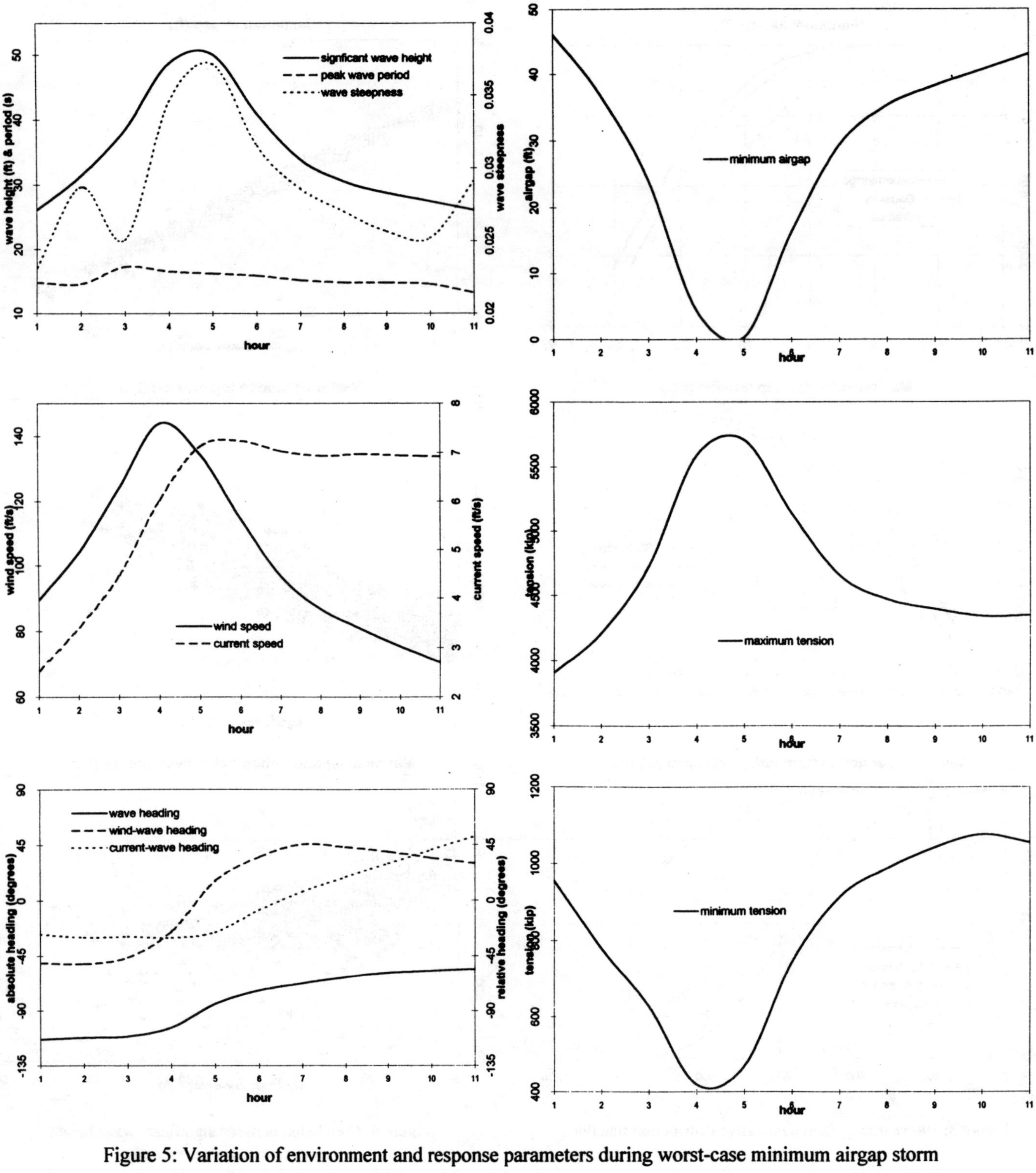

Figure 5: Variation of environment and response parameters during worst-case minimum airgap storm

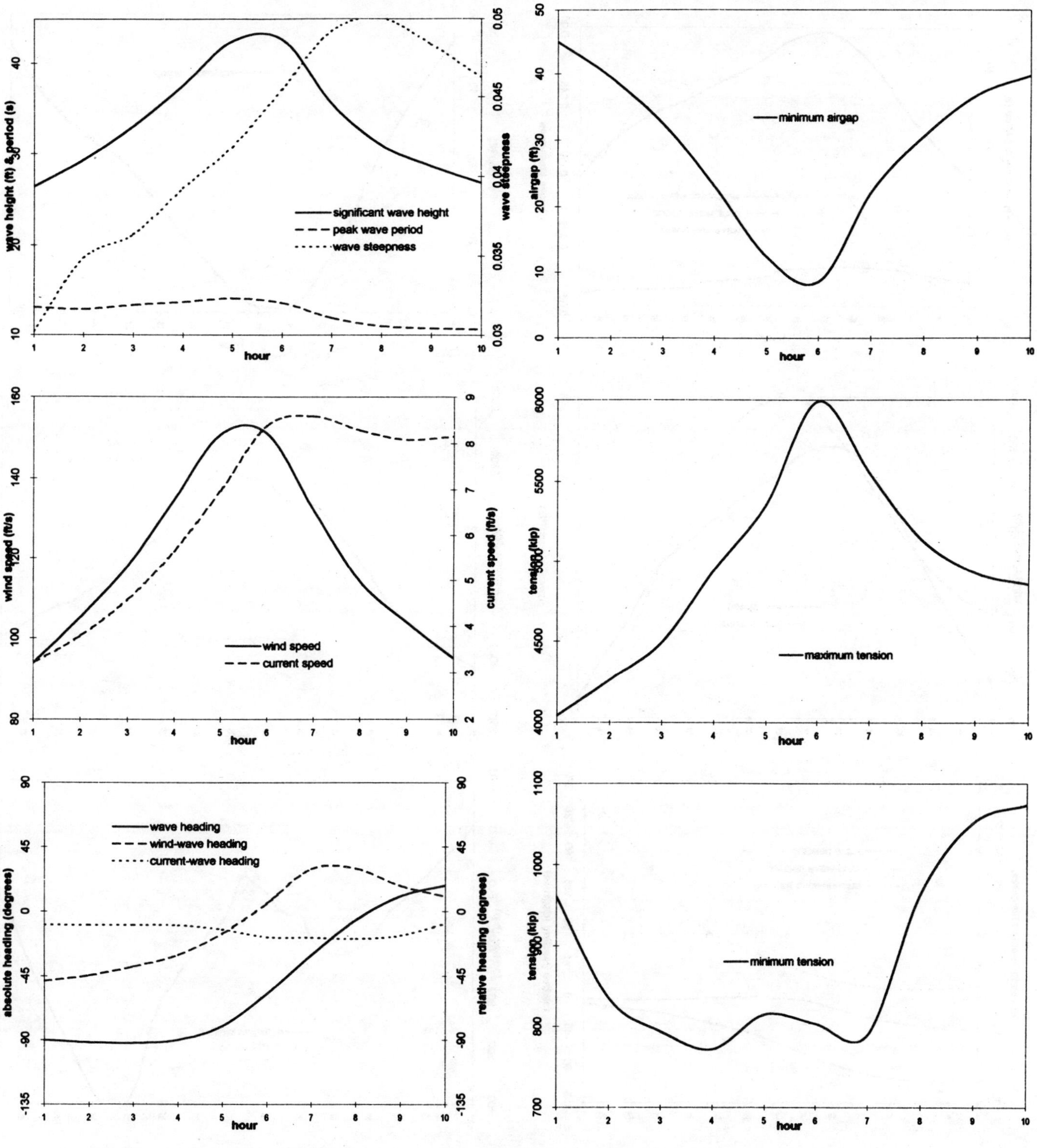

Figure 6: Variation of environment and response parameters during worst-case maximum tension storm

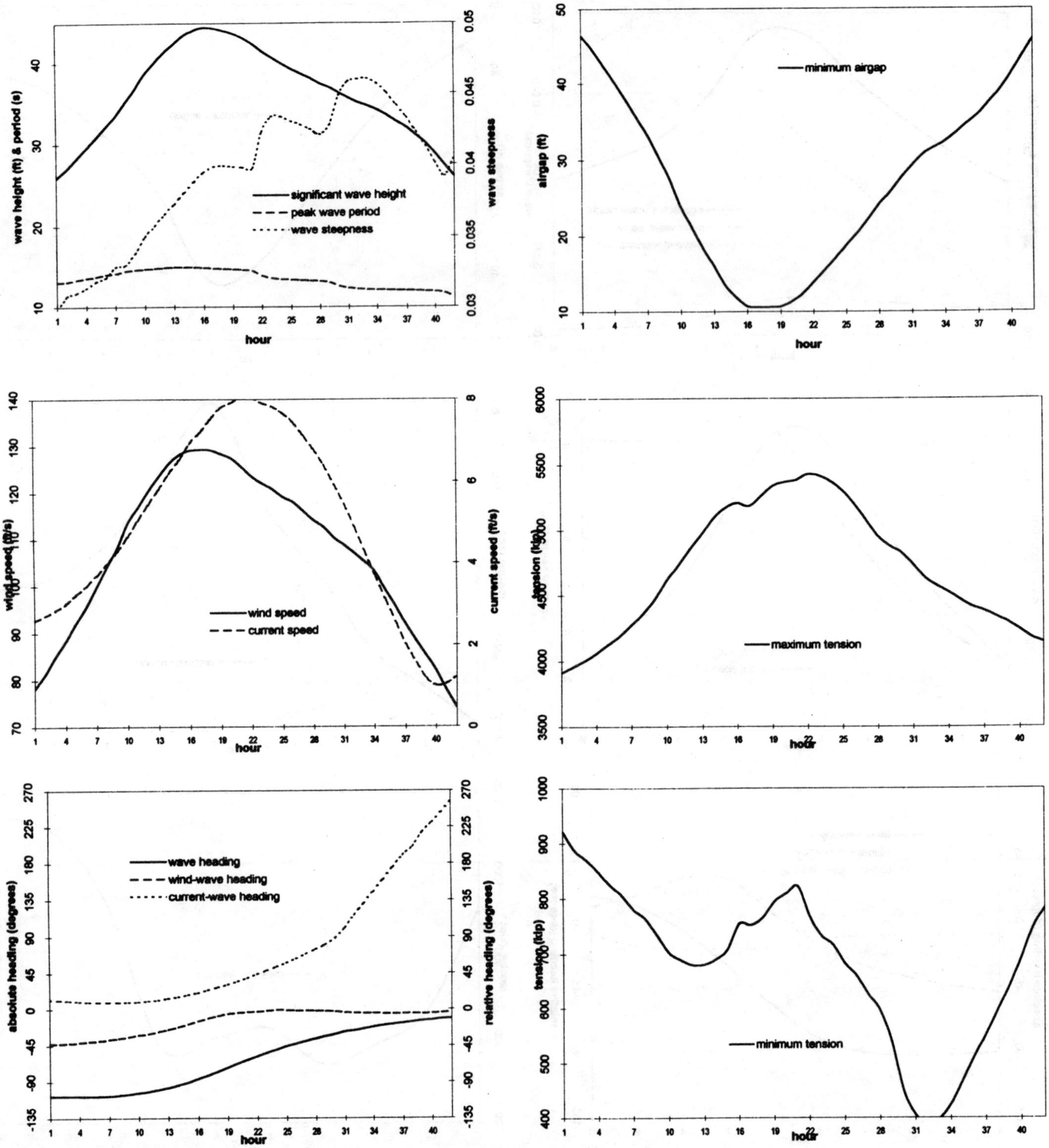

Figure 7: Variation of environment and response parameters during worst-case minimum tension storm

Proceedings of the Seventh (1997) International Offshore and Polar Engineering Conference
Honolulu, USA, May 25-30, 1997

ISBN 1-880653-28-1 (Set); ISBN 1-880653-29-X (Vol. I)

An Analysis Procedure for the Tendons of a Tension Leg Platform

Luiz Henrique Moraes Alves
PETROBRAS Research & Development Center

Breno Pinheiro Jacob and Gilberto Bruno Ellwanger
COPPE/UFRJ

Rio de Janeiro, Brazil

ABSTRACT

This work presents the results of a sequence of dynamic analyses of the tendons of a tension leg platform (TLP). The first step consists in stochastic time domain analyses of the platform, where environmental loads are evaluated using alternatively the Diffraction theory and the Morison approach. The results are the motions of the hull and the tension at the top of the tendons, which are subsequently used as input for a time domain analysis of a tendon, modeled with nonlinear space frame finite elements. The main results of the tendon analysis, axial force and bending moments, may then be employed for the structural verification and evaluation of fatigue damage of the tendon. This sequence of analyses may then comprise an analysis and design procedure for the tendons of a TLP. The application studied is a TLP considered for oil exploration in Brazilian deep waters, where the marine current is important not only for the static offset of the floating structure, but also in the generation of bending effects on the tendons. Thus, the precise consideration of the nonlinear behavior of the tendons becomes of crucial importance.

INTRODUCTION

The recent identification of huge oil fields located in deep waters has increased the demand for structural systems to support exploration platforms in water depths greater than 1000 meters. The Tension Leg Platform (TLP) (Figure 1) is one of the alternatives considered nowadays; it is essentially a floating, semisubmersible structure, moored by vertical tethers or tendons tensioned by excess buoyancy. The TLP is in fact an evolution of the concept of semisubmersible platforms; by restraining the vertical movements it allows dry well completion, that facilitates control and interventions in the wells, avoiding the use of dynamic positioning ships that are needed with wet completion where the Christmas trees remain at the seabottom. Another important advantage of the TLP concept is that, by avoiding the catenary mooring lines employed in semisubmersible platforms, it allows a simpler, "cleaner" bottom arrangement.

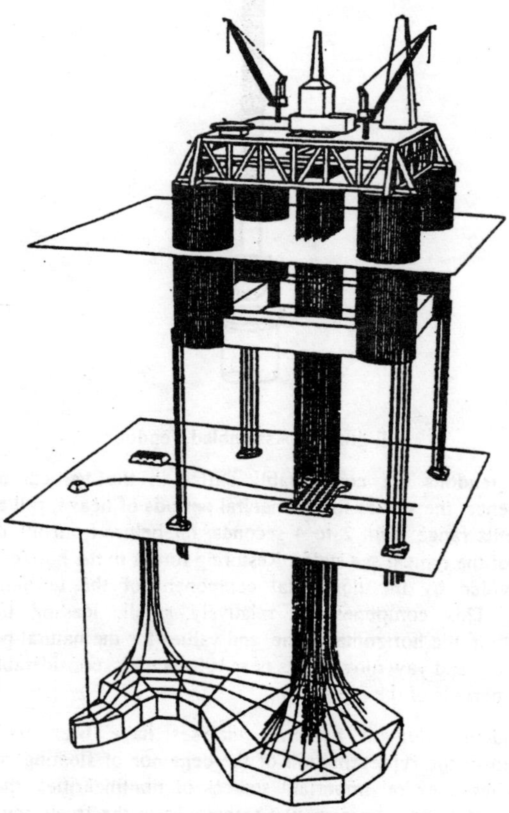

Figure 1 - Schematic view of production system with a TLP (ASME, 1992)

The tendons (Figure 2) of a TLP are, in general, sealed non-pressurized tubes, that may be fabricated in a single segment or in interconnected segments with length of about 70 meters. Neutral buoyancy is required to allow feasible transportation and verticalization of single-segment tendons. Such tendons may be employed in water depths not greater than 500 meters, since, in deeper waters, the consideration of the hydrostatic pressure in the design of the tendon leads to small values for the diameter-to-thickness ratio, and neutral buoyancy cannot be attained.

Figure 2 - Assembled Tendon

The tendons are considerably stiffer in the vertical plane; in consequence, the values for the natural periods of heave, roll and pitch movements range from 2 to 4 seconds, far below from the dominant periods of the typical sea states. Restoring forces in the horizontal plane are provided by the horizontal component of the tension on the tendons. This component is relatively small, leading to higher flexibility in the horizontal plane, and values for the natural periods of surge, sway and yaw movements near 100 seconds, considerably higher than the periods of the sea states.

Frequency domain dynamic analyses have been traditionally employed in the representation of the behavior of floating structures. Nevertheless, several important sources of nonlinearities, that would require approximate linearization techniques in the frequency domain, are present in the behavior of a TLP. These sources of nonlinearities include: geometric effects due to large displacements; the drag term of the Morison equation, which is proportional to the square of the relative fluid-structure velocity; the variation of the free surface of the waves; changes in the position of the hull in relation to the waves; variation of the tension of the tendons. These nonlinear effects could be more easily and appropriately treated by time domain dynamic analysis.

This work, therefore, considers the application of time domain dynamic analysis in the study of the tendons of a TLP for the environmental conditions of Campos Basin, Southeastern Brazil. The TLP is designed for a water level of 938 m, with a production plant of 100.000 bpd. The platform movements are first calculated by the TDSIM6 program (Paulling, 1992), with an uncoupled model where the tendons are represented by non-linear, weightless springs. The resulting time histories of the horizontal movements and top tensions are then imposed in the analysis of a finite-element model of the tendons by the ANFLEX program (Mourelle, Casaprima and Jacob, 1995). Such uncoupled model is known to present good results for water depths not greater than 1000 meters (Sircar, 1993).

The representation of the environmental loadings include: waves (employing the Morison theory or a hybrid Morison/Diffraction approach), current and wind (static force plus wind spectra). Random wave loads are represented by the two-parameter Pierson-Moskowitz spectrum (Barltrop and Adams, 1991); the variable character of the wind is considered through the API spectrum (1991). The tendon is modeled with geometric nonlinear spatial beam elements. The boundary conditions at the ends of the tendon are represented by rotational springs, that simulate the flexible joints.

Parametric studies were performed in order to assess: **i)** the influence of the consideration of the diffraction theory in the analysis of the TLP movements over the resulting forces along the tendon, and **ii)** The relative participation of wave, current and wind on the TLP movements and the tension on tendon top.

The main results of the dynamic analysis of the tendons are: tension and bending moments along the tendon; horizontal reaction force on the load transmission element at the bottom and top of the tendon; top and bottom tendon angles. These results are employed for the structural verification according to the API-RP-2T code (1987); the verification of the minimum effective tension at the bottom of the tendon, in order to avoid its disconnection from the seabottom device; and, finally, for the fatigue analysis and calculation of total damage and fatigue life.

TIME DOMAIN ANALYSIS OF THE TLP

The procedure for the movement analysis of the TLP in the time domain considers the equations of large amplitude, rigid-body movements (Meirovich, 1970). All nonlinear effects due to large displacements and environmental loads mentioned in the Introduction of this work are taken into account. The time integration of these equations employs the fourth-order Runge-Kutta predictor-corrector method.

In the calculation of wave and current loadings, the Morison formulation can be safely employed whenever the diameter D of the cylindrical members that comprise the platform may be considered small when compared to the wave length λ ; this is true when $\lambda/D > 5$. Otherwise, the diffraction theory should be applied. In this latter case, the calculation of the fluid forces is performed employing the results of WAMIT program, that generates an input file for the analysis program containing added mass coefficients, drift coefficients, and potential damping coefficients. As a result, a hybrid model is employed in the time domain, obtaining simultaneously the first and second order movements due to wave, current and wind acting together.

The analyses are performed for the *in-situ* condition of the considered tension leg platform (Figure 3), with four circular columns, four rectangular pontoons, and three tendons per column. The main characteristics of the TLP are summarized in Table 1.

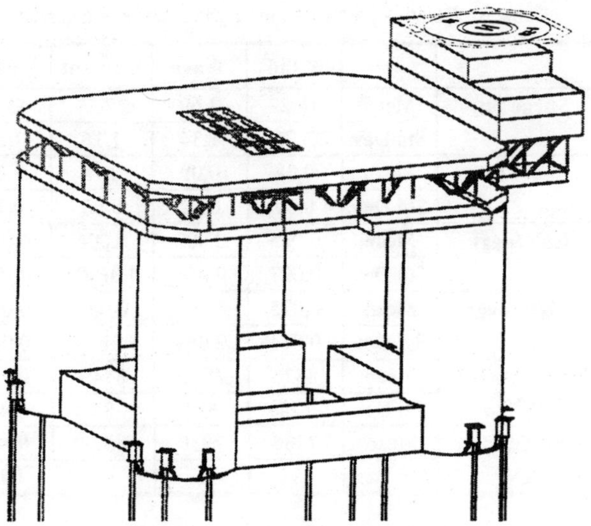

Figure 3 - Schematic view of TLP

Table 1 - TLP Characteristics

Displacement		41287.00 tf
Mass		31567 t
Column Diameter		18.20 m
Pontoon Dimensions		9.45x8.40 m
Distance between Columns		51.80 m
Deck Dimensions		70.0x70.0 m
Draft		28.00 m
G.C. Elev. Relative to Column Base		29.60 m
Natural Periods:	Surge	123.0 s
	Heave	3.3 s
	Pitch	3.0 s
	Sway	123.0 s
	Roll	3.0 s
	Yaw	98.0 s

The operation draft of the TLP in its initial position is 28.0m. Preliminary analyses indicated that the mean set-down of the platform is about 2.0 meters. The diffraction analysis that generates an input file for the TLP movement analysis considered a mean draft of 29.0m.

The drag coefficients of the Morison equation are the proposed by DNV (1990): Cd = 0.7 for circular section columns, Cd = 2.0 for rectangular section pontoons. When the diffraction theory is not considered, the following values are taken for the added mass coefficients: Cm = 2.0 for the columns; and, according to the direction of the fluid flow on the pontoons: Cm = 2.02 or Cm = 2.22 .

Wind forces acting on the hull and deck are given by (API, 1991):

$$Fv = \frac{1}{2}\rho_{ar} V_v^{\ 2} A_{ww}$$

where ρ_a is the specific mass of air, V_w is the total wind velocity (V_w = mean 1-hour velocity + fluctuating component), and A_{ww} is a factor representing the sum of the exposed areas multiplied by the respective drag coefficients. For wind incidences of 0 and 45 degrees, the following values are taken, respectively: A_{ww} = 2312 m^2 and A_{ww} = 2867 m^2 .

In order to obtain the maximum forces on the tendons, the analyses presented in this work consider the least favorable environmental condition, where wave, wind and current are assumed to be simultaneously acting in the SW direction, diagonal in relation to the platform. The verification of the minimum effective tension at the bottom of the tendon is performed considering that current is not acting. It should be observed that other environmental conditions should also be considered in an actual design procedure for the tendons; the sequence of analysis with the condition considered here is intended to illustrate such procedures.

The environmental data considered for the analyses, for a recurrence time of 100 years, are the following: **Wave**: significant height H_s = 7.8 m ; mean value of the ascending zero periods T_z = 11.5 s; **Wind**: V_w(10 min) = 37.22 m/s, acting in an elevation of 10 meters above average sea level; **Current**: Velocity profile ranging from 1.34 m/s at the surface, to 0.54 m/s at a water depth of 915 m, and 0 m/s at seabottom; **Tide**: Since the reference level of the TLP is the average sea level, the tide variation in relation to the initial condition is taken as 1.0 m up and 0.9 m down. Crack-growth of 70 mm on the tendon thickness is considered, down to a water depth of 50 meters.

Representation of the Tendons

Remembering that, in an uncoupled analysis of the TLP, the tendons are represented by springs, some approximations must be considered in order to take into account the mass of the tendons, and also the hydrodynamic loadings acting on the tendons.

In particular, the consideration of the current forces acting on the tendons is important for the environmental conditions on the Campos Basin, where the polygonal current profile presents high values of current velocity. This fact makes the current on the tendons to be the cause of considerable part of the horizontal drift of the TLP, as will be demonstrated by the parametric analyses presented in the following sections.

Here we consider that one third of the mass of the tendons is concentrated at the points where the tendons connect to the hull, following recommendations of the API-RP-2T code (1987).

The drag forces due to the current on the tendons are very significant and cannot be omitted. They are represented by introducing and additional force on the hull, calibrated by increasing the drag coefficient of the pontoons on the horizontal direction, following a procedure outlined in (Alves, 1996). The increase of the drag coefficient affects the viscous damping in the horizontal directions; this should not alter the maximum tendon tensions, that are more sensitive to the vertical damping.

Analysis Results

The nonlinear time domain dynamic analysis of the platforms movements is then performed, adopting a time step value of 0.1s and a total simulation time of 1200s. This latter value is conservatively selected considering Chou (1983), that states that 100 simulation cycles are sufficient to generate the necessary data for the prediction of the extreme forces of the tendon. Assuming that the tendon responds with the period of the waves (aprox. 9.4s), 940s simulation time would be enough.

The sea spectrum is converted in a sea elevation signal, considering that the sea is represented by a finite number of sinusoidal waves (Chakrabarti, 1987).

Parametric studies are performed to assess the influence of the consideration of the diffraction theory, and the relative participation of wave, current and wind on the TLP movements and the tension on tendon top. The main results of the analyses are time histories of the platform movements and of the forces on the tendons.

Tables 2, 3 and 4 present values of the statistical parameters (mean and standard deviation) calculated from the results of the analysis with and without the consideration of the diffraction theory. The comparison shows that diffraction effects are not very important in this case; with extreme waves for which the ratio λ/D is equal to 7.6 (for λ = 138 m and D = 18.2 m). In particular, it can be observed that the standard deviation of the surge motion is 32% higher when diffraction effects are considered. This is due to the more accurate consideration of the slow drift term due to the waves, that is provided by the diffraction theory.

Table 2 - Surge and Heave Movements

Wave Forces	Surge		Heave	
	Mean (m)	Std Dev. (m)	Mean (m)	Std Dev. (m)
Morison	42,38	1,21	1,87	0,11
Difr.+Morison	43,85	1,60	2,00	0,14

Table 3 - Roll and Pitch Movements

Wave Forces	Roll		Pitch	
	Mean (deg)	Std Dev. (deg)	Mean (deg)	Std Dev. (deg)
Morison	0,0326	0,0385	0,0338	0,0469
Difr.+Morison	0,0347	0,0363	0,0358	0,0465

Table 4 - Tension on Tendons

Wave Forces	Max. Tension		Min. Tension	
	Mean (kN)	Std. Dev. (kN)	Mean (kN)	Std. Dev. (kN)
Morison	10200,0	736,0	9000,0	693.0
Difr.+Morison	10400.0	745,0	9080,0	683,0

In order to evaluate the relative importance of the environmental loadings, parametric analyses are performed where wind, wave and current act isolately. The results are summarized in Table 5, from which we can observe that:

- Current is responsible for about 60% of the horizontal offset of the TLP, while the wind is responsible for 35%; also, current causes the higher static variation of the tensions, while waves generate most of the dynamic part of the tensions.
- The amplitude of the horizontal movements due to wind is greater than the corresponding to the waves. Wind is also the responsible for most of the pitch and roll mean rotation, while dynamic terms are dominated by the waves.

As it can be expected due to the nonlinearity of the problem, the sum of the average of the results for each isolated environmental loading is not equal to the result of the analysis considering every loading acting simultaneously.

Table 5 - Relative participation of wind, wave and current

		Wind	**Wave**	**Current**	**All**
Surge (m)	Mean	16,22	2,69	27,31	43,85
	Std Dev	2,75	1,34	1,16	1,59
Heave (m)	Mean	0,28	0,09	0,78	2,00
	Std Dev	0,09	0,02	0,06	0,14
Roll (deg)	Mean	0,025	0,002	0,006	0,035
	Std Dev	0,027	0,043	4,0E-04	0,036
Pitch (deg)	Mean	0,025	0,002	0,006	0,036
	Std Dev	0,014	0,043	6,0E-04	0,047
Max. Tension (kN)	Mean	8620	8010	8740	10400
	Std Dev	324	694	82	745
Min. Tension (kN)	Mean	7760	7930	8520	9080
	Std Dev	318	722	45	683

NONLINEAR DYNAMIC ANALYSIS OF THE TENDON IN THE TIME DOMAIN

The nonlinear dynamic analysis of the tendon in the time domain is performed, employing the ANFLEX program (Mourelle, Casaprima and Jacob, 1995). This program is based on the Finite Element Method, performing static and dynamic analyses where the nonlinear effects that characterize the behavior of the tendon are taken into account. This is accomplished by an incremental-iterative algorithm based in the standard or modified versions of the Newton-Raphson method (Bathe, 1982), where the tangent stiffness matrix is updated after each iteration, or kept constant during the iteration cycle and updated only in the beginning of a load (or time) step.

The differential equations of movement are integrated in time by the Hilber-Hughes-Taylor (1977) operator, implicit and unconditionally stable. The HHT operator combines the advantages of the traditional "trapezoidal rule" with a kind of numerical damping that does not interfere on the low frequency modes, where the ratio $\Delta t / T$ is lower than 0.1 , and grows exponentially for the high frequency modes. This improves the quality of the dynamic response, since it works as a filter for these latter modes, that do not have a significant participation on the response.

Calculation of hydrodynamic forces induced by wave and current on the tendon is performed in three stages: **i)** fluid velocities and accelerations are determined by an appropriate wave theory, and added to the current velocities, **ii)** the Morison formulation is employed to determine values of distributed loadings on the elements that discretize the tendon, and **iii)** Equivalent nodal loads are calculated. The Morison formulation is valid, since the dimensions of the cross-section of the tendon are small when compared to the wave lengths. The drag coefficient of the Morison formula is obtained according to Vandiver (1994), considering the amplification of this coefficient due to the oscillations of the tendon due to vortex-shedding effects.

The main characteristics of the tendons are summarized in Table 6. The tendon selected for the analyses is the one that presents the higher tension values obtained from the TLP movement analysis.

Table 6 - Characteristics of the Tendons

Diameter	0.711m (28")
Area of Steel	0.06777 m^2
Length	907.0 m
Yield Stress	490 Mpa
Added Mass Coefficient	2.0
Thickness	0.03175m (1.25")
Inertia	0.00392 m^4
Axial Stiffness	15325.2 kN/m
Pre-Tension	7946.1 kN
Drag Coefficient	1.7

In some situations the tendons of a TLP can have relatively small diameter values, as in the Hutton platform, where d≈0.25m. In such cases the bending effects can be safely disregarded. However, with larger tendons such as the one considered in this work, bending effects must be considered by employing frame elements in their discretization. Therefore, the tendon is discretized with nonlinear space frame elements based in the co-rotational formulation (Mourelle, 1993; Mourelle et al., 1995). This element is a powerful tool for the geometric nonlinear analysis, since it allows finite rotations in space and present good convergence characteristics.

The discretization of the tendon (Figure 4) comprises an uniform mesh with 184 elements with length of 5 meters. The flexible joints at the ends of the tendon are represented by generalized scalar elements, with rotational stiffness coefficients K_{rot} = 4200 kNm/rad.

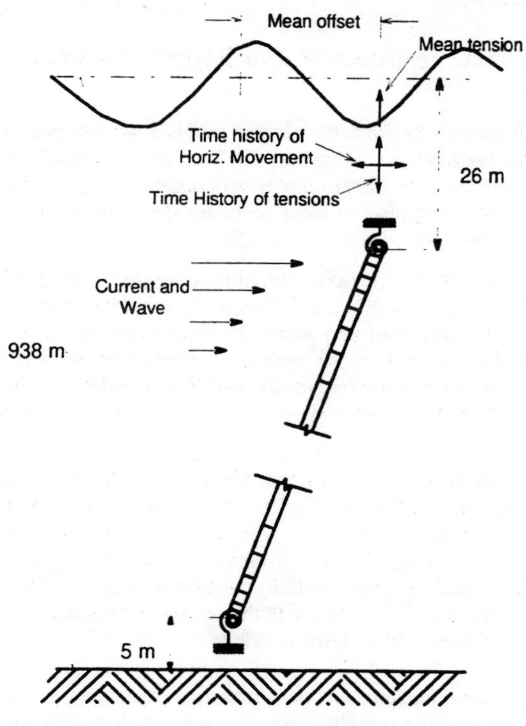

Figure 4 - Discretization and Loadings on the Tendon

In an uncoupled analysis, two alternatives may be considered to impose, on the top of the tendon, the results from the TLP movement analysis: **a)** Imposing horizontal and vertical movements, or **b)** horizontal movements and top tension. This latter alternative is the preferred (Fylling and Larsen 1984), since the former does not allow the accurate estimation of the axial force of the tendon. This is due to the fact that the tendon axial stiffness that is considered in the TLP movement analysis (k=EA/L, Figure 5) is larger than the stiffness that results from the analysis of the tendon itself, where the curvature of the tendon (that reduces the stiffness value) is considered.

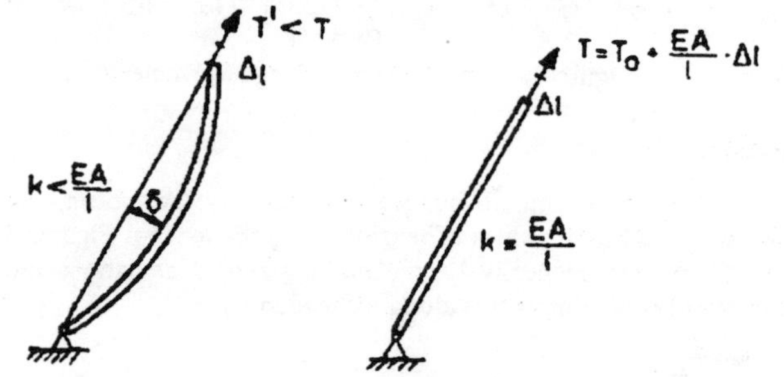

Figure 5 - Tendon Stiffnesses

Analysis Strategy

The analysis of the tendon is performed in two stages:

- A static analysis is performed first, where the axial tensions and horizontal displacements that correspond to the static offset are imposed at the top of the tendon. The static components of the environmental loadings are also applied, including self-weight, buoyancy and current. Figure 6 presents the sequence of application of these loadings. The percentage of the axial tension that corresponds to pre-tension of the tendon is applied first, in order to assure the stability of the system.

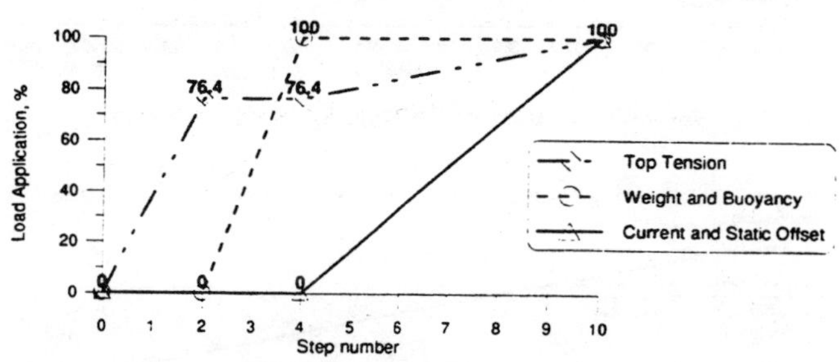

Figure 6 - Application of Loadings in the Static Analysis

- The dynamic analysis in the time domain is then performed; static loadings applied in the first stage are maintained, and time-histories of the dynamic components of horizontal movements and top tension are applied. These components correspond to the values obtained from the TLP movement analysis after 150 seconds of integration time, when the steady-state response had been reached (Figure 7).

As a result of this strategy, the CPU time required for the analysis is considerably reduced, since the transient that would occur if the static loads were applied incrementally in the dynamic analysis is eliminated.

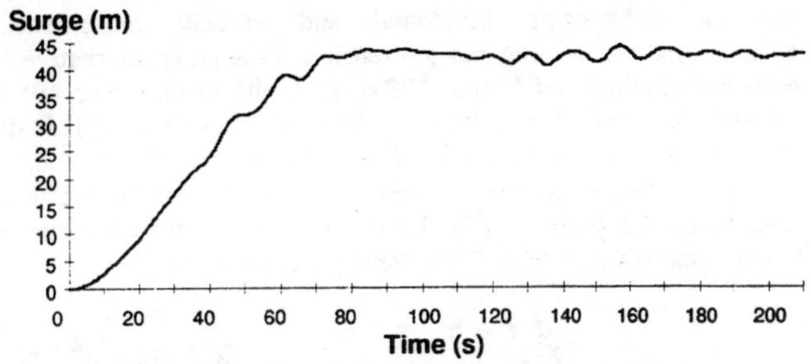

Figure 7 - Time History of Surge Movements

Results

Figures 8, 9 and 10 present, respectively, the mean, standard deviation, and total values of tension along the tendon. Figures 11, 12 and 13 present, respectively, the mean, standard deviation, and total values of bending moments along the tendon.

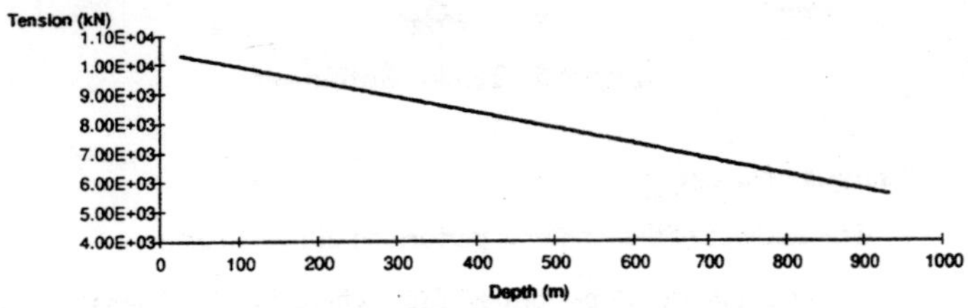

Figure 8 - Mean Tension on Tendon

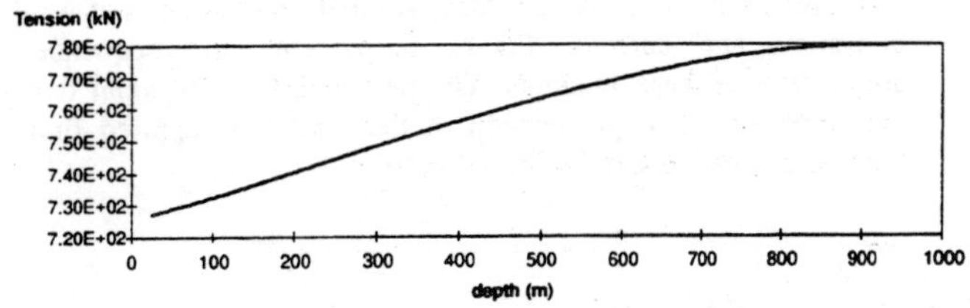

Figure 9 - Tension on Tendon, Standard Deviation

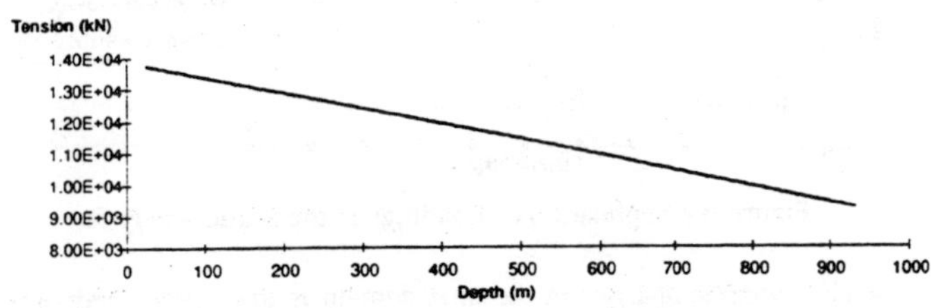

Figure 10 - Total Tension on Tendon

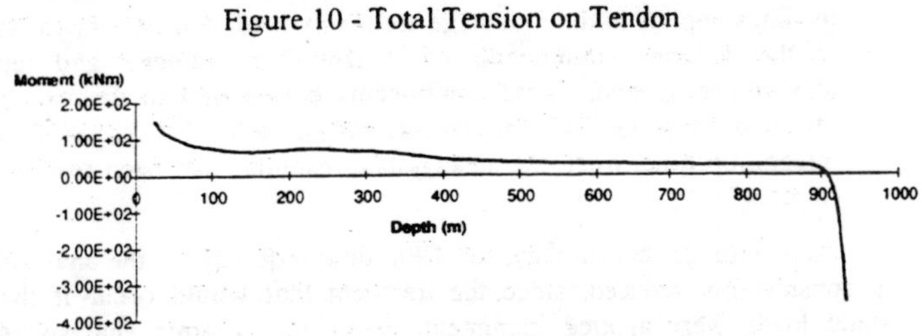

Figure 11- Mean Bending Moment on Tendon

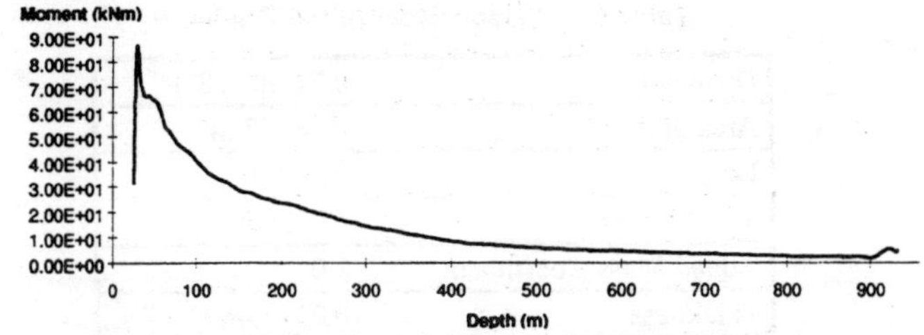

Figure 12 - Bending Moment on Tendon, Standard Deviation

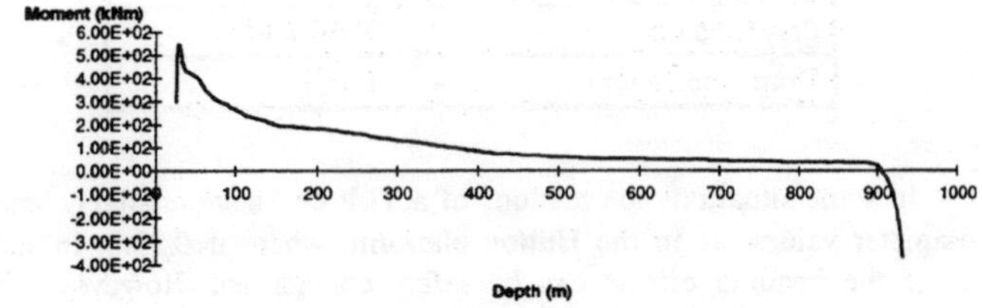

Figure 13 - Total Bending Moment on Tendon

Finally, in order to assess the influence of the consideration of the diffraction effects in the TLP movement analysis that generated the movements and tension imposed on the top of the tendon, another set of analyses of the tendon was performed, now employing results from a TLP movement analysis without diffraction effects. We observed that the difference between both results is small, more noticeable only for the mean tensions, that are lower when only the Morison approach is employed. This is due to the slow drift term of the waves, which is only considered by the Diffraction theory. Diffraction effects are expected to be more important for the environmental conditions considered in the fatigue analyses.

STRUCTURAL VERIFICATION AND FATIGUE ANALYSIS OF THE TENDONS

With the results from the tendon analysis we can proceed to its structural verification, following the Von Mises equivalent tension criteria, and the API (1991) interaction equation, thus demonstrating that the tendon complies to these codes for the extreme environmental condition considered.

We also verify the axial stability criterion, that involves the verification of the minimum effective tension at the bottom end of the tendon. This value must be positive, in order to assure that the tendon will not disconnect from its seabottom connection device. For this criterion the extreme environmental condition considers that there is no current acting on the tendon; this leads to a lower value of mean tension.

Next, the fatigue analysis of the tendon in the frequency domain is performed from transfer functions for the vertical movement on the top of the tendon, obtained from the composition of the transfer functions of the roll, pitch and heave movements (Chou, Ghosh and Huang, 1983) that resulted from a TLP movement analysis. The transfer function is then used together with the sea states from the wave scatter diagram of Campos Basin (Alves, 1996).

The determination of the structural damage is performed considering that the response spectrum presents a narrow bandwidth; therefore, the Rayleigh distribution (Price and Bishop, 1974) can be employed to obtain the stress range density function. The Wirsching correction (Wirsching and Light, 1980) is also employed.

The C curve from the DNV (1984) code is the S-N curve employed to obtain the allowable number of cycles. Two stress concentration factors are applied to the stress range amplitudes: the first factor, $SCF_1 = 1.30$, is determined from an expression given in the DNV (1984) code, to consider the misalignment of the tubes. The second factor takes into account other sources of uncertainty in the analysis, and is taken as $SCF_2 = 1.27$ (Alves, 1996).

As a result of the analysis for the considered sea states, we observed that the response bandwidth is always smaller than 0.8, and the mean period of the response is close to the dominant sea period.

Assuming only the Rayleigh distribution for the response spectrum, the obtained value for the estimated fatigue life of a tendon is 1025 years. When the Wirsching correction is applied, the fatigue life changed to 1235 years, an increase of 21%, which demonstrates that the Rayleigh distribution leads to conservative values. The consideration of the endurance limit of the S-N curve results in still increased values, 3703 and 4406 respectively without and with the Wirsching correction. It is observed that the API-RP-2T code recommends that the tendons present a fatigue life greater than 10 times the operational life of the TLP (which is 20 years for the platform considered here).

Finally, we should mention that this calculated damage refers to the first order movements of the TLP. Other sources of damage should also be considered, including the second order movements (springing effects), and vortex-shedding effects. Preliminary studies have suggested that this latter effect does not cause significant damage for the environmental conditions of the Campos basin.

FINAL REMARKS

The analysis of the tendons of tension-leg platforms has been the subject of many investigation activities in the last years, due to the increasing utilization of the TLP concept for the exploration of oil fields in deep waters. This work has, therefore, presented a sequence of analysis that define an analysis, design and verification procedure for the tendons of a TLP, considering the environmental conditions particular to the Brazilian seas. Of course, in the application of this procedure to real-life design activities, other possible loading situations should be taken as the extreme, operation and damaged conditions, as recommend the API-RP-2T code.

Other issues related to the design of the tendons of a TLP should also be considered, including fatigue calculations for the tendon connections, and consideration of the hardening of the flexible joints when dynamically excited with small angles.

Further developments in the definition of an analysis procedure of the tendons of a TLP should consider the incorporation of the results of studies on advanced topics related to the nonlinear dynamic behavior of this class of platforms, including: incorporation of high-frequency effects (*springing* and *ringing*) in the fatigue calculation, and the vortex-shedding effects and its influence on the determination of the drag coefficient of the Morison formulae.

Finally, the validity of an uncoupled model for water depths greater than 1000 meters should also be assessed, by comparing its results with those obtained from a fully coupled analysis of the hull-tendon assembly.

ACKNOWLEDGMENTS

The authors would like to acknowledge the support provided by Petrobras, that included the permission for the publication of some preliminary results of the design of a TLP. In particular, we would like to acknowledge the contribution of the CENPES/PETROBRAS TLP design staff that contributed with the accomplishment of this work.

REFERENCES

Alves, L.H.M. (1996), *Dynamic Analysis of the Tendons of a Tension Leg Platform*, M.Sc. Thesis (in Portuguese), COPPE/UFRJ, Civil Engineering Department.

American Society of Mechanical Engineers (ASME) (1992), *TLP Design Technology Seminar*, Houston, Texas, USA.

American Petroleum Institute (API) (1987), *API-RP-2T Recommended Practice for Planning, Designing and Constructing Tension Leg Platforms.*

American Petroleum Institute (API) (1991), *Recommended Practice for Planning, Designing and Constructing Fixed Offshore Platforms, Working Stress Design*, Twentieth Edition.

Aveleda, A.A. (1996), *Data Generation and Signal Processing in Offshore Engineering*, M.Sc. Thesis (in Portuguese), COPPE/UFRJ, Civil Engineering Department.

Barltrop, N.D.P. and Adams, A.J. (1991), *Dynamics of Fixed Marine Structures*, Butterworth-Heinemann Ltd.

Bathe, K.J. (1982) *Finite Element Procedures in Engineering Analysis*, Prentice Hall.

Chakrabarti, S.K. (1987), *Hydrodynamics of Offshore Structures*, Computational Mechanics Publications.

Chou, F.S.F., Ghosh, S., Huang, E.W. (1983), "Conceptual Design Process of a Tension Leg Platform", *SNAME Transactions*, Vol. 91.

Det Norske Veritas (DNV) (1990), *Rules for Classification of Mobile Offshore Structures*, Part 3.

Det Norske Veritas (DNV) (1984), *Classification Notes N-30.2, Fatigue Strength Analysis for Mobile Offshore Units.*

Faltinsen, O.M. (1990), *Sea Loads on Ships and Offshore Structures*, Cambridge University Press.

Fylling, I.J. and Larsen, C.M. (1984) "Nonlinear Static and Dynamic Behavior of Tension Leg Platform Tethers", *Procs. of the 3th Int. Conf. on Offshore Mechanics and Arctic Engineering - OMAE*, pp. 551-555, Houston, Texas/USA.

Hilber, H.M., Hughes, T.J.R., Taylor, R.L. (1977), "Improved Numerical Dissipation for Time Integration Algorithms in Structural Dynamics", *Earthquake Engineering and Structural Dynamics*, vol. 5

Meirovitch, L. (1970), *Methods of Analytical Dynamics*, McGraw-Hill Book Company.

Mourelle, M.M. (1993), *Dynamic Analysis of Structural Systems composed by Risers and Anchor Lines*, D.Sc. Thesis (in Portuguese), COPPE/UFRJ, Civil Engineering Department

Mourelle, M.M., Casaprima E.G., Jacob B.P. (1995), "ANFLEX - Computational System for Flexible and Rigid Riser Analysis" in: F.L.L.B. Carneiro et alii (eds), *International Offshore Engineering*, pp. 441-458, John Wiley & Sons, Chichester/New York.

Paulling, J.R. (1992), *TDSIM6 Time Domain Platform Motion Simulation with Six Degrees of Freedom, Theory and User Guide*

Price, W.G. and Bishop, R.E.D. (1974), *Probabilistic Theory of Ship Dynamics*, John Wiley & Sons.

Wirsching, P.H. and Light, M.C. (1980), "Fatigue under wide band random process", *Journal of the Structural Division - ASCE*, July 1980

Sircar, S., Kleinhans, J.W., Prasad, J. (1993), "Impact of Coupled Analysis on Global Performance of Deep Water TLP's", *OTC 7145*

Vandiver, J.K. (1994), *SHEAR7 Theoretical Manual - Program for Vortex Induced Vibration Response Prediction of Beams or Cables with Linearly Varying Tension in Non-Uniform Flow or Uniform Flow*, MIT Press.

Proceedings of the Seventh (1997) International Offshore and Polar Engineering Conference
Honolulu, USA, May 25-30, 1997

ISBN 1-880653-28-1 (Set); ISBN 1-880653-29-X (Vol. I)

UOE-Formed Tendon Pipes for Deep Water Oil & Gas Exploration in the Gulf of Mexico

Tomoaki Ikeda, Kazushi Ohnishi, Akio Yamamoto, Makoto Nagase and Izumi Takeuchi
Sumitomo Metal Industries, Ltd.
Kashima, Japan

James D. Smith
Shell Offshore Inc.
Houston, TX, USA

ABSTRACT

This paper describes the development, pre-qualification testing and production properties of heavy wall-thickness TLP tendon pipes for the TLP projects in the Gulf of Mexico, X-60, 26" O.D. x 1.300"W.T. (AUGER), X-70, 28"O.D.X1.200"W.T. (MARS, RAM/POWELL), X-65, 32"O.D. X 1.500"W.T. (URSA).

The AUGER TLP, installed in 2,860 ft of water in 1994, was the first step-wise development of deep water Gulf of Mexico discoveries. This success in subsea field exploitation accelerated the consecutive deep water TLP construction ,MARS, RAM/POWELL, and URSA, was based on the lessons obtained. The thick-walled tendon pipes were also developed in this stream with the TLP construction projects.

The tendon pipes for AUGER TLP was the first thread, manufactured by UOE process using proprietary TMCP-practiced steel plate with double-submerged arc seam welding (DSAW). The development of tendon pipes for MARS, RAM/POWELL, and URSA TLP's went on stream. Although the steel grade or wall thickness was increased more than for AUGER tendons, its excellent weldability with low carbon equivalent had to be maintained in order to keep the fabrication cost equivalent to the application of no-preheat easy welding methods. From these points of view, the same control ranges in chemical compositions were applied for all grades and thicknesses but with some modification in the TMCP parameters and reinforcement in the UOE press machine for higher grade and heavier wall thickness.

The evaluation test results and production history show the superb weldability and excellent fracture toughness for every tendon pipe and within tight production tolerance. This also

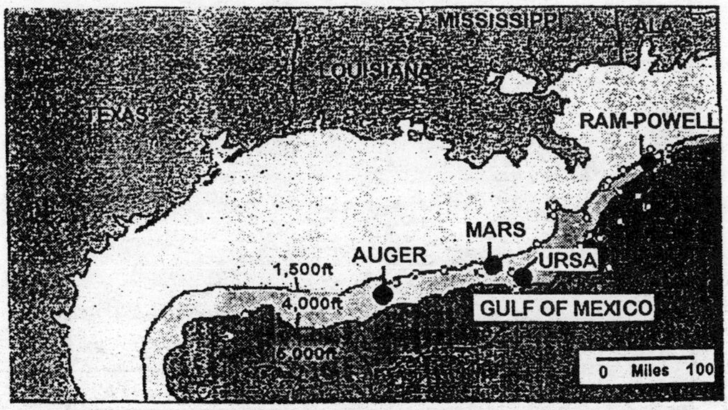

Fig.1 Location of Shell TLP's in the Gulf of Mexico

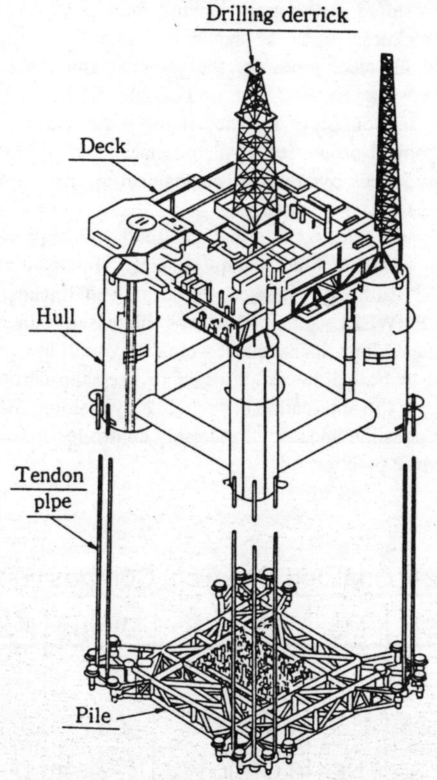

Fig.2 General structure of TLP

Table 1 TLP Projects in the Gulf of Mexico

Project	AUGER	MARS	RAMPOWELL	URSA
Water Depth	2,860ft	2,933ft	3,220ft	4,102ft
Steel Supply for Tendon Pipe	5,300mt	5,520mt	6,000mt	14,000mt
Pipe Size	26"O.D.X1.300" W.T.	28"O.D.X1.200" W.T.	28"O.D.X1.200" W.T.	32"O.D.X1.500" W.T.
Pipe Grade	X-60	X-70	X-70	X-65
Production Start	1994	1996	1998	2000

enabled the rapid consecutive construction TLP's in the Gulf of Mexico setting world records. For the success of all these TLP construction projects were a result of the concentrated efforts of TLP design engineers, construction fabrication contractor, and the steel maker.

1. INTRODUCTION

In the Gulf of Mexico, new construction of offshore structures is rapidly progressing based on expectation of a world-wide shortage of oil in the beginning of the 21st century, resulting from a sharp increase in the demand for oil in China and India.

Nowadays, major oil companies are planning to continue investing heavily in deep water exploration and development in the Gulf of Mexico. This trend was triggered by the success of AUGER and MARS projects. Shell Offshore Inc. has also been constructing huge tension leg platforms (TLPs), RAMPOWELL and URSA in the Gulf of Mexico. These are to set the new world record of water exploration depth. Also, enlargement of the TLPs has resulted in the great demand for higher grade and heavier wall thickness pipes, as shown in Table 1. Sumitomo Metal supplied all steel pipes for the projects and the supply quantity is increasing year after year, or TLP after TLP.

At first, the tendon pipes for the Auger project, which had excellent mechanical properties, HAZ toughness and weldability, have been produced by thermo mechanical control process (TMCP), in 1991.

And TMCP was selected due to their good weldability, with some modified stringent quality requirements. In other words, by using TMCP, although the steel grade or wall thickness for MARS, RAMPOWELL and URSA was increased more than that of AUGER tendons, its excellent weldability with low carbon equivalent had to be maintained in order to keep the fabrication cost equivalent to the application of preheat easy welding method. This was only possible because of precisely controlled innovative proprietary TMCP practice.

Table 2 Qualified Chemical Composition

C	Si	Mn	P	S	Others	Pcm
0.04 /0.08	0.05 /0.30	1.34 /1.64	≦ 0.013	≦ 0.005	Cu-Ni -Cr-Nb-Ti	0.14 /0.19

Secondly, the chemical composition and the steel making process were critically examined and optimized to ensure the highest possible strength level with adequate fracture toughness. In addition, high heat affected zone (HAZ) toughness, has also been kept within the same control ranges in chemical compositions for all projects.

This paper mainly describes the development, pre-qualification testing, and production properties of the URSA, X65 grade 32"O.D. X 1.500"W.T. steel pipes. The heavy wall thickness URSA pipes were produced by reinforcing TMCP facilities and UOE press machine.

2. METALLURGICAL DESIGN

One of the most important qualities for the offshore structural steel is to ensure safety of the welded structure through sufficiently high toughness in heat affected zone (HAZ). Lowering carbon content is effective in improving fracture toughness of welded joints, and TMCP enables us to achieve high strength and high toughness of plates with lower carbon content through controlled rolling and accelerated cooling.

In particular, the following three points are considered in manufacturing of the URSA pipes :

1) High critical HAZ CTOD value at +4° C at high and low heat input / inter pass temperature combination
2) Low temperature toughness of base metal (Charpy transition temperature of less than or equal to that of AUGER pipe)
3) Good weldability to enable low preheat temperature

<u>Chemical Composition</u>

As mentioned before, HAZ toughness should be considered as the first step in chemical design. And the mechanized automatic gas-metal arc girth welding method has been widely employed in construction of TLP, owing to its high efficiency. However, this welding method tends to raise the HAZ hardness due to its low heat input welding.

Fig. 3 Microstructure of URSA Pipe (at mid-thickness)

In the case of multi-pass welding, HAZ toughness will deteriorate due to a local brittle zone (LBZ). Therefore, Pcm values were controlled to be at the low necessary for good weldability without preheating.

In addition, brittle fracture initiation properties due to the LBZ have been evaluated with the CTOD test. CTOD properties in HAZ are impaired by the increase of the initial austenite grain size and by the increase of the fraction of M-A constituent. The M-A constituent is especially harmful because it has a tendency to nucleate micro-cracks easily. The M-A constituent is strongly affected by C content. So C content should be lowered. Also, Cu-Ni were added to improve strength with less impairment of weldability and fracture toughness.

Moreover, to achieve the specified performances of base metal and welded joints, a precisely controlled small amount of Nb-Ti microalloying elements was added on the basis of recent research into the toughness improvement resulting from the precipitation effect of Nb-Ti complex compounds.

Low C-Cu-Ni-Nb-Ti microalloyed steel was developed considering the properties of the base metal, and basis of controlling LBZ in the HAZ.

The control range of chemical composition of the developed steel is shown in Table 2. And the same range is applied to all projects, AUGER, MARS, RAM-POWELL and URSA, in spite of variation of steel grade and thickness.

Table 3 Manufacturing Process

Project	AUGER	MARS.RAMPOWELL	URSA
Pipe Size	26"O.D.X1.300"W.T.	28"O.D.X1.200"W.T.	32"O.D.X1.500"W.T.
Pipe Grade	X-60	X-70	X-65
Plate Rolling	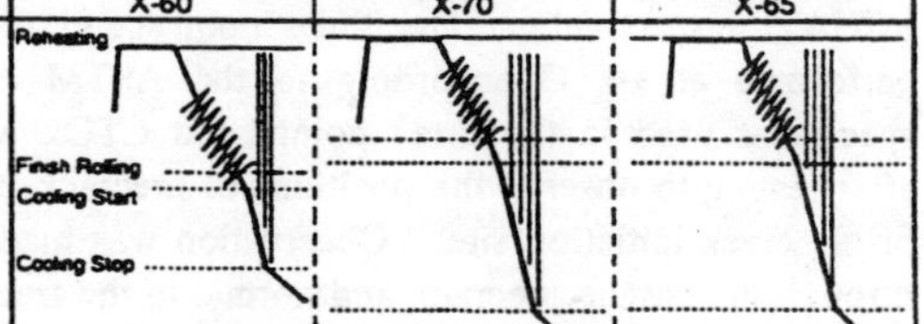		
Pipe Forming	C press→U press→O press→Welding→ Expanding→X-ray→Dimension check		

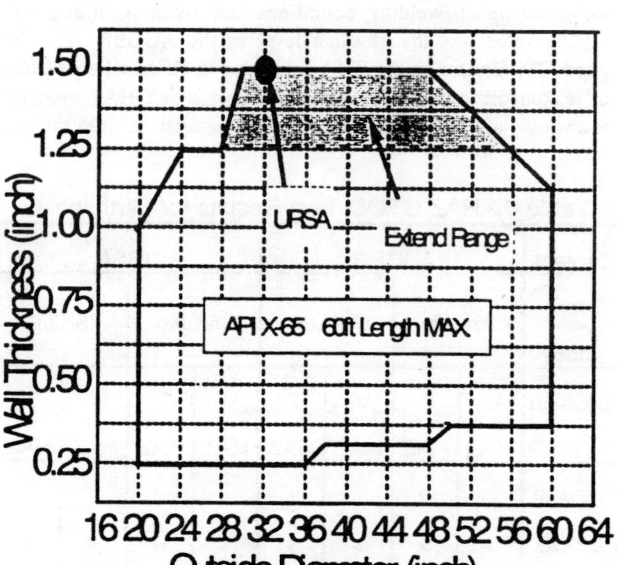

Fig.4 Available Size Range of UOE mill

Microstructure

In the case of heavy wall thickness plate with low carbon, there tends to be coarsened grain size in the mid-thickness portion and a deteriorate on toughness. However, in the base metal of URSA pipe, the fine formed bainite microstructure was mainly observed at the center thickness as shown in Fig. 3, even though it was a heavy wall thickness pipe. This microstructure was obtained by reduction at non-recrystallized austenite region, followed by cooling from this temperature range. This microstructure resulted in excellent toughness of the base metal in the center portion.

3. MANUFACTURING PROCESS

The manufacturing process for the steel plate is shown in table 3. In the steel making process , impurity contents such as P ,S and H_2 gas were sufficiently reduced to low levels. The steel was continuously cast into 300 mm thick slabs. In the plate rolling process, the slab reheated to high temperature was controlled rolled mainly in the non-recrystallized austenite region and then accelerated cooled to the appropriate temperature range. The rolling and accelerated cooling conditions were adjusted in accordance with the steel grade or wall thickness for AUGER, MARS, RAMPOWELL and URSA projects . In the case of AUGER pipes, the following TMCP conceptual factors were controlled to obtain specified strength and toughness of base metal:

1) The slabs were reheated at high temperature to dissolve Nb-Ti complex compounds

2) The lower finishing rolling temperature and higher accelerated cooling stop temperature were adopted, because of the relatively lower grade and lesser thickness.

The AUGER pipes, which were manufactured by the following the above procedure, had excellent base metal properties. However, in the case of the other project pipes, steel grade and /

Table 4 Pre-Qualification Test Items for Plate and Pipe

Project			AUGER					URSA				
Plate	Thickness Position	Direction	Test Results									
Tensile Test (API strip specimen)	1/2t	T	YS (ksi)	TS (ksi)	YR (%)	El (%)	RA (%)	YS (ksi)	TS (ksi)	YR (%)	.El (%)	RA (%)
			67	82	82	31	76	69	87	79	30	79
Charpy Impact Test			vTrs = -73°C					vTrs = -83°C				
Drop Weight Test			NDT Temp. = -50°C					NDT Temp. = -70°C				
Pipe	Thickness Position	Direction	Results									
			YS (ksi)	TS (ksi)	YR (%)	El (%)	RA (%)	YS (ksi)	TS (ksi)	YR (%)	El (%)	RA (%)
Tensile Test (B.M)	Full	T	67	88	76	48	64	73	89	82	56	77
	1/2t	L	73	85	86	29	69	73	85	85	27	82
All weld	1/2t	L	93	101	-	25	71	88	95	-	29	75
Charpy (W.M)	1/2t	T	vE-15.6°C = 163ft-lb					vE-15.6°C = 142ft-lb				
(HAZ)			vE-5.6°C = 185ft-lb					vE-5.6°C = 266ft-lb				
Drop Weight Test	1/2t	T	-23.4°C No break					-23.4°C No break				
CTOD Test (B.M)	1/2t	L	+3°C>1.411mm					+4°C>1.465mm				

or wall thickness are more severe. This is the reason why we needed to change TMCP parameters. In order to increase base metal strength, the finishing temperature and cooling start temperature were raised, and cooling stop temperature was decreased in comparison with those of AUGER. In general, this changes of TMCP parameters mentioned above would result in the deterioration of base metal toughness. However, rolling condition before accelerated cooling, such as reduction ratio per one pass rolling, total reduction amount in the non-recrystallized region and so on, was changed at the same time. As this result, base metal toughness of URSA pipe has become better than that of AUGER pipe.

By the way, in order to achieve this wide range of accelerated cooled steels, the capacity of TMCP facilities has been expanded. The water supply for TMCP facilities has been increased to insure the cooling rates. The improvement of mechanical properties, the decrease of mechanical properties dispersion, and improvement of plate flatness were expected from the increased capacity of the TMCP facilities. As a result of the cooling rate increase with new TMCP facilities, all project pipes could be manufactured with the same chemical composition.

In the pipe forming process, the available range of UOE mill is shown in Fig. 4. The present press machine could not form the tendon pipes for URSA because of high strength (X-65), heavy wall thickness (1.50") and length (60ft-length). To deal with this, we have reinforced the press machines in UOE mill to make possible higher grade and heavier wall thickness pipe production. In order to manufacture the heavy wall thickness pipe, the U-press power had to be increased and the other press machines (C-press and Expander) were slightly modified. In particular, the rocker-arm of the U-press machine was altered to be able to bend the steel with increased bending moment. On the other hand, the oil pressure of each press machine was not altered in order to save construction costs and to delivery of the pipes as quick as possible. As shown in Fig. 4, after this reinforcement, the available range was extended, and the URSA pipes could be manufactured.

Table 5 Pre-Qualification Girth Welding Conditions.

Welding Method	GMAW	SAW
Groove Shape	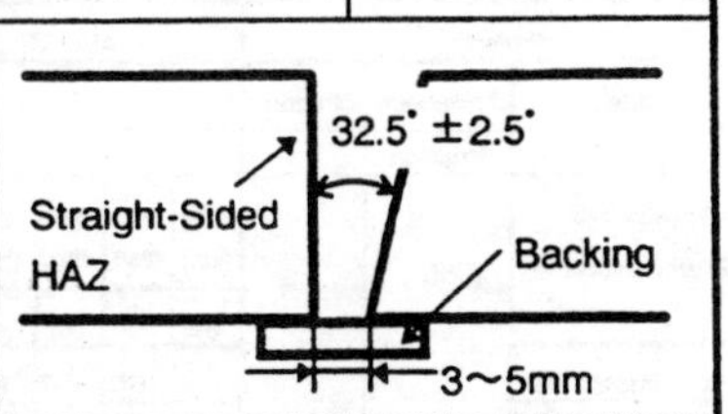	
Heat Input	25KJ/in	75KJ/in
Welding Material	SM-1NT (1.2 φ) (80%Ar+20%CO_2)	W36(4 φ)+BL55
Preheat Temperature	Ambient Temp.	149° C (300° F)
Interpass Temp.	21-35° C (70-95° F)	249-263° C (480-505° F)

4. PRE-QUALIFICATION(PQ)

Mechanical Properties of Base Metal

Pre-qualification (PQ) test for base metal was performed on plate and pipe for Auger and URSA project.

Drop weight test (DWT) , in addition to Charpy impact test, was performed for plate in order to determine nil-ductility temperature and Charpy transition temperature, respectively. The base-metal PQ test results are shown in Table 4. The strength of the URSA pipes satisfied the requirements of API X65 grade. Moreover, the steel plate had good toughness equivalent to that of AUGER plate.

HAZ toughness

The performance of the girth weld is the greatest concern for the welded-type tendon pipes. To evaluate this performance, girth welded joints were investigated at a lower heat input (25kJ/in) by GMAW and a higher heat input (75kJ/in) by SAW. Table 5 shows the groove shapes and the welding conditions. Controlled interpass temperature at ambient (≦38° C) was coupled to the 25kJ/in heat input to generate maximum cooling rate in HAZ. Full thickness CTOD tests were performed to examine the HAZ toughness at the vertical straight bevel side. The notch, which was machined along the straight fusion line of the K-bevel and fatigue-precracked before testing, was located at THAZ (Transformed-HAZ : that is along the fusion line) and VHAZ (Visible-HAZ : that is along outer line of visible HAZ). THAZ and VHAZ were analogous to API RP2Z's CGHAZ and SCHAZ, respectively. The three point bend CTOD test was performed at +4° C according to the ASTM E 1290. The specimens used in the three point bend CTOD were sectioned after testing to observe the positions of precracked notch tip and brittle crack initiation site. Observation was made on the plane through the fatigue precrack and normal to the fracture surface of the three point bend CTOD test specimen. Twelve specimens were targeted for THAZ and VHAZ, respectively. The CTOD test results was shown in Table 6 in comparison to AUGER pipe with URSA pipe. The steel for the URSA TLP had good HAZ toughness in all welding conditions and notch positions. The CTOD value was the at same level as the AUGER one. The good CTOD value in the THAZ was due to chemical equivalence of it composition for AUGER pipes, because CGHAZ toughness is strongly affected by steel chemical composition. On the other

Table 6 HAZ CTOD Test Results for Girth weld.

Project	AUGER				URSA			
Heat input	25KJ/in		75KJ/in		25KJ/in		75KJ/in	
position	V-HAZ	T-HAZ	V-HAZ	T-HAZ	V-HAZ	T-HAZ	V-HAZ	T-HAZ
Critical CTOD Value (mm)	> 1.307	> 1.356	> 1.338	> 1.345	> 1.004	> 0.930	> 1.463	> 1.054

hand, the toughness in the VHAZ was strongly affected by the toughness of base metal and it has been proved that the toughness increase in base metal due to grain refining is effective in improving the SCHAZ toughness. The toughness of the base metal of URSA pipe was at the same level as that of AUGER as a result of adjusting TMCP condition. These levels were substantially higher than the specified requirement, indicating a significant margin of safety.

5.WELDABILITY

The test methods are shown in Table 7. No cold cracking was observed on both plates, which were designed with lower Pcm value, in spite of artificially high diffusible hydrogen levels. This test resulted that the design of chemical composition was emphasized to enable preheating-free welding.

Table 7 CTS Cracking Test.

Welding Conditions	
(1) Welding Process	: SMAW (Shielded Metal Arc Welding)
(2) Filler Metal	: 4.0mm dia.
(3) Position	: Flat (1G)
(4) Preheat	: None
(5) Post weld Heat	: None
(6) Welding Technique	: Single Pass Single Electrode
(7) Joint	: T-Joint
(8) Welding Parameter	
Current	: 180A
Voltage	: 23V
Speed	: 23cm/min
Heat Input	: 1.08KJ/mm (27.4KJ/in)
(9) Diffusible Hydrogen Measurements	: 10.54(ml/100g)

6. PRODUCTION RESULTS

Finally, SMI's recently mass-produced tendon pipes in the Gulf of Mexico were introduced. Fig.5 shows the production histograms of the chemical composition in ladle analysis for 25 heats of AUGER and 37 heats of URSA, respectively. Because the same control range in the chemical composition was applied for both pipes, the actually achieved ranges were within a extremely narrow spread. The carbon contents were low, in the 0.06 ~ 0.07% range, in order to keep lower Pcm values of 0.16 ~ 0.17% to maintain good HAZ toughness and superior weldability. Attention should be paid to the low phosphorus and low sulfur content to ensure the steels soundness. Phosphorus content didn't exceed 0.013%, and sulfur content was less than 0.002%.

Figs. 6 and 7 show the production histograms of the mechanical properties for AUGER and URSA plates.

Although the URSA plates were of heavier wall thickness than those of AUGER plates, the yield strength and tensile strength range of URSA plates were controlled to within the same range as the AUGER plates with little variation. This was due to the strict control of TMCP conditions. As a result, high strength equivalent to X70 grade and good toughness were obtained for URSA plates and uniformity of mechanical properties were also demonstrated in mass-production.

Additionally, production roundness measurement results for pipe were shown in Fig.8. The production roundness of URSA pipes range was the same as for AUGER pipes within a narrow range. The UOE pipe forming could be performed under strict dimension control within narrow ranges through reinforcement of U-press.

All of the production histograms exhibit the high-quality steel properties of the product.

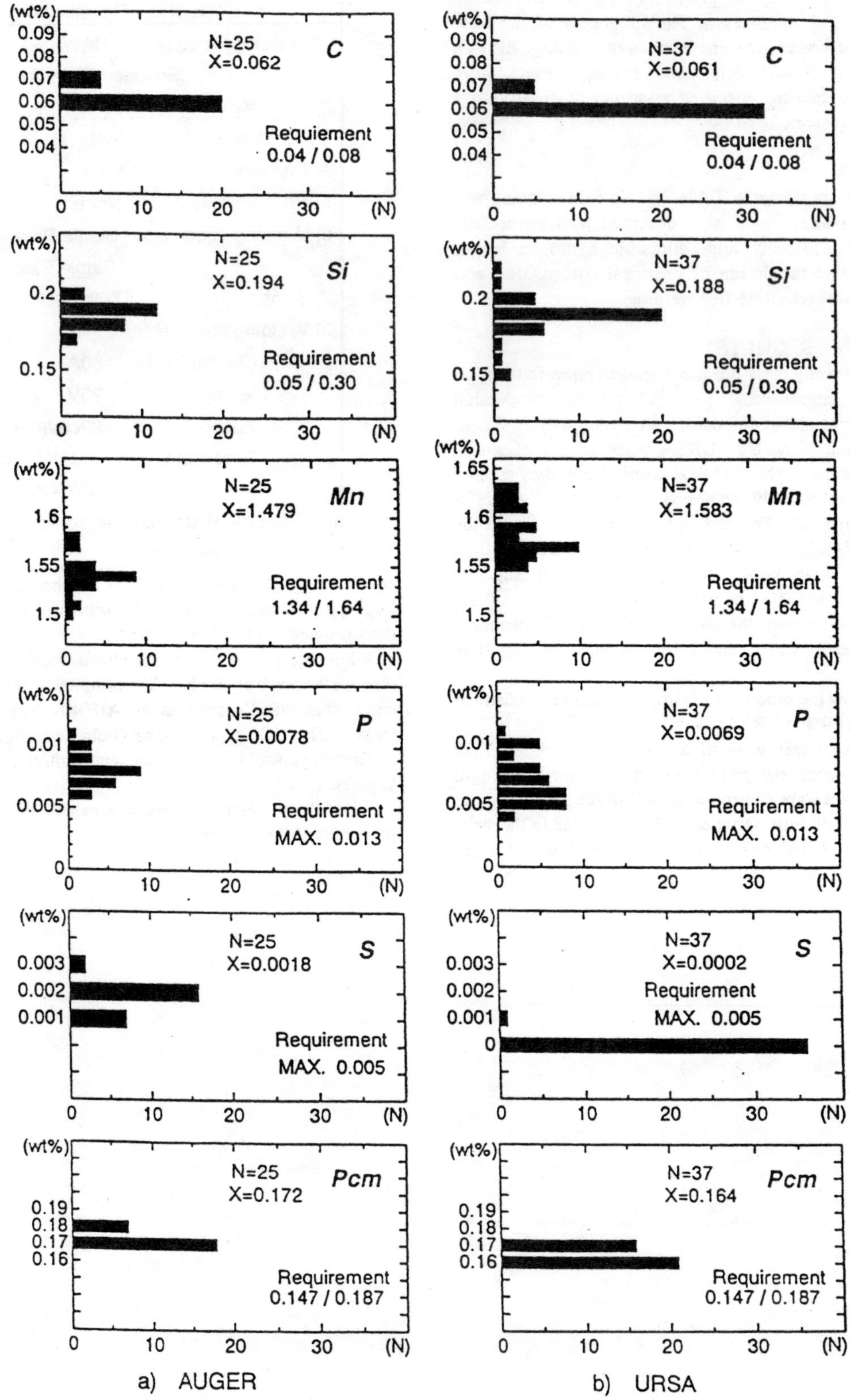

Fig 5 Chemical Composition of Mass Production

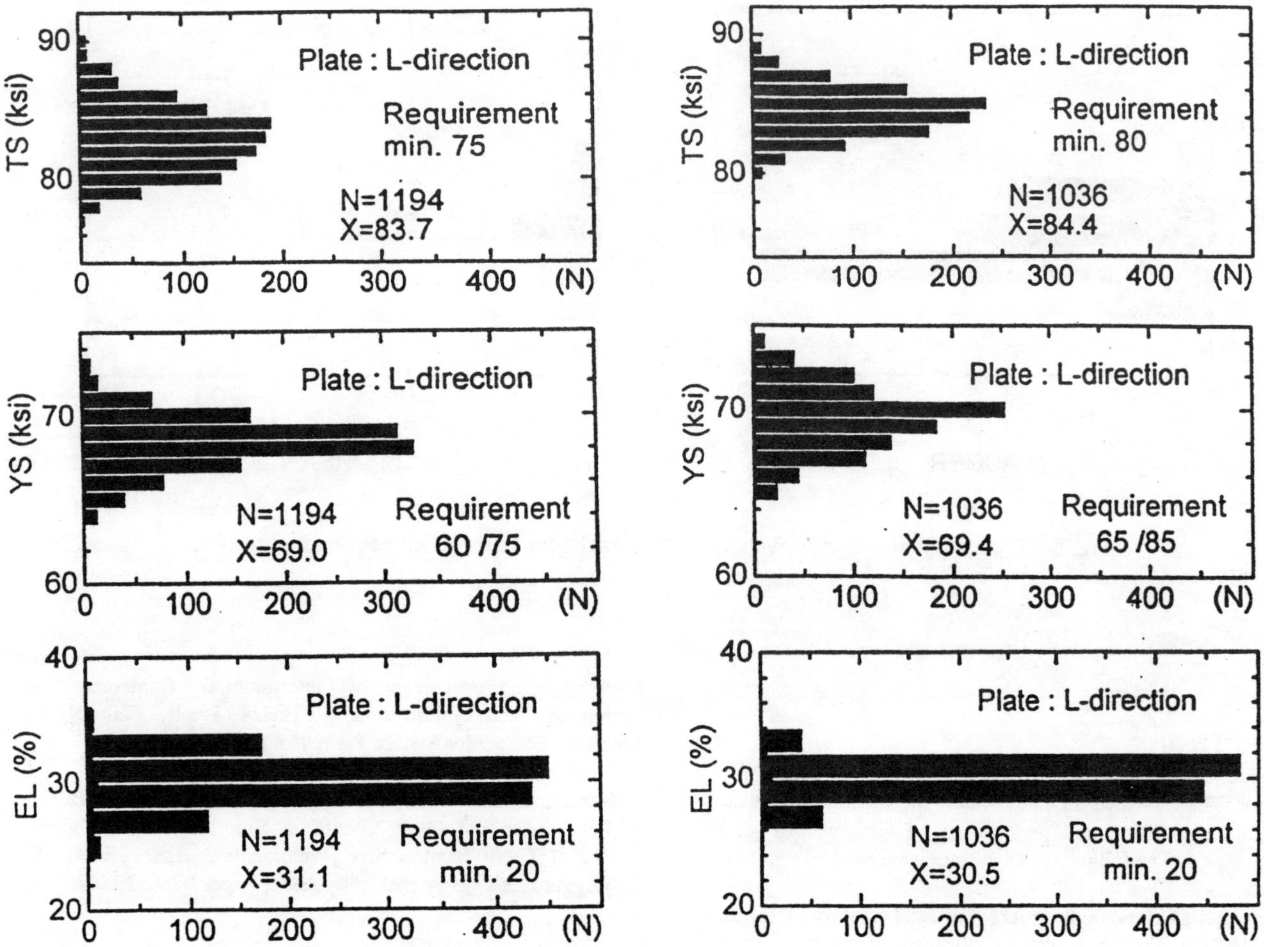

a) AUGER b) URSA

Fig 6 Tensile Test Data for Plate

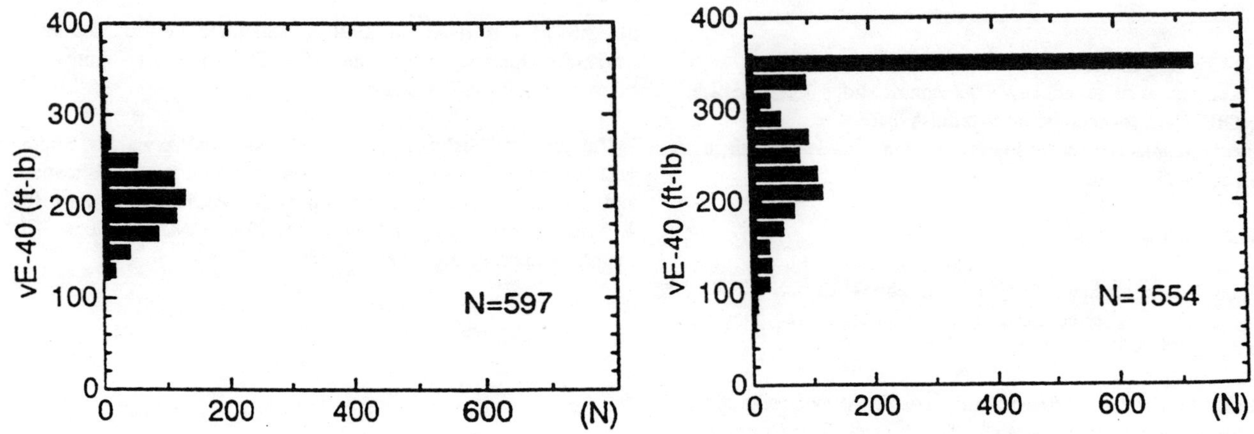

a) AUGER b) URSA

Fig. 7 Charpy Impact Test Data for Plate

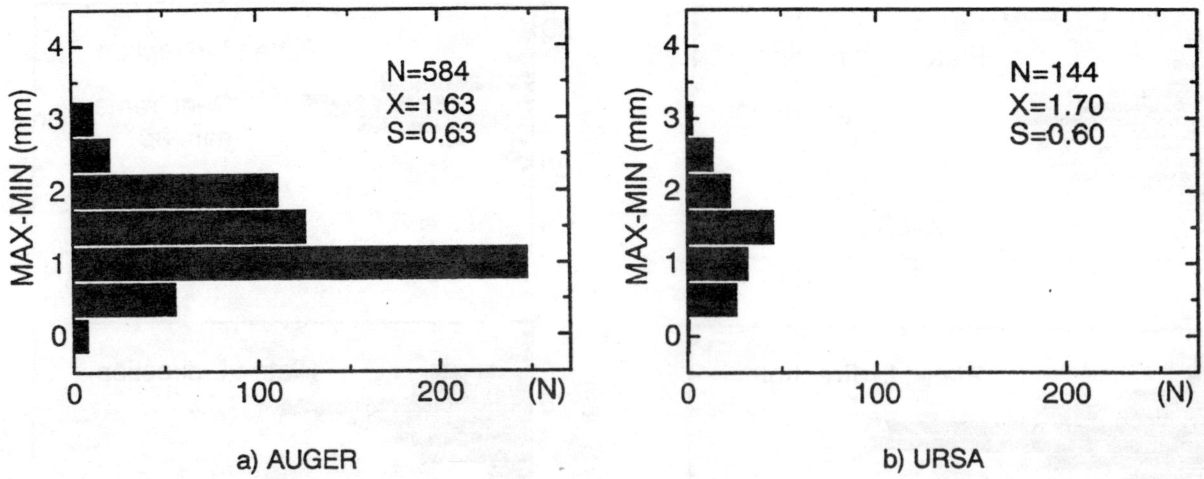

Fig. 8 Production dimensional Measurement Results for Pipe (Out-of-roundness)

7. CONCLUSION

In order to develop high strength and heavy wall tendon pipe used for TLP installed in deep water, manufacturing process, chemical composition and performance of girth welded joints were studied. The following conclusions were reached.

1) Various high strength up to API X70 and heavy wall thickness up to 1.5"WT UOE pipe can be manufactured with the same chemical composition by adjusting TMCP conditions and UOE facilities. Based on this development, tether pipes were successfully mass-produced within a strict control range of mechanical properties of the base metal.

2) Extremely low Pcm level of the steels allowed the implementation of lower preheat welding with resultant cost saving

8. ACKNOWLEDGMENT

The authors wish to acknowledge contributions by SHELL OFFSHORE INC. for permission to publish these here.

A Special thanks is directed to every person who worked for in the Gulf of Mexico project.

9. REFERENCES

1) Fukada, Y. and Komizo, Y (1992). "Study on Critical CTOD Property of Heat Affected Zone in C-Mn Microalloyed Steel", IIW Doc. IX-1664-92, SC-IX-A

2) Roberts, W. (1983). "Recent Innovation in Alloy Design and Processing of Microalloyed Steels" HSLA Steels Technology & Application, Proceedings of International Conference on Technology and Applications of HSLA Steels, Philadelphia, American Society for Metals, Paper No. 8306-072

3) Suzuki, S., and Weatherly, G. C., (1985). "Characterization of Precipitates and Grain Grouth in Simulated HAZ Thermal Cycles of Ti-Nb Bearing Steel Weldments", HSLA Steels'85, Beijing, proceedings published by ASM, Paper No.8520-014

4) Suzuki, S., Weathery, G. C., and Houghton, D. C., (1987). "The Response of Carbo-Nitride Particles in HSLA Steels to Weld Thermal Cycles", Acta Metallurgica, Vol.35, No. 2, pp.341-352

5) SHELL (1995). "Supplement 8, SHELL OFFSHORE INC., Material Quality Level 1, Carbon and Low Alloy Steel , Seamless-Marine -Riser-Tubular Requirements 50-100 ksi SMYS 0.40"-0.75" wall Thickness."

6) ASTM STANDARDS (1996). "Standard Test Method for Crack-Tip Opening Displacement (CTOD) Fracture Toughness Measurement". ASTM E1290

7) Takeuchi, I., Nishimoto, K., Nagse, M., Nishizawa, M., Smith, J. D. and Bowen K. G. (1995). "Development, Pre-Qualification, and Production History of 60 ksi UOE Steel Tendon Pipe for AUGER Tension leg Plat Form". 1995 OMAE-Volume III, materials Engineering ASME 1995, pp347-355

Proceedings of the Seventh (1997) International Offshore and Polar Engineering Conference
Honolulu, USA, May 25-30, 1997

ISBN 1-880653-28-1 (Set); ISBN 1-880653-29-X (Vol. I)

Response Prediction of a TLP in Cyclonic Storm Conditions

S.A. Higgins and J.B. Hinwood
Australian Maritime Engineering CRC Ltd.
Melbourne, Australia

K.P. Thiagarajan
Australian Maritime Engineering CRC Ltd.
Launceston, Australia

ABSTRACT

A six degree of freedom frequency domain model of a Tension Leg Platform (TLP), which includes the effects of non-collinear wind, wave and current, is developed to determine the response and sensitivities to environmental input. The model includes the steady forcing effects due to currents and waves, wind and wave drift. The dynamic effects include Morison type forces due to inertia and drag, spatially correlated wind and slowly varying wave drift effects. Attention is given to the position dependent mooring stiffness incorporating an iterative procedure for the determination of mean static offset position as well as the linearised mooring stiffness at the displaced location. The linearised effects of body motion, current and wave particle velocities are considered for drag determination.

This model is applied to predict the response of a TLP to a sample hindcast cyclonic storm off the North West Shelf of Australia. Storm parameters used in the response model include significant wave height, spectral peak period, wave direction, storm current and direction, wind speed and direction. Results show that peak response does not necessarily coincide with the maximum sea state, but rather, different motions have different phasing with the input condition.

1 INTRODUCTION

In the preliminary design stage of offshore structures there is a requirement to understand the relationships between the site specific metocean environment and the global structural response. With this understanding, a structure's main geometric configuration may be optimised to achieve an acceptable design; for example, limiting responses may be maximum overturning moment for a rigid steel jacket or maximum static and dynamic offset for a compliant structure. This process may be relatively simple if the response is governed by a single environmental parameter, or may be quite complex for a structure sensitive to a multi-dimensional metocean input. This paper concerns the latter case and is directed towards a formulation of response surface prediction for a Tension Leg Platform (Figure 1) situated in a deep water cyclonic environment. This work aims to develop a model which could enable a designer to implement a 6 degrees-of-freedom (6 DOF) response model to predict the response to specified metocean criteria. The model is to be used in a procedure for screening preliminary designs, and hence must be capable of producing results rapidly without a significant loss of accuracy.

Offshore oil and gas production is frequently located in regions of intense cyclonic (or hurricane) activity such as the Gulf of Mexico or Australia's North West Shelf (NWS). A tropical cyclone is characterised by an intense atmospheric low pressure system on the order of 950 hPa or less and wind speeds of up to and sometimes exceeding 50 m/s. These atmospheric conditions may extend over 100 km in diameter. The moving cyclone causes rapidly changing fetch lengths and directions due to the spatially and temporally changing wind strength. This mechanism creates a wave field which could be multi-directional, fetch-limited, and uncorrelated in direction with the instantaneous local wind.

Few in-situ data exist for cyclones in the NWS and so hindcast studies are usually undertaken to predict wind, wave and current time series. This research uses results of a hindcast study from a series of cyclones during the period 1960-90. The model used in the hindcast provides simultaneous values of wind speed, wave conditions and current velocity, and relative headings. Any changes in mean water level due to atmospheric pressure and wind stress are not included. Figure 2 provides metocean particulars of a specific storm used in this study. The wind and wave conditions can be described as a sea state 9 (Beaufort 10). As can be seen, current speed is reasonably constant, while the wind speed has a 50% variation between maximum and minimum. The waves have a roughly constant peak period, with the significant wave height varying over 50%. Waves and current are initially along the same heading, while the wind is roughly 30 degrees off. About 10 hours into the storm the current heading diverges from the wind and wave directions with a shift of more than 50 deg at the end of the simulation.

The TLP has been well documented for its response to first order wave, second order wave, dynamic wind as well as the steady force contributions from current, wind and second order wave. Various authors (Li 1988, Vickery 1988, Mak 1990 and Incecik 1993) have given highly complex or simplified analyses for various degrees of freedom in frequency and time domain numerical simulations. Only a small number (eg. Mak, 1990; Li, 1988) have presented a full 6 DOF model which could simulate TLP response for an entire cyclonic storm with its associated non-collinear metocean input. The run time of these models would be prohibitive for preliminary design screening, or in the simplified case of Incecik (1993), exclusion of coupling between the degrees of freedom, both in forcing and response, may not produce the required accuracy.

The work of Li (1988) appears to be the most comprehensive among published 6 DOF models incorporating nonlinearities in both the frequency and time domain solutions. The method involves a full boundary element diffraction solution which also includes the quadratic transfer functions for potential slow drift forcing. Viscous wave and current loads, which contribute to mean and slowly varying force excitation, are derived in terms of Hermite polynomials up to the free surface. Li (1988) quotes that it takes about 45 minutes CPU run time on a VAX 8500 for his time domain solution (4 hours of time history) and four minutes for the frequency domain solution. These times do not include the time taken to compute the diffraction force or drift quadratic transfer functions. These long computation times make Li's (1988) models unattractive for feasibility and preliminary design studies, where a range of alternative design configurations must be rapidly and economically screened. A similar criticism can be made of Mak's (1990) full time domain model. Thus there is a need for simplified cost-effective models that can be used to screen TLP designs for regions that are exposed to cyclone forcing.

In order to meet this need a new model was created as part of this research. To simplify code development and to minimise execution times, the model is written as a set of MATLAB procedures. The model will form part of a design evaluation system which recognises response constraints and can perform trade offs or 'optimisations'. The model is required to predict the surge, sway and heave motions, tether tensions and air gap of a generic TLP with non-collinear wind, current and wave forcing. The strategy is not to analyse every nonlinear aspect of the response but to predict the overall trends for comparative studies. Thus, high frequency nonlinear phenomena such as springing and ringing are not considered

Following a description of the metocean environment, the response model is described below. Wave forcing is described by a linearised Morison equation accounting for the effects of relative water particle and structural motion. Wind loading is based on a two dimensional coherence function approach. Nonlinearities in the mooring stiffness are solved by a Newton-Raphson iterative solution for the mean offset position where a linearisation is sought for the solution of response. The linear equation of motion is solved in the frequency domain, incorporating an iteration in the linearised viscous damping matrix. Other damping contributions include structural Rayleigh (Clough and Penzien, 1975) and mean wind aerodynamic effects.

2 ENVIRONMENT DESCRIPTION

WAVE

As discussed previously, a cyclone has rapidly changing fetch lengths and directions, hence special attention should be given to the selection of a suitable wave spectrum. JONSWAP wave spectra have been fitted to rapidly changing wind fields in hurricanes (Whalen and Ochi 1978) but there exists a large range for the five parameters describing the shape and energy content. For the purposes of this paper, the JONSWAP spectrum given in Faltinsen (1995) is used. The input variables T_p and H_s are directly obtained from the hindcast output. For simplicity, a 2D wave field is considered in this model.

WIND

Very limited data are available on the specific characteristics of tropical cyclones over water. Up to 50 metres elevation, the atmospheric boundary layer is strongly influenced by the turbulence generated by the roughness of the sea surface. Above this height in cyclonic conditions there is strong convective activity whose properties are relatively unknown.

The mean wind speed over the sea as a function of elevation is approximated by the logarithmic law, as derived for over land:

$$\overline{U}(z) = \frac{1}{k} u_* \ln \frac{z}{z_0} \tag{1}$$

where $\overline{U}(z)$ is the mean velocity at height z, u_* is the friction velocity, k is the von Karman's constant and z_0 is roughness length . The surface drag coefficient C_{10} is defined as

$$C_{10} = \left(\frac{u_*}{\overline{U}_{10}} \right)^2 \tag{2}$$

There is much conjecture as to the selection of C_{10} as it is a function of the sea state and atmospheric stability. Over land C_{10} is nearly independent of $\overline{U}(z)$ but over the ocean, higher winds create higher waves and thus the drag coefficient is larger (Melbourne and Blackman 1982). For the North West Shelf region of Australia, Bureau of Meteorology data exist for a number of cyclones on the North West Cape near Exmouth, Western Australia (Wilson 1979a, b). Mean and gust speed measurements were made from US Naval Communications Tower at heights of 9, 60, 191, 279 and 300 metres which was situated on flat salt bush covered land 5 km from the coast. Data from Wilson (1979a, b) suggest $\overline{U}(z)$ below 60 metres can be fitted to a logarithmic profile, while above 60 meters the mean wind speed was approximately constant. The following relations for extreme design conditions apply:

$$\begin{aligned} \overline{U}(z) &= \overline{U}_{10} 0.285 \ln(z/0.3) && \text{for } z \le 60 \text{ m} \\ \overline{U}(z) &= 1.5\, \overline{U}_{10} && \text{for } z > 60 \text{ m} \end{aligned} \tag{3}$$

The distribution of the along-wind rms velocity, $\sigma_u(z)$, with height is defined by the local turbulence intensity,

$$I_u(z) = \frac{\sigma_u(z)}{\overline{U}(z)} \tag{4}$$

The single sided auto-spectral density of velocity fluctuations in the along wind direction is given by Ochi and Shin(1988). From measured wind data over the ocean at various locations around the world they proposed an empirical formulation of the auto spectral density, G_u:-

$$
G_u(f_*) = \begin{cases} 583 f_* & \text{for } 0 \le f_* \le 0.003 \\ \dfrac{420 f_*^{0.7}}{(1+f_*^{0.35})^{11.5}} & \text{for } 0.003 \le f_* \le 0.1 \\ \dfrac{838 f_*}{(1+f_*^{0.35})^{11.5}} & \text{for } f_* \ge 0.1 \end{cases} \qquad (5)
$$

where the dimensionless frequency $f_* = fz/\overline{U}_z$, and $G_u(f_*) = f G_u(f)/u_*^2$.

The applicability of this formulation for tropical cyclones would have to be questioned as the most severe wind conditions analysed were for the Norwegian Sea at a height of 110 m and mean wind variation between 2 - 36 m/s. For the data collected in cyclonic regions the maximum mean wind speed is less than 12 m/s with maximum height of 29 meters. This would not seem appropriate for a typical tension leg structure which extends at least 100 m above mean water level and could be exposed to cyclonic winds greater than 50 m/s. This spectrum does however contain the largest energy content at the TLP surge natural frequencies when compared with all other spectra given by Ochi and Shin (1988). Application of this spectrum to the present structure is therefore considered to be the most conservative option.

The frequency dependent correlation of wind speed between two locations, i and j, is defined by a coherence function. To the authors' knowledge there is no published information on coherence function for cyclonic conditions. The root coherence defined by Kareem (1985) is adopted for across wind separations and is defined as,

$$
\gamma_{u_{ij}}(\Delta r, f) = \exp\left(-\left((C_x \Delta x)^2 + (C_y \Delta y)^2\right)^{0.5} f / \overline{U}_m\right) \qquad (6)
$$

where $C_x = 10$ and $C_y = 16$, and $\Delta r = (\Delta x, \Delta y)$ is the separation distance. $\overline{U}_m$ is the averaged mean wind speed between i and j. The cross-spectral density can then be obtained as.

$$
G_{u_{ij}}(f) = \gamma_{u_{ij}}(\Delta r, f)\left(G_{u_{ii}}(f) G_{u_{jj}}(f)\right)^{0.5} P_{u_{ij}} \qquad (7)
$$

The imaginary part of the cross spectrum is usually small and is ignored.

CURRENT

Currents modelled in the hindcast storm simulation include contributions from wind induced shear on the water surface and the surge induced currents from the action of atmospheric pressure. The tidal contribution has not been included for the present study. Over the computational interval of one hour, the current is considered constant in magnitude and direction. Based on published expressions for the depth velocity current profile for a draft of 32m in approximately 500m of water (Kokkinowrachos 1987), the current is assumed to be constant over all elements of the main structure.

3 DYNAMIC LINEARISED SYSTEM MODEL

Firstly it would be beneficial to clarify the model's aims and assumptions. These are:

AIMS

- To predict the global response for a TLP due to 3-D multiple load mechanism from wind, wave and current for an entire cyclonic storm data base
- To model hydrodynamic characteristics that contribute to maximum structural response.
- To be simple to implement, inexpensive and capable of running on a standard desktop PC using a readily available commercial language, eg. MATLAB.
- To be user-friendly, and be applicable to a range of TLP examples with simple user input.

ASSUMPTIONS

- Deep water (543 m)
- Linear wave theory up to mean water line
- Longcrested wave
- Linearised drag
- Linearised equations of motion
- Diffraction effects are not significant and Morison's equation is valid.

The assumptions are based on a method given by Patel (1989). Morison's equation is considered applicable to a transparent structure with characteristic dimensions less than a fifth of the wave length. For a typical column diameter of 18 metres and a wave period of less than 7.6 sec. diffraction analysis would be required. However for the cyclonic data base used in this research the minimum $T_P = 7.9$ sec. with a mean of 10.8 sec. and therefore Morison's equation is assumed to be valid.

The general equation of motion of a TLP at its mean offset position can be stated as

$$
\mathbf{M\ddot{x}} + \mathbf{B\dot{x}} + \mathbf{Kx} = \mathbf{F(t)} \qquad (8)
$$

where

$\mathbf{M} = \mathbf{m} + \mathbf{A}$
$\mathbf{B} = \mathbf{B_{VL}} + \mathbf{B_A} + \mathbf{B_S}$
$\mathbf{K} = \mathbf{K_M} + \mathbf{K_H}$
$\mathbf{m}$ = Structural mass matrix
$\mathbf{A}$ = Hydrodynamic added mass matrix
$\mathbf{B_{VL}}$ = Linearised viscous damping matrix
$\mathbf{B_A}$ = Aerodynamic damping matrix
$\mathbf{B_S}$ = Structural damping matrix
$\mathbf{K_M}$ = 1st order mooring stiffness matrix at mean offset
$\mathbf{K_H}$ = Hydrostatic stiffness
$\mathbf{F(t)}$ = 1st order dynamic wave and current load + 2nd order dynamic wave load + dynamic wind load
$\mathbf{\ddot{x}}, \mathbf{\dot{x}}, \mathbf{x}$ = Acceleration, velocity, displacement in terms of O'X1X2X3

Figure 1 details the coordinate systems used. The equations of motion are solved in terms of the O'X1X2X3 system, which is considered fixed at the platform's centre line mean offset position. OXYZ axes represent the platform's initial still water space fixed coordinate system which also serves as a reference for the wind, wave and current coordinate systems.

The mass matrix is formulated in terms of the body fixed coordinate system O'X1X2X3 and combines the structural and hydrodynamic mass terms. Using the discretisation of the pontoons and columns, a numerical summation is made and transformed to body-fixed axes. The method relies on strip theory and ignores any structure-wave interactions, which would become significant as its transparency diminishes. For the simulation shown in this study the pontoons and columns are subdivided into ten elements each.

The program takes into account the contributions to the damping matrix due to hydrodynamic viscous, aerodynamic and structural energy dissipation. The viscous damping matrix is derived from the structural velocity using the linearisation coefficient term in equation (10). It is applied to all below mean water line structural elements excluding the tethers. The structural damping is evaluated from the stiffness matrix calculated at the mean offset position and is based on Rayleigh's method (eg. Clough and Penzien, 1975). Two percent damping ratio is assumed. The aerodynamic damping matrix is given by equation (20) below.

Le Boulluec et al. (1994) have investigated the contributions of slow drift damping on a Hutton TLP example. They included the effects of viscous damping on the pontoons and tethers and wave drift damping on the columns. A severe ($H_S = 15.5$m and $T_p =$ 15.8 sec.) and a mild ($H_S = 5$m and $T_p =$ 11 sec.) sea state were considered, with the authors concluding that wave drift damping played only a minor role in the slow drift motions and was comparable with the tether damping. Drag forces on the pontoons provided the major contribution. For the present study the damping due to tethers and wave drift have been neglected.

Based on the method described in Patel (1989), the hydrostatic and mooring stiffnesses are determined at the exact mean position of the structure with respect to body fixed coordinate system. The current method does take into account the nonlinear variation in restoring force but the stiffness is calculated to first order at the offset position.

FIRST ORDER WAVE AND CURRENT LOADING

This study uses linear wave theory evaluated up to the mean water line. The structure's cylindrical pontoons and columns below the mean water line are discretised into a user defined number of segments. The wave and structure kinematics are calculated at each element centroid based on a combination of Morison equation and dynamic pressure at the column bases (eg. Faltinsen. 1995).

The wave and current hydrodynamic load on a slender cylindrical element can be expressed by the vectorial form of the Morison equation (eg. Faltinsen. 1995).

$$d\mathbf{F}(t) = \rho d\forall \dot{\mathbf{u}}_n + C_a \rho d\forall (\dot{\mathbf{u}}_n - \ddot{\mathbf{r}}_n) + \frac{1}{2}\rho C_d dA\,|\mathbf{U}_n + \mathbf{v}_n|(\mathbf{U}_n + \mathbf{v}_n) \tag{9}$$

where $\dot{\mathbf{u}}_n$ and $\ddot{\mathbf{r}}_n$ are the fluid particle and structural element acceleration (subscript refers to the element axis normal); $\mathbf{U}_n$ is the current velocity; $\mathbf{u}_n$ and $\dot{\mathbf{r}}_n$ are the fluid particle and structural element velocity normal to element axis. and $\mathbf{v}_n = \mathbf{u}_n - \dot{\mathbf{r}}_n$ is the relative fluid particle velocity; C_a and C_d are the added mass and drag coefficients. The linearised form of the drag term in Equation (9) can be expressed as

$$dF^{DL} = \frac{1}{2}\rho dAC_d(\mathbf{L_1 v_n} + \mathbf{L_2 U_n}) \tag{10}$$

where $\mathbf{L_1}$ and $\mathbf{L_2}$ for this study are the linearisation matrices given by Leira (1987). The force and moment contribution of each element about the body fixed coordinates O'X1X2X3 are summed to provide the resultant force and moments.

In the frequency domain the forces due to drag, inertia and dynamic pressure all contribute to the force cross-spectral density for m and n degrees of freedom. Summing over all segments (i and j) for the relevant mode, we obtain

$$G_{F_{mn}}^{wave}(\omega) = \sum_{i=1}^{nelements}\sum_{j=1}^{nelements}\begin{pmatrix} G_{F_{mn}}^{drag}(\omega) + G_{F_{mn}}^{inertia}(\omega) + G_{F_{mn}}^{pressure}(\omega) + \\ 2G_{F_{mn}}^{drag/inertia}(\omega) + 2G_{F_{mn}}^{drag/pressure}(\omega) + \\ 2G_{F_{mn}}^{inertia/pressure}(\omega) \end{pmatrix} \tag{11}$$

The last three terms on the right hand side represent cross terms between various forcing components.

STEADY AND SLOWLY VARYING DRIFT FORCE

The TLP is by design compliant in the surge, sway and yaw degrees of freedom with natural periods of oscillation of order 100 seconds in these modes. Therefore consideration must be given to all relevant steady and slowly varying forcing especially that which extends to the surge, sway and yaw natural frequencies. This study includes the steady and slowly varying drift contribution from the potential flow effects of wave elevation and velocity head. The steady viscous drift due to wave and current is included in Equation (10), while the slowly varying component below the mean waterline is introduced by a method similar to Burns (1983). The following are not included for the sake of simplicity: body motion and rotational inertia contributions; viscous steady and slowly varying column free surface loads; and slowly varying effects due to viscous contributions in the splash zone (with and without current). Based on Li (1988) and Chakrabarti (1990), contributions due to these effects are minor compared to the included effects.

For a regular wave, the equation for the steady wave elevation and velocity head potential forcing on a surface piercing cylinder in deep water arising from the first order velocity potential can be written (Chakrabarti 1990).

$$\overline{F}_1 + \overline{F}_2 = \frac{2\rho g a^2 D}{\pi^2 (kR)^3}\sum_{n=0}^{\infty}[1 - \frac{n(n+1)}{(kR)^2}]^2 \frac{1}{A_n A_{n+1}} \tag{12}$$

where

$A_n(kR) = J_n'^2(kR) + Y_n'^2(kR)$

$\overline{F}_1, \overline{F}_2$ are the wave elevation and velocity head steady forces,

$a =$ wave amplitude, $D =$ cylinder diameter (2R), $k =$ wave number.

For the case of a four column array in a two dimensional irregular wave field, the approximations of Pinkster (1975) can be applied to obtain the mean and slowly varying wave force. For a single surface-piercing column the mean potential drift force is,

$$\bar{F}_{pd} = \rho g \int_0^{\infty} G_\eta(\omega) C_{FD}^2(\omega) d\omega \qquad (13)$$

where C_{FD} is the drift force coefficient. To take into account the column spacing for the evaluation of the potential drift force, Pinkster's method applicable to a two hull structure is adopted. For a given platform orientation with respect to the wave field two equidistant column pairs are selected. The slowly varying autospectral density of the force is obtained as:

$$G_{Fpd}(\mu) = 2\rho^2 g^2 \int_0^{\infty} G_\eta(\omega) G_\eta(\omega+\mu) \left[C_{FD}'^2(\omega+\mu) \left[\begin{array}{l} \cos(\frac{(\omega+\mu)^2 - \omega^2}{g} \cdot \frac{l_1}{2}) \\ + \cos(\frac{(\omega+\mu)^2 - \omega^2}{g} \cdot \frac{l_2}{2}) \end{array} \right] \right]^2 d\omega \qquad (14)$$

where C_{FD}' is the drift force coefficient for two cylinders, l_1 and l_2 are the column pair spacings in the direction of the wave.

Burns (1983) describes a simplified technique for including the wave/current viscous slowly varying force below the mean water line applied to the columns of a tension leg platform. The assumption is that in a regular wave the steady force due to wave with a specified current is proportional to the wave amplitude squared. In the present response model, a viscous drift coefficient is calculated based on current velocity, and amplitude a and frequency ω of the wave. The viscous drift force is obtained by integration of a Morison's type drag term from the base of a column to the mean water line, and averaging over a wave cycle. Thus,

$$\bar{F}_{vd} = \frac{\rho}{2} D_c C_d \int_0^{\frac{2\pi}{\omega}} dt \int_{-d}^{0} |\mathbf{U}_n + \mathbf{v}_n| (\mathbf{U}_n + \mathbf{v}_n) dz \qquad (15)$$

$$\bar{F}_{vd} = \frac{1}{2} \rho g C_{VD}^2 a^2 \qquad (16)$$

The viscous drift coefficient, C_{VD}, is then squared and added to the $C_{FD}'^2$, and applied to equation (14). A similar method is also described by Kobayashi et al. (1985).

STEADY AND DYNAMIC WIND FORCE

A quasi-steady approach to the wind loading on a structure can be applied if the characteristic dimension is much less than the significant wave lengths of the fluctuating wind, viz. $fD/\bar{U} << 1$ (Davenport 1979). This formulation was used in the wind loading analysis for line-like structures. The slender body quasi-steady analysis can also be extended to larger bodies by replacing a structure by slender elements. Taking into consideration the mean wind speeds and frequency range of significant fluctuating wind energy it is estimated that the characteristic body dimension should be less than 20 metres for use of the quasi-steady approach.

The aerodynamic load on the ith elemental area is the component normal to its surface,

$$\mathbf{f}_{w_i} = \frac{1}{2} \rho_a C_{w_i} \mathbf{U}_{w_i}^2(t) A_i \qquad (17)$$

where

$\mathbf{U}_{w_i} = \bar{\mathbf{U}}_{w_i} + \mathbf{u}'_{w_i} - \dot{\mathbf{r}}_{w_i}$

$\bar{\mathbf{U}}_{w_i}$ = mean along-wind velocity, (10 min mean)

$\mathbf{u}'_{w_i}$ = time varying along wind velocity

$\dot{\mathbf{r}}_{w_i}$ = structural velocity.

Expanding gives,

$$\mathbf{f}_{w_i} = \frac{1}{2} \rho_a C_{w_i} (\bar{\mathbf{U}}_{w_i}^2 + 2\bar{\mathbf{U}}_{w_i} \mathbf{u}'_{w_i} - 2\bar{\mathbf{U}}_{w_i} \dot{\mathbf{r}}_{w_i}) A_i \qquad (18)$$

where the first term on the right is the mean load followed by the fluctuating load and then finally the aerodynamic damping term. The structural and fluctuating wind velocity squared terms are small compared with the mean wind velocity and so are neglected. The vertical and lateral wind velocity fluctuations are also ignored due to their insignificant contribution.

In Equation (18), the second term on the right hand side is the dynamic forcing term. The global force vector is then expressed in terms of the mean offset axes O'X1X2X3 as

$$\mathbf{F}_W(t) = \rho_a \sum_{i=1}^{nelements} C_{w_i} A_i \mathbf{T}_i \bar{\mathbf{U}}_{w_i} \mathbf{u}'_{w_i} \qquad (19)$$

where $\mathbf{T}_i$ matrix transforms the normal force on the element to forces and moments defined at O'X1X2X3.

Similarly, the third term in Equation (18), provides for the determination of the aerodynamic damping matrix. Following Li (1988), this is

$$\mathbf{B}_A = \rho C_W \sum_{i=1}^{nelements} A_i \bar{\mathbf{U}}_{w_i} \mathbf{T}_i \mathbf{T}_i^T \qquad (20)$$

Equation (19) is transformed into the frequency domain to obtain the cross-spectral density for m and n degrees of freedom,

$$G_{F_{mn}}^{W}(\omega) = \rho_a^2 \sum_{i=1}^{nelements} \sum_{j=1}^{nelements} C_{w_i} A_i \mathbf{T}_{i_m} \bar{\mathbf{U}}_{w_i} C_{w_j} A_j \mathbf{T}_{j_n} \bar{\mathbf{U}}_{w_j} G_{u_{ij}}(\omega) \qquad (21)$$

4 RESPONSE SOLUTION

In the frequency domain the solution of the response cross spectral density is evaluated by

$$\mathbf{G_X}(\omega) = \mathbf{H}^*(\omega) \mathbf{G_F}(\omega) \mathbf{H}^T(\omega) \qquad (22)$$

$\mathbf{G_F}(\omega)$ is the cross-spectral density matrix of total contributions of first order wave and current, slowly varying wind and slowly varying potential and viscous drift forces by addition of Equations 11, 14 and 21. The transfer function of the dynamic system is simply given by

$$\mathbf{H}(\omega) = \left[\mathbf{K} - \omega^2 \mathbf{M} - i\omega \mathbf{B}\right]^{-1} \qquad (23)$$

QUASI-STATIC ITERATION

The solution of TLP dynamic and static response is based on a Newton-Raphson iteration of the nonlinear restoring force of the mooring and a general first order approximation of the mooring stiffness matrix at any defined offset position. At the still water position the equation of motion is solved, inclusive of its own iteration sequence to optimise the linearisation of the damping matrix until convergence of response solution, and mean wind, wave and current loads are determined. Using the initial stiffness matrix at still water, the new offset position is determined. At this new displaced position, the equation of motion is again solved iteratively and the nonlinear restoring force of the moorings and the updated mean external load are evaluated. Both the restoring and updated mean load are added as vectors to predict a new displacement based on the estimation of the stiffness matrix. The entire process is repeated until the new displacement is within a pre-determined error limit. The last calculated cross-spectral displacement response (dynamic and mean static) is then used to determine the individual tether tensions and platform position for the sea state simulated.

VERIFICATION OF MODEL

Published data are very limited for comparison with the model developed in this research. Li (1988) presents two 6 DOF simulations, one of which has the following environmental input:

	Spectrum	*Data*	*Heading*
Waves	JONSWAP	f_0=0.08 Hz, γ=3.3, α=0.0061	22.5 deg
Wind	Kareem	U_{wind}=17 m/s @ 10m	0 deg
Current		$U_{current}$=1 m/s	35 deg

The structural data used for Li's (1988) simulation are identical to the particulars given in Table 1. The present model compares well with Li's (1988) findings in both the surge and sway modes, for the wave and low frequency motion. The RMS values of the response fall within 10 percent of the results from Li's (1988) complex fully nonlinear model. In these modes, the motion response is dominated by the damping matrix, which in turn has the most significant contribution from viscous dissipation.

The mean offset results obtained from the present model are compared with Li's results below:

Mean Offset	*Li (1988)*	*Present model*
Surge	7.6 m	7.1 m
Sway	5.8 m	1.8 m
Heave	-0.088 m	-0.050 m

There is good agreement in the predicted mean surge offset (a difference of less than 7%). The results are widely different in sway and heave offsets. The final horizontal position of the TLP based on Li's mean offsets, is at an angular coordinate of 37 deg with respect to OX-axis. In contrast, the present model predicts the structure's final position at an angle of 14 deg. Considering that the wind wave and current headings are 0, 22.5 and 35 degrees respectively, the present results appear to be consistent. Differences in mean offsets also contribute to disagreements in heave and pitch motions in the wave frequency region. However, both models agree in wave frequency roll and yaw motions, to within 2 and 6 percent respectively. A comparison of the surge response due to potential slow drift was made with the work of Li (1988). Over the wave spectral peak frequency range 0.08 - 0.12Hz agreement was within 5 % but the results outside this frequency under predict the response with the present model.

5 STORM SIMULATIONS

As mentioned in the introduction, the present hindcast storm data base includes a series of storms for which a response surface is to be developed for the specific location. As an example of the response output, a typical cyclonic event has been chosen (Figure 2). Figure 2 shows the build up and decay of wind speed and H_S with a maximum sea state occurring around the 10 hour mark. The extreme sea state condition lags that of the wind which is as expected. The directional information indicates a strong correlation between wave and wind while the current diverges after the maximum sea state. It should be noted that the initial surge orientation of the platform was set in line with the initial wind.

Figure 3 shows the response spectra at the H_S peak (approximately 10 hours). The dominant excitation at this time is that of sway with the low frequency response dominating. The low frequency yaw response shows the influence of the spatially correlated wind loading acting on the topsides.

Figure 4 illustrates the trajectory of the mean position of the platform centre line in terms of the initial still water coordinate system OXYZ. The effect of changing fetch direction is clearly visible indicating that the maximum mean excursion may not occur at the maximum sea state, but rather at a range of combinations including all the environmental parameters. Figures 5 and 6 give a detailed description of the dynamic response, and it is evident, in the surge and sway modes, that the peaks do not necessarily occur at the maximum sea state. In fact, every response has its own unique phasing with the cyclone.

The mean tension variation in each tether is shown in Figure 7. T_1 and T_2 represent the up wind mean tension, while T_3 and T_4 correspond to the down wind positions at simulation time start. The model correctly predicts a higher mean tension and transfer of loading for T_4 and T_2 as the storm progresses from a surge to sway dominant loading.

6 CONCLUSIONS

As part of a project to develop an integrated procedure for preliminary design of offshore facilities, a model to predict the response of a TLP has been developed. The model includes independent wind, wave and current vectors and contributions to the damping matrix from viscous, aerodynamic and structural energy dissipation, with relevant terms evaluated at the position of displacement under mean loads. The model has been shown to reproduce adequately the responses predicted by other published models while yielding results with much shorter computation time. The proposed simplified method currently takes approximately 6 minutes on a desktop PC per sea state.

Simulation of the response for a conceptual location on the North West Shelf, Australia during the passage of a tropical cyclone has confirmed the capability of the model. The simulation demonstrated that all of the responses change significantly, as the metocean variables change during the passage of the cyclone, making the selection of the critical design cases very difficult. This factor clearly indicates the necessity to simulate an entire cyclonic passage to correctly predict the phasing between the responses, both dynamic and static, to locate maximum and minimum response conditions.

Acknowledgments
The first author's PhD work was funded by an Australian Postgraduate Award (Industry) at Monash University. Top-up Scholarship provided by the Australian Maritime Engineering CRC Ltd. is acknowledged.

References

Le Boulluec, M., Le Buhan, P., Chen, X. B., Deleuil, G., Foulhoux, L. Molin, B. and Villeger, F., (1994), "Recent Advances on the Slow Drift Damping of Offshore Structures", *7 th International Conference on the Behaviour of Offshore Structures*, MIT, pp 9 - 30.

Burns, G. E., (1983), "Calculating Viscous Drift of a Tension Leg Platform", *Proceedings 2 nd Off Mech Arc Eng*, Houston, pp 22 - 30.

Chakrabarti, S. K., (1990), *Nonlinear Methods in Offshore Engineering*, Elsevier Science Pub., Amsterdam.

Clough, R. W., and Penzien, J., (1975), *Dynamics of Structures*, McGraw-Hill Pub. New York.

Davenport, A. G, (1977) "The Prediction of the Response of Structures to Gusty Wind", In *Safety of Structures Under Dynamic Loading*, Report Norwegian Inst. Tech., Trondheim, Vol. 1, pp 257-84.

Faltinsen, O. M., (1995), *Sea Loads on Ships and Offshore Structures*, Cambridge University Press, London.

Incecik, A., (1993), "The Effect of Non-collinear Wave, Wind and Current Loading on Dynamic Response of a Tension Leg Platform", *Proceedings 12th Off Mech Arc Eng*, pp 17 - 24.

Kareem, A., (1985), "Wind Induced Response Analysis of Tension Leg Platforms", *J. Structural Engng*, ASCE, Vol. 111, pp. 37-55.

Kobayashi, M., Shimada, K. and Fujihara, T., (1985), "Study on Dynamic Responses of a TLP in Waves", *Proceedings Fourth Off Mech Arc Eng*, pp 29 - 35.

Kokkinowrachos, K., (1987), "Hydrodynamic Analysis of Offshore Fixed Platforms", *Offshore Platforms and Pipelines*, (Mazurkiewicz, B.K. ed.), Trans Tech Publications, Clausthal - Zellerfeld Pub.

Leira, B. J., (1987), "Multidimensional Stochastic Linearisation of Drag Forces", *Appl. Ocean Res.*, Vol. 9, No. 3, pp 150 - 162.

Li, Y-S., 1988. "Stochastic Response of a Tension Leg Platform to Wind and Wave Fields". *Ph.D. Thesis*, Univ. of Houston.

Mak, T. T-Y., 1990. "Dynamics of Tension Leg Platform in Random Directional Wind-wave Fields Co-existing with a Uniform Current", *M.E.Sc. Thesis*, Univ. of Melbourne.

Melbourne, W. H. and Blackman, D. R. , (1982), "Wind Turbulence Over Seas in Tropical Cyclones", *Proc 18 th. Int Conf Coastal Eng*, Vol 2, pp 370-393.

Ochi, M. K. and Shin, Y. S., (1988), "Wind Turbulent Spectra for Design Consideration of Offshore Structures", Paper No. 5736, *Proceedings OTC*, Houston, pp 461-467.

Patel, M. H., (1989), *Dynamics of Offshore Structures*, Butterworth & Co. London.

Pinkster, J. A., (1975), "Low Frequency Phenomena Associated with Vessels Moored at Sea", *Petroleum Trans.* AIME, pp 487 - 494.

Vickery, P. J., (1988), "Wind and Wave Loads on a Tension Leg Platform: Theory and Experiment", *Ph.D. Thesis*, Univ. of Western Ontario.

Whalen, J. E. and Ochi, M. K., (1978), "Variability of Wave Spectral Shapes Associated with Hurricanes", *Proceedings 10th OTC*, Houston, pp 1515-1523.

Wilson, K. J., (1979a), "Wind Observations from an Instrumented Tower During Tropical Cyclone Karen, 1977", *12th Technical Conference on Hurricanes and Tropical Meteorology*, New Orleans.

Wilson, K. J., (1979b), "Characteristics of Subcloud Layer Wind Structure in Tropical Cyclones", *International Conference on Tropical Cyclones*, Perth.

No. of vertical columns	4
Column diameter	18 m
Column draft	31.9 m
Column spacing	66.2 m
No. of horizontal pontoons	4
Pontoon diameter	9 m
Plane (OX_1X_3) wind projected area	5320 m^2
Plane (OX_2X_3) wind projected area	5350 m^2
Total displacement	28509 t
Total tether preload	16640 t
Centre of gravity of platform	+6 m on X_3 axis
No. of tethers	4
Total tether stiffness	3.028 E08 N/m
Moment of Inertia (Roll)	3.09 E10 kg-m^2
Moment of Inertia (Pitch)	3.09 E10 kg- m^2
Moment of Inertia (Yaw)	2.98 E10 kg- m^2
Inertia Coefficient (column)	2.0
Inertia Coefficient (pontoon)	2.0
Drag Coefficient (column)	0.8
Drag Coefficient (pontoon)	0.8
Aerodynamic drag coefficient	
lattice structures	0.7
main deck	1.4
columns	1.0

Table 1: Particulars of a Generic TLP (Li, 1988).

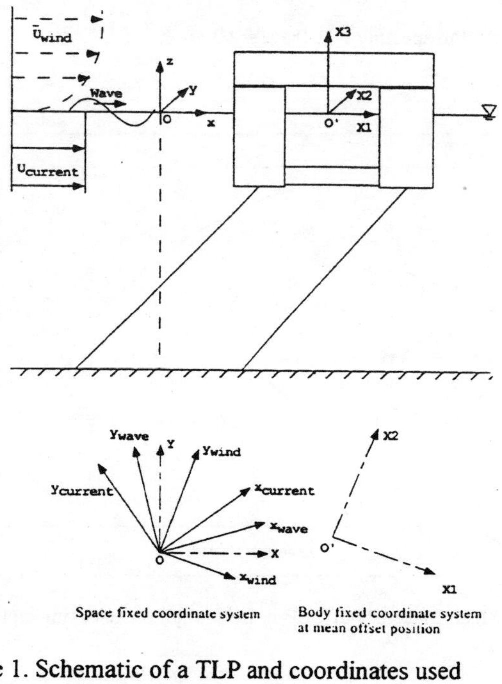

Figure 1. Schematic of a TLP and coordinates used

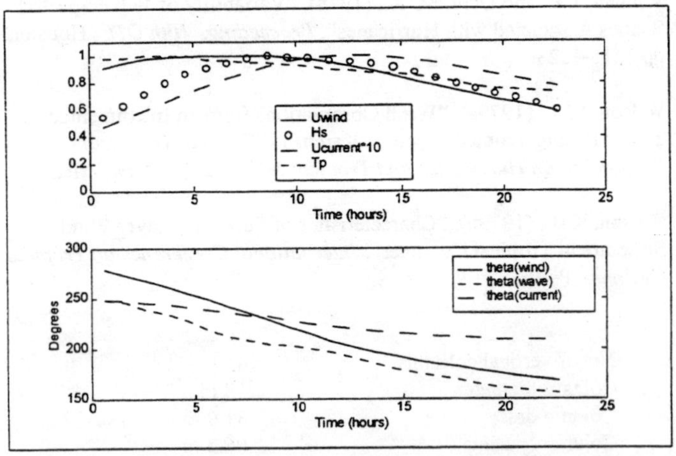

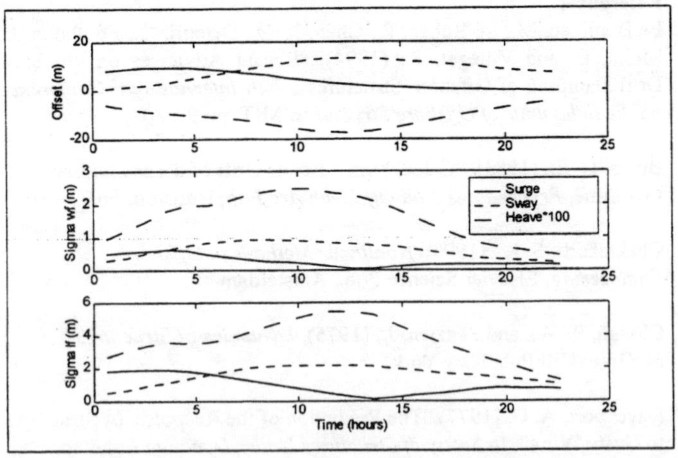

Figure 2. Sample cyclonic nondimensional wind, wave and current conditions and their associated relative headings.

Figure 5. Mean offsets and standard deviations for sample storm. Sigma wf- rms response in wave frequency (w>0.2 rads/s); Sigma lf- rms response in low frequency (w<0.2 rads/s).

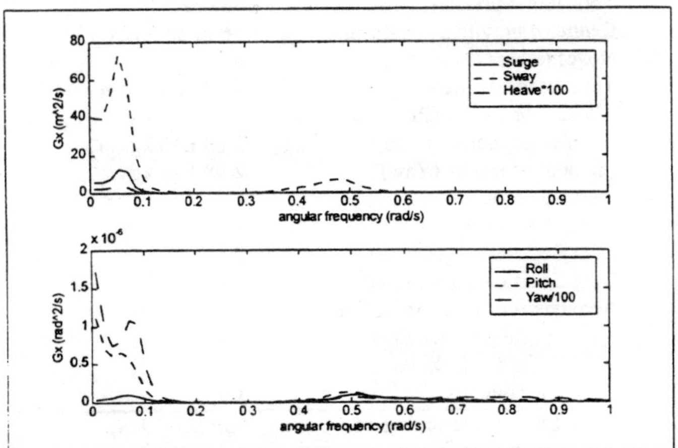

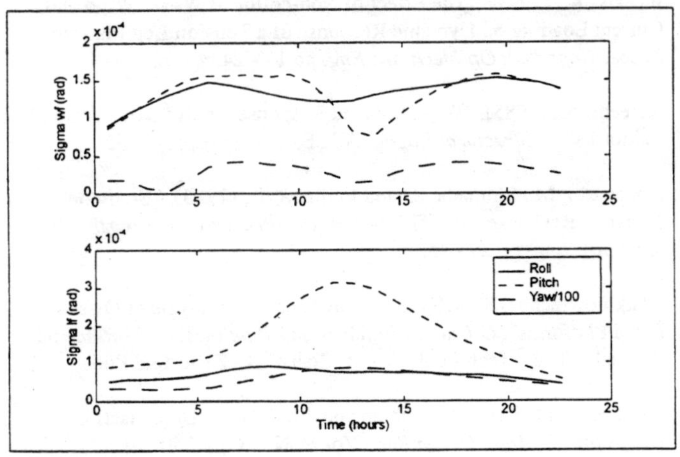

Figure 3. Motion spectra of TLP at t = 10 hours.

Figure 6. Standard deviations of TLP rotational motions for the storm duration.

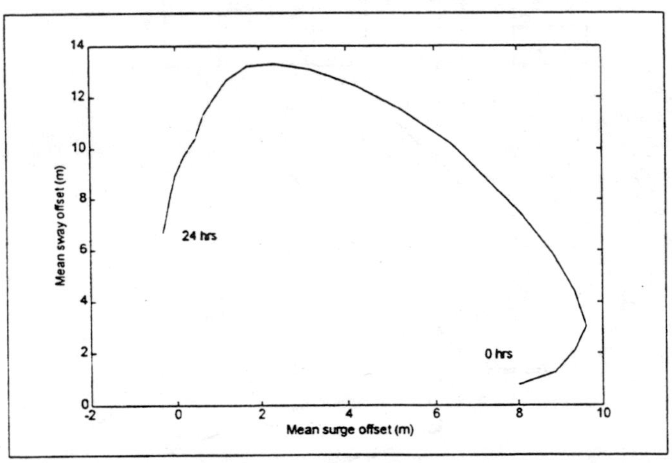

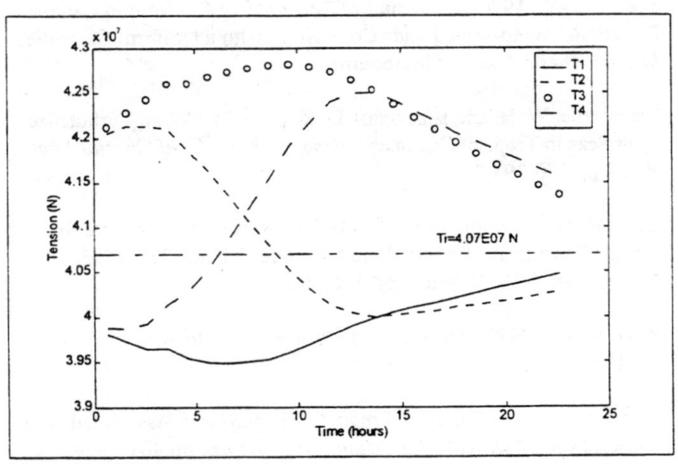

Figure 4. Mean trajectory of TLP in horizontal plane during storm.

Figure 7. Mean tether tensions for storm duration.

Proceedings of the Seventh (1997) International Offshore and Polar Engineering Conference
Honolulu, USA, May 25-30, 1997

ISBN 1-880653-28-1 (Set); ISBN 1-880653-29-X (Vol. I)

An Investigation of Extreme Wave Behaviour Around a Model TLP

A.D. Arnott and C.A. Greated
University of Edinburgh, Edinburgh, United Kingdom
A. Incecik
University of Newcastle-on-Tyne, United Kingdom
A. McLeary
University of Glasgow, Glasgow, United Kingdom

Abstract

This paper presents further results from an investigation into the response of TLP's in extreme waves. Using a scale model of the *Heidrun* TLP in a wave basin, the wave behaviour around the model in regular waves of various frequencies, heights and approach angles was recorded using wave gauges and video cameras. These revealed considerable wave runup on the columns, wave-height amplification under the deck and waterspouts. A selection of the results obtained are presented here.

Key Words:

TLP's, extreme waves, wave amplification, runup & jetting.

Introduction

As waves pass a cylinder, partially-immersed vertically in the water, the cylinder will, to a lesser or greater extent, experience 'run up' and 'draw down' around its circumference. By this, the maximum heights and depths that the water reaches to on the surface of the cylinder are in excess of the crest and trough amplitudes respectively of the undisturbed wave train. In addition the cylinder reflects a proportion of the wave energy which then meets and interferes with the approaching crests. Where this interference is constructive, the wave steepens and its velocity field becomes increasingly non-linear. Many investigations, for example by Hallermeier (1977) or Tørum (1989), have been concerned with the implications for the prediction of forces on such single, surface-piercing columns.

However, offshore platforms usually incorporate multiple-column geometries. Hydrodynamic interference between the columns results in intensification of the behaviour mentioned above, both in front of the platform and under its deck. These have important implications for the design of the air gap between platform and water, which is governed by the maximum local wave amplitude. For a TLP or semi-submersible, the design of the airgap is critical due to weight and stability considerations, but the possibility of wave impact (slamming) on the underside of the deck must also be minimised. This is complicated by the inverted-pendulum motion of a TLP which gives rise to "set down" of the platform relative to the water surface, thus reducing the airgap. Although slight, this should also be accounted for in the choice of airgap.

Investigations using TLP configurations by Owen & Padilla Perez (1987), Eatock-Taylor & Sincock (1989) and Niedzwecki & Huston (1992) found that the local wave heights could be amplified by a factor of up to 2 in regular waves and by approximately 10% in random seas. For a given wave period, the ratio of runup on a forward leg to that on an aft leg was found to increase with decreasing column spacing. Additionally with decreasing column spacing, the maximum amplitude of upwelling under the deck and the wave period at which the maximum occurred, were found to increase. An investigation by Bruce & Easson (1997) into local amplifications of wave height between pairs of cylinders, found jets of water erupting from the surface between the cylinders. To investigate whether this particular phenomenon should be considered in the design of offshore platforms, the investigation is currently being extended to semi-submersible platforms in association with the present authors.

Natvig (1994) has shown that ringing forces are affected by vertical shifts in a TLP's centre of gravity, *ergo* the choice of the design air gap will, to some extent, affect the transient forces acting on a TLP. An experimental investigation by Incecik *et al* (1996), studied these higher-order forces and the wave conditions which produced them, using a scale model of the *Heidrun* TLP. Also observed during these tests however, were significant amplifications of wave height under the deck and "jetting" upstream of the model. Therefore further experiments were performed to examine these phenomena more closely, results from which are presented in this paper.

Experimental Arrangement

The model TLP

The model TLP used in the experiments was a 1:98.4 scale model of the *Heidrun* TLP. The model, shown in figures 1(a) to (c), incorporates 4 circular cylinders for its supporting columns, scale replicas of the tether balconies being incorporated at their bases. Connecting the bases of adjacent legs are pontoons of rectangular cross-section. The deck is made from a sheet of plywood, 25 mm thick, although the airgap is not scaled, that of the model being less than that of the full size platform. Stiffening for the model structure is provided by a steel-tube frame running throughout. Although not detailed here, the model is equipped with load cells and strain gauges for the measurement of local and global forces in order to investigate ringing. In addition one column is transparent to facilitate velocity measurements using Particle Image Velocimetry, whilst another column is used for tuft-flow visualisation. These measurements of forces and fluid velocities have already been performed in both regular waves and Pierson-Moskowitz Spectra, samples of these having been presented by McLeary *et al* (1996). However a complete presentation of all the results from the investigation is given in the final report to the project sponsors (Incecik *et al*, 1996).

The wide, wave tank at the University of Edinburgh

The wave tank used was the Wide Wave Tank at Edinburgh University, shown in figures 2 & 3. The working area of the basin is 7 m in length, 25 m in width, with a water depth of 1.2 m. Waves are generated by 75 paddles, of the absorbing-flap type (hinged at their bases), arrayed side by side along one of the long sides of the basin. The generation of waves is based on superposition of wave components. This allows random seas to be modelled, in addition to sinusoidal waves propagating at various angles to the line of the paddles. Control of the wave paddles is achieved via computer programs run on a PC-486.

One of the short sides of the basin is constructed from glass panels, giving a view under the water surface. Along the remaining 2 sides are passive wave absorbers. Walkways exist above the wave makers and the wave absorbers opposite.

Installation of the model TLP

The model was held rigidly in the water from above such that it was square to the line of paddles, 5 m from them, as shown in figures 2 & 3. The figures also show the Cartesian coordinate system O*XYZ*, the origin O being on the still-water line at the centre of the square formed by the legs. The positive *X*-direction is towards the paddles, the positive *Y*-direction is normal away from the glass wall and the positive Z-direction is normal, upwards from the still water surface. For convenience the 2 legs closest to the paddles are referred to as forelegs, those furthest from the paddles as aft legs. The 2 legs nearest the glass wall in figure 2 are referred to as "starboard" legs, those furthest from the glass wall, "port" legs.

Apparatus for the measurement of surface elevation

Wave probes

Four capacitance-type wave probes, positioned as shown in figure 4, were used to measure the surface elevation between the legs. The accuracy of the probes in measuring the wave height was ±1 mm, due to the calibration procedure. The probes were connected to the interface and PC that operated the wave paddles; the programs to generate the wave patterns also controlled the sampling of the probes.

Video camera recordings

To obtain records on the wave amplification away from the gauges, a high-speed video camera and a miniature *CCD* camera in a waterproof housing were used. Both cameras recorded images onto video tape, the former at 200 frames per second (*fps*), the latter at 25 *fps*. The latter was used to obtain images from close to the waterline where it was impractical to use the high-speed camera. However the resolution of the high-speed camera was quite low and focusing was made more difficult by being performed on only a quarter of the area available on a monitor screen. Circumferential lines were drawn on the 2 starboard legs, parallel to the still-water line at intervals $\Delta Z = \pm 20$ mm, above and below it, to a height/depth of $Z = \pm 180$ mm. Angular position around the circumferences of the legs, was shown by vertical lines marked at intervals of 15° around the circumferences of the legs.

Experimental procedure

The wave gauges were calibrated at the start of each day on which they were used for data measurements. Wave frequencies used for the sets of measurements ranged from 0.7 Hz to 1.3 Hz and the peak-to-trough heights of the waves, from 7 cm to 13 cm.

Each particular set of wave conditions was run past the model 5 times, the approach angle of the waves being changed each time. These approach angles were: 0° (head on), -15°, -30°, +15° & +30°. A period of 18 s was left after starting the wave paddles before sampling of the wave gauges commenced. The gauges were sampled for 16 s, after which the paddles were brought to rest again; the motion of the paddles was ramped up and down. The starting point of the video recordings was often arbitrary however, dependant on the availability of an assistant to operate the camera whilst the experimenter was operating the PC. After sampling, a period of time was left for the surface of the water to become still to the naked eye, before the measurement run was repeated.

Wave elevation measurements were also carried out in irregular waves using the Pierson-Moskowitz Spectrum. Due to space limitations however, it is felt that better justice to the work would be served by their presentation in a future paper. In addition, some diffraction-theory calculations have been performed at Oxford University for this TLP model, in the regular wave conditions presented here. However no comparisons have yet been drawn between the calculations and experiments. Once these have been performed though, it is anticipated that both sets of data will form the subjects of future papers.

Discussion of results

Selected wave-elevation/time series are presented, accompanied by sketches of the relevant wave direction, followed by images grabbed from individual frames of the video recordings. In the figures of the gauge records, the traces for the undisturbed waves are omitted for clarity, but the relevant peak-to-trough heights are included in the captions.

Figure 5 presents a sketch of the model and gauges in a head sea (0°). Figure 6(a) shows wave-elevation/time traces for a head sea of 1.0 Hz waves of 7 cm peak-to-trough height, the Fourier transforms of the traces being shown in figure 6(b). Harmonics of the wave frequency are evident at all 4 gauge positions, these being believed to be associated with the occurrence of ringing and springing. However amplification of the wave crest is only significant for gauges 1 & 2. The height of runup on the leg is about twice the undisturbed crest amplitude, the wave elevation just in front of the centre of the model being slightly greater still. The gauge records are confirmed by video evidence in figure 15.

As the undisturbed wave height at 1.0 Hz was increased, the runup and upwelling at the forward legs increased, the latter also becoming significant at gauges 3 & 4. In 10 cm waves (figure 7) the wave amplification just in front of the forward 2 legs is again greater than the runup on the leg (gauge 1). The video image in figure 16 shows the existence of a significant upwelling, midway between the 2 front legs, which reaches an amplitude of approximately 85% of the height of the deck above the still-water surface.

In steeper waves (1 Hz, 13 cm), the wave elevations at gauges 1 & 2 in figure 8 seem identical at first sight. However figure 17 shows that the jet between the 2 forward legs has increased in volume and height such that it impacts the deck of the model. As such the peaks of the traces for gauge 2 are not valid as the jet has overtopped the gauge. Also the maximum runup on the forward legs occurs about 30° around them, towards their inboard sides, not on the centrelines of the legs. The trough amplitude between the forward legs is about 20% more than the draw-down on the fronts of the legs, which was corroborated by video evidence.

Wave gauge results from 1.3 Hz, 7 cm waves approaching from 0°, are shown in figure 9. Interestingly, gauges 1 & 2 show between 20 - 30% less runup and wave amplification than in 1 Hz waves of the same height. Under the deck in figure 18, the wave crest as it passes between the legs takes on a "devil's horns" appearance.

Increasing the wave height to 10 cm, at this frequency and angle (figure 10), a significant increase in wave elevation was measured only at gauge 1, *i.e.* runup. Also, in general, the crest amplitudes measured by all 4 gauges were smaller than in 1 Hz waves of this height (figure 7). The gauges alone do not show the full picture however. Figure 19 shows a video frame of the front of the model for the wave conditions in figure 10. In this a large jet can be seen erupting from the surface, to a height significantly higher than the air gap. This jet is upstream of gauge 2, where the wave activity is much less severe. Runup on the forward legs was observed to reach the deck and other jets were also in evidence under the centre of the deck (figure 20). Runup was also significant on the rear legs as shown by figure 21.

Changing the angle of approach of the waves produced interesting results too. Figures 11 & 13 show the wave directions relative to the model for -30° & +30° waves respectively, whilst figures 12 & 14 shows the gauge records for 1.0 Hz, 13 cm waves approaching from the 2 respective angles. In these cases, the interference becomes more complex as the line of symmetry is removed and reflections from the "downstream" legs are at different phases to each other. For example, in figures 12 & 14, the wave amplifications at gauge 3 are approximately equal to those for gauge 2 (and also the runup at gauge 1 in figure 12). Again the 4 gauges alone show a limited picture and a sample of the surface behaviour under the deck for the waves in figure 14 is presented in figure 22. Here significant upwelling and splashing is evident, which reaches virtually to the deck height of the model.

It has not been reported openly whether similar behaviour has also been observed in tests of larger-scale models of *Heidrun*, or even the full scale platform during operation. Also, bearing in mind that the airgap of the model is less than that of the full scale platform, the paper is not intended to imply that the deck of *Heidrun* is subject to forces from jetting or runup. However the fact that these effects are so pronounced around this configuration indicates that they should be considered during the design of future TLP's and warrants further investigation. Additional experiments, to enable a 3-dimensional picture of the wave surface around the model to be built up, would be extremely useful in the ongoing development of models to predict the wave behaviour under the deck of a platform.

Conclusions

The research indicates that, in addition to significant runup on the legs, the large dimensions of this platform cause large waterspouts to erupt from the surface in steep waves. It is advised that this should be considered in the design of airgaps and deck overhangs for TLPs.

The greatest height of runup measured was twice the crest amplitude of the undisturbed wave.

At the wave frequencies tested, the jetting phenomena begins to occur as the wave height increases from 7 to 10 cm. Jetting also occurs under the deck, in oblique wave seas, due to interference between the legs.

Also, for a constant wave frequency, jetting becomes more severe as the peak-to-trough height of the waves is increased.

Acknowledgement

These results were obtained as part of the managed programme of research: Uncertainties in Loads on Offshore Structures (1994-96). This was sponsored by the UK's Engineering & Physical Sciences Research Council, through the Marine Technology Directorate Ltd and jointly funded with: Aker Engineering a.s., Amoco (UK) , BP, Brown & Root Ltd., Exxon, UK Health & Safety Executive, Shell UK, Den Norske Stats Oljeselskap a.s. and Texaco Britain Ltd.. The authors gratefully acknowledge the financial support and guidance from the sponsors.

The authors would also like to thank Jennifer Perrins and Professor Rodney Eatock-Taylor of the Department of Engineering Science, Oxford University, for performing the diffraction-theory calculations, mentioned earlier in the paper, around the TLP model.

References

Bruce T & Easson WJ (1997). Wave enhancement due to structure blockage in semi-submersibles. *Proceedings of the 7th International Offshore and Polar Engineering Conference*, 25-30 May, Honolulu, USA.

Eatock-Taylor R & Sincock P (1989). Wave upwelling effects in TLP and semi-submersible structures. *Ocean Engineering*, **16**, pp 281-306.

Hallermeier RJ (1977). Non-linear flow of wave crests past a thin pile. *ASCE Journal of Waterways, Harbours and Coastal Engineering*, **102**, pp 365-377.

Incecik A, Faulkner D, McLeary A, Greated CA, Easson WJ & Arnott AD (1996). Experiments on the case study Tension Leg Platform in extreme waves and currents. Final report under Project D1 of the UK managed programme: *Uncertainties in Loads on Offshore Structures*.

McLeary A, Cornut S, Incecik A, Arnott AD & Greated CA (1996). Wave-drift and viscous-damping forces acting on TLPs: some experimental results. *Proceedings of the 6th International Offshore and Polar Engineering Conference*, 26-31 May, Los Angeles, USA, **3**, pp 445-454.

Natvig B (1994). A proposed ringing analysis model for higher-order tether response. *Proceedings of 4th International Offshore and Polar Engineering Conference*, Osaka, Japan, **1**, pp 40-51.

Niedzwecki J & Huston J (1992). Wave interaction with Tension-Leg Platforms. *Ocean Engineering*, **19**, pp 21-37.

Niedzwecki J & Rijken O (1994). Characterising some aspects of stochastic wave-structure interactions. *Stochastic Dynamics and Reliability of Non-linear Ocean Systems*, **77**, pp 109-118.

Owen DG & Padilla Perez IA (1987). Experimental investigation of upwelling and interaction effects within blocked structures. Final report, volume 1, under *Programme of Research into the Dynamics of Compliant Structures*, 1985-87.

Tørum A (1989). Wave forces on pile in surface zone. *ASCE Journal of Waterways, Harbours and Coastal Engineering*, **115**, pp 547-565.

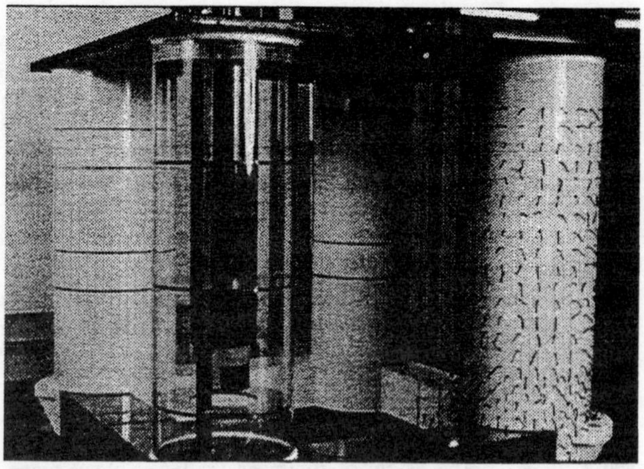

Figure 1(a): Photograph of completed model (1:98.4) of *Heidrun*.

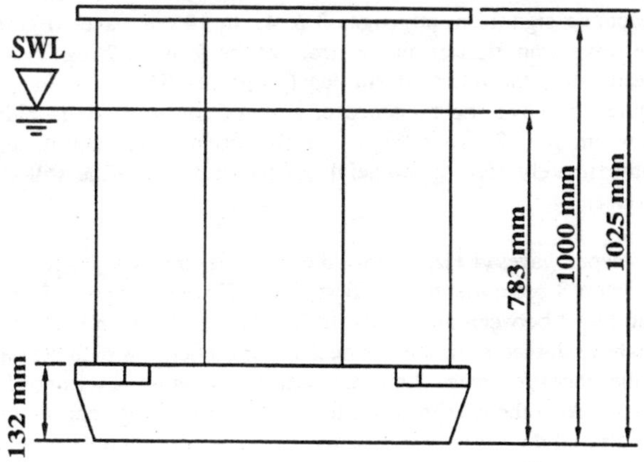

Figure 1(b): Side view of *Heidrun* model.

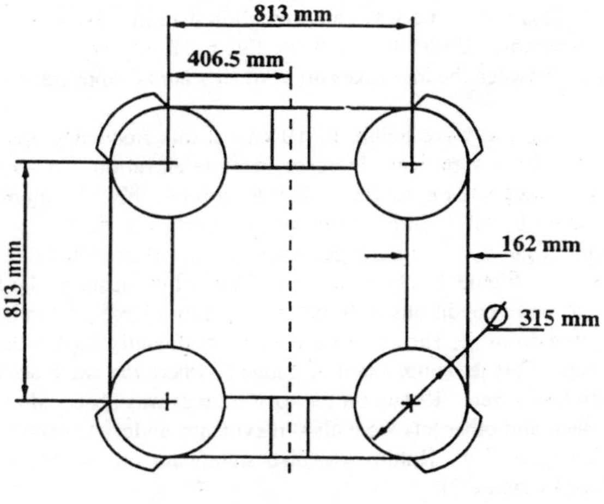

Figure 1(c): Plan view of *Heidrun* model.

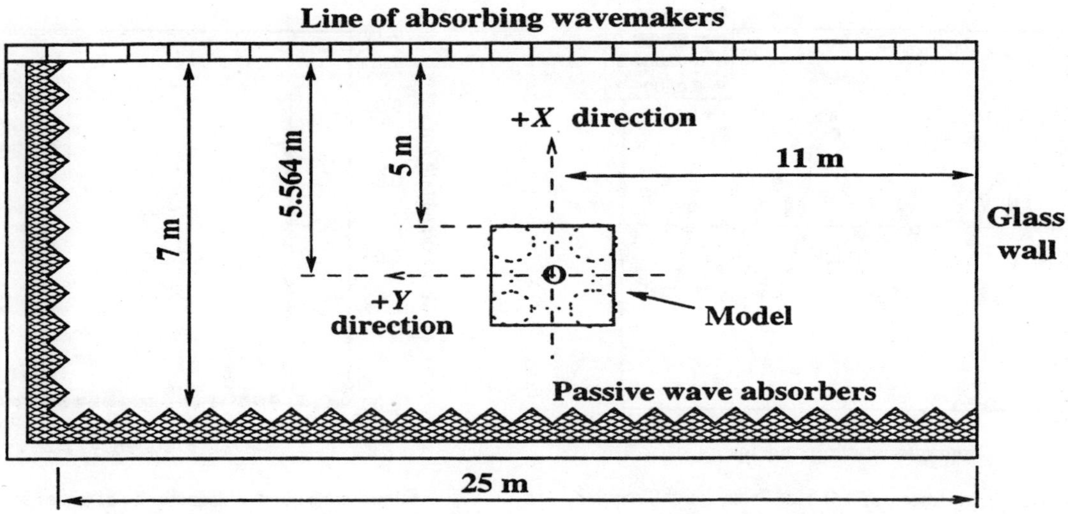

Figure 2: Plan view of *Heidrun* model in the wide wave basin at The University of Edinburgh.

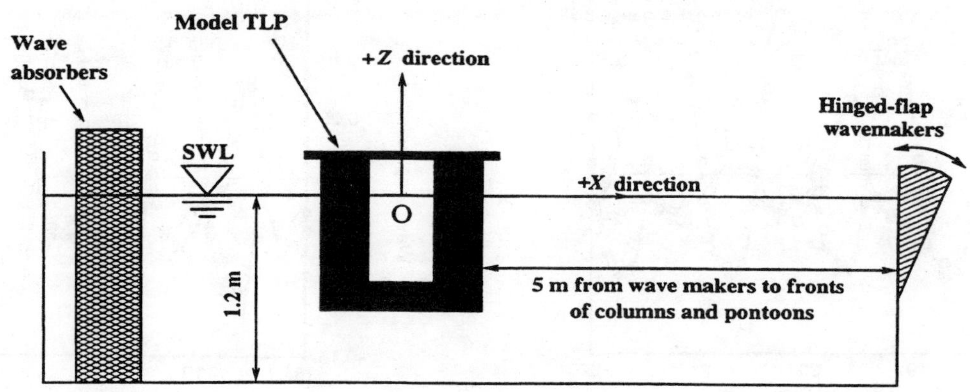

Figure 3: Side view of *Heidrun* model in the wide wave basin at The University of Edinburgh (looking through glass wall in figure 2).

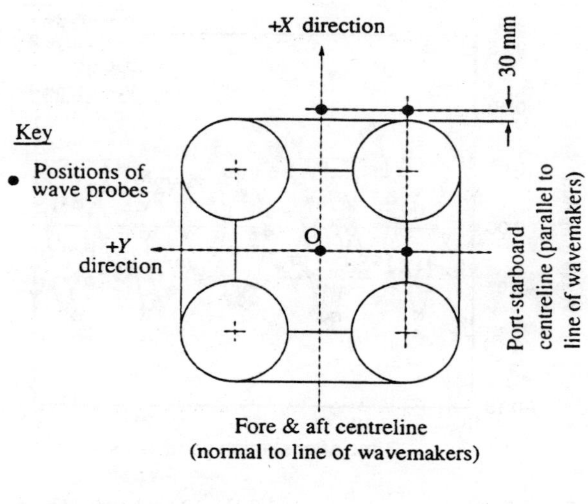

Figure 4: Locations of wave gauges around the model.

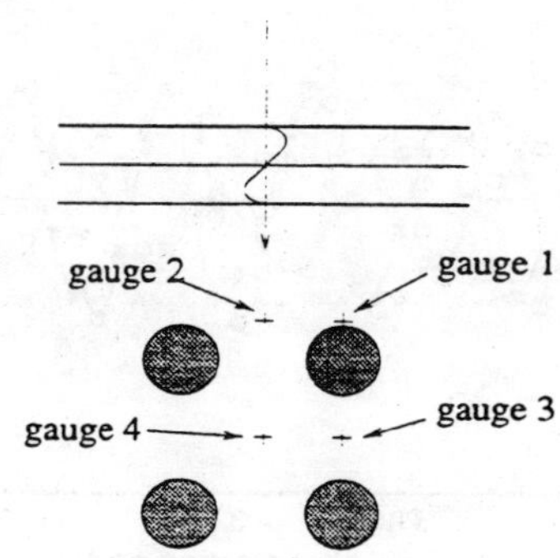

Figure 5: Sketch of wave direction in a head (0°) sea.

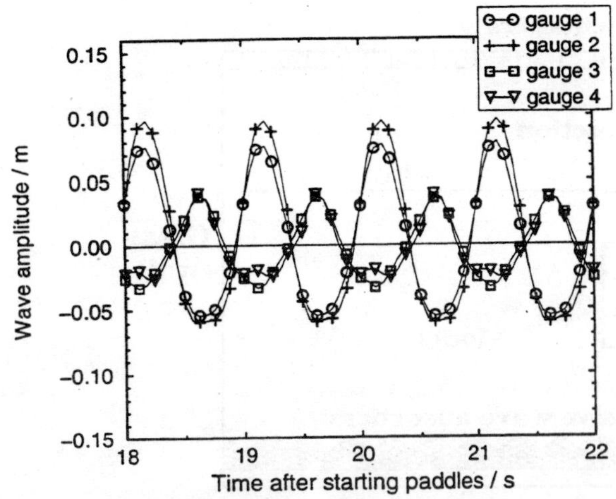

Figure 6(a): Wave gauge traces in 1.0 Hz, 7 cm height regular waves from head on (0°) to the model.

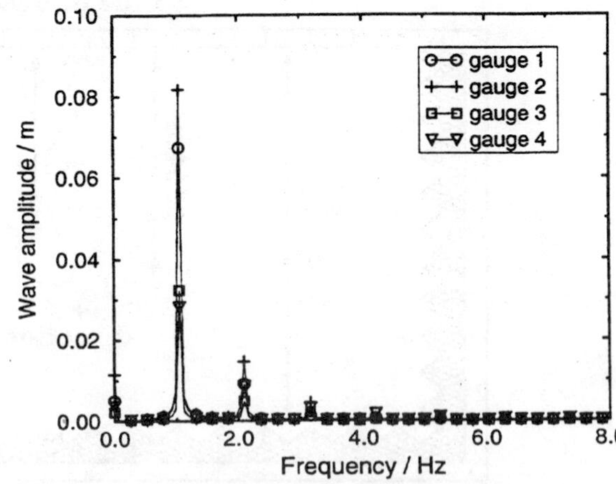

Figure 6(b): Fourier transforms of traces in figure 6(a).

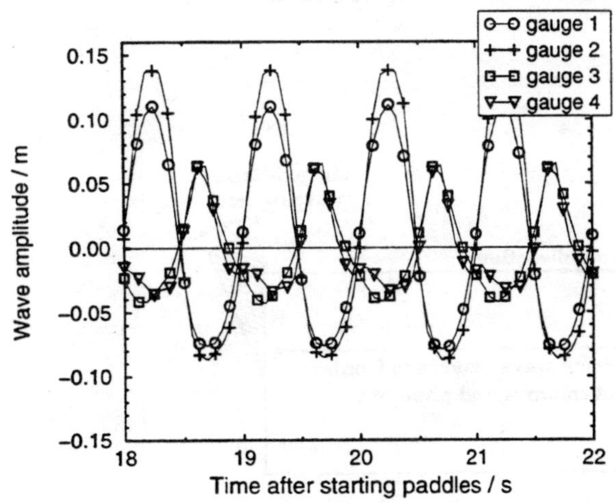

Figure 7: Wave gauge traces in 1.0 Hz, 10 cm height regular waves from head on (0°) to the model.

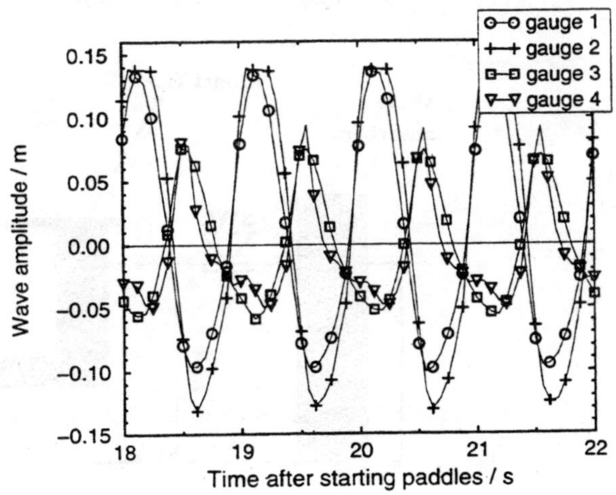

Figure 8: Wave gauge traces in 1.0 Hz, 13 cm height regular waves from head on (0°) to the model.

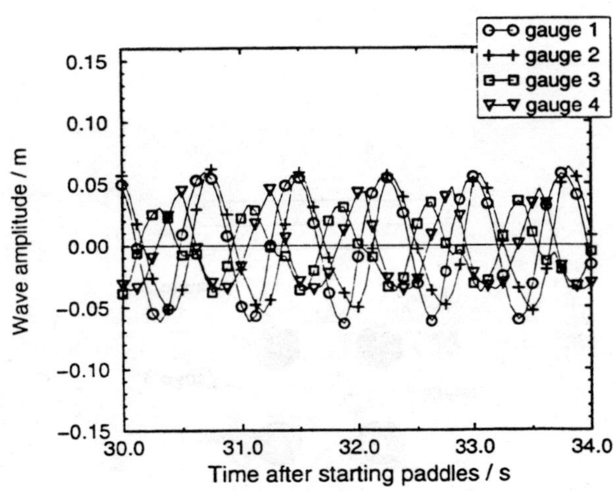

Figure 9: Wave gauge traces in 1.3 Hz, 7 cm height regular waves from head on (0°) to the model.

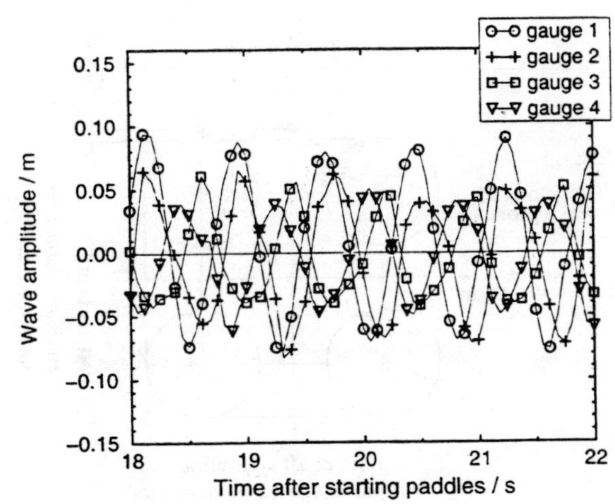

Figure 10: Wave gauge traces in 1.3 Hz, 10 cm height regular waves from head on (0°) to the model.

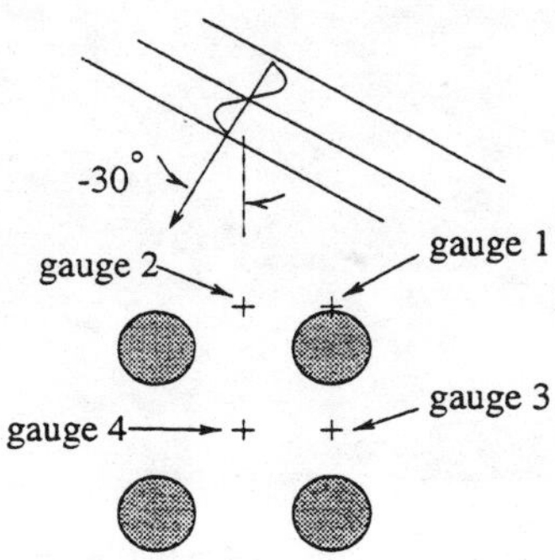

Figure 11: Sketch of regular waves approaching the model from -30°.

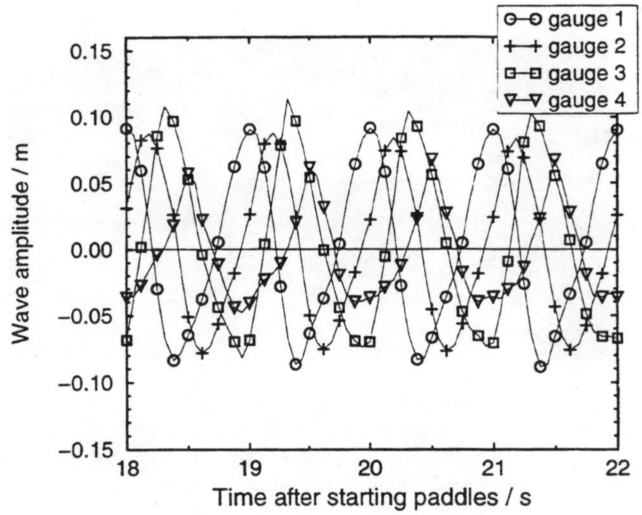

Figure 12: Wave gauge traces in 1.0 Hz, 13 cm height regular waves from -30° to the model as in figure 11.

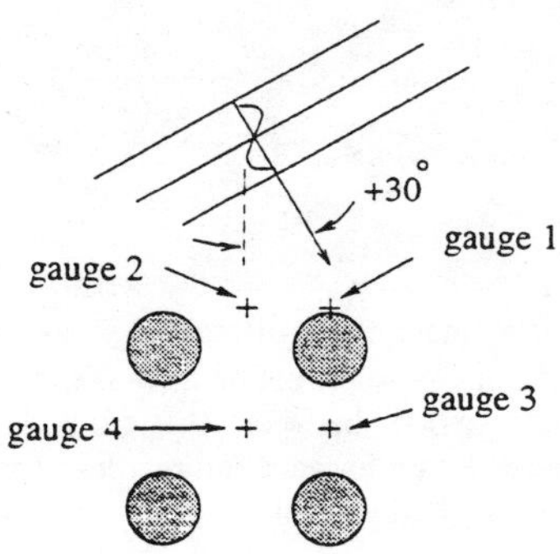

Figure 13: Sketch of regular waves approaching the model from +30°.

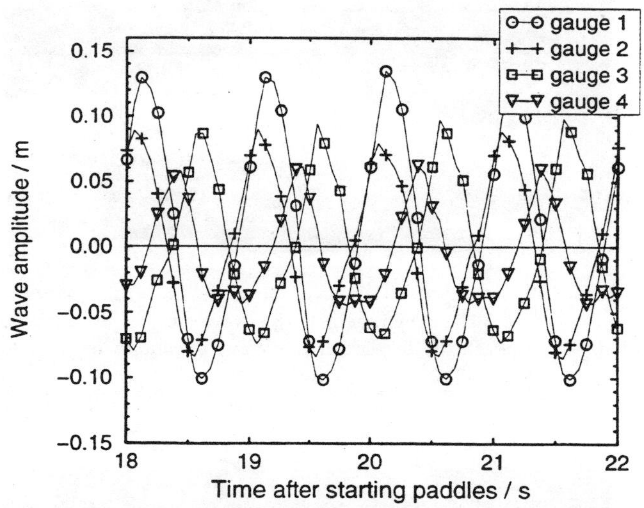

Figure 14: Wave gauge traces in 1.0 Hz, 13 cm height regular waves from +30° to the model as in figure 13.

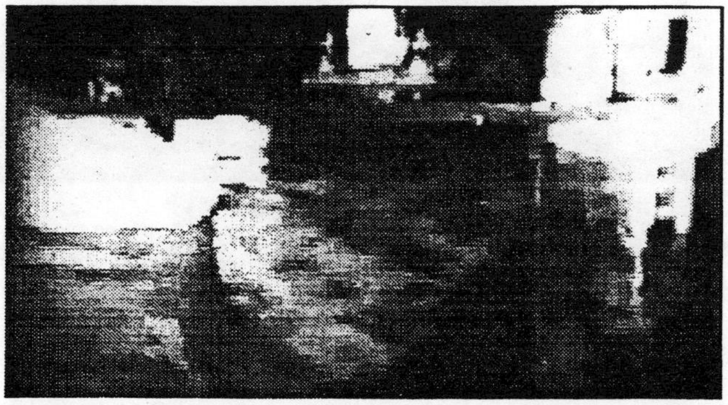

Figure 15: Video still of wave crest meeting forward legs in regular waves of 1.0 Hz, 7 cm height, approaching from 0° (into page).

Figure 16: Video still of the view under the deck, as a wave crest meets the forward legs, in regular waves of 1.0 Hz, 10 cm height, from 0° (into page). Here the jetting is significant, but does not quite reach to the deck.

Figure 17: Video still of the view under the deck, as a wave crest meets the forward legs, in regular waves of 1.0 Hz, 13 cm height, approaching from 0°. Here the jetting is of sufficient height to impact the deck.

Figure 18: Video still under in regular waves of 1.3 Hz, 7 cm height, approaching from 0° (into page). A wave crest is passing between the rear 2 legs, the next wave is in the foreground (slightly out of focus).

Figure 19: Video still upstream of the model, in regular waves of 1.3 Hz, 10 cm height, as a wave crest is just about to reach the forward legs. Wave direction:- head on to model (0°, *i.e.* into page).

Figure 20: View under deck in regular waves of 1.3 Hz, 10 cm height. Wave direction from 0° (into page). The wave has passed between the front legs. Interference between the forward and rear legs has produced a water jet which impacts the deck.

Figure 21: View under deck in regular waves of 1.3 Hz, 10 cm height. Wave direction from 0° (into page). The wave crest is passing between the rear legs showing that runup can also be significant there.

Figure 22: View under deck from the starboard side of the model in waves of 1.0 Hz, 13 cm height, approaching the model from +30°. Wave direction right to left in frame.

Proceedings of the Seventh (1997) International Offshore and Polar Engineering Conference
Honolulu, USA, May 25-30, 1997

ISBN 1-880653-28-1 (Set); ISBN 1-880653-29-X (Vol. I)

Time Domain Simulation of a TLP's Response Due to Laboratory-Generated Asymmetric Irregular Waves

J. Zou, Y. Xu and C.H. Kim
Texas A&M University
College Station, TX, USA

KEY WORDS: mean offset and setdown, slow drift motion, springing and ringing, second-order wave forces, tendon dynamics, stretching formula, Morison equation.

ABSTRACT

The nonlinear response of a coupled ISSC-TLP system in a design sea state is simulated by TIOSTAMU. The low-, wave- and high-frequency responses are systematically investigated and statistical analyses of the TLP response are performed.

The objective of this study is to investigate the effects of numerically simulated and experimentally measured forces on the nonlinear responses including both springing and ringing. The present study reveals that springing is due to weak asymmetric waves while ringing due to strong asymmetric waves.

INTRODUCTION

The TLP system is one of the best designs for oil production in deep water, since the construction cost does not increase dramatically with increasing depth of water, and the resonant motions are detuned from the frequencies of dominant wave energy spectra. However the low- and high-frequency forces excite the low- and high-frequency resonant motions, while the mean forces create the offset and setdown. These are associated with the key design parameters, i.e. air gap, tether fatigue and extreme loading. Therefore we need to develop reliable and sound techniques for the dynamic analysis of TLPs.

The waves, wind and current are the important environmental elements to yield the loads on the TLP system. However the present investigation considers only the wave loads in order to compare the effects of theoretical and experimental wave loads.

The low- and high-frequency wave forces are due to the nonlinear interactions of the linear responses. The second-order diffraction theory (Molin, 1979; Kim and Yue, 1989, 1990; Chau and Eatock Taylor, 1992) gives the interaction terms up to the second-order.

Third-order diffraction theory has recently been proposed by Faltinsen et al (FNV, 1995) and Malenica and Molin (MM, 1995) to compute ringing wave loads. Rainey (1995) argues that the FNV assumption of wave amplitude and cylinder radius being of the same order could lead to divergence of the Stokes' expansion. Malenica and Molin (1995) point out that these 3rd-order theories (FNV and MM) give 50% different results from each other even at a very low wave slope of 0.035. Krokstad et al. (1996) find that there exist large overpredictions of high-frequency horizontal forces by the FNV on a vertical cylinder.

The slender-body theory (Rainey, 1989, 1995) assumes that the structure is slender relative to the wavelength. A comparison of the diffraction and slender body theories (Eatock Taylor et al, 1992) shows the occurrence of disparity for shorter waves.

In contrast to the above studies, the present study uses the measured forces and moments of the ISSC-TLP due to the laboratory waves. In addition, a method of local distortion has recently been developed to generate strong asymmetric irregular waves. The large front-concave asymmetric wave in the irregular wave train can produce the ringing clearly.

STRONG AND WEAK ASYMMETRIC WAVES

The theoretical wave by Longuet-Higgins & Cokelet (1976) is highly nonlinear and asymmetric. The crest of the wave has a distinctly concave front and convex rear. Thus the wave may be called strong asymmetric wave. Waves which do not have the property are weak asymmetric waves. The transient waves are strong asymmetric (Kim et al, 1992; Zou and Kim, 1996), while the conventional laboratory waves are usually weak asymmetric. Examples of these waves are shown in Figs. 21 (weak asymmetric) and 22 (strong asymmetric).

A strong asymmetric wave group similar to the transient wave has to be in the irregular wave train. Funke et al (1982) proposed a distortion technique to generate asymmetric irregular waves. A similar approach is used herein to locally distort the largest wave group in the time domain keeping the energy spectrum unaltered.

Myrhaug et al (1984) analyzed field data and found that the vertical asymmetry parameter λ (Eq 9) varied from 1.2 to 2.1. This indicates the fact that there are wind driven strong asymmetric waves in the real ocean.

NUMERICAL RANDOM WAVES AND THREE DESIGN SEAS

Zhao (1996) developed a random wave generation technique which can produce arbitrarily long time series without repeating. Three design waves (Table 1) are employed in this study, where H_s, T_p, γ and λ represent significant wave height, peak period, overshoot parameter and vertical asymmetry parameter (Eq 9), respectively.

Table 1. Parameters of the Design Waves

Run Case No.	H_s (m)	T_p (s)	γ	λ
1 (Numerical)	12.5	12	2.0	None
2 (Measured)	12.5	12	2.0	None
3 (Measured)	13.8	9	6.5	2.0

Applying the filtered spectrum method (Zhao, 1996) we generate numerically a random wave of significant wave height 12.5 m (Case 1). The same parameter of JONSWAP spectrum is used to generate the irregular wave in the wave tank (Case 2). The wave amplitude spectra of Cases 1 and 2 are shown in Figs. 1 and 2 respectively. These wave amplitude spectra are similar to each other except for the significant energy shifting in Case 2, from the primary wave spectrum, into the low- and high-frequency regions due to the nonlinear interactions of physical waves. In Table 2, one finds that the numerical wave is very close to Gaussian, while the measured deviates slightly from Gaussian.

The wave of Case 3 is generated in the wave tank using the local time domain distortion method as mentioned in the preceding section.

Table 2. Statistics of Waves, Case 1 and Case 2

	Gaussian Wave	Case 1	Case 2
Mean (m)	0.0	-0.0013	-0.128
RMS (m)	3.125	3.160	2.98
Skewness	0	-0.096	0.195
Kurtosis	3.0	2.9	3.01

THE THEORETICAL WAVE LOADS IN TIME DOMAIN

The first- and second-order frequency response functions for the first- and second-order forces and moments are computed by using the higher-order boundary element method (Liu et al., 1991, 1995). These are Fourier transformed to the first- and second-order impulse response functions, which are used in the following Volterra model.

$$y(t) = \int_{-\infty}^{\infty} g_1(\tau)x(t-\tau)d\tau + \int_{-\infty}^{\infty}\int_{-\infty}^{\infty} g_2(\tau_1,\tau_2)x(t-\tau_1)x(t-\tau_2)d\tau_1 d\tau_2 \tag{1}$$

$$g_1(\tau) = \frac{1}{2\pi}\int_{-\infty}^{\infty} \hat{G}_1(\omega)e^{-i\omega\tau}d\omega \tag{2}$$

$$g_2(\tau_1,\tau_2) = \frac{1}{(2\pi)^2}\int_{-\infty}^{\infty}\int_{-\infty}^{\infty} \hat{G}_2(\omega_1,\omega_2) e^{-i(\omega_1\tau_1+\omega_2\tau_2)}d\omega_1 d\omega_2 \tag{3}$$

where x(t) and y(t) denote input wave elevation and output wave loads respectively: $g_1(\tau)$ and $g_2(\tau_1,\tau_2)$ are the first- and second-order impulse response functions respectively, while $\hat{G}_1(\omega)$ and $\hat{G}_2(\omega_1,\omega_2)$ the first- and second-order transfer functions for the forces and moments.

Eq 1, 2 and 3 are used to compute wave loads on the platform, while the linear Morison equation is applied to calculate the wave loads on the tendons, where C_m and C_d are 2.0 and 1.0 respectively.

EMPIRICAL MODEL FOR HORIZONTAL FORCE AND MOMENT

A model has been developed for prediction of horizontal forces on a vertical truncated cylinder (Kim et al., 1996 A & B). Given the wave elevation time series, the horizontal particle velocities are predicted by a stretching formula. Thus using Morison equation one can compute the force and moment of the cylinder.

The complex amplitude spectrum of the measured wave elevation time history has been used to reproduce the measured wave record exactly on the sampling interval. This is represented by

$$\eta(x,t) = R_e \sum_{j=1}^{n} a_j \exp\left[i\left(k_j x - \omega_j t + \phi_j\right)\right] \tag{4}$$

where η is wave elevation at x and a_j, ω_j, k_j and ϕ_j are amplitude, frequency, wave number and phase angle of the jth component, respectively. It is assumed that the nonlinear irregular wave elevation is transferred linearly to the nonlinear irregular horizontal velocity in the form:

$$u(x,z,t) = R_e \sum_{j=1}^{n} a_j \exp\left[i\left(k_j x - \omega_j t + \phi_j\right)\right] G_u(z,\omega_j), \quad z \le 0 \tag{5}$$

with

$$G_u(z,\omega_j) = \frac{g k_j}{\omega_j}\frac{\cosh k_j(z+h)}{\cosh k_j h} \tag{6}$$

$$\frac{\omega_j^2}{g} = k_j \tanh(k_j h) \tag{7}$$

where G_u is the linear transfer function (LTF) for the horizontal velocity at z, and h is the water depth.

Use of Eq 6 makes the velocity at the MWL diverge. Wheeler (1970) introduced a stretching method to avoid the divergence and determine the horizontal particle velocity in the wave crest. Wheeler's stretching model is improper for the prediction of kinematics in a steep wave crest. Perturbation models for the wave kinematics may be adequate for the waves of small slope. As the slope becomes larger, the model may diverge or converge very slowly. In order to have the stretching model simple and accurate for arbitrary wave slope, a so-called local stretching model is proposed, which is different from the previous work (Kim et al., 1996A),

$$z_e = a z_a^3 + b z_a^2 + c z_a + d \qquad \text{for } -h \le z_a \le H_c \tag{8}$$

with

$$\begin{aligned} a &= [(-h+H_c)+k(h+H_c)]/(h+H_c)^3 \\ b &= [-2(h^2-hH_c+H_c^2)-k(h+H_c)(H_c-2h)]/(h+H_c)^3 \\ c &= [H_c(H_c^2-H_ch+4h^2)+kh(h+H_c)(h-2H_c)]/(h+H_c)^3 \\ d &= -\{h^2H_c[k(h+H_c)+2H_c]\}/(h+H_c)^3 \\ k &= (2.00-\lambda)H_t/H \\ \lambda &= T_f/T_r, \text{ if } H_c/H_t \text{ is less than 1.0, then } \lambda=1.0; \\ & \quad \text{if } \lambda \text{ is larger than 1.95, then } \lambda=1.95 \end{aligned} \tag{9}$$

where z_e is the effective vertical coordinate ($-h \le z_e \le 0$), whereas z_a the actual vertical coordinate ($-h \le z_a \le H_c$). Eq 8 is the mapping equation between z_e and z_a. Substituting the effective coordinate Eq 8 into Eq 6, one obtains the LTF G_u as a function of the actual coordinate. H_c, H_t and H represent the wave crest, trough and total height at the referred wave cycle. h is water depth and k and λ are asymmetry parameters. T_f and T_r indicate the rise and fall time of the wave in the crest region. H_c/H or H_t/H and T_f/T_r represent the horizontal asymmetry about the MWL and vertical asymmetry about vertical axis through wave crest respectively.

The above Eqs are used in the fully nonlinear Morison equation for computing the forces and moments, where C_M (1.95) and C_D (0.9) are assumed to be uniformly distributed along the cylinder axis.

MEASUREMENT AND ESTIMATION OF FORCES AND MOMENTS

The wave tank is 37 m long, 0.91m wide and 1.22 m deep. The TLP model consists of two of structure A and one of structure B. Structure A consists of two columns and one pontoon which are paralleled to the wave direction while structure B consists of two pontoons which are normal to the wave direction. The horizontal and vertical forces of these structures are measured and the moments are estimated using the measured forces and empirical prediction model.

Fig. 3 shows an example of horizontal force prediction of a vertical truncated cylinder (16.4 cm diameter and 30 cm draft) in a strong asymmetric transient wave of a nominal slope 0.28.

DYNAMIC EQUATION OF COUPLED TLP SYSTEM

Applying both kinematic and dynamic constraint conditions at the bottom and top joints of the tendons, the six-degree coupled dynamic equations are rewritten as

$$M\ddot{U} + C\dot{U} + K\Delta U = F \tag{10}$$

where U is the response motion,

$$M = \begin{bmatrix} M_V + T_I^T M_{II} T_I & 0 \\ 0 & M_{JJ} \end{bmatrix} \tag{11}$$

$$C = \begin{bmatrix} C_V + T_I^T C_{II} T_I & T_I^T C_{IJ} \\ C_{JI} T_I & C_{JJ} \end{bmatrix} \tag{12}$$

$$K = \begin{bmatrix} T_I^T K_{II} T_I & T_I^T K_{IJ} \\ K_{JI} T_I & K_{JJ} \end{bmatrix} \tag{13}$$

and

$$F = \begin{Bmatrix} F_f + F_{PT} \\ F_J \end{Bmatrix} \tag{14}$$

where M_V denotes the virtual mass of the platform, M_{II} the mass of interface elements, M_{JJ} the mass of the other elements; T_I indicates the connection matrix, T_I^T the transpose matrix of T_I. The indexes in [C] and [K] denote similar meanings. C_{IJ} (C_{JI}) and K_{IJ} (K_{JI}) represent coupled damping and stiffness respectively. It is assumed that no coupled damping exists. F_f denotes wave vector forces obtained from either the theory or measured data. F_{PT} are the interaction vector forces exerted by the platform on the connecting elements, while F_J the vector forces on the remainder of the elements.

Eq 10 can be condensed by removing the redundant DOF. The time incremental equations are solved by the updated Lagrangian method. Theoretical damping estimations at the low- and high-resonant frequency regions are still under developed. The damping coefficients used herein are the experimental data due to Huse (1990) and Thiagarjan and Troesch (1993).

NUMERICAL RESULTS AND DISCUSSIONS

The computations of ISSC-TLP response in the three different head seas, Case 1, 2 and 3 are systematically analyzed. The detail particulars of the ISSC-TLP can been found in Kim et al. (1994).

Since there are resonant frequencies in the low- and high-frequency regions, it is essential to investigate the platform motions and tether tensions in three different frequency regions, as defined in Table 3. Due to the limited paper size, the heaving and pitching motions are not discussed. Three hours long simulations are desirable for the statistics of the slow drift motion and ringing. Since the measured time length is about 27 min, the statistical meaning of the data is limited to a certain degree.

Table 3. Definition of Three Frequency Regions

	ω (rad/s)
Low -frequency region	$\omega \leq 0.25$
Wave -frequency region	$0.25 < \omega \leq 3.0$
High -frequency region	$3.0 < \omega \leq 4.0$

LOW-FREQUENCY RESPONSES

The slowly varying motions and tensions due to the theoretical and measured waves (Cases 1 and 2) are compared as shown in Figs. 4 and 5. The statistics of the motions are given in Table 4, where one sees that the mean offset and the SD of Case 1 are considerably larger than those of Case 2.

Table 4. Statistics of Slowly Varying Surge Motion

	Case 1	Case 2	Abs Difference
Mean (m)	4.74	3.81	0.93
Max. (including mean, m)	29.6	27.6	2.0
Min. (including mean, m)	-23.0	-17.0	6.0
Standard deviation SD (m)	10.94	8.89	2.05
Skewness	-0.055	0.072	
Kurtosis	2.19	2.66	

It is postulated that the overestimations are due to the overpredictions of slow drift forces. However in Table 5, one sees that the mean and SD of surging forces of Case 2 are larger than those of Case 1. Therefore the overestimations of the offset and SD are not caused by the overpredicted slow drift forces. It is difficult to give a physical explanation at the time of writing, but there exist differences of, the statistical nonlinear parameters, skewness and kurtosis.

Table 5. Statistics of Slow Drift Force

	Case 1	Case 2	Abs Difference
Mean (N)	9.05E5	1.19E6	2.85E5
Max. (including mean, N)	5.29E6	7.47E6	2.18E6
Min. (including mean, N)	-3.81E6	-2.62E6	1.19E6
Standard deviation SD (N)	1.01E6	1.48E6	4.7E5
Skewness	0.163	0.881	0.718
Kurtosis	6.66	4.96	1.70

The tendon tension of element # 10 (Kim et al., 1994) due to the waves Cases 1 and 2 are presented by the amplitude spectra as shown in Figs. 6 and 7. A comparison of these shows that the tensions due to the measured forces are much larger than those of the theoretical wave forces.

The filtered slowly varying tensions due to the two waves are shown in Figs. 8 and 9. The mean of wave frequency tendon forces is put to zero to display the tension forces clearly. The mean and SD of slow-varying tensions due to the theoretical wave are slightly larger than those of Case 2, as given in Table 6.

Table 6. Statistics of Slowly Varying Tension

	Case 1	Case 2	Abs Difference
Mean (N)	4.03E7	3.95E7	8.0E5
Max. (including mean, N)	4.37E7	4.28E7	9.0E5
Min. (including mean, N)	3.82E7	3.63E7	1.9E6
Standard deviation SD (N)	1.22E6	8.62E5	3.58E5
Skewness	0.59	0.56	
Kurtosis	2.51	4.6	

WAVE-FREQUENCY RESPONSES

Similarly the surging motions and tensions due to the two waves (case 1 and 2) in the wave-frequency region are analyzed. The surging motions of the two waves are shown in Figs. 4 and 5. The SDs are about the same but the max. and min. values are much different (Table 7).

The tension amplitude spectra in the wave-frequency regions are plotted in Figs. 6 and 7, while the corresponding filtered time series in Figs. 8 and 9. It is evident from these figures and Table 8 that the SD of tensions of Case 1 is much smaller than those of Case 2.

Table 7. Statistics of Surge in Wave-Frequency Region

	Case 1	Case 2	Abs Difference
Max. (removed mean, m)	3.6	3.86	0.26
Min. (removed mean, m)	-3.02	-3.37	0.35
Standard deviation SD (m)	1.03	0.93	0.1
Skewness	-0.011	-0.032	
Kurtosis	2.98	3.86	

Table 8. Statistics of Tension in Wave-Frequency Region

	Case 1	Case 2	Abs Difference
Max. (removed mean, N)	4.54E6	1.11E7	6.56E6
Min. (removed mean, N)	-6.82E6	-1.07E7	3.88E6
Standard deviation SD (N)	1.51E6	3.09E6	1.58E6
Skewness	-0.342	0.022	
Kurtosis	3.08	3.23	

HIGH-FREQUENCY RESPONSES

The springing and ringing are due to the resonant pitching and heaving motions at high frequencies. The typical ratios of max. tension to the SD vary from 4 to 6 for the springing, and above 6 for the ringing. There are a few different definitions of the criteria.

To investigate the ringing phenomenon, we employ a third wave train, Case 3 (Table 1), which contains a strong asymmetric wave crest that has concave front and convex rear as shown in Fig. 22.

The tension amplitude spectra of element #10 of Cases 1, 2 and 3, are shown in Figs. 6, 7 and 10, while the corresponding filtered time series are illustrated in Figs. 11, 12 and 13. One sees in Table 9 that the max. tensions of Case 1, 2 and 3 are 0.95 MN, 5.16 MN and 11.0 MN, respectively.

The statistics of the TT and HFT are compared in Table 9. The HFTs are all extracted means. It is interesting to note that the SD of HFT of Case 1 is about 20% of that of Case 2 and 11% of Case 3. This implies that tensions in high frequency region are significantly underestimated by the second-order diffraction theory.

Table 9. Statistics of Total Tension (TT) and High Frequency Tension (HFT)

	Case 1		Case 2		Case 3	
	TT	HFT	TT	HFT	TT	HFT
Mean (N)	4.03E7	3.6E-3	3.95E7	1.25E-2	3.92E7	3.28E-3
Max*	4.8E7	9.5E5	5.19E7	5.16E6	5.57E7	1.1E7
Min *	3.35E7	-9.7E5	2.81E7	-5.16E6	2.0E7	-1.1E7
Variance (N^2)	3.8E12	3.16E11	1.12E13	8.74E11	1.47E13	2.35E12
SD** (N)	1.95E6	1.78E5	3.35E6	9.35E5	3.83E6	1.53E6
Skewness	0.2554	1.8E-8	9.11E-2	1.84E-8	-5.06E-3	-1.7E-8
Kurtosis	2.9843	6.5156	3.2748	5.7307	3.9944	6.6672

*including mean, N
**Standard Deviation

The filtered ratios of the tensions to the SD due to the three waves are shown in Figs. 14, 15 and 16. It is found that springing occurs in Cases 1 and 2, while ringing does in Case 3.

The probability distributions of the TT and HFT of the three waves are shown in Figs. 17, 18 and 19. It is seen that the TTs of Case 1 and 2 are about Gaussian, while that of Case 3 and all of the HFTs are clearly non-Gaussian. It has to be noted that the probability of the ringing has to be taken from more than three hours long data. The ringing occurs, according to field observations (Natvig, 1994), once in more than three hours. The reason we have the ringing in the wave of Case 3 is that the largest wave group in the wave train is artificially distorted to make a strong asymmetric wave crest. That may be generated in the long and huge wave tank.

SPRINGING AND RINGING WAVES

It is found in the preceding section that the strong asymmetric wave of Case 3 produces the ringing, while those of Cases 1 and 2 generate the springing. The largest HFTs are generated at the time of about 511 s in Case 1 (Fig. 11), 822 s in Case 2 (Fig. 12) and 1396 s in case 3 (Fig. 13). The waves causing these tensions are demonstrated in Figs. 20, 21 and 22.

A comparison of the two waves in Figs. 20 and 21 shows that the nonlinearity of the numerical wave is much weaker than that of the laboratory-generated wave. This implies that the numerical wave method is valid for the seas less than moderate.

It is seen that the largest waves of Cases 2 and 3 (Figs. 21 and 22) are much different, i.e. the asymmetry of Case 3 is much larger than that of Case 2. The wave of Case 3 has the crest with a concave front and a convex rear, while that of Case 2 has convex front and convex rear. Thus the former is strongly asymmetric, while the latter is weakly asymmetric. It is concluded that the strong asymmetric wave is the ringing wave, whereas the weak asymmetric wave is the springing wave.

CONCLUDING REMARKS

The results are summarized as follows:

The slow drift motion of Case 1 (theoretical forces) is larger than that of Case 2 (measured forces); the slow-varying tension of Case 1 is larger than that of Case 2; the SD of surging motion of Case 1 is almost the same as that of Case 2, in the wave-frequency region; the tension of case 1 is much smaller than that of Case 2, in the wave-frequency region; the tension of Case 1 is 20% of that of Case 2 and 11% of that of Case 3, in the high frequency region.

Springing is generated in the waves of Cases 1 and 2, while ringing is generated in the wave of Case 3. The strong asymmetric wave of Case 3 is the ringing wave, while the large weak asymmetric waves of Cases 1 and 2 are the springing waves.

The max. tensions of Case 1, 2 and 3 are 0.95 MN, 5.16 MN and 11.0 MN, respectively, while the ratios of these to the SDs are 5.3, 5.5 and 7.3. It is incomplete to use only the ratio of extreme tension to SD as a criterion of springing and ringing, because the absolute HFTs may be much different even the ratios are about the same.

The numerical wave generation methods (Case 1) are only good for less than moderate seas. An arbitrary application of the method to storm seas would cause significant errors. The second-order diffraction theory is applicable for the low or moderate seas if extended carefully. A consistent higher-order diffraction theory for high seas or strong nonlinear waves is currently not available.

It is recommended to produce simulations much longer than 27 minutes to have more realistic statistical analysis.

ACKNOWLEDGMENT

This work was cosponsored by ABS and Koje Samsung Maritime Research Institute. The TIOSTAMU was supported by AMOCO, ABS, BPE, and El-Dorado Engineering. The authors are very grateful to the sponsors for their supports of the research and permission to publish the paper.

REFERENCES

Chau, FP and Eatock Taylor, R (1992). "Second-order Wave Diffraction by a Vertical Cylinder" *J Fluid Mech*, Vol 242, pp 571-599.

Eatock Taylor, R, Rainey, RCT and Dai, DN (1992). "Nonlinear Hydrodynamic Analysis of TLPs in Waves: Slender Body and Diffraction Theories Compared," *Proc 8th BOSS'92*, London, pp 569-583.

Faltinsen, OM , Newman, JN and Vinje, T (1995). "Nonlinear Wave Loads on a Slender Vertical Cylinder," *J Fluid Mech*, Vol 289, pp 178-198.

Funke, ER, Mansand, EPD and Barthel, V (1982). "Generation of 'Random' Waves with Wave Asymmetries," *Proc 18th Conf on Coastal Engr*, Cape Town, South Africa.

Huse, E (1990). "Resonant Heave Damping of Tension Leg Platforms," *Proc of 22nd OTC*, OTC 6317, pp 431-436.

Kim, CH, Kim, MH, Liu, YH and Zhao, CT (1994). "Time Domain Simulation of Nonlinear Response of a Coupled TLP System," *Int J Offshore and Polar Engr*, ISOPE, 4(4), pp. 284-291.

Kim, CH, Randall, RE, Boo, SY and Krafft, MJ (1992). "Kinematics of 2-D Transient Water Waves Using Laser Doppler Anemometry," *J Waterway, Port, Coastal and Ocean Engr*, ASCE, Vol 118, pp 147-165.

Kim, CH, Xu, Y, Zou, J and Won, YS (1996A). "A Model for Weak Impacting Force on Vertical Truncated Cylinder due to Steep Asymmetric Wave," *Proc 6th Int Offshore and Polar Engr Conf*, Los Angeles, Vol 3, pp 215-220.

Kim, CH, Xu, Y and Zou, J (1996B). "Impact and Non-Impact on Vertical Truncated Cylinder Due to Strong and Weak Asymmetric Wave," will appear in *Int J Offshore and Polar Engr*, June, 1997.

Kim, CH, Zhao, CT and Zou, J (1995). "Springing and Ringing Due to Nonlinear Waves on a Coupled TLP," *Proc 5th ISOPE*, Vol 1, pp 83-89.

Kim, CH, Zhao, CT, Zou, J and Xu, Y (1997). "Springing and Ringing Due to Laboratory-Generated Asymmetric Waves," *Int J Offshore and Polar Engrg*, Vol 7, No 1.

Kim, MH and Yue, DKP (1989). "The Complete Second-order Diffraction Solution for an Axisymmetric Body, Part 1, Monochromatic Model by Incident Waves," *J Fluid Mech*, Vol 200, pp 235-264.

Kim, MH and Yue, DKP (1990). "The Complete Second-order Diffraction Solution for an Axisymmetric Body, Part 2, Bichromatic Incident Waves and Body Motions," *J Fluid Mech*, Vol 211, pp 571-593.

Krokstad, JR, Stansberg, CT, Nestegard, A and Marthinsen, T (1996). "A New Non-slender Ringing Load Approach Verified against Experiments," *OMAE*, Vol I, part A, pp 371-387.

Liu, YH, Kim, CH and Lu, XS (1991). "Comparison of Higher-order Boundary Element and Constant Panel Method for Hydrodynamic Loadings," *Int J Offshore and Polar Engrg*, Vol 1, No 1, pp 8-17.

Liu, YH, Kim, MH and Kim, CH (1995). "The Computation of Second-order Mean and Double Frequency Wave Loads on a Compliant TLP by HOBEM," *Int J Offshore and Polar Engrg*, Vol 5, pp 111-119.

Longuet-Higgins, MS and Cokelet, ED (1976). "The Deformation of Steep Surface Waves on Water-1. A Numerical Method of Computation," *Proc Roy Soc Lond*, A350, pp 1-26.

Malenica, S and Molin, B (1995). "Third-harmonic Wave Diffraction by a vertical Cylinder," *J Fluid Mech*, Vol 302, pp 203-229.

Molin, B (1979). "Second-order Diffraction Loads upon Three Dimensional Bodies," *Appl Ocean Research*, Vol 1, pp 197-202.

Myrhang, D and Kjeldsen, SP (1984). "Parametric Modeling of Joint Probability Density Distribution for Steepness and Asymmetry in Deep Water Waves," *Applied Ocean Research*, Vol. 6, No. 4, pp. 207-220.

Natvig, BJ (1994). "A Proposed Ringing Analysis Model for Higher Order Tether Response," *Proc Int Offshore and Polar Engr Conf*, Osaka, ISOPE, Vol 1, pp. 40-51.

Rainey, RCT (1989). "A New Equation for Calculating Wave Loads on Offshore Structures," *J Fluid Mech*, Vol 204, pp 295-324.

Rainey, RCT (1995). "Slender-body Expressions for the Wave Load on Offshore Structures," *Proc R Soc Lond A*, Vol 450, pp 391-416.

Thiagarjan, KP and Troesch, AW (1993). "Hydrodynamic Damping Estimation and Scaling for Tension Leg Platforms," *Proc 12th OMAE*, Vol 1, pp 473-479.

Wheeler, JD (1970). "Method for Calculating Forces Produced by Irregular Waves," *J Pet Tech*, March, pp 359-367.

Zhao, CT (1996). "Theoretical Investigation of Springing-Ringing Problems in Tension Leg Platforms," Ph.D. dissertation in Ocean Engr, TAMU, May.

Zou, J and Kim, CH (1995). "Extreme Wave Kinematics and Impact Loads on a Fixed Truncated Circular Cylinder," *Proc 5th ISOPE*, Vol 3, pp 216-225.

Zou, J and Kim, CH (1996). "Experimental Study of Impacting Wave Force on Vertical Truncated Cylinder," *Int J Offshore and Polar Engr*, Vol 6, No 4, pp 291-293.

Zou, J and Kim, CH (1996). "Generation of Strong Asymmetric Irregular Wave with JONSWAP Spectrum," Progress Report No. 1, submitted to ABS and Samsung.

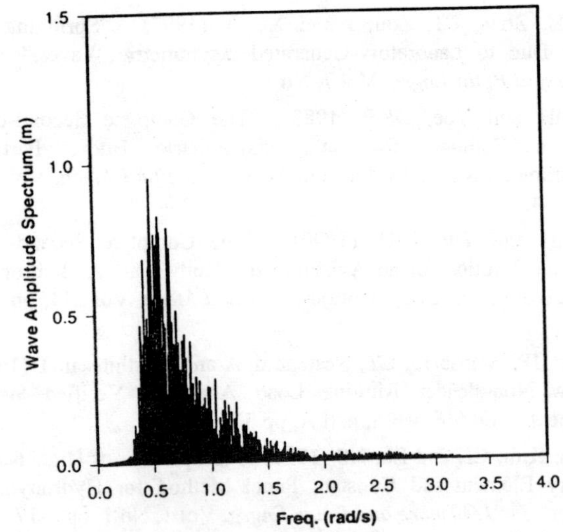

Fig. 1 Wave amplitude spectrum (case 1)

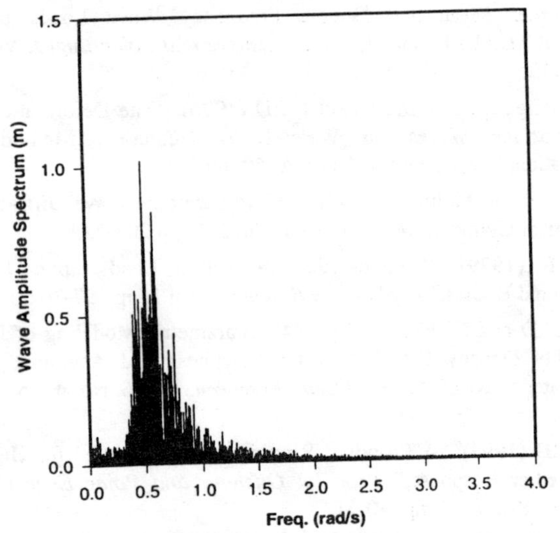

Fig. 2 Wave amplitude spectrum (case 2)

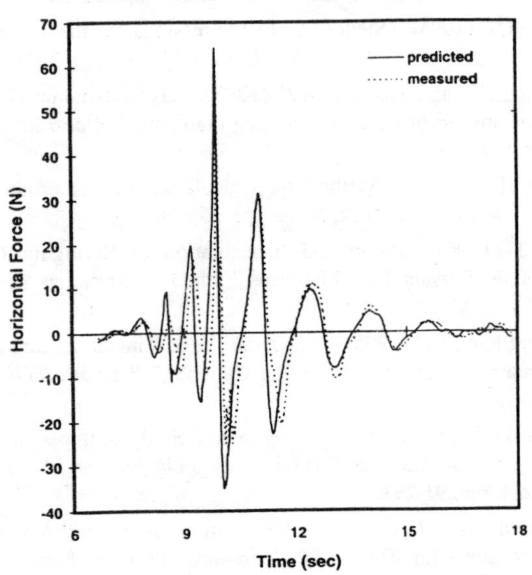

Fig. 3 Comparison of horizontal force on vertical truncated circular cylinder

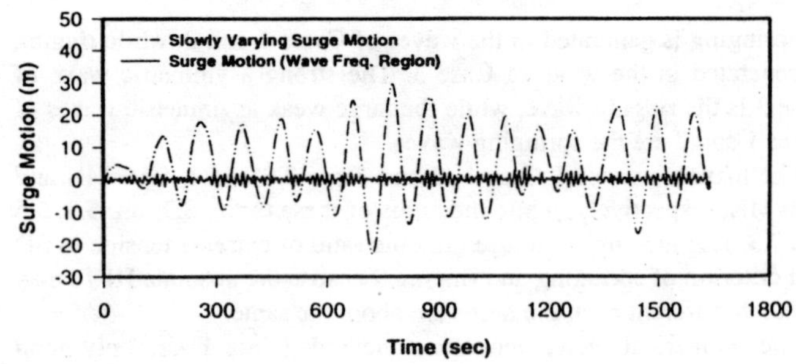

Fig. 4 Surge motion (case 1)

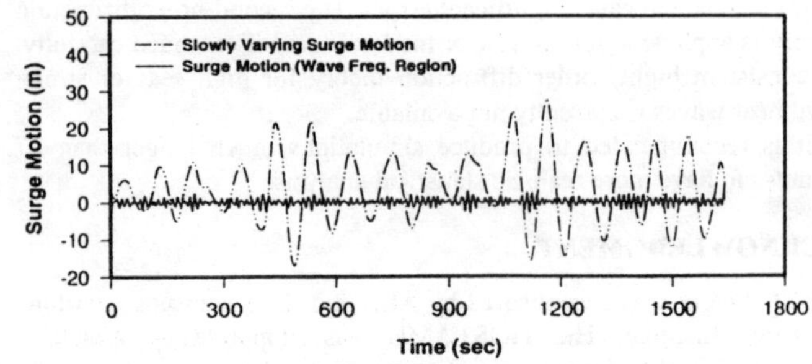

Fig. 5 Surge motion (case 2)

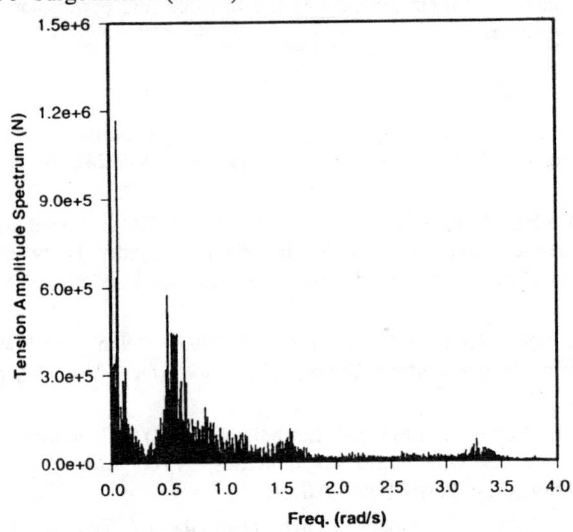

Fig. 6 Tension amplitude spectrum (case 1)

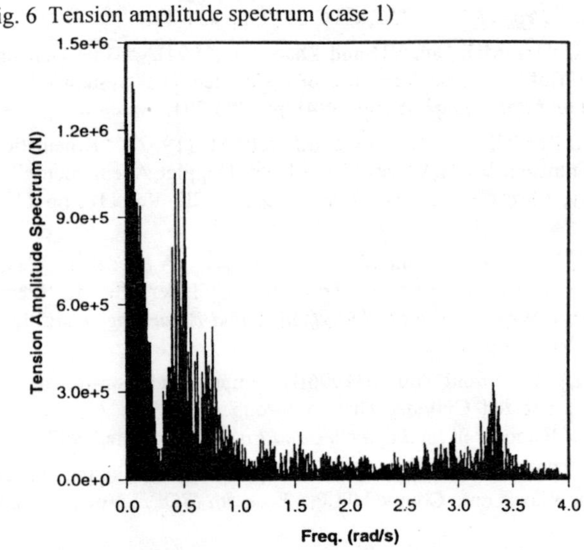

Fig. 7 Tension amplitude spectrum (case 2)

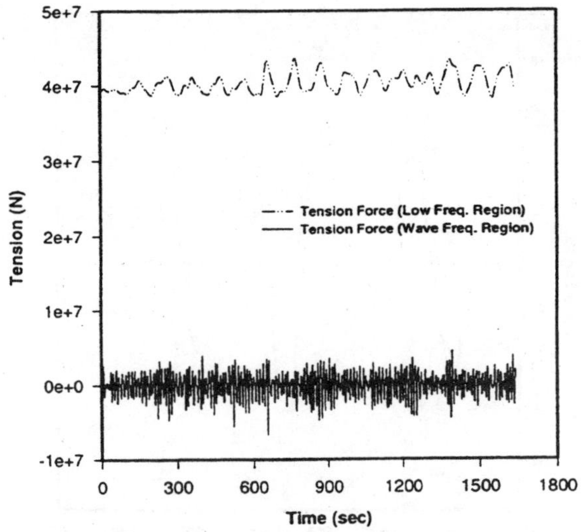

Fig. 8 Low and wave frequency forces (case 1)

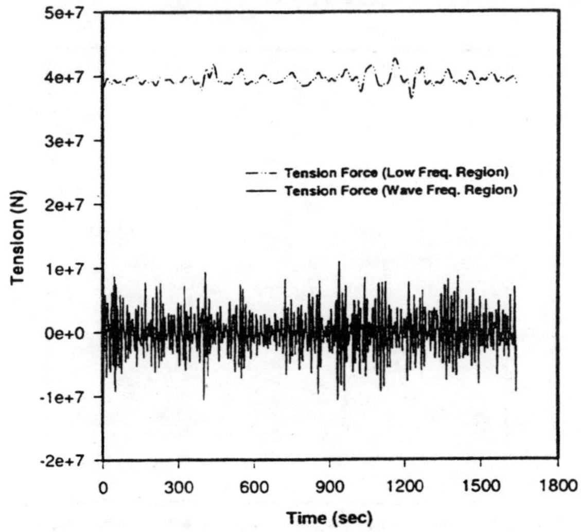

Fig. 9 Low and wave frequency forces (case 2)

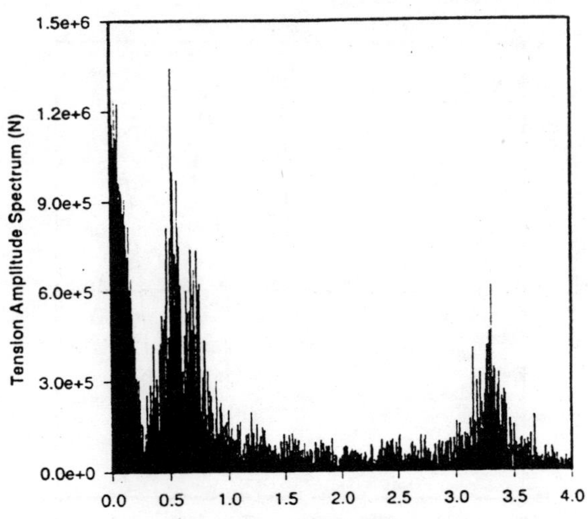

Fig. 10 Tension amplitude spectrum (case 3)

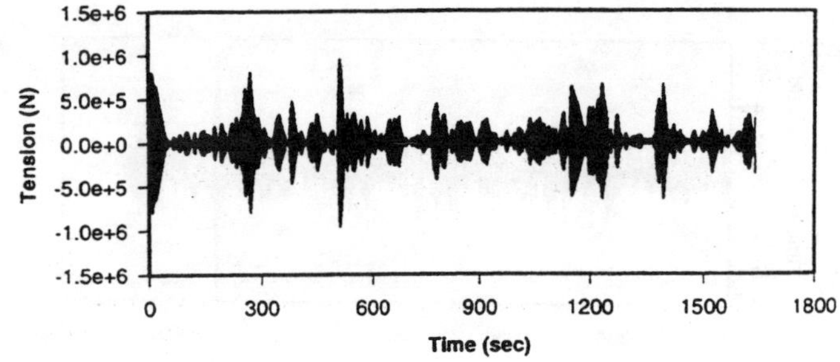

Fig. 11 Filtered tension force (3.0-4.0 rad/s, case 1)

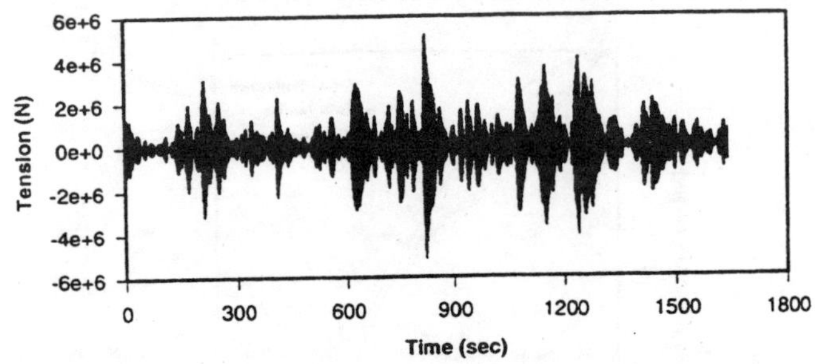

Fig. 12 Filtered tension force (3.0-4.0 rad/s, case 2)

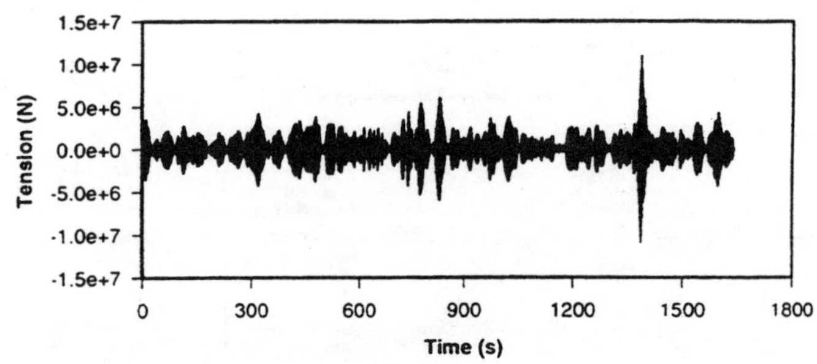

Fig. 13 Filtered tension force (3.0-4.0 rad/s, case 3)

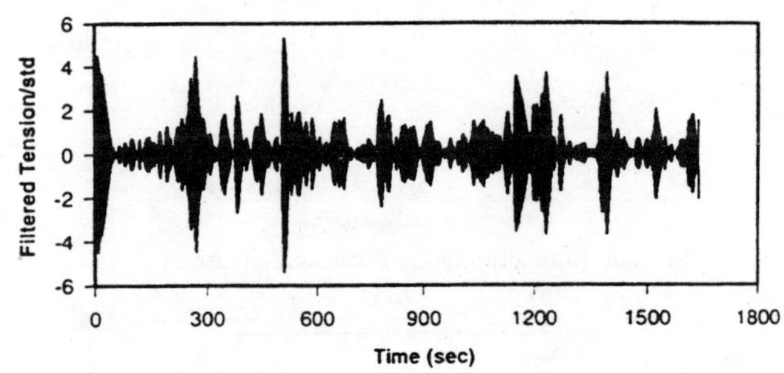

Fig. 14 Filtered tension/standard deviation (case 1)

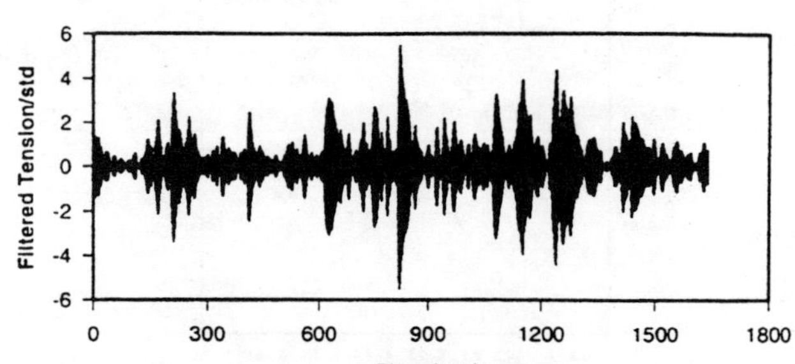

Fig. 15 Filtered tension/standard deviation (case 2)

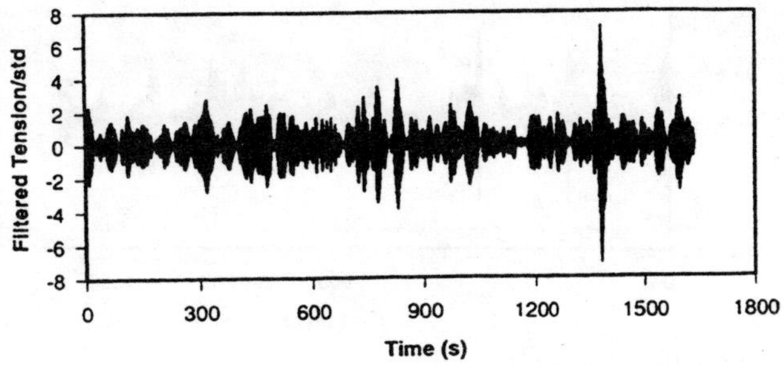

Fig. 16 Filtered tension/standard deviation (case 3)

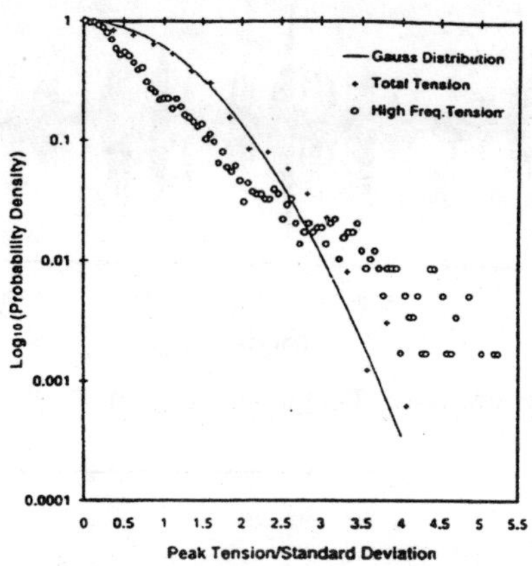

Fig. 17 Tension force probability distribution (case 1)

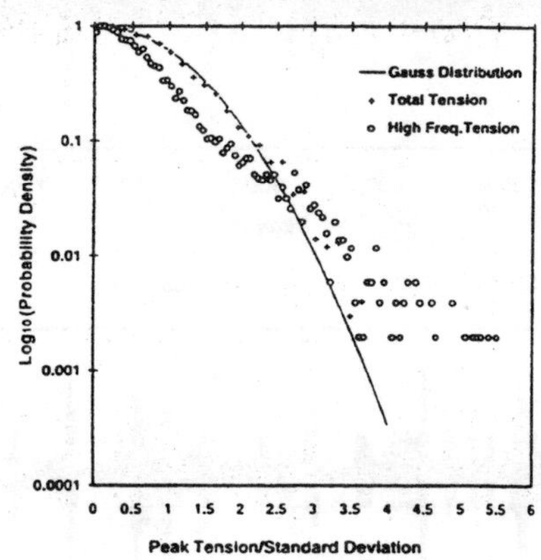

Fig. 18 Tension force probability distribution (case 2)

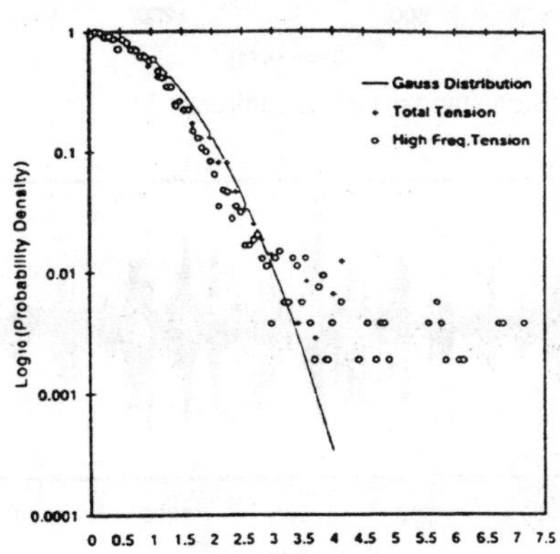

Fig. 19 Tension force probability distribution (case 3)

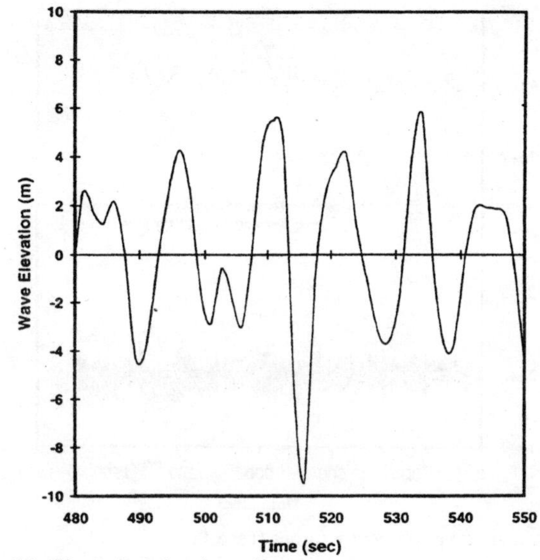

Fig. 20 Wave elevation (case 1)

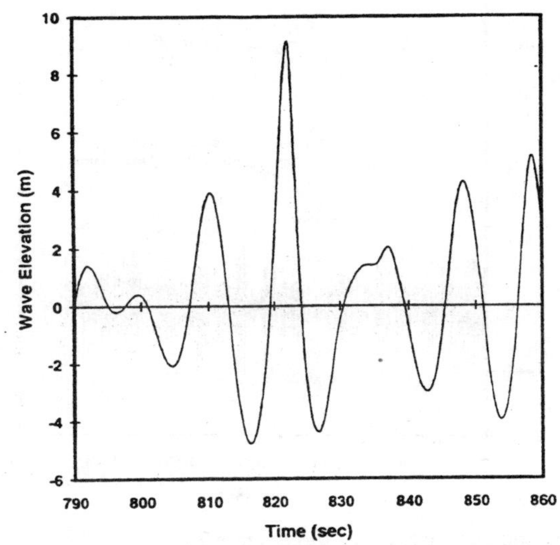

Fig. 21 Wave elevation (case 2)

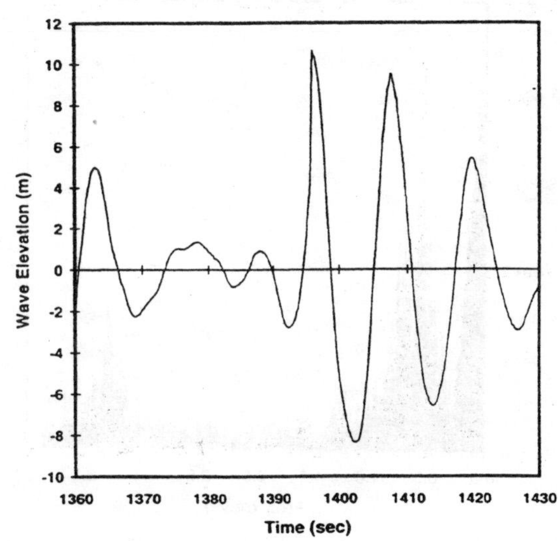

Fig. 22 Wave elevation (case 3)

Proceedings of the Seventh (1997) International Offshore and Polar Engineering Conference
Honolulu, USA, May 25-30, 1997

ISBN 1-880653-28-1 (Set); ISBN 1-880653-29-X (Vol. I)

Dragged Surge Motion on Tension Leg Platform Tethers

Hsien Hua Lee, Pei-Wen Wang and Chung-Pan Lee
National Sun Yat-Sen University
Kaohsiung, Taiwan, China

ABSTRACT

This paper presents an analytical solution for the tethers of the tension leg platform system when the platform subjected to the wave-induced surge motion and the flow-induced drag motion. Along with the analysis the coupling problem of a two dimensional tension leg platform interacting with a monochromatic linear wave train in an inviscid and incompressible fluid is also being taken into account. In the interaction problem for the tension leg platform system the scattering problem and radiation problem are firstly solved separately and then combined together to resolve for all unknowns. The dynamic behavior of the tethers was further solved based on these solutions.

KEY WORDS: Tension leg platform, surge motion, tethers, structure wave interaction

INTRODUCTION

Tension leg platforms (TLPs) are usually used as working station for oil exploration in deep water. Lately, the application on this structural system is even wider such as the floating break-water system and the fish-farming cage system. The TLPs are constituted of a semi-submerged structure and pre-tensioned tethers anchored to the ocean bed. The difficulty in the dynamic analysis of the TLPs is due to the nonlinearity of the TLP motion and the boundary conditions on the free surface and on the structural bottom (Garrison, 1974). In the problem the linear boundary value problem is incorporated into a scattering problem and a radiation problem (Sarpkaya and Isaacson, 1981). This problem was solved numerically and experimentally by Yamamoto *et al.* (1982) and numerically and analytically by Lee and Lee (1991, 1993), in which the surge motion of platform with pre-tensioned tethers was calculated. However, the elasticity of the tethers was only implied and the motion of tethers was also simplified as on-line rigid-body motion as the top platform.

For the fish-farming cage systems with tension leg as anchoring devices the tether's behavior will be crucial concerning about the space inside the cage and the cage safety during severe storm conditions. Therefore, in this study based on Lee and Lee's (1993) results an analytical solution for the tethers of the tension leg platform system was derived when subjected to flow-induced drag motion and the surge motion of top platform induced from incident waves. Firstly, the scattering problem is solved and the results are used to calculate the forcing function for the radiation problem and then both solutions were used for the solution of the tethers. In this preliminary analysis the interaction between the tether motion and the platform motion was ignored. However, the surge motion of the platform was taken into account when the equation of the motion for the tether was derived and solved.

GENERAL WAVE THEORY

For the inviscid and incompressible fluid and irrotational flow, a single-valued velocity potential ϕ can be defined as

$$u = -\nabla\phi \tag{1}$$

where u is the flow velocity, ∇ is the gradient operator. The velocity potential satisfies the Laplace equation:

$$\nabla^2\phi = 0 \tag{2}$$

and the Bernoulli equation

$$-\frac{\partial\phi}{\partial t}+\frac{1}{2}\nabla\phi\nabla\phi+\frac{p}{\rho}+gz=0 \tag{3}$$

in the flow field, where p is the pressure and ρ is the water density. The nonlinear term in equation (3) is neglected when the linear small amplitude wave is assumed.

A two-dimensional tension leg platform interacting with a monochromatic small amplitude wave propagating in the $+x$-direction was considered here as shown in Fig. 1. The wave form and the associated velocity potential are given accordingly as,

$$\eta_i = -iA_i e^{-(K_1 x + i\sigma t)} \tag{4}$$

and

$$\phi_i = \frac{A_i g}{\sigma} \frac{\cos[K_1(z+h)]}{\cos(K_1 h)} e^{-(K_1 x + i\sigma t)} \tag{5}$$

where A_i is the wave amplitude, g is the gravitational constant, h is the water depth, $\sigma = (2\pi/T)$ is the angular frequency, T is the period, and $K_1 = -ik,\ k = (2\pi/L)$ is the wave number, L is the wave length. K_1 satisfies the dispersion relation

$$\sigma^2 = gK_1 \tan(K_1 h) \tag{6}$$

TENSION LEG PLATFORM SYSTEM

In the platform system the motion of the structure induced by the small amplitude incident wave is assumed to be small. The structural and wave induced motion can be solved from the imposed boundary problem. Because the linearity of the problem, the problem can be incorporated into a scattering and a radiation problem (Black et al., 1971). The wave forces calculated from the scattering problem provide the force function in the radiation problem, and the forced oscillation then generate outgoing waves.

A tension leg platform system was illustrated as Fig. 1, where the flow field is divided into three regions with two artificial boundaries at $x = -b$ and $x = b$. In region I, the total velocity potential ϕ_I consists of incident waves ϕ_i, scattered waves ϕ_{Is}, and the radiated waves ϕ_{Iw}. In region II and III, the total velocity potential ϕ_{II} and ϕ_{III} consists of scattered waves ϕ_{IIs} and ϕ_{IIIs}, and the radiated waves ϕ_{IIw} and ϕ_{IIIw}. The subscript s denotes the scattering problem and w denotes the radiation (wave making) problem. All the velocity potentials satisfy the Laplace equation. Furthermore, Sommerfeld's radiation condition is satisfied at the infinity of region I and III to secure unique solutions.

$$\lim_{x \to \pm\infty}\left[\frac{\partial \phi}{\partial x} \pm \frac{1}{C_w}\frac{\partial \phi}{\partial t}\right] = 0 \tag{7}$$

Scattering Problem

In the scattering problem, the incident wave is considered to be diffracted by a fixed structure. The corresponding boundary value problem was also shown in Fig. 1. Applying the method of the separation of variables, matching the horizontal boundary conditions in each region, and applying Sommerfeld's condition to regions I and III, the corresponding surface elevation and velocity potential can be found.

⇒In region I :

$$\phi_{Is} = \sum_{j=1}^{\infty} \frac{A_{Isj} g}{\sigma} \frac{\cos[K_j(Z+h)]}{\cos(K_j h)} e^{[K_j(x+b) - i\sigma t]} \tag{8}$$

and

$$\eta_{Is} = -i\sum_{j=1}^{\infty} A_{Isj} e^{[K_j(x+b) - i\sigma t]} \tag{9}$$

⇒In region III:

$$\phi_{IIIs} = \sum_{j=1}^{\infty} \frac{A_{IIIsj} g}{\sigma} \frac{\cos[K_j(Z+h)]}{\cos(K_j h)} e^{-[K_j(x-b) + i\sigma t]} \tag{10}$$

and

$$\eta_{IIIs} = -i\sum_{j=1}^{\infty} A_{IIIsj} e^{-[K_j(x-b) + i\sigma t]} \tag{11}$$

where the eigenvalues K_j can be solved from the dispersion equation

$$\sigma^2 = gK_j \tan(K_j h) \tag{12}$$

with

$$K_1 = -ik,\ for\ j = 1; \qquad (2j-3)\frac{\pi}{2} < K_j h < (j-1)\pi \tag{13}$$

⇒In region II:

$$\phi_{IIs} = \frac{ig}{\sigma}[(A_{IIsP1}\frac{x}{b} + A_{IIsN1})\cos K_{II1}(Z+h) + \sum_{j=2}^{\infty}(-1)^{j-1} \sum_{j=2}^{\infty}(A_{IIsPj}e^{-K_{IIj}(x+b)} + A_{IIsNj}e^{K_{IIj}(x+b)})\cos K_{IIj}(Z+h)]e^{-i\sigma t} \tag{14}$$

where the eigenvalues K_{IIj} can be solved from the dispersion equation

$$K_{IIj} = \frac{(j-1)\pi}{h-d}, j \ge 1 \tag{15}$$

The series of the four unknowns $A_{Isj}, A_{IIsPj}, A_{IIIsj}, A_{IIsNj}$ can further be solved from the equations derived from the two artificial boundaries of region II.

Radiation Problem

In the radiation problem the structure is considered to be forced in motion by the wave forces induced by incident waves and scattered waves. The corresponding boundary value problem is illustrated in Fig. 2. The displacement of the dragged surge motion is given by

$$\xi = Se^{-i\sigma t} \tag{16}$$

where S is the unknown amplitude of the surge motion.

By means of the method similar to the scattering problem, the solutions of the radiation problem are written as follows.

⇒In region I :

$$\phi_{Iw} = \sum_{j=1}^{\infty} \frac{A_{Iwj} g}{\sigma} \frac{\cos[K_j(Z+h)]}{\cos(K_j h)} e^{[K_j(x+b) - i\sigma t]} \tag{17}$$

$$\eta_{Iw} = -i\sum_{j=1}^{\infty} A_{Iwj} e^{[K_j(x+b) - i\sigma t]} \tag{18}$$

⇒In region III:

$$\phi_{IIIw}=\sum_{j=1}^{\infty}\frac{A_{IIIwj}g}{\sigma}\frac{\cos[K_j(Z+h)]}{\cos(K_jh)}e^{-[K_j(x-b)+i\sigma t]} \tag{19}$$

$$\eta_{IIIw}=-i\sum_{j=1}^{\infty}A_{IIIwj}e^{-[K_j(x-b)+i\sigma t]} \tag{20}$$

⇒In region **II**:

$$\phi_{IIw}=\frac{ig}{\sigma}[(A_{IIwP1}\frac{x}{b}+A_{IIwN1})\cos K_{II1}(Z+h)+\sum_{j=2}^{\infty}(-1)^{j-1}(A_{IIwPj}e^{-K_{IIj}(x+b)}+A_{IIwNj}e^{K_{IIj}(x+b)})\cos K_{IIj}(Z+h)]e^{-i\sigma t} \tag{21}$$

Again the four series of the unknown amplitudes of waves $A_{Iwj}, A_{IIwPj}, A_{IIwNj}, A_{IIIwj}$ can be solved from the four equations derived from the two vertical boundaries. To solve for the unknown amplitude S of the surge motion as shown in the kinematic boundary conditions, the equation of the surge motion of the structure was requested.

Motion of the Platform

The equation of motion of the structure can be derived from Newton's second law, and written as

$$M\frac{d^2\xi}{dt^2}=F_{wx}-F_{Tx} \tag{22}$$

where M is the mass of the structure, F_{wx} is the horizontal wave force acting on the two vertical sides of the structure, and F_{Tx} is the horizontal component of the static tension force applied by the tension legs. The wave force F_{wx} is the integration of the total hydrodynamic pressure over the vertical surfaces of the structure at x= -b and x= b (from the linearized Bernoulli equation).

$$F_{wx}=\rho\int_{-d}^{0}\left(\frac{\partial\phi_I}{\partial t}\Big|_{x=-b}-\frac{\partial\phi_{III}}{\partial t}\Big|_{x=b}\right)dz \tag{23}$$

The static tension force is assumed proportional to the displacement of the surge motion ξ, as

$$F_{Tx}=f\xi \tag{24}$$

where $f=2\rho gb^{(d-d_0)}/_{(h-d)}$, $d_0=M/_{(2\rho b)}$, d and d_0 are the drafts of he structure before and after applying the pre-tension force. The lynamic tension in an order of ξ^2 is neglected here.

Substituting Equation (23) and (24) into Equation (22), the equation of motion can be rewritten in the form as following:

$$M\frac{d^2\xi}{dt^2}-\rho\int_{-d}^{0}\left(\frac{\partial\phi_{Iw}}{\partial t}\Big|_{x=-b}-\frac{\partial\phi_{IIIw}}{\partial t}\Big|_{x=b}\right)dz+f\xi=\rho\int_{-d}^{0}\left(\frac{\partial(\phi_i+\phi_{Is})}{\partial t}\Big|_{x=-b}-\frac{\partial\phi_{IIIs}}{\partial t}\Big|_{x=b}\right)dz \tag{25}$$

The fifth equation which relates to the amplitude of the surge motion may be obtained by solving Equation (25) as

$$\frac{2}{bQ}B_1A_{IIwP1}+\frac{1}{Q}\sum_{j=2}^{\infty}(-1)^{j-1}K_{IIj}\left(1+e^{-2K_{IIj}b}\right)B_j\left(-A_{IIwPj}+A_{IIwNj}\right)-\frac{i}{Q}\left[\sum_{\beta=1}^{\infty}\frac{\langle Z_\beta^0\rangle}{\cos(K_\beta h)}\left(A_{I\!w\beta}-A_{IIIw\beta}\right)+\frac{\langle Z_1^0\rangle}{\cos(K_1h)}e^{K_1b}A_i\right]=\frac{\sigma^2}{g}S \tag{26}$$

where

$$Q=\frac{1}{\rho}\left[\frac{f}{\sigma^2}-M-2\rho\sum_{\gamma=1}^{\infty}\left(C_\gamma\langle Z_\gamma^0\rangle\right)\right] \tag{27}$$

EQUATION OF MOTION FOR TETHERS

The equation of motion for the pre-tension tether subjected to the wave induced forces and surge motion of top platform can be written as

$$\rho\frac{\partial^2\chi}{\partial t^2}+C\frac{\partial\chi}{\partial t}-f\frac{\partial^2\chi}{\partial z^2}=F(z,t) \tag{28}$$

where χ is the horizontal deflection of the tether; f is the pre-tension force and C is the damping coefficient of unit length. The unit length wave force is obtained through Morison's equation (Newman, 1977) for small body accounted for the relative structural motions as

$$F(z,t)=0.5\rho_wDC_d|u-\dot{\chi}|(u-\dot{\chi})+0.25\pi\rho_wD^2C_m(u-\dot{\chi})+0.25\pi\rho_wD^2\ddot{\chi} \tag{29}$$

where ρ_w is the density of water; D is the diameter of tether; C_d and C_m are the drag coefficient and added mass coefficient respectively while $\dot{\chi}$ and $\ddot{\chi}$ are the velocity and acceleration of the tether. Substituting into the equation of motion, Equation (28) is rewritten in the form.

$$\left[\rho+0.25\pi\rho_wD^2(C_m-1)\right]\frac{\partial^2\chi}{\partial t^2}+C\frac{\partial\chi}{\partial t}-f\frac{\partial^2\chi}{\partial z^2}=0.25\pi\rho_wD^2C_m\dot{u}+0.5\rho_wDC_d|u-\dot{\chi}|(u-\dot{\chi}) \tag{30}$$

There exists a nonlinear term in the force function, and it is replaced by a linear drag according to Lorent's hypothesis of equivalent work (Sollitt and Cross,1972) by assuming that the linear drag consumes the same energy in a wave period as that of the nonlinear drag. The linear drag coefficient Rd^* is defined as

$$Rd^*=\frac{\int_0^T\left[0.5\rho_wDC_d|u-\dot{\chi}|(u-\dot{\chi})\right](u-\dot{\chi})dt}{\int_0^T\rho_wD\sigma(u-\dot{\chi})(u-\dot{\chi})dt} \tag{31}$$

After applying this linear drag coefficient Equation (30) can be rewritten as

$$\left[\rho+0.25\pi\rho_wD^2(C_m-1)\right]\frac{\partial^2\chi}{\partial t^2}+\left(C+Rd^*\rho_wD\sigma\right)\frac{\partial\chi}{\partial t}-f\frac{\partial^2\chi}{\partial z^2}=0.25\pi\rho_wD^2C_m\dot{u}+Rd^*\rho_wD\sigma u \tag{32}$$

where the velocity is obtained from the potential, thus $u=-\frac{\partial\phi_{II}}{\partial x}$ and $\dot{u}=\frac{\partial u}{\partial t}$. Substituting u and $\dot{u}$ into Equation (32) we have

$$\begin{aligned}&\left[\rho+0.25\pi\rho_w D^2(C_m-1)\right]\frac{\partial^2\chi}{\partial t^2}+\left(C+Rd^*\rho_w D\sigma\right)\frac{\partial\chi}{\partial t}-f\frac{\partial^2\chi}{\partial z^2}\\&=-\left(0.25\pi\rho_w D^2C_m g+iRd^*g\rho_w D\right)\Big\{\frac{A_{IIs/wP1}}{b}\cos K_{II1}(z+h)\\&+\sum_{j=2}^{\infty}(-1)^{j-1}\left(-K_{IIj}A_{IIs/wPj}+K_{IIs/wNj}e^{-2bK_{IIj}}\right)\cos K_{IIj}(z+h)\Big\}e^{-i\sigma t}\end{aligned}\tag{33}$$

The motion of the tether at the bottom of the structure will be same as the motion of top platform, that was assumed to be harmonic with amplitude S and frequency σ, as $\chi(-d,t)=Se^{-i\sigma t}$, and at the sea bed the motion is restrained thus, $\chi(h,0)=0$

Assuming that the motion of the tether is in the same harmonic form as the top platform $\chi(z,t)=Z(z)e^{-i\sigma t}$, and substituted in the equation of motion , a reduced equation is given by

$$\begin{aligned}&f\frac{d^2Z}{dz^2}+\left\{\left[\rho+0.25\pi\rho_w D^2(C_m-1)\right]\sigma^2+\left(C+Rd^*\rho_w D\sigma\right)(i\sigma)\right\}Z(z)\\&=\left(0.25\pi\rho_w D^2C_m g+iRd^*g\rho_w D\right)\Big\{\frac{A_{IIs/wP1}}{b}\cos K_{II1}(z+h)\\&+\sum_{j=2}^{\infty}(-1)^{j-1}\left(-K_{IIj}A_{IIs/wPj}+K_{IIs/wNj}e^{-2bK_{IIj}}\right)\cos K_{IIj}(z+h)\Big\}\end{aligned}\tag{34}$$

with reduced boundary conditions.

$$\begin{cases}Z(-d)=S\\Z(-h)=0\end{cases}\tag{35}$$

Now Equation (34) is an ordinary differential equation and ready to be solved in the following infinite series form

$$Z(z)=C_1e^{az}+C_2e^{-az}+\sum_{j=1}^{\infty}a_j\cos K_{IIj}(z+h)\tag{36}$$

where C_1, C_2, a, and a_j are given as

$$\begin{aligned}C_1=&\frac{1}{e^{a(h-d)}-e^{-a(h-d)}}\Big\{e^{ah}\left[S-\sum_{j=1}^{\infty}a_j\cos K_{IIj}(h-d)\right]\\&+e^{ad}\sum_{j=1}^{\infty}a_j\Big\}\end{aligned}\tag{37}$$

$$\begin{aligned}C_2=&\frac{1}{e^{a(h-d)}-e^{-a(h-d)}}\Big\{e^{-ah}\left[\sum_{j=1}^{\infty}a_j\cos K_{IIj}(h-d)-S\right]\\&-e^{-ad}\sum_{j=1}^{\infty}a_j\Big\}\end{aligned}\tag{38}$$

$$a=\pm\sqrt{\frac{1}{f}\left\{\sigma^2\left[\rho+0.25\pi\rho_w D^2(C_m-1)\right]+i\sigma\left(C+Rd^*\rho_w D\sigma\right)\right\}}\tag{39}$$

$$a_j=\frac{\left(0.25\pi\rho_w D^2g+i\rho_w DRd^*g\right)(-1)^{j-1}\left(K_{IIj}A_{IIs/wPj}+K_{IIj}A_{IIs/wNj}e^{-2bK_{IIj}}\right)}{-f(K_{IIj})^2+\sigma^2\left[\rho+0.25\pi\rho_w D^2(C_m-1)\right]+i\sigma\left(C+Rd^*\rho_w D\sigma\right)}\tag{40}$$

NUMERICAL RESULTS AND DISCUSSION

To demonstrate the analytical solution, the numerical analysis was carried out for the profile of the tethers when the periods of incident waves and the water depth were varied. The periods adopted for the incident waves in the numerical demonstration were 1-second, 5-second and 10-second, while the depths shown in the solution were 10 m, 20 m and 30 m. As follows are the numerical results shown in two groups.

Tether Profile Corresponding to Wave Periods

Shown as Fig. 3, 4 and 5 are the displacement profiles for the tether in 10 m deep water subjected to 1-second, 5-second and 10-second incident waves respectively. In the same water depth when the period of the incident waves is lengthened the number of nodes staying put in the tether vibration is reduced. The vibration amplitude during the 5-second waves is relatively larger than 1-second and 10-second waves because 5-second is close to the resonant period about 7 seconds for the system.

Fig. 6, and 7 illustrate the displacement profiles for the tether in the water of 20 m depth subjected to 5-second and 10-second incident waves respectively. Again it shows the correspondence between the wave period and the vibration mode, while corresponding to the lengthened wave period the displacement amplitude of the tether is increased.

Sketched in Fig. 8 is the displacement profile for the tether in 30 m deep water subjected to 5-second period of waves. Observed from Fig. 4, 6, and Fig. 8, for tethers located in different water depth but subjected to waves with the same period, the number of nodes staying put in the tether vibration is increased corresponding to the increase of the water depth.

Time History of Tether Corresponding to Water Depth

Shown in Fig. 9 is the time history for tethers located in 10 m deep water subjected to 5-second incident waves, where corresponding to five positions along the water depth the relative motion could reach as large as 5 m. Fig. 10 shows the time history for tethers located in 20 m deep water subjected to 10-second incident waves, in which the relative motion between two different positions is relatively small. It shows that the relative displacement between two positions in the tether could be amplified by certain combinations of the water depth and the wave period.

CONCLUSIONS

An analytical solution was derived and presented as an infinite series form for the dynamic behavior of tethers utilized for the tension leg platform structural system. The solution was also demonstrated in the numerical examples. As was illustrated in the figures and discussed for the results, the vibration profile for the tether was highly dependent on the period of the incident waves and the water depth while wave height, structural dimensions and the tether properties were held constant. The motion of the tether generally follow the motion of the platform except for the vibration amplitude, which is varied corresponding to the period of the incident waves. A small displacement in the platform motion does not necessarily induce a small amplitude for the tether vibration, which depends on the wave periods and water depth. Corresponding to the increase of the water depth, the maximum amplitude of the tether vibration is increased when subjected to incident waves with period ranged from 1~15 seconds. The relative displacement between two positions (depths) in the tether could be amplified by certain combinations of the water depth and the wave period.

ACKNOWLEDGMENT

This study was partially supported by National Science Council, Taiwan under grant No. 86-2611-E-110-009. It is gratefully acknowledged here.

REFERENCE

Black, JL, Mei, CC and Bray, MCG (1971). "Radiation and scattering of water waves by rigid bodies," *J. Fluid Mechanics*, Vol 46.

Garrison, CJ (1974). "Dynamic response of floating bodies," OTC 2067.

Lee, C-P and Lee, J-F (1991). "Interaction between waves and tension leg platform," *Engnrg. Mechanics Conf. on Mechanics Computing in 1990s and Beyond*, Columbus, Ohio.

Lee, C-P and Lee, J-F (1993). "Wave induced surge motion of a tension leg structure," *Ocean Engineering*, Vol 20, No.20.

Newman, JN (1977). *Marine Hydrodynamics*. MIT Press, Mass.

Sarpkya, T and Isaacson, M (1981). *Mechanics of wave forces on offshore structures*, Van Nostrand Reinhold, New York.

Solliitt, CK and Cross, RH (1972). "Wave transmission through permeable breakwaters," *Proceeding of 13th ICCE*, ASCE.

Yamamoto, T, Yoshida, A and Ijima, T (1982). "Dynamics of elastically moored floating objects," In *Dynamics Analysis of Offshore Structures*, edited by Kirk, C.L. CML, Southampton.

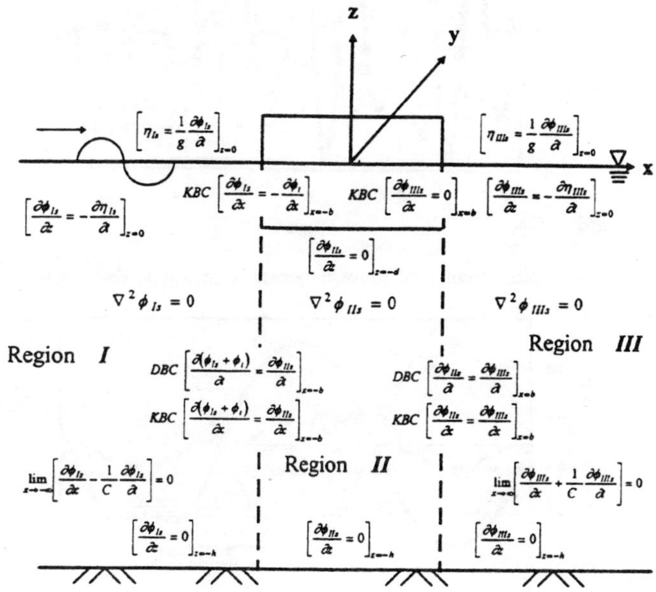

Fig. 1 Illustration for the scattering problem

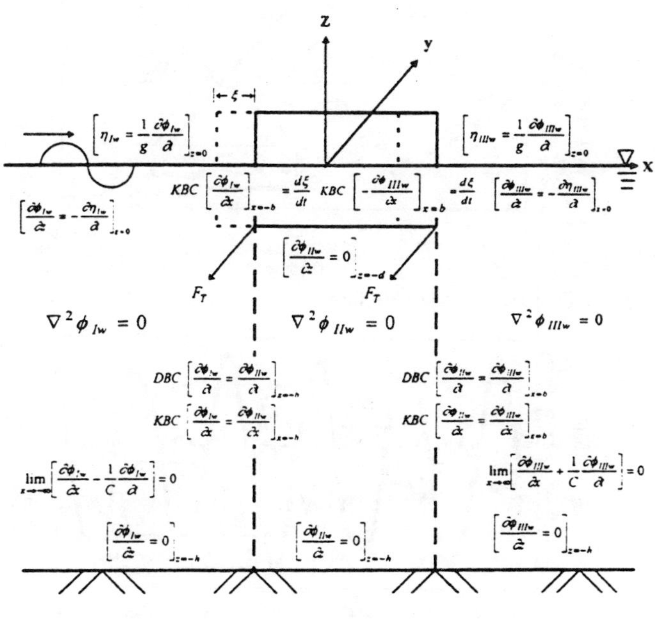

Fig. 2 Illustration for the radiation problem

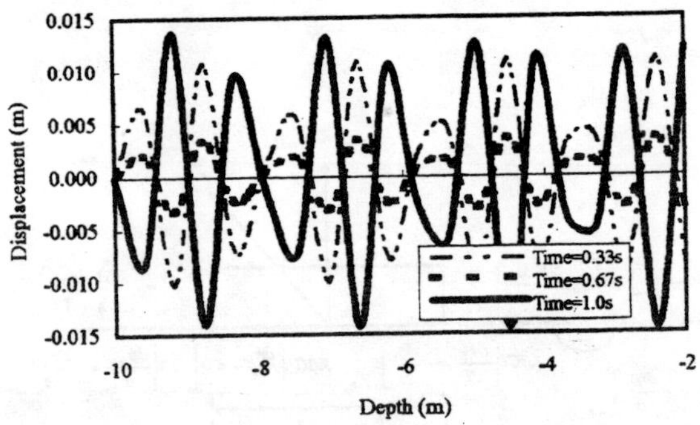

Fig. 3 Displacement profile for tether in 10 m water, 1 sec. waves

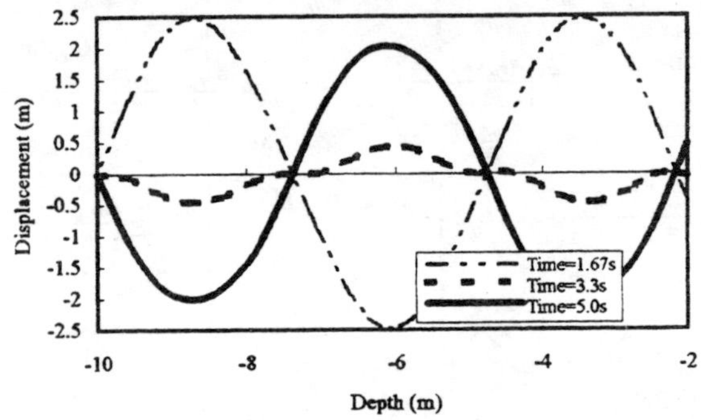

Fig. 4 Displacement profile for tether in 10 m water, 5 sec. waves

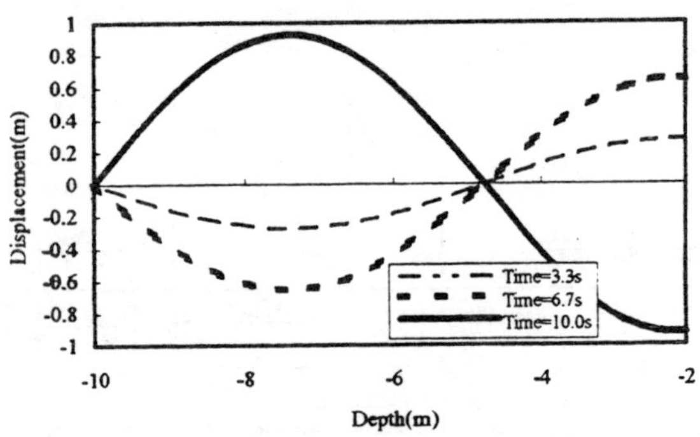

Fig. 5 Displacement profile for tether in 10 m water, 10 sec. waves

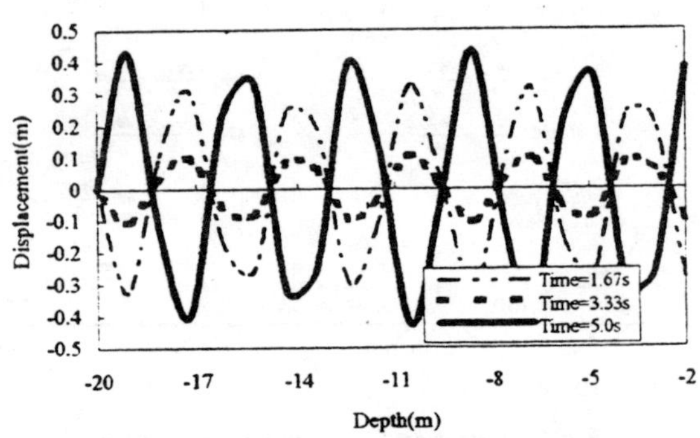

Fig.6 Displacement profile for tether in 20 m water, 5 sec. waves

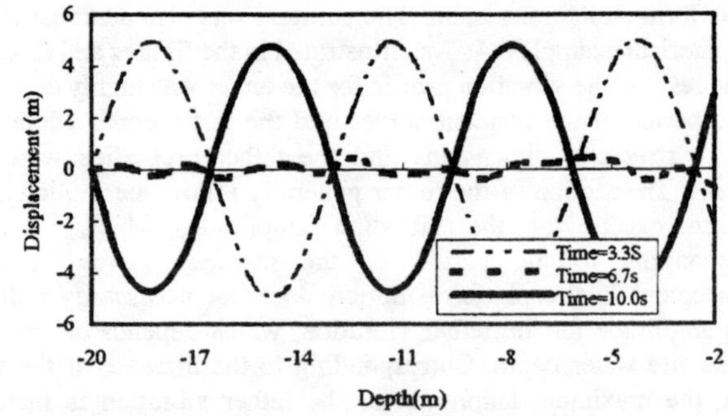

Fig. 7 Displacement profile for tether in 20 m water, 10 sec. waves

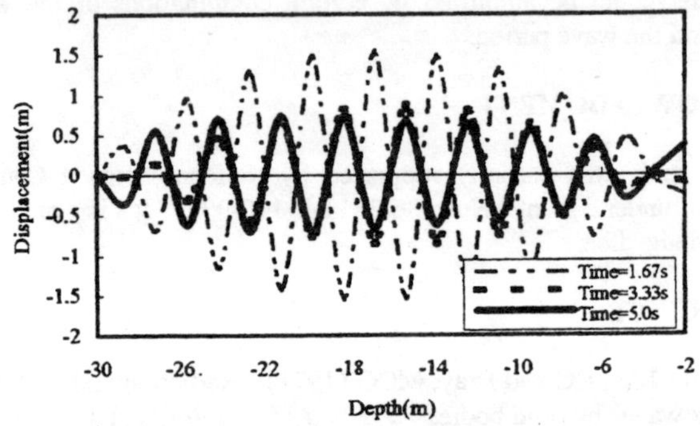

Fig. 8 Displacement profile for tether in 30 m water, 5 sec. waves

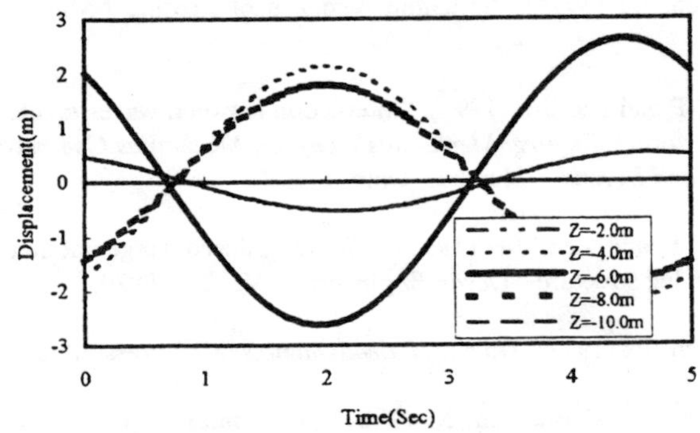

Fig. 9 Time history for tether in 10 m water, 5 sec. waves

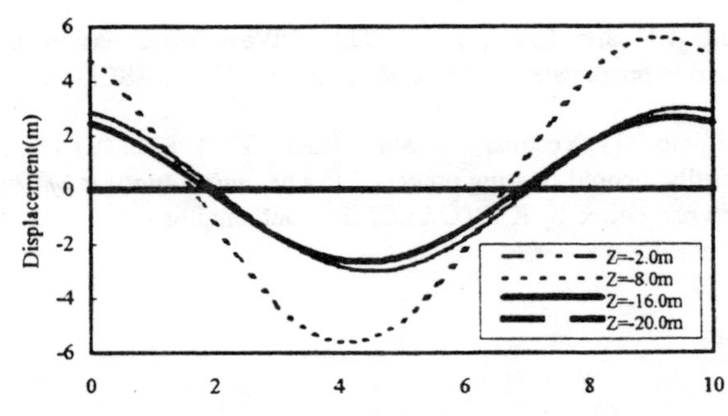

Fig. 10 Time history for tether in 20 m water, 10 sec. waves

Proceedings of the Seventh (1997) International Offshore and Polar Engineering Conference
Honolulu, USA, May 25-30, 1997

ISBN 1-880653-28-1 (Set); ISBN 1-880653-29-X (Vol. I)

Interaction of Waves and a Porous Tension Leg Platform with an Impermeable Top Layer

Chung-Pan Lee and Wen-Kuei Ker
National Sun Yat-Sen University
Kaohsiung, Taiwan, China

ABSTRACT

An analytical solution has been developed here to quantify the flow field and the surge motion of a porous tension leg platform induced by linear waves. The porous layer of the TLP is considered to be anisotropic but homogeneous. The nonlinear form drag in the porous layer is replaced by a linear drag according to Lorentz's hypothesis of equivalent work. The convergence of the series solution is verified. The dependence of the flow field and the surge motion of the platform, as well as its resonant frequency, on wave and structure properties has been studied.

KEY WORDS: TLP, porous, resonance, scattering, radiation, surge.

INTRODUCTION

Tension leg platforms (TLPs) have been used in deep water environment as working stations for oil exploration and for OTEC (IOA, 1990) and as floating breakwaters (Harms, 1979). TLP's desirable capability of being reusable, transportable, and rapidly deployable calls for great attention on the related studies (Lee et al., 1986; ISOPE-97). A typical tension leg platform usually includes a floating body and is anchored on the sea floor by pre-tensioned legs. The floating body may be comprised of group of cylinders and a top of working platform. In the study, the group of cylinders is simulated by a layer of anisotropic but homogeneous porous structure and the working platform by an impermeable layer. This is also similar to a porous floating breakwater capped with a pontoon-like floater.

Study on the behavior of a porous tension-leg floating structure and the flow field induced by incident waves might be difficult mainly due to the presence of the nonlinearity in the boundary value problem, the turbulent flow in the porous structure, and the nonlinear drag on the legs. Simplified models for the surge motion of an impermeable structure with linearly elastic, pre-tensioned legs have usually been considered (Lee and Lee, 1991; Yamamoto et al., 1982; Mei, 1978; Lee and Lee, 1993). Lee (1994) further applied the hypothesis of Lorentz's equivalent work (Sollitt and Cross, 1972) to solve analytically the boundary value problem with the nonlinear drag on the legs. However, only experimental studies has been done on a floating breakwater containing porous layers (Harms, 1979).

In the present study, the hypothesis of Lorentz's equivalent work is further applied to the flow field in the porous structure. The flow field in the anisotropic but homogeneous porous medium can be proved to be irrotational under a periodic small motion (Lee, 1987). Velocity potentials of the corresponding boundary value problem are then solved analytically after incorporating into a scattering and a radiation problem by the method of the separation variables. In the flow field containing the porous structure, the range of each complex root of the dispersion equation is given by an explicit equation (Ker, 1997). The complex eigenvalues of the potentials are then solved by Ward's or Newton's methods. The analytic solution has been accommodated by a coded computer program. The convergence of the series solution is verified and theoretical behaviors of the dragged surge motion of the tension leg structure and the flow properties are illustrated for typical wave and structure conditions.

BOUNDARY VALUE PROBLEMS

Potential Theory

Consider a flow field within an anisotropic but homogeneous porous structure under the action of a small amplitude incident wave train. The fluid is assumed to be incompressible. The flow field can be proved to be irrotational if the induced motion is small and periodic and initially irrotational. In the flow field, a single-valued velocity potential exists and can be defined by (Lee, 1987)

$$u_j = a_j^2 \frac{\partial \phi}{\partial x_j} \tag{1}$$

$$a_j^2 = \left(\frac{i}{f_j - is_j} \right), \qquad j=1, 2 \tag{1a}$$

where the subscript j represents the variables in x- or z-direction, respectively, as it is equal to 1 or 2, s is the inertial force coefficient given in Morrison's equation, and f is the linear friction drag coefficient according to the hypothesis of Lorentz's equivalent work, which is given by

$$f_j = \frac{\varepsilon_j}{\sigma}\left\{\frac{\nu}{K_{pj}} + \frac{C_{fj}\varepsilon_j}{K_{pj}^{1/2}} \frac{\int_V \int_{t_0}^{t_0+T} |u_j|^3 \, dtdV}{\int_V \int_{t_0}^{t_0+T} |u_j|^2 \, dtdV}\right\} \quad (2)$$

where C_{fj} is a dimensionless turbulent coefficient, ν is the kinematic viscosity of the fluid, ε_j is the directional areal porosity, and K_{pj} the intrinsic permeability of the porous media. For specific porous media, they can be evaluated a priori from standard tests or from empirical expressions and be applied when the wave length is long with respect to wave amplitude and media grain size (Ward, 1964; Dinoy, 1971; Sollitt and Cross, 1972).

The velocity potential satisfies the modified Laplace equation derived from the continuity equation

$$\sum_{j=1}^{2}\left[a_j^2 \frac{\partial^2 \phi}{\partial x_j^2}\right] = 0 \quad (3)$$

and the linearized Bernoulli equation

$$-\frac{\partial \phi}{\partial t} + \frac{P}{\rho} + gz = 0 \quad (4)$$

which is integrated from the equation of motion. It is noted that Eqs.(1), (3) and (4) reduce to those in clear water where an inviscid fluid is assumed (Lee, 1987).

Scattering and Radiation Problems

In the study, a two-dimensional TLP capped with an impermeable top layer is interacting with a monochromatic small amplitude wave train propagating in the +x direction, as shown in Fig. 1. The velocity potential of the incident wave is given by

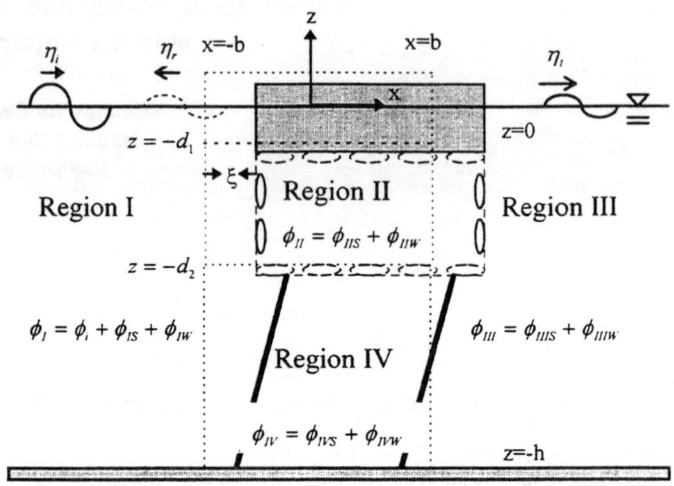

Fig. 1. The definition sketch

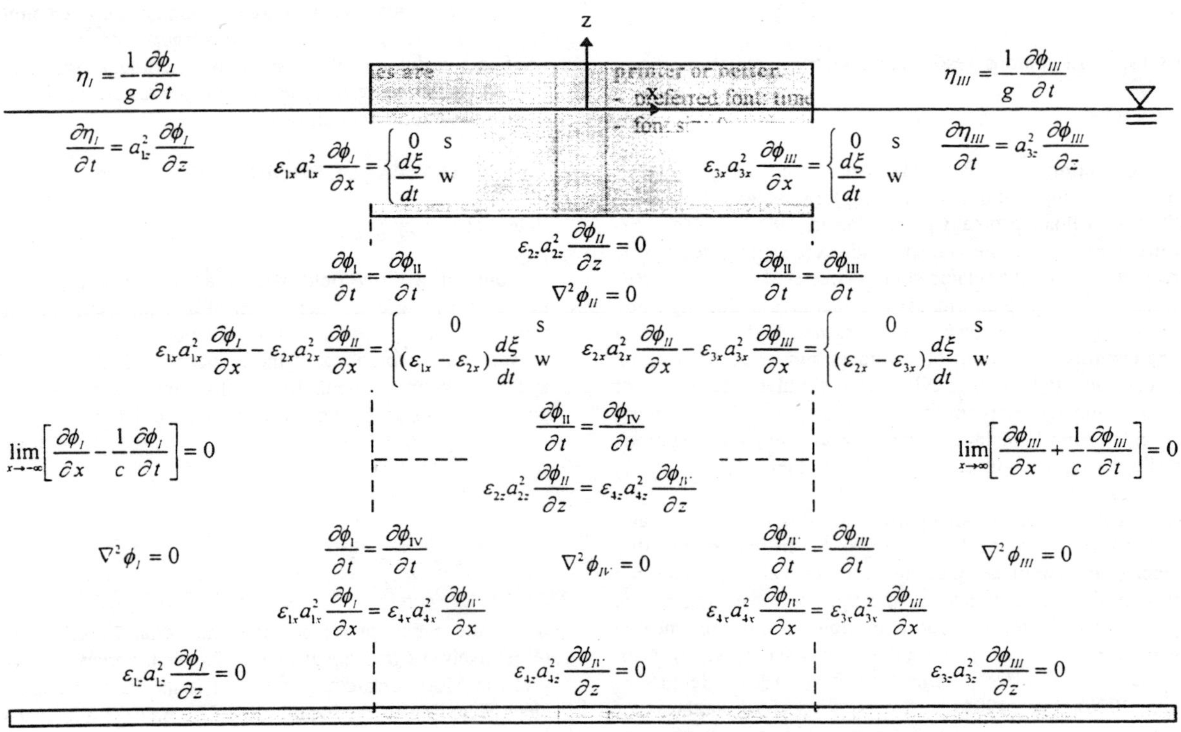

Fig. 2. Boundary value problems of scattering and radiation problems
s: scattering problem, w: radiation problem

$$\phi_i = \frac{A_i g}{\sigma}\frac{\cos K_1(h+z)}{\cos K_1 h}e^{-(K_1 x + i\sigma t)} \tag{5}$$

where h is the water depth, g is the gravitational constant, A_i is the wave amplitude, σ = (2π/T) is the angular frequency and T the period, and K_1= -ik, k = (2π/L) is the wave number and L the wave length. K_1 satisfies the dispersion relation

$$\sigma^2 = -gK_1 \tan(K_1 h) \tag{6}$$

The motion of the structure induced by the small amplitude incident waves is assumed to be small. Because of the linearity of the problem, the wave-structure interaction problem can be incorporated into a scattering (or diffraction) problem and a radiation (or wave-maker) problem. The wave forces calculated from the scattering problem provide the forcing function in the radiation problem, and the forced oscillation then generates out-going waves.

The flow field is first divided into three columns with two artificial boundaries at x = -b and x= b. In the second column, there are two layers named Region II, which is the porous layer in the study, and Region IV, respectively. In the first column, also named Region I, -∞ <x < -b, the total velocity potential ϕ_I consists of incident waves ϕ_i, scattered waves ϕ_{IS}, and radiated waves ϕ_{IW}. In the second column, -b < x < b and -h < z < $-d_1$, and the third column, also named Region III, b < x < ∞, the total velocity potentials ϕ_{II}, ϕ_{IV}, and ϕ_{III} consist of both scattered, ϕ_{IIS}, ϕ_{IVS}, and ϕ_{IIIS}, and radiated waves, ϕ_{IIW}, ϕ_{IVW}, and ϕ_{IIIW}, as shown in Figure 1, where the subscript s denotes the scattering problem and w the wave-maker (i.e. radiation) problem. The corresponding boundary value problems of the scattering and wave-maker (radiation) problems are given in Fig. 2, where the displacement of the surge motion with unknown amplitude S is given by $\xi = Se^{-i\sigma t}$.

ANALYTICAL SOLUTIONS

Applying the method of separation of variables, matching the boundary conditions on the horizontal boundaries of each region, and applying the Sommerfeld's radiation condition to Regions I and III, the velocity potentials of the scattering problem, with subscript s, and the wave-maker (radiation) problem, with subscript w, can be found as follows (Ker, 1997):

In Region I:

$$\phi_{IS/W} = \sum_{n=1}^{\infty}\frac{A_{IS/Wn} g}{\sigma}\frac{\cos K_n(h+z)}{\cos K_n h}e^{K_n(x+b) - i\sigma t} \tag{7}$$

In Region III:

$$\phi_{IIIS/W} = \sum_{n=1}^{\infty}\frac{A_{IIIS/Wn} g}{\sigma}\frac{\cos K_n(h+z)}{\cos K_n h}e^{-K_n(x-b) - i\sigma t} \tag{8}$$

where the eigenvalues K_n can be solved from the dispersion equation

$$\sigma^2 = -gK_n \tan(K_n h) \tag{9}$$

with

$$K_1 = -ik_1, \text{ for n=1; } (2n-3)\pi/2 < K_n h < (n-1)\pi, \text{ for } n \geq 2. \tag{10}$$

In Region II:

$$\phi_{IIS/W} = \frac{g}{\sigma}\left[\sum_{n=1}^{\infty} X_{IIS/Wn}(x) Z_{IIS/Wn}(z)\right]e^{-i\sigma t} \tag{11}$$

with

$$X_{IIS/Wn}(x) = \begin{cases} \frac{A_{IIS/W1}}{b}x + B_{IIS/W1} & for \quad n=1 \\ A_{IIS/W}e^{K_{IIn}x} + B_{IIS/W}e^{-K_{IIn}x} & for \quad n\geq 2 \end{cases} \tag{12}$$

$$Z_{IIS/Wn}(z) = \begin{cases} 1 & for \quad n=1 \\ \cos\frac{K_{IIn}}{a_{2xz}}(d_2+z) + iQ_{2n}\sin\frac{K_{IIn}}{a_{2xz}}(d_2+z) & for \quad n\geq 2 \end{cases} \tag{13}$$

In Region IV:

$$\phi_{IVS/W} = \frac{g}{\sigma}\left[\sum_{n=1}^{\infty} X_{IVS/Wn}(x) Z_{IVS/Wn}(z)\right]e^{-i\sigma t} \tag{14}$$

with

$$X_{IVS/Wn}(x) = \begin{cases} \frac{A_{IVS/W1}}{b}x + B_{IVS/W1} & for \quad n=1 \\ A_{IVS/Wn}e^{K_{IIn}x} + B_{IVS/Wn}e^{-K_{IIn}x} & for \quad n\geq 2 \end{cases} \tag{15}$$

$$Z_{IVS/Wn}(z) = \begin{cases} 1 & for \quad n=1 \\ \cos\frac{K_{IVn}}{a_{4xz}}(h+z) & for \quad n\geq 2 \end{cases} \tag{16}$$

The eigenvalues K_{IIn} are given by the dispersion equation

$$\tan\frac{K_{IIn}}{a_{2xz}}(d_2-d_1) = -\frac{\varepsilon_{4z}a_{4x}a_{4z}K_{IVn}}{\varepsilon_{2z}a_{2x}a_{2z}K_{IIn}}\tan\frac{K_{IVn}}{a_{4xz}}(h-d_2) \tag{17}$$

where

$$iQ_{2n} = -\frac{\varepsilon_{4z}a_{4x}a_{4z}K_{IVn}}{\varepsilon_{2z}a_{2x}a_{2z}K_{IIn}}\tan\frac{K_{IVn}}{a_{4xz}}(h-d_2) \tag{18}$$

Furthermore, in the second column with multiple layers, eigen-values and unknown coefficients of the x-dependent term in Regions II and IV can be related to each other by the following equations

$$\varepsilon_{2x}a_{2x}^2K_{IIn}^2 = \varepsilon_{4x}a_{4x}^2K_{IVn}^2 \tag{19}$$

and

$$A_{IVn} = \frac{1}{2}\left[\left(\frac{K_{IVn}}{K_{IIn}}+1\right)A_{IIn} + \left(\frac{K_{IVn}}{K_{IIn}}-1\right)B_{IIn}\right]\frac{\frac{\sinh K_{IIn}b}{\sinh K_{IVn}b}}{\cos\frac{K_{IVn}}{a_{4xz}}(h-d_2)} \tag{20a}$$

$$B_{IVn} = \frac{1}{2}\left[\left(\frac{K_{IVn}}{K_{IIn}}-1\right)A_{IIn} + \left(\frac{K_{IVn}}{K_{IIn}}+1\right)B_{IIn}\right]\frac{\frac{\sinh K_{IIn}b}{\sinh K_{IVn}b}}{\cos\frac{K_{IVn}}{a_{4xz}}(h-d_2)} \tag{20b}$$

Equations (19) and (20) were obtained from the continuity of the horizontal mass flux across the ends of the horizontal boundary between Regions II and IV at the vertical boundaries x=-b and x=b (Lee, 1987; Ker, 1997), as shown in Figure 2. It is noted that Equation (20) reduces the number of the unknown coefficients of each wave mode to only two in the second column.

Four equations derived from the four boundary conditions on the two vertical boundaries of the middle column can be used to determine the series of unknowns in the solution. The four equations can be obtained from the dynamic and kinematic boundary conditions at x =-b and x=b,respectively. Orthogonality of the eigenfunctions in each column has been shown to exist in the interval from the impermeable sea bed to the free surface (Lee, 1987) or to the impermeable bottom of the capped layer (Ker, 1997). Thus four equations can then be obtained,for m≥1,

$$\left[(1-2\delta_{m1})E1(m)\right]e^{-K_{IIm}b}A_{IIS/Wm} + E1(m)e^{K_{IIm}b}B_{IIS/Wm} - \sum_{n=1}^{\infty}E2(n,m)A_{IS/Wn}$$

$$= \begin{cases} E2(1,m)A_i e^{K_1 b} & for \quad "s" \\ 0 & for \quad "w" \end{cases} \tag{21}$$

$$\frac{Z_n Z_m(m,m)}{\cos K_m h}K_m A_{\mathrm{I}sm} + \left[\varepsilon_{2x}a_{2x}^2\frac{CS_{nm}(1,m)}{b} + \varepsilon_{4x}a_{4x}^2\frac{Z_{4n}Z_m(1,m)}{b}\right]A_{IIS/W1}$$

$$+\sum_{n=2}^{\infty}\left[\varepsilon_{2x}a_{2x}^2 K_{IIn}CS_{nm}(n,m)e^{-K_{IIn}b} + \varepsilon_{4x}a_{4x}^2 K_{IVn}(\alpha\gamma - \beta\gamma')Z_{4n}Z_m(n,m)\right]A_{IIS/Wn}$$

$$+\sum_{n=2}^{\infty}\left[-\varepsilon_{2x}a_{2x}^2 K_{\mathrm{IIn}}CS_{nm}(n,m)e^{K_{IIn}b} + \varepsilon_{4x}a_{4x}^2 K_{\mathrm{IVn}}(\beta\gamma - \alpha\gamma')Z_{4n}Z_m(n,m)\right]B_{IIS/Wn}$$

$$= \begin{cases} \dfrac{Z_n Z_m(m,m)}{\cos K_m h}K_m e^{K_1 b}A_i\delta_{1m} & for \quad "s" \\ \left[ZD_2^0(m) - \varepsilon_{2x}ZD_2^1(m)\right]\dfrac{i\sigma^2 S}{g} & for \quad "w" \end{cases} \tag{22}$$

$$E1(m)e^{K_{IIm}b}A_{IIS/Wm} + E1(m)e^{-K_{IIm}b}B_{IIS/Wm} - \sum_{n=1}^{\infty}E2(n,m)A_{IIIS/Wn}$$

$$= 0 \quad for \ "s" \ and \ "w" \tag{23}$$

and

$$-\frac{Z_n Z_m(m,m)}{\cos K_m h}K_m A_{IIIS/Wm} + \left[\varepsilon_{2x}a_{2x}^2\frac{CS_{nm}(1,m)}{b} + \varepsilon_{4x}a_{4x}^2\frac{Z_{4n}Z_m(1,m)}{b}\right]A_{IIS/W1}$$

$$+\sum_{n=2}^{\infty}\left[\varepsilon_{2x}a_{2x}^2 K_{IIn}CS_{nm}(n,m)e^{K_{IIn}b} + \varepsilon_{4x}a_{4x}^2 K_{IVn}(\alpha\gamma' - \beta\gamma)Z_{4n}Z_m(n,m)\right]A_{IIS/Wn}$$

$$+\sum_{n=2}^{\infty}\left[-\varepsilon_{2x}a_{2x}^2 K_{\mathrm{IIn}}CS_{nm}(n,m)e^{-K_{IIn}b} + \varepsilon_{4x}a_{4x}^2 K_{\mathrm{IVn}}(\beta\gamma' - \alpha\gamma)Z_{4n}Z_m(n,m)\right]B_{IIS/Wn}$$

$$= \begin{cases} 0 & for \quad "s" \\ \left[ZD_2^0(m) - \varepsilon_{2x}ZD_2^1(m)\right]\dfrac{i\sigma^2 S}{g} & for \quad "w" \end{cases} \tag{24}$$

where δ is the Kronecker delta, and the other notations used in the above equations are defined in the Appendix.

Equations (21)-(24) can then be solved for the four series of the unknowns A_{ISn}, A_{IISn}, A_{IIISn}, and A_{IVSn} in the scattering problem and substituted into the corresponding equations to calculate the flow properties. On the other hand, a fifth equation is required to combine with the four equations to resolve the five sets of unknowns including the surge amplitude S in the radiation problem.

The fifth equation can be obtained from the equation of motion derived from Newton's second law and written as $M\,d^2\xi/dt^2 = F_{wx} - F_{Tx}$ where M is the mass of the structure, F_{wx} is the horizontal wave force acting on the both sides of the structure, and F_{Tx} is the horizontal component of the static (or the pre-tensioned) tension force applied by the tension legs. The static tension is given by $F_{Tx} = f_0\xi$ where

$$f_0 = 2\rho g b(d_2 - d_0)/(h - d_2) \tag{25a}$$

$$d_0 = \varepsilon(d_2 - d_1) + M/2\rho b \tag{25b}$$

In Eq. (25), ε is the volumetric porosity, d_0 and d_2 are the drafts of the structure before and after applying the pre-tension force, respectively, and $d_2 - d_1$ is the thickness of the porous layer. On the other hand, the wave force F_{Wx} is simply the integration of the total hydrodynamic pressure over the vertical surfaces of the structure at x = -b and x = b, by using the linearized Bernoulli equation, Eq.(3) with $\phi_{\mathrm{I}} = \phi_i + \phi_{\mathrm{IS}} + \phi_{\mathrm{IW}}$ and $\phi_{\mathrm{III}} = \phi_{\mathrm{IIIS}} + \phi_{\mathrm{IIIW}}$. Solving the equation of motion therefore results in

$$\sum_{n=1}^{\infty}\left[(1-\varepsilon_{2x})\frac{ZD_2^1(n)}{\cos K_n h} + \frac{ZD_1^0(n)}{\cos K_n h}\right](A_{IWn} - A_{IIIWn}) - \frac{i\sigma^2}{\rho g}\left[\frac{f_0}{\sigma^2} - M\right]S$$

$$= -\left\{\left[\frac{ZD_2^1(n)}{\cos K_1 h} + \frac{ZD_1^0(n)}{\cos K_1 h}\right]e^{K_1 b}A_i + \sum_{j=1}^{\infty}\left[(1-\varepsilon_{2x})\frac{ZD_2^1(n)}{\cos K_n h} + \frac{ZD_1^0(n)}{\cos K_n h}\right](A_{ISn} - A_{IIISn})\right\}$$

....(26)

The five sets of unknowns A_{IWn}, A_{IIWn}, A_{IIIWn}, A_{IVWn}, and S can then be solved from the five sets of equations, Eqs.(21) to (24), and (26), and the related flow properties be computed from their corresponding equations.

ANALYTICAL BEHAVIOR

The computation of the final solutions of the wave-structure interaction includes two parts: first solving the matrix equation obtained from Equations (21)-(24) for the scattering problem in which the incident wave conditions are given. This solution is then incorporating into Equations (21)-(24), and (26) to solve for the final solution of the radiation problem.

For a given set of wave conditions, structure geometry, and media properties, the computation begins with the determination of the eigenvalues in each column. To initiate the solution, a value of the linear friction drag in the porous layer is assumed. Then the eigenvalues of appropriate number in each column are solved numerically from the dispersion equation (combined with the relationship between eigenvalues in different layers) by either the Newton or Ward (Carnahan et al., 1969) methods.

The eigenvalues are used to enumerate the coefficients in Eqs.(21)-(24), and (26). The resulting coefficients related to the unknowns constitute the coefficient matrix, and the coefficients related to the incident wave in the scattering problem (the scattering waves in the radiation problem) constitute the right hand side of this matrix equation. The matrix equation is then solved by the LU decomposition method.

After the unknowns are solved, the fluid particle velocity and pressure at any position can be computed from the corresponding equations. The real parts of the velocities are extracted and substituted into Eq.(2) to calculate linear friction drags. The computation is complete when the difference of the linear friction drags found in two consecutive procedures is within the required accuracy, e.g. 1 % in this paper, otherwise, the above procedures will be repeated. The integrations in Eq.(2) are performed numerically by the trapezoidal rule. A Fortran program has been developed to perform the computation described in this section.

The solutions given in the previous sections involve infinite series. This is because the eigenfunctions in neighboring flow fields are in general not orthogonal (Lee, 1994). In practical computation, only the first N modes of waves will be taken and matrix equations of 4N × 4N for the scattering problem and (4N+1) × (4N+1) for the radiation problem will be constructed from the corresponding equations. In order to secure the existence of the series solutions, the convergence of the series should be verified. It has been shown that the flow properties converge to constant values as the number of the wave modes increases to about 12 (Ker, 1997).

Illustration of The Analytical Solutions

The dependence of wave reflection coefficient K_r, transmission coefficient K_t, and wave-induced structure surge motion abs(S/Ai) on incident wave conditions, structure properties, and linear friction drag will be demonstrated in this section. Structure properties include mass M, width b, drafts d_1 and d_2 of the porous layer. Dependence is shown as one of the above parameter is varying while others are held constant. Unless specifying in the figures, the test conditions are given as h = 10m, Ai = 0.3m, M = 1000Kg, b = 2m, d_1= 0.5m, d_2 = 2.2m, $f_x = f_z = 1.0$, $\varepsilon_x = \varepsilon_z = 0.5$, and $\varepsilon = 0.75$.

The dependence of K_r, K_t, and abs(S/Ai) on structure mass M has been shown in Fig. 3. As in Eq.(25), d_0 increases as M increases, and the pre-tension force of the tension legs f_0, which is also the stiffness of the TLP, is inversely proportional to the structure mass for a specific structure geometry. This results in an increasing on K_r as the structure mass decreases, which is equivalent to the increase of the stiffness of the oscillating system. Furthermore, the resonance frequency is shifting to a lower frequency as the structure mass is increasing or as the corresponding stiffness is decreasing. The TLP becomes transparent to long waves and corresponding to higher K_t values as expected.

In Fig. 4, the results for varying structure width with a fixed mass are shown. The result shows a reverse trend as that shown in Fig. 3. In this case, K_r decreases and the resonance frequency shifts to a higher frequency as the structure width increases. This is because the effect of

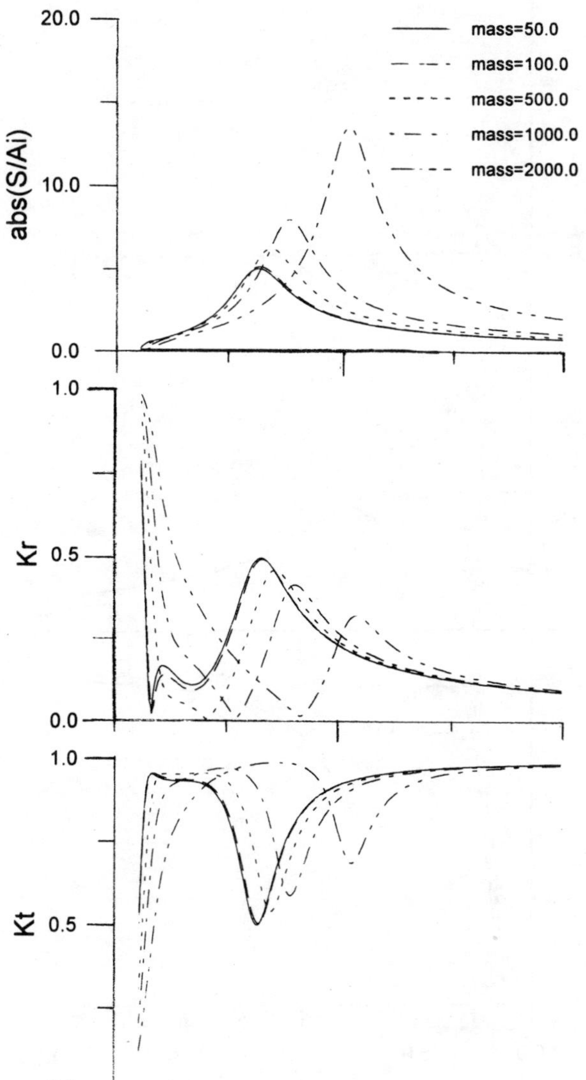

Fig. 3. Dependence of K_r, K_t, and $abs(S/A_i)$ on structure mass with fixed linear friction drags.

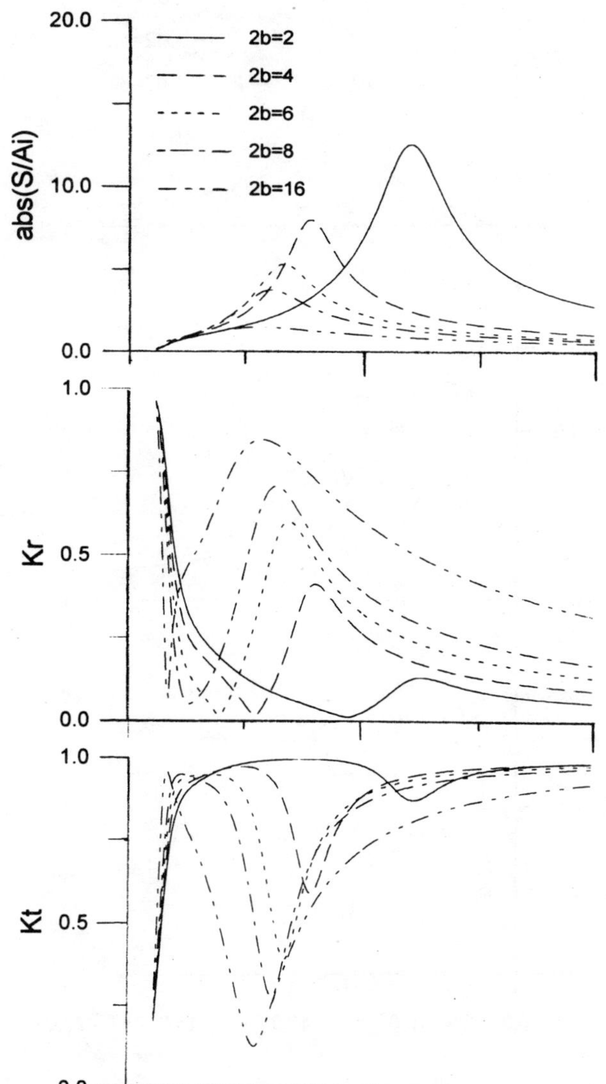

Fig. 4. Dependence of K_r, K_t, and $abs(S/A_i)$ on structure width with fixed linear friction drags.

the width on the structure stiffness is just opposite to that of the mass as given in Eq.(25). Thus, the increase of stiffness dominates in determining the resonance frequency even thought it has been shown that a larger body will result in a larger effective added mass (Lee, 1994). Similar trend as varying structure width can be found as shown in Fig. 5 for different drafts d_2, which has similar effect on the structure stiffness as the width does. It is noted that the resonance amplitude is greatly reduced as the pre-tension force is large in all above cases.

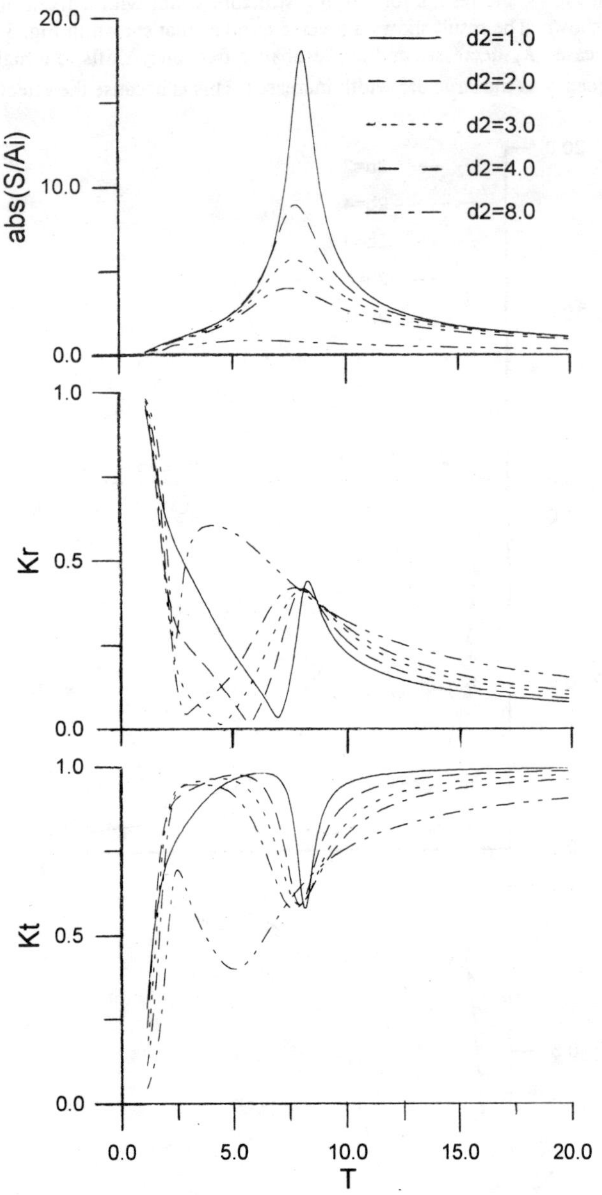

Fig. 5. Dependence of K_r, K_t, and $abs(S/A_i)$ on structure draft with fixed linear friction drags.

The dependence on the thickness of the porous layer has been shown in Fig. 6, in which the draft d_2 is taken as 4.0m. As the value of d_1 increases from 0.01m to 3.99m, d_0 will decrease and f_0 increase correspondingly according to Eq.(25). Additionally, the total damping with fixed linear friction drags will then decrease as the thickness of the porous layer decreases. These two effects shift the resonance to a higher frequency for the cases of $d_1 = 0.3$m and 0.5m. However, it can be shown that the virtual mass is inversely proportional to the porosity of the porous layer by substituting Eqs. (22) and (24) into the equation of motion (Ker, 1997). Since an increase on the virtual mass will shift the resonance to a lower frequency, this effect for a thinner porous layer seems to overcome the previous two effects and results in shifting the resonance to a lower frequency (larger period) for the cases of $d_1 = 2.0$m and 3.99m as shown in Fig. 6.

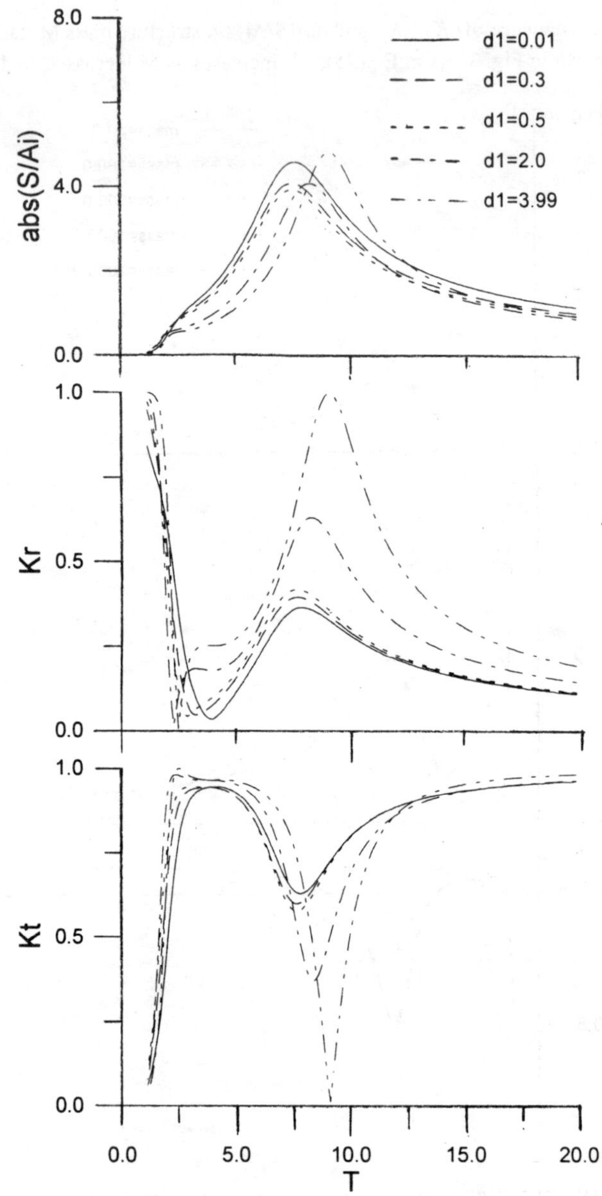

Fig. 6. Dependence of K_r, K_t, and $abs(S/A_i)$ on the thickness of the porous layer with fixed linear friction drags.

It is well known that a resonance will shift to lower frequencies in a damped linear oscillation if the system damping increases. However, as shown in Fig. 7 for varying the interior linear friction drags, it is found that the resonance slightly shifts to a higher frequency as the drag changes from 0.1 to 0.3 and then to lower frequencies as the drag further increases to 0.5, 1.0, and 5.0. This can be understood because the damping of the oscillation system includes the energy dissipation damping caused by the friction drag and the radiation damping caused by the induced propagating waves which carry energy away from the oscillation system. Thus, the above result may simply imply that the radiated wave energy of the case of 0.1 is greater than that of 0.3. In Fig. 7, it is also noted that the amplitude of the TLP motion decreases first as the friction drag in the porous layer increases from 0.1 to 1.0 and then increases as the drag further increases to 5.0. This is because the flow going to the porous layer may become small as the resistance force in the layer is large and the force is then to force the structure to oscillate positively.

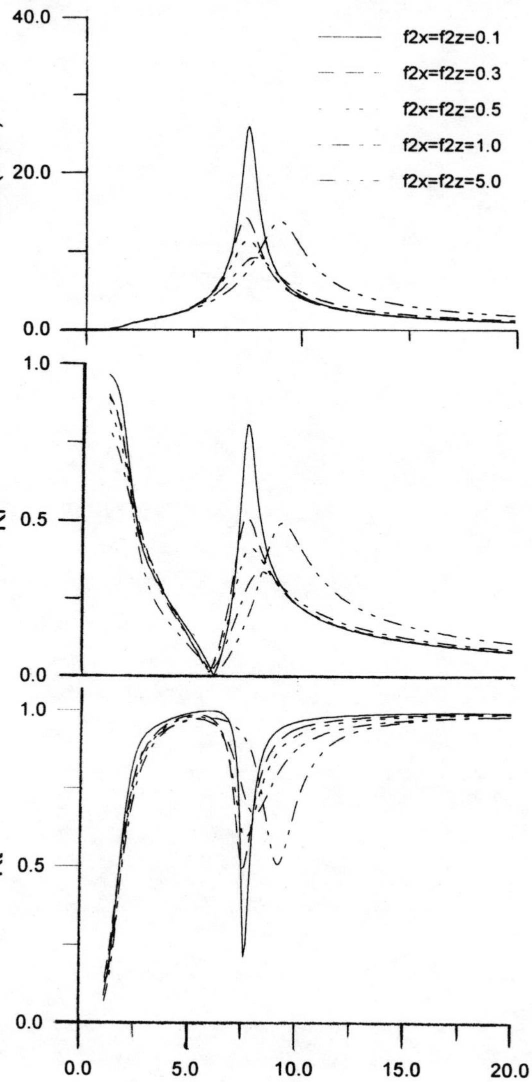

Fig. 7. Dependence of K_r, K_t, and $abs(S/A_i)$ on linear friction drags.

CONCLUSIONS

An analytical solution has been developed to describe the wave-induced flow field and the surge motion of a porous TLP capped with an impermeable layer. The porous layer may contain anisotropic but homogeneous media and the interior nonlinear turbulent drag has been replaced by a linear drag according to Lorentz's hypothesis of equivalent work. The corresponding boundary value problem is then incorporated into scattering and radiation problems and solved analytically to obtain series solutions in terms of the linear incident wave conditions and structure properties. The convergence of the series solutions is verified for typical structure and wave conditions.

The dependence of wave reflection, transmission, as well as structure surge motion on incident wave conditions, structure properties, and linear friction drag has been demonstrated. Structure properties include its mass, width, drafts, thickness, and porosity of the porous layer. The results show that the drags change the behavior of the floating structure significantly, especially as the resonance occurs, and, consequently, the corresponding flow field.

APPENDIX

$$Z_n Z_m(n,m) = \int_{-h}^{0} \cos(K_n h)\cos(K_m h)\,dz \tag{27}$$

$$Z_{2n} Z_{2m}(n,m) = \int_{-d_2}^{-d_1} \left[\cos\frac{K_{IIn}}{a_{2xz}}(d_2+z) + iQ_{2n}\sin\frac{K_{IIn}}{a_{2xz}}(d_2+z)\right] \left[\cos\frac{K_{IIm}}{a_{2xz}}(d_2+z) + iQ_{2m}\sin\frac{K_{IIm}}{a_{2xz}}(d_2+z)\right]dz \tag{28}$$

$$Z_{4n} Z_{4m}(n,m) = \int_{-h}^{-d_2} \cos\frac{K_{IVn}}{a_{4xz}}(h+z)\cos\frac{K_{IVm}}{a_{4xz}}(h+z)\,dz \tag{29}$$

$$Z_{2n} Z_m C(n,m) = \int_{-d_2}^{-d_1} \cos\frac{K_{IIn}}{a_{2xz}}(d_2+z)\cos K_m(h+z)\,dz \tag{30}$$

$$Z_{2n} Z_m S(n,m) = \int_{-d_2}^{-d_1} \sin\frac{K_{IIn}}{a_{2xz}}(d_2+z)\cos K_m(h+z)\,dz \tag{31}$$

$$Z_{4n} Z_m(n,m) = \int_{-h}^{-d_2} \cos\frac{K_{IVn}}{a_{4xz}}(h+z)\cos K_m(h+z)\,dz \tag{32}$$

$$CS_{nm}(n,m) = Z_{2n} Z_m C(n,m) + iQ_{2n} Z_{2n} Z_m S(n,m) \tag{33}$$

$$E1(m) = \frac{\varepsilon_{2z}}{\varepsilon_{2x}} Z_n Z_m(m,m) + \frac{\varepsilon_{4z}}{\varepsilon_{4x}} \frac{Z_{4n} Z_{4m}(m,m)}{\left[\cos\frac{K_{IVm}}{a_{4xz}}(h-d_2)\right]^2} \tag{34}$$

$$E2(n,m) = \frac{\varepsilon_{2z}}{\varepsilon_{2x}} \frac{CS_{nm}(m,n)}{\cos K_n h} + \frac{\varepsilon_{4z}}{\varepsilon_{4x}} \frac{Z_{4n} Z_m(m,n)}{\cos K_n h} \frac{1}{\cos\frac{K_{IVm}}{a_{4xz}}(h-d_2)} \tag{35}$$

$$ZD_2^1(m) = \int_{-d_2}^{-d_1} \cos(K_m h)\,dz \tag{36}$$

$$ZD_1^0(m) = \int_{-d_1}^{0} \cos(K_m h)\,dz \tag{37}$$

$$ZD_2^0(m) = \int_{-d_2}^{0} \cos(K_m h)\,dz \tag{38}$$

ACKNOWLEDGEMENT

This research was sponsored by National Science Council, Taiwan, under Grant No. NSC86-2611-E-110-006.

REFERENCE

Carnahan, B, Luther, HA, and Wilkes, JO (1969). Applied Numeri-cal Methods. John Wiley & Sons, Inc., New York, USA.

Dinoy, AA (1971). "Friction Factor and Reynolds Number Relationship in Flow through Porous Media," *ME Thesis*, AIT, Bangkok, Thailand.

Harms, VW (1979). "Design Criteria for Floating Tire Breakwaters," *J. of The Waterway, Port, Coastal and Ocean Division*, Vol. 105, No. WW2, 149-170.

IOA NEWSLETTER (1990). "British Engineer Proposed A New CWP Design," *International OTEC/DOWA Association*, Vol.1 No.3.

ISOPE-97 (1997). *Bulletin of 7th International OFFSHORE AND POLAR ENGINEERING CONFERENCE*, Hawaii, USA, May 25-30, 1997.

Ker, WK (1997). "Wave Interaction with Porous Tension Leg Platform," *Master Thesis*, Department of Marine Environment, National Sun Yat-sen University, Kaohsiung, Taiwan.

Lee, CP, Sollitt, CK, McDougal, WG, and Perry, TJ (1986). "Mechanically Coupled Flap Type Breakwaters: Theory and Experi-ment," *Ocean Structural Dynamic Symposium '86*, Corvallis, Oregon, USA.

Lee, CP (1987). "Wave Interaction with Permeable Structures," *Ph.D. dissertation*, Civil Engineering Department, Oregon State University, Corvallis, Oregon, USA.

Lee, CP, and Lee, JF (1991). "Interaction Between Waves and Tension Leg Platform," *Engineering Mechanics conference on Mechanics Computing in 1990's and Beyond*, 1991, May 20-22, Columbus, Ohio, USA. (ASCE).

Lee, CP, and Lee, JF (1993). "Wave-induced Surge Motion of A Tension Leg Structure," *J. of Ocean Engineering*, Vol. 20, No 2, 171-186.

Lee, CP (1994). "Dragged Surge Motion of A Tension Leg Structure," *J. of Ocean Engineering*, Vol. 21, No. 3, 311-328.

Mei, CC (1978). "Numerical Methods in Water Wave Diffraction and Radiation," *Ann. Rev. Fluid Mech.* 10, 393.

Sollitt, CK, and Cross, RH (1972). "Wave Transmission Through Permeable Breakwaters," *Proceeding of 13th ICCE*, ASCE, 1827-1846.

Ward, JC (1964). "Turbulent Flow in Porous Media," *J. of the Hydraulics Division*, ASCE.

Yamamoto, T, Yoshida, A, and Ijima, T (1982). "Dynamics of Elastically Moored Floating Objects," In DYNAMIC ANALYSIS OF OFFSHORE STRUCTURES, Kirk, C.L. (ed.), pp. 106-113. CML, Southampton.

Proceedings of the Seventh (1997) International Offshore and Polar Engineering Conference
Honolulu, USA, May 25-30, 1997

ISBN 1-880653-28-1 (Set); ISBN 1-880653-29-X (Vol. I)

First- and Second-Order Wave Forces on a Large Spar Platform

Hyungsuk Yoo, David C. Weggel, Edward J. Powers and Jose M. Roesset
University of Texas at Austin
Austin, TX, USA

Abstract

The first- and second-order wave forces acting in the surge mode of a spar platform are computed and compared using two different approaches. The first approach relies on second-order diffraction theory to directly compute wave forces on the spar. The second approach rests upon using higher-order statistical signal processing to determine a second-order orthogonal Volterra-like model relating wave excitation and surge response. This model is used to decompose the observed surge motion into its first- and second-order components, then first- and second-order force spectra are estimated using a linear deconvolution technique. The comparisons are carried out for both experimental (model test) and numerically generated time series data.

I. Introduction

The design and analysis of compliant offshore structures such as spar platforms require, among other issues, an accurate determination of the hydrodynamic forces due to wave action accounting for nonlinear effects. Spars are typically designed to possess natural frequencies that are below the dominant incident wave frequencies. Detailed design and analysis considerations for spar platforms can be found in [Santeen 76], [Glanvi 91], and [Horton 92] for example. The status of spar platforms is reported in [Halkyar 96].

Both first- and second-order forces are exerted on the structure by the incident waves. The second-order force is proportional to the square of the incident wave elevation, and contains some very low-frequency components due to difference frequency generation associated with square-law mixing effects. Some of these very low frequencies lie within the surge motion resonant bandwidth. Although these forces are very weak, they lead to large-amplitude resonant motions in surge, because the system is very lightly damped. Such large-amplitude resonant motions are of major concern because of the resultant loads on the mooring lines. For this reason, the determination of the second-order forces is of importance to obtain reasonable estimates of spar responses. However, this is a difficult *experimental* problem because the second-order forces are typically more than an order of magnitude smaller than the first-order forces.

In this paper we compute first-order (linear) and second-order (quadratic) wave force spectra for a spar platform subjected to a random incident wave. Two methods of achieving this goal are compared. In the first method we directly compute wave forces on the spar using second-order diffraction theory. Wave-to-force linear and quadratic transfer functions (LTF's and QTF's) are computed and the input wave spectrum is multiplied by the transfer functions to produce first- and second-order force spectra. In the second method we utilize higher-order statistical signal processing (HOSSP) techniques to process model test data in order to determine a second-order, orthogonal, Volterra-like, frequency-domain model relating wave excitation to spar response. The Volterra model is then utilized to decompose the spar response into its constituent first-order (linear) and second-order (quadratic) components. Using a linear model relating the force and response of the spar, we then deconvolve the first- and second-order responses to obtain first- and second-order wave force spectra, respectively.

The input (wave elevation) and output (spar response) time series were generated both experimentally and numerically. The experiments were conducted at the Offshore Technology Research Center's wave basin on a 1:55 scale model of a large, deep-draft spar platform.

Comparison of the two methods will help to: 1) check the validity of second-order diffraction theory and spar modeling assumptions, 2) illustrate the application of higher-order statistical signal processing to the estimation of first- and second-order wave forces from excitation and motion-response time series data, and 3) provide insight into possibly desirable changes in the experimental protocol of model basin experiments designed to elucidate the higher-order nonlinear characteristics of offshore structures in random seas.

In section II, we describe the experimental setup and wave environment. In section III, the prediction of forces and responses using diffraction theory is reviewed. In section IV, the methodology of recovering second-order forces using higher-order statistical signal processing is explained. In sections V and VI, the results of the two approaches are compared and discussed for experimental and simulation data, respectively. Section VII, concludes this paper.

Diameter [ft]	133.0
Draft [ft]	650.0
Mass (with entrapped water) [slugs]	1.78E+07
Center of gravity below waterline [ft]	347.7
Mooring point below waterline [ft]	346.6
Radius of gyration wrt waterline [ft]	402.8
Mooring stiffness (0 to 45.0 ft offset) [kip/ft]	13.1
Mooring stiffness (45.0 to 85.3 ft offset) [kip/ft]	27.8
Surge natural frequency [rad/sec] / period [sec]	0.019 / 332
Pitch natural frequency [rad/sec] / period [sec]	0.094 / 66.8
Heave natural frequency [rad/sec] / period [sec]	0.216 / 29.0

TABLE I. Large Spar Specifications

II. Experimental Setup

Experiments were performed on a large spar platform model at the Offshore Technology Research Center's (OTRC) wave basin under the direction of Deep Oil Technology (DOT). The spar tests were part of a joint industry project (JIP) to gain a better understanding of spar platform behavior under wind, wave and current loading. The spar was a 1:55 scale model and tested in a prototype water depth of $3000ft$. It was tautly moored to the basin bottom. Free oscillation tests were performed to experimentally determine the natural frequencies of the spar in all response modes. Table I contains the large spar's specifications.

The spar was subjected to unidirectional irregular waves that represented the 100-year survival condition. Waves were generated to represent a JONSWAP spectrum defined by a significant wave height $H_s = 43.0ft$, a peak period of $T_p = 14.0sec$, and a peak enhancement factor $\gamma = 2.0$. Fig. 1 (a) shows the incident wave time series, and Fig. 2 is the corresponding power spectrum on a semilog scale. Fig. 1 (b) and (c) are time series plots of the surge and pitch responses recorded at an optical tracking target attached to the spar and located $178.8ft$ (prototype scale) above the mean water level (MWL). Here we observe that the surge and pitch responses are closely coupled, and their dominant frequencies are much lower than that of the input wave. This is confirmed by comparing the input wave, surge response, and pitch response power spectra in Figs. 2, 3, and 4, respectively. The data were collected at a sampling frequency 2.6968 Hz (16.95 [rad/sec]) for 3600 sec for a total of 9709 data points. The input wave spectrum has the most power between 0.3 and 0.7 [rad/sec] and is negligible after 1.5 [rad/sec], thus for the higher-order spectral estimation we sub-sampled every fifth point, to yield a sampling rate of 0.5394 Hz (3.39 [rad/sec]). All dimensions, frequencies, and other experimental parameters are for prototype scale.

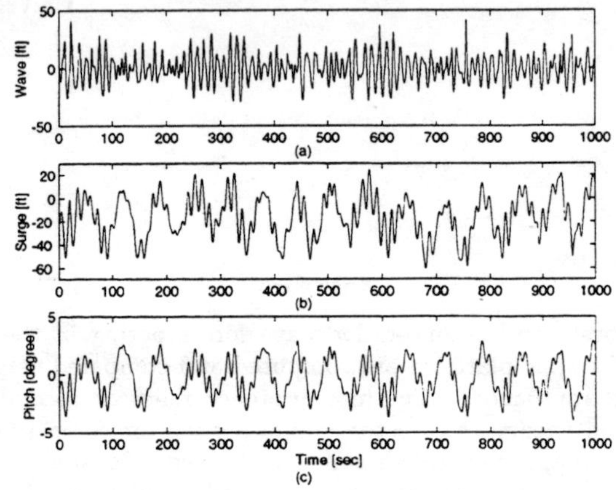

Fig. 1. Wave, surge, and pitch time series for first 1000 seconds

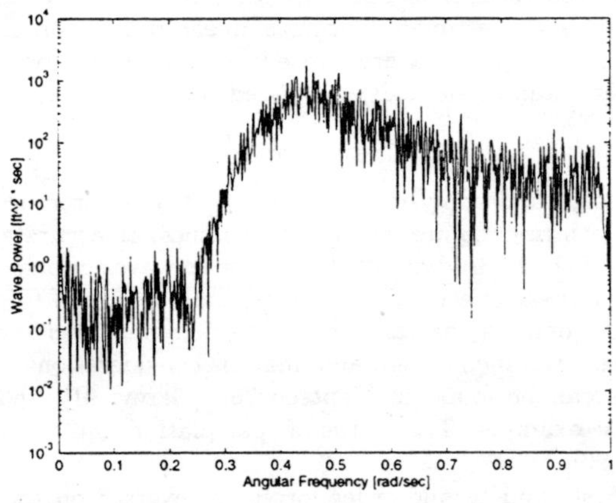

Fig. 2. Input wave power spectrum of experimental data

III. Prediction of Forces and Responses via Diffraction

Force and response predictions were numerically computed in the frequency domain using second-order diffraction/radiation theory. The actual wave time series was FFT'ed and used as input into the computed wave-to-force linear and quadratic transfer functions (LTF's and QTF's). The resulting forces were used to drive a linear equation of motion of the structure. Force spectra were also computed showing the component first- and second-order force contributions. The force LTF's and QTF's were computed using WAMIT[MIT 94], a well-known diffraction/radiation computer program developed at MIT which is used extensively by the offshore industry.

Complete second-order diffraction/radiation analyses produce first-order (linear) complex forces, linear radiation coefficients in the form of added mass and added damping coefficients, and second-order (nonlinear) complex forces at sum- and difference-frequencies of the input wave frequencies. Force linear transfer functions (force LTF's) provide a complex force quantity (or force modulus and phase) for a given wave amplitude (and frequency). Quadratic forces (second-order forces) are obtained from force quadratic transfer functions (force QTF'S) and are a function of two interacting waves. The change in hydrostatic force due to wave passage is included in the diffraction forces.

An extensive presentation of force LTF's and radiation coef-

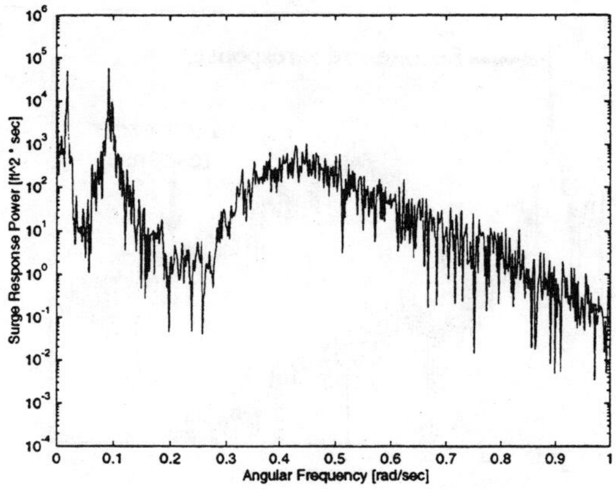

Fig. 3. Surge response power spectrum of experimental data

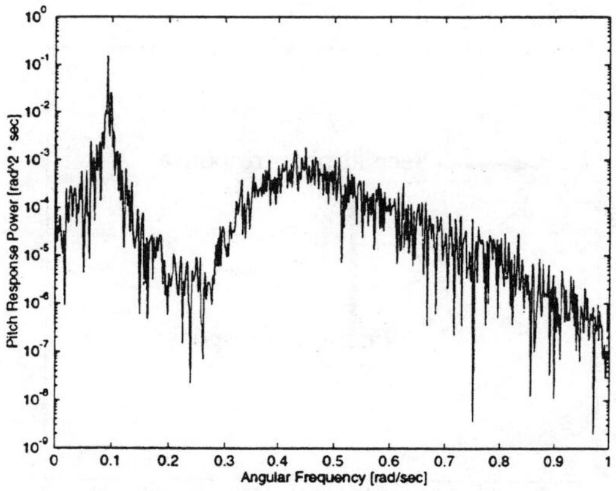

Fig. 4. Pitch response power spectrum of experimental data

ficients with approximate formulae is given in [Weggel 94] for truncated cylinders of several aspect ratios. A variety of surge, pitch, and heave force QTF's are plotted on the bi-frequency plane for two cylinder aspect ratios in [Weggel 97]. Fig. 5 shows the predicted first- and second-order surge force spectra for the large spar. These spectra cannot be compared directly with experimental results because second-order forces are very difficult to measure directly. As a result, this plot is compared with forces recovered using HOSSP (to be discussed in Sec. IV). In the response predictions linearized drag damping was included in the dynamic equation of motion (EOM); its formulation is discussed in [Weggel 97].

Surge and pitch response power spectra are shown in Figures 6 and 7, respectively. For surge and pitch the first-order (linear) response predominates for frequencies $\omega \geq 0.25[rad/sec]$ while the second-order response predominates for $0 \leq \omega < 0.25[rad/sec]$. The predicted surge spectrum shows two peaks: one at the surge natural frequency and one at the pitch natural frequency due to the strong coupling between these response modes. A discrepancy between the experimental and predicted responses can be seen at the surge natural frequency; the theoretical model overpredicts the response contribution at that frequency (Fig. 6 and Fig. 3). The assumptions of linear mooring stiffness and linearized drag damping may be responsible for this over-prediction. The surge response at the pitch natural frequency seems to be reasonably predicted. As seen in Figures 7 and 4, the predicted pitch response is in excellent agreement with the experimental data. There is only one peak in this case corresponding to the pitch natural frequency.

IV. Recovery of Second-Order Force via Higher-Order Statistical Signal Processing (HOSSP)

The surge and pitch responses of a spar exhibit a strong low-frequency response where the input wave power is very weak, as illustrated in Figs. 2, 3, and 4. Since these low-frequency responses are associated with the generation of higher-order wave forces, a second-order Volterra model is a good choice to model the excitation-response relationship. The excitation consists of time series records of wave elevation not wave force, since wave forces are much more difficult to measure. The discrete frequency-domain second-order Volterra model is as follows,

$$\begin{aligned} Y[m] &= Y_L[m] + Y_Q[m] + \epsilon[m] \\ &= H_L[m]X[m] \\ &+ \sum_p \sum_q H_Q[p,q]X[p]X[q]\delta[p+q-m] \\ &+ \epsilon[m] \end{aligned} \tag{1}$$

where m denotes the discrete frequency, $X[m]$ is the wave elevation, $Y[m]$ is the response, $H_L[m]$ is the linear transfer function, and $H_Q[p,q]$ is the quadratic transfer function of the system. $Y_L[m]$ is the output due to the linear transfer function, $Y_Q[m]$ is the sum of the outputs due to the quadratic mixing effects, $\epsilon[m]$ is the modeling error, and $\delta[p+q-m]$ represents the frequency selection rule, i.e., for given response frequency m, only those frequency pairs (p,q) which satisfy $m = p + q$ contribute to the response at m.

Although the model is nonlinear in terms of the input, it is linear in terms of the transfer function coefficients. One can solve[Nam 94] for these using cross-spectra of input and output up to third-order and auto-spectra of input up to fourth-order for a general nonGaussian input. For the nonGaussian case (which this case is, because of nonlinear wave interactions which occur before the waves impinge on the spar), the linear output Y_L and the quadratic output Y_Q are not orthogonal to each other. This results in interference spectra which prevent one from decomposing the response power spectrum into its constituent linear and quadratic components. An orthogonalized Volterra-like model[Im 96] based on the modified Gram-Schmidt method is therefore used. Once the response has been decomposed into its constituent first- and second-order components using the orthogonal model, deconvolution techniques can be applied using the linear equation of motion (which relates wave force and surge/pitch response), to recover the linear and quadratic wave force spectra. Specific details of this calculation may be found in [Yoo 96].

V. Results Using Experimental Input

Diffraction: The diffraction predicted surge force spectrum due to the actual experimental wave profile is shown in Fig. 5. Each point in the wave elevation time series was used producing a 9708-point FFT of complex wave amplitudes. The amplitudes were input into the wave-to-force LTF's and QTF's. The first-order force dominates at frequencies above $\omega = 0.25[rad/sec]$. In this region second-order forces are approximately two to four orders of magnitude smaller. For frequencies below $\omega = 0.25[rad/sec]$, however, the second-order forces overtake the first-order forces. As a result, these forces are expected to contribute the most to the slowly-varying surge response.

As discussed earlier, Fig. 6 is the predicted surge response power spectra based on the actual experimental wave pro-

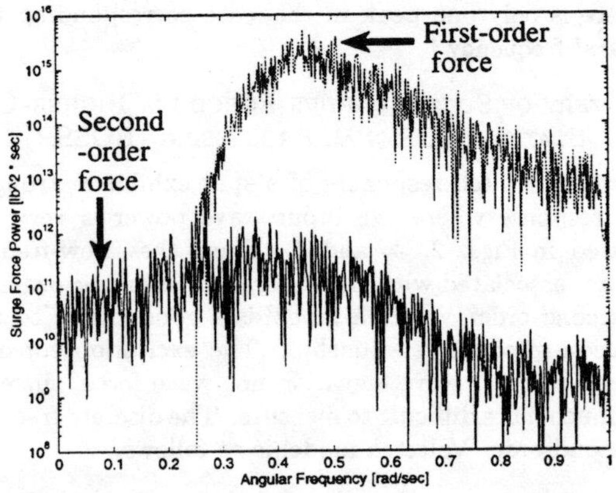

Fig. 5. Diffraction: predicted surge force power spectra using experimental wave profile

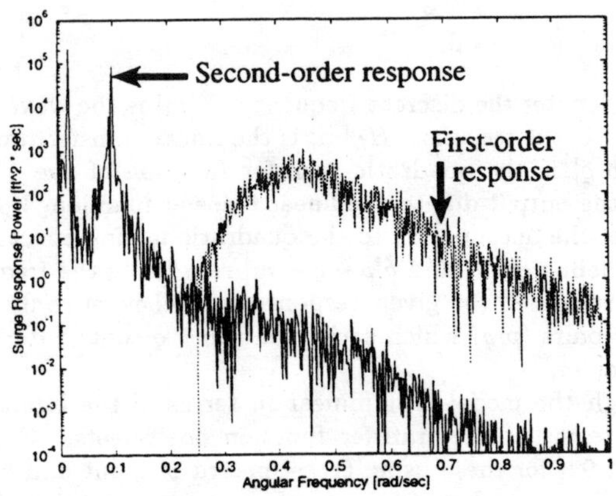

Fig. 6. Diffraction: predicted surge response power spectra using experimental wave profile

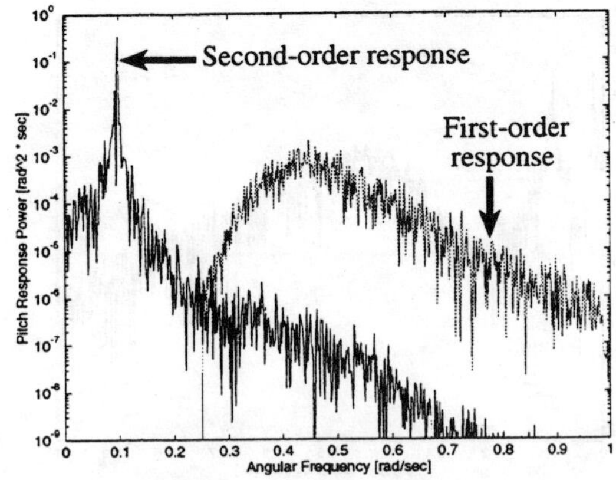

Fig. 7. Diffraction: predicted pitch response power spectra using experimental wave profile

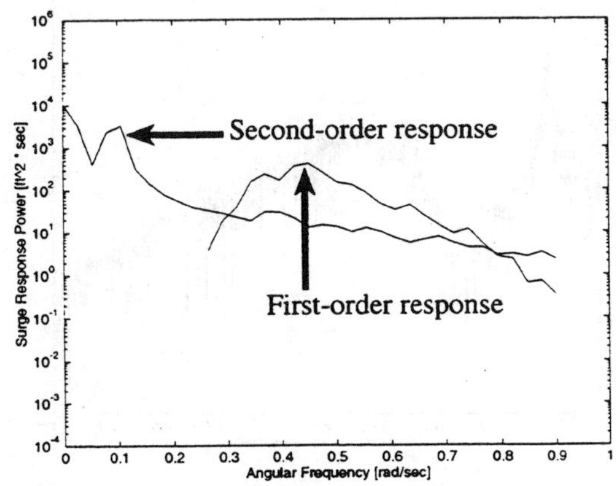

Fig. 8. HOSSP: decomposed surge response power spectra of experimental data

file. In the plot, first-order responses dominate again for $\omega \geq 0.25[rad/sec]$; then they drop off rapidly for frequencies below this value. Thus the first-order response is cut-off at $\omega = 0.25[rad/sec]$. Response contributions below this frequency are due predominantly to second-order difference-frequency forces. In this low-frequency region the spectrum shows a peak at the surge natural frequency ($\omega_{n_1} = 0.019[rad/sec]$) and another at the pitch natural frequency ($\omega_{n_5} = 0.094[rad/sec]$) because of the strong coupling between these two modes. The fact that these peaks are higher than the values at other frequencies classifies the spar as a resonant-dominated structure.

The diffraction predicted pitch response power spectrum decomposition is shown in Fig. 7.

Higher-order statistical signal processing: For the recovery of wave forces using higher-order statistical signal processing (HOSSP), we employed 128-point discrete Fourier transforms over 75 segments, which is the maximum number allowed for the total 9709 points. Fig. 8 shows the experimentally observed surge response power spectra decomposed, via a Volterra model, into its first- and second-order components. Comparison with Fig. 3 indicates that the surge natural frequency is embedded in the peak at the dc (zero-frequency) component indicating that a 128-point discretization is not fine enough to resolve the surge natural frequency. Fig. 9 shows the surge forces (first-order and second-order) recovered following the procedure outlined in Sec. IV and described in detail in [Yoo 96]. The recovered linear force spectrum between 0.3 and 0.7 [rad/sec], where most of the wave energy exists, agrees with that of Fig. 5 (diffraction-predicted) relatively well. The maximum of both results are between 10^{15} and 10^{16}. But the second-order force spectrum at the surge natural frequency is about an order of magnitude larger than that of Fig. 5.

A definitive comparison of force spectra predicted by diffraction (Fig. 5) and estimated by HOSSP (Fig. 9) is not possible for the following reasons. The HOSSP results are characterized by low variance but poor frequency resolution, while the diffraction predictions are characterized by very fine frequency resolution but high variance. With respect to this latter point, note that the "band" characterizing the second-order force in Fig. 5 is approximately an order of magnitude wide. Thus we conclude that, in order to make a more informative comparison of force spectra predicted by diffraction and estimated from HOSSP, we need force spectra estimates of comparable frequency resolution and variance. To test this conclusion we conducted a numerical experiment as described in the next section. The advantage of such experiments lies in the fact that long data records can be generated relatively inexpensively. The availability of such long records enables us to improve the frequency resolution of the HOSSP force spectra estimates. To reduce the variability of

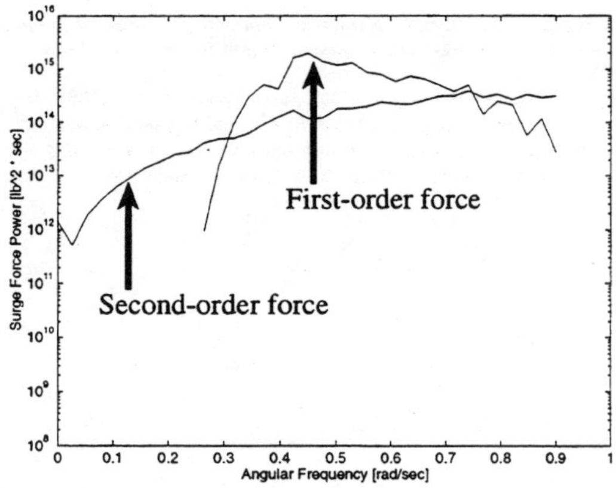

Fig. 9. HOSSP: recovered surge force power spectra of experimental data

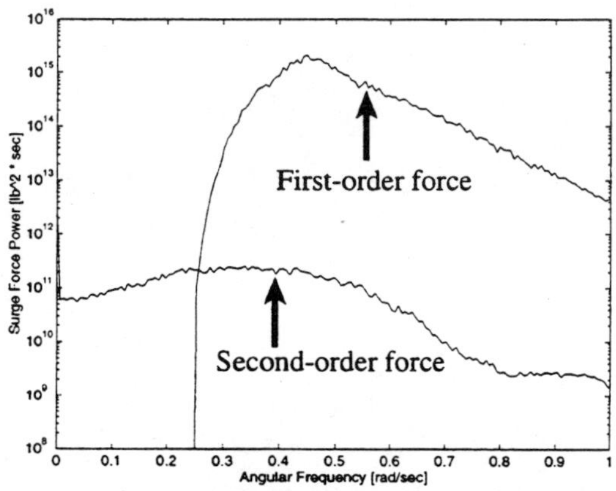

Fig. 10. Diffraction: averaged surge force power spectra predicted with numerically generated data (simulation)

the diffraction estimates and to render the frequency resolution comparable to the new HOSSP results, we frequency average over adjacent frequency bins. Specific details are provided in the next section.

VI. Results Using Numerically Simulated Input

The simulation was based on the JONSWAP spectrum with $H_s = 43.0ft$, a peak period of $T_p = 14.0sec$, and a peak enhancement factor $\gamma = 2.0$. Since the JONSWAP input spectrum has most of its energy around 0.5 [rad/sec], and decreases as frequency increases, we can approximate it by restricting the frequency to about 1.5 [rad/sec], say $\pi/2$. This approximation determines the critical sampling frequency to be $\omega_s = \pi[rad/sec]$ ($f_s = 0.5[Hz]$), or equivalently the sampling interval is $t_s = 2[sec]$. "Noisy" simulation was used where both wave amplitudes and phases were taken as random variables [Ude 96]. The resulting complex incident wave components were input into the wave-to-force LTF's and QTF's to produce force spectra.

The wave excitation time series and response data consisted of 262,144($= 2^{18}$) points extending over a 524,288 sec duration. The number of data points is twice the minimum number of data points required for second-order Volterra transfer function estimation using 512($= 2^9$)-point DFTs. As pointed out below, we note that the 512-point DTF is the minimum for differentiating the surge natural frequency from the dc (zero frequency) component.

Fig. 10 shows the *frequency-averaged* first- and second-order

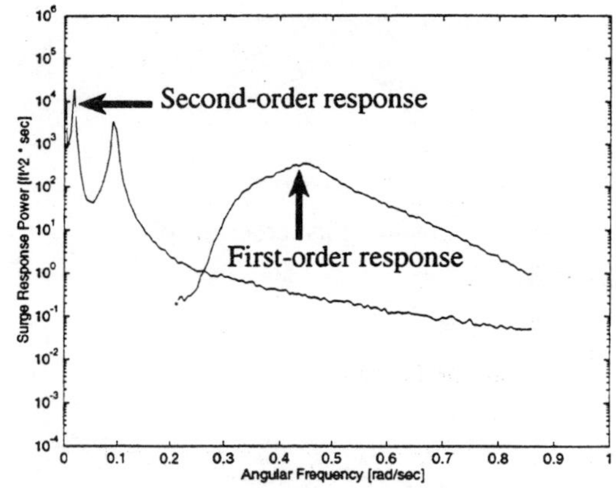

Fig. 11. HOSSP: decomposed surge response power spectra of numerically generated data (simulation)

surge force spectra determined via diffraction. The averaging was performed for two purposes. The first is to reduce the variance of the estimation and the second is to provide comparable frequency resolution to facilitate direct comparison with the HOSSP result. The bin size (frequency resolution) is adjusted by averaging approximately to the bin size of the 512-point DFT which is 0.0061 [rad/sec]. Note that the bin size 0.0061 [rad/sec] can differentiate the surge natural frequency 0.019 [rad/sec] from dc (zero frequency) component, while the bin size of 256-point DFT can not.

Fig. 11 shows the decomposed surge response power spectra of the numerically generated data using 512-point Fourier transforms in the HOSSP approach.

Fig. 12 shows the HOSSP recovered surge force power spectra corresponding to the responses given in Fig. 11. Compared to Fig. 10, the first-order force for frequencies greater than 0.25 [rad/sec] is in good agreement. Note that in the vicinity of surge natural frequency, the diffraction and HOSSP second-order force spectra are also in rather good agreement. However, as the frequency increases the HOSSP second-order force spectra in Fig. 12 increases somewhat faster than the second-order force spectra estimated from diffraction in Fig. 10. Possible explanations include spectral leakage and/or energy in the surge response spectra due to coupling with pitch motion, where this additional energy may not have been fully compensated by the values utilized in the mass, stiffness, and damping matrices in the equation of motion. Above 0.4 [rad/sec] the diffraction predicted second-order force (Fig. 10) drops off with frequency since sum-frequency interactions are not included. This does not happen in the HOSSP estimate (Fig. 12) because the HOSSP approach does include sum-frequency interactions.

VII. Discussion

The experimental determination of second-order forces is a very difficult problem because such forces are often orders of magnitude smaller than the corresponding first-order forces. In this paper and reference [Yoo 96] a methodology has been presented, based on higher-order statistical signal processing (HOSSP) and Volterra models, whereby such second-order forces can be estimated from the wave-excitation structural-response time-series data. The results are compared with the first- and second-order forces predicted by appropriate diffraction codes. The results indicate that when the first- and second-order force spectra are compared using comparable variance and frequency resolution,

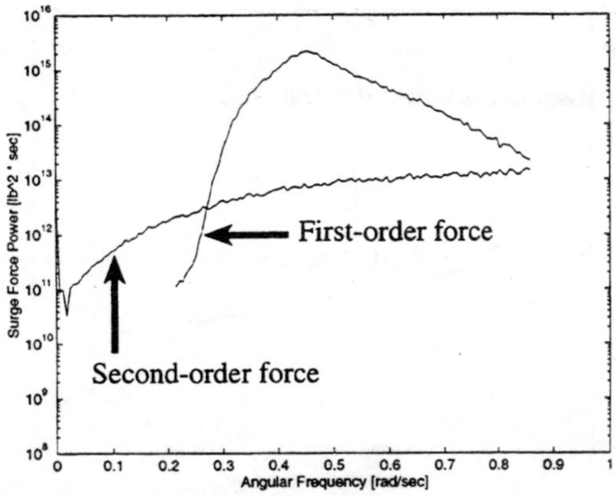

Fig. 12. HOSSP: recovered surge force power spectra of numerically generated data (simulation)

the first-order force spectra agree very well, and the second-order force spectra agree quite well in the vicinity of surge natural frequency. We are currently investigating why the second-order force spectrum estimated via HOSSP increases somewhat more rapidly with frequency (just above surge resonance) than the diffraction predicted second-order force. We are also using the HOSSP force spectra and wave input spectra to determine wave-force LTF's and QTF's, which can then be directly compared to those predicted by diffraction theory.

ACKNOWLEDGMENTS

This work is supported in part by NSF Grant CDR-8721512 through the Offshore Technology Research Center. The orthogonal Volterra-like model methodology was developed principally under the auspices of the Joint Services Electronic Program AFOSR Contract F-49620-95-C-0045 and Department of Navy Grant N00014-92-J-1046. The United States Government is authorized to reproduce and distribute reprints for governmental purposes notwithstanding any copyright notation thereon.

REFERENCES

[Glanvi 91] Glanville, RS, Paulling, JR, Halkyard, JE, and Lehtinen, TJ (1991). "Analysis of the Spar Floating Drilling Production and Storage Structure," *Proceedings of the 1991 Offshore Technology Conference*, OTC 6701, Vol.4, pp.57–68, Houston, Texas.

[Halkyar 96] Halkyard, JE (1996). "Status of Spar Platform for Deepwater Production Systems," *Proceedings of the 6th International Offshore and Polar Engineering Conference*, Vol I,ISOPE-96, pp.262-272, Los Angeles, California.

[Horton 92] Horton, EE, and Halkyard, JE (1992). "A Spar Platform for Developing Deep Water Oil Field," *MTS '92*, pp.998–1005, Marine Technology Society, Washington, D.C.

[Im 96] Im, S, and Powers, EJ (1996). "A Sparse Third-order Orthogonal Frequency-domain Volterra-like Model," *J. Franklin Inst.*, Vol.333(B), No.3, pp.385–412. (Special Issue on Higher-Order Statistics)

[MIT 94] MIT (1994). *WAMIT Version 5.1s User Manual*, Department of Ocean Engineering, MIT.

[Nam 94] Nam, SW, and Powers, EJ (1994). "Application of Higher-Order Spectral Analysis to Cubically Nonlinear System Identification," *IEEE Trans. Signal Processing*, Vol.42, No.7, pp.1746–1765.

[Santeen 76] van Santen, JA, and de Werk, K (1976). "On the Typical Qualities of Spar Type Structures for Initial or Permanent Field Development," *Proceedings of the 1976 Offshore Technology Conference*, OTC 2716, Vol.3, pp.1105–1118, Houston, Texas.

[Ude 96] Ude, TC, Kumar, S, and Winterstein, SR (1996) *TFPOP2.1: Stochastic Response Analysis of Floating Structures Under Wind, Current, and Second-Order Wave Loads*, Report No. RMS-18, Reliability of Marine Structures Program, Civil Engineering Department, Stanford University.

[Weggel 94] Weggel, DC (1994). *Hydrodynamic Forces on Truncated Cylinders*, Report No. CDR-8721512, Offshore Technology Research Center, The University of Texas at Austin.

[Weggel 97] Weggel, DC (1997). *Nonlinear Dynamic Response of Large Diameter Offshore Structures*, Ph.D. Dissertation, The University of Texas at Austin, Austin, TX.

[Yoo 96] Yoo, H, Powers, EJ, Weggel, DC, and Roesset, JM (1996). "Recovery of Second-Order Force via Higher-Order Statistical Signal Processing," *Proceedings of 7th ICSPAT (International Conference on Signal Processing Applications & Technology)*, Vol.2, pp.1234–1238, Boston, Massachusetts.

Proceedings of the Seventh (1997) International Offshore and Polar Engineering Conference
Honolulu, USA, May 25-30, 1997

ISBN 1-880653-28-1 (Set); ISBN 1-880653-29-X (Vol. I)

Numerical Study of the Hydrodynamic Forces on Huge Floating Structures in Waves and Ocean Currents

Yoshiyuki Inoue, Xuangang Zhang and Shigeru Tabeta
Yokohama National University
Yokohama, Japan

ABSTRACT

A linearized analysis of the hydrodynamic forces of the huge floating structure in waves is presented in this paper. The hydroelastic interactions of the elastic deformations and the hydrodynamic forces of a floating airport are discussed. The motion responses and drift forces are investigated by the present method. Interactions of the huge floating structure and ocean currents are also important to estimate environmental forces. Some numerical simulations by means of multilevel model are carried out to discuss interactions of ocean currents and the huge floating structure in a bay.

KEY WORDS: Huge Floating Structures, Hydrodynamic Forces, Elastic Deformations, Separated Region Method, Multilevel Model

INTRODUCTION

The world population grows rapidly now which increases the demand for land and space. There will be a need in the future for floating airports mainly due to the lack of adequate land space near coast lines where there are densely populated metropolitan areas. Although a lot of projects about ocean airports are being discussed in Japan, the application of reclamation system in deeper sea is very difficult to realize because of the necessity of the mass sand for the filling of the land. Long period of the construction and environmental impact are also obvious defects of this system. On the contrary, an airport with floating system clearly has advantages in some points. Especially, it is considered to give less damage to the environment than reclaimed one. Therefore, matters for investigation in designing a floating airport system are frequently discussed recently (Inoue and et al., 1995).

The huge floating structure like an airport requires special response analysis techniques, because it has a very small vertical dimension compared to its longitudinal and lateral dimensions. The elastic deformations due to the dynamic loads of waves may be of the order that cannot be neglected in comparison with the motion as rigid body. The hydroelastic interactions that the elastic deformations and the hydrodynamic forces interact with each other should be considered in the predictions. Further, when a normal size floating structure is constructed in the sea, current forces acting on the floating body can be estimated by means of many empirical methods. On the other hand, there is no empirical data for estimating current forces of huge floating structures such as a floating airport that has not been built in the world. However, current forces acting on floating structures are important in design stage.

A linearized analysis of the hydrodynamic forces of the floating structure in waves is presented in this paper. The three-dimensional hydroelastic analysis of the huge floating structure requires significant computation resources in comparison with other available method. It is very difficult to investigate hydroelastic responses thoroughly by three-dimensional method for a typical floating airport. The present study is restricted in two-dimensional. The separated region method is applied to calculate the hydrodynamic forces in this study. Ijima (1972) used the separated region method to analyze the motion responses of moored and freely floating rectangular bodies in waves at finite water depth. The calculated motion responses and transmission coefficients by this method have been proved to be in good agreement with the experimental data. In present study, this method is applied to the hydroelastic analysis to calculate the hydrodynamic forces of huge floating structures in waves. To get the elastic response of the structure, it is assumed that the elastic structure is divided into many small rigid blocks. The continuous response of the elastic structure is expressed approximately as the motion responses of these blocks. The effect of bending rigidity to motion responses is considered as the restoring forces in rigid motion equations of each divided block. The hydrodynamic and elastic problems can be solved simultaneously.

Further, some numerical simulations are carried out to discuss interactions of currents and a huge floating structure in a bay. A multilevel model is used in order to describe current field considering the influence of the floating structure. Current forces acting on the structure are estimated as well by the numerical simulations.

MATHEMATICAL FORMULATION

Heave responses of the floating elastic structure in waves

In the coordinate system chosen to describe the problem (see Fig.1), y is the horizontal coordinate that is coincident with the undisturbed free-surface, and z is positive pointing upwards. The floating structure has the breadth B and the draft d. The structure in consideration is divided into a finite number (M) of blocks having the length b_j and each block is denoted by the index, j, as described in Fig.1. The center of gravity of block j is (y_{jo}, z_{jo}). The elastic response of entire floating structure is approximately taken by the rigid responses of M blocks. It is assumed that the mooring system is such that is allowed significant motions only in the vertical direction. This assumption is reasonable when the floating structure has a very small vertical dimension compared to its longitudinal and lateral dimensions. The heave amplitude of block j is denoted as $Z_{j,a}$.

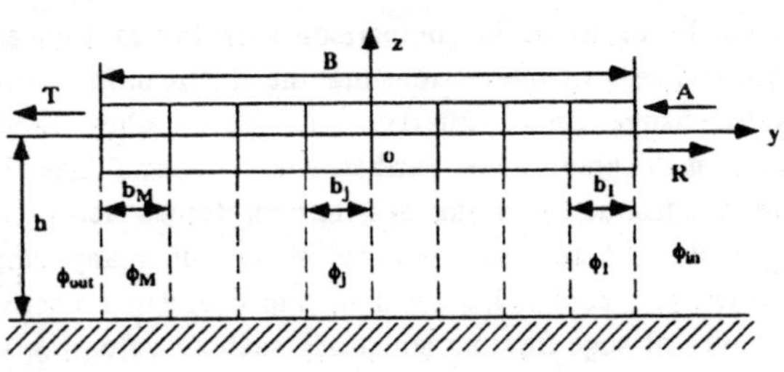

Fig. 1 Schematic sketch of the coordinate system

The fluid is assumed to be incompressible, inviscid and the flow irrotational. Following the assumption of small amplitude waves and motions, the velocity potential $\phi e^{i\omega t}$ (expressed as $\phi_{in}e^{i\omega t}$, $\phi_{out}e^{i\omega t}$, $\phi_j e^{i\omega t}$ in the fluid domains) must satisfy the following conditions:

$$\frac{\partial^2\phi}{\partial y^2}+\frac{\partial^2\phi}{\partial z^2}=0 \qquad \text{Laplace's equation} \tag{1}$$

$$\frac{\partial\phi}{\partial z}=0 \qquad \text{on z=-h, sea bottom condition} \tag{2}$$

$$\frac{\partial\phi}{\partial z}-\frac{\omega^2}{g}\cdot\phi=0 \qquad \text{on z=0, free surface condition} \tag{3}$$

$$\frac{\partial\phi_j}{\partial z}=i\omega\cdot Z_{j,a} \qquad \text{on z=-d, bottom condition of block j} \tag{4}$$

$$\frac{\partial\phi_{in}}{\partial y}+ik\phi_{in}\rightarrow\infty \qquad as\ y\rightarrow+\infty \qquad \text{radiation condition} \tag{5}$$

$$\frac{\partial\phi_{out}}{\partial y}-ik\phi_{out}\rightarrow\infty \qquad as\ y\rightarrow-\infty \qquad \text{radiation condition} \tag{6}$$

According to the velocity potential expressions of one rigid floating body (Ijima and et al., 1972; Nojiri, 1975) the solutions of ϕ_{in}, ϕ_{out}, ϕ_j can be expressed as follows:

$$\phi_{in}=[Ae^{ik(y-\frac{B}{2})}+Re^{-ik(y-\frac{B}{2})}]\cdot\frac{\cosh k(z+h)}{\cosh kh}+\sum_{n=1}^{\infty}C_n e^{-k_n(y-\frac{B}{2})}\cdot\frac{\cos k_n(z+h)}{\cos k_n h} \tag{7}$$

$$\phi_{out}=Te^{ik(y+\frac{B}{2})}\cdot\frac{\cosh k(z+h)}{\cosh kh}+\sum_{n=1}^{\infty}G_n e^{k_n(y+\frac{B}{2})}\cdot\frac{\cos k_n(z+h)}{\cos k_n h} \tag{8}$$

$$\phi_j=\sum_{s=0}^{\infty}\{[D_{j,s}\frac{\cosh\bar{S}(y-y_{jo})}{\cosh SB_j}+E_{j,s}\frac{\sinh\bar{S}(y-y_{jo})}{\sinh SB_j}]\cdot\cos\bar{S}(z+d)+4i(-1)^s\cdot\frac{h}{b_j}\cdot Z_{j,a}\omega h\cdot\cos\mu_{j,s}(y-y_{jo})\cdot\frac{\cosh\mu_{j,s}(z+h)}{(\mu_{j,s}h)^2\sinh\overline{\mu_{j,s}}h}\} \tag{9}$$

where

$$q=\frac{d}{h},\quad \bar{S}=\frac{s\pi}{(1-q)h},\quad SB_j=\frac{\bar{S}b_j}{2} \tag{10}$$

$$\mu_{j,s}=\frac{(2s+1)\pi}{b_j},\quad \overline{\mu_{j,s}}=(1-q)\mu_{j,s} \tag{11}$$

In above expressions, A, R, T, C_n, G_n, $D_{j,s}$, $E_{j,s}$ are all the complex constraints. A is representing the incident waves, R, the reflected waves, T, the transmitted waves. C_n and G_n are representing non-propagating wave modes, $D_{j,s}$ and $E_{j,s}$, the arbitrary coefficients. ω is wave frequency, k and k_n are eigenvalues decided by the following equations:

$$\frac{\omega^2}{g}-k\tanh kh=0,\quad \frac{\omega^2}{g}+k_n\tan k_n h=0 \tag{12}$$

The wave transmission and reflection coefficients of the floating structure are expressed as follows:

$$K_t=\left|\frac{T}{A}\right|,\qquad K_r=\left|\frac{R}{A}\right| \tag{13}$$

For the elastic analyses of the floating structure, the structure is considered as a two-dimensional uniform beam on elastic foundation and the slope of the deflection curve is assumed small enough to neglect the rotatory inertia in present study. The bending rigidity is considered as not a function of displacement. Considering the hydroelastic interactions of the elastic deformations and the hydrodynamic forces, the equation of vertical response for each divided floating block can be expressed as,

$$m_j\frac{d^2z_j}{dt^2}=F_{W,j}-m_jg-F_{Ej}\qquad j=1,2,......M \tag{14}$$

where

m_j: mass of the divided floating block j

$F_{W,j}$: wave exciting force acting on the block j

F_{Ej}: restoring force by elastic response of the block j

When elastic responses of the floating structure are calculated, the structure is divided into M blocks. The structure between adjoining blocks' gravity centers is modeled by a beam element with bending rigidity EI_j. This flexible beam is assumed that both ends of the floating structure are free. To get hydrodynamic forces, however, the slopes of each divided block are small enough to satisfy equation (9).

For the block j, the vertical response equation is expressed as follows:

$$m_1 \frac{d^2 z_1}{dt^2} = \int_{\frac{B}{2}-b_1}^{\frac{B}{2}} p_1 dy - m_1 g - F_{E,1} \qquad (15)$$

$$m_j \frac{d^2 z_j}{dt^2} = \int_{\frac{B}{2}-\sum_{i=1}^{j} b_i}^{\frac{B}{2}-\sum_{i=1}^{j-1} b_i} p_j\, dy - m_j g - F_{Ej} \qquad j = 2,3,......M \qquad (16)$$

z_j can be expressed by the amplitude of motion responses as follows:

$$z_j = z_{jo} + Z_{j,a} e^{i\omega t} \qquad j = 1,2,......M \qquad (17)$$

p_j is calculated using the Bernoulli's equation, as,

$$p_j = -\rho \frac{\partial(\phi_j e^{i\omega t})}{\partial t} - \rho g(-d + Z_{j,a} e^{i\omega t}) \qquad j = 1,2,......M \qquad (18)$$

According to the assumptions stated before, F_{Ej} (j=1,2,......M) of every block can be expressed by the following equations:

$$F_{E,1} = \frac{3EI_1}{\left(\frac{b_1}{2}+\frac{b_2}{2}\right)^3} \cdot (z_1 - z_2) \qquad (19)$$

$$F_{E,2} = \frac{3EI_1}{\left(\frac{b_2}{2}+\frac{b_1}{2}\right)^3} \cdot (z_2 - z_1) + \frac{12EI_2}{\left(\frac{b_2}{2}+\frac{b_3}{2}\right)^3} \cdot (z_2 - z_3) \qquad (20)$$

$$F_{E,j} = \frac{12EI_{j-1}}{\left(\frac{b_j}{2}+\frac{b_{j-1}}{2}\right)^3} \cdot (z_j - z_{j-1}) + \frac{12EI_j}{\left(\frac{b_j}{2}+\frac{b_{j+1}}{2}\right)^3} \cdot (z_j - z_{j+1})$$

$$j = 3,4,......M-2 \qquad (21)$$

$$F_{E,M-1} = \frac{12EI_{M-2}}{\left(\frac{b_{M-1}}{2}+\frac{b_{M-2}}{2}\right)^3} \cdot (z_{M-1} - z_{M-2})$$

$$+ \frac{3EI_{M-1}}{\left(\frac{b_{M-1}}{2}+\frac{b_M}{2}\right)^3} \cdot (z_{M-1} - z_M) \qquad (22)$$

$$F_{E,M} = \frac{3EI_{M-1}}{\left(\frac{b_M}{2}+\frac{b_{M-1}}{2}\right)^3} \cdot (z_M - z_{M-1}) \qquad (23)$$

According to the continuity conditions for velocities and pressures in the separated regions, the velocity potentials in the fluid domains should satisfy the following equations:

$$\frac{\partial \phi_{in}}{\partial y} = 0 \qquad 0 \geq z \geq -d,\ y = \frac{B}{2} \qquad (24)$$

$$\frac{\partial \phi_{in}}{\partial y} = \frac{\partial \phi_1}{\partial y},\ \phi_{in} = \phi_1 \qquad -d \geq z \geq -h,\ y = \frac{B}{2} \qquad (25)$$

$$\frac{\partial \phi_{out}}{\partial y} = 0 \qquad 0 \geq z \geq -d,\ y = -\frac{B}{2} \qquad (26)$$

$$\frac{\partial \phi_{out}}{\partial y} = \frac{\partial \phi_M}{\partial y},\ \phi_{out} = \phi_M \qquad -d \geq z \geq -h,\ y = -\frac{B}{2} \qquad (27)$$

$$\frac{\partial \phi_j}{\partial y} = \frac{\partial \phi_{j+1}}{\partial y},\ \phi_j = \phi_{j+1} \qquad -d \geq z \geq -h,\ y = \frac{B}{2} - \sum_{i=1}^{j} b_i$$

$$j = 1,2,......M-1 \qquad (28)$$

All the unknown constraints in the velocity potentials and heave amplitudes $Z_{j,a}$ are decided by solving the equations (24)~(28) and (15)~(16) simultaneously. The elastic responses of floating structure can be obtained by the proposed method.

The wave drift force of the floating elastic structure

Modifying Maruo's theory of wave drift force in deep water, the drift force in shallow water is calculated by the following formula:

$$F_d = \frac{1}{2}\rho g\left(1 + \frac{2kh}{\sinh 2kh}\right)\zeta_r^2 \qquad (29)$$

The drift force coefficient is expressed as follows:

$$C_{Fd} = \frac{F_d}{\frac{1}{2}\rho g \zeta_a^2} = \left(1 + \frac{2kh}{\sinh 2kh}\right) \cdot K_r^2 \qquad (30)$$

ζ_a and ζ_r are the amplitudes of incident and reflected waves of the floating structure. For the rigid floating body, the numerical results are in good agreement with the experimental data. In present study, the drift force of the huge elastic structure is calculated by the same method. If the reflection coefficients are decided, the drift force coefficients can be calculated by the formula (30).

Interactions of currents and the floating elastic structure

The estimation of tidal current force acting on the huge floating structure is important in designing a mooring system and others. However, there is no established method to estimate tidal force in the huge floating structures. In the present study, the current force is estimated from the numerical calculation with multilevel model, which is the modified method by Tabeta and Inoue (1995).

When the floating structure is divided into a finite number (M) of mesh blocks described in Fig.2, the static equilibrium of forces

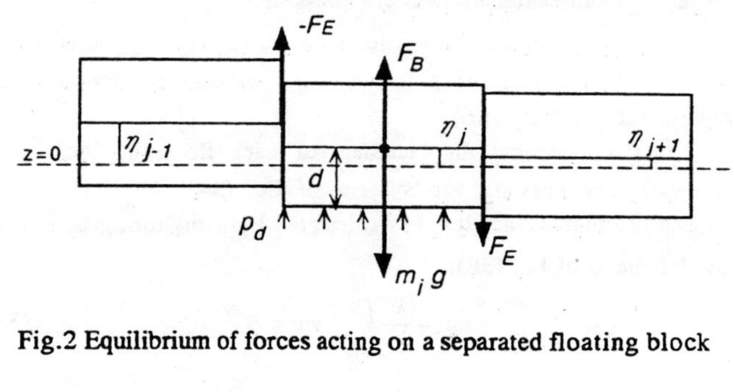

Fig.2 Equilibrium of forces acting on a separated floating block

acting on the block j is,

$$m_j g - F_{Bj} + F_{Ej} = 0 \tag{31}$$

The buoyancy force $F_{B,j}$, is described as,

$$F_{Bj} = p_{dj} A_{wj} \tag{32}$$

p_{dj} is the hydrostatic pressure at the bottom of each block, A_{wj} is the bottom area of block j. The restoring force due to the rigidity of the floating structure is also assumed to be proportional to the difference of the vertical displacement of adjacent blocks and described by using the equivalent damping coefficients. In this section the vertical displacement of block j is expressed as η_j. Now, the virtual surface level ζ_j is introduced as follows;

$$p_{dj} = \rho g(\zeta_j + d - \eta_j) \tag{33}$$

Using ζ_j instead of p_{dj}, Eq. (31) derives,

$$\rho g(\zeta_j - \eta_j) A_{wj} + F_{Ej} = 0 \tag{34}$$

If the virtual surface level ζ_j is known, the vertical displacement of each block is derived by solving the system of above equation.

Suppose u and v are two horizontal velocity components, w is the vertical velocity in the fluid domain. The following assumptions are applied to take the huge floating structure into the multilevel model. The subscript j for the index of partitioned block will be omitted in the expressions below.

$$w = 0 \quad at\ z = \eta - d \tag{35}$$

In the fluid domain where the floating structure exists, horizontal velocity components below the floating structure are calculated as follows in order to satisfy the following equation derived by integrating the equation of continuity vertically using Eq. (35):

$$\frac{\partial}{\partial x}\int_{-h}^{\eta-d} u dz + \frac{\partial}{\partial y}\int_{-h}^{\eta-d} v dz = 0 \tag{36}$$

1) Tentative virtual surface level $\dot{\zeta}$ is assumed to be equal to the value of the previous time step.

2) Tentative pressure $\dot{p}$ at each grid point is calculated by

$$\dot{p} = \rho g(\dot{\zeta} - d) + p_a \tag{37}$$

where p_a denotes the atmospheric pressure.

3) Tentative horizontal velocities $\dot{u}$, $\dot{v}$ are calculated by horizontal momentum equations using the tentative pressure $\dot{p}$. These are not always satisfied Eq. (36).

4) Tentative vertical displacement of each floating block $\dot{\eta}$ is calculated by solving the system of Eq. (34).

5) The residual surface level is calculated by substituting $\dot{u}$, $\dot{v}$, $\dot{\eta}$ to the left hand of Eq. (36).

$$\frac{\Delta\dot{\zeta}}{\Delta t} = -\frac{\partial}{\partial x}\int_{-h}^{\dot{\eta}-d} \dot{u} dz - \frac{\partial}{\partial y}\int_{-h}^{\dot{\eta}-d} \dot{v} dz = 0 \tag{38}$$

6) The tentative virtual surface level is corrected by:

$$\alpha\rho g\Delta\dot{\zeta} \rightarrow \Delta\dot{\zeta} \tag{39}$$

where α is correction factor.

7) The procedures from 2) to 6) are iterated until the maximum value of residual surface level in the domain of floating structure $\Delta\dot{\zeta}_{max}$ becomes small enough.

8) If the iteration converges, the horizontal velocity components and vertical displacement of each block are determined:

$$u^{(n+1)} = \dot{u},\quad v^{(n+1)} = \dot{v},\quad \eta^{(n+1)} = \dot{\eta} \tag{40}$$

It is assumed that the force F due to tidal current acting on the floating structure is composed of two components, pressure force F_p and frictional force F_f.

$$F = F_p + F_f \tag{41}$$

In the present method, the pressure resistance force F_p is calculated by integrating the hydrostatic pressure over the surface of the structure, in which the change of current speed is considered.

$$F_p = \int_S p\mathbf{n} ds \tag{42}$$

where S is the surface of the floating structure and **n** is the inward normal unit vector of S. The frictional resistance force is estimated by using the frictional coefficient C_f:

$$F_f = \int_{S_b} C_f \frac{1}{2}\rho |v_b| v_b dS_b \tag{43}$$

where $\mathbf{v}_b$ is the horizontal velocity vector at the layer just below the floating structure. S_b is the bottom surface of the floating structure. As the frictional coefficient C_f, Prandtl-Schlichting formula is used:

$$C_f = \frac{0.455}{\log Re^{2.58}},\qquad Re = \frac{|v_b|L}{\upsilon} \tag{44}$$

where L is the length of the floating structure in the direction of the current and υ is the coefficient of kinematic viscosity.

NUMERICAL CALCULATIONS

Validation of the computed results

In the previous section an outline of the theoretical aspects and the computational method has been given. In this section some results of other method or experimental data are compared with the results of computations on the method of the previous section.

First, numerical results of heave responses by the present hydroelastic method are compared with the results of the floating elastic model studied by Nakai and et al. (1995). The principal dimensions of the model are listed in Table1. Fig. 3 (a)-(d) are the amplitudes of vertical displacement for the floating elastic model in different wavelengths. The theoretical calculated results and experimental data are shown in these figures for comparisons. In Nakai and et al.'s calculation, they used the boundary element method combined with finite element method considering the effect of the elasticity. The experiment has been performed for two-dimensional model structure. In order to perform the experiment in two-dimensional state, the channel of the wave tank has been divided into two channels. In present calculation, the elastic body is divided 11 equal blocks. Generally the present calculated results have fairly well agreement with the experimental data except in some positions. The agreement in smaller wavelengths is better than it in larger wavelengths. The three-dimensional effect may be inevitable in the experiment for larger wavelengths.

Table 1 Principal dimensions of experimental model

Dimensions	Elastic Model
Length L(m)	10.0
Width B(m)	0.50
Depth D(m)	0.076
Draft d(m)	0.017
Displacement W/B(kg/m)	167.2
Bending Stiffness EI(kg•m^2)	192.0
Water Depth h(m)	1.10

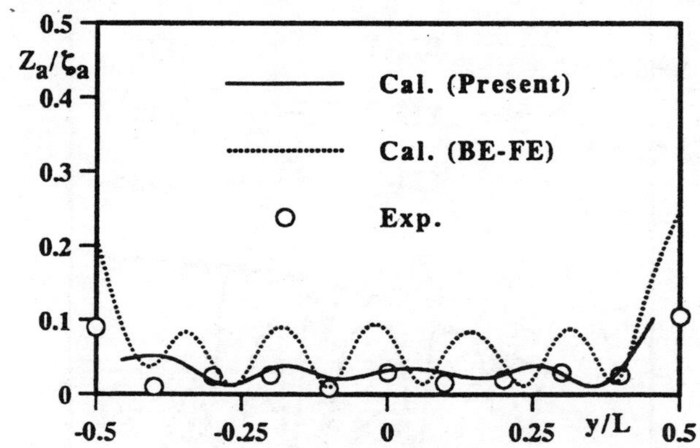

(a) L/λ=17.90

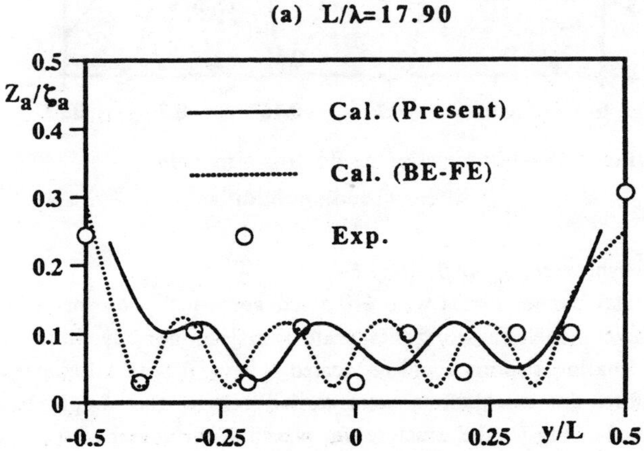

(b) L/λ=10.07

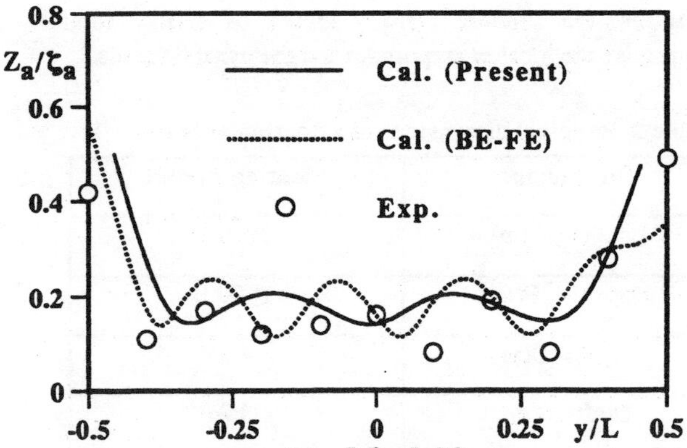

(c) L/λ=5.22

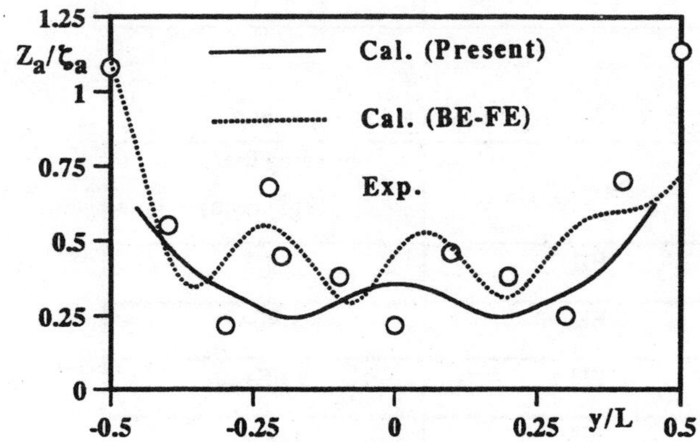

(d) L/λ=2.47

Fig. 3 The heave amplitudes of vertical displacement responses for elastic model

Elastic responses and wave drift forces of a floating airport

The floating airport recently discussed in Japan is considered as a good example of the huge floating structure. It was taken as the analysis model in this study. Many authors proposed the dimensions of the floating airport differently. The principal dimensions of the pontoon type floating airport used in present study are listed in Table 2. Considering the floating airport is set along the coastline, a two-dimensional model is constructed by a lateral section of the airport in present analyses. The airport is divided into 11 equal blocks of length 90.91m each. In calculations the bending rigidity is shown in Table 3. A zero value of bending rigidity means that the blocks of the airport can move freely and independently. Only the fluid forces effect the heave amplitudes of each block. Fig.4 (a)–(c) are heave amplitudes of three blocks (Block 1, 6, 11) of the airport in different bending rigidity conditions. It is seen that the rigidity of the structure has great influence in the hydroelastic responses of the floating airport. For smaller values of EI (such as EI1 and EI2) and shorter wavelengths, only the weather side of the airport (Block 1) is affected by waves, and the motions decline from the weather side to inwards. Generally the heave amplitudes decrease with the increase of the rigidity of the structure in longer wavelengths, but the influence is quite different in shorter wavelengths. For the case of EI5, the amplitudes that are as the responses of rigid body are very small values near zero. Although the deflections are increased with the introduction of the rigidity, it helps to reduce the dynamic loads on the bottom of the floating airport. According to design conditions, the suitable rigidity should be decided carefully to compromise the motion amplitudes and the dynamic loads.

Table 2 Principal dimensions of a floating airport

Dimensions	Floating Airport
Length L(m)	3000.0
Breadth B(m)	1000.0
Depth D(m)	4.5
Draft d(m)	3.0
Water Depth h(m)	20.0

Table 3 Bending rigidities of a floating airport

Symbol	Bending Rigidity ($kgf\cdot m^2/m$)
EI1	0.0
EI2	1.0E+9
EI3	1.0E+10
EI4	1.0E+11
EI5	1.0E+13

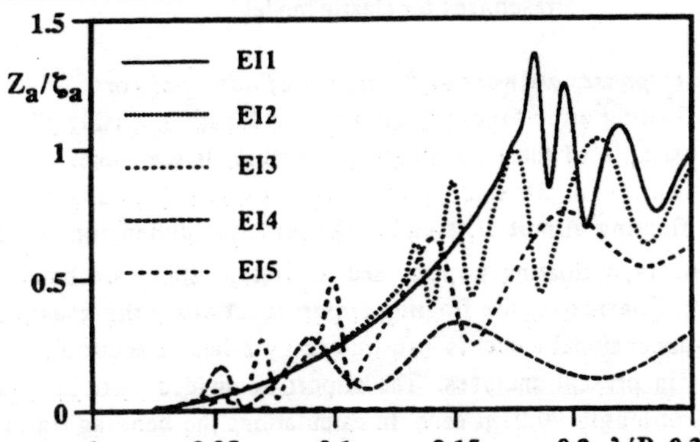

(a) Block 1

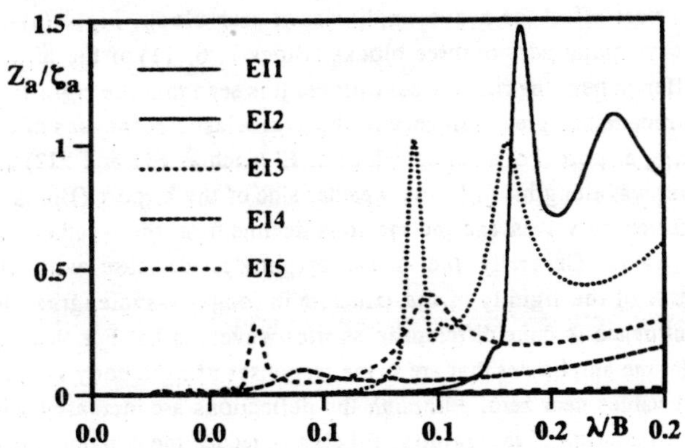

(b) Block 6

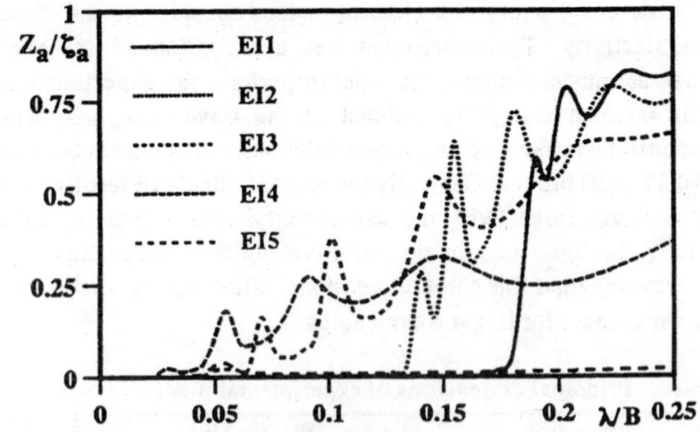

(c) Block 11

Fig. 4 The heave amplitudes of the floating airport in different bending rigidities

The wave drift forces of the floating airport with different bending rigidities are shown in Fig.5. The shallow water effect plays an important role for the drift forces in wide range of variable wavelengths for rigid body. For elastic body, the bending rigidity is important for the drift forces of the floating airport. The drift forces change clearly according to the variation of the bending rigidity. The drift forces of the floating airport decrease with the introduction of the rigidity.

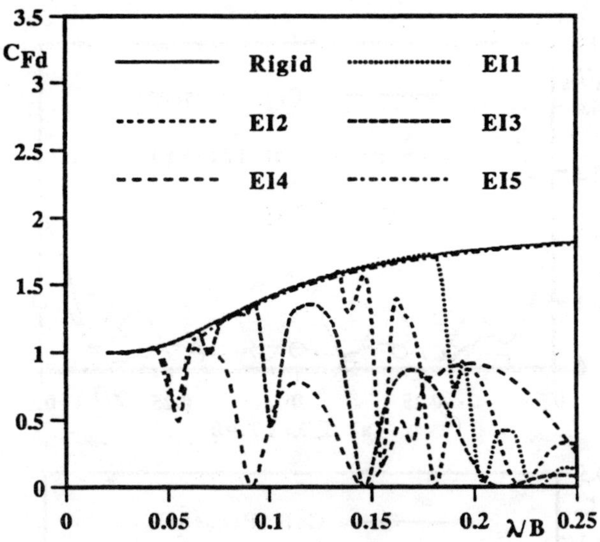

Fig. 5 The drift forces of the floating airport in different bending rigidities

The current forces of the floating airport

The tidal current forces were estimated numerically by means of multilevel model. The horizontal scale of major inner bays in Japan where floating airports could be located is several tens' kilometers. Therefore, the calculations were carried out in the simple bay described in Fig.6. The structure was placed at the center of the bay that is 30km long, 15km wide and has constant depth of 20m. The tidal wave given at the open boundary had the amplitude of 0.5m and the period of 12 hours. The horizontal mesh size was taken to be 1500m and the number of vertical levels is five. The neighborhood domain of the floating airport was analyzed by finer mesh whose size is 500m. The Coriolis force was ignored. The calculation was carried out for five tidal periods (60 hours) and discussion was done about the result of the last one cycle. Further, restoring force due to the rigidity of the airport was ignored because the structure is supposed to deform along the shape of the surface elevation in the waves.

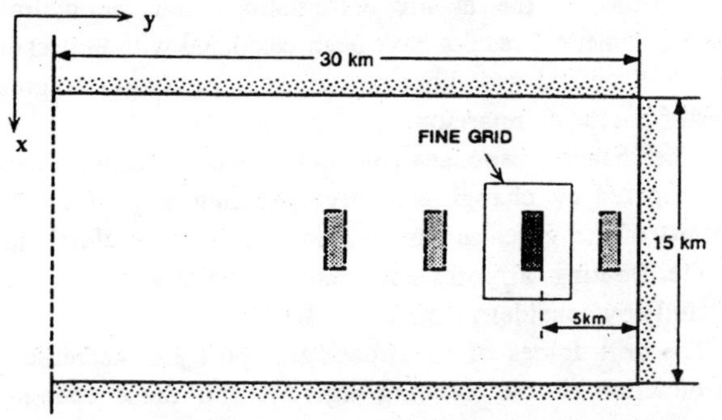

Fig.6 The rectangular basin and positions of the floating airport used for calculation of current forces

Fig.7 shows the time variation of pressure force, frictional force, and total force in the longitudinal direction of the bay (the positive values denote the force toward the inner bay). The phase of the pressure force is different from that of the frictional force whose phase is almost the same as that of the velocity. It indicates that the rate of the component due to the gradient of sea level, whose phase is almost the same as that of the sea level variation, in the pressure force should be significantly large. Because the gradient of the sea level due to tide is usually very small, the force due to it can be ignored when the size of the structure is not very large. However, when the huge floating structure is considered, the effect of the gradient of tidal level should be considered.

Fig.8 shows the maximum values of the pressure force, frictional force, and total force due to the position of the floating airport. Because there are phase lags between the pressure force and frictional force, the simple sum of two does not give the total force. The force acting on the floating airport becomes large when it is located offshore, because the maximum value of the current velocity and the surface gradient becomes larger, as shown in Fig.9 and Fig.10.

Next, the forces due to wind-driven currents are also calculated by the same method. The calculations were carried out under the constant wind that blows in the direction from the entrance to the interior of the bay. Fig.11 shows the results under the operational condition at 60 hours after the beginning of the calculation when the currents reach almost steady state. The surface elevation and its gradient become larger at the interior part of the bay because of wind set up. Then the pressure force mainly due to the surface gradient works in the direction from the interior to the entrance of the bay and it becomes larger at the interior part of the bay, as shown in Fig.12. On the other hand, the direction of the frictional force is from the entrance to the interior because currents flow in that direction near the surface. When the floating structure is located offshore, the total force acting on the floating structure works in the direction toward the interior part of the bay because the frictional component is dominant. The magnitude of the total force becomes small at the interior part of the bay because the pressure component and frictional component cancel each other.

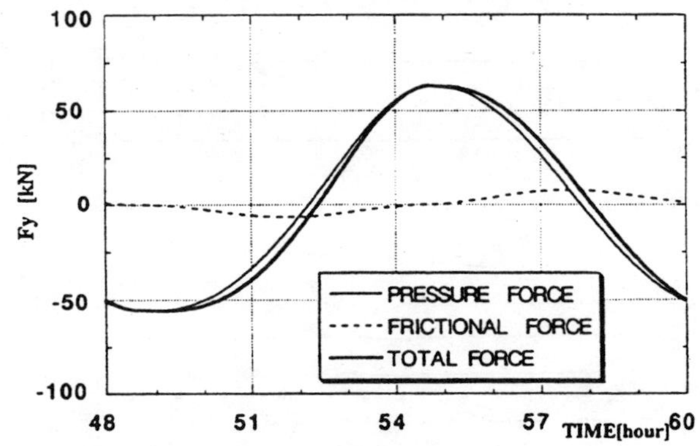

Fig. 7 Time variation of tidal current force

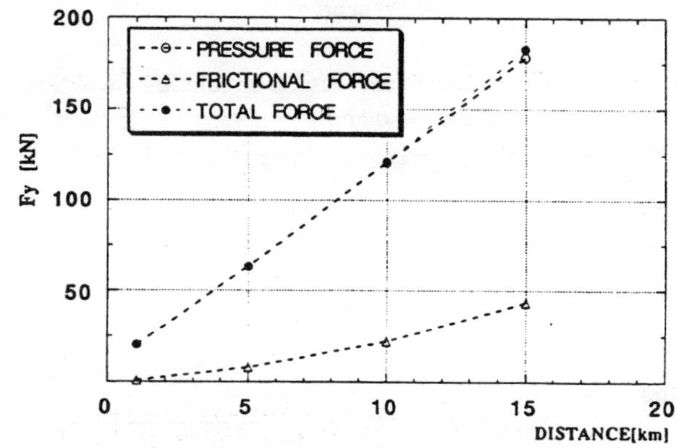

Fig. 8 Maximum values of tidal force at various locations in the bay

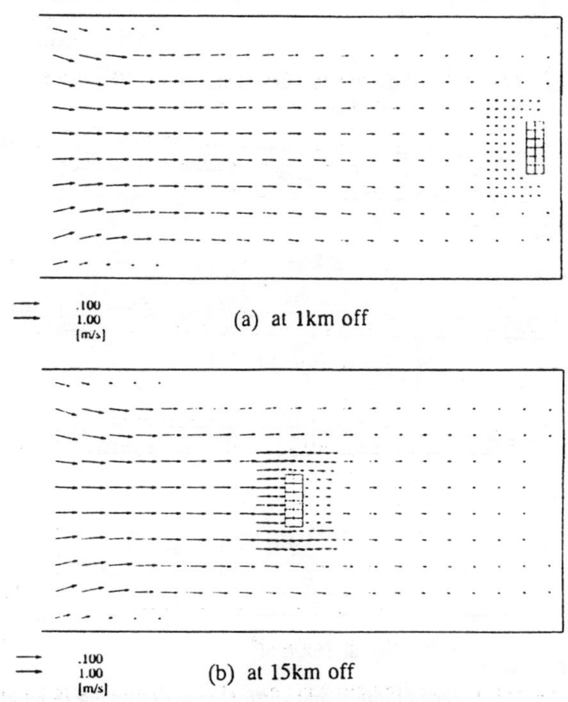

Fig. 9 The surface current velocity around the airport at t=60 hours

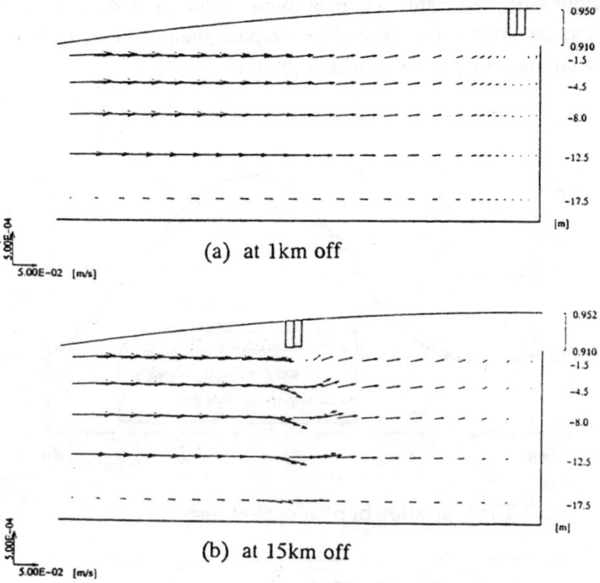

(a) at 1km off

(b) at 15km off

Fig. 10 The surface gradient and vertical circulation due to tidal wave at t=60 hours

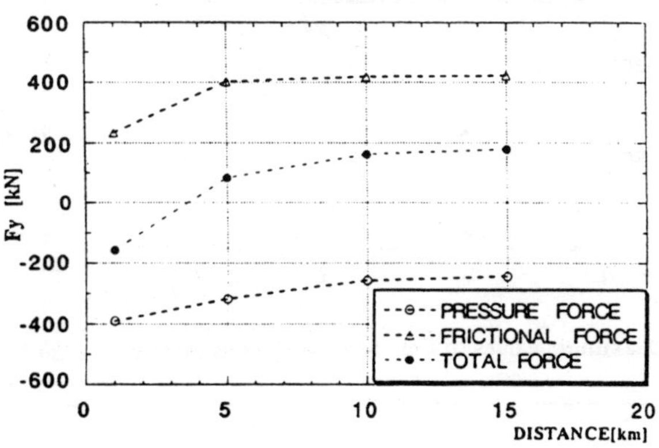

Fig. 11 Forces by wind-driven current at various locations in the bay

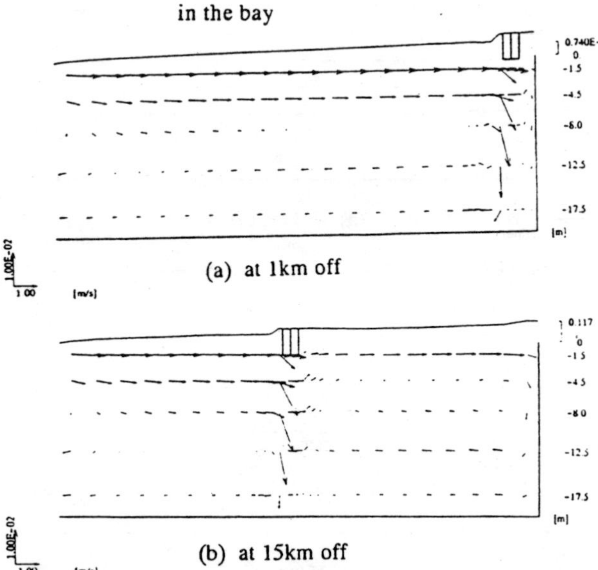

(a) at 1km off

(b) at 15km off

Fig.12 The surface gradient and vertical circulation due to wind at t=60 hours

CONCLUSIONS

In this study, the hydrodynamic forces acting on the huge floating structure and the hydroelastic responses in waves were investigated. The main conclusions obtained in this study can be summarized as follows:

1. The hydroelastic responses of the huge floating structure can be predicted by the present study when considering the hydroelastic interactions of the elastic deformations and the hydrodynamic forces. Numerical results have been compared with the results based on other method and experimental data. Fairly good agreement is found from the comparison.
2. The heave responses of the floating airport have been investigated by changing relative parameters. It is seen that the rigidity of the structure has great influence in the elastic responses of the floating airport. The bending rigidity should be decided carefully by considering many conditions.
3. The drift forces of the floating airport can decrease with the introduction of the bending rigidity, and these change greatly according to the variation of the bending rigidity.
4. The current forces acting on the huge floating structure were estimated by means of numerical simulation with multilevel model. The result indicated that the force by the gradient of tidal level ignored in the conventional estimation might be taken into account for the huge floating structure.

ACKNOWLEDGMENT

This research was supported by Grand-in-Aid for Scientific Research from the Ministry of Education, Science and Culture of Japan.

REFERENCES

Ijima, T, Tabuchi, M, and Yumura, Y (1972). "Scattering of Surface Waves and the Motions of a Rectangular Body by Waves in Finite Water Depth," *Proc. of The Japan Society of Civil Engineers*, No.202, pp. 33-48

Inoue, Y, Arai, M, Tabeta, S, Nakazawa, K, Zhang, X, and Takei, Y (1995). "Dynamic Behavior of a Floating Airport and Its Effects on Ocean Current," *Proc. of the 5th International Offshore and Polar Engineering Conference*, pp. 406-413

Nakai, K, Sekita, K, Oota, H, Utsunomiya, T, and Watanabe, E (1995). "Experiments and Analyses on Responses of Very Large Floating Structures in Waves," *Proc. of the 13th Ocean Engineering Symposium, The Society of Naval Architects of Japan*, pp. 209-216

Nojiri, N (1975). "Dynamics Characteristics of Large Floating Bodies Moored in Shallow Waters," *Ishikawajima-Harima Engineering Review*, Vol.15, No.4, pp. 485-494

Tabeta, S, and Inoue, Y (1995). "Interaction of Ocean Current and a Huge Floating Structure in Restricted Sea," *Proc. of International Conference on Technology for Marine Environment Preservation*, Tokyo, Vol.1, pp. 505-512

Proceedings of the Seventh (1997) International Offshore and Polar Engineering Conference
Honolulu, USA, May 25-30, 1997

ISBN 1-880653-28-1 (Set); ISBN 1-880653-29-X (Vol. I)

A B-Spline Galerkin Scheme for Computing Wave Forces on a Floating Very Large Elastic Plate

Masashi Kashiwagi
Research Institute for Applied Mechanics, Kyushu University
Fukuoka, Japan

ABSTRACT

An effective calculation scheme is described for computing wave forces on an elastic plate in the regime of very short wavelengths. The scheme is featured in utilizing bi-cubic B-spline functions for representing the unknowns and a Galerkin method for satisfying the body boundary condition. Obtained results of first-order wave forces are quite satisfactory in terms of good accuracy with relatively less computation time and fewer unknowns. Wave-induced hydroelastic responses are computed by a mode-expansion method and convergence of the results with increasing the number of mode shapes is confirmed. Also computed is the second-order drift force in both shallow and deep waters, with elastic-deflection effects taken into account.

KEY WORDS: Very large floating structure, Hydroelastic responses, Drift force, B-spline function, Galerkin method

INTRODUCTION

Recently in Japan, a floating airport is considered and its safety and performance in waves are intensely studied. The configuration primarily considered is of barge type, and its size will be of order of 5 km long, 1 km wide, and a few meters in depth. Therefore this kind of structure is featured in that: (1) the wavelength of practical interest is very small compared to the horizontal dimensions of the structure; and (2) hydroelastic responses are more important than the rigid-body motions due to relatively small flexural rigidity.

To take account of these features, several analytical techniques have been proposed. Among them, probably the most common one is the mode-expansion method, representing the structural deflection by a set of 'generalized' mode-shape functions. One of the issues in this method is that the wave forces corresponding to each of specified mode shapes must be accurately computed even for very short wavelengths of order of $1/50 \sim 1/100$ of the structure length.

Since the structure under consideration can be approximated by a zero-draft rectangular plate on the free surface, the analyses have been based mainly on the pressure-distribution method; e.g. Yamashita (1979), Ikoma et al. (1995), Yago (1995), and Takaki et al. (1996). However, the obtained results so far are not accurate in short wavelength regime, because the integral equation appearing in the method is discretized with a limited number of panels, and on each panel the unknown pressure is represented by a constant value. If this traditional zero-th order panel method is used for the present analysis, the number of unknowns must be $O(10^4)$ for wavelengths of interest, and thus the computation time will be far from practical.

In this paper, a new solution scheme is presented for the integral equation in the pressure-distribution method. The scheme employs bi-cubic B-spline functions for representing the unknown pressure and a Galerkin method for converting the integral equation into a linear system of simultaneouns equations. The Galerkin method is expected to increase the accuracy but, as a trade-off, is said to increase the computation time. In the present case, however, we can use 'relative similarity relations' in evaluating the influence coefficients in the matrix; which drastically reduces the computation time. Singular integrals originating from a Rankine source multiplied by B-spline functions are evaluated analytically, which also contributes to the achievement of good accuracy.

Excellent performance of the scheme is demonstrated through numerical checks of the energy-conservation principle, the Haskind relation, and convergence of the results with increasing the number of panels. Satisfactory results are obtained up to $L/\lambda = 50$ (L: structure length, λ: wavelength), with feasible computation time and the number of unknowns for routine use.

In the latter half of the paper, numerical examples of the diffraction pressure and elastic responses are shown for head and oblique waves. Furthermore, with all contributions of elastic deformation taken into account, the second-order drift force in head waves is computed up to the regime of very short wavelengths. Shallow-water effects on the pressure distribution, elastic responses, and drift force are also discussed.

MATHEMATICAL FORMULATION

Cartesian coordinates are defined with $z = 0$ as the plane of the undisturbed free surface and $z = h$ as the horizontal sea bottom. The incident regular wave comes from the negative x-axis with incident angle β.

The fluid is assumed inviscid and incompressible with irrotational motion, which permits the introduction of the velocity potential. Considering time harmonic motions, we can write the velocity potential as

$$\Phi(x,y,z,t) = \mathrm{Re}\left[\,\phi(x,y,z)\,e^{i\omega t}\,\right] \tag{1}$$

where ω is the circular frequency of the incident wave, and Re means only the real part should be taken. With this understanding, the analysis will proceed in terms of the complex quantities dependent only on spatial coordinates (x,y,z).

Assuming that the problem can be linearized, we write the velocity potential ϕ, pressure p, and vertical displacement of the structure w in the following decomposed form:

$$\phi = i\omega a\{\phi_I + \phi_S\} + \sum_j i\omega X_j\,\phi_j \tag{2}$$

$$p = \rho g\left[\,a\{p_I + p_S\} + \sum_j X_j\,p_j\,\right] \tag{3}$$

$$w = a\{\zeta_I + \zeta_S\} + \sum_j X_j\,\zeta_j \tag{4}$$

where a is the amplitude of incident wave, ρ the fluid density, and g the gravitational acceleration.

Suffix I represents quantities related to the incident wave, S the scattering component, and j the radiation component of j-th mode of motion. In the definition of mode indices, not only conventional rigid-body motions, but also a set of 'general' modes to be used for representing elastic deformations are included. X_j denotes the complex amplitude of normalized mode functions, ζ_j, which will be described later.

The structure under consideration is of rectangle in the plan view with length L and width B, and the draft is regarded as substantially zero because of very small value relative to L and B.

In the analyses to follow, the length dimensions are nondimensionalized in terms of $L/2$, and thus the structure exists in the region of $|x| \le 1$ and $|y| \le b \equiv B/L$ on $z=0$.

Hydrodynamically, the disturbance due to the presence of structure can be expressed by the pressure applied on the free surface. Then the dynamic and kinematic free-surface boundary conditions are given by

$$p_j = K\phi_j + \zeta_j\ ,\quad \frac{\partial \phi_j}{\partial z} = \zeta_j \quad \text{on } z=0 \tag{5}$$

where $K=\omega^2/g$, and $p_j = 0$ outside of the structure.

Since the velocity potential can be given by the convolution integral of the unknown pressure p_j and the Green function satisfying (5) with $p_j = 0$, it is of relative ease to show that the integral equation for the unknown pressure takes the form (Yamashita, 1977)

$$p_j(x,y) - K\iint_{S_H} p_j(\xi,\eta)G(x-\xi,y-\eta,0)\,d\xi d\eta = \zeta_j(x,y) \tag{6}$$

where S_H denotes the bottom of the structure situated on $z=0$ and $G(x,y,z)$ the Green function, which is given for the finite-depth case in the form

$$G(x,y,z) = \frac{1}{\pi}\sum_{n=1}^{\infty} C_n \frac{\cos k_n(z-h)}{\cos k_n h} K_0(k_n R) + \frac{i}{2} C_0 \frac{\cosh k_0(z-h)}{\cosh k_0 h} H_0^{(2)}(k_0 R) \tag{7}$$

where

$$\left.\begin{aligned} C_0 = \frac{k_0^2}{K + h(k_0^2 - K^2)}\ ,\ C_n = \frac{k_n^2}{K - h(k_n^2 + K^2)} \\ k_0 \tanh k_0 h = K\ ,\ k_n \tan k_n h = -K \end{aligned}\right\} \tag{8}$$

K_0 and $H_0^{(2)}$ in (7) are the 2nd kind of modified Bessel function and Hankel function, respectively, with $R = \sqrt{x^2+y^2}$.

In actual computations, the Green function is evaluated by use of the method developed by Seto (1992).

The right-hand side of (6) is the vertical displacement of the j-th mode, which will be specified as below.

In the diffraction problem ($j=S$), $\zeta_I + \zeta_S = 0$ and thus the vertical displacement due to scattering is given by

$$\begin{aligned} \zeta_S(x,y) &= -\zeta_I(x,y) \\ &= -\exp\{-ik_0(x\cos\beta + y\sin\beta)\} \end{aligned} \tag{9}$$

In the extended radiation problem ($j = 1,2,\cdots$), the vertical displacement is expressed in terms of an appropriate set of mathematical mode shapes. Here we write

$$\sum_j X_j\,\zeta_j(x,y) = \sum_{m=0}^{\infty}\sum_{n=0}^{\infty} X_{mn} u_m(x) v_n(y) \tag{10}$$

where the mode functions in the x- and y-axes, $u_m(x)$ and $v_n(y)$ respectively, are the natural modes for bending of a uniform beam with free ends. Specifically $u_m(x)$ can be written as (Newman, 1994; Wu et al., 1995)

$$\left.\begin{aligned} u_0(x) &= \frac{1}{2} \\ u_{2m}(x) &= \frac{1}{2}\left[\frac{\cos\kappa_{2m}x}{\cos\kappa_{2m}} + \frac{\cosh\kappa_{2m}x}{\cosh\kappa_{2m}}\right] \end{aligned}\right\} \tag{11}$$

$$\left.\begin{aligned} u_1(x) &= \frac{\sqrt{3}}{2}x \\ u_{2m+1}(x) &= \frac{1}{2}\left[\frac{\sin\kappa_{2m+1}x}{\sin\kappa_{2m+1}} + \frac{\sinh\kappa_{2m+1}x}{\sinh\kappa_{2m+1}}\right] \end{aligned}\right\} \tag{12}$$

where the factors κ_m denote the positive real roots of the equation

$$(-1)^m \tan\kappa_m + \tanh\kappa_m = 0 \tag{13}$$

These functions are orthogonal, satisfying

$$\int_{-1}^{1} u_m(x)u_k(x)\,dx = \frac{1}{2}\,\delta_{mk} \tag{14}$$

for all m and k, where δ_{mk} is the Kronecker delta, equal to 1 when $m=k$ and otherwise zero.

$v_n(y)$ can be also written in a similar form, with x replaced by y/b in the right-hand sides of (11) and (12).

Summing up the above, the right-hand side of (6) is given as $\zeta_j(x,y) = u_m(x)v_n(y)$. Depending on the combination of odd and even numbers of m and n, these are categorized into four types, as follows:

1) $\zeta_j = u_{2m+1}(x)v_{2n}(y)$; which is odd in x and even in y, and referred to as $F(X)$ type.
2) $\zeta_j = u_{2m}(x)v_{2n+1}(y)$; which is even in x and odd in y, and referred to as $F(Y)$ type.
3) $\zeta_j = u_{2m}(x)v_{2n}(y)$; which is even in both x and y, and referred to as $F(Z)$ type.
4) $\zeta_j = u_{2m+1}(x)v_{2n+1}(y)$; which is odd in both x and y, and referred to as $F(N)$ type.

In numerical computations, the mode numbers in these types are defined as in Table 1. Among these, only three modes correspond to the conventional rigid-body motions: u_1v_0 (pitch), u_0v_1 (roll), and u_0v_0 (heave).

Equation (9) for the diffraction problem can also be categorized into the same four types as those stated above.

Table 1 Definition of mode numbers

mode number	(m,n)	$F(X)$	$F(Y)$	$F(Z)$	$F(N)$
1	(0, 0)	u_1v_0	u_0v_1	u_0v_0	u_1v_1
2	(1, 0)	u_3v_0	u_2v_1	u_2v_0	u_3v_1
3	(0, 1)	u_1v_2	u_0v_3	u_0v_2	u_1v_3
4	(1, 1)	u_3v_2	u_2v_3	u_2v_2	u_3v_3
5	(2, 0)	u_5v_0	u_4v_1	u_4v_0	u_5v_1
6	(2, 1)	u_5v_2	u_4v_3	u_4v_2	u_5v_3
7	(0, 2)	u_1v_4	u_0v_5	u_0v_4	u_1v_5
8	(1, 2)	u_3v_4	u_2v_5	u_2v_4	u_3v_5
9	(2, 2)	u_5v_4	u_4v_5	u_4v_4	u_5v_5
⋮	⋮	⋮	⋮	⋮	⋮

Therefore, taking advantage of the fact that the pressure has the same symmetries as those of mode shapes, the unknowns in the integral equation (6) can be confined to the first quadrant ($x > 0$ and $y > 0$) of the structure.

NUMERICAL COMPUTATION METHOD

To solve the integral equation (6) with fewer unknowns even for very short wavelengths, the unknown pressure is represented by use of bi-cubic B-spline functions, in the form

$$p(x,y) = \sum_{k=0}^{NX+2} \sum_{\ell=0}^{NY+2} \alpha_{k\ell}\, B_k(x) B_\ell(y) \tag{15}$$

Here $B_k(x)$ and $B_\ell(y)$ are normalized cubic B-spline functions, which can be obtained by Boor-Cox's recursion formula (Sakurai et al., 1982).

NX and NY are the number of panel division in the x- and y-directions, respectively. Since one cubic spline function extends its influence over four panels, the number of total unknowns, $\alpha_{k\ell}$ in (15), is $(NX+3)*(NY+3)$.

To determine these unknowns with good accuracy, a Galerkin scheme is employed. That is, after substituting (15) in (6), both sides of (6) are multiplied by $B_p(x)B_q(y)$, where $p = 0 \sim NX+2$ and $q = 0 \sim NY+2$, and integrated over the plate S_H.

This procedure gives a linear system of simultaneous equations, in the form

$$\sum_{k=0}^{NX+2} \sum_{\ell=0}^{NY+2} \alpha_{k\ell} \left[\mathcal{L}^{(1)}_{pq,k\ell} + \mathcal{L}^{(2)}_{pq,k\ell} \right] = \mathcal{R}_{pq} \tag{16}$$

where $p = 0 \sim NX+2$, $q = 0 \sim NY+2$.

The influence coefficients in this matrix are defined as

$$\mathcal{L}^{(1)}_{pq,k\ell} = \iint_{S_H} B_p(x)B_q(y)\, B_k(x)B_\ell(y)\, dxdy \tag{17}$$

$$\mathcal{L}^{(2)}_{pq,k\ell} = \iint_{S_H} B_p(x)B_q(y) \Big[\iint_{S_H} B_k(\xi)B_\ell(\eta) \times G(x-\xi, y-\eta, 0)\, d\xi d\eta \Big]\, dxdy \tag{18}$$

$$\mathcal{R}_{pq} = \iint_{S_H} B_p(x)B_q(y)\, u_m(x)v_n(y)\, dxdy \tag{19}$$

Among these, (17) and (19) are relatively easy to evaluate, because these integrands are in fact simple products of independent functions of x and y, respectively. For these integrals, a Clenshaw-Curtis quadrature is used in the x- and y-directions independently, with absolute error less than 10^{-7} required.

The most time-consuming part is the integral in (18). However, when the discretization is made into panels of equal size, the amount of computations can be drastically diminished by taking advantage of 'relative similarity relations' in the integral; a more detailed explanation for that can be found in Kashiwagi (1996).

For integrals with respect to (ξ,η) and (x,y) in (18), the Gauss quadrature is applied with $6 \times 6 = 36$ points used as integration points within one panel, which is found to give very precise results.

The above-mentioned treatment is valid for the regular part of the integrand. When the field point is located near or within the integration panel, a special treatment is required for singularities due to functions of $1/R$ and $\ln R$ included in the Green function.

In the present case, since the bi-cubic B-spline function is used as a weight function in the Galerkin scheme, the following singular integrals must be evaluated:

$$\psi_{mn} \equiv \iint_{\Delta S} \frac{(\xi-x)^m(\eta-y)^n}{R}\, d\xi d\eta \tag{20}$$

$$\varphi_{mn} \equiv \iint_{\Delta S} (\xi-x)^m(\eta-y)^n \ln R\, d\xi d\eta \tag{21}$$

where m and n are integers between 0 and 3, and ΔS denotes an elementary panel of rectangle.

In this paper, these are analytically integrated, by applying the procedure described in Newman (1986) for evaluating higher-order polynomial distributions of sources.

FIRST-ORDER PRESSURE FORCE

The computation of pressure force in the extended radiation problem including elastic modes can be summarized in a similar form to the conventional rigid-body analysis, as follows:

$$\frac{F_i}{\rho g a (L/2)^2} = \sum_j \left[K\{ A'_{ij} - i B'_{ij} \} - C'_{ij} \right] \left(\frac{X_j}{a} \right) \tag{22}$$

where

$$A'_{ij} - i B'_{ij} = -\frac{1}{K} \iint_{S_H} (p_j - \zeta_j)\, \zeta_i\, dxdy \tag{23}$$

$$C'_{ij} = \iint_{S_H} \zeta_j \zeta_i\, dxdy = \frac{b}{4} \delta_{ij} \tag{24}$$

Here indices i and j can take on any values among not only rigid-body modes but also extended elastic modes. A'_{ij}, B'_{ij}, and C'_{ij} are 'generalized' added mass, damping coefficient, and hydrostatic restoring force coefficient, respectively, given here in nondimensional form. Equation (24) has been obtained by using (10) and (14).

In the diffraction problem, $p_I = 0$ on $z = 0$ and thus the wave-exciting force in the i-th direction can be computed by

$$\frac{E_i}{\rho g a (L/2)^2} \equiv E_i' = -\iint_{S_H} p_S \, \zeta_i \, dx dy \tag{25}$$

Numerical accuracy of above computations can be confirmed by checking various hydrodynamic relations which are theoretically proven. In this paper two different relations are considered: one is the energy-conservation principle associated with the damping force, and the other is the Haskind relation between the radiation and diffraction problems.

Omitting derivations, final results can be expressed in the form

$$B_{jj}' = \frac{1}{4\pi} C_0 \int_0^{2\pi} | H_j(k_0, \theta) |^2 d\theta \tag{26}$$

$$E_j' = H_j(k_0, \beta + \pi) \tag{27}$$

where

$$H_j(k_0, \theta) = \iint_{S_H} p_j(\xi, \eta) \, e^{ik_0(\xi \cos\theta + \eta \sin\theta)} \, d\xi d\eta \tag{28}$$

is defined as the Kochin function.

SECOND-ORDER DRIFT FORCE

It is known that the steady drift force can be computed only from quadratic terms of first-order quantities. In the present case, the x-component of the contribution from the bottom of the plate can be evaluated from

$$F_D^{(1)} = \frac{1}{4} \iint_{S_H} \{ p(x,y) \frac{\partial}{\partial x} \overline{w(x,y)} + \overline{p(x,y)} \frac{\partial}{\partial x} w(x,y) \} \, dx dy \tag{29}$$

where the overbar means the complex conjugate.

Substituting (4) and (6) for $w(x,y)$ and (3) for $p(x,y)$ into (29) and transforming further with the energy-conservation principle, we can obtain the expression

$$\frac{F_D^{(1)}}{\rho g a^2 (L/2)} = \frac{1}{4} \int_{-b}^{b} \Big[| p(x,y) |^2 \Big]_{-1}^{1} dy + \frac{K}{8\pi} k_0 C_0 \int_0^{2\pi} |A(k_0,\theta)|^2 \Big(\cos\beta - \cos\theta \Big) \, d\theta \tag{30}$$

where $\quad A(k_0,\theta) = H_S(k_0,\theta) + \sum_j \Big(\frac{X_j}{a}\Big) H_j(k_0,\theta) \quad (31)$

is the wave amplitude function comprised of all components of the Kochin function, and $[\quad]_{-1}^{1}$ in the first term on the right-hand side of (30) denotes the difference between the values at $x = 1$ and -1.

Note that $F_D^{(1)} = 0$ in the diffraction problem, because $w(x,y) = 0$ is satisfied as the boundary condition in (9). (What is happening in fact is that the first and second terms on the right-hand side of (30) cancel out each other.)

Since this is not the case, there must be a missing term; the pressure integral between $z = 0$ and actual wave elevation on the vertical walls at $x = \pm 1$ is also of second order. Let this contribution be denoted by $F_D^{(2)}$. Then it can be shown that $F_D^{(2)}$ is the same in form and opposite in sign to the first term on the right-hand side of (30).

In summary, the wave drift force given by $F_D^{(1)}$ plus $F_D^{(2)}$ takes the form

$$\frac{F_D}{\rho g a^2 (L/2)} = \frac{K}{8\pi} k_0 C_0 \int_0^{2\pi} |A(k_0,\theta)|^2 \Big(\cos\beta - \cos\theta \Big) \, d\theta \tag{32}$$

which is consistent with the result derived from Maruo's theory (Maruo, 1960).

ELASTIC RESPONSES IN WAVES

To compute $A(k_0,\theta)$ from (31), the complex amplitude of j-th mode shape, X_j/a, must be known; which will be given by solving the vibration equation of an elastic plate. An outline of its solution method is briefly described below.

Firstly, the vibration equation of an elastic plate is given as

$$-m\omega^2 w(x,y) + D\Big\{ \frac{\partial^4}{\partial x^4} + 2\frac{\partial^4}{\partial x^2 \partial y^2} + \frac{\partial^4}{\partial y^4} \Big\} w(x,y) = -p(x,y) \tag{33}$$

where m is the distribution of mass, equal to M/LB in the case of uniform distribution (M being the total mass), and D is the flexural rigidity.

Substituting (3) and (4) into (33) and writing the result in nondimensional form, we obtain

$$-K M' \Lambda \sum_j \Big(\frac{X_j}{a}\Big) \zeta_j(x,y) + D' \sum_j \Big(\frac{X_j}{a}\Big) \Big\{ \frac{\partial^4}{\partial x^4} + 2\frac{\partial^4}{\partial x^2 \partial y^2} + \frac{\partial^4}{\partial y^4} \Big\} \zeta_j(x,y) = -p_S(x,y) - \sum_j \Big(\frac{X_j}{a}\Big) p_j(x,y) \tag{34}$$

where $M' = M/\rho LBd$, $\Lambda = 2d/L$, $D' = D/\rho g(L/2)^4$, with d being the draft of structure. In obtaining (34), the relations of $p_I = 0$ and $\zeta_I + \zeta_S = 0$ on $z = 0$ have been used.

Multiplying (34) by $\zeta_i(x,y) = u_k(x) v_\ell(y)$ (the relation of mode indices among k, ℓ, and i is the same as in Table 1) and integrating over the bottom of the structure, we obtain a linear set of motion equations

$$\sum_j \Big(\frac{X_j}{a}\Big) \Big[-K\Big(M_{ij} + A_{ij}' - i B_{ij}' \Big) + C_{ij}' + D_{ij} \Big] = E_i' \tag{35}$$

where

$$M_{ij} = M'\Lambda \iint_{S_H} \zeta_i \, \zeta_j \, dx dy = M'\Lambda \frac{b}{4} \delta_{ij} \tag{36}$$

$$D_{ij} = D' \iint_{S_H} \zeta_i \Big(\frac{\partial^4}{\partial x^4} + 2\frac{\partial^4}{\partial x^2 \partial y^2} + \frac{\partial^4}{\partial y^4} \Big) \zeta_j \, dx dy \tag{37}$$

Here M_{ij} is called the mass matrix and D_{ij} the stiffness matrix which is the restoring force coefficient due to structural rigidity. A_{ij}', B_{ij}', C_{ij}', and E_i' in (35) are defined in (23)~(25).

One important issue in this mode-expansion method is the free-end boundary conditions along the periphery of the structure. In fact, each of the adopted mode functions, $\zeta_j(x,y) = u_m(x) v_n(y)$, does not satisfy the appropriate free-end conditions. However, as suggested in Newman (1994), the end conditions can be taken into account in transforming the stiffness matrix by partial integrations.

It can be shown that the integrated terms appearing after integrating (37) twice by parts must be zero by the end conditions and the final result is of the form

$$D_{ij} = D' \iint_{S_H} \Big\{ \frac{\partial^2 \zeta_i}{\partial x^2} \frac{\partial^2 \zeta_j}{\partial x^2} + \frac{\partial^2 \zeta_i}{\partial x^2} \frac{\partial^2 \zeta_j}{\partial y^2} + \frac{\partial^2 \zeta_i}{\partial y^2} \frac{\partial^2 \zeta_j}{\partial x^2} + \frac{\partial^2 \zeta_i}{\partial y^2} \frac{\partial^2 \zeta_j}{\partial y^2} \Big\} \, dx dy \tag{38}$$

These integrals can be evaluated analytically, by utilizing orthogonality relation (14) and the 4-th order differential equation satisfied by elementary mode functions ζ_i and ζ_j.

RESULTS AND DISCUSSIONS

Numerical accuracy and computation time

Computations were performed for a rectangular plate of $L/B = 5$. The discretization of the plate into panels is made such that the first quadrant ($x > 0$ and $y > 0$) is subdivided into NX in the x-axis and NY in the y-axis with ratio $NX/NY = 5$, meaning that each panel is a square.

Convergence of the results with increasing the number of panels is shown in Figs. 1 and 2, where Fig. 1 shows the added mass in heave mode ($\zeta_j = u_0(x)v_0(y)$ in Table 1) and Fig. 2 the damping coefficient in the same heave mode. The water depth is taken as infinity in this case, and the abscissa is the length ratio L/λ

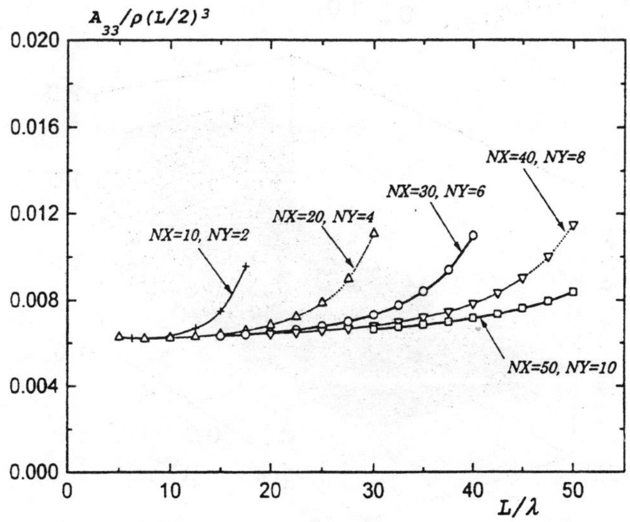

Fig. 1 Convergence of heave added mass for the plate of $L/B = 5$, in deep water

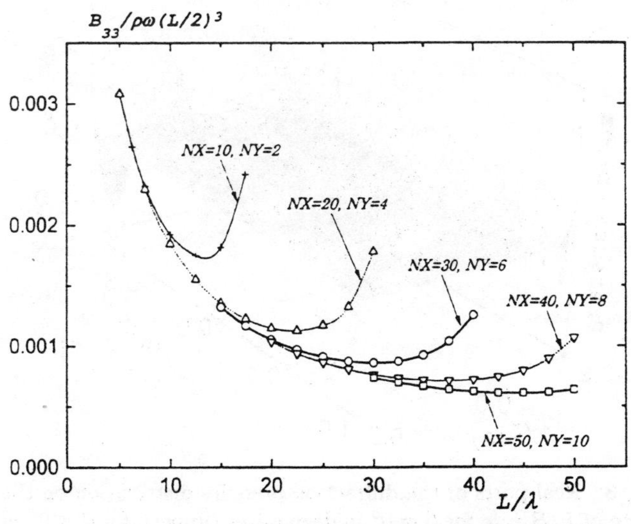

Fig. 2 Convergence of heave damping coefficient for the plate of $L/B = 5$, in deep water

Table 2 Relative numerical error in the energy conservation principle, listed in percentage ($NX = 40, NY = 8, L/\lambda = 45, h = \infty$)

MODE	BX(I,I)	BY(I,I)	BZ(I,I)	BN(I,I)
1	.2227E-01	.1363E-01	.2505E-01	.1330E-01
2	.2189E-01	.1239E-01	.1886E-01	.1301E-01
3	.2784E-02	.1170E-02	.1010E-01	.4330E-02
4	.2280E-02	.4132E-02	.3198E-02	.4178E-02
5	.2068E-01	.1342E-01	.2298E-01	.1263E-01
6	.2018E-02	.4629E-02	.1112E-02	.3293E-02
7	.5100E-02	.4719E-03	.6977E-03	.3421E-02
8	.5151E-02	.4239E-02	.5964E-02	.3460E-02
9	.4536E-02	.2332E-02	.4328E-02	.2234E-02
10	.1985E-01	.1177E-01	.1827E-01	.1227E-01
11	.9950E-03	.2454E-02	.2263E-02	.2821E-02
12	.3865E-02	.2049E-02	.4611E-02	.1258E-02
13	.3102E-02	.2718E-02	.9917E-02	.1117E-02
14	.2565E-02	.1865E-02	.2021E-02	.1467E-02
15	.4295E-02	.8927E-03	.3059E-02	.3270E-03
16	.5915E-02	.1047E-03	.5317E-02	.6188E-03
17	.1833E-01	.1245E-01	.2014E-01	.1180E-01
18	.1058E-03	.3354E-02	.3817E-04	.2941E-02
19	.3582E-02	.2958E-03	.3300E-02	.1900E-03
20	.7494E-02	.8798E-03	.6425E-02	.1233E-02
21	.2638E-02	.3325E-02	.8399E-04	.4133E-02
22	.2698E-02	.3287E-02	.2854E-02	.3404E-02
23	.1430E-02	.3313E-02	.2055E-02	.2427E-02
24	.7174E-03	.2087E-02	.1323E-02	.1779E-02
25	.2679E-03	.1618E-02	.4669E-03	.1239E-02

between structure length L and wavelength λ.

It can be clearly seen that the results approach smoothly converged values as the number of panels increases. Rough estimate from these results indicates that reliable results may be obtained, provided $L/\lambda < 0.8 * NX$ is satisfied.

Considering that the computed results are very smooth, there seem to be no phenomena of irregular frequencies at least within the computed range of L/λ.

Another thing to be noted is that all the results plotted in Figs. 1 and 2 satisfy quite accurately the energy-conservation principle and the Haskind relation, even when the results are a little away from converged values.

One typical example for that is shown in Table 2, which shows the relative error in the energy-conservation principle, checked for $L/\lambda = 45$ in deep water, with $NX = 40$ and $NY = 8$. Surprisingly the error is within 0.03% for all specified mode shapes. Excellent agreement of this order was also found in the Haskind relation and for other numbers of panels.

This fact implies that the numerical check based only on the energy-conservation principle and/or the Haskind relation may be not sufficient to ensure the accuracy of numerical results.

Table 3 Computation time vs. numbers of panels

(NX, NY)	Number of unknowns	Average CPU time*
(10, 2)	65	20 *sec*
(20, 4)	161	4 *min* 15 *sec*
(30, 6)	297	20 *min* 05 *sec*
(40, 8)	473	62 *min* 10 *sec*
(50, 10)	689	153 *min* 30 *sec*

* *by EWS HP 9000 series/model 735*

Not only the accuracy, but also the computation time and the number of unknowns are of great interest. All computations in this paper have been implemented using an engineering workstation of HP 9000 series / model 735, and the developed computer program is not fully optimized. Nevertheless, as shown in Table 3, the average computation time per one wavelength, including numerical checks discussed above, was found comparatively little.

Elastic motions and drift force in waves

In the preceding subsection, it was elucidated that the converged results could be obtained by use of panels satisfying $L/\lambda < 0.8 * NX$. With enough panels to meet this requirement, wave-induced motions of the structure and the wave drift force were computed.

As an illustration, computations have been performed with the data shown in Table 4, which were selected as a model of floating airport under consideration.

Table 4 Numerical data for computations of elastic motions

$L \times B \times d = 5000\ (m) \times 1000\ (m) \times 5\ (m)$
$h = 50\ (m)$ or *infinity*
$M' = M/\rho LBd = 1.0$
$D' = D/\rho g(L/2)^4 = 0.6153 \times 10^{-5}$

The number of mode functions defined in Table 1 was taken equal to 256 at maximum for each of four different types of motion pattern (1024 modes in total), which may be sufficient to obtain fully converged results.

Some examples are shown in Fig. 3 for the diffraction pressure and in Fig. 4 for the resultant elastic deformation. These are computed at the frequency corresponding to $L/\lambda_\infty = 20$ and show only the real part (that is, a picture taken at $t = 0$). In both Fig. 3 and Fig. 4, the upper one is the result for head wave in deep water, the middle one is for head wave in shallow water of $h = 50$ m, and the lower one is for oblique wave of $\beta = 30°$ in deep water.

We can see that the diffraction pressure in head wave is almost zero inside and lee side of the plate and sharply increases near the weather side with small oscillations along the rim of both sides parallel to the direction of wave propagation.

On the other hand, the elastic deformation is large even inside the plate, especially in the fore half near the weather side, which is markedly different from the distribution of diffraction pressure. The dominant wavelength of elastic deformation is much longer than that of incident wave. In fact, as explained by Ohkusu and Nanba (1996), the wavelength of elastic deformation is determined by the dispersion relation modified due to the flexural rigidity of a structure located on $z = 0$.

In shallow water, the wavelength of incoming wave becomes short relative to the infinite-depth case, which results in the shorter wavelength of the pressure oscillation. Another shallow-water effect is to constrain the flow horizontally under the plate, increasing the magnitude of the pressure and the depth of penetration of dominant peak pressure.

Due to these effects on the pressure, the typical wavelength of elastic deformation in shallow water is somewhat shorter than that in deep water (that is, the slope of elastic deformation becomes steeper).

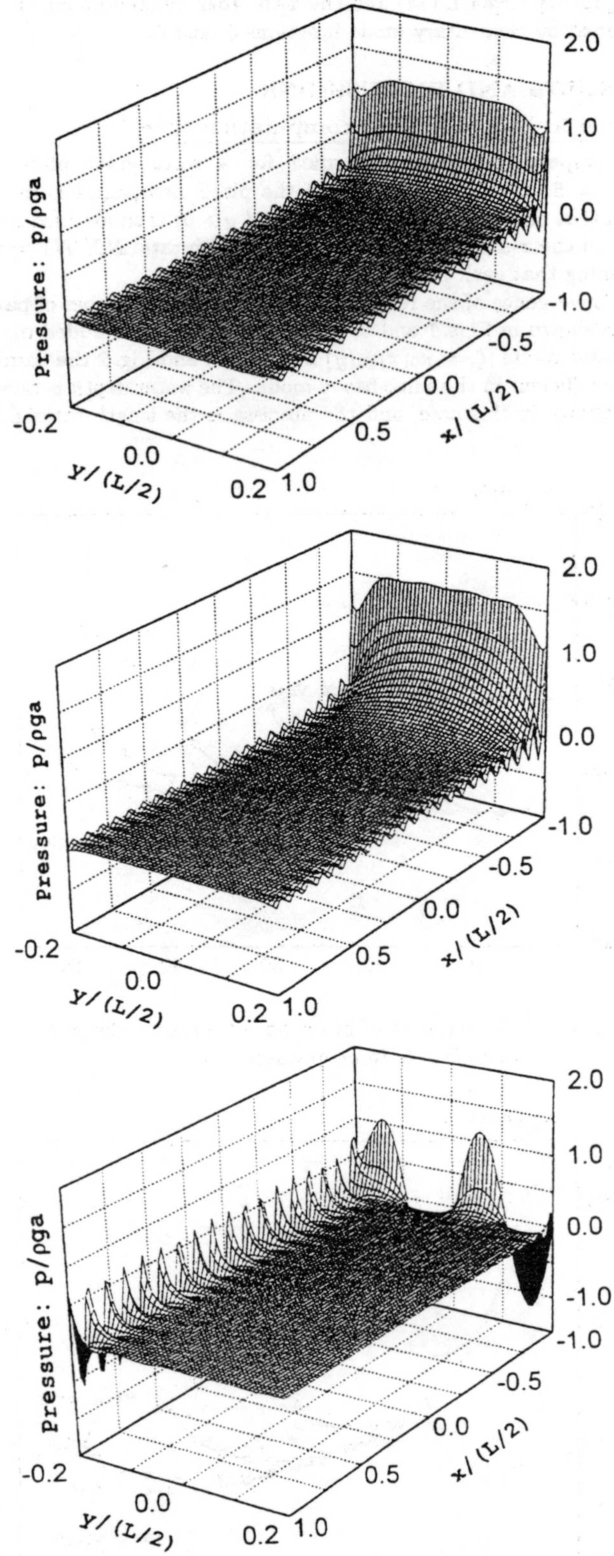

Fig. 3: Real parts of the diffraction pressure distribution on the plate of $L/B = 5$; for $\beta = 0°$ in deep water (upper), for $\beta = 0°$ in shallow water of $h/\lambda_\infty = 1/5$ (middle), and for $\beta = 30°$ in deep water (lower). $L/\lambda_\infty = 20$ in all cases.

In oblique wave, the oscillation pattern of both the diffraction pressure and elastic deformation is considerably different from the head-wave case. The pressure is oscillatory along the upwave side, in phase with the incident wave, and negligible elsewhere. On the other hand, the elastic deformation is small except in the bow part (upwave end) of the plate.

Figure 5 shows the computed drift force in deep water for both cases of motion fixed (diffraction problem) and motion free (including all contributions of elastic mode shapes). The results are plotted in nondimensional form of $F_D/\rho g a^2(B/2)$; that is, the denominator is the incident-wave energy per breadth of the plate, and therefore 1.0 of the ordinate corresponds to the perfect reflection of the incident wave from the bow part.

We can see that the results converge with increasing the number of panels and, in the diffraction problem, they approach the limiting value of perfect reflection with decreasing the wavelength. In the case of motion free, the drift force becomes small

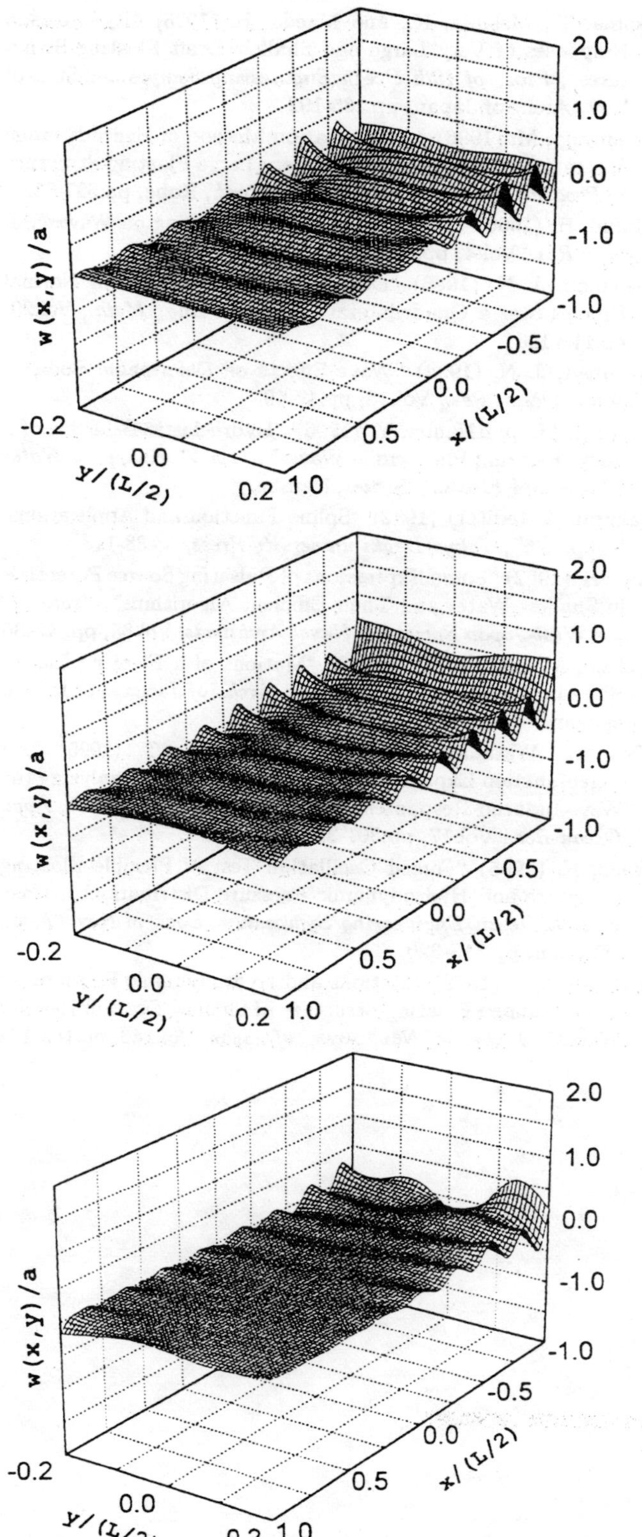

Fig. 4: Real parts of the elastic deformation of the plate of $L/B = 5$; for $\beta = 0°$ in deep water (upper), for $\beta = 0°$ in shallow water of $h/\lambda_\infty = 1/5$ (middle), and for $\beta = 30°$ in deep water (lower). $L/\lambda_\infty = 20$ in all cases.

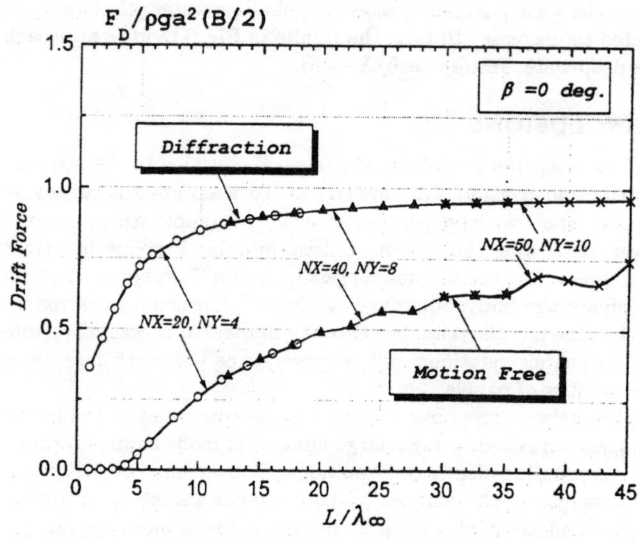

Fig. 5 Wave drift force acting on the plate of $L/B = 5$ in deep water ($\beta = 0°$)

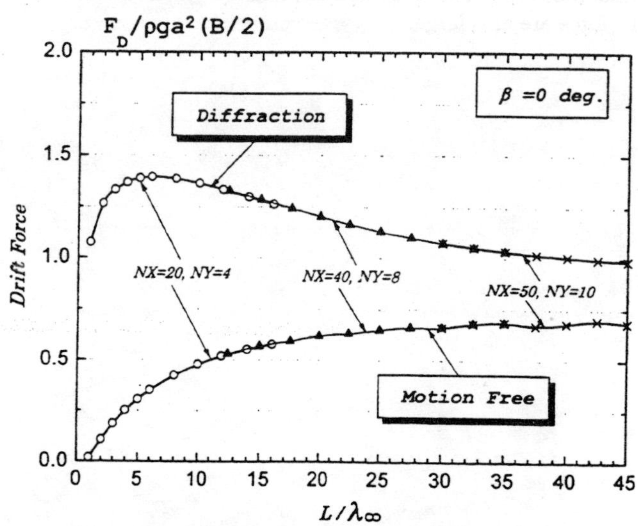

Fig. 6 Wave drift force acting on the plate of $L/B = 5$ in shallow water of $h = 50$ m ($\beta = 0°$)

compared to the case of motion fixed. This result makes us envisage that a part of the incident wave does not reflect but propagates beneath the elastic plate.

As the wavelength decreases, small fluctuation can be observed, which is hopefully physically relevant phenomenon associated with interactions between complicated elastic motions and ambient waves.

Corresponding results in the shallow-water case are shown in Fig. 6. The water depth is constant and equal to 50 m for all computed frequencies, and therefore the effects of shallow water are more prominent in the longer wavelength region.

In the diffraction problem, the nondimensional value of the drift force is larger than 1.0, implying that the flow is strongly constrained horizontally and the extent of wave reflection is greater than the breadth of a structure. In the case of elastic motions allowed, the drift force becomes small due to the penetration of incident wave beneath the elastic plate, but still larger than the corresponding value in deep water.

As the wavelength decreases, the shallow-water effects are expected to decrease. In fact, the results in Fig. 6 tend to approach the deep-water results as $L/\lambda \rightarrow \infty$.

CONCLUSIONS

A new scheme for solving the integral equation in the pressure distribution method was described. To keep good accuracy in the very short wavelength regime with less computation time and fewer unknowns, the scheme utilizes bi-cubic B-spline functions for representing the unknown pressure and a Galerkin method for satisfying the body boundary condition. Excellent performance of the scheme was validated through numerical checks of various hydrodynamic relations and a convergence test with increasing the number of panels.

Hydroelastic responses of a plate were computed by the mode-expansion method, using a large number of mode-shape functions for representing elastic deformations. Convergence of the results with increasing the number of mode shapes was also confirmed.

With effects of these elastic responses taken into account, the wave drift force was computed for very short wavelengths up to $L/\lambda = 45$. It was shown that the flexibility of a plate diminishes considerably the drift force and the shallow-water effects on the drift force are also large.

REFERENCES

Ikoma, T., Masuda, K., and Maeda, H. (1995) "Hydroelastic Responses of Very Large and Shallow Draft Floating Structures", *Proc. of 13th Ocean Engineering Symposium*, Soc. of Nav. Arch. of Japan, pp.185-192

Kashiwagi, M. (1996) "A Calculation Method of Hydrodynamic Forces on a Shallow-Draft and Very Large Floating Structure ", *Proc. of Techno-Ocean '96 Int. Conf.*, Kobe, pp.627-632

Maruo, H. (1960) "The Drift of a Body Floating on Waves", *J. Ship Res.*, Vol.4, pp.1-10

Newman, J. N. (1986) "Distributions of Sources and Normal Dipoles over a Quadrilateral Panel", *J. Eng. Math.*, Vol.20, pp.113-126

Newman, J. N. (1994) "Wave Effects on Deformable Bodies", *Appl. Ocean Res.*, Vol.16, pp.47-59

Ohkusu, M. and Nanba, Y. (1996) "Hydroelastic Behavior of a Large Floating Platform in Waves", *11th Workshop on Water Waves and Floating Bodies*, Hamburg

Sakurai, A. (editor) (1982) "Spline Function and Applications: Chapter 6", *Tokyo Denki university Press*, pp.88-118

Seto, H. (1992) "Some Expressions of Pulsating Source Potentials in Shallow Water and Their Efficient Algorithms", *Trans. of the West-Japan Society of Naval Architects*, Vol.85, pp. 25-36

Takaki, M. and Gu, X. (1996) "Motions of a Floating Elastic Plate in Waves", *J. Soc. of Nav. Arch. of Japan*, Vol.180, to appear

Wu, C., Watanabe, E., and Utsunomiya, T. (1995) "An Eigenfunction Expansion-Matching Method for Analyzing the Wave-Induced Responses of an Elastic Floating Plate", *Appl. Ocean Res.*, Vol.17, pp.301-310

Yago, K. (1995) "Forced Oscillation Test of Flexible Floating Structure and Hydrodynamic Pressure Distributions", *Proc. of 13th Ocean Engineering Symposium*, Soc. of Nav. Arch. of Japan, pp.313-320

Yamashita, S. (1979) "Motions and Hydrodynamic Pressures of a Box-Shaped Floating Structure of Shallow Draft in Regular Waves", *J. Soc. of Nav. Arch. of Japan*, Vol.146, pp.165-172

Proceedings of the Seventh (1997) International Offshore and Polar Engineering Conference
Honolulu, USA, May 25-30, 1997

ISBN 1-880653-28-1 (Set); ISBN 1-880653-29-X (Vol. I)

Three-Dimensional Slamming of a Distorted Plate

Ken Takagi
Osaka University
Osaka, Japan

ABSTRACT

A theoretical and numerical approach to the distorted plate penetrating the calm water surface is investigated as a flow model of a blunt body slamming in rough seas and the effect of air trapping is also investigated in this paper. This theory is based on the variational formulation in which the displacement potential is obtained instead of the velocity potential.

Numerical results show that the maximum impact force becomes smaller and smaller as the steepness of distortion increases, and the duration of the impact becomes longer as the steepness increases. In order to clarify the reduction effect of the roughness of water surface, a diagram of the maximum impact force with variation of the deadrise angle of the wedge is depicted. This diagram shows that the reduction effect seems significant in the region of small deadrise angle. Other numerical results show that the trapped air makes the impact force stronger and the reduction of the maximum impact force is mainly brought about by the three-dimensional effect of the roughness of water surface when the water surface is rough enough to guarantee the formation of a cavity in advance. It is considered that the effect of the trapped air becomes dominant for the reduction of the maximum impact force as the steepness of surface waves decrease.

Summarizing the numerical results, a new scenario is implied in that the reduction effect of the maximum impact force acting on a blunt body is mainly brought about by the three-dimensional effect of the roughness of water surface when the body falls into rough seas, and when the water surface is calm the effect of the trapped air becomes dominant as indicated in previous experimental works.

KEY WORDS : Three-Dimensional Slamming, Air Trapping, Variational Formulation

INTRODUCTION

'Slamming' has been widely studied in the literature through the last decades. Most studies treated slamming as a two-dimensional problem since the slender body theory for the ship motion insists that the slamming problem is solved as a part of the inner problem which is usually two-dimensional.

The three-dimensional slamming theory studied first by Chuang (1969). He investigated the slamming of cone-shaped bodies. Watanabe (1986) studied the pressure distribution on the inclined flat rectangular bottom of a ship penetrating the water surface with the three-dimensional flat plate formulation. Recently, Toyama (1996) gave a theoretical approach to the slamming problem of three-dimensional bodies. However, their research object is limited to the simple shaped-body penetrating the calm water surface.

When a large ship experiences slamming, the condition of sea is rough. High swells induce large ship motions and, as a result, slamming occurs at the bow of the ship. In these cases, the surface of the sea water is not smooth, since the displacement of the water surface is described by the summation of many component waves which include very short waves and have different directions of propagation. Therefore, the three-dimensional effect due to short waves is expected and is investigated in this study.

The theoretical approaches to the slamming problem are based on Wagner's theory (Wagner, 1932) which was improved by Armand and Cointe (1986) and Watanabe (1986) using the matched asymptotic expansion method. According to their theory, a linearized slamming formulation is obtained as the first order outer problem. In the flame work of the linearized slamming formulation, the deformation of water surface is identical with the distortion of the bottom plate of a ship. Therefore, in this paper, a theoretical and numerical

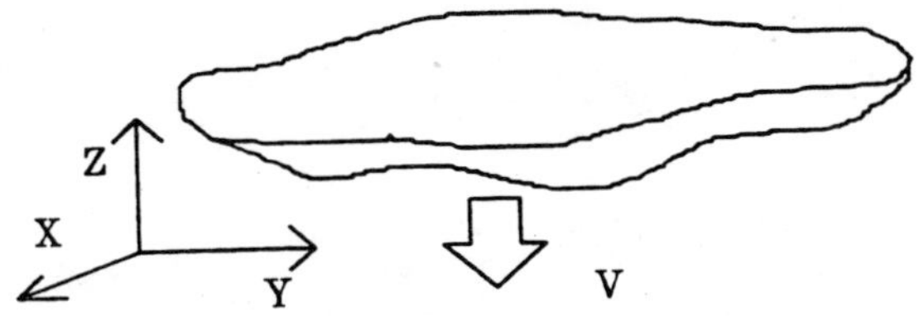

Fig. 1 Coordinate system.

approach to a distorted plate penetrating calm water surface is investigated as a flow model of the above mentioned slammings in rough seas.

When a body bottom of which has very small deadrise angle approaches the water surface, the air which is pushed ahead of the body cannot escape completely from the gap between the bottom and the water surface and, as a result, a cavity filled with the trapped air may be formed. However, it is offen the case that the process of the cavity formation is unstable and depends very much on small variation of parameters such as attack angle, fine geometry of the bottom near the edge and so on. Korobkin (1996) considered the falling of a two-dimensional blunt body with a shallow depression at its top to the free surface of an ideal incompressible liquid. In this case that the cavity formation is guaranteed in advance. Similarly, the falling of the distorted plate to the water surface guarantees the formation of the cavity, however it is more difficult to judge the air trapping of the three-dimensional flow than that of the two-dimensional case. Therefore, the effect of the air trapping is also investigated in this paper.

THEORY

The theory is based on the matched asymptotic expansion method and the perturbation parameter is preferred to be the ratio between the vertical scale of the distortion of the plate and the horizontal scale of it. The problem is composed of the outer problem and the inner problem. The outer problem is obtained by employing the following nondimensionalization for the space coordinate and the time.

$$(x, y, z) = \left(\frac{X}{L}, \frac{Y}{L}, \frac{Z}{L}\right) \quad , \quad t = \frac{TV}{H}. \tag{1}$$

Where, L is a typical horizontal scale of the distortion of the plate, H is a typical vertical scale of the distortion of the plate and V is the drop speed of the plate. Fig. 1 shows the definition of the coordinate system.

The inner problem is defined at the singular point of the outer problem and that point appears at the intersection between the surface of the plate and the free surface. However, since Armand and Cointe (1986) pointed out that the contribution of the inner solution to the impact force is a higher order and is negligible, only the outer problem is discussed in the present paper.

It is assumed that the fluid motion is inviscid, irrotational and incompressible. The velocity potential Φ exists and it is convenient to employ the following nondimensionalization for the manipulation of equations appearing in the outer problem.

$$\phi = \frac{\Phi}{LV} = O(1). \tag{2}$$

The flow is characterized by the following two parameters.

$$\frac{H}{L} = \varepsilon \ll 1 \quad , \quad F_n = \frac{V}{\sqrt{gL}} = O(1) \tag{3}$$

Where, ε is the perturbation parameter as stated previously and F_n is Froude Number of the flow.

Without cavity case

When the plate touches the water surface, the intersection between the plate surface and the water surface seems as a line on the plane of $z = 0$. If this line makes a closed loop, the air would be trapped between the plate and the water, and a cavity would be formed. On the contrary, when the loop is not closed the cavity is not formed under the plate. This is the case treated in this section.

It is assumed that the air pressure on the free surface is zero, however it is expected that the air pressure in the loop is higher than that in the outside because of the viscous effect of the air flow at the opening of the loop when the opening of the loop is very small.

The governing equation of the fluid is Laplace's equation from the basic assumptions mentioned above. The first order approximation of the free surface condition and the plate surface condition are represented respectively as follows:

$$\nabla^2 \phi = 0 \qquad for \qquad z < 0, \tag{4}$$

$$\frac{\partial \phi}{\partial z} = -1 \qquad for \qquad z = 0, \quad (x, y) \in \Omega_B, \tag{5}$$

$$\frac{\partial \phi}{\partial t} = 0 \qquad for \qquad z = 0, \quad (x, y) \in \Omega_F, \tag{6}$$

$$\frac{\partial \eta}{\partial t} - \frac{\partial \phi}{\partial z} = 0 \qquad for \qquad z = 0, \quad (x, y) \in \Omega_F. \tag{7}$$

Where, Ω_B denotes the projection of the wetted surface of the plate on the plane $z = 0$, Ω_F denotes the projection of the free surface on the plane $z = 0$ and η is the nondimensional free surface displacement.

The boundary value problem defined by (4) $\sim$ (7) is reduced to an integral equation by applying the Green's theorem. However, it is difficult to obtain an accurate solution of the integral equation, since the gradient of the velocity potential at the intersection between the area Ω_B and the area Ω_F is unbounded.

Howison et al. (1991) presented a variational formulation which is originally suggested by Korobkin (1982) and indicated that the boundary value problem is easily solved by this algorithm. According to the spirit of the so-called

'Baiocchi transformation' in the theory of variational inequalities, a displacement potential is defined as follows:

$$\varphi(x,y,z,t) = \int_0^t \phi(x,y,z,\tau)d\tau. \tag{8}$$

Substituting this into (4) $\sim$ (7), it is found that the governing equation of the displacement potential is also Laplace's equation and the dynamic free surface condition (6) becomes as follows:

$$\varphi(x,y,0,t) = 0 \qquad for \qquad t < \gamma(x,y). \tag{9}$$

Where, $t = \gamma(x,y)$ denotes the shape of Ω_B. Similarly, the kinematic free surface condition (7) reduced to the following form.

$$\frac{\partial\varphi}{\partial z} = \eta \qquad for \qquad t < \gamma. \tag{10}$$

Combining the kinematic free surface condition (7) with the body surface condition (5), the following boundary condition is obtained.

$$\frac{\partial\varphi}{\partial z} = b(x,y) - t \qquad for \qquad t > \gamma. \tag{11}$$

Where, $b(x,y) = B(X,Y)/H$ is the nondimensional shape of the plate and the dimensional shape of the plate at $T = 0$ is represented as $Z = B(X,Y)$.

The great advantage of working with the displacement potential instead of the velocity potential is that the gradient of the displacement is bounded on $t = \gamma$ and so the displacement potential satisfies the 'complementarity problem'.

$$\nabla^2\varphi = 0 \qquad z < 0 \tag{12}$$

$$\varphi\left(\frac{\partial\varphi}{\partial z} + t - b\right) = 0 \qquad on \quad z = 0 \tag{13}$$

$$\text{with} \quad \varphi \leq 0 \;, \quad \frac{\partial\varphi}{\partial z} + t - b \geq 0 \qquad on \quad z = 0 \tag{14}$$

This provides both a mathematical basis for the present formulation and a minimization algorithm for numerical calculations. In concrete term, the (14) is numerically solved by the SOR method, and this will be explained later.

The first order impact pressure is obtained by differentiating the velocity potential respect to the time. Therefore, the integration of the displacement potential on the body surface gives the double time integral of the impact force.

$$F(t) = \int_{\Omega_B} \varphi(x,y,0,t)ds. \tag{15}$$

The impact force is given as a second-order differentiation of $F(t)$.

$$f = \frac{d^2F}{dt^2}. \tag{16}$$

With cavity case

When the intersection loop mentioned above is closed, a cavity is formed between the plate surface and the water surface. First of all, the nondimensionalization of the pressure is defined.

$$p = \frac{P}{\rho_w V^2}. \tag{17}$$

Where, ρ_w is the density of the water.

It is assumed that the pressure inside of the cavity starts increasing just after the loop is closed. Since the transformation of the air is adiabatic, the pressure inside of the cavity is represented as follows:

$$p(t) = \beta\left(\frac{1}{\nu(t)^n} - 1\right). \tag{18}$$

Where, n is the ratio of the two specific heats of the gas and its value is 1.4 for the air, $\nu(t) = N(t)/N_0$ is the nondimensional cavity volume, $N(t)$ is the instantaneous cavity volume and N_0 is the cavity volume on the time just after the cavity is formed. The parameter β represents the scale effect of the air pressure and is defined by

$$\beta = \frac{1}{n}\frac{\rho_a}{\rho_w}\left(\frac{c_0}{V}\right)^2\frac{H}{L}. \tag{19}$$

Where, ρ_a is the density of the air and c_0 is the speed of sound.

Applying this pressure, the dynamic free surface condition in the cavity is given.

$$\frac{\partial\phi}{\partial t} = -\beta\left(\frac{1}{\nu(t)^n} - 1\right) \qquad for \qquad z = 0 \quad (x,y) \in \Omega_C. \tag{20}$$

Where, Ω_C denotes the projection of the cavity on the plane $z = 0$. Integrating both R.H.S. and L.H.S. of (20) and combining with (9), the free surface condition for the displacement potential is obtained.

$$\varphi(x,y,0,t) = -q(x,y,t) \qquad for \qquad t < \gamma(x,y). \tag{21}$$

Where, the function $q(t)$ is given by

$$\frac{d^2q}{dt^2} = \beta\left(\frac{1}{\nu(t)^n} - 1\right) \quad for \quad (x,y) \in \Omega_C, \tag{22}$$

$$q(x,y,t) = 0 \quad for \quad (x,y) \in \Omega_F. \tag{23}$$

When the displacement potential satisfies the free surface condition (21), the complementary problem is represented as follows:

$$(\varphi + q)\left(\frac{\partial\varphi}{\partial z} + t - b\right) = 0 \qquad on \quad z = 0, \tag{24}$$

$$\text{with} \quad \varphi \leq -q \;, \quad \frac{\partial\varphi}{\partial z} + t - b \leq 0 \qquad on \quad z = 0. \tag{25}$$

Solving this problem numerically, the instantaneous displacement potential is obtained. The with cavity case requires longer computational time than that of the without cavity case, since the time increment should be small enough to guaranty the accurate evaluation of the function q(x,y,t) by the double integral in time.

NUMERICAL METHOD

Since the displacement potential is governed by Laplace's equation, the following integral equation is obtained by applying Green's theorem to the fluid domain.

$$\frac{\partial \varphi(x,y,z,t)}{\partial z} = -\frac{1}{2\pi}\int_{\Omega_B+\Omega_C} \varphi(x,y,0,t)\frac{\partial^2}{\partial z \partial z'}\left(\frac{1}{r}\right) ds, \quad (26)$$

$$\text{where} \quad r^2 = (x-x')^2 + (y-y')^2 + (z-z')^2.$$

To discretize the above mentioned integral equation, the displacement potential is represented by the quadratic B-spline function.

$$\varphi(x,y,0,t) = \sum_i \varphi_i S[(x-x_i)/\Delta x] S[(y-y_i)/\Delta y]. \quad (27)$$

Where, the B-spline function $S[x]$ is defined as follows:

$$S[x] = \begin{cases} \frac{3}{4} - x^2 & 0 < |x| < \frac{1}{2} \\ \frac{1}{2}\left(|x| - \frac{3}{2}\right)^2 & \frac{1}{2} < |x| < \frac{3}{2} \end{cases}. \quad (28)$$

Substituting (27) into the integral equation (26), the system of linear equations for the evaluation of the unknowns φ_i is obtained.

$$\frac{\partial \varphi(x_i, y_i, 0, t)}{\partial z} = \sum_j \Gamma_{ij}\varphi_j \quad (29)$$

Where, the coefficient Γ_{ij} is obtained by performing the analytical integration over the certain panel. This system of linear equations is solved by the SOR method so as to satisfy the inequalities (14) or (25). The Runge-Kutta method is applied to integrate (22) in time and the time increment is chosen to be $\Delta t = 0.005 \sim 0.02$ according to the scale of the distortion.

In order to validate the numerical procedure, a corn entry problem is solved as an example. In this calculation, the shape of the plate is represented by the following equation.

$$f(x,y) = x^2 + y^2 \quad (30)$$

Fig. 2 shows a comparison between numerical results of the displacement potential on a line y=0, z=0 and the analytical solution. The shape of Ω_F is a square whose side length is 2, and the number of panels is 49. The numerical results are in very good agreement with the analytical result in spite of the small number of panels. This results confirms that the present numerical procedure is accurate enough, even if the number of panels is small. It is expected that the accuracy of the numerical result depends on the shape of the plate. Therefore, smaller panels ($\Delta x = 0.105, \Delta y = 0.105$ and the number of panels is 361.) are used for main calculations shown in the next section.

RESULTS AND DISCUSSIONS

In the main calculations of this study, plate shapes of A, B and C are used. The shapes A and B are employed for the with out cavity case and the shape C is applied to the without cavity case.

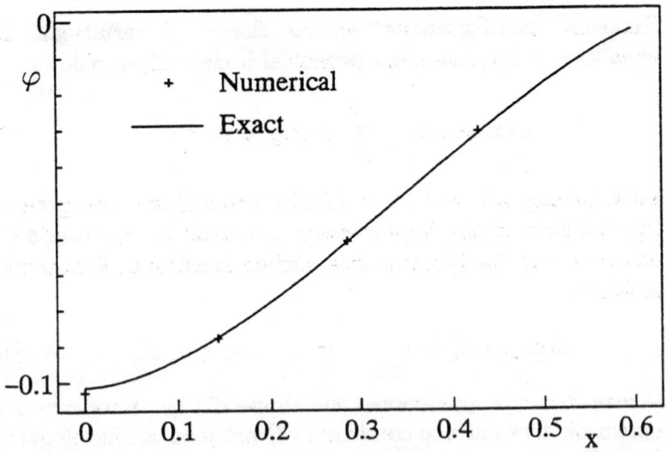

Fig. 2 Comparison between numerical results and the analytical solution of the displacement potential for slamming of a cone-shaped body.

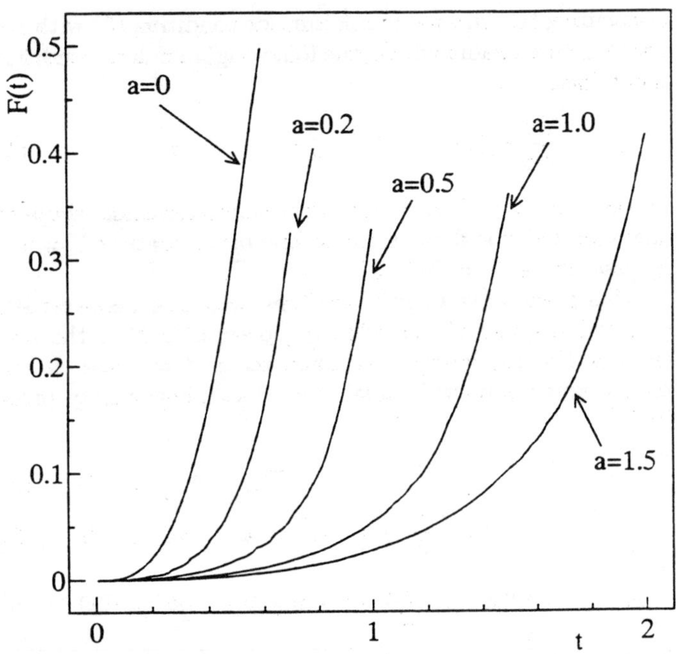

Fig. 3 Comparison of time histories of the function $F(t)$ in the case of shape A.

Without cavity case

The shape A is represented by

$$f(x,y) = \begin{cases} |x| - \frac{a}{2}\left(\cos\frac{5\pi}{2}x + \cos\frac{5\pi}{2}y\right) + a & |x| \le \frac{\pi}{4}, \\ \infty & |x| > \frac{\pi}{4}. \end{cases} \quad (31)$$

This shape corresponds to the case where a wedge-shaped body whose breadth is $\frac{\pi}{2}$ falls to the bi-directional waves whose nondimensional steepness is $\frac{4}{5}a$. It is also noted that the top of the wedge touches the crest of waves at $t = 0$ in this model. Since the two-dimensional solution of the wedge entry problem is well known, this solution is utilized for the part of the present solution to improve the numerical accuracy. This formulation is described in Appendix.

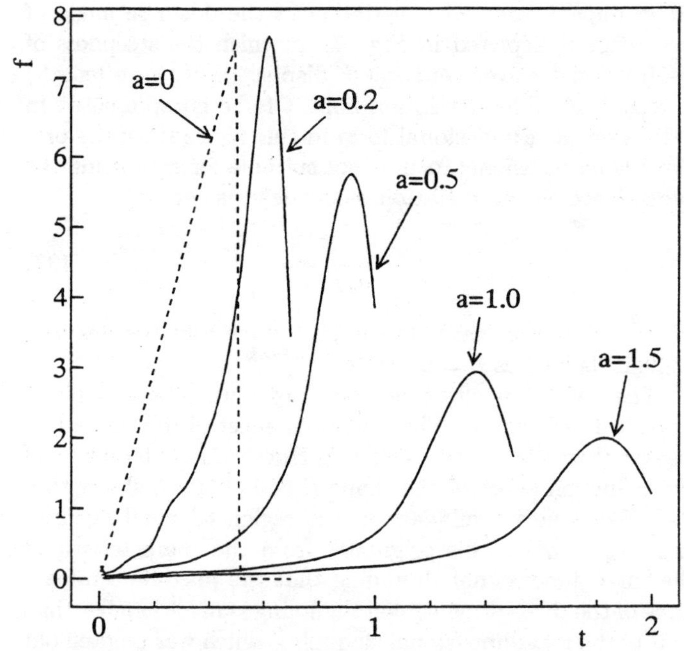

Fig. 4 Comparison of time histories of the impact force with variation of the steepness of bi-directional waves in the case of shape A.

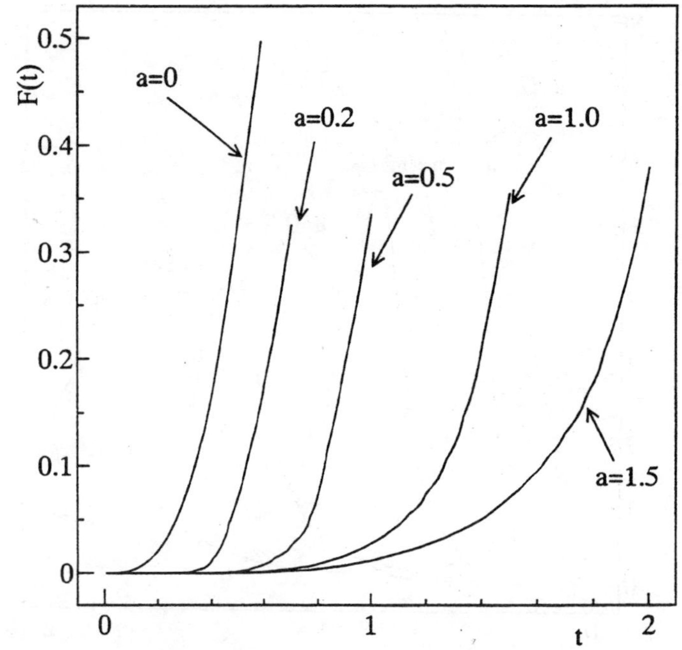

Fig. 5 Comparison of time histories of the function $F(t)$ in the case of shape B.

Fig. 3 shows the comparison of time histories of the function $F(t)$ which gives the double integration of the impact force. The physical meaning of the function $F(t)$ is not obvious since the function $F(t)$ corresponds to the integration of the impulse acting on the bottom of the plate. However, at least, it can be stated that the larger value of $F(t)$ indicates the stronger force and impulse acting on the bottom of the plate.

In order to clear the physical meaning of the numerical results, we try to obtain the impact force acting on the bottom of the plate by differentiating the function $F(t)$ twice respect to the time. When the numerical differentiation, for example the central difference, is applied to the differentiation of the function $F(t)$ the obtained impact force has a very big fluctuations, since the numerical result of the gradient of the function $F(t)$ is not continuos.

From the theoretical point of view, the function $F(t)$ should be, at least, C^2 class in time. So, the reason why the numerical result of the gradient of the function $F(t)$ is not continuos is investigated by carrying out a precise test computation. Then, it is found that the discontinuity happens when one of collocation points in the calculation of inequality (14) is located on the boundary between the domain Ω_B and the domain Ω_F. If the panel size is infinitely small, the transition of the collocation point from the domain Ω_B to Ω_F would be smooth and, as a result, the gradient of the function $F(t)$ would be continuos. But, it is impossible to use infinitesimal panel in the actual computation. Thus, a numerical filter is applied to reduce the numerical fluctuation of the time history of the impact force due to the numerical discontinuity of the function $F(t)$.

The disadvantage of applying a numerical filter is that the result depends on the intensity of the filter. Therefore, in the present study, the intensity of the filter is determined so that the maximum impact force in the case of a=0.2 is as same value as that in the case of a=0 which is obtained exactly by an analytical method. This procedure decreases the numerical accuracy of the displacement potential formulation with the theory of variational inequalities and should be improved in the future work.

Fig. 4 shows comparison of the time histories of the impact force acting on the bottom of plate with variation of the steepness of bi-directional waves. It is evident from this figure that the maximum impact force becomes smaller and smaller as the steepness of bi-directional waves increase and the duration of the impact becomes longer as the steepness increases.

The shape B is represented by

$$f(x,y) = \begin{cases} |x| + \dfrac{a}{2}\left(\cos\frac{5\pi}{2}x + \cos\frac{5\pi}{2}y\right) + a & |x| \le \frac{\pi}{4}, \\ \infty & |x| > \frac{\pi}{4}. \end{cases} \quad (32)$$

The situation corresponding to the shape B is almost as same as that of the shape A, and the only difference is that the center of the wedge falls to the trough of the bi-directional waves and the side of the wedge touches the water surface earlier than the top of the wedge.

Fig. 5 shows the comparison of time histories of the function $F(t)$ for the shape B. This figure shows that the value of $F(t)$ becomes smaller and smaller as the wave steepness increases. Fig. 6 shows a comparison of the time histories of the impact force with variation of the steepness of bi-directional waves for the case of shape B. The impact force is obtained by applying the numerical filter whose intensity is as same as that in the case of shape A. It is also evident in

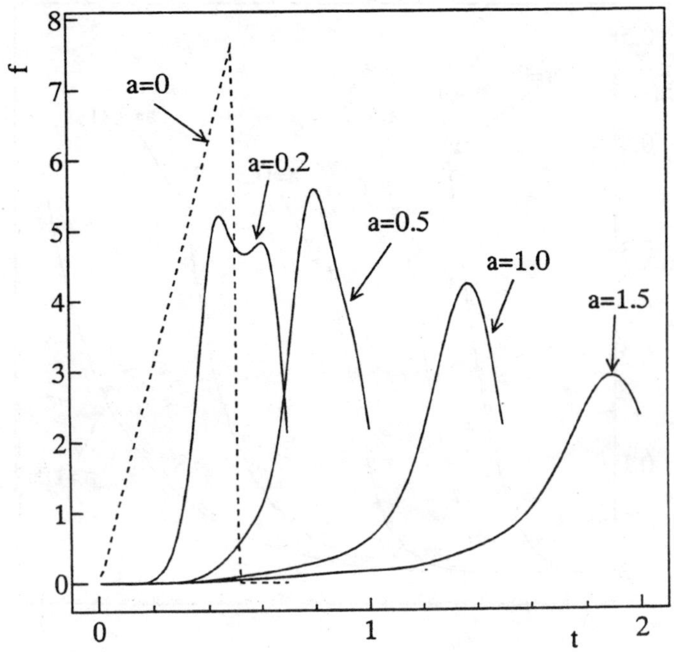

Fig. 6 Comparison of time histories of the impact force with variation of the steepness of bi-directional waves in the case of shape B.

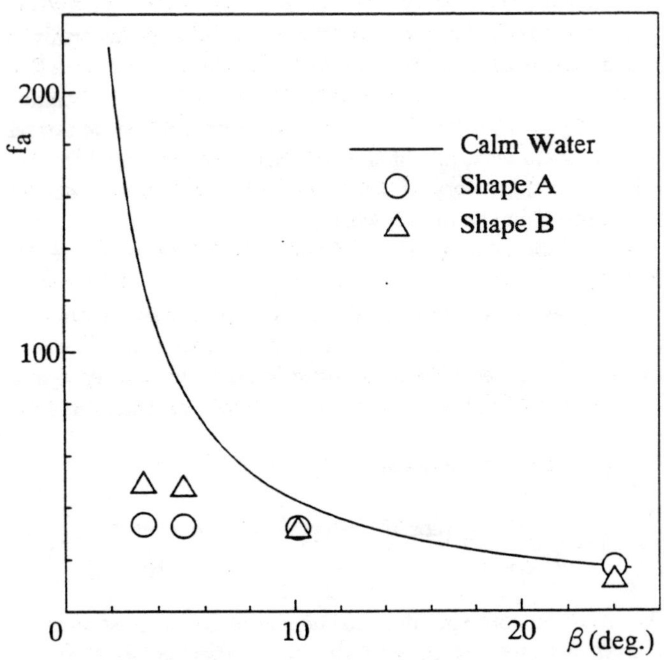

Fig. 7 Diagram of the maximum impact force with variation of the deadrise angle of the wedge-shaped body.

the case of shape B that the maximum impact force becomes smaller and smaller as the steepness of bi-directional waves increase and the duration of the impact becomes longer as the steepness increases.

From these numerical results, it seems that the roughness of water surface reduces the impact force acting on a wedge-shaped body. In order to clarify the reduction effect of the roughness of water surface, a diagram of the maximum impact force with variation of the deadrise angle of the wedge is depicted in Fig. 7, in which the steepness of bi-directional waves (maximum displacement/wave length) is kept 1/14. The maximum impact force is represented in a different nondimensional form in this figure since the previous nondimensional form is not suitable for expressing the dependence on the variation of the deadrise angle.

$$f_a = \frac{f}{\varepsilon} = \frac{\mathcal{F}}{\rho_w L^2 V^2} \tag{33}$$

Where, $\mathcal{F}$ is the maximum impact force and the deadrise angle is defined as $\beta = \arctan \varepsilon$.

The reduction effect on the maximum impact force is evident in this figure. The reduction effect of the shape B is weaker than that of the shape A. However, the tendency of the reduction effect of the shape B and shape A shows that the effect seems significant in the region of small deadrise angle ($\beta < 5°$). This originates from the characteristic of the three-dimensional slamming that the maximum impact force of the three-dimensional slamming is much smaller than that of the two-dimensional slamming which was pointed out by Toyama (1996).

With cavity case

The shape C is represented by

$$f(x,y) = \frac{1}{2}\{g(x) + g(y)\} \tag{34}$$

$$\text{where,} \quad g(x) = \begin{cases} \cos 7.854x & |x| \le 0.6, \\ 9.09\left(x^4 - 0.6^4\right) & |x| > 0.6. \end{cases} \tag{35}$$

This shape corresponds to the case where a flat-bottom body falls to bi-directional waves. It is easy to expect the intersection line of this model which surrounds the cavity formed between the bottom surface and the water surface, since it is apparent that the intersection line is located between crests of bi-directional waves.

Fig. 8 shows the comparison of time histories of the function $F(t)$ for the shape C. This figure shows that the value of $F(t)$ becomes larger as the coefficient β increases. Fig. 9 shows a comparison of time histories of the impact force with variation of the coefficient β. The impact force is obtained by applying the numerical filter whose intensity is as same as that in the case of shape A.

It is evident that the maximum impact force becomes larger as the coefficient β increases and the duration of the impact becomes shorter as β increase. From the definition of the coefficient β, the larger value of β means the stronger effect of the trapped air. Therefore, it can be steated that the trapped air makes the impact force stronger and the reduction of the maximum impact force is mainly brought about by the three-dimensional effect of the roughness of water surface. However, it should be noted that this statement is limited for the case where the water surface is rough enough to guarantee the formation of a cavity in advance, and it is considered that the effect of the trapped air becomes dominant for the reduction of the maximum impact force as the

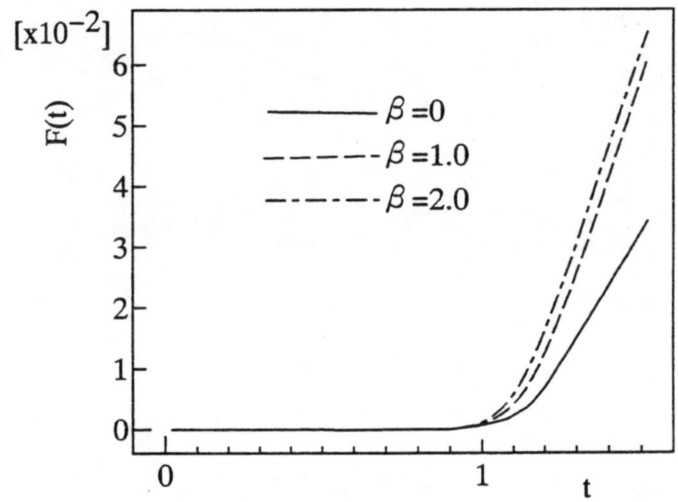

Fig. 8 Comparison of time histories of the function $F(t)$ in the case of shape C.

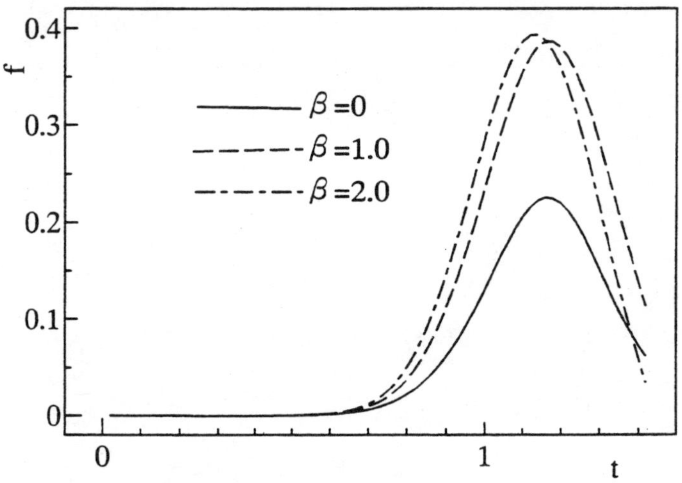

Fig. 9 Comparison of time histories of the impact force with variation of the scale effect of trapped air.

steepness of surface waves decrease.

Summarizing the numerical results, a new scenario is implied in that the reduction effect of the maximum impact force acting on a blunt body is mainly brought about by the three-dimensional effect of the roughness of water surface when the body falls into rough seas, and when the water surface is calm the effect of the trapped air becomes dominant as indicated in previous experimental works (for example Chuang (1967)).

CONCLUSIONS

A theoretical and numerical approach to the distorted plate penetrating the calm water surface is investigated as a flow model of a blunt body slamming in rough seas. Since the falling of the distorted plate to the water surface guarantees the formation of the cavity in advance, it is easy to treat the air trapping between the body surface and the water surface theoretically. Therefore, the effect of air trapping is also investigated. The theory is based on the variational formulation in which the displacement potential is obtained instead of the velocity potential.

Numerical results of the 'Without Cavity Case' show that the maximum impact force becomes smaller and smaller as the steepness of surface waves increase and the duration of the impact becomes longer as the steepness of surface waves increase. In order to clarify the reduction effect of the roughness of water surface, a diagram of the maximum impact force with variation of the deadrise angle of the wedge is depicted. This diagram shows that the reduction effect seems significant in the region of small deadrise angle. Numerical results of the 'With Cavity Case' show that the trapped air makes the impact force stronger and the reduction of the maximum impact force is mainly brought about by the three-dimensional effect of the roughness of water surface when the water surface is rough enough to guarantee the formation of a cavity in advance, and it is considered that the effect of the trapped air becomes dominant for the reduction of the maximum impact force as the decrease of the steepness of surface waves.

Summarizing the numerical results, a new scenario is implied in that the reduction effect of the maximum impact force acting on a blunt body is mainly brought about by the three-dimensional effect of the roughness of water surface when the body falls to rough seas, and when the water surface is calm the effect of the trapped air becomes dominant as mentioned in previous experimental works.

ACKNOWLEDGMENT

The author gratefully acknowledge the support of the Grant-in-Aid for Scientific Research from the Ministry of Education (Grant No. 08651095).

REFERENCES

Armand, JL and Cointe, R (1986). "Hydrodynamic Impact Analysis of a Cylinder," Proc of 5th Int OMAE Symposium, Tokyo, Vol 1, pp250-256.

Chuang, SL (1967). "Experiments on Slamming of Wedge-Shaped Bodies with Variable Deadrise Angle," JSR, Vol 11, pp190-198.

Chuang, SL (1969). "Theoretical Investigations on Slamming of Cone-Shaped Bodies," JSR, Vol 13, pp276-283.

Howison, SD, Ockendon, JR and Wilson, SK (1991). "Incompressible Water-Entry Problem at Small Deadrise Angles," J Fluid Mech, Vol 222, pp215-230.

Korobkin, A (1982). "Formulation of Penetration Problem as a Variational Inequality," Din Sploshnoi Sredy, Vol 58 pp73-79.

Korobkin, A (1996). "Entry Problem for Body with Attached Cavity," Proc of 11th Int Workshop on Water Waves and Floating Bodies, Hamburg.

Toyama, Y (1996). "Flat Plate Approximation in the three-Dimensional Slamming," J Soc Naval Arch Japan, Vol 179, pp271-279.

Wagner, H (1932). "Über Stoss und Gleitvergänge an der

Oberfläche von Flüssigkeiten," Zeitschr für Angewendte Mathematik und Mechanik, Bind 12, Heft 4, pp193-215.
Watanabe, I (1986). "Analytical Expression of Hydrodynamic Impact Pressure by Matched Asymptotic Expansion Technique," Trans West-Japan Soc Naval Arch, Vol 71, pp77-85.
Watanabe, I (1986). "On Hydrodynamic Impact Pressure Acting upon Flat Bottomed Ships," J Soc Naval Arch Japan, Vol 159, pp193-202.

APPENDIX

The solution of the two-dimensional wedge entry problem is well known and its displacement potential is represented as follows:

$$\varphi_0(x,y,0,t) = \begin{cases} -\frac{1}{2}\left(t\sqrt{c^2t^2-x^2} - \frac{x^2}{c}\ln\frac{x}{ct-\sqrt{c^2t^2-x^2}}\right) & t \le \frac{l}{c}, \\ -\frac{1}{2}\left(t\sqrt{c^2t^2-x^2} - \frac{x^2}{c}\ln\frac{x}{ct-\sqrt{c^2t^2-x^2}}\right) & \\ \qquad -\left(t-\frac{l}{c}\right)\sqrt{l^2-x^2} & t > \frac{l}{c}. \end{cases} \tag{36}$$

Where, $c = \pi/2$ and l is the nondimensional half breadth of the wedge. The gradient of φ_0 in z-direction is given as

$$\frac{\partial\varphi(x,y,0,t)}{\partial z} = \begin{cases} -t+x & t \le \frac{l}{c}\ ,\quad x \le ct \\ \frac{x}{c}\arcsin\frac{ct}{x} - t & t \le \frac{l}{c}\ ,\quad x > ct \\ -t+x & t > \frac{l}{c}\ ,\quad x \le l \\ \frac{x}{c}\arcsin\frac{l}{x} & \\ \qquad +\left(t-\frac{l}{c}\right)\frac{x}{\sqrt{x^2-l^2}} - t & t > \frac{l}{c}\ ,\quad x > l \end{cases} \tag{37}$$

Substituting (36) and (37) into boundary conditions (5) ~ (7) and after some manipulations, the following complementary problem is obtained.

$$\nabla^2\varphi = 0 \qquad z < 0 \tag{38}$$

$$(\varphi+\varphi_0)\left(\frac{\partial\varphi}{\partial z} + \frac{\partial\varphi_0}{\partial z} + t - b\right) = 0 \quad on \quad z = 0 \tag{39}$$

$$\text{with}\quad \varphi+\varphi_0 \le 0,\quad \frac{\partial\varphi}{\partial z}\frac{\partial\varphi_0}{\partial z} + t - b \ge 0 \quad on \quad z = 0 \tag{40}$$

Solving this problem by the same numerical procedure, the displacement potential for the cases of shape A and B is obtained.

Proceedings of the Seventh (1997) International Offshore and Polar Engineering Conference
Honolulu, USA, May 25-30, 1997

ISBN 1-880653-28-1 (Set); ISBN 1-880653-29-X (Vol. I)

Dynamic Wave Interaction Analysis of Floating Elastic Plates

Tadaaki Tanabe, M. Nishimura and Tetsuo Kawamura
Nagoya University
Nagoya, Japan

ABSTRACT

Three-D finite element formulation of wave-structure interaction analysis considering elastic deformation has been presented. Linearized potential which is the addition of scattering wave potential and radiation potential has been directly coupled with the elastic deformation so that they give the equilibrium of forces at the boundary of wave and plates and to satisfy the compatibility condition. The infinite boundary condition is satisfied by hybrid method that connects the numerical solution with analytical solution at certain boundary. In this way, quite efficient and consistent finite element solution algorism is obtained. Moreover, the method to reduce the degree of freedoms of the total system is presented taking into account the fact that the nodal points of wave potential need not be the same as the nodal point of displacement of elastic plates.

KEY WORDS: Three D analysis, Elastic deformation, Hybrid method, Finite element method, Reduction of freedom, Floating airport, Mooring.

INTRODUCTION

Analysis of large floating plates made by either concrete or steel has been a focus of structural engineers since the publicizing of construction plan of offshore air port sometime back in Japan. Since then, we have started to formulate the numerical solution scheme to analyze plate bending coupled with wave forces and a part of the results have been already published several years ago in Japanese in the numerical analysis community of concrete structures in Japan (Nishimura, 1990). Recently a number of research results on the same topic were made public (Wu, 1996)(Meylan, 1996)(Newman, 1994). Among them, the paper by Wu and et.al. (1996) may be most representatively referred to, since it treats exactly the same topic of long elastic floating plate. The method in the paper (Wu, 1996) is considered to give the results of accuracy of the same level with the authors formulation when the number of the modal degree is increased. However, for the analysis of much wider variety of configuration of the floating body, we need an analysis method that have the capacity applicable to any kind of configuration without loss of accuracy. It seems there still exists a quite large field to be investigated for the purpose. In the paper, we will introduce the fundamental solution scheme by directly giving the boundary condition for liquid and plate boundary and enforcing the continuity of the potential and its gradient at the cylindrical numerical calculation boundary by hybrid method. It is shown that the solution scheme make the analysis applicable for quite large varieties of structures in the consistent manner.

GOVERNING EQUATIONS IN 3-D DOMAIN AND FINITE ELEMENT FORMULATION

Derivation of the Finite Element Formulation

Let consider the 3-D wave domain with constant depth as shown in Fig. 1 where fluid is assumed to be incompressible, inviscid and irrotational in which a floating elastic plate exists. The total velocity potential $\varphi(t)$ is expressed as the addition of incident wave potential $\varphi^I(t)$ and the potential φ^E, i.e.,

$$\varphi = (\varphi^I + \varphi^E)e^{-i\omega t} \tag{1}$$

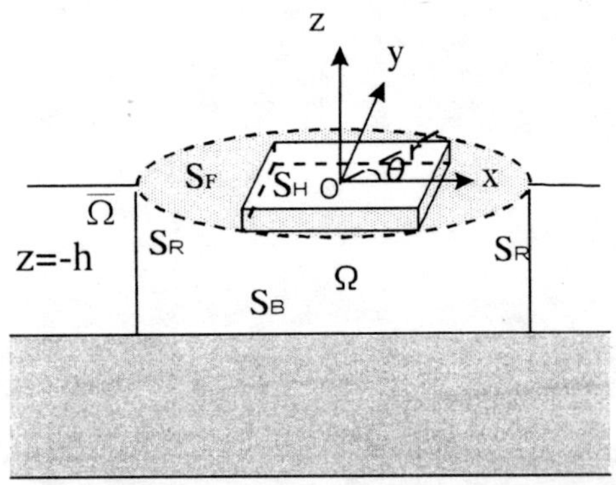

Fig.1 Water and a Plate Field to Be Solved (Cylindrical domain implies the area where numerical solution is sought)

in which incident potential $\varphi^I(t)$ is written as

$$\varphi^I(t) = \frac{-igH\cosh k_0(z+h)}{2\omega\cosh k_0 h}\exp(-ik_0 r\cos(\theta-\theta_I)) \qquad (2)$$

where g: gravity acceleration, H: wave amplitude, k_0: wave number, h: water depth. Three components r, θ, z are cylindrical coordinates. The notation, φ^E denotes the potential other than the potential of incident wave. In the traditional sense, this is the addition of the scattering wave potential and the radiation potential. Since the problem is linear field, we have the boundary value problem as follows in terms of φ^E,

$$\nabla^2\varphi^E = 0 \qquad \text{in } \Omega \qquad (3)$$

with boundary conditions that

$$\left.\begin{aligned} &\frac{\partial\varphi^E}{\partial z} - \frac{\omega^2}{g}\varphi^E = 0 && \text{on } S_F, \\ &\frac{\partial\varphi^E}{\partial n} = 0 && \text{on } S_B, \end{aligned}\right\} \qquad (4)$$

$$\left.\begin{aligned} &\frac{\partial\varphi^E}{\partial n} = -\frac{\partial\varphi^I}{\partial n} + \frac{\partial w}{\partial t} && \text{on } SH_1, \\ &\frac{\partial\varphi^E}{\partial n} = -\frac{\partial\varphi^I}{\partial n} + \frac{\partial u}{\partial t} && \text{on } SH_2, \\ &\frac{\partial\varphi^E}{\partial n} = -\frac{\partial\varphi^I}{\partial n} + \frac{\partial v}{\partial t} && \text{on } SH_2, \end{aligned}\right\} \qquad (5)$$

$$\sqrt{k_0 r}\left(\frac{\partial}{\partial r} - ik_0\right)\varphi^E \to 0 \qquad as \qquad k_0 r \to \infty$$

in which u,v,w denote displacements of a plate to the X,Y and Z directions and n denotes the normal directional coordinate to the plate surface. The notation, Ω denotes the domain to be solved, S_F, surface wave boundary, S_B, bottom of sea bed, and SH_1, SH_2, SH_3 denote the bottom surface, side surfaces to the X direction and to the Y direction respectively of floating body. It is readily noted that ϕ^E is reduced to the scattering potential when $w = u = v = 0$, examining Eq.(5)

In regards to a floating plate, the governing differential equation in regards to the deflection to the Z direction is written as

$$\rho_c\frac{\partial^2 w}{\partial t^2} + D\nabla^4 w + \rho_w\left(\frac{\partial\varphi^I_{(x,z)}}{\partial t} + \frac{\partial\varphi^E_{(x,z)}}{\partial t}\right) + \rho_w g w + k_m w = 0 \qquad (6)$$

in which w: the vertical displacement of a plate, D: rigidity of a plate, ρ_c: density of concrete, ρ_w: density of water, k_m: rigidity of mooring system, g: gravity acceleration. The boundary condition is written as

$$M_n = EI\frac{\partial^2 w}{\partial n^2} = 0, \quad M_{nt} = \frac{\partial^2 w}{\partial n\partial t} = 0, \quad Q_n = 0 \qquad (7)$$

In the similar way, the governing equation in regards to the displacements to the X and to the Y directions are written respectively as

$$\left.\begin{aligned} &\rho_c D_H\frac{\partial^2 u}{\partial t^2} + \frac{E_x}{1-\nu^2}D_H\left(\frac{\partial^2 u}{\partial x^2} + \nu\frac{\partial^2 v}{\partial x\partial y}\right) - Q_x = 0 \\ &\text{and} \\ &\rho_c D_H\frac{\partial^2 v}{\partial t^2} + \frac{E_y}{1-\nu^2}D_H\left(\frac{\partial^2 v}{\partial y^2} + \nu\frac{\partial^2 u}{\partial x\partial y}\right) - Q_y = 0 \end{aligned}\right\} \qquad (8)$$

in which D_H denotes the total depth of a plate.

Boundary condition for these equation are written as

$$\begin{aligned} &x = \pm\frac{l_l}{2}, \quad Q_x = \int_0^{D_H}\sigma_x dz = \mp\int_0^{D_H}\left(-\frac{\partial\varphi}{\partial t} + \rho z\right)dz \\ &y = \pm\frac{l_w}{2}, \quad Q_y = \int_0^{D_H}\sigma_y dz = \mp\int_0^{D_H}\left(-\frac{\partial\varphi}{\partial t} + \rho z\right)dz \end{aligned} \qquad (9)$$

Now we try to solve the coupled differential equation of Eq.(3), Eq.(6), Eq.(8) with boundary conditions of Eq.(5),Eq.(7) and Eq.(9)

However, before we come to the solution scheme, it is necessary to tackle with infinite boundary. Following the hybrid method, let separate the domain to be solved in two parts as shown in Fig.1, i.e., one, the cylindrical domain where a plate exists and where numerical solution is sought by appropriate numerical method, and the domain outside of the cylinder where analytical solution is applied to satisfy the infinite boundary condition. We enforce the continuity of potential and its gradient as $\varphi^E = \overline{\varphi^E}$ and $\frac{\partial\varphi^E}{\partial r} = \frac{\partial\overline{\varphi^E}}{\partial r}$ at the boundary in which φ^E is the potential in the numerical domain i.e., inside the boundary and $\overline{\varphi^E}$ is the potential of analytical domain i.e., outside of the boundary. Taking the domain of cylindrical boundaries between them, the analytical solution which satisfy the governing equation outside of the cylindrical domain is written in 3-D domain as

$$\begin{aligned} \overline{\varphi^E} &= \sum_{m=0}^{\infty}(\alpha_{m0}\cos m\theta + \beta_{m0}\sin m\theta)\cosh\kappa_0(z+h)H^{(1)}_{m(\kappa_0 r)} \\ &+ \sum_{n=1}^{\infty}\sum_{m=0}^{\infty}(\alpha_{mn}\cos m\theta + \beta_{mn}\sin m\theta)\cos\kappa_n(z+h)K_{m(\kappa_n r)} \end{aligned} \qquad (10)$$

in which $H^{(1)}_{m(\kappa_0 \gamma)}$ and $K_{m(\kappa_n \gamma)}$ are the Hankel function of the first kind and the Bessel function of the second kind respectively. Eq.(10) are rewritten using normalized factor such as Eq.(11) and these unknown values are determined in the solution process.

$$\alpha H^{(1)}_{m(\kappa_0 \gamma)} = \left[\alpha_{mn} P_{(m,\kappa_0 \gamma)}\right] \frac{H^{(1)}_{m(\kappa_0 \gamma)}}{P_{(m,\kappa_0 \gamma)}}$$

$$\alpha K_{m(\kappa_n \gamma)} = \left[\alpha_{mn} Q_{(m,\kappa_0 \gamma)}\right] \frac{K_{m(\kappa_n \gamma)}}{Q_{(m,\kappa_0 \gamma)}} \tag{11}$$

where

$$P_{(m,\kappa_0,\gamma_0)} = \begin{cases} -i\sqrt{\dfrac{2}{\pi \tan\alpha}} e^{m(\alpha - \tanh\alpha)} & \left[\dfrac{\kappa_0 \gamma_0}{m} < 1\right] \\ -\dfrac{1}{3\pi}\Gamma\left(\dfrac{1}{3}\right)\left(e^{\frac{5\pi i}{6}} + i\right)\left(\dfrac{6}{m}\right)^{\frac{1}{3}} & \left[\dfrac{\kappa_0 \gamma_0}{m} = 1\right] \\ -e^{\frac{3\pi i}{4}}\sqrt{\dfrac{2}{\pi m \tan\alpha}} e^{im(\tan\alpha - \alpha)} & \left[\dfrac{\kappa_0 \gamma_0}{m} = 1\right] \end{cases} \tag{12}$$

$$Q_{(m,\kappa_0,\gamma_0)} = \sqrt{\frac{\pi}{2\kappa_0 \gamma_0}} e^{\kappa_0 \gamma_0} \sum_{j=0}^{\infty} \frac{(m,j)}{(2\kappa_0 \gamma_0)^j}$$

and

$$(m,j) = \frac{\Gamma\left(m + j + \frac{1}{2}\right)}{j!\Gamma\left(m - j + \frac{1}{2}\right)} \tag{13}$$

With all these preparations, we seek the weak form for the governing differential equation, Eq.(3), with boundary conditions of Eq.(4) and Eq.(5).

$$\iiint_{\Omega} \nabla^2 \varphi^E \delta\varphi^E dv - \iint_{S_F} \left(\frac{\partial \varphi^E}{\partial z} - \frac{\omega^2}{g}\varphi^E\right)\delta\varphi^E ds + \iint_{S_B} \left(\frac{\partial \varphi^E}{\partial z}\right)\delta\varphi^E ds +$$
$$\iint_{S_{H2}} \left(\frac{\partial \varphi^E}{\partial n} + \frac{\partial \varphi^I}{\partial n} - \frac{\partial w}{\partial t}\right)\delta\varphi^E ds - \iint_{S_{H2}} \left(\frac{\partial \varphi^E}{\partial n} + \frac{\partial \varphi^I}{\partial n} + m\frac{\partial u}{\partial t}\right)\delta\varphi^E ds +$$
$$\iint_{S_{H2}} \left(\frac{\partial \varphi^E}{\partial n} + \frac{\partial \varphi^I}{\partial n} + m\frac{\partial v}{\partial t}\right)\delta\varphi^E ds - \iint_{S_R} \left(\frac{\partial \varphi^E}{\partial r} - \frac{\partial \overline{\varphi^E}}{\partial r}\right)\delta\varphi^E ds -$$
$$\iint_{S_R} (\varphi^E - \overline{\varphi^E})\frac{\partial \delta\varphi^E}{\partial r} ds = 0 \tag{14}$$

In Eq.(14), the last two terms imply that the velocity and its gradient is continuous along the cylindrical boundary between the numerical domain and analytical domain. It should also be noted that the weight for the continuity conditions is taken as $\dfrac{\partial \delta\varphi^E}{\partial r}$ to maintain the symmetry of the coefficients matrix of K_{LC} and K_{CL}.

By applying the Greens first theorem and adding terms in the same integration point, the following equation is derived.

$$\iiint_{\Omega} \nabla \delta\varphi^E \nabla \varphi^E dv + \iint_{S_F} \delta\varphi^E \left(-\frac{\omega^2}{g}\varphi^E\right) + \iint_{S_H} \delta\varphi^E \frac{\partial \varphi^I}{\partial n} dS +$$
$$\iint_{S_H} \delta\varphi^E \left(-\frac{\partial w}{\partial t}\right) ds + \iint_{S_{H2}} \delta\varphi^E \left(m\frac{\partial u}{\partial t}\right) ds + \iint_{S_{H2}} \delta\varphi^E \left(m\frac{\partial v}{\partial t}\right) ds +$$
$$\iint_{S_R} \frac{\partial \varphi^E}{\partial r} \delta\varphi^E dS + \iint_{S_R} \frac{\partial \delta\varphi^E}{\partial r}(-\overline{\varphi^E}) dS = 0 \tag{15}$$

Applying appropriate interpolation function for ϕ^E, μ, w, u and v, we obtain the next linear equations between each nodal freedoms where μ denotes the normalized coefficients of the each terms of series of Eq.(10)

$$K_{LL}\varphi + K_{LC}\mu + K_{LW}w + K_{LU}u + K_{LV}v + f_L = 0 \tag{16}$$

and

$$K_{CL}\varphi + K_{CC}\mu = 0 \tag{17}$$

For the governing equation of the plate, i.e. Eq.(6), the weak form is written as

$$\iint_S \delta w\left\{ \rho_c \frac{\partial^2 w}{\partial t^2} + D\nabla^4 w + \rho_w\left(\frac{\partial \varphi^I_{(x,z)}}{\partial t} + \frac{\partial \varphi^E_{(x,z)}}{\partial t}\right) + \rho_w g w + k_m w\right\} ds = 0 \tag{18}$$

This equation is reduced to the following equation also using Greens theorem and applying the boundary condition of Eq.(7)

$$\iint_S D\left(\frac{\partial^2 \delta w}{\partial x^2}\frac{\partial w^2}{\partial x^2} + 2\frac{\partial^2 \delta w}{\partial x \partial y}\frac{\partial^2 w}{\partial x \partial y} + \frac{\partial^2 \delta w}{\partial y^2}\frac{\partial w^2}{\partial y^2}\right) dxdy +$$
$$\iint_S \left(\rho_c \omega^2 \delta w w + \rho_w \delta w\left(\frac{\partial \varphi^I_{(x,z)}}{\partial t} + \frac{\partial \varphi^E_{(x,z)}}{\partial t}\right) + \rho_w g \delta w w + k_m \delta w w\right) dxdy = 0 \tag{19}$$

Using appropriate interpolation function for W, above equation is reduced to

$$(K_W + K_{WR} + K_{WM} + M_W)w + K_{WL}\varphi^E + f_W = 0 \tag{20}$$

Similarly for the X direction and for the Y directions, the weak forms of Eq.(8) are written as

$$\iint_{SH} \delta u\left\{\left(\rho_c D_H \frac{\partial^2 u}{\partial t^2} + \frac{E}{1-\nu^2} D_H\left(\frac{\partial^2 u}{\partial x^2} + \nu\frac{\partial^2 v}{\partial y^2}\right) - Q\right\} dxdy = 0$$
$$\iint_{SH} \delta v\left\{\left(\rho_c D_H \frac{\partial^2 v}{\partial t^2} + \frac{E}{1-\nu^2} D_H\left(\frac{\partial^2 v}{\partial y^2} + \nu\frac{\partial^2 u}{\partial x^2}\right) - Q\right\} dxdy = 0 \tag{21}$$

These are again reduced to the equations of the following form

$$K_{UU}u + K_{UV}v + K_{UL}\varphi + f_U = 0$$
$$K_{VV}v + K_{VU}u + K_{VL}\varphi + f_V = 0 \tag{22}$$

In this way all together, the governing differential equations are reduced to the linear equations of unknown nodal values of φ^E, w, u, v, and μ.

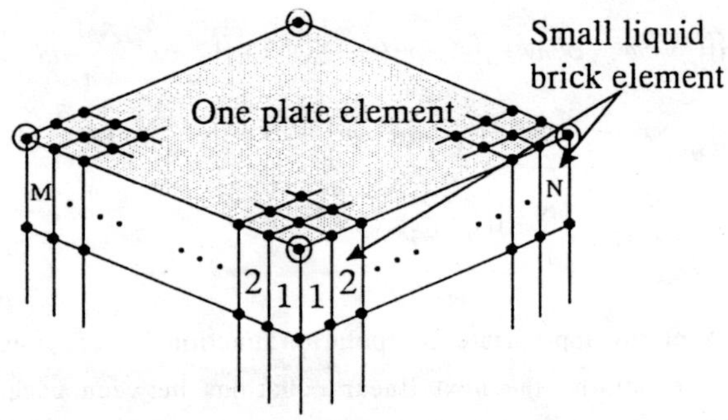

Fig.2 An Example of Discretization
for a Plate and Liquid

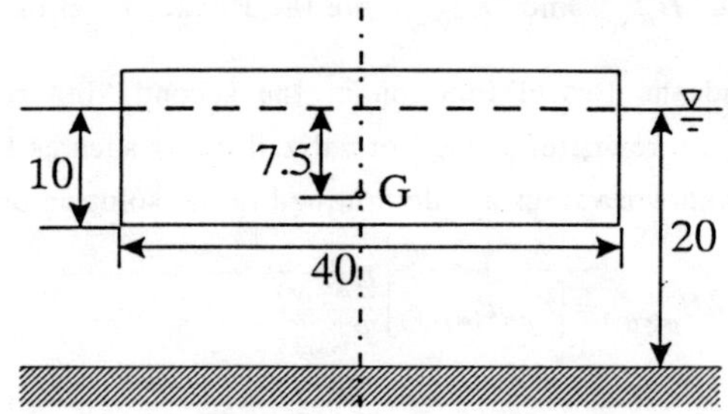

Fig.3 Rectangular Rigid Body Calculated (by m)

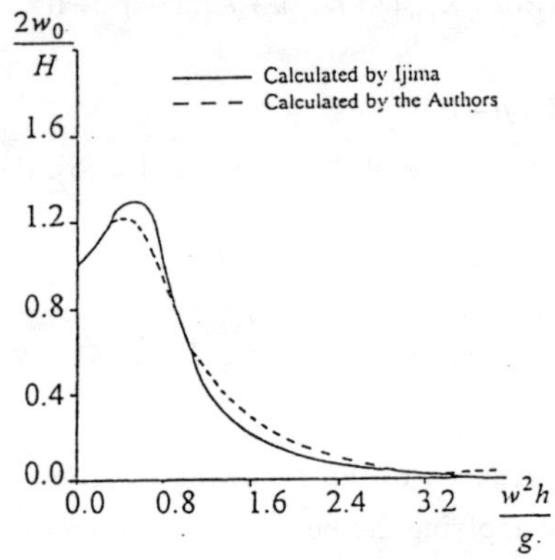

Fig.4 Heaving Amplitude

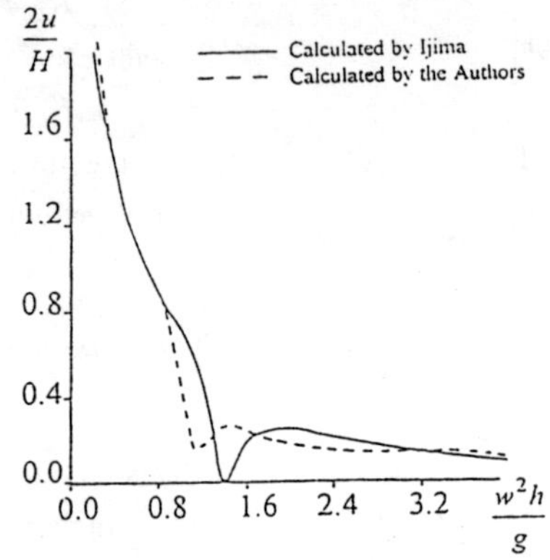

Fig.5 Swaying Amplitude

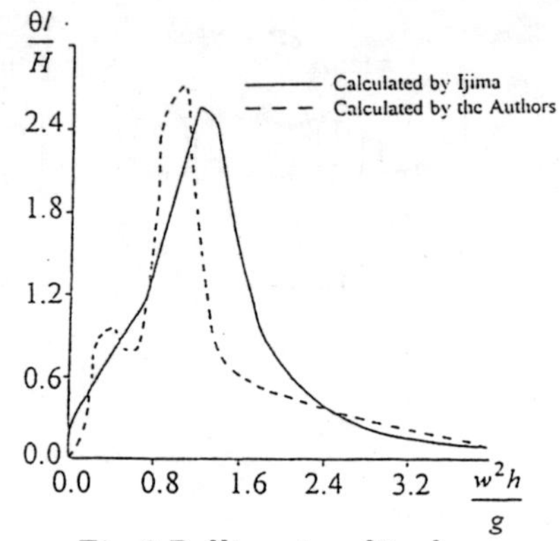

Fig.6 Rolling Amplitude

$$\begin{bmatrix} K_{LL} & K_{LW} & K_{LU} & K_{LV} & K_{LC} \\ K_{WL} & (K_W + K_{WR} + K_{WM} + M_W) & 0 & 0 & 0 \\ K_{UL} & 0 & K_{UU} & K_{UV} & 0 \\ K_{VL} & 0 & K_{VU} & K_{VV} & 0 \\ K_{CL} & 0 & 0 & 0 & K_{CC} \end{bmatrix} \begin{Bmatrix} \varphi^E \\ w \\ u \\ v \\ \mu \end{Bmatrix} = \begin{Bmatrix} -f_L \\ -f_W \\ -f_U \\ -f_V \\ 0 \end{Bmatrix} \tag{23}$$

Now the problem is only to solve the complex linear equations above.

Reduction of The Number of Nodal Freedoms

It is obvious that the number of degrees of freedom increases to the quite number to analyze the floating plate which length is several kilometers if we keep the strategy of maintaining the same nodal points in the plate as well as liquid elements. For a plate, we need not discretize it to far smaller elements than the deforming configuration except for the place of stress concentration while for liquid element discretization should be at least smaller than a quarter of the wave length. Therefore for this situation, the formulation to incorporate the plate element which have several attached liquid element may be utilized as shown in Fig.2. In this case, one plate nodal point relates to several liquid elements. In this situation, K_{LW} is slightly modified as

$$\begin{aligned} \iint_e \delta\varphi(i\omega) w ds &= \delta\varphi_1(i\omega) \iint N_1^l N^w ds_1 w^w + \\ &\quad \delta\varphi_2(i\omega) \iint N_2^l N^w ds_2 w^w + \cdots + \delta\varphi_n(i\omega) \iint N_n^l N^w ds_n w^w \\ &= \delta\varphi^T K_{LW} w \end{aligned} \tag{24}$$

The proof of the accuracy instead of the reduction of freedom is shown elsewhere (Tanaka, 1993).

Comparison with The Existing Solution for a Rectangular Rigid Body

After deriving the formulation, a way to checked its accuracy may be to compare the results with the one already published and recognized as correct. For this purpose, let take the solution example done by Ijima (1972) for the 2-D rectangular structure of 10m submerged depth and 40m long with the center of gravity of 7.5m from the water surface and in the shallow water of the depth of 20m as shown in the Fig.3. The floating body is rigid and has no mooring. The present solution scheme enables to calculate the rigid body oscillation by simply giving quite high rigidity values to the floating structure.

The Figs. 4, 5, and 6 show the calculated results of both Ijima's solution with solid line and the present research by dotted

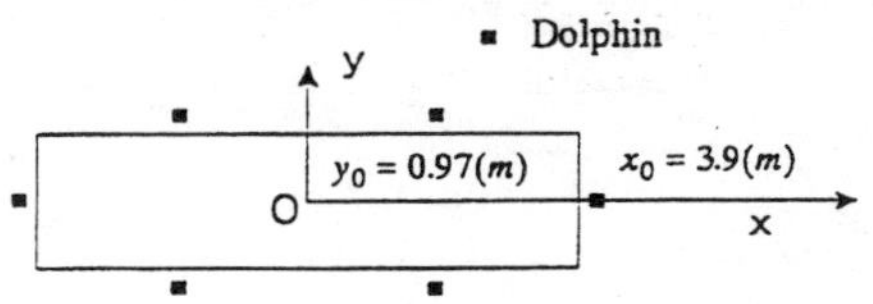

Fig.7 Schematic Diagram of the Experimental Specimen

Table-1 Parameters of the floating structure model

		Model	Real structure
Tank length B		7.80m	390m
Width L		1.94m	97m
Height H_e		0.52m	26m
Weight	Ballast	13720N(1.4tf)	1.72×10^9N(1.75×10^5tf)
	Half loaded	42140N(4.3tf)	5.24×10^9N(5.35×10^5tf)
Submerging depth	Ballast	9.0cm	4.5m
	Half loaded	28.3cm	14.2m
Height of the center of gravity (from the water surface)	Ballast	21.5cm	10.8m
	Half loaded	−11.7cm	−5.9m

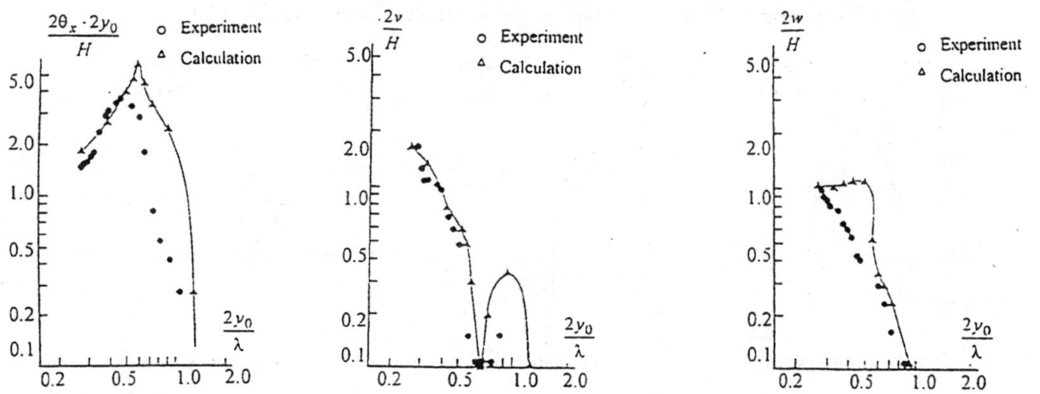

(a) Rolling Angle (b) Sway Displacement (c)Heaving Displacement

Fig.8 Comparison with Calculation for the Case of a Half Full-load and Non Mooring

line for the normalized heaving amplitude, swaying amplitude and rolling amplitude respectively in which H denotes the incident wave amplitude as shown in Eq.(2), and θ denotes the rotation angle. By comparison, we notice that there exist slight differences between the two, specially in the value of predominant period. The reason of this is considered partly due to the formulation in the side of a plate to treat the wave effects in the average as shown in the Eq.(19). This effect becomes eventually very small when the plate become long. As a whole, it looks to give a good enough solution with this coarse mesh discretization.

COMPARISON WITH THE TEST BY TAKAYAMA (1980)

Another example is the comparison with the experiments. Takayama has carried out the experiment simulating the oil storage floating tank moored to the elastic spring as shown in Fig.7. The dimension of the specimen is 7.8m long 1.94m wide and 0.52m high box, while incident wave characteristics are 2.0cm height and the angle of incident is 90 degree and their period is changed to several levels. The scaled down dimension of the test specimen are shown in Table 1,in comparison with the dimension of prototype. Test results for the free mooring condition are shown in Fig.8 and Fig.9 where solid circles and blank circles show the experimental results while solid triangle or blank triangle with solid lines show the calculation. Fig.8 shows the normalized rolling angle, sway displacement and heaving displacement respectively for the half full-loaded condition. Fig.9 shows the same amounts for ballistic condition. You may be able to notice that the overall trend can be represented well by the calculation though there exist some discrepancy in the range of small wave length. It should be remembered here again that the rigidity of the structure is taken quite high to express rigid body condition of the structure. For the moored condition, experimental results are compared with the calculation in the Fig.10 and Fig.11. Here again, you may be able to notice that the overall trend of the experiments are well captured by the calculation. However, it should be mentioned that in this case the spring constants are reduced to the half of the experimentally used mooring system since the gap between the box specimen and the dolphin exist and full elasticity is not assured and the calculation just modified the spring constant to take it into account. We do not get into deep into the discussion any further since this is only to show the capability of the solution scheme.

COMPARATIVE STUDY OF A PLATE WITH 1KM LENGTH

The dynamic behavior of a plate, such as displacements,

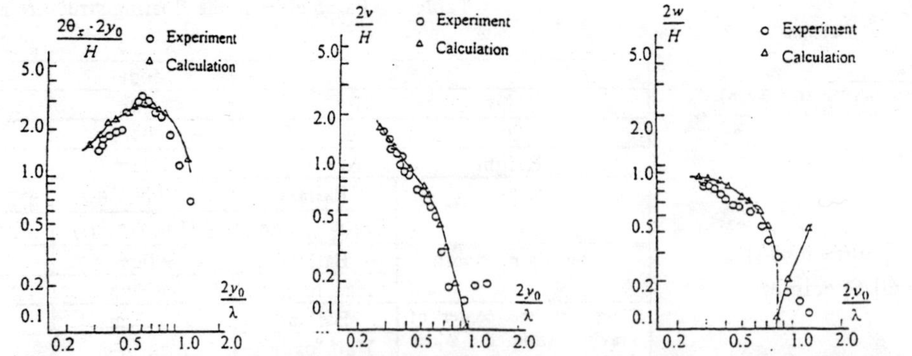

(a) Rolling Angle (b) Sway Displacement (c)Heaving Displacement

Fig.9 Comparison with Calculation for Ballistic Case with Non Mooring

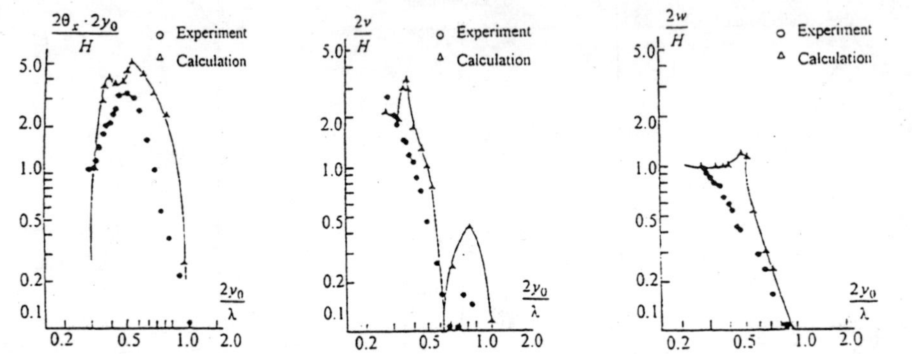

(a) Rolling Angle (b) Sway Displacement (c)Heaving Displacement

Fig.10 Comparison with Calculation for the Case of a Half Full-load and Mooring

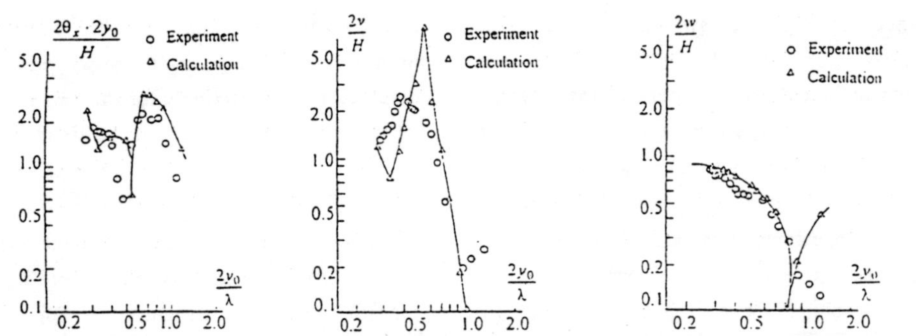

(a) Rolling Angle (b) Sway Displacement (c)Heaving Displacement

Fig.11 Comparison with Calculation for Ballistic Case with Mooring

magnitude of bending moment and shear force are surely affected by the relative rigidity of the section and the assumption that the floating body remains as rigid can not be applied any further to those very long structures. To show those characteristics, the numerical calculation of the floating plates with several length levels up to 1000m was performed.

The section treated is shown in Fig.12, and its sectional rigidity $EI = 5.0 \times 10^8 tf \cdot m^2 / m$. The incident wave is the following nature that amplitude is 4.6m and the period is 9.6 second, and the wave length is 115m. The wave length is the one that had been adopted as the design wave at the construction of Kansai International Airport in the sea off Osaka City, Japan.

In the beginning , Fig.13 shows the deflection pattern for non mooring case which is dependent on the rigidity of the section and in case of $EI = 5.0 \times 10^8 tf \cdot m^2 / m$ oscillation looks to have three to five peak deflection points while in the case of $EI = 5.0 \times 10^{10} tf \cdot m^2 / m$, it is reduced to two showing clearly that the amplitude of oscillation is also dependent of the rigidity of the section.

Figs.14 and 15 show the vertical displacements and horizontal displacements of a plate without mooring. From these figures, you will notice that the rigidity does not affect the amplitude until the length is 200 meters. However, beyond that length amplitude, bending moment and shear forces etc. are affected by the rigidity

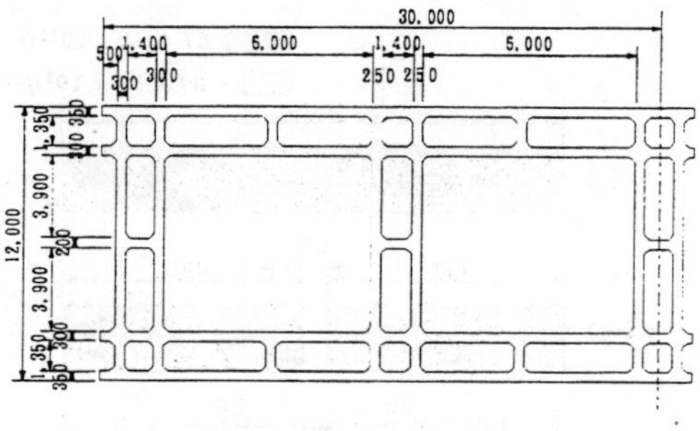

Fig.12 The Unit Section of a Calculated Floating Plate (by mm)

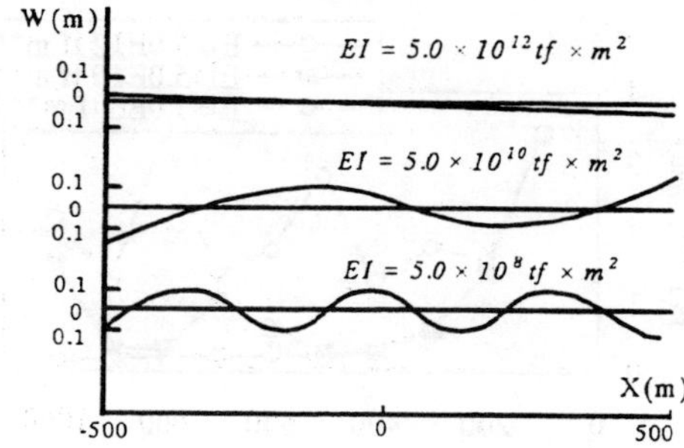

Fig.13 The Oscillating Configuration of a Plate

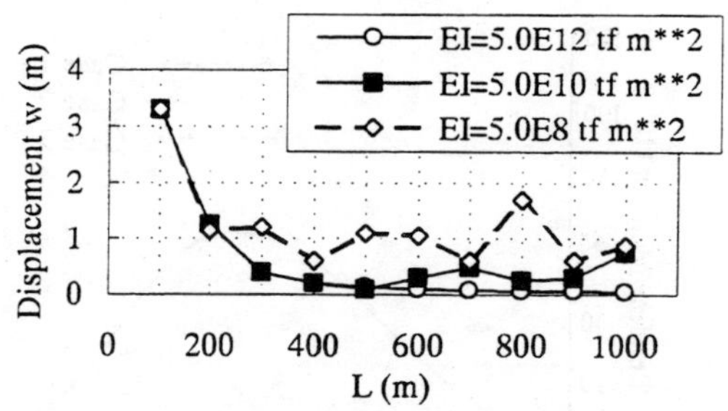

Fig.14 The Calculated Vertical Displacement for Non Moored Case

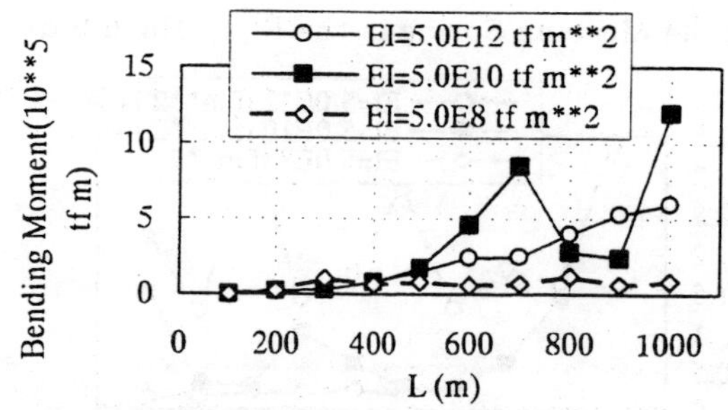

Fig.16 Bending Moment of the Section for Non Moored Case

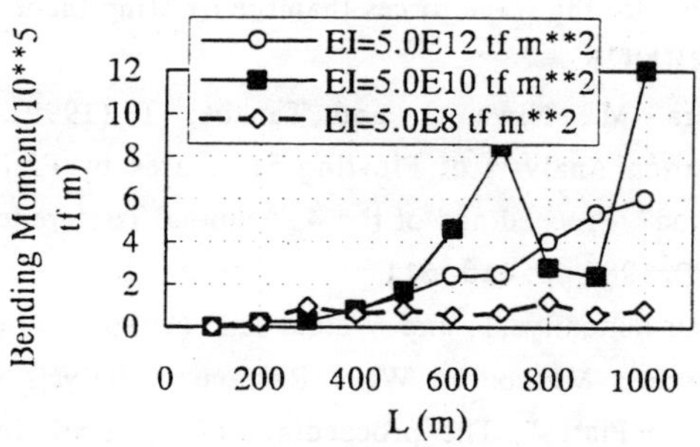

Fig.16 Bending Moment of the Section for Non Moored Case

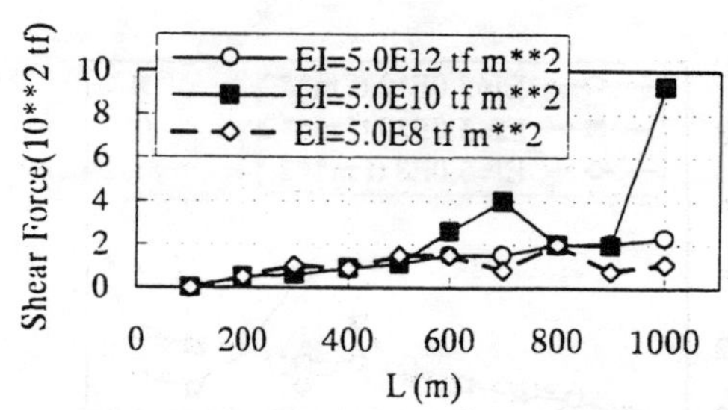

Fig.17 Shear Force of the Section for Non Moored Case

quite amount and it is concluded that the plates deflection surely be incorporated in the calculation, otherwise it is impossible to obtain the correct sectional forces. In Figs.16, and 17 the bending moments as well as shear forces are shown for non mooring case to recognize the effect of the length of a plate and the rigidity. It looks that the plate with the rigidity of $EI = 5.0 \times 10^8 tf \cdot m^2 / m$ and the rigidity of $EI = 5.0 \times 10^{12} tf \cdot m^2 / m$ gives smooth transition of those sectional forces with the increase of length of the body , however, in the case of a plate with $EI = 5.0 \times 10^{10} tf \cdot m^2 / m$, the oscillating curves are obtained.

Similar curves are obtained for moored case. In Fig.18, the mooring forces toward vertical direction and toward the horizontal direction are shown. General trend that those forces oscillate according to the length of the floating plate are also shown. All these values affect directly the determination of sectional dimension of floating structure as well as the design of mooring system.

As an additional information, it should be mentioned that the calculation method enables to calculate any type of floating structure and the results for the structure which is connected by the joint in several places and affected by the same incident wave of Kansai Air Port ,are shown in Fig.19.

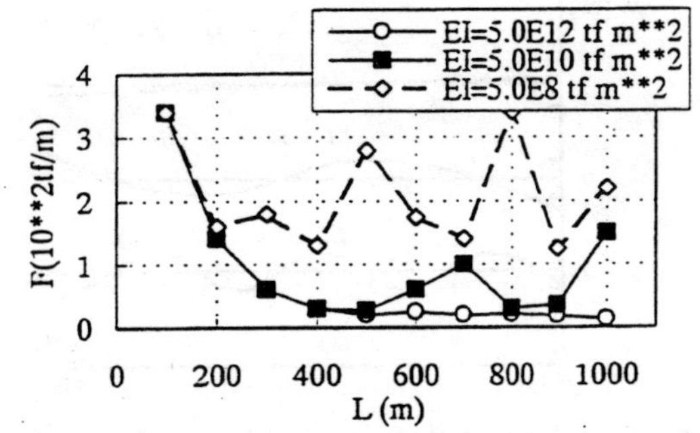

(a) For the Mooring Spring in the Right Hand Side

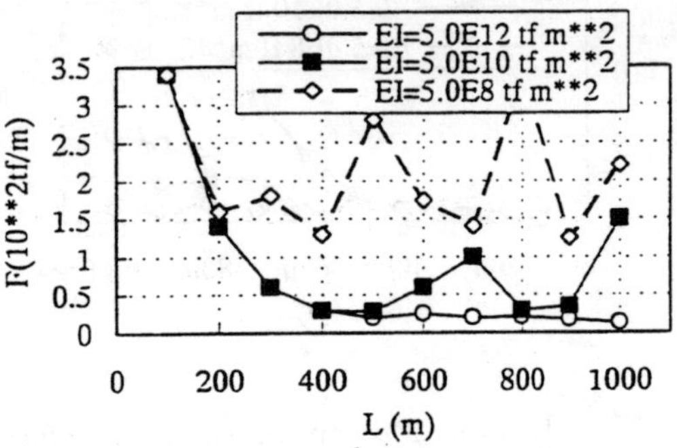

(b) For the Mooring Spring in the Left Hand Side

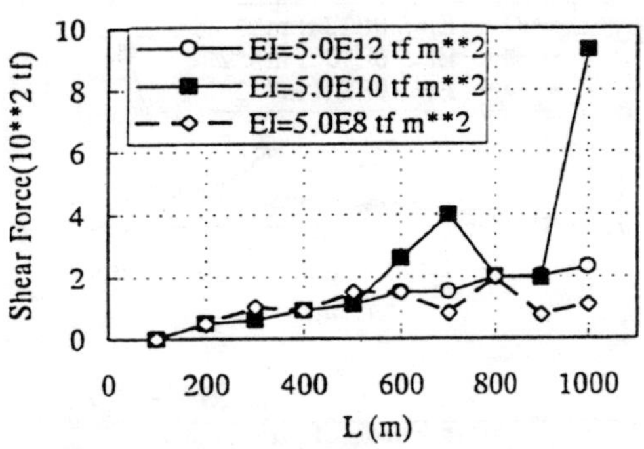

(c) Horizontal Mooring Force

Fig.18 Mooring Forces (Incident wave approaches from right-hand side)

CONCLUSION

Formulation to solve the coupled wave forces and elastic deformation analysis is presented. The differences of plate responses according to the rigidity of the plate are shown when the plate become large. It is clear that there exist a limit of the length and the rigidity of a plate that can be treated as rigid body as usually done in the normal floating body analysis. In the case of $EI = 10^8 kgf / cm^2$, the limit of the length of floating is found to be about 200m and beyond that, it is necessary to calculate response considering the deformation of the plate. The present formulation gives much wider possibility to analyze floating body which is deformable for the wave forces than the existing theories.

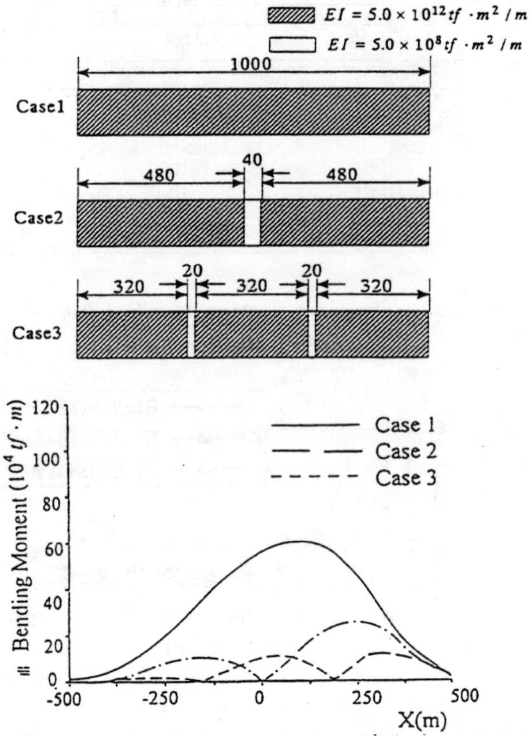

Fig.19 Bending Moment for a Plate Connected with Hinge

REFERENCE:

Nishimura, M., Niwa, J. and Tanabe, T (1990). "Flexural Vibration Analysis of Floating Structures by Finite Element Method", Proceedings of the 45th annual conference of JSCE, Vol. 2, 1988,pp. 610- 611.

Wu,C., Utshunomiya,T. and Watanabe,E (1996). "Application of Galerkin's Method in Wave Response Analysis of Flexible Floating Plates", The proceedings of the sixth international offshore and polar engineering conference, Vol. 3, pp.307-314.

Meylan,M.H. (1996). "The Wave Response of Floating Thin Plates of Shallow Draft by a Variational Method", The proceedings of the sixth international offshore and polar engineering conference, Vol. 3, pp.328-334.

Newman, J. N. (1994). "Wave Effects on Deformable Bodies, Applied Ocean Research", Vol.16, pp.47-59.

Tanaka,S. (1993). "Analysis of Wave-Structure Interactive Oscillation of a Large-Sized Floating RC Structures", Bachelor Thesis submitted to Nagoya Univ..

Ijima,T., Tabuchi,Y. and Yamura,Y.(1972). "Scattering of Surface Waves and The Motions of A Rectangular Body by Waves in Finite Water Depth", Proc. of JSCE, No.202, pp.33-48.

Takayama,T., Nagi,N., Kikuchi,O. and Moroishi,K.(1980). "Motions and Mooring Forces o a Rectangular Floating Body in Wave (1st Report), Rep. of P.H.R.I,Vol.19, No.3, pp.71-104.

Proceedings of the Seventh (1997) International Offshore and Polar Engineering Conference
Honolulu, USA, May 25-30, 1997

ISBN 1-880653-28-1 (Set); ISBN 1-880653-29-X (Vol. I)

Shallow Water Effect on Hydrodynamic Coefficients of Very Large Floating Structures

Shin-ichi Aoki
Toyohashi University of Technology
Toyohashi, Japan

ABSTRACT

An analytical formulation for wave radiation and diffraction problems due to very large floating structures is carried out. Floating structures are modelled as a one-dimensional thin beam placed on the water surface. The modal analysis for beam vibration and the eigenfunction expansion method are used for solving the boundary-value problem. Numerical calculations of hydrodynamic coefficients and wave exciting forces show significant shallow water effects especially on added masses in the elastic deformation modes.

KEYWORDS: offshore structure, very large floating structure, shallow water effect, hydrodynamic force, added mass, damping coefficient

INTRODUCTION

Constructing very large floating structures in coastal shallow water area is recently considered in Japan as an alternative of offshore reclaimed islands such as Kansai Airport. There are however some technical problems to overcome. One of them is the estimation of shallow water effects on hydrodynamic forces in the elastic deformation modes of the structures. Since the ratio of horizontal scale (structure length) to vertical scale (water depth) may be in the order of 10^2, the effects of sea bottom become more significant than ordinary floating structures. Thus the topographical change of the bottom should be considered in estimating hydrodynamic forces and also material conditions such as muddy bottom possibly influence the hydrodynamic forces in some cases.

This paper focuses on the effect of water depth on hydrodynamic forces of barge type very large floating structures. We first develop the theoretical formulation using the modal analysis and the eigenfunction expansion method for solving interaction problems between beam vibration and water waves and then discuss the shallow water effects based on the numerical results. This paper is an extension of the author's previous paper in Japanese (Aoki, 1996). Wu et al. (1995) proposed similar formulation to the author's and discussed the response of an elastic plate.

THEORETICAL FORMULATION

Modal Analysis and Natural Modes of Beam Vibration

The modal analysis is first applied to a one-dimensional beam of length $2l$ with free ends resting on an elastic foundation as shown in Fig. 1. The beam is assumed very thin so that we can apply Euler - Bernoulli theory. The elastic foundation represents restoring forces due to hydrostatic pressure. Let vertical displacement of the beam and external force per unit length be $\xi(x,t)$ and $f(x,t)$, respectively. The mass per unit length m and the flexural stiffness EI of the beam and the spring constant of the elastic foundation K are assumed constant over the beam. The differential equation for the deflection function $\xi(x,t)$ reduces to

$$m\frac{\partial^2 \xi}{\partial t^2} + EI\frac{\partial^4 \xi}{\partial x^4} + K\xi = f(x,t) \tag{1}$$

To obtain the natural modes of vibration, we solve the free vibration problem. Using the method of separation of variables, we write

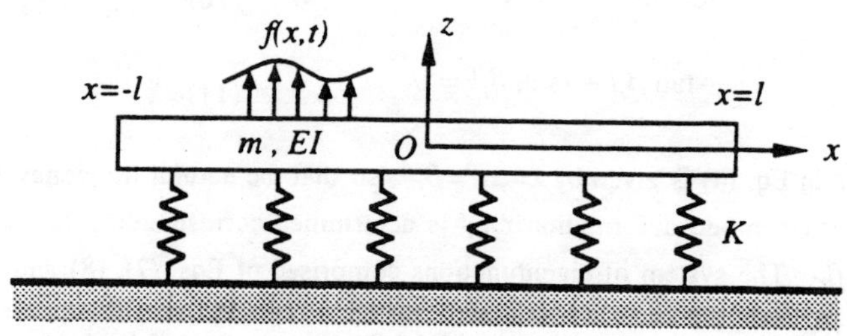

Fig. 1 An elastic beam on elastic foundation

$$\xi(x,t)=G(x)T(t) \tag{2}$$

where $G(x)$ and $T(t)$ are independent functions of x and t and represent beam deformation and its time variation, respectively. Substituting Eq. (2) into Eq. (1) with $f(x,t)=0$, we obtain

$$-\frac{1}{T(t)}\frac{d^2T(t)}{dt^2}=\frac{EI}{m}\frac{1}{G(x)}\frac{d^4G(x)}{dx^4}+\frac{K}{m} \tag{3}$$

This equation has solutions only when both sides of the equation are constant. Letting this constant be $\omega_n{}^2$, we have the following equation for $G(x)$:

$$d^4G/dx^4-\varepsilon G=0 \tag{4}$$

where $\varepsilon=(m\omega_n{}^2-K)/EI$ (>0). The general solutions of Eq. (4) that satisfy the boundary conditions at both ends:

$$d^2G/dx^2=0\,,\ d^3G/dx^3=0 \quad at\ x=\pm l \tag{5}$$

are

$$G(x)=\sum_{r=0}^{\infty}\left\{C_r^P P_r(x)+C_r^Q Q_r(x)\right\} \tag{6}$$

where $C_r{}^P$ and $C_r{}^Q$ are constants. $P_r(x)$ and $Q_r(x)$ are eigenfunctions for free vibration and expressed as

$$P_0=1/\sqrt{2}\quad,\quad Q_0=\sqrt{3/2}\cdot x/l \tag{7}$$

$$P_r=\sqrt{2}\left(\frac{\cos\alpha_r x}{\cos\alpha_r l}+\frac{\cosh\alpha_r x}{\cosh\alpha_r l}\right)\qquad r=1,2,\cdots \tag{8}$$

$$Q_r=\sqrt{2}\left(\frac{\sin\beta_r x}{\sin\beta_r l}+\frac{\sinh\beta_r x}{\sinh\beta_r l}\right)\qquad r=1,2,\cdots \tag{9}$$

α_r and β_r are eigenvalues determined by the characteristic equations:

$$\tan\alpha_r l+\tanh\alpha_r l=0 \tag{10}$$

$$-\tan\beta_r l+\tanh\beta_r l=0 \tag{11}$$

ε in Eq. (4) is given by $\varepsilon=\alpha_r{}^4=\beta_r{}^4$, so that the natural frequency ω_n of rth mode deformation ω_{nr} is determined corresponding to α_r or β_r. The system of eigenfunctions comprised of Eqs. (7), (8) and (9) are orthogonal and normalized as the integration of the square of each function over the interval $[-l,l]$ gives a value of l. As seen in Eqs. (7) - (9), P_0 and Q_0 correspond to heave and pitch modes of rigid body motion respectively, and P_r (r=1,2,..) and Q_r (r=1,2,..) represent symmetric and asymmetric modes of elastic deformation, respectively. Figure 2 shows the shapes of these modes.

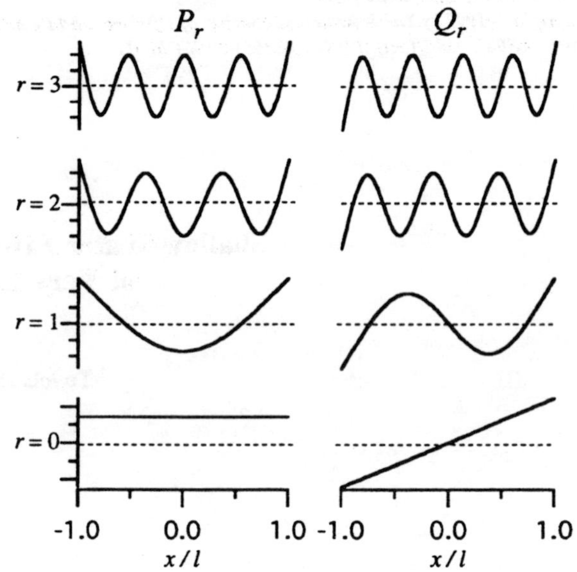

Fig. 2 Mode shapes of deformation (P_r: symmetric, Q_r: asymmetric)

Equations of Vibration and Associated Hydrodynamic Coefficients

As in Fig. 3, we consider a thin beam with the length of $2l$ vibrating harmonically in time on the water surface. Writing the system of eigenfunctions $\{P_r\ ,Q_r\ ;r=0,1,2,..\}$ simply as $\{Y_r\ ;r=0,1,2,..\}$, deflection of the beam is expressed as

$$\xi(x,t)=\sum_{r=0}^{\infty}Y_r(x)T_r(t) \tag{12}$$

from the expansion theorem (e.g. Meirovitch, 1967), where $T_r(t)$ denotes time variation of the magnitude of rth mode of vibration. Substituting Eq. (12) into Eq. (1) we obtain

$$\sum_{r=0}^{\infty}\left[mY_r\frac{d^2T_r}{dt^2}+EI\frac{d^4Y_r}{dx^4}T_r+KY_rT_r\right]=f_E(x,t)+f_R(x,t) \tag{13}$$

where we divide the external force f into two components f_E and f_R. f_E is the wave exciting force acting on a fixed beam and f_R is the radiation force due to beam vibration.

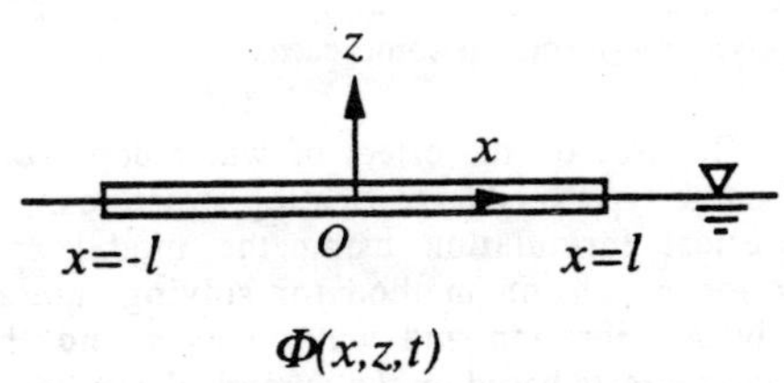

Fig. 3 A thin beam vibrating on water surface

Using complex amplitude T_{or} we express $T_r(t)$ as

$$T_r(t)=\mathrm{Re}[T_{or}e^{-i\omega t}] \tag{14}$$

where ω is angular frequency. Within the linear theory f_R also varies harmonically in time and any harmonic function can be expressed by using sine and cosine type functions: d^2T_r/dt^2 and dT_r/dt.

$$f_R(x,t) = -\sum_{r=0}^{\infty}\left[\mu(x)\frac{d^2T_r}{dt^2} + \lambda(x)\frac{dT_r}{dt}\right] \tag{15}$$

where $\mu(x)$ and $\lambda(x)$ are unknown functions. Substituting Eq. (15) into Eq. (13) we obtain

$$\sum_{r=0}^{\infty}\left[\{mY_r + \mu_r(x)\}\frac{d^2T_r}{dt^2} + \lambda_r(x)\frac{dT_r}{dt} + EI\frac{d^4Y_r}{dx^4}T_r + KY_rT_r\right] = f_E \tag{16}$$

We next multiply the both sides of this equation by $Y_s(x)$ (s=0,1,2, ...) and then integrate them over the interval $[-l, l]$. Owing to the orthogonality of the eigenfunctions, we have

$$m\frac{d^2T_s}{dt^2} + \sum_{r=0}^{\infty}\left[\frac{1}{l}\int_{-l}^{l}\mu_r(x)Y_s(x)dx\cdot\frac{d^2T_r}{dt^2} + \frac{1}{l}\int_{-l}^{l}\lambda_r(x)Y_s(x)dx\cdot\frac{dT_r}{dt}\right] + m\omega_{ns}^2T_s = \frac{1}{l}\int_{-l}^{l}f_E(x,t)Y_s(x)dx \quad (s = 0,1,2,...) \tag{17}$$

This gives the expression of added mass M_{sr} and damping coefficient N_{sr} in sth mode due to the vibration in rth mode:

$$M_{sr} = \frac{1}{l}\int_{-l}^{l}\mu_r(x)Y_s(x)dx \;,\; N_{sr} = \frac{1}{l}\int_{-l}^{l}\lambda_r(x)Y_s(x)dx \tag{18}$$

Next we express the velocity potential $\Phi(x,z,t)$ due to beam vibration as

$$\Phi(x,z,t) = \mathrm{Re}\left[\phi(x,z)e^{-i\omega t}\right] = \mathrm{Re}\left[\sum_{r=0}^{\infty}\phi_r(x,z)(-i\omega)T_{or}e^{-i\omega t}\right] \tag{19}$$

where $\phi_r(x,z)$ is the radiation potential due to vibration in rth mode normalized by the complex amplitude of the vibration $-i\omega T_{or}$. The radiation force $f_R(x,t)$ is expressed as

$$f_R(x,t) = -\rho\left.\frac{\partial\Phi}{\partial t}\right|_{z=0} = \mathrm{Re}\left[i\rho\omega\phi(x,0)e^{-i\omega t}\right]$$

$$= \mathrm{Re}\left[-\rho\sum_{r=0}^{\infty}\left\{\mathrm{Re}(\phi_r)(-\omega^2T_{or}) + \omega\,\mathrm{Im}(\phi_r)(-i\omega T_{or})\right\}e^{-i\omega t}\right]$$

$$= -\sum_{r=0}^{\infty}\left[\rho\,\mathrm{Re}[\phi_r(x,0)]\frac{d^2T_r}{dt^2} + \rho\omega\,\mathrm{Im}[\phi_r(x,0)]\frac{dT_r}{dt}\right] \tag{20}$$

Comparing this equation with Eq. (15) we can define $\mu(x)$ and $\lambda(x)$ and obtain the final expressions of M_{sr} and N_{sr}

$$M_{sr} = \frac{\rho}{l}\int_{-l}^{l}\mathrm{Re}[\phi_r(x,0)]Y_s(x)dx \tag{21}$$

$$N_{sr} = \frac{\rho\omega}{l}\int_{-l}^{l}\mathrm{Im}[\phi_r(x,0)]Y_s(x)dx \tag{22}$$

Assuming the wave exciting force f_E to be in the form:

$$f_E(x,t) = \mathrm{Re}\left[f_{oE}(x)e^{-i\omega t}\right] \tag{23}$$

the right hand side of Eq. (17) can be expressed as

$$F_{Es}(t) = \mathrm{Re}\left[F_{oEs}e^{-i\omega t}\right] = \mathrm{Re}\left[\frac{1}{l}\int_{-l}^{l}f_{oE}(x)Y_s(x)dx\cdot e^{-i\omega t}\right] \tag{24}$$

where F_{OEs} is the complex amplitude of the exciting force in the sth mode of vibration. The equation of vibration finally reduces to

$$m\frac{d^2T_s}{dt^2} + \sum_{r=0}^{\infty}\left[M_{sr}\frac{d^2T_r}{dt^2} + N_{sr}\frac{dT_r}{dt}\right] + m\omega_{ns}^2T_s = F_{Es}(t) \tag{25}$$

Boundary Value Problem and its Solution

a) Radiation problem

We here show the method to obtain the radiation potential ϕ_r due to harmonic vibration of the thin beam set on the water surface. The boundary condition underneath the beam is

$$\partial\xi/\partial t = \partial\Phi/\partial z \qquad at\ z = 0 \tag{26}$$

Considering Eqs. (12), (14) and (19), Eq. (26) yields the boundary condition for ϕ_r :

$$\partial\phi_r/\partial z = Y_r \qquad at\ z = 0 \tag{27}$$

To obtain solutions of Laplace equation that satisfy the above condition, the free surface and bottom conditions, we use the eigenfunction expansion method. As shown in Fig. 4, we divide the region of water of depth h into three subregions: (-), (o) and (+) and let the radiation potentials corresponding to the regions be ϕ_{r-}, ϕ_{ro} and ϕ_{r+}. ϕ_{r-} and ϕ_{r+} are expressed as

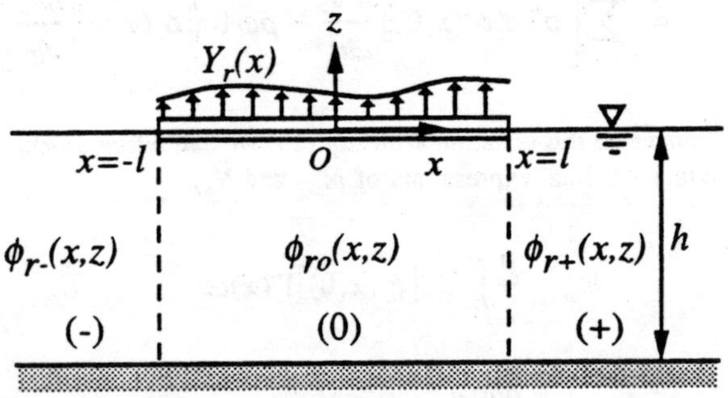

Fig. 4 Divided water region and velocity potentials

$$\phi_{r\pm} = A^{\pm} e^{\pm ik(x\mp l)} \frac{\cosh k(z+h)}{\cosh kh} + \sum_{n=1}^{\infty} B_n^{\pm} e^{\mp k_n (x\pm l)} \frac{\cos k_n (z+h)}{\cos k_n h} \tag{28}$$

where $A^{\pm}$ and $B_n^{\pm}$ (n = 1, 2, ...) are complex unknown coefficients. k and k_n are eigenvalues that satisfy the equations:

$$\omega^2 h / g = kh \tanh kh = -k_n h \tan k_n h \tag{29}$$

The radiation potential in the region (o) is expressed as

$$\phi_{ro} = \frac{1}{h}\left\{ C_0 + D_0 \frac{x}{l} - \int\left(\int Y_r(x)dx\right)dx \right\} + \frac{2}{h}\sum_{m=1}^{\infty}\left[C_m \frac{\cosh \mu_m x}{\cosh \mu_m l} + D_m \frac{\sinh \mu_m x}{\sinh \mu_m l} + \frac{(-1)^m}{\mu_m}\left\{ \cosh \mu_m x \int \sinh \mu_m x \cdot Y_r(x)dx - \sinh \mu_m x \int \cosh \mu_m x \cdot Y_r(x)dx \right\}\right] \cos \mu_m (z+h) \tag{30}$$

by determining an inhomogeneous solution that satisfies Eq. (27) using the Fourier cosine transform, where C_m and D_m (m = 1, 2, ...) are complex unknown coefficients and $\mu_m = m\pi/h$.

To determine the unknown coefficients in Eqs. (28) and (30), we use the continuity conditions of pressure and velocity at the imaginary boundaries that separate the regions:

$$\phi_{r\pm}\big|_{x=\pm l} = \phi_{ro}\big|_{x=\pm l} \ , \quad \partial\phi_{r\pm}/\partial x\big|_{x=\pm l} = \partial\phi_{ro}/\partial x\big|_{x=\pm l} \tag{31}$$

According to Yoshida et al. (1990), we make up a set of linear algebraic equations for the unknown coefficients applying Eq. (31) to the reference points taken on the imaginary boundaries as the number of them becoms the same as that of the unknowns. Substituting the radiation potential solved into Eqs. (21) and (22) and integrating them, we obtain the added mass M_{sr} and damping coefficient N_{sr}, where the symmetric relations $M_{sr}=M_{rs}$ and $N_{sr}=N_{rs}$ are satisfied. Furthermore, symmetric and asymmetric modes of vibration expressed by P_r and Q_r are fully independent of each other and no coupling effects exist between these modes.

b) Diffraction problem

Similar analysis to the radiation problem can be applied to the diffraction problem except that the boundary condition underneath the beam becomes $\partial\Phi/\partial z=0$ instead of Eq. (26) and we add the incident wave component to the velocity potential in either region (-) or (+). The coefficient matrix of the simultaneous equations is the same as in the radiation problem. The complex amplitude of the wave exciting force F_{oEs} is given by Eq. (24).

RESULTS

Required Number of Eigenfunctions in Velocity Potentials

We first discuss the required number of eigenfunctions in series form of the velocity potentials in Eqs. (28) and (30). We let the number of eigenfunctions be n both in the equations and investigate the property of convergence of the solution with increasing n, where the number of reference points on the imaginary boundaries becomes n+1. Figure 5 shows the variations of added mass $M_{sr}{}^P$ for symmetric modes, where the added masses are normalized by that for n=30. We find that a large number of eigenfunctions are necessary in the calculation for higher modes of vibration.

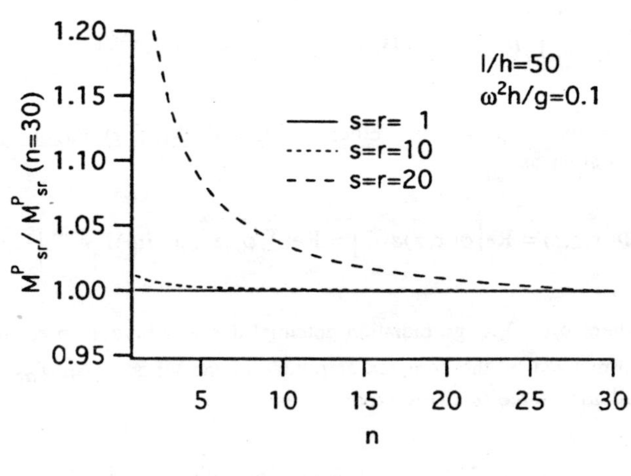

Fig. 5 Influence of the number of eigenfunctions on convergence of solution

Shallow Water Effects on Hydrodynamic Coefficients

The hydrodynamic coefficients are functions of beam length $2l$, water depth h and period of vibration $T=2\pi/\omega$. To make difference clear we take simulation cases for very long (l=1000m) and short (l=10m) beams. Figure 6 shows the added masses and damping coefficients for symmetric modes as functions of water depth, where the subscripts $s=r=0$ indicates the heave motion as a rigid body. The variation in water depth gives rapid increase or decrease in the added masses in shallow water (h<20m), whereas the damping coefficients varies gradually with varying water depth. Furthermore, three response curves of added mass for different T agree with each other showing no influence of vibration period.

In Fig. 7 the hydrodynamic coefficients are plotted as functions of vibration period taking water depth h as a parameter. We find little effect of vibration period on the added masses and conversely a small difference in water depth gives great difference in the added masses. In the figures of damping coefficients, we notice both effects of vibration period and water depth. The reason why the dependencies of added mass and damping coefficient on vibration period and water depth show different characteristics may have relation to fluid motion under the beam. In shallower water the stream lines of fluid motion under a long beam induced by its vibration becomes horizontal like that of shallow water waves for any modes or periods of vibration. The added masses that have close relations to standing wave components thus depends only on water depth. On the other hand, the damping coefficients are functions of wave energy propagation velocity c_g of radiating waves from the beam, which depends on wave period as long as the ratio of water depth to wavelength is not very small.

Figure 8 shows the results for the short beam (l=10m). We find the effects of vibration period even on the added masses and less effects of water depth on damping coefficients than in Fig. 7. We have shown the results only for the symmetric modes of vibration. The hydrodynamic coefficients in asymmetric modes have similar characteristics to the symmetric modes.

Shallow Water Effects on Wave Exciting Force

Generally the response curves of the wave exciting force in the ith mode takes similar forms to that of the diagonal term of damping coefficients in the same ith mode. As in the figures of damping coefficients in Fig. 7, the exciting forces become large in the range of shorter wave period for higher modes of vibration and the dependence on the water depth is not as great as added mass.

CONCLUSIONS

The shallow water effects on hydrodynamic forces acting on very large floating structures are discussed based on numerical simulations in which the modal analysis and the eigenfunction expansion method are applied to the interaction problem between water waves and a thin beam placed on the water surface. Significant shallow water effects are found especially on the added mass.

ACKNOWLEDGMENTS

Financial support from Fundamental Research Developing Association for Shipbuilding and Offshore is gratefully acknowledged.

REFERENCES

Aoki, S. (1996): Shallow water effect on hydrodynamic coefficients of very large floating structures, Proc. of Coastal Eng., JSCE, vol.43, No.1, pp.926-930 (in Japanese).

Meirovitch L. (1967): Analytical Methods in Vibrations, Macmillan Publishing Co., Inc., 555p.

Wu C., E. Watanabe and T. Utsunomiya (1995): An eigenfunction expansion-matching method for analyzing the wave-induced responses of an elastic floating plate, Applied Ocean Research, vol. 17, No.5, pp.301-310.

Yoshida, A., H. Kojima and Y. Tsurumoto (1990): A collocation method of matched eigenfunction expansions on the boundary-value problem of wave-structure interactions, Proc. JSCE, 417/II-13, pp.265-274 (in Japanese)

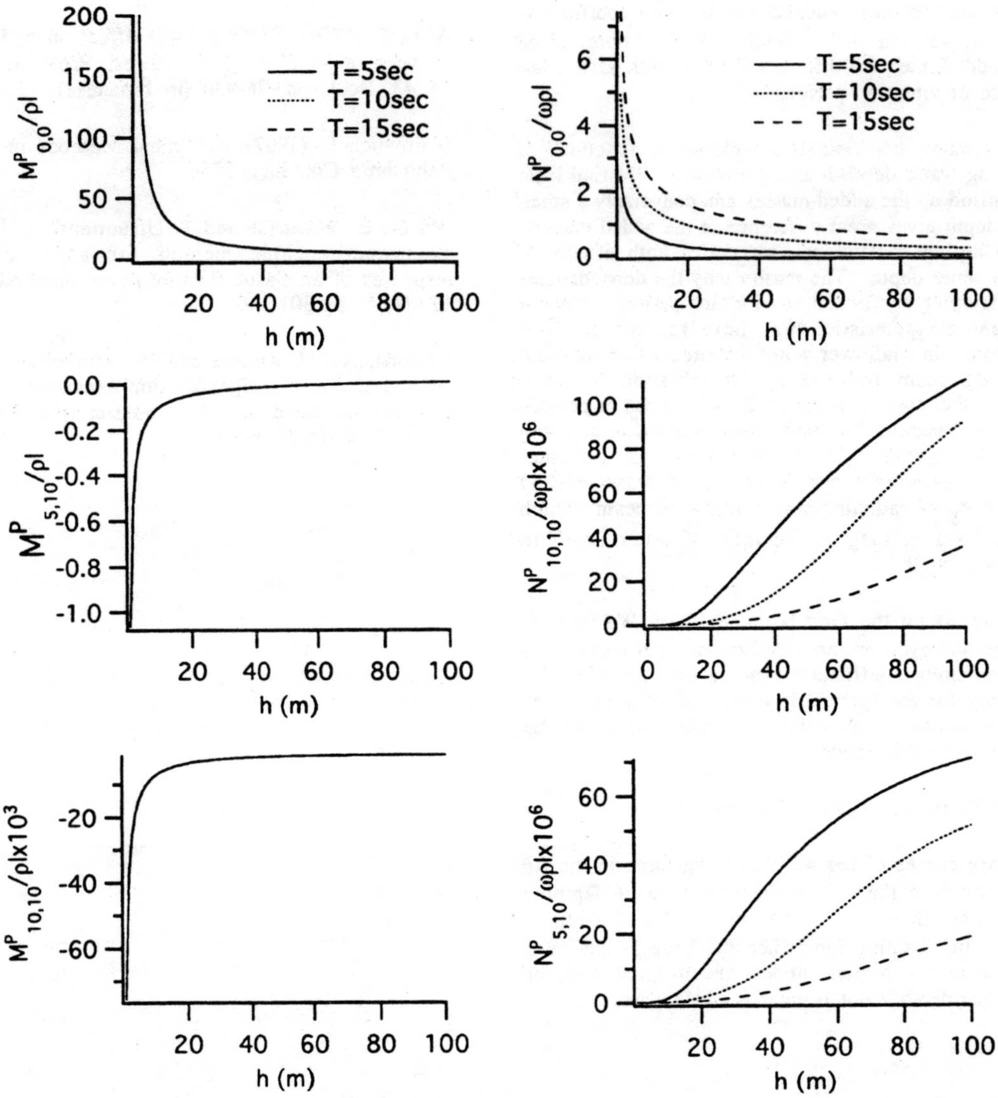

Fig. 6 Added mass and damping coefficient as functions of water depth (symmetric modes, l=1000m)

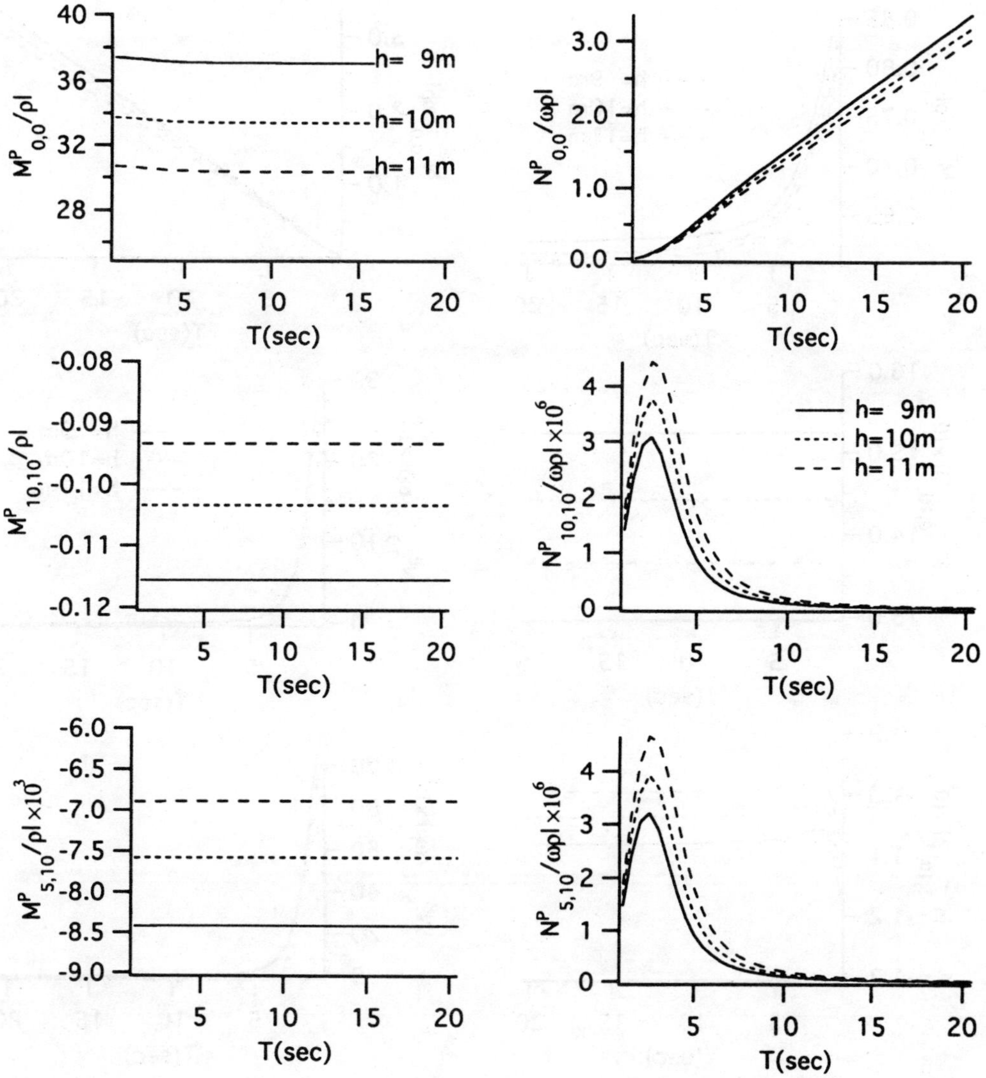

Fig. 7 Added mass and damping coefficient as functions of vibration period (symmetric modes, l=1000m)

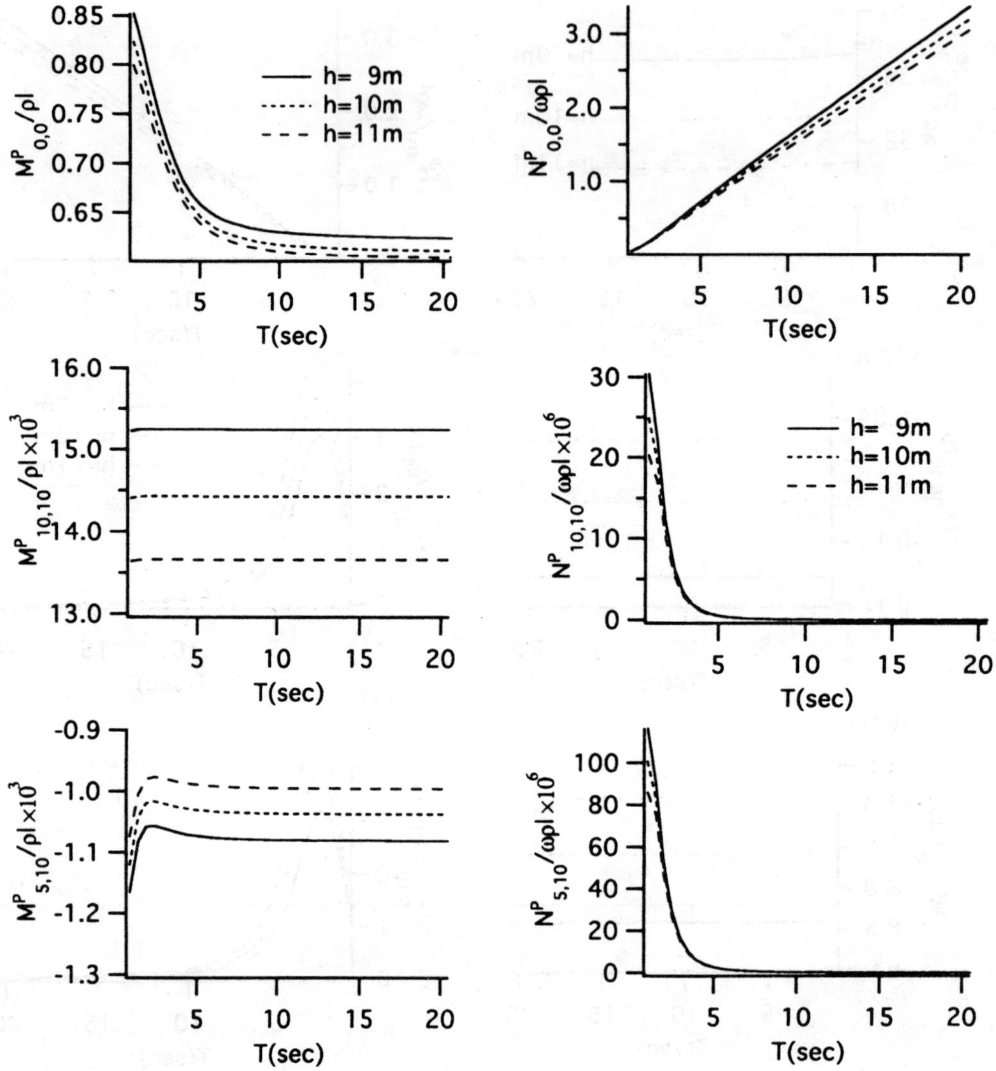

Fig. 8 Added mass and damping coefficient as functions of vibration perio (symmetric modes, l=10m)

Proceedings of the Seventh (1997) International Offshore and Polar Engineering Conference
Honolulu, USA, May 25-30, 1997

ISBN 1-880653-28-1 (Set); ISBN 1-880653-29-X (Vol. I)

Effects of Seaquake on Offshore Floating Structure

Kouji Mochizuki
Mitsui Engineering Shipbuilding Co.
Tokyo, Japan

Hajime Kihara, Ken Takagi and Shigeru Naito
Osaka University
Osaka, Japan

ABSTRACT

Seaquakes are generally considered the phenomena that a ship or a floating structure are shocked by the seismic waves which are propagated through the seawater. The effects of seaquakes on offshore floating structures are studied. Both effects concerning the compressibility of fluid and the free surface are discussed. The dynamic pressure acting on floating structures by seaquakes is investigated. The resonance is the important role of the effects of seaquakes on offshore floating structures. Numerically, boundary value problems are solved using the Green function method. After computations of the radiation hydrodynamic force and the vertical exciting force by seaquake, the frequency response of structure's heaving motions is evaluated.

KEYWORD : Seaquake, Compressibility of fluid, Resonance, Frequency of primary wave, Floating structure

1. INTRODUCTION

In an effort to create much needed space, coastal developments such as the reclamation of shallow beach front areas and the creation of off shore artificial islands have been evolved in Japan. These developments are beneficial in areas where population density is concentrated and outward expansion is impractical due to surrounding mountainous terrain. The land reclamation has been the general policy to date, however a recent growing realization of the clear possible long term negative effects upon the coastal ocean environment of such land reclamation projects has given impetus to interest in offshore floating structures.

The Great Hanshin Earthquake of 1995 which struck the Kansai area encouraged this movement toward the development of floating structures. In the artificial islands off Kobe, the massive earthquake caused the liquefaction of reclaimed lands and the quay wall built by the caisson method led to serious damage. However it is important for floating pier structures to survive the earthquake with a significant lack of damage. Additionally, there are no reports that any ships in Kobe bay area received by the earthquake. The lack of damage to ships is partly because the earthquake did not induce tsunami waves and seaquakes. Offshore floating structures are designed with quake-proof ability in mind, however the effects of the vertical oscillations on such offshore floating structures remains unclear. Seaquakes are a phenomenon in which a ship or a offshore floating structure receive a shock caused by the seismic waves which are propagated through the seawater. It is deemed necessary to investigate the effects of vertical oscillations caused by earthquakes on offshore floating structure as vertical oscillations are dominant. In situations where a floating structure is moored at a quay wall utilizing a tight mooring system, there is a possibility that the floating structure will suffer from the effects of horizontal oscillations earthquake induces. [5)]

This paper describes the effects of vertical oscillations caused by seaquakes on offshore floating structures. Whether the compressibility in seawater and the free surface waves are to be considered significant or not is shown here. The hydrodynamic forces and the frequency response of the heaving motions are evaluated. The resonance phenomenon plays an important role considering the effects of seaquake on offshore floating structures. It is also confirmed that there exist some resonance modes in seaquakes. In numerical work, the Green function method is employed to solve the problem.

2. SEAQUAKE

2.1 TSUNAMI AND SEAQUAKE

There are tsunami waves and seaquakes as ocean phenomena induced by an earthquake. An earthquake causes a sea-bed deformation such as an upheaval or a subsidence in a very short time. Consequently the disturbances in the seawater propagate all around as long surface waves and grow extremely-developed waves called tsunami near a coast.

On the other hand, a structure on the free surface suffers the strong shock from an earthquake, just as it suffers the shock on the ground. This is because the sea-bed vibrations by an earthquake reach the free surface through the seawater as longitudinal (compressive) waves. According to some reports, the feeling looks as if a ship struck a rock. Such a phenomenon is called seaquake, which is distinguished from tsunami.

2.2 OSCILLATION OF SEA-BED

It is usually known that the seismic waves are classified into two kinds of waves according to their properties. One is called P-wave (primary wave) which is observed for the initial duration of the vibrations and both amplitude and period are relatively small. P-wave is a longitudinal wave and oscillates in parallel with the propagative direction. The other is called S-wave (secondary wave) which is observed after P-wave and its amplitude is larger. S-wave is a transverse wave and oscillates vertically to the propagative direction. The incident direction of seismic waves is nearly perpendicular to the sea-bed because the propagative velocity becomes more slowly with coming close to the sea-bed. Therefore, the vertical oscillations are caused by P-wave and the horizontal oscillations are by S- wave.

3. FORMULATION OF SEAQUAKE

3.1 VELOCITY POTENTIAL OF SEAQUAKE

The compressibility of fluid has been generally neglected considering the mathematical formulation of seaquakes, partly because the propagative velocity of longitudinal waves is fast[4),7)]. However, this assumption is inadequate to explain phenomena of the past reports about seaquakes[2),3)]. While the frequency of P-wave is considered high, its band width is wide, and besides the dominant frequency is not so high. Suppose that the following effects are valid in this section.

(i) The effect of compressible of fluid

(ii) The effect of surface wave

Assuming the fluid is inviscid and irrotational, its motion is specified by the velocity potential. Table.1 shows considering assumptions and the description of each velocity potential.

Table 1: Velocity potentials of seaquake

	Surface wave	No surface wave
Compressible	ϕ_1	ϕ_2
Incompressible	ϕ_3	ϕ_4

3.2 GOVERNING EQUATION AND BOUNDARY CONDITIONS

Figure 1 shows the mathematical model of a seaquake and its coordinate system. The wave source is part of the sea-bed region with length $2a$ at the water depth $-h$, whose vertical

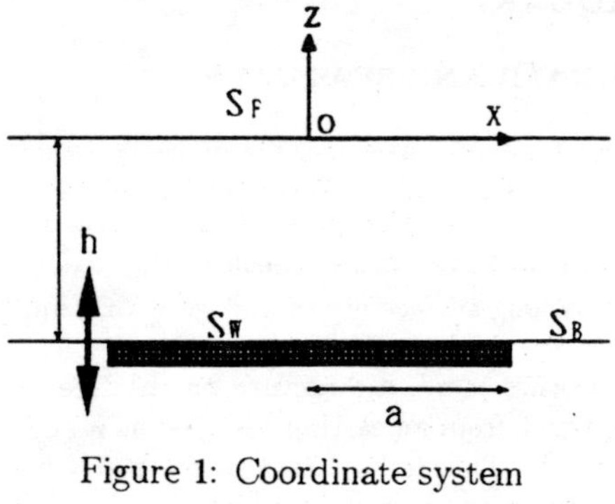

Figure 1: Coordinate system

oscillation is harmonic. The boundary value problems to be solved are described in Table 2. Governing equations and free surface conditions in this table are as follows :

Helmholtz equation [Eq.(1)]

$$\nabla^2\phi_l + K^2\phi_l = 0 \tag{1}$$

Laplace equation [Eq.(2)]

$$\nabla^2\phi_l = 0 \tag{2}$$

Free surface condition [Eq.(3)]

$$\frac{\partial\phi_l}{\partial z} - \nu\phi_l = 0 \tag{3}$$

Free surface condition [Eq.(4)]

$$\phi_l = 0 \tag{4}$$

where ω is angular frequency, $K = (\omega/c)$ is the wave number, c (=1500m/s) is the underwater sound velocity, $\nu = \omega^2/g$, g is the gravitational acceleration and v_n is the normal velocity of the wave source surface.

Table 2: Governing equations and boundary conditions

	ϕ_1	ϕ_2	ϕ_3	ϕ_4
Governing equation (in Ω)	Eq.(1)		Eq.(2)	
Free surface condition (on S_F)	Eq.(3)	Eq.(4)	Eq.(3)	Eq.(4)
Sea-bed condition (on S_B)	$\partial\phi_l/\partial z = 0$			
Wave source condition (on S_W)	$\partial\phi_l/\partial z = v_n$			
Radiation condition				

S_F:Free surface, S_B:Sea-bed, S_W:wave source surface

3.3 GREEN FUNCTIONS

We introduce Green functions G_l ($l = 1 \sim 4$) which satisfy every boundary condition except for the boundary condition on the wave source and are expressed using following eigen functions [1),6)].

$$G_1(x,z,\xi,\eta) = \frac{-i}{2\lambda_0}\varphi_0(z)\varphi_0(\eta)e^{-i\lambda_0|x-\xi|} + \sum_{n=1}^{s}\frac{-i}{2\lambda_n}\varphi_n(z)\varphi_n(\eta)e^{-i\lambda_n|x-\xi|} + \sum_{n=s+1}^{\infty}\frac{1}{2\lambda_n}\varphi_n(z)\varphi_n(\eta)e^{-\lambda_n|x-\xi|} \tag{5}$$

$$G_2(x,z,\xi,\eta) = \sum_{n=1}^{s}\frac{-i}{2\lambda'_n}\varphi'_n(z)\varphi'_n(\eta)e^{-i\lambda'_n|x-\xi|} + \sum_{n=s+1}^{\infty}\frac{1}{2\lambda'_n}\varphi'_n(z)\varphi'_n(\eta)e^{-\lambda'_n|x-\xi|} \tag{6}$$

$$G_3(x,z,\xi,\eta) = \frac{-i}{2\lambda_0}\varphi_0(z)\varphi_0(\eta)e^{-ik_0|x-\xi|} + \sum_{n=1}^{\infty}\frac{1}{2\lambda_n}\varphi_n(z)\varphi_n(\eta)e^{-k_n|x-\xi|} \tag{7}$$

$$G_4(x,z,\xi,\eta) = \sum_{n=1}^{\infty}\frac{1}{2\lambda'_n}\varphi'_n(z)\varphi'_n(\eta)e^{-k'_n|x-\xi|} \tag{8}$$

$$
\varphi_n(z) = \begin{cases} \sqrt{\dfrac{2k_0{}^2}{h(k_0{}^2-\nu^2)+\nu}}\dfrac{\cosh k_0(z+h)}{\cosh k_0 h} \\ (\text{for} \quad n=0) \\ \sqrt{\dfrac{2k_n{}^2}{h(k_n{}^2+\nu^2)-\nu}}\dfrac{\cos k_n(z+h)}{\cos k_n h} \\ (\text{for} \quad n=1,2\cdots) \end{cases} \tag{9}
$$

$$
\lambda_n = \begin{cases} \sqrt{K^2+k_0^2} \\ (\text{for} \quad n=0) \\ \sqrt{K^2-k_n^2} \\ (\text{for} \quad n=1,2\cdots s) \\ \sqrt{k_n^2-K^2} \\ (\text{for} \quad n=s+1,s+2\cdots) \end{cases} \tag{10}
$$

$$
\varphi_n'(z) = \sqrt{\frac{2}{h}}\sin k_n' z \quad (\text{for} \quad n=1,2\cdots) \tag{11}
$$

$$
\lambda_n' = \begin{cases} \sqrt{K^2-k_n'^2} \\ (\text{for} \quad n=1,2,3\cdots s) \\ \sqrt{k_n'^2-K^2} \\ (\text{for} \quad n=s+1,s+2\cdots) \end{cases} \tag{12}
$$

where k_0 and k_n $(n=1,2\cdots)$ are positive real numbers obtained by solving $k_0\tanh k_0 h=\nu$, $k_n\tan k_n h=-\nu$ $(k_1<k_2<\cdots)$, respectively, $k_n'=(2n-1)\pi/2h$ $(n=1,2\cdots)$, and s is the maximum value of natural numbers satisfying k_n, $k_n'<K$.

3.4 APPROACH BY INTEGRAL EQUATION

Applying Green's theorem to these boundary value problems, boundary integral equations are obtained. Assuming the position of wave sources is on the sea-bed, the velocity potentials ϕ_l are expressed with the following integral form.

$$
\phi_l(P) = \int_{S_W} v_n(Q)G_l(P,Q)ds, \quad (l=1\sim 4) \tag{13}
$$

where $P(x,z)$is a field point, $Q(\xi,\eta)$is a source point. The velocity potentials $\phi_l(l=1\sim 4)$ can be obtained by analytical integration in equation (13) here, because it is not an integral equation any more. However, in case of diffraction and radiation problems which are dealt with at second stage, an integral equation must be solved.

3.5 DYNAMIC PRESSURES BY SEAQUAKE

The dynamic pressures p_l $(l=1\sim 4)$ can be written by using the linearized Bernoulli's equation.

$$
p_l(x,z) = -i\rho\omega\phi_l(x,z), \quad (l=1\sim 4) \tag{14}
$$

where ρ is the density of the fluid.
The dynamic pressures in seawater are computed about three cases. In all calculations the half breadth of wave source is 500 [m] and the water depth is 100[m], i.e., these ratio a/h is 5, referring to the Izu-Hanto-Toho-Oki Earthquake in 1980 [8]

CASE(1)

The acceleration amplitude A_0 of the sea-bed is constant over a range of every frequency and the whole wave source region oscillates uniformly. Figure 2 shows frequency response amplitude function of dynamic pressures p_l $(l=1\sim 4)$ at the observation point (x/h= 5.5, z/h= −0.05). The vertical axis is nondimensional parameter $|p_l|/\rho A_0 D$, where D is the distance from the free surface to the observation point. The horizontal axis is nondimensional frequency $Kh(=\omega h/c)$. Each dynamic pressure value is almost equal near $Kh=\pi/8$. Over a range less than $\pi/8$, it clearly shows differences between the cases (p_1, p_3) and the cases (p_2, p_4). The former cases take account of the surface waves effect, and the latter cases do not. Therefore, the effect of surface waves is dominant over a range of low frequencies. As the frequency gets higher, the dynamic pressures (p_3, p_4) in incompressible fluid are constant, while the dynamic pressures (p_1, p_2) in compressible fluid become bigger. Therefore, the effect of the compressibility is no more negligible over a range of high frequencies.

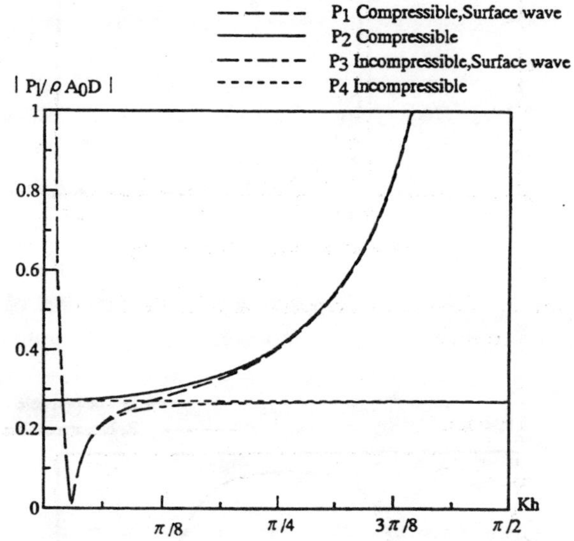

Figure 2: Frequency response amplitude function of dynamic pressure ($x/h=5.5$, $z/h=-0.05$)

Figure 3 shows dynamic pressures at two different observation points, where the horizontal distance from the center of the wave source is different. Observation points are located at (x/h= 0.0, z/h= −0.05), (x/h = 10.0, z/h= −0.05) respectively. Dynamic pressure p_4 is constant value, while dynamic pressure p_2 has sharp peak values at the same frequencies $Kh=\pi/2$ and $3\pi/2$ in each figure. This is due to the resonance for the longitudinal wave between the surface of sea and sea-bed, where the longitudinal wave pattern is node at sea-bed and is antinode at the free surface. According to sailors who encountered seaquake, it is reported as follows, [8] "We felt very strong shock as if the ship was running ashore or received a bomb" and "We encountered seaquake several times after an interval". The resonance enables us to explain these facts and it implies the necessity to take account of the compressibility. Furthermore, figure 4 shows the propagative characteristic of dynamic pressure, where the horizontal axis is the distance x/h and the wave source locates in $|x/h|\leq 5$. The depth of the observation points are all constant and two cases of Kh=$2\pi/5$, $2\pi/3$ are computed. The dynamic pressure p_4 is constant above the wave source and is decayed gradually as it moves away from the wave source. The dynamic pressure p_2 propagates without decay in case of Kh=$2\pi/3$. The latter result has the same tendency as the fact that seaquake propagates even at a long distance from the wave source [8]. The propagative characteristic of seaquakes has relations closely with the compressibility of fluid. The dynamic pressure p_2 dose not propagate a long away off in case of Kh=$2\pi/3$ even considering the compressibility of fluid. Adding to the analytical interpretation, the pressure p_2 can be written as follow :

$$p_2(x,z) = \sum_{n=1}^{s} \frac{-\rho\omega}{2\lambda'_n}\varphi'_n(z)\varphi'_n(\eta)e^{\mp i\lambda'_n x}\int_{-a}^{a} v_n(x)e^{\pm i\lambda'_n\xi}d\xi + \sum_{n=s+1}^{\infty}\frac{-i\rho\omega}{2\lambda'_n}\varphi'_n(z)\varphi'_n(\eta)e^{\mp\lambda'_n x}\int_{-a}^{a} v_n(x)e^{\pm\lambda'_n\xi}d\xi \quad (15)$$

$$(x \geq |a|)$$

where $v_n(x) = A_0/i\omega$. The first term of the right hand in above equation is the progressive wave's component, and satisfies the condition : $Kh - (2n-1)\pi/2 > 0$. So the progressive wave's component is not induced in case of Kh=$2\pi/3$.

CASE(2)

Here the acceleration amplitude is $-A_0$ in $-a \leq x \leq 0$, and A_0 in $0 \leq x \leq A_0$. It is assumed that earthquake fault is located at $x = 0$ and oscillatory phase is different in two region. Figure 5 shows the propagation of the dynamic pressures p_2, p_4 at z/h= -0.05. It is found that the dynamic pressures are different by π in phase and are anti-symmetry with respect to $x = 0$. Moreover, the maximum value of the dynamic pressure is accord with the results in figure 4.

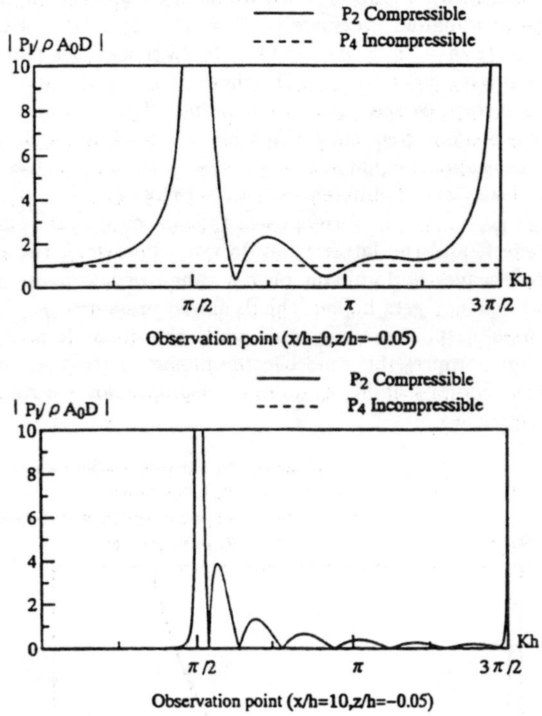

Figure 3: Frequency response amplitude function of dynamic pressure

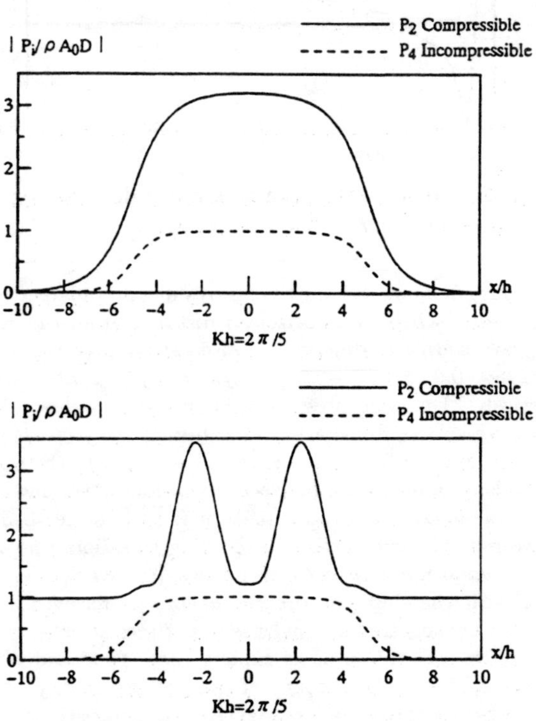

Figure 4: Propagation of dynamic pressure

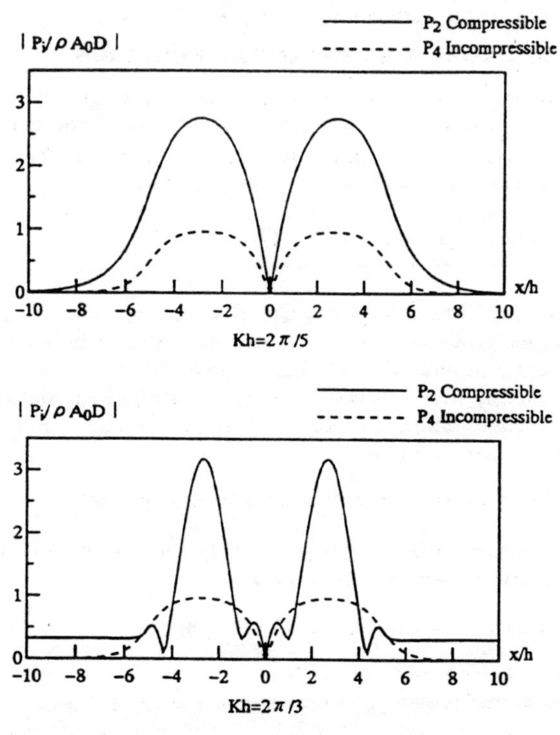

Figure 5: Propagation of dynamic pressure

CASE(3)

The displacement amplitude Z_0 of the sea-bed is constant over a range of every frequency and the whole wave source region oscillates uniformly. Let us consider the free surface displacement here. Figure 6 shows frequency response amplitude function of the surface wave at the observation point (x/h= 10.0, z/h= −0.05). The vertical axis is the nondimensional value η_w/Z_0 of the surface wave, where η_w is the amplitude of the surface wave. The surface wave is induceded only over a range of low frequency (Kh=0∼$3\pi/80$), just as it was confirmed in figure 2.

From above all studies, the assumption of the compressibility of fluid is necessary to explain seaquake phenomenon. The effect of the compressibility is not negligible as the frequency get higher, while the effect of the free surface is negligible in such a frequency domain. Therefore these assumptions is used to estimate the hydrodynamic forces acting on the floating structure in next stage.

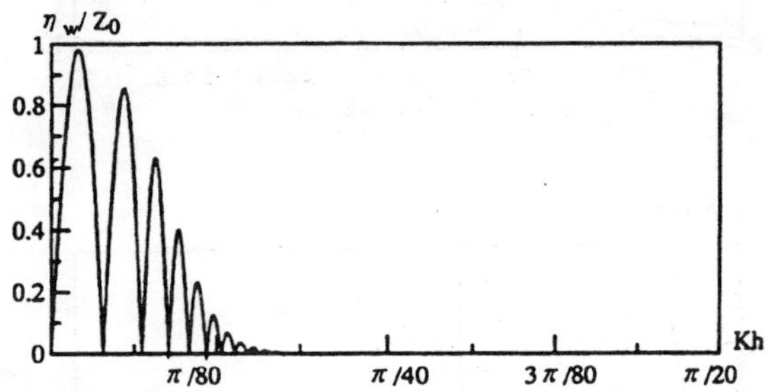

Figure 6: Frequency response amplitude function of surface wave

4. MOTION OF THE FLOATING STRUCTURES BY SEAQUAKE

In this section, we study two problems to consider the motion of the floating structure. One is a diffraction problem, in which the exciting force by seaquake acting on a fixed structure on the free surface is estimated. The other is a radiation problem, in which the hydrodynamic forces such as added mass force and damping force are obtained by forced oscillations.

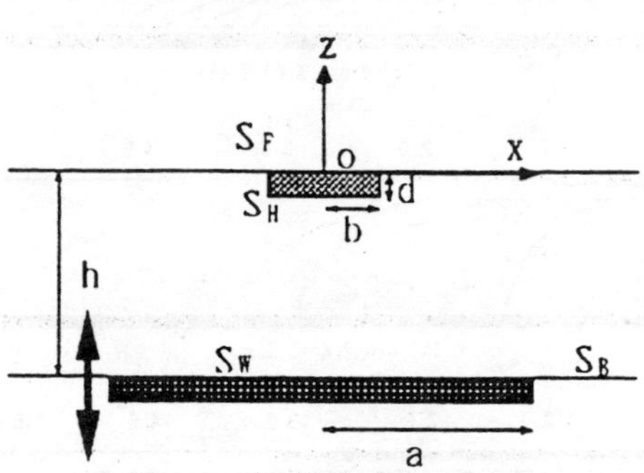

Figure 7: Coordinate system

4.1 DIFFRACTION PROBLEM

A floating structure with half breadth b and draft d is fixed on the free surface as shown in figure 7. Supposing the vertical harmonic oscillation of the wave source with half breadth a at water depth h and the velocity potential $\phi_d e^{i\omega t}$, the boundary value problem to be solved is as follows.

$$\left.\begin{array}{ll} \nabla^2\phi_d + K^2\phi_d = 0 & \\ \phi_d = 0 & on \quad S_F \\ \dfrac{\partial\phi_d}{\partial n} = 0 & on \quad S_H \\ \dfrac{\partial\phi_d}{\partial z} = 0 & on \quad S_B \\ \dfrac{\partial\phi_d}{\partial n} = v_n & on \quad S_W \end{array}\right\} \tag{16}$$

where S_H is the body surface of the floating structure. Applying Green's theorem to the problem, the following boundary integral equation is obtained, where G_2 is Green function in equation (6).

$$C\phi_d(P) + \int_{S_H} \phi_d(Q)\frac{\partial G_2(P,Q)}{\partial n}ds = \int_{S_W} v_n(Q)G_2(P,Q)ds \tag{17}$$

where $P(x,z)$ is a field point and $Q(\xi,\eta)$ is a source point. C is 1 when $P(x,z)$ is in the wave fields and is 1/2 when $P(x,z)$ is on the smooth boundary surface. This integral equation can be solved numerically by discretizing the boundary. The velocity potential is assumed to be constant on each discretized element.

The dynamic pressure p_d acting on the floating structure can be written in the same form as equation (14).

$$p_d(x,z) = -i\rho\omega\phi_d(x,z) \tag{18}$$

The vertical exciting force f_d acting on the floating structure is shown with the integral form of the dynamic pressure p_d.

$$f_d = \int_{-b}^{b} p_d(x,-d)dx \tag{19}$$

4.2 RADIATION PROBLEM

Let the floating structure forced to oscillate vertically in the still water as shown in figure 8. Supposing the forced oscillation is harmonic, the displacement of the structure and the velocity potential are written in forms of $z = z_o e^{i\omega t}$ and $\phi_r e^{i\omega t}$ respectively. The boundary value problem to be solved is

$$\left.\begin{array}{ll} \nabla^2\phi_r + K^2\phi_r = 0 \quad in \quad \Omega & \\ \phi_r = 0 & on \quad S_F \\ \dfrac{\partial\phi_r}{\partial n} = V_n & on \quad S_H \\ \dfrac{\partial\phi_r}{\partial z} = 0 & on \quad S_B \end{array}\right\} \tag{20}$$

where V_n is the normal velocity on the body surface. The velocity potential ϕ_r are obtained by solving the integral equation in the similar way with the diffraction problem, and the dynamic pressure p_r and the vertical exciting force f_r acting on the floating structure can be expressed as follows:

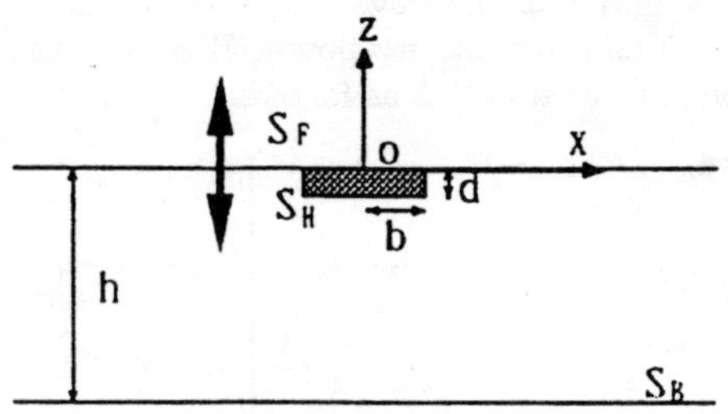

Figure 8: Coordinate system

$$p_r(x,z) = -i\rho\omega\phi_r(x,z) \tag{21}$$

$$f_r = \int_{-b}^{b} p_r(x,-d)dx \tag{22}$$

The force f_r is a complex value and its real part and imaginary part corresponds to added mass m_h and damping N_h respectively.

$$m_h = \frac{\mathrm{Re}[f_r]}{z_o\omega^2} \tag{23}$$

$$N_h = -\frac{\mathrm{Im}[f_r]}{z_o\omega} \tag{24}$$

where Re[ ] is a real part, and Im[ ] is a imaginary part.

4.3 EQUATION OF HEAVING MOTIONS

The motion equation for heaving mode is expressed by using equations (19),(23),(24).

$$(M + m_h)\ddot{z} + N_h\dot{z} + \rho gBz = f_d \cdot e^{i\omega t} \tag{25}$$

where M is the mass of the floating structure, $B(=2b)$ is the breadth. Frequency response function of heaving motions can be obtained by solving equation (25).

4.4 NUMERICAL RESULTS AND DISCUSSIONS

To study the effect of the structure's breadth, calculations are carried out in the three cases, i.e. breadth-depth ratio is (a)$B/h = 0.4$, (b)$B/h = 2.0$ and (c)$B/h = 6.0$, respectively. The acceleration amplitude of the wave source is constant at A_0 and the whole wave source region oscillates uniformly, and $a/h = 5.0$, $d/h = 0.05$ are constant.

4.4.1 DIFFRACTION PROBLEM

Figure 9 shows frequency response amplitude function of the vertical exciting force. Only in case of (b)$B/h = 2$ the calculation of $a/h = 5$ is added to the calculation of $a/h = 0.5$ to investigate the effect of the wave source's breadth. There are sharp peaks at a certain frequency as seen in figure 3. This is also due to the resonance for the longitudinal wave in the fluid. However, the remarkable point is that the resonance frequencies change according to the structure's breadth. So as it is difficult to estimate the resonance frequencies, we examined the resonance condition by calculating the dynamic pressure in fluid. Figure 10 are the pressure contour about the first resonance which occurs near the dominant frequency of P-wave.

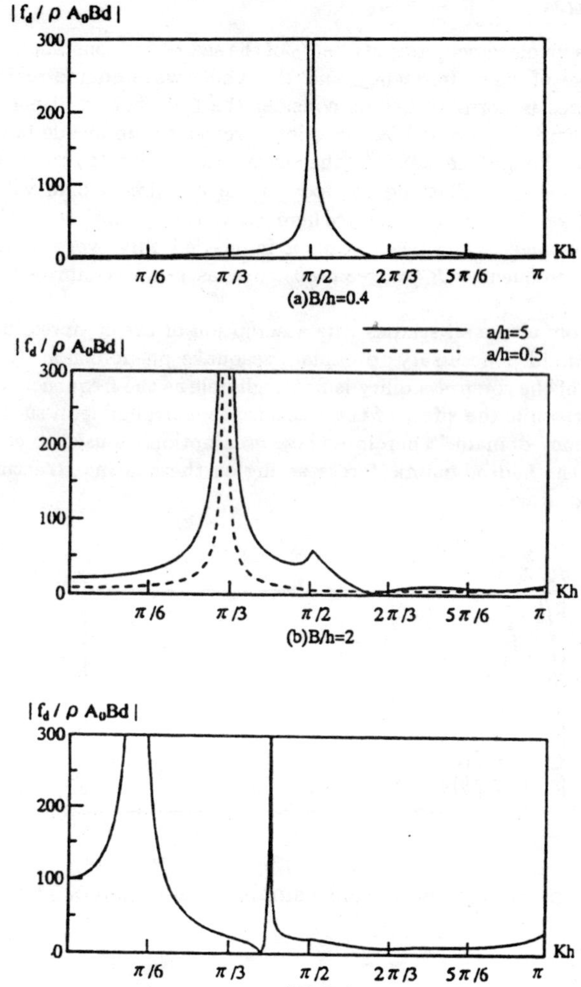

Figure 9: Frequency response amplitude function of vertical exciting force

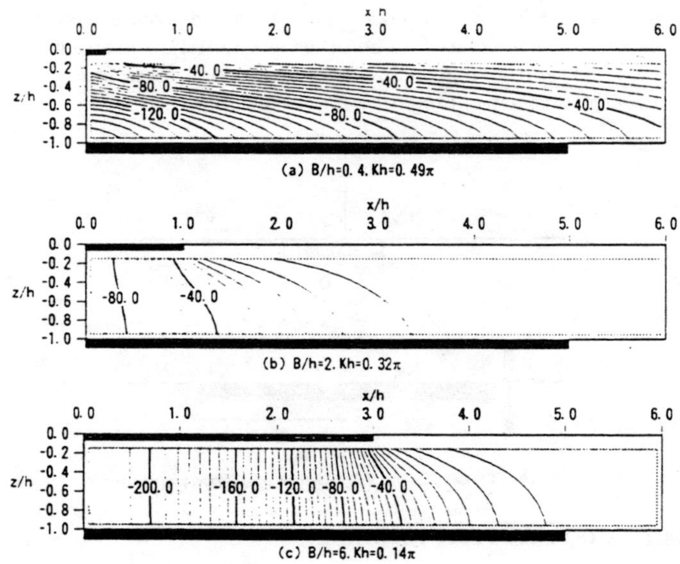

Figure 10: Dynamic pressure contour

In this figure the dynamic pressure is expressed by nondimension parameter $|p_l|/\rho A_0 h$ and is described by the domain of $0 \leq x/h \leq 6$ because of z-axis symmetrical property. The each resonance mode is a little different among three calculations. The longitudinal wave pattern is a node at sea-bed and is an anti-node at the surface of sea in case of (a)$B/h = 0.4$, and it is node respectively at both ends of the floating structure in case of (c)$B/h = 6.0$, and the case of (b)$B/h = 2.0$ is in transition process from (a) to (c). When the breadth is small, the resonance in vertical direction occurs and it shows the peak value at frequencies satisfying $Kh = (2n-1)\pi/2$ $(n = 1, 2 \cdots)$. When the breadth get larger, the resonance in the horizontal direction occurs and it shows the peak at frequencies satisfying $Kh = (2n-1)\pi h/B$ $(n = 1, 2 \cdots)$. So it means that not only the depth but the breadth are important parameters to predict the resonance frequencies of the vertical exciting force acting on the floating structure. It is supposed that there is the resonance between the structure's bottom and the sea-bed besides above resonance modes. This phenomenon appears at the frequency satisfying $Kh = (2n-1)\pi h/(h-d)$ $(n = 1, 2 \cdots)$ as the second resonance. Figure 9(b) shows that the vertical exciting forces are nearly in proportion to the wave source's breadth, while the resonance frequencies are independent of the effect its breadth.

4.2.2 RADIATION PROBLEM

Figure 11 shows the frequency response characteristics of added mass coefficient $m_h/\rho b^2$. Added mass coefficient is constant over a range of low frequency because of the assumption of no surface waves, while the resonance is caused at the same frequency as in case of the vertical exciting force. Figure 12 shows the frequency response characteristics of damping coefficient $N_h/\rho\omega b^2$. Damping coefficient exists only in case that Kh is more than $\pi/2$, which is the critical frequency to induced the first order's progressive component(the longitudinal wave), because damping depends on the progressive waves.

4.4.3 RESPONSE CHARACTERISTICS OF VERTICAL ACCELERATION OF STRUCTURES

Figure 13 shows frequency response amplitude function of vertical acceleration, in which the vertical axis is expressed by nondimensional value $\ddot{z} / A_0$. The vertical acceleration gives almost equal value independent of the structure's breadth over a range of the low frequency except for the usual resonance. Moreover, the resonance is caused between the free surface and the sea-bed at $Kh = \pi/2$ in common with three cases. However, the resonance in the direction of the structure's breadth dose not occur even in case of $Kh = 6.0$. This is because the resonance results in the cancellation between the added mass force and the vertical exciting force. The seaquake's impulsive phenomenon is induced by the resonance for the longitudinal wave between the free surface and sea-bed. These results are obtained with the assumption that the floating structure is the rectangular rigid-body. In case of a mega floating structure, another approach is needed such as considering the elasticity.

5. CONCLUSIONS

As the coastal developments are evolved, a floating structure tends to be larger in the future. In this paper, the effects of seaquakes are studied for the design of offshore floating structures. The results are summarized as follows :

1. The longitudinal waves are induced by the vertical oscillation of the sea-bed, and propagate through the seawater. Consequently seaquakes are caused by the resonance of longitudinal waves. Therefore the effect of the compressibility of fluid cannot be ignored.
2. The resonance frequencies of the vertical exciting forces acting on floating structures are affected by the breadth. The resonance occurs between the free surface and the sea-bed when the floating structure is small, and does between both ends of the structure when the floating structure is large.
3. Concerning the added mass force and the vertical exciting force, the resonance is caused at the same frequencies and their frequencies change according to the breadth. On the other hand, the damping depends on whether the progressive wave's component of the longitudinal waves occurs or not.
4. The floating structure's motion increase more only at the frequencies at which the resonance is caused between the free surface and the sea-bed. This is because the resonance results in the cancellation between the added mass force and the vertical exciting force.

References

[1] Isshiki, H. and Maeda, H. (1975). "The Formulae for a Pulsating Source in a Fluid with a Free Surface and the Methods of Their Numerical Evaluations", Journal of the Kansai Society of Naval Architects, Japan, Vol.157, pp.73-82.

[2] Kiyokawa,T. and Inada, H. (1989). "On the mechanism of seaquakes", Proc. of Coastal Engineering, JSCE., Vol.36, pp.734-738.

[3] Kiyokawa, T. (1995). "Numerical Prediction For Responce Motion of Floating Structures Subjected to Seaquake", Proceedings of the Fifth(1995) International Offshore and polar Engineering Conference , Vol III, pp.429-434.

[4] Baba, E. (1978). "A study on the effect of seaquakes on a floating body", Journal of Society of Naval Architects of Japan, Vol.162, pp.90-98

[5] Hamamura, K (1987)."Can earthquake excitations be escape on free surface?-investigation of seaquakes- ", FUNE-NO-KAGAKU, ,Vol.40, pp58-61

[6] Higo, Y. et.al.(1992). "A Study of the Two-dimensional Underwater Sound Field Radiated from Vibrating Bodies in Shallow Water, Journal of Society of Naval Architects of Japan, Vol.172, pp.539-549.

[7] Matsuoka, K. (1988). "On quake-proof ability of floating structures", Bulletin of the Society of Naval Architects of Japan, No.706, Vol.40, pp214-220.

[8] Mogi,K. and Mochizuki,H. (1980)."Observation of High Frequency Seismic Waves by a Hydrophone just above the Focal Region of the 1980 Izu-Hanto -Toho-Oki Earthquake and a Precise Estimate of Location of the Submarine Earthquake Fault", Bulletin of Earthquake Research Institute, University of Tokyo, Vol.55, pp.1017-1041.

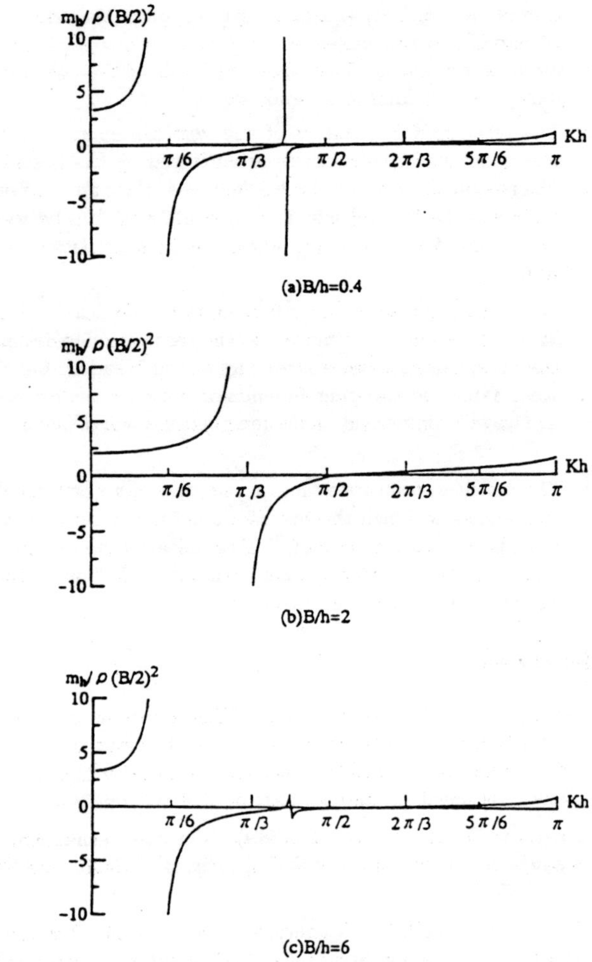

Figure 11: Heaving added mass coefficient of floating structure

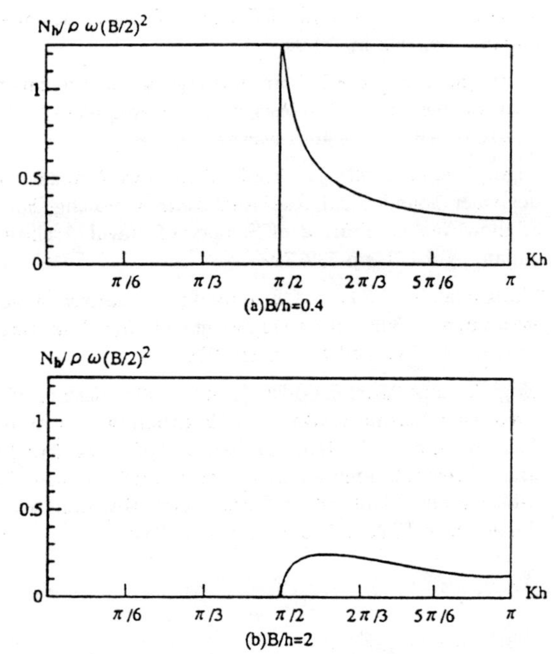

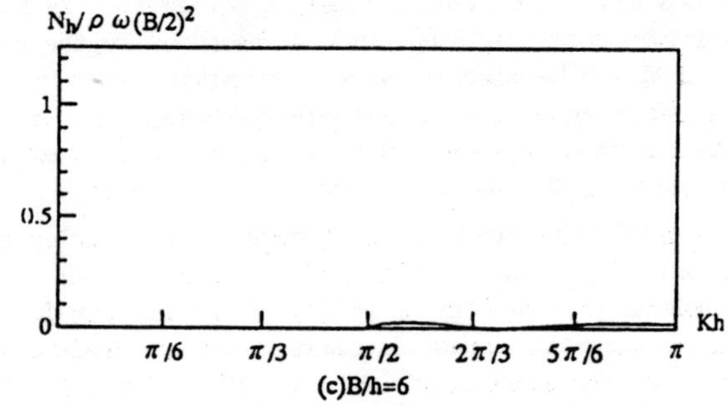

Figure 12: Heaving damping coefficient of floating structure

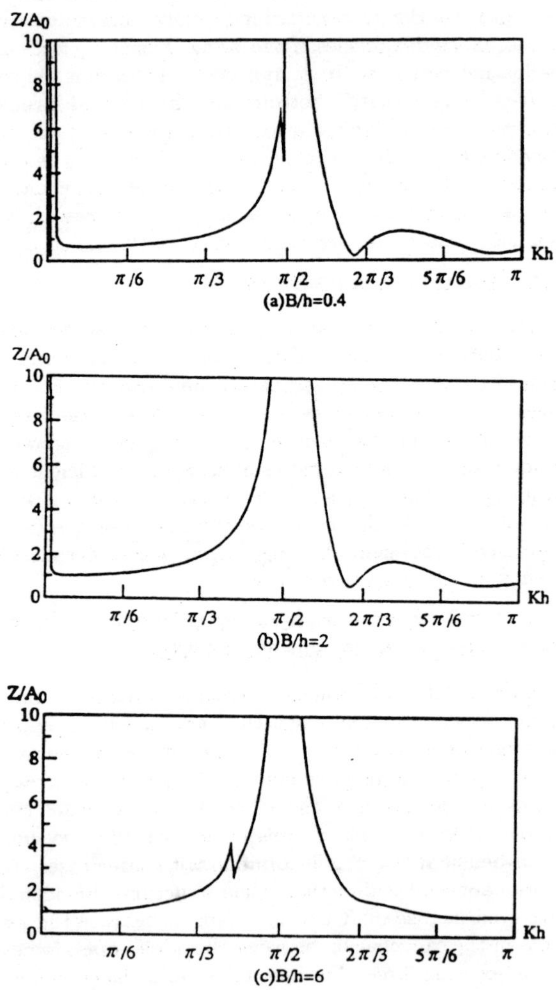

Figure 13: Frequency response amplitude function of vertical acceleration

Proceedings of the Seventh (1997) International Offshore and Polar Engineering Conference
Honolulu, USA, May 25-30, 1997

ISBN 1-880653-28-1 (Set); ISBN 1-880653-29-X (Vol. I)

Dynamic Response of a Very Large Floating Structure with Active Pneumatic Control

T. Hirayama and N. Ma
Yokohama National University
Yokohama, Japan

ABSTRACT

A pneumatic type active control apparatus to reduce wave-induced motions was developed and applied to very large floating structures (VLFS) in this study. Active control of motions and deflections of structure are achieved by controlling air pressure inside the air tanks individually which are attached to the bottom of column footings. The effectiveness of this control system has been investigated through the model experiments and theoretical discussions. The heave and deflections of structure can be reduced successively especially at low frequencies. The applicability of developed system are discussed with considerations of pre-pressure, positions and parameters of control.

KEY WORDS: Attitude Control, Pneumatic Pressure, Rigid Motion, Deflection, Static Recovery, Dynamic Motion Reduction

1 INTRODUCTION

The utilization of ocean space by using large floating structure will tend to deeper water depth in future. However, waves in such offshore area have significant values both in height and length and could cause unexpectedly large motions and deflections although the structure may be moored in certain ways. Of cause as a result of the displacement due to waves or other external influences, the operation and safety will be affected unexpectedly. Considering a floating airport, such as semisubmersible type which consists of multiple removable units proposed by Hirayama et al (1995a, 1995b) for the aim of long durability, it is necessary to keep on its operations (landing and take-off) conforming the requirements of airport when replacing units. Thus in order to achieve the attitude control and to ensure the safety of structure, an static and dynamic control system is indispensable. On the other hand, from the viewpoint of the safety of the structure and its mooring system, it is desirable for it to be equipped with control apparatus to suppress the motion while considering the structure subjected to an extreme wave such as tsunami or seismic impacts.

Regarding the active control of the motion of offshore structure, a few recent studies can be found. For examples, Ando et al (1994, 1995) carried out the model tests of motion reduction by controlling the movable floats underwater. Yamashita (1992) presented a method to reduce the heave of offshore structure by using the controlled piston amid the columns in experiments. Comparing to these mechanical control techniques, few studies which concerned the use of air pressure. Blood (1996) proposed a pneumatically supported platform and studied the hydrostatic stiffness, air movement. Patel et al (1987) studied the applications of pneumatic compliances to control air pressure inside air tanks which are attached to columns at waterline of semisubmersible actively. Air pressure utilization brings many advantages, such as the few pollution effects upon the environment because its cleanness, relatively cheap to build the control apparatus and easy to maintain. Hirayama et al (1996) have developed a pneumatic pressure control apparatus for floating structures by controlling air pressure inside air columns (air tank) which are attached to the bottom of column footings.

In this study, we apply it to a semisubmersible type VLFS to suppress not only its rigid motions but also deflections. The objective of this paper is to present a preliminary investigation on the characteristics of the newly developed control system. Numerical investigations into hydroelastic response are also shown and compared with the results of measured ones. The static recovery tests from inclined condition and the active attitude control tests were carried out in still water and in waves of various periods. Finally, the applicability of proposed pneumatic type active control are discussed.

2 PNEUMATIC ACTIVE CONTROL

2.1 VLFS Model and Measurement

Model experiments were carried out in the towing tank of Yokohama National University by using semisubmersible type VLFS model. This model is supposed for a medium scale airport which consists of 36 removable units. Each unit is compromised of 4 columns with footings arrayed squarely. For the sake of simplicity, the elasticity of structure is simulated by connecting these unit with elastic beams (aluminum) each other in transverse and longitudinal directions at upper deck.

The model and positions of measurements of deflections and bending moments are shown in Fig.1. Hence, Zn represents the vertical displacement of point-n, M_1, M_2, M_3 correspond to bending moment at three different positions as shown in Fig.1 respectively. The deflection were detected by using a non-touch type optical sensing system which as shown in figure as LED and Cameras. The model were pneumatically controlled at 12 units by controlling air mass trapped inside the underwater air column (bottom open air tank) as shown in Fig.1. Linear spring mooring system was adopted in experiments to prevent only the motion in horizontal plane. The principal dimensions of model, measured and calculated natural periods of motions and elastic deformation modes are shown in Table 1. The calculations of natural periods were based on the wet mode analysis.

2.2 Apparatus, Algorithm of Control

Two compressors outside of model were used to provide the necessary air pressure in fore and rear part of model respectively. Four columns of each unit are linked together, so that every unit are controlled individually. The description of control of each unit is shown in Fig.2. In the experiment, the mass of air in air column was not directly controllable. However by controlling the air flow between the compressor and the air columns, the mass of air trapped inside the air column can be controlled through the on/off control of electromagnetic valves and this result in a change of buoyancy of supporting column. Air flows into air column (flow in) and outside to atmosphere (flow out) is controlled with the open / close interval of electromagnetic valves and hand made pistons as shown in Fig.2. These provide three conditions named "in", "hold", "out" to each unit, where "hold" means there is neither flow-in nor flow-out from air column - no change of air mass. The calculation of control command is performed based on the traditional P-D control algorithm on personal computer as controller. The block diagram of control is shown in Fig.3 and this is same for the air columns of one unit. The flow chart of electromagnetic valve control is shown in Fig.4. Applying this control to all controlled units (see Fig.1) individually, the different vertical control force at each unit can be generated consequently, thus the motions in different modes can be reduced.

The motion reduction is done in effect by controlling the speed of air flow, in other words, mass flow rate ($\dot{m}_i$) associated with each air column for the proposed system. Generally, $\dot{m}_i$ will be expressed as the equation of the state of the system as

$$\dot{m}_i = f(\mathbf{X}, \dot{\mathbf{X}}, \mathbf{E}) \tag{1}$$

where **X** is the state vector and **E** is external force vector. The state vector contains

$$\mathbf{X} = \{\mathbf{Z}^T, \dot{\mathbf{Z}}^T, \mathbf{m}^T\} \tag{2}$$

where Z is the displacement vector. **m** is the vector containing the mass of air in each column.

Generally, if we assume the control to be linear then the equation of state of the system can be written as,

$$\dot{m}_i = \mathbf{g}_{1i}{}^T \mathbf{X}(t) + \mathbf{g}_{2i}{}^T \dot{\mathbf{X}}(t) + \mathbf{r}_i{}^T \mathbf{E}(t) \tag{3}$$

where g_{1i}, g_{2i} and r_i are constant vectors associated with i-th air column. The third term in eqn. (3), i.e. the contribution of external force, is generally uncertain and this could be accounted by controlling only the displacement and rate with appropriate gains. The first and second term, the gains of P-D control, are determined through preparatory tests and the frequency independent values are used in experiments.

2.3 Description of Experiments

In general, the selection of air compressor should be made with considerations of the necessary capacity of air, required speed of air flow and adaptability with the period of motion etc. For the sake of simplicity, two small compressors with maximum initial pressure of 9.9 kgf/cm^2 (6 kgf/cm^2 used in experiment) and capacity of 65 liters were chosen for the experiments. Air mass flow into air column has non-linearity with respect to the valve open length (Hirayama et al.(1996)) and results in bad adaptability of control especially at initial stage of flow-in, to overcome this difficulty we selected the suitable initial air pressure and control gains through a number of tests. Because of insufficient information on the air actuator, the control gains were determined by pre-experiments in still water, hence the empirical data at certain frequency where the heave motion was successfully suppressed had been applied for other frequencies.

Since the aim of the proposed control is to provide of both attitude recovery control and dynamic motion control, these two kinds of experiments were conducted. The former one was conducted mainly in still water and later one was conducted in regular waves of various periods.

In addition, the frequency response characteristics were also investigated through forced oscillation tests by using this air actuator. In this case, the control input variables are the pseudo vertical displacement generated by a frequency generator.

3 ANALYSES OF MOTION AND DEFLECTION

As for the response prediction of VLFS, special analysis techniques are needed due to the presence of flexible deformation modes in addition to the usual rigid motion modes. e.g. hydrodynamic force and elastic deformation must be determined simultaneously. Hirayama et al (1995b) and Ma et al (1996) has developed a numerical method based on the linear hydroelasticity theory by using the mode superposition method. The hydrodynamic forces are evaluated in three dimensions by using sink source method and the structural analysis is executed by using modal analysis based on FEM. From the comparison in Table 1, the analyzed natural periods show very good coincidences with experimental ones.

3.1 Mode Superposition Method

According to the mode superposition method, the motion equation of principal coordinate of r-th mode (p_r) is expressed in following form.

$$(a_{rk} + A_{rk})\ddot{p}_r + (b_{rk} + B_{rk})\dot{p}_r + (c_{rk} + C_{rk})p_r = F_r e^{i\omega t} \quad (r, k = 1,2,3\ldots) \tag{4}$$

where a_{rk}, b_{rk}, c_{rk} are structural mass, damping and stiffness coefficients respectively. A_{rk}, B_{rk}, C_{rk} are added mass, fluid damping and hydrostatic stiffness respectively. F_r is wave exciting force in principal coordinate.

Having the principal coordinate of each mode obtained, the displacement (u) could be predicted consequently by aggregating those eigenmodes as follows.

$$\mathbf{u}(x,y,z,t) = \sum_{r=1}^{m} \mathbf{u}_r(x,y,z)\, p_r(t) \tag{5}$$

where $\mathbf{u}_r$ is the mode vector of r-th mode.

3.2 Motions of Pneumatically Controlled VLFS

Taking the air's effects on hydrostatic stiffness and varying exciting forces into accounts (see Patel et al (1987), for example), the motion equation of a pneumatically controlled VLFS could be expressed as follows, for instance for heave in time domain.

$$(M + a_{33})\ddot{Z} + B_{33}\dot{Z} + (C_{33} + QC_0^{(a)})Z + QC_t^{(a)}e^{i\omega t} = (F_3 + F_3^{(a)}e^{i\varepsilon_a})e^{i\omega t} \tag{6}$$

where M is mass of the structure, $C_0^{(a)}$, $C_t^{(a)}$ is the constant and time varying pneumatic stiffness respectively. Q is the air constant determined by amount of air trapped inside of air column and temperature etc. $F_3^{(a)}$ is time varying exciting force induced by air, ε_a is the phase lag of pneumatic force.

Furthermore, if the air perturbation is small then $C_t^{(a)}$ could be assumed to be negligible. eqn.(6) may be reproduced in frequency domain assuming the harmonic oscillation.

$$\left\{C_{33} + QC_0^{(a)} - \omega^2(M + a_{33}) + i\,\omega B_{33}\right\}Z = F_3 + F_3^{(a)} \tag{7}$$

By solving equation (7), the heave motion can be determined. However this requires the information of air stiffness and air-induced exciting force in detail.

4 RESULTS AND DISCUSSIONS

4.1 Attitude Control and Forced Oscillation

For the purpose of reduction of the down time when the VLFS is supposed to be used as an airport in inclined conditions due to lack of unit for replacement and others, it becomes indispensable to recover the structure from heeled, trimmed or bent conditions. Fig.5 shows the measured deflection in unit removed condition (two units on centerline at fore part) and recovered them with only fore 4 units. Although a slight sinking was observed in recovered deflection, the fore end which had a deflection of about -5 cm (sinking) were restored satisfactorily so that the structure returned to be flat again.

The valve actions of 3 units of port side (A1, G1, L1: see Fig.1) and the simultaneously measured responses of deflections (vertical displacement, Z1, Z7, Z16), inclinations (pitch and roll denoted as θ_y, θ_x respectively) and bending moments are shown in Fig.6 for a static recovery from trimmed and bent condition due a weight loaded on aft end (unit L2). The deflection Z16 was about -2 cm and this was recovered within about 4 seconds. It can also be observed that the valve action of L1 unit was turned to "in" position for the same time, but short iterations between "hold" and "out" position for center unit and very few actions for the fore unit. The deflections, inclination and moment had been well restored except the deflection Z1 and the moment M3.

In order to clarify the frequency response characteristics of dynamic control, forced oscillation tests were carried out by using the external transient signal as feedback signal for the input of the controller. The signal for each air column is different in phase each other so that the desired oscillation mode can be achieved.

As an example, the time histories of pitch mode forced oscillation is shown in Fig.7. It can be recognized the valves at fore and aft end unit repeat on/off actions responded to external signal, inclination (output of clinometer) and vertical displacement at fore and center show reasonable response due to excited pitch motion. The Fourier spectrum of input signal and transfer functions of inclination and vertical displacements are shown in Fig.8. It can be concluded that the pitch motion was excited precisely up to frequency of 5.0 rad/s by this apparatus, therefore the proposed control is considered to have effects at relatively low frequencies. The forced oscillation of bending mode has also been carried out but the results are omitted here because of limited space. The desired deformation shape was obtained successfully and the frequency response characteristics is similar to rigid motion.

4.2 Motion Suppression in Waves

It is important to understand the characteristics of wave-induced motion and deflection of VLFS in order to achieve effective control of them. The predictions of hydroelastic response were done by the method aforementioned, the experimental results were obtained from tests in transient water waves. The transfer function of heave motion without control is shown in Fig.9, the experimental data plotted in figure is the measurement of Z-deflection at center (Z7) of elastic model. The discrepancies at short wave periods lower than 1.0 sec is due to the elastic deformation effects which are not included in calculation of heave. However, the response operators and the phases due to long waves are simple than those for short waves, this implies that the controls at these frequencies is easier to achieve.

The response amplitudes of deflections in Z-direction at Z2 (fore) and Z7 (center) are shown in Fig.10. The calculations agree with measured values fairly good including resonance at natural frequencies of elastic deformation mode which are drastically different to the usual rigid structure. The deflection at fore-end show peaks at 1st and 2nd bending mode frequencies which correspond to 2-nodes and 3-nodes bending modes. On the other hand, responses at center show peak only at 1st mode frequency because this point corresponds to the node of 2nd bending mode.

Controls of response at high frequencies where the structure has remarkably large displacement due to its resonance to waves were also attempted, however due to the shortage of available energy needed to achieve controller's commands, the controls in high frequency (normal wave period) had less effects. As expected, because of the compressibility of air, the system has a limit of its adaptability at high frequency, thus the investigations are restricted for relatively long periods. Comparison of deflections in regular waves with and without pneumatic control is shown in Fig.10 for period = 7.0 sec and in Fig.11 for period = 10.0 sec. From the timehistories of displacements, it is confirmed that the vertical displacements at three points of structure has been successively suppressed by the pneumatic control. According to the transfer functions (left lower in figures) which are analyzed with respect to the different wave of each experiment, the response at incident wave frequency was approximately reduced to 55% and 65% for 7.0 sec and 10.0 sec respectively. In fact, the 2-nodes bending mode and unstable pitch motions were observed in case of control, these could be explained as the effects of non-linear, transient excitation of air and waves and should be investigated more accurately.

5 CONCLUDING REMARKS

A pneumatic type active control apparatus for VLFS is developed by changing the additional buoyancy in supporting columns and its applicability has been examined by experiments in model scale. Through the examinations in still water and regular long waves, the proposed control system is considerably available not only for the attitude control but also for the use as a motion compensator in long waves. In addition, the possibility of the control system as a forced oscillator of both rigid motion and elastic deformation modes has been shown.

The potential of the proposed control system exists and the effectiveness can be enhanced through further improvements of apparatus considering the available energy to achieve an adaptive control. Also the control algorithm and optimization of control should be studied further. The application of the proposed system to a prototype VLFS can be realized enough by utilizing other adaptive control together, and further investigations with special considerations of scale effects are needed in future.

ACKNOWLEDGMENTS

The authors would like to express their gratitude to Mr. K. Miyakawa and Mr. T. Takayama, the staffs of our laboratory as well as the students, for their efforts in conduction of this experiment and analysis works.

REFERENCES

Ando, H, and Kato, S (1994). "Research on Active Controlled Floating Offshore Structure (Part 2)," *Proc. of 64 General Meeting of Ship Res. Inst. of Ministry of Transportation of Japan*, pp 90-94 (in Japanese).

Ando, H (1995). "Experiments on Active Control of Floating Body in Waves," *Proc. of 66 General Meeting of Ship Res. Inst. of Ministry of Transportation of Japan*, pp 248-249 (in Japanese).

Blood, H (1996). "Model Tests of a Pneumatically Stabilized Platform," *Proc. of Int. Workshop on Very Large Floating Structures, Hayama, Japan*, pp 12.1-12.8.

Hirayama, T, and Ma, N (1995a). "Response Characteristics of a Long Life Type Floating Airport in Waves (1st report)," *Journal of Soc. of Naval Architects of Japan*, Vol. 177, pp 219-230 (in Japanese).

Hirayama, T, Ma, N, and Nishio, FO (1995b). "Response Characteristics of a Long Life Type Floating Airport in Waves (2nd report)," *Journal of the Soc. of Naval Architects of Japan*, Vol. 178, pp 225-236 (in Japanese).

Hirayama, T, Takayama, T, Miyakawa, K, and Ma, N (1996). "Development of Active Attitude Control System for Floating Structure by Using Air Pressure," *Journal of Kansai Soc. of Naval Architects of Japan*, No. 226, pp 99-105 (in Japanese).

Ma, N, Hirayama, T, and Sato, N (1996). "Comparative Study of Hydroelastic Response of Semisubmersible Type and Pontoon Type Very large Floating Structures in Waves," *Journal of Soc. of Naval Architects of Japan*, Vol. 180, pp 353-363 (in Japanese).

Patel, MH, and Witz, JA (1987). "On Active Control of Marine Vehicles with Pneumatic Compliances," *International Shipbuilding Progress*, Vol.34, No.395, pp 134-144.

Yamashita, S (1992). "An Experimental Investigation of Active Control of Floating Body," *Proc. of 11th Ocean Engineering Symposium of Soc. of Naval Architects of Japan*, pp 271-277 (in Japanese).

Table 1 **Principal** Dimensions and Natural Periods (exp. and cal.)

	Model (1/256)		Full Scale
Length (L)	7.10 m		1817.6 m
Breadth (B)	1.70 m		435.2 m
Depth (D)	0.36 m		92.2 m
Draft (d)	0.115 m		29.4 m
Displacement (W)	267.4 kg		4,486,228 ton
KG	0.268 m		68.48 m
GM_T	1.0 m		256.07 m
GM_L	21.562 m		5519.87 m
Radius of Gyration			
Roll (K_{xx})	0.528 m		135.12 m
Pitch (K_{yy})	2.517 m		644.22 m
Yaw (K_{zz})	2.955 m		756.43 m
Bending Rigidity (EI)	992.1 Nm^2		$1.09 \times 10^{15} Nm^2$
Natural Frequency	T_{exp} (sec)	T_{cal} (sec)	T_{cal} (sec)
Heave	1.72	1.54	24.64
Roll	1.84	1.76	28.16
Pitch	1.62	1.68	26.88
1st Bending	1.18	1.16	18.56
2nd Bending	0.61	0.59	9.44
3rd Bending	0.32	0.32	5.12

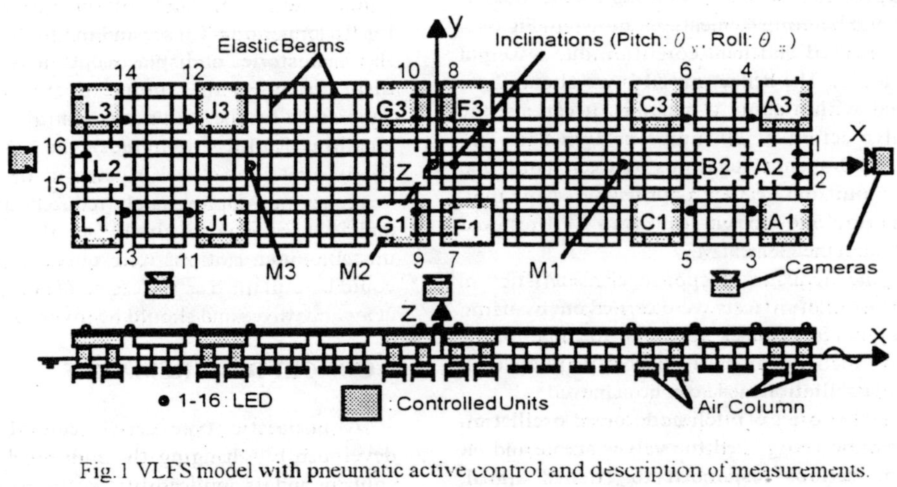

Fig. 1 VLFS model with pneumatic active control and description of measurements.

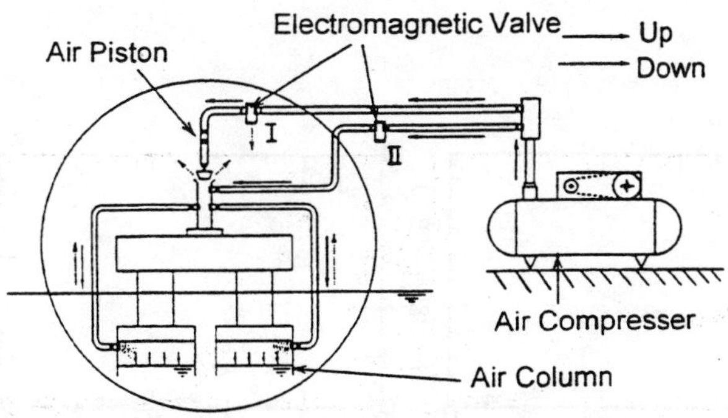

Fig.2 Actuator of pneumatic type active control on unit of VLFS

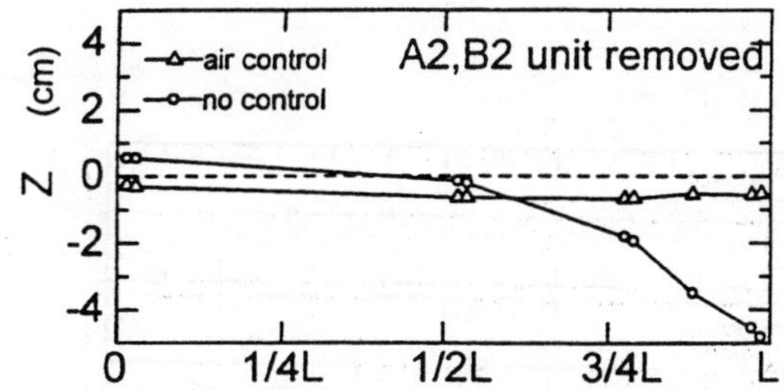

Fig.5 Static deflection distribution due to remove of units and recovery by air control

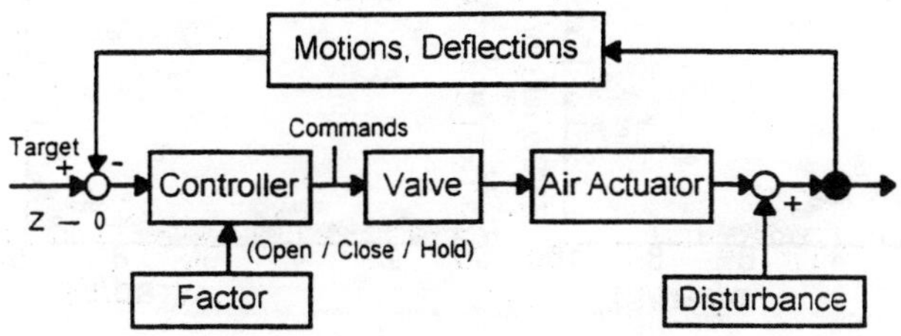

Fig.3 Block diagram of air valve control

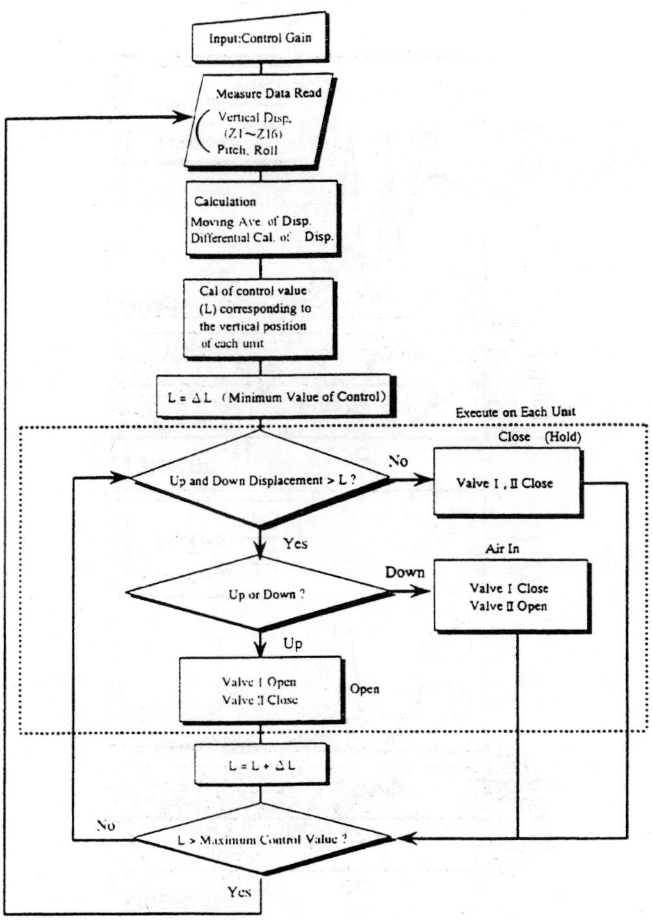

Fig.4 Control flow of electromagnetic air valves

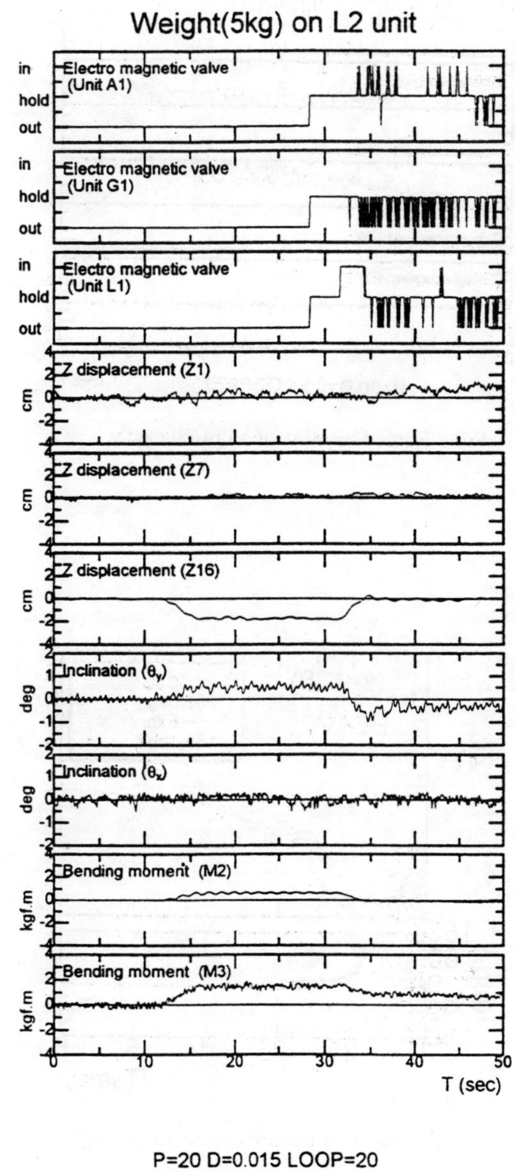

Fig.6 A recovery from inclined condition due to static load

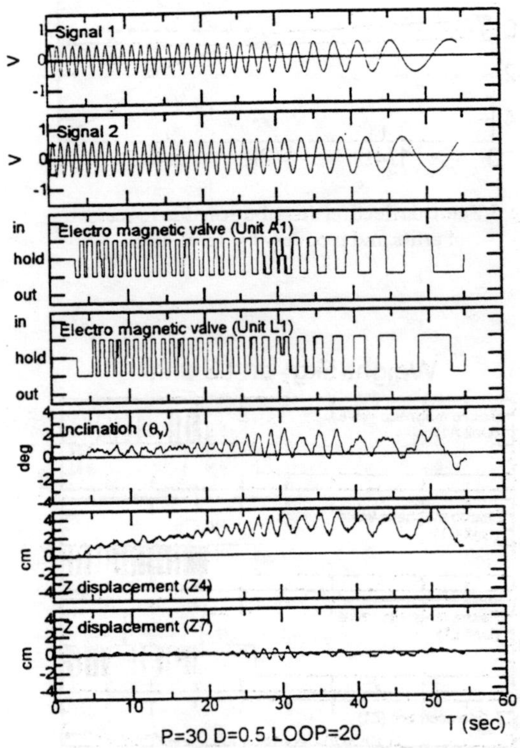

Fig.7 Forced oscillation in pitch mode

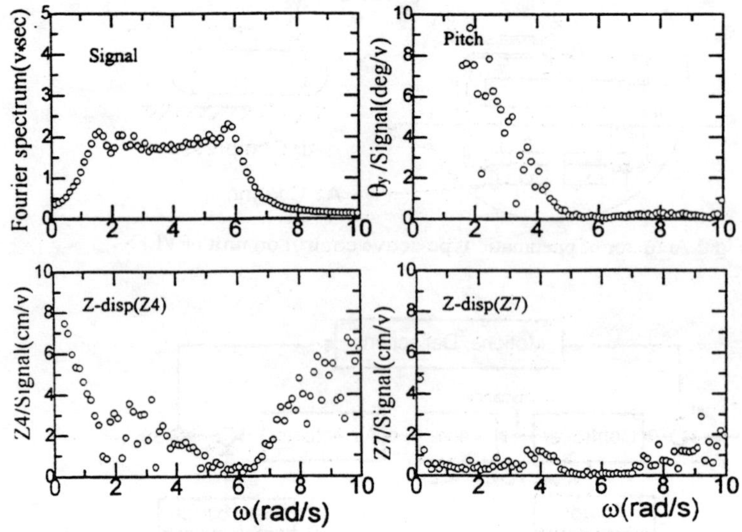

Fig.8 Frequency response function of forced pitch

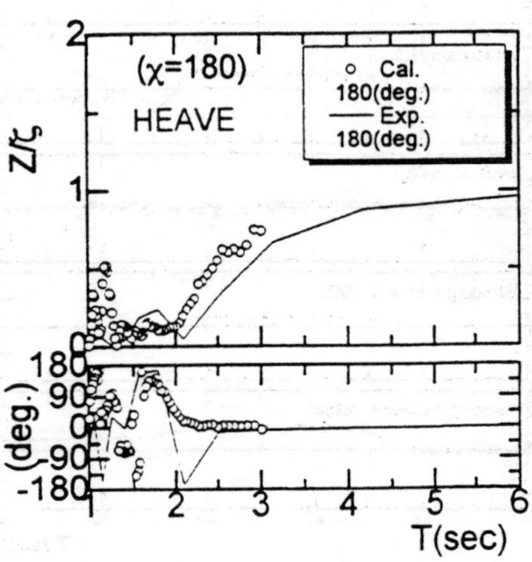

Fig.9 Transfer function of heave in head seas

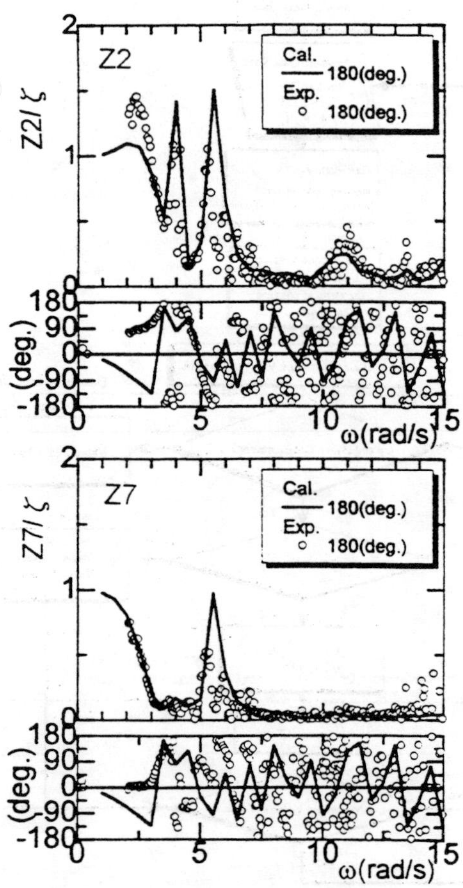

Fig.10 Transfer functions of deflections in head seas

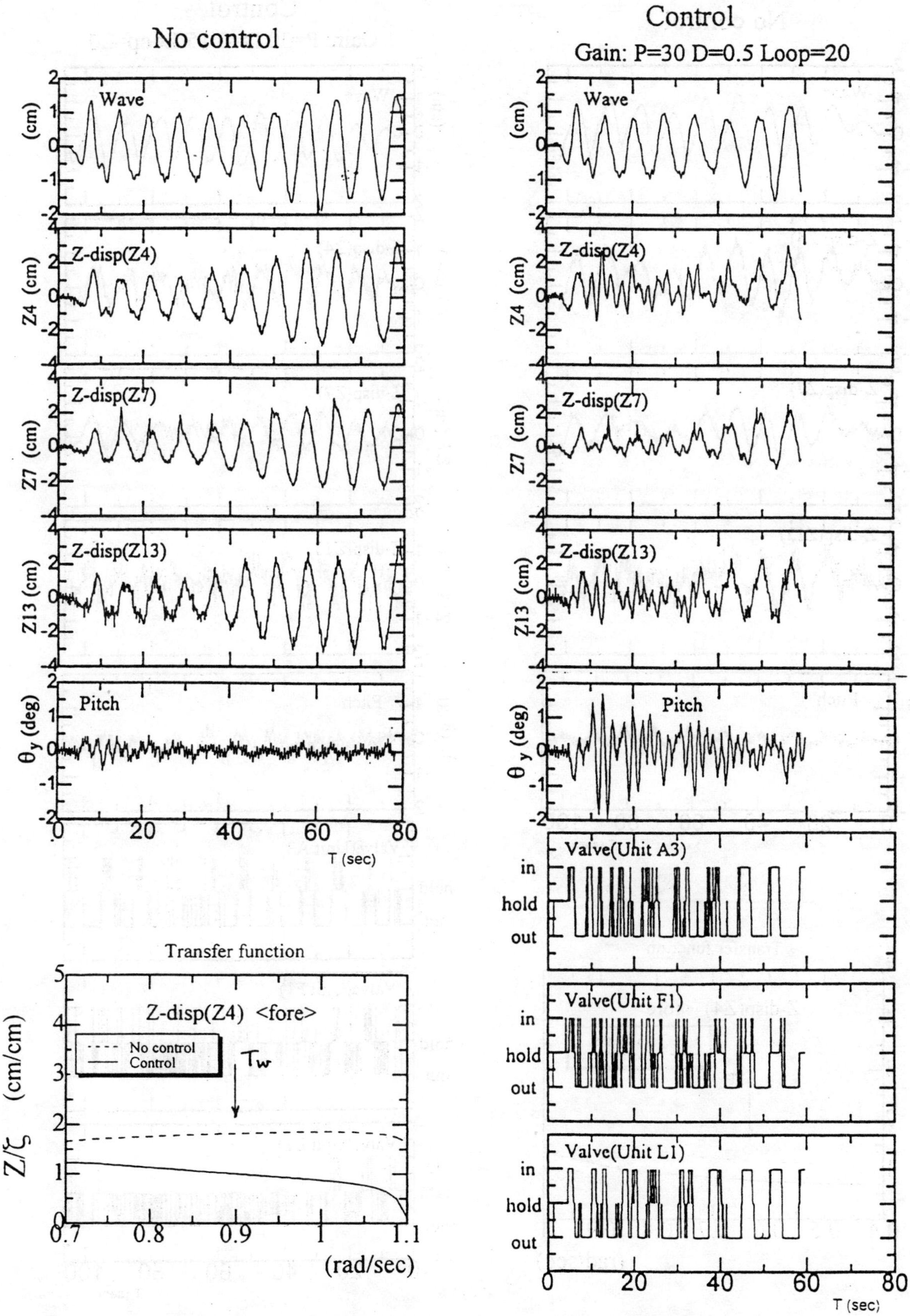

Fig. 11 Responses in regular waves with and without control (wave period=7.0 sec)

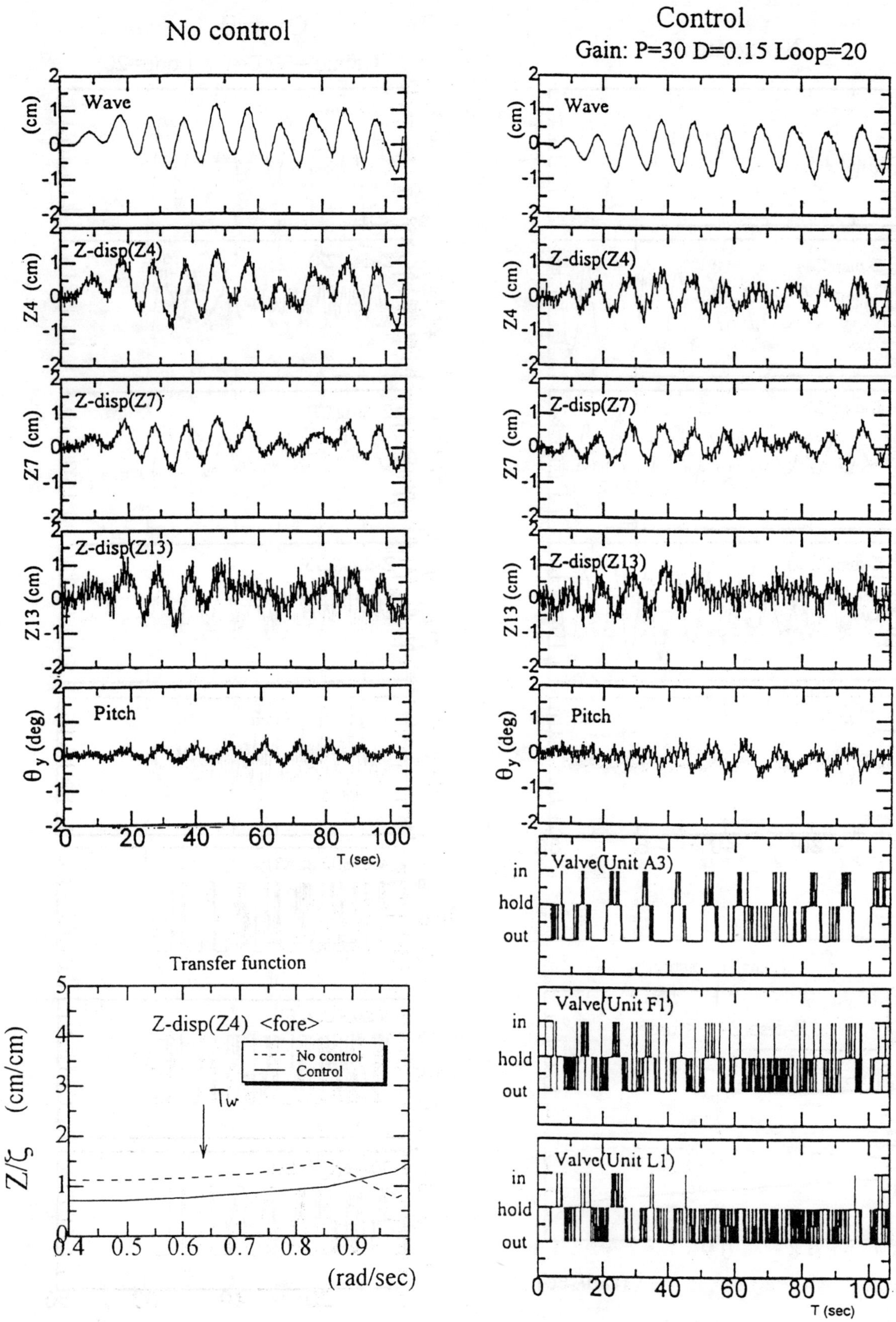

Fig. 12 Responses in regular waves with and without control (wave period=10.0 sec)

Proceedings of the Seventh (1997) International Offshore and Polar Engineering Conference
Honolulu, USA, May 25-30, 1997

ISBN 1-880653-28-1 (Set); ISBN 1-880653-29-X (Vol. I)

Wave-Induced Motions of a Very Large Floating Structure in Quartering Seas

Mikio Takaki and Xiechong Gu
Hiroshima University
Higashi-Hiroshima, Japan

ABSTRACT

This paper deals with three-dimensional responses of huge floating structure in head and quartering seas. Huge mat-like floating structure is considered as an elastic plate. First we estimate dry-eigenmodes of the completely free plate by employing finite element method. Then, hydrodynamic pressures exerted on it are evaluated by combining pressure distribution method and the dry-eigenmodes. Wave-induced motions of the plate are expressed as a superposition of the dry-eigenmodes, and the principal coordinates are finally determined by solving a set of equations of motion. Present modal analysis method is verified through comparisons of calculated amplitudes of vertical motions and instantaneous shapes of a plate model in head seas with available measured data. They show reasonable agreements. Next, we apply this method to a large floating module of $300m \times 60m$, and investigate its motions and disturbance waves around it in head and quartering seas. Finally, response amplitude operators of its first 10 modes in all heading waves with various wave lengths are shown.

KEY WORDS: Wave-induced Motions, Mat-like Floating Structure, Hydroelasticity, Modal Analysis Method

INTRODUCTION

Recently, a number of studies have been made for predicting the behavior of wave-induced motions of very large floating structure (Takaki, 1995a, 1995b; Kagemoto and Yue, 1994; Lee and Webster, 1994; Mamidipudi and Webster, 1994; Hirayama, 1994; Ertekin and Wang, 1994; Maeda, 1995; Murai, 1995). Mat-like floating structure has been proposed to be a candidate of a floating airport. This kind of structure has dimensions of a few kilometers long, some hundred meters wide, while its depth is around ten meters. This large length to depth ratio makes the structure very flexible, and the natural frequencies of elastic vibrations are significantly reduced. In addition to that, incident wave length becomes relatively short, thus elastic vibrations are remarkably induced by wave excitations and flexible deformations will be comparable with the reduced rigid motion responses. Thereby, elasticity of this sort of structure also must be taken into account. So far, this kind of study has mainly dealt with the motions of a long body like a beam except for the one due to Mamidipudi and Webster(1994) in head seas. They have proposed a method of evaluating three-dimensional motions of mat-like floating structure by using the discretized panel method. However, there are very few studies dealing with motions of mat-like structure in quartering seas.

In this paper, we develop the estimating method dealing with three-dimensional motions of mat-like structure in quartering seas by using modal method. The present analysis method is verified through comparisons with available measured data(Murai, 1995). Then we apply this method to the large floating module of $300m \times 60m$, and investigate motions of it and disturbance waves around it in head and quartering seas. Finally, the response amplitude operators of the first 10 modes in all heading waves with various wave lengths are shown.

FORMULATION

Mat-like floating structure will be treated as a plate in this study. We hereafter focus our attention on the floating elastic plate. A Cartesian coordinate system (x, y, z) with upward z-axis is used, and the origin o is set on the center of the plate as shown in Fig. 1. A long-crested regular wave of small amplitude incomes from direction of angle χ. Draft of the plate is very small comparing with its length and width. This allows us to employ pressure distribution method (Yamashita, 1979) to evaluate hydrodynamic pressures on the plate. Thereby horizontal motions (surge, sway and yaw) are excluded in this study. Wave-induced motions of the plate are coupled with the hydrodynamic pressures.

Motion Equations of Structure

For the sake of generality, the elastic plate is assumed to be orthotropic which obeys the following constitutive law(Timoshenko

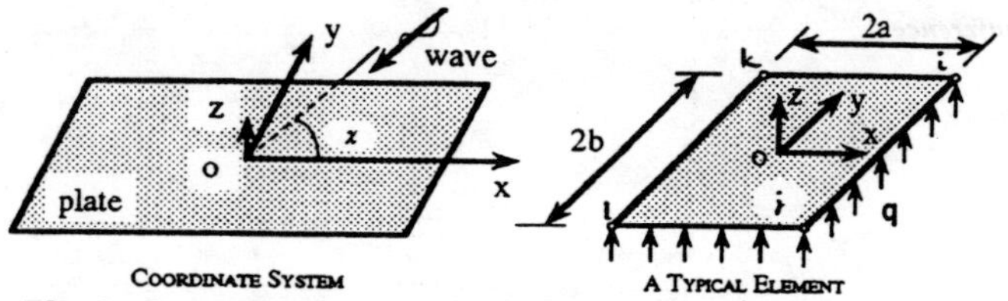

Fig.1 Coordinate System and A Typical Element

and Woinowsky-Krieger, 1959)

$$\begin{pmatrix} M_x \\ M_y \\ M_{xy} \end{pmatrix} = \begin{pmatrix} D_x & D_1 & 0 \\ D_1 & D_y & 0 \\ 0 & 0 & D_{xy} \end{pmatrix} \begin{pmatrix} -\dfrac{\partial^2 w}{\partial x^2} \\ -\dfrac{\partial^2 w}{\partial y^2} \\ 2\dfrac{\partial^2 w}{\partial x \partial y} \end{pmatrix} \quad (1)$$

where M_x, M_y and M_{xy} are bending moments and twisting moment, $w(x, y)$ deflections and D_x, D_y, D_1, D_{xy} rigidities of the plate. Motion equations of such a completely free plate can be derived by employing finite element discretization. The plate is divided into rectangular elements with 4 nodal points. At each point three degrees of freedom, vertical movement w, angular movements θ_x and θ_y, are permitted. Small angular displacements θ_x and θ_y are approximately equal to the slopes respect to y- and x-directions as below

$$\theta_x = \frac{\partial w}{\partial y}, \quad (2)$$

$$\theta_y = -\frac{\partial w}{\partial x}. \quad (3)$$

A typical element of length $2a$ and width $2b$ is depicted in Fig. 1. A polynomial with 12 parameters due to Zienkiewicz and Taylor(1991) is introduced as an interpolation function of deflection of the element

$$w = Q\alpha, \quad (4)$$

where

$$Q = (1, x, y, x^2, xy, y^2, x^3, x^2y, xy^2, y^3, x^3y, xy^3), \quad (5)$$

$$\alpha = (\alpha_1, \alpha_2, \alpha_3, \cdots, \alpha_{10}, \alpha_{11}, \alpha_{12})^T. \quad (6)$$

The coefficients $\alpha_i (i = 1, \cdots, 12)$ depend linearly on nodal displacement vector

$$\delta_e = (w_i, \theta_{xi}, \theta_{yi}, w_j, \theta_{xj}, \cdots, \theta_{yk}, w_l, \theta_{xl}, \theta_{yl})^T \quad (7)$$

of the element. Deflection w is thus interpolated by δ_e as follows

$$w = N\delta_e \quad (8)$$

where N is known as shape function which is a row vector of 12 polynomial functions. By applying the principle of virtual work to the element, stiffness matrix K_e of the element, corresponding mass matrix M_e and nodal force vector f_e due to constant load q distributed on the element are derived as follows

$$K_e = \frac{1}{60ab} L (D_x K_1 + D_y K_2 + D_1 K_3 + D_{xy} K_4) L, \quad (9)$$

$$M_e = \frac{\rho_s ab}{6300} L M_g L, \quad (10)$$

$$f_e = \frac{1}{3} abq (3, -b, a, 3, b, a, 3, -b, -a, 3, b, -a)^T, \quad (11)$$

and the motion equations of the element is written as

$$K_e \delta_e + M_e \frac{\partial^2 \delta_e}{\partial t^2} = f_e. \quad (12)$$

Matrices in these expressions, L, K_1, K_2, K_3, K_4 and M_g, can be found from Zienkiewicz(1991) and Washitsu(1983). Furthermore, ρ_s represents mass per unit area of the plate. We assemble all the element's mass matrix, stiffness matrix and nodal force vector, and get motion equations of the plate as follows

$$K\delta + M\frac{\partial^2 \delta}{\partial t^2} = f, \quad (13)$$

where K and M are stiffness matrix and mass matrix respectively, and δ is nodal displacement vector of the plate and f is the nodal force vector due to the piecewise constant load q. Eigenmodes d_j and eigenvalues ω_j^2 are estimated by solving the following eigenvalue problem

$$Kd_j - \omega_j^2 M d_j = 0. \quad (14)$$

Tensor expressions will be adopted in this paper without mention, but the second term in Eq. (14) shoud not be recognized as the summation convention. By introducing a matrix λ_{ij}

$$\lambda_{ij} = \begin{cases} \omega_j^2 & i = j \\ 0 & i \neq j \end{cases}, \quad (15)$$

equation (14) can be rewritten as the following form

$$K_{ik} d_{kj} - M_{ik} d_{kl} \lambda_{lj} = 0. \quad (16)$$

Eigenmodes are usually normalized with respect to the mass matrix as follows

$$d_{ki} M_{kl} d_{lj} = \delta_{ij}, \quad (17)$$

$$d_{ki} K_{kl} d_{lj} = \lambda_{ij}. \quad (18)$$

As an example some eigenmodes of an isotropic square plate are evaluated by solving the above eigenvalue problem. Weaver, Timoshenko and Young(1990) represented natural frequency ω_j of the j-th mode in the following non-dimensional form

$$\alpha_j = \omega_j a_0^2 \sqrt{\frac{\rho_s}{D}}, \quad (19)$$

where a_0 and D are side length and bending rigidity of the plate. Table 1 shows that our numerical values of α_j for the lowest 3 modes agree well with Timoshenko's results(Weaver, 1991).

Table 1 α_j of A Completely Free Square Plate

Nodal Line	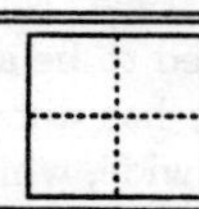	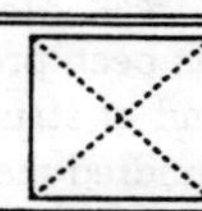	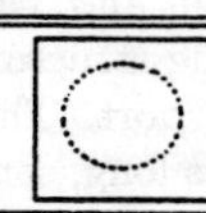
present computation	14.13	20.40	23.90
Timoshenko's result	14.10	20.56	23.91

Hydrodynamic Pressures

We assume that fluid is incompressible and inviscid, and flow is irrotational. Thus velocity potential $\Phi(x, y, z, t)$ exists in the fluid domain Ω, and can be expressed as the sum of the incident potential ϕ_I, the diffraction potential ϕ_D and the radiation potential ϕ_R as follows

$$\Phi(x, y, z; t) = i\omega\{\phi_I(x, y, z) + \phi_D(x, y, z) + \phi_R(x, y, z)\}e^{i\omega t} \quad (20)$$

where

$$\phi_I = \frac{\zeta_A}{\kappa} e^{\kappa z + i\kappa(x\cos\chi + y\sin\chi)}$$

ω; incident wave frequency,
κ; wave number$(=\omega^2/g)$,
ζ_A; incident wave amplitude,
g; gravitational acceleration.

We define pressure function $p(x, y)$ and vertical motion $\zeta(x, y)$ without the term of time dependency as follows

$$\left.\begin{array}{ll} P(x, y; t) = -\rho g p(x, y)e^{i\omega t}, & \text{on } S_h \\ Z(x, y; t) = \zeta(x, y)e^{i\omega t}, & \text{on } S_h \end{array}\right\}, \quad (21)$$

where ρ is fluid density, and S_h is mean wetted body surface at $z = 0$. Potential function $\phi(x, y, z)$, which is ϕ_D or ϕ_R, is governed by three-dimensional Laplace equation [L]. It has to satisfy a linearized free surface condition [F], a hull surface condition [H], a bottom condition [B], and a radiation condition [R]. They can be expressed as follows

$$\left.\begin{array}{lll} [L] & \nabla^2\phi(x, y, z) = 0, & \text{in } \Omega \\ [F] & \kappa\phi - \phi_z = 0, & \text{on } S_f \\ [H] & \phi_z = V_z, & \text{on } S_h \\ [B] & \phi_z \to 0, & z \to -\infty \end{array}\right\}, \quad (22)$$

where S_f is the mean free surface $z = 0$, and subscript denotes partial differentiation. The hull surface condition [H] for radiation potential ϕ_R and diffraction potential ϕ_D is written as

$$\left.\begin{array}{l} \dfrac{\partial\phi_R}{\partial z} = \zeta(x, y) \\ \dfrac{\partial\phi_D}{\partial z} = -\dfrac{\partial\phi_I}{\partial z} = -\kappa\phi_I \end{array}\right\} \quad \text{on } S_h. \quad (23)$$

The pressure function $p(x, y)$ is evaluated by the following equation

$$p = \phi_z - \kappa\phi. \quad (24)$$

By introducing a Green function $G(x, y, z; x', y', z')$ which satisfies Eq. (22) except for [H], $\phi(x, y, z)$ can be expressed as the following integration

$$\phi(x, y, z) = -\iint_{S_h} p(x', y')G(x, y, z; x', y', 0)dx'dy'. \quad (25)$$

Substituting Eq. (25) into Eq. (24), we get the following integral equation with respect to the pressure function $p(x, y)$ as follows

$$\phi_z(x, y, 0) = p(x, y) - \kappa\iint_{S_h} p(x', y')G(x, y, 0; x', y', 0)dx'dy', \quad (26)$$

where Green function $G(x, y, 0; x', y', 0)$ is the same as the one used by Yamashita(1979).

We apply constant panel method to solve this equation for the evaluation of pressure function $p(x, y)$. First, S_h is divided into panels and centroids of the panels are selected as collocation points. Then pressure p_j on j-th panel is set to be a constant. The following simultaneous equations are thus resulted

$$\phi_{zi} = C'_{ij}p_j \quad (27)$$

where C'_{ij} is evaluated by the following integration

$$C'_{ij} = \delta_{ij} - \kappa\iint_{\Delta S_{hj}} G(x_i, y_i, 0; x', y', 0)dx'dy' \quad (28)$$

δ_{ij} is the Kronecker delta

$$\delta_{ij} = \begin{cases} 1, & i = j \\ 0, & i \neq j \end{cases}. \quad (29)$$

We can get pressure p_j from Eq. (27) by using matrix C_{ij}, the inverse of C'_{ij}, as follows

$$p_i = C_{ij}\phi_{zj}. \quad (30)$$

We get the corresponding pressure function p_D and p_R by substituting Eq. (23) into Eq. (30). Thereby we get hydrodynamic pressure on i-th panel in the form

$$P_i = \rho g C_{ij}(\kappa\phi_{Ij} - \zeta_j)e^{i\omega t}. \quad (31)$$

Wave-Body Interactions

To calculate wave-induced motions of a plate, the finite elements are managed to coincide with the panels, and deflection $w(x, y)$ of the plate corresponds to the vertical motion $\zeta(x, y)$. Then the nodal force vector f due to piecewise constant hydrodynamic pressures, Eq. (31), can be expressed as

$$f_i = \rho g F_{ij}(\kappa\phi_{Ij} - w_j)e^{i\omega t}, \quad (32)$$

where matrix F_{ij} is a transform of matrix C_{ij} in Eq. (30) by using transformation Eq. (11), and ϕ_{Ij} and w_j are the incident potential ϕ_I and deflection w evaluated at the centroid of j-th panel. Motions δ_i due to wave-body interactions is also harmonic and can be written as

$$\delta_i = u_i e^{i\omega t}, \quad (33)$$

where displacement vector u_i can be expressed as a superposition of eigenmodes of the plate as follows

$$u_i = d_{ij}q_j. \quad (34)$$

Substituting Eq. (32), (33) and (34) to Eq. (13) the following equations for determining unknown principal coordinates q_j are obtained

$$(K_{ik}d_{kj} - \omega^2 M_{ik}d_{kj} + \rho g F_{ik}w_{kj})q_j = \rho g\kappa F_{ik}\phi_{Ik}, \quad (35)$$

where w_{ij} is deflection of j-th mode at the centroid of i-th panel. Multiplying the above equation with d_{im} and taking into account the orthogonality in Eq. (17) and (18), we can get a set of equations as follows

$$\{\lambda_{mj} - \omega^2\delta_{mj} + \rho g d_{im}F_{ik}w_{kj}\}q_j = \rho g d_{im}F_{ik}\kappa\phi_{Ik}. \quad (36)$$

Thus, principal coordinates q_j are determined by solving the above equations, and then wave-induced motions Eq. (33), (34) are obtained.

Table 2 Principal Particulars of the Plate Model

Length	2.0m
Width	0.5m
Thickness of Plate	0.005m
Draft of Body	0.05m
E (Young's modulus)	3.0×10^8 (Kgf/m^2)
υ (Poisson's ratio)	0.3

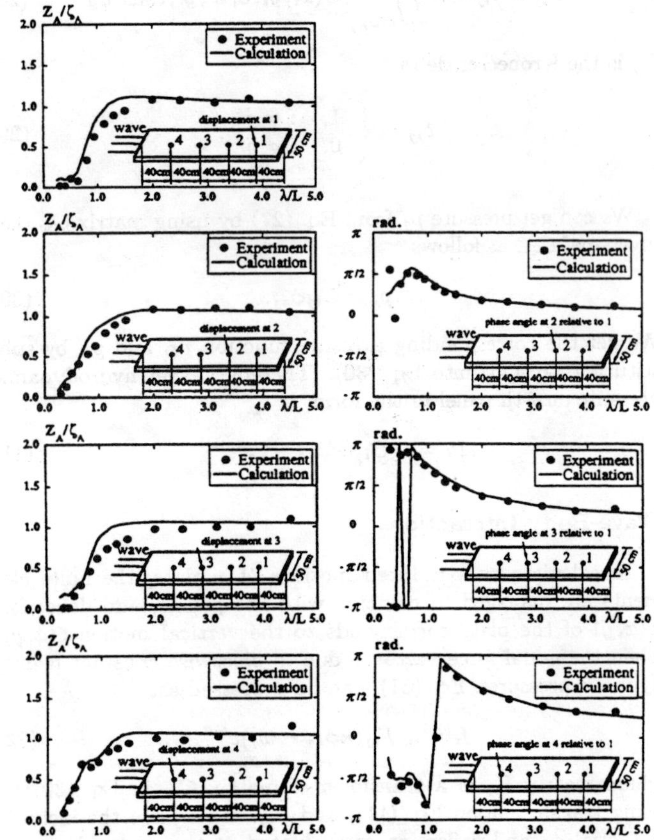

Fig.2 Comparisons of Amplitudes of Vertical Displacement at locations of point 1, 2, 3, 4, and Their Phase Angles Relative to Point 1 with Measured Data (Murai, 1995).

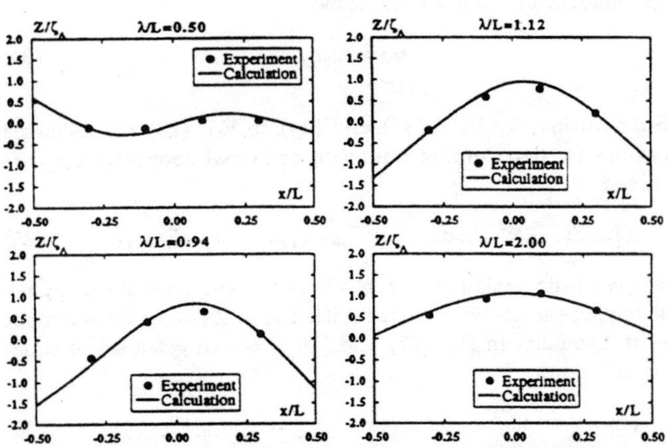

Fig.3 Comparison of Instantaneouse Shapes of the Plate Model with Measured Data (Murai, 1995).

NUMERICAL RESULTS

Validation

Murai et al(1995) carried out an experiment for a plate model. Its principal particulars are listed in Table 2, where Poisson's ratio of the plate is added by the authors. In calculation the plate model is divided into 40×10 panels, and the first 20 eigenmodes are evaluated and brought into calculation for determining wave-induced motions. Natural frequency of two nodal line eigenmode by calculation is $0.6990H_z$. Fig. 2 shows comparisons of amplitudes and phase angles of vertical motion at point 1, 2, 3, 4 with the measured data(Murai, 1995), in which phase angles are relative to point 1. They show good agreements. Instantaneous shape of the plate centerline are recorded by measuring vertical displacements at point 1, 2, 3, 4 in the experiment. Fig. 3 shows that calculated shapes agree well with measured points.

Table 3 Principal Particulars of the Module

Length	300m
Width	60m
Thickness	5m
Draft	0.72m
Moment of inertia of cross-section	0.406(m^4/m)
E (Young's modulus)	2.06×10^{11}(N/m^2)
υ (Poisson's ratio)	0.3

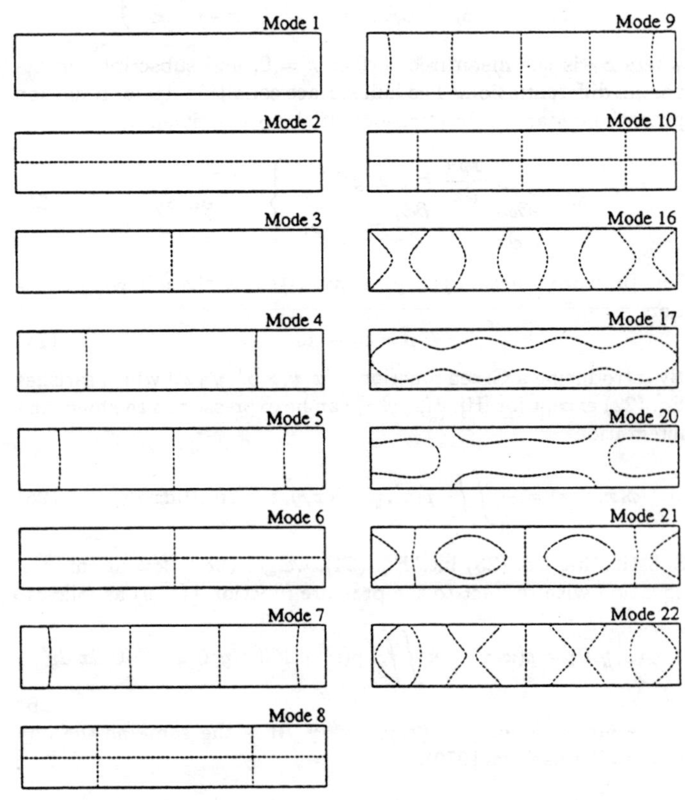

Fig.4 Nodal Lines of Several Modes of the Module

Table 4 Natural Frequencies of Several Eigenmodes of the Module

mode number	1	2	3	4	5
frequency (Hz)	Heave	Roll	Pitch	0.43	1.18
mode number	6	7	8	9	10
frequency (Hz)	1.30	2.33	2.66	3.87	4.15
mode number	16	17	20	21	22
frequency (Hz)	10.69	11.27	12.43	13.67	14.28

Motions Performance of A Module

Next we investigate motions of a large floating module ($300m \times 60m$) (Shin, 1995). The principal particulars of the module are shown in Table 3. We divide the module into 60×12 panels, and the first 60 eigenmodes are evaluated for determining its motions. Fig. 4 shows nodal lines of the first 10 eigenmodes and 16-, 17-, 20-, 21-, 22-th mode. Table 4 are the corresponding natural frequencies. The first 3 modes correspond to the rigid modes, heave, roll and pitch, and the 4-th mode is the bending mode of 2 nodal lines, the first elastic mode. While 6-, 8- and 10-th modes are twisting modes which have a very important role in oblique seas. The symmetric modes, 16-, 17-, 20-, 21-, 22-th modes, are specially shown here to emphasize that their nodal lines are different from those of beam.

It is worth to pay more attention to the value of natural frequency of the first elastic mode, denoted as f_4 here. f_4 of the module is equal to $0.43H_z$ as shown in Table 4. If a $2000m \times 300m$ floating airport is constructed by connecting this kind of modules, the value of f_4 will be greatly reduced to $0.0096H_z$, and thereby it will be very flexible. However, for conventional ship it is much larger than the one of the floating module. For example, a destroyer has the value of $2.12H_z(13.3rad/s)$(Bishop and Price, 1979). A character number, $\tilde{f} = f_4\sqrt{L}$, can be defined based on the law of similarity(Lee and Webster, 1994), which describe the flexibility of structure. If a $2m \times 0.3m$ plate is used as an experimental model of the $2000m \times 300m$ floating airport, the value of f_4 for the model will be $0.30H_z$ due to the law of similarity. It is much less than that of the plate model mentioned in section 3.1.

Fig. 5 shows motions of the module at wave length ratio $\lambda/L = 0.25$ in head seas. More than 10 to 15 panels are required in one wave length for reasonable evaluation of pressure function (Yago, 1996). In the present computation, it contains 15 panels. Disturbance wave profiles around the module which include the diffraction and radiation waves are shown in Fig. 6. The motions in quartering seas of incident angle 135° at the same wave length ratio are calculated, and are shown in Fig. 7. We can see that motions are no more symmetric with respect to the centerline. Furthermore, we can see from Fig. 8 that amplitude of vertical motion at a point around $x/L = 0.3$ is small. In fact, point $x/L = 0.333, y/L = -0.033$ is almost a nodal point where amplitude of vertical motion is equal to 0.0029. Disturbance wave profiles around the module are shown in Fig. 9, and hydrodynamic pressures are given in Fig. 10. Fig. 11 shows the effects of the elasticity on heaving, rolling and pitching amplitudes at wave length ratio $\lambda/L = 0.25$ in various wave headings. The elastic deformations are calculated by using 60 modes. In Fig. 11 the results as a rigid body do not show significant discrepancies with those as an elastic body.

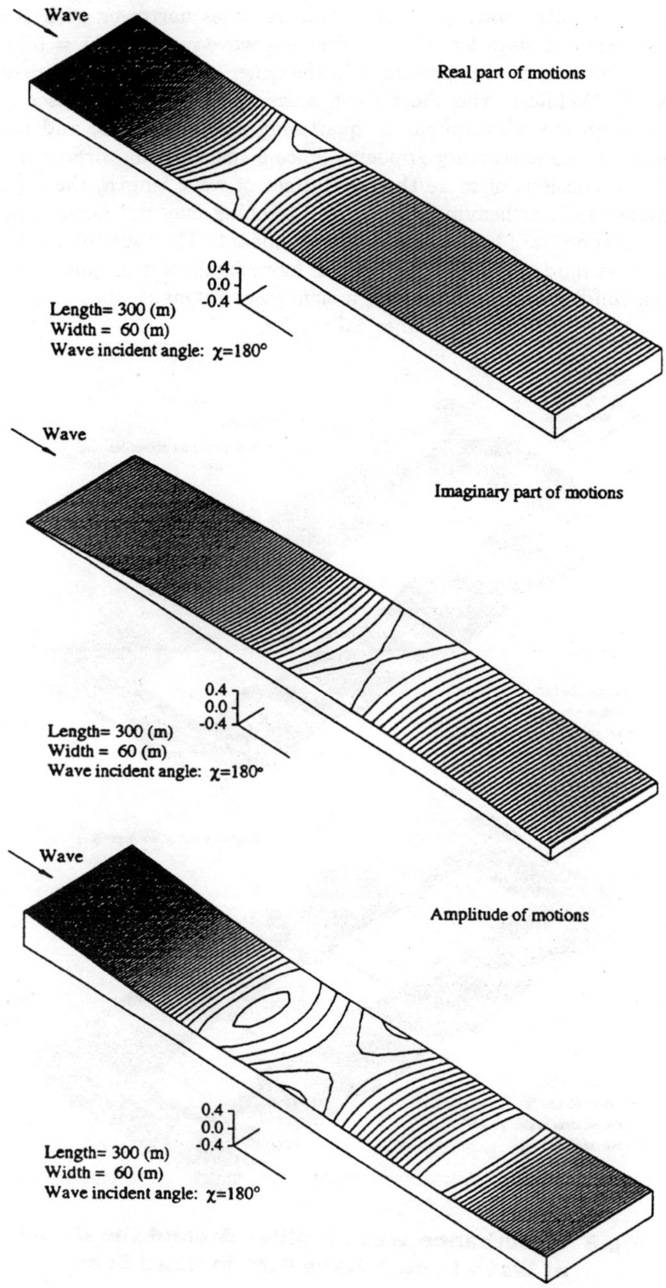

Fig.5 Motions of the Module at Wave Length Ratio 0.25 in Head Seas

Fig. 12 shows response amplitude operators (RAOS)(Bishop and Price, 1979) of the first 10 modes, in which the maximum deflections of each mode are taken to be one unit. We can see from the figures that the response amplitude operators become smaller with the higher modes. It means that the present modal analysis method can evaluate deflections of the module by using the responses of a few lower modes. In the next place, the significant responses of each mode appear in a limit range of wave conditions. For instances, the responses of higher modes appear in the shorter wave lengths, which becomes narrower with the reduction of wave lengths. In the long wave range ($L/\lambda = 0.1$), only heaving mode appears, and the other modes do not appear at all. While in the short wave range ($L/\lambda = 6$), responses of the higher modes appear in quartering sea conditions, and the heaving mode scarcely appears. Since a real floating airport has the dimensions of more than 100 times of wave length, the rigid modes such as heaving, rolling and pitching may not occur, and the responses of higher modes are significant. The twisting modes such as mode-6, -8 and -10 become more significant in quartering sea conditions. Furthermore, a significant resonant phenomenon does not occur in the numerical results.

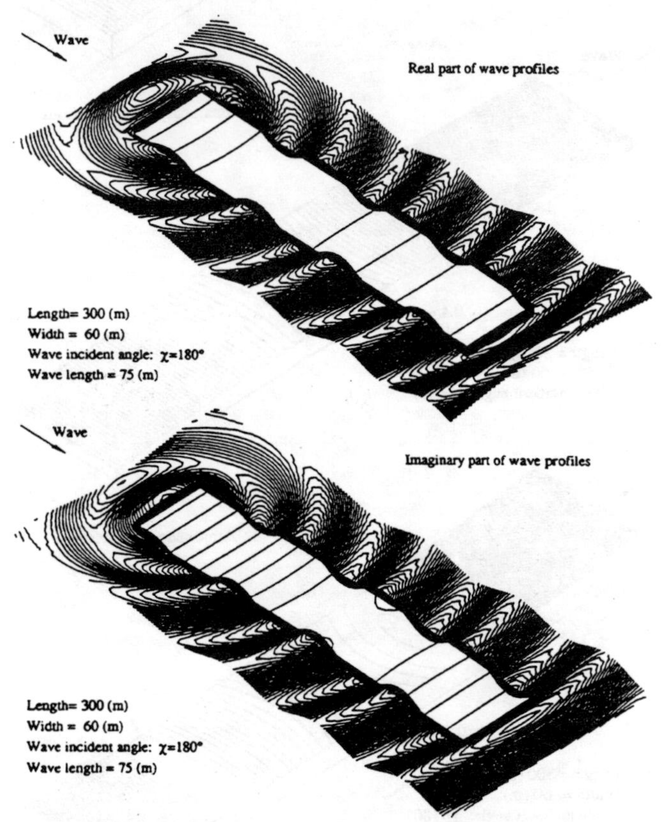

Fig.6 Disturbance Wave Profiles Around the Module at Wave Length Ratio 0.25 in Head Seas

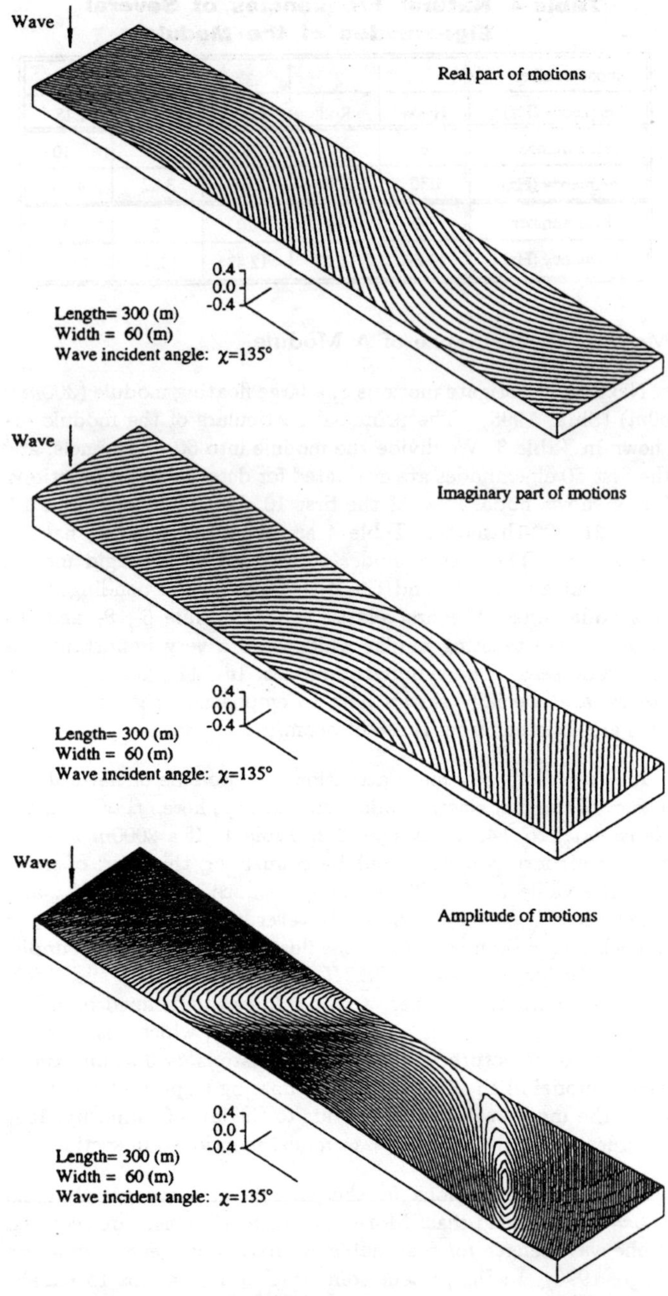

Fig.7 Motions of the Module at Wave Length Ratio 0.25 in Quartering Seas of 135 °.

CONCLUSIONS

In this paper we developed the plate modal method which combines pressure distribution method in hydrodynamics with finite element method in elasticity. It makes the prediction of wave-induced motions of very large floating structure possible in any heading waves.

The main conclusions obtained in this study can be summarized as follows;

(1) The accuracy of present method is verified by comparisons of calculated vertical displacements and instantaneous shapes of the $2m \times 0.5m$ plate model in head seas with available measured data, which show reasonable agreements.

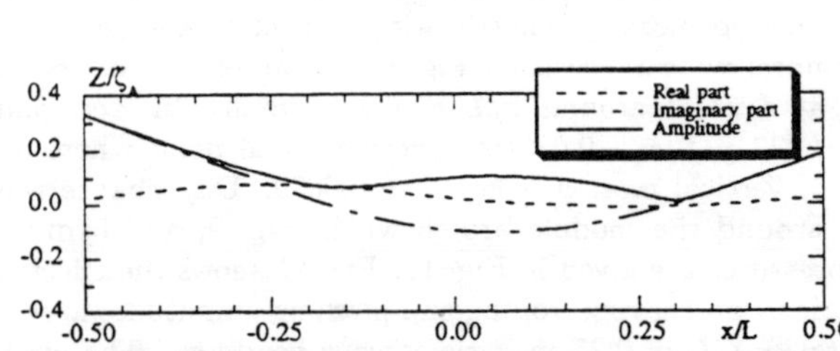

Fig.8 Displacement of the Module Centerline at Wave Length Ratio 0.25 in Quartering Seas of 135 °.

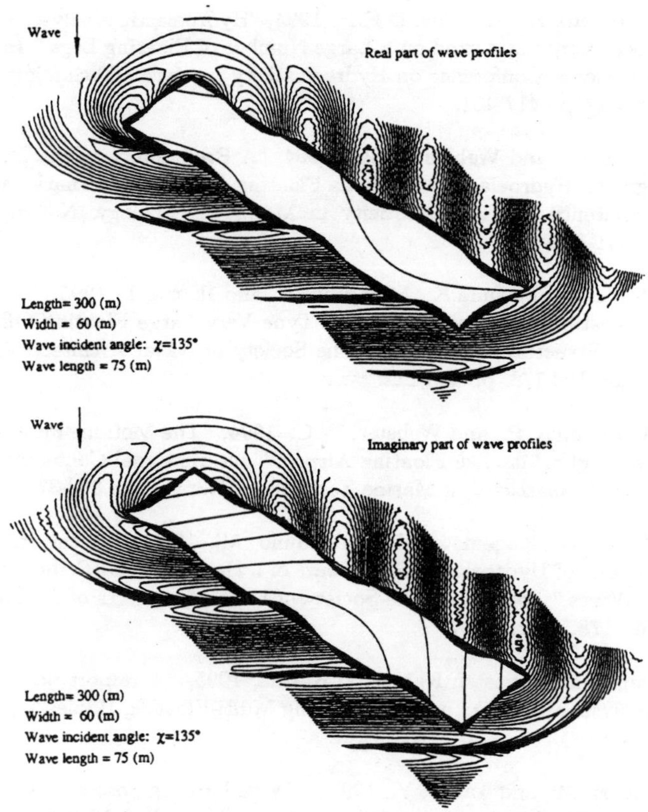

Fig. 9 Disturbance Wave Profiles Around the Module at Wave Length Ratio 0.25 in Quartering Seas of 135 °.

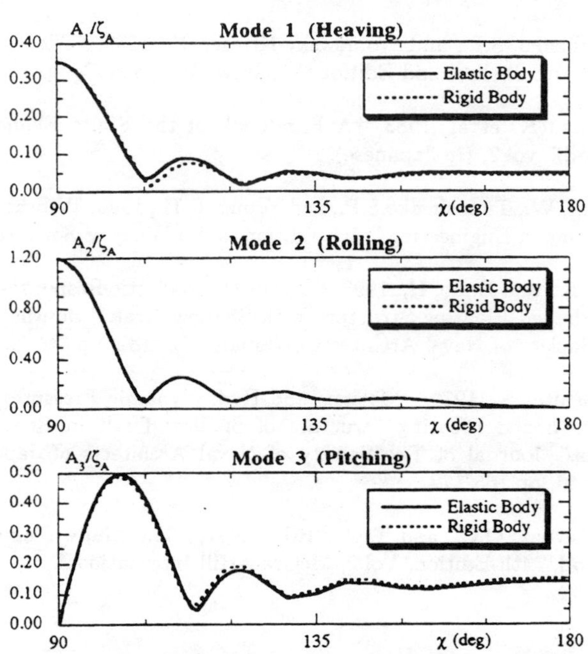

Fig. 11 Effect of Elasticity on the Heaving, the Pitching and the Rolling Amplitudes at Wave Length Ratio 0.25.

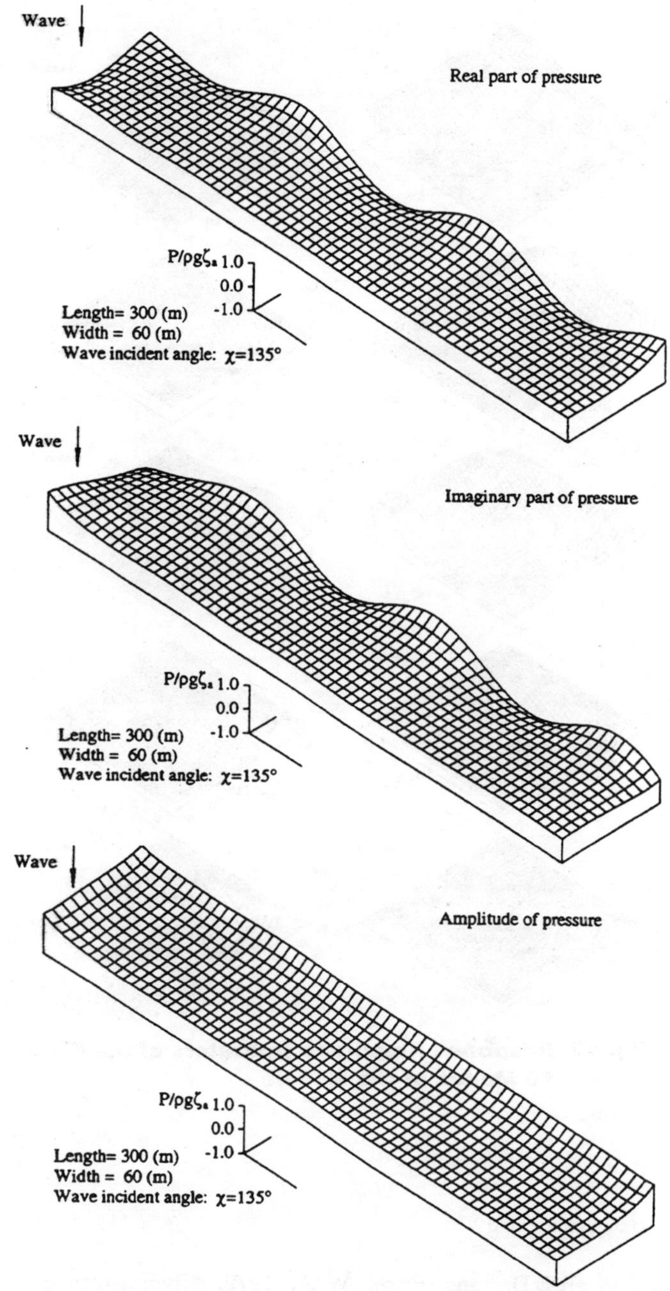

Fig.10 Pressures on the Bottom of the Module at Wave Length Ratio 0.25 in Quartering Seas of 135 °.

(2) The involvement of twisting modes makes the prediction of motion in oblique seas more reasonable. Even in head seas, prediction by present method are still show some difference from that by the beam modal method.

(3) We propose the character number $\tilde{f}$ related to natural frequency of the first elastic mode. The value of $\tilde{f}$ indicates the flexibility of plate.

(4) It is made clear that the twisting modes appear in the motions of the large module of $300m \times 60m$ in quartering seas, and significant resonance phenomena of it scarcely appear in waves.

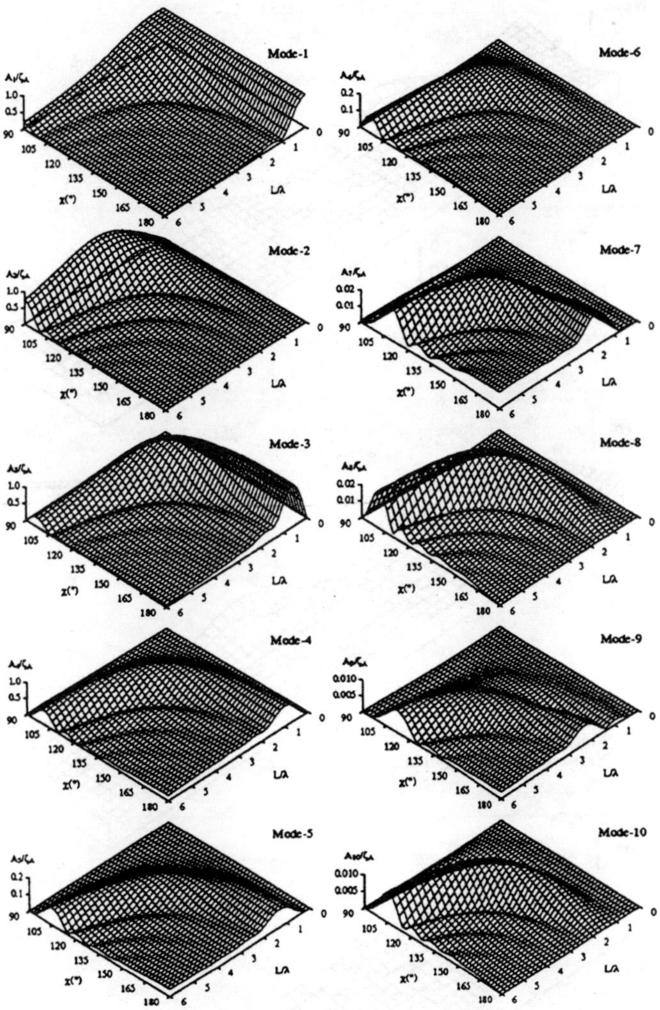

Fig. 12 Response Amplitude Operators of the First 10 Modes of the Module

REFERENCES

Bishop, R.E.D., and Price, W.G., 1979, "Hydroelastisity of Ships," Cambridge University Press, pp.175.

Ertekin, R.C., and Wang, S.Q., 1994, "Hydroelastic Response of a Floating Runway," International Conference on Hydroelasticity in Marine Technology, Norway, pp.389-400.

Hirayama, T., Ma, N., and Ueno, S., 1994, "Influence of Flexibility on the Motions and Deflections of an Airport-oriented Floating Long Offshore Structure," International Conference on Hydroelasticity in Marine Technology, Norway, pp.377-387.

Kagemoto, H., and Yue, D.K.P., 1994, "Hydroelastic Analyses of a Structure Supported on a Large Number of Floating Legs," International Conference on Hydroelasticity in Marine Technology, Norway, pp.417-431.

Lee, S.W., and Webster, W.C., 1994, "A Preliminary to the Design of a Hydroelastic Model of a Floating Airport," International Conference on Hydroelasticity in Marine Technology, Norway, pp.351-362.

Maeda, H., Masuda,K., Miyajima, S., and Ikoma, I., 1995, "Hydroelastic Responses of Pontoon Type Very Large Floating Offshore Structure," Journal of The Society of Naval Architects of Japan, Vol.178, pp.203-212.

Mamidipudi, P., and Webster, W.C., 1994, "The Motions Performance of a Mat-like Floating Airport," International Conference on Hydroelasticity in Marine Technology, Norway, pp.363-375.

Murai, M., Kagemoto, H., and Fujino, M., 1995, "On the Predictions of Hydroelastic Behaviours of a Huge Floating Structure in Waves," Journal of The Society of Naval Architects of Japan, Vol. 178, pp.213-224.

Shin Nihon Seitetsu Kabushiki Kaisha, 1995, "A Report on the Analysis of Connecting Force among Multi-Floating Bodies," (in Japanese).

Takaki, M., and Tango, Y., 1995a, "Wave Drifting Forces on Very Large Floating Structures," International Journal of Offshore and Polar Engineering, Vol. 5, No. 3, pp.204-211.

Takaki, M., Lin, X., and Higo, Y., 1995b, "Effect of Connecting Condition on Motions of Huge Floating Structures in Waves," Transactions of West-Japan Society of Naval Architects, No. 89, pp.111-121.

Timoshenko, S.P., and Woinowsky-Krieger, S., 1959, "Theory of Plates and Shells," 2nd Edition, McGraw-Hill Book Company.

Washitsu K. et al, 1983, "A Handbook of the Finite Element Method," vol.2, (in Japanese).

Weaver, W., Timoshenko,S.P., and Young, D.H., 1990, "Vibration Problems in Engineering," 5th Edition, John Wiley & Sons, Inc.

Yago, K., and Endo, H., 1996, "On the Hydroelastic Responses of Box-Shaped Floating Structure with Shallow Draft," Journal of The Society of Naval Architects of Japan, Vol. 180, pp.341-352.

Yamashita, S., 1979, "Motions and Hydrodynamic Pressures of a Box-shaped Floating Structure of Shallow Draft in Regular Waves," Journal of The Society of Naval Architects of Japan, Vol. 146, pp.165-172.

Zienkiewicz, O.C., and Taylor R.L., 1991, "The Finite Element Method," 4th Edition, Vol.2, McGraw-Hill International.

Proceedings of the Seventh (1997) International Offshore and Polar Engineering Conference
Honolulu, USA, May 25-30, 1997

ISBN 1-880653-28-1 (Set); ISBN 1-880653-29-X (Vol. I)

Hybrid Dynamic Analysis of Large Tension Leg Floating Structures Using Plate Elements

Takuji Hamamoto and Akinori Suzuki
Musashi Institute of Technology
Tokyo, Japan

Ken-ichi Fujita
Yokogawa Bridge Corporation
Chiba, Japan

ABSTRACT

This paper presents a three dimensional hydroelastic formulation of the wet mode free vibration of large tension leg floating structures which consist of multiple barge-type modules by a BEM-FEM hybrid method, taking account of module and connector flexibilities. Tension legs can be attached to any nodal point on the water-structure interface. The sea water is modeled by BEM using four node quadrilateral isoparametric (linear) elements. The floating structure is modeled by FEM using four node quadrilateral isoparametric (plate) elements which can account for flexural as well as shear deformation of both modules and connectors. Only the water-structure interface is discretized by boundary elements. Wet mode free vibration analysis is carried out by an iterative method because added mass terms are frequency dependent. Numerical examples are presented to show the variation in wet mode frequencies and mode shapes of module linked floating structures due to irregular configuration as well as connector flexibility. Moreover, numerical results are compared with published experimental results to discuss the efficiency and accuracy of the proposed method.

KEYWORDS : Module linked floating structure, BEM-FEM hybrid analysis, Tension leg mooring, Wet mode analysis, Plate finite element, Linear boundary element

INTRODUCTION

Module linked large floating structures are preferable structural systems from a view point of fabrication, transportation and deployment of each module. However, an overall structural deformation becomes significant because of small thickness relative to large expanse of the structures. In shallow water, breakwaters can be constructed around the floating structures to eliminate or mitigate wave forces. In deep water, however, the construction of breakwaters is not a good solution from economical and environmental points of view. One of the promising methods to construct large floating structures in deep water is the application of tension leg mooring system. The tension leg mooring can stiffen the overall structural system and considerably change its dynamic characteristics to reduce structural deformation.

The hydroelastic analysis of module linked floating structures has been studied from both analytical and experimental points of view. Maeda *et al.* (1979) have studied the one dimensional dynamic behavior of floating structures which consist of rigid modules with rigid or pin connectors for regular waves by a strip method. Kagemoto and Yue (1987) have studied the dynamic behavior of large floating structures which consist of multiple vertical cylinders connected by elastic springs, taking account of multiple diffraction among cylinders. Riggs and Ertekin (1993) have studied the three dimensional hydroelastic analysis of module linked floating structures subjected to regular waves with arbitrary angle by using FEM for structures and a strip method for water. Takaki and Tango (1995) have studied the wave drifting forces acting on a multiple module linked floating structure which is connected by rigid or pin connectors by a three dimensional panel method. So far, however, a BEM-FEM hybrid method has not yet extensively used for large floating structures. Taylor and Waite (1978) have studied the three dimensional hydroelastic response of cantilevered and freely floating cylinders subjected to regular waves. Jades and Mathews (1990) have studied the elastic response of a spherical shell and cantilevered plate in the acoustic media. Liu *et al.* (1991) have investigated the relative merits of the higher-order boundary elements and constant panel methods for calculating hydrodynamic loadings.

The authors have already formulated the hydroelastic response of module linked large floating structures by a three dimensional BEM-FEM hybrid analysis (Hamamoto and Fujita, 1995, 1996). In the analysis, the structure-water interface is discretized by constant boundary elements. On the other hand, the floating structure is discretized by eight node brick or four node plate elements. As a result, the compatibility condition of structure and water is not strictly satisfied because nodal points of boundary and finite elements do not coincide on the structure-water interface.

In this study, the combination of linear boundary elements and four node quadrilateral (plate) finite elements is used to satisfy the compatibility condition completely. The study is restricted to a wet mode free vibration problem. Numerical examples are presented to show the variations in wet mode frequencies and mode shapes of module linked floating structures due to irregular configuration as well as connector flexibility. Moreover, numerical results are compared with published experimental results which were obtained by experimental modal analysis of a small scaled model of module linked floating structure(Endo *et al.*,1995) to discuss the efficiency and accuracy of the BEM-FEM hybrid method.

ANALYTICAL MODEL AND ASSUMPTIONS

The analytical model of a module linked floating structure is shown in Fig.1. The structural system is composed of multiple flexible modules and connectors. The floating structure is anchored by tension legs to the sea bed. Each module has a square plan. Connectors of finite width are placed between modules. In the figure, a is the length of module, b is the width of connector, h is the thickness of module, d is the water depth, $\bar{d}$ is the distance from the bottom of the floating structure to the sea bed, Ω is the fluid domain, Γ_f is the free water surface, Γ_s is the sea bed, and Γ_υ is the structure-water interface. The global coordinate system x-y-z is defined in such a way that the z-axis is directed upward.

The following assumptions are introduced in the wet mode free vibration analysis of module linked floating structures.

1. The structural motion is governed by a linear vibration theory.
2. The structure is elastic and of uniform thickness.
3. The structure is vertically guided without friction along the sides.
4. The sea water is inviscid, incompressible and irrotational.

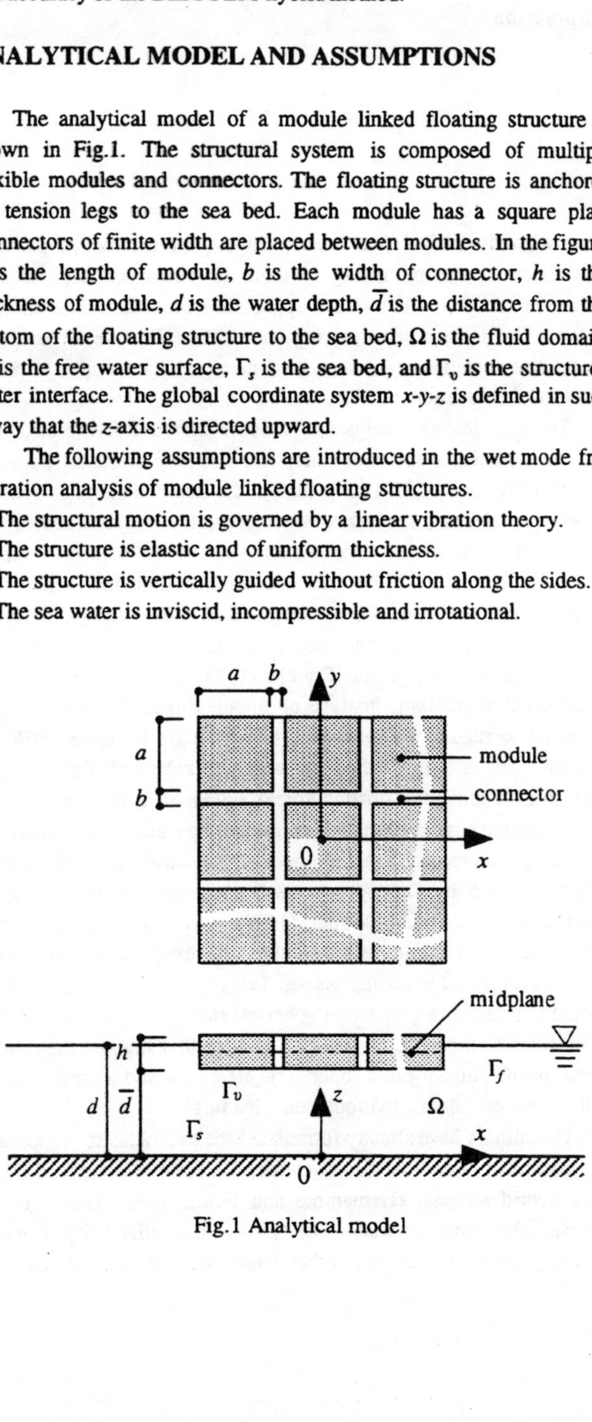

Fig.1 Analytical model

5. The sea water is of constant depth and extends to infinity.
6. The tension legs are elastically anchored to the sea bed and transfer only vertical force.
7. The mass effect of tension legs is ignored.
8. The effect of wetted vertical faces is negligibly small in comparison with that of horizontal lower surface.
9. Only added mass effect is considered, *i.e.*, hydrodynamic radiation effect is ignored.

BEM MODEL FOR FLUID DOMAIN

Boundary value problem

The whole fluid domain is governed by the velocity potential function of radiated wave,

$$\Phi(x,y,z,t) = \phi(x,y,z)e^{-i\omega t}, \tag{1}$$

where ϕ is the spatial function of Φ.

When a floating structure is moving and vibrating with harmonic motion in still water, the radiated wave potential ϕ satisfies the following three dimensional Laplace equation and boundary conditions.

Laplace equation :

$$\nabla^2\phi = 0 \qquad -\infty < x,y < +\infty,\ 0 \le z \le d, \tag{2a}$$

Free surface condition :

$$\frac{\partial\phi}{\partial z} = \frac{\omega^2}{g}\phi \qquad at\ z = d, \tag{2b}$$

Sea bed condition :

$$\frac{\partial\phi}{\partial z} = 0 \qquad at\ z = 0, \tag{2c}$$

Structure-water interface condition :

$$\frac{\partial\phi}{\partial z} = -i\omega\bar{u}_z \qquad at\ z = \bar{d}, \tag{2d}$$

Sommerfeld's radiation condition :

$$\lim_{r\to\infty}\sqrt{r}\left(\frac{\partial\phi}{\partial r} - ik\phi\right) = 0 \qquad at\ r \to 0, \tag{2e}$$

where $\nabla^2 = \partial^2/\partial x^2 + \partial^2/\partial y^2 + \partial^2/\partial z^2$, $r = \sqrt{x^2+y^2}$ is the distance from the origin of the global coordinate system, g is the gravitational acceleration and $\bar{u}_z$ is the amplitude of structural vertical displacement. Eq.(2d) represents the continuity of structural motion and water particle motion at the structure-water interface.

Boundary integral equation

The above boundary value problem can be solved by using the second Green's formula,

$$\iiint_\Omega\left(\phi\nabla^2 G - G\nabla^2\phi\right)d\Omega = \iint_\Gamma\left(\phi\frac{\partial G}{\partial n} - G\frac{\partial\phi}{\partial n}\right)d\Gamma, \tag{3}$$

where G is the Green function, n is the outward normal unit vector and Γ denotes the total boundary surface. In this study, the Green function satisfies the following three dimensional Laplace equation and boundary conditions (Kiyokawa *et al.*, 1983).

Laplace equation :

$$\nabla^2 G = -4\pi\delta(P,Q) \qquad in\ \Omega, \tag{4a}$$

Free surface condition :

$$\frac{\partial G}{\partial z} = \frac{\omega^2}{g} G \qquad on\ \Gamma_f, \tag{4b}$$

Sea bed condition :

$$\frac{\partial G}{\partial z} = 0 \qquad on\ \Gamma_s, \tag{4c}$$

Sommerfeld's radiation condition :

$$\lim_{r\to\infty} \sqrt{r}\left(\frac{\partial G}{\partial r} - ikG\right) = 0 \qquad at\ r\to\infty, \tag{4d}$$

where $\delta(P,Q)$ is the Dirac delta function, $P(x, y, z)$ is the point of application of the delta function, and $Q(\xi, \eta, \zeta)$ is the point under consideration. The Green function which satisfies Eqs.(4a) - (4d) is given by (John, 1950)

$$G(P,G) = 2\pi i C_0 \cosh kz \cosh k\zeta H_0^{(1)}(kR) + 4\sum_{j=1}^{\infty} C_j \cos k_j z \cos k_j \zeta K_0(k_j R), \tag{5}$$

where $R = \sqrt{(x-\xi)^2 + (y-\eta)^2}$ is the horizontal distance from P to Q, $H_0^{(1)}(kR)$ is the Hankel function of zero order of the first kind, $K_0(k_j R)$ is the modified Bessel function of zero order of the second kind, and coefficients C_0 and C_j are given by

$$C_0 = \frac{k^2 - v^2}{d(k^2 - v^2) + v}, \quad C_j = \frac{k_j^2 + v^2}{d(k_j^2 + v^2) - v}, \tag{6}$$

where $v = \omega^2/g$. The wave numbers k and k_j satisfy the following transcendental equation :

$$\omega^2 = kg \tanh kd = -k_j g \tan k_j d. \tag{7}$$

Incorporating Eq.(5) into Eq.(3) yields the following integral equation,

$$4\pi\phi(P) + \iint_{\Gamma_\upsilon} \left\{\phi(Q)\frac{\partial}{\partial n}G(P,Q) - G(P,Q)\frac{\partial\phi(Q)}{\partial n}\right\} d\Gamma = 0, \tag{8}$$

where P is a point in the fluid domain Ω and Q is a point on the boundary Γ_υ. In Eq.(8), the Green function G and its derivative $\partial G/\partial n$ have singularities when P coincides with Q on the boundary Γ_υ. To avoid the singularity in the integral operation, the boundary point is assumed to be at the center of the sphere and afterwards the radius is reduced to zero. As a result, the following boundary integral equation is obtained for a point P on the boundary Γ_υ,

$$c(P)\phi(P) + \iint_{\Gamma_\upsilon} \left\{\phi(Q)\frac{\partial}{\partial z}G(P,Q) - G(P,Q)\frac{\partial\phi(Q)}{\partial z}\right\} d\Gamma = 0, \tag{9}$$

where $c(P)$ is a constant which depends on whether P is located on the smooth or non-smooth boundary. In Eq.(9), the normal derivative is replaced by the derivative with respect to z.

Boundary element formulation

The boundary integral equation, Eq.(9), is applied to only the structure-water interface Γ_υ. To discretize by a boundary element method, the boundary Γ_υ is divided into E boundary elements. If the velocity potential ϕ is assumed to vary linearly on each element, the boundary Γ_υ can be discretized by 4-node quadrilateral isoparametric (liner) elements. The velocity potential and its derivative, ϕ and, $\partial\phi/\partial z$, may be described in terms of interpolation functions,

$$\phi = \sum_{i=1}^{4} N_i \phi_i = \{N\}^T\{\phi_0\}, \tag{10a}$$

$$\frac{\partial\phi}{\partial z} = \sum_{i=1}^{4} N_i \frac{\partial\phi_i}{\partial z} = \{N\}^T\left\{\frac{\partial\phi_0}{\partial z}\right\}, \tag{10b}$$

where $\{N\}$ is the interpolation function vector (4×1) and $\{\phi_0\}$ is the nodal velocity potential vector (4×1). Each element of $\{N\}$ is given by

$$N_1 = \frac{1}{4}(1-\xi)(1-\eta),\ N_2 = \frac{1}{4}(1+\xi)(1-\eta),$$
$$N_3 = \frac{1}{4}(1+\xi)(1+\eta),\ N_4 = \frac{1}{4}(1-\xi)(1+\eta). \tag{11}$$

Incorporating Eqs.(10a,b) into Eq.(9) and assembling the contribution from other elements yield the discretized integral equation at a nodal point P_i on the structure-water interface,

$$c(P_i)\phi(P_i) + \sum_{e=1}^{E}\left[\iint_{\Gamma_e}\{N\}^T\frac{\partial G(P_i,Q)}{\partial z}d\Gamma\{\phi_0\} - \iint_{\Gamma_e}\{N\}^T G(P_i,Q)d\Gamma\left\{\frac{\partial\phi_0}{\partial z}\right\}\right] = 0, \tag{12}$$

where E is the total number of elements, and $c(P_i)$ is 2π when P_i is located on the surface, 3π on the edge, and 3.5π at the vertex, respectively. Assembling Eq.(12) with respect to all nodal points yields the following algebraic equation,

$$[H]\{\phi\} - [G]\left\{\frac{\partial\phi}{\partial z}\right\} = \{0\}, \tag{13}$$

where the element vectors $\{h\}$(4×1) and $\{g\}$(4×1) of the global matrices $[H]$ (N_P×N_P) and $[G]$ (N_P×N_P) are given by

$$\{h\}^T = \iint_{\Gamma_e}\{N\}^T\frac{\partial G(P_i,Q)}{\partial z}d\Gamma, \tag{14a}$$

$$\{g\}^T = \iint_{\Gamma_e}\{N\}^T G(P_i,Q)d\Gamma, \tag{14b}$$

and

$$\{\phi\}^T = \{\phi_1, \phi_2, \cdots, \phi_{N_P}\}, \tag{14c}$$

$$\left\{\frac{\partial\phi}{\partial z}\right\}^T = \{-i\omega\bar{u}_{z1}, -i\omega\bar{u}_{z2}, \cdots, -i\omega\bar{u}_{zN_P}\}, \tag{14d}$$

where the subscript N_P is the total number of nodal points of boundary elements. The integral calculation in $\{h\}$ and $\{g\}$ can be performed by a 4-point Gauss quadrature rule when P_i and Q are located on different elements respectively. If P_i and Q are located on the same element, the integration in $\{h\}$ and $\{g\}$ becomes singular due to the nature of the Green function. The Green function can be considered to nearly behave as $1/r$ in the same element, *i.e.*,

$$G(P_i,Q) \approx \frac{1}{r}, \tag{15}$$

where $r = \sqrt{(x-\xi)^2 + (y-\eta)^2 + (z-\zeta)^2}$. The integration of $\{h\}$

becomes 0 because of $\partial G/\partial z = 0$ on the structure-water interface. On the other hand, the element of $\{g\}$ can be written as

$$g_i = \iint_{\Gamma_e} N_i(Q)\frac{1}{r}d\Gamma$$
$$= \iint_{\Gamma_e} (N_i(Q)-1)\frac{1}{r}d\Gamma + \iint_{\Gamma_e} \frac{1}{r}d\Gamma, \tag{16}$$

Due to the nature of interpolation functions, the first term of Eq.(16) vanishes and only the second term remains (Matsui and Kato, 1988). Consequently, Eq(16) can be calculated in explicit form as (Garrison, 1973)

$$g_i = \iint_{\Gamma_e} \frac{1}{r}d\Gamma$$
$$= 2\sqrt{\frac{\Delta S_i}{b}}\left\{\ln\left(b+\sqrt{b^2+1}+b\ln\left[\frac{1+\sqrt{b^2+1}}{b}\right]\right)\right\}, \tag{17}$$

where ΔS_i is the area and b is the aspect ratio of the i-th element.

FEM MODEL FOR FLOATING STRUCTURE

Module linked floating structures are discretized by 4-node quadrilateral isoparametric plate elements which can account for flexural as well as shear deformation of both modules and connectors. The element displacement and rotation components may be described in terms of interpolation functions as

$$u_z = \sum_{i=1}^{4} N_i u_{zi} = \{N\}^T\{u_{z0}\}, \tag{18a}$$

$$\theta_x = \sum_{i=1}^{4} N_i \theta_{xi} = \{N\}^T\{\theta_{x0}\}, \tag{18b}$$

$$\theta_y = \sum_{i=1}^{4} N_i \theta_{yi} = \{N\}^T\{\theta_{y0}\}, \tag{18c}$$

where $\{N\}$ is the interpolation function vector which is represented by Eq.(11), $\{u_{z0}\}$ is the z-directional nodal displacement vector (4 ×1); $\{\theta_{x0}\}$ and $\{\theta_{y0}\}$ are the nodal rotation vectors(4×1) with respect to x-and y-axes, respectively.

The equation of motion for a global system of floating structure can be written as

$$[M_S]\{\ddot{U}\}+[C_S]\{\dot{U}\}+([K_S]+[K_T])\{U\}=\{F\}, \tag{19}$$

where $[M_S]$, $[C_S]$ and $[K_S]$ are the global structural mass, damping and stiffness matrices, respectively, $[K_T]$ is the global stiffness matrix of tension leg mooring and $\{F\}$ is the global fluid force vector acting on the bottom surface of the structure, and $\{U\}$ is the global displacement vector of the structure which has three components at the i-th nodal point, *i.e.*,

$$\{U_i\} = \{u_{zi}, \theta_{xi}, \theta_{yi}\}^T \qquad (i = 1,2,\cdots N_p) \tag{20}$$

The element matrices of the global matrices $[M_S]$ are given by

$$[m_z] = \rho_s h \iint \{N\}^T\{N\}|J|d\xi d\eta, \tag{21a}$$

$$[m_{\theta x}] = \rho_s I_x \iint \{N\}^T\{N\}|J|d\xi d\eta, \tag{21b}$$

$$[m_{\theta y}] = \rho_s I_y \iint \{N\}^T\{N\}|J|d\xi d\eta, \tag{21c}$$

where $[m_z]$ is the element mass matrix corresponding to the z-directional motion, $[m_{\theta x}]$ and $[m_{\theta y}]$ are the element mass matrices corresponding to the rotational motions with respect to x- and y-axes, respectively, ρ_s is the mass density of the structure, and I_x and I_y are the rotational inertia with respect to x-and y-axes, respectively.

The element stiffness matrix $[k_s]$ of the global stiffness matrix $[K_s]$ is represented by the sum of bending and shear stiffnesses.

$$[k_s] = \iint [B_b]^T[D_b][B_b]|J|d\xi d\eta + \iint [B_s]^T[D_s][B_s]|J|d\xi d\eta, \tag{22}$$

where the first and second terms of the right hand side represent the bending and shear stiffnesses, respectively, $[B_b]$ (3×12) and $[B_s]$ (2 ×12) are the strain-displacement relationship matrices corresponding to the bending and shear deformation, respectively, $[D_b]$ and $[D_s]$ are the stress-strain relationship matrices corresponding to the bending and shear deformation, respectively, and given as

$$[D_b] = \frac{Eh^3}{12(1-\nu^2)}\begin{bmatrix}1 & \nu & 0\\ \nu & 1 & 0\\ 0 & 0 & (1-\nu)/2\end{bmatrix}, \tag{23a}$$

$$[D_s] = \frac{5Eh}{12(1+\nu)}\begin{bmatrix}1 & 0\\ 0 & 1\end{bmatrix}, \tag{23b}$$

where E and ν are Young's modulus and Poisson's ratio of the structure, respectively.

The structural damping is assumed to follow the Rayleigh damping assumption as follows:

$$[C_S] = \alpha[M_S] + \beta[K_S], \tag{24}$$

where coefficients α and β are given by

$$\begin{Bmatrix}\alpha\\ \beta\end{Bmatrix} = 2\begin{bmatrix}1/\omega_1 & \omega_1\\ 1/\omega_2 & \omega_2\end{bmatrix}^{-1}\begin{Bmatrix}h_1\\ h_2\end{Bmatrix}, \tag{25}$$

where ω_i and h_i are the i-th natural circular frequency and damping ratio, respectively.

Tension legs can be attached to any nodal point on the structure-water interface. Each element at nodal points attached to tension legs in the global stiffness matrix $[K_T]$ is given by

$$k_T = \frac{A_T E_T}{l}, \tag{26}$$

where A_T and E_T are the sectional area and Young's modulus, respectively, and l is the length of each tension leg.

COUPLING OF BEM AND FEM

Fluid coupled equation of motion is formulated for module linked floating structures by coupling BEM and FEM. To combine BEM and FEM, the continuity of structure and water particle motions have to be assured on the structure-water interface. In the formulation, the continuity condition can be completely satisfied at all interface nodal points on the structure-water interface.

The element nodal fluid force vector at the structure-water interface is given by the following equation,

$$\{f\} = \iint_{\Gamma_e} \{N\}^T p_r |J| d\xi d\eta \cdot e^{-i\omega t}, \tag{27}$$

where p_r is the hydrodynamic pressure induced by structural motion in still water and given as

$$p_r = i\omega\rho_w\{N\}^T\{\phi\} - \rho_w g\{N\}^T\{\bar{u}_z\},\tag{28}$$

where ρ_W is the mass density of sea water. Incorporating Eq.(28) into Eq.(27) yields the following equation,

$$\{f\} = \iint_{\Gamma_e}\{N\}^T\left(i\omega\rho_w[N]^T\{\phi\} - \rho_w g[N]^T\{\bar{u}_z\}\right)|J|d\xi d\eta\cdot e^{-i\omega t}.\tag{29}$$

The fluid force may be divided into two components which are in phase with structural acceleration and velocity, respectively. The element velocity potential function vector $\{\phi\}$ may be written as

$$\{\phi\} = [\lambda]\{-i\omega\bar{u}_z\},\tag{30}$$

where $[\lambda]$ is the element matrix of the global matrix $[H]^{-1}[G]$ and can be represented as

$$[\lambda] = \mathrm{Re}[\lambda] + i\,\mathrm{Im}[\lambda],\tag{31}$$

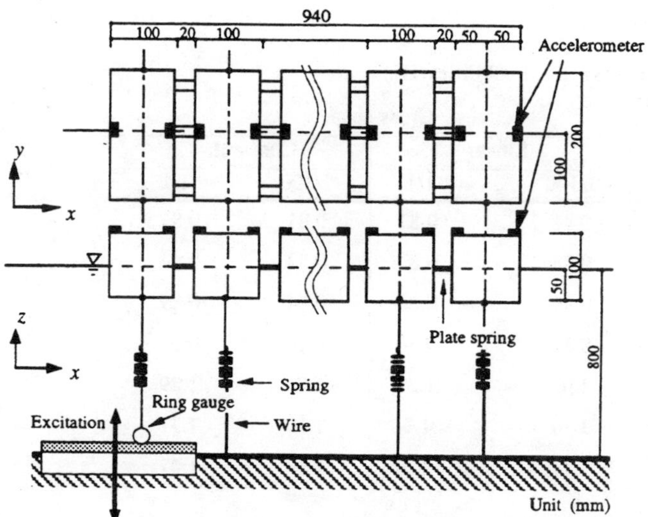

Fig.2 Experimental models

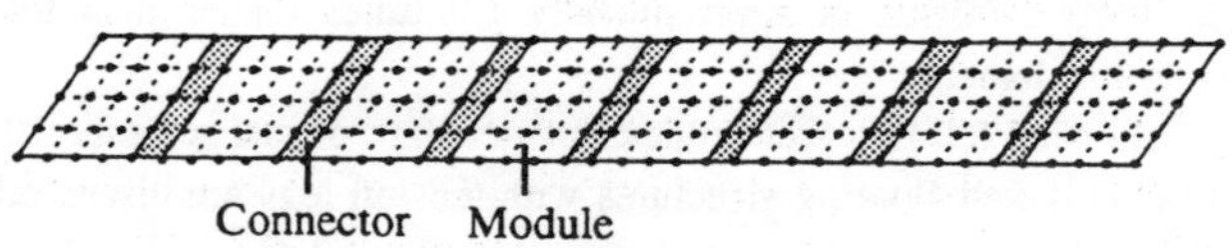

Fig.3 Finite element model

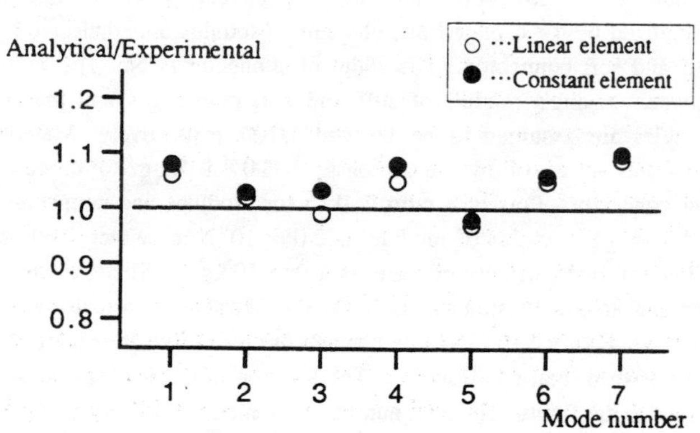

Fig.4 Ratio of analytical and experimental wet mode frequencies

where Re $[\lambda]$ and Im $[\lambda]$ are the real and imaginary parts of $[\lambda]$, respectively. Incorporating Eqs.(30) and (31) into Eq.(29), the element fluid force vector is given by

$$\{f\} = \iint_{\Gamma_e}\{N\}\left(\omega^2\rho_w\{N\}^T \mathrm{Re}[\lambda]\{\bar{u}_z\}\right.$$

$$\left. + i\omega\rho_w\{N\}^T \mathrm{Im}[\lambda]\{\bar{u}_z\} - \rho_w g\{N\}^T\{\bar{u}_z\}\right)|J|d\xi d\eta\cdot e^{-i\omega t}$$

$$= -[m_w]\{\ddot{u}_w\} - [c_w]\{\dot{u}_w\} - [k_w]\{u_w\},\tag{32}$$

where $[m_w]$, $[c_w]$ and $[k_w]$ are the element added mass, added damping and added stiffness matrices, respectively, and given as

$$[m_w] = \rho_w\iint_{\Gamma_e}\{N\}\{N\}^T \mathrm{Re}[\lambda]|J|d\xi d\eta,\tag{33a}$$

$$[c_w] = \rho_w\omega\iint_{\Gamma_e}\{N\}\{N\}^T \mathrm{Im}[\lambda]|J|d\xi d\eta,\tag{34b}$$

$$[k_w] = \rho_w g\iint_{\Gamma_e}\{N\}\{N\}^T[\lambda]|J|d\xi d\eta.\tag{34c}$$

The global fluid force vector $\{F\}$ may be obtained by assembling Eq.(32) with respect to all interface elements. Incorporating $\{F\}$ into Eq.(19) and neglecting the structural and radiation damping matrices yield the fluid coupled equation of motion for the undamped free vibration of module linked floating structures,

$$([M_S]+[M_W])\{\ddot{U}\} + ([K_S]+[K_T]+[K_W])\{U\} = \{0\},\tag{34}$$

where $[M_W]$ and $[K_W]$ are the global matrices of the element matrices $[m_w]$ and $[k_w]$, respectively. The frequency equation can be obtained as

$$\left|[K_S]+[K_T]+[K_W] - \bar{\omega}^2([M_S]+[M_W])\right| = 0,\tag{35}$$

where $\bar{\omega}$ is the wet mode circular frequency. Eq.(35) can be solved by an iterative method because the added mass matrix $[M_W]$ is frequency dependent. 4 to 5 iterations are necessary to achieve the enough convergence for linear elements, while 7 to 8 iterations for constant elements.

NUMERICAL EXAMPLES AND DISCUSSION

To show the effectiveness of the BEM-FEM hybrid formulation, numerical results are compared with experimental results which were

Table 1 Wet mode frequencies for linear and constant boundary elements (Hz)

Mode	Experimental	Analytical	
		Constant	Linear
1	1.28	1.39	1.36
2	1.37	1.41	1.41
3	1.49	1.54	1.47
4	1.57	1.70	1.66
5	1.81	1.78	1.76
6	2.32	2.47	2.45
7	2.93	3.24	3.21

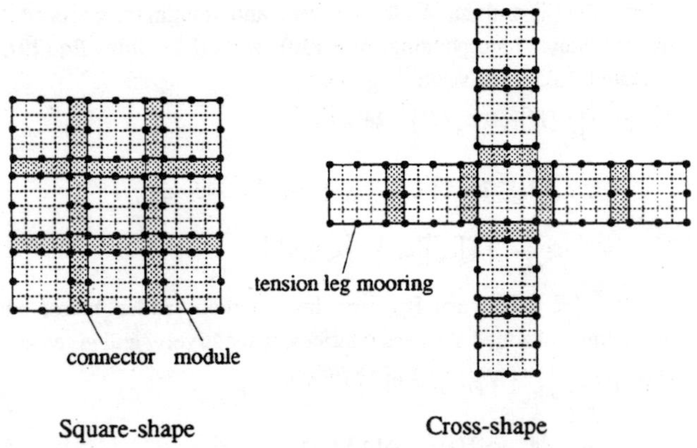

Fig.5 Finite element discretizations for square and cross-shapes

Table 2 CPU time for linear and constant elements(sec)

	Linear	Constant
Number of elements	156	156
Number of nodal points	200	156
CPU time	1.52×10^4	8.13×10^3

Table 3 Wet mode frequencies for different connector flexibilities (Hz)

Mode	Square-shape				Cross-shape			
	Linear		Constant		Linear		Constant	
	1/100	1/1	1/100	1/1	1/100	1/1	1/100	1/1
1	1.01	1.01	1.01	1.01	0.81	0.81	0.81	0.81
2	1.06	1.07	1.07	1.08	0.83	0.83	0.83	0.84
3	1.13	1.15	1.14	1.16	0.84	0.88	0.84	0.89
4	1.24	1.51	1.25	1.51	0.88	0.96	0.88	0.97
5	1.25	1.52	1.26	1.52	1.02	1.28	1.02	1.29
6	1.36	1.62	1.38	1.63	1.10	1.30	1.10	1.31
7	1.48	2.12	1.49	2.12	1.17	1.46	1.17	1.47

obtained by an experimental modal analysis for a small scaled model of module linked floating structure without tension legs. The dimensions of experimental model are shown in Fig.2. The size of each module is 0.1m×0.2m×0.1m. The width of connector is 0.02m. Mass density of each module is 3.72×10^2kg/m^3 and connector stiffness is 2.85×10^{-2}Nm^2. Water depth is 0.8m and mass density of water is 1.06×10^3kg/m^3. The finite element model corresponding to the experimental model is shown in Fig.3. The total number of finite elements are 156 which includes 28 connector elements. In the analytical model, the equivalent mass density of connector is 7.0kg/m^3 and Young's modulus of connector is 1.71×10^3N/m^2. In this study, the highest mode to be predicted is assumed to be the 6th mode. Of course, higher modes can be predicted by using much more fine mesh.

Table 1 shows the analytical and experimental values of wet mode natural frequencies without tension legs. Figure 4 shows the ratio of analytical and experimental wet mode frequencies. The difference between wet mode frequencies calculated by using linear and constant boundary elements is relatively small, although the results of linear elements have a slightly better agreement with experimental values than those of constant elements. The first six wet mode frequencies have a relatively good agreement with experimental values. Table 2 shows the CPU time for linear and constant elements. The calculation is performed by SUN workstation 10. The CPU time of linear elements is approximately 1.9 times longer than that of constant elements.

In what follows, wet mode characteristics of real scale models of module linked floating structures with tension legs are discussed for different configurations and connector flexibilities. Two different shapes with square and cross plans are considered. Both shapes are composed of 9 modules. The length, width and thickness of each module are 100m,100m and 5m, respectively. Each module is discretized by 4×4 plate finite elements. Modules are connected by stiff and soft connectors. The width of connector is 5m. The ratios between Young's moduli of stiff and soft connectors and that of modules are assumed to be 1/1 and 1/100, respectively. Material properties are as follows: mass density is 5.0×10^2kg/m^3 for modules and connectors, Poisson's ratio is 0.15 for modules and connectors, and Young's modulus of modules is 2.06×10^{10}N/m^2. Water depth is 200m and mass density of water is 1.06×10^3kg/m^3. The total cross sectional area of tension legs is 1/10000 of the area of strucutre-water interface. Figure 5 shows finite element discretizations for square and cross-shaped floating structures. The location of tension legs is also shown in the figure. The total number of elements is 196 including 52 connector elements for square-shaped model and 176 including 32 connector elements for cross-shaped model.

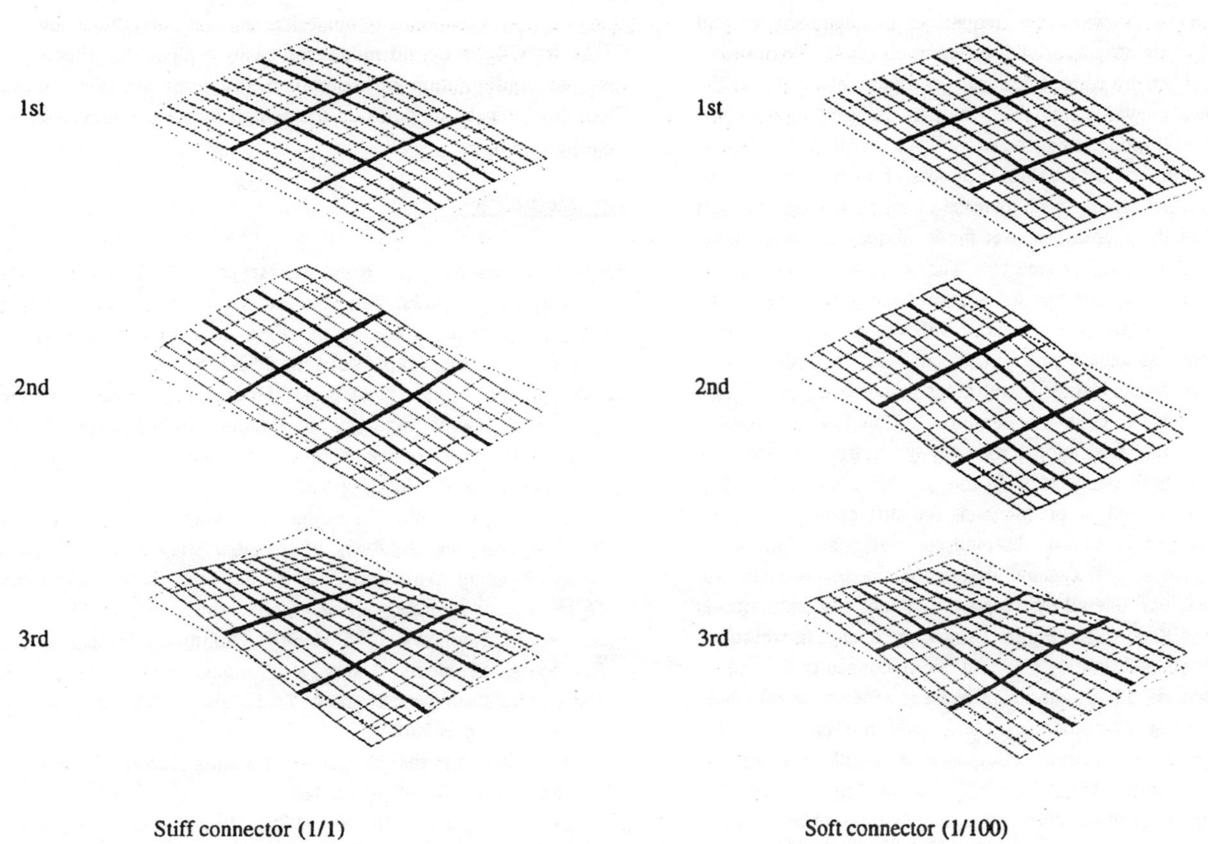

Fig.6 Wet mode shapes of square-shaped floating structure for different connector flexiblities

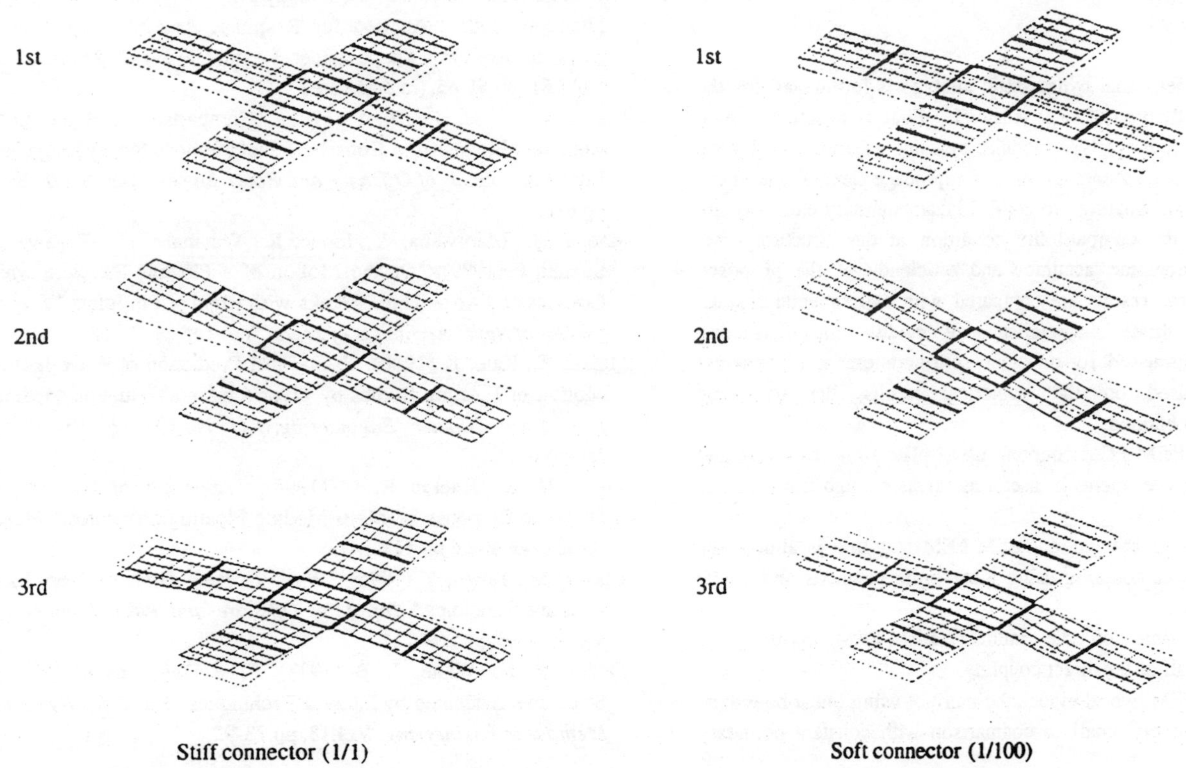

Fig.7 Wet mode shapes of cross-shaped floating structure for different connector flexiblities

Table 3 shows the wet mode frequencies of square-shaped and cross-shaped floating structures with different connector flexibilities, which are calculated by using linear and constant elements. In the first two modes in which rigid body motion is predominant, the difference in wet mode frequencies is very small for different connector flexibilities. In the higher modes, however, wet mode frequencies increase rapidly with increase of mode number for stiff connector, while the increase in wet mode frequencies with mode number is small for soft connector. The increase in wet mode frequencies is more remarkable for square-shape than cross-shape. This may be due to the larger overall stiffness of module linked floating structures associated with mutual constraints of modules.

Figure 6 shows the change in wet mode shapes of square-shaped floating structure due to connector flexibility for the first three modes of vibration. In the first mode, rigid body motion of heave is predominant for both connectors. In the second mode, rigid body motion of pitch or roll is predominant for stiff connector, while flexural deformation is clearly observed for soft connector. In the third mode, torsional and flexural deformation is predominant for both connectors. The overall deformation becomes discontinuous at connectors for soft connector. Figure 7 shows the change in wet mode shapes of cross-shaped floating structure due to connector flexibility. In the first mode, rigid body motion of heave is predominant for both connectors. In the second mode, rigid body motion of pitch or roll is predominant. However, flexural deformation is slightly observed for stiff connector, while significant deformation appears for soft connector. In the third mode, the combination of flexural deformation with both concave and convex shapes is observed for both connectors.

CONCLUSIONS

A three dimensional hydroelastic analysis is formulated for the wet mode free vibration of large tension leg floating structures which consist of multiple barge-type modules by using a BEM-FEM hybrid model. Plate finite elements are used to reproduce flexural and shear deformation of the floating structure. Linear boundary elements are used to assure the compatibility condition at the structure-water interface. To show the accuracy and efficiency of the proposed method, numerical results are compared with experimental results. Moreover, wet mode characteristics of module linked floating structures are discussed for different configurations and connector flexibilities. Based on the numerical results, the following conclusions are obtained,

1. The indirect BEM-FEM coupling using plate finite elements and constant boundary elements is useful as far as enough fine mesh is used.
2. The accuracy of the indirect BEM-FEM coupling is almost the same as the direct coupling using plate finite elements and linear boundary elements.
3. The CPU time of the indirect BEM-FEM coupling is approximately half of the direct coupling.
4. In the BEM-FEM hybrid model, the merit of using linear boundary elements is relatively small in comparison with constant boundary elements from viewpoints of numerical accuracy and efficiency.
5. The BEM-FEM hybrid model can easily evaluate the effects of the irregular configuration of assembled structures and the connector flexibility between modules on the overall dynamic behavior of large floating structures.

REFERENCES

Endo, R., Hamamoto, T., Kato, T., Hayashi, T., Tosaka, N (1995). "Experimental Modal Analysis on Unit-Linked Large Floating Structure Models," *Proc. of the Fifth International Offshore and Polar Engineering Conf.*, Vol.1, pp.153-158.

Hamamoto, T., Fujita. K (1995). "Three Dimensional BEM-FEM Coupled Dynamic Analysis of Module-Linked Large Floating Structures," *Proc. of the Fifth International Offshore and Polar Engineering Conf.*, Vol.3, pp.392-399.

Hamamoto, T., Hayashi, T., Fujita. K (1996). "3D Coupled BEM-FEM Hydroelastic Analysis of Irregular shaped, Module-Linked Large Floating Structures," *Proc. of the Sixth International Offshore and Polar Engineering Conf.*, Vol.1, pp.362-369.

Jeans, A. R., Mathews, C. I (1990). "Solution of Fluid-Structure Interaction Problems Using a Coupled Finite Element and Variational Boundary Element Technique," *J. Acoust. Soc. Am.*, Vol.88, No.5, p.2459-2466.

John, F (1950). "On the Motion of Floating Bodies," *Comm. Pure and Appl. Math.*, Vol.3, pp.45-101.

Kagemoto, H., Yue, D. K. P (1987). "Wave-Induced Motions of Multiple Floating Bodies," *J. of the Society of Naval Architects of Japan*, Vol.161, pp.152-158.

Kiyokawa, T., Ohyama, T., Kobayashi, H (1983). "A Green's Function Method Applied for Response Analysis of a Floating Body of Arbitrary Shape under Regular Waves," *Trans. JSCE*, Vol.161, pp.55-65, (in Japanese).

Liu, Y. H., C. H., Lu, X. S (1991). "Comparison of Higher-Order Boundary Element and Constant Panel Methods for Hydrodynamic Loadings," *Int. J. of Offshore and Polar Engineering*, Vol.1, No.1, pp.8-17.

Maeda, H., Maruyama, S., Inoue, R., Watanabe, K., Togawa, S., Suzuki, F (1979). "On the Motion of a Floating Structure which Consists of Two or Three Blocks with Rigid or Pin Joints," *J. of the Society of Naval Architects of Japan*, Vol.145, pp.71-78.

Matsui, T., Kato, K (1988). "Numerical Prediction of Wave-Induced Motion of Floating Bodies by Hybrid Integral-Equation Method," *J. of Struct. Constr. Engineering, AIJ*, No.393. pp.165-175, (in Japanese).

Riggs, H. R., Ertekin R. C (1993). "Approximate Methods for Dynamic Response of Multi-Module Floating Structures," *Marine Structures*, Vol.6 pp.117-141.

Takaki, M., Tango, Y (1995). "Wave Drifting Force on Very Large Floating Structures," *Int. J. of Offshore and Polar Engineering*, Vol.5, No.3, pp.204-211.

Taylor, R. E., Waite, J. B (1978). "The Dynamics of Offshore Structures Evaluated by Integral Techniques," *Int. J. for Numerical Methods in Engineering*, Vol.13, pp.73-92.

Proceedings of the Seventh (1997) International Offshore and Polar Engineering Conference
Honolulu, USA, May 25-30, 1997

ISBN 1-880653-28-1 (Set); ISBN 1-880653-29-X (Vol. I)

The Wave Forcing of a Three Dimensional Floating Elastic Body

Michael H. Meylan
University of Auckland
Auckland, New Zealand

ABSTRACT

A systematic method for solving the linear hydroelastic problem of the wave forcing of a floating body is presented. The variational equation for a linear deformable body and the boundary element equation for a floating body subject to a linear wave force are combined. From this a variational equation which is expressed entirely in terms of the body displacement is obtained. We use this variational equation to derive simple formula for a coupled finite element - boundary element solution expressed entirely in terms of the finite element nodal displacements.

Key Words. Hydroelastic, finite element, boundary element, wave forcing.

INTRODUCTION

We are concerned with the hydroelastic problem of the wave induced motion of a floating elastic body. In general the solution depends on solving two problems simultaneously, the linear wave forcing problem for a floating body and the forced response problem for an elastic body. The solution of the wave forcing problem for a floating body by the boundary element method (which began with the pioneering work of John, 1949 and 1950) has become standard. If the geometry of the elastic body is assumed to be sufficiently simple, e.g. one dimensional (a beam), then the forced elastic response problem is trivial. Perhaps for this reason the solution method to many hydroelastic wave forcing problems has been a simple modification of the solution method for the wave forcing of a rigid body, e.g. Newman's (1994) solution for the wave forcing of a beam in a fluid. This method involves a solution expansion in the free modes of motion which are given trivially for a rigid body or a beam.

While a modal expansion method can be applied to higher dimensional elastic bodies, e.g. the solution by Wang et. al. (1995) for a two dimensional elastic body (a thin plate), considerable computational effort is required to solve for the modes. For two and three dimensional elastic bodies the forced elastic response problem is as difficult as the wave forcing problem and a solution method should be developed which reflects this. In general the forced elastic response problem is best solved numerically by a method derived from a variational equation, e.g. the finite element method. In fact the finite element method was the method used by Wang et. al. (1995) to solve for the modes. A further problem with the modal expansion method is that it requires a solution of the radiation problem for each mode.

Meylan (1996) also solved the problem of wave forcing of a floating thin plate but this solution was derived from a variational equation for the fluid - thin plate system. This had the advantage of solving the problem directly and required only a single radiation problem to be solved. The solution which Meylan (1996) presented was based on a Rayleigh-Ritz method using global polynomials. This solution method is not as numerically stable as a finite element method and it did not exploit the discretisation, which is required to solve the boundary element problem, in the solution of the variational equation.

In this paper we derive a variational equation for a floating three dimensional elastic body subject to wave forcing assuming small displacements (linearity). This equation is then used to derive a numerical solution using a coupled finite element - boundary element method given entirely in terms of the generalized displacements associated with the finite element method. The solution method requires only a single radiation problem to be solved, has the numerical stability of the finite element method and uses the same discretisation to solve both the radiation and elastic body problem. The basic layout of the paper is as follows. In section 2 we derive the variational equation of motion for a linear deformable body which is float-

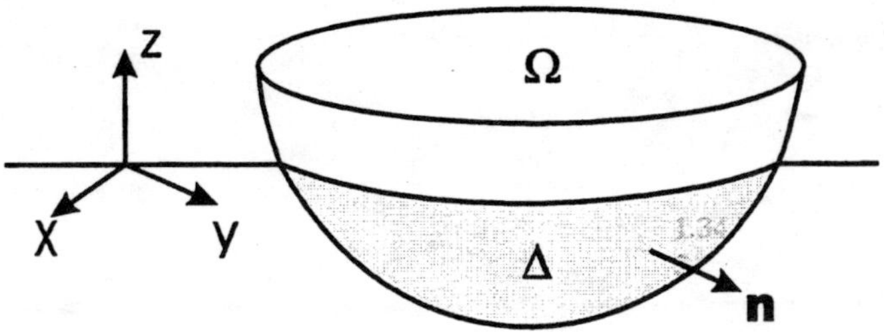

Figure 1: A schematic diagram of the floating deformable body.

ing in a fluid expressing the fluid forcing term as a function of the velocity potential of the fluid. In section 3 we use the boundary element equations for linear waves in a fluid to derive an expression for the velocity potential over the wetted body surface in terms of the input wave potential and the outward surface normal derivative of the potential. In section 4 we introduce the operator $\mathbf{P}^{\Delta}$ which maps the body displacement to the outward surface normal derivative of the potential (by continuity). This gives our desired variational equation of motion in terms of the body motion alone. We then use this equation in section 5 to derive a finite element - boundary element solution for a smooth three dimensional elastic body (for the case of a straight forward implementation of both methods).

A VARIATIONAL EQUATION FOR A DEFORMABLE BODY IN A FLUID

In this section we derive a variational equation for a deformable body in a fluid assuming small displacements (i.e. the problem is linear). The geometry is shown in Fig. 1, the fluid occupies the region $z < 0$. We denote the body by Ω, the body surface by $\partial\Omega$ and the wetted surface by $\Delta \subset \partial\Omega$. The deformation of the body at any moment in time we will denote by $\bar{\mathbf{u}} \in \mathcal{C}^{\infty}(\Omega, \mathbb{R}^3)$ where $\mathcal{C}^{\infty}(\Omega, \mathbb{R}^3)$ is the set of all infinitely differentiable functions from Ω to $\mathbb{R}^3$ where $\mathbb{R}$ is the set of real numbers. The bar we will use to denote quantities which are a function of time. In what follows we shall consider only forcing of a single frequency ω and from the linearity it follows that all time varying quantities are proportional to $\Re(e^{-i\omega t} \times$ complex quantity) where $\Re$ denotes the real part. Time independent quantities will be unbared, e.g. $\bar{\mathbf{u}} = \Re\left(e^{-i\omega t}\mathbf{u}\right)$ where $\mathbf{u}$ is the complex body displacement, $\mathbf{u} \in \mathcal{C}^{\infty}(\Omega, \mathbb{C}^3)$ where $\mathbb{C}$ is the set of complex numbers.

We identify the elements of $\mathbf{u}$, which we denote by u_i, such that $\mathbf{u} = (u_1, u_2, u_3) = (u_x, u_y, u_z)$ (here the subscript refers to the direction of the element not the derivative). We denote the outward surface normal by $\mathbf{n} \in \mathcal{C}^{\infty}(\partial\Omega, \mathbb{R}^3)$ (Fig. 1) which we again represent in element form as $\mathbf{n} = (n_1, n_2, n_3) = (n_x, n_y, n_z)$. Note that through out we make the assumption of small displacements and therefore we do not distinguish the reference state of the body Ω and the deformed state Ω'.

The principal of virtual work for a three dimensional deformable body gives us the following variational equation of motion (Fung, 1965)

$$\int_{\Omega} \bar{\sigma}_{ij}\delta\bar{\varepsilon}_{ij}dV - \int_{\Omega}\left(\bar{B}_i - \rho\frac{\partial^2\bar{u}_i}{\partial t^2}\right)\delta\bar{u}_i dV = \int_{\Delta}\bar{T}_i\delta\bar{u}_i dA \quad (1)$$

where $\bar{\sigma}_{ij}$ is the stress, $\bar{\varepsilon}_{ij}$ is the strain, $\bar{B}_i$ is the body force, ρ is the solid density and $\bar{T}_i$ is the applied surface force per unit area. In Equation 1 and through out the paper (unless stated otherwise) we use the Einstein summation convention, summing over any repeated indices. We assume that the body force $\bar{B}_i$ is due to gravity and hence a constant with respect to time. This force is exactly balanced by a constant hydrostatic pressure (the Archimedes' force) which determines the reference state of the body in the fluid. From now on we shall ignore these static forces and any corresponding displacements they induce (since we are considering a linear problem the static solution may be included if required by simple addition). Restricting Equation 1 to the quantities which are proportional to $e^{-i\omega t}$ we obtain

$$\int_{V} \sigma_{ij}\delta\varepsilon_{ij}dV - \omega^2\int_{V}\rho u_i\delta u_i dV = \int_{\Delta}T_i\delta u_i dA \quad (2)$$

where the quantities in Equation 2 are now complex. The surface traction T_i is due entirely to the fluid pressure which is given by the linearized Bernoulli's equation

$$\bar{p} = -\rho_w g z - \rho_w\frac{\partial\bar{\phi}}{\partial t} \quad (3)$$

where $\bar{p}$ is the fluid pressure, ρ_w is the water density and $\bar{\phi}$ is the velocity potential of the fluid. Ignoring the static component of the pressure and considering only the quantity p which is proportional to $e^{-i\omega t}$ Equation 3 becomes

$$p = -\rho_w g u_3 + i\omega\rho_w\phi \quad (4)$$

and the surface force per unit area is

$$T_i = (\rho_w g u_3 - i\omega\rho_w\phi)\, n_i \quad (5)$$

where n_i is the appropriate component of the outward surface normal. Substituting Equation 5 into Equation 1 we obtain the following variational equation for a three-dimensional deformable body subject to wave forcing

$$\int_{\Omega} \sigma_{ij}\delta\varepsilon_{ij}dV - \omega^2\int_{\Omega}\rho u_i\delta u_i dV = \int_{\Delta}(\rho_w g u_3 - i\omega\rho_w\phi)\, n_i\delta u_i dA. \quad (6)$$

WAVE FORCING OF A BODY IN A FLUID

In this section we derive an equation relating the velocity potential and its normal derivative over the wetted body surface when the body is subject to wave forcing. In keeping with our desire for simplicity we consider a fluid of infinite depth, the finite depth case being a straight forward modification of the theory presented here (John, 1949 and 1950, Wehausen and Laitone, 1960 and Sarpkaya and Isaccson, 1981). The linearized boundary value problem for the fluid velocity potential

$\bar{\phi}$ is

$$\left.\begin{aligned} \nabla^2\bar{\phi} &= 0, & -\infty < z < 0, \\ \frac{\partial \bar{\phi}}{\partial z} &= 0, & z \to -\infty, \\ \rho_w\left(g\frac{\partial \bar{\phi}}{\partial z} + \frac{\partial^2 \bar{\phi}}{\partial t^2}\right) &= -\frac{\partial \bar{p}}{\partial t}, & z = 0, \end{aligned}\right\} \tag{7}$$

where $\bar{p}$ is the fluid pressure, ρ_w is the density of the water and g is the acceleration due to gravity. We also require appropriate conditions to be met as $R \to \infty$ (where R is the "distance from the body", "$R = \sqrt{x^2 + y^2}$"). In what follows we consider forcing form a linear small amplitude wave of a single frequency ω so we again consider complex amplitudes, i.e. $\bar{\phi}(x, z, t) = \Re\left(\phi(x, z)e^{-i\omega t}\right)$ etc. The boundary value problem (Equation 7) becomes (assuming that the surface pressure is a constant),

$$\left.\begin{aligned} \nabla^2\phi &= 0, & -\infty < z < 0, \\ \frac{\partial \phi}{\partial z} &= 0, & z \to -\infty, \\ \frac{\partial \phi}{\partial z} - k\phi &= 0, & z = 0, \end{aligned}\right\} \tag{8}$$

where k is the wave number ($gk = \omega^2$). The condition as $R \to \infty$ is the standard Sommerfield radiation condition (Sarpkaya and Isaccson, 1981)

$$\sqrt{R}\left(\frac{\partial}{\partial R} - ik\right)(\phi - \phi_{\mathrm{i}}) = 0, \text{ as } R \to \infty. \tag{9}$$

where ϕ_{i} is the far field input wave potential, a unit amplitude wave travelling in the θ direction, where θ is the angle from the x-axis,

$$\phi_{\mathrm{i}}(x, y, z) = -\frac{i\omega}{k}e^{ik(x\cos\theta + y\sin\theta)}e^{kz}. \tag{10}$$

The standard solution method to the linear wave problem is to transform the boundary value problem into an integral equation by an appropriate Green's function and then to seek a solution of the integral equation (the boundary element method). Performing just such a transformation Equations 8 and 9 become (John, 1949 and 1950),

$$\beta\phi(\mathbf{x}) = \phi_{\mathrm{i}}(\mathbf{x}) - \int_{\Delta} \left(G^n(\mathbf{x};\mathbf{y})\phi(\mathbf{y}) - G(\mathbf{x};\mathbf{y})\phi^n(\mathbf{y})\right) dS_{\mathbf{y}}. \tag{11}$$

In Equation 11 ϕ^n is the outward surface normal derivative of the potential and G is the Greens function which satisfies

$$\left.\begin{aligned} \nabla^2 G(\mathbf{x};\mathbf{y}) &= \delta(\mathbf{x};\mathbf{y}), & -\infty < z < 0, \\ \frac{\partial G}{\partial z} &= 0, & z \to -\infty, \\ \frac{\partial G}{\partial z} - kG &= 0, & z = 0, \end{aligned}\right\} \tag{12}$$

plus appropriate radiation conditions and G^n is the outward surface normal derivative of G. By the outward surface normal we mean the outward surface normal with respect to the body not the fluid. It is usual in boundary element formulations to consider the outward surface normal with respect to the fluid but this convention would contradict the definition of the outward surface normal given in section 2 so through out this paper the outward surface normal will be with respect to the body. β is given by (assuming the body surface is smooth)

$$\beta = \begin{cases} \frac{1}{2}, & \mathbf{x} \in \boldsymbol{\Delta}, \\ 1, & \mathbf{x} \notin \boldsymbol{\Delta}. \end{cases}$$

G as defined by Equation 12 may be easily calculated numerically and expressions are given in John (1949 and 1950), Wehausen and Laitone (1960) and Sarpkaya and Isaccson (1981). In our derivation of Equation 11 we have assumed that the wave frequency does not correspond to the so called irregular frequencies (John, 1950 of Sarpkaya and Isaccson, 1981). These frequencies, which correspond to eigenvalues of the Dirichlet problem for the floating body geometry, do not in general pose serious problems to the solution since they are isolated points.

We write Equation 11 in operator notation assuming that $\mathbf{x} \in \boldsymbol{\Delta}$,

$$\frac{1}{2}\phi = \phi_{\mathrm{i}} - \mathbf{G}^n\phi + \mathbf{G}\phi^n \tag{13}$$

where ϕ and $\phi^n \in \mathcal{C}^\infty(\Delta, \mathbb{C})$ and $\mathbf{G}^n$ and $\mathbf{G}$ are bounded linear operators on the space $\mathcal{C}^\infty(\Delta, \mathbb{C})$, i.e. $\mathbf{G}^n$ and $\mathbf{G} \in \mathcal{L}(\mathcal{C}^\infty(\Delta, \mathbb{C}), \mathcal{C}^\infty(\Delta, \mathbb{C}))$. Since we are only considering the space defined by Δ we consider ϕ and ϕ^n as two functions related by Equation 13. In the boundary element solution of Equation 11 or Equation 13 the operators $\mathbf{G}^n$ and $\mathbf{G}$ become matrices. We solve Equation 13 formally as

$$\phi = \left(\frac{1}{2} + \mathbf{G}^n\right)^{-1}(\phi_{\mathrm{i}} + \mathbf{G}\phi^n) \tag{14}$$

(where $\frac{1}{2}$ denotes $\frac{1}{2}\times$ the appropriate identity operator). Solving the radiation problem by constructing these operators, although apparently complicating the solution, means that the radiation problem needs only be solved once.

COUPLING THE FLUID AND DEFORMABLE BODY EQUATIONS

In this section we construct an operator $\mathbf{P}^\Delta$ which maps the body displacement $\mathbf{u}$ to the normal derivative of the velocity potential over the wetted surface of the body ϕ^n. From the definition of the velocity potential $\bar{\phi}$ we know that $\bar{\phi}^n$ represents the velocity of the fluid in the direction of the outward surface normal. Therefore we must have

$$\bar{\phi}^n(\mathbf{x}) = \frac{\partial \bar{u}_i(\mathbf{x})}{\partial t} n_i(\mathbf{x}), \ \mathbf{x} \in \Delta \tag{15}$$

which gives us (from the $e^{-i\omega t}$ proportionality)

$$\phi^n(\mathbf{x}) = -i\omega u_i(\mathbf{x}) n_i(\mathbf{x}), \ \mathbf{x} \in \Delta. \tag{16}$$

Mathematically we introduce the following linear map $\mathbf{P}^\Delta \in \mathcal{L}\left(\mathcal{C}^\infty(\Omega, \mathbb{C}^3), \mathcal{C}^\infty(\Delta, \mathbb{C})\right)$ such that $\mathbf{P}^\Delta(\mathbf{u}) = \phi^n$ where the definition of $\mathbf{P}^\Delta$ is given by Equation 16, i.e.

$$\mathbf{P}^\Delta(\mathbf{u})(\mathbf{x}) = -i\omega u_i(\mathbf{x}) n_i(\mathbf{x}), \ \mathbf{x} \in \Delta. \tag{17}$$

Equation 6 becomes

$$\int_\Omega \sigma_{ij}\delta\varepsilon_{ij} dV - \omega^2 \int_\Omega \rho u_i \delta u_i dV \tag{18}$$

$$= \int_{\Delta} \left(\rho_w g u_3 - i\omega\rho_w \left(\frac{1}{2} + \mathbf{G}^n\right)^{-1} \left(\phi_i + \mathbf{G}P^{\Delta}(\mathbf{u})\right)\right) n_i \delta u_i dA \tag{19}$$

which is a variational equation for a floating deformable body expressed entirely in terms of the body displacement.

FINITE ELEMENT - BOUNDARY ELEMENT COUPLING FOR A SMOOTH 3-D ELASTIC BODY

The mathematical work in deriving Equation 19 was to allow us to develop a coupled finite element - boundary element method to allow simple numerical solution of the floating elastic body problem. In this section we consider a smooth elastic body and use the most straight forward finite element and boundary element approach. It should be obvious that the solution we present can easily be extended if required, e.g. a body with corners can be considered by an alteration in the equations linking the surface elements, nodal points and nodal displacements.

Finite Element

We approximate the displacement by its value at a number of nodes which we denote by $\vec{u}$, the column matrix of nodal displacements. Assuming that the body is elastic we can approximate the various integrals in Equation 19 by (Petyt, 1990)

$$\int_{\Omega} \sigma_{ij}\delta\varepsilon_{ij} dV \approx \frac{1}{2}\delta\left(\vec{u}^T \mathbf{K}\vec{u}\right), \tag{20}$$

$$\omega^2 \int_{\Omega} \rho u_i \delta u_i dV \approx \omega^2 \frac{1}{2}\delta\left(\vec{u}^T \mathbf{M}\vec{u}\right) \tag{21}$$

and

$$\int_{\Delta} \left(\rho_w g u_3 - i\omega\rho_w \left(\frac{1}{2} + \mathbf{G}^n\right)^{-1} \left(\phi_i + \mathbf{G}P^{\Delta}(\mathbf{u})\right)\right) n_i \delta u_i dA \approx \vec{f}\delta\vec{u}, \tag{22}$$

where $\mathbf{K}$ is the stiffness matrix, $\mathbf{M}$ is the inertia matrix, $\vec{f}$ is the column matrix of equivalent nodal forces and ^{T} denotes transpose. Equations 20, 21 and 22 give us the following standard finite element equation

$$\mathbf{K}\vec{u} - \omega^2 \mathbf{M}\vec{u} = \vec{f}. \tag{23}$$

The dimension of $\vec{u}$ we denote by N, the elements of $\vec{u}$ we denote by $\vec{u}_i$, $i \in I_N$ where $I_N = \{1, ..., N\}$. For each $\vec{u}_i$ we denote the associated nodal position by $\mathbf{x}_i \in \Omega$ (there will be three displacements for every node) and the unit vector in the direction of positive displacement by $\hat{u}_i \in \mathbb{R}^3$. We identify those displacements which correspond to nodal points on the surface of the body which we denote by $I_\Delta \subset I_N$ such that $\mathbf{x}_i \in \Delta$ if $i \in I_\Delta$. The set of separate surface nodes we denote by $\mathbf{x}_i^{\Delta}$, $i \in I_M$ where $I_M = \{1, ..., M\}$. We introduce an operator $\Sigma : I_\Delta \longrightarrow I_M$ such that if $i \in I_\Delta$ then $\mathbf{x}_i = \mathbf{x}_{\Sigma(i)}^{\Delta}$ (i.e. Σ maps the surface nodes in the $\mathbf{x}_i$ representation to the surface nodes in the $\mathbf{x}_i^{\Delta}$ representation, Σ is not one to one but "three to one").

Although it is possible to use higher order methods to calculate the nodal force we present results for a simple constant panel method, i.e. for each surface node we associate an element of surface area and assume that the fluid pressure is a constant over this area. In notation, for each $\mathbf{x}_i^{\Delta}$ we associate an element of surface area Δ_i such that $\mathbf{x}_i^{\Delta} \in \Delta_i$, $\bigcup_{i=1}^{M} \Delta_i = \Delta$ and $\Delta_i \bigcap \Delta_j = \emptyset$ (the empty set) if $i \neq j$. The elements of $\vec{f}$ are therefore given by

$$\vec{f}_i = \begin{cases} 0, \ i \notin I_\Delta, \\ -p(\mathbf{x}_i)\left(\mathbf{n}(\mathbf{x}_i) . \hat{u}_i\right) \mu(\Delta_{\Sigma(i)}), \ i \in I_\Delta, \ (\text{N.S.}) \end{cases} \tag{24}$$

where $p(\mathbf{x}_i)$ is the complex fluid pressure at the point $\mathbf{x}_i$, $\mathbf{n}(\mathbf{x}_i)$ is the outward surface normal at $\mathbf{x}_i$, . denotes the usual vector dot product, μ denotes the surface area and $n.s.$ stands for no summation indicating that the Einstein summation convention does not apply (note that if $i \in I_\Delta$, $\mathbf{x}_{\Sigma(i)}^{\Delta} = \mathbf{x}_i$).

Boundary Element

For each surface node $\mathbf{x}_i^{\Delta}$ we wish to determine a value of ϕ and ϕ^n, i.e. we represent the potential and its normal derivative by the column matrices $\vec{\phi}$ and $\vec{\phi}^n$ of dimension M, where the elements $\vec{\phi}_i$ and $\vec{\phi}_i^n$ are the values at the $\mathbf{x}_i^{\Delta}$ node. $\mathbf{G}$ and $\mathbf{G}^n$ become $M \times M$ matrices whose elements are given by

$$G_{ij} = \int_{\Delta_j} G\left(\mathbf{x}_i^{\Delta}, \mathbf{y}\right) dS_{\mathbf{y}} \ \text{ and } \ G_{ij}^n = \int_{\Delta_j} G^n\left(\mathbf{x}_i^{\Delta}, \mathbf{y}\right) dS_{\mathbf{y}} \tag{25}$$

(in general these integrals are calculated by further approximations). The operator $\mathbf{P}^{\Delta}$ becomes an $M \times N$ matrix (such that $\vec{\phi}^n = \mathbf{P}^{\Delta}\vec{u}$) whose elements are given by

$$P_{ij}^{\Delta} = \begin{cases} 0, \ j \notin I_\Delta, \\ 0, \ j \in I_\Delta \text{ and } i \neq \Sigma(j), \\ -i\omega\left(\mathbf{n}(\mathbf{x}_j).\hat{u}_j\right), \ j \in I_\Delta \text{ and } i = \Sigma(j). \ (\text{N.S.}) \end{cases} \tag{26}$$

The surface pressure is

$$\vec{p} = -\rho_w g \mathbf{D}\vec{u} + i\omega\rho_w \left(\frac{1}{2} + \mathbf{G}^n\right)^{-1} \left(\vec{\phi}_i + \mathbf{G}\mathbf{P}^{\Delta}\vec{u}\right) \tag{27}$$

where $\mathbf{D}$ is a $M \times N$ matrix whose elements, from Equation 24, are given by

$$D_{ij} = \begin{cases} 0, \ j \notin I_\Delta, \\ 0, \ j \in I_\Delta \text{ and } i \neq \Sigma(j), \\ \hat{u}_j.\hat{k}, \ j \in I_\Delta \text{ and } i = \Sigma(j), \end{cases} \tag{28}$$

where $\hat{k}$ is the unit vector in the z direction. We now introduce a further operator $\mathbf{P}^{\Omega}$, an $N \times M$ matrix such that $\vec{f} = \mathbf{P}^{\Omega}\vec{p}$, whose elements are given by

$$P_{ij}^{\Omega} = \begin{cases} 0, \ i \notin I_\Delta, \\ 0, \ i \in I_\Delta \text{ and } j \neq \Sigma(i), \\ -\left(\mathbf{n}(\mathbf{x}_i).\hat{u}_i\right)\mu(\Delta_{\Sigma(i)}) \ i \in I_\Delta \text{ and } j = \Sigma(i). \ (\text{N.S.}) \end{cases} \tag{29}$$

Therefore if we introduce the following $N \times N$ matrix

$$\mathbf{W} = \mathbf{P}^{\Omega}\left\{-\rho_w g\mathbf{D} + i\omega\rho_w\left(\frac{1}{2} + \mathbf{G}^n\right)^{-1} \mathbf{G}\mathbf{P}^{\Delta}\right\} \tag{30}$$

and the following $N \times 1$ column matrix

$$\mathbf{c} = \mathbf{P}^{\Omega}\left\{i\omega\rho_w\left(\frac{1}{2}+\mathbf{G}^n\right)^{-1}\vec{\phi}_{\rm i}\right\} \qquad (31)$$

and combine Equations 23, 24, 27, 30 and 31 we obtain

$$\mathbf{K}\vec{u} - \omega^2\mathbf{M}\vec{u} = \mathbf{W}\vec{u}+\mathbf{c} \qquad (32)$$

a simple matrix equation for the displacement $\vec{u}$. All the matrices which are required to calculate Equation 32 are either sparse matrices of dimension N or full matrices of dimension M.

CONCLUSIONS

A systematic method for solving linear hydroelastic problems involving wave forcing (and with small modification other forcing regimes) for a floating body has been presented. The first stage in the solution method was to combine the variational equation for an elastic body and the boundary element equation for a floating body. This was accomplished straight forwardly by the introduction of an operator $\mathbf{P}^{\Delta}$ which maps the elastic body displacement to the outward surface normal derivative of the fluid velocity potential over the wetted body surface. This gave a variational equation which was expressed entirely in terms of the body displacement. The second stage of the solution method was to use this variational equation to derive simple formula for a coupled finite element - boundary element solution, expressed entirely in terms of the finite element nodal displacements.

REFERENCES

John, F, (1949). "On the Motion of Floating Bodies I," *Commun Pure Appl Math*, Vol 2, pp 13–57.

John, F, (1950). "On the Motion of Floating Bodies II," *Commun Pure Appl Math*, Vol 3, pp 45–101.

Fung, Y. C., (1965). *Foundations of Solid Mechanics*, Prentice-Hall Inc., New Jersey, 525pp.

Meylan, M. H., (1996). "The Wave Response of Floating Thin Plates of Shallow Draft by a Variational Method", *Int. Offshore and Polar Eng. Conf.*, Vol 3, pp 328-334.

Newman, J.N., (1994). "Wave Effects on Deformable Bodies, *Applied Ocean Research*, Vol 16, pp 47-59.

Petyt, M, (1990). *Introduction to Finite Element Vibration Analysis*, Cambridge University Press, 558pp.

Sarpkaya, T, and Isaacson, M, (1981). *Mechanics of Wave Forces on Offshore Structures*, Van Nostrand Reinhold, pp 381–472.

Wang, S, Ertekin, RC, van Stiphout, ATFM, and Ferier, PGP, (1995). "Hydroelastic-Response Analysis of a Box-Like Floating Airport of Shallow Draft", *Proc 5th Int Offshore and Polar Engin Conf*, Vol 1, pp 145–152.

Wehausen, JV, and Laitone, EV, (1960). "Surface Waves," in *Handbuch der Physik*,

ed. S Flugge, Springer-Verlag, Berlin, Vol 9, pp 446-778.

APPENDIX FEM - BEM solution for a Floating Thin Plate of Shallow Draft

In this appendix we present a finite element (FEM) - boundary element (BEM) solution for a floating thin plate of shallow draft (a floating thin plate whose submergence is assumed negligible). This problem has already been treated by Wang et. al. (1995) and by Meylan (1996) and many of the relevant details can be found in these papers. The solution presented in these papers we believe to be inferior to the one which we present here and this is the reason for the inclusion of this appendix. Several simplifications to the formulation already presented for a three dimensional body are possible for this geometry as will be shown.

For the thin plate $\Omega = \Delta$. In the finite element solution for a thin plate we represent the generalized displacement by $\vec{w}$, whose elements we represent by $\vec{w}_i$, $i \in I_N$ where $I_N = \{1, ..., N\}$ and for each $\vec{w}_i$ we denote the associated nodal position by $\mathbf{x}_i$. Instead of having three displacements for each node as in the three dimensional elastic body we will now have one displacement and two derivatives specified at each node. In analogous fashion to what was done previously we introduce a subset $I_\Delta \subset I_N$ such that for each $i \in I_\Delta$, $\vec{w}_i$ represents a displacement (assumed to be in the positive z direction). The set of separate surface nodes we denote by $\mathbf{x}_i^{\Delta}$, $i \in I_M$ where $I_M = \{1, ..., M\}$. We introduce an operator $\Sigma : I_\Delta \to I_M$ such that if $i \in I_\Delta$ then $\mathbf{x}_i = \mathbf{x}_{\Sigma(i)}^{\Delta}$ (Σ is now one to one).

The finite element equation is exactly analogous to Equation 23 (Petyt 1990)

$$\mathbf{K}\vec{w} - \omega^2\mathbf{M}\vec{w} = \vec{f} \qquad (33)$$

where $\mathbf{K}$ and $\mathbf{M}$ are the stiffness and inertia matrices respectively for a thin plate. Again for each $\mathbf{x}_i^{\Delta}$ we associate an area of surface area Δ_i such that $\mathbf{x}_i^{\Delta} \in \Delta_i$, $\bigcup_{i=1}^{M} \Delta_i = \Delta$ and $\Delta_i \bigcap \Delta_j = \emptyset$ if $i \neq j$. The elements of $\vec{f}$ are (since $\mathbf{n}\left(\mathbf{x}_i^{\Delta}\right) = -\hat{k}$)

$$\vec{f}_i = \begin{cases} 0, \; i \notin I_\Delta, \\ p(\mathbf{x}_i)\mu(\Delta_{\Sigma(i)}), \; i \in I_\Delta. \; (\text{N.S.}) \end{cases} \qquad (34)$$

Again we represent the potential and its normal derivative by column matrices $\vec{\phi}$ and $\vec{\phi}^n$ of dimension M. Equation 11 is altered due to the shallow draft assumption and becomes

$$\phi(\mathbf{x}) = \phi_{\rm i}(\mathbf{x}) + \int_{\Delta} G(\mathbf{x};\mathbf{y})\left(k\phi(\mathbf{y}) + \phi^n(\mathbf{y})\right) dS_{\mathbf{y}} \qquad (35)$$

and the expression for G is simplified (Meylan, 1996). $\mathbf{G}$ becomes an $M \times M$ matrix whose elements are given by

$$G_{ij} = \int_{\Delta_j} G\left(\mathbf{x}_i^{\Delta}, \mathbf{y}\right) dS_{\mathbf{y}}. \qquad (36)$$

The elements of $\mathbf{P}^{\Delta}$ are

$$P_{ij}^{\Delta} = \begin{cases} 0, \; j \notin I_{\Delta}, \\ 0, \; j \in I_{\Delta} \text{ and } i \neq \Sigma(j), \\ i\omega, \; j \in I_{\Delta} \text{ and } i = \Sigma(j), \end{cases} \tag{37}$$

the surface pressure is

$$\vec{p} = -\rho_w g \mathbf{D}\vec{u} + i\omega\rho_w \left(1 - k\mathbf{G}\right)^{-1} \left(\vec{\phi}_{\mathrm{i}} + \mathbf{G}\mathbf{P}^{\Delta}\vec{u}\right) \tag{38}$$

and the elements of $\mathbf{D}$ are

$$D_{ij} = \begin{cases} 0, \; j \notin I_{\Delta}, \\ 0, \; j \in I_{\Delta} \text{ and } i \neq \Sigma(j), \\ 1, \; j \in I_{\Delta} \text{ and } i = \Sigma(j). \end{cases} \tag{39}$$

The elements of $\mathbf{P}^{\Omega}$ are

$$P_{ij}^{\Omega} = \begin{cases} 0, \; i \notin I_{\Delta}, \\ 0, \; i \in I_{\Delta} \text{ and } j \neq \Sigma(i), \\ \mu(\Delta_{\Sigma(i)}), \; i \in I_{\Delta} \text{ and } j = \Sigma(i), \end{cases} \tag{40}$$

the matrices $\mathbf{W}$ and $\mathbf{c}$ are

$$\mathbf{W} = \mathbf{P}^{\Omega} \left\{ -\rho_w g \mathbf{D} + i\omega\rho_w \left(1 - k\mathbf{G}\right)^{-1} \mathbf{G}\mathbf{P}^{\Delta} \right\} \tag{41}$$

$$\mathbf{c} = \mathbf{P}^{\Omega} \left\{ i\omega\rho_w \left(1 - k\mathbf{G}\right)^{-1} \vec{\phi}_{\mathrm{i}} \right\} \tag{42}$$

and Equation 33 becomes

$$\mathbf{K}\vec{w} - \omega^2 \mathbf{M}\vec{w} = \mathbf{W}\vec{w} + \mathbf{c} \tag{43}$$

as before.

Proceedings of the Seventh (1997) International Offshore and Polar Engineering Conference
Honolulu, USA, May 25-30, 1997

ISBN 1-880653-28-1 (Set); ISBN 1-880653-29-X (Vol. I)

Experimental Study for the Behavior of a Huge Floating Structure

H.S. Shin and D.J. Yum
Hyundai Heavy Industries Co., Ltd.
Ulsan, Korea

ABSTRACT

In the present paper, experimental investigations are performed to find out the motion characteristics of a huge floating offshore structure moored in regular and irregular waves. A floating type offshore structure with dimension of 5 km by 1 km is proposed as a concept design for the offshore airport at a Korean offshore site. Model tests are performed for the 1/10 part of the whole structure. Motions at 16 deck points above footings are traced using the optical measuring system and bending moments are measured using strain gages. Transfer functions for vertical motions and bending moments are obtained. Elastic mode shapes are found by drawing 10 snapshots of the model during one wave period. In addition to the motion tests, the forced oscillation tests are performed to find out wet mode shapes of the elastic structure. The model is connected to the oscillation device at its center. Motions at deck positions and vertical forces at connecting point are measured. Through the present experimental studies, motion characteristics for the huge floating offshore structure are found and good reference data to verify the theoretical calculation method are obtained.

KEY WORDS: Huge Floating Offshore Structure, Hydroelasticity, Dynamic Behavior, Mode Shape, Model Test

INTRODUCTION

Feasibility studies for a new concept of offshore structure are now carried out in Japan after a long preparation since the Kansai airport project. Through theoretical analyses, model tests and real sea tests, many kinds of uncertainties in motion responses and designs may be clarified. However, it may still be unclear how huge floating offshore structures in the order of kilometers in length behave really in an ocean environment of wave, current and wind until a real structure is constructed in the sea.

After Bishop and Price(1979) studied dynamic responses of hydroelastic bodies, Yoshida et al.(1992), Riggs and Ertekin(1993) and Kagemoto and Yue(1994) tried to find efficient numerical methods for the same problem of multi-module floating structures. Hamamoto and Hayashi(1996), Yoon and Choi(1996) made efforts to solve hydroelastic behaviors using FEM or modal analysis method for elastic responses and BEM for the flow problem in a coupled manner. These methods look still too ideal to cover the whole structure because too long calculation time and too large computer capacity is required. In order to reduce calculation time, Maeda et al.(1996) adopted the pressure distribution method instead of BEM and coupled with the modal analysis method. They found good agreements with model tests performed for a pontoon type very large floating structure. Murai et al.(1996) solved the flow field using the area division method and coupled with the modal analysis method. Hirayama et al.(1996) reduced the calculation time in BEM using symmetric relation and performed various model tests for a medium size floating airport in intact and damage condition. Yoshida et al.(1996) investigated the effect of interaction between underwater bodies on the design of a large floating offshore structure. However, there are not rigorous numerical schemes to calculate the dynamic responses of very large floating structures.

It is almost impossible to perform normal model tests for the whole dimension of a huge floating offshore structure because there exist some physical limitations such as the model basin size and wave making capacity. If the test model is fabricated for the whole dimension of a structure with very large scale ratio and the incident wave height becomes too small, then the generated incident waves cannot be regarded as normal gravity waves(Lee and Webster, 1994). Lee and Webster surveyed the model test for a floating airport and proposed many valuable guidelines.

The multi-point tracking system is necessary to measure the dynamic behaviors of hydroelastic structures. Maeda et al.(1996) used the image processing system to trace marked targets on the model. The image processing of this system is not real time but is performed after model tests finish. Shin and Yum(1996) used the optical measuring system to trace LED(Light Emitting Diode) targets installed on model deck. The setting of camera angle and sensitivity is important. Lee and Webster(1994) proposed an alternative measurement scheme, which allows a large number of slopes on the model.

A huge floating type offshore structure with dimensions of 5 km by

1 km is proposed as a concept design for the offshore airport at a Korean offshore site(Shin et al., 1995). In the present paper, model tests performed for a part of this structure are described (see Photo). The model size is determined so that the hydroelastic characteristics of the structure is properly shown. The model is fabricated with the scale ratio 100.

Model tests are conducted according to Shin and Yum(1996). An optical motion measuring system is used to measure the motions at multi-points of the elastic model. Twelve LED targets are set up on the 8 points of model deck just above the footing. Model tests are performed twice for one test condition switching the positions of LED targets from the fore half part to the aft half part of the model. The mode shapes of the whole model can be obtained by combining two test results. Bending moments at 5 deck points are measured using the strain gauges.

Forced oscillation tests are performed for the heave mode. Elastic deformations and reaction forces on the oscillation device are measured. By towing the model at very low speeds, drag forces on the model are measured.

MODEL TEST

Model

A floating type offshore structure with dimensions of 5 km in length and 1 km in width is proposed as a concept design for the offshore airport at a Korean offshore site. Considering the size of the model basin, model tests are performed for a 1/10 part cut from the whole structure. It is assumed that the elastic property is homogeneous all over the deck. The scale ratio for the local model dimension is 100. It is aimed at to see physical phenomena close to natural reality.

As shown in Fig. 1, the model has dimensions of 4.65 m (between footing centers) in length, 0.9 m in breadth and 95 mm in draft. Total 224 footings are used as underwater supporting bodies. They constitute 7 by 32 footing array. The footings are attached to the longitudinal plate deck of aluminum alloy. The footings are made with the polyurethane foam. The longitudinal bending stiffness of the structure is realized with strips of the aluminum plate according to the similitude law. The longitudinal bending stiffness EI for the model is 213.36 kg-m^2. Then the width and thickness of the plate are 50 mm and 10.1 mm respectively. The section modulus is assumed as uniform along the whole deck structure and the transverse stiffness is not considered. However, deck plates are connected each other using 3 light transverse bars so as not to be free in lateral direction. Model is assumed to be moored to the carriage with 2 linear springs of spring constant 1.786 kg/m at fore and aft end.

The positions of LED targets are switched from the fore half part of the deck to the aft half part as in Fig. 2. Duplicate tests for each regular wave are carried out to find hydroelastic deformation for the whole structure. Bending moments are measured using strain gauges installed on 5 points of the center deck plate.

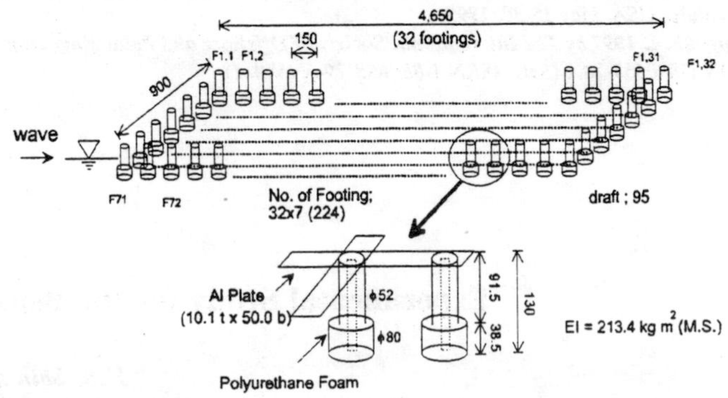

Fig. 1 Model of a Huge Floating Offshore Structure

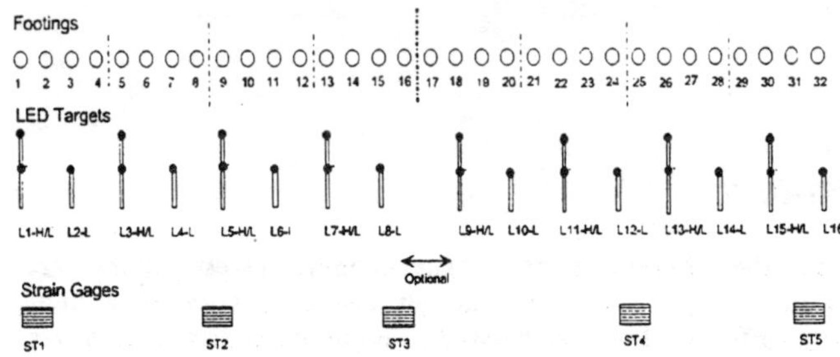

Fig. 2 Set-up of Experiment

Measurement

Fifteen regular waves are generated, length of which are multiples of the distance between adjacent footings varying from 2 to 62. The model wave height is 2.0 cm. The model is moored with 2 linear springs at 50 m off the wave maker. Irregular waves for a sea state are generated according to the ITTC wave spectrum. The significant wave height for the sea state is 1.6 cm and the mean period is 1.12 sec.

An optical position sensor system is used to measure the elastic behaviors of the model. The sensitivity and scope of the system limit the relative positions of the lens and LED targets and the number of traced LED targets. Scale factors for the LED motions are found by calculating the ratio of the distance between adjacent LEDs and measured signal voltages for the stationary LEDs. Four additional LED targets are used as shown in Fig. 2 to find the scale factors for the vertical motions. The sampling rate is 40 Hz.

Forced heave oscillation tests are performed for the 15 frequencies of regular wave tests. The heave amplitude is fixed to 0.5 cm for all oscillation frequencies. Motions at 8 points of the deck and reaction forces from the model are measured. The reaction force is measured by a 3-component load cell, through which the model and the oscillation equipment is connected.

Drag forces on the model are measured by towing the model at low speeds. The same measurement device with the forced oscillation test is used to measure horizontal forces. These horizontal forces are considered to be equivalent to current forces with an assumption that towing speeds are equivalent to current speeds. Towing speeds are varied in the range of 0.1 m/sec and 0.25 m/sec.

Results and Discussions

RAOs of vertical motions at some LED positions are shown in Fig. 3. Measured raw data are filtered to pass between 0.1 and 2.0 Hz and fitted to sinusoidal curves through the regression analysis. As expected, the largest vertical motions are found at fore part of the model against incident waves. The motions become smaller in the middle part and become larger again in the aft part. At most LED positions except both end parts of the model, vertical motion amplitudes are close to incident wave amplitudes in long waves and close to zero amplitudes in short waves. Watching the intermediate range of wave length, 0.760 Hz and 1.020 Hz are found as peak wave frequencies.

RAOs of bending moments at 5 deck points are shown in Fig. 4. The trends of bending moments are opposite to the motion trends. Bending moments are small at both end parts of the model and large in the middle part. Distinctive peak values are found at 0.579 Hz for all measuring points of the deck.

Hydroelastic deformations are shown in Fig. 5 for some incident wave frequencies. Ten snapshot pictures for the model test during one wave period are displayed. The time steps are identified by the symbols. The left side of the figure shows the hydroelastic deformations of the fore half part and the right side shows ones of the aft half part of the model. As expected from the Fig. 3 and Fig. 4, large vertical motions are found at fore and aft part of the model and very calm motions in the middle part. In long waves, the deformed shapes of the model show combination of the heave and the first mode of the elastic deformation. In short waves, model shapes look more complicated and it is not easy to definitely derive out any special mode shapes.

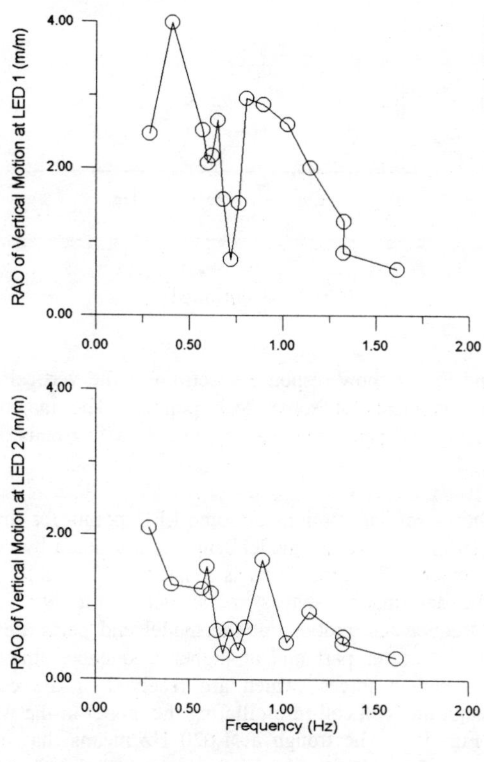

Fig. 3 RAOs of Vertical Motions at LED Points

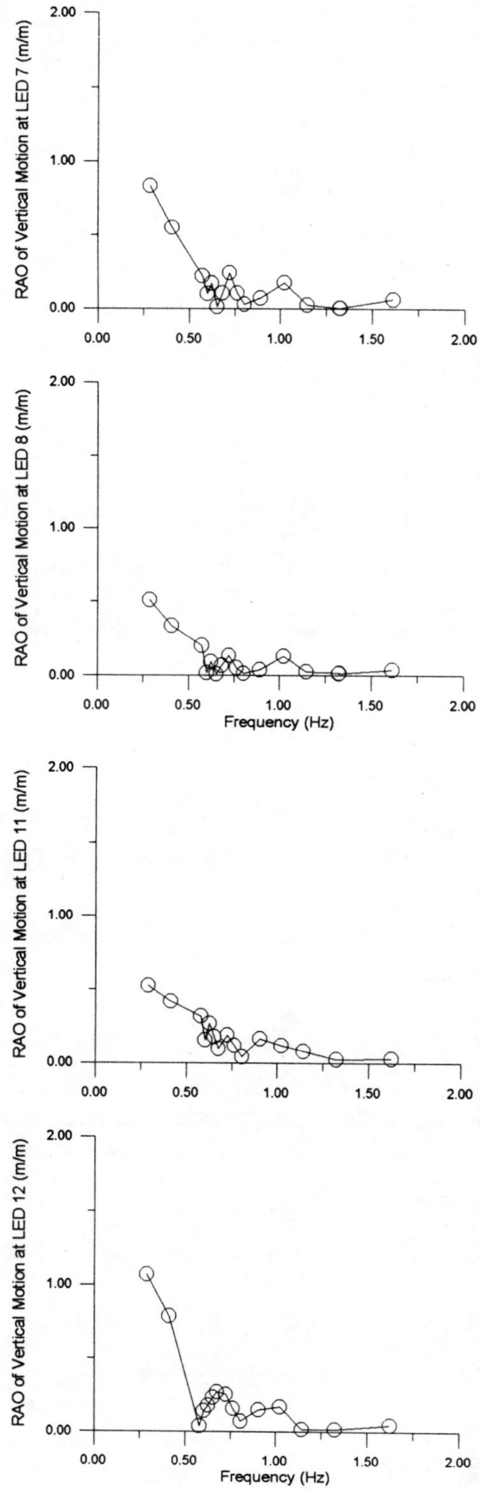

Fig. 3 (Continued)

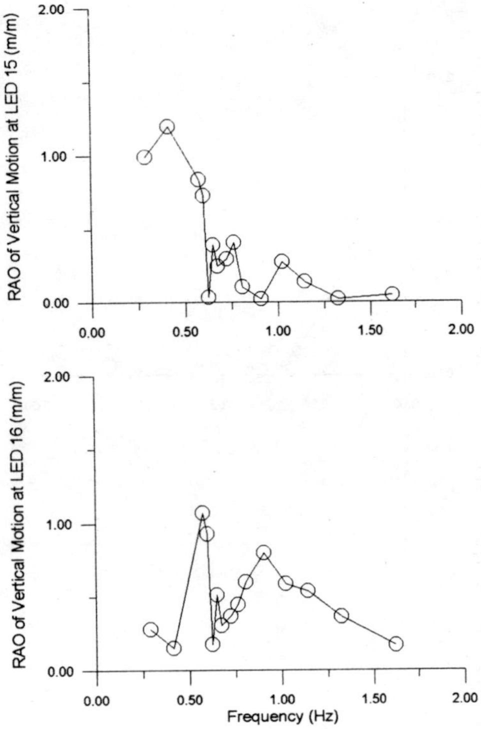

Fig. 3 (Continued)

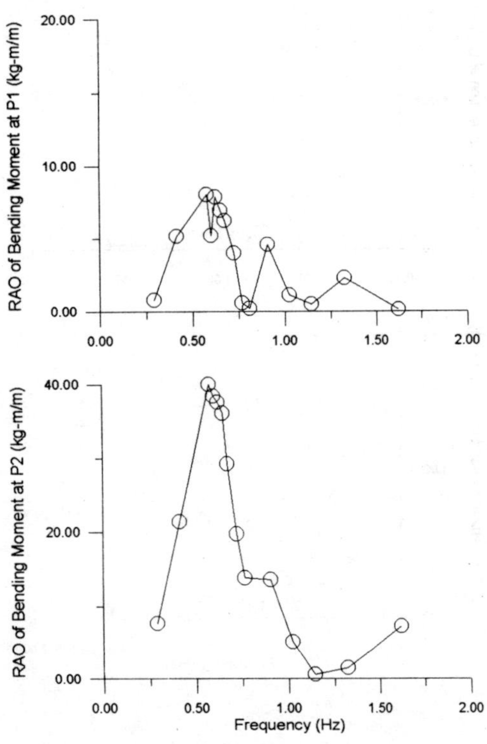

Fig. 4 RAOs of Bending Moments at Deck Points

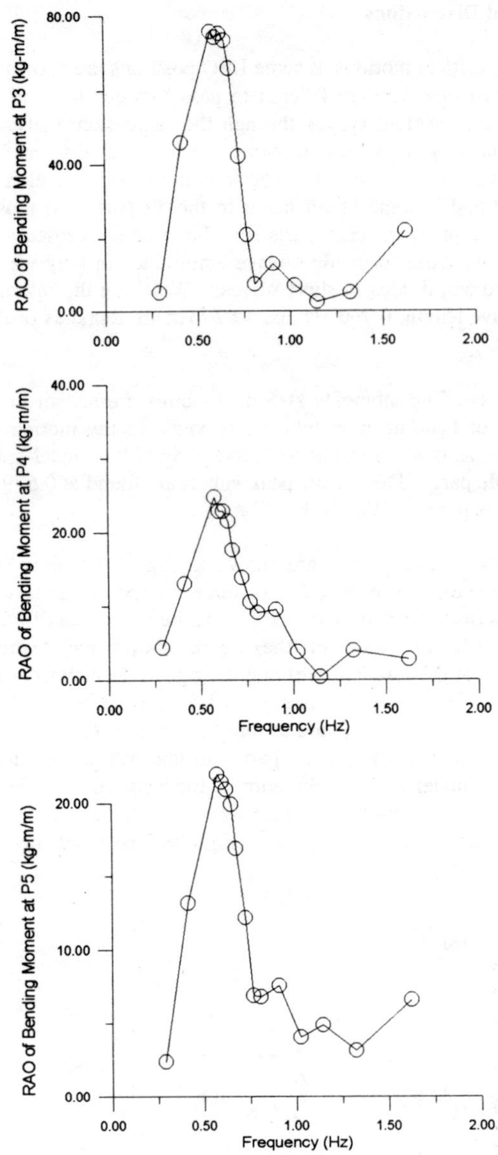

Fig. 4 (Continued)

Fig. 6 and Fig. 7 show response spectrum of the vertical motions and bending moments at some deck points. The magnitude of spectral density and peak frequencies coincide with trends of Fig. 3 through Fig. 5.

Fig. 8 shows vertical motions at some LED points for the forced heave oscillations, where the model center is oscillated by the heave oscillation device. Peak frequency is found at 1.317 Hz for all LED points. The deformation shapes are shown in Fig. 9. At lower oscillation frequencies, motions at the model end parts are smaller than the model center part and at higher frequencies the trend is reversed. Reaction forces, which are received at the oscillation device through the load cell in oscillating the model in the water, are shown in Fig. 10. The trough at 1.020 Hz means that minimum force is required to activate the heave motion of the model at this frequency. The trough frequency is regarded as the natural frequency for the heave mode of the model motion. Considering the peak reaction force and large motion at the model end parts at the frequency of 1.317 Hz, this frequency seems close to the natural frequency for the first mode of the elastic deformation. However, further studies are necessary about how to perform forced oscillations in higher mode and how to evaluate response motions and reaction forces in the relation to natural frequencies.

Fig. 11 shows current forces. It is found that current forces are proportional to the square of current speed. The drag coefficient for the model is estimated as 0.0666, which is non-dimensionalized by the square of the flow velocity, water density and the projected area of the model. Here the projected area is the total area of all submerged footings. The estimated drag coefficients look lower than generally used values for a single footing of the circular section. It is due to the shield effects.

CONCLUSIONS

From the present experimental studies, following conclusions are drawn.

1) Hydroelastic characteristics of a huge floating offshore structure are successfully investigated through model tests for the 1/10 part of the structure, which is conceptually designed as an offshore airport.

2) The elasticity of the designed structure is stiff enough not to give large deformation except at the fore and aft part of the model. Hence no problems are found for landing and take-off of airplanes if the middle part of the structure is used for those purposes. Large bending moments are found in the middle part of the model and safe structural design is required for the part.

3) Through the forced oscillation test, the hydroelastic behaviors are investigated, which will be the most important problem in the very large floating offshore structure. Further studies in the theoretical calculation and experiment are necessary.

4) Large shield effects for multiple underwater body arrays are found in the current force test.

ACKNOWLEDGMENT

This research was supported by the Ministry of Science and Technology, Korea. The present work is a part of the joint research project between the University of Ulsan, Korea and Ship Research Institute, Ministry of Transport, Japan.

REFERENCES

Bishop, RED, and Price, WG (1979). "*Hydroelasticity of Ships*," Cambridge University Press

Riggs, HR, and Ertekin, RC (1993). "Approximate Methods for Dynamic Response of Multi-module Floating Structures," J. of Marine Structures, Vol. 6, pp117-141

Hamamoto, T, and Hayashi, T (1996). "3D BEM-FEM Coupled Hydroelastic Analysis of Irregular Shaped, Module Linked Large Floating Structures," ISOPE 96, Vol. I, pp362-369

Hirayama, T, Ma, N, and Nishio, FO (1996). "Response Characteristics of a Long Type Floating Offshore Airport in Waves," J. of SNAJ, Vol. 178, pp225-236, in Japanese

Kagemoto, H, and Yue, DKP (1994). "Hydroelastic Analyses of a Structure Supported on a Large Number of Floating Legs," Proc. of Hydroelasticity in Marine Technology, pp417-431

Lee, SW, and Webster, WC (1994). "A Preliminary to the Design of a Hydroelastic Model of a Floating Airport," Proc. of Hydroelasticity in Marine Technology, pp351-362

Maeda, H, Masuda, K, Miyajima, S, and Ikoma, T (1996). "Hydroelastic Responses of Pontoon Type Very Large Floating Offshore Structure," J. of SNAJ, Vol. 178, pp203-212, in Japanese

Murai, M, Kagemoto, H, and Fujino, M (1996). "On the Predictions of Hydroelastic Behaviours of a Huge Floating Structure in Waves," J. of SNAJ, Vol. 178, pp213-224, in Japanese

Shin, HS, and Yum, DJ (1996). "Elastic Behaviors of a Huge Floating Structure," ISOPE 96 Vol. I, pp347-353

Yoon, JD, and Choi, HS (1996). "A Study on the Hydroelastic Behavior of Large Floating Structures," Korea-Japan Joint Workshop on Ship and Marine Hydrodynamics 96, pp176-182

Yoshida, K, Goo, JS, Kobayashi, K, and Suzuki H (1992). "Wave Exciting Forces and Dynamic Responses of Multi-unit Floating Structures in Waves," J. of SNAJ, Vol. 172, pp69-80, in Japanese

Yoshida, K, Suzuki, H, Iijima, K, and Oka, N (1996). "Analyses of Response of Large Floating Structures in Waves and Hydrodynamic Interaction Effects on their Design," J. of SNAJ, Vol. 178, pp297-304, in Japanese

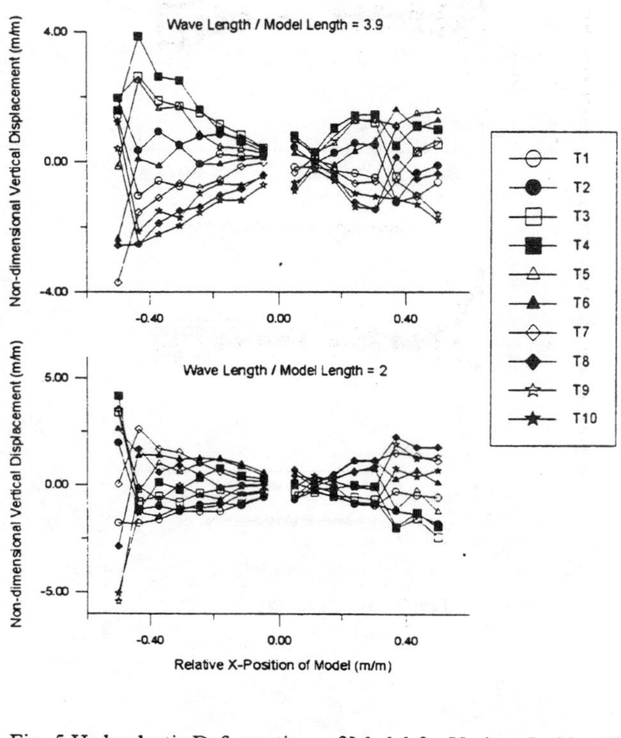

Fig. 5 Hydroelastic Deformations of Model for Various Incident Wave Lengths; X position is non-dimensionalized by the model length and backward is +, Y axis is non-dimensionalized by the incident wave amplitude

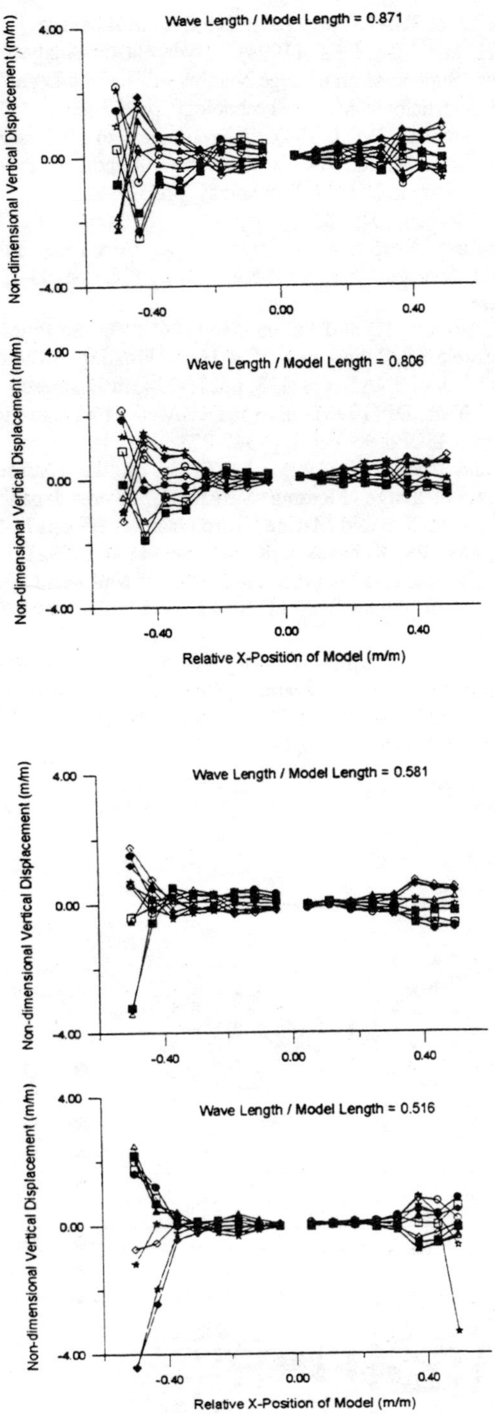

Fig. 5 (Continued)

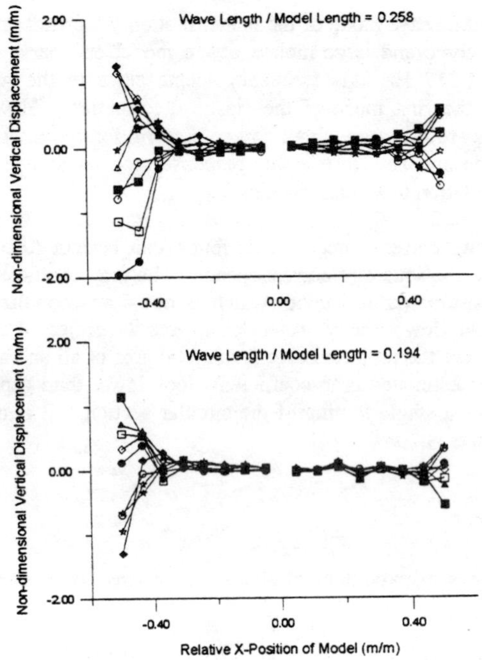

Fig. 5 (Continued)

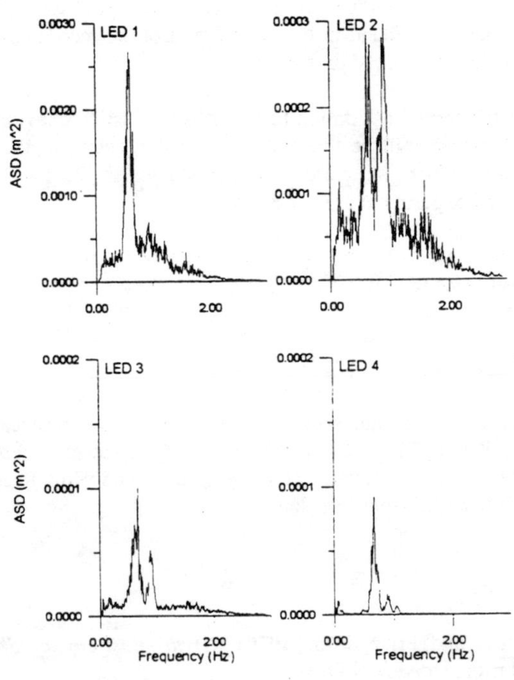

Fig. 6 Spectrum of Vertical Displacement at LEDs

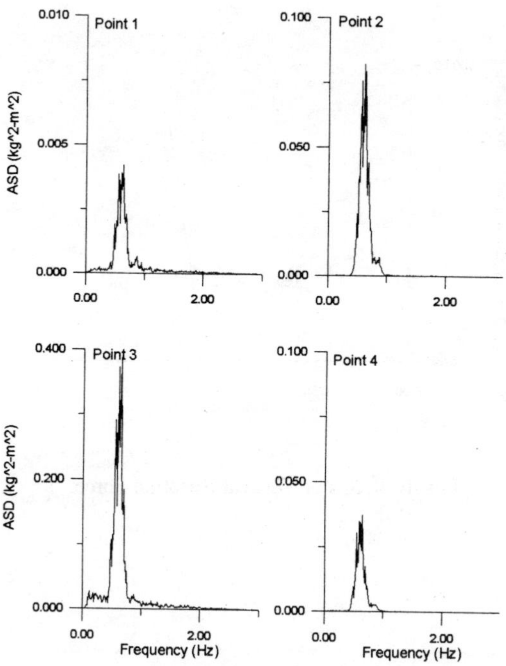

Fig. 7 Spectrum of Bending moment

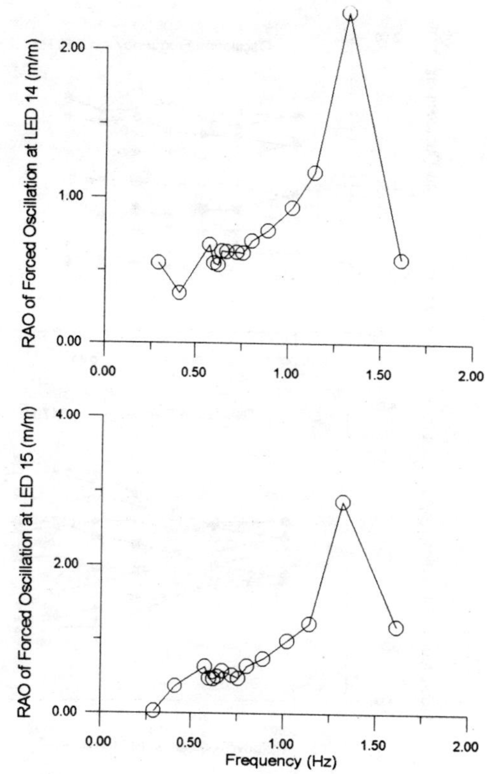

Fig. 8 (Continued)

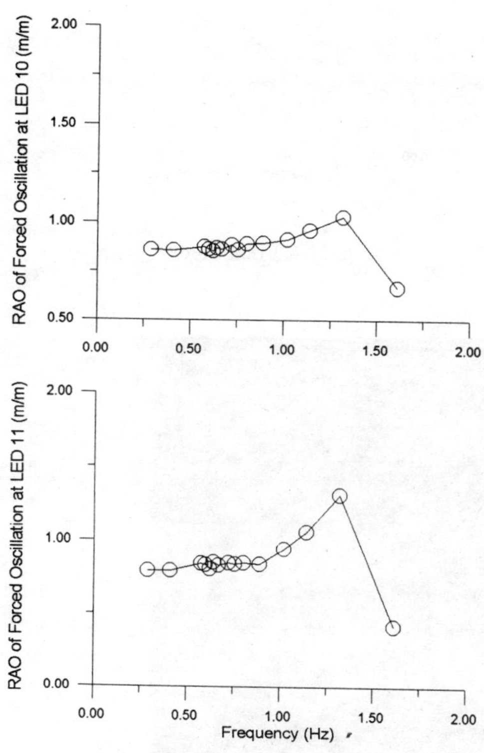

Fig. 8 Response RAOs for Forced Oscillation

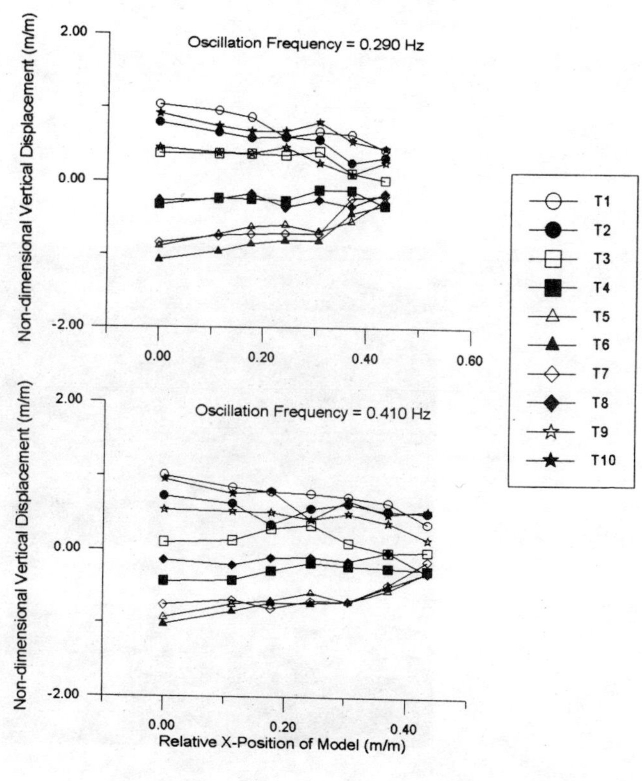

Fig. 9 Response Shapes for Forced Oscillations

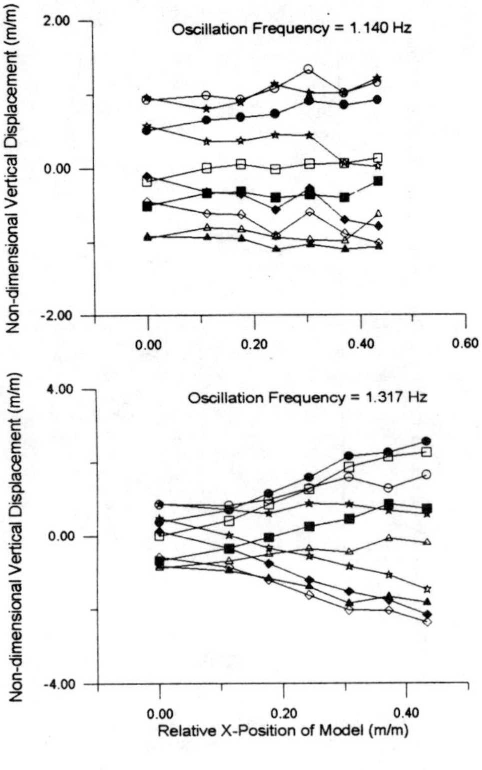

Fig. 9 (Continued)

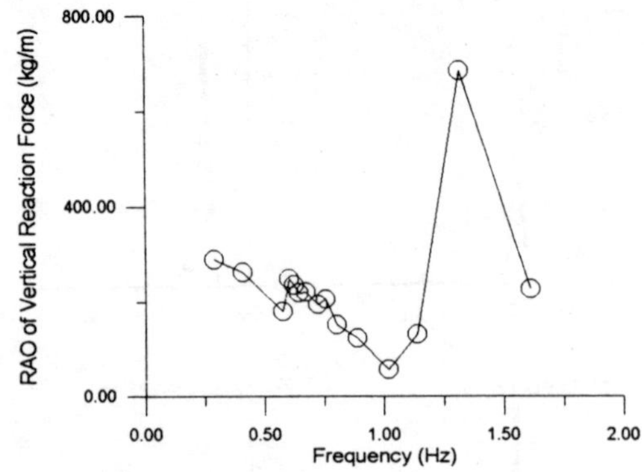

Fig. 10 RAOs of Vertical Reaction Force

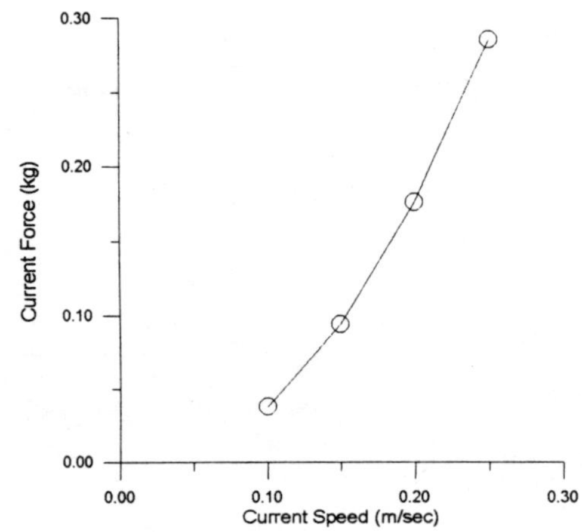

Fig. 11 Current Forces

Photo Model Test for an Offshore Airport

Proceedings of the Seventh (1997) International Offshore and Polar Engineering Conference
Honolulu, USA, May 25-30, 1997

ISBN 1-880653-28-1 (Set); ISBN 1-880653-29-X (Vol. I)

Study on the Inverse Analysis Method for Cathodic Protection Problems

M. Iwata, Q. Wu, Y. Huang and H. Yajima
Hiroshima University
Higashi-Hiroshima, Japan

ABSTRACT

Aiming at the cathodic protection monitoring technique for large scale floating structures, the pseudo-noise inverse analysis method for the estimation of the electropotential distribution on the surface of floating structures based on the measured potential data in sea water is proposed.

Considering the fact that the errors in the measurement are unavoidable in practice, the unknown pseudo-noise variables are introduced into the simultaneous algebraic equations obtained by the boundary element method for the cathodic protection problem, so that the errors induced by the ill-posed matrix in the inverse analysis procedure are controlled successfully, and therefore a rational potential distribution on the surface of the floating structure can be estimated from the measured potential data in sea water.

A series of simulating calculations for the floating structure model by changing the noises in the simulated input potential data in sea water show good stability of the inverse solutions with the proposed method, and it may be concluded that the potential distribution on the surface of the floating structure can be reliably estimated from the measured potential data in sea water when the measurement error is less than 10 mV.

KEY WORDS: Inverse analysis, Cathodic protection, Ill-posed matrix, Pseudo-noise

INTRODUCTION

For large scale floating structure such as a marine floating air port, covering method with some anti-corrosion materials such as titanium(Tadokoro et al 1992) and cathodic protection technique(John et al 1977) can be employed to prevent its corrosion in the severe corroding environment, so that a service life of 100 years is expected.

Based on the developed cathodic protection techniques for general marine structures, it is possible to make a reasonable design of the cathodic protection system for the submerged part of a large scale floating structure. During the long service life, however, the protection state of the floating structure surface will surely change due to many factors, such as the damage of the sacrificial anodes, marine growth on structural surface, etc.. In order to keep watching on the changes of cathodic protection state all the time, researchers have been developing a cathodic protection monitoring technique, which aims at electric potential distribution on structural surface according to the measured data of the electric potential field, so as to make a reasonable maintenance plan for the cathodic protection system.

Generally speaking, when the robot is employed to measure the electric potential fields formed by the cathodic protection system, it is very difficult to measure the potential distribution just on the surface of the floating structure, but only potential values at certain positions apart from the structural bottom surface, for example 1m or 50cm, are measured. Therefore, the method for estimating potential distribution on the floating structure surface based on the potential data in the sea water measured by robots becomes necessary in the cathodic protection monitoring technique for large scale floating structures.

In the numerical analysis of the cathodic protection problem, the simultaneous algebraic equations for the cathodic protection electric field can be formed conveniently by using the boundary element method(Brebbia 1978, Iwata et al 1990 and Huang et al 1990). When the simultaneous algebraic equations are directly employed for the inverse analysis of the cathodic protection problem of a large scale floating structure, in which the potential distribution on the surface of the floating structure becomes unknown and the measured potential data in sea water are taken as inputs, a real solution of the potential distribution on the surface of the floating structure can not be obtained because of the ill-posed matrix and the measurement errors.

In this paper, in the inverse analysis procedure based on the above-said simultaneous algebraic equations, unknown pseudo-noise variables for all the measured potential data in sea water are introduced to compensate the measurement errors so that the inverse solution errors induced by the ill-posed matrix are controlled successfully, and therefore a rational potential distribution on the surface of the floating structure can be estimated from the measured potential data in sea water.

A series of simulating calculations of floating structure model under cathodic protection by changing the noises in the simulated measured potential data in sea water are carried out. The investigated results show that the inverse solutions by the proposed method are very stable for the possible measurement errors in input data, and it may be concluded that the real potential distribution on the surface of the floating structure can be reliably estimated from the measured potential data in sea water when the measurement error is less than 10 mV.

BASIC EQUATIONS FOR CATHODIC PROTECTION PROBLEM

Mathematical Model of Cathodic Protection Problem

As for the system of cathodic protection, when the specific resistance doesn't change everywhere in the electrolyte domain, the governing equations in the domain and on the boundary are as follows(Iwata et al 1990 and Huang et al 1990),

$$\begin{cases} \frac{1}{\rho}\nabla^2\Phi = 0 & \text{in}\,\Omega \\ q = -\frac{1}{\rho}\frac{\partial \Phi}{\partial n} & \text{on}\,S \end{cases} \tag{1}$$

where ∇^2 is the Laplace operator, ρ is the specific resistance, Φ is the electropotential, q is the electric current density, S is the boundary, Ω is the electrolyte domain and n is the outward normal on the boundary S. In a general cathodic protection electric field, the corresponding boundary conditions for the forward problems as shown in Fig.1(a) are classified as follows,

$$\begin{cases} \Phi = \bar{\Phi} & \text{on } S_1 \\ q = \bar{q} & \text{on } S_2 \\ q = f(\Phi) & \text{on } S_3 \end{cases} \tag{2}$$

where $\bar{\Phi}$ and $\bar{q}$ are the prescribed Φ and q on the boundaries S_1 and S_2, S_3 is the anodic or cathodic boundary and f(Φ) is the anodic or cathodic polarization characteristic.

For the inverse problem, however, the corresponding boundary conditions as shown in Fig.1(b) are not only S_1,S_2 and S_3 but also S_4 and S_5 as follows(Aoki et al 1991).

$$\begin{cases} \Phi, q = \text{known} & \text{on}\,S_4 \\ \Phi, q = \text{unknown} & \text{on}\,S_5 \end{cases} \tag{3}$$

In the inverse analysis for the cathodic protection problem of a large scale floating structure, the measured potential data in sea water should also be taken as known conditions expressed as follows,

$$\Phi = \bar{\Phi}_{in} \tag{4}$$

where $\bar{\Phi}_{in}$ is the measured potential in sea water.

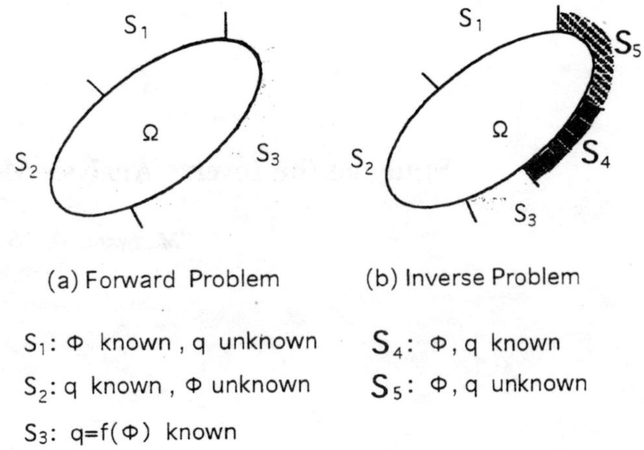

Fig.1 Boundary Kinds for the Forward Problem and Inverse Problem

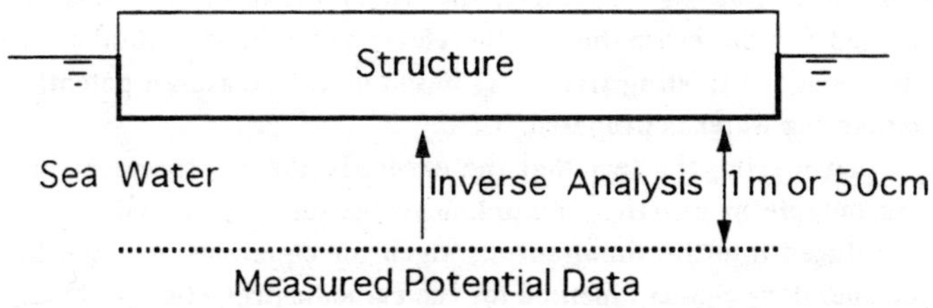

Fig.2 Model for the Potential Monitoring System

In this paper, studying on the potential distribution on the surface of the floating structure as shown in Fig.2, the following mathematical equation for the estimation of the potential distribution on the surface of the floating structure is used.

$$\begin{cases} \nabla^2\Phi = 0 & \text{in sea water } \Omega \\ q = -\frac{1}{\rho}\frac{\partial \Phi}{\partial n} & \text{on } S \\ \Phi = \bar{\Phi}_{in} & \text{in sea water } \Omega \end{cases} \tag{5}$$

Simultaneous Algebraic Equations for Inverse Analysis

According to the boundary element method(Brebbia 1978, Iwata et al 1990 and Huang et al 1990), the following system of algebraic equations can be obtained,

$$[H]\{\Phi_b\}=[G]\{q_b\} \quad \text{for the boundary} \tag{6}$$

and,

$$[G_{in}]\{q_b\}-[H_{in}]\{\Phi_b\}=\{\bar{\Phi}_{in}\} \quad \text{for inner points} \tag{7}$$

where Φ_b and q_b are the vectors consisting of the boundary values of potential and electric current density on the nodes for the boundary element method, and the matrices [G], [H],$[G_{in}]$ and $[H_{in}]$ are the known coefficient matrices depending on the geometric relationship between each other of the points and the elements. In order to solve the above equations, the Φ 's and q's are classified as,

$$\{m\}=\text{the vector consists of knowns in } \{\Phi_b\} \text{ and } \{q_b\} \tag{8}$$

and,

$$\{d\}=\text{the vector of } \{\bar{\Phi}_{in}\} \text{ and knowns in } \{\Phi_b\} \text{ and } \{q_b\} \tag{9}$$

and then the following Eq.(10) can be obtained from Eqs.(6) and (7).

$$[M_{HG}]_1\{m\}=[M_{HG}]_2\{d\} \tag{10}$$

As for the forward problem, matrices $[M_{HG}]_1$ and $[M_{HG}]_2$ are the new matrices obtained from Eq.(6), and the number of the unknowns is just equal to the number of the equations after substituting the boundary conditions into Eq.(10) and the matrix $[M_{HG}]_1$ is well-posed. Therefore, the solution of the forward problem can be obtained as Eq.(11).

$$\{m\}=[M_{HG}]_1^{-1}[M_{HG}]_2\{d\} \tag{11}$$

For the inverse problem, where the matrices $[M_{HG}]_1$ and $[M_{HG}]_2$ are obtained from Eqs.(6) and (7), however, the matrix $[M_{HG}]_1$ is not always just determined so that the pseudo-solution should be written as Eq.(12),

$$\{m\}=[M_{HG}]_1^{-g}[M_{HG}]_2\{d\} \tag{12}$$

where $[M_{HG}]_1^{-g}$ is the generalized least square inverse matrix(William 1989) of matrix $[M_{HG}]_1$ and which is expressed as Eq.(13).

$$[M_{HG}]_1^{-g}=([M_{HG}]_1^T[M_{HG}]_1)^{-1}[M_{HG}]_1^T \tag{13}$$

Defining a matrix [A] by Eq.(14),

$$[A]=[M_{HG}]_1^{-g}[M_{HG}]_2 \tag{14}$$

Eq.(12) can be rewritten as Eq.(15).

$$\{m\}=[A]\{d\} \tag{15}$$

For the case of the floating structure under cathodic protection, {d} is the vector consisting of measured potential data in the sea water and {m} is the vector consisting of potential values to be estimated on the surface of the floating structure.

INVERSE ANALYSIS WITH PSEUDO-NOISE

Since the errors in the measured potential data {d} in sea water can not be avoided and the matrix $[M_{HG}]_1$ is usually ill-posed, the errors in the measured potential data {d} will be amplified into {m} badly. Therefore, in order to get the satisfying solution {m} from Eq.(15), the unknown pseudo-noise(Huang et al 1996 and Wu et al 1997) is introduced for all the measured potential data {d} to compensate the measurement errors so as to overcome the above problem. In the analysis of floating structures under cathodic protection, the pseudo-noises can be thought as the errors in the measured potential data {d} in sea water and then the following expression can be obtained,

$$\{d\}_T=\{d\}_M+\{d\}_E \tag{16}$$

where $\{d\}_T$ is the true potential data in the sea water (unknown), $\{d\}_M$ is the measured potential data in the sea water(known) and $\{d\}_E$ is the pseudo-noises or the errors in the measurement. And then Eq.(15) should be rewritten as follows.

$$\{m\}=[A](\{d\}_M+\{d\}_E) \tag{17}$$

And now, the problem boils down to calculating the values of $\{d\}_E$. Therefore, the following optimization model is used to evaluate $\{d\}_E$

$$\min \quad \sum |d_E|$$

subject to:

$$\begin{cases} [F]\{m\} \geq 0. \\ \{m\} = [A](\{d\}_M + \{d\}_E) \end{cases} \tag{18}$$

where matrix [F] is called the constraint matrix, which expresses the physical conditions encountered in the problem under research and for example, on the anode, $q \geq 0$.

At last, Eq.(17) is used to estimate the unknowns {m} in the inverse analysis model.

NUMERICAL ANALYSIS

In order to show the practicability of the proposed inverse analysis method in the cathodic protection monitoring technique for large scale floating structures, simulating calculations for the floating structure model are carried out.

The floating structure model under cathodic protection for the simulating calculations is shown in Fig.3. In this model, the structural surface is prescribed as the heavily coated area, the old coated area and the bared area. The polarization characteristics for the old coated area and bared area of the structural surface and that of the anode are shown in Fig.4.

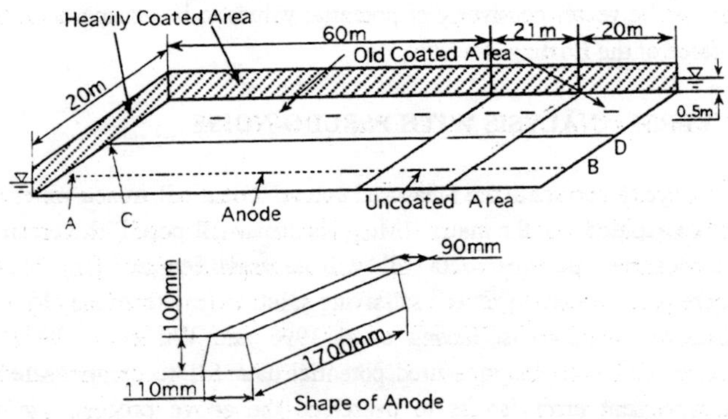

Fig.3 Floating Structure Model for Simulating Analysis

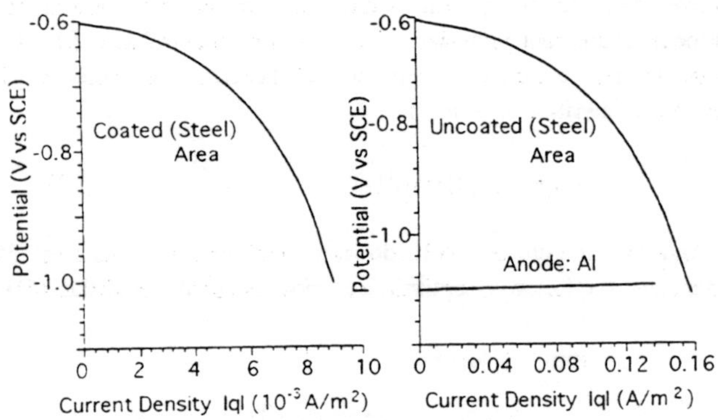

Fig.4 The Polarization Characteristics Used in the Forward Analysis of the Floating Structure Model

For the simulating calculations with the proposed inverse analysis method, the corresponding forward analysis for the floating structure model as shown in Fig.3 is firstly carried out, in which the polarization characteristics shown in Fig.4 are used. And through Eq.(19), the input potential data in sea water for the inverse analysis are then generated by adding certain noises to the analytical solutions of the potential in sea water from the forward analysis.

$$\Phi_M = \Phi_F + 2b(r-0.5) \qquad (19)$$

where Φ_M and Φ_F are the simulated input potential data in sea water and the analytical solutions of forward analysis, r is the random number between 1 and 0 so as to simulate the error distribution in the measurement and b is the amplitude of the added noises.

In order to investigate the stability of the inverse solution with the proposed method, a series of simulating calculations by changing the noises in the simulated input potential data in sea water are performed, and the average absolute error Ce of the inverse solutions of the potential distribution on the surface of the floating structure model comparing with the analytical potential distribution on the surface of the floating structure model by forward analysis is evaluated by Eq.(20),

$$Ce = \sum_{i=1}^{n} |\Phi_{inv} - \Phi_{for}| / n \qquad (20)$$

where n is the number of the nodes on the cathodic area and Φ_{inv} and Φ_{for} stand for the inverse solution and the forward solution(analytical solution) respectively.

Generally speaking, the larger the noises in the input potential data in sea water are, the worse the inverse solutions of the potential distribution on the surface of the floating structure become. In order to investigate the sensitivity of the inverse solution with the proposed inverse analysis method to the noise in the measurement, the inverse analyses are carried out by selecting b in Eq.(19) as 1, 5, 10 and 20mV respectively. In this numerical simulation of the above model, the number of the boundary nodes is 483 and the number of simulating measuring points is 516. Fig.5 shows the simulated input potential data in sea water for the inverse analyses.

For four cases, the inverse solutions of the potential distribution on the surface of the floating structure model are compared with the analytical solutions by forward analysis in Fig.6, and the average absolute errors Ce defined by Eq.(20) are shown in Table 1. It can be said that the error of the inverse solution with the proposed inverse analysis method is not so sensitive to the noise in measurement, and in practice, it is possible to estimate the real potential distribution on the surface of floating structure based on the measured potential data in sea water with the proposed inverse analysis method.

CONCLUSION

Aiming at the cathodic protection monitoring technique for large scale floating structures, the pseudo-noise inverse analysis method for the estimation of the electropotential distribution on the surface of floating structures based on the measured potential data in sea water is proposed.

Considering the fact that the errors in the measurement are unavoidable in practice, the unknown pseudo-noises are introduced into the simultaneous algebraic equations obtained by the boundary element method for the cathodic protection problem, so that the errors induced by the ill-posed matrix in the inverse analysis procedure are controlled successfully, and therefore a rational potential distribution on the surface of the floating structure can be estimated from the measured potential data in sea water.

A series of simulating calculations for the floating structure model by changing the noises in the simulated input potential data in sea water show good stability of the inverse solutions with the proposed method, and it may be concluded that the potential distribution on the surface of the floating structure can be reliably estimated from the

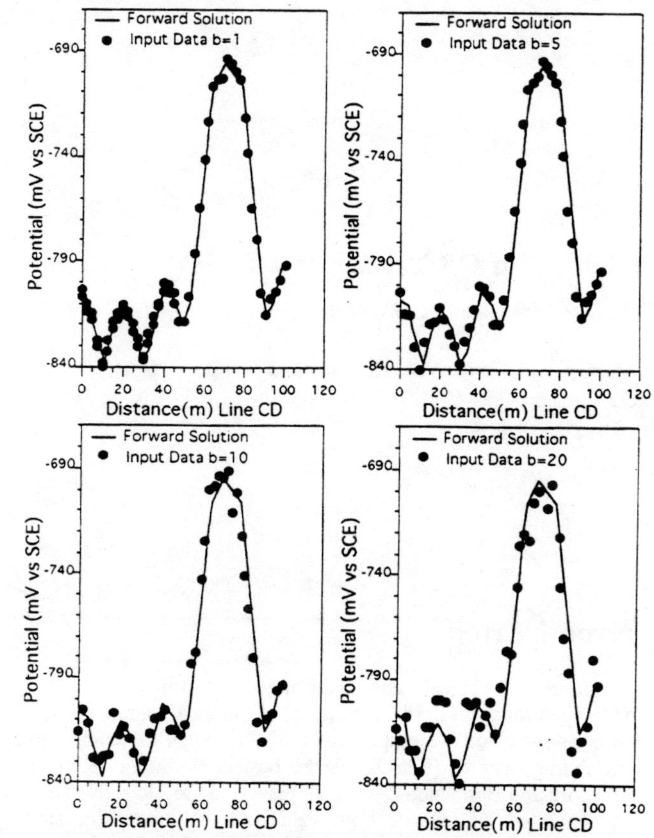

Fig.5 Comparison of the Forward Solution of the Measuring Points (50cm under the Bottom of Floating Structure) and Input Data for Inverse Analysis of Different Amplitude of Added Noise

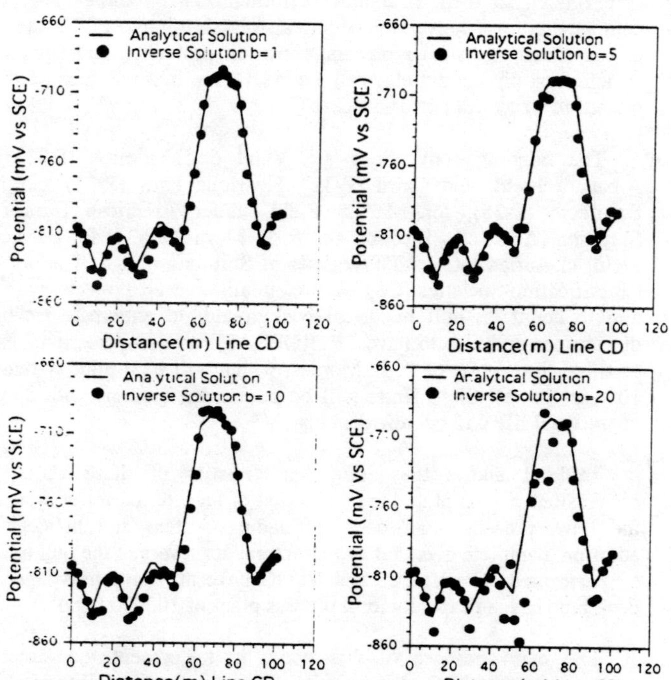

Fig.6 Comparison between the Analytical Solution and the Inverse Solution of the Potential Distribution on the Surface of the Floating Structure (along Line CD)

Table 1 The Average Absolute Error Ce (mV)

Amplitude of the Noise b	1	5	10	20
Average Absolute error Ce	4.0	5.1	7.2	11.1

measured potential data in sea water when the measurement error is less than 10 mV.

REFERENCES

Aoki S. and Urai Y.(1991), "Inverse Analysis of Galvanic Corrosion Problems Using Clustering Techniques", Material, Vol.40.

Brebbia C.A.(1978), The Boundary Element Method in Engineering, Pentch Press.

Huang Y., Iwata M., and Jin Z.L.(1990), "Numerical Analysis of Electropotential Distribution on the Surface of Marine Structure under Cathodic Protection," Journal of the Society of Naval Architects of Japan, Vol.168.

Huang Y., Wu Q., and Iwata M.(1996), "A Study on the Inverse Analysis Method for Cathodic Protection Problems: Report 1 An Inverse Estimation Method of Potential Distribution," Journal of the Society of Naval Architects of Japan, Vol.180.

Iwata M. and Huang Y.(1990), "Numerical Analysis of the electropotential distribution on the surface of metal structure under cathodic protection," Boundary Elements 12, Vol.1.

John O.M. Bockri and Amulya K.N. Reddy(1977), Modern electrochemistry.

Tadokoro Y., Honma K., Nagatani T., Hitoshi S., Yoshida K., Yamaya K. and Ito S.,(1992) "Development of Super Anti-Corrosion Method for Steel Structures Using Titanium," Nippon Steel Technical Report,No.344.

William M.(1989), Geophysical Data Analysis: Discrete Inverse Theory,Academic Press.

Wu Q., Iwata M. and Huang Y.(1997),"Study on the Application of Pseudo-noise and Boundary Element Method in Inverse Problems of Potential Fields," Memoirs of the Faculty of Engineering of Hiroshima University. Vol 45.

Proceedings of the Seventh (1997) International Offshore and Polar Engineering Conference
Honolulu, USA, May 25-30, 1997

ISBN 1-880653-28-1 (Set); ISBN 1-880653-29-X (Vol. I)

Structural Requirements for the Conversion of an Oil Tanker to a FPSO as a Permanent System

Marina Barbosa Fachetti, Deborah Martinez de Mattos and Carlos Alberto Bardanachvili
PETROBRAS R&D Center
Rio de Janeiro, Brazil

ABSTRACT

In the near future PETROBRAS will install, at least, five FPSOs at Campos Basin - Brazil, to exploit its giant fields, Marlim, Albacora and Barracuda Fields, considering for each Unit an oil processing around 100000 bopd and a 20-year operational life without docking. The conversion of its own oil tankers, older than 20 years, into FPSOs has been adopted, as a result of a Technical and Economic Feasibility Study.

In order to guarantee a suitable performance of the Unit during the whole life, special requirements were established for structural verification of the FPSO, taking into account peculiar conversion aspects, such as previous fatigue damage as an oil tanker, thickness corrosion rates, Brazilian environmental and new cargo loading conditions.

The main requirements, concerning yielding, buckling and fatigue criteria, can be gathered in four groups. First of all, a verification of the longitudinal hull-girder strength has to be performed with updated loading conditions, taking into account the loads introduced due to the new function, such as turret and process plant loads. The cargo region, the most stressed area of the hull, has also to be verified, specially in relation to fatigue analysis and new hydrodynamic forces. Regarding the load transmission between the hull and the turret at the bow region, a refined structural model and analysis of the turret region has to be carried out. New heavy structures (process plant deck, helideck, flare, cranes, etc.) are also focused.

The structural requirements are presented and also compared with those established by Classification Societies.

KEY WORDS: FPSO, ship conversion, structural requirements, structural analysis

INTRODUCTION

Recent Technical and Economic Feasibility Studies (TEFS) developed for giant fields discovered at Campos Basin, offshore Brazil, have pointed out the employment of Floating Production Storage & Offloading systems (FPSOs) as the best exploitation option. Marlim, Albacora and Barracuda fields will receive up to 1998 five Units based on converted oil tankers from PETROBRAS fleet. The strategy of converting existing oil tankers into FPSOs was based on oil anticipation, lower investment cost and the availability of oil tankers, resulting in a clear economic advantage over a new building unit. On the other hand, these oil tankers older than 20 years need additional care related to their structural performance. The three major cost components for a conversion of an existing ship into a FPSO, apart of new systems, are steel renewals, corrosion protection and equipment (machinery, piping and electric) overhauls, the first two being directly related to structural performance.

The ongoing conversions are Vidal de Negreiros (P-31), for Albacora Field, and Cairu (P-32), Henrique Dias (P-33) and José Bonifácio (P-35), for Marlim Field, under American Bureau of Shipping (ABS) classification and P. P. Moraes (P-34) for Barracuda Field, classified by Lloyd's Register of Shipping (LRS). The original Classification Societies (CS) were maintained in all projects, with the aim to count on hull historical background, to anticipate technical discussions and also to have PETROBRAS Basic Engineering Design certified. Exception of P. P. Moraes, which will be a pilot system for 10-year life, all these Units will be permanent systems with 20-year operational life without any docking.

Table 1 shows the main characteristics of these Units. All conversions contemplate the installation of an internal turret system at the bow, process plant deck, offloading system and helideck. In addition, complete overhaul and structural surveyors of the hull have to be performed. The VLCCs and VLOO have no limitations regarding deck area or load capacity for a process plant of 100000 bopd.

The main purpose of this paper is to present the structural requirements established in order to guarantee the feasibility of old oil tankers, complementing those issued by Classification Societies. A summary of CS Rules is presented, and also a brief explanation on our design premises for structural verification.

TABLE 1 - PETROBRAS FPSOs Main Particulars

UNIT	P-34	P-32	P-31	P-33	P-35
Field	Barracuda	Marlim	Albacora	Marlim	Marlim
Water depth (m)	840	160	330	780	860
Production plant (bopd)	45 000	-	100 000	50 000	100 000
Treatment plant (bopd)	-	100 000	200 000	50 000	100 000
Deadweight (dwt)	55 000	280 000	280 000	280 000	270 000
Class	Class P	VLCC	VLCC	VLCC	VLOO
LBP (m)	231.10	320.00	320.00	320.00	320.00
Breadth molded (m)	26.00	54.50	54.50	54.50	54.50
Depth molded (m)	16.87	27.80	27.80	27.80	28.00
Ship construct. date	1959	1974	1974	1978	1974
Expected first oil	Jun./97	May/97	Sept./97	Apr./98	May/98
Upper bearing diameter (m)	13.6	5	9.8	≈ 7.5	≈ 7.5
Number of risers	34	8	27	35	47
Classification Society	Lloyds	ABS	ABS	ABS	ABS

CLASSIFICATION SOCIETIES RULES FOR FPSOs

The Rules for FPSO Classing issued by the two CS that are contracted for our ongoing conversions are listed below. They are also employed as a design guide.

- *Guide for Building and Classing Floating Production and Storage Systems* (FPSO Guide) by ABS;
- *Provisional Rules for Floating Production Installations at a Fixed Location* - Part 9 by LRS.

ABS Rules state that the design and construction of **new hull** to be used as a FPSO should comply with all the applicable requirements of the Steel Vessel Rules (*Rules for Building and Classing Steel Vessels*) and in particular with section 5/2 "Vessel Intended to Carry Oil in Bulk". Global and local strength of the FPSO and also the fatigue life of structural details should be evaluated. Two approaches may be adopted, one based on a finite element method (FEM) analysis for a three-tank model, already implemented in the ABS SafeHull System for Tankers program and the other using the Dynamic Load Approach (DLA) with a complete FEM model with allowable Von Mises stresses limited to 85% of the material yield.

FPSOs originated from **converted hulls** receive by ABS an individual treatment, not specifically described in the Rules, where service history, age and condition of the existing vessel are considered. The requirements for conversion are dealt case by case.

LRS Rules also allow the new hull scantlings to be determined by either direct calculation or empirical formulae, as furnished in the Rules and Regulations for the Classification of Ships- Parts 3 & 4. The Structural Design Assessment procedure (SDA), implemented in LRS Shipright program, is mandatory for new oil tankers with lengths over 190 m and recommended when the owner wants to obtain more confidence in the ship.

Additionally to Classification Society Rules, regulations applicable to territorial waters must also be taken into account. Brazilian Administration follows the IMO requirements stated in SOLAS/MODU Code and in the amendments to the MARPOL Convention which does not impose the use of double hull ships for FPSOs.

Normally all these Rules are related to new building FPSO and do not contemplate all requirements for conversions. In the case of fatigue requirements, they do not mention how to estimate the fatigue damage over the previous life. Even in the case of a new hull, these Rules present some gaps related to the design of specific structure systems like process plant deck and turret system.

PETROBRAS DESIGN PREMISES

During the last 20 years PETROBRAS has been using conversion of drilling semi-submersibles into floating production units, at the beginning as early production systems and more recently as definitive production systems, achieving a considerable background on this area. However, the operation of P.P. Moraes moored to a CALM-Yoke system in Albacora Field from 1987 to 1993 was the only experience on FPSO.

Regarding the ongoing FPSO conversions, the generation of BID documentation was performed in parallel, taking into account our previous experience on semi-submersibles, as well as, the experience on operation of oil tankers. The most difficult task in a conversion work is how to deal with the existing parts and at this point both naval architecture and structural subjects are the great stars of the problem and our challenge to be overcome.

As mentioned before, the required life for all permanent systems has been defined as 20 years without docking. Pilot systems have variable requirements, and a 10-year life has been adopted for P. P. Moraes FPSO. Fatigue life, painting system and cathodic protection system should be compatible with the service life defined for the FPSO.

The FPSO, including mooring systems, will be designed to withstand the centenary environmental condition of the location, except some systems which operations are limited to specific environmental conditions, such as pull-in winches and offloading systems. Table 2 presents the maximum centenary, annual, pull-in, offloading and beam sea wave heights and associated periods, wind velocity and surface current velocity for Campos Basin Fields. Brazilian environmental data have some peculiarities regarding current velocity and directionality. The intensity is high throughout the depth and the directionality at the surface is only lightly related to wind approach and both of them vary according to different Campos Basin Fields, as shown in Table 3. It is important mentioning that the simultaneity of events among wave, wind and current is not yet available, but great engineering efforts have been employed in order to improve the techniques for environmental data treatment and consequently to establish adequate Safety Factors (SFs). The current peculiarity also helps to make it hard.

TABLE 2 - Maximum Environmental Conditions at Campos Basin (Marlim Field)

PARAMETER	MAXIMUM ENVIRONMENTAL CONDITIONS				
	Centenary	Annual	Beam Sea	Off-loading	Pull-in
Significant wave height (H_s) (m)	7.6	5.1	4.5	3.5	2.0
Zero up-crossing period (T_z) (s)	9.2	7.9	8.0	8.0	7.0
Wind velocity* (m/s)	37.22	21.22	10	21.22	18
Surface current velocity (m/s)	1.75	1.42	0	1.42	1.42

* The wind values are referred to 10 m above mean sea level and 10 min duration.

TABLE 3 - Vertical Profile of Extreme Centenary Current for Marlim and Albacora Fields

MARLIM *			ALBACORA **		
Depth (m)	Velocity (m/s)	Direction	Depth (m)	Velocity (m/s)	Direction
Surface	1.75	SE	Surface	2.06	S
50	1.61	SE	60	1.73	S
100	1.26	SE	100	1.65	S
140	0.82	E	150	1.25	S
230	0.76	E	250	0.86	S
340	0.79	E	300	0.72	S
750	0.61	NE	-	-	-
Bottom	0	-	Bottom	0	-

* At José Bonifácio FPSO location
** At Vidal de Negreiros FPSO location

For all FPSOs, the weathervane capability is furnished by turret location at the bow, without any help of thrusters, which means that relevant crossed environmental conditions could reach the Unit. In terms of structural and naval verifications the motions analyses should be performed considering wind and wave approaches from the same direction and current approach varying from 0° to 45° with regard to the former.

Regarding our operational experience, an additional environmental condition was introduced, related to the swell condition daily observed in Brazilian coast. Waves coming from South Atlantic, may have a very regular shape and heights in the range of 2.5 m to 4.5 m, which correspond to conditions smaller than the 1-year storm. Twice a year, when wind and currents have low intensities, leading to a non-preferential ship heading, the swell may reach the Unit transversally, causing a beam sea condition. Although wave heights are not as high as the centenary case, roll motions and accelerations may govern some design parameters, such as bearing sizing and production plant performance. Also a damage condition, corresponding to a trim of 10° and a heel of 22.5° is to be considered.

The Unit structural elements, i.e., hull, turret, process plant deck, flare boom etc., have to be designed and verified using the yielding, buckling and fatigue criteria according to the CS Rules, complemented by PETROBRAS design premises. The structural elements have to be verified or designed for the most unfavorable combination, position and direction of the static plus adequate environmental loading conditions. In the following items some particularities are described.

Material Steel & Design Areas

Steel materials for hull reinforcements should be the same of those used in the existing ship. New structures should be designed using preferably steel with yielding stress over or equal to 355 N/mm². In order to prevent the lamellar tearing effect, steel with Z quality (strength through the thickness) should be used in locations with stresses in plate transversal direction which should be avoided by employing special details.

FPSO areas are classified taking into account the importance of a structural member to the overall integrity of the Unit and safety. This area classification is made for the inspection planning of structural components and welds, for material selection, depending on design temperature, and to guide the structural design.

The steel structure is divided into three main design areas:

Primary Areas: structural elements essential to the overall FPSO integrity, e. g., midship structure and turret shaft;

Special Areas: regions of the Primary Structure that are in the way of critical load transfer joint, critical stress concentrations or critical dynamic forces, e. g., bearing support and chain table;

Secondary Areas: structural elements of minor importance, which failure is unlikely to affect overall Unit integrity.

It should be emphasized that material grades selection provides the required level of structure performance. However, for a conversion, this assumption is not always correct as most of the steel quality used in ships is grade A that does not comply with Charpy requirements. In this case the engineering judgment and experience may interfere in the final material grade selection. In such circumstances, if the stress levels are very low and are not fatigue sensitive the grade material requirement could be updated.

Corrosion Margins

The definition of hull plating replacement is based on the results of periodical Survey of Hull of the Steel Vessel Rules in conjunction with the renewal requirements, considering the corrosion effects. Normally, these values are in compliance with the survey frequency of an oil tanker, at maximum period of 5 years.

In order to reflect an average overall corrosion wastage for 20 years in service, assuming good maintenance schedules and an effective system for coating protection in ballast tanks, minimum corrosion margins were established and have to be considered in addition to the renewal thickness fixed by the CS for different part of the hull. For instance in the case of deck plating a corrosion margin equal to 1.3 mm (0.065 mm/year X 20 years) for 20 years has been added to CS renewal thickness. Besides considering for plating replacement criteria, these values must also be adopted when verifying hull strength, for which net scantlings of the hull should be considered.

Fatigue Requirements

All of these conversions are using oil tankers older than 20 years which are almost at the end of their operational life as ships and will operate for more 20 years as FPSOs. So, a procedure is settled in order to assure the previous statement in relation to fatigue. The main aspect of concern refers to the previous life of the ship. It would be a hard design task to reproduce the history of the ship trips, including the seastates it encountered over its life as an oil tanker. A better way to follow is to override this drawback and get information from inspection of the ship during the conversion work. Feedback from surveys shows that the main source of plate replacement is corrosion effect. As fatigue is characterized by crack growth over the service period of the structure, results from a set of inspected joints with no fatigue crack detected after 20 years are important information that could even feed a reliability analysis, to help on establishing proper safety factors (SF) for the conversion design.

Ship maintenance workers and the CS inspection surveyor have a relevant idea of the real state of the ship structure, from previous surveys, and can choose the best inspection method to be employed. Fatigue inspection, using non-destructive tests (NDT), of all structural details would be expensive and probably not necessary. A close-up visual inspection of almost all details, together with NDT inspection of identified important ones seems to be an adequate procedure to be followed, complemented, if necessary, by crack tip opening displacement (CTOD) tests. Welded connections of longitudinals to webs or transverse bulkheads, by means of brackets or not, are critical points to be checked. The selected set of joints to be inspected should be improved after the results from previous sets, according to whether any flaws has been detected or not.

The ship was originally designed with an SF for fatigue equal to unity, associated to a design life of 20 years. However, it was implicit that the unit would be docked at intervals of years defined by the CS, inspected at different levels of detail, and, eventually, repaired. Although the FPSO is supposed to be inspected over its service life at specified intervals of time, decreasing its storage capacity during inspection, it will be very difficult to carry out any necessary repair. So, based on engineering judgment and our exeperience, it has been demanded an SF of unity for converted ships to be applied to the previous life fatigue damage, under unrestricted services, and the SFs in Table 4 for the expected damage at the new life as a FPSO, taking advantage of less severe seastates conditions of Campos Basin. The calculated fatigue life, obtained from the total damage, is then compared to the design required service life, around 20 years.

TABLE 4 - Safety factors for fatigue analysis

STRUCTURAL COMPONENTS	SAFETY FACTOR (SF)
Primary elements that transfer load from turret to hull	2.0
Connection between hull and production plant	2.0
Others primary elements of the hull	1.0
Turret/hull interface	4.0
Flare boom joints and connection to the hull	4.0
Non acessible areas	10.0

Fatigue analysis may be carried out through Simplified Fatigue or Full Stochastic Approaches (5). In the first approach, a Weibull distribution is assumed for the long term distribution of stress or force ranges, for which the parameters must be calibrated at representative hot spot or load points. Long term waves data are supplied by the scatter diagram of Hs and Tz, for Campos Basin. However, as no correlation of wind and waves with current have been available on data, some conservative or reasonably simplified assumptions have to be applied to heading directionality. The variating draft during loading/unloading of tanks should be treated as additional long term data, requiring a set of models for motion analyses.

MAIN STRUCTURAL ITEMS DESIGN

Longitudinal Hull Girder Strength

The longitudinal hull girder strength should be verified according to the Vessel Rules of the Classification Society, by calculation of the still water bending moments and shear forces for all envisaged loading patterns, considered at Trim & Stability Booklet and adding the dynamic wave forces. For offloading operations, all loading pattern must comply with trim limitations between 0 m towards bow and 3 m towards stern in normal conditions and between 2 m towards bow and 6 m towards stern in inspection conditions.

As an alternative to CS wave forces formulations, the direct calculation of wave loading acting on the FPSO may also be used, considering a full stochastic analysis or a design wave approach in order to evaluate the maximum wave bending moment and shear forces along the length of the vessel. In this case the centenary site environmental seastate should be considered and the hydrodynamic analysis should be performed based on a 3-D radiation-diffraction method.

Envelope curves for the still-water bending moment and shear forces, taking into account all loading patterns are to be provided. In the case of performing a direct calculation, these envelope curves are to consider the combined wave and still-water forces.

In Vessel Rules, which focus the design of oil tankers to travel around the world, experiencing a wide variety of environmental conditions, the wave loading is a function of the dimension of the vessel. In contrast, a FPSO will generally operate at one location for its whole life with all intermediate stages of cargo, loading/unloading sequences, that may impose higher stresses. In addition, the lightweight will be updated considering all modifications occurred in the ship, such as inclusion of process plant, turret system, helideck and flare. Adopting the Brazilian seastates data, lower wave bending moments and shear forces would be achieved, even with centenary condition, compared to using Vessel Rules values, which may be used to compensate any increase in the still water forces due to loading condition modifications. Nevertheless, this approach does not take automatically into account the Safety Factors included in the Vessel Rules formulations and a previous agreement with the CS must be done.

As the vessel tanks will be inspected during operation, without docking, an additional care should be taken referred to inspection loading conditions in order to maximize cargo capacity and be compatible with the design allowable stresses. According CS, the tanks should be in a condition of "free for men" which means that some adjacent tanks should also be empty and clean, possibly affecting the acting still-water bending moment.

For the longitudinal strength verification, the cross sections properties along the length of the hull are to be calculated considering the actual measured scantlings or the latest CS Reassessment scantlings, taking into account corrosion margins for the service period.

Cargo Region

As mentioned previously Vessel Rules do not consider obligatory a Strength Assessment of the Cargo Region for conversions. However, PETROBRAS documentation required this analysis, performed in compliance with ABS or LRS procedures, established for new hulls, which covers the three criteria: yielding, buckling and fatigue, in order to assure the integrity of the hull structure. The strength analysis, again to be performed using the actual measured or latest CS reassessment scantlings, will help to have more information about structural behavior and critical fatigue details, in conjunction with survey reports.

In addition, as cargo and ballast tanks will operate, most part of the time, partially full, longitudinal and transverse bulkheads scantlings should be verified against the dynamic forces due to sloshing effects.

The critical part of the hull for fatigue is located amidships, for which the simplified approach may be adequate as CS Rules specify values for the Weibull shape parameter for groups of structural details. However, modifications related to the previously mentioned SFs have to be considered to account for previous and new lives.

Turret Region

The turret region comprises the turret shaft, the turret/hull interface and upper turret deck structure. The turret shaft is the interface structure that effectively transmits loads from mooring lines, risers and upper turret deck to the FPSO main structure, through bearing assemblies.

The design loading conditions for the turret shaft and turret supports, generally, correspond to the worst condition from mooring and risers analysis for bearings dimensioning. In addition to those loads, the inertia forces due to FPSO motions, wind, gravity, buoyancy, FPSO deformation and fluid effects loads also have to be considered acting on the turret region.

The bearing supports, turret moonpool and all hull reinforcements take part of turret/hull interface. It is worth mentioning that loads distribution trough turret region is considerably affected by turret diameter dimension compared to ship breadth and depth and loading level, which may define the number of bearing levels and respective types as well as the proper structural model to be adopted. Apart of this, care should be taken relating to deformations and dimensional tolerances of the structure in order to be in accordance with manufacturer bearing limitations.

The stresses acting on the structure may be evaluated by a rational linear frequency domain analysis, comprising the calculation of hydrodynamic wave pressures on the hull through radiation-diffraction theory, then transferred to a 3-D structural FEM model of the region of interest. The refinement of the mesh should be compatible for the desired analysis: yielding, buckling or fatigue.

Regarding fatigue prediction, the simplified fatigue approach have been proved to be suitable for old and new structures around the turret region. The input to FEM model are the most probable maximum values in 20 years obtained from the long term Weibull distribution of force and moment ranges acting on envisaged points, for instance, the bearings, used for stresses evaluation. The shape parameter, in conjunction with the proper SN curve of each detail, define the allowable stresses. As the distributions are obtained by long-term summation of non-linear motion responses of the turret moored system, the parameters must be calculated for the specific system. The procedure involves the determination of short-term response functions of forces to a set of fatigue loading conditions, which are associated to the main environmental parameters, current, wind and waves, and to different drafts. The long-term data for Campos Basin, together with an assumed correlation among the parameters is then introduced in the analysis.

Good results from inspection of few parts of the structure around the turret located at the bow region, normally reinforced for collisions, would allow for a zero damage assumption for previous life, as no fatigue flaws are expected during this period.

New Heavy Structures

The process plant deck, flare boom, helideck and crane supports are some examples of new heavy structures which have some peculiarities in the design procedures. The hull deformations due to still water and wave bending moments as well as FPSO motions are important imposed loads to be considered in those structures, besides functional loads.

Care should be taken relating to hull deformations in process plant decks, which length may reach up to one third (1/3) of the total length of the ship, as these deformations play an important role during design phase. In order to reduce this effect and to avoid the necessity of an integrated hull and process plant deck analysis, the CS requires that the process plant deck should be segmented in parts with at maximum one tenth (1/10) of the total length of the ship. However, we consider that in any case the hull deformation has to be applied in the structure analyses.

The connection with the hull should be extensively investigated with the aim of properly transmit the involved loads as well as for fatigue verification. Deck supports have to be checked against fatigue both due to motions, mainly roll motions, and to hull deformation. For the last case, care should be taken when designing simulated hinges through reduction of inertia in longitudinal direction, as the stress concentration factors may make the flexible connection not adequate for fatigue.

It is important mentioning that the pipe-rack dimensioning, normally located above the process plant deck, will considerably be affected by hull deformations and motions. In addition, for equipment supports, attention should be paid to high accelerations, also imposed by motions, besides equipment vibration.

As a result of layout study complying safety requirement, the best flare location was defined to be at the bow of the ship nearby turret system, using an inclined boom. According to radiation analysis considering the emergency gas flow rate, the boom length could reach values up to 60 m, which represents an additional concern related to high accelerations.

For the fatigue analysis of the flare boom, cyclic forces come from wind dynamics and also from ship motions. A criterion was defined in which a safety factor of 4 is required for the fatigue damage obtained from wind dynamics, to account for both cases.

CONCLUSIONS

The FPSO is a combination of ship traditions and mobile offshore units. The success of this combination depends on how we can extract the best from all activities. This statement is valid for the Owners, shipbuilders and Classification Societies.

On the other hand, the numerous interfaces between ship, turret and process plant may interfere in the design development, requiring an intensive interface control as well as a detailed technical specification in order to not generate several interpretations of the performance requirements.

In the recent BID of unit P-37 for Marlim Field, PETROBRAS did not offer an oil tanker to be converted into FPSO. In addition, the ongoing experience of the design and construction of the five FPSOs, have led to implement some additional requirements, such as exclusion of ship existing piping and steam system. Considering the required service life of 20 years, the ship to be offered to the conversion must not have its first ship register dated before 1973.

PETROBRAS complementary requirements, beyond those issued by CS, concerned to structural subject proved to be a right decision. However some pending questions have to be improved and a clear distinction between the obligatory and desirable items must be done, with the purpose of not generating contractual misunderstandings.

REFERENCES

DNV (Det Norske Veritas) (1984) "Fatigue Strength Analysis for Mobile Offshore Units, " *Classification Notes, No. 30.2.*

Fachetti, MB, Figueiredo, JA, Mattos, DM (1996). "Floating production systems: The PETROBRAS approach," *11th Annual FPS Conference '96*, London.

Markatos, PA, Mahmood, M (1996) "A rational approach to assess the safety and structural requirements for site specific operation of FPSO's," *FPS '96 Conference*, IIR Limited, London.

Mendonça, CE, Fachetti, MB and Mattos, DM (1996). "PETROBRAS challenges in FPSO design," *FPSO Tech '96*, Oslo.

Souza, MJ (1996). "Process plant deck for FPSO - Vessel motions and deformations influence in structural analysis," *MSc Thesis Rio de Janeiro Federal University (UFRJ)*. (in portuguese)

Proceedings of the Seventh (1997) International Offshore and Polar Engineering Conference
Honolulu, USA, May 25-30, 1997

ISBN 1-880653-28-1 (Set); ISBN 1-880653-29-X (Vol. I)

Process Decks for FPSOs - Effects of Ship Motions and Deformation on Structural Analysis

Marcos José de Souza
Petrobras R&D Center
Rio de Janeiro, Brazil

Breno Pinheiro Jacob and Gilberto Bruno Ellwanger
Federal University of Rio de Janeiro
Rio de Janeiro, Brazil

ABSTRACT

This work deals with the influence of ship motion and deformation on the structural design of decks to support production plants in floating production and storage offloading units (FPSOs). These decks are space frames with lengths ranging from 30 to 100 meters and a variable number of rigid supports linked to the ship hull. The decks and ship hull are modeled together for naval and structural analysis. A global coupled analysis is performed considering the deck and ship structures; sea loads are evaluated by means of probabilistic methods.

KEY WORDS

Process decks; FPSOs; structural analysis; ship motion; ship deformation.

INTRODUCTION

Floating Production, Storage and Offloading units - **FPSOs** (Figures 1 and 2) have been recently established as attractive alternatives for the exploitation of offshore oil fields. This fact is due to the versatility and economicity of this new concept, which is especially suited for deep water fields located where a pipeline network is not available, with limited recoverable reserves, and with a small number of wells. Actually, with the increasing development of this new technology, the use of FPSOs has been shown to be adequate and reliable also for the exploitation of larger oil fields.

FPSOs are usually based on barges or existing tankers, converted to its new function. The main advantages of this concept are[1]: **a)** Integration, in one system, of all functions needed for production, storage and oil offloading to a shuttle tanker; **b)** Flexibility in the choice of the type of ship or tanker to be employed; **c)** Ease of installation and relocation; **d)** Large working area, contributing to an easy arrangement of production equipments and facilities; **e)** Simplicity of construction, with several qualified shipyards all over the world; **f)** Tanks with large capacity for storage of the oil production; **g)** Relatively low costs and time lag between design and production phases, especially in the case of converted tankers.

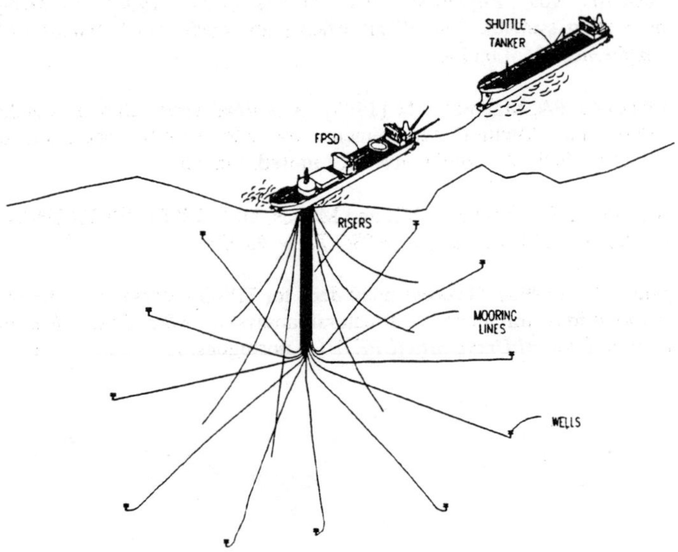

Figure 1 - General View of a FPSO unit

The first known application of a FPSO, in 1977, was located in the Castellon field in Spain[2]. Since then more than 40 fields have been developed in about 15 countries, mostly in the current decade. The first application in Brazil was the process ship P.P. Moraes, installed in the Garoupa field in 1979 and operating until 1993 in the Albacora field[3]. Presently a new process plant is being installed in this ship, which will be relocated at a water depth of 840m in the Barracuda Field. During the next years four more tankers will be converted to

FPSOs, all of them currently belong to the FRONAPE fleet. They have been used for more than 20 years, and present high operation costs that inhibit their continued use for oil transportation.

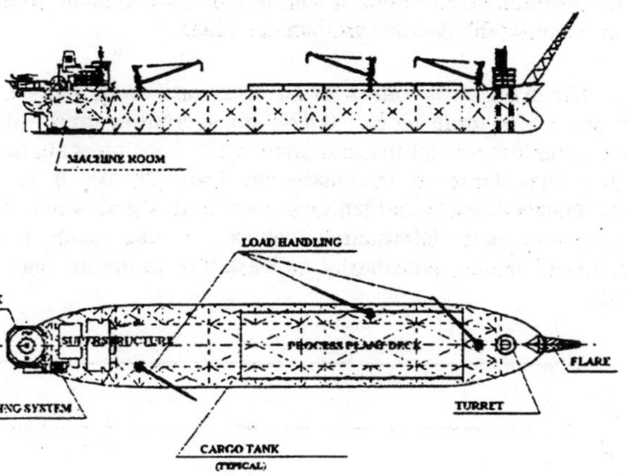

Figure 2 - Schematic representation of a FPSO

The increasing demand for this concept, therefore, motivates the studies presented in this work[4], that include results of coupled analyses for the structural design of decks to support FPSOs. These decks are space frames with lengths ranging from 30 to 100 meters, with a variable number of rigid supports linked to the ship hull. The deck and ship hull are modeled together for naval and structural analysis. Global, coupled analyses are performed considering the deck and ship structures. Sea loads are evaluated by means of probabilistic methods. Results for four different drafts will be compared. Finally, an analysis is performed specifically to check the recommendation of the ABS code[5] regarding the maximum length of one module of a deck.

MODELS

The importance of a coupled analysis encompassing the process decks and the naval and structural model of the ship resides in the precise consideration of the influence of the ship movement and deformation in the design of the process decks. The analyses presented here will consider the maximum resistance of the structural elements of the deck, not including fatigue effects.

The dimensions of the ship considered in the analyses are: length - 320 - meters; width - 54.5 meters; height - 27.8 meters. The ship is represented by two distinct models - a naval and a structural model (Figures 3 and 4). In the former model the geometry of the hull is represented by the strip theory from cross-sections taken along the length of the ship. The structural model is a space frame with a grid topology at ship depth mid level. The frames that support the decks are modeled as space frames, being included in the global model.

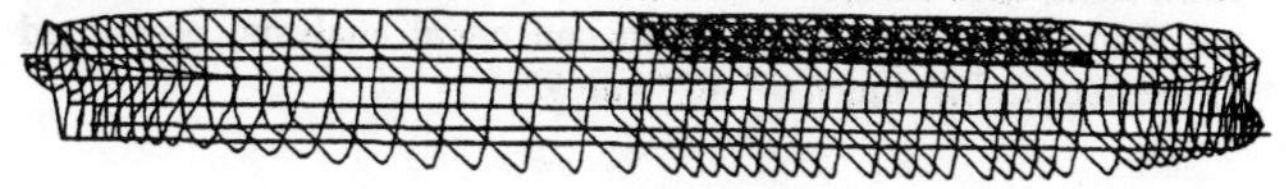

Figure 3 - Naval Model of the Ship-Decks Assembly

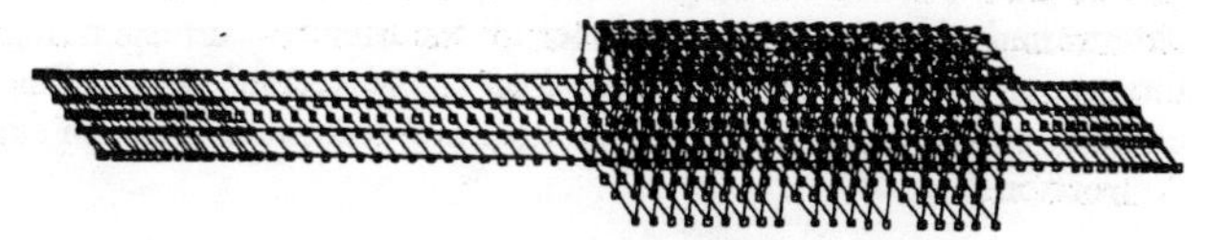

Figure 4 - Structural Model of the Ship-Decks Assembly

LOADINGS

Self-weight and ship loadings are calculated and applied to the structure as distributed loads along its length, and also as concentrated loads. The weight of oil and other fluids is added to the model as loadings on the longitudinal beams. For each wave with unit height and given period, the corresponding hydrostatic and hydrodynamic forces are calculated and applied automatically on the structural nodes of the ship that are associated to each section of the naval model. The inertial loadings are calculated from the masses and movements for each wave and period.

ANALYSES

The ship-decks assembly was first analysed in a static condition, submitted to a loading case (referenced as FRQMEAN) comprised by all vertical loadings and hydrostatic thrust. In this condition the program MOSES[6] automatically determines the equilibrium of the assembly, defining a draft for the ship as a function of the loading

As a result of a dynamic analysis in the frequency domain, the program generates one load case that corresponds to the mean position of the model, and two response amplitude operators ("RAO") load cases for each heading and period; one for real part and one for the imaginary part of the response. The ISSC spectra are considered for the maximum responses, with significant height of 7.8 m, $\cos^2$ spreading type and average periods varying from 5 to 12 seconds at intervals of 0.5 seconds.

The maximum responses are conservatively combined with the FRQMEAN load case, in order to maximize the forces at the element ends. This is accomplished by adding the absolute values and adopting the signs of the FRQMEAN case. The case with reversed FRQMEAN signs is also considered.

The ship-deck system is analysed to four distinct drafts: 7m, 13m, 21m and transport. Inertial forces and fluid pressures acting on the ship hull are obtained from one naval analysis, performed with three distinct headings: 0°, 45° and 90°.

The maximum forces acting on each bar are determined through a spectral calculation for fifteen unidirectional sea states, with significant height of 7.8 m for the 0° and 45° cases and of 3.5 m for the 90° case. The mean periods vary from 5 to 12 seconds on all considered directions, according to the recommendations of the DNV code[7], in order to accurately ascertain the maximum response of the system. Each one of these sea states is combined with a RAO to generate an internal force spectrum response. The maximum values are determined considering the number of occurrences corresponding to a three-hour storm, and mean period of the 10-second waves. For each element of one analysis 45 stress ratios are determined (15 sea states x 3 directions).

The results will be presented initially for the draft of 21 m. Next, follows comparisons of the results for all four considered drafts.

RESULTS FOR 21-M DRAFT

Figure 5 presents the results of the coupled analysis for the fourth deck, for a draft of 21 meters, in terms of the distribution of the number of bars for each interval of maximum stress ratios.

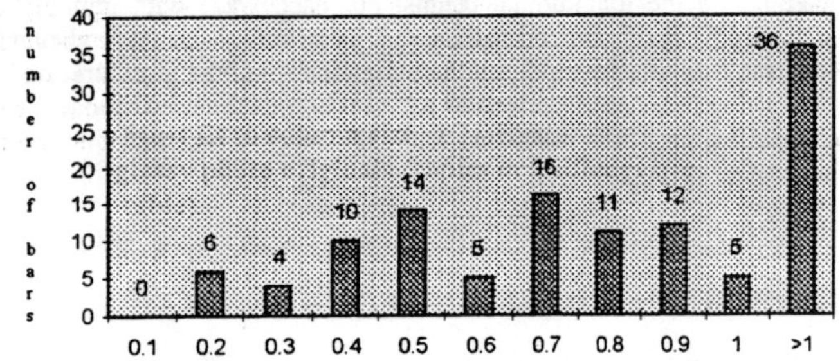

Figure 5 - Distribution of the maximum stress ratios (deck 4)

Table 1 shows the distribution of the number of dimensioned bars for each load direction. It can be observed that the loading case that determines the dimensioning of the majority of the bars is that of the heading at 45°. The relative importance of the heading at 90° is very small, since it determined the dimensioning of only two bars.

Table 1

Direction	Deck 1	Deck 2	Deck 3	Deck 4	TOTAL
0°	21	17	14	8	**60**
45°	52	102	104	110	**368**
90°	0	0	1	1	**2**
TOTAL	73	119	119	119	**430**

The mean stress ratios for this case are: deck 1 - 0.745; deck 2 - 0.764; deck 3 - 0.665; deck 4 - 0.760 .

COMPARISON BETWEEN DRAFTS

The results of the comparative analysis of the four drafts are presented in Figure 6 for the four decks, in terms of the distribution of the maximum stress ratios. It can be observed that the transportation draft is noticeably less severe than the others.

The mean stress ratios for all decks are: draft 7 m - 0.710; draft 13 m - 0.678; draft 21 m - 0.694; transportation draft - 0.613 .It is interesting to note that the minimum draft (7 m) presents mean values higher than those of the maximum draft (21 m). It is, therefore, recommended that more than one draft be analyzed when determining acceleration and deformation values for use with a simplified uncoupled model, considering at least the minimum and maximum drafts.

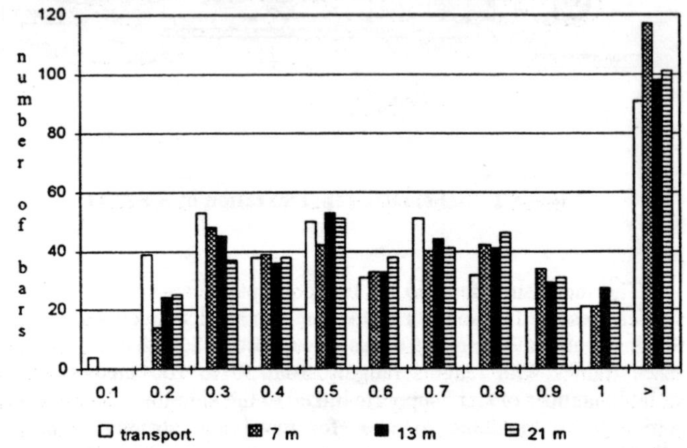

Figure 6 - Distribution of the maximum stress ratios

Considering the influence of the ship linear accelerations in the structural design, we observe that the mean stress ratio obtained without these accelerations is .505 for the 21m draft. This represents that more than 40% of the mean stress ratio is due to the linear accelerations imposed to the ship.

Returning to the dimensioning load cases, Table 2 shows that the heading at 45° still is more important. The observed increase of the number of bars dimensioned for the 90° case and transportation draft is due to the fact that, for this draft, we consider the annual wave for the 0° and 45° cases, and not the centenary wave.

Table 2

Direction	7 m	13 m	21 m	Transport.
0°	58	81	60	60
45°	372	347	368	262
90°	2	2	2	108

VERIFICATION OF THE RECOMMENDATION OF THE ABS

The American Bureau of Shipping - ABS, according to [5], recommends that the length of each module that comprises a process deck does not exceed 10% of the total length of the ship. In order to verify the validity of this recommendation, a comparative analysis is performed employing the coupled model at the draft of 21 m. All four decks are rigidly connected in this analysis. Figure 7 compares the results (in terms of the distribution of the maximum stress ratios for the four decks) with those obtained by the analysis presented in the previous section, where the decks are disconnected.

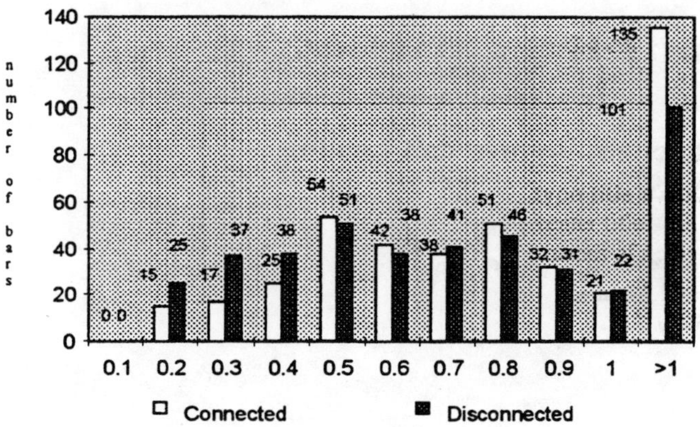

Figure 7 - Maximum Stress Ratios for Connected and Disconnected Decks

Table 3 compares the distribution of the number of dimensioned bars for each load direction, for both "disconnected" and "connected" analyses. It can be observed that the heading 90° is still unimportant for the design. There is, however, a noticeable increase on the number of dimensioned bars by the heading 0° with the "connected" model, from about 14% to about 30% of the total number of bars. This demonstrates the influence of the ship bending deformation on the design of the decks for this case.

Table 3

	deck 1		deck 2		deck 3		deck 4	
direction	dis	con	dis	con	dis	con	dis	con
0°	21	33	17	50	14	31	8	16
45°	52	40	102	69	104	88	110	102
90°	0	0	0	0	1	0	1	1

Table 4 compares the mean stress ratios for both models. The stress ratios for the connected models are always higher. The increase of the stress ratios is more noticeable in decks 1 and 3. Although this increase is not dramatic for all decks, it can be observed that the continuity of the decks is prejudicial to the overall structural behavior of the decks.

Table 4 - Mean Stress Ratios

	deck 1	deck 2	deck 3	deck 4
disconn.	0.691	0.689	0.599	0.808
connected	0.822	0.693	0.700	0.817
variation	+19.0%	+0.6%	+16.9%	+1.1%

FINAL REMARKS AND CONCLUSIONS

The results presented in this work allow some important conclusions to be drawn, regarding the design of process decks for FPSOs. The first conclusion is related to the choice of the significant wave height for the 0° and 45° directions. Employing the same value for both directions leads to the dimensioning of the majority of the bars (more than 80% of the total number) by the heading at 45°. Additional studies must, therefore, be performed in order to verify this assumption, and to possibly determine a reduction of the value of the significant wave height for the 45° direction; or else, a reduction on the heading angle when considering the centenary wave to, say, 22.5°.

Regarding the comparative analyses between the four draft values (7 m, 13 m, 21 m and transport), it has been observed that the results for the maximum draft, which is usually the one employed in an analysis, are not necessarily the most critical. It is recommended, therefore, that more than one draft be considered (at least both minimum and maximum), when determining acceleration and deformation values for use with a simplified uncoupled model.

Finally, the general increase on the stress ratios obtained in the analysis with all decks rigidly connected helped to demonstrate the relevance of the recommendation of the ABS, concerning the maximum length of each module that comprises a process deck. Therefore, this recommendation should, in general, be enforced, unless in special cases where a detailed coupled analysis demonstrates that continuous decks can be employed.

ACKNOWLEDGMENTS

The authors would like to thank Petrobras for clearance to publish this work.

REFERENCES

[1] Cannavacciuolo, A., *FPSO Technology Under Development* - 8Th Deep Offshore Technology, Rio de Janeiro, October 1995.

[2] Key J.W., *History of FPSOs* - Sname, FPSO Technology Symposium, 1993.

[3] Carneiro, P.R.B., *Barracuda Field: New Record for Turret Moored FPSOs* - 8th Deep Offshore Technology, Rio de Janeiro, October 1995.

[4] Souza, M.J., *Process Decks for Floating Production, Storage and Offloading Units - Effects of Ship Motions and Deformations on Structural Analysis* (in Portuguese), M.Sc. Thesis, COPPE/UFRJ, Civil Engineering Department, Rio de Janeiro, 1996.

[5] American Bureau of Shipping (ABS), *Rules for Building and Classing Steel Vessels Part 3 Hull Construction and Equipment*, 1994.

[6] Ultramarine, Inc., *Reference Manual for Moses - Multi Operational Structural Engineering Simulator*, 1994.

[7] Det Norske Veritas (DNV), Classification Notes N-30.5 - *Environmental Conditions and Environmental Loads* - March, 1991.

Proceedings of the Seventh (1997) International Offshore and Polar Engineering Conference
Honolulu, USA, May 25-30, 1997

ISBN 1-880653-28-1 (Set); ISBN 1-880653-29-X (Vol. I)

An Efficient System for Floating Production, Storage and Loading in Ice-Covered Sea

A. Bech, C.M. Bech, K. Syvertsen and T. Vinje
Advanced Production and Loading
Arendal, Norway

ABSTRACT

The feasibility of using an STL system for offshore loading of oil in icy waters is discussed. It is shown that the system is easily adaptable to icy conditions. Since the STL system is easily disconnectable there will be an economical trade off between investments in mooring system, investments in oil-storage capacity, and investment in icebreaking assistance, specially in waters with heavy ridge formation.

KEYWORDS
Offshore Loading, Ice, Floating Production, Submerged Turret

INTRODUCTION

Exploration drilling in the Russian part of the Barents Sea and the Pechora Sea has been quite promising, and plans are under way for development of several oil and gas fields, for instance the Prirazlomnoye field. A critical factor in the development of the oil fields is the export system. Because of the complete lack of infrastructure, and the long distances, export of LPG and oil by tankers to western Europe definitely is an alternative. There are several critical aspects in this context, such as: the offshore loading system, transmission pipelines, the ice management system and suitable tankers.

The present paper will concentrate on the terminal for offshore loading. There definitely are several alternatives: a Single Point Mooring (SPM), a CALM or SALM, direct offloading from a GBS, and a Submerged Turret Loading system (STL). To narrow the scope down further we will concentrate on the STL system in the present discussion. The use of the Submerged Turret Production (STP) in ice will also be briefely discussed.

Chapter 3 will, in some detail, describe the STL system. Chapter 4 describes the ice conditions in the Pechora Sea, which indicates what ice conditions a STL/STP system has to operate in. Chapter 5 describes the results of two series of model tests in ice, which yields the actual forces the STL system has to deal with when exposed to ice. Finally, Chapter 6 describes the modifications that have to made to the STL system to meet the requirements when operating in ice. The consequences for STP will also be commented on in this chapter.

DESCRIPTION OF THE STL/STP SYSTEM

The basic concept is a loading buoy connected to a vessel with an integrated turret. Mooring lines and the loading riser are connected to the buoy. When in operation the STL buoy is pulled into a mating recess in the ship hull and locked. When the vessel is locked to the buoy, the vessel can freely weather-vane. Figure 1 shows the STL Buoy in connected and disconnected condition.

The STL system comprises a subsea STL system and a shipborne STL ship equipment. The subsea system includes :

- the STL buoy with an integrated turret
- a pick-up system
- mooring lines
- riser

The shipborne equipment comprises :

- a mating cone with a locking arrangement
- a loading manifold connected to the tankers cargo piping system or a multi path swivel
- a loading and access compartments
- a pull-in winch and guiding system

The main body of the STL Buoy is built as a welded steel structure consisting of several independent buoyancy chambers. Typical measures for the complete buoy are a maximum diameter of about 7.5 meters and a height between 7 and 12 meters. Variation in size is due to different requirements at various application. The applications may vary from loading terminals in shallow water, to Floating Production Storage and Offloading (FPSO) applications in harsh environment and in deep water.

The interface geometry between the buoys and the mating recess in the tankers is identical for all the various buoy sizes, i.e. the mating recess in the vessel and the locking collar on the buoy are similar for all STL systems. The STL buoy geometry is thus standardized allowing any

STL vessel to connect to any of the installed STL buoys. The geometry is kept identical also for the buoys built for STP application.

The STL Buoy is shown in Figure 2. The figure shows the turret, built as a central vertical shaft with mooring line connections arranged in a turret table at the upper end of the "turret tank".. A lower radial bearing carries the major part of the horizontal mooring forces on the turret, and an axial bearing in the upper part of the shaft carries the vertical forces. The bearings are design with good margins for long term operation. E.g. at the Heidrun STL Buoys the bearing design life is 25 years.

The turret shaft has a central opening allowing an "oil pipe" to be placed inside the shaft. The oil pipe has in its lower end a riser connection which can include a flexible element. The flexible element is introduced to reduce the bending forces in the riser end termination. The upper end of the oil pipe has an arrangement with an in-line swivel and a valve unit of the "dry break" type, as used for standard offshore loading operations in the North Sea. The oil pipe can be replaced without any assisting vessel, a task requiring less than 24 hours. For STP applications there are multiple oil pipes, as well as umbilicals, passing through the swivel. The inner turret shaft diameter will for this purpose be increased. Maximum opening within the limitations of the standard STL/STP Buoy is about 2.7 meter diameter.

The STL mooring system is one of the vital parts in the STL system. For each installation a careful design is made with basis in advanced computer analysis and model tests. The mooring line configuration will depend on several factors, i.e. water depth, environmental conditions, operational philosophy, vessel size, etc.. The configuration for moderate and shallow water depths is a system comprising both wire and chain segments. Wire segments will always be found in the connection to the STL Buoy, thus avoiding high buoyancy requirement to the STL Buoy. In deeper water it has been necessary to introduce Mooring Line Buoyancy Elements (MLBE). Typical for water-depth in excess of 250 meters is a system with two wire segments, one buoyancy element and one chain segment. (See Figure 3). The "bottom" part of the mooring line is the anchor. The STL systems in operation are installed with either pile anchors, drag anchors or suction anchors. The latest installations include suction anchors.

In parallel with the construction and installation of STL systems, a continuous development of the mooring system design is ongoing. In addition to optimisation of mooring systems at North Sea water depths, focus has also been given to installations in areas with other environmental conditions and depths. Mooring system designs have been made for water depths ranging from 25 to 1500 meters. Several model tests have been completed in water depths ranging from 40 meters to 900 meters confirming that the design approach gives feasible and cost effective mooring system both in shallow and deep waters. It has been demonstrated, both by station keeping analyses and by model tests, that the operational limits for the STL system are high. The system can, with a suitably designed mooring system, be operated safely in sea states corresponding to the most severe conditions in the North Sea.

The STL system offers solutions with a high flexibility with respect to operational requirements and with respect to cost. For "low cost operations" at marginal fields both the mooring system and the general arrangement are effectively reduced to a competitive level. The system may be used either as basis for new-built vessels or for the conversion of standard tankers. The size of the mating recess and the loading compartments is small. The conversion of a tanker thus only requires moderate modifications including only standard yard operations.

The STL systems have been continuously developed for various type of applications ranging from the simple shallow water terminals to sophisticated deep water FPSO units. Some of the applications of the STL/STP system being delivered by APL, or being in order at present, are listed below.

- Direct Shuttle Loading (DSL) permits loading of processed crude oil directly onto shuttle tankers without the requirement for oil storage at the field. Two STL loading buoys located adjacent to the production installation allows continuous loading. A DSL system is installed at the Heidrun Field (Conoco). 3 dedicated tankers (newbuilds) operate at the field. The production rate is 250.000 bpd.

- STL as a Floating Storage Unit (FSU) application has been selected for the Fulmar Field (Shell/Esso), the Yme Field (Statoil/Saga) and the Njord Field (Norsk Hydro). The Fulmar and Yme FSU's are converted standard tankers (Aframax and Suezmax). The Njord FSU is a purpose built barge with Aframax size. The mooring systems for the three FSU's are designed for operation in harsh North Sea environments. The STL system allows the FSU units to be disconnected for service and maintenance. A safe rapid disconnection can also be made if the environmental conditions exceed the design conditions or in an emergency situation. The Fulmar FSU has been in operation since 1993. The Yme STL Buoy was installed in 1995, with first oil delivered early 1996. The Njord FSU will be in operation in 1997.

- A Submerged Turret Production (STP) has been developed for the Lufeng Field in the South China Sea. The system is designed to be disconnected during typhoons, but is capable of withstanding the 25 year soliton condition and "cold wave" condition. The system will be installed during the Summer of 1997.

ICE CONDITIONS IN THE PECHORA SEA

For application of STL/STP system in the Arctic it is necessary to get an impression of the ice conditions it has to operate in. One of the actual areas for development at the moment is the Pechora Sea.

The information on the ice conditions in the Pechora Sea is taken from Løset & al (1997). The ice conditions in the area are in general characterized by first year ice, and by icebergs being practically non-existent. The icy period lasts for about 185 days in the western part of the area, and 240 days (in the extreme for 300 days) in the eastern part. The most extensive ice cover is observed in March-April, when 10/10 of the area is covered.

A typical value for the extreme thickness of level ice in the area is 1.6 m, with rafted ice thickness of about 2.5 m. Characteristic values for extreme ridges is a sail height of 4 m, a keel depth of 15 m, and more, with a consolidation depth of about 3.5 m. The drift speed of ice in the area may get as high as 1 m/s, and is mostly due to tidal effects. The flexural strength of the level ice basically varies between 300 kPa and 500 kPa.

MODEL TESTS OF THE STL/STP SYSTEM IN ICE

Good mathematical models for calculation of oscillatory forces on vessels in ice are scarce, and model tests form the most reliable tool for prediction of these forces. This is the reason for performing model tests of the STL system in ice at this stage. Model tests have been performed at the Kværner Masa-Yards ice tank in Helsinki, Finland, and at the Hamburg Ship Model Basin in Germany. In both cases tests were run in level ice and in ridges. Both model basins use real ice for their tests, but introduce different techniques for making the strength of the ice scale properly.

At the Masa-Yard tank the ice is basically pushed against the structure to simulate ice drift, at the Hamburg tank a carriage is moved along the basin, and the ice is stationary. Accordingly, the model, and thus the mooring system, remains stationary at the Kværner Masa-Yard, and in Hamburg the mooring system has to be placed on a false bottom moving with the carriage, as shown in Figure 4. In Hamburg a surface piercing column, connecting the false bottom and the carriage thus has to be introduced. This column is situated in the wake of the vessel.

At the Kværner Masa-Yard the model scale was 1:48.13, and at the Hamburg model basin 1:36. Further more, the models were of different vessels in the two test series. They differed both in terms of full scale length and width, as well as bow form. Both vessels had a bow form of an "ice-going tanker" though. For both test series the "water depth" was 30 m, and the draft of the vessel 15 m. Both mooring systems were eight legged systems. The Kværner Masa-Yard mooring was stiff (but formed a catenary), while the Hamburg basin, to a certain extent, had modelled the flexibility of the lines. With these differences in the modelling one should expect a certain deviation of the test results.

Model Tests in Level Ice

One test run with the vessel head on in level ice was run in both basins under nearly equal conditions, which made it possible to compare results. The comparison is given in Table 1.

As is clearly seen, the results are consistent, and thus lead to a certain confidence in the results from both facilities.

The test referred to above is the only one run with the vessel head on in level ice at the Kværner Masa-Yard. At the Hamburg Model Basin, on the other hand, three series of tests were run with different ice thickness and different ice drift speeds. Unfortunately, during one of the test series (the 4000 series) the test conditions were not perfect, and the results are thus not totally reliable, at least not for two of the three runs. The conditions for the third one were not too bad, and the results are referred to below, but must be regarded with a certain suspicion. The results for the three series of tests are given in Table 2.

The mean force in the direction of the ice drift is the result in the table which is least dependent on local variations in the ice conditions along the basin. The extreme force and, to an even greater extent, the maximum line tension are dependent on both the variation in ice conditions along the track and on the conditions normal to the track.

Lindquist (1989) has developed a semi-empirical model for the (mean) forces acting on a vessel running in level ice. According to this model the dependence of the mean force on the drift velocity is linear. Linear regression analyses of the results from the 1000 and the 2000 test series respectively were performed. With only three points available from each series of runs the results from an regression analysis are not particularly reliable, but the results showed that the relative error of any one of the six points was less than 1.5%, which is promising.

According to Lindquist's model the dependence on the flexural strength and the thickness of the ice should be linear with respect to the flexural strength, and at most quadratic with respect to the ice thickness. The data do definitely not support these approximations. A regression analysis, not taking the dependency on the flexural strength into account (utilising test runs 1030, 2030 and 4040, all with a drift velocity of 0.75 m/s) yields the formula:

$$F_E \approx 1544 \cdot h^{2.82}$$

where F_E is given in kN and the ice thickness, h, in m. The measured and estimated values are compared in Table 3.

F_E seems to fit the data quite well, which should not be surprising in view of the fact that only three points have been used for the regression analysis.

Introducing the assumption that the mean force is written in the form:

$$F_E \approx K \sigma_F \cdot h^{\nu}$$

yield a value for ν of 3.8 and relative error of F_E varying between -18% and + 18% for the three data. From these analyses one can only draw the following conclusion:s

- The test data indicate that the mean force depends on the cube of the thickness of the ice sheet, rather than on the square.
- The scatter of the test data is that large, and the amount of data is that small, that no more specific conclusions can be drawn.

A corresponding analysis for the maximum line tension leads to the following approximation:

$$L_M \approx 1907 \cdot h^{1.59}$$

where again L_M is in kN and h in m. Even though the exponent deviates considerably from the one found for the mean force, the data seemingly fit quite well, as is shown in Table 4.

An extrapolation of L_M, according to the formula given above, to an ice thickness of 2 m yields an extreme line tension of about 6000 kN. An utilisation of the lines of 0.6 relative to minimum breaking load, which is in accordance with the API (1995) recommendations, indicates a wire rope of a moderate diameter of 108 mm for an ice thickness of 2 m.

Tests in Ridges

Tests were performed in ridges both at the Kværner Masa-Yard and at the Hamburg Ship Model Basin. At both test facilities the height of the ridges were about 18 m, with a consolidation depth of 3-4 m. The drift velocity, on the other hand, was different for the two sets of tests. One of the tests series at the Hamburg Ship Model Basin was run with an unrealistically high strength of the ice (flexural strength of 2700 kPa), and has been disregarded in the following. The results for the remaining four tests are given in Table 5.

The flexural strength referred to in the table is the strength of the ice outside the ridge, and thus only acts as an indicator for the actual strength of the ridge. The data from the Hamburg tests reflect that the force depends on the flexural strength, and actually is proportional to it. This is not directly contradicted by the tests at Kværner Masa-Yard, with the two tests run in ice with flexural strengths being rather close to each other.

From the tests in Hamburg it was found that the extreme line tension was between 44% and 47% of the extreme force on the turret. So, estimating the maximum line tension to be about half of the maximum force seems reasonable. Further more, the flexural strength given in the design basis for the Pechora Sea is between 300 kPa and 500 kPa., which is much lower than what was experienced during the tests. It would thus be a reasonable assumption that a maximum line tension of 13,000 kN would be experienced in a ridge of the size indicated above. With an utilisation of 0.6 (API) this corresponds to a minimum breaking load of the lines of some 22,000 kN. This is acceptable for an eight legged system with double wire lines, each with a diameter of 114 mm.

SPECIAL CONSIDERATIONS REGARDING THE STL/STP SYSTEM IN ICE

For the STL system proposed for operation in ice in shallow water, considerations have been made to avoid problems with broken ice that may be present below the bottom of the tanker. In addition to modifications to the buoy itself, modifications also have to be made to the locking system, the mooring line connections, and to the riser connection. Due to shallow water the idle buoy has to rest on the seabed a pedestal also has to be added to the bottom of the buoy.

Locking system

As shown in Figure 2, the upper part of the buoy is equipped with an integrated locking collar. This collar is for North Sea operation designed for a vertical load of about 15.000 kN, and the locking unit on board the vessel contains two locking shafts. This system allows the buoy to be released with loads acting on the shafts. When the buoy is in position inside the turret recess, the locking system is secured with a mechanical over centre mechanism. For operation in ice both the locking collar and the locking shafts will be strengthened, and certain modifications will be made to the locking mechanism, ensuring safe release when extreme forces are acting on the buoy. The main modification is to add a bearing on the supporting clamps along the span of the locking shafts (this is at present a steel to steel contact), which will permit load-transfer when the vertical force exceeds 8,000 kN. In addition the over centre mechanism will be modified to allow release under full loading on the shaft.

STL Buoy hull

For ice operation added strength to, the otherwise robust, body of the buoy will be introduced where required. The turret shaft and turret shaft bearing system is, in the basic design, built for typical load levels of 15,000 kN (dynamic load) both horizontal and vertically. The bearing system will be upgraded to carry loads in excess of 20,000 kN.

As will be discussed in more detail below, the lower part of the turret shaft will be integrated with a buoyant structure. This design is selected to obtain a better transfer of forces into the turret system and to reduce the total forces in the locking system. The added buoyancy will also give an increased flexibility with respect to the mooring system design. Heavier mooring lines, or mooring lines with increased pretension, can be used without increasing the requirements to the pull-in force from the traction-winch during mating of the buoy.

STL Buoy pedestal

An STL Buoy designed for operation in shallow water is equipped with a pedestal that allows the idle buoy to rest on the seabed. The design of this pedestal will depend on several factors, such as: soil conditions, submerged weight of the buoy, sideways forces from current and waves, etc.. For the operation in ice, the pedestal will be designed as a protective structure for the riser connection.

The pedestal design is a cylindrical structure with an open recess to one side. The riser connection and the riser connecting pipe will be located in this recess. The pedestal is either built as a transparent structure with respect to current and waves, or as a plate-coverd structure. In the first case the hydrodynamic forces are reduced compared with the second. In the case of the transparent design ice lumps may enter the framework, but their size will depend on the framework spacing. The plate-covered structure gives a better protection against ice (but will be exposed to higher hydrodymnamic forces).

Buoyancy of the STL Buoy

The buoyancy of the STL Buoy will be tuned for various operational modes:

1. During pick-up and mating, the buoyancy must be large enough to allow use of a standard traction winch for pick-up.
2. The STL buoy, with mooring lines connected, will totally have a negative buoyancy, i.e. the buoy will rest on the seabed. The resting force on the seabed should be relatively small, but still large enough to avoid the buoy to move horizontally when exposed to current and waves. Tilting is not regarded as a problem due to the strong restoring effect of the chain segments closest to the buoy.

The difference between the force against seabed and the pull-in force (amounting to about 600 - 800 kN) is due to the added weight from the mooring lines when lifted from the seabed during pull-in.

Protection of the mooring lines

The mooring lines are connected to padeyes in the outer circumference of the "turret tank". (see Figure 2). To avoid damage to the mooring lines from ice in this area, a chain segment will be introduced closest to the buoy. The mooring chain is a robust element which will not be damaged by ice.

Riser protection

The riser is the most sensitive element of the system. There are two alternatives for protection of the riser. The first is to build the upper section of the riser as a robust steel pipe. The second alternative is to extend the turret pedestal by some distance, and thus lowering the location of the riser connecting.

Steel pipe as upper riser segment

To use a steel pipe as upper riser segment will give a robustness against ice contact. With an arrangement as shown in Figure 6 the flexible part of the riser will start at some distance below the riser connecting point. By lowering the flexible riser, the risk of interference with ice is reduced. The riser connection to the turret is via one (or possibly two) flexible elements (of e.g. OSI type). The flexible element will also reduce impact forces from ice on the steel pipe.

Extended turret pedestal

The second alternative is to use an extended pedestal. The riser connection will then be in the lower part of this pedestal as shown in Figure 5. The riser connecting point is then lowered to a location where the risk for interference with ice is negligible. A flexible element is introduced in addition to the bending stiffener for the connection to the turret. This flexible element will allow the buoy to rotate during disconnection.

Remotely Operated Pickup (ROP)

The pick-up system has to be improved for operation in ice; since no pick-up buoy is now permitted on the surface. For this purpose the Remotely Operated Pickup System (ROP) is designed.

The ROP is basically a ROV with main purpose to locate the STL Buoy, and lock-on to it. The ROP is connected to the STL vessel via the pull-in line. This, in turn, means that the pull-in of the STL Buoy can be done directly after lock-on of the ROP to the buoy.

A prototype has been built and tested in calm water. Offshore testing at the Heidrun field will take place during the spring of 1997.

There are several considerations that have to be made when applying the ROP in ice. The most wounderable component of the system is the ROP umbilical. To avoid infliction of damages to the umbilical the following precautions will be taken:

- To avoid direct contact with the ice, and potential damage to the umbilical, it will be protected externally by a riser carcass or a steel web.

- A quick release mechanism for the umbilical at the ROP-end will ensure that the umbilical can be disconnected and hauled in once the ROP is connected to the STL Buoy. This will minimize the exposure of the umbilical to ice. To ensure the possibility of an emergency disconnection of the ROP from the STL Buoy after the umbilical has been hauled in, a system for remote disconnection will be introduced.

- In principle all forces acting on the ROP will be transferred to the vessel via the pull-in line. In case the umbilical is taking some of this load, or the umbilical is directly exposed to forces from ice, the quick release mechanism on the ROP will automatically be activated, and the umbilical will be hauled in. The ROP has to be hauled in and the umbilical reinstalled before a new approach, and a connection to the STL Buoy, takes place.

- In the unexpected event that an umbilical is damaged by ice, the ROP has to be hauled in for repair. A quick release mechanism at both ends of the umbilical ensures a simple and fast replacement of the damaged umbilical by a spare stored aboard the vessel. The time spent on the replacement should not be significantly longer than the time for reinstallation of an umbilical released because of ice forces acting on it.

The ROP hull has to be designed to withstand direct loads from ice. The external cover will have a minimum of appendices and has to have a rounded slender form to avoid being locked within the ice. The pull-in line has to be strong enough to withstand ice forces on the buoy during pull-in.

The pull-in line can easily be exposed to ice. To avoid wear of a polyester pull-in line it could be covered by a steel web on the outside, or better, the polyester line could be replaced by a steel wire rope.

The STP system in ice

At a first glance the STP system does not differ much from the STL system, which to a certain extent is correct. One of the important difference between the two systems is the number of risers/umbilicals, with several for the STP system, versus the single riser for the STL system. The number of risers/umbilicals (usually four or more) creates requirement for a much more complex swivel than is the case for the STL system. This swivel is situated inside the STP vessel, and is automatically protected from ice. The requirement for quick release, which is a more important issue for STP system in ice than in open water, has been met by the recent development of a swivel with quick release couplings.

Another difference between the two systems is the much higher internal pressure in the STP risers than in the STL riser. This does not introduce any fundamental difference between the two systems regarding the approach to protecting the risers against ice, only makes it more important to protect the risers; the principals for protection of the risers is the same.

All in all, the STP system can be adapted to operation in icy waters just as easily as the STL system.

CONCLUSIONS

It is clear from the discussion given above that the STL system can easily be adapted to operation in icy waters:

- The forces from drifting level ice and non-extreme ridges can easily be absorbed by mooring systems formed by standard size mooring lines made out of wires and chain. Ice management is required for the STL system in areas with heavy formation of large ridges.
- Protection of the risers from ice is done partly by special design of the STL Buoy to protect the upper part of the riser and partly by redesigning the part of the riser closest to the Buoy. Protection of the mooring lines is done simply by introducing chain in the parts of the lines that are exposed to ice.
- The STL Buoy and its locking mechanisms inside the STL vessel must be strengthened, which is a straight forward engineering task.
- The pick-up of the STL Buoy in ice is performed by a Remote Operated Pick-up (ROP), which is a ROV launched from the STL vessel with a pull-in line and connects to the Buoy.

- The STL vessel operating in ice has to be designed for the task. At present there is development of different kinds of such vessels going on, for instance the Double Acting Tanker (DAT) from Kærner Masa Yard. It is worth noticing that the standardization of the STL Buoys allows the fields to be serviced by standard tankers during the non-icy season.

Table 1 : Comparison of test results in level ice from the Hamburg Ship Model Basin and from Kværner Masa-Yard.

	Hamburg Ship Model Basin	Kværner Masa-Yard
Model Scale	1:36	1:48.13
Ship's width (full scale)	39 m	40 m
Vessel Draft	15 m	15 m
Ship's length	235 m	263 m
"Water Depth"	30 m	30 m
Ice Drift Velocity	0.75 m/s	0.75 m/s
Ice Thickness	1.0 m	1.0 m
Flexural Strength	900 kPa	900 kPa
Total Force in x-Direction		
Mean Value	1520 kN	1240 kN
Standard Deviation	760 kN	---
Maximum	4080 kN	4030 kN

Table 2 : Test results with vessel head on in level ice.

Test No.	Ice Thickness	Flexural Strength	Speed	Mean Force in x-Dir.	Max. Force in x-Dir.	Max. Line Tension
1010	1.0 m	1260 kPa	0.3 m/s	960 kN	3090 kN	1620 kN
1020	1.0 m	1260 kPa	0.45 m/s	1150 kN	2850 kN	1570 kN
1030	1.0 m	1260 kPa	0.75 m/s	1520 kN	4080 kN	1920 kN
2010	1.5 m	720 kPa	0.3 m/s	3960 kN	7900 kN	3410 kN
2020	1.5 m	720 kPa	0.45 m/s	4210 kN	7200 kN	3210 kN
2030	1.5 m	720 kPa	0.75 m/s	5010 kN	8000 kN	3510 kN
4040	1.6 m	900 kPa	0.75 m/s	5540 kN	12500 kN	4140 kN

Table 3 : Results from regression analysis for the mean force

h	1 m	1.5 m	1.6 m
F	1520 kN	5010 kN	5540 kN
F_E	1544 kN	4844 kN	5811 kN
$(F_E - F) / F$	+ 1.6%	- 3.3%	+ 4.9%

Table 4 : Results from regression analysis of the extreme line tension

h	1 m	1.5 m	1.6 m
L	1920 kN	3510 kN	4140 kN
L_M	1907 kN	3634 kN	4026 kN
$(L_M - L) / L$	- 0.7%	+ 3.5%	- 2.8%

Table 5 : Results from tests in ridges at the two model basins

Basin	Hamburg Ship Model		Kværner Masa-Yard	
Height of Ridge	17 m	18 m	17 m	20 m
Drift Velocity	0.3 m/s	0.3 m/s	0.75 m/s	0.75 m/s
Flexural Strength	1260 kPa	900 kPa	1080 kPa	1180 kPa
Maximum Force	38,000 kN	27,500 kN	28,000 kN	28,000 kN

REFERENCES

API (1995): Recommended Practice of Design and Analysis of Seakeeping Systems for Floating Structures, API RP 2SK, 1995

Lindquist, G. (1989): "A straight forward method for calculation of ice resistance of ships", *10th Int. Conf. on Port and Ocean Eng.* (POAC), Luleå, 1989

S.Løset, K. Shkhinek, P. Strass, O.T. Gudmestad, E.B. Michalenko and T. Kärnä (1997): "Ice Conditions in the Barents and Kara Sea", *16th OMAE Conf.*, Yokohama, 1997.

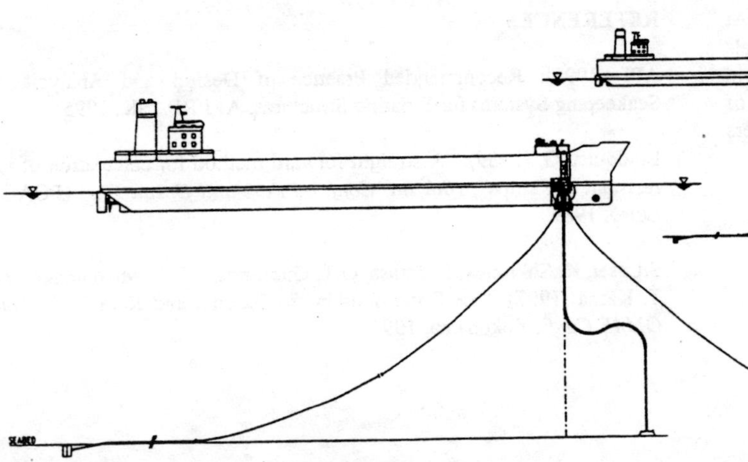

Figure 1. The STL System. Connected and Disconnected

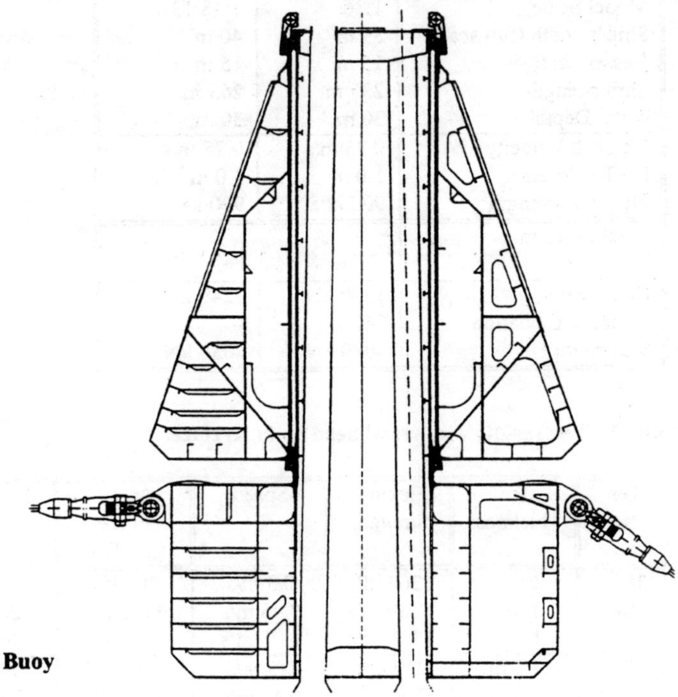

Figure 2. The STL/STP Buoy

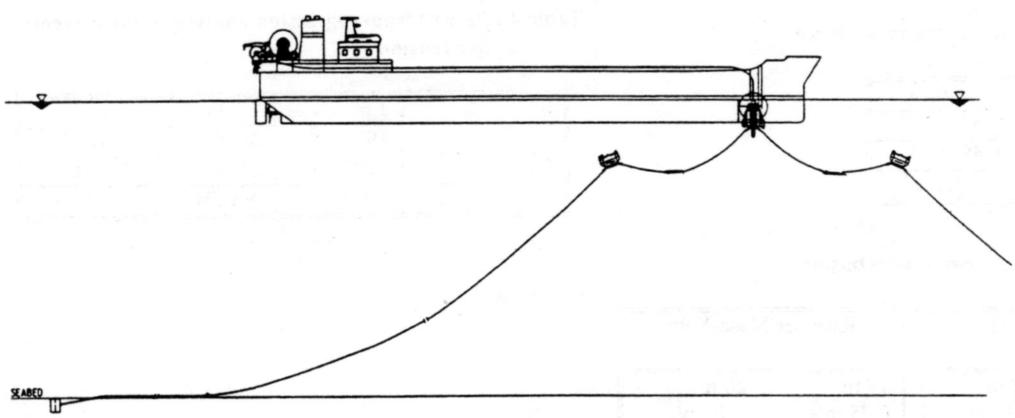

Figure 3. The STL/STP mooring for water depths over 250 m.

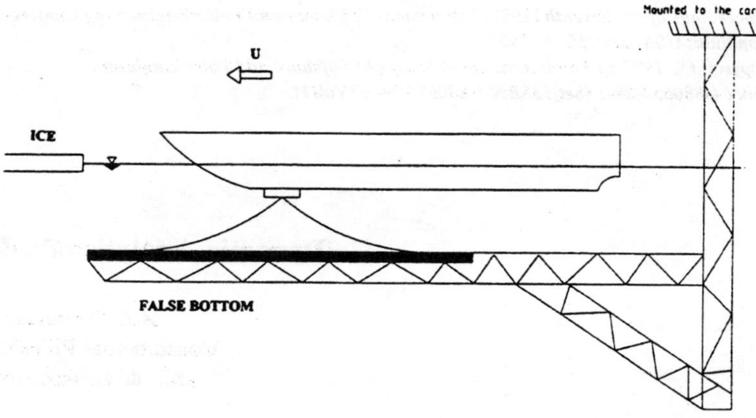

Figure 4. Arrangement of false bottom at Hamburg Ship Model Basin

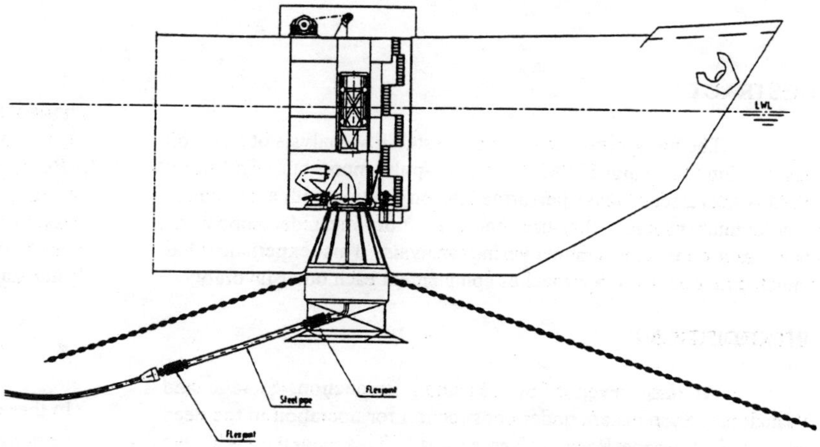

Figure 5. The STL Buoy, including the upper part of the riser and mooring lines, when connected to the tanker.

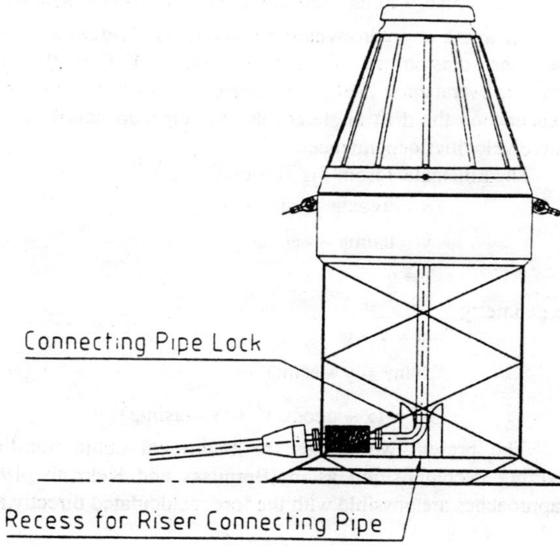

Figure 6. The extended turret pedestal.

Proceedings of the Seventh (1997) International Offshore and Polar Engineering Conference
Honolulu, USA, May 25-30, 1997

ISBN 1-880653-28-1 (Set); ISBN 1-880653-29-X (Vol. I)

Dynamic Analysis of a FPSO System

A.C. Fernandes
Consultant at PETROBRAS
Rio de Janeiro, Brazil

S. Sphaier
Federal University of Rio de Janeiro
Rio de Janeiro, Brazil

ABSTRACT

The work presents a linearized stability analysis of ships on Single Point Moorings (SPM). The fixed-pole, mono-buoy and turret SPM is addressed. This is performed in conjunction with a non-linear time-domain program that uses the cubic non-linear decomposition from traditional ship maneuvering analysis. This experience has indicated that the two approaches complement each other favorably.

INTRODUCTION

At least seven FPSO (Floating Production, Storage and Offloading) Systems are under construction for operation in the deep waters of Campos Basin offshore Brazil. This, together with the conventional mono-buoy network for oil distribution along the remaining Brazilian Coast is enough motivation to discuss the theories to explain the horizontal plane slow Dynamics of large ships on a Single Point Mooring (SPM). See (Carneiro, 1995) for a Campos Basin case description.

Besides the application of commercially available softwares, PETROBRAS has been developing with key Brazilian Universities and Consultancy, new programs and new theoretical approaches. Besides that, a crew training program is under development with the recently operational CIAGA Ship Maneuvering Simulator. For general purposes, model testing, full-scale measurements are also been made.

The work here presented is inside this context, trying to combine two aspects: the time-domain simulation and the linear dynamic stability assessment. In what follows, the equations of motion are presented and three SPM types: the fixed-pole, the mono-buoy and the turret are addressed subsequently. In the end conclusion are presented, indicating a methodology to be applied in design and analysis.

EQUATIONS OF MOTION

The equations of motion (1), (2) and (3) that represent the horizontal behavior of a ship attached to a SPM fixed-pole are given below and refer to Fig. 1. They agree in several references such as (Bernitsas and Kekridis, 1985). Here, only a constant direction and intensity current is the only environment component. This is usual in basic dynamic stability analysis, whose intention is to discover the most critical parameters. The wind and waves may be considered in later stages.

$$(m - X_{\dot{u}})\dot{u} - mrv = X(u,v,r) + T\cos\omega \qquad (1)$$

$$(m - Y_{\dot{v}})\dot{v} + mru - Y_{\dot{r}}\dot{r} = Y(u,v,r) - T\sin\omega \qquad (2)$$

$$-N_{\dot{v}}\dot{v} + (I_z - N_{\dot{r}})\dot{r} = N(u,v,r) - Ta\sin\omega \qquad (3)$$

In these equations, m is the mass of the vessel; u, v, r are the relative velocity of the vessel with respect to the water; $X_{\dot{u}}$ and $Y_{\dot{v}}$ are the surge and sway added masses; $N_{\dot{r}}$ is the yaw added inertia; $Y_{\dot{r}}$ and $N_{\dot{v}}$ are the coupled yaw-sway and sway-yaw added masses; a is the distance between the hawser fairlead and the center of gravity of the vessel, both at the center plane and the ship center line; $X(u,v,r)$, $Y(u,v,r)$ and $N(u,v,r)$ are the hydrodynamic forces that will be expressed in terms of hydrodynamic derivatives (H-derivatives); T expresses the tension in the SPM hawser. Differently from (Fernandes and Aratanha, 1996), the current forces on the vessel which depend on the drift angle are left to come up naturally from the relative velocities dependence.

The kinematics from Fig. 1 yields

$$\dot{x} = u\cos\psi - v\sin\psi - U$$

$$\dot{y} = u\sin\psi - v\cos\psi$$

$$r = \dot{\psi}\,.$$

and the geometry

$$\omega = \gamma + \psi$$

$$\ell\sin\gamma = y + a\sin\psi$$

$$\ell^2 = (x + a\cos\psi)^2 + (y + a\sin\psi)^2$$

The present work uses the traditional cubic non-linear maneuvering decomposition as in (Bernitsas and Kekridis, 1985). Other approaches are possible with the forces calculated directly as in

Other approaches are possible with the forces calculated directly as in (Marin, 1994) and even some that uses the slow maneuveing coefficients as in (Nishimoto, Brinati and Fucatu, 1995). For any approach however, the state-of-the-art requires several empirical coefficients that are easily not obtained. For super tankers, or VLCCs (Very Large Crude Carrier), on the other hand, the open literature seems generous, as in (Abkowitz, 1981) and (Marin, 1993). The cubic decomposition requires the following H-derivatives as in (Bernitsas and Kekridis, 1985).

$$X = X_u u + \frac{1}{2}X_{uu}u^2 + \frac{1}{6}X_{uuu}u^3 + \frac{1}{2}X_{vv}v^2 + \frac{1}{2}X_{rr}r^2 + \frac{1}{2}X_{vvu}v^2u + \frac{1}{2}X_{rru}r^2u + \frac{1}{2}X_{vr}vr + X_{vru}vru \quad (4)$$

$$Y = Y_u u + \frac{1}{2}Y_{uu}u^2 + Y_v v + \frac{1}{6}Y_{vvv}v^3 + \frac{1}{2}Y_{vrr}vr^2 + Y_{vu}vu + \frac{1}{2}Y_{vuu}vu^2 + Y_r r + \frac{1}{6}Y_{rrr}r^3 + \frac{1}{2}Y_{rvv}rv^2 + Y_{ru}ru + \frac{1}{2}Y_{uru}ru^2 \quad (5)$$

$$N = N_u u + \frac{1}{2}N_{uu}u^2 + N_v v + \frac{1}{6}N_{vvv}v^3 + \frac{1}{2}N_{vrr}vr^2 + N_{vu}vu + \frac{1}{2}N_{vuu}vu^2 + N_r r + \frac{1}{6}N_{rrr}r^3 + \frac{1}{2}N_{rvv}rv^2 + N_{ru}ru + \frac{1}{2}N_{uru}ru^2 \quad (6)$$

where the indices represent partial derivatives.

A non-linear time-domain program has been programmed for the solution of the above equations. This program uses a Runge-Kutta fourth order integration scheme.

SPM - FIXED POLE

The classical linearized stability analysis requires the linearized equations of motion. For the symmetric case (ship and environment) it can be shown from Eq. (1), (2) and (3) and kinematics properties that the longitudinal equation uncouples from the lateral motion equations that remain coupled. These last ones may be written as

$$(m - Y_{\dot v})\ddot y - Y_{\dot r}\ddot\psi - Y_v\dot y + (Y_{\dot v}U + Y_r)\dot\psi + \frac{T}{\ell}y + [Y_vU + T(1+\frac{a}{\ell})]\psi = 0 \quad (7)$$

$$-N_{\dot v}\ddot y + (I_z - N_{\dot r})\ddot\psi - N_v\dot y + (N_{\dot v}U - N_r)\dot\psi + T\frac{a}{\ell}y + [N_vU + Ta(1+\frac{a}{\ell})]\psi = 0 \quad (8)$$

These equation includes some inertial terms (proportional to U in the velocity coefficients) that have been neglected in (Fernandes and Aratanha, 1996). The H-derivatives with respect to v may be expressed in terms of derivatives of the drift angles of the external current forces as discussed in the last reference. Equations in the form (7) and (8) are however more usual. It is important to realize that these equations are implied in (Bernitsas and Kekridis, 1985) and in (Garza-Rios and Bernitsas, 1995). The former studies the Ship Towing problem, while the latter studies the more general Spread Mooring case. Due to direct application in real cases as cited in the Introduction, the present work focus the attention on the SPM case, with the linear stability receiving a closer look as shown next.

From (7) and (8) the following characteristic equation may be written:

$$a_0\sigma^4 + a_1\sigma^3 + a_2\sigma^2 + a_3\sigma + a_4 = 0 \quad (9)$$

where

$$a_0 = (m - Y_{\dot v})(I_z - N_{\dot r}) - Y_{\dot r}N_{\dot v} \quad (10)$$

$$a_1 = -(m - Y_{\dot v})N_r - (I_z - N_{\dot r})Y_v - Y_{\dot r}N_v - Y_rN_{\dot v} + mUN_{\dot v} \quad (11)$$

$$a_2 = a_{20} + a_{21}T \quad (12)$$

$$a_{20} = Y_vN_r - Y_rN_v + mUN_v \quad (13)$$

$$a_{21} = a_{210} + \frac{a_{211}}{\ell / a} \quad (14)$$

$$a_{210} = (m - Y_{\dot v})a + N_{\dot v} \quad (15)$$

$$a_{211} = (m - Y_{\dot v})a + (I_z - N_{\dot r})/a + Y_{\dot r} + N_{\dot v} \quad (16)$$

$$a_3 = a_{31}T \quad (17)$$

$$a_{31} = a_{310} + \frac{a_{311}}{\ell / a} \quad (18)$$

$$a_{310} = -Y_va + N_v \quad (19)$$

$$a_{311} = -Y_{\dot v}U + N_{\dot v}U/a - Y_va - N_r/a + Y_r + N_v \quad (20)$$

$$a_4 = a_{41}T \quad (21)$$

$$a_{41} = \frac{a_{411}}{\ell / a} \quad (22)$$

$$a_{411} = (-Y_va + N_v)U/a \quad (23)$$

In the last coefficients the dependence on ℓ / a and T has been separated in order to verify the consequences in the stability criteria. The present work applies the Routh-Hurwitz necessary and sufficient (NS) criteria as shown in (Kuo, 1975). These conditions require that the Hurwitz determinants should all be greater than zero:

NS1:

$$D_1 = a_1 > 0 \quad (24)$$

NS2:

$$D_2 = a_1a_2 - a_0a_3 > 0 \quad (25)$$

NS3:

$$D_3 = D_2a_3 - a_1^2a_4 > 0 \quad (26)$$

NS4:

$$D_4 = D_3a_4 > 0 \quad (27)$$

There are also necessary conditions, requiring that all the coefficients of Eq. (9) to have the same sign, but when they do not coincide with an NS condition they are weaker.

For a conventional ship the NS1 is usually verified, since the virtual mass terms are large and positive and Y_v and N_r are large and negative (see Eq. (11)). On the other hand, provided NS3 is verified, the condition NS4 is fulfilled when $a_4 > 0$. This is a Static Condition that may be obtained by applying the Virtual Work Principle (Fernandes and Aratanha, 1996) and result in the following inequality:

$$a > \frac{N_v}{Y_v} \quad (28)$$

The other most significant condition is NS3 that may be considered a Dynamic Condition. The three independent variables other than the H-derivatives are T, ℓ and a. However, due to operational reasons, usually the hawser fairlead is fixed at the bow, so that,

$$\frac{a}{L} \cong 0.5$$

Hence the present work tries to express the Dynamic Stability Condition in terms of T and ℓ, constructing the T- ℓ map.

From equations above, NS3 may be written as

NS3:

$$\alpha T+\beta>0 \tag{29}$$

where

$$\alpha = A+\frac{B}{\ell / a}+\frac{C}{(\ell / a)^2} \tag{30}$$

$$\beta = D+\frac{E}{\ell / a} \tag{31}$$

$$A = a_1 a_{210} a_{310} - a_0 a_{310}^2 \tag{32}$$

$$B = a_1 a_{210} a_{311} + a_1 a_{211} a_{310} - 2a_0 a_{310} a_{311} \tag{33}$$

$$C = a_1 a_{211} a_{311} - a_0 a_{411}^2 \tag{34}$$

$$D = a_1 a_{20} a_{310} \tag{35}$$

$$E = a_1 a_{20} a_{311} - a_1^2 a_{411} \tag{36}$$

If, as usual,

$$\alpha>0 \tag{37}$$

then, for stability

NS3:

$$T>-\frac{\beta}{\alpha} \tag{38}$$

If

$$\alpha<0 \tag{39}$$

then, any T>0 will lead to stability, or in other words NS3 is not active.

The same analysis has been made in the context of ship towing by (Bernitsas and Kekridis, 1985).

Using coefficients from a VLCC (Abkowitz, 1981) and the model 80 from (Eda and Crane, 1995), one may construct Fig. 2 the T- ℓ map. In this figure, the variables are made non-dimensional by L (ship length), T (ship draft), ρ (water density) and U as explained in the figure captions. This figure indicates the dynamic stability limit of each ship. It is also important to assess the static stability (28).

From (Abkowitz, 1981), in non-dimensional terms as in the Fig. 2, $Y_v = -0.386$ and $N_v = -0.155$ leading to the following condition for static stability

$$\frac{a}{L} > 0.402$$

On the other hand from (Eda and Crane, 1965), the model 80, with $Y_v = -0.354$ and $N_v = -0.0864$, requires

$$\frac{a}{L} > 0.244$$

This implies that the static stability requirement is stronger for the VLCC.

For Dynamic Stability that difference is not significant for smaller ℓ / a as shown in Fig. 2. However, the requirement for larger ℓ / a are again stronger for the VLCC. Therefore, the VLCC is also dynamically less stable.

The construction of the T- ℓ map is an easy task since only the knowledge of the linear H-derivatives are required and algebraic relations. A point above the limiting curve of a specific ship means stability. Another ship means other limit curve. It is important, on the other hand, to understand how the curves relates with time-domain results for the SPM analysis. For that reason, the more stable model 80 (Eda and Crane, 1965) is studied a little closer with the cited time-domain program. The results of the processing of three conditions are indicated in three T- ℓ maps, in Fig. (3), (4) and (5). Each curve corresponds to a different X_{uu} resistance coefficients (see Eq. (4)). In these figures, the variation of the hawser tension in the last range of the time series simulation is included for two nominal cable length (50 m and 150 m). The cable length variation is small leading to an almost vertical trace.

Observing Fig. (3), (4) and (5), for the cable with 50 m length one may notice that as the resistance coefficients increase, the limit cycle becomes less open, eventually becoming a point in Fig. (5), consistently inside the stable region. For 150 m length cable, despite less open the limit cycle is still the behavior. In this case, the average tension, again consistently, is inside the unstable region.

The conclusion is probably that there is no hope to get a stable situation if the average is with the unstable region of the T- ℓ map.

SPM - MONO-BUOY

As the water depth is increased, a fixed-pole tends to be replaced by compliant structures and one of them is the mono-buoy, usually with conventional mooring (catenary). These mono-buoy may be as stiff as the hawser.

However, as the water depth continues to increase, it is not possible to keep the horizontal restoring capability since due to excessive weight a large portion of the catenary is replaced by steel cables. In this case, the mono-buoy has to displace much more horizontally to yield the same required force.

Together with equations (1), (2) and (3), the following equations of motion, considering the mono-buoy as a material point (mass m_b) should be considered:

$$(m_b - X_{b\dot{u}})\ddot{x}_b = X_b(u_b, v_b) + T\cos\gamma + R\sin\beta \tag{40}$$

$$(m_b - Y_{b\dot{v}})\ddot{y}_b = Y_b(u_b, v_b) - T\sin\gamma - R\sin\beta \tag{41}$$

where the same nomenclature of Fig. 1 is kept with respect to point O now assumed to be the mono-buoy position. The force R is the restoring force due to the mono-buoy displacement assumed to be always pointing to the static equilibrium and making an angle β with the horizontal and measured like the angle γ. The hydrodynamic force due to the environment and relative motion on the mono-buoy are X_b and Y_b.

In principle, there is a necessity of two more degrees of freedom. A simplification though is possible since the mono-buoy mass is much smaller than the ship mass that enters into the equations (40) and (41) via T. It is also reasonable to assume that X_b and Y_b are small again compared to the ship inertia and resistance. These considerations lead to

$$T \cong R \tag{42}$$

that is the hawser tension is transmitted entirely to the mono-buoy which characterizes a two spring in series.

However, one should not forget that the mono-buoy displacement may be high and may influence significantly the dynamic stability. Hence a new equivalent tension (T_e) will have to be considered with an equivalent length (ℓ_e), such that

$$\ell_e = \ell_h + \ell_b \tag{43}$$

where the right hand side subscripts are referred to the hawser and mono-buoy respectively.

Assuming that both the hawser and the mono-buoy restoring characteristics may be represented by a sequence of straight lines, it is not difficult to conclude that the equivalent spring will also be represented in this manner, with the equivalent spring constant (k_e) such that

$$\frac{1}{k_e} = \frac{1}{k_h} + \frac{1}{k_b} \tag{44}$$

valid for well defined intervals related to the intervals where the hawser (k_h) and the mono-buoy (k_b) spring constants are defined.

With these considerations, T- ℓ maps like the one shown in Fig. 2 may still be used, provided the T_e and (important!) ℓ_e are used.

SPM - TURRET

For the Turret (see Fig. 6), there is no cable but with correct definition of T, the equations of motion (1), (2) and (3) may still be used, once it is assumed a linear vertically axisymmetric mooring restoring force, such that

$$T = kd \tag{45}$$

Here $d \geq 0$ is the distance between the turret static equilibrium position and the instantaneous dynamic position. The Fig. 6 geometry leads to

$$T\sin\omega = k(y\cos\psi - x\sin\psi) \tag{45}$$

$$T\cos\omega = k(-x\cos\psi - y\sin\psi - a) \tag{46}$$

that together with (2) and (3) will lead again to a longitudinally uncoupled equations for the symmetric case;

$$(m - Y_{\dot v})\ddot y - Y_{\dot r}\ddot\psi - Y_v\dot y + (Y_{\dot v}U + Y_r)\dot\psi + ky + [Y_vU + ka]\psi = 0 \tag{48}$$

$$-N_{\dot v}\ddot y + (I_z - N_{\dot r})\ddot\psi - N_v\dot y + (N_{\dot v}U - N_r)\dot\psi + kay + [N_vU + ka^2]\psi = 0 \tag{49}$$

Again, different from (Fernandes and Aratanha, 1996), here, the inertial terms proportional to the velocity are not neglected and the derivative with respect to the drift angle are replaced by the derivatives with respect to the relative velocities. The same characteristic equation (9) may be used for the stability analysis, together with the coefficients a_0 (Eq. (10)), a_1 (11), a_2 (12) and a_{20} (13). For the Turret specific case,

$$a_{21} = (I_z - N_{\dot r}) + (Y_{\dot r} + N_{\dot v})a + (m - Y_{\dot v})a^2 \tag{50}$$

$$a_3 = a_{31}k \tag{51}$$

$$a_{31} = (N_{\dot v}U - N_r) + (-Y_{\dot v}U + Y_r + N_v)a - Y_va^2 \tag{52}$$

$$a_4 = a_{41}k \tag{53}$$

$$a_{41} = (-Y_va + N_v)U / a \tag{54}$$

Analogously as before, the NS Routh-Hurwitz criteria are applied. The results are again expressed by (24), (25), (26) and (27). Note the Static Condition is, as expected, corresponding to $a_4 > 0$, the same for the SPM - fixed pole/mono-buoy case, inequality (28). The predominant Dynamic Condition is NS3 and has to be verified for large a. In Fig. 7, using the (Abkowitz, 1981) VLCC coefficients a k-a map is constructed, indicating the minimum k for dynamic stability. The equation that express k in terms of a is similar to (37) so that

$$k > -\frac{B}{A} \tag{55}$$

since, also in this case usually, $A > 0$. The variables in Eq. (55) are such that

$$A = a_{31}(a_1a_{21} - a_0a_{31}) \tag{56}$$

$$B = a_1(a_{20}a_{31} - a_1a_{41}) \tag{57}$$

where the most recent definitions of the coefficients for the turret should be used. The right hand side of Eq. (55) will lead to a polynomial ratio (second degree over fourth degree) in terms of a. A concrete case is observed in Fig. 7.

For a fixed a, Fig. 7 yields the limiting k for stability. For a given k and a, the SPM in a turret will be stable for currents such that

$$U < \sqrt{\frac{k}{k_{lim}\frac{1}{2}\rho T}} \tag{58}$$

Considering for instance k=15kN/m and a/L=0.45, that seems typical for deep water FPSOs with VLCCs, and estimating $k_{lim} \cong 0.15$ from Fig. 7, the conclusion, using (58) is that the system is stable for current velocities up to 3 m/s.

A Figure like 7 may be used in conjunction with time domain programs producing figures similar to (3), (4) and (5). Although already available this is not performed here for conciseness.

CONCLUSIONS

The SPM analysis should not be made only with time domain programs. The linear Routh-Hurwitz criteria since straight forward, based only on algebraic calculations can always be performed. For SPM with hawser on fixed-pole or mono-buoys, the T- ℓ map seems to be the right frame work for stability assessment, since normally, the hawser fairlead is fixed at the bow.

Whether the average tension is located in a stable T- ℓ region or not seems to be a valuable information for a time-series interpretation.

It may also be concluded that the mono-buoy may use the fixed-pole results provided both equivalent tension and length are used.

For the SPM-turret, the k-a map is helpful for design and analysis, yielding elements to choose the minimum mooring line k for a stable operation in a certain environment.

Finally, based on order of magnitude of the examples and the observed consistency, it may be concluded that the cubic maneuvering model used in the computer program may be considered a valuable option with immediate application to the FPSOs that are being installed in Campos Basin this year.

ACKNOWLEDGMENTS

Naturally, PETROBRAS should be greatly acknowledged for providing so many opportunities. Eng. Sylvio H. S. C. da Silva is remembered for the direct incentive received from him within the effort to train PETROBRAS crew in the Brazilian Navy CIAGA Simulator.

REFERENCES

Abkowitz,M.A., 1981, "Measurement of Hydrodynamic Characteristics from Ship Maneuvering Trials by System Identification", Trans. SNAME, Vol. 89, pp 283-318.

Bernitsas,M.M. and Kekridis,N.S., 1985, "Simulation and Stability of Ship Towing", Int. Shipbuilding. Progress, Vol. 32, No. 369, May.

Carneiro,P.R.B., 1995 "Barracuda Field: New Records for Turret Moored FPSO", Deep Offshore Technology Conference, Rio de Janeiro, Nov..

Eda,H. and Crane Jr.,L., 1965 "Steering Characteristics of Ships in Calm Water and Waves", Trans. SNAME, Vol. 73, pp 135-177.

Fernandes,A.C. and Aratanha,M., 1996, "Classical Assessment to the Single Point and Turret Dynamic Stability Problem", OMAE, International Conference of Offshore Mechanics and Artic Engineering, Florence.

Garza-Rios,L.O. and Bernitsas,M.M., 1995, "Analytical Expressions of the Bifurcation Boundaries for Symmetric Spread Mooring Systems", *Applied Ocean Research*, Vol. 17, pp 325-341.

Kuo,C.B., 1975, "Automatic Control Systems", Print.-Hall,.

Marin, 1993, "Prediction of Wind and Current Loads on VLCC's. An Update", Marin R. z011524-2-OE; to OCIMF; Oct..

Marin, 1994, "TERMSIM II. User's Guide", Marin, Oct..

Nishimoto, K., Brinati, H.L. and Fucatu, C.H., 1995, "Analysis of Single Point Moored Tanker Using Manouvering Hydrodynamic Model", OMAE, International Conference of Offshore Mechanics and Artic Engineering, Copenhagen.

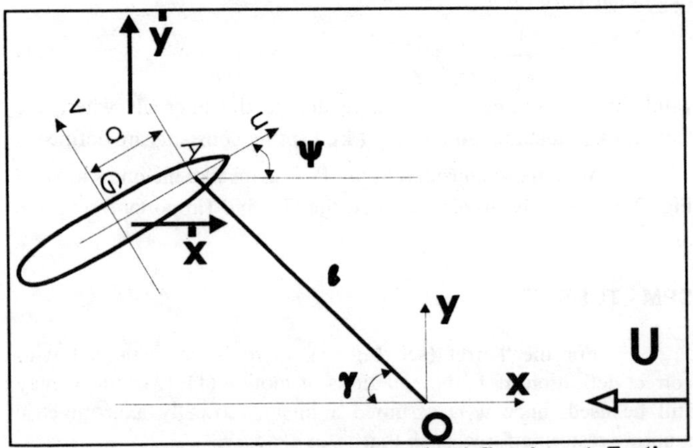

Figure 1. SPM reference systems: (x,y) fixed w.r.t. Earth; center of gravity (G) absolute acceleration denoted by $\dot{x}$ and $\dot{y}$; the ship motion u,v,r are relative to the water.

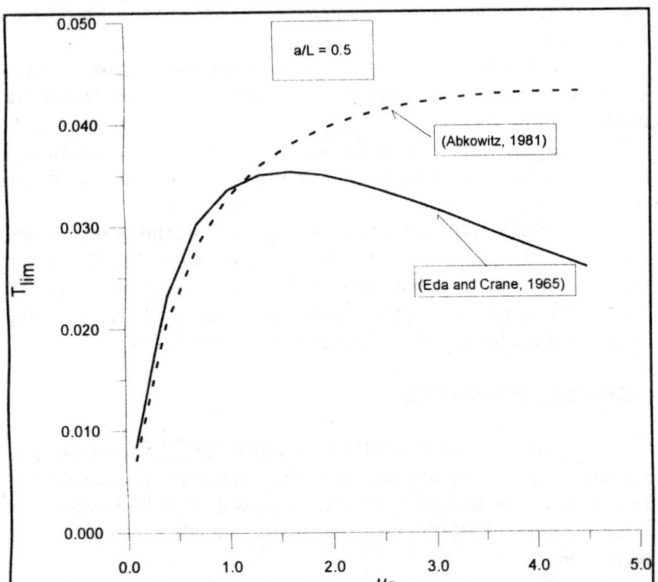

Figure 2 Dynamic Stability Criteria for a fixed-pole SPM configuration with the (Eda and Crane, 1965) model 80 tanker and the (Abkowitz, 1981) super tanker; fair-lead at ship bow; T_{lim} is the value made non-dimensional with $\frac{1}{2}\rho LTU^2$.

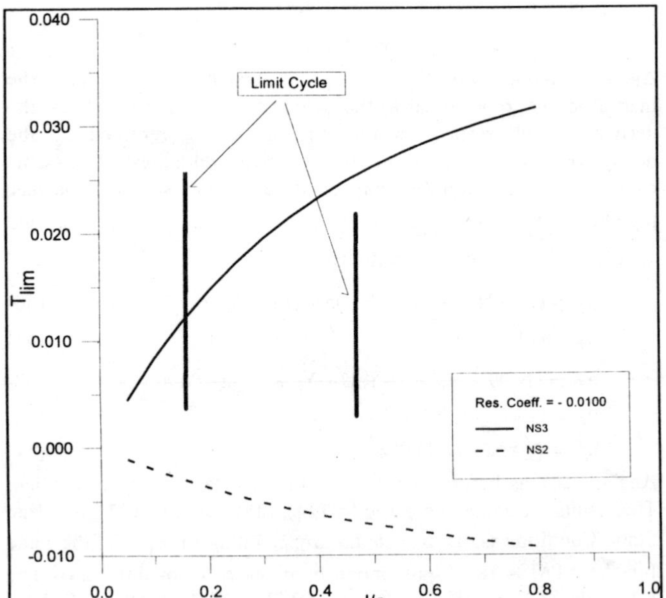

Figure 3 Traces on the T-l map for lower resistance hull for (Eda and Crane, 1965) model 80; T_{lim} is the value made non-dimensional with $\frac{1}{2}\rho LTU^2$.

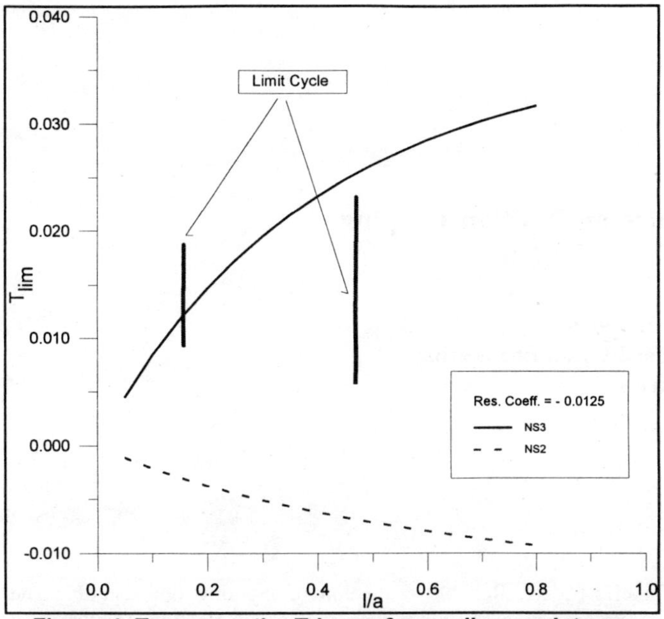

Figure 4. Traces on the T-l map for medium resistance hull for (Eda and Crane, 1965) model 80; T_{lim} is the value made non-dimensional with $\frac{1}{2}\rho LTU^2$.

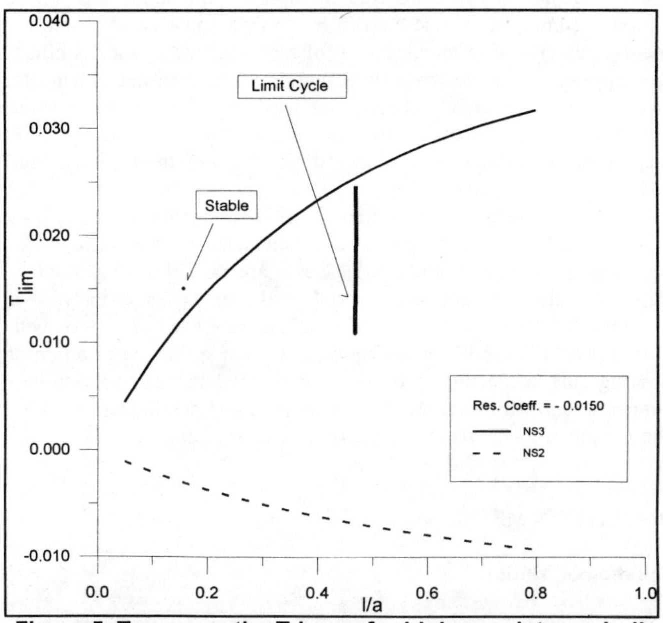

Figure 5. Traces on the T-l map for higher resistance hull for (Eda and Crane, 1965) Tanker; T_{lim} is the value made non-dimensional with $\frac{1}{2}\rho LTU^2$.

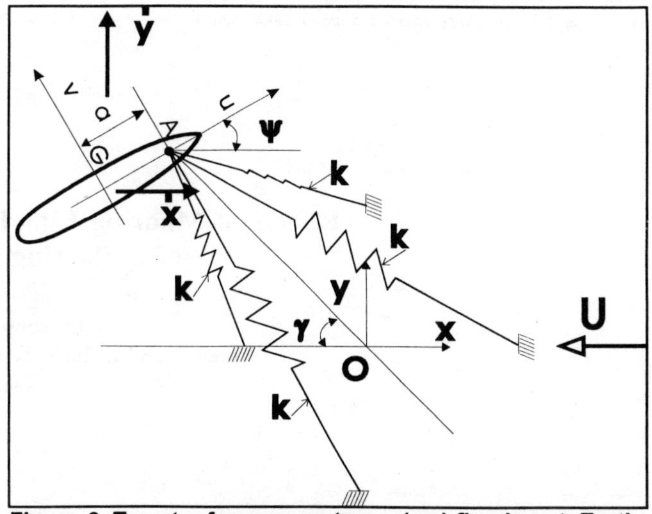

Figure 6. Turret reference systems: (x,y) fixed w.r.t. Earth; center of gravity (G) absolute acceleration denoted by $\dot{x}$ and $\dot{y}$; the ship motion u,v,r are relative to the water. O is the Turret position when no environment forces are present.

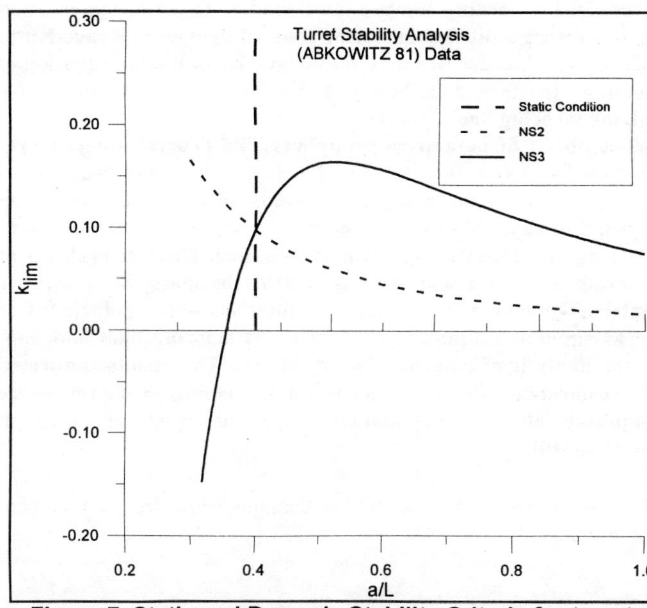

Figure 7. Static and Dynamic Stability Criteria for turret SPM configuration with the (Abkowitz, 1981) super tanker;

$$k_{lim} = \frac{k}{\frac{1}{2}\rho TU^2}.$$

Proceedings of the Seventh (1997) International Offshore and Polar Engineering Conference
Honolulu, USA, May 25-30, 1997

ISBN 1-880653-28-1 (Set); ISBN 1-880653-29-X (Vol. I)

Effects of Mooring Line Dynamics on Position Keeping of a Floating Production System

Sup Hong and Sa Y. Hong
Korea Research Institute of Ships and Ocean Engineering
Taejon, Korea

ABSTRACT

In order to investigate the effects of mooring lines damping on slow drift motion of floating vessel, a parametric study on the coupled dynamics of floating structure and its mooring system was carried out. Hydrodynamic forces on floating structure of ship type were estimated by 3-D panel method. Mooring catenaries were discretized according lumped-mass model. Two time integration schemes were combined in solving coupled dynamics: Runge-Kutta method of 4th order was used for integration of motion equation of floating structure and Newmark method was implemented for solving mooring lines dynamics.

The object of parametric study was TPS (Turret Moored Production System) with a tanker of 124,400 ton, which was one of standard models for comparative study in FPS 2000 workshop (1989, Norway). Mooring system was designed for two different mooring stiffnesses using catenary equation. Contact problem of mooring lines with seafloor was modeled by mass-spring-damper system. Three kinds of non-linear simulations were conducted: free decay simulation, quasi-static analysis of mooring lines and non-linear analysis of mooring lines dynamics. The results confirmed the remarkable effects of mooring lines damping on the resonance amplitude of slow drift motion of moored vessel depending on mooring stiffness.

KEY WORDS : TPS, mooring line damping, slow drift motion, contact problem

INTRODUCTION

As the water depth increases in installation of Floating Production Systems, dynamic effect of mooring system on behaviors of floating unit is growing up. In particular, FPS of ship type such as turret moored tanker is affected in slow drift motion amplitude to large extent by mooring line damping [ITTC, 1990]. The damping mechanism of mooring line is essentially nonlinear and consists mainly of viscous drag on the mooring lines moving relative to water particles and bottom effects, etc. Reliable evaluation of mooring line damping and accurate prediction of slow motion response of floating structure is of great significance for economical and effective design of FPS mooring system.

In order to investigate effects of mooring line damping on the slow drift motion of FPS, a ship type vessel was chosen as object model and a series of analyses is conducted in motion simulation. The vessel motion is simulated in time domain by solving equation of motion of floating vessel having coupled term of mooring line dynamics effect. Mooring line is modeled by a finite number of lumped-mass elements. Two different numerical schemes are used to calculate the coupled motion of moored floating structure: Runge-Kutta method for time integration of equation of motion and Newmark method for dynamic analysis of mooring line.

Free decay simulation of moored vessel is conducted to find out the natural period of surge motion and to estimate the mooring line damping. Coupled motion simulations in waves are carried out by conducting quasi-static mooring analysis and nonlinear dynamic analysis of mooring lines, respectively. The simulation results are compared with motion simulation results using equivalent linear spring. The portion of mooring line contacting with seafloor is modeled by mass-spring-damper system which accounts for the normal interaction from seafloor. The change of touch-down point is taken into account.

SIMULATION METHOD

Equation of Motion

Hong & Hong (1996) proposed a simulation method solving slow drift motion as well as fast motion of column structured platform, i.e. DDF(Deep Draft Floater) considering coupled effects of mooring dynamics, in which viscous drag forces on columns were estimated by Morison equation. In this paper we have neglected the viscous drag on vessel hull and use the following equation of motion:

$$\left\{m_{ij} + a_{ij}(\infty)\right\}\ddot{X}(t) + \int_0^t K_{ij}(\tau)\dot{X}(t-\tau)d\tau + CX(t) = F(t) \qquad (1)$$

where K_{ij} represents the memory effects due to free surface, m_{ij} +

$a_{ij}(\infty)$ the mass and added mass matrix, C the hydrostatic coefficient matrix and $F(t)$ the forcing vector. The forcing vector $F(t)$ consists of wave exciting force, mean and slow drift force, and mooring lines restoring force. Slow drift force is estimated following Newman's approximation(1974). $X(t)$ is the motion displacement vector composed of slow motion and fast motion.

Mooring Line Analysis

A mooring line shows strongly nonlinear characteristics in its dynamic behaviors. It becomes even more nonlinear in coupled dynamics with moored vessel in irregular waves. In Eq. (1) the mooring effect on vessel motion takes part as an external force, i.e. top tensions of mooring lines.

The principal role of mooring tension is a restoring reaction force against the motion of vessel. The restoring forces to be obtained by the transformation of mooring line top tensions depend on the *displacements* of vessel motion, and are caused by the change of mooring line configurations according to the displacements of mooring line top position. That is the reason why static analysis of mooring line provides a good estimation of mooring stiffnesses. In the meanwhile, dynamic analysis of mooring line is indispensable in estimation of slow motion of moored vessel, in particular for surge motion of turret moored tankers, because of importance of mooring line damping [Wichers & Huijsmans, 1990]. In spite that dynamic effects of mooring lines on the fast motion of vessel are negligible due to the order of magnitude of wave exciting loads on the vessel and the incomparably small mass of mooring lines relative to the mass of vessel, the mooring line damping is likely to be increased by fast vertical motion of vessel when combined with slow surge motion [Dercksen et al., 1992]. A TPS vessel of ship type has only a negligible potential damping for slow drift motion.

For that reason a fully nonlinear dynamic analysis of mooring lines is inevitable in order for turret moored tankers to evaluate the effects of mooring lines damping on the slow motion amplitude. Non-linear dynamic characteristics of mooring line is due to non-linear viscous drag on mooring line, geometric non-linearity of mooring line, non-linear material property and contact of mooring line with seafloor, etc.

A non-linear dynamic analysis method [Hong, S., 1992], in which the viscous drag and the geometric non-linearity of mooring line are taken into account and incremental-iterative time integration scheme using Newmark's 2 parameter method is adopted for discretized lumed-mass model, was employed for analysis of slow drift motion of a DDF(Deep Draft Floater) model [Hong & Hong, 1996].

In this paper, the method for non-linear analysis of mooring line used for motion simulation for DDF is updated by considering the touch-down effect of mooring line by a contact model consisting of mass, spring and damper.

Contact Model of Mooring Line

In order to simplify the complex contact problem of mooring line with seafloor we have adopted simple spring-mass-damper system for interactions on the seafloor[van den Boom et al., 1987]. The mooring line segment lying on the seafloor is supported by springs and dampers in normal direction.

The normal reaction force from the seafloor can be expressed as

$$F_{c_r} = -k_c(d+y) - C_c\dot{y} \qquad (2)$$

where d is water depth, k_c the spring constant, y and $\dot{y}$ are vertical coordinate of lumped mass node positive upwards from mean water line and its time derivative, respectively. C_c is critical damping coefficient.

Defining the penetration depth of mooring line, δ beneath the mud line, the non-linear spring constant is determined by $k_c = W/\delta$ for $y \le -d$ and $k_c = 0$ for $y > -d$, where W is lumped mass of mooring line. Critical damping coefficient, $C_c = 2\sqrt{k_c(m+a)}$, where $m+a$ is the lumped virtual mass of mooring line, was employed in order to suppress impulsive reaction forces from bottom spring at the contacting moment.

Numerical Integration Method

Eq. (1) is solved in time domain by using Runge-Kutta method of fourth order, whereas the top tensions in mooring lines are obtained by means of Newmark's 2 parameter method. Runge-Kutta method is applied for the transformed formulation of Eq. (1)

$$\begin{aligned}\ddot{X}(t) &= -\left[M_{ij}\right]^{-1}\left\{\int_0^t K_{ij}(\tau)\dot{X}(t-\tau)d\tau + \left[C_{ij}\right]X(t) - F(t)\right\} \\ &= f(\tau,\dot{X},\ddot{X})\end{aligned} \qquad (3)$$

Mooring lines top tensions included in the forcing term F(t) on the right-hand side of Eq. (3) depend upon the instantaneous positions of fairleads due to the vessel motion. The fairlead position at each time step plays a role of boundary condition for the mooring line dynamics. Newmark integration is conducted with the half time step of the motion simulation.

NUMERICAL SIMULATIONS AND DISCUSSIONS

Simulation Object Model

A tanker of 124,400 ton, which was one of standard models for comparative study in FPS 2000 workshop [Nielsen et al. 1990], was chosen for numerical application. Main particulars of the object tanker are shown in Table 1. Using 3-D code DIFFRAN [Hong, D.C. et al., 1987] hydrodynamic coefficients and wave exciting forces are calculated. Fig. 1 shows added mass, damping, retardation function, wave drift force and wave exciting force for surge mode. Each terms in the figure are expressed in nondimensionalized forms defined as below.

$$a_{11} = a_{11}^{'}/r\nabla$$
$$b_{11} = b_{11}^{'}/r\nabla\sqrt{g/L}$$
$$K_{11} = K_{11}^{'}/r\nabla\sqrt{g/L}$$
$$\overline{F}_x = \overline{F}_x/rgA^2L$$

Where $a_{11}^{'}$ and $b_{11}^{'}$ mean the added mass and damping, $K_{11}^{'}$ the memory function, $\overline{F}_x$ the mean drift force, respectively. L is the characteristic length, 41m. It is found that negative drift force exists in the calculation. Drift force was calculated following near filed method using 3-D source and doublet distribution method [Hong, D.C., 1987]. Panel discretization of the model tanker was represented in Fig. 2. Slow drift force was generated with frequency interval of 0.025 rad/sec.

Mooring Systems of TPS

Catenary mooring was adopted to design two mooring systems having different mooring stiffnesses, (1) taut mooring: k_{11}= 38.2419 ton/m, k_{22} = 32.3271 ton/m, (2) soft mooring: k_{11}= 9.5605 ton/m, k_{22} = 8.0816 ton/m. Spread mooring with 16 lines was chosen for water depth of 500 m. Mooring depth of 485.2 m was taken in calculation by considering draft and penetration depth of 0.2m. Table 2 shows the initial static data of single mooring line obtained by means of catenary stiffnesses equations for a stud-link chain of 182 kg/m in water.

Suspended segment of each mooring line was discretized into 30 lumped mass elements and the length of segment part on the seafloor was varied in the range of 150 - 500m. Initial equilibrium states of discretized mooring lines were obtained by static analysis. Static pull simulation of turret showed that the surge mooring stiffness generated by discrete mooring line model is lower than the initially designed value, i.e. (1) k_{11}= 31.68 ton/m for taut mooring, (2) k_{11}= 8.3 for soft mooring. Fig. 3 and Fig. 4 represent mooring system layouts designed with 16 lines for soft mooring case.

Four kinds of simulation in view point of mooring system,

- Simulation with linear mooring spring (LS)
- Free decay simulation (FD), Free decay simulation (FDW)
- Static mooring simulation (SS)
- Nonlinear dynamic simulation (ND)

were performed for two mooring stiffnesses,

- Taut mooring (TM)
- Soft mooring (SM)

in two irregular waves,

- Operation condition (SP1: $H_{1/3}$=6.5m, T_z=8.5sec)
- Survival condition (SP2: $H_{1/3}$=15.5m, T_z=13.5sec)

In order to investigate the effect of mooring line damping separately, wave drift damping as well as hull damping was not considered in the all simulations.

Table 1 Main particulars of TPS tanker

Items			Value
Length	LPP	(m)	230
Breadth	B	(m)	41
Draft	T	(m)	15
Displacement	∇	(m^3)	121386
Center of buoyancy	(x, y, z)	(m)	(-4.1, 0., 7.34)
Center of gravity	(x, y, z)	(m)	(-4.1, 0., 0.)
Center of turret	-	(m)	20
Radii of gyration	r_{55}	(m)	57.5
	r_{35}	(m)	-4.1

Table 2 Initial static characteristics of single mooring line

Items	Unit	Taut	Soft
Mass	(kg/m)	209.2	209.2
Weight in water	(kgf/m)	182.	182.
Mooring depth	(m)	485.2	485.2
Pre-tension	(ton)	299.0543	157.5405
Suspended length	(m)	1165.804	777.539
Horizontal span	(m)	1026.132	556.583
k_{11}	(ton/m)	3.3847	0.8462
k_{33}	(ton/m)	0.8428	0.5321
k_{13}	(ton/m)	1.4087	0.5280

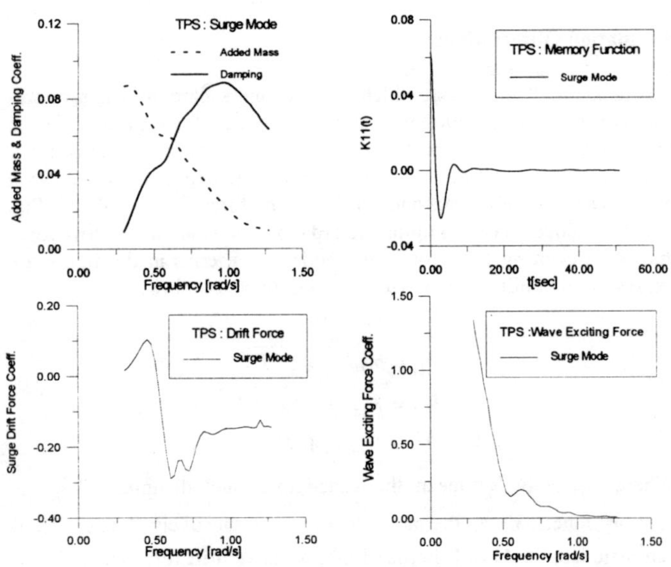

Fig. 1 Hydrodynamic characteristics of TPS tanker by DIFFRAN

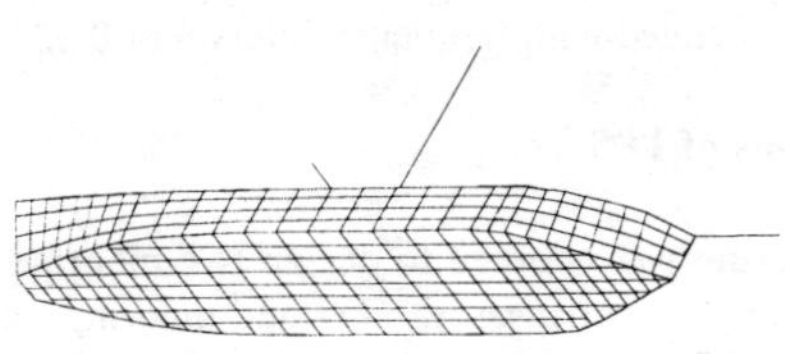

Fig. 2 Panel discretization of TPS tanker

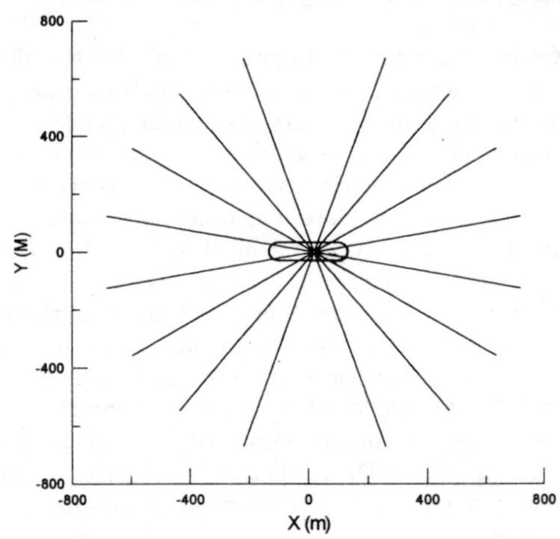

Fig. 3 Mooring lines layout of soft mooring

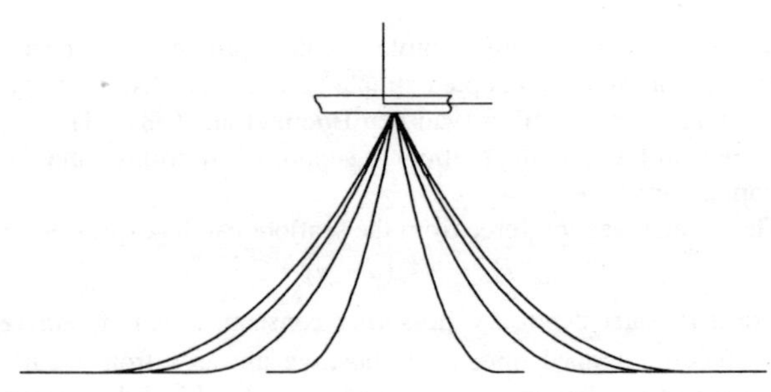

Fig. 4 Initial equilibrium state of soft mooring condition

Time interval was varied from 0.05 sec to 0.1 sec depending upon sea state. Simulations were carried out for real time of 3,600 seconds. The Morison equation was employed for estimation of hydrodynamic forces of mooring lines. Drag coefficient of 1.0 and mass coefficient of 2.0 were used. A slight numerical damping was introduced by using the Newmark damping parameter, $\gamma = 0.6$, and $\beta = 4.0$.

The results of free decay simulations in calm water are shown in Fig. 5. Three different free decay simulations were conducted by nonlinear static, nonlinear dynamic analyses of mooring lines, and free oscillation simulation with linear mooring springs. Surge displacement vs. mooring force curves indicating the hysteresis due to viscous drag on mooring lines are displayed in Fig. 6. Fig. 7 shows the surge responses of TPS tanker with equivalent linear springs. The results from static mooring simulations are given in Fig. 8. The effects of mooring lines dynamics on the surge motion of TPS vessel are represented in Fig. 9 - Fig. 10.

As shown in Fig. 5, it was found by the free decay simulations that the natural periods of slow surge motion of TPS tanker with taut and soft linear mooring springs are about 116.2 sec and 232.4 sec, respectively. However, the natural surge periods were found by static mooring simulations as for two discrete mooring systems to be about 127.5 sec. and 255 sec., respectively. Since the hull damping was ignored in the simulation, the results showed a free oscillation maintaining its initial amplitude in case of simple linear spring and static simulation of mooring lines. The dynamic simulation of mooring lines, however, shows a remarkable decay of surge oscillation amplitude. Herewith, the significance of mooring line damping on slow surge motion is clearly to be seen.

The mooring damping in the case of taut mooring is higher than that in the soft mooring as shown in Fig. 5b. Rapid motion causes more damping than slow motion. The damping ratio estimated from logarithmic decrement analysis was about 5% of critical damping, which is about 5 - 20 times of the hull damping in calm water [Wichers 1988].

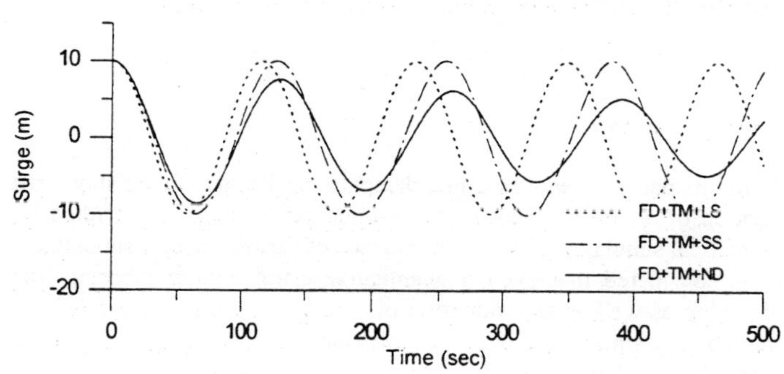

Fig. 5a Free decay simulations for taut mooring

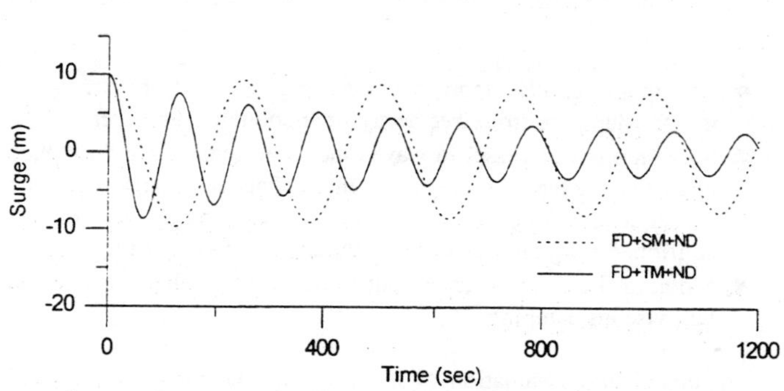

Fig. 5b Free decay simulation by nonlinear dynamic analysis of mooring lines

Fig. 6 represents the relation between surge motion and mooring restoring force for the case of Fig. 5b. In Fig. 7, a typical beating response can be seen, which is caused by slight difference of the excitation frequency from the resonance frequency. The difference between excitation period and natural period was 9.46 seconds for the case of linear spring. The resonant surge motion in Fig. 8 is due to the close excitation frequency to the natural frequency, the period difference was only 1.84 seconds.

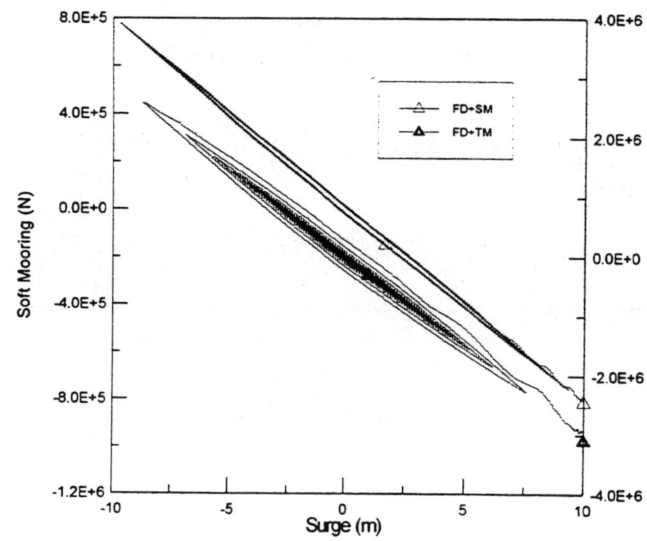

Fig. 6 Surge displacement vs. mooring force curves from free decay simulation: *mooring line damping*

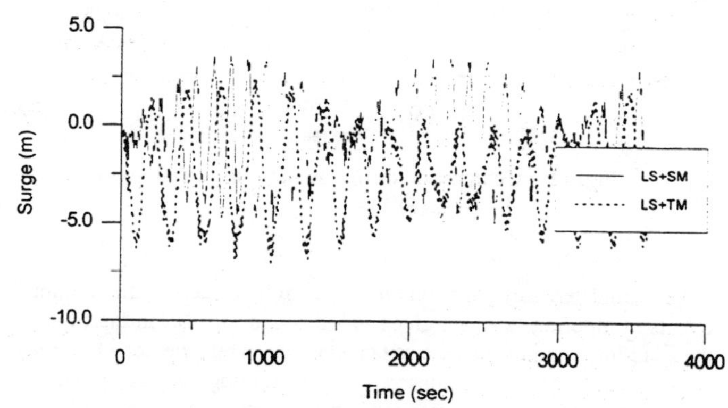

Fig. 7 Surge motions of TPS tanker with linear springs at SP1

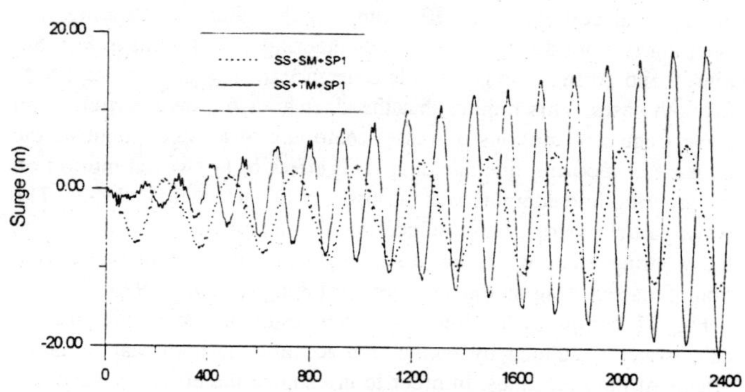

Fig. 8 Surge motions of TPS tanker by static mooring analyses

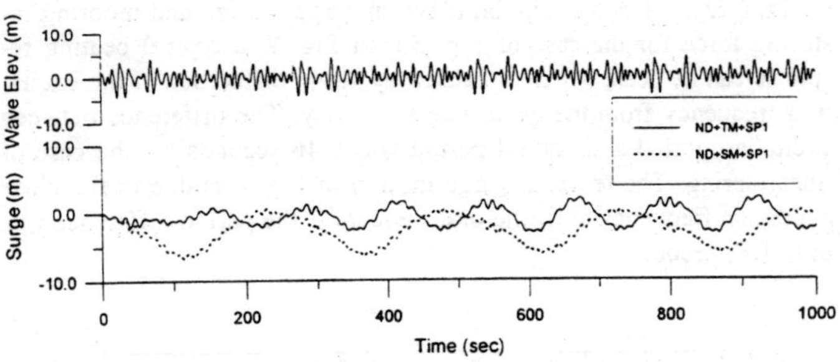

Fig. 9 Surge motions of TPS tanker at SP1

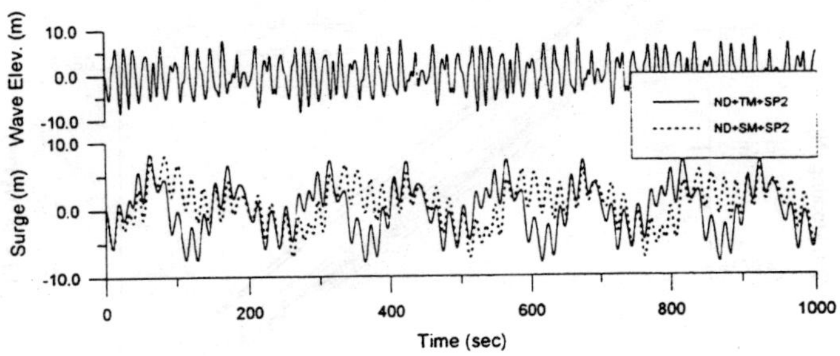

Fig. 10 Surge motions of TPS tanker at SP2

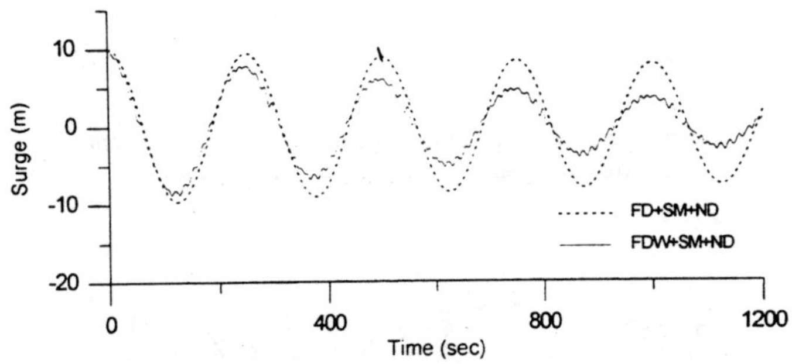

Fig. 11 Comparison of free oscillations in calm water and in waves (SP1)

The simulated surge motions in waves by considering the dynamic effects of mooring lines are given in Fig. 9 and Fig. 10. In the figures, the solid line denotes the case of taut mooring, while the dotted line the soft mooring case. As a consequence of mooring line damping, the steady state of the surge motion is reached immediately comparing to Fig. 8. Further, the amplitude of slow motion is bounded to 5m even in the survival condition of SP2. Although the effect of mooring line damping is more distinguished in taut mooring condition as in Fig. 5b, Figs. 9 and 10 show no noticeable difference in slow motion amplitude for both cases. This implies that the kinematics of wave particles and wave frequency motions of vessel contribute to a large extent to the increase of mooring line damping. This effect of fast vessel motion on the mooring line damping was discussed in Dercksen et al. (1992). The wave particle velocity can induce low frequency viscous damping on the mooring lines, when the Morison equation is used for the estimation of viscous drag on the mooring line [Hong & Hong, 1996].

Fig. 11 compares the free oscillation result in waves with that in calm water simulated by taking into account the nonlinear dynamic effects of mooring lines. In order to investigate the effect of wave frequency components, the mean and slow drift forces were excluded in this simulation. A dramatic increase of mooring line damping can be clearly recognized herein.

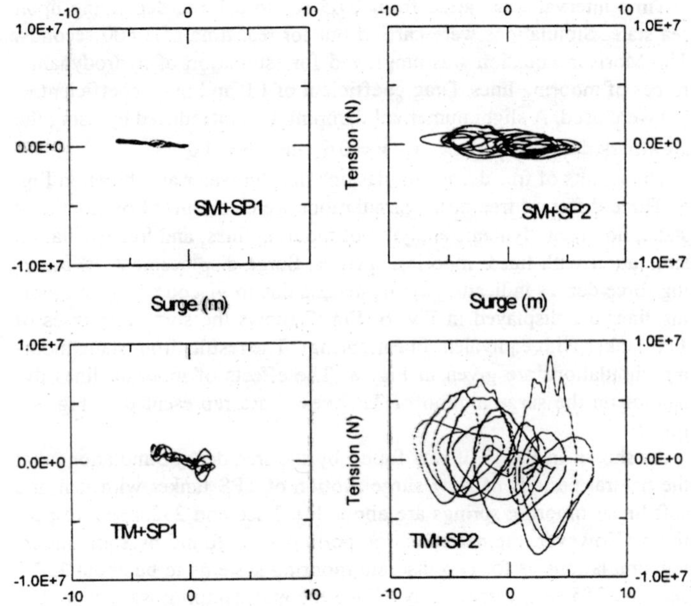

Fig. 12 Mooring restoring force vs. surge displacement

Fig. 12 shows the surge displacement vs. mooring restoring force curves obtained by simulation of nonlinear mooring dynamics in waves with respect to the mooring conditions.

The mooring line contact model with bottom adopted in this paper provided a good simulation results for the mooring lines aligned in the direction of external loads. However, the fluctuation of dynamic tensions might be overestimated because of not considering the lateral interaction between bottom and mooring line in this study.

CONCLUSIONS

In this paper the mooring line damping mechanism on the slow drift motion of a TPS (Turret-moored Production System) was studied by numerical simulations. As a consequence of various motion simulations it is considered that even a quantitative estimation of mooring line damping as well as an evaluation of qualitative characteristics will be possible by simulation, if sufficient parametric calculations and thereupon the simulation model improvement are conducted.

The results of motion simulations for the TPS object model justify the following concluding remarks on the effects of mooring line damping.

- Taut mooring rather than soft mooring gives rise to higher mooring line damping from free decay simulation in calm water.
- Surge motion of vessel in waves shows no remarkable difference upon the change of mooring stiffness. This is understood as the consequence of the increase of mooring damping by wave effects confirmed from the free decay simulation in waves of Fig. 11.
- Mooring damping is turned out to be incomparably greater than hull viscous damping.

For quantitative evaluation of mooring line damping a robust and physically acceptable contact model of mooring line with seafloor will be required.

ACKNOWLEDGMENT

The authors are in debt to appreciate Korea Ministry of Science and Technology (MOST) for the full support of this study.

REFERENCES

van den Boom, J. H. H., Dekker, J. N. and Elsaecker, A. W. (1987), "Dynamic Aspects of Offshore Riser and Mooring Concepts", Proc. of Ocean Technology Conference, paper 5531.

Dercksen, A., Huijsmans, R. H. M. and Wichers, J. E. W. (1992), "An Improved Method for Calculating the Contribution of Hydrodynamic Chain Damping on Low-Frequency Vessel Motion", Proc. Offshore Technology Conference, OTC paper 6967.

FPS 2000 COMPARATIVE STUDY Workshop (1989), "Presentation of Data", Hydro, Norway.

Hong, D. C. (1987), "On the Improved Green Integral Equation Applied to the Water Wave Radiation-Diffraction Problem", J. Soc. Naval Architects of Korea, Vol. 24, No. 1.

Hong, D. C. et al. (1987), "Development of Design Technology for a Semi-Submersible Drilling Rig(I)", KIMM Report No. UCN076-929.D.

Hong, Sa Y. et al. (1995), "Development of Dynamic Analysis Techniques for the Behaviors of Large Floating Structures" (in Korean), KIMM Report No. UCN337-1868.D.

Hong, Sa Y. and Hong, Sup (1996), "Motion Simulation of Floating Structure Coupled with Mooring Lines", Proc. 6th Int. Offshore & Polar Eng. Conf., Los Angeles, May.

Hong, Sup (1992), "Three-Dimensional Non-linear Static and Dynamic Analysis of Flexible Risers in Seaway", Doctoral Dissertation (in German), Tech. Univ. of Aachen

Newman, J. N. (1974), "Second order, slowly varying forces on vessels in irregular waves", Int. Symposium on the Dynamics of Marine Vehicles and Structures in Waves, London.

Nielsen, F. G., Herfjord, K. and Loeken, A. (1990), "Floating production system in waves. Results from a comparative study on hydrodynamic coefficients, wave forces and motion responses", FPS2000 Floating production systems for oil and gas, A HTNF Research Programme, Vol. 1, 2, 3, Hydro.

Report of Ocean Engineering Committee (1990), ITTC, Vol. 1.

Wichers, J. E. W. (1988), "A Simulation Model for a Single Point Moored Tanker", Publication No, 797, MARIN, The Netherlands.

Wichers, J. E. W. and Huijsmans, R. H. M. (1990), "The Contribution of Hydrodynamic Damping Induced by Mooring Chains on Low-Frequency Vessel Motions", Proc. Offshore Technology Conference, OTC paper 6218.

Proceedings of the Seventh (1997) International Offshore and Polar Engineering Conference
Honolulu, USA, May 25-30, 1997

ISBN 1-880653-28-1 (Set); ISBN 1-880653-29-X (Vol. I)

Disconnectable Concrete FPSO for Arctic Conditions

Loic des Déserts, Nicolas Parsloe and François Thiébaud
DORIS Engineering
Paris, France

ABSTRACT

The cost effectiveness and technical feasibility of Floating Production, Storage and Offloading (FPSO) systems in regions of harsh seastate environments has been proven in recent years. The Arctic regions challenge further the designer of such systems not only with harsh sea states but also with extreme temperatures, the presence of icebergs and pack ice. In order to avoid unacceptably high design loads on moorings, risers and the hull, and hence reduce costs, it is desirable to be able to disconnect the FPSO from its risers and moorings once conditions exceed certain criteria. The Health, Safety and Environmental aspects of production in iceberg infested waters leads to the requirement for emergency disconnection. The present paper shows the feasibility and highlights the advantages of a disconnectable turret-moored Concrete FPSO in such circumstances. The aspects of reconnection, damage to risers and moorings, and production downtime are addressed. The advantages that are inherent in a concrete vessel in terms of motion behaviour, thermal properties and ice impact resistance are put forward.

Key Words: FPSO, Concrete, Arctic, Moorings, Disconnection

1. Introduction

In recent years DORIS Engineering have invested a great deal of effort in the development of concrete hulled FPSOs for harsh environments. Designs have been developed for areas such as the West of Shetland (Harris, 1996). The advantages in general of FPSOs installed in such areas are well known, indeed the FPSO is often the only technically and economically feasible solution in areas where no infrastructure is present.

DORIS Engineering have produced a design for a concrete hulled FPSO which is fully viable, cost competitive, and which presents some unique benefits; for example :

- Competitive in terms of cost and schedule compared to a traditional steel hull
- Improved motion performance due to higher inertia relative to a steel hull
- Durability of concrete makes for a long life with minimal maintenance and inspection requirements
- No fatigue problems
- Thermal insulation provided by the concrete reduces or eliminates the need for heating of stored oil even for a single bottomed hull, particularly important in cold waters
- Concrete hull does not need ballasting in the light ship condition, which reduces in-hull systems, maintenance and sea-water discharge
- The greater stability of a concrete ship gives it a greater flexibility with respect to topsides loads, and the position of its supports, as well as an ability to accommodate easily future extensions

Further development work has been performed for Arctic areas such as the Grand Banks off Newfoundland, Canada. The particular aspects of this area that must be taken into consideration are the relatively shallow water depth (90m), the severe storm conditions and the presence of pack ice and icebergs.

The Health and Safety regulations in such areas lead to the requirement that FPSO can be disconnected from its moorings and risers in order to avoid collision with an iceberg. A collision could give rise not only to damage to the production system but also to pollution from broken risers. Different types of disconnection are defined such as a controlled disconnect or an emergency disconnect.

This paper describes a concrete hulled FPSO that meets all criteria for Arctic conditions, for the example of the Grand Banks, offshore Newfoundland, Canada, and then addresses the philosophy of disconnection and its impact on capital and operating costs.

2. Design of Concrete FPSO

The basic design of the concrete FPSO meets the following production requirements:

- 125 000 bopd production leading to a topsides weight of up to 15 000 t in operating conditions
- 850 000 bbl storage

- 5 500 m^3/h offloading rate direct to shuttle tanker in tandem
- 14 risers
- Accommodation for 65 persons

This leads to the following characteristics of the hull

• Overall length	280m
• Breadth	55m
• Depth	28m
• Fully loaded draft	20m
• Light ship draft	13m
• Freeboard	from 15m to 8m
• Concrete volume	62 200 m^3
• Steel weight:	
Reinforcement	20 060 t
Prestress	5 070 t

The storage volume is derived from a study of the offloading possibilities. For the Grand Banks area three possible locations for offloading were considered : Newfoundland, Halifax (Nova Scotia) and Philadelphia. The storage on the FPSO must be greater than the volume of the shuttle tanker plus the volume produced during waiting on weather time. For most of the year the waiting on weather time is comparable to North Sea values. During the ice season, consideration must be given to the presence of pack ice, which could interfere with the connection between the shuttle and the FPSO, and of icebergs, which could cause the whole operation to be delayed if icebergs are found within the warning zones (see section 6). An average weather window of 72 hours was considered for the determination of storage volume. It should be noted that if there is a risk of iceberg impact the production will be shut down and so no production volume is required.

The concrete hull is a rectangular box, rounded at the bow (figure 1). It is made of vertical walls and flat slabs for the roof and keel. At the bow the hull is fitted with a 13m high bulwark, giving a minimum height of 21m above the waterline when fully loaded, which gives good protection from green water. The hull consists of 6 main cargo cells, a moonpool in which the turret is situated and 32 peripheral cells which provide buoyancy and protection of the cargo cells from external impact, as shown in figure 2. The length of the cargo cells is defined so as to avoid resonance in the sloshing of liquids within the compartments. The calculation is based on the Bureau Veritas rules. The dimensions of the compartment are determined such that the natural frequencies of liquid movements are higher than the frequencies of vessel motions by a sufficient margin.
The length of the external buoyancy cells is determined from marine stability requirements under damaged conditions where two adjacent compartments are assumed fully invaded. Four trim tanks located in pairs fore and aft are supplied in order to provide a trimming capacity. Two slop tanks are located at mid-ship.

The accommodation is located aft. All the topsides equipment is located between the accommodation and the turret, as shown in figure 3. The turret is situated 30m from the bow (11% of the length). This position allows passive weathervaning of the ship which reduces the environmental loading of the hull. This reduces the motions of the ship thereby allowing production to continue during storms, and reduces the bending induced in the hull. Weathervaning is particularly important in Arctic conditions where loads from even low concentrations of drifting pack ice are to be minimised. Indeed a turret placed at the mid length of the ship would not be able to weathervane without assistance.

Thrusters for the dynamic positioning system are provided fore and aft. This system is required for the FPSO to manoeuvre when disconnected from its moorings and during the reconnection procedure. It can also be used to limit excursions during large storms thereby reducing loads in the mooring lines. This system is sized so as to keep the FPSO on station with moorings disconnected during the 1 year storm (Hs = 10.9 m, Tp = 14.1s, $\gamma = 1.7$).

The use of a single bottom in Arctic waters gives rise to the question of heating the stored oil. A study has been performed for the heat loss accounting for the single bottom, the single roof and double walls with an oil storage temperature of 65°C and an external air and sea temperature of -2°C. The study considered the heat transfer between the oil and the surrounding air filled compartments, the sea beneath the hull and the air above the roof. The study considered convection in the oil and in the air in the buoyancy tanks. The maximum heat flow was calculated and used to determine the oil temperature variation with time. The study found that even after 20 days shut down of production the temperature of the stored oil is sufficient to avoid wax deposits, meaning that heating coils within the tanks are not required. After 20 days the oil temperature dropped to 55°C, still well above the assumed wax formation temperature of 46°C.

Thicknesses of the concrete hull are in the range 0.45m to 0.65m.

These thicknesses are quite standard and permit the walls to be slipformed. This technique is preferred because of the quality of the finished product and the speed of construction. Indeed the repetitive nature of the structure is designed for rapid construction.

The outer hull of a concrete FPSO is designed to resist external water pressures acting on the wall of an empty buoyancy cell. The wall is designed as a plate stiffened at 4.0 to 5.0m intervals. The resulting thickness of wall is sufficient to resist pack ice pressures, of the order of 4 kPa over 4.5m height across the whole span between stiffeners, without any damage. A small increase in thickness in exposed areas, such as the bow, would enable the hull to resist higher pressures such as those from iceberg impact. The required hull thickness would be determined from the size of iceberg that should be resisted . The current thickness is sufficient to resist growler impacts, modelled as a 5 MN load applied on a 1 m^2 area at the midspan. This is similar to design loads used on GBS projects in the Grand Banks area. The inherent abrasion resistance of concrete means that even local damage under an iceberg impact will be minimal.

3. Design of Mooring System

The environmental conditions on the Grand Banks are quite severe. The 100 year storm used in determining the extreme motions of the ship and in the design of the moorings is :

- Hs = 15.4m, Tp = 16.7s JONSWAP, gamma = 1.3, 3 hours duration, heading 10°
- water depth = 90m.
- current = 0.91 m/s, constant with depth, heading -45°
- wind (1 hr @ 10m) = 77 knots, heading 10°

The study of motions and the mooring design was based on a diffraction analysis using the DIODORE program. The main results of the hydrodynamic analysis are given in table 1.

motion	Natural period	maximum amplitude at CoG
surge		± 5.29 m
sway		± 1.47 m
heave	12.4s	± 5.38 m
roll	13.7s	± 1.74°
pitch	11.9s	± 7.90°

Table 1. Results of Hydrodynamic analyses.

The design of the mooring system also accounts for slow drift effects which are summarised in table 2.

Motion	Mean	Maximum	Min
surge (m)	-24.79	-8.47	-44.64
sway (m)	-0.66	6.27	-9.79
heave (m)	0.00	5.44	-5.50
roll (°)	0.00	2.04	-2.04
pitch (°)	0.00	6.86	-7.21
yaw (°)	0.56	6.21	-2.95

Table 2. Slow drift motions

The mooring system is composed of 12 lines arranged in three groups of four at 120° intervals around the turret. A 4° separation between individual lines is provided to avoid interference. The complete mooring system is attached to a buoy in the centre of the base of the turret. The risers are accommodated in two groups between the sets of mooring lines as shown in figure 4. They are not attached to the mooring buoy.

The design of the mooring system is performed using the ARIANE mooring analysis software. A safety factor of 1.85 is taken on the line tension obtained from a full dynamic analysis, which is more severe than a factor of 2.0 on a quasi-static analysis. The required line configuration is:

- A 40t anchor
- 800m 6" chain grade 4
- 80m 6" chain grade 4 with3 x 80m 6" chain as a sinker
- 90m 6¼" chain grade 4
- connecting chain within turret 6¼"

The situation that governs the design of the moorings is a combination of slow drift surge motions towards the stern combining with heave and pitch motions that submerge the bow of the vessel.

4. Design of Riser system

The design of the riser system commenced with a screening of the different types of configuration suitable for shallow waters with a vessel undergoing large motions. The supplementary requirement was to allow the system to be laid down onto the sea floor in a controlled manner during the disconnection procedure. In this way the damage to the risers would be minimal and so replacements would not be needed at each disconnection.

The risers would be connected to the turret independently of the mooring system so that the two systems can be disconnected one after the other.

Five configurations of riser were studied :

- Free Hanging catenary risers draped to the sea floor
- Lazy Wave with distributed buoyancy providers supporting the riser which is draped to the seafloor
- Steep Wave with distributed buoyancy providers supporting the riser which is connected to a seafloor riser base with a bending stiffener
- Steep S consisting of a catenary section from the vessel to a submerged buoy then a tensioned section from the buoy to a riser base
- Lazy S consisting of a catenary section from the vessel to a submerged buoy then a draped section to the sea floor. The buoy is tethered to a dead weight on the sea floor

The selection of the best configuration was made according to the following criteria :

- the ability of the configuration to follow the motions of the vessel in the design storm
- the ability of the configuration to accommodate the motions of the vessel with respect to induced tension and curvatures especially the minimum radius of curvature in the riser and maximum angle of upper termination
- ease of initial installation
- ability to be laid down in a controlled manner
- ability to be recovered from the sea floor and be reconnected to the FPSO with a minimum of external assistance

The comparative study led to the retention of the Lazy S configuration. A dynamic analysis of one typical riser was then performed to determine the riser geometry. The criteria of upper termination tension, upper termination angle, minimum radius of curvature were checked. The resulting geometry was found to be feasible.

The riser connection to the turret is made separate from the moorings so that the disconnection can take place without clashes. The risers will be connected at the upper part of the turret, with a bend restrictor at the lower flange of the turret. The connectors will be chosen so as to limit spillage to the environment even in the emergency disconnect event.

5. Disconnection and Reconnection of Mooring and Riser Systems

The design of the turret and disconnection systems conform to the following criteria :

- Ability to disconnect in four hours or less
- Ability to perform an emergency disconnect
- Ability to reconnect risers from the FPSO with minimal assistance
- Minimise damage to risers during disconnection
- Minimise damage to equipment left behind after sailaway

The system that has been devised uses the following procedure during an emergency disconnect:

- Decision to disconnect
- Hold position with DP
- Emergency shut down of subsea equipment
- Emergency shut down of production equipment
- Release moorings
- Release risers
- Move off station

This emergency disconnect procedure will not attempt to accurately position or orient the vessel but will try to limit motions so that the disconnection will not fail. The moorings on their buoy are released first to be sure that the vessel can move off station, even if there are problems to release cleanly the risers. Since no control of riser laydown is performed it should be assumed that they will be damaged and need to be replaced.

The system that has been devised uses the following procedure during a controlled disconnect:

- Decision to disconnect
- Shut down wells
- Shut down production
- Flush risers and flowlines
- Manoeuvre with DP over prepared sea floor
- Controlled descent of moorings
- Manoeuvre with DP to guide laydown of first set of risers, sinking buoy simultaneously
- Release of first set of risers
- Manoeuvre with DP to guide laydown of second set of risers, sinking buoy simultaneously
- Release of second set of risers
- Move off station

This procedure lays down all equipment between the vessel and the sea floor in order to minimise damage from the passage of an iceberg.

The design of the system allows for the reconnection to be performed in seas of at least 2.1m significant wave height. The Dynamic Positioning (DP) system is used to manoeuvre the vessel during reconnection. The risers are reconnected first. Retrieval of risers is performed using ROVs and winches. An anchor handling vessel is assumed to be called up during the disconnection period to assist with the reconnection of moorings. The time estimated for the reconnection of the risers and moorings is less than one day.

6. Philosophy for Disconnection

The reliability of a disconnectable turret has been demonstrated in various situations around the world (for example Forest-Dodelin, 1992, Gruy, 1993). An application in Arctic zones has not been made to this date.

The philosophy for disconnection has a great impact on the design of the turret, mooring system and riser system. Two distinct categories of disconnection are defined : an emergency disconnection, where limited damage to parts of the system is acceptable, and a controlled disconnection, where no damage occurs thereby allowing production to restart as soon as the FPSO is reconnected.

The environmental events are reviewed to determine which type of disconnection will be required. It should be borne in mind that a minimum of an emergency disconnect system will be required for Health and Safety reasons.

The first event that is considered is an impending impact with an iceberg (due to a failure in the early warning system) or a drifting vessel. In this case an emergency disconnection with the aim of saving lives and/or the production vessel is the only possible solution. For the purposes of this design a disconnection in 5 to 10 minutes was assumed.

The second event considered is the case of an approaching iceberg. There is an average period of about 3½ months a year when icebergs can be seen on the Grand Banks. Statistically, a few icebergs each year could pass within 20 nautical miles of a typical field. The actual route of an iceberg is unpredictable since it is dependent on the current and wind conditions which are unlikely to be collinear. In order to minimise the risk of impact the disconnection procedure should commence as soon as an iceberg of a given size enters a warning zone of a set radius around the vessel. The size of the zone will depend on the time required for : the safe shutdown of production and subsea systems, flushing of flowlines, disconnection of risers and disconnection of moorings. The sailaway of the vessel would then occur with the iceberg still at a safe distance from the vessel, say one nautical mile. It has been assessed that all the above operations could be performed within four hours. With an average speed of an iceberg at 0.6m/s this gives a warning zone of radius 5.67 nautical miles. Any iceberg entering this zone, whatever its trajectory, would initiate the disconnection procedure. An early warning system based on radar is recommended because of restricted visibility due to the common presence of fog on the Grand Banks. Currently available radar systems can observe large icebergs at 200 nautical miles and bergy bits (which present only about 5m² cross section) at 20 nautical miles. For this case a controlled disconnection system is required. Alternatively an emergency disconnection system can be envisaged for a reduced warning circle, say of radius one nautical mile. The smaller the warning circle the more unlikely it is that a disconnection is required since less icebergs will enter the zone. The fact that the concrete FPSO can be designed to resist iceberg impacts up to a given size means that the number of disconnections over the life of the field would be further reduced, since disconnection would be initiated only for the larger icebergs entering the warning zone.

The third case considered is sea ice. Sea ice can occur between December and July and is very unpredictable in coverage. Some areas of the Grand Banks can be free of sea ice for years at a time, yet significantly covered for weeks when sea ice occurs. The sea ice tends to be relatively weak ice that will crush against the hull. With the weathervaning of the vessel the head-on loads due to sea ice are small relative to wave loads, so sea ice will not govern mooring and riser design. However in order to be able to navigate in the ice floe, a disconnection criterion may be set when a certain percentage of sea is covered with ice, say 40% to 60% of the sea surface. This will correspond to several weeks duration per year in the worst years. In contrast, a coverage of 100% will only occur a few times over the life of the field. The required disconnection system will therefore depend on whether navigation in the ice floe is required (controlled disconnect) or whether the FPSO is to be kept on station for all coverages of sea ice.

The fourth possible cause for disconnection is during severe storms. Having invested a great deal in a a disconnectable system, it is tempting to extend the number of occasions in which it will be used. If production is to be suspended during major storms it may be worthwhile to disconnect in order to reduce mooring dimensions. However the prediction of the severity of a storm is not easy and the risk is that too many disconnections are effected. It is more cost effective to design the system to resist the most severe storms without disconnection.

In all of the above circumstances the relative costs of the turret, the moorings and the replacement of damaged equipment need to be assessed in order to refine the disconnection philosophy. For example the cost of a turret that provides only emergency disconnection will be much cheaper than that for a turret providing for frequent disconnections.

The probability of occurrence of the above events has to be assessed in order to decide on the final philosophy. If it is shown that the likelihood that an iceberg approaches within one mile of the vessel is very small it may be acceptable to provide only for emergency disconnection and to replace all damaged equipment.

7. Production Downtime

The costs associated with the disconnection are not only capital costs, but also operational costs. The most important of these costs is that associated with lost production.

The downtime associated with the disconnection philosophy is of five types :

- Duration of disconnection procedures. During a controlled disconnect a total of 4 hours is assumed.
- Duration of FPSO off station. In the case of a drifting iceberg a minimum of 4 hours is required for the iceberg to leave the warning zone. In the case of sea ice the pack may take days even weeks to pass by.
- Waiting on weather for reconnection. It is desirable to design the system so that the reconnection can be performed in relatively severe conditions since this will reduce the waiting on weather time. The design reconnection seastate will depend on the environmental conditions during the iceberg season. For the Grand Banks area the waiting on weather can be up to three days.
- Reconnection time. The reconnection procedure is quite straightforward with minimal subsea intervention, which leads to a total reconnection time for both moorings and risers at less than one day. It should be noted that the use of a DP system allows production start-up to commence once risers are connected even if the moorings are still disconnected.
- Duration of start up of production. This will include all aspects of the production from wells through flowlines and risers to the process equipment. If the risers and flowlines have been flushed or if they are full but the disconnection period is of a short duration, the production can start up within a few hours. However, if filled risers and flowlines are left unattended for a long period in Arctic waters, waxing may occur. This may imply the replacement of all flowlines in the most unfavourable scenario, leading to significant production downtime as well as the capital cost of replacement. The design of the flowlines should take into account their insulating properties to cater for this event.

The minimum downtime is therefore dependent more on the weather than on the physical disconnection and reconnection processes. It is clear that a disconnection philosophy that reduces the number of disconnections throughout the life of the field will minimise lost production.

8. Conclusions

The design of a disconnectable turret moored FPSO for Arctic waters has been found feasible. The cost of providing a system that allows frequent disconnections has been found to be high, especially for the turret. Frequent disconnections also imply a loss of production. On the other hand a system that supplies only emergency disconnection will have lower initial CAPEX and less production downtime, but may incur the cost of replacement of equipment.

A great advantage of the concrete hull is that it can easily resist impact pressures due to pack ice. The total loading of the vessel is reduced by weathervaning which can be actively assisted by the DP system. Indeed the sizing of the DP system and moorings can be such that disconnections are not necessary for pack ice.

In the case of icebergs, the warning circle defined for a controlled disconnection covers an area of 350 km^2. If the warning circle radius is reduced to 1 nautical mile the area is 10 km^2. The probability of an iceberg entering this smaller warning circle is significantly reduced meaning that an emergency disconnect due to an iceberg will be very rare. Further reductions in the number of disconnections can be achieved if the hull is designed to resist iceberg impacts.

The choice of disconnection philosophy should be made taking into account both cost and risk factors. There will be a cost trade-off between an expensive disconnectable turret and a less expensive turret associated with possible replacement of equipment. The cost of lost production should be included. The degree of risk that the operator is willing to take will influence any decision. The comparison of systems should be performed by means of a risk analysis and should account for life-of-field costs

9 References

Forest-Dodelin, D. (1992), "The disconnectable mooring system of the FPSO vessel at Huizhou, South China Sea, lives up to its promises", OTC 6972.

Gruy,RH et al (1993), "Design & Construction of a Disconnectable Turret Mooring for a FSO in South China Sea", SNAME FPSO Symposium, (1993).

Harris, RJS et al. (1996), "Concrete FPSO", OTC 8047.

10 Figures

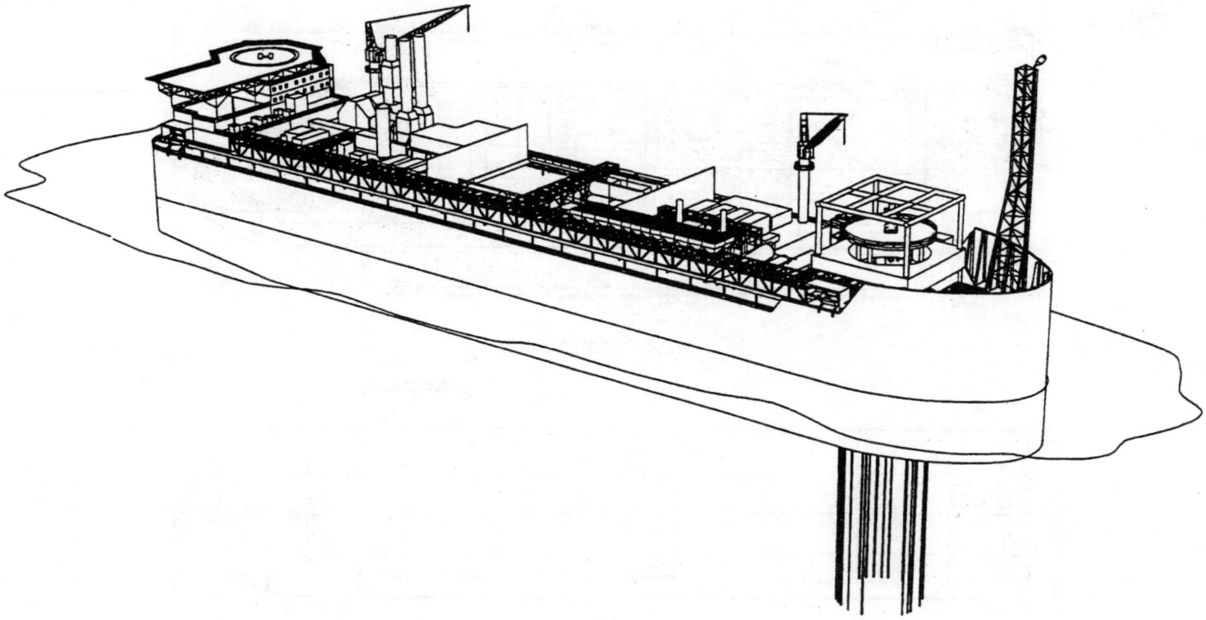

Figure 1 General view of the Concrete FPSO

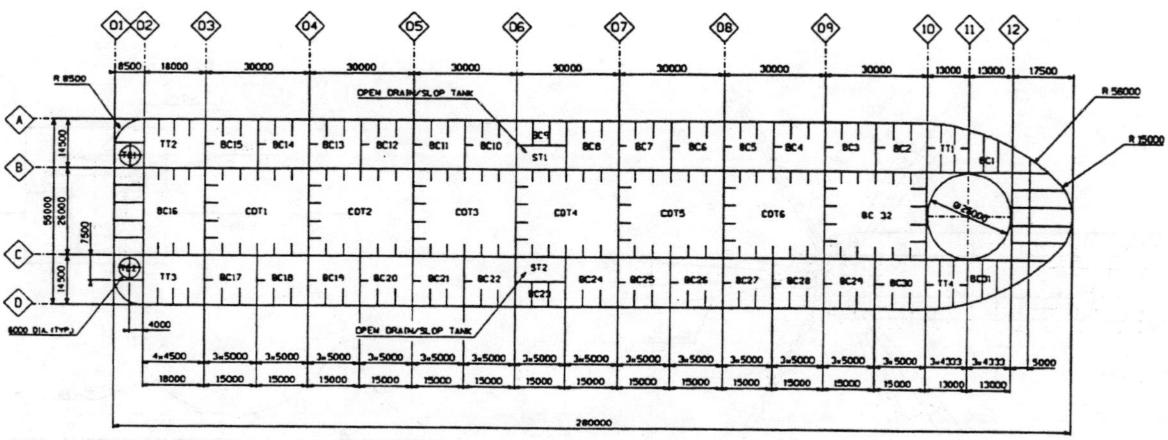

Figure 2 Compartment Layout

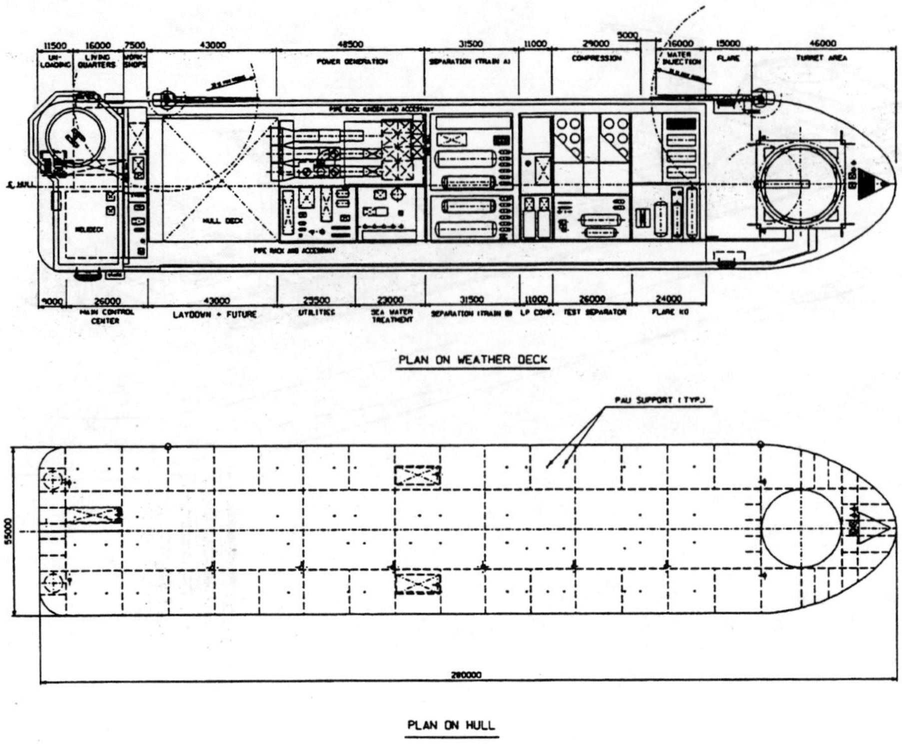

Figure 3 Plan view of the Concrete FPSO

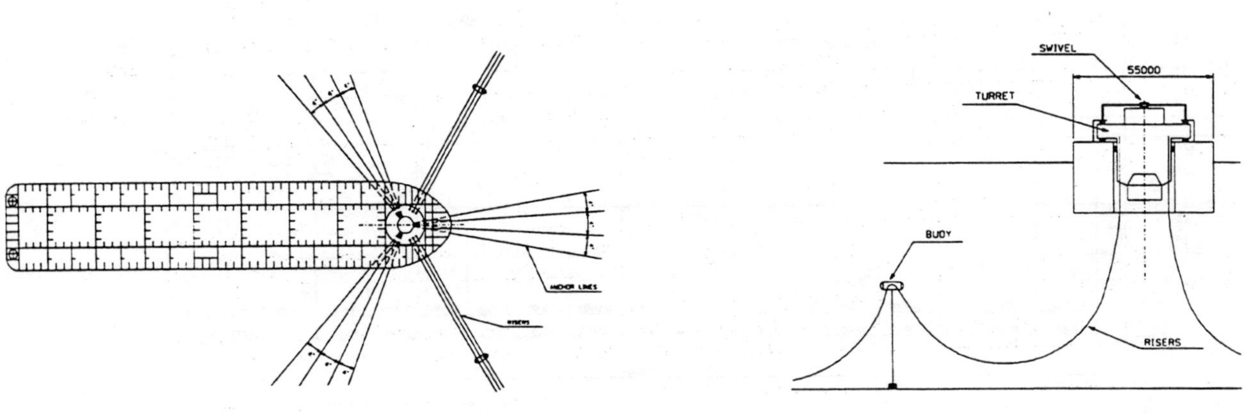

Figure 4 Mooring and Riser system

Proceedings of the Seventh (1997) International Offshore and Polar Engineering Conference
Honolulu, USA, May 25-30, 1997

ISBN 1-880653-28-1 (Set); ISBN 1-880653-29-X (Vol. I)

Quasi-Linear Model of a Tanker in a Turret Configuration

C.A. Martins and M.R. Martins
Escola Politécnica, USP, Sao Paulo, Brazil
A.J.P. Leite
PETROBRAS, Rio de Janeiro, Brazil
J.A.P. Aranha
Escola Politécnica, USP, Sao Paulo, Brazil

ABSTRACT

This paper describes a quasi-linear model that allows one to make a frequency domain analysis of a tanker in a turret configuration. The model takes profit of recently developed expressions for the wave drift damping, for the low frequencies forces and moment in the horizontal plane and of an algebraic approximation for the dynamic tension in the mooring lines. It incorporates also the damping due to wind and ocean current.

The only hydrodynamic information needed for the analysis are the standard ones: the force and moment coefficients for unitary wind and current velocities, the added mass at zero frequency, the RAO of the floating system and the drift force coefficients for regular waves.

The results of this analysis are the steady, quasi-steady and total offsets off the moored unit together with the components of tension in the mooring lines as a function of its pre-tension. With these results one can choose the "best" pre-tension to fit some design criteria.

In particular, a turret configuration is studied as a function of the pre-tension in the mooring lines and of the angle between the ocean current and wind/waves.

KEYWORDS: Mooring, turret

INTRODUCTION

The mooring system of a floating production unit is designed to withstand environmental loads due to wind, ocean current and waves. In doing so it introduces restoring forces and moments in the horizontal plane that are small when compared with the gravitational restoring ones.

The natural frequencies for the motion in the horizontal plane are low but comparable with the low frequencies that appear at second order in an irregular sea. So the resonant response is usually large, since the damping mechanisms are small in this range of frequencies. Thus the large amplitude low frequency oscillation (slow drift) must be added to the static displacement in order to obtain the maximum offset of the floating unit.

The design of a mooring system is not an easy task by several reasons. Firstly it depends on the restoring forces and moments provided by the system, that are nonlinear functions of the displacements, and the second order wave forces are difficult to compute in the strict sense. Furthermore, a precise estimation of the damping factor is essential in the analysis of the resonant response (slow drift) although this takes a job, specially in this case, where the damping is supposedly small and different mechanisms should be accounted for: the viscous damping on the hull and mooring lines and the "wave drift damping" (see Wichers(1982)).

Strictly, the problem should be solved in the time domain but the solution is difficult. For an assumed mooring system, one must compute the low frequencies second order wave forces influenced by the relative velocity between the medium (ocean current) and the floating system and then estimate the maximum offset of the unit in an irregular sea. Once this is done, the static tension in each line can be determined and must be added to the dynamic tension caused by the first order motion of the floating unit, again a difficult task, one must then check if the global offset is within the bounds defined by the design criteria and if the lines can withhold the ensuing total tension (static plus dynamic); if either the obtained offset or tension are larger than the stipulated ones, one must redefine the mooring system and repeat the analysis; if both variables are within the desired bounds the proposed mooring system is acceptable, although not necessarily the most suitable one.

One can certainly relax some of the basic assumptions in order to solve the problem in the frequency domain, an option that can produce a reasonable good estimate of the final answer in a much shorter time. Some of the difficulties, however, remain; the estimation of the low frequency force spectrum continues to be cumbersome and the analysis of the influence of the forward speed on the drift force ("wave drift damping") is as difficult as before. Further simplifications are needed and some of them have become very popular in the field. "Newman's approximation" for the slow drift force spectrum (see Newman (1974)) is almost always assumed, although different forms of the this approximation are proposed by different authors without any deeper discussion about the relative merits of each one. Furthermore, recent numerical computations of the second order forces seem to indicate that this approximation is not as good as it was thought by the community, although the numerical computation itself may be wrong.

With respect to the "wave drift damping" the situation is even worse. Until recently the only known approximation was a heuristic expression proposed by Wichers (1982) that is known to be inexact, the actual "wave drift damping" being in general much larger.

Some of these difficulties have been overcome lately. In fact, Aranha (1994) has derived an exact expression for the "wave drift damping" in a two dimensional problem that depends only on the standard drift force coefficient for a harmonic wave; this expression has been further generalized in Aranha (1996) for all three horizontal degrees of freedom in the three dimensional problem. Also Aranha & Fernandes (1994) have shown that some of the so called "Newman's approximation" are indeed very precise in the low frequency regime. Furthermore, they showed that the ***exact*** second order slow drift force spectrum is ***flat*** (it has a "white noise" like behavior) in low frequency, its value being determined by a simple integral of the drift force coefficient in harmonic waves multiplied by the sea spectrum. This last result not only simplifies the determination of the low frequency force spectrum but also indicates that there is little hope of detuning the floating unit natural frequencies, by an eventual change in the mooring system, from the low frequency wave excitation.

These analytic results, coupled with an algebraic expression for the dynamic tension in the line derived formerly (see Aranha et al. (1993)), make it possible to develop a software aimed to define, in a relatively fast and precise way, the most suitable mooring line system within a certain possible family of systems. In particular, if one chooses the pre-tension in one line as a control parameter, one can determine the most suitable pre-tension such that the mooring system fits the desired design requirements: the minimum offset compatible with the maximum possible tension in the most loaded line.

In a previous work Martins et al. (1995) treated the problem with an unidimensional approach that considers all the environmental loads acting in the same direction and the motion of the floating unit restricted to that direction. The new version of the software eliminated these restrictions treating the problem as essentially bidimensional.

This paper briefly describes the mathematical models and presents some results for a tanker in a turret configuration

MATHEMATICAL MODEL

The problem is solved by a four-step algorithm. In the first step the steady forces are applied on the floating unit in a incremental manner and the static equilibrium position and the static tensions in the mooring lines are obtained. In the second step the slow drift displacements are also applied resulting in a new static configuration and quasi-steady forces in the lines. In the next step the dynamic problem is solved by linearizing the motion around the static position resulting in dynamic displacements of the unit. The only nonlinear term considered in this phase is the damping in still water. In the last step the dynamic tensions in the mooring lines are calculated using the algebraic expression proposed in Aranha et al. (1993).

The loads that influence the floating unit motion are from several natures: ocean current, wind, waves and the relative motion between the unit and water. The mathematical model must take into account all of these effects. In the following, each one of them is presented in a brief manner

Current forces and moment

If U_C is the current velocity and α_C the angle between current direction and the axis of the floating unit, using the notation of Wichers (1987) the longitudinal force X_C, the lateral force Y_C and the moment N_C can be expressed by

$$\begin{aligned} X_C(\alpha_C) &= \frac{1}{2}\rho TLC_{1C}(\alpha_C)U_C^2 \\ Y_C(\alpha_C) &= \frac{1}{2}\rho TLC_{2C}(\alpha_C)U_C^2 \\ N_C(\alpha_C) &= \frac{1}{2}\rho TLC_{6C}(\alpha_C)U_C^2 \end{aligned} \quad , \quad (1)$$

where ρ is the water density, L the length and T the draft of the floating unit and C_{1C}, C_{2C} and C_{6C} are the hydrodynamic coefficients of the current loads.

When the unit moves these loads change as the angle α_C changes. If $\dot{x}$ and $\dot{y}$ are the components of the velocity of the unit and $\dot{\alpha}$ is its angular velocity, the new forces and moment $\tilde{X}_C$, $\tilde{Y}_C$ and $\tilde{N}_C$ can be expressed as, considering only the linear terms in $\dot{x}$, $\dot{y}$ and $\dot{\alpha}$,

$$\begin{Bmatrix} \tilde{X}_C \\ \tilde{Y}_C \\ \tilde{N}_C \end{Bmatrix} = \begin{Bmatrix} X_C \\ Y_C \\ N_C \end{Bmatrix} - \mathbf{B}^{(C)} \cdot \begin{Bmatrix} \dot{x} \\ \dot{y} \\ \dot{\alpha} \end{Bmatrix} - \mathbf{K}^{(C)} \cdot \begin{Bmatrix} x \\ y \\ \alpha \end{Bmatrix} \quad , \quad (2)$$

where

$$\mathbf{B}^{(C)} = \rho LTU_c \begin{Bmatrix} C_{1C}(\alpha_C) \\ C_{2C}(\alpha_C) \\ LC_{6C}(\alpha_C) \end{Bmatrix} \times \begin{bmatrix} \cos\alpha_C & \sin\alpha_c & 0 \end{bmatrix} + \frac{1}{2}\rho LTU_c \begin{Bmatrix} C'_{1C}(\alpha_C) \\ C'_{2C}(\alpha_C) \\ LC'_{6C}(\alpha_C) \end{Bmatrix} \times \begin{bmatrix} -\sin\alpha_C & \cos\alpha_C & 0 \end{bmatrix} \quad (3)$$

and

$$\mathbf{K}^{(C)} = \frac{1}{2}\rho LTU_C^2 \begin{Bmatrix} C'_{1C}(\alpha_C) \\ C'_{2C}(\alpha_C) \\ LC'_{6C}(\alpha_C) \end{Bmatrix} \times \begin{bmatrix} 0 & 0 & 1 \end{bmatrix} \quad . \quad (4)$$

The coefficients C'_{1C}, C'_{2C} and C'_{6C} are the derivatives of the hydrodynamic coefficients C_{1C}, C_{2C} and C_{6C} in respect of incidence angle. If the latter coefficients are expressed by means of Fourier series, these derivatives can be easily calculated.

The first parcel in equation (2) is steady and must be incorporated in the loads applied to obtain the static equilibrium configuration. The other two must appear in the dynamic equation.

Wind forces and moment

The expressions for the forces and moment due to the wind are very similar to the ones for the ocean current. If U_W is the wind velocity and α_W the angle between the wind velocity and the axis of the floating unit, using the notation of Wichers (1987) the longitudinal force X_W, the lateral force Y_W and the moment N_W can be expressed by

$$\begin{aligned} X_W(\alpha_W) &= \frac{1}{2}\rho_a A_T C_{1W}(\alpha_W)U_W^2 \\ Y_W(\alpha_W) &= \frac{1}{2}\rho_a A_L C_{2W}(\alpha_W)U_W^2 \\ N_W(\alpha_W) &= \frac{1}{2}\rho_a A_L C_{6W}(\alpha_W)U_W^2 \end{aligned} \quad , \quad (5)$$

where ρ_a is the air density, A_T the exposed longitudinal area, A_L the exposed lateral area and C_{1W}, C_{2W} and C_{6W} are the hydrodynamic coefficients of the wind loads.

When the unit moves these efforts change as the angle α_W changes. If $\dot{x}$ and $\dot{y}$ are the components of the unit velocity and $\dot{\alpha}$ is its angular velocity, the new forces and moment $\tilde{X}_W$, $\tilde{Y}_W$ and $\tilde{N}_W$ can be expressed as, considering only the linear terms in $\dot{X}$, $\dot{y}$ and $\dot{\alpha}$,

$$\begin{Bmatrix}\tilde{X}_W\\ \tilde{Y}_W\\ \tilde{N}_W\end{Bmatrix}=\begin{Bmatrix}X_W\\ Y_W\\ N_W\end{Bmatrix}-\mathbf{B}^{(W)}\cdot\begin{Bmatrix}\dot{x}\\ \dot{y}\\ \dot{\alpha}\end{Bmatrix}-\mathbf{K}^{(W)}\cdot\begin{Bmatrix}x\\ y\\ \alpha\end{Bmatrix} \quad , \qquad (6)$$

where

$$\mathbf{B}^{(W)}=\rho U_W\begin{Bmatrix}A_T C_{1W}(\alpha_W)\\ A_L C_{2W}(\alpha_W)\\ A_L L C_{6W}(\alpha_W)\end{Bmatrix}\times\begin{bmatrix}\cos\alpha_W & \sin\alpha_W & 0\end{bmatrix}+ \frac{1}{2}\rho U_W\begin{Bmatrix}A_T C'_{1W}(\alpha_W)\\ A_L C'_{2W}(\alpha_W)\\ A_L L C'_{6W}(\alpha_W)\end{Bmatrix}\times\begin{bmatrix}-\sin\alpha_W & \cos\alpha_W & 0\end{bmatrix} \qquad (7)$$

and

$$\mathbf{K}^{(W)}=\frac{1}{2}\rho U_W^2\begin{Bmatrix}A_T C'_{1W}(\alpha_W)\\ A_L C'_{2W}(\alpha_W)\\ A_L L.C'_{6W}(\alpha_W)\end{Bmatrix}\times\begin{bmatrix}0 & 0 & 1\end{bmatrix} \qquad (8)$$

The coefficients C'_{1W}, C'_{2W} and C'_{6W} are the derivatives of the hydrodynamic coefficients C_{1W}, C_{2W} and C_{6W} in respect of incidence angle. If the latter coefficients are expressed by means of Fourier series, these derivatives can be easily calculated.

The first parcel in equation (6) is steady and must be incorporated in the loads applied to obtain the static equilibrium configuration. The other two must appear in the dynamic equation.

Wave drift damping

For an harmonic sea with unitary amplitude, the steady wave forces can be represented by the vector of generalized forces:

$$\mathbf{D}_0(\omega;\alpha_O)=D_{x,0}(\omega;\alpha_O)\mathbf{i}+D_{y,0}(\omega;\alpha_O)\mathbf{j}+N_{z,0}(\omega;\alpha_O)\mathbf{k} \quad , \qquad (9)$$

where $D_{x,0}(\omega;\alpha_O)$ is the surge steady force, $D_{y,0}(\omega;\alpha_O)$ is the sway steady force and $N_{z,0}(\omega;\alpha_O)$ is the yaw steady moment. These loads are functions of the wave frequency ω and the angle α_O between the wave direction and the floating unit axis. They are determined in the solution of the hydrodynamic linear problem in the frequency domain and are supposed to be know in this analysis.

Is there is an ocean current with velocity $U_{R,C}$ and direction $\alpha_{R,C}$ relative to the body, the steady wave forces are altered. Aranha (1996) has shown that ignoring terms quadratic in velocity the new values can be expressed exactly. Let $\alpha_{O,C}$ be the relative angle between the directions of wave and current. Then

$$\alpha_{O,C}=\alpha_O-\alpha_{R,C} \qquad (10)$$

Defining the *encounter frequency* ω_E as

$$\omega_E=\left[1-\frac{U_{R,C}}{c}\cos\alpha_{O,C}\right]\omega \quad , \qquad (11)$$

where c is the wave velocity

$$c=\frac{g}{\omega} \qquad (12)$$

and the *refraction angle* α_R as

$$\alpha_R=\alpha_O+2\frac{U_{R,C}}{c}\sin\alpha_{O,C} \quad , \qquad (13)$$

according Aranha (1996) the steady wave forces influenced by the ocean current can be expressed by

$$\mathbf{D}_U(\omega;\alpha_O)=\left[1-4\frac{U_C}{c}\cos\alpha_{O,C}\right]\cdot\mathbf{D}_O(\omega_E;\alpha_O) \qquad (14)$$

Expanding this expression in Taylor's series, ignoring the terms quadratic in velocity and defining the auxiliary vectors

$$\mathbf{b}_W=4\mathbf{D}_0(\omega;\alpha_O)+\omega\frac{\partial\mathbf{D}_0}{\partial\omega}(\omega;\alpha_O) \qquad \mathbf{b}_R=-2\frac{\partial\mathbf{D}_0}{\partial\alpha_O}(\omega;\alpha_O) \qquad (15)$$

leads to

$$\mathbf{D}_U(\omega;\alpha_O)=\mathbf{D}_0(\omega;\alpha_O)+U_C\mathbf{B}_H^{(WD)}\cdot\begin{Bmatrix}\cos\alpha_C\\ \sin\alpha_C\\ 0\end{Bmatrix} -\mathbf{B}_H^{(WD)}\cdot\begin{Bmatrix}\dot{x}\\ \dot{y}\\ \dot{\alpha}\end{Bmatrix}-\mathbf{K}^{(WD)}\cdot\begin{Bmatrix}x\\ y\\ \alpha\end{Bmatrix} \qquad (16)$$

where

$$\mathbf{B}_H^{(WD)}(\omega)=\frac{1}{c}\cdot\left[\mathbf{I}_{(3,3)}\middle|\mathbf{I}_{(3,3)}\right]_{(3,6)} \cdot\begin{bmatrix}\cos\alpha_O\cdot\mathbf{b}_W & \sin\alpha_O\cdot\mathbf{b}_W & \mathbf{0}\\ \sin\alpha_O\cdot\mathbf{b}_R & -\cos\alpha_O\cdot\mathbf{b}_R & \mathbf{0}\end{bmatrix} \qquad (17)$$

and

$$\mathbf{K}_H^{(WD)}(\omega)=\frac{1}{2}\begin{bmatrix}\mathbf{0} & \mathbf{0} & \mathbf{b}_R\end{bmatrix}_{(3,3)} \quad . \qquad (18)$$

In a random sea, if $S(\omega)$ is its spectrum, the stationary force, influenced by the ocean current and the body movement, becomes

$$\hat{\mathbf{D}}_U(\alpha_O)=\hat{\mathbf{D}}_0(\alpha_O)+U_C\mathbf{B}^{(WD)}(\alpha_O)\cdot\begin{Bmatrix}\cos\alpha_C\\ \sin\alpha_C\\ 0\end{Bmatrix} -\mathbf{B}^{(WD)}(\alpha_O)\cdot\begin{Bmatrix}\dot{x}\\ \dot{y}\\ \dot{\alpha}\end{Bmatrix}-\mathbf{K}^{(WD)}(\alpha_O)\cdot\begin{Bmatrix}x\\ y\\ \alpha\end{Bmatrix} \quad , \qquad (19)$$

where

$$\hat{\mathbf{D}}_0(\alpha_0) = 2\int_0^\infty S(\omega)\cdot \mathbf{D}_0(\omega;\alpha_0)d\omega$$

$$\mathbf{B}^{(WD)}(\alpha_0) = 2\int_0^\infty S(\omega)\cdot \mathbf{B}_H^{(WD)}(\omega;\alpha_0)d\omega \tag{20}$$

$$\mathbf{K}^{(WD)}(\alpha_0) = 2\int_0^\infty S(\omega)\cdot \mathbf{K}_H^{(WD)}(\omega;\alpha_0)d\omega$$

Low frequency oscillatory forces

The coupling of the several harmonic components of an random sea results in low frequency random forces that excite the floating unit in its natural periods in the horizontal plane. Let $S_{FX}(\mu;\alpha_0)$ be the surge force spectrum, $S_{FY}(\mu;\alpha_0)$ the sway force spectrum and $S_{NZ}(\mu;\alpha_0)$ the yaw moment spectrum in low frequency, with μ being the "difference frequency". In the frequency range of interest it can be shown that this spectra are essentially white noise, see Aranha & Fernandes (1994). So

$$S_{FX}(\mu;\alpha_0) \cong S_{FX} = 8\int_0^\infty S^2(\omega)\cdot D_{x,0}^2(\omega;\alpha_0)d\omega$$

$$S_{FY}(\mu;\alpha_0) \cong S_{FY} = 8\int_0^\infty S^2(\omega)\cdot D_{y,0}^2(\omega;\alpha_0)d\omega \tag{21}$$

$$S_{NZ}(\mu;\alpha_0) \cong S_{NZ} = 8\int_0^\infty S^2(\omega)\cdot N_{z,0}^2(\omega;\alpha_0)d\omega$$

Damping in still water

The viscous damping due to the motion of the hull in still water is non-linear. If ω_x, ω_y and ω_z are the natural frequencies in surge, sway and yaw and σ_x, σ_y and σ_z are the rms of the motions in those directions, disregarding the coupling terms such damping can be expressed as

$$\mathbf{B}_N(\sigma) = \begin{bmatrix} B_{11}^{(N)} & 0 & 0 \\ 0 & B_{22}^{(N)} & 0 \\ 0 & 0 & B_{66}^{(N)} \end{bmatrix} \tag{22}$$

where

$$B_{11}^{(N)} = \frac{8}{3\pi}\sqrt{2}\sigma_x\omega_x\left[\frac{1}{2}\rho TLC_{1c}(0^\circ)\right]$$

$$B_{22}^{(N)} = \frac{8}{3\pi}\sqrt{2}\sigma_y\omega_y\left[\frac{1}{2}\rho TLC_{1c}(90^\circ)\right] \tag{23}$$

$$B_{66}^{(N)} = \frac{8}{3\pi}\sqrt{2}\sigma_z\omega_z\left[\frac{1}{16}\rho TL^4C_{1c}(90^\circ)\right]$$

Dynamic Equation

In the quasi-linear approach, the motion of the floating unit around its equilibrium position, is modeled by the equation

$$\mathbf{M}\ddot{\mathbf{q}} + \mathbf{B}\dot{\mathbf{q}} + \mathbf{K}\mathbf{q} = \mathbf{F} \tag{24}$$

where **q** is a vector the elements of which are the displacements and rotation in the horizontal plane, **M** is the *mass matrix*, **B** is the *damping matrix*, **K** is the *stiffness matrix* and **F** is a vector that contains the components of low frequency excitation

$$\mathbf{F} = \begin{Bmatrix} F_x(t) \\ F_y(t) \\ N_z(t) \end{Bmatrix} . \tag{25}$$

Each one of these matrices are described in the following.

Mass matrix

If M is the mass of the floating unit, M_{11} and M_{22} the added mass in surge and sway, I is rotational inertia in yaw and M_{66} the added inertia in the horizontal plane, disregarding the coupling terms, the mass matrix **M** has the form

$$\mathbf{M} = \begin{bmatrix} M+M_{11} & 0 & 0 \\ 0 & M+M_{22} & 0 \\ 0 & 0 & I+M_{66} \end{bmatrix} \tag{26}$$

Damping matrix

The damping matrix **B** is the sum of a linear parcel $\mathbf{B}_L$ and a non-linear one $\mathbf{B}_N(\sigma)$ due to the damping in still water (see equations (23) and (22)):

$$\mathbf{B} = \mathbf{B}_L + \mathbf{B}_N(\sigma) \tag{27}$$

The linear damping matrix $\mathbf{B}_L$ is itself another sum that incorporates the different mechanisms that contribute to the linear damping:

$$\mathbf{B}_L = \mathbf{B}^{(C)} + \mathbf{B}^{(W)} + \mathbf{B}^{(WD)} \tag{28}$$

where $\mathbf{B}^{(C)}$ is the contribution of current (equation (3)), $\mathbf{B}^{(W)}$ is the damping due to the wind (equation (7)) and $\mathbf{B}^{(WD)}$ is the wave-damping (equation (17)).

Stiffness matrix

The stiffness matrix can be assembled by the sum

$$\mathbf{K} = \mathbf{K}^{(M)} + \mathbf{K}^{(C)} + \mathbf{K}^{(W)} + \mathbf{K}^{(WD)} \tag{29}$$

where $\mathbf{K}^{(M)}$ takes into account the restoring forces of the mooring system, $\mathbf{K}^{(C)}$ is the contribution of current (equation (4)), $\mathbf{K}^{(W)}$ is the parcel of stiffness due to the wind loads (equation (8)) and $\mathbf{K}^{(WD)}$ is due to the wave-damping (equation (18)).

Response spectrum

From equation (24) we can define the *transfer matrix*

$$\mathbf{H}(\omega;\sigma) = \left[H_{jk}(\omega;\sigma)\right] = \left[-\omega^2\mathbf{M} + i\omega\mathbf{B}(\sigma) + \mathbf{K}\right]^{-1} \tag{30}$$

and introducing the matrix

$$\mathbf{H}^{(2)} = \left[\left|H_{jk}(\omega;\sigma)\right|^2\right] \tag{31}$$

then the response spectrum is

$$\begin{Bmatrix} S_x(\omega) \\ S_y(\omega) \\ S_\alpha(\omega) \end{Bmatrix} = \mathbf{H}^2 \times \begin{Bmatrix} S_{FX}(\omega) \\ S_{FY}(\omega) \\ S_{NZ}(\omega) \end{Bmatrix} \tag{32}$$

NUMERICAL RESULTS

In this section a study case is presented to exemplify the method. The unit chosen was a FPSO moored in a turret configuration which is operating at Campos basin on a 1000 meters water depth.

Floating unit data

The unit main characteristics are:

total length:	330 m;
draft conditions:	17 %;
maximum displacement:	280000 tons;
added mass in surge:	2496 tons;
added mass in sway:	34869 tons;

The turret position is 150 m from the middle of the ship.

Environmental data

The environmental data chosen for the analysis was the 10-year worst conditions in Campos basin:

Wind velocity:	29.23 m/s;
Current velocity:	1.60 m/s;
Significant wave height:	7.80 m;
Zero up-crossing period:	11.50 s;

It was assumed that current, wind and velocity data are equal in all directions.

Mooring system characteristics

The mooring lines considered for the analysis had three different segments the characteristics of which are described on Table 1. The mooring pattern is composed by 8 lines settled in angles: 0, 45, 90, 135, 180, 225, 270 and 315 degrees counterclockwise from the north.

Table 1 - Characteristics of mooring lines

	lower(near the anchor)	middle	upper
Material	Chain grade 4	Wire	Chain grade 4
Diameter (m)	0.203	0.109	0.160
Length (m)	1350	1250	150
Weight in sea (kN/m)	1.729	0.421	1.364
Elastic Modules (MN)	793.88	536.790	620.68

Results

The analysis done was a parametric study in the pre-tension of the mooring lines. It was assumed that the pre-tension was equal of all the lines varying in the range 800 and 4000 kN. For some of these values the results presented were for a system in a taut-leg configuration.

In a turret system, the floating unit is free to rotate in the horizontal plane to achieve the equilibrium position. The first results presented aim to discuss how the heading of the unit varies with the angle between current and wind-waves. The heading of the unit was measured counterclockwise from the North to the its axis and the heading of the environmental loads and the heading of environmental agents counterclockwise from the North to the direction they go, see Figure 1.

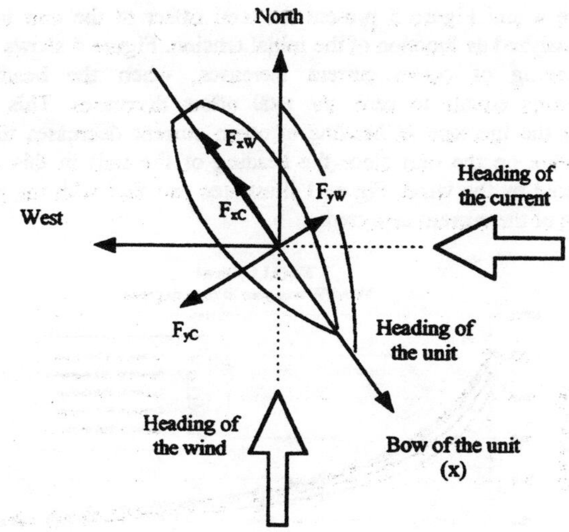

Figure 1 - Effect of the heading of the current

Figure 2 shows how the heading of the unit varies when the heading of current is equal to zero and the heading of the wind/waves varies between zero and 90 degrees. Figure 3 shows the same results in the case the heading of the wind/waves is zero and the heading of the ocean current varies.

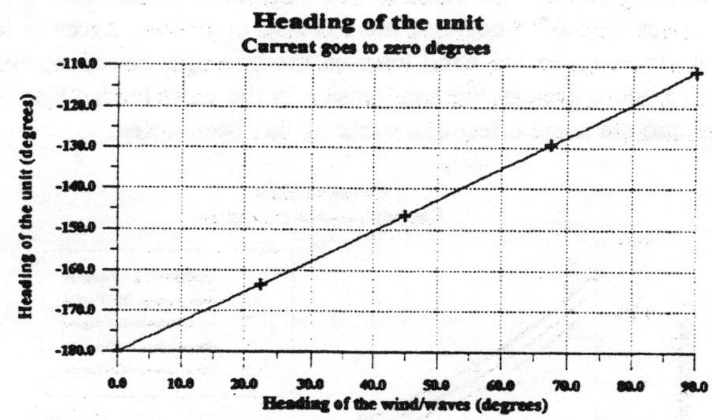

Figure 2 - Heading of the unit (current to 0 degrees)

These figures show that the behavior is almost linear with the angle of the heading that varies. In the case studied the heading of the unit presents a great dependence of the heading of the wind due the draft condition. With low draft, the wind forces are greater than the current forces due to the great projected area emerged.

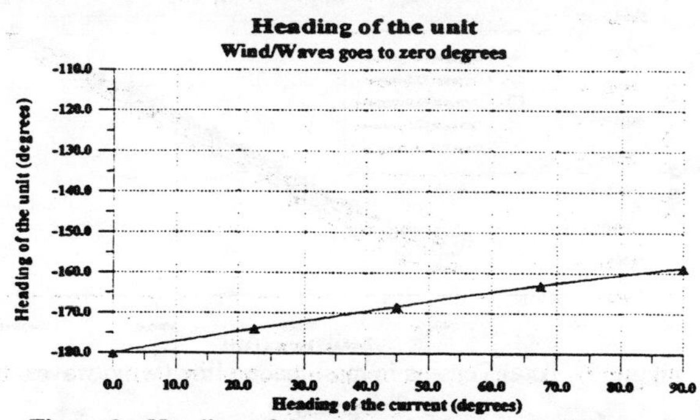

Figure 3 - Heading of the unit (wind/waves to 0 degrees)

Figure 4 and Figure 5 present the total offset of the unit for both cases analyzed as function of the initial tension. Figure 4 shows that as the heading of ocean current increases, when the heading of wind/waves equals to zero, the total offset decreases. This occurs because the increase in heading of ocean current decreases the total solicitation on the unit since the heading of the unit in this case is dominated by the wind. Figure 1 illustrates this fact with the greatest heading of the current analyzed.

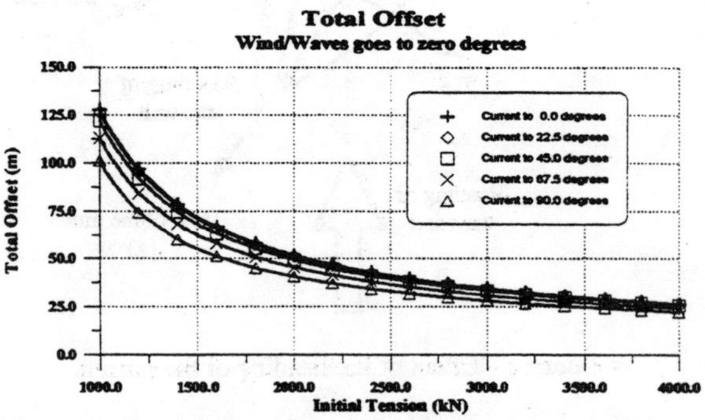

Figure 4 - Total Offset of the unit (wind/waves to 0 degrees)

Figure 5 presents the behavior of the total offset in the other case. Now the effect of the current forces do not decrease the total offset with the increase in heading of the wind, this effect is sensible only with heading of the wind equal to 67.5 degrees. For the heading of the wind lesser than 67.5 degrees, the increase in current forces is lesser than the increase in the wind force on the opposite direction. Finally, Figures 5 and 6 present the total tension in the most loaded line of the system and the same effect of the current has been noted.

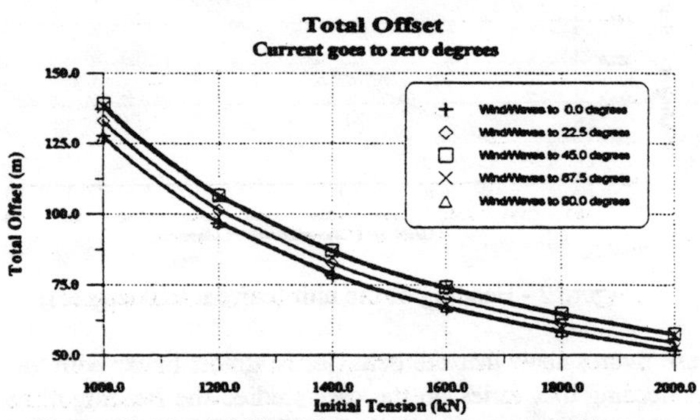

Figure 5 - Total Offset of the unit (current to 0 degrees)

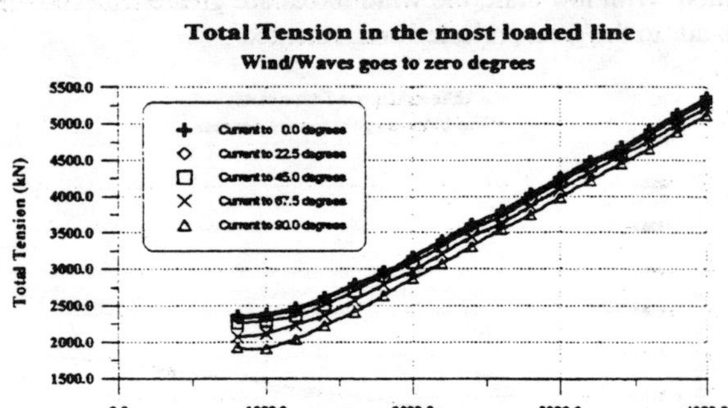

Figure 6 - Total Tension in most loaded line (wind/waves to 0 degrees)

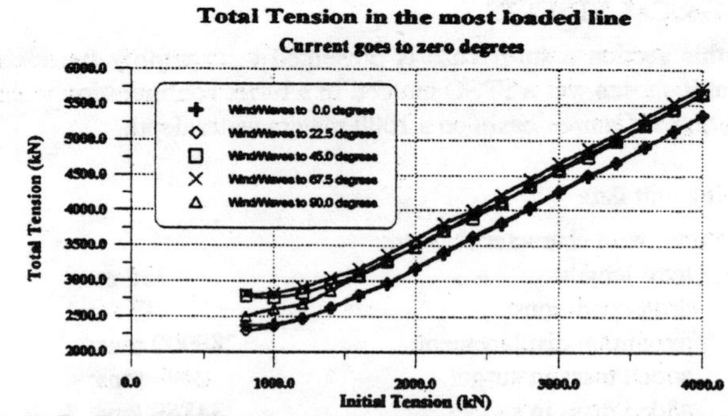

Figure 7 - Total Tension in most loaded line (current to 0 degrees)

CONCLUSION

Although the problem is somewhat complex, the quasi-linear approach make its solution feasible. It was possible to implement the model in a computer program which obtains the solution for a large set of pre-tension values in few minutes. That program revealed to be a suitable tool for mooring system design.

REFERENCES

Aranha, J. A. P. & Pesce, C. P. & Martins, C. A. & Andrade, B. L. R., (1993). "Mechanics of submerged cables: asymptotic solution and dynamic tension", *3rd International Offshore & Polar Eng. Conference*, Singapore, Jun.6-11, 1993, vol.2, pp.345-356.

Aranha, J. A. P. (1994). "A formula for wave damping in the drift of a floating body", *J. Fluid Mech.*, Vol.275, pp.147-155.

Aranha, J. A. P. (1996). "Second order horizontal steady forces and moment on a floating body with small forward speed", *J. Fluid Mech.*, Vol. 312, pp. 39-54.

Aranha, J. A. P. & Fernandes, A. C. (1994). "On the second order slow drift force spectrum of a floating body", *Applied Ocean Research*, Vol 17, pp 311-313.

Crandall, S. H. & Mark, W. D. (1963). "Random Vibration", *Academic Press.*

Faltinsen, O. (1994). "Wave and current induced motions of floating production systems", *Applied Ocean Research*, vol 15, N.6, pp 351-370.

Kim, M. H. & Yue, D. K. P. (1990). "The complete second order diffraction solutions for an axisymmetric body. Part2. Bichromatic incident waves and body motions", *J. Fluid Mech.*, vol.211, pp.557-593.

Newman,J.N. (1974). "Second order slowly varying forces on vessels in irregular waves", *Proc. Int. Symp on Dynamics of Marine Vehicles*, London, 1974.

Wichers,J.E.W. (1982). "On the low-frequency surge motion of vessels moored in high seas", *14 Annual OTC*, Houston, Texas.

Wichers,J.E.W. (1987). "A Simulation Model for a Single Point Moored Tanker", Publication N.797, *Maritime Research Institute Netherlands*, Wageningen.

Proceedings of the Seventh (1997) International Offshore and Polar Engineering Conference
Honolulu, USA, May 25-30, 1997

ISBN 1-880653-28-1 (Set); ISBN 1-880653-29-X (Vol. I)

Offshore Single Point Mooring Systems for Import of Hazardous Liquid Cargoes Offshore Southern California

Aaron M. Salancy
NASSCO
San Diego, CA, USA
Robert G. Bea
University of California at Berkeley
Berkeley, CA, USA

ABSTRACT

The goal of the study summarized in this paper was the determination of feasibility of single point mooring systems (SPMS) for use as deepwater ports for the import of hazardous liquid cargoes offshore southern California. The use of deepwater ports is advocated because it has been determined by the U.S. Coast Guard that they represent the least risky form of crude oil import, lessening the likelihood of occurrence and environmental impact severity of accidents. Two configurations of SPMS were examined as deepwater ports in this study: catenary anchor leg mooring (CALM) and single anchor leg mooring (SALM). Two sites for these systems were chosen offshore southern California by the California State Lands commission : El Segundo and Morro Bay. The study examined the environmental conditions at both sites, developed analytical models with which to evaluate the suitability of SPMS to these environmental conditions, determined the reliability of the systems by use of state-of-the-art reliability methods, and evaluated the feasibility of the systems by comparing reliability to system costs. The results of this study indicate that SPMS for offshore southern California conditions are feasible and do not require major technological developments to allow such systems to be designed, constructed, and operated. Use of these systems should lower the number of accidents due to hazardous liquid cargo import, as well as reduce the impact of those accidents which do occur.

KEY WORDS: mooring, single point mooring systems, deep water ports, oil transportation, reliability, risk analysis

INTRODUCTION

The purpose of this study was to perform an evaluation of the reliability and feasibility of deepwater ports, specifically those consisting of a single point mooring system (SPMS) and support equipment for tanker discharge, for offshore southern California. These deepwater ports could serve as discharge ports for tankers delivering crude oil from Valdez, Alaska, or other supply points, to southern California. Two locations along the California coast, El Segundo and Morro Bay, were studied as potential locations for these facilities. These two locations were specified by the California State Lands Commission. The water depth for both facilities was proposed as one thousand feet.

This study investigated the feasibility of deepwater ports at the two locations proposed by the California State Lands Commission. This investigation required examining the existing types of SPMS, evaluating how they would function in the offshore southern California environment, determining what configurations would adequately withstand the environmental conditions while performing satisfactorily, and establishing the financial and technical feasibility of the resulting configurations. The determination of feasibility was based on an assessment of the reliability characteristics of each proposed system and the costs to build and operate the system.

Two facility systems were chosen for detailed examination. Each facility was to be capable of servicing tankers up to roughly very large crude carrier (VLCC) size (loosely defined as 200,000 to 275,000 dead weight tons), with the tankers requiring no major modification to use the facility. Each facility would also meet all major relevant requirements and guidelines for an offshore installation of this type. It was desirable that the facilities should not require major leaps in technology from that currently existing in any component, or in installation, maintenance, or regular operation. This was considered necessary to insure that reasonable reliability, feasibility, and cost estimates could be obtained. Above all else, the facility should be capable of rapid, safe disconnection in deteriorating sea states and be capable of surviving intact the 100 year storm and seismic conditions with a sufficiently high probability of success. Only once these requirements were met by a system would the financial feasibility analysis be conducted for that system.

The scope of this study included: the SPMS, the tankers which are expected to use the facility, all equipment necessary for connection and discharge of the tankers, and the piping system to transfer oil from the facility to the shore. The pipeline to shore itself was not a main focus of this study, however, and the shoreside facilities have not been examined. These aspects of the systems should be given attention in the future, as they will heavily impact system feasibility by their effect on facility cost.

SPMS TYPES

AN SPMS is a mooring which allows a ship to weather vane around the mooring, thus minimizing the environmental loads on the system by allowing the moored ship to head into the prevailing weather. The SPMS also provides the interface between the tanker and the pipeline for the discharge of crude oil. The primary SPMS in placed in operation to date include: catenary anchor leg mooring (CALM), single anchor leg mooring (SALM), turret mooring, and fixed, articulated loading or catenary articulated tower. After a review of the merits of these alternative SPMS for the proposed application offshore California, the CALM and SALM systems were choosen for further study.

SPMS have been used successfully in many applications around the world in many different conditions. The are more than 500 SPMS in operation around the world. The challenge posed in this study for the use of SPMS is the specified water depth of one thousand ft. and the offshore California oceanographic and seismic conditions. This is not an unprecedented depth for the use of SPMS, as the Marlim Field catenary anchor leg mooring (CALM) off the coast of Brazil is located in approximately 1312 ft. of water (Hwang and Bensimon, 1990). However, the environmental conditions offshore California, including seismic activity, are more severe than those encountered by most SPMS. Even so, the design of a SPMS for use in one thousand ft. of water off the California coast should require no major breakthrough developments.

STUDY SITES & CONDITIONS

Two sites along the southern California coast were identified by California State Lands Commission as prospective SPMS facility sites : El Segundo and Morro Bay (Fig. 1). The specified water depth of 1000 ft. gives a distance offshore ranging from 6 to 12 miles at both sites, for a variety of specific locations.

The weather along the southern California coast is primarily a product of extra-tropical storms originating in the Northeast Pacific during winter (Stevens, 1977). Other weather phenomena such as tropical storms, thunderstorms, tornadoes and waterspouts are rare at best.

The main environmental components of wind, waves, current and seismic activity are summarized in Table1. The environmental conditions are presented in a probability framework, which will allow for rapid integration with the methods used to determine probability of failure due to environmental loadings and effects. This method is based on the concept of "return period". The return period is the mean elapsed time expected between occurrences of an event. These values are based on past statistical data and extrapolation.

Wind in this region is primarily from the northwest, circulating around the Pacific High, and varies from winter to summer. The most notable exception to this trend are the Santa Ana winds, generated inland and blowing out over the coast. However, given the distance offshore of these facilities, the effects of this type of wind can be ignored. Another phenomenon of this region is the Catalina Eddy. This eddy causes recurvature of winds locally near the coast, but, for this study, it can also be considered insignificant (Stevens, 1977).

The wind speed can be expected to vary throughout the region under consideration, but estimates for the area should prove sufficient for this study. Values for wind speed were researched from several sources (Intersea Research Corporation, 1974; Stevens, 1977) and are given by return period in Table 2.

The primary wave direction offshore southern California is north to northwesterly, with little seasonal variation. Values for maximum wave height were researched from several sources (Intersea Research Corporation, 1974; Department of Navigation and Ocean Development, 1977; Stevens, 1977; API, 1989) and are given by return period in Table 1. It should be noted that for these values, maximum wave height is related to significant wave height by Eqn. 1 (Stevens, 1977):

$$H_{max} = 1.86\, H_S \tag{1}$$

Fig. 1 : Site Locations

Table 1 : Environmental Conditions by Return Period

Return Period (years)	Wind Velocity (@ 10m) (knots)	Surface Current Velocity (knots)	Expected Maximum Wave Height (ft.)	Peak Vertical Ground Accel. (g's)
2	39	1.2	29	0.01
10	55	1.7	37	0.05
100	72	2.2	46	0.18
1000	88	2.7	54	0.40

Currents in this region are usually relatively small. It is normal practice to estimate current speed by a combination of two components. The first component is shear force imparted by wind. The second component consists of tidal flows and currents arising from the topography of the ocean floor. Surface currents are usually wind-driven, while subsurface currents are driven by geostrophic factors and tides.

In this study, the surface current speed will be taken as 4% of the steady wind speed, subsurface currents at mid-depth will be taken as one-half surface current speed, and near-bottom currents will be taken as one-third surface current speed (Stevens, 1977). This approximation includes the small effect of tides. The resulting values for current speed correlate well with those found in other sources (Intersea Research Corporation, 1974). Values for maximum current are given by return period in Table 1.

For the analyses done later in this paper involving environmental loads, it is necessary to know the primary direction of wind, wave and current. Since only extreme conditions are examined, directionality need only be known for these conditions.

In this region, storms are usually from the northwest. In a large storm, wind and wave direction will tend to coincide, especially when short wind gusts are discounted. Surface current will also follow the direction of the wind, as discussed in the previous section. Therefore, the environmental components will be assumed to act unidirectionally. There are some flaws to this assumption, as some directional spreading is inevitable and subsurface currents are ignored. However, directional spreading can be addressed, and it is shown later that subsurface currents have little effect on the analytical models. Therefore, unidirectionality of wind, wave and currents was assumed.

The southern California area has a relatively high degree of seismic activity. Offshore structures are usually analyzed for seismic safety by examining the effects of local activity as well as distant, more severe activity. The seismicity of a specific region, however, is highly variable, depending on local and distant fault positions. Since the area of interest in this study is located in deep water, fault positions are relatively unknown, and an accurate portrayal of seismicity in the region is not possible. Therefore, values from a study of the entire offshore southern California area have been used for peak ground accelerations by return period (Bea, 1992). These values are given in Table 1.

The features of the ocean floor of both sites were evaluated by examination of nautical charts prepared by the National Ocean Service (U.S. Department of Commerce, 1991). The purpose of this examination was to determine the slope of the floor in these locations and discover which locations would be unfavorable due to excessive slope, which may indicate a likelihood of sliding.

At El Segundo, the slope of the floor was found to vary from 12.8 degrees to 2.2 degrees at 1000 ft. depth, with the slopes growing steeper to the south. The only feature of note in this area is the Santa Monica Canyon, which is not especially deep. The slopes at Morro Bay were found to be less severe, ranging from 1.9 degrees to 0.9 degrees.

Soils were investigated for the purpose of calculating anchor and anchor pile holding power, as well as determining the possibility of scour, sliding and other ocean floor phenomena which may have an effect on the facilities. The soils in this region consist of "clayey silt" to a depth of approximately 15 ft. below the sea floor, and "stiff, silty clay" below this level to a depth of approximately 200 ft.. The former type of soil has an undrained shear strength of approximately 1.5 kilopounds per square foot, while the latter soil type has an undrained shear strength of approximately 2.0 kilopounds per square foot (Woodward-Clyde, 1984).

Scour, slumping and sliding are all possible in this area, but risks should be lower in areas with shallower floor slopes (Intersea Research Corporation, 1974). Silt soils, such as those at the sea floor, have a low resistance to scour. Clay soils have a lower susceptibility to scour, however, so while some scour may occur, it will not be severe, due to the presence of the stiff clay soils below the silt soils (Woodward-Clyde, 1983).

Liquefaction is probably a greater concern. This phenomenon involves the sudden loss of soil strength due to ground shaking or wave effects, and could have a severe effect on the holding power of pile anchors. The top soil type, "clayey silts" will be expected to experience some loss in strength, on the order of 15%, but soil strength loss below 50 ft. should be negligible (Woodward-Clyde, 1978). Therefore, if piles extend more than 50 ft. below the sea floor, they should be relatively safe from liquefaction risks.

DESIGN REQUIREMENTS

An initial criteria for this study was the requirement that both facilities meet all relevant rules and guidelines governing safety. In this study, the primary guidelines are the ABS factors of safety on mooring lines, anchors and fatigue, and the API "watch circle" guideline governing maximum buoy offset based on product riser type (Jones, 1992; API, 1991).

A factor of safety (F.S.) of 2.0 is required by ABS on mooring lines for floating production systems examined by quasi-static analysis in the intact condition (Jones, 1992). The factor of safety is defined as the ratio of the capacity to the 100-year load, as calculated by either quasi-static or dynamic analysis. This factor of safety drops to 1.6 for the damaged condition, which refers to one mooring line broken. These and all other applicable ABS Factors of Safety are given in Table 2 (Jones, 1992).

Table 2 : Applicable ABS Factor *of* Safety Requirements

Component / Analysis	Holding Power	F.S. : Intact Condition	F.S.: Damaged Condition
Mooring Lines			
Quasi-static	Tension	2.0	1.6
Dynamic	Tension	1.67	1.33
Foundation			
Anchors			
Quasi-static	Tension	2.0	1.5
Dynamic	Tension	1.5	1.1
Piles			
Quasi-static	Vertical	2.0	1.5
	Horizontal	1.5	1.3
Dynamic	Vertical	1.5	1.25
	Horizontal	1.5	1.1
Fatigue			
Quasi-static	Tension	3.0	NA
Dynamic	Tension	3.0	NA

As Table 2 shows, factors of safety for dynamic analysis are lower than those for quasi-static. The methods of analysis employed in some parts of this study (line tensions and anchor loads for the CALM system) are considered dynamic, while the remainder of the analyses are considered quasi-static.

The watch circle specified by API states that the maximum horizontal buoy offset from calm water position under the maximum design conditions (taken as the 100-year storm in this study) for systems employing flexible production risers in deep water (2000 ft. to 3000 ft.) must not exceed 15% of the water depth. For shallow water (below 300 ft.) the criteria is 15% to 25% offset (API, 1991). In this case, "flexible product riser" refers to any hose or pipe falling from a buoy in a catenary shape or attached to a vertical leg system such as a SALM. Therefore, a maximum offset of 15% of water depth (150 ft. in this study) will be used as a design guideline.

Limits on operational sea states are another requirement that will be placed on these facilities. Requirements on operational sea states for the facility will be set by the Facility Manager, in conjunction with the Coast Guard. Establishing these operational constraints is necessary because both systems are disconnectable, and require disconnection in order to be adequately resilient to severe storm conditions. In operation, the tanker captain and the facility pilot must decide when to stop cargo transfer and disconnect from the SPMS due to adverse or expected adverse environmental conditions.

CALM

A catenary anchor leg mooring (CALM) system consists of a set of anchored catenary legs arranged in a radial pattern around a large buoy, with some type of flow line from the sea floor to the buoy for transporting liquids (Fig. 1). This is the most common type of SPMS. A CALM system derives its restoring force to offset from the tension in the anchor legs due to the weight of the legs and an initial pretension.

The disadvantages of these systems are : disconnection requires substantial amounts of time (except for emergency disconnection) while weather may be deteriorating; operation is limited by the seakeeping ability of the vessels assisting in line recovery; and tankers and other vessels can come into contact with the catenary legs, causing damage to the legs. A problem which this system shares with all other systems not involving dedicated vessels is the presence of floating hoses on the water surface. Two sections of approximately 1000 foot long hose will be floating on the surface when the facility is not in operation. These hoses are vulnerable to damage from passing vessels. This is why it is necessary to establish zones for the facility that are free of most maritime traffic and are constantly monitored for stray vessels (LOOP, 1992).

The final design of the CALM for this facility is illustrated in Fig. 2. This design is the result of iteration between the analytical models and the reliability and feasibility analyses performed during this study.

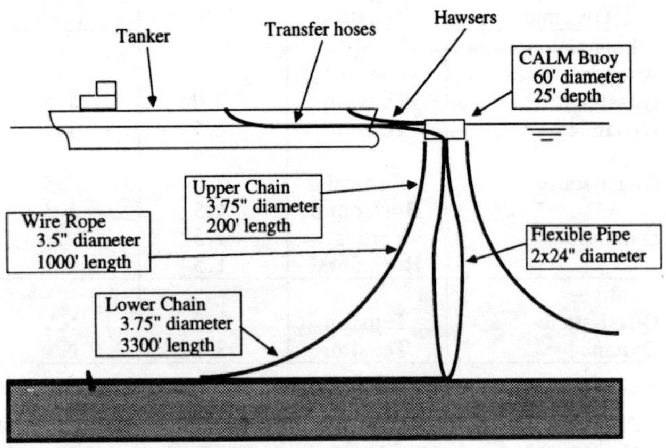

Fig. 2: Final CALM Design

The two main components of this system are the buoy and the anchor legs. The buoy is 60 ft. in diameter, 25 ft. in depth and weighs 271 Long Tons (LT). There are 8 anchor legs, arranged in a 45 degree spread. Each leg has three components : an upper chain section, a wire rope section and a lower chain section. The upper section is made of chain for tensioning and wear purposes, while the lower section is chain for bottom-abrasion purposes. The lower section also adds to the system holding power. Wire rope is used in the supported section of the leg because it gives a higher restoring force for a given pretension, due to its lower weight when compared with chain (API, 1991).

SALM

A single anchor leg mooring (SALM) system differs from a CALM in that it has only one anchor leg, which is vertical and highly tensioned. A SALM derives its restoring force from this single tensioned leg by the horizontal component of the leg tension when angularly displaced, as well as the added buoyancy from the buoy resulting from horizontal displacement. A simple description of the physics of the system would be to describe it as an inverted pendulum. A SALM consists of a vertical buoyant riser, a foundation, a pretensioned leg from the sea floor to the riser, and a flow line from the sea floor to the surface. A SALM has the same connection possibilities as a CALM system, which, in this study, means that only hawser connections are under consideration. The foundation is typically of the pile anchor type in deep water depths.

There is some variation in SALM designs in the nature of the tensioned leg employed : chains, wire rope, tubular risers and articulated risers have all been used, along with various combinations of these elements. In this study, wire rope was initially investigated as the anchor leg, but loadings and resulting tensions required switching to an articulated tubular riser.

A SALM has the advantages over a CALM of being more forgiving of collisions, due to the nature of its restoring force, and has a lower likelihood of contact and entanglement problems as the leg is located directly beneath the buoy.

A SALM system has many drawbacks, however. They are more complicated and expensive than a comparable CALM system. They have not yet been proven in this water depth. Maintenance can be a problem, as a SALM fluid swivel is considered to be a high-maintenance item, and is usually located at the sea floor. And, like the CALM, the floating hoses on the surface are vulnerable to damage.

The final SALM design for this study is illustrated in Fig. 3. As with the CALM, this design is the result of iteration between the analytical models and the reliability and feasibility analyses performed during this study . The SALM buoy is 15 ft. in diameter, 70 ft. in depth and weighs 47.4 LT. The tubular riser is divided into three sections. The sections are, from top to bottom, 300 ft., 350 ft. and 320 ft. in length. All four are four feet in outer diameter. The wall thickness varies by section due to the differences in hydrostatic pressures. The uppermost section has 0.5 in. thick walls, the middle section has 0.75 in. thick walls and the lower section has 1.0 in. thick walls. In this design, the fluid swivel is located in the bottom of the buoy, which makes it easily accessible. The articulated leg is divided into three sections, each of which is slightly buoyant prior to installation and made negatively buoyant during installation. The transfer hoses connect to the bottom of the SALM buoy.

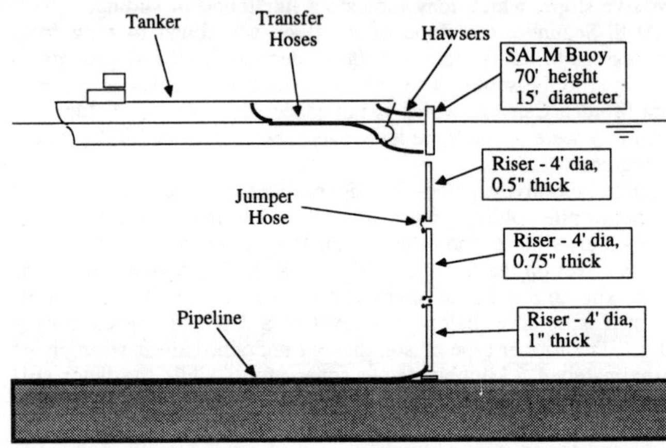

Fig. 3: Final SALM Design

ASSOCIATED COMPONENTS

The elements of each facility under consideration in this study are: the SPMS, including buoy, anchor legs and anchors or anchor piles; the crude oil tanker discharging cargo to the facility; the pair of hawsers to hold the tanker in position while it is using the facility; two tending vessels to assist the tanker in the use of the facility; two transfer hoses for the transfer of crude oil from the tanker to the product riser(s); the product riser(s) to the sea floor; and the pipeline itself, running from the sea floor to the shore-side facility. It should be noted that intermediate pumping stations will be required, but these were outside the scope of this study.

The San Diego class of crude oil tanker was examined in this study, as this class is considered typical of the Valdez, Alaska to southern California trade. The particulars of this tanker class can be found in Table 3.

Table 3 : San Diego Class Tanker Particulars

L. B. P. (ft.)	Beam (ft.)	Draft (ft.)	Lightship (LT)	Full Load (LT)
915	166	59.3	30,000	188,500

A typical rate of discharge for this size of tanker is 80,000 to 100,000 barrels per hour. This gives a time for discharge of approximately 17 hours for an entire cargo load of crude oil.

Product risers -- the flow line(s) used to transfer crude oil from the moored tanker to the ocean floor pipeline -- can be of three types : hoses, rigid pipe or flexible pipe. Hose will be used for the SALM system, as the product riser is attached to the anchor leg and need not support itself. Flexible pipe will be used for the CALM system. The flexible pipe will be free to hang in a catenary curve from the buoy to the sea floor.

Flexible pipe is preferred over rigid pipe because flexible pipe allows for larger relative motions and does not require heave compensation equipment. Flexible pipe consists of seven or more separate layers of material. These layers may be bonded or unbonded, although unbonded pipes are becoming the standard. The layers are typically (from inside out) : a steel carcass, a plastic sheet, a layer of wound steel wire, a flat steel carcass, an anti-friction plastic sheet, armor layers and an external plastic sheet. A typical value for minimum radius of curvature of flexible pipe is ten times pipe outer diameter (Sødahl, 1991).

Flexible pipe is well-proven in offshore applications. It has been used in water depths up to 850 meters. Over 2500 kilometers total of flexible pipe have been installed to date, with approximately half of that total presently over ten years old (Coutarel, 1992).

The design of the pipeline necessary for these facilities was considered to be outside the scope of this study. However, some important points concerning pipeline design for offshore southern California were found, and these are summarized here to give some feel for the difficulties involved in pipeline design.

Laying and servicing pipeline in 1,000 ft. of water is within the reach of today's technology. However, the equipment required is not readily available on the West Coast, and would probably need to be delivered from the Gulf of Mexico, the North Sea, or Brazil. This would have a substantial impact on the installation and maintenance costs of the pipeline. Of the environmental conditions present at the locations examined in this study, only one significantly affects the design of sub-sea pipeline. This is the seismic activity of the southern California region (California Coastal Commission, 1978).

Seismic events can cause three possible actions : soil liquefaction, elastic ground waves, and inelastic, permanent ground movement. The issue of soil liquefaction has already been mentioned. Elastic ground waves typically have peak-to-peak lengths of several miles, and amplitudes of in. or fractions of in.. These waves will not have a significant effect on ocean floor pipeline. Inelastic, permanent ground movement occurs along faults during seismic events, and may be either horizontal or vertical, with rupture regions up to several hundred ft. in length. These ruptures are a major source of concern in pipeline design. Locating pipelines clear of faults is the best design solution, but this is often not feasible, as faults in deep water are difficult to locate. When it is known that a section of pipeline must cross a fault, the pipeline may be reinforced at this section or lifted off the ocean floor by bents. If the latter course of action is taken, the area must be prohibited to commercial fishing and any drag-type operations (California Coastal Commission, 1978).

Federal regulations specify that offshore pipelines for the transfer of liquids must be buried. Burial eliminates the possibility of damage from anchors and other ocean floor equipment. However, burial is very expensive, especially for large diameter pipelines such as those considered in this study. Of course, pipelines should not be buried in regions where they cross known faults. Burial can also be counter-productive in areas where liquefaction is likely (California Coastal Commission, 1978).

Two vessels will be required for normal operations. One of these vessels will handle the hawsers during connection while the other retrieves the hoses. The primary mooring vessel must be sufficiently large and powerful to assist the tanker in adverse sea conditions. This vessel will be a standby/towing vessel of 60 to 80 meters length. It should be capable of operating in all weather conditions under which operation of the facility is to be conducted, with an added margin for emergencies. It will have hose handling capability of 10 meters by 22 meters with deck containment and drainage for crude oil. Wooden clad will be provided for servicing floating hoses. It must be highly maneuverable, with twin-ducted propellers, twin rudders and transverse thrust units. It should be equipped with two towing wires of approximate strength to the bollard pull, two towing win. (or a dual winch) and all necessary wire pennants and ancillary towing equipment. The required bollard pull will be based on the vessel being able to assist the tanker in the following environmental conditions : 35 knot wind, 1 knot adverse current and 3.5 meter significant wave height. This should give a required bollard pull of at least 60 tonnes. The vessel should also have fire-fighting capability. This vessel would be a 155 foot tractor tug with twin 7,300 horsepower (hp) engines, employing Voith Schneider propulsion.

The secondary mooring vessel is a line handling vessel which is smaller in size and does not require the same bollard pull capacity or sea-keeping capability. This vessel will pass the hawser messenger lines to the tanker in connection. This vessel will be launches of 85 foot length, with twin 1,200 hp engines.

Two hawsers will be used to connect the tanker to the SPMS buoy. The use of two hawsers for safety through redundancy is considered standard. The hawsers for use with both facilities examined in this study will be 200 ft. long, 21 in. diameter nylon ropes with chafe chain attachments at both ends.

A tanker using the facility will be guided to the SPMS by the vessel traffic controller (VTC) of the facility, and will be escorted by the primary mooring vessel, to insure against possible damage resulting from a loss of power by the tanker and subsequent drifting in the area of the facility. Once the tanker reaches the facility, the primary mooring vessel will be responsible for keeping the floating hoses clear of the tanker, while the secondary mooring vessel passes the hawser messenger lines to the tanker, connecting the tanker to the SPM buoy. The primary vessel then assists in connection with the floating hoses to the tanker manifold. Discharge can begin once the hoses are attached. The primary vessel maintains a stern tow on the tanker during operations to avoid sudden swings by the tanker which can result in very high transient loadings on the hawsers. Disconnection is handled in a similar fashion, removing the hoses from the manifold, lowering them into the water and then disconnecting the hawser lines from the tanker. Total time for connection, discharge and disconnection for both facilities is estimated to be 19 hours for San Diego class tankers.

Disconnection states must be determined for both systems. Disconnection will be carried out to avoid sea states which result in excessive loadings on the system for the attached condition. There will also be a margin on this disconnection sea state, to allow for worsening of weather during disconnection. Normal disconnection is estimated to take approximately one hour, including disconnection of the hoses and lowering the hoses into the water, but emergency disconnection should take no more than five minutes. This emergency disconnection involves the use of cargo pump emergency trips, and is not recommended for non-emergency use. Normally 20 to 30 minutes notice is recommended before hose disconnection is performed.

The determination of disconnection while in operation will be made by the captain of the tanker in conjunction with the pilot/facility manager. General guidelines for disconnection will be set by the facility in conjunction with the Coast Guard.

STEADY ENVIRONMENTAL FORCES

Three environmental components act as relatively steady forces. These are : wind force, current force and mean wave drift force. Each is described in detail in the following sections. These forces are combined and applied to the facility to produce a steady offset, to which oscillatory offsets are added.

Wind

The effects of wind on the SPMS and other facility components have been evaluated by the conventional drag Eqn. (Simiu, 1978). Each above-water facility component (buoy, tanker hull and tanker superstructure) has been treated separately, with its own wind velocity based on centered elevation. Wind velocities were given previously for a 10 meter elevation above still water. These velocities must be adjusted to the component centered elevation velocity by Eqn. 2 for each component (Simiu, 1978).

$$V_Z = V_{10}\left(\frac{Z}{10m}\right)^{0.125} \tag{2}$$

This velocity is then modified by gust duration as in Eqn. 3 (Bea, 1993). The velocities given previously are for a one-minute duration gust. It was decided that a three minute gust duration is appropriate for calculating steady force, due to the period of low frequency motions, following the example of Hunter, et al, for evaluating wind loadings on moorings and vessels (Hunter, 1993).

$$V_t = V_{1min}\left[1+\frac{1}{2}ln\left(\frac{1min}{t}\right)\left(0.00535+0.00042V_{1min}\right)^{1/2}\right] \tag{3}$$

The steady wind force is then calculated from Eqn. 4 with this adjusted wind velocity.

$$F_{Wind} = \frac{1}{2}\rho_{Air}C_S V_{tZ}|V_{tZ}|A \tag{4}$$

The wind shape coefficient is considered typical of marine systems with relatively solid shapes (Bea, 1993). The total steady wind force can be found in Table 4.

Mean Wave Drift

The effect of waves on the facility in question have been separated into three components: first order wave motions (wave frequency motions), second order wave motions (low frequency motions) and a mean wave drift force (API, 1987). While model tests would be a superior measure, this approach is considered to be an adequate approximation. The mean wave drift force is considered to be a steady force component, while the first and second order motions are considered oscillating motions.

Mean wave drift force was calculated based on Eqn. 5 (Le Tirant, 1990) for the tanker in the head-seas condition. It can be seen from the Eqn. that wave drift is roughly proportional to the square of the wave amplitude, and for a constant amplitude, the mean wave drift increases as wave period decreases.

$$F_{Bow,\ drift} = 0.13\ C_{Drift}B_B{}^2LH_S^2 \tag{5}$$

The average drift coefficient for tankers is 0.05 (Le Tirant, 1990). Significant wave height is determined by Eqn. 2.1. For the buoy and the riser, the mean wave drift force is calculated by Eqn. 4.5 (Le Tirant, 1990). The drift coefficient is taken as 1175 Ns2m-4.

$$F_{Mean\ drift} = C_{Df}D_{Buoy}{}^2\left(\frac{H_S}{T_S}\right)^2 \tag{6}$$

The resulting mean wave drift force can be found in Table 4.

Current

The current forces have been evaluated by the drag Eqn. for the buoys and risers (Bea, 1993) and by the API guidelines for the tanker (API, 1991) with depth effects linearized. Eqn. 7 gives the steady current force on the buoys and the risers.

$$F_{Current,\ buoy} = \frac{1}{2}\rho_{SW}C_D V_C|V_C|A \tag{7}$$

The drag coefficient was taken as 1.0 for the buoys and 0.9 for the risers (Bea, 1993). The steady current force on a tanker for the head-seas condition is given by Eqn. 8 (API, 1991).

$$F_{Current,\ Ship} = 0.016\ SV_C^2 \tag{8}$$

The wetted surface was calculated by Eqn. 9 (Lewis, 1988).

$$S = C_{WS}\left(\Delta_{Vol}L\right)^{1/2} \tag{9}$$

The wetted surface coefficient was taken as 2.7, a typical value for tankers of the San Diego class size (Lewis, 1988). The current force can be found in Table 4.

Table 4 : Steady Environmental Forces, Two-Year Return Period Conditions

System Component	Wind Force (Long Tons)	Mean Wave Drift Force (LT)	Current Force (LT)
SALM buoy	0.60	0.33	1.18
CALM buoy	0.76	5.36	1.65
Tanker, lightship, superstructure	12.25	---	---
Tanker, full load, superstructure	11.53	---	---
Tanker hull, lightship	11.82	9.12	0.86
Tanker hull, full load	4.37	9.12	2.16
Flexible Pipe	---	0.01	1.24
Hoses	---	0.04	2.06

OSCILLATING ENVIRONMENTAL MOTIONS

Two of the environmental components can be treated as oscillating components. These are first order (wave frequency) motions and second order (low frequency) motions, both due to wave action. For the connected condition, the tanker was considered to generate the governing motion, and buoy motions were ignored. The motions were combined by adding the maximum wave frequency motions to the significant low frequency motions (API, 1987).

First Order Wave Motions

Wave frequency motions were determined by the use of a ship motions program, SEAWAY (Journée, 1992). SEAWAY is a PC-based ship motions program using ordinary and modified strip theory method. It calculates wave-induced loads and motions with six degrees of freedom. SEAWAY can simulate moorings as well, by adding up to six linear springs to the model.

SEAWAY was used to model the San Diego class tanker and both the CALM and SALM buoys. Springs were then added to these models to simulate the mooring restoring force. Although the mooring restoring force is not perfectly linear, for small offsets this is a relatively good approximation. These models were tested against the given wave events for various return periods. The offsets due to first order motions are given in Table 5.

Second Order Wave Motions

Low frequency motions have been determined by use of the API curves (API, 1987). These curves were generated for drill ships in the 400 foot to 540 foot range, but with the given correction factors for vessels outside this length range, these curves should be an adequate approximation for tankers. A separate set of curves for semisubmersible hulls was used for low frequency motions of the SPMS buoys. The offsets due to second order motions are given in Table 5. Total motions were determined by adding the maximum wave frequency motions to the root mean square low frequency motions.

Table 5: Oscillating Motions, Two-Year Return Period Conditions

Condition	Wave Frequency Motions (ft.)	Low Frequency Motions (ft.)
Tanker, full load	2.4	2.9
Tanker, lightship	2.7	2.9
CALM buoy	6.2	11.4
SALM buoy	7.2	8.6

SEISMIC ACTIVITY

Seismic motions were examined for the SALM system only. This is because seismic motions will effect only systems with significant vertical stiffness. The SALM must have a high vertical stiffness due to its method of providing restoring force, but the CALM system does not require a high vertical stiffness.

The SALM system was modeled as a spring. PCNSPEC (Mahin, 1983), an earthquake analysis program, was used to determine peak motions. PCNSPEC is an inelastic response program for viscously damped single-degree-of-freedom systems. The SALM system was tested against the El Centro earthquake, scaled to have the peak ground accelerations given in Table 2 for the various return period events. The resulting offsets are given in Table 6. As the values in the table indicate, seismic motions are not large, especially considering the anchor leg's 1000 foot length.

Table 6 : SALM Vertical Seismic Offsets

Return Period years	Maximum Relative Vertical Offset (ft.)
2	0.0123
10	0.0617
100	0.2217
1000	0.4922

FATIGUE

Fatigue is defined as the degradation of component characteristics (such as strength or stiffness) due to cyclic straining and stressing. In this application, the cycling is due to wave action. Cycling can also result from operations (cargo pumping, operation of other equipment), construction (installation, transport to installation, launching) or other environmental components (thermal changes, wind, current, earthquakes) (Bea, 1990). However, wave cycling is considered to be the dominant source of cyclic loading in this study.

Fatigue effects are calculated for the chain, wire rope and connections in the CALM system, and for the riser and articulations in the SALM system. These components are considered to be the most likely components to fail due to fatigue.

The calculation of fatigue "load" is done in a different manner than for other environmental loads. Since fatigue failure is a result of cumulative damage over a (usually) long period of time, it is more appropriate to determine a mean fatigue life. This mean fatigue life is the expected life of the component or structure before failure due to cyclic fatigue occurs. The mean fatigue life is calculated by Eqn. 10 (Bea, 1990).

$$T_{fL} = \frac{\Delta K}{B^m \Omega} \tag{10}$$

The accumulated fatigue damage parameter is set equal to 1.0 for fatigue failure. The stress range model error parameter is taken as 0.80, a standard value for the marine environment (Bea, 1990). The values for the negative slope and the log life intercept are taken from the S-N curve for a particular component (API, 1989; API, 1991; Bea, 1990). These values are given in Table 7.

Table7 : Component Fatigue Characteristics

Component	m	K (cycles)
Wire Rope	4.09	1.30 x 10^{10}
Chain	3.36	4.60 x 10^9
Connections	3.74	1.79 x 10^{10}
Tubular Riser	4.38	1.50 x 10^{12}
Articulations	3.74	1.79 x 10^{10}

These values for K are biased, however. This bias is a result of the standard practice of offsetting S-N curves by two standard deviations for design guidelines. This bias is removed by Eqn. 11, which, in effect, adds the two standard deviations back to K.

$$B_K = \frac{\overline{K}_{True}}{\overline{K}_{Nominal}} = \frac{exp\left[ln\left(\overline{K}_{Nominal}\right) + 2\sigma_{lnK}\right]}{\overline{K}_{Nominal}} \tag{11}$$

The natural log of the standard deviation of the fatigue life is calculated by Eqn. 12.

$$\sigma_{lnT} = \left[ln(1+C_\Delta^2)(1+C_K^2)(1+C_B^2)^{m^2}\right]^{1/2} \tag{12}$$

where : $C_\Delta = 0.3$, $C_K = 0.73$, and $C_B = 0.5$. The values given for the various coefficients of variation are considered typical for marine systems (Bea, 1990).

The unbiased values for K can be found in Table 7. The stress range parameter is calculated by Eqn. 13.

$$\Omega = \lambda(m) f_0 S_m^m \left[lnN_T\right]^{-m/\varepsilon} \Gamma\left[\frac{m}{\varepsilon} + 1\right] \tag{13}$$

The rainflow correction and epsilon are both taken as 1.0, considered a typical value for both variables for marine systems (Bea, 1990). The design period was taken as 100 years, and the number of cycles was based on this length of time and an average wave period of 13 seconds. The average frequency of the stresses was taken as the inverse of the period of the waves.

The fatigue design stress range is calculated in Eqn. 14.

$$S_{fD} = \left[\frac{KH_{fD}^{\alpha m}}{T_{SF} Y_0}\right]^{1/m} \tag{14}$$

where

$$T_{SF} = F_{SF} T_{DP} \tag{15}$$

and

$$Y_0 = \frac{N_T}{T_{DP}} H_{fD} \left(lnN_T\right)^{-m\alpha/\varepsilon} \Gamma\left(1 + \frac{\alpha m}{\varepsilon}\right) \tag{16}$$

The fatigue life safety factor is typically 3.0 for marine systems (Bea, 1990). This value is related to the nominal design stress range by the stress concentration factor, as given in Eqn. 14. This nominal design stress range is then used as the largest stress value in Eqn. 11. These relations result in the mean fatigue lives given in Table 8.

Table 8 : Summary of Mean Fatigue Life

Component	Unbiased K value	Mean Fatigue Life (years)
Wire Rope	$4.8x10^{10}$	66,820
Chain	$1.7x10^{10}$	25,460
Connections	$6.6x10^{10}$	42,072
Tubular Riser	$5.5x10^{12}$	98,034
Articulations	$6.6x10^{10}$	42,072

CALM ANALYTICAL MODEL

The details of the analytical model of a CALM are not discussed in depth here. The reader is referred to Ma (1994) for a detailed explanation of the model used in this study. After several design iterations, it was decided to use a system pretension of 23.6 LT for the CALM. The line tensions and anchor loads are summarized in Table 9 by return period.

The holding power of drag-embedment anchors needed to be determined for the CALM system (API, 1991). The chain section of the anchor leg lying on the sea floor also contributes to the holding power of the system, as described by Eqn. 17.

$$P_C = f_{CO} L_C W_C \tag{17}$$

Table 9 : CALM Line Tensions and Anchor Loads

	Line Tension (LT)	Anchor Load (LT)
CALM Buoy		
2-year conditions	31.7	15.6
10-year conditions	36.6	19.6
100-year conditions	42.9	25.4
Buoy with Lightship, 2-year conditions	45.5	26.3

The coefficient of friction between chain and the ocean bottom is taken as 0.7 (API, 1991). The unit weight of 3.75 in. chain is 132 pounds per foot. Approximately 2000 ft. of chain will be in contact with the ocean bottom during extreme conditions. This results in a chain holding power of 82.5 LT.

The anchor was then selected based on anchor loads and the given chain holding power. Bruce FFTS anchors of 6.7 LT weight (15 kilopounds) and 192 LT holding power (430 kilopounds) were selected, giving a total anchor leg holding power of 274 LT.

SALM ANALYTICAL MODEL

A SALM derives its restoring force from the tension in its leg (Fig. 4). This leg may be a chain, wire rope or solid riser. The choice depends on the tension required and the specifics of the application. In this study it was decided that a solid riser would be necessary to support the tension required. Articulations were necessary in the riser due to the water depth of 1000 ft..

There are two components involved in SALM restoring force: the horizontal component of the leg tension and the added buoyancy due to offset which acts to increase leg tension. The design inputs for a SALM are : buoy size, riser size, number of riser sections, weight of buoy and risers, and the dimensions of the anchor pile. The anchor leg of a SALM is given a pretension so that it will develop an adequate restoring force. In this design, a pretension of 300 LT was decided upon.

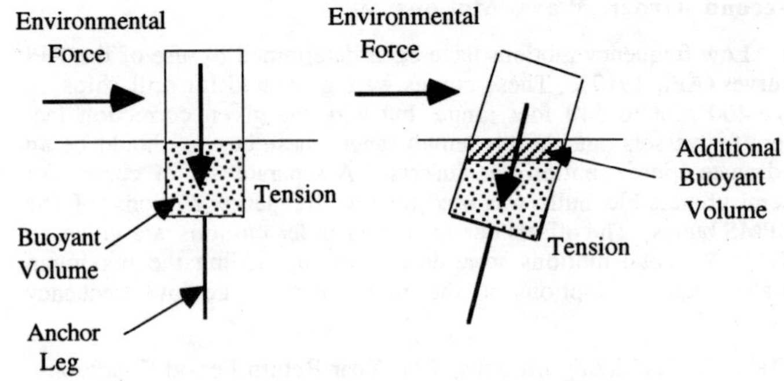

Fig. 4: SALM Restoring Force Schematic

With system dimensions, weights and pretension decided, the SALM could be modeled analytically. The horizontal environmental force acting on the system must be offset by the horizontal component of the system tension, which is a function of the system pretension and the added buoyancy due to offset, which are both functions of the angular displacement of the anchor leg. The line tensions for storms and seismic events are given in Table 10.

Table 10: SALM Storm and Seismic Leg Tensions

	Leg Tension, Storm (LT)	Leg Tension, Seismic (LT)
SALM Buoy		
2-year conditions	275	272.8
10-year conditions	278.3	273.8
100-year conditions	284.5	277.3
Buoy with Full-load ship, 2-year conditions	300.5	200.3
Buoy with Lightship, 2-year conditions	311.7	200.3

The determination of anchor pile characteristics were based on the leg tensions, as theses act directly on the anchor. The anchor load is considered to be strictly vertical, due to the small angular displacement values involved (usually under 10 degrees for extreme conditions). It was decided, based upon the reliability analysis to have a target pull-out capacity of 2,000 LT. The ultimate pull-out capacity is given by Eqn. 18 (API, 1989).

$$Q_P = f_{SC} A_S + W_P \tag{18}$$

The unit skin friction capacity is determined by Eqn. 19 (API, 1989).

$$f_{SC} = \alpha_P c_{USS} \tag{19}$$

The undrained shear strength was determined to be 2.0 kilopounds per foot. The dimensionless pile factor (a) is taken as 1.0 (API, 1989).

From these relations, the length of a pile can be estimated. Wall thickness was based upon API criteria for minimum wall thickness, as given in Eqn. 18 (API, 1989).

$$T_{WT} = 0.25 + \frac{D_P}{100} \tag{20}$$

With these criteria, a pile was selected of 140 ft. length, 60 in. diameter and 1 in. wall thickness. This gives a pull-out load of approximately 2,000 LT.

PROBABILITY OF FAILURE

Failure is defined as the breakage of one or more legs of the SPMS, insufficient holding power developed by the anchors or anchor pile, major damage to the risers / hoses, or damage to the buoy which would halt operations. The total probability of failure is the sum of the annual chances of these events occurring, as given in Eqn. 21 (neglecting small cross-product terms).

$$P_{f,\ Total} = P_{f,\ Mooring\ Legs} + P_{f,\ Anchors} + P_{f,\ Risers} \quad (21)$$

The probability of failure of the facilities under consideration is dependent upon four relatively independent failure hazards. Failure may be due to storm loadings, seismic loadings, cyclic fatigue loadings, or human and organizational error (HOE). The total annual probability of failure is expressed in Eqn. 22 (again neglecting small cross-product terms).

$$P_{f,total} = P_{f,storms} + P_{f,seismic} + P_{f,fatigue} + P_{f,HOE} \quad (22)$$

The method of computation of each one of these components is examined in depth in the following sections.

Component Probability of Failure

The procedure for calculating probabilities of failure for storm loadings, seismic loadings and cumulative fatigue damage is outlined in this section. The analyses are based on a Lognormal - Lognormal relationship between component loadings and component capacities, whether loadings are due to storms, seismic events or fatigue degradation. This type of relationship is considered typical of large marine systems analyzed over a long time period (Bea, 1990). The mean values of load (line tensions, anchor loadings and fatigue damage) and capacity (line strength, anchor holding power and mean fatigue life) are known, either by calculation or from manufacturer's specifications. Variances of these properties are considered, as well as possible correlation between load and capacity.

The probability of failure for a log normal system is determined by Eqn. (23).

$$P_f = 1 - \Phi(\beta) \quad (23)$$

In some cases, values for the safety index fall outside those tabulated in the standard normal distribution table. In these cases, an approximation due to Abramowitz (Melchers, 1987) is used. This approximation was found to be a superior fit for the values of beta encountered in this study.

The safety index, ß, is defined in Eqn. 24 for components exposed to lognormally distributed loadings due to storms or seismic events. This Eqn. relates the median load to the mean capacity, with the effect of variations and correlations included.

$$\beta = \frac{ln\left(\frac{R_{50}}{S_{50}}\right)}{\sqrt{\sigma_{lnR}^2 + \sigma_{lnS}^2 - 2\,\rho_{RS}\sigma_{lnR}\sigma_{lnS}}} \quad (24)$$

The safety index is defined by Eqn. 25 for cyclic fatigue degradation. This fatigue analysis differs from the environmental analysis in that it is based on a time frame instead of a peak loadings. This means that capacity will be replaced by mean fatigue life and load will be replaced by accumulated fatigue damage, as in Eqn. 25, where the mean fatigue life is that calculated previousloy.

$$\beta = \frac{ln\left(\frac{T_{fL}}{T_{SL}}\right)}{\sigma_{lnT}} \quad (25)$$

The components of both peak load-based and time-based relations are discussed in the following sections.

Load and Capacity

The mean loadings used in the peak load analysis are the 2-year return period loadings determined earlier for storms and seismic events, and the mean fatigue life for the seismic analysis. The capacities are the ultimate limit state capacities given by the manufacturer, which are listed in Table 11 (Avallone, 1987; Bureau Veritas, 1980).

Variance and Deviation

There are two types of uncertainty in structures analyzed by reliability methods. Type I uncertainties refer to natural or inherent randomness, such as peak environmental conditions. This type of uncertainty cannot be controlled. Type II uncertainties refer to modeling uncertainties. This type of uncertainty includes uncertainties in computations of forces, uncertainties in measurement and uncertainties due to limited data sets. This type of uncertainty is systematic. It is also information sensitive and can be reduced by acquiring additional information, whether the information is research, inspection or quality control / assurance (Bea, 1992).

For distributions with Type I uncertainties, the variance of the distribution can be measured by two parameters. The coefficient of variation is a normalized measure of the variability of a parameter. The standard deviation is a measure of dispersion or variability of a distribution, as is the natural log standard deviation. The relations between these parameters are given in Eqns. 26 and 27.

$$V_X = \frac{\sigma_X}{X} \quad (26)$$

$$\sigma_{lnX} = \sqrt{ln\left(1 + V_X^2\right)} \quad (27)$$

However, these relations require more information than has been generated to this point. Therefore, Eqn. 5.8 is more suitable for the calculation of Type I uncertainties, as loads for the 2-year and 100-year return periods are known. Eqn. 28 calculates the coefficient of variation from the 2-year return period and the 100-year return period loads.

$$\sigma_{lnX} = \frac{ln\left(X_{99}/X_{50}\right)}{2.33} \quad (28)$$

Systems were analyzed for the disconnected case to determine loading variance, as the systems should never be connected for the 100-year condition. This gave the variances listed in Table 11 for environmental Type I uncertainty. The values for capacity Type I uncertainty (Yong, 1991; Bea, 1990) are considered typical for marine system components. The anchor / anchor pile variation is based on the soil type (Bea, 1990). These uncertainties are listed in Table 11 It should be noted that the systems were designed so that the connections are the most likely element of each system to fail. This was done because the connections are the easiest component of each system to maintenance and replace. Their failure should also cause the least amount of damage. However, for the SALM system, nearly any failure will be a serious one, as its single load-path allows for no redundancy.

Type II uncertainties are more difficult to determine, as they must be based on historical analysis of analytical methods. Therefore, representative values were taken from existing literature for this study. For storm loadings on marine structures, the Type II variation has been estimated as 0.07 to 0.11 (Bea, 1992; Nikolaidis, 1992). A value of 0.10 was taken for this study. For seismic loadings, literature Type II variation ranged from 0.0 to 0.31 (Bea, 1992; Nikolaidis, 1992). A value of 0.10 was selected for this study, based on the modeling tool used (PCNSPEC) and the small effect of seismic loadings in line tensions. Type II uncertainties in fatigue analysis were discussed earlier.

Table 11 : Component Reliability Characteristics

Component	Capacity (LT)	Variance
CALM		
Wire rope	447	0.10
Chain	321	0.10
Connections	100	0.10
Anchors	274	0.40
SALM		
Articulations	600	0.10
Riser	700	0.10
Anchor Pile	2000	0.40

Table 12 : Environmental Variance

System	Leg Loads (Storm)	Leg Loads (Seismic)	Anchor Loads (Storm)
CALM	0.115	---	0.186
SALM	0.015	0.007	0.015

The Type I and II variances are combined into a single variance by Eqn. 29. This relation is valid for systems which have independence between load and capacity. This independence is discussed in the next section.

$$V_X = \sqrt{V_S^2 + V_R^2 + V_{II}^2} \tag{29}$$

Correlation

The correlation coefficient expresses how strongly two variables are related. A value of 1.0 indicates perfect correlation, while a value of 0.0 indicates complete independence. The correlation coefficient can be negative as well, up to a value of -1.0, which indicates perfect negative correlation. There are three types of correlation which required examination in this study : correlation between load and capacity, correlation between capacities of different components, and correlation between failure modes of different components.

The determination of load to capacity correlation is best carried out by model testing, as it is very difficult to determine analytically. There may be some small positive correlation in these systems due to larger wave heights (higher loads) encountering more structure (higher capacity), but this may be offset by slight changes in environmental directionality. Therefore, it was assumed that there would be no correlation between load and capacity.

The correlation between capacities of different elements is likely to be very high, due to similarities in design and manufacture, and has been taken as 1.0.

System Probability of Failure Calculations

The system probability of failure is determined by calculating component probabilities of failure from the given relations and reducing the systems into series and parallel elements. The CALM system is composed of eight catenary anchor legs, which are modeled as series-loaded subsystems. These subsystems are combined to form an eight element parallel system. The SALM system is much simpler, being composed of one tensioned leg modeled as a series system.

One new attribute of a system must be considered in the calculation of system reliability. Correlation can exist between system components, due to their manufacture and due to their modes of failure. An estimate of system failure mode correlation based on relative uncertainties attributed to Cornell (Bea, 1990) is used here (Eqn. 30).

$$\rho_{FM} = \frac{V_S^2}{V_R^2 + V_S^2} \tag{30}$$

For series elements, the system reliability is calculated by Eqn. 31. This Eqn. is valid for systems with perfect element-to-element correlation. Although this assumption concerning correlation may not be completely true here, it is a good approximation. As has been pointed out elsewhere (Bea, 1990), the probability of failure of a system is well approximated by the probability of failure of the most-likely-to-fail element in the system.

$$P_{f,\,system} = max(P_{f,\,element}) \tag{31}$$

For parallel elements, the system reliability is calculated by Eqn. 32. This equation is valid for normally distributed identical parallel elements.

$$\beta_S = \beta_E \sqrt{\frac{N}{1+\rho_E(N-1)}} \tag{32}$$

It should be noted that failure of one leg is essentially system failure for the CALM system (it has already been defined as failure for the SALM). Failure of one leg of the CALM will halt operation. Therefore, "system failure" for the CALM refers to one leg broken or severely damaged.

Failure due to Storm Loadings

The probability of failure due to storms must cover two separate cases : the SPMS alone and the SPMS with an attached tanker. Loads in the system will be much higher with a tanker present. The facility will have guidelines governing when a tanker using the facility should disconnect from the SPMS to reduce loads. Therefore, the probability of failure due to storm loadings can be expressed as given in Eqn. 33.

$$P_{f,storms} = \left(P_{f,storms}|disc\right)\left(1-P_{conn}\right) + \left(P_{f,storms}|conn\right)\left(P_{conn}\right) \tag{33}$$

The percent of time in which a tanker is using the facility is an economic decision, affected only slightly by environmental conditions and facility downtime. The downtime of the facility is expected to be very small, according to gathered data. Typical downtime values for systems operating for one year or more range from 0% to 3.2% (Key, 1993). For this study, three San Diego class tankers will use the facility per week. Given the 17 hour discharge time of this tanker class and the estimated 2 hour connection and disconnection time, the facility will be in use approximately 31% of the time, allowing for a 3% system downtime.

The probability of failure while disconnected is a relatively straightforward calculation, comparing disconnected loadings with capacities, as in Eqn. 24. All data for this calculation have already been determined, and the resulting probability of failure can be found in Table 13. Failure while connected is a more difficult calculation, due to the question of what sea states will be encountered before disconnection will occur. Disconnection can be ordered to occur at the appearance of a certain sea state, but the actual occurrence of disconnection depends upon perception of the sea state by the captain and pilot, and weather deterioration which may occur before disconnection can be completed. Therefore, it is necessary to model disconnection as a probability distribution in relation to sea state. The probability of failure while connected can therefore be expressed as in Eqn. 34.

$$P_{f,storms}|conn = \sum_{all\ sea\ states}\left[(P_f|conn, sea\ state)(P_{sea\ state})(P_{conn}|sea\ state)\right] \tag{34}$$

The probability of a given sea state being the annual maximum encountered is a function of the annual maximum wave height distribution. The probability of connection in a given sea state is based upon assumed behavior and the uncertainty associated with identifying sea states and weather deterioration. The values for these probabilities are given in Table 13. The probability of connection in a given sea state is based on an instruction to disconnect to avoid operation (and connection) during conditions with wind exceeding 32.5 knots or waves exceeding 25.5 ft. maximum, or a 13.7 foot significant wave height. The distribution is Lognormal over the spectrum of sea states. The distribution is based on the mean disconnection state being the 25.5 ft. maximum wave height state, with a 5% chance of being connected at the 29.5 ft. maximum wave height sea state.

Table 13 : Probability of Connection by Sea State

Plsea state	Wave Height (ft.)	Wind Velocity (knots)	Current Velocity (knots)	Probability of Connection
10	21	26.5	0.8	0.90
10	23.5	30	0.9	0.70
10	25.5	32.5	1.02	0.50
10	27	35.5	1.1	0.25
10	28.5	38	1.18	0.10
10	29.5	40.5	1.28	0.05
10	31	43.5	1.35	0.01
10	32.5	46.5	1.45	0.002
20	37	55	1.7	0.0005

The probability of failure for a given sea state while connected was then calculated based on Eqn. 23, with the mean loading based on the mean wave height, wind speed and current speed for a given sea state, and variance as previously determined. The probabilities of failure are then summed over all sea states to produce a total probability of failure for the SPMS while connected to a tanker. The results of this calculation are given in Table 14.

Reliability for Storm Loadings

System reliability is calculated for each facility based on the component probabilities of failure, as explained earlier. The probabilities of failure are combined with the probabilities of given sea states being encountered and connection during that sea state, as in Eqn. 33. The probabilities of failure are then summed over all sea states, as in Eqn. 34. The resulting values for probability of failure are given in Table 14 for the case of storm loadings.

Table 14 : Probability of Failure due to Storm Loadings

System State	Percent Time	CALM P_f	SALM P_f
Connected	31	$6.7x10^{-7}$	$2.4x10^{-7}$
Disconnected	69	$1.6x10^{-11}$	$3.4x10^{-7}$
TOTAL	100	$2.1x10^{-7}$	$3.0x10^{-7}$

Failure due to Seismic Loadings

The probability of failure due to seismic loadings is calculated in the same manner as the storm loadings for the unconnected facility. However, seismic loadings are considered to be significant for the SALM system only, as the CALM system has relatively little vertical stiffness. Seismic loadings were determined earlier, while capacities are given in Table 11. Variations in load were determined earlier. Correlation between load and capacity is again assumed to be equal to 0.0.

It will be noted that the variance on seismic loadings is lower than that of storm loadings, which may seem counter-intuitive. However, seismic loadings on the SALM system are of relatively low magnitude, and tend to be overshadowed by the constant pretension force of the system.

System reliability for seismic loadings is similar to that for storm loadings. Since there is only one load path, the system is a series one. As in other system reliability sections, the system probability of failure of a series system is well approximated by the probability of failure of the most-likely-to-fail component.

The probability of failure of the SALM system due to seismic loadings is given in Table16.

Failure due to Cyclic Fatigue Loadings

Fatigue reliability is characterized by mean fatigue life (analogous to capacity), service life (analogous to load) and a deviation on the fatigue life, as in Eqn. 15. Mean fatigue life was determined earlier, while service life is a design decision. The natural log of the standard deviation of the fatigue life is calculated by Eqn. 11.

The resulting natural log standard deviation is found to be 1.53. It should be pointed out that this variance is very high in comparison with other variances calculated in this section.

The fatigue probabilities of failure are given in Table 15. It can be seen that the large mean fatigue lives are offset by the large variance calculated in Eqn. 35, giving probabilities of failure similar to those for other environmental components.

This figure serves to illustrate the concept of preventative maintenance. By periodically replacing the elements of a system which are likely to suffer fatigue damage, the probability of failure is lowered.

Table 15 : Component Probability of Failure due to Fatigue

Component	Probability of Failure
CALM	
Chain	$2.4x10^{-7}$
Wire Rope	$5.8x10^{-8}$
Connections	$1.2x10^{-7}$
SALM	
Articulations	$3.3x10^{-8}$
	$1.2x10^{-7}$

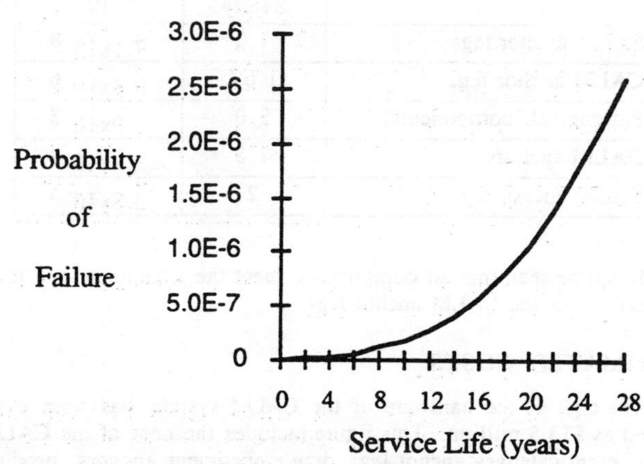

Fig. 5 : Fatigue Reliability versus Service Life

The system effect of fatigue reliability is similar to that for other types of system reliability with high correlation. The most likely to fail element's probability of failure is taken as the system probability of failure.

Facility Probability of Failure

The probabilities of failure calculated in the previous sections are summarized in Table 16. These probabilities are the annual probabilities of failure due to each loading case for each facility. The total probability of failure for each facility is the sum of these individual probabilities of failure for each facility. This total probability of failure assumes independence between the individual probabilities of failure. Although values for probability of failure have not been calculated for human and organiation errors (HOE), it should be noted that these probabilities of failure may be very significant.

Table 16 : Facility Probabilities of Failure

Failure Component	CALM	SALM
Storm Loadings	$2.1x10^{-7}$	$3.0x10^{-7}$
Seismic Loadings	---	$4.6x10^{-6}$
Fatigue Degradation	$2.4x10^{-7}$	$1.2x10^{-7}$
TOTAL	$4.5x10^{-7}$	$5.0x10^{-6}$

Acceptable Reliability

An acceptable reliability can be determined from the factors of safety given in Table 2. This is done by equating the factor of safety (FS) to the 100-year load and the mean capacity, as in Eqn. 36 (Jones, 1992; Bea, 1990).

$$FS = \frac{R_{99}}{S_{50}} \tag{36}$$

Eqn. 26 can be manipulated to give Eqn. 37, relating the safety index to the deviations and factor of safety.

$$\beta_{acceptable} = \frac{2.33\sigma_{lnS} + ln(FS)}{\sigma_{ln}} \tag{37}$$

Table 17 presents the applicable factors of safety and the calculated target probabilities of failure (for the intact condition only).

Table 17 : Acceptable Reliabilities

System Component	Factor of Safety	Acceptable P_f
SALM anchor legs	2.0	$7.1x10^{-8}$
CALM anchor legs	1.67	$9.8x10^{-6}$
Fatigue (all components)	3.0	$1.0x10^{-3}$
CALM anchors	1.5	$3.2x10^{-4}$
SALM anchor leg	2	$3.8x10^{-5}$

It can be seen that all components meet the acceptable reliability except for the SALM anchor legs.

ESTIMATED COSTS

The cost of the hardware of the CALM system has been estimated as $13.5 million. This figure includes the cost of the CALM buoy, eight catenary anchor legs, drag embedment anchors, product risers, hawsers and transfer hoses. It does not include the cost of the pipeline or any onshore facilities, including vessel traffic control. This cost estimate was based upon the existing Marlim CALM facility offshore Brazil (Hwang and Bensimon, 1990).

The cost of the hardware of the SALM system has been estimated as $12.5 to $14 million. This price is more uncertain than that of the CALM, as no SALM systems have been constructed for this water depth to date.

For the both the CALM and the SALM system, the total cost of engineering, certifying, study management, construction and installation supervision, and transportation has been estimated as $1.25 million.

The installation of the CALM system is expected to take 20 to 25 days. The installation is estimated to cost from $3 to $5 million. This high cost is due in part to the necessity of bringing equipment to the site from the Gulf of Mexico, as deepwater equipment is not readily available on the West Coast.

The installation of the SALM system cannot be estimated as accurately. It is expected to be more expensive, due to the necessity of driving the anchor pile at the sea floor. An estimate of $5 million was used in this study.

The total facility costs are given in Table 18. These total costs do not include the pipeline or any onshore facilities. The costs also do not include vessel traffic control. The entry "other" refers to contingencies, insurance and overheads.

Table 18 : Total Initial Facility Costs

Cost Component	CALM Facility (millions of $)	SALM Facility (millions of $)
Hardware	13.5	12.5 to 14
Development & Engineering	1.25	1.25
Installation	3 to 5	5
Other	1.5	1.5
TOTAL	19.25 to 21.25	20.25 to 21.75

These figures indicate that the initial cost of the two systems are comparable. However, there is a greater degree of uncertainty regarding the SALM estimates, as no comparable system exists.

Table 19 outlines the cost of operation for the tending vessels which will be used at the facilities. The total yearly cost is based on three days of operation per week, following the earlier assumption that the facility will be visited by three tankers per week, each taking approximately 19 hours to connect, discharge and disconnect. No spot chartering has been assumed in the cost estimate.

Table 19 : Annual Operation Costs

Activity	Annual Estimated Cost
Yearly contract (vessel and crew, per tending vessel)	$2.4 million
Fuel (on a consumption basis)	1000 gallons per day @ $0.50 per gallon
Spot Charter	$6500 per day (plus fuel consumption)
TOTAL ANNUAL COST	$4.96 million

The cost of tanker charter is approximately $40,000 per day.

Preventative maintenance will be carried out on the hawsers used to connect tankers to the SPM buoys. These hawsers can have severe fatigue problems, and they can cause serious problems when they break by snapping back. It is common practice to replace them often -- LOOP uses only 20% of the calculated fatigue life before replacing hawsers. Hawsers will be replaced every six months at the facilities to avoid fatigue problems. The cost of hawsers was not available.

CONCLUSIONS

It has been shown by the U. S. Coast Guard that deepwater ports are a viable way to reduce the environmental risks arising from the import of crude oil. This study has shown that two types of single point moorings can be used -- without stretching today's technology -- as deepwater ports offshore southern California.

The CALM system developed during this study is a relatively simple design. The engineering is not complex and the system does not require custom-built components other than the CALM buoy itself. The CALM proved to be relatively robust in reliability analysis, due to its inherent safety by redundancy -- failure of one anchor leg will not entail catastrophic failure for the system.

The SALM system was also a relatively straightforward design. However, it requires more complex engineering and more custom construction of components. It also requires much more attention for reliability, as the system is not inherently robust -- failure in one leg is a catastrophic failure. The system also required a very high pretension to provide adequate restoring force to meet API guidelines concerning system offset. This high pretension dictated the need for the unusual components. It also necessitated the high vertical stiffness of the SALM also means that more attention must be paid to seismic effects.

The study also showed that disconnection criteria are very important in reliability based design of systems which disconnect to avoid extreme environmental conditions. The greater probabilities of failure associated while connected strongly affect total system reliability.

In summary, there are no technical barriers to the use of SPMS for deepwater ports offshore southern California in 1,000 ft. of water. However, a CALM system appears to be much simpler than a SALM, with a greater degree of control of risk in the design.

ACKNOWLEDGMENTS

Many people have been instrumental in the development of this paper. We would like to thank Professor William C. Webster for his contribution to the direction of this research. We would also like to thank Dr. Wei Ma who was responsible for a parallel effort to develop analytical models for CALM moorings.

This study was made possible by the support provided by LOOP, IMODCO, and Chevron Petroleum Technology company and Chevron Shipping Company. We would like to thank Dale G. Rollins of the Louisiana Offshore Oil Port for taking the time to show us around the LOOP facility, as well as for answering many questions at other times concerning single point moorings and deepwater ports. We would also like to thank Michael J. LeBlanc and Wayne E. Lolan of LOOP for their input and assistance.

M. Steven Mostarda of IMODCO was of great help in the systems design of the SPMS, as well as in providing helpful background on the systems. We would like to thank Capt. A. F. Fantauzzi of the Chevron Shipping Company and Christopher P. Whatley of the Chevron Petroleum Technology Company for their help in understanding the challenges involved in design and operation of facilities of this type.

We would also like to thank the following for their contribution to the success of this research : Jay Phelps of the California State Lands Commission, Paul Palo of the Naval Civil Engineering Laboratory, Dr. Malcolm Sharples of Noble, Denton and Associates, Inc., and Gavin Cleator of Bruce Anchor Limited.

This paper is funded in part by a grant from the National Sea Grant College Program, National Oceanic and Atmospheric Administration, U. S. Department of Commerce, under grant number NA36RG0537, study number R/OE-26 through California Sea Grant College, and in part by the California State Resources Agency. The views expressed herein are those of the authors and do not necessarily reflect the views of NOAA or any of its sub-agencies. The U. S. Government is authorized to reproduce and distribute for governmental purposes.

REFERENCES

American Petroleum Institute. 1987. Analysis of Spread Mooring Systems for Floating Drilling Units. API RP 2P. Washington, DC.

American Petroleum Institute. 1989. Recommended Practice for Planning, Designing and Constructing Fixed Offshore Platforms. API RP 2A. Washington, DC.

American Petroleum Institute. 1991. Draft Recommended Practice for Design, Analysis and Maintenance of Mooring for Floating Production Systems. API RP 2FP1. Washington, DC.

Anaturk, A. R., Tromans, P. S., van Hazendonk, H. C., Sluis, C. M., and Otter, A. 1992. Drag Force on Cylinders Oscillating at Small Amplitude : A New Model. Journal of Offshore Mechanics and Arctic Engineering. Volume 114.

Avallone, Eugene A. and Baumeister, Theodore III. 1987. Marks' Standard Handbook for Mechanical Engineers, Ninth Edition. McGraw-Hill, Inc.

Bea, R.G. 1990. Reliability Based Design Criteria for Coastal and Ocean Structures. National Committee on Coastal and Ocean Engineering. Institution of Engineers, Austrailia.

Bea, R.G. 1992. Re-qualification of Platforms : Offshore California & Alaska. Department of Civil Engineering and Department of Naval Architecture, University of California at Berkeley.

Bea, R.G. 1993. Course reader for NA 205B : Wind & Wave Forces on Marine Structures. University of California at Berkeley.

Berman, M.Y., Birrell, N. D., Irick, J. T., Lee, G. C., Rubin, M. and Utt, M. E. 1990. The Role of the API Committee on Standardization of Offshore Structures. Proceedings of Offshore Technology Conference: OTC 6206. Houston, TX.

Bruen, F.J., Gordon, R. B., and Vyas, Y. K. 1991. Reliability of a Deepwater Gulf of Mexico FPS Spread Mooring. OMAE, Volume II, Safety and Reliability. ASME.

Bureau Veritas. 1980. Rules and Regulations for the Construction and Classification of Steel Vessels, Offshore Platforms and Submersibles : Characteristics and Inspection of Materials. Bureau Veritas, Paris, France.

California Coastal Commission. 1978. Study of Pipeline Feasibility for Offshore LNG Terminal Sites. San Francisco, California.

Chryssostomidis, Chryssostomos and Connor, Jerome J. 1983. Behaviour of Off-Shore Structures, Volume 2. Hemisphere Publishing Corporation, NY, New York.

Chopra, Anil K. 1980. Dynamics of Structures. Earthquake Engineering Research Institute. Berkeley, California.

Cornell, C. Allin. 1987. Offshore Structural Systems Reliability : A Report to the Joint Industry Study. C. Allin Cornell, Inc.

Coutarel, A. 1992. Flexible Pipe Installation in Deep and Very Deep Waters. Proceedings of the Marine Technology Centre, International Seminar on Recent Research and Development of Flexible Pipe Technology. Trondheim, Norway.

Cramer, E.H. 1992. Fatigue Reliability of Continuous Welded Structures. Dr. Engineering Dissertation. Department of Naval Architecture. University of California at Berkeley.

de Kat, Jan O. and Wichers, Johan E. W. 1991. Behavior of a Moored Ship in Unsteady Current, Wind, and Waves. Marine Technology. Society of Naval Architects and Marine Engineers. September 1991.

Department of Navigation and Ocean Development, Sacramento, California. 1977. Deep-Water Wave Statistics for the California Coast, Station 5.

Department of Navigation and Ocean Development, Sacramento, California. 1977. Deep-Water Wave Statistics for the California Coast, Station 6.

Det norske Veritas. 1991. Structural Reliability Methods. Det norske Veritas, Division Ship and Offshore. Norway.

Donley, M. G. and Spanos, P. D. 1992. Stochastic Response of a Tension Leg Platform to Viscous and Potential Drift Forces. OMAE, Volume II, Safety and Reliability. ASME.

D'Souza, Richard, Dove, Peter S., Hervey, Donald G., and Hardin, Donald J. 1992. The Design and Installation of Efficient Deepwater Permanent Moorings. Marine Technology. Society of Naval Architects and Marine Engineers. Volume 29, No. 1.

Froberg, Carl-Erik. 1985. Numerical Mathematics : Theory and Computer Applications. Benjamin/Cummings Publishing Company, Inc. Menlo Park, California.

Guenard, Yves Francois. 1984. Application of System Reliability Analysis of Offshore Structures. John A. Blume Earthquake Engineering Center, Department of Civil Engineering. Stanford University.

Hanna, S., Dove, P. G. S., and Breese, G. B. H. 1988. The Design, Procurement and Installation of Placid Oil Company's Green Canyon Floating Production Platform Mooring System. Proceedings of Offshore Mechanics and Arctic Engineering. Houston, TX.

Huang, E. W., Beynet, P. A., Chen, Jing Hui, Li, Jian Xun, and Lu, Zi Guang. 1993. FPSO Model Tests and Analytical Correlation. Proceedings of Offshore Technology Conference: OTC 7146. Houston, TX.

Hunter, K.C., Froning, J.P. and DeSouza, P.M.F.M. 1993. BHP's Experience with FPSO's. SNAME FPSO Technology Symposium. Houston, Texas.

Huse, E. and Matsumoto, K. 1989. Mooring Line Damping Due to First- and Second-Order Vessel Motion. Proceedings of Offshore Technology Conference: OTC 6137. Houston, TX.

Hwang, Yuh-Lin and Bensimon, Luiz Fernando Callado. 1990. Design Analysis for the Marlim Deepwater Calm System. Society of Petroleum Engineers, SPE Latin American Petroleum Engineering Conference, October 14-19. SPE 21122.

Intersea Research Corporation Research Corporation. 1974. Observed Winds, Uncompensated Hindcast Waves and Calculated Currents Offshore Southern California. Prepared for Shell Oil Company. La Jolla, California.

Jiao, Guoyang. 1992. Limit State Design for Flexible Pipes. Marine Structures 5 : 431-454. Elsevier Science Publishers Ltd., England. Great Britain.

Jones, D. E., and McIntyre, S. R. 1992. A Classification Society's Approach and Procedures Utilized in Assuring Compliance of Floating Production Systems to Applicable Codes, Standards and Regulatory Requirements. Seventh International Conference on Floating Production Systems. Oslo, Norway. American Bureau of Shipping.

Journée, J.M.J. 1992. SEAWAY - DELFT; User Manual of Release 4.00. Delft University of Technology, Ship Hydromechanics Laboratory. Delft, The Netherlands.

Kentosh, J. M., Muscettola, M., and Ziliotto, F. 1988. Design Criteria for Permanent Mooring of Offshore FPSUs : Mooring of a 138,000 DWT Tanker in the Sicily Channel. Seventh International Conference on Offshore Mechanics and Arctic Engineering, Houston, Texas.

Key, Joe W. 1993. History of FPSO's. SNAME FPSO Technology Symposium. Houston, Texas.

Kwan, C.T. 1991. Design Practice for Mooring of Floating Production Systems. Marine Technology, Society of Naval Architects and Marine Engineers. Volume 28, No. 1.

Le Tirant, Pierre and Meunier, Jacques. 1990. Anchoring of Floating Structures. Scientific and Technical Information Ltd. Oxford, UK.

Lewis, Edward V. 1988. Principles of Naval Architecture, Volume I, Second Revision. The Society of Naval Architects and Marine Engineers. Jersey City, New Jersey.

Lou, Jack Y. K. and Choi, Han S. 1990. Nonlinear Response of an Articulated Offshore Loading Platform. OTRC Report No. 6/90-A-2-100. Ocean Engineering Program, Texas A&M University.

Louisiana Offshore Oil Port. 1992. LOOP Operations Manual, Volume 7.

Lundgren, H., Sand, Stig E., and Kirkegaard, J. 1983. Drift Forces and Damping in Natural Sea States. Proceedings of the Third International Conference on the Behaviour of Off-Shore Structures. Massachusetts Institute of Technology, Cambridge, Massachusetts.

Luo, Yong and Ahilan, R.V. 1991. Mooring Safety Assessment Using Reliability Techniques. Proceedings of Offshore Technology Conference: OTC 6783. Houston, TX.

Ma, Wei. 1994. An Analytical Approach to the Dynamics of Cables. Report to Sea Grant Program University of California at Berkeley.

Mahin, S. A. and Lin, J. 1983. Construction of Inelastic Response Spectra for Single-Degree-of-Freedom Systems : Computer Program and Applications. Report number UCB/EERC-83/17, Earthquake Engineering Resource Center, College of Engineering, University of California at Berkeley.

McIntyre, S. R. and Jones, D. E. 1992. Classification and Certification of Floating Production Systems Utilized in Asian/Australasian Waters. Proceedings of the Ninth Offshore South East Asia Conference, Singapore. OSEA 92178.

Melchers, R. E. 1987. Structural Reliability : Analysis and Prediction.. The Camelot Press, Southampton, Great Britain.

Moore, W. H. 1993. Management of Human and Organizational Error in Operations of Marine Systems. Dr. Engineering Dissertation. Department of Naval Architecture. University of California at Berkeley.

Nikolaidis, E. and Kaplan, P. 1992. Uncertainties in Stress Analyses on Marine Structures. International Shipbuilding Progress, Volume 39, Delft University Press.

Peuker, M., Kokkinowrachos, K., and Giese, K. 1987. Development of Flexible Risers for Floating Offshore Production. Proceedings of Offshore Technology Conference: OTC 5469. Houston, Texas.

Saevik, Svein. 1992. On Stresses and Fatigue in Flexible Pipes. Department of Marine Structures, The Norwegian Institute of Technology, University of Trondheim, Norway.

Seelig, William N., Kriebel, David and Headland, John. 1992. Broadside Current Forces on Moored Ships. Proceedings of the International Conference of Civil Engineering in the Oceans V. American Society of Civil Engineers. New York, New York.

Simiu, Emil and Scanlan, Robert H. 1978. Wind Effects on Structures. John Wiley & Sons, Inc. NY, New York.

Sødahl, Nils. 1991. Methods for Design and Analysis of Flexible Risers. The Norwegian Institute of Technology. Trondheim, Norway.

Sterling, G. H. 1992. The Development of Oil & Gas in the Deepwater, Gulf of Mexico. Shell Offshore Incorporated. Ocean Engineering Seminar, University of California at Berkeley, January 22.

Stevens, P. M. 1977. Environmental Design Data for the Southern California OCS Region. The Aerospace Corporation. El Segundo, California.

Tein, Y.S.D., Lohr, C. J., Bruen, F. J., and Huang, K.. 1989. A Probabilistic Dynamic and Fatigue Analysis Procedure for Mooring Systems. Proceedings of Offshore Technology Conference: OTC 6036. Houston, TX.

U.S. Department of Commerce, National Oceanic and Atmospheric Administration. 1991. Nautical Chart 18700, Point Conception to Point Sur. Washington, D.C.

U.S. Department of Commerce, National Oceanic and Atmospheric Administration. 1991. Nautical Chart 18744, Santa Monica Bay. Washington, D.C.

U.S. Department of Transportation, Coast Guard. 1993. Deepwater Ports Study. Office of Marine Safety, Security and Environmental Protection. Washington, D.C.

Woodward-Clyde Consultants. 1978. Preliminary Geotechnical Evaluation of Potential Island and Offshore California LNG Import Terminal Sites. Orange, California.

Woodward-Clyde Oceaneering. 1983. Soil and Foundation Investigation, Jackup Rig Sites, Leases 3120 and 3242, Santa Barbara Channel, California. Houston, Texas.

Woodward-Clyde Oceaneering. 1984. Geotechnical Investigation : Boring 308N Coal Oil Point. Houston, Texas.

Yong, Luo and Ahilan, R.V. 1991. Mooring Safety Assessment Using Reliability Techniques. Proceedings of Offshore Technology Conference: . Houston, TX.

Proceedings of the Seventh (1997) International Offshore and Polar Engineering Conference
Honolulu, USA, May 25-30, 1997

ISBN 1-880653-28-1 (Set); ISBN 1-880653-29-X (Vol. I)

Dynamics of Moored Tankers - SPM and Turret

K. Nishimoto, H.L. Brinati and C.H. Fucatu
University of São Paulo
São Paulo, Brazil

ABSTRACT

Single Point Mooring (SPM) systems are largely used as a tanker terminal for loading and offloading purpose.Turret_moored tanker have been conceived by the oil industry as a floating production storage and offloading (FPSO) system. Both systems when subjected to the environment forces may present dynamic instability. The present paper is concerned with the dynamic analysis of these two types of mooring systems with focus on the ship motions and hawser (or lines) tension.

The paper presents a description of the mathematical model used to represent the dynamics of the mooring systems. Particular emphasis is given to the slow ship dynamics in which a manoeuvering model is used to represent the damping forces. Simulation of the mathematical model is used to evaluate the dynamic behaviour of the mooring systems. It is firstly analysed the performance of a ballasted tanker moored to a SPM in shallow water. The effect of the wave forces on the ship motions and hawser tension is investigated. Later on it is studied the dynamical behaviour of a turret-moored tanker. The influence of the turret position and wave angle of incidence upon the system stability is investigated.

INTRODUCTION

The dynamic behaviour of two different concepts of mooring systems is analysed in this paper. The first one is the Single Point Mooring (SPM) system, which is widely used as a tanker terminal. Petrobrás, the brazilian oil company, has been operating this type of tanker terminal for a long time in the south coast of the country. The company's experience shows a good performance of SPM systems for tankers up to 200 kDWT. Recently, Petrobrás, motivated by the economical advantage of employing VLCC for delivering oil in that region, ordered a feasibility study for berthing a 280 kDWT tanker at a SPM system. We have been working in this project and part of this paper is dedicated to present the results of this study.

The other type of mooring system focused in the paper is the turret one, which is being seen by Petrobrás as an attractive option for a floating production, storage and offloading system since it allows the company to convert old tankers into FPSO systems. The main advantage of the turret-moored tanker configuration is its low cost-performance ratio when compared to other solutions. Another advantage of the turret system is its capability of keeping the ship aligned with the direction of the environmental forces.

The determination of the dynamic behaviour of ship mooring systems requires a very detailed and complex modelling of the forces and moments acting on the ship. SPM system, Spread Mooring System (SMS) and the turret-moored ships have been studied by several authors - Wichers (1991, 1994); Obokata (1987); Bernitsas (1995). We have already in a previous paper - Nishimoto et al. (1995) - investigated the problem of dynamic instability of SPM system subjected to the action of wind and current. In that study a new formulation was introduced to deal with the SPM dynamics. The formulation is based on a manoeuvering model developed by Takashina (1986, 1990) for low speed ship motion. The results of simulation tests using this model were in good agreement with those presented by Wichers (1991) and Obokata (1987) for SPM system.

The utilization of this mathematical model was also very useful to explain the unstable behaviour of a 280 kDWT tanker moored to a single buoy IMODCO in the brazilian south coast; in that case the application of other commercial softwares did not foresee the system instability.

In the present paper the stability analysis is done taking into account also the wave effects. As an illustration it is studied the technical feasibility of berthing a VLCC at a SPM terminal. The mathematical models used in the SPM analysis were modified to deal with a turret-moored-tanker. We included also in the ship dynamics formulation the effect of wave drift damping. The influence of the turret position and the environment forces on the dynamic performance of the system is investigated.

The mathematical models of the ship mooring system are presented in the next section. They consist of the ship model, which includes the hydrodynamic forces, the mooring lines and the wave force models. The two following sections are dedicated to analyse the dynamic behaviour of the SPM and turret systems, respectively. Finally, the main conclusions of the study are pointed out.

MATHEMATICAL MODELS

The mathematical models required to represent the motion of a moored tanker are presented in this section. Firstly it is shown the equations of motion for the horizontal plane slow ship dynamics. Afterwards we describe the formulation for the wave forces and mooring lines.

Equations of Motion for the slow ship dynamics

Figure 1 depicts two coordinate systems used to represent the motion of a moored ship:
- the OXYZ axis system fixed on the earth (inertial system);
- the Gxyz axis fixed on the ship with origin at the ship center of gravity.

Applying Newton's law of motion, the following differential equations describing the relationship between the motion variables, referred to the Gxyz system, and the external forces may be obtained:

$$\begin{aligned} M(\dot{u} - rv) &= F_{X\,ext} \\ M(\dot{v} + ru) &= F_{Y\,ext} \\ I_z\dot{r} &= F_{N\,ext} \end{aligned} \tag{1}$$

where M is the ship mass, I_z is the moment of inertia about the Gz axis, u and v are, respectively, the surge and sway velocities, r is the yaw angular velocity, $\dot{u}$, $\dot{v}$ and $\dot{r}$ are the time derivatives of u, v and r, respectively; $F_{X\,ext}$, $F_{Y\,ext}$ and $F_{N\,ext}$ are the excitations forces in x and y directions, and moment about the z axis.

The external forces and moment for each ship may be expressed in terms of the different factors that affect the ship dynamics, as for instance for $F_{X\,ext}$:

$$F_{X\,ext} = F_{X\,i} + F_{X\,h} + F_{X\,w} + F_{X\,c} + F_{X\,wv} + F_{X\,m} \qquad (2)$$

where $F_{X\,i}$ and $F_{X\,h}$, are the hull inertia and damping forces, respectively, $F_{X\,w}$, $F_{X\,c}$ and $F_{X\,wv}$ are the wind, current and wave forces, respectively, $F_{X\,m}$ is the mooring force.

In order to complete the mathematical model describing the horizontal ship motion one should include the kinematics relationships:

$$\begin{aligned} \dot{X} &= u\cos\psi - v\,sen\psi \\ \dot{Y} &= v\cos\psi + u\,sen\psi \\ \dot{\Psi} &= r \end{aligned} \qquad (3)$$

where ψ is the ship heading angle.

Thus, the mathematical model which represents the motion of the FPSO and SPM system in the horizontal plane consists of 3 differential equations.

Hydrodynamic Models

In a previous paper (Nishimoto et al., 1995) the authors used two different formulations to model the hydrodynamic forces (inertia and damping) which act on the ship hull: the first one is a classical model used by several authors like Wichers (1991), Obokata (1987); the other one is based on a manoeuvering model, specially developed to represent low speed ship motion. A comparison between the results given by the two formulations was presented there.

The second model is used in the present study because we believe it gives a better representation of the hull damping forces. For deep water condition these forces are given by (Takashina, 1986):

$$\begin{aligned} X_H &= 0.5\rho LDU^2(X'_u u' + X'_{vr} v' r') \\ Y_H &= 0.5\rho LDU^2(Y'_v v' + Y'_{vvv} v'^3 + Y'_{vvvvv} v'^5 + Y'_{ur} u' r' + Y'_{ur|r|} u' r' |r'| + Y'_{v|r|} v' |r'| \\ N_H &= 0.5\rho L^2 DU^2(N'_v v' + N'_{uv} u' v' + N'_{vvv} v'^3 + N'_{uvvv} u' v'^3 + N'_r r' + N'_{r|r|} r' |r'| + \\ &+ N'_{uv|r|} u' v' |r'| + N'_{vvr} v'^2 r' \end{aligned} \qquad (4)$$

where ρ is the water density, L is the ship length, D is the ship draft; u', v' and r' are the nondimensional surge, sway and yaw velocities: $u' = u / U$; $v' = v / U$; $r' = rL / U$; $U = (u^2 + v^2)^{0.5}$.

The coefficients in equation (4) are the nondimensional hydrodynamic derivatives, obtained from captive model tests. The coefficients determined by Takashina for a particular ship can be used for another ship provided they have a geometric similarity. Otherwise it is required to carry out a set of test with a scale model.

To take into account the shallow water effects, Takashina (1990) introduced in his formulation a correction function to evaluate the ship hydrodynamic derivatives at a water depth W_d based on the values of the derivatives at infinite water depth; this function was also included in our model.

Two modifications were made in Takashina's model to deal with the dynamics of mooring systems. Firstly, the dimensionless coefficients of equation (4) were dimensionalized for digital simulation purpose. The other modification was done in the surge damping term. Initially, in order to increase the damping force which was small compared with that given by Obokata formulation, we had changed $X_u u$ for $Cd_x u^2$, but there was no significant change in the model response. Later on, we reintroduced the term $X_u u$; the coefficient X_u was evaluated from the results of a full scale test performed in Santos basin with a 30 kDWT tanker.

The current forces are taken into account in the model by substituting the absolute surge and sway velocities by the corresponding relative ones.

Mooring Line Model

The mooring forces that act on the ship are computed using a simple mathematical model. It is not considered the hawser dynamics; the hawser model is given by its force-elongation curve (stiffness characteristic).

For the turret system a simplification was made to model the line forces. It was assumed that all lines are connected to a imaginary point on the ship bottom at the longitudinal plane of symmetry; each line is modelled by an appropriate stiffness curve.

Wave forces

The wave forces on the ship hull are evaluated considering separately the mean wave drift forces, the slow varying (second order) drift forces and the high frequency (first order) forces.

Mean wave force. The mean wave force acting on the ship hull is evaluated considering the procedure described below:

-the quadratic transfer functions of wave drift forces for a regular wave are evaluated for different angles of incidence, χ, and different frequencies, ω, within a specified sea spectrum (ITTC); considering only the terms that influence ship motion in the horizontal plane, situated on the diagonal of the matrix of quadratic transfer functions, we obtain the following terms: $X_{dm}(\chi, \omega)$, $Y_{dm}(\chi, \omega)$ and $N_{dm}(\chi, \omega)$. A 3D source distribution method is used to compute the quadratic transfer functions.

- the mean wave forces and moment acting on the ship for each angle of incidence, $X_{Dm}(\chi_I)$, $Y_{Dm}(\chi_I)$ and $N_{Dm}(\chi_I)$, can be computed through:

$$\begin{aligned} X_{Dm}(\chi_I) &= 2\int_0^\infty S(\omega) X_{Dm}(\chi_I, \omega)\, d\omega \\ Y_{Dm}(\chi_I) &= 2\int_0^\infty S(\omega) Y_{Dm}(\chi_I, \omega)\, d\omega \\ N_{Dm}(\chi_I) &= 2\int_0^\infty S(\omega) N_{Dm}(\chi_I, \omega)\, d\omega \end{aligned} \qquad (5)$$

where $S(\omega)$ is the considered sea spectrum.

Slow varying drift forces. In a random sea, besides the mean drift forces, a body is subjected to low frequency oscillating drift forces. These forces, that have frequencies equal the sum and difference of two regular wave frequencies, are usually of low amplitude but they may have significant effect when their frequencies are near the system natural frequencies. In this condition a moored ship may exhibit oscillating motion of high amplitudes.

The slow drift forces spectra, S_{XL}, S_{YL}, S_{NL}, may be computed from the cross-correlation of the sea spectrum and the quadratic transfer function, as shown below:

$$\begin{aligned} S_{XL}(\chi, \mu) &= 8\int_0^\infty S(\omega) S(\omega + \mu) \left[X_{Dm}\left(\chi, \omega + \frac{\mu}{2}\right)\right]^2 d\omega \\ S_{YL}(\chi, \mu) &= 8\int_0^\infty S(\omega) S(\omega + \mu) \left[Y_{Dm}\left(\chi, \omega + \frac{\mu}{2}\right)\right]^2 d\omega \\ S_{NL}(\chi, \mu) &= 8\int_0^\infty S(\omega) S(\omega + \mu) \left[N_{Dm}\left(\chi, \omega + \frac{\mu}{2}\right)\right]^2 d\omega \end{aligned} \qquad (6)$$

where $\mu = \omega_1 - \omega_2$ is the frequency difference.

In equations (6) we used, as proposed by Aranha & Fernandes (1994), μ equal zero.

Once the slow drift spectra has been obtained, we may evaluate the correspondent time varying forces, F_{DLX}, F_{DLY}, F_{DLN}, using the inverse Fourier transform.

The mean wave drift forces and the slow varying drift forces are introduced in equation (1) in order to compute the second order ship motion in the horizontal plane.

First order wave forces. The first order wave forces give rise to the high frequency ship motions. In the study of ship mooring systems, it is usually assumed that the heave and pitch motions are the most important high frequency ship motions. We did not include in our model the high frequency components of surge, sway and yaw motions which would require the utilization of memory functions. It must be pointed out that this simplification is valid since the amplitude of the horizontal plane motions produced by the first order wave forces on the moored ship are about ten times smaller than the motion amplitudes caused by the second order ones, whose frequencies are near the natural frequency of the mooring systems.

Therefore the high frequency ship motion is computed only for heave and pitch, as it is shown below:

$$(m+m_z)\ddot{z}+b\dot{z}+c_{zz}z+b_{z\theta}\ddot{\theta}+b_{z\theta}\dot{\theta}+c_{z\theta}\theta = F_z(\chi,\omega)e^{-j\omega t}+F_{zm}$$
$$(I+I_\theta)\ddot{\theta}+b_{\theta\theta}\dot{\theta}+c_{\theta\theta}\theta+m_{\theta z}\ddot{z}+b_{\theta z}\dot{z}+c_{\theta z}z = M_\theta(\chi,\omega)e^{-j\omega t}+M_{\theta m} \quad (7)$$

where z and $\theta$are, respectively, the heave and pitch displacement of the ship centre of gravity; m_z is the added mass in heave; I_θ is the added moment of inertia in pitch; b_{zz}, $b_{\theta\theta}$, $b_{z\theta}$ and $b_{\theta z}$ are potential damping coefficients; c_{zz}, $c_{\theta\theta}$, $c_{z\theta}$ and $c_{\theta z}$ are hydrostatic restoring coefficients; F_{zm} and $M_{\theta m}$ are the vertical component and the pitch moment of the mooring force respectively.

The hydrodynamic coefficients and the excitation forces that appear in equations (7) are evaluated using a 3D source distribution method. It was assumed that the added mass and damping coefficients are frequency independent.

The first order wave forces spectra are computed as described below:

$$S_{F_z}(\chi,\omega) = S(\omega)(F_z(\chi,\omega) / h)^2$$
$$S_{F_\theta}(\chi,\omega) = S(\omega)(M_\theta(\chi,\omega) / h)^2 \quad (8)$$

where S_{F_z} and S_{F_θ} are the first order wave heave force and pitch moment spectra, respectively; $F_z(\chi,\omega)/h$ and $M_\theta(\chi,\omega)/h$ are the heave force and pitch moment transfer functions, respectively, for a wave of amplitude h.

Wave drift damping. Description of the phenomenon and formulation of computation methods have been presented recently - Wichers(1994), Grue(1996) and Aranha(1994). In the last paper it was presented a procedure that takes into account the current effect. The damping force is derived considering the influence of the current velocity on the wave encounter frequency (Doppler effect). In the present paper we adopted this method, so the wave drift damping matrix $\mathbf{B}_H^{(wd)}$ is given by:

$$\mathbf{B}_H^{(wd)}(\omega) = \frac{1}{c}\left[\mathbf{I}_{(3,3)}|\mathbf{I}_{(3,3)}\right]_{(3,6)} \times \begin{bmatrix} cos\alpha_o\mathbf{b}_w & sin\alpha_o\mathbf{b}_w & \mathbf{0} \\ sin\alpha_o\mathbf{b}_r & cos\alpha_o\mathbf{b}_r & \mathbf{0} \end{bmatrix}_{(6,3)} \quad (9)$$

where :

a_o is the wave direction relatively to ship longitudinal axis

$$\mathbf{b}_w = 4.\mathbf{D}_0(\omega,\alpha_0)+\omega\frac{\partial\mathbf{D}_0}{\partial\omega}(\omega,\alpha_0)$$

$$\mathbf{b}_r = -2.\frac{\partial\mathbf{D}_0}{\partial\omega}(\omega,\alpha_0)$$

$\mathbf{I} = identity\ matrix$

in which $D_0(w, a_0)$ is the vector of slow drift forces:

$$D_o(\omega,\alpha_o) = D_{xo}(\omega,\alpha_o)\vec{i}+D_{yo}(\omega,\alpha_o)\vec{j}+D_{zo}(\omega,\alpha_o)\vec{k}$$

The matrix of wave drift damping for a random sea is given by:

$$\mathbf{B}^{(wd)}(\alpha_o) = 2\int_0^\infty S(\omega).\mathbf{B}_H^{(wd)}(\omega,\alpha_o)d\omega \quad (10)$$

Finally, the matrix of wave drift damping under the influence of the current is given by:

$$F_{wd} = \mathrm{Vc}.\mathbf{B}^{(wd)}(\alpha_o).\begin{bmatrix} cos\alpha_o \\ sin\alpha_o \\ 0 \end{bmatrix} - \mathbf{B}^{(wd)}(\alpha_o).\begin{bmatrix} \dot{x}_1 \\ \dot{x}_2 \\ \dot{x}_6 \end{bmatrix} \quad (11)$$

In this way the wave drift damping corresponds to the variation of the mean wave drift force, which may be positive or negative, depending on the wave and current directions.

Wind forces

Wind forces acting on the ship hull are modelled using traditional formulation whose coefficients are based on model tests. We adopted in our study the OCIMF experimental curves.

ANALYSIS OF SPM SYSTEM

The SPM analysis is divided in two parts. In the first one it is studied the effect of the wave forces on the mooring system behaviour. In the second one it is investigated the technical feasibility of berthing a specified tanker in a SPM terminal.

SPM Data

The system under consideration is a 280 kDWT tanker moored by a bow hawser to a CALM system terminal. The main particulars of this ship are presented in Table 1 for two loading conditions. Three different curves for the system stiffness were considered, as it is shown in Figure 2. The system stiffness is the combination of the buoy lines and hawser stiffness. The curve elast-cab is the one that better fits to the monobuoy characteristics.

Table 1 -The characteristics of the tanker

Designation	Symbol	Unit(SI)	Loaded	Ballasted
Lenght	L	m	327.00	327.00
Breadth	B	m	57.24	57.24
Draft	D	m	20.96	10.5
Depth	H	m	32.00	32.00
Displacement volume	∇	m^3	310.32	146.34
Mass	M	Kg	3.15 10^8	1.50 10^8
Radius of gyration	k11	m	14.90	15.01
	k22		75.80	77.5
	k33		80.09	83.67
Wind area of seperstructure	A_L	m^2	922	922
	A_t		853	853
Added Mass	a11	Kg	3.00 10^7	7.40 10^6
	a22	Kg	1.11 10^8	6.55 10^7
	a66	Kg*m^2	1.05 10^{12}	5.00 10^{11}
	a26	Kg*m	-1.43 10^9	-1.48 10^9

Analysis of Wave Effects

In order to have a better understanding of the wave effects on the SPM performance , and particularly on the hawser tension, the second order and the first order forces were included sequentially in the simulation model.

It was assumed in the present analysis that wind and current have the same direction with a incidence angle of 180 degrees relatively to earth fixed system. It was adopted a Pierson-Moskowit sea spectrum, with a 2.5 m significant wave height .The wave direction is taken as a variable whose effects we want to evaluate.

Second order forces. Since it was selected a low energy sea spectrum, it is not expected that the second order wave forces have a large effect on the mooring system dynamics, particularly when the wind and current speeds are large. We want to investigate, however, the influence of the wave direction. Several simulation tests were carried out varying the wave angle of incidence as well as the wind and current velocities. The simulation results indicated, as it could be expected, that for angle of incidence zero (stern waves) there is a reduction of the longitudinal force on the ship and, therefore the hawser mean tension decreases; depending on the magnitude of wind and current speeds it may occur system instability - oscillatory ship motions and alternating hawser tension. On the other hand, with bow waves (180 degrees) there is an increase in the hawser mean tension which may lead the system to a better stability condition. Figure 3 shows the time series of ship motions and hawser tension when the system is subjected to moderate wind and current speeds, 12 and 1 m/s, respectively, for three different conditions. It may be seen that the amplitudes of ship motion are larger for stern waves but the hawser tension are not smaller in this condition because, although the alternating force may be smaller, the mean hawser tension is larger. Simulation results indicate that for larger wind and current speeds there is no influence of wave forces on the system response.

Figure 4 presents the system behaviour for wave angle of incidence equal to 45 and 90 degrees, with the same values of wind and current speeds. In these cases it is much more clear the stabilizing effect of the wave force. The ship assumes a new heading, around minus 30 degrees, the mean hawser tensions are higher but there is practically no difference in peak hawser tension due to the second order wave forces.

First Order Forces. The first order ship motions, particularly heave and pitch, induced by the waves, may cause large oscillatory forces on the hawser, which for stability purpose is specified with a small length. In this case the vertical motion of the ship bow may become an important factor influencing the hawser tension. The effect of the first order wave forces is illustrated in Figure 5, which shows the time series of ship motions and hawser tension. It may be seen that the inclusion of the high frequency forces produces a small effect on the ship motion but has a great influence on the hawser tension. It should be pointed out that the small difference in the ship motion is due to the variation in mooring force caused by the heave and pitch motions.

When the hawser is new, with its specified elastic properties, it may absorb the first order bow motion. However, as the hawser becomes old, it looses its elasticity and gets stiffer. When this occurs, the hawser can not absorb the vertical ship motions and there may be a failure. The catenary configuration of the buoy mooring lines is an important factor in the definition of the system stiffness. For small water depths the lines elasticity tends to be small.

In the present study it was investigated how the system stiffness may contribute to attenuate the effects of the first order ship motion on the hawser tension. Three differents curves were adopted to represent the system stiffness:

- Wichers' curve, reproduced from Wichers (1994), corresponds to the most elastic system;
- the catenary stiffness curve takes into account only the buoy lines elasticity, assuming that the hawser stiffness is infinity;
- the elast. cab. curve is obtained by the series association of two springs which represent the buoy lines and the hawser.

It was adopted in this analysis wind and current speeds equal to 18 and 1.5 m/s, respectively, and a wave angle of incidence equal to 45 degrees; under these environment conditions the ship motion in the horizontal plane is relatively stable, and the hawser mean tension is high. Figure 6 shows the time series of the hawser tension for the 3 mooring stiffness. It may be seen that the alternating hawser tension is relatively small when the system stiffness is modelled by Wichers or elast. cab. curves. However, when the catenary curve is used the alternating hawser tension is extremely large. This critical system behaviour is due to the combined effect of a large hawser mean tension and a very stiff mooring system. In this condition, although the system is stable, a small ship motion may generate high hawser tension.

An illustrative example. The mathematical models and the correspondent code were used to investigate the technical feasibility of mooring a large tanker of 280 kDWT to a given terminal. The approach used in this investigation was to identify the critical environment conditions for a given hawser length. The acceptability criterion is based on the maximum hawser tension (peak tension). It was assumed that it is feasible to moore the ship to the monobuoy if the maximum tension does not exceed a given value specified by the hawser manufacturer. For the type of hawser (nylon) used, the maximum allowable tension is 2200 kN which corresponds nearly to 30% of its ultimate strength. Considering the inaccuracies of the mathematical models, we adopted as the limiting value a maximum tension of 1800 kN.

Several simulations tests were carried out to define the limiting environment conditions for each hawser length, varying the wind and current speeds and the wave angle of incidence. A first evaluation of the wave effects may be seen comparing the results shown in Figure 7. It was adopted in the simulations wind and current angle of incidence equal to 180 degrees; the wave angle of incidence is 0 degree. It may be noticed that the inclusion of wave effects produces an elevation of the tension surfaces for the two hawser lengths.

Figure 8 presents the results for the angles of incidence equal to 45 and 90 degrees. It is interesting to observ that in these cases the hawser tension increases, in comparison to those shown in Figure 7, for the short hawser and decrease for the large hawser. As it was explained before, for these angles of incidence the second order wave forces lead the ship to a static equilibrium position in which the mean tension is high. Therefore, the first order motion produces large maximum tensions for the shorter hawser.

Taking into account the results shown in Figures 7 and 8 we may conclude that if a 75 m long hawser is used, it is feasible to moore the 280 kDWT tanker for wind and current speeds up to 18 and 1.5 m/s, respectively, and a sea spectrum with significant wave height equal to 2.5 m. .

ANALYSIS OF THE TURRET SYSTEM

The analysis of the turret-moored tanker was divided in two parts. In the first one it is studied the effect of the turret position on the dynamic behaviour of the system subjected to wind and current forces. In the second part the wave forces are included in the models and their effects on the tanker motion and on the line forces are investigated.

Data of the turret-tanker system

The ship used in this analysis is a 200kDWT tanker whose main characteristics are given in Table 2. The turret mooring system consists of 6 equally spaced chain line, as shown in Figure 9. All the lines have the same properties, as presented on Table 3. The above ship and turret data were drawn from Wichers (1994). From the same reference it was taken the stiffness curve of the mooring lines. The dynamic behaviour of the mooring system was investigated for water depth to draught ratio equal to 4.4, for which the stiffness curve was obtained.

Table 2 - Particulars of tanker

Designation		Symbol	Unit	Value
Length		LPP	m	260
Breadth		B	m	44.5
Depth		D	m	23.8
Draft (even keel)		T	m	16.18
Displacement volume		∇	m^3	1.532 E8
Projected area	longit.	ALS	m^2	600
	frontal	ATS	m^2	500
Centre of gravity above base		KG	m	10.00
Centre of gravity aft FP		LCB	m	0.004506
Metacentric height	longit.	GM_l	m	300.31
	transv.	GM_t	m	7.913
Longitudinal radius of gyration		kyy	m	65
Transverse radius of giration		kxx	m	14.6
Yaw radius of giration		kzz	m	65
Added mass	surge	A11	kg	9.11 E6
	sway	A22	kg	1.36 E8
	heave	A33	kg	1.49 E8
	roll	A44	kg	8.14 E9
	pitch	A55	$kg.m^2$	6.43 E11
	yaw	A66	$kg.m^2$	5.43 E11
	sway and yaw	A26	kg.m	-6.06 E8
	heave and pitch	A35	kg.m	3.27 E8

Table 3- Main particulars of mooring arrangement

Designation	Symbol	Unit	Value
Number of chains	-	-	6
Chain diameter	D	inch	5
Length of each chain	L	m	700
Chain table diameter	d	m	7.22
Chain mass in air	M	$kfgs^2/m^2$	37.61
Chain weight in air	W_a	kgf/m	369
Chain weight in water	W	kgf/m	321
Longitudinal stiffness	EA	tf	1.45E5
Pre-tension angle	α	deg	60.0
Pre-tension	P	tf	40.83
Chain table above sea bottom	H	m	63.6

Effect of the turret position on the system stability

The assessment of the turret position is one of the most important steps in the mooring system design. It may be concluded from the analysis of a simplified model that the system stability depends on the parameter $x_p=a/L$, where a is the distance from the turret to midship and L is the ship length. When x_p is small the system is unstable.

The effect of the turret position was investigated by digital simulation of the mathematical model. All the tests were carried out considering only the effects of wind and current , which have the same direction (180°); the values assumed for the wind and current velocities are 20.5 and 1.5 m/s, respectively.

The test results (ship motion in the horizontal plane and lines tension), for 3 different values of a, are presented in Figure 10. It may be seen that for $a = 0$ the system is unstable in the original position of equilibrium, but the ship reaches a new condition of equilibrium, with a ship heading of about 90 degrees; in this condition the ship is exposed to beam wind and current and after 11000 seconds the system become unstable (oscillatory sway motion). On the other hand, for values of a equal to 90 and 155 m, the ship exhibits small oscillatory motions, with constant amplitudes, which is typical of a limit cycle response.

Effect of waves on the turret system behaviour

It is presented in this section the analysis of the performance of the turret-moored tanker under the combined action of wind, current and waves. It is adopted wind and current angles of incidence equal to 180 degrees, and a Pierson-Moskowit sea spectrum with significant height equal to 10 m and mean period of 9 sec. It is assumed that the turret is located 120 m forward the midship.

Several simulation tests were carried out varying the wave angle of incidence. The test results indicated that with the inclusion of wave forces there is a considerable increase in the lines tension. Figure 11 shows the variation of the maximum tension in the most loaded line, as a function of wind and current velocities, for two wave angles of incidence (180 and 60 degrees). The lines tension, which remain below 700 kN when the tanker is subjected only to the action of wind and current, reaches values around 3500 kN with a wave angle of incidence equal to 60 degrees, and even higher values (above 10000 kN) when the angle of incidence is 180 degrees. In this last condition, the line tension is dramatically large exceeding the ultimate strength for current speeds above 2 knots.

Figures 12a and 12b present the time series of ship motions and lines tension, respectively, for two combinations of wind and current velocities. Although the ship exhibits large motions in the horizontal plane, it may be seen that the larger tension peaks occur when the pitch and heave motions have the higher amplitudes.

CONCLUSION

Two types of mooring systems were analysed in this paper - the SPM system and the turret-moored system. The analysis of the SPM was focused on the evaluation of the wave effects on the system stability. It was shown that, considering the worst direction of wind and current, the wave angle of incidence has a relevant effect on the system behaviour. It was also shown that the hawser tension depends on the first order motion induced by waves and, if catenary stiffness curve is used to model the hawser elasticity, the alternating tension may be very large. It was presented a procedure which allows one to define for a given ship the limiting environmental conditions for a selected hawser length.

In the study of the turret-moored tanker it was shown that the system stability depends on the turret position and a better system performance is obtained when the turret is located relatively far from midship. The analysis of the combined effect of wind, current and waves indicated that for the specified sea spectrum there is a great effect of the wave forces; depending on the wave direction, large ship motions and very large line tensions may occur.

ACKNOWLEDGEMENT

The authors express their gratitude to Eng André P Leite from Petrobrás; Dr A. C. Fernandes, consultant of Petrobrás and Prof J. A. P. Aranha from USP for their valuable comments and useful contribution. Part of the results were obtained from a joint research project developed with Petrobrás and the authors express their appreciation to the company for giving us the opportunity to test the methodology presented here.

The gratitude is extended to Mr Leo W. Yamauchi, João Paulo Matsuura and Fausto R. Alves, undergraduate students of USP, for their cooperation in software development and preparation of this paper.

REFERENCES

Aranha & Fernandes (1994): "On the second order low frequency force spectrum", Applied Ocean Research.

Aranha, J. A. P. (1994): "A formula for 'wave damping' in the drift of a floating body", I. Of Fluid Mechanics, vol 272, pp 147-155;

Bernitsas, M.M., Garza-Rios L. O. (1995): "Effect of Mooring Line Arrangement on the Dynamics of Spread Mooring Systems"; Offshore Mechanics and Arctic Engineering OMAE - 1995; Volume 1 - Part B

Grue, J & Palm, E, (1996): "Wave drift damping of floating bodies in slow yaw-motion", to appear;

Hsu & Blenkam (1970): "Analysis of peak mooring forces caused by slow vessel drift oscillations in random waves", Proc. 2nd OTC.

Marthinsen, T., (1983): "Calculation os slowly varying drift force". Appl. Ocean Res., 5, 141-144.

Newman , In., (1974), Second order slowly varying forces on vessels in irregular waves, Proc. Int. Symp. On Dynamics of Marine Vehicle and Structure, UCL.

Nishimoto, K., Brinati, H. L., Fucatu, C. H. (1995): "Analysis of Single Point Moored Tanker Using Manoeuvering Hydrodynamic Model". Offshore Mechanics and Artic Engineering OMAE - 1995; Volume 1 - Part B

Obokata, J. (1987): "On the basic design of single point mooring systems (1st report)", Journal of the Society of Naval architects of Japan, vol161, june 1987.

Oil Companies International Marine Forum (1977), "Prediction of Wind and Current Loads on VLCCs", OCIMF, 1977 (first version).

Pinkster, J. A. (1976): "Low frequency second order wave forces on vessels moored at sea", Proc. 11th Symp. of Naval Hydrodynamics, U.C.L.

Takashina, J. (1986): "Ship Maneuvering Motion due to Tug boats and its Mathematical Model", Jornal of the society of Naval Architects of Japan, vol. 160, Dec. 1986.

Takashina, J., Hirano, M. (1990): "Ship Maneuvering Motion by Tugs in Deep and Shallow Water", MARIN&ICSM90, June 1990.

Wichers, J.E.W., Dercksen, A. (1994): "Investigation Into Scale Effect on Motions and Mooring Forces of a Turret-Moored Tankers"OTC paper N 7444, 1994.

Wichers, J.E.W (1991) The Behaviour of a Moored Tanker in Unsteady Current, Wind and Waves, Marine Tecnology - September 1991.

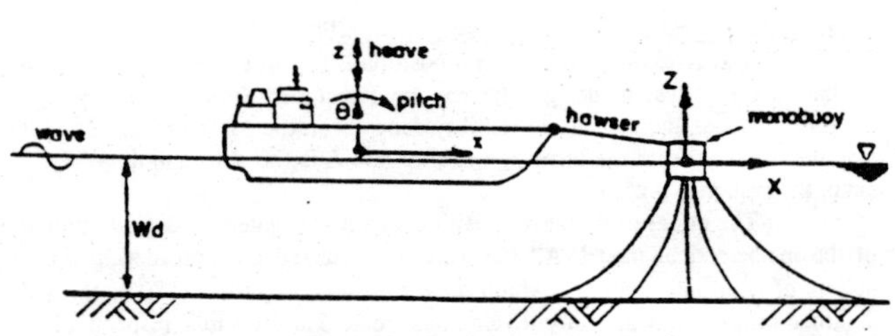

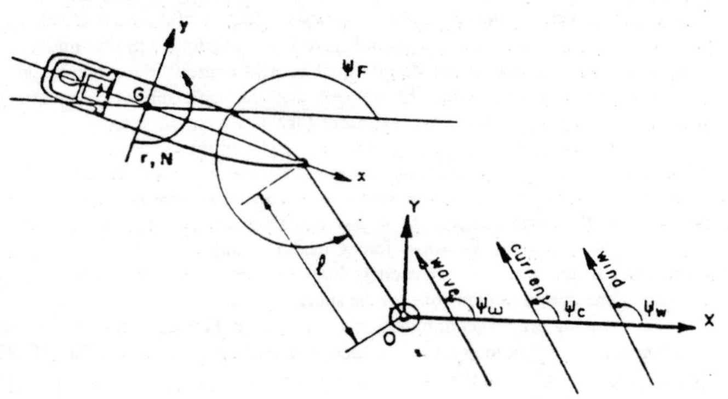

Figure 1 - Coordinate systems

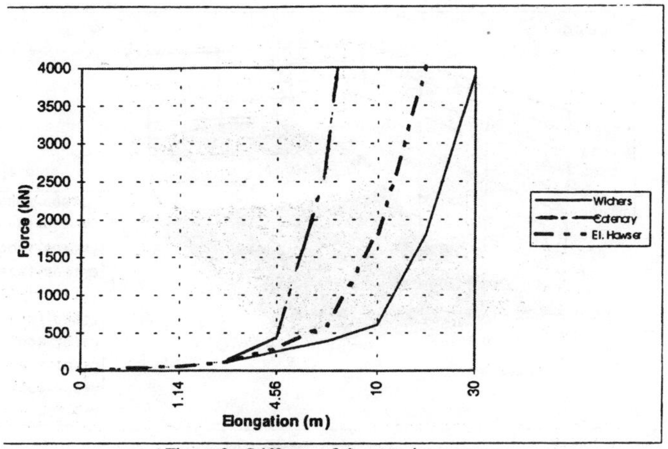

Figure 2 - Stiffness of the mooring system

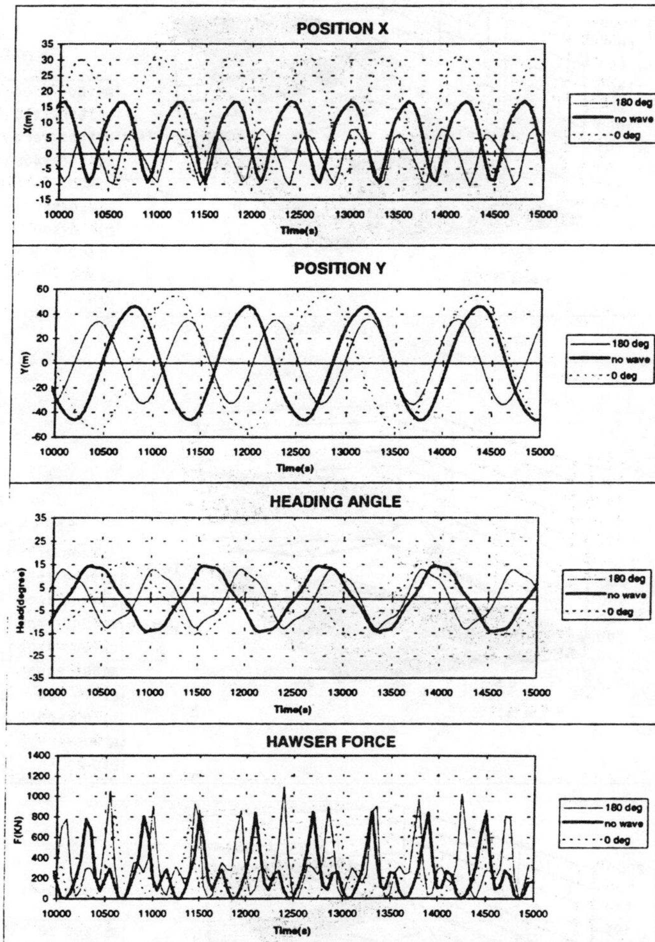

Ballast T=6s C=1(m/s) El. Hawser
H=2.5m V=12(m/s) L=75m

Fig. 3 - Effect of the wave angle of incidence on the SPM system

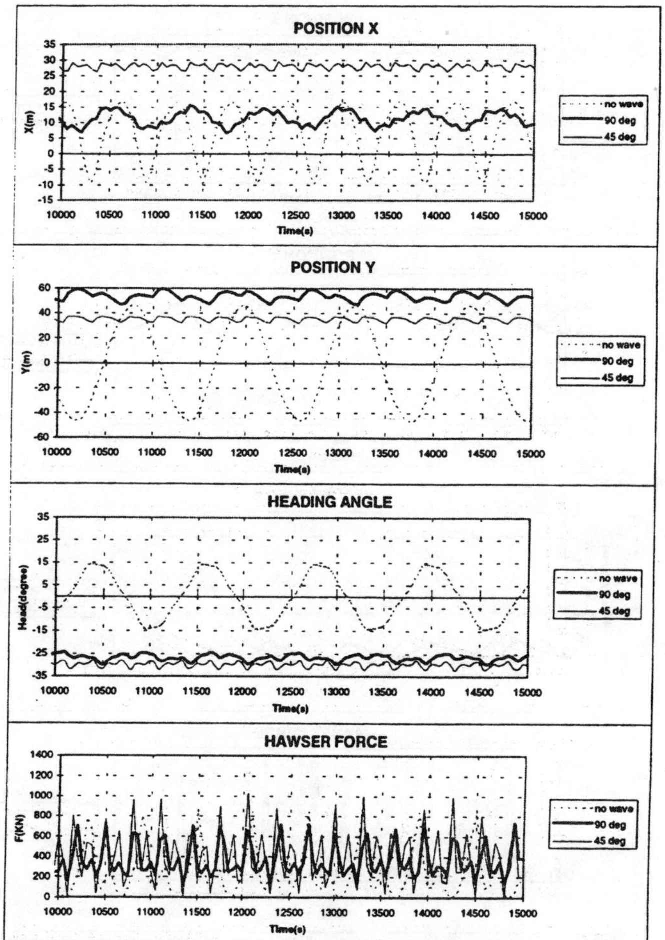

Ballast T=6s C=1(m/s) El. Hawser
H=2.5m V=12(m/s) L=75m

Fig. 4 - Effect of the wave angle of incidence on the SPM system

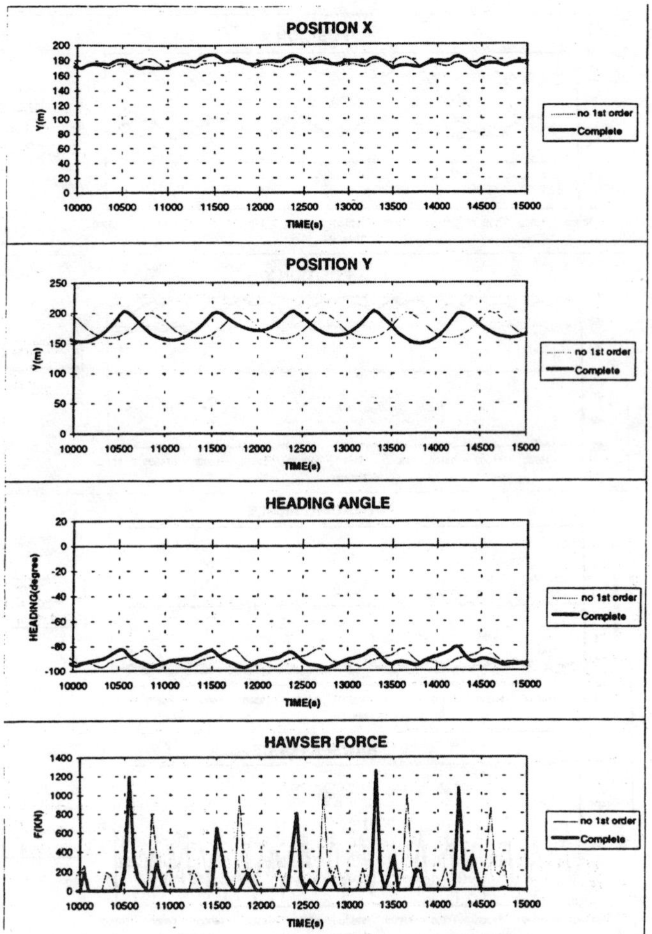

Ballast T=6s C=0.5(m/s) Aw=45 deg El. Hawser
H=2.5m V=0(m/s) L=40m

Fig. 5 - Effect of first order wave forces on the system behaviour

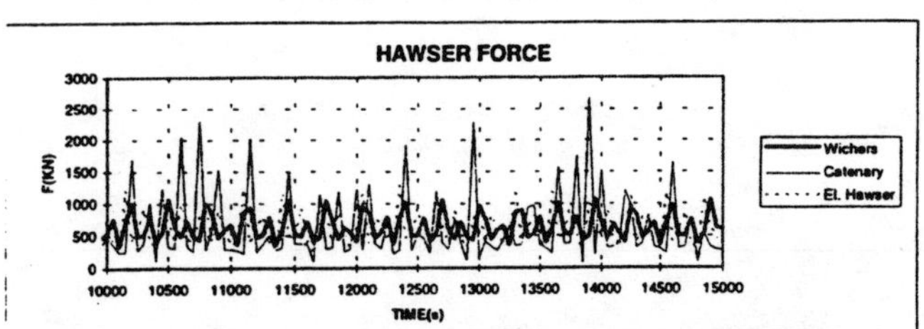

Ballast T=6s C=1.5(m/s) Aw=90 degrees
H=2.5m V=18(m/s) L=40m Takashina

Fig. 6 - Influence of the system elasticity on the hawser tension

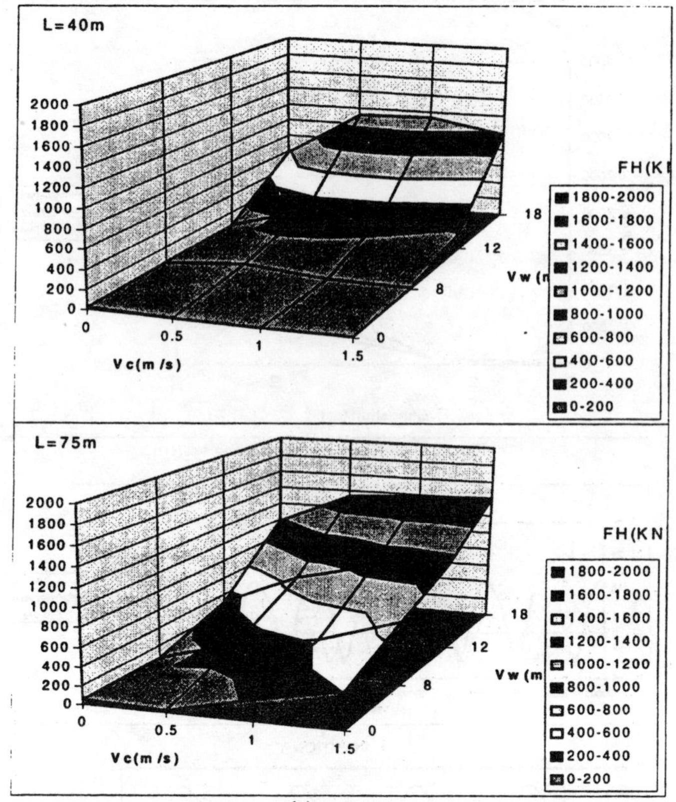

(a) no waves

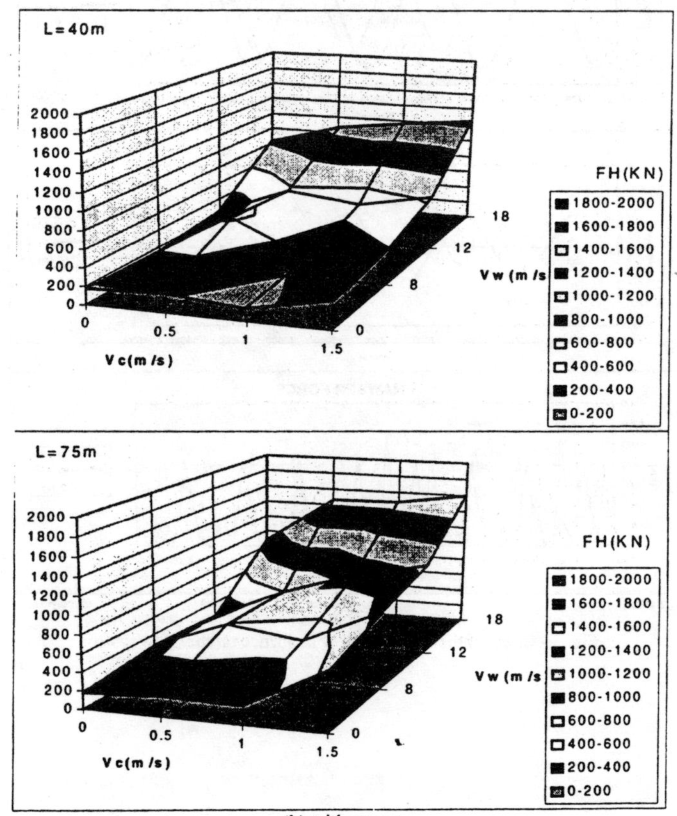

(b) with waves

Figure 7 - Effect of environment forces on the hawser tension

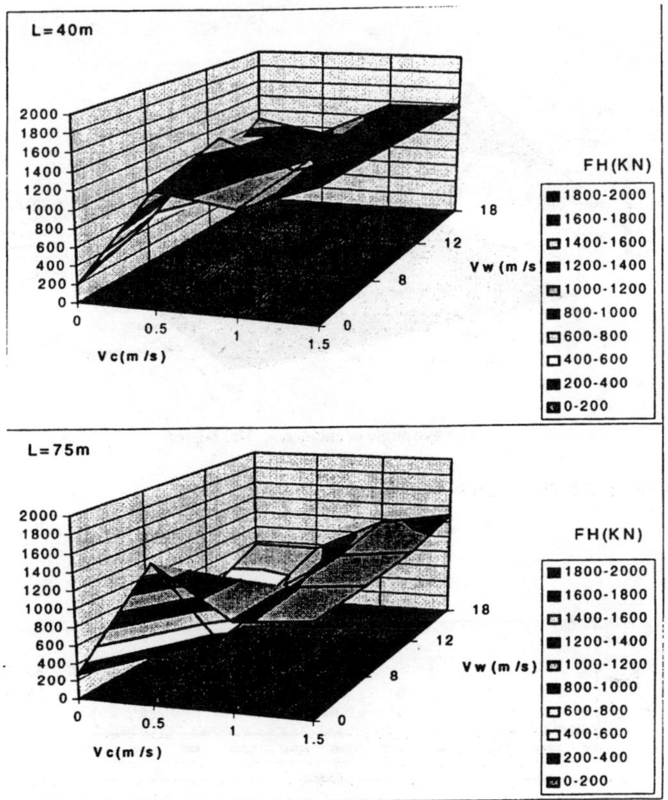

(a) 45 degrees

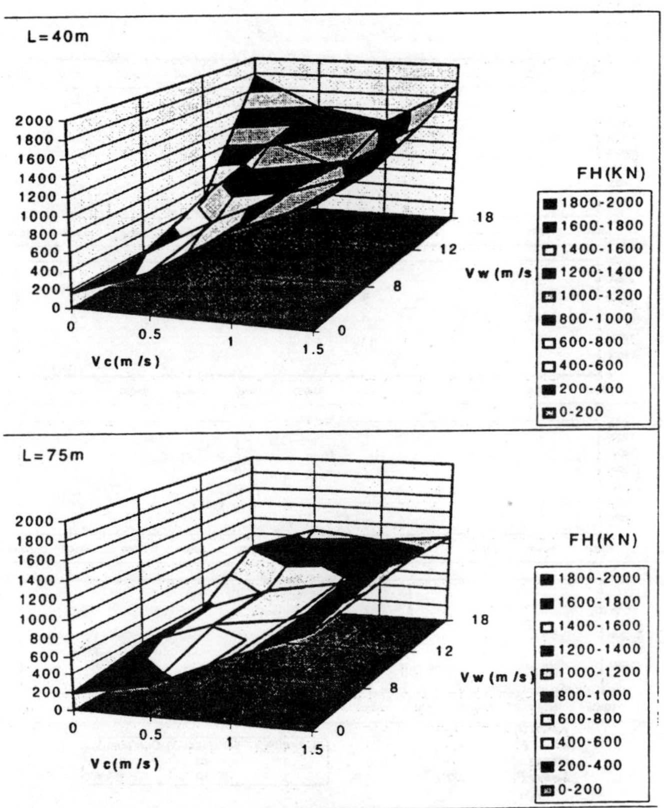

(b) 90 degrees

Figure 8 - Effect of the wave angle of incidence on the hawser tension

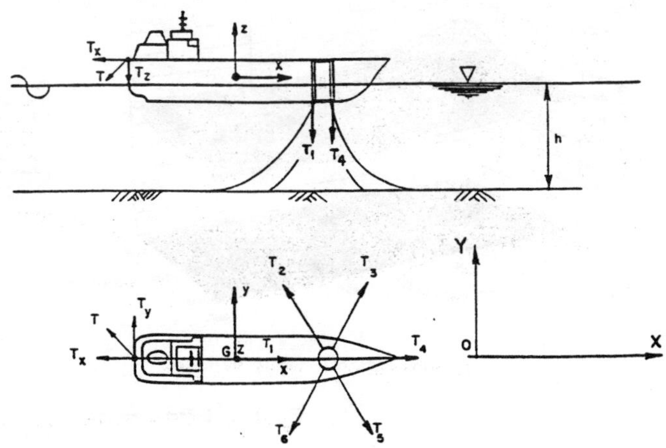

Figure 9 - Geometry of the Turret-moored tanker and coordinate systems

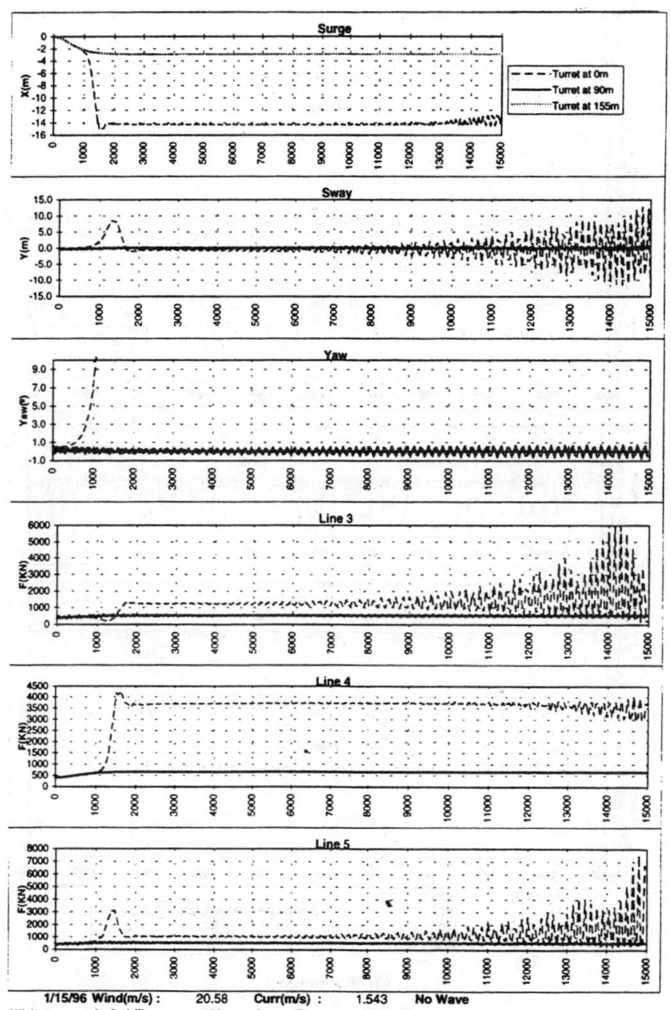

Figure 10 - Effect of turret position on the dynamic behaviour of the system

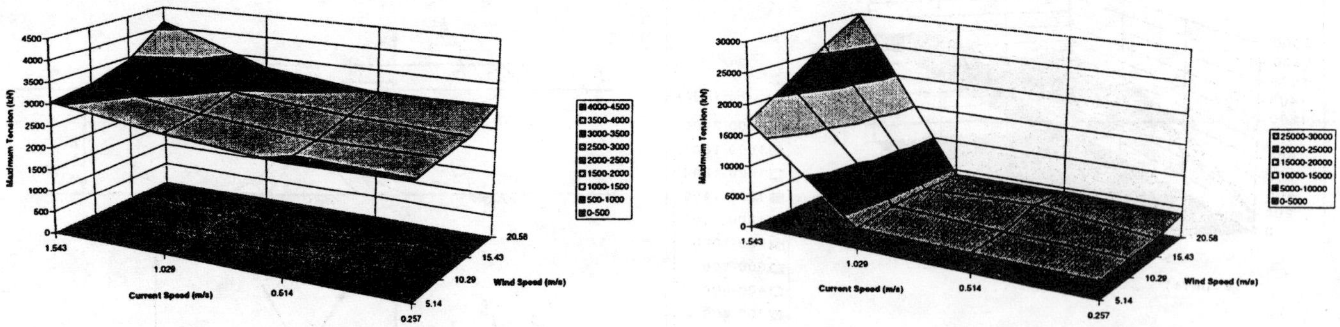

(a) wave angle of incidence: 60 degrees

(b) wave angle of incidence: 180 degrees

Figure 11 - Effect of environmental forces on the tension of the most loaded line

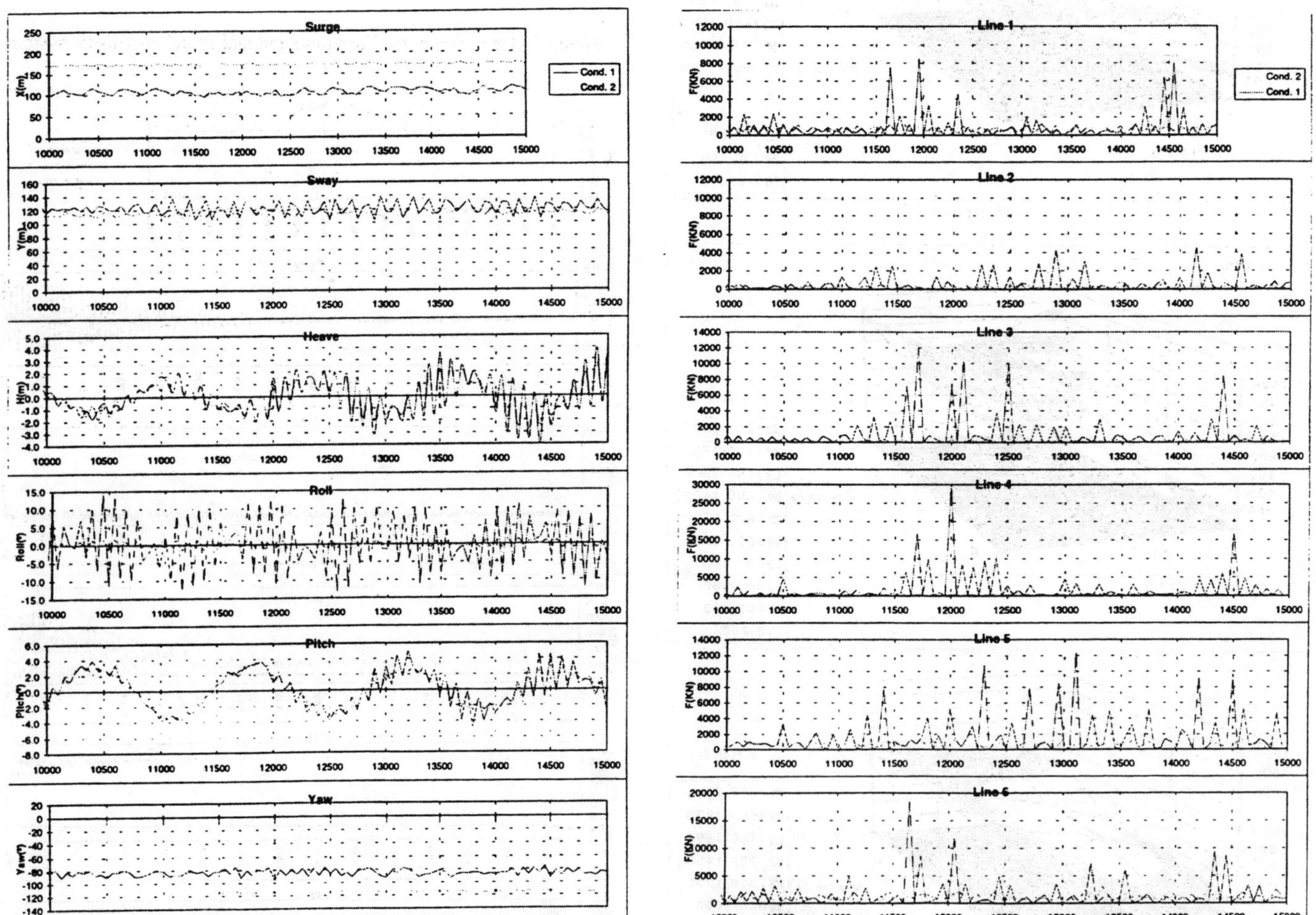

Awind (°) : 180.0
Ac (°) : 180.0
Wave (m) 10
Awave (°) 60.0
T(s) 9

Condition	Wind (knots)	Curr (knots)
1	40	3
2	10	1

(a) ship motions

Awind (°) : 180.0
Ac (°) : 180.0
Wave (m) 10
Awave (°) 60.0
T(s) 9

Condition	Wind (knots)	Curr (knots)
1	40	3
2	10	1

(b) lines tension

Figure 12 - Behaviour of the turret-moored tanker-ship motions

Proceedings of the Seventh (1997) International Offshore and Polar Engineering Conference
Honolulu, USA, May 25-30, 1997

ISBN 1-880653-28-1 (Set); ISBN 1-880653-29-X (Vol. I)

Barge Mounting Technique for Installation of Large Offshore Bridge Structure

Y.T. Kim
Inha University
Incheon, Korea
Chul H. Jo
Hyundai Heavy Industries Co.
Ulsan, Korea

ABSTRACT

Most offshore structures have been installed with assistance of derrick crane vessels. However as the size of offshore structure increases, the barge mounting method (sometime called as IFO technique but it will be called as BM method in this paper for convenience) has been introduced and applied. And this method has proved that without heavy crane vessels, large offshore structures can be installed safely and economically. The key of the BM method is the control of barge motion during docking and mating operations. Also to absorb impact loads and to help with load transfer during mating between substructure and deck, several damping mechanisms have been developed in the past.

This method is being planned for the application of a bridge installation in Korea because it is found that the BM technique is not only attractive to heavy offshore structures but also can be effectively applied to the bridge installation. The most critical area of concern is the safe docking and mating operation of the bridge being mounted on high level piers. To overcome these difficulties, a strand jacking system is applied and introduced for the installation of bridges on piers. Docking can be assisted by tug boats and mooring lines secured to the existing substructures. Mating will be achieved by a jacking system together with a ballasting of transportation barge. Since the motion of barge is also very critical in this operation, the acceptable sea states and time are to be carefully investigated.

This paper describes the technical considerations and installation procedures illustrating detail sequence from docking to mating operation of a 4,500 ton bridge construction in Busan, Korea and also introduces the engineering analysis of the BM method for the specified bridge installation.

KEY WORDS

Barge Mounting, Bridge Structures, IFO Method

INTRODUCTION

A BM method has many advantages over to a conventional lifting mode. For a crane lifting, due to the limited crane capacity and boom length, the dimension control of the structure is very important. Also during the design stage, the detail considerations should be made according to the derrick barge specifications and limitations. The conventional deck has higher density due to a compact design and arrangement considering the available lifting capacity and boom reach of selected derrick vessel. Also the lifting aids such as padeyes, slings, rigging platforms, structural reinforcement, etc. can be neglected in the BM scheme. This method does not only save installation cost but also can achieve a remarkable schedule improvement as well. That is why this method is very attractive especially for large offshore structure installation.

The BM method will be applied to the offshore bridge installation on the high level piers in Korea. The wide spacing between piers implicates the restriction of fender and bumper application and also make it more difficult to secure the barge to external forces. The location is subjected to the relatively high tide and this determines the mating time. Also to accommodate the height of piers, elevation towers are to be built on the transportation barge to support the bridge structure and also a jacking system is to be applied to lift and lower the structure. With the specific site conditions and job requirements, detail mooring and barge motion analyses were performed together with a study on the vertical and lateral loads transfer.

A large truss bridge top side weighing about 4,500 ton is to be loaded out onto a floating transportation barge equipped with four elevating towers. The barge with the structure will be towed to the site having relatively low center of gravity to be more stable during sailing. As the barge reaches near to the location, the structure will be lifted up by strand jacking system of the elevating towers. Then the barge will be maneuvering in between piers. After the barge is secured by various mooring lines and two tug boats, the bridge top side will be lowered by the jacking system and ballasting of the transportation barge. A strand jacking system improves significantly the stability of barge during transportation by lowering the center of gravity and provide a safe and quick load transfer during mating operation.

PROJECT DESCRIPTION

Figure 1 shown a schematic diagram of "Kwangan three-span truss bridge" installation. The bridge of 360m long is a part of the Kwangan Urban Highway near Kwangan beach area in Busan, Korea. Kwangan bridge project involves fabrication, transportation and installation of two (2) three-span continuous truss bridges which are being built at the fabrication yard of Hyundai Heavy Industries Co. in Ulsan, Korea. As fabrication is done, it will be loaded onto the HDB-1006 and be towed out to the site near to "Sooyong Bay" in Busan. The bridge will be divided into three (3) blocks. Each block will be installed by the BM method with assistance of jacking system without conventional floating derrick cranes. The main particulars of the bridge block and barge are as follows:

Bridge Block

- L x B x H	=	132m x 20m x 15m
- block weight	=	4000 tons
- support/lift frame weight	=	500 tons
- total lift weight	=	4500 tons
- pier elevation	=	25-32m from C.D.

Barge

- name	=	HDB-1006
- L x B x D	=	168.3m x 47m x 9.5m
- draft MLD	=	6.6 m
- gross tonnage	=	20,000 tons
- net tonnage	=	16,400 tons
- dead weight	=	34,000 tons
- ballast tank capacity	=	57073 m^3
- ballast pumps	=	one centrifugal pump 4000 m^3/hr x 20 m head two centrifugal pump 500 m^3/hr x 20 m head

INSTALLATION METHODOLOGY

The site conditions and the period of installation are very important factors to consider in the BM method. Also the mooring plus fender systems are critical to secure the barge against the external forces during mating operation. Particularly, the strand jack system that will be applied for the bridge installation for Kwangan bridge allows a lower center of gravity than that of the final location during transportation which provides with safer and more stable states. Also this system will increase the weather window for mating operation due to a fast lowering and load transferring capacity.

The main equipment required for the BM method are as listed below;

- one transportation and ballasting barge
- lifting towers with bridge support and lift frame
- towing tugs with anchor handling tug
- primary mooring winch & anchoring system and secondary mooring winch & line system
- four tugger winch & line system
- strand-jack lift & lowering system
- power generation
- ballasting system
- control & monitoring system
- bumper & guide system on piers

SYSTEM DESCRIPTION

1. Strand-Jacking System (Up and Down)

The jacking system is one of the key components for the installation of bridge or similar structure over to a high level piers. The hydraulic jacking system should allow the reverse operation as well. Also a fast lowering function should be equipped with a computerized monitoring and control system. The system should be designed to maintain an even load distribution and requires the following provisions:

- capability to continue a mating operation even for a failure of one hydraulic power unit
- capability to allow a rapid jack-down prior to barge demobilization after load transfer
- pulling capability at high pressure in order to retract the support and guide structure in case of wedging
- capability to operate in an overload for a short time for a failure of one jack of the each group
- capability to continue a lowering operation with a damage to hydraulic pipes or hydraulic connections for a power failure
- capability to operate a lifting up or lowering for a failure of electronic control system

2. Computerized Monitoring and Control System

The monitoring of whole operations requires an access to information relative to the barge and to the jacking equipment. Also, the knowledge of the environmental conditions is essential to a decision making process at each step of the operation.

This will imply the monitoring of the following information:

- all data relative to the operation of the hydraulic system; pressure, flow rate, etc.,
- strand-jack extension and acting load at each support
- total weight and center of gravity of the bridge
- draft, heel, trim, and overall center of gravity of the barge
- environmental conditions (wave, swell, wind, tide and current)
- weather forecast for coming hours

- barge positioning system (GDPS) and CCTV video camera to visualize the mating operation

- mooring lines tensions

The above information and data should be continuously monitored and accessed to the control station which collects them. These information will be the basis of a decision making process for docking and mating operation. The analysis and interpretation will be done by the assistance of computer programs which predict the barge motions, external forces, mooring forces, etc. The control system is designed to accommodate any failure mode and is equipped with a switching function from automatic to manual during jacking up and down operation. The system is facilitated with an alarm function to announce any malfunction and abnormal situations during operation. Also back up and contingency procedures for each step are to be developed to cope with unexpected situations and conditions.

3. Mooring System.

The mooring system should have enough resistance against barge motions and external environmental conditions. The primary mooring system consists of six steel wires with Flipper Delta anchors stretching out to various locations to support mostly the transverse and longitudinal motions of the barge. The secondary mooring system is composite lines which are made with polypropylene ropes and steel wires connected to sheaves and bits of the footing piers.

The pre-tensioned lines are defined to provide a precise vessel positioning by securing the barge against external forces. Possible resonance effects should be eliminated and confirmed by a low frequency analysis considering the wave period and mooring characteristics.

Winches, sheaves, bollards and miscellaneous fittings are designed and sized to satisfy all possible configurations as the barge being moved in and out and ballasted. The barge draft would vary by about 1.2 meters during the ballast and de-ballast operations.

4. Ballasting System

The ballasting system is designed to ballast or de-ballast the barge to control the trim, heel at any time during docking and mating operation. During mating process, a ballasting would be done to assist the lowering of the bridge together with jacking system. After a load transfer, ballasting operation would be required in order to keep the clearance between bridge and the elevating towers for evacuation of the site.

5. Bumper/Guide System/Fender on Pier

Protection of piers and structures during docking and un-docking-locking is required. Depending on the size of barge and dynamic behaviors, the bumper size and type could differ. Even though it is not easy to predict the exact impact force, some type of guideline should be given to bumper design and selection. The impact force was estimated under the condition that the barge is moored by six mooring lines during docking. The external forces were taken from the available data of the region at the time of installation during last 20 years. Also very simple calculation of estimating the barge impact force to bumper can be made based on the energy transfer concept. For this project, rectangular timbers were attached on the bridge outside and bumpers/guides equipped with rubber fender (elastomer material) on the piers are arranged. The fender size and type should be decided depending on the vessel impact magnitude and barge motions and energy absorption.

6. Communications

The communication system is vital to the success of the BM operation. To achieve this goal, the minimum of the below equipment will be utilized:

- VHF radio communication with all involved personnel on barge/pier and assistance tugs

- satellite telephone and telefax to keep permanent liaison with main office, supplier and weather forecast service

INSTALLATION PROCEDURE

1. Tow to Site

Towing operation from Ulsan to the site in Busan off Kwangan beach by tugs will be commenced under daily weather forecast after the completion of the load-out the tests of jacking system. One tug will tow the barge with structure and two AHT's (anchor handling tug) at the site will assist on the spreading of anchors and positioning before and after docking.

2. Arrive at Site

After arrival at the site around 200m off piers, the barge will be maneuvered by tugs and reached 100m apart to the piers. First, two bow anchors will be dropped and two primary bow anchor wires to be payed out by AHT. Then primary stern anchors will be placed. The other anchors will be placed as per the anchor pattern. The anchor patterns should be decided to accommodate various positions of barge during docking and un-docking operation. Also to protect existing subsea facilities and offshore structures, the anchor patterns should be thoroughly reviewed and approved among concerning parities. Also two anchor handling tug boats are to assist on the maneuvering of barge and secure the barge during mating process together with mooring lines to piers. The ballasting system will be used to control the barge trim and heel during the mooring and docking operation. The anchor winches will adjust the minor location of barge and the mooring tension.

3. Barge Docking Preparation

The following items are to be checked and be ready prior to docking:

- radio and communication system

- stand jacking system for mating by lowering

- ballast system

- mooring winches and power units

- tidal range

- tug boats

- weather forecast

- crawler crane

- fenders on piers and bridge

- working platforms on piers and equipment and personnel

- checking list

As the transpiration barge arrives at about 200 m off piers, tow bow anchors dropped at about 115 meter off piers and tow other bow anchor with mooring wires will be payed out by a AHT. The barge will travel to reach about 30 meter off piers.

Four side mooring lines will be connected to piers. Four side mooring winches pull the lines with tug boat assistance and release four bow anchor lines. The barge reaches about 10 meter off piers and makes a stop. Two stern mooring line/anchors are to be payed out by a AHT. The bridge will be lifted up by jacking system to reach the required elevation.

4. Bridge Lift-Up Operation

After the preparatory works are done and the checking results prove that everything is acceptable, the bridge lifting up will be commenced by activating the stand-jack system to the elevation required to place the bridge on the piers. The decision of lifting time should be made considering the entire period of mating operation since once the bridge is lifted up and the docking is made, it is very difficult to evacuate the site due to a severe weather condition which is not suitable for the BM installation.

5. Bridge Mating

Once the barge is in the mating position and it satisfies the mating conditions, the bridge structure being lifted up about 2.5 meter off the top of piers will be lowered. The structure will be carefully lowered by the strand jacking system and the ballasting of the barge. This operation should be started during the falling tide. This operation is very critical and all relating parties should be communicating closely and each step should be checked and monitored.

6. Barge Un-docking

After the load transfer, lowering jacking system and ballasting the barge will continue until it reaches the un-docking clearance between the elevating towers and the bridge. The anchor winches will wind in the anchoring wires to advance for an un-docking . The anchors will be retrieved by AHT and will be fixed to be towed away by a tug boat.

MODELING AND ANALYSIS

The following analyses have been performed for the BM method:

- structural analysis to check the strength of the barge

- Barge stability analysis of each time frame during transportation, bridge lifting up operation, docking, mating (load transfer) operation

- Mooring analysis with the worst environmental conditions during the BM operation

- Estimation of impact loads to pier to determine the fender size and type

1. Modeling

The barge was modeled with 3-D diffraction elements as shown in Figure 9. The origin is located at the bow, keel and centerline. The x-axis is from bow to stern, with the y-axis to starboard and z-axis upward. The centerline of the bridge was located at 93.5 meter from the bow. The geometry of the towers and frames were set accordingly. The mooring lines were broken down into 2 groups, anchor lines and taut lines. The taut lines are from the barge to the footing back and back to the barge. The anchor lines include the rest of the mooring system. The properties for chain and wire are as shown in Table 1. Figure 10 shows the modeled barge being loaded with bridge structures.

2. Loading Conditions

For the computer analysis, six (6) loading conditions have been selected and used for modeling as summarized in Table 2.

3. Environmental Conditions/Criteria

After a complete study of environmental conditions (wave, wind and currents) which were collected for years in the regions, the conditions specified in Table 3 were selected for analyses.

4. Docking and Mating Analysis Results

Based on the selected environmental conditions, the external forces acting on the barge during standby, docking and mating modes were obtained as summarized in Table 4.

5. Towing Motion Analysis Results

The maximum accelerations at the bridge center of gravity during tow operation were estimated in Table 5.

CONCLUSION

The BM method can be applied not only to the installation of large offshore structures but also to the bridge construction. The bridge installation by a BM technique requires many areas to study prior to the selection of barge, mooring system, pier size and type, etc. considering the nature of the site and environmental conditions. Due to the height of piers, the bridge structure should be lifted up before docking process by a jacking system. For a better stability during transportation, the center of gravity of the structure can be lowered by placing it below than the designed elevation. The quick load transferring is very critical in this operation as well. The application of a strand jacking system for bridge structures can achieve these needs by lifting and lowering the structure on barge. Since the dynamic behavior of barge is very important to the BM operation, various analyses for each step should be intensively performed such as vessel stability and dynamic analyses, mooring study, docking and mating simulations. The coordination among involving parties are critical in BM operation together with continuos monitoring system. The BM technique has many advantage over to the ordinary crane lifting method since it can reduce construction time, transportation and installation cost and does not limit the structure configuration. The BM technique for the Kwangan truss bridge construction was introduced in this paper and the analyses proved that a BM technique is also acceptable for the application of this type of structure.

REFERENCE

William David Hartell and S. Michael Beattie, "Integrated, Float-over Design Considerations", Offshore Technology Conference, OTC 8119, May 1996.

Birdwell, J.R. and Jordan, J.E., "Mobile Bay Project Float-on Deck Installation" Offshore Technology Conference, OTC 7499, May 1994.

White, G.J., Ramzan F.A., "Offshore Installation of an Integrated Deck onto a Preinstalled Jacket", Offshore Technology Conference, OTC 5260, May 1986.

NOTE

Due to the restriction of paper quantity, the details will be illustrated and explained during presentation.

Table 1. Mooring System Properties

Description	Diameter (mm)	AE (m-tons)	Length (m)	Breaking (m-tons)	Winch (tons)
Chain	68	36300	220	406	
Wire 1	58	26400	283	202	75
Wire 2	46	16600	300	127	50
Wire 3	32	8040	Varies	61	30

Table 2. Loading Conditions

Draft (m)	Displacement (tons)	KG (total) (m)	Bridge Weight (tons)	Bridge KG (m)
3.50	25250	13..30	4500	42.0
2.85	20250	14.67	4500	42.0
4.40	31950	11.47	4500	42.0
3.50	25250	15.12	4500	50.0
2.85	20250	16.93	4500	50.0
4.40	31950	12.90	4500	50.0

Table 3. Selected Environmental Conditions

Description	Maximum Docking/Mating Operation Condition	Maximum Design Condition (Standby Operation)	Towing Condition
Significant Wave Height (m)	0.3 (head and quartering)	1.5 (head and quartering)	1.5 (head and quartering)
Mean Period (sec)	4-5	6-7	6-7
Current (cm/sec)	20	30	.
Wind Velocity (knots)	20	40	40
Towing Speed (knots)	.	.	4

Table 4. Environmental Force

Operation Condition	Environmental Force (m-tons)	Bow	Beam	Quartering
Maximum Standby	Wind	104.5	47.3	101.2
	Current	0.6	15.7	10.9
	Mean Wave Drift	4.1	17.0	17.5
	TOTAL	**109.2**	**80.0**	**129.6**
Maximum Docking/Mating	Wind	26.1	11.8	25.3
	Current	0.2	3.8	2.6
	Mean Wave Drift	1.9	11.3	11.5
	TOTAL	**28.2**	**26.9**	**39.4**

Table 5. Maximum Acceleration During Tow at 4 knots

Significant Wave Height (m)	Tow Speed (knot)	Heave Acceleration (m/sec^2)	Roll Acceleration (m/sec^2)	Pitch Acceleration (m/sec^2)
1.0	4	1.07 g	0.11 g	0.02 g
1.5	4	1.16 g	0.16 g	0.04 g
2.5	4	1.27 g	0.27 g	0.06 g

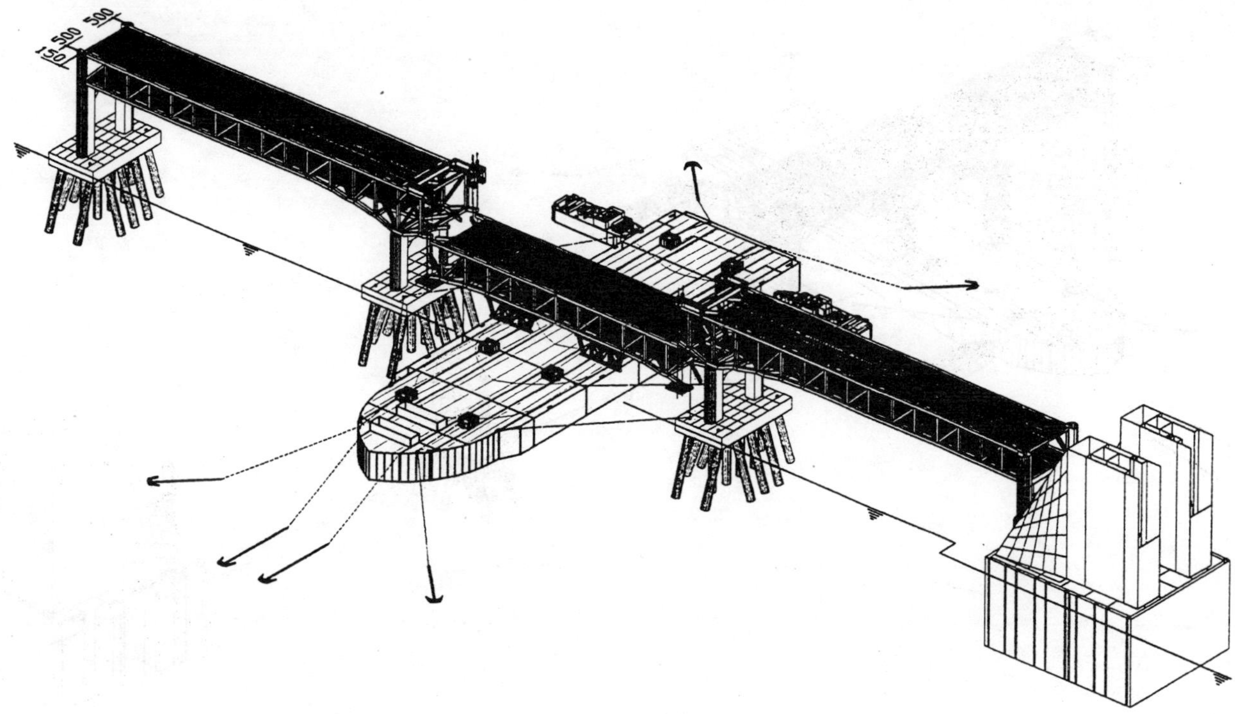

Figure 1. Bridge Installation Sequence (Continued) – STEP 3

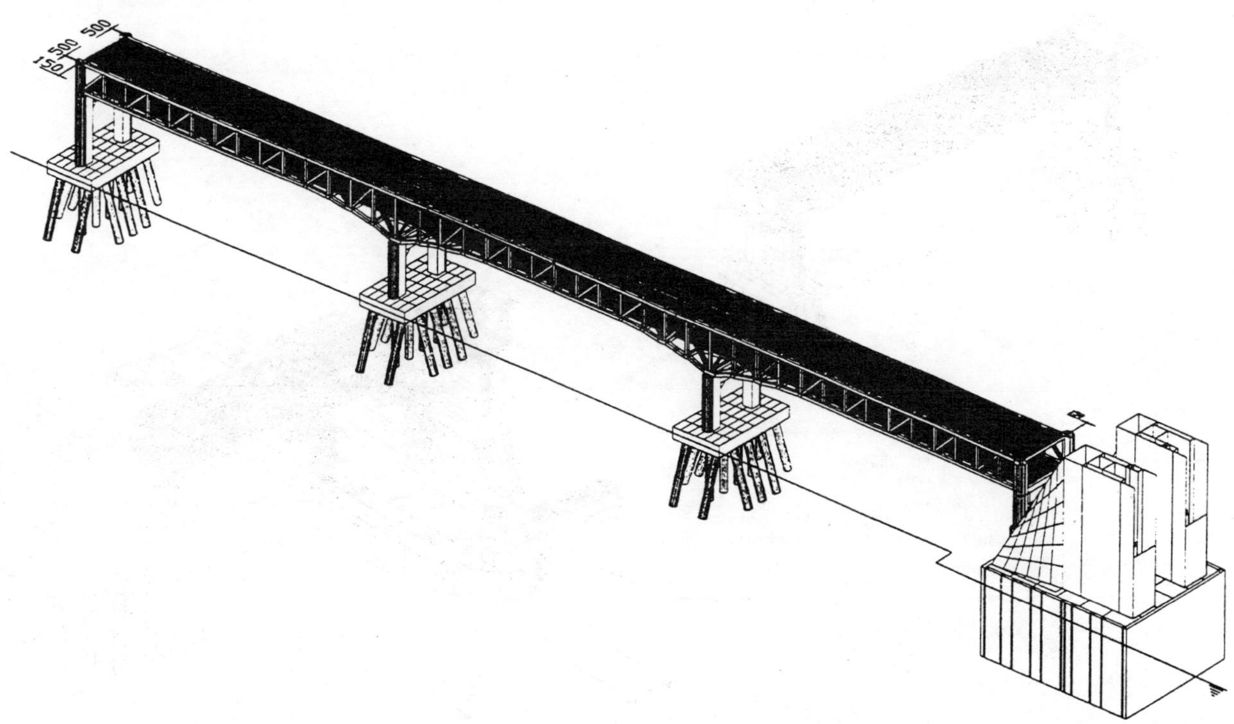

Figure 1. Bridge Installation Sequence (Continued) – STEP 4

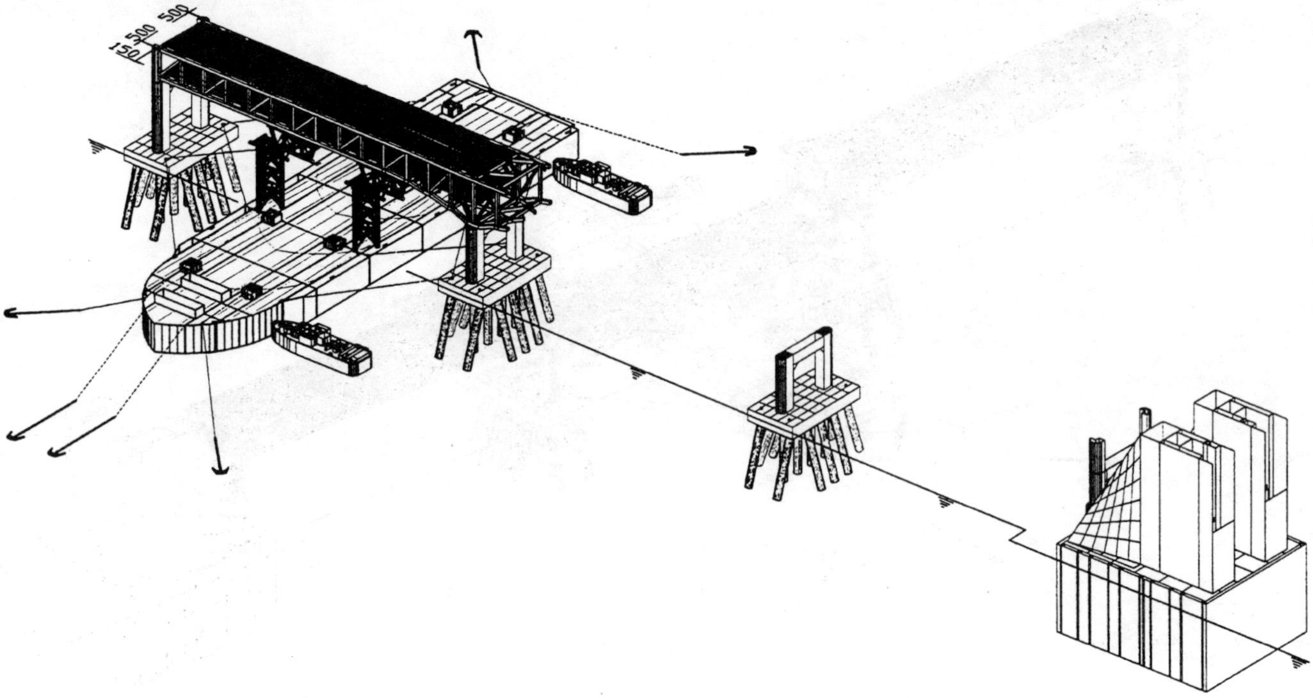

Figure 1. Bridge Installation Sequence – STEP 1

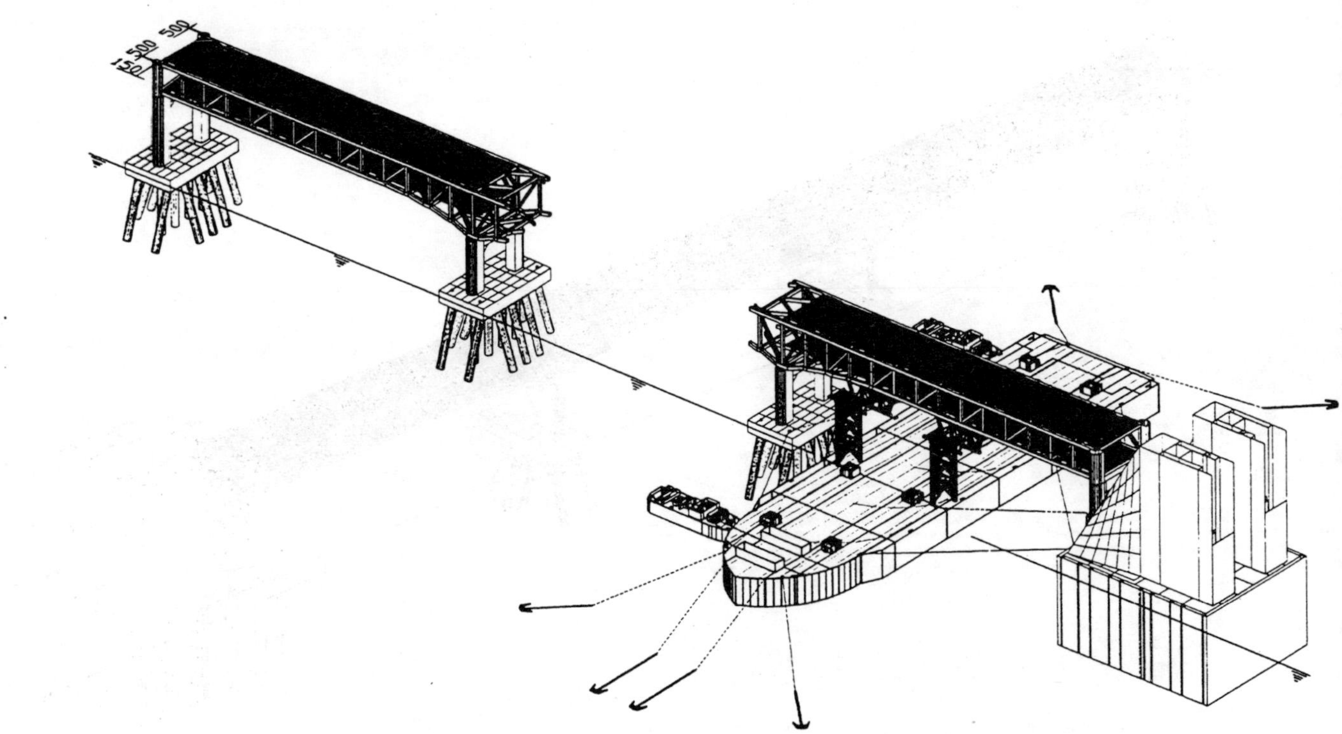

Figure 1. Bridge Installation Sequence (Continued) – STEP 2

Proceedings of the Seventh (1997) International Offshore and Polar Engineering Conference
Honolulu, USA, May 25-30, 1997

ISBN 1-880653-28-1 (Set); ISBN 1-880653-29-X (Vol. I)

Harald Valve Station - a Danish Subsea Installation in the North Sea

Nils Gjersøe Fog
LICengineering A/S

Poul Henrik Jønsson
Mærsk Olie og Gas AS

Esbjerg, Denmark

ABSTRACT

The Harald Valve Station was installed in the autumn 1996. The valve station is located close to the Harald A platform in the Danish part of the North Sea and comprises main safety valves for the pipelines between Harald, Svend and Tyra fields. The valve station includes manually or ROV operated valves for the Harald 24" and 16" gas/oil export pipelines. The design of structure and layout for valves, piping and adjacent expansion spools are discussed. New concepts were applied to ensure an open layout of the structure reducing seabed scour around and within the installation while maintaining full overtrawl capability. The overall weight of approximately 80 tonnes is significantly lower than found for existing similar designs, enabling installation from a diving support vessel in one operation including the valve spools. The main structural design considerations were loads imposed by impact or hooking from fishing gear. The design analysis of structure and piping, for functional and environmental loads as well as accidental loads is described.

KEY WORDS: subsea installations, subsea structural design, subsea valves, pipelines, wave generated seabed scour, diver/ROV operations.

INTRODUCTION

The Harald Valve Station is located approximately 250 m south of the Harald A platform.

The main function of the valve station is to provide a safety shut down capability for the 24" and 16" pipelines carrying gas and oil produced at the Harald field, approximately 80 km south towards the Tyra facilities.

The valve station is placed between the main exit lines for the pipelines from Harald to Svend and Tyra fields. It is tied-in to the pipelines by z-shaped expansion spools. The pipeline approach layout is shown on Fig. 1. The valve station general arrangement is shown on Fig. 2.

The valve station contains a check valve and a ball valve for each of the two pipelines. The main safety valves are the check valves placed on the upstream side of the valve station. The ball valves are placed on the downstream side. Their main purpose is to enable service being carried out on check valves without flooding the main pipelines.

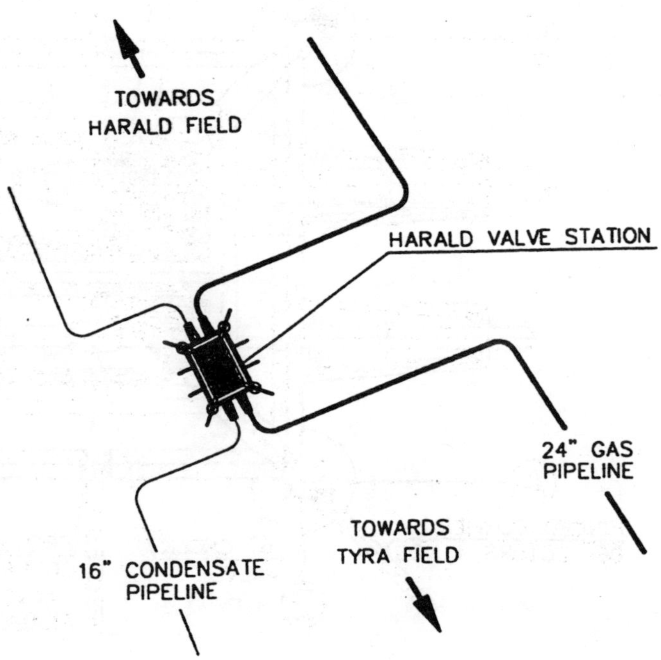

Fig. 1 Layout of pipelines/spools at the Harald Valve Station.

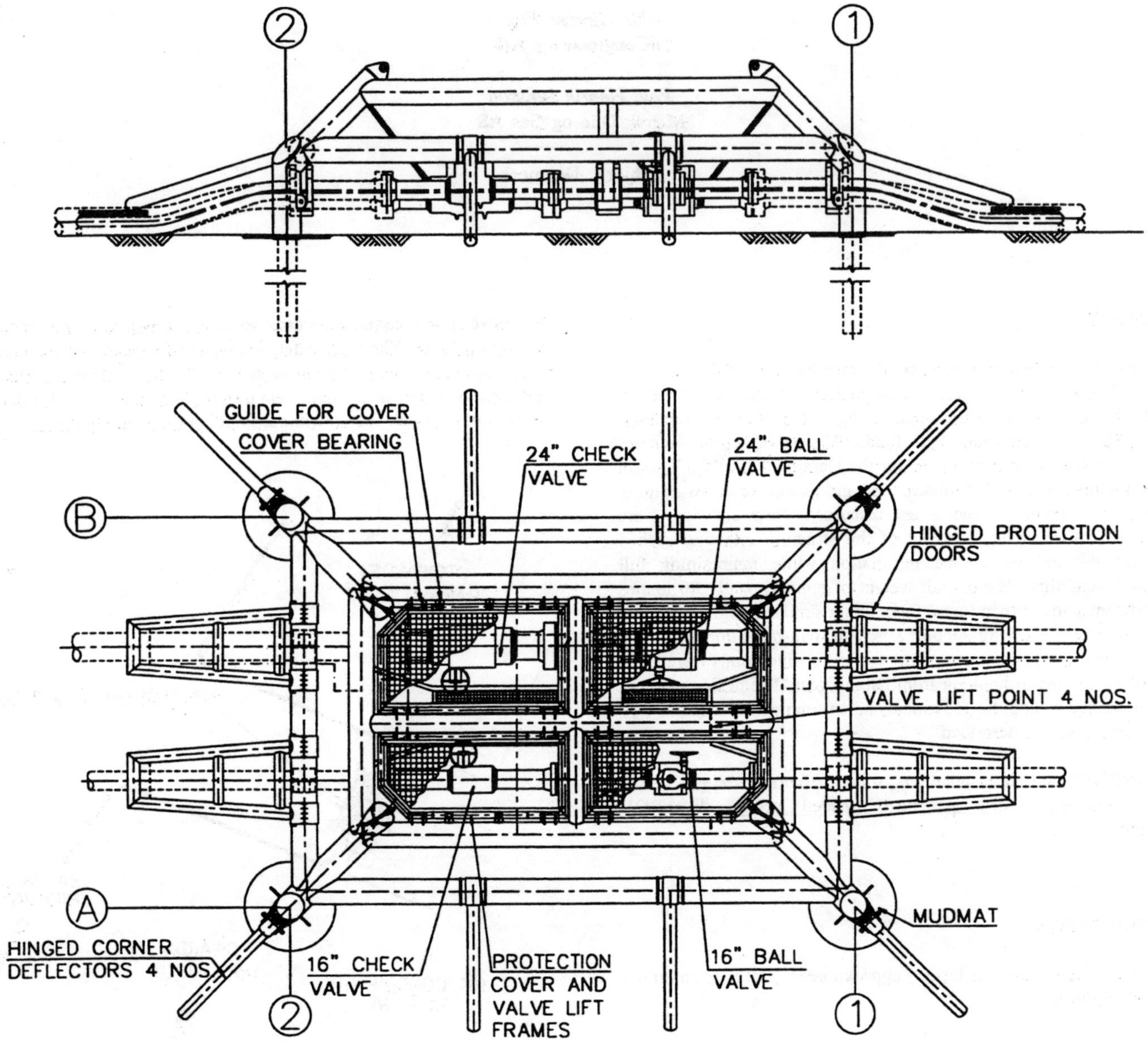

Fig. 2 Valve station general arrangement

DESIGN PHILOSOPHY

The valve station is designed to protect and support the valves on the 24" and 16" pipelines from Harald A. The main design considerations for the valve station layout have been focused on the following issues:

- overtrawl capability
- low weight, installation in one operation from a diving support vessel
- minimisation of scour
- easy access to valves
- possibility for lift out/replacement of check valves

Traditionally, subsea valve stations have been designed, based on a support frame resting on the seabed. This will result in a compact solution with a high overall weight. New concepts have been employed in the design of the Harald Valve Station in order to satisfy the above design criteria. The design was further based on good experience with open structure design in the North Sea - primarily from the Regnar subsea wellhead completion, put in production in 1993, ref. /1/ and from test results during development of the Gannet B. field, ref. /2/. The design has resulted in an open, strong support and protection structure, piled into the seabed. The valve station covers an area of approximately 17 x 13 m on the seabed when installed. It appears as a low profile with inclined corner members.

The four corner legs have been designed with the upper part angled and the lower part vertical, thereby pile sleeves will form an integrated part of the main structure.

Protection against damage by fishing gear overtrawling is a major consideration in the North Sea. Substantial trawl boards and bottom dragged trawls are used, see ref. /3/, /4/ and /5/. Trawl board deflectors are employed at the corner legs and at each main side of the structure. The deflectors will ensure overtrawl capability by guiding the tow lines and trawl boards over the structure. Thereby fishing gear is stopped or deflected before it enters into the structure, hitting a valve or other vulnerable items. Deflectors are hinged to the main structure. The deflectors will adjust to the actual seabed level depending on the scour development.

The overall weight of the valve station including valve spools is approximately 80 tonnes, sufficiently low for the structure and valve spools to be installed in one operation from a medium size diving support vessel. The structure has four main lifteyes built into the upper corner members. The valve station including valves can be installed as a whole by lifting from these four points, see Fig. 2.

The layout of the structure has been chosen as open as possible. In this way scour is minimised and access to valves is improved. A work platform is positioned between the two valve spools to be used by divers for operation and inspection of valves.

The valve station has four roof panels, one above each valve. The roof panels will protect valves against dropped objects.

The valve spools and tie-in spools are clamp supported from the main structure above. This is part of the general philosophy of maintaining an open layout of the valve station with few scour generating objects at seabed level. A stop clamp placed centrally in the structure lock each spool for axial movements. The tie-in spools are connected to the valve station main tubulars by bolted clamps at each end of the valve station.

Each valve spool is divided into two flanged sections, a check valve and a ball valve spool. The tie-in spool flanges are placed protected inside the valve station. The short check valve spools can be brought out by lifting through the roof. In this operation the check valve will be hanging in straps below the roof panel above, utilising roof cover as a lifting/guide frame. No clamps are placed on the check valve spool in order to ease a possible check valve retrieval operation. The lifting operation is shown in Fig. 3.

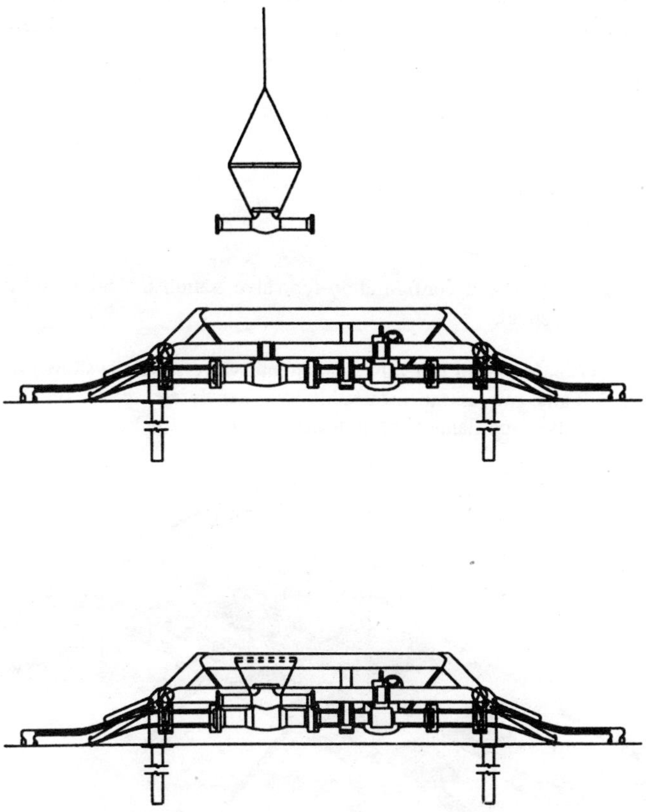

Fig. 3 Replacement of check valve by lift out through roof

Tubular hinged doors are placed above each tie-in spool at the entrance to the valve station in order to protect the flowlines against dropped objects. Deflectors and doors are rotated to be in an upright position during load out and installation, as they are attached by loose fitting clamps directly around the main horizontal valve station tubulars. Space requirements during sea transportation and installation will thereby be minimised. The load-out configuration is shown in Fig. 4.

Rock cover has been applied for protection of the pipeline and tie-in spools at the upstream side of the valve station towards the Harald A platform.

Downstream of the valve station, valve spools must be allowed to take up substantial pipeline expansion. The z-shaped expansion spools are therefore covered by concrete mattresses, allowing pipeline expansion to be taken up by expansion spool movements below the mattresses.

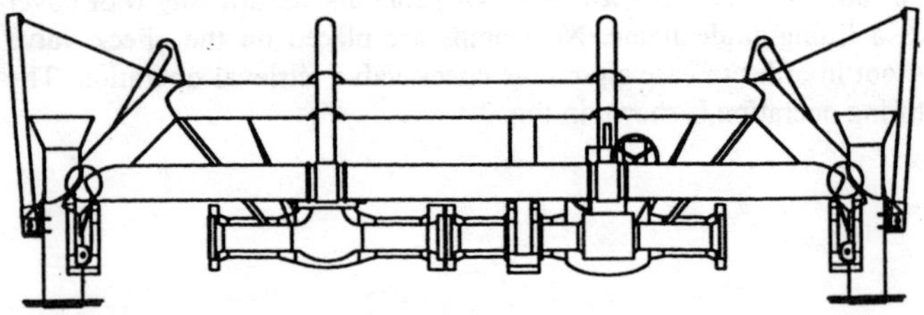

Fig. 4 Load-out configuration for valve station including valve spools.

A 3D visualisation of the valve station and valve spools is shown in Fig. 5. Only two of the detachable roof panels are shown on the figure. Valves and flanges are protected within the structure.

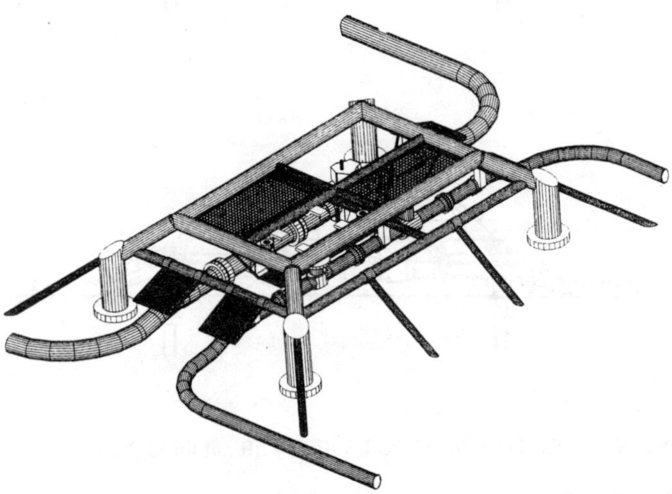

Fig. 5 3D View of valve station and valve spools

STRUCTURAL LAYOUT OF VALVE STATION

The valve station is designed as a welded structure, mainly composed of large welded standard tubulars. The main structure measures approximately 14 x 9 x 3.5 m. The general arrangement of the valve station is shown in Fig. 2.

The structure is piled by 16" piles through each corner member to a depth of approximately 10 m. The piles are grouted. When piles are driven they are fully placed hidden inside the leg members, thereby not generating fishing gear snagging points.

The four corner legs have circular mudmats for initial support before piles are driven and grouted. The layout of pile, corner member and mudmat is shown in Fig. 6.

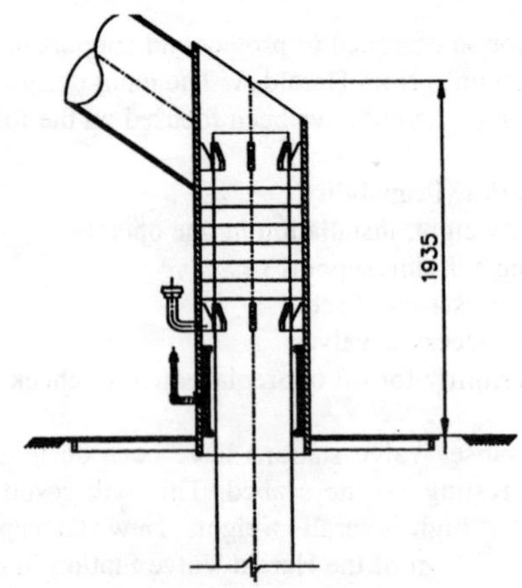

Fig. 6 Layout of corner leg, mudmat and pile

The roof covers are bolted to the roof tubulars. The roof covers are grated and designed to resist impact from dropped objects.

Trawl board deflectors and hinged doors are designed to a lower structural strength than the main structure attachment points. If deflectors are damaged by fishing gear, replacement will be possible because the main structure remains intact.

VALVES, OPERATION AND ROV ACCESS

Each valve spool includes a subsea check valve and a manually operated subsea ball valve. The valve spools are elevated approximately 1 m above the seabed by swan necks on the tie-in spools.

The check valves will automatically close in the event of loss of pressure in the upstream pipeline. The check valves are connected to the ROV docking panel at one side of the valve station by hydraulic hoses. The valves can therefore be operated hydraulically by ROV intervention. The valves are designed for operational pigging by intelligent pigs. For this purpose, the valves can be locked in open position. The layout of the ROV docking panel is depicted in Fig. 7.

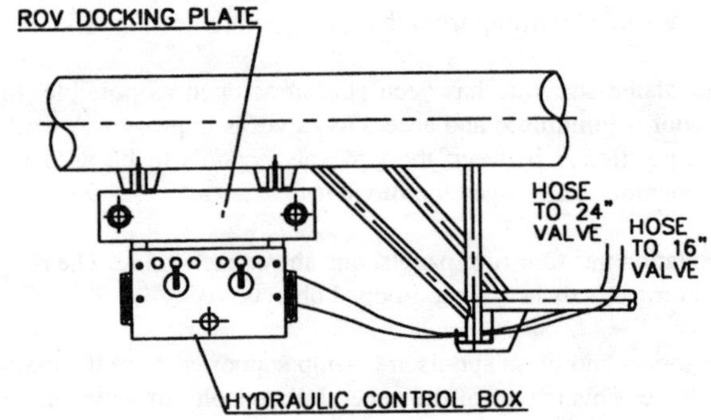

Fig. 7 ROV docking panel

The ball valves are manually operated by divers from the working platform between the valve spools.

When closing the ball valves, service can be carried out on the check valves including a possibility for replacement of a check valve.

SCOUR CONSIDERATIONS

Scour is a major consideration in the Danish sector of the North Sea and must therefore be accounted for in the design of subsea installations. Scour generates due to the sandy seabed composition in combination with large wave and current velocities. The water depth in the Danish sector is in the range 35-65 m. This means that large velocities are experienced also near seabed level. Objects placed on the seabed may generate scour holes. Scour hole development will also depend on the size and geometry of the object placed on the seabed. Large objects may generate holes of several metres depth.

Scour considerations have been an important parameter in the valve station design. The design has been based on investigations of scour mechanism, see ref. /6/ - /12/, as well as previous subsea structure design in the Danish Sector - e.g. the Regnar wellhead protection structure, ref. /1/.

In order to minimise the scour development for a structure placed on the seabed, an open design without plated sections should be used. The global scour anticipated is in the order 0.1 m extending 2-5 m away from the structure. The scour from the horizontal members, the pipelines and valves will cause a lowering of the seabed in the central part of the valve station approximately 0.4 m deep. Around the corner legs scour holes of approximately 1.0 m depth are anticipated to develop.
Concrete mattresses are used for spool protection south of the Harald valve station. The concrete mattress stability and resistance to scour development has been secured by installation of shingles deposits at specific locations.

DESIGN ANALYSIS

The design analysis has been carried out mainly focusing on the following loads:

- fishing gear, impact and hooking analysis
- dropped objects
- functional and environmental loads during operation
- loads during sea transportation and installation

The structural design analysis was carried out using a finite element model for the main structure including piles and valves. Specific items as roof panels, hinged doors and deflectors were checked by separate calculations.

The structural analyses were carried out for a design life of 30 years. For accidental loads a plastic design criterion was used - i.e. the structure may experience plastic deformation, but shall keep its integrity and maintain full functionality.

The protection structure was designed according to Danish Standard, DS 449, whereas the piping structural analysis was carried out according to the DNV rules for submarine pipelines.

The analysis of the protection structure was coupled with the structural analyses of the tie-in spools, using clamps as interface points.

The functional and environmental design loads were found to be low. No lock-on to vortex shedding was found. Max. stresses during normal operation including loads from expansion spools were found to be around 60 MPa for the structure and 90 MPa for the piles.

The accidental loads were found to be the main design parameters, especially considering hooking of the structure by fishing gear. The structure is designed to deflect fishing gear. In spite of this, the structure can resist the full hooking load - i.e. a load of 100 metric tonnes. Denting of main structural tubulars was also considered.

The impact/hooking analysis was carried out for an impact load of a 2.7 tonnes trawl board, dragged at a speed of 7 knots. A number of load cases for the hooking event were defined and analysed. The max. stress level in the structure was found to be around 300 MPa. Max. forces in the piles were found to be around 800 kN. Max. indentation of main tubulars was found to be 7 mm. Local high stresses were found for impact analysis at secondary members - i.e. deflectors and attachments. Deflector attachments were designed to act as weak links in a local hooking event.

The dropped object analysis was carried out for a 5 tonnes load with a footprint of 2 x 2 m. The roof panel grating was designed to take up dropped object impact load as membrane forces.

The piles and structure are designed to accommodate a max. scour of 2 m at the corner legs.

INSTALLATION OF VALVE STATION

The valve station was loaded out on the deck of a medium size diving support vessel. The load out configuration is shown in Fig. 4. The valve station was installed after lay-down of the 16" and 24" pipelines passing the valve station location. Straight flanged lay- spools were included at the valve station location when installing the pipelines. The straight lay-spools were removed before tie-in to valve station, see spool layout depicted in Fig. 1.

The valve station including valve spools was installed in one operation utilising the vessel crane. The valve station was lifted by a single point lift from the four main lifteyes. During sea transportation and installation each valve spool was hanging from the valve spool stop clamp in combination with slings to the roof structure.

After tie-in to the pipelines, valve spools are supported by three clamps each. The guide clamps at each end of the valve station were prefitted to the z-shaped expansion spools during installation. The guide clamps were bolted to the valve station tubular structure by divers during installation. Adjustment possibility was built into the clamp design to ease installation of spools and clamps.

Deflectors and doors were hinged in an upright position during load out and installation, thereby minimising space requirements. After installation on the seabed, the trawl board deflectors and the tie-in spool protection doors were rotated to their final positions.

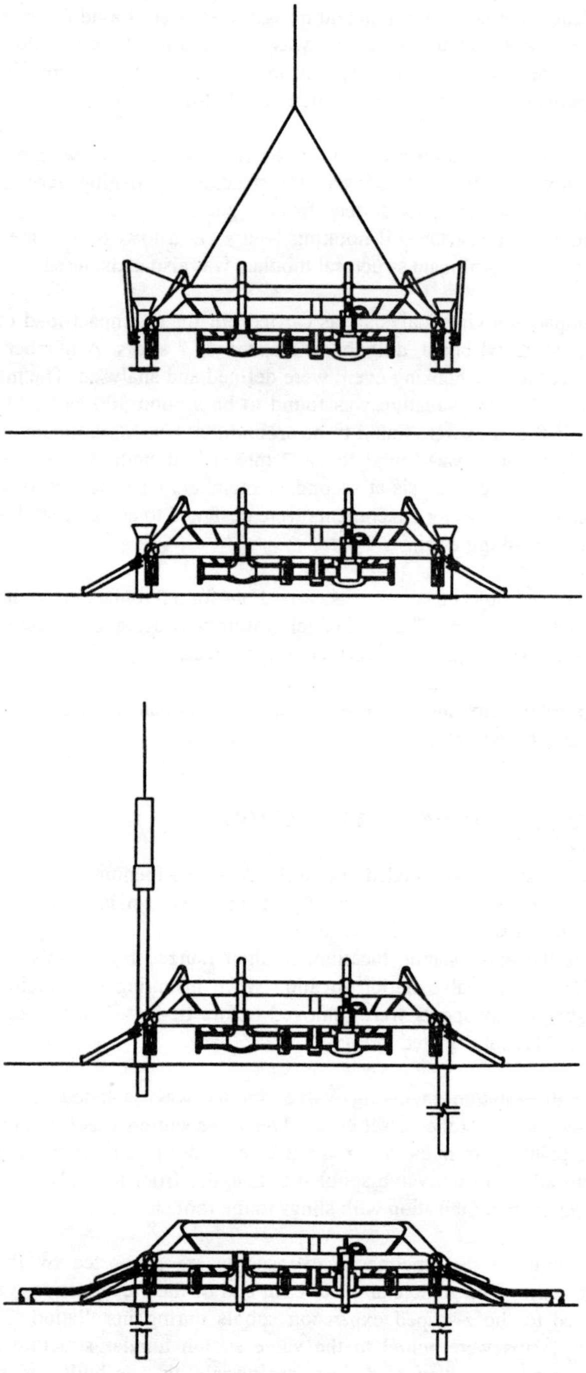

Fig. 8 Installation sequence

The structure was piled by one 16" pile at each corner leg using a follower for the drive operation in order to terminate the pile fully within the corner member, see Fig. 6. Fittings for a levelling system were included on each leg enabling levelling of the structure to be carried out in relation to the pile driving operation.

The installation sequence is depicted on Fig. 8 . The top picture shows lowering of the structure including valve spools by a single point lift from the DSV heave compensated crane. Corner deflectors are then unfolded, piling and levelling is carried out by driving piles internally through each of the four legs. Finally, spools are tied in and remaining deflectors and the hinged doors are rotated to their final position.

MAINTENANCE

The valve station is coated with a paint system. The valve station structure is further protected by sacrificial anodes to a design life of 30 years.

Inspection/maintenance on valves can be carried out by utilising the divers platform placed between the valve spools. Check valves can be replaced or brought out for overhaul by lift through the roof structure as shown in Fig. 3.

ACKNOWLEDGEMENT

The authors wish to thank the management of Mærsk Olie og Gas AS, Texaco Denmark Inc. and Shell Olie og Gasudvinding, Danmark BV, for permission to present this work.

REFERENCES

/1/ Regnar Subsea Protection Frame, Design Report (1993). Doc. No. 105.3.142.040.01 Rev. 1, Mærsk Olie og Gas AS, by RLD Joint Venture, 1993

/2/ Heal, D. and Naughton, L. (1991). Wellhead Protection at Gannet B., Underwater Technology, Vol. 4, 1991

/3/ Bottom Trawling Interaction with Submarine Pipelines (1991). Mærsk Olie og Gas AS, DOC. No. 141.4 , by RLD Joint Venture, 1991

/4/ Guijt, J and Horenberg, J.A.G. (1987). Recent Investigations Concerning the Effect of Bottom Trawl Gear Crossings on Submarine Pipelines Integrity, OTC 5616, 1987

/5/ Moshagen, H. and Kjeldsen, S.P. (1980). Fishing Gear Loads and Effects on Submarine Pipelines, OTC 3782, Houston, 1980

/6/ Sumer, B. M. and Fredsøe, J. (1992). Scour around Vertical Piles in Waves. ASCE, J. of Waterway, Port, Coastal and Ocean Engineering, Vol. 118, No. 1, pp 15-31

/7/ Sumer, B. M. and Fredsøe, J. (1990). Scour below Pipelines in Waves. ASCE, J. of Waterway, Port, Coastal and Ocean Engineering, Vol. 116, No. 3, pp 307-323

/8/ Sumer, B. M. and Fredsøe, J. (1992). Time Scale of Scour around a Vertical Pile. Proc. of the Second International and Polar Engineering Conference, Vol. III, pp 308-315

/9/ Sumer, B. M. and Fredsøe, J. (1993). Influence of Cross Section on Wave Scour around Piles. ASCE, J. of Waterway, Port, Coastal and Ocean Engineering, vol. 119, No. 5, pp 477-495

/10/ Angus, N.M. and Moore, R.L. (1982). Scour Repair Methods in the Southern North Sea, Paper OTC 4410, 14th Annual OTC in Houston, Texas, May 3-6, 1982

/11/ SISS (1992). Sea Bottom Instability around small Structures. Erodible Bed Laboratory Tests (Phase 1). Final Report, Text and Drawings. DHI and Snamprogetti, Contract INGE91/SP/03060, June 1992

/12/ Breusers, H.N.C. and Raudkivi, A.J., (1991). Scouring. Hyd. Structures Design Manual 2, IAHR, Balkema, Rotte

Proceedings of the Seventh (1997) International Offshore and Polar Engineering Conference
Honolulu, USA, May 25-30, 1997

ISBN 1-880653-28-1 (Set); ISBN 1-880653-29-X (Vol. I)

Study of Oilfield Produced-Water Pre-Separation Hydrocyclone

Jie He
Daqing Petroleum Institute, Anda, China
Fuzhou Dong and Zengrui Li
Daqing Petroleum Administration, Daqing, China
Minghu Jiang
Daqing Petroleum Institute, Anda, China

ABSTRACT

Hydrocyclone used on pre-separation of the high-water-cut oilfield produced-water was explored according to the needs of oilfield. After researching and analyzing on its flow field properties, we found that the short circuit flow would be increased dramatically if enlarging the diameter of the upflow tube. In order to enhance the pre-separating efficiency, the length of the upflow finder must be required. At the same time, the relationship of the upflow diameter and the downflow diameter was proposed on the view of experiments, and based on which, a sample hydrocyclone was designed and made. Its oilfield tests show that the pre-separating efficiency can reach 99%. The pre-separation hydro-cyclones can be spread and applied on oilfield production.

KEY WORDS: Hydrocyclone, pre-separation, short circuit flow, split ratio

NOMENCLATURE

C	oil-cut (PPM or mg/l)
D	diameter (mm)
F	split ratio= $\frac{Q_u}{Q_i}$ (%)
P	pressure (MPa)
ΔP	pressure drop (MPa)
Q	flowrate (m^3/h)
Suffixes	
d	downflow (oil depleted stream)
i	inlet (mixed stream)
u	upflow (oil enriched stream)

INTRODUCTION

In China, many oilfields have stepped into middle or high-water-cut exploration period. The average water-cut exceeds 80%, and the disposing of oilfield produced-water has become a significant problem. Due to the increasing of water-cut in the oilwells produced-fluid, the surface construction scale was increased, the investment and the cost of crude oil were raised. So the manufacture of high efficiency oil-water separation device becomes an important task in exploring oilfields economically. And the study of oilfield produced-water pre-separation hydrocyclone(OPPH) has been proposed based on this situation.

The requirements to the OPPH are to decrease the water-cut of produced-water from 80% to below 30%(in upflow), and the oil-cut of downflow water is around 1%, so as to link up with the following two industrial processes, i.e. de-water of crude oil and sewage-disposal. The OPPH must decrease the water-cut of upflow to a certain level, but also control the oil-cut of downflow in a certain range, which makes the manufacture of the OPPH become very difficult.

From the view of the produced-water property, when its water-cut is around 80% and the temperature is 40℃ approximately, it is in the state of "oil-in-water" generally, i.e. water is the continuous phase and the oil disperse. So the OPPH in use for above produced-fluid is still de-oil hydrocyclone. Certainly, its structural parameters obviously differ from today conventional ones. First, the conventional de-oil hydrocyclone is mainly used for disposing the sewage whose oil-cut is far lower than 1% and its upflowrate(or split ratio F) is only 1%~2%. The upflow diameter D_u is also very small, the smallest is only about 2 mm. While the OPPH should remove more than 20% fluid through upflow, thus the split ratio F must be larger than the inlet oil-cut C_i so as to separate crude oil from the produced-fluid and remove it by upflow. With the increasing of upflowrate Q_u and the relatively decreasing of downflowrate Q_d, the upflow diameter D_u, the downflow diameter D_d and some other structural parameters must be changed appropriately. At the same time, the change of the flow field in hydrocyclone and the effects on separating properties would be studied after changing these structural parameters.

We measured and analyzed the flow field in the OPPH with the Laser Doppler Velometer(LDV), and specific measure to weaken short circuit flow was proposed. Based on experiments, the relationship of upflow diameter D_u, downflow diameter D_d, split ratio F and other parameters was proposed, and according to the relationship model machine was made. The oilfield tests indicate that its properties has achieved the desired results completely.

FLOW FIELD PROPERTIES OF THE OPPH

In order to determine the reasonable geometric structure of the OPPH, its flow field properties were studied.

Short Circuit Flow

As enlarging the upflow diameter of hydrocyclones, a part of fluid which entering the OPPH through inlets will directly flow up to the upflow without separation(Fig. 1). And with the increasing of the upflow diameter, the intensity of the SCF will be strengthened. Through transparent sample hydrocyclone, the existence of SCF can be obviously observed on the test rig. But it's regrettable that we couldn't measure its velocity quantitatively.

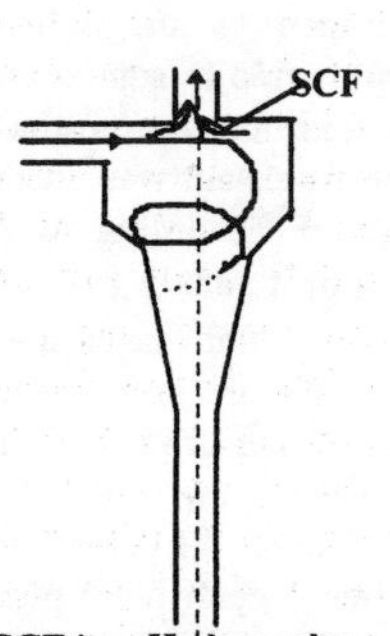

Fig. 1 SCF in a Hydrocyclone

SCF is one of the important factors affecting the separating effects of the OPPH. With the arising of SCF, not only is the water-cut of upflow increased, but also disturbances to the oil-water mixed fluid are produced and makes the rotating velocity of fluid inside vortex chamber weakened.

Analysis of Pressure Property

The state of pressure field changed also, while enlarging the upflow diameter of hydrocyclone. We did some experiments on a hydrocyclone with 38mm main diameter. In the case of not changing other geometric parameters, the upflow diameter D_u are 2mm, 3mm, 4mm, 6mm and 8mm respectively. During the experiments, the inlet pressure P_i was maintained with 0.32 MPa, while the upflow pressure P_u was raised from 0.015 MPa to 0.02 MPa, and the downflow pressure D_d decreased slightly from 0.12 MPa to 0.09 MPa, i.e. pressure drop ΔP between the inlet and the downflow increased, the energy lose raised to some extent. The reason may be that the energy lose promoting the upflow out of hydrocyclone increased and those promoting the downflow decreased, due to the increasing of upflowrate and the raising of upflow pressure, so that the rotating intensity in the separating segment of hydrocyclone would be weakened relatively and affect the separating effect. So in view of the OPPH, increasing the inlet pressure appropriately P_i may be necessary to maintain higher separating efficiency.

Change of Split Ratio F

In order to determine the relationship between upflow diameter D_u and inlet oil-cut C_i (i.e. split ratio F), we measured and obtained two groups of curves, F- D_u and F- P_i (Fig. 2 and Fig. 3).

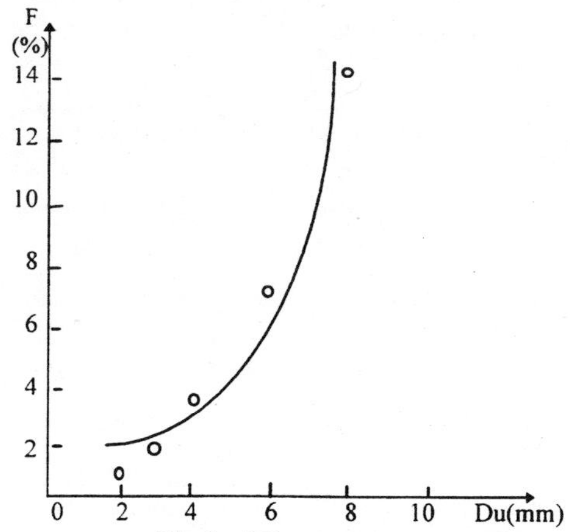

Fig. 2 F-Du curve

We can see from Fig. 2 that the split ratio F increased sharply with the increasing of D_u, so the effective method to enhance the split ratio of a hydrocyclone is to enlarge its upflow diameter. But as mentioned above, with the increasing of D_u, the SCF will be strengthened obviously and the diameter of oil core will be augmented, so much water without being separated will be removed from upflow tube, which could affect the separating efficiency.

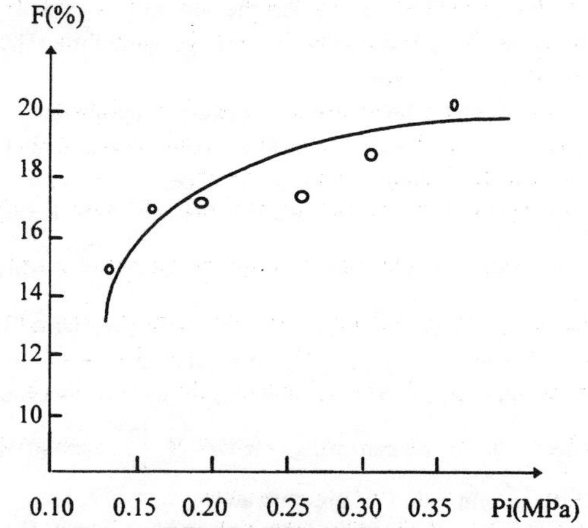

Fig. 3 F-Pi curve

Fig. 3 shows that the split ratio F increased obviously with the increasing of inlet pressure P_i when other conditions are maintained. And as stated above, raising the inlet pressure appropriately is beneficial to enhance the separating efficiency. So we determined the OPPH's new parameters with enlarging upflow diameter D_u and raising inlet pressure P_i during its manufacture.

RELATIONSHIP BETWEEN UPFLOW DIAMETER D_u AND DOWNFLOW DIAMETER D_d

In accordance with analyzing for the flow field property of the OPPH, at the same time of enlarging its upflow diameter, the downflow diameter should be decreased appropriately so as to be beneficial to enhance the separating efficiency. We accomplished a series of experiments in order to obtain the reasonable relationship between D_u and D_d.

First we utilized the relation curve of split ratio F and upflow diameter D_u in Fig. 2 to determine in advance the size of D_u, and then did a series of experiments on the hydrocyclone, D=38 mm, under different upflow diameters, and the following empirical equation was obtained:

$$Q_i = 11.1 D_u^{0.003} \Delta P^{0.43} \quad (1)$$

This formula reflects the relationship of inlet flowrate Q_i, pressure drop ΔP and upflow diameter D_u of hydrocyclone. Where, Q_i is the inlet flowrate, m^3/h, D_u is the upflow diameter, mm, ΔP is the pressure drop between inlet pressure and downflow pressure, MPa.

For a certain hydrocyclone, the relationship between its inlet flowrate Q_i and inlet pressure P_i is determined. And with the definition of Q_i and P_i, ΔP can be determined, so D_u can be calculated.

Many experiences show that the split ratio F should be determined according to the inlet oil-cut C_i. The reasonable F should be $1.2\sim1.3\,C_i$, i.e.

$$F=(1.2\sim1.3)C_i \quad (2)$$

Furthermore, we obtained another empirical formula of F, D_u, D_d and Q_i based on the equations (1) and (2):

$$F = 0.71\left(\frac{D_u}{D_d}\right)^{2.2}\sqrt{Q_i} \quad (3)$$

Where, D_u and D_d are diameters of upflow and downflow, mm, Q_i is flowrate, m^3/h.

From the formula (1) we can see that the power of D_u is only 0.003, i.e. the variations of D_u affect on the flowrate Q_i quite little. The changing of Q_i depends on ΔP mainly.

The proportional relationship between upflowrate Q_u and downflowrate Q_d (i.e. the split ratio F) is mainly changed after enlarging the upflow diameter, which is just what we hope.

From the equation (3) we can see that the split ratio F changes with $\left(\frac{D_u}{D_d}\right)^{2.2}$, the value of split ratio F would be increased sharply with the increasing of D_u or decreasing of D_d, that is to say, the split ratio F is sensitive to the ratio of D_u to D_d, while the change of Q_i has little effect on the split ratio F. The equation (1) shows that the change of D_u hardly affects Q_i. So we can change the size of $\frac{D_u}{D_d}$ appropriately to suit for the needs to split ratio F of pre-separation.

The above empirical equations have been verified repeatedly, the errors are in the range of 5%, so they can be applied to engineering practice.

OILFIELD TESTS

Based on the theoretical analysis and the indoor experiments, we finished the oilfield tests on a certain scale. The design flowrate of oilfield test hydrocyclone is $1000\,m^3/d$, the design pressure is 1MPa. The tests were accomplished in an oil-transfer-station of Daqing Oilfield, and the average water-cut of produced-fluid in this station is 85% approximately.

The purposes of this test were to verify the actual application condition of the OPPH in oilfield, and to know whether or not it could achieve the desired requirements when it is used to finish pre-separation of oilfield produced-water. The requirements for the test were that the water-cut of upflow was around 30%, so as to be directly transported to de-water technological process without other intermediate links, at the same time the oil-cut in downflow would be decreased to about 1%, and then it was disposed again so that its oil-cut can reach about 1000 PPM which makes it can be directly transported to the sewage-disposing station to be purified further. This subject is being in progress now.

Although no other power device was added during the tests, the existing output pump (centrifugal pump) of the oil-transfer-station was utilized to increase the pressure, so the test pressure was higher. Most tests were accomplished at the pressure of 0.6~0.8 MPa. The test results are listed in Tab. 1.

Tab. 1 Oilfield test results

	Before separating					After separating	
	Inlet flowrate Qi (m^3/h)	Inlet pressure Pi (MPa)	Inlet water-cut 1-Ci (%)	Upflow-rate Qu (m^3/h)	Split ratio F (%)	Upflow water-cut 1-Cu (%)	Downflow oil-cut Cd (mg/l)
1	45	0.80	85.7	8.6	20.0	31.2	2640
2	48	0.76	78.7	9.0	22.8	23.0	10437
3	48	0.78	77.2	12.0	25.0	20.0	14092
4	45	0.68	87.9	9.0	20.0	26.5	7852
5	45	0.71	81.7	12.0	26.7	26.5	4200
6	49	0.83	90.3	8.0	16.1	22.0	11528
7	59	0.80	90.7	6.0	10.2	28.0	8080
8	42	0.70	82.0	8.5	20.2	33.1	16307
9	48	0.76	91.3	3.0	9.3	24.0	15358

During the tests, the mixed fluid processed by hydrocyclone was pumped by the output pump, so the oil in the produced-water was emulsified to some extent, which caused the oil-cut in downflow raising. In addition, the produced-fluid often changed during the production and the water-cut was unstable. It was very difficult to timely adjust the split ratio F according to the oil-cut of the inlet fluid, and the upflowrate couldn't timely suit for the inlet oil-cut C_i, which was also one of the reasons that caused the unstability of upflow water-cut and downflow oil-cut. We are now further researching how to timely and exactly measure the oil-cut so as to control the operating parameters of hydrocyclone.

During the tests we also found that putting appropriate de-emulsifying agent into the produced-fluid was beneficial to the separation. Particularly when a centrifugal pump was used to be the pressure-raising device, the emulsification increased, so putting appropriate de-emulsifying agent and operating the pump in moderate speed would also obtain better separating results.

From Tab.1 we can see that, in general, if the relationship between the inlet oil-cut C_i and the split ratio F was adjusted appropriately, it could completely achieve to control the water-cut of upflow to below 30%, and the oil-cut of downflow to about 1% (i.e. around 10 000 mg/l). Sometimes the oil-cut of downflow was higher, according to the test observation, which was due to the putting amount of de-emulsifying agent was not appropriate so that intensified the emulsification. The oilfield test results fully prove that the study of the OPPH has achieved the desired purposes, the OPPH can be applied to engineering practice.

CONCLUSIONS

1. Due to the enlargement of the upflow diameter, the SCF is intensified so that the structure of upflow tube should be adjusted appropriately, e.g. to extend the upflow tube into the vortex chamber of hydrocyclone, so as to prevent happening of the SCF.

2. Reasonably determine the split ratio F according to the inlet oil-cut C_i may be an important step to enhance the separating efficiency. Based on experiments, F equals to 1.2~1.3 C_i may be reasonable.

3. It has been verified to be correct to determine D_u and D_d of the OPPH with the equations (1) and (2), which can be applied to actual engineering.

4. The oilfield tests indicate that the structure of the OPPH is reasonable. When the inlet oil-cut is 80%~90%, the upflow water-cut can be decreased to below 30%, and the downflow oil-cut to 1% approximately after separation. The OPPH can be applied to the engineering practice.

Proceedings of the Seventh (1997) International Offshore and Polar Engineering Conference
Honolulu, USA, May 25-30, 1997

ISBN 1-880653-28-1 (Set); ISBN 1-880653-29-X (Vol. I)

Fuzzy Syllogistic Inspection Planning Paradigm for Offshore Structures

Hari B. Kanegaonkar
Aker Maritime, Aker Offshore Partner
Stavanger, Norway

Abstract

Planning in-service inspection of offshore structures is a complex task. Several methodologies have been developed in the past which range from use of totally qualitative to highly quantitative involving the system reliability and redundancy evaluations. The decision making process needs to take into account several parameters. The best available inputs from many of these parameters - more often than not, are linguistic descriptions.

The important factor in decision making for inspection planning is development of prognosis. By its very nature we are dealing with fuzzy system rather than a probabilistic system.

The calculus of fuzzy rules provides a technique to deal with imprecise dependencies. The fuzzy calculus captures the human intuitive expert knowledge. A method using fuzzy logic concepts is proposed here for inspection planning for topsides of offshore structures.

The important parameters which affect the decisions in the inspection plan are identified. A knowledge data base is generated for each item through pertinent questionnaire to collect information regarding all the parameters. The information is concised into simple linguistic codes. Fuzzy IF...THEN rules/principles are developed and codes from all the parameters are combined using the sup-star compositions to develop the structural condition prognosis using linguistic variables. The prognosis is combined with the inspection technique/frequency and the failure consequence of the inspection on structural integrity and serviceability. The fuzzy inference engine is developed which consists of combinations of fuzzy IF...THEN rules to obtain inspection plan with associated risk levels.

The method has been applied to two structures in the north sea. The inspection plans developed based on the fuzzy logic concepts show that significant cost savings can be achieved by using this new technique.

Introduction

Planning in-service inspection of offshore structures is a complex task. Several methodologies have been developed in the past which range from use of totally qualitative to highly quantitative involving the detailed redundancy and system reliability evaluations. (R.J.van Foeken et al, 1994, Tang , M.X. et al, 1994). The philosophy of inspection planning is to keep the consequences of the possible non-detected/non-detectable deterioration at a level whereby the overall structural and functional behaviour has sufficient reliability while minimising the efforts for inspection and repairs to detectable/detected deterioration

The decision making process for scheduling has to take into account several uncertain parameters. The best available input from many of these parameters more often than not are linguistic descriptions. The approximate nature of the inputs from many of these parameters can be used advantageously by using logic which considers the linguistic variables. In general, with the use of this logic, the nature of assessments necessary to arrive at proper action would save time and detail compared to the conventional decision analysis. The assessments could be imprecise or inaccurate with wider bounds leading to simple and faster assessments without jeopardising the robustness of the actions. The logic of the evaluation process corresponds closely to the human reasoning. The procedures and recommendations of a fuzzy logic system could use linguistic variables.

The rigorous calculations for redundancy and its relationship to the system reliability which have been used by jacket inspection planning experts are questionable if duplicated for the topside structures. The pushover analysis performed for jackets utilises the environmental design load as a basis and these loads are increased gradually to obtain the ultimate load capacity, reserve and residual strengths and redundancy in the system. The same approach cannot be used for the topside for obvious reasons. The major environmental loads for the topside come from the wind loading. The ultimate capacity of the topside for the wind loads is too high for any meaningful estimation of redundancy. Moreover the 90% of the stresses in the topside members are due to the topside systems (equipment/bulk loads on the members). Some of the joints in the concrete gravity topside are subjected to fatigue loading from the waves on the shafts. The different topside system weights cannot be increased in the same proportion as the systems are independent. The various configuration of the load levels in different systems will produce different failure patterns It is difficult to capture the failure mode leading to global minimum ultimate load. Thus the evaluation of redundancy reliability relationship will be much more uncertain for the topsides compared to the jacket structure.

Basics of Fuzzy Logic System

Fuzzy logic is a logic of approximate reasoning which may be viewed as a generalisation and extension of multi-valued logic. Since the fuzzy logic is constructed from fuzzy IF--THEN rules, linguistic information is easily incorporated. The basic configuration of fuzzy logic system with fuzzifier and defuzzifier is (Wang, 1992) shown in Fig.1. The fuzzifier maps crisp points in U to fuzzy sets in U and the defuzzifier maps fuzzy sets in V to crisp points in V

Fuzzy Set Let U be a collection of objects $U=R^n$ and be called the universe of discourse. A fuzzy set F in U is characterised by a membership function $\mu_F : U \rightarrow [0,1]$ with μ_F the grade of membership of $u \in U$ in fuzzy set F. A fuzzy set may be viewed as a generalisation of the concept of an ordinary set (i.e. crisp set) whose membership function takes only two values {0,1}. In a fuzzy set transition between membership and non-membership is gradual rather than abrupt.
The support of a fuzzy set F is the crisp set For all points $u \in U$ such that $\mu_F(u) \rangle 0$

The centre of a fuzzy set F is the point(s) $u \in U$ at which $\mu_F(u)$ achieves its maximum value.

If the support of a fuzzy set F is a single point U at which $\mu_F = 1$ the F is called a fuzzy singleton.

Linguistic Variables If a value of a variable can be expressed by words in natural language, the variable is defined as a linguistic variable. The words are usually labels of fuzzy sets. A linguistic variable can have numbers as its value as well. e.g. Fatigue Life at a location in a structure can be considered as a linguistic variable which can take "Low" , "Medium" , "High", etc., as its values.

Hedges The linguistic descriptions can be intensified by using various hedges such as "very", "more or less" etc. Let F be a fuzzy set in U then very F is defined as a fuzzy set in U with membership function as defined below

$$\mu_{veryF}(u) = \mu_F^2(u) \qquad (1)$$

and more or less F is a fuzzy set in U with member ship function given by

$$\mu_{more.or.lessF}(u) = \sqrt{\mu_F(u)} \qquad (2)$$

Fuzzy Rule Base A fuzzy rule base consists of a collection of fuzzy IF..THEN rules in following forms.

$$R^l : IF\ x_1\ is\ F_1^l\ and \ldots\ldots x_n\ is\ F_n^l,\ THEN.y.is.G^l$$

where $F_i^l .and. G^l$ are fuzzy sets in $u_i \subset R.and.V \subset R$ respectively and

$$\underline{x} = (x_1 \ldots\ldots x_n)^T \in (U_1 \ldots\ldots U_n).$$

$$y \in V$$

are linguistic variables with l = 1,......M and $\underline{x}$ is the input and y is the output.

The fuzzy rule base is the core of the fuzzy logic system. Fuzzy IF..THEN rules in the form shown above provide a convenient frame work for human experts to express their domain knowledge. The fuzzy rule base shown above is general enough to include all types of linguistic information. There are two main methods to obtain the fuzzy rule base (1) provided by human experts (2) using training algorithm based on measured data. When the rules are provided by the human experts, the membership functions for F_i^l and G^l are specified by the experts since these functions form the integral part of the expert knowledge. If the rules are determined by the numerical data, then the functional form for μ_{Fi} and μ_G^l are determined by most commonly used functional forms such as Gaussian, triangular or trapezoidal, etc.

Fuzzy Inference Engine
In fuzzy inference engine, fuzzy logic principles are used to combine the fuzzy rule base into a mapping from fuzzy sets in $U = U_1 \times \ldots\ldots\ldots U_n$ to fuzzy set V A fuzzy IF..THEN rule is interpreted as a fuzzy implication $F_1^l \times \ldots\ldots\times F_n^l \rightarrow G^l$ in UxV

Let a fuzzy set A' in U be the input to the fuzzy inference engine then each fuzzy set B' in V using sup-star composition is

$$\mu_{B'}(y) = Sup.x \in U\left[\mu_{F_1^l \times \ldots F_n^l \rightarrow G^l}(x,y) * \mu_{A'}(x)\right] \qquad (3)$$

The '*' in above can be replaced by different interpretations such as sup-min, sup-product, sup-maxmin, sup-Boolean etc. (Wang,1994)

DIVISION OF INPUT-OUTPUT INTO FUZZY SPACES

The parameters shown in the Fig. 2 are basic input parameters, intermediate output from the basic parameters and constraints together resulting into the inspection plan.
The parameter spaces are divided into fuzzy regions with membership functions. The expert knowledge required to establish these are a result of answers to a questionnaire prepared for each parameter. The questionnaire and the possible answers to the questions with explanations wherever found necessary are provided below to establish an expert knowledge data base to arrive at the values and membership functions for the fuzzy sets with linguistic variables.

DFI RESUME PARAMETER (DFIR)

The DFI resume provides the knowledge data base of the structure during design, fabrication and installation stages. The data base is suitable to select partial list of important items for inspection. To define the DFI resume parameter membership function, the information from the resume is systematically obtained through the pertinent questions. The answers to following questions affect the value of the linguistic variable as well as the membership function.

Q.1 : Is deviation from the design premises reported ?
Q.1a : Does the deviation affect this location ?
Q.1b: What is the severity level of the effect ?
A. : 1. Very Low 2. Low 3. Medium 4. High 5. Very High
Q.2 : Is the location reported as highly utilised ?
Q.2a : What is the level of utilisation ?
A. : 1. Very Low 2. Low 3. Medium 4. High 5. Very High
Q.3 : Is the location reported as with low fatigue life ?
Q.3a : What is the level of fatigue life ?
A. : 1. Very Low 2. Low 3. Medium 4. High 5. Very High
Q.4 : Is the location reported as with weld defect ?
Q.4a : What is the type of the defect ?
A. : 1. Lack of fusion 2. Incomplete penetration 3. Slag inclusion 4. Lamellar tearing 5.Solidification cracks 6. Hydrogen cracking 7. Undercuts

This is to find out whether any type of the defect which has significant effect on fatigue life has been introduced during fabrication.

Q.4b : Will the weld defect grow into a crack ?
Q.4c : What is the possibility of crack growth ?
A. : 1. Unlikely 2. Less possibility 3. Possible 4. High possible 5. Very High possibility

These are heuristic arguments based upon information and expertise as an aid to decision making.

Q.5 : Is the material at the location reported as defective ?
Q.5a : What is the type of material defect ?
Q.5b : Will the material defect grow into crack ?
Q.5c : What is the possibility of crack growth ?
A. : 1. Unlikely 2. Less possibility 3. Possible 4. High possibility 5. Very High possibility
Q. 6 : Is any damage reported during installation ?
Q. 6a : Will the damage grow in time ?
Q.6b : What is the possibility of damage growth ?
A. : 1. Unlikely 2. Less possibility 3. Possible 4. High possibility 5. Very High possibility
Q. 6c :What is the severity of the damage regarding capacity reduction?
A. : 1. Very Low 2. Low 3. Medium 4. High 5. Very High

STATIC UTILISATION PARAMETER (SU)

The static utilisation value and the corresponding membership function is based on usages for the operating and extreme ultimate limit load conditions.

The values and the membership functions are given based on the answers to questions related to the usages in DFI which provide information regarding the uncertainties involved in the estimation of usages.

The parameter is given linguistic value based on the usages such as shown in Fig.3

DYNAMIC UTILISATION PARAMETER (DU)

The codes for the dynamic usages are based on whether the equipment based or environmental loads. The high dynamic utilisation can be either due to high stresses with relatively less occurrences or smaller stresses with high level of occurrence leading to fatigue problems. The parameter values and the membership functions are based on following questions.

Q.1 : Do dynamic stresses result from reciprocating equipment ?
Q.2 : Do dynamic stresses result from centrifugal equipment ?
Q.3 : Is structural frequency close to operating frequency ?
Q.3 : Do dynamic stresses result from environmental forces ?
Q.4 : What is the operating dynamic stress range?

FATIGUE LIFE PARAMETER (FL)

Fatigue life is an important parameter in the determination of the inspection type and frequency of the inspection for the detail. Theoretical background for these calculations must be considered as an important sub-parameter since large uncertainties can be created by the approach adopted for fatigue life estimates.

Q.1 : Is method for fatigue life estimate conservative ?

The parameter is given linguistic value based on the fatigue life such as shown in Fig.4.

STRUCTURAL REDUNDANCY PARAMETER (SR)

Q.1 : Is this part of a cantilever structure ?
Q.2:Are members in one plane / two planes / multiple planes available for redistribution of stresses?
Q.3 : According to engineering judgement is redistribution of stresses possible ?

The answers to above questions are useful in providing the values for the structural redundancy parameter as answers to the questions presented below

Q.4 What is the possibility of progressive collapse if this member fails?
A. : 1. Unlikely 2. Low possibility 3. Possibility 4.High Possibility 5. Very High Possibility
Q.5 : What is the importance of the member for serviceability ?
A. : 1. Negligible 2. Low 3. Medium 4. High 5. Very High

The parameter will be dependent upon the answers to above questions and their combinations. The membership function is assumed based on singleton fuzzifier assumption.

STRUCTURAL CONDITION RECORD PARAMETER (SCR)

Q.1 : Is major defect affecting the structural integrity found during inspection at this location ?
Q.2 : Is fatigue crack found during the inspection ?
Q.3 : Is weld failure found during inspection ?
Q.4 : Are laminations and cracks found during inspections ?
Q.5 : Are signs of material yielding found during inspection ?
Q.6 : Are excessive deformations found during inspection ?
Q.7 : Is localised deformation found during the inspection ?
Q.8 : What is the level of corrosion ?
A. 1. Negligible 2. Low 3. Medium 4. High 5 Very High
Q.9 : Is reduction in material thickness due to corrosion found during the inspection ?

Based on the answers to above questions, following questions are answered.

Q.10.: What is the structural condition with regard to fatigue cracking at this location?
Q.11.: What is the structural condition with regard to high stress cracking, large deformation, yield signs at this location?
Q.12.: What is the structural condition with regard to corrosion at this location?

The questions 10 to 12 are answered as below.

A.: 1.Very Poor 2. Poor 3.Medium 4.Good 5. Very Good

The membership functions for the parameter values can be based on singleton fuzzifier assumption.

HISTORY INSPECTION TYPE PARAMETER (HIT)

Q.1 : What is the method of inspection used ?

HISTORY INSPECTION FREQUENCY PARAMETER (HIF)

Q.1 :What is the frequency of inspection for this item.

HISTORY FAILURE PARAMETER (HF)

Q.1 : Has this or similar element failed in the past ?
Q.2 : Does fatigue life suggest strong possibility of element failure ?
Q.3 : Are redundancy code and failure of the element correlated ?
Q.4 : What is the inspection method for the failed element ?
Q.5 : What is the inspection frequency for the failed element ?
Q.6 : Does structural condition record indicate failure ?
Q.7 : Was the fault detected in advance ?
Q.8 : Does prognosis parameter suggest strong possibility of failure ?
Q.9 : What were the consequences of failure ?
A. : 1. Negligible 2. Low 3. Medium 4. High 5. Catastrophic
Q.10 : Was repair performed at this location ?

In case of no failure history for similar element the inspection planning will not have any input from the similar element which either might supplement or contradict the current values for this item and hence the type and frequency of inspection to be based on the current values of this item alone.
This parameter is introduced to draw upon the experience of the inspection of similar items in the past to plan failure preventive checks.

HISTORY REPAIR PARAMETER (HR)

Q.1 : Has repair been performed at the location during previous inspection interval ?
Q.2 : Has failure of the repair occurred ?
Q.3 : Has repair at similar locations been performed ?
Q.4 : Has the effect of repair been included in the new codes ?
Q.5 : Does the prognosis including the effect of repair conform to inspection records ?

This code includes the information regarding the changed status of the structure in relevant parameter values. The repair would restore the structure to the original state. The questions for the parameter are directed at collecting the information regarding the efficiency of the repair. The membership function can be based on the answers to above questions. If the structure has gained its original strength, then the membership function is 1.0, if no repair is performed, membership function is 0.0

CURRENT PERIOD CHANGES PARAMETER (CPC)

Q.7.:.Will the structure condition prognosis parameter value membership function change due to changes in static usage, dynamic usage, fatigue life, redundancy, member importance , etc. ?

This parameter is to take into account the changes made in the structure during the operating period of the long-term inspection plan before the plan is revised. The answers to above questions may change the categorisation and membership functions of the relevant input parameters.

FUTURE PERIOD CHANGES PARAMETER (FPC)

This parameter is to take into account the changes to be made in the structure during the operating period of the long-term inspection plan after the plan is revised. The answers to above questions may change the categorisation and membership functions of the relevant input parameters.

STRUCTURAL CONDITION PROGNOSIS PARAMETER (SCP)

Q.1 : Has defect been found at this location ?
Q.2 : What will be effect of defect find at the location for integrity of the structure ?
Q.3.: What will be effect of defect find at the location for serviceability of the structure ?
Q.4 : What is the consequence of the defect being not repaired ?
Q.5 : What is the possibility of progressive collapse in case of defect ?
Q.6 : What is the possibility of catastrophic collapse ?
Q.7 : What is the possibility of large deformations ?
Q.8 : What is the possibility of fatigue failure ?
Q.9 : What is possibility of crack propagation ?
Q.10 : What is possibility of crack initiation ?
Q.11 : What is the possibility of reduction in thickness due to corrosion acceleration ?
Q.12 : What is the possibility of leakage ?
Answers to questions 2 through 12 are given by following values
1. Unlikely 2. Low possibility 3. Possibility 4. High Possibility 5. Very High Possibility

The answers to questions for structural condition prognosis code should be based on the input from DFI, Static Utilisation, Dynamic Utilisation, Fatigue Life, Structural Redundancy, History Repair, Current and Future Changes, Structural Condition Record, History Inspection Type and Frequency. This is an important intermediate output from the input parameters The structural condition prognosis is a result of combination of fuzzy parameters which reflect the knowledge of the structure. The fuzzy rule base to establish this parameter are developed and explanation for this is provided in the section on Generation of Fuzzy Rule Base and Inference Engine.

CONSEQUENCE PARAMETER (CONSQ)

Q.1 : What will be the consequence for integrity and serviceability if damage is not detected through GVI/CVI/NDT ?
Q.2: What is the consequence of inspection failure on integrity and serviceability?
Answers to these questions are given in linguistic values as 1.Negligible 2. Low severity 3. Medium severity 4. High severity 5. Very High Severity

This parameter is a combination of prognosis, expert knowledge on type of inspection required to detect the predicted condition and experience with reliability of performance on site during inspection.

INSPECTION CHANGES PARAMETER (IC)

Q.1 : What change in the inspection method is necessary ?
A. : Increase Detail or Decrease Detail or No Change
Q.2 : What change in the inspection frequency is necessary ?
A. : Increase or Decrease or No Change

Based upon the SCP parameter and CONSQ parameter the increase or decrease in the details of inspection and the frequency can be decided. In this case the combinations of the codes should result in consequences between negligible and the acceptable risk level such as low or medium.
If the prognosis is such that the consequences after GVI for not detecting the damage are between negligible and low severity the GVI can be chosen hence no necessity to increase the detail of inspection. If the prognosis is such that there is high possibility detecting the damage through GVI but the consequence of failure is high, instead of increasing the detail of inspection, frequency can be increased to achieve low consequences of not detecting the damage.

ACCESSIBILITY (ACC)

Open easy access given as "Easy "code. Some work necessary to get the access is given "Medium "code. Significant work necessary to obtain the access is given the code "Difficult" and where access is not possible "No Access" is reported
Accessibility is considered as a constraint on increasing the detail or the frequency of the inspection. With this constraint by decreasing the inspection detail or the frequency if the consequence parameter is still within the required acceptable risk, inspection efforts can be significantly reduced.

CERTIFICATION REQUIREMENTS (CERT)

Any specific requirements regarding the type and frequency of inspection can be prescribed under this parameter. This is the second constraint which must be followed while preparing the revised long-term inspection plans.

The above parameters are combined using fuzzy IF..THEN rules. The generation of fuzzy rule base and application of fuzzy logic concepts to combine these rules is explained in following section.

GENERATION OF FUZZY RULE BASE AND INFERENCE ENGINE

A fuzzy rule base consists of a collection of fuzzy IF..THEN rules in following forms.

$$R^{l} : IF\ x_1\ is\ F_1^{l}\ and.......x_n\ is\ F_n^{l},\ THEN.y.is.G^{l}$$

where $F_i^{l}.and.G^{l}$. are fuzzy sets in $u_i \subset R.and.V \subset R$ respectively

and $\underline{x} = \left(x_1 \cdots\cdots x_n\right)^T \in \left(U_1 \cdots\cdots U_n\right).and$

$y \in V$

are linguistic variables with l = 1,......M and $\underline{x}$ is the input and y is the output.

The input parameters are DFIR, SU, DU, FL, SR, SCR, HIT. HIF, HF, HR, CPC, FPC, with intermediate out-put parameter SCP. The answers to different question in the above mentioned parameters are used to develop IF..AND..OR...THEN rules to obtain the prognosis at the different locations. Several such rules are established which cannot be included in here due to lack of space, however, typical examples and how these are developed is explained in the following..

A typical rule lookup table for a pair of input-output parameters is shown in Table 1. This table shows the rules for structural condition prognosis SCR(fatigue failure) as a combination of Fatigue life (FL) and Structure condition record (SCR (fatigue crack)). It will be seen that more importance is given to the relationship between the inspection records compared to the fatigue life estimation. Similar look-up tables are made for different rules and degrees for each rule are assigned.

The relationship between individual fuzzy parameters and the SCP (Structural Condition Prognosis Parameter) are denoted by an importance factor representing the strength of the ties by a number in the range 0 to 1.0 The closer is the value to 1.0 stronger is the link between the two.

When two or more rules must be jointly satisfied, the output is evaluated with respect to each component of the input separately and assigned the importance score equal to the lowest/highest of the scores of the individual rules.

A typical multi-input multi-output fuzzy rule is as shown below.

Rule
IF DFI(premise) is "very low" and DFI (SU) is "high" DFI(FL) is "High" DFI(defect) is "very low" SU is "High" and SR(Collapse) is "Low possibility" and SCR (fat crack) is "Very good" SCR(stress) is "Good" and SCR(corrosion) is "Poor" and HIT is GVI and HIF is "Standard" and HF is (0...1.0) and HR is "(0...1.0), **THEN SCP(deform) is "Low possibility", SCP(fatigue failure) is "Unlikely" SCP(crack initiation) "Low Possibility"**
The rules are based on the expert knowledge similar to the guide for inspection provided in Reference 2. (DnV, 1980).

The rules can also be written in different format as

Rule

SCP (fatigue failure) is "Unlikely or Low Possibility" unless (DFI(severity) is "Medium to Very High" and/or DFI(FL) is "Medium" or DFI(defect) is "Possible" and/or FL is "Medium" SCR(fatigue crack) is "Poor" or (HIT is "GVI" and HF is 1.0" and FL is "Medium") or CPC(dynamic utilisation) is "Increased" or CPC (fatigue life) "Decreased" or FPC(dynamic utilisation) is "Increased" or FPC (fatigue life) "Decreased"

The number of rules made according to the second representation will be less compared to the first representation.

When the SCP parameter has a value from "Low Possibility" to "Very High Possibility", the item is considered as a candidate which needs inspection. This depends upon the acceptable risk level in the CONSQ parameter. For all such candidate items, the following procedure is adopted to establish the inspection requirements.

The fuzzy sets of Structural Condition Prognosis (SCP) are then combined with the Structural redundancy(SR), History of Inspection Type(HIT), History of Inspection Frequency (HIF) with Consequence (CONSQ) to obtain the changes required in the inspection plan The fuzzy relations are combined using Sup star Composition as shown in Eq.4.

$$\mu_{R\circ S}(u,w) = \sup v \in V\left[\mu_R(u,v) * \mu_S(v,w)\right] \quad (4)$$

Considering the output from the set of inputs to be fuzzy implication, the interpretation can be based on Boolean rule of fuzzy implication (Weiss,J.J. and Donnell, M.L.,1979)
This rule follows the material implication by using fuzzy union for $\dot{+}$ in propositional calculus (Wang,1994)

$$\mu_{A\to B}(u,v) = \mu_{\bar{A}}(u) \dot{+} \mu_B(v) \quad (5)$$

resulting in

$$\mu_{A\to B}(u,v) = \mu_{\bar{A}\dot{+}B} = \max(1-\mu_A, \mu_B) \quad (6)$$

The inspection plan is then subject to conditions on accessibility and the requirement from the certifying authority.

A simplified solution for the above considering fuzzy parameters, e.g. DFI, SU., FL SR, SCP, CONSQ has been carried out with other remaining parameters considered as fuzzy singletons i.e. with membership function 1.0.

The method has been applied for revisions of inspection plans for topsides of two gravity based offshore structures in the North Sea. The inspection plan of one structure was based on conventional procedure. The structure had 120 significant items to be inspected during four year period. Using the fuzzy logic based inspection planning method, the number of inspection points were reduced to 80 points during four year period. In addition to the reduced number of items to be inspected, the inspection types and details were down graded resulting in reduced inspection costs. The inspections for the second structure were planned based on the reliability based method. The inspection plan had 77 items to be inspected in four year period. The number of items to be inspected were reduced to 42 items over the same period for the same acceptable risk level. It has been seen that rational method of decision making significantly reduces the inspection costs

CONCLUSIONS

A rational method for decision making for inspection planning based on expert knowledge and fuzzy logic principles is proposed. The expert knowledge data base is developed by answering a questionnaire related to each parameter affecting the inspection plan. The Knowledge-space is divided into fuzzy subspaces. The fuzzy IF..THEN rules are developed and combined using Sup-star composition. Due to the nested nature of fuzzy rules, Boolean rule of fuzzy implications is seen to be more appropriate.

The method has been applied to revise the inspection plans for two topsides of offshore platforms in the North Sea. This rational method of decision making reduced considerably the number of inspection items necessary to assure adequate safety levels.

REFERENCES

1.Blockley, D.I.(1979) The Nature of Structural Design and Safety
2.Det Norske Veritas, (1980) Rules for the Design Construction and Inspection of Offshore Structures, Appendix I In service Inspection
3. Dubois, D and Prade, Henri, (1979) Outline of Fuzzy set Theory : An Introduction, in Advances in Fuzzy Set Theory and Applications, Ed. Gupta,M.M. at al, North Holland Publishing Company
4.Foeken, R.J. et al, (1994), Fracture Mechanics Limit States for Reassessment and Maintenance of Fixed Offshore Structures, *4th Int. Offshore and Polar Engineering Conference, Osaka, Japan*
5.Hunter,A.B.,(1991)Developments in Artificial Intelligence Reasoning in Artificial Intelligence in Engineering, Ed. Winstanley G., John Wiley & Sons Ltd
6.Kosko Bart, (1992) Neural Networks and Fuzzy Systems, Prentice Hall
7.Pedrycz, Witold,(1989) Fuzzy Control and Fuzzy Systems, John Wiley & Co ,
8.Tang, M.X. et al, (1994) Use of Knowledge Based Systems for Practical Reliability Analysis Based Inspection and Maintenance Planning for Offshore Structures, *4th Int.Offshore and Polar Engineering Conference, Osaka, Japan*
9 Wang, Li-Xin, (1994) Adaptive Fuzzy Systems and Control, PTR Prentice Hall
10.Weiss, J.J. and Donnel, M.L., (1979) A General Purpose Policy Capturing Device Using Fuzzy Production Rules, in Advances in Fuzzy Set Theory and Applications, Ed. Gupta,M.M. at al, North Holland Publishing Company
11.Chen,S.M.(1996) A new Method for Evaluating Weapon Systems Using Fuzzy Set Theory, IEEE Transactions on Systems, Man, and Cybernetics, Part A : Systems and Humans, Vol. 26 No.4, pp 493-497

SCR \ FL	V.Low	Low	Medium	High	V. High
V. Poor	V.High	High	High	High	High
Poor	High	High	High	Medium	Medium
Medium	High	High	Medium	Low	Low
Good	Medium	Medium	Low	Low	Unlikely
V. Good	Medium	Medium	Low	Unlikely	Unlikely

Table 1. Lookup table for SCP(fatigue failure) with FL and SCR as input parameters

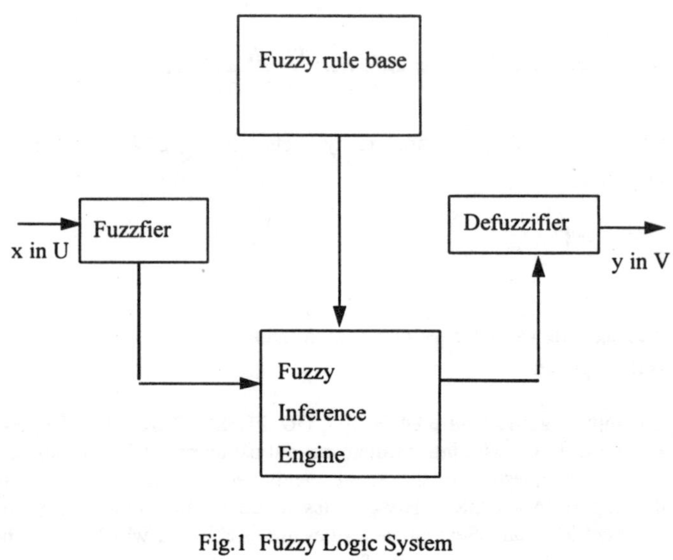

Fig.1 Fuzzy Logic System

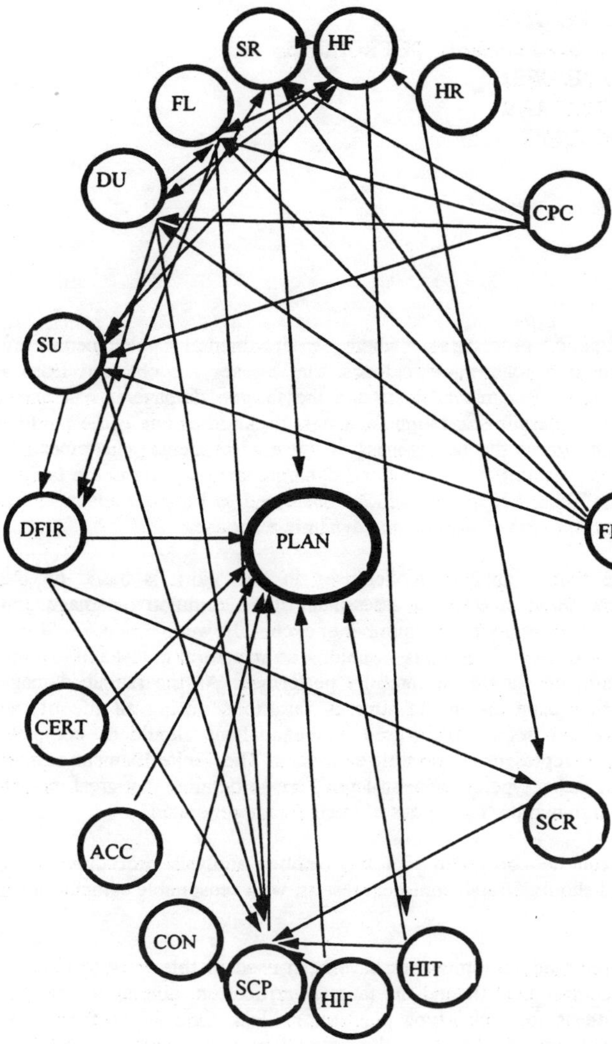

Fig. 2 Parameters affecting inspection plan

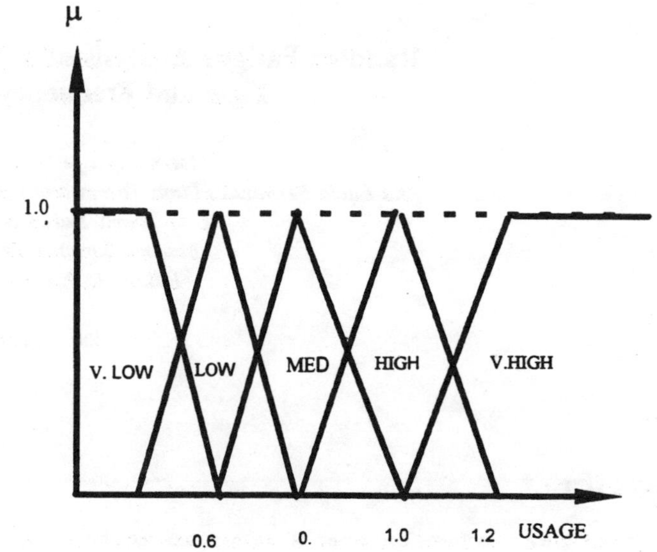

Fig. 3 Fuzzy membership functions for Static Usage (SU) parameter

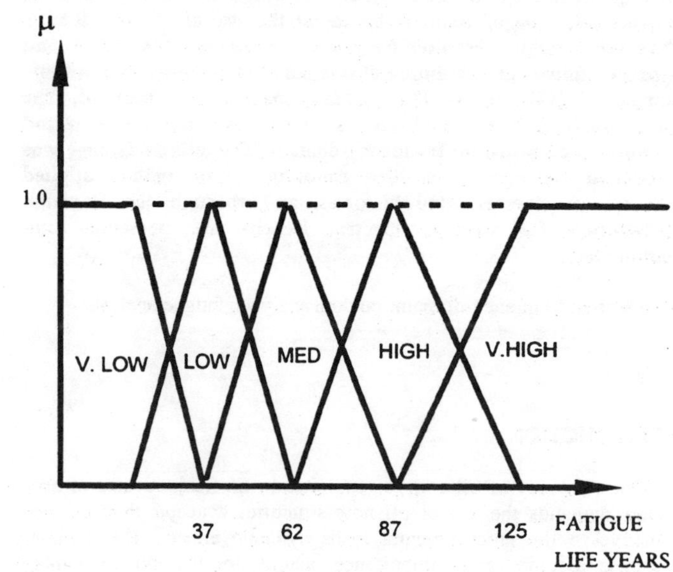

Fig. 4 Fuzzy membership functions for Fatigue Life (FL) parameter

Proceedings of the Seventh (1997) International Offshore and Polar Engineering Conference
Honolulu, USA, May 25-30, 1997

ISBN 1-880653-28-1 (Set); ISBN 1-880653-29-X (Vol. I)

Random Fatigue Analysis of a Jacket Type Offshore Structure: Time and Frequency-Domain Approaches

Marcos Queija de Siqueira, COPPE/UFRJ
Ana Lucia Fernandes Lima Torres and Carlos Alberto Bardanachvili, PETROBRAS
Luis Volnei Sudati Sagrilo, COPPE/UFRJ
Marina Barbosa Fachetti, PETROBRAS
Edison C.P. de Lima, COPPE/UFRJ

Rio de Janeiro, Brazil

ABSTRACT

For offshore structures composed of welded steel members, the use of structural random analysis procedures are recommended in order to conveniently represent the environment load and the structure's response. In this paper, some joints of a jacket type offshore platform were studied in order to determine the fatigue damage by means of three structural random dynamic analysis approaches. One of them is based on the use of stresses RAO's obtained through a dynamic frequency-domain analysis. The second approach allows the obtaining of stresses time-histories from a time-domain analysis, where the sea-state spectra are simulated. The third approach is a hybrid one, as the sea-state are simulated and response is obtained on frequency-domain. The fatigue damage was calculated through a rainflow counting cycles method applied directly on stresses time histories, and through the Rayleigh distribution for stresses spectra. Results are presented and commented.

Key Words: frequency-domain; random analysis; fatigue analysis

INTRODUCTION

The tendency of oil and gas exploitation on fields located at deep waters demands the use of offshore structures concepts that are very sensitive to the environmental loads dynamic effects. The dynamic response assumes great importance, mainly for the fatigue damage verification.

Offshore structures are subjected to several types of loads during their life that may be static or time-varying ones. When installed, the action of environmental phenomena like wind, current and sea waves, that have a time-varying random behavior, generates stresses fluctuations on the material. In the case of the welded joints of jacket type platforms, this process contributes to the fissures development and propagation, possibly leading to the collapse for fatigue damage.

Two basic different approaches are generally used by designers in order to perform the fatigue verification: a) One of them is based on deterministic procedures, where environmental phenomena are represented by deterministic loads, the stresses are obtained from a deterministic dynamic analysis and the fatigue damage is calculated through a deterministic fatigue analysis. b) Another one is the random approach, where the randomness of the environmental phenomena is taken into account in the structural dynamic analysis and in the fatigue analysis. The random approach is considered to be more adequate for the representation of loading and dynamic response.

The fatigue analysis model used in this work is based on the Palmgren-Miner rule for the assessment of the cumulative damage, and the S-N curve to access the number of cycles allowed. This procedure is sensitive to stress amplitudes variations, so structural model and loading representation should be carefully performed. As the fatigue damage calculation depends on the stresses variations during all life of the structure, the set of loads used in the analysis should be complete enough to represent all possible situations. The Finite Element meshes used for jacket type offshore structures are so detailed that great models with a high number of degrees of freedom are generated.

It can be concluded that the random analysis procedure to be adopted should furnish confident results, with reasonable computational costs.

Three random approaches have been used in this work to perform the structural analysis and the fatigue verification, aiming at verifying their fitness for jacket-type platform models. One is the frequency-domain approach, the other the time-domain approach and finally a hybrid method.

The first one is a classic frequency-domain approach where the stresses spectra for each sea-state have been calculated using the RAO of member forces built from the structural transfer function (Fachetti et al., 1992). It is the faster approach, but ignores the non-linear phenomena near the free surface, without considering the relative fluid-structure velocity. The nonlinearities in Morison's equation are not completely taken into account. The random frequency-domain analysis is more attractive due to the lower computer time consuming, but it demands the previous linearization of eventual existing nonlinearities, both in the applied loading and in the structural model. This restriction limits the range of application of this analysis, and demands care in its use.

In the time-domain approach the sea-state spectra were treated by a time-simulation method, so fluid nonlinearities and fluid-structure interaction were represented. The dynamic analysis was carried out in time-domain by means of a direct integration method (ANLIDE. 1995; Mourelle and Gonzalez, 1991). The time-domain random analysis is considered to be the more appropriate to be used due to the possibility of representing the existing nonlinearities of the model. Structural nonlinearities, drag forces, fluid-structure relative velocity, or sea surface level variations may be taken into account. The main drawback of this approach in structural design is the great computer time necessary for the analysis of large structures.

The hybrid method used is named "Dual Method" (ADEP. 1987; Lima et al., 1991; Lima et al., 1993; Siqueira et al., 1995; Videiro and Lima, 1991). Random wave loads were treated by a time-simulation method, so fluid nonlinearities and fluid-structure interaction were represented. The structural response was obtained by a frequency-domain analysis with a mode superposition method, from the Fourier transforming of loading. The main goal of this method is the computational efficiency, which enables the analysis of large structures complete models. The time-frequency hybrid methods attempt to join the advantages of both the frequency-domain and the time-domain approaches.

These three random approaches were used to the evaluation of the fatigue damage and the lifetime of some joints of a jacket platform model similar to those installed in Campos Basin, Brazil. The random fatigue analysis was performed using the stresses time-histories, stresses spectra or stresses RAO, obtained from time-domain or frequency-domain analyses.

All these analysis were performed by means of PETROBRAS's in-house computer softwares, developed and implemented as part of projects from "CENPES-The Research and Development Center of PETROBRAS S.A." with "COPPE/UFRJ-The Engineering Post-Graduating Coordination of the Federal University of Rio de Janeiro".

RANDOM FATIGUE ANALYSIS PHASES

The phases for the fatigue analysis may be summarized as: a) the load definition and the corresponding percentage of occurrence, obtained from a wave scatter diagram; b) the random dynamic analysis of the structure, for member end forces calculation. The analyses were performed in time-domain, frequency-domain and by means of a hybrid method; c) evaluation of the hot-spot stresses around the joint section circumference, from the member end forces associated to stress concentration factors; d) the random fatigue analysis and fatigue damage calculation.

STRUCTURE'S DYNAMIC RANDOM ANALYSIS

Hydrodynamic Loading

Besides the static loads, the main environmental loads that act on offshore structures are due to the action of hydrodynamic load caused by ocean waves and currents, and wind loads. Only hydrodynamic loads were focused in this work.

In a long-term sense, the wave load is a nonstationary process that can be represented as a summation of stationary processes called sea-states, defined by their wave amplitudes spectral density functions, or spectra, the significant wave height, Hs, and the mean zero upcrossing period, Tz. Within each sea-state the sea surface elevation is assumed to be a Gaussian stationary process (Borgman, 1969: Chakrabarti, 1987).

This is the short-term description of wave load. The long-term description of acting load may be obtained from the wave scatter diagram. Each situation, or load condition, is related to a percentage of occurrence. In this work the modified Pierson-Moskowitz spectrum was adopted for all sea-states:

$$S(\omega) = \frac{A}{\omega^5}\exp\left(-\frac{B}{\omega^4}\right) \quad (1)$$

where:

$$B = \frac{1}{\pi}\left(\frac{2\pi}{T_z}\right)^4 \quad (2)$$

$$A = \frac{B}{4}H_s^2 \quad (3)$$

T_z is the mean zero crossing period

H_s is the significant wave height.

Frequency-Domain Approach

The PETROBRAS's in-house software ADEP (1987) was used for this analysis. It is an oriented language software that allows static and dynamic, linear and nonlinear analyses of offshore structures subjected to all possible kinds of loadings. It is composed of several modules of analyses, based on the Finite Element Method. In this work, the deterministic dynamic frequency-domain module, named ADEP/DINA, was used.

The structural amplitude response is considered to be proportional to the wave amplitude, as given by:

$$S_R(w) = [RAO(w)]^2 S_\eta(w) \quad (4)$$

where RAO(ω) is the RAO function, $S_R(\omega)$ is response spectrum and S_η is the wave spectrum.

The RAO function is the ratio between the response amplitude and the wave amplitude. It was determined by applying individual regular waves, with unitary height and period associated to the division of the sea-state spectrum on the structure, and calculating the resulting response.

For each applied wave, loads were generated by Morison's equation in time-domain. So, for each individual wave, free-surface variation and drag forces were represented with the wave height following the wave steepness relation. These loads were converted to the frequency-domain through a Fourier Transform. The structural response was determined by a frequency-domain analysis associated to a Mode Superposition Method with the Static Correction Technique (Torres, 1984). RAO ordinates for each wave frequency are obtained from the standard deviation of the stress response.

Time-Domain Approach

For each load condition or sea-state associated to a percentage of occurrence, a time-domain structural analysis was performed in order to determine time-histories of member end forces.

The PETROBRAS's in-house software ANLIDE (1995) was used to perform this analysis. It is a non-linear time-domain analysis software. The hydrodynamic load is calculated with the Morison's equation, using particle velocities and accelerations time-histories obtained with the linear Airy's theory from the time simulation of sea surface elevation spectrum. Above the still water level, a vertical extrapolation technique is used. Free surface fluctuation and fluid-structure relative velocity are represented. Dynamic response is obtained by the Newmark time-step integration algorithm. ANLIDE allows the modeling of nonlinear geometric effects on beam elements and nonlinear material behavior in the soil-structure interaction. The member end forces results are generated as time histories.

In the simulation process, the sea elevation time-history is represented by the summation of a finite number of harmonic waves associated to random phases, obtained from the spectrum discretization into intervals:

$$\eta(x,t) = \sum_{n=1}^{N} a_n \cos(k_n x - \varpi_n t + \phi_n) \qquad (5)$$

where ϖ_n is a frequency and ϖ_N has a sufficient high value such that $S(\omega) \cong 0$ for $\omega \rangle \omega_N$.

Let $\varpi_n = (\omega_n - \omega_{n-1})/2$ and the quantity "a_n" is given by:

$$a_n = \sqrt{2S(\varpi_n)\Delta\omega_n} \qquad (6)$$

in which $\Delta\omega_n = \omega_n - \omega_{n-1}$

The phase angles ϕ_n are independent random variables uniformly distributed over the interval $(0, 2\pi)$, and k_n is the wave number.

The random response of the structure was obtained by a random time-domain analysis, that generated time-histories of member end forces for the fatigue program.

Hybrid approach

The PETROBRAS's in-house software ADIFI (ADIFI. 1995; Lima et al., 1991) was used to perform this analysis. It is a dual time-frequency random analysis system. The hydrodynamic load is calculated with the Morison's equation, using particle velocities and accelerations time-histories obtained with the linear Airy's theory from the time simulation of sea surface elevation spectrum. Free surface fluctuation is considered and fluid-structure relative velocity is taken into account by means of an iterative technique. A previous nonlinear static analysis may be performed to determine the mean displaced position. Load is converted to the frequency-domain and response is obtained by means of a frequency-domain analysis with mode superposition method, with the use of the Ritz-Wilson vectors basis (Siqueira et al., 1995). The member end forces results are generated in the form of Fourier components (real and imaginary parts), as forces spectra.

As in ANLIDE, the sea elevation time-history is represented by the summation of a finite number of harmonic waves associated to random phases, obtained from the spectrum discretization into intervals:

The random response of the structure was obtained by a linear frequency-domain analysis, that generated spectra of member end forces for the fatigue program.

RANDOM FATIGUE ANALYSIS

The random fatigue analysis was performed by means of PETROBRAS's in-house software, the "POSFAL-Random Fatigue Analysis" (POSFAL. 1995; Torres et al., 1995), that aims at calculating fatigue damage and lifetime of welded steel tubular joints.

The structural model is represented by a Finite Element mesh. The input data to POSFAL is the same as the analysis software input data, reducing the possibility of introducing errors. The final displaced position of the structure may be considered.

The load conditions are defined related to their percentage of occurrence.

The Stress Concentration Factors must be furnished, as they are not calculated in the program. There is an interface with the program ADEP which contains formulas for SCFs, like those of Kuang (Kuang et al., 1975) and Effythimous (1988). The SCF may be defined associated to the different load incidence directions. The Effythimous formulas were used in this work for SCF calculation, using the ADEP software (1994).

Depending on the random dynamic analysis performed, member end forces are generated as time-histories, spectra of the random processes, or RAOs. The POSFAL software offers the possibility of using member end forces given as time-histories, spectra or RAO, in global or local reference system, performing the adequate rotation to joint's plane reference system. The stresses components are calculated in 8 points around the joint section.

Hot-Spot Stresses

Frequency-Domain Approach

Hot spot stress RAO's were calculated from ADEP-DINA deterministic frequency-domain analysis results for each regular wave. Stresses due to member end forces, increased by the stress concentration factors, were combined taking into account the Fourier response components. RAO ordinates were then calculated from the standard-deviation of the hot spot stress time histories for each frequency:

$$S_i(t) = SCF_x^i \frac{F_x^*(\omega_j,t)}{A} \pm SCF_y^i \frac{M_y^*(\omega_j,t)}{Iy} z_i \pm SCF_z^i \frac{M_z^*(\omega_j,t)}{Iz} y_i \qquad (7)$$

where i=1,...,8 stands for the cross section points; j=1,...,N refer to RAO wave frequencies; $F_x^*(t,w_j)$, $M_y^*(t,w_j)$ and $M_z^*(t,w_j)$ are the Fourier response components of member axial force, in-plane and out of plane bending moments; A is the cross section area, Iy and Iz are the inertia moments related to section axes y and z; SCF_x, SCF_y and SCF_z are the stress concentration factors and y_i, z_i are the point distances to section axes y and z.

Time-Domain Approach

The hot-spot stresses time-histories, *S(t)*, were generated based on the results of the ANLIDE's time-domain analysis, whose resultant member end forces were increased by means of the stress concentration factors:

$$S^i(t) = SCF_x^i \frac{Fx(t)}{A} \pm SCF_y \frac{My(t)}{Iy} z_i \pm SCF_z \frac{Mz(t)}{Iz} y_i \qquad (8)$$

where i=1,..,8 stands for the cross section points; $F_x(t)$, $M_y(t)$ and $M_z(t)$ are the time-histories of member axial force, in-plane bending and out-of-plane bending moments.

The technical bibliography in the area of offshore structural random analysis recommends, for a convenient statistical representation of fatigue damage, the use of a set of stress time-histories obtained from different simulations, or realizations, of each sea-state, instead of only one. Nevertheless, in common practice of design, this is difficult to perform due to the great computer time consuming , so only one simulation for each sea-state was adopted (Torres, 1993).

Hybrid Approach

In this case, frequency components of stresses were calculated in 8 points around the joint cross section, using the following expression:

$$S_i^*(\omega_j) = SCF_x^i \frac{F_x^*(\omega_j)}{A} \pm SCF_y^i \frac{M_y^*(\omega_j)}{Iy} z_i \pm SCF_z^i \frac{M_z^*(\omega_j)}{Iz} y_i \quad (9)$$

where sign * is used to identify the complex form, i=1,..,8 stands for the cross section points; j=1,..(number of frequency components) stands for the frequency discretization of spectral density function; F_x^* , M_y^* and M_z^* are the frequency components for member axial force, in-plane bending and out-of-plane bending moments, respectively.

The stress spectral density function is obtained from the stress frequency components as:

$$Sp^i(w_j) = \frac{\overline{S}^i(w_j) S^i(w_j)}{2\Delta w_j} \quad (10)$$

where the bar stands for the conjugate complex and $\Delta\omega_j$ is the frequency interval associated to the frequency ω_i.

Damage Calculation

The S-N curves model was used for the calculation of the expected fatigue life of tubular structural joints. The fatigue behavior of a material described by an S-N curve is assumed to be of the form:

$$NS^m = K \quad (11)$$

where S is a stress range, N is the number of cycles to failure, m and K are material constants obtained from experimental tests. The S-N curves of DNV (1982) and API (1989) are implemented, and there is the possibility of giving a curve defined by points.

The hot-spot stress can be considered as a random continuous process composed of stresses cycles, each of them associated to a sub-life that contributes to the total life. In order to calculate fatigue life, the linear cumulative fatigue damage law known as "Miner's rule" is assumed. In this model each stress cycle, S, induces a damage D defined as:

$$D = \frac{1}{N} = \frac{S^m}{K} \quad (12)$$

and the total damage caused by the stress process is given by:

$$D_T = \sum_{i=1}^{n} D_i = \frac{1}{K} \sum_{i=1}^{n} S_i^m \quad (13)$$

where the index i stands for each stress cycle in the time-varying stress process; S is a stress range, N is the number of cycles, m and K are material constants for S-N curve. The fatigue failure is expected to occur when $D_T = 1$. The major aspect in Miner's rule is to identify each stress cycle for a time-varying stress process.

From the long term description, the percentage of occurrence of each load condition is used. Total damage is obtained from the summation of each damage related to a sea-state, with the associated probability of occurrence, γ, resulting:

$$D_T = \frac{1}{K} \sum_{j=1}^{M} \sum_{i=1}^{n_j} S_{ij}^m \gamma_j \quad (14)$$

where S_{ij} is the i-th stress cycle associated to the j-th load condition. The expected fatigue life is assumed to be the inverse of the fatigue damage.

Cycles Counting

For the time-domain approach, being disposable the stresses time-histories, the rainflow algorithm (Wirshing and Light, 1980; Wirshing and Shehata, 1977) was used in order to identify and count each stress cycle. So, for each sea-state, damage was calculated by the equation (13).

For the frequency-domain and hybrid approaches, stresses spectra are generated. Depending on the process, it is said to be narrow-banded or wide-banded.

For a Gaussian narrow-banded process, the probability distribution of its amplitudes follows a Rayleigh distribution (Newland, 1975), given by:

$$p(a) = \frac{a}{m_0} \exp\left(-\frac{1}{2}\left(\frac{a}{m_0}\right)^2\right) \quad (15)$$

where a stands amplitude and m_0 is the variance of the process.

Assuming that the stress time-history constitutes a narrow-banded Gaussian process S(t), the fatigue damage for a particular load condition can be calculated as algorithm (Wirshing and Light, 1980; Wirshing and Shehata, 1977):

$$D = \frac{n}{K} 2^K \int_0^\infty a^K p(a) da = \frac{n}{K} (2\sqrt{2m_0})^m \Gamma(\frac{m}{2}+1) \quad (16)$$

where $n = f_o T$ is the expected number of stress cycles within the time duration T of the load condition being considered, $p(a)$ is defined in equation (14) and $\Gamma(.)$ is the Gamma function. The total fatigue damage can be rewritten as:

$$D_T = \frac{T}{K} (2\sqrt{2})^m \Gamma(\frac{m}{2}+1) \sum_{i=1}^{M} f_{0i} \gamma_i \left(\sqrt{m_{0i}}\right)^m \quad (17)$$

where the index $i=1,2,...,M$ stands for each particular load condition used to represent the long-term process, and γ_i is the corresponding percentage of occurrence.

Already based on the assumption of narrow-bandness, instead of using equation (16) for damage calculation, numerical integration of Rayleigh distribution is performed in order to assess the number of cycles. In this case, the threshold limit in S-N curves is considered.

In the case of a wide-banded process, a time-simulation can be applied and the rainflow counting cycles technique is used. Another way, aiming at maintaining the simplicity of predicting the fatigue damage for a narrow-band Gaussian process, the Wirshing and Light empirical formula may be used, relating the actual fatigue damage caused by a wide-band process to the damage obtained assuming an equivalent narrow-band Gaussian process with the same variance and zero upcrossing frequency, that is:

$$D = \lambda(\varepsilon, m) D_{NB} \qquad (18)$$

where ε is the spectral width parameter, m is the S-N curve parameter, D_{NB} is the damage assuming narrow-band hypothesis and $\lambda(\varepsilon,m)$ is the correction factor for a wide-banded process given by

$$\lambda(\varepsilon, m) = a(m) + (1 - a(m))(1-\varepsilon)^{b(m)} \qquad (19)$$

where $a(m) = 0.926 - 0.033m$ and $b(m) = 1587m - 2.323$.

The following techniques are available in POSFAL for evaluating fatigue life:
- narrow-band assumption with equation (16);
- narrow-band assumption with numerical integration Rayleigh distribution;
- narrow-band assumption with the band correction factor from equation (18);
- rainflow technique applied directly on the original stress time-histories;
- rainflow technique applied on the simulated stress time histories from the spectrum. The fatigue damage is taken as the average damage of the set of records. The simulated stress time-history is obtained from the following expression:

$$S^i(t) = \sum_{j=1}^{NFREQ} (2Sp^i(\omega_j)\Delta\omega_j)^{1/2} \cos(\omega_j t + \phi_j) \qquad (20)$$

where ϕ_j is a random phase angle and the other variables have been previously defined.

APPLICATION

The objective of this application is to present the fatigue damage calculation through the three random dynamic analysis approaches presented previously, with the use of the described PETROBRAS in-house softwares.

The fatigue verification was performed at some joints of a jacket type platform model, similar to those installed in Campos Basin, Brazil. The model main characteristics are described on Table 1 and Fig. 1. Structural and added hydrodynamic masses were calculated automatically by the softwares, and applied at model joints. Soil-structure interaction was represented by linear springs attached to the basis of the structure. The 10 first natural periods of the structure are presented on Table 2.

Table 1
Model Characteristics

number of joints	702
number of space frame elements	812
critical damping : first/second modes	2%
Rayleigh coefficients	α=0.0595 β=0.0067
water depth	88.7 m

Table 2
Natural Frequencies and Periods

MODE	FREQUENCY (rad/s)	PERIOD (s)	MODE	FREQUENCY (rad/s)	PERIOD (s)
1	2.922	2.150	6	13.989	0.449
2	3.372	1.864	7	14.102	0.445
3	6.442	0.975	8	15.403	0.408
4	9.144	0.687	9	16.537	0.380
5	12.520	0.502	10	16.973	0.370

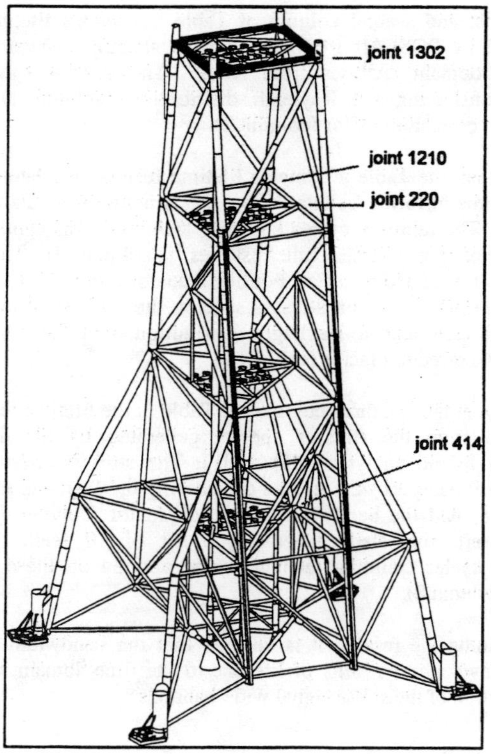

Figure 1
Jacket Model

The frequency-domain approach used doesn't allow the representation of current loads, so only wave loads were considered, although both the time-domain and hybrid methods may represent current action. ANLIDE and ADIFI softwares allow the nonlinear dynamic analysis but the linear beam element was used in order to performs the comparison to the frequency-domain procedure. As the structure doesn't present significant nonlinearities the linear analysis is more faster than the nonlinear one. Fluid-structure relative velocity was not considered, either.

For a real project, several sea-states should be analyzed, but for this work, in order to reduce computer time, the structure was subjected to the action of the four sea-states presented on Table 3. The joints and members analyzed are located as described in Table 4 and located in Fig. 1:

Table 3- Sea-States

SEA-STATE	Hs (m)	Tz (s)	OCCUR. (%)
1	1.22	3	0.007
2	1.65	5	0.578
3	2.1	7	0.361
4	2.87	9	0.054

Table 4 - Joints Location

JOINT	BRACE	CHORD	LEVEL (M)
1210	1209	237	-16.40
220	237	245	-16.40
414	438	2413	-63.50
1302	1302	2210	8.325

In Fig. 2, a window of the hot-spot stress time-history at a section point of joint 1210, brace 1209, associated to the sea-state 3 is presented. It was generated using the member end forces generated from the dynamic time-domain analysis, from ANLIDE. The associated spectrum obtained from a FFT treatment is presented on Fig. 3.

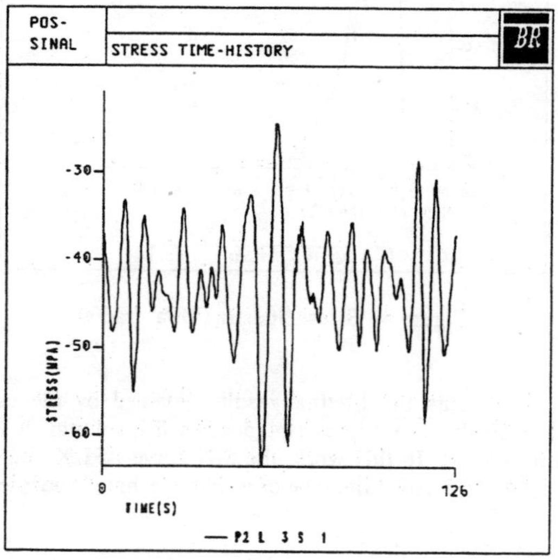

Figure 2- Stress Time-History (Mpa)

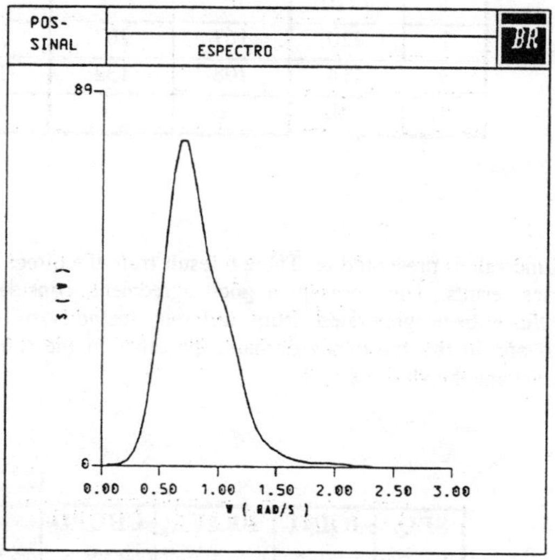

Figure 3 - Stress Spectra (MPa^2 /rad/s)

The hot-spot stress spectrum for the same joint, calculated from the spectra forces generated in the dual time-frequency dynamic analysis from ADIFI, is also presented on Fig. 4.

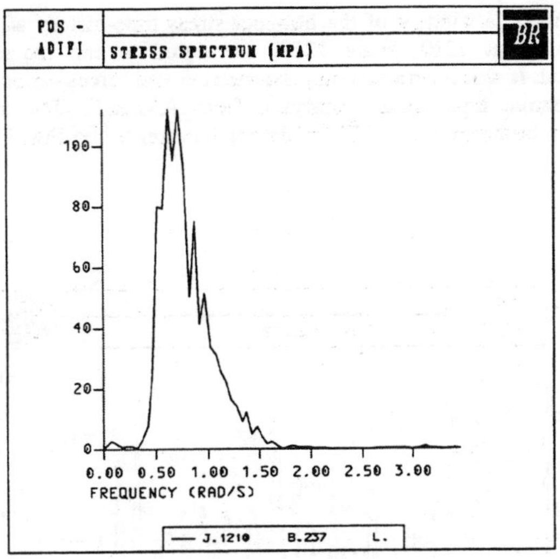

Figure 4 - Stress Spectra (MPa^2 /rad/s)

Table 5 presents the lifetime results obtained by the use of the different methods. The superscript denotes the column of the table, referred in the text. In this work, the S-N curve API X' was used, in order to take into account the type of welding technique adopted.

The first and second column of Table 5 contains lifetime results calculated by POSFAL from the stresses spectra, generated by the frequency-domain analyses from ADEP. The stresses spectra were treated considering the Rayleigh distribution (column 1) and the bandwidth correction factor (column 2).

After that, the Table 5 contains lifetime results calculated from the stresses time-histories generated by the time-domain analyses from ANLIDE. The rainflow cycles counting technique was applied on the original hot-spot stresses time-histories (column 3). These time-histories were treated by a Fast Fourier Transform in order to obtain the stresses spectra. Based on stresses spectra, the number of cycles were calculated assuming a Rayleigh distribution (column 4), and the bandwidth correction factor (column 5).

Finally, at the last three columns of Table 5, the lifetime results were calculated from the stresses spectra generated by the dual time-frequency hybrid analysis from ADIFI. The stresses spectra coming from ADIFI analyses were treated considering the Rayleigh distribution (column 6) and the bandwidth correction factor (column 7). So, the spectra were simulated, generating a set of 10 realizations. The Rainflow cycles counting technique was applied on those simulated signals (column 8).

Examining the results, it is verified that the bandwidth correction factor allows closer results of lifetime to the time-domain, due to the consideration of the stress signal wide-bandness.

Table 5 - Lifetime: Frequency-Domain X Time-Domain X Hybrid Method

		LIFETIME (YEARS) FROM							
		FREQ-DOMAIN		TIME-DOMAIN			HYBRID		
SEQ.	JOINT	RAYL.[1]	RAYL.[2]	ORIG.[3]	RAYL.[4]	RAYL.[5]	RAYL.[6]	RAYL.[7]	SIMUL.[8]
1	1210	63	77	89	84	105	80	99	108
2	220	167	207	257	319	398	138	172	229
3	414	108	132	163	145	181	215	269	471
4	1302	42	53	28	22	28	28	34	35

Lifetime values presented on Table 6 result from the direct treatment of stresses results. They present a good agreement, considering that stresses have been generated from different methods of structural analyses, one in the frequency-domain, the other in the time-domain and the last one the dual method.

The reasons that influences the differences found are the different way of sea-state spectra division and the sensibility of S-N curves, that are exponential, so any difference in stresses results in great differences in the number of cycles allowed.

Table 6

				LIFETIME (YEARS)		
SEQ.	JOINT	BRACE	CHORD	FREQUENCY-DOMAIN/ RAYLEIGH WITH CORRECTION	TIME-DOMAIN/ RAINFLOW	HYBRID/ RAYLEIGH WITH CORRECTION
1	1210	1209	237	77	89	99
2	220	237	245	207	257	172
3	414	438	2413	132	163	269
4	1302	1302	2210	53	28	34

CONCLUSIONS

In order to attain confident structural response, new methods of analysis are necessary, that allow to conveniently consider the acting load characteristics and structure's behavior. The use of integrated software facilitates the designers to access the information they need. The procedures implemented in the softwares used in this work aim at reaching these objectives, as they contemplate different types of structural response and procedures of fatigue damage calculation.

The linear frequency-domain approach may be a useful tool if nonlinearities are not significant. In the case of representing loading nonlinearities, hybrid methods may furnish better results with lower CPU time compared to the time-domain procedure.

REFERENCES

1. "ADEP-User's Manual" (1987). PETROBRAS/CENPES/DIPREX/ SEDEM.

2. "ADIFI: Random Dynamic Analysis of Fixed Platforms - User's Manual" (1995). PETROBRAS/CENPES/DIPREX/SEDEM (in Portuguese).

3. "ANLIDE: Nonlinear Analysis of Offshore Platforms Considering Soil-Structure Interaction - User's Manual" (1995). PETROBRAS/ CENPES/DIPREX/SEDEM (in Portuguese).

4. API (American Petroleum Institute) "Recommended Practice for Planning and Constructing Fixed Offshore Platforms", API RP2A, (1989).

5. Borgman,L.E (1969) "Ocean Wave Simulation for Engineering Design", *Journal of the Waterways and Harbors Division, Proceedings of the American Society of Civil Engineers*, vol. 55, no WWS.

6. Chakrabarti,S.K. (1987) "Hydrodynamics of Offshore Structures", Springer-Verlag, Berlin

7. DnV (Det Norske Veritas) "Rules for Design, Construction and Inspection of Offshore Structures, Steel Structures" (1982).

8. Efthymiou,M. (1988). "Development of SCF Formulae and Generalized Influence Functions for Use in Fatigue Analysis", *OTJ'88 - Recent Developments in Tubular Joints Technology*.

9. Fachetti,M.B., Bardanachvili,C.A., Ayabe,C. (1992). "FADAL-User's and Theoretical Manual", PETROBRAS/CENPES/DIPREX/ SEPRE.

10. Kuang,J.G, Potvin,A.B, Leick, R.D. (1975). "Stress Concentration in Tubular Joints", *Offshore Technology Conference*, no 2205.

11. Lima,E.C.P., Sagrilo,L.V.S. and Siqueira,M.Q. (1991). "The ADIFI Theoretical Manual", Report to PETROBRAS/CENPES/DIPREX/ SEDEM (In Portuguese).

12. Lima,E.C.P.; Torres,A.L.F.L; Sagrilo,L.V.S.; Siqueira,M.Q.; Videiro,P.M. (1993). "Dual Time-Frequency Random Analysis of a Compliant Tower", *Proceedings of the Symposium on Offshore Mechanics and Arctic Engineering (OMAE)*, Edinburgh, Scotland.

13. Mourelle,M.M., Gonzalez,E.C., (1991). "ANFLEX Program - Utilization Course" (in Portuguese).

14. Newland, D.E. (1975). "An Introduction to Random Vibrations and Spectral Analysis", London, Longman Group Limited.

15. "POSFAL: Random Fatigue Analysis - User's Manual" (1995). PETROBRAS / CENPES / DIPREX/SEDEM (in Portuguese).

16. "Procedures for the Calculation of Efthymioy's SCF in ADEP System" (1994). COPPETEC Report for PETROBRAS/CENPES/ DIPREX/SEDEM (in Portuguese).

17. Siqueira,M.Q.; Sagrilo,L.V.S.; Lima,E.C.P. (1995) "Ritz Method for Dynamic Analysis of Structures Submitted to Random Loads", *Brazil Offshore'95*.

18. Torres, A.L.F.L. (1984). "Fatigue Analysis of Steel Welded Structures Subjected to Dynamic Loading", M.Sc. Thesis, COPPE/UFRJ (in Portuguese).

19. Torres, A.L.F.L. (1993). "Random Dynamic Analysis Applied to Offshore Platforms", D.Sc. Thesis, COPPE/UFRJ (in Portuguese).

20. Torres, A.L.F.L., Sagrilo.L.V.S., Siqueira,M.Q., Lima,E.C.P. (1995) "A Procedure for Random Fatigue Analysis of Offshore Structures", *Brazil Offshore'95*.

21. Videiro, P.M., Lima,E.C.P. (1991). "Direct Integration of the Steady-State Response Using the Ritz-Wilson Method with Improved Basis", *Brazil Offshore'91*, vol. 8, 332-349.

22. Wirshing,P.H. and Light,M.C. (1980). "Fatigue Under Wide-Banded Random Stresses", Journal of Structural Division, 106, ST7:1593-1607.

23. Wirshing,P.H. and Shehata, A.M. (1977). "Fatigue Under Wide-banded Random Stresses Using the Rainflow Method", *Journal of Engineering Materials and Technology*, 99, 3:205-211.

Proceedings of the Seventh (1997) International Offshore and Polar Engineering Conference
Honolulu, USA, May 25-30, 1997

ISBN 1-880653-28-1 (Set); ISBN 1-880653-29-X (Vol. I)

Study of the Multi-Function Removable Concrete Platform

Yu-pu Song and Yi Wei
Dalian University of Technology
Dalian, China

ABSTRACT

The paper presents a kind of multi-function removable concrete platform suitable for marginal oil fields in the Bo Hai of China. Type selection and structure optimum of the platform are finished. The temperature distribution of the oil tank is calculated. The study provides the theory basis of platform structure design. The application of the new type of concrete platform should reduce the cost of opening up offshore oil fields, and promote the development of China offshore oil fields.

KEY WORDS: multi-function removable concrete platform, marginal oil field, type selection, structure optimum, temperature distribution.

INTRODUCTION

With the development of offshore oil fields, it is not easy to find the large high-yield oil field. Some medium and small sized oil fields can't be opened up for the high cost of development. The offshore oil fields being developed in China are mainly medium and small sized oil fields with yearly production less than 1000,000 tons. Therefore, to reduce the cost of engineering facilities in the development offshore oil fields is an important problem of China.

It is an effective way in the development of marginal fields to use the concrete platform. Some studies (Yupu Song, 1991) about the fixed concrete platform lying on hard soil layer have been finished. But the study about the removable concrete platform to be used on soft soil layer is seldom found. Presently, a kind of multi-function removable concrete platform suitable for offshore oil fields in the Bo Hai is developed by Dalian University of Technology and China Bo Hai Offshore Oil Corporation. In this paper, analysis and calculations about platform optimum and temperature distribution are given. The study provides the theory basis of platform structure design.

STYLE SELECTION

The concrete platforms generally have three functions such as drilling, production and storage, but several self-elevating drill ships owned by Bo Hai Offshore Oil Corporation are enough to develop Bo Hai fields, and most marginal fields had predrilled well. Therefore, to simplify facilities on the deck, the concrete platform in the paper has only two functions such as production and storage.

The type of the concrete platform should be selected according to the condition of the Bo Hai oil fields.

Firstly, the ports and the shipways are shallow, so the draught of the platform can not be deep. Secondly, offshore oil fields in the Bo Hai are mainly marginal fields with little reserves and short productive life. For above two reasons, the platform should be refloated and reused after the oil field has been produced, the type of platform is that the bottom is concrete cassion to store oil and the upside is steel leg column. The type not only brings material characteristics into play, but also makes platform satisfy the requirements such as shallow draught, being suitable for different deep water and removable.

The surface of the sea floor is silt layer with very low bearing capacity, so how to construct gravity concrete platform on this soft foundation is an important problem. To solve the problem, there are three ways as follows. The first two ways are to reduce the basement stress, such as to use large bottom area caisson or to use separate foundation. Another way is to sink the foundation into the soil layer with higher bearing capacity. For the the tank storage is not large, the caisson shouldn't be made too big. In this paper, the last two ways are used: the platform has three separate foundations and the tanks sink into better soil layer. During sinking tanks, silt is flushed into mud and drawn out. After tanks touch the better soil layer, sand should be filed in the conical place under tanks. The last platform type is shown in Fig. 1.

STRUCTURE OPTIMUM

Optimum Model

The structure optimum of the platform is a constraint nonlinear programming problem. The general form can be written as:

$$\begin{cases} \min & f(\vec{x}) \qquad \vec{x}=x_1,x_2,\cdots,x_n \\ \text{s.t.} & g_j(\vec{x}) \geqslant 0 \qquad j=1,2,\cdots,m \\ & x_{imin} \leqslant x_i \leqslant x_{imax} \qquad i=1,2,\cdots,n \end{cases} \tag{1}$$

In the paper, the cost of the platform structure is the objective function, the cost is only the cost of structure material which doesn't include the cost of installation and top facilities. :

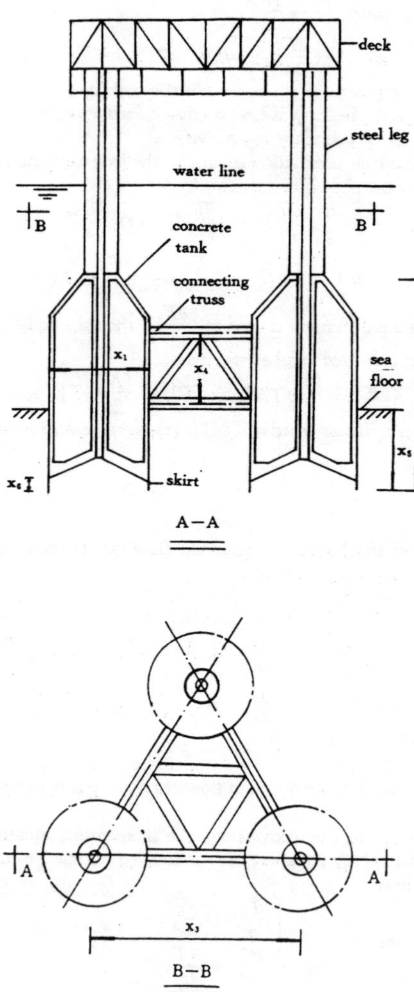

Fig. 1 The type of concrete platform

$$f(\bar{x}) = \alpha_c W_c + \alpha_s W_s \tag{2}$$

where α_c, α_s are unit cost of concrete and steel respectively. W_c, W_s are weight of concrete and steel respectively.

The six optimum variates are: the diameter of tank x_1, the height of tank x_2, the distance between the centers of tanks x_3, the high of the connecting truss x_4, the depth of tank sinking x_5, the height of the skirt x_6, which are shown in Fig. 1..

Considering eight constraint conditions as follows:

1. the storage of tanks: $V \geq [V]$ (3)
2. the draught of the platform: $H_p \leq H_W - H_F$ (4)
3. initial metacentric height: $h \geq GM$ (5)
4. shear stress resisting slide: $\tau_d \geq \tau_{mob}$ (6)
5. moment resisting overturning: $M_W \leq k_0 M_q$ (7)
6. allowable bearing capacity of the foundation: $q_d \geq \frac{F_v \cdot r_f}{A'}$ (8)
7. the truss stress: $r_f \sigma_i \leq \frac{f}{r_m}$ (9)
8. the refloating capacity: $P_F \geq r_F P_V$ (10)

where $[V]$ is allowable storage. H_W is depth of the shipway. H_F is margin depth. GM is allowable initial metacentric height. F_V is vertical load. A' is equivalent foundantion area. f is tensile and compressive strength of steel. P_V is vertical load when the platform refloats. k_0, r_f, r_m, r_F are safety coefficient.

Optimum Method

The function and constraint inequality of platform optimum are complex. It is difficult to get their analytic solutions. Therefore we use ALM (Augmented Lagrange Method) which needn't derivation calculation.

In ALM the object function $f(\bar{x})$ is changed to the augmented lagrange function as:

$$M(\bar{x},\bar{z},\bar{\lambda}) = f(\bar{x}) - \sum_{j=1}^{m} \lambda_j (g_j(\bar{x}) - z_j^2) + \frac{c}{2} \sum_{j=1}^{m} (g_j(\bar{x}) - z_j^2)^2 \tag{11}$$

and min $\overline{M}(\bar{x},\bar{z},\bar{\lambda})$ is equiralent to $minL(\bar{x},\bar{\lambda})$ (12)

$$\text{where } L(\bar{x},\bar{\lambda}) = f(\bar{x}) + \frac{1}{2c} \sum_{j=1}^{m} \{[0, \lambda_j - cg_j(\bar{x})]^2 - \lambda_j^2\} \tag{13}$$

Giving a big enough penalty factor c, to every λ_j find the minimal value of $L(\bar{x},\bar{\lambda})$ using PCD (Powell Conjugate Directions Method). After each iterative step, correct the value of λ_j untill $\lambda_j^{k+1} \to \lambda_j^k$. Then the last $f(\bar{x})$ and $\bar{x}$ are the optimal valve. The iterative function of λ_j is:

$$\lambda_j^{k+1} = \max[0, \lambda_j^k - cg_j(\bar{x}^k)] \tag{14}$$

To ensure convergence and reduce the number of iteration, some measures are used as follows:

1. To the wave load calculation which takes most time, quasi-static method is used. When the changes of dimensions of the tank are little, use the value of wave load before changing. After each iterative step, the value must be corrected.
2. Because the magnitude of the constraint conditions are very different, different c_j are given to different constraints to make the penalties to the constraints similar. During the iteration c_j increases progressively.
3. If the limits of optimum variates are changed to inequality constraints, the dimensionality of the problem should be increased. In this paper, a direct method that make the variates which are out of the limits lie on the limits is used. This method changes the direction of search, but it is the best method according to the results of caculations.
4. In the real engineering, the dimensions of the platform should be modular. In this paper, the convergence criteria is determined according to the modulus of variates. so iterations between the modutuses are avoided.

Through using above measures, the convergency of optimum is improved, and the time of caculation is shortened about 70%.

Optimum Result

Structure optimum is finished to a Bo Hai offshore oil field BZ28—1. When the allowable metacentric height GM = 0. 5m, the result of structure optimum is: $f(\bar{x}) = 664 \times 10^4$RMB¥, $x_1 = 10.0$m, $x_2 = 19.5$m, $x_3 = 15.6$m, $x_4 = 6.2$, $x_5 = 7.1$m, $x_6 = 1.1$m. The real initial metacentric height $h = 0.5$m. This shows that the constraint condition of initial metacentric height is the control constraint condition to BZ28 —1. To different depth of water z_w, the relations between the optimum results of the object function $f(\bar{x})$ and the allowable metacentric height GM are shown in Fig. 2.

From Fig. 2, it is seen that if the depth of water is deep or the value of *GM* is great, the effect of the value of *GM* is great. Different organizations give different values of *GM* shown in Table 1.

Table 1 The Valus of *GM* Given by Different Organizations

Organization	DNV(1982)	NMD(1981)	BV(1975)
GM(m)	Towing: 1. 0 Temporary state: 0. 3	0. 3	0. 3
Organization	USCG(1978)	CMC(1987)	CBSI(1992)
GM(m)	0. 05	0. 3	0. 5

To the new type platform, it is suggested that when towing platform, $GM=1.0$m, and when submerging at temporary state, $GM=0.5$m.

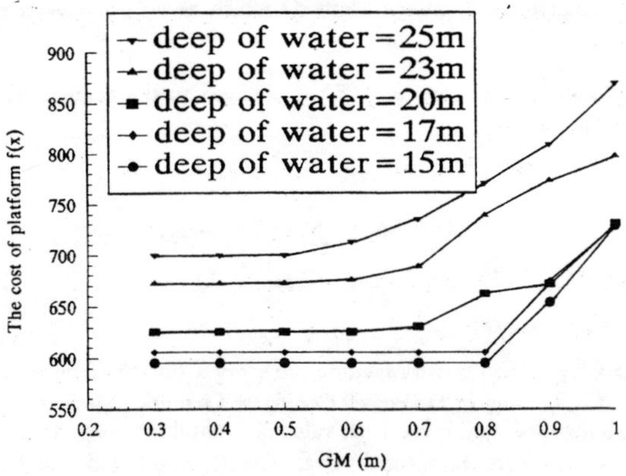

Fig. 2 The relations between $f(x)$ and GM to different deep of water

CALCULATION OF TEMPERATURE DISTRIBUTION

The crude oil of Bo.Hai fields is mostly high solidifiction point oil. The initial temperature of the oil input the tank is about 50℃. The lowest water temperature in winter is −1.7℃. If the temperature difference between two sides of the tank wall is directly given as $\Delta T=50$℃, the temperature stress in the tank wall should be too large to bear. From the experiment of oil-water displacement finished by Dalian University of Technology (Lu, 1995), the solidstate oil layer which adheres to the tank wall reduces the temperature difference as a thermal insulating layer. Therefore, how to give a reasonable value of the temperature difference of tank wall is an important problem to design.

The temperature distribution calculation of the tank is a heat conduction, convection and phase change coupled axisymmetrical heat transfer problem. The paper provids a kind of increment-iteration method. The measure is shown as follows:

Both the tank wall and solidstate oil layer are named solid phase region. The crude oil is named liquid phase region.

To each time step Δt_i:
1. From the temperature of liquid phase centre region T_{fi-1} and the boundary temperature of solid phase region T_{bi-1} got on last step, heat transfer coefficient h_i and temperature distribution in boundary layer T'_i may be calculated through convection calculation.
2. According to the heat conduction theory, the temperature distribution of solid phase region T'_{si} can be got.
3. The thickness of the solidstate oil layer S_i and the temperature distribution of solid phase region T_{si} may be calculated with the phase change theory. According to the whole heat equalization, the temperature of liquid phase centre region T_{fi} can be got.
4. The new head conduction element mesh should be given according to S_i.
5. To repeat 1~4 untill the last time step.

The methods to solve the problems of heat conduction, convection and phase change are respectively discussed as follows.

Heat Conduction

FEM is used to solve the heat conduction problem. To the axisymmetrial system, the heat conduction equation is:

$$\frac{\partial T}{\partial t}=\frac{k}{\rho C_P}\left(\frac{\partial^2 T}{\partial r^2}+\frac{1}{r}\frac{\partial T}{\partial r}+\frac{\partial^2 T}{\partial x^2}\right)+q_v \qquad (15)$$

initial condition: $T(x,r,t_0)=T_0(x,r)$ (16)

B.C. first kind: $T(x,r,t)|_{s\in w_1}=T_1(x,r,t)$ (17)

B.C. sencond kind: $-k\frac{\partial T}{\partial n}|_{s\in w_2}=q_s$ (18)

B.C. third kind: $-k\frac{\partial T}{\partial n}|_{s\in w_3}=h(T-T_f)$ (19)

where T is temperature. t is time. k is thermal conductivity. e is density. C_p is specific heat. q_v is heat produced form element. w_1, w_2, w_3 are three kinds of boundary respectively.

To resolve the heat conduction equation, the extreme value of the following functional should be calculated.

$$I[T(r,x)]=\iint_D\left\{\frac{kr}{2}\left[\left(\frac{\partial T}{\partial x}\right)^2+\left(\frac{\partial T}{\partial r}\right)^2\right]+r\rho c\frac{\partial T}{\partial t}T-q_v rT\right\}dxdr$$
$$+\int_{w_3}h\left(\frac{1}{2}T^2-T_fT\right)dA+\int_{w_2}q_s rTdA \qquad (20)$$

Backward finite-difference is used to $\{\frac{\partial T}{\partial t}\}_t$, then the equation of transient temperature field can be writen as:

$$\left([K]+\frac{[N]}{\Delta t}\right)\{T\}_t=\{P\}_t+\frac{[N]}{\Delta t}\{T\}_{t-\Delta t} \qquad (21)$$

where $[K]$ is stiffness matrix. $[N]$ is transient temperature variation matrix.

Convection

The axisymmetrical system convection flow can be described by the following conservation equations:

$$\frac{1}{r}\frac{\partial}{\partial r}(rv_r)+\frac{\partial vx}{\partial x}=0 \qquad (22)$$

$$\rho\left(v_r\frac{\partial v_r}{\partial r}+v_x\frac{\partial v_r}{\partial x}\right)=-\frac{\partial p}{\partial r}+\mu\frac{\partial}{\partial r}\left(\frac{1}{r}\frac{\partial}{\partial r}(rv_r)\right)+\mu\frac{\partial^2 v_r}{\partial x^2} \qquad (23)$$

$$\rho\left(v_r\frac{\partial v_x}{\partial r}+v_x\frac{\partial v_x}{\partial x}\right)=-\frac{\partial p}{\partial x}+\mu\frac{1}{r}\frac{\partial}{\partial r}\left(r\frac{\partial v_x}{\partial r}\right)+\mu\frac{\partial^2 v_x}{\partial x^2}+\rho g \qquad (24)$$

$$\rho C_p\left(v_r\frac{\partial T}{\partial r}+v_x\frac{\partial T}{\partial r}\right)=k\frac{1}{r}\frac{\partial T}{\partial r}\Big)+k\frac{\partial^2 T}{\partial x^2} \qquad (25)$$

where v_x, v_r are x and r direction flow velocity. p is pressure. μ is viscosity.

To the problem of free convection, the momentum equations (23), (24) and the energy equation (25) is coupled. Some assumptions are given as follows.

$$\begin{cases}\frac{\partial^2 v_x}{\partial r^2}\gg\frac{\partial^2 v_x}{\partial x^2},\\ \frac{\partial^2 T}{\partial r^2}\gg\frac{\partial^2 T}{\partial x^2},\\ \frac{\partial v_r}{\partial r}\gg\frac{v_r}{r}\end{cases} \qquad (26)$$

Then the Eq. 22 to Eq. 25 can be changed to the free convection boundary-layer equations:

$$\frac{\partial v_r}{\partial r}+\frac{\partial v_x}{\partial x}=0 \qquad (27)$$

$$v_r\frac{\partial v_x}{\partial r}+v_x\frac{\partial v_x}{\partial x}=\nu\frac{1}{r}\frac{\partial}{\partial r}\left(r\frac{\partial v_r}{\partial r}\right)+\beta g(T_f-T_b) \qquad (28)$$

$$v_r\frac{\partial T}{\partial r}+v_x\frac{\partial T}{\partial x}=\alpha\frac{1}{r}\frac{\partial}{\partial r}\left(r\frac{\partial T}{\partial r}\right) \qquad (29)$$

where ν is momentum diffusivity. β is inflation coefficient. α is thermal diffusivity.

Boundery conditions:

when $r=R$: $v_x=v_r=0$, $T=t_b$ (30)

when $r=R-\delta$: $v_x=0$, $T=T_f$, $\frac{\partial v_x}{\partial r}=\frac{\partial T}{\partial r}=0$ (31)

where δ is thickness of the boundary-layer.

To introduce similarity variables as follows.

$$\eta=cx^{-1/4}r,\ f=\frac{\psi}{4\nu cx^{3/4}},\ \theta=\frac{T-T_f}{T_b-T_f} \qquad (32)$$

where $c=[\beta g(T_f-T_b)/(4\nu^2)]^{\frac{1}{4}}$, $v_x=\frac{\partial\psi}{\partial r}$, $v_r=-\frac{\partial\psi}{\partial x}$ (33)

where ψ is the stream function.

Utilizing Eq. 32 and Eq. 33 we can transform Eq. 27 to Eq. 31 into:

$$f'''+3ff''+1/\eta\cdot f''-(2f')^2+\theta=0 \qquad (34)$$
$$\theta''+3Prf\theta'+1/\eta\cdot\theta'=0 \qquad (35)$$

Boudary conditions:

when $\eta=\eta_1=cx^{1/4}R$: $f=f'=0$, $\theta=1$ (36)

when $\eta=\eta_2=cx^{-1/4}(R-\delta)$: $f'=f''=\theta=\theta'=0$ (37)

where $f' = df/d\eta$, $f'' = \frac{d^2 f}{d\eta^2}$, $\theta' = d\theta/d\eta$ (38)

The higher differential equations can be solved with the N—s iterative method (Nachtseim and Swigert, 1965).

Phase Change

It is assumed that specific heat C_p and thermal conductivity k of crude oil and solidstate oil layer are same. Through the calculation of convection, the temperature distribution of the crude oil near the solid phase region T_{i1} can be got. Firstly, neglecting the effect of the solidifiction heat, the thickness increment of the solidstate oil layer during time $\triangle t_i$ is obtained. Secondly, assuming that real thickness increment is $\triangle s_{i1}$ and giving of the solidifiction heat of $\triangle s_{i1}$ thickness of solidstate oil, the oil tempreture distribution T_{i2} is calculated with heat conduction theory. Then the thickness increment of the solidstate oil $\triangle s_{i2}$ can be got. If $\triangle s_{i2} \neq \triangle s_{i1}$, the $\triangle s_{i2}$ should be modified. To repeat the iteration until $\triangle s_{ik} \rightarrow \triangle s_{i-1}$. Then $\triangle s_{ik}$ is the real thickness increment of solidstate oil layer during time $\triangle t_i$.

Result of Calculation

In this paper, the calculation of oil tank temperature distribution to crude oil pruduced in Daqing oil field is finished. The physical property is given as follows, density $\rho = 0.85 \times 10^3 kg/m^3$, solidifiction point $T_m = 35$℃, thermal conductivity $k = 0.1337 w/m \cdot$ ℃, momentum diffusivity $\nu = 36.1 \times 10^{-6} m^2/s$, specific heat $C_p = 2.093 \times 10^3 J/kg \cdot$ ℃, solidifiction heat $\lambda = 2.198 \times 10^5 J/kg \cdot$ ℃.

The density of concrete $\rho_c = 2.4 \times 10^3 kg/m^3$, the specific heat $C_{pc} = 1.084 \times 10^3 J/kg \cdot$ ℃. The thermal conductivity of concrete $k = 2.0 w/m \cdot$ ℃.

The analysis model of the tank is shown in Fig. 3. The outside of tank wall should satisfy first kind of boundary condition: $T = T_{sw} = 0$℃. The initial oil temperature $T_{f_0} = 50$℃. From the analysis of the experiment by Dalian University of Technology (Lu, 1995), it is found that the heat transfering through the oil-water interface is about three percent of the total heat transfer from the tank. So the axis of symmetry and oil-water interface should satisfty thermal insulation boundary condition.

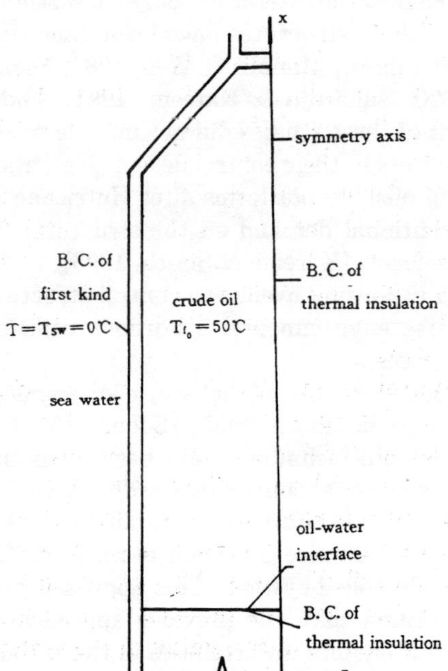

Fig. 3 Heat transfer analysis modle of the tank

The initial average heat transfer coefficient $\bar{h}_0 = 31.5 w/m^2 \cdot$ ℃.

The maximal temperature difference distribution along tank wall is shown in Fig. 4.

From Fig. 4. it is found that except the top part of tank, the temperature difference of other parts of tanks is less than 20℃. Therefore, it is suggested that the temperature difference of tank wall is 25℃, and the top of tank should take thermal insulation measure.

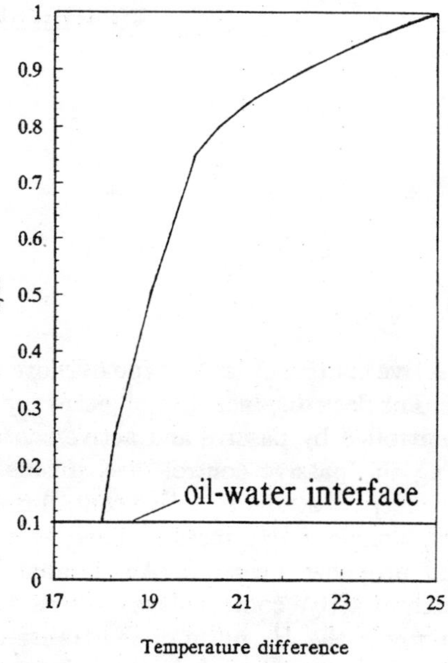

Fig. 4 Temperature difference of tank wall

CONCLUSION

1. The type of the platform selected in this paper is an effective way to reduce the cost of engineering facilities and develop the marginal oil fields.
2. The value of GM has a great effect on the cost of the platform, so it is important to give a proper value of GM to the new type platform.
3. The solidstate oil layer which adheres to the tank wall reduces the temperature difference as a thermal insulating layer.

REFERENCES

Burean Veritas (1975). "Rules and Regulations for the Construction and Classification of Offshore Platforms".

China Bureau of Ship Inspection (1992). "Code for the Construction and Classification of Mobile Offshore Platform".

China Ministry of Communication (1987). "Code for Harbour Engineering Technology".

Det Norske Veritas (1982). "Rules for Classification of Mobile Offshore Units".

Nachtseim P. R. and Swigert P. (1965). "Satisfaction of Asymptotic Boundary Conditions in Numerical Solution of Systems of Nonlinear Equations of Boundary-Layers Tape". NASA TN d-3004.

Norwegian Muritine Directorate (1981). "Regulations for Mobile Drilling Platforms with installations".

Pei-qiong Lu (1995). "The Feasibility Study of Crude Oil Storage in Underwater Rig Mat". China Offshore Platform.

U. S. C. G Department of Transportation (1978). "Requirements for Mobile Offshore Drilling Units".

Yu-pu Song and Jian Wang (1992). "Finite Element Method for Design of Reinforced Concrete Offshore Platforms". China Ocean Engineering, Vol. 6, No. 1, pp. 27—36.

Proceedings of the Seventh (1997) International Offshore and Polar Engineering Conference
Honolulu, USA, May 25-30, 1997

ISBN 1-880653-28-1 (Set); ISBN 1-880653-29-X (Vol. I)

Structural Control of Offshore Platforms

J. Suhardjo and A. Kareem
University of Notre Dame
Notre Dame, IN, USA

Abstract

In this study, active control of jacket type offshore platforms is investigated. The deck displacement of jacket type offshore platforms is controlled by passive and active control mechanisms. Among the passive control mechanisms, a tuned mass damper concept is studied. Active control mechanisms considered here include active mass damper, active tendon mechanism and propeller thruster. An optimal frequency domain approach to active control of wave-excited platforms is proposed in which the H_2 norm of the transfer function from the external disturbance to the regulated output is minimized. The drag force is evaluated using JONSWAP wave spectrum. The influence of nonlinearity in drag force is retained by expressing the nonlinear force components in terms of higher order convolutions of the water particle velocities. The performance with and without control devices under different sea states is examined. It is demonstrated that the control devices are useful in reducing the displacements response of jacket type offshore platforms, especially when the wave forces are concentrated at frequencies close to the natural frequencies of the platform. With increase in water depth the natural frequencies of jacket type platforms fall closer to the dominant wave frequencies. Addition of control devices for deep water applications will ensure not only a reduction in global response, but also in localized effects, such as fatigue of welded joints.

Introduction

In order to tap untouched resources of oil in deep waters or in hostile environment there is a growing need for improving the design and construction practice of drilling and production platforms. It is essential that these platforms can withstand the action of wind, waves, and earthquakes both for normal operating and extreme conditions. Conventional gravity type platforms in deeper water approach natural frequencies that are within the range of occurrence of everyday wave heights resulting in excessive fatigue related damage. A number of design modifications have been introduced that range from a host of structural systems, e.g., guyed towers and tension leg platforms, to utilization of control devices in platforms.

There are two types of systems to control structural response, i.e., passive and active. Passive control devices utilize the dynamics of the structure itself to dissipate energy. An example of passive control devices which can be used to control an offshore platform is the tune mass damper (TMD) which has been used extensively for controlling tall buildings under wind excitations (Petersen, 1980). However, passive control devices can only control the response of a structure up to a certain limit and are generally tuned to one of the natural frequencies of the structure. For the control of structures under large excitations, active control has shown considerable promise in recent years.

In civil engineering literature, only a few studies on active control of offshore structures have been done (Sirlin & others, 1986; Reinhorn, Manolis & Wen, 1987; Yoshida, Suzuki & Oka, 1990; Suhardjo & Kareem, 1991; Abdel-Rohman, 1996). Most of these studies did not include modeling of the wave load effects in their control design. Recent emphasis on the safety of offshore platforms after Hurricane Andrew has imposed additional demand on the structural resistance of offshore platforms (Kareem & Smith, 1993). Structural control offers a promising avenue to control structural response under adverse environmental conditions utilizing auxiliary damping devices.

Most control design methods in civil engineering literature have been in time domain (Soong, 1988). Currently, frequency domain techniques have been introduced for civil engineering structural control (Suhardjo, Spencer, Jr. & Kareem, 1992). In this approach, a stabilizing, optimal control is designed to minimize a certain norm of a transfer function of the controlled system. This approach provides more physical meaning than the previous approaches, it is very flexible, and it enables uncertainties in the system models to be included in the control design.

Although acceleration is the most practically measurable response of a structure (Reinhorn, Manolis & Wen, 1987), in most of current civil engineering structural control studies have been based on the assumption that the states of a structure, *i.e.*, the displacements and velocities, are measurable. Using frequency domain control design techniques, the measurement of any structural response can be easily implemented in the control design. It is also easy to specify any structural response to be controlled. This is an example of the flexibility of the frequency domain control design techniques.

In this study, the feasibility of active control of the jacket type offshore platforms will be investigated. A frequency domain control method, the H_2 control, will be used. The displacements of the decks of jacket type offshore platforms will be controlled by passive and active control mechanisms.

Theoretical Background

In this section, first some theoretical background concerning frequency domain based H_2 control and its solution procedure are outlined. It is followed by state space representation of wave load effects. An example jacket type platform is used to illustrate the procedure and to evaluate the performance of different systems.

State Space Modeling of Wave Load Effects

In control theory, it is a common practice to represent the equations of motion of a system in a state space description. This representation is general enough to make it readily applicable in numerical algorithms for analysis and synthesis. A state vector is a vector of smallest dimension which is needed to express the equations of motion in a state space representation. Typically in structural control it is a $2n$-dimensional vector of displacements and velocities for an n-degree-of-freedom structure. Similarly, the external loading needs to be expressed in the state space framework.

In many cases, spectral properties of environmental loadings on a structure are known and can be modeled. By including loading models in the control design, a better control of the structure can be obtained with little additional effort. A feedback-feedforward control design method where a stochastic model of the external disturbance was included in the control design was presented by Suhardjo, Spencer, Jr. & Sain, 1990. In this case, the external disturbance was an earthquake ground acceleration. The approach taken was to augment the equations of motion of the structural system with an appropriate model of earthquake excitation based on filtering a Gaussian white noise process. The augmented equations of motion were used to determine a control which utilized both feedback and feedforward compensation. The feedback loop incorporated measurements of the response of the structure into the control law. The information from both the structure and the earthquake excitation was utilized in the feedforward control law with an observer designed to estimate the states of the external excitation model based upon the base acceleration measurements. In the following discussion, a state space representation for wave load effects is presented.

Power Spectral Density Functions of Wave Forces

The hydrodynamic wave loads acting on a typical offshore platform consisting of small diameter members is given by the Morison equation which can be explained as given below:

$$F_{wave}(t) = F_{inertia}(t) + F_{linear_drag}(t) +$$

$$F_{current}(t) + F_{nonlinear_drag}(t) \qquad (1)$$

where $F_{inertia}$ is the inertia force, F_{linear_drag} is the linear part of wave drag force, $F_{current}$ is the force created by the current, and $F_{nonlinear_drag}$ is the nonlinear part of the wave drag force. Each component of the force is expressed in terms of the wave height fluctuation. Accordingly, their corresponding spectral density functions can be related.

One of the most commonly used spectral density function of wave surface elevations is the JONSWAP (Joint North Sea Wave Project) spectrum (Chakrabarti, 1984). By introducing $\hat{\omega} = \frac{\omega}{\omega_0}$ where ω_0 is the peak spectral frequency of the wave surface elevations, the JONSWAP spectrum with a sharpness magnification factor γ can be written in a dimensionless form

$$\hat{S}_{\eta\eta}(\hat{\omega}) = \frac{S_{\eta\eta}(\omega)}{S_{\eta\eta}(\omega_0)} = exp(1.25)\hat{\omega}^{-5}exp[-1.25\hat{\omega}^{-4}]\gamma^{\mu-1} \quad (2)$$

where $\mu = exp\left[-\frac{1}{2\sigma^2}(1-\hat{\omega})^2\right]$ with $\sigma = 0.07 for \hat{\omega} \leq 1$ and $\sigma = 0.09 for \hat{\omega} > 1$. This spectrum can be approximated into a form which is suitable for a state space representation (Spanos, 1986) as

$$\hat{S}_{\eta\eta}(\hat{\omega}) \approx \hat{S}_A = \frac{G\hat{\omega}^4}{[(\hat{\omega}^2-k_1)^2+(c_1\hat{\omega})^2][(\hat{\omega}^2-k_2)^2+(c_2\hat{\omega})^2]} \quad (3)$$

where the parameters G, k_1, k_2, c_1, and c_2 are determined by a least square algorithm. Note that Eq. 3 represents a single input single output (SISO) system.

The spectral density function of horizontal water particle velocities at any water level z is related to the spectral density function of wave surface elevations via

$$S_{vv}(\omega, z) = |H_{v\eta}(j\omega, z)|^2 S_{\eta\eta}(\omega). \qquad (4)$$

The magnitude of the transfer function from wave surface elevation to horizontal water particle velocities is

$$|H_{v\eta}(j\omega, z)|^2 = \omega^2\Lambda(\omega, z)(1 + cos\kappa b(z))S_{\eta\eta}(\omega) \qquad (5)$$

$$\Lambda(\omega, z) = \frac{cosh^2(\kappa z)}{sinh^2(\kappa d)}$$

where $j = \sqrt{-1}$, $\kappa = \omega^2/g$ is the wave number, g is the gravitation acceleration, $b(z)$ is the horizontal distance the two legs of the platform at level z, and d is the water depth.

The cross spectral density function of the wave forces per unit depth of the platform between water levels z_m and z_n is

$$S_{q_m,q_n}(\omega, z_m, z_n) =$$

$$S_i(\omega, z_m, z_n)+S_I(\omega, z_m, z_n)+S_{II}(\omega, z_m, z_n)+S_{III}(\omega, z_m, z_n) \quad (6)$$

where the spectral density function of the inertia force is

$$S_i(\omega, z_m, z_n) = k_I(z_m)k_I(z_n)S_{a_m a_n}(\omega), \quad (7)$$

the spectral density function of the linear part of the wave drag force is

$$S_I(\omega, z_m, z_n) =$$

$$k_D(z_m)k_D(z_n)B(\gamma_c(z_m))B(\gamma_c(z_n))\sigma_{v_m}\sigma_{v_n}S_{v_m v_n}(\omega), \quad (8)$$

the spectral density function of the nonlinear wave force can be expressed in terms of a quadratic and cubic representation, $S_{II}(\omega)$ and $S_{III}(\omega)$, given below

$$S_{II}(\omega, z_m, z_n) =$$

$$2k_D(z_m)k_D(z_n)\Phi(\gamma_c(z_m))\Phi(\gamma_c(z_n))S^{*2}_{v_m v_n}(\omega), \quad (9)$$

$$S_{III}(\omega, z_m, z_n) =$$

$$\frac{8}{3}\frac{k_D(z_m)k_D(z_n)}{\sigma_{v_m}\sigma_{v_n}}\phi(\gamma_c(z_m))\phi(\gamma_c(z_n))S^{*3}_{v_m v_n}(\omega), \quad (10)$$

$$k_I(z) = \frac{\pi}{4}\rho C_M D^2(z),$$

$$k_D(z) = \frac{1}{2}\rho C_D D(z),$$

$$B(\gamma_c(z)) = 2\gamma_c(z)\Phi(\gamma_c(z)) + 4\phi(\gamma_c(z)),$$

$$\gamma_c(z) = \frac{U_c}{\sigma_u(z)},$$

$$\Phi(\gamma_c(z)) = 2\int_0^{\gamma_c(z)} \phi(t)dt,$$

$$\phi(\gamma_c(z)) = \frac{1}{\sqrt{2\pi}}\exp -\frac{\gamma_c^2(z)}{2},$$

the spectrum of wave particle acceleration

$$S_{a_m a_n}(\omega) = \omega^2 S_{v_m v_n}(\omega) \quad (11)$$

with ρ the density of the water, C_D drag coefficient, C_M inertia coefficient, $D(z)$ characteristic diameter of the platform leg at water level z, U_c uniform water current, $\sigma_u(z)$ standard deviation of wave particle velocity at water level z, σ_{v_m} is the standard deviation of the horizontal wave velocities at water level z_m, and the convolution S^{*n} is defined as

$$S^{*n}(\omega) \triangleq \int_{-\infty}^{\infty} S(g)^{*(n-1)}S(\omega - g)dg. \quad (12)$$

Notice that for a multi input multi output (MIMO) wave force model, $\sigma_{v_m}\sigma_{v_n}$ is an element of the covariance matrix of the horizontal water particle velocities. Note that in the absence of current, the S_{II} component of the wave force can be neglected.

In modeling the wave forces, we will use the spectral density matrix factorization approach (Davis, 1963). In this approach, a random process with a spectral density matrix $S(\omega)$ is modeled as the output of a filter with a unit intensity white noise process as an input. In Laplace domain (by substituting s for $j\omega$ or $-s^2$ for ω^2), this spectral density matrix is factorized as

$$S(s) = H(-s)H^*(s) \quad (13)$$

where H^* is the transpose of the complex conjugate of H. Using the definition of spectral density function, it can easily been shown that if the input of $H(s)$ is a unit intensity white noise, then the output of $H(s)$ will have a spectral density of $S(s)$. Thus, $H(s)$ is the filter used to model the random process with spectral density function $S(s)$. By representing $H(s)$ in the state space, this filter is readily used for structural control design.

In a jacket type offshore platform model used in this study, the wave forces will enter the through five levels. Hence we will need to construct a wave forces model that have one input, *i.e.*, a unit intensity white noise, and five outputs, *i.e.*, the wave forces on the legs of the platform at different sea levels. To simplify the procedure, this modeling will be done in three steps. The first step is modeling the wave surface elevations, the second step is modeling the transfer function from wave surface elevations to horizontal wave velocities, and the third step is using the results of the first two steps to obtain the final model. By doing this in three steps, the models in the first two steps are SISO models which transform the spectral matrix factorization processes into simple algebraic operations. In the first step, based on Eq. 13, a SISO model with a unit intensity white noise process as an input and wave surface elevations as an output is obtained. In the second step, five SISO models with wave elevations as input and horizontal water particle velocities at five levels of the platform legs as output are obtained separately. Note that this step can be done in the same manner as the first step because

$$|H_{v\eta}(jw)|^2 = H_{v\eta}(-jw)H^*_{v\eta}(jw) \quad (14)$$

which is the same form as Eq. 13. The third step which transfers the five horizontal water particle velocities into five wave forces on the legs of the platform, involves only constants.

The modeling procedure in the first step of $H_{\eta w}$ and second step of $H_{v\eta}$ is equivalent to the modeling of the filter for S_{vv}. By Eq. 13,

$$S_{vv}(s) = H_{vw}(-s)H^*_{vw}(s). \quad (15)$$

Substituting $H_{vw}(s) = H_{v\eta}(s)H_{\eta w}(s)$ into Eq. 15,

$$S_{vv}(s) = H_{v\eta}(-s)H_{\eta w}(-s)H^*_{v\eta}(s)H^*_{\eta w}(s). \quad (16)$$

The resulting full model from a unit intensity white noise process to wave forces on the platform is shown in Figure 1. In this figure, w is the unit intensity white noise process, η is the wave surface elevations, v is the horizontal water particle velocities, and q is the wave forces per unit depth of the platform.

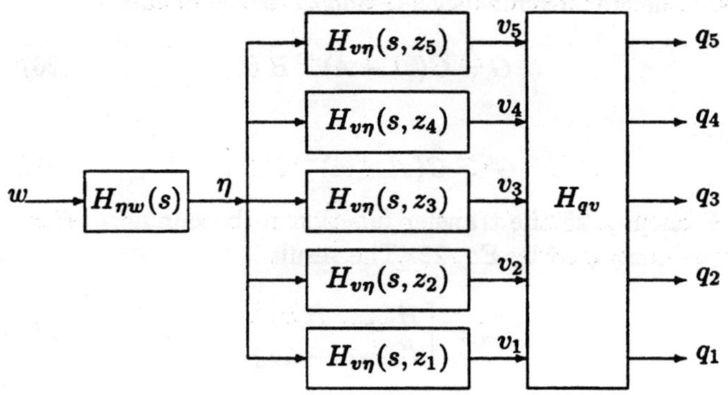

Figure 1: A full model of wave forces on offshore platforms.

The filters for S_i, S_{II}, and S_{III} can be obtained in a similar way to the filter for S_I. For the filter of S_i, the power spectral density function of the of water particle acceleration $S_{a_m a_n}$ is needed. This function can be obtained from $S_{v_m v_n}$ by Eq. 11 which in Laplace domain is

$$S_{a_m a_n}(s) = -s^2 S_{v_m v_n}(s). \tag{17}$$

The filter for $S_{a_m a_n}$ can then be obtained in the same way as the first step in obtaining the filter of $S_{v_m v_n}$. From that, the filter for S_i can be obtained following the second and third steps in obtaining S_I. For the filters of S_{II} and S_{III}, the power spectral density functions $S^{*2}_{v_m v_n}$ and $S^{*3}_{v_m v_n}$ are needed. These functions can be numerically obtained from the convolution of $Sv_m v_n$ as described in Eq. 12. The results are pairs of numbers describing $S^{*2}_{v_m v_n}$ and $S^{*3}_{v_m v_n}$. By some curve-fit processes, these numerical results are then presented in the forms of ratios of polynomials which can be factorized as in Eq. 13. From these ratios of polynomials, the filters for $S^{*2}_{v_m v_n}$ and $S^{*3}_{v_m v_n}$ are obtained. From there, the processes of getting the the filters of S_{II} and S_{III} are straight forward.

Frequency Domain Control

Using a block diagram description, the structural control problem can be depicted as in Figure 2, where y is the measured output vector, z is the regulated output vector, u is the control input vector, and d is the exogenous input vector. The regulated output vector z may consist of any combination of states of the system and components of the control input vector u. By appropriate choice of elements of z, different control design objectives can be included in the problem formulation. Weighting functions can be added to elements of z to determine the frequency range where each element of z is to be minimized. The "system" in Figure 2 can then contain the structure plus filters and weighting functions in frequency domain.

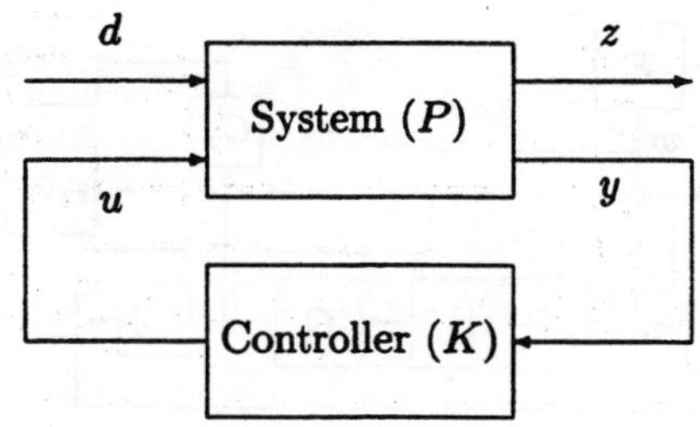

Figure 2: Basic structural control system.

The task here is to design a stabilizing controller in such a way that a norm or "size" of the transfer function from disturbance d to regulated output z is minimized (Suhardjo, 1990). The control design method which minimizes this H_2 norm is termed the H_2 control.

The freedom of the control designer to choose elements of z and frequency dependent weighting functions makes this approach very flexible.

Let us denote the "system" in Figure 2 by P and the "controller" by K. Partitioning P into its components,

$$P = \begin{bmatrix} P_{zd} & P_{zu} \\ P_{yd} & P_{yu} \end{bmatrix} \tag{18}$$

we find that

$$y = P_{yd}d + P_{yu}u = P_{yd}d + P_{yu}Ky \tag{19}$$

and

$$z = P_{zd}d + P_{zu}u. \tag{20}$$

Note that in Eq. 18, a matrix with square brackets is used instead of a matrix with parentheses. This notation is used to emphasize that the elements of the matrix are not constant matrices, but transfer functions.

Rearranging Eq. 19 and substituting into Eq. 20, we obtain

$$z = [P_{zd} + P_{zu}K(I - P_{yu}K)^{-1}P_{yd}]d$$

or

$$z = H_{zd}d \tag{21}$$

where

$$H_{zd} = P_{zd} + P_{zu}K(I - P_{yu}K)^{-1}P_{yd}. \tag{22}$$

H_{zd} is the transfer function from d to z that needs to be minimized.

To illustrate this approach, consider the problem of controlling an offshore platform under wave loading. One possible block diagram set up for P is depicted in Figure 3.

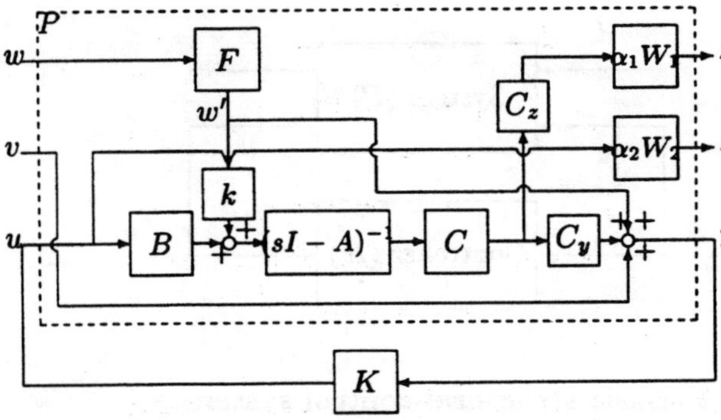

Figure 3: A control problem for offshore platform under wave loading.

In this figure, A, B, and C are matrices which forms the state-space representation the equations of motion of the platform (with $D = 0$) and the block F represents the wave force model. C_z and C_y are matrices needed to extract parts of the system output to form regulated and measured output vectors. The input of F is a vector of unit intensity white noise processes, w and the output of F is a vector of wind forces, w'. Note that in general the control force u and the wave forces w' enter the system differently. For example, the control force comes into the system via the active control device as indicated by the matrix B, while wave forces produce loadings on every lumped mass of the platform. The external excitation d consists of two elements: w and v, or

$$d = \begin{pmatrix} w \\ v \end{pmatrix} \tag{23}$$

where w is a white noise input vector to the wave force model filter and v is the measurement noise. The regulated output z consists of z_1 and z_2 *i.e.*,

$$z = \begin{pmatrix} z_1 \\ z_2 \end{pmatrix} \tag{24}$$

where z_1 is the weighted structural response and z_2 is the weighted control force.

W_1 and W_2 are frequency dependent weighting functions, v is the measurement noise, α_1, α_2 are scalar multipliers for W_1, W_2, respectively, and k is a scalar which expresses a preference in minimizing the transfer function from d to z versus minimizing the transfer function from v to z. If k is taken as a large value, then the effect of measurement noise is neglected in the control design.

The objective of the control design is to determine a controller K such that a weighted norm of the transfer function from d to z is minimized. With this setup, the matrix P expressed in the form of Eq. 18 is given by

$$P = \begin{bmatrix} P_{z_1w} & P_{z_1v} & P_{z_1u} \\ P_{z_2w} & P_{z_2v} & P_{z_2u} \\ P_{yw} & P_{yv} & P_{yu} \end{bmatrix} =$$

$$\begin{bmatrix} \alpha_1 W_1 C_z G'k & 0 & \alpha_1 W_1 C_z G \\ 0 & 0 & \alpha_2 W_2 \\ C_y G'k + F & I & C_y G \end{bmatrix}. \tag{25}$$

where the transfer functions G and G' are defined as

$$G \triangleq C(sI - A)^{-1}B \tag{26}$$

and

$$G' \triangleq C(sI - A)^{-1}F. \tag{27}$$

From Eq. 25 the transfer function to be minimize, H_{zd}, can be computed by Eq. 22. The result is

$$H_{zd} = \begin{bmatrix} H_{z_1w} & H_{z_1v} \\ H_{z_2w} & H_{z_2v} \end{bmatrix} \tag{28}$$

where

$$H_{zd}(11) = \alpha_1 W_1 (C_z G'k + C_z GK(I - C_y GK)^{-1} C_y G'k)$$

$$H_{zd}(12) = \alpha_1 W_1 C_z GK(I - C_y GK)^{-1}$$

$$H_{zd}(21) = \alpha_2 W_2 K(I - C_y GK)^{-1}(C_y G'k + F)$$

$$H_{zd}(22) = \alpha_2 W_2 K(I - C_y GK)^{-1}.$$

Increasing $\alpha_1 W_1$ in a frequency range while choosing a large value of k causes the transfer function from the white noise excitation vector w to the weighted structural response z_1 to be minimized more. This will help improve the performance of the system at those frequencies. Increasing α_2 in a frequency range while choosing a large value of k causes the transfer function from the white noise excitation w to the weighted control force z_2 to be minimized more. This will help limit the control forces. Increasing one of the weightings in a frequency range decreases the importance of the other weighting in that frequency range. Hence, we are faced with the problem of tradeoff between different design objectives.

H$_2$ Control Solution Procedure

As suggested by the name, H_2 control design is a control design which searches for a stabilizing controller while minimizing the H_2 norm of H_{zd}, $\|H_{zd}\|_2$. The H_2 solution procedure is based on the state-space realization of the system P in Figure 2 in the form

$$\dot{x} = Ax + B_1 d + B_2 u \tag{29}$$

$$z = C_1 x + D_{12} u \tag{30}$$

$$y = C_2 x + D_{21} d + D_{22} u. \tag{31}$$

According to Chiang & Safonov, 1988, the H_2 norm optimal control problem is equivalent to a conventional LQG optimal control problem of the system

$$\dot{x}(t) = Ax(t) + B_2 u(t) + w(t), \tag{32}$$

$$y(t) = C_2 x(t) + v(t) \tag{33}$$

with a performance index

$$J = \mathrm{E}\left[\int_0^\infty \begin{pmatrix} x(t) & u(t) \end{pmatrix} \begin{pmatrix} Q & X \\ X^T & R \end{pmatrix} \begin{pmatrix} x(t) \\ u(t) \end{pmatrix} dt\right]$$

$$= \mathrm{E}\left[\int_0^\infty \begin{pmatrix} x^T & u^T \end{pmatrix} \begin{pmatrix} C_1^T \\ D_{12}^T \end{pmatrix} \begin{pmatrix} C_1 & D_{12} \end{pmatrix} \begin{pmatrix} x \\ u \end{pmatrix} dt\right] \quad (34)$$

and the correlated white plant noise w and white measurement noise v have the joint correlation function

$$E\left[\begin{pmatrix} w(t) \\ v(\tau) \end{pmatrix} \begin{pmatrix} w(t) & v(\tau) \end{pmatrix}^T\right] =$$

$$\begin{pmatrix} S_w & X_f \\ X_f^T & S_v \end{pmatrix} \delta(t-\tau) = \begin{pmatrix} B_1 B_1^T & B_1 D_{21}^T \\ D_{21} B_1^T & I \end{pmatrix} \delta(t-\tau). \quad (35)$$

The final H_2 optimal controller is the transfer function

$$\dot{x}_{cp} = A_{cp} x_{cp} + B_{cp} y\,; \quad u = C_{cp} x_{cp} + D_{cp} y \quad (36)$$

where

$$A_{cp} = A - K_f C_2 - B_2 K_c + K_f D_{22} K_c, \quad (37)$$

$$B_{cp} = K_f, \quad (38)$$

$$C_{cp} = K_c, \quad (39)$$

and

$$D_{cp} = 0 \quad (40)$$

where K_c is the control gain matrix and K_f is the Kalman-Bucy Filter gain matrix, both are obtained by solving respective Riccati equations.

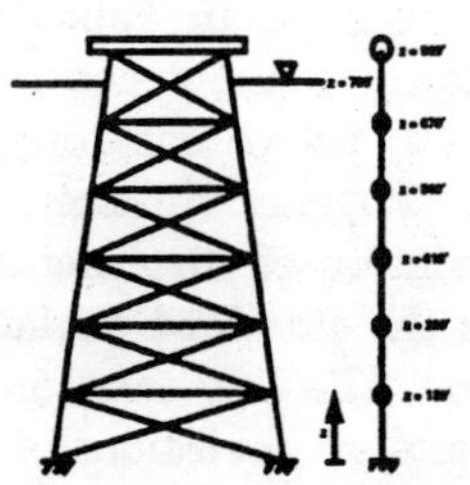

Figure 4: Elevation of jacket type offshore platforms used in numerical examples.

Example

A jacket type offshore platform is modeled as six lumped masses, five of them are under the mean sea level. The depth of the sea is 700 feet. The lumped masses are $2 \times 10^5, 1.8 \times 10^5, 1.6 \times 10^5, 1.4 \times 10^5, 1.2 \times 10^5$, and 1.0×10^6 slugs located at $z = 150, 280, 410, 540, 670$, and 800 feet above the sea bed, respectively. The distances between the legs of the platforms below the mean sea level at the locations of the lumped masses are $b(z) = 165$, 148.75, 132.5, and 116.25 feet, respectively. The corresponding equivalent characteristic diameters of the legs are $D(z) = 45$, 40, 35, 30, and 25 feet, respectively. The elevation of the platform is shown in Figure 4.

For completeness, other properties of the platform are listed below. The first six natural frequencies of the platform are 0.2, 0.98171, 1.9527, 2.7209, 3.2802, and 3.8471 cycles/sec (Hz), respectively. The corresponding modal damping ratios are 0.02, 0.49704, 0.866, 0.11579, 0.13705, and 0.15859, respectively.

The sea state is defined by the JONSWAP wave spectrum with peak spectral frequency $\omega_0 = 0.35$ rad/s or period of 18 Hz, and sharpness magnification factor $\gamma = 3.78$.

The following parameters are obtained based on the charts in Spanos, 1986: $G = 50$, $k_1 = 1$, $k_2 = 4$, $c_1 = 0.15$, and $c_2 = 3.5$. To obtain a significant wave height of 15 m, a value of $G = 50$ has been chosen. Because this process involves a SISO system, numerator and denominator of the spectral density function can be treated separately. Details are omitted here, the reader is referred to Suhardjo & Kareem, 1991.

Recall from Eqs. 5 and that the magnitude of the transfer function from wave surface elevations to horizontal water particle velocities is

$$|H_{v\eta}(j\omega, z)|^2 = \omega^2 \Lambda(\omega, z)(1 + cos\kappa b(z))$$

with

$$\Lambda(\omega, z) =$$

$$\frac{cosh^2(\kappa z)}{sinh^2(\kappa d)} = \frac{1 + \frac{1}{2}(\kappa z)^2 + \frac{1}{24}(\kappa z)^4 + \frac{1}{720}(\kappa z)^6 + \ldots}{1 + \frac{1}{6}(\kappa d)^3 + \frac{1}{120}(\kappa d)^5 + \frac{1}{5040}(\kappa d)^7 + \ldots}.$$

Also

$$(1 + cos\kappa b) = 2 - \frac{1}{2}(\kappa b)^2 + \frac{1}{24}(\kappa b)^4 - \frac{1}{720}(\kappa b)^6 + \ldots.$$

Let us denote the numerator of $|H_{v\eta}|^2$ by N_H and its denominator by D_H. Substituting $\kappa = \omega^2/g$, $\omega^2 = -s^2$, and truncating up to the sixth power of s, we obtain

$$N_H(s, z) = \frac{2z^2 - b^2}{2g^2} s^6 + 2\, s^2$$

and

$$D_H(s) = \frac{d^3}{6g^3} s^6 - 1.$$

Now, using the same procedure as in the first step, $N_H(s, z)$ and $D_H(s, z)$ can be easily factorized to obtain the numerator and denominator of the model filter $H_{v\eta}(s, z)$.

The third step is to "factorize" a constant matrix S with individual element

$$\frac{8}{\pi} k_D(z_m) k_D(z_n) \sigma_{v_m} \sigma_{v_n}$$

and to connect the result to $H_{v\eta}$ and $H_{\eta w}$. Note that $\sigma_{v_m}\sigma_{v_n}$ is an element of the covariance matrix of the horizontal water particle velocities. The "factorization" here is to find a matrix H_{qv} such that

$$S = H_{qv} H_{qv}^T$$

which simplifies into

$$H_{qv} = S^{\frac{1}{2}}$$

where H_{qv} is a constant matrix of real numbers due to the fact that S is symmetric. This operation can be done using various available matrix decomposition algorithms.

The filter of the wave particle acceleration spectrum is

$$H_a(s) = s\,\frac{N_{H_{\eta w}}(s)}{D_{H_{\eta w}}(s)}\,\frac{N_{H_{v\eta}}(s)}{D_{H_{v\eta}}(s)}.$$

This wave particle acceleration filter will then be used to calculate S_i in Eq. 7 following the three steps described above.

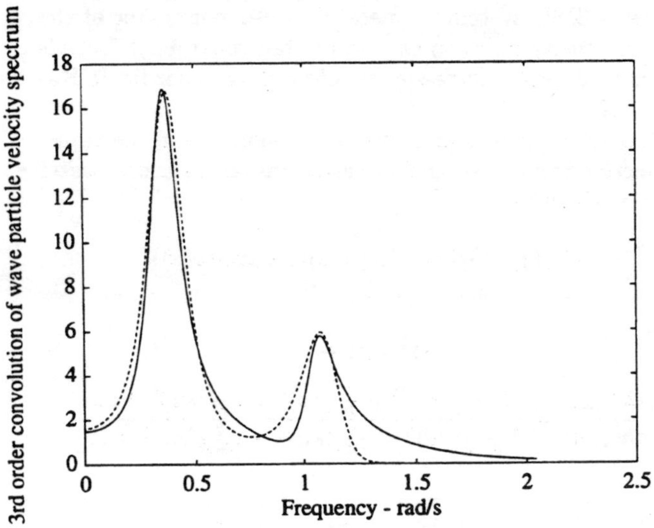

—— Exact third order convolution wave particle velocity spectrum
.... Curve-fitted third order convolution wave particle velocity spectrum

Figure 5: Third order convolution wave particle velocity spectra.

The third order convolution of wave particle velocity spectra are calculated by Eq. 12 from the first and second order convolution wave particle velocity spectra. These spectra are curve-fitted by functions that can be factorized according to Eq. 13. For example, the third order convolution of wave particle velocity spectrum for the wave with period of 18 s that enters the platform at its 5th lumped mass is curve-fitted by a function

$$S_u^{*3}(s) = \frac{N_{S_u^{*3}}(s)}{D_{S_u^{*3}}(s)} \tag{41}$$

where

$$N_{S_u^{*3}}(s) = 0.01$$

and

$$D_{S_u^{*3}}(s) = s^{24} + 1.0476 \times 10^1 s^{22} + 4.9357 \times 10^1 s^{20} +$$

$$1.3717 \times 10^2 s^{18} + 2.4798 \times 10^2 s^{16} + 3.0324 \times 10^2 s^{14} +$$

$$2.5286 \times 10^2 s^{12} + 1.4173 \times 10^2 s^{10} + 5.1760 \times 10^1 s^8 +$$

$$1.1923 \times 10^1 s^6 + 1.7372 s^4 + 1.4821 \times 10^{-1} s^2 + 6.1326 \times 10^{-3}$$

Both the exact and the curve-fitted spectra are plotted in Figure 5.

Based on the spectrum on Eq. 41, the third order convolution wave particle velocity filter is obtained as

$$F_{uIII}(s) = \frac{N_{F_{uIII}}(s)}{D_{F_{uIII}}(s)}$$

where

$$N_{F_{uIII}}(s) = 2.1794 \times 10^{-2}$$

and

$$D_{F_{uIII}}(s) = s^{12} + 1.9656 s^{11} + 7.1982 s^{10} + 9.7126 s^9 +$$

$$1.8081 \times 10^1 s^8 + 1.7157 \times 10^1 s^7 + 2.0043 \times 10^1 s^6 +$$

$$1.3004 \times 10^1 s^5 + 9.7087 s^4 + 3.9011 s^3 + 1.6588 s^2 +$$

$$3.2358 \times 10^{-1} s + 7.0769 \times 10^{-2}.$$

This third order convolution wave particle velocity filter will then be used to calculate S_{III} in Eq. 10.

In this study, currents are omitted which eliminates the contribution of S_{II}.

The block diagram description for this control problem is given in Figure 3. In this figure, the matrix C_z is such that the regulated output z_1 is the weighted displacement of the platform deck and C_y is such that the measured output y is the accelerations of all the lumped masses of the platform.

The numerical results for various cases of control of Platforms A and B are given in Tables 1 and 2. These cases include the platforms without control, with passive control by TMD, and with active control by an AMD, a tendon mechanism, or a propeller thruster. In these tables, σ_d is the standard deviation of displacement of the top mass in cm, and $\sigma_{d_{MD}}$ is the standard deviation of displacement of the mass damper. The last entry in these tables, σ_u (kN) represents the standard deviations of control force generated by the active control device.

Table 1: Standard deviations of displacements, and control force. Wave period = 18 s.

	$\sigma_{d_6}(cm)$	$\sigma_{d_{MD}}(cm)$	$\sigma_u(cm)$
Case I	1.9236E1	—	—
Case II	1.7151E1	7.3883E1	—
Case III	1.4345E1	2.3101E2	3.2291E2
Case IV	1.3507E1	—	3.9293E3
Case V	1.4000E1	—	7.6260E2
Case VI	1.3685E1	6.0375E1	3.8310E3
Case VII	1.3637E1	5.9009E1	9.2715E2

Case I represents the platform with no external control device. Case II represents the platform with a TMD installed on the platform deck. The design of the optimum TMD is based on the design formulae in Kareem, 1983. The properties of the TMD are: mass 172 tons (1% of the generalized mass associated with the first mode of the platform), stiffness 268 kN/m, and damping coefficient 20 tons/s. The eigen frequency of the damper is 0.199 Hz which is 99.3 % of the first natural frequency of the platform.

Table 2: Standard deviations of displacements, and control force. Wave period = 6 s.

	$\sigma_{d_6}(cm)$	$\sigma_{d_{MD}}(cm)$	$\sigma_u(cm)$
Case I	6.4129E0	—	—
Case II	4.7863E0	3.1443E1	—
Case III	2.0849E0	9.1357E1	9.3222E1
Case IV	2.0653E0	—	8.1844E2
Case V	2.1355E0	—	1.7679E2
Case VI	2.0532E0	1.7294E1	7.6264E2
Case VII	1.9528E0	1.6447E0	1.8552E2

Case III represents the platform with an AMD. With structural control, we want to reduce the structural response in the frequency range where the structure is sensitive to disturbance, *i.e.*, at low frequencies. At high frequencies where the structure is often not sensitive to disturbance, we want to "roll off" or lower the control because the control is not effective. Hence for minimization of structural responses, a weighting function which has a large magnitude at low frequencies and rolls off at high frequencies is needed. One class of functions that satisfies these requirements is in the form of

$$\left(\frac{a}{s+a}\right)^n \quad ; \quad n = 1, 2, 3, \ldots.$$

Here, a is termed the "corner frequency" *i.e.*, the frequency where the function "starts" to roll-off. The slope of the roll-off depends on n; a larger n yields steeper slope. For this example, the weighting function is

$$W_1 = \frac{a}{s+a}. \tag{42}$$

The corner frequency $a = 4$ rad/s is chosen throughout these numerical studies. For minimization of control forces, a weighting function which has a small magnitude at low frequencies and increases at high frequencies is needed. For this example, we will use

$$W_2 = \frac{s+0.1}{1 \times 10^{-6}\, s + 0.1}. \tag{43}$$

Case IV represents the platform with a tendon mechanism between its 5th and 6th lumped masses. Case V represents the platform with a propeller thruster on its 5th lumped mass. Case VI represents the platform with a combination of a TMD and a tendon mechanism between its 5th and 6th lumped masses. Case VII represents the platform with a combination of a TMD and a propeller thruster on its 5th lumped mass.

We can see from Table 1 that the response of the platform is reduced by the application of TMD and AMD reduces it further. However, for some cases of AMD, the displacements of the mass damper are too large to be practical for control of jacket type offshore platforms. These displacements can be reduced by using larger mass for the mass damper, but this solution might not be practical. Using only a tendon below the deck of the platform, a similar reductions in platform displacements are achieved with a significantly larger control force. Using only a propeller thruster just below the sea water level, the same reductions of platform displacements are achieved by a significantly smaller control force. In one case in Table 1, the control force produced by the propeller thruster is even smaller than the control force generated by the tendon of the AMD. Using a TMD and a tendon below the deck of the platform, the control force is smaller than just using a tendon, and the displacement of the damper mass is smaller than the displacement of the AMD. Hence it can be seen here that the tendon mechanism helps to reduce the travel distance of the AMD. Hence using the combination of TMD and tendon mechanism, we can afford a smaller mass ratio for the AMD. Using a TMD and a propeller thruster just below the sea water level, the control force is even smaller than the previous case. Hence this combination is probably the best for control of jacket the type offshore platforms considered here.

Conclusions

This paper examines the feasibility of controlling a jacket type offshore platforms using passive/active auxiliary devices. An optimal frequency domain control method based on minimization of H_2 norm of the system transfer function is used. The JONSWAP wave spectra has been employed to model the wave surface elevation, from which the wave force spectra was obtained, including contributions of the nonlinear effects. This wave force spectra was then factorized to give a linear filter for expressing wave load effects in a state space format which is essential for the control procedure utilized here. A numerical study, where the displacements of the decks of platforms were controlled by passive and active control mechanisms, have been conducted. The passive control mechanism consisted of a TMD, whereas the active control devices considered included an AMD, active tendon mechanism, propeller thrusters, and a combination of these.

The size of the control forces and the reduction in the platform response during different sea states were the primary factors taken into consideration in this feasibility study. This study suggests that the combined TMD and propeller thruster mechanism were the most effective means to control the jacket type offshore platforms studied here. For the sea states in which the wave forces were concentrated at frequencies near the dominant frequencies of the platform, the response of the platform was reduced up to about 75% of the uncontrolled platform. This study suggests that by including a single or a combination of auxiliary active/passive damping devices, the platform response can be reduced significantly, thus improving both serviceability performance and fatigue life.

Aknowledgement

The support for this work was provided in part by the National Science Foundation Grants BCS 90-96274, CMS 95-

03779, Office of Naval Research Grant, N 00014-93-1-0761, and several oil companies. The authors are also very grateful to Dr. B. F. Spencer, Jr. for his input.

References

Abdel-Rohman, M. (1996). Structural control of a steel jacket platform. *Structural Engineering & Mechanics*, **4**(2), 125–138.

Chakrabarti, S. K. (1984). On the formulation of JONSWAP spectrum. *Applied Ocean Research*, **6**, 175–.

Chiang, R. Y. & Safonov, M. G. (1988). *Robust–Control Toolbox*. The MathWorks, South Natick, Massachusetts.

Davis, M. C. (1963). Factoring the spectral matrix. *IEEE Transactions on Automatic Control*, **8**(4), 296–305.

Kareem, A. (1983). Mitigation of wind induced motion of tall buildings. *Journal of Wind Engineering and Industrial Aerodynamics*, **11**, 273–284.

Kareem, A. & Smith, C. (1993). Performance of offshore platforms in hurricane andrew. In Reinhorn, A. M. (Ed.), *Proceedings of Hurricanes of 1992: Andrew & Iniki, One Year Later*, ASCE, New York.

Petersen, N. R. (1980). Design of large scale tuned mass dampers. In Leipholz, H. H. (Ed.), *Structural Control*, pages 581–598, North-Holland, Amsterdam.

Reinhorn, A. M., Manolis, G. D., & Wen, C. Y. (1987). Active control of inelastic structures. *Journal of Engineering Mechanics, ASCE*, **113**(3), 315–333.

Sirlin, S., *et al.* (1986). Active control of floating structures. *Journal of Engineering Mechanics, ASCE*, **112**(9), 947–965.

Soong, T. T. (1988). State-of-the-art review: Active control in civil engineering. *Engineering Structures*, **10**, 74–84.

Spanos, P.-T. D. (1986). Filter approaches to wave kinematics approximation. *Applied Ocean Research*, **8**(1), 2–7.

Suhardjo, J. (1990). *Frequency Domain Techniques for Control of Civil Engineering Structures with Some Robustness Considerations.* PhD dissertation, University of Notre Dame, Department of Civil Engineering.

Suhardjo, J. & Kareem, A. (1991). Feasibility of active control of offshore platform. Technical Report NDCE 91-004, *Dept. of Civil Engrg & Geological Sciences*, University of Notre Dame, Notre Dame, Indiana.

Suhardjo, J., Spencer, Jr., B. F., & Kareem, A. (1992). Frequency domain optimal control of wind excited buildings. *Journal of Engineering Mechanics, ASCE*, **118**(12), 2463–2481.

Suhardjo, J., Spencer, Jr., B. F., & Sain, M. K. (1990). Feedback-feedforward control of structures under seismic excitation. *Structural Safety*, **8**, 69–89.

Yoshida, K., Suzuki, H., & Oka, N. (1990). Control of dynamic responses of towerlike offshore structures in waves. *Transactions of the ASME*, **112**, 14–20.

Proceedings of the Seventh (1997) International Offshore and Polar Engineering Conference
Honolulu, USA, May 25-30, 1997

ISBN 1-880653-28-1 (Set); ISBN 1-880653-29-X (Vol. I)

Lifting Eye Design

Edwin P. Russo and John N. Crisp
University of New Orleans, New Orleans, LA, USA
William W. St. Cyr
National Aeronautics and Space Administration, Stennis Space Center, MS, USA
Engin A. Egeseli
Avondale Industries, Avondale, LA, USA

ABSTRACT

This paper is an extension of earlier works by the authors on the design and analysis of lifting eyes (padeyes). The emphasis is on providing formulas that can be easily incorporated into computer programs. The paper describes the philosophy of analysis for the design of lifting eyes. A systematic ordered approach to the design using simple beam theory is presented. Typical modes of failure for this connection associated with direct pin action on the padeye are bearing, pullout, and tension tearing. In the body of the lifting eye shear, tension, and bending stresses occur. A semi-graphical solution is also presented for quick order-of-magnitude hand calculations.

Key Words: Lifting eye, padeye, connection

NOMENCLATURE

- A_b - Minimum pin bearing area, in^2 (mm^2)
- A_c - Area of hypothetical column, in^2 (mm^2)
- A_s - Area of pin shear pullout, in^2 (mm^2)
- A_{t1} - Area of net tearing section at pin hole, in^2 (mm^2)
- A_{t2} - Area of net tearing section at pin hole, in^2 (mm^2)
- a - Distance from upper edge of padeye to center of pinhole, in (mm)
- b - Distance from center of pin hole to attachment structure, in (mm)
- C_C - Slenderness ratio transition point
- c - Size of cheek plate fillet weld, in (mm)
- D - Diameter of pin hole, in (mm)
- D_P - Diameter of pin, in (mm)
- E - Modulus of elasticity, ksi (MPa)
- e - Edge distance at pin hole, in (mm)
- F - Load factor
- F_a - Allowable axial stress, ksi (MPa)
- F_y - Minimum specified yield stress, ksi (MPa)
- h - Total length of plate at any cross section, in (mm)
- I_c - Moment of inertia of hypothetical column, in^4 (mm^4)
- I_1 - Moment of inertia of main plate, cheek plates in section and acting stiffeners for in-plane bending, in^4 (mm^4)
- I_2 - Moment of inertia of main plate, cheek plates in section, and acting stiffeners for out-of-plane bending, in^4 (mm^4)
- K - Effective length factor of hypothetical column
- L - Main plate attachment length, in (mm)
- L_{t1} - Length of plate tearing section, in (mm)
- l - Length of hypothetical column, in (mm)
- M_1 - In-plane bending moment, in-kips (N•mm)
- M_2 - Out-of-plane bending moment, in-kips (N•mm)
- n - Number of padeyes carrying the load
- P - Dynamic design load, kips (N)
- P_t - Transverse load on padeye, kips (N)
- R - Radius of main plate, in (mm)
- R_e - Effective radius of main plate, in (mm)
- r - Outside radius of cheek plate, in (mm)
- r_e - Effective radius of cheek plates, in (mm)
- r_g - Radius of gyration of hypothetical column, in (mm)
- T - Thickness of main plate, in (mm)
- t - Thickness of cheek plate, in (mm)
- W - Total weight of lift, kips (N)
- w - Size of main plate fillet weld, in (mm)
- w_e - Effective maximum width of plate for hypothetical column in compression, in (mm)
- x - Distance from center of pin hole along X axis, in (mm) (Fig. 1)
- x' - Distance from center of pin hole to neutral axis of section, in (mm) (Fig. 6)
- y - Distance from center of pin hole along Y axis, in (mm)
- z - Distance from center of pin hole along Z axis, in (mm)
- z' - Distance from center of pin hole to cross section being analyzed, in (mm) (Fig. 6)
- α - Angle sling makes with the horizontal, degrees
- β - Angle padeye attachment section makes with load, degrees
- ϕ - Angle between vertical and attachment section, degrees
- γ - Angle of taper, degrees
- γ_1 - Angle of right taper of padeye, degrees
- γ_2 - Angle of left taper of padeye, degrees
- σ_1 - Stress due to combined bending and axial load, ksi (MPa)
- σ_w - Nominal weld electrode tensile strength, ksi (MPa)
- τ_1 - In-plane shear stress in main plate, ksi (MPa)
- τ_2 - Out-of-plane shear stress in main plate, ksi (MPa)

INTRODUCTION

This paper expands the work presented in earlier works by Russo and Hall (1977) and Russo, Gelpi, Cayll, and Garic (1992) on the design of lifting eyes (padeyes). The method presented is based on simple beam theory. The philosophy of analysis for the design of the connection is presented in a systematic ordered approach. The formulas can be easily incorporated into computer programs such as the one that the authors are developing for the offshore/shipyard industry through the Gulf Coast Region Maritime Technology Center at the University of New Orleans. A semi-graphical solution is also presented for quick order-of-magnitude hand calculations.

Typical modes of failure for a padeye due to direct pin action on the padeye are bearing, pullout, and tension tearing. In the body of the padeye shear, compression, tension and bending stresses occur. Stresses associated with these various modes of failure must be determined. These stresses should be compared to the appropriate allowable stresses (e.g. the AISC Manual of Steel Construction). Although the methods and ideas given in this paper are adequate for most padeyes, proper engineering judgement should always be used and final design should be reviewed by a professional engineer.

DESIGN LOAD

The dynamic design load, P, for a lifting eye (padeye) is a function of the number of eyes to be used for the lift, n, a load factor, F, the total weight to be lifted, W, and the angle, α, between the horizontal and the lifting sling (see Figure 1). Thus

$$P = F\frac{W}{n\sin(\alpha)} \qquad (1)$$

The load factor F should be chosen with proper judgement to include such factors as impact, possible unequal lengths of slings and clevises that could result in unsymmetrical loading on the lifting eyes, etc.

SIDE LOADS

In addition to the main load (Figure 1) the effect of side loads and torsion must be considered in the design. The American Petroleum Institute (1989), paragraph 2.4.2a, recommends using five percent (5%) of the main design load as a side load. This load should be applied at the center of the pinhole. This transverse load may be resisted by properly designed gusset plates or stiffeners. Torsion about the Z axis (Figure 1) is caused by an out-of-plane force component and a pin hole being offset from the neutral axis. This torsion can be resisted by appropriately spaced stiffeners as well. The degree to which stiffeners share the total load with the main plate can be estimated by including all of the elements in a given cross section in the properties of that cross section. However, care should be exercised in selecting the proper height for the gussets so as not to interfere with the sling shackles.

PRELIMINARY SIZING CALCULATIONS

A preliminary estimate for the thickness of a padeye plate can be based on the pin bearing stress. Using the AISC Manual of Steel Construction (1989) allowable of $0.9F_y$ for bearing stress on the projected area of the pin (paragraph J8), the minimum thickness of the padeye at the pin hole is given by

$$T + 2t \geq \frac{P}{0.9\,F_y D_P} \qquad (2)$$

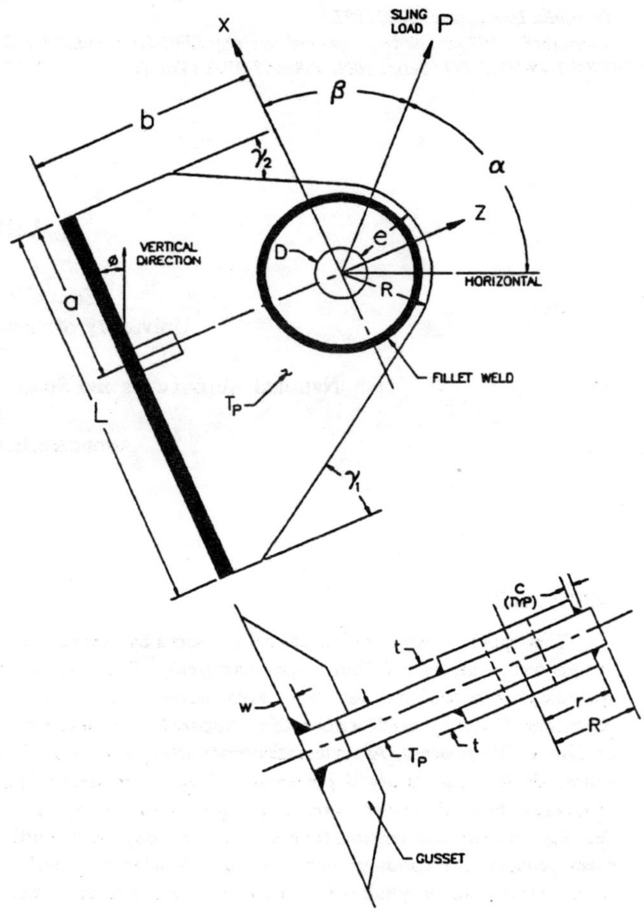

Figure 1. Padeye Geometry and Nomenclature

Load P	Pin Diameter, D_P		Inside Clearance*	
[M.Tons]	[in.]	[mm]	[in.]	[mm]
200	4⅛	105	5.9	150
300	5¼	133	7.3	185
400	6¼	160	8.7	220
500	7	180	9.8	250
600	7¾	200	10.8	275
700	8½	215	11.8	300
1000	10⅝	270	15	380

* Note: Shackle inside clearance limits maximum padeye thickness, T + 2t.

Table 1. Nominal Shackle Pin Diameters and Shackle Inside Clearances

where D_P is the diameter of the pin, T is the thickness of the main plate, and t is the thickness of the cheek plate if cheek plates are deemed necessary. Typical pin diameters are shown in Table 1.

Quite often the section of the padeye along the attachment point to the support structure is the most critical section. Therefore an analysis of this section of the padeye will give an estimate of the length of plate required (Russo and Hall, 1977). For ductile materials, the two criteria generally accepted for determining yielding are the Maximum Shear-Stress (Tresca) Criterion and the Distortional Energy Density (von Mises) Criterion. Since the maximum shear stress criterion is slightly more conservative than the von Mises criterion, and computationally simple to use, the estimate of the minimum length of the plate, L, given below is based on this criterion. Taking the yield strength in shear to be one-half the yield strength in tension, and the maximum allowable shear stress equal to 60% of the yield shear stress, the design allowable maximum shear stress is 0.6 [½ F_y] = 0.3 F_y.

By analyzing an element at the neutral axis of this section the length of the plate, L, must satisfy

$$L \ge \frac{P}{0.6F_yT}\sqrt{1+8\cos^2(\beta)} \qquad (3)$$

The angle β is determined from the angle of lift, α, and the vertical angle, ϕ, as shown in Figure 1. That is

$$\beta + \alpha = \phi + 90^\circ \qquad or \qquad \beta = 90^\circ + \phi - \alpha \qquad (4)$$

An element at the end of this attachment section is subjected to combined axial tension and bending stresses. Limiting the maximum shear stress in this area to $0.3F_y$, the length L must satisfy the following equation for $0 \le \beta \le 90^\circ$

$$L \ge -\frac{P\sin(\beta)}{0.6F_yT} + \sqrt{\left(\frac{P\sin(\beta)}{0.6F_yT}\right)^2 + \frac{6P[b\cos(\beta)+a\sin(\beta)]}{0.6F_yT}} \qquad (5)$$

where T is the thickness of the padeye main plate (Figure 1). If a side load equal to 5% of the main design load is included, the above equation becomes

$$L \ge -\frac{P[\sin(\beta)+0.15\,b/T]}{0.6F_yT} + \sqrt{\left(\frac{P[\sin(\beta)+0.15\,b/T]}{0.6F_yT}\right)^2 + \frac{6P[b\cos(\beta)+a\sin(\beta)]}{0.6F_yT}} \qquad (6)$$

The larger value of L predicted by equations (3) and (5) [or (6)] can be used as a first estimate of the cross section; however, other elements throughout the plate should also be analyzed using the analysis given in subsequent sections.

Using equation (2) to eliminate the plate thickness from equations (3) and (5), the minimum padeye length, L, can be expressed as a function of a, b, t and β as shown in Figures 2a, 2b, and 2c.

Comparison of Figures 2a, 2b, and 2c clearly indicate were the maximum shear stress occurs. Without cheek plates, the maximum shear stress occurs at the neutral axis only for small β and stubby padeyes (i.e. small b). The addition of cheek plates permits the use of a thinner main plate and results in higher stresses at the attachment of the padeye to the main structure, which leads to longer padeyes. For example, with $b = 2D_P$ and t = T/2 (Figure 2b), equation (3) sets the minimum length for β up to approximately 45°, after which equation (5) establishes the minimum length.

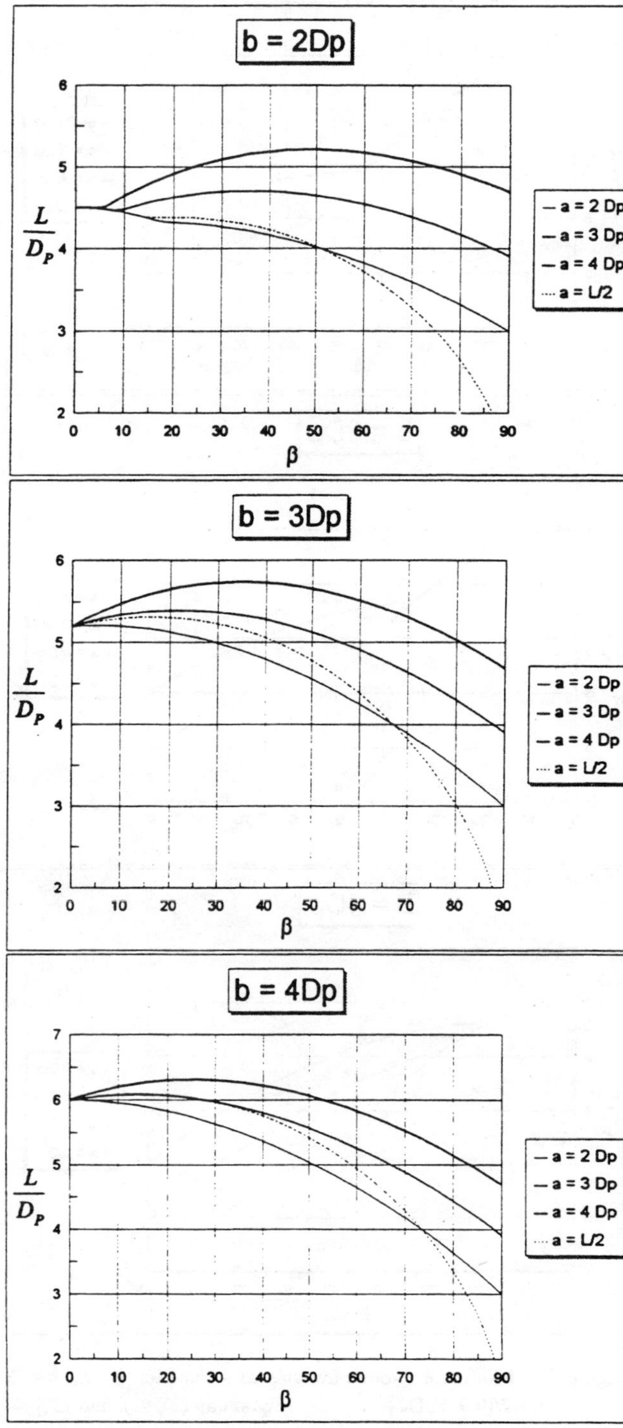

Figure 2a. Minimum Padeye Lengths as a Function β for t = 0 (i.e., no cheek plates).
T = P/[0.9 $F_y D_P$] [Equations (2), (3), and (5)]

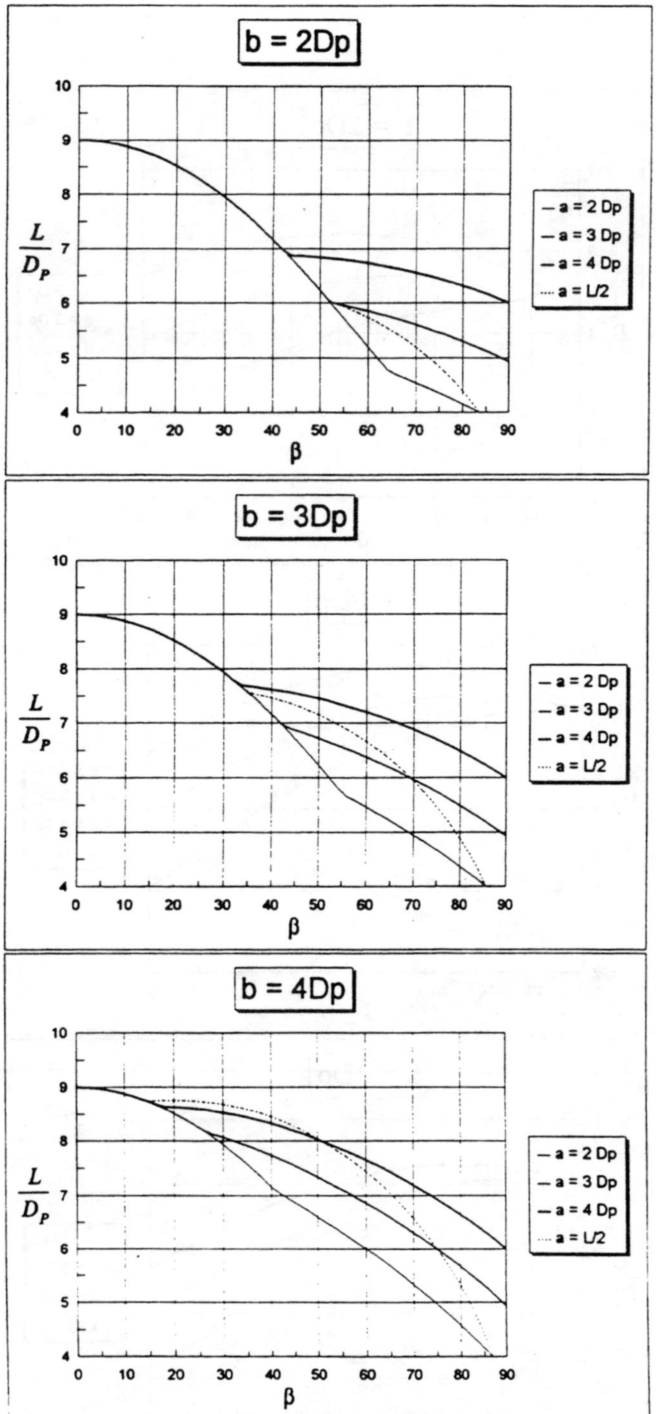

Figure 2b. Minimum Padeye Lengths as a Function β for t = T/2.
T = ½P/[0.9 F_y D_P] [Equations (2), (3), and (5)]

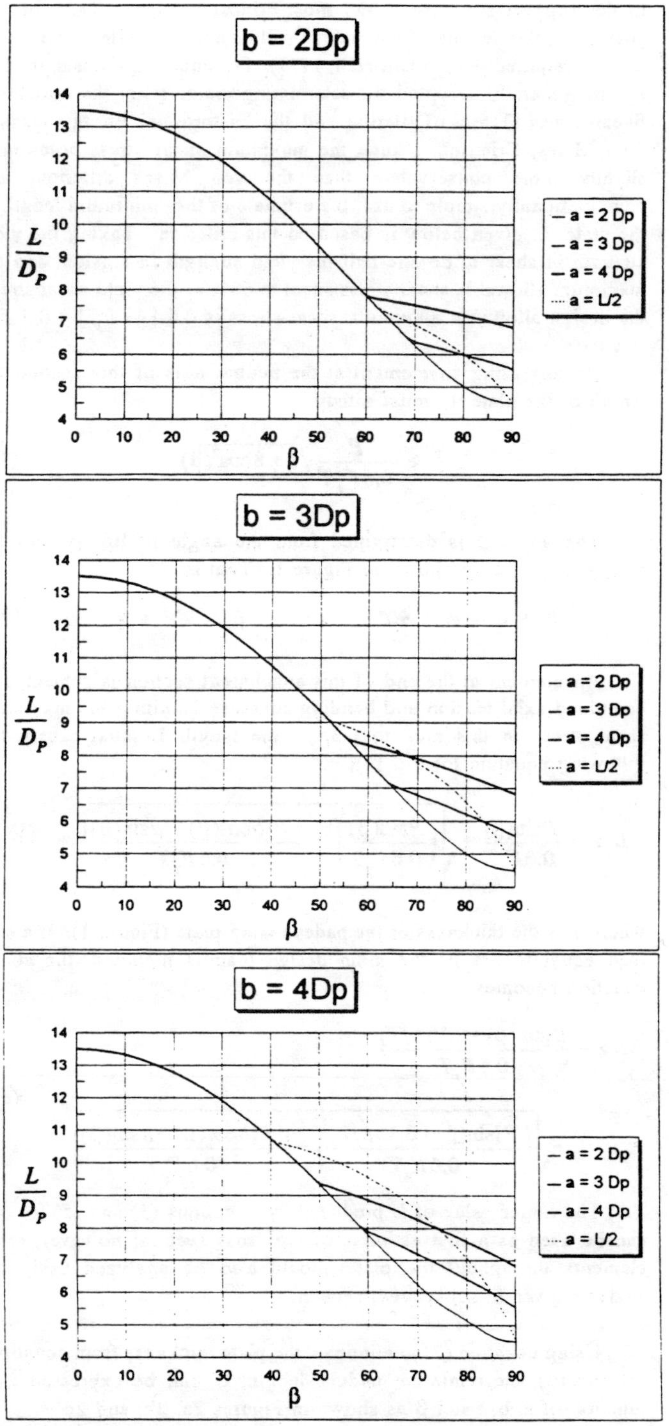

Figure 2c. Minimum Padeye Lengths as a Function β for t = T.
T = ⅓P/[0.9 F_y D_P] [Equations (2), (3), and (5)]

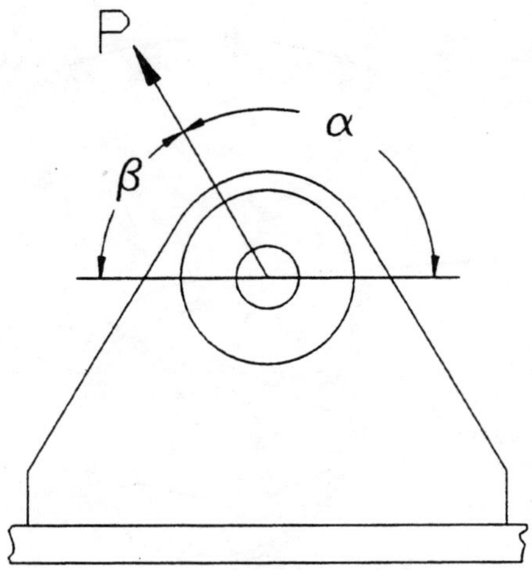

Figure 3. Symmetric Padeye. $a = L/2$, $\gamma_1 = \gamma_2$

For a symmetric padeyes such as shown in Figure 3, the *a* dimension will be L/2, and the minimum length can be determined by using the dashed line on Figure 2a, 2b, or 2c.

The maximum shear stress failure criteria results in a larger length, L, for the padeye than would be obtained using the maximum normal stress criteria of failure, particularly for stubby padeyes, i.e., b small, and small angle β. If the maximum normal stress criteria is used, with the maximum stress limited to $0.6F_y$, equation (3) becomes (Russo and Hall, 1977)

$$L \geq \frac{0.5P}{0.6F_y T}\left[\sin(\beta) + \sqrt{1 + 8\cos^2(\beta)}\right] \tag{3'}$$

The minimum length based on the stress at the end of the attachment section (equation (5) or (6)) will be the same for both the maximum shear stress and the maximum normal stress failure criteria.

PLATE THICKNESS

The thickness of the lifting eye main plate, T, is determined by the stress and tearing associated with the pin/padeye interaction.

PIN BEARING

The area required to support the bearing stress of the pin on the plate is based on the projected area of the pin (Figure 4)

$$A_b = (T + 2t)D_P \geq \frac{P}{0.9\,F_y} \tag{7}$$

The minimum area requirement shown in equation (7) is based on the allowable bearing stress as defined in the AISC Manual of Steel Construction (1989). This allowable considers high bearing stress concentrations (Section C-J8). The total thickness of the padeye at the pin hole, T + 2t, is then given by

$$T + 2t = \frac{A_b}{D_P} \geq \frac{P}{0.9F_y D_P} \tag{8}$$

If the thickness, T + 2t, is too large to be practical, it may be feasible to use multiple plate padeyes. The use of cheek plates around the hole in order to sustain the bearing stresses is the designer's choice. If they are deemed unnecessary, then set t = 0 in the equations.

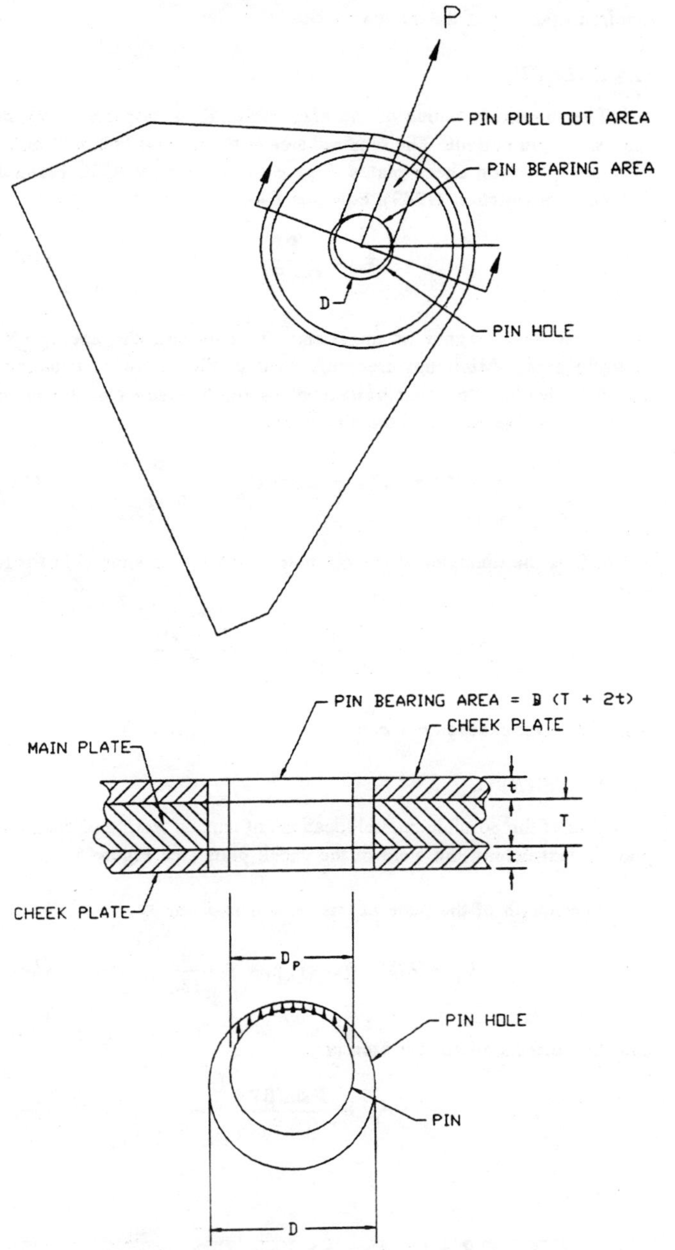

Figure 4. Pin Bearing and Pin Pullout Areas

CHEEK PLATES

The radius of the cheek plates, *r*, is to be chosen by the designer. A reasonable value for *r* is one to two times the pin diameter, D_P. If cheek plates are used then the fillet weld size, c, is given by

$$c \geq \frac{Pt}{1.33\,r\,(T + 2t)\,\sigma_w} \tag{9}$$

where r is the radius of the cheek plate and σ_w is the nominal weld electrode tensile strength. Note that the AISC Manual of Steel

Construction (1989), Table J2.4, lists requirements on minimum and maximum fillet weld size as a function of plate thickness. Also, care should be taken to assure that the cheek plates do not interfere with the required opening of the sling shackles.

PIN PULLOUT

The minimum radius of the main plate, R, is determined by an analysis of pin pullout. The required area to resist shear failure through the main plate and cheek plates (Figure 4) is given by AISC Manual of Steel Construction (1989), equation F4-1

$$A_s \geq \frac{P}{0.4\,F_y} \tag{10}$$

There is a degree of uncertainty in choosing the appropriate shearing area. Minimum areas are used in the following equation, therefore, leading to conservative values for the radius of the main plate, R, and the radius of the cheek plate, r,

$$A_s = 2(2r - D)t + (2R - D)T \geq \frac{P}{0.4F_y} \tag{11}$$

where D is the diameter of the pin hole. Solving equation (11) for R gives

$$R \geq \frac{D}{2} + \frac{P}{2(0.4F_yT)} - \frac{(2r - D)t}{T} \tag{12}$$

where R must exceed $r + c$.

PLATE TEARING

One of the possible critical locations of tension tearing of the main plate is just behind the weld of the cheek plate (see Figure 5).

The length of the plate at this section is given by

$$L_{t1} = 2R + (r + c)(\gamma_1 + \gamma_2)\frac{\pi}{180} \tag{13}$$

and the corresponding tear area is

$$A_{t1} \geq \frac{P\sin(\beta)}{0.6\,F_y} \tag{14}$$

where

$$A_{t1} = L_{t1}T = \left[2R + (r + c)(\gamma_1 + \gamma_2)\frac{\pi}{180}\right]T \geq \frac{P\sin(\beta)}{0.6F_y} \tag{15}$$

Another possible location of critical tension tearing is at the location of the hole. The AISC Manual of Steel Construction (1989), paragraph D3.2, gives two limitations on the edge distances for the sides and in front of the hole for pin connected plates. The first restriction is limited by the hole diameter. It requires that the maximum edge distance, *e*, used in calculations shall be (Figure 5)

$$e \leq 0.8\,D \tag{16}$$

The second restriction is limited by the thickness of the plate(s) at the pin hole. It requires that the maximum edge distance used in calculations be

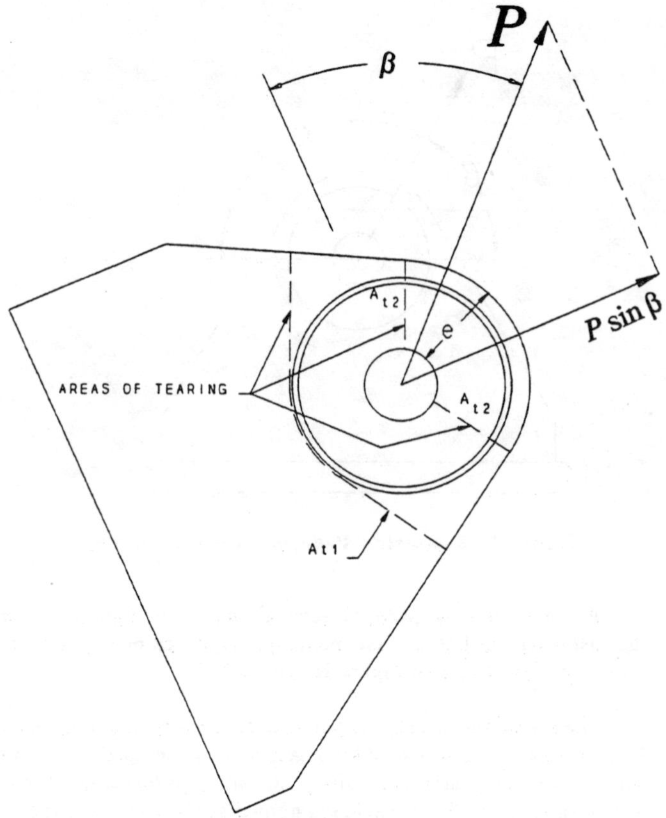

Figure 5. Plate Tearing Areas

$$e \leq 4T \tag{17}$$

By using the lesser of the edge distances from the previous two equations effective radii for the main plate and cheek plates may be computed as the smaller of

$$R_e = R \qquad or \qquad R_e = e + \frac{D}{2} \tag{18}$$

and

$$r_e = r \qquad or \qquad r_e = R_e \tag{19}$$

Using the appropriate effective radii, the net section area (A_{t2}) through the hole (Figure 5) is calculated by:

$$A_{t2} = T(2R_e - D) + 2t(2r_e - D) \geq \frac{P\sin(\beta)}{0.6F_y} \tag{20}$$

PLATE LENGTH

The plate length may be determined by satisfying the stress equations given in the following sections. It is suggested that a computer program be utilized to determine the stresses at number of points along the attachment section. Then increment to the next section outward from the attachment to a new section where, for example, the tapers change the overall depth of the section, and then check several points along this section. Repeat this process until the entire padeye has been analyzed.

AXIAL AND BENDING STRESSES

The ability of a padeye to withstand stresses caused by bending combined with axial tension or compression must be analyzed. It should be noted that a portion of the lifting eye may have a uniform cross section and a portion may be tapered along either or both edges (Figure 6). The worst case combination of the above stresses can exist in any one of these portions of the padeye; hence, it is not adequate to check only one cross section. It should be noted that the bending stresses increase with the pin hole offset distance *b*. Thus it may be desirable to make this dimension small. However caution should be exercised to make sure that there is clearance for the shackle.

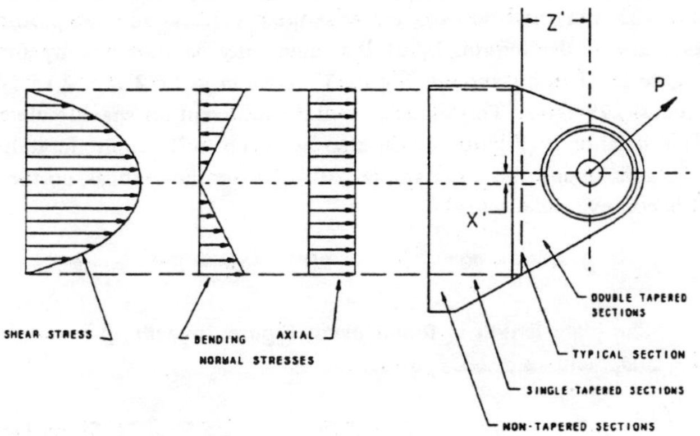

Figure 6. Typical Stress Distributions

The in-plane and out-of-plane bending moments exerted on a section located a distance z (Figure 6) from the pin hole are given by

$$M_1 = Pz'\cos(\beta) - Px'\sin(\beta) \qquad (21)$$

and

$$M_2 = P_t z' \qquad (22)$$

where x' and z' are as shown on Figure 6 and P_t is the transverse (i.e., side) load. The American Petroleum Institute (1989), paragraph 2.4.2a, recommends 5% of the sling load P as a suitable side load.

The elemental normal stresses which are distributed linearly along a section may be expressed for points along the section by

$$\sigma_1 = \frac{M_1(x + x')}{I_1} + \frac{M_2 y}{I_2} + \frac{P\sin(\beta)}{hT} \qquad (23)$$

where h is the length of the cross section under consideration.

SHEAR STRESS

The shear stress in the non-tapered portion of the padeye may be computed by the classical parabolic stress distribution formula; however, the shear stress in the tapered section of the lifting eye is much more complicated. An equation estimating the shear stress in a tapered beam undergoing axial and bending forces has been derived by Russo and is given in Gelpi, Cayll, Coyne and Russo (1990) and Russo and Garic (1992) and can be used for the tapered portion of the lifting eye. This equation is

$$\tau_1 = \frac{1.5P\cos(\beta)}{Th} - \frac{6Px^2\cos(\beta)}{Th^3} - \frac{3M_1\tan(\gamma)}{Th^2} + \frac{36M_1x^2\tan(\gamma)}{Th^4} + \frac{2Px\sin(\beta)\tan(\gamma)}{Th^2} \qquad (24)$$

where h is the length of the cross section under consideration and γ is the taper angle for the respective side.

The average out-of-plane stress is given by

$$\tau_2 = \frac{P_t}{Th} \qquad (25)$$

The above stresses should satisfy the following unity checks

$$\frac{\sigma_1}{2F_a} + \sqrt{\left(\frac{\sigma_1}{2F_a}\right)^2 + \left(\frac{\tau_1}{0.6F_y}\right)^2 + \left(\frac{\tau_2}{0.6F_y}\right)^2} \le 1 \qquad (26)$$

and

$$\sqrt{\left(\frac{\sigma_1}{0.8F_y}\right)^2 + \left(\frac{\tau_1}{0.4F_y}\right)^2 + \left(\frac{\tau_2}{0.4F_y}\right)^2} \le 1 \qquad (27)$$

COMPRESSION

When a portion of the padeye undergoes an axial compressive stress then that portion should be treated as a hypothetical column with an appropriate Kl/r_g. The effective width (w_e) of the column is taken to be that portion of the padeye plate cross section that is undergoing compressive stresses. For unstiffened elements under compression, the AISC Manual of Steel Construction (1989), paragraph B5.1, suggests a width of

$$w_e \le \frac{127\,T}{\sqrt{F_y}} \qquad (28)$$

The length (l) of the column is the distance from the cross section under question to the padeye attachment structure. The effective length factor (K) can be assigned a value of unity if the padeye has stiffeners and can be taken conservatively to be 2.1 if there are no stiffeners (AISC, 1989, paragraph C-C2).

The radius of gyration is given by

$$r_g = \sqrt{\frac{I_c}{w_e T}} \qquad (29)$$

The allowable stress for the compressive loading with Kl/r_g greater than C_c is given by

$$F_a = \frac{12\pi^2 E}{23(Kl/r_g)^2} \qquad (30)$$

If Kl/r_g is less than C_c the allowable stress is reduced to

$$F_a = \frac{\left[1 - \frac{(Kl/r_g)^2}{2C_c^2}\right]F_y}{\frac{5}{3} + \frac{3(Kl/r_g)}{8C_c} - \frac{(Kl/r_g)^3}{8C_c^3}} \tag{31}$$

where

$$C_c = \sqrt{\frac{2\pi^2 E}{F_y}} \tag{32}$$

MAIN PLATE WELD

If only fillet welds are used to attach the padeye to the support structure, the minimum main plate fillet weld size may be expressed as

$$w = \frac{2.35T\sqrt{\sigma_1^2 + \tau_1^2 + \tau_2^2}}{\sigma_w} \tag{33}$$

SAMPLE DESIGN ANALYSIS

Design a padeye that can be used to lift 500 tons (454,000 kg). The load will be lifted from the four corners of the structure (i.e. n = 4) and the load impact factor, F, should be 2. Use ASTM A36 steel for the padeye material [F_y = 36 ksi (248 MPa)]. Assume that the slings are sized for a lifting angle, α, of 50° and the general configuration is as shown in Figure 7.

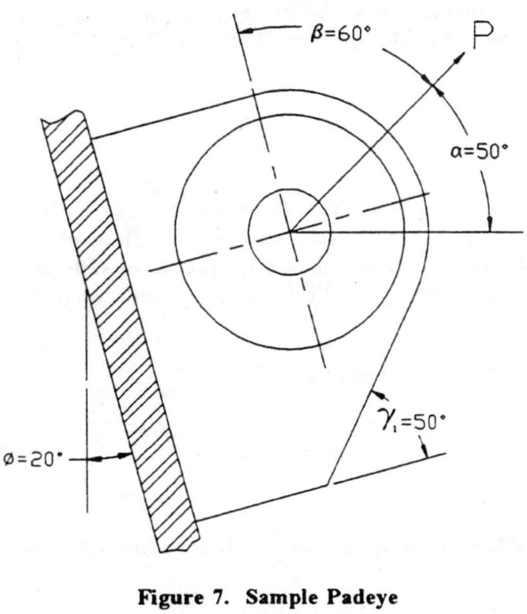

Figure 7. Sample Padeye

For these parameters the design load is (equation (1))

$$P = F\frac{W}{n\sin(\alpha)} = (2)\frac{2(500)}{(4)\sin(45^o)} = 707 \ kips \quad (3\ MN)$$

The 707 kip (3 MN) load calls for a 350 ton shackle, and from Table 1, the pin diameter, D_P, will be approximately 5½ inches (140 mm). Consult a manufacturer's catalog for more precise dimensions.

The thickness required to sustain the pin bearing is estimated from equation (2) to be

$$T + 2t \geq \frac{P}{0.9F_y D_P} = \frac{707}{(0.9)(36)(5.5)} \approx 4 \ inches \quad (102\ mm)$$

Two 1 inch (25 mm) thick cheek plates along with a 2 inch (50 mm) main plate will be used. Thus T = 2 inches (50 mm) and t = 1 inch (25 mm) will be used in subsequent calculations. A preliminary estimate of the length, L, of the plate may be obtained by utilizing Figure 2. The distance b (Figure 1) is taken to be $2D_P$ = 2 (5.5) = 11 inches (279 mm). This distance will be sufficient for shackle clearance. The distance a (Figure 1) will also be taken to be approximately $2D_P$. The attachment section is sloped 20°, ϕ, from the vertical, so the angle β becomes (equation (4))

$$\beta = \phi + 90^o - \alpha = 20^o + 90^o - 50^o = 60^o$$

The plate length is found using Figure 2b with β = beta = 60°, b = 2 D_P and a = 2 D_P to be

$$\frac{L}{D_P} \approx 5.2 \quad or \quad L = 5.2D_P = 5.2(5.5) = 28.6 \ inches \ ($$

Using equations (3) and (5), the minimum length must satisfy L ≥ 5.155 D_P = 28.4 inches (720 mm) and L ≥ 4.868 D_P = 26.8 inches (680 mm), respectively. If the maximum normal stress criterion is used to establish minimum lengths instead of the maximum shear stress criterion, then it can be shown that the minimum length must satisfy L ≥ 3.866 D_P = 21.3 inches (540 mm) and L ≥ 4.868 D_P = 26.8 inches (680 mm).

The radius of the cheek plates will be taken to be 1.5 times D_P (i.e., r = 8.25 inches [210 mm]). The minimum fillet weld size for attaching the cheek plates is computed with equation (9) using an electrode tensile strength σ_w = 70 ksi (483 MPa).

$$c \geq \frac{Pt}{1.33r(T+2t)\sigma_w} = \frac{(707)(1)}{1.33(8.25)(4)(70)} = 0.23 \ in \quad (5.9\ mm)$$

Table J2.4 of the AISC Manual of Steel Construction (1989) specifies a minimum fillet weld size of 5/16" (8 mm) for plates over 3/4" (19 mm) thick. Consequently a 5/16" (8 mm) fillet weld would be the minimum acceptable size, i.e., c = 5/16" (8 mm).

The diameter of the pin hole, D, will be taken to be 0.5 inches (12.7 mm) greater than the pin (approximately 10% larger) to provide sufficient clearance for easy assembly. Therefore D = D_P + 0.5 = 5.5 + 0.5 = 6 inches (152 mm).

[Note: AISC paragraph D3.2 states (in part): "For pin-connected members in which the pin is expected to provide for relative movement between connected parts while under full load, the diameter of the pin hole shall not be more than 1/32 in. (0.8 mm) greater than the diameter of the pin."]

The radius of the main plate, R, may now be determined by utilizing equation (12)

$$R \geq \frac{D}{2} + \frac{P}{2(0.4F_yT)} - \frac{(2r-D)t}{T}$$

$$= \frac{6}{2} + \frac{707}{(2)(0.4)(36)(2)} - \frac{[2(8.25)-6](1)}{2}$$

$$= 10 \ inches \quad (254 \ mm)$$

The tear areas must now be checked. The first tear area is computed using equation (15)

$$A_{t1} = L_{t1}T = \left[2R + (r+c)(\gamma_1 + \gamma_2)\frac{\pi}{180}\right]T$$

$$= \left[2(10) + \left(8.25 + \frac{5}{16}\right)(50+0)\frac{\pi}{180}\right](2) = 55 \ in^2 \quad (35500 \ mm^2)$$

where the taper angles, γ_1 and γ_2, have been taken to be 50° and 0°, respectively (Figure 7).

In order to check the second tear area the proper edge distance, *e*, must, must be established. The edge distance, *e*, must be the lesser of the following two calculations (equation (16))

$$e \leq 0.8D = 0.8(6) = 4.8 \ inches \quad (122 \ mm)$$

or (equation (17))

$$e \leq 4T = 4(2) = 8.0 \ inches \quad (203 \ mm)$$

thus e = 4.8 inches (122 mm) will be used in subsequent calculations. The effective radii for the main plate and cheek plates is the smaller of (equations (18))

$$R_e = R = 10 \ inches \quad (254 \ mm)$$

$$\text{or} \quad R_e = e + \frac{D}{2} = 4.8 + \frac{6}{2} = 7.8 \ inches \quad (198 \ mm)$$

and (equation (19))

$$r_e = r = 8.25 \ inches \quad (210 \ mm)$$

$$\text{or} \quad r_e = R_e = 7.8 \ inches \quad (198 \ mm)$$

Thus r_e = 7.8 inches [198 mm] and R_e = 7.8 inches (198 mm) will be utilized in subsequent calculations. From equation (20) the net section area is

$$A_{t2} = T(2R_e - D) + 2t(2r_e - D)$$

$$= (2)[2(7.8)-(6)] + 2(1)[2(7.8)-(6)]$$

$$= 38 \ in^2 \quad (24500 \ mm^2)$$

Areas A_{t1} and A_{t2} must be greater than

$$A_t \geq \frac{P\sin\beta}{0.6F_y} = \frac{707(0.866)}{0.6(36)} = 28 \ in^2 \quad (18300 \ mm^2)$$

Using the above dimensions other critical points on the plate should be checked.

SUMMARY AND CONCLUSIONS

The paper presents a philosophy of analysis and provides suitable equations that can be used to design padeyes. The design procedure utilizes simple beam theory to predict the appropriate stresses.

ACKNOWLEDGEMENTS

This paper is an extension of the work done by S. Gelpi and D. Cayll of Chevron Research and Technology Corporation and E. Russo, M. Coyne and G. Garic of the University of New Orleans. The authors acknowledge the support given by the NASA John C. Stennis Space Center's University Affairs Office, the Propulsion Test Directorate, and the National Research Council's Research Associateship Program.

REFERENCES

American Institute of Steel Construction, (1989), Manual of Steel Construction, Allowable Stress Design,

American Petroleum Institute, (1989), Recommended Practice for Planning, Design, and Constructing Fixed Offshore Platforms, RP 2A,

Gelpi, S. L., (1976-1990), Personal notes and calculations,

Gelpi, S.L., Cayll, D.R., Coyne, M.J. and Russo, E.P., (1990), PADEYE, Computer Program Description and User Manual, Version 2, Chevron Research and Technology Co., New Orleans, Louisiana, October,

Russo, E.P. and Hall, R.A., (1977), "Systematic Approach to Lifting Eye Design," Ocean Resources Engineering, November, pp. 33-38,

Russo, E.P., Gelpi, S.L., Cayll, D.R. and Garic, G., (1992), "Padeye Design and Analysis," Proceedings of the Second International Offshore and Polar Engineering Conference, San Francisco, June 14-19, pp. 326-330,

Russo, E.P. and Garic, G., (1992), "Shear Stress Distribution in Symmetrically Tapered Cantilever Beams," ASCE Journal of Structural Engineering, vol. 118, No. 11, November, pp 3243-3249.

Proceedings of the Seventh (1997) International Offshore and Polar Engineering Conference
Honolulu, USA, May 25-30, 1997

ISBN 1-880653-28-1 (Set); ISBN 1-880653-29-X (Vol. I)

The Environmental Consequences of Large-Scale Research in the Deep Ocean

Hjalmar Thiel
Alfred Wegener Institute for Polar and Marine Research
Bremerhaven, Germany

Anthony L. Rice
Southampton Oceanography Centre
Southampton, United Kingdom

ABSTRACT

The European Union called for proposals for "the assessment of any possible risk likely to affect the marine environment in association with research, monitoring and surveying in marine sciences and technologies". An assessment was made by a small group of deep-sea ecologists, stating that normal scale research would impact the environment to a negligible extent. Future requirements for ecological deep-ocean research related to the use of the deep for waste disposal and mining of metal-rich ores were also presented. Experimental large-scale research for commercial-scale impact assessment as well as pre-commercial pilot operations are discussed. It is concluded that studies of this type are still of an order of magnitude that is acceptable and unlikely to cause significant impacts.

KEY WORDS: Deep sea, environment, risk through research, mining, waste disposal.

INTRODUCTION

Serious research in the open ocean began in the 1870s with the circumnavigation of the British naval surveying vessel HMS Challenger. The sample and data gathering techniques used at that time were, of course, very different from those used today. But the scale of sampling and the broad approach adopted during the Challenger Expedition were not dramatically different from those of the late twentieth century. Admittedly, the spatial sampling regime during the Challenger and many subsequent research cruises, in which individual sampling localities were separated by tens or even hundreds of kilometres, is not typical of modern cruises. In the last thirty or forty years, the increasing realisation that the deep ocean is much more variable, physically, chemically, geologically and biologically, and at a wide range of spatial and temporal scales, has led to multiple sampling within restricted areas and repeated visits to the same locality at different times. Nevertheless, because of the small size of individual samples in relation to the size of the environment studied, the environmental impact of this conventional research must be trivial. For example, the total area of seafloor sampled "destructively" by all the gears used by oceanographers in the 125 years since the Challenger, trawls, dredges, corers and grabs, amounts to no more than a few square kilometres, a tiny fraction of one percent of the more than 300 million square kilometres of the deep ocean. But in recent years, oceanographers have begun to question whether this 'traditional' approach is adequate to answer some of the questions about the oceans that are now being posed. As we have already noted, even to understand how the natural system functions it is clear that more intense spatial and temporal sampling and data gathering is necessary. However, a much more difficult problem is that of predicting the likely environmental consequences of anthropogenic impacts. The consequences of general impacts such as global warming or the build-up of atmospheric carbon dioxide are, of course, of great potential importance. Detecting or predicting them may require very precise data collection and analysis, but not significantly different from the techniques used in conventional research. In contrast, predicting the impacts to be expected from direct human intervention in the deep-sea, such as metal extraction or sludge disposal, by extrapolation from conventional research results, including square metre-scale recolonisation experiments, is difficult if not impossible. Accordingly, some deep-sea scientists believe that in order to make such predictions it may be necessary to conduct much larger scale experiments.

One such experiment, DISCOL (Disturbance and re-colonisation experiment), is an attempt to simulate at least some of the probable effects of manganese nodule mining. In this experiment (Schriever et al. 1997) some 10 square kilometres of the seafloor at a depth of 4150m in the southeastern Pacific was physically disturbed by ploughing, using a modified farm harrow, and the resulting recovery of the area has been monitored at intervals ever since. Although the data obtained during this project will certainly be valuable, their relevance to the much larger scale physical disturbance to be expected from a commercial operation, of one square kilometre of seafloor destruction every day for up to twenty years, is questionable. Moreover, even if they were, DISCOL simulates only part of the potential impact of commercial mining and does not, for instance, address the effects of the inevitable large mud plume which would be produced. In order to address this impact further, large-scale experiments have been conducted by blowing large amounts of sediment into the water column, observing their resettlement and its influence on the fauna (Fukushima 1995, Kaneko et al. 1997, Trueblood & Ozturgut 1997). These experimental studies influenced several square kilometres of seafloor. In other possible activity areas, either of resource extraction or waste disposal, no impact experiments have yet been conducted. A recent desk study, RISKER (Environmental risk from large-scale ecological research in the deep sea. A desk study), funded by the European Union under the MAST II programme, has recently investigated this general area.

In 1993 the European Union published a call for proposals (93/C203/19) asking for "the assessment of any possible risk likely to affect the marine environment in association with research, monitoring and surveying in marine sciences and technologies". Funded under this programme, RISKER's overall objective was to identify the types and scales of research that would be needed to predict the impact of a range of deep-sea activities and to asses the likely environmental impact, and therefore acceptability, of the research itself. As a result of the RISKER study we suggest that future scales of ecological research will range from corer, grab and trawl samples for the benthos and water bottle, pump and plankton net sampling for the mid-water realm to the experimental creation of large-scale impacts for the evaluation of commercial uses of the deep sea. These potential industrial activities include both resource extraction and waste disposal and the assessment of their environmental impacts will require research at a variety of scales.

Scales of environmental deep-sea research

Possible deep sea exploitable resources include manganese nodules and crusts, phosphorites and massive sulphides, and metaliferous muds. Although the scales of operations and the technologies needed to recover these resources vary, they share a number of common features. They would all involve severe physical disruption of the sea floor over relatively large areas, and they would all also result in extensive mud plumes in the water column. While plume phenomena can be modelled to some extent using available data, we suggest that in order to assess the likely environmental consequences it will also be necessary to conduct and monitor relatively large-scale in situ simulations. The result from a single experiment may provide data relevant to a range of different activities, but it may be necessary to conduct several large-scale experiments in order to cover the full range of possible impacts.

The use of the deep ocean for the disposal of waste is rather different. The wastes considered range from sewage sludges and dredge spoils, through radioactive wastes and munitions, to large structures such as redundant offshore oil and gas installations, and finally to industrial excess carbon dioxide in liquid or frozen form. Each of these categories would probably involve the use of unique technologies, and would certainly result in quite different impacts depending upon the consistency and volume of the waste as well as its toxicity. Because of these differences, and varying socio-political attitudes towards them, we suggest that the types of research required to investigate their environmental impacts would also have to be different.

In some cases, for instance the disposal of radioactive waste, a realistic in situ experiment in the current socio-political climate would be unthinkable, any more than would renewed deep sea disposals of such waste. Instead, we suggest that the existing emplacements of such wastes should be regarded as if they were scientific experiments, and they should be monitored using current conventional research techniques, or even ones yet to be developed, but at similar scales to those traditionally used in the past close-up observations of the storage containers and targeted sampling in their immediate neighbourhood seems essential. The most appropriate technique would be the deployment of remotely controlled vehicles (ROVs) or manned submersibles. The results could certainly be of considerable purely scientific interest, but might also be relevant to other anthropogenic impacts in, for example, elucidating transport rates using the radioactive wastes as tracers. In the context of RISKER, the environmental impacts resulting from by such research activities would remain rather limited, although ROVs and submersibles may impact several square meters of sea floor by working close to it.

A similar situation applies to munitions dumping, but in this case dumping localities have been kept confidential because of their highly political nature. However, oceanographers should have access to these sites to inspect and measure the impacts on the environment. Nevertheless, the research techniques needed would be of the conventional scale and therefore, we believe, would have no significant and unacceptable environmental impact.

Recent events have placed the disposal of large structures in much the same situation as radioactive waste and munitions disposal. The events surrounding the BRENT SPAR suggest that any attempt within the foreseeable future to carry out a similar operation, either commercially or "experimentally", would encounter very strong opposition. We therefore suggest that in order to investigate the possible impacts of such events, unlikely though they may be, it would be necessary to study existing natural or man made analogues. These range from natural oil and gas seeps, through hydrothermal vents, to existing (or future) shipwrecks. Again, the research techniques needed, either currently available or to be developed in the future, would be at the traditional research scales, including the use of ROVs and submersibles, and therefore would again have no significant environmental impact themselves.

Whereas for these deep ocean disposals environmental impacts can be deduced from earlier experience or disasters, this is not the case for other potential candidates for deep ocean storage: sewage sludge, dredge spoils, and carbon dioxide. Sewage sludge from the city of New York has been dumped at the so-called Dumpsite 106 in the western North Atlantic Ocean at a depth of 2000m. The settlement of the particulate matter, its transport and distribution on and along the seafloor have been monitored and the results published (e.g. Van Dover et al. 1992, Takizawa et al. 1993, Bothner et al. 1994). This experience may help to predict abyssal disposal to some extent, but the true effects cannot be extrapolated from this continental slope situation.

Therefore, the large-scale experimental approach is a *conditio sine qua non*.. Only such experiments will allow appropriate environmental impact assessments. Techniques of transport, deep-sea emplacement, and the monitoring programme are likely to be similar for sewage sludge and dredge spoils, and the same large-scale experimental design may be sufficient for both types of waste materials. In contrast, the introduction of carbon dioxide into deep water layers or into sea floor depressions will need specific research efforts.

But for both situations a stepwise scaled approach is necessary to evaluate different ecological levels and to avoid excessive costs. For evaluating the effects of carbon dioxide at the species, population, and community levels, four steps should be involved: laboratory tests on the reactions of individual organisms to increased CO_2 concentrations under aquarium (1) and mesocosm (2) conditions; under natural conditions with a medium-scale experiment contaminating a fjord system for a restricted period (3); and finally the ultimate large-scale experiment by disposing of a large volume of carbon dioxide into the deep ocean (4). Steps 3 and 4 should be based on extensive modelling of the oceanographic conditions to design the experiment at scales warranting results with predictive values for economic scale disposals. The steps suggested for sewage sludge and dredge spoil should also involve four stages. The first of these would be a pre-experimental and design phase in which a prospective disposal site and a control site would be surveyed, the disposal technique selected and some preliminary modelling work undertaken. The second phase would be a relatively small in situ experiment in which the equivalent of a single load of a bulk carrier would be emplaced within an area with a radius of about 1 kilometre and monitored for about one year. If the impact at this stage is acceptable the third phase would involve the emplacement of about 1 million m^3 of waste in a similarly restricted area over a period of about one year and then monitored for a further year. Again, if the impact at this stage is still acceptable the fourth phase would involve annual emplacements of about one million tonnes each year for ten years.

The potential impacts of deep-sea mining must be evaluated by large-scale mining or simulation experiments. Techniques for mining of manganese nodules and metalliferous muds (Thiel et al. 1991) are available, although if and when mining starts in a few decades, more effective methods may be employed. Prepilot Mining Tests (PPMTs) have been conducted already in the late seventies, and environmental studies accompanied these tests. The results led to the new approach of large-scale experiments (Thiel 1991) already mentioned. For mining other ore types no techniques have been designed, but suitable methods may be deduced from those already known. However, the resulting environmental impacts are likely to vary according to the ore type and local ocean conditions. Therefore, specific large-scale experiments must be foreseen in association with each industrial intrusion into the deep sea. Moreover, the mining of phosphorites and the associated environmental impacts must be considered separately from the other ores since they occur in relatively shallow areas of the deep sea and on continental shelves, specifically in upwelling areas. Plumes created by mining activities, particularly in these regions, might be transported to large distances by the prevailing currents, and they might reach the highly productive surface waters or interact with fish stocks and their larvae.

Careful impact predictions are of importance for all these mining activities. Only large-scale experiments, which allow extrapolation of the results to commercial-scale mining, will provide adequate information.

Is large-scale research in the deep sea acceptable?

We believe that conventional deep sea research has not had and will not have an unacceptable impact either on the bottom or in mid-water. The disturbance by conventional sampling, observational or recording methods remains so small in relation to the large areas and volumes available that even the slow pace of life in the deep sea will repair any damage in relatively short times. "Normal"-scale biological deep ocean research is therefore excluded from further consideration.

However, deep-sea ecologists are faced with new dimensions of their research as man penetrates the most remote and inaccessible space on Earth. Ethically there appears to be no difference between exploitation of terrestrial or extraterrestrial realms and under water space. Nevertheless, it is justifiable to ask whether large-scale research approaches would exert unacceptable risks for sustainable developments. The new experimental approaches have already involved deep sea areas of 10 or more square kilometres. Such sites have received an additional sediment cover of a few millimetres from material disturbed in their immediate neighbourhood or from some distance upstream, but still from the same general ecological setting. In other places the sediment surface has been severely mixed and turned over, and the benthos community has certainly been strongly shifted to lower standing stocks and different dominance ratios. Although reestablishment of communities does occur, this takes considerable time (Bluhm & Thiel 1996, Borowski & Thiel 1996). In the case of severe habitat change, for example the burial of manganese nodules below the sediment surface, hard substrate epi- and crevice fauna cannot return, but their ecological space will be occupied by species of the softbottom subcommunity, now spreading over the total area.

No specific criteria are available to discern between acceptable and unacceptable environmental impacts. This poses problems in environments which are much better known than the deep sea. For this little known space some rather general principles may be summarised:

- Persistent adverse effects on regional biodiversity should not be accepted.
- No species should be driven to extinction.
- Ecological processes should not be impaired. This pertains specifically to those processes with climatic relevance and the transport of toxic substances.
- Living resources should not be adversely effected.
- Interference with other ocean uses should not occur.
- Risks for keystone communities and/or processes and even to human safety must be excluded.

Large scale experiments may be undertaken as purely scientific investigations, or they may be conducted in conjunction with industrial developments and tests. In deep ocean mining the Pilot Mining Operation (PMO) phase is the first developmental step which, probably under reduced scales, will lift larger masses of ore and tailings to the surface. Whereas the scientific experiments causing of the order of 10 square kilometres of disturbance may easily meet the above criteria of acceptability, this may not be true for a PMO. During this phase, large masses of material will be transported from the seafloor to the ocean surface and the tailings will be discharged at a depth to be determined according to the local conditions. Although the techniques to be employed are not known, the extent of this pre-commercial phase should also meet the criteria. The duration of such PMOs would be limited to a few months, and probably will be executed at only about one fifth of the scale of subsequent commercial mining. It seems likely, therefore, that the acceptability criteria will be met by such action. However, oceanographers of all disciplines must carefully monitor such PMO events to ensure that the environmental impacts, and therefore acceptability, of commercial scale mining can be predicted accurately.

Analogous to the PMO concept for mining, the RISKER study introduces the term Pilot Disposal Opertion (PDO). As indicated above, a stepwise experimental design should be the approach to waste disposal with the acceptability at each stage demonstrated unequivocally before proceeding to the next stage.

In summary, the RISKER study concludes that traditional scale research activities and the large-scale scientific experiments, as conducted recently, meet the environmental acceptability criteria. Because of their limited extent in space and time, it is believed that PMO and PDO commercial tests would also be acceptable although the specific techniques which are likely to be used for mining and disposal cannot yet be predicted. Reconsiderations in the early phases of any future developments will be essential before decisions are made to proceed with PMOs and PDOs.

REFERENCES

Bluhm, H and Thiel, H (1996). "Photographic and video surveys for large scale animal and seafloor surface charting aiming at ecological characterization of habitats and communities," *Proceedings of the International seiminar on deep sea-bed mining technology*, Oct. 18 - 20, 1996, COMRA, Peking, PR China, pp C15-C23.

Borowski, C and Thiel, H (1996). "Indicator taxa for dominant deep-sea animal groups in their importance for comparative biodiversity and environmental studies," *Proceedings of the Inernational seminar on deep sea-bed mining technology*, Oct. 18 - 20, 1996, COMRA, Peking, PR China, pp C35 -C43.

Bothner, MH, Takada, H, Knight, IT, Hill, RT, Butman, B, Farrington, JW, Colwell, RR and Grassle, JF (1994). "Sewage contamination in sediments beneath a deep-ocean dump site off New York," *Mar. Environm. Res.*, 38, pp 43 - 59.

Fukushima, T (1995). "Overview "Japan Deep-sea Impact Experiment = JET," *Proceed. First ISOPE Ocean Mining Symposium*, Tsukuba, Japan, pp 47-53.

Kaneko, T, Maejima, Y, and Teishima, H (1997). "The abundance and vertical distribution of abyssal benthic fauna in the deep-sea Japan Environmental Impact Experiment," *Proceed, Seventh Intern. Offshore Polar Engineering Conference*, Honolulu, Hawaii, in print.

Schriever, G, Ahnert, A, Bluhm, H, Borowski, C, and Thiel, H (1997). "Results of the large-scale deep-sea environmental impact study DISCOL during eight years of investigation," *Proceed. Seventh Intern. Offshore Polar Engineering Conference*, Honolulu, Hawaii, in print.

Takizawa, M, Straube, WL, Hill, RT, and Colwell, RR (1993). "Near bottom pelagic bacteria at a deep-water sewage sludge disposal site", *Appl. Environm. Microbiol.*, pp 3406-3410.

Thiel, H, Foell, EJ, and Schriever, G (1991). "Potential environmental effects of deep seabed mining," *Ber. Zentrum Meeres. u. Klimaforsch., Universitat Hamburg*, 26, pp 243.

Thiel, H (1991). "From MESEDA to DISCOL: a new approach to deep-sea mining risk assessments." *Mar. Mining*, 10, pp 369-386.

Trueblood, DD, and Ozturgut, E (1997). "The Benthic Impact Experiment: A Study of the Ecological Impacts of Deep Seabed Mining on Abyssal Benthic Communities," *Proceed. Seventh Intern. Offshore Polar Engineering Conference*, Honolulu, Hawaii, in print.

Van Dover, CL, Grassle, JF, Fry, B, Garritt, RH, and Starczak, VR (1992). "Stable isotopic evidence for entry of sewage-derived organic material into a deep-sea food web," *Nature*, 360, pp 153-156.

Proceedings of the Seventh (1997) International Offshore and Polar Engineering Conference
Honolulu, USA, May 25-30, 1997

ISBN 1-880653-28-1 (Set); ISBN 1-880653-29-X (Vol. I)

Results of the Large Scale Deep-Sea Environmental Impact Study DISCOL During Eight Years of Investigation

Gerd Schriever and Ahmed Ahnert
BIOLAB Forschungsinstitut
Hohenwestedt, Germany

Hartmut Bluhm, Christian Borowski and H. Thiel
Alfred-Wegener-Institut fur Polar- und Meeresforschung
Bremerhaven, Germany

ABSTRACT

DISCOL, the first large scale impact experiment in the deep sea, was initiated in 1988 together with the Federal Ministry of Science and Technology of Germany. After severe experimental disturbance of the seabed in a manganese nodule field in early 1989, the recolonization of the benthos in the DISCOL Experimental Area (DEA) was monitored over a seven year period and compared with the baseline information on the community gained immediately before the impact. Four randomly sampled post impact data sets were collected during RV SONNE-cruises SO 61 (immediately after the impact), SO 64 (six months later), SO 77 (after three years), and SO 106 (after seven years). Due to the impact, the abundances of all faunal taxa decreased significantly and, except for the bacteria, did not reach the values of undisturbed sediment half a year later. Three years after the impact, densities of the major faunal groups (mega-, macro-, and meiofauna) significantly exceeded those determined for the baseline study. The sample processing for the fourth post impact investigation is still in progress, but some initial results are presented herein.

Interpretation of the animal density numbers has provided information about the start and progression of the recolonization process. Statistical analyses within major animal groups, supported by taxonomic identifications, have revealed that quantitative recovery of animal densities does not necessarily imply the recovery of the faunal composition to conditions similar to an undisturbed community: The Polychaeta demonstrated aberration in their biodiversity and in the composition of their functional groups for three years. Indicator taxa for deep-sea risk assessment studies are also suggested.

KEYWORDS: environmental assessment, recolonization, deep sea, manganese nodules, ocean mining, indicator taxa

INTRODUCTION

The DISCOL (DISturbance and reCOLonization experiment in a manganese nodule area of the Southeastern Pacific) study included a long-term, large-scale DISturbance and reCOLonization experiment in

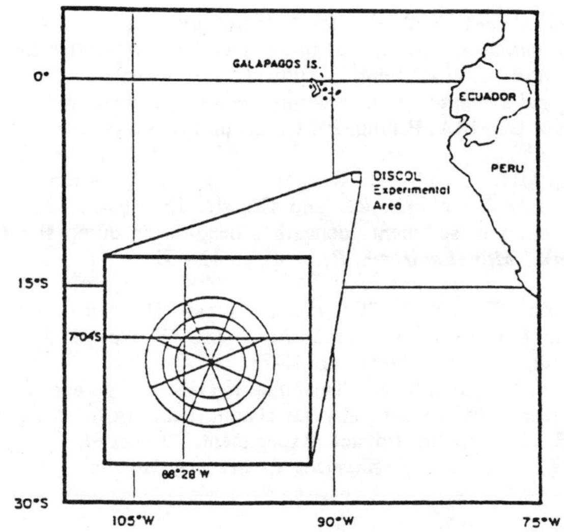

Fig. 1: The DISCOL Experimental Area (DEA) in the southeastern Pacific Ocean of Peru

the abyssal Southeast Pacific Ocean (Fig. 1). It was devised to achieve a better understanding of the rate, sequence, and direction of benthic community development after disturbance in the deep sea, similar to that expected to result from industrial usage of the deep-sea floor, e.g. mineral mining. The DISCOL-project, funded for a five years period (1988-1993) by the German Bundesministerium für Forschung und Technologie, is now succeeded by the ECOBENT (Investigation of the benthic fauna in the abyssal ecosystem of the south-east Pacific Ocean) investigations as part of the interdisciplinary ATESEPP (Auswirkungen technischer Eingriffe in das Ökosystem der Tiefsee im Süd-Ost-Pazifik vor Peru), funded by the German Bundesministerium für Bildung, Forschung, Wissenschaft und Technologie from 1995 to

1998. DISCOL field work has been in progress since February 1989, when the disturbance experiment was initiated during a two-months cruise aboard RV SONNE. Recolonization of the benthic fauna in the DISCOL Experimental Area (DEA) was monitored and compared with baseline information on the community which we obtained before the impact.

Four randomly sampled post impact data sets were collected during RV SONNE-cruises SO 61 (immediately after the impact), SO 64 (six months later), SO 77 (after three years), and SO 106 (after seven years). In this paper we report on the recolonization processes of mega-macro-, and meiofauna, as observed during three DISCOL cruises, and present first results from the fourth post impact study ECOBENT conducted seven years after disturbance.

METHODS

After site selection and baseline assessments, we marked out a circular experimental area, covering approximately 10.8km². This shape ensured optimal disturbance with our specially constructed disturbing device which we called the "plow-harrow" (Thiel and Schriever, 1989). By towing the disturber on diametrical courses through the DEA, we created the heaviest disturbances in central areas whereas the periphery was less severely impacted. For sampling purposes, we partitioned the DEA into eight pie-shaped sectors positioned along the main geographical directions. Additionally, we subdivided the area using concentric circles, with the inner circle having a radius of 1000m and a second circle at a radius of 1350m, leaving a width of 500m for the outer ring region. The positioning of the ship and the sampling and observation equipment was controlled with transponder navigation.

Megafauna was monitored by a towed photo-/videosystem OFOS (Ocean Floor Observation System) during all DISCOL and ECOBENT cruises. The distance OFOS and the sea floor was kept at approximately three meters (Bluhm, 1993, Bluhm et al., 1995). Image analysis provided information about the diversity and density of the megafauna (Bluhm, 1994; Bluhm et al., 1995; Bluhm and Thiel, lecture at the International Seminar on Deep Sea-bed Mining Technology, Oct. 18-20 1996, Beijing). Macrofauna (>500µm) was sampled by a box corer (KG, sample size: 50 x 50cm). For meiofauna we used the multiple corer (MC, sample size: 10 x 70.8cm²). Sediment sampling was planned to be conducted in the central circular area and in the outer ring area, representing the regions of maximum and minimum disturbances, respectively. The intermediately impacted middle ring area served as a buffer zone for experimental seperation and was not sampled. Due to time constraints, we did not examine each sector in the same level of intensity, but randomly selected five sample sectors in each of the two target areas. For navigational reasons, sampling positions were located at the midline in each sample sector, i. e. 500m and 1600m away from the DEA center, but usual navigation deviations resulted in a random distribution of the samples around the target position. Faunal material was obtained from three box corer samples and one multiple-corer sample at each selected location. We compared "disturbed samples", which we obtained from the plow harrow tracks, with "undisturbed" ones from areas between the tracks. This rather formal sampling scheme was applied during the pre-impact study and the three post-impact investigations during DISCOL cruises. During the later DISCOL cruises we also sampled an unaffected reference site three nautical miles to the south of the DEA .

In order to minimize our sampling effort during ECOBENT, we selected specific sample areas with high track densities, as well as undisturbed areas inside the DEA and at the reference site, instead of repeating the DISCOL scheme. This was achieved by equiping the box and the multiple corers with a video camera for the visual control during targeted sampling. Since many statistical tests require equal sized (statistical) samples, we decided to obtain a minimum of 8 box corer samples for macrofauna and 2 multi-corer samples (with at least 2 cores each) for meiofauna analyses from each of the "disturbed DEA", "undisturbed DEA", and "reference site" categories.

Detailed descriptions of all four cruises including the gear and methods are given in Thiel and Schriever (1989), Schriever (1990), Schriever and Thiel (1992), and Schriever et al. (1996).

RESULTS

Environment

Heavy impacts were created within and adjoining the plow tracks, which covered approx. 20% of the DEA. The semi-liquid uppermost sediment material was removed from the surface, and the exposed clayish sediments from deeper layers initially formed sharp edged track contours. The sediment cloud created by the disturber passage had resettled and blanketed large areas of the DEA with various thicknesses (Bluhm et al., 1995). Soft surface sediments, transported by water currents, progressively refilled the artificial sea bed depressions along the plow-harrow tracks and smoothened their relief with passing time. As soon as three years after impact, the thickness of the semi-liquid upper layer in some disturbed sediment samples exceeded natural conditions (Borowski, 1996). However, the tracks remained clearly recognizable for the entire period of investigation (Bluhm et al., 1995; Schriever et al., 1996). In the sediment samples, the disturbance impact was recognizable by the presence of lighter colored carbonate-rich clay and uneven sample surfaces, even after seven years had passed.

Megafauna

Images from the baseline study (Thiel and Schriever, 1989) and from later observations which were conducted outside of the DEA represent the undisturbed situation of the animal community. The megafauna densities in these undisturbed areas remained on a more or less constant level during the entire period of investigation. This indicated that, apart from the experimental impact, no detectable ecological factors had affected the megafauna.

The plowing destroyed or buried the majority of the megafauna in the tracks (Fig. 2). Half a year later, highly motile animals had repopulated these areas (Bluhm et al., 1995). Three years after the impact, megafauna densities in the tracks reached the level of the baseline study. However, the diversity was still low, and highly mobile animals like Ophiuroidea, Holothurioidea, and Crustacea (mainly *Probeebei mirabilis*) were dominant. Less motile animals, e.g. Lophenteropneusta, Asteroidea, and Actiniaria, occured only in the image material from the ECOBENT expedition in 1992. At this time the animal densities had declined to levels lower than those of the baseline study. The plowing had buried the manganese nodules and thus had in effect removed the hard substrate from within the disturber tracks. Since hard bottom animals such as Porifera and Crinoidea did not survive this procedure, and were also unable to recolonize the nodule free areas again, the megafauna in the tracks changed from the typically mixed to a pure soft bottom assemblage.

In the between track areas of the DEA, the megafauna densities remained on the baseline level during the early post-impact

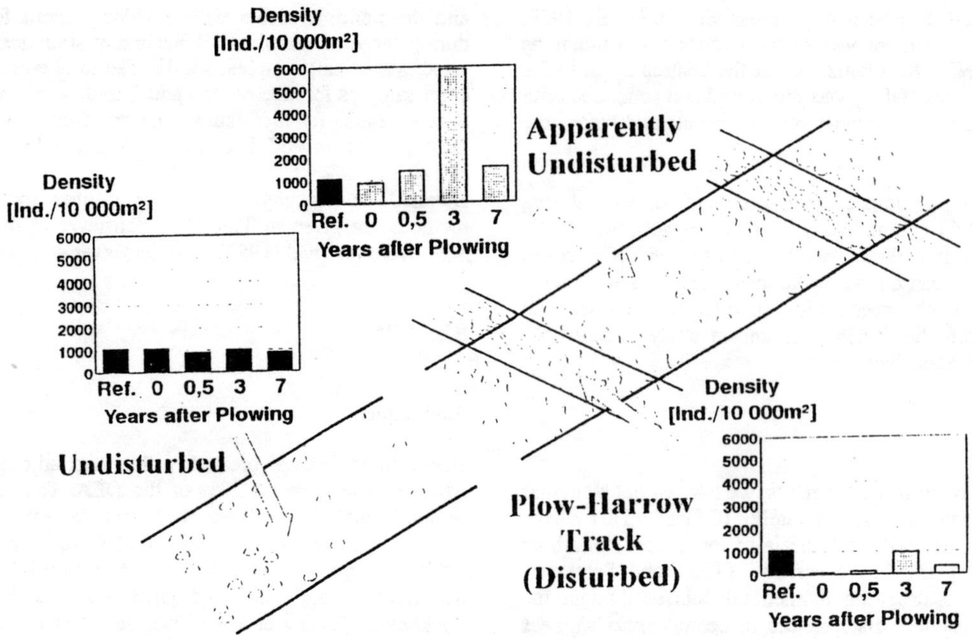

Fig. 2: Recolonization of the megafauna in the DEA between 1989 and 1996. Undisturbed: baseline data (= Ref.) and areas outside the DEA. Apparently undisturbed: Areas within the DEA, which were not directly influenced by the disturber, but showed different degrees of resedimentation. Disturbed: Plow-harrow tracks. Filled bars represent data from the baseline study.

investigations. Three years after the impact, however, the abundances, especially of highly mobile soft bottom animals, increased remarkably (Bluhm et al., 1995). We conclude, that increased numbers of smaller animals (see below) may have provided enhanced food availability.

First results from the ECOBENT expedition seem to show further declines in megafauna densities, compared to the previous cruise. Additional analyses of the available image material is expected to provide more detailed information.

Macrofauna

Polychaeta, Tanaidacea, Isopoda, and Bivalvia dominated the macrofauna and together included more than 90% of all animals in this size class in the baseline and refererence samples. Immediately after the plowing disturbances, these groups were significantly reduced in disturbed samples, and macrofauna density decreased to 38.6% of that in undisturbed sediments (Fig. 3).

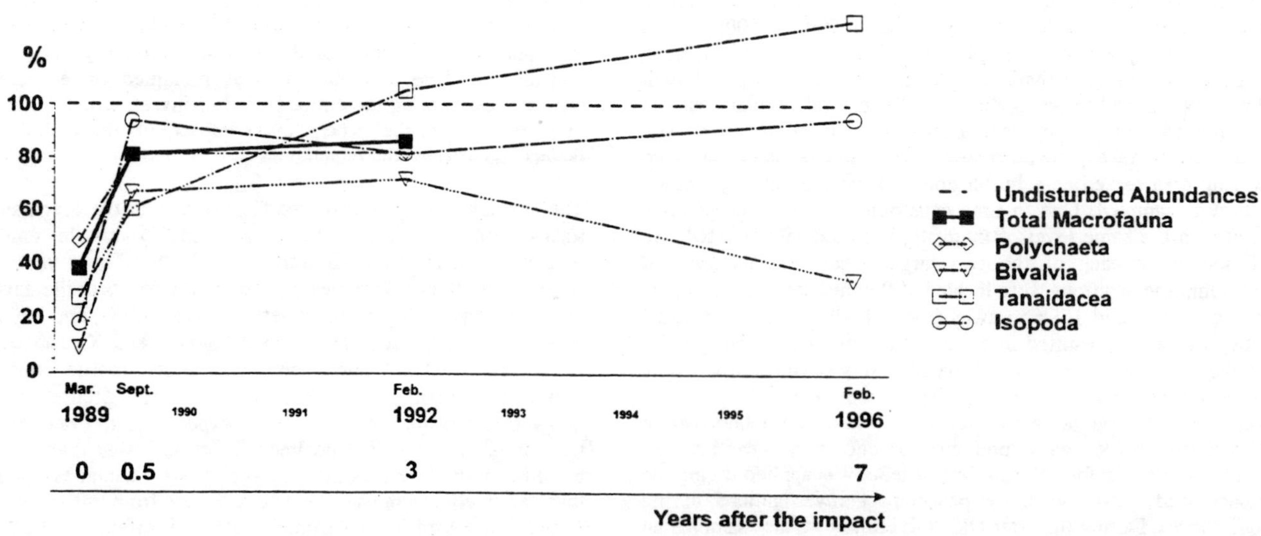

Fig. 3: Development of the relative macrofauna abundances in disturbed sediments during a seven years period of investigation. Abundances of the total macrofauna and of the four dominant taxa are displayed as percentages of undisturbed values for each time of observation. Undisturbed values are given as hatched 100% line. The symbols indicate mean values of all disturbed samples from the respective expedition (0 years: N=7, 0,5 years: n=9, 3 years: n=9, 7 years: n=4). Seven years data for the Polycheata and the total macrofauna are not yet available.

The intensity of the impact on the different groups can be regarded as a function of the average dwelling depths in the sediment. Shallow-dwelling Bivalvia were almost extirpated (9.8%), whereas the impact had a lesser effect on the deeper-dwelling Polychaeta (48.6%). However, a less harmful effect on the Polychaeta may also reflect the high motility of these animals. This feature can have facilitated rapid initial recolonization of free habitat resources in the plow tracks from the unplowed between-track areas.

The recovery started rapidly. Only half a year after impact, the macrofauna densities in the disturbed samples rose to 81.8% of the undisturbed values, and for the following years, further leveling between the different sample types was expected. Three years after the impact, however, balanced abundances between disturbed and undisturbed samples were observed only for the Tanaidacea, whereas the recovery of all other taxa stagnated or, in case of the Isopoda, even regressed (Fig. 3). At this time, the macrofauna depth distribution in disturbed samples had significantly shifted into deeper sediment layers. We found very high animal densities in those samples which corresponded to tracks refilled with semi-liquid material (Borowski, 1996).

Seven years after the impact, disturbed abundances equalled undisturbed levels for the Isopoda, but for the Tanaidacea and the Bivalvia, the disturbed values still displayed strong oscillations. The polychaete data analyses of the most recent observation are still in progress, and those of the other taxa (and therefore also for the whole macrofauna) still remain incomplete. However, we interpret the oscillations of single taxa as a symptom of persisting disturbance in the affected sediment areas.

Quantitative recovery of the abundance data for entire major taxonomic groups does not necessarily imply recovery of the faunal composition to conditions similar to an undisturbed community. Figure 4 illustrates aberrations of the polychaete species diversity in disturbed sediments extending over at least three years. While the rarefaction curves (Hurlbert, 1971) of undisturbed samples all lie close together and display high species diversity, the curves of the disturbed group are more dispersed. Some of them fall below the 95% confidence margin of the undisturbed group and thus indicate significantly lower species diversities. We obtained similar results even on the higher taxonomic level of polychaete families (Borowski and Thiel, 1996). At the same time, the abundance of subsurface deposit feeding polychaetes was still significantly reduced in disturbed samples ($p<0.05$, t-test). High proportions of this functional group are generally considered asan indicator for stable successional stages in recolonization studies of disturbed macrofauna communities (Rhoads and Boyer, 1982; Probert, 1984). These data illustrate that, at least for a three year period, the macrofaunal composition in the impacted areas was far from reaching the state of an undisturbed community. Further processing of the samples of the recent expedition will reveal if this can still be recognized after seven years.

Meiofauna

Foraminifera, Nematoda, and Harpacticoida were determined to investigate the recolonization process of the meiofauna. Immediately after the impact, the densities of these three groups decreased to approx. 50% of the baseline study. Half a year later, this process continued for the Nematoda and the Harpacticoida, while Foraminifera densities slightly increased (Fig. 5). We assume that our experiment had disturbed the food chain for the Nematoda and Harpacticoida so that even more animals continued to die because of decreased food availability (Schriever et al., 1991).

Three years after the impact, the situation had changed. The densities of the Nematoda had increased to levels more than twice those of the baseline study, while Harpacticoida had increased by more than 60%. Such effects are regularly observed in recolonization processes and also known from shallow water areas and terrestrial ecosystems (Pearson and Rosenberg 1978, Rumohr, 1980). However, in contrast to

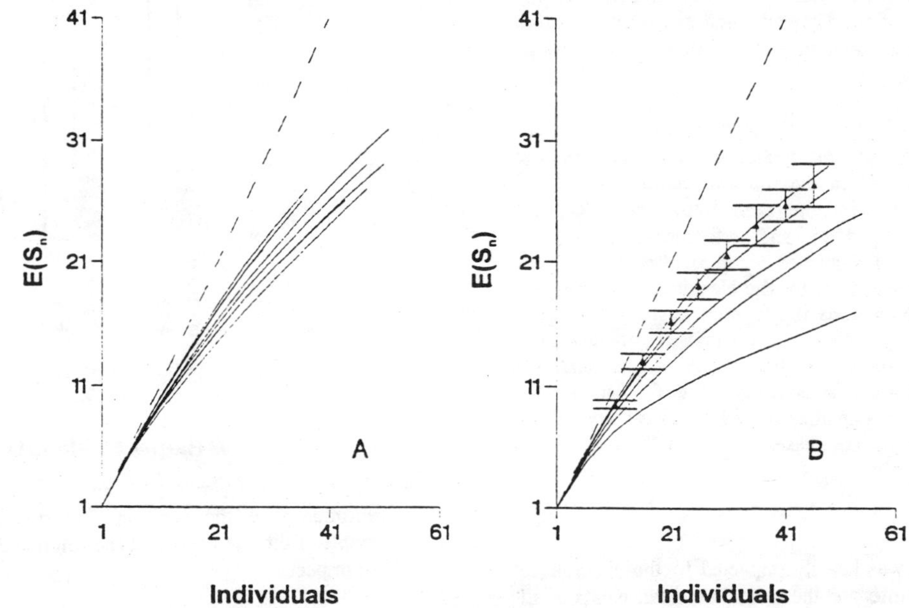

Fig. 4: Hurlbert-rarefaction curves of selected box core samples obtained three years after the impact, on the basis of 78 polychaete species (>500μm). A: eight undisturbed samples; B: solid curves represent six disturbed samples; the dotted curve is the calculated mean for the undisturbed group (A) with 95% confidence limits. The theoretical value E(Sn) represents the number of expected species per number of individuals. Steep slopes indicate high species diversity. The hatched lines represent theoretical maximum diversity, where each specimen in a sample represents another species.

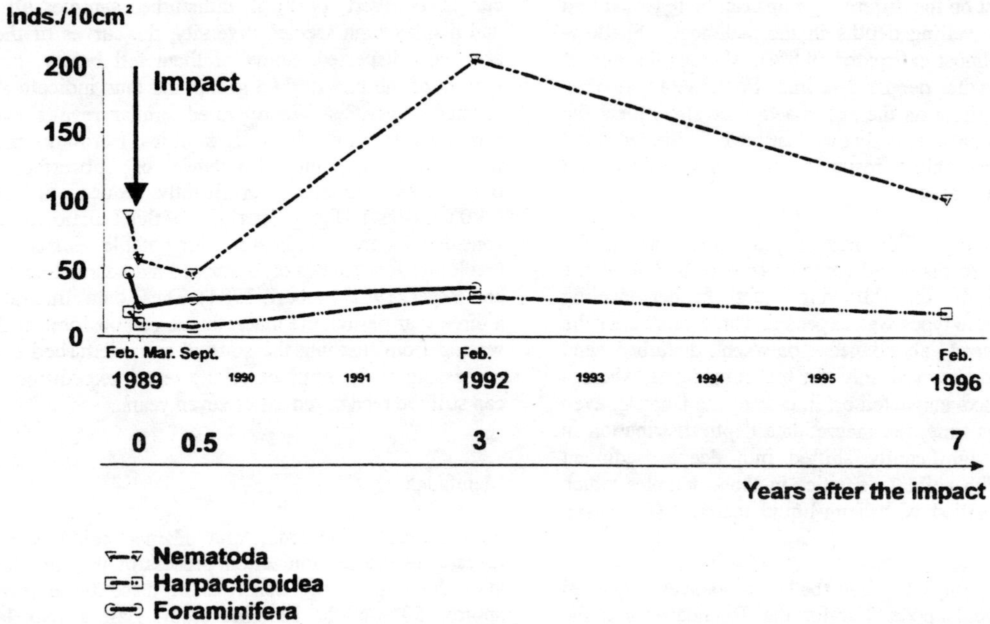

Fig. 5: Time series of major meiofauna taxa abundances (individuals/10cm^2) in the DEA from 1989 to 1996. The bold arrow indicates the time of the impact.

the megafauna, meiofauna densities had also increased at the reference site. We assume that this was due to an additional food input from the euphotic zone. This must have occured equally effectively as well at the disturbed sites, and we consequently are not able to distinguish between the recolonization and food input arguments.

Preliminary results from our recent sampling, seven years after the impact, show again decreased densities of the Nematoda and Harpacticoida. Nematoda densities were still 50% higher than the baseline values, while those of the Harpacticoida exceeded baseline levels by 15%. At the reference station, the densities of both groups nearly equalled those of the baseline study.

Faunal analyses of seperate depth intervals from selected sediment cores revealed that 70-95% of the metazoan meiofauna in the undisturbed DISCOL area dwell in the top two centimeters of the sediment. This is consistent with the data from other deep-sea investigations (Bussau et al., 1995 and references therein). A comparison of undisturbed and disturbed samples after seven years shows inverse abundance relationships for the Nematoda and Harpacticoida: The density of the Nematoda is higher in disturbed samples compared to undisturbed ones, whereas the opposite situation was observed for the Harpacticoida. At the reference stations we determined intermediate densities for both taxa (Fig. 6). Samples for the evaluation of Foraminifera densities from ECOBENT are still in the processing and data development phase.

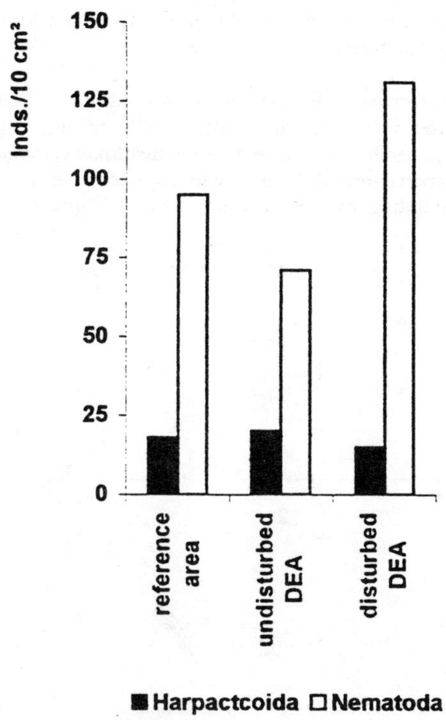

Fig. 6: Densities of the Nematoda and Harpacticoida in the upper six centimeters of the sediment, in disturbed and undisturbed samples from the DEA and the reference site, seven years after the impact.

DISCUSSION

The fauna of all size groups was heavily impacted by the disturbance within the DEA in 1989. We interpret the animal abundance data of all four post-impact cruises as showing that the Foraminifera were the first colonizers in a recovery process that started rapidly after the disturbance. This process generally resembled recolonization in shallow water areas, but it progressed at a strongly reduced rate. As we had expected, seven years after the impact, the density data of the soft bottom community still did not indicate a completed reestablishment of the fauna to conditions which resemble the undisturbed situation of the baseline study. However, we must admit that our results may be somewhat biased by the limited knowledge available on seasonal and long-term variations of faunal densities that are normally found in the deep sea.

The densities of Nematoda and Harpacticoida in the upper two centimeters of the sediment showed significant differences between disturbed and undisturbed samples. The plowing impact had resuspended the very soft top-layer sediments from the plow tracks, and the created plume was subsequently dislocated by water currents. Vane shear strength measurements in the plow tracks demonstrated altered structures of the top sediment layers down to depths of 9-11cm (Grupe and Langner, 1996). In these areas, some clayish material of deeper origin remained exposed at the sediment surface, whereas depressions in plow tracks trapped soft surface material transported by near-bed water currents. Within a seven year period of investigation, this process formed an irregular mosaic pattern with „hard" (high vane shear strength) and „soft" (low vane shear strength) sediment patches that we expect to exist for considerable time, extending at least over some decades. Clayish sediments obviously hampered the recolonization of harpacticoids. We assume that mainly burrowing Nematoda (Bussau, 1993), supported by "burrowing" and very small Harpacticoida, were able to invade the disturbed areas. Their bioturbation possibly prepared the clayish sediment for other colonizers. Soft bottom dwelling Harpacticoida, however, were nearly excluded from early recolonization. Our findings contrast with the data presented by Kaneko et al. (1995) from the JET program (Japanese Deep Sea Impact Experiment) in which the abundances of the Nematoda also decreased subsequent to an experimental disturbance impact. But, in comparison to the DISCOL data, Nematoda abundances remained on a low level, even one year later. The densities of the Harpacticoida, however, were not directly impacted, but had significantly increased again after one year.

From our sediment fauna data, we conclude that the recovery progress of the semi-liquid sediment surface layer is an important factor for the pace of the recolonization process of burrowing infauna. The texture and the food content of the top sediment layers seem to control recolonization success. The reestablishment of the top sediment layer apparently is strongly influenced by material transport with near-seabed water currents. The recovery processes of physically disturbed benthic communities probably can vary with different hydrodynamic regimes and may need much longer time spans in oceanic regions with weaker bottom currents than at the DISCOL location (Borowski, 1996).

Taxonomic groups which are available in sufficient quantities from the sediment samples or from video observations and for which species can be determined (or at least distinguished) gain special importance as target taxa for comparative studies on disturbed deep-sea communities. Within our DISCOL material, the taxa Polychaeta, Tanaidacea, Isopoda (Borowski and Thiel, lecture at the International Seminar Deep Sea-bed Mining Technology, Oct. 18-20 1996, Beijing), Nematoda, Holothurioidea, as well as the Harpacticoida on generic level, meet these requirements. Based on our investigations we recommend that these groups be selected as target taxa in other similar recolonization and mining impact studies. However, different animal groups may be more important in other deep-sea regions. Obviously, intensive baseline information about various benthos size groups is required from other deep-sea areas of potential commercial interest before further large-scale experiments or monitoring programs associated with pilot mining operations and other planned commercial activities are designed.

The results of the two large scale experiments DISCOL and JET are not directly comparable. The aims of DISCOL and the Japanese JET-study are as different as the way and methods of creating the impacts. DISCOL focused on the recolonization of a large scale physical sea-bed disturbance, comparable to mining collector tracks, whereas JET (and other BIE-studies) investigates the impact of a resettled sediment plume on the fauna. The DISCOL disturbance modified the structure of the benthic environment from semi-liquid (very low vane shear strength) to hard surface sediments (high vane shear strength), whereas in JET the sediment structure in principle remained unchanged, except that a few millimeter thick, resedimented soft layer was added on top. The variable thickness of this layer, the grain size distribution and the degree of particle packing are the determining factors for the penetration of oxygen and food materials into deeper horizons, as well as for the ability of the fauna to migrate upwards. The results of both experiments are complementary to another and increase our knowledge about possible impacts of future mineral mining upon the deep sea ecosystem.

ACKNOWLEDGEMENTS

We thank the crews of FS „ SONNE" and all persons who helped to collect and process the samples, and Eric Foell for streamlining the English. The ECOBENT-project is funded by the German Bundesminister für Bildung, Wissenschaft, Forschung und Technologie (Ministry for Education, Science, Research, and Technology) under contract number 03 G 0106 A. The authors are responsible for the contents of the paper.

REFERENCES

Bluhm, H. 1993: Effects of deepsea mining for manganese nodules on the abyssal megabenthic community. Proceedings, 25th Annual Offshore Technology Conference, Houston Texas, May 3-6, 1993. OTC Paper 7134: 521-529.

Bluhm, H. 1994: Comparison of megabenthic communities in abyssal manganese nodule sites of the Northeastern and Southeastern Pacific Ocean. Aquatic Conservation: Marine and Freshwater Ecosystems, 4: 187 - 201.

Bluhm, H., G. Schriever and H. Thiel 1995: Megabenthic recolonization in an experimentally disturbed abyssal manganese nodule area. Marine Georesources and Geotechnology, 13, 393-416.

Bluhm, H. and H. Thiel 1996: Photographic and video surveys for large scale animal and seafloor surface charting aiming at ecological characterization of habitats and communities. Printed papers of the International seminar on deep sea-bed mining technology, Oct. 18-20, 1996, COMRA, Beijing, PR China: C15-C23.

Borowski, C. 1996: Taxonomische und ökologische Untersuchungen an der sedimentbewohnenden Tiefsee-Makrofauna eines Manganknollenfeldes im Peru-Becken (Äquatorialer Ostpazifik) unter besonderer Berücksichtigung der Polychaeta. Ergebnisse aus dem Tiefsee-Umweltschutzprojekt DISCOL. Thesis Univ. Hamburg,. FB Biol., Shaker Verlag, Aachen: 321 pp.

Borowski, C. and H. Thiel 1996: Indicator taxa for dominant deep-sea animal groups in their importance for comparative and environmental studies. Printed papers of the International Seminar on Deep Sea-bed Mining Technology, Oct. 18-20, 1996, COMRA, Beijing, PR China: C35-C43.

Bussau, C. 1993: Taxonomische und ökologische Untersuchungen an Nematoden des Peru-Beckens. Thesis Univ. Kiel: 621 pp.

Bussau, C., G. Schriever and H. Thiel 1995: Evaluation of abyssal metazoan meiofauna from a manganese nodule area of the eastern South Pacific Ocean. Vie et Milieu 45 (1): 39-48.

Grupe, B. and W. Langner 1996: MEPARSED. In: G. Schriever, A. Koschinsky, and H. Bluhm: Cruise report ATESEPP - Auswirkungen technischer Eingriffe in das Ökosystem der Tiefsee im Süd-Ost-Pazifik vor Peru. Berichte aus dem Zentrum für Meeres- und Klimaforschung, E, 11: 113-129.

Hurlbert, S. H. 1971: The nonconcept of species diversity: a critique and alternative parameters. Ecology, 52 (4): 577-586.

Kaneko (Sato), T., K. Ogawa and T. Fukushima 1995: Preliminary results of meiofauna and bacteria abundance in an environmental impact experiment. Proceedings of the first (1995) ISOPE Ocean mining Symposium, Tsukuba, Japan: 181-186.

Pearson, T.H., & R. Rosenberg, 1978: Macrobenthic succession in relation to organic enrichment and pollution of the marine environment. Oceanogr. Mar. Biol. Annu. Rev., 16: 229 - 311.

Probert, P. K. 1984: Disturbance, sediment stability, and trophic structure of soft-bottom communities. J. Mar. Res., 42: 893-921.

Rhoads, D. C. and L. F. Boyer 1982: The effects of marine benthos on physical properties of sediments: A successional perspective. In: McCall, P. L. and M. S. S. Tevesz (editors): Animal-sediment relations. Plenum Press, New-York: 3-52.

Rumohr, H., 1980: Der "Benthosgarten" in der Kieler Bucht - Experimente zur Bodentierökologie. Reports SFB 95, Bd. 55.

Schriever, G. 1990: Cruise-Report DISCOL 2, SONNE 64. Ber. Zentrum für Meeres- u. Klimaforsch. Univ. Hamburg, 6: 51 pp.

Schriever, G. and H. Thiel 1992: Cruise Report DISCOL 3, SONNE cruise 77. Ber. Zentrum für Meeres- u. Klimaforsch. Univ. Hamburg E, 2: 59 pp.

Schriever, G., C. Bussau and H. Thiel 1991: DISCOL - precautionary environmental impact studies for future manganese nodule mining and first results on meiofauna abundance. Proc. Adv. Mar. Tech. Conf. 4: 47-57.

Schriever, G., A. Koschinsky and H. Bluhm 1996: Cruise report ATESEPP - Auswirkungen technischer Eingriffe in das Ökosystem der Teifsee im Süd-Ost-Pazifik vor Peru. Berichte aus dem Zentrum für Meeres- und Klimaforschung, E, 11: 195 pp.

Thiel, H. and G. Schriever 1989: Cruise-Report DISCOL 1, SONNE - Cruise 61, with contributions by C. Borowski, C. Bussau, D. Hansen, J. Melles, J. Post, K. Steinkamp and K. Watson. Ber. Zentrum Meeres- u. Klimaforsch. Univ. Hamburg, 3: 91 pp.

DISCOL-Publication No. 51

Proceedings of the Seventh (1997) International Offshore and Polar Engineering Conference
Honolulu, USA, May 25-30, 1997

ISBN 1-880653-28-1 (Set); ISBN 1-880653-29-X (Vol. I)

Experiments on the Influence of Technical Activities in the Deep-Sea on Heavy Metal Cycles

Andrea Koschinsky
Freie Universität Berlin, Berlin, Germany
Hans Gerber
Technische Universität Berlin, Berlin, Germany
Axel Szemeitat
Technische Universität Clausthal, Clausthal-Z., Germany

ABSTRACT

Our projects within the German research group for deep-sea environmental protection (TUSCH) investigate biogeochemical cycles in the bottom water - deep-sea floor interface and its possible reactions to technical activities, especially binding and release of heavy metals. During research cruise SO 106 disturbance experiments with previously undisturbed sediment were performed for the first time; a Maxicorer, a benthic experimental chamber and a sampling and measuring system had been developed for these experiments. After stirring up the upper centimeters of the sediment an immediate release of pore water constituents like nutrients and heavy metals was observed. Fast metal sorption in the suspension directly after the pore water release proved that the suspended particles represent effective scavengers for the released metals. Sorption was followed by a certain metal mobilization and enhanced microbial activity after a few days. Only in the presence of reducing or organically complexing agents critical amounts of toxic heavy metals like Cu, Pb, Tl, and Cd were mobilized.

At present, the laboratory experiments are continued and the necessity for in-situ experiments is discussed.

KEY WORDS: deep-sea ecosystem, anthropogenic activities, environmental protection, heavy metals, metal release, metal adsorption, deep-sea technology

INTRODUCTION: OBJECTIVES OF THE EXPERIMENTS

The investigations presented here are part of the activities of the German TUSCH cooperative project, which was established to investigate the problems that might arise from a future commercial utilization and exploitation of the deep-sea (see for example Thiel and Forschungsverbund TUSCH, 1995; Schriever et al., 1996). The goal is to obtain the necessary knowledge of the deep-sea ecosystem in its status quo and its reactions on anthropogenic disturbances. We concentrate on the biogeochemical cycles in the bottom water - deep-sea floor interface and on binding and release of heavy metals with possible toxic effects for the ecosystem. A disturbance at the seafloor and the redistribution of the surface sediment and other precipitates like Mn-Fe oxide phases creates new reactive boundaries between particles and dissolved phases, releases pore water and mobile sediment constituents into the near-bottom water, and may change physico-chemical parameters like pH and redox potential (Fig. 1). Decomposition of resuspended organic material and microbial activity might be promoted.

These possible changes in the chemical environment exert an influence on the heavy metal distribution between solid and dissolved phases and on the bioavailability and toxicity of the metals. Laboratory investigations on the association of heavy metals with solid phases and the mobilization by changes in pH, Eh, and chemical compounds have been carried out in the predecessor projects. It was shown that a change of the redox environment and release of reducing agents would be the critical factor and could cause significant dissolution of mineral carrier phases (especially MnO_2 and FeOOH) accompanied by mobilization of associated toxic heavy metals.

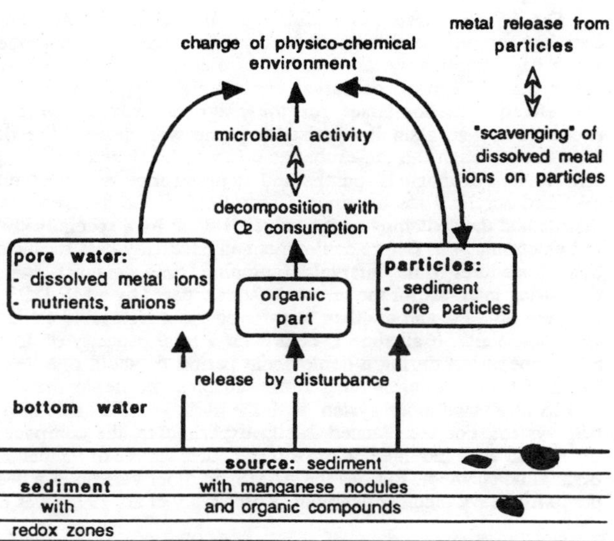

Fig. 1 Possible reactions in the bottom water - deep-sea sediment system when the upper sediment layer is resuspended into the bottom water layer

To obtain information whether such significant changes in environmental conditions could be caused by heavy sea bed disturbance, in situ and laboratory disturbance and observation experiments were planned for the SO 106 cruise in the Peru Basin which took place in 1996.

As this type of experiments had not been carried out before, special new equipment had to be developed: a Maxicorer for sampling undisturbed sediment cores and a benthic experimental chamber with a water sampling and measuring system. It was intended to place the benthic experimental chamber on the seafloor, disturb the upper sediment layer in the chamber with a stirrer, and measure the changes of pH, Eh and dissolved oxygen and take water, particle and heavy metal samples with an active measuring and sampling system (ASMS) within a week. Because of an unexpected incident during the first deployment of the benthic chamber, the planned in-situ experiments mostly had to be substituted by laboratory experiments with sediment cores recovered with the Maxicorer and the Multicorer.

TECHNICAL DEVELOPMENT AND EQUIPMENT

No appropriate equipment was available for this new concept of geochemical disturbance experiments with emphasis on heavy metal cycles. To handle the very low metal concentrations in seawater and avoid severe contamination problems with regular sampling systems, a special sampling and measuring system was built. For the planned in-situ experiments, a metal-free benthic chamber with a stirring mechanism was required. The reason for developing the Maxicorer for taking deep-sea sediment samples was to combine and improve the functions of existing tools like Boxcorer or Multicorer i. e. to get big and completly enclosed samples with undisturbed sediment including soil-water interface and bottom water. These samples were the precondition for laboratory experiments as close as possible to in-situ conditions.

ASMS (Active Sampling and Measuring System)

The ASMS (Active Sampling and Measuring System; see Schriever et al., 1996) was constructed and used for the following sampling and measuring actions: a) measuring of pH, redox potential, dissolved oxygen concentration and temperature in widely selectable intervals (0.1 to 10 minutes, b) sampling of water in 12 PTFE 2 liter sampling bags for anion and organic analyses at selectable times and flow rates (0 to 2000 ccm/min), c) metal enrichment on 12 ion-exchange columns (Chelex 100) for trace heavy metal analyses at selectable times and low flow rates (about 15 ccm/min) and d) particle collection with two-step prefilters.

The ASMS consists of a sampling unit (energy supply, controller box, bag-sampling and ion-exchange system) and a sensor-probe. The water bag sampling system consists of a stepping-motor driven peristaltic pump, 12 magnetic valves to close the bags, a flush valve and two electric junction-boxes for the magnetic valves. For the ion-exchange system a smaller peristaltic pump was chosen. The flow of seawater through the ion-exchange columns is steered by a magnetic turn valve. All magnets, pumps and stepping motors are mounted in oil-filled acrylic glass housings. Controlling of the different stepping motors and the switching of the valves is done by a specially customized electronic unit. Before deployment all necessary data for the operating procedures time intervals (pumping time, magnet switching, measuring intervals for the sensors) are calculated by a MS-DOS based program on a PC and are then transferred via a releasable cable to the electronic unit. 56 alkaline D-cells with a total capacity of 1500 Wh allow operations during a deployment period of about one week. An electropolished stainless steel frame enables the deployment of the ASMS as a stand-alone system with the hydrographic cable or a mooring system. For the planned in-situ experiments, the components of the ASMS were mounted on the top and outer sphere of the benthic experimental chamber. Besides the ASMS, an older version of a deep-sea pump for heavy metal enrichment (Koschinsky et al., 1996) was used.

Benthic Experimental Chamber

The design of the autonomous benthic experimental chamber consists of the following main components (Fig. 2): a) reaction chamber with stirring mechanism and equipment, b) buoyancy glass spheres with stand, c) ballast lip (ballast weight), and d) main frame. The reaction chamber is a welded cylindrical aluminium vessel with a total volume of approximately 1 cubic meter. On the bottom side the disc area is open, on top there are some openings for the floating lids, for the instalment of the sensor probe, for the sampling hoses of the ASMS and for the driving unit of the stirring mechanism. In the interior of the reaction chamber a software controlled propeller with a diameter of 40 mm is linked to the driving unit. The lids of the four sediment traps inside can be closed by software controlled electromagnets. The surface inside of the chamber is coated with E-CTFE. All other components in the interior of the chamber are made of PTFE, to guarantee a chemically inert environment.

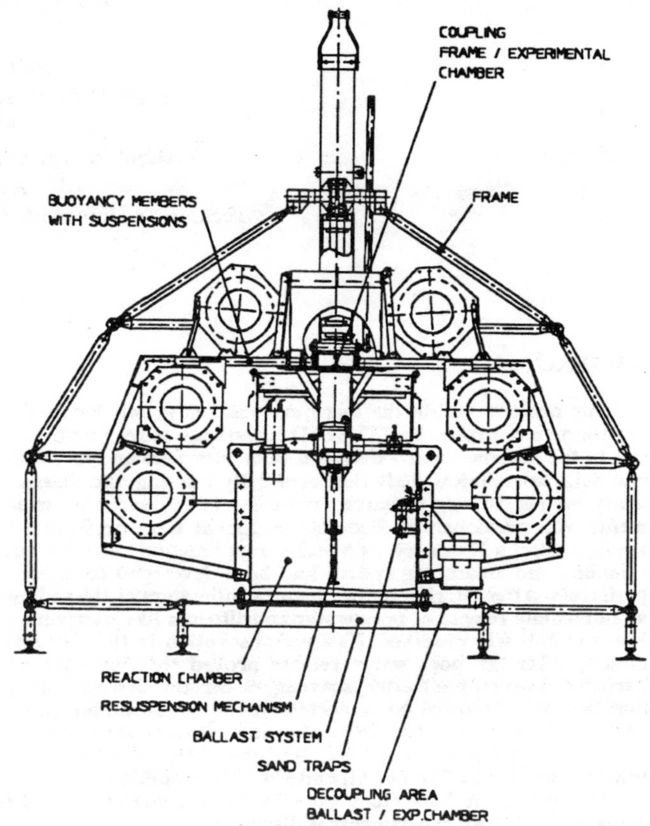

Fig. 2 Technical sketch of the benthic experimental chamber

All further equipment, like power supplies, control device, parts of the sampling system, electromagnets, release mechanism and transponder are appropriately mounted to the outer ribs of the reaction chamber. The sampling bags of the ASMS are positioned concentrically on top of the chamber. The power supply for the stirring mechanism, for the electromagnets and for the sediment traps is realized by the use of a lead/acid battery. For buoyancy purposes 21 plastic covered glass spheres of Nautilus Marine Service Ltd. are positioned on three levels around the chamber providing a good trim. Each of them has a buoyancy of 270 N. The main frame is an octagonal, pyramid shaped steel construction which measures 4.3x4.3x4.2 m. It is made of premanufactured tube elements. This frame contains a damping unit that allows a defined soft settling of the chamber onto the sediment surface after the frame has touched the seafloor. After settling, the chamber with buoyancy device and lip of ballast is decoupled from the main frame. The frame is lifted back to the ship. During the entire operation on the sea floor the reaction chamber stands on the lip of ballast, which penetrates the soil nearly with its entire height of 190 mm. For recovery of the system the lip of ballast has to be released. The mechanical release can be activated through the use of three independent mechanisms: a) electromagnet (general steering), b) burn-wire (autonomous steering and power supply), and c) transponder (acoustic release from the mother ship).

The benthic chamber was first tested successfully in the Baltic Sea in 1995. Before testing it on cruise SO 106 in the Peru Basin, the buoyancy spheres were lowered to 4000 m below sea level to perform a pressure test under pressure vonditions of the planned operation. In addition a combined floating and sinking test of the complete chamber system was performed. All tests were finished sucessfully. Then a deep-sea test on the rope was performed to check all its components under deep-sea conditions (Fig. 3). Therefore, all components which could be detached via the release device were secured by short ropes to the chamber and the main frame. The chamber was lowered until contact with the sea floor at 4170 m water depth and was left there for fifteen minutes. All steering devices were programmed in such a manner that the most relevant data were stored to be checked after the mission. The three systems for releasing the lip of ballast were also activated. Even though all buoyancy glass spheres had been checked under deep-sea conditions before (see above), they imploded for unknown reasons during this test and damaged the complete system to that extent that no sensor data or samples could be recovered. Any further usage of the benthic chamber during the SO 106 cruise was impossible.

Fig. 3 Deployment of the benthic experimental chamber with ASMS, buoyancy system, and frame, for the deep-sea test on R/V Sonne during cruise SO 106

Maxicorer

The general concept (see also Gerber et al., 1996) is the following: The three liners, which are the heart of the Maxicorer containing the samples after the operation, were arranged in a triangle around a central supporting unit also housing the closing mechanisms. This supporting structure was mounted to a main frame using a cardanic suspension, which guarantees its safe, stable and vertical position on the sea floor. Penetration of the liners was realized by a guided and damped vertical movement of the unit liner/support-structure relative to the position fixed main frame after touch down on the sea floor.

The liners have a diameter of 300 mm and a height of 600 mm. Tubes of acrylic glass are embedded into a steel construction. After penetration into the soil the liners are closed tightly through use of a closing lever and a top lid. While heaving the Maxicorer, steel cables which are connected with the deep sea cable force the lever to move. The magnetic lock is opened during that procedure and the liners are closed as described above. The main frame is manufactured from steel tubes and was later galvanized. The total height of the system is 4.1 m during lowering and 5.34 m after taking the samples and during the heaving procedure. The entire operation of the Maxicorer deployment is shown in Fig. 4: 1) lowering the Maxicorer, touch down of the frame and relieving the deep sea cable, 2) damped lowering of the support tube - liner - unit, disconnecting the gear from the supporting tube by use of the release mechanism and sediment penetration of the liner, 3) lifting the deep sea cable to close the liner as well as the top lids and continued lifting of the liner, and 4) lifting until the ropes have tightened and heaving back onboard.

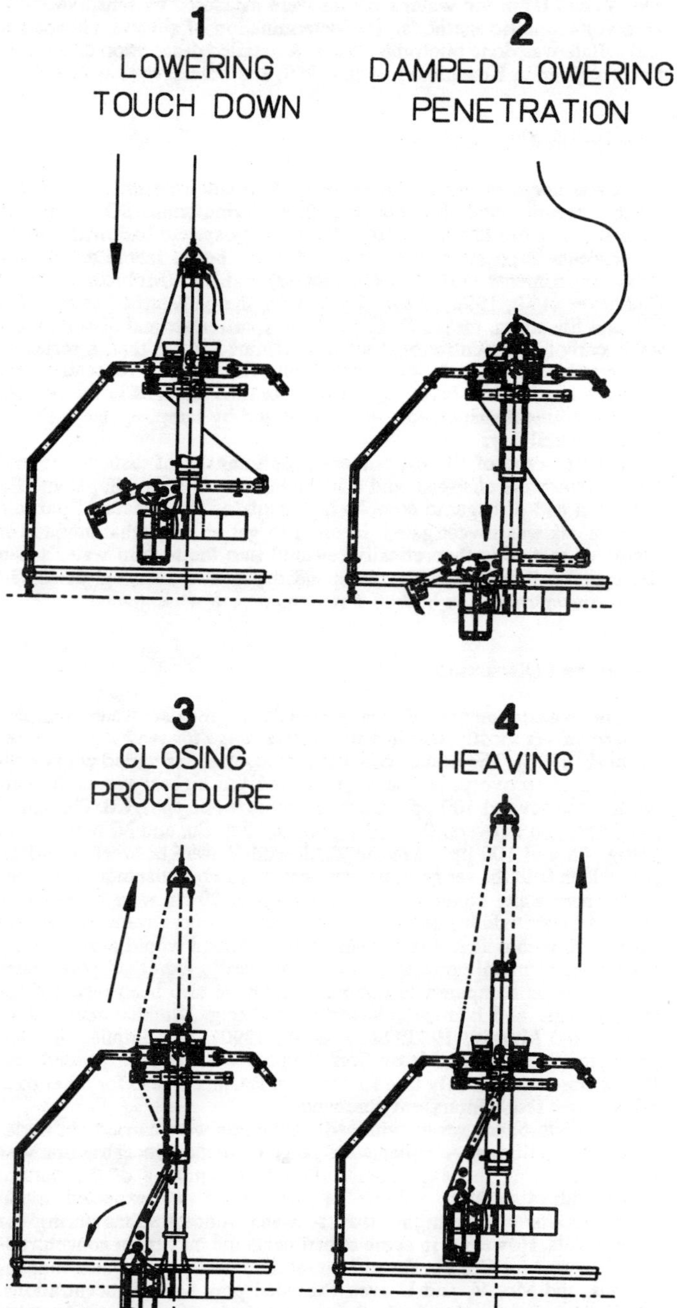

Fig. 4 Deployment and sampling process of the Maxicorer

The Maxicorer was first tested in the Baltic Sea in October 1995. Later in spring 1996 during research cruise SO 106 it was deployed 15 times on R/V Sonne between 3,500 and 4,300 m water depth. The Maxicorer operated successfully which could be deduced from the state of the samples including the undisturbed surface, the boundary layer and water temperature in the liners which was still close to deep-sea temperature after recovery.

THE GEOCHEMICAL EXPERIMENTS

Analytical Methods

Dissolved heavy metals (Mn, Fe, Co, Ni, Zn, Cd, Pb, Cu, Ti, Tl, Mo, V, and U) in the water samples were measured by sensitive stripping voltammetric methods. The determination of silicate, phosphate, and sulfate was done photometrically. A detailed description of the methods is given in Koschinsky et al. (1996) and Schriever et al. (1996).

Experimental Procedures

It was intended to run three one-week in-situ experiments with the benthic experimental chamber (Fig. 2, 3) during cruise SO 106 in the Peru Basin. After the failure of the buoyancy system had made ithese experiments impossible, we conducted a number of laboratory disturbance experiments with MX (Maxicorer) and MC (Multicorer) liners (Schriever et al., 1996). First, the data for the undisturbed sample (T, pH, and Eh, anion, metal, POC and DOC (particulate and dissolved organic carbon) concentrations) were determined. After that, a series of disturbance experiments was carried out. The upper 5-10 centimeters of the brown surface layer were stirred and the changes in the bottom water - sediment suspension were monitored by sampling and measuring for several days.

During some of the experiments, the intensity of disturbance and temperature were changed, and also the influence of a slight pH and Eh reduction and of organic complexing agents on the release of particle components was investigated. In order to get an idea what amount of dissolved metals is theoretically released into the bottom water when the upper sediment layer is resuspended, we also analyzed dissolved metal concentrations in bottom water and pore water extracts.

Results and Discussion

The measurements of heavy metals in regular water samples showed values mostly within a reasonable range for seawater (Schriever et al., 1996). The metal pore water concentrations varied with depth and area of recovery. Mn and Fe showed the highest variations between 1 and several 100 ppb. Co was very low (<1 ppb), Cd, Pb, and U mostly ranged between 0.1 and 5 ppb, and Zn, Cu, and Ni were found between 5 and 100 ppb. Measured Mo and V were between 5 and 10 ppb, which is in the range of normal seawater . For most metals the measured pore water concentrations in the upper 20 cm were at least one order of magnitude higher than open ocean or bottom water concentrations. This means that a resuspension of sediment would also increase bottom water metal concentrations temporarily. Increased pore water concentrations compared to bottom water have also been reported for other pelagic and hemipelagic sediments (e.g. Klinkhammer, 1980; Sawlan and Murray, 1983; Shaw et al., 1990). This implies that the principle of fast metal release from the pore water of resuspended surface sediment is not only true for the Peru Basin, but also for other oxic and suboxic sedimentary environments.

Ten disturbance series with sediment cores were carried out under different conditions. A rather wide range of metal concentrations was measured during the experiments. Due to the mixing of the bottom water with a smaller volume of pore water we expected metal concentrations a little higher than seawater concentrations during the experiments. However, in some experiments the maximum concentrations measured for Cd, Pb, Zn and Cu reached nearly pore water concentrations, and Mo, V, and U even exceeded pore water concentrations. This indicates a certain metal release not only from pore water, but also from solid phases. Initial metal release was mostly followed by a fast decrease, and after a few days we often found dissolved metal concentrations to increase slightly again.

Fig. 5 shows the typcial metal gradients of a laboratory experiment, in which seawater containing dissolved heavy metals in the concentration range of pore water was mixed with ferromanganese ore particles and the adsorption/release processes were monitored for 7 days. The concentration courses support the idea that besides simple pore water release followed by adsorption on marine particles, also a certain release from the particles is induced after a couple of days. However, no striking release of metals was evident in all experiments under unchanged conditions or with slight decrease of pH (by adding diluted HCl). Only when ascorbic acid (as reducing agent) or EDTA (as complexing agent) were added in the range 0.0001-0.001 Mol/l during the disturbances, dissolution processes releasing up to several hundreds of ppb of heavy metals were observed. This artifical addition of chemical compounds was done to simulate stronger environmental changes, e.g. caused by the deposition of waste material and tailings in the deep-sea.

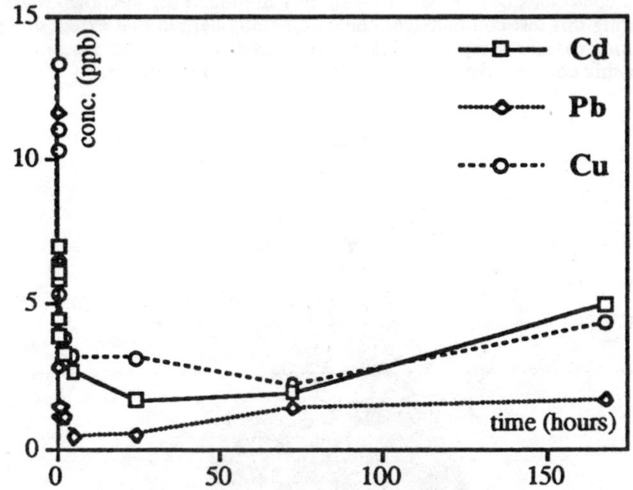

Fig. 5 Metal concentrations during an adsorption experiment with dissolved heavy metals and ferromanganese crust particels (size 63-125 µm, concentration 1 g/l) in seawater

The samples of the disturbance experiments were also analyzed microbiologically and biogeochemically for particulate and dissolved organic carbon as well as for particulate amino acids. Cell counting gave increased numbers of bacterial cells in the upper sediment layers with progression of the experiment, i.e. with time and number of disturbances. Parallel to this, concentrations of particulate organic carbon, total nitrogen, amino acids and hexosamines increased which could be due to the enhanced bacterial biomass production. Decreasing C/N ratios and changes in the spectral distribution of individual amino acids support that the additional organic matter is of freshly produced bacterial origin (see Haake et al., 1993). We assume that the slow release of heavy metals after a few days of the disturbance experiment is related to enhanced microbial acitivity with production of fresh organic matter.

In one disturbance series, we tried to evaluate the influence of temperature on the metal release, which is a very important question concerning the applicability of laboratory experiments on in-situ conditions. One disturbance series was carried out in the sunlight around 30°C, another one in the dark at 4°C, and the third series was carried out in the dark and cool as well, however, with a stronger and deeper disturbance of the upper sediment layer. The dissolved metal distributions of the three series (Fig. 6) show that the stronger disturbance in the cool/dark experiment released on average somewhat higher metal concentrations than the cool/dark experiment with a normal disturbance. The released metal concentrations in the experiment conducted in sunlight was significantly higher for most metals, especially Zn, Cu, Mn, Ni, and also V and Mo. This clearly shows that temperature has an

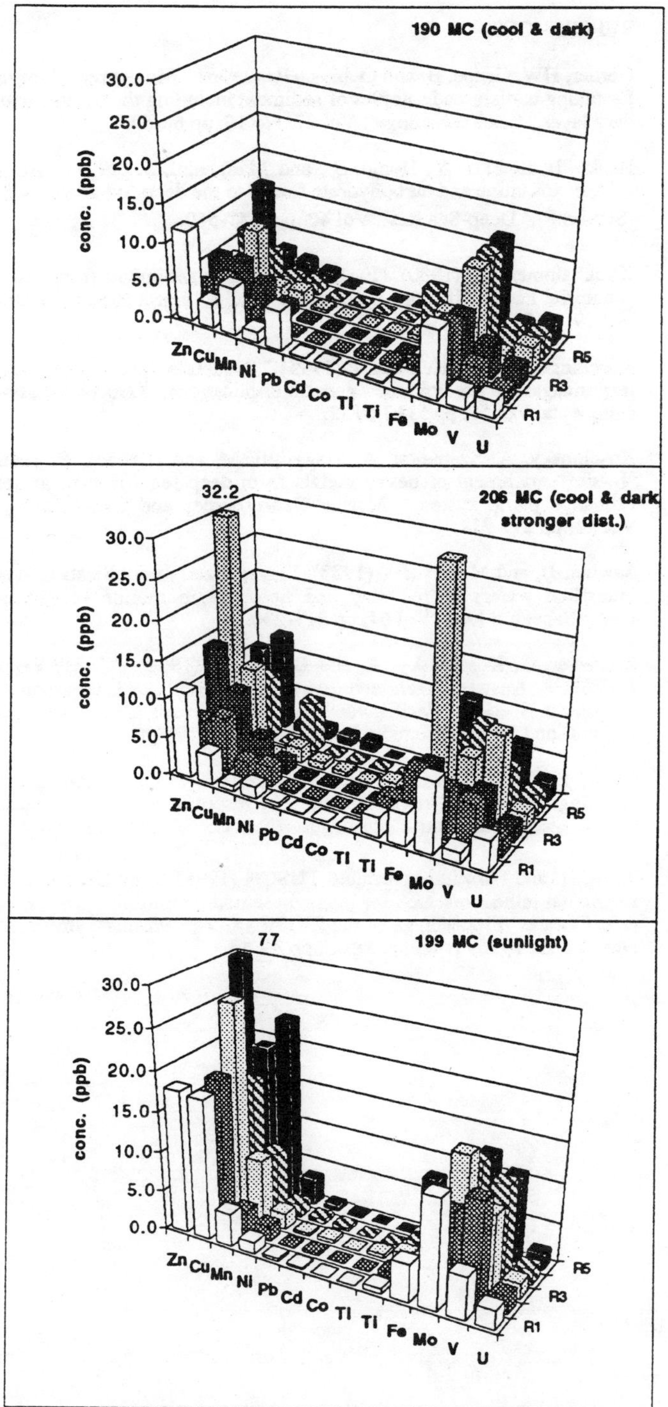

190 MC:	R1 undisturbed R2 after 1st disturbance R3 before 2nd disturbance R4 after 2nd disturbance R5 after 3rd disturbance R6 2 days after 3rd disturbance	206 MC:	R1 undisturbed R2 after 1st disturbance R3 before 2nd disturbance R4 after 2nd disturbance R5 before 3rd disturbance R6 after 3rd disturbance
199 MC:	R1 undisturbed R2 after 1st disturbance R3 after 2nd disturbance		R4 after 3rd disturbance R5 2 days after 3rd disturbance

Fig. 6 Metal concentrations measured in the bottom water during three different disturbance series of sediment cores

important influence on the metal cycle in this system, which is probably also related to microbial activity.

Fig. 7 shows the pH recordings of a disturbance experiment during which liner 1 was disturbed only once, and liner 2 four times. The pH decreased about 0.2 units after every disturbance and then slowly increased again, which means release of H+ ions from pore water or desorption of H+ from particles caused by the disturbance, followed by resorption or consumption of H+ due to a chemical reaction. Repeated disturbance caused repeated pH decrease. The nutrient-type anions (Fig. 7) silicate and phosphate show a clear release after the first disturbance and more silicate is released with every additional disturbance. This is in agreement with 5-10 times higher silicate concentrations in pore water compared to seawater (data in Schriever et al., 1996).

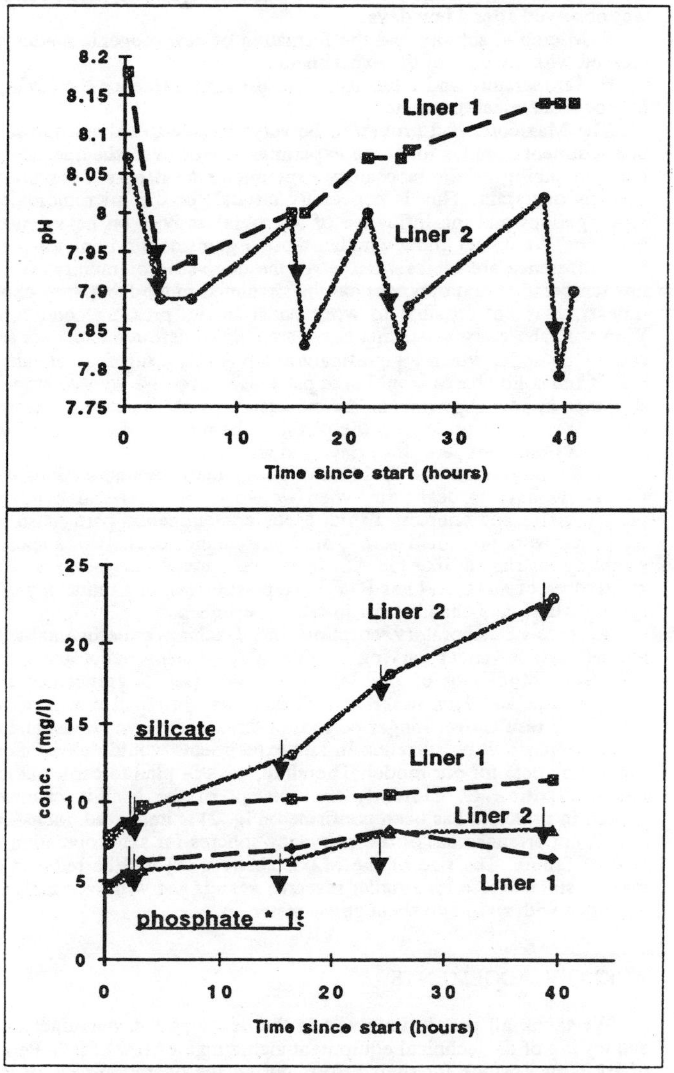

Fig. 7 Variations of pH and dissolved silicate and phosphate concentrations during a sediment disturbance experiment
Liner 1: one disturbance at the beginning of the experiment
Liner 2: four disturbances during the experiment
(each arrow represents one disturbance with a stirrer)

SUMMARY AND CONCLUSIONS

Our geochemical experiments with deep-sea sediment cores showed that the principal reaction process of dissolved chemical compounds after a disturbance of the upper sediment layer is the following:

1. Due to the higher concentrations in pore waters compared to bottom water, dissolved pore water components like nutrients and heavy metals are released, and pH decreases slightly.
2. Release is followed by fast adsorption of most released heavy metals on suspended sediment or Mn/Fe oxide particles, if the physicochemical environment remains unchanged.
3. If the environment is changed to more reducing conditions or strongly complexing organic compounds are present, further heavy metal release instead of sorption is induced.
4. Also in unchanged oxic environments a certain metal release was observed after a few days.
5. Microbial activity and the formation of new biogenic matter increased with the time of the experiment.
6. Temperature and intensity of disturbance influence the strength of biogeochemical reactions.

The Maxicorer had proven to be very suitable to deliver undisturbed sediment samples for these experiments. However, the question of the applicability of the laboratory experiments on deep-sea conditions remains uncertain. This is especially critical for the microbiological investigations and the influence of microbial activity on heavy metal reactions. We do not know whether those organisms surviving the pressure difference are representative for the deep-sea community. While low temperature and darkness can be simulated in the laboratory experiments, it is not possible to work under in-situ pressure conditions. With our laboratory experiments we are able to define parameters and conditions under which a significant and possibly dangerous metal release from solid phases would take place. However, we want to emphasize that in-situ experiments are necessary to measure real changes under deep-sea conditions in the physico-chemical environment of the deep-sea floor over periods of days and weeks.

A second point is the problem of heavy metal contaminations that we always have to deal with when we work with conventional sampling systems and onboard a ship. Metal contamination partly also affected the work presented here. This confirms the necessity of a special sampling instrument like the ASMS for trace metal analyses, which is mostly free of metal and has PTFE and plastic sampling units, and of a specially designed chamber for in-situ experiments.

At present, laboratory sorption and leaching experiments with marine particles under varying conditions (also using radiotracers) are continued. Modelling of our laboratory data (see Segschneider and Sündermann, and Jankowski and Zielke, this vol.) will enable us to extend our results over longer periods of time, over larger areas, and to consider transport by currents. In-situ experiments would deliver more supporting data for our model. Therefore, we still plan to carry out the in-situ experiments. Currently the concept of the benthic chamber (which in principle has been confirmed; Fig. 2) is improved, including a different arrangement of the buoyancy spheres for safer operation in case of failure. The size of the Maxicorer (Fig. 4) will be reduced to make it suitable also for smaller research vessels and will eventually be equipped with an in-situ shear stress meter.

ACKNOWLEDGEMENTS

We thank all people involved in the development, manufacturing and testing of the technical equipment including the crew of R/V Penck of BMS in Rostock for their efforts during the operation tests of the Maxicorer and the benthic chamber with ASMS in the Baltic Sea in fall 1995, and the crew of R/V Sonne for the skilled handling of our equipment on cruise SO 106. Furthermore, we acknowledge the work of all people who contributed to the scientific results presented here, especially the scientific crew of SO 106 and other people involved in the TUSCH projects. The projects of the TUSCH group are funded by the "Bundesministerium für Bildung, Wissenschaft, Forschung und Technologie" (here: project no's. 03G0106F, 03G0106G). The manuscript has been improved substantially by the comments of H. Thiel and I. König.

REFERENCES

Gerber, HW, Grupe, B, and Oebius, HU (1996). „Maxicorer - A device for taking undisturbed samples of sediment including the benthic boundary layer," Sea Technology, Vol 37, No 10, pp 66-69.

Haake, B, Ittekkot ,V, Honjo, S , and Manganini, S. (1993). "Amino acid, hexosamine and carbohydrate fluxes to the deep Subarctic Pacific (Station P)," Deep-Sea Res., Vol 40, pp. 547-560.

Klinkhammer, GP (1980). "Early diagenesis in sediments from eastern equatorial Pacific. II. Pore water metal results," Earth Planet Sci Lett, Vol 49, pp 81-101.

Koschinsky, A, and Halbach, P (1995). "Sequential leaching of marine ferromanganese precipitates - genetic implications," Geochim Cosmochim Acta, Vol 59, pp 5113-5132.

Koschinsky, A, Szemeitat, A, Maggiulli, M, and Halbach, P (1996). "In-situ enrichment of heavy metals from deep-sea water by an ion-exchange pump system," Marine Georesources and Geotechnology, Vol 14, pp 297-314.

Sawlan, JJ, and Murray, JW (1983). "Trace metal remobilization in the interstitial waters of red clay and hemipelagic marine sediments," Earth Planet Sci Lett, Vol 64, pp 213-230.

Schriever, G, Koschinsky, A, and Bluhm, H. (1996). "Cruise Report ATESEPP: Auswirkungen technischer Eingriffe in das Ökosystem der Tiefsee im Sued-Ost-Pazifik vor Peru," Berichte aus dem Zentrum für Meeres- und Klimaforschung, Reihe E, Vol 11, 195 pp.

Shaw, TJ, Gieskes, JM, and Jahnke, RA (1990). "Early diagenesis in differing depositional environments: The response of transition metals in pore water," Geochim Cosmochim Acta, Vol 54, pp 1233-1246.

Thiel, H, and Forschungsverbund TUSCH (1995). "The German environmental impact research for manganese nodule mining in the SE Pacific Ocean, " Proceedings of the ISOPE - Ocean Mining Symposium, Nov. 21-22, 1995, Tsukuba, Japan, pp 39-45.

Proceedings of the Seventh (1997) International Offshore and Polar Engineering Conference
Honolulu, USA, May 25-30, 1997

ISBN 1-880653-28-1 (Set); ISBN 1-880653-29-X (Vol. I)

Data Support for Modelling of Deep-Sea Mining Impacts

J.A. Jankowski and W. Zielke
University of Hannover
Hannover, Germany

Abstract

The paper critically reviews the presently available experimental data from various tests and experiments connected with the deep-sea mining issue with regard to their feasibility in supporting and validating the developed numerical models. Numerical modelling is applied mainly to predetermining the plume development and seafloor blanketing caused by various sediment discharges. The paper describes processes included in these models and discusses the experimental acquisition of needed model parameters. The existing models and their validation are shortly reviewed and parameters essential to operate and validate them are pointed out. Recommendations for further field studies are given in order to improve the quality of model forecasts.

Keywords

Deep-sea mining, environmental impact assessment, hydrodynamic numerical modelling, data acquisition and assimilation, model verification, validation.

Introduction

Although deep-sea mining does not seem to be profitable at present due to the market situation, the mining technology is being further developed in many countries. Among deep-sea deposits, the manganese nodules are of greatest interest. They form flat fields at the bottom at depths of 4000–6000 m, partially covered by or buried under sediments. As field tests have shown, deep sea mining is technologically feasible (Bischoff and Piper, 1979; Halbach et al., 1988; Padan, 1990). Mining will inevitably be accompanied by serious changes of the seafloor and generate plumes at different depths. The plumes will consist mainly of fine bottom sediments, nodule fines, or fragments, dissolved effluents, and biota debris. Depending on the mining technology, the discharges will take place in different depths and with various intensities. The greatest discharge will appear at the bottom during nodule harvesting, or as a consequence of mining tailings introduction (Thiel, 1991).

Numerical modelling is applied mainly to determine plume developments and seafloor blanketing caused by various sediment discharges in order to predict their environmental impact. However, the reliability of the present numerical models depends on the quality of field data.

At present, we have a stagnation phase in the deep-sea mining issue. This presents a chance for research work on precautionary deep-sea environmental studies, and for consultation in technological methods ensuring the least possible impact before the start of commercial mining. This also implies that the environmental impact assessment has to be formulated without impact examples providing sufficient information, which makes it a difficult task. A characteristic feature of the more recent field experiments is the generation of artificial impacts by different methods. These impacts are small, when compared to the planned recovery on an industrial scale.

Since the field experiments are limited, there exists a need to provide numerical models which allow evaluation and extrapolation to the dimensions to be expected in future commercial mining. The models require data, which are assimilated in different ways, namely as:

1. data to formulate basic assumptions in the models,
2. input data to operate the models,
3. verification data,
4. validation data.

For deep-sea mining impact modelling the specially designed field studies play the most important role among different oceanographic data sources.

Goal of numerical modelling

Modelling concerned with the assessment of the environmental impact of deep-sea mining is mainly applied to forecasting the discharged material plumes, which appear at different depths, and seafloor blanketing caused by settling particles. The impacts to be estimated describe the exposition of marine organisms to large amounts of particles (sediments, ore particles, biota) and other effluents (e.g. dissolved heavy metals) in concentrations exceeding ambient oceanic ones, as well as their settling. Basically, it is necessary to estimate how long the plume will persist before it dilutes and reaches the ambient concentration level or settles at the bottom, and which new sediment coverage will result. The period in which the surface plumes hinder the penetration of the solar radiaton has to be estimated, too.

The probability of occurrence, the recovery rate of organisms, the consequences and overall significance of these influences are the object of present research. Deep-sea ecology, as well as the response of the deep-sea organisms to anthropogenic activities has not yet been understood sufficiently (Thiel et al., 1991). The links between the description of the impact in physicochemical categories and effects on the biota are not clearly known.

There are no well-defined environmental standards defining the limits of excess concentration and its persistence time over a given bottom area, as well as of the resulting deposition thickness.

Required model parameters

The models describing transport of discharged sediment and/or other effluents may differ greatly according to the basic assumptions of the conceptual model. The assumptions mainly depict the scope of the physical phenomena, which are included in the model and their parametrization. For the modelling process, these physical mechanisms and relevant effects of the model environment must be formulated in terms of model variables. Models are reliable, when they are verified, validated and their errors can be quantified.

The data for modelling are input, verification and validation data. The input data describe the initial conditions, configuration or properties of the modelled system (e.g. the geometry, fluid and sediment properties, etc.), as well as parameters needed to run and control the model (e.g. the boundary conditions).

Model verification shows whether the mathematically formulated problem is properly posed and solved. It is done by tests of the model performance and accuracy for well-defined parameter ranges, which together cover the model application domain. For this purpose, we use analytical solutions or data sets for selected physical phenomena included in the model.

The sensitivity analysis is a part of the model calibration, helping to assess the response of the model to parameter variations. It determines the parameters crucial for the accuracy of the model, and help to formulate guidelines for the model application. Parameter studies which compare the results with measurements are also an important step of model calibration. Calibration is usually understood as adjustment of model parameters to a given application in order to obtain improved results.

Validation is based on tests, which are designed to find out, how closely the output of a verified and possibly calibrated model describes the real environment. During validation the model results should be compared to the field data which have not been used in the calibration process.

The main part of all models discussed in this paper is a mathematical description of the sediment transport, based on the transport equation with its appropriate boundary and initial conditions (McLean, 1985). The main physical phenomena controlling plume spreading are: velocity field, processes in the bottom or surface boundary layer, density effects, temperature and salinity profiles, bottom topography, different discharge characteristics, sediment settling velocity, cohesive properties of sediments, scavenging by external particles (e.g. marine snow), sediment erosion and deposition processes.

In the following we attempt to present a list of *input parameters* necessary for a three-dimensional and time-dependent model applied to the deep-sea mining sediment discharges, which includes all relevant physical phenomena.

1. **Current velocity field.** The current measurements are applied to:

   (a) detect the most important hydrodynamic phenomena (tides, inertial waves, eddies, bottom boundary layer) for given transport scales,

   (b) produce typical current scenarios,

   (c) or incorporate them directly into the models.

   Additionally, these measurements are used for formulation of time-dependent boundary conditions. Measurements should be intended on obtaining the velocity field in space and time resolutions appropriate for a given model scale. Special attention must be paid to the characterization of the turbulent bottom or surface boundary layer (bottom roughness, velocity profile, wind influence). Global circulation features enhancing vertical transport, as topographic influences, up- and downwelling, must be also addressed.

2. The **turbulent viscosities and diffusivities** with their spatial and temporal variabilities. They are obtained from the current measurements using statistical methods, or from spreading of easily detectable substances from known sources. Special attention must be paid to the vertical diffusivity, a factor which can balance the settling velocity of the sediment.

3. **Bathymetry** of the area in sufficient resolution for a given model scale is needed in order to take into account topographic influences.

4. **Profiles of salinity, temperature, and turbidity** provide the ambient, normal particle concentrations and density field. They are especially important for modelling the surface discharges, where the thermocline is a boundary between two areas with different hydrodynamical characteristics.

5. **Characteristics of various discharges** (discharge rate and form) must be carefully quantified taking into consideration the mining technology (Ozturgut et al., 1981). Depending on the mining technology, the sediment mass available for discharge at the bottom limits the maximum discharge rates from all possible sources.

6. **Density effects.** The initial concentration and density of the discharge, which deviate from the surrounding waters are important for characterization of density effects. These effects are not only due to suspended sediment, but also due to discharge salinities and temperatures, differing from the surrounding water. Density currents generate near-bottom, flat distribution of the bottom plume in the initial stages after discharge, causing an additional settling effect. The stratification caused by the suspended sediment dampens the vertical mixing (Jankowski et al., 1994).

7. The **sediment settling velocity in situ** is the most complex parameter. The particle size distribution of the sediments from the mining areas shows that most of the

particles have diameters smaller than 60μm, so that the cohesive forces between the particles cannot be neglected (McCave, 1984, Klein, 1993). Settling velocities for the plume particles, interacting with each other, flocculating and breaking up, range from 10^{-3} m/s to 10^{-7} m/s and require special attention. In the models, the settling velocity is treated using the following methods:

(a) to assume the settling velocity to be constant in time and space, and equal to a mean settling velocity of the sediment spectrum (non-cohesive case);

(b) to use a number of sediment classes characterized by settling velocities without interactions between various classes (non-cohesive case);

(c) to use an empirical formula to treat flocculation and break-up, with the settling velocity as a function of concentration and turbulence characteristics (cohesive case);

(d) to use sediment classes and to account for interactions between them (cohesive case).

In order to apply the non-cohesive cases (a) or (b), it is usually assumed that the information provided by the particle diameter and density distributions is satisfactory to obtain the settling velocity. The problems connected with shape, aggregation level and porosity of the particles are dealt with using empirical parameters. The assumption of constant settling velocity implies that the particle diameter distribution remains constant in time. In the case (b), the different sediment classes settle independently and the mean settling velocity diminishes with the age of the plume, as the larger particles settle out.

In order to take into account the cohesive sediment properties influencing the mean sediment settling velocity, the empirical formulations, case (c), are widely used in estuaries and shelf regions. They parametrize the processes of aggregation and break-up of particles using suspended sediment concentration and turbulent shear (McCave, 1984; Dyer, 1989; van Leussen, 1994; Malcherek, 1994). Theoretically considered, the break-up phenomena are insignificant in the low energetic ocean bottom boundary layer. Differential settling dominates other flocculation mechanisms as Brownian motion and turbulent shear. In this case an empirical model is appropriate, in which the settling velocity is a function of the sediment concentration only (Jankowski et al., 1995). The possible flocculation effects may, therefore, accelerate the deposition. Unfortunately, data covering the cohesive properties of the deep-sea sediments in situ are presently too scarce to allow the application of this formulation in a non-speculative way (McCave and Gross, 1991). Laboratory experiments using sediments from the equatorial Pacific mining areas have shown that flocculation effects are significant in the plumes at concentrations above 100 mg/l (Ozturgut and Lavelle, 1986). These higher concentrations, under which flocculation effects are stronger, are found in the vicinity of the source. Flocculation between plume particles may be insignificant in the diluted plumes far away from the source. Scavenging by rapidly sinking external particles (marine snow) may be more important in this case.

It is difficult, but feasible to describe sediment settling, using a flocculation model based on a spectrum of interacting, i.e. cohesive sediment classes, case (d), (Hill and ~~Nowell,~~ 1995). The most serious problems encountered are uncertainities in interaction rates.

Since it is very difficult to estimate the parameters describing the settling velocity in situ, most estimations are based on laboratory experiments. In situ measurements are still very scarce (McCave and Gross, 1991; Spinrad et al., 1989). There are also difficulties in estimating the role of the organic substances, and interactions between suspended particles and living organisms.

8. The **scavenging rate of the plume particles by external particles** is found by observing the vertical flux of the ambient particles appearing naturally in the ocean, as marine snow. These large, amorphous particles sweep large ranges of the water column, collecting finer particles underway and transporting them downwards. If the mining plume drifts in areas swept by these particles, scavenging may be an effective agent in removal of the diluted plumes consisting of slowly settling fine particles (Lavelle, 1987).

9. **Erosion and deposition** processes can be described without difficulty in the low-energetic boundary layer, typical for the regions of deep-sea mining. The small velocity magnitudes (Bischoff and Piper, 1979; Klein, 1993) allow assumptions that no erosion occurs and that each plume particle, which reaches the bottom, actually deposits. If mining takes place in regions of stronger currents, the formal description of erosion and deposition processes must be introduced. They are parametrized by the bottom shear stress, the critical stress for the erosion, the resuspension rate and the critical stress for deposition. Very little is known about these parameters relevant for the deep-sea sediments (McCave and Gross, 1991).

For model *validation*, data on the model output must be collected in order to compare them with the model results. They are:

1. **Transient, three-dimensional suspended sediment concentration field**. It is usually obtained by nephelometer observations from fixed and variable positions, and water probes sampling.

2. **Current measurements** correlated with the concentration data.

3. **Deposition fluxes**, measured by means of sediment traps and by geochemical methods (bottom sediment sampling).

4. **In-situ particle characteristics** in different plume dispersion stages (plume age) are essential for validation of settling velocity formulations.

Model data acquisition in conducted experiments

From the very beginning of the ocean mining investigations, concern has been expressed about the potential environmental impact due to the exploitation of deep-sea deposits. Tests and experiments related to deep-sea mining started in the early seventies (Thiel et al., 1991). This section concentrates exclusively on the experiments associated with the manganese nodule mining, in which most field work is under way. It does not consider tests with other marine deposits, such as hydrothermal metalliferous sediment deposits in the Red Sea (Amann, 1985; Thiel,

Table 1. Model-relevant deep sea mining tests and experiments.

Name	Date	Organization	Position, Depth	Action
OMI mining test, DOMES, Deep Ocean Mining Environmental Study	March-May 1978	Ocean Management Inc. (OMI), USA, cooper. with National Oceanic and Atmospheric Administration (NOAA)	9°N, 151°W, DOMES Site A, 5100 m	102 h collector work, 900 t nodules recovered
OMA mining test, DOMES	Oct.-Nov. 1978	Ocean Mining Associates (OMA), USA, cooperation NOAA	15°N, 126°W, DOMES Site C, 4300 m	18h collector work, 500 t nodules recovered
DISCOL Disturbance and Recolonization Experiment	1: Feb. 1989 2: Sept. 1989 3: Jan. 1992 4: Feb. 1996	TUSCH research group, Germany, BMBF–sponsored	7°04′S, 88°28′W, Peru Basin, 4150 m	78 plow-harrow tows, 10 km^2 bottom disturbed, recolonization observation
BIE Benthic Impact Experiment	I. 1991/1992 II. July-Sep. 1993	NOAA, USA, cooperation with CGGE (Yuzhmorgeologiya), Russia	12°56′N, 128°36′W, Clarion-Clipperton Fracture Zone, 4800 m	98h benthic disturber action, 1450 t sediment resuspended
JET Japan Deep Sea Impact Experiment	Aug.-Sep. 1994	MMAJ (Metal Mining Agency of Japan), cooper. with CGGE, Russia	9°15′N, 146°15′W, Japanese mining claim, 5300 m	20.5h benthic disturber action, 352 t sediment resuspended
IOM experiment	Summer 1995	InterOceanMetal (IOM) East-European	Clarion-Clipperton Fracture Zone, IOM claim	reports expected

1991). Basic information on the field studies discussed in this section is shown in Table 1.

It is difficult to compare the deep-sea mining field studies, because they were conducted independently and with different methods. The most striking differences are found in the discharge characteristics, including real mining devices in action as well as artificial disturbances. However, there are some common features in the global procedure followed by all these experiments. The main procedure can be divided into the following stages: pre-impact studies, impact monitoring, immediate post-impact survey, and long-term post-impact observations. In this section we describe field studies with an overall aim to find out their strong and weak points.

DOMES project and OMI/OMA tests. In 1975-80 National Oceanic and Atmospheric Administration of USA (NOAA) coordinated the multidisciplinary project DOMES (Deep Ocean Mining Evironmental Study) intended to: (1) establish environmental baselines in three reference areas in the Clarion–Clipperton Fracture Zone (CCFZ), (2) create a database for environmental guidelines, (3) observe actual impacts during mining tests, and (4) develop predictive capabilities for determining the environmental hazards of nodule mining (Bischoff and Piper, 1979). The main result was a classification of the environmental problems according to their importance. Few effects were recognized to be most probable causes for severe negative environmental impacts: (1) the impact on benthic biota due to seafloor destruction by the mining collectors, (2) the impact due to seafloor blanketing at a distance from the mining area, and (3) impact on the upper ocean biota due to the surface discharge plume. Future research needs were pointed out on: (1) the impact of the benthic plume and (2) the trace metal uptake by organisms (Padan, 1990).

During the DOMES project (1978) two successful and so far unique pre-pilot mining tests by Ocean Mining Inc. (OMI) andby Ocean Mining Associates (OMA) were monitored (Ozturgut et al.,1981; Lavelle et al., 1982). The nodule recovery rate was about one order of magnitude lower than during planned commercial production. The observations included monitoring of the benthic and surface plumes.

The parameters describing the OMI test bottom plume were measured in water samples, as well as by nephelometers and current meters moored at fixed different levels above the bottom, providing vertical profiles of velocity and turbidity. The collector system discharge was not exactly known. Only approximations based on optical observations of the deposition at a distance from the collector tracks were available (photograph analyses). As evidence for density currents, the optically detectable deposition, which was found as far as 50 m upstream from the collector tracks was pointed out. Eventually, a computer model was used to obtain the parameters which could not be measured during the test, with special attention to the settling velocity and near-bottom vertical diffusivity (Lavelle et al., 1981). The current varied strongly in the test area. The most striking result was the high mean settling velocity, which was about 10^{-3} m/s, i.e. much higher than expected, with a broad settling velocity spectrum. The large settling velocity caused massive deposition (estimated 90%) in the first 100 m away from the device tracks. The residual plume excess concentrations, measured in the plume 5–8 days of age and transported about 20 km away of the source, were 10 μg/L. The tests showed the inherent problems met when monitoring of the narrow benthic plume by meters in fixed positions. The purely optical analysis of the bottom deposition was a particularly weak point of these observations.

The monitoring of the surface plume was performed during

both tests. The modelling made use of the OMI test data only (Ozturgut et al., 1981; Lavelle et al., 1982). The characteristics of the surface discharge were exactly known, in contrast to the benthic ones. Again, the plume measurements using nephelometer transects at 5 m depth (horizontal), water samples (vertical) and particle chemistry (Fe-Mn) indicated that the discharged material settled more rapidly than it had been expected for pelagic clays, i.e. $6 \cdot 10^{-4}$ m/s. Plumes older than 37 h could not be detected any more in the surface mixed layer. It was concluded from these observations and laboratory experiments that flocculation could play a role in settling of the plume. The density effects were excluded, because the discharge came from a moving vessel, and was comparatively small. Density differences were in the range of 1%. During the tests the velocity profiles were not measured, so only pre-impact measurements were available. Some effects, e.g. those due to the significant velocity directional shear, remained unclear, especially with regard to particle settling below the pycnocline.

Both tests were on a smaller scale than the potential commercial mining, and regarded as too limited to allow an extrapolation to the industrial full-scale exploitation (Ozturgut et al., 1981). Monitoring of the plumes concentrated on selected aspects, and complete information was never gathered. Actually, the computer models were applied to obtain the unmeasured parameters. However, observations provided valuable information on the impacts to be caused by deep-sea mining, because mining devices under real conditions were used.

DISCOL. The mining tests in 1978 were the only two mining tests in real conditions ever made, no further tests have been conducted due to the fading interest of the industry. So there was no chance to repeat or improve the measurements in situ with the presence of mining devices in action. The characteristic aim of most of the post-DOMES experiments was to generate an artificial bottom disturbance or resuspension without applying mining technology.

The first group to follow this pattern was the German interdisciplinary TUSCH (German: *Tiefseeumweltschutz*, deep sea environmental protection) research group. They concentrate their experimental activities (Disturbance and Recolonization Experiment (DISCOL)) in the DISCOL Experimental Area (DEA) in the Peru Basin in the Southeast Pacific Ocean, in the vicinity of a potential German mining area (Thiel and Schriever, 1990; Thiel, 1995; Schriever et al., 1996). The main purpose of the study is to evaluate the reaction of organisms to seafloor disturbances. It started in 1989 with baseline pre-impact investigations of the benthic community. In order to obtain an impact similar to mining activities, the bottom was disturbed in a circular area of 3.5 km in diameter using a towed 8 m wide plow-harrow. The area was traversed 78 times, tilling about 20% of the area. All the nodules were buried. The rest of the area was blanketed by resedimentation with varying thickness. This is not exactly the kind of disturbance to be expected by mining, as the nodules were not collected but buried and the bottom was not squeezed by the collector weight. Nevertheless, the disturbance effect was achieved in an area of 10 km^2, as was confirmed by post-impact optical observations. A feature of DISCOL is that no scheduled observations of the plume in suspension and its deposition were made.

Both pre- and post impact observations followed the same pattern: bottom sediment sampling, photographic and video documentation, as well as CTD, hydrographic and sedimentological measurements. These activities were accompanied by long-term current profile measurements for over two years (Klein, 1993). In 1996 they were repeated for a month with higher time and space resolutions.

The modelling in the TUSCH group profits mainly from the basic support data collected during the cruises, i.e. the current measurements, bathymetric surveys, sediment characteristics and geochemical aspects. As mentioned above, no validation data are collected, with the exception of current measurements. The main objectives are collecting input data for modelling and biological evaluation.

It was planned to revisit the DEA every two years and to observe the subsequent recolonization by benthic community, and to continue the baseline studies. The area was actually revisited 6 months after the disturbance, then in 1992, and later in early 1996 (Schriever et al., 1996). The preliminary recolonization experiment results were summarized by Thiel, (1995) and Schriever (1995).

Although a discharge experiment was considered during preparations for the 1996 DISCOL cruise, it was not scheduled. The new experimental idea was to create a small experimental field at the bottom, in a closed chamber, where sedimentological and geochemical experiments could be carried out in situ and under control. Although the experiment failed, this method to circumvent the prohibitive difficulties in monitoring bottom plumes may bring good results in estimating the in situ cohesive sediment properties and geochemical coefficients (Schriever et al., 1996, Koschinsky et al., this issue).

BIE. The next disturbance experiment, Benthic Impact Experiment (BIE), was organized by NOAA in cooperation with Russia's Central Marine Geological and Geophysical Expedition (CGGE). The intention of the experiment was "to simulate the environmental effects of sediment resuspension by deep seabed mining operations and to assess the environmental impact on the deep-sea benthos" (Trueblood, 1993). The main effects to be produced were sediment burial and food resource dilution. After two attempts, the BIE-II experiment was conducted in Summer 1993 in an area in the CCFZ, where background data had been collected previously.

As a characteristic feature of the experiment, the sediment was resuspended by a special device, a *benthic disturber* (Deep Sea Sediment Resuspension System, DSSRS-II) (Brockett and Richards, 1994; Trueblood, 1993). The towed disturber is designed to fluidize, lift and discharge a slurry of bottom sediment and to blanket an area of the sea floor in a manner as can be expected during mining activites. Again, as in DISCOL, no nodules were removed from the bottom. The main impact was due to the deposition and destruction caused by the device sleds. Discharge parameters are given in Table 2. As a result, a deposition with gradual reduction of the thickness at increasing distance from the tracks was achieved in an area of about 2 km^2, ranging from 10 to 1 mm.

For measuring the extent of the deposition 18 sediment traps situated 2 mab (m above bottom) were used. They were placed on both sides of the towing area, accounting for possible current reverse, and in three rows, approximately: 50 m, 150 m and 400 m from the tow zone. Two current meters with nephelometers were deployed 2 mab with sediment traps 5 mab on both sides of the towing zone. During the operation 3 sediment traps and 1 current meter mooring were recovered in order to monitor the progress of the sediment plume dispersion.

The sediment trap contents indicated a rapid deposition north of the tracks. The nephelometers gave distinct signals indicating the passes of the disturber. The photographic data confirmed that the heaviest deposition was within 50 m from the tracks. It was concluded that a portion of the sediment deposited quickly due to the near-bottom sediment-laden density flow. The influence of the changing bottom topography around the tow zone was also made responsible for the quick deposition. Additional CTD casts and radionuclide analyses of sediment cores were used to map the sediment plume and its redeposition pattern. The CTD casts were unsuccessful, due to difficulties to locate the plume properly, but it was hoped that radionuclide measurements could bring more information. The box corer sampling was carried out throughout the area before and after the disturbance in order to conduct biological and sediment grain size analyses.

In Summer 1994 recolonization by the benthic organisms and the reestablishment of the sediment structure was observed, and current meter moorings were collected.

The deposition monitoring was appropriate to find out about the redeposition pattern. However, the sediment traps were located comparatively high (2 mab), so that the load in the densest, near-bottom plume parts went probably undetected. Since the experiment was restricted to the resulting redeposition, monitoring of the concentration and distribution of the bottom plume was not sufficiently taken into account. The nephelometer signals indicated that the near-source plume consisted of isolated clouds or bands. The nephelometers were also located at 2 mab, so no information on vertical plume profile is available. Probably, the main difference between the disturber discharge and a real collector plume lies in the temporal development of the sediment cloud and its deposition. The disturber worked discontinuously in an irregular way for total 98 hours during 19 days of deployment. The envisioned collectors produce a continuous discharge, and follow a pattern at the bottom avoiding coming back to the already mined areals.

No attempt was made to follow the plume at some distance from the tracks. There were uncertainities regarding the source strength, since only estimates from samples collected at the outlet of the discharge pipe were available.

JET. In summer 1994 the Metal Mining Agency of Japan (MMAJ) has conducted Japan Deep Sea Impact Experiment (JET) in cooperation with CGGE and NOAA in the Japanese mining claim in the CCFZ, at a depth of 5300 m depth in an abyssal valley with comparatively smooth topography. In this area MMAJ had been collecting baseline data since 1991. The objectives and methodology were similar to those used in BIE experiment (MMAJ, 1994; Fukushima, 1995).

The benthic disturber uf the US BIE experiment was used again. The disturber was towed in alternate directions (SW-NE) only 19 times from 50 planned. Two parallel tracks were laid in order to make sure that a region of heavy deposition was located between them. The discharge parameters are given in Table 2. The discharged dry sediment mass was estimated by analyzing the slurry samples from the rossette sampler at the top of the discharge pipe (Barnett and Yamauchi, 1995). These samples were used for estimating the sediment settling velocity. Before the experiment, sediment samples were collected, and the currents were measured during 76 days at 5 and 50 mab, with an additional sediment trap at 30 mab.

Throughout the experiment numerous mooring systems were

Table 2. A brief comparison BIE/JET.

Parameter	BIE	JET
mean site depth	4800 m	5300 m
disturber deployment	19 d	14 d
disturber action	98 h	20.5 h
towing area	150×3000m	two tracks 150×2000m
number of tows	49	19
discharge rate	125 L/s	125 L/s
discharge height	5 m	4 m
wet sediment discharge vol. (after 24h settling)	4888 m^3	2475 m^3
dry weight discharged	1450 t	352 t
aimed deposition area	ca. 2 km^2	ca. 2 km^2(?)
maximum deposition thickness	10 mm	1.9 mm
sediment traps	18+4	12+2
current meter moorings during experiment	2	5
pre/post observations current meter moorings	2/4	2/1
nephelometer deployed/failed	2/0	4/2
pre/post sediment sampling stations	3/7	13/14

deployed to monitor the current and sedimentation conditions (sediment traps with two tubes and nephelometers) around the disturber tracks. Two systems were deployed at some distance from the area in order to observe the current near the bottom, and the current assumed to be unaffected by the bottom topography. Two current meter systems with transmissometers and sediment traps were located NW, and towards the ends of the device tracks, in order to describe exactly the conditions in these places, where the greatest disturbance occured (turning points of the device). Two other systems with sediment traps were also equipped with current meters and transmissometers, and were placed centrally on the both sides of the tracks. This array was completed by ten sediment traps at 2 mab, placed very close to the disturber tracks.

During the experiment (after 11 tows) two moorings were recovered to check the current, indicating a significant current reversal during the experiment. To clarify the reason of the current reversals, the post-disturbance survey was continued for over a year with one mooring consisting of 6 current meters at different water depths and two sediment traps.

The results from the numerous sediment traps were interpolated in the entire experimental area, yielding blanketing thicknesses up to 1.91 mm. In addition, different geo- and radiochemical estimations of the blanketing thickness were made.

Pre- and post-disturbance surveys using sediment sampling and an underwater camera with CTD sensors were carried out in order to produce a photographical documentation of the bottom, and estimate the bottom blanketing thickness.

Most of the comments regarding the BIE apply to the JET as well. The strong side of this experiment was the description of the sediment deposition at the bottom together with the near-bottom currents. By concentrating, as BIE, on documenting the deposition, the suspended plume was observed rudimentarily. Therefore, the JET yields no additional information on the plume drifting away from the experimental area. Information concerning the bottom boundary layer characteristics is much better than during BIE (moorings in different places provided measurements at various heights above the bottom), but no information on the vertical structure of the plume is available. Again, no continuous plume was produced. Due to technical problems the discharge could be carried on for no longer than total 20.5 h (discontinuously) during 14 days.

Synthesis. A simple quality estimate on the model data acquisition in the four discussed experiments is given in Table 3. During the mining tests, efforts were made to monitor not only the bottom blanketing, but also the plume in its different stages. The later experiments neglected the drifting plume. These experiments concerned only two aspects of the impact, i.e. bottom destruction and plume deposition at the bottom. Direct comparison between the disturbances caused by a plow-harrow or the disturber to the impact caused by real mining devices cannot be made. The main difference is that the nodules in the experiments were not removed from the bottom. Other obvious differences are due to scale, spatial and temporal development of the produced discharges. The discharge components may be also different, because in the case of real mining the separation of nodules from surrounding sediments is a process which probably increases heavy metal mobilization, and introduces also nodule fines into the water column. The experiments did not consider the possible effects caused by the diluted plumes drifting away from the source, the modellers are left alone with theoretical concepts. Already in the DOMES conclusions this aspect was pointed out as a field for future research. Although the resuspension experiments give a chance to estimate the sediment settling velocity in situ, and to clear the uncertainities connected with the density effects, these aspects were not considered.

Table 3. Data acquisition in the field studies

Parameter	DOMES	DISCOL	BIE	JET
long-term current	+	+	+	+
current during exp.	+/−	+	+	++
BBL characteristics	−/+	−/+	−/+	+/−
current vert. resol.	+/−	+/−	−/+	+/−
current hor. resol.	−	−	+	+
diffusivity	−	+/−	+/−	+/−
bathymetry	−/+	+	++	++
salinity, temp., turb.	+/−	+	+	+
scavenging	−	−	−	−
discharge charact.	+/−	−	+	+
deposition	−/+	−/+	+	++
sed. particle charact.	+	+/−	+	+
settling vel. in lab.	+	+/−	+	+
settling vel. in situ	−	−	−	−
cohesivity in lab.	−/+	−/+	−	−
cohesivity in situ	−	−	−	−
heavy metal mobil.	−	+/−	−	−

++ good, + satisfactory, +/− partly usable
−/+ unsatisfactory, − not addressed

Existing models and their validation

A number of computer models applied to deep-sea material transport are available. They were applied to the natural sediment transport phenomena, as well as to the environmental impact assessment of human activities (e.g. Marietta and Simmons, 1988; Gross and Dade, 1991).

The first attempts to model the deep sea mining plumes were made by Hess and Hess (1976) and Ichiye and Carnes (1977). Further, Lavelle et al. (1981) developed an analytical model for the bottom plume and for the surface discharge (Lavelle and Ozturgut, 1981), based on an analytical solution of the sediment transport equation in a uniform velocity field. Actually, the analytical model of the bottom plume was used to estimate the unmeasured parameters during the OMI test. Although the authors were aware that the chances of verification were limited, they tried to extrapolate the test results to industrial scale mining in both bottom and surface plume cases (Ozturgut et al., 1981; Lavelle et al., 1982).

A few years later, Lavelle reanalyzed the problem using a two-dimensional numerical finite-difference model, in order to include the effects of the bottom boundary layer, particle scavenging by marine snow, and new settling velocity laboratory analyses (Ozturgut and Lavelle, 1986; Lavelle, 1987).

Taguchi et al. (1995) developed a three-dimensional finite-difference model applied to the JET experiment. The current measurements were incorporated directly into the model yielding the velocity field. For model validation well-documented resedimentation data from the comparatively small JET area were used, and the effective settling velocity was compared with laboratory values.

Within the TUSCH group, a three-dimensional, finite element, mesoscale model (Jankowski et al. 1994, 1995) and a large-scale model with finite-differences (Segschneider and Sündermann 1995, 1996, Zielke et al. 1995) were developed. The TUSCH modelling efforts concentrate on parameter sensivity studies and scenario computations. The input data used came mainly from the DISCOL experimental area. The main problem encountered is connected with missing validation data. The large-scale hydrodynamics can only be compared with known patterns of global circulation (Segschneider and Sündermann 1995, Zielke et al. 1995). The group intends to formulate an extrapolation to the industrial recovery level. Both models include dispersion of particle-reactive effluents (as heavy metals), with input data from systematic DISCOL experimental campaigns (Segschneider and Sündermann, this issue, Koschinsky et al., this issue).

The existing models are listed chronologically in Table 4. While the early models were specifically dedicated to deep-sea mining discharges alone, the latest are adjustments of generic, complex models for this particular application. There is a tendency to include all relevant physical phenomena.

As can be expected, most of the models related to deep-sea mining are rather strictly connected with particular experimental activities. Although most of these models can be adapted to different mining areas, they were applied for a given experimental area only. They differ also greatly in basic assumptions as to the conceptual model and the included physical phenomena, as well as the application domain scale. While the models are

Table 4. Existing models of the deep sea mining discharge impact.

Name	Year	Type, Method	Data	Comments
Hess and Hess	1976	analytical		surface plume
Ichiye and Carnes	1977	analytical		surface plume
Lavelle et al.	1981	analytical	OMI/OMA	bottom plume and deposition
Lavelle et al.	1981	analytical	OMI/OMA	surface plume
Lavelle	1987	numerical, finite differences	OMI/OMA, settling velocity measurements	2D, bottom plume, < 20 km
Jankowski et al.	1993-5	numerical, finite element	DISCOL	3D, mesoscale, bottom plume and deposition
Segschneider et al.	1993-5	numerical, finite differences	DISCOL	3D, large-scale, geostrophic, both plumes, deposition
Nakata (unpubl.)	1994	numerical, finite differences	BIE	3D, for BIE area, bottom plume
Taguchi et al.	1995	numerical, finite differences	JET	3D, for JET area, bottom plume and deposition

usually well verified using measurements or analytical solutions (in order to test if the effects of the included physical phenomena are reproduced properly), their validations were not possible.

The sensitivity studies with the models point out that the crucial parameter for the model reliability and accuracy is the settling velocity of the cohesive deep-sea sediments. Most of the studies use constant mean sediment settling velocities. Tests with a broad range of constant settling velocities are performed in order to detect the inaccuracies. A theoretical flocculation model was discussed, but not applied, due to the data situation (Jankowski et al., 1995). Other domains, where uncertainities affecting model reliability exist, are listed under conclusions. In contrast, the reproduction of the hydrodynamics with simple bottom boundary layer descriptions and topographic influences is usually satisfactory.

Conclusions

1. Numerical models are quite advanced with regard to the physical processes which govern the sediment plume resulting from deep sea mining. Recent models include real bathymetry, variability of currents, bottom boundary layer, density changes induced by suspended sediment and resulting currents. Even flocculation and scavenging have been included, in spite of the fact that they have been only scarcely investigated in the deep sea.

2. There is a lack of input data to operate the models and an even greater lack of calibration and validation data. Therefore a formal process of model calibration and succeeding validation is not possible at the present time.

3. It seems that present models are capable of reliable prediction of the deposition in the experimental areas, on the experimental scale, provided sufficient input data are provided.

4. Sensitivity studies with a model allow the selection of those parameters which are crucial for its predictive capability. In addition to the current data, the most important parameter is the settling velocity of the cohesive deep-sea sediments.

5. The uncertainties in model parameter values, influencing model reliability and accuracy and defining future research requirements, concern the modelling of the following phenomena:
   (a) cohesive sediment properties including sediment settling velocity,
   (b) exact discharge characteristics, especially for the near-bottom plume,
   (c) scavenging of the plume particles by external particles (as marine snow),
   (d) density currents and stratification induced by suspended sediment in the vicinity of the source,
   (e) the bottom boundary layer (shear-induced mixing),
   (f) the role of effluents, such as heavy metals,
   (g) global circulation features enhancing vertical transport.

6. The available data sets are insufficient for the validation of the predictions of the plume development and sedimentation at a distance from its source, i.e. in the mesoscale (< 500 km), where a major impact is to be expected, and in the large-scale (> 500 km), where also long-term effects may occur. No validation of the models on these scales has been performed.

7. Past experiments such as DISCOL, BIE and JET, as well as the pre-pilot mining operations (PPMT), as the OMI and OMA tests in 1978, were limited in scale. They allowed determination of only some of the parameters needed by models. Actually, some parameters can be obtained even without generating any impact. It seems that only appropriate monitoring of pilot mining operations (PMO) may bring a chance to estimate the full scope of needed parameters. These operations are envisioned as longer (a few months) tests of commercial mining. The PMO's may also allow efficient collection of the validation data.

Recommendations

1. The validation of impact modelling can be conducted only during appropriately monitored longer pilot mining opera-

tions in future. The most challenging task (from the experimental point of view) is to provide the models with input and validation data in appropriate spatial and temporal scales and resolutions. The transient, three-dimensional dispersion data together with the current and deposition flux measurements are needed for the model validation. Additionally, information on in-situ particle characteristics in different transport stages are essential for the modelling of the settling velocity.

2. The experimental and model activities should be concentrated in reference areas in the regions of potential deep seabed mining, where the baseline data are systematically sampled and where the future mining tests will probably be carried out. Models of appropriate resolution and transport scales should be adapted or developed for these regions, using all available data, so that they are suitable for predictions of the future mining impacts.
3. The challenge remains to extend modelling of physical processes to ecological modelling.
4. At present, research groups work independently in different areas, with almost identical subjects and goals. Cooperation should be increased.

References

Amann, H., 1985. Development of ocean mining in the Red Sea. *Marine Mining*, **5**(2), 103–116.

Barnett, B. and Yamauchi, H., 1995. Deep sea sediment resuspension system used for the Japan Deep Sea Impact Experiment. In: *Proceedings of the First (1995) ISOPE Ocean Mining Symposium, Tsukuba, Japan*, 175–179.

Bischoff, J., and Piper, D., Eds., 1979. Marine Geology and Oceanography of the Pacific Manganese Nodule Province, Plenum Press. 839 pp.

Brockett, T. and Richards, C.Z., 1994. Deepsea mining simulator for environmental impact studies. *Sea Technology*, 1994(8), 77–82.

Dyer, K., 1989. Sediment processes in estuaries: future research requirements. *Journal of Geophysical Research*, **C10**, 14327–14332.

Fukushima, T., 1995. Overview Japan Deep-Sea Impact Experiment = JET. In *Proceedings of the First (1995) ISOPE Ocean Mining Symposium, Tsukuba, Japan*, ISOPE, pp. 47–53.

Gross, T., and Dade, W., 1991. Suspended sediment storm modeling. *Marine Geology*, **99**(3/4), 343–360.

Halbach, P., Friedrich, G., and Stackelberg, U., v., Eds., 1988. The manganese nodule belt of the Pacific Ocean, Enke Verlag, Stuttgart. 254 pp.

Hess, W., and Hess, W., 1976. An advection-diffusion model of the DOMES turbidity plumes. Tech. Rep. ERL 382 - OD 13, NOAA.

Hill, P., and Nowell, A., 1995. Comparison of two models of aggregation in continental-shelf bottom boundary layers. *Journal of Geophysical Research*, **100**(C11), 22,749–22,763.

Ichiye, T., and Carnes, M., 1977. Modelling of sediment dispersion during deep ocean mining operations. In *9th Annual Offshore Technology Conference in Houston*, pp. 421–432. Paper OTC 2778.

Jankowski, J., Malcherek, A., and Zielke, W., 1994. Numerical modeling of sediment transport processes caused by deep sea mining discharges. In *Proceedings of the OCEANS 94 Conference, Brest*, vol. III, IEEE/SEE, pp. 269–276.

Jankowski, J., Malcherek, A., and Zielke, W., 1996. Numerical modelling of suspended sediment due to deep-sea mining. *Journal of Geophysical Research*, **101**(C2), 3545–3560.

Klein, H., 1993. Near-bottom currents in the deep Peru Basin, DISCOL Experimental Area. *Deutsche Hydrographische Zeitschrift*, **45**, 31–42

Lavelle, J., 1987. Effects of boundary layer structure and macroparticle scavenging on the benthic deposition of fine sediment resuspended during nodule mining. Final report to noaa office of ocean and coastal resource management, NOAA Pacific Marine Environmental Laboratory, Seattle. 39 pp.

Lavelle, J., and Ozturgut, E., 1981. Dispersion of deep-sea mining particulates and their effect on light in ocean surface layers. *Marine Mining*, **3**(1/2), 185.

Lavelle, J., Ozturgut, E., Baker, E., and Swift, S., 1982. Discharge and surface plume measurements during manganese nodule mining tests in the North Equatorial Pacific. *Marine Environmental Research*, **7**, 51–70.

Lavelle, J., Ozturgut, E., Swift, S., and Ericson, B., 1981. Dispersal and resedimentation of the benthic plume from deep-sea mining operations: a model with calibration. *Marine Mining*, **3**(1/2), 59.

van Leussen, W., 1994. Estuarine macroflocs and their role in fine-grained sediment transport. PhD thesis, University of Utrecht. Utrecht, Netherlands.

Malcherek, A., 1994. Numerical modelling of cohesive settling velocities. *Int. J. of Sediment Research*, **9**, 97–106. special issue Dec. 1994.

Marietta, M., and Simmons, W., 1988. Dispersal of radionuclides in the oceans: models, data sets and regional descriptions. *Feasibility of disposal of high-level radioactive waste into the seabed.* Nuclear Energy Agency, Paris. 462 pp.

McCave, I., 1984. Size-spectra and aggregation of suspended particles in the deep ocean. *Deep-Sea Research*, **31**.

McCave, I., and Gross, T., 1991. In-situ measurements of particle settling velocity in the deep sea. *Marine Geology*, **99**(3/4), 403–412.

McLean, S. R., 1985. Theoretical modelling of deep ocean sediment transport. *Marine Geology*, **66**, 243–265.

MMAJ, 1994. Japan Deep Sea Impact Experiment (JET), August–September in 1994, R/V Yuzhmorgeologiya, Cruise Report, MMAJ, (Metal Mining Agency of Japan), 83 pp.

Monin, A., and Ozmidov, R., 1985. Turbulence in the ocean. D. Reidel Publishing Co., Dordrecht, Boston, Lancaster. 247 p.

Ozturgut, E., and Lavelle, J., 1986. Settling analysis of fine sediments in salt water at concentrations low enough to preclude flocculation. *Marine Geology*, **69**, 353–362.

Ozturgut, E., Lavelle, J., and Erickson, B., 1981. Estimated discharge characteristics of a commercial nodule mining operation. *Marine Mining*, **3**(1/2), 1–13.

Padan, J., 1990. Commercial recovery of deep seabed manganese nodules: twenty years of accomplishments. *Marine Mining*, **9**, 87–103.

Schriever, G., 1995. DISCOL- disturbance and recolonization experiment of a manganese nodule area of the southeastern Pacific. In *Proceedings of the First (1995) ISOPE Ocean Mining Symposium, Tsukuba, Japan*, ISOPE, pp. 163–166.

Schriever, G., Koschinsky, A., Bluhm, H., 1996. Cruise Report ATESEPP (Impact of potential technical interventions on the deep-sea ecosystem of the southeast Pacific off Peru) Report Nr. 11, Zentrum für Meeres- und Klimaforschung der Universität Hamburg, Reihe E: Hydrobiologie und Fischereiwissenschaft, pp. 195.

Segschneider, J., and Sündermann, J., 1995. Large scale transport of suspended matter. *EOS*, **75**(44).

Segschneider, J. and Sündermann, J., 1996, Large-scale transport of suspended matter released by anthropogenic activities, submitted to *Submitted to Journal of Marine Systems.*

Spinrad, R., Bartz, R., and Kitchen, J., 1989. In situ measurements of marine particle settling velocity and size distributions ising the Remote Optical Settling Tube. *Journal of Geophysical Research*, **94**(C1), 931–938.

Taguchi, K., Nakata, K., Aoki, S., and Kubota, M., 1995. Environmental study on the deep-sea mining of manganese nodules in the Northeastern Tropical Pacific. In *Proceedings of the First (1995) ISOPE Ocean Mining Symposium, Tsukuba, Japan*, ISOPE, pp. 167–174.

Thiel, H., 1991. From MESEDA to DISCOL: a new approach to deep-sea mining risk assessments. *Marine Mining*, **10**, 369–386.

Thiel, H., 1995. The German environmental impact research for manganese nodule mining in the SE Pacific Ocean. In *Proceedings of the First (1995) ISOPE Ocean Mining Symposium, Tsukuba, Japan*, ISOPE, pp. 39–45.

Thiel, H., Foell, E., and Schriever, G., 1991. Potential Environmental Effects of Deep Sea Mining. Report Nr. 26, Zentrum für Meeres- und Klimaforschung der Universität Hamburg. 243 pp.

Thiel, H., and Schriever, G., 1990. Deep-sea mining, environmental impact and the DISCOL project. *AMBIO*, **19**(5), 245–250.

Trueblood, D., 1993. US Cruise report for BIE II. NOAA Technical Memorandum NOS OCRM 4, NOAA, National Ocean Service.

Zielke, W., Jankowski, J., Sündermann, J., and Segschneider, J., 1995. Numerical modeling of sediment transport caused by deep sea mining. In *Proceedings of the First (1995) ISOPE Ocean Mining Symposium, Tsukuba, Japan*, ISOPE, pp. 157–161.

Proceedings of the Seventh (1997) International Offshore and Polar Engineering Conference
Honolulu, USA, May 25-30, 1997

ISBN 1-880653-28-1 (Set); ISBN 1-880653-29-X (Vol. I)

Large Scale Transport of Particle Reactive Tracers - Numerical Simulations

J. Segschneider and J. Sündermann
University of Hamburg
Hamburg, Germany

ABSTRACT

The spreading of particle reactive tracers that disperse from a point source in the eastern tropical Pacific is simulated with a global scale Lagrangian transport model. In the application that is presented here, the mining of manganese nodules and the related release of elements like Mn, Pb, Mo, and Cu into the water column are the focus of the study. Observations in the German mining claim in the eastern equatorial Pacific (EEP) show, that concentrations of these elements in the water samples taken close to experimentally stirred-up sediment are about one order of magnitude higher than in undisturbed areas. This rise in concentration can be traced back to the stirring up of pore water. Concentrations will be even higher if manganese nodules are crushed during the process of mining. As a result from the model experiment, three dimensional, time dependent concentrations of the trace elements are calculated. As it is known that hydrographic parameters in the upper EEP show strong interannual variations (El Niño Southern Oscillation cycles) disturbances of the environment caused by the release of tailings into the upper ocean may also vary in response to changed hydrographic conditions. To study these variations, experiments are performed with both El Niño and La Niña boundary conditions.

KEY WORDS

Deep-sea mining, eastern tropical Pacific, particle dynamics, numerical transport modelling, environmental protection, large scale oceanography

INTRODUCTION

Mining of manganese nodules and ore-containing oozes from the bottom of the deep sea may be harmful for the benthic ecosystem (Ozturgut et al., 1981; Thiel et al., 1991). Organisms may be destroyed by the collecting systems or burried by additional sedimentation. Bacteria that are attatched to the stirred-up sediment consume oxygen (Suess, 1980) which reduces the amount of oxygen available for other organisms. For the release of tailings into the upper ocean, the strong interannual variation of hydrographic conditions in the eastern tropical Pacific is important. To investigate the consequences of deep-sea mining for the benthic ecosystem, the interdisciplinary TUSCH-research (Tiefseeumweltschutz) group has been formed. Modelling efforts so far have been restricted to simulate transport paths of the resuspended sediment at the ocean bottom and suspended matter discharged into the near surface ocean (Jankowski et al., 1996; Zielke et al., 1995).

In recent experiments that were performed in the DISCOL area (disturbance and recolonization area, $\approx 88^oW, 7^oS$) enhanced concentrations of several elements such as Mn, Pb, Mo, and Cu have been observed after a stirring-up of the uppermost centimeters of sediment (Schriever et al., 1996). It has been found in previous numerical studies (Zielke et al., 1995) that fine grained particles drift several thousands of kilometers. To study the dispersion of e.g. Mn and Pb though, the particle reactive behaviour of theses elements has to be taken into account, whereas other heavy metals like Mo are only slowly scavenged from the water column. Adsorption of Mn and Pb onto organic macroflocs is thought to be important for the vertical transport of the trace elements (Hunter, 1983; Honeyman et al., 1988). In the numerical model presented here, particle fluxes are calculated with the Hamburg Ocean Carbon Cycle Model (HAMOCC; Maier-Reimer, 1993). To study the effect of scavenging, we applied the large scale model used in previous studies (Zielke et al., 1995) to the transport of dissolved constituents including simplified particle dynamics. In the following sections, the single components of the model and results from two model experiments will be described.

THE OCEAN CIRCULATION MODEL

The hydrographic model that provides the ocean currents in our study is the large scale geostrophic ocean model described in Maier-Reimer et al. (1993) with an enhanced vertical resolution (22 instead of 11 layers). The model works on a global domain with a horizontal resolution of $3.5^o \times 3.5^o$ on an Arakawa E-grid (Arakawa and Lamb, 1977). The model output consists of three dimensional velocity fields, temperature and salinity as well as ice cover and convective overturning. The coarse resolution allows long integration times. Thus it is possible to run the model long enough for the deep ocean currents to equilibrate without relaxation to hydrographic data. To investigate interannual variability of the hydrographic conditions in the EEP, the model is forced with transient boundary conditions, i.e. monthly means of the European Center for Medium Range Weather Forcasting (ECMWF) 1000 mb windstress as described in Winguth et al. (1994). These data extend from 1979 to 1994.

In Fig. 1 the velocity field is shown for an El Niño year (1983) and a La Niña year (1988). It can be seen, that even in the deep layers there is large interannual variability in the simulated ocean currents in response to the variations in the wind forcing. Close to the DEA (DISCOL experimental area) the current reverses from westerly directions in January 1983 to southeasterly directions in January 1988. The velocity is up to several cm/s.
Variations, even reversels of the equatorial surface currents and the equatorial undercurrent have been observed during El Niño events (McPhaden and Hayes, 1990).

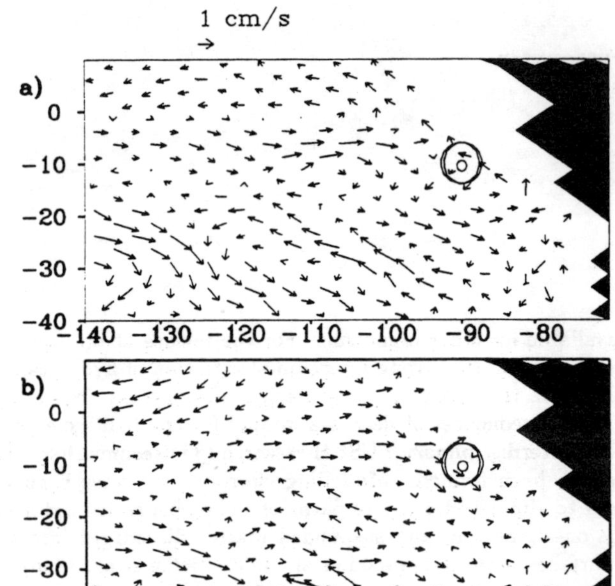

Fig. 1 Horizontal velocities in the two lowermost model layers in the vicinity of the assumed source (marked with a circle). a) January 1983, 3187 m, b) January 1988, 3187 m, c) January 1983, 3683 m, d) January 1988, 3683 m. The velocity scale is on top of a).

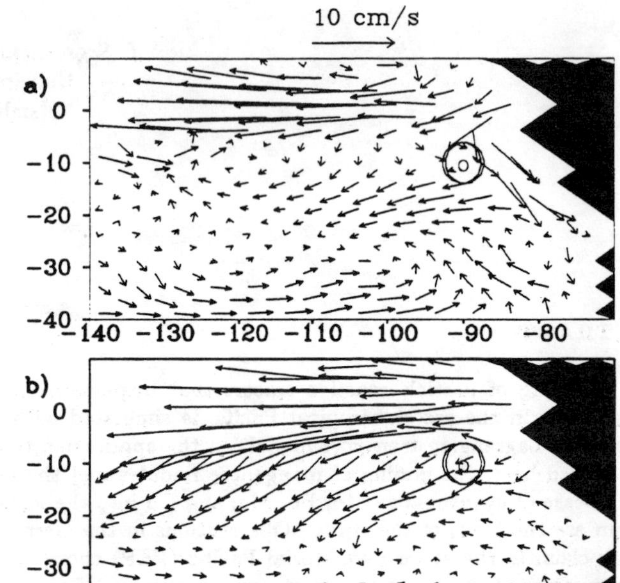

Fig. 2 Horizontal velocities in the two uppermost model layers in the vicinity of the assumed source (marked with a circle). a) January 1983, 25 m, b) January 1988, 25 m, c) January 1983, 80 m, d) January 1988, 80 m. The velocity scale is on top of a).

The computed horizontal velocities are up to several tens of cm/s in the equatorial area, but no more than 5 cm/s in the DEA. During the El Niño year 1983, trade winds were blowing weaker than normally. Consequently, computed currents in the two uppermost layers are slower in 1983 than in 1988, were anomalous strong trade winds were observed. While the surface current in the DEA is directed towards the South American coast in 1983 (Fig. 2a) it is directed westward in 1988 (Fig. 2b). Extensions of the equatorial undercurrent in the DEA, on the other hand, are roughly the same in the two years. Thus, La Niña years are more suitable for surface discharge, if any, than El Niño years.

THE TRANSPORT MODEL

To simulate the transport and diffusion of a dynamically passive tracer, the advection-diffusion equation, i.e.

$$\frac{\partial q}{\partial t} + \sum_i v_i \frac{\partial q}{\partial r_i} - \sum_i \frac{\partial}{\partial r_i}(A_i \frac{\partial q}{\partial r_i}) = S \tag{1}$$

has to be solved, where q is the quantity of the passive tracer, v_i the components of the three-dimensional current vector, r_i the components of the three dimensional location vector, A_i the exchange coefficients and S the source term of the passive tracer. Here, we use a Lagrangian solution technique for eq. (1) to avoid the well known numerical problems of an Eulerian treatment which arise especially in the case of strong gradients of passive tracers (e.g. Gerdes et al., 1991). Furthermore, Lagrangian models allow the direct simulation of transport paths and the computation of residence times. The size distribution of resuspended particles can easily be taken into account as each Lagrangian particle can have different properties while in Eulerian models usually mean settling velocities are prescribed. Lagrangian transport models have been applied extensively to suspended matter transport in coastal seas such as the North Sea (Maier-Reimer, 1973; Maier-Reimer and Sündermann, 1981; Sündermann, 1993, 1994). Here, the use of the Lagrangian scheme is extended to the global ocean and to particle reactive elements. Trajectories for an arbitrary number of Lagrangian tracers are calculated based on the three dimensional velocities of the circulation model with a fourth order Runge Kutta scheme that allows a relatively long timestep of 30 days.

For solving the advection-diffusion equation the examined system can be divided into two subsystems that are separated by the time scale of their fluctuations (Einstein; 1905, Taylor, 1921; Hasselmann, 1976). The advecting currents of the circulation model represent the slowly varying part, whereas the diffusive term, that is formed by randomly distributed and fast changing 'turbulent' velocity contributions, describes the short scale subsystem. For such a system, the probability p to find a particle at location r at time t is determined by a Fokker-Planck type advection-diffusion equation (e.g. Hasselmann, 1976),

$$\frac{\partial p}{\partial t} + \sum_i \frac{\partial}{\partial r_i}(\overline{v}_i p) - \sum_i \frac{\partial}{\partial r_i}(A \frac{\partial p}{\partial r_i}) = 0 \tag{2}$$

where $\overline{v}_i$ are the components of the mean velocity over the ensemble of particles and A is a diffusion coefficient. While investigating Brownian motion, Einstein (1905) showed that A can be expressed as $\frac{1}{2}\frac{\partial \sigma^2}{\partial t}$, which in the case of stationary turbulence ($\frac{\partial < v' >}{\partial t} = 0$, where $< v' >$ is the ensemble average of the turbulent velocity v'), reduces to $\frac{\sigma^2}{2t}$. The variance σ^2 of the displacements caused by the small scale velocities is given by

$$\sigma^2(t) = \frac{1}{N}\sum_{n=1}^{N}(r_{i,n}(t) - \overline{r_i}(t))^2 \tag{3}$$

where N is the number of particles and $\overline{r_i}$ the mean position of the particle ensemble. The position r_i of a particle along its trajectory at time T is determined by its initial location r_{i_0}, and by the displacements due to advection and turbulence:

$$r_i(T) = r_{i_0} + \sum_{t=0}^{T} \overline{v_i}(r_i,t)\Delta t + \sum_{t=0}^{T} b_i v_i'(r_i,t)\Delta t \tag{4}$$

where the small scale velocity fluctuations v_i' are taken randomly out of the interval [-0.5,+0.5] and scaled by a bandwidth b_i. The total diffusion of the particle cloud is then determined by relative diffusion, i.e. the randomly fluctuating small scale velocities, and the dispersion of the cloud that is caused by the velocity shear of the large scale circulation, i.e. the ocean model's currents. As only the dispersion of passive tracers is calculated with the model, no feedback mechanisms between the two subsystems have to be considered. The velocity fluctuations in the diffusive term in eq. (4) are top-hat distributed in our model. It can be shown by use of the central limit theorem that for a large number of realizations (i.e., numerical timesteps) the resulting distribution of the particles in space is Gaussian, where 'large' is on the order of ten (see, e.g., Hammersley and Handscomb, 1964). The central limit theorem theoretically requires an infinite number of particles, but it has been shown that it works for particle numbers in the order of 10^4 in practical applications (Maier-Reimer, 1973).

The dependence of the exchange coefficient on the maximum diffusion velocity v'_{imax}, is given by (e.g. Csanady, 1973)

$$A_i = \frac{1}{6}\, v'^2_{imax}\, \Delta t \tag{5}$$

with

$$v'_{imax} = \frac{b_i\, \Delta x_i}{\Delta t} \tag{6}$$

where Δx_i are the distance between two grid points in the i-th direction, and Δt the model's timestep. The relation between exchange coefficient and bandwidth then reads

$$A_i = \frac{1}{6}\frac{b_i^2\, \Delta x_i^2}{\Delta t} \tag{7}$$

The approach of Taylor (1921) leads to $A_i = v_i' \cdot T_{Li}$ for long time integrations ($T >> T_L$), where T_L is the Lagrangian integral time scale (also called decorrelation time), i.e. the time after which subsequent random movements of a particle are statistically independend. In our model this is the case between two timesteps, i.e. T_L = 30 days.

THE SCAVENGING MODEL

Particle reactive elements like Mn and Pb are scavenged from the water column to the ocean bottom by particles. Naturally, these particles either originate from the ocean's surface (biogenic and terrigenous) or the ocean bottom (hydrothermal vents and resuspension of sediment by bottom currrents). Here, we investigate the role of biogenic particles. Particles of biogenic origin have settling velocities in the order of 100 m/day (Bacon and Anderson, 1982). The production varies with space and time, depending on the availability of nutrients and solar energy in the euphotic zone. Surface production rates of calciumcarbonate (Fig.3), silicate and particulate organic carbon are calculated with the Hamburg oceanic carbon cycle model (Maier Reimer, 1993). With increasing depth, these particles are dissolved and the number of particles and hence the scavenging probability decreases (Honjo et al., 1982). The scale depth for disaggregation of the biogenic particles in the carbon cycle model is 2000 m for calciumcarbonate, 12000 m for silicate and 1000 m for particulate organic carbon. We assume a scavenging probability p_{scav} of 1/1000 in the area of maximal production. Wether or not a Lagrangian particle is 'scavenged' is then determined by a random number ν [0,1], and the actual particle flux F in relation to the maximal particle flux F_{max}:

$$\frac{F}{F_{max}} \cdot p_{scav} = p$$

If ν is smaller than p, the Lagrangian tracer is assumed to be scavenged and its settling velocity is increased to that of the biogenic particles ($1 \times 10^{-4}\ m/s$).

Hydrothermal vent waters are heavily loaded with particles, as can be observed by attenuation measurements. From these 'hydrothermal' particles, additional potential to scavenge dissolved manganese exists. In the DEA however, which is well away from the East Pacific Rise, the influence of hydrothermal vents can be neglected. Other sources of particles which are not included in the model yet are the input of terrigenious debris via the sea surface and the resuspension of sediment at the ocean bottom.

Scavenging rates depends not only on the particle concentration, but also on the adsorption rates of the single elements. A two dimensional model for the dispersion of manganese from a point source is described by Lavelle (1992). For the removal of dissolved manganese from the water column Lavelle assumes a scavenging time scale of 360 days for microorganisms and of 60 days for macroaggregates. The corresponding remobilization time scales are 180 days and 6 days, respectively. We use the same adsorption rates on fine and large particles for our calculations. For the other elements like Pb and Cu, adsorption rates yet have to be determined. The adsorption rates of the different elements also vary with respect to the species of particles. Lead prefers calciumcarbonate as scavenging agent (Broecker and Peng, 1982), whereas manganese is presumanbly scavenged by terrestrial alumino-silicates (von Damm, 1995) and encapsuled bacteria (Cowen et al., 1986) instead. Experiments to determine the adsorption rates of various elements are currently performed by A. Koschinsky at the FU Berlin.

Fig. 3 Annual calcium carbonate production in $g\ C/m^2$ as calculated with the HAMOCC model (Maier-Reimer, 1993).

Resuspension of resettled sediment caused by bottom currents is not included in the model, as the observed bottom currents in the DEA are rather slow (Klein, 1993).

ESTIMATION OF THE INITIAL DISTURBANCE

Initial concentrations of manganese and other elements close to mining activities yet have to be estimated, as no field data are available. Consequently, the calculated concentrations of the particle reactive tracer are shown as % of the initial concentration. This approach allows to re-calibrate predicted estimates when source rates are known.

The geographical location of the source in the model is at $90^o W, 10^o S$. The near bottom source is located at 500 meters above bottom, the near surface source at 50 m below the surface. The additional particle concentration, based on a near bottom source of 50,000 t per day and a near surface source of 4,000 t per day (Thiel et al., 1991) is 6170 μg/l for the bottom release and 494 μg/l for the near surface discharge of tailings. The observed size distribution of the particles in the upper centimeter of sediment in the source region is shown in Tab. 1.

class	f [%]	d [μm]	m[g]	w_s[m/s]	t_{100} [d]
a	14	100	$5.51 \cdot 10^{-7}$	$1.6 \cdot 10^{-4}$	7.2
b	3	60	$1.21 \cdot 10^{-7}$	$7.8 \cdot 10^{-5}$	14.8
c	7	40	$3.63 \cdot 10^{-8}$	$4.2 \cdot 10^{-5}$	27.5
d	36	20	$4.69 \cdot 10^{-9}$	$1.5 \cdot 10^{-5}$	77.1
e	20	4	$6.20 \cdot 10^{-11}$	$1.0 \cdot 10^{-6}$	1157
f	20	0.75	$7.80 \cdot 10^{-13}$	$1.0 \cdot 10^{-7}$	11574.1

Tab.1 : Class names, frequency f, mass m per particle and settling velocity w_s depending on the particle diameter d (McCave, 1975) and the time t_{100} required to sink 100 m for six sediment classes of the uppermost centimeter of sediment in the DEA (Klein, 1993).

The particulate phase of the released tracer is distributed over the particle spectrum or has zero settling velocity if it exists in its dissolved phase. Experiments performed by Koschinsky (Schriever et al., 1996) show, that most of the elements Mn, Fe, Zn, Cu, and Ni in the disturbed suspension exist in a colloidal or particulate phase. Therefore, we assume that 90% of the released tracer is particle bound and only 10% is dissolved. The total release of the tracer per time step (i.e., one month) is normalized to one.

EXPERIMENT SETUP AND RESULTS

One hundred Lagrangian tracers per month are started at 90° W, 10° S over a period of 20 years, resulting in a total release of 24000 tracers. The climatologic currents of the OGCM are used to advect the Lagrangian tracers at the current state of work. Experiments with El Niño and La Niña conditions are planned for the near future. According to Okubo (1971) and the grid length of the ocean model, which is about 500 km, the horizontal exchange coefficient is set to 1000 m^2/s. The Lagrangian tracers have the following properties: mass and settling velocity according to Tab. 1, and a loading with a passive tracer in arbitrary units. The Lagrangian tracers are scavenged by biogenic particles as described in the scavenging model section.

Results are shown after 20 model years as percentage of the initial concentration. These concentrations are determined by counting the Lagrangian tracers in boxes and summing-up their loadings devided by the volume of the box. If the calculated position of the particle is deeper or equal to the water depth, the loading of the particle is added to the bottom concentration of the respective box. Afterwards, the particle is removed from the computation.

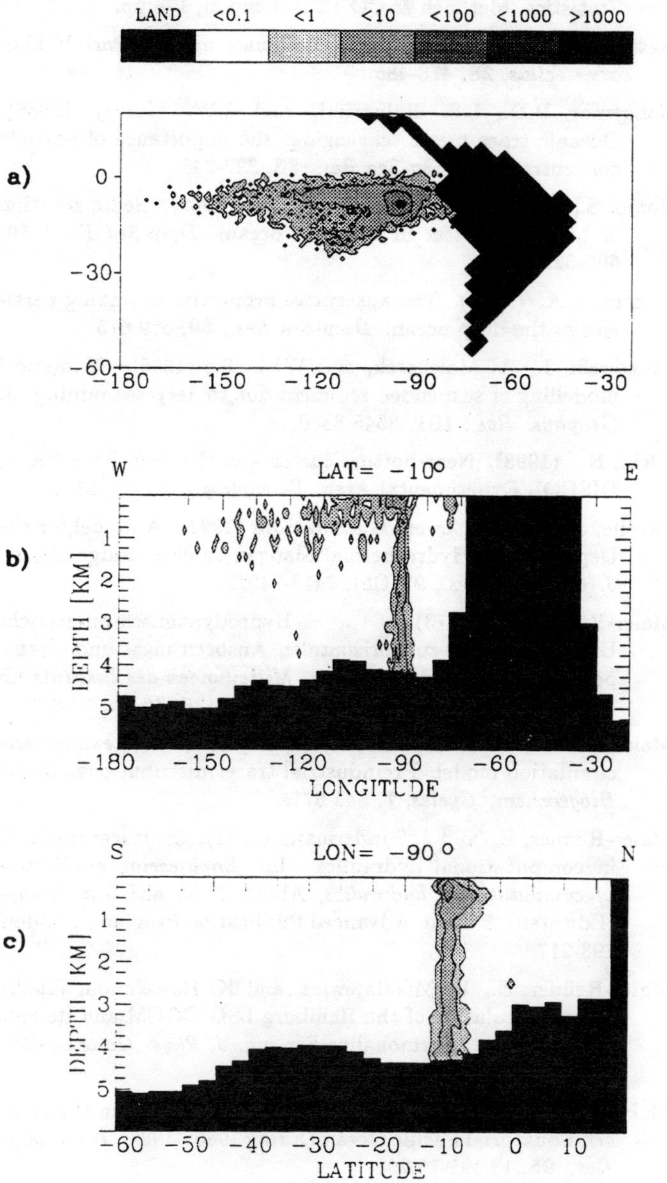

Fig. 4 Surface source: Concentration of passive tracer as % of initial concentration at the end of the integration time. a) vertically averaged, b) zonal section along 10° S, and c) meridional section along 90° W.

The computed tracer concentration after 20 model years, that is at the assumed end of the mining activities, is shown in Fig. 4. The main direction of the spreading is to the west. The major part of the tracers, however, is scavenged by the rapidly settling of large particles and sinks to the ocean bottom within a few hundred kilometers of the source. The zonal section (Fig. 4 b) shows that slow settling fine particles drift as far as 5000 km and more. Small amounts of the tracer are advected to the east by extensions of the equatorial undercurrent, but the concentrations close to the south-american coast are less than 10 % of the initial concentration.

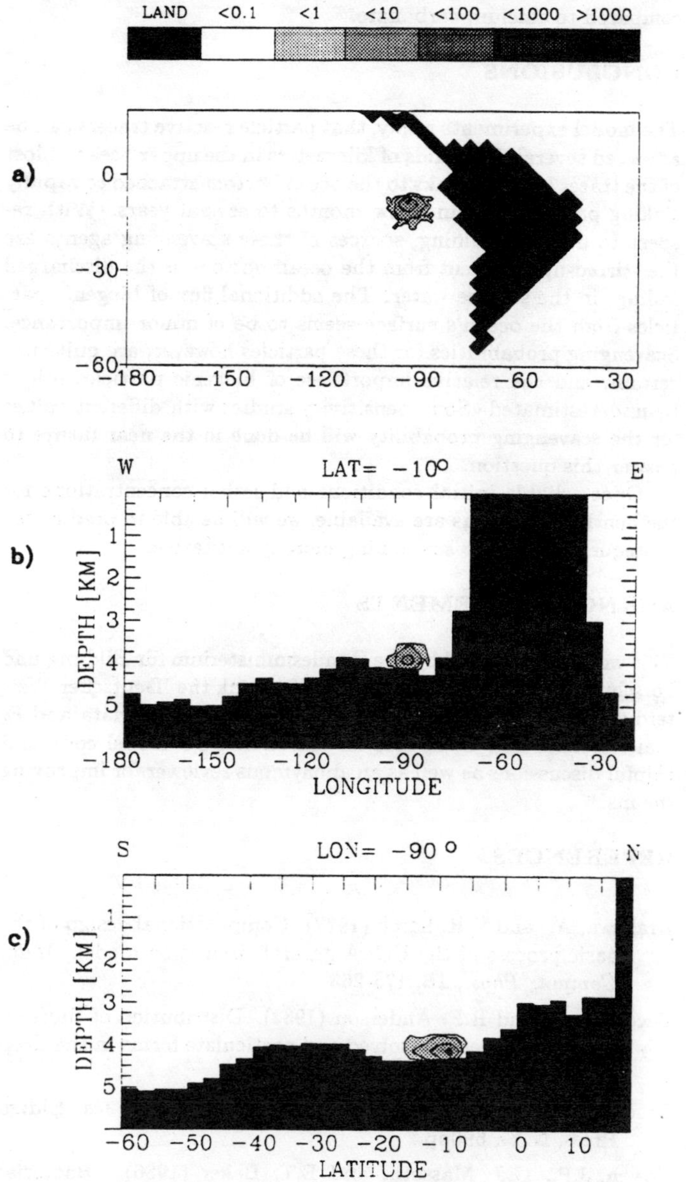

Fig. 5 Bottom source: Concentration of passive tracer as % of initial concentration at the end of the integration time. a) vertically averaged, b) zonal section along 10° S, and c) meridional section along 90° W.

Computed concentrations for the bottom source show, that due to the slower deep-sea currents and the shorter settling time of all particles, the affected area is much smaller than for the discharge of tailings. Consequently, resettling rates are higher for the near bottom source. At the ocean bottom, which is at 4100 m depth in the DEA, the stirred-up and resettling sediment scavenges the dissolved tracer from the water column. Calcium carbonate is dissolved in the upper 3000 m of the ocean and thus no calcareous particles occur at the bottom source depth. Surface production of silicate is confined to a relatively small tongue centered at the equator. Thus the concentration of silicious particles in the DEA is low in spite of the larger scale depth for dissolution of silicate compared to calcium carbonate.

CONCLUSIONS

The model experiments imply, that particle reactive tracers can be advected several thousands of kilometers in the upper ocean. Most of the tracer however sinks to the ocean bottom attached to rapidly sinking particles within a few months to several years. With respect to deep sea mining, sources of these scavenging agents are the stirred-up sediment from the ocean bottom or the discharged tailings in the surface water. The additional flux of biogenic particles from the ocean's surface seems to be of minor importance. Scavenging probabilities for these particles however, are quite uncertain, and the relative importance of biogenic particles might be underestimated. Some sensitivity studies with different values for the scavenging probability will be done in the near future to answer this question.

Once reliable initial conditions and lethal concentrations for the benthic organisms are available, we will be able to predict the consequences of deep-sea mining more quantitative.

ACKNOWLEDGEMENTS

This work is supported by the Bundesministerium für Bildung und Forschung under grant 03G0106D. We thank the 'Deutscher Wetterdienst' for processing and providing the ECMWF-data and E. Maier-Reimer and C. Heinze for the HAMOCC model code and helpful discussions as well as an anonymous reviewer for improving the ms.

REFERENCES

Arakawa, A. and V.R. Lamb (1977). Computational design of the basic process of the UCLA general circulation model. *Meth. Comput. Phys.*, **16**, 173-263.

Bacon, M.P. and R.F. Anderson (1982). Distribution of thorium isotopes between dissolved and particulate forms in the deep sea. *J. Geophys. Res.*, **87**, 2045-2056.

Broecker, W.S. and T.H. Peng (1982). *Tracers in the sea.* Eldigo Press, N.Y., 690pp.

Cowen, J.P., G.J. Massoth, and E.T. Baker (1986). Bacterial scavenging of Mn and Fe in the mid- to far-field hydrothermal particle plume. *Nature*, **322**, 169-171.

Csanady, G.T. (1973). *Turbulent Diffusion in the Environment.* D. Reidel Publishing Company, Dordrecht-Boston, 248 pp.

v. Damm, K.L., (1995). Controls on the Chemistry and Temporal Variability of Seafloor Hydrothermal Fluids. *Seafloor Hydrothermal Systems: Physical, Chemical, Biological, and Geological Interactions.* S.E. Humphries, R.A. Zierenberg, L.S. Mullineaux, R.E. Thomsen (eds). *Geophysical Monograph Series,* bf 91, 222-245.

Einstein, A. (1905). Über die von der molekularkinetischen Theorie der Wärme geforderte Bewegung von in ruhenden Flüssigkeiten suspendierten Teilchen. *Ann. Phys.*, **IV**, 549-560.

Gerdes, R., C. Köberle, and J. Willebrand (1991). The influence of numerical advection schemes on the results of ocean general circulation models. *Climate Dyn.*, **5**, 211-226.

Hammersley, J.M. and D.C. Handscomb (1964). *Monte Carlo Methods.* Methuen's Monographs on Applied Probability and Statistics, Methuen & CO LTD, London, 178 pp.

Hasselmann, K. (1976). Stochastic climate models, Part I: Theory. *Tellus*, **28**, 473-485.

Honeyman, B.D., L.S. Balistrieri, and J.W. Murray (1988). Oceanic trace metal scavenging: the importance of particle concentration. *Deep-Sea Res.*, **35**, 227-246.

Honjo, S., S.J. Manganini, and J.J. Cole (1982). Sedimentation of biogenic matter in the deep ocean. *Deep-Sea Res.*, **29**, 609-625.

Hunter, K.A. (1983). The adsorptive properties of sinking particles in the deep ocean. *Deep-Sea Res.*, **30**, 669-675.

Jankowski, J., A. Malcherek, and W. Zielke (1996). Numerical modelling of suspended sediment due to deep-sea mining. *J. Geophys. Res.*, **101**, 3545-3560.

Klein, H., (1993). Near-bottom currents in the deep Peru Basin, DISCOL Experimental Area. *D. Hydrogr. Z.*, **45**, 31-42.

Lavelle, J.W., J.P. Cowen, G.J. Massoth (1992). A Model for the Deposition of Hydrothermal Manganese Near Ridge Crests. *J. Geophys. Res.*, **97**(C5), 7413-7427.

Maier-Reimer, E. (1973) Hydrodynamisch-numerische Untersuchungen zu horizontalen Ausbreitungs- und Transportvorgängen in der Nordsee. *Mitteilungen des Instituts für Meereskunde der Universität Hamburg*, **26**, 56 pp.

Maier-Reimer, E. (1993). Geochemical cycles in an ocean general circulation model: Preindustrial tracer distributions. *Global Biogeochem. Cycles*, **7**, 645-677.

Maier-Reimer, E. and J. Sündermann (1981). On tracer methods in computational hydraulics. In: *Engineering application of computational hydraulics*, Abott, M.B. and J.A. Cunge (Editoren), Pitman Advanced Publication Program, London, 198-217.

Maier-Reimer, E., U. Mikolajewicz, and K. Hasselmann (1993). Mean Circulation of the Hamburg LSG OGCM and its Sensitivity to the Thermohaline Forcing. *J. Phys. Oceanogr*, **23**, 731-757.

McPhaden, M.P. and S.P. Hayes (1990). Variability in the Eastern Equatorial Pacific Ocean During 1986 - 1988. *J. Geophys. Res.*, **95**, 13,195-13,208.

Okubo, A. (1971). Oceanic diffusion diagrams. *Deep-Sea Res.*, **18**, 789-802.

Ozturgut, E., J.W. Lavelle, and R.E. Burns (1981). Impacts of Manganese Nodule Mining on the Environment: Results from Pilot-Scale Mining Tests in the North Equatorial Pacific. *Marine Environmental Pollution 2, Dumping and Mining* R.A. Geyer (ed), Elsevier Oceanography Series, **27b**, 437-474.

Suess, E. (1980). Particulate organic carbon flux in the oceans–surface productivity and oxygen utilization. *Nature,* **288**, 260-263.

Schriever, G. , A. Koschinsky, and H. Bluhm (1996). Cruise Report ATESEPP: Auswirkungen technischer Eingriffe in das Ökosystem der Tiefsee im Sd-Ost-Pazifik vor Peru. *Berichte aus dem Zentrum für Meeres- und Klimaforschung, Reihe E*, **11**, 195pp.

Sündermann, J. (1993) Suspended Particulate Matter in the North Sea: Field Observations and Modell Simulations. *Philos. Trans. Roy. Soc. London (A)*. **343**, 423-430.

Sündermann, J. (1994) (Editor) *Circulation and Contaminent Fluxes in the North Sea.* Springer Verlag Heidelberg, 654 pp.

Taylor, G.I. (1921). Diffusion by Continuous Movements. *Proceedings of the London Mathematical Society*, **2** (XX), in: The scientific papers of Sir Geofr. Ingman Taylor, Vol II, *Meteorology, Oceanography and Turbulent Flow*, 1960, Cambridge, 172-184.

Thiel, H., E. Foell, and G. Schriever (1991). *Potential Environmental Effects of Deep Sea Mining.* Report Nr. 26, Zentrum für Meeres- und Klimaforschung der Universität Hamburg, 243pp.

Winguth, A.M.E. , M. Heimann, K.D. Kurz, E. Maier-Reimer, U. Mikolajewicz, J. Segschneider (1994). El Niño-southern oscillation related fluctuations of the marine carbon cycle. *Glob. Biogeochem. Cyc.*, **8**(1), 39-63.

Zielke, W. , J.A. Jankowski, J. Sündermann, and J. Segschneider (1995). Numerical Modeling of Sediment Transport Caused by Deep-Sea Mining. *Proceedings of the ISOPE-Ocean Mining Symposium.* 157-162.

Proceedings of the Seventh (1997) International Offshore and Polar Engineering Conference
Honolulu, USA, May 25-30, 1997

ISBN 1-880653-28-1 (Set); ISBN 1-880653-29-X (Vol. I)

Use of Marine Mineral Tailings for Aggregate and Agricultural Applications

John C. Wiltshire
University of Hawaii
Honolulu, HI, USA

ABSTRACT

The disposal of tailings from a marine mining operation could be its greatest environmental problem. Recent work at the University of Hawaii has shown that manganese tailings can be used for a wide range of construction applications. For large tailings volume use however, agriculture and aggregate utilization offer the best potential. Tailings were used in an agriforestry experiment growing Koa, a native Hawaiian hardwood and as fine grained aggregate in concrete. The results in both cases were positive and are presented in this paper. The tailings stimulated the growth of Koa in comparison to strict controls. The tailings added useful properties to the concrete including reducing porosity and increasing strength. The manganese tailings may also give the concrete some resistance to biofouling.

KEY WORDS: Manganese, tailings, marine minerals, agriforestry, concrete

INTRODUCTION

Three factors have significantly increased the likelihood that Pacific manganese nodule and crust resources will be mined at some point in the twenty-first century. These are i) the legal clarification provided by the Law of the Sea Treaty now in full effect, ii) the rapid economic growth in East Asia and iii) the sharp rise in world demand for cobalt for new high tech uses such as batteries for electric cars and high tensile strength super alloys. There are also political and national policy considerations which are driving national programs in several Asian countries. The Chinese Ocean Minerals Research and Development Association (COMRA) has announced plans to begin mining manganese nodules early in the next century. The Chinese government has a U.N. granted manganese nodule lease just a few hundred miles outside the Hawaiian EEZ. COMRA has also recently signed a cooperative agreement with the State of Hawaii for the cooperative investigation of issues related to this industry (December, 1996). The Chinese have inquired of the State of Hawaii as to the possibility of processing manganese nodules on the Island of Hawaii. Korean groups have also announced plans to continue their major mining development work with a view to mine nodules around the year 2010. Other active programs are occurring in India and Japan. This activity will ultimately result in a deep sea mining operation for manganese nodules or crusts (Markussen, 1990).

Much work remains to be done before such an operation could take place in an environmentally sound, sustainable way. Current studies (DOI, 1990) show 80% of the environmental effect of a marine minerals industry is associated with ore processing, particularly with waste (tailings) disposal. Numerous options exist for nodule and crust tailings disposal (Wiltshire, 1993; NOAA, 1981; DOI, 1990). These include ocean dumping, nearshore disposal through an ocean outfall, on land tailings ponds storage and disposal in injection wells. However, all of these potential disposal schemes have significant environmental drawbacks. More sophisticated management of the tailings from a marine minerals industry will be a coming environmental requirement.

As a tailings management strategy, a better alternative than disposal would be the utilization of this waste material. Beneficial uses for marine manganese tailings have previously been documented by Lay *et al*. (1996), Wiltshire (1993,1995), El Swaify and Chromec (1985) and Osaki *et al*. (1987). Initial work has indicated that the tailings have some very useful properties. Current ongoing research at the University of Hawaii indicates that tailings from proposed hydrometallurgical processing of manganese nodules or crusts can be made into useful construction products. These include tiles, ceramics, coatings, resin castings for plumbing fixtures, additives to strengthen concrete, stack gas cleaners, rust-proofing and antibiofouling coverings. We have had special success with the tailings as a filler for plastics, as a glaze additive, as a soil amendment, and as an anti-termite and anti-rust coating. The economics of such usage appears favorable (Loudat *et al*, 1994). This approach is an innovative strategy for tailings management and reduces potential environmental damage associated with marine mining.

A significant constraint encountered in our initial research effort was the lack of manganese tailings. We obtained twenty-five tons of Clarion-Clipperton nodules from Kennecott Copper and used a bench scale sulfuric acid leach system to make tailings. Work by the U.S. Bureau of Mines (DOI, 1990) has shown that acid leach tailings of manganese nodules and manganese crusts are virtually identical. These tailings had to be made in small quantities, limiting the experimentation that could be done and greatly increasing expenses. With the help of the

Broken Hill Proprietary Company (BHP) we determined that tailings from their largest manganese mine at Groote Eylandt, Northern Australia were virtually identical to leach nodule and crust tailings both chemically and in grain size. Considerable work was done to quantify the properties of the tailings and to insure that for all of the types of work we were undertaking there was no significant difference between the BHP tailings and the nodule tailings. This work was done in association with the Department of Mining and Metallurgical Engineering at the Technical University of Nova Scotia and is presented in Lay *et al.* (1996). The result was not surprising as the Groote Eylandt deposit is a marine manganese crust deposit which is now uplifted on land on an island a few tens of feet out of the water (William A. Scheel, BHP Mine Manager, Groote Eylandt, personal communication). BHP, one of the world's largest mining companies with 50,000 employees and $20 billion in annual sales is interested in this project and has shipped 3 tons of tailings to the project. This allowed sufficient quantities of tailings for full experimentation to document the wide range of the beneficial uses.

However, it is necessary to focus on very large volume uses of tailings to be certain to be able to use all the waste produced. For this reason, more work is being focused in the areas of agricultural applications and aggregate for concrete. As aggregate, the tailings appear to slightly increase concrete strength raising compression strength to 5000 psi from a standard 3500 psi. The tailings also appear to reduce concrete porosity which can be a very useful characteristic in the marine environment or in pond liners. For applications requiring increased grain size the tailings can be pelletized. In much of the South Pacific, there is a major aggregate shortage. Coral reefs are being dredged up to supply aggregate for the construction industry. In itself, this practice is very environmentally destructive. A significant new source of aggregate could be supplied from tailings. The use of tailings, in time, could be a major environmental asset to planned development. In Hawaii, which has over the years been referred to as a future processing site for marine minerals, the government is especially concerned over any tailings disposal issues related to an industry such as marine mineral processing. It appears that tailings may be able to be used to rehabilitate old lava land and turn it into useful agricultural land, particularly for agriforestry. Ongoing research is looking at the apparent stimulation of the growth of a native hardwood, Koa, by the addition of tailings to poor native soil or crushed lava cinders. Koa is highly prized for woodworking. This offers the possibility of developing a whole new tailings industry ancillary to a manganese nodule or crust processing industry, providing new jobs as well as removing the problems of waste disposal.

This paper outlines the results of a series of experiments aimed at agriforestry uses of the tailings as well as experiments trying to quantify the role of the tailings as aggregate in concrete. Manganese tailings do have a series of unusual properties which must be highlighted before the agricultural and aggregate applications can be fully appreciated.

MANGANESE TAILINGS PROPERTIES

Two basic routes exist for processing manganese nodules or crusts: acid leaching and smelting. Numerous processes have been devised for each. The tailings of an acid leach process after drying are a fine grained black powder. A smelting process produces smelter slag which is blocky and often inert. It can be used as road ballast if it is sufficiently inert to resist rainwater leach into aquifers. Generally smelter slag is not the disposal problem that acid leach tailings are. For this reason and the fact that acid leach processes are less energy intensive and therefore potentially favored in Pacific Island areas, this study focused solely on the acid leach manganese tailings. Many terrestrial manganese mines, such as BHP's Groote Eylandt Mine, beneficiate and separate the manganese size fractions by physical processes that do not involve acid. After extensive testing, we found the fine grained tailings produced by such processes to be very analogous to the marine mineral acid leach tailings after the acid residues are neutralized. As these analogous terrestrial tailings are abundant and consistent in composition between batches they were the focus of this study.

Chemical and physical property tests of the tailings used in this study, fine grained manganese tailings from the BHP tailings ponds at their Groote Eylandt Manganese Mine, were done at the University of Hawaii as well as by BHP and Micromeritics Corporation (for particle size distribution and gas absorption). Similar detailed tailing studies have been done for a range of tailings by Haynes *et al.* (1985a, 1985b, 1982). Summary data are presented in Table 1 for chemical composition and Table 2 for other properties. Basically the tailings are small angular manganese dioxide grains intermixed with clay and small amounts of iron oxide and silica. They pack well because of their angularity but have a high surface area.

Table 1: Chemical Composition

Oxide	Percentage	Elemental Percentages	%
MnO_2	36.0	Mn	22.6
SiO_2	28.1	Si	13.1
Al_2O_3	20.3	Al	10.7
FeO	5.6	Fe	4.4
BaO	0.60	Ba	0.54
TiO_2	0.48	Ti	0.29
K_2O	0.3	K	0.25
MgO	0.14	Mg	0.08
SrO	0.04	Sr	0.03
CaO	0.04	Ca	0.02
Na_2O	~0.5 (estimate)		
H_2O	~8.0 (estimate)		

Table 2: General Chemical Properties of the Tailings

Specific Gravity	3.46
Melting Range	1240-1285°C
Ph	5.5 (in water suspension)
Grain Size Range	0.1 to 300 microns
Median Grain Size	3 microns, 50 % of sample mass was between 0.7 and 9 microns
Mineralogy	36% pyrolusite, 51% silicates (largely kaolinite), trace quartz
Magnetic Properties	Pyrolusite (paramagnetic), Kaolinite (non-magnetic)
Hardness	6 on the Moh's scale
Particle Surface Area	very high, 26.9 m^2/g
Average Pore Diameter	146 angstroms
Gas Absorption	0.30 cm^3/g absorption of CO gas
Electrical Properties	Resistor at all temperatures
Solubility	Insoluble in water
Appearance	Gray-black , fine grained powder, no odor
Toxicity	Non-toxic (standard EPA type toxicity tests run by BHP)

SOIL AMENDMENT EXPERIMENT

The agricultural potential of ferromanganese tailings was investigated by El Swaify and Chromec (1985). Even though these ferromanganese tailings were of marine origin,[1] their characteristics are similar to the Groote Eylandt tailings. There were several primary conclusions of the El Swaify study related to the

[1] Due to their marine origin, crust or nodules tailings or have a high salinity level if not thoroughly washed. This is not true of Groote Eylandt manganese tailings given their terrestrial origin.

agricultural potential of the tailings. First, with proper management, normal plant growth(corn was used) is possible using tailings as the growing medium. To achieve normal growth, however, large initial inputs of lime[2] and major nutrients, particularly phosphorous, are required. Thereafter, much smaller maintenance additions should be sufficient to sustain such growth. Additionally, to further enhance growth, micronutrient additions may also be required or additions of manure or compost.[3]

The second primary conclusion of the El Swaify and Chromec study relates to the use of tailings as a soil amendment. Tailings were added to four different soils at a rate of 0-20 percent by weight. The impact of the tailings on: particle size, aggregate stability, dispersion, slaking, swelling potential and water retention were measured on the four soil types. The results are shown in Table 3.

Table 3: Tailings impact on soil characteristics

Soil	Aggregation	Aggregate Stability	Dispersion	Slaking	Swelling Potential	Water Retention
Lualualei	promotes	increase	no effect	none	decreased	no effect
Ewa	promotes	increase	no effect	none	decreased	increase
Hilo	reduces	no effect	no effect	none	no effect	decrease
Kapapala	weak	increase	no effect	at 20%	decreased	decrease

Table Notes: Lualualei is a very fine, montmorillonitic, isohyperthermic family of Typic Chromustert dominated by montmorillonite which exhibits a high CEC (cation exchange capacity) and shrink potential. Ewa is a fine kaolintic, isohyperthermic family of Aridic Haplustoll with a high content of hallosyte and is less reactive than the Lualualei. Hilo is a thixotropic isohyperthermic family of Typic Hydrandept formed under constant moisture conditions on layers of volcanic ash that have weathered into amorphous aluminosilicates, complexed with organic matter, which dehydrate irreversibly. Kapapala is a medial isomesic family of Typic Dystrandept and is less than 60% volcanic ash, not thixotropic and is dominated by amorphous materials.

The tailings interact most strongly with montmorillonitic soils (2:1 clay minerals) significantly increasing the amount of the silt size secondary particles and reducing the swelling potential. This appears to be related to the charge characteristics of the soil (El Swaify and Chromec, 1985). Impacts of tailings on other soil characteristics for the different soils used were either not evident or more weakly expressed.

The experiment undertaken here investigated the agricultural potential of Groote Eylandt tailings as a soil amendment. To that end we devised an experiment to determine the impact of the tailings soil application on the growth of Koa trees. The site of the experiment is Wood Valley which is located in the Ka'u district of the Island of Hawaii. The site elevation is approximately 670 meters, annual rainfall is approximately 254 cm per year. Wood Valley is adjacent to Kapapala. Thus, its soils and the impact of tailings added to Wood Valley soil as an amendment would be expected to be most similar to the Kapapala soil tested in the earlier El Swaify experiments. Our experiment does not, however, measure the impact of tailings applications on the soils, but rather, the experiment measures impact on the growth of Koa trees.

Experiment Details

On June 14, 1995, in preparation for field trials, various levels of Groote Eylandt tailings and amendments were mixed with soil located on site. The different levels of tailings and amendments used are shown in Table 4 on a ton per acre basis. Sets 1 and 2 are controls. Large soil and tailings clumps were broken into their smallest aggregate pieces before mixing the two.

Table 4: Tailings and Amendment Levels

	Tons per Acre			
	KMg	P-Rock	Calcium	Tailings
Four Pot Set - 1	0.0	0.0	0.0	0.0
Four Pot Set - 2	0.5	0.5	3.0	0.0
Four Pot Set - 3	0.5	0.5	3.0	7.5
Four Pot Set - 4	0.5	0.5	3.0	15.5
Four Pot Set - 5	0.5	0.5	3.0	25.0
Four Pot Set - 6	0.5	0.5	3.0	33.0

Table Notes: KMg is potassium/magnesium, P-Rock is granulated phosphate rock, and calcium is crushed coral ($CaCo_3$).

The amendment levels are based on applications made by regional Koa growers. The alternative tailings levels were chosen to obtain an adequate range of growth responses to tailings application.

An additional four treatments were also devised using mixtures of soil and tailings only. This trial was conducted to determine if high concentrations of tailings have any toxic effects on Koa growth. The ratios of soil to tailings used is shown in Table 5.

Table 5: Soil and tailings ratios in no amendment trial

	Soil to Tailings Ratio
Four Pot Set - 7	19 to 1
Four Pot Set - 8	9 to 1
Four Pot Set - 9	4 to 1
Four Pot Set - 0	3 to 2

Measurements

Koa seedling growth measurements were made on January 13, 1996, 6 months after planting and April 22, 1996, 10 months after planting or 3 months after the first measurement. Measurements made were stem diameter (mm), height (mm) and number of leaves which was made for the first measurement date only. The experiment was terminated after the second measurement as the Koa trees had become root bound.

[2] The tailings used by El Swaify and Chromec (1985) were from a high temperature-pressure sulfuric acid leach process. Thus, their pH would be very low (pH 2-3) and consequently require high lime inputs to raise the pH to levels allowing normal plant growth. Groote Eylandt slimes would not require such remediation to support plant growth since their pH is 5.5, a pH well within the pH bounds required for normal plant growth. Another noteworthy point related to pH is that optimum tailings pH was found to be about 5. Poor growth occurred at higher pH due to the reduced availability of micronutrients, especially zinc.

[3] Adding chicken manure to the tailings in addition to the basal fertilizer applications tripled yields. This was thought to have been due to an increase in pH (to about 6), coupled with an improvement in the K:Ca+Mg balance of the plants. The compost addition also improved phosphate uptake and may have increased the availability of micronutrients through the effects of chelation. This result and the high initial nodule tailings phosphorous requirement, suggest that an inexpensive source of organic matter, such as sewage sludge, might be used to partially meet the phosphorous requirement, and, depending on the type of sewage sludge, fulfill the lime requirement.

No Amendments

Figure 1 shows the relationship between tailings as a percentage of the total soil mixture and the various growth measures for the 2 measurement dates. Koa tree growth for each measure and measurement date shows a dramatic positive response to the addition of the tailings (5%). This increased growth rate is also noted for the 10% addition but decreases for the 20% and 40% tailings mixtures. This pattern remained the same for the two measurement dates. When the pots were examined the tailings in the higher percentage range had clumped, making the soil surface very hard in these pots. This likely inhibited water and nutrient absorption and resulted in less stimulation of growth than may have occurred had the soil been broken up (the higher tailings concentration plants are behaving like the plants at no tailings addition). We believe it is this physical soil parameter that is causing the decrease in Koa growth stimulation at the higher tailings concentrations. In any case, there is no evidence for tailings toxicity even at our highest tailings concentration in the soil.

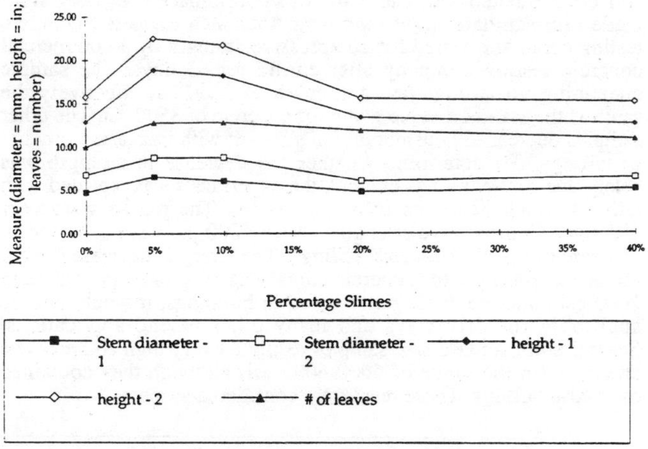

Figure 1: Growth of Koa seedlings in soil with increasing percentages of tailings added and no other amendments as noted in table 5. The graph indicates stimulation of Koa growth at lower percentages of tailings added but not at higher percentages. There is no evidence of tailings toxicity.

In summary, Koa seedling growth with the tailings across all the measures is higher than the control until a concentration of 20 percent tailings is reached at which point growth is about the same. Even though Figure 1 shows a positive growth response to tailings addition to the soil at a 5 percent level, the difference from the zero percent tailings control is not statistically overwhelming.[4]

With Amendments

Figure 2 shows the relationship between tailings application on a ton per acre basis and the various growth measures when amendments are added. Koa seedling growth for each measure shows a surprising initial negative response to the addition of very small amounts of tailings when other soil amendments are added to the soil/tailings mix. Relative to the control (*i.e.* amendments but no tailings) for measurement 2, the difference is statistically significant.[5] Nonetheless, this is an anomalous result given Koa growth at higher tailings levels. In that it only represents four pots, other artifacts of pot placement, watering or variation in seedlings could easily explain this result (e.g. two sickly plants out of the four bringing the average down).

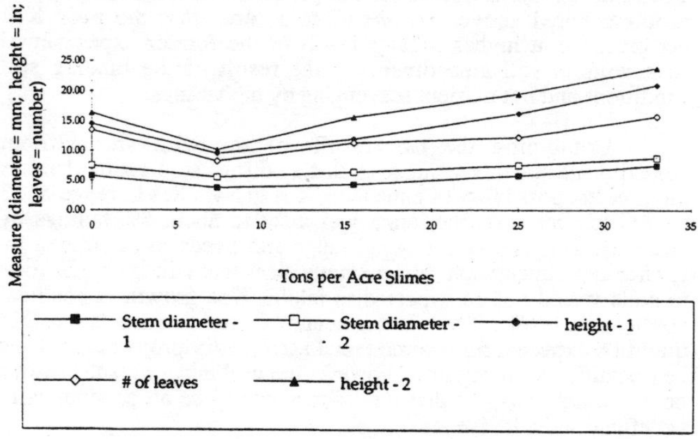

Figure 2: Growth of Koa seedlings with increasing application of tailings to the soil where other amendments have already been added as noted in table 4. This shows a clear response of increased Koa growth with increased tailings applications

Koa growth steadily increases with higher additions of tailings, until the two highest levels of tailings application levels are attained (*i.e.* 25 and 33 tons per acre). The growth measures are markedly higher for these higher tailings application levels. Nonetheless as a statistical caveat, given the small sample size, the differences from the control are statistically significant in only one instance.[6] This statistically significant height difference between the control and the second height measurement suggests an increasing Koa growth response from the tailings as the trees age given that this result did not appear until the second measurement was made.

Discussion

For the "no amendment" situation the following was observed. An initial positive growth response occurred with the addition of tailings which subsequently dampened at higher tailings application levels. Such a result is not surprising given the fact that the tailings were observed to compact into a 'hard pan' soil at the higher concentrations which would be expected to lower plant nutrient and moisture availability. An alternate explanation to this would be that the tailings because of their large surface area and scavenging nature, bind plant nutrients that would otherwise be available, hence reducing growth. If this latter explanation were true one would expect that the higher tailings to soil ratios with no amendment addition would lead to poorer Koa growth. Consistent with both of these explanations is the fact that for measurement 2, Koa tree height and diameter growth were slightly less than the control for the higher tailings application amounts (*i.e.* 20 and 40%). This was not the case at the time measurement 1 was made. Even though none of these differences for measurement 1 or 2 are statistically significant, this result suggests that in pot bound plants, with time, higher levels of tailings applications may appear to dampen Koa growth.

When amendments were added to the soil/tailings mix, the low tailings addition growth response was negative which became much greater than the control (*i.e.* positive) at the higher tailings application levels. This difference in growth from the control does not become positive until higher tailings application levels. This suggests that the tailings are not only not binding plant nutrients at these tailings application levels when amendments are added, but

[4] The highest level of significance measured (*i.e.* p-value) was 0.23.

[5] The measured level of significance (*i.e.* p-value) was 0.002 for the difference in height between the control and measurement 2 height for 7.5 tons of slimes and 0.058 for the difference in diameter between the control and measurement 2 diameter for 15.0 tons of slimes.

[6] The measured level of significance (*i.e.* p-value) was 0.009 for the difference in height between the control and measurement 2 height for 33.0 tons of slimes per acre.

they are creating some soil condition that leads to positive growth of Koa. As the overall tailings levels were lower in the case of the amendment experiment, we did not get the soil clumping/'hard pan' problem noted above. We therefore assume that the poor Koa performance at higher tailings levels in the former experimental case without soil amendments is the result of the binding soil conditions and not nutrient scavenging by the tailings.

Combining the two results of the with and without amendments situations, suggests that a tailings application that has the greatest possibility of enhancing Koa growth lies in range of 33 to 500 ton per acre (the latter equivalent to about 5% tailings by volume). This is a rather broad range and needs to be refined by further experimentation. It is certainly clear that tailings application to soils would not be expected to inhibit Koa growth at practical application levels. The exact amount of growth stimulation that might be expected for a commercial agriforestry project cannot yet be quantified with certainty, however the preliminary results shown here certainly suggest that the tailings would be an economically beneficial soil additive.

Suggested Future Research

The initial very economically significant results documented by these Koa experiments should be followed up by several larger, more focused studies. Such studies should include both pot and field components.

Pot Experiments

Future pot experimentation is necessary to validate and possibly expand on the conclusions drawn in the above discussion. It should be conducted wherein the following occurs:

1. Pre-selection of uniform Koa seedlings for transplanting to test pots to reduce the measurement impacts of high Koa genetic variability. High Koa genetic variability causes large variances which may have partially masked the differences between the various controls and tailings treatments in the first round of experiments.
2. A reduced treatment set, perhaps 5 tailings concentrations, with more trials (pots) per treatment, perhaps 10 instead of four. Treatments would be different tailings applications ranging from 5 to 50 tons per acre (or whatever would be considered a practical amendment limit) with the standard amendment application for Koa.

Field Experiments

After the results of the second set of pot experiments, a Koa field experiment using tailings at different application levels would be undertaken for confirmation of the initial results over a longer period of time and in field conditions similar to those experienced in agriforestry. The objective of such experimentation would be not only to confirm the tailings impacts on Koa growth but secondly, in the long-term, to assess the tailings impact on Koa wood color and possible wood termite resistance due to uptake of Mn into the wood due to higher soil concentrations.

In the pot experiments done in the first round of work only one seedling was grown per pot. The seedlings were selected from groups of plants collected in the wild and grown to a viable size for transplantation in trays. This process coupled with the high genetic variability of Koa introduced considerable statistical variability into the experiment. In the follow-up phase of field experiments a much larger number of seedlings will be used with a narrower range of tailings additions to give better statistical control.

AGGREGATE FOR CONCRETE

The American Concrete Institute (1990) has done a considerable amount of work on admixtures to concrete. This includes the successful incorporation of superplasticizers to increase concrete durability and blast furnace slag to increase strength and weight. Specialty concretes for the marine environment have been given a lot of study. Another area of major ongoing research is the texturing of concrete pavements particularly to provide tough skid-resistant roadways. This is done both by sculpturing the concrete with grooves and also by adding gritty material to the concrete. Although mineral tailings other than blast furnace slag were not used directly in this work, the approaches and testing methods are applicable.

Our concrete work began by testing varying mixtures of Portland cement, coarse sand and tailings. Mixtures were made with 0-60% tailings in the concrete. Two sets of experiments were conducted (see figure 3). The first involved nodule tailings which were not fully neutralized after acid leach processing. These tailings had been washed and had a pH of approximately 4. They were made into standard eight inch long, four inch diameter, concrete testing cores and tested for compressive strength by a commercial concrete testing company after curing for 33 days. The sample containing no tailings had a strength of 3460 psi effectively the same as the standard value given for concrete of 3500 psi. The other samples decreased considerably in strength with increasing amount of tailings. To determine whether the decrease in strength was solely due to the tailings content, the experiment was repeated with tailings which had been fully neutralized. The results were very different. Compressive strengths above 4000 psi were achieved in concrete containing 20-25% tailings. The strength decreased fairly rapidly to 1000 psi for concrete containing 50% tailings. Although 1000 psi concrete could not be used in buildings, it would still be applicable for driveways and many other paving applications. Several non-standard size samples exhibited very high compressive strengths (in the range of 5000-6000 psi) although they contained over 40% tailings. These results are considered suspect.

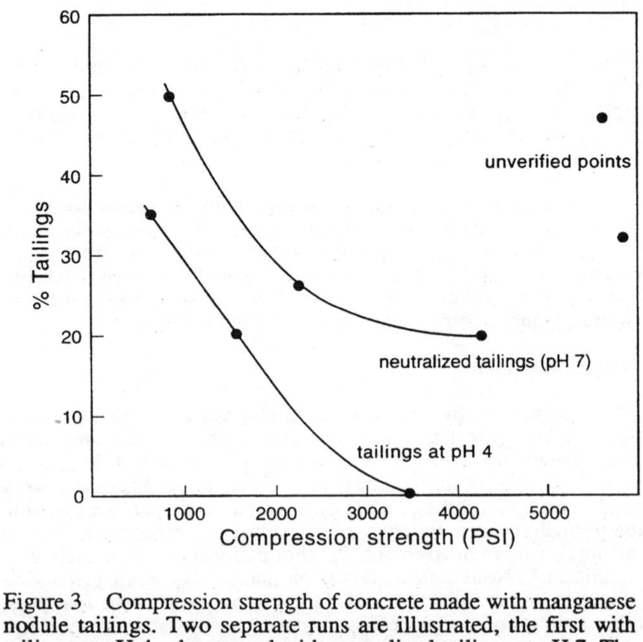

Figure 3 Compression strength of concrete made with manganese nodule tailings. Two separate runs are illustrated, the first with tailings at pH 4, the second with neutralized tailings at pH 7. The results clearly show that tailings must be neutralized and when this is done, concrete strengths in excess of the standard 3500 psi can be achieved by tailings additions of about 20%.

In addition to increased strength the tailings appear to give the concrete several other interesting properties. The fine grained nature of the tailings appears to make the concrete more moldable and bubble-free. This was demonstrated in two ways. First, a moldability test was performed using a 12 inch by 4 inch latex rubber mold of an ornamental Japanese fish. The mold was made with considerable attention to fine detail. This fine detail was picked up in a casting using concrete containing 30% tailings but not by the standard concrete. Further, the ferromanganese concrete gave a much smoother bubble-free surface. The surface textures were compared in another test by examining cut surfaces of the ferromanganese tailings concrete and a standard precast concrete brick. The ferromanganese surface had bubble pits over less than 2% of its surface. These pits ranged in size from pinholes to 2 mm in diameter averaging about 500 microns. By contrast, the precast brick had pits covering over 20% of its surface ranging to 8 mm in diameter and averaging about 2 mm in diameter. The difference was very marked. The precast brick had a rough surface, the ferromanganese concrete a very smooth surface. In addition, the tailings give the ferromanganese concrete a considerably greater density than regular concrete in that iron and manganese are being substituted for the less dense silica and aluminum of sand and the ferromanganese concrete has less than one tenth the bubbles found in regular concrete.

Greater density and bubble reduction are two particularly important properties for concrete to be used in the marine environment or in freeze-thaw situations. In the marine environment, wave action compresses air into the pores and pits of the concrete which over time breaks the concrete down. The same action results from water freezing and expanding in these pores. The lower number of pores subjected to this action the longer the concrete will endure. There is evidence to indicate that ferromanganese surfaces repel the growth of organisms (Department of the Interior, 1990). If this can be confirmed, it may be that ferromanganese concrete would not be covered as quickly with algae or encrusting organisms which would offer advantages for outfall pipes as well as many other marine and terrestrial structures.

When the physical properties of the tailings are related to the characteristics of the resulting concrete it becomes clear what is happening. The tailings have a specific gravity of 3.46, much higher than normal silica sand. They are making the concrete denser. The tailings have a very high surface area which rapidly absorbs water. This makes the concrete dry faster. It also causes the fine clay grains in the tailings to expand to fill voids in the concrete causing the concrete to have lower pore space and better moldability. The greater density, reduced pore space and better grain linking, caused by this void filling, in turn translate into higher compression strength. Evidently this advantage begins to be lost above an addition of about twenty percent tailings to the concrete mix when the lack of coarse grain aggregate becomes important presumably after all the small voids have already been filled by the tailings.

Clearly there are considerable avenues for further research on the properties of ferromanganese tailings concrete. Initial indications of increased compressive strength, superior moldability, higher density and lower porosity give reason to believe that particularly for specialty concretes the addition of the ferromanganese waste is imparting very economically desirable characteristics.

CONCLUSIONS

Initial work in agricultural applications and aggregate in concrete indicates considerable promise for large volume usage of manganese tailings from future manganese nodule or crust mining operations in the twenty-first century. These first results are preliminary and must be extended and repeated before completely definitive conclusions can be drawn. Initial economic studies indicate that these proposed uses should be economic and able to use most if not all of the tailings derived from an acid leach crust or nodule operation. This is exciting news if it is correct. It means that tailings need not be by definition a waste and a cost for an ocean mining operation but could provide a byproduct industry that might contribute to the overall economic viability of ocean mining. However, before any future ocean miner would commit to this innovative kind of tailings management, there would have to be a large scale test of the concept that demonstrated feasibility in real world economic terms. This would need to done before there were large volumes of nodule tailings. In fact, these demonstrations would need to be completed before the nodule processing plant and its disposal systems were designed. Fortunately, the tailings of one terrestrial manganese mine are sufficiently similar to nodule tailings to allow a test. It should be possible, over the next decade to experiment with the Groote Eylandt tailings and optimize a system for their utilization. Once successfully commercially demonstrated at this mine, the manganese tailings problem will be solved for the future ocean miner.

ACKNOWLEDGMENTS

This paper is funded in part by a grant from the National Oceanic and Atmospheric Administration, project #R/MC-1 which is sponsored by the University of Hawaii Sea Grant College Program, SOEST, under Institutional Grant No. NA36RG0507 from NOAA Office of Sea Grant, Department of Commerce. The views expressed herein are those of the author and do not necessarily reflect the views of NOAA or any of its sub-agencies. This is UNIHI-SEAGRANT - CP-97-02. This research has also been partially supported by the Department of the Interior's Mineral Institute Program administered by the U.S. Minerals Management Service through the Generic Mineral Technology Center for Marine Minerals under grant number G1115128. The Broken Hill Proprietary Company (BHP) also contributed critical funding. This work is State of Hawaii Ocean Resources Branch Contribution Number 130.

REFERENCES

American Concrete Institute. 1990. Manual of Concrete Practice. ACI, Detroit. MI.

Aplin, C.L. and Argall, George O. 1973. Tailings Disposal Today. Proceedings of the First International Tailings Symposium held 31 October-3 November 1972 in Tucson. Miller Freeman Publications. San Francisco, 861 p.

El Swaify S. and W. Chromec. 1985. The Agricultural Potential Of Manganese Nodule Waste Material. in P. Humphrey (ed.) Marine Mining: A New Beginning. State of Hawaii, Department of Planning and Economic Development, Honolulu, 319 pp.

Haynes B. and S. Law. 1982. Predicted Characteristics of Waste Materials from the Processing of Manganese Nodules. BU Mines IC 8904, 10 pp.

Haynes B. *et al.* 1985a. Characterization of Waste Materials from Manganese Nodule Processing, in P. Humphrey (ed.) Marine Mining: A New Beginning. State of Hawaii, Department of Planning and Economic Development, Honolulu, 319 pp.

Haynes B. , D. Barron, G. Kramer, R. Maeda and M. Magyar. 1985b. Laboratory Processing and Characterization of Waste Materials From Manganese Nodules. Bureau of Mines Report of Investigations, RI 8938. 16pp.

Johnson, C.J. 1990. Future Technical and Economic Dimensions of Deep-Sea Mining. Materials and Society v.14 p.209-223.

Loudat, T.A. 1992. An Economic Analysis of Manganese Crust Mining. Pacific Congress on Marine Science and Technology Abstracts. PACON International, Honolulu.

Markussen, J.M. 1990. Commercial Exploitation of Polymetallic Nodules: When, Why, Who and How. Materials and Society . v.14 p.397-413.

National Oceanic and Atmospheric Administration (NOAA). 1981. Deep Seabed Mining: Final Programmatic Environmental Impact Statement. U.S. Department of Commerce. Washington. 295 pp.

National Oceanic and Atmospheric Administration (NOAA). 1991. Deep Seabed Mining: Report to Congress. U.S. Department of Commerce. Washington. 14 pp.

Osaki, K., Hamada, Y. and Ichijo, M. 1987. Basic Study for Production of Light Weight Aggregate from Leaching Residue of Manganese Nodules, Journal of the Mining and Metallurgical Institute of Japan. vol. 103, no. 1197, 793-778.

U.S. Department of the Interior (DOI), Minerals Management Service. 1990. Final EIS: Proposed Marine Mineral Lease Sale in the Exclusive Economic Zone Adjacent to Hawaii and Johnston Island . State of Hawaii, Department of Business and Economic Development and Tourism (publisher). Honolulu.

Wiltshire, J. 1991. Application of Ferromanganese Marine Mineral Tailings in Concrete and Ceramics. Proceedings of the 22nd Underwater Mining Institute. September 29-October 2, 1991. Marine Minerals Technology Center. Honolulu.

Wiltshire, J. 1992. Beneficial Uses of Ferromanganese Marine Mineral Tailings. Pacific Congress on Marine Science and Technology Abstracts. PACON International. Honolulu.

Woolsey J. R., 1992, Compliance Possibilities and Industrial Opportunities for Coal-Burning Utilities Using Offshore Mineral Commodities for Flue Gas Desulfurization. Marine Minerals Technology Center Continental Shelf Division Technical Progress Report , University of Mississippi, Oxford, pp.17-22 .

Proceedings of the Seventh (1997) International Offshore and Polar Engineering Conference
Honolulu, USA, May 25-30, 1997

ISBN 1-880653-28-1 (Set); ISBN 1-880653-29-X (Vol. I)

The Abundance and Vertical Distribution of Abyssal Benthic Fauna in the Japan Deep-Sea Impact Experiment

Tomomi (Sato) Kaneko and Yoriko Maejima
Marine Biological Research Institute of Japan Co.
Sapporo and Tokyo, Japan

Hirohiko Teishima
Metal Mining Agency of Japan
Tokyo, Japan

ABSTRACT

The Japan Deep-Sea Impact Experiment, abbreviated "JET", was undertaken to study resedimentation in a deep-sea environment. In 1994, 13 samples before the artificial disturbance (JET1) and 14 samples following the disturbance (JET2) were collected in the eastern central Pacific. One year later, 12 samples (JET3) were collected in the same area. The abundances and vertical distributions of meiofauna and bacteria were studied using those samples. The average abundance of nematodes significantly decreased in the JET2, while those of copepods and bacteria significantly increased in the JET3. Different sediment depths contained significantly different abundances of each faunal component.

KEY WORDS: resedimentation, meiofauna, nematodes, copepods, bacteria, abundance, vertical distribution, JET samples.

INTRODUCTION

In the late 1970s, explorations related to mining manganese nodules from the deep sea were initiated by international, industrial, and governmental consortia. Pre-pilot mining tests were conducted in 1978-79 and follow-up research to monitor the test mining tracks was conducted in 1983, in the Pacific Ocean. In those explorations, many studies were reported on the deep sea benthic fauna (Hecker et al., 1979; Jumars, 1981; Spiess et al., 1987). However, until the mid-1980s, there were few fundamental information on the deep sea meiofauna of the Pacific Ocean (Thiel, 1975; Snider et al., 1984; Shirayama, 1984a, 1984b), because most reports focus on macrofauna.

In the last ten years, large scale experiments to disturb the sea floor and to create artificial resedimentation have been conducted in the Pacific Ocean. In relation to those experiment, there were many fundamental studies to evaluate future potential mining impacts on both the deep sea macrofauna and meiofauna (Schriever et al., 1991, 1992; Thiel et al., 1992; Wilson, 192; Bussau et al., 1995).

One of those experiments, the Japan Deep-Sea Impact Experiment (JET project), was commenced in 1994 in the Eastern Central Pacific, by the Metal Mining Agency of Japan (MMAJ). An aim of this project is to examine long-term effects of artificial resedimentation on deep sea meiofauna and bacteria. During the artificial disturbance in the 1994 research cruise, the direction of the bottom current and the volume of the flux were monitored. Bottom samples were collected before and immediately after the disturbance (named JET1 and JET2 samples respectively) in 1994. And two subsequent research cruises in 1995 and 1996 collected the samples (JET3 and JET4 samples respectively) from the same area. As an update, this paper compares the abundances and vertical distributions of meiofauna and bacteria from the samples collected prior to the disturbance (JET1), just after the disturbance (JET2) and one year later (JET3).

MATERIALS AND METHODS

JET1 and JET2 samples were collected during two cruises on board of the Russian R/V Yuzhmorgeologia during August and September, 1994. JET3 samples were collected during September - October (1995) on board of the R/V Hakurei-maru No.2 of the Metal Mining Agency of Japan.

The experimental region was established in the center of a shallow trough located in the Eastern Central Pacific, approximately 5,300 m depth at 146°15′-16′W, 9°14′-15′N. During the 1994 research cruise, a plum of resuspended bottom sediments was generated by a benthic disturber, known as D2 (Brockett et al. 1994; Barnett et al, 1995), along two parallel tracks approximately 2 000 m long and 150 m wide, with a distance of about 100 m. For baseline studies, thirteen sampling stations for JET1 were selected at random from 704 grid elements dividing the experimental region. Fourteen JET2 stations were established in the surrounding area of the towing tracks to investigate the resedimentation. Twelve JET3 stations were arranged along two lines that fall at right angles to the towing zones where heavy resedimentation was expected based on the current direction and particle flux data collected in 1994 research. All sampling stations are shown in Figure 1.

Bottom samples were taken using a SMBA-multiple corer (Barnett et al., 1984; Thiel et al, 1989) with eight plastic sampling tubes and an attached deep sea camera. Each sampling tube takes undisturbed sediment samples (70.8 cm^2) and water samples overlaying the sediment. Details of the sampling method and sample treatments are described in the cruise report (MMAJ, 1994) and Kaneko et al (1995). Briefly, the overlaying water was aspirated retaining the top water (upper 2 cm from the sediment). After that, cylindrical subcores (6.15 cm²) were extracted from each tube sample for biological and chemical studies. The sediment of the subcore was sliced into six layers, down to 3 cm depth as follows: 0-0.25 including the top-water, 0.25-0.50, 0.50-0.75, 0.75-1.0 1.0-2.0, and 2.0-3.0. The sliced sediment samples for biological studies were immediately preserved with formalin. The meiofauna and bacteria samples were treated and evaluated separately with different procedures. The meiofauna samples were sieved through two different mesh sizes (300 μmm and 30 μmm). Organisms that passed through the 300 μmm mesh and remained in the 30 μmm mesh were stained with Rose Bengal, classified into faunal component of metazoan meiofauna (except foraminifers) and were counted under a microscope. All meiobfauna samples were sieved, but in this study we only analyzed 44 subcores collected from ten JET1 stations, 34 subcores from ten JET2 stations and 48 subcores from twelve JET3 stations. Bacteria were counted under an epifluorescent microscope after staining the samples with DAPI, using one subcore sample collected from twelve JET1 stations, eleven JET2 stations and twelve JET3 stations, respectively.

The current direction and flux data were collected by current meters (Aanderaa RCM8) and sediment traps (NICHIYU GIKEN KOGYO Model-SMD13W-6000 and single sample sediment trap owed by NOAA)(MMAJ, 1994b).

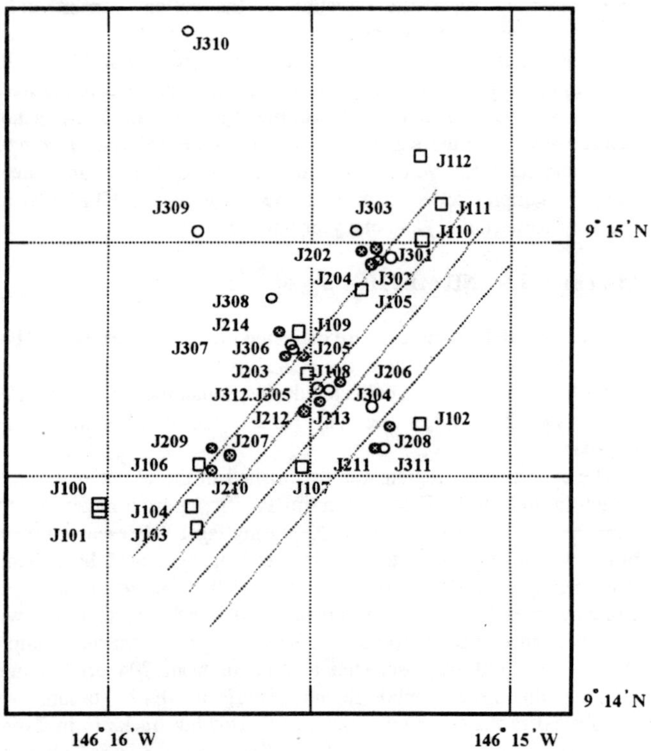

Fig. 1 Location of sampling stations and towing zones. Square, closed-circle and open-circle denote the JET1, JET2 and JET3 stations, respectively.

RESULTS

Abundances of metazoan meiofauna and bacteria

Based on individual subcores, the abundances of metazoan meiofauna and bacteria at each station are given in Table 1, and shown in Figures 2. Figure 3 depicts the major taxonomic components of the metazoan meiofaunal community. The average abundance of metazoan meiofauna based on all JET1 samples (101.0 ± 55.9 inds./10cm² x 3cm) was similar to those of other deep-sea studies at similar depths (Shirayama, 1984a; Schriever et al., 1991; Foell et al., 1992; Vincx et al., 1994). The average abundance of bacteria in JET1 samples (1.1E+09) also fell within the range of other abundances presented for the deep-sea environment (Demming, 1992; Foell et al., 1992). Nematodes were by far the most abundant of the metazoan meiofaunal community and constituted about 90 % of total meiofauna in the JET1 samples, while copepods (harpacticoids and nauplii) were the second most abundant. These two faunal components accounted for more than 95 % of total meiofauna in all JET samples. Dominant and subdominant components in the JET area (except foraminifers) were the same as those of other deep sea studies (Gage & Tyler, 1991; Schriever et al., 1991). However, copepods and other taxa (such as ostracods, tardigrades, and kinorhynchs) were rather rare in the JET area compared to other deep-sea areas.

The t-test results, using data from subcore samples (44, 35 and 48 subcores of JET1, JET2 and JET3, respectively for the metazoan meiofauna analyses, 12, 11 and 12 subcores for the bacteria analyses), are given in Table 2. The average abundances of metazoan meiofauna in the JET1, 2 and 3 samples were statistically compared using t-tests. JET1 had significantly higher abundances with the JET2 approximately half those of the JET1 ($P<0.001$), however there was no significant difference between the metazoan meiofauna abundances in the JET1 and JET3 ($P=0.09$). The nematode abundance in JET2 was significantly less than JET1 ($P<0.001$), but there was no significant difference between the copepod abundances in these samples ($P=0.23$). Therefore, the decrease of metazoan meiofauna in JET2 is due to a decrease in nematodes. No significant difference between the metazoan meiofauna abundances in the JET1 and JET3 does not necessarily indicate a recovery in the abundance of nematodes. The nematode abundance in the JET 3 compared to the JET1 had a significantly lower density ($P<0.01$), and the copepod abundance was significantly higher ($P<0.001$). The percentage of copepods in JET3 was approximately twice as much as that in JET1. For bacteria, the average abundance was not significantly different ($P=0.08$) between JET1 and JET2, but the average abundance in the JET3 was approximately ten times greater than that of the JET1 and JET2 ($P<0.001$). From these results, it is apparent that the community structure of metazoan meiofauna had changed after the disturbance.

Vertical distributions of metazoan meiofauna and bacteria

In the three year baseline study prior to the experiment, more than 95 % of the total metazoan meiofauna were distributed in the upper 5 cm of the sediments, about 90 % were located in the upper 3 cm, with most organisms concentrated close to the sediment surface (MMAJ, 1993, 1994). This distribution pattern (i.e. organisms decreasing in number with increasing sediment depth) is similar to trends reported in previous studies (Snider et al., 1984; Shirayama, 1984b, 1989; Schriever et al., 1991). Resulting from those baseline studies, it was only necessary to analyze the upper 3 cm of sediment for each JET sample. The average abundances and the cumulative percentage curves of abundance with respect to depths in sediment are given in Table 3, and illustrated in Figures 4 and 5. Cumulative percentage curves indicate that most organisms were concentrated the upper 1 cm of sediment in (50 % to 80 %). The percentage of copepods in the upper layer of sediment was higher than that of nematodes and bacteria.

The average abundances of metazoan meiofauna and bacteria in each sediment layer were statistically compared among the JET1, JET2 and JET3 samples. Although there were significant differences in the nematode abundances among the three JET samples, the results of Comparisons in each sediment layer showed a similar density at the 1-

2 and 2-3 cm layers in the JET1 and JET3. Whereas the nematode abundances in all sediment layers of the JET2 had significantly lower densities than JET1 ($P<0.001$ in all sediment layers). JET3 samples had significantly greater abundances of copepods than JET1 ($P<0.001$) only in the upper 1 cm sediment layer, whereas bacteria were significantly more abundant at each depth ($P<0.001$) for the same samples. Copepod abundances of the first 0.25 cm layer in the JET2 samples and bacterial abundances of the upper two layers (0-0.5 cm depth in sediment) were lower than those of JET1 samples, although they were not significantly different ($P<0.2$). Vertical distributions in the sediment profile indicated that the extended effects of disturbance on the average abundance were different for each faunal component, but changes in abundance of total fauna were greater in the upper layers of the sediment.

Table 1 Average abundance of metazoan meiofauna and bacteria at each JET1, JET2, JET3 and JET4 station. Station numbers of each JET survey are not equivalent to each others. (meiofauna: No. of individuals; bacteria: No. of cells; both per 10 cm^2 and 3 cm sediment depth).

JET1

	J100	J101	J102	J103	J104	J105	J106	J107	J108	J109	J110	J111	J112	AVE.
Nematoda		86.1	114.2	112.4		113.7	107.2	117.0	77.5	39.0	82.9	43.3		94.6
Harpacticoida		1.6	8.9	1.6		3.5	3.7	1.4	0.5	0.0	7.6	1.6		2.8
Nauplius larva		0.5	6.1	4.3		5.1	2.4	0.8	1.4	0.0	9.2	0.0		3.0
Others		1.6	2.4	0.8		1.1	0.0	0.3	0.8	0.0	0.5	0.0		0.6
total		89.9	131.6	119.2		123.5	113.3	119.4	80.2	39.0	100.2	45.0		101.0
S.D.(N)		15.9(3)	47.6(4)	35.8(6)		62.3(6)	21.9(4)	64.5(6)	62.1(6)	10.5(3)	50.2(3)	9.4(3)		55.9(44)
Bacteria	1.1E+09		1.0E+09	5.4E+08	8.9E+08	1.7E+09	6.7E+08	1.2E+09	1.6E+09	1.5E+09	2.2E+09	4.0E+08	6.6E+08	1.1E+09

JET2

	J201	J202	J203	J204	J205	J206	J207	J208	J209	J211	J212	J213	J214	AVE.
Nematoda	48.3	24.9	24.4	36.3	49.8				59.0	29.0	13.8	89.4	35.7	39.3
Harpacticoida	1.6	5.4	2.7	1.1	2.2				1.6	3.2	0.4	3.2	1.1	2.2
Nauplius larva	5.7	3.5	0.5	0.5	3.8				2.2	1.4	0.8	2.7	1.1	2.1
Others	1.6	0.0	0.5	0.0	0.0				1.1	0.3	0.4	0.0	0.5	0.6
total	57.3	33.9	28.2	37.9	55.8				63.9	33.9	15.4	95.3	38.5	44.2
S.D.(N)	17.6(4)	3.3(3)	3.8(3)	8.4(3)	10.7(3)				9.8(3)	13.8(6)	5.2(4)	27.6(3)	4.1(3)	25.0(35)
Bacteria	1.3E+09	9.7E+08	6.1E+08	5.2E+08	3.4E+08	5.4E+08	1.5E+09	7.0E+08	4.3E+08		1.5E+08		1.2E+09	7.5E+08

JET3

	J301	J302	J303	J304	J305	J306	J307	J308	J309	J310	J311	J312	AVE.
Nematoda	85.3	81.2	59.7	59.3	78.4	101.2	75.6	45.9	76.4	80.8	54.4	48.7	69.6
Harpacticoida	12.6	9.7	6.5	4.1	4.5	10.2	6.9	0.8	9.7	7.3	6.5	6.1	7.0
Nauplius larva	3.7	3.7	3.2	2.8	2.4	13.0	8.9	2.8	11.0	5.7	5.7	5.7	5.7
Others	4.1	3.7	2.0	0.8	1.2	6.9	2.4	0.4	1.2	2.4	0.8	2.8	2.2
total	105.6	98.3	71.5	67.0	86.5	131.2	93.8	50.0	98.3	96.3	67.4	63.4	84.5
S.D.(N)	45.6(4)	20.1(4)	21.6(4)	43.7(4)	34.2(4)	40.3(4)	22.7(4)	31.9(4)	14.4(4)	23.2(4)	5.4(4)	11.5(4)	33.0(48)
Bacteria	1.5E+10	1.7E+10	1.1E+10	9.2E+09	6.4E+09	9.5E+09	1.2E+10	5.3E+09	5.6E+09	1.1E+10	1.6E+10	7.2E+09	1.0E+10

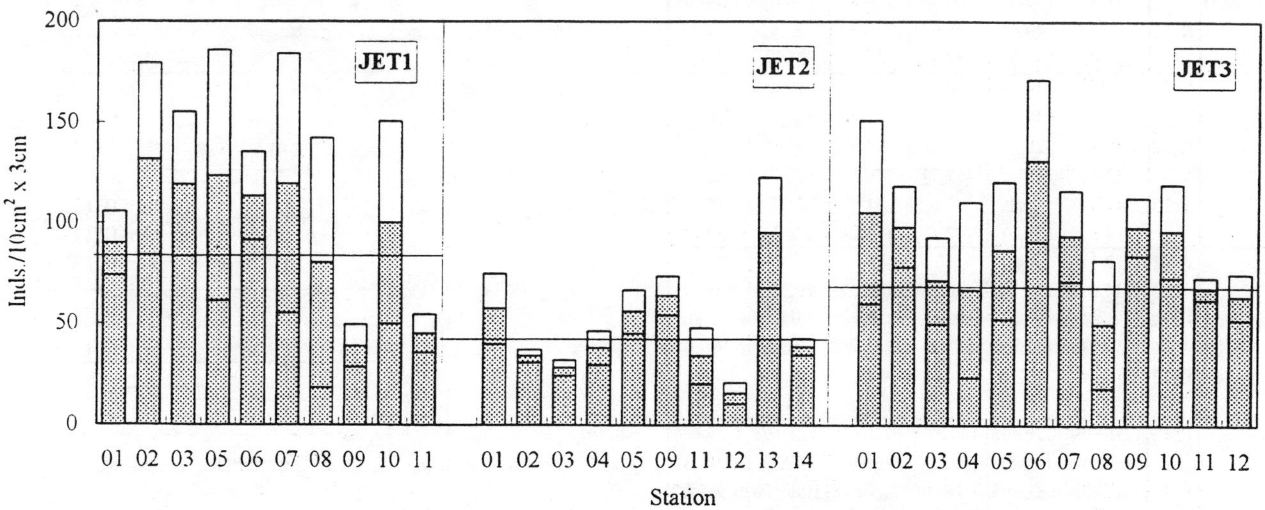

Fig.2 Average abundance of metazoan meiofauna and its standard diviation at each station. Lines indicate the average abundance based on all subcore samples of each JET survey.

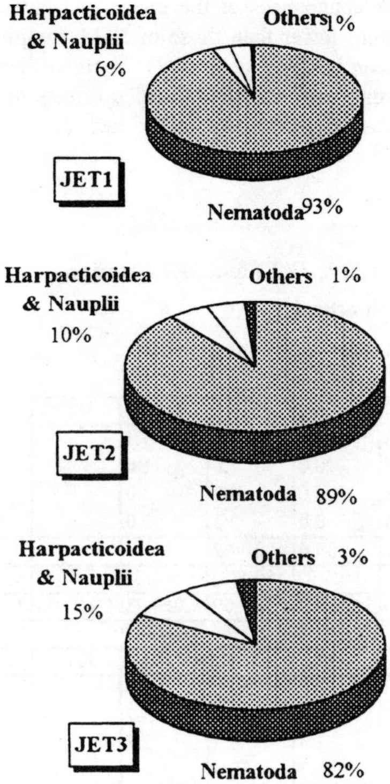

Fig.3 Faunal composition of metazoan meiofauna based on average abundances.

Table 2 Significant differences in abundances between JET1 and JET2, JET3 samples using t-test.

* Significant value (P<0.01)

	JET1-JET2		JET1-JET3	
Metazoan meiofauna	8.95E-08	*	0.0876	
Nematoda	5.95E-08	*	0.0065	*
Copepoda	0.2304		8.96E-05	*
Bacteria	0.0805		7.98E-06	*

Table 3 Average abundances (cumulative %) in JET1, JET2 and JET3, organized by sediment layers. meiofuna: No. of individuals; bacteria: No. of Cells; both per 10cm^3

Sediment Layers		0-0.25	0.25-0.5	0.5-0.75	0.75-1.0	1.0-2.0	2.0-3.0
Average of Metazoan meiofauna	JET1	115.4	52.9	50.2	41.8	20.5	15.7
		(28.5)	(41.5)	(53.9)	(64.3)	(84.5)	(100.0)
	JET2	55.3	31.9	18.9	18.0	8.7	4.4
		(31.3)	(49.4)	(60.1)	(70.3)	(90.0)	(100.0)
	JET3	60.3	49.6	49.6	36.7	23.4	12.1
		(17.8)	(32.5)	(47.2)	(58.0)	(85.7)	(100.0)
Nematoda	JET1	104.0	49.3	48.3	39.1	19.4	15.0
		(27.5)	(40.5)	(53.3)	(63.6)	(84.2)	(100.0)
	JET2	48.8	28.4	16.7	15.4	7.9	4.0
		(31.1)	(49.1)	(59.7)	(69.5)	(89.7)	(100.0)
	JET3	41.4	38.2	37.5	31.0	21.5	11.0
		(14.9)	(28.6)	(42.1)	(53.2)	(84.1)	(100.0)
Copepoda	JET1	10.3	3.4	1.9	2.1	1.0	0.4
		(44.0)	(58.5)	(66.7)	(75.5)	(92.5)	(100.0)
	JET2	6.1	3.0	2.0	2.6	0.6	0.3
		(34.7)	(51.6)	(63.2)	(77.9)	(92.6)	(100.0)
	JET3	15.6	10.2	10.6	5.1	1.5	0.7
		(30.8)	(50.9)	(71.8)	(82.0)	(94.1)	(100.0)
Bacteria	JET1	1.5E+09	8.5E+08	4.5E+08	3.2E+08	2.2E+08	1.2E+08
		(33.5)	(52.5)	(62.5)	(69.6)	(89.3)	(100.0)
	JET2	8.2E+08	4.3E+08	3.7E+08	2.7E+08	1.8E+08	9.9E+07
		(27.3)	(41.6)	(53.9)	(62.9)	(86.8)	(100.0)
	JET3	5.9E+09	7.7E+09	5.1E+09	4.6E+09	2.3E+09	2.3E+09
		(14.0)	(32.5)	(44.5)	(55.6)	(78.0)	(100.0)

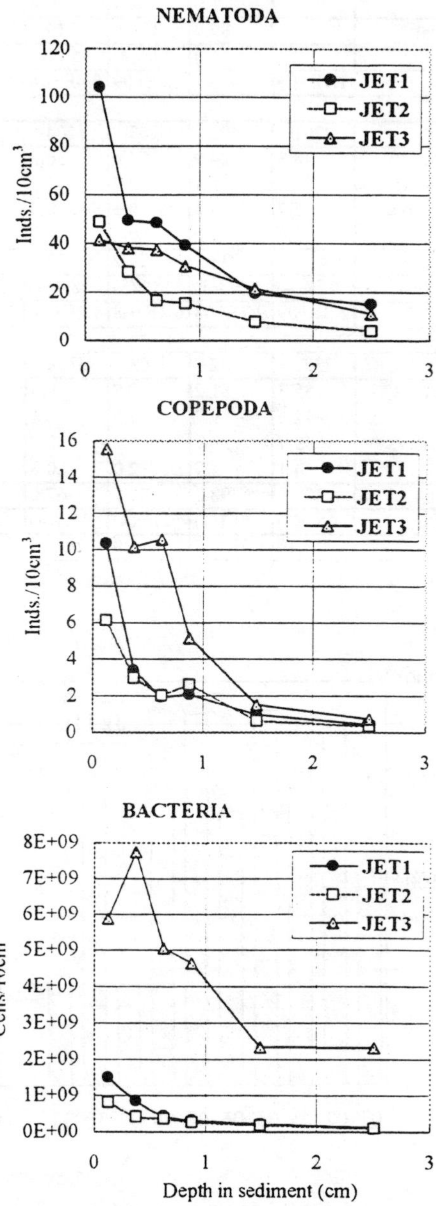

Fig.4 Vertical distribution in the sediment based on average abundances of meiofauna and bacteria.

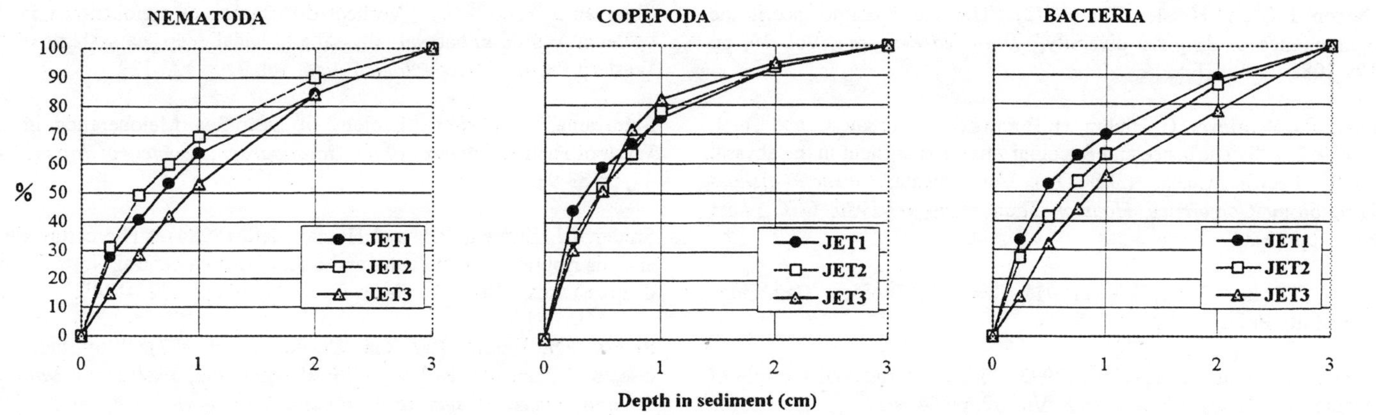

Fig.5 Cumulative percentage curves of average abundances of meiofauna and bacteria drawn with respect to depth in sediment.

DISCUSSION

It is thought that the artificial resedimentation cause the burial of the food resource and the habitat for the benthic organisms, so that, the resedimentation can be expected to affect on benthic organisms, especially sessile organisms or those inhabiting the upper sediment layers, as they have low mobility and depend on detritus for food (Dayton et al., 1972; Grassle et al., 1973; Hessler et al., 1974, Jumars, 1981; Wilsson, 1992). According to this hypothesis, the effects of resedimentation may be stronger on meiofauna and bacteria which is what we found in the JET area. Furthermore, it was shown that there was a different time scale in response for each meiofaunal component. The average abundance of nematodes collected straight after the disturbance (JET2) had significantly decreased in comparison to that taken prior to it (JET1). This decrease still persisted one year later as shown by the JET3 samples. The copepod abundance did not show any differences initially after the disturbance, but had significantly increased in the following year (JET3 samples). Changes in bacterial abundances were similar to those of the copepods. We assume the differences in response time among nematodes, copepods and bacteria are attributable to differences in their mobility, food resources available after the disturbance and in their life history, but now these relationships can't be examine.

The only other study to report the effects of bottom disturbance on the meiofauna is that of the DISCOL study in the Southeast Pacific, although the bottom disturbance was created by the different way each other. The response time of each nematodes, copepods and bacteria in the JET study were markedly different to the DISCOL study (viz: Schriever et al., 1991, 1995, Foell et al., 1992). However, as demonstrated commonly in both the JET and DISCOL studies, we consider that the abundance of meiofauna decreases after the impact, but moves towards recovery.

CONCLUSION

In this paper, we present an update of the JET experiment that is assessing the effects of resedimentation on the abundances and vertical distributions of metazoan meiofauna and bacteria after an artificial disturbance. By comparing samples collected before and after the disturbance, the abundance of metazoan meiofauna significantly decreased immediately after the impact (JET2 samples) but one year later seemed to be recovering to similar levels than those prior to the impact. In the most dominant taxa of the metazoan meiofauna and bacteria, immediately after the disturbance, there was a significant decrease in nematode abundance, but no significant difference in the copepods and bacteria. In samples collected one year after the disturbance, the lower numbers of nematodes persisted, but there was an obvious increase in numbers of copepods and bacteria. These significant differences in abundances were seen most clearly in the upper layers of the sediment. Furthermore, the trend of recovery towards abundance levels prior to the disturbance were apparent in the deeper layers of sediment. It is concluded that each component of the deep-sea meiofauna and the bacteria react in different ways through time and space to environmental impacts.

ACKNOWLEDGMENTS

This project was funded by the Metal Mining Agency of Japan (MMAJ). The authors would like to thank Mr. J Watanabe of the Marine Biological Research Institute of Japan, for sorting and counting the specimens. The present study was supported by many people who supervised the JET project and members of National Institute for Resources and Environment (NIRE) and MMAJ.

REFERENCES

Barnett, B, (1995). "Deep sea sediment resuspension system used for the Japan Deep sea Impact Experiment," *Proc 1st Ocean Mining Symp,* ISOPE, pp. 175-179.

Barnett, PRO, Watson, J. and Connelly C, (1984). "A multiple corer for taking virtually undisturbed samples from shelf, bathyal and abyssal sediments," *Oceanologica Acta*, vol. 7(4), pp. 339-408.

Brockett, T, and Richards, CZ, (1994). "Deep sea mining simulator for environmental impact studies," *Sea Technology*, pp 77-81.

Bussau, C, Schriever, G, and Thiel, H, (1995). "Evaluation of abyssal metazoan meiofauna from a manganese nodule area of the eastern South Pacific Ocean," *Vie et Milieu*, 45(1), pp. 39-48.

Deming, JW, and Yager, PL, (1992). "Natural bacterial assemblages in deep-sea sediment," In *Deep-sea food chains and the global carbon cycle*, ed. Rowe GT, and Pariente V, pp 11-27, Kluwer Academic Publishers.

Dayton, PK, and Hessler, RR, (1972). "The role of disturbance in the maintenance of deep-sea diversity," *Deep-Sea Research*, Vol. 19, pp 199-208.

Foell, EJ, Schriever, G, Bluhm, H, Borowski, C, Bussau, C, and Thiel, H, (1992). "Disturbance and Recolonization Experiment in the abyssal South Pacific Ocean (DISCOL): An Update," *Proc Offshore Technology Conference, Houston Texas, May 4-7,1990*, OTC paper 6805, pp 25-34.

Gage, JD, and Tyler, PA, (1991). *Deep-Sea Biology*, Cambridge University Press.

Grassle, JF, and Sanders, HL, (1973). "Life histories and the role of disturbance," *Deep-Sea Research*, Vol. 20, pp 643-659.

Hecker, B, and Paul, AZ, (1979). "Abyssal community structure of the benthic in fauna of Eastern Equatorial Pacific: DOMES sites A, B, and C," In *marine geology and oceanography of the Pacific manganese nodule province*, ed. Biscaoff JL, and Piper, DZ, pp 83-112, New York: Plinum Press.

Hessler, RR, and Jumars, PA, (1974). "Abyssal community analysis from replicate box cores in the Central North Pacific," *Deep-Sea Research*, Vol. 21, pp 185-209.

Jumars, PA, (1981). "Limits in predicting and detecting benthic communities responses to manganese nodule mining," *Marine Mining*, Vol. 3, pp 213-229.

Kaneko ST, Ogawa, K, and Fukushima, T, (1985). "Preliminary results of meiofauna and bacteria abundance in an environmental impact experiment," *Proc 1st Ocean Mining Symp,*ISOPE, pp 181-186.

MMAJ, (1993). *Report on the Environmental Research Impact for Manganese Nodule Development*, Metal Mining Agency of Japan.

MMAJ, (1994a). *Report on the Environmental Research Impact for Manganese Nodule Development*, Metal Mining Agency of Japan.

MMAJ, (1994b). *Cruise report of Japan Deep Sea Impact Experiment (JET)*, Metal Mining Agency of Japan.

Schriever, G, Bussau, C, and Thiel, H, (1991). "DISCOL-precautionary environmental impact studies for future manganese nodule mining and first results on meiofauna abundance," *Proc Advanced Marine Technology Conf*, Vol. 4, pp 47-57.

Schriever, G, Bluhn, H, Boetius, A, Borowski, C, Bussau, C, and Thiel H, (1992). "DISCOL-precautionary large scale environmental impact studies for future polymetallic nodule mining from the deep sea," *Proceedings XV. Congreso Mundial de Minetia, Madrid, Spain*, pp 1311-1319.

Schriever, G, (1995). "DISCOL - Disturbance and Recolonization Experiment of a Manganese Nodule Mining Area of the Southeastern Pacific," *Proc 1st Ocean Mining Symp,*ISOPE, pp 163-166.

Shirayama, Y, (1984a). "The abundance of deep-sea meiobenthos in the western Pacific in relation to environmental factors," *Oceanologica Acta*, vol. 7, pp 113-122.

Shirayama, Y, (1984b). "Vertical distribution of meiobenthos in the sediment profile in bathyal, abyssal and hadal deep sea systems of the Western Pacific," *Oceanologica Acta*, vol. 7, pp 123-129.

Shirayama, Y, (1989). "Ecology of Deep-Sea Meiobenthos in the Western Pacific," *Journal of the Oceanographic Society of Japan*, Vol. 45, pp 83-93.

Snider, LJ, Burnett, BB, and Hessler, RR, (1984). "The composition and distribution of meiofauna and nanobiota in a central North Pacific deep-sea area," *Deep-Sea Research*, Vol. 31A, pp 1225-1249.

Spiess, FN, Hessler RR, and Weydert, M, (1987). "Environmental effects of deep sea dredging," *Final report prepared under National Oceanic and Atomospheric Administration Contract Number 83-SAC-00659.*

Thiel, H, (1975). "The size structure of the deep-see benthos," *Internationale Revue der Ges. Hydrobiologie*, Vol. 60(5), pp 575-606.

Thiel, H, and Schriever, G, (1989). "Cruise-Report DISCOL 1, SONNE-Cruise 61," *Ber. Zentrum Meeres- u. Klimaforsch, Univ. Hamburg*, Nr. 3. 91 pp.

Thiel, H, Bluhm, H, Borowski, C, Bussau, C, Gooday, AJ, Maybury, C, and Schriever, G, (1992). "The impact of mining on deep sea organisms. The DISCOL-Project," *Ocean Challenge*, 3(1), pp 40-46.

Vincx, M, Bett, BJ, Dinet, A, Ferrero, T, Gooday, AJ, Lambshead, PJD Pfannkuche, O, Soltwedel, T, and Vanreusel, A, (1994). "Meiobenthos of the Deep Northeast Atlantic," *Advances in Marine Biology*, Vol. 30, pp 3-88.

Wilson, GDF, (1992). "Biological evaluation of a preservational reserve area. Faunal data and comparative analysis," *Prepared under National Oceanic and Atomospheric Administration Ocean Minerals and Energy office Contract Number 50-DSNC-90-00108.*

Proceedings of the Seventh (1997) International Offshore and Polar Engineering Conference
Honolulu, USA, May 25-30, 1997

ISBN 1-880653-28-1 (Set); ISBN 1-880653-29-X (Vol. I)

The Benthic Impact Experiment: A Study of the Ecological Impacts of Deep Seabed Mining on Abyssal Benthic Communities

Dwight D. Trueblood and Erdogan Ozturgut
National Oceanic and Atmospheric Administration
Silver Spring, MD, USA

ABSTRACT

The Benthic Impact Experiment (BIE) is designed to address the effects of sediment redeposition prior to the commencement of commercial mining in the deep sea. The experiment is being conducted in collaboration with Russian scientists from the Yuzhmorgeologiya Association aboard the R/V YUZHMORGEOLOGIYA. Sediment resuspension and subsequent redeposition during manganese-nodule mining is predicted to be the primary impact on benthic communities living on the abyssal seafloor. Sediment redeposition could adversely affect abyssal benthic communities through entombment, spatio-temporal changes in recruitment patterns, or through starvation caused by food dilution and obstructed feeding.

The Deep Sea Sediment Resuspension System was built by the National Oceanic and Atmospheric Administration (NOAA) to study the effects of sediment redeposition on the deep-sea benthos. Initial trials and experiments with the Deep Sea Sediment Resuspension System were conducted in 1991 and 1992. However the sediment redeposition results were not satisfactory; a second generation Deep Sea Sediment Resuspension System (Disturber) was manufactured and tested in 1993. The new Disturber fluidizes the sediments in the top layer, lifts the sediment slurry and discharges it 5 m above the seafloor.

In July 1993 NOAA started the BIE with baseline studies which included deployment of a transponder net and current meters, and side scan sonar/camera surveys in the study area. In August 1993 the baseline current meters were recovered, baseline box core samples were randomly collected from the study area, and sediment traps and current meters with transmissometers were deployed.

Following the baseline activities the 4900-m deep experimental area was blanketed with varying thicknesses of sediment using the Disturber. In the BIE, the Disturber was towed 49 times through a 150-m-x-3,000-m tow zone resuspending 4,000 m^3 of wet sediment. Northerly currents at the time of disturbance transported the bulk of resuspended sediment to the north as indicated by the sediment trap and transmissometer data. Results indicate that the discharged sediment did not travel far, settling quickly as a sediment-laden fluid flow.

After towing, samples were randomly collected in the area of heaviest resedimentation to determine the impacts on macro- and meiofauna immediately after towing and one year later in 1994. Meiofuanal analyses indicate that nematodes exhibit a significant decrease in abundance in the sediment redeposition area 9 months after the initiation of the BIE. Of the 71 macrofaunal families analyzed, only Sabellid polychaetes and Macrostylid isopods exhibite a significant treatment affect that could be attributed to the sediment redeposition. Overall species diversity remained unaffected by sediment redeposition.

KEY WORDS: Deep Sea, Benthic Impact Experiment, Mining, Benthos

INTRODUCTION

In July 1993 the National Oceanic and Atmospheric Administration's (NOAA) Ocean Minerals and Energy Division conducted a study called the Benthic Impact Experiment (BIE) (Trueblood and Ozturgurt, 1994). The BIE is the first successful deep sea disturbance experiment conducted by the United States that is designed to assess the environmental effects of sediment redeposition on the deep-sea benthos. Sediment redeposition is predicted to adversely affect abyssal benthic communities through burial, spatio-temporal changes in recruitment patterns, or through starvation caused by food dilution and obstructed feeding (Jumars, 1981). The BIE is designed to address these environmental concerns raised by NOAA's Deep Ocean Mining Environmental Study (DOMES) conducted in 1977-1978 and the recommendations of the

1984 National Research Council (NRC, 1984).

The action of the mining collector as it moves across the seafloor creates a benthic plume. The exact impact of increased sedimentation rates and particulate loads on the benthic fauna in the mining area of the Clarion-Clipperton Fracture Zone (CCFZ) is not known at present. Sediment deposited in the near-field from the benthic plume might form a sedimentary blanket entombing the benthic fauna. Fine sedimentary particles from the plume may also be carried some distance from the mine site by bottom currents, settling and diluting the food supply of the far-field benthic fauna. The action of the nodule collector as it moves across the destroys the benthic fauna inside the track and smothers those between the tracks. This unavoidable effect is clearly adverse, but should be mitigated somewhat by "islands" of undisturbed seafloor that are created by unmineable portions of the mine sites and, to a lesser extent, the difficulty of executing precisely adjacent mining tracks.

DISCOL was the first large scale experiment conducted in a deep-sea manganese nodule zone. In this experiment a large area was plowed over and the biological impacts of such a disturbance were monitored over 6 years (Schriever et al, 1991).

Following NOAA's BIE, Japan, and a multinational mining consortium InterOceanMetal (IOM) located in Poland have conducted similar studies using NOAA's Disturber. The Metal Mining Agency of Japan (MMAJ) conducted their benthic impact study called JET in their mining claim area in 1994 and collected benthic samples one and two years after disturbance in 1995 and 1996. IOM conducted their BIE with the Disturber in 1995 and will be making the first biological monitoring cruise in 1997. The government of India in collaboration with Russia will be conducting their BIE in the Indian mining claim area in the Indian Ocean in May/June 1997. Similarly the Republic of Korea's KORDI plan to start their BIE with baseline studies in 1997 and with the Disturber in 1998.

In this paper we present the BIE concept, methodologies used, and preliminary biological results for 1993 and 1994.

METHODS

Study Site

The experimental study area is located at approximately 128° 35' W, 12° 55' N at 4800 m depth in the Clarion-Clipperton Fracture Zone (CCFZ) of the Northern Equatorial Pacific Ocean. This part of the deep seabed manganese mining province was chosen for conducting the BIE project because of it's proximity to three US mining consortium claims. Multibeam bathymetry reveals a "hill and valley" topography with a 200-m relief that is very typical of the US mining claims located in this part of the CCFZ. The manganese nodule coverage in the study area is about 15%. Two years of current meter data from 1989 to 1991 in this area reveal a mean boundary layer current speed 5 m above the seafloor of 3.2 cm/s (maximum speed = 13.4 cm/s) with a semidiurnal tidal component velocity of 1.4 cm/s. One of the notable features of the deep-sea circulation in this area is mesoscale eddies with periods of weeks to months. The passage of mesoscale eddies seems to correlate well with the reversal of the bottom currents in the study area.

Deep Sea Sediment Re-suspension System

The Deep Sea Sediment Re-suspension System (Disturber) was conceived by NOAA and Japanese scientists, and designed and built by Sound Ocean Systems, Inc., Seattle, WA (Brockett, 1994). The Disturber was the critical piece of equipment used to generate the sediment redeposition

Fig. 1 Photograph of the Deep Sea Sediment Re-suspension System configuration used to create the sediment redeposition impact area in the BIE.

environment on the 4800-m deep abyssal plain (Fig. 1). Fluidized sediments are lifted through a riser pipe to a discharge height of 5 m. To assess the sediment concentration of the discharge plume, a rosette sampler was attached to the top of the riser pipe when desired (Fig. 2).

Experimental Design

Russian scientists performed the initial baseline studies by deploying five transponders and two current meters and conducting six side scan sonar and camera transects. Before the start of towing the two current meters were recovered and five 0.25-m^2 UNSEL-type box cores were randomly collected from the study area.

Following the baseline sampling, 18 sediment trap and 2 current meter moorings were deployed to map the extent of the sediment redeposition created by the Disturber. The sediment traps consisted of 1-m long PVC cylinders with a cross-sectional area of 80 cm^2. Two replicate cylinders were mounted on each mooring so that the mouth of the sediment traps was positioned 2 m above the sea floor. Due to mesoscale eddies observed that reverse the bottom currents in the study area, the sediment traps were arrayed symmetrically in three rows on each side of the tow zone (Fig. 3). The first two rows of sediment traps, consisting of four sediment trap moorings each, were deployed with in 50 m on either side of the tow zone. The second two rows of sediment traps in the array, consisting of three sediment trap moorings each, were deployed approximately 150 m on each side of the tow zone. The third two rows of sediment traps, consisting of two moorings each, were deployed approximately 400 m from the tow zone. One mooring on the northern and southern outer edge of the mooring array had Aanderra current meters and Sea Tech transmissometers (25 cm path length) deployed 2 m above the seafloor, and sediment traps deployed 5 m above the seafloor. The amount of sediment collected in one replicate trap from each mooring was determined by centrifuging on board and then weighing on shore; the contents of the other replicate trap was saved, filtered through a 0.2-µm filter, and weighed.

Immediately following mooring recovery in 1993, seven box core samples were randomly collected for community analysis from the sediment redeposition zone. In 1994, four treatment and five control box core and multicorer samples were similarly collected.

Sample Processing

All box core samples were subdivided into 25 100-cm^2 subcores. Two subcores were randomly chosen for assessing meiofaunal community analysis, and the remaining subcores used for macrofaunal community analysis. All macrofaunal samples were processed using 500-µm sieves. All meiofaunal samples were processed using 45-µm sieves. Macrofaunal subcores were sectioned into top 2 cm, 2-5 cm and 5-10 cm fractions and preserved using the methods outlined in Fauchald 1977, and identified to the species level. All meiofaunal subcores were sectioned into 0.5 cm fractions over the top 3 cm of each subcore and preserved and processed following the methods of Pfannkuche and Thiel (1988) and the dominant taxa, harpacticoid copepods and nematodes, have been identified to the species level.

Statistical Analysis

Macrofaunal samples from the BIE-2 were analyzed in two ways, using a fixed-effects, two-way Analysis of Variance (ANOVA), and using Sanders-Hulbert rarefaction analysis (Sanders 1968, Hurlbert 1971). Deep sea samples characteristically have large numbers of species but have a small number if individuals within any given species. Therefore the ANOVA analysis was conducted using family-level

Fig. 2 Photograph of the Deep Sea Sediment Re-suspension System with the rosette sampler mounted to sample the sediment discharge plume. This configuration was used to assess the sediment concentration of the discharge plume.

taxanomic groupings to determine if there were significant experimental treatment or time effects. For this analysis, the data were log(x+1) transformed, and the alpha levels conservatively adjusted for the proper number of comparisons using Bonferrioni's technique (Miller 1966). Significant changes in species diversity between the different treatments was determined by comparing Sanders-Hulbert rarefaction diversity curves (Sanders 1968, Hurlbert 1971) using jackknifed 95% confidence limits (Heltshe and Forrester, 1985). Meiofauna abundances were analyzed using the Wilcoxon Rank Sum Test (Conover, 1980).

RESULTS AND DISCUSSION

Simulating Mining Disturbance

The main objective of the BIE was to create a sediment plume that would blanket a 1-to-2-km^2 area of the seafloor immediately adjacent to the tow zone. Our goal was to create a redeposition area decreasing in thickness away from the tow zone. The transponder array and acoustic navigation enabled us to keep all of the Disturber tows within a polygon 3,000-m long and 150-m wide (Fig. 3). This accuracy was needed in order to avoid damaging the sediment trap moorings deployed next to the tow zone and to maintain the integrity of the adjacent deposition area. In the BIE we were able to successfully complete 49 Disturber tows in 19 days within the 150-m-x-3,000-m polygon (Fig. 3). Based on 41 rossette samples collected over the 49 tows (mean = 33.3 g/l dry wt.), we estimate a total of 4,000 m^3 of wet sediment was resuspended.

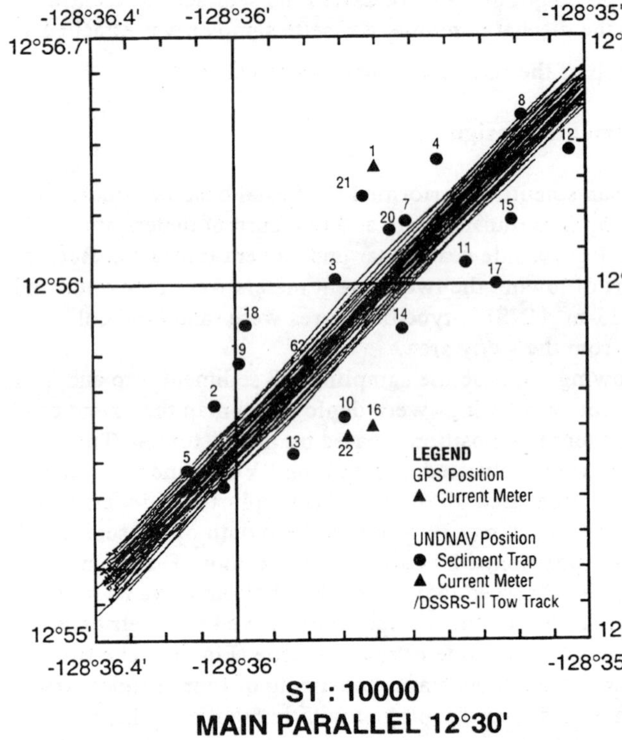

Fig. 3 The locations of the 18 sediment traps, 2 current meters and 49 Disturber tow tracks in the BIE study area.

Sediment Re-deposition

The mean distribution of sediment collected as a function of distance from the tow zone ranged over two orders of magnitude (24 to 1,094 mg), with the highest mean values occurring in the sediment traps positioned within 50-m north of the tow zone. The average amount of material within the traps drops off rapidly from 1,094 mg 50-m north of the tow zone to 360 mg 300-m north of the tow zone; south of the tow zone the average amount of material drops from 252 mg to 24 mg over the same distance. This indicates that the majority of the sediment slurry discharged from the Disturber settled rapidly to the north. This is substantiated by the current meter data collected during the Disturber towing operations and by the bottom photographs taken after towing showing a band of heaviest deposition to the north of the tow zone. The bottom currents 2 m above the seafloor were to the west-northwest with a mean speed of 3.9 cm/s. The transmissometer record shows distinct minima corresponding to each Disturber pass through the tow zone at 4-5 h lags (Fig. 4). The transmission minima correspond to 2-3 mg of suspended material.

An accurate mapping of the resedimentation layer thickness distribution remains allusive. The sediment traps were deployed to provide an estimate of the spatial distribution of the redeposited sediment layer thickness. The maximum redeposited sediment thickness calculated from the quantity of sediment collected in the traps is less than 1 mm, whereas the photographs of the seafloor taken after disturbance show that near the tow zone the nodules were buried under 1-2 cm of redeposited sediments. This difference in thickness estimated from the sediment trap data and photographs are likely due to one or combination of : (1) depth variations caused by bathymetry between the Disturber and the sediment traps, (2) spreading of the discharge from the Disturber as a density flow under the traps, (3) poor collection efficiency of the traps.

Meiofauna

Comparison of treatment and control samples collected immediately after the disturbance in 1993 (Table 1) show no significant difference in nematoda abundance (W=13, p = 0.2). A similar comparison of samples taken 9 months after the disturbance in 1994 indicate a significant decrease (W=30, p = 0.02) in nematoda abundance in the disturbed treatment zone (Table 1). Total harpacticoid copepod abundances on both dates showed no significant differences between the treatment and control samples. Species identification of nematodes and harpacticoid have just been

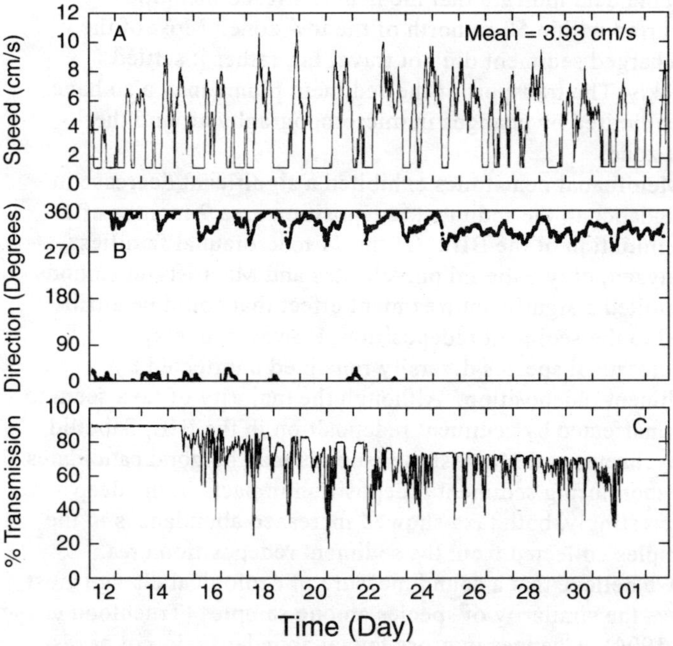

Fig. 4 Current meter data 500 m north of the BIE tow zone. The transmissometer data shows the plume moving northward during the experimental period, August 14 through September 1, 1996.

Table 1. BIE Nematoda abundance / 10 cm^2 in the top 2 cm of the sediments.

	1993		1994	
	Control	Treatment	Control	Treatment
	60.8	168.0	67.0	58.2
	123.2	118.4	58.3	38.7
	182.3	182.8	86.3	52.0
	105.5	294.4	78.2	50.7
				52.4
Mean	117.9	190.9	72.5	50.4
SD	50.3	74.3	12.3	7.2

Macrofauna

Of the 71 families of macrofauna found in the BIE study area, 3 taxa have significant treatment effects, 4 taxa have significant time effects and 4 taxa have significant treatment-x-time interaction effects (Table 2). Sabellid polychaetes and Macrostylid isopods show marked treatment effects, increasing in abundance in the area impacted by sediment redeposition (Fig. 5). Both of these taxa rank in the top 10 for overall abundance at the BIE site (Table 2). The significant treatment effect observed for amphinomid polychaetes is due primarily to the increase in abundance of amphinomids in the control treatment 9 months following the sediment redeposition (Fig. 5) and is also why a significant treatment-x-time interaction is observed (Table 2). Capitellid polychaetes show a significant decrease in abundance over the course of the BIE experiment and appears to be due to natural temporal variability. The tanaids Anarthruridae and Typhlotanaidae and the bivalve Kelliellidae exhibit significant treatment-x-time interactions, indicating only that there is an inverse relationship in the change in these taxa's abundance in the treatment and control areas over the duration of the experiment (Fig. 5). Overall species diversity remained unaffected in the BIE study area by sediment redeposition (Fig. 6).

CONCLUSIONS

The BIE is a deep-sea disturbance experiment designed to assess the effects of sediment redeposition on the deep-sea benthos prior to the commencement of commercial mining. In the BIE we towed the Disturber on the 4800-m deep abyssal seafloor through a narrow polygon (150-m wide), resuspending 4,000 m^3 of wet sediment to create a resedimentation regime that can be reliably sampled. The

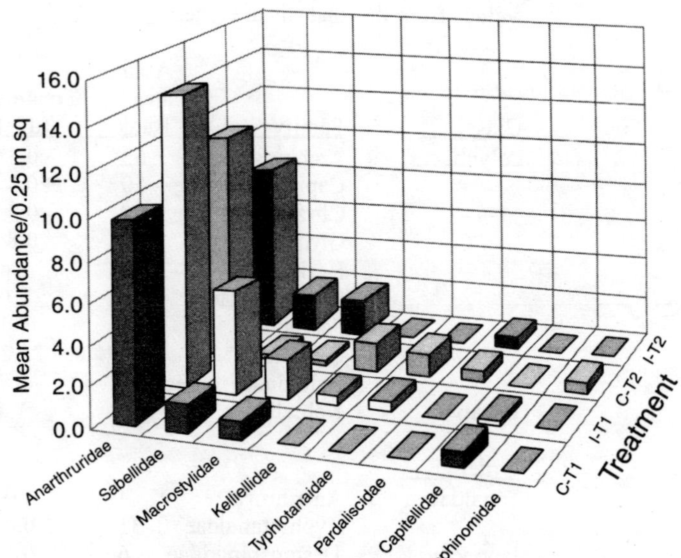

Fig. 5 Taxa that had significant treatment or time effects during the BIE. C-T1 : 1993 pre-impact control samples, C-T2: 1994 post-impact control samples, I-T1: 1993 treatment samples, I-T2: 1994 treatment samples.

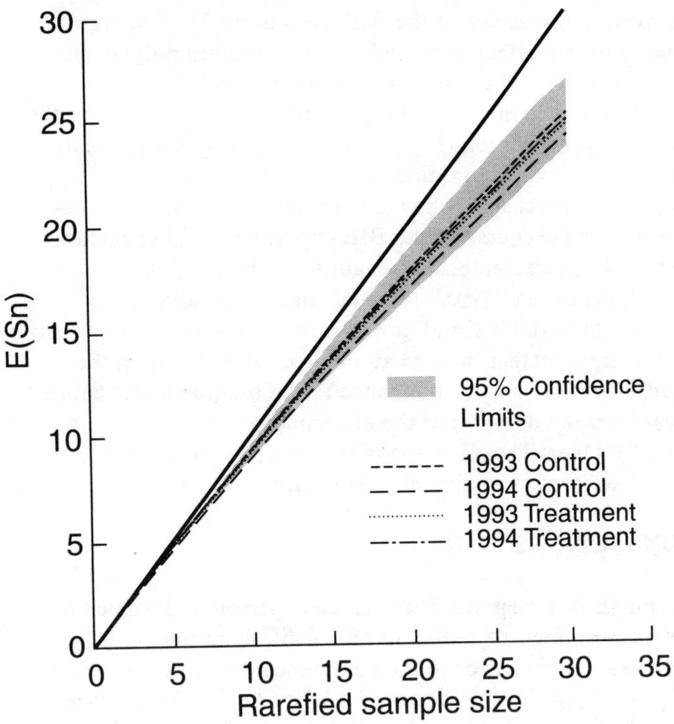

Fig. 6 Sanders-Hurlbert rarefaction diversity curves and jackknifed 95% confidence limits for the BIE samples.

sediment trap, current meter, transmissometer, and photographic data indicate that the heaviest resedimentation occurred within 50-m north of the tow zone. Most of the discharged sediment did not travel far, rather it settled quickly. The transport of the sediment plume may also have been limited by changes in microtopography around the tow zone.

Meiofuanal nematodes exhibited a significant decrease in abundance in the sediment redeposition area 9 months after the initiation of the BIE. Of the 71 macrofaunal families analyzed, only Sabellid polychaetes and Macrostylid isopods exhibited a significant treatment effect that could be attributed to the sediment redeposition. However, overall macrofaunal species diversity remained unaffected by sediment redeposition. Although the majority of taxa seem to be unaffected by sediment redeposition in the BIE, Sabellid polychaetes and Macrostylid isopods may be good candidates for monitoring sediment redeposition impacts in the deep sea. Interestingly, both taxa showed increased abundances in the samples collected from the sediment redeposition area.

We believe that a sound measure of ecological change must asses the similarity of species among samples (Trueblood et al, 1994). Changes in gross faunal abundance do not necessarily measure the health of a system nor does it provide adequate information about the nature of ecological change. Although shifts in faunal composition within a community have been identified as valuable early warning indicators, more rigorous multivariate statistical methodologies need to be applied in order to understand the fundamental nature of the observed ecological change.

Table 2. ANOVA results for selected taxa from the BIE-2 resedimentation experiment. Mean square values from the analysis are listed.

Class	Family	Abundance Rank	Treatment (df=1)	Time (df=1)	Treatment x Time Interaction (df=1)	Error (df=17)
Polychaeta	Amphinomidae	63	0.3*	0.3*	0.3*	0.1
	Capitellidae	50	0.3	0.7*	0.3	0.1
	Cirratulidae	4	0.2	0.2	0.3	2.7
	Glyceridae	12	0.8	0.2	0.4	1.0
	Hesionidae	10	2.4	4.2	0.0	1.8
	Lacydoniidae	8	2.0	0.1	0.1	1.6
	Paraonidae	2	0.1	12.2	1.0	11.7
	Sabellidae	7	25.2**	17.0*	4.4	2.5
	Spionidae	3	0.7	19.8	5.9	6.0
Bivalvia	Kelliellidae	30	1.4	1.4	4.7**	0.5
Order						
Tanaidacea	Anarthruridae	1	4.0	26.4	70.5*	16.3
	Typhlotanaidae	31	0.4	0.4	2.6*	0.5
Isopoda	Desmosomatidae	6	0.1	2.4	0.5	4.0
	Haploniscidae	11	0.9	3.1	0.0	2.4
	Macrostylidae	9	15.6***	1.6	0.2	1.0
	Nannoniscidae	5	8.9	1.1	0.1	5.8
Amphipoda	Paradaliscidae	53	0.0	1.0*	0.0	0.1

* a=0.05, ** a=0.01, ***a=0.005

REFERENCES

Brockett, T (1994). "Deep-sea mining simulations for the Benthic Impact Experiment - an update," *Proc Mar Tech Soc*, Washington, DC.

Conover,WJ (1980). "Practical Nonparametric Statistics, 2nd ed.," *Wiley*, New York, NY.

Heltshe, JR, and Forrester, NE (1985). "Statistical evaluation of the jackknife estimate of diversity when using quadrat samples," *Ecol*, Vol 66, pp 107-111.

Hurlbert, SH (1971). "The nonconcept of species diversity: a critique and alternative parameters," *Ecol*, Vol 52, pp 577-586.

Jumars, PA (1981). "Limits in predicting and detecting benthic community responses to manganese nodule mining," *Mar Mining*, Vol 3, pp 213-229.

Miller, RG (1966). "Simultaneous statistical inference," *McGraw-Hill*, New York, NY.

National Research Council (NRC) (1984). "Deep seabed stable reference areas," *National Academy Press*, Washington, DC.

Pfannkuche, O, and Thiel, H (1988). "Sample processing, **In:** Introduction to the study of meiofauna, RP Higgins and H Thiel, eds," *Smithsonian Institution Press*, Washington, DC.

Sanders, HL (1968). "Marine benthic diversity: a comparative study," *Am Nat*, Vol 102, pp 243-282.

Sato, KT, Ogawa, K, and Fukushima, T (1995). "Preliminary results of meiofauna and bacteria abundance in an environmental impact experiment," *Proct Ocean Mining Symp*, *ISOPE*, Tsukuba, pp 181-186.

Schriever, G, Bussau, C, and Thiel, H (1991). "Discol-precautionary environmental impact studies for future manganese nodule mining and first results on meiofauna abundance," *Proc Advanced Mar Tech Conf*, Vol 4, pp 44-57.

Trueblood, DD, Gallagher, ED, and Gould, DM (1994). "The three stages of seasonal succession on the Savin Hill Cove mudflat, Boston Harbor," *Limnol.Oceanogr*, Vol 39(6), pp 1440-1454.

Trueblood, DD, and Ozturgut, E (1994). "The Benthic Impact Experiment (BIE): A study of environmental impacts of manganese-nodule mining on the abyssal seafloor," *Proc Mar Tech Soc*, Washington, DC.

Proceedings of the Seventh (1997) International Offshore and Polar Engineering Conference
Honolulu, USA, May 25-30, 1997

ISBN 1-880653-28-1 (Set); ISBN 1-880653-29-X (Vol. I)

Benthic Environmental Baseline Investigations in the Manganese Nodule Area of the Central Indian Basin

Rahul Sharma, B. Nagender Nath, S.M. Gupta and Z.A. Ansari
National Institute of Oceanography
Goa, India

ABSTRACT

In order to exploit the manganese nodule deposits in future, an assessment of the environmental impact due to potential mining activity, has been undertaken in the Central Indian Basin. Under this programme, seabed surveys in five selected areas have been conducted to study the bathymetry, nodule distribution and benthic fauna. Each of the areas, having dimensions of 10x10 nautical miles, has a total relief of 160-200 m with variable undulations (30-60 m) and few elevated features (50-100 m). The sites with relatively flat topography have higher sediment thickness (50-100 m) than those with uneven topography (20-35 m).

Nodules in areas of thicker sediment cover have low abundances (32 out of 39 locations: <3 kg/sq.m, average abundance: 1-2 kg/sq.m), as compared to those with relatively thin sediment cover, (11 out of 25 locations: 5-12 kg/sq.m, and average abundance: 3.5-5.0 kg/sq.m). Almost all nodules appear to be buried under these sediments. Nodule exposure ranges between 0 and 2%, except in one station with 50% coverage.

The sediment underlying the nodules is predominantly radiolarian ooze with occasional burrows (mainly thalassinoid type). Out of the 22 cores studied for benthic activity, five displayed pronounced layered structure in the top (0-15 cm) layer and a comparatively coherent sub-layer (below 15 cm). In three cores, evidence of partial induration/lithification has been noticed, leading to hard ground formation.

Quantitative estimates of deep-sea benthos recorded from core samples indicate moderately rich fauna. The meiofauna density ranges from 3-45/10 cm^2, dominated by Nematoda. Similarly, the macrofaunal density ranges from 2-16/0.25 m^2 and was dominated by Polychaeta (>50%). The ratio of macro- to meiofauna in the total population was 1:500.

KEY WORDS: Nodules, benthic environment, baseline studies, Indian Ocean.

INTRODUCTION

Several studies have been carried out in the Pacific Ocean, such as DISCOL (German), BIE (American), and JET (Japanese). All of them have focussed on selecting a test site and monitoring large-scale experimental pre- and post-disturbance environmental conditions (Foell, 1990; Trueblood, 1994; Fukushima, 1995).

In order to evaluate the potential environmental impact due to nodule mining, a multidisciplinary program, named as the 'Indian Deepsea Environment Experiment' (INDEX) was initiated in 1995 in the nodule area of Central Indian Basin.

In this program, baseline studies have been carried out for collecting information on various parameters of the seafloor in the Central Indian Basin. This information will be used to identify potential test and reference areas as well as to assess the existing conditions on the seafloor.

MATERIALS AND METHODS

Bathymetric maps generated from hydrosweep surveys in the Central Indian Basin were used to select five study areas (Fig. 1). Each one covers an area of 10x10 nautical miles. Their co-ordinates are as follows:

T1 : 10°00' - 10°10'S, 75°10' - 75°20' E
R1 : 10°00' - 10°10'S, 75°35' - 75°45' E
A1 : 10°00' - 10°10'S, 75°55' - 76°05' E
T2 : 12°15' - 12°25'S, 75°45' - 75°55' E
R2 : 11°55' - 12°05'S, 76°00' - 76°10' E

Four cruises were conducted onboard RV *A.A. Sidorenko* during April-May, 1996 and September-October, 1996 for the collection of benthic baseline data. The study areas were surveyed by echosounding, sub-bottom profiling, as well as by box corers and Okean grabs with camera for nodule and sediment samples (Fig. 2). The photographs (Fig. 3) were evaluated for nodule coverage and seafloor features, such as sediment cover, nodule exposure and biological activity.

From the core samples, the overlying water was siphoned off and the interface water was collected with syringes in small plastic bottles. Four large PVC tubes with a diameter of 15 cm were inserted into the box and the sediment was retrieved for porewater studies Tubes with 10 cm diameter were inserted for geochemical, micropaleontological, sedimentological and geotechnical studies. All the sediment cores were uniformly sectioned at 0-2 cm, 2-4 cm, 4-6 cm, 6-8 cm, 8-10 cm, 10-15 cm and 15-20, 20-25, 25-30 cm intervals for porewater studies.

Macrobenthic (>0.5 mm in size) samples from 29 grabs and 16 box cores and meiobenthic samples from 16 box cores were collected during these cruises. A sub-core (diameter 4.5 cm, lenth 10 cm) was taken from box-cores after removing the supernatent water and sub-sampled at 2 cm intervals. The sediment was sieved through a mesh size of 0.5 mm and 0.062 mm for separating meiobenthos from macrobenthos. All samples were preserved in Rose- Bengal formalin (5%) buffered solution for further analysis. Sorting and counting of the samples were completed under binocular microscope.

RESULTS

(a) Bathymetry

The echosounding records were digitised and merged with ship's navigation data, and depth contours were plotted. In general, the depths in the study areas range between 5120m and 5400m, each having a specific range, average and standard deviation (Table 1). In the northern areas (T1, R1 and A1), the topography is gently undulating with reliefs of 30-40 m, and a few knolly features of 50-100 m relief. The undulations in the southern areas (T2 and R2) are higher (50-60 m), more abundant and have more peaks.

The morphometric analyses show lower average slope angles in longitudinal (1-1.4 degrees) as well as latitudinal (0.68-0.93 degrees) directions in the northern areas, compared to those in the southern areas (1.67-1.76 degrees in longitudinal, and 1.02- 1.36 degrees in latitudinal direction). The average vertical difference between 1 minute positions in northern areas is lower (12-22.6m) than in the southern areas (14.2-30.5m) in longitudinal as well as latitudinal directions. The average number of peaks (>10 m) in the northern areas are also less (2.9-4.7), as compared to southern areas (4-7.8).

Whereas the depths do not vary drastically between different areas the average slopes, vertical differences as well as the number of peaks between digitised points at 1 minute interval, are lower for the northern areas as compared to southern areas, demonstrating that the northern areas are less rugged than the southern ones.

(b) Nodule distribution

Exposed nodules are limited in the northern areas, with most of the locations showing occurrence of no or few nodules (0-2% coverage) on the sediment surface. Similarly, most of the locations in the southern areas also show no or very few (0-2% coverage) nodules except for one location in T2 (~50% coverage).

The overall nodule distribution in all areas (Table 2) shows that the northern areas have relatively lower nodule abundances (average= 1-2 kg/sq.m) compared to the southern areas (average= 3.7-5 kg/sq.m). Most of the locations in the northern areas (32 out of 39) have <3 kg/sq.m whereas the southern areas have 14 out of 24 locations with >3 kg/sq.m abundance.

The locations with higher abundance (>3 kg/sq.m) show the occurrence of exposed nodules in the photographs, whereas the remaining locations show no nodules exposed on the seabed. Hence, whereas the burial of nodules is controlled by the sediment cover, their exposure also depends on the quantitity or abundance of nodules available at these locations.

(c) Sediment thickness

The photographs in all the areas invariably show the occurrence of sediments on the seabed. The complete to almost complete penetration of the camera trigger weight into the sediment indicates the soft, watery nature of the sediment layer, which is probably responsible for burial (or non-exposure) of many of the nodules occurring in these areas.

The sediment thickness observed from subbottom records is higher in the northern areas than the southern areas. In the north generally the sediment thickness is between 25 and 60 m, which is reduced to almost 0 m on the peaks of abyssal hills, and increases to 100 m in flat regions and negative reliefs. Whereas the southern areas have sediment thicknesses generally between 20-40 m, with a maximum of 50 m and many locations with <20 m, due to undulating topography.

(d) Sediment characteristics:

In all cores from areas T1, R1, and A1, the top 0-10 cm sediment is either moderate brown (5YR 3/4 of rock colour chart) or moderate yellowish brown (10YR 5/4) in colour followed by 10- 20 cm of light to yellowish brown coloured sediments. While the top layer (0-10 cm) is always uniform, the second layer (10-20 cm) has mottled features with light intercalations, probably due to the benthic mixing and bioturbation (Fig. 4a,b,c). At places, streaks of light grayish or olive greenish intercalations were noticed in deeper layers (20-30 cm) indicating microenvironments of low oxygen levels. The sediments are mostly silty clay to clayey silt in size (Rao and Nath, 1988).

Three cores each were collected from Areas R2 and T2. The cores recovered from Area R2 contain a homogenous layer of moderate brown (5YR 3/4) varying in thickness between 7 and 9 cm with grayish orange (10YR 7/4) intercalations. One core (# 145) composed of 27 cm of grayish orange coloured sediment as second layer with intercalations and mottling and a distinct grayish band at 23 cm. The cores recovered from area T2 are similar to the ones from R2; all the three have three units viz: Unit I- moderate brown, Unit II- mixed zone and Unit III- grayish orange specks and lenses in brown matrix. One of the cores (# 151) had open burrows of 0.5-1 cm diameter on the surface.

All the porewater components were higher in concentration than the overlying water. The porewater profiles reflect typical deep-sea sedimentary environment. High nitrate values, low phosphate concentrations and high sediment Eh indicate the redox conditions within the top 15-20 cm with suboxic conditions in the deeper zones with higher mottling. Porewater silica values are very high due to the dissolution of radiolarian tests which is the typical sediment type in the present area (Nath et al.1989). A semi-indurated incipient pelagic hard ground was encountered at location # 172 characterised by several open empty burrow holes over gently hummocky relief (Fig. 4d). A number of grayish coloured lumps resembling oolites, larger than 1 mm in size were also recovered from the mottled layer (Fig. 4e). These lumps indicate the processes of synsedimentary lithification leading to the formation of incipient pelagic hard ground (Gupta, 1995).

(e) **Megabenthos and bioturbation**:

Traces of megabenthos was observed in many of the photographs in the form of reworked sediments, bioturbation, as well as fecal casts, pellets, trail marks and Thalassinoides type burrows of benthic organisms (Fig. 4e). A single polychaete specimen in dead condition was recovered from location #180, which may be responsible for sediment mixing and burrowing in the general area. Some of the prominent features observed in photographs are circular fecal coils, locomotion trails. Observations of megabenthic activity associated with nodules from the Central Indian Ocean have been reported earlier by Sharma and Rao, (1982).

(f) **Benthos**

The preliminary results indicate the dominance of Polychaetes and Nematodes in macro and meiofauna respectively. The fauna is moderately rich, comprising 6 macrofaunal and 3 meiofaunal taxa. The population density varies from 3-45/10 cm^2 for meiofauna and 8-64/m^2 for macrofauna. Among the meiofauna taxa, Nematodes with a 100% occurrence were observed, followed by the Harpacticoid copepods. Harpacticoid larvae were also recorded in some samples, their average density varying significantly between the sites.

The vertical distribution of meiofauna in different sediment layers shows that the Nematodes were regular in their occurrence at various depths, although the density decreases with increasing depth. The Harpacticoid copepods on the other hand, were confined to the top 5 cm. This pattern was however, not consistent. It is important to mention that meiofauna samples, used here, were collected from box corers which also enclosed the supernatent, thick layer of water..

Among the macrofauna, Polychaetes were the most important group, followed by peracarid Crustaceans (Amphipods, Tanaidacea), Bryozoa and Rhizopoda. The abundance of macrofauna varied within a modest range of 5-15%, as compared to metazoan meiofauna.

The biomass varied significantly due to the presence of large specimens of crustacea at some locations. No significant differences in the depth distribution of macrofauna were observed. Based on few benthic samples, Neyman et al. (1973) reported low benthic biomass in the abyssal region of the tropical Indian Ocean. According to them the present study area falls into the oligotrophic zone. Occurrence of rich benthic fauna associated with manganese nodules has been reported (Parulekar et. al. 1982).

CONCLUSIONS

Evaluation of echosounding, sub-bottom profiling, sampling and photography data has revealed the baseline conditions of the benthic environment in selected areas. The bathymetric variations, sediment thickness and nodule concentrations will be used for selection of the Test and Reference Areas for the 'Indian Deep-Sea Environmental Experiment (INDEX)'. The sediment characteristics as well as benthos data will constitute reference information for the environmental impact assessment.

ACKNOWLEDGEMENTS

The authors acknowledge the support of Dr. E. Desa, Director and Mr. R.R. Nair, Deputy Director, NIO, during the course of the study. The officers, scientists and crew of RV A A Sidorenko are thanked for data collection. Hydrosweep maps were processed by Dr. V.N. Kodagali. The study was carried out under the 'PMN-EIA studies in Central Indian Basin', funded by the Department of Ocean Development, Government of India.

References

Foell, E.J.; Thiel H.; Schriever G. (1990). "DISCOL: A long-term, large-scale, disturbance-recolonization experiment in the Abyssal Eastern Tropical South Pacific Ocean," *22nd Offshore Technology Conference*, Houston, OTC 6328, pp 497-503.

Fukushima T. (1995). "Overview - Japan deep-sea impact experiment: JET," *Proceedings of ISOPE - Ocean Mining Symposium*, Tsukuba, Japan, pp 47-53.

Gupta S.M. (1995). "Palaeogene hardgrounds and associated intraclast lag deposits as the substrates of ferromanganese crusts and nuclei of nodules: inferences of abyssal currents in the Central Indian Ocean Basin". *Journal of Palaeontological society of India*, 40, pp77-86.

Nath B.N., Rao V.P. and Becker K.P. (1989). "Geochemical evidence of terrigenous influence in deepsea sediments upto 8°S in the Central Indian Basin." *Marine Geology*, 87, pp 301-313.

Neymann, A.A., Sokolova, M.N., Vinogradova, N.G. and Pasternak, F.A. (1973). "Some patterns in the distribution of bottom fauna in the Indian Ocean," In: *The Biology of the Indian Ocean (Ecological Studies 3)*, B. Zeitschel, editor, Springer - Verlag, Berlin, pp 467 - 473.

Parulekar A.H.; Harkantra S.N.; Ansari Z.A.; Matondkar S.G.P. (1982). "Abyssal benthos of the Central Indian Ocean", *Deep- Sea Research*, 29, pp 1531-1537.

Rao V.P., Nath B.N. (1988). "Nature, distribution and origin of clay minerals in grain size fractions of sediments from manganese nodules field, Central Indian Ocean," *Indian Journal of Marine Sciences*, 17, pp 202-207.

Sharma R.; Rao A.S. (1992). "Geological factors associated with megabenthic activity in the Central Indian Basin." *Deep-Sea Research*. 39, pp 705-713.

Trueblood D.D. (1993). "US cruise report for BIE II-cruise 1". *NOAA Technical Memorandum NOS OCRM 4*, pp 51.

Table 1: MORPHOMETRIC CHARACTERISTICS OF THE SEAFLOOR IN THE STUDY AREAS (at 1 min interval from echosounding records)

Area	No. of data Points	Depth Range	Average Depth	SD	Average Slope		Average Vertical Difference		Average No. of Peaks	
	(n)	(m)	(m)		Longi-tude	Lati-tude	Longi-tude	Lati-tude	Longi-tude	Lati-tude
T1	1666	5150-5340	5217	34	1.07	0.78	19.3	22.6	4.0	2.9
R1	1857	5200-5400	5330	37	1.40	0.93	19.6	20.4	4.7	3.1
A1	1739	5180-5400	5327	25	1.00	0.68	16.2	12.0	4.3	2.9
T2	1581	5120-5320	5217	41	1.76	1.02	30.5	20.8	5.8	4.0
R2	3035	5200-5380	5297	32	1.67	1.36	20.4	14.2	7.8	7.8

Note : Long = longitudinal direction
Lat = latitudinal direction

Table 2: NODULE DISTRIBUTION CHARACTERISTICS IN THE STUDY AREAS (Number of samples per weight class)

Area	No.of	Frequency of abundance (kg/sq.m)						Average
	data points	0	<1	1-3	3-5	5-10	>10	abundance
T1	13	2	5	2	2	2	-	1.94
R1	13	4	7	-	1	1	-	1.14
A1	13	4	5	1	-	2	1	2.1
T2	13	3	4	-	1	4	1	3.41
R2	12	1	1	2	2	5	1	5.05

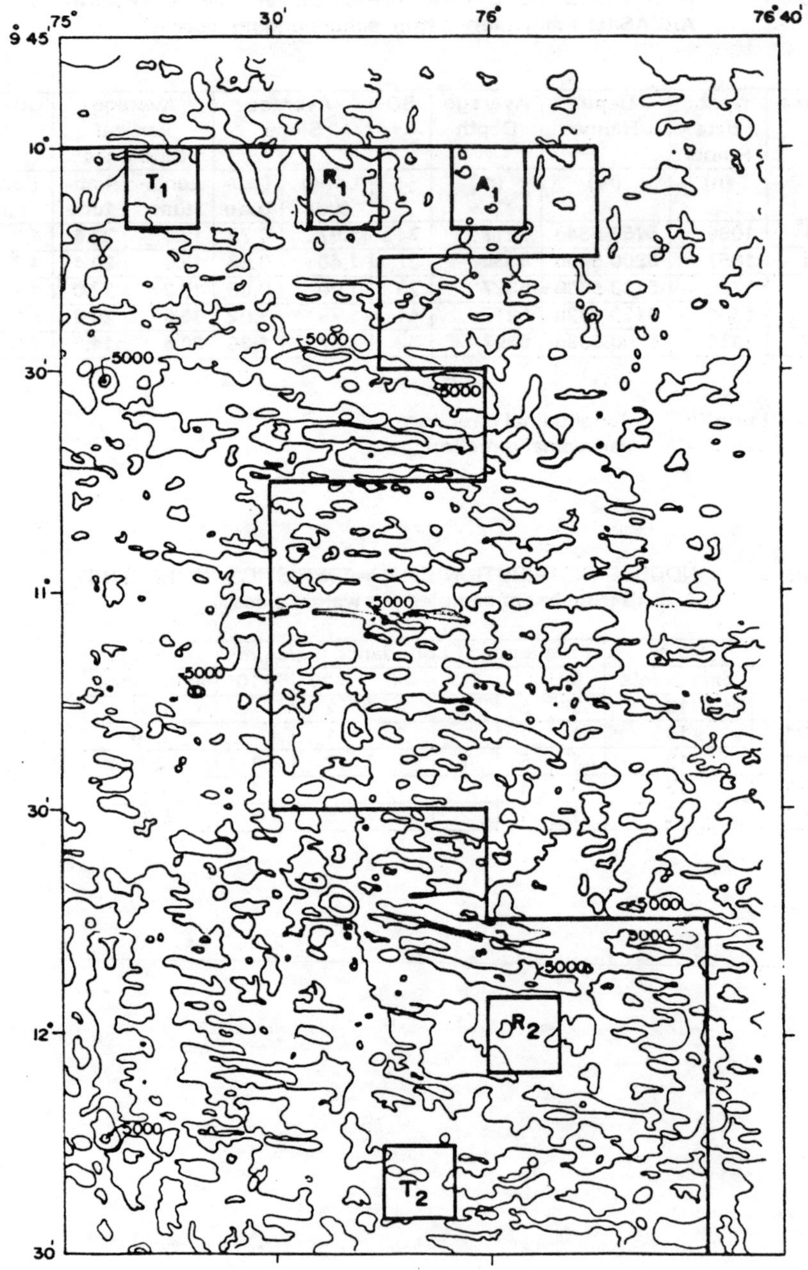

Fig.1 Locations of study areas.

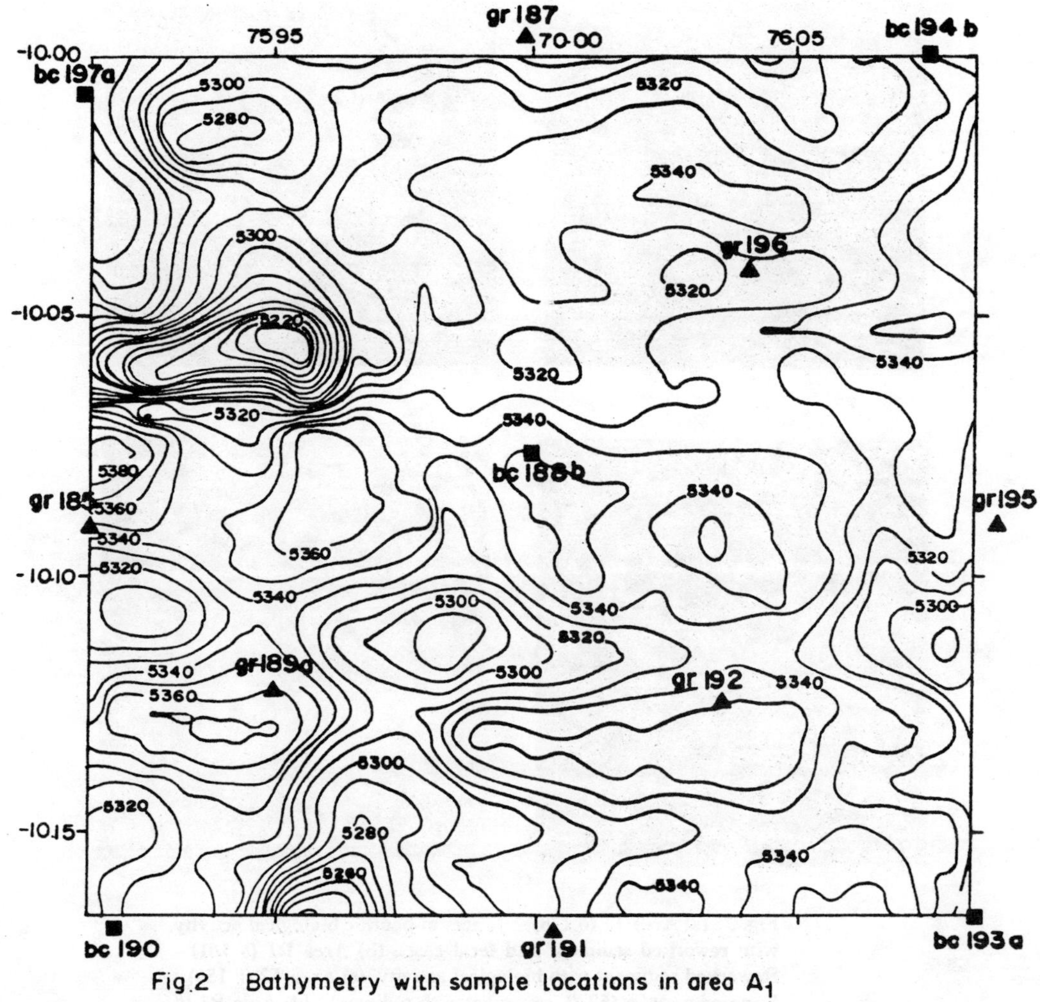

Fig.2 Bathymetry with sample locations in area A_1

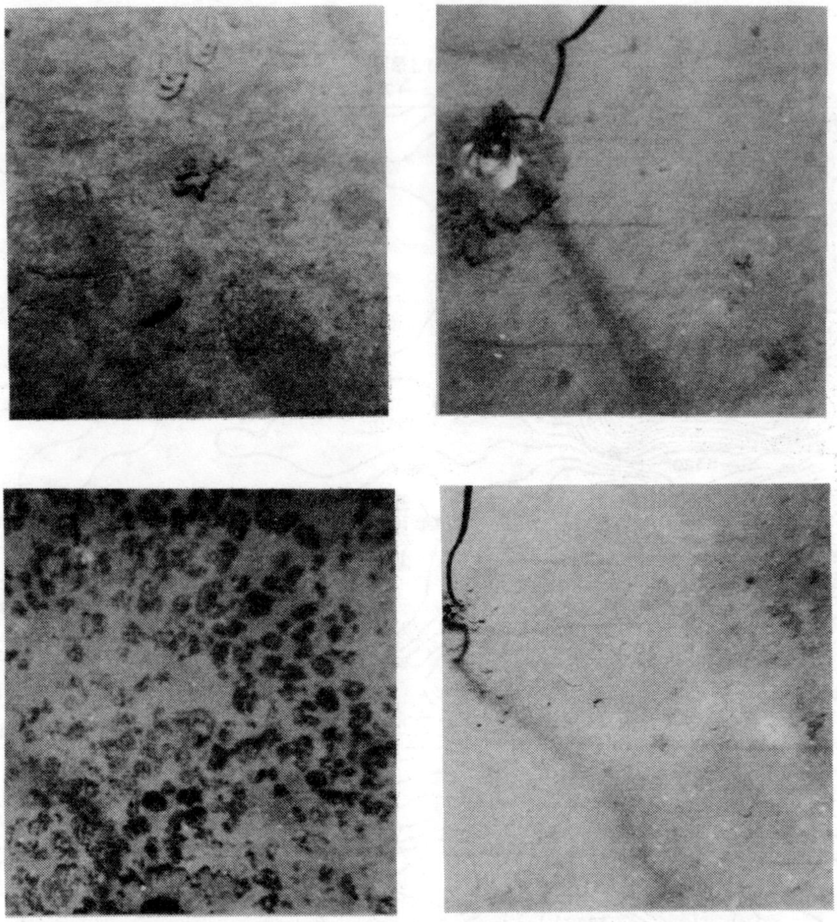

Fig. 3 : (a) Area T1 (# 178) - Traces of benthic biological activity with reworked sediment and fecal casts. (b) Area R1 (# 161) - Reworked sediment with biological activity. (c) Area T2 (# 156) - Exposed nodules (50 % coverage) with sediment. (d) Area R2 (# 138) - Reworked sediment with burrow openings and fecal casts.

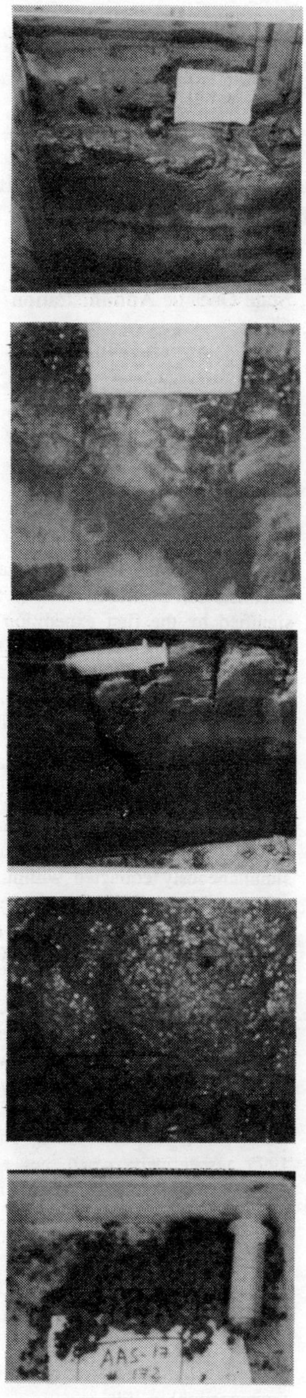

(a) Vertical section of box core (type - Reineck) AAS, # 159b showing top 10 cm homogenous brown buff coloured pelagic ooze overlying the partially mottled grayish brown coloured sub-surface layer in 10-20 cm sediment depth.

(b) Vertical section of core AAS # 160a, with top 10 cm homogenous, 10-20 cm mottled gray brown, 20-30 cm brown pelagic clay with gray coloured lenses. An elliptical lensoid burrow of unknown origin in the lower left corner suggests intense reworking of the sediment.

(c) Vertical section of another core with typical stratification and a gently curving vertical hollow (empty structure) of thallassinod type of burrows. Plastic syringe is 2.5 cm in diameter. Burrow diameter is 1-1.3 cm. Mottling seems to be the result of bioturbation.

(d) Surface view of the box core AAS #172 with hummocky micro relief with micro depressions filled with slurry at sediment water interface associated with manganese nodules, several burrow openings are visible on the hummocky mound.

(e) Coarse grannular fraction >1 mm, with lumpy aggregates of diagenetically lithified intraclasts suggesting formation of an incipient pelagic hardground encountered in core AAS # 172.

Fig. 4 Sedimentry structures and bioturbations feature observed in box core samples.

Proceedings of the Seventh (1997) International Offshore and Polar Engineering Conference
Honolulu, USA, May 25-30, 1997

ISBN 1-880653-28-1 (Set); ISBN 1-880653-29-X (Vol. I)

Chinese Environmental Investigation Related to Potential Impact of Deep-Sea Mining

Huai-Yang Zhou
Second Institute of Oceanography, State Oceanic Administration
Hangzhou, China

ABSTRACT

It is known that China Ocean Mineral Resources R & D Association (COMRA), a "pioneer investor" registered by UN Seabed Authority in 1991 and a national organizer and coordinator of activities concerned with oceanic mineral resources research and development, is undertaking a 15 year program (1991-2005) of exploration and exploitation on manganese nodules in "invested area" in northeast Pacific. During the time, COMRA would fulfill the obligations required as a pioneer investor and make a contribution to the ocean environment and eco-system protection.

Although the environmental investigation could be traced in early 80' when Chinese scientists began their geological and geophysical exploration on manganese nodules in Pacific, the well-planed environmental work was carried out in the last 5 year (1991-1995). A lot of environmental investigation have been carried out during this period and therefore some basic information on meteorology, physical oceanography, chemical oceanography, biology have been collected in the "pioneer area".

COMRA will continue and enlarge its scale of environmental research related to potential environmental impact of manganese nodules mining in registered area during next 10 years. An environmental baseline study project, to be conducted between 1996 to 2000, termed "Natural Variability of Baseline (NaVaBa)" (Tab.2) is reasonable planned by COMRA.

KEY WORDS: COMRA, Environmental investigation, impact experiment, baseline data, NaVaBa project

As a "pioneer investor" registered by the UN Seabed Authority in 1991 and a national organizer and coordinator of activities concerned with oceanic mineral resources research and development, the China Ocean Mineral Resources R & D Association (COMRA) is known to undertake a 15 year program (1991-2005) of exploration and exploitation on manganese nodules in the "invested area" of the Northeast Pacific. Meanwhile, COMRA would fulfill the obligations required as a pioneer investor and make a contribution to the ocean environment and ecosystem protection.

The environmental investigation in Pacific Ocean by Chinese scientists can be traced back to the early 80ies of this century that is signified by the first expedition of the State Oceanic Administration (SOA) in 1983 when Chinese scientists began their geological and geophysical exploration on manganese nodules . Before 1990, besides of the first cruise mentioned above, about 8 voyages were carried out by the scientists from the SOA (1985-1986, 1987, 1988, 1990) with RV Xiang-Yang-Hong 16 and from the Ministry of Geology & Mineral Resources (MGMR, 1986, 1987, 1988, 1989) with RV Hai-Yang 4, respectively. Approximately 2,000,000 km^2 in the Eastern and Central Pacific Ocean were geophysically, geologically and geochemically investigated. During this time, some environmental data were simultaneously collected within meteorology, physical oceanography, chemical oceanography and biology. This phase is termed the first period of oceanic environmental investigation that is related to potential mining of manganese nodules in China.

When COMRA was founded in 1990 and became the "pioneer investor" in 1991, being the representative of China, the Chinese oceanic environmental studies, together with other projects, went into its second period. During the first five years (1991-1995), the environmental baseline studies were arranged in all of the 4 expeditions, by RVs Hang-Yang 4 in 1992, 1993, Xiang-Yang-Hong 9 in 1994 and Da-Yang 1 in 1995, respectively, when they explored the Chinese "invest area". A complete set of environmental baseline data, compiled in Tab.1 was collected and investigated from the "invest area" by means of CTD, ADCP, Seabeam (Multibeam), Deep-Tow (Sonar + Video & Photo Cameras), and Boxcorer.

Besides, an NS transect investigation on the hydrology, chemistry and geology in the eastern part of the Chinese "invest area" was cooperatively carried out by COMRA and IOM (Interoceanmetal Joint Organization) in 1994.

After its first five year (1991-1995) project, COMRA will continue and enlarge its scale of environmental research related to potential environmental impact of manganese nodules mining in the registered area during the next 10 years. COMRA is now planning the environmental work for the next years, and the following factors are considered:

Because the environmental work in the ocean is of international and world-wide interest, the results and experiences collected by other scientific groups could be based as the references for the planning.

Table 1. The items of investigated baseline data in the Chinese 'invest area" from 1991 to 1995

Meteorological Observations	temperature, pressure and humidity cloud cover and visibility wind and cyclone conditions
Current and Wave Measurements	speed and direction of currents, wave height
Physical Characters of Seawater	temperature salinity density
Chemical Characters of Seawater and Pore Water in Surface Sediment	pH, Eh, dissolved oxygen nutrients heavy metals
Biology and Biochemistry	mega- and macrobenthos phyto- and zooplankton chlorophyll-a and primary production
Geological Characters	geomorphological characters geological and geochemical characters of surface sediment geological and geochemical characters of manganese nodules

Besides considering the health of humans, the safety of ships and mining machinery, the main aims of such environmental research and/or environmental impact studies which are guided by the "precautionary principle" is to understand and recognize the communities in the deep sea as well as the recovery capacity of different ecosystems after the non-natural disturbance through mining. Furthermore, it could help to develop the mining and processing technologies by avoiding or minimizing potentially harmful impacts, that is, to support the development of "clean" technologies.

Since the beginning of the large scale exploration of manganese nodules in the 60ies of this century, some industrialized countries and international consortia have made much effort and carried out a series of environmental research projects, such as "The Blake Plateau Mining Test Study" (USA, 1970), "The Bermuda Rise Study" (USA, 1972), "The Continuous Line Bucket (CLB) Mining Test Study" (Japan-USA-France, 1972), "The Deep Ocean Mining Environmental Study (DOMES)" (USA, 1975-1980), "The ECHO-1 Expedition" (USA, 1983), "The Acute Mortality Experiment" (USA, 1987), "The QUAGMIRE II Expedition " (USA, 1990), "The DISCOL Project and Expeditions" (Germany, 1989-1994), and its prolongation "The ECOBENT Project and Expedition" (Germany, 1996), "The Benthic Impact Experiment (BIE)" (USA, 1993; Japan, 1994; IOM, 1995), although some of them concentrated their emphasis on functions and the safety of mining machinery.

Particularly from the comprehensive studies of DOMES, most of the scientists recognized that negative environmental impacts of deep-sea mining might be especially associated with two broad depth zones:

- at the seafloor within and downstream of the mining tracks due to the action of the collector system, and
- in the water column near and below the depth where tailings and other wastes are disposed of

Afterwards, impact experiments such as DISCOL, BIE and JET focused mainly on the environment near the seafloor. Their approach, first realized by DISCOL, was the large-scale disturbance experiment, based on the results from former studies, that experimental large-scale disturbances would allow the extrapolation to full-scale mining impacts.

It is well known that the results of impact experiments could be significant and meaningful only under the presupposition of understanding the natural baseline of the ecosystem potentially damaged and its natural changes. However, some problems as follows should be noticed:

1. Until now, most of the oceanologists recognized that, although a significant amount of basic and applied research work has been done in the past decades and some of them are still in progress, the current knowledge and understanding of the deep-sea environment and ecology are nevertheless much limited.

- It is common that a surprisingly high quantity of new species is found by almost every expedition in the recent decades.
- It is recognized that the biological community in the deep sea is one of the most diverse on earth.
- The natural changes or variabilities of the ecosystem temporally and spatially, demonstrated by several researchers in recent expeditions, could be unexpected large.

As a result, the baseline data collected in short time and small scale is not sufficient to be taken as the natural standard used in some risk assessments.

2. It could be recognized by, examing the different sampling gears and procedures, methods of sub-sampling, mesh sizes of sieves and selection of monitoring indicators which used by different scientific groups that, there are no "international standard techniques, equipment and format" for measuring and reporting environmental parameters and, therefore, their results must be compared with care.

3. Because of complicated reasons, commercial mining of manganese nodules is estimated to commence at least 30 years later. Of course, this provides a time-gap and temporary opportunity to carry out some well-planned environmental studies including impact experiments, and it would be useful to develop appropriate techniques and strategies for the monitoring of impacts during long-term PMOs that will create the largest environmental disturbances prior to the initiation of commercial mining.

However, on the other hand, because the technology and procedures for mining and processing manganese nodules are still under development, they cannot currently provide the references of operational parameters necessary for ultimate impact evaluation. That is, the technique to mine the manganese nodules in 30 years or in the later future is still not exactly known. This raises the question whether the results, especially those of impact experiments currently being done (referring to the nowadays deep-sea mining technology), will still be valid at that time. The speed of high-technology development and the progressively consciousness of environment protection, make it difficult to imagine, for example, that we apply the environmental assessment data collected in the 60ies or even before, to plan our industrial activities now in the 90ies. The environmental impacts could be so dependent upon and may largely differ between the various techniques employed. Of course, the environmental data collected until today have taught us much about the ocean and the mining impacts, although they must be used with care in the later impact assessments.

Consequently, an environmental baseline study project, to be conducted between 1996 to 2000, termed "Natural Variability of Baseline (NaVaBa)" (Tab.2) is planned by COMRA.

Table 2. Short outline of the "NaVaBa" project:

Expeditions	two times partly repeated in same area and same season in 1997 and 1998, respectively
Area	the Chinese invest area and reference areas (2 plots with 15-20 km² representatively in western and eastern Chinese registered area, respectively)
Investigation	collection of biological baseline data collection of hydrological baseline data collection of chemical baseline data collection of geological baseline data

Through the planned investigation, and combined with the data collected by previous Chinese cruises and also other scientific groups abroad, NaVaBa intends to make a contribute to the understanding the natural changes and variabilities of the environmental and ecological system in the deep sea of Chinese "invest area". Such a set of data is considered to become a reliable baseline as a natural standard for artificial impacts.

REFERENCES

COMRA (1996). "Report on Environmental Investigation in the 'Pioneer Area' from 1991 to 1995". Beijing. (in Chinese and in press).

Jin, J.-C, and Zhou, N (1996). "Basic Results and Prospects for COMRA's Activities in the Area". International Seminar on Deep Sea-bed Mining Technology. Beijing, October.

NOAA (1981). "Deep Seabed Mining. Final Programmatic Environmental Impact Statement". (Vol. 1 and Vol. 2) NOAA/OOME, Washington, D. C.

Thiel, H, Foell, E, J, Schriever, G (1991). "Potential Environmental Effects of Deep Seabed Mining". Umweltforschungsplan des Bundesministers fuer Umwelt, Naturschutz und Reaktorsicherheit. Berichte aus dem Zentrum fuer Meeres- und Klimaforschung der Universitaet Hamburg 26.

Thiel, H (1995) "The German Environmental Impact Research for Manganese Nodule Mining in the SE Pacific Ocean". Proceedings of the ISOPE - Ocean Mining Symposium, Nov. 21-22, 1995, Tsukuba, Japan.

Tkatchenko, G, G, and et al. (1996) "A Comparative Analysis of Results Provided by Comprehensive Studies on Reference Environmental Transects in Pioneer Areas of COMRA and IOM". International Seminar on Deep Sea-bed Mining Technology. Beijing, October.

Zhou, H,-Y (1996) "Natural Variability of Baseline (NaVaBa) Studies in Deep Sea--a challenge and chance for international cooperation". International Seminar on Deep Sea-bed Mining Technology. Beijing, October.

Proceedings of the Seventh (1997) International Offshore and Polar Engineering Conference
Honolulu, USA, May 25-30, 1997

ISBN 1-880653-28-1 (Set); ISBN 1-880653-29-X (Vol. I)

Direct Ocean Storage of CO_2: Options and Obstacles

Peter M. Haugan
Geophysical Institute, University of Bergen
Bergen, Norway

ABSTRACT

Separation and liquefaction of CO_2 from power plants and large industries with subsequent direct injection at large depths in the ocean may give a significant contribution towards reducing the build-up of CO_2 in the atmosphere. There is a range of technical options for transporting CO_2 to large depths, each with different cost, efficiency and environmental effects. The major obstacle may be the environmental effects associated with acidification of the seawater, or public reactions to such use of the oceans to help solving global climate change problems.

KEY WORDS: carbon dioxide. ocean environment, acidity, climate, injection.

INTRODUCTION

The main aim of this paper is to describe some options for disposing of CO_2 in the oceans, and what environmental problems that need to be considered in conjunction with direct ocean storage. I start with an overview of the motivation for considering this technology at all, realising that many readers may find it hard to believe that anybody would ever seriously consider it. End-of-pipe solutions like separating CO_2 from the exhaust gas of power plants and disposing of it somewhere, perhaps far away in the deep ocean, seem like costly last-resort options. However, research results from several groups over the past several years have shown that fossil fuel may be competitive with other energy sources even if the penalty of separation, transport and injection costs have to be paid.

Non-technical factors like the public perception of the seriousness of the climate change issue and the feelings aroused by using the oceans as some kind of waste bin, may in the end determine whether direct ocean storage will ever happen in one form or another. Meanwhile, scientists and engineers should work on the various aspects and provide more firm estimates of the efficiency, cost and possibly negative environmental effects of the various options. When the decision has to be made on whether to implement direct ocean storage, more knowledge will be needed, and it will take some time to develop that knowledge.

Following a broad overview of the CO_2 problem and other options which are available to deal with the problem, I describe below the oceanic sink of CO_2 and also some biological methods that have been proposed for enhancing that sink. However, physical methods, i.e. direct storage, seem to have a greater potential, and are the main focus of the paper. A number of technical options are discussed, followed by a discussion in turn of their efficiency, costs and environmental impact. Subsequently, public perception and international conventions are briefly touched upon, as well as future research needs, before the concluding section.

THE CO_2 PROBLEM

The atmospheric CO_2 content is now about 25 % higher than pre-industrial levels. Fossil fuel use is expected to give at least a doubling of the atmospheric CO_2 content in the next century, maybe much more. The increasing CO_2 levels in the atmosphere are viewed by many as perhaps the most serious global environmental problem facing mankind. If irreversible changes in the global climate system are triggered by the increased greenhouse effect caused by CO_2 and other gases whose natural cycles have been perturbed, conditions for future generations may deteriorate.

In order to cope with these problems, a wide range of actions are already being implemented. Improving energy efficiency, fuel switching to less carbon intensive fuels, e.g. from coal and oil to natural gas, and other no-regret options seem to enable some developed countries to stabilise or slightly reduce their emissions. However, due to the slowness of natural carbon sinks, global emissions would have to be reduced by about 60 % in order to stabilise the atmospheric CO_2 content (Houghton et al., 1995). This is totally unrealistic. Rapidly developing countries like China seem unable to decouple economic growth from energy use and have a very rapid increase in emissions. If the CO_2 problem is taken seriously, it is clear that more drastic actions will be required.

A carbon tax is now enforced on emissions from Norwegian offshore activities. In order to avoid venting of 1 million tonnes of CO_2 per year to the atmosphere, naturally occurring CO_2 in the

Sleipner Vest natural gas field has since fall of 1996 been separated from the produced gas stream and injected in an aquifer about 1000m below the seabed in the North Sea. Similar operations at much larger rates are being planned elsewhere. In regions such as the North Sea, there are aquifers with capacity to store large amounts of CO_2, not only from produced gas, but also from burning of fossil fuel.

Aquifers and oceans seem to be the only storage locations with sufficient capacity to handle a significant fraction of the global CO_2 emissions. However, aquifers on land, in particular in populated areas, are not so attractive for safety reasons, and offshore aquifers are not so widespread or well known. It is therefore anticipated that some countries will consider acceleration of the oceanic sink of CO_2 as a climate technology that may help them meet their international obligations and national objectives of reducing emissions.

THE ROLE OF THE OCEANS IN THE CARBON CYCLE

The potential contribution of CO_2 emissions to climate change, has motivated increased research on the global carbon cycle during the last decades (Siegenthaler and Sarmiento, 1993). The major anthropogenic influences on fluxes and reservoirs have occurred during the past 200 years driven by the needs of a growing world population. Requirement for food led to changes in land use with considerable net carbon releases primarily in the 19th and the first part of the 20th century, and requirements for energy, primarily in the last 50 years, led to accelerating mining and burning of coal, oil and natural gas, with projected continued increases well into the 21st century.

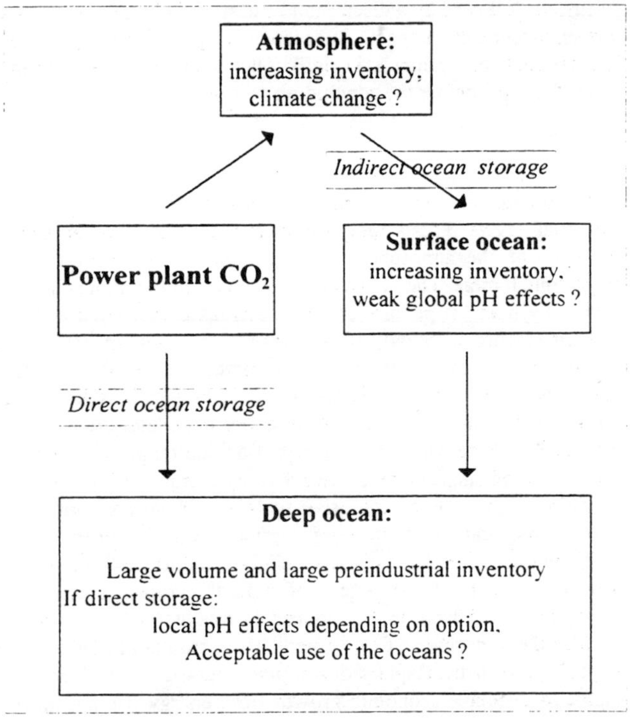

CO_2 is unusual compared to other atmospheric gases because about 98.5 % of the total atmosphere-ocean inventory is in the ocean. This is primarily due to the fact that CO_2 reacts with water to form carbonate and bicarbonate ions. It implies that the fractional increase of carbon content in the ocean since the start of the industrial revolution is much smaller than in the atmosphere even if the absolute increases are of the same order of magnitude.

Deep water contains more carbon than surface water. Photosynthesis in the near surface ocean fixes carbon which is recycled or transported to depth in particulate or dissolved form. Cooling of surface water allows carbon uptake because of the increase in solubility with decreasing temperature. Cold water sinks to great depths at high latitudes and forms deep water in the world oceans. This deep water is subsequently enriched in carbon due to the biological pump.

Emissions of CO_2 to the atmosphere is accompanied by uptake across the air-sea interface. The oceanic sink of anthropogenic carbon is presently estimated at 2 Giga tonnes of carbon per year (GtC yr^{-1}, 1 Gt = 10^9 tonnes = 1 Pg = 10^{15} g), or about one third of the total emissions. Due to the slow ventilation of deep ocean water masses, most of the anthropogenic carbon in the oceans is in near surface waters, while very little change has so far occurred at depth (Quay et al. 1992). CO_2 emitted to the atmosphere is equilibrated with surface water on a time scale of a few years. Uptake of CO_2 disturbs the chemical equilibrium between dissolved CO_2, carbonate, and bicarbonate and leads to increased acidity, i.e. reduced pH. The increase in atmospheric CO_2 since the start of the industrial revolution has already decreased surface water pH by about 0.1 pH unit (Haugan and Drange, 1996).

Calculations based on carbonate ion abundance indicate that the deep oceans can absorb at least 1200 GtC without significantly changing the chemical balance of seawater (Cole et al. 1995). This is about 200 times the present annual emissions. This input will result in reductions of sea-water pH by a few tenths and increase the area of sea floor that is in contact with water undersaturated with calcite.

Dissolution and reaction of sediment calcite will increase the capacity of the ocean, possibly enough to accommodate 2000 to 6000 GtC. This latter process is slow. However, even neglecting dissolution of sediments, it is clear that a significant fraction of the emissions during the next few centuries could be stored in the ocean. In the short term, 30-50 % of the emissions are taken up by the oceans. In a perspective of a thousand years or more this fraction will by natural processes increase to around 90 %. Are there ways of accelerating the oceanic sink so that the potentially devastating transient peak in atmospheric CO_2 can be avoided?

BIOLOGICAL OR PHYSICAL METHODS?

Proposals have been made to stimulate biological fixation by intentional addition of macro- (P, N, Si) or micro-nutrients (Fe), for increased transport of carbon out of the euphotic zone to sedimentation in the deep ocean. However, quite apart from the practical problems with such schemes, 3-D global modelling studies have shown that the efficiency may not be so good.

In the case of iron (Fe) enrichment, Sarmiento and Orr (1991) showed that even if the enrichment led to total macro-nutrient depletion in the surface water in the entire Southern Ocean, increases in atmospheric CO_2 could only be delayed and not counteracted. Furthermore, the effect would depend on continued supply of iron and lead to other undesirable consequences such as oxygen depletion in parts of the Indian Ocean and reduction of low latitude new production.

Concerning increasing the algae production in surface waters by addition of nutrients, Orr and Sarmiento (1992) showed that harvest and burning of the (macro-)algae as a substitute for fossil fuel was necessary to achieve a significant effect. Just relying on strengthening the biological pump with increased sedimentation would be inefficient.

Harvesting of organic matter from the sea such as seaweed or kelp, and subsequent burning of products based on this material on the other hand would create a fast subcycle of the global carbon cycle where the CO_2 released to the atmosphere during burning, would again be fixed in oceanic new production. This is quite analogous to using terrestrial renewable energy sources such as wood. In the case of the oceans it requires both large fertiliser production and development of large scale harvesting systems, neither of which seem feasible.

Marchetti (1977) suggested to dispose CO_2 from European power plants directly into the water flowing from the Mediterranean across the sill in the Gibraltar Strait and spreading at about 1000m depth in the Atlantic. In the past 5-6 years, variations of this idea of direct storage (physical method), and different ways of capturing CO_2 from exhaust gas, has been investigated by researchers in several countries, notably in Japan, North America and Europe.

DIRECT STORAGE OPTIONS

The different options rely on several physical properties of CO_2. Transport requirements, and also many separation methods, dictate that the CO_2 will be delivered in dense or «liquid» phase, normally by pressurising to 50-100 bar for delivery through pipelines, but possibly also by cooling which allows Liquefied Petroleum Gas (LPG) type ship transport with less stringent pressure requirements on the CO_2 containers.

If CO_2 is released in the oceans at depths less than 4-500m, it will vaporise. Rising CO_2 bubbles will dissolve in seawater during ascent. If released at low rates, the bubbles may dissolve completely before reaching the sea surface (Haugan and Drange, 1992). However, except at special sites where persistent downwards currents are found, near surface water is normally in vigorous contact with the atmosphere. A considerable fraction of the CO_2 would then be expected to be vented to the atmosphere. Transport to greater depths is therefore normally considered necessary.

When CO_2 is solved in seawater, the density of the combined solution increases relative to normal seawater. This has motivated the idea of intentionally creating dense gravity currents which would transport CO_2-enriched water along the sloping sea floor (Haugan and Drange, 1992, Drange et al., 1993). Transport along a sloping seafloor is considerably more efficient than releasing in the open water column since inmixing of ambient water is reduced. The vertical penetration depth is determined primarily by the slope, bottom roughness and ambient stratification and may be of the order of a few hundred meters. The starting point for such a gravity current could be a specially designed dissolution chamber (Adams et al., 1995). Another technique (the «GLAD» system) exploits the buoyancy of CO_2 gas for efficient production of a dense solution by using gas lift in an inverted U-tube (Kajishima et al., 1995).

Liquid CO_2 is less dense than seawater for depths less than 3000m. Release of liquid CO_2 from a fixed point in the 600-3000m depth range is expected to create a rising droplet plume above the release depth. However, since solution of CO_2 in seawater increases the water density, the plume will peel off enriched water (Liro et al., 1992, Haugan et al., 1995), bringing CO_2-enriched water finally down below the release depth. Liquid CO_2 release is further complicated by the possible formation of hydrate clathrates at the interface between CO_2 and seawater. Hydrate shells on droplets will impair the solution of CO_2 in seawater (Hirai et al., 1996) and may affect dynamical droplet coalescence and break-up.

Release from a moving ship by sprinkling through a towed pipe has also been considered (Nakashiki et al., 1995). With a perforated pipe of several kilometer length, CO_2 may be spread out in small droplets over a wide depth interval. The wake behind the towed pipe contributes to dilution so that droplet-droplet interaction is reduced and some of the more complicated plume effects are avoided. Effects of hydrates on single small droplets are more easily predictable and controllable than similar effects on droplet plumes.

Creating dense dry ice blocks and dropping them from ships has even been tried in practice (Nakashiki et al., 1991). This option would spread the CO_2 in the water column in a similar way to sprinkling. However, the energy penalty of creating dry ice and problems with handling of such blocks makes this option unrealistic and impractical for large scale application. The same can be said about dropping torpedos or penetrators containing liquid and/or solid CO_2 for burial in the seafloor (Murray et al., 1996).

Finally, there is the option of creating a «lake» of liquid CO_2 in depressions at the sea floor. Liquid CO_2 is more dense than seawater at depths greater than 3000m. At depths greater than 3700m, pure CO_2 is also more dense than CO_2-saturated water which is expected to form at the interface between the pure CO_2 and the water above. Thus a stable stratification is obtained. However, even in this case, leakage of CO_2 to the water above is expected, so the lake will only delay the solution of CO_2 into the water column and not be a permanent repository shielded from the water above. This option could be served both by moving ships to a floating platform and by pipelines along the seafloor.

EFFECTS ON ATMOSPHERIC CO_2

Release in the upper 4-500m will generally not give acceptable residence times in the ocean. Although this method may be used in special locations, we concentrate here on options which may be applied more widely. As a first guess, the deeper the release, the longer the time until any of the stored CO_2 reaches the sea surface. However, this will depend strongly on the large scale ocean circulation pattern as well as the local intensity of vertical mixing, which is again governed by factors such as the water column density stratification and bottom topography (Haugan and Drange, 1995).

Three dimensional ocean circulation modelling studies have shown that the expected rate of partial outgassing of stored CO_2 to the atmosphere may be strongly dependent on location. In some cases there is little to be gained from increasing the injection depth from say 1000m to 2000m. Site specific studies are required to clarify which fraction of the stored CO_2 will return to the atmosphere on different time scales. This is further complicated by the uncertainty in the assumed background atmospheric CO_2 scenarios over the next few centuries (Bacastow and Dewey, 1996). Generally, however, by injection below the main thermocline in the oceans, i.e. below about 1000m, and taking into account local oceanographic conditions in the design of the operation, it is possible to sequester most of the injected CO_2 on a time scale comparable to the expected atmospheric peak in CO_2, i.e. over several hundred years and up to about a thousand years. On time scales of several thousand years, deep ocean circulation, if it stays approximately as today, will anyway distribute anthropogenic CO_2 between the oceanic and atmospheric reservoirs so that most will reside in the deep ocean.

COSTS

The costs of separation, transport and storage are presently dominated by the costs of capture of CO_2 from the fossil fuel power plant flue gases. A whole range of different techniques have been developed. Presently, the total costs of capture, transport and storage, would increase the price of electricity with somewhere between 20 and 100 % depending on the type of plant (Haugen and Eide, 1996). However methods under development and incorporating the need for capture into plant design may bring down the costs substantially. The requirements for capture (and transport), and thus the costs, are normally similar for underground storage and ocean storage.

The relatively low costs of transport and injection facilities compared to that of capture is the reason why limited efforts have gone into detailed cost estimation of these elements. There is a very significant economy of scale associated with pipeline transportation of CO_2 (Skovholt, 1993). Ship transport will normally have problems with handling CO_2 at a pressure of 50 bars or more. If cryogenic CO_2 separation methods which deliver liquid CO_2 at low pressure can be developed, there is a possibility that power plants close to shore can combine favourably with ship transport giving a low cost total installation and operation.

Based on the cost estimates, direct ocean storage may become attractive to power plant operators in developed countries if a carbon tax at a moderate level, say 50 US $/tonne of CO_2, is introduced. This is the present Norwegian offshore CO_2 tax. In developing countries, one could imagine that the additional costs of a CO_2 separation and storage scheme would be paid by the developed world, to allow developing countries to expand their power sector, perhaps based on domestic fossil fuel resources, while minimising the negative impact of CO_2 emissions on the global environment.

ENVIRONMENTAL IMPACTS

For dense gravity currents there is a need for high concentration of CO_2 in order to obtain the necessary excess density. This inevitably leads to strongly reduced pH, down to a pH of 3-4 inside the dense plume. This would wipe out most if not all forms of marine life at and near the sea floor in the area.

Open ocean droplet plumes created from fixed point releases are tall (maybe more than 100m) and narrow (approx. 10m radius), depending on release mode and ambient stratification (Haugan et al., 1995). Inside the plume, pH may go down to 5-6. This is unacceptable to most forms of marine life if they are exposed for a long time. However, mobile species are expected to avoid the area. The problem is then how large perturbations in pH that occur in a larger area downstream of the release point and how long time mainly passively drifting species will stay in the different pH regimes. This has been modelled in detail by Herzog et al. (1996), showing that limited lethal effects are to be expected. Less is known experimentally about possible sublethal effects (Ormerod and Angel, 1996).

Storage in depressions in the seafloor («lake mode») is expected to give a temporary delay in the release to the water column. Unless the depression is very deep and/or has extremely steep sidewalls, CO_2-enriched water will move away from the release site in a layer with relatively high dissolved CO_2 concentrations. The added density due to dissolution of CO_2 keeps the water from mixing upwards. The CO_2-enriched water will inevitably wipe out benthic life close to the site. How far away strong impacts can be expected and how harmful this impact is to essential functions of large marine ecosystems is poorly known at present.

Release at middepths (1000-2000m) from a moving ship or by other technical solutions which can ensure dilution, is probably the most environment-friendly way to perform ocean storage. Even from a purely ocean environmental protection point of view, and without considering the climate change aspects of emissions, this type of release may well turn out to be preferred in comparison to indirect near surface storage via the atmosphere (Haugan and Drange, 1996). It is the moving ship option which most properly can be viewed as a true acceleration of the uptake from the atmosphere. It bypasses the euphotic zone where most of the biological production, including that which is important for food supply from the oceans, occurs, and also avoids high CO_2 concentrations near the seafloor where there is another maximum in biological activity and species diversity.

PUBLIC PERCEPTION

Increasing the content of CO_2 in the oceans will be viewed by many as pollution, even if CO_2 is already present in the ocean in large quantities. Direct storage may be viewed as improper use of the oceans, even by people who may accept invasion of similar amounts of CO_2 to the oceans via the atmosphere. There will always be some uncertainties about the possible environmental impacts of a direct storage scheme, in particular long term effects. Why not give it up?

Research has shown that seemingly environment-friendly options, such as the moving ship option, do exist. If these options are further developed, and it can be shown that the perturbations in ocean carbon content and pH due to direct storage is comparable to natural variability in some location (Haugan and Drange, 1996), rational practise would be to allow such storage. There will always be uncertainties about environmental effects of CO_2 both in the atmosphere, on land and in the oceans, as there will be uncertainties about effects of other ways to provide energy supply. If it can be scientifically established that some form of direct ocean storage is essentially harmless, we are almost obliged to allow it.

Environmental impact assessments and monitoring schemes would have to be established based on principles similar to those used by coastal states for various kinds of outlets. Since CO_2 storage will be long term, may be expected to transcend national boundaries and even take place in international waters, a complicated set of partially overlapping international conventions will have to be examined in order to ensure that they are not violated. In this process, inconsistencies and legal obstacles may occur. One can only hope that if a scientifically sound way to handle CO_2 is developed, it will not be stopped by irrational reactions from the public. On the other hand, irrational overreactions from the public to warnings about global climate change should not be allowed to lead to irresponsible, large scale storage in the oceans.

THE MOVING SHIP OPTION

The moving ship option may be the best available or practicable technology from an ocean environmental point of view. It is attractive because of its flexibility concerning location. Recent engineering studies (Masahiko Ozaki, Mitsubishi Heavy Industries, pers. comm. 1996) indicate that it is technically feasible and that costs are not necessarily prohibitive. Even if it may be more costly than other available options in some application, there may be a requirement to evaluate it as an alternative for environmental reasons. However, the concentration distribution in this case has not been much studied. The vertical rise of CO_2 released from a moving ship may be similar to that from a fixed point plume if the release rates are high or the ship speed is slow. If release rates are low and ship speed is high, single droplet dynamics may prevail, determined mainly by the effective droplet size and the extent of hydrate formation.

Consider a ship which is to inject a nominal rate of Q = 100 kg CO_2 s^{-1} = 2.2 x 10^3 mol s^{-1}. This rate corresponds to a power plant with output between 500 and 1000 MW depending on the type of plant and its efficiency. The towed pipe may have a diameter of 400 mm (16 "). If the wake produced by the pipe has a width W [m], the ship speed is V [ms^{-1}], and the vertical depth interval over which the CO_2 is dissolved is H [m], the average carbon concentration perturbation in the wake behind the pipe is ΔC [mol m^{-3}]:

$$\Delta C = Q/(WHV)$$

If W is 2 m, H is 200 m and V is 1 ms^{-1}, then C is 5.5 mol m^{-3}. The perturbation is reduced to 0.22 mol m^{-3} if H is 1000 m and V is 5 ms^{-1}, but is increased to 22 mol m^{-3} if H is 100 m and V is 0.5 ms^{-1}. Since the natural inorganic carbon concentration in seawater is of the order 2 mol m^{-3}, these perturbations are of the order 0.1 to 10 times the natural level. This is a considerable perturbation. It is obvious then that mixing induced within the wake itself is not sufficient to achieve acceptable dilution. Even if the release takes place at rates which imply droplet plume dynamics, the mixing induced by the buoyancy of a droplet plume is unable to reduce the concentrations below these levels (Haugan et al, 1995).

We have to rely on ambient lateral diffusion to reduce the peak concentration behind the ship. Okubo (1971) synthesised available data from dye experiments on time scales from one hour to one month into a correlation for variance σ^2 [m^2] in a radially symmetric distribution as

$$\sigma^2 = 1.08 \times 10^{-6}\, t^{2.34}$$

where t [s] is the time since release. In the present case, concentration is reduced only by diffusion in the transverse direction normal to the ship heading. A length scale Y [m] for the transverse spreading can then be defined by

$$Y = \sqrt{3}\,\sigma = 0.18 \times 10^{-4}\, t^{1.17}$$

If the correlation is valid also for time scales less than one hour, it indicates widening of the patch from the instantaneous half width of 1 m to about 25 m in one hour and to about 1 km in 24 hours, i.e. dilution by a factor of order 10^3 during 24 hours after the ship has passed. With the range of ship speeds and dissolution heights discussed above, this brings the average perturbation in carbon concentration down to 10^{-4} to 10^{-2} of the natural concentration within 24 hours.

Based on the correlations of Herzog et al (1996) for mortality of marine organisms from exposure to reduced pH, and a gradually diluted perturbation according to the above very simplified and tentative calculation, zero mortality seems within reach for the moving ship option. However, there are several questions remaining, including how to engineer the pipe release so that the CO_2 is dissolved efficiently in the wake, whether ambient lateral diffusion in the area works as indicated by the above application of Okubos (1971) correlation somewhat beyond its empirical basis, and whether especially sensitive organisms present in the area, could suffer more than expected from the general impact correlations of Herzog et al (1996) or whether significant sublethal effects may occur from this relatively brief exposure to reduced pH.

FUTURE RESEARCH AND DEVELOPMENT

On the technical side, there is a need to better understand CO_2 droplet plume dynamics which is quite complicated due to hydrate clathrate effects and the increase in water density due to solution of CO_2. Theoretical and laboratory scale research is ongoing and short term, small scale field experiments are contemplated in order to obtain better knowledge about the horizontal and vertical spread of dissolved CO_2 around an injection site, and interaction with ambient oceanographic conditions. Results from such investigations would in turn be the basis for engineering design and environmental impact assessment of fixed point release from some candidate sites.

Costs are intimately linked to CO_2 capture and transport which suggests that system studies including all these elements may be more fruitful than optimising each component separately. If new power plants are the target, new designs suitable for capture and transport may be developed. With a given CO_2 tax or total allowed cost, a range of options may be technically feasible, but not all of them will be environmentally acceptable.

From an environmental point-of-view it seems that the flexible moving ship option is the one which stands the best chances of demonstrating minimal impact and thus being acceptable. As will be evident from the discussion above, this option has not been very much studied. Research needs are similar to those of fixed point release for the near field dissolution process. Improved knowledge of ambient lateral diffusion is required for the far field.

For all options, biological impact needs to be better understood. Laboratory experiments on effects of increased CO_2 on individual species are still very few (Ormerod and Angel, 1996) and should be executed also to understand effects of CO_2 invading from the atmosphere. Long term (few years) experiments on the ecosystem level may be performed in semi-enclosed deep fjord systems (more than 500m deep) which simulate the open ocean but has a somewhat simpler ecosystem and better possibilities for monitoring.

CONCLUSIONS

We are still in the early days of research on options for direct ocean storage of CO_2. New ideas and combinations of old ones may lead to options which will actually be implemented. There is a range of partial technical solutions and a need for site specific studies to decide which are feasible in each case. Ultimately, the public perception of the global climate change issue as well as the issue of using the global oceans to assimilate some of our CO_2 «waste» may dictate where the CO_2 will go in the future. Direct storage in the oceans, bypassing the temporary storage in the atmosphere is doable now at a certain cost. Whether the environmental impacts and costs of each of the options are acceptable remains to be seen.

ACKNOWLEDGEMENTS

Numerous colleagues and friends in different parts of the world have contributed to the knowledge base on direct ocean storage. The International Energy Agency Greenhouse Gas R&D

Programme has helped identify critical issues. The preparation of this paper has received financial support from Statoil a.s. Thanks to Olav Kårstad of Statoil for always keeping an eye on ocean storage schemes even if underground storage may have been closer to implementation locally. The perspective and views expressed in this review are on behalf of myself only.

REFERENCES

Adams, EE, D Golumb, XY Zhang and HJ Herzog (1995). "Confined release of CO_2 into shallow seawater," pp 153-164, in: *Direct Ocean Disposal of Carbon Dioxide*, N Handa and T Ohsumi (eds), Terra Scientific Publishing Co, Tokyo.

Bacastow, RB and RK Dewey (1996). "Effectiveness of CO_2 sequestration in the post-industrial ocean," *Energy Conversion Management* Vol 37, pp 1079-1086.

Cole, KH, GR Stegen and D Spencer (1995). "The capacity of the deep oceans to absorb carbon dioxide," pp 143-152, in: *Direct Ocean Disposal of Carbon Dioxide*, N Handa and T Ohsumi (eds), Terra Scientific Publishing Co, Tokyo.

Drange, H, G Alendal and PM Haugan (1993). "A bottom gravity current model for CO_2-enriched seawater," *Energy Conversion Management* Vol 34, pp 1065-1072.

Haugan, PM and H Drange (1992). "Sequestration of CO_2 in the deep ocean by shallow injection," *Nature* Vol 357, pp 318-320.

Haugan, PM and H Drange (1995). "Disposal options in view of ocean circulation," pp 123-141 in: *Direct Ocean Disposal of Carbon Dioxide*, N Handa and T Ohsumi (eds), Terra Scientific Publishing Co. Tokyo.

Haugan PM and H Drange (1996). "Effects of CO_2 on the ocean environment." *Energy Conversion Management* Vol 37, pp 1019-1022.

Haugan PM, F Thorkildsen and G Alendal (1995). "Dissolution of CO_2 in the ocean." *Energy Conversion Management* Vol 36, pp 461-466.

Haugen, HA and LI Eide (1996). "CO_2 capture and disposal: The realism of large scale scenarios." *Energy Conversion Management* Vol 37, pp 1061-1066.

Herzog, HJ, EE Adams, D Auerbach and J Caulfield (1996). "Environmental impacts of ocean disposal of CO_2." *Energy Conversion Management* Vol 37, pp 999-1005.

Hirai, S, K Okazaki, N Araki, H Yazawa, H Ito and K Hijikata (1996). "Transport phenomena of liquid CO_2 in pressurized water flow with clathrate-hydrate at the interface." *Energy Conversion Management* Vol 37, pp 1073-1078.

Houghton, JT, LG Meira Filho, J Bruce, H Lee, BA Callander, E Haites, N Harris and K Maskell, eds. (1995). "*Climate Change 1994. Radiative Forcing of Climate Change and an Evaluation of the IPCC IS92 Emission Scenario.*" Cambridge University Press.

Kajishima, T, T Saito, R Nagaosa and H Hatano (1995). "A gas-lift system for CO_2 release into shallow water," *Energy Conversion Management* Vol 36, pp 467-470.

Liro, CR, EE Adams and HJ Herzog (1992). "Modelling the release of CO_2 in the deep ocean," *Energy Conversion Management* Vol 34, pp 1065-1072.

Marchetti, C 1977. "On geoengineering and the CO_2 problem," *Climate Change* Vol 1, pp 59-68.

Murray, CN, L Visintini, G Bidoglio and B Henry (1996). "Permanent storage of carbon dioxide in the marine environment: The solid CO_2 penetrator," *Energy Conversion Management* Vol 37, pp 1067-1072.

Nakashiki, N, T Ohsumi and K Shitashima (1991). "Sequestering of CO_2 in the deep ocean, fall velocity and dissolution rate of solid CO_2 in the ocean," Central Research Institute of Electric Power Industry (CRIEPI) report EU91003, Tokyo.

Nakashiki, N, T Ohsumi and N Katano (1995). "Technical view on CO_2 transportation onto the deep ocean floor and dispersion at intermediate depths," pp 183-193, in: *Direct Ocean Disposal of Carbon Dioxide*, N Handa and T Ohsumi (eds), Terra Scientific Publishing Co, Tokyo.

Okubo, A (1971). "Oceanic diffusion diagrams", *Deep-Sea Research* Vol 18, pp 789-802.

Ormerod, W and M Angel, eds. (1996). "*Ocean storage of carbon dioxide: Workshop 2 - Environmental Impact,*" IEA Greenhouse Gas R&D Programme, Cheltenham. 131pp.

Orr., JC and JL Sarmiento (1992). "Potential of marine macroalgae as a sink for CO_2: Constraints from a 3-D general circulation model of the global ocean." *Water, Air and Soil Pollution* No 64, pp 405 - 421.

Quay, PD, B Tilbrook and CS Wong (1992). "Oceanic uptake of fossil fuel CO_2: Carbon-13 evidence." *Science* Vol 256, pp 74-79.

Sarmiento, JL and JC Orr (1991). "Three-dimensional simulations of the impact of Southern Ocean nutrient depletion on atmospheric CO_2 and ocean chemistry," *Limnol. Oceanogr.* Vol 36 No 8, pp 1921-1950.

Siegenthaler, U and JL Sarmiento (1993). "Atmospheric carbon dioxide and the ocean." *Nature*, Vol 365, pp 119-125.

Skovholt, O (1993). "CO_2 transportation system". *Energy Conversion Management* Vol 34, pp 1095-1103.

Proceedings of the Seventh (1997) International Offshore and Polar Engineering Conference
Honolulu, USA, May 25-30, 1997

ISBN 1-880653-28-1 (Set); ISBN 1-880653-29-X (Vol. I)

Simulation of Deep Ocean Relocation of Dredged Material

Philip J. Valent and David K. Young
Naval Research Laboratory
Stennis Space Center, MS, USA

Andrew W. Palowitch
Oceaneering International, Inc.
Upper Marlboro, MD, USA

ABSTRACT

Each year, 300×10^6 m^3 (400×10^6 yd^3) of sediment are dredged in the United States of which 3 to 5 percent is polluted and cannot be used beneficially or returned to the environment. Disposal of this polluted dredged material is a national problem. One possible solution under study is to relocate polluted dredged material to the seafloor of the abyssal ocean, considered to include those depths greater than 3000 meters. A project sponsored by the Defense Advanced Research Projects Agency focused on physics-based simulations of processes involved in this waste management option.

SimDOR, which stands for Simulation of Deep Ocean Relocation, concentrated primarily on sites of the abyssal seafloor of the North Atlantic Ocean and considered dredged material coming from the eastern seaboard of the United States. Results from the SimDOR project are also generally applicable to west coast and Gulf of Mexico ports.

The SimDOR project took advantage of recent advances in numerical modeling capabilities, simulation concepts, and visualization technology. Concepts and analytical models were developed and integrated to enable simulation of the entire process from dredging through long-term impact on the abyssal environment, allowing engineers, scientists, environmentalists, and the public an opportunity to study and understand all aspects of this option. Systems to dredge sediments, load containers, transport them to the ocean site, and ensure their safe descent to the abyssal seafloor were designed, simulated, and evaluated. Hydrodynamic, geochemical, and biological effects of the relocation process on scales ranging from the time of disposal of dredged material on the abyssal seafloor to years, tens of years, and thousands of years in the future were simulated and visualized. It is concluded that deep ocean relocation of polluted, dredged material offers a potentially economically sound and environmentally safe alternative to land-based waste isolation.

Key Words – Dredging, Dredged Materials, Numerical Modeling, Environmental Effects, Abyssal Seafloor

THE PROBLEM

With increased international trade and continued growth of port cities in the United States, keeping vital shipping ports dredged and continuously operating is of national importance. Each year, 300×10^6 m^3 (400×10^6 yd^3) of sedimentary materials are dredged of which 3 to 5 percent is contaminated and cannot be used beneficially or returned to the environment near dredging sites. Finding a cost-efficient, environmentally safe, and politically acceptable alternative to land-based disposal of dredged material is becoming increasingly important to the welfare of our economy, our nation, and our national defense (NRC, 1985, 1989).

Contaminated sediments in our nation's coastal waterways are a reservoir of toxins which can be passed through food webs potentially to man via bioaccumulation pathways. Removal of the contaminated sediments from harbors would eliminate this continued source of toxic material from the nation's productive coastal waters.

By law (Marine Protection, Research, and Sanctuaries Act, 1972, Public Law 92-532), contaminated dredged materials—after removal from the harbor or waterway floor—either have to be isolated from the environment or chemically altered, rendering them safe for disposal (EPA, 1991). Remediation of contaminated dredged materials, though technically feasible, is very expensive and is currently being tested and used only on a limited basis. Alternative solutions, such as shipping contaminated dredged materials to inland sites remote from human habitation, are currently practiced.

The number of available disposal sites is steadily declining due to environmental, sociological, and political constraints. Rising landfill costs severely limit land-based disposal methods. The primary problem is to minimize environmental impacts upon humans, and to our air, water, and food supplies by assuring that contaminants are isolated. But even if we can continue the expense of operating well-managed, contaminated waste material landfills, how much longer can we continue to consume valuable land space by this practice?

A SOLUTION

A practical solution to this problem is to relocate contaminated marine dredged material to the seafloor of the abyssal ocean. The abyssal ocean encompasses those areas where the water is deeper than 3000 meters. These isolated ocean regions contain most of the world's water and cover over half of the surface of the globe.

SimDOR, a project sponsored by the Defense Advanced Research Projects Agency and conducted by the Naval Research Laboratory, focused on model simulations of the relocation of contaminated dredged material to the abyssal ocean. SimDOR, which stands for Simulation of Deep Ocean Relocation, concentrated primarily on sites on the abyssal seafloor of the North Atlantic Ocean and on dredged material coming from the eastern seaboard of the United States. Results from the SimDOR project are also generally applicable to west coast and Gulf of Mexico ports, as reported from an earlier study assessing suitability of deep-water sites within 1600 km (1000 nautical miles) of the continental United States for disposal of wastes (Young and Valent, 1996). The SimDOR project took advantage of recent advances in numerical modeling capabilities, simulation concepts, and visualization technology. Concepts and analytical models were developed and integrated to enable simulation of the dredging, transport, and isolation of dredged material in containers, ensuring that the contained material passes through the upper layer of the ocean without detrimentally affecting organisms in this productive region (Fig. 1).

CONTAINMENT AND DELIVERY SYSTEM

The containers, fabricated from geosynthetic fabrics of woven polyester or polypropylene fibers, have been used internationally in a range of beach replenishment, erosion prevention, island creation, and sediment disposal applications. The strength and durability of these fabrics assure containment of sediments in marine environments for decades (Valent et al., 1996).

A system to dredge sediments, load the geosynthetic fabric containers, transport them to the ocean site, and ensure their safe descent to the seafloor has been proposed and predictably modeled. During the dredging and disposal operation, potentially polluted dredged materials must be contained. A specially designed clamshell dredge completely encapsules underwater sediments so that the amount of sediment released into the water at the dredging site is minimized. While maintaining the efficiency of commercial clamshell dredge operations, the total containment clamshell dredge head minimizes leakage while the excavator arm moves out of the water and over to the storage hopper.

Large debris is removed from the dredged sediment in the storage hopper with separator screens and grinder rollers. Transfer augers lead up from the bottom of the storage hopper to control flow rates during sediment distribution to the transporter. The totally enclosed auger pipes prevent the possible spread of pollutants from the sediments during container loading. Once on the transporter, a set of five distribution hoppers feed the twenty geosynthetic fabric containers which are positioned in segmented cells. Each container is filled with roughly 600 m^3 (800 yd^3) of sediment through a top central spout that is closed when loading is complete.

The transporter is then pushed by an ocean-going tug (or towed in Sea State 5 and higher) to the open ocean disposal site in an integrated tug-barge configuration. Physics-based dynamic modeling, used to test the transporter for stability in various sea states, indicates that for increased safety the transporter would have to be towed instead of pushed in sea states greater than four. The design process considered not only the vessel's design, but also the economic and operational viability of the entire process.

Using satellite positioning and ocean current measurement techniques, the transporter locates a release point, such that the container will free fall to a predetermined area on the abyssal seafloor. The containers are sequentially released from their individual cells. Discrete element and numerical flow modeling techniques were used to assess risk during container exit, descent, and seafloor impact. Stresses within the container fabric during these periods rapidly change, potentially risking container rupture. The prediction of high stress areas assists in developing deployment hardware and operational procedures that ensure container integrity. During its descent, the container reaches a steady state downward velocity of about 5 meters per second. Upon reaching the bottom, the container settles into a slumped shape. Please note, the primary purpose of the container is to isolate the dredged material from the productive, near-surface water; should the container rupture upon impact on the seafloor, the primary function of the container would still be served.

THE ABYSSAL ENVIRONMENT

Once the containers are deposited on the seafloor, what is their long term environmental impact upon the abyssal seafloor site? Many parts of the abyssal seafloor, far from continental land masses, are geologically stable and are not subject to frequent disturbances characteristic of shallower water and land masses which are exposed to the atmosphere (reviewed by Gage and Tyler, 1991). The lack of seafloor relief on the abyssal plains is contrasted with the steep slopes and roughness of continental margins.

As one progresses seaward from continental margins, the abyssal seafloor typically becomes progressively smoother without underwater mountains or hills. Abyssal plains are covered by a fine, thin layer of pelagic or hemipelagic sediment. These large, nearly flat abyssal plains are deposits of mud, silt, and dead marine organisms that have accumulated gradually over millions of years. In the abyssal ocean there is no sunlight, the waters are constantly below 4°C, and the pressure is several hundred times greater than at the sea surface.

SITE SELECTION

With so many variables, how does one find suitable sites for isolating materials on the abyssal seafloor? To answer this question, a site selection model was developed at the Naval Research Laboratory (NRL) to quantitatively compare the suitability of abyssal seafloor sites for waste isolation (Fleischer, 1995). Areas were divided into 1-degree squares of latitude-longitude. Scorable factors were chosen that apply to the biologic, geologic, oceanographic, meteorologic, and anthropogenic criteria of significance to siting for waste isolation. Areas were excluded beyond 1600 km of the

Fig. 1 The deep ocean relocation process for dredged materials involves the removal of contaminated sediments from the harbor (lower left), the processing and loading of the sediment onto an ocean transporter (top), and the transport of the sediment to the abyssal seafloor in geosynthetic fabric containers (lower right).

coastline of the U.S., within foreign economic zones, and shallower than 3000-m depth. The entry of factors into the computer model and their weighting and scoring was carefully designed to maximize an effective environmental isolation plan.

The most favorable 1-degree squares in the North Atlantic for the isolation scenario are in the Hatteras Abyssal Plain (Fig. 2). The characteristics of these areas are a nearly flat, uniform, muddy bottom, tranquil water over the seafloor, and long distances from other human activities such as shipping lanes and submarine cables.

ENVIRONMENTAL SIMULATIONS

Simulations of the North Atlantic ocean circulation, conducted with the NRL basin-scale, multilayer model, show that average current

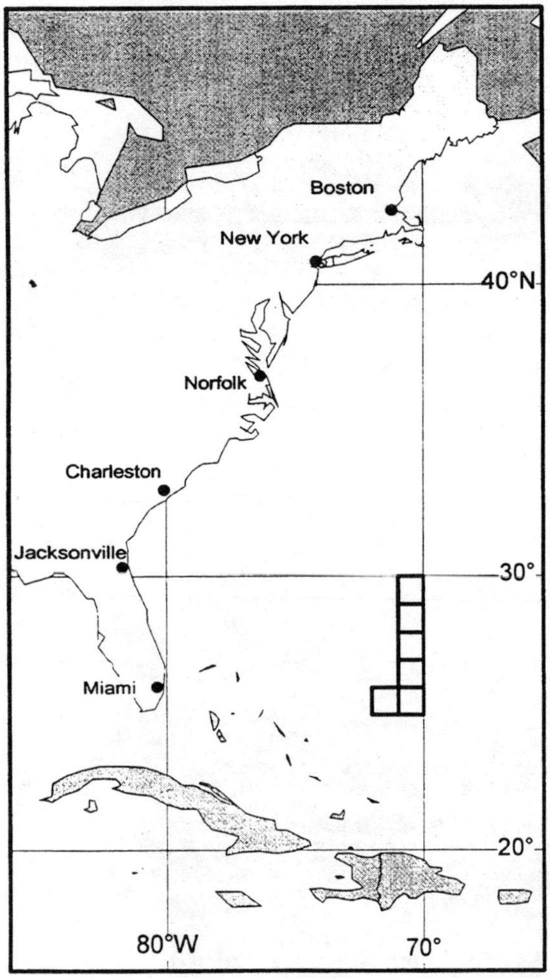

Fig. 2 Most favorable areas (1° latitude-longitude squares) for isolation of contaminated dredged material within 1600 km (1000 nautical miles) of the U.S. East Coast.

velocities and kinetic energies of the mean flow of abyssal currents at the selected sites are very low (Gallacher, 1995). This is part of a worldwide circulation pattern that cycles water from the surface through abyssal depths and back again on the time scale of centuries. Water that flows through the western North Atlantic in the region of the Hatteras Abyssal Plain may not come to the surface of the ocean for hundreds of years.

High-resolution computer simulations show that effects from impact of the geosynthetic fabric containers with the seafloor will remain local to the site (Gallacher, in press). Although sizeable plumes may result from this impact, the low levels of turbulent energy ensure that the dispersion of suspended and dissolved material from seafloor and containers (if ruptured) will be minimal.

Numerical simulations have been conducted to predict the rates and distributions of deposition and remineralization of a 1-year disposal of 8.5×10^6 m^3 of dredged material on the abyssal seafloor (Jahnke, 1995). As the organic matter within the deposit decomposes, oxygen and sulfate in the underlying sediments are depleted. Simulation results indicate that the oxygen-depleted zone does not significantly spread laterally into the surrounding sediments, but rather is essentially confined to the sediments directly below the deposit. Additionally, there is no significant anoxia of the overlying water affecting areas adjacent to or downstream from the site of deposition. Return to geochemical equilibrium conditions may take as long as thousands to tens-of-thousands of years. This should be no surprise because most turbidite deposits, which are natural analogs of the relocation process, have not yet reached geochemical equilibrium despite being laid down 10,000 years ago.

Ten-year numerical simulations of an abyssal food chain have been run to determine effects of increased organic matter on the organisms (Rowe, 1995). At sites where animals are not buried, model results show increased growth rates. The food chain model was also used to simulate contaminant transport by incorporating polycyclic aromatic hydrocarbons (PAHs) into simulations. Results from these simulations suggest that the one hypothetical pathway for PAHs to leave abyssal depths is by transport in the yolks of eggs of certain deep-sea invertebrates and fishes. The potential for measurable biological transport of contaminants by this pathway, however, is negligible (Rowe, 1995).

CONCLUSIONS

These results show that potential environmental impacts resulting from relocating contaminated dredged material to the abyssal seafloor would be localized at the disposal site chosen by a site selection model to optimize isolation. The local changes that would occur have to be balanced by the immediate detrimental effects occurring in many of our harbor areas posed by the *status quo*. Research and development for finding new and innovative alternatives to the disposal of contaminated dredged materials must continue to determine safe, economical, and dependable isolation methods for the growing problem of managing these materials. The Deep Ocean Relocation program offers the potential for an economically sound and environmentally safe alternative to land-based contaminated dredged material isolation, which can help reduce the amount deposited on an ever-decreasing amount of available land.

ACKNOWLEDGMENTS

The authors thank the participants of the SimDOR Project, some of whom are referenced in this paper, for sharing results of their work summarized in this paper. We acknowledge the help of many authorities—too numerous to mention here—in academia, government, and industry, who were consulted and gave freely their wealth of knowledge and experience. We appreciate funding of this project by the Defense Advanced Research Projects Agency, Program Element 0603226E, Dr. Allan Steinhardt, Program Manager, NRL Contribution Number NRL/PP/7401--96-0025.

REFERENCES

EPA (1991). "Evaluation of Dredged Material Proposed for Ocean Disposal (Testing Manual)," EPA-503/8-91/001 (U.S. Environmental Protection Agency, Washington, D.C.), 196 p.

Fleischer, P (1995). "Site Selection Model," In: PJ Valent and DK Young (Editors), *Abyssal Seafloor Waste Isolation: Environmental Report,* NRL/MR/7401--95-7576, Naval Research Laboratory, Stennis Space Center, MS, pp 133–188.

Gage, JD and Tyler, PA (1991). *"Deep-Sea Biology: A Natural History of Organisms at the Deep-Sea Floor,"* Cambridge Univ. Press, 504 p.

Gallacher, PC (in press). "Hydrodynamical Dispersion of Dredged Materials Sequestered on Abyssal Plains," *J Mar Sys.*

Gallacher, PC (1995). "Physical Oceanographic Modeling," In: PJ Valent and DK Young (Editors), *Abyssal Seafloor Waste Isolation: Environmental Report,* NRL/MR/7401--95-7576, Naval Research Laboratory, Stennis Space Center, MS, pp 254–276.

Jahnke, RA (1995). "Predictions of Near- and Far-Field Effects Resulting from Introduction of Wastes to Abyssal Environments," In: PJ Valent and DK Young (Editors), *Abyssal Seafloor Waste Isolation: Environmental Report,* NRL/MR/7401--95-7576, Naval Research Laboratory, Stennis Space Center, MS, pp 326–351.

NRC (1985). "Dredging Coastal Ports, An Assessment of the Issues," Marine Board, National Research Council (National Academy Press, Washington, D.C.), 212 p.

NRC (1989). "Contaminated Marine Sediments—Assessment and Remediation," Committee on Contaminated Marine Sediments, Marine Board and Commission on Engineering and Technical Systems (National Academy Press, Washington, D.C.), 493 p.

Rowe, GT (1995). "Numerical Simulations of Deep-Sea Food Chains with an Estimate of Potential Export of Contaminants from Abyssal Waste Isolation," In: PJ Valent and DK Young (Editors), *Abyssal Seafloor Waste Isolation: Environmental Report,* NRL/MR/7401--95-7576, Naval Research Laboratory, Stennis Space Center, MS, pp 356–366.

Valent, PJ, Young, DK, and Palowitch, AW (1996). "Technical and Economic Assessment of Waste Isolation on the Abyssal Seafloor," *Proc Oceanology International 96* (Spearhead Exhibitions, Ltd. Publishers, Surrey, UK), Vol. 1, pp 305-316.

Young, DK and Valent, PJ (1996). "Environmental Assessment of Waste Isolation on the Abyssal Seafloor," *Proc Oceanology International 96* (Spearhead Exhibitions, Ltd. Publishers, Surrey, UK), Vol. 1, pp 317-326.

Proceedings of the Seventh (1997) International Offshore and Polar Engineering Conference
Honolulu, USA, May 25-30, 1997

ISBN 1-880653-28-1 (Set); ISBN 1-880653-29-X (Vol. I)

Numerical Modeling of the Settling Processes of Dredged Material Disposed in Open Waters

Qin Jiang and Hiroshi Kunisu
Penta-Ocean Institute of Technology
Nasugun, Tochigi, Japan

Akira Watanabe
University of Tokyo
Tokyo, Japan

ABSTRACT

Laboratory experiments were conducted to observe the settling behaviors of ocean disposal material from barges. A vertical 2-D numerical model was then developed for predicting the short-term fate of discharged material after an individual disposal operation. In the proposed model, the released material is separated into the descending material core and the suspended particle cloud with each possessing different characteristics. The ambient fluid motion and the transport-dispersion of disposal material are calculated from the Navier-Stokes equations and mass transport equation by using the Marker and Cell technique applied on boundary fitted grids. Comparisons with experimental data indicate that the proposed model can give satisfactory results.

KEY WORDS: Ocean disposal material, ambient fluid motion, descending material core, suspended particle cloud, transport-dispersion, Marker and Cell technique, Boundary Fit Method.

1. INTRODUCTION

Ocean disposal operations of dredged material or soil are usually encountered in the dredging of navigation channels and harbors, or in the construction of breakwaters as well as offshore man-made islands. Therefore, prediction of the settling processes of disposed material in open waters, such as the descending and transportation-dispersion of the material, is of practical importance for both engineering management and environmental conservation.

Over the past few decades, experimental and theoretical studies have been carried out to investigate the movement of ocean disposal material during the initial stage after dumping from a barge. Ogawa and Takeuchi (1969) conducted laboratory experiments on the dispersion of dumped sand from a hopper-barge and derived analytical formulas for evaluating the settling velocity, the dispersion width and the distribution of dumped sand deposited on the bottom. Koh and Chang (1973) developed a mathematical model for predicting the short-term fate of ocean disposal material from barges, in which the total process was divided into three phases including the convective descent, dynamic collapse, and long-term passive transport-dispersion of the material. With this model, instantaneous changes in the disposal body size and suspended sediment concentration can be obtained. Major modifications of the Koh-Chang model were made by Brandsma and Divoky (1976). Field evaluations by Bokuniewicz *et al.* (1978) and physical model tests by Johnson *et al.* (1993) clarified this three step-process. Through experiments, Murota et al. (1988) investigated the solid-buoyant clouds formed after the release of soil from barges and analyzed the cloud motion by using thermal theory.

A common deficiency of the above numerical models is the inadequacy of their representation of the descending process in real disposal operation, e.g. treating the disposal from a barge as a single spherical or hemispherical cloud descending through the water column. Such an assumption prohibits the accurate simulation of suspended sediment concentration in water column.

On the other hand, Horie (1989) solved the material descending problem by introducing the Marker and Cell technique into his numerical model to calculate the ambient fluid flow and the motion of the released material. However, the poor accuracy of his model is unavoidable due to the static pressure distribution and incompressible fluid assumptions. Oda *et al.* (1993,1994) developed a numerical model (DEMAC) by combining the Marker and Cell technique and the Discrete Element Method, in which effects of the relative motion between the material particles and surrounding waters have been taken into account. Using this model, they analyzed the dispersion behavior and the deposition configuration of dumped particles in waters. The DEMAC model is only appropriate for the simulation of the descending coarse material because the suspended motion of the stripped fine particles is not considered. This prohibits its applications to the practical problems where the stripping of material from the released material core is significant and the accurate evaluation of suspended sediment concentration is necessary.

Even though various models on the subject have been developed, there is no general method for practical applications. In addition, the settling processes of ocean disposal material are not fully understood. Therefore, the purpose of this study is focused on gaining additional insight into the physical phenomenon of the problem. With the knowledge obtained from the experiment, a vertical two dimensional numerical model was constructed to simulate the short-term movement of ocean disposal material released from barges. In the proposed model, the released material is divided into the descending material core with high density and the suspended particle cloud formed around the material core. The Navier-Stokes equations and mass transport equation are used to compute the motion of the descending material core, as well as, the surrounding water and the transport dispersion of the suspended particle cloud. The Marker and Cell technique applied on a boundary fitted grid was used in the numerical model. It is verified that good agreement between the calculation results and experimental data is achieved.

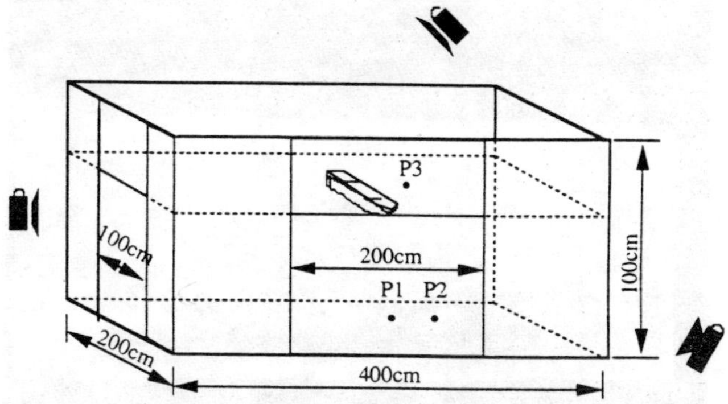

Fig. 1. Sketch of experimental setup

2. EXPERIMENT AND RESULTS

2.1 Laboratory experiment

Laboratory experiments were carried out to clarify the physical phenomenon of dredged material placement from barges. The test facility is a flume of dimensions 4.0m long by 2.0m wide by 1.0m high at a model-to-prototype scale of 1:100. On the front and left side walls of the flume, two viewing windows (2.0m by 1.0m and 1.0m by 1.0m) were built for observing the experiment. The disposal vessel is a hopper barge which was constructed on the basis of an actual design at the same scale as the flume. Its disposal volume is 2000 cubic centimeters. The disposal material used for the testing is dredged material taken from a real seabed. Material characteristics are listed in Table 1.

Table 1: Characteristics of the disposal material

Specific Gravity	2.75
Grain Size	$1.0 \sim 60.0\ \mu m$
Median Diameter	$10.0\ \mu m$
Chemical Composition	SiO_2 ($\geq 90.0\%$) Al_2O_3 ($\leq 3.0\%$) Fe_2O_3 ($\leq 1.0\%$)

Five physical tests were made in which the water content ratio of the sample was 145%, and the water depths varied from 20.0, 30.0, 40.0, 50.0 to 70.0 centimeters. As noted, the volume of each dump was 2000.0 cubic centimeters, and the releasing operation was nearly instantaneous.

All tests were conducted in static and unstratified waters, and the disposal vessel was placed at the center of the flume and kept stationary. Each test was videotaped from each of the two viewing windows. The movement of the disposal material was recorded during the time after its release from the vessel to the final resting stage. Still photos from above the water surface, from the front, and from the left side windows were taken at every two seconds. The water level fluctuation, current velocity and sediment concentration were measured at several points as shown in Fig. 1.

2.2 Experimental results

Figures 2 through 4 are still photos which give front views of the descending disposal material. In this case, the water depth is 50.0cm and the photos shown were taken at 2, 10 and 20 seconds after the disposal operation. An example of an overhead view is shown in Fig. 5. (The grid size in the photos is 10cm in length by 10cm in height)

Fig. 2. Still photo viewed from the front (at Time=2s)

These figures verified that the discharged material from the hopper barge vessel into open waters experiences the following processes.

(1) **Descending stage** – The disposal material falls through the water column due to gravity. During this period, an ambient fluid current is induced and the fine sediment particles on the outer edge of the material body are entrained into the ambient water by resistant shear forces.

(2) **Dynamic collapse** – This happens when the falling material cloud impacts the bottom. In this case, the descending momentum of the material will be transmitted into horizontal forces. In addition, a large number of sediment particles may be forced into suspension by the intensive impact forces.

(3) **Horizontal spreading** – After the impacts of the material cloud on the bottom, the sediment particles possess a large amount of energy which may drive them into horizontal spreading. Turbulent effects become a predominate

Fig. 3. Still photo viewed from the front (at Time=10s)

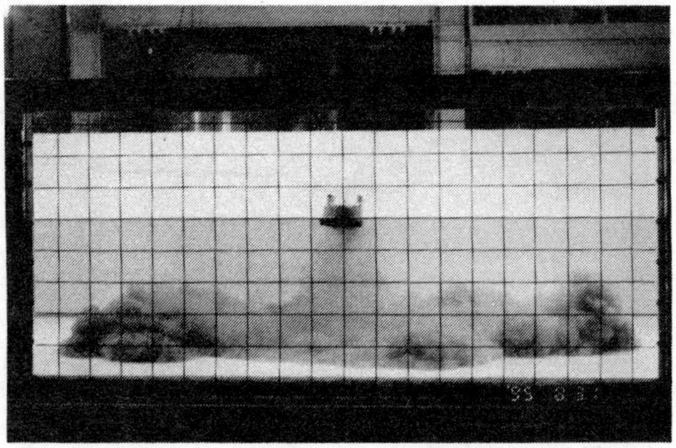

Fig. 4. Still photo viewed from the front (at Time=20s)

Fig. 5. Still photo viewed from the overhead (at Time=20s)

factor in this stage. When the surge energy is dissipated or exhausted, the disposal material deposits or settles to the bottom.

(4) **Transport dispersion** – Under the influence of ambient fluid currents and the propagating turbulent surges, the suspended sediment particles which are detached from the main body of the material are transported passively in the water column.

From these experimental results, it can be concluded that after an individual disposal operation, the placement of dredged material in open waters generally follows a two phase process, (a) the descending process of high density material core under the influence of gravity, which includes rapid movement through the water column, the bottom impact, and horizontal spreading and deposition on the bottom, (b) the passive transport-dispersion of the suspended particle cloud by the ambient currents and turbulent. The movement of the descending material core results in the primary driving forces for the ambient fluid motion.

3. NUMERICAL MODEL

To simulate the settling processes of ocean disposal material from barges, a vertical 2-D numerical model has been developed, considering the different characteristics of the two phases discussed above. The movement of the descending material core and the ambient fluid motion are computed from the Navier-Stokes equations, and the transport-dispersion of the suspended particle cloud is calculated from a mass transport equation in the proposed numerical model.

3.1 Governing equations

In reality, the placement of disposal material in open waters is a typical solid-liquid phase problem. By neglecting the relative motion between particles of the material and surrounding waters, the problem is simply treated as a single phase compressible fluid flow. This assumption is valid if the water content of the dredged material is not too low. As a result, the ambient fluid motion induced by the ocean disposal material from barges can be computed from the Navier-Stokes equations for unsteady flow and the mass conservation equation.

The equations for two-dimensional compressible flow in a vertical physical plane are written as follows:

$$\frac{\partial \rho}{\partial t} + \frac{\partial \rho u}{\partial x} + \frac{\partial \rho v}{\partial z} = 0 \tag{1}$$

$$\frac{\partial \rho u}{\partial t} + u\frac{\partial \rho u}{\partial x} + v\frac{\partial \rho u}{\partial z} = -\frac{\partial p}{\partial x} + \frac{\partial}{\partial x}\left(2\mu\frac{\partial u}{\partial x}\right) + \frac{\partial}{\partial z}\left[\mu\left(\frac{\partial u}{\partial z} + \frac{\partial v}{\partial x}\right)\right] \tag{2}$$

$$\frac{\partial \rho v}{\partial t} + u\frac{\partial \rho v}{\partial x} + v\frac{\partial \rho v}{\partial z} = -\frac{\partial p}{\partial z} - \rho g + \frac{\partial}{\partial x}\left[\mu\left(\frac{\partial u}{\partial z} + \frac{\partial v}{\partial x}\right)\right] + \frac{\partial}{\partial z}\left(2\mu\frac{\partial v}{\partial z}\right) \tag{3}$$

in which, u and v are the velocity components in the horizontal x and vertical z directions, p is the pressure, μ is the viscosity.

The transport-dispersion of the suspended particle cloud is governed by the mass transport equation,

$$\frac{\partial \rho C}{\partial t} + u\frac{\partial \rho C}{\partial x} + v\frac{\partial \rho C}{\partial z} = \frac{\partial}{\partial x}\left(D_x\frac{\partial C}{\partial x}\right) + \frac{\partial}{\partial z}\left(D_z\frac{\partial v}{\partial z}\right) + \omega_0\frac{\partial \rho C}{\partial z} \quad (4)$$

where, C is the suspended sediment concentration in the water column, ω_0 is the settling velocity of suspended sediment particle, and D_x and D_z are the diffusion coefficients in the x and z directions.

In the above equations, ρ is the density of the combined water-material fluid which is calculated by,

$$\rho = \rho_w + \Delta\rho_m + \Delta\rho_c \quad (5)$$

in which, ρ_w is the density of water, $\Delta\rho_m$ denotes the increased density by the descending material core and $\Delta\rho_c$ stands for that due to the suspended particles.

The density of the combined fluid varies in space and time corresponding with the movement of the disposal material. Its gradient is the primary driving force for the ambient fluid motion.

3.2 Boundary conditions

The boundary conditions for u, v, p and C in the physical domain are as follows.

1). At the water surface, the suface position must be determined at each time step with its own boundary conditions as follows,

$$\frac{\partial u}{\partial z} = 0, \quad v = \frac{\partial h}{\partial t} + u\frac{\partial h}{\partial x}, \quad p = 0, \quad \frac{\partial C}{\partial z} = 0 \quad (6)$$

2). At the bottom, the non-slip boundary conditions are used, i.e.,

$$u = v = 0, \quad \partial C/\partial z = 0, \quad \partial p/\partial z = -\rho g \quad (7)$$

3). At the ends of the calculation domain, the boundary conditions for a fixed wall are applied,

$$u = v = 0, \quad \partial C/\partial x = 0, \quad \partial p/\partial x = 0, \quad \partial h/\partial x = 0 \quad (8)$$

4. COMPUTATION METHOD

4.1 Coordinate transformation

For the present problem, the upper surface is a moving boundary which changes in space and time. As a result, the generated grid for numerical simulation is also a moving system in space and time. It is difficult to solve the governing equations in such a moving and curvilinear grid. Therefore, the Boundary Fit Method is introduced in the numerical simulation, i.e. to map the moving grid to a fixed and linear one through a "sigma-stretched" coordinate transformation as in Fig. 6.

The relationships between the physical domain (x, z, t) and the computational domain (ξ, η, τ) are expressed as follows:

$$\xi = x, \quad \eta = \frac{z}{h(x,t)}, \quad \tau = t \quad (9)$$

where, $h(x,t)$ is the total water depth.

With Eq. (9), the transformed form of governing equations, Eqs.(1) through (4), in the computational domain are then derived as,

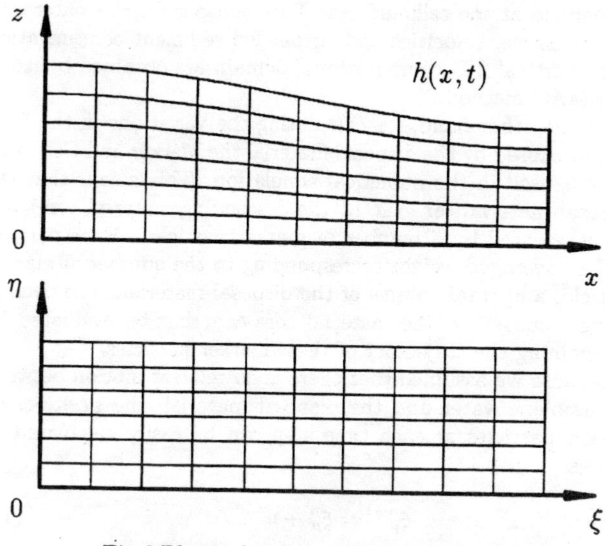

Fig.6 Physical and computational domains

$$\frac{\partial \rho h}{\partial t} + \frac{\partial \rho h u}{\partial \xi} + \frac{\partial}{\partial \eta}\left(\rho v - \eta\frac{\partial h}{\partial \xi}\rho u - \eta\frac{\partial h}{\partial \tau}\rho\right) = 0 \quad (10)$$

$$\begin{aligned}\frac{\partial \rho u}{\partial \tau} + u\frac{\partial \rho u}{\partial \xi} + \left(v - \eta\frac{\partial h}{\partial \xi}u - \eta\frac{\partial h}{\partial \tau}\right)\frac{\partial \rho u}{\partial \eta} = -\frac{\partial p}{\partial \xi} \\ + \frac{\eta}{h}\frac{\partial h}{\partial \xi}\frac{\partial p}{\partial \eta} + \left(\frac{\partial}{\partial \xi} + \frac{\eta}{h}\frac{\partial h}{\partial \xi}\frac{\partial}{\partial \eta}\right)\left(2\mu\frac{\partial u}{\partial \xi}\right) \\ - \left(\frac{2\eta}{h}\frac{\partial h}{\partial \xi}\frac{\partial}{\partial \xi} + \frac{1}{h^2}\frac{\partial}{\partial \eta}\right)\left(2\mu\frac{\partial u}{\partial \eta}\right) + \frac{1}{h}\frac{\partial}{\partial \eta}\left(\mu\frac{\partial v}{\partial \xi}\right) \\ - \frac{\eta}{h^2}\frac{\partial h}{\partial \xi}\frac{\partial}{\partial \eta}\left(\mu\frac{\partial v}{\partial \eta}\right) + 2\mu\frac{\eta}{h}\frac{\partial^2 h}{\partial \xi^2}\frac{\partial u}{\partial \xi} - \mu\frac{1}{h^2}\frac{\partial h}{\partial \xi}\frac{\partial v}{\partial \eta}\end{aligned} \quad (11)$$

$$\begin{aligned}\frac{\partial \rho v}{\partial \tau} + u\frac{\partial \rho v}{\partial \xi} + \left(v - \eta\frac{\partial h}{\partial \xi}u - \eta\frac{\partial h}{\partial \tau}\right)\frac{\partial \rho v}{\partial \eta} = -\frac{1}{h}\frac{\partial p}{\partial \eta} - \rho g \\ + \frac{\mu}{h}\left(\frac{2}{h}\frac{\partial}{\partial \eta} - \eta\frac{\partial h}{\partial \xi}\frac{\partial}{\partial \xi}\right)\frac{\partial v}{\partial \eta} + \left(1. - \frac{\eta}{h}\frac{\partial h}{\partial \xi}\right)\frac{\partial}{\partial \xi}\left(\mu\frac{\partial v}{\partial \eta}\right) \\ - \left(\frac{\eta}{h^2}\frac{\partial h}{\partial \xi}\frac{\partial}{\partial \eta} + \frac{1}{h}\frac{\partial}{\partial \xi}\right)\left(2\mu\frac{\partial u}{\partial \eta}\right) - \frac{\mu}{h^2}\frac{\partial h}{\partial \xi}\frac{\partial v}{\partial \eta}\end{aligned} \quad (12)$$

$$\begin{aligned}\frac{\partial \rho C}{\partial \tau} + u\frac{\partial \rho C}{\partial \xi} + \frac{1}{h}\left(v - \eta\frac{\partial h}{\partial \xi}u - \eta\frac{\partial h}{\partial \tau} - \omega_0\right)\frac{\partial \rho C}{\partial \eta} = \\ \frac{D_x\eta}{h}\left(\frac{h}{\eta}\frac{\partial^2 C}{\partial \xi^2} - 2\frac{\partial h}{\partial \xi}\frac{\partial^2 C}{\partial \xi \partial \eta} - \frac{\partial^2 h}{\partial \xi^2}\frac{\partial C}{\partial \eta}\right) + \frac{D_z}{h}\frac{\partial^2 C}{\partial \eta^2}\end{aligned} \quad (13)$$

Similarly, the boundary conditions must be transformed in the computation domain.

4.2 Numerical Solution Method

The transformed equations and corresponding boundary conditions are solved numerically by the finite deference method. The numerical calculation is carried out on a staggered grid, in which p, C and ρ are defined at the center of each cell and u, v are defined at the cell surfaces. The solutions for the water surface, pressure, velocities and suspended sediment concentration in the vertical 2-D computational domain are obtained by using the SMAC method.

On the other hand, for calculating the variations in the fluid density caused by the disposal material the Marker and Cell technique is used in the numerical simulation. This means that the disposal material (or exactly, the descending material core) will be represented by a number of marker particles. Each particle is of an averaged weight corresponding to the number of marker particles and total volume of the disposal material. The density changes caused by the material core can thus be evaluated by determining the trajectory of these marker particles.

Because we assumed that there is no relative motion between the ambient water and the disposal material, the positions of marker particles at each time step can be easily calculated as follows.

$$\xi_p^{n+1} = \xi_p^n + u^{n+1}\Delta t \tag{14}$$

$$\eta_p^{n+1} = \eta_p^n + \frac{\Delta t}{h}\left(v^{n+1} - \eta\frac{\partial h}{\partial x}u^{n+1}\right) \tag{15}$$

Therefore, assuming the ambient density is uniform, the change in the fluid density of each computation cell is given by,

$$\Delta\rho_{i,j} = \Delta\rho^n_{m(i,j)} + \Delta\rho^n_{c(i,j)} = \frac{(1-\alpha)W_p N^n_{i,j}}{\Delta\xi\Delta\eta} + C_{i,j}{}^n \tag{16}$$

In Eqs.(14) through (16), ξ_p and η_p denote the coordinates of marker particles, α stands for the volume ratio of the suspended particle to the total discharged material, W_p ($= \rho_m V_m / \sum N_{i,j}$, V_m and ρ_m: the total volumn and the density of disposal material, respectively) is the weight per particle, $N_{i,j}$ is the number of particles in the (i,j) cell, Δt is the time increment $\Delta\xi$ and $\Delta\eta$ are the grid lengths in the ξ and η directions, $C_{i,j}$ is the suspended sediment concentration in the (i,j) cell and the superscript indicates the time level.

With the distribution of fluid density determined at each time step, the ambient fluid motion and the movement of disposal material can then be determined from Eqs. (10) – (13). Finally, the calculated u, v, p and C in the computation domain can be transformed back to their real positions in the physical domain by Eq.(9).

5. CALCULATION RESULTS

To verify the above-described numerical model, computed results are compared with those from the experiments under the conditions shown in Table 2.

Table 2: Computation conditions

Water Depth	Flume Length	Material Volume	Water content	Volume Ratio	Particle Number
h_0(cm)	L(cm)	V_m(cm^3)	W(%)	α(%)	N
50.0	400.0	2000.0	145.0	3.0	600

In the computation, the grid lengths in the horizontal x and vertical z directions were chosen as 4.0 and 1.0 centimeters, respectively. Marker particles are initially arranged at 2 by 2 central surface grids, representing the descending material core which is of $(1-\alpha)$ percent of the total released material. The residual α percent of the released material is distributed around the marker particles as initial concentration of suspended material cloud. It should be noted that simplification was made here because the stripping of material into the water column is a dynamic process and thus the formation of suspended material changes temporally during the settling process. Whereas, α in this model is only a ratio of the amount of suspended particles produced in the whole settling process to that of the released material.

Figures 7 – 9 are the calculated distributions of local velocity and suspended sediment concentration corresponding to the experimental results shown in Figs. 1 – 3. Circles in Figs. 7(b), 8(b) and 9(b) indicate marker particles of the descending material core. Figures 10 and 11 give comparisons of the calculated and measured horizontal velocity and suspended sediment concentration at the fixed points P1 (x=220cm, z=45cm) and P2 (x=260cm, z=45cm) indicated in Fig. 7, respectively. It is seen that general features of the settling processes of material disposed in open waters have been reproduced well by the proposed numerical model, except for a little discrcepancies between calculated and measured temporal changes in the horizontal velocity and suspended sediment concentration as seen in Fig. 10 and 11.

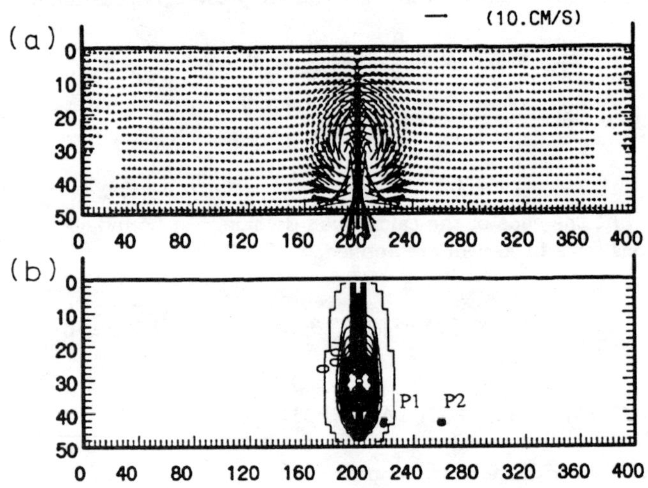

Fig. 7. Calculated distributions of velocity and concentration(Time=2s)

6. CONCLUSIONS

Through laboratory experiments, it was indicated that the settling processes of the ocean disposal material can be classified into two phases, the descention of a high-density material core and the transport of suspended fine particle cloud. On this basis, a vertical 2-D numerical model was developed by combining the Navier-Stokes equations with the Marker and Cell method along with the convective-diffusion equation. The former is used to simulate the movement of the descending material core as well as the ambient fluid motion, the later is aimed at representing

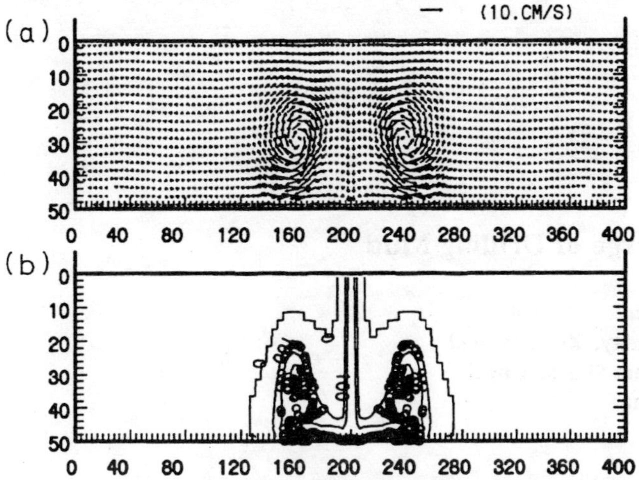

Fig. 8. Calculated distributions of velocity and concentration(Time=10s)

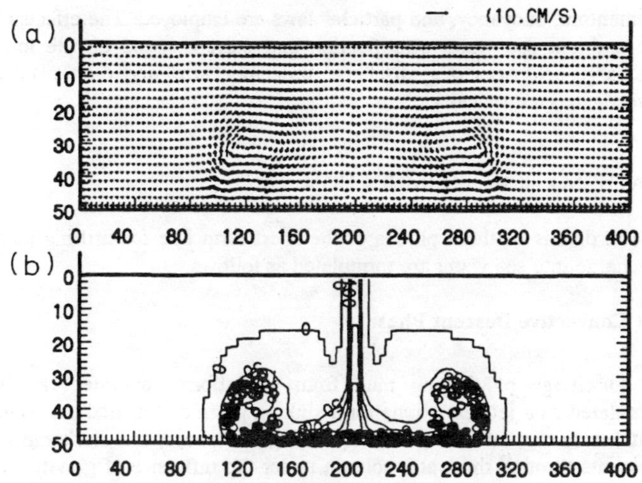

Fig. 9. Calculate distributions of velocity and concentration(Time=20s)

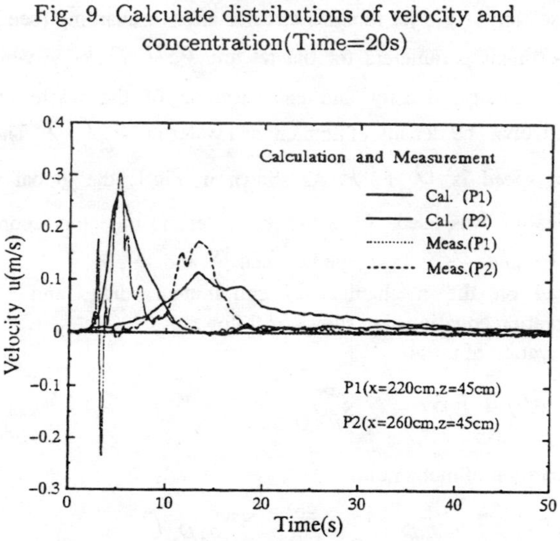

Fig. 10. Comparisons of velocities(at P1,P2)

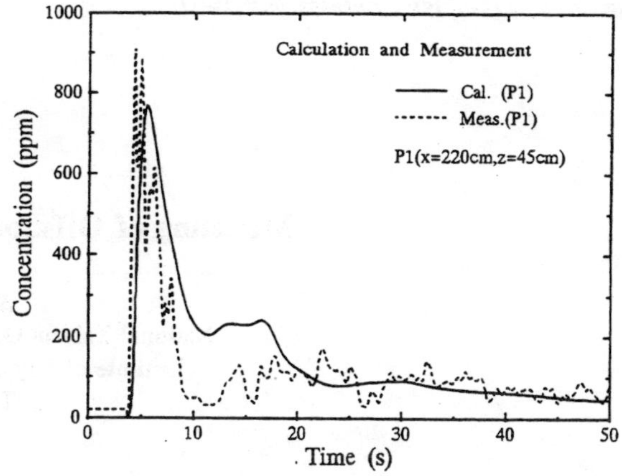

Fig. 11. Conparison of concentration(at P1)

the transportation of suspended fine particles which are detached from the main body of the disposed material. The proposed numerical model is able to give reasonable results for the physical processes.

ACKNOWLEDGMENT

The authors would like to thank K. Murakami and H. Tsuruya, the Port and Harbor Research Institute of Japan, for their valuable discussions.

REFERENCES

Brandsma, M. G. and D. J. Divoky (1976): Development of models for prediction of short-term fate of dredged material discharged in the estuarine environment, Report No. D–76–5, tetra Tech., Pasadena, California.

Horie, T. (1983): MAC model for prediction of short-term fate of dumped material from barge, Technical Note of The Port and Harbor Research Institute, No.471, pp.1–34.

Johnson, B. H. et al. (1993): Physical model tests of dredged material disposal from a split-hull barge and multiple bin vessel, Technical Report DRP-93-1, US Army Engineering Waterways Experiment Station, Washington, DC.

Oda, K., T. Shigematsu and H. Kanno (1993): Numerical study of settling and dispersion behaviors of particles in flows with DEMAC Method, Proc. of Coastal Eng., JSCE, Vol. 40, pp.951–955 (in japanese)

Oda, K and T. Shigematsu (1994): Development of a numerical simulation method for predicting the settling behavior and deposition configuration of soil dumped into waters, Proc. 24rd International Conference in Coastal Engineering, Vol.3, pp.3305–3319.

Yauchi, E. and K. Katoh (1992): Dispersion processes and the settlement pattern of mud dumped in oceans, Proc. of 23rd International Conference In Coastal Engineering, Vol.3, pp.2735–2744.

Proceedings of the Seventh (1997) International Offshore and Polar Engineering Conference
Honolulu, USA, May 25-30, 1997

ISBN 1-880653-28-1 (Set); ISBN 1-880653-29-X (Vol. I)

Modeling of Offshore Discharge of Drilling Mud

Bao-Shi Shiau
National Taiwan Ocean University, Keelung, and
Institute of Physics, Academia Sinica, Taipei
Taiwan, China

ABSTRACT

This paper presents the results of a development of mathematical model that simulates the fate of drilling mud discharged into the sea water. The mathematical model simulates the short term fate of the offshore petroleum exploration drilling mud discharging into the sea water by three phases: jet convective descent phase, dynamic collapse phase, and encounter sea bottom phase. The model calculated centerline trajectory of the drilling mud discharge is close to the laboratory measurements. The effects the ambient water density stratification and the cross currents on the discharge trajectory, size, and dilution are investigated.

KEY WORDS: Drilling mud discharge, jet convective descent, dynamic collapse, stratification, cross current

INTRODUCTION

Since the marine environment protection consciousness arising, people pay more attention on the pollution caused by the offshore discharge of drilling mud. Realistic response to these attentions requires description of the processes influencing the fate of discharge mud. The fate of drilling mud on the ocean bottom can be categories as short-term fate and long-term fate. The long-term fate processes begin after initial decomposition occurs and the time period for the short term rages from minutes to hours. The long term time period ranges from weeks to years.

Koh and Chang (1973) had proposed the mathematical model for barged ocean disposal of waste. Brandsma and Divoky (1976) had developed model for predicting the short term fate of dredged material discharge in the estuarine environment.

The behaviors of the drilling mud discharge from the offshore petroleum exploration platform are similar to those of dredged material dumping. The main purpose of this study is to develop a numerical model to simulate the short-term behavior of drilling mud discharges from offshore platform. The short term fate of the drilling mud discharging into the sea water are described by three phases : (1) jet convective descent phase, (2) dynamic collapse phase, and (3) encounter sea bottom phase. In each phase, the conservation of mass, momentum, buoyancy, and particles laws are employed. The effects of the ambient sea water stratification and current speed on the mud discharge behaviors (e.g. discharge trajectory, size, and dilution) are investigated numerically.

MATHEMATICAL FORMULATION

The details of three phases of the short term fate for drilling mud discharge into sea water are formulated as follows.

Jet Convective Descent Phase

Discharge of drilling mud from a platform to water can be simulated as a jet. Equations for a sinking jet in a stratified sea water with cross current is formulated. During this phase, the discharge jet descents through the water column under the influence of gravity and rapidly entrains seawater from surrounding area. It is assumed that the cross section of the jet remains circular after discharging (see Fig.1). The pertinent parameters for the jet are b, U, ρ, C_{si} which are radius, velocity, density and concentration of the waste material respectively. The density of ambient sea water is $\rho_a(y)$. The cross current speed is $U_a(y)$. As shown in Fig.1, the global system coordination was used. s is the jet centerline trajectory coordinate. θ_1 、θ_2 and θ_3 is the angle between s and x, y, z.

Based on the mechanism of entrainment, drag, and settling, conservation equations are shown as follows.

Conservation of mass

$$\frac{d}{ds}(\rho\forall) = E\rho_a - \sum_i s_i \rho_{si} \tag{1}$$

Conservation of momentum

$$\frac{d}{ds}\vec{M} = F\vec{j} + E\rho_a\vec{U}_a - \vec{F}_D - \sum_i s_i\rho_{si}\vec{U} \tag{2}$$

Conservation of buoyancy

discharge pipe diameter: D
cross current: Ua

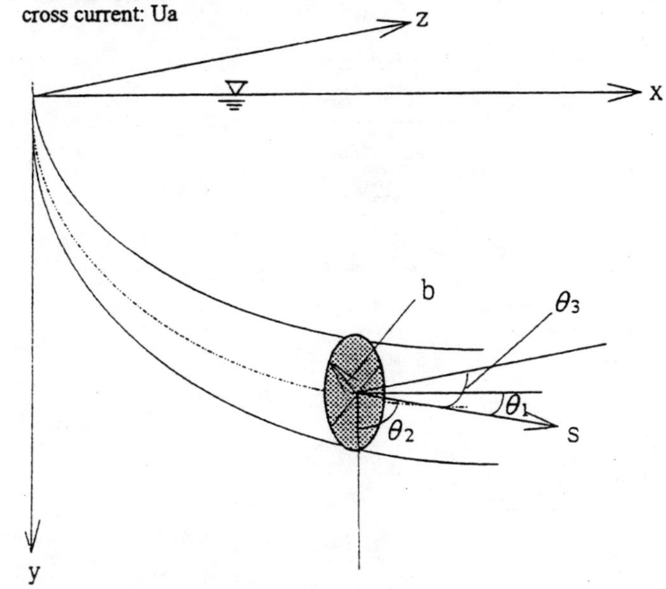

Fig.1 Schematic diagram of the jet

$$\frac{d}{ds}B = E\Delta f - \sum_i s_i \Delta_{si} \quad (3)$$

Conservation of particles

$$\frac{d}{ds}p_i = -s_i \quad (4)$$

Where $\forall = \pi b^2 U$: the volume flux; E : entrainment; s_i : particles descent; $\vec{M}$: momentum flux; F : buoyancy force per unit length; j : vector in normal direction; $\vec{F}_D$: drag force; B: buoyancy flux; Δf : density difference between ambient free surface and position s; Δs_i : density difference between ambient free surface and solid particle; p_i : solid particles volume flux.

The trajectory of the jet centerline can be obtained through the following equations.

$$\frac{dx}{ds} = \cos\theta_1 \quad (5.a)$$

$$\frac{dy}{ds} = \cos\theta_2 \quad (5.b)$$

$$\frac{dz}{ds} = \cos\theta_3$$
(5.c)

Dynamic Collapse Phase

As the jet is far downstream from the nozzle, it is no longer behaves like a jet. The jet tends to collapse vertically and spread out horizontally seeking a hydrostatic equilibrium in the ambient density gradient. The cross section is assumed to have a shape of ellipse.

$$\frac{y'^2}{a^2} + \frac{x'^2}{b^2} = 1 \quad (6)$$

where y', x' are coordinates with the origin fixed on the centroid of the cross section. a, b are minor and major axes of the ellipse and are function of time. The ambient density distribution is

$$\rho_a(y) = (\rho + \Delta\rho)\{1 - \varepsilon(y)y'\} \quad (7)$$

where $\Delta\rho$ is the cloud density difference at y'. $\varepsilon(y)$ is the ambient sea water density gradient.

$$\varepsilon = -\frac{1}{\rho_{a0}}\left(\frac{d\rho_a}{dy}\right) \quad (8)$$

where ρ_{a0} is sea water density of free surface, y is sea water depth and is upward with positive. As the ambient fluid was entrained, it makes the cloud density change.

$$\rho^*(y', x', t) = \rho\{1 - \gamma_1 \tfrac{a_0}{a}\varepsilon(y)y'\} \quad (9)$$

where γ_1 is the distribution constant. And $0<\gamma_1<1$.

The conservation equations are derived as follows.

Conservation of mass

$$\frac{d}{dt}(\rho\forall) = E\rho_a - \sum_i s_i \rho_{si} \quad (10)$$

Conservation of momentum

$$\frac{d}{dt}(\vec{M}) = F\vec{j} + E\rho_a\vec{U}_a - \vec{D} - \sum_i s_i \rho_{si}\vec{U} \quad (11)$$

Conservation of buoyancy

$$\frac{d}{dt}(B) = E\Delta f - \sum_i s_i \Delta_{si}$$
(12)

Conservation of particles

$$\frac{d}{dt}p_i = -s_i \quad (13)$$

where $\vec{D}$: drag force.

The trajectory of the waste cloud can be found by the following equation

$$\frac{dx}{dt} = u \quad (14.a)$$

$$\frac{dy}{dt} = v \quad (14.b)$$

$$\frac{dz}{dt} = w \quad (14.c)$$

Sea Bottom Encounter Phase

The waste material jet can reach the sea bottom and spread out there, if the ambient density stratification is not strong enough to keep the vertical descent of the jet somewhere above the bottom.

In sea bottom encounter phase, the cross section of the jet is assumed to have a half elliptical shape. The sea bed is assumed to be horizontal. The jet plume is allowed to move as an entity with respect to the ambient fluid and the sea bed, while it collapses vertically and spread out horizontally. The conservation equations are similar as those in dynamic collapse phase except conservation of momentum which is

$$\frac{d}{dt}(\vec{M}) = \vec{Fj} + E\rho_a \vec{U}_a - \vec{D} - \sum_i s_i \rho_{si} \vec{U} - \vec{F}_F$$

(15)

where $\vec{F}_F$: sea bed friction force.

NUMERICAL CALCULATION

Basic Assumptions

The basic assumptions for the numerical model calculation are as follows.

(1). The sea bed is horizontal, i.e. the water depth is constant.

(2). The open water area is considered.

(3). The particles contained in the drilling mud discharge is cohesive.

(4). The particles settling velocity as suggested by Ariathurai et al. (1977) is

$$w_s = \begin{cases} 0.00052m/sec, & c < 0.025 \\ 7.13\times 10^{-6} \times c^{4/3} m/sec, if & 0.025 \le c \le 0.3 \\ 0.014m/sec, & c > 0.3 \end{cases} \quad (16)$$

where c is the solid particle cohesive coefficient.

(5). The liquid limit and void ratio of the waste cloud are not considered.

Numerical Scheme And Procedure

The governing equations described in the mathematical formulation section are ordinary differential equations with the following types

$$\frac{dy_i}{dx_i} = F_i \quad (17)$$

where y_i are dependent variables; x_i are independent variables; and Fi are given values. As the proper initial conditions are given, the fourth order Runge-Kutta method can be applied to solve these ordinary differential equations step by step. In general, the numerical scheme is stable.

The numerical calculation begins from the jet descent phase with the input of the initial conditions. The flow chart for the numerical calculation is shown in Fig.2.

RESULTS AND DISCUSSION

The governing equation systems for the short term fate of the drilling mud discharge into the sea water with cross current are derived and formulated in accordance with their physical phenomenon. Then, the numerical model is established with the fourth order Runge-Kutta scheme. The validity of the present model is tested with other researcher's laboratory data. And the effects of the ambient seawater density stratification and the currents on the discharge behaviors are investigated from the numerical calculation results and discussed as follows.

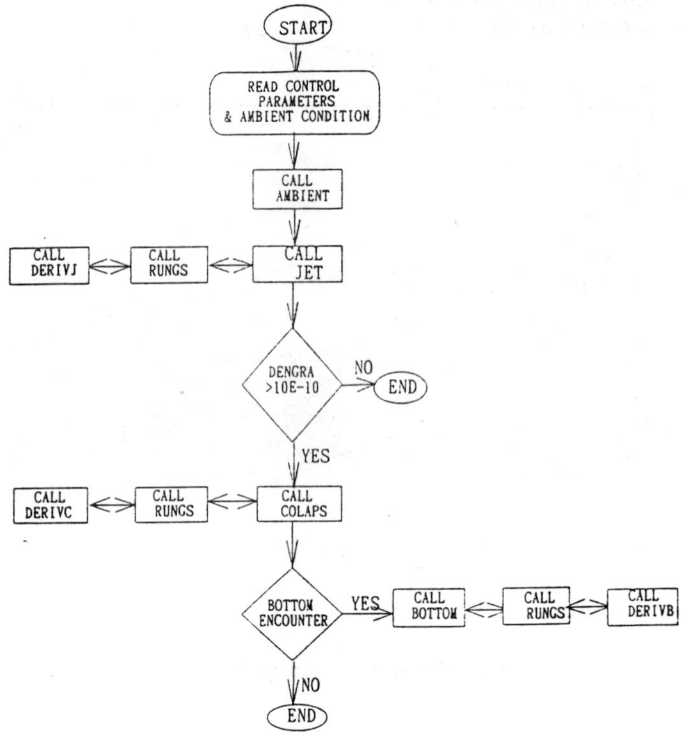

Fig.2 Flow chart of the numerical model

Verification Of The Model

Fig. 3 shows the comparison of the calculated discharge centerline trajectory with the laboratory data of Davis et al. (1983). The results of this study yield good agreement with the experimental result in the jet descent phase. In the dynamic collapse phase, there is a little deviation between the numerical and laboratory data. But in summary, they are close.

Ambient Water Density Stratification Effect

Fig. 4 shows the discharge trajectory variations for different ambient sea water density gradients. For zero density gradient, the discharge goes down and encounters the sea bottom. When the density gradient of ambient water increases, the buoyancy force increases. So the discharge does not hit the bottom and its trajectory is closer to the water surface as the density gradient increases.

The discharge dilution as the function of time for different sea water density gradients are plotted in Fig. 5. It is shown that the dilution becomes large as time increases. Also the discharge dilution becomes large when the density gradient decreases.

The variations of the discharge size for various ambient sea water density gradients are shown in Fig. 6. For non-stratification case, the discharge grows as the time increases. For stratification ambient water,

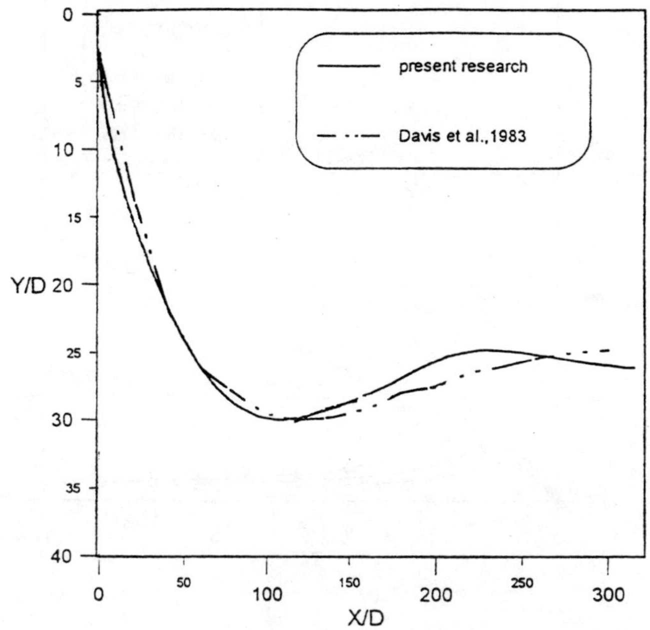

Fig.3 Comparison of the discharge trajectory with the experiment

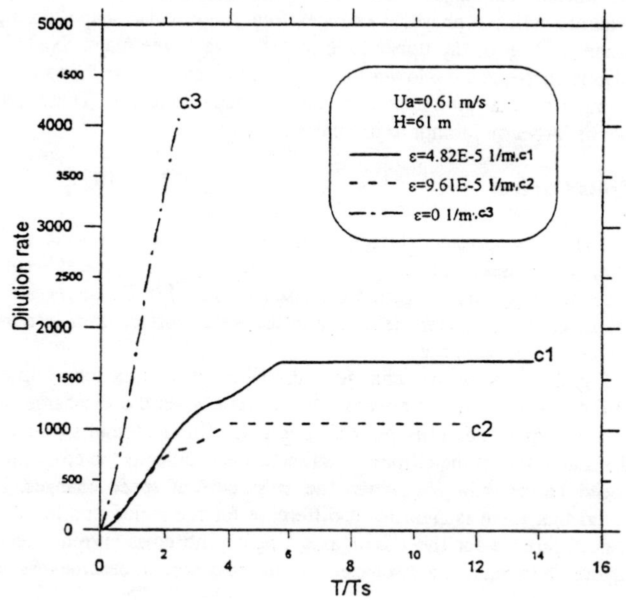

Fig.5 Discharge dilution as the function of time for different ambient water density gradients

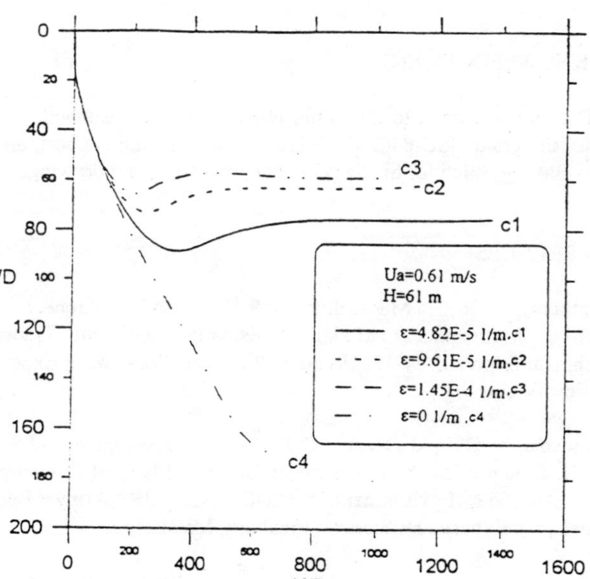

Fig.4 Discharge trajectory variations for different ambient water density gradients

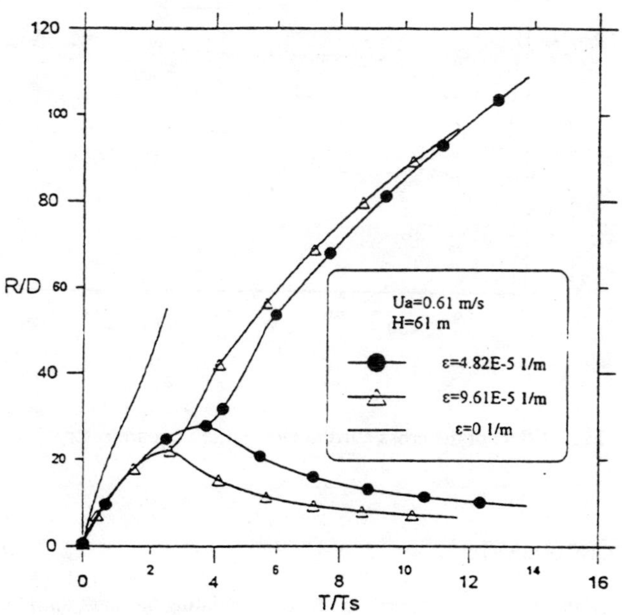

Fig.6 Discharge size variations as the function of time for different ambient water density gradients

discharge grows as the time increases. For stratification ambient water, the discharge at the jet descent phase has a circular cross section. At the dynamic collapse phase, it changes from the circular shape to elliptic shape. In Fig. 6, the upper lobe of the curve is the major axis of the elliptic shape, and the lower lobe is the minor axis. Fig.6 shows that the stronger stratification makes the elliptic shape more oval. That is major axis is larger and minor axis smaller.

Cross Current

The cross current evidently deflects the discharge trajectory. This effect can be seen in Fig. 7. The larger the cross current speed becomes, the more trajectory deflection the discharge has. Fig. 7 also reveals that the discharge trajectory is closer to the water surface with increasing the cross current speed.

Fig. 8 shows the size change of discharge as the function of time for different cross current speeds. As the time goes the discharge cross section expands and becomes elliptic shape finally. Results reveal that the major axis of the elliptic cross section is smaller as the cross current speed becomes larger. When the cross current speed changes, it is noted that there is significant difference for the major axis length but less difference for the minor axis length. The cross current speed is favorable to make the discharge elliptic cross section become less oval.

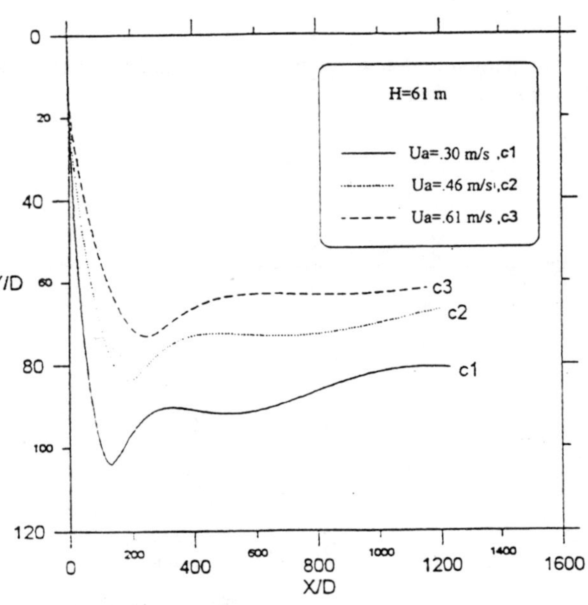

Fig.7 Effect of the cross current on the discharge trajectory

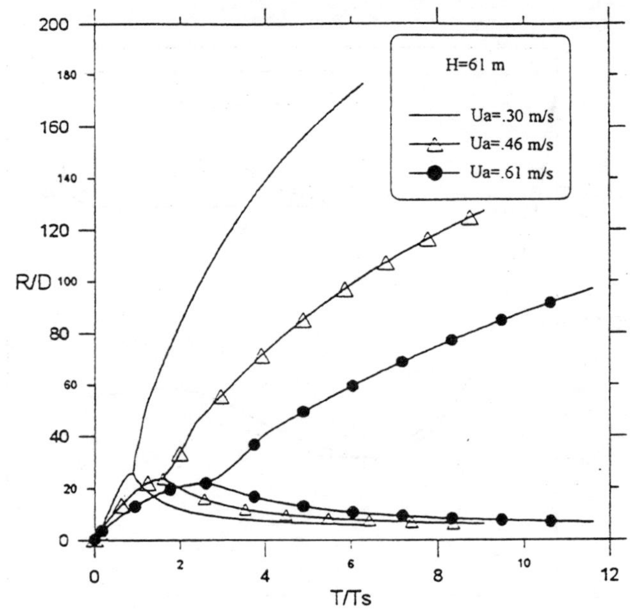

Fig.8 Discharge size variations as the function of time for different cross current speeds

CONCLUSION

In this study, a numerical model for simulating the short term fate of the offshore drilling mud discharge is developed and used to investigate numerically for the ambient environment conditions affecting the discharge behaviors. The results of numerical model prediction calculations is close to the measured results of Davis et al. (1983). Numerical investigation results show that a stronger sea water stratification will inhibit the discharge to penetrate downward the water and make the shape of the discharge cross section become more oval. The cross current deflects the discharge trajectory. The trajectory is closer to the water surface as the cross current speed becomes larger. The major axis of the elliptic discharge cross section will be longer when the cross current speed changes to smaller.

ACKNOWLEDGEMENT

The author wishes to thank the National Science Council, Taiwan, under the grant NSC-8402611-E019-007 for funding support, and Mr. Wen-Shyang Shieh for his help in data processing and drawings.

REFERENCES

Ariathurai, R., MacArthur, R.C., and Krone, R.B., (1977), "Mathematical Model of Estuarial Sediment Transport," Technical Report D-77-12, US Army Engineer. Waterways Experiment Station, Vicksburg

Brandsma, M.G., and Divoky, D.J. (1976). "Development of Models for Prediction of Short-term Fate of Dredged Material Discharged in the Estuarine Environment." CR-D-76-5, US Army Engineer Waterways Experiment Station, Vickburg, Mississippi

Davis, L.R., Cavola, R.G., and Mohebbi, B., (1983), "An Experimental Investigation of Drilling Fluid Disposal in a Flowing Stratified Environment," Department of Mechanical Engineering, Oregon State University

Koh, R.C.Y., and Chang, Y.C. (1973). "Mathematical Model for Barged Ocean Disposal of Waste." Environmental Protection Technology Series, EPA 660/2-73-029, US Environmental Protection Agency, Washington D.C.

Proceedings of the Seventh (1997) International Offshore and Polar Engineering Conference
Honolulu, USA, May 25-30, 1997

ISBN 1-880653-28-1 (Set); ISBN 1-880653-29-X (Vol. I)

Establishing a Simulation Strategy for the Dispersion of Drill-Cuttings

L. Carles and I. Bryden
The Robert Gordon University
Aberdeen, United Kingdom

ABSTRACT

In order to fully ascertain the impact of drilling developments, it is necessary to determine accurately the dispersion of the drilling discharge into the sea. The development of a dispersion model would enable the prior determination of the different concentrations of drill-cuttings (of any size) deposited on the seabed or the thickness of the cuttings pile (if present) on the seafloor. Some models already exist and could be improved to achieve a more accurate model, which would predict the behaviour of even fine particulates into the water column and their deposition on the seafloor.

KEY WORDS: Simulation, dispersion, drilling wastes.

INTRODUCTION

In 1995, the operational wastes permitted to be released into the marine environment were: produced water, water-based drilling muds (WBM), cuttings (from both WBM and oil-based drilling muds (OBM) applications), ballast, bilge, and storage water, sanitary and food wastes, deck drainage, produced sand, well treatment fluids, cooling water, desalination water, and other wastes and residues.

The predominant environmental concerns over offshore oil and gas operations have centred on the impacts of hydrocarbons on the benthic environment. The effects of discharges of particulate drill wastes - drill muds and cuttings- typically have been viewed within the framework of their effects on benthic communities, either through physical impacts (burial, smothering) or chemical toxicities. A considerable amount of research has been conducted on the environmental fate and effects of operational drilling wastes, particularly in the United States and in Europe. Of particular concern have been the potential impacts of those wastes released in the largest quantities, primarily produced water, water-based drilling mud, and cuttings.

Drilling muds are circulated downhole to lubricate the bit, to return cuttings to the surface and to control well pressure. The drill-cuttings, once lifted to the surface, are separated from the drilling mud via the solids control system. The drilling mud is recirculated and the cuttings are discharged to sea (Environmental Consultancy Services Ltd, 1993). The release of the drilling discharge occurs, after relevant treatment, from a vertical caisson submerged at different depths depending on the platform. Usually, there is no pressure added and the drilling discharge falls into the sea under gravity. Sometimes, drill mud and cuttings are discharged near the sea surface to ensure maximum dispersion in the strong surface tidal currents and minimise potential for cuttings pile formation.

Drill cuttings contain rock fragments contaminated with drilling muds. The contamination can include hydrocarbons as well as other products which are potentially hazardous, like bentonite, to the benthic community. When cuttings and associated discharges are released into the water column, they are subjected to a wide variety of mixing and transport processes. In order to determine the areas potentially at risk from drilling operations and prevent possible conflicts between different users of the sea, it is necessary to understand the complex physics of these processes to be able to predict the behaviour of the released particulate matter. The principal uncertainties relate to the potential for flocculation and aggregation of particles during the release and subsequent settling processes. The behaviour of relatively large drill-cuttings is well-defined but that of the fine particulates is still unprecisely described, mainly because of the difficulty in quantifying the processes of aggregation and turbulence.

This paper presents, firstly, a brief summary of the physical processes involved in the dispersion of the particulate matter in the water column and its deposition onto the seabed. Secondly, the water column is divided into several different zones in which specific physical processes significantly influence the mixing processes and for which boundary conditions are described. Then, a scheduled methodology for the completion of the project of study is planned, from a theoretical and experimental point of view.

PHYSICAL PROCESSES INVOLVED IN THE DISPERSION

Initially, the physical processes that have to be studied will depend on whether the discharge is from a point above or below the sea surface. For the purpose of this paper, the situation where drilling discharge is released from below the sea surface will be more thoroughly considered. In effect, in the North Sea, the discharge of drilling fluids usually occurs from a submerged point. Drilling discharge in this context means drilling muds and drill-cuttings together.

Several factors will affect the final dispersion of the particulate wastes. They are concerned with the particulates and the marine environment receiving the discharge, and can be summarised as follows:

For the particulates:
- initial velocity at the exit of the orifice;
- density (i.e. mass and volume);
- type of rock;
- shape and size.

For the marine environment receiving the discharge:
- water density;
- water depth;
- tidal and residual currents characteristics;
- turbulence (at different scales) characteristics;
- meteorological conditions;
- morphology of the seabed (i.e. its topography and nature of the sea floor).

Some other parameters like the orientation and the shape of the exit orifice, the presence of solid boundaries (e.g. coastal boundaries or platform legs) will also affect the behaviour of the discharged fluid.

After the discharge, the particulates will go through different processes like initial dilution, flocculation (which will change the density of the drilling discharge and therefore, the whole process of settlement), dispersion, suspension, settlement, turbulence and diffusion. Once deposited onto the seabed, the particles can also be resuspended and redeposited. In order to make the presentation of these processes as clear as possible, a brief definition of each process is given below:

- Initial Dilution: the primary process that allows the discharged fluid to mix with the ambient fluid. This process is totally driven by the difference in density of the two fluids or the shear stresses between a jet and the ambient environment.

- Flocculation: the formation of small masses formed in a fluid through attraction between fine suspended particles.
- Suspension: the process resulting in a mixture of fine, nonsettling particles of any solid within a liquid or a gas (McGraw-Hill, 1994).
- Settlement: the gradual downward movement of a particle.
- Deposition: the laying, placing, or throwing down of any material; specifically, the constructive process of accumulation (McGraw-Hill, 1994).

In this paper, the term "settlement" is used more precisely to describe the downward movement of a particle while the term "deposition" is used to describe the final placement of that particle on the seabed.

- Turbulence: the motion of fluids in which local velocities and pressures fluctuate irregularly, in a random manner (McGraw-Hill, 1994).

- Dispersion: the distribution of finely divided particles in a medium (McGraw-Hill, 1994).

- Diffusion: the migration of atoms or molecules due to their random thermal motions. The process is very slow in solids, faster in liquids (depending a lot on viscosity) and rapid in gases (Macmillan, 1986).

The terms "dispersion" and "diffusion" can sometimes be interchangeable. A distinction between these two terms can be defined by using "diffusion" at a molecular scale only. Another approach to distinguish these two terms is to consider that "dispersion" is the combination of diffusion and advection.

The importance of each of these processes on the dispersion of the drilling wastes has yet to be fully quantified. For example, experimental tests using a perspex tank will be conducted to determine the influence of flocculation on the cuttings settling speed. Later, experimental tests to define the influence of turbulence and flocculation together on the cuttings settling speed will also be performed. Flocculation starts occurring within the caisson where sea water is used to entrain the drilling wastes in their chute and to make sure that the caisson keeps clear. According to Postma (1967, quoted by Muschenheim and Milligan, 1996), the process of flocculation can increase the settling speed of fine particles by up to two orders of magnitude. The increased settling speed serves to retain fines from drilling wastes in the benthic boundary layer near the discharge point, and hinders dispersal.

According to Gordon et al (1995), model results are very sensitive to the values taken for settling speed (i.e. big difference between using 1cm/sec and 2cm/sec).

Hence it is crucial to accurately determine the cuttings fall velocities. Few mathematical models have already been developed, taking into account the shape, the size and the density of the drill-cuttings. The biggest uncertainties come from the definition of the cuttings shape. The cuttings, depending on the drilling processes (type of bits and muds used for drilling, rate of penetration, rock composition...etc.), can be of various shapes. Typical shapes, that can be considered as being acceptable approximations, are: spheres, disks and flatplates. The process of flocculation and turbulence also influence the cuttings shape. For example, an individual clay particle can be reasonably assumed to have a flatplate shape, while, after flocculation, a sphere may be more accurate. So, the shape parameter in the final model should be a function of the drilling processes, the type of rock penetrated and the rate of flocculation. The shape of drill-cuttings will, therefore, not be an easy parameter to quantify.

Once the physical processes are defined and their influence on the dispersion of the drilling wastes into the marine environment is determined, then, the greatest problem remains their mathematical quantification. One way to achieve this can be to define their influence on the density of the particulate matter or on the settling speed. For example, there are models existing to define more or less accurately the changes in particles density due to the process of flocculation. So, each process influences the dispersion of the drilling wastes in a specific way. It can change the distance between particles (like initial dilution or flocculation), the particles density (like flocculation or deflocculation- that may be caused by turbulence), the shape of the particle (like flocculation or turbulence)...etc. The accurate definition of the influence of each process is crucial to realistically quantifying this influence.

One way of doing so is to define which parameter of the dispersion is significantly affected by the each process, and then develop mathematical models (as analytically as possible) to quantify those processes. Once that is satisfactorily achieved then simple submodels can been created for each process and therefore, they can be applied to the adequate zones that are described in the next section of this paper.

DEFINITION OF THE DIFFERENT ZONES AND SIGNIFICANT PROCESSES

The aim of this section is to initially describe a methodology for the accurate modelling of the marine environment and then for the dispersion of the drilling wastes introduced into this complex environment. In order to accurately model the marine environment in which the drilling fluid is released, it is necessary to divide the water column into several different zones (modified from Carles, 1996). These zones have been carefully defined, depending on the physical processes involved in the dispersion, keeping in mind the mathematical dimension for the definition of those zones.

Firstly, three different horizontal layers have been defined (see Figure. 1):
- surface layer: influenced by meteorological and tidal conditions. In this layer, wave action, tidal currents, thermocline and pycnocline have to be taken into account in the changes of the general current speed profiles;
- transition layer. This, which is effectively a transition layer, may not exist depending upon the overlapping of the two other layer;
- bottom layer: influenced by the friction between water flow and the seabed (considered to be "waterproof" for the purpose of this paper).

The thickness of this layer depends on the velocity of the marine currents, the water depth and the morphology of the seabed.

Secondly, three range related "fields" can be assumed:
- Near field, where coarse and fine cuttings get separated by buoyancy effects.

Here, the initial kinetic energy is dissipated in the turbulent mixing of the jet with the surrounding fluid, while the buoyancy force (positive or negative) drives the discharge vertically upwards or downwards producing further mixing and dilution with the surrounding fluid. The term "jet" is used in this paper to describe those cases where relative kinetic energy is still present while the term "plume" is used to describe the cases where buoyancy is the dominant driving force on the discharge movement. This field is characterised by "convective descent".
- Mid field. In this region, the collapse phase begins when the plume encounters its level of neutral buoyancy (or the sea bottom), at which point the descent is retarded and horizontal spreading dominates. Generally speaking, the collapse phase continues until the material is spread because of turbulent diffusion and not anymore because of dynamic mechanisms. This field is characterised by "dynamic collapse".
- Far field. Only relatively fine particles in small concentrations can reach this region, where ambient conditions drive the motion of the wastes. This field is characterised by "passive diffusion".

Up to 9 relevant zones, related to the fields and layers, can be defined in terms of mathematical and physical boundaries. These boundaries are described in the next section.

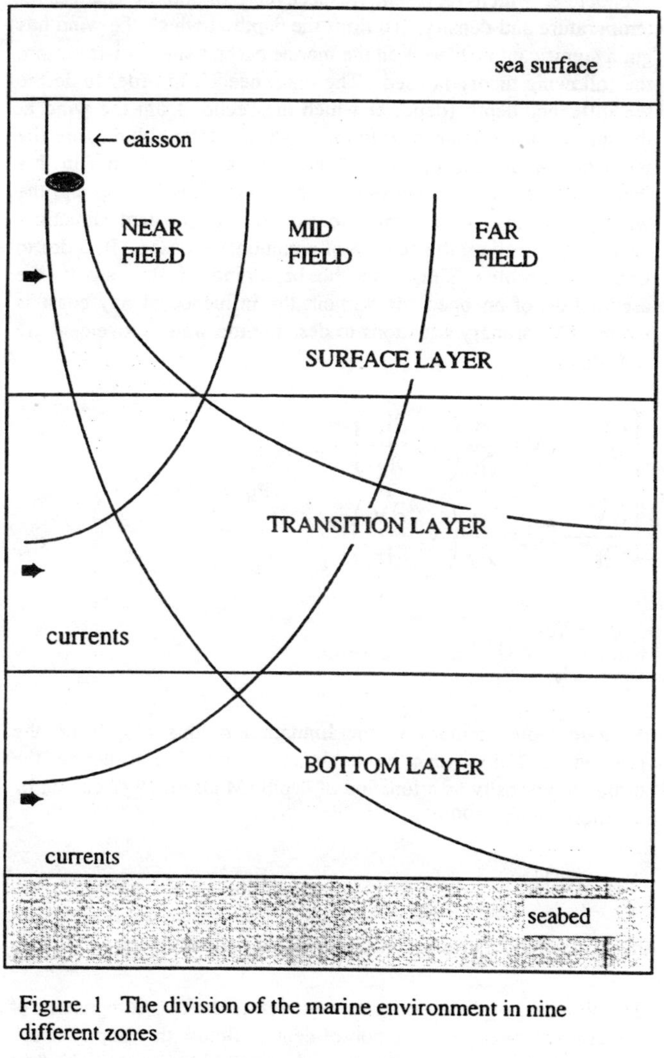

Figure. 1 The division of the marine environment in nine different zones

(modified from Carles, 1996).

Once the different zones have been defined in terms of physical processes, the most difficult problem is to describe them in terms of mathematical boundary conditions. This can be achieved by using existing theories and developing them for this particular case.

DETERMINATION OF ZONE BOUNDARIES

This section deals with identification of the boundaries between the layers and fields and subsequent division of the environment into the possible 9 zones. Submodels will be used in each zone, each of which will have horizontal and range related boundaries.

Division of the Water Column into Layers

The first horizontal zone is influenced by the waves and then, by the conditions at the surface of the sea. These conditions are themselves directly influenced by meteorological phenomenon such as wind, storms, rain, warm weather...etc. In this first "layer", not only wave action has to be considered but also the thermocline and pycnocline, which represent, respectively, important changes in temperature and density. To limit the depth at which the wind has got a significant influence on the marine current speed distribution, the following theory is used. The input needed in order to define the influence depth (depth at which the "echo" from the wind at the surface of the sea can be felt through the water column) are the wind (at the surface of the sea) intensity and direction. In this theory, the sea water is considered as a homogeneous fluid and the wind as having a uniform intensity and a constant direction, initially blowing on the surface of a motionless ocean. In order to keep the problem simple at the beginning of this study, the assumption of an open sea without the influence of any coast is made. The primary equations to describe the water movement are as follows:

$$\begin{cases} \dfrac{\partial U}{\partial t} + fV = \dfrac{\partial}{\partial z}\left(v_t \dfrac{\partial U}{\partial z}\right) \\ \dfrac{\partial V}{\partial t} - fU = \dfrac{\partial}{\partial z}\left(v_t \dfrac{\partial V}{\partial z}\right) \end{cases} \qquad \text{(Eq. 1)}$$

with: $$\frac{\partial W}{\partial z} = 0 \qquad \text{(Eq. 2)}$$

To gain more accuracy in the limitation of this first layer, the previous equations can be analytically solved considering the turbulent viscosity as a function of depth (Madsen, 1977 quoted by Verdier-Bonnet, 1996):

$$v_t = \kappa_a U z \qquad \text{(Eq. 3)}$$

κ_a is the Von Karman constant and U the shear velocity.

The solutions of these equations will define the two components of the current speed as a function of depth. Below this depth, it can be considered that the influence of the wind blowing at the surface is negligible on the current speed distribution. Tidal conditions may also influenced the current speed profiles throughout the water column. In this paper, tidal movement is assumed to be periodic. The influence of waves upon the dispersion in the upper layer must, of course, be quantified. This is done by using the traditional approach of considering all but the most energetic sea states as a gaussian random process and applying linear wave theory.

The second layer is a transition layer where the influence of any processes at the surface or the bottom of the sea can be considered as negligible. Here the conditions are less turbulent. In the case of shallow waters and strong winds combined, this transition may not exist, as the two others layers may overlap each other. Another very important factor of influence on the current speed profiles (intensity and direction) is the presence of coastal boundaries. This problem is far from being easy to solve and will probably be taken into account at a later stage in the development program.

The third layer is a benthic boundary layer. Here, the shear stress between the water flow and the seabed has to be quantified. The shear stress is mainly caused by friction between currents and the seabed and, in the case of very shallow waters and relatively strong currents, by the wave action on the seabed. Here, it has been considered as reasonable to assume the Coriolis effects as negligible. So, in the final model, the water depth will be of great importance as it will determine the height (and even presence!) of each layers. The choice of the grid for the computational model is also very important. For example, in the first horizontal layer, the grid should be finer at the top where many changes in the parameters occur while it can be less fine at the bottom where the changes can be assumed to be less drastic. The reverse is the case in the third layer.

In order to quantify the shear stress at the bottom, the tidal influence is described as the propagation of a sinusoidal wave with the following characteristic equations:

$$\begin{cases} g\dfrac{\partial \eta}{\partial x} = \omega a_x \cos(\omega t + \varphi_x) \\ g\dfrac{\partial \eta}{\partial y} = \omega a_y \cos(\omega t + \varphi_y) \end{cases} \qquad \text{(Eq. 4)}$$

$\omega = \dfrac{2\pi}{T}$ is the frequency of the wave, a_x and a_y are the amplitudes on x and y axis, while φ_x and φ_y are the corresponding phases.

When shallow waters and strong currents are considered, the fluid within the whole water column is well-mixed (due to a high level of turbulence), so that the fluid can be considered as homogeneous. In the case of deeper waters, the assumption of a stratified fluid is necessary, as buoyancy effects decrease turbulent movement (Verdier-Bonnet, 1996). In the North Sea, the stratification of the water column is considered to be relevant from a depth of 50m with tidal currents of 0.50m/s (Davies, 1993). In waters deeper than 50m, the fluid can still be considered as homogeneous in the benthic boundary layer because the high turbulence, due to the friction between currents and the seabed, effectively mixes the ambient fluid.

Range Divisions

The near field is characterised by the highly turbulent motion of the drilling fluid released into the sea water, by a high level of flocculation and by a high rate of initial dilution. The parameters influencing the dispersion of the drilling wastes are numerous and the significance of their influence needs to be checked before trying to establish mathematical models. In the near field, for example, the shape, diameter and orientation of the caisson has got a noticeable influence on the jet. In this particular zone, the jet is supposed to be well-mixed because of high level of turbulence, so the fluid can be considered as homogeneous. The concentration method is used here. Eulerian equations are used, therefore, to describe the movement of the particulate matter. One way of simply, but with an acceptable accuracy, limiting this zone, is to determine a threshold in the particle concentration above which it is not acceptable anymore to speak in terms of concentrations but in terms of individual particles.

Then, in the mid field, Lagrangian equations are used to describe the trajectory of each individual particle. Equations to pass from Eulerian equations onto Lagrangian ones are well-known. These equations give the initial position of the particle while it enters one cell of the grid and, after a certain time, the ultimate position when the particle exits that cell to enter the next one.

In the far field, Lagrangian equations are still used. Only very fine particles reach this region where they follow the movement of the ambient fluid.

General Lagrangian equations to describe the movement of particles can be as follows (Ryhming, 1988):

$$dF = \frac{\partial F}{\partial t}dt + \frac{\partial F}{\partial x}dx + \frac{\partial F}{\partial y}dy + \frac{\partial F}{\partial z}dz \qquad \text{(Eq. 5)}$$

and therefore,

$$\frac{dF}{dt} = \frac{DF}{DT} = \frac{\partial F}{\partial t} + \nabla F \cdot \frac{dr}{dt} = \frac{\partial F}{\partial t} + v \cdot \nabla F \qquad \text{(Eq. 6)}$$

here, the vector "r" represents a function of Lagrangian coordinates.

CONCLUSIONS

The general methodology presented in this paper is effectively the understanding, definition and quantification of the physical processes affecting the dispersion of the drilling wastes released into the marine environment. For each physical process, a submodel is created, which will be used in the relevant zone. Each zone is defined in terms of horizontal as well as "range related" boundary conditions, which means that marine effects, as well as dispersive effects are taken into account in each zone.

The logic of this methodology is to model first the marine environment without the release of any drilling wastes, then to model each physical process affecting the studied dispersion and then, finally, to put all these data together with the characteristics of the drilling wastes in order to define the trajectory of the particles. Once the particles trajectory is known, it is "simple" to determine the concentrations of the particulate matter deposited on the seabed and therefore the thickness of the cuttings pile (if present).

ACKNOWLEDGMENTS

The authors would like to thank the Robert Gordon University for the opportunity to participate in the studies outlined in the paper, and also the different institutions which kindly contribute in the provision of data.

REFERENCES

Carles L. (1996). "Simulating the Dispersion of Offshore Drill-Cuttings", *Institute of Offshore Engineering, Heriot-Watt University* (Edinburgh), pp 68-74.

Davies A.M. (1993). "Numerical Problems in Simulating Tidal Flows with a Frictional-Velocity-Dependent Eddy Viscosity and the Influence of Stratification", *Int. J. Num. Meth. Fluids*, 16.

Environmental Consultancy Services Ltd (1993). "South Coast Exploration Drilling Programme 1993: Environmental Review", *Elf Entreprise*, pp 6, 7.

Gordon et al (1995). "Prediction of Potential Benthic Impact Zones Produced by the Discharge of Operational Wastes from Drilling Platforms", *Internal Review of PERD Projects on the Fate and Effects of Operational Drilling Wastes* (23 January 1995).

Macmillan (1986). "Macmillan Dictionary of Physics", *Macmillan Press Ltd*, pp 77.

McGraw-Hill (1994). "Dictionary of Scientific and Technical Terms", Fifth Edition, *McGraw-Hill Inc.*, pp 544, 592, 1964, 2083.

Muschenheim D.K. and Milligan T.G. (1996). "Flocculation and Accumulation of Fine Drilling Particulates on the Scotian Shelf (Canada)", *Marine Pollution Bulletin*, July 1996.

Ryhming I.L. (1988). "Dynamique des Fluides", *Presses Polytechniques Romandes*, pp 16, 17.

Verdier-Bonnet C. (1996). "Modelisation des Ecoulements Cotiers Stratifies a Surface Libre et Parametrisation de la Turbulence. Application a la Dissemination Larvaire.", *University of Toulon et du Var* (Toulon, France), pp 75, 81.

Proceedings of the Seventh (1997) International Offshore and Polar Engineering Conference
Honolulu, USA, May 25-30, 1997

ISBN 1-880653-28-1 (Set); ISBN 1-880653-29-X (Vol. I)

Infiltration Control Cover Technology Demonstration at Marine Corps Base Hawaii - Preliminary Results of Water Balance Monitoring

Leslie Karr
Naval Facilities Engineering Service Center, Port Hueneme, CA, USA
T.E. Hakonson
Colorado State University, Fort Collins, CO, USA
Bryan Harre
Naval Facilities Engineering Service Center, Port Hueneme, CA, USA

Abstract

This paper presents preliminary results from a landfill capping demonstration conducted by the Naval Facilities Engineering Service Center (NFESC) at Marine Corps Base Hawaii (MCBH). Specific details about construction and sampling methods used in the study are not presented here as they appear elsewhere (Hakonson, 1994 and 1996). This paper summarizes some of the data that has been collected on precipitation, runoff, and percolation from the MCBH covers beginning November 1995 through July 1996. Data from this analysis were also compared against corresponding data for a clay cap obtained from Environmental Protection Agency's (EPA's) HELP, Version 3, water balance model (EPA, 1994). Preliminary results from the runoff enhancing structures (RES) designs appear encouraging in that they are producing results in line with the technical theory behind them.

Keywords

Landfill, Capping, Water Harvesting, Water Balance, Infiltration

BACKGROUND

Surface covers to control water infiltration to waste buried in landfills serve as an attractive remediation alternative for most of the Navy's hazardous and sanitary waste landfills. Surface covers can effectively manage the human and ecological risks of a site, along with offering easy maintenance and lower costs than other remediation options. Conventional wisdom would suggest that landfill capping technology is well developed as evidenced by the availability of EPA guidance (EPA, 1989) for designing and constructing what has become known as the Resource Conservation and Recovery Act (RCRA) cap and its several variations. In practice, however, very little field testing of the RCRA cap, or any other design, has been done to evaluate the long term effectiveness of various covers in limiting infiltration of water into the waste.

EPA encourages the using alternative cap designs if they can be shown to be "equivalent" to the RCRA cap (EPA, 1989). Demonstrating "equivalency" presumably must involve the use EPA's Hydrologic Evaluation Leachate Potential (HELP) model since few side-by-side comparisons of capping technologies have been done.

The NFESC studied several capping designs at MCBH. These large scale lysimeters are used to perform a mass balance on the precipitation received at the site. The objective of the study was to evaluate three alternative covers including: a 20% water harvesting structure; a 40% water harvesting structure; and a control plot, as shown in Figure 1. The equivalency of the three MCBH covers to a modified RCRA clay cap was also evaluated by comparing the data with that predicted with Version 3 of the Hydrologic Evaluation of Landfill Performance, or HELP3 model (EPA, 1994).

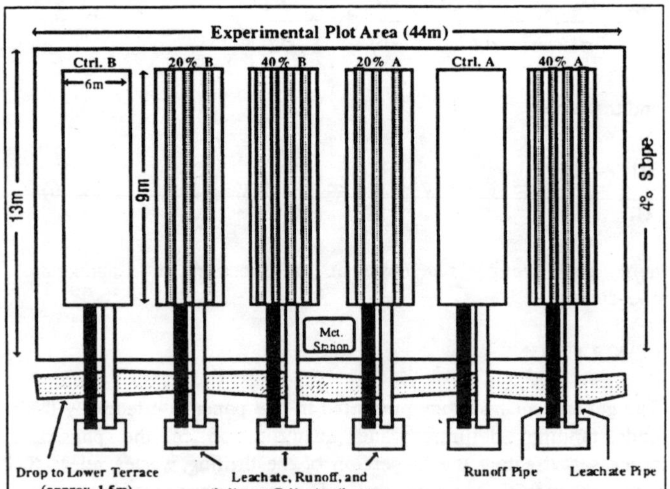

Figure 1. Plot Layout

---

*Ms. Karr and Mr. Harre are from the Naval Facilities Engineering Service Center, Port Hueneme, CA., and Dr. Hakonson is from the Colorado State University, Fort Collins, CO.

TECHNICAL APPROACH

Most capping technologies use design features that control one or more of the processes governing the fate of precipitation falling on the landfill. The fate of meteoric water falling on the landfill is often referred to as the water balance of the site. A simplified representation of water balance describes surface runoff and one-dimensional movement of water in the soil profile to the plant rooting depth. For net rates and amounts, the water balance equation is:

$$\delta S/dt = (P - R - ET - L)/dt \quad \text{(Eq. 1)}$$

where:

δS = Change in soil moisture storage
P = Precipitation
R = Runoff
ET = Evapotranspiration
L = Percolation below the root zone
t = Unit of time used in solving the equation

Units of depth, or cm, are used in applying the equation in this paper.

The concept of water balance, relative to design of landfill covers, accounts for the strong interactions between the various terms of Equation 1. For example, a reduction or elimination of the runoff term (R), increases infiltration of water into the soil, resulting in increased soil moisture storage followed by an increase in ET and/or percolation. Likewise, an enhancement of runoff must be followed by reduction in the amount of precipitation that infiltrates into the soil with a potential reduction in ET and/or percolation.

Because the maximum amount of soil moisture that can be removed by ET has an upper limit that is defined by climate and other local conditions, a cover is needed for sites where the amount of water that infiltrates into the soil exceeds the potential for ET to remove it. This need especially applies to humid sites that receive more than 75 cm of precipitation per year or to more xeric sites where snow is an important source of annual precipitation.

There are several possible ways to manage soil moisture in a landfill cover including the use of hydraulic or capillary barriers to laterally divert percolating water away from the waste environment or methods to limit infiltration of precipitation into the cover soil (Hakonson et al. 1990; Nyhan et al., 1990; Hakonson et al., 1992). In either case, vegetation plays a key role in the ability of the cover to limit percolation by removing large amounts of stored soil moisture via transpiration. The study at MCBH focuses on controlling infiltration by enhancing surface runoff as a means of reducing the amount of precipitation available for infiltration into the cover soil. The theory behind this design is to limit percolation of water through the cap to levels that the vegetation growing on the cover can remove via transpiration (Figure 2). If the concept is viable, it could greatly simplify the construction and cost of covers where it can be used because the technology can be applied to an existing landfill cover, where it is easily repairable, and involves a minimum of materials, equipment, and labor.

Some of the pioneering work on infiltration control landfill covers in the United States was conducted by Dr. R. K. Schulz, University of California, Berkley, with Nuclear Regulatory Commission funding (Schulz, et al., 1990). However, the technical basis for this means of water management was understood by ancient Hebrew farmers who developed and used infiltration control techniques to support a flourishing agriculture 8,000 years ago in the Negev Desert (Evenari, et al., 1961).

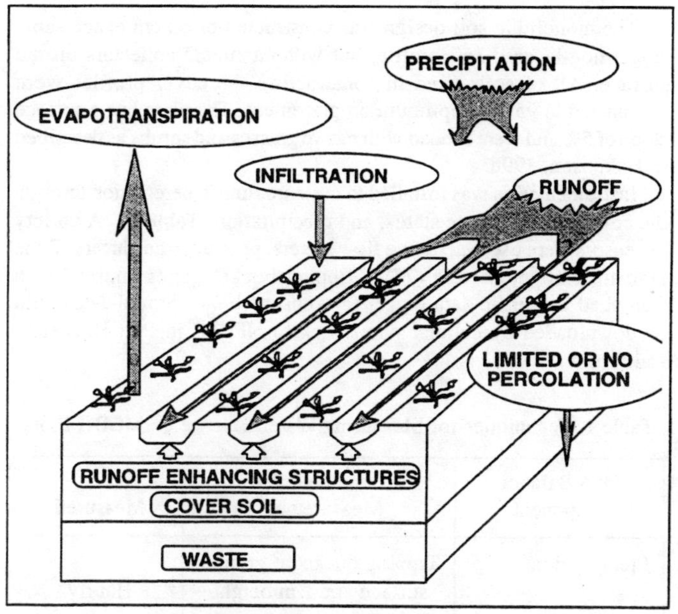

Figure 2. Infiltration Control Cover

METHODS AND MATERIALS

Six plots, 6 m x 9 m long (Hakonson, 1994), were constructed to measure all of the water balance terms in Equation 1, except evapotranspiration. The latter was derived by solving Equation 1. The cap designs to limit infiltration consisted of replicates of a non-layered soil profile, 60-cm thick, with two levels of RES on the cover surface (Figure 1). Initial calculations, based on an average MCBH precipitation of 99 cm/yr, suggested that RES installed on about 20 and 40% of the cover surface should be sufficient to reduce percolation through the cover to a very low level. RES were constructed from 12-cm wide rain gutter placed parallel to the slope on the ground surface. Figure 3 provides a more detailed layout of one of the test plots.

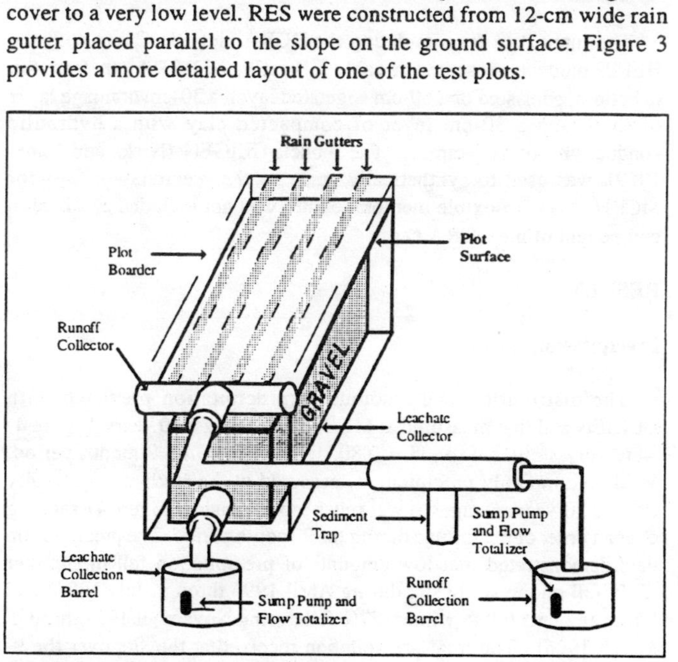

Figure 3. Individual Test Plot Diagram.

The monolithic soil design was constructed of 60 cm of the same single non-layered soil profile, but without runoff collectors on the surface. All of soils used in constructing the cover profiles were compacted to 95% of optimum on placement. The plots had a surface slope of 5% and were seeded with native grasses and shrubs as described in Hakonson, 1996.

Instrumentation was installed to measure runoff, percolation through the cover, soil moisture status, and precipitation (Table 1). A variety of sensors were used including flow meters, pressure transducers, Time Domain Reflectometers (TDR),tipping bucket gages coupled to a Campbell Scientific data logger (Hakonson, 1994). Stored data could be downloaded by cellular phone link to NFESC in Port Hueneme, California.

Table 1. Techniques for Measuring Water Balance on MCBH Plots

Water Balance Component	Method of Measurement	Measured
Precipitation	Tipping bucket at ground surface and 1 m height	Hourly
Soil Moisture	4 TDR probes/plot	Hourly
Runoff Transducers	Flow meters/pressure	Every 15 min.
Leachate Transducers	Flow meters/mressure	Every 15 min.
Evapotranspiration	Solve water balance equation	---
Sediment	Total collection	Every 3 mos

The input data for simulating the RCRA, subtitle C cap with the HELP3 model is presented in Table 2. Briefly, the RCRA cap, from top to bottom, consisted of a 60-cm vegetated layer, a 30-cm drainage layer of sand, and a 30-cm layer of compacted clay with a hydraulic conductivity of 10^{-7} cm/sec. The model, CLIGEN (Nicks and Lane, 1989), was used to synthetically generate the precipitation files for MCBH. Also, a flexible membrane liner was not included as a design component of the RCRA cap.

RESULTS

Precipitation Data

The distribution and amounts of precipitation predicted with CLIGEN and that measured at MCBH over the last 30 years (Figure 4) were very similar (i. e., $r^2 = 0.80$, 1, 7 df) over the 9-month period. While the monthly precipitation measured at the study site generally tracked 30-year averages, it was much more variable (Figure 4), ranging over an order of magnitude during the 9-month period. The precipitation data demonstrated that low amounts of precipitation fall in summer (23% fell on the study site during April 1996 through July 1996) and larger amounts fall in winter (77% fell during November 1995 through March 1996). The total precipitation received at the site over the 9-month period was about 66 cm. This is about 85% of the 77 cm/yr that is the 30-year average at MCBH and about 84% of the 79 cm predicted by CLIGEN.

Table 2. HELP3 Input Data Used in Simulating the EPA RCRA Cap for MCBH Conditions (Station Latitude = 21.33 degrees)

Layer	Amount of Measured Runoff
Layer 1- Vertical Percolation Layer	
SCS Runoff Curve Number	80.50
Evaporative Zone Depth	0.61 meters
Thickness	0.61 meters
Porosity	0.4640vol/vol
Field Capacity	0.3100 vol/vol
Wilting Point	0.1870 vol/vol
Initial Soil Water Content	0.2719 vol/vol
Effective Sat. Hyd. Cond.	6.4 x 10^{-5} cm/sec
Layer 2- Lateral Drainage Layer	
Thickness	0.305 meters
Porosity	0.3970 vol/vol
Field Capacity	0.0320 vol/vol
Wilting Point	0.0130 vol/vol
Initial Soil Water Content	0.0328 vol/vol
Effective Sat. Hyd. Cond	0.3 cm/sec
Slope	4.00%
Drainage Length	63.4 meters
Layer 3- Barrier Soil Liner	
Thickness	0.305 meters
Porosity	0.4270 vol/vol
Field Capacity	0.4180 vol/vol
Wilting Poin	0.3670 vol/vol
Initial Soil Water Content	0.4270 vol/vol
Effective Sat. Hyd. Cond.	1 x 10^{-7} cm/sec
Evapotranspiration and Weather Data	
Maximum Leaf Area Index	5.00
Start of Growing Season (Julian Date)	0
End of Growing Season (Julian Date)	367
Average Annual Wind Speed	18.83 km/h

Precipitation file for MCBH was generated using CLIGEN (Nicks and Lane, 1989.)

Runoff Data

Monthly runoff (based on plot pair averages) is plotted in Figure 5 for November 1995 through July 1996. Statistical tests (Students t-test for paired data) for differences in runoff between cover treatments were all significant at $p \leq 0.05$. Results clearly show that runoff enhancement has occurred on those plots treated with RES compared to the untreated soil surface.

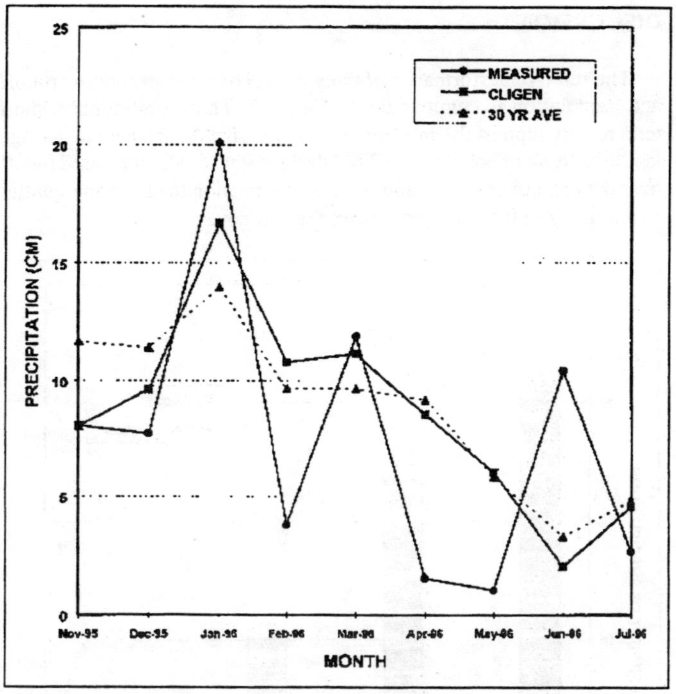

Figure 4. Measured, 30-year average, and HELP3 predicted monthly precipitation at MCBH.

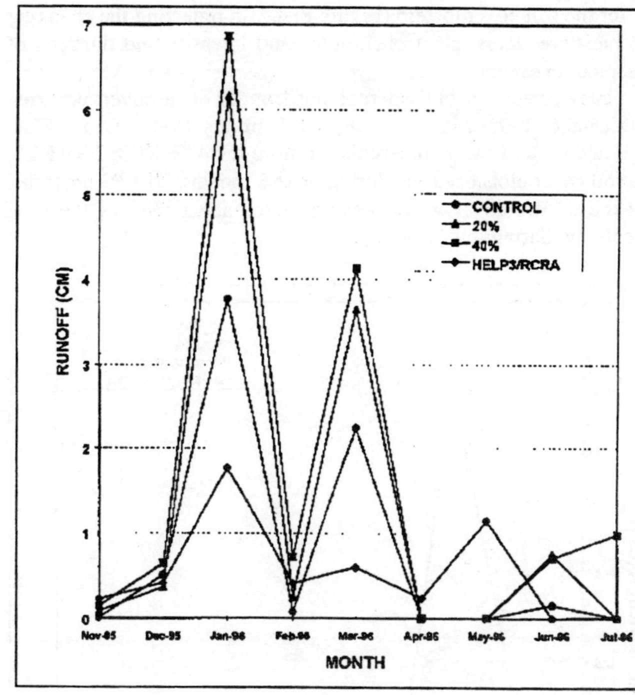

Figure 5. Average monthly runoff from various covers at MCBH.

Additionally, the amount of runoff generated was related to the level of runoff enhancement incorporated into the cover design. In general, the untreated soil cover design had the least runoff, the 20% enhancement was intermediate, and the 40% enhancement had the most.

Runoff distribution over the 9-month period (Figure 6) paralleled that observed for precipitation: 88 to 98% was generated during November 1995 through March 1996 and 2 to 12% during April through July 1996 (Table 3). The total amount of runoff was strongly related to the cover type with the 40% and 20% runoff enhancement designs producing 2.1 times (14.2 cm) and 1.7 times (11.3 cm) more runoff than the 6.8 cm produced by the monolithic soil cover.

The RCRA cap, based on HELP3 simulations, was estimated to produce 4.8 cm of runoff. Over the 9-month period, the relative amount of runoff averaged 21, 17, 10, and 6 from the 40% RES, 20% RES, soil, and RCRA designs, respectively. On a month-to-month basis, the relative amount of runoff generated by the various covers (Figure 6) was as high as 35% and as low as zero reflecting the influence of changing soil moisture status, vegetation phenology, and duration and intensity of individual rain storms.

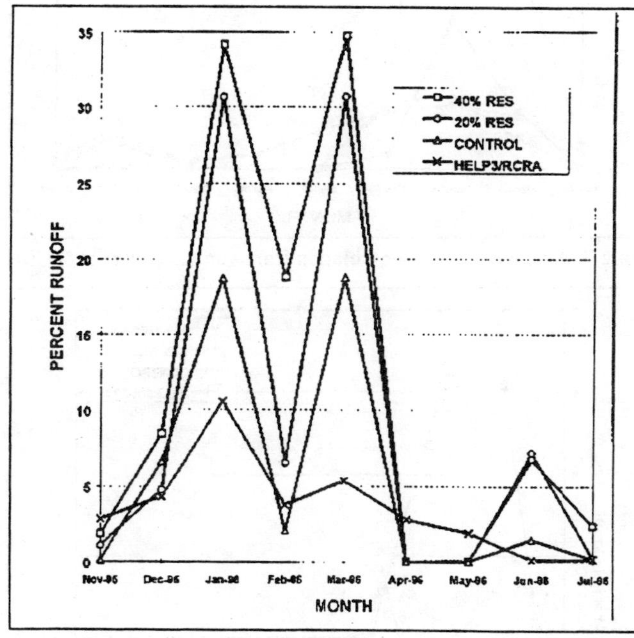

Figure 6. Relative amount of runoff from various covers at MCBH.

Table 3. Distribution of Precipitation and Runoff from Various Cover Designs by Season at MCBH

Cover Design	% Runoff Apr - Jul 1996	% Runoff Nov 1995 - Mar 1996
Precipitation	23.0	77.0
Soil Cover	2.3	97.3
20% RES	6.7	93.3
40% RES	12.0	88.0

Percolation

Percolation from both infiltration control covers and that estimated for the RCRA cover was significantly lower ($p = 0.03$) than that produced from the soil cover (Figure 7). In contrast, differences in percolation between the two RES covers were not significantly different ($p = 0.35$). Additionally, differences in percolation rates from the RES covers were also not significantly different ($p = 0.25$ and $p = 0.48$) from those predicted by HELP3 for the RCRA cover.

Percolation through all covers (Figure 7) averaged 0.69% (40% RES), 0.83% (20% RES), 3.0% (soil cover), and 0.71% (HELP3/RCRA cover) of the total precipitation that fell during the 9-month period. On a month-to-month basis, the relative percolation varied from a high of

6% for the soil cover to zero (Figure 8), again reflecting the changing soil moisture status, plant phenology, and intensity and duration of precipitation events.

A large percentage of the percolation from all of the covers occurred in December 1995, January 1996, and February 1996. About 80%, 89%, and 91% of the total percolation from the 40% RES, 20% RES, and soil cover plots occurred during these 3 months. HELP3 predicted that the RCRA cover would have produce about 68% of the total percolation during this 3-month period.

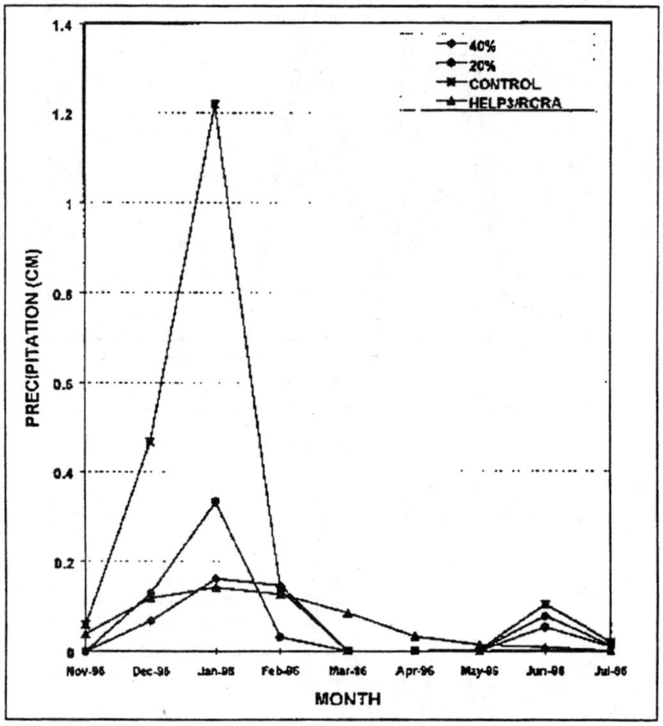

Figure 7. Average monthly percolation from various covers at MCBH.

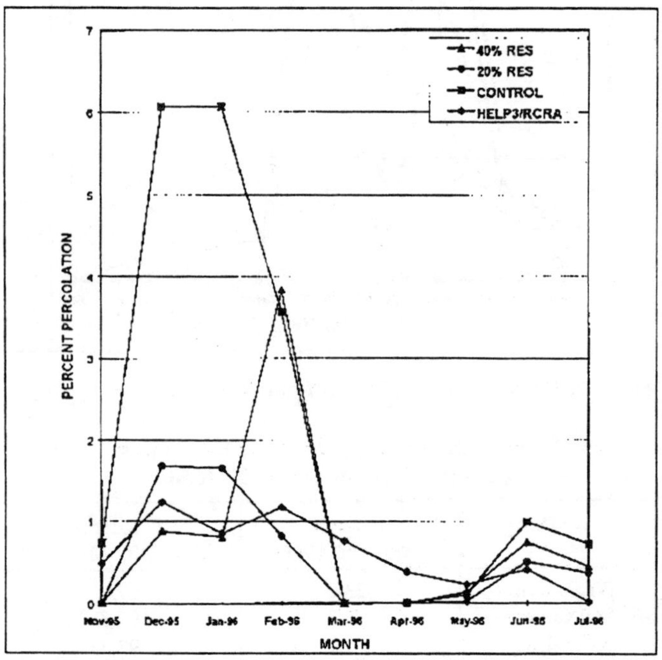

Figure 8. Relative amount of percolation from various covers at MCBH.

DISCUSSION

The relative performance of the various covers in generating runoff and percolation is summarized in Figure 9. These preliminary, short term results support the concept of using runoff enhancement to manage landfill site water balance. The RES designs statistically increased runoff over the soil cover design and reduced percolation to about one-quarter to one-third of that measured from the soil cover.

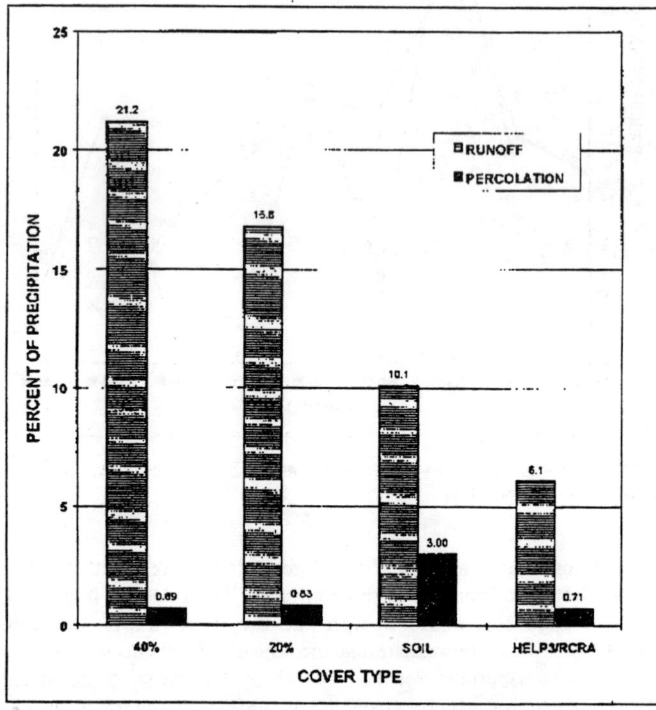

Figure 9. Relative amount of runoff and percolation produced from runoff enchancement, soil, and HELP3 simulated RCRA covers at MCBH for November 1995 through July 1996.

Results also demonstrate that the hydrologic response of the MCBH covers depends on the season and a related variable — the amount of precipitation falling during a particular month. Most of the runoff and percolation was associated with a few months in winter when most of the precipitation occurs. Additionally, months with threshold precipitation above about 10 cm, generated most of the runoff or percolation. Summer months generally received less than 10 cm and produced little of the runoff and percolation measured on the plots.

Based on a comparison of the predicted hydrologic performance of a modified RCRA cover and the two RES covers, these preliminary results suggest that they are about equivalent in their ability to limit percolation. While the RES covers rely on limiting infiltration, the RCRA cpver relies on lateral diversion of soil moisture in a drainage layer to prevent percolation. In the HELP3 simulation of the RCRA dover, about 15% of the precipitation was predicted as lateral flow from the drainage layer.

SUMMARY

The preliminary results from the RES covers appear encouraging in that they are producing results in line with the technical theory behind them. Should the performance characteristics of the RES covers that have been observed thus far be validated with further monitoring data, these covers would offer a simple and inexpensive alternative for interim stabilization or closure of landfills, particularly in more humid sites.

ACKNOWLEDGMENTS

Funding for this project was provided by the Naval Facilities Engineering Services Center, Port Hueneme, California. The authors wish to thank Mr. David Garcia, Ms. Doris Tong, Ms. Barbara Nelson, and Mr. Ian Stewart of NFESC, who contributed significantly to the data collection and analysis phases of this study. The on-site support of the Environmental Restoration Group at MCBH, particularly that of Mr. James Abbott, is also gratefully acknowledged.

REFERENCES

EPA (1989). "Final Covers on Hazardous Waste Landfills and Surface Impoundments," EPA/530-SW-89-047, July 1989.

EPA (1994). "The Hydrologic Evaluation of Landfill Performance (HELP) Model," EPA/600/R-94/168a.

Evenari, M., L. Shanan, N. Tadmor, and Y. Aharoni, 1961. "Ancient Agriculture in the Negev. *Science* 133(3457):979-996.

Hakonson, T. E., L. J. Lane, J. W. Nyhan, F. J. Barnes, and G. L. DePoorter (1990). "Trench cover systems for manipulating water balance on low-level radioactive waste sites." M. S. Bedinger and P. R. Stevens, (eds.), pp 73-80, in: *Safe Disposal of Radionuclides in Low-Level Radioactive Waste Repository Sites: Low Level Radioactive Waste Disposal Workshop*, USGS Circ.- 1036.

Hakonson, T. E., L. J. Lane, and E. P. Springer (1992). "Biotic and Abiotic Processes," Reith, C. C. and Thomson, B. M. (eds.) in *Deserts as Dumps, The Disposal of Hazardous Materials in Arid Ecosystems*, University of New Mexico Press, ISBN 0-8263-1297-7.

Hakonson, T. E. 1994. "Infiltration Control Cover Technology Demonstration at the U. S. Navy Marine Corp Base Hawaii - Test Plan and Field Procedures," Los Alamos National Laboratory Report, LAUR-94-2836.

Hakonson, T. E. 1996. "Infiltration Control Cover Technology Demonstration at the Marine Corps Base Hawaii - Project Objectives, Approach, and Preliminary Results of Vegetation Sampling in 1995," Colorado State University Report, CSU-DRHS-MCBH-96-1.

Nicks, A. D. and L. J. Lane. 1989. "Weather Generator." *In USDA-Water Erosion Prediction Project: Hillslope Profile Model Documentation*. L. J. Lane and M. A. Nearing (eds.) NSERL Report No. 2, USDA-ARS National Soil Erosion Research Laboratory, Purdue University, West Lafayette, IN.

Nyhan, J. W., T. E. Hakonson, and B. J. Drennon (1990). "A Water Balance Study of Two Landfill Cover Designs for Semiarid Regions," *Journal of Environmental Quality*, Vol 19, No. 2, pp. 281-288.

Schulz, R. K., R. W. Ridky, and E. O'Donnel. 1990. "Control of water infiltration into near surface LLW disposal units," NUREGCR-4918V4.

Proceedings of the Seventh (1997) International Offshore and Polar Engineering Conference
Honolulu, USA, May 25-30, 1997

ISBN 1-880653-28-1 (Set); ISBN 1-880653-29-X (Vol. I)

Curtains on Board for Preventing Spilt Oil Diffusion

Masayoshi Kubo
Kobe University of Mercantile Marine, Kobe, Japan
Kenji Asaki and Kazusei Yamamoto
Marine Technical College, Ashiya, Japan
Tsuyoshi Oki
Shibata Industrial Co., Ltd., Akashi, Japan

ABSTRACT

Oil pollution caused by casualties is becoming a serious public problem. Even if a large quantity of oil spills from the ship, there are no effective countermeasures that seamen can take by themselves. Most of oil spills at the early stage after the accident. In addition, no other support can be expected just after the accident because rough weather condition often continues for several days. Therefore, some effective devices which can be operated by only seamen should be boarded. In this paper, the authors propose a new device to prevent spilt oil diffusion. It encloses a hull with curtains whose depth is nearly equal to the full load draft of the ship and sinkers are mounted on the foot of the curtain to be stabilized in the water. Through some two-dimensional model experiments, it is shown that the device is better than an oil boom in performance to prevent spilt oil diffusion. Furthermore, it becomes clear on the basis of a full-scale experiment that the most practical setting method of the device is to handle the rolled curtain in the shape of a rope.

1. INTRODUCTION

The conservation of an oceanic environment is one of the most important obligations for seamen. Then, they should make endeavors to prevent oil spilling out of their ship and should always prepare effective countermeasures taken in the accident. However, there are no effective countermeasures taken by seamen when a large quantity of oil spills.

In this paper, firstly, the authors propose a new device which is designed so that a seaman can operate effectively to prevent spilt oil diffusion in several minutes just after an accident. Secondly, the effects of the device are discussed on the basis of some results of two-dimensional model experiments. Finally, a practical operation method of the device is presented with a result of a full scale experiment carried out on an actual ship.

That is to say, the purpose of this paper is to introduce the new device to take an effective countermeasure for preventing spilt oil diffusion and to show the practicability of the device through some experiments.

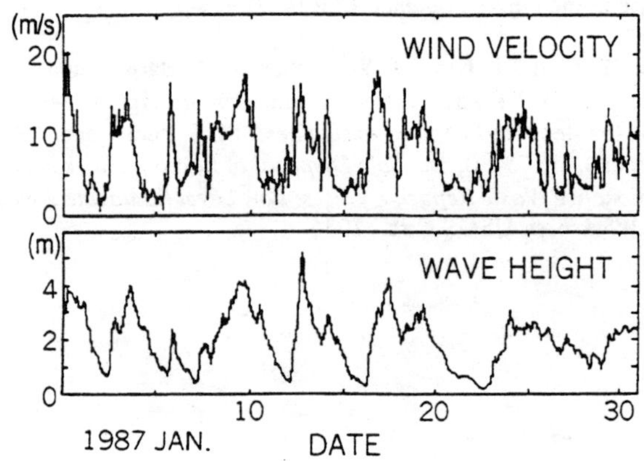

Fig.1 Wind velocity and wave height in Niigata offing of the Japan Sea

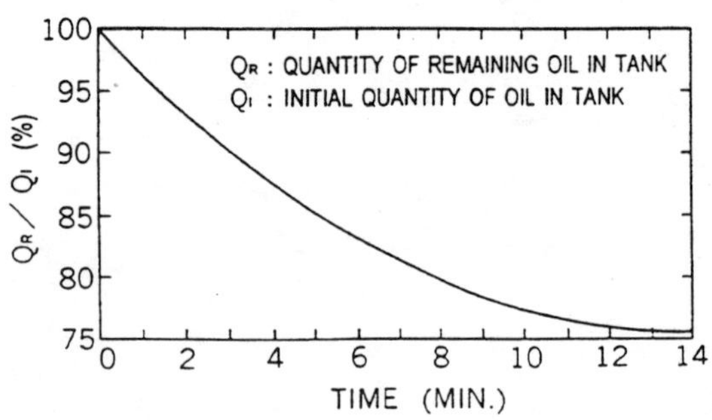

Fig.2 Quantity of remaining oil in ship's tank. Calculated by the simple Bernoulli's theorem. Oil spills through a hole of $1m^2$ at a tank bottom of 100,000DWT tanker.

2. NEW DEVICE TO PREVENT SPILT OIL DIFFUSION

Fig.1 shows wind velocity and wave heights in Niigata offing of the Japan Sea, where the accident of the Juriana occurred. Wave heights are 3m or more for a quarter of a month, and 2m or more for a half of a month. Wind velocity is 15m/s or more for a one fifth of a month, and 10m/s or more for a half of a month.

An existing typical device to prevent spilt oil diffusion is an oil boom, which needs some other support such as a jolly boat to spread it on the sea surface. Even if a device has sufficient performance to prevent spilt oil diffusion, if the device needs some other support to be set, it is impossible to use the device under the heavy weather condition as shown above because the support can not be expected.

Fig.2 shows quantity of remaining oil in a ship's tank after beginning of oil spill, which is calculated by the simple Bernoulli's theorem. It can be seen from this figure that most of oil completely spills in a brief period of about 14 minutes after the accident. Consequently, to take an effective countermeasure to prevent spilt oil diffusion, a ship should have a device which can be operated without any other support just after the beginning of oil spill.

Then, the authors propose a new device shown in Fig.3 to make up for the fault of an oil boom. This device encloses a hull with thin curtains and contains the spilt oil within the region formed with the hull and the curtain. The curtain is stored on board so that the normal operation of the ship is never disturbed. Once oil spills or the risk happens, the curtain is set in several minutes by a fairly simple handling. The advantages of this device are as follows.

(1) Spillage of contained oil over the curtain due to the wave can be prevented, because the curtain can be set close to the hull and the head of the curtain can be positioned higher above the water surface.
(2) When the flow around the hull can be controlled as a result of the proper form of the curtain, spillage of contained oil under the foot of the curtain can be prevented.
(3) This device is useful for both an initial countermeasure just after an accident and a preventive one before the oil spills, because it needs no other support and can be set in several minutes.
(4) The hull construction is hardly required to be changed, because the device is compactly stored outside of the hull. Therefore, storing the device does not influence the amount of the freight. Furthermore, it can be mounted on all the kind of ships having the space available to be stored.

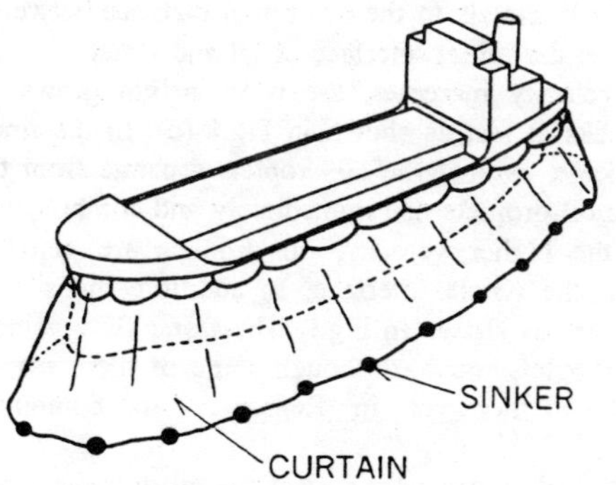

Fig.3 Curtain for preventing spilt oil diffusion

3. PERFORMANCE OF THE CURTAIN TO PREVENT SPILT OIL DIFFUSION

On the design of the device, it is important to know the behavior of contained oil and the performance of the device. It is guessed that the behavior of oil enclosed with the curtain including catamaran or outrigger type booms is different from it in an oil boom, because the proposed device prevents spilt oil diffusion by a pair of two structures such as a hull and the curtain while most oil booms are used with a single float. Therefore, the standards to evaluate the performance of an oil boom can not be applied with the device. Then, it is necessary to examine new standards against the proposed device. In this section, the behavior of contained oil will be presented through some model experiments and the standard of the device will be discussed.

3.1 Experimental apparatus and method

In this paper, the behavior of spilt oil under two dimensional steady flow are investigated. From preliminary experiments, the shapes of flexible curtains and a flow pattern are observed as shown in Fig.4. Fig.5 shows the experimental apparatus. A two-dimensional hull model and rigid curtains are fixed in the circular tank. In the following experiments, rigid curtains, which are made of acrylic boards of 5mm thickness, are used to hold the shapes of curtains steady and the shapes are replaced as shown in Fig. 5-(a) or Fig.5-(b). The hull model has a center tank to store oil and a small opening with a plug at the bottom. Pressure on the oil surface is always equal to the atmospheric pressure because there is no ceiling in the tank. The depth of water is 25cm and the ship's draft is 11cm through all experiments in this section.

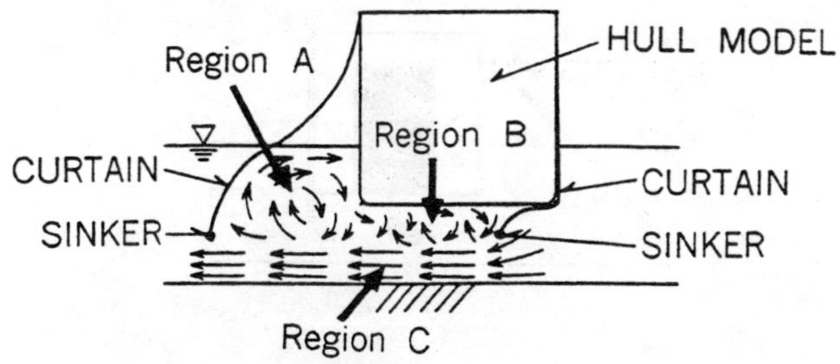

Fig.4 Typical shape of flexible curtain in flowing water and flow pattern around hull and curtains. Observed by tracer method.

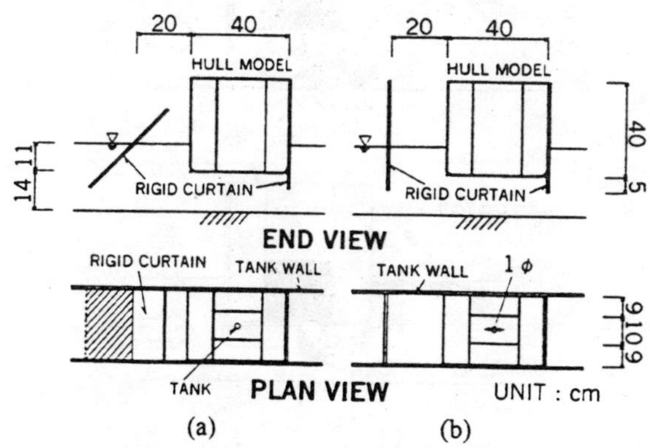

Fig. 5 Arrangement of experimental apparatus in two-dimensional model experiments

Table 1 Characteristics of test oil

Items	Test Oil A	Test Oil B
Oil Density ρ (15 ℃, g/cm^3)	0.873	0.884
Kinematic Viscosity ν (50 ℃, cSt)	2.77	8.75

3.2 Behavior of oil just after spill

Fig. 6 shows the behavior of oil from the beginning of spill to the end. Elapsed times from the beginning are indicated in the figures.

As soon as the bottom plug is opened, the cloud-shape oil spouts mightily to outboard and collides with the bottom of the circular tank and is thrown into the downstream. Next, after the oil becomes a lot of droplets of various sizes, they rise to the water surface in "Region A" or to the region right under the hull bottom as they are drawing the circular arc and rotating clockwise. In this stage, the majority of the oil droplets stay in "Region A" or "Region B", while only some of small oil droplets in "Region C" pass under the downstream curtain.

Oil droplets rising in "Region A" are thrown to the downstream largely and the maximum throwing distance is observed just after the spill begining. It is the reason that the oil droplets are influenced easily by the flow through "Region C", because the oil mightily spouts owing to the high level of the oil in the tank in the early stage of spill and the oil droplets reach "Region C" deeply.

On the other hand, the oil rising in "Region B" tends to stay near the upstream curtain as shown in Fig.6-(c). This tendency appears more remarkably when flowing velocity is fast. In the experiment under the about 15.7cm/sec of flowing velocity, almost all spilt oil stays near the curtain in the early stage. Since the curtain is rigid and fixed in this experiment, the oil never goes into the region between the hull and the upstream curtain. However, this phenomenon indicates a possibility that the oil goes into the region because a flexible curtain is used in an actual device.

As shown in Fig.6-(d), small interfacial waves always exist at the interface of water and oil in "Region B". Although some oil droplets separate from the interface, no droplet generated at this interface passes under the curtain. The waves move from the upstream to the downstream and the wave amplitude grows as quantity of oil in "Region B" increases. Most of oil in "Region B" still moves and rises in "Region A" by the influence of this wave motion after the oil spillage is completed. Finally, about 88% of the whole spilt oil is contained in "Region A". Within the limits of these experiments, the results indicate that the performance of the device largely depends on the limit of containing oil within "Region A", because almost all spilt oil tends to stay in the region.

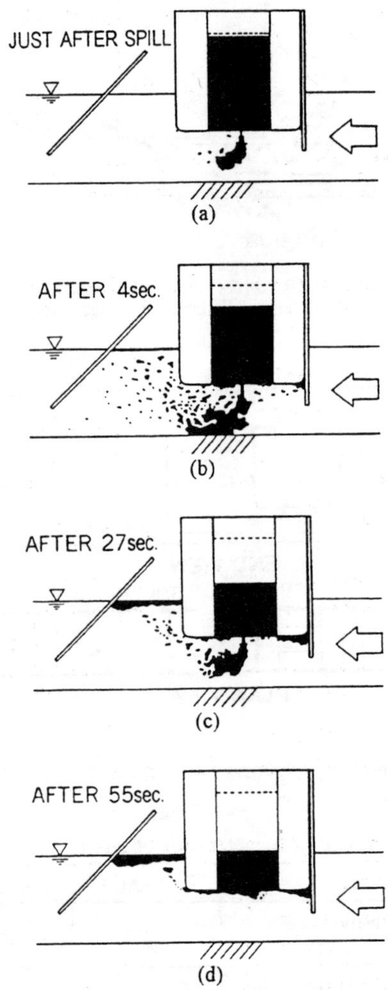

Fig.6 Behavior of spilt oil from the beginning of the spill to the end. Flowing velocity : 9.4cm/s. Quantity of oil : 2000cm^3. Test oil A.

3.3 Behavior of oil contained in "Region A"

Fig. 7 to 9 show the difference of oil behavior in "Region A" for various velocities of flowing water. Every experiment is carried out under the same initial condition which quantity of oil is 2000cm^3 and the initial thickness of the oil layer is 3.4cm.

In the early stage of the velocity increasing, small intefacial waves with high frequency are generated at the interface as shown in Fig.7. The every interfacial wave height is almost the same and the oil layer thickness is also almost uniform over the whole interface. Next, as the velocity increases, the oil is swept to the hull owing to the whirlpool shown in Fig.4 and the oil layer thickness near the hull increases as shown in Fig.8-(a). This phenomenon appears remarkably when the quantity of oil is fairly little. This state disappears in a very short time and then big interfacial wave moving downstream appears. Fig.8 shows this movement. When the interfacial wave reaches the downstream curtain, it returns to the initial state again as shown in Fig.8-(a). Thereafter, the interface repeats the same movement of period of 4 to 5 second though some are irregular. In addition, the wave length nearly equals to the horizontal distance between the hull and the curtain at the initial interface of oil and water.

As the velocity increases, the wave height grows. Moreover wave crest breaks at last as shown in Fig.8-(d). In the first stage of the wave breakage, even if the oil droplets separate from the broken wave crest, the oil droplets rise immediately and combine with the oil layer. Under the higher velocity, oil droplets are generated more frequently over the whole interface. In addition, the size of an oil droplet gets larger as shown in Fig.9. The faster flow velocity is, the wider the oil droplets move. Although some of them reach "Region B", a large part of them stay in "Region A" and combine with the former oil layer.

As mentioned above, only part of the oil droplets escape out of the region, even so continuous spillage of oil under the curtain is not observed through these experiments.

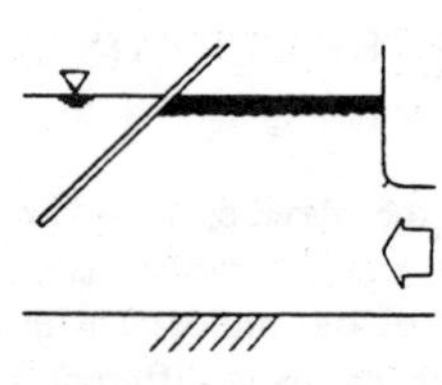

Fig.7 Initial behavior of oil layer in "Region A". Flowing velocity : 8.6cm/s. Quantity of oil : 2000cm^3. Test oil A.

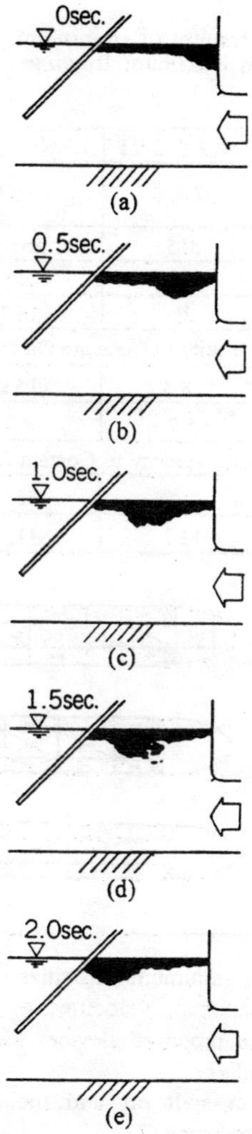

Fig. 8 Behavior of oil layer in "Region A". Flowing velocity : 18.2cm/s. Quantity of oil : 2000cm^3. Test oil A.

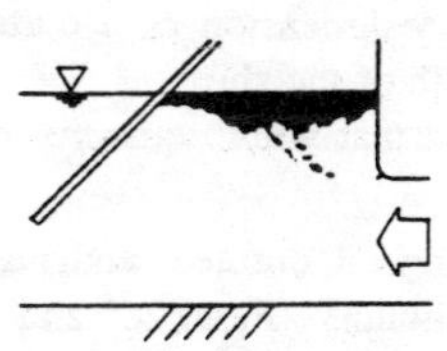

Fig. 9 Separation of oil droplets from oil layer in "Region A". Flowing velocity : 25.8cm/s. Quantity of oil : 2000cm^3. Test Oil A.

3.4 Generating of oil droplets and oil containing performance

From the experimental results, it can be expected that a standard to evaluate the performance of the device will be examined by understanding a generating mechanism of oil droplets in "Region A". That is, some oil droplets intermittently pass under the foot of the curtain in the first stage of oil spill, whereas this phenomenon never occurs under the condition that no oil droplets exist in "Region "A.

As mentioned above, the oil droplets are generated as the interfacial wave height at the interface of oil and water grows. Therefore, let us try a simple quantitative analysis on the basis of the measured wave height at the interface.

3.4.1 Dimensional analysis

On performance of an oil boom, it was reported that the Froude number given by Eq.(1) had close relation to the phenomenon that oil pass under an oil boom caused by high velocity flow,

$$Fa = \frac{V}{\sqrt{g(1-\sigma)}\sqrt{S}} \quad , \qquad (1)$$

where, V = velocity of flowing water, g = acceleration of gravity force, S = area (volume in three demensions) of oil region, $\sigma = \rho_o / \rho_w$, where, ρ_o = oil density and ρ_w = water density.

A mono-hull oil boom prevents spilt oil diffusion by one body structure, while the proposed device does that by two structures, a hull and a curtain. Therefore, it is estimated that the influence of the flow on oil is different. Consequently, Eq.(1) cannot be applied to the phenomena in this device. Then, a new non-dimensional number will be found as follows.

It is assumed the behavior of oil in "Region A" is governed mainly by Fi, force of inertia of water or oil, Fb, buoyancy of oil, and Fv, viscous force of oil. Each force can be expressed by following equations,

$$\boxed{Fi = \rho\, l^3 \frac{l}{t^2} = \rho\, l^2 v^2} \quad , \qquad (2)$$

$$\boxed{Fb = \Delta\rho\, g\, l^3} \quad , \qquad (3)$$

$$\boxed{Fv = \mu\, l^2 \frac{v}{l}} \quad , \qquad (4)$$

where, ρ = density of oil or water, l = length, t = time, v = velocity, μ = coefficient of viscosity, $\Delta\rho = \rho_w - \rho_o$, an equation within a rectangle means a representative quantity. In expression by Π theorem, two non-dimensional numbers shown by following equations can be obtained,

$$\boxed{\pi_1 = \frac{Fi}{Fb} = \frac{\rho v^2}{\Delta\rho\, g\, l}} \quad , \qquad (5)$$

$$\boxed{\pi_2 = \frac{Fi}{Fv} = \frac{\rho l v}{\mu} = \frac{l v}{\nu}} \quad , \qquad (6)$$

where, ν = kinematics viscosity.

Next, let us discuss both representative quantity of l and of v. It is inferred that the behavior of oil in "Region A" is influenced chiefly by the movement of the whirlpool in the same region, which is assumed to depend on the following factors; δod : thickness of oil

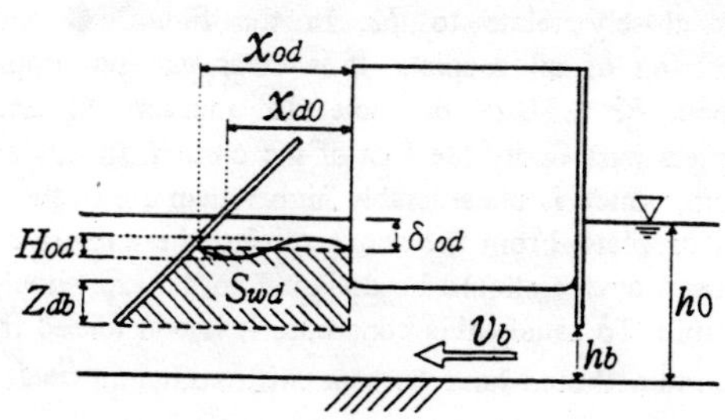

Fig. 10 Parameter in dimensional analysis

layer, Zdb : depth of the downstream curtain under the ship bottom, $xd0$: horizontal distance between broadside and the curtain at the water plane, a shape of the curtain and v_b : the velocity in "Region C". The four factors from the first among these are concerned with Swd, an area of vertical section of "Region A" under the oil layer. Thus, it is appropriate to use $\sqrt{Swd}$ as a representative quantity of length, and v_b as a representative quantity of velocity. Therefore, Eqs.(5) and (6) are rewritten in Eqs.(7) and (8) respectively,

$$Fc = \frac{\rho v_b^2}{\Delta \rho g \sqrt{Swd}} , \qquad (7)$$

$$Rc = \frac{\sqrt{Swd} \cdot v_b}{\nu} , \qquad (8)$$

As is evident from Eq.(5) or Eq.(6), Fc and Rc are non-dimensional numbers based on Froude's law and Reynold's law, respectively. In general, it is impossible to satisfy both laws in model scaling of an experiment under liquid of the same quality as a full scale experiment. In the case of an oil boom, Fa is obtained by neglecting a fluid viscosity and is applied in high velocity flow. In the case of the proposed new device as well as an oil boom, it is assumed that a fluid viscosity can be neglected, because oil droplets are generated in high velocity flow. Thus, in this paper, Rc is ignored and only Fc is used as the non-dimensional number having closerelation to the phenomenon.

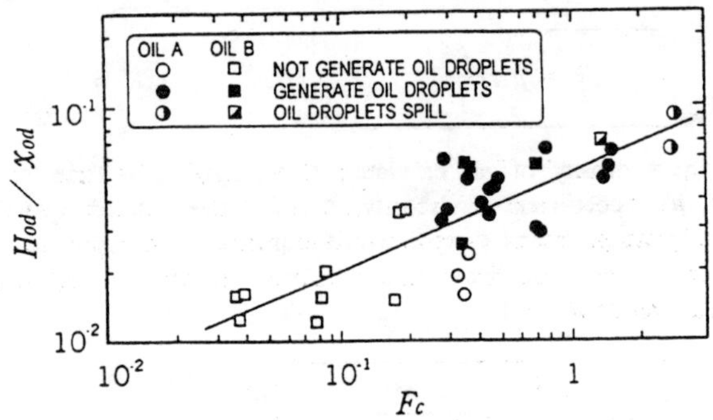

Fig. 11 Relation between interfacial wave height Hod and Froude number Fc.

3.4.2 Limit of generating oil droplets.

The relations between the non-dimensional heights of interfacial wave and Fc are shown in Fig.11. Where, Xod is horizontal distance between the broadside and the curtain at the interface of water and oil. The experimental values are scattered a little because the velocity distributions of flowing water on a vertical section of the circular tank is not uniform. However, it can be explained that the interfacial wave heights closely relate to Fc. In the figure, ● and ■ indicate generating of oil droplets. It is clear that oil droplets are generated when Fc is 0.26 or more. In addition, ◑ and ◪ mean oil droplets pass under the foot of the curtain. In this case, Fc is 1.36 or more, which is considerably higher than Fc in the limit of generating oil droplets. From the above results, the authors conclude that the proposed device should be designed so that Fc should be as lower as possible. To satisfy this condition, it is considered from Eq. (7) that the curtain should have a shape with sufficient Swd to any v_b.

Table 2 Calculation results of minimum flowing velocity to generate oil droplets and maximum flowing velocity to contain oil within proposed device

Items	280,000 DWT	150,000 DWT	80,000 DWT
Quantity of Spilt Oil (m^3)	29,400	16,600	12,000
Length over all (m)	315	263.3	230
Breadth (m)	57	44.5	42
Draft (m)	21	16.7	12.19
Minimum Flowing Velocity to Generate Oil Droplets (Fc=0.26)			
v_b (knots)	18.3	15.9	15.0
v_0 (knots)	5.7	4.9	5.3
Maximum Flowing Velocity to Contain Oil (Fc=1.36)			
v_b (knots)	41.9	36.5	34.4
v_0 (knots)	13.1	11.3	12.1

280,000 DWT
150,000 DWT
80,000 DWT

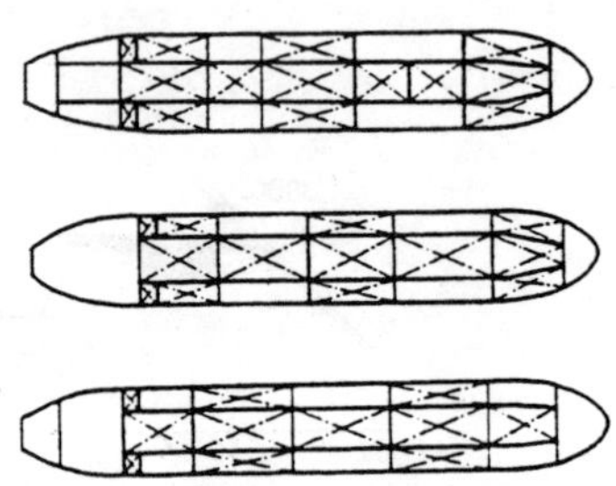

Table 2 shows the minimum velocities to generate oil droplets in "Region A" and maximum velocities to contain oil under the assumption of using the proposed device. Each value is calculated under the following condition.

- The quantity of spilt oil and the ship's dimensions are quoted from the reference (7).
- Each ship's form is replaced by a box shape of the same principal dimensions.
- The shapes of the curtains are the same as them shown in Fig.4-(a) and every dimension of a curtain is scaled up on the basis of the breadth of the ship.
- It was assumed that whole quantity of spilt oil is contained in "Region A".

v_b when generating oil droplets and passing under the curtain are obtained by substituting Fc=0.26 and Fc=1.36 into Eq.(7), respectively. Furthermore, v_0 is obtained from Eq.(9) under the condition that a depth under the foot of the curtain is 1.8Zdb,

$$v_0 = v_b \cdot h_b / h_0 , \qquad (9)$$

where, h_0 : depth of water, h_b : depth under the foot of the upstream curtain. Since a flow of water is accelerated under the ship's bottom when whole quantity of flowing water pass through the region, v_b gets considerably higher than v_0. However, it is estimated in actual grounding that an oil droplet is generated only under higher flowing velocity than that shown in Table 2 because the flow is separated not only vertically but also horizontally under three-dimensional flow. In the case of an oil boom, it was reported that an oil droplet is generated and spills under about 1 knot or more. Even if the difference of the experimental condition is considered, it is clear from Table 2 that the proposed device is more effective than an oil boom.

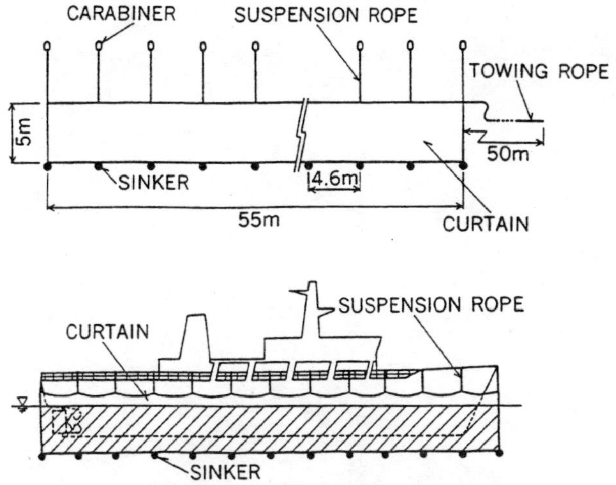

Fig. 12 Experimental apparatus in full-scale experiment

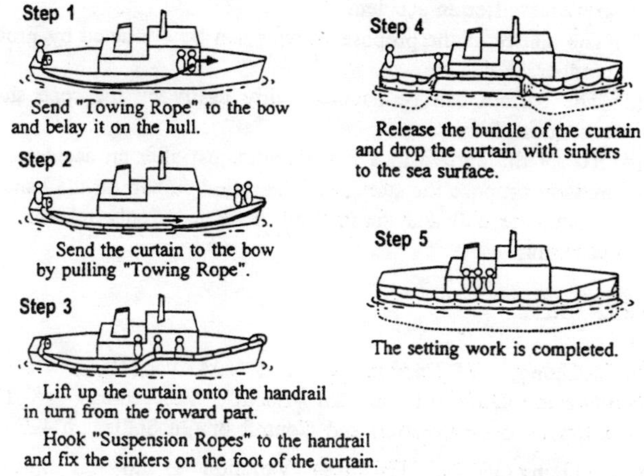

Fig. 13 Procedure of setting curtain in full-scale experiment

4. PRACTICAL METHOD TO SET THE CURTAIN

To investigate a practical method of setting the curtain, some experiments on an actual ship are carried out. As mentioned earlier, a device to prevent spilt oil diffusion should be set as soon as possible. Thus, the experimental result in which the curtain is set for the shortest time will be mentioned.

The model curtain is sketched in Fig.12. The curtain of half length is set only along a broadside of the hull in the experiment although a curtain encloses whole surroundings of a hull in an actual device. Hence, the setting time in the experiment should be doubled in actual use of the device. The experiment is carried out by 3 men. Sea state is smooth and light breeze is blowing.

Fig.13 shows the procedure of the experiment. To improve portability and storing ability of the curtain, it is rolled in a rope shape of 5cm in diameter, is bundled by bands at 1.5m intervals and is wound onto a reel in advance.

In the first step, three workers carry the curtain onto the poop deck which has enough space to work safely. Next, a thin rope called "Towing Rope", whose after end is connected with the fore end of the bundled curtain, is sent to the bow along the broadside by two workers, "A" and "B", and then the fore end of the rope is belayed on the hull. In the third step, the curtain is put out the outboard by "Worker C", and is sent gradually to the bow by pulling the rope.

As shown in Fig.12, the curtain is suspended with several thin ropes called "Suspension Rope". In the initial condition, the top of each rope is joined to the origin of the next rope. In the forth step, the curtain is lifted up onto the handrail by hauling in each "Suspension Rope" in turn from the forward part. Moreover, each top of the rope is detached from the next rope and hooked to the handrail by "Worker B".

To sink the curtain quickly into the water and to prevent rising of the curtain due to the influence of wave and tide, several sinkers are fixed on the foot of the curtain. In the fifth step, sinkers, which have already been arranged at each proper position on the deck, are fixed by all workers. After every sinker has been fixed completely on the curtain, bands of the curtain are released. Finally, by dropping the curtain with sinkers to the sea surface, the set of the curtain is completed. The time required from the beginning of the experiment to the end is about 15 minutes.

From the experiment, it becomes evident that the handling a rolled curtain in a rope shape leads to the efficient setting work although the curtain has a large area and length. Furthermore, the above method is practical in the light of the time required and the non-support. However, there are some risk on the safety in the work under rough weather condition because a worker is required to lean out of a gunwale when he lifts up the curtain or pulls "Towing Rope".

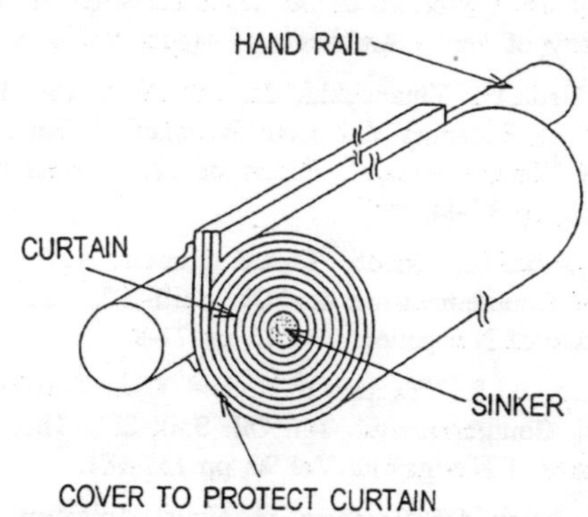

Fig. 14 The most practical storage of curtain to be set rapidly and safely

Thus, a more practical and safer method is proposed. As shown in Fig.14, a rolled curtain with sinkers is wrapped in "Protection Cover" and always mounted on the handrail or the ship's side around the hull. On using this storage, the curtain can be set by only releasing the fastening. Therefore, not only safe and quick work can be expected under rough weather condition, but also highly trained skill is not required in the setting work.

5. CONCLUSIONS

To prevent spilt oil diffusion from a ship's damage, a new device, which encloses a hull with curtains, was proposed. Furthermore, the performance of the device was investigated by model experiments and the setting work ability was evaluated by field experiments on an actual ship. The obtained results are summarized as follows:

(1) To take an effective countermeasure against oil spill, a ship

should have a new device which can be operated by only the crew just after an accident.

(2) The ability of the proposed device can be evaluated by Froude number, *Fc*.

(3) The curtain can be handled fairly easily by the rope shape curtain in the setting.

(4) To set the curtain in a few minutes just after an accident, the authors propose the storage of the rolled curtain on the handrail around the hull and the method to set it by only releasing the fastening.

REFERENCE

(1) J.S.Chung, G.R.Cunningham and J.M.Evon,(1973), "Design Parameter Study of an Oil Boom", Proccedings of 1973 Conference on Prevention and Control of Oil Spills, pp 427-439.

(2) J.S.Chung,(1977), "Thickness Variation of an Oil Spill on Waves", Proceedings of Cavitation and Polyphase Flow Forum-1977, pp15-18.(2) Emori,I. (1985), "Theory and Application of Model Experiments", Gihodo,pp18-39.

(3) Tamiya,S., Kato,H., Watanabe,Y. and Komura,T. (1974), "A Treatment of the Equibrium of Oil Layer on Water Flow", Journal of the Society of Naval Architects of Japan, Vol 135, pp 71-80.

(4) Ueda,K., Ueda,Y., Ymanouchi, H. and Watanabe, K. (1985), "Prevention of Escaping Oil from Mechanical Barriers in the Presence of Water Currents", Papers of Ship Research Institute, Vol.22, No.5, pp 33-44.

(5) Kubo,M., Asaki,K., Onoda, T. and Shibata, A.,(1991), "Basic Reserch on Countermeasure for Oil Spill- I ", The Journal of Japan Institute of Navigation, Vol 85, pp 71-81.

(6) Kubo,M., Asaki,K., Yamamoto,K. and Oki, T.,(1994), "Basic Reserch on Countermeasure for Oil Spill-III", The Journal of Japan Institute of Navigation, Vol 90, pp 157-184.

(7) The Association for Research of Naval Architects of Japan, (1985), "Investigation and Research on large quantity of oil spill from ship", Report No.186R, pp18-39.

(8) Kubo,M., Asaki,K., Yamamoto,K. and Oki, T.,(1995), "Basic Reserch on Countermeasure for Oil Spill-V", The Journal of Japan Institute of Navigation, Vol 94, pp 281-289.

(9) Emori,I.(1985),"Theory and Application of Model Experiments",Gihodo,pp18-39.

Proceedings of the Seventh (1997) International Offshore and Polar Engineering Conference
Honolulu, USA, May 25-30, 1997

ISBN 1-880653-28-1 (Set); ISBN 1-880653-29-X (Vol. I)

In Situ Anaerobic Bioremediation of Fuel Contaminated Ground Water at Seal Beach, CA

Martin Reinhard, Gary D. Hopkins and Eva Orwin
Stanford University
Stanford, CA, USA

Carmen A. Lebron
Naval Facilities Engineering Service Center
Port Hueneme, CA, USA

ABSTRACT

Benzene, toluene, ethylbenzene, *m*-xylene and *o*-xylene (BTEX) have been shown to transform within the anaerobic zone of a historic gasoline spill site. In a series of slug tests, ground water was augmented in situ with different electron acceptors and BTEX compounds and incubated for 60 to 90 days. Under denitrifying conditions, transformation was complete for toluene, ethylbenzene, *m*-xylene and*o*-xylene whereas benzene was stable. Under sulfate-reducing conditions, toluene and *m*- and *o*-xylene were completely transformed whereas ethylbenzene and benzene removals were very slow. Under methanogenic conditions, toluene, m-xylene and o-xylene were completely removed and benzene and ethylbenzene were partially removed.

KEY WORDS

BTEX, Transformation, Bioremediation, Anaerobic, Sulfate Reducing, Nitrate Reducing, Methanogenic.

INTRODUCTION

Since refining and storage facilities of fuel hydrocarbons are often located in coastal regions, groundwater contamination with fuel hydrocarbon compounds is a frequently encountered problem. Remediation of such sites poses significant technical challenges since the quantities of fuel spilled and the areas affected can be large. Interest in intrinsic biotransformation processes has been increasing since it may be less intrusive and more cost effective than conventional remediation technologies such as pump-and-treat. Intrinsic bioremediation is defined as the innate capability of indigenous microbial communities to biodegrade or transform pollutants into benign products without human intervention. Intrinsic bioremediation of fuel hydrocarbon plumes may be aerobic at the fringes of the plume where the supply of oxygen is rapid, and anaerobic inside the plume where oxygen is depleted. The focus of this study was on anaerobic biotransformation processes. Because anaerobic biotransformation of hydrocarbons is still poorly understood, rates are difficult to predict .

Benzene, toluene, ethylbenzene, *o*-,*m*- and *p*-xylene (BTEX) combined constitute over 50 wt% of the water soluble fraction of unleaded gasoline. (Coleman et al., 1984) Laboratory and field research has shown that the biooxidation of aromatics creates a sequence of characteristic geochemical reaction regimes. If oxygen is present, then it will be utilized first, followed by nitrate, sulfate, and carbon dioxide. (Biological oxidation of BTEX compounds with Fe(III) as the electron acceptor was not observed at this site and was not studied in the slug tests.) Microcosm studies have reported aromatic hydrocarbon degradation under denitrifying (e.g. Zeyer et al., 1986, Major et al., 1988, Hutchins et al., 1991, Barbaro et al., 1992), sulfate-reducing (e.g., Beller et al., 1992, Edwards et al., 1992, Edwards and Grbic-Galic, 1992), and methanogenic (Grbic-Galic and Vogel, 1987) conditions. The reported studies indicate that when the BTEX compounds are present as a mixture, the degradation rates tend to be sequential, with toluene most readily degraded under both nitrate- and sulfidogenic-conditions. Benzene was often found to be resistant except in a study by Edwards and Grbic-Galic, (1992) under presumed sulfidogenic-conditions.

There is a significant need to demonstrate that results of small laboratory studies can be extrapolated to the field. So far, few attempts have been made to demonstrate intrinsic anaerobic bioremediation at the field scale under controlled conditions and then only under denitrifying conditions (Patrick et al., 1985 and 1986, Barbaro et al., 1992, Hutchins et al., 1991). This paper summarizes the results of controlled release experiments designed to evaluate anaerobic BTEX biotransformation in situ under nitrate- and sulfate-reducing and methanogenic conditions. Results of the unamended experiments were reported elsewhere (Reinhard et al., 1996 and 1997).

CONTROLLED RELEASE EXPERIMENTS

Controlled release experiments were conducted within the anaerobic plume of a weathered gasoline plume site. The controlled release experiments consisted of slug tests in which approximately 1,000 L of pretreated groundwater was slowly injected into the anaerobic test zone. The slug was augmented with bromide salt as the tracer, salts to readjust the ionic strength, electron acceptors such as nitrate and sulfate as needed, and trace quantities of BTEX compounds. Contaminant behavior was studied by comparing the concentration - time profile of the reactant with that of the conservative tracer and with model predictions. The inner test zone (the slug) was surrounded by a concentric buffer of medium with the same composition as the reaction zone, but without the organics and the tracer. Thus, constant reaction conditions were maintained even during the period of decreasing tracer and organics concentration.

The site hydrogeology is characterized by a small hydraulic gradient, and the plume appears to spread radially. The aquifer consists of sandy/silty alluvial and coastal deposits with some clay lenses (Schroeder, 1991). Inside the plume, sulfidogenic and methanogenic conditions existed in wells as evidenced by the occurrence of black iron sulfide precipitate and methane gas in the headspace of the wells. Outside the gasoline plume, the groundwater was aerobic and the background sulfate, nitrate and bromide concentrations were approximately 85 mg/L, 5 mg/L, and 0.6 mg/L, respectively. Such high sulfate concentrations are expected to be relatively typical for ground waters near the coast due to the impact of the marine environment.

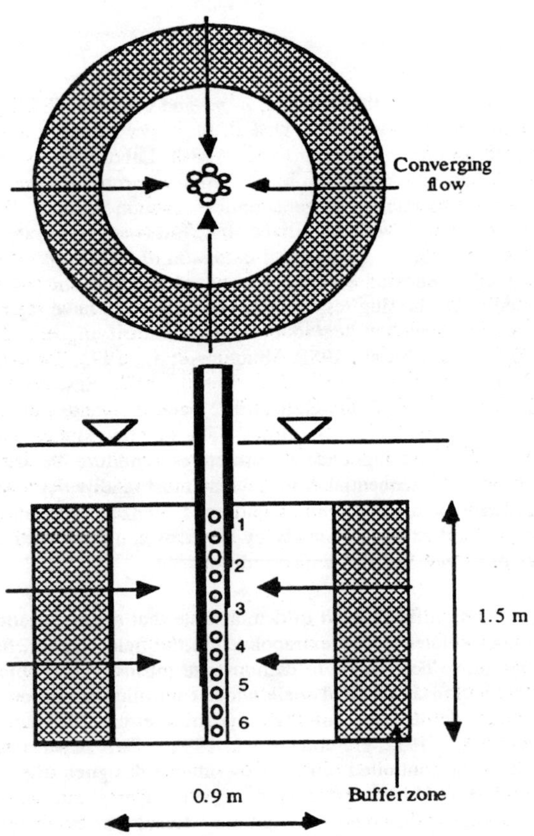

FIGURE 1: Schematic of idealized test zone and sampling device (not to scale). Dimensions given are approximate assuming a slug of 1 m^3. Top: view of converging flow. Bottom: Cross section of test zone. The circles indicate the depths of the individual sampling tubes.

The injection/extraction well used was screened over 1.524 m depth starting from 30.5cm below the water table as shown schematically in Figure 1. The well was instrumented with multilevel samplers consisting of a central 6.35mm stainless steel tube capped at the lower end, and six individual tubes reaching to different depths. Further details are given elsewhere (Reinhard et al., 1997). Samples collected from these multilevel samplers were numbered 1 through 6, with 1 representing the uppermost point and 6 representing the deepest point. Samples collected through the central 6.35mm tube were considered a composite sample. The data discussed here were obtained using the composite samples.

Water used for the designed injection experiments was extracted from a well designated as EO and treated by activated carbon adsorption (GAC) for organics removal and ion exchange columns as discussed elsewhere (Reinhard et al., 1997). Residual dissolved oxygen in the injection water was removed by passing the water through a helium purged gas stripping tower. Chloride salts of calcium, magnesium, and sodium were added back to the injection water to readjust the ionic strength to values similar to the average value measured in the native groundwater. The concentrations of the injection water for the two experiments, the nitrate and sulfate reducing experiments (designated as EO4 and EO6), are indicated in Table 1. For injection, the bromide spiked water was metered into the well at an average flow rate of 1.5 L/min. to minimize pressure mounding during the injection. Sampling was the main driving force for the reversed flow during the "pull back" phase and thus was discontinuous. At weekly or biweekly intervals, 60 L of water was extracted from the composite and each multilevel sampling tube and sent to Stanford University for analysis of bromide, sulfate, nitrate, and BTEX compounds. Approximately 8 h after the injection, a set of samples was taken without purging to determine the concentration in the injection water. These water samples were taken from within the well casing but without contacting the aquifer. Thereafter, the samples were collected at regular intervals with sufficient flushing of the well borehole to produce water from outside the well casing.

TABLE 1. Experimental Conditions for the Nitrate- (EO4) and Sulfate-Reducing (EO6) and Methanogenic (EO8) Experiments.

	Experiment		
	Nitrate Reducing (EO4)	Sulfate Reducing (EO6)	Methanogenic (EO8)
Test Volume (L)	900	750	710
Buffer Volume(L)	948	735	2700
Bromide (mg/L)	58	56	-54
Nitrate (mg/L)	230	0	0
Sulfate (mg/L)	2	45	0.63
Benzene (μg/L)	241	255	268
Toluene (μg/L)	286	259	289
Ethylbenzene	286	296	322
m-Xylene	211	227	218
o-Xylene	250	277	298

A total of ten slug tests (designated as EO1 through EO10) were conducted. The data from these experiments have been summarized by Reinhard et al., (1996 and 1997). During the first three experiments (EO1 through EO3), the BTEX contaminants originated from the contaminated aquifer. Thereafter, BTEX concentrations were too low and BTEX compounds were amended to the injection water at concentrations between 200 and 300 μg/L. Under unamended conditions, i.e. during incubation with ground water, removals were observed only for toluene and the sum of m- and p-xylene. The other BTEX compounds remained stable (Reinhard et al., 1996). A model for the evaluation of the slug test data has been presented by Reinhard et al., (1997).

RESULTS AND DISCUSSION

A model was developed that considered advective-dispersive flow coupled with a zero-order reaction term (Reinhard et al., 1995). Radial advective flow in the direction of the extraction wells was assumed to occur only during sampling. The dispersivity and the regional groundwater flow velocity was obtained by fitting the tracer response curve to the bromide data. The tracer response curves were reproducible in successive tracer tests, indicating that the hydraulic conditions were stable. The longitudinal dispersivity and the regional groundwater velocity were estimated as 5 cm and 0.7 cm/day, respectively (Reinhard et al., 1995).

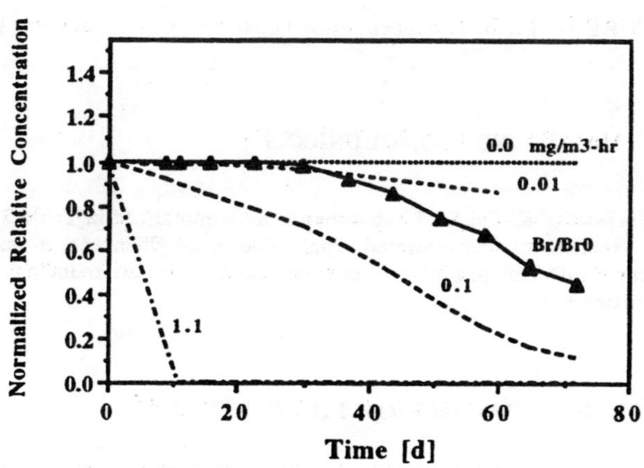

FIGURE 2: Normalized relative concentration of reactive solutes with different zero-order reaction rates (in mg/m3-hr) and tracer concentration normalized to the initial concentration (Br/Br_o).

Figure 2 shows the predicted responses of transforming solutes along with the measured concentration-time profile for the normalized conservative tracer (Br/Br_o). Rates were selected to approximately match results observed in the controlled release experiments. The Br/Br_o data was taken from the nitrate (EO4) experiment and is included to indicate breakthrough of the buffer. For the first 30 days, the tracer concentration remained constant indicating that undiluted slug water was withdrawn. Then the Br/Br_o ratio gradually decreased as the buffer began breaking through and the slug water was diluted. The nitrate experiment was terminated when the Br/Br_o value was below 0.5. In calculating the concentration responses for reactive solutes, zero-order kinetics was assumed. The experimental conditions and zero-order rates (K_b = 0, 0.01, 0.1, and 1.1 mg/m^3-hr) were selected to approximately match the observed BTEX responses in the nitrate and sulfate experiments discussed below. The indicated concentrations are normalized to the initial concentrations of the respective solute (C/C_o) and relative to the normalized tracer concentration (Br/Br_o). Thus, the concentrations shown represent the normalized relative concentrations given by ratio $C_r = (C/C_o)/(Br/Br_o)$. Dividing C/C_o by Br/Br_o accounts for the effects of dispersion.

The model predicts a linear decrease of C_r for compounds that are completely transformed during the undiluted period (the first 30 days). For compounds that degrade more slowly, the C_r decreases linearly during extraction of the undiluted slug. As the dilution becomes more and more significant, C_r decreases faster. For compounds that transform very slowly, the effect of breakthrough on C_r is less pronounced: i.e. for the compound which transforms at 0.01 mg/m^3-hr, the slope of C_r remains relatively small even when the slug is diluted by more than 50 percent. For non-transforming solutes such as bromide (K_b = 0.0 mg/m^3-hr), C_r remains constant at 1. This applies to both the Br tracer and solutes that sorb but do not transform. Thus, the shape of the curve is not influenced by sorption as long as it does not influence the rate of transformation.

Figure 3 shows an analogous plot for the aromatics tested in the first nitrate experiment (EO4). The data shown represents the composite samples. The initial nitrate concentration was 230 mg/L, in excess of the theoretical demand. The nitrate concentrations remained relatively constant during the first 38 days and then decreased at some levels (data not shown). The organics showed a distinct behavior. Toluene, *m*-xylene, and ethylbenzene decreased rapidly and were completely removed by day 30; *o*-xylene decreased more slowly but by the end of the experiment it had also disappeared. Benzene increased initially, probably due to desorption from the aquifer material, but then remained stable. Rapid removal of toluene, ethylbenzene, and *m*-xylene and slow removal of *o*-xylene were also observed in laboratory experiments. The persistence of benzene was consistent with results from laboratory experiments using Seal Beach solids (Ball and Reinhard, 1996).

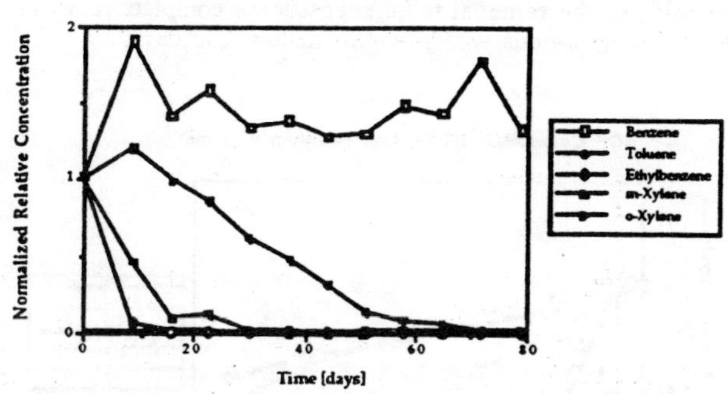

FIGURE 3: Denitrification conditions: Normalized concentrations (C/C_o) of benzene, toluene, ethylbenzene, and *m*- and *o*-xylene.

Figure 4 shows the results of the first controlled release experiment under sulfate-reducing conditions. This experiment lasted 60 days. The removal of aromatics was also sequential, as in the nitrate-reducing experiment, with toluene, o- and m-xylene being removed first and relatively rapidly. Ethylbenzene and benzene were more resistant to degradation. Benzene removal was not evident in the composite, but appeared to be significant at individual levels (Reinhard et al., 1997). The observed slow removal is consistent with results obtained in the laboratory (Edwards and Grbic-Galic, 1992) but should be confirmed in longer term experiments. The fact that toluene was degraded in the test zone without a lag time under presumed sulfidogenic conditions immediately after the denitrifying experiment, indicates that sulfidogenic bacteria remain active even after the exposure to high concentrations of nitrate (230 mg/L). The observed removals are in qualitative agreement with results obtained from laboratory studies (Edwards et al., 1992, Ball and Reinhard, 1996) with few exceptions. In some microcosm studies, toluene appeared to inhibit the degradation of m-xylene, which commenced only after toluene was completely removed. In the field, this inhibitory effect was not observed although toluene was the fastest compound to degrade.

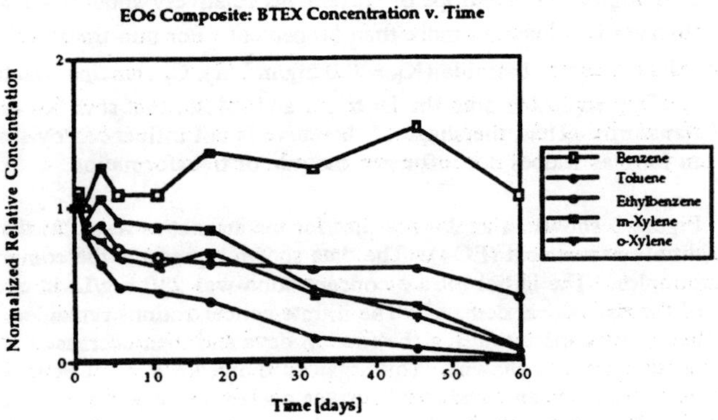

FIGURE 4: Sulfate-reducing conditions: Normalized concentrations (C/C_O) of benzene, toluene, ethylbenzene, and *m*- and *o*-xylene (Composite Data).

Figure 5 shows the removal of the BTEX compounds under methanogenic conditions. Toluene, m-xylene and o-xylene were removed in less than 30 days. Benzene and ethylbenzene were removed slowly at nearly the same constant rate. After 73 days, approximately 50 % of the initial concentration was removed. Extrapolating the removal trend suggests the complete removal of these two compounds within approximately 150 days.

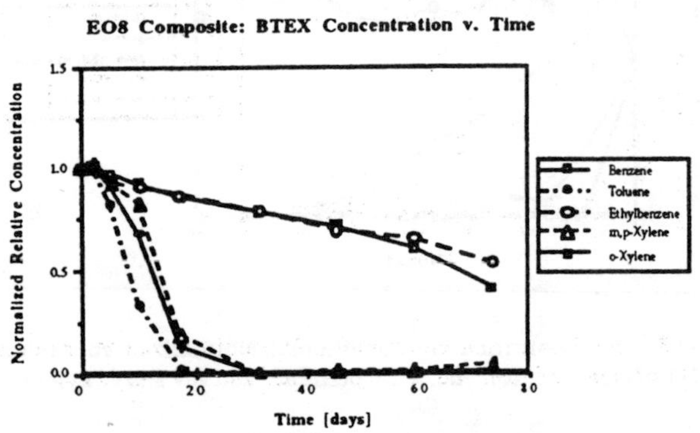

FIGURE 5: Methanogenic conditions: Normalized concentrations (C/C_O) of benzene, toluene, ethylbenzene, and *m*- and *o*-xylene.

However, there are no laboratory results available to corroborate this finding. The formation of methane in the composite and at the different levels is indicated in Figure 6 confirming methanogenic conditions in the test zone. Mass balance considerations indicate that only a small fraction of the methane originates from the BTEX compounds (Reinhard et al., 1996).

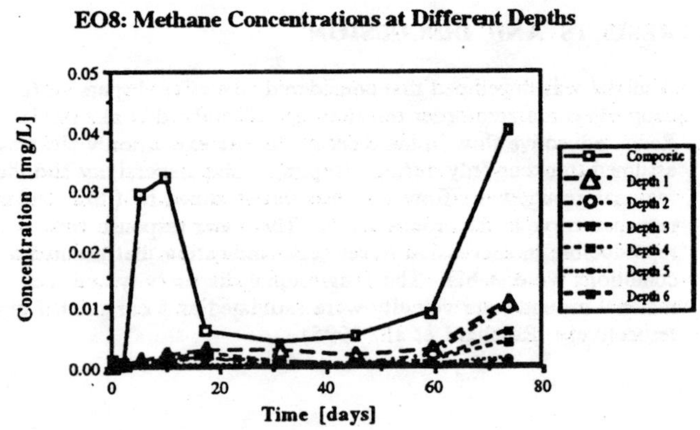

FIGURE 6: Methane concentrations observed in Experiment EO8.

SUMMARY AND CONCLUSIONS

The results of all the field experiments are summarized in Table 2. The experiments demonstrated rapid toluene, ethylbenzene, *o*- and *m*-xylene transformation in situ under nitrate- and sulfate-reducing conditions.

Table 2. SUMMARY OF FIELD RESULTS

Compound	Nitrate-Reducing	Sulfate-Reducing	Methanogenic
Benzene	not observed	very slow, variable from level to level	slow
Toluene	fast	fast	fast
E-benzene	fast	slow	same as benzene
o-Xylene	slow	fast	fast
m-Xylene	fast	fast	fast
p-Xylene	-	-	slow

Key:

very slow: less than 50% removal over 60 to 90 days;

slow: more than 50% removal over 60 to 90 days;

fast: complete removal in less than 30 days;

- : not tested

Benzene removal appeared significant only at individual levels and needs to be investigated in longer-term experiments. In the absence of nitrate and sulfate, methanogenic conditions developed as indicated by the formation of methane. Under methanogenic conditions, all BTEX compounds tested were removed although at significantly different rates. The observed field data obtained in this study lend credence to bioremediation schemes that consider intrinsic anaerobic removal processes.

ACKNOWLEDGMENTS

The author thanks the U.S. Department of Navy, Naval Facilities Engineering Service Center, Port Hueneme, California for their support under a grant No. R-815738-01 through the U.S. Environmental Protection Agency supported Western Region Hazardous Substance Research Center. Also thanks to the Chevron Research and Technology Company. The contents of this study do not necessarily represent the views of these organizations. We thank Lew Semprini, Oregon State University for suggesting the slug test methodology and Dale Lorenzana for technical assistance and Larry Vitale, Special Projects Division, California Regional Water Quality Control Board, Santa Ana Division for his support.

REFERENCES

Ball, H.A., M. Reinhard, E. Orwin, H.F. Ridgway, and D. Phipps (1993). Factors Influencing the Anaerobic Biodegradation of Gasoline Hydrocarbons by Groundwater Bacteria at Seal Beach, CA. Orange County Water District, Fountain Valley, CA.

Ball, H.A. and M. Reinhard (1996). "Monoaromatic Hydrocarbon Transformation under Anaerobic Conditions at Seal Beach, California: Laboratory Studies," *Environ.Tox.Chem.*, Vol. 15, pp. 114-122.

Barbaro, J.R., J.F. Barker, L.A. Lemon, and C.I. Mayfield (1992). "Biotransformation of BTEX Under Anaerobic, Denitrifying Conditions: Field and Laboratory Observations." *J Cont. Hydrol,* Vol. 11, pp. 245-272.

Beller, H.R., D. Grbic-Galic and M. Reinhard (1992). "Microbial Degradation of Toluene under Sulfate-Reducing Conditions and the Influence of Iron on the Process." *Appl. Environ. Microbiol.* Vol. 58, No 3, pp. 786-793.

Coleman, W.E., J.W.Munch, R.P.Streicher, H.P.Ringhand, and F.C.Kopfler. (1984). "The identification and measurement of components in gasoline, kerosene, and No.2 fuel oil that partition into the aqueous phase after mixing." *Arch. Environ. Contam. Toxicol.* Vol. 13, pp.171-178.

Edwards, E.A., Wills, L.E., Reinhard, M. and Grbic-Galic, D. (1992). "Anaerobic Degradation of Toluene and Xylene by Aquifer Microorganisms under Sulfate-Reducing Conditions." *Appl. Environ. Microbiol.* Vol. 58, No3, pp. 794-800.

Edwards, E.A., and Grbic-Galic, D. (1992). "Complete Mineralization of Benzene by Aquifer Microorganisms under Strictly Anaerobic Conditions." *Appl. Environ. Microbiol.* Vol. 58, No.8, pp. 2663-2666.

Grbic-Galic, D. and Vogel, T.M. (1987). "Transformation of Toluene and Benzene by Mixed Methanogenic Cultures." *Appl. Environ. Microbiol.* Vol. 53, No. 2, pp. 254-260.

Hutchins, S.R., G.W. Sewell, D.A. Kovacs, and G.A. Smith. (1991). "Biodegradation of Aromatic Hydrocarbons by Aquifer Microorganisms Under Denitrifying Conditions." *Environ. Sci. Technol.* Vol. 25, No 1, pp. 68-76.

Major, D.W., C.I.Mayfield, J.F.Barker. (1988). "Biotransformation of benzene by denitrification in aquifer sand." *Ground Water,* Vol. 26, pp. 8-14.

Patrick, G.C., C.J. Ptacek, R.W. Gillham, J.F. Barker, J.A. Cherry, D.W. Major, C.I. Mayfield, and R.D. Dickhout. (1985). The Behavior of Soluble Petroleum Product Derived Hydrocarbons in Groundwater. Pet. Assoc. Conserv. Can. Environ., Ottawa, Ont., PACE Phase I, Rep. No. 85-3, pp.70.

Patrick, G.C., Barker, J.F., Gillham, R.W., Mayfield, C.I., and Major, D. (1986). The Behavior of Soluble Petroleum Product Derived Hydrocarbons in Groundwater. Pet. Assoc. Conserv. Can. Environ., Ottawa, Ont., PACE Phase II, Rep. No. 86-1, pp. 59.

Reinhard, M., S. Shang, P.K. Kitanidis, E. Orwin, G.D. Hopkins, C.A. Lebron. (1997). "In Situ BTEX Biotransformation Under Intrinsic and Enhanced Nitrate- and Sulfate-Reducing Conditions," *Environ. Sci. Technol.*, Vol. 31, No. 1, pp. 28-36.

Reinhard, M., Shang, S., Kitanidis, P.K., Orwin, E., Hopkins, G., Lebron, C.A. (1995). Measurement of in situ BTEX Biotransformation Rates in an Anaerobic Gasoline Plume under Different Electron-Accepting Conditions. Technical Report WRC-6; Western Region Hazardous Substance Research Center; Stanford, CA, 1995.

Schroeder, R.A. (1991). Delineation of a Hydrocarbon (Weathered Gasoline) Plume in Shallow Deposits at the U.S. Naval Weapons Station, Seal Beach, California. U.S. Geological Survey.

Zeyer, J., E.P. Kuhn, and R.P. Schwarzenbach. (1986). "Rapid Microbial Mineralization of Toluene and 1,3-Dimethylbenzene in the Absence of Molecular Oxygen." *Appl. Environ. Microbiol.* Vol. 52, No. 4, pp. 944-947.

Proceedings of the Seventh (1997) International Offshore and Polar Engineering Conference
Honolulu, USA, May 25-30, 1997

ISBN 1-880653-28-1 (Set); ISBN 1-880653-29-X (Vol. I)

Laser Heterodyne Imaging for Shallow Water Surveillance

M.L. Riaziat, R.X. Cao and L.C. Wong
Varian Research Center
Palo Alto, CA, USA

R. Bourke
Oceanit Laboratories
Honolulu, HI, USA

Abstract

A laser imaging system was designed and fabricated for optimum performance in highly turbid (less than one meter visibility) coastal, river, and lake waters. The system uses coherence gating and heterodyne detection. The laser source is a compact 683 nm semiconductor laser with an external cavity, modified to improve its coherence properties. The power consumption of the laser is 5W. Imaging was done at the fixed range of two meters in a laboratory tank. The water was made turbid by adding concentrated constituents of brackish natural waters of Kaelepulu Pond in Honolulu. Images of low contrast, high contrast and metallic objects were obtained in waters with attenuation lengths as low as 18 cm (visibility of less than 50 cm). The resolution obtained was approximately 2 mm at the range of 11 attenuation lengths.

Key Words: imaging, under-water, turbid-water, laser, heterodyne

Terminology

The optical power in a beam of light propagating a distance L through water is attenuated exponentially according to Beer-Lambert Law:

$$P = P_0 e^{-(\sigma_s+\sigma_a)L} \qquad (1)$$

where σ_s and σ_a are scattering and absorption coefficients respectively. The inverse of these coefficients are the *scattering length* $L_s = 1/\sigma_s$ and the *absorption length* $L_a = 1/\sigma_a$. Attenuation of a beam of light through water is due to both scattering and absorption. *Attenuation length (AL)* is defined as $AL = 1/(\sigma_s + \sigma_a)$. These numbers are defined for monochromatic light and have a strong wavelength dependence.

NTU or Nephelometric Turbidity Unit is proportional to the amount of light scattered at 90 degrees from a standard illuminating beam. NTU is not a unique measure of scattering in water. Its utility is only in comparing turbidities for any known water constituents and distribution of particles. In experiments reported here NTU can be converted to attenuation length using Figure 3.

Visibility as used here is another measure of water turbidity. It is defined as the maximum distance from the observer at which a standard sized high contrast object (Secchi Disk) can be seen. Visibility is affected by ambient illumination. Daylight illumination is assumed here.

Introduction

The approach to under water optical imaging needs to be tailored to the particular water type of interest. An imaging system that has satisfactory performance in deep ocean may not necessarily function well in turbid coastal waters. Our imaging system was designed specifically for highly turbid waters with visibility of less than one meter. Besides striving to achieve the highest resolution at the longest range, we also placed some emphasis on simplicity, compactness, and reliability of the system as well as low power consumption. These characteristics will allow the system to be operated as an autonomous unit.

The simplest under water imaging systems are cameras using floodlight illumination. They are capable of distinguishing objects up to 2 attenuation lengths (AL) away from the camera in "mildly scattering" sea water [Kulp, et. al., 1993]. In these waters, it is the backscattered light in the volume ahead of the object of interest (common volume) that limits detectability. The reception of this backscattered light by the detector is minimized if laser illumination and synchronous scanning are used. The range increases to approximately 3 to 5 AL [Caimi, 1995] in mildly scattering sea water. In "strongly scattering" waters on the other hand, the entire approach should be reexamined. The wavelength of light can be optimized to give far less scattering. Also, the path taken by light may involve random direction changes through multiple scattering. In this case, reduction of the common volume is helpful, but range improvement beyond that may be needed. Time gating or coherence gating can be used to enhance resolution at increased range [McLean, et. al., 1995]. The system we describe here uses laser illumination and coherence gating, and operates at a wavelength that is more suitable for imaging in the presence of strong scattering.

Water Type Selection

We designed the imaging system to operate in turbid coastal waters where visibility is between 0.5 to 0.8 meters. To simulate coastal water conditions, our tests were conducted in the laboratory using a water tank with anti-reflection coated optical windows. A concentrated Standard Turbidity Substance (STS) was prepared using the natural sediments of a brackish water Hawaiian pond. The particle size distribution of the STS is shown in Figure 1. The turbidity units (NTU) used to characterize the water in our experiments represent the level of dilution of the STS.

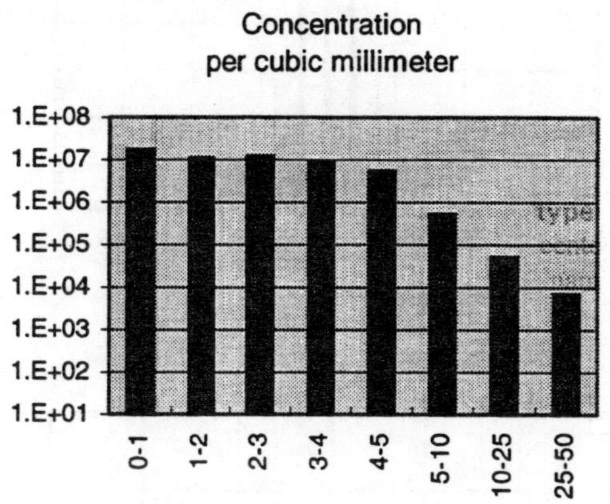

Figure 1. Particle size distribution (in microns) in Honolulu's Kaelepulu pond water from which the standard turbidity substance was prepared for laboratory use.

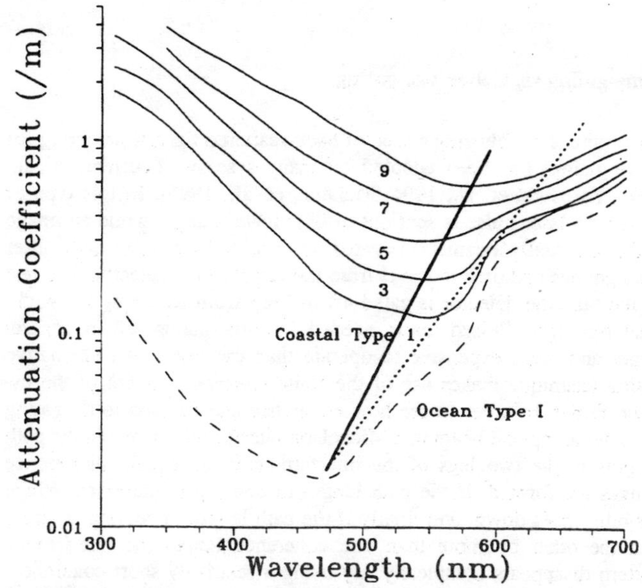

Figure 2. The spectrum of downward irradiance of light in various types of ocean water [Jerlov, 1976]. The wavelength of minimum attenuation shifts to longer wavelengths at higher turbidities. The dotted line shows the more appropriate wavelengths for imaging that take advantage of absorption.

Wavelength Selection

Imaging through turbid water is limited by blurring rather than optical attenuation. In other words, the limiting factor is the scattering length rather than the absorption length (Eqn. 1). For this reason it is not justified to use green or blue green light with absorption lengths exceeding 25 meters when the scattering length is below one meter. Instead it is advantageous to use longer optical wavelengths in order to reduce scattering. Scattering has a strong wavelength dependence: $(a/\lambda)^4$ in case of Rayleigh scattering. For example, deep red light of 700 nm wavelength encounters a Rayleigh scattering cross-section that is lower by a factor of 4 from green light. Scattering due to larger particles outside of the Rayleigh regime is also reduced at longer wavelengths although at a slower rate. Figure (2) shows how the minimum attenuation of light shifts to longer wavelengths in higher turbidity waters. One might suggest that the best operation wavelength at a given water quality is the wavelength of minimum attenuation. This is in fact the case for relatively clear water. However, in highly turbid water it is advantageous to operate at still longer wavelengths where water absorption is higher. Increased absorption discriminates against highly scattered light that does not take the shortest available geometrical path. This helps to sharpen the image and to reduce the background noise.

Based on these arguments it was found that the optimum operation wavelength when water visibility is less than a meter lies close to 700nm. We searched the availability of laser diodes with sufficient power and appropriate coherence characteristics, and found the SDL 7630 diode laser to closely match most of the requirements. This laser operates at 680 nm wavelength, has a nominal output power of 500 mW, and uses 5 watts of electrical power. This laser was chosen for the imaging system and the operation wavelength was fixed at 680 nm which is deep red. Figure (3) shows measured attenuation of deep red and green light as a function of water turbidity.

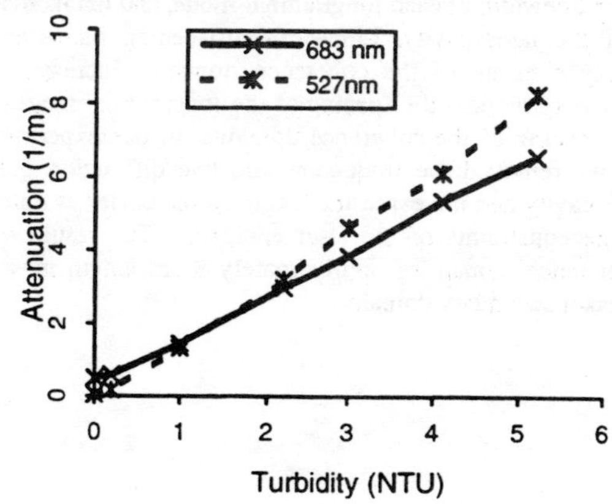

Figure 3. Measured attenuation of red (683 nm) and green (527 nm) light as a function of turbidity. Concentrated samples of water constituents from Kaelepulu pond were used to create the desired turbidity in the laboratory.

Time-gating vs. Coherence-gating

To minimize the blurring effect of back-scattered light, time-gating (or range-gating) has been adopted by many systems [Fournier, et. al., 1994, Klepsvik, et. al., 1994, McLean, et. al., 1995]. In this type of imaging a laser pulse is sent out to illuminate a target while an image detector (usually a camera) is gated on to record the image only after the light has had time to travel from the target to the detector. The rest of the time the detector is gated off to keep from recording the back-scattered light. Pulsed lasers needed for time-gating are in general larger and more expensive to operate than cw lasers. An alternative gating technique makes use of the finite *coherence length* of the cw laser output. In order to see how coherence can be used to do gating consider an optical beam in a Michelson interferometer. When the path lengths in the two legs of the interferometer are equal, interference fringes are formed. If the path length in one leg is increased, fringe visibility goes down, and finally if the path length in one leg is larger that the other by more than one coherence length the interference pattern disappears completely. By using a relatively short coherence-length laser a gating effect similar to time-gating can be achieved. Optical interference occurs with light that takes the desired path length, but not with scattered light that deviates from this path. This is the approach that we adopted in order to use the more compact and less expensive cw lasers. The actual gating is done by heterodyne detection itself and no extra switching is needed. In fact, with heterodyne detection, the source laser can be switched between a pulsed laser and a short coherence-length cw laser without any other modification to the system.

Coherence Length of the Laser

A laser with multiple longitudinal modes does not have a well defined coherence length. Instead, it has multiple domains of coherence or "fringe visibility" in interference experiments [Alaruri 1993]. Figure (4) shows a typical external cavity laser diode output spectrum and its corresponding coherence characteristics as a function of distance, which is the Fourier transform of the output spectrum. Since it is desirable for the laser to have a single coherence domain, it is possible to lower the effect of the secondary domains. This can be done by (1) broadening the linewidth of each longitudinal mode, and (2) increasing the length of the laser cavity. Linewidth broadening increases the envelope decay constant of the coherence domains. Increasing the length of the cavity reduces the spacing of the longitudinal modes and increases the spacing of the coherence domains. In our experimental arrangement we removed the frequency selective diffraction grating from the laser cavity and increased the length of the cavity as much as allowed by the constraints of the laser enclosure. The result was a dominant coherence domain of approximately 3 cm length in water, and much weaker secondary domains.

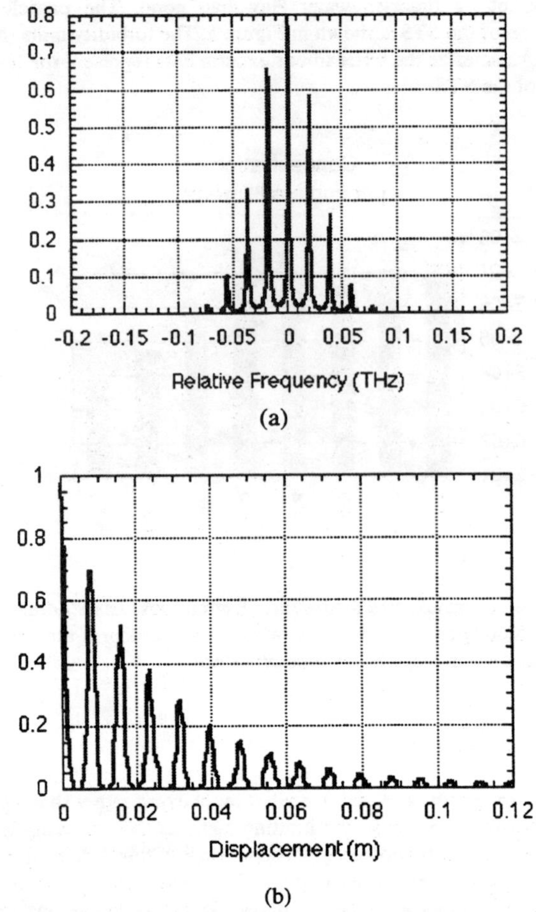

Figure 4. A typical laser diode output spectrum (a) and the corresponding fringe visibility vs. path length difference in a Michelson interferometer (b).

Balanced Heterodyne Detection

In heterodyne detection the received optical signal is mixed with a reference beam (LO) at an offset frequency, and the difference frequency (IF) is detected. The mixing is done in a 50% power divider (3-dB coupler) with two outputs terminated in photodetectors. If the power division is equal and the divider is lossless the two IF outputs will be 180 degrees out of phase. The two outputs are combined in a pair of balanced detectors followed by a transimpedance amplifier. An important advantage of the balanced detection is that is suppresses the noise due to laser power fluctuations.

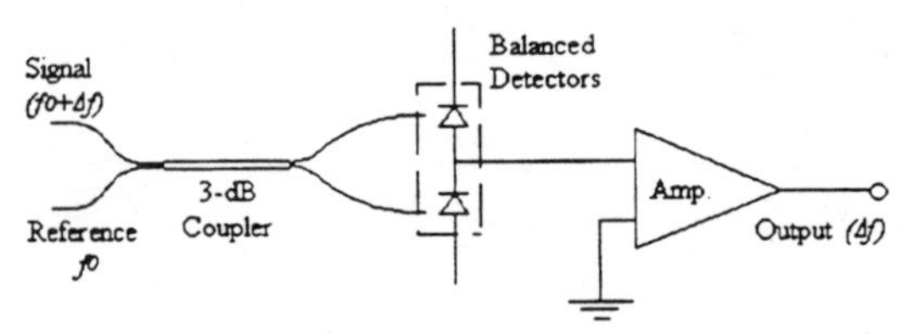

Figure 5. The schematic diagram of a balanced heterodyne detection system.

Heterodyne detection is superior to direct detection from a signal to noise point of view. It is capable of achieving at least twice the signal-to-noise ratio than shot-noise limited direct detection using the same bandwidth. This factor becomes much greater if the detector's thermal noise, dark current, or background radiation are significant [Pratt 69, Riaziat 96 Okoshi, 82, Abbas, et. al., 85, Yariv, 85]. Since heterodyne signal is the geometric mean of the test signal and the reference signal, it is capable of "amplifying" the signal above thermal noise and dark current, by boosting the reference signal. This capability of always operating in the shot-noise limited regime does not exist with direct detection. Responding to the geometric mean of the test signal and the LO translates into increased dynamic range for heterodyne detection. Also, the narrow detection bandwidth inherent in heterodyne, essentially eliminates the problem of background radiation.

Experimental Results

The schematic diagram of the experimental system is shown in Figure 6. The laser source is a tunable semiconductor laser diode with an output wavelength of 683 nm. Maximum measured output power is 420 mW. The dimensions of the laser are 17X16X3 inches, and its power requirement is 2.3 A at 2.2 V. The output of the laser is injected into a single mode optical fiber directional coupler which branches off 5 percent of the power into a reference beam used as the heterodyne local oscillator signal. The main beam passes though a Bragg cell and gets upshifted in frequency by 80 MHz and then impinges on the test target through a camera lens. The reflected light is collected by another camera lens and injected into a 3 dB directional coupler. The reference beam is frequency upshifted by 110 MHz using a second Bragg cell. It then passes through a variable delay, and is mixed with the signal beam in the 3-dB single mode optical fiber coupler. The path length of the reference beam is adjusted to be equal to the optical path of the main beam through the water.

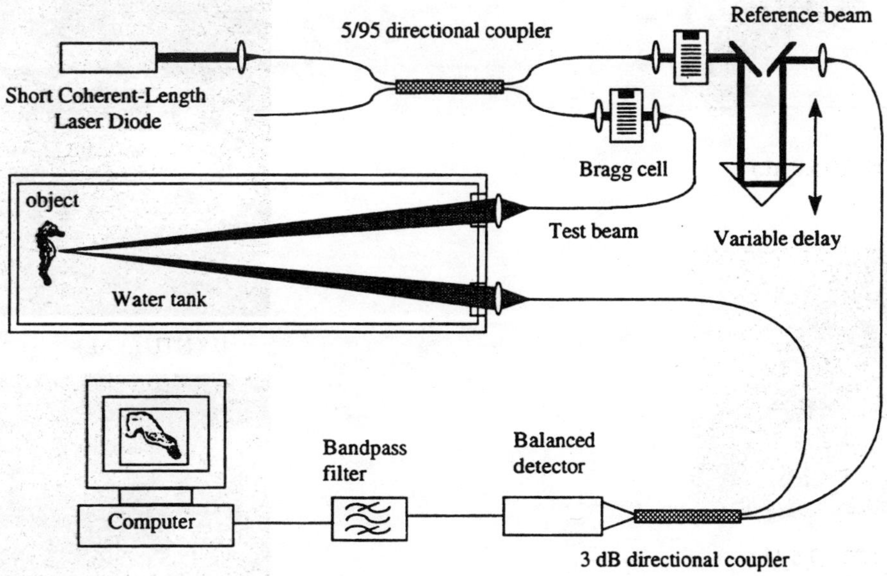

Figure 6. The schematic diagram of Varian's heterodyne coherence gated laser imaging system.

A beat frequency of 30 MHz is generated in a balanced photodetector / preamplifier arrangement followed by a narrowband detector (spectrum analyzer). The signal beam and the reference beam are both frequency shifted in order to avoid electrical and optical interference.

The 30 MHz beat frequency is observed if light received from the test object has not lost its coherence with the reference beam. In this way, the heterodyne system performs range gating, and discriminates against the incoherent light and light that is scattered away from the incident beam. The desired range is set by the variable delay line.

Figure 7. Shows images obtained in the laboratory at various water turbidity levels. The target was a rubber stamp with dimensions of 1.0 by 1.0 inch. The rubber stamp was chosen as a representative target that has no highly reflective features. The low contrast of the rubber stamp puts realistic constraints on the imaging system. The system has no laser beam scanning capability, instead the target was scanned by an automated 2-axis mechanical motion controller and was illuminated by a fixed laser beam.

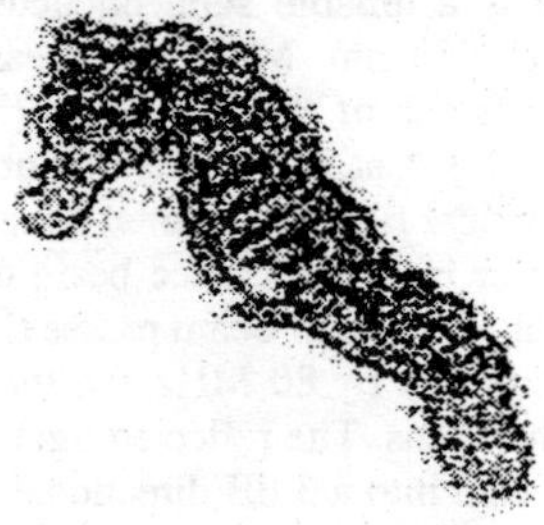

0.88 NTU (2.4 AL)

2.0 NTU (5.8 AL)

3.0 NTU (7.6 AL)

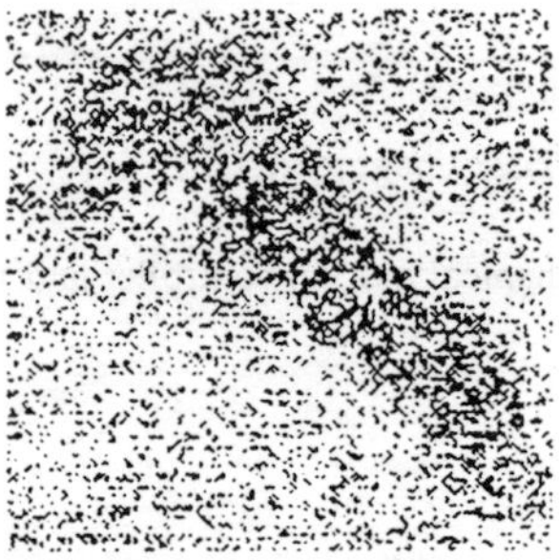

4.0 NTU (10.2 AL)

Figure 7. Scanned two-dimensional images of a rubber stamp target at the range of two meters in waters with various turbidities. The size of the target is 1.0X1.0 inch. The images are 125X100 pixel each with no image enhancement.

The target range was 2 meters, and the angle between the projection beam and reception beam was 12 degrees. The data acquisition was done by a computer controlled spectrum analyzer. The data acquisition software was designed on a LabView graphic software package. At the turbidity of 4.0 NTU the visibility in the water we used was 50 cm and the attenuation length was 20 cm.

0.8 NTU (2 AL)

3.6 NTU (9 AL)

Figure 8. A highly reflective object and its image in clear and turbid water. Note that at 3.6 NTU the portion of the field that reflects the light directly toward the detector is very clear while the rest of the field is dark.

Figure 8 shows the image of a highly reflective object (coin) at two different turbidities. The areas of the object that reflect light directly toward the detector are distinguished clearly even in high turbidity, while the other areas of the object remain dark.

The third target used in the imaging experiments is an AIIM standard resolution pattern made by Applied Images, Inc. It consists of black stripes on a white background. The stripe dimensions imaged here are 0.7, 1.1 and 2.0 millimeters. Figure 9 shows the images of the resolution pattern in various turbidity waters.

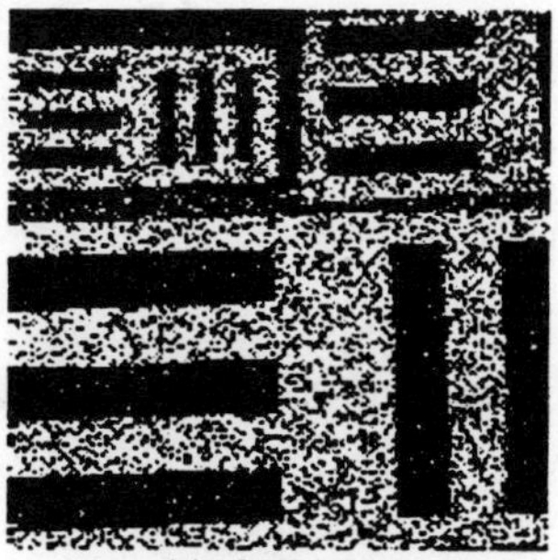

(a) clear water

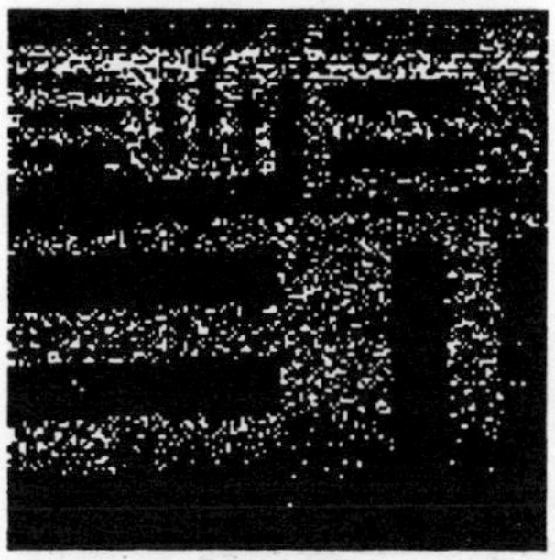

(b) 11 AL

Figure 9. Images of the resolution target in (a) clear water and (b) in water with a turbidity of 4.2 NTU. The stripe dimensions are 0.7, 1.1 and 2.0 millimeters. Note that the imaging range at the turbidity of 4.2 NTU is 11 attenuation lengths.

Conclusion

A heterodyne coherence-gated system for imaging in turbid coastal waters was designed and a prototype was fabricated. Images were obtained with a resolution of 2 millimeters at the range of 11 attenuation lengths. The system is compact and has low power consumption. The power requirement for the laser is five watts. The system is particularly suitable for imaging in waters with less than one meter of contrast visibility.

Acknowledgments

This work was funded by the National Defense Center of Excellence for Research in Ocean Sciences (CEROS). CEROS is part of the Natural energy laboratory of Hawaii Authority (NELHA), an agency of the Department of Business, Economic Development & Tourism, State of Hawaii. CEROS is funded by the Defense Advanced Research Projects Agency (DARPA) through grants to NELHA. This report does not necessarily reflect the position or policy of the Government, and no official endorsement should be inferred.

References

Abbas, GL, Chan, VWS, and Yee, TK (1985) "A Dual -Detector Optical Heterodyne Receiver for Local Oscillator Noise Suppression", Vol. LT-3, No. 5, pp. 1110-1122.

Alaruri, S (1993) "Experimental Method for Determining the Coherence Length of CW Lasers Using a Michelson Interferometer", Journal of Laser Applications, Vol. 5, No. 2, pp. 33-36.

Caimi, F(1995) "Technical Challenges and Recent Developments in Underwater Imaging", SPIE Vol. 2383, pp. 408-418.

Fournier, GR, Bonnier, D, Forand, JL (1994) "LUCIE-Laser Enhanced Underwater Camera", Sea Technology, pp. 55-59.

Jerlov, NG (1976) *Marine Optics.* Elsevier Scientific Publishing Co., New York.

Klepsvik, JO, Bjarnar, M, and Brosstad, PO (1994), "A Novel Laser Radar System for Subsea Inspection and Mapping", IEEE-Oceans 94, pp. II-700-705.

Kulp, TJ, Garvis, D, Kennedy, R, Salmon, T, and Cooper, K (1993) "Development and Testing of a Synchronous-Scanning Underwater Imaging System Capable of Rapid Two-Dimensional Frame Imaging", Applied Optics, Vol. 32, No. 19, pp. 3520-3530.

McLean, EA, Burris, HR, and Strand, MP (1995) "Short-Pulse Range-Gated Optical Imaging in Turbid Water", Applied Optics, Vol. 34, No. 21, pp. 4343-4351.

Okoshi, T (1982), "Heterodyne and Coherent Optical Fiber Communications", IEEE Trans. MTT-30, No. 8, pp. 1138-1148.

Park, H, Chodorow, M, and Kompfner, R (1981) "High Resolution Optical Ranging System", Applied Optics, Vol. 24, No. 14, pp. 2389-2394.

Pratt, WK (1969) *Laser Communication Systems,* John Wiley and Sons, Inc.

Riaziat, ML (1996) *Introduction to High Speed Electronics and Optoelectronics*, John Wiley and Sons, Inc., New York.

Swartz, BA (1994) "Laser Range Gate Underwater Imaging Advances", IEEE-Oceans 94, pp. II-722-727.

Yariv, A (1985) *Optical Electronics* 3rd Ed., Holt, Rinehart and Winston, Inc.

Yoo, KM, Liu, F, and Alfano, RR (1991) "Imaging Through a Scattering Wall Using Absorption", Optics Letters, Vol. 16, No. 14, pp. 1068-1070.

Proceedings of the Seventh (1997) International Offshore and Polar Engineering Conference
Honolulu, USA, May 25-30, 1997

ISBN 1-880653-28-1 (Set); ISBN 1-880653-29-X (Vol. I)

An Impressive Reduction in False Diffusion for Multidimensional Phenomena

Takaaki Shigematsu and Kazuki Oda
Osaka City University
Osaka, Japan

Masahide Takeda
Toa Construction Co.
Japan

ABSTRACT

In calculating the convection terms in conservation equations, upwind differences are sometimes used. It is pointed out, however, that objections based on the appearance of a false or numerical diffusion often arise especially in the presence of flow oblique to the grid lines. In this paper the improvement of the discretization method of the convection terms is proposed and some calculations are performed to validate the present method under a simple condition. Moreover, the present method is applied to the diffusion problem of muddy water around a sheet barrier.

KEY WORDS: False Diffusion, Convective Transport, Upwind Difference, Numerical Calculation

INTRODUCTION

In order to discuss the influence of human developments upon the environment, correct predictions of the transport of substances, e.g. dissolved oxygen, suspended sediment, oil leakage from a ship, etc., involving fluid flow is required. The transport of the substances is usually predicted by numerically solving the convective diffusion equation. The numerical solution of the convection-dominated transport equation is plagued with several problems including instability, unrealistic oscillations and numerical diffusion.

In approximating the convective diffusion equation, upwind differences are sometimes used because numerical stability and physical reasonableness can be obtained. But the upwind scheme suffers from severe false diffusion when the flow direction is not parallel to the grid direction as described by Patanker(1980). deVahl Davis and Mallinson(1972) indicated that the false diffusion of a first-order upwind scheme in the case of a square grid is maximum when $\alpha = 45^\circ$ (α =angle between velocity vector and co-ordinate axes).

In this paper, first, the concept of a first-order upwind scheme is described and the reason why false diffusion caused from the angle between the flow direction and the grid axes occurs is presented. Then, an improved first order upwind method without damaging simplicity of the upwind scheme is developed. A simple calculation is performed and achievement of the reduction of the false diffusion by using the presented method for the severest condition is indicated. Finally, the presented method is applied to the simulation of the diffusion problems around a sheet barrier.

CONCEPT OF THE UPWIND SCHEME

The convective terms of the convection diffusion equations represent the transporting of the substance upstream to downstream by fluid flow. In discretizing the convective terms, the flow direction at a cell should be taken into consideration. The estimate of the convective terms for the convection diffusion equation in one dimension by a first order upwind scheme is

$$\frac{\partial uC}{\partial x} = \frac{1}{2\Delta x}[(u_E + |u_E|)C_i + (u_E - |u_E|)C_{i+1} - (u_W + |u_W|)C_{i-1} - (u_W - |u_W|)C_i] \quad (1)$$

where u is flow velocity and C is concentration of a substance (see **Fig.1**). According to eq.(1) the concentration flux for inflow and outflow at each face of the cell can be calculated reasonably with direction in which a substance is to be transported.

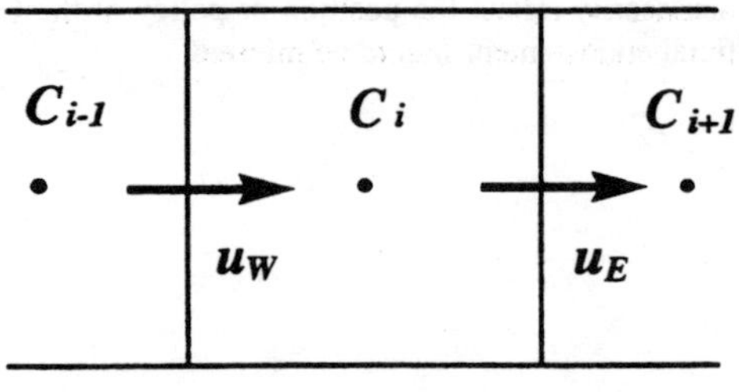

Fig. 1 First Order Upwind Scheme

But in applying the upwind scheme for multidimensional problems the transport direction becomes uncertain. For a pure con-

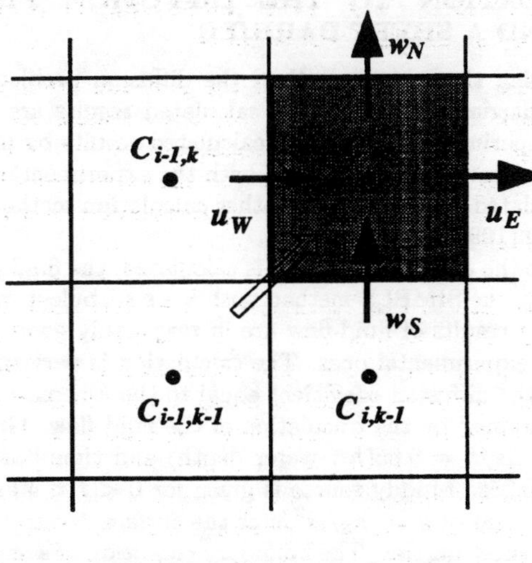

Fig. 2 Two Dimensional Upwind Scheme

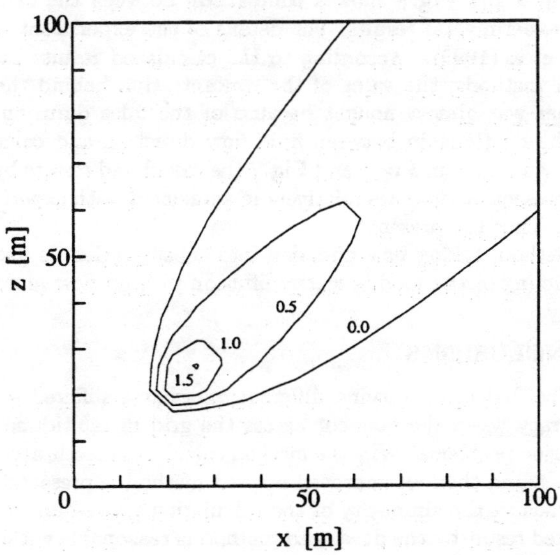

Fig. 3 Calculated Result by First Order Upwind Scheme

vection problem in two dimension, the discretized equation based on the upwind scheme is

$$
\begin{aligned}
&\frac{C^{n+1} - C^{n}}{\Delta t} \\
&+ \frac{1}{2\Delta x}[(u_E + |u_E|)C_{i,k} + (u_E - |u_E|)C_{i+1,k} \\
&- (u_W + |u_W|)C_{i-1,k} - (u_W - |u_W|)C_{i,k}] \\
&+ \frac{1}{2\Delta z}[(w_N + |w_N|)C_{i,k} + (w_N - |w_N|)C_{i,k+1} \\
&- (w_S + |w_S|)C_{i,k-1} - (w_S - |w_S|)C_{i,k}] = 0 \qquad (2)
\end{aligned}
$$

where u and w are the flow velocity components and u_E, u_W, u_N, u_S are defined as shown in **Fig.2**. For the sake of convenience the discussion for $u = w > 0$ will be developed here. In this case the direction of fluid flow is represented by $\Longrightarrow$ in **Fig.2**. Namely the cell $(i-1, k-1)$ is physical upstream cell of (i, k).

The transport of a substance along the stream direction, however, can never be calculated directly by eq.(2) because eq.(2) doesn't include $C_{i-1,k-1}$. By eq.(2) the inlet fluxes from the cell $(i-1,k)$ and the cell $(i,k-1)$ are calculated, which means that a substance in the cell $(i-1, k-1)$ is transported to the cell $(i, k-1)$ and $(i-1, k)$ and then transported to the cell (i, k). This calculation process causes the false diffusion in multidimensional problems.

A calculated result of the pure convective diffusion equation by the upwind difference is shown in **Fig. 3**. The calculation for the case where the source is located in the cell including $x = z = 22.5$ m is performed under $u = w = 1$ m/s, $\Delta x = \Delta z = 5$ m (grid 20 × 20) and the emission source ratio $q = 1\text{g/m}^2/\text{s}$. It is found from **Fig. 3** that the concentration profile becomes wider and the value becomes smaller with transporting downstream due to the false diffusion.

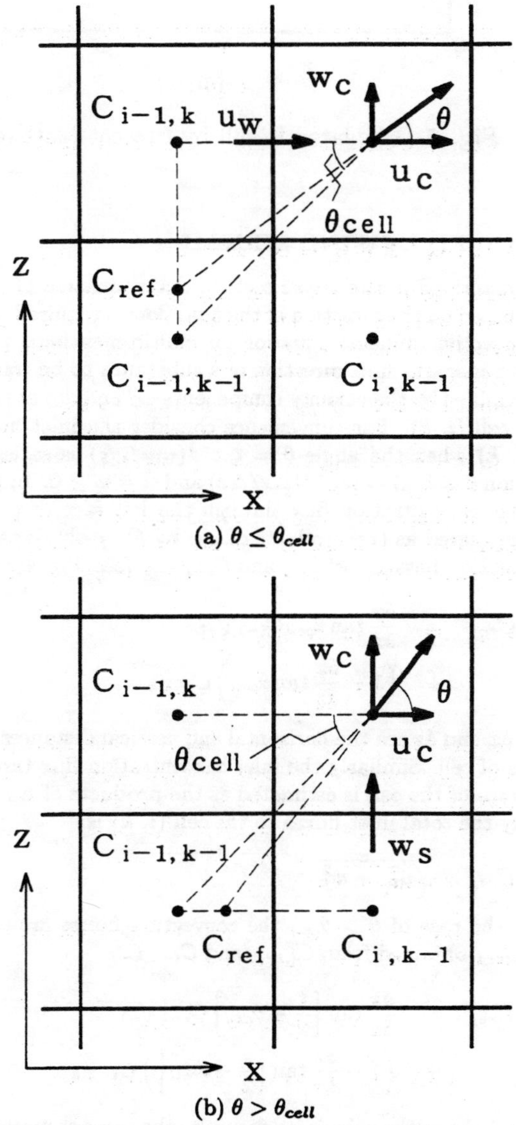

Fig. 4 Definition of Transport Direction of a Substance

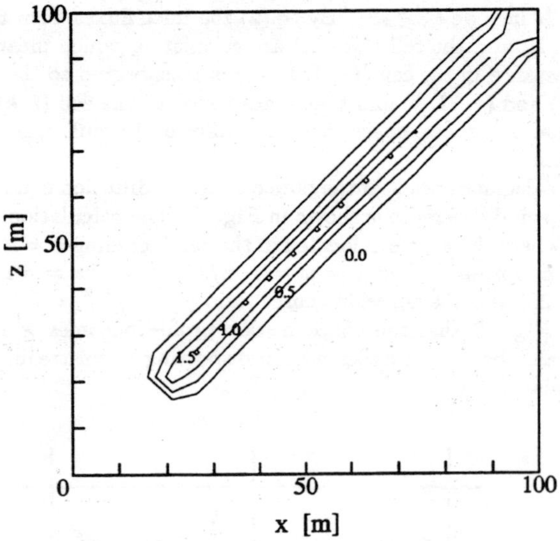

Fig. 5 Calculated Result by Present Method

IMPROVED UPWIND SCHEME

As described in the above section, determination of upstream nodes based on the direction of the fluid flow is required in solving the convective diffusion equation for multidimensional problems. In this paper, the flow direction of a substance to be transported is determined by the velocity components u_C and w_C at the center of the cell $(i,\ k)$. For convenience consider the inlet flux to the cell $(i,\ k)$ when the angle $\theta = \tan^{-1}(u_W/w_S)$ is satisfied with condition $\theta \leq \theta_{cell} = \tan^{-1}(\Delta z/\Delta x)$ and $u = w \geq 0$. In this case, the inlet concentration flux through the left face of the cell $(i,\ k)$ is estimated as the products of u_W by C_{ref} obtained by liner interpolation between $C_{i-1,k}$ and $C_{i-1,k-1}$ (see **Fig.4**(a)).

$$\begin{aligned} C_{ref} &= \frac{\delta x}{\delta z}\tan\theta_{cell}C_{i-1,k-1} \\ &+ \left(1-\frac{\delta x}{\delta z}\tan\theta_{cell}\right)C_{i-1,k} \end{aligned} \quad (3)$$

where δx and δz are the horizontal and vertical distances between centers of cell. Similarly the inlet concentration flux through the down face of the cell is estimated as the products of w_S by C_{ref}. Namely the total inlet fluxes to the cell $(i,\ k)$ is

$$C_{ref} \times \sqrt{u_W^2 + w_S^2} \quad (4)$$

For the case of $\theta > \theta_{cell}$ the convection fluxes are estimated from C_{ref} obtained from $C_{i,k-1}$ and $C_{i-1,k-1}$.

$$\begin{aligned} C_{ref} &= \frac{\delta z}{\delta x}\tan\left[\frac{\pi}{2}-\theta_{cell}\right]C_{i-1,k-1} \\ &+ \left(1-\frac{\delta z}{\delta x}\tan\left[\frac{\pi}{2}-\theta_{cell}\right]\right)C_{i-1,k} \end{aligned} \quad (5)$$

Fig.5 shows the calculated result by the present method under the same calculation condition as the first-order upwind scheme above. From **Fig.5** it is found that a substance is transported diagonally along the flow direction without numerical diffusion.

It is noted that mass is perfectly conserved by the proposed method.

APPLICATION TO THE DIFFUSION PROBLEM AROUND A SHEET BARRIER

By using the present method the diffusion problems around a sheet barrier are solved and calculated results are compared to the experimental ones. The calculated results by using other calculation methods don't agree with the experimental ones. For the calculated results by using other calculation methods refer to Oda et al.(1994).

Before the diffusion equation is calculated, the fluid flow is calculated by the SIMPLE method with $k-\varepsilon$ turbulent model. The calculated results of fluid flow are in reasonably good agreement with the experimental ones. The calculation is performed as the value of the diffusion coefficient equal to the kinematic eddy viscosity obtained by the calculation of the fluid flow. Grid scale is $\Delta x/h = \Delta z/h = 0.05(h:$ water depth) and time iteration step is $\Delta t = 0.25$s. Muddy source is given for $0 < t < 40$ s with the emission ratio of $q = 1\text{kg/m}^2/\text{s}$ at the surface 1 meter upstream from the sheet barrier. The boundary condition of a sheet barrier is given based on the empirical transformed shape in flow.

Fig.6 and **Fig.7** show a comparison between the calculated and experimental results. For details of the experiment refer to Oda et al.(1993). According to the calculated results by using other methods, the value of the concentration behind the sheet barrier was almost nought because of the false diffusion based on the relationship between fluid flow direction and calculation grid. As shown in **Fig.6** and **Fig.7** the calculated results by using the present method are relatively in agreement with experimental ones. Thus the present method, taking flow direction into consideration, is useful for predicting of the muddy water diffusion in fluid flow more accurately.

CONCLUSIONS

The first order upwind differencing scheme suffered a loss in accuracy when the flow cut across the grid in multidimensional diffusion problems. Why the method calculates inaccurately is described and then an improved upwind method is presented without loosing the simplicity of the calculation procedure. The calculated result by the developed method is reasonable without the false diffusion even for the case where the worst angle between flow direction and grid line exists. Furthermore the effectiveness of the developed method is illustrated by comparison between the calculated and experimental results of a muddy water transport around a sheet barrier.

REFERENCES

de Vahl Davis, G. and Mallinson, G.D. (1972) : False Diffusion in Numerical FLuid Mechanics, Univ. of New South Wales, School of Mech. and Ind. Eng. Rept. 1972/FMT/1.

Oda K., T. Shigematsu, T. Kobayashi and S. Yamase (1993) : Vertically 2-D Experiment on the Diffusion of Suspended Soil around a sheet barrier, JCEE, pp.976-980 (in Japanese).

Oda K., T. Shigematsu and T. Kobayashi (1994) : Numerical Simulation of 2-D Diffusion Equation around a Thin Body, Proc. of the Annual Conf., JSCE, CS49 (in Japanese).

Patanker S.V. (1980):Numerical Heat Transfer and Fluid Flow, Hemisphere Publishing Co. 197p.

Raithby G.D. (1976): A Critical Evaluation of Upstream Differencing Applied to Problems Involving Fluid Flow, Com. Meth. Appl. Mech. Eng. 9, pp.75-103.

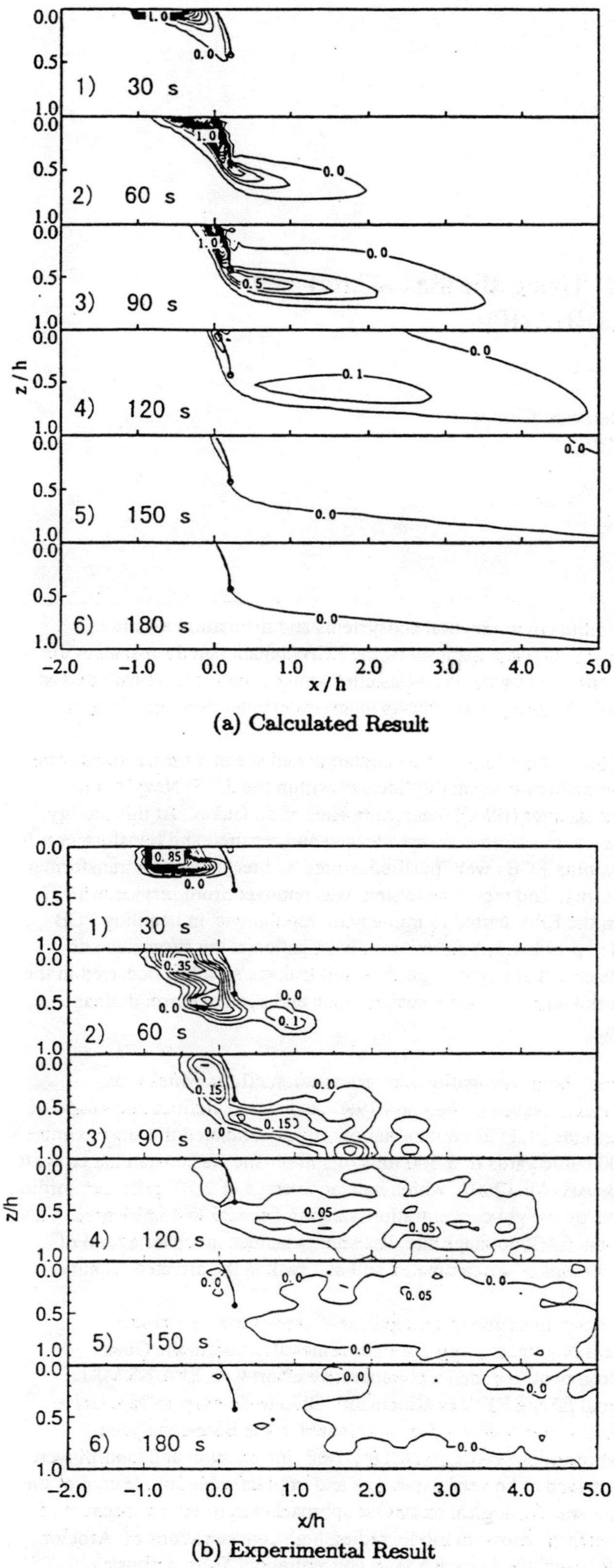

(a) Calculated Result

(b) Experimental Result

Fig. 6 Comparison of Calculated Results with Experimental One (Hanging Type).

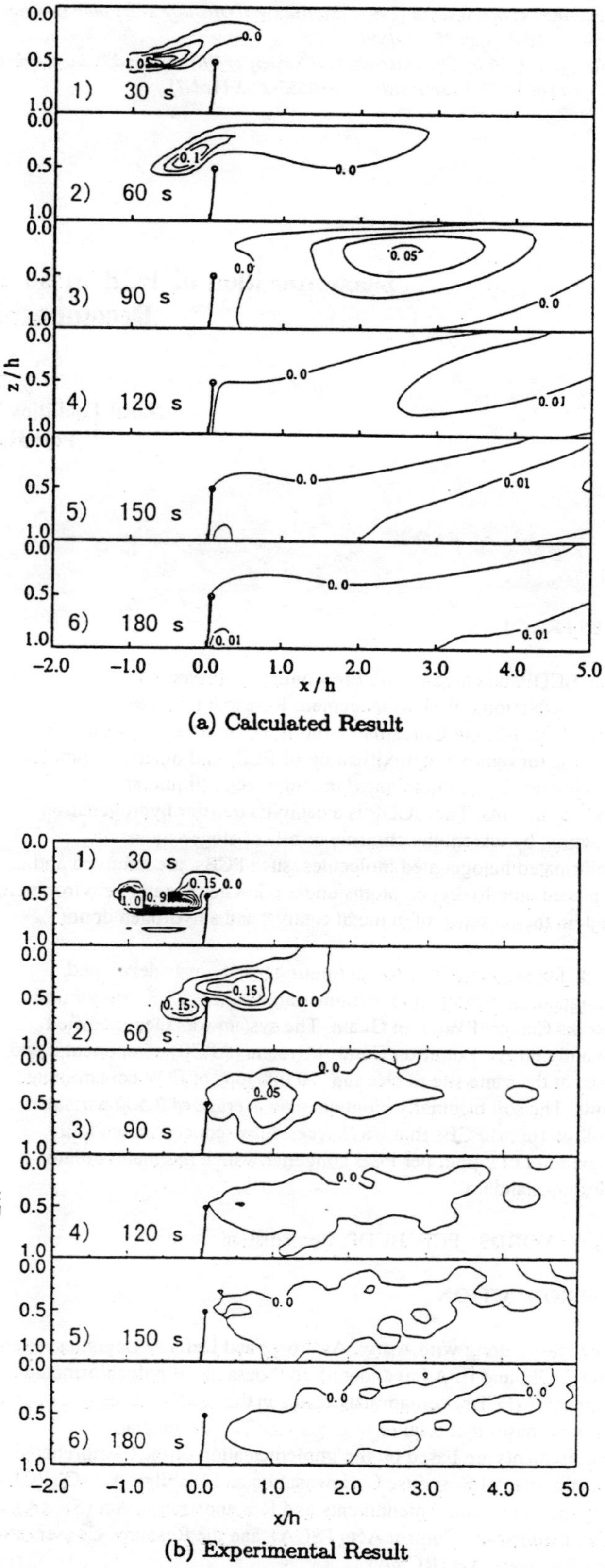

(a) Calculated Result

(b) Experimental Result

Fig. 7 Comparison of Calculated Results with Experimental One (Standing Type).

Proceedings of the Seventh (1997) International Offshore and Polar Engineering Conference
Honolulu, USA, May 25-30, 1997

ISBN 1-880653-28-1 (Set); ISBN 1-880653-29-X (Vol. I)

Decomposition of PCB in Soils/Sediments Using the Base-Catalyzed Decomposition Process (BCDP)

D.B. Chan
Naval Facilities Engineering Service Center
Port Hueneme, CA, USA

ABSTRACT

The BCDP, patented by the Environmental Protection Agency/National Risk Management Research Laboratory (EPA/NRMRL) in Cincinnati Ohio is a cost-effective treatment process for removal/detoxification of PCBs and other halogenated compounds in a contaminated medium, either liquid or soils/sediments. The BCDP is a catalytic transfer hydrogenation reaction, by which, the chlorine or other halogen atoms on chlorinated/halogenated molecules, such PCBs, are removed and replaced with hydrogen atoms under a low-temperature environment and in the presence of a metal catalyst and a hydrogen donor.

A BCDP system of 1-2 ton per hour capacity was developed, designed, fabricated, and demonstrated at the U. S. Navy Public Works Center (PWC) in Guam. The system was then modified (mainly on Air Pollution Control System (APCS) components) and used at the same site to cleanup 10,000 tons of PCB contaminated soils. The soil originally contained an average of 2,500 parts per million (ppm) PCBs that was successfully reduced down to a maximum of 2 ppm per PCB congener, which has met regulator's cleanup standard.

KEY WORDS PCB BCDP Remediation

INTRODUCTION

The Navy, along with Army, Air Force and Defense Logistics Agency and DOE, and EPA have identified thousands of polychlorinated biphenyls (PCBs) contaminated sites in the United States and overseas military bases that require remediation. These remediation requirements are based on the implementation of the Comprehensive Environmental Response Compensation and Liability Act (CERCLA), and the Superfund Amendments and Reauthorization Act (SARA), the Toxic Substance Control Act (TSCA), and the Resource Conservation and Recovery Act (RCRA).

PCBs are known to have significant effect to human health , especially when its impurities/contaminants dioxins and furans are co-existing , and to the environmental ecosystems and the natural resources. Removal and safe disposal of the PCB contaminated substances are thus enforced by the laws (as cited above) and it has constituted as one of the top priority tasks in environmental cleanup program.

The Navy identified a PCB contaminated site at a former transformer repair and reworks facility located within the U. S. Navy Public Works Center (PWC) near Apra Harbor, in Guam. At this facility, electrical transformers were cleaned and repaired and transformer oils containing PCBs were purified, stored and recycled. The transformer oil storage and recycling system was removed from service in 1977 when the EPA started to implement regulations in handling PCB related problems. A follow-on investigation at the suspected site confirmed that a significant PCB oil leakage/spill had occurred in the rework facility's nearby surface zone and a major storm drainage ditch.

Several site investigation/characterization efforts which were conducted between 1983 and 1990 had clearly defined the scope of magnitude of PCB contamination. It was estimated that approximately 4,000 cubic yards or 5,500 tons of soil on-site was contaminated with PCB (Aroclor 1260) , which had an average of 2,500 parts per million (ppm) by weight concentration, ranging from 25 to 45,860 ppm. This original 5,500 tonnage site was later expanded to cover a total of 10,000 tons of contaminated soil and rock in the affected vicinity.

The Navy in search of an applicable, cost-effective, on-site remediation technology for PCB removal in soil in the Guam environment, initiated a collaborative effort with EPA/NRMRL (formal EPA/RREL) in Cincinnati, Ohio to develop an innovative alternative technology that culminated in the Base-Catalyzed Decomposition Process (BCDP). Both incineration and landfill were determined to be very expensive and impracticable for cleanup of the Guam site. Biological treatment approach was ruled out because of its ineffectiveness in biodegrading high concentrations of Aroclor 1260., specially in such a high concentration. Thus, a physico-chemical process, BCDP, became economically attractive to be developed, demonstrated and ultimately being used for remediating the Guam site.

BCDP TECHNOLOGY DESCRIPTION

Catalytic Transfer Hydrogenation

Hydrogenation of an unsaturated carbon structure using hydrogen gas and a metal catalyst is a reaction well known to the chemical industry. The possibility of achieving reduction with the aid of an organic molecule that functions as the hydrogen donor in the presence of a catalyst is a process known as catalytic transfer hydrogenation.

In the hydrogenation reaction, the chlorine atoms on chlorinated molecules are removed and replaced by hydrogen atoms. It is illustrated in an equation as shown below:

$$\text{H-donor} + \text{chlorinated compounds} \xrightarrow[> 300\,^\circ\text{C}\ (572\,^\circ\text{F})]{\text{base catalyst}}$$

$$\text{donor} + \text{dechlorinated compounds} (+\text{H}) + \text{salt (chlorides)}$$

In principle, the donor compounds can be any organic compounds whose oxidation potential is sufficiently low that hydrogen transfer can occur under relatively mild (such as in a low temperature) conditions. This principle is now successfully applied to dehalogenate many toxic environmental contaminants, such as PCBs, dioxins and furans, PCP, DDT, lindane, and many other halogenated compounds of insecticides, herbicides.

The Rotary Kiln Reactor (RKR)

The rotary kiln reactor (RKR) (see Figure 1) provides a first stage treatment of PCB contaminated soil/solids. In case of treating sediments, it must be dehydrated before preceding with the RKR operation. The processed (effluent) product quality meets EPA's TSCA permit standard of 2 parts per million (ppm) per PCB congener. The cleaned soil, exiting from the RKR can be returned to the site from where it was excavated.

In the processing (see Figure 2), the contaminated soil is first screened, sized with a rock crusher and stockpiled for feeding. This is followed by feeding it to a pugmill where a measured quantity of catalyst, sodium bicarbonate ($NaHCO_3$), is uniformly mixed with stockpiled soil (in the amount of 10 % by weight) before being fed to the RKR, in a continuous operation mode. The PCB-laden soil is then heated to 332° C (630° F) for an hour inside the RKR. By this treatment, the PCBs in the soil are either chemically decomposed and/or thermally desorbed. As such, the PCB content in the soil is reduced to less than 2 ppm PCB per congener. By using GC/MS (EPA Standard Procedure 8080 and 8280) to analyze processed soil samples, it had revealed that the RKR process had achieved reduction of PCB congeners (shown in an analytical chromatography) from about 35 peaks (congeners) in the feed down to 4 or 5 peaks in the product soil, and each congener species generally had less than 0.5 ppm in concentration. That was equivalent to having less than 2 to 2.5 ppm total PCB in the product soil.

Generation and Collection of Residues Emitted from the RKR

Residues generated from the RKR consist of dust (particulate), water vapor, and gases. These residues contain residual PCBs and organics from the soil and they must be captured and treated for a safe final disposal.

As shown in Figure 2, the RKR off-gases (carrying the residues) pass through a cyclone, and a baghouse where all the particulates (larger than one micron in size) are removed. The residual vapor, gases and fines are passed through a venturi-(wet) scrubber, then through a demister and passed through carbon adsorbers. The cleaned/purified gases which are free of PCBs are then emitted to the atmosphere. The scrubber liquid (which was not used for recycling) is mixed with activated carbon for removal of remaining organics including PCBs, and then pass through a filter press where all the carbon loaded with the organics are separated (in filter cake form). The filtrate from the filter press is pumped through an activated carbon purifier to complete removal of any residual PCB to less than one part per billion (ppb) in the liquid effluent, making it feasible to be discharged into the on-site sewerage system.
Balances of the residues including particulates, spent carbons, filter cake and some condensate collected are treated on-site with a second stage process reactor, called a Stirred Tank Reactor (STR).

The Stirred Tank Reactor (STR)

The STR, operated in a batch mode is charged with all the residues, dust (particulates) collected from the cyclone and baghouse; precipitate and skimmer from the scrubber liquid settling tank, and the "spent" carbon from exhaust gas treatment, filter cake, and filtrate treatment, and condensate resulting from gas/vapor emission control. In the STR processing, sodium hydroxide (NaOH, in the amount of 2.5 times of PCB load by weight), a high boiling point oil, such as bunker oil #6 (in the amount of 1 to 1 of total residues by volume), a hydrogen donor (in the amount of 1 to 1 of PCB load by weight) and a proprietary catalyst (in the amount of 1 % of residue weight) are added to PCB containing residues. The mixture is continuously stirred while it is being heated to a temperature of 350° C or 662° F and is held at that temperature for 3 hours. Nitrogen gas is continuously pumped into the STR in a very low quantity during the treatment to avoid any potential fire hazard. All PCBs in the residues are destroyed in the STR treatment. The bunker oil is then reclaimed for reuse (about 80 % by volume) in the subsequent STR operation. The accumulated oily waste resulted from the STR operation is being solidified with asphalt and disposed at the processed product soil disposal site as a surface cover to the product soil pile built. The gases emitted from the STR are purified with an activated carbon adsorber, before its being discharged into the air. The treated vapor condensate that contained less than 1 ppb of PCBs, is acceptable for discharging it to the sewer. The resulting spent carbons are either mixed with feed soil and treated in the RKR or treated in the following STR batch operation.

THE BCDP SYSTEM

The BCDP System Description

The BCDP system consisted of a number of equipment that were integrated at the project site. They are described briefly in the following :

1. Feed Preparation Equipment
The excavated soil is crushed in the Eagle AR-125G portable crushing plant to prepare the feed particle to be less than 1.5 inches (or 3.8 cm) size. This plant is capable of crushing up to 20 ton/hr of material. Then

the crushed feed is mixed with 10% by weight sodium bicarbonate in a Kolberg Model 50 portable twin pugmill mixing plant. This plant is also capable of processing up to 20 ton/hr of material. A week's worth of feed could easily be prepared in an 8-hr shift to match with the designed 1-2 ton/hr RKR processing load. A Kolberg Series 12 Electric Feed Conveyor is used to transport the soil/sodium bicarbonate mixture to the inlet of the RKR.

2. The Rotary Kiln Reactor (RKR)
The centerpiece of the BCDP soil treatment system, the RKR, used in Guam is an indirectly fired rotary kiln reactor built by Heyl Patterson, see Figure 1. As described previously the RKR was designed to process 1 to 2 tons per hour of soil (maximum 20% by weight moisture content) and operate continuously for several months without the need for a maintenance shutdown. The RKR includes a 3 feet diameter by 38 feet long (or 0.91 meter dia. @ 11.6 meter long) rotating shell constructed of mild steel, and a furnace surrounding the shell for approximately 29 ft (or 8.8 m) long.. The furnace has 14 Hauck high-velocity burners (ignited with propane gas fire to burn #2 diesel fuel) and can output thermal heat up to 7 million Btu/hr. The soil is heated as it slowly flows through the RKR (at about 1 RPM with a 3-degree tilt). The soil resident time in the RKR is about one hour.

Figure 1. The Rotary Kiln Reactor (RKR)

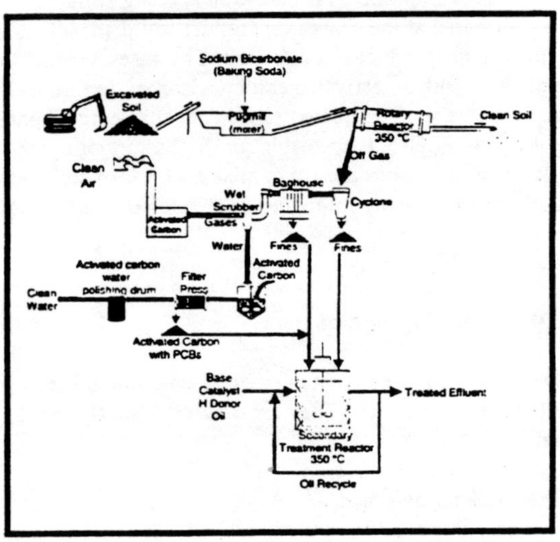

Figure 2. Basic Base-Catalyzed Decomposition Process (BCDP) System Flow Diagram.

3. The Residues (Offgas) Collection and Treatment System
The residues collection and treatment system are where the RKR operation products stripped off from the soil are being collected for further treatment in some supporting processors and ultimately in the STR, the second stage of BCDP processing. An Induced-Draft (ID) fan in the supporting processors maintains a slight vacuum in the RKR, cyclone, baghouse and wet scrubber to prevent any gas, vapor and particulates leakage/release to the environment.

Particulates are removed by the cyclone, baghouse, and wet scrubber. The wet scrubber also condenses water vapor, biomass oils from the soil and any thermally desorbed PCB. A Flex-Kleen Model 16" FILOS cyclone, a Flex-Kleen Model 58-BVBS-36, No. III baghouse containing thirty-six 58 inches (or 1.5 m) Nomex bags and a conventional wet scrubber were used for the collection. An activated carbon canister contained 300 pounds (or 140 kg) of carbon (two operated in series, Nixtox Model N-1500) was used to finally purify the exhaustion gases, prior to its being discharged into the atmosphere. The water used in the venturi scrubber is collected in the base of the unit where it is pumped to a combined carbon and filter press treatment. A SRC media pack system was used for oil/water separator.

4. Improved Residue Collection System (also called Air Pollution Control System (APCS))
Substantial modification/rebuild was made to the original design of residues collection system. It was deemed necessary for having this innovative design in order to operate the BCDP system as a production system instead of a technology demonstration. The modification was aimed to achieve mainly for (a) eliminating the potential of generating dioxins, (b) producing less residues, (c) meanwhile effectively collecting dust/particulates in the submicron range, and (d) generating a cleaner gas emission. The modified system included a steam injection, used as a sweeping gas in the RKR, a multiclone, a quench and a Wet Electrostatic Precipitator (WESP), a primary condenser, a chiller condenser, a High Efficiency Mist Eliminator (HEME) and a larger capacity ID fan and carbon treatment unit for exhausting gas. The modified system has performed so effectively that the gas emission not only met but also exceeded the EPA's PCB Destruction Removal Efficiency (DRE) six nine standard requirements, which is primarily set for incineration.

5. The Stirred Tank Reactor (STR)
As described previously, the STR was designed to treat the particulates collected by the cyclone and baghouse, the organic condensate/sludge collected in the oil/water separator, and the spent carbons from offgas cleaning and scrubber water treatment. The STR was a 400-gallon batch kettle-type reactor heated with an external electric heater surrounding the vessel. As shown in Figure 3, the small amount gas stream exited from the condenser was polished through activated carbon before it is vented to the air. After the reaction has been completed, the reactor was cooled down (operated as a batch system), and the oily residue was transferred to a waste oil storage tank. The treated slurry was allowed to settled for 24 hours so that clarified oil was separated for reuse. From laboratory testing, it was estimated that approximately 80% (by volume) of the oil could be reclaimed for reuse in the subsequent STR runs.

6. Automated Data Acquisition
Many of the critical process parameters during operation of the BCDP system were monitored and recorded with automated data acquisition equipment. They included parameters such as temperatures, tank levels, flowrates etc. Many automated system operation controls and

critical alarm hardware were also added in the production operation time.

THE BCDP SYSTEM OPERATION

Technology Demonstration

In late August 1992, the BCDP system was shipped to PWC Guam and delivered to the demonstration site. Two natural disasters, Typhoon Omar (150 MPH wind) on August 28, 1992 and a major earthquake (8.1 on the Richter scale) on August 8, 1993 had caused the technology demonstration to be delayed until February 1994.

Figure 3. The Stirred Tank Reactor (STR)

Two series of test runs, one in February and the other in November of same year, were conducted for the technology demonstration. The results for each of the test runs are summarized and presented respectively in Tables 1 & 2. Table 1 shows the first BCDP treatment of 15 tons of PCB contaminated soil from the ditch. We used the RKR only, which did achieve the removal of PCB in the soil to meet the EPA standard. The data also shows that most of the desorbed PCB was deposited and concentrated in the cyclone and baghouse dust/particulates. In a rough material balance calculation, as shown in Table 1, there was about 29% unaccounted PCBs presumably being decomposed by BCDP inside the RKR. The figures displayed in the water treatment indicate the effectiveness of carbon treatment. That is, the effluent becomes acceptable for sewer discharge. Sample analysis did indicate however, that there was a very trace amount of dioxins and furans in the RKR feed soil but, there was none in the treated soil. Also as noted in the table, an elevated level of the dioxins and furans was found in the cyclone and baghouse dust, as well as the PCB concentration. It apparently was a result of the stripped off and concentration mechanism, but not through a "new production" inside the RKR because, the low temperature and lack of oxygen in the RKR operation would not have generating them.

Table 2 shows the second series test run results of non-stop treatment operation of about 45 tons PCB contaminated soil. It continued to show that the RKR processing had consistently produced a clean soil. Again, it indicates the presence of elevated concentration of PCB in the cyclone and baghouse particulates sample. This was primarily due to condensation of oily droplets (rich in PCB content) which was generated from the soil treatment in RKR, and condensed within the baghouse. The second stage carbon treatment appears to experience a carbon adsorption break-through difficulty that caused the effluent to contain more than 1 ppb PCB , the sewer discharge limit. This problem was easily solved by using a new set of carbon bed. The observed concentrations or ratios of PCBs, dioxins and furans in the feed, product, and the dust/particulates from cyclone and baghouse during the second series test run were very similar to results of the first test run conducted in February.

The STR was tested during the second series of RKR run. Results verified the chemistry principle used, that the STR was capable in completely treating the PCBs, dioxins and furans containing in the feed. A second series test run of the STR was conducted in May 1995, after some minor system modifications. A new hydrogen donor and one new reactant were introduced during the test run. The particulates, spent carbons and liquid condensate collected from previous RKR operation were homogenized and fed to the second stage's STR treatment. Appropriate chemical reactants were added to residues feed and was heated to a temperature of 670° F, holding for 3 hours while the mixture was continuously stirred. Samples were drawn for analysis when the STR was cooled down to ambient temperature. Three typical run results are tabulated in Table 3, see below. The results reveal also completeness in treating PCB by the STR operation.

Production Operation

The BCDP system in operation for production (site cleanup) was conducted in two periods; the first period during July 1995 to September 1995 and the second period during April 1996 to January 1997 . The first period called rapid start initiated in July 1995 used only partial equipment of modified Air Pollution Control System (APCS). Later, from September 1995 until February 1996 the remaining APCS equipment was received in Guam and installed. Thus, the second period called full scale production operation was started in April 1996 and completed in January 1997. As briefed previously, the APCS modification was deemed necessary and it has since accomplished those previously mentioned four major goals of ; (1) it ensures non-production of dioxins (PCB DRE was 7 nines and the dioxin/furan toxic equivalent was 0.18 nanogram per cubic meter) , (2) it effectively collects submicron particulates including oil droplets, (3) it produces cleaner air emission gases and (4) it reduces total emission rate of the residues from RKR operation.

During both production operation periods, only the RKR and APCS were operated. The STR that would have required as much modification as the APCS to the RKR was abandoned for use. Thus, because of the site specific conditions, especially the low volume of residues production by the RKR (less than 0.4 % by volume), it was determined that it would be more cost effective to ship the residues back to the mainland for final disposal .

Table 1. BCDP Treatment Results for First Test Series (STR was not tested)

Run 1 & 2, performed respectively on February 5, 10 tons/10 hours & February 10, 1994, 5.7 tons/5.25 hours)

A. Soil Treatment		B. Water Treatment		C. Rough Material Balances		
SAMPLE POINT	PCBs CONCENTRATION, ppm *	SAMPLE	PCBs CONCENTRATION	STREAM	TOTAL STREAM wt (lbs)	PCBs Conc./mass, ppm/lbs
Feed	**Run #1,** 630 +/- 100, (10% $NaHCO_3$) Run #2, 1,300 +/- 100, (5% $NaHCO_3$)	Before Treatment	1,054 ppm	Feed	20,000	630/13
Product	< 0.5, (0.0049 - 0.3)	After filter press	129 ppb	Cyclone	65	20,000/1.3
Cyclone	20,017	After first stage carbon	4.64 ppb	Baghouse	22	18,000/0.4
Baghouse	17,913	After second stage carbon	< 0.5 ppb	Scrubber water	6,000	1,100/6.6
Offgas	ND **			Carbon filter press	100	8,900/0.89
				Uncounte d PCBs		/3.81 or 29 %

*** Used Method 680; ** ND = No Detectable PCBs.**

Notes: (1) Sample analysis was done by the Southwest Research Institute (SWRI)
(2) Dioxin, very trace amount in feed; none in product; and elevated levels in cyclone and baghouse dust/particles
(3) ppm = parts per million (mg/kg), ppb = parts per billion (μg/kg), 1 ton = 2,000 lbs, 1 lb = 2.2 kg

Table 2 BCDP Test Results for Second Test Series (STR Tested)

Run 3 (November 19 - 21, 1994, 44.411 tons/36 hours)

A. Soil Treatment		B. Water Treatment	
Sample Point	**PCBs Concentration, ppm ***	**Sample**	**PCBs Concentration**
Feed	1,360 (Avg.)	Before treatment	12 ppm
Product	0.2 (Avg.)	After filter press	2.6 ppm
Cyclone	15,742	After 1st stage carbon	3 ppb
Baghouse	109,331	After 2nd stage carbon	1.75 ppb
Offgas	ND **	Filter cake	48,000 ppm

* Used Method 680.; ** ND = Undetermined

Note : (1) Sample analysis was done by the Southwest Research Institute (SWRI)
(2) ppm = parts per million (mg/kg), ppb = parts per billion (μg/kg), 1 ton = 2,000 lbs, 1 lb = 2.2 kg

Table 3 STR Test Results (batch mode of operation)

Test Run No.	Before Treatment (Total PCB Conc. in ppm)	After STR Treatment (Each PCB Congener Conc. in ppm)
1	850	non-detect*
2	25,000	non-detect*
3	42,000	non-detect*

* Note : (1) PCB detection limit was set at 0.5 ppm for each congener
(2) ppm = parts per million (mg/kg)

To date, the BCDP system using the modified RKR (using steam as sweep gas to carry residues out of the reactor) along with the new series of APCS equipment has successfully and cost-effectively removed and cleaned (up) all on-site PCB contaminated soil , a total of 10,000 tons . An estimated 40 tons of "residues" collected is to be shipped back to U. S. for final disposal, likely incineration and landfill.

TOTAL TREATMENT COST

A total annual treatment cost for using BCDP system is estimated and illustrated in Table 4, which includes both fixed capital investment depreciation/amortization and operating cost. Based on assumptions made, a total annual treatment cost (including RKR and STR two-stage processing) is projected to be $2.75M/year for the 1.25 ton/hour BCDP system that has 10 years life. The unit cost for treating projected 7,700 ton/year is estimated at **$360/ton** . This estimate does not take into account the costs of obtaining any regulatory permits that may be required for treatment operations at other location than Guam. Approximately 62% of the total treatment costs are from labor costs (operation supervision, and overhead). Therefore, efforts to reduce the treatment costs should be focused on reducing labor requirements e. g. incorporate more automated systems for materials handling and equipment operation.

CONCLUSIONS

1. A continuous, 1 to 2 tons/hour Base - Catalyzed Decomposition Process (BCDP) System was successfully demonstrated and used in cleanup PCB contaminated soil, a total of 10,000 tons, at the U. S. Navy PWC in Guam. The System has produced a product soil that contains less than 2 ppm per resolvable PCB congener from original 2,500 ppm average concentration in the contaminated soil.

2. Use of granular activated carbon (GAC) to treat process scrubber water was effective in reducing the PCB concentration to less than 1 ppb, which meets sewer discharge standard.

3. An innovative Air Pollution Control System designed to replace original equipment used in technology demonstration, was successfully used in the production operation. It was much more effective in removing and concentrating the contaminants. It achieved the goals of collecting much less residues , producing cleaner air emission, effectively collecting submicron particulate and oil droplet , and eliminating any potential in generating dioxins and furans. PCB Destruction Removal Efficiency (DRE) reached seven 9s, and the dioxin and furan Toxic Equivalent (TEQ) was 0.18 nanogram per cubic meters.

4. An economic evaluation was performed to estimate the BCDP treatment cost. The evaluation has derived the unit cost to be approximately $360 per ton of treating PCB contaminated soil. This unit cost has proved that the BCDP treatment is a cost-effective innovative alternative technology.

ACKNOWLEDGEMENTS

The project is totally funded by the Naval Facilities Engineering Command (HQ & Pacific EFD). EPA/NRMRL provided the technology support, in particular Charles Rogers and Al Kornel of BCD Group. NFESC Remedial Action Contractor (RAC), DOE/PNNL & Battelle Columbus and IT Corporation conducted respectively technology demonstration and production operation at PWC, Guam. They are all gratefully acknowledged here.

Table 4. BCDP Treatment Cost Estimate Summary.

Item	Unit (b)	Unit Cost, $	Number of Units Per Year	Annual Cost, $/yr
Chemical				
$NaHCO_3$	lb	0.21	1,390,000	290,000
Granular activated carbon (GAC)	lb	1.50	39,000	59,000
LW-110	gal	2.00	5,300	11,000
NaOH (solid)	lb	0.50	9,600	4,800
Hydrogen donor	lb	0.40	7,700	3,100
Catalyst	lb	0.39	2,600	1,000
Disposal/Recycle Fee	gal	3.00	14,000	42,000
Subtotal				**410,000**
Utilities				
Fuel Oil	gal	1.20	55,000	66,000
Electricity	kWh	0.10	290,000	29,000
Subtotal				**95,000**
Labor				
Supervisors	hr	30	7,200	220,000
Operators	hr	25	29,000	730,000
Subtotal				**950,000**
Supervision (20% of labor)				190,000
Offsite Analysis	PCB Analysis	500	100	50,000
Maintenance Supplies 5% of FCI[(a)]/yr				151,000
Operating Supplies 1% of FCI[(a)]/yr				30,000
Plant Overhead Costs, 60% of labor				570,000
Depreciation, 10% of FCI[(a)]/yr				301,000
Total Treatment Cost per year				**2,750,000**

(a) FCI, Fixed Capital Investment (i.e., capital necessary for equipment, installation, and all auxiliaries for complete process operation) for 1.25 ton/hr BCD system is $3,010,000.

The Unit Cost = $2,750,000 / 7,700 tons (per year) = $ 357.14/ton

(b) 1 lb = 2.2 kg, 1 ton = 2,000 lbs, 1 gal = 3.785 l, hr = hour; $ = U.S. dollar,

Proceedings of the Seventh (1997) International Offshore and Polar Engineering Conference
Honolulu, USA, May 25-30, 1997

ISBN 1-880653-28-1 (Set); ISBN 1-880653-29-X (Vol. I)

Red Hill Fuel Tank Leak Detection

William Major, Leslie Karr, George Warren and John Norbutas
Naval Facilities Engineering Service Center
Port Hueneme, CA, USA

Abstract

The Naval Facilities Engineering Service Center (NFESC) is currently tasked with a Research, Development, Test and Evaluation (RDT&E) project for a leak detection system for the large tanks in the Red Hill Underground Fuel Storage Facility near Pearl Harbor, Hawaii. Initial calibration and demonstration tests have been conducted on a prototype version of the system installed in an underground storage tank (UST) at the Naval Air Station North Island, San Diego. The Red Hill Facility system is scheduled for calibration, demonstration and certification testing in the summer of 1997.

Keywords

Leak detection; fuel storage; underground storage tanks; finite element modeling; low range differential pressure system

BACKGROUND

The Red Hill Fuel Tank Farm has twenty USTs, each with a nominal capacity of 47,700 m^3 (300,000 barrels). Tank capacities actually vary from 45,345 to 48,143 m^3. The twenty tanks were constructed in two rows of ten each, as shown in Figure 1. The facility was constructed by tunneling into a volcanic rock ridge. The tanks are separated by 30.5 meters of undisturbed rock. The tops of the tanks range from 30.5 meters to 53.4 meters below the surface, while the bottoms are all above sea level. The original construction of the tanks, with upper and lower access tunnels, underground pump house, fuel pier and three miles of tunnel with piping and narrow gauge railroad tracks, was completed in 1943. Since then, individual tanks have been taken "off-line" for modification, modernization or maintenance, but the fuel farm has been in continuous service. These tanks are the largest of their type in the world. In recognition of the uniqueness and immense size of the facility, the American Society of Civil Engineers recently designated it as a historical monument.

Figure 1. Red Hill Fuel Storage Facility.

An elevation view of a typical tank is depicted in Figure 2. The tank is a vertical right circular cylinder with upper and lower domed ends. The internal diameter of the tank is 30.5 meters, and the overall internal height is 76.2 meters. The two domed ends are near hemispherical. The upper dome is made of six frustums of cone sections,

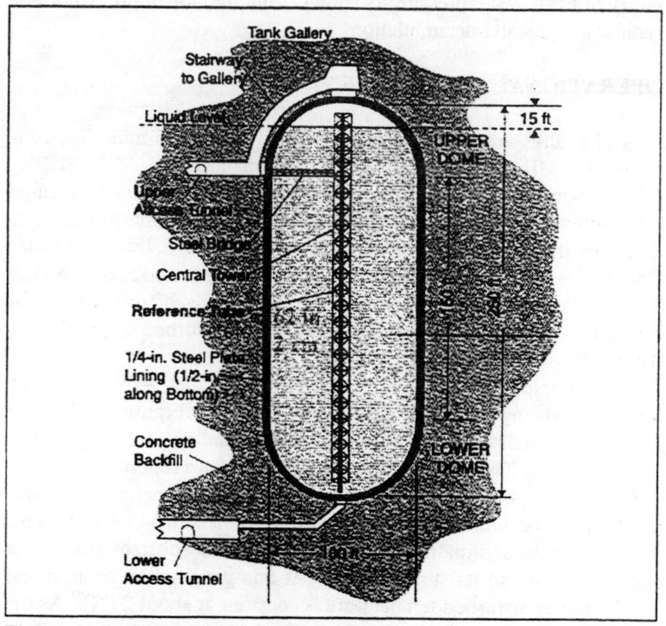

Figure 2. Typical tank elevation.

and the lower dome is made of four frustum sections. The flat top of the tank is about 7.9 meters in diameter. The flat bottom of the tank is approximately 6.1 meters in diameter. During construction, the butt-welded, 6.4-mm-thick steel plate tank lining served as the inside forms for the 1.2- to 2.4-meter-thick concrete tank walls (12.8-mm steel plate lining was used for the bottom domes). The steel lining is anchored to the concrete tank walls.

An open-frame steel elevator tower in the center of the tank, which was used during construction, remains in each tank. Access to the tower is provided from the upper access tunnel through an 2.4-meter-diameter flanged hatch near the 61.0-meter level of the tank. An open grate catwalk leads to the tower. Inspection access at the top of the tank is provided by a narrow stairway/ladder from the upper access tunnel to a gallery above the top of the tank, where a manhole leads into the top of the tank. Each tank has two receive/issue pipes, varying in size from 400 to 800 mm in diameter, an 200-mm slop drain line, and several smaller pipes that all run from the lower access tunnel over and up through the flat bottom of the tank. All active pipe lines are double valved, some motorized and some manual.

EXISTING TANK GAUGING SYSTEM

Tank inventory monitoring (tank gauging) is currently a volume system based on measuring the fuel surface level in the tanks. A guyed and counterbalanced stainless steel float is connected to an electromechanical device located in the gallery above each tank. The electrical output signal is transmitted to the Gauger Room in the lower access tunnel, where it can be displayed and recorded. The system can monitor the liquid level with a resolution of 0.3 mm for the full usable height range of each of the twenty tanks. The volume of fuel in the tanks is determined from the level readings with the use of the tank strapping tables. A fuel temperature measurement system and a fuel sample piping system provide data to determine the fuel standard volume. Platinum resistance temperature detector (RTD) probes, with an accuracy of ±5/9°C, are located adjacent to the center tower at tank elevations of 23, 41, and 61 meters (measured from the bottom of the tank). Piping is installed in each of the tanks to provide a means to draw fuel samples from 2.4-, 18.3-, 36.6- and 53.3-meter elevations. Sample density analyses are performed by the Fuel Department Laboratory of FISC. Samples are also drawn from the slop drain to monitor sediment and water accumulation.

OPERATIONAL ENVIRONMENT

The tanks are currently being used to store JP-5 and F76, conforming to MIL-T-5624 and MIL-F-16884, respectively. MIL-HDBK-200 supplements the specifications. The specifications allow for a range of acceptable values for various physical and chemical properties. For example, the specific gravity of JP-5 may vary from 0.7883 to 0.8448. The facility uses a first-in/first-out system of receipt and issue. A tank can be filled over a period of time with fuel received from more than one source. Once filled, a tank can remain undisturbed for more than 18 months.

A sample time history of the fuel level in Tank #15, using data from the existing tank gauging system, is shown in Figure 3. The tank had been partially filled, with a steady level reading of 26.288 meters for eight days. Then on February 21, 1995, fuel at a reported 20°C was transferred from a ship, filling the tank to the 71.606-meter level. Over the next 44 days, the level in the tank rose more than 111.8 mm. This is not an unusual situation. The incoming fuel is substantially cooler than the tank and its surroundings. At this geographic location and depth, the undisturbed temperature is constant at about 27°C. As the fuel warms in the tank, it expands and the level in the tank rises. This

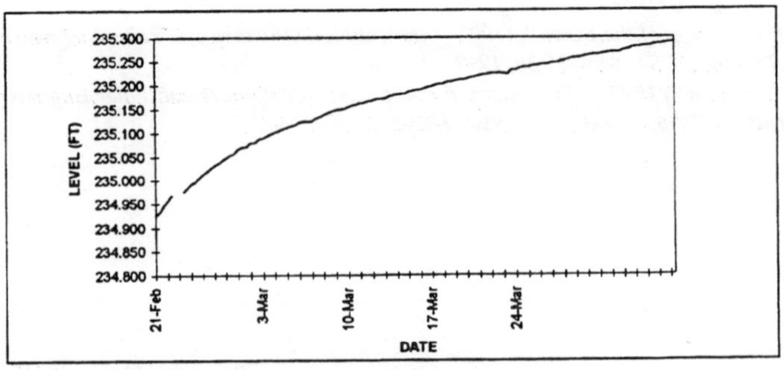

Figure 3. Tank #15 gauge data time history.

rise in level corresponds to an overall average temperature increase of the fuel of 0.9°C. The very large amount of fuel, together with the thermal properties of the fuel, tank and surroundings, results in very long times to approach thermal equilibrium. The early part of the curve, rising at a rate of 9.1 mm per day, indicates an overall average temperature increase of 0.0031°C per hour from an overall bulk heat transfer rate of 232,000 kJ per hour. The latter part of the curve, which is rising at a rate of 0.9 mm per day, indicates an overall average temperature increase rate of 0.00029°C per hour from an overall bulk heat transfer rate of 22,150 kJ per hour.

A thermister-based instrumentation system was used to obtain fuel temperature profile data. A cable assembly with thermister probes attached at 3-meter intervals was lowered into the tanks from their galleries. Probe resistance data were uploaded via a switching and digital metering system to a PC-based field data processor, where they were recorded. A sample temperature profile plot is shown in Figure 4. The plot reveals the cooler, denser fuel at the bottom of the tank and the warmer fuel above, with a prominent interface at about the 24-meter level.

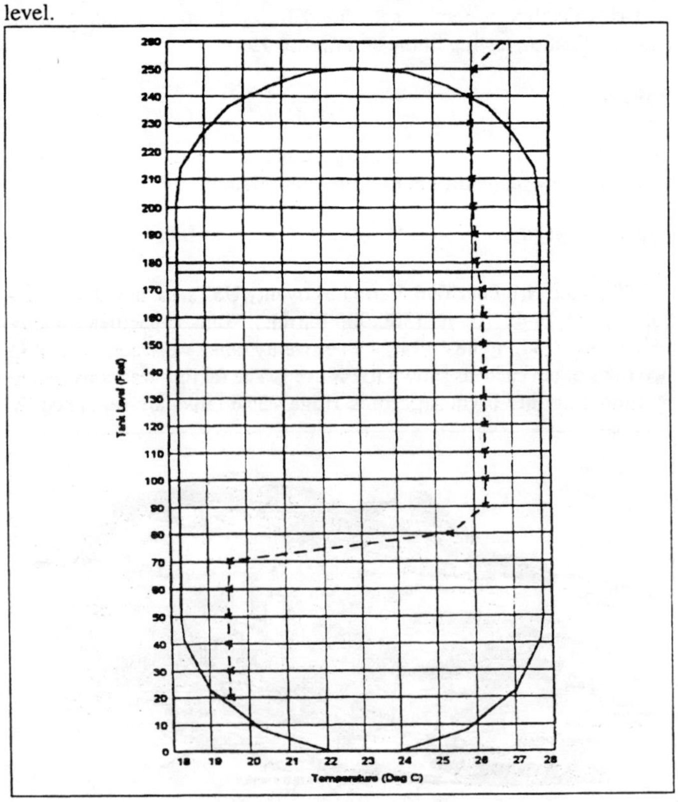

Figure 4. Tank #13 measured temperature versus elevation.

Table 1. Material Properties

Material	Dynamic Viscosity (Pa-sec)	Heat Capacity (J/kg-°C)	Thermal Conductivity (J/sec-m-°C)	Volume Expansion (°C^{-1})	Mass Density (kg m^{-3})	Gravity Acceleration (m sec^{-2})	Surface Tension (N m^{-1})
JP-5	0.0013	1940.0	0.11	99.0 E-05	809	9.8	0.023
Concrete	-	664.0	1.8	0.99 E-05	2319	9.8	-
Air	1.9 E-05	1015.0	0.025	360.0 E-05	1.0	9.8	-
Water	0.0014	4192.0	0.59	21.6 E-05	995	9.8	0.073

COMPUTER MODEL

A finite element model computer analysis of a simple axisymmetric tank model was conducted using the ADINA fluids/thermal program module. Material properties used in the model are shown in Table 1. All case runs had the tank and surrounding rock initially at 27°C. Case runs were made with the fuel initially at a lower temperature, and with the tank partially filled with fuel at 27°C and partially with fuel at a lower temperature. Sample results are shown in Figures 5, 6 and 7. Figure 5 shows the temperature distribution after 60 hours for a case run with the bottom 15.2 meters of the tank filled with JP-5 at 24°C and the next 45.7 meters with JP-5 at 27°C. Figure 6 is a plot of convection currents for the same case at 1 hour after start. Figure 7 is a plot of the overall average temperature of the fuel in the tank as a function of time. The constant fuel temperature across horizontal planes, the non-uniform vertical temperature gradient, the presence of convection currents, and the long times to approach thermal equilibrium are all consistent with reported field observations.

TECHNOLOGY SELECTION

The Environmental Protection Agency (EPA) recognizes a variety of leak detection methods for underground storage tanks. It has published guidelines for their selection and use as well as standardized test procedures for evaluation and certification of systems. Criteria (for smaller USTs) are written in terms of detecting leaks of a fraction of a gallon per hour within 48 hours. Both volume-based and mass-based tank gauging/tank tightness leak detection methods are candidates for use in the Red Hill tanks.

An error budget analysis, summarized in Table 2, was conducted to evaluate alternative technology methods for leak detection for the Red Hill tanks. The two approaches considered were volume-based and mass-based. A volume-based approach involves measuring fuel level, temperature, and density, and reading the tank strapping tables. Tank strapping tables, listed as TST in Table 2, are look-up tables that relate level and gross volume. A mass-based approach involves measuring hydrostatic and atmospheric pressure and density and reading the tank strapping tables. For the analysis, the technology groups were divided into (1) existing at Red Hill, (2) current technology, and (3) technology improvements. All resultant volumetric random errors are shown as gallons of fuel to permit direct comparison.

To provide the accuracy and precision needed for leak detection with a volume-based approach, such as the existing tank gauging system, would require elaborate fuel temperature monitoring and compensation. A nominally full tank with 47,700 m^3 is filled to the 71.6-meter level. At this level the tank diameter is about 22 meters. For this amount

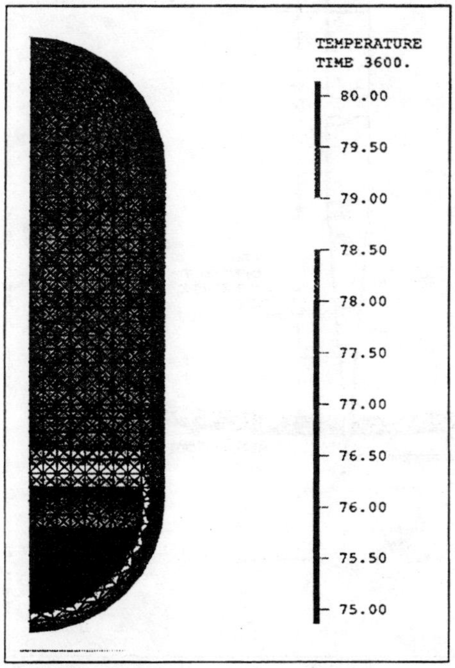

Figure 5. Computer model -- temperature plot.

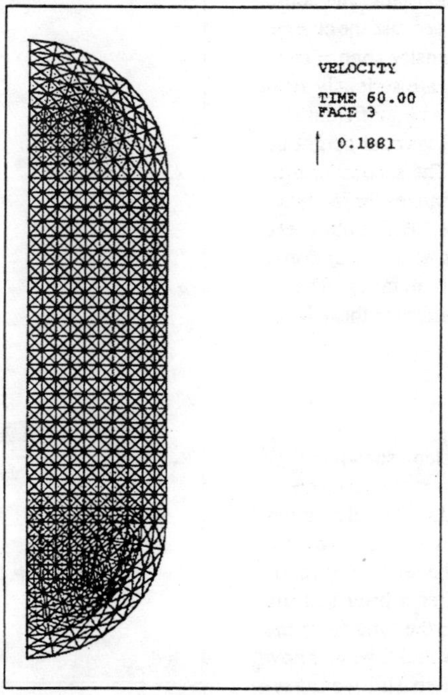

Figure 6. Computer model -- convection currents.

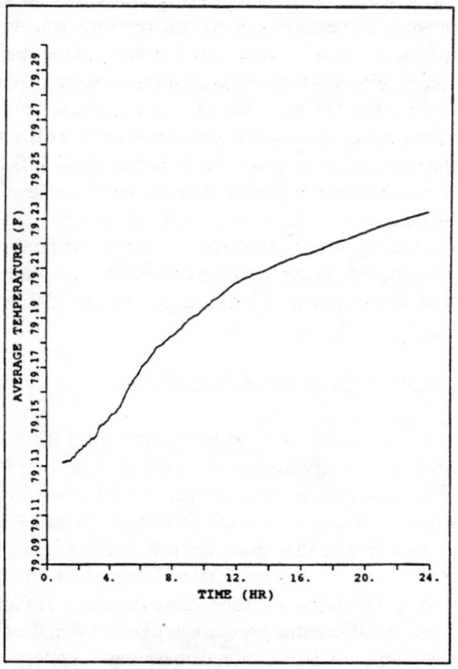

Figure 7. Computer model -- average tank temperature versus time.

Table 2. Random Error Budget for Large Bulk Storage Tank Leak Detection for Red Hill Facility

Measured Quantity	Existing at Red Hill		Current Technology		Technological Improvements (Transducer and/or Signal Processing)	
	Measurement Random Error	Resultant Volumetric Error	Measurement Random Error	Resultant Volumetric Error	Measurement Random Error	Resultant Volumetric Error
Level-Based System						
Level	±1.5875 mm	±157 m³	±1.0 mm	±1.0 m³	±0.18 mm	±0.19 m³
Temp	±0.55°C	±31.5 m³	±0.055°C	±3.15 m³	±0.013°C	±0.72 m³
TST	±0.0017 m²	±0.14 m³	±0.0017 m²	±0.14 m³	±0.00019 m²	±0.015 m³
Temp Profile	±0.33°C	±18.9 m³	unknown	unknown	TBD	TBD
API	±0.1°API	±0.65 m³	±0.1°API	±0.66 m³	±0.024°API	±0.16 m³
Mass-Based System						
Hyd. Pressure	N/A	N/A	±0.006 to ±0.06 kPa	±0.84 to ±8.45 m³	±0.0014 to ±0.014 kPa	±0.2 to 1.73 m³
Atm Pressure	N/A	N/A	±0 to ±0.25 kPa	±0 to +31.9 m³	TBD	TBD
TST	±0.0017 m²	±0.14 m³	±0.0017 m²	±0.14 m³	±0.00019 m²	±0.015 m³
Avg Density	N/A	N/A	±0.000014 kg/m³	±0.015 m³	±0.0000026 kg/m³	±0.004 m³

of fuel, a mere 0.055°C rise in temperature results in a volume expansion of 2.62 m³. For a tank throat diameter of 22 meters, this corresponds to a rise in level of 7 mm. However, both the computer model and the field measurements show that while the temperature is fairly uniform across any elevation, the vertical profile can be quite nonuniform and unsteady. The problem, then, is not only how accurately the temperature can be measured, but how to obtain an accurate temperature distribution.

The mass-based system approach was selected for the Red Hill tanks because of its inherent compensation for the effect of temperature changes on the fuel. For a uniform cross section tank, the static pressure at the bottom of the tank remains unchanged because the change in height (head) is offset by a change in the fuel density (ρgh = constant). However, conventional mass-based systems are seriously limited by two factors. The first is a pressure sensor with sufficient dynamic range to measure level changes over the full vertical height of the tank lacks the precision to detect small leaks. The second limitation is the need to compensate for amplified level changes during thermal expansion due to the shape of the tank. At Red Hill, both these limitations would substantially impact the precision of a conventional mass-based system because of the size and shape of the tanks. Therefore, development efforts focused on methods to minimize these limitations.

LRDP SYSTEM DESCRIPTION

A Low Range Differential Pressure (LRDP) system, shown in Figure 8, was configured to significantly improve pressure transducer precision and to inherently compensate for thermally induced level changes due to tank shape. The key component in the LRDP system is a vertical reference tube that spans the full usable height of the tank. The reference tube is filled and drained with the tank. After a brief settling period, the reference tube valve is closed, isolating the tube from the tank. A differential pressure gauge between the tank and tube will now read zero. A conventional mass-based system at Red Hill would require a 0- to 586-kPa full-range pressure transducer. The LRDP system requires less than a 0- to 6.9-kPa full scale and, therefore, provides

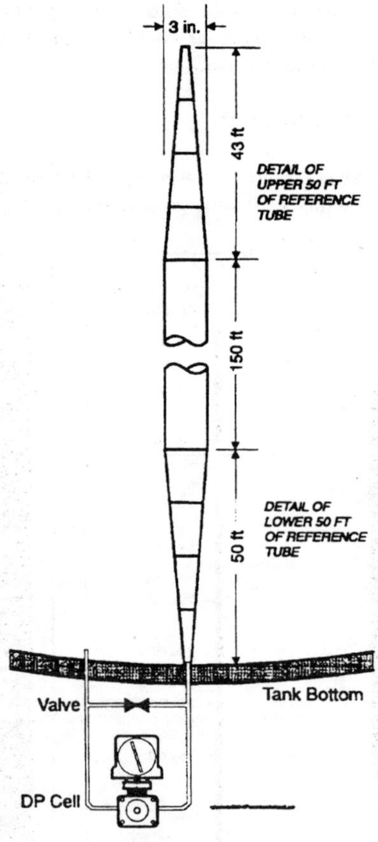

Figure 8. LRDP system concept.

a 100X improvement in transducer precision. Except for the motorized valve, there are no moving parts, which results in a highly reliable and robust system. The reference tube is a fixed stainless steel tube that is expected to last the life of the tank. The increased precision realized from the reference tube allows the use of an off-the-shelf industrial grade differential pressure transducer. Additionally, the LRDP system is self-calibrating since, after fueling operations, the pressure differential is known and is set at zero.

The Red Hill tank temperature profiling efforts demonstrated that the temperature change over location and time is variable and dynamic. Temperature-induced changes in the reference tube must therefore locally mimic the tank changes. To accomplish this, the reference tube must be located inside the tank and be shaped to match the tank. When properly matched, temperature-induced changes at any elevation will contribute equally in the tank and in the tube. The result is essentially a zero change in the differential pressure between the reference tube and the tank due to temperature (inherent passive temperature effect compensation).

PROTOTYPE INSTALLATIONS

A prototype LRDP system was installed at the Red Hill Underground Fuel Storage Facility in Tank #16 in July 1996. The installation has both an external and an internal system. The external system version has the differential pressure transducer located in the lower access tunnel, near the reference tube fill valve. The internal system version has the differential pressure transducer in an enclosure located at the bottom of the tank. Both system versions use the same reference tube. The reference tube is 76.2 mm in diameter throughout the midsection of the tank. The upper and lower sections of the tube are tapered to match the tank dome ends. The tube was welded into 6.1-meter sections and then welded together in place in the tank. In addition, gauge pressure transducers were installed at the bottom of the tank, the bottom of the reference tube, and in the slop line.

Both the internal and the external differential pressure transmitters are Rosemount model 3051CD capacitive cell type transducers, with a 4- to 20-mA dc loop output signal. For a span setting of 0 to 12.7 mm inches of water, the transmitters have an accuracy of ±0.275%. The temperatures of the sensors are monitored with RTDs to provide compensation data, if required. The sensor circuits are all protected by Stahl Intrinsic Safety Barriers. Sensor data are uploaded via a Data Acquisition and Control Unit to a PC-based logger running LabView for Windows, where they are recorded in ASCII format. Laboratory bench experiments with the differential pressure transmitter indicate that the instrumentation is capable of detecting leak rates of less than on 0.0038 m^3 per hour. The calibration, demonstration, and certification testing of the prototype LRDP leak detection system are planned for the summer of 1997, following completion of the scheduled tank refurbishment.

Another prototype version of the LRDP system has also been installed in Tank #1006 at Naval Air Station North Island. This shallow-buried UST, with about 1.2 meters of sand cover, measures approximately 27 meters in diameter by 4 meters in height and has a capacity of 95,400 m^3. Since there is no below tank access for this UST, an internal LRDP system is being used with the pressure sensor in an enclosure located on the bottom of the tank. The reference tube is considerably shorter and of constant cross section to match the tank geometry. The signal conditioning and data acquisition and control are similar to the Red Hill installation. System checkout, calibration, development of operating procedures, and initial evaluation and certification testing were started in January 1997.

SYSTEM TESTING

Initial system checkout and calibration of the prototype LRDP systems are being conducted to determine the effects of actual tank environment. Figure 9 is a plot of the differential pressure gauge data, converted to gallons of fuel, from an early test run at the NASNI installation. The linear regression plot indicates a leak rate of 0.0015 m^3 of fuel per hour. The effects of the unsteady state heat transfer, convection currents, evaporation and condensation in the tank all need to be evaluated in addition to the dynamics introduced during receipt and issue from the tank. Operating procedures, such as securing valves and tank venting, can impact required settling time before valid leak data can be collected.

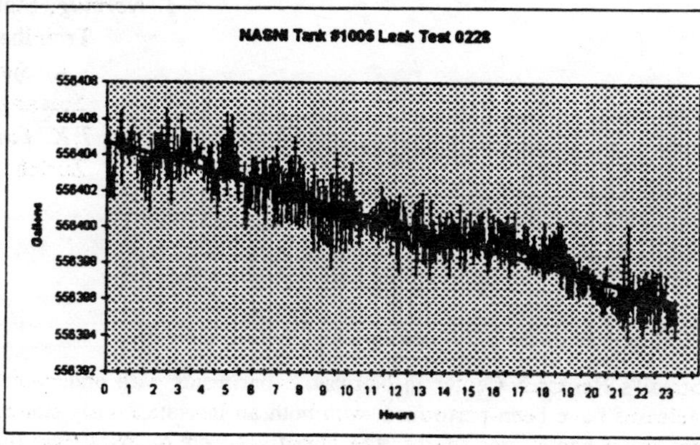

Figure 9. Tank lead test data and regression plot.

FUTURE PLANS

Third party evaluation and certification testing will be conducted on both the Red Hill and North Island LRDP installations. The test and evaluation protocol being used for these systems, "Alternative Test Procedures for Evaluating Leak Detection Methods: Evaluation of Bulk Field Constructed Tanks," was recently accepted by the National Workgroup on Leak Detection Evaluations. Coordination of detailed test procedures is underway with EPA, California State Water Resources Board, and Hawaii Department of Health personnel. The test protocol requires twelve test runs, each of 48 hours duration. Three tests each are to be run at induced leak rates of 0X, 0.5X, 1X, and 2X of the target threshold leak rate. The total volume of product removed from the tank during each test period will be used to determine the actual induced rate. The procedure for the statistical calculations for determining the actual system threshold in cubic meters per hour to obtain a 95% probability of detection and a 5% probability of false alarm is described in the EPA protocol for volumetric systems: "Standard Test Procedures for Evaluating Leak Detection Methods: Automatic Tank Gauging Systems."

ACKNOWLEDGMENTS

The authors would like to recognize the contributions of Vista Research, Inc., and Asteroid Corporation in this program.

Proceedings of the Seventh (1997) International Offshore and Polar Engineering Conference
Honolulu, USA, May 25-30, 1997

ISBN 1-880653-28-1 (Set); ISBN 1-880653-29-X (Vol. I)

Surface Flow and Gas Dispersion from a Subsea Release of Natural Gas

T. Engebretsen, SINTEF Energy
Trondheim, Norway
T. Northug, Statoil Research Center
Trondheim, Norway
K. Sjøen, Statoil
Stavanger, Norway
T.K. Fanneløp, ETH
Zürich, Switzerland

ABSTRACT

At Statoil's Research Center in Norway, experiments with underwater gas releases have been performed, with both an instantaneously started source and a continuous source. The objectives were to investigate the surface currents near the bubble plume, and the gas dispersion above the surface. The currents can cause stability problems for vessels or offshore platforms, high concentrations of natural gas can result in fires or explosions for releases occurring below or near offshore installations. The experiments were performed in a basin of 7 m depth with volumetric flow rates ranging from 0.083 Nm^3/s to 0.75 Nm^3/s, corresponding in scale to that expected from a full pipeline rupture or blowouts from shallow pockets. Measurements of water velocities and void fractions were performed in the vertical bubble plume and in the horizontal surface flow. The gas concentration above the water surface was monitored by means of a novel measuring technique; for these experiments, a mixture of helium and air was used to obtain the same density as natural gas. The velocity profile in the surface flow was found to be different from that observed by others; the profile depth was found to be decreasing with increasing distance from the center of the bubble plume, whereas previous work indicate an increase of profile depth due to entrainment. The horizontal transport of gas (bubbles) was found to be greater than expected; the gas was dispersed into the atmosphere from an area corresponding to about six times the area of the central boil region. In the present experiments the release rates were significantly higher than in previous investigations, and we observed a fountain of significant height above the reference surface. The return flow from this fountain penetrated deep into the water. Its interaction with the horizontal flow from the bubble plume, resulted in a significant transport of water and bubbles deep in the horizontal layer. On comparing different release rates, the gas concentration profiles above the surface appeared to be of similar shape, but the rise velocity of the gas increased with increasing release rate. Some progress has also been made in correlating the gas movements above the surface to the parameters of the bubble plume.

KEY WORDS: Blow-out, Safety, Bubble-plume, Fire hazard.

INTRODUCTION

From experience we know that large releases of natural gas under water are low-probability events. However, a blowout of some size can be a serious event and is strongly unwanted due to risks related to safety, environmental impact and costs. It is a primary objective for the oil production and exploration industry to prevent blowouts of oil and gas. Large resources are used to develop designs, equipment and operational procedures to build and maintain safe systems. If a blowout situation occurs a major effort is still required to reduce the number of casualties, to reduce the damage to rig and equipment and to ensure minimal environmental damage. To meet these requirements we need more knowledge about the physics of subsea gas releases. In 1993 a co-operation agreement was signed between the Norwegian oil companies Hydro and Statoil to start a research project on sudden releases of natural gas under water. The primary objective was to verify existing theories presented by Fanneløp and Sjøen (1980) and Bettelini and Fanneløp (1993), and to describe the interaction of a steady bubble plume with the ocean surface in order to predict the concentrations of hydrocarbon gas in the atmosphere above the plume. This is a problem in need of predictive methods. It has been reported that ships and other types of floating installations have lost buoyancy and sunk when engulfed by a gas plume. Based on present knowledge, we are more concerned about high concentrations of hydrocarbon gases in the air that can lead to explosion and fire, than about loss of buoyancy. The "boiling activity" on the surface manifested by a high (transient) fountain and strong radial currents, represent new and unexpected hazards.

Experiments have been carried out in a 7 meter deep basin at Statoil's Research center in Trondheim with gas release rates ranging from 0.083 to 0.750 Nm^3/s. The instrumentation was located both below and above the water surface. Video cameras were used in order to record plume shape, bubble size, fountain height and rise time as function of release rate. The data obtained are compared with existing experimental data and results from previous theoretical work.

We have also access to interesting information and reports from full scale events in the ocean, where uncontrolled blowouts have taken place. However, the available information from these events are based on qualitative observations, and give a rather poor basis for comparison with theory or laboratory experiments. Some accidental gas releases are listed in Table 1. Not many of these types of accidents are reported, and the release rates, when given, are often underestimated. Sometimes drilling ships are known to have disappeared, most likely caused by underwater gas plumes. In addition to fire and explosion hazards, ships and floating offshore installations could experience major stability problems due to the strong currents in the surface region; water velocities of the order of 10 m/s are realistic, leading to anchor forces in excess of the design limits.

A schematic description of a steady-source bubble-plume is given in Figure 1 (left); the plume fluctuates with large horizontal excursions (wandering), due to large scale turbulence. Superimposed on this wandering, we have smaller fluctuations due to turbulence with smaller length scales. A time-averaged description of the plume is also given in the figure (right); this representation of the plume will give a wider plume. Velocity profiles are also sketched; for the time-averaged case the velocity profile will be wider and lower, and with no ocean currents present, the profile will be axisymmetric in contrast to that of the real (instantaneous) case. Description of the dynamic behavior is not discussed much in earlier publications on experiments with bubble plumes, and hardly any theory exists. The present study will deal, mainly, with time-averaged quantities; however, for dynamic loading of vessels and offshore platforms, the dynamic behavior will be of importance; both the initial (transient) effects associated with the starting plume and the effects of turbulence. Consequently, the present study also covers, in part, the dynamic behavior of the plume with resulting vertical and horizontal currents; this work will be completed in the near future. An interesting observation made both in the laboratory and in full-scale events in the ocean, is the presence of a high, pulsating fountain of water and gas above the plume's "boil area".

Table 1 Accidental gas releases.

Location/year	Ocean depth (m)	Comments
Dubai (Papa and Stone, 1977)	55	Shallow pocket, gas fire above the plume one week after start of release.
Gulf of Mexico (IXTOC I), 1979 (Hamilton, 1986)	Approx. 40	Gas fire above the plume.
West Vanguard, North Sea (NOU, 1986)	223	Gas fire above the plume.
Asia (no open ref.)	126	Large gas release due to failure of geological structure below the seabed.

PREVIOUS WORK AND THEORETICAL BACKGROUND

Previous studies of steady-source bubble-plumes

Figure 1 (right) gives a time-averaged representation of a bubble-plume. The figure shows a conical-shaped plume, before transition to a horizontal surface flow. The model indicates a virtual origin for the source to compensate for the jet region in the lower part of the plume. In the plume region, it is assumed that buoyancy is the only force driving the plume. For deep-water releases, the difference in depth between real and virtual source will most often be negligible.

Previous work has shown that time-averaged plumes and jets have velocity profiles which are very close to Gaussian; this is also the case for the void profile (the density profile is given in Fig. 1; the void profile will be similar to the velocity profile, but narrower). Several workers have investigated this problem experimentally; those most important for the present study are listed in Table 2.

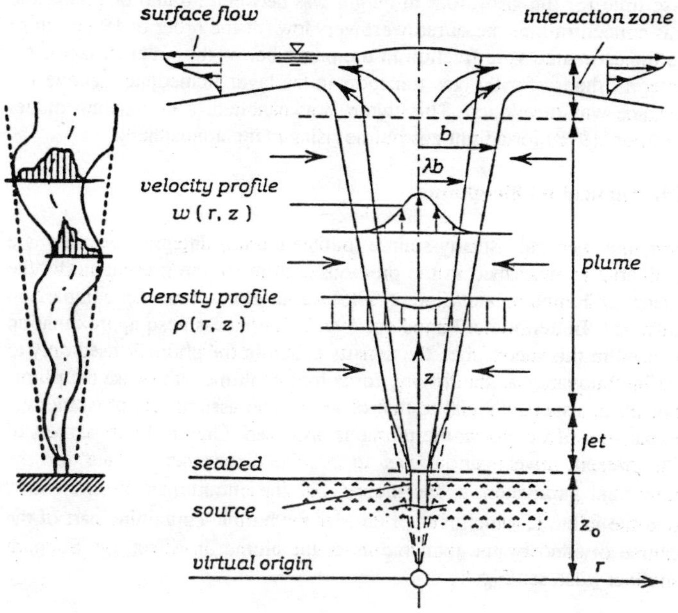

Figure 1 Schematic descriptions of a steady-state bubble-plume; snapshot (left) and time-averaged representation (right) of a bubble-plume with surface flow (from Fanneløp and Sjøen, 1980). "b" is the plume radius (Gaussian), "λb" is the radius of the void profile, i.e., the bubble-containing (visible) part of the plume.

Table 2 Bubble plumes; previous work.

Reference	Water depth (m)	Release rate (Nm^3/s)
Topham (not published)	23 60	0.06-0.65 0.30-0.40
Fanneløp and Sjøen, 1980	9.9	0.005-0.022
Milgram and Van Houten, 1982	4.0	0.00021-0.0023
Milgram, 1983	50	0.024-0.59
Løes and Fanneløp, 1989	50	0.6-1.3
Present study	6.9	0.083-0.75

Referring to Table 2: in Topham's, Fanneløp/Sjøen's and Milgram's experiments velocity measurements were performed in the bubble-plume to investigate the development of the plume as it approached the water surface. Fanneløp/Sjøen's and Milgram's experiments showed that the velocity profiles were well represented by Gaussian fits, as is the case also for light gas plumes. The entrainment coefficient, which is an empirical parameter expressing the horizontal transport of water into the vertical plume, was found to increase with release rate and with distance from the source. To model these flows, integral theories based on Gaussian or top-hat profiles were developed. In Fanneløp/Sjøen's study the solution was found to be insensitive to variations in the entrainment coefficient, so it could be treated as constant, whereas Milgram expressed this parameter as a function of a bubble Froude number. In Milgram/Van Houten's study void fraction measurements were also performed; the width of the void profile of the vertical plume was found to be about 80% of the velocity profile. In Løes/Fanneløp's study the rise time for the initial bubble cloud ("cap") from seabed to water surface was measured, and the hydrocarbon gas concentration in air was monitored at four different horizontal positions. One gauge was located above the centerline and the remaining three gauges located symmetrically around the plume center. All measuring points were located 0.5 m above the water surface. The gas released was methane, with a density relative to air of 0.55. Even with this high buoyancy, the rise time for the gas to 0.5 m height was between 30 and 60 s, and the gas concentrations measured were very low (of the order of 1%, or close to the estimated void fraction in the plume below the water surface). An integral theory for the gas transport in the layer immediately above the surface was developed. This theory was matched to an existing model (Scorer, 1978) for a light gas plume rising in the atmosphere.

The vertical bubble-plume

For the vertical, steady-source bubble-plume, integral theories are available as described in the previous section. In the present study, the theory of Fanneløp and Sjøen (1980) was used as a basis for comparison (also ref. Bettelini and Fanneløp, 1993). The Boussinesq approximation is used in this theory; i.e., the density deficit in the plume is used only to define buoyancy as the driving force for the plume, otherwise the plume density is assumed equal to that of water. The assumption of isothermal expansion of the gas in the plume is also used. One of the objectives of the present investigation was to provide experimental data for the empirical parameters in the theory, like the entrainment coefficient, α and the value λ, which is the radius of the bubble-containing part of the plume divided by the total radius of the plume, in effect, the Schmidt number (also see Fig. 1).

Surface flow

Very little theoretical work is available for predicting the surface flow from a vertical bubble-plume. The theoretical basis, for comparisons with the present experiments, is taken from the work of Fanneløp and Sjøen (1980). Their work predicts the decay of velocity and the increase in depth as a function of the radial distance from the plume, assuming Gaussian or top-hat profiles. An entrainment coefficient of $\beta = 0.06$ for the surface flow was found to give the best agreement with experimental data.

Gas transport above the bubble-plume

As for the surface flow, little theoretical information is available for the gas transport above a bubble-plume. Løes and Fanneløp (1989) developed a global theory for the light gas plume above the bubble-plume, based on the conservation of mass and momentum within a control volume. A schematic representation of the flow above the water surface is given in Figure 2.

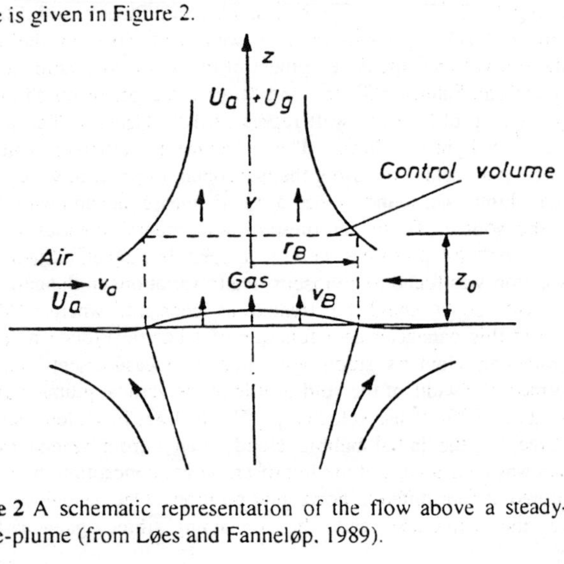

Figure 2 A schematic representation of the flow above a steady-state bubble-plume (from Løes and Fanneløp, 1989).

An expression for the gas concentration above the bubble-plume as a function of the release rate and the "boil area" was found. The predicted increase in gas concentration with increasing release rate at a fixed water depth, was found to agree well with the experimental results.

EXPERIMENTAL DETAILS

The basin and the gas release system

The experiments were conducted in fresh water, in a rectangular basin of 7 m depth and with a surface area of 6 x 9 m. Video recordings were taken at several positions, both under and over the water surface. A high-speed video camera was used for recordings of bubble shapes, sizes and movements of the plume. The gases released were either air, or a mixture of 40% helium in air, defined by the respective partial pressures at 7 bar total overpressure. Pure air was used when only parameters below the water surface were to be investigated. When gas concentration measurements were of interest, the helium-air mixture was used to simulate the buoyancy of natural gas. A tracer gas (ethylene) was added to the helium-air mixture for the concentration measurements above the water surface. The gas to be released was contained in a vessel of approx. 5 m^3 volume. A pipeline was connected between this vessel and the point of release at the bottom center of the basin. To ensure a constant flow rate of gas during the release, a pressure regulator was mounted in the pipeline to keep the downstream pressure constant at a preset value, depending on the desired flow rate. The release valve was of a piston type (rapidly acting, because the present study was also concerned with the initial transients). The gas was released in the vertical direction, and just downstream of the exit a circular disk was located to reduce the vertical momentum of the initial jet. This disk, in turn, was enclosed by a wider tube to turn the horizontal flow from the disk to the vertical direction. With this geometry, the fluctuations in the gas flow were substantially reduced, and the length of the jet region was minimized. This was very important in the present investigation; the maximum water depth was only 7 m, and without this momentum trap the jet phase with its large fluctuations would extend 1-1.5 m above the point of release. With our limited depth it was necessary to have a fully developed buoyancy-dominated plume as close to the release point as possible, and the experimental conditions should also be repeatable.

Three different release rates were used: 0.050, 0.100 and 0.450 m^3/s (nominal values) referred to the state at the point of release, or: 0.083, 0.17 and 0.75 Nm^3/s (the highest flow rate was used only in a few experiments).

Diagnostics and experimental program

Table 3 Instrumentation used in the present investigation.

Parameter measured	Measuring technique
Gas flow rate	Vortex flowmeter with pressure and temperature compensation, type: Yokogawa YF104, with electronics for compensation: Yokogawa STLD-301
Vertical and horizontal water currents	Lightweight turbine flow meters, type: Höntzsch ZS 16 GFA, 9 mm rotor diameter
Void fraction in the vertical plume and in the surface flow	Conductivity probes, in-house type (50x100 mm electrodes with 50 mm gap), with electronics: DHI 102E, and compensation probes (in-house)
Gas concentration	UVIC (Ultra Violet Ionization Chamber), Enviro Systems Ltd.
Plume widths and fountain heights	Video cameras (different types; Betacam and High-8 systems
Bubble shapes and movements	High-speed camera (NAC HSW 500) and strobelight

Table 3 gives an overview of the instrumentation used. The response time for the turbine flowmeters was approx. 0.3 s, for the conductivity probes practically negligible. The gas analyzers were of a novel type, UVIC (see Tab. 3), with a response time of 0.2 s and with sub-ppm accuracy (manufacturer's data). For measurements in the vertical plume, five turbine flowmeters and five conductivity probes were mounted on a horizontal beam, at five radial positions. The beam could be moved to different vertical positions, measurements were performed in three vertical positions. The beam was narrow and wing-shaped, and did not perturb the flow significantly. The instruments were mounted on vertical rods attached to the beam, 0.4 m below the beam to avoid measuring in a perturbed flow. Measurements in the horizontal surface flow were performed in a similar manner with the turbine flowmeters and conductivity probes turned 90^0 relative to their directions in the plume. Four of each kind of probes were mounted on a vertical beam to measure the velocity and void profiles at two radial positions from the plume center. The gas concentration was measured by the UVIC gas analyzers in 6 positions; 4 radial positions from the plume center and at a fixed height of 1.0 m above the water surface to avoid water splashes. For the highest release rate the inlets of the UVICs were lifted to 1.5 m due to the higher fountain. Above the plume center, 3 additional UVICs were located with a spacing of 0.5 m to determine the vertical movements of the light gas in air. The measurements were recorded on a digital recorder with a sampling rate of 100 Hz, due to the rapid response of the conductivity probes. The recorder interfaced with a PC, and the data acquisition system was based on LabView (by National Instruments Inc.).

In the experiments with only underwater measurements, the duration of the release was limited to 20 s to eliminate the influence of re-circulating flow in the basin (the rise time for the initial bubble cap, formed immediately after opening of the valve, was about 5 s). For the gas concentration measurements, the release time was 3 minutes to obtain steady-state conditions in the flow above the water surface. The experiments were repeated several times in order to produce good statistical data (ensemble averaging).

EXPERIMENTAL RESULTS AND DISCUSSION

Vertical plume

The steady-state bubble-plume is characterized by large fluctuations, due to turbulence with a wide variety of length scales and intensities, from sub-bubble-level to large amplitude fluctuations reminiscent of von Kàrmàn vortices. Figure 3 gives typical raw data. The vertical water velocities are given for one elevation of the instrument traverse, at different radial positions from the centerline. As expected, the highest velocity was measured along the centerline, and it decreased with increasing radial distance from the center. All velocity measurements show the plume's dynamic behavior, due to it's wandering in the horizontal direction and to turbulence of different scales superimposed. The record of the central velocity had a dominating frequency of about 0.25 Hz, i.e., the most low-frequent and high-amplitude fluctuation had a period of about 4 s close to the water surface.

Figure 4 shows time-averaged velocity and void profiles at the three different elevations. The profiles are given as Gaussian fits. The averaged experimental data deviated very little from this fit, which is typical for steady-state plumes (Fanneløp and Sjøen, 1980; Milgram, 1983). The figure shows how the profiles grow wider and lower due to entrainment, as the flow approaches the surface.

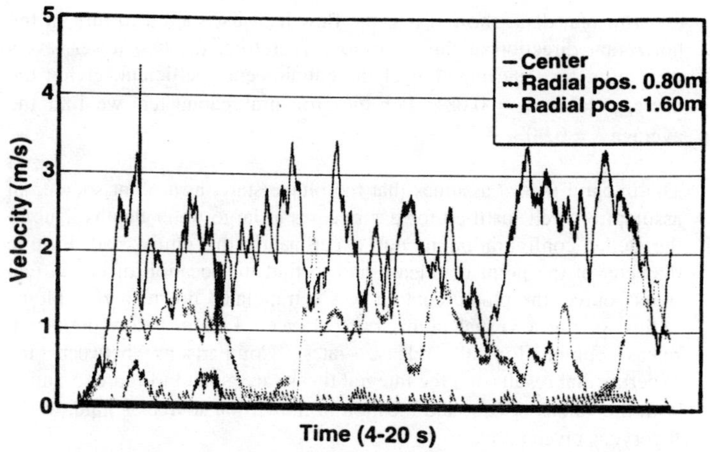

Figure 3 Measurements in the vertical bubble plume; water velocities at different radial positions from the centerline. Elevation: 5.9 m from the release point, or 1.0 m below the water surface. Radial pos.: 0, 0.8 and 1.6 m from the plume center. Release rate: 0.17 Nm^3/s. The highest velocities are readings at the centerline.

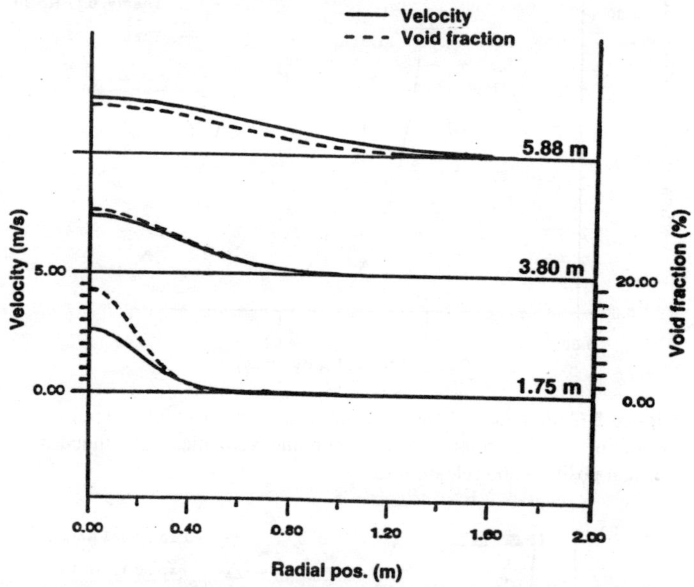

Figure 4 Time averaged velocity and void profiles (the average of 5 repeated experiments), at three different elevations from the release point. Release rate: 0.17 Nm^3/s.

Within the large variation of release rates in the present experiments, all profiles (as in Fig. 4) could be well represented by Gaussian fits, and the profile widths were approximately the same for the different release rates. With reference to the "Previous work" section, it is known that the development of the plume radius with increasing distance from the point of release, can be used to determine the entrainment coefficient α for the plume, assuming its value to be constant. This assumption gives a reasonably good agreement between integral theory and experimental data, because the solution is rather insensitive to minor variations in the entrainment coefficient (Fanneløp and Sjøen, 1980; Bettelini and Fanneløp, 1993). The plume width ratio λ is the radius of the void profile divided by the radius of the velocity profile. This structural parameter gives information about the width of the bubble-containing part of the plume, compared to its total width. The value of λ can be derived from the measured profiles. The upper instrument position was left out in calculating the entrainment coefficient; analysis of the experimental data, showed that the flow had just started to turn in the horizontal direction at this elevation. Therefore, the two lower levels were used for determination of the entrainment coefficient, giving the average value $\alpha = 0.081$. For the structural parameter, we find the average $\lambda = 0.90$.

The integral theory assumes that the plume starts as a point source, an assumption well justified for gas releases at large water depths. Due to the outlet configuration in our experiments, the plume had a finite diameter at the point of release and to find the location of our virtual point source, the plume contour was extrapolated down to zero plume radius, giving a virtual point source located 1.20 m below the actual release point for all release rates. Comparisons between the experimental results and the integral theory are given in Figures 5 and 6, with the values of α, λ and location of the virtual source as input to the theory, as given above.

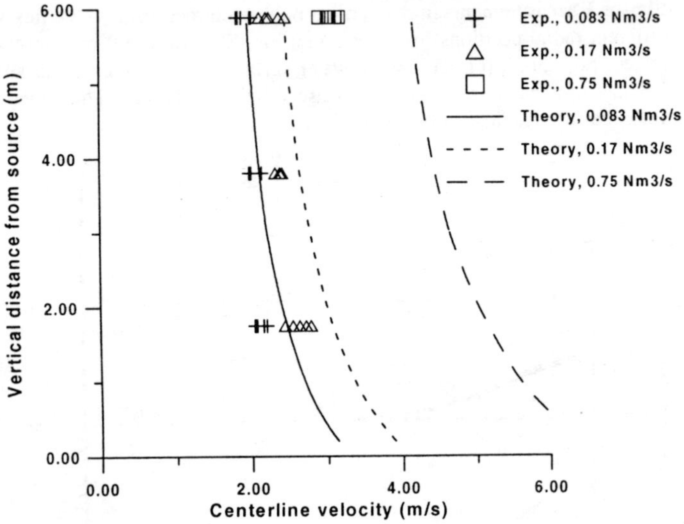

Figure 5 Comparisons between experimental results and integral theory (Bettelini and Fanneløp, 1993); centerline velocities as a function of vertical position and release rate.

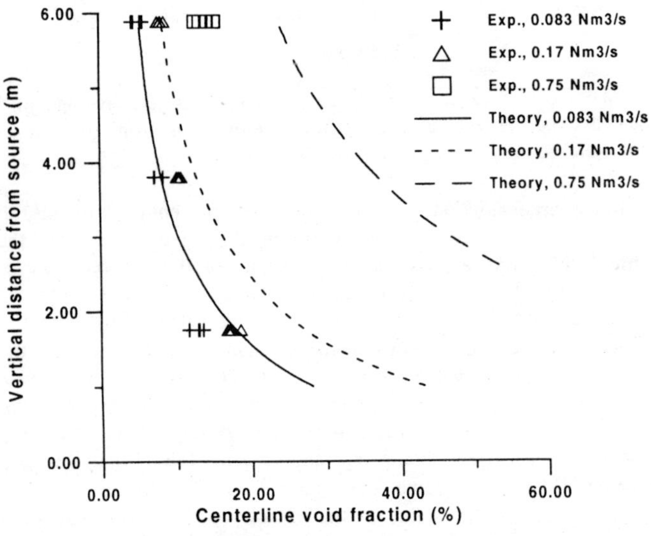

Figure 6 Comparisons between experimental results and integral theory (Bettelini and Fanneløp, 1993); centerline void fraction as a function of vertical position and release rate.

Figures 5 and 6 show that good agreement is obtained, as long as the void fraction does not exceed approx. 10%. The upper limit of validity of a theory based on the Boussinesq assumption is not easily established, but the present integral theory has been assumed to be valid for void fractions below 10-15% (Fanneløp, 1993 and onwards). This is now verified by the results of the present investigation. For most accidental subsea releases, values as high as 10% will be realistic; but accidents have been reported, where enormous amounts of gas have been released (seabed cave-in, blowouts from shallow pockets) where the void fraction by far must have exceeded 10% (sources classified). The limitation of the present integral theory is primarily due to the Boussinesq approximation as noted. The limitation for a bubbly-flow regime in a plume, is believed to represent a void fraction in excess of 30%; the predicted void fractions above this level, shown in Fig. 6, represent cases for which the theory does not apply. This explains the discrepancies between theory and experiments for the larger gas flow rate, 0.75 Nm^3/s.

Surface flow and fountain height

The nature of the velocity and void measurements in the surface flow, was similar to that observed in the vertical plume. The major difference was an even more cyclic nature of the void record, showing an intermittent gas transport in the surface flow. The fluctuations of the velocity records of the surface flow , with the lowest frequency and highest amplitude, had a period of about 4 s; i.e., the same as for the fluctuations in the vertical flow. An example of the time-averaged data for the horizontal flow is given in Figure 7.

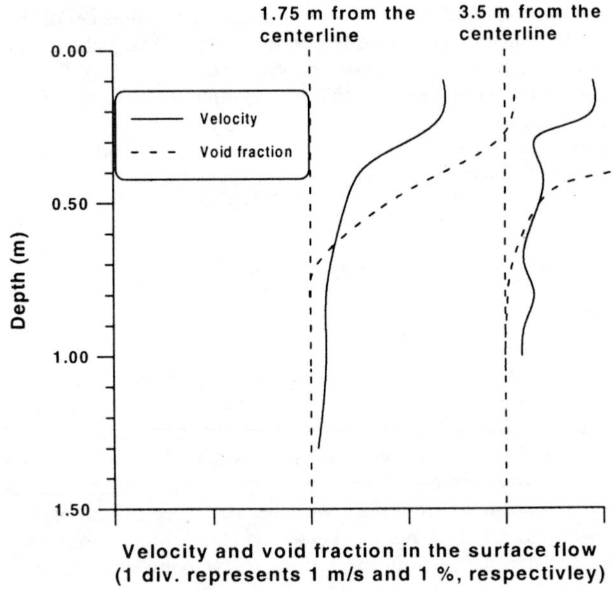

Figure 7 Velocity and void profiles in the horizontal flow (curve fits to the experimental data), at two different radial positions from the plume center. Release rate: 0.17 Nm^3/s.

With reference to Figure 7, the measurements show a relatively deep transport of bubbles in the horizontal flow, about 0.5 m below the surface. For the high release rate of 0.75 Nm^3/s, the bubble transport was deeper than 1.0 m; for this release rate measurements were performed only at radial position 3.5 m. This subsurface transport was rather unexpected, as we expected nearly all gas to be released in the central "boil area", with the exception of the small amount contained in the smallest bubbles. Fig. 7 also shows that the profile depth gradually is reduced with increasing distance; this is contradictory to the results of previous work at 5.5 m and 9.9 m water depth, where the depth of the velocity profile was found to increase outwards to approx. 10 plume radii from the center (Fanneløp and Sjøen, 1980). Our radial positions represent 2.3 and 4.6 plume radii, which is well within this region. The most likely explanation for the different behavior in our experiments, is the much higher release rate at a comparable water depth. Our minimum release rate was 0.083 Nm^3/s at 7 m depth, whereas the maximum rate in the previous study was 0.022 Nm^3/s at 9.9 m depth. In the previous study the water surface was only slightly elevated during the release, whereas in our study a fountain was always present. The return flow from the fountain penetrates deep below the water surface and interferes with the radial flow from the plume, as it is deflected from the surface.

The fountain is in itself of particular interest. From blowouts in the sea, fountain heights up to 70 m above the sea level have been reported (information restricted). However; these heights are most probably water splashes, and not at all representative as averaged fountain heights. But still, they represent interesting observations. A sample of fountain heights observed in the present study are given in Table 4.

Table 4 Observed fountain heights. Water depth: 6,9 meter.

Release rate (Nm3 /s)	Fountain heights, Initially (m above water surface)	Fountain heights, maximum (m above water surface)
0.17	0.3	0.65
0.72	0.45	1.25

The initial fountain heights were measured immediately after the leading bubble-cap front first penetrated the water surface. The maximum fountain heights were observed over a period of 15 seconds in the transient phase. The most extreme water splash was registered in connection with the experiments with the UVICs. The main units of the UVICs were placed 2 meters above the water surface and one of the UVICs was once hit by a water splash.

The velocity and void profiles (like those in Fig. 7) were used to calculate the gas transport in the outward moving layer below the water surface; Table 5 gives the results as a percentage of the total release rate at two radial positions. There is one uncertainty in these calculations. The return flow from the fountain acts as a small waterfall, penetrating the surface. Some air will be entrained in this process, but at present, we have no information about the amount. Air entrained in the boiling area is neglected in the further discussion.

Table 5 Gas transport in the horizontal surface flow as a percentage of the release rate.

Release rate Nm^3/s	Gas transport at radial distance 1.75 m (% of release rate)	Gas transport at radial distance 3.50 m (% of release rate)
0.083	44	37
0.17	66	56
0.75	no measurements	62

Table 5 shows that the horizontal transport of gas, as a percentage of the release rate, increases slightly with release rate. This is reasonable because the fountain height will increase, resulting in a deeper outward-moving layer and thus a deeper transport of bubbles in the radial direction. The gas transport also decreases with increasing distance, which is reasonable because some gas will rise to the surface between the two radial positions.

Gas concentrations above the water surface

Figure 8 shows experimental data from the UVIC gas analyzers mounted in the horizontal plane, one meter above the water surface.

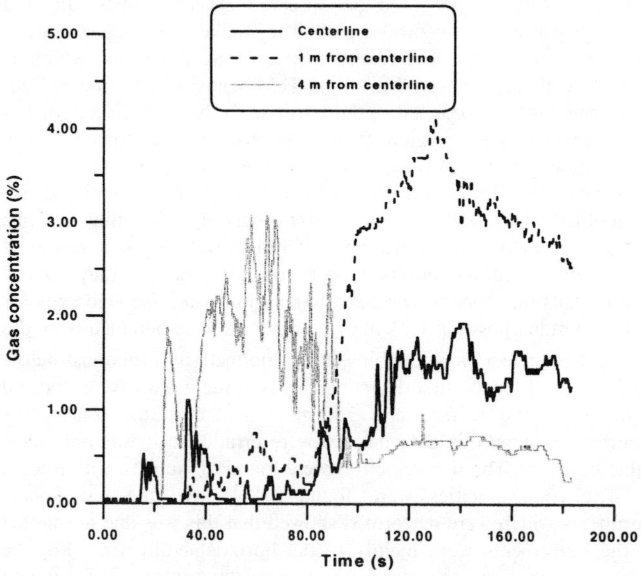

Figure 8 Measurements of gas concentrations at 3 radial positions from the plume centerline, 1.0 m above the water surface. Release rate: 0.083 Nm^3/s. Time zero refers to the opening of the release valve at the basin bed.

As described previously, our investigation covered both the initial transient of the release and the steady-state condition, and the present paper deals with both situations. Figure 8 shows both the initial response and the steady-state situation after the time of release. Typical for all measurements, was relatively high gas concentrations close to the walls of the basin, the low concentrations above the central region at early times, and a transition to low concentrations along the walls and higher in the central region in the steady-state. For the highest release rate it was possible to measure only the initial transient due to the limited supply volume, so this highest rate is not discussed further. As can be seen in Fig. 8, the gas concentration measurements show strong fluctuations. Time-averaged gas concentration profiles in the steady-state situation, are given in Figure 9. The void profiles in the bubble-plume, 1.0 m below the surface, are also given for comparison.

The gas concentration profiles in Figure 9 are polynomial fits drawn through the time averaged experimental data. The gas concentration profiles are much wider than the void profiles. The maximum values are measured as peaks averaged over 1 s. As can be seen from Fig. 9, the highest concentrations observed is about 50% higher than the average values. The high maximum concentrations measured in radial positions 3.0 and 4.0 m, can be explained as release of gas bubbles at the basin wall. It can be seen from Fig. 9 that the measured gas concentration in air is much lower than the void fraction measured 1 m below the water surface. The concentrations, in terms of percent of volume, have been compared in previous work. It has been proposed, as a rule of thumb, that the concentration just above the surface is approximately the same as the void fraction just below the surface (Løes and Fanneløp, 1989).

For both release rates shown, the concentration was highest at the radial position 1.0 m, and somewhat lower above the plume center. Radial position 1.0 m is just outside the fountain; in this region there was an intense "boil", the large amount of gas was expected to be released here.

From 1.0 m outwards, the concentration gradually decreased before, once again, it increased close to the walls of the basin. (Radial position 4.0 m was 0.5 m from the side wall). For visualization, a few experiments were performed with a layer of smoke on the water surface, and here a flow of air could be observed at early times, directed downwards towards the central region. The smoke was swept sideways towards the side walls, and forced to rise along the walls, which is consistent with the measured profiles. This behavior was unexpected, because the gas was expected initially to rise in the form of a light gas plume above the central region. Pictures published from blowouts show this flow configuration. Another surprising observation, was that the gas concentration profiles did not change significantly when the release rate was doubled. This will be discussed further. By mounting the gas analyzers vertically, with a spacing of 0.5 m above the plume center, it was possible to follow puffs of peak concentrations as they moved upwards. This information was used to determine the rise velocities for the gas. At radial position 1.75 m, rise velocities were obtained from the times for first detection of bubbles at the conductivity probe just below the surface, and for the first detection of gas at the gas analyzer, located 1.0 m above the surface and 2.0 m from the plume center (the measurements from this gas analyzer were erratic, so it was only used for this purpose). The rise velocities were found to increase with release rate. The rise velocities were found to be low, the few smoke experiments which were performed showed that this was due to the fact that the movements were mainly in the horizontal direction. For the steady-state situation, the maximum average concentration was about 3% (ref. Fig. 17). Løes and Fanneløp (1989) observed rise velocities of about 0.02-0.008 m/s, which was surprisingly slow; however, their maximum average gas concentration was approx. 1%, which gives significantly less buoyancy than in the present investigation. For comparison with their experimental results, the rise velocities of the present experiments are plotted as a function of the average centerline void fraction close to the surface in Figure 10. Motivation: the gas concentration near the surface is related to the fraction of gas (and the buoyancy) in the plume, and therefore influences the rise velocity.

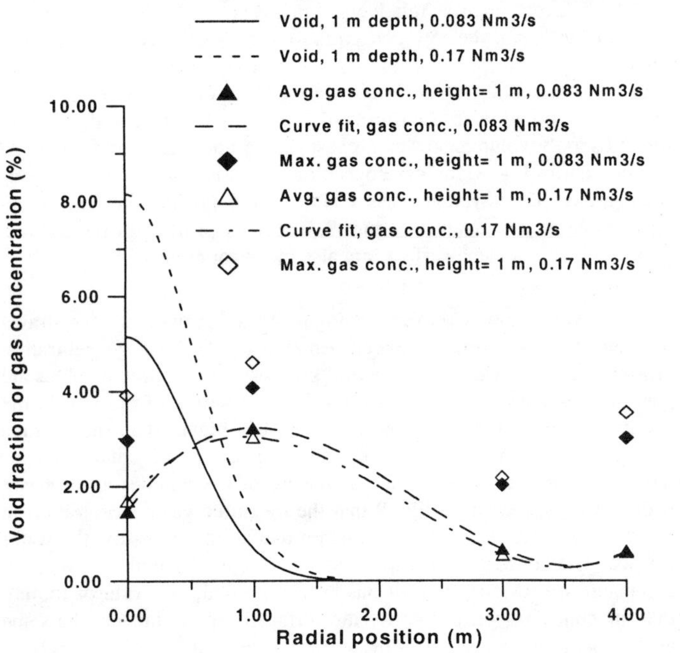

Figure 9 Gas concentration profiles measured 1.0 m above the surface, bubble-plume void fraction measured 1.0 m below the surface. Bubble-plume profiles are Gaussian fits, gas concentration profiles are polynomial fits.

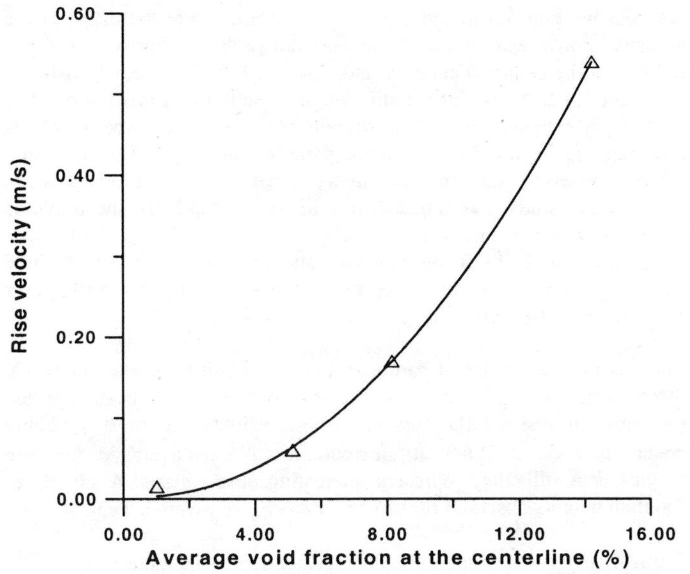

Figure 10 Average rise velocities for the gas immediately above the surface, as a function of the centerline void fraction just below the surface. The three data points at the right-hand side represent the average of the present experiments. The curve is valid for natural gas. The lower, left point is from previous work at 50 m depth and gas concentration measurements 0.5 m above the surface (Løes and Fanneløp, 1989); here the void fraction is estimated, and not measured.

The polynomial curve fit given in Fig. 10 represents the following expression:

$$v_{rise} = -0.0030\varepsilon + 0.0029\varepsilon^2, \qquad (1)$$

where: ε is the void fraction (volume fraction).

Figure 10 shows that the rise velocity is well approximated by a second-degree function. By using the rise velocities given in Fig. 10 (i.e., the vertical component) and the gas concentration profiles, the gas transport in the vertical direction could be estimated. The gas transport profiles are given in Figure 11; the gas transport is defined as the gas concentration multiplied with the rise velocity, i.e., like a local volumetric flow. The gas transport profiles for the bubble-plume are also given in the figure; here, the void-fraction is multiplied with the rise velocity for the bubbles, which is the water velocity plus the slip velocity. The slip velocity is the rise velocity for a bubble relative to the water velocity, which is very close to 0.30 m/s for bubble diameters ranging from about 1.5 to 15 mm (Whalley, 1987).

Fig. 11 explains why the concentration profile did not change with doubled release rate; the concentration is the same, but the rise velocity is approximately doubled so that the gas transport profile is doubled (approximately). The total amount of gas measured by the gas analyzers, is given as a fraction of the released amount of gas in Table 6. For the horizontal transport of gas below the water surface, the gas transport profiles were also integrated and the results are already presented (Tab. 5).

Tables 5 and 6 show that approximately 50% of the gas is transported in the horizontal flow below the surface, all the way to the side walls, whereas about 50% is released to the atmosphere between the plume center and the side walls, mainly in the central areas. The gas transported under the water surface, was most likely released as the horizontal flow was deflected by the side walls, i.e., just outside the last measuring position at 4 m; this is supported by the few smoke experiments, where a vertical flow with relatively high velocity could be observed at the rim of the basin. By adding the gas transport below the surface to that above the surface for the 0.17 Nm^3/s release rate, the total exceeds 100%; this is most likely due to entrainment of air where the return flow from the fountain penetrates the surface. The contribution due to air entrainment is expected to increase with the release rate, Tab. 6 gives an indication of this behavior (gas concentrations were not measured in the horizontal bubble flow).

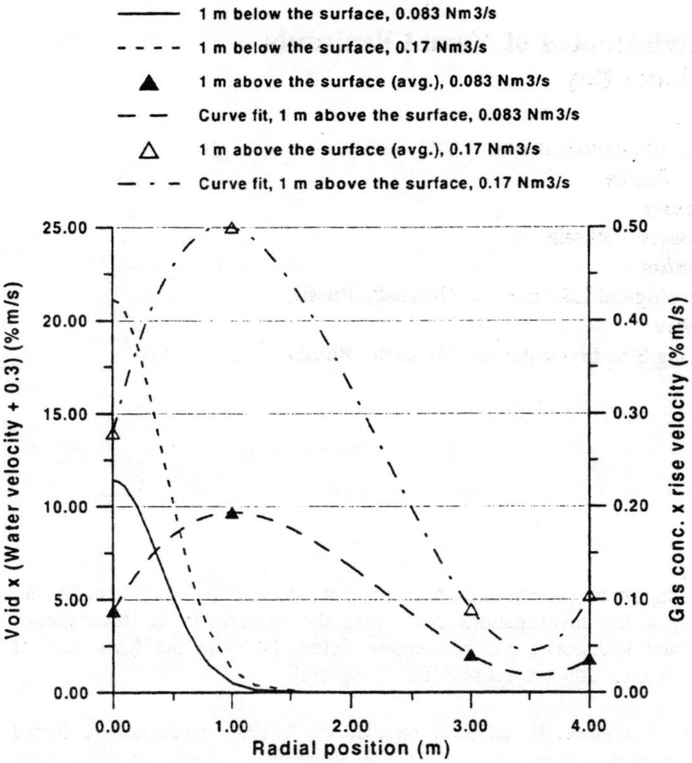

Figure 11 Gas transport profiles, polynomial fits; local volumetric flow rates of gas, 1.0 m below and 1.0 m above the water surface.

Table 6 Gas transport in the vertical flow above the water surface, as a percentage of the release rate.

Release rate (Nm^3/s)	Gas transport above the surface (% of released gas)
0.083	45
0.17	55

CONCLUSIONS

The vertical, steady state bubble-plume is well predicted by the existing integral theory model given by Bettelini and Fanneløp 1993, provided that the centreline void fraction does not exceed approximately 10%. By combinations of release rates and depths resulting in high void fractions in the surface region, fountains of significant heights are established above the reference surface. Previous experiments have most often been performed with relatively low void fractions. This gives a more or less negligible fountain, and little interaction with the surface flow. We observed a deep horizontal flow in the radial direction from the plume centre. The flow from the fountain penetrates the deflected surface flow and contributes to a deeper velocity profile. The boiling area on the surface was observed to be much larger than the theoretical radius of the bubble plume just below the water surface. This can also be explained by the flow interactions between the return flow from the fountain and the opposing plume flow.

The volume concentration of light gas in the air above the plume is observed to be significantly lower than the void fraction in the plume centre just below the surface, in contradiction to a known practical "rule of thumb". Also, the gas concentration profile is much wider and the rise velocity is much lower than expected. A correlation is given to predict the rise velocity of natural gas in air as a function of void fraction in the centreline of the plume close to the surface. It is observed that the gas concentration recordings are dynamic, with measured maximum values about 50% higher than the average values, in spite of the calm conditions in the laboratory. In the field, due to the environmental effects (wind, temperature gradients), it will be more difficult to quantify the hazards related to explosion and fire, because much higher ratios of peak to mean concentrations can be expected. The transport of gas above the bubble-plume is a topic in need of further investigations, and efforts in this direction will be continued.

The theoretical analyses in the present investigation give additional information about bubble plumes, over and above that available in existing theories. At present, work is in progress where all processes involved are modelled numerically by a CFD-code. Integral theories are also being developed to provide simplified methods to predict the plume's and surface flow's dynamic behavior. The objectives for this activity are to be able to predict both

- the level of gas concentration and the hazards related to explosions and fire in the vicinity of the plume,
- and the hazards related to fountain height and surface flow and their interactions with ships and floating installations in the area influenced by the plume.

REFERENCES

Bettelini, M. S. G. and Fanneløp, T. K. (1993): "Underwater Plume From an Instantaneously Started Source", Applied Ocean Research 0141-1187/93.

Fanneløp, T. K. (1993 and onwards): private communication. Professor, Institute of Fluid Dynamics, ETH, Switzerland.

Fanneløp, T. K. and Sjøen, K. (1980): "Hydrodynamics of Underwater Blowouts". AIAA, Paper No. 80-0219.

Hamilton, A. (1986): "Oil - The Price of Power", ISBN 820210891-8.

Løes, M. and Fanneløp, T. K. (1989): "Concentration Measurements Above an Underwater Release of Natural Gas", SPE Drilling Eng., 171-8.

Milgram, J. H. (1983): "Mean Flow in Round Bubble Plumes", J. Fluid Mech., 133, 345-76.

Milgram, J. H. and Van Houten, R. J. (1982): "Plumes From Subsea Well Blowouts", Conference on the Behavior of Off-shore Structures, 659-684.

NOU Report No. 16 (1986): "West Vanguard - Rapporten" (in Norwegian).

Papa, G. and Stone, A. D. (1977): "Cleaning up after a blowout", Ocean Industry, Oct. -77, 35-38.

Scorer, R.S. (1978): "Environmental Aerodynamics", Ellis Harwood Ltd./John Wiley and Sons, New York City.

Topham, D. R.: "Measurements of Deepset Bubble Plumes", Institute of Ocean Sciences, Sidney, Canada (internal report).

Proceedings of the Seventh (1997) International Offshore and Polar Engineering Conference
Honolulu, USA, May 25-30, 1997

ISBN 1-880653-28-1 (Set); ISBN 1-880653-29-X (Vol. I)

Specialized Information System on Environment of Yamal Peninsula and Baydaratskaya Bay

G.E. Odisharia, VNIIGAS, Moskovskaja obl.
Leninsky raion, Russia
A.S. Tsvetsinsky
JSC Eco-System, Moscow, Russia
N.N. Mikhailov
All Russian Research Institute of Hydrometeorological Information, Obninsk, Russia
G.I. Dubikov
Production and Research Institute of Engineering Site Investigation, Moscow, Russia

ABSTRACT

The paper discusses the specialized information system (SIS-Yamal) that is to form the basis for providing the project of the Russian Joint-Stock Company (RJSC) "Gasprom" with environmental data and information . This project is designed for the exploration of Yamal gas fields and gas transportation to European Russia and Western Europe. The paper analyzes the state of the art in this field, objectives of the SIS-Yamal development, application domain of the system, concept of its creation and its architecture. It is emphasized that currently the system is at the initial development stage. However, the design concept, being the basis of the system, made it possible to fulfil, during the first year, one of the most labor consuming tasks consisting in collection of environmental observations (factographical, textual, and spatial) obtained from research on the Yamal project in 1990-96.

Using single methodology and descriptive means, the materials of observations and investigations began to be digitized and included in the archived data bank of the SIS-Yamal. The creation of the archived data bank is considered as the first step in the system's development which solves the problem of saving data and preparing them to be further used.

KEY WORD: Environment, data, information, database, data bank, informational system, subject-oriented tasks, dataware, exploration, gas condensate fields, Yamal Peninsula, Kara sea, Baydaratskaya bay

INTRODUCTION

Timely and efficient provision of data and information (hereafter referred to as dataware) on the environment conditions of Yamal Peninsula and adjacent area of the Kara Sea is one of the most important prerequisites to carry out successfully the scientific and practical investigations concerning the exploration of Bovanenkovo and Kharasavei gas condensate fields (GCF) in the framework of "Yamal", the project of RJSC "Gasprom".

Therefore, in 1996, a number of Russian organizations started preparing data sets on the hydrometeorological, geocryological, ecological and other aspects of environment which is considered a critical component of dataware in research and production activities around GCF.

Nevertheless, given a swift spread of information over the world, dataware is interpreted in a wider sense than merely preparing and using systematized data on computer-compatible media (Kopylov, 1995). For a long time it had been believed that the amount and quality of knowledge ("Artificial intellect", 1990) about the objects of environmental, social and industrial spheres depended directly on the amount of data describing the objects' properties. Thus the major research principle for these objects consisted in implementing the statement "a lot of data imply a lot of information". However, it may also happen that very large amounts of data do not yield new knowledge. This situation can be interpreted as "a syndrome of rich data and poor information", the reality of this situation being repeatedly mentioned by a number of authors (Ward et al., 1986, Lyubarsky, 1990). In this connection, of particular importance here is the transformation of data on the object into information on them.

The use of integrated scheme for acquisition, accumulation and complex analysis of various flows of data currently represent a major principle of new information technologies for research and exploration of the environmental objects (Valkman and Suvorov, 1983, Fedra, 1996). New information technologies (Pospelov, 1983, Suvorov, 1996) along with data and information, cover knowledge and mathematical models ("Algorithmic modelling...", 1992). Much attention is given to a basic part of information technologies which consists of a regular sequence of actions and operations with data to achieve the objective.

Thus, the future development of dataware for the Yamal project are associated with creation of the specialized information system (hereafter referred to *as SIS-Yamal*). This system aimed at realization of the acquisition, accumulation and transformation of environmental data in order to obtain information required for designing, constructing and maintaining gas production and transportation facilities.

The development of this system has been started by the organizations and institutes of Russia involved in the environmental section of the Yamal project in 1996. In order to preserve and systematize unique and expensive data derived from the previous five-year work on the project, a top-priority task was posed that consisted in creating the archived data bank which represents the specialized *SIS-Yamal* data bank maintaining the data sets on the environment of the region under study (Fig.1) with the help of single formatting and documentation means. This data bank is scheduled to be created in 1998. In 1996, concurrent with the formation of the archived data bank on the environment of the region under study, a set of documents was compiled aiming at the development of *SIS-Yamal* in 1997-2000 as a special task of the Yamal project.

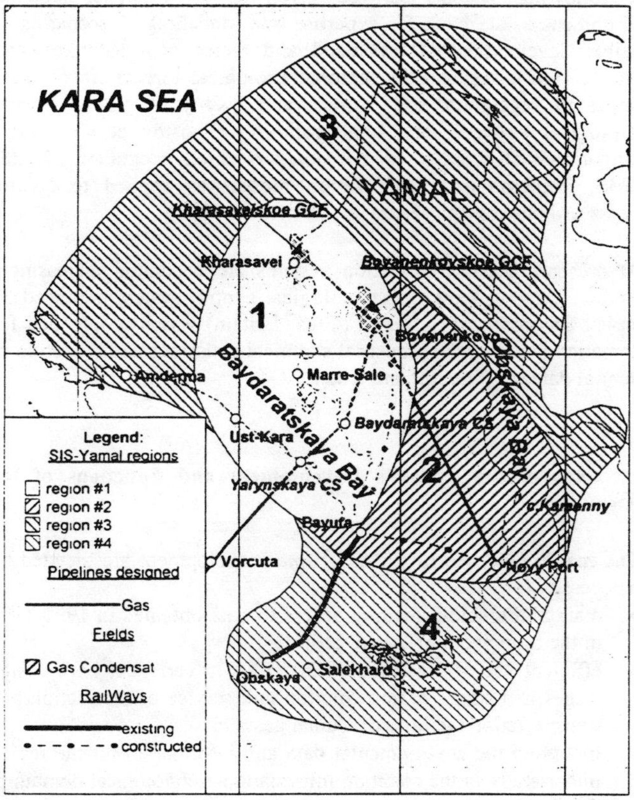

Fig.1 Areas covered on the successive stages of *SIS Yamal* implementation. 1: 1996-97; 2: 1997-98; 3: 1998-99; 4: 1999-2000.

The present paper deals with major aspects of the system design and presents the results of its development.

STATE OF THE ART: BRIEF INFORMATION

Environmental data derived from the Yamal project

Regular observations of the environment in Yamal Peninsula and adjacent area of the Kara Sea started in the first decade of the present century (Baydaratskaya Bay Environmental Conditions, 1997). Extensive multi-discipline expedition studies were performed in 1990-1996 in the framework of the Yamal project which produced a large amount (about 1000 Mbytes) of environmental data for the region under study. Observations, summaries, modelling results and calculations are both held at customer- organizations (All Russian Scientific Research Institute of Natural Gases and Gas Technology - VNIIGAS, Nadym Branch of Gasprom) and at participant-organizations (Joint Stock Company "Eco-System" - Eco-System, Production and Research Institute of Engineering Site Investigation - PNIIIS, Institute of Problems of North Area Exploration of Siberian Division of Russian Academy of Science - IPOS, Institute of Ecology, Flora and Fauna of Ural Division of Russian Academy of Science, Arctic Marine Engineering-Geological Expedition-AMIGE), as well as in a number of other institutions of Russia (All Russian Research Institute of Hydrometeorological Information - World Data Centre - RIHMI-WDC, Murmansk Marine Biological Institute of Russian Academy of Science - MMBI, Arctic and Antarctic Research Institute-AARI, Moscow State University, Regional Centre "Monitoring of Arctic").

Data are mainly held in tables, books, maps, research and technical reports, as well as on computer-compatible media in non-standard formats. Table 1 presents major information about environmental data for Yamal Peninsula and adjacent area of the Kara Sea. The data derived from research programs of the Yamal project make up about 50% of the entire information holding on the environment of this region. As seen in the table, the part of data were digitized in the prosess of *SYS-Yamal* development in 1996. On the whole, the present capabilities of dataware in the project are restricted to (mostly manual) preparation and provision of selected information on environmental and other objects or provision of reports of researchers experienced in conducting research in the region under study.

Major Tendencies in Information Technologies of Environmental Study

At present, the development of information technologies for the application domain treated as a sphere of *SIS-Yamal* activities is oriented to the preparation of global environmental data bases and banks and the development of information systems with a data bank as one of their elements.

By the present time, a number of Russian institutions have created data bases and data banks for different aspects of the environment. The following developments can be given as an example: informational-reference hydrometeorological data banks like "Oceanography: Seas of Russia", "Meteorology", etc. ("Regional oceanographic data banks", 1988, "Operational activities..., 1990). "Ground waters" data base, bank of geological and economical information for oil and gas. "Oil and gas of Western Siberia: prediction, search and exploration" data bank.

However, it should be noted that since data bases and data banks in both Russia and the former USSR were being created in different years by using various software, hardware, information and language means, most of the developments proved to be incompatible.

Table No 1

General Characteristics of Environmental Data for Yamal Peninsula And Adjacent Area of the Kara Sea Including Baydaratskaya Bay

No	Types of observations and description of materials	Volume (Mbyte) 1	2	3
	MAPS			
1.	**Topographic charts of Yamal Peninsula** Scale: 1:1000000, 1:100000, 1:25000, 1:5000	600	-	200
2.	**Bathymetric charts of Kara Sea parts adjacent to Yamal Peninsula. Scale:** 1:1000000, 1:500000, 1:200000, 1:100000)	35	-	3
	Total Number of Charts:	**635**	**-**	**203**
	YAMAL PENINSULA			
1	**Engineering and geological conditions** (observations of composition, chemistry and thermics of ground; results of geocryological analysis and regioning around Bovanenkovo and Kharasavei GCF in 1972-94)	640	435	293 (78)
2	**Hydrology of surface waters** (hydrographic and hydrochemical observations, observations of ground erosion in 1989-95)	265	224	21
3	**Biota and Soil** (observations of soil, vegetation and animals in 1962-95)	120	95	22
4	**Social and economic conditions** (social, economic, medical, ethnographic and archeological investigations around Bovanenkovo and Kharasavei GCF in 1975-95)	75	60	5
	Overall data on Yamal Peninsula:	**1100**	**814**	**341**
	BAYDARATSKAYA BAY			
1	**Engineering and geocryological conditions** (observations of composition, chemistry and thermics of ground; results of geocryological analysis and regioning in the transition area, etc. in 1972-94)	90	67	7
2	**Lythodynamic conditions** (observations of sediment dynamics and caving in 1980-94)	26	22	13
3	**Hydrometeorological conditions** (observations of 10 hydrometeorological stations in 1933-95 and of more than 100 RV cruises in 1921-95, ship hydrometeorological observations in 1900-95, observations of currents in 1965-93, etc.)	55	5	24
4	**Ice conditions** (observations of characteristics and physical and mechanical ice properties from aircrafts, helicopters, satellites and caterpillar sled trains; survey of ground ice and furrows in the transition area, etc. in 1925-94)	135	50	8
5	**Pollution of marine and land objects** (cruise observations in 1985-95)	8	3	1
6	**Biota** (hydrobiological observations of 13 RV cruises in 1934-95)	2	1	1
	Overall data on Baydaratskaya Bay:	**316**	**148**	**54**
	TOTAL:	**2051**	**962**	**598**

Comments:

1 - All known data
2 - Data obtained under «Yamal» Program
3 - Data included into Archive Data Base of *SIS-Yamal*

In western countries, similar works aimed at creating centralized data archives providing for standardization, archiving and application of data based on mass-memory soft- and hardware and on the access through computer networks. In the USA, the most representative systems of centralized data archiving belong to two largest environment-related organizations: NOAA (NOAA central..., 1993, Soracco, 1995) and NASA (Olsen et al., 1991).

A special feature of the current period of the information systems (IS) development is in the fact that the instrumental IS-supporting software is being most actively developed in the form of geoinformation systems (GIS) and data base management systems (DBMS). In Russia, there are now 100 organizations and firms (Miller, 1994) providing software of instrumental GIS, such as Arc/Info, Map/Info and GeoDraw/GeoGraph, which, when combined with powerful DBMSs (Oracle, SQL Microsoft, etc.), allow rather sophisticated IS to be created.

Against the background of intensive development of instrumental IS-supporting means, there are extremely few examples of specialized IS for the environment which are capable of implementing the integrated technology of data accumulation, management and processing by inexperienced end users. However, these works seem to be expanded. The following systems can be given as an example: regional environmental data bank for expertise and modelling of scenarios of possible development of urbanized and recreational (Shitikov and Kostina, 1992); system of a regional ecological forecast to estimate changes in vegetation, soil, storage and quality of ground waters, hydrological network, and total pollution (Shustov et al., 1993); information system based on the object-oriented integration of GIS, DBMS, imitation and optimization models as applied to natural process and phenomena (Fedra and Jamieson, 1993).

At present, "alert" information systems are becoming increasingly abundant. Among these is the ecological control system produced by "Implex Environment Systems" (Great Britain) which is intended for automatic monitoring of industrial chemical emissions and reduction of potential damage (Qureshi and Khar, 1994).

Objectives of the System's Development and Functions of the System

The main objectives of the *SIS-Yamal* development are dictated by the necessity for:

- maintaining and providing access to data obtained in 1990-1996 in the framework of Yamal project;
- efficient use of the obtained materials to carry out both design stages and stages of construction and service of production and transportation facilities of Yamal gas;
- including the environmental data and information for the region under study in the common information-technological domain of RJSC «Gasprom».

Exploration of resources of Bovanenkovo and Kharasavei GCF will undoubtedly require efficient transformation of observations and modelling results, as well as expert estimates on the environment into adequate and diversified information to elaborate economical and environmental measures and to make management decisions at different stages of GCF exploration. The *SIS-Yamal* purpose consists in implementing the process of transformation (Fig.2) by using the integrated technological scheme of acquisition, accumulation, processing, storage, analysis, interpretation and provision of environmental data to users.

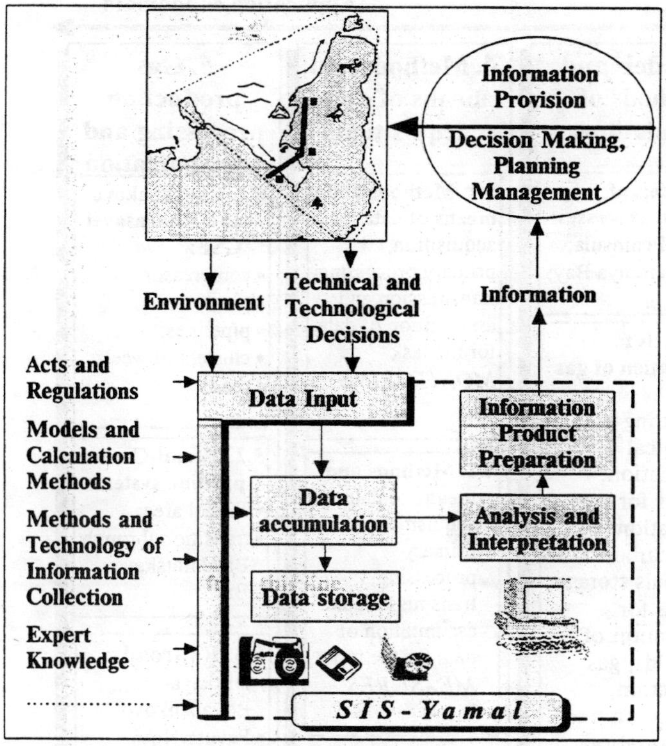

Fig.2 Role and place of *SIS-Yamal* in information providing for tasks of design, building and exploitation of sites of Yamal gas production and transport as well as for tasks of monitoring of environment and social conditions of Yamal Peninsula and adjected part of the Kara Sea.

Major subject-oriented tasks to be solved by using *SIS-Yamal* are determined by the following high-priority sectors of dataware for design, construction and service of gas production and transportation facilities:

- provision of data and information by requests as hard copies or in the electronic form (hereafter referred to as *REQUESTS*);
- obtaining environmental characteristics (including extreme) and assessment of the background conditions of the environment (*BACKGROUND*);
- environment impact assesment (*IMPACT*);
- elaborating measures to reduce environmental impacts during construction and maintenance of industrial facilities (*MEASURES*);
- monitoring of the current environmental conditions (*MONITORING*);
- management of reliable and safety maintenance (*MANAGEMENT*).

APPLICATION DOMAIN OF THE SYSTEM

Application domain (AD) of the *SIS-Yamal* is defined by above-mentioned sectors of dataware and consists of several classes of AD objects (table 2):

- *"Environment"* - data and information on environmental conditions of Yamal Peninsula and adjacent part of the Kara Sea;
- *"Sites of gas production and transport"* - technology-engineering characteristics of sites of gas extraction, processing and transport;
- *"Models and methods of calculation of environmental characteristics"* - models and methods of calculation of environmental characteristics for the region of investigation;
- *"Legal and regulatory acts"* - legal and regulatory information, necessary to make decisions on appropriate issues at different phases of site exploration;
- *"Tools and methods of data collection"* - methods and techniques of collection, primary processing and dissemination of current data on conditions of environmental and industrial objects to be used for operational solution of problems mentioned above.

Each class of AD objects determines appropriate data and information sets to be included into the *SIS-Yamal*, i.e. subject-oriented sections of the system (table 2). At that the AD boundaries of the system (i.e. subject-oriented sections) may be expanded due to necessity of new tasks to be performed, change of geographical orientation of the system etc. For instance, the class of AD objects of the system "Environment" may be expanded by inclusion of the new subject-oriented section "Geological characteristics of gas deposits".

CONCEPT OF THE SYSTEM DESIGN

Due to several reasons (insufficient formalization of application domain, complexity of tasks, difficulties in application of standard means under conditions of discipline variety of large data flows etc.) the system being designed should have unique properties. Since ready-made software-technological solution which makes it possible to realize dataware of considered scale is currently not available, justified selection of the effective concept of the *SIS-Yamal* design becomes of special importance.

Considering the facts mentioned above, the concept of the *SIS-Yamal* design should be based on the following basic principles:

- *compositional approach*, meaning isolation of structural elements of the system (components, subsystems, blocks) by their application domain as well as functions with respect to their interconnection and interrelation (technological, technical methodological, organizational) and dividing of the system development into appropriate steps;
- *open character of the system*, meaning its ability to expand the range of subject-oriented tasks by means of both inclusion of new classes of application domain objects and adaptation to other (besides Yamal) regions;
- *unity of realization tools*, meaning use of unified standards of information, methodological, hardware and software tools, being obtained by co-ordination and unification of data description means, classifiers and codifiers, data exchange protocols and formats, specifications of basic computational platforms, peripheral equipment and software etc.

Table 2

Application domain of *SIS-Yamal*

Application subdomain →

Subject oriented sections ↓

1. Environment and social conditions	2. Acts, regulations and instructions	3. Models and methods of calculations	4. Methods and means of data acquisition	5. Gas production processing and transportation
1.1. Yamal Peninsula • maps • hydrometeorological conditions; • engineering-geological and geocryological conditions; • geomorphology of land; • hydrochemistry of land waters and geochemistry; • biology; • social and economical conditions; •	**2.1. Legal information** • federal; • territorial.	**3.1. Models of natural processes:** • Yamal Peninsula; • Baydaratskaya Bay; •	**4.1. Methods and means** of data acquisition, primary processing, transmission and assimilition of data for the task *MONITORING*	**5.1. Bovanenkovo and Kharasavei GSF:** • compressor plant; • pipelines; • clusters of weels; • roads; • ....
1.2. Baydaratskaya Bay • maps • hydrometeorological conditions; • ice conditions; • engineering-geological and geocryological conditions; • geomorphology of bottom and coast; • hydrochemistry (including water pollution); • hydrobiology; •	**2.2. Regulatory documentation:** • documents of environmental protection and monitoring organizations; • standards (national and international); • regulatory documents referring to construction ; • regulatory documents of supervision bodies referring to construction; • repartamental regulatory documents; •	**3.2.Models for calculation of gas fields:** • processing of geological information; • models for calculation hydrocarbon raw materials storage; • models for exploration of gas fields, gas production, preparation and transportation; •	**4.2. Methods and means** of data acquisition, primary processing, transmission and assimilation of data for the task *MEASURES*	**5.2. Yamal-Center pipeline system:** • Yamal area; • transition through Baydaratskaya Bay
		3.3. Auxiliary means: • Standard methods of information processing; • Specialized software.	4.3.	**5.3. Railroads:** • Obskaya-Bovanenkovo • Payuta-Novy Port.
				5.4. ....

BASIC STRUCTURAL ELEMENTS OF THE SYSTEM

By analogy with the adopted approach for representation of a data model (Revunkov et al.,1992, Martin,1984a, Martin,1984b), three levels of abstraction are used to define the *SIS-Yamal* structure:

- *Physical level* reflecting views of a system programmer, developing basic elements of the system. At this level the *SIS-Yamal* consists of different components (research-methodological, informational etc.).
- *Realisation level* reflecting views of an applied programmer, using components of the system to develop tools for implementations of its functions. At this level the system is defined as a set of functional subsystems, based on combinations of its components. Functional subsystems may be divided into functional blocks as subsystem elements, realizing definite operations with data.
- *External level* reflecting views of a system user. At this level the *SIS-Yamal* is considered as a set of *SIS-Yamal* workstations designed to handle data within the boundaries of its functional subsystems. A workstation represents a part of the system visible for a user and is supported by a single functional block or several of them.

Components of the System

On the physical level the *SIS-Yamal* consists of the following basic elements:

- research-methodological component, determining the mode and the subject orientation of the *SIS-Yamal* operation;
- information component, which consists of the information holding in the form of archived databases of different data types and integrated database as well as tools of unification of data presentation in the system;
- linguistic component, representing tools to support the unified set of terms, notions and information languages used for data description and manipulation;
- software-technological component, which consists of DBMS (Oracle type), GIS (Arc/Info, Map/Info type), software package to support an expert system, calculation and modelling programs, realizing information technologies in the system;
- hardware component, representing computer and periphery for operation of the *SIS-Yamal;*

- organizational-legal component, ensuring personnel (foundation of a special structural division responsible for support and use of the system) and organizational-legal status of the system within the RJSC "Gasprom" structure, including rules of interaction with other information systems.

Functional Subsystems and Blocks

Subject-oriented tasks considered above are executed by the following functional subsystems of *the SIS-Yamal*: subsystem of archived banks, subsystem of integrated data bank and the subsystem of problem-oriented applications. Peculiarities of interaction of subsystems are shown at Fig.3.

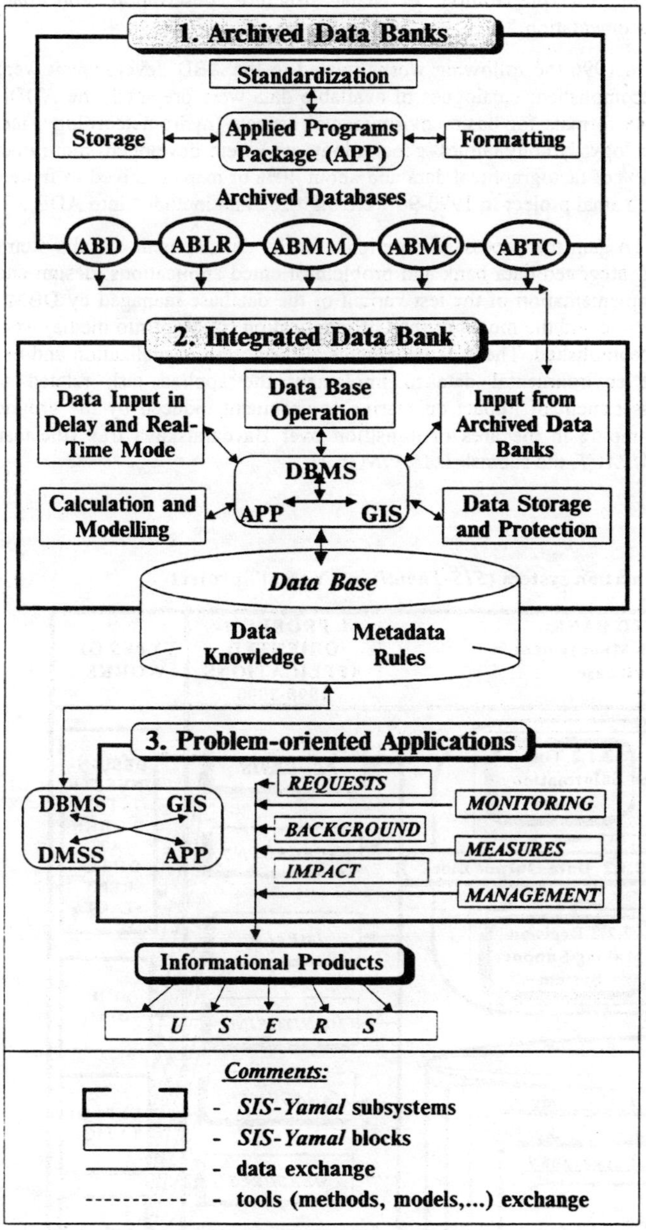

Fig.3 Functional structure of *SIS-Yamal*

Subsystem of archived banks (AB) includes the following:

- archived data bank of environmental conditions of Yamal Peninsula and adjacent part of the Kara Sea (ABD);
- archived bank of legal and regulatory information (ABLR);
- archived bank of models and methods of calculation of environmental conditions characteristics for the region of investigation (ABMM);
- archived bank of methods and tools for collection of environmental data (ABMC);
- archived bank of technical and technology-engineering characteristics of sites of gas extraction, processing and transport (ABTC).

Development of archived banks is generated by the need to overcome initial informational and technological incompatibilities of data sets, methodological and software-technological tools for data collection, processing and analysis by bringing them to the unified technical, language and linguistic media. AB have their own importance as an effective and reliable tool of long-term data storage.

Subsystem of integrated data bank includes the following:

- database, i.e. complex-organized set of interconnected data, information, knowledge, calculation scenarios etc., which integrates the archived database in accordance with current subject orientation of the system on the whole;
- software-technological tools for data entry, database management, calculation and modelling, supported by instrumental DBMS and GIS subsystems.

The subsystem is designed to fulfil definite operations related to loading of data, coming from the subsystem of AB; registration and management of data and results of calculation and modelling; preparation of observation data and results of calculation and modelling to implement applied tasks and to store the results of implementation. From user viewpoint the subsystem will consist of several workstations such as Operator and Administrator of integrated database.

The subsystem of problem-oriented applications is designed for the following:

- analysis of data and information to implement such applied tasks as *BACKGROUND, IMPACT, MONITORING, MEASURES* and *MANAGEMENT*;
- preparation of output information products of the system in the form of hard and electronic copies (materials for Technical-Economical Ideas, Technical-Economical Justifications, Technical Projects, Working Documents, consultative materials, bulletins, historical-reference materials, atlases etc.) in accordance with the required formats;
- dissemination of information products according to the list of system users and requests sent by regular and electronic mail.

The subsystem consists of functional blocks to fulfil separate applied tasks and generating of information products. Workstations of the subsystem may be located at the central (organization-customer) and local (region of investigation) levels.

THE SEQUENCE OF DEVELOPMENT OF THE SYSTEM AND ITS CURRENT STATUS

The general approach to development of the *SIS-Yamal* is shown in table 3, where the block-scheme is given to illustrate the content and the structure of the system combined with information on the sequence and dates of preparation of its basic elements. The system is developed in several steps:

- the first step - preliminary works, i.e. accumulation of data for all classes of the application domain of the system and preparation of the subsystem of archived banks to ensure initial conditions for development of the system itself and storage of accumulated data;
- the second and the third steps: a) design of the system, which includes the following: preparation of specifications and various design documentation for the structural elements of the system; development of the basic components of the system - informational, linguistic, software-technological etc.; b) creation on the basis of the components, of the subsystem of integrated data bank to accumulate and operate the complex database of the system using DBMS- and GIS-technologies;
- the fourth step - creation of the subsystem of problem-oriented applications, intended for analysis and interpretation of data required to implement such applied tasks as *BACKGROUND, IMPACT, MONITORING, MEASURES* and *MANAGEMENT* and to obtain information products for making technical, technological and environmental decisions related to exploration of Yamal gas fields and its transport.

In fact the development of the *SIS-Yamal* has been started only since 1996. Currently implementation of initial step is in progress, i.e. the archived data bank (ADB) containing environmental data for Yamal Peninsula and Baydaratskaya Bay is developed. The ADB will define to a great extent the ability of the system to ensure dataware for the "Yamal" project.

The ADB is developed on the basis of uniform language-informational and software-technological tools used to prepare factographical, textual and spatial data on PC media. These data cover all environmental characteristics of the region under consideration

To enclose factographical data into the archived database the specialized formatting system in the form of archived data description language (ADDL) has been developed. The software system SHIP (textual data) and the archived database management system ADBMS (factographical data) are used for software-technological support of the ADB development. These language means and software tools provide the compatibility with existing environment data bases. For instance, user can convert the data from ADDL media to standard DBMS formats and vice versa using special functions of ADBMS.

To prepare the archived database section on spatial data, instrumental geoinformational systems Map/Info and Arc/Info are used. The unified technology of the ADB development is based on specially designed detailed methodological recommendations containing requirements and rules for preparation of data structure description and data documentation.

In 1996 the following works related to the ABD development were accomplished: catalogues of available data were prepared; the ADDL data formats for basic environmental aspects (hydrometeorology, ice, ecology, lythodynamics, geocryology etc.) were developed; more than 50% of factographical data and about 40% of maps received in frames of Yamal project in 1990-96 were digitized and included into ADB.

To gain the necessary experience in the development of subsystems of integrated data bank and problem-oriented applications, design and implementation of the test variant of the database managed by DBMS Oracle and the model of the GIS-application (in Map/Info media) were accomplished. These developments allow access, visualization and use of environmental data to implement the applied task related to assessment of impact on marine environment induced by the ground damping in the area of transition over Baydaratskaya Bay (the task *IMPACT*, the subtask *DAMPING*).

Table 3.

Succession of development and use of specialized information system (*SIS-Yamal*) in "Yamal" project

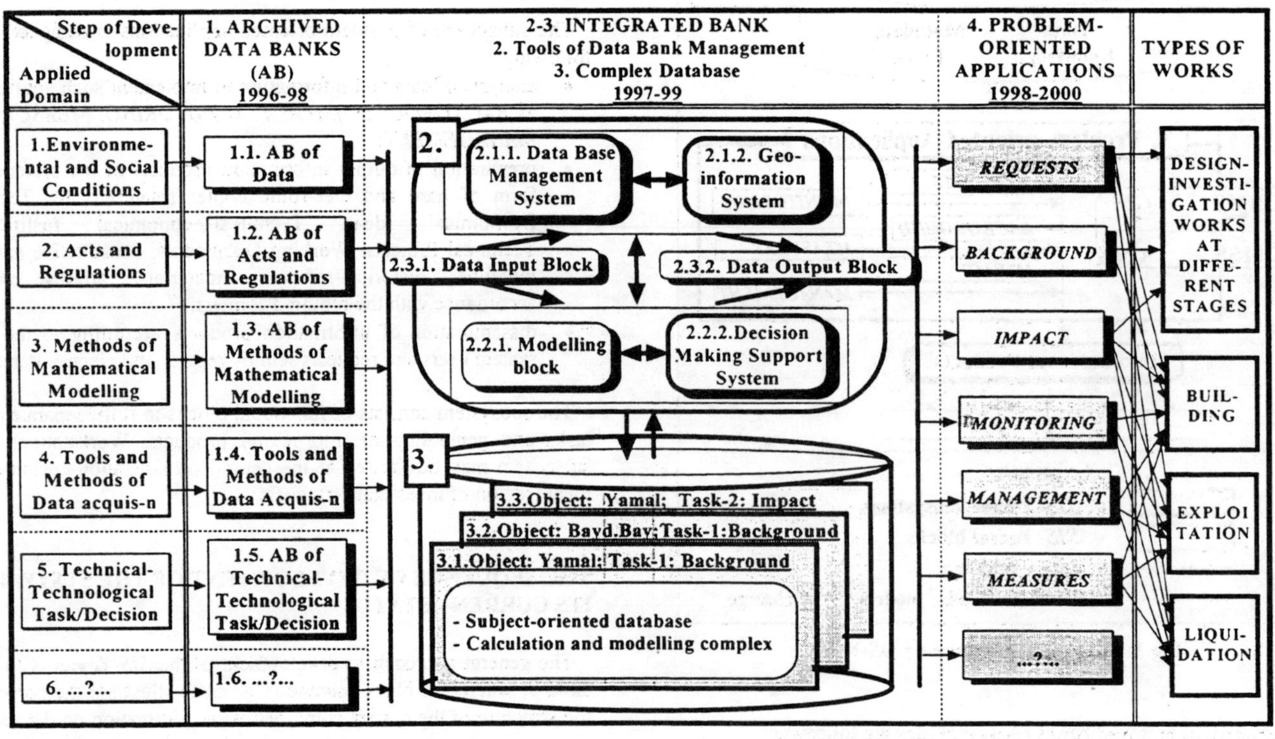

CONCLUSIONS

As might be seen from the concept of the *SIS-Yamal* design which was considered above, it has the following basic peculiarities:

- compositional approach realized by isolation of components, subsystems andblocks of the system and by dividing the system development into several steps;
- isolation of the step related to creation of the archived data banks in order to save ("data rescue") and prepare for subsequent use the data of observations and investigations, accumulated in the process of Yamal project implementation;
- possibility to determine priorities in development of the system with regard to importance of some subject-oriented tasks;
- from the beginning of the system development application of unified decisions and unified methodological and software-technological tools for various types of heterogeneous data.

The concept of *SIS-Yamal* design is realized in practice through development of the archived data bank containing data on natural conditions of the region under consideration. The successful accomplishment of this work will create th information basis to subsequently develop the system elements of a higher level.

REFERENCES

Kopylov, VA (1995). "Problems of informatization and legislation in the field of information", *RTI,* Ser.2, No 6 (In Russian).

"Artificial intellect" (1990). Book 2. "Models and methods", Guide/ed. Pospelov D.A., Moscow: Radio i svyaz (In Russian).

Ward RC, Loftic JC, McBridge (1986). "The "data rich but information - poor" syndrome in water quality monitoring", *Environmental Monitoring,* No 10.

Lyubarsky YY (1990). "Intellectual environment systems", Moscow: Nauka (In Russian).

Valkman YP, Suvorov AI (1983). "Principles of building integrated systems of database models", *Programmnye producty and systemy,* No 3 (In Russian).

Fedra K (1996). "Integrated environmental information systems: from data to information", *Report on NATO Advanced Research Workshop "Integrated Environmental Data Management Systems",* Izmir, Turkey, September 1966.

Pospelov GS (1983). "Artificial intellect: modern information environmenty", *Vestnik AN SSSR,*- No 6 (In Russian).

Suvorov., (1996). "On the terms "technology", "information technology" and "new information technology"", Programmnye producty and systemy.- No 3 (In Russian).

"Algorithmic modelling. Instrumental tools and models" (1992), ed. Ivanischev VV, SPb: SPII RAN (In Russian).

Baydaratskaya Bay Environmental Conditions. The Basic Results of the 1990-1996 Study for Pipeline Yamal-Center Crossing Design, 1997, Moscow, Geos (In Russian).

"Regional oceanographic data banks" (1988), Ed. Lamanov VI, Moscow, Gidrometizdat (In Russian).

"Operational activities of Goskomgidromet of the USSR in relation with obtaining, collection, processing and dissemination of hydrometeorological and geologogeophysical information" (1990), Leningrad, Gidrometizdat (In Russian).

"NOAA central library catalog of CD-ROM holdings", 1993.

Soracco M (1995). "NOAA Satellite information system (NOAASIS)", *Earth System Monitor,* V 6,- No 2.

Olsen LM, Closs JW, Vanderpool J, Humberson W (1991), "NASA's Climate Data System", Primer Version 4.0.

Miller SA (1994). "Review of Russian market of software for geoinformation technologies", *Annual review "Software-hardware, digitized material holdings, services and legal regulatory base of geoinformatics in Russia and CIS"*

Shitikov VK, Kostina NV, (1992). "Information system for ecological expertize of urbanization and recreation territories", *Ecological fundamentals of optimization of urbanization and recreation environment: Abstracts of papers of international meeting, Togliatti, 30 May - 2 June 1991 - Part 1,* Togliatti.

Shustov MV, Polyansky YV, Biktimirov TZ, Rosenberg GS (1993). "Database "Region-Ulyanovsk" as a basis of effective use of nature in Ulyanovsk region", *Ecological problems of basin of large rivers: Abstracts of international conference,* Togliatti, 6-10 September 1993 - Togliatti (In Russian).

Fedra K, Jamieson DG (1993). "An object-oriented approach to model integration: a river basin information system example", *In: Kovar K. and Nachtnebel H.P.[eds.]: IANS Publ.-,* No 235

Qureshi MA, Khar S (1994). "Environmental pollution information system using remote sensing and GIS: report on training programme", *J.Sci. and Ind. Res.* 53, No 4.

Revunkov GI et al. (1992). "Data and knowledge bases and banks", Moscow: Vysh. Shk (In Russian).

Martin J (1984a). "Planning of computerized systems development", Moscow, Finansy i statistika (In Russian).

Martin J (1984b). "Organization of databases in computer systems development", Moscow, Finansy i statistika, 1984 (In Russian).

Proceedings of the Seventh (1997) International Offshore and Polar Engineering Conference
Honolulu, USA, May 25-30, 1997

ISBN 1-880653-28-1 (Set); ISBN 1-880653-29-X (Vol. I)

Environmental Impact Assessment for the Baydaratskaya Bay Crossing of the Yamal - West Europe Gas Pipeline System

Alexander V. Beloshapkov, Vladimir B. Leschinsky and Alexander S. Tsvetsinsky
EPS Eco-System
Moscow, Russia

Konstantin A. Kashunin
JSC Petergaz BV
Moscow, Russia

ABSTRACT

The Baydaratskaya Bay crossing in the Kara Sea is a part of the Yamal - West Europe gas system project. It is the first experience of the large-diameter gas pipeline engineering in the Russian Arctic seas. The Environmental Impact Assessment of the construction for the first two lines of the crossing is based on the results of six-year scientific program of Baydaratskaya Bay environmental studies. Laws, rules and regulations applicable to Environmental Impact Assessment in the Russian Federation are considered. The sources and types of possible impact during construction and operation stages are shown.

KEY WORDS: Environment, impact, assessment, pipeline, underwater crossing, ecological examination.

INTRODUCTION

The development of the Bovanenkovskoye gas field on the Yamal Peninsula in Northwest Siberia by JSC Gazprom, the Russian Gas company, requires a gas transport system to be laid across the Baydaratskaya Bay in the Kara Sea. The underwater crossing of main gas pipeline will connect the Yamal gas fields and onshore pipelines in the Yamal Peninsula with the Ural - West Europe pipeline system. The length of the offshore part of the crossing is about 67 km (Fig. 1), the maximal depth reaches 22 m. The Baydaratskaya Bay underwater crossing will consist of 8 parallel lines of 1220 mm gas pipelines. Detailed engineering has been completed for the first two lines. It was carried out by the Joint Stock Company Petergaz BV.

The environmental conditions of the crossing region are rigorous. The ice free period in the Bay is generally about 70 days. Storms with intense waves, currents and surges are a common occurrence during the ice free period. The shallow zone of the crossing is covered by ice in the winter, while drifting ice dominates in the central part of the Bay. The environmental studies were carried out by the engineering production society (EPS) Eco-System during 1990-95 within the framework of JSC Gazprom program. It included 10 offshore and 15 onshore expeditions including oceanographic, side scan, shallow seismic and biological surveys of the area, as well as geological and sediment

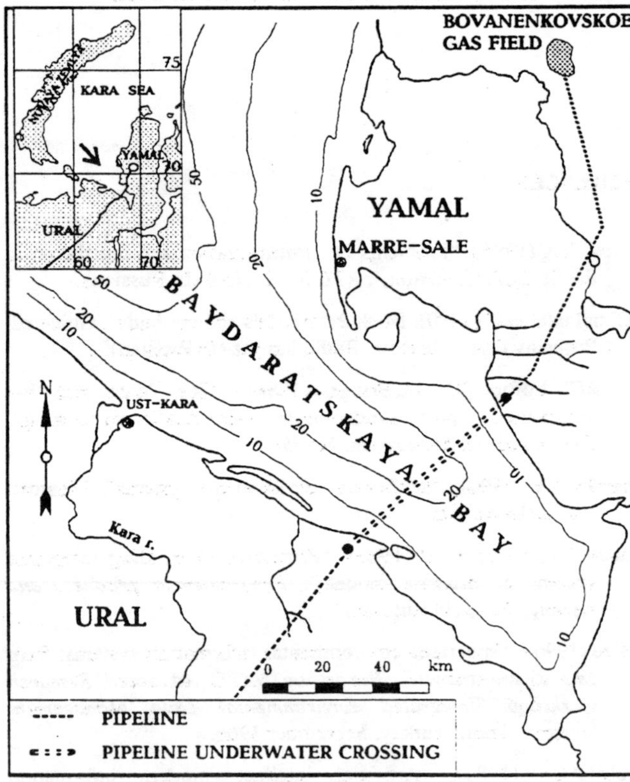

Fig. 1 Location map

deposit probing. Data from several satellites were collected and analyzed as well. Some models and calculation methods for evaluation of the crossing construction impact on the environment of the Baydaratskaya Bay were worked out and adjusted for the region conditions. Specialized information system is considered in the paper by Odisharia et al (1997).

This paper describes requirements of environmental protection legislation and regulations in the Russian Federation and some results obtained by EPS Eco-System and Petergaz BV in the framework of the preparation of the Environmental Impact Assessment (EIA) at the stage of feasibility study report.

CHARACTERISTICS OF CROSSING CONSTRUCTION

The design of the crossing provides optimal technology for construction in different sectors (Fig. 2). Seven sectors were selected as follows:

- Yamal and Ural onshore sectors;
- Yamal and Ural shore crossing sectors - to 1.5 m sea depth offshore;
- Yamal and Ural shallow sectors - from 1.5 m to 13.8 m sea depth;
- Deep-water sector.

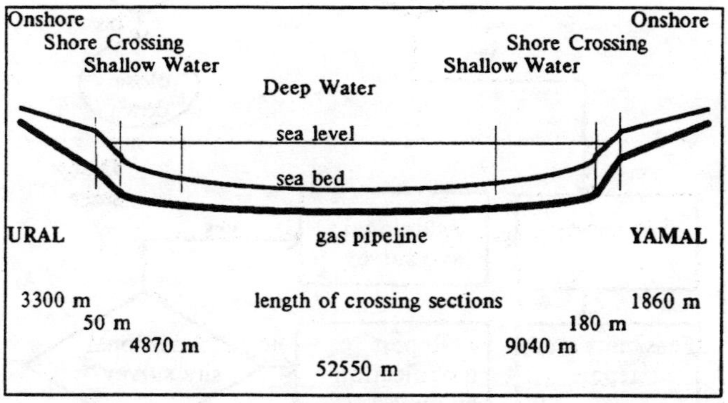

Fig. 2 Demarcation of Crossing Sections

The construction of the onshore sectors will be carried out by underground pipeline laying. The construction of the Yamal and Ural shore crossing sectors will be accomplished by construction of temporary berms, seawalls and cofferdams (Beloshapkova et al, 1996).

The crossing construction in the shallow sectors will consist of operations including dredging of underwater trenches and access channels for laybarges, pipe laying, and backfilling of trenches.

The pipeline will be installed on the sea bottom from the laybarge in the deep water sector. Being laid it will be post-trenched with a special device and will be backfilled by a self-propelled cutter suction dredgers using suction pipe. It is planned to carry out the first phase of crossing construction during three years.

LAWS, RULES AND REGULATIONS APPLICABLE TO ENVIRONMENTAL IMPACT ASSESSMENT IN RUSSIAN FEDERATION

Environmental protection law (19.12.1991; Amendments 21.02.1992 and 02.06.1993)

This law covers the environmental issues generally. Water pollution control, rules and regulations concerning waste water discharge into coastal water are considered. The law defines principles of ecological safety at different stages of construction work.

Earth depths law (21.02.1992; Amendment 03.03.1995)

The law lays the basis for geological study, rational use and protection of earth depths in the Russian Federation. State control for these activities is described in Art.3. Art.4 defines payment principle for exploitation of earth depths within local borders.

Continental shelf law (25.10.1995)

The Russian Federation executes the jurisdiction in relation to scientific sea investigations; protection sea environment due to exploitation of living resources, waste water discharge into coastal waters; laying and exploration of underwater cables and pipelines.

Ecological examination law (23.11.1995)

The purpose of this law is stated (Art.1) as to establish the correspondence of planning activity to ecological requirements. There are two types of ecological examination in the Russian Federation: state and public (Art.4). The law is obligatory in planning and controlling of land use, regulations on buildings and constructions, protection of forests and green areas, etc.

The EIA preparation at the stage of the feasibility study report of the Baydaratskaya Bay underwater crossing project is carried out with the use of the mentioned laws and by-laws, as "Instructions on ecological substantiation of economical and other activities" (Instructions.., 1995) and departmental rules and regulations e.g. manual "Bases of ecological safety" (Bases..., 1996), compulsory for branch structures of a fuel and energy complex of the Russian Federation. JSC Gazprom regulations (Environmental protection regulation..., 1994) are compulsory for oil and gas industrial complexes, located on the territory of the Yamal-Nenez national region.

These regulations define a set of ecological norms, requirements and restrictions during investigation, designing, constructions and operation of the branch enterprises and recultivation of disturbed grounds in the course and upon completion of construction and operation in the northern regions of Tyumen area in view of environmental specific features and traditional economic activities of the local population. These regulations are agreed with Yamal-Nenez regional committee for environmental protection and rational use of natural resources and are compulsory for use in territory of the region.

The design documentation concerning environmental safety in the Baydaratskaya Bay is a complex problem having some features in scientific and in methodical aspects. There are no regulations and experience in EIA use for underwater pipelines construction and operation in the Arctic seas in the Russian Federation, but anyway some possible available experience with EIA at similar projects was taken into consideration. Some recommendations of onshore environmental safety manuals worked out by E&P Forum were taken into account, as well.

ENVIRONMENTAL IMPACT ASSESSEMENT PROCEDURE IN THE RUSSIAN FEDERATION

The Environmental Impact Assessment in the Russian Federation maintains a certain procedure where it starts with the account of the ecological requirements in the preparation of the decisions on social and economical development. Its objective is to find out ecological, social, economic and other consequences of economic activity which may bring a critical threat to the environment and society as well. The EIA procedure in the Russian Federation comprises two independent stages:

- Environment Impact Assessment report preparation;
- State Ecological Examination of the planned activity projects.

According to the Standards of the Russian Federation (Instruction on the procedure..., 1995, Procedure of development..., 1995) EIA procedure is obligatory at the stage of the investment substantiation and at the stage of feasibility study report. Some aspects of EIA shall be included into the pre-design documentation at the stage of business proposition study. The positive decision of the local authorities on the business proposition is necessary for the onset of the investment substantiation study. The EIA procedure results in the decision of state ecological examination bodies on the permissibility or inadmissibility of the possible impact on the environment (Fig. 3).

In the Russian Federation, as well as in the majority of developed countries, the "permissive principle" of ecological examination (Rule of ecological..., 1996) is accepted. The state ecological examination is carried out with the use of independent experts. Depending on the significance of the project the examination shall be carried out at the federal level or at level of the Federation subject. The final decision about permissibility or inadmissibility of the project according to the results of state ecological examination will be made by the federal or local bodies consequently.

The technique of EIA preparation in the Russian Federation and developed countries, working onshore and offshore oil and gas deposits of the Arctic is different. In these countries the technique of EIA preparation is based on the principle of «normative regulation» of the project safety (Regulations..., 1990). An estimation of possible impact on environment in terms of some ecological risk is considered here within the framework of the management of project risk on the whole. The decision making on the impact permissibility will start with risk estimation of established limits of influence on environment components. In the Russian Federation the technique of EIA is based on an "obligatory principle" that is the principle of compatibility with numerous environmental specifications. The decision making on planned activity possibility will start with comparison of levels of the project impact on environmental components with the allowable levels regulated by the environmental specifications. The evaluations of the risk as a part of EIA technique are obligatory but there are no regulations for risk level in the Russian Federation.

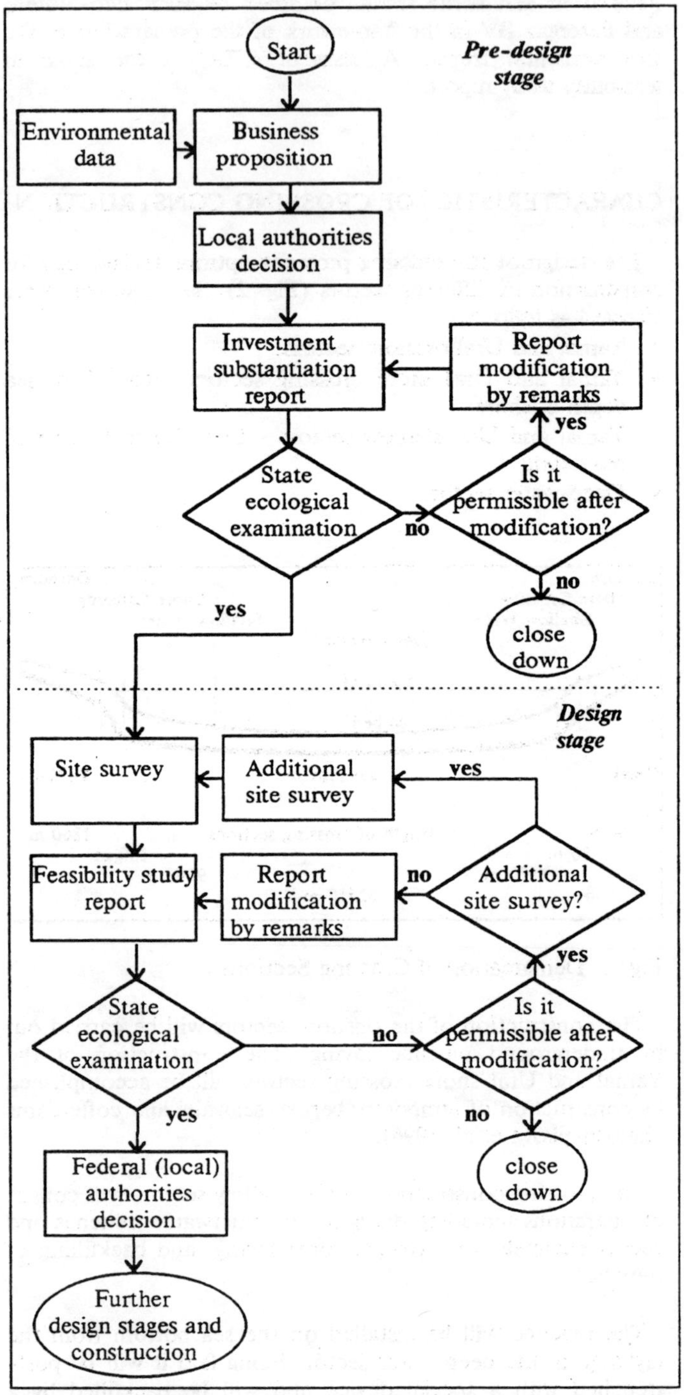

Fig. 3 Phases of Environmental Impact Assessment in Russian Federation

POSSIBLE SOURCES AND TYPES OF IMPACT

Considering construction of the underwater crossing on the basis of the environmental safety requirements, experience of work and analysis of fulfilled researches, it is possible to underline sources and types of impact on the environment. The following groups of the impact due to construction and operation of the crossing are singled out:

- the atmosphere (air);
- the hydrosphere (water) of the Baydaratskaya Bay;
- the lithosphere (sediments and permafrost);
- the biosphere (plants and animals);
- personnel.

Sources and types of impact on environment during construction and operation of the pipeline in a large measure are defined by design features, technology and organization of construction, but also by operational regime of the pipeline.

At the crossing construction stage sources of impact on the environment will be connected with:

- technological processes on crossing construction with the use of various engineering tools;
- technological processes on hydraulic test of pipelines.

At the crossing operation stage the impact on the environment will be connected with possible thermal and acoustical impact and possible failures.

The impact of the crossing construction on environment is complex and can differ significantly in the deep-water, Yamal and Ural shallow and shore crossing sectors (Tab. 1). The following types of construction will have most significant influence on the environment:

- digging of trenches, temporary roads construction in the onshore sections;
- construction of seagulls and cofferdams in the shore crossing sectors;
- dredging of underwater trenches and access channels in the shallow sectors, excavation of sand from the sea sand pits;
- pipeline post-trenching in the deep-water sector;
- backfilling of underwater trenches and access channels in the deep-water and shallow sectors, dumping of dredged soils.

Table 1: Possible impact of gas pipeline construction on the environment at different sectors of the crossing

	Crossing sectors			
	onshore	shore crossing	shallow	deep water
atmosphere	insignificant	insignificant	insignificant	insignificant
hydrosphere	weak	insignificant	permissible	permissible
litosphere	permissible	permissible	permissible	permissible
biota	permissible	weak	permissible	permissible
personnel	insignificant	insignificant	insignificant	insignificant

These sources of impact will differ by:

- space position (ground, water, underwater), that is defined by site, on which work will be carried out;
- space forms (point, connected with location of possible pollution sources; linear, connected with construction of a linear part of crossing; and area).

The main types of impact will also differ on a number of aspects, in particular, by:

Type of a impact:

- mechanical impact (digging, dredging, post-trenching and backfilling);
- chemical pollution (ejection dumping of polluting and other substances in gas, liquid and firm condition);
- physical impact (noise, vibration, electromagnetic and thermal impact);
- use of natural resources (water for technological and drinking needs, use of land for construction, trenches and etc.).

Substances imposed in the environment, their phase condition and composition:

- ship and construction installation smoke discharge- into the atmosphere;
- chemical reagents dissolved in water, used for hydraulic test - into the water;
- fuel and oil - into water and on ground;
- ground particles at dredging and damping - into water environment;
- firm construction wastes - on the ground.

Resources taken from the environment owing to direct or indirect impact:

- the ground (constant and temporary removals);
- biological resources (reduction due to construction).

In emergency the construction stage can be aggravated by:

- Fuel spills and leakages;
- Non controllable ejection of inhibited waters at hydraulic tests (in case of pipeline destruction).

At normal operation of underwater crossing there will be some thermal impact of transported gas on the geological environment (Tab. 2). As a result a seasonal thawing and frozen layer form on the shores of the Bay, mostly consisting of permafrost, around the pipeline. There are possible natural gas accidental explosions in the water and in the atmosphere on the pipeline crossing.

Table 2: Possible impact of gas pipeline normal operation on the environment at different sectors of the crossing

	Crossing sectors			
	onshore	shore crossing	shallow	deep water
atmosphere	insignificant	insignificant	insignificant	insignificant
hydrosphere	weak	insignificant	insignificant	insignificant
litosphere	weak	insignificant	insignificant	insignificant
biota	weak	insignificant	insignificant	insignificant
personnel	insignificant	insignificant	insignificant	insignificant

Besides mentioned problems Petergaz BV and EPS Eco-System develop the following directions in the framework of the Environmental Impact Assessment for the Baydaratskaya Bay underwater crossing at feasibility study report preparation:

- evaluations of the existing anthropogenic impact on the environment of the Bay and shores; the levels of different groups of contamination concentrations in the air, water, soils, biota;
- impact assessment on the air, water, permafrost, shores, biota and human health due to underwater crossing construction, normal operation and possible accident situations;
- environmental risk estimations and evaluation of economic losses due to possible ejection to the atmosphere, spills and leakages into the water;
- evaluations of efficiency for proposed environment protection plans at all phases of construction and operation of the crossing, including biota reproduction plans;
- evaluations of possible changes in social and economical conditions of native population due to crossing construction.

CONCLUSION

The obtained results show that the possible impact on the environment due to construction and operation of underwater crossing will have local character. It is possible to make a conclusion about the construction permissibility under the project requirements fulfillment and realization of the necessary environment control measures.

REFERENCES

Bases of ecological safety for a fuel and energy complex (designing, construction, operation). *Authorised by Fuel&Energy Ministry of Russian Federation* 10.06.1996 (in Russian).

Beloshapkova S, Beloshapkov A and Leont'ev I (1996) «Sediment Transport Changes Caused by Underwater Pipeline in Kara Sea», *Proc. 6th Int. Offshore and Polar Eng. Conf.*, ISOPE, Los-Angeles, Vol. 2, pp. 105-110.

Environmental protection regulation of oil and gas deposits development in the northern regions of Tyumen area. *TyumenNIIgiprogas*, 1994 (in Russian).

Instructions on ecological substantiation of economical and other activities. *Authorised by Russian Ministry of Nature* 29.12.1995, No 539 (in Russian).

Instruction on the procedure of development, coordination, approval and composition of design documentation for construction of enterprises, buildings and structures. *GP CENTRINVESTproekt*, Russia, 1995.

Odisharia GE, Tsvetsinsky AS, Mikhailov NN, Dubikov GI (1997) "Specialized Information System on Environment of Yamal Peninsula and Baydaratskaya Bay", *Proc. 7th Int. Offshore and Polar Eng. Conf.*, ISOPE, Honolulu, in press.

Procedure of development, coordination, approval and composition of feasibility studies for construction, of enterprises, buildings and structures. *GP CENTRINVESTproekt*, Russia, 1995.

Regulations concerning pipeline systems in the petroleum activities (1990), Norwegian petroleum Directorate.

Rule of ecological examination order. *Authorised by Government of Russian Federation* 11.06.1996 No 698 (in Russian).

Proceedings of the Seventh (1997) International Offshore and Polar Engineering Conference
Honolulu, USA, May 25-30, 1997

ISBN 1-880653-28-1 (Set); ISBN 1-880653-29-X (Vol. I)

Numerical Simulation of Flow and Density Field of Kagoshima Bay in the Summer

Yusaku Kyozuka
Kyushu University, Kasuga, Japan
Changhong Hu
Kyushu University, Fukuoka, Japan
Nobuyuki Iwakiri,
City office of Kagoshima, Kagoshima, Japan

ABSTRACT

A multi-level model is applied to the simulation of tidal currents and density field of Kagoshima Bay in summer. Tidal currents are calculated for tidal level of twelve hours(M_2) given at the mouth of the bay. Heat flux is given by a constant value at sea surface, neglecting the diurnal changes. Numerical simulations are conducted in time marching from an initial state of the variables using a finite difference method. Time averaged values of 25th period of tidal oscillation from the initial state are taken as steady solutions. Distributions of the residual current, temperature, salinity and density thus obtained are compared with available observations.

As an application of the present method, tidal simulations are presented for the past topography of Kagoshima Bay before the eruption in 1914, by which East Sakurajima Strait were blocked with lava. Distributions of temperature and salinity before the eruption were compared with those of present bay. Effects of East Sakurajima Strait on the physical environment are presented in graphics.

KEY WORDS: Tidal Flow, Density Field, Numerical Simulation, Multi-Level Model, Kagoshima Bay

INTRODUCTION

Kagoshima Bay is a narrow bay, $75km$ along the North-South and $20km$ along the East-West. It is located at Southern edge of Kyushu. The size is almost same as Tokyo Bay but the topographic and the oceanographic characteristics are quite different. There is Sakurajima volcanic island at the interior of the bay and there is an oceanic basin of $400m$ in depth at the center of the bay. Before the great eruption of Sakurajima in 1914, there was a strait of $400m$ breadth and $70m$ depth.

Since fish was found polluted by mercury in 1973 and the cultured fish were sometimes damaged by red tide, many investigations concerning the circulation of water and mass-transport in the bay have been done. For those investigations, ocean measurements and sampling of marine lives were mainly done, but the numerical simulation approach might bring useful information about flow and density field of the bay.

In this study, we have developed a multi-level model for the analysis of the flow and density field of Kagoshima Bay and the validation of the numerical results is discussed by comparing with the field measurements. Finally, as an application of this approach, the flow and density field of the bay with East-Sakurajima Strait before the great eruption of Sakurajima, are simulated and are compared with the present bay.

NUMERICAL MODEL

Formulation of the problem

We consider the flow and density field in a bay, where tide, wind and density anomaly are taken into account as driving forces.

Assuming f-plane, Boussinesq and hydrostatic approximations, the equation of continuity is given by:

$$\frac{\partial u}{\partial x} + \frac{\partial v}{\partial y} + \frac{\partial w}{\partial z} = 0 \tag{1}$$

Momentum equations are written as:

$$\frac{Du}{Dt} - fv = -\frac{1}{\rho_0}\frac{\partial p}{\partial x} + A_M \nabla_h u + \frac{\partial}{\partial z}(K_M \frac{\partial u}{\partial z}) \tag{2}$$

$$\frac{Du}{Dt} + fu = -\frac{1}{\rho_0}\frac{\partial p}{\partial y} + A_M \nabla_h v + \frac{\partial}{\partial z}(K_M \frac{\partial v}{\partial z}) \tag{3}$$

$$0 = -\frac{1}{\rho}\frac{\partial p}{\partial z} - g \tag{4}$$

Here, the operators $\frac{D}{Dt}$ and ∇_h are:

$$\frac{D}{Dt} = \frac{\partial}{\partial t} + u\frac{\partial}{\partial x} + v\frac{\partial}{\partial y} + w\frac{\partial}{\partial z} \tag{5}$$

$$\nabla_h = \frac{\partial^2}{\partial x^2} + \frac{\partial^2}{\partial y^2} \tag{6}$$

In these equations, f is the Coriolis parameter, ρ is density of water, ρ_0 is reference density of water, p is the pressure, g is acceleration of gravity, and A_M and K_M are the eddy viscosity in horizontal and vertical motions.

Integrating (4) in z from the bottom($z = H$) to the sea surface($z = \eta$), the pressure is obtained by:

$$p = p_a + \rho g(\eta - z) \tag{7}$$

where p_a is the atmospheric pressure.

The surface elevation is calculated by the following equation which is obtained by integrating (1) from the bottom to the sea surface and using the boundary conditions at both ends:

$$\frac{\partial \eta}{\partial t} + \frac{\partial}{\partial x}\left(\int_H^\eta u dz\right) + \frac{\partial}{\partial y}\left(\int_H^\eta v dz\right) = 0 \tag{8}$$

The density current is driven by the density difference, which is largely determined by the water temperature($T(°C)$) and the salinity($S(‰)$). We used the following formula by Mamaev(1968):

$$\begin{aligned} \rho &= 1000 + \sigma \\ &= 1028.152 - 0.0135T - 0.004691T^2 \\ &\quad +(0.802 - 0.002T)(S - 35.0) \end{aligned} \tag{9}$$

T and S are calculated by the following diffusion equations:

$$\frac{DT}{Dt} = A_C \nabla_h T + \frac{\partial}{\partial z}\left(K_C \frac{\partial T}{\partial z}\right) + Q_T \tag{10}$$

$$\frac{DS}{Dt} = A_C \nabla_h S + \frac{\partial}{\partial z}\left(K_C \frac{\partial S}{\partial z}\right) - \frac{RS}{V_R} + Q_S \tag{11}$$

where A_C and K_C denote eddy diffusivity in horizontal and vertical directions, R is the flux of river discharge, V_R is the volume of river discharge and Q_T and Q_S denote the thermal and salinity fluxes.

Boundary conditions

Boundary conditions for the fluid motion are given by no-slip condition as follows:

$$K_M \frac{\partial u}{\partial z} = \frac{\tau_x}{\rho}, \quad K_M \frac{\partial v}{\partial z} = \frac{\tau_y}{\rho} \tag{12}$$

The shear stress on the sea bottom is given as follows:

$$\left.\begin{aligned} \tau_x &= \gamma_b^2 \rho u_b \sqrt{u_b^2 + v_b^2} \\ \tau_y &= \gamma_b^2 \rho v_b \sqrt{u_b^2 + v_b^2} \end{aligned}\right\} \tag{13}$$

at the sea surface,

$$\left.\begin{aligned} \tau_x &= C_d \rho_a u_w \sqrt{u_w^2 + v_w^2} \\ \tau_y &= C_d \rho_a v_w \sqrt{u_w^2 + v_w^2} \end{aligned}\right\} \tag{14}$$

where γ_b^2 is friction coefficient at the sea bottom; C_d is friction coefficient at the sea surface; (u_b, v_b) is velocity vector at the sea bottom; (u_w, v_w) is wind velocity; and ρ_a is air density.

The boundary conditions for T are derived from the thermal exchanges at each boundary. Usually, the adiabatic condition is applied at the coastline and the sea bottom. Thermal exchanges are used at sea surface and open boundaries. At the sea surface, the following four components are taken into account: solar radiation(Q_s), long wave radiation(Q_b), the latent heat transport by evaporation(Q_e) and the sensible heat transport(Q_c). Total thermal exchang at the sea surface is given as:

$$Q_T = Q_s - (Q_b + Q_c + Q_e) \tag{15}$$

where Q_s, Q_b, Q_c and Q_e are calculated by:

$$\begin{aligned} Q_s &= (1 - \alpha_s)Q_{s'} \\ &= 0.93(1 - 0.67n)Q_{s0} \end{aligned} \tag{16}$$

$$\begin{aligned} Q_b &= 1.32 \times 10^{-12}\theta_w^4\{0.49 - 0.066\sqrt{e(T_a)}\} \\ &\quad \times(1 - 0.65n^2) + 5.27 \times 10^{-12}\theta_w^3 \\ &\quad \times(T_w - T_a) \quad [cal/m^2 \cdot s] \end{aligned} \tag{17}$$

$$\begin{aligned} Q_c &= K_c\{T_w - T_a\} \\ &= \rho_a C_{pa} \frac{u_*^2}{W_h(z)}\{T_w - T_a\} \end{aligned} \tag{18}$$

$$\begin{aligned} Q_e &= Q_c/B \\ &= Q_c/\{0.66\frac{T_w - T_a}{e(T_w) - e(T_a)}\} \end{aligned} \tag{19}$$

where

α_s	:	Albedo at sea surface, (= 0.07)
$Q_{s'}$	:	Solar radiation at sea surface
n	:	Cloudiness, ($0 < n < 1$)
Q_{s0}	:	Solar radiation under clear sky
θ_w	:	Water temperature in °K
$e(T_a)$	:	Vapor pressure at temperature T_a
T_a	:	Atmospheric temperature at the sea surface
T_w	:	Water temperature at the sea surface
K_c	:	Turbulent thermal conductivity
C_{pa}	:	Atmospheric specific heat
u_*	:	Atmospheric friction velocity
$W_h(z)$	:	Mean wind velocity at z
B	:	Bowen ratio

Flux of salinity at sea surface Q_S, is given as follows:

$$Q_S = S(E_v - P_r) \tag{20}$$

where

E_v	:	Evaporation of water at the sea surface
P_r	:	Precipitation at the sea surface

The boundary conditions for T and S at the open boundary are given by the zero-gradient for outflow and by the observed values for inflow.

NUMERICAL METHOD

Topography and physics of Kagoshima Bay

Kagoshima Bay is a narrow bay in North-South as shown in Fig.1. Its maximum depth is approximately $250m$. The bay is classified into three areas which are: the bay mouth (of $100m$ depth), the center of bay (of $200m$ depth) and the interior of the bay (of $100m$ depth). The interior of the bay is connected with the center of the bay through West-Sakurajima Strait (of $40m$ depth).

Wind over the bay usually ranges from 3 to $3.5m/s$ throughout the year, and the direction of wind is most frequently from

North-West, except for summer. There is no main direction of wind in summer. Water temperature of the bay is almost the same as air temperature during the period from May to September, but is higher than air temperature during other months.

Fresh water is mostly supplied by the rain and the rivers. There are no big rivers, except near the interior of the bay. Ten rivers are considered in this study. Discharge rate of the rivers is taken as year-averaged value as shown in Table 1([1]).

There is no measured data concerning the evaporation of sea water. Therefore, it is estimated from water temperature, air temperature and wind([1]). Precipitation is measured at the Kagoshima weather station. A maximum value of 493mm was observed in June. From those data, monthly flux of fresh water through sea-surface are estimated to be 426mm/month in June and 245mm/month in July([1]).

Numerical cases

Three cases are calculated for August as follow:

(Cal. 1) Heating at sea-surface, discharge from rivers and tidal level(M_2) are considered.

(Cal. 2) In addition to (Cal. 1), the evapolation and precipitation at sea-surface are considered.

(Cal. 3) Simulation using the original topography before closure of East-Sakurajima Strait.

The mesh size is ($1km \times 1km$) in horizontal plane. Five levels ($0, 20, 40, 90, 150m$, bottom) are used in vertical direction. Initial conditions are given by constant values as follows: flow velocity($0m/s$), salinity(34.5‰), and water temperature(14.5°C). The numerical method used here is similar to that of Hu and Kyozuka(1996).

Eddy viscosity(K_M) and eddy diffusivity(K_C) in the horizontal plane are set to equal and constant values as : $A_M = A_C = 50m^2/s$. In the vertical direction, they are given by the stratification function as follow (Tabeta and Fujino(1994)):

$$K_M = K_{M0}(1 + 5.2R_i)^{-1} \quad (21)$$

$$K_C = K_{C0}(1 + \frac{10}{3}R_i)^{-1.5} \quad (22)$$

where K_{M0} and K_{C0} denote values in non-stratified water, and R_i is the Richardson number defined as follow:

$$R_i = -\frac{g(\frac{\partial \rho}{\partial z})}{\rho(\frac{\partial U}{\partial z})^2} \quad (23)$$

Primary parameters used in this study are shown in Table 2, and the water temperature and salinity at the open boundary are given in Table 3.

RESULTS AND DISCUSSIONS

Fig.2 shows time evolutions of water temperature and salinity at the center of the bay(St.6). They diverge from the upper level, values with progression of time. For the lowest level, the change is very small. Constant values are not reached even after 300 hours (25 tidal periods).

Numerical results of residual currents at levels 1 and 3 in (Cal.1) are shown in Fig.3 and they agree well with the measurements by No.10 Headquarters of Sea Safety Agency(1977). Residual currents near sea-surface flow out from the bay(Fig.3). Also, the clockwise circulation at the center of the bay and the flow pattern of the interior of the bay in Fig.3 agree with the observations shown in Fig.4. These results indicate that the numerical method can accurately simulate the tidal currents and the residual currents of the bay.

In Fig.5, the horizontal distributions of salinity and water temperature at top level in (Cal.1) are shown. The observed distribution of salinity is generally higher at the mouth of the bay and lower at interior of the bay, but the calculations contradict that behavior. This may be caused by the initial value of salinity inside the bay and the value of salinity at open boundary. Additionally, the precipitation of rain may be important in determining salinity distribution. This effect is considered in (Cal.2). Unlike salinity, the distribution of water temperature agrees well with the obserbation in the bay.

In Fig.6, the vertical distributions of salinity and temperature are shown in the plane along the line $X_1 - X_2$ of Fig.1. The stratifications are clearly visible. The interior of the bay is connected to the center through the shallow West Sakurajima Strait. Thus, the interior forms an indepentent water structure.

Two calculations(Cal.1 and 2) of the vertical distributions of temperature, salinity and density at the interior(St.1) of the bay are comapered with observations by Takahashi(1977) in Fig.7. Calculated temperature distributions agree with observation very well. For the salinity, (Cal.2) gives better result than (Cal.1). This indicates that precipitation effect is important in the interior of the bay.

Similar comparisons are shown in Fig.8 for the point(St.6) in the center of the bay. Temperatures at sea-surface and deep water agree with the observation, but the shapes of the distributions are different. Although calculated distributions change their gradients slowly for 200m in depth, observated distributions change rapidly within 100m from the surface. This may be caused by the vertical eddy viscosity, eddy diffusivity, number of levels and the computational time. Similar discussions apply for the salinity and density, where observations show typical stratified distributions. These results show that the numerical method accurately simulate the temperature and density of the bay, although there are some problems in salinity.

Next, as an application of the present method, the flow and density fields of the bay before the eruption of Sakurajima were examined. The original East-Sakurajima Strait was 400m wide and 70m deep, but we use a strait of ($1km \times 28m$) in the calculations. Fig.9 shows the residual currents at levels 1 and 2. Comparison with Fig.3 shows that the residual currents at East-Sakurajima Strait flow out at level 1 but flow in at level 2, and the circulation observed in the interior of the present bay disappeared by the original topography.

Fig.10 shows the differences of the water temperature, salinity and density at level 1 between (Ca.1) and (Cal.3). The effects of the closure of East-Sakurajima Strait appear at both straits and the interior of the bay. A large difference in temperature appears in small area near both straits. Fig.11 shows the differences at level 2. Fig.10 shows a wide area where temperature is higher by more than 1°C in the interior of the bay. This may be due to the warm water which flows through the lower level from the center of the bay. The above numerical results indicate that the East-Sakurajima Strait has potential to improve the water quality of the interior of the bay. However, further study is needed to examine the impact on the ecosystem.

CONCLUSION

1. The present method accurately simulates the tidal cuurents and the residual currents of Kagoshima Bay.
2. The vertical distribution of water temperature by the present method agrees well with the observation in summer, but it does not agree with observed salinity values. Including precipitation improves the results. Therefore, it may be concluded that the effect of precipitation on the salinity distribution is relatively large in Kagoshima Bay.
3. The tidal current and the residual current in the interior of the Kagoshima Bay before the closure of the East-Sakurajima Strait might be quite different from those of the present bay. More than $1°C$ increse in water tempeature will occur over a wide area in the interior of the bay, if the strait exists. Further study is needed to examine the impact on the water quality and the ecosystem in the bay.

References

[1] The Oceanographical Society of Japan, (1985). Coastal Oceanography of Japanese Islands, Tokai University Press.

[2] Mamaev, O.I.(1968). A Simplified Relationship between Density, Temperature and Salinity of Sea Water, Bull. Akad. Sci. USSR. Geophy. Ser. 2.

[3] Takahashi, Y.(1977). Hydraulics of Kagoshima Bay, Bull. on Coastal Oceanography, Vol 14, No 1·2.

[4] Nakata, K. (N.Yokoyama Edit.), (1993). Marine Environmental Simulation, Hakua shobou.

[5] Hu, C.H. and Kyozuka, Y.(1996). Tidal Simulation of a Bay with a Very Large Floating Structure Using Multi-Level Model, Trans. West-Japan Soc. Naval Arch., Vol 91.

[6] Tabeta, S. and Fujino, M.(1994). Numerical Simulation of Currents and Density Variation in Tokyo Bay by means of Multi-Level Model, J. Soc. Naval Arch. Japan, Vol 176.

[7] No.10 Headquarters of Sea Safety Agency, (1977). Tidal current of Kagoshima Bay.

Table 1 Discharge rate of rivers

Nagatagawa	1.5 (m^3/s)
Koutukigawa	4.5
Omoigawa	1.6
Beppugawa	4.5
Amikakegawa	2.0
Amorigawa	18.0
Kenkougawa	1.6
Honjyogawa	1.7
Uenogawa	2.5
Ogawa	6.8

Table 2 Parameters used in simulations

Horizontal mesh size	Δx ,Δy	1km
Number of levels	NZ (Δz)	5(0,-20,-40,-90,-150m,bottom)
Time step	Δt	5 sec
Coriolis parameter	f	$7.5 \times 10^{-5} s^{-1}$
Gravity acceleration	g	$9.8m/s^2$
Bottom frictional coefficient	γ_b	0.0026
Reference density of water	ρ_0	$1025kg/m^3$
Density of air	ρ_a	$1.226kg/m^3$
Horizontal eddy viscosity	A_M	$50m^2/s$
Vertical eddy diffusivity	A_C	$50m^2/s$
Vertical eddy viscosity	K_{M0}	$1 \times 10^{-4} m^2/s$
Vertical eddy diffusivity	K_{C0}	$1 \times 10^{-4} m^2/s$
Tidal level at open boundary	η_1	$0.77m(M_2)$
Solar radiation under clear sky	Q_{so}	$751 cal/cm^2 day$
Quantity of cloud	n	0.5
Temperature	T_a	$27°C$
Relative humidity	f_a	80 %

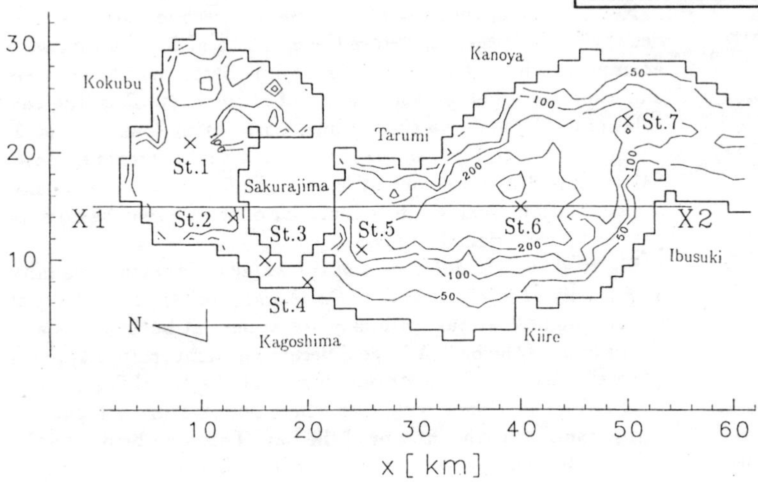

Fig.1 Contour map of Kagoshima Bay

Table 3 Values at open boundary

	Top level	2nd level	3rd level
Salinity($°/_{oo}$)	32.5	33.1	34.3
Temperature($°C$)	27.5	25.0	19.1

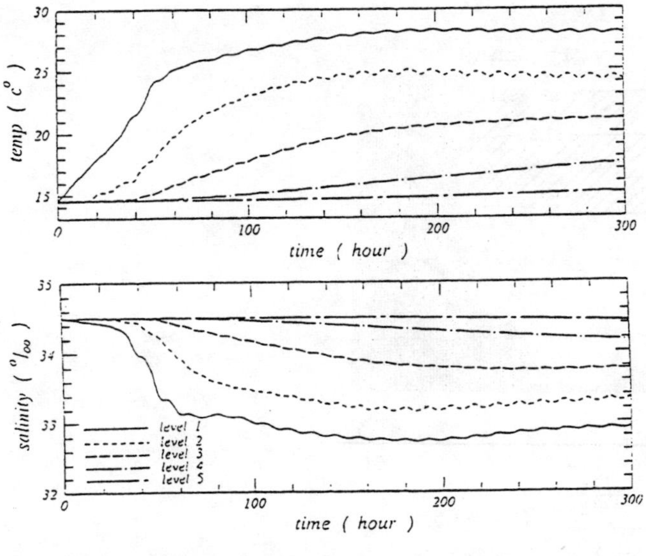

Fig.2 Time evolution of temperature and salinity at the center of the bay (St.6)

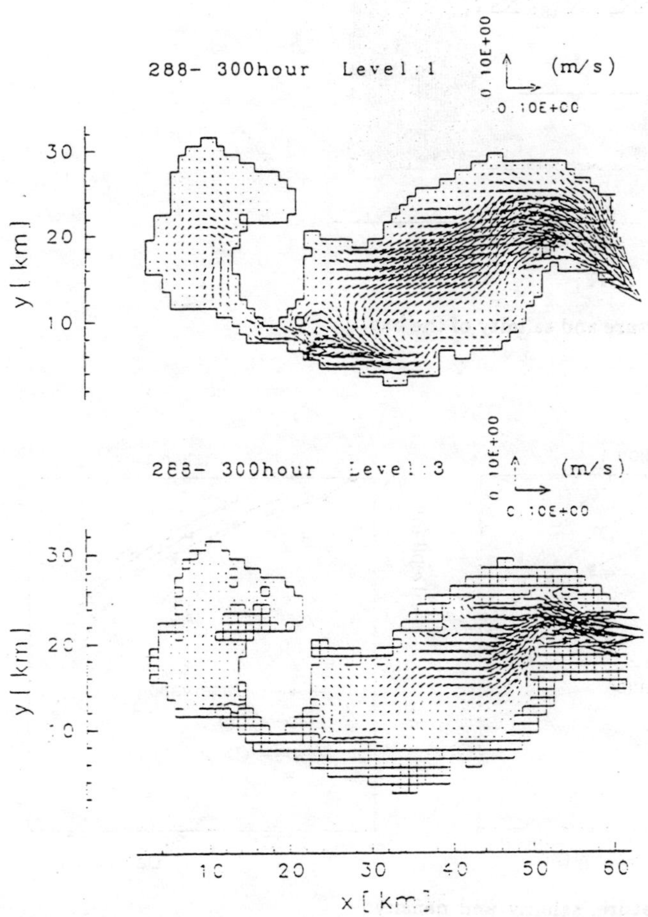

Fig.3 Calculated residual current of the bay (Level 1 and 3, Cal.1)

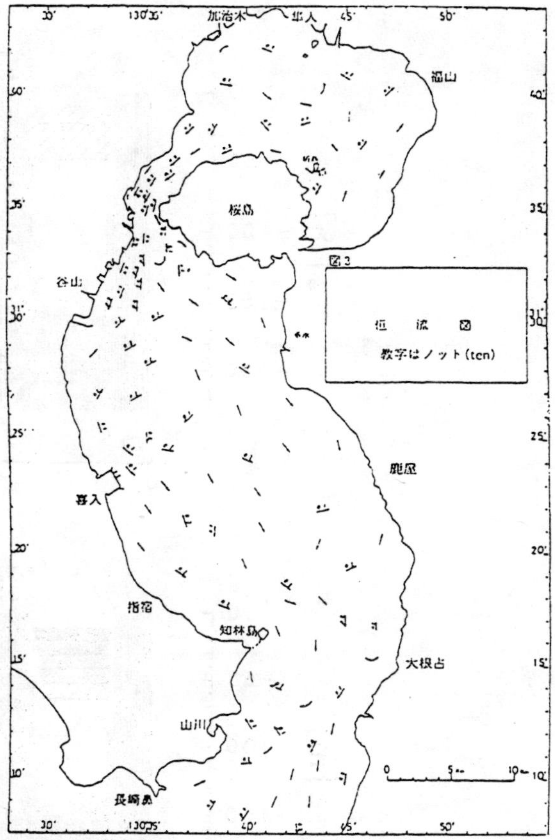

Fig.4 Observed residual current at upper surface of the bay[7]

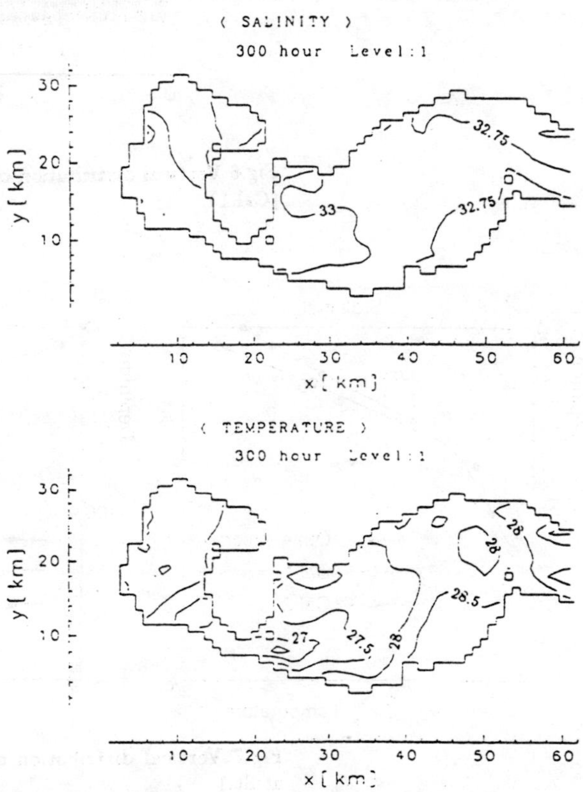

Fig.5 Horizontal distribution of temperature and salinity of the bay (Cal.1)

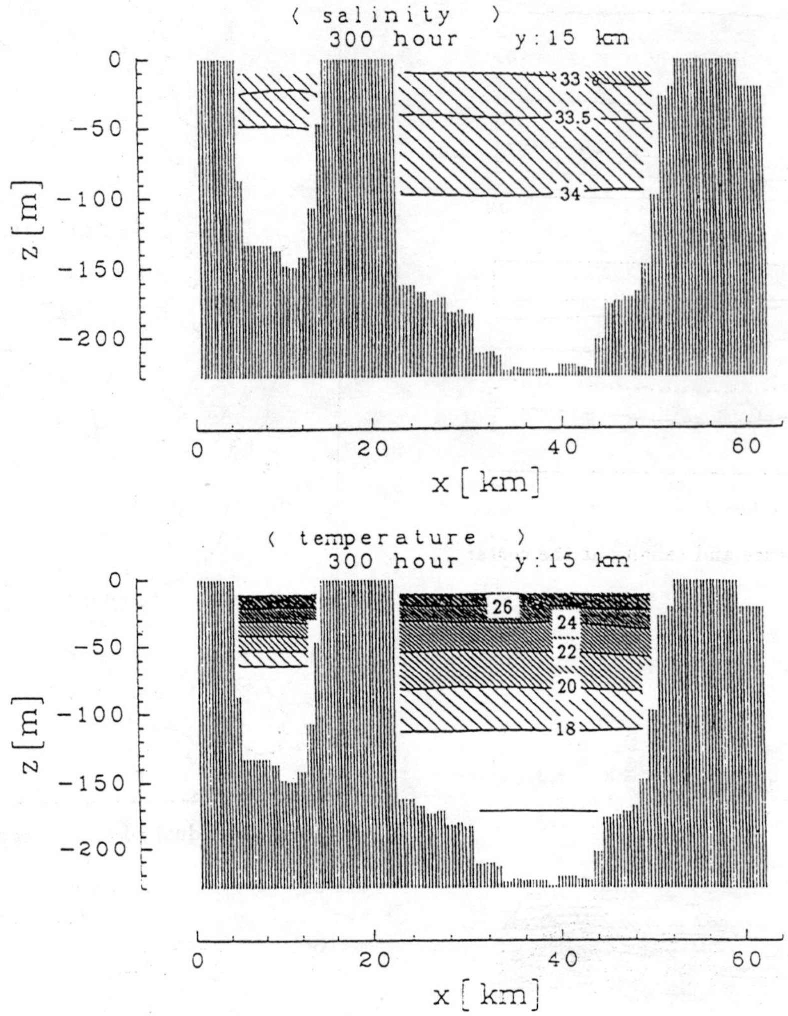

Fig.6 Vertical distribution of temperature and salinity of the bay (Cal.1)

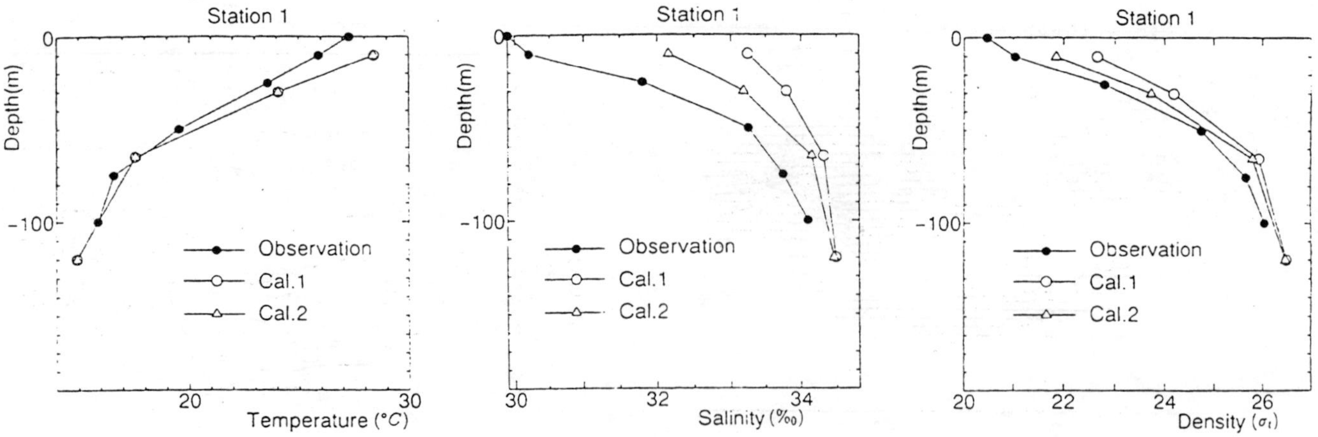

Fig.7 Vertical distribution of temperature, salinity and density at St.1

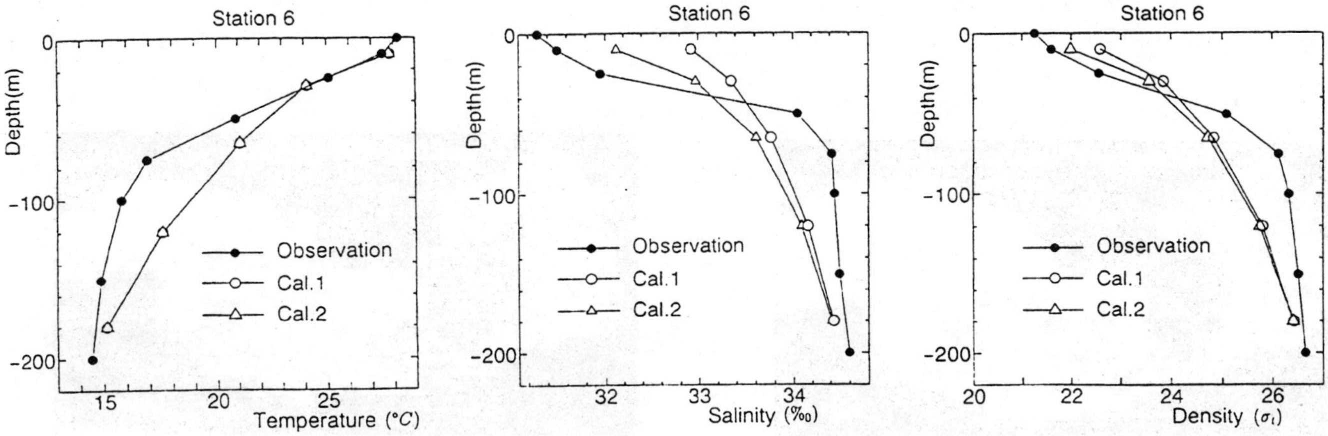

Fig.8 Vertical distribution of temperature, salinity and density at St.6

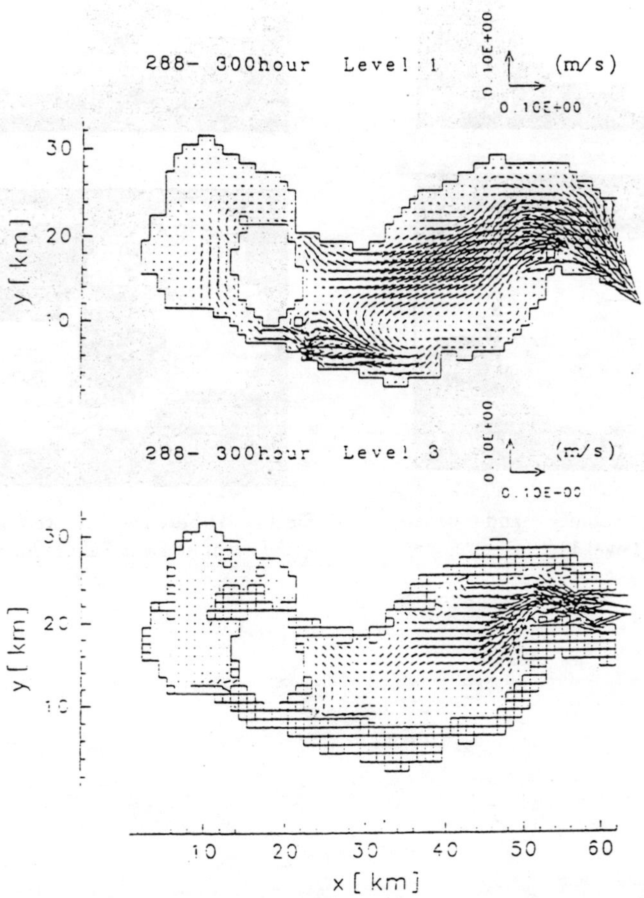

Fig.9 Calculated residual current with East-Sakurajima Strait (Cal.3)

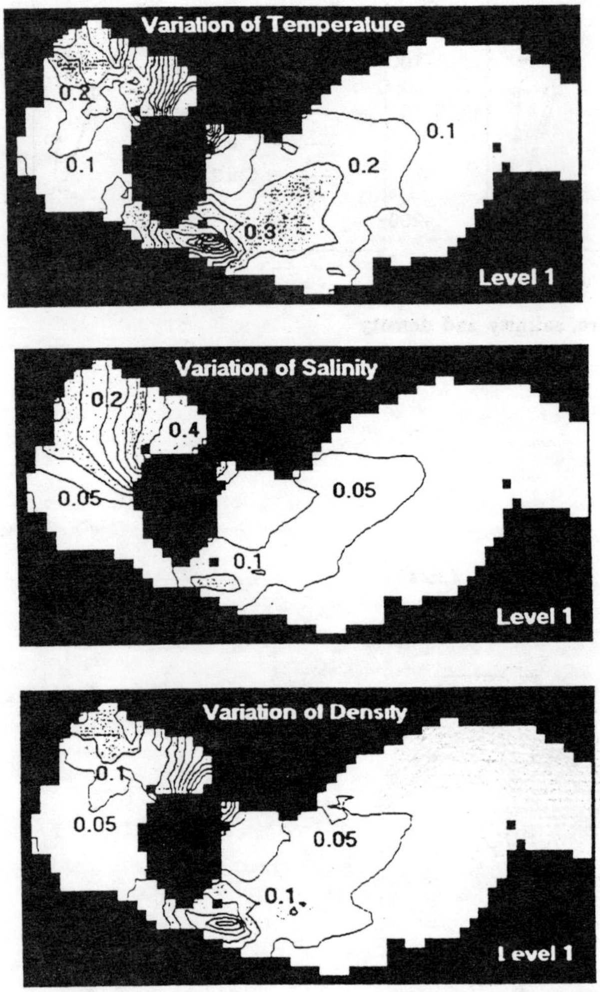

Fig.10 Difference of temperature, salinity and density with/without East Sakurajima Strait (Level 1)

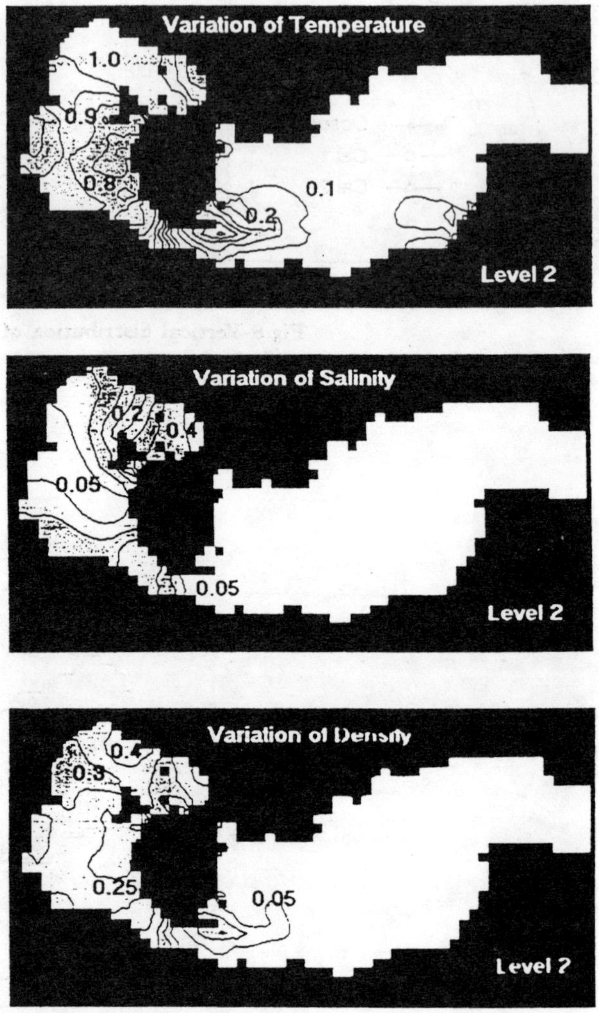

Fig.11 Difference of temperature, salinity and density with/without East Sakurajima Strait (Level 2)

Proceedings of the Seventh (1997) International Offshore and Polar Engineering Conference
Honolulu, USA, May 25-30, 1997

ISBN 1-880653-28-1 (Set); ISBN 1-880653-29-X (Vol. I)

On Validation of Multi-Level Method for Simulation of Flow Around a Mega-Float: Comparison of Numerical Calculation with Model Experiment

Hideyuki Omori
Akishima Laboratories (Mitsui Zosen), Akishima, Japan
Yusaku Kyozuka
Kyushu University, Kasuga, Japan
Changhong Hu
Kyushu University, Fukuoka, Japan
Hiroyuki Nakagawa and Masanori Kobayashi
Akishima Laboratories (Mitsui Zosen), Akishima, Japan

ABSTRACT

A multi-level model for the calculation of tidal flow and material diffusion of a bay has been developed to simulate ocean environmental problems. Recently, it has been extended to the problem of a bay with a very large floating structure or a mega-float. However, the validation of the application is required before the model can be generalized, because the multi-level model is based on a quasi-three dimensional method with hydrostatic pressure approximation.

In this paper, results of a model experiment are compared with those of numerical calculations and the applicability of multi-level model is discussed. In the model experiments, the flow around a towed pontoon was measured. The draft was changed to investigate the effect of draft/water depth ratio, and the mass-flux below a pontoon was estimated from the measured velocity profile. Then measured distributions of velocity and mass-flux below the pontoon are compared with calculations by a multi-level model in which the existence of a mega-float is taken into account.

KEY WORDS: Mega-Float, Multi-Level model, Numerical simulation, Laboratory experiment

INTRODUCTION

There are ideas to use a very large floating structure or a mega-float for utilization of coastal area such as off-shore city, floating airport etc. To realize these ideas, a verification experiment of a very large floating structure or a mega-float settled in coastal area or a bay has being done by setting Mega-float model (Length 300m, Width 60m and Draft 0.5m) in Oppama(Yokosuka,Japan) since 1995. There the researchers of T.R.A.M.(Technical Research Association of Mega-float) have been observing ocean environment changes caused by settlement of Mega-float model. However, some new problems arise because the assumed floating-body size is so huge and can be up to a few kilometers. Kyozuka et al (1995) stated that one of the problems concern the environmental impacts on the surrounding sea area due to setting a mega-float in coastal area or a bay.This problem should be answered by public environmental-assessment. Though it is said that physical, chemical and biological elements and factors of environment concerns complicate to the environmental effect assessment, the changes of a flow pattern in bay is important as one of factors controlling basic feature of many kinds of diffusion and circulation of matters.

A multi-level model for the calculation of tidal flow and diffusion of a bay has been developed and used for the simulation of ocean environmental problems. Recently, it is extended to the problem of a bay with a very large floating structure or a mega-float.

However, validation of the application is required before the method can be generalized, because the multi-level model is based on a quasi-three dimensional method with hydrostatic pressure approximation and does not consider the existence of a mega-float.

EXPERIMENT

The experiment was carried out in a towing tank at Akishima Laboratories(Mitsui Zosen)inc. . Fig.1 shows the three cases of the experiment and measuring positions of fluid velocity. The measuring positions, in a horizontal plane, were 5 points in longitudinal direction and 3 points in lateral direction. In a vertical plane, the measuring positions are at intervals of 0.05m and 0.10m, at water depth of 0.3m and 0.9m, respectively. Considering symmetry of the pontoon, up-stream and down-stream measurements were performed by a towing car back and forth without moving the current meter. An electro-magnetic 2-dimensional current meter was used to measure fluid velocity using an average recording duration of 20 seconds.

Assuming the real length of the pontoon vessel is 300m and according to similarity law used in hydraulical experiment of civil-environmental engineering we have a comparison between the model test and a real situation as in Table.1. The scale of horizontal plane and vertical plane shown in Table.1 are different to satisfy flow and diffusion process. From Table.1, flow around the pontoon could be three-dimensional in both cases A and B owing to the large water depth and the large draft, so it would not be suitable to use multi-level model. These will be discussed below.

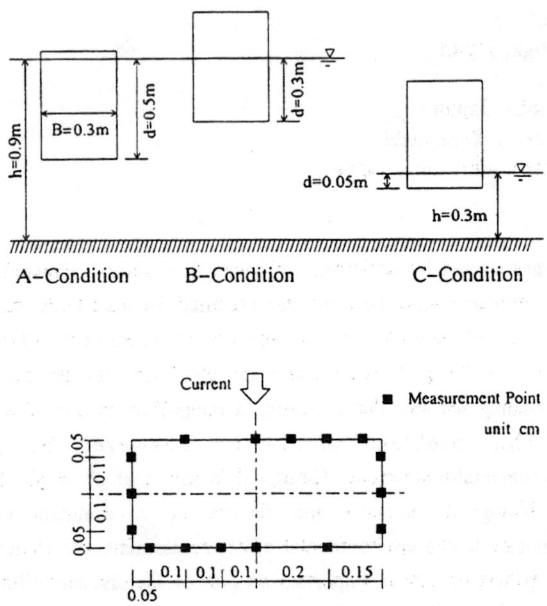

Fig.1 Model conditions and measured points

NUMERICAL SIMULATION

The level model is a way to calculate three dimensional flow by dividing the calculating field vertically to some levels and calculate the two dimensional flow in each level by considering three dimensional effect to the two dimensional flow calculation so it is called a quasi-three dimensional model. MAC method is often use to calculate the flow around a floating structures and it would be extended to three dimension with fully three dimensional assumption. Though MAC method needs very small calculating mesh and calculating time step, calculating a huge field such as a bay by MAC method is not practical at present because of the capacity of memories of computer and calculation speed.

Horie et al (1983) described the features of two models of level model. One is a multi-level model which describes the vertical movement of fluid by vertical velocity, and the other is a layer-level model which describes the vertical movement of fluid by vertical movement of level boundary. By this feature layer-level model is not suit for calculation of the diffusion such as salinity, temperature etc., and calculation often becomes unstable when the changes of the sea bed is large. So at present a multi-level model is considered suitable practical numerical method for the problem of some environmental assessment of the mega-float.

A multi-level model is based on quasi-three dimensional method in which the vertical velocity is assumed to be smaller than horizontal velocity, so basically it is not suitable to calculate the flow around floating structures. However, the multi-level model can give good approximation if the velocity of fluid is small so that three-dimensional flow is restricted to the area near the corners of structures, such as pontoon type mega-float planed.

Table.1 Scale ratio of the model

condition	Notation	Length (m)	Breadth (m)	Draft (m)	Depth of water (m)	Current (m/s)	Depth of water / Draft
A				0.5 / 28.4	0.9 / 51.2		1.8
B	model / full-scale	0.3 / 300	0.3 / 128.6	0.3 / 17.1		0.1 / 0.754	3.0
C				0.05 / 2.8	0.3 / 17.0		6.0
	(Scale)	(S_0)	(S_0)	($S_0^{2/3}$)	($S_0^{2/3}$)	($S_0^{1/3}$)	

BASIC EQUATIONS

Curvature of the earth can be ignored if the horizontal scale of the calculating sea area of tidal current is below several kirometers (approximation of f-plane). Using the hydrostatic pressure approximation the basic equation of tidal analysis in a sea area without off-shore structures is obtained as follows.

The continuity equation is as follows:

$$\frac{\partial u}{\partial x}+\frac{\partial v}{\partial y}+\frac{\partial w}{\partial z}=0 \tag{1}$$

The momentum equations in x,y,z direction are as follows:

$$\frac{\partial u}{\partial t}+u\frac{\partial u}{\partial x}+v\frac{\partial u}{\partial y}+w\frac{\partial u}{\partial z}$$

$$=-\frac{1}{\rho}\frac{\partial p}{\partial x}+A_M\left(\frac{\partial^2 u}{\partial x^2}+\frac{\partial^2 u}{\partial y^2}\right)+\frac{\partial}{\partial z}\left(K_M\frac{\partial u}{\partial z}\right)+fv \tag{2}$$

$$\frac{\partial v}{\partial t}+u\frac{\partial v}{\partial x}+v\frac{\partial v}{\partial y}+w\frac{\partial v}{\partial z}$$

$$=-\frac{1}{\rho}\frac{\partial p}{\partial y}+A_M\left(\frac{\partial^2 v}{\partial x^2}+\frac{\partial^2 v}{\partial y^2}\right)+\frac{\partial}{\partial z}\left(K_M\frac{\partial v}{\partial z}\right)-fu \tag{3}$$

$$0=-\frac{1}{\rho}\frac{\partial p}{\partial z}-g \tag{4}$$

where (u,v,w) stands for the fluid velocity ;(t) the time ,(ρ) the typical density of sea water,(P) the pressure, (A_M) and (K_M) the horizontal and the vertical eddy viscosity, respectively; (f) the Coriolis parameter and (g) the gravitational acceleration.

By integrating equation (4) in the vertical direction, the pressure at depth z is obtained.

$$p=p_a+\rho g(\eta-z) \tag{5}$$

where (η) denotes the surface elevation, (P_a) the atmospheric pressure.

Integrating equation (1) from the sea bottom($Z=-h$) to free surface ($z=\eta$),the following equation can obtained, which is used to determine the surface elevation.

$$\frac{\partial \eta}{\partial t}+\frac{\partial}{\partial x}\left(\int_{-h}^{\eta}udz\right)+\frac{\partial}{\partial y}\left(\int_{-h}^{\eta}vdz\right)=0 \tag{6}$$

NUMERICAL SIMULATION METHOD

Numerical studies of tidal currents in a sea area or a bay with a very large floating structure have been reported by Inoue et al (1994) and Hu et al (1996). In this paper Hu's method was applied. In Hu's method, the fluid field is divided into two areas ,the area with free surface and the area beneath mega-float. For the second area the pressure was obtained by a two dimensional Poisson equation which was derived from the equations (2) and (3).

For discretization of the basic equation a staggered mesh is used. An explicit leap-frog scheme was used for the time and a central difference scheme for the space. An up-wind scheme of the first order precision was used for the convection term.

In this calculation the area beneath the floating structure is divided into 4 to 6 levels. A slip condition was used for the structure and the sea bottom boundaries. Horizontal mesh size was $\Delta x=\Delta y=0.1$m, and vertical mesh were;

1) conditions A and B : $\Delta z=0.10$m , 9 levels
2) condition C : $\Delta z=0.05$m, 6 levels

Eddy viscosity was $A_M=K_M=0.01\text{m}^2/\text{s}$ in all cases of the numerical calculation. For each calculation, the current velocity was increased from zero to the steady velocity at the beginning. The time step was $\Delta t=0.005$sec.

COMPARISON OF EXPERIMENT AND CALCULATION

FLOW PATTERN

Fig.2(a) shows the calculated velocity vectors of A-condition at the section of $y=0$. Like this example, the calculation results of a multi-level model can only give a laminar flow without vertex shed from the bottom edge of the pontoon .

Fig.2(b) shows a comparison of the measured and calculated horizontal velocity u at front and rear edges of the pontoon for the A-condition. These are the results of the center section of pontoon, however the results of the side section of pontoon almost give the same flow pattern.

For the results of up-stream side, measured velocity agree with calculated velocity. For the results of down-stream side, however, the measured velocity shows a reversed flow near the bottom of pontoon which indicates a existence of large vortex in that area, while the calculated velocity shows a similar trend to the up-stream side.

Fig.2(c) shows a comparison of the vertical velocity w of A-condition. The large downward velocity was measured in experiment in up-stream side, while the calculated velocity is generally small. The reason for monotonous change of measured velocity in down-stream side is supposed that the measuring point is inside the wake.

These results show that if the draft of pontoon is deep like A-condition, the assumption of a multi-level model is not correct and especially the model can not simulate the vortex shedding at the bottom corner of a pontoon. However, it can give us some useful qualitative information.

Fig.3(a) shows the calculated velocity vectors of B-condition. It shows almost the same pattern of A-condition. The velocity vector at a position little far from pontoon shows that the horizontal velocity is small near the surface and large near the sea bottom.

It can be considered that this is the characteristic flow pattern of the case that pontoon floating in shallow water, and measured data confirms this tendency.

Fig.3(b) shows a comparison of the measured and calculated horizontal velocity u of B-condition. In this case, calculated velocity in the up-stream side agrees well with measured velocity. In down-stream side, the experimental velocity shows a flow toward upstream due to the effect of vortex shedding. Except for the points near the pontoon bottom, calculated velocity compare well to the measured velocity.

Fig.3(c) shows the comparison of the vertical velocity w of B-condition. The downward velocity of calculation agrees with the measured velocity very well in up-stream side. In down-stream side, the calculation shows upward velocity ,but measured data shows downward velocity. This experimental result is also the effect of three dimensional flow and the vortex shedding from the bottom edge of the pontoon. Comparing calculated velocity of up-stream side and of down-stream side, the absolute velocity of them are not symmetry. This is indicates a three-dimensional effect in calculation.

Fig.4(a) shows calculated velocity vectors of C-condition. This case, as shown in table.1,is followed to the condition of the verification experiment of mega-float in Oppama(Tokyo-bay). This figure shows that only the velocity near the front and rear ends of the pontoon have some vertical velocity, and at the other points the vertical velocity are very small. Therefore, in this case the approximation by multi-level model is supposed to be satisfied .

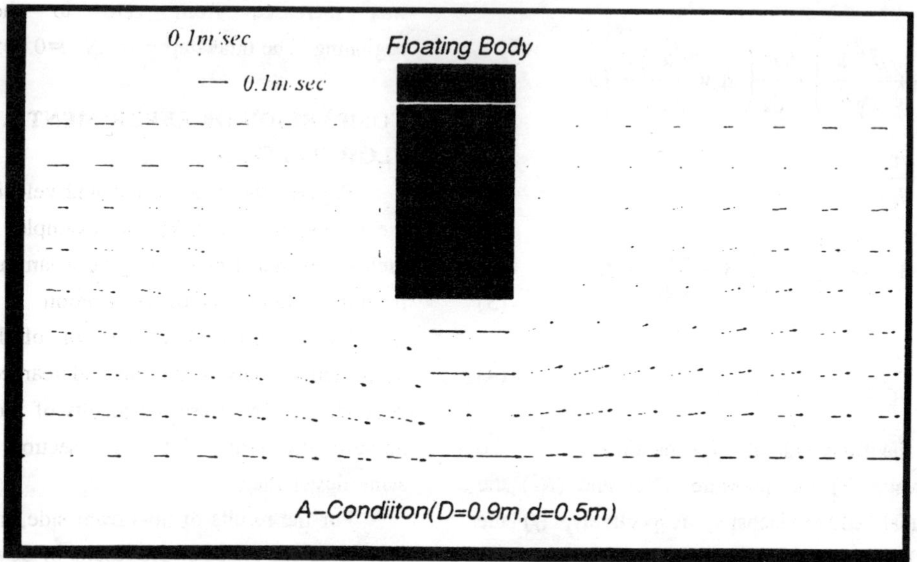

Fig.2 (a) Velocity vectors at $y = 0$ in condition (A)

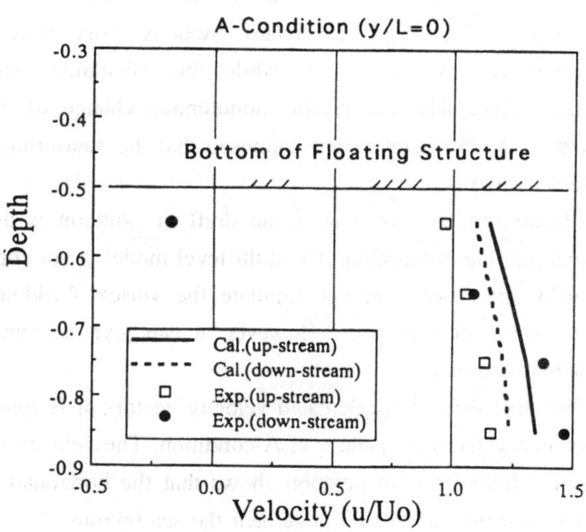

Fig.2 (b) Distribution of horizontal velocity at $y = 0$ in condition (A)

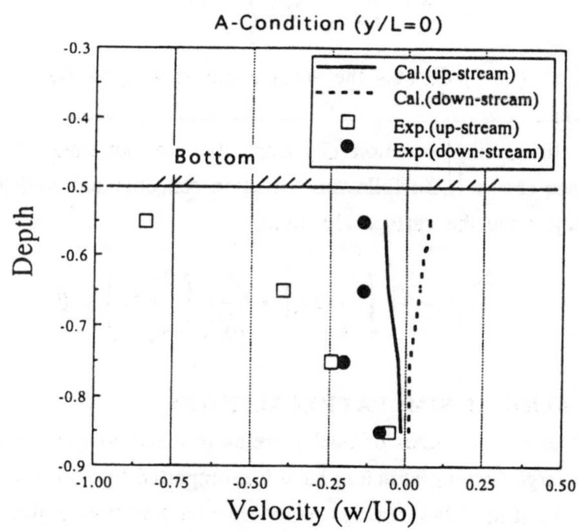

Fig.2 (c) Distribution of vertical velocity at $y = 0$ in condition (A)

Fig.4(b) shows a comparison of the measured and calculated horizontal velocity u of C-condition. In this figure, the calculated velocity in the up-stream side agrees quantitatively with the measured velocity well. However, in down-stream side, calculated velocity is little smaller than up-stream side as similar to the other cases, but the measured velocity is 10 to 20% larger in up-stream side than in down-stream side. It is necessary to check the continuity of mass flux under pontoon, and this is discussed later.

Fig.4(c) shows a comparison of vertical velocity w of C-condition. In this figure, the measured velocity near bottom of pontoon is larger than calculated velocity in up-stream side, and the comparison of vertical velocity w is not as good as horizontal velocity u. However, in the down-stream side, measured velocity agrees with calculated velocity very well. These results show that the application of the multi-level model is most satisfied in C-condition, and calculated velocity agree generally with measured data well.

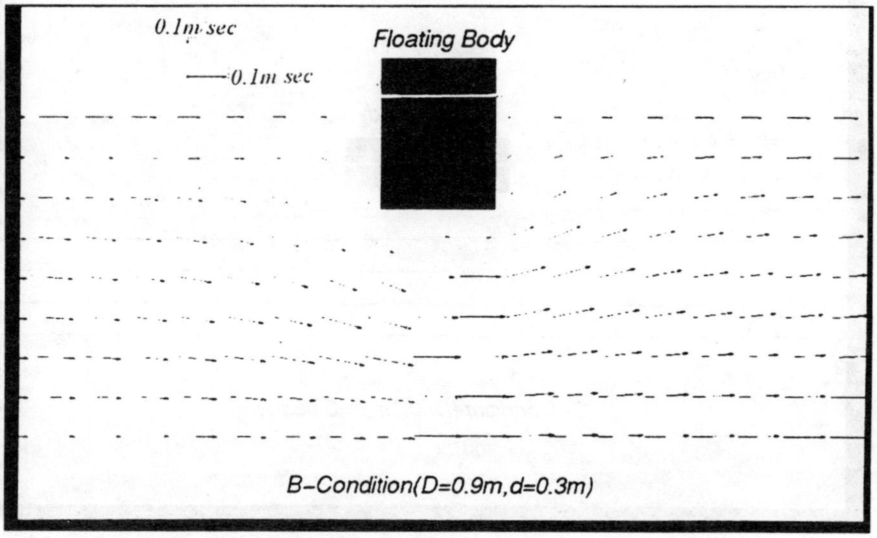

Fig.3 (a) Velocity vectors at $y = 0$ in condition (B)

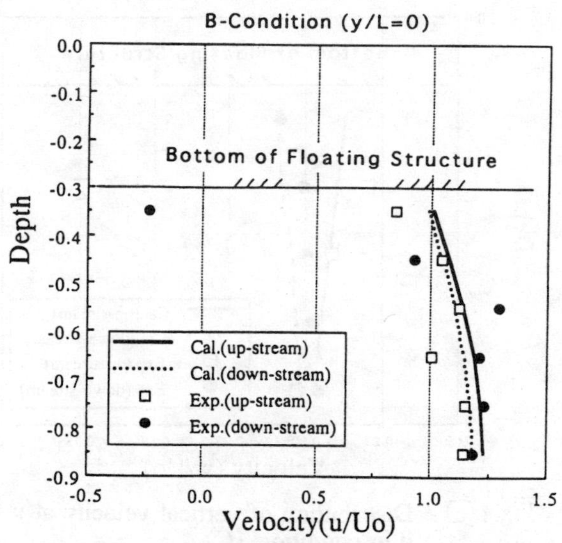

Fig.3 (b) Distribution of horizontal velocity at $y = 0$ in condition (B)

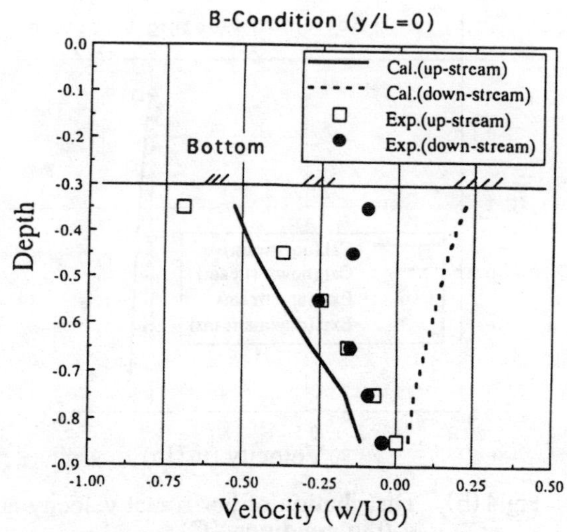

Fig.3 (c) Distribution of vertical velocity at $y = 0$ in condition (B)

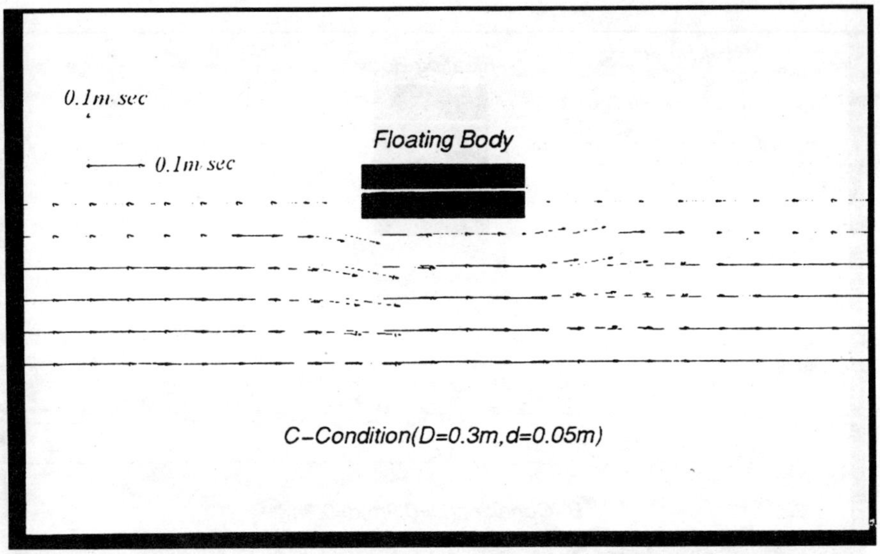

Fig.4 (a) Velocity vectors at $y = 0$ in condition (C)

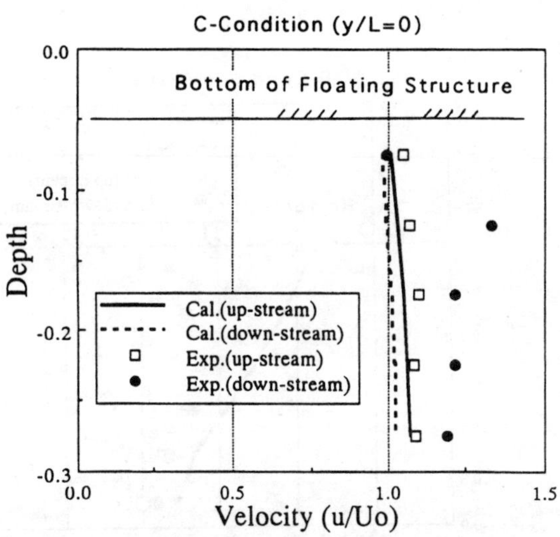

Fig.4 (b) Distribution of horizontal velocity at $y = 0$ in condition (C)

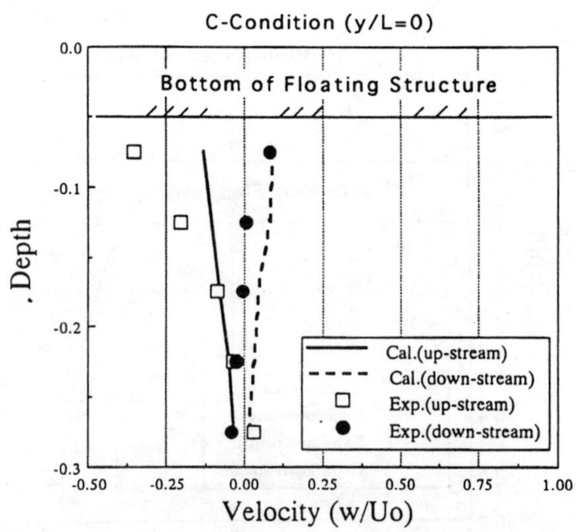

Fig.4 (c) Distribution of vertical velocity at $y = 0$ in condition (C)

MASS FLUX BENEATH PONTOON

Fig.5 shows the distribution of the mass flux beneath the pontoon. The mass flux excluded constant velocity is evaluated by equation ;

$$q_x(y)=\frac{1}{U_0(h-d)}\int_{-h}^{-d}\{u(y,z)-U_0\}dz$$
$$q_y(x)=\frac{1}{U_0(h-d)}\int_{-h}^{-d}v(y,z)dz \qquad (7)$$

The results of experiments A- and B-condition give a similar pattern with the numerical calculation. Calculated value of average velocity in main flow direction q_x shows much larger than measured value at flow-in and flow-out side. Along both front and rear side of the pontoon, the calculated value of q_x is almost constant, but the experimental value becomes smallest in center of pontoon. The tendency of q_y along the sides of pontoon is compared well between the calculated results and the measured results, but only the measured results are much larger. In the calculation, slip condition was used at the bottom of pontoon, and this may be a part of the reason for this result.

The result of C-condition is different from those of A- and B-condition. Measured value of both q_x and q_y is larger than calculated value, especially q_x in the down-steam side is much larger than other measuring points. This result is inconsistent with the law of conservation of mass. As the constant velocity U_0=0.1m/sec is small, the accuracy of measurement of the velocity is not supposed to be good and this may be an important reason for the result. However, the velocity profile and direction of flow-in and flow-out agree generally.

CONCLUSION

Considering the object for simulation, stability of calculation in large changes of sea bed ,capacity of memories of computer and calculation time, layer-level model or MAC method would not satisfy requirements of practical calculation of tidal flow and material diffusion of a bay with a very large floating structure or a mega-float, so at present a multi-level model is considered suitable practical numerical method for the problem of some environmental assessment of the mega-float.

However the precision of this method for the problem is not clear, as multi-level model is based on a quasi-three dimensional method with a hydrostatic pressure approximation in z-direction.

This paper is attempting to solve this problem. The following conclusion were obtained from the comparison between the results of numerical calculation and model experiment.

In case of a very large draft pontoon (h(depth)/d(draft) < 1.8), the agreement of results of calculation and model experiment is not necessarily good because the three dimensional effect becomes server in the area near the pontoon. When h/d is larger than 3, the agreement of results between calculation and model experiment gets better, even the velocity u is comparatively fast (in real field it is approximately u=0.8m/s). However, even in large draft conditions(h/d<1.8) the comparison between the calculation velocity and measured velocity of the flow in the up-stream side or the distribution of vertical velocity caused by shallow water effect and the mass flux beneath a mega float is generally good.

In larger scale field, the vortex shedding at the bottom edge of the pontoon assumed to be much smaller. So it is confirmed that a multi-level model is applicable for the simulation of the flow and the material diffusion of a bay with a very large floating structure or a mega-float.

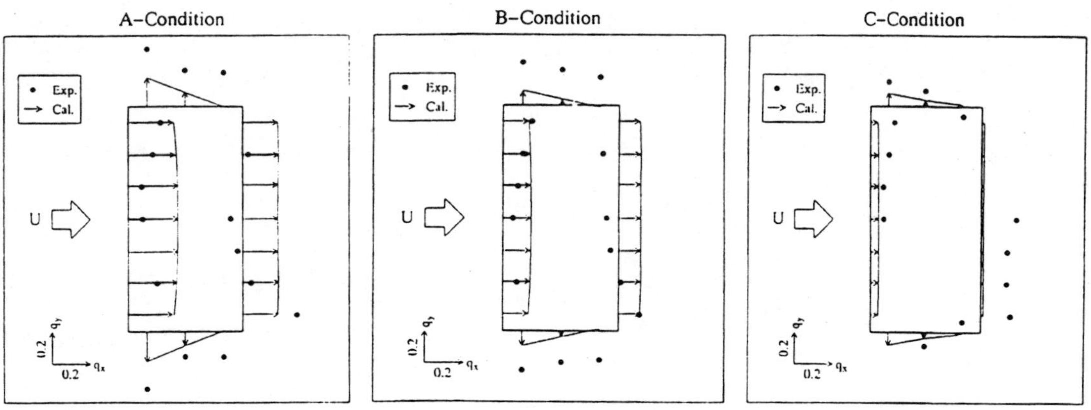

Fig.5 Distribution of flux below model

Acknowledgment

We sincerely appreciate Tsuyoshi Honma(Akishima Laboratories(Mitsui Zosen)inc., Japan), for his corporation to put this research in operation, make his assistance in the experiment and in the analysis of the results.

References

Kyozuka, Y, Tabeta, S, Nakagawa, H and Masuda, K (1995). "Huge Ocean Structure and Marine Environmental Assessment," *13th Ocean Engineering Symposium* ,Vol.13 , pp 21 - 30

Horie, T, and Hosokawa, Y (1983). "Water-quality Controlling Factors in an Eutrophication Model(Japanese)," *REPORT OF THE HORBOUR RESEARCH INSTITUTE*,VOL,22 NO.3, pp.159-205

Hu, C, and Kyozuka, Y (1996). "Tidal Simulation of a Bay with Very Large Floating Structure Using Multi-Level Model," *Transactions of the West-Japan Society of Naval Architects* , Vol.91 , pp 51 - 59

Inoue, Y, and others (1994). "Feasibility Design of a Floating Airport and Investigation into its Effect on the Environment -Effect of a Floating Airport on Ocean Current," *JOURNAL of The Society of Naval Architects of Japan* , Vol.176 , pp 75 - 81

Proceedings of the Seventh (1997) International Offshore and Polar Engineering Conference
Honolulu, USA, May 25-30, 1997

ISBN 1-880653-28-1 (Set); ISBN 1-880653-29-X (Vol. I)

Response to Nutrient Addition in Tropical Nearshore Waters, Bays and Estuaries

Hans J. Krock
University of Hawaii at Manoa
Honolulu, HI, USA

ABSTRACT

Data gathered at five locations in the tropical Pacific are used to determine the kinetic constants relating nutrient concentrations and residence times to chlorophyll-a concentrations. The nutrient concentrations were found to be log-normally distributed and the residence times could be determined either from a salt balance or from measurements of advection or eddy diffusivity. The eddy diffusion coefficient was related to vertical and horizontal restrictions. Total Nitrogen (TN) or Total Phosphorus (TP) were more applicable indicators of nutrient limitation than the conventional nitrate or orthophosphate. This is indicative of the symbiotic relationship of bacterial and algal plankton in real-world systems. The techniques developed can be used to define mass emission rates of TN or TP that will result in a desired level of chlorophyll-a. They are also useful in optimizing the location and design of wastewater outfalls.

Key Words: Nutrients, Kinetics, Residence Time, Water Quality, Wastewater, Standards.

INTRODUCTION

The general kinetic relationship describing enzymatic activities on substrates was first discovered early in this century (Michaelis and Menten, 1913). Each enzymatic reaction is with a specific substrate making this a well defined system whose rate can be described mathematically using a relatively simple relationship. That relationship involves two variables; the reaction rate (μ) and the substrate concentration (S), and two kinetic constants; the maximum reaction rate ($\hat{\mu}$) and the half saturation constant (K_s). The maximum reaction rate would occur when the substrate concentration is so high (mathematically infinite) that every enzyme molecule instantly finds another substrate molecule as soon as it completes its reaction with the previous substrate molecule. The half saturation constant is the value of the substrate concentration at which the reaction rate is half the maximum reaction rate. The Michaelis-Menten relationship is:

$$\mu = \frac{\hat{\mu} S}{K_s + S} \tag{1}$$

Each enzymatic reaction in a living cell deals with a different substrate and has different kinetic constants. These reactions occur in series and in parallel, but all have the same general mathematical form. Given the great complexity of all of the reactions involved in a living cell, it took real insight for Monod to propose that the growth rate of a bacterial culture can be described by the enzyme reaction rate equation of Michaelis-Menten if one considers that one of these enzyme reactions must be the slowest and hence "limiting" to the whole process (Monod, 1949).

This required a re-definition of μ to signify growth rate and S to be the concentration of the limiting substrate. Subsequently, this relationship was applied by biological scientists to determine the growth kinetics of photosynthetic and bacterial organisms with all manner of substrates in monocultural chemostats. This relationship was also used by biochemical engineers to design bioreactors. Sanitary engineers were bold enough to apply the relationship to the multicultural activated sludge process used in sewage treatment. In this case, the substratum is also complex, namely all the organic material susceptible to breakdown under aerobic conditions, expressed as Biochemical Oxygen Demand (BOD). Although far removed from the specific enzymatic reaction and specific substrate that might be controlling the overall growth rate, the relationship (Eq. 1) is accurate enough to support the design of sewage treatment plants.

It has also been noted that the time related growth response is exponential when the limiting substrate is added to the chemostat or the bioreactor or the activated sludge plant. The equation for such exponential growth is:

$$C_t = C_o e^{\mu t} \tag{2}$$

The initial concentration of the organism or population (C_o) increases with time (t) and the growth rate (μ) to the higher concentration (C_t).

The effect of competitive or non-competitive inhibitory substances on enzymatic reactions can also be described by Michaelis-Menten kinetics if the concentration of the inhibitory or toxic substance is treated mathematically as inverse substrate (Krock, 1971).

All of these observations in controlled environments indicate that the more complex biological systems of the natural environment could possibly also be described by Michaelis-Menten kinetics and exponential growth. The problem then becomes one of identifying the limiting substrate (or finding a parameter that is proportional to the limiting substrate) and of determining the time available for a growth (or decay) response to take place.

STUDY LOCATIONS

The real world systems that are included in this study are all located in the tropical Pacific and have similar water column biosystems of bacterial and algal plankton, zooplankton and higher trophic level organisms. They have similar temperature, salinity and incident sunlight conditions. They differ in nutrient concentrations, in biomass and chlorophyll-a concentrations, in transport and mixing characteristics and in residence times.

The first location studied was the Hawaii Kai Marina on the Hawaiian Island of Oahu. This body of water is an expanded shallow tidal pond. The marina is well mixed by the wind but the principal exchange mechanism with the ocean is the tide. The Hawaii Kai Marina has relatively high nutrient levels (Sunn, Low, Tom and Hara Inc., 1974).

Kaneohe Bay on the windward side of Oahu is a relatively large embayment for which nutrient and chlorophyll-a data were available with and without the addition of a significant amount of domestic wastewater (Sunn, Low, Tom and Hara Inc., 1976).

The Kona Coast of the Island of Hawaii is unusual in that there is essentially no surface runoff because of the high permeability of the geologically new lava. Consequently, there is proportionally little input of phosphorus to the nearshore waters relative to the input of nitrogen, which travels with the groundwater (Krock, 1986). Also, a considerable amount of water quality data exists for these waters because of the on-going studies for Ocean Thermal Energy Conversion (OTEC) and related industry.

The Ala Wai Canal on Oahu is a nutrient rich drainage canal and sediment basin which exhibits the classical two layer flow of an estuary.

The approach developed in this study was also applied successfully to predict the chlorophyll-a concentration of the Barbers Point Deep Draft Harbor on Oahu and to describe the effect of the discharges from the tuna canneries on Pago Pago Harbor on the Island of Tutuila in American Samoa.

RESIDENCE TIME

In order to utilize Equation (2) to determine the net growth rate, it is necessary to determine the time available for growth to take place. The methods developed, by chemical engineers in the design of reactor vessels and of environmental engineers in the description of secondary treatment plants and tertiary pond systems, are applicable to the real-world systems analyzed in this study. The average residence time in a reaction vessel, whether it is plug flow or a Constant Flow Stirred Tank Reactor (CFSTR), or anything in between, is the volume divided by the flow rate. The most workable approach to analyzing the exchange characteristics of the various water bodies was to develop appropriate box models. These box models took into account stratification, hydraulic control sections, eddy diffusivity, and advection. Where there was a definable quantity of fresh water input (Kaneohe Bay and the Ala Wai Canal), it was possible to use a salt balance to define the saline water flow rate.

Advection, due to tidal currents and wind, was measured using current meters located at the hydraulic control sections (Hawaii Kai Marina and Kaneohe Bay).

In those areas without significant advection, the principal transport mechanism was eddy diffusivity. This was measured using the spreading rate of a dye patch as recorded by timed aerial photographs (Hawaii Kai Marina, Kaneohe Bay, Kona Coast, and Pago Pago Harbor). As has been noted by many other researchers, the rate of increase of the coefficient of eddy diffusivity with scale is limited by boundaries. This means that mixing is limited in shallow, or stratified or narrow water bodies and the Richardson's Law 4/3 exponent that is frequently assumed in computer models is not applicable.

The residence time in a straight channel or perpendicular to an open coast is inversely related to the eddy diffusivity and directly to distance.

With the use of these techniques, the residence times of the various areas of each water body could be determined with respect to the transition zone to the ocean. A plot was then made relating the residence time (arithmetic scale) and the chlorophyll-a concentration (log scale). The slope of the line corresponded to the exponential net growth rate as defined by integrating Equation (2).

It was noted that the chlorophyll-a concentration data was log-normally distributed and that the best descriptor of the central tendency of that distribution is the geometric mean or the median.

The net growth rates observed in these real world systems ranged from a low of 0.06 per day off the Kona Coast to a high of 0.30 per day in the Ala Wai Canal. Hawaii Kai Marina had a net growth rate of 0.25 per day. The net growth rate in Kaneohe Bay changed from 0.09 per day without sewage effluent addition to 0.12 per day with that addition.

MICHAELIS-MENTEN KINETICS

The kinetic constants in Equation (1) can be defined by inverting the equation to result in a form that is a straight line in a plot of the reciprocal net growth rate versus the reciprocal limiting substrate concentration. That inverted equation is:

$$\frac{1}{\mu} = \frac{1}{\hat{\mu}} + \left(\frac{K_s}{\hat{\mu}}\right)\left(\frac{1}{S}\right) \tag{3}$$

In a plot of this equation $\frac{1}{\hat{\mu}}$ is the intercept of the $\frac{1}{\mu}$ axis, while $\frac{K_s}{\hat{\mu}}$ is the slope of the line.

Figure 1. Nutrient Limited Net Growth in Tropical Saline Waters for Phosphorus

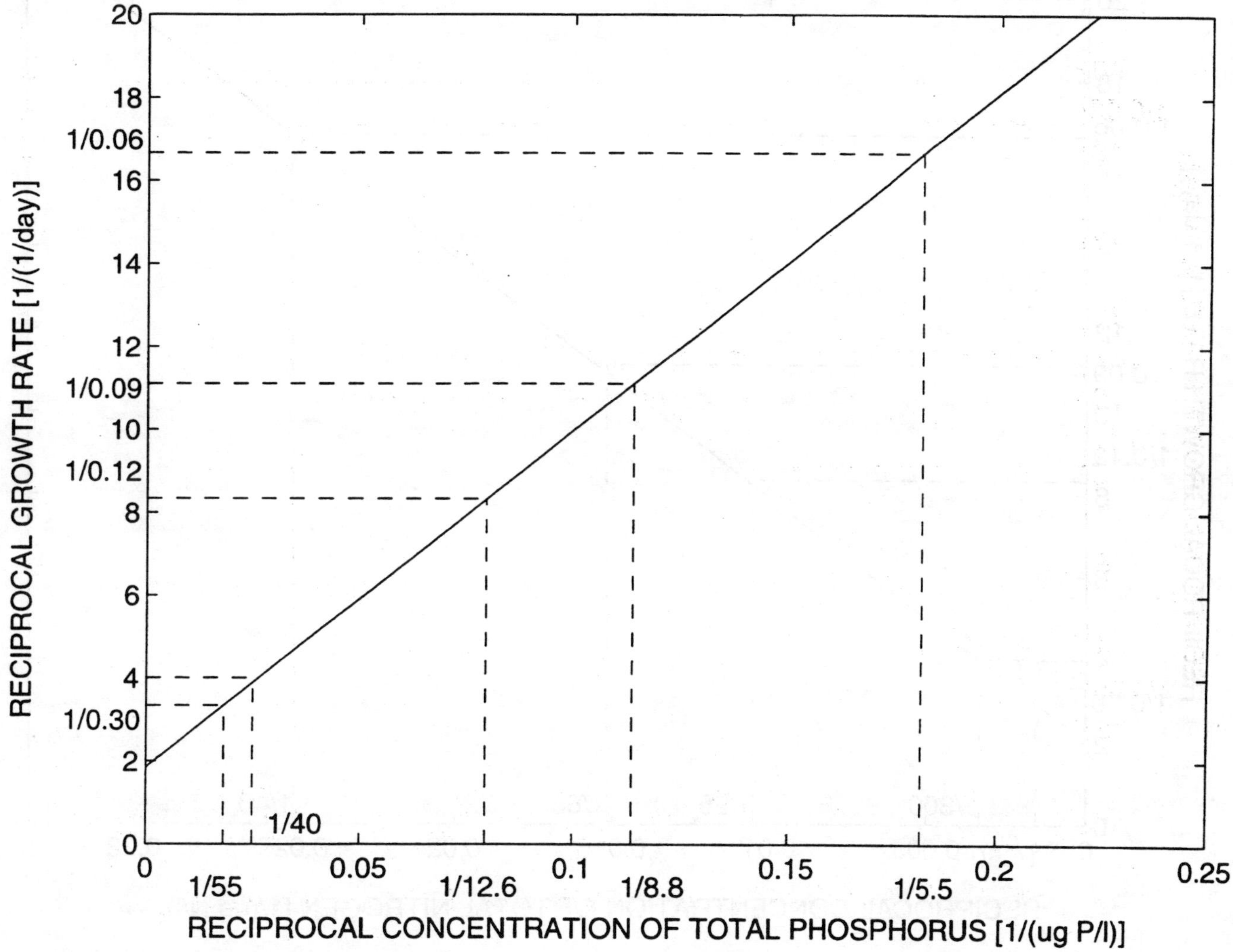

What remains to be done is to find an appropriate definition of the appropriate limiting substrate in this complex biological system. Following the experience of biological scientists with monoculture chemostats, the concentrations of inorganic nutrients (NO_3^-, NH_3, and orthophosphate) were plotted. The result was not successful in that the higher growth rates were often paired with lower inorganic nutrient concentrations. However, when a plot is made of Equation (3), using either Total Phosphorus (TP) or Total Nitrogen (TN) as the limiting substrate, very satisfactory straight lines result. The plot for the geometric mean of Total Phosphorus is shown in Figure 1, while that for the geometric mean of Total Nitrogen is in Figure 2. The data used in these plots is given in Table 1.

The kinetic constants obtained from Figures 1 and 2 are given in Table 2.

Table 1. Median Nutrient Concentrations and Net Growth Rates

LOCATION	µ PER DAY	TP µg/ℓ	TN µg/ℓ
Off Kona	0.06	5.5	40
Kaneohe Bay w/o Sewage	0.09	8.8	66
Kaneohe Bay w/ Sewage	0.12	12.6	95
Hawaii Kai Marina	0.25	40	300
Ala Wai Canal	0.30	55	650

Figure 2. Nutrient Limited Net Growth in Tropical Saline Waters for Nitrogen

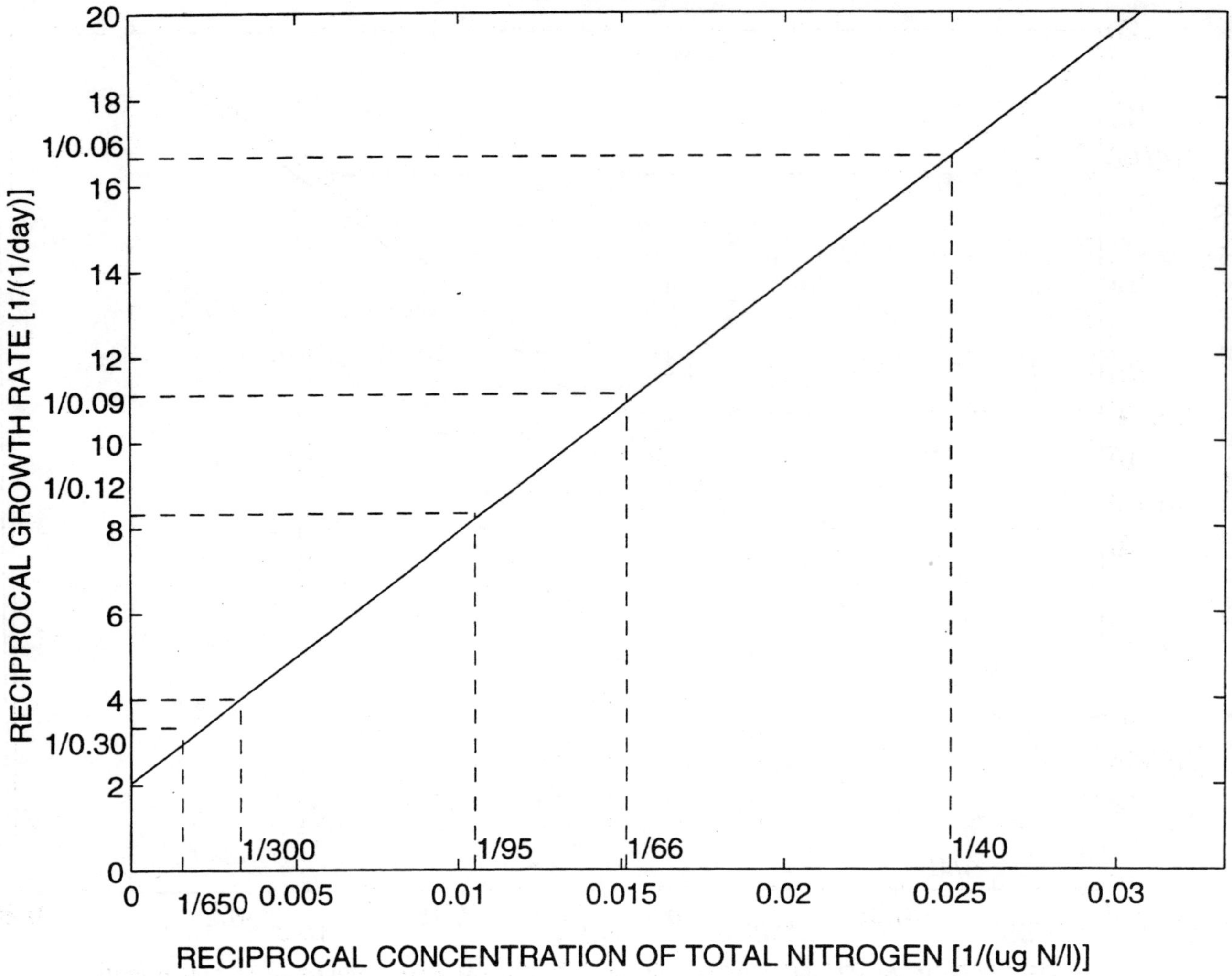

Table 2. Kinetic Constants

LIMITING SUBSTRATE	$\hat{\mu}$ PER DAY	K_s $\mu g/\ell$
Total Phosphorus	0.50	40
Total Nitrogen	0.50	300

DISCUSSION

The growth rates that are observed for marine phytoplankton under controlled conditions in pure culture chemostats or in well engineered monoculture bioreactors are often in the range of 1 or 2 per day. Consequently, the net growth rates measured in real world systems appear to be too low—especially the maximum net growth rate ($\hat{\mu}$). However, it must be remembered that these are 'net' growth rates which take into account the affects of predation, competition, night-time light limitation, toxic substances, settling out, and other inhibiting conditions.

The use of TP or TN as the limiting substrate for the net growth rate of phytoplankton is not readily obvious to biological scientists who try to model the complex interactions of real-world systems by finding the kinetic constants for each trophic level (or possibly each species) and then somehow adding the results. That approach has met with very limited success—except for highly controlled bioreactors. Of specific concern in this approach is the assumption that the measured values of inorganic nutrients in the water accurately reflect what is available to the phytoplankton. This assumption misses the close interaction between bacterial plankton and phytoplankton. This symbiotic relationship processes nitrogen and phosphorus in all its forms and consequently it is reasonable to consider TP and TN as better measures of what is available to the real-world biological system.

It is perhaps also not surprising that the analysis of the five real-world systems considered in this study resulted in very close outcomes for both TP and TN as candidate limiting substrates. The molar ratio of the half saturation constants for TN to TP is very close to 16, the ratio found in living organic material. Also, the sources of

TP and TN in the systems studied are mostly organic materials having this ratio.

APPLICATIONS

The results of this study can be used to predict the water quality of new projects such as marinas or harbors. The analytical approach can also be used to establish site specific water quality standards and to determine the level of mass emission rate control that would be required to meet such standards.

ACKNOWLEDGEMENTS

Kaiser-AETNA supported the study of the Hawaii Kai Marina. The U.S. Army Corps of Engineers supported the studies of Kaneohe Bay and Pago Pago Harbor. The State of Hawaii, Department of Transportation supported the study of the Barbers Point Harbor. The County of Hawaii supported the study off Kona. Data for the Ala Wai Canal was from a high school student project supported by the NSF and from the Oceanic Institute.

REFERENCES

Krock, H.J. (1971). "Bioassays of Lower Trophic Levels," Volume VI of "A Study of Toxicity and Biostimulation in San Francisco Bay-Delta Waters," SERL Report No. 71-8, University of California at Berkeley.

Krock, H.J. (1986). Unpublished Measurements Taken Off Kona For: Sea Engineering, Inc., Honolulu, Hawaii.

M & E Pacific, Inc. (1978). "Barbers Point Deep Draft Harbor EIS," for the State of Hawaii Department of Transportation.

M & E Pacific, Inc. (1979). "Baseline Water Quality Study for American Samoa," for the U.S. Army Engineering Division, Honolulu, Hawaii.

Michaelis, L. and Menten, M.L. (1913). "Die Kinetik Der Invertinwirkung," Biochem.Z. 49:333.

Monod, J. (1949). "The Growth of Bacterial Cultures," Annual Review of Microbiol. 3:371.

Oceanic Institute, Inc. (1992). Prepublication Data of Water Quality in the Ala Wai Canal, Honolulu, Hawaii.

Sunn, Low, Tom & Hara, Inc. (1974). "Investigation of Hawaii Kai Marina Waters," for Kaiser-AETNA, Honolulu, Hawaii.

Sunn, Low, Tom & Hara, Inc. (1976). "Kaneohe Bay Water Resources Data Evaluation," for the U.S. Army Engineering Division, Honolulu, Hawaii.

Proceedings of the Seventh (1997) International Offshore and Polar Engineering Conference
Honolulu, USA, May 25-30, 1997

ISBN 1-880653-28-1 (Set); ISBN 1-880653-29-X (Vol. I)

Compliance Methodology for Lognormal Populations

Ross S. Tanimoto
City and County of Honolulu

Hans J. Krock
University of Hawaii at Manoa

Honolulu, HI, USA

ABSTRACT

The City and County of Honolulu, Department of Wastewater Management (WWM) is a publicly owned treatment works discharging roughly 115 mgd of treated wastewater into the marine environment. The State of Hawaii, Department of Health (DOH) and the Environmental Protection Agency (EPA) regulate the discharge of treated effluent in compliance with the Clean Water Act and Hawaii Revised Statues through applicable discharge permits. The existing statistically based state water quality standards consider criteria having potential impacts to the local marine ecosystem. For these reasons, the City and County of Honolulu has established a statistical methodology to show compliance with applicable standards. The methodology is unique in that it assumes a lognormal distribution of several environmental parameters and reliably describes environmental processes. This paper presents the statistical methodology and identifies the benefits of the method as a diagnostic tool for assessing the water quality around the discharge points.

KEY WORDS: Water quality, lognormal distribution.

INTRODUCTION

The City and County of Honolulu, Department of Wastewater Management (WWM) is a publicly owned treatment works discharging roughly 115mgd of treated wastewater into the marine environment using four wastewater treatment facilities. Approximately 100 mgd is discharged at a depth of roughly 200 feet after primary treatment from two deep ocean outfalls. The remaining 15 mgd is discharged after secondary treatment at a depth of roughly 100 feet from two shallow marine outfalls. The State of Hawaii, Department of Health (DOH) and the Environmental Protection Agency (EPA) regulate the discharge of treated effluent in compliance with the Clean Water Act and the Hawaii Revised Statues through applicable discharge permits. The existing statistically based Water Quality Standards for the State of Hawaii (State WQS) consider water quality standards relative to impacts to local marine ecosystem. For these reasons, the City and County of Honolulu has established a statistical methodology to show compliance with these standards, specifically Hawaii Administrative Rules, Title 11, Chapter 54. The methodology is unique in that it assumes a lognormal distribution of several environmental parameters and reliably describes environmental processes. The State of Hawaii has adopted marine receiving water quality standards for the following parameters: total nitrogen, ammonia nitrogen, nitrate + nitrite, total phosphorus, chlorophyll-a, and turbidity. Each parameter has three criteria: (1) Geometric mean, (2) "Not to exceed the given value more than ten percent of the time," and (3) "Not to exceed the given value more than two percent of the time." The City and County of Honolulu, has developed a methodology to demonstrate compliance with these standards, and in the process found the method to be a valuable tool for assessing the water quality condition of the receiving waters.

METHODOLOGY

The methodology is based on the assumption that the data follow a lognormal distribution. Testing of the three compliance categories (i.e., geometric mean, not to exceed 10% and not to exceed 2%) for any parameter is done by using a graphical method on lognormal probability paper.

Instructions on the method to test compliance include the following steps.

a. In column (1) of the worksheet, number the rows from 1 to n, where n is the number of data points to be used in the compliance determination. See Table 1 for a sample worksheet.

b. List the data for a particular parameter (e.g., turbidity) in column (2). The data should include all applicable stations in the area of interest. Control station data should not be included in the analysis. Instead, a separate analyses should be done for comparison. The data used should be taken over a time period of at least one year. The size of the data base should exceed thirty (30) data points, or $n > 30$. Larger data bases are suggested to increase statistical confidence.

c. Order the data in column (2) from smallest to largest into column (3).

d. Take the logarithms of the ordered data from column (3) and list in column (4). It does not matter what type of logarithm is used provided consistency is maintained. The natural logarithm is specified for consistency.

e. Find the mean and sample standard deviation of the natural logarithm values in column (4).

f. To calculate the geometric mean (or 50% value) concentration, take the antilog of the mean natural logarithm value from step "e" above.

Multiply and divide the geometric mean value by the antilog of the sample standard deviation from step "e" to define the concentrations associated with 84% and 16% respectively.

Plot these three points [(GM or 50%), (84%), and (16%)] on Log-Probability paper. A line drawn through the three points will be straight on Log-Probability paper. Identify this line as the *calculated line*. Plot, on the same graph, the three points (or criteria for 50%, 90%, 98%) from the State WQS for the parameter in question, connecting the points with a straight line. Identify the resulting line as the *standard line*. Compare both lines. If the *calculated line* is entirely below the *standard line*, the water quality parameters for the 50%, 90%, and 98% criteria have not been exceeded. In other words, comparison of both lines is the method used to determine exceedences of the standard, not an individual data point. From the *calculated line*, determine the parameter values corresponding to the 90% and 98% probability. Compare these values and the GM with the appropriate State WQS criteria for the parameter in question.

g. Plot all data points on Log-Probability paper, use the ranked data (column (3)) as the abscissa (y-axis) and the probability (column (5)) as the ordinate (x-axis). The purpose of plotting the data is to validate the lognormal distribution. If the plotted data points approximate the *calculated line* on Log-Probability paper, then the lognormal distribution assumption is verified. Other distributions may be applied; however, the assumption of this methodology is a lognormal distribution.

h. If the plotted data confirms a lognormal distribution, the results from step (f) above are valid.

i. If the plotted data does not correspond to the *calculated line* (e.g. confirmation of the lognormal distribution is not validated), several events could be occurring. There could be something wrong with the data or another process may be occurring in the receiving waters. Lastly, but not necessarily finally, the process may not be lognormal distributed, and therefore, compliance cannot be determined.

TABLE 1
SAMPLE WORKSHEET

I	Listing of raw data as received	Data, xi ranked smallest to largest	Log of the ranked data	Probability (percent) 100*(I-1/2)/n [%]
1	y_1	x_1	$\ln(x_1)$	$(1-\frac{1}{2})/n$
2	y_2	x_2	$\ln(x_2)$	$(2-\frac{1}{2})/n$
:	:	:	:	:
n	y_n	x_n	$\ln(x_n)$	$(n-\frac{1}{2})/n$

where n = the number of data points.

The above analyses should be done on an annual (calendar year) basis, using a sample size of roughly thirty (30) data points. If thirty data points are not available during a given monitoring year, the data obtained during the previous monitoring year can be included in the compliance data base. The procedure should be continued until a compliance data base of thirty or more data points is achieved. An analyses which includes at least five (5) years of data should also be done at the same time as the annual analyses. The purpose of this analyses is to identify possible long term impacts.

APPLICATION

The method outlined herein has been applied to measurements taken over a period of more than five years in the discharge and dilution areas of the City ad County of Honolulu wastewater discharges. The method has been incorporated in the computer data system and has proven to be a useful tool in checking compliance with the State Water Quality Criteria. Experience has shown that in every case where a data set did not follow a log-normal distribution, the data was found to be invalid or another source of input was operating. Such input included heavy non-point discharges or (on one occasion) sea bird droppings.

The method using log-normal distributions has proven to be superior to the previous use of single number standards or of a regular normal distribution because a log-normal distribution accurately describes the natural variation of the water quality parameters under consideration.

ACKNOWLEDGEMENT

The Department of Wastewater Management, City and County of Honolulu provided the monitoring data from their four marine outfalls.

REFERENCES

State of Hawaii (1992). "Hawaii Administrative Rules, Title 11, Chapter 54, Department of Health Water Quality Standards.

State of Hawaii (1977). "An Ecosystem Approach to Water Quality Standards - Report of the Technical Committee on Water Quality Standards," Technical Report No. 1, Honolulu, Hawaii.

Proceedings of the Seventh (1997) International Offshore and Polar Engineering Conference
Honolulu, USA, May 25-30, 1997

ISBN 1-880653-28-1 (Set); ISBN 1-880653-29-X (Vol. I)

Static and Dynamic Properties of Marine Soils

Shamsher Prakash
University of Missouri-Rolla
Rolla, MO, USA

Vijay K. Puri
Southern Illinois University
Carbondale, IL, USA

ABSTRACT

Static and dynamic soil properties of marine soils are discussed. The effect of oil contamination on static geotechnical properties of near shore soils is also reviewed. Methods commonly used for determination of dynamic soil properties are discussed along with typical test data.

KEY WORDS: Static, and dynamic soil properties, soil, contamination marine soils

INTRODUCTION

Offshore structures are designed for a combination of static and dynamic loads. In order to ensure safe design and performance of such sturctures, the soil properties must also be determined for these load combinations.

Goetechnical properties of soil for static and dynamic loads are reviewed. Also the effect of oil contamination in near shore soils is described.

STATIC SOIL PROPERTIES

The static soil properties of marine soils are discussed here by considering four sedimentary environments of southern California; continental borderland (Bouma et al., 1972; Gershowetz et al., 1972), namely; (1) the Continental Shelf (0 to 100 m), (2) Basin Floor (700 to 100 m); (3) Basin Slope (100 to 800 m), and (4) Submarine Canyon Fan (700 to 800 m). These properties are listed in Table 1.

It may be observed from Table 1, that the shelf sediments are characterized by greater shear strengths, grain size and bulk densities, and lower water content, specific gravity, porosity and compression index. The deeper water basin and fan sediments, on the other hand, are characterized by greater water content, specific gravity, porosity, compression index and lower shear strengths, grain sizes and bulk densities. Slope sediments are transitional in these properties. The average specific gravity for Californian Borderland sediments is 2.76 which is greater than values of onland specific gravities of about 2.65.

TABLE 1 - Physical Properties for Sedimentary Environments of Continental Borderland (Bouma et al., 1972)

PROPERTY	SHELF	SLOPE	BASIN	SUBMARINE FAN	OVERALL AVERAGE
Median grain size, microns	58.6	37.8	5.6	8.5	38.4
Specific gravity	2.740	2.795	2.775	2.741	2.760
Salt water content, %	51.5	97.6	136.4	143.4	82.5
Unit dry weight, kN/m^3	12.0	8.6	6.5	6.0	9.7
Unit submerged weight, kN/m^3	7.5	5.5	4.0	3.7	6.2
Porosity%	57.7	72.5	78.8	79.0	69.3
Cohesion, kN/m^2 (at 5 kN/m^2 confining)	7.5	5.8	4.8	5.7	6.4
Compression index, C_c	0.112	0.557	0.988	1.036	0.481

NORTH ATLANTIC and NORTH PACIFIC SEA BASINS

The North Atlantic and North Pacific ocean basins are two major ocean basins that differ quite a bit in size and topography. The Atlantic basin is basically divided into two halves by the Mid-Atlantic Ridge, with similar topography on the east and west sides of the ridge, whereas the Pacific basin has numerous ridges, trenches, island chains, scarps and seamounts all of which influence the deposition of the sediments. The average depth of the Pacific is also greater than that of the Atlantic which also influences the sediment distribution (Keller, 1969). The main sediment types in these two classes are: (1) fluvial-marine (sand-silt) representing the coarse fraction (larger than 0.016 mm); (2) fluvial-marine (silt-clay) the finer fraction (smaller than 0.016 mm) of material derived from terrestrial drainage: (3) "red clay" the inorganic pelagic clays which are usually chocolate brown, but vary in color; (4) calcareous ooze, the sediment material from various planktonic animals and plants; (5) calcareous sand and silt that consist of shell fragments and coralline debris of sand and silt size particles; and (6) siliceous oozes.

The average shear strength values for both the basins range from 3.5 to 18.0 kN/m^2 for the upper 1.0 m of the sea-floor. Shear strength of 7.0 to 11.0 kN/m^2 seem to dominate the North Atlantic and can be attributed to the calcareous deposits. Values smaller than 3.5 kN/m^2 can often be found around the coastal areas of both the Atlantic and Pacific basin, where local drainage or current conditions strongly affect the deposition of sediments.

Water content values in the North Atlantic range from 30% to 170% but are commonly between 50% and 100%. Relatively high water contents are found in the basin as a result of the influence of local drainage and current conditions.

In the North Pacific, water contents vary from 50% to 375%, but are most commonly 100% to 200%. These higher water contents can be attributed to the "red clay" deposits in the Pacific.

Sediment densities of 14.8 to 17.1 kN/m^3 make up a major portion of the Atlantic sea-floor. Sediment densities of 12.4 to 14.8 kN/m^3 cover extensive regions of the Pacific sea-floor.

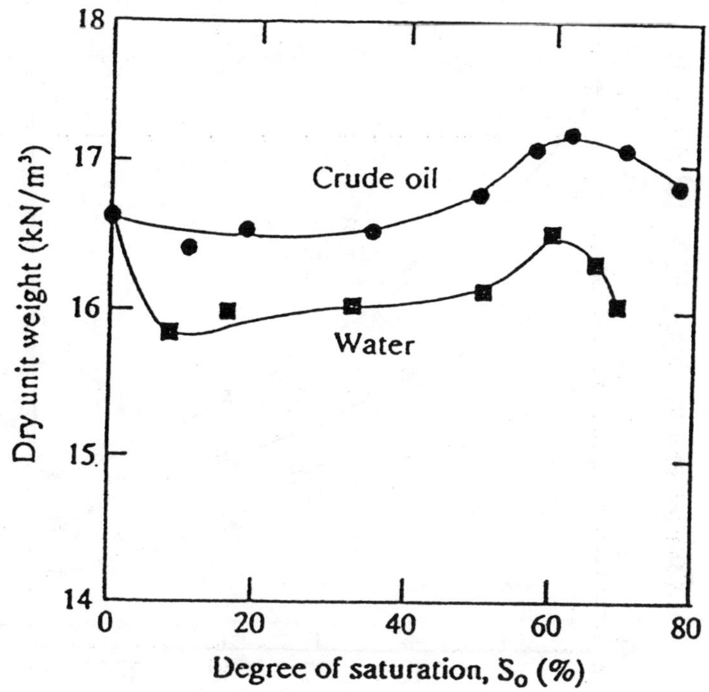

Figure 1 Plot of dry unit weight versus degree of saturation (Puri et al. 1991)

ENGINEERING PROPERTIES OF OIL CONTAMINATED SOILS

Accidental off-shore oil spills may cause not only environmental and ecological problems; they may also have an adverse effect on the mechanical properties of soils especially the near shore soils. Recent investigations have shown that the mechanical properties of soils are influenced to a considerable extent by chemical contamination Anderson (1982), Lukas (1972), Kumpaley and Ishola (1985), Sherard et al. (1972), and Shridharan et al (1981). Cook et al. (1992) and Puri et al. (1993) investigated the effect of crude oil contamination on the compaction, shear strength, one-dimensional compression characteristics of a poorly graded river sand.

Compaction behavior with water and crude oil as the pore fluid is compared in Fig. 1. The maximum dry unit weight achieved with water as the pore fluid is about 16.45 kN/m^3 when the degree of saturation is about 62% (water content = 13.45%). The maximum dry unit weight obtained under the same compactive effort is about 6% higher when crude oil is used for compaction as compared to water. This is due probably to the fact that oil is more effective in reducing the friction between the sand grains thus helping them to occupy closer configuration resulting in higher dry unit weight.

The angle of internal friction ϕ (based on total stress) also decreases due to the presence of oil. The variation of the angle of friction (total stress) with degree of oil saturation is shown in Fig. 2. For tests at a relative density (D_r) of 85%, the angle of friction is about 41° for dry sand and its value decreases to about 30° when the degree of oil saturation increases to about 19.5%. Within the range of parameters studied, the percentage of decrease in the values of angle of friction ranged from 17.6% to 25% as compared to its value at the same relative density in the dry state. The results on one-dimensional compression tests showed that the presence of oil in the soil mass had an adverse effect on the compression characteristics of sand. Figure 3 shows the variation of constrained modulus E_c with a degree of oil saturation at predetermined strain levels. The value of E_c decreases with an increase in degree of oil saturation. More work is needed to understand the behavior of soils with different type of chemical spills.

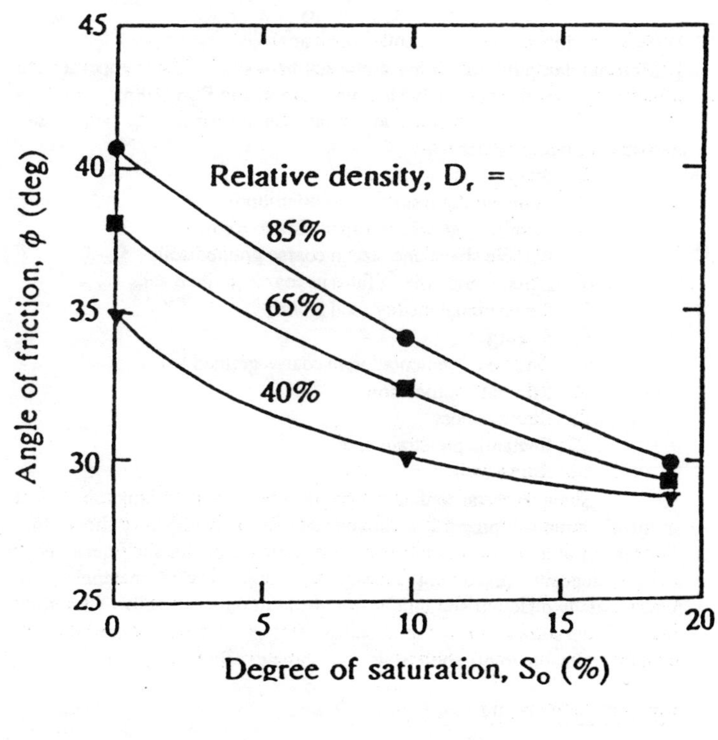

Figure 2 Variation of angle of friction versus degree of oil saturation (Puri, et al. 1993)

DYNAMIC PROPERTIES OF MARINE SOILS

Seismicity studies of the earth have shown that many areas of the ocean are seismically active, and a high percentage of all earthquakes occur in the circumpacific belt including the continental shelf (Chandrasekaran and Krishna 1976).

The dynamic loads are also imposed by sea waves, and sea storms. In order to analyze the dynamic-soil structure interaction effects of offshore structure, the dynamic soil-properties are required. The dynamic properties of interest are:

1) Strength of soil under dynamic loads

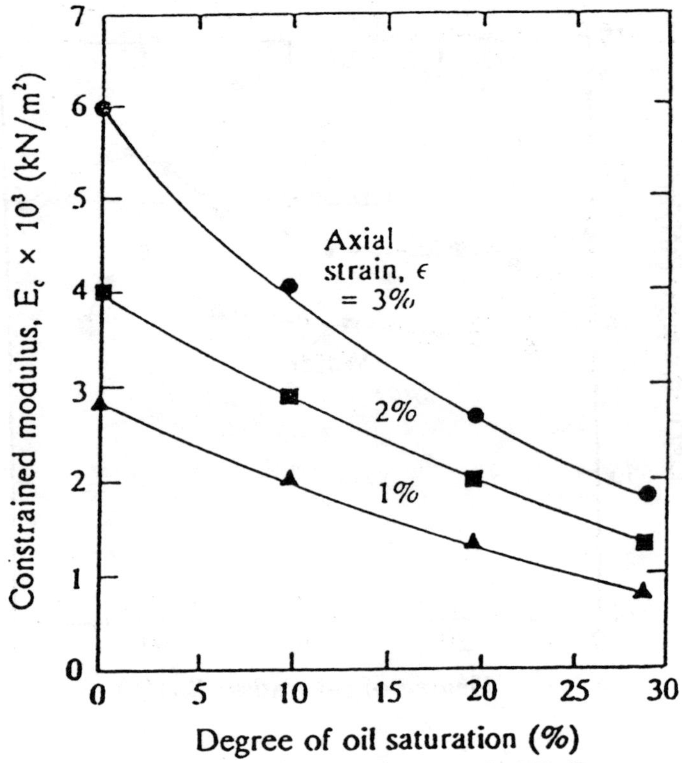

Figure 3 Constrained modulus versus degeree of oil saturation (Cook et al, 1992)

2) Shear modulus of soil for displacement analysis
3) Material damping due to hysteretic soil behavior. These properties are affected by several factors listed below (Prakash and Puri 1988).

1. Type of soil and simple soil properties e.g. void ratio, consistency, water content etc.
2. Strain level
3. Initial and pulsating stress conditions
4. Confining pressure (mean effective stress)
5. Particle shape and size in coarse grained soils
6. Percent and type of fines in coarse grained soils
7. Depositional history (soil fabric)
8. Ageing
9. Degree of cementation in coarse grained soils
10. Mineral composition
11. Stress history
12. Dynamic pre-strain and
13. Stress path

Strain level and confining pressures have the most important effect on the dynamic soil properties. Chaney and Fang (1983) have shown that there is no indication that physical characteristics of the shelf deposits or pelagic deposits (excluding calcareous and siliceous materials) are fundamentally different and unique compared to on-shore soils. Therefore, most of the studies on dynamic soil properties of the on-shore soils are equally applicable to off-shore soils with some exceptions.

EFFECT OF STRAIN

The behavior of any soil under cyclic shear loading depends strongly on the level of cyclic strain induced in the soil. At very small strains the stress-strain response may be linear ie $G \approx G_{max}$. The upper limit of this range, which has been called "elastic threshold" is about 10^{-5} in sands and about 10^{-4} in normally consolidated clays of Plasticity Index, PI ≈ 50. At small strains the stress-strain response is relatively linear, the value of G is only slightly lower than G_{max}, the material damping (ζ) is low, and there is little or no stress-strain degradation or hardening with number of cycles. The upper limit of this range, which has been variously called "threshold strain" or "plastic threshold" is about 3 to 5 x 10^{-5} for gravel, near 10^{-4} for sands, and on the order of 10^{-3} for normally consolidated clays of high plasticity. At intermediate strains above the plastic threshold the stress-strain response becomes strongly nonlinear, with much higher material damping and with considerable stiffness degradation or hardening caused by the cyclic loading (Dobry et al. 1989).

At large strains on the order of 1 percent (10^{-2}) to several percent, G is only a small fraction of G_{max}; the stress-strain response can change very dramatically in as little as one or two cycles; and the cyclic stress-strain behavior is controlled by factors other than G_{max}. These factors include mainly (1) the degree of saturation of the soil, (2) its contractive or dilative character, and (3) its monotonic and cyclic shear strength characteristics. Therefore, in this strain range, both G/G_{max} and ζ must be specified for site response analyses considering mainly the large strain behavior of the soil, including its monotonic shear strength (with due consideration to strain rate effects), rather than by a simple extrapolation of (or fitting an analytical model to) G/G_{max} and ζ curves determined at smaller strains.

The current state of the practice for critical facilities includes obtaining G_{max} from both in situ measurements and laboratory determinations on small specimens, and the cyclic properties at larger strains such as G/G_{max} and ζ from the laboratory. For less critical structures and more limited budgets, very often some or all cyclic soil properties are estimated rather than obtained from measurements. This is done using our accumulated knowledge on the factors controlling G_{max}, G/G_{max}, ζ, as well as published correlations between these cyclic properties and soil type, degree of saturation, effective confining pressure, index properties, penetration resistance, relative density, unconfined compressive strength, and plasticity index (Dobry et al. 1989). A brief summary of this data has been included.

Effect of Confining Pressure. Several investigations have developed empirical relations for G_{max} in terms of simple soil properties as: Hardin and Black (1969)

$$G_{max} = 1230(OCR)^k \frac{(2.973-e)^2}{1+e}(\bar{\sigma}_o)^{\frac{1}{2}} \quad \textbf{(1)}$$

where

G_{max} = Maximum value of G in psi at $\gamma = 10^{-6}$
OCR = Over consolidation ratio
k = A factor depending upon PI of clays
e = Void ratio

$$\bar{\sigma}_o = \frac{\bar{\sigma}_1+\bar{\sigma}_2+\bar{\sigma}_3}{3}$$, average mean effective confining pressure in psi

Eq. 1 is a general equation for all types of soils e.g. sand and clays. Several other similar relations have been developed by other investigators, which show that $G_{max} \propto (\bar{\sigma}_o)^{½}$ (Prakash 1981, Prakash and Puri 1988). It will thus been seen that G_{max} is a function of $(\bar{\sigma}_o)^{½}$. If it is argued that G also depends upon all the factors which control G_{max}, it may be concluded that G is also a function of $(\bar{\sigma}_o)^{½}$

FIELD METHODS

Some of the field methods used for determining the shear modulus and/or damping are.

1. Cross-Borehole Wave Propagation Test
2. Up-Hole or Down-Hole Wave Propagation Test
3. Surface Wave Propagation Test

4. Vertical Footing Resonance Test
5. Horizontal Footing Resonance Test
6. Spectral analysis of Surface Waves (SASW)
7. Field Cone Test (Piezocone)
8. Back Calculation from Seismic Arrays
9. Seismic Cone
10. Standard Penetration Test

It is seen that most of the in situ methods are seismic techniques (crosshole, downhole or uphole, seismic cone, and SASW), used mainly to determine G_{max} by measuring V_s. Similar seismic methods are used to determine V_p also at very small strains (Prakash and Kumar (1994), Prakash and Puri (1988)). There is a scarcity of in situ methods to reliably measure G at intermediate and large strains and ζ at any strain. These development would be very desirable due to (i) the difficulty of sampling and testing some soils (saturated sands, and gravels), and (ii) the uncertainty in the lab results because of sample disturbance in all soils. A brief description of seismic cone and SASW techniques is given here. For other tests see Dobry et al (1989), Prakash and Sharma (1990), and Chang, et al (1991).

<u>Seismic Cone</u>. The seismic cone has been used by several investigators (Dobry et al. 1989). This method provides information on V_s and G_{max} in addition to regular static cone penetration resistance. The seismic cone test configuration is shown in Figure 4a. Basically, a cone penetrometer containing a triaxial receiver system is statically penetrated into the soil deposit. Then downhole seismic tests are conducted by exciting sources on the ground surface in the vicinity of the cone. Wave velocities and moduli are inferred from the travel times of the waves between source and receiver. Information on soil strength is obtained from the cone penetrometer results. Figure 4b shows a comparison between results from downhole seismic cone and crosshole tests for clay at a site in Norway.

The seismic cone method offers the capability of providing information on low strain behavior and strength characteristic at the same location. The limitations of the method include the inherent problems of cone penetrometer testing, that is, its lack of ability to test coarse-grained (gravel-type) soil deposits. Campanella et al. (1986) have shown good comparison of seismic cone data with crosshole data.

<u>Spectral analysis of Surface Waves (SASW)</u>. Stokoe and Nazarian (1985) have developed a surface wave method for monitoring G_{max} of the soil with depth without the use of boreholes. In this technique, two vertical transducers are placed on the ground surface at equal distances from an imaginary center line as shown in Fig. 5. A vertical impulse is then generated on the ground surface at a distance which for the near receiver is approximately equal to the distance between receivers (Fig. 5). Surface waves of the Rayleigh wave type are monitored as they propagate past the two transducers.

The depth of soil sampled with the SASW method is a function of the wave length generated, with low frequency waves sampling greater depths. This is related to the dispersion of surface waves in actual soil deposits, and it forms the basis of the SASW method (Rix et al. 1990). Dispersion means that the propagation velocity of surface waves is representative of the material stiffness over depths where there is a significant particle motion. For example, wave which has a wave length λ_1, less than the thickness of the top layer in the profile shown in Fig. 6 will exhibit a propagation velocity which is only dependent on the stiffness of that top layer (Fig. 6). On the other hand, the velocity of a surface wave which has a longer wave length $\lambda_2 > \lambda_1$, and thus has particle motions in all of these layers will be influenced by the stiffness of all layers (Fig. 6). Thus, by measuring the surface wave velocity over a wide range of wave lengths (frequencies), it is possible to assess the stiffness of the layers over a range of depths. The basic relationship between wave velocity, frequency, and wavelength is:

$$V_R = f \cdot \lambda \quad (2)$$

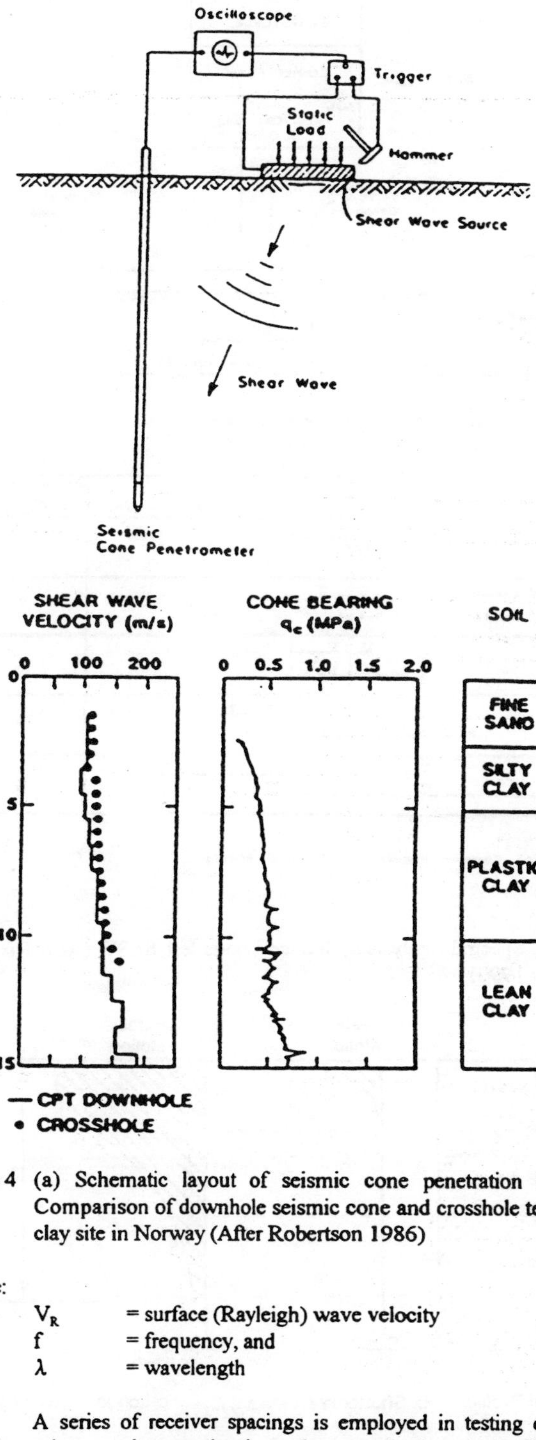

Figure 4 (a) Schematic layout of seismic cone penetration test, (b) Comparison of downhole seismic cone and crosshole tests for a clay site in Norway (After Robertson 1986)

Where:

V_R	= surface (Rayleigh) wave velocity
f	= frequency, and
λ	= wavelength

A series of receiver spacings is employed in testing one site. Initially, a close receiver spacing is used to sample the near-surface soils. Then the receiver spacings are increased about the imaginary center line (Fig 5), so that deeper soils are sampled. For each receiver spacing the source is located first to the left and then to the right of the receivers. With this approach, average values of wave propagation between the receivers are collected in an attempt to minimize problems with inhomogeneities between them. In this manner, a composite dispersion curve (V_R versus λ) is developed from all receiver spacings.

Once the dispersion curve has been developed for a site, it must be inverted. Inversion is the process of calculating the shear wave velocity profile from the dispersion curve. A theoretical dispersion curve is calculated

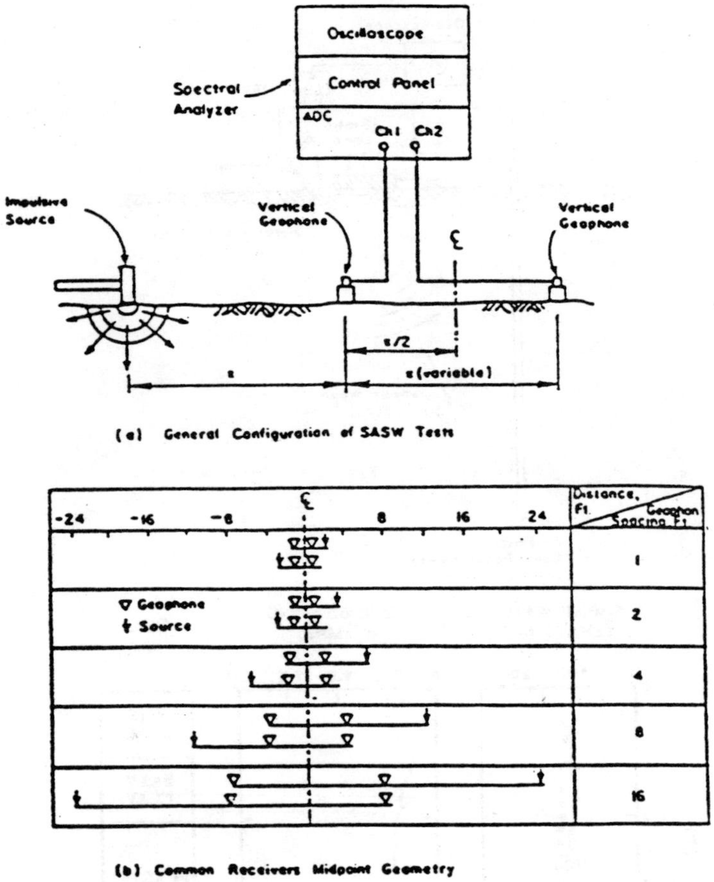

Figure 5 Spectral Analysis of Surface Waves test, SASW (After Rix et al. 1990)

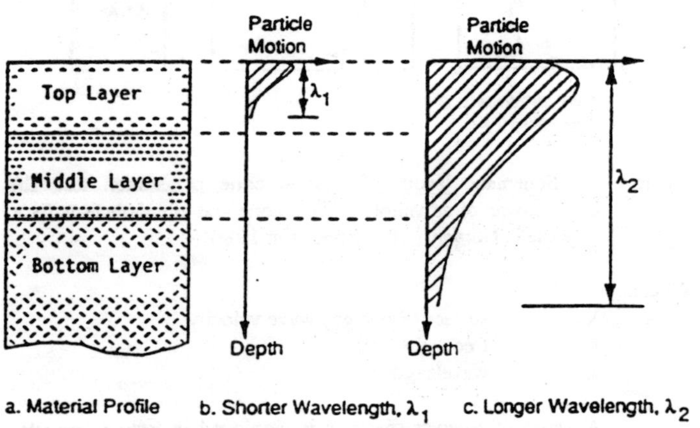

Figure 6 Approximate distribution of vertical particle motion with depth for two surface waves of different wavelength (After Rix et al. 1990)

from an assumed velocity profile and is then compared to the field dispersion curve. The assumed velocity profile contains a sufficiently large number of sublayers to define the in situ variation in stiffness with depth. The theoretical curve is calculated using a modified Haskell-Thompson matrix algorithm (Thompson, 1950; Haskell, 1953; Nazarian, 1984). The shear wave velocities and thicknesses of the sublayers in the assumed profile are adjusted by trial and error until a satisfactory match is obtained, at which time the final profile is assumed to represent the in situ profile. The computer

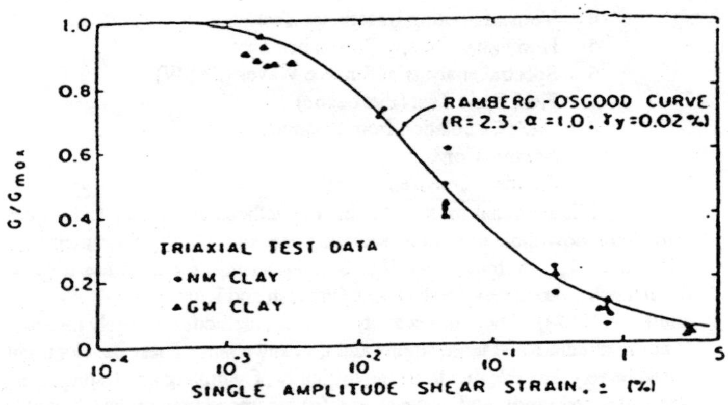

Figure 7 Modulus degradation with shear strain, NC and UC clays (After Singh and Gardner, 1979)

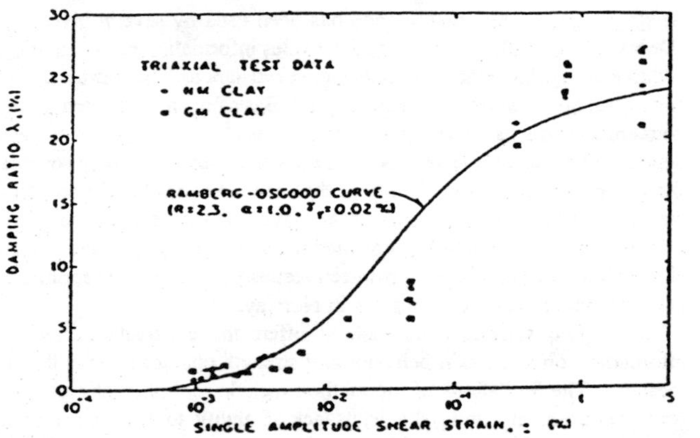

Figure 8 Damping variation with cyclic shear strain on NC and OC clays (After Singh and Gardner, 1979)

programs used for this SASW inversion continue to be developed, and their availability is presently (1996) quite limited.

Stokoe and Nazarian (1985) and Rix and Stokoe (1990) report good correlation between results of SASW and crosshole measurements. However, Sirles (1987, 1988) reports some discrepancies in another case study. This seems to occur in layered soils with inclined boundaries or in heterogeneous soil deposits.

LABORATORY TESTS

The following laboratory tests are performed to determine dynamic shear modulus and material damping of soils:

1. Resonant column tests
2. Cyclic triaxial test
3. Cyclic simple shear test

Resonant column test is a low strain test while cyclic triaxial and simple shear tests are intermediate to large strain tests. Since the soil parameters cannot be determined from one test at all strain levels, the results must be reported at a common confining stress.

Properties of typical marine soils will now be described.

DYNAMIC PROPERTIES OF TYPICAL SOILS

Houston and Herrmann (1979) reported test data on 9 different soils from Atlantic, Pacific and San Francisco Bay.

Stokoe and Lodde (1978) have studied normalized shear modulus degradation with shear strain of the San Francisco Bay Mud. Singh and Gardner (1979) reported tests on normally loaded and over consolidated marine clays. A plot of G/G_{max} versus shear strain (5% to 10^{-4}%, 10^{-4}% = 10^{-6}) is shown in Fig. 7. Variation of damping factor ζ versus shear strain is plotted in Fig. 8.

Calculation from Back Array. Efforts are being directed to monitor ground motion data during earthquakes. Also soil response with depth up to 17m, has been monitored during similar events (Chang et al. 1991). The properties of soil have been back calculated at appropriate strain. G_{max} values have been determined by wave propagation test method. The G/G_{max} versus strain have been plotted in Fig. 9. (Dobry et al. 1989).

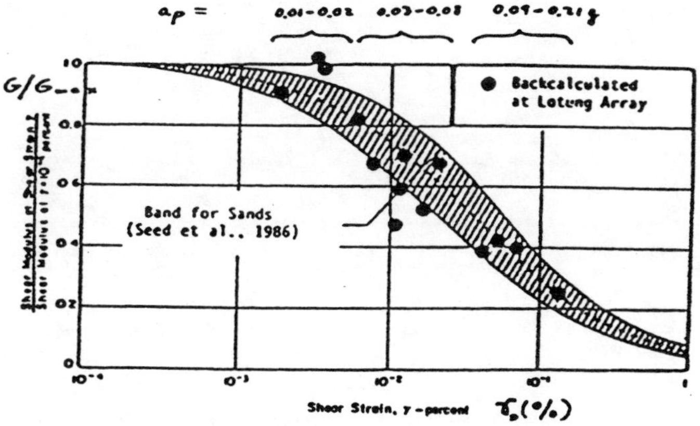

Figure 9 G/G_{max} versus strain, back calculated at Lotung array and band for sands (After Dobry et al, 1989)

Simulation of Ocean Bottom. The test results described above are primarily for on-shore conditions. Luke et al. (1993) conducted experiments to investigate the dominant seismic wave motion at the surface of a homogeneous material overlain by water. Ocean bottom was simulated by a 0.4 m (1.25 ft) freshly placed concrete slab laid over a mature concrete slab in a lake, overlaid by 0.4 m (1.25 ft) water standing over the fresh concrete. To permit testing over a large range of stiffness SASW and crosshole measurement were made as the concrete cured. The primary finding was that the predominant wave motion at the surface of a homogeneous material overlain by deep water has a velocity very close to that of a Rayleigh wave velocity on the surface of the same material overlaid by air (Prakash and Kumar 1994).

CONCLUSIONS

Foundation design for offshore structures and near shore structures needs an appropriate knowledge of static and cyclic characteristics of loads and how the soils behave under these loads. Oil spills are likely to have an adverse effect on the performance of near shore structures. More information in this regard is needed. Wave and earthquake loads induce cyclic shear stresses in the foundation soils. Stresses are generated not only by the structure resting on the soil but also due to free field strains. Earthquake loads are applied in a more concentrated time interval and at higher frequencies than those due to storm waves. Also, the load transfer for these two situations differs. Load transfer occurs from superstructure to foundation for waves and from foundation to the superstructure for earthquake loading. These differences in load characteristics may cause differences in soil response only. Cyclic loading of soils leads to development of cyclic and permanent strains, excess porewater pressures and possible decrease in soil strength.

REFERENCES

Anderson, D (1982), "Does Landfill Leachate Make Clay Liner More Permeable", Civil Eng, ASCE, September, pp 66-69.

Bouma, A., Sweet, W., Dunlap, W. (1972) "Comparison of Geological & Engineering Parameters of Marine Sediments", CTC 1514, Vol. 1.

Campanella, R.G., Robertson, P.K. and Gillespie, D. (1986), "Seismic Cone Penetration Test", Geotechnical Special Publication No. 6, (use of Insitu Tests in Geotechnical Engineering), editor, S.P. Clemence, p 116-130.

Chandrasekaran, A.R. and Krishna, J. (1976), "Elements of Earthquake Engineering", Sarita Prakashan, Merut India.

Chaney, R.C. and Fang, M.Y. (1983), "Static and dynamic Properties of Marine Sediments: State of the Art", Marine Geotechnology and Nearshore/Offshore Structures, ASTM STP 923, pp 74-111.

Chang, C.Y., Mok, C.M. Power, M.S., Tang, Y.K. Tang, H.T. and Stepp, J.C. (1991), "Development of Shear Modulus Reduction Curves Based on Lotung Downhole Ground Motion Data", Proceedings, Second International Conference on Recent Advances in Geotechnical Earthquake Engineering and Soil Dynamics, Vol. 1, March 11-15, St. Louis, Missouri, Paper No. 1.44, pp 111-118.

Cook, E.E., Puri, V.K. and Shin, E.C. (1992), "Geotechnical Characteristics of Crude Oil Contaminated Sands", Proceeding of Second International Offshore and Polar Engineering Conference, San Francisco, June 14-19, Vol I, pp 384-387.

Dobry, R., Bhatia, S.K., Cramer, C.H., Henke, R., Higgins, C., Jamiolkowski, M., Ko, H.Y., Ladd, R.S., Prakash, S., Saada, A., and Sirles, P.D., (1989), "Low and High Strain Cyclic Material Properties", Ch3-Proc. NSF/EPRI Workshop on Dynamic Soil Properties and Soil Characterization, Vol I pp 3-1 to 3-100.

Gorshowitz, R.; Liao, J.; Harris, F. (1972) "Engineering Properties of Marine Sediments - Gulf of Paria, Trinidad, W.I."; CTC 1650, Vol. 2.

Hardin, B.O. and Black, W.L. (1968), "Vibration Modulus of Normally Consolidated Clay", Journal of the Soil Mechanics and Foundation Division, ASCE, Vol. 94, March pp 358-369.

Hardin, B.O. and Black, W.L. (1969), "Closure to Vibration Modulus of Normally Consolidated Clay", Journal of the Soil Mechanics and Foundation Division, ASCE, Vol. 95, November, pp 1531-1537.

Haskell, N.A. (1953), "The Dispersion of Surface Waves on Multilayered Media," Bulletin of Seismological Society of America, Vol. 43, No. 1, pp. 17-34.

Houston, W.N. and Herrmann, G. (1979), "Undrained Strength of Marine Soils Under Long-term Cyclic Loading", Soil Dynamics in the Marine Environemnt, ASCE, Boston.

Keller, G.H. (1969) "Engineering Properties of Some Sea-Floor Deposits," Journal of Soil Mech. And Found. Division, ASCE, Vol 95, NO SM1, pp 1379-1392.

Kumapley, NK and Ishola, A (1985), "The Effect of Chemical Contamination on Soil Strength," Proc 12th Int Conf Soil Mech Found Eng, San Francisco, Vol 3, pp 1199-1201.

Lukas, RG and Gnaedinger, RJ Jr. (1972), "Settlement Due to Chemical Attack of Soils," Proc ASCE Specialty Conference on Performance of Earth and Earth Supported Structures, Purdue University, Lafayette, Indiana, pp 1087-1104.

Luke, B.A., Stokoe II, K.H., Wright, S.G., Roessel, J.M. and Byungsik, L. (1993), "Experimental Investigations of Surface Wave Velocities in Simulated Ocean Bottoms", Offshore Technology Conference, Vol. 1, pp 287-297.

Nazarian, S (1984), "Insitu Determination of elastic Moduli of Soil Deposits and Pavement Systems by SASW Method" Ph.D. Disertation, Univ. of Texas-Austin.

Prakash, S. (1981), "Soil Dynamics" McGraw Hill Book Co, New York, Reprinted by S.P. Foundation, Rolla, 1991).

Prakash, S. And Puri, V.K. (1988), "*Foundations for Machines, Analysis and Design*", John Wiley and Sons, New York, NY.

Prakash, S. And Sharma, H.D. (1990), "*Pile Foundations in Engineering Practice*", John Wiley and Sons, New York, NY.

Prakash, S. And Kumar, S. (1994), "On Dynamic Properties for Marine Environment", Proceedings of the Fourth International Offshore and Polar Engineering Conference, Osaka, Japan, April 10-15, Vol I, pp 404-412.

Puri, V.K., Das, B.M., Cook, E.E. and Shin, E.C. (1993), "Geotechnical Properties of Crude Oil Contaminated Sand", ASTM STP 1221 pp--.

Rix, G.J., Bay, J.A. and Stokoe, K.H. (1990), "Use of Seismic Tests to Access Insitu the Properties of Portland Cement Concrete During Curing", Presented at the 1990 Annual Meeting of the Transportation Research Board.

Sherard, J.L, Decker, R.S. and Ryker, N.L. (1972), "Piping in Earth Dams of Dispersive Clays," Proc. ASCE Specialty Conference on Performance of Earth and Earth Supported Structures, Purdue University, Lafayette, Indiana, pp 589-626.

Shridharan, A., Nagraj, T.S. and Suvapulliah, P.V. (1981), "Heaving of Soil Due to Acid Contamination," Proc. 10th Int Conf Soil Mech Found Eng, Stockholm, Vol 2, pp 383-386.

Singh, R.D. and Gardner, W.S. (1979), "Characterization of Dynamic Properties of Gulf of Alaska Clays", Soil dynamics in the Marine Environment, ASCE, Boston.

Sirles, P.C. (1987), "Shear Wave Velocity and Altenu Measurements in Liquefiable Soil Deposits in the Tru Meadows, Reno, Navada", M.S. thesis, University Nevada, Reno.

Sirles, P.C. (1988), "Evaluation of SASW and Crosshole Vel Measurements Before and After Dynamic Compaction Jackson Lake Dam, Wyoming", U.S. Bureau Reclamation Report (Unpublished).

Stokoe II, K.H. and Nazarian, S. (1985), "Use of Rayleigh Waves in Liquefaction Studies", in "Measurement and Use of Shear Wave Velocity for Evaluating Dynamic Properties", Proceedings, Geotechnical Enginee Division, Editor, R.D. Woods, ASCE, pp 1-17.

Stokoe II, K.H. and Lodde, P.F. (1978), "Dynamic Respon San Francisco Bay Mud", Proceeding of the A Geotechnical Engineering Division Specialty Confer Earthquake Engineering & Soil Dynamics, June 19 Pasadena, CA.

Thompson, W.T. (1950), "Transmission of elastic Waves through a Stratified Solid Medium," Journal of Applied Physics, Vol. 21, pp. 89-93.

Proceedings of the Seventh (1997) International Offshore and Polar Engineering Conference
Honolulu, USA, May 25-30, 1997

ISBN 1-880653-28-1 (Set); ISBN 1-880653-29-X (Vol. I)

Consolidation Analysis of Sensitive Clay by Elasto-Viscoplastic Two-Fractions Mixture Model

Nobuharu Abe
Osaka University
Osaka, Japan

ABSTRACT

In this paper, an elasto-viscoplastic constitutive model is proposed to predict the one-dimensional consolidation behavior of natural sensitive clay deposits. In order to describe the yielding behavior of sensitive clays the proposed model is derived being based on the theory of two-fractions mixture and the concept of damage. The applicability of the model is examined by comparing calculated results with laboratory and in-situ behavior of natural sensitive clays, and it is confirmed that the catastrophic yielding of sensitive clays including strain-softening can be expressed by the proposed elasto-viscoplastic two-fractions mixture model.

KEY WORDS

Sensitive clay, Consolidation, Elasto-viscoplasticity, Two-fractions mixture, Damage, Time-dependency, Yielding, Strain-softening

INTRODUCTION

Sensitive clays are encountered mostly in the coastal regions of many countries. Increased use of coastal lands for urbanization and industrialization has posed special problems of settlement and stability for sensitive clay deposits. However, the prediction methods for settlement and stability of sensitive clays has not been established yet, because most of sensitive clays have complicated compressibility characteristics including time-dependency and sudden yielding behavior. Therefore, it is very important to model the compression property of sensitive clays and to predict the settlement and stability of sensitive clay deposits.

Fig. 1 schematically shows the void ratio vs. logarithm of vertical effective stress curve of a natural sensitive clay. The compression behavior of a sensitive clay for a stress increment spanning the yield stress can be recognized to have three zones (Nagaraj et al., 1990) ; (1) Zone-I (from σ_{zo} up to σ_{zy}), (2) Zone-II (between σ_{zy} and σ_{zt}), (3) Zone-III (beyond σ_{zt}). σ_{zo} , σ_{zy} and σ_{zt} are the initial stress, the yield stress and the transition stress, respectively. In zone-I, sensitive clays are aged and stiffened due to the cementation bonds, so the compression of sensitive clays is very small. Beyond the yield stress σ_{zy} , there is sudden compression of relatively high magnitude. The yield stress is an effect of the clay structure, that is, internal bonding, therefore, yielding behavior of sensitive clays can be explained in terms of the deterioration of internal bonding or the collapse of clay structure. The yielding behavior in zone-II is dependent on the degradation rate of clay structure. In zone-III, where the clay structure is completely broken, the behavior of sensitive clays is similar to that of remolded clays.

In the following sections, first, the one-dimensional elasto-viscoplastic model for sensitive clays is proposed being based on the above-mentioned conceptual model. The theory of two-fractions mixture and the concept of damage are employed for numerical modeling. And then, the applicability of the proposed model is investigated by comparing calculated results with laboratory and in-situ behavior of sensitive clays. Finally, from the results of numerical simulation and in-situ measurement, it is confirmed that the generation of pore water pressure of sensitive clays during one-dimensional consolidation under the constant total stress is caused by strain-softening behavior.

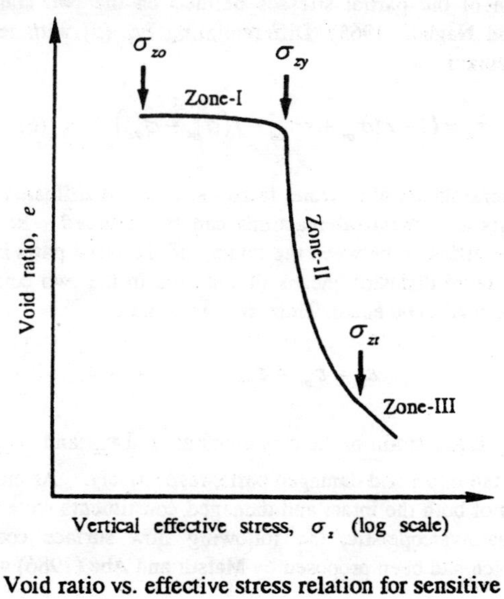

Fig. 1 Void ratio vs. effective stress relation for sensitive clay

ELASTO-VISCOPLASTIC TWO-FRACTIONS MIXTURE MODEL

The above-mentioned compressibility characteristics of sensitive clays were modeled based on the theory of two-fractions mixture and the concept of damage (Hirao, et al., 1991; Abe, et al., 1992). It was assumed that the compression behavior of sensitive clays is decomposed in two constituents, intact (non-damaged) and damaged (Frantziskonis and Desai, 1987). During the loading process, intact constituent is progressively transformed to damaged one due to increasing deformation to express the gradual deterioration of sensitive clay structure.

We consider a clay element of volume V which consists of the intact fraction of volume V_n and the damaged fraction of volume V_d. The average stresses σ_{zn} and σ_{zd} in the intact and damaged fractions are expressed as follows:

$$\sigma_{zn} = \frac{1}{V_n}\int_{V_n} \bar{\sigma}_{zn} dV_n \qquad (1)$$

$$\sigma_{zd} = \frac{1}{V_d}\int_{V_d} \bar{\sigma}_{zd} dV_d \qquad (2)$$

where $\bar{\sigma}_{zn}$ and $\bar{\sigma}_{zd}$ are internal stresses acting in the intact and damaged fractions, respectively. The average homogeneous stress σ_z in the total volume V is defined as:

$$\sigma_z = \frac{1}{V}\left(\int_{V_n} \bar{\sigma}_{zn} dV_n + \int_{V_d} \bar{\sigma}_{zd} dV_d\right) \qquad (3)$$

From the definition, the damage parameter r is expressed as follows:

$$r = V_d/V \qquad (4)$$

Substituting Eqs. (1), (2) and (4) into Eq. (3), we have

$$\sigma_z = (1-r)\sigma_{zn} + r\sigma_{zd} \qquad (5)$$

The relation in Eq. (5) states that the average stress of mixture is the summation of the partial stresses defined on the two components (Green and Naghdi, 1965). Differentiating Eq. (5) with respect to time, we obtain

$$\dot{\sigma}_z = (1-r)\dot{\sigma}_{zn} + r\dot{\sigma}_{zd} - \dot{r}(\sigma_{zn} - \sigma_{zd}) \qquad (6)$$

In the general theory of mixture, factors such as the diffusion between constituents and chemical reactions can be included. In the case herein, the diffusion between the intact and damaged parts is absent. The absence of diffusion means that strains in the two components can be assumed to be equal. Thus, we can write

$$\varepsilon_z = \varepsilon_{zn} = \varepsilon_{zd} \qquad (7)$$

where ε_z is the strain of the clay element, and ε_{zn} and ε_{zd} are the strains of the intact and damaged parts, respectively. As mechanical properties of both the intact and damaged constituents were assumed to be elasto-viscoplastic, the following flow surface constitutive model which had been proposed by Matsui and Abe (1986) was used.

$$F = \mu \ln\left[\frac{1}{\delta}\left[\left\{1-\exp\left(-\frac{\delta}{\mu}\dot{\varepsilon}^{v}_{zr}t\right)\right\}\exp\left(\frac{f}{\mu}\right) + \delta\exp\left(-\frac{\delta}{\mu}\dot{\varepsilon}^{v}_{zr}t\right)\right]\right] - \varepsilon^{vp}_z = 0 \qquad (8)$$

$$f = \frac{\lambda - \kappa}{1+e_0}\ln\left(\frac{\sigma_z}{\sigma_{z0}}\right) \qquad (9)$$

where λ and κ are compression and swelling indices; μ, δ and $\dot{\varepsilon}^{v}_{zr}$ are viscous parameters of sensitive clays; ε^{vp} is viscoplastic strain; and e_o and σ_{zo} are initial values of void ratio and average homogeneous stress. When δ is zero, flow function in Eq. (8) is simplified as follows:

$$F = \mu \ln\left[1 + \left(\frac{\dot{\varepsilon}^{v}_{zr}}{\mu}t\right)\exp\left(\frac{f}{\mu}\right)\right] - \varepsilon^{vp}_z = 0 \qquad (10)$$

Elasto-viscoplastic constitutive relations for the intact and damaged components are obtained from the flow surface theory of viscoplasticity, together with the consistency condition as follows:

$$\dot{\sigma}_{zn} = \left[\frac{\kappa_n}{1+e_0}\frac{1}{\sigma_{zn}} + \frac{\partial F_n}{\partial \sigma_{zn}}\right]^{-1}\left(\dot{\varepsilon}_{zn} - \frac{\partial F_n}{\partial t}\right) = D^{ev}_n\left(\dot{\varepsilon}_{zn} - \frac{\partial F_n}{\partial t}\right) \qquad (11)$$

$$\dot{\sigma}_{zd} = \left[\frac{\kappa_d}{1+e_0}\frac{1}{\sigma_{zd}} + \frac{\partial F_d}{\partial \sigma_{zd}}\right]^{-1}\left(\dot{\varepsilon}_{zd} - \frac{\partial F_d}{\partial t}\right) = D^{ev}_d\left(\dot{\varepsilon}_{zd} - \frac{\partial F_d}{\partial t}\right) \qquad (12)$$

where F_n and F_d are flow functions for the intact and damaged components, and D^{ev}_n and D^{ev}_d are elasto-viscoplastic constitutive matrices for two components.
Substituting Eqs. (11) and (12) into Eq. (6) , we can obtain the constiitutive relation of the elasto-viscoplastic two-fractions mixture model.

$$\dot{\sigma}_z = D^{ev}\dot{\varepsilon}_z - L \qquad (13)$$

$$D^{ev} = (1-r)D^{ev}_n + rD^{ev}_d \qquad (14)$$

$$L = (1-r)D^{ev}_n\frac{\partial F_n}{\partial t} + rD^{ev}_d\frac{\partial F_d}{\partial t} + \dot{r}(\sigma_{zn} - \sigma_{zd}) \qquad (15)$$

For the completion of formulation, the evolution law for the damage variable r needs to be defined. In this study, the damage variable r is expressed as a function of history of viscoplastic strain such that

$$r = r\left(\varepsilon^{vp}_z\right) \qquad (16)$$

$$\text{where} \quad \varepsilon^{vp}_z = \int_0^t \dot{\varepsilon}^{vp}_z dt \qquad (17)$$

Specifically, the following equation is employed:

$$r = r_u\left[1 - \exp\left(-h\left\langle\varepsilon^{vp}_z - k\right\rangle^m\right)\right] \qquad (18)$$

where r_u is the ultimate value of the damage variable r ; h and m are parameters for the damage variables; k is the threshold value of viscoplastic strain at which the damage occurs. The symbol $\langle * \rangle$ is Macauley bracket. Thus, the damage variable r is zero in low viscoplastic strain region $\left(\varepsilon_z^{vp} \le k\right)$, therefore, sensitive clays express the compressibility characteristics in zone I.

The proposed elasto-viscoplastic two-fractions mixture model has nine material parameters. Material parameters λ, κ, μ and $\dot{\varepsilon}_{zr}^{v}$ for the intact and damaged fractions can be obtained from the results of the consolidation tests in Zone-I and Zone-III. However, parameters δ and m, k, h and r_u for the damage variable have to be estimated by applying the curve fitting method to the laboratory or in-situ behavior of sensitive clays.

CONSOLIDATION ANALYSIS OF SENSITIVE CLAY

In this section, the applicability of the proposed elasto-viscoplastic two-fractions mixture model is examined through numerical simulations for the yield behavior of sensitive clays by the finite element method.

Oedometer test of Olga clay

Mesri (1985) performed the oedometer test on Olga clay, which had highly sensitive structure, and observed that the excess pore water pressure generated during consolidation after yielding of the clay. In this test, the specimen was loaded in one step from 83.6 kPa to 120 kPa so as to span the consolidation yield stress 95 kPa determined by the multi-stage loading test. This oedometer test of the high sensitive clay was simulated , using the proposed model (Abe et al., 1994). The material parameters listed in Table 1 were determined based on the void ratio vs. vertical effective stress curve in the reference (Mesri, 1985). In this analysis, the coefficient of permeability of Olga clay was assumed to be dependent of void ratio. In the finite element analysis the specimen with a height of 1.88cm is subdivided into 10 one-dimensional elements.

Fig. 2 shows the comparison of the calculated and the experimental results for the vertical strain and excess pore water pressure behavior. It is seen from the figure that, as a whole, the calculated results agree with the observed ones, although a little bit difference between these two can be recognized especially in the rate of initial dissipation of pore water pressure. The calculated effective stress vs. strain curves of three elements during consolidation are shown in Fig. 3. They have the steep slopes near the yield stress. In particular, the effective stress near the impermeable bottom of specimen (element No. 10) decreases after the sudden yielding. The generation of pore water pressure at the bottom of specimen results from this strain-softening behavior.

Embankment test in Berthierville

Kabbaj et al.(1988) reported the in-situ behavior of a test embankment on a soft sensitive clay layer 3.2 m thick at Berthierville in Canada. In this site, the long-term behavior of settlements and pore water pressures were measured and the in-situ vertical effective stress-strain curves of the sublayer were defined. The in-situ and the laboratory vertical effective stress-strain curves were very different; in particular, the stress softening was observed in-situ.

The consolidation behavior of this sensitive clay layer was analyzed, using the proposed model (Abe and Hashimoto, 1993). Though Berthierville test fill is a circular embankment with a diameter of 24m and a height of 2.4m, the consolidation can be considered to be one-dimensional from the condition that the ratio of the embankment

Table 1 Material parameters on Olga clay

Material parameters	Olga clay	
	Intact fraction	Damaged fraction
Compression index, λ	0.46	1.3
Swelling index, κ	0.024	0.08
Coefficient of secondary compression, μ	0.006	0.03
Reference strain rate, $\dot{\varepsilon}_{zr}^{v}$ (min^{-1})	1.0×10^{-6}	4.5×10^{-5}
Internal restraint strain rate, δ	1.0×10^{-4}	1.0×10^{-4}
Parameter for damage variable, m	1.05	
Parameter for damage variable, k	0.0	
Parameter for damage variable, h	90	
Parameter for damage variable, r_u	1.0	
Coefficient of permeability, k_{z0} (cm/min)	1.8×10^{-5}	
Permeability change index, C_k	0.7	

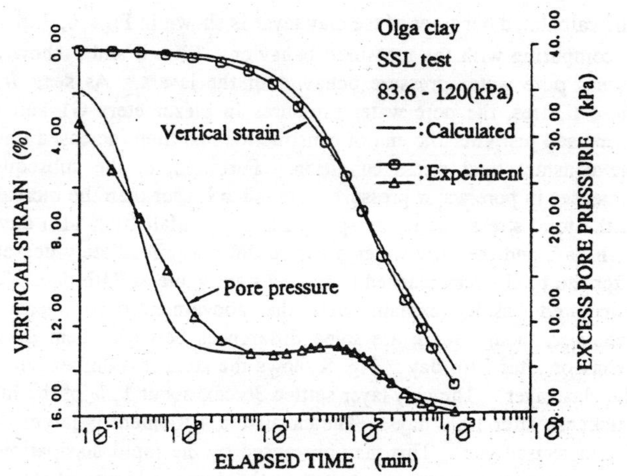

Fig. 2 Developments of strain and pore pressure with elapsed time on Olga clay

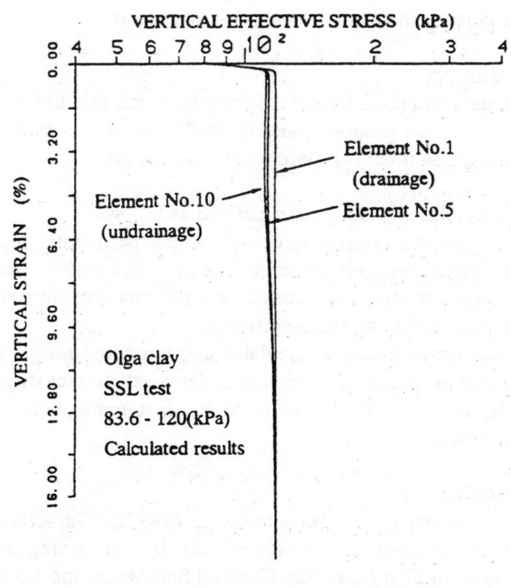

Fig. 3 Calculated effective stress vs. strain relations on Olga clay

dimension to the thickness of the clay layer is 9. Both the top and the base of clay layer were assumed to be drained. The initial distributions of the void ratio, the coefficient of permeability and the vertical effective stress within the clay layer were determined as shown in Fig. 4 from the geotechnical profile in the reference (Samson and Garneau, 1973). While most of material parameters, summarized in Table 2, were determined from the geotechnical profile in the references, parameters for damage variables were estimated by using the curve fitting method to in-situ pore water pressure behavior (see Fig. 7). In this analysis, the coefficient of permeability of Berthierville clay was assumed to be dependent of void ratio. The variation in applied total stress with time at the mid-depth of the clay layer is shown in Fig. 5. The applied total stress increased due to the spell of heavy rainfall , though it kept the constant value after the completion of construction. In finite element analysis, the clay layer 3.2m thick is subdivided into 32 one-dimensional elements.

The calculated behavior of the clay layer is shown in Figs. 6, 7, 8 and 9, comparing with the measured behavior. Figs. 6 and 7 show the excess pore water pressure behavior of the layers. As seen from these figures, the pore water pressures in piezometers G1 and G2 decreased just after the end of construction and then increased under the constant total stress condition. Furthermore, the subsequent increases in pore water pressure were much higher than the increases of the total stress due to the spell of heavy rainfall from 11th day to 24th day, and the pore water pressure did not immediately decrease after the total stress reached to the constant value at 24th day. The calculated results explain well the above-mentioned measured behavior, though there are some differences between them on the behavior after 24th day. Fig. 8 shows the compression behavior of the clay layer. The clay layer settled 36 cm, about 11% of the layer thickness after 1000 days. The calculated settlement is larger than the measured one. This may be caused by the rapid dissipation of the calculated pore water pressure in the upper sublayer of the clay layer. Fig. 9 shows the calculated vertical stress-strain curves in the upper and mid sublayers of the clay layer. The in-situ vertical effective stress-strain curve is also shown in Fig. 9. Both effective stress behaviors agree well. Therefore, it is elucidated that the strain-softening behavior after yielding causes the generation of pore water pressure in the natural sensitive clay deposit.

CONCLUSIONS

In this paper, an elasto-viscoplastic two-fractions mixture model for sensitive clays was proposed, and its applicability was examined by experimental and in-situ behavior of sensitive clays.

The main conclusions can be summarized as follows:

(1) As the simulation results agree well with experimental and in-situ behavior of sensitive clays, therefore, the proposed elasto-viscoplastic two-fractions mixture model can express the catastrophic yielding of sensitive clays including strain-softening.

(2) It is confirmed from the calculated and measured results that the generation of the pore water pressure under constant total stress in the natural deposit can be explained by the strain-softening behavior after yielding of sensitive clays.

REFERENCES

Abe, N., Yoshida, K. and Hashimoto, K. (1992). "One-dimensional consolidation analysis of sensitive clay by interacting continua model," Proc. of 27th Japan Nat. Conf. on Soil Mech. and Found. Eng. pp375-378 (in Japanese).

Abe, N. and Hashimoto, K. (1993). "One-dimensional consolidation analysis of natural clay deposits by two-fractions mixture model," Proc. of 48th Annual Conf. of Japan Society of Civil Engrg, Vol.3, pp982-983 (in Japanese).

Abe, N., Hashimoto, K and Yasui, T. (1994). "One-dimensional consolidation analysis of natural clay deposits by two-fractions mixture model (II)," Proc. of 49th Annual Conf. of Japan Society of Civil Engrg, Vol.3, pp258-259 (in Japanese).

Table 2 Material parameters on Berthierville clay

Material parameters	Berthierville clay	
	Intact fraction	Damaged fraction
Compression index, λ	0.032	0.297
Swelling index, κ	0.01	0.031
Coefficient of secondary compression, μ	0.003	0.014
Reference strain rate, $\dot{\varepsilon}^{\ast}_{v}$ (day^{-1})	1.3×10^{-4}	1.0×10^{-2}
Internal restraint strain rate, δ	-	-
Parameter for damage variable, m	1.05	
Parameter for damage variable, k	5.0×10^{-4}	
Parameter for damage variable, h	120	
Parameter for damage variable, r_u	1.0	
Coefficient of permeability, k_w (m/day)	5.1×10^{-2}	
Permeability change index, C_k	0.6	

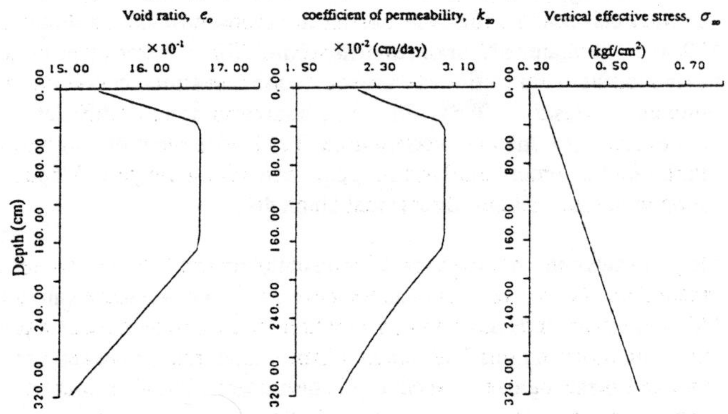

Fig. 4 Initial Distributions of void ratio, coefficient of permeability and effective stress within clay layer (1 kgf/cm^2 =98 kPa)

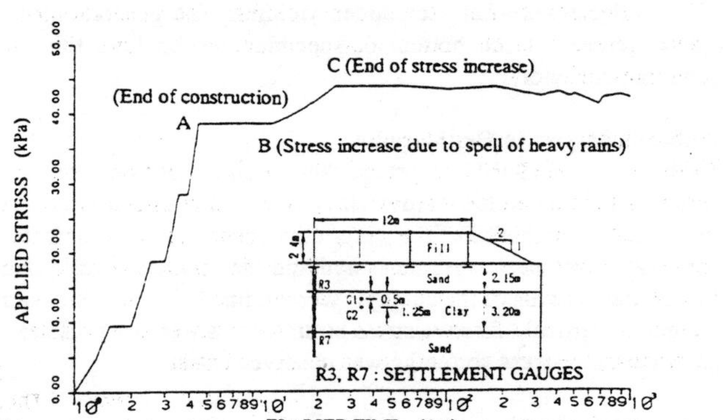

Fig. 5 Variation in applied total stress at mid-depth of clay layer with elapsed time

Frantziskonis, G. and Desai, C.S. (1987). "Elastoplastic model with damage for strain softening geomaterials," Acta Mechanica, 68, pp151-170.

Green, A.E. and Naghdi, P.M.(1965). "A dynamical theory of interacting continua," Int. J. Engrg Sci., Vol.3, pp231-241.

Hirao, K., Abe, N., Matsui, T. and Hashimoto, K. (1991). "One-dimensional elasto-viscoplastic strain-softening model for clays," Proc. of 46th Annual Conf. of Japan Society of Civil Engrg, Vol.3, pp390-391 (in Japanese).

Kabbaj, M., Tavenas, F. and Leroueil, S.(1988). "In situ and laboratory stress-strain relationships," Geotechnique, 38 (1), pp83-100.

Matsui, T. and Abe, N. (1986). "Flow surface model of viscoplasticity for normally consolidated clay," Proc. 2nd Int. Symp. Num. Models Geomech., Ghent, pp157-164.

Mesri, G.(1985). "Written discussion," Proc. 10th Int. Conf. on Soil Mech. and Found.Eng. Stockholm,Vol.4, p659.

Nagaraj, T. S., Murthy, B.R.S., Vatsala, A. and Joshi, R.C. (1990). "Analysis of compressibility of sensitive soils," J. Geotech. Engrg, ASCE, 116 (1), pp105-118.

Samson, L. and Garneau, R. (1973). "Settlement performance of two embankments on deep compressive soils," Can. Geotech. J. 20 (2), pp211-266.

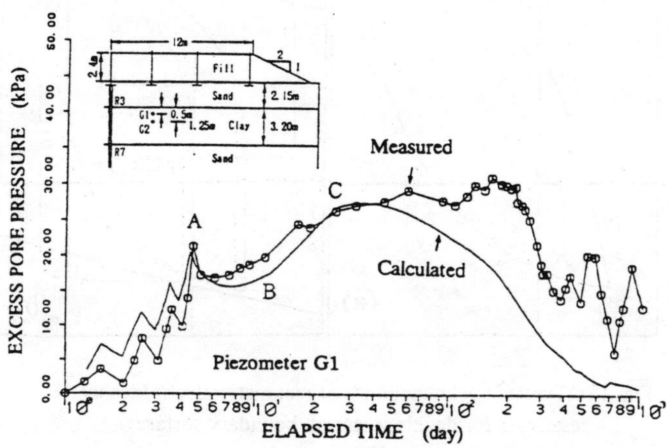

Fig. 6 Variation in pore water pressure in piezometer G1 with elapsed time

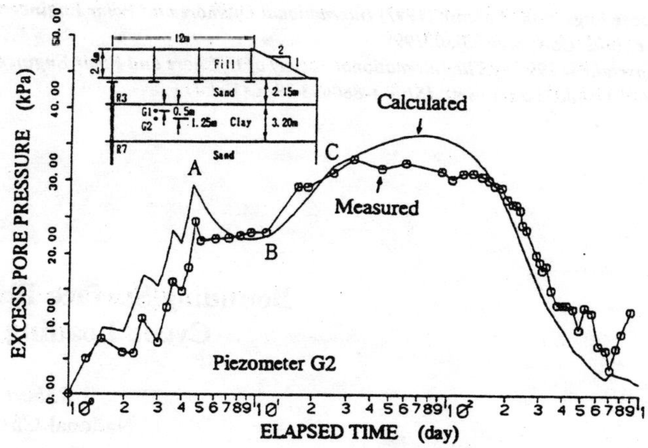

Fig. 7 Variation in pore water pressure in piezometer G2 with elapsed time

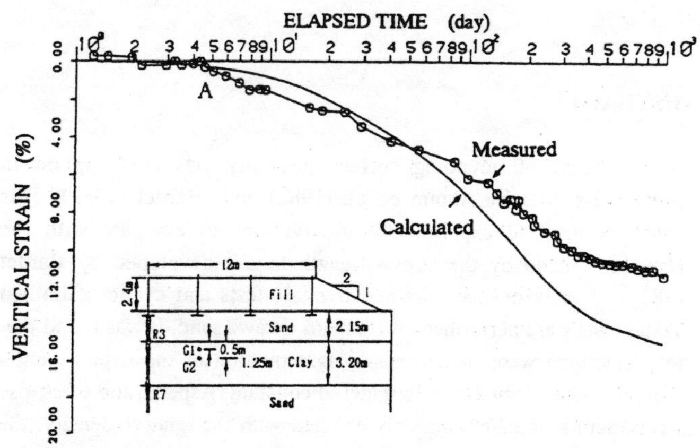

Fig. 8 Evolution of settlement of clay layer with elapsed time

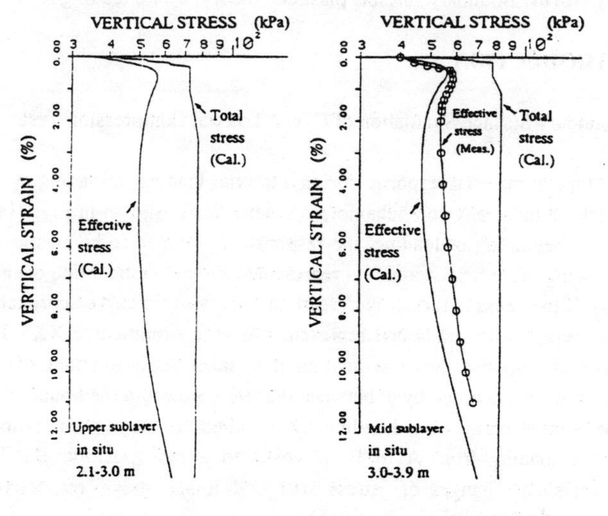

Fig. 9 Stress vs. strain curves obtained from simulations and in-situ measurement

Proceedings of the Seventh (1997) International Offshore and Polar Engineering Conference
Honolulu, USA, May 25-30, 1997

ISBN 1-880653-28-1 (Set); ISBN 1-880653-29-X (Vol. I)

Bounding Surface Plasticity Theory Applied to Cyclic Loading on Cohesionless Soil

Robert Yun-Pin Chin
National Cheng Kung University
Tainan, Taiwan, China

ABSTRACT

The concept of bounding surface plasticity was first applied to cohesionless soils by Aboim et. al. (1982) and Bardet (1983). This paper presents the experimental observations to compare with the predictions made by the stress-strength theory developed by Bardet (1983). A series of static drained triaxial tests and cyclic undrained triaxial tests were performed on standard Ottawa sand. It was found that the predictions were in reasonable agreement with the experi-mental laboratory tests. However, the material constant, Aspect ratio of ellipse in contracting domain, is mainly effected with the relative density, the mathematical relationship of the material constant, to the relative density has been developed in this paper.

Key words: Boundary surface, plasticity theory, cyclic loading

INTRODUCTION

Boundary Surface Simulation of Cyclic Triaxial Compression Test

A typical material response during a triaxial loading-unloading cycle, which includes any soil behavior, is schematically shown in Fig. 1(a). At the beginning of loading, the response is assumed to be elastic. At removing the elastic strain, it is represented by the vertical line of Fig. 1(b). If the stress is further increased, the stress-strain curve approaches and merges with the bound represented by the straight line XX'. The slope of response curve at any point is taken as a function of the distance AA', denoted by d, between the stress state and the bound XX'. The point a" on the bounding line XX' is called the image point. During plastic loading from A to B it moves to a new position B'. The infinitesimal changes of stress state and image stress, respectively denoted $d\sigma$ $d\sigma$ and $d\bar{\sigma}$ are related by

$$d\sigma / s = d\bar{\sigma} / s_b \qquad (1)$$

where S is the slope of stress-strain curve at A, and S_b the slope ofbound XX' at A'. During unloading the image point is selected on theother line YY', and all the ideas defined for loading apply to describe the unloading response.

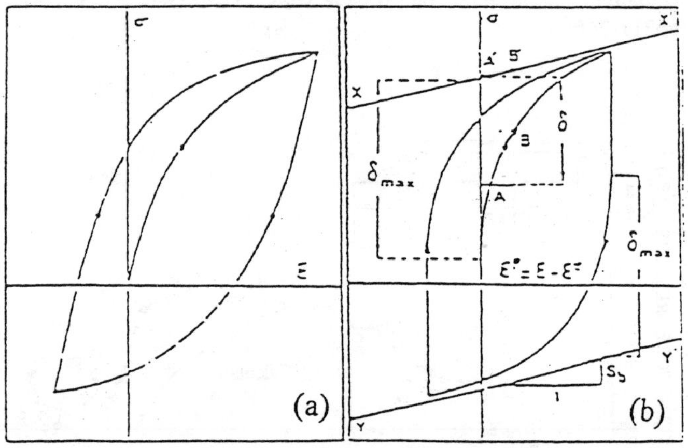

Fig. 1 (a)Typical cyclic response of real material; (b) Idealized responses for development of boundary surface plasticity.(Bardet, 1983)

Definition of Two-Dimensional Stress Space

The stress parameters used in the analysis of triaxial test results are

$$p = (\sigma_1' + 2\sigma_3')/3, \quad q = \sigma_1' - \sigma_3' \qquad (2a)$$

$$\varepsilon_v = \varepsilon_1 + 2\varepsilon_3, \quad \varepsilon_r = 2(\varepsilon_1 - \varepsilon_3)/3 \qquad (2b)$$

where p = the mean effective principal stress; q = the effective deviatoric stress; σ_1' and σ_3' denote the axial and lateral stress, respectively; ε_v = the volumetric strain, and ε_q = the deviatoric strain; ε_1 and ε_3 denote the axial and lateral strain, respectively. However it is difficult to monitor directly the lateral strain, the value of the deviatoric strain, ε_q is determined by the relationship, $\varepsilon_r = 3\varepsilon_1/2$, which hold valid for undrained triaxial compression test condition. The incremental work dW which is generally defined as

$$dW = p d\varepsilon_v + q d\varepsilon_q \tag{3}$$

The radial mapping of any stress state (p, q) to the image stress $(\bar{p}, \bar{q})$ on the bounding surface are

$$f(\bar{p}, \bar{q}, \varepsilon_v^p) = 0; \quad \bar{p} = \alpha p \text{ and } \bar{q} = \alpha q \tag{4a}$$

$$f(\alpha p, \alpha q, \varepsilon_v^p) = 0 \tag{4b}$$

where α denote the function of no-vector. Then the distance δ which separates the image point from the stress state is expressed by the relationship

$$\delta = \left[(\bar{p} - p)^2 + (\bar{q} - q)^2\right]^{1/2} \tag{5}$$

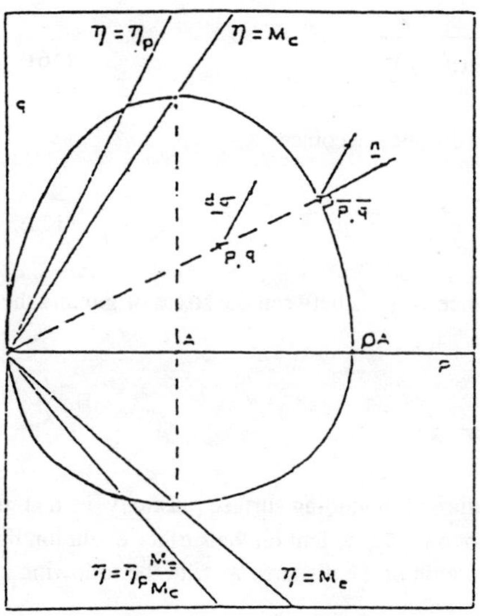

Fig. 2 Selected boundary surface in the $p-q$ space. (Bardet, 1983)

As shown in Fig. 2, the plastic flow direction is collinear to the unit vector normal to the bounding surface at the image stress, with components η_p, η_q

$$\eta_p = \frac{1}{g}\frac{\partial f}{\partial p} \text{ and } \eta_q = \frac{1}{g}\frac{\partial f}{\partial q} \tag{6a}$$

where the partial derivatives with respect to p and q are evaluated at the image stress, and g is expressed by the relationship

$$g = \left[(\frac{\partial f}{\partial p})^2 + (\frac{\partial f}{\partial q})^2\right]^{1/2} \tag{6b}$$

The plastic modulus H, which generalizes the slope S of Fig. 1(a), depends upon δ and should satisfy the following conditions

$$H = +\infty \text{ if } \delta > \delta_{max} \tag{7a}$$

$$H = H_b \text{ if } \delta = 0 \tag{7b}$$

where H is the value of the plastic modulus at the image stress, which corresponds to the slope S of Fig. 1(b). The plastic strain increment $(d\varepsilon_v^p, d\varepsilon_q^p)$ corresponding to a stress increment (dp, dq) is

$$d\varepsilon_v^p = \left\langle \frac{1}{H}(\eta_p dp + \eta_q dq) \right\rangle \eta_p \tag{8a}$$

$$d\varepsilon_q^p = \left\langle \frac{1}{H}(\eta_p dp + \eta_q dq) \right\rangle \eta_q \tag{8b}$$

where

$$\langle x \rangle = x \text{ if } x > 0 \ ; \ \langle x \rangle = 0 \text{ if } x \leq 0 \tag{8c}$$

As shown in Fig. 2, plastic flow takes place if the stress increment translated at the image stress points outwards to the bounding surface. The plastic strain develops when the stress state is strictly inside the bounding surface. The Eq.1 of triaxial stress is generalized into the following relation

$$(\eta_p dp + \eta_q dq)/H = (\eta_p d\bar{p} + \eta_q d\bar{q})/H_b \tag{9a}$$

where (dp, dq) and $(d\bar{p}, d\bar{q})$ are respectively the increments of stress state and image stress during plastic flow. H is calculated by forcing the image stress to remain on the bounding surface, which is equivalent to apply the consistency condition of conventional plasticity at the image stress

$$H_b = -(\frac{\partial f}{\partial \varepsilon_v^p})(\frac{\partial f}{\partial p})(\frac{1}{g^2}) \tag{9b}$$

$$d\varepsilon_v = d\varepsilon_v^e + d\varepsilon_v^p \text{ and } d\varepsilon_q = d\varepsilon_q^e + d\varepsilon_q^p \tag{9c}$$

where the increment of elastic strain $(d\varepsilon_v^e, d\varepsilon_q^e)$ is related to the bulk modulus B, and the shear modulus, G as

$$d\varepsilon_v^e = dp/B \text{ and } d\varepsilon_q^e = dq/3G \tag{10}$$

BOUNDARY SURFACE PLASTICITY TO SAND BEHAVIOR

Elastic Contribution

For simplicity, it is assumed that the Poisson's ratio μ is constant, and the bulk modulus, B is proportional to the mean effective pressure p as

$$B = B_o\, p \qquad (11a)$$

with B as elastic materials constants, the shear modulus G which is related to the bulk modulus through a constant Poisson ratio μ is therefore also pressure dependent

$$G = \frac{3(1-2\mu)}{2(1+\mu)} \qquad (11b)$$

Definition of Boundary Surface

The bounding surface is the ellipse shown in Fig. 2.

$$(\rho - A)^2 - q^2 / M^2 - A^2 = 0 \qquad (12)$$

where M is the slope of the critical and characteristic states in the space. M is equal to M_c or M_e respectively, depending upon whether q is positive or negative. A is the coordinate of the ellipse summit on p-axis. The image point on the bounding surface is:

$$\bar{p} = \gamma A \text{ and } \bar{q} = \gamma \eta A \qquad (13a)$$

where $\eta = q/p$; $\gamma = \dfrac{2}{1+x^2}$ and $x = \eta / M$. From Eq. 5, the distance between image and stress state is

$$\delta = (\gamma A - p)\sqrt{1+\eta^2} \qquad (13b)$$

According to Eqs. 6 and 12, the unit vector normal to the bounding surface has for components η_p and η_q is

$$\eta_p = M(\gamma - 1)/g, \quad \eta_q = \gamma x / q \qquad (14a)$$

$$\text{where } g = \sqrt{M^2(\gamma-1)^2 + \gamma^2 x^2} \qquad (14b)$$

The aspect ratio, ρ of the bounding surface is related to the plastic flow direction. A strong stress-dilatancy applicable to dense sands is obtained for values of ρ between 1.5 and 2.0, and value between 2.0 and 3.0 for loose sands.

Plastic Modulus at Image Point

It is assumed that stress increments at stress point, where (dp, dq), and at image stress, $(d\bar{p}, d\bar{q})$, give the same plastic strain increment, which, according to Eq. 9a, is equivalent to

$$(\eta_p\, dp + \eta_q\, dq)/H = (\eta_p\, d\bar{p} + \eta_q\, d\bar{q})/H_b \qquad (15a)$$

and then

$$\begin{aligned} H_b &= (\eta_p\, d\bar{p} + \eta_q\, d\bar{q})H / (\eta_p\, dp + \eta_q\, dq) \\ &= (\eta_p\, d\bar{p} + \eta_q\, d\bar{q})\eta_p / d\varepsilon_v^p \end{aligned} \qquad (15b)$$

By substitution Eq.15b into Eq.14, the plastic modulus H_b becomes

$$H_b = \frac{1+e_o}{\lambda - \kappa}\frac{A}{g^2} M^2(\gamma-1)[\gamma + \rho(\rho-2)] \qquad (15c)$$

when $H_b > 0$, (strain hardening); $H_b < 0$, (strain softening) and $H_b = 0$, (critical state)

Plastic Modulus at Stress State

The distance δ which separates the current stress state and its corresponding image point on the bounding surface, is

$$\delta = \sqrt{(\bar{p} - \rho)^2 + (\bar{q} - \rho)^2} \qquad (16)$$

From Eq.13, this distance becomes

$$\delta = (\gamma A - \rho)\sqrt{1+\eta^2} \qquad (17a)$$

The largest distance δ_{max} is between the stress origin and the image point

$$\delta_{max} = \gamma A \sqrt{1+\eta^2} \qquad (17b)$$

A specific feature of bounding surface plasticity is that the plastic modulus H is not only dependent on the surface evolution but is also an adjustable function of the distance δ with the following requirements;

$$H = +\infty \text{ if } \delta \geq \delta_{max} \;;\; H = H_b \text{ if } \delta = 0 \qquad (18a)$$

The restrictions imposed by Eq. 15 ensure that the strains are elastic far from the bounding surface, and that the current stress state will move with the bounding surface when it reaches it. the following form is proposed for the plastic modulus H:

$$H = H_b + \frac{1+e_o}{\lambda - \kappa}\frac{\delta h_0 p}{\delta_{max} - \delta}\left[\frac{\eta_p}{M_c} x\right] \qquad (18b)$$

where η_p, h_0 = two material constants. The term which is added to H_b in Eq.15 is a decreasing positive function of δ. Where δ exceeds δ_{max}, H becomes very large, and the response is purely elastic. The distance dependence can be simplified

$$\frac{\delta}{\delta_{max}-\delta}=\frac{\gamma A}{\rho}-1 \qquad (19)$$

TESTING PROGRAM AND PROCEDURES

Test Materials and Testing Program

Experimental program were performed on Ottawa sand, having a specific gravity of solids, G_s = 2.64. The sand for the test has a grain size distribution ranging from 0.85 to 0.16 mm. and the uniformity coefficient, C_u = 2.0. The mean grain size, D_{50} = 0.42 mm. The maximum dry density, $\gamma_{d\,max}$ = 1.83 Mg/m^3 , and minimum dry density, $\gamma_{d\,min}$ = 1.56 Mg/m^3 .

Static Triaxial Compression Tests

The test specimens had dimensions of 100 mm. in height and 50 mm. in diameter, Its were prepared directly on triaxial cell pedestal. For preparing a specimen, a pre-weighed amount of air-dried sand mixed with about 8% of water contents at an appropriately density was poured into a membrane lined mold in five equal layers with zero height of drop. Slight leveling was required to achieve appropriate densities of test specimens. A small negative pressure is applied to the pore water to maintain the stability of the specimen while the mold is removed prior to theapplication of a confined cell pressure. All specimens were saturated by using the carbon dioxide and a back pressure of about 0.2 MPa to assure that pore water pressure para- meter B-value equaled or exceeded 0.95 and became saturated.

The consolidated drained and undrained triaxial compression or extension tests with pore water pressure measurements were performed at the relative densities of 15% with the effective confining pressures in the range from 100 to 800 kPa and the rate of stress control at 10 kPa per minute . All specimens were initially subjected to isotropic consolidated pressure under an appropriate effective confining pressures, through a mercury pot system to the fluid in the cell. This pressure was maintained for at least 2.0 hours to ensure full consolidation.

During isotropic consolidation, the volume change of the specimens was measured to correct the specimens deformation. To cause the specimens results to shear failure, an axial deviator stress was applied through a vertical ram. Two pressure transducers were used to measure the axial deviator stress and pore water pressure. An LVDT displacement transducer was used to measure the axial displacement during the application of the deviator stress. For consolidated drained test, the volume change indicator fixed into the drainage line was use to measure the volume change of the test specimen. During consolidated drained test, the drainage connection was kept closed. The pore water pressure was measured by using pressure transducers.

Cyclic Triaxial Compression Tests

Cyclic triaxial compression test were carried out by using the CKC automated triaxial testing system. The cylindrical specimens mesuring 75 mm in diameter and 150 mm in height were tested at the initial relative density of 15 %. The test specimens were prepared according to same procedure of static triaxial compression tests, all specimens were saturated to assure that pore water pressure parameter B-value equaled or exceeded 0.95. During cyclic triaxial compression test, the specimens were performed without drainage and the electric signal or frequency and magnitude of loading programed by a Computer was applied to an electro-pneumatic transducers which controlled the pneumatic amplifiers for the application of loads. To begin the test, an appropriate cyclic deviator stress was applied with a sine wave pulse at a low frequency of 1/60 Hz. During each test, cyclic load, axial strain and pore water pressure were monitored with electronic transducers and recorded by PC computer.

EXPERIMENTAT RESULTS VERSUS THEORIAL PREDICATION

Drained Triaxial Compression and Extension Test

For the drained compression and extension tests were performed on saturated Ottawa sand, the numerical simulations compared to the experimental results are shown in Fig. 3 . The relationship between effective deviatoric stress and axial strain of drained triaxial compre- ssion tests in Fig. 3 (a) indicated that it predicats well at the lower effective pressures of100 and 200 kPa, However, it is not situable at the higher effective pressures of 400 and 800 kPa due to the harder grain on Ottawa silica sand. . Fig. 3(b) shows the predicated and experimental results for drained triaxial extension tests.,The effect of the effective deviatoric stress on axial strain is situable for the lager axial stain.

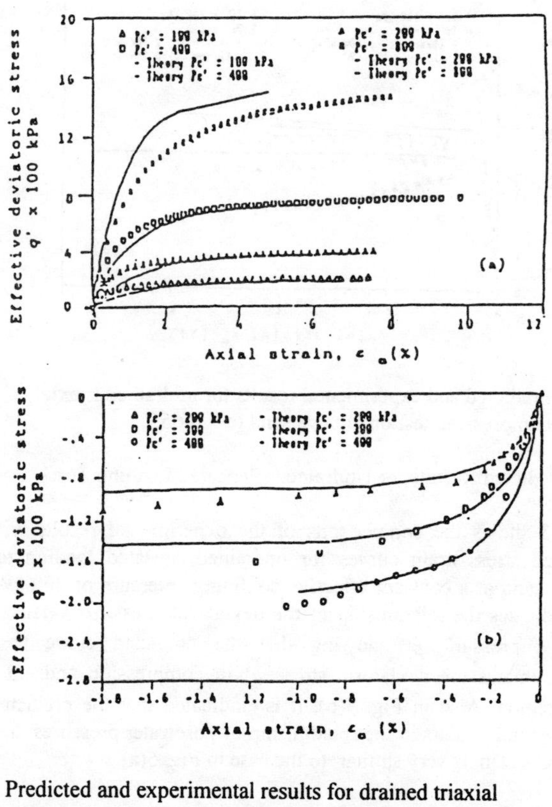

Fig. 3 Predicted and experimental results for drained triaxial compression and extensipn test on Ottawa sand (e = 0.655)

For the undrained triaxial compression tests with constant confining pressure ranging from 100 to 800 kPa, the effective stress paths of the specimens are plotted in Fig. 4(a).The shape and position of the effective stress paths in this plot will depend on the magnitude of the preshear consolidation pressure. It is noted that the initial portion of the effective stress paths for all the specimens are nearly parallel to the q - axis. For all the tests at the low consolidation pressure of 100, 200 and 400 kPa, the predicted and experimental results for compression and extension side are in good well agreement. However, for the specimens sheared at pre-shear condition pressure of 800 kPa, the effective stress paths deviate towards the origin with increasing p. For the speimens in the apparent confining pressure, plots of the normalized stress ratio, η (q/p) against shear strain are shown in Fig. 4(b). It is evident that the stress ratio against shear strain relationship is dependent on the effective confined pressures, and it seems that the distance between those curves becomes larger as the stress ratio, η increases.

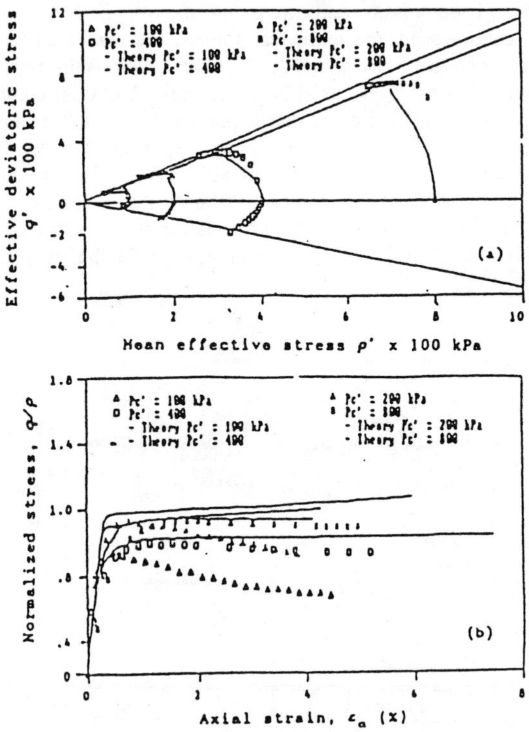

Fig. 4 Predicted and experimental results for undrained triaxial compression test on Ottawa sand (e = 0.655)

The Predicted Results for Undrained Repeated Loading Test

Fig. 5 shows the comparisons of the experimental results with the predicted stress-strain curves for undrained repeated loading test on Ottawa sand at a constant effective confining pressure of 200 kPa. Fig. 5(a) indicates the relationship of the deviatoric stress, q' versus mean effective pressure p' showing that the predicted curve tends to underestimates the deviatoric stress q' in compression and extension stress region. Also in Fig. 5(b), it is indicated that the predicted and experimental results in the relationship of porewater pressures. u versus the axial strain, is very similar to the case in Fig. 5(a).

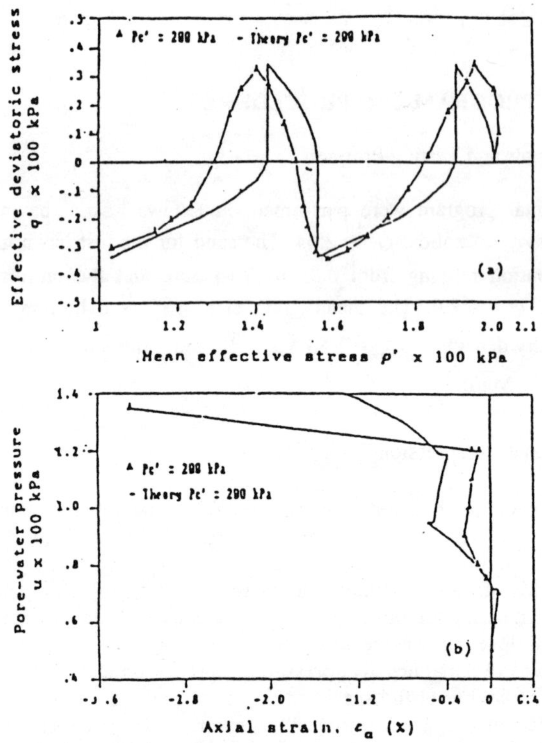

Fig. 5 Predicted and experimental results for undrained repeated loading test on Ottawa sand (e = 0.655)

Determination of Material Constant

The value of λ and κ can be related to the commomly known quantities such as compression index and swelling index. λ value is defined as the slope of virgin loading line on $e - \ln p'$ plot, κ value is defined as the slope of unloading, swelling, curve on the same plot. Another constant M deals with the frictional properties of sand, and is equal to the stress ratio, $\eta = p/q$ at critical state. The test results were presented in a previous paper (Chin, 1987), With λ = 0.615, κ = 0.00982 and M_c = 1.150, M_s = 0.60. The material constants to specify Ottawa sand are summarized in Table 1.

From the previous predicted results, the stress incremental relations (Eq. 9c) have been specified for the values of Table 1, the values of ellipse aspect ratios, ρ is one of the important influential factors to predict in constitutive modeling. Fig. 6 shows the effective stress paths on Ottawa sand of various ellipse aspect ratios, ρ value at a constant confining pressure of 400 kPa, it is indicates that the model constant of ellipse aspect ratios, ρ will injure the accuracy in model's prediction.

Table 1 Material Constant for Ottawa Sand

Consolidated pressure , kPa	p	100 200 300 400 800
Ellipse aspect ratio	ρ	2.60 2.20 1.95 1.85 1.50
Elastic swelling modulus	k	0.00982
Elastic poission's ratio	μ	0..30
Critical e^P at p = 100 kPa	Γ_p	0.619
Slope of critical state line in e - ln p plane	λ	0.615
Slope of characteristic state in p - q plans in compression q ≥ 0	M_c	1.15
in tension q < 0	M_s	0.60
Maximum peak stress ratio	η_p	1.20
Plastic modulus	h_o	1.00

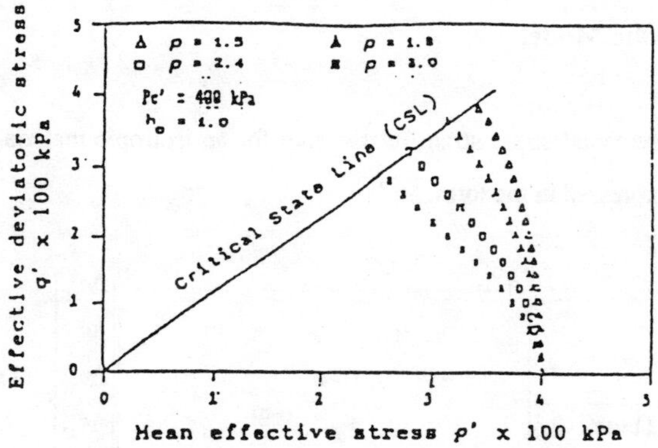

Fig. 6 The stress paths at various aspect ratios ρ value

Method of Determination of Ellipse Ratios ρ

From the above mentioned, the values of elipse aspect ratio, ρ, is one of the important factor to predict in constitutive modeling are showed in Fig. 7, the relative density for each specimen is found to increase with decrease in the ellipse aspect ratios, ρ. At any particular density of sand, the ellipse aspect ratios, ρ, is expressed as the following equation for the accuracy in model's prediction;

$$\rho = 2.68(D_r)^2 - 4.48(D_r) + 2.75 \quad (20)$$

where, D_r is the value of relative density in (%).

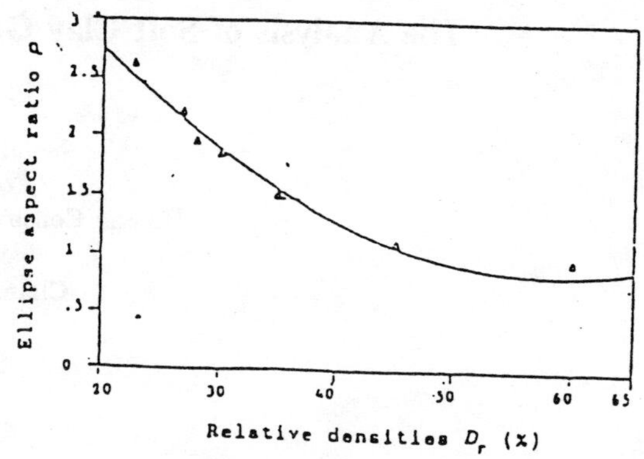

Fig. 7 The relationship of aspect ratios ρ value and relative density D_r on Ottawa sand (e = 0.655).

CONCLUSIONS

To investigate the bounding surface plasticity model to static and cyclic sand behaviour proposed by Bardet (1983) various kinds of triaxial tests were carried out on Ottawa sand. The proposed Bardet's sand model can be applied in geotechnical problems. Simplicity has been a major consideration in the development of this model and has been dictated by the practicality of future applications and by limitations on the available experimental results. Bounding surface plasticity has provided a simple and flexible theoretical framework to assemble the proposed model. The comparison of the model predic-ions with various experimental results indicate good performance of the model along monotonic and cyclic triaxial tests. However, this model which has been completely formulated in the $p-q$ space can only simulate sand behaviour during triaxial tests.

ACKNOWLEDGMENTS

This study was performed as part of research grant by the National Science Committee, R.O.C., NSC75-0410-E006-02. The support is gratefully acknowledged.

REFERENCES:

Aboim, CA and Roth, WH (1982). "Bounding Surface Plasticity Applied to Cyclic Loading of Sand, " *Proc. of the international Symposium on Numerical Models in Geomechanics*, Zurick,.

Bardet, J P (1983) . "Application of Plasticity Theory to Sand Behavior, a New Sand Model," *PH.D. thesis, The California Insti tute of Technology.*

Chin, YP (1987) . " Evaluation of Elastic-Plastic Model under Low Stress State on Cohesionless Soil, "*The Second Conference of Geo-tcchnical Engineering*, R.O.C. (in Chinese)

Proceedings of the Seventh (1997) International Offshore and Polar Engineering Conference
Honolulu, USA, May 25-30, 1997

ISBN 1-880653-28-1 (Set); ISBN 1-880653-29-X (Vol. I)

The Analysis of Soft Clay Ground Behavior Using Spline Method

Song Lee
Seoul City University
Koou-Ho Hwang
Doosan Construction & Engineering Co.
Soo-Sam Kim
Chung-Ang University

Seoul, Korea

Abstract

The Duncan's hyperbolic model is one of the most wide-spread models. That provides a simple governing equation encompassing the most important characteristics of soil stress-strain behavior. The values of the hyperbolic parameters can be obtained easily from conventional triaxial compression.

But, this model is inappropriate to express the plastic behavior of soil. In this study a new model to explain the softening behavior of overconsolidated soil is suggested using the functions involving the OCR.

***Keyword** : hyperbolic model, nonsymmetric parabolic model, OCR*

Introduction

It is possible to approximate the nonlinear inelastic soil behavior in stress analyses of soils, however, it is necessary to describe the stress-strain behavior of the soil in quantitative terms.

The hyperbolic stress-strain relationships were developed for use in nonlinear incremental analysis of soil deformations. In each increment of such analyses, the stress-strain behavior of the soil is treated as being linear, and the relationship between the stress and the strain is assumed to be governed by the generalized Hooke's Law of elastic deformations. [2), 3), 6), 7)]

Hyperbolic Model

The incremental stress-strain relationship for an isotropic material may be expressed in the form: [1), 2)]

$$\begin{Bmatrix}\Delta\sigma_{11}\\ \Delta\sigma_{22}\\ \Delta\sigma_{33}\\ \Delta\sigma_{12}\\ \Delta\sigma_{23}\\ \Delta\sigma_{13}\end{Bmatrix}=\frac{E}{(1+\nu)(1-2\nu)}\begin{bmatrix}(1-\nu) & \nu & \nu & & & \\ & (1-\nu) & & & 0 & \\ & & (1-\nu) & & & \\ & & & \frac{(1-2\nu)}{2} & & \\ & \text{Symmetric} & & & \frac{(1-2\nu)}{2} & \\ & & & & & \frac{(1-2\nu)}{2}\end{bmatrix}\begin{Bmatrix}\Delta\varepsilon_{11}\\ \Delta\varepsilon_{22}\\ \Delta\varepsilon_{33}\\ \Delta\varepsilon_{12}\\ \Delta\varepsilon_{23}\\ \Delta\varepsilon_{13}\end{Bmatrix} \quad (1)$$

in which $\Delta\sigma$: stress increments

$\Delta\varepsilon$: strain increments

E : Young's modulus

ν : Poisson's ratio

By varying the values of Young's modulus and Poisson's ratio as the stresses within the soil vary, it is possible to use Eq. 1, to model the three important characteristics of the stress-strain behavior of soils, namely, nonlinearity, stress-dependency, and inelasticity. The equation of this hyperbola is :

$$(\sigma_1 - \sigma_3) = \frac{\varepsilon}{\dfrac{1}{E_i} + \dfrac{\varepsilon}{(\sigma_1 - \sigma_3)_{ult}}} \qquad (2)$$

where E_i is the initial tangent modulus or the initial slope of the stress-strain curve and $(\sigma_1 - \sigma_3)_{ult}$ is the asymptotic value of stress difference which is closely related to the strength of the soil. The value of $(\sigma_1 - \sigma_3)_{ult}$ is always greater than the stress difference at failure for the soil, as shown in Fig. 1. [6] (Duncan, J.M. and Chang, C.Y., 1970)

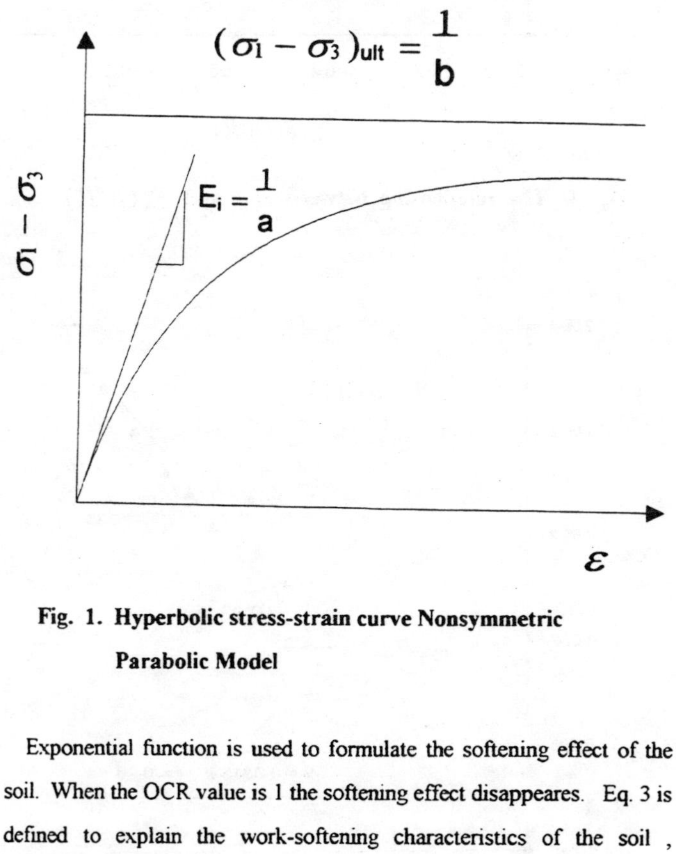

Fig. 1. Hyperbolic stress-strain curve Nonsymmetric Parabolic Model

Exponential function is used to formulate the softening effect of the soil. When the OCR value is 1 the softening effect disappeares. Eq. 3 is defined to explain the work-softening characteristics of the soil , R(OCR), which approaches zero as the OCR value reaches 1. Fig. 2 shows this relationship.

$$R(OCR) = EXP\left[-\left|\frac{1}{0.98 - OCR}\right|\right] \qquad (3)$$

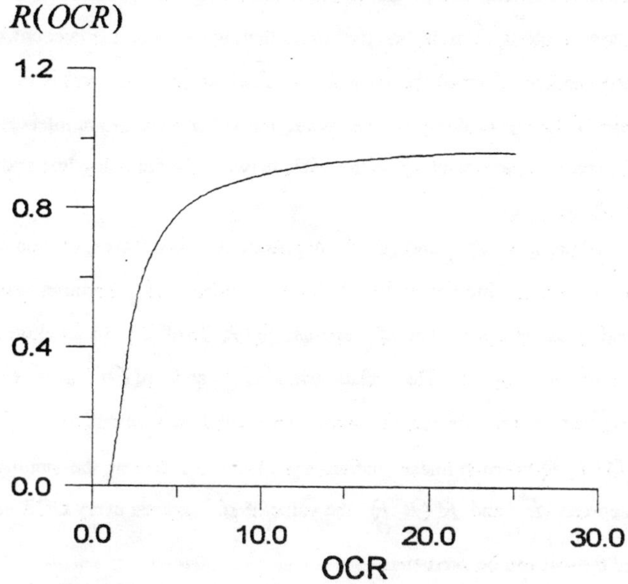

Fig. 2. The relationship of OCR value and exponential function

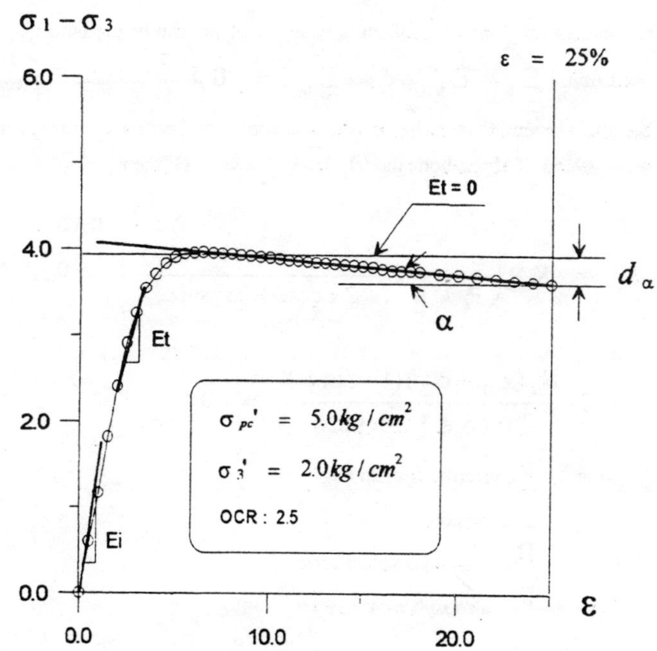

Fig. 3. Stress-strain relationship of the soil of OCR : 2.5

To predict the softening behavior of the soil, two parameters, d_α and α are defined, as shown in Fig. 6. The first parameter d_α is the distance between two points, one at which the line at 25% strain and the line to prolong straight line part of softening of the soil is encountered; and another at which the vertical line which strain of the soil is at 25% and the line to prolong the peak point of soil horizontally are intersected. The second parameter α is the angle between the softening line and the horizontal line.

To determine d_α and α, the results of the triaxial compression tests were plotted. Increasing the OCR value higher, d_α becomes greater and α the steeper. The d_α against $R(OCR)$ of Eq. 3 was plotted as shown in Fig. 4. This relationship d_α and $R(OCR)$ is called exponential transformation. After exponential transformation, d_α and $R(OCR)$ become linear relationship. Therefore, finding the inclination between d_α and $R(OCR)$, the value of d_α against every OCR value of the soil can be predicted.

The inclination, coefficient A is defined in Fig. 4 . And the inclination, coefficient B is defined in Fig. 5

After determing coefficient A and B, if OCR value of the soil being known, two parameters, d_α and α are obtained. Using two parameters, d_α and α, softening behavior of soil can be predicted.

Redefining E_t as E_t', and set $E_t' = 0$ in the ultimate strength, because tangential modulus in ultimate condition becomes zero. From the equation of Hyperbolic model, let $E_t' = 0$, then

$$E_t' = K P_a \left(\frac{\sigma_3}{P_a}\right)^n \left[1 - \frac{R_f(\sigma_1 - \sigma_3)(1-\sin\phi)}{2c\cos\phi + 2\sigma_3\sin\phi} \right]^2 = 0 \quad (4)$$

$$1 - \frac{R_f(\sigma_1 - \sigma_3)(1-\sin\phi)}{2c\cos\phi + 2\sigma_3\sin\phi} = 0 \quad (5)$$

where, ϕ : internal friction angle

c : cohesion

R_f : constant failure ratio

P_a : atmospheric pressure

A constant failure ratio R_f is defined as $(\sigma_1 - \sigma_3)_f = R_f(\sigma_1 - \sigma_3)_{ult}$ and R_f is typically between 0.6 to 0.9 for most soils.[6] (Duncan, J.M. and Chang, C.Y., 1970)

Eq. 5 becomes

$$R_f(\sigma_1 - \sigma_3)(1-\sin\phi) = 2c\cos\phi + 2\sigma_3\sin\phi \quad (6)$$

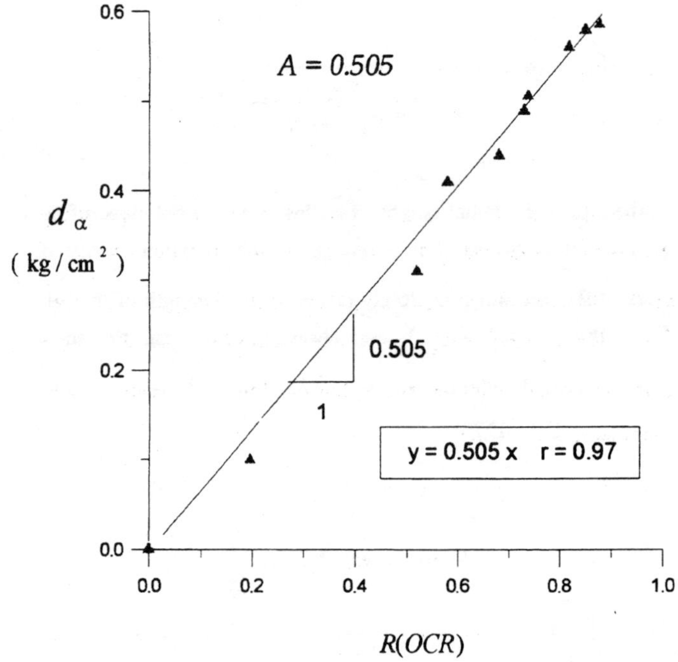

Fig. 4. The relationship between d_α and $R(OCR)$

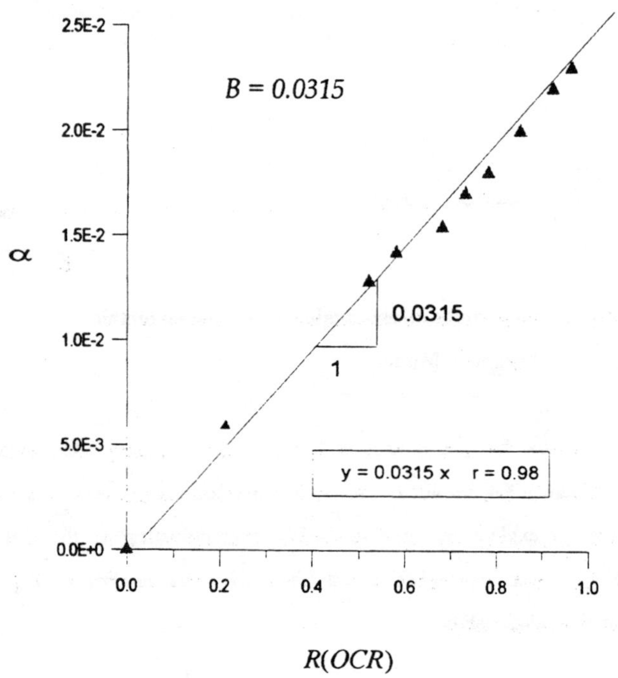

Fig. 5. The relationship between α and $R(OCR)$

Then Eq. 6,

$$(\sigma_1-\sigma_3) = \frac{2c\cos\phi+2\sigma_3\sin\phi}{R_f(1-\sin\phi)} \tag{7}$$

Deviator stress $(\sigma_1-\sigma_3)$ from Eq. 7 means ultimate strength, $(\sigma_1-\sigma_3)_f$, for ultimate condition.
After ultimate strength $(\sigma_1-\sigma_3)_f$ is determined, draw the line to pass $(\sigma_1-\sigma_3)_f$ horizontally.

And draw the line again, which is away from ultimate strength line at distance d_α, and whose angle is α. The matching point is obtained from two lines intersected on ultimated strength as shown in Fig. 6.

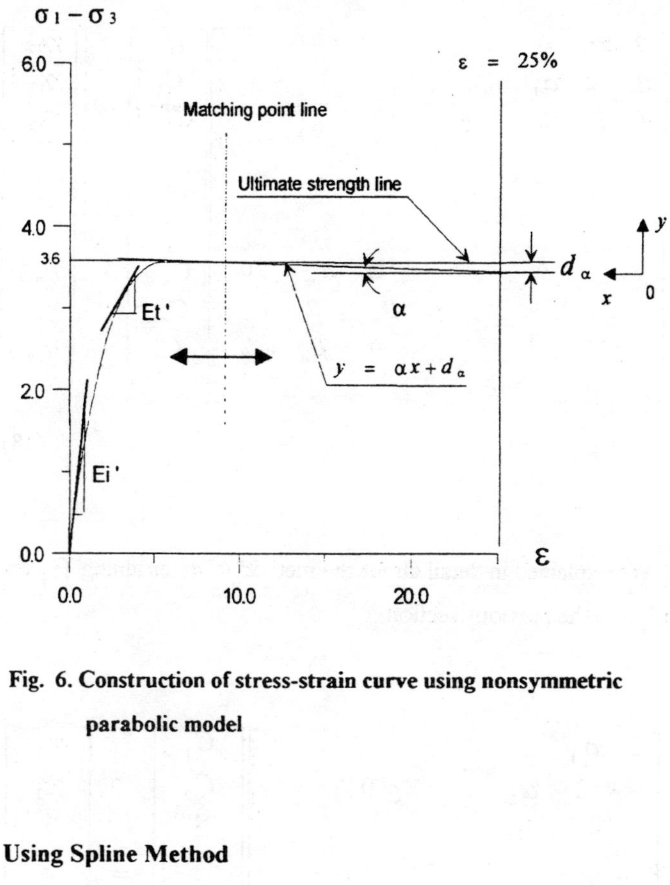

Fig. 6. Construction of stress-strain curve using nonsymmetric parabolic model

Using Spline Method

Splines and their mathematical properties is not described herein. However, a cubic spline is described and utilized. [4), 5)]

Let the subintervals of Fig. 7 be called I_1 to I_n, so that $I_i = (\varepsilon_{i-1}, \varepsilon_i)$.

Also define

$$\Delta\varepsilon_i = \varepsilon_i - \varepsilon_{i-1} \tag{8}$$

and the subintervals need not be of equal length.
Let the spline function on the segment I_i be $S_i(\varepsilon)$, and then $S''_i(\varepsilon)$ assume

$$S''_i(\varepsilon) = C_{i-1}\frac{\varepsilon_i-\varepsilon}{\Delta\varepsilon_i}+C_i\frac{\varepsilon-\varepsilon_{i-1}}{\Delta\varepsilon_i} \tag{9}$$

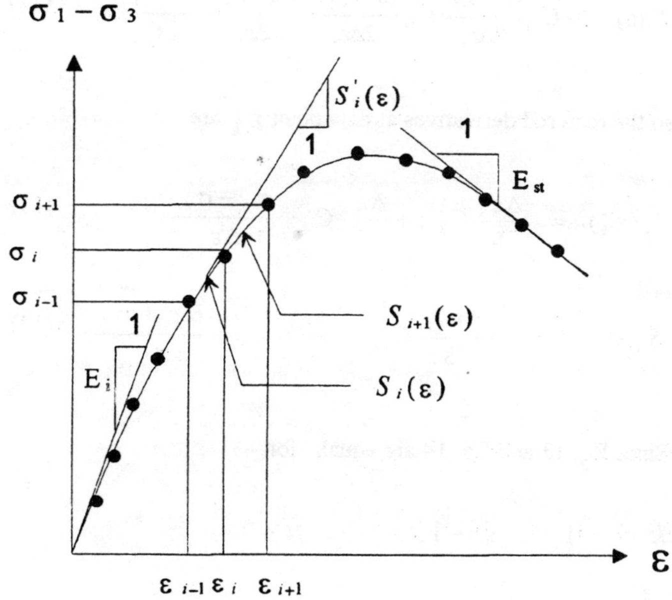

Fig. 7. Spline curve to stress-strain data

Integrating Eq. 9 twice to obtain the spline segments and then yields

$$S_i(\varepsilon) = C_{i-1}\frac{(\varepsilon_i-\varepsilon)^3}{6\Delta\varepsilon_i}+C_i\frac{(\varepsilon-\varepsilon_{i-1})^3}{6\Delta\varepsilon_i}+d_i(\varepsilon_i-\varepsilon)+e_i(\varepsilon-\varepsilon_{i-1}) \tag{10}$$

where, C_i, d, e are arbitrary constants

From Fig. 7, we must have $S_i(\varepsilon_{i-1}) = \sigma_{i-1}$ and $S_i(\varepsilon_i) = \sigma_i$. These conditions fix d_i and e_i, and lead to $S_i(\varepsilon)$

$$S_i(\varepsilon) = C_{i-1}\frac{(\varepsilon_i-\varepsilon)^3}{6\Delta\varepsilon_i}+C_i\frac{(\varepsilon-\varepsilon_{i-1})^3}{6\Delta\varepsilon_i}+\left(\sigma_{i-1}-\frac{C_{i-1}\Delta\varepsilon_i^2}{6}\right)\frac{\varepsilon_i-\varepsilon}{\Delta\varepsilon_i}$$

$$+\left(\sigma_i-\frac{C_i\Delta\varepsilon_i^2}{6}\right)\frac{\varepsilon-\varepsilon_{i-1}}{\Delta\varepsilon_i} \qquad (11)$$

as may be existed between ε_{i-1} and ε_i.

Differentiating Eq. 11,

$$S_i'(\varepsilon) = -C_{i-1}\frac{(\varepsilon_i-\varepsilon)^2}{2\Delta\varepsilon_i}+C_i\frac{(\varepsilon-\varepsilon_{i-1})^2}{2\Delta\varepsilon_i}+\frac{\sigma_i-\sigma_{i-1}}{\Delta\varepsilon_i}-\frac{C_i-C_{i-1}}{6}\Delta\varepsilon_i \qquad (12)$$

so the required derivatives at data point ε_k are

$$S_k'(\varepsilon_k^-) = \frac{\Delta\varepsilon_k}{6}C_{k-1}+\frac{\Delta\varepsilon_k}{3}C_k+\frac{\sigma_k-\sigma_{k-1}}{\Delta\varepsilon_k} \qquad (13)$$

and

$$S_{k+1}'(\varepsilon_k^+) = -\frac{\Delta\varepsilon_{k+1}}{3}C_k-\frac{\Delta\varepsilon_{k+1}}{6}C_{k+1}+\frac{\sigma_{k+1}-\sigma_k}{\Delta\varepsilon_{k+1}} \qquad (14)$$

Since Eq. 13 and Eq. 14 are equal, for

$k = 1, \ldots, n-1,$

$$\frac{\Delta\varepsilon_k}{6}C_{k-1}+\frac{\Delta\varepsilon_k+\Delta\varepsilon_{k+1}}{3}C_k+\frac{\Delta\varepsilon_{k+1}}{6}C_{k+1} = \frac{\sigma_{k+1}-\sigma_k}{\Delta\varepsilon_{k+1}}-\frac{\sigma_k-\sigma_{k-1}}{\Delta\varepsilon_k} \qquad (15)$$

Let

$$\alpha_i = \frac{\Delta\varepsilon_{i+1}}{\Delta\varepsilon_i+\Delta\varepsilon_{i+1}}$$

$$\beta_i = 1-\alpha_i = \frac{\Delta\varepsilon_i}{\Delta\varepsilon_i+\Delta\varepsilon_{i+1}}$$

$$\gamma_i = \frac{6}{\Delta\varepsilon_i+\Delta\varepsilon_{i+1}}\left(\frac{\sigma_{i+1}-\sigma_i}{\Delta\varepsilon_{i+1}}-\frac{\sigma_i-\sigma_{i-1}}{\Delta\varepsilon_i}\right) \qquad (16)$$

for $i = 1, \ldots, n-1$.

Eq. 15 can then be be rewritten, still for $i = 1, \ldots, n-1$, as

$$\beta_i C_{i-1}+2C_i+\alpha_i C_{i+1} = \gamma_i \qquad (17)$$

Then the system can be rewritten in the matrix form as Eq. 18. The values of initial point and ending point are each $i = 0$ and n. However, C_0 and C_n are not determined. In Desai's paper, setting $C_0 = C_n = 0$, which is assumed a natural spline, and solves the underdetermined condition. But in this paper, from Fig. 7 and Eq. 12, defining $S_1'(\varepsilon_1) = E_i$ and $S_n'(\varepsilon_n) = E_{st}$, and solve the two underdetermined condition of Eq. 18, and setting strain space equal, then become Eq. 19.

$$\begin{bmatrix} 2 & \alpha_0 & 0 & & & & \\ \beta_1 & 2 & \alpha_1 & & 0 & & \\ 0 & \beta_2 & 2 & & & & \\ & & & \ddots & & & \\ & & 0 & & 2 & \alpha_{n-2} & 0 \\ & & & & \beta_{n-1} & 2 & \alpha_{n-1} \\ & & & & 0 & \beta_n & 2 \end{bmatrix} \begin{bmatrix} C_0 \\ C_1 \\ C_2 \\ \vdots \\ C_{n-2} \\ C_{n-1} \\ C_n \end{bmatrix} = \begin{bmatrix} \gamma_0 \\ \gamma_1 \\ \gamma_2 \\ \vdots \\ \gamma_{n-2} \\ \gamma_{n-1} \\ \gamma_n \end{bmatrix} \qquad (18)$$

We explained in detail about the method of determining E_i and E_{st} in the previous section.

$$\begin{bmatrix} 2 & \alpha_1 & & & \\ & 2 & \alpha_2 & 0 & \\ & & \ddots & \ddots & \\ & & & \ddots & \alpha_{n-3} \\ & Sym & & 2 & \alpha_{n-2} \\ & & & & 2 \end{bmatrix} \begin{bmatrix} C_1 \\ C_2 \\ \vdots \\ C_{n-2} \\ C_{n-1} \end{bmatrix} = \begin{bmatrix} \gamma_1 \\ \gamma_2 \\ \vdots \\ \gamma_{n-1} \\ \gamma_{n-2} \end{bmatrix} \qquad (19)$$

nonsymmetry parabolic model represented the softening behavior after failure, but hyperbolic model did perfectly plastic behavior.

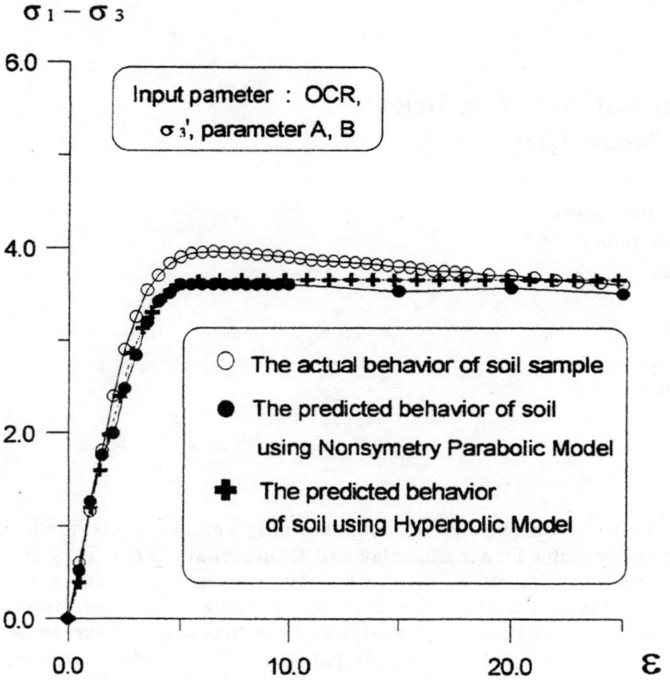

Fig. 8. The results of the analysis of two models

Fig 8. is shown that behaviors of the soil of three cases are compared, hyperbolic model, nonsymmetric parabolic model and actual behavior of soil. The nonsymetry parabolic model appears softening behavior such as the soil sample after failure, but hyperbolic model does not appear softening behavior.

Conclusion

To explain softening effect of soil three parameters are defined, that are, $R(OCR)$, d_α and α. Function $R(OCR)$ is the function to be varied according to the OCR value as shown in Fig. 2.

We can obtain two parameters d_α, α from triaxial compression tests and then predict the behavior of the soil, interpolating with the spline function as shown in Fig. 7.

We could find that the behavior to be predicted using

References

1. Bardet, J. P. And Choucair, W., "A Linearized Integration Technique for Incremental Constitutive Equations," Int. Journal for Numerical Analytical Methods in Geomechanics, Vol. 15(1991), pp. 1 - 19.
2. Bishop, A.W., "The Strength of Soils as Engineering Materials," Geotechnique, Vol. 16(1996), pp. 91 - 130.
3. Chakrabarty, J., Theory of Plasticity, McGraw-Hill, 1987
4. Desai, C. S., "Nonlinear Analysis Using Spline Functions," J. Soil Mech. Found. Div., ASCE, Vol. 97, No. SM 10, Oct. 1971, pp. 1461 - 1480.
5. Desai, C. S., and Christian, J. T. (Eds.), "Numerical Methods in Geotechnical Engineering," McGraw-Hill Book Company, New York, 1977.
6. Duncan, J.M., and Chang, C.Y., "Nonlinear Analysis of Stress and Strain in Soils," J. Soil Mech. Found. Div., ASCE, Vol. 96, No. SM 5, Sept. 1970, pp.1627 - 1653.
7. Hill, R., The Mathematical Theory of Plasticity, Oxford University Press, London, 1950.
8 Roscoe, K. H., Schofield, A., and Thurairajah, A., "Yielding of Clays in State Wetter than Critical," Geotechnique, Vol. 13, No. 3, 1963, pp. 211 - 240.

Proceedings of the Seventh (1997) International Offshore and Polar Engineering Conference
Honolulu, USA, May 25-30, 1997

ISBN 1-880653-28-1 (Set); ISBN 1-880653-29-X (Vol. I)

Major Effect of Grain Size Distribution to Coefficient of Permeability of Marine Clay

Mitsuharu Fukuda and Seiji Suwa
Osaka Soil Test Laboratory
Osaka, Japan

Takao Uno
Gifu University
Gifu, Japan

ABSTRACT

Coefficiency of permeability of clay has a tendency to take an important role in the consolidation analysis of clay. The permeability properties of clay, however, seems to remain a relatively minor investigated field. One of a few findings in this field of study is the existence of a linearity between the logarithm of coefficient of permeability and void ratio.

Therefore, for the purpose of further investigation of the permeability properties of clay, a relationship between coefficient of permeability and grain size distribution curve was focused as one of the study themes.

This paper first presents the new concept developed by authors for clarifying soils based on a grain size distribution curve. It is defined as the classification radius method. Applying this method, with a sand sample used in this paper shows that good relationship between permeability coefficient of sand and grain size distribution curve is detected. Furthermore, it will be shown that the soil classification method, when used on the study of permeability properties of clay, can derive coefficient of permeability of clay closed to the value obtained by a consolidation test. Finally, this paper will show the figure which represents a relationship of depth and coefficient of permeability including sand stratum and clay. The figure gives us a lot of information on the permeability properties of ground.

KEY WORDS: clay, grain size distribution curve, coefficient of permeability, classification

INTRODUCTION

Recently, the coefficient of permeability of a marine clay has come to take an important role in a nonlinear consolidation analysis (Imai,1988, Yoshikuni, et al. 1990). However, comparing to the volume of findings on a permeability properties of sand and gravel strictly concerning with grain size distribution curves(Uno, et al. 1993), in general, a physical meaning of the permeability of clay hasn't been summarized yet(Takane et al. 1985, Imai et al. 1989). Therefore, this study will proceed to define the role of a grain size distribution curve as a principal factor which affects the variance of the permeability properties of a marine clay.

There are some methods used for obtaining a grain size distribution curve available for a marine clay soil (Muta, et al. 1976). They are, however inadequate as they are only able to figure out samll differnces in the grain size distribution curves from each other. Therefore, exact grain size analysis of clay has been proven to be one of the most difficult tests. In general, it is concluded that grain size distribution curves of clay are doubtful and not useful for solving geotechnical problems. Nevertheless, there have been some findings concerned with the relationship between engineering properties of clay and the grain size distribution curve. One of them is that the content less than the particle size of 2 μm governs the consistency properties of clay, this is called the activity, and is proposed by Skempton (Skempton, 1953). Another finding has been that a liquid limit tends to increase when subjected to the increasing of fine particles(Fukuda, et al. 1995).
Therefore, this study of the permeability of clay was carried out by emphasizing the role of the grain size distribution curve.

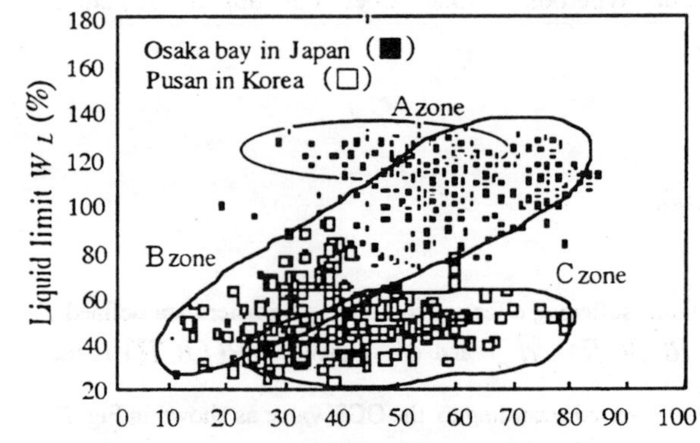

Fig.1 Relationship between liquid limit and clay contents finer than 0.05mm in diameter

RELATIONSHIP BETWEEN CONSISTENCY PROPERTIES OF CLAY AND GRAIN SIZE DISTRIBUTION CURVES

Fig.1 shows the liquid limit characteristics which clearly supports the existence of the linear relationship between the liquid limits and clay contents of samples. They were collected at the Osaka bay area in Japan and at Pusan in Korea. Data included in the B zone and C zone, which contain the results of the samples taken in both countries, tends to change following the linear relationship between both factors. Whereas, data in A zone tend to keep a constant liquid limit independent of clay contents. This property seems to be curious. However, further investigation made it clear that the test belonging to A zone was carried out by the same company, and the testing was badly proceeded. Therefore, the linear relationship between liquid limit and clay contents is the propertiy.

Fig.2 shows the activity of clay comparing the samples in Japan with one in Pusan. There appears to be differences of proportional characteristics between both samples. In each group of samples, however, the linear relationships between the activity and clay contents can be observed like broken line except several data.

BASIC EQUATIONS FOR PERMEABILITY OF SAND

Authors have derived the equations for estimating coefficient of permeability using distribution the grain size characteristics of coarse

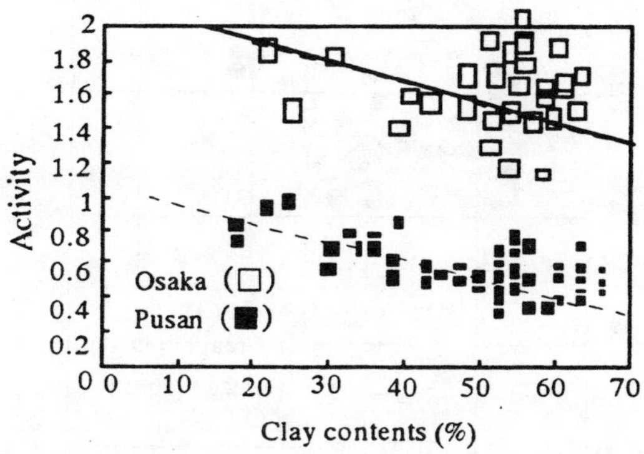

Fig.2 Relationship between activity and clay contents finer than 0.02mm in diameter

sand as are followed.

$$h = \frac{0.3d_{50}}{\exp\left\{0.5(0.484 + 0.420\ln Uc)^2\right\}}\frac{e}{Gs} \tag{1}$$

$$Uc = \frac{d_{60}}{d_{10}} \tag{2}$$

$$\log k = 2.87(1 + \log h) \tag{3}$$

Where k is coefficient of permeability (cm/s), h is equivalent to an imaginable average thickness of void surrounding each particle (mm). d_{10}, d_{50} and d_{60} are 10%,50% and 60% particle size subject to a grain size distribution curve respectively and their unit is (mm). Uc means coefficient of uniformity. Compared with many test results, these equations have proved to be effective for estimating the coefficient of permeability of sand and gravel.

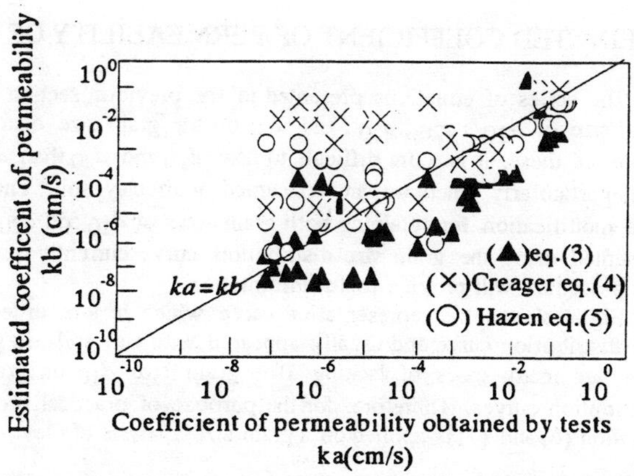

Fig.3 Comparison between the experimental data with the permeability estimated

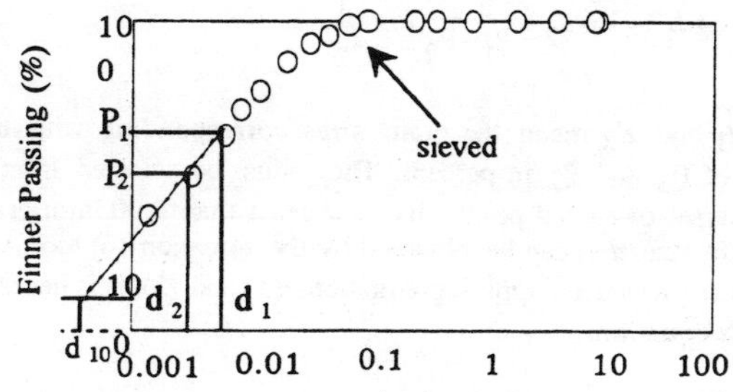

Fig.4 Sehmatic instruction for estimating the 10% grain size d_{10} on grain size distribution curve

Among the simplified equations presented for predicting coefficient of permeability based on grain size distribution curve, two famous formulas will be investigated for the comparison study. Creager's equation and Hazen's one are familiar with the practical engineers. Two equations are shown as followings.

Creager's equation

$$k = 0.359 d_{20}^{2.327} \tag{4}$$

Hazen's equation

$$k = 100 d_e^2 \tag{5}$$

where, d_{20} is the particle size corresponding with 20% of grain size distribution curve and de is called the effective size, which is generally defined as equivalent to the diameter of d_{10} previously shown.

Fig.3 shows the variable properties of the relationship between estimated values derived by those equations and the experimental data. As clearly detected, the coefficient of permeability predicted by the equations (3) tends to scatter within the narrow band away from the ka=kb line which means that the predicted value is equal to the experimental value. While, the values derived by Creager's formula and Hazen's one scatter more than 2 orders away from the experimental data.

Therefore, within the limitation of using sand, it has been proven that the sensitivity of practical application of equation (3) is better than both other equations, Creager's and Hazen's.

PREDICTED COEFFICIENT OF PERMEABILITY OF CLAY

The series of equations presented in the previous section require grain size of d_{10} , d_{20},, d_{50},, and d_{60} on the grain size distribution curve. Of these, it is more difficult to take d_{10} and d_{20} than d_{50} and d_{60}, particularly when we are concerned with clay soil. Therefore, here, modification for obtaining both grain sizes of d_{10} and d_{20} will be presented using the grain size distribution curve currently used, and this procedure will provide useful information.

Fig .4 shows the representative curve which is a modified grain size distribution curve and usually appeared with clay soils. In general, there are many cases of lacking 10% grain size d_{10} on grain size distribution curves. Therefore, for the purpose of practical procedure, equation (6) and (7) is applied on a grain size analysis of clay.

$$\log d_{20} = \log d_2 - \frac{P_2-20}{P_1-P_2}\log\frac{d_1}{d_2} \qquad (6)$$

$$\log d_{10} = \log d_2 - \frac{P_2-10}{P_1-P_2}\log\frac{d_1}{d_2} \qquad (7)$$

where d_1 and d_2 mean the grain sizes corresponding with the finer passing of P_1 and P_2 in percent. They must be selected from among the larger set of sieved points that is at least than 0.003mm grain size. 20% grain size d_{20} can be obtained by the equation (6) too, Although d_{20} isn't requested on applying equations (1) and (3), it is necessary for Creager's equation.

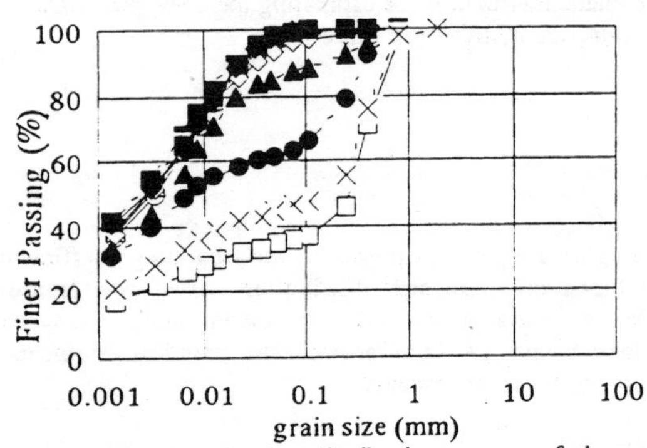

Fig.5 Group of grain size distribution curves of clay soil used for analyzing

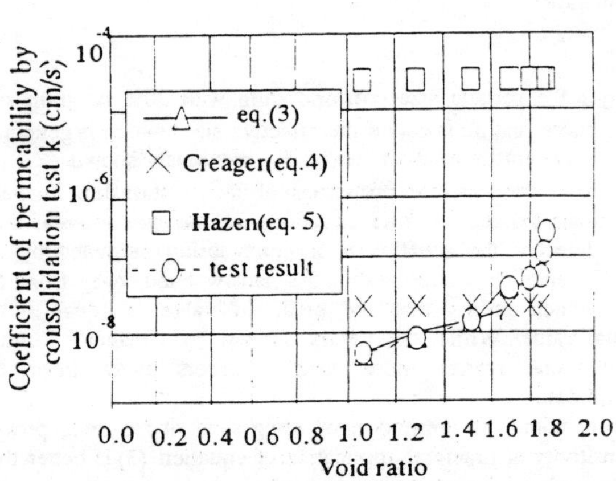

Fig. 6 Comparison between predicted permeability and consolidation test result (1)

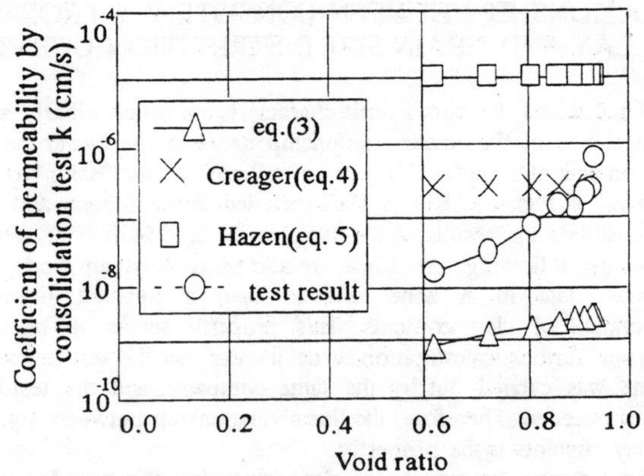

Fig. 7 Comparison between predicted permeability and consolidation test result (2)

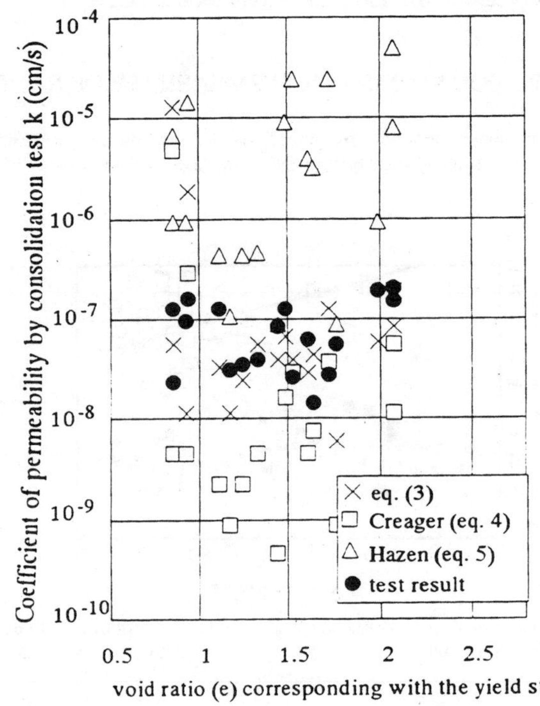

Fig.8 Coefficient of permeability properties corresponding with the yield stress in consolidation test

Fig.5 describes the group of the grain size distribution curves of clay used for analyzing. They are representative examples of grain size distribution curves obtained from around the Osaka bay area in Japan.

Fig.6 and Fig.7 show the representative examples of results on applying the above equations to predict coefficient of permeability. The coefficient of permeability taken by the test is known the consolidated coefficient of permeability, because it is generally derived from the coefficient of consolidation obtained by curve fitting to the consolidated settlement time curve. Void ratio (e) included in Fig.6 and 7 are ones corresponded with the coefficient of permeability in the consolidation test result. In the case of application of equation (3), these corresponding void ratios are used.

Good agreement is found between the predicted coefficient of permeability using equation (3) and the experimental data included in the normally consolidated zone. A large gap, however, is detected between both values in the over consolidated zone. Whereas, when comparing with the relationship which appeared in the results by equation (3), the predicted values derived from Creager's and Hazen's equations largely scatter. One of the major reasons is that the two equations don't include void ratio as a governing factor.

The fitness properties between the coefficient of permeability predicted by equation (3) and void ratio in the over consolidated zone is different from the ones in the normalized zone. But this special aspect remained in a problem on the next stage study. When restricting to the permeability properties within the normalized zone, the relationship between a permeability and void ratio can be symbolized by a tendency of a relationship between the logarithms of permeability and void ratio obtained at the consolidated yield stress.

Fig.8 shows the summarized data derived from consolidation tests and the set of estimated coefficients of permeability. This figure proves clearly that the value predicted using equation (3) fits the test results, while, there is the large discrepancy between the test results and the predicted value by Creager's and Hazen's equations, because the coefficient of permeability derived by Creager's and Hazen's equation keeps constant against a void ratio changing.

However, there are two different types of fitness in the application results of equation (3). This phenomena is associated with a type of grain size distribution curves. For examples, good agreement is bounded with the type of a grain size distribution curve smoothly changing, however, large discrepancy is found in data bounded with

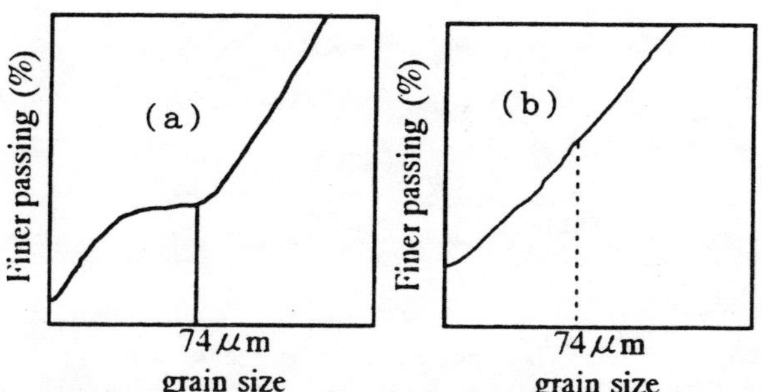

Fig.9 Relationship between type of distribution curves and sensitivity of estimation

the grain size distribution curves which have a staircase-like portion. The two types of distribution curves are shown in Fig.9 (a) and (b) respectively.

It is said that the coefficent of permeability of clay soil and void ratio and logarithms of the coefficient of permeability have linearity properties. These characteristics are applied on an elast-plastic analysis using FEM.

Besides of Creager's and Hazen's equation, another characteristics detected is presented that the function of $e^3/(1+e)$ and the coefficient of permeability make linear relationship. Although rare practical application of a model, capillary tube model is promoted to some formulas for estimating a coefficient of permeability. Of these, Taylor's equation is shown as representative and includes not only grain size but also void ratio as shown in equation (8).

$$k = D_S^{\,2}\frac{\gamma_w}{\eta_w}\frac{e^3}{1+e}C \tag{8}$$

where, D_S is the representative grain size, C constant value, γ_w density of water, and η_w viscosity of water. Operating logarithms of both sides, equation (8) is changed into equation (9).

$$\log k = 2\log D_S + \log\left[\frac{e^3}{1+e}\right] + \log\left[\frac{\gamma_w}{\eta_w}\cdot C\right] \tag{9}$$

On the basis of equation(9), logarithms of permeability is proportional to the logarithms of a grain size and the function composed of void ratio. Devising equation (11) into two parts, equation (3) can be rearranged as follows.

$$h = CI\cdot\frac{e}{G_s} \tag{10}$$

$$CI = \frac{0.3d_{50}}{\exp\left\{0.5\left(0.484+0.420\ln U_c\right)^2\right\}} \tag{11}$$

$$\log k = 2.87\log(CI) + 2.87\log(e) + 2.87(1-\log G_s) \tag{12}$$

where CI is defined as classification radius.

Comparing equation (12) with equation (9), the first term of both equations are concerns with a representative grain size, the second with void ratio, and the third constant value. Both equations are classified in the similar form. Moreover, the constant values of the first term in the right hand side are 2.87 and 2.0respectively, which are similar values. And the ratio of $\left\{e^3/(1+e)\right\}$ to $e^{2.87}$ ranges 0.3 to 0.6 within the void ratio of 0.4 to 3.0. Therefore, equation (3) has the same properties with the Taylor's equations.

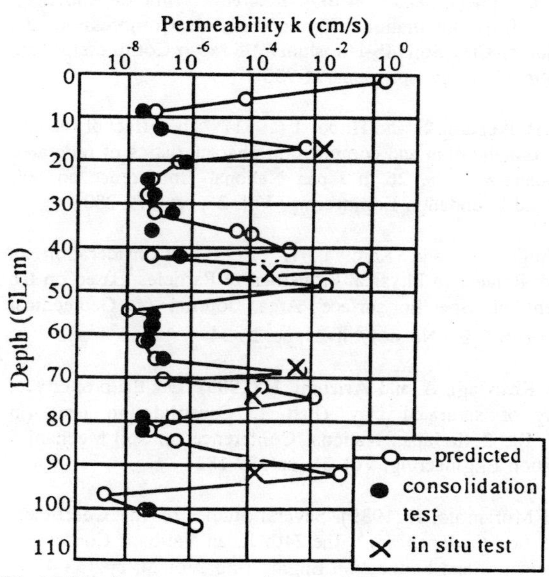

Fig.10 Distribution pattern of permeability in depth

APPLICATION EXAMPLE OF CLASSIFICATION RADIUS

Fig.10 shows the comparison between the predicted permeability based on the equation (3) and the permeability test results. The permeability test contains in situ test which applied to the sand layers and the consolidation tests with clay samples. The distribution patterns of the permeability in the depth direction taken by the two methods are similar to each other. That is, permeability characteristics could be correctly obtained on the chart including permeability of various kinds of soils, especially clay or clayey soils. Then, the classification radius method will take an important role, because it can be applied not only on sand soils but also on clay soils.

CONCLUSIONS

In this paper, good relationship between experimental results and estimated values of the coefficient of permeability could be pointed to have a gap variance of both values within 1 orders, if grain size distribution curve smoothly changes. But if shape pattern of distribution curve is figured like staircase, a discrepancy between experimental value and estimated one will be scattered over the range of 2 orders, although proportional properties between both factors are similar to each other.

Creager's equation and Hazen's one, which are equations used in dam construction field, have a tendency to present a estimated value away from experimental results. Therefore, permeability properties of clay couldn't be traced by using these equations.

Another viewpoint presented in this paper is that coefficient of permeability of clay is so strictly characterized that the logarithm of coefficient of permeability is conducted linearly changing with logarithm of void ratio. The equation presented by authors can be applied for the properties of permeability with the classification radius of soils.

REFERENCES

Imai, G (1995). "Analytical examinations of foundations to formulate consolidation phenomena with inherent time-dependence," Proceedings of the International Symposium on Compression and Consolidation of Clay Soils-IS-Hiroshima '95 Japan Compression and Consolidation of Clayey Soils, pp.891-935

Yoshikuni, H, Ikegami, S, and Hirao, T (1991). "The effect of loading rate on the consolidation and compression characteristics of reiterated clay (in Japanese)," The 26 th Japan National Conference on Soil Mechanics and Foundation Engineering, Vol. 26, pp. 279-280.

Uno, T, Sugii, T, and Kamiya, K (1993). "Consideration of Permeability Related to Physical Properties of Particles Based on the Measurement of Specific surface Area, Journal of Geotecnical Engineering (J.S.C.E.) No. 469/ III-23, pp. 25-34

Takane, H, Kasiwagi, A, and Arizumi, K (1985)."On the property of permeability of saturated clay (Part 1. Consolidation test) (in Japanese)," The 21th Japan National Conference on soil Mechanics and Foundation Engineering, Vol..21, pp.179-182

Imai,G, and Muramoto, K (1989)."Several Studies of The Coefficient of Consolidation (in Japanese)," The 24th Japan National Conference on Soil Mechanics and Foundation Engineering, Vol.24, pp.433-436

Muta, A, and Watanabe, S (1976). "Correct grain size distribution curve of Kanto loam and clay (On the some problems related to sedimentation in liquid grain analysis (in Japanese), The 20th Colloid Chemistry Conference, pp.81-82

Skempton, A.W. (1953). "The Colloidal "Aetivity" of Clays," Proc. 3rd. IC SM, Vol.1, pp.57-61.

Fukuda, M, Suwa, S, Akai, T, and Hamada, T (1995). "Consistency properties of clays sampled in Korea and in Osaka (in Japanese)," The Japanese Geotechnical Society "Symposium on the Soil Classification", pp.197-204.

Proceedings of the Seventh (1997) International Offshore and Polar Engineering Conference
Honolulu, USA, May 25-30, 1997

ISBN 1-880653-28-1 (Set); ISBN 1-880653-29-X (Vol. I)

A Numerical Analysis on the Deformation of Decomposed Granite Soil Foundation Using Elasto-Plastic Model

Bang-woong Shin, Kwang-dong Lee, Se-wook Oh and Tae-wook Cheong
Chung-buk National University
Cheongju, Korea

Abstract

This paper aimed at predicting the behaviour of deformation of decomposed granite soil foundation using elasto-plastic constitutive model proposed by Lade. Sample soil is weathered granite soil that be founded at Cheong-ju district. To determine soil parameters, isotropic consolidation tests and cylinderical triaxial compression tests are carried out, at this time it was maintained compaction energy of 95% on the laboratory testing in order to regulate fracture effect by virtue of the compaction.

In the deformation analysis of foundation, it was vertified by the model load test through centrifuge test that maintained same stress condition in situ as being laboratory test that deformation value of foundation was predicted.

Basic physical properties is determined by geotchnical tests, and for application of constitutive model with multiple intersecting yeild surface, it keep constant compaction energy and then determined soil parameters(14) needed in constitutive model. The centrifuge test was carried out for behaviour analysis of foundation.

1. Introduction

Offshore reclamation is important at increase the effect of land use. Soil used of reclamation is weathered granite soil that carried into situ out of inland. Weathered granite soil is geotechnical sample that came in contacted easily a kind of soil is various and now semipermanent weathered processing, and this soil is different according the component or ground composition as a mass granite size distribution, specific gravity etc,. As geotechnical property is different according to degree of weathering, water content etc., analysis and predict of foundation deformation of reclaimed ground introduced stress-strain theory are more difficult. To analysis or predict foundation deformation of reclaimed ground, it have to used the relevant constitutive laws introducing geotechnical property or boundary condition in site.

Since elasto-plastic model is presented by Drucker et al in the latter of 1950s, then Cam-clay model proposed by the Roscoe et al and constitutive model of Cap model etc. that founded on continuous failure concept of soil has been proposed. Elasto-plastic constitutive model proposed by Lade[1,2,3] is new elasto-plastic work hardening model with two yield surface containing work hardening rule and plastic potential surface that are presented by yield surface and hardening softening process upon plastic compression.[4] Generally, this model can predicts foundation behavior very well, also it can presents stress-strain theories more than anything else. Soil parameters using in the Lade model is determined from conventional triaxial compression test(CTC), but it has not been putted to practical use. because non-associative flow rule that plastic potential function agree not with yield surface is applied, and process of determination of soil parameters is difficult. As weathered soils behave with characters of sand-clayey following degree of weathering, it is accurately determined parameters for application of virtual foundation as well as theoretical consideration of character of geotechnical engineering.

Sample that used in this paper is granite weathered soils distributing the cheong-ju district. For informationized construction, basic physical properties is determined and fracture effect is maintained constantly keeping compaction energy 95% over. For determination of soil parameters, isotropic consolidation test and CTC is carried out. In the analysis of foundation deformation, predict value of foundation deformation is verified comparing numerical analysis with model loading test by centrifuge testing.

2. Constitution model of multiple intersecting surface by Lade

2.1 Summary of model

For the purpose of modelling the stress-strain behavior of soils by elasto-plastic theory, the total strain increment ($d\varepsilon_{ij}$), and divide into an elastic component ($d\varepsilon^e_{ij}$) a plastic collapse component($d\varepsilon^c_{ij}$) and a plastic expansive component($d\varepsilon^p_{ij}$). These strain component are calculated separately, the elastic strain by Hook's law and the plastic collapse strain and the plastic expansive strain by plastic stress-strain theories that involve, respectively, a cap-type yield surface and a conical yield surface with apex at the origin of the stress space.[5,6,7,8)]

2.2 Elastic strains

The elastic strain increments with are recoverable upon unloading, are calculated from Hook's law, using the unloading-reloading modulus defined ad:

$$Eur = kur \cdot P_a \cdot (\sigma_c/P_a)^n \tag{2.1}$$

In equation (2.1) pais atmospheric pressure expressed in the same units as *Eur* and σ_3 *kur, n* are soil parameters.

Also the value of Poisson's ratio has been found for the elastic parts of unloading reloading in triaxial compression test the value is used in the following equation

$$\nu = \frac{1}{2}\left(1 - \frac{\Delta\varepsilon_v}{\Delta\varepsilon_1}\right) \tag{2.2}$$

2.3 Plastic expansive strains

The yield surface for the plastic expansive strains is equaled the following equation with the plastic work.

$$f_p = f_p'(\sigma) - f_p''(W_p) \tag{2.3}$$

$$\text{where, } f_p' = \left(\frac{I_1^3}{I_3} - 27\right) \cdot \left(\frac{P_a}{I_1}\right)^m \tag{2.4}$$

$$f_p' = \eta_1 \text{ (in failure)} \tag{2.5}$$

$$\text{where, } I_1 = \sigma_x + \sigma_y + \sigma_z \tag{2.6}$$

$$I_3 = \sigma_x \cdot \sigma_y \cdot \sigma_z + \tau_{xy} \cdot \tau_{yz} \cdot \tau_{zx} + \tau_{yx} \cdot \tau_{zy} \cdot \tau_{xz} - (\sigma_x \cdot \tau_{yz} \cdot \tau_{zy} + \sigma_y \cdot \tau_{zx} \cdot \tau_{xz} + \sigma_z \cdot \tau_{xy} \cdot \tau_{yx}) \tag{2.7}$$

η_1 and m are soil parameters that by determined.

The plastic expansive strains is determined for soil using non-associative flow rule and the plastic potential function is modelled on the yield function as follows.

$$g_p = I_1^3 - \left(27 + \eta_2 \cdot \left(\frac{P_a}{I_1}\right)^m\right) \cdot I_3 \tag{2.8}$$

where η_2 constant for given values of *fp* and the confining pressure. The work-hardening and work-softening low employed with the concial yield surface is expressed as

$$f_p'' = a \cdot e^{-b \cdot W_p}\left(\frac{W_p}{P_a}\right)^{\frac{1}{q}} \qquad q>0 \tag{2.9}$$

where, a, b, c are constants for given value of the confining pressure.

2.4 Plastic collapse strains

The yield for the plastic collapse strains with the plastic work is expressed as

$$f_c = f_c'(\sigma) - f_c''(W_c) \tag{2.10}$$

$$\text{where, } f_c' = I_1^2 + 2I_2 \tag{2.11}$$

$$\text{where, } I_2 = \tau_{xy} \cdot \tau_{yx} + \tau_{yx} \cdot \tau_{zy} + \tau_{zx} \cdot \tau_{xz} - (\sigma_x \cdot \sigma_y + \sigma_y \cdot \sigma_z + \sigma_z \cdot \sigma_x) \tag{2.12}$$

The plastic potential function is identical to the yield function ;

$$g_c = I_1^2 + 2I_2 \tag{2.13}$$

Thus, the normality condition used for this surface implies an associative flow rule, The work-hardening law used with the spherical yield surface is expressed as;

$$f_c'' = P_a^2 \cdot \left(\frac{1}{C}\right)^{\frac{1}{p}} \cdot \left(\frac{W_c}{P_a}\right)^{\frac{1}{p}} \tag{2.14}$$

where, C and P are constants.

3. Experimental results for decomposed granite soil

3.1 physical tests

To abtained basic physical properties for decomposed granite soil in Cheong-ju district, it is carried out EDS analysis, grain size distribution, compaction test, specific gravity, consolidation tests, permeability test. Sample for decomposed granite soil that used tests is gathered on condition that disturbed at the first weathering zone that the outer layer is eliminated. Water content in site is 9.23% and shape had on rock structure and grain is sand, some is clayey. Sample is made by A-1 compaction method with compaction energy over 95%. Basic physical properties is as follows.

Table 3.1 Physical Properties of Decomposed Granite Soil

Specific Gravity	Water Content in Site(%)	Plastic Index (PI)	USCS	γ_{dmax} (t/m^3)	OMC (%)	K
2.661	9.23	NP	SM	1.97	16	9.62 x 10^{-5} cm/sec

Fig. 3.1 is grain size distribution curve, passing quantity of 0.074 sieve is 18%, Cu is 31.5 and Cc is 2.57.

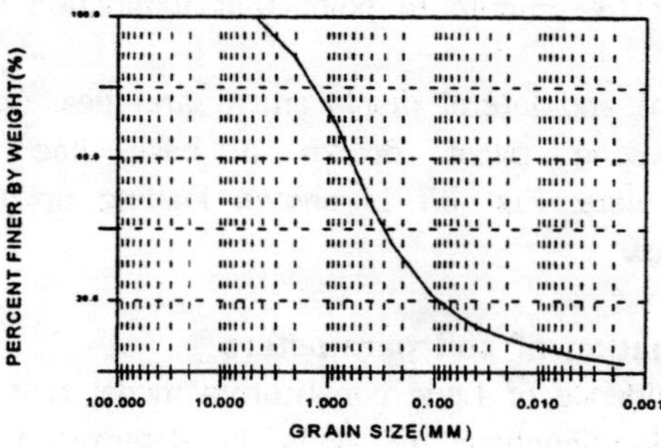

Fig. 3.1 Grain Size Distribution of Decomposed Granite Soil

Table 3.2 shows the mineral analysis of compaction granite soil.

Table 3.2 EDS Analysis

Element	Weight(%)	Atomic(%)*	Precision 2sigma	K-Ratio**
Na	0.00	0.00	0.00	0.0000
Al	20.32	22.79	0.13	0.2191
Si	60.43	65.12	0.20	0.5528
K	6.56	5.08	0.08	0.0710
Ca	1.18	0.89	0.04	0.0135
Ti	0.71	0.45	0.04	0.0083
Fe	7.91	4.29	0.11	0.0998
Cu	2.89	1.38	0.10	0.0354

Total : 100

* : Atomic percent is normalized to 100

** : k-ratio = k-ratio x R

where, R = reference(standard)/reference(sample)

3.2 Isotropic consolidation test

Sample made by A-1 compaction method with compaction energy over 95%, is applied according as confing stress(σ_3) is 0.5, 1, 2, 3, 4kg/cm^2. And then it is removed confining stress with interval of 24 hours respectively. Fig. 3.2 shows result of isotropic consolidation test with volume change of conventional triaxial specimen.

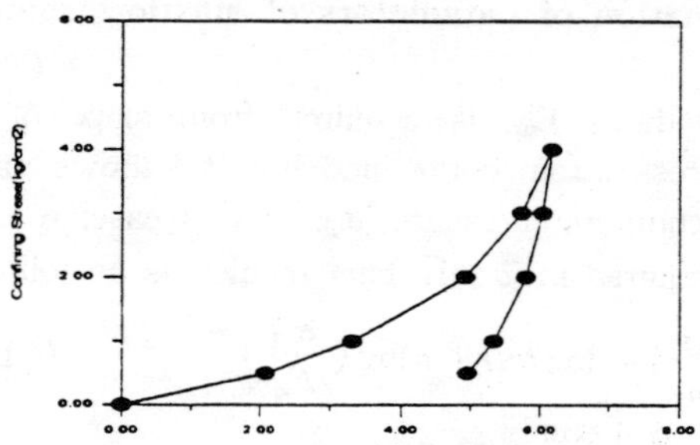

Fig. 3.2 Isotropic Consolidation Test with Volume Change of Conventional Triaxial Specimen

3.3 Triaxial compression test

Sample used in conventional triaxial compression test is that diameter is 50mm and height is 123mm. Since it was saturated for 72 hours, then isotropic consolidated for 24 hours, undrained shear tests is carried out by 0.4mm/min through condition of stain control of 0.5, 1, 2, 3kg/cm^2. Fig. 3.3 is results of conventional triaxial compression test and shows relationship of axial stain vs deviator stress, and Fig. 3.4 shows that normalized each undrained shear tests.

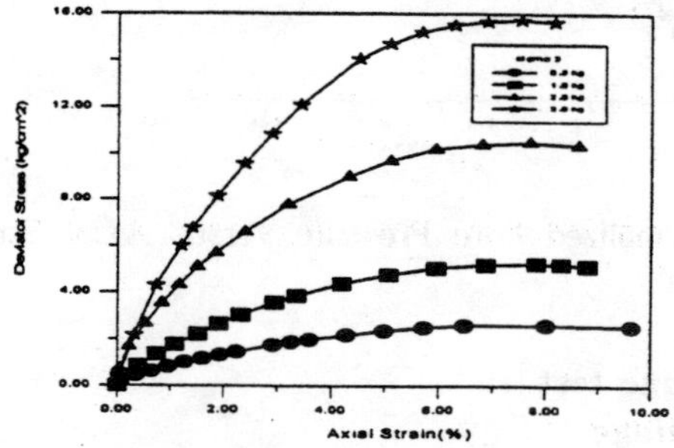

Fig. 3.3 Undrained Compression Tests with Deviator Stress versus Axial Strain for CTC

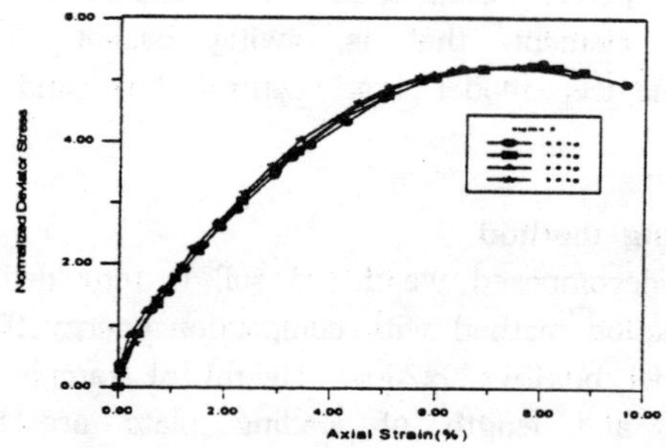

Fig. 3.4 Normalized Deviator Stress versus Axial Strain of CTC

Fig. 3.5 is results of CTC that is axial stain vs pore pressure and Fig. 3.6 is normalized each pore pressure.

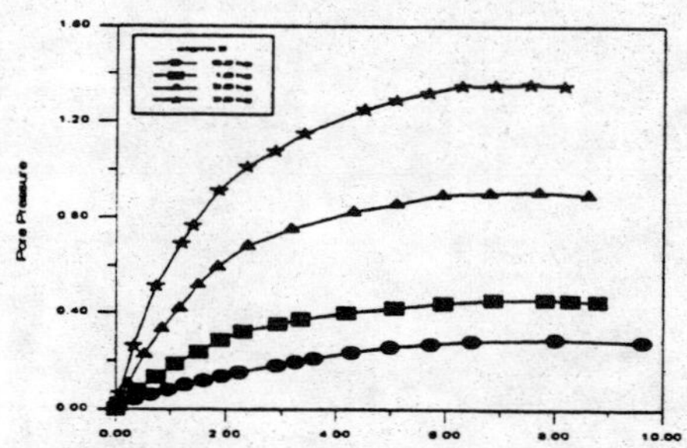

Fig. 3.5 Undrained Compression Tests with Pore Pressure versus Axial Strain for CTC

As result of CTC, it was found out that axial strain is existed the interval of 3~6% along confining pressure in failure. Also degree of manifestation of pore pressure is

50% of confining pressure, axial stress and pore pressure normalized are shown that curve is uniform generally.

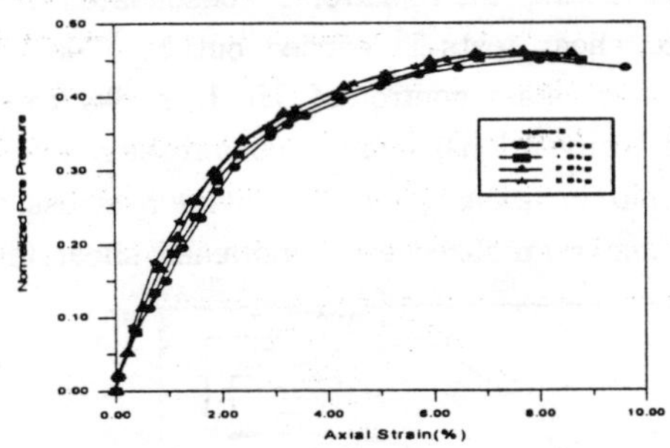

Fig. 3.6 Normalized Pore Pressure versus Axial Strain for CTC

3.4 Centrifuge test

3.4.1 Centrifuge

Basic principle of centrifuge test apparatus is to revive stress condition that induced in virtual foundation to manifest centrifugal force though rotation on uniform velocity property. Centrifuge test apparatus has 5 composition element; that is, swing basket, slip ring, recording device, model box, control box and loading system.

3.4.2 Testing method

Sample decomposed weathered soil is remolded by the A-1 compaction method with compaction energy 95% over in the model box(51x30x20㎤). Height of sample is 20㎝ and width and length of loading plate are 8, 20㎝, respectively. And the loading apparatus is settled at the center of upper. Testing method is that since swing velocity is raised from 1g to 25g-level, then loaded velocity(0.4mm/min) through condition of strain control. It

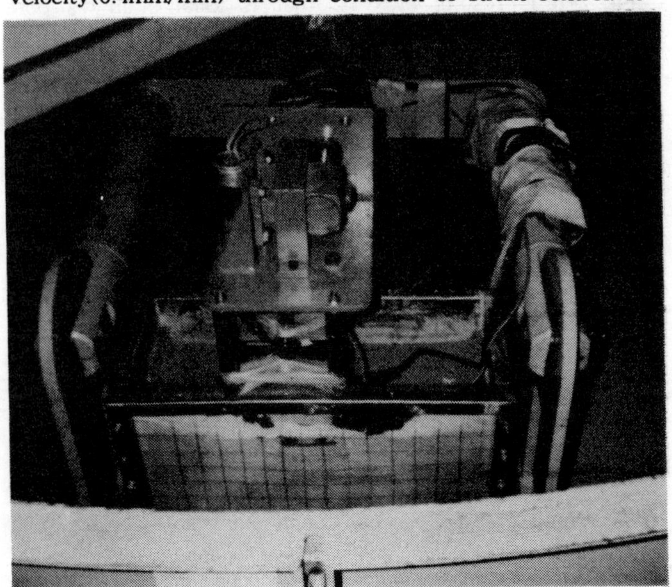

Fig. 3.7 Photo of load system for centrifuge test

is tested for 180 minute to point that foundation stress is declined.

Settlement and lateral displacement are measured using chinese drawing paper drawn a base line on the transparent plate. Fig. 3.7 is shown loading apparatus in the model box.

4. Determination of soil parameters

The confidence of Lade constitutive model that selected for foundation analysis of soils is determined on the predict capacity of stress-strain theories. As this predict capacity is determined on accuracy of parameters for constitutive model, computer program used linear and non-linear regression analysis is used for procedure of determination of parameters from the results of laboratory test. Fig. 4.1 show flow chart for procedure of determination.

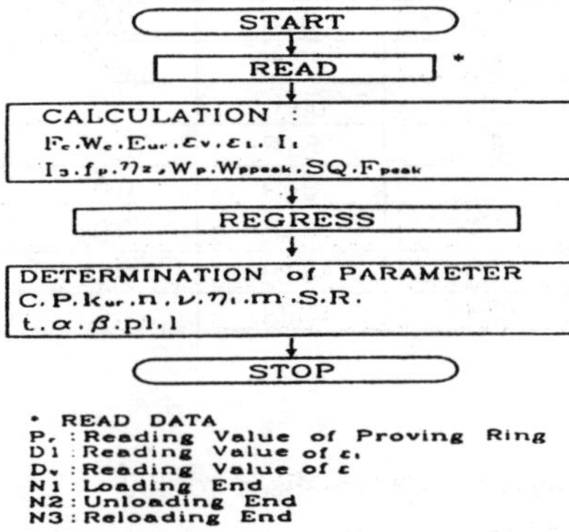

Fig. 4.1 Flow chart

For the determination of parameters, isotropic consolidation test and CTC is performed(confining pressure is 0.5, 1, 2, 3kg/㎠, respectively). It is determined that elastic stains is K_{ur}, n, ν, plastic collapse strains is c, p, plastic expansive strains is R, S, t, α, β, pl, l and failure criterion is η_1, m.

4.1 Determination of parameters of elastic strains, K_{ur}, n and ν

Elastic modulus E_{ur}, is acquired from slope of initial loading in stress-strain curve, and Fig. 4.2 shows variation of E_{ur} with confining pressure σ_3. In this case, n is slope and K_{ur} is acquired in σ_3=1. This relation is as following.

$$\log\left(\frac{Eur}{P_a}\right) = \log kur + n\log\left(\frac{\sigma_3}{P_a}\right) \tag{4.1}$$

양변에 log를 제거하면,

$$Eur = kur \cdot P_a \cdot \left(\frac{\sigma_3}{P_a}\right)^n \tag{4.2}$$

K_{ur} is 285 and n is 0.967, as shown as Fig. 4.2. Poisson's ratio ν not related with σ_3. And because ν is

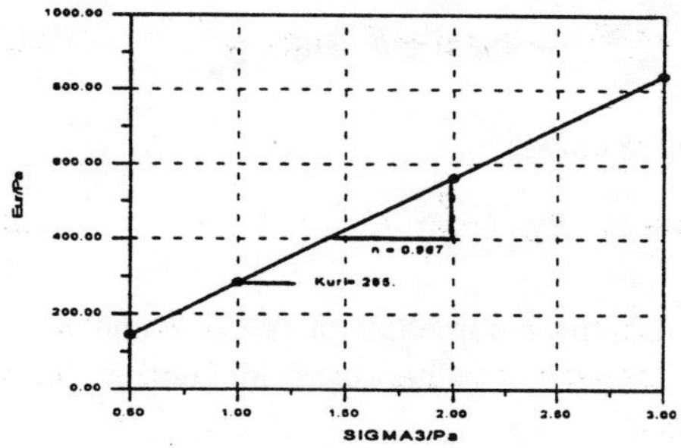

Fig. 4.2 Variation of elastic modulus with confining pressure for granite soil

constant, mean poisson's ratio is 0.25 from result of triaxial test.

4.2 Determination of parameters c and p, plastic collapse strains

plastic collapse work is determined by using results of isotropic compression test.

$$W_c = \int \sigma_3 \cdot d\varepsilon_v^c \qquad (4.3)$$

Where, $d\varepsilon_v^c$ is determined by using strains abtained from isotropic compression and f_c for isotropic compression is determined from the following equation.

$$f_c = 3\sigma_3^2 \qquad (4.4)$$

Fig. 4.3 shows relationship of Wc and fc, and the equation is as following.

$$\log\left(\frac{W_c}{P_a}\right) = \log c + p \cdot \log\left(\frac{f_c}{P_a^2}\right) \qquad (4.5)$$

If log is removed,

$$W_c = c \cdot P_a\left(\frac{f_c}{P_a^2}\right)^p \qquad (4.6)$$

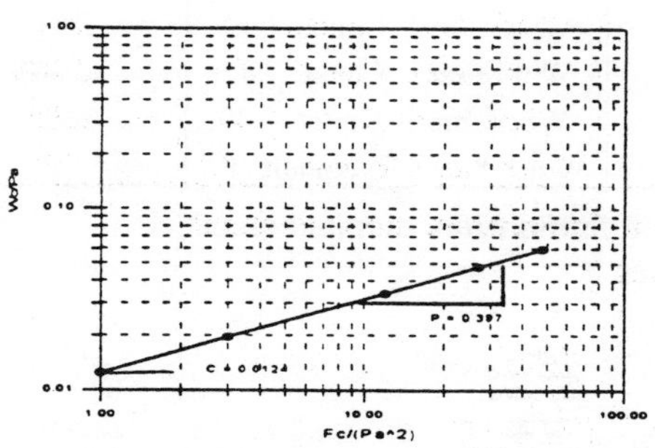

Fig. 4.3 Relation between plastic collapse work W_c and the value of f_c

c is an intercept value $\left(\frac{f_c}{P_a^2}\right) = 1$, p is slope, as shown as Fig. 4.3. Where, c=0.0124 and p=0.379.

4.3 Determination of parameters for plastic expansion stains

4.3.1 Determination of η_1 and m

Fig. 4.4 shows variation of $\frac{I_1^3}{I_3} - 27$ with $\frac{P_a}{I_1}$ calculated by using stress in failure from results of undrained triaxial compression test.

In Fig. 4.4, the variation of $\frac{I_1^3}{I_3} - 27$ with $\frac{P_a}{I_1}$ is a straight line and is as following.

$$\log \eta_1 = \log(I_1^3/I_3 - 27) + m \cdot \log(P_a/I_1) \qquad (4.7)$$

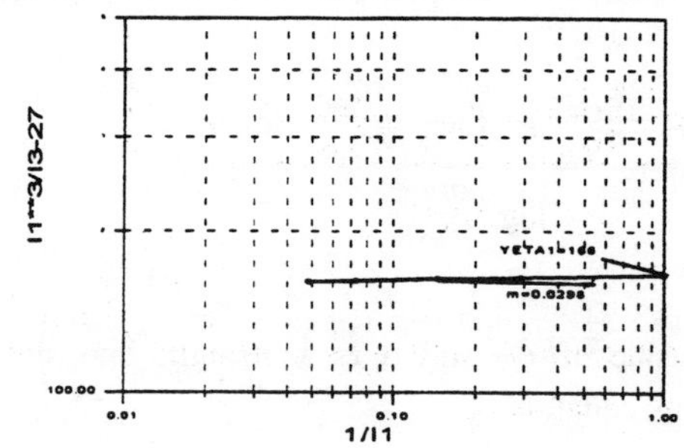

Fig. 4.4 Determination of the value of η_1 and m in failure criterion

If log is removed,

$$\eta_1 = \left(\frac{I_1^3}{I_3} - 27\right) \cdot \left(\frac{P_a}{I_1}\right)^m \qquad (4.8)$$

Where, $\eta_1 = \frac{I_1^3}{I_3} - 27$ in $\frac{P_a}{I_1} = 1$, m is slope of this line.

Determined η_1 is 166 and m is 0.0298.

4.3.2 Determination of S, R, t (η_1-f_p)

Fig. 4.5 shows variation of f_p with η_2, using undrained triaxial compression test.

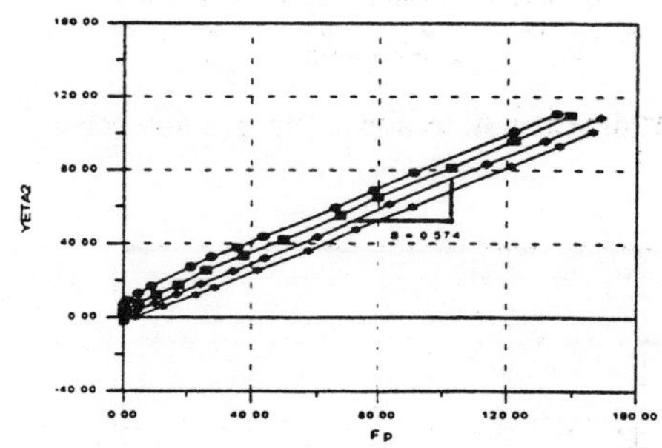

Fig. 4.5 Variation of η_2 with f_p for granite soil

For η_2, the equation is as follows.

$$\eta_2 = \frac{3 \cdot (1+\nu^p) \cdot I_1^2 - 27 \cdot \sigma_3(\sigma_1 + \nu^p \cdot \sigma_3)}{\left(\frac{p_a}{I_1}\right)^m \cdot \left[\sigma_3(\sigma_1 + \nu^p \cdot \sigma_3) - \left(\frac{I_3}{I_1}\right) \cdot m \cdot (1+\nu^p)\right]} \qquad (4.9)$$

Variation of η_2 with f_p shows straight line and relationship of η_2, f_p and σ_3 is as follows.

$$\eta_2 = S \cdot f_p \cdot R\sqrt{\frac{\sigma_3}{P_a}} + t \qquad (4.10)$$

Where, S (=0.574) is slope of straight line. R (=-0.237) and t (=5.8) is slope and 절편, respectively, as shown as Fig. 4.6..

4.3.3 Determination of α, β, pl, l (f_p-W_p)

Fig. 4.7 shows relationship of q and σ_3 from results of undrained triaxial compression test. q is presented as following.

$$q = \frac{\log\left(\frac{W_{ppeak}}{W_{p60}}\right) - \left(1 - \frac{W_{p60}}{W_{ppeak}}\right)\log e}{\log\left(\frac{\eta_1}{f_{p60}}\right)} \qquad (4.11)$$

Relationship of σ_3 and q is a straight line and present as following equation.

$$q = \alpha + \beta \cdot \left(\frac{\sigma_3}{P_a}\right) \qquad (4.12)$$

Where, α=3.82 and β= -0.349.

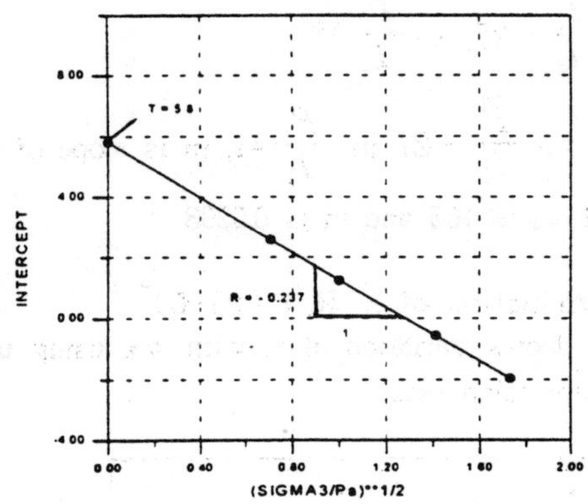

Fig. 4.6 Determination of R and t for granite soil

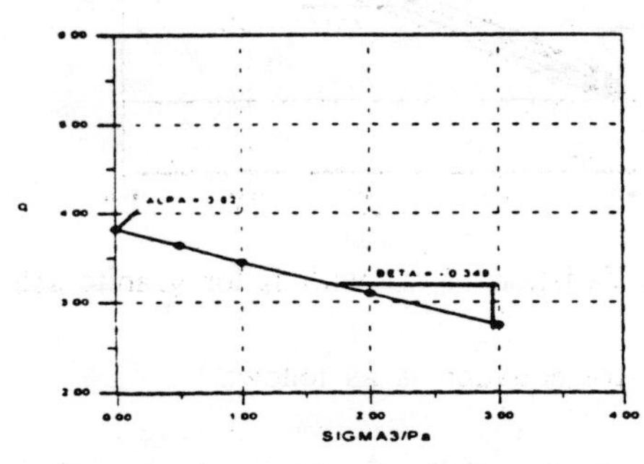

Fig. 4.7 Variation of q with confining pressure σ_3

Fig. 4.8 shows relationship of Wppeak (plastic expansion day) and σ_3 (confining pressure) in failure.

$$\log\left(\frac{W_{ppeak}}{P_a}\right) = \log pl + l \cdot \log\left(\frac{\sigma_3}{P_a}\right) \qquad (4.13)$$

If log is removed,

$$W_{ppeak} = pl \cdot P_a \cdot \left(\frac{\sigma_3}{P_a}\right)^l \qquad (4.14)$$

In Fig. 4.8, the parameters pl is 0.328 and L is 0.2. The number of parameter values used in Lade is 14, as shown as Table 4.1.

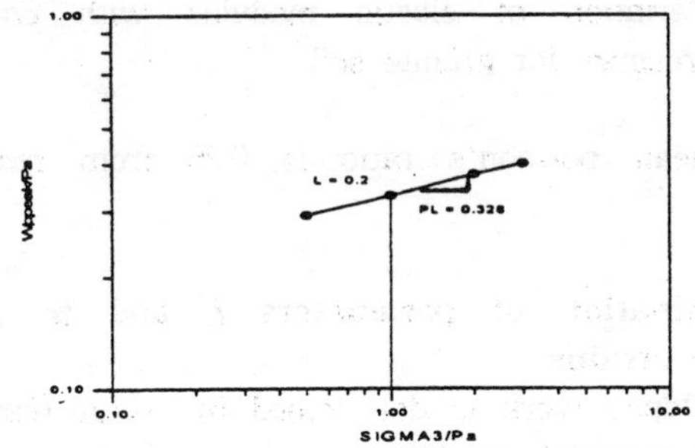

Fig. 4.8 Variation of W_{ppeak} with confining pressure σ_3

Table 4.1 Summary of determined parameter values

Strain Component	Soil Parameter	Value
Elastic	Modulus No. (kur)	285
	Exponent (n)	0.967
	Poisson's Ratio (ν)	0.25
Plastic Collapse	Collapse Modulus (c)	0.0124
	Collapse Const. (p)	0.397
Failure Criterion	Yield Const. (η_1)	166
	Yield Exponent (m)	0.0298
Plastic Expansive	pl.Potent Const. (R)	-0.237
	pl.Ponent Const. (S)	0.574
	pl.Potent Const. (t)	5.8
	Work-hard. Const. (α)	3.82
	Work-hard. Const. (β)	-0.349
	Work-hard. Const. (pl)	0.328
	Work-hard. Exponent. (l)	0.2

5. Analysis of foundation displacement

5.1 Flow chart

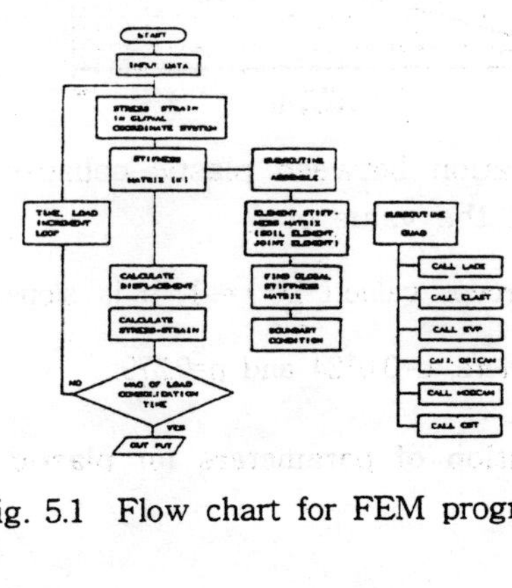

Fig. 5.1 Flow chart for FEM program

5.2 FEM and results of test

5.2.1 Imput data

For analysis of behavior in elasto-plastic foundation, it is carried out centrifuge test, then it is analyzed with FEM through following boundary condition of model box.

1) Fig. 5.2 is FEM analysis model for foundation in model box. The number of element and node is 56, 72, respectively.
2) Boundary condition of foundation is that upper layer is free and both sides is fixed, drain condition is that upper layer is drain and lower layer is undrain.
3) The constitutive equation used in numerical analysis is applied Lade's constitute model for total foundation.
4) Soil foundation is regarded as K_0 consolidation layer.

5.2.2 Consideration of numerical analysis results

Soil parameters, required in Lade's constitute model with multiple intersecting yield surface, is determined by conventional triaxial compression test. And Using the result, it is compared result of numerical analysis with centrifuge test for undrained behavior.

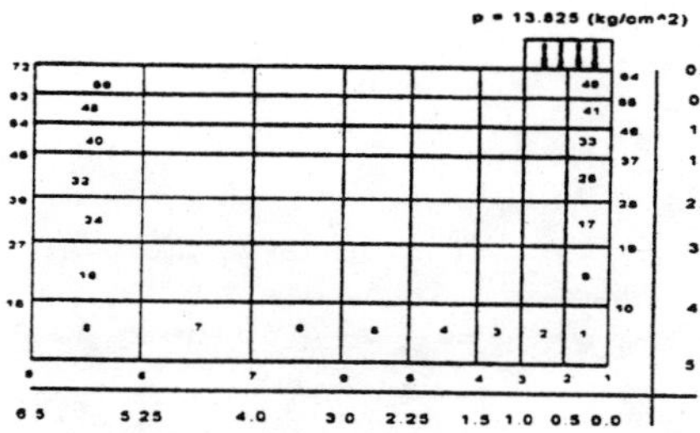

Fig. 5.2 FEM Grid

1) The ultimate bearing capacity

The ultimate bearing capacity of both value of measurement and predict are 13.83kg/cm^2. Fig. 5.3 presents the condition between local shear failure and general shear failure. The measurement value presents similarity as the predict value.

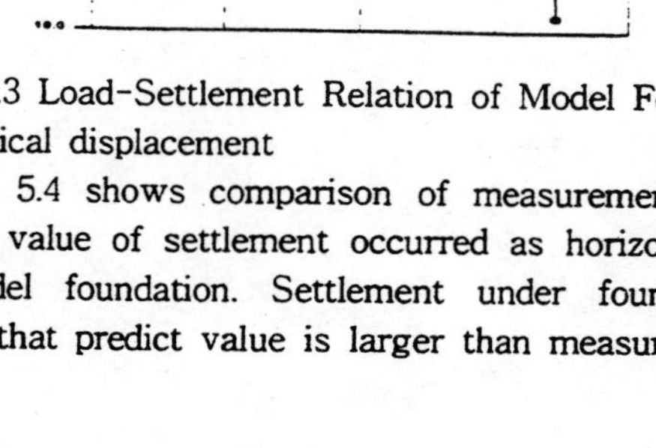

Fig. 5.3 Load-Settlement Relation of Model Foundation

2) Vertical displacement

Fig. 5.4 shows comparison of measurement value and predict value of settlement occurred as horizontal distance in model foundation. Settlement under foundation plate shows that predict value is larger than measurement value, measurement value is horizontal and predict value is curve, in maximum settlement. it is because is calculated each joint in FEM analysis.

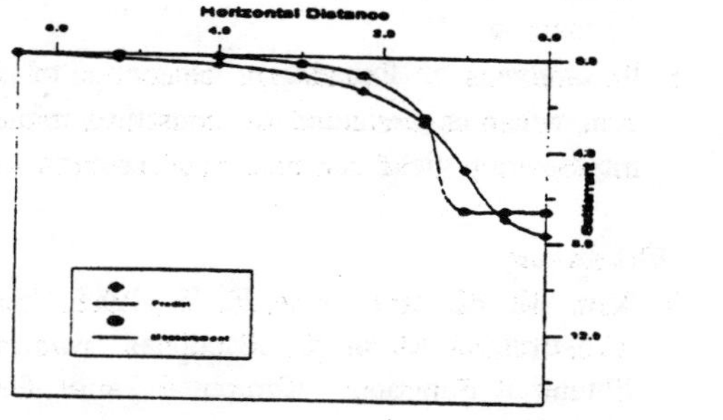

Fig. 5.4 Vertical Displacement for Model Foundation

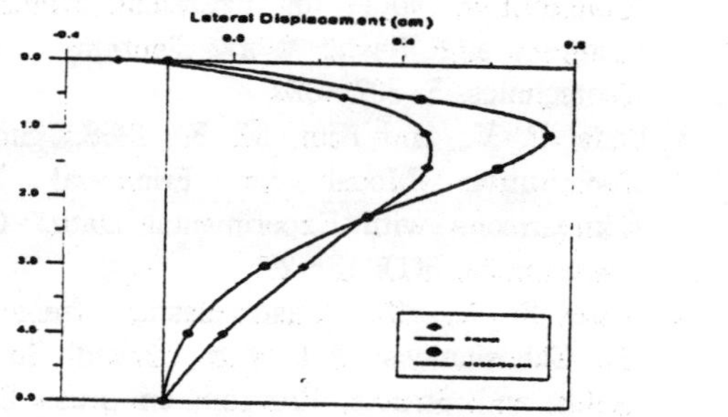

Fig. 5.5 Lateral Displacement at the Edge of Model Foundation

3) Lateral Displacement

Fig. 5.5 shows, when failure load is 13.83kg/㎠, lateral displacement in the point 1m away from the center of loading plate. Measurement value and maximum lateral displacement appear in 1B depth of width of loading plate. And measurement value is somewhat large. General displacement(measurement displacement and predict displacement) enters upon a similar aspect.

6. Conclusion

The sample used in this paper is granite weathered soils distributing the cheong-ju district. Basic physical properties is determined by many test. And for application of constitute model with multiple intersecting yield surface, it keep constant compaction energy and then determined soil parameter(14) needed in constitutive model by isotropic consolidation test and CTC. The centrifuge test was tested for behaviour analysis of foundation and acquired the conclusion as following.

1. From conventional triaxial compression test, vertical strains in failure increases and presents 3~6% as confining pressure increases.
2. For compaction granite weathered soils, η_2 is 166 and C, R, β, l unacquired in clay is 0.0124, -0.237, -0.349, 0.2, respectively.
3. Failure of foundation generates in 13.83kg/㎠ and the failure is influenced to 2B depth under loading plate. so

it shows that the limit of plastic region extends to 2B depth under loading plate.

4. In model loading test, settlement is that predict value larger than measurement value but lateral displacement is reverse.
5. In analysis of foundation behaviour of deposed granite soil, when is predicted by constitute model with multiple intersecting yield surface, measurement value is good.

Reference

1. Kim. M. K., and Lade, P. V., 1988, "Single Hardening Constitutive Model for Frictional Materials, I, Plastic Potential Function." Computers and Geotechnics, 5(4), 307~324.
2. Lade. P. V., and Kim, M. K., 1988, "Single Hardening Constitutive Model for Frictional Materials, II, Yield Criterion and Plastic Work Contours " Computers and Geotechnics, 5, 307~324.
3. Lade. P. V., and Kim, M. K., 1988, "Single Hardening Constitutive Model for Frictional Materials, III, Comparisons with Experimental Data." Computers and Geotechnics, 6(1), 13~29.
4. Lade. P. V., 1977,"Elasto-plastic Stress-Strain Theory for Cohesionless Soil with Curved Surfaces". Int. J. Solids and Structs, Program on press Inc. New York, Vol.13, pp.311~323.
5. Lade. P. V., 1978, "Prediction of Undrained Behaviour of Sand". Journal of the Geotechnical Engineering Division, ASCE, Vol. 104, No. GT6, pp.721~735.
6. Lade. P. V. and Duncan, J. M., 1973, "Cubical Triaxial Test on Cohesionless Soil", J. Soil Mech. Found. Div., ASCE, Vol. 99, No. SM10, Oct. pp. 793~812.
7. Lade. P. V., 1974, "Three-Dimensional Behavior and Parameter Evaluation of an Elasto-plastic Soil model" Geomechanical Modeling in Engineering.
8. Lade. P. V. and Nelson. R. B., 1984, "In Crementalization Procedure for Elasto-plastic Constitute Model with Multiple Intersecting Yield Surface", International Journal for Numerical and Analytical Methods in Geomechanics, Vol. 8. pp. 311~323.
9. Lade. P. V., "Stress-Strain Theory for Normally Consolidation Clay.
10. Lade. P. V., 1979, "Stress-Strain Theory for Normally Consolidated clay" Proc. of 3rd Int. Conference on Numerical method in Geomechanics.
11. Chistian. J. T., 1968, "Undrained Stress Distributions by Numerical Method", Journal of the Soil Mechanics and Foundations, Division, ASCE, Vol. 94, No. SM6, PP. 1333~1345.
12. Lade. P. V., 1979, "Stress-Strain Theory for Normally Consolidated Clay", Proc. of 3rd Int. Conference on Numerical Method in Geomechanics.
13. Lee. G. D., 1994, "A Numerical Analysis on the Deformation of Clayey Foundation Using Elasto-plastic Model", Chonnam Uni.

Proceedings of the Seventh (1997) International Offshore and Polar Engineering Conference
Honolulu, USA, May 25-30, 1997

ISBN 1-880653-28-1 (Set); ISBN 1-880653-29-X (Vol. I)

Comparison of Bathymmetry and Consolidation Settlement of Dredge Mound in Long Island Sound

H.G. Brandes
University of Hawaii
Honolulu, USA

A.J. Silva
University of Rhode Island
Narragansett, RI, USA

ABSTRACT

Self-weight consolidation of a capped dredge disposal mound in Central Long Island Sound is compared to periodic bathymmetric surveys. The properties of the dredged materials and the seabed were determined from a geotechnical characterization program conducted on samples obtained at various stages during mound construction and beyond. Predictions from two modeling approaches are in good overall agreement with the bathymmetry. This finding agrees with other monitoring studies that suggest that little or no erosion has taken place and that all the deposited sediments are well contained in their present state. Calculations indicate that nearly three years after construction of the mound well over 90% of primary consolidation has been completed.

KEY WORDS: Consolidation, dredging, modeling, bathymmetry, capping, mounds, coring.

INTRODUCTION

The Central Long Island Sound (CLIS) disposal site, located some 5.6 miles south of New Haven in about 19 m of water, serves as one of New England's primary shallow water repositories of dredged spoils (Fig. 1). This site has received significant amounts of material from recent dredging operations of harbors and navigation channels in Long Island Sound, much of which is unacceptably-contaminated (UC). Upon placement at the disposal site, the spoils are typically covered with an uncontaminated cap to isolate them from the overlying water column. In 1977 the New England Division (NED) of the US Army Corps of Engineers initiated the Disposal Area Monitoring System (DAMOS) program to monitor the long-term physical, chemical and biological effects of capped dredge disposal mounds in shallow water. Of particular interest to this study is assessing the physical integrity of the UC dredge and cap system over time by characterizing the material geotechnically, estimating mound consolidation, and correlating the predictions with the results from periodic bathymmetric surveys. Preliminary bathymmetric monitoring and calculations of nearby mounds had indicated substantial settlement (Silva et al., 1994a; Uchytil, 1992; Brandes et al., 1991), although the available data base was not sufficient to warrant firm conclusions on the physical integrity of these mounds.

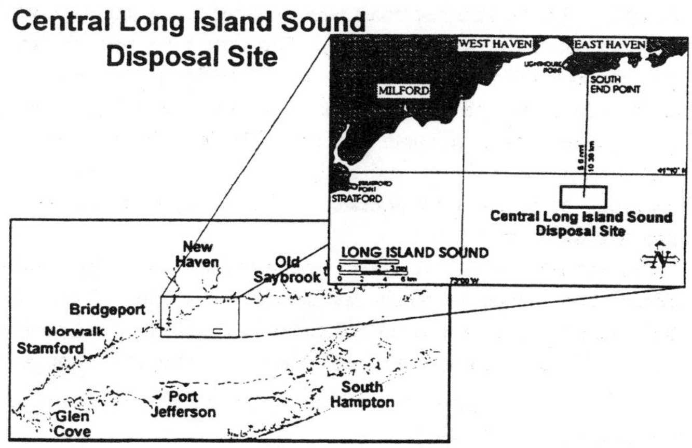

Fig. 1 CLIS Disposal Site (Figure courtesy of SAIC).

We focus on a recently completed (1993-94) dredge mound located at the center of a ring of previous mounds (Fig. 2) near the northeast corner of the CLIS site. The material was dredged from the main New Haven Harbor navigation channel, with smaller amounts coming from nearby marinas and other permit projects (Fredette, 1994). The bulk of the UC dredge material came from the upper reaches of New Haven Harbor (590,200 m^3), whereas the cap (569,300 m^3) consisted of sediments dredged from the outer harbor and the other smaller projects. Construction of the mound lasted from October 1993 through January of 1994. Careful planning of this project allowed for periodic high resolution bathymmetric surveys and detailed geotechnical sampling, providing a unique opportunity to track and model settlement of the sediments from a time prior to initiation of material placement, through the various stages of mound construction, and beyond.

FIELD SAMPLING AND LABORATORY TESTING

Geotechnical field sampling (Silva et al., 1994b) included obtaining large-diameter (10.2 cm) gravity cores of the base, UC dredge, and cap materials at five different stages of the project: September 21, 1993 (pre-disposal, 6 cores), November 10, 1993 (mid-disposal, 7 cores), March 15, 1994 (upon disposal completion, 7 cores), July 18, 1994 (4 months post-disposal, 7 cores), and August 28, 1995 (17 months post-disposal, 7 cores). In addition, 9 grab samples of the upper sediments were obtained between the September and November surveys, and another 9 grab samples between the November and March surveys. Bathymmetric surveys were conducted each time gravity cores were collected and during July of 1996. The undisturbed cores were obtained using the URI/MGL Large-diameter Gravity Corer (LGC). This system has been designed to minimize sediment disturbance while obtaining representative samples large enough for consolidation and strength testing. The 34 gravity cores ranged in length from 69 to 278 cm. All the cores were sealed with o-ring plugs and refrigerated until processing. The locations of the cores are indicated in Fig. 2, superimposed on the pre-disposal bathymmetry.

In the laboratory, the cores were processed using the Multi-Sensor Core Logger to obtain non-destructive profiles of sediment bulk density using a gamma ray source and scintillation counter system. Density was also measured directly on the opened cores with

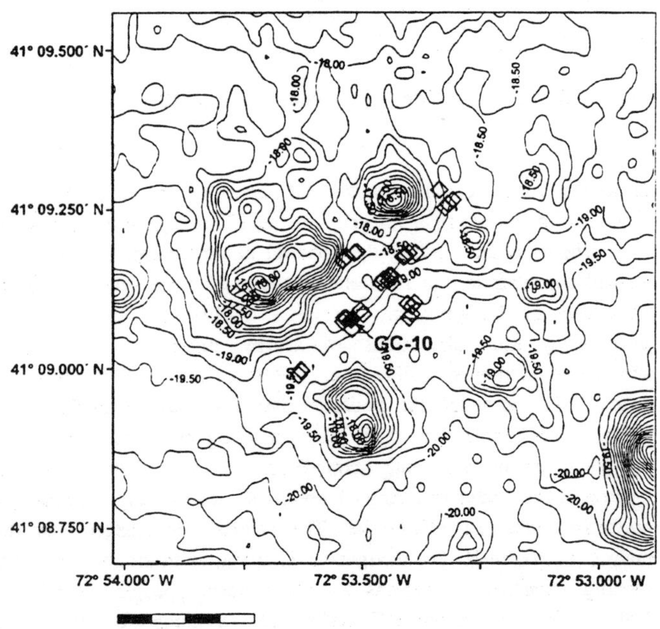

Fig. 2 Core locations and pre-disposal bathymmetry (Figure courtesy of SAIC).

a constant volume plug sampler, and calculated from the water contents assuming full saturation. Following logging, the cores were extruded vertically or split horizontally to examine the nature and distribution of the sediments, to make measurements of shear strength, and to take subsamples for further laboratory testing. Subsampling consisted of taking water content samples and constant-volume plug samples to determine the variations in density and water content throughout the mound sediments. Shear strength was measured using a motorized vane shear device outfitted with a sensitive torque transducer and a 2.54×2.54 cm blade. In addition, 78 thin-walled undisturbed subsamples were taken for consolidation testing.

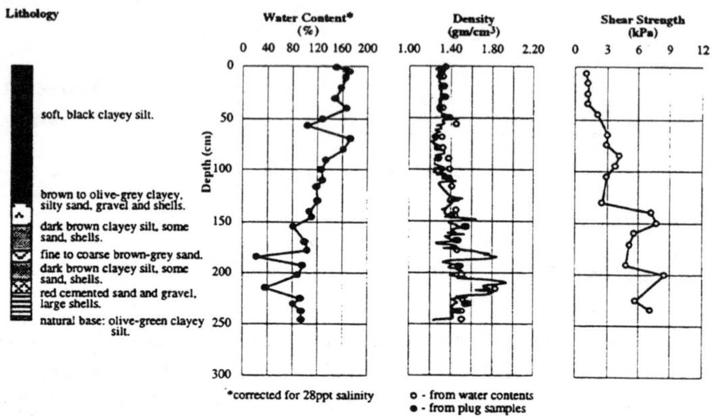

Fig 3 Profile of physical parameters - core GC-10.

A typical vertical sediment profile is shown in Fig. 3 for core GC-10, taken during the August 1995 survey (Fig. 2). The upper 130 cm of the core consist of a soft black clayey silt, of which approximately the top 50 cm constitute cap sediments. The similarity between inner and outer New Haven Harbor sediments makes distinguishing the interface between cap and UC sediments difficult. A decrease in water content with depth throughout this layer suggests that at least some consolidation has taken place during the first 17 months following completion of the mound. The depth interval from 130 cm to 222 cm shows a rapid succession of silt, sand and gravel layers with sharp changes in physical properties. It is possible that some of the lower sections of this interval are actually old dredge material from the nearby mounds. The uniform olive-green clayey silt below 222 cm constitutes the natural seabed. This clayey silt extends to a depth of 10 m and is followed by 25 m of permeable sand and gravel (Poindexter-Rollings, 1990).

Physical property tests were conducted in the laboratory to determine Atterberg limits, grain size distribution and specific gravity of the different sediments (Silva et al., 1994c). The bulk of the cap and the UC sediments throughout the mound have similar, relatively uniform properties, consisting of a black, soft, organic, clayey silt with high water contents and relatively low permeabilities. Typical plastic limits range between 50 and 60 and liquid limits between 120 and 130. The remainder of the UC and old dredge spoils have highly variable properties, ranging from fine clay to large gravel and cobble-size rocks. The natural base sediments are very uniform and consist of an olive-green clayey silt with traces of small shell fragments and sand. Water content was relatively uniform at near 100%, which is close to the average liquid limit of 99. The average plastic limit was 39.

NUMERICAL COMPUTATIONS

Two computer codes, CONSOL and PCDDF89, were used to predict the primary consolidation behavior of the newly deposited materials and the natural base sediments below (Silva et al., 1994d). The basic premise underlying the analysis is that changes in mound configuration are primarily due to void ratio changes as pore water is expelled during consolidation. Both programs assume one-dimensional conditions, i.e. deformations and fluid flow are restricted to the vertical direction. This is a reasonable simplification for semi-infinite, horizontally stratified deposits, and applies to the present case given that the new mound is some 200 times wider than it is high. The analysis also assumes that the materials are fully saturated at all times.

The program CONSOL solves for the total settlement and time rate of consolidation of sediment layers using a finite difference implementation of Terzaghi's one-dimensional consolidation equation:

$$\frac{\partial u}{\partial t} = c_v \frac{\partial^2 u}{\partial z^2} \qquad (1)$$

where:

u = excess pore water pressure
t = time
c_v = coefficient of consolidation
z = vertical space coordinate

Additional assumptions inherent in Terzaghi's approach are: a) the sediment particles and the pore water are incompressible, b) Darcy's law is valid, c) the coefficient of permeability (k) is constant, and d) the coefficient of compressibility (a_v) is constant. In theory these assumptions limit the usefulness of CONSOL to small-displacement, small-strain problems where the effects of material properties and geometric non-linearities can be neglected. However, a first order approximation can be achieved by using an average slope for the $e\text{-}log\sigma'$ relationship over the appropriate stress range. With this approach a reasonably good estimate of the compression of the mound and base materials can be made.

Whereas CONSOL was not specifically developed to solve for the problem of self-weight consolidation, the computer program PCDDF89, developed by the US Army Corps of Engineers, incorporates material and geometric non-linearities that are important for large-displacement analyses of thick deposits (Stark, 1991). The mathematical model used in PCDDF89 is based on the finite strain theory of consolidation developed by Cargill (1982). The governing equation, developed by Gibson et al. (1967), is given by:

$$\left(\frac{\gamma_s}{\gamma_w}-1\right)\frac{d}{de}\left[\frac{k(e)}{1+e}\right]\frac{\partial e}{\partial z}+\frac{\partial}{\partial z}\left[\frac{k(e)}{\gamma_w(1+e)}\frac{d\sigma'}{de}\frac{de}{dz}\right]+\frac{\partial e}{\partial t}=0 \qquad (2)$$

where:

γ_s = unit weight of solids
γ_w = unit weight of water
e = void ratio
$k(e)$ = coefficient of soil permeability
z = vertical material coordinate
σ' = effective stress
t = time

The above equation provides for the effects of: a) self-weight consolidation, b) permeability varying with void ratio, c) non-linear void ratio versus effective stress relationship, and d) large strain deformations. The solution technique employs an explicit finite difference scheme and, for input data, requires tables of void ratio versus permeability and void ratio versus stress (derived from laboratory testing), as well as initial and boundary conditions. Both CONSOL and PCDDF89 iterate within each time step until either convergence or a maximum number of iterations have been reached.

MATERIAL PARAMETERS

A settlement analysis requires characterizing the consolidation properties of the sediments, specifying initial stress and void ratio conditions, and describing the time-load history of the problem. The consolidation characteristics of the base, contaminated dredge, and cap materials are based on the results of six undisturbed standard Soiltest consolidation tests, two reconstituted large-diameter consolidation tests (conducted in a 10.2 cm Proctor mold), and seven undisturbed and reconstituted back-pressured, constant-rate-of-deformation consolidation (CRD) tests. Reconstituted tests were conducted to simulate consolidation of the initially disturbed dredge materials, which presumably underwent significant remolding and mixing with water during dredging, transportation, and placement at the disposal site.

The program CONSOL uses a constant, the compression index C_c (slope of effective stress versus log of void ratio relationship), to characterize the compression behavior, whereas the PCDDF89 program requires a tabulated description of the non-linear void ratio versus stress relationship over the range of operating stresses. The time-dependent behavior is characterized by a constant value of the coefficient of consolidation c_v in the case of CONSOL, and the full void ratio versus permeability relationship in the case of PCDDF89.

Compressibility results for UC dredge sediments are shown in Fig. 4. Included are the results from three reconstituted tests and one undisturbed one. The interpreted trendline, used as input for PCDDF89, favors the results from the large-diameter reconstituted test in the low stress range since it is believed that this type of test simulates the early in situ consolidation behavior the best. The trendline reflects the non-linearity typical at low stresses during

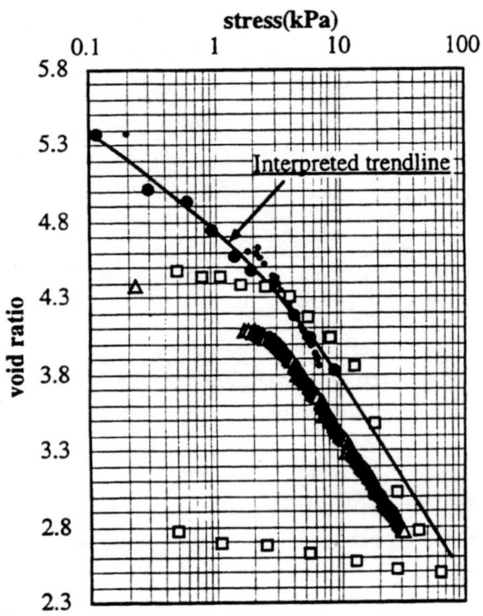

• Reconstituted, remolded CRDtest, center

● Reconstituted, remolded large-diameter test, 100m S

Δ Reconstituted, remolded CRD test, 200m E

□ 2.0" undisturbed, conventional test, 26-30cm depth, near core GC-10

Fig 4 Compressibility - UC dredge material.

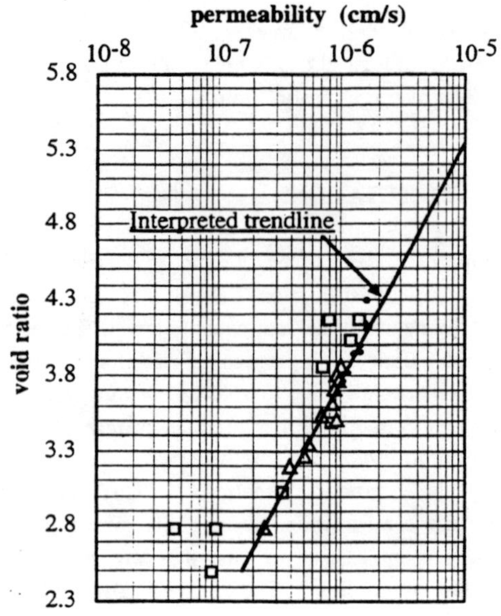

• Reconstituted, remolded CRDtest, center

Δ Reconstituted, remolded CRD test, 200m E

□ 2.0" undisturbed, conventional test, 26-30cm depth, near core GC-10

Fig 5 Permeability - UC dredge material.

virgin compression. Since the self-weight field stresses are typically small, an average C_c of 0.80 was selected from the low stress range in Fig. 4 as input for CONSOL. Permeability results for the UC dredge material shown in Fig. 5 clearly indicate that permeability is not a constant but decreases with void ratio as the material densifies during consolidation. The interpreted trendline is extrapolated to higher void ratios to characterize the early consolidation behavior. The coefficient of consolidation, determined from the time-dependent data at various applied stress stages during the consolidation tests, indicates that c_v decreases somewhat with void ratio (Fig. 6). As input for CONSOL we assume an average value of 0.45×10^{-3} cm²/s. Similar consolidation and permeability data was determined for the cap and the natural base sediments.

For the CONSOL program, initial stress conditions in the UC dredge and cap materials were assumed as being constant throughout, at 0.1 kPa. This value was chosen after careful consideration of the data from the cores taken during surveys prior to, during, and immediately after mound completion. However, none of these cores are thought to accurately represent the initial conditions, given the actual complex construction sequence and the fact that the recovered sediments had already experienced some consolidation by the time they were tested in the laboratory. The initial stress of 0.1 kPa reflects our best estimate based on the laboratory consolidation test results and the water content and density determinations from the recovered cores. By extrapolating the compressibility results in Figure 4 to 0.1 kPa, we can approximate the initial void ratio of the UC dredge material to be 5.4. Similarly, the initial void ratio of the cap is 4.4. The base material was assumed to be normally consolidated with a uniform density of 1.45 gm/cm³ (see Fig. 3). As part of the analysis, the PCDDF89 program evaluates the initial stress distributions under self-weight conditions. For PCDDF89, the specific gravities of the base material (2.66) and the dredge material (2.57) are also required as input parameters.

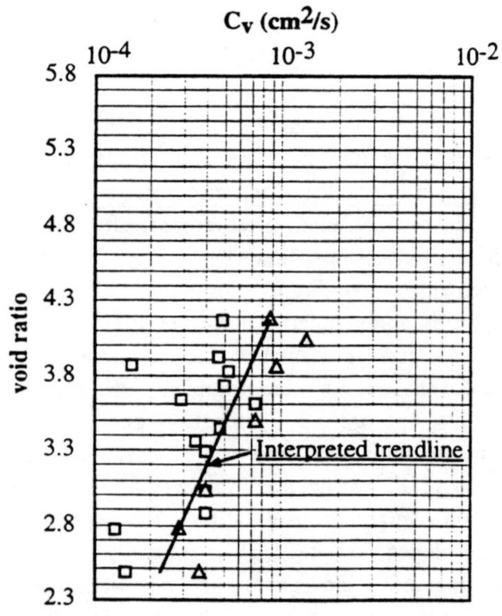

Δ Reconstituted, remolded CRD test, 200m E

□ 2.0" undisturbed, conventional test, 26-30cm depth, near core GC-10

Fig 6 Coefficient of consolidation - UC dredge material.

LOADING HISTORY AND MODELING APPROACH

Dredging of New Haven Harbor and disposal at the CLIS site occurred between the beginning of October 1993 and the end of January 1994 (Charles and Inglin, 1994). Most of the UC dredge material (460,100 m^3) came from the navigation channel in the inner harbor and was deposited during the month of October at the NHAV93 buoy, which marks the center of the new mound. A smaller amount of UC dredge material from the navigation channel and from five nearby marine terminals (130,100 m^3) was deposited southwest of the buoy during the months of November and December. Capping operations began in the middle of November and lasted through the end of January. The cap material (569,300 m^3) was spread more or less evenly across the new mound to achieve the stated goal of a 0.5 to 1.0 m cap (Fredette, 1994). Thus the new mound received a total of 1,159,500 m^3 of sediment.

The actual sequence of events is more complicated than that described above, which makes accurate modeling of the construction process difficult. As a first approximation, we assume that all dredge material was deposited at once, at the center of the new mound, on October 17 (t = 0 days). To estimate the height of the resulting mound, we further assume that it has a conical shape with a diameter of 590 m, estimated from the November 93 bathymmetric survey. Using these dimensions, the height of the apex becomes 5.0 m. Cap material was added during the months of November, December, and January at selected locations, chosen in an attempt to achieve uniform coverage of the UC dredge material. The thickness of the cap is estimated from the bathymmetric survey of March 94 and the volume of cap sediments to be an average 0.5 m. Since the cap thickness is only 10% of the UC dredge layer, we further assume that all the sediments forming the new mound were deposited simultaneously and instantaneously at t = 0. The calculations described are for the center of the mound (near the NHAV93 buoy) where we estimate the thicknesses of the foundation base, dredge and cap layers to be 10, 5 and 0.5 m respectively. Further, drainage takes place through the top of the mound and through the bottom of the base layer.

Although PCDDF89 was designed for multiple layered sediments, the available code crashed unexpectedly for the case when dissimilar materials were specified for the cap and the UC dredge layers. Results were only obtained for cases with a single dredge layer and the base. Since the cap cover was much thinner than the UC dredge layer, and the bulk of the sediment in both layers had similar properties, the PCDDF89 analysis was conducted using a single dredge thickness of 5.5 m having the material properties shown in Figs. 4 and 5. With these parameters PCDDF89 selects a starting void ratio of 5.4 and proceeds to calculate primary consolidation settlement in the dredged mound and the underlying base. The program CONSOL can handle dissimilar material properties but is not set up to solve the problem of self-weight consolidation directly. To simulate this process we specify a uniform initial void ratio of 5.4 (and a corresponding stress of 0.1 kPa) throughout the cap and UC dredge layers and a final stress corresponding to a normally consolidated state of stress resulting from an average unit weight of 14.2 kN/m^3 measured in the cores.

COMPARISON BETWEEN PREDICTIONS AND BATHYMMETRY

Bathymmetric surveys were conducted using Science Applications International Corporation's Integrated Navigation and Data Acquisition System. This system incorporates real-time navigation and positioning (±3 m) by triangulation with two shore stations along the Connecticut coast in the horizontal control of North American Datum of 1927 (Morris, 1996). Individual depths were recorded to an accuracy of 3.0 cm by a narrow beam system operating at 208 kHz. When individual errors from the positioning system, vertical motion of the survey vessel, changes in sound velocity through the water column, the slope of the bottom, and tidal corrections are accounted for, the system is able to reliably detect changes in water depth of 0.20 m or greater from survey to survey (Morris, 1996a). The bathymmetric data was analyzed using SAIC's Hydrographic Data Analysis System to construct the bathymmetric charts.

A comparison of settlement predictions from the two codes for the first 1500 days (4.1 years) is shown in Fig. 7. As expected, the rate of settlement decreases exponentially with time. However, the rate of settlement is somewhat different for the two formulations. PCDDF89 predicts a faster initial rate of settlement than CONSOL, but a smaller rate after approximately 2 years. PCDDF89 allows estimating the settlement in the UC dredge-cap and base layers

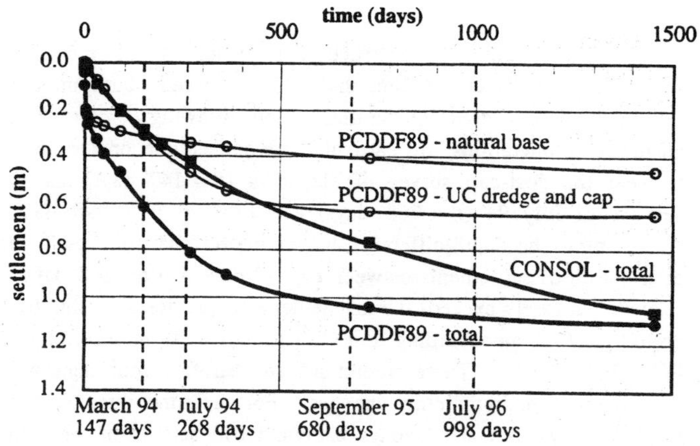

Fig 7 Numerical settlement predictions.

separately (Fig. 7), which indicates that after an initial period of about 150 days the majority of the settlement occurs in the 10 m base layer as opposed to the dredge mound. The total ultimate settlement that can be expected from primary consolidation is 1.70 m according to CONSOL and 1.17 m according to PCDDF89. These differences reflect a combination of contrasting formulations, input material properties, and modeling approaches. Although it is difficult to determine which of these factors has the most significant effect, it should be noted that PCDDF89 uses a more advanced, and presumably more accurate, formulation. According to PCDDF89 nearly 95% of the total ultimate settlement had occurred by the time of the July 96 survey (t = 998 days), whereas according to CONSOL only 54% had occurred at that time.

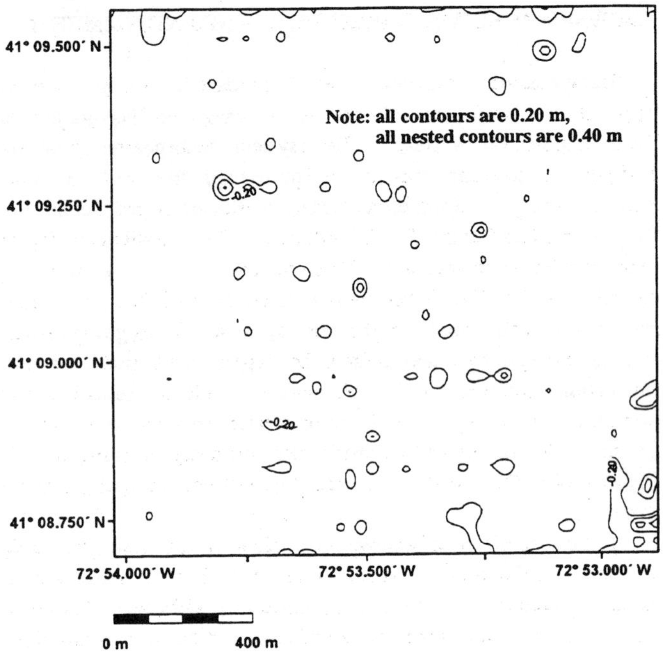

Fig 8 March 1994 to July 1994 bathymmetric changes (Figure courtesy of SAIC).

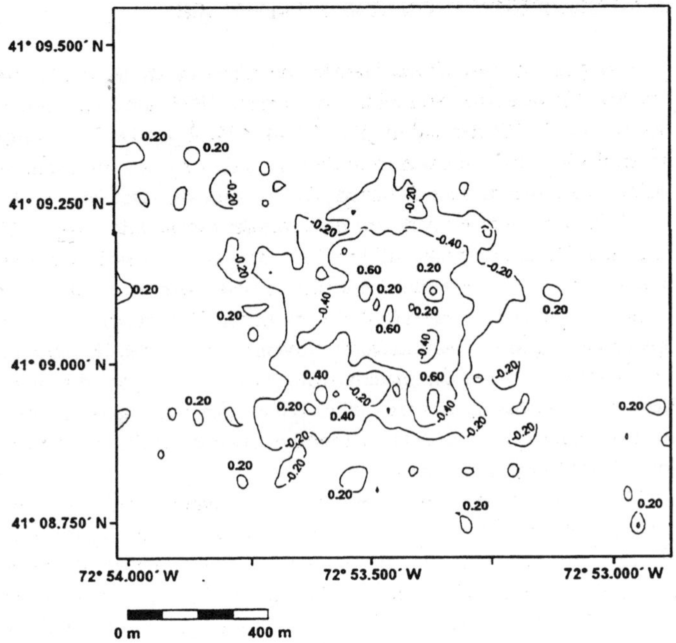

Fig 9 March 1994 to July 1996 bathymmetric changes (Figure courtesy of SAIC).

Given the complex construction sequence of the mound, a comparison between predictions and observed bathymmetric settlements is best done for the time period following completion of all dredging operations. In particular, we concentrate on the period between the post-cap survey of March 94 (t = 147 days) and the surveys of July 94 (t = 268 days) and July 96 (t = 998 days). According to the depth-difference bathymmetric chart in Fig. 8, the distribution of settlements between $t = 147$ days and $t = 268$ days is uneven, but rarely exceeds 0.2 m. Settlements predicted for this time period are 0.13 m according to CONSOL and 0.20 m according to PCDDF89 (Fig. 7). These predictions are therefore consistent with the observed bathymmetric changes. For the time interval from t = 147 days to t = 998 days bathymmetric settlement near the center is at least 0.40 m, and likely higher at the apex of the mound (Fig. 9). Predictions according to PCDDF89 are 0.44 m and according to CONSOL 0.58 m.

SUMMARY

The majority of the observed settlements can be accounted for by one-dimensional consolidation processes, which tends to support the overall modeling approach and the set of assumed material parameters. This conclusion is further supported by periodic sediment-profile photography and bottom current studies conducted by SAIC (Morris, 1996b). However, two possible sources of error need to be kept in mind. The 0.20 m bathymmetric reliability level limits correlation with predictions to settlements of this order. Much more accurate monitoring of settlements is highly desirable. Also, the complex mound construction sequence makes selecting the initial conditions for the analysis problematic. A three-dimensional model that incorporates a better description of the true time-load sequence of events would obviously result in more accurate predictions. Given the fine-grained nature of many of the sediments, future calculations should also include the effects of secondary compression. Initial estimates indicate that creep can add up to 10% or more to the total settlements (Silva et al., 1994a). Nevertheless, the predictions presented above are thought to be reliable enough to make useful settlement predictions, particularly for long times after the initial stages of the consolidation process.

Based on our calculations, the 1993-1994 dredge mound constructed at the CLIS site had already undergone well over 50% of the total estimated primary settlement as of September 1996. According to the more reliable predictions from the PCDDF89 code, less than 0.10 m of primary settlement can be expected in the future. Finally, it is worth noting that for this particular mound the largest portion of the settlement occurs in the base materials, and not in the dredge mound.

ACKNOWLEDGMENTS

We wish to thank the New England Division of the US Army Corps of Engineers, Waltham, MA, and Science Applications International Corporation, Newport, RI, for their support. In particular, our thanks go to John Morris (for help with the bathymmetry data) and Tom Fredette.

REFERENCES

Brandes, HG, Silva, AJ, and Fredette, T (1991). "Offshore Disposal of Dredge Materials.," *Maritimes*, University of Rhode Island, Graduate School of Oceanography, November, pp 12-14.

Cargill, KW (1982). "Procedures for prediction of consolidation in soft fine-grained dredged material," *Technical Report D-83-1*, US Army Engineer Waterways Experiment Station, Vicksburg, MS.

Charles, J, and Inglin, D (1994). "Monitoring Surveys of the New Haven Capping Project, 1993-1994," *SAIC Report No. 319*, Science Applications International Corporation, Newport, RI.

Fredette, TJ (1994). "Disposal Site Capping Management: New Haven Harbor," *Dredging '94, Second International Conference on Dredging and Dredged Material Placement*, Lake Buena Vista, FL, Vol 2, pp 1142-1151.

Gibson, RE, England, GL, and Hussey, MJL (1967). "The Theory of One-Dimensional Consolidation of Saturated Clays, I. Finite Non-Linear Consolidation of Thin Homogeneous Layers," *Geotechnique*, Vol 17, pp 261-273.

Morris, JT (1996a). "Monitoring Cruise at the Central Long Island Sound Disposal Site, July 1996, Data Report," *SAIC Report No. 3XX*, Science Applications International Corporation, Newport, RI.

Morris, JT (1996b). "DAMOS Site Management Plans," *SAIC Technical Report No. 365*, Science Applications International Corporation, Newport, RI.

Poindexter-Rollings, ME (1990). "Methodology for Analysis of Subaqueous Sediment Mounds," *Technical Report D-90-2*, US Army Engineer Waterways Experiment Station, Vicksburg, MS.

Silva, AJ, Brandes, HG, Uchytil, CJ, Fredette, TJ, and Carey, D. (1994a). "Geotechnical Analysis of Capped Dredge Material Mounds," *Dredging '94, Second International Conference on Dredging and Dredged Material Placement*, Lake Buena Vista, FL, Vol 1, pp 410-419.

Silva, AJ, Brandes, HG, and Veyera GE (1994b). "Geotechnical Coring and Core Processing Summary for DAMOS/CLIS New Haven Harbor Maintenance Project," *Technical Report* to Science Applications International Corporation, Newport, RI, October 1994.

Silva, AJ, Brandes, HG, Veyera, GE, and Ag, A (1994c). "Geotechnical Laboratory Testing Summary," *Technical Report* to Science Applications International Corporation, Newport, RI, October 1994.

Silva, AJ, Brandes, HG, Veyera, GE, and Ag, A (1994d). "Preliminary Consolidation Analysis of a Newly Completed (1993-1994) Dredge Mound at the CLIS Disposal Site," *Technical Report* to Science Applications International Corporation, Newport, RI, November 1994.

Stark, TD (1991). "Program Documentation and User's Guide: PCDDF89, Primary Consolidation and Desiccation of Dredged Fill," *Instruction Report D-91-1*, US Army Corps of Engineers Dredging Operations Technical Support Program.

Uchytil, CJ (1992). "Geotechnical Analysis of Dredge Mounds," *M.S. Thesis*, Department of Ocean Engineering, University of Rhode Island, Narragansett, RI.

Proceedings of the Seventh (1997) International Offshore and Polar Engineering Conference
Honolulu, USA, May 25-30, 1997

ISBN 1-880653-28-1 (Set); ISBN 1-880653-29-X (Vol. I)

A Case Study on the High Pressure Injection Method for the Quaywall Construction

B.S. Chun, Hanyang University, Seoul, Korea
Y.S. Chae, Suwon University, Suwon, Korea
H.K. Yoo, Hanyang University, Ansan, Korea
D.K. Chung, Seongha Geology Co., Ltd., Seoul, Korea
Y.S. Kim, Hyundai Engineering & Construction Co., Ltd., Seoul, Korea
Y.W. Nam, LG Construction Co., Seoul, Korea
S.J. Oh, Hanyang University, Seoul, Korea

ABSTRACT

The construction of the quaywall over a 40m thick layer of marine clay required the soil stabilization by hydraulic filling, soil improvement by preloading and paper drain. But, the filling processs prior to the construction of the wall was suspended by excessive ground deformation and slope failure. A high pressure injection method(i.e., JSP, Soilmech : double tube method) was then used to reduce the ground deformation and to eliminate the successive slope failure due to the dredging operations of the front side of the wall and backfilling.

The slope stability analysis and numerical analysis by FEM were performed using the soil parameters before and after the grout injection. By the high pressure injection method, the soft ground was improved very much, the safety factor of slope stability was increased and the ground deformation was decreased.

KEY WORDS : FEM, harbor, high pressure injection, improvement, marine clay, stability, strength.

INTRODUCTION

New construction or expansion of airport and harbor facilities is in active progress for the efficient mobilization program of the increasing export-and-import materials to be accomplished in a successful manner. This paper presents a case study for the expansion of existing harbor facilities located in Ulsan, Korea, where the subsurface formation consists of soft marine clay, which is carried out in link with the magnification process of natural infrastructures. A high pressure injection method was applied to eliminate the possibilities of the slope failure and heaving associated with the soft ground treatment required for the construction of harbor facilities. Analysis for the slope stability and ground deformation were performed using the soil properties determined from laboratory and in-situ tests. However, the required locations of ground improvement were determined from the results of stability analysis for the section treated with the high pressure injection technique. The degree of improvement of the uniaxial compressive strength in relation to cement contents was investigated.

FIELD CONDITION

Soil Condition

The variation of soil layer is very complex and erratic, and the formation is composed of fill, 40m thick marine clay(CH), lower marine deposit(SP, SM) and weathered rock from the ground surface. The average N-Value(obtained from Stadanrd Penetration Test) of very soft marine clay at the front side 7 quaywalls was estimated at 0 to 5 and 5 to 8 at the back side of the wall indicating the medium stiff clay layer. The N-Value at deeper zone are 6 to 7 at the front side and 13 to 14 at the back side. Fig. 1 shows the plan view of the job site and the boring locations, and the geological cross section of boreholes(MH17～MH18, MH21～MH20) is shown in Fig. 2.

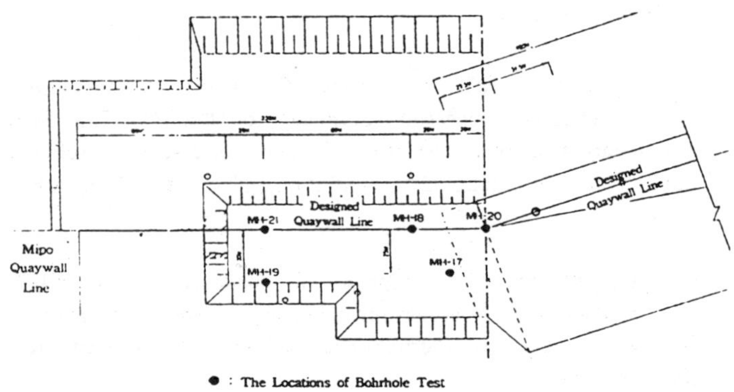

Fig. 1 Plan view and borehole locations

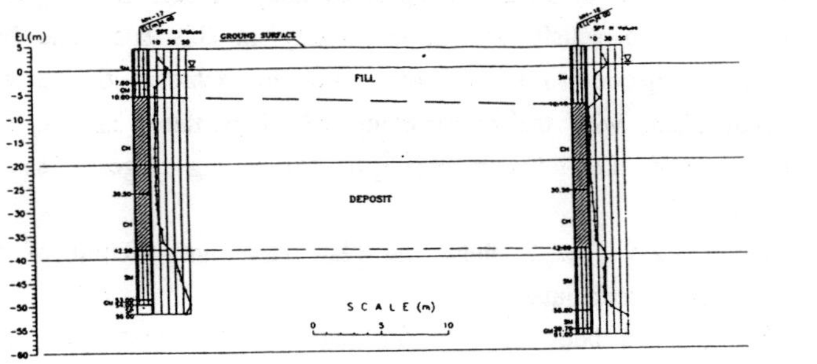

a) MH17~MH18

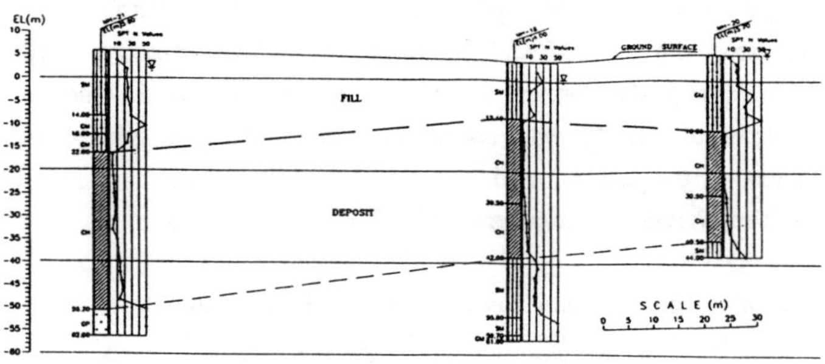

b) MH21~MH20

Fig. 2 Geological cross section

The results of laboratory test for the marine clay sample taken at different depths of MH17, MH18 and MH20 are summarized in Table 1. Fig. 3 and Fig. 4 present the results of the N-Value, vane shear test, cone penetration test and unconfined compression test for the marine clay layer across the boring locations.

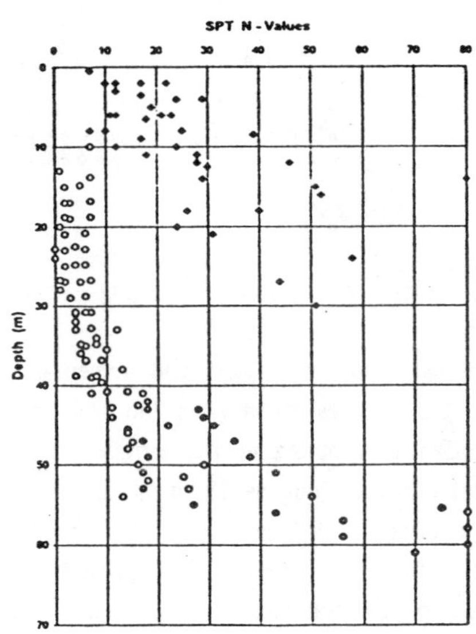

◆Filled Soil ○Deposit Soil

Fig. 3 Distribution of N-value

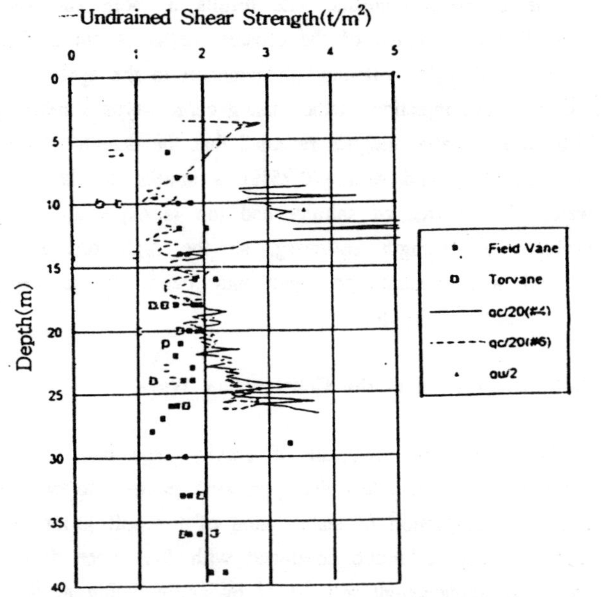

Fig. 4 Results of vane shear, cone penetration and unconfined compression test

Table 1 The results of laboratory test of marine clay

Elevation (m)	Natural moisture content (%)	Liquid limits (%)	Plastic index	Compression index	Undrained strength (kg/m²)	void ratio
-8.0	66.7	83.5	48.7	-	0.33	1.93
-10.0	69.9	84.6	47.5	0.689	0.27	1.85
-12.0	70.9	89.6	53.9	0.936	0.18	1.99
-14.0	74.0	89.2	54.3	-	0.21	2.00
-16.0	75.2	87.1	51.7	0.748	0.19	2.06
-18.0	74.2	88.5	53.3	-	0.23	1.94
-20.0	80.0	92.0	56.7	0.831	0.21	2.07
-22.0	79.5	93.5	56.2	0.766	0.21	2.09
-24.0	77.0	89.6	53.1	-	0.28	2.12
-28.0	74.3	90.1	50.5	-	0.4	2.08
-30.0	65.1	82.4	47.9	-	0.4	1.77
-32.0	57.7	76.4	44.5	-	0.4	1.53
-34.0	51.2	64.6	34.3	-	0.55	1.40

Note) USCS : CH, Specific gravity : 2.67, Passing #200 seive : 93.4~99.0%, γ_t=1.56t/m³

The organic matter contained in the marine clay was found to be 12.6% through the TGA technique and was indicated as 13.2%

by the chemical analysis. The results of both analysis do not show that the content of the organic matter is not so high as to cause the strength softening phenomenon in the application of the high pressure injection method using cement-type solidifying agent. The ground water analysis revealed that the density of Cl^-(0.45%), SO_4^{2-}(241ppm) and salinity(0.75%) is similar to that of the sea water. The degree of salinity and ion is expected to cause the problem of strength softening, so the high pressure injection method with cement-type agent may cause the problem in the ground of the job site.

Brief Description of the Construction Work

The 40m long quaywall of gravity type and the remaining 180m quaywall of piled pier type were planned to be constructed after the completion of stabilization of the soft ground using the staged filling technique combined with the paper drain method, using the replacement method of hydraulic filling at the sea and constructing the harbor at the ground. Heaving of the marine clay was observed, causing the slope failure(refer to Fig. 1 and Fig. 2). Accordingly, a severe damage for the adjacent quaywall was expected in case of continuous process of preloading. So the high pressure injection method was applied to accomplish a mightier ground improvement.

Fig. 5 The view of marine clay heaved above sea level

APPLICATION OF HIGH PRESSURE INJECTION METHOD

Due considerations should be given to the purpose, geological condition, construction condition and cost in the choice of a proper method of the ground improvement..

Different from the case where a solidifying method is used to improve the bearing capacity against the vertical loading, a method capable of making an grouted body of soil in the ground is required to satisfy the slope stability in this site. The high pressure injection method was used to achieve the ground improvement, with the agent compared of portland cement, slag cement and ESCA(Earth Strength Controlling Agent) as an admixture.

Fig. 6 and Fig. 7 show the plan view and section of the improvement technique.

The grouted body of 1.0m in diameter was constructed and the space betweeen the piles was 1m.

Soil cement column in marine clay was formed to increase the slope stability. Jumbo Special Pattern(J.S.P.) with double rod was applied with SOILMECH equipment of 500bar capacity. Portland cement and slag cement were used as grout material. And as an admixture, ESCA was added to reduce the strength softening effect of the cement used.

The chemical components of ESCA is as follows.

Table 2 The chemical components of ESCA (Unit:%)

	N.S.C[1)]	Micro Silica	$FeSO_4$	$FeCl_2$	$CaCl_2$	$CaSO_4$	beside
ESCA	20~50	15~20	0~30	0~30	10~30	0~10	0~5

1) N.S.C : Naphthalene Sulfonate-Sodafornaldehyde Condensation

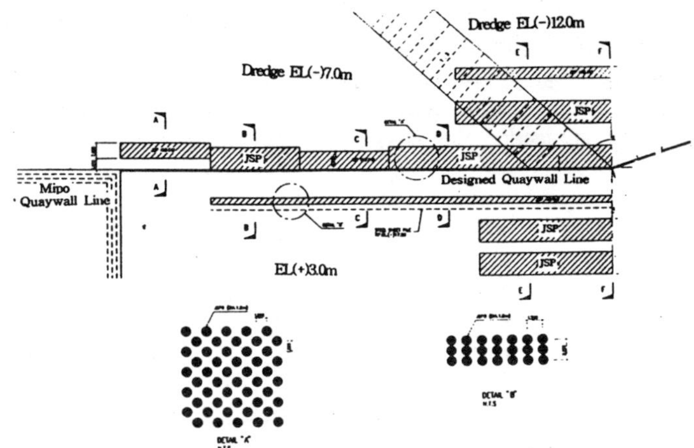

Fig. 6 Plan view of grouting by high pressure injection method

The unconfined compression test and cement-content test were performed on the samples taken from the soil cement column to measure the effect of grouting in a quantitive manner.

The samples cured as long as 28 to 94 days were prepared dry before the unconfined compression test. The samples were 5cm in diameter and 10cm in height. The amount of cement was measured based on the formation of 1m length of the soil cement column. Approximately 401kg of cement per 1m length of column(451kg for the case of C-448) was used and the injection pressure was controlled to 200, 400, and 500bar. The unconfined compression strengths of the improved soil cement columns associated with the depth and curing time were summarized in Table 3.

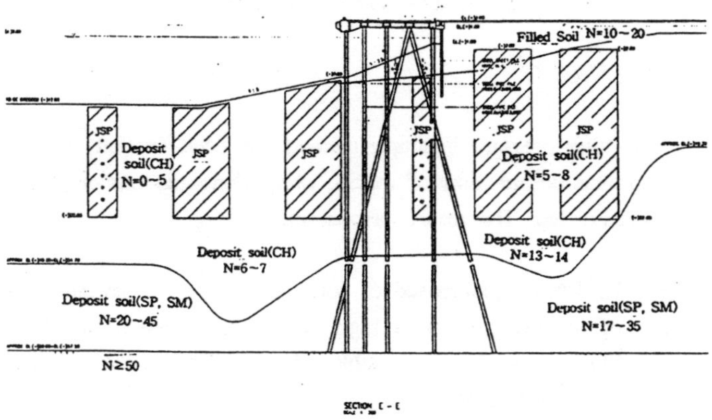

Fig. 7 Cross section of grouting by high pressure injection method

Table 3 The unconfined compression strength of improved marine clay samples

Location	Depth of sampling (m)	Curing days	Type of cement	Injection pressure (bar)	Usage of cement (kg/m)	Contents of cement (%)	Compression strength (kg/cm²)
E-105	EL-17.5	40	Portland	400	401	31.4	32.59
E-98	EL-10.5	50	Portland	200	401	10.6	0.8
E-99	EL-15~29.5	50	Portland	200	401	10.1~23.8	-
C-118	EL-22	94	Portland	200	401	-	27.7
E-105	EL-28	82	Portland	400	401	-	21.9
C-380	EL-15.0	28	Slag	400	401	-	15.50
C-380	EL-29.0	28	Slag	400	401	-	16.44
C-380	EL-30	76	Slag	400	401	-	70.8
C-448	EL-22	55	Slag	400	451	-	20.8
F-229	EL-27	37	Slag	400	401	-	20.8
F-156	EL-25	57	Slag	400	401	-	97.0
F-56	EL-5.0	43	Slag	500	401	-	40.4
F-56	EL-10.0	43	Slag	500	401	-	52.0
F-56	EL-20.0	43	Slag	500	401	-	56.1
F-56	EL-26.0	34	Slag	500	401	-	73.9
F-56	EL-32.0	43	Slag	500	401	-	35.5
E-58	EL-30	52	Slag+ESCA	400	401	-	63.1

In general, there were many kinds of salts like chlorides, sulfates and organic matters in the marine clay ground. These gave bad effects on the cement hydrates and eroded strength of the cement grouting. Above all, Cl^- and SO_4^{2-} in the ground badly affected cement hydrates. The erosion by Cl^- occurred at the early curing stage of hydration. Cl^- reacted on $Ca(OH)_2$ among the cement hydrates resulted in producing soluble $CaCl_2$ which speeded up to make the porous matrix of cement grouting hydrates.

The erosion mechanism of SO_4^{2-} was to make ettringite which was produced by the reaction of SO_4^{2-} between calcium aluminate hydrates. That is, the transformation of monosulfer aluminate into ettringite which accelerate the expanded failure of cement grouting hydrates.

The contents of SO_4^{2-}, Cl^- and organic matters which caused the strength inferiority of cement grouting were highly measured from the chemical analysis, X-Ray diffraction analysis, TGA and SEM-photographing. In the application of the high pressure injection method, the comparison of portland cement with slag cement as a grout for the application in the marine clay indicates that slag cement has more strength than portland cement even though it has more irregular strength and lack of early strength than portland cement. The effect of the unconfined compression strength was dependent on the injection pressure rather than the amount of cement. So, in the case of F-56 spot grouted up to EL-48.3m with the injection pressure 500bar, and the unconfined compressive strength of the improved column was 35.5~73.9kg/cm².

The unconfined compression strength of the sample from E-58 where ESCA was used was 63.5kg/cm², which proved a considerable strengh increase.

Thus, although it depends on the kind of cement, the injection pressure and curing day were the dominant factors over the unconfined compression strength of the improved column. In cases of E-98, and E-99, the samples were not molded and tested 0.8kg/cm² in strength because the contents of cement examined were 10.1~23.8%. It was analyzed that the reasons of such low strengths of improved column were not vertically injected and sampled. Also, the general discharge of soil was not performed and cements were overflown under the construction of high pressure injection.

STABILITY ANALYSIS

Analysis of Slope Stability

The analysis of slope stability was done in the E-E section by PCSTABL5M. In the analysis section, an uniform load of 2t/m was applied in the back of quaywall considering the equipment load under the construction and the traffic load after construction. The property of the improved soil was represented by the equivalent cohesion of the improved columns to the unimproved soil. The underground water level was the same as the sea level. Janbu's simplified method was adopted for the analysis. The E-E section for the analysis and the minimum safety factor are shown in Fig. 8.

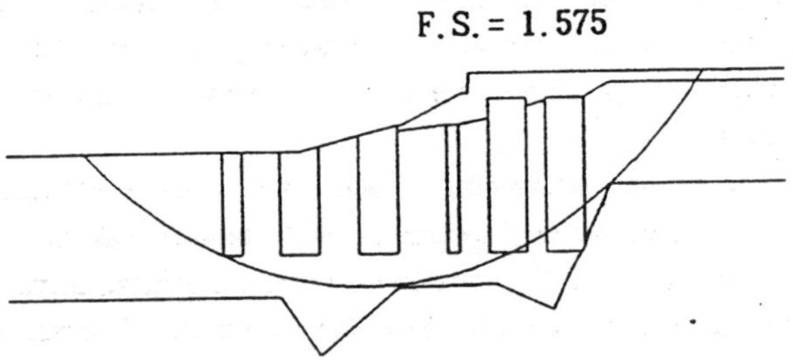

Fig. 8 Safety factor and failure surface of E-E section

Owing to the ground improvement by the high pressure injection method, the minimum safety factor 1.575 satisfied the Specification of Harbor Facilities Design Manual in Korea.

Analysis of Ground Deformation

As stated before, the section was safe as far as the slope stability was concerned. But the slope stability analysis based on the limit equilibrium could not deal with stress-strain relations. Continued deformation was expected during and after construction, so the pattern of future ground deformation was investigated through the deformation analysis. FEM code, SIGMA /W was used, and the ground was modeled by Modified Cam-Clay Model.

The soil cement column was modeled to be an elastic element. The properties obtained from the ground invesitgation, which were used in the analysis, are summerized in Table 4.

Table 4 The properties of marine clay (E-E section)

	O.C.R	ν	λ	κ	Γ	M	Pc' (t/m^2)
back of quaywall	1	0.4	0.293	0.029	2.657	1.027	8
front of quaywall	1	0.4	0.341	0.040	2.907	1.027	6

Based on the results of deformation analysis, the settlement at the back of quaywall was about 75cm on the completion of the ground deformation development, and the heave of ground was expected to be as much as 40cm in the area of no treatment. The amount of heave estimated in the grouted area was 10cm, indicating that the high pressure injection method was effective to control the ground deformation. A vertical displacement of the area at which the quaywall was located was about 10cm, while the horizontal displacement was 6 to 12cm. The horizontal displacement at the upper part was larger than that of the lower part.

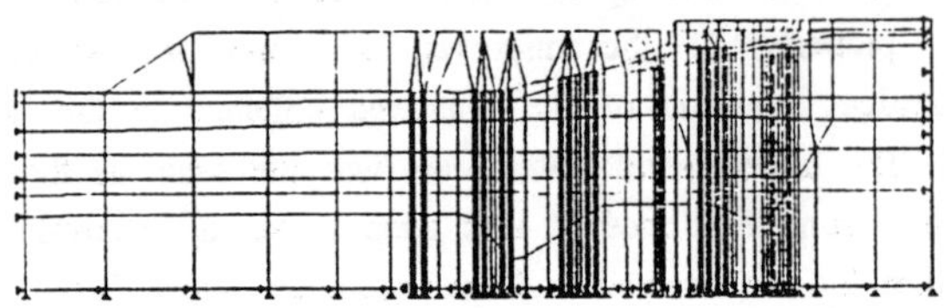

Fig. 9 Cross section for FEM analysis

CONCLUSION

The high pressure injection method has been applied to control the ground deformation of marine clay where the harbor facilities are constructed. The results are as follows:

(1) The job site composed of soft marine clay where the content of organic matters is high and a good amount of Cl^- and SO_4^{2-}, may give a possible adverse-effect in the application of cement grout.

(2) The degree of early strength development was less prominent in slag cement case than that in portland cement one, depending on the kinds of grout material, and the slag cement case gives a good indication of improvement with longer curing time, even though erratic. Futhermore, with the addition of ESCA, a considerable degree of strength increase was obtained.

(3) Injection pressure have a greater effect on the improvement of strength of the soil cement column than the amount of cement used

(4) The results of slope stability show that the ground is safe with consideration of harbor facilities. With the results of FEM analysis of ground deformation, the high pressure injection method would improve the ground, thereby reducing the ground deformation, even though a little amount, and is inevitable for the soft ground.

REFERENCES

Chun, B.S. (1996a). "A Study on Soil Improvement Effect under Poor Ground Condition", Journal of the Korean Geotechnical Society, Vol. 12, No. 2, pp 115-130.

Chun, B.S. (1996b). "The Report for Ground Improvement with High Pressure Injection Method in Quaywall Foundation in Ulsan", Construction Research Institute of Hanyang Univ., pp 146-148.

Chun, B.S. (1996c). "A Case Study on the Countermeasure of Sliding Failure in Marine Clay by Jet Grouting", Proceedings of the '96 Conference of KSCE, Vol. III, pp 531-534.

GEO-SLOPE (1995). "User's Guide SIGMA/W", GEO-SLOPE International Ltd., Canada, pp 8.59-8.70.

Shroff, A.V. (1993). "Grouting Technology in Tunnelling and Dam Construction", A.A. BALKEMA, pp 13-16.

Proceedings of the Seventh (1997) International Offshore and Polar Engineering Conference
Honolulu, USA, May 25-30, 1997

ISBN 1-880653-28-1 (Set); ISBN 1-880653-29-X (Vol. I)

Liquefaction Potential in Coastal Areas of Proposed Kaohsiung Deep-Sea Harbor, Taiwan

Shen-Chi Chen
Public Construction Commission, Taiwan, China
Gwan-Shong Wang
Dept. of Kaohsiung Mass Rapid Transit, Taiwan, China
Yun-Pin Chin
National Cheng Kung University, Tainan, Taiwan, China

ABSTRACT

This Paper presents some of the soil dynamics consideration used in evaluating liquefaction potential in coastal areas of proposed 20-meters deep deep-sea harbor of the Kaohsiung port at southwestern corner island of Taiwan. The subsurface soils beneath the sea floor, along some parts of the alignment, consist of fine sands which may be susceptible to possible earthquake of magnitude 7.5 on the Richter scale. It was concluded after performing simplified analyses, that, liquefaction was not a potential problem.

INTRODUCTION

The Kaohsiung Harbor located at longitude 120° 15′ east and latitude 22° 37′ north, extends as long as 12.0 km from the first entrance in the northwest to the second entrance in the southeast. The Kaohsiung Harbor has a total water area of 12.7 km^2. It enjoys brisk seaborne traffic and is now the biggest international harbor in Taiwan. It is one of the largest container transhipment center in the Far East. The location of proposed 20-meters deep eep-sea harbor is selected around right side of the second entrance around the coasted area about 4-km in length. The layout of the Kaohsiung Harbor is shown in Figure 1.

EXPERIMENTAL PROGRAM

Subsurface Exploration Program

The subsurface exploration program consisted of the drilling of 6 borings at the maximum depth of 50-meter below the ground surface along the alignment of proposed 20-meters deep deep-sea harbor. The approximate locations of the borings are shown in Figure1. There were four boreholes (BH-1 to BH-4) which were located on land along sea-shore; and borehole, BH-5 and BH-6 were located at shallow water within shore area. Hydraulic rotary diamond drilling rig was used for the drilling works. During drilling, the standard penetration test were conducted at 1.5-m of internal depth. Undisturbed soil samples within soft clayey layer or loose sand deposit were taken at selected depths by using the tri-tube core barrel. All undisturbed soil samples were stored in sampling container which is kept low temperature with some pieces of dry ice. Therefore, these undisturbed soil samples were kept on less disturbed situation. Figure 2 illustrates the tri-tube core barrel. The simplified soil profile based on the boring records is shown in Figure 3.

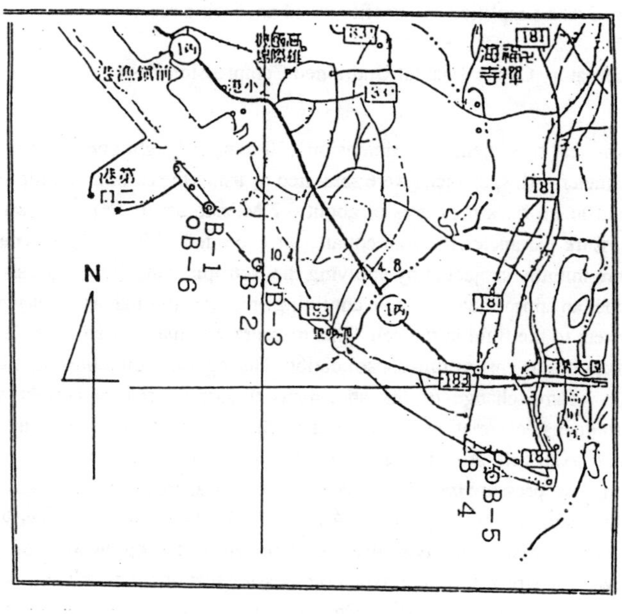

Fig. 1 - The layout of the Kaohsiung harbor

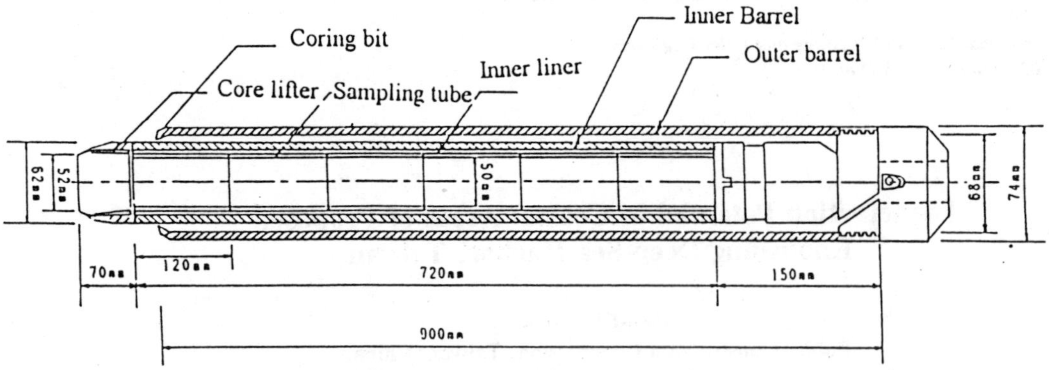

Fig. 2 The tri-tube core barrel (based on Chin, 1993)

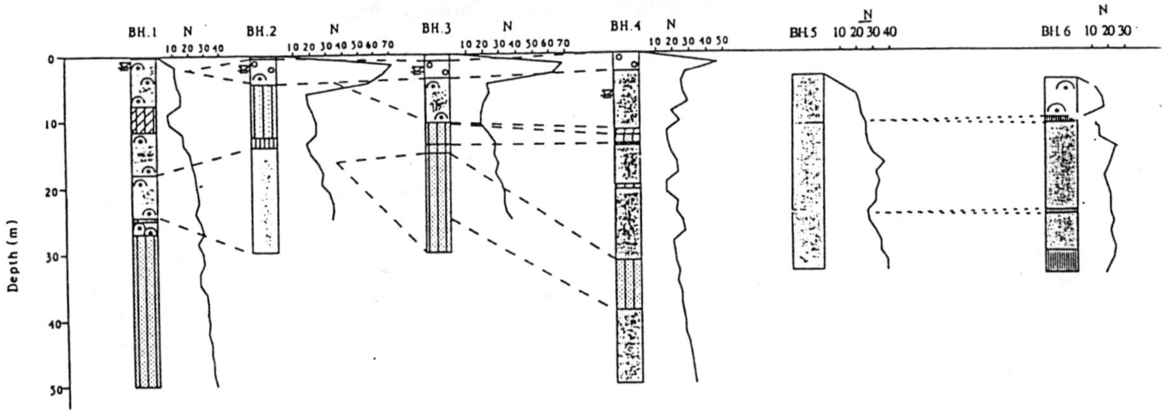

Fig. 3 The simplified soil profile based on boring records

Laboratory Cyclic Triaxial Undrained Compression Tests

The test specimens had dimension 150 mm. in height and 75 mm. in diameter. All specimens were saturated by using the carbon dioxide and a sufficient back pressures of about 0.2 MPa to assure that pore water pressure parameter B-value equaled or exceeded 0.95. All specimens were initially subjected by applying the isotropic consolidated pressure under an appropriate effective confining pressures through a mercury pot system to the fluid in the cell. This pressure was maintained for at least 2.0 hours to ensure full consolidation. During isotropic consolidation, the volume change of the specimens was measured to correct the specimens deformation. To cause the shear failure of the specimens in the triaxial cell, an axial deviatoric stress, is applied through a vertical ram. Two pressure transducers were used to measure the axial deviatoric stress, and pore water pressure. An LVDT displacement transducer was used to measure the axial displacement during the application of the deviatoric stress. Cyclic triaxial compression test on the specimens were performed without drainage. Computer programmed electric signal or frequency and magnitude of loading was applied to an electro-pneumatic transducers which controlled pneumatic amplifiers for the application of load. To begin the test, the frictionless piston was connected to the load cell, an appropriate cyclic deviatoric stress without drainage was applied with a sine wave pulse at a low frequency of 1/60 Hz. During each test, cyclic load, axial strain and pore water pressure were continuously monitored with electronic transducers and recorded by PC computer.

The cyclic triaxial tests were performed at various magnitudes of cyclic shear stress while the magnitude of initial static shear stress was held constant. The specimens were then tested by subjecting them to cyclic loading which ranged from 5-50 cycles till the pore water pressure just became equal to the confining pressure indicating that the effective stress momentarily dropped to zero. Based on Seed. (1966), this constitutes the initial liquefaction.

EVALUATION OF THE LIQUEFACTION POTENTIAL OF SAND DEPOSIT

Catalog of Earthquake in Taiwan

Taiwan is located at the plate boundary of the Philippine Sea and the Eurasian continental plate. These two plates have collided with each other since late Miocene. This plate collision has caused the passive continental margin to be deformed and has resulted in the rapid uplifting of the Central Range (Figure 4).

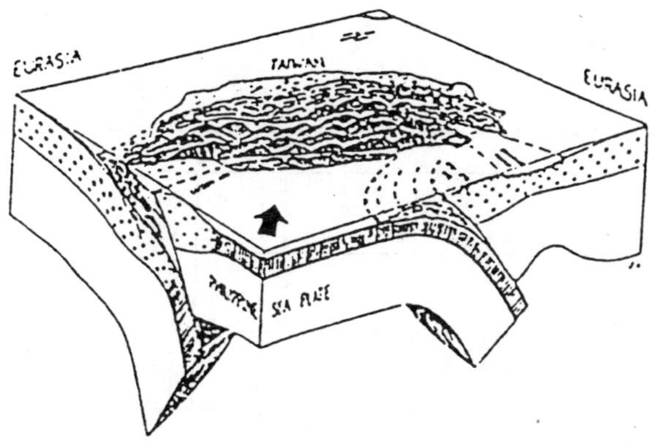

Fig. 4 Lithospheric plate structure in and around Taiwan
(after HO, 1982)

Liquefaction Analysis From Standard Penetration Resistance in Field

Evaluation of the liquefaction potential of sand deposits based upon field observations of past performance during earthquakes have been widely studied (Das, 1993). Seed (1976) has suggested the use of a semi-empirical procedure for estimating the earthquake-induced liquefaction potential ground deposit due to cyclic stress applications. The procedure determined by the Standard Penetration Test (SPT) and cyclic stress ratio, which is an indirect measure of ground shaking. Accordingly, a design chart, (Fig. 5), present cyclic stress ratio, τ_h / σ_v likely to cause liquefaction in the field as a function of SPT N values corrected to an overburden pressure of 1 ton / ft^2 (95.76 kPa) i.e. N'. Figure 5 shows the lower bounds of correlation curve causing liquefaction for an earthquake magnitude M = 7.5 in the field.

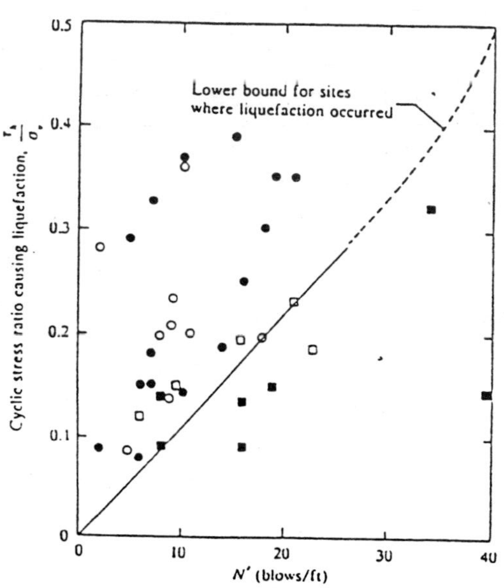

Fig. 5 Correlation between τ_h / σ_v and SPT N values
(Seed, 1979, and Das, 1993)

However, the above correction charts cannot be used with confidence in the field, because they do not take into consideration the magnitude of the earthquake and the duration of shaking (Das, 1993). Excellent agreement is observed when the SPT values of N' and the corresponding, $(\tau_h / \sigma_v)_{field}$ with the lower-bound curves for an earthquake magnitude M = 6, 7.5 and 8.25 shown in Figure 6, which can be used for determination of the liquefaction of the liquefaction potential in the field (Das, 1993). The corrected N' value are expressed that

$$N' = C_N N_F \qquad (1)$$

where N_F = field standard penetration tests values, C_N = correction factor to convert to an effective overburden pressure, σ'_v of 1 ton / ft^2 (95.76 kPa), that can be expressed as

$$C_N = \sqrt{1/\sigma'_v} \qquad (1a)$$

where σ'_v is in tons per square foot or kilogram per square centimeter.

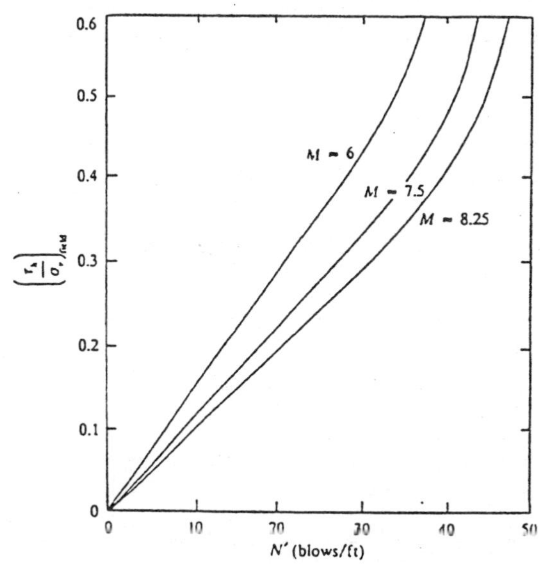

Fig. 6 Variation of $(\tau_h / \sigma_v)_{field}$ with N' and M
(after Seed, 1979 and Das, 1993)

From the results of field investigation as shown in Figure 3, it can be seen that the subsurface soils beneath the sea floor along some parts of the alignment consist of fine sands which may be susceptible to liquefaction for earthquakes of magnitude 7.5 or more on the Richter scale.

Liquefaction analysis from standard penetration resistance, SPT for the boreholes Nos. BH-1 to BH-4 inclusive in this study were indicated in Figure 7. From these analysis, the results showed that the safety of factor for sand deposit almost more than 1.2 against possible liquefaction failure which may be caused by an earthquake of magnitude 7.5 on the Richer scale. liquefaction were not a potential problem.

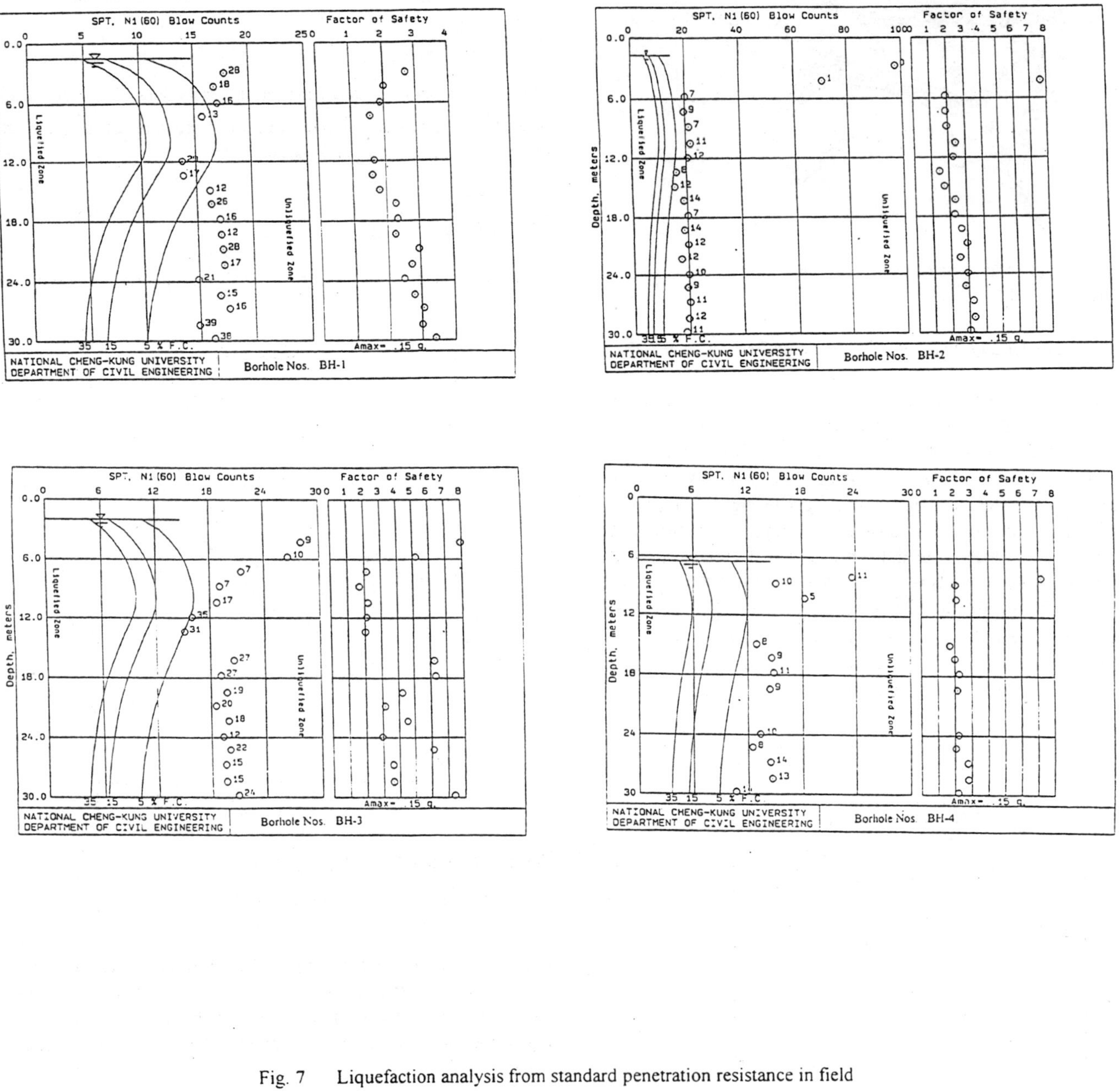

Fig. 7 Liquefaction analysis from standard penetration resistance in field

Table 1 Factor of safety against liquefaction failure for cyclic undrained triaxial compression test

BH-1	Depth m	σ_o kPa	u kPa	$\bar{\sigma}_o$ kPa	γ_d	A_{max} g	τ_{ave} kPa	$\tau_{cy(eq)}$ kPa	C_t	$\tau_{cy(field)}$ kPa	F.S.	Remark
B1T3	5.00 - 5.70	105.3	37.67	67.61	0.920	0.15	9.44	24.4 24.2	0.94	22.94 22.75	2.43 2.41	UD Remold
B1T6	11.00 - 11.70	222.3	96.53	125.77	0.830	0.15	17.98	39.5 32.2	0.94	37.13 30.27	2.06 1.68	UD Remold
B1T7	12.50 - 13.20	254.5	111.25	143.25	0.807	0.15	20.03	48.8 27.4	0.94	45.87 25.76	2.29 1.29	UD Remold

BH-2	Depth m	σ_o kPa	u kPa	$\bar{\sigma}_o$ kPa	γ_d	A_{max} g	τ_{ave} kPa	$\tau_{cy(eq)}$ kPa	C_t	$\tau_{cy(field)}$ kPa	F.S.	Remark
B2T1	9.50 - 10.00	185.9	79.36	106.54	0.852	0.15	15.44	30.6 27.5	0.94	28.76 25.85	1.86 1.67	UD Remold
BA2T1	10.00 - 10.70	194.7	84.27	110.43	0.845	0.15	16.04	33.1 43.1	0.94	31.11 40.51	1.94 2.53	UD Remold
BA2T2	11.00 - 11.75	213.2	94.08	119.12	0.829	0.15	17.24	30.7 29.0	0.94	28.86 27.26	1.67 1.58	UD Remold

BH-3	Depth m	σ_o kPa	u kPa	$\bar{\sigma}_o$ kPa	γ_d	A_{max} g	τ_{ave} kPa	$\tau_{cy(eq)}$ kPa	C_t	$\tau_{cy(field)}$ kPa	F.S.	Remark
B3T1	9.50 - 10.20	187.3	76.71	110.59	0.852	0.15	15.56	24.0 37.0	0.94	22.56 34.78	1.45 2.23	UD Remold
B3T2	11.00 - 11.70	217.0	91.43	125.57	0.830	0.15	17.55	34.5 26.3	0.94	32.43 24.72	1.84 1.41	UD Remold
B3T4	14.00 - 14.70	277.0	120.86	156.14	0.785	0.15	21.19	35.7 32.8	0.94	33.56 30.83	1.58 1.45	UD Remold

BH-4	Depth m	σ_o kPa	u kPa	$\bar{\sigma}_o$ kPa	γ_d	A_{max} g	τ_{ave} kPa	$\tau_{cy(eq)}$ kPa	C_t	$\tau_{cy(field)}$ kPa	F.S.	Remark
B4T2	10.80 - 11.50	222.4	45.57	176.83	0.833	0.15	18.05	56.0 43.3	0.94	52.64 40.70	2.92 2.25	UD Remold
B4T4	14.0 - 14.70	280.3	76.93	203.37	0.785	0.15	21.45	52.2 40.5	0.94	49.07 38.07	2.29 1.77	UD Remold
B4T5	18.50 - 19.20	364.7	121.03	243.67	0.717	0.15	25.50	57.2 49.5	0.94	53.77 46.53	2.11 1.82	UD Remold

Computation of The Average Shear Stress in The Sand Deposit

A ground response analysis was performed (Seed, 1976) to evaluate the stress history at various depths and to determine how the pore pressure dissipates during an earthquake and after earthquake motion has ceased. The shear stresses developed at any point during an earthquake is primarily due to the upward propagation of shear waves in the deposit. If the soil column of soil height *h* behaved as a rigid body and the maximum ground surface acceleration was A_{max}, the average equivalent uniform shear stress induced by a given earthquake in the soil deposit, (τ_{av}) is obtained from simplified expression as (Seed, 1979; Das, 1993)

$$\tau_{av} = 0.65 \frac{A_{max}}{g} \sigma_0 \gamma_d \qquad (2)$$

in which $0.65A_{max}$ = the average acceleration associated with the number of significant cycles of earthquake shaking; g = the acceleration of gravity (= 980 gals); σ_o = the total overburden pressure at the depth under consideration; γ_d = a stress reduction factor of dynamic shear stress to account for the deformation of the ground varying from a value of 1.0 at the ground surface to a value of 0.9 at a depth of 9.1 m, The cyclic shear stress to cause liquefaction can be given as (Seed, 1976, and Das, 1993). The factor of safety (F.S.) can then be estimated as

$$\tau_L = \sigma_o' \left[\frac{\sigma_{dp}}{2\sigma_3} \right] C_t \qquad (2a)$$

in which σ_o' = the effective overburden stress; σ_{dp} = the single amplitude cyclic axial stress, σ_3 = the initial effective confining pressure applied to the specimen, $\sigma_{dp} / 2\sigma_3$ = the stress ratio causing liquefaction in the cyclic triaxial compression test; and C_t = a factor varying from 0.57 to 0.68 depending upon the relative density of the soil (Seed, 1976).

$$F.S. = \frac{C_t \sigma_{dp} / 2\sigma_3}{\tau_{av} / \sigma_o} \qquad (2b)$$

The value of C_t = 0.57 was used in this study for the appropriate relative density of about 35 %. The subsurface soils beneath the sea floor, along some parts of the alignment, consist of fine sands which may be susceptible to liquefaction due to earthquakes of magnitude 7.5 or more on the Richter scale. The maximum ground accelerations (A_{max}) and an equivalent number of significant stress cycles N were taken as 0.15g and 20 cycles, respectively. These values were used in subsequent analyses of liquefaction potential. From the results of Cyclic undrained triaxial compression tests on undisturbed sand samples obtained from boreholes, BH-1 to BH-4 inclusive, and remold samples were molded after each undisturbed sample had been completed for cyclic compression test. The factor safety against liquefaction failure for cyclic undrained triaxial compression test are summarized in Table 1. It is seen from Table 1 that the factor of safely against liquefaction failure is also more than 1.2, and compared with the results of liquefaction analysis from standard penetration resistance in field is very closed. Possible liquefcction failure which may be caused by an earthquake of magnitude 7.5 on the Richer scale were not a potential problem.

SUMMARY AND CONCLUSIONS

The results of standard penetration tests obtained during field from exploration boreholes and laboratory cyclic undrained triaxial compression tests on undisturbed and remolded specimens were used to evaluate the shearing resistance for the Kaohsiung Harbor of proposed 20-meters depth deep-sea harbor project against seismic reaction. Based on the results of this investigation, the following conclusions can be drawn

1. Liquefaction analysis from standard penetration resistance, SPT for the boreholes Nos. BH-1 to BH-4 inclusive, showed that the safety of factor for sand deposit almost more than 1.2 against possible liquefaction failure which may be caused by an earthquake of magnitude 7.5 on the Richer scale. liquefaction were not a potential problem.

2. The factor safety against liquefaction failure for cyclic undrained triaxial compression test are summarized. It is seen that the factor of safely against liquefaction failure is also more than 1.2, and compared with the results of liquefaction analysis from standard penetration resistance in field is very closed. Possible liquefaction failure which may be caused by an earthquake of magnitude 7.5 on the Richer scale were not a potential problem.

REFERENCE

Academia-Sinica (Dec.1989). "Catalog of Earthquake in Taiwan". *Institute of Earth Sciences,* R.O.C.

Chin, YP (1993). "The Study of Ground Improvement Technique for Deep-sea Harbor," *Research Report, 93-41-221, National Cheng Kung University,* Tainan, Taiwan, R.O.C.

Das, BM (1993). "Principles of Soil Dynamics,". *PWS Publishing Co., New York,* pp. 396-458.

Ho, CS (1982). "Tectonic Evolution of Taiwan," *Ministry of Economic Affairs,* R.O.C.

Seed, HB and Lee, KL (1966). "Liquefaction of Saturated Sand During Cyclic Loading," *J. Soil Mech. Found Div., ASCE,* Vol. 92, No. SM6, pp. 105-134.

Seed, HB (1976). "Evaluation of Soil Liquefaction Effects on Level Ground During Earthquake," *paper presented at The ASCE Annual Convention,* Philadelphia.

Proceedings of the Seventh (1997) International Offshore and Polar Engineering Conference
Honolulu, USA, May 25-30, 1997

ISBN 1-880653-28-1 (Set); ISBN 1-880653-29-X (Vol. I)

Three Dimensional Liquefaction Analysis of Reclaimed Island

F. Oka, A. Yashima and M. Sugito
Gifu University
Gifu, Japan

Y. Taguchi
Taisei Corporation
Nagoya, Japan

ABSTRACT

A three-dimensional effective-stress based liquefaction analysis of Port Island, a reclaimed island, was carried out in order to understand the basic behavior of the man-made island subjected to liquefaction. The reclaimed island and the surrounding seabed subsoil were modeled with the elasto-plastic constitutive model, in which the non-linear kinematic hardening rule and the generalized flow rule were incorporated. In the FEM analysis, the brick elements were used. Analytical results were compared with the borehole array observation records in Port Island obtained during the 1995 Hyogoken-Nambu Earthquake. The analytical results is in good agreement with the array records irrespective of the rather simplified modeling.

KEY WORDS: liquefaction, reclaimed ground, FEM, effective stress, elasto-plasticity, earthquake

INTRODUCTION

Recently, large scale structures with important facilities have been constructed on the reclaimed islands. Significant problems such as liquefaction, consolidation settlement, etc. are existing. Especially, the liquefaction phenomena will immediately lose bearing capacity of the structure, and will bring large displacements and settlements with significant structural damages. It is important to clarify the mechanism of liquefaction and to carry out any appropriate countermeasures for the reclaimed ground.

In the present study, a three-dimensional effective-stress based liquefaction analysis was carried out in order to understand the basic behavior of the reclaimed island subjected to liquefactions. Port Island and the surrounding ground in Kobe city, Japan were modeled with a three-dimensional finite element mesh. In Port Island, the borehole array observation records were obtained during the 1995 Hyogoken-Nambu Earthquake. It can be seen from the array observation records that the liquefaction phenomena occurred in the reclaimed ground. The analytical results are compared with the array observation records. The wide-range behavior of the reclaimed island and its surrounding ground are also discussed in the present study.

ANALYTICAL METHOD

The constitutive model in the present study is constructed with the concept of the non-linear kinematic hardening rule which had originally been used in the field of metal plasticity. Oka et al. (1992) derived a cyclic elasto-plastic constitutive model for sand. Tateishi et al. (1995) incorporated a new stress dilatancy relationship and cumulative plastic strain dependent characteristic of the shear modulus to modify the original model. The constitutive model is formulated under the three dimensional stress conditions. The validity of the constitutive model has been verified by the experimental evidences from the hollow cylindrical torsional shear tests under various stress conditions.

Then this constitutive model has been incorporated into a coupled finite element -finite difference (FEM-FDM) numerical method for the liquefaction analysis of a fluid-saturated ground. Using a **u-p** (displacement of the solid phase and pore water pressure of the fluid phase) formulation, the numerical method was developed (Oka et al., 1994). The finite element method is used for the spatial discretization of the equilibrium equation, while the finite difference method is used for the spatial discretization of the continuity equation. Newmark's β-method is used for the time discretization of both equations. The applicability of the proposed method for the actual ground (including model ground by centrifuge test) had been already verified by previous studies (e.g. Taguchi et al., 1996).

GROUND MODEL AND BOUNDARY CONDITIONS

The borehole array record in Port Island obtained during the 1995 Hyogoken-Nambu Earthquake was used for the analytical input motion. Three directional acceleration records at GL-83m and ground conditions surveyed around the borehole location were used in the present study. The material parameters used in the numerical analysis are summarized in Table 1. These are obtained from previous studies by Taguchi et al. (1996).

The reclaimed island is modeled as a rectangular parallelepiped

Table 1 Cross-section and material parameters of the analytical model.

Depth (GL m)	Soil type <No.>	Unit weight tf/m3 (SPT)	Void ratio e	shear wave Velocity Vs m/sec	Poisson ratio ν	Coefficient of permeability m/sec	G_0/σ'_{m0}	Phase transformation angle degree (stress ratio M_m^*)	Failure angle degree (stress ratio M_f^*)	liquefaction strength N_c=10	liquefaction strength N_c=30	Plastic modulus parameters upper G^P_{max}, mid. G^P_{min}, lower C_f	Dilatancy parameters upper: D_0^*, lower: n	Reference strain upper: γ^{E*}_{DAr}, lower: γ^{P*}_{DAr}	bulk modulus of fluid K^f kPa
0.0 ▲ ~ -2.4 ▽	<1> sandy gravel (reclaimed)	2.0 (5.2)	0.60	140	0.25	—	3350	—	—	—	—	66500 1300 50000	0.0 —	∞ ∞	2.0×10^4
-2.4 ~ -5.0	<2> sandy gravel (reclaimed)	2.0 (5.2)	0.60	180	0.25	3.0×10^{-5}	1700	22.4 (0.71)	25.2 (0.81)	0.22	0.17	6500 650 2000	0.6 2.8	0.005% 0.05%	2.0×10^4
-5.0 ~ -12.6	<3> sandy gravel (reclaimed)	2.0 (6.5)	0.60	195	0.25	3.0×10^{-5}	1050	23.5 (0.75)	26.4 (0.85)	0.22	0.17	5100 510 2000	0.8 2.8	0.005% 0.10%	2.0×10^5
-12.6 ~ -19.0 (-16m)▲	<4> sand with gravel (reclaimed)	2.0 (6.5)	0.60	220	0.25	3.0×10^{-5}	820	23.5 (0.75)	26.4 (0.85)	0.22	0.17	4300 430 2000	0.8 2.7	0.005% 0.10%	2.0×10^5
-19.0 ~ -27.0	<5> alluvial clay	1.7 (3.5)	1.50	180	0.30	2.0×10^{-6}	350	23.4 (0.74)	23.4 (0.74)	—	—	12100 240 10000	0.0 —	∞ ∞	2.0×10^6
-27.0 ~ -33.0 ▲(-32m)	<6> alluvial sand	2.0 (13.5)	0.60	245	0.25	2.0×10^{-5}	615	28.0 (0.91)	31.4 (1.03)	0.40	0.30	8200 820 2000	0.8 3.0	0.02% 0.10%	2.0×10^6
-33.0 ~ -50.0	<7> diluvial sand with gravel	2.0 (36.5)	0.50	305	0.25	1.0×10^{-5}	700	35.0 (1.16)	42.0 (1.41)	0.45	0.35	9900 990 2000	0.8 2.5	0.02% 0.10%	2.0×10^6
-50.0 ~ -61.0	<8> diluvial sand	2.0 (61.9)	0.50	350	0.25	1.0×10^{-5}	680	35.0 (1.16)	46.6 (1.57)	0.60	0.40	11000 1100 2000	0.5 3.0	0.02% 0.10%	2.0×10^6
-61.0 ~ -79.0	<9> diluvial clay	1.75 (11.7)	1.20	303	0.30	1.0×10^{-6}	365	30.3 (0.99)	30.3 (0.99)	—	—	11400 230 10000	0.0 —	∞ ∞	2.0×10^6
-79.0 ~ -83.0 ▲	<10> diluvial sand with gravel	2.0 (68.0)	0.50	320	0.25	1.0×10^{-5}	410	35.0 (1.16)	46.6 (1.57)	0.60	0.40	11000 1100 2000	0.6 2.3	0.05% 0.10%	2.0×10^6

Note 1. ▲ accelerometer 2. K_0=0.5 3. OCR ratio···1.5 (alluvial and diluvial deposit), 1.2(reclaimed ground)

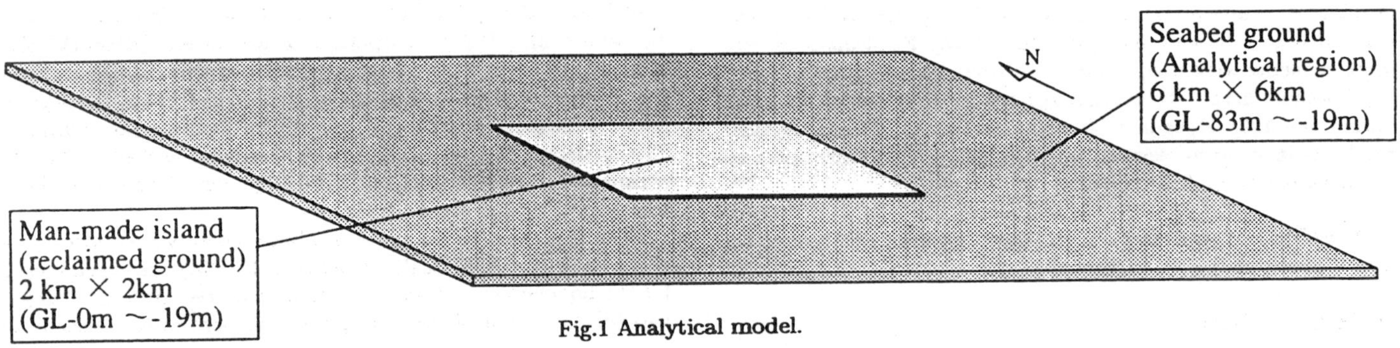

Fig.1 Analytical model.

(2km wide, 2km long and 19m deep), whose volume was nearly equivalent to that of the first stage reclamation in Port Island. The ground from GL-19m to GL-83m was modeled as natural deposit layers, 6km wide and 6km long. Ground model used in the analysis is shown in Fig.1. All ground layers were assumed as level layers. The boundary displacement conditions are free except the bottom surface. The ground water level (GL-2.4m) in the reclaimed island and the seabed surface (GL-19m) were drained boundaries, while the bottom and lateral surface were undrained boundaries. About 600 brick elements were used for the modeling, which were rather flat and simplified shape. The inclination of soil layer, localization of soil properties, and the variation of input seismic motion at a depth were not considered. Although ground shapes are rather simplified, it could be possible to obtain an approximate behavior of the reclaimed island and its surroundings. For example, it is possible qualitatively to compare the behavior at the edge of the reclaimed island with that at the center of the reclaimed island.

ANALYTICAL RESULTS AND DISCUSSIONS

Analytical results were compared with the borehole array records. Fig.2 shows the comparison between the analytical results and the observed records of the response acceleration time histories. As for the horizontal accelerations, the amplitudes as well as the phases were almost reproduced in the simulation. However, as shown in Fig.2, high frequency component was lost in the analysis. One of the reason may be the simplification of the FEM mesh. As for the vertical accelerations, on the other hand, the amplitudes as well as the phases in the simulation are well reproduced, except at GL-16m. In the present study, in order to take into account the condition of partial saturation for the reclaimed ground and the compressibility of ground water due to soluted air or air bubbles (Yashima et al., 1995), the bulk modulus of the fluid phase was decreased to express the increase in compressibility.

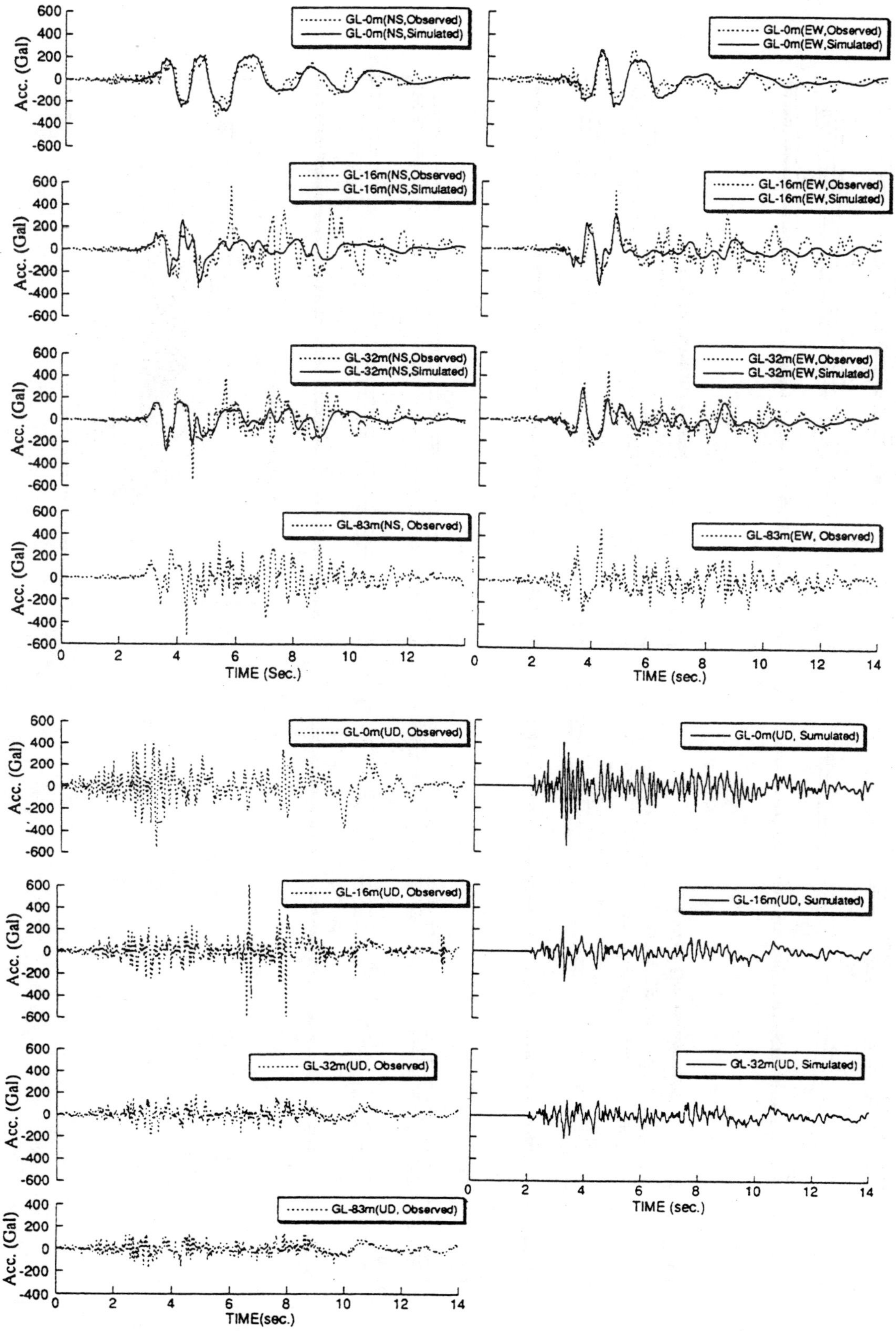

Fig.2 Time histories of horizontal and vertical acceleration (observed and analyzed).

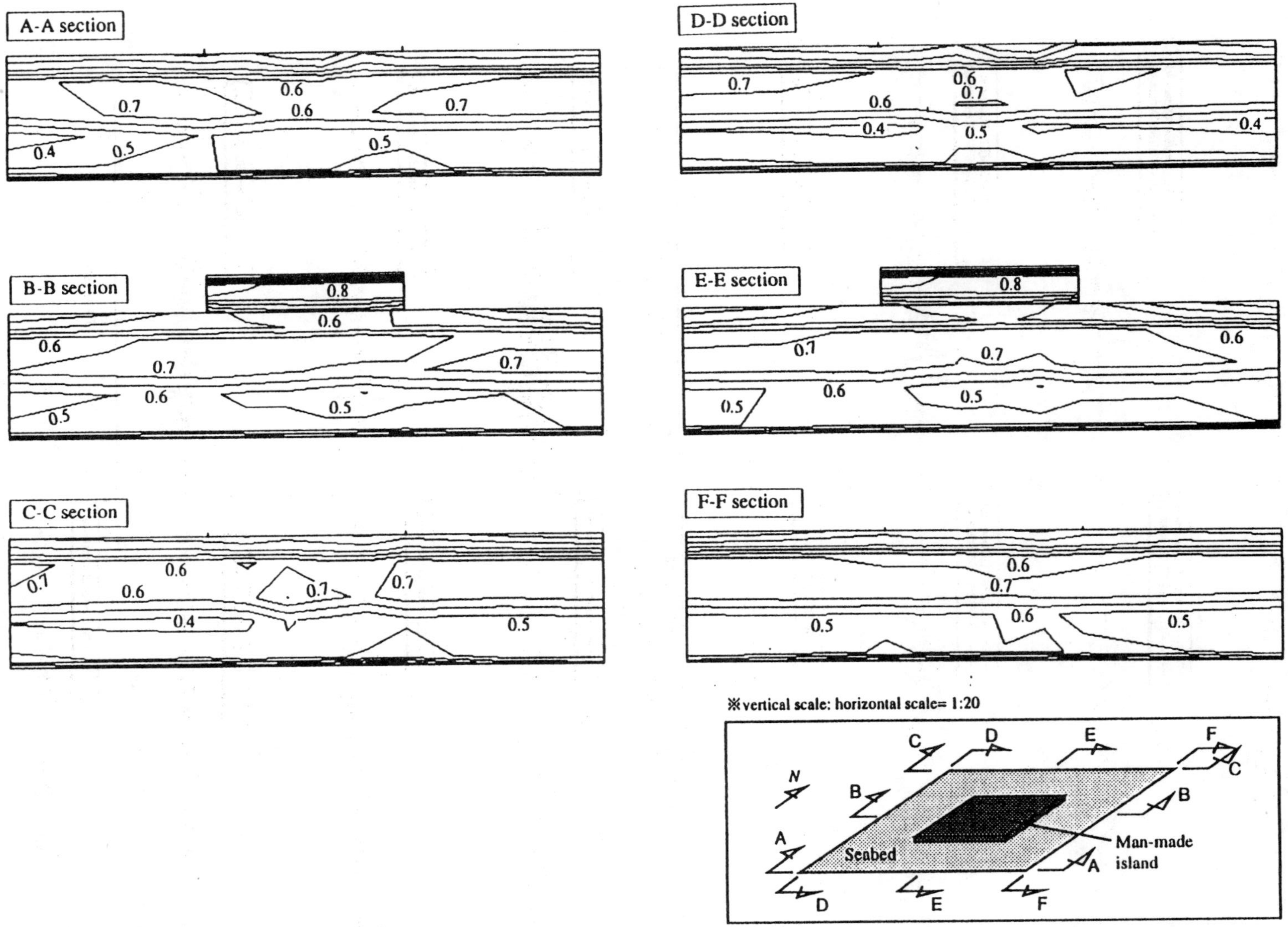

Fig.3 Distribution of the excess pore water pressure ratio at the end of shaking.

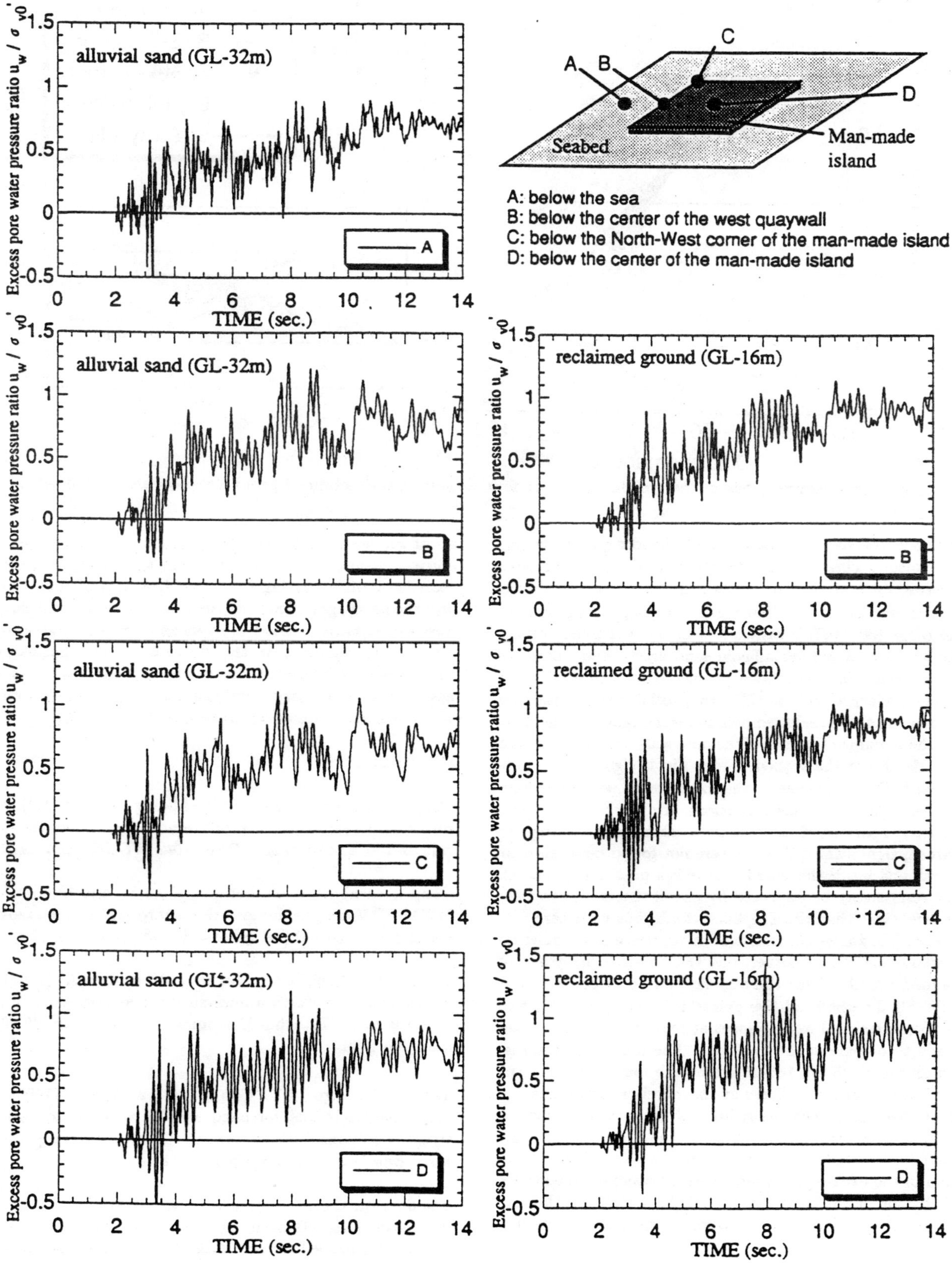

Fig.4 Time histories of excess pore water pressure ratio at selected points.

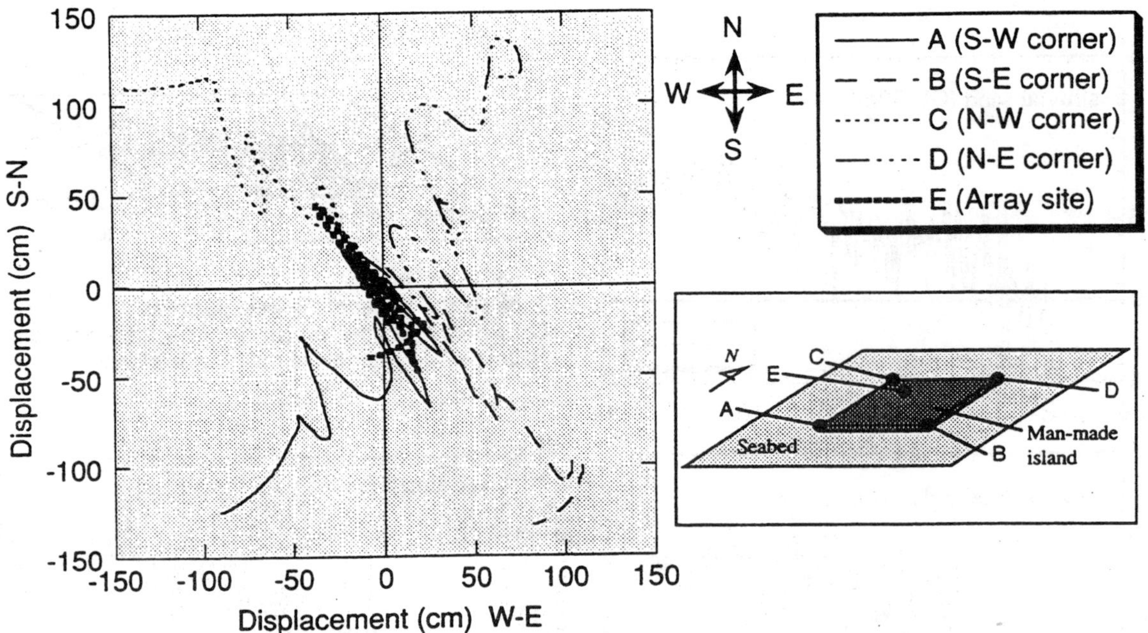

Fig.5 Displacement orbits of the ground surface at four corners of the reclaimed ground and borehole array site.

The behavior of the reclaimed island including surrounding ground are also discussed. The distribution of the excess pore water pressure ratio at the end of shaking (t=14 sec.) is shown in Fig.3. It is seen from Fig.3 that higher excess pore water pressure ratio range (over 0.7) is distributed in the reclaimed ground below the water level and the alluvial sand layer. Time histories of excess pore water pressure ratios for selected points at GL-32m (alluvial sand layer) as well as GL-16m (Reclaimed ground) are also shown in Fig.4. The selected points are the center of the west quaywall, the north-west corner of the quaywall, the center of the reclaimed island and the seabed. From Fig.4, significant differences in the build-up process of the excess pore water pressure could not be obtained at these locations. In the present study, since the lateral surface of the reclaimed island was assumed to be undrained, significant difference were not found between near the quaywall and inside the island. There is a possibility that the numerical results may be improved by use of the more practical drainage conditions than the simplified conditions used here.

Fig.5 shows the analytical results of the displacement orbits on the ground surface at four corners and the borehole array site of the reclaimed island. Since the seismic motion is predominant to the NW-SE direction, all five points in Fig.5 are shaking to that direction. Residual displacement at four corners were, however, bound to outside of the quaywall. The maximum residual displacements in Fig.5 are 100-150cm at the four corners of the reclaimed island and about 50cm at the borehole array site. In the present modeling, since the rigidity of the quaywall was not considered, the accuracy of the value of the displacement may be restricted. In addition, the consolidation settlement due to dissipation of the excess pore water pressure was not considered. Actual behavior of the reclaimed island during the earthquake must be understood with a combination of such several mechanisms.

CONCLUDING REMARKS

A three-dimensional effective stress-based liquefaction analysis of the wide-range reclaimed island was carried out. The analytical results was compared with the borehole vertical array records obtained during the 1995 Hyogoken-Nambu Earthquake, and the good agreement was obtained. It was found that the proposed analytical method is applicable to such wide-range three-dimensional analysis. In the future work, it is necessary to take more practical boundary conditions and effective mesh size. The appropriate boundary conditions and finer mesh configuration will bring about more accurate analytical results.

REFERENCES

Oka,F., Yashima,A., Kato,M. and Sekiguchi,K. (1992), "A constitutive model for sand based on the non-linear kinematic hardening rule and its application," *Proc. 10th WCEE*, pp.2529-2534.

Oka,F., Yashima,A., Shibata,T., Kato,M. and Uzuoka,R. (1994), "FEM-FDM coupled liquefaction analysis of a porous soil using an elasto-plastic model," *Applied Scientific Research*, 52, pp.209-245.

Taguchi,Y., Tateishi,A., Oka,F. and Yashima,A. (1996), "Three-dimensional liquefaction analysis method and array record simulation in Great Hanshin Earthquake," *Proc. 11th WCEE*, Pergamon, No.1042.

Tateishi,A., Taguchi,Y., Oka,F. and Yashima,A. (1995), "A cyclic elasto-plastic model for sand and its application under various stress conditions," *Proc. 1st Int. Conf. on Earthquake Geotechnical Eng.*, IS-Tokyo'95, pp.399-404.

Yashima,A., Oka,F., Taguchi,Y. and Tateishi,A. (1995), "Three dimensional liquefaction analysis considering the compressibility of fluid phase," *Proc. 40th Symp. Geotechnical Eng.*, JGS, pp.257-264 (in Japanese).

Proceedings of the Seventh (1997) International Offshore and Polar Engineering Conference
Honolulu, USA, May 25-30, 1997

ISBN 1-880653-28-1 (Set); ISBN 1-880653-29-X (Vol. I)

Evaluation of Liquefaction Resistance of Sand by Its Maximum Shear Modulus

Yao-Chung Chen
National Taiwan Institute of Technology
Taipei, Taiwan, China

ABSTRACT

This research performed undrained cyclic shear tests on remolded specimens of Taipei sand by a hollow cylindrical torsional shear apparatus to obtained their liquefaction resistance. The maximum shear moduli were obtained by resonant column tests. Remolded samples with different fines contents were prepared. The influences of fines content on the relationship between maximum shear modulus and liquefaction resistance were studied. To obtain a unique relationship between maximum shear modulus and liquefaction resistance, the maximum shear modulus is to be normalized by the effective confining pressure and a fabric factor. Under the conditions of same maximum shear modulus, liquefaction resistance decreases with increasing fines content.

KEY WORDS: Liquefaction resistance, shear modulus, sand, torsional shear test

INTRODUCTION

Liquefaction generally includes all phenomena involving excessive deformations as a result of transient or repeated disturbance of saturated granular soils. Initial liquefaction indicates a condition where, during the course of cyclic stress applications, the residual pore water pressure on completion of any full stress cycle becomes equal to the applied confining pressure. Cyclic mobility represents a condition in which cyclic stress applications develop a peak cyclic pore pressure ratio of 100% and subsequent cyclic stress applications cause limited strains to develop because the soil dilates during the deformation. In this paper liquefaction resistance is defined as the cyclic stress ratio required to cause soil to reach initial liquefaction at a certain number of cyclic loading.

The liquefaction of a site can be determined by the following approaches. The first approach is to assume that the factors of stress history and density, which control liquefaction resistance, also affect penetration resistance. The blow count obtained by the standard penetration test or the cone tip resistance is related to the liquefaction resistance of the deposit by means of empirical equations based on penetrometer measurements at sites that did (or did not) liquefy, supplemented by the results of large-scale liquefaction tests. Such correlation obviously depends on the general applicability of penetration data obtained at relatively few sites. This method is generally believed to be conservative.

The second approach is to recover undisturbed, from the problem deposit, specimens that could be subjected to actual liquefaction tests. However, conventional sampling techniques will produce significant disturbance of clean sand specimens and the true liquefaction resistance will be considerably underestimated in laboratory tests.

Another approach is to correlate the state of laboratory specimens with in-situ soil characteristics using some appropriate index property. The maximum shear modulus of sandy soils, measured at a strain amplitude less than 10^{-6}, is influenced by many of the variables that influence liquefaction, such as mean effective confining stress, void ratio, stress history, over-consolidation ratio, sample preparation, soil fabric and soil type (Hardin, 1978, Mulilis et al, 1977). Therefore, once the relationship between maximum shear modulus and liquefaction resistance is established, the in-situ liquefaction resistance can be directly predicted by measuring the shear wave velocities.

This paper will discuss the applicability and limitations of using maximum shear modulus to evaluate the liquefaction resistance of sand.

LITERATURE REVIEWS

De Alba et al (1984) performed cyclic triaxial tests together with measurements of elastic wave velocities to establish the relationship between liquefaction resistance and shear- or compressive-wave velocities in saturated sand. The test results for six types of sands showed that a characteristic relationship between elastic-wave velocities and liquefaction resistance could be established for each sand.

Tokimatsu et al (1986) performed cyclic triaxial tests on sands with different densities and stress histories to investigate the relationship between maximum shear modulus and liquefaction resistance. Test results showed that a good correlation could be established for a given soil under a given confining pressure.

Chen and Chen (1991) performed resonant column tests and cyclic torsional shear tests on Taipei Sand to study the effects of sample preparation, relative density, effective confining pressure, stress history, and fines content on the relationship between liquefaction resistance and maximum shear modulus. Test results showed that a good relationship could be established for different relative densities and sample preparation methods under a given effective confining pressure and fines content, as shown in Fig. 1. However, if specimens are over-consolidated or liquefied previously, the relationship between maximum shear modulus and liquefaction resistance will be different, as shown in Fig. 2.

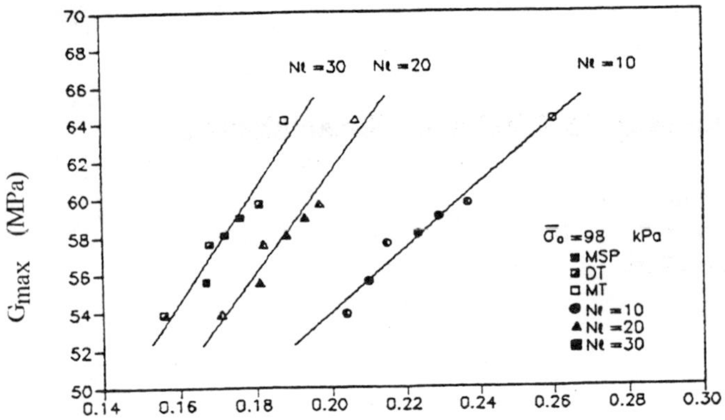

Fig. 1 Relationship between cyclic stress ratio to cause initial liquefaction and maximum shear modulus for Taipei sand with 10% fines content (MSP, multiple sieving pluviation; DT, dry tamping; MT, moist tamping) (Chen and Chen, 1991)

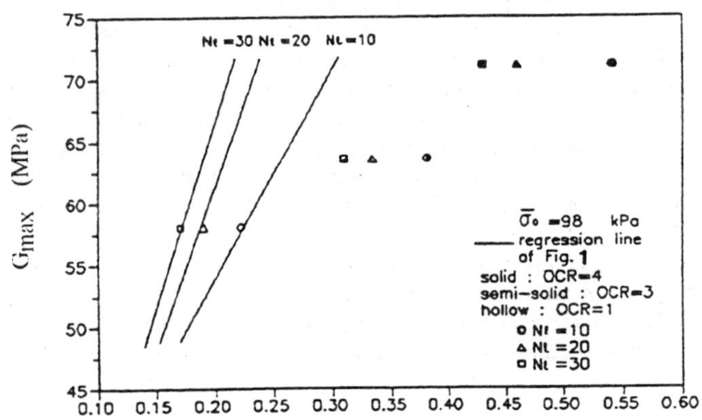

Fig. 2 Influence of over-consolidation on the relationship between cyclic stress ratio to cause initial liquefaction and maximum shear modulus (Chen and Chen, 1991)

Chen and Lee (1994) investigated the relationship between maximum shear modulus and liquefaction resistance of Lanyang sand. They found that a linear relationship exists between maximum shear modulus and liquefaction resistance if the maximum shear modulus is modified for different confining pressures. However, for liquefied specimens the relationship is different from that of virgin specimens (Fig. 3).

Robertson et al (1992) studied the liquefaction potential for sand using seismic cone penetration test. They proposed to use normalized shear-wave velocities to evaluate the critical stress ratio causing liquefaction. They suggested that shear-wave velocities should be normalized by the overburden pressure to the power of 0.25.

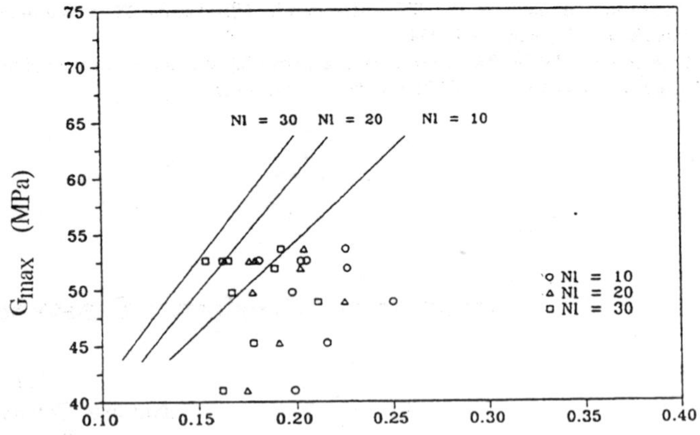

Fig. 3 Influences of liquefaction on the relationship between shear modulus and stress ratio, symbols are liquefied samples (Chen and Lee, 1994)

According to the previous research, there exists a relationship between maximum shear modulus and liquefaction resistance under certain confining pressure for a specific soil type. Tokimatsu and Uchida (1990) proposed a set of normalization procedures that could eliminate the effects of different confining pressures and soil types. After the normalization of maximum shear modulus, a unique relationship exists between normalized shear modulus and liquefaction resistance. The normalization procedures are described as follows.

Maximum shear modulus (G_{max}) can be expressed as a function of the mean effective confining pressure (σ_m') and void ratio (e) in the following equation (Hardin, 1978):

$$G_{max} = AF(e)(\sigma_m')^n \qquad (1)$$

in which, $\sigma_m'=1/3(\sigma_1'+\sigma_2'+\sigma_3')$, σ_1', σ_2' and σ_3' are the effective principal stresses. Parameters A and n are constants. Parameter A reflects the soil fabric induced by various stress histories. F(e) is a function of void ratio and has the following form depending on the shapes of soil particles:

$$\text{For angular shape of particles, } F(e) = (2.97-e)^2/(1+e) \qquad (2)$$

$$\text{For rounded shape of particles, } F(e) = (2.17-e)^2/(1+e) \qquad (3)$$

Because liquefaction resistance is the ratio of cyclic shear stress and effective confining pressure, it is generally unaffected by confining pressure (Peacock and Seed, 1968). From equation (1), it is shown that maximum shear modulus is a function of confining pressure. So to eliminate the effects of confining pressure, maximum shear modulus should be normalized in the following way:

$$G_1 = G_{max}/(\sigma_m')^n \qquad (4)$$

Although test results show that n value is generally about 1/2, Tokimatsu and Uchida (1990) found that maximum shear modulus normalized by n=2/3 is better than n=1/2. They thought the possible reason to be that the liquefaction resistance is slightly decreasing instead of being constant when the confining pressure increases.

To eliminate the effects of soil types, maximum shear modulus should be normalized in the following way:

$$G_n = G_1 / [F(e_{min})] \tag{5}$$

Substitute equations (1) and (4) into (5):

$$G_n = A\,[F(e) / F(e_{min})] \tag{6}$$

in which, e_{min} is the minimum void ratio. The ratio $[F(e) / F(e_{min})]$ reflects the influence of soil density, it equals one when $e = e_{min}$, and decreases as void ratio increases. By the above procedures, normalized shear modulus would have unique relationship with liquefaction resistance. This paper will examine the suitability of these normalization procedures.

TEST PROGRAM

Test Material and Equipment

The soil used in this study is Taipei sand. The soil particles are sub-angular. Natural soil has fines contents between about 8 to 14%. This study used several types of remolded samples with fines content ranging from 2 to 20%.

The test equipment used are resonant column test apparatus and torsional shear test apparatus. Resonant column tests were performed to obtain the maximum shear modulus of soil at a strain amplitude less than 10^{-6}. Torsional shear tests were performed to obtain the liquefaction resistance of soil.

Test Conditions

Specimens of different relative densities were prepared either by dry tamping, moist tamping, or multiple sieving pluviation methods. The soil was poured into the mold in five equal layers to control the density and uniformity. Specimens were circulated with CO_2 and de-aired water and subjected to a back pressure of 200 kPa to ensure good saturation. Specimens were consolidated at three kinds of effective confining pressures (0.5, 1, and 1.5 kg/cm^2) to study the effects of confining pressure on the relationship between maximum shear modulus and liquefaction resistance.

RESULTS AND DISCUSSIONS

Results of Taipei Sand

The maximum shear modulus of Taipei sand with different fines contents could be expressed in the following equations:

$$\text{F.C.} = 2\%,\quad G_{max} = 245(2.97-e)^2/(1+e)\,(\sigma_m')^{0.48} \tag{7}$$

$$\text{F.C.} = 5\%,\quad G_{max} = 235(2.97-e)^2/(1+e)\,(\sigma_m')^{0.47} \tag{8}$$

$$\text{F.C.} = 10\%,\quad G_{max} = 216(2.97-e)^2/(1+e)\,(\sigma_m')^{0.38} \tag{9}$$

$$\text{F.C.} = 15\%,\quad G_{max} = 216(2.97-e)^2/(1+e)\,(\sigma_m')^{0.45} \tag{10}$$

$$\text{F.C.} = 20\%,\quad G_{max} = 220(2.97-e)^2/(1+e)\,(\sigma_m')^{0.46} \tag{11}$$

in equations(7) to (11), the units of G_{max} and σ_m' are kgf/cm^2.

Fig. 4 shows the results of liquefaction tests of samples prepared by dry tamping method with fines contents equal to 10 and 15 %. It is shown in the figure that samples with higher fines content will have lower liquefaction resistance. The liquefaction resistance of Taipei sand with fines contents 2, 5, 10, 15, and 20%, at confining pressure 1 kgf/cm^2, is plotted versus maximum shear modulus in Fig. 5. The liquefaction resistance in this figure is the cyclic stress ratio needed to cause initial liquefaction in 10 cycles. From the figure, it is shown that liquefaction resistance increases with increasing maximum shear modulus in a slightly non-linear way, but using linear regression will still have good correlation. The relationship between liquefaction resistance and maximum shear modulus will shift to the right-hand side as the fine content increases. Under the conditions of equal maximum shear modulus, samples with higher fines contents will have lower liquefaction resistance.

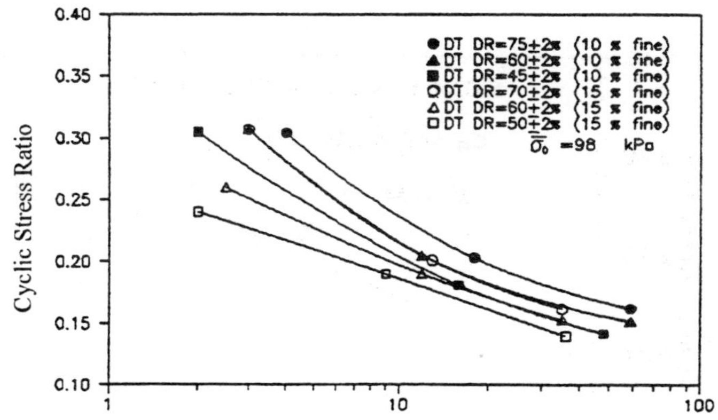

Fig. 4 Cyclic stress ratios to cause initial liquefaction versus cyclic numbers (fines contents=10 & 15%; DT, dry tamping method)

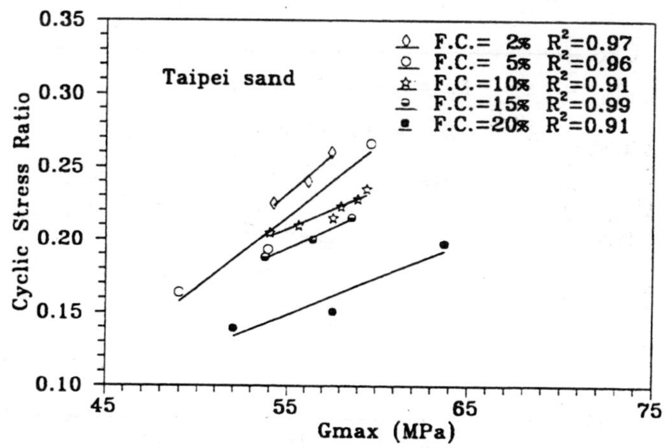

Fig. 5 Maximum shear modulus versus cyclic stress ratio to cause initial liquefaction at 10 cycles for Taipei sand with different fines contents (R^2, correlation)

Normalized Relationship for Taipei Sand

Tokimatsu and Uchida (1990) suggested to use equation (6) for the normalization of maximum shear modulus and recommended the n value to be 2/3. In this study, the maximum shear moduli of Taipei sand for various fines contents were normalized according to equation (6). Fig. 6 shows the relationship between normalized shear modulus and liquefaction resistance for n=1/2. The correlation coefficient R^2 equals 0.54. Fig. 7 shows the relationship between normalized shear modulus and liquefaction resistance for n=2/3. The correlation coefficient R^2 equals 0.27. From Figures 6 and 7, it is shown that normalization by n=1/2 is better than n=2/3. This is different as compared with Tokimatsu and Uchida (1990). From equations (7) to (11), it is shown that the values of n for Taipei sand with different fines contents are about 0.4 to 0.5, so the results are better for normalization by n=1/2 than by 2/3.

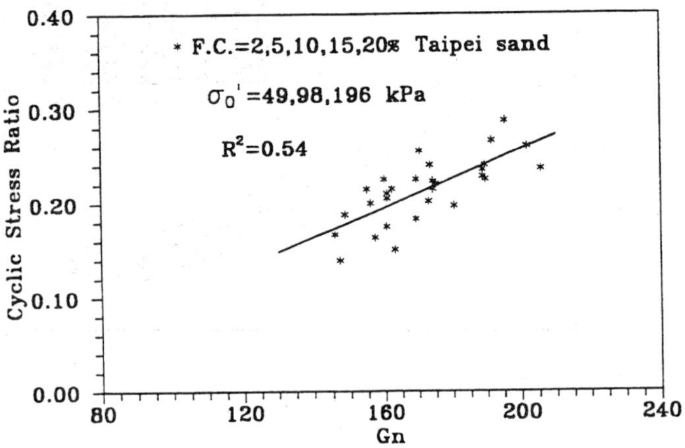

Fig. 6 Normalized shear modulus versus liquefaction resistance for Taipei sand (n=1/2)

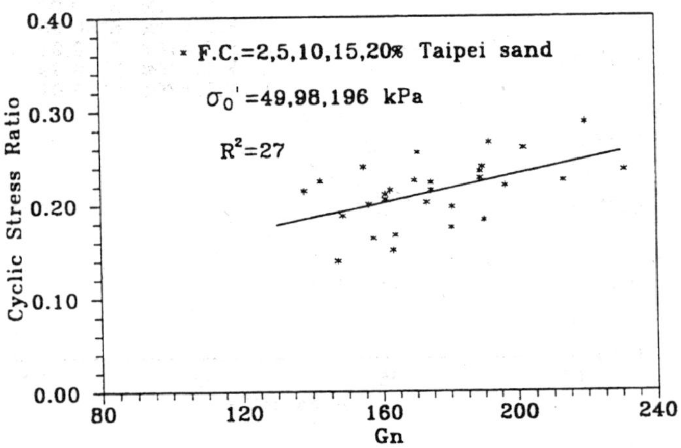

Fig. 7 Normalized shear modulus versus liquefaction resistance for Taipei sand (n=2/3)

From the original ideas of the normalization procedures, if liquefaction resistance is affected by confining pressure then maximum shear modulus should be normalized by the individual n value for specific soil type to obtain the best results. In Fig. 8, the maximum shear moduli were normalized by using the individual n values from equations (7) to (11) and were plotted versus liquefaction resistance. The correlation coefficient R^2 equals 0.59, which is slightly better than n=1/2.

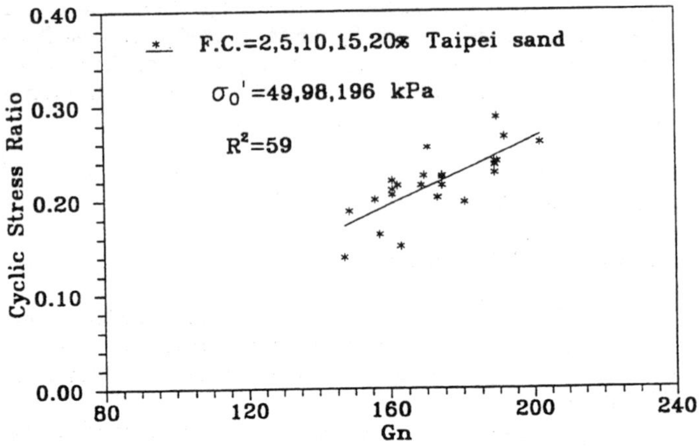

Fig. 8 Normalized shear modulus versus liquefaction resistance for Taipei sand (with individual n values)

The test results of Taipei sand were plotted together with Tokimatsu and Uchida's test results in Fig. 9. It is shown that different relationships between normalized shear modulus and liquefaction resistance are observed. After examination of the normalization procedures, it is found that the differences come from the different functions of void ratio used in Equation (6). The particle shapes are rounded for Tokimatsu and Uchida's tests, while in this study they are sub-angular. If the test results of Taipei sand are normalized by using the void ratio function of rounded particles, that is equation (3), they will compare quite well with Tokimatsu and Uchida's results, as shown in Fig. 10.

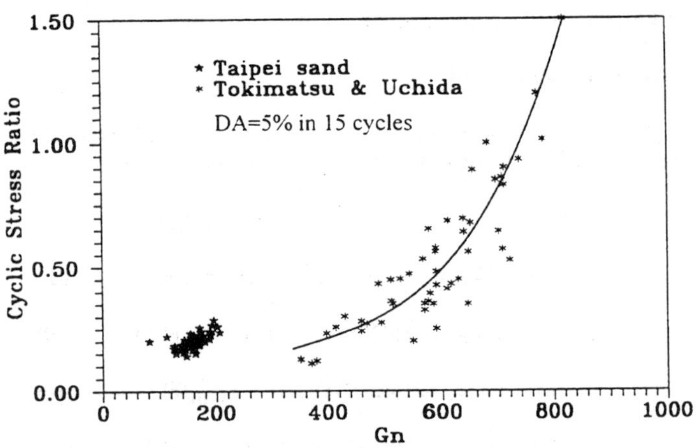

Fig. 9 Comparisons with Tokimatsu and Uchida's test results using different void ratio function

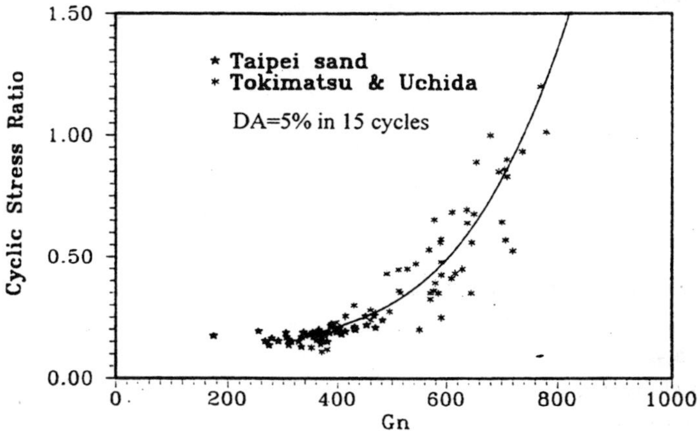

Fig. 10 Comparisons with Tokimatsu and Uchida's test results using same void ratio function (DA=5% in 15 cycles)

Effects of anisotropic consolidation

In the field conditions, soils are consolidated under different vertical and horizontal stresses. Previous tests were performed under isotropic consolidation. It is important to know the effects of anisotropic consolidation on the relationship between cyclic stress ratio and maximum shear modulus. The maximum shear modulus of Taipei sand is not much influenced by the anisotropic consolidation. For stress ratio (vertical to horizontal) less than 3 the effects are less than 10%. The effects of anisotropic consolidation on the liquefaction resistance can be normalized by defining the cyclic stress ratio as the ratio between the cyclic shear stress and the mean effective confining pressure (Chen and Laih, 1995). A series tests were performed to investigate this problem. Samples were consolidated under different stress ratios first. Then, they were subjected to undrained cyclic shear tests after consolidation. The relationship between normalized shear modulus and cyclic stress ratio for anisotropically-consolidated samples is shown in Fig. 11. It is shown that anisotropic consolidation has little effects on the relationship between cyclic stress ratio and normalized shear modulus.

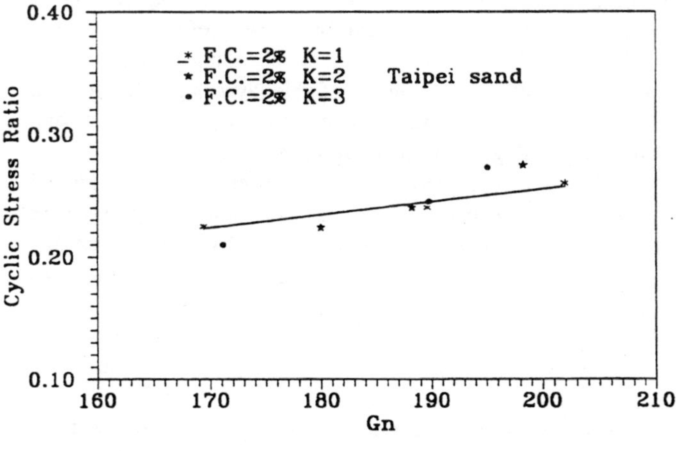

Fig. 11 Effects of anisotropic consolidation on the relationship between cyclic stress ratio and normalized shear modulus

CONCLUSIONS

(1) Considering the test results of Taipei sand, liquefaction resistance will decrease with increasing fines contents under the conditions of same maximum shear modulus.
(2) The relationship between normalized shear modulus and liquefaction resistance will be better if the maximum shear modulus is normalized by the individual n value for specific soil type.
(3) The void ratio functions used in the normalization procedure are different for rounded and angular particles. This will result in different normalized relationship for different particle shapes.

ACKNOLEGEMENT

This work is financially supported by the National Science Council, Republic of China, under Grant No. NSC 84-2211-E011-025.

REFERENCES

Chen, YC, and Chen, ML (1991). "Evaluation of Liquefaction Potential by Shear Modulus of Sand," *Journal of the Chinese Institute of Civil and Hydraulic Engineering*, Vol 3, No 3.

Chen, YC, and Laih, YF (1995). "Relationship between Maximum Shear Modulus and Liquefaction Resistance of Sand under Anisotropic Consolidation," *Journal of the Chinese Institute of Civil and Hydraulic Engineering*, Vol 7, No 2.

Chen, YC, and Lee, CG (1994). "Evaluation of Liquefaction Resistance of Sand by Maximum Shear Modulus," *Journal of the Chinese Institute of Engineers*, Vo 17, No 5.

De Alba, P, Baldwin, K, Janoo, V., Roe, G, and Celikkol, B (1984). "Elastic-Wave Velocities and Liquefaction Potential," *Geotechnical Testing Journal, GTJODJ*, Vol 7, No 2.

Hardin, BO (1978). "The Nature of Stress-Strain Behavior of Soil", *Proceeding of the ASCE, Geotechnical Engineering Division Specialty Conference*, Vol 1.

Mulilis, JP, Seed, HB, Chan, CK, Mitchell, JK, and Aluranandan, K (1977). "Effects of Sample Preparation on Sand Liquefaction," *Journal of the Geotechnical Engineering Division, ASCE*, Vol 103, No GT2.

Peacock, WH, and Seed, HB (1968). "Sand Liquefaction Under Cyclic Loading Simple Shear Conditions," *Journal of the Soil Mechanics and Foundations Division, ASCE*, Vol 94, No SM3.

Robertson, PK, Woeller, DJ, and Finn, WDL (1992). "Seismic Cone Penetration Test for Evaluating Liquefaction Potential Under Cyclic Loading," *Canadian geotechnical Journal*, Vol 29.

Tokimatsu K, and Uchida A (1990). "Correlation between Liquefaction Resistance and Shear Wave Velocity", *Soils and Foundations* , Vol 30, No 2.

Tokimatsu, K, Yamazaki, T, and Yoshimi, Y (1986). "Soil Liquefaction Evaluation by Elastic Shear Moduli," *Soils and Foundations* , Vol 26, No 1.

Proceedings of the Seventh (1997) International Offshore and Polar Engineering Conference
Honolulu, USA, May 25-30, 1997

ISBN 1-880653-28-1 (Set); ISBN 1-880653-29-X (Vol. I)

Liquefaction Behavior of Sandy Soil Sandwiched by Clay Layers

Tej B.S. Pradhan
Yokohama National University
Yokohama, Japan

ABSTRACT

It has been attempted to clarify experimentally the mechanism of liquefaction behavior of a clean sand sandwiched by reconstituted marine clay layers. Post liquefaction settlement behavior of such a composite ground has also been investigated. A series of undrained cyclic triaxial tests have been performed on composite soil specimens with sand layer sandwiched by clay layers. A special technique was used to measure the pore pressure developed at the middle sand layer. When the excess pore pressure at the sand layer has exceeded about 95% of the effective confining pressure, the cyclic loading was terminated and allowed to drain from either the sand layer or clay layers. The effect of the drainage layer on the settlement of the composite ground has been investigated. It has been found that the liquefaction strength of sand layer is not much affected by the presence of clay layer irrespective of its thickness. However the local strains induced in sand and clay layers are affected by the thickness of clay layer.

KEY WORDS: Sand, Clay, Composite ground, Liquefaction, Post Liquefaction Settlement, Drainage layers

INTRODUCTION

In general a natural ground is composed of alternating layers of sand, silt or clays. Analytical and experimental studies have shown that liquefaction of sand could occur at some depth below the ground surface (Lee and Seed, 1967). After the subsurface sand layer is liquefied, the excess pore water pressure therein is dissipated through the overlying soil layer which is not liquefied. From the case histories of the ground liquefaction, it is reported that the damage at the surface was less due to the existence of a thick layer of non-liquefiable soil like clay overlaying a liquefied sand layer (for example, Ishihara et al., 1985).

Yoshimi et al. (1973) has proposed a numerical method to calculate the pore water pressure development in the surface non-liquefied soil overlying a completely liquefied sand stratum. They have concluded that if the compressibility of the surface soil is one order of magnitude smaller than that of the liquefied soil, the maximum pore pressure in the surface soil is not significantly affected by the initial pore pressure in the surface soil and also by the relative thickness of the surface soil relative to the liquefied soil.

Ishihara (1985) has pointed out the significance of the thickness of non-liquefiable clay layer overlying a liquefied sand layer on the ground damage. Ishihara et al. (1985, 1993), based on a vast amount of data obtained from the Niigata earthquake (1964), the Nihonkai-chubu earthquake (1983), the Tozan earthquake (1976) and the Luzon earthquake (1990), has proposed a boundary line separating the surface ground damage like sand boil, structure damage etc., due to the liquefaction of sand layer which is overlaid by non-liquefied clay layer. According to them, no significant surface damage will occur if the thickness of surface non-liquefiable layer is greater than 3m irrespective of the thickness of the underlying liquefied sand layer for a peak ground acceleration of 200 gal.

Another severe problem is the settlement of the ground surface after the dissipation of the accumulated excess pore water pressure. Some research specifically focused on the settlement of sandy ground after liquefaction due to earthquakes can be found (for example, Tokimatsu et al., 1987; Ishihara et al., 1992 among others). However, research on the effect of the drainage layer in case of composite grounds on settlement behavior has not been done so far.

From the above mentioned facts, it seems necessary to investigate the effects of the existence of surface clay layer on the liquefaction behavior of underlying sand layer. This study is focused on following two points: (a) liquefaction behavior of sand layer sandwiched by clay layers and (b) the effect of drainage layer on the settlement behavior of the composite ground associated with the liquefaction of underlying sand layer due to the dissipation of accumulated pore water pressure.

TESTING MATERIALS

The grading curves of the sand and the clay used is shown in Fig. 1. The basic properties of the soils used to fabricate the composite specimen are as follows.

Sand: A Japanese standard sand called Toyoura sand was used. The physical properties are: grain density 2.64 Mg/m^3, maximum void ratio 0.997, minimum void ratio 0.605, mean particle size 0.16 mm and zero percent of fines less than 75 μm.

Clay: A dredged clay from Tokyo bay was passed through a 75 μm sieve and used as clay sample. The physical properties are: grain density 2.65 Mg/m^3, Liquid limit 75.4%, Plasticity index 37.

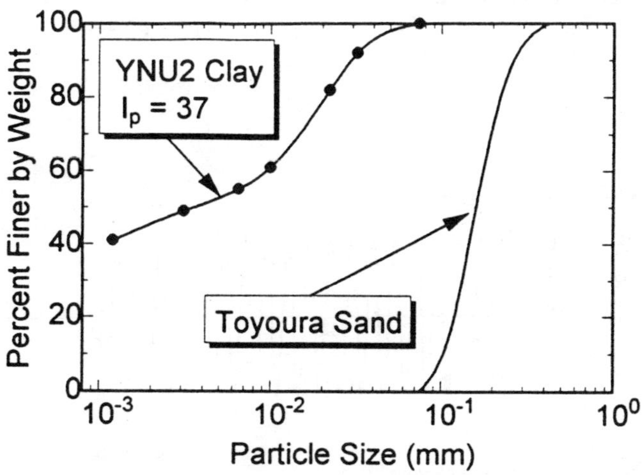

Fig.1 Grading curves of Clay and Sand

Photo -1 A composite specimen

TESTING METHOD AND TESTING CONDITIONS

The sand specimens were made by air pluviation method in a mold (5 cm in inner diameter) so as to obtain a relatively loose state with a relative density of (50 $\pm$3) %. Deaired water was percolated so as to make it moist but not fully saturated. Partial saturation state was found to be adequate so as not to overcome excess volume expansion of the specimen during freezing. Then it was frozen one dimensionally so as not to

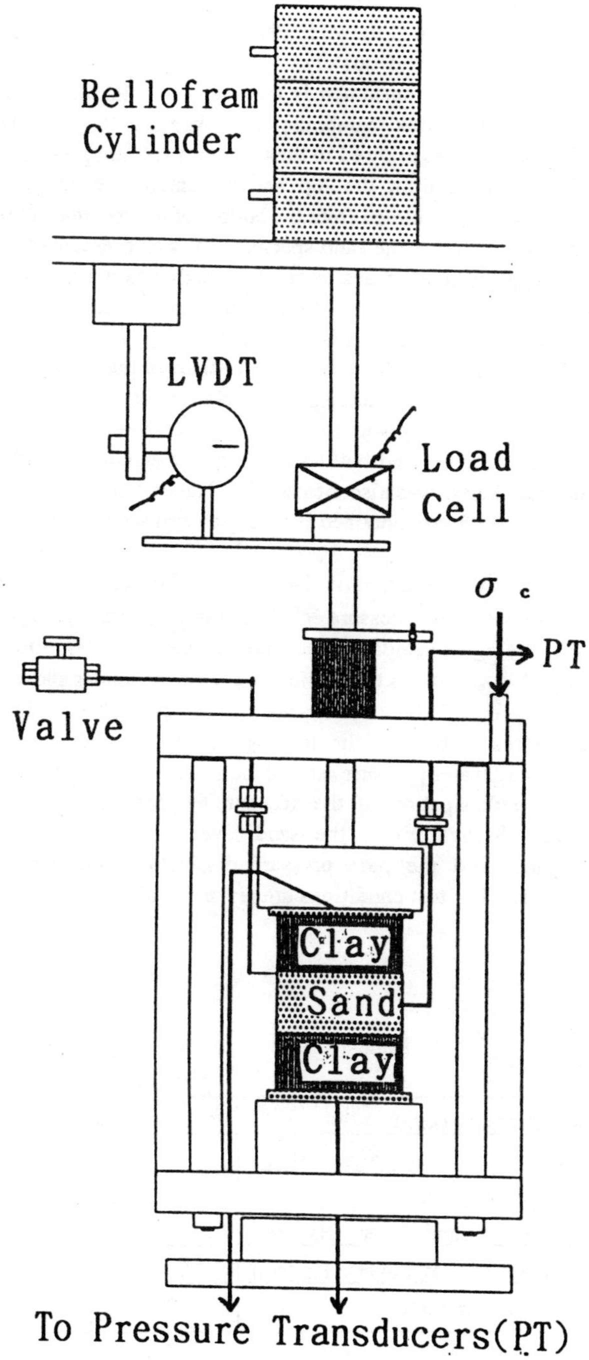

Fig.2 Triaxial apparatus and the specimen

disturb the structure of the specimen.

The clay sample was first thoroughly mixed at a water content of twice the liquid limit and then one dimensionally preconsolidated in a mold (30 cm in inner diameter and 35 cm in height) at a vertical pressure of 50 kPa. After preconsolidation was completely finished after about 10 days, the sample was trimmed to a size of diameter 5 cm and a specified length according to the testing condition.

The frozen sand specimen trimmed to a specified length was sandwiched between clay specimens so that the total height of the specimen was 10 cm. A typical photograph of the composite specimen is shown in Photo-1. The height Hs of the sand layer is varied adequately.

The composite specimen is then set in a triaxial cell and wrapped with a latex membrane of thickness 0.25 mm. The sand specimen was allowed to thaw under a vacuum pressure of 20 kPa applied from the top and bottom of the specimen. Deaired water percolated from the bottom of the specimen to the top so as to saturate the sand specimen. A back pressure of 200 kPa is applied in all cases to ensure a higher saturation of the composite specimen. A triaxial loading system and the condition of the specimen is illustrated in Fig. 2. A technique was used to measure the pore water pressure at the periphery of the midheight of the specimen as depicted in Fig. 2. The details of this method is given in Pradhan et al.(1995). The pore pressure at the top and bottom of the specimen corresppondng to the clay layers was also measured. The Bishop's B-value of more than 0.96 was obtained both at clay and sand layers in all cases.

After the specimen is isotropically consolidated to an effective confining pressure σ_c' of 100 kPa, undrained cyclic loading using sinusoidal waves under a constant stress ratio SR (σ_d /2 σ_c': σ_d is the cyclic deviator stress) under a loading frequency of 0.1 Hz. Tests were carried out under different cyclic stress ratios. Cyclic loading was terminated when an excess pore water pressure ratio of the sand layer exceeded 95%. Then the dissipation of the accumulated pore pressure was allowed through either the sand layer or clay layer. The settlement and the pore pressure dissipation behavior was monitored. The test conditions are given in Table-1.

Table-1 Test conditions

Back Pressure (kPa)	200			
Effective Confining Pressure (kPa)	100			
Loading Frequency (Hz)	0.1			
Thicknes of Sand Layer Hs (cm)	0	3	5	10
Thickness of Clay Layer Hc (cm)	10	7	5	0
Drainage from the Layer	Clay		Sand	

TEST RESULTS AND DISCUSSIONS

1. Behavior during undrained cyclic loading:

A typical test result for a specimen with 5 cm thickness of sand layer is shown in Fig. 3. In the figure, time histories of stress ratio SR, total axial strain ε and excess pore water pressure ratio $\Delta u/\sigma_c$' at the sand layer and clay layers are shown. From Fig. 3(c), it can be noticed that the pore pressures at the top and bottom clay layers are almost the same while the pore pressure at the sand layer is quite higher compared to that in clay layers. If $\Delta u/\sigma_c$' at sand layer is too high, the migration of excess pore pressure seemed to occur through the inner side of membrane since the membrane gets loose when the effective confining pressure becomes very low. The test was terminated before such phenomena was observed. It is quite interesting to note that the excess pore pressure ratio in the clay layer is only about 20% even though the ratio is more than 95% in the sand layer for a certain SR.

The pore pressure in the clay layer was measured at the top and bottom and hence it can be surmised that the pore pressure in clay at the vicinity of sand layer may be slightly larger. However it is quite difficult to measure this pore pressure.

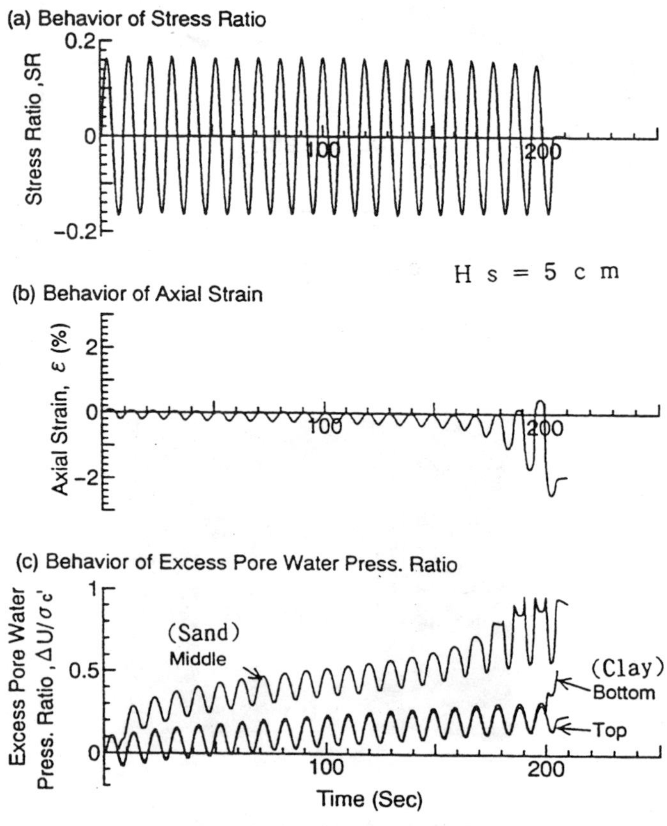

Fig.3 Typical test results

The relationship between the stress ratio SR and the number of cycles to induce a total double amplitude axial strain $\varepsilon_{DA,t}$ of 1% and 2% are shown in Figs. 4(a), (b) for specimens with different thickness Hs of sand layers. Following points could be indicated from these figures. (a) Undrained cyclic strength of clay specimen is larger than that of sand and composite specimens. This is quite understandable since relatively loose sand specimen was used. (b) Undrained cyclic strength of composite specimen is almost the same as that of sand specimen irrespective of the thickness of sand layer. This observation suggests that the liquefaction behavior of sandwiched sand layer is not affected by the overlying or underlying clay layer. (c) The total double amplitude axial strain of the composite specimen is small even though the excess pore pressure ratio in the sandwiched sand layer is more than 95%. However, it can be surmised that the local strains in individual layers are different since the accumulated excess pore pressures are different.

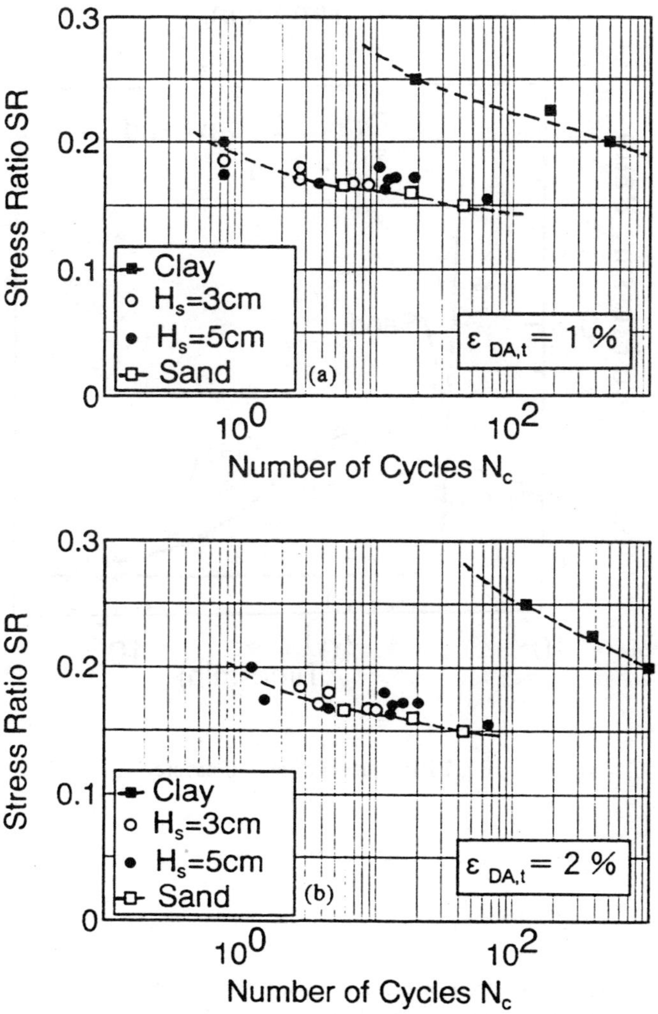

Fig.4 Stress ratio vs loading cycles

Fig. 5 shows the relationship between the accumulated excess pore pressure ratio $\Delta u/\sigma_c'$ and induced double amplitude axial strain for clay specimen only (no sand layer) subjected to different stress ratios. There exists an unique relationship irrespective of the applied stress ratio and can be expressed by a hyperbolic function as Eqn. 1.

$$\varepsilon_{DA,C}(\%) = (\Delta u/\sigma_c')_{Clay} / \{1-(\Delta u/\sigma_c')_{Clay}\} \quad \text{------ (1)}$$

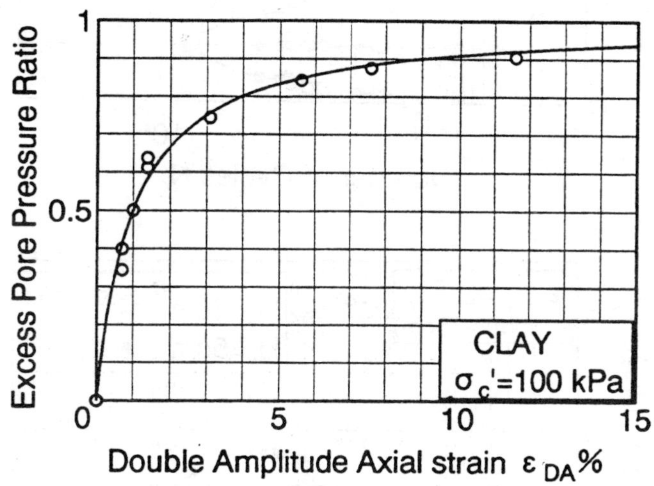

Fig.5 $\Delta u/\sigma_c'$ vs ε_{DA} for clay

Here $\varepsilon_{DA,C}$ is the double amplitude axial strain and $(\Delta u/\sigma_c')_{Clay}$ is the accumulated excess pore pressure ratio for the clay specimen only. This relationship is used to calculate the local strain $\varepsilon_{DA,S}$ induced in sand layer by taking into account the compatibility condition of the total deformation of each layer. Figs. 6 (a) and (b) show the local double amplitude axial strain of each layer in the composite specimens plotted against number of cyclic loading Nc for a given total double amplitude axial strain $\varepsilon_{DA,t}$. The data points for higher Nc corresponds to lower values of applied stress ratio SR. Following points could be noticed from the figures. (a) The local strain in sand layer is extremely higher than that in clay layer. This may be due to the difference in the accumulated pore pressure. (b) The local strain in the clay layer is independent of sand layer. However, the local strain in sand layer is larger when the thickness of sand layer is smaller for a given number of loading cycle. (c) If the level of applied stress ratio is small or at larger number of loading cycles, there is a tendency that the composite specimen deforms rather uniformly.

2. Settlement behavior due to the dissipation of pore pressure.

Settlement behavior due to the dissipation of accumulated pore pressure from either sand layer or clay layer has been

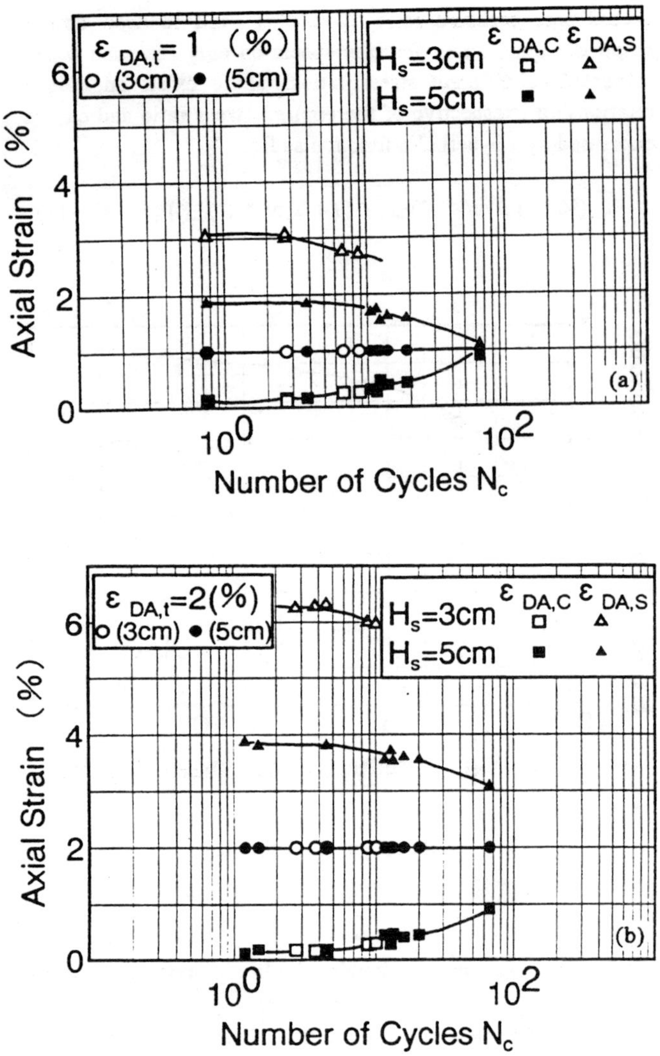

Fig.6 Axial strain vs loading cycles

investigated. Figs. 7(a) and (b) show respectively the dissipation process of pore pressure with time when the drainage layer is either clay or sand . When the drainage is allowed from clay layer, the accumulated pore pressure in the sand layer has to dissipate through the clay layer and hence the time for the full dissipation depends on the thickness of clay layer. When the drainage is allowed from the sand layer, it is interesting to note that the pore pressure of the clay layer is increased at first and then start to dissipate. When the cyclic loading is terminated, the migration of pore pressure occurs from sand layer to clay layer resulting in the increase in pore pressure in clay layer at the vicinity of sand layer. Another important point to be noticed is that the time for the full dissipation of pore pressure is almost independent of the drainage layer if the accumulated pore pressure in each layer is the same.

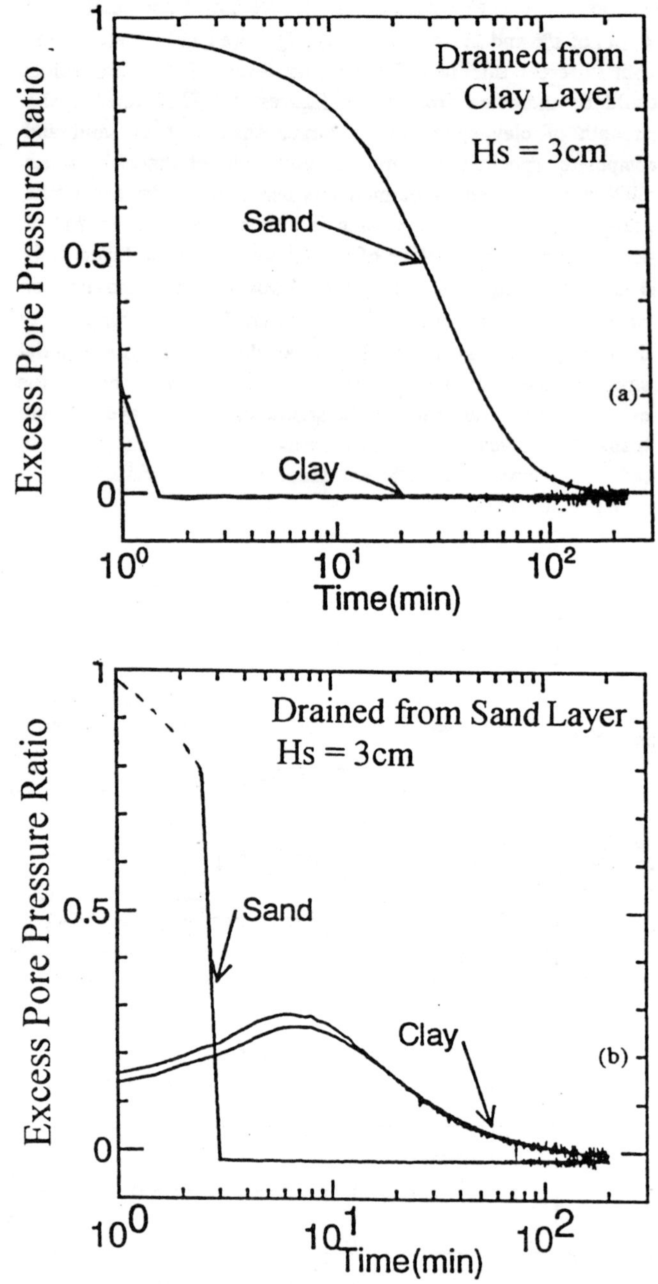

Fig.7 Dissipation of pore pressure

The axial strain measured during the dissipation of pore pressure is plotted against time in Figs. 8(a) and (b). It is obvious that when the drainage is allowed from sand layer, a large instaneous settlement occurs at the beginning due to the fast dissipation of pore pressure in sand layer. The final settlements shown in Figs. 8(a) and (b) are different because the induced total double amplitude axial strains are different in two cases.

Fig. 9 shows the relationship between the accumulated excess pore pressure ratio ($\Delta u/\sigma_c'$) in clay layer and the axial strain induced due to the dissipation of the excess pore pressure for different specimens. The excess pore pressure ratio in sand layer is greater than 95% in all cases. It can be observed that some unique relationship seems to exist in case of clay specimen only, however, no clear relationship exist for composite specimens and sand specimens. Fig. 10 shows the relationship between the total double amplitude axial strain $\varepsilon_{DA,t}$ after the cyclic loading is terminated and the associated axial strain due to the dissipation of the accumulated pore pressure.

Following points could be noticed from the figure. (a) A positive correlation can be seen between the total axial strain due to the dissipation of pore pressure and the induced $\varepsilon_{DA,t}$ after the cyclic loading for all kinds of specimens. (b) Axial strain is larger at smaller $\varepsilon_{DA,t}$ in case of clay specimen and vice versa for sand specimen. The axial strains for composite specimens lie in between them. (c) No significant difference was observed in axial strain due to the thickness of the sandwiched sand layer.

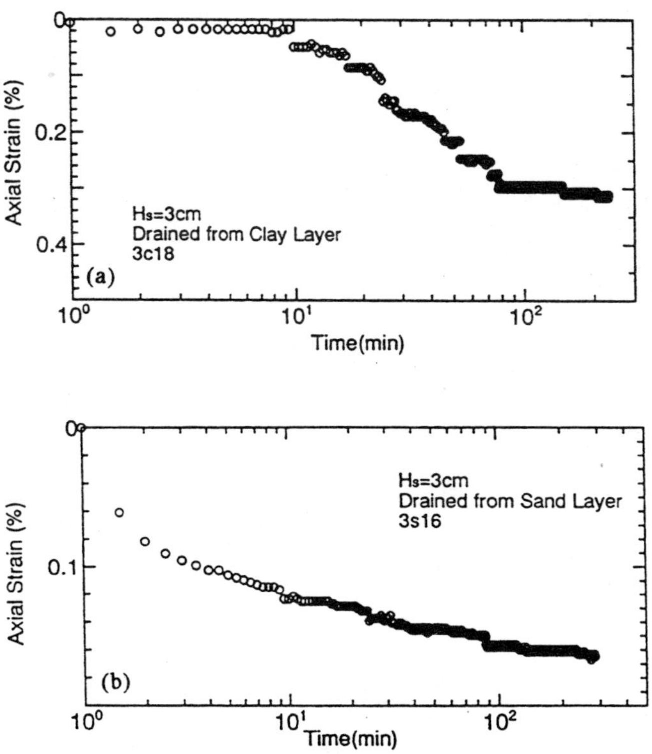

Fig.8 Settlement behavior

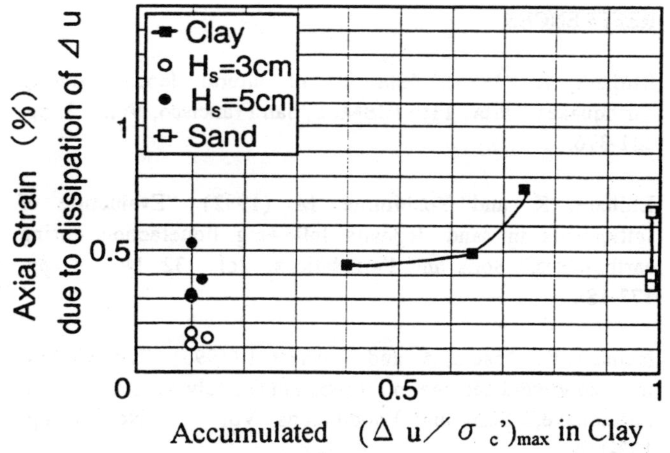

Fig.9 Axial strain vs $\Delta u/\sigma_c'$

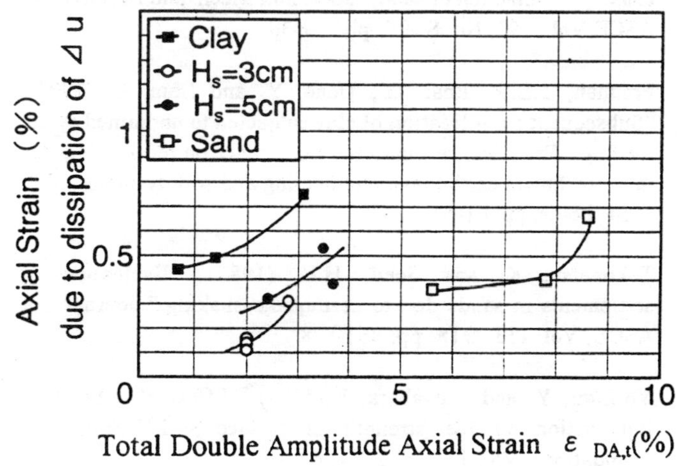

Fig.10 Axial strain vs $\Delta u/\sigma_c'$

CONCLUSIONS

Following conclusions can be drawn from the limited number of tests carried out in the present research.

1. The liquefaction strength of sand layer is not much affected by the presence of clay layer and its thickness. However, the local strains induced in sand layer is extremely higher than that in clay layer and is affected by the thickness of sandwiched sand layer.

2. The settlement behavior is affected by the drainage condition. However, the time for the full dissipation of pore pressure is almost independent of the drainage layer if the accumulated pore pressure in each layer is the same.

3. A positive correlation can be seen between the total axial strain due to the dissipation of excess pore water pressure and the induced total double amplitude axial strain after the cyclic loading.

REFERENCES

Ishihara, K. (1985). "Stability of natural deposits during earthquakes," Proc. 11th ICSMFE, San Francisco, Vol. 2, pp. 321-376.

Ishihara, K. and Yoshimine, M. (1992). "Evaluation of settlements in sand deposits following liquefaction during earthquakes," Soils and Foundations, Vol. 32, No. 1, pp. 173-188.

Ishihara, K., Acacio, A. and Towhata, I (1993). "Liquefaction induced ground damage in Dagupan in the July 16, 1990, Luzon Earthquake," Soils and Foundations, Vol. 33, No. 1, pp. 133-154.

Lee, K.L. and Seed, H.B. (1967). "Cyclic stress conditions causing liquefaction of sand," Jour. Soil Mech. and Found. Div., ASCE Vol. 90, No. SM 5, pp. 1-30.

Pradhan, T.B.S., Imai, G., Ueno, Y. and Dam, L. (1995). "Subsequent consolidation of clay subjected to undrained cyclic loading," Proc. 3rd International Conference on recent advances in geotechnical earthquake engineering and soil dynamics, Vol. 1, St. Louis, pp. 61-64.

Tokimatsu, K. and Seed, H.B. (1987). "Evaluation of settlements in sands due to earthquake shaking," Journal of ASCE, Vol. 113, GT8, pp. 861-878.

Yoshimi, Y. and Kuwabara, F (1973). "Effect of subsurface liquefaction on the strength of surface soil," Soils and Foundations, Vol. 13, No. 2, pp. 67-81.

Proceedings of the Seventh (1997) International Offshore and Polar Engineering Conference
Honolulu, USA, May 25-30, 1997

ISBN 1-880653-28-1 (Set); ISBN 1-880653-29-X (Vol. I)

Behavior of Gravity Type Quay Wall During Earthquake Observed in Model Shaking Table Test

K. Miura and E. Kohama
Hokkaido University, Sapporo, Japan
S. Kurita
Hokkaido Development Bureau, Sapporo, Japan
N. Ohtsuka
North Japan Port Consultants Co., Ltd., Sapporo, Japan
N. Yoshida
Sato Kogyo Co. Ltd., Tokyo, Japan

ABSTRACT

The mechanism of the damage to gravity type quay walls caused by the liquefaction of the backfill ground during the earthquake is made clear. A series of model shaking table tests were conducted in order to obtain the behavior of the caisson type quay wall during the earthquake. The comparative examination showed that the movement of the caisson were quite different whether liquefaction occurred in the backfill ground or not. The fluctuating earth pressure on the caisson has a tendency to suppress the movement of the caisson when liquefaction does not occur at the backfill. On the other hand, the fluctuating component of earth pressure and the inertia force coincide in phase angle when liquefaction occurs in the backfill ground.

KEY WORDS: Gravity Type Quay Wall, Earthquake, Liquefaction, Inertia Force, Earth Pressure, Shaking Table

INTRODUCTION

Significant damages were caused by liquefaction in the port and harbor facilities, especially those in reclaimed lands, during the recent big earthquakes in Japan. The functions of quay walls were frequently lost severely when the gravity type quay wall caissons were displaced toward sea and/or subsided. It looks as if the damage was enhanced by the occurrence of the liquefaction of the backfill (e.g., Japanese Geotechnical Society, 1994; Inagaki, et al., 1966; Kamon, et al., 1996). Close examinations on the damages to quay walls in Hokkaido Island, Japan, during three recent big earthquakes, i.e., 1993 Kushiro-oki, 1993 Hokkaido Nansei-oki and 1994 Hokkaido Toho-oki earthquakes, clarified that the degrees of the damage are strongly related to the occurrence of the liquefaction in the backfill ground (Japanese Geotechnical Society, 1994; Hokkaido Development Bureau, 1996). A clear proof is seen by looking at the damage to treated quay walls. Remedial measures against liquefaction have been positively conducted in Hokkaido Island after the 1983 Nihonkai-chubu earthquake, and the quay walls whose backfill ground was improved by a combination of sand the compaction pile and gravel drain methods survived these earthquakes with slight damage, whereas quay walls whose backfill was not treated sometimes suffered seriously.

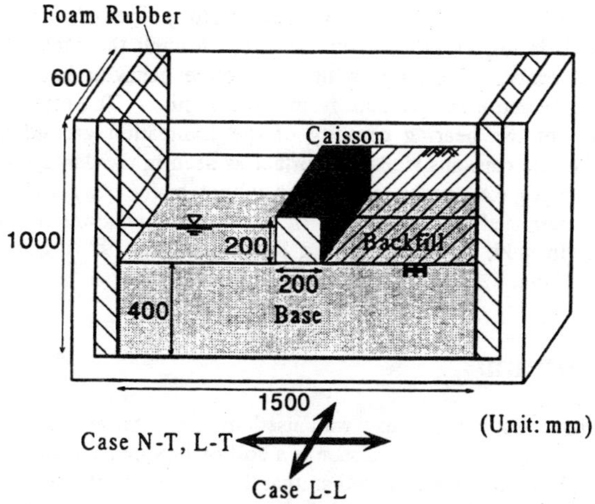

Fig. 1(a)

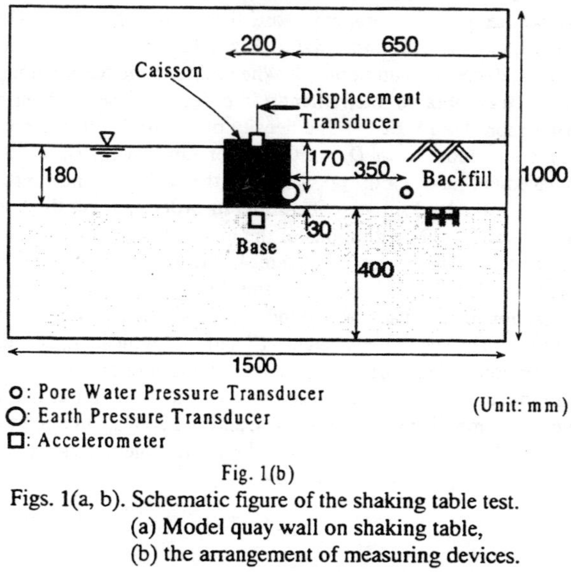

Fig. 1(b)

Figs. 1(a, b). Schematic figure of the shaking table test.
(a) Model quay wall on shaking table,
(b) the arrangement of measuring devices.

Table 1. Physical properties of siliceous sand used

Grain density ρ_s	Mean diameter D_{50}	Uniformity coefficient, U_c	Maximum density, ρ_{dmax}	Minimum density, ρ_{dmax}
2.717 g/cm^3	0.18 *mm*	1.82	1.61 g/cm^3	1.255 g/cm^3

Table 2. Test cases

Test case	Relative density, *Dr*	Liquefaction	Vibrated direction
N-T	72%	No Liquefaction	Transverse
L-T	31%	Liquefaction	Transverse
L-L	38%	Liquefaction	Longitudinal

The causes of the damage to quay walls are classified into

- Inertia force of the caisson
- Earth pressure acting on the back of the quay wall
- Reduction of the bearing capacity of the foundation ground

Inertia force is a body force induced by the vibration through the foundation, which acts in the opposite direction to the acceleration. Both static and fluctuating components of the earth pressure acts on the back wall of the caisson. The fluctuation of the earth pressure occurs from the interactive behavior between the vibrations of the caisson and the backfill. In the present study, the behavior of the quay wall during vibration is investigated through the shaking table tests and analysis focusing on the interactive behavior between the inertia force and earth pressure from the backfill. The effect of the reduction of the bearing capacity of the foundation ground to the damage to the caissons is also important as seen in the damage during the1995 Hyogoken-nanbu earthquake where many quay walls were severely damaged by liquefaction of the sand foundations as well as backfill (Inagaki, et al., 1966; and Kamon, et al., 1996), but is not discussed here.

TEST METHOD

Model caisson type quay wall used in the shaking table tests is shown in Fig. 1(a). The caisson is a hollow box made of steel plates, in which lead balls were stuffed so that the total mass of the caisson is 50kg and averaged density is 2.2t/m^3. Both the base underlying the caisson and the backfill ground were made with the same siliceous sand whose physical properties are listed in Table 1. The sand ground was deposited under water (Kiku, 1993) in order to obtain the sufficient degree of saturation. When making the foundation ground, the base was vibrated intermittently in order to possess a high relative density more than 90%. The density of the back fill was controlled to be either loose with D_r = 30-40% or dense with D_r = 70-80 %. Displacement and earth pressure on the caisson and pore water pressure in the backfill were measured in order to grasp the behaviors of quay walls. The locations of the measuring devices are shown in Fig. 1(b). Three test cases are listed in Table 2. The first characters "L" and "N" in case name indicate the occurrence of the liquefaction in the backfill ground: "L" for liquefy and "N" for not liquefy. The second characters "T" and "L" indicate the direction of the vibration: "T" for transverse to the direction of caisson and "L" for longitudinal. The input sinusoidal wave acceleration at the shaking table is common for all the test cases, and is 220 Gals in amplitude and 0.4 second in period. Details of the testing method is described in Kohama, et al. (1996) and Oka (1996).

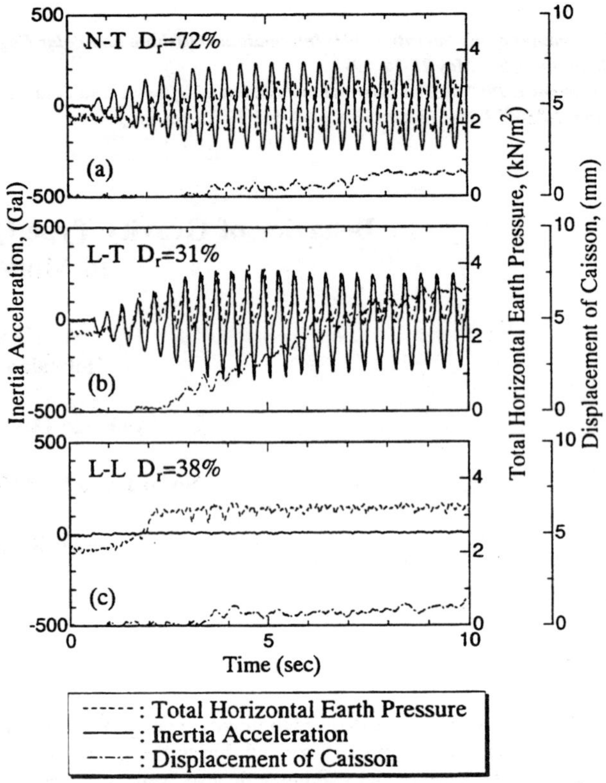

Figs. 2(a-c). Time histories of acceleration, earth pressure and displacement measured in;
(a) Case N-T, (b)Case L-T, (c) Case L-L.

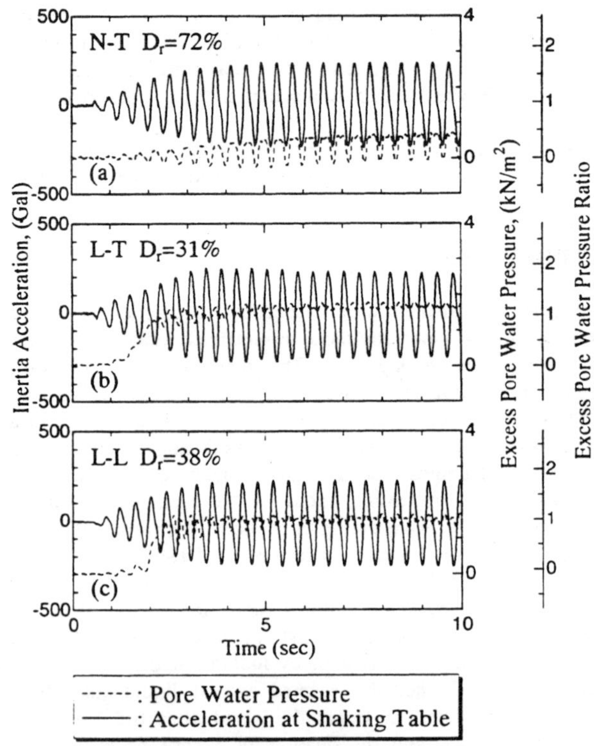

Figs. 3(a-c). Time histories of acceleration and pore water pressure in backfill measured in the test.
(a) Case N-T, (b)Case L-T、 (c) Case L-L.

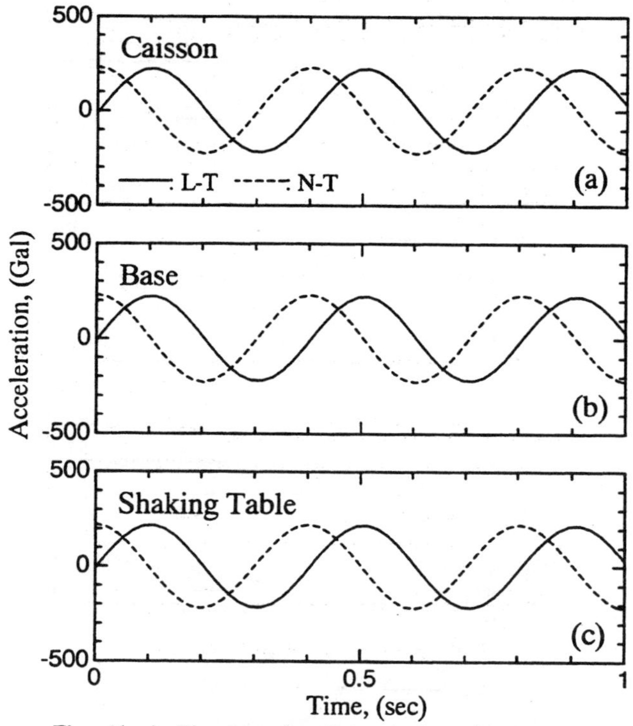

Figs. 4(a-c). Time histories of measured accelerations.
(a) at top of caisson, (b) at base beneath caisson, (c) on shaking table.

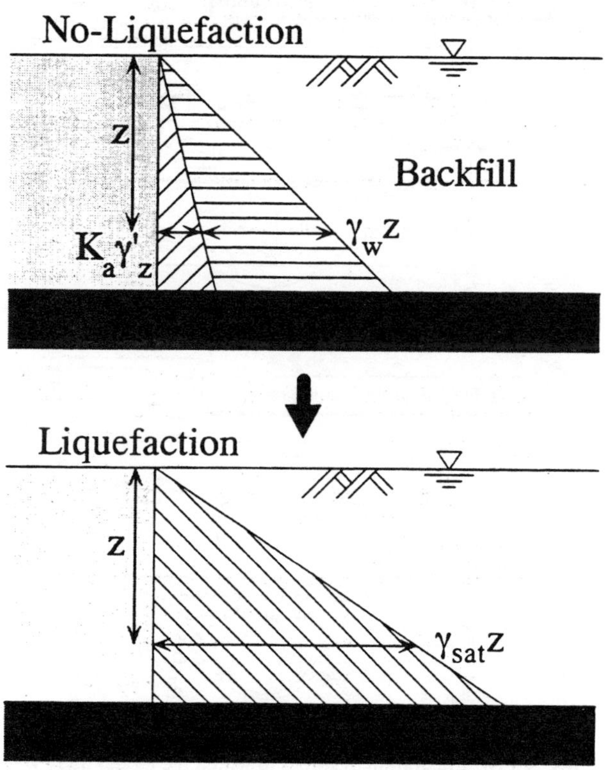

Fig. 5. Rise in static component of earth pressure due to loss of effective stress by liquefaction.

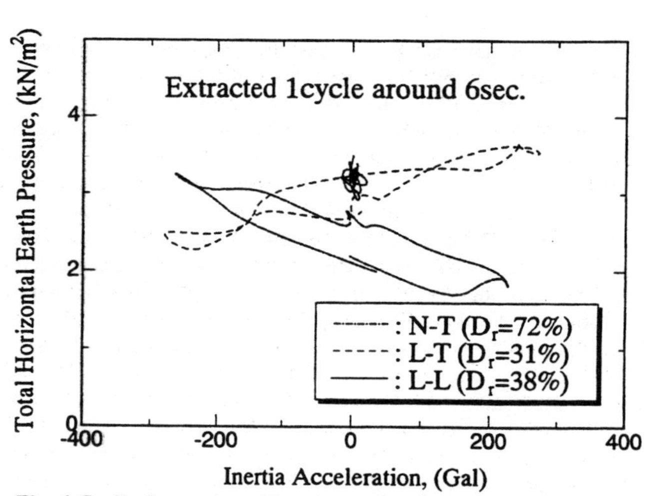

Fig. 6. Cyclic fluctuation of inertia acceleration measured in cases N-T, L-T and L-L

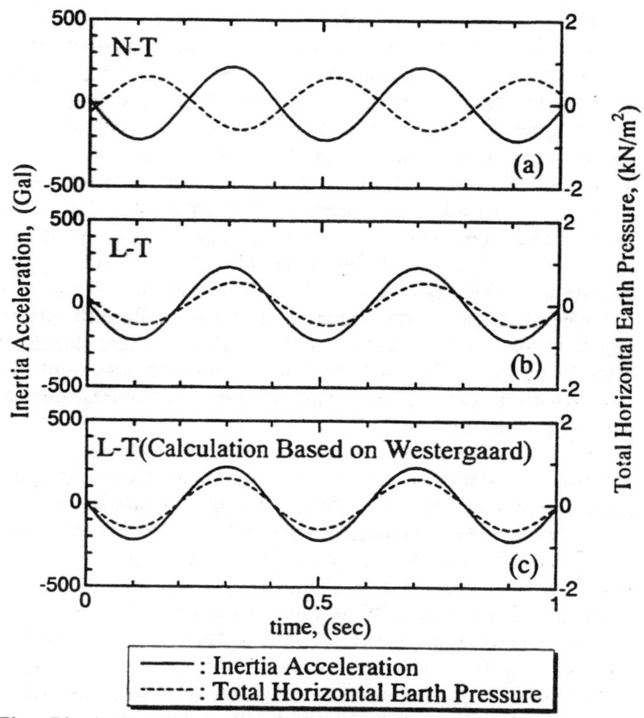

Figs. 7(a-c). Dominant components of inertia acceleration and earth pressure analyzed from time histories;
(a) measured in case N-T, (b) measured in case L-T,
(c) calculated for case L-T based on Westergaard's formula.

TEST RESULTS AND DISCUSSION

Shown in Figs. 2 and 3 are time histories of the total horizontal earth pressure, acceleration and displacement of the caisson. Acceleration was not detected in case L-L because direction of vibration is perpendicular to the direction of measurement. The earth pressure also did not fluctuate, although changes was observed from static to liquefied state. Time histories of the acceleration at top of the caisson, at the base beneath the caisson and on the shaking table are shown in Fig. 4, where only predominant component (frequency of 2.50Hz) was extracted by using Fourier Analysis. All the wave forms in different cases and at different locations are almost the same to each other; phase difference between cases N-T and L-T was intentionally set only for the demonstration. This suggests that the base is sufficiently rigid and resonance between the base and the caisson does not occur. Actually accumulation of the excess pore water pressure was negligibly small in the foundation ground.

As shown in Fig. 3, the backfill ground liquefied in both cases L-T and L-L; the excess pore water pressure increased resulting in the perfect loss of the effective stress within 3 seconds from the start of the vibration. On the other hand, in case N-T where backfill did not liquefy, pore water pressure fluctuated but only a half of the initial effective stress was lost at maximum in the backfill. The caisson moved significantly only in case L-T, but the movement or slide ceased at the time of the termination of input acceleration. The stability of the caisson was evaluated by assuming the friction angle between the caisson and the base to be of 35 degrees. The safety factor F_s against sliding was 2.65 under ordinary condition under which both earth pressure and water pressure at rest works. It was 0.70 under the earthquake condition under which inertia force and earth pressure under horizontal seismic intensity of k_h = 220/980 were considered, where 220 and 980 are maximum horizontal acceleration and acceleration of gravity, respectively. The earth pressure increase more when liquefaction occurs and effective stress losses in the backfill ground as shown in Fig. 5, in which case F_s is 0.62. The caisson could slide during the earthquake even in case N-T (F_s = 0.70) based on the stability analysis. The displacement of the caisson was, however, small to be only 3 mm or 2% of the height of the caisson. This indicates that both inertia force and liquefaction of backfill both are needed to cause instability of the caisson. Comparison of the behaviors between cases L-T and L-L suggests that the fluctuation component of the earth pressure of liquefied backfill plays an important role for the instability of the caisson. These experimental results agrees with the trend of the actual damage to quay walls during the earthquake; damage was significant when the liquefaction in the backfill ground occurred.

The relationships between the inertia acceleration and earth pressures are plotted in Fig. 6 for 1 cycle after 6 seconds from the beginning of the vibration. Since the input vibration wave is a little apart from the perfect sinusoidal wave and the deformation properties of backfill and foundation grounds are not linear, the loops shown in the figure are not complete elliptic shape. In case L-L, neither inertia acceleration or earth pressure fluctuated; only a rise in earth pressure caused by the occurrence of liquefaction was recognized (see Figs. 2(c) and 5). The same amount of the rise of excess pore water pressure is also found in case L-T. On the other hand, in cases N-T and L-T, both inertia acceleration and earth pressure oscillated, but the directions of the loops are different from each other. The predominant components of the inertia acceleration and the fluctuating earth pressure are shown in Figs. 7(a, b). The directions of the inertia acceleration and fluctuating component of the earth pressure are almost opposite in case N-T, whereas almost the same in case L-T. The cause why instability of the caisson occurred only in case L-T is explained from this contrast in the direction of inertia acceleration relative to earth pressure.

The forces acting on the caisson during the vibration are illustrated in Fig. 8, where horizontal components of the forces are concerned. The four cases shown in the figure have the following characteristics:

Case A - Figure 8(a) shows the inertia force acting on the caisson; the existence of the sea water and backfill ground are neglected. The outward inertia force ΔF_i becomes maximum at state (A).

Case B - Figure 8(b) shows the inertia force, static components of

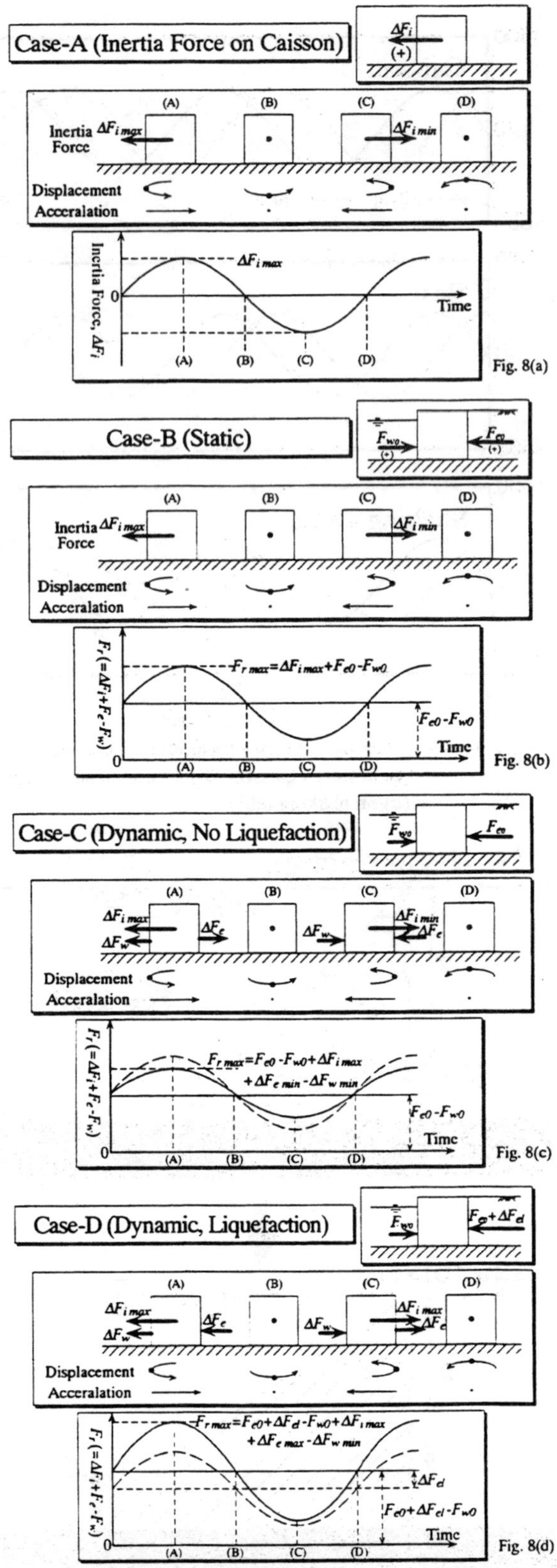

Figs. 8(a-d). Illustration of forces to which the model quay wall is subjected; (a) inertia forces, (b) static components of water and earth pressures, (c) dynamic component in the no-liquefaction case, (d) dynamic component in the liquefaction case.

sea water pressure and backfill earth pressure; fluctuating components of the pressures are neglected. Since the static earth pressure F_{eo} including pore water pressure is larger then water pressure F_{wo}, maximum resultant force $F_{r\,max}$ increases by $(F_{eo}$-$F_{wo})$ from Case A.

Case C - Figure 8(c) shows the case of no-liquefaction; both fluctuating components of earth pressure and water pressure are considered. Based on the test results shown in Figs.6 and 7(a), the fluctuating earth pressure ΔF_e works in the opposite direction to the inertia force ΔF_i. Therefore, the maximum resultant force $F_{r\,max}$ reduces and the caisson becomes stable compared with Case B.

Case D - Figure 8(d) shows the case of liquefaction; both fluctuating components of earth pressure and water pressure are considered, too. Since the test results shown in Figs. 6 and 7(b) show the increase in static component of earth pressure and the coincidence of the directions of inertia force and fluctuating component of earth pressure, the maximum resultant force $F_{r\,max}$ rises and the caisson becomes more unstable than Case B, which is contrary to Case C.

ANALYSIS OF EARTH PRESSURE

If perfect liquefaction occurs, soils lose their shear stiffness and strength completely and behaves like a liquid. When evaluating the earth pressure acting from the liquefied backfill to the quaywall in this stage, the back fill can be modeled to be a liquid. The closed form formula for dynamic water pressure on dams was derived by Westergaard (1933), where a set of wave equations were solved under 2-dimensional condition for the structure with upright back wall for the reservoir subjected to the sinusoidal base oscillation. The dynamic water pressure is evaluated to be

$$p = \frac{8\alpha\rho h}{\pi^2}\sum\frac{1}{n^2 c_n}\left(\sin\frac{n\pi}{2h}y\right)\exp\left(\frac{2i\pi t}{T}\right), \quad c_n = \sqrt{1-\frac{16h^2\rho}{n^2 k T^2}} \qquad (1)$$

whose abbreviated formula is

$$p = \frac{3\alpha\rho}{4}\sqrt{hy} \qquad (2)$$

where α is an amplitude of the base sinusoidal acceleration wave, T is wave period and h is the height of the dam. Parameters ρ and k are mass density and bulk modulus of the reserved water, respectively. This equation can be used for evaluating the earth pressure when liquefaction occurs in the backfill ground because the upright condition of the caisson is satisfied in Case L-T as shown in Fig. 4. Figure 7(c) shows calculated fluctuating earth pressure from Eq. (1), in which ρ is set to be $1.96\text{x}10^3\ kg/m^3$ and k is $0.933\text{x}10^5\ kN/m^2$, which are typical values suggested in JSCE (1994).

Both the measured and calculated behaviors in Figs. 7(b) and (c) show that fluctuating component is the same in phase angle with the inertia acceleration. Westergaard (1933) noticed this point and emphasized that fluctuation of the water pressure during the earthquake may notably reduce the stability of the dam and, therefore, it is an important factor in the earthquake resistant design. As shown in these figures, measured fluctuating earth pressure is in good agreement with the calculated one.

Shown in Fig. 9 is the explanation of the resistant forces acting on the caisson at the most unstable state in cases N-T and L-T. The fluctuating earth pressure was evaluated from the measurement shown in Fig. 7(a) for case N-T where the backfill was not liquefied, whereas that was calculated from Eq. (1) for Case L-T where the backfill was liquefied. The angle of resultant force from the vertical direction is much larger in the liquefaction case than in the no liquefaction case. This is attributed to the coincidence of the phase angle of inertia force and fluctuating earth pressure when the backfill is liquefied; this tendency is in accordance with the measured sliding of the caisson shown in Fig. 2.

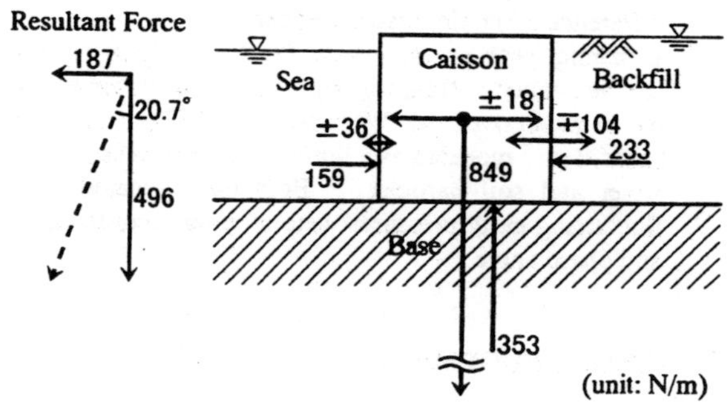

Fig. 9(a)

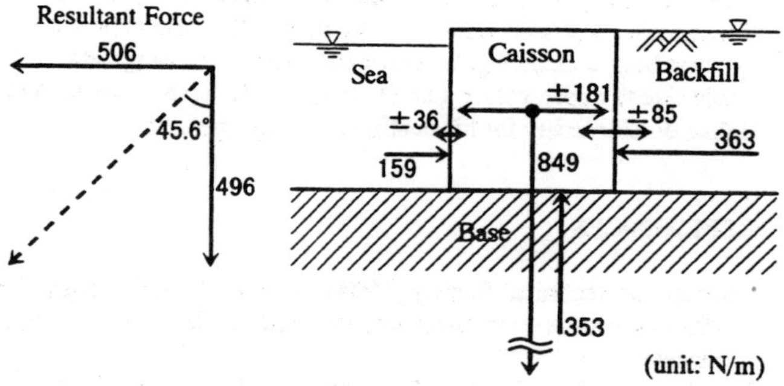

Fig. 9(b)

Figs. 9(a, b). Forces acting on model quay wall during earthquake; (a) in case N-T (no-liquefaction), (b) in case L-T (liquefaction).

CONCLUDING REMARKS

A series of model shaking table tests were conducted in order to investigate the mechanism of the damage to gravity type quay walls during earthquakes. The following conclusions were obtained from the comparative examination of the observed behaviors.

- The sliding of the caisson during the vibration is largely enhanced by the occurrence of the liquefaction in the backfill ground. The caisson became unstable when vibrated in the transversal direction causing the liquefaction of the backfill, in which three components, i.e., the increase in static component of the earth pressure induced by the liquefaction, fluctuation of the earth pressure from liquefied backfill and inertia force, cooperate together to slide the caisson.
- If liquefaction does not occur in the backfill, the fluctuating component of the earth pressure works in opposite to the inertia force, whereas the fluctuating component of the earth pressure and inertia force are same in the direction with no phase angle

difference when liquefaction occurs.

- Dynamic earth pressure with fluctuation from liquefied backfill can be evaluated from the formula for the dynamic water pressure on dams derived by Westergaard (1933), where the liquefied backfill was modeled as a liquid with mass density as a mixture of water and soil particles. Both the measured and calculated fluctuating earth pressures were in good accordance in amplitude and phase angle.

ACKENOWLEDGEMENT

The authors are grateful to Messrs. Takahiko Sasajima and Sadamitsu Akeda of Hokkaido Development Bureau for their supports. They are also grateful to Prof. Susumu Yasuda for his advice on the shaking table test, and to Messrs. Nobukatsu Oka and Kazuma Harada for their help conducting the experiment and processing data. They would like to express their sincere thanks to Professor Shosuke Toki of Hokkaido University for his continuous encouragement.

REFERENCES

Japanese Geotechnical Society (1994). "Report of the Damages during 1993 Kushiro-oki and Notohanto-oki Earthquakes." pp.175-182. (in Japanese)

Igarashi, H., Iai, S., Sugano, T., Yamazaki, H. and Inatomi, T. (1996). "Performance of Caisson Type Quay Walls at Kobe Port." Special Issue of Soils and Foundations, pp.119-136.

Kamon, M., Wako, T., Isemura, K., Sawa, K., Mimura, M., Tateyama, K. and Kobayashi, S. (1996). "Geotechnical Disasters on the Waterfront." Special Issue of Soils and Foundations, pp.137-147.

Westergaard, H. M. (1933). "Water Pressures on Dams during Earthquakes, Trans. ASCE, Vol.98, pp.418-432.

Hokkaido Development Bureau (1996). "Report on the Improvement of the Seismic Resistance of Port Facilities.", Cold Region Port and Harbor Research Center. (in Japanese)

Oka, N. (1996). "A Study on the Stability of Gravity Type Quay Wall by Shaking Table Tests," Bachelor Thesis, Hokkaido University. (in Japanese)

Kiku, H. (1993). "A Study on the Permanent Displacement of Liquefied Ground," Thesis for the Partial Fulfillment for the Degree of Doctor of Engineering , Kitakyushu Institute of Technology. (in Japanese)

Konama, E., Miura, K., Oka, N., Sasajima, T. and Ohtsuka, N. (1996). "Shaking Table Experiments on the behavior of Gravity Type Quay Wall during Earthquake." Proc., 31st Annual Meeting of Japanese Geotechnical Society, pp.1199-1200. (in Japanese)

Japan Society of Civil Engineers (1994). "Coastal Wave, - state of the art of the analytical methods of interactions among waves, structures and grounds -," pp.481-483. (in Japanese)

Yoshimi, Y. (1985). "Liquefaction of Sand Grounds", Tsuchi-to-kiso, Japanese Geotechnical Society, Vol.33, No.8, pp.3-8. (in Japanese)

Proceedings of the Seventh (1997) International Offshore and Polar Engineering Conference
Honolulu, USA, May 25-30, 1997

ISBN 1-880653-28-1 (Set); ISBN 1-880653-29-X (Vol. I)

Seabed Response to Wave Loading

Q.S. Yang
Hong Kong University of Science and Technology
Hong Kong

H.B. Poorooshasb
Concordia University
Montreal, Canada

ABSTRACT

The stability and deformation of a seabed during ocean wave represents an important environmental factor affecting the safety of coastal and offshore structures. The seabed deposit is considered as a saturated porous medium for which two-phase coupled governing equations are derived. A constitutive model based on the non-associative bounding surface plasticity is adopted, which is capable of simulating marine sand behaviour under diverse loading conditions, in particular, rotations of the principal stress direction occurred typically under wave loading. The coupled governing equations and the marine sand model have been implemented into a displacement-pore pressure coupled, nonlinear, finite element program to analyze the seabed response to wave loading. The numerical results show that the coefficient of the permeability has a great influence on the susceptibility of the seabed to liquefy. This integrated approach is expected to provide a better understanding of wave-induced seabed failure mechanism and to further the state of the art in coastal and offshore foundation design.

KEY WORDS: sand, plasticity, liquefaction, seabed, wave loading

INTRODUCTION

The phenomenon of wave-induced seabed liquefaction has been recognized in recent years. It may occur in sand or silt deposits, thereby exerting damaging influences on the engineering installations placed nearby, for example, the flotation of pipelines (Christian, 1974), loss of lateral resistance of piles (Finn and Martin, 1979). The stability and deformation of the seabed under storm waves are important considerations in the design of marine pipelines, anchors and gravity base or pile supported offshore platforms. Problems associated with wave-induced seabed response have been studied extensively. Various methods including empirical, analytical and numerical procedures for predicting the response of marine deposits have been proposed. Traditionally, the seabed was treated as linear elastic or an equivalent linear elastic medium. Henkel (1970) was one of the first to carry out a total stress analysis for seabed under wave loads following the classical procedure of limit equilibrium slope stability analysis. Analytical solutions of effective stress analysis for two-phase coupled response of a poro-elastic bed to water wave were provided by Yamamoto (1978) applying the Biot theory of consolidation. In the finite element analysis of wave-induced seabed response, an assumption of elastic behavior of soils is popular (e.g. Dunham et al., 1971; Gatmiri, 1990). However, nonlinear, anisotropic and hysteric behaviours of soil are apparent from field data and laboratory tests.

Some simplified procedures, which basically were modifications of those for seismic response analyses (Seed and Idriss, 1971), have been proposed for wave-induced liquefaction analysis (e.g. Seed and Rahman, 1978; Finn et al., 1983). However, it is worth noting that certain differences exist between earthquake excitation and wave loading although both are cyclic in nature. Ocean wave periods and duration are considerably longer than the earthquake periods and duration. Drainage conditions become very important. The undrained assumption for seismic response is no longer valid for wave response analysis. On the other hand, the dynamic terms may be entirely negligible for a typical storm wave with a period in seconds (Zienkiewicz and Bettess, 1982; Yang, 1990). Nevertheless, ocean wave loadings are imposed at the surface of the seafloor as a pressure load rather than earthquake excitation at lower boundaries as base shaking. The ground motion developed in soil deposits during an earthquake is dominated by the upward propagation of shear waves. However, the seabed soil under wave loading is stressed in a more complicated manner, characterized by a continuous changing magnitudes and directions of the principal stresses.

This paper presents the standing wave-induced response of seabed deposits. A non-associative bounding surface plasticity constitutive model proposed by Yang (1996) is used in this analysis. The build-up of pore water pressure and the accumulation of permanent soil deformations are predicted by the numerical analysis. Results from the coupled nonlinear analysis show that seabed soil liquefaction may indeed occur for certain loose deposits under ocean wave loading.

FIELD EQUATIONS FOR SATURATED POROUS MEDIA

The classical early work of Terzaghi (1943) on soil consolidation was followed by the more explicit work of Biot (1955). The saturated soil could be considered as a two-phase porous medium consisting of solid particles and water in the pores with the following basic aasumptions:

1. The skeleton of the solid particles is governed by a soil constitutive model based on the non-associative bounding surface plasticity (Yang, 1996) in terms of the effective stress.
2. The pores are saturated with water slightly compressible. Water flow follows Darcy's law.
3. The load exerted on the seabed surface due to the standing waves is of a sinusoidal nature with fairly long wave period. It is rational to omit the dynamic terms from the equilibrium equation.

Based on above assumptions, the coupled governing equations could be derived and written in an incremental matrix form which is more suitable for dealing with the inelastic soil behaviors,

$$d\sigma = d\sigma' + \mathbf{m}\ \mathbf{dp}$$
$$\mathbf{d\varepsilon} = \mathbf{L}\ \mathbf{du}$$
$$d\sigma' = \mathbf{D}\ \mathbf{d\varepsilon} \qquad (1)$$
$$\mathbf{L^T}\ \mathbf{d\sigma} = \mathbf{0}$$
$$\nabla^T((\mathbf{k}/\gamma_w)(\nabla p - \rho_w \mathbf{g})) + \mathbf{m}^T\dot{\varepsilon} = \dot{p}\, n / K_w$$

where $\sigma^T = [\sigma_{11}, \sigma_{22}, \sigma_{33}, \sigma_{12}, \sigma_{23}, \sigma_{31}]$ is the stress vector, $\varepsilon^T = [\varepsilon_{11}, \varepsilon_{22}, \varepsilon_{33}, \varepsilon_{12}, \varepsilon_{23}, \varepsilon_{31}]$ is the strain vector, $\mathbf{u}^T = [u_1, u_2, u_3]$ is the displacement vector, $\mathbf{D}$ is a 6 x 6 matrix of material stiffness coefficients, $\mathbf{m}^T = [1, 1, 1, 0, 0, 0]$, and $\mathbf{L}$ is a 6 x 3 matrix operator equivalent to the small strain definition. The first equation in (1) states the Terzaghi's principle of effective stress where σ is the total stress, σ' is the effective stress and p, the pore water pressure. The fourth equation in (1) is the equilibrium of the total stress which includes the evolution of the pore water pressure. The last equation in (1) represents the transient seepage of the pore water, coupled with the deformation of the solid skeleton where $\mathbf{k} = [k_1, k_2, k_3, 0, 0, 0]$ is the Darcy's permeability vector and for isotropic case $k_1 = k_2 = k_3 = k$ and $\mathbf{k} = \mathbf{m}^T k\ \mathbf{m}$. ρ_w is the density of water, $\mathbf{g} = [g_1, g_2, g_3]$ is the gravitational acceleration vector, n is the soil prosity and K_w is the compressibility of water.

CONSTITUTIVE MODEL FOR MARINE SAND

In generalized plasticity formulation (Zienkiewicz and Mroz, 1984), a unit vector $\mathbf{n}$ in the effective stress space is required to determine both loading and unloading. The incremental stress-strain relations could be written in compact matrix notation in which the subscripts l and u refer to loading and unloading respectively,

$$d\sigma' = \mathbf{D}_l\ d\varepsilon \qquad \text{if } \mathbf{n}^T d\sigma' > 0 \text{ loading} \qquad (2)$$

$$d\sigma' = \mathbf{D}_u\ d\varepsilon \qquad \text{if } \mathbf{n}^T d\sigma' < 0 \text{ unloading} \qquad (3)$$

where the tangent matrices are related (Zienkiewicz and Mroz, 1984) by

$$(\mathbf{D}_l)^{-1} = (\mathbf{D}^e)^{-1} + \mathbf{n}_{gl}\, \mathbf{n}^T / H_l \qquad (4)$$

$$(\mathbf{D}_u)^{-1} = (\mathbf{D}^e)^{-1} + \mathbf{n}_{gu}\, \mathbf{n}^T / H_u \qquad (5)$$

in which H_l and H_u are scalar parameters known as loading and unloading plastic moduli. $\mathbf{D}^e$ is the elastic matrix which ensures the uniqueness of strain increments for $\mathbf{n}^T d\sigma' = 0$,

$$d\varepsilon = (\mathbf{D}^e)^{-1} d\sigma' \qquad (6)$$

If the unit vectors $\mathbf{n}_{gl}$ and $\mathbf{n}_{gu}$ differ from the loading/unloading vector $\mathbf{n}$ then the flow is non-associative and will lead to asymmetric tangent matrices $\mathbf{D}_l$ and $\mathbf{D}_u$. The material behaviour can be fully described by $\mathbf{D}^e$, $\mathbf{n}$, $\mathbf{n}_{gl/u}$ and $H_{l/u}$. By straightforward algebraic manipulation, equation (4) and (5) can be converted into the so-called elasto-plastic matrix,

$$\mathbf{D}_{l/u} = \mathbf{D}^e - \frac{\mathbf{D}^e \mathbf{n}_{gl/u}\, \mathbf{n}^T \mathbf{D}^e}{H_{l/u} + \mathbf{n}^T \mathbf{D}^e \mathbf{n}_{gl/u}} \qquad (7)$$

The constitutive equations in terms of $\mathbf{D}^e$, $\mathbf{n}$, $\mathbf{n}_{gl/u}$, $H_{l/u}$ for this marine sand model are specified by a bounding surface (Dafalias and Popov, 1975) incorporated with a plastic potential (Poorooshasb, et al., 1966). A set of stress invariant (r, θ, Z) is used, which could be related to the conventional stress invariant (I_1, J_2, J_3) as,

$$r = (2 J_2)^{1/2}$$
$$\theta = \sin^{-1}(-3\sqrt{3}\, J_3 / (2 J_2^{3/2}))/3, \quad -\pi/6 \le \theta \le \pi/6$$
$$Z = I_1 / \sqrt{3} \qquad (8)$$

where I_1 is the first invariant of the effective stress tensor σ'_{ij}, while J_2 and J_3 are the second and third invariant of the stress deviation tensor s_{ij},

$$I_1 = \sigma'_{kk} \quad , \quad s_{ij} = \sigma'_{ij} - \sigma'_{kk}\delta_{ij}/3$$
$$J_2 = s_{ij} s_{ij}/2, \quad J_3 = s_{ij} s_{jk} s_{ki}/3 \qquad (9)$$

in which r and θ give the true value on the octahedral plane ($\theta = -\pi/6$ represents triaxial compression and $\theta = \pi/6$ is for triaxial extension) and z gives the true value along the diagonal of the principal effective stress space, as shown in Figure 1.

The bounding surface (see Figure 1) is specified by

$$F = F(r, \theta, Z, e^*) = r - \eta(e^*)\, g(\theta)\, Z = 0 \qquad (10)$$

in which e* is a measure of the total plastic distortion given by the expression

$$e^* = \int de^*,$$
$$de^* = (de^P_{ij}\, de^P_{ij})^{1/2} \qquad (11)$$
$$de^P_{ij} = d\varepsilon^P_{ij} - d\varepsilon^P_{kk}\,\delta_{ij}/3$$

where de^P_{ij} is the incremental plastic deviatoric strain. The function $\eta(e^*)$ is defined in terms of plastic material parameters A and η_f which is related to the angle of internal friction ϕ_f,

$$\eta(e^*) = \eta_f[1 - \exp(-e^*/A)], \qquad (12)$$

$$\eta_f = 2\sqrt{2}\, \sin\phi_f / (3 - \sin\phi_f) \qquad (13)$$

by which the size of the bounding surface is determined. The shape of the bounding surface is a cone having straight meridians and a noncircular cross section with its apex at the origin and its axis along the diagonal of the stress space. The cross section of the bounding surface on the octahedral plane is describe by the function g(θ). An ellipse formulation proposed by William and Wande (1975), as shown in Figure 2, is adopted to quantify the function g(θ). In this model, $g_{max} = g(-\pi/6) = 1$ is chosen for triaxial compression and $g_{min} = g(\pi/6) = (3 - \sin\phi_f)/(3 + \sin\phi_f)$ for triaxial extension to consistent with the Mohr-Coulomb failure criterion.

The plastic potential is defined by

$$G = G(r, \theta, Z, Z_0) = r + M_{cr}\, g(\theta)\, Z \ln(Z/Z_0) = 0 \qquad (14)$$

$$M_{cr} = 2\sqrt{2}\, \sin\phi_{cr} / (3 - \sin\phi_{cr}) \qquad (15)$$

where M_{cr} is another plastic material parameter. The cone surface $\eta = r/Zg(\theta) = M_{cr}$, on which no plastic volumetric strain increment occurs, is synonymous with 'the critical state line' (CSL) or 'the phase transformation line' by Ishihara, et al. (1975).

The state of the soil skeleton is defined by the effective stress σ' and the plastic internal variable e*, which is used to account for the past loading history. If the stress point σ' is on the bounding surface, i.e. $F(\sigma', e^*) = 0$, it is called virgin loading. The bounding surface is identified as a yield surface as in the classical plasticity. The loading/unloading unit vector is given by the unit normal vector of the bounding surface. If $\mathbf{n}^T d\sigma' > 0$, the loading flow vector $\mathbf{n}_{gl}$ is determined by the unit normal vector of the plastic potential G(σ'). Imposing the consistency condition dF = 0, the loading plastic modulus H_l is obtained,

$$H_l = \frac{r}{A}\left(\frac{\eta_f}{\eta} - 1\right)\left[\left(dev\frac{\partial G}{\partial \sigma'}\right)^T\left(dev\frac{\partial G}{\partial \sigma'}\right)\right]^{1/2}$$

$$/\left[\left(\frac{\partial F}{\partial \sigma'}\right)^T\left(\frac{\partial F}{\partial \sigma'}\right)\right]^{1/2}\left[\left(\frac{\partial G}{\partial \sigma'}\right)^T\left(\frac{\partial G}{\partial \sigma'}\right)\right]^{1/2}$$

with $\eta = \eta(e^*) = r / (g(\theta) Z)$ (16)

If neutural loading, $\mathbf{n}^T d\sigma' = 0$ or unloading, $\mathbf{n}^T d\sigma' < 0$ takes place on the bounding surface, pure elastic response is assumed,

$$H_u = \infty, \quad \mathbf{D}_u = \mathbf{D}^e \tag{17}$$

If the stress point σ' has indeed moved inside the bounding surface, i..e. $F(\sigma', e^*) < 0$, a stress reversal process is initiated. The bounding surface acts as a prestress memory surface. $\mathbf{n}$, $\mathbf{n}_{gl/u}$ and $H_{l/u}$ are defined in terms of a conjugate surface $F_c(\sigma', e^*_c) = 0$, which passes through the current stress point σ' and homologous to the bounding surface. The unit normal vector, $\mathbf{n}_c^T$, of the conjugate surface acts as the loading/unloading unit vector. If $\mathbf{n}_c^T d\sigma' > 0$, it is a stress reversal loading and the loading flow vector $\mathbf{n}_{gl}$ is given by the unit normal vector of the conjugate potential $G_c(\sigma')$, which also passes through the current stress point σ'. If $\mathbf{n}_c^T d\sigma' = 0$, it is pure elastic response with H_u and $\mathbf{D}_u$ given by (17). If $\mathbf{n}_c^T d\sigma' < 0$, it is a stress reversal unloading. The unloading flow vector $\mathbf{n}_{gu}$ is related to the loading flow vector $\mathbf{n}_{gl}$ in a reflecting manner as shown for the triaxial configuration in Figure 3,

$$\mathbf{n}_{gu}(n^v_{gu}, n^s_{gu}) = (-n^v_{gl}, -n^s_{gl}) \quad \text{if } \eta = r/Zg(\theta) > M_{cr}$$

$$\mathbf{n}_{gu}(n^v_{gu}, n^s_{gu}) = (n^v_{gl}, -n^s_{gl}) \quad \text{if } \eta = r/Zg(\theta) < M_{cr} \tag{18}$$

where superscript s refers to the deviatoric component in the octahedral plane and v stands for the hydrostatic component along the space diagonal.

To specify the loading/unloading plastic moduli $H_{l/u}$ for the stress reversal process, an 'image' point on the bounding surface is defined by a 'deviatoric radial' mapping rule. The image point is the intersection of the bounding surface and a radial on the octahedral plane passing through the current stress point. For loading, the image point and the current point are located on the same side, and for unloading, they are on the opposite sides. This could be expressed by the distance δ, as shown in Figure 3,

$\delta = \underline{ac}$ if loading

$\delta = \underline{ad}$ if unloading (19)

$\delta_0 = \underline{ac} + \underline{ad}$ (see Figure 3)

The loading /unloading plastic moduli $H_{l/u}$ is determined by an interpolation rule:

$$\frac{1}{H_{l/u}} = \frac{1}{H_B}\left(1 - \frac{\delta}{\delta_0}\right)^{\gamma} \tag{20}$$

where γ is a material parameter and H_B is the plastic loading moduli at the image point. It may be noted that although H_l and H_u are expressed similarly in equation (20), in fact they are different since the image points, and thus H_B and δ are not the same for stress reversal loading and unloading.

There are total seven material parameters, e_{in}, κ, ν (three for elastic) and $\eta_f(\phi_f)$, $M_{cr}(\phi_{cr})$, A, γ (four for plastic) in this model. The elastic parameters are required to determine $\mathbf{D}^e$. In this study, Poisson's ratio ν is taken as constant and the bulk modulus K is assumed to depend on the effective mean stress σ'_m as,

$$K = p_a \frac{1 + e_{in}}{\kappa}\left(\frac{\sigma'_m}{p_a}\right)^{1/2} \tag{21}$$

where p_a is the atmospheric pressure for the nondimensionalization of the constant κ, and e_{in} is the initial void ratio.

SEABED RESPONSE TO STANDING WAVES

A computer program STWAVE was developed, in which the proposed material constitutive model (Yang, 1996) is implemented. According to the linear wave theory, the water pressure exerted on the sea bottom due to the standing wave would be of a sinusoidal nature. It is sufficient, by the argument of symmetry, to consider only a half-wave length L/2 as shown in Figure 4. Plane strain is assumed.

The initial conditions are

$\sigma'_{yy} = (\gamma_s - \gamma_w) y$, $\sigma'_{xy} = 0$, $\sigma'_{xx} = k_0 s'_{yy}$, $p = \gamma_w (d + y)$, @ t= 0;

and the boundary conditions are

$\sigma'_{yy} = 0$, $\sigma'_{xy} = 0$, $p = (H/2)\gamma_w \cos(2\pi x/L)\sin(2\pi t/T)$, @ y= 0;

$\partial p / \partial y = 0$, $u_x = 0$, $u_y = 0$, @ y= h;

$\partial p / \partial x = 0$, $u_x = 0$, @ x= 0;

$\partial p / \partial x = 0$, $u_x = 0$, @ x= L/2.

where k_0 is the coefficient of the earth pressure at rest, H/2 and T are the amplitude and the period of the wave, respectively. The finite element mesh used in this analysis is shown in Figure 5, which shows the nodal numbering, element numbering and the position of Gauss integration points within an element. For the particular example cited here, the material properties correspond to data for loose Fuji river sand tested by Ishihara and Tatsuoka (1975) and the following wave load specifications, geometrical and material parameters were employed:

L/2 = 80 m	h = 40 m	d = 20 m	H/2 =6.62 m
γ_w = 9.8 kN/m^3	γ_s = 19 kN/m^3	k_0 = 0.5	T = 100 sec
n/K_w = 10^{-7} m^2/kN	e_{in} = 0.85	κ = 0.008	ν = 0.2
ϕ_f = 27°	ϕ_{cr} = 27°	A = 0.0022	γ = 5.0
k = 10^{-1} and 10^{-3} cm/sec			

The computation results are presented at locations of nodes and Gauss integration points. Thus, a curve labeled E2G1 is showing the numerical results obtained at Gauss point No. 1 of Element No. 2. The graphs shown in Figure 6 represent the variation of the pore water pressure. It may be noted that in the case of marine sand with a higher permeability, 10^{-1} cm/sec in this example, the variation of the pore water pressure essentially oscillates about a constant mean value. However, this is not the case of marine sand with a relatively lower permeability, 10^{-3} cm/sec in this example. The pore water pressure oscillates about a mean value which is increasing with time. This build up of the pore water pressure is detrimental to the integrity of the system.

Figure 7 shows the variation of the mean effective stress with time. When the permeability of sand is higher (10^{-1} cm/sec), the surface wave loading appears to have little effect on the variation of the mean effective stress, i.e. the water just enters and leaves the pores of the soil without affecting the stress regime to any extent. The situation is quite different for sand with lower permeability (10^{-3} cm/sec). Curves indicate a general tendency for the mean effective stress to decrease in magnitude as the number of loading cycles increase. This reduction of the mean effective stress is clearly an indication of the loss of the soil shear strength, and can lead to seabed failure. The mean effective stresses at the Gauss points E2G2 and E2G1 approach zero only after ten cycles of surface wave loading. The seabed soil at these locations will behave like a liquid with no intrinsic shear strength and the phenomenon is generally referred as liquefaction.

Figure 8 gives what may be called as a stress path followed the wave loading for sand element located at the Gauss points E8G3 and E8G4 with a lower permeability (10^{-3} cm/sec in this case). They show that the decrease in the vertical effective stress is associated with the increase in the pore water pressure and the decrease in the mean effective stress. The stress path migrates towards the origin which shows that the condition for liquefaction is indeed developed.

In Figure 9, the vertical displacement at Node 37 is plotted against time, which is closest to the initial liquefied point E8G4. For sand with a higher permeability (10^{-1} cm/sec), a small but finite amount of densification of the sand is experienced. This is, of course, expected since plastic deformation is always irreversible. However, it is directly contrast to the commonly held view that the liquefaction potential is a function, only, of the sand density. For sand with a lower permeability (10^{-3} cm/sec), the initial soil densification is followed by a heaven (upward movement shown by negative numbers) of the deposit. It shows that seabed instability due to liquefaction is likely to take place. These results demonstrate the importance of the sand permeability on the susceptibility of the seabed to liquefy.

CONCLUSION

The ocean wave-induced response of seabed deposits and liquefaction hazards have received considerable attention in marine geotechnical engineering. Massive movement of seabed soil and the collapse of the waterfront structure foundations during storm waves is the direct results of soil liquefaction. Although an assumption of elastic seabed soil was adopted for study the ocean wave induced seabed response, it is possible to develop a more rigorous analysis procedure, incorporated with a valid elastoplastic soil constitutive model, as presented in this study.

It is a well known fact that the relative density of a sand plays a very important role in its liquefaction potential. The behavior of marine sand under wave loading is different from that under fully drained or undrained condition. The example cited here clearly demonstrate that the influence of the coefficient of permeability on the susceptibility of the marine sand to liquefy. For loose sand with a lower value of permeability, seabed liquefaction is likely to take place. However, for loose sand with a relatively higher value of permeability, the seabed is safe.

It should be noted that inclusion of the structures in the analysis would make the problem much more complicated. A free field solution may provide some valuable information for design of the offshore structure foundations. Further research on the post liquefied soil behavior is needed to allow the calculations to be pursued after a localized liquefaction and possibly up to the global liquefaction failure of the seabed deposits.

REFERENCES

1. Biot, M.A. (1955). Theory of elasticity and consolidation for a porous anisotropic solid, Journal of Applied Physics, Vol. 26, pp. 182–185.
2. Christian, J.T., Taylor, P.K., Yen, J.K.C. and David, R.E. (1974). Large diameter underwater pipeline for nuclear power plant designed against soil liquefaction, Proceedings, Sixth Annual Offshore Technology Conference, Houston, Taxas, Paper No. OTC 2094, pp. 597-606.
3. Dafalias, Y.F. and Popov, E.P. (1975). A model of nonlinearly hardening materials for complex loading, Acta Mechanica, Vol. 21, pp. 173-192.
4. Dunham, R.S., Olson, R.E. and Reese, L.C. (1971). Analysis of a potential sea floor slide at Block 70, Gulf of Mexico, Report to Shell Oil Company, offshore Division Construction, Southern E & P Region, ODC Report 27.
5. Finn, W. D.L. and Martin G.R. (1979). "Analysis of piled foundations for offshore structures under wave and earthquake loading", Proceedings, paper No.38, Second International Conference on Behavior of Off shore Structures, London, England.
6. Finn, W. D. L., Siddharthan, R. and Martin, G.R. (1983). "Response of seafloor to ocean waves", J. Geotechnical Engineering, ASCE, Vol. 109, GT4, pp.556-572.
7. Gatmiri, B. (1990). "A simplified finite element analysis of wave induced effective stress and pore pressures in permeable seabeds", Geotechnique, Vol. 40, No.1, pp.15–30.
8. Henkel, D.J. (1970). "The role of waves in causing submarine landslides", Geotech. 20(1), pp. 75–80.
9. Ishihara, K., Tatsuoka, F. and Yasuda, S. (1975). "Undrained deformation and liquefaction of sand under cyclic stress", Soils and Foundations, 15(1), pp.29-44.
10. Poorooshasb, H.B., Holubec, I. and Sherbourne, A.N. (1966). "Yielding and flow of sand in triaxial compression, Part 1", Can. Geotech. J., 3, pp. 179-190.
11. Seed, H.B. and Idriss, I. M. (1971). "Simplified procedure for evaluating soil liquefaction potential", J. Soil Mechanics and Foundation, ASCE, Vol. 97, No. SM9, pp.1249-1273.
12. Seed, H. B. and Rahman, M. S. (1978). "Wave–induced pore pressure in relation to ocean floor stability of cohesionless soils", Marine Geotechnology, Vol. 3, No. 2, pp.123–150.
13. Terzaghi, K. (1943). Theoretical Soil Mechanics, Wiley, New York.
14. Williams, K. T. and Wanda, E. P. (1975). "Constitutive models for the triaxial behavior of concrete", Proc. Int. Ass. Bridge Sstruct. Engng., Vol. 19, pp. 1-30.
15. Yamamoto, T. (1978). "Seabed instability from waves", Proc. 10th Annual Offshore Technology Conferences, paper 3262 1, Houston, Texas, 1819-1824.
16. Yang, Q.S. (1990). "Wave induced response of sea floor deposits: A simple model for sands and nonlinear analysis by FEM", Ph.D. thesis, Concordia University, Montreal, Canada.
17. Yang, Q.S. (1996). "A non-associative bounding surface plasticity model for marine sands", Proceedings of the 6th Int. Offshore and Polar Engineering Conference, Los Angeles, USA, Vol. 1, pp.377-384.
18. Zienkiewicz, O.C. and Bettess P. (1982). "Soils and other saturated media under transient, dynamic conditions; General formulation and the validity of various simplifying assumptions", International Symposium on Soils Under Cyclic and Transient Loading, G.N. Pande and O.C. Zienkiewicz, eds., John Wiley and Sons, Inc., London, U.K., pp.1–16.
19. Zienkiewicz, O.C. and Mroz, Z. (1984). Generalized plasticity formulation and applications to geomechanics, in Mechanics of Engineering Materials, Editor, Desai, C.S. and Gallagher, R.H., Wiley, Ch. 33.

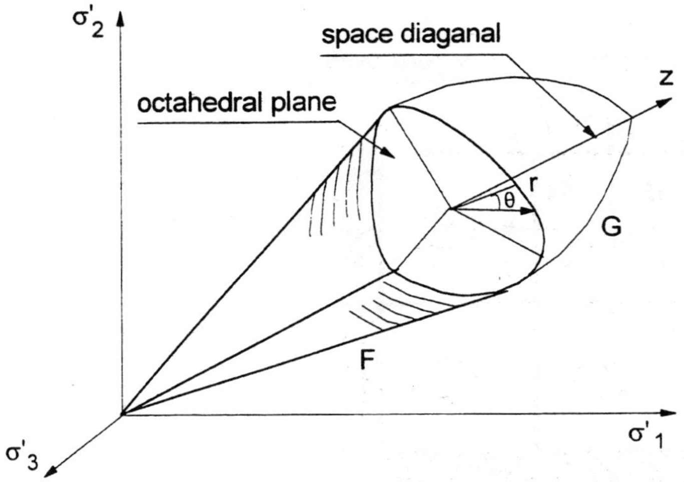

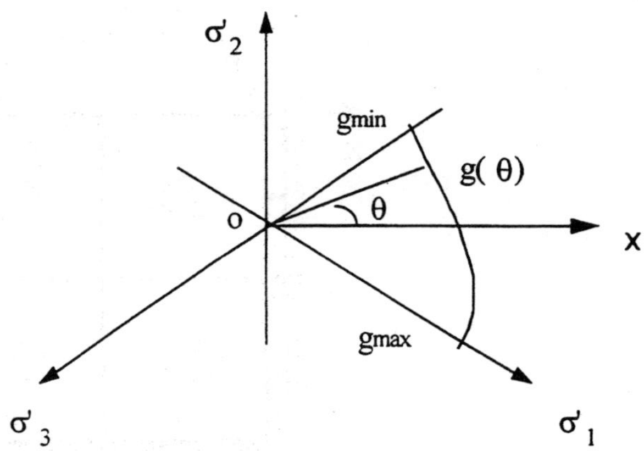

Figure 2. Elliptic Trace of Function g(θ).

Figure 1. Expression of r, θ, Z in the Principal Effective Stress Space.

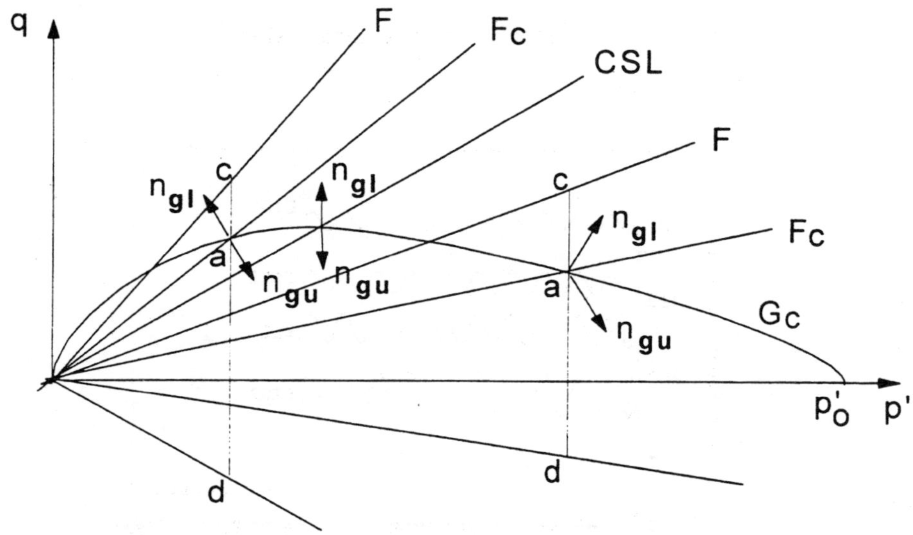

Figure 3. Model in Triaxial Configuration.

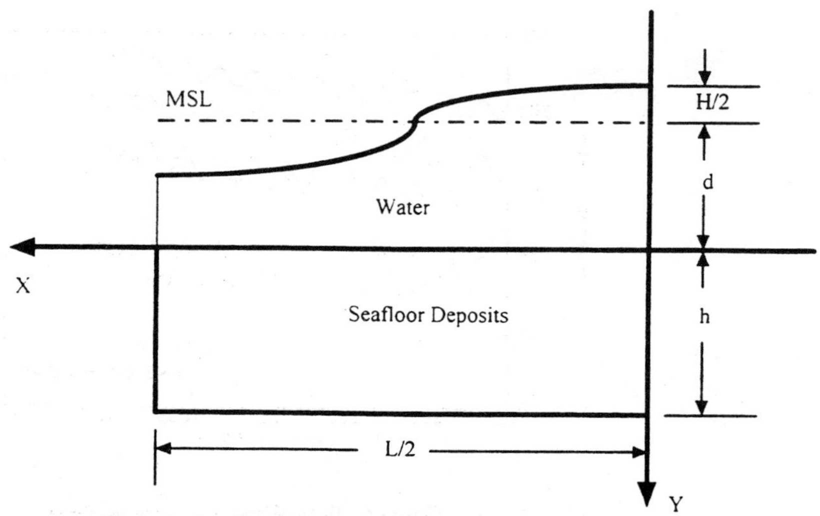

Figure 4. Standing Wave Load on the Seabed.

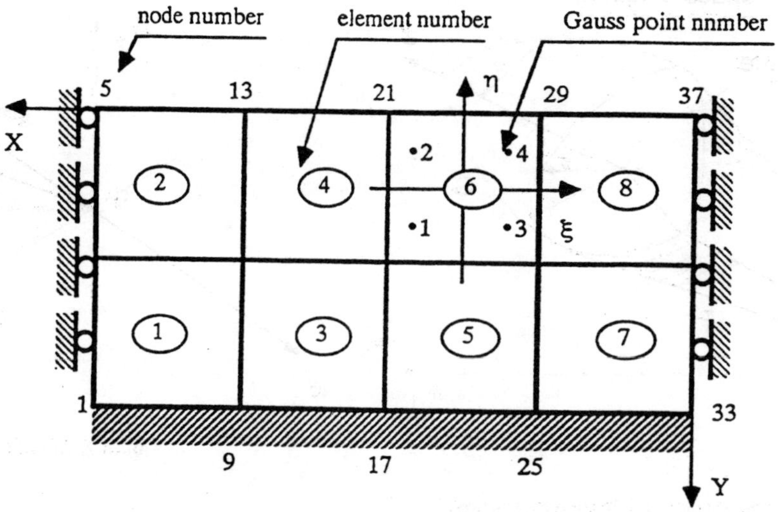

Figure 5. The Finite Element Mesh.

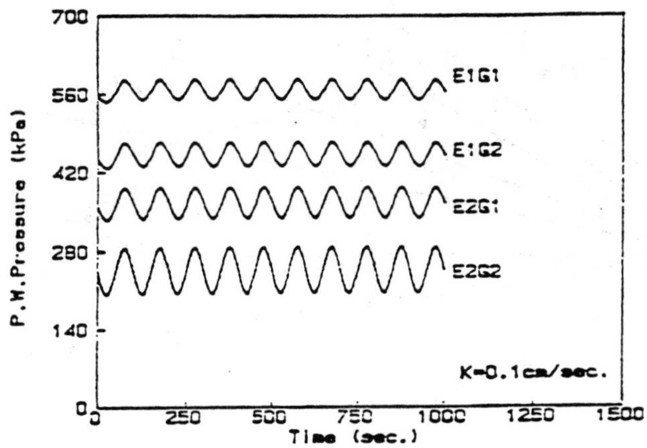

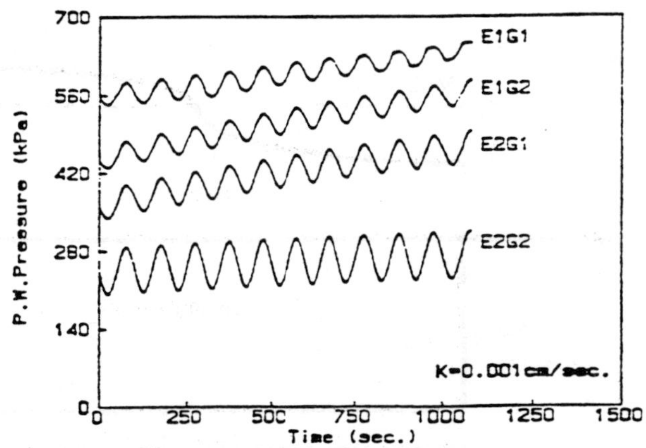

Figure 6. Pore Water Pressure vs Time.

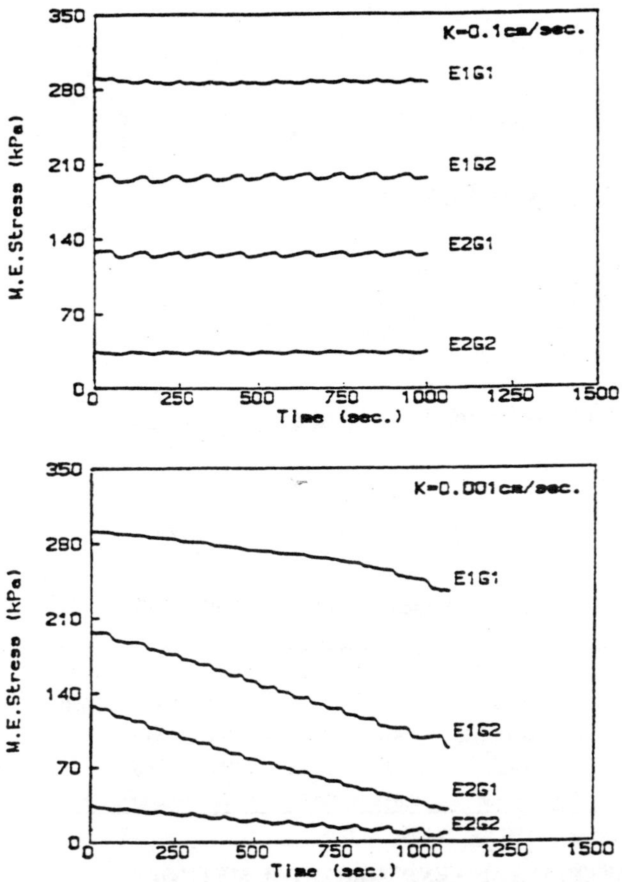

Figure 7. Mean Effective Stress vs Time.

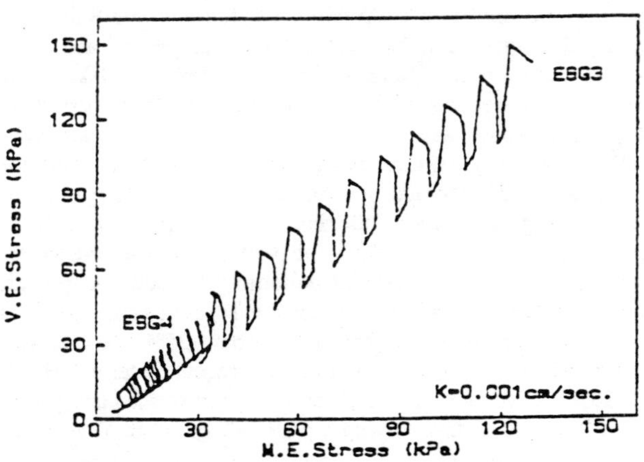

Figure 8. Vertical Effective Stress vs Mean Effective Stress.

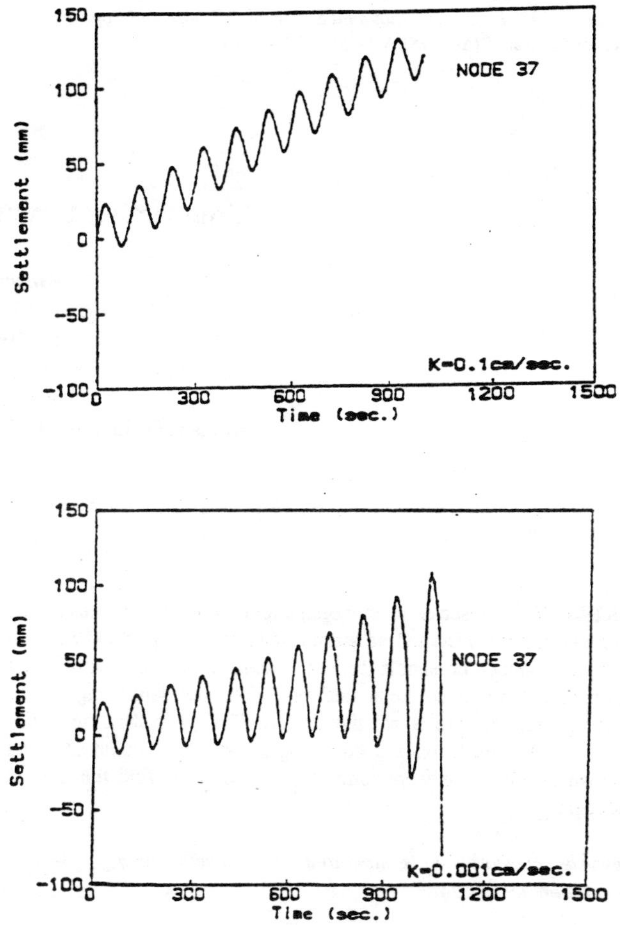

Figure 9. Vertical Displacement vs Time.

Proceedings of the Seventh (1997) International Offshore and Polar Engineering Conference
Honolulu, USA, May 25-30, 1997

ISBN 1-880653-28-1 (Set); ISBN 1-880653-29-X (Vol. I)

Liquefaction Strengths of Crushable Soils

Masayuki Hyodo
Yamaguchi University, Ube, Japan
Adrian F.L. Hyde
University of Bradford, Bradford, United Kingdom
Noritaka Aramaki
Kumamoto Institute of Technology, Kumamoto, Japan

ABSTRACT: A series of monotonic and cyclic triaxial tests was performed on two decomposed granite soils and volcanic soil Shirasu which are typical crushable soils in Japan. Both contractive and dilative monotonic shear behaviour were observed depending on the confining stress and initial relative density. The cyclic shear strengths were also dependent on the confining stress. The monotonic and cyclic shear behaviour were compared each other to find the mutual relationship.

Keywords: *crushable soils, monotonic shear, cyclic shear, liquefaction, steady state*

INTRODUCTION

In 1995 during the great Hanshin earthquake damage occurred in areas such as the Kobe Port Island and Rokko Island, both land areas were reclaimed from the sea using a local decomposed granite soil, Masado, as a fill material. Serious damage occurred to harbour facilities and coastal defences. Masado is a residual granite soil widely distributed in the western parts of Japan. It was recognized as a high quality fill material. Since the Niigata earthquake liquefaction studies have concentrated on hard grained sands found in coastal and estuarine areas. However, in many cities and industrial areas decomposed granite soils have been used as fill materials and now it seems they could be problematical under earthquake conditions. Volcanic soils are widely distributed in Japan (Ishihara et al., 1996, Miura et al., 1995) and among these Shirasu is commonly found in large areas of Southern Kyushu. This material is also often used as a fill (Umehara et al., 1975). Given that both of the above mentioned soils are crushable materials it is important to investigate the liquefaction properties of crushable soils. The authors have conducted monotonic and cyclic triaxial tests on two types of crushable soils; Masado (Hyodo et al., 1991) and Shirasu (Okabayashi et al., 1994). The effect of both confining pressure and relative densities on the shearing properties of these soils has been considered. The research presented here was performed in order to understand the effect of soil compressibility and crushability on the monotonic and cyclic undrained shear strength of soils. In addition, an attempt has been made to relate monotonic and cyclic shear strengths.

Table 1 Physical proprerties for soils tested

	G_s	e_{max}	e_{min}	U_c	L_i (%)
Ube Masado	2.615	1.243	0.728	4.772	2.690
Hiroshima Masado	2.610	1.394	0.784	5.819	3.731
Shirasu	2.489	1.494	0.775	29.50	–
Toyoura sand	2.643	0.973	0.635	1.200	–

MATERIALS AND EXPERIMENTAL METHODS

Physical Properties of Materials

In this study we used two crushable soils, Masado and Shirasu to investigate the correlation between shearing properties and other mechanical characteristics. Masado soil for the tests was taken from two locations, an area east of Hiroshima and Ube. From now these two soils will be known as Ube Masado and Hiroshima Masado. They are both geologically similar being weathered granite from the Hiroshima facies. The Shirasu was sampled from a weathered deposit at Airagun in Kagoshima prefecture. Finally, for comparison purposes, Toyoura silica sand was also tested. The particle size distributions are shown in Fig. 1. The coefficient of uniformity for both the Masado soils and the Shirasu is greater than for the Toyoura sand. The fines content of the Shirasu (< 0.074mm) was about 35% but was non-plastic, containing the same materials as the parent material. (Hyodo et al.,1980). The Masado, on the other hand, was washed and passed through a 2mm sieve, thus having a particle size range, 0.074mm - 2mm. Physical properties for each soil are shown in Table 1. The specific gravity of the individual pumice grains of Shirasu is very low, due to intraparticle occluded voids. The maximum and minimum void ratios for both the Masado soils and Shirasu was greater than that for Toyoura sand. This was particularly so for the maximum void ratio. Masado has various degrees of weathering leading to a variation in mechanical properties. The degree of weathering can be shown by the ignition loss, in Table 1, which increases with the degree of weathering (Murata et al., 1987). So for example, based on this, the Hiroshima Masado is more weathered than the Ube Masado.

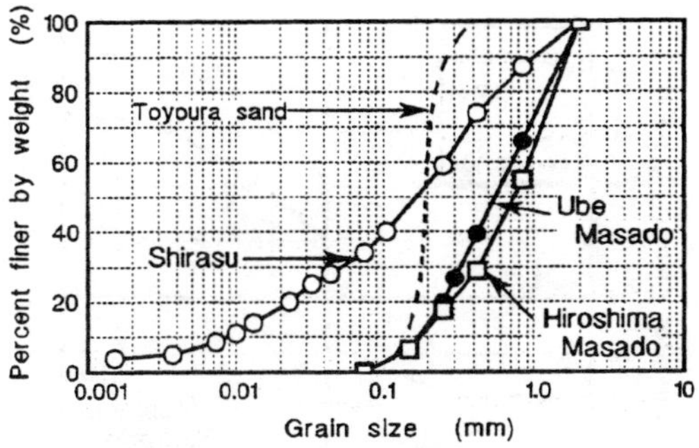

Fig.1 Grain size distribution curves for soils tested

Experimental Method

The dimensions of the triaxial specimen were 50mm diameter and 110mm height. Masado specimens were air pluviated to the same height producing an initial relative density Dr_i of 60% for Hiroshima Masado and 50% for Ube Masado. The initial void ratios were quite similar. However, because Shirasu has a high fines content and varying specific gravity of individual grains, air pluviation and water pluviation was not used. Instead, the required mass of material was weighed out and a loose sample was prepared by rodding the material into the mould. Following this, the required relative density was achieved by tapping the mould until the volume had reduced sufficiently. The relative densities of 50% loose and 90% dense were tested for this material. In order to produce full saturation the samples were first flooded with CO_2 gas which was then displaced by de-aired water and a back pressure of 100 kPa was applied to produce a B value greater than 0.96. After the application of the required cell pressure, monotonic and cyclic undrained triaxial tests were performed. Monotonic triaxial tests were performed in both compression and extension under constant cell pressure conditions and an axial strain rate of 0.1% per minute. Cyclic tests were performed under stress controlled conditions with a sinusoidal load being applied by a pneumatic actuator at a frequency of 0.1 Hz. Polished stainless steel end platens with a diameter of 60mm were used together with a silicone grease and a membrane were used to reduce end friction. The top and bottom platens each had a 5mm diameter porous stone at the centre for drainage and pore water pressure measurement.

Stress Parameters

The stress invariants p' and q used, have been defined for triaxial conditions as mean principal effective stress, $p' = (\sigma_a' + 2\sigma_r')/3$ and deviator stress $q = \sigma_a - \sigma_r$. In this paper the effective consolidation stress is distinguished by using the notation σ_c' and the amplitude of cyclic deviator stress is given by σ_d.

ISOTROPIC COMPRESSION CHARACTERISTICS

The results for the isotropic compression tests on each of the sands are shown in Fig. 2. In this figure the hatched area contains the effective consolidation pressure used for the shear tests. The initial relative densities shown for the Masado and Shirasu soils are the same as that used for the shear tests. For the stress range considered it can be seen that even loose Toyoura sand, Dr_i = 40% is relatively compressible and its curve lies well to the right of the crushable soils. In this figure, the yield stress, p_y, conventionally defined as the intersection of the straight line section of the e-log p curve, has been marked for each soil. The yield point for the Toyoura silica sand is an order of magnitude

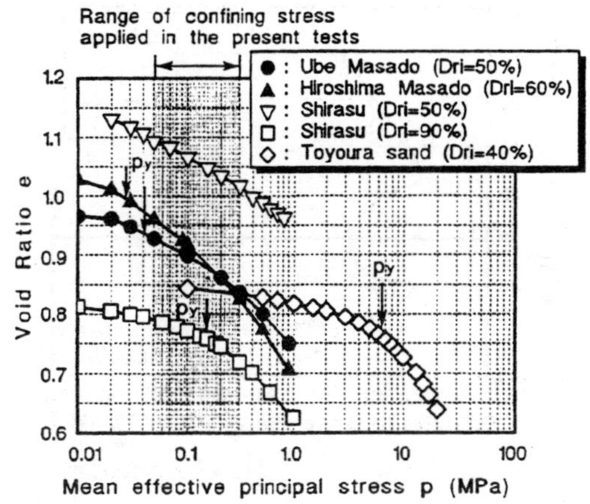

Fig.2 Isotropically compression curves for each sample

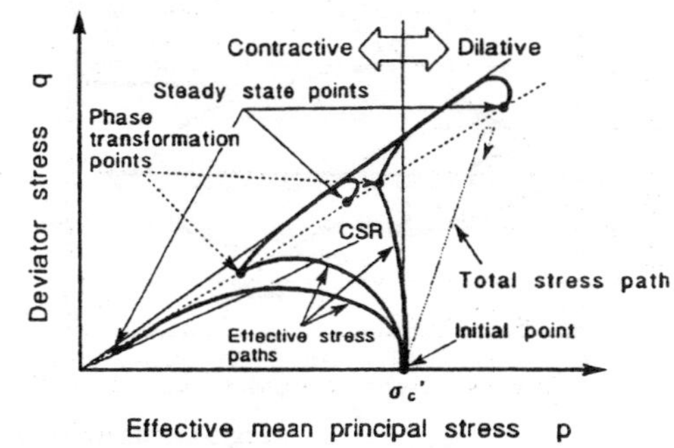

Fig.3 Schematic diagram of effective stress path for undrained monotonic shear tests

greater than that for crushable soils, being 6 MPa. In the hatched working stress region the compressibility of the silica sand is negligible. This is why in previous studies the normalized liquefaction resistance, $\sigma_d/2\sigma_c'$, has not apparently been affected by confining pressure (Ishihara., 1994). On the other hand, the yield points for the Masado were at particularly low pressures. Similar behaviour was observed for dense Shirasu with the yield point occurring in this zone and although no apparent yield point was noted for the loose material, a marked decrease was noted for the loose Shirasu in this zone. It can be seen that the crushability of the soil grains is an important factor in the compression of both Masado and Shirasu.

UNDRAINED MONOTONIC SHEAR PROPERTIES

Fig. 3 shows sets of possible effective and total stress paths for undrained shear tests. The lowest effective stress path is for loose sands and demonstrates purely contractive behaviour until the steady state point is reached on the far left of the diagram (Vaid et al., 1985, 1989, Castro et al., 1975, 1977). Dense sands, however, pass through a phase transformation after an initial contractive period and dilate to a steady state point. In this figure, if the final steady state is at a mean principal effective stress greater than the initial confining stress, then the sample is defined as "dilative". Otherwise, it is "contractive". In order to investigate the effect of confining pressure on the undrained shear properties of the crushable soils, undrained monotonic compression and extension tests were performed at various cell pressures.

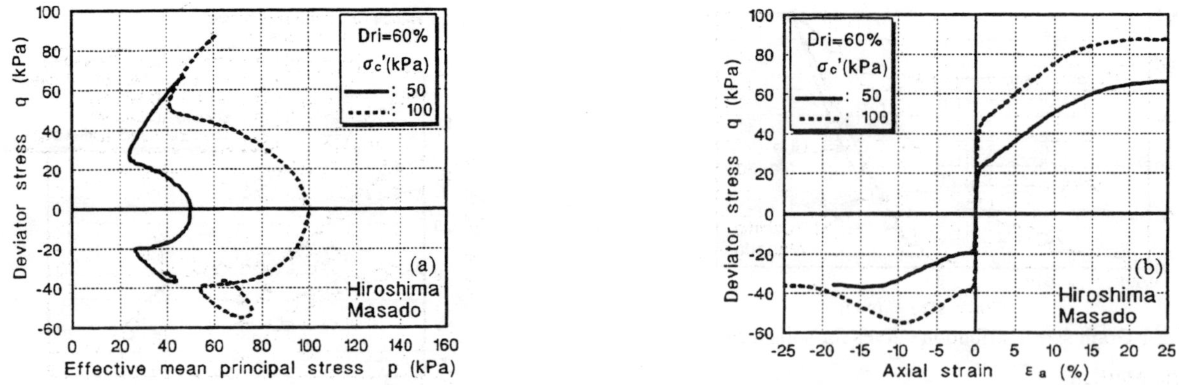

Fig.4 Undrained monotonic triaxial behaviour of Hiroshima Masado

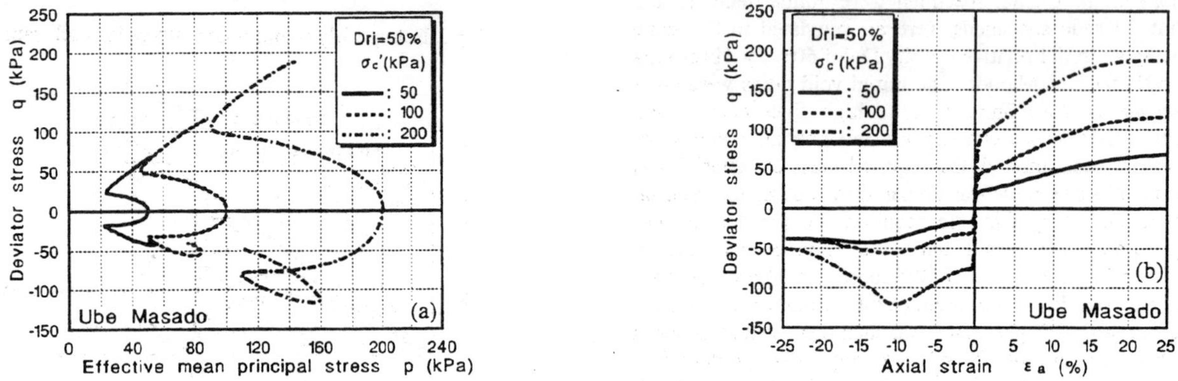

Fig.5 Undrained monotonic triaxial behaviour of Ube Masado

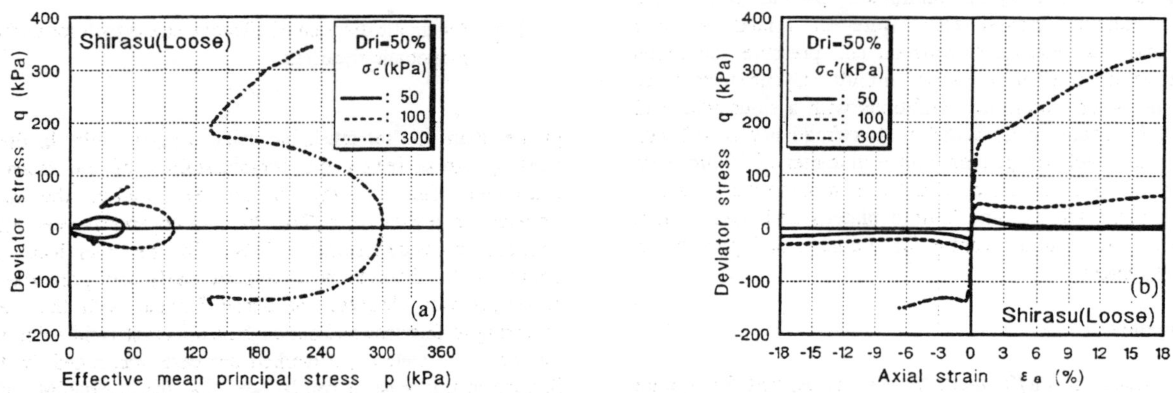

Fig.6 Undrained monotonic triaxial behaviour of loose Shirasu

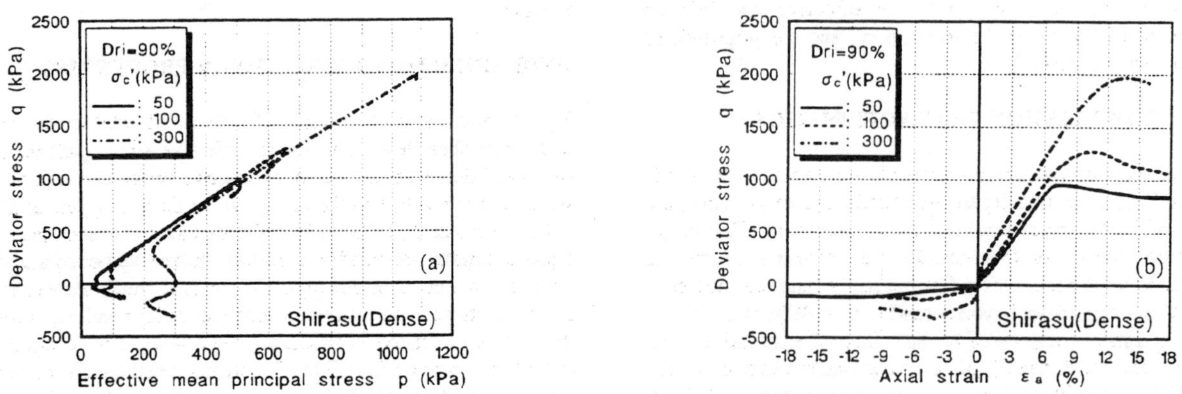

Fig.7 Undrained monotonic triaxial behaviour of dense Shirasu

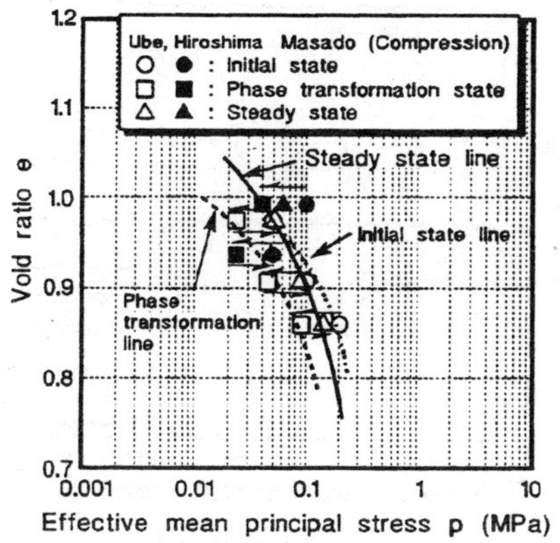

Fig.8 State diagram for Hiroshima and Ube Masado

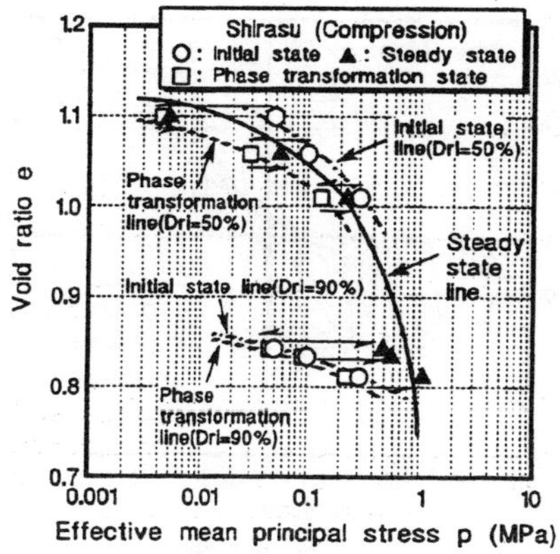

Fig.9 State diagram for Shirasu

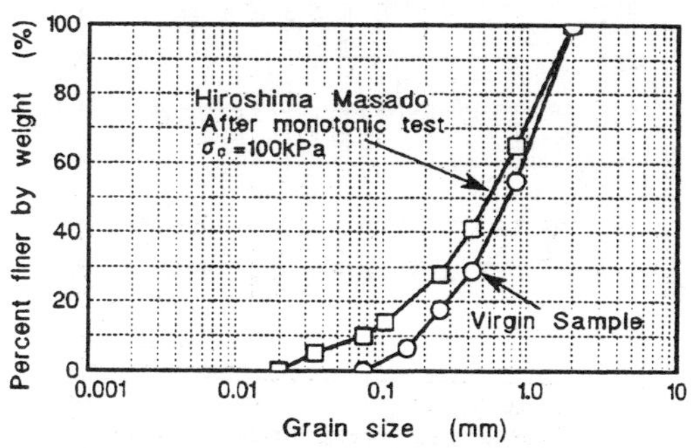

Fig.10 Grain size distribution curves for Masado in pre- and post-triaxial test

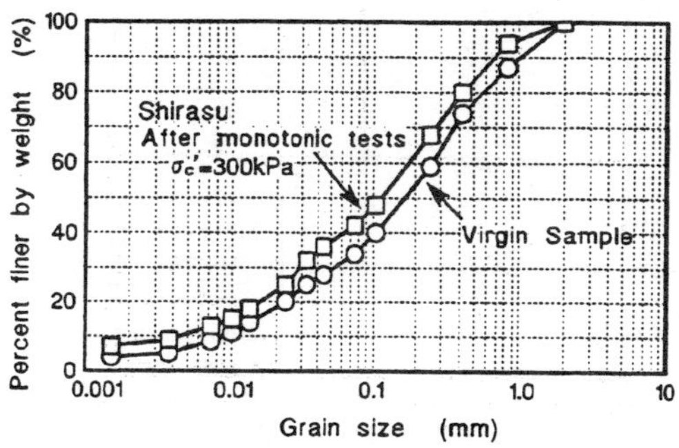

Fig.11 Grain size distribution curves for Shirasu in pre- and post-triaxial test

The cell pressures used were 50 and 100 kPa for Hiroshima Masado; 50, 100 and 200 kPa for Ube Masado and 50, 100 and 300 kPa for loose and dense Shirasu. Figs. 4(a) and 5(a) show the effective stress paths for Hiroshima and Ube Masado respectively. The effective stress paths for these materials are similar, each showing an initial contraction followed by a point of phase transformation and then increasing dilatancy. However, each of the stress paths terminated before any net dilation occurred. Figs. 4(b) and 5(b) show the deviator stress versus the axial strain. In the case of these materials, it can be observed that very little axial strain occurred up to the point of phase transformation. After this, axial compressive strains rapidly increases to values in excess of 20% at which point it is considered that steady state conditions were achieved. Although there was an increase in sample strength with increasing confining pressure, the shape of each stress-strain curve remained similar.

Figs. 6 (a) , (b) and 7 (a) , (b) show the stress paths and stress-strain curves for loose (Dr_i = 50%) and dense (Dr_i = 90%) Shirasu, respectively. Examination of Fig. 6(b) shows marked strain softening for confining pressures of 50 and 100 kPa. In particular at 50 kPa only contractive behaviour was observed (Fig. 6(a)) with the sample achieving a steady state without any phase transformation occurring. At 100 kPa a slight strain hardening was observed to occur after the phase transformation point and shortly after this the sample reached the steady state conditions on the contractive side. At 300 kPa there is no strain softening and indeed after the phase transformation point strain hardening occurs. It is clear for loose Shirasu that the behaviour is much affected by the level of confining pressure. In contrast, the dense Shirasu reached the phase transformation point at a very early stage and went on to develop strain hardening behaviour.

Figs. 8 and 9 show the void ratio-effective stress relationships at the initial state, phase transformation and steady state conditions from the compression side shear tests on Masado and Shirasu. Although similar results were obtained in extension tests, at large strains necking occurred and so the steady state data was not considered reliable in this region. In these figures the steady state line is unique for a given soil regardless of initial relative density, void ratio and confining pressure. The steady state line for Shirasu was particularly curved. Both Ube and Hiroshima Masado soils are shown in the same figure. Although their initial states are slightly different their overall behaviour was very similar, each having almost the same steady state lines. The initial confining pressure shown by the initial state line was wholly located on the right of the steady state condition indicating contractile behaviour during shear. The normal yield stress, p_y for these materials was in each case less than the initial confining pressure indicating that the soils were well within their crushing region. This is typical of crushable materials such as Masado and Shirasu.

Castro (1975) cautioned that such soils, where the initial state was on the right of the steady state line, could be subject to classical liquefaction, meaning unstable softening or contractive behaviour, if subjected to cyclic loading. In order to evaluate the degree of crushing after shearing for Masado and Shirasu, samples were sieved before and after testing. Figs. 10 and 11 show typical results for each material selected at two different confining pressures. Examination of the size distribution curves for each material indicate an increase in the proportion of fines in each case. It can be seen that particle breakage has occurred across the full range of particle sizes. In particular for the Masado it is thought that the breakdown of loosely

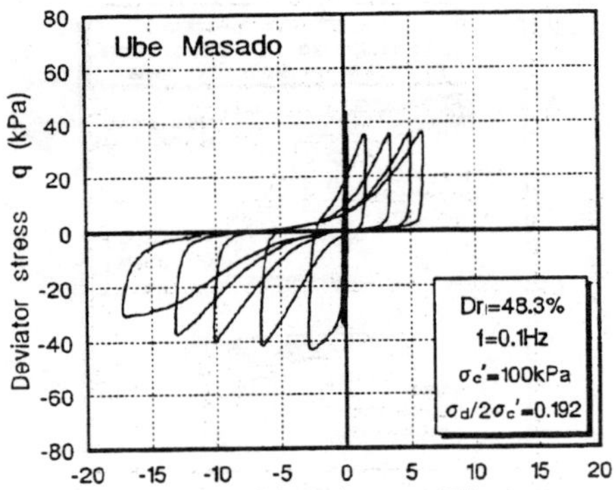

(a) Ube masado

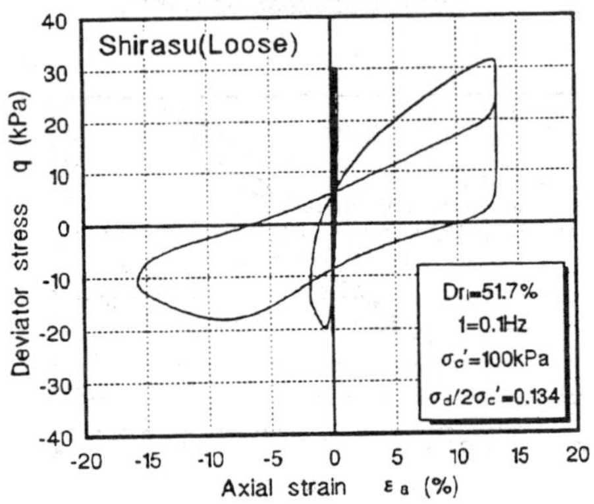

(b) loose Shirasu

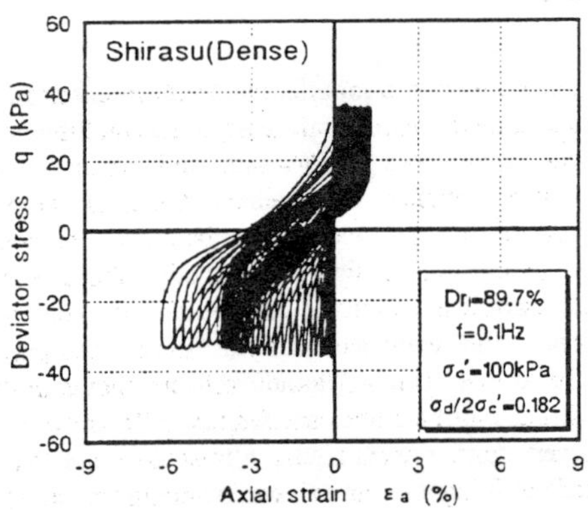

(c) dense Shirasu

Fig.12 Relationship of deviator stress against axial strain in undrained cyclic triaxial tests

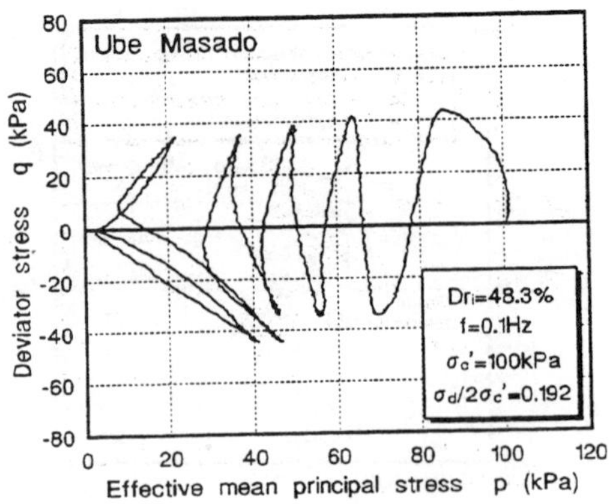

(a) Ube masado

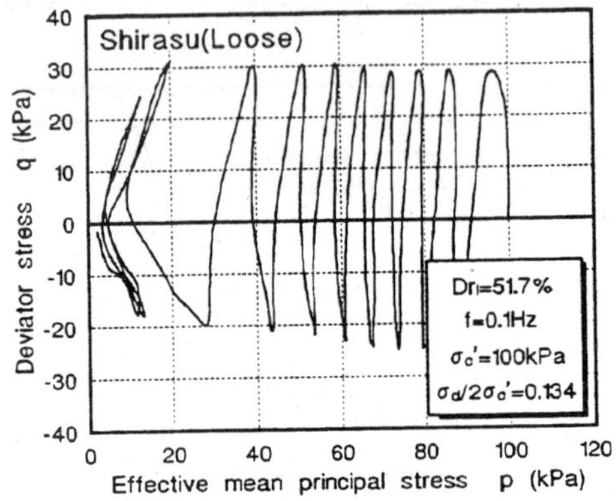

(b) loose Shirasu

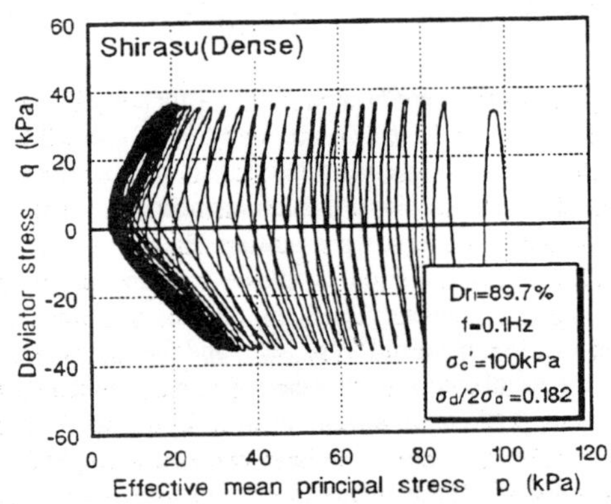

(c) dense Shirasu

Fig.13 Effective stress path in undrained cyclic triaxial tests

cemented aggregated particles has also produced a larger quantity of finer particles.

UNDRAINED CYCLIC SHEAR DEFORMATION

Undrained cyclic shear tests were performed on Masado and Shirasu at the same confining pressures as for the monotonic tests. Figs. 12 (a), (b) and (c) show typical cyclic stress-strain plots for samples tested at 100 kPa. The behaviour described in Fig. 12 (a) for Ube Masado is also typical of the Hiroshima Masado. It should be noted that large axial strain only began to occur after the cyclic stress path reached the phase transformation point. Although incremental development of axial strain occurred on the compression side, the behaviour was markedly anisotropic and much larger strains occurred in extension. In Fig. 12 (b) for the loose Shirasu once again large axial strains only develop after the phase transformation point. In this case however large symmetrical strains develop very rapidly compared with Masado soils. The initial state condition for this material is situated further to the right of the steady state line leading to the expectation of more marked flow deformation behaviour under cyclic loading.

For hard grained silica sands such an initial state representing a very loose sand would be quite unusual. However for crushable soils which are often also angular this kind of initial state is easily achieved. The corresponding stress paths for each of the materials are shown in Figs. 13 (a) - (c). Figs. 13 (a) and (b) for Masado and loose Shirasu show

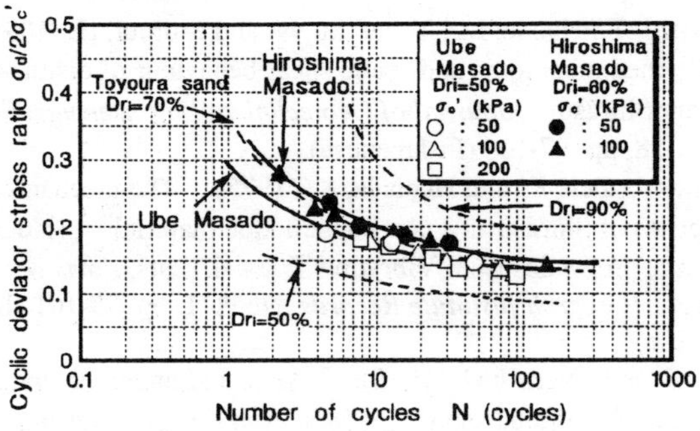

Fig.14 Cyclic shear strength curves for Hiroshima and Ube Masado

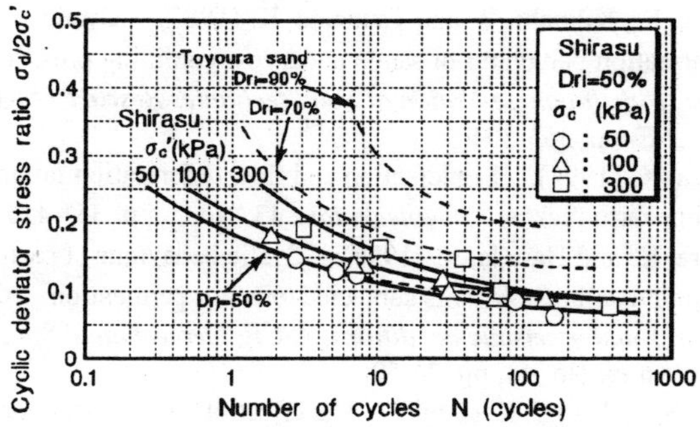

Fig.15 Cyclic shear strength curves for loose Shirasu

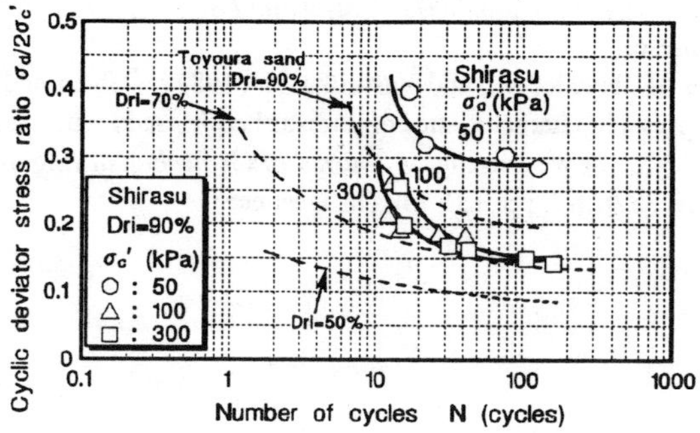

Fig.16 Cyclic shear strength curves for dense Shirasu

the development of liquefaction for these materials after very few cycles when the effective confining pressure becomes zero. On the other hand dense Shirasu exhibits typical cyclic mobility behaviour where the mean principal effective stress does not quite become zero. Examination of Figs. 12 (a), (b) and 13 (a) , (b) show that as p' approaches zero the average slope of the stress - strain loop decreases. On the other hand in Fig. 13 (c), it can be seen that during the final cyclic mobility phase of the behaviour a residual p' remains and hence the slope of the cyclic stress - strain curves, although decreasing slightly, remains at a higher value.

Defining the liquefaction failure as a double amplitude strain DA = 5%, then figures 14 -16 show the plots of stress ratio $\sigma_d/2\sigma_c'$, against number of cycles, N, to cause liquefaction failure. The figures show data for the crushable soils over a range of confining pressures and for comparison purposes include the data for Toyoura sand at relative densities of 50%, 70% and 90%. These latter lines are for a single confining pressure of 100 kPa, however , it has been shown that there is no effect due to confining pressure on silica sand over this range of relative densities (Ishihara et al., 1983, Yunoki et al., 1982). In these figures the cyclic strength of Hiroshima Masado is slightly greater than that for Ube Masado and neither was affected by confining pressure. It can also be seen that for the corresponding monotonic stress paths in Figs. 4 and 5 are similar as are the phase transformation strengths for different cell pressures. On the other hand for loose and dense Shirasu as shown in Figs. 15 and 16 there is a clear confining pressure dependency. However the cyclic strength of loose Shirasu increased with increasing confining pressure while for the dense material the opposite tendency is observed. Similarly opposite tendencies can also be observed in the monotonic stress paths in Figs. 6 and 7.

CONCLUSIONS

This study has examined the cyclic behaviour of soils of varying crushability characteristics in relation to basic parameters determined from monotonic undrained stress paths. The conclusions can be summarized as follows:

1) The slope of the isotropic compression line in the working stress zone for Toyoura sand was very small. On the other hand the Masado and Shirasu soils exhibited grain crushing and yielding behaviour and hence a much steeper compression line in the same stress range.

2) The Masado soils exhibited dilative behaviour after the phase transformation point but still reached the steady state on the contractive side.

3) Monotonic undrained shear behaviour of loose Shirasu changes from strain softening to strain hardening with increasing confining stress. This is the opposite of what is normally expected.

4) The monotonic undrained shear behaviour of dense Shirasu behaved dilatively at every cell pressure and finally ended on the steady state line in dilative state.

5) The steady state line for Masado and Shirasu was shown to be a unique curve regardless of confining pressure.

6) Undrained cyclic shear behaviour of Masado and loose Shirasu liquefaction occurred suddenly after the phase transformation point was reached in extension. On the other hand for dense Shirasu cyclic strains gradually increased until a state of cyclic mobility was achieved.

7) Both Hiroshima and Ube Masado had a unique cyclic strength curve regardless of confining pressure. In loose Shirasu the cyclic strength increased with increasing confining pressure while for dense Shirasu the opposite was true.

REFERENCES

Castro, G. (1975): "Liqeufaction and cyclic mobility of saturated sand," *Journal of Geotechnical Engineering, ASCE*, **101**, No. GT6, pp. 551-569.

Castro, G. and Poulos, S. J. (1977): "Factors affecting liquefaction and cyclic mobility" , *Journal of Geotechnical Engineering, ASCE*, **103**, GT6, pp. 501-516.

Hyodo, M., Yamanouchi, T. and Sakai, A. (1980): "The effect of particle size distribution on the liquefaction characteristics of Shirasu" . *Proc. 15th Japanese National Soil Mechanics Research Meeting, JSSMFE*, pp. 581-584 (In Japanese).

Hyodo, M., Yasufuku, N., Murata, H., Yamamoto, Y. and Marayama, H. (1991): "The effect of confining pressure on the liquefaction characteristics of Masado" , *Proc. 26th Japanese National Soil Mechanics Research Meeting, JSSMFE*, **1**, pp. 841-842 (In Japanese).

Ishihara, K., Kikuchi, Y. and Tsutsumi, K. (1983): "Cyclic deformation behaviour of sands under low confining pressure", *Proc. 18th Japanese Nation al Soil Mechanics Research Meeting, JSSMFE*, pp. 353-354.

Ishihara, K. (1994): "Liquefaction and flow deformation failure during earthquakes", *Geotechnique* **43**, No. 3, pp. 351-415.

Ishihara, K. and Harada, K. (1996): "Cyclic behaviour of partially saturated collapsible soils subjected to water permeation", *Ground failures under seismic conditions, ASCE, Geotechnical Special Publication* No. 44, pp. 34-50.

Miura, S. and Yagi, K. (1995): "The effect of stress history on the dynamic characteristics of a course grained volcanic soil", *Proc. of Symposium on the properties of volcanic ash soil related to design and construction. Japanese Geotechnical Society*, pp.229-236, (In Japanese).

Murata, H., Hyodo, M., and Yasufuku, N. (1987): "Shear and compression characteristics of undisturbed Masado with respect to the degree of weathering", *Journal of Geotechnical Engineering, JSCE*, 382/III-7, pp. 131-140 (In Japanese).

Okabayashi, T., Hyodo, M., Yasufuku, N. and Murata, H. (1994): "Undrained monotonic and cyclic shear behaviour of disturbed primary Shirasu", *Journal of Geotechnical Engineering, JSCE*, 499/III-28, pp. 97-106 (In Japanese).

Umehara, Y., Zen, K. and Hamada, K. (1975): "The mechanical properties of Shirasu as a port construction material", (No 5), Dynamic properties from Vibration Triaxial Tests, *Ports & Harbour Research Instutute Records*, No. 211, pp. 79-101 (In Japanese).

Vaid Y. P. and Chern J. C. (1985): "Cyclic and monotoic undrained response of saturated sand", *Advances in the art of testing soils under cyclic conditions, ASCE convention*, pp.120-147.

Vaid, Y. P., Chung, E. K. F. and Kuerbis R. H. (1989): "Preshearing and undrained response of sand, *Soils and Foundations*, **29**, No. 1, pp. 49-61.

Yunoki, Y., Ishihara, K. and Seki, M., Yoneda, Y. (1982): "The effect of the initial effective confining pressure on cyclic triaxial shear behaviour of dense sand", *Proc. 17th Japanese National Soil Mechanics Research Meeting, JSSMFE*, pp. 1649-1652.

Proceedings of the Seventh (1997) International Offshore and Polar Engineering Conference
Honolulu, USA, May 25-30, 1997

ISBN 1-880653-28-1 (Set); ISBN 1-880653-29-X (Vol. I)

Developments in Prefabricated Vertical Drain Enhanced Soil Flushing

J.D. Quaranta, M.A. Gabr, E.E. Cook and D. Szabo
West Virginia University
Morgantown, WV, USA

ABSTRACT
The innovative application of prefabricated vertical drains (PVDs) for enhancement of the *in situ* soil flushing remediation technique is presented. Testing was performed to evaluate the internal well resistance as a function of flowrate of the PVDs operating under vacuum pressures. The results indicated an approximate linear relationship between the PVD flowrate versus the total vacuum head as well as the head loss versus flowrate. The zone of influence was characterized by testing the PVDs under vacuum conditions in soils consisting of 100% sand and 80% sand/20% kaolinite mixture. Results of the areal extent of the PVD zone of influence were quantified and showed relationships between the 100% sand specimens and the 80% sand/20% kaolinite specimens. Pilot-scale tracer flushing tests demonstrated the feasibility of using PVDs for flushing in the 100% sand and the 80/20 soil. The tracer testing demonstrated that longer flushing times were necessary to achieve 80% clean-up levels as the soil hydraulic conductivity decreased when the kaolin was added to the sand.

Key Words: prefabricated vertical drain, soil flushing, geosynthetic, vacuum enhanced consolidation

INTRODUCTION
Soil flushing is a treatment technology used to remove contaminants from the subsurface with the aid of solutions. The technology uses drilled injection and extraction wells located within the site's contaminated zone. The injection wells are the source point for introducing the flushing solution into the subsurface soil. The remediation process occurs as the solution permeats through the contaminated soil and desorbing the soil-bound contaminants into solution. The extraction wells are used to recover the contaminated groundwater solution for subsequent treatment.

This technology is limited to sites having coarse-grained soils (i.e., coarse sands) with hydraulic conductivity ranging 10^{-1} cm/s to 10^{-2}cm/s. In the case of fine-grain fraction soils with hydraulic conductivities less than 10^{-3}cm/s, conventional soil flushing is not always effective for remediation of contaminated soils and the use of pump and treat systems on long term basis may be required. This is because the extraction of the contaminant from the soil and movement of the flushing solution are restricted due to the lower permeability and adsorption characteristics of the fine-grain soils.

In the area of geotechnical engineering prefabricated vertical drains (PVDs) are routinely used for soil improvement in consolidation of soft soils. The PVDs shorten the drainage path for removal of the pore water by either applying a surface surcharge to squeeze the soil, or by applying a vacuum to the PVDs causing a hydraulic gradient promoting pore water movement toward the drain. PVDs are a composite system of an inner core and an outer filter jacket. The core is typically constructed of extruded, polypropylene and the filter jacket is typically a durable, non-woven polypropylene geotextile, refer to Figure 1.

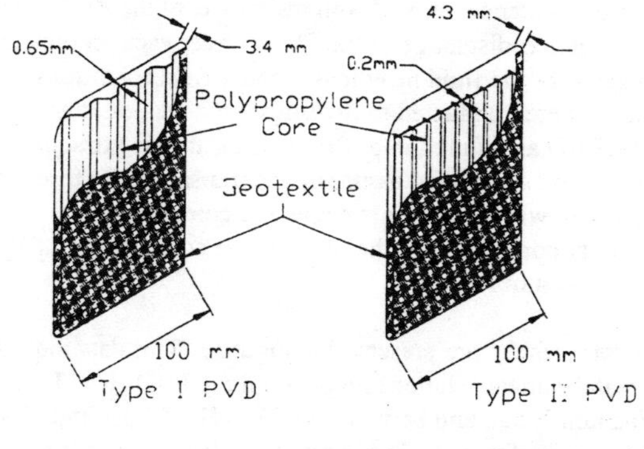

Figure 1 - Prefabricated Vertical Drains (Types I&II)

The idea behind using the PVD system in a soil flushing scheme is to shorten the drainage path of the groundwater flow and promote subsurface liquid movement for expediting the soil flushing process. The PVDs in this case are used for injection of a flushing solution into the *in situ* soil concurrent with an extraction process for removal of the contaminated solution. The flushing process occurs while the soil is in a saturated condition. This diminishes the potential for inducing a volumetric change in the soil thereby reducing the hydraulic conductivity and increasing the flushing time. The concept of this technology is illustrated in Figure 2.

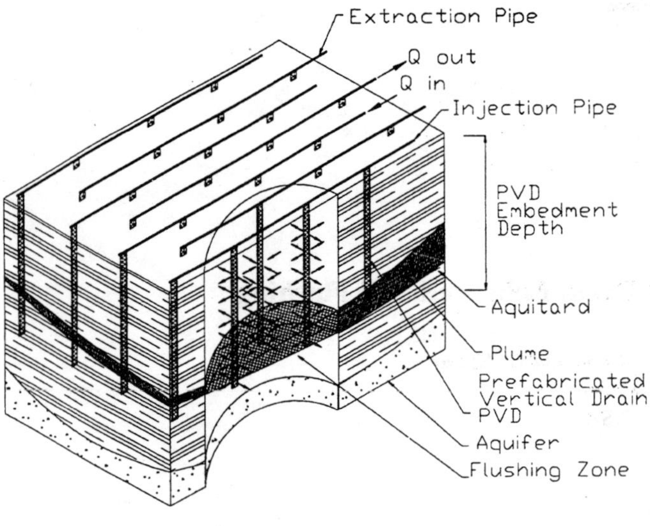

Figure 2 PVD Soil Flushing Concept

The PVD soil flushing technology is applicable to coastal or marine environments in two ways. The first is for removal of contaminants from existing soils and dredged materials. Secondly, the PVD technology may be used as a technique for re-introducing electrolytes to the existing soil pore fluid for the purpose of increasing soil strength properties.

This paper presents an overview of three key research areas of importance for determining the feasibility of PVDs for enhanced soil flushing: 1) well resistance, 2) characterization of the PVD zone of influence, and 3) contaminant recovery pilot-scale tests to investigate the vacuum/injection feasibility.

WELL RESISTANCE

The internal resistance, termed well resistance, of the PVDs effects the vertical extraction discharge of the PVDs under vacuum pressures. In order to quantify the fluid head losses under field conditions a series of experiments were conducted to investigate the well resistance effects within the PVD as a function of flow rate during an extraction process. These tests were conducted assuming the transmissivity through the jacket (parallel with the core) is negligible compared to the transmissivity component in the plane of the core. The two PVD types tested are shown in Figure 1.

Experimental results are presented in Figure 3. This data indicates an approximately linear relationship between the PVD Head Loss and Total Vacuum Head, and between the PVD Head Loss and Flowrate for all of the tests performed. The hydraulic radius (cross sectional area of flow/wetted perimeter) for the PVDs was calculated as: Type I= 25.6 mm. and Type II=46.4 mm. The Type II PVD has a larger hydraulic radius by approximately 45%. The larger hydraulic radius of the Type II PVD is credited with providing a 2.8 times higher flowrate at approximately the same operating vacuum as compared to Type I PVD. The Type II PVD consistently showed data supporting its superiority in producing higher flowrates at lower applied vacuum head and with less head loss than the Type I PVD.

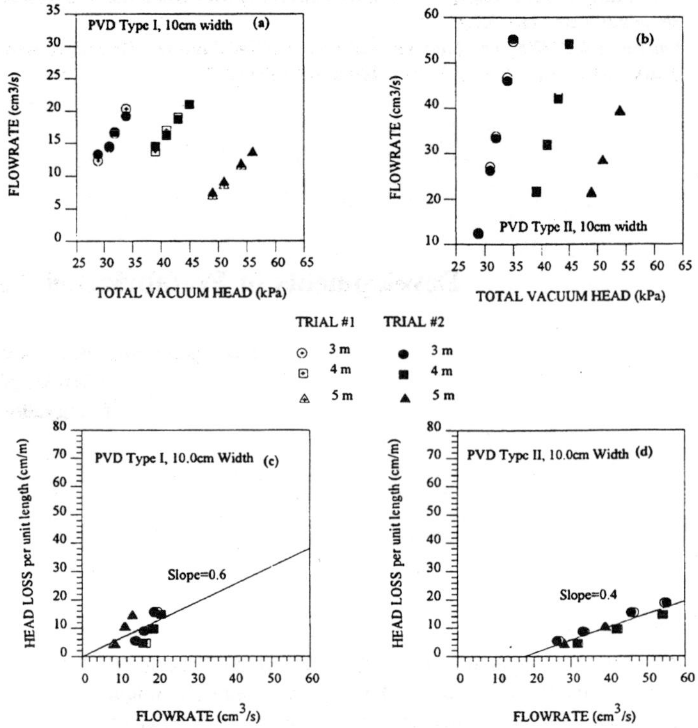

Figure 3 - PVD Well Resistance
a) PVD Flowrate vs Total Head, Type I, b)PVD Flowrate vs Total Head, Type II, c) PVD HeadLoss per unit length vs Flowrate, Type I, d) PVD HeadLoss per unit length vs Flowrate, Type II

ZONE OF INFLUENCE

The zone of influence addresses the areal coverage within which the PVD is effective in influencing the subsurface liquid movement under vacuum or pressure conditions. The tests provide insight into the necessary PVD spacing and configuration requirements. The experiments were carried out using cylindrical cells measuring 0.3m in diameter, refer to Figures 4 a&b.

The zone of influence was characterized in two study soils: a 100% Ottawa sand, and an 80% Ottawa sand/20% Kaolinite mix. Water surface levels were obtained through piezometer readings during the experiments. The zone of influence of the drains is quantified by comparing the total hydraulic head of the subsurface liquid versus time in the study soils and was determined as 2 times the minimum radial distance where changes in the water surface were considered not to be significant. The Type II PVD was used for all testing.

In the 100% sand tests, no clear Zone of Influence was developed. For the 80/20 specimens, the drawdown profiles are shown in Figure 5 a&b and corresponding surface contours in Figure 5c. A lower hydraulic conductivity for the 80/20 specimen is credited with development of the PVD Zone of Influence.

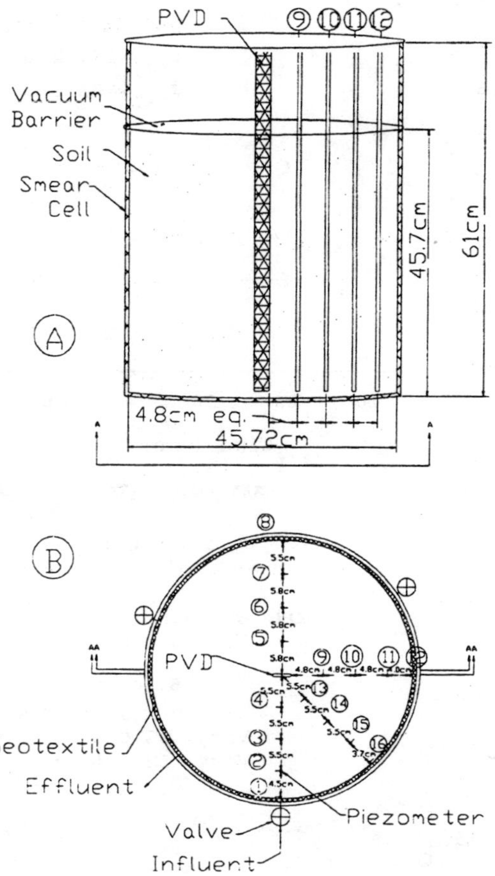

Figure 4 a) Smear Cell Elevation, B) Plan

The PVD Equivalent Diameter is the relationship of the rectangular PVD to an equal size cylinder. For this testing the PVD specimens Equivalent Diameter was taken as the circumference/π, and calculated as 1.9 cm. The PVD's Zone of Influence was taken to be 10 cm for the 2.5 kPa vacuum and 15 cm for the 15 kPa vacuum. The relationship between the Zone of Influence divided by the Equivalent Diameter was 5.3 for the 2.5 kPa and 8 for the 15 kPa.

CONTAMINANT RECOVERY

Experiments to evaluate the extraction rates and the feasibility of the injection vacuum scheme were performed using Contaminant Recovery Cells (CRCs) measuring 0.59m x 0.59m x 0.76m (width/depth/length), refer to Figure 6. This work entails testing the PVD system in a fluid flushing scenario at the pilot-scale.

The study soils consisted of a 100% Ottawa sand and a mixture of 80% Ottawa sand and 20% Kaolin. Tracer tests using a sodium chloride solute were performed to quantify tracer migration rates and extents within the cells during a flushing operation. The CRC tests are designed to obtain data which will assist in verifying fluid flushing feasibility and development of optimum PVD field configurations.

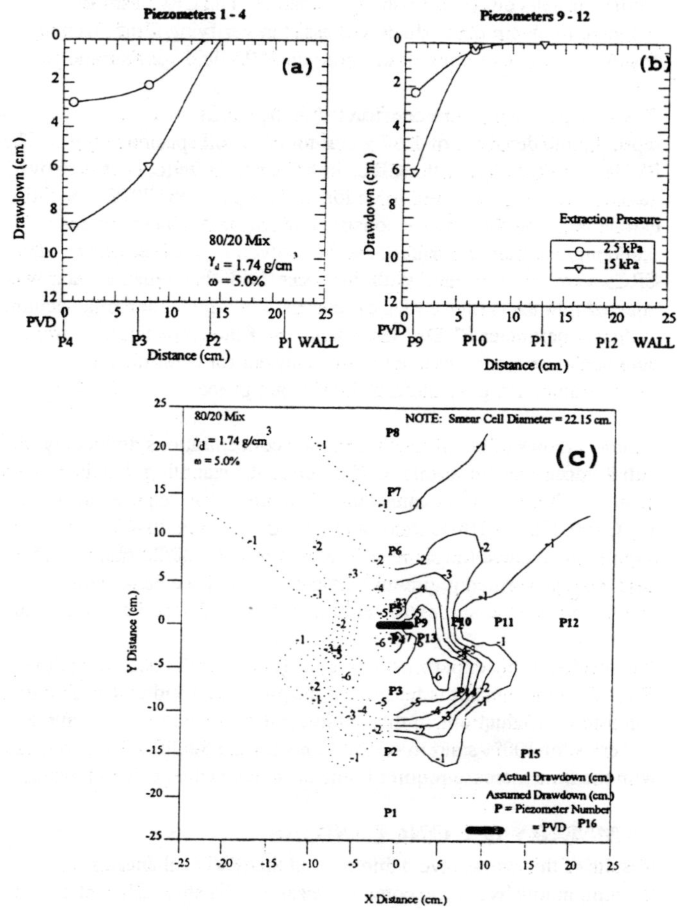

Figure 5 - Drawdown Profiles 80/20 Specimen
a) Drawdown Profile Piezometer 1-4

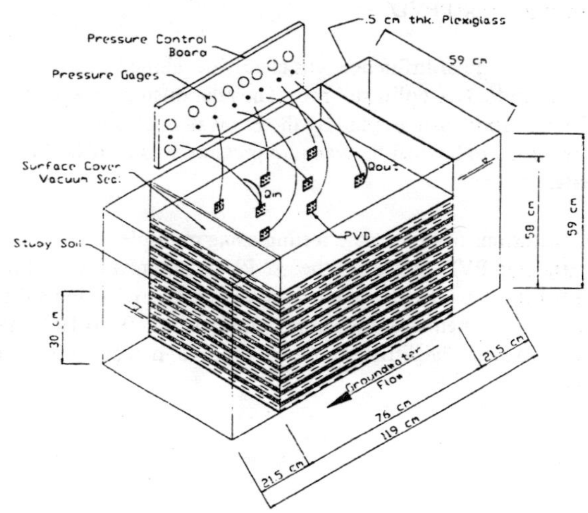

Figure 6 Contaminant Recovery Cell

The study soils consisted of a 100% Ottawa sand and a mixture of 80% Ottawa sand and 20% Kaolin. Tracer tests using a sodium chloride solute were performed to quantify tracer migration rates and extents within the cells during a flushing operation. The CRC tests are designed to obtain data which will assist in verifying fluid flushing feasibility and development of optimum PVD field configurations.

The soil specimens were constructed in the CRCs to achieve reproducible densities of 1.67 g/cm^3 for both soil specimen types. The PVDs were installed in the CRCs in off-set rows with the center row having two PVDs for fluid injection and six perimeter PVDs for fluid extraction. The fluid tracer consisted of prepared concentrations of 1,000 mg/l sodium chloride (NaCl) dissolved in distilled water. The CRCs were first saturated with the tracer fluid, then potable water was injected into the center PVDs concurrent with application of a vacuum to the six perimeter PVDs. Grab samples of the extracted fluid were obtained at set time increments and analyzed for chloride ion concentration using a standard chloride ion probe.

Figure 7 shows plots of the chloride ion concentrations divided by the initial tracer concentrations (C/Co) versus the sampling time increment. Results in Figure 7a indicates that 40 minutes were required to achieve a C/Co=0.2 in the 100% sand sample, and in Figure 7b 40 hours were required to achieve the same criterion limit in the 80/20 sample. The tests were performed maintaining a constant balance between the volume of injected and extracted fluids to minimize soil consolidation.

The results of this experimentation have successfully shown that the PVD flushing concept is feasible at the pilot-scale and under laboratory conditions. Reductions in the soil hydraulic conductivities from 6.8 x 10^{-2} cm/s for 100% sand to 4 x 10^{-5} cm/s for the 80/20 soil are consistent with the time increase required time to achieve equal clean-up levels.

SUMMARY & CONCLUSIONS

Results of this work have demonstrated that PVD enhanced soil flushing in low hydraulic conductive soils is feasible. Characterization of the PVD's internal resistance shows linear relationships between the discharge head loss and applied air vacuum, and with the discharge head loss and discharge flowrate. The Type II PVD having the larger hydraulic radius is credited with proving a 2.8 times higher flowrate than the Type I PVD.

A PVD Zone of Influence was clearly developed and maintained in 80% sand/20% kaolin soil mix. The relationship between the PVD Zone of Influence divided by the Equivalent Diameter was maintained between 5 to 8 when the vacuum pressure was increased from 2.5 to 15 kPa.

Conclusions of pilot-scale testing using multiple injection and extraction PVDs indicates the use PVDs for enhanced soil flushing is feasible in low hydraulic conductive fine-grain fraction soils. Decreases in soil specimen hydraulic conductivity from 10^{-2} to 10^{-5} cm/s resulted in equivalent NaCl tracer removal times to increase from 40 minutes to 40 hours.

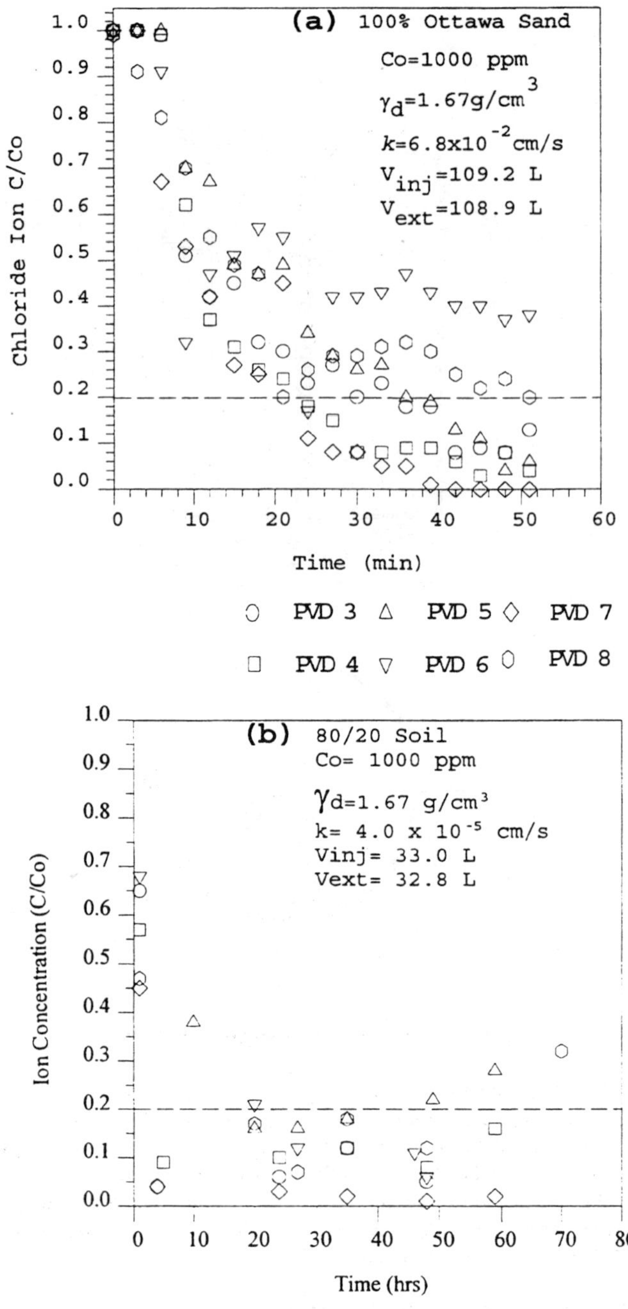

Figure 7 Chloride Ion C/Co vs Time
a) 100% Sand, b) 80/20 Mix

ACKNOWLEDGMENTS

This study was supported by the US Department of Energy, Morgantown Energy Technology Center (METC). The authors would like to thank Dr. Rodney Geisbrecht, METC Project Manager, for his interest and support in this work. The authors appreciate the engineering assistance and in-kind support provided by Mr. Walter van Woudenberg and Mr. James M. Cramer of the Nilex Corp. Several undergraduate students assisted in this study and their help is greatly appreciated.

Proceedings of the Seventh (1997) International Offshore and Polar Engineering Conference
Honolulu, USA, May 25-30, 1997

ISBN 1-880653-28-1 (Set); ISBN 1-880653-29-X (Vol. I)

Evaluation of Ground Densification by Dynamic Compaction Using SASW Method

Dong-Soo Kim, Hyung-Choon Park and Gi-Chul Kweon
Korea Advanced Institute of Science and Technology
Taejon, Korea

Sung-In Kim
Technology Institute Department DASAN Consultants
Seoul, Korea

ABSTRACT

In order to evaluate the quality and depth of ground densification by dynamic compaction, standard penetration test and/or cone penetration test are frequently used in Korea. Because these evaluation techniques are intrusive and one-point tests, they require a substantial cost and time to evaluate a large area. The Spectral-Analysis-of-Surface-Waves(SASW) method which is non-intrusive, has a great potential for rapid determination of effectiveness of ground improvement. This method consists of making field measurements of surface wave velocity at various frequencies and determining shear wave velocity (V_S) profile through a inversion process. In this paper, the quality and extend of ground improvement by dynamic compaction was evaluated using SASW method. The shear wave velocity profiles were determined using SASW method before and after compactions. Resonant column (RC) tests were performed using reconstituted specimens of various densities and confinements, and shear wave velocity-void ratio-confinement (V_S-e-σ_0) relationship of the site was determined. The variations in void ratio with depth before and after dynamic compactions were evaluated by combining Vs profiles determined by SASW method with V_S-e-σ_0 relationship determined by RC test. Based on this study, the feasibility of using SASW method in the application of assessing the quality and depth of ground improvement was verified.

KEY WORDS: SASW method, resonant column test, dynamic compaction, surface wave, shear wave velocity, dispersion curve, ground improvement

INTRODUCTION

As Korean economy is rapidly growing, a number of infrastructure systems are being constructed. Because of the limited land, most of the projects require a large-scale reclamation and the following ground improvement. As one of the ground improvement techniques, dynamic compaction is frequently used on granular soils. The procedure involves systematically dropping a large steel or concrete tamper from heights onto a grid spacing to densify the underlying soils. In order to evaluate the quality and depth of ground densification due to dynamic compaction, standard penetration test (SPT) and/or cone penetration test (CPT) are performed before and after dynamic compaction, and the variation in relative density is correlated empirically with SPT N-values and/or cone tip resistance (Bowles, 1988). However, both methods are intrusive and one point tests which require a substantial time to evaluate a large area. Because the relative density is empirically correlated with the penetration resistance, quality evaluation is quite dependent on the operation technique and soil type.

The Spectral-Analysis-of-Surface-Waves (SASW) method has a great potential for rapid determination of layer thickness and small-strain shear wave velocity (V_S) of soil deposits (Nazarian and Stokoe II, 1985, 1986). The SASW method is nonintrusive and nondestructive. It is based on the dispersive characteristics of surface waves. This method consists of making field measurements of surface wave velocity at various frequencies and determining Vs profile through a inversion process. The shear wave velocity can be related with the maximum shear modulus of granular soils which are mostly influenced by void ratio and confinement (Hardin, 1978). In order to obtain a

V_S-void ratio-confinement relationship, resonant column (RC) test has been widely used (Drnevich, 1985; Kim, 1995). The quality of ground improvement can be evaluated using a V_S-void ratio-confinement relationship determined by laboratory RC test.

In this paper, the quality and extent of ground improvement by dynamic compaction was evaluated using SASW method. At a reclamation site in the western coastal area, the shear wave velocity profiles were determined using SASW method before and after compaction. RC test was performed using reconstituted specimens of various densities and confinement, and V_S-void ratio-confinement correlation was determined. The void ratio profiles before and after compactions were determined using the developed correlation.

DYNAMIC COMPACTION

Dynamic compaction is a ground improvement method on granular soils by dropping a heavy steel or a concrete weight repeatedly at the same spot from 10 to 40 m height. Compaction is usually done systematically, in a rectangular or triangular pattern. The spacing between the impact points depends mainly on the depth of compressible layer, the permeability, and the location of ground water table. Deep craters with depth of 1 to 2 m are formed by the heavy tamping, and these are filled with sand after each pass. The depth of compaction, d, can generally be estimated from the following empirical correlation (Leonards et al., 1980; Mayne et al., 1984):

$$d=\alpha\sqrt{WH} \tag{1}$$

where W is the mass of the pounder in tons, H is the dropping height in meters, and α is a empirical coefficient ranging from 0.3 to 0.8.

In order to assess the effectiveness of the compaction, penetration tests such as standard penetration test (SPT) and cone penetration test (CPT) are mostly used, and Menard pressuremeter test (PMT) is sometimes used in Korea. The penetration resistance such as N value in SPT or tip resistance in CPT is usually converted to the relative density using the empirical correlations or these values are directly used in the foundation design. However, the current evaluation techniques are intrusive and/or require a borehole. When the site contains large particles such as boulder or gravel, test results are not reliable. Since these methods are one point testings, they require a large number of tests and thus a substantial time to evaluate a large area.

SPECTRAL ANALYSIS OF SURFACE WAVES (SASW) METHOD

Overview

The spectral analysis of surface waves (SASW) method is an in-situ, seismic method for determining the shear wave velocity (or shear modulus) profile of the site. Because shear wave velocity and shear modulus are mainly dependent on the effective stress and density, this method is useful in estimating the changes in soil strength and density resulting from dynamic compaction. The SASW method is non-intrusive and non-destructive.

When the surface is excited by an impulse load, the following three types of stress waves are transmitted away in an elastic half space: compression(P) wave, shear(S) wave, and surface(R) wave. Because approximately two-thirds of energy is transmitted away by the surface wave and surface wave decays much more slowly with distance than P and S waves, it can be assumed that at large distances (relative to the wave length) from the source, the energy contained in the body waves (P and S) is insignificant compared with that transmitted by the surface waves.

Most of the surface wave energy exists within one wave length of depth. In layered media, the velocity propagation of surface wave depends on the frequency (or, wavelength) of the wave because waves of different wave lengths sample different parts of the layered medium. For example, high-frequency (short wavelength) waves propagate only in the near-surface layers. Low-frequency waves with longer wave lengths propagate through the near-surface layers and also deeper layers. Therefore, by using surface waves over a wide range of frequencies, one can effectively sample different portions of the material profile. The SASW method is based on this dispersive characteristics of surface waves.

Field Testing and Data Reduction Procedures

The general coniguration of receivers, source, and recording equipment in the SASW test is shown in Fig. 1. Two vertical receivers were placed on the ground at an equal distance from a fixed centerline. Impact source was used to apply vertical excitation in the line with the two receivers at a distance, D, away from the near receiver. A FFT signal analyzer was used to record the receiver signals and then to transform them into the frequency domain. A reverse test was also performed with a source on the opposite side of the receiver array. Testing continues by progressively moving the receivers away from the fixed centerline. Typical source and receiver array used in this study is shown in Fig. 2.

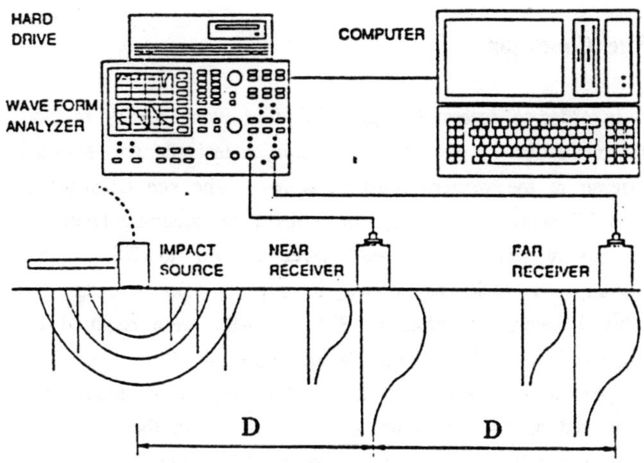

Fig 1. General Configuration of the SASW Testing Setup

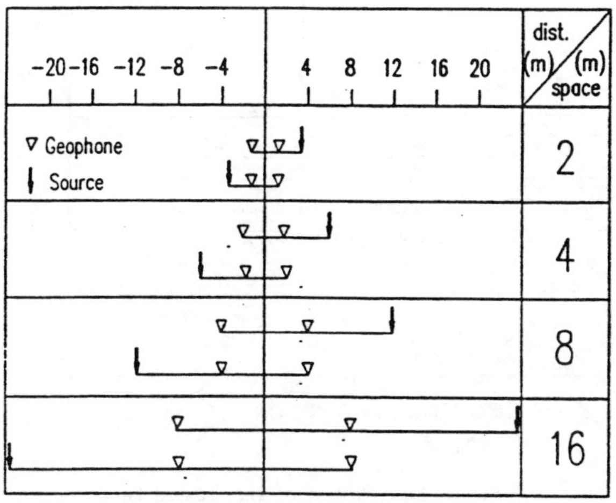

Fig 2. Typical Source and Receiver Array in This Study.

A phase information of the cross-power spectrum, which represents the phase difference between two receiver signals as a function of frequency, was obtained as shown in Fig. 3.

From the cross power spectrum, time delay between receivers is obtained for each frequency by:

$$t(f) = \theta_{xy}(f) / 2\pi f \quad (2)$$

where $\theta_{xy}(f)$ is the phase shift of the cross power spectrum in radian and the frequency, f, is in cycles/second. The surface wave phase velocity, V_R, is then calculated using:

$$V_R(f) = D \;/\; t(f) \quad (3)$$

where D is the distance between two receivers. The corresponding wavelength of the surface wave, λ_R is

$$\lambda_R(f) = V_R(f)/f \quad (4)$$

These calculations yield an experimental dispersion curve (V_R versus λ_R) for the receiver spacing.

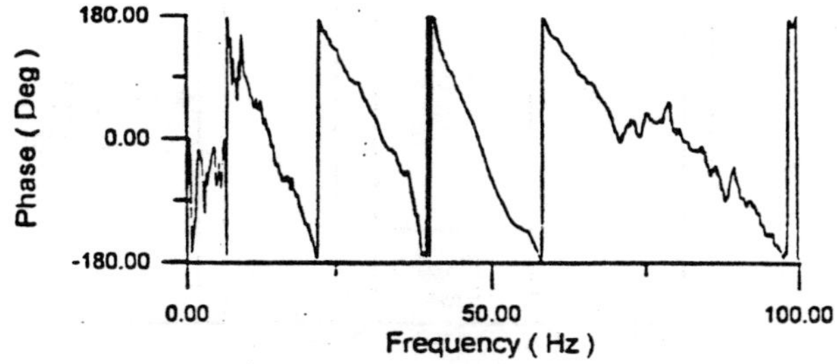

Fig 3. Typical Phase Information of the Cross-Power Spectrum

A shear wave velocity profile for each testing array is obtained through an iterative process matching the experimental dispersion curve to the theoretical dispersion curve. The theoretical dispersion curve was calculated for the assumed horizontal layered profile and compared with the experimental one. The assumed shear wave velocity and layer thicknesses in the profile were adjusted until satisfactory agreement between two dispersion curves was obtained. The shear wave velocity profile for the final theoretical dispersion curve was assumed to represent the actual profile of the site. The theoretical dispersion curve was calculated using a computer program of WinSASW developed at the University of Texas at Austin(1992).

RESONANT COLUMN TEST

In order to estimate the in situ void ratio using the shear wave velocity profile determined by SASW method, the laboratory resonant column (RC) test was performed using reconstituted specimens of various void ratios. The shear wave velocity (or, shear modulus) - void ratio - confining pressure relationship of the tested site was determined based on RC test results.

The RC testing equipment of the torsional fixed-free type was used. The bottom end of the specimen was rigidly fixed against rotation at the base pedestal while the top (free end) is connected to a drive system that is used to excite and monitor torsional motion, as shown in Fig. 4a. The basic operational principle is to vibrate the cylindrical specimen in first-mode torsional motion. Once first mode is established, measurements of the resonant frequency and amplitude of vibration are made as shown in Fig. 4b. These measurements are then combined with equipment characteristics and specimen size to calculate the

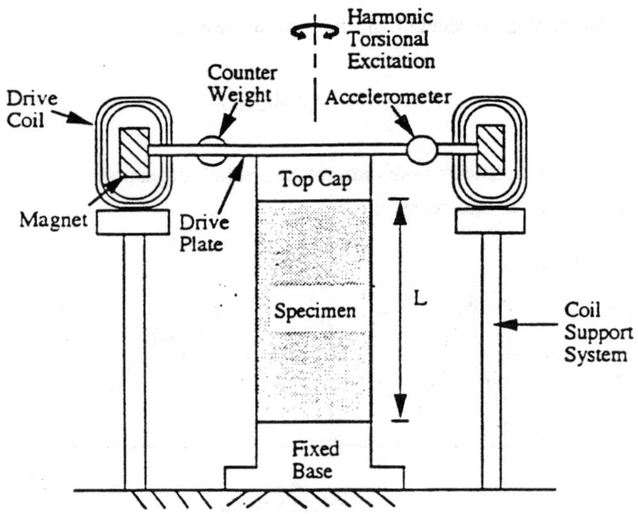

(a) Schematic Configuration of RC Test

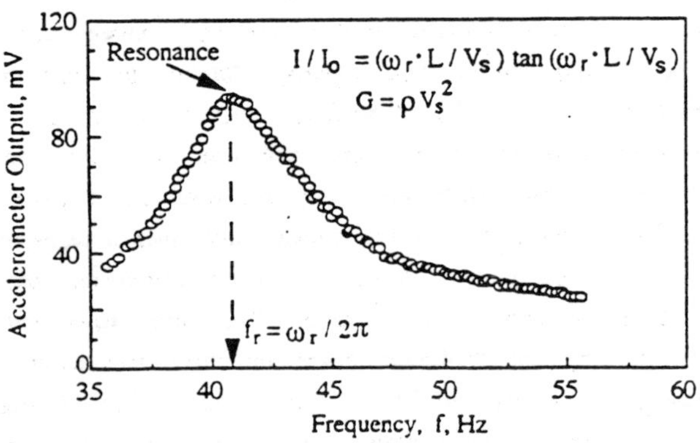

(b) Associated Frequency Response Curve

Fig 4. Schematic Configuration of RC Test and an Associated Frequency Response Curve

shear wave velocity (Vs), shear modulus (G), and shearing strain amplitude (γ). The basic data-reduction equation is expressed based on elastic wave propagation as follows:

$$I/I_0 = \omega L/V_s \tan(\omega L/V_s) \tag{5}$$

where I and I_0 are the moment of inertia of specimen and driving system, respectively, ω is the natural frequency of system and L is the length of specimen. Typical specimen size used in this study is 50mm in diameter and 100mm in height.

Once the value of shear modulus is determined, the shear modulus of the specimen can be calculated from:

$$G = \rho V_s^2 \tag{6}$$

where ρ is the mass density.

FIELD APPLICATION

Site Condition

In order to assess the applicability of using SASW method in the evaluation of ground densification, a test site was selected at Daesan in the western coast in Korea. The site consisted of a land fill layer of 1.5 to 2.0 m, a hydraulic dredging layer of 5.0 to 6.0 m, an alluvial sand layer of 7.0 to 8.0 m and a decomposed rock, as shown in Fig. 5. Both dredging and alluvial layers, classified as SM and ML, were required to be improved to build a large storage tank. A 20-ton tamper was dropped at a height of 30 m, and the depth of compaction was estimated as 14.7 m using α value of 0.6 in the design stage. The compaction was continued to the 4th tamping and a typical grid pattern of compaction is shown in Fig. 6.

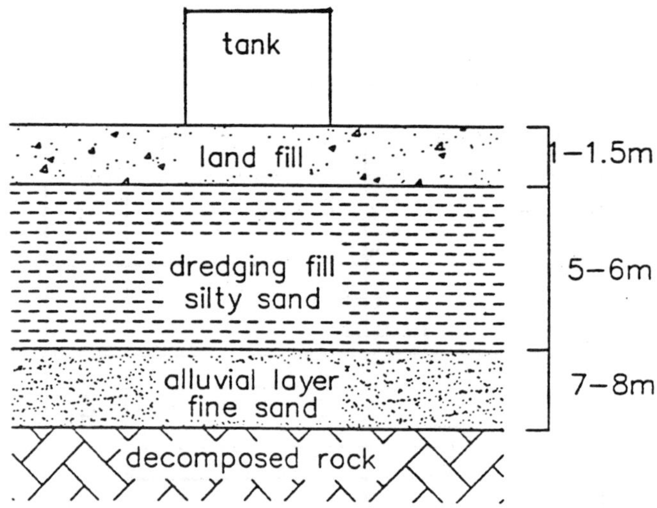

Fig 5. Soil Profile of the Tested Site

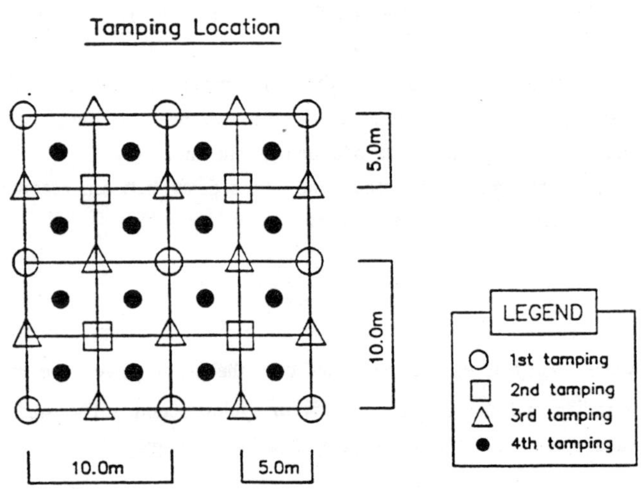

Fig 6. Typical Tamping Pattern of Compaction

SASW Test Results

SASW tests were performed before and after dynamic compactions. Vertical velocity transducers (Mark Products Model L-4) with a natural frequency of 2 Hz were used as receivers. The arrangement of receivers which is called the common receiver midpoint geometry, was used as shown in Fig.2. At a receiver spacing up to 8 m, a sludge hammer was employed as a source and for receiver spacings greater than 8 m a large weight used for dynamic compaction was employed. A signal analyzer (Hewlett Packard 35665A) was used to receive time domain signals, and calculate the phase information of cross power spectrum and the coherence function.

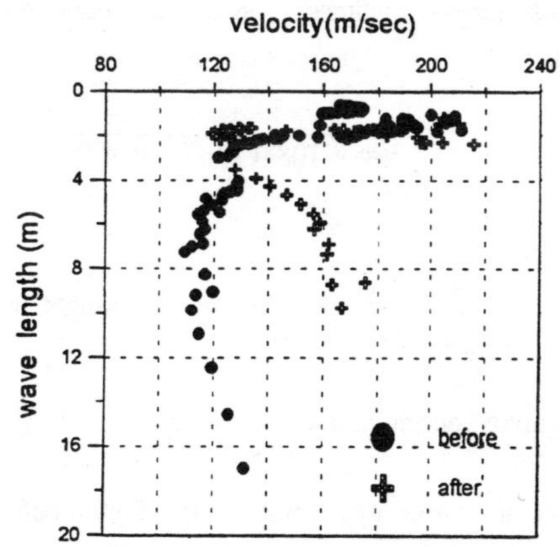

Fig 7. Experimental Dispersion Curves before and after Dynamic Compaction

The dispersion curves of the site determined by the series of cross power spectrums are shown in Fig. 7. The two dispersion curves are plotted together to allow a comparison of the curve obtained prior to dynamic compaction with the curve measured after compaction. Although it is usually difficult to draw detailed conclusions about the site profile from dispersion curves, it is possible to make a qualitative information based on a comparison of dispersion curves before and after. The phase velocity of near surface materials corresponding wave length of about 3.0m or less appears to be unchanged or a little less, because the top layer is usually disturbed by the heavy tamping. In the wavelengths below 3.0m, the phase velocity prior to compaction ranges from 110 to 130 m/sec, and the phase velocity after compaction ranges from 120 to 170 m/sec, which increased about 40 m/sec due to compaction.

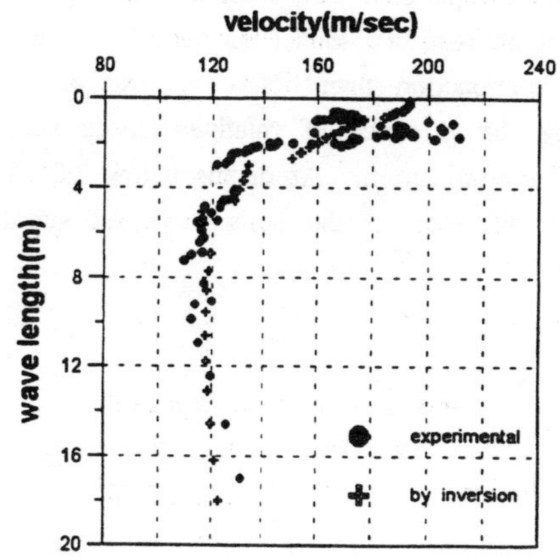

(a) before

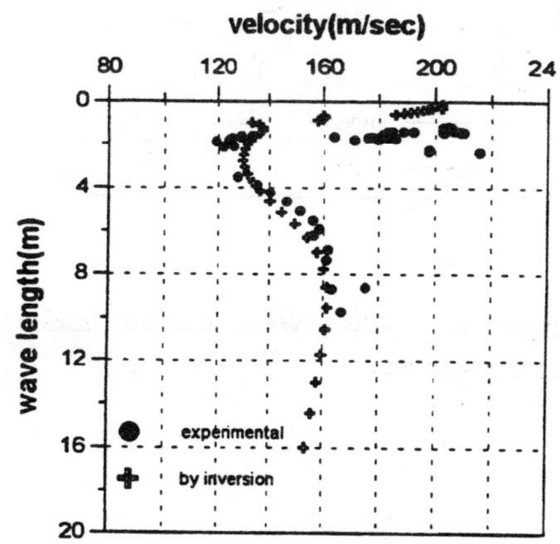

(b) after

Fig 8. Comparison of Experimental and Theoretical Dispersion Curves

The composite profile of the site was determined by the inversion of the dispersion curve using the computer program of WinSASW. The layering of the site and shear wave velocity of each layer were assumed based on experimental dispersion curves. The Poisson's ratio and unit weight of soil, which are known to little influence the results (Nazarian and Stokoe, 1985, 1986), were assumed to equal 0.25 and 1.7 t/m^3. The theoretical dispersion curve was calculated for the assumed layering and shear wave velocities, and compared with experimental dispersion curve. Unless satisfactory agreement was achieved, the assumed layering and properties were varied and repeated this forward modeling. The final comparisons between experimental and theoretical dispersion curves before and after compactions are shown in Fig. 8.

The shear wave velocity profiles were determined before and

after dynamic compactions and compared in Fig. 9. The results indicated that near-surface soils (less than 2.0-m depth) were less sti ff after compaction than before compaction. This can be explained by the presence of relatively loose back fill in the hole left after compaction. At depths below 2.0 m, shear wave velocity (or stiffness) of the soils increased significantly after compaction.

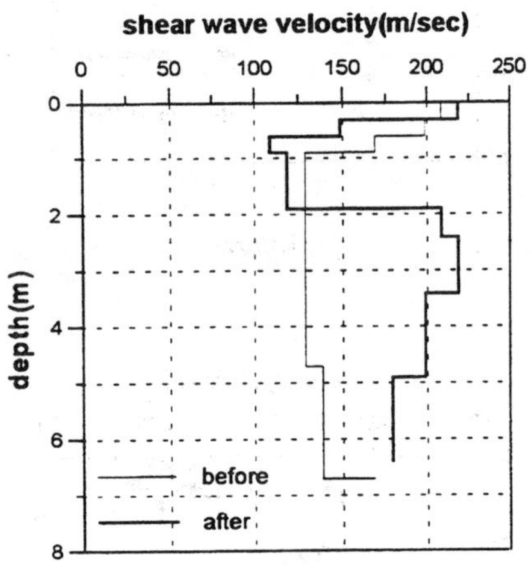

Fig 9. Comparison of Shear Wave Velocity Profiles before and after Compaction

RC Test Results

Shear wave velocity of sands is significantly influenced by the confining pressure and void ratio. The shear wave velocity-confining pressure-void ratio relationship was determined by RC test with samples brought from the field. The specimen was prepared by pluviation method and the void ratio of the specimen was varied by changing nozzle size of the system.

Table 1. Summary of RC Test Result

Confining Pressure (kPa)	Void Ratio : 0.76		Void Ratio : 0.88		Void Ratio : 1.03	
	V_s (m/sec)	G_{max} (MPa)	V_s (m/sec)	G_{max} (MPa)	V_s (m/sec)	G_{max} (MPa)
19.6	159.9	38.46	152.5	31.34	136.5	24.37
39.2	205.9	63.82	200.3	54.25	172.5	38.95
78.4	253.6	96.78	248.5	83.54	216.3	61.13

The shear wave velocities and the small-strain shear moduli (G_{max}) of reconstituted specimens of various void ratios of 0.76, 0.88, and 1.03, were determined at confining pressures of 19.6, 39.2, and 78.4 kPa and tabulated in Table 1.

Hardin (1978) proposed that a estimation of small-strain shear modulus (G_{max}) could be made from:

$$G_{max} = \frac{A}{0.3+0.7e^2}\left(\frac{\sigma_0}{P_a}\right)^n P_a \qquad (7)$$

where A = dimensionless coefficient, e = void ratio, σ_0 = mean effective stress, P_a = atmospheric pressure, and n = slope in the plot of log G_{max} versus log σ_0

Using the data shown in Table 1, the coefficients of A and n were calculated as 795 and 0.68, respectively, and the shear wave velocity - void ratio - confining pressure relationship was determined as follows:

$$e = \sqrt{1084(1/\alpha)} - 0.429 \qquad (8)$$

where $\alpha = \dfrac{V_s^2 \cdot (\frac{\gamma_t}{g})}{(\frac{\sigma_0}{P_a})^{0.68} \cdot P_a}$, γ_t = unit weight of soil

and g = gravity acceleration.

Variations in field void ratio with depth before and after dynamic compactions were estimated as shown in Fig. 10, by using eq. 8 and the shear wave velocity profils determined by SASW method. As shown in the figure, the quality and the depth of dynamic compaction can be assessed by combining the in situ SASW test and the laboratory RC test, and this clearly indicates the feasibility of applying the developed method for the quality evaluation of ground improvement on sands. Direct comparison between the field and laboratory test results is quite difficult because of the variability of site condition, different testing technique, aging effect of in situ soils, and so on. In order to improve the quality and applicability of the developed technique, further research is presently underway.

CONCLUSIONS

The quality and extent of ground densification by dynamic compaction was evaluated using SASW method. RC test was also employed to evaluate the shear wave velocity-void ratio-confinement relationship. Following conclusions can be drawn from this study:

1. The feasibility of using SASW method in the application of assessing the quality and depth of ground improvement by dynamic compaction was verified.

2. Compared with conventional evaluation techniques such as SPT and CPT, SASW method which is nonintrusive and spatial testing can be an alternative economical method to evaluate the ground improvement in a large area.

3. The variations in void ratio with depth before and after dynamic compaction were evaluated by combining the shear wave velocity profiles determined by SASW method with the shear wave velocity-void ratio-confinement relationship determined by laboratory RC column test.

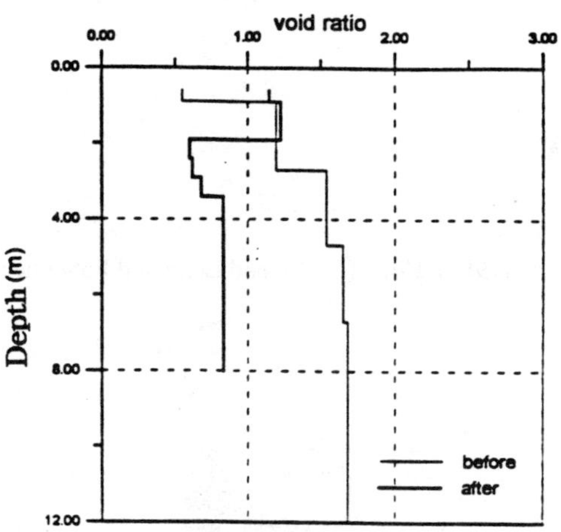

Fig 10. Comparison of In-Situ Void Ratio before and after Compaction

ACKNOWLEGEMENT

This work was supported by the Construction and Transportation Technology R&D Project (RND-96-0007). The authors express their appreciation for this support. In addition, the authors thank Professor Kenneth H. Stokoe II at the University of Texas who allowed to use of WinSASW.

REFERENCES

Bowles, E.J.(1988). "Foundation Analysis and Design," McGraw -Hill Book Company, 4th Edition, pp 141-142.

Drenvich, V. P.(1985). "Recent Developments in Resonant Column Testing," Richart Commemorative Lectures, ASCE Annual Meeting, Detroit MI, ASCE. pp79-107.

Hardin(1978). "The Nature of Stress-Strain Behavior of Soils," proceedings of the Earthquake and Soil Dynamics Conference, ASCE, Padasadena, Ca., June, 19-21, Vol I, pp 3-90.

Kim, D. S.(1995). "Deformational Charcteristics of Dry Sand Using Resonant Column / Torsional Shear Testing Equipment," J. of Korean Geotechnical Society Vol 11, No 1, pp 101-112.

Leonards, G.A., Cutter, W.A. and Holtz, R.D. (1980). "Dynamic Compaction of Granular Soils," J. of Geotechnical Engineering, ASCE, Vol 106, GT1, pp 35-44.

Mayne, P.W., Jones, J.S and Dumas, J.C. (1984). "Ground Response to Dynamic Compaction," J. of Geotechnical Engineering, Vol 110, No 6, pp 757-774.

Nazarian and Stokoe II (1986). "In Situ Determination of Elastic Moduli of Pavement Systems by Spectral-Analysis-of-Surface-Waves Method(Theoretical Aspects)," FHWA,Report No. FHWA /TX-87/46+437-2.

Nazarian and Stokoe II (1985). "In Situ Determination of Elastic Moduli of Pavement Systems by Spectral-Analysis-of-Surface-Waves Method(Practical Aspects)," FHWA, Report No. FHWA /TX-86/136+368-1F.

User's Guide of WinSASW (1992), Data Reduction and Analysis Program for SASW test, Univ. of Texas at Austin.

Proceedings of the Seventh (1997) International Offshore and Polar Engineering Conference
Honolulu, USA, May 25-30, 1997

ISBN 1-880653-28-1 (Set); ISBN 1-880653-29-X (Vol. I)

Prefabricated Vertical Drains in Geotechnical Engineering: State of the Art Review

M.A. Gabr, J.D. Quaranta and E.E. Cook
West Virginia University
Morgantown, WV, USA

D.T. Mooney
University of Arizona, USA

ABSTRACT
A theoretical and practical review of the utilization of prefabricated vertical drains (PVDs) in geotechnical engineering is presented. Recommended spacing of the PVDs and associated zone of influence are presented along with the general boundary theory normally utilized in soil consolidation analysis. Issues related to the installation process and the disturbance of the surrounding soil are discussed. Work reported in the literature on the effect of well resistance and smear is presented. Geotechnical projects dealing with the use of PVDs for coastal applications are reviewed. Case histories reported by several researchers on the stabilization of coastal dredge material, dissipation of pore pressure in the foundation, and the drainage of sludge material are discussed. Finally, recent developments dealing with the premise of using the PVDs for environmental remediation of contaminated soils is presented.

KEY WORDS: Consolidation, Flushing, Hydraulic Conductivity, Clay, Radial, Sediments, Vacuum.

INTRODUCTION

Coastal and harbor construction activities often encounter contaminated sediments and dredged soft material. The use of the Prefabricated Vertical Drains (PVDs) for a range of applications in offshore engineering is gaining wide acceptance. In addition to the current usage in soil improvement through accelerated consolidation and vacuum assisted consolidation, as reported by Jacob et al. (1994), an innovative development for using PVDs in environmental soil flushing was presented by Gabr et al. (1995) and Bowders and Gabr (1996).

The PVDs, also referred to as wick or strip drains, were originally developed as a substitute for the commonly used sand drains. With the advent of material science, these drains are manufactured from synthetic polymers such as polyproplene and high density polyethylene. PVDs are normally manufactured with a corrugated or channeled synthetic core enclosed by a geotextile filter shown schematically in Figure 1. Installation rates reported in the literature are on the order of 0.1 to 0.3 m per second excluding equipment mobilization and setup time. The PVDs have been used extensively in the past for expedient consolidation of low permeability soils under surface surcharge as, for example, described by Hansbo and Torstensson (1977), Hansbo(1983), Fellenius and Castonguay (1985), Koerner et al. (1986), Holtz (1987), and Long and Fontaine (1992).

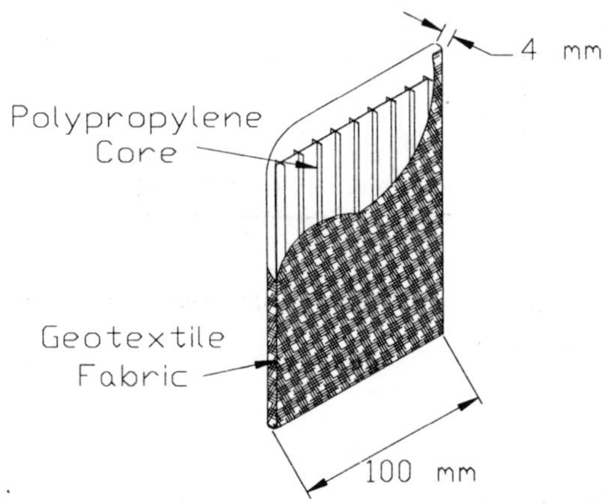

Figure 1 Prefabricated Vertical Drain (PVD)

Utilizing the PVDs in soil improvement is based on inducing a hydraulic gradient, via surface surcharge or vacuum potential, to promote the expulsion of water from the soil voids. As the PVDs are installed on a relatively close spacing, the rate of water drainage is increased due to the decreased drainage path. The state of the art review presented herein includes theoretical description of flow mechanisms and associated volume changes for vertical drains as presented by Terzaghi (1943), Barron (1948) and subsequently by Hansbo (1979,1981). Approximations of these theories as apropos to the design and analysis of PVDs installations are reviewed and presented.

PVDs IN CONSOLIDATION THEORY

Design and analysis of PVDs for accelerated consolidation is largely based on Terzaghi's (1943) radial consolidation theory. Richart (1959) presented an extensive review of these theories and provided a discussion of numerical procedures to solve the various consolidation equations. Barron (1948) and Hansbo (1979, 1981) contributed further to the work in radial consolidation for sand drains and later its application to PVDs.

As a system of PVDs is installed and surcharged, see Figure 2, the rate of consolidation will be increased due to the decreased drainage path with the longest drainage path being one-half of the PVDs spacing (S/2). Compared to the case without the PVDs where the drainage path will be equal to one-half the height of the compressible layer (H/2), for a double drained system. The governing equation of consolidation in this case can be expressed as:

$$\frac{\delta u}{\delta t} = c_r\left(\frac{\delta^2 u}{\delta r^2} + \frac{1}{r}\frac{\delta u}{\delta r}\right) + c_v\left(\frac{\delta^2 u}{\delta z^2}\right) \qquad (1)$$

where c_r and c_v = coefficients of consolidation in radial and vertical directions, respectively, and u = excess pore pressure, r = radial distance from PVD, and Z = depth.

The PVD has a zone of influence which is defined by the spacing. This zone was originally given the shape of a square for square spacing patterns and a hexagon for triangular or staggered patterns. Replacing the polygonal zone of influence with a right circular cylinder of equal area simplified the problem by making it axisymmetric. The diameter of the zone of influence (De) is defined by 1.13 times the spacing for square patterns and 1.05 for triangular patterns. This assumption has been shown to introduce diminutive errors into the calculations (Terzaghi, 1943). General boundary conditions for solving equation 1 are that the outer boundary of the zone of influence is an impermeable membrane, i.e., a no flow boundary, the lower surface is a no flow boundary, and excess water escapes only through the drain or through the top of the zone of influence.

Barron (1948) considered two extreme cases of free strain and equal vertical strain during the radial consolidation process. In the free strain case, differential settlements are assumed to occur with no arching in the load fill and the applied load is therefore uniformly distributed over the surface of the zone of influence. Equal vertical strain implies that as the consolidation process occurs horizontal planes within the consolidating soil mass remain horizontal. This is affected by the arching of the load fill acting to redistribute the load such that all vertical strains are equal. This solution was developed for the ideal case and the case with well smear and well resistance; however, the differences in the solution to the two cases is negligible as presented by Hansbo (1979). The equal vertical strain case is simpler and observations of field settlement supported the assumption of equal vertical strain (Hansbo, 1979), therefore the equal strain case is usually assumed in the design and analysis of sand drain and PVD installations. Figure 3 shows the average rate of consolidation for both cases with curves designated as no arching referring to the case of free strain.

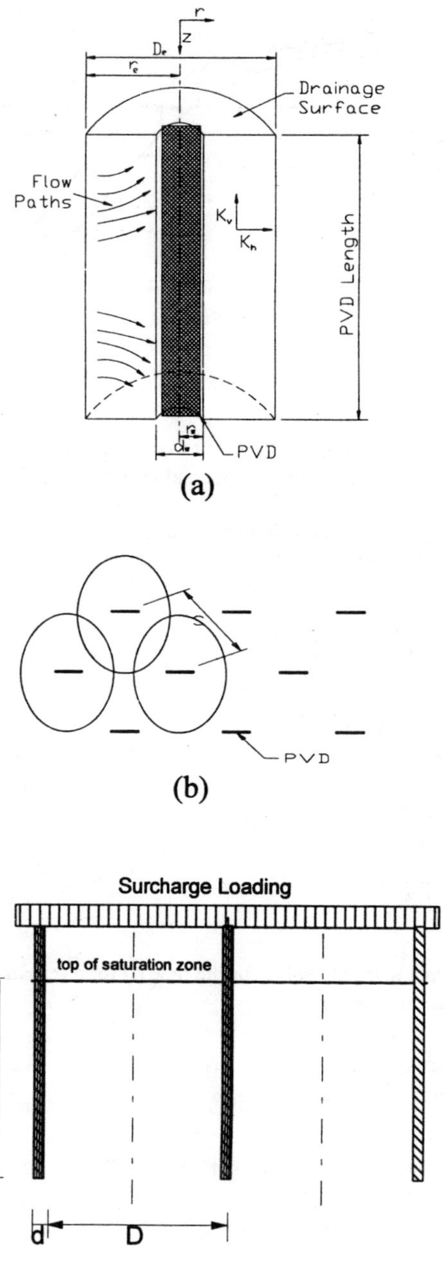

Figure 2. PVDs Layout and Field Configuration:
a) Zone of Influence, b) Installation Grid, and
c) Schematic of Profile with Surcharge Loading

THEORETICAL DEVELOPMENTS

Through some slight modifications and approximations to the above theories are apropos to the design and analysis of PVD installations. Hansbo (1979) published a succinct presentation of these adaptations which has become the preferable design methodology in use today (Koerner, 1990). In 1981 Hansbo published more rigorous derivations of the solutions found in his 1979 work. Hansbo (1979) uses only the equal strain theory of Barron (1948) citing the small differences in their results and the agreement of field observations with the equal strain theory.

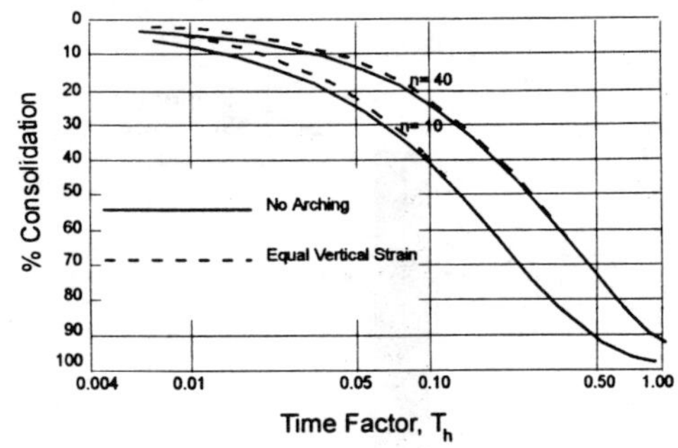

Figure 3 - Time Factor and Percent Consolidation for Arching Versus Equal Strain Analysis, after Hansbo (1979)

The average degree of consolidation (as opposed to the average excess pore pressure) due to radial flow only is given by Hansbo (1979) as:

$$U_h = 1 - e^{\frac{-8T_h}{\mu}} \qquad (2)$$

$\mu = f(D_e/d)$, defined below, and
T_h=Dimmensionless time factor for radial/horizontal drainage

Recalling that:

$$T_h = \frac{c_h t}{D^2}$$

it is a simple matter to solve for the time (t) for a certain degree of consolidation to occur as:

$$t = \frac{D^2\mu}{8c_h}\ln\left(\frac{1}{1 - U_h}\right) \qquad (3)$$

Based on Barron's (1948) work, Hansbo (1979) approximates the μ term as:

$$\mu \cong \frac{n^2}{n^2-1}\left(\ln(n) - 0.75 + \frac{1}{n^2}\right) \qquad (4)$$

Equations 2, 3 and 4 are for the ideal case. Graphical solutions are given in Figure 4 of equation 2 which enable the user to directly obtain the maximum drain spacing to achieve a given degree of consolidation.

PVD EQUIVALENT DIAMETER

A major consideration in the application of radial consolidation theory for PVDs was its shape. The PVDs typically consist of geotextiles wrapped around plastic drainage cores 75 to 100 mm wide and 3 to 10 mm thick. The drains are typically placed in a triangular or square grid pattern with 1 to 4 m spacing. Kjellman (1948) stated that the circumference of the PVDs was more critical than the diameter, i.e., the cross-sectional area, for accelerated consolidation. Results of an experimental study by Singh and Hattab (1979) tended to verify Kjellman's assumption. Using different shaped mandrels in examining the effects of installation and spacing they found that the highest rates of consolidation were obtained using a closed, cross-shaped mandrel. Although they used a mandrel with a higher circumference (star shaped) the resulting angularity of this shape caused more disturbance thereby reducing the rate of consolidation below that obtained with the cross-shaped mandrel.

Consequently equivalent diameter (d) for PVDs are based on equivalent circumferences and not necessarily on cross-sectional area. A typical

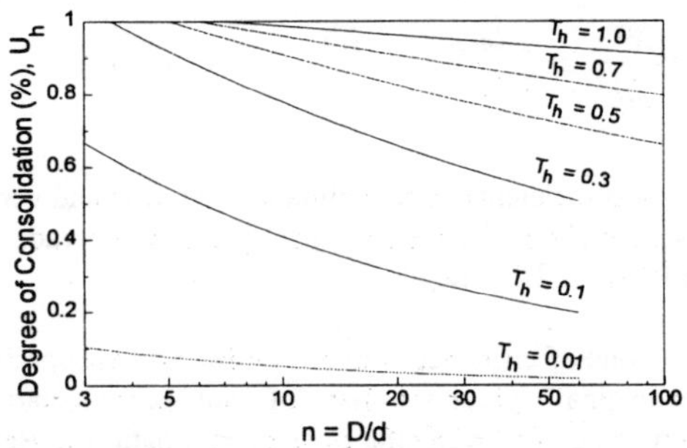

Figure 4. Degree of Consolidation as a Function of PVDs Spacing

PVD having a thickness of 4 mm and width of 100 mm has a circumference of 208 mm with an equivalent drain diameter of 66 mm (circumference/π). However, it should be noted that estimating (d) based on the circumference does not allow for the throttle which occurs as flow lines come together rapidly in an area such as at the edge of a PVD. Atkinson and Eldred (1981) reported that the Delft Laboratory of Soil Mechanics in the Netherlands proposed a further reduction of (π/4) to d in order to take such an effect into account.

SMEAR EFFECT

Regardless of the installation process, some disturbance of the surrounding soil will occur due to the mandrel installation. As stated previously this disturbance is often termed "smear". A classic example is the case of a varved clay in which the varves consist of a clay overlying a coarser layer, e.g., silt. During the mandrel installation the clay will be "smeared" down over the more permeable silt layers creating a zone of low permeability adjacent to the well (Barron, 1948).

Obviously the larger the mandrel used in the installation process the larger is the zone of disturbance with the circular, hollow, closed-ended mandrel considered to create the largest zone of disturbance. Therefore, in order to prevent excess disturbance, the size of the mandrel for PVD installation should be as close as possible to the size of the PVD itself.

The effects of smear can be accounted for by either using a smaller diameter PVD for calculations, or assuming a smeared zone of radius (r_s) and reduced permeability (k_c') to exist around the PVD. For PVD installation design and analysis, a reduced equivalent diameter of 50 mm has been found to agree with observations as reported by Hansbo (1979) and Davies and Humpheson (1981). Using a reduced equivalent diameter (and assuming no well resistance) the solutions for the ideal case can be used without further modification. When considering a smeared zone to exist around the well the μ term in Eq. 2 should be replaced with μ_s:

$$\mu_s \cong \ln\left(\frac{n}{s}\right) + \frac{k_c}{k_c'}\ln(s) - 0.75 + \pi z(2l - z)\left(\frac{k_c}{q_w}\right)\left\{1 - \frac{\frac{k_c}{k_c'} - 1}{\left(\frac{k_c}{k_c'}\right)\left(\frac{n}{s}\right)^2}\right\} \tag{5}$$

where: k_c = permeability of the undisturbed soil, k_c'= reduced permeability of the smear zone
$s = r_s/r_{eq}$,
l = length of the PVD (½ length when the PVD is double draining),
q_w = discharge capacity of the PVD.

Unfortunately the values for s and k_c' are somewhat debatable. Values of s between 1.5 and 3 have been reported in the literature. Casagrande and Poulos' (1969) work for driven sand drains results in an s value of 1.4. The range for k_c/k_c' is however less well defined (Hansbo, 1979). Barron (1948) assumed a value of k_c/k_c' to be 0.1. Later, Casagrande and Poulos (1969) concluded that the permeability in the smeared zone could be up to 3 orders of magnitude lower than in the undisturbed zone. Hansbo (1979) stated that the best agreement with Equ. 5 has been by using the reduced equivalent diameter method.

WELL RESISTANCE

Resistance to flow would occur in the PVD resulting in head losses that are particularly significant for large flow rates and/or when the diameter of the drain is small. Unfortunately filter requirements, to prevent clogging of the drains by fines, precludes the use of very coarse material in the drains. The use of very thin geotextiles as filters on PVD's help to eliminate some of this resistance. In 1948 Barron had only developed solutions for the equal strain case, for well resistance. In them he considered both the presence and absence of a peripheral smear zone at the well. In both cases he assumed the vertical permeability of the consolidating soil to be zero, however, ($\delta u/\delta z \neq 0$).

The modifications for well resistance and smear are similar to those of Barron (1948). Considering well resistance only, the term μ In Equ. 2 should be replace with μ_r:

$$\mu_r = \ln(n) - 0.75 + \pi z(2l - z)\frac{k_c}{q_w} \tag{6}$$

l = length of drain, ½ length when the well is double draining,
k_c = permeability of the, undisturbed soil, and
q_w = discharge capacity of the well.

In the case of PVDs when the ratio k_c/q_w is less than about 3000 m^2 the effect of well resistance cannot be ignored (Hansbo,1979). Determination of the discharge capacity is preferably accomplished by full scale field tests or alternatively by laboratory tests which attempt to closely model field conditions including soil type, *in situ* stresses, hydraulic gradients and PVD type as presented by Hansbo (1979) and Hansbo, Jamiolkowski, and Lok (1981).

VACUUM-ASSISTED CONSOLIDATION

Vacuum enhanced consolidation using PVDs was first proposed by Kjellman in the 1950s followed by the Corps of Engineers who subsequently investigated the feasibility of vacuum enhanced soil consolidation in the 1960s. Vacuum enhancement of the PVDs is used as a cost effective alternative to surcharging the soils for inducing pre-consolidation. The vacuum enhancement promotes a reduction in the soil pore pressure at a constant total stress. This is in lieu of increasing the soil's total stress from the application of a traditional surcharge fill. Cost savings have been observed using vacuum enhanced consolidation in France, where the applied vacuum system replaced an equivalent 4.5 meters of surcharge height (Cognon, et.al, 1994).

The significant limitation to using vacuum enhancement is the amount of the vacuum pressure that can be applied. Water undergoes cavitation at a negative atmosphere gage pressure, which corresponds to approximately 101 kPa. For 100% system efficiency, this theoretically limits the on-land use of vacuum enhanced soil water removal to a depth of approximately 10.33 meter (Cognon, et.al, 1994). Cognon, et. al, (1994) reported that the foremost practical problem encountered in the field where PVDs have been used for vacuum-assisted consolidation, was maintaining a maximum vacuum efficiency of 75 percent. Cognon listed the following considerations governing the effectiveness and economy of vacuum-assisted consolidation: 1) integrity of the surface cover membrane, 2) perimeter seal between the ground and membrane, 3) soil stratification having permeable seams within the consolidating soil, and 4) the *in situ* groundwater depth.

COASTAL APPLICATIONS

Application of PVDs to ground improvements of coastal and offshore dredged materials have been reported by Cognon, et al. (1994), Jacob et al. (1994), and Sarkar et al. (1988). In Cognon and Jacob's applications, the PVDs were used as part of the site ground modifications for improving a hydraulic fill site used for pier construction. The PVDs were installed in the hydraulic fill and were coupled with a vacuum-assisted consolidation operation to replace the conventional surface surcharge technique. The researchers reported success with the PVD application but noted several operational difficulties with maintaining vacuum efficiency because of the on-land application. In the Sarkar et al. (1988) research, the PVDs were used to accelerate consolidation of clayey soils for a wharf platform and storage area. A total of approximately 2 million linear feet of PVDs were installed for the project. The construction preload in the PVD area successfully preconsolidated the clayey subsoils for the design load condition.

PVDs FOR SOIL FLUSHING

Soil flushing is an established environmental remediation technology that is used to flush-out contaminated subsurface soils with the aid of flushing solutions. The technology is based on using injection and extraction wells that are strategically located within the contaminated zone. A flushing solution is injected into the subsurface under positive pressure through the injection well gallery. As the solution permeates through the contaminated soil it desorbes and mobilizes the contaminants towards the extraction wells which are then used to collect the contaminated soil pore water where it is removed for subsequent treatment. In the case of soils with a fine-grain fraction, i.e. clays and silts, both the extraction of the contaminant from the soil and movement of the flushing solution are less than that observed for the relatively inert, coarse grained soils. In the former case, both the extraction of the contaminant from the soil and movement of the flushing solution are restricted due to the low permeability and the desorbtion/adsorbtion characteristics of the fine-grain fraction soils.

Conventional soil flushing may not be effective in this case and the use of "pump and treat systems" on a long term basis may often be required

A PVD-vacuum system has been demonstrated by Gabr et. al, (1995) to successfully and expediently remove pore water from soils of relatively low hydraulic conductivity. On-going research has been looking into using PVDs to enhance the soil flushing and use them as an *in situ* environmental remediation technique. The application of PVDs for flushing tracer fluids has been demonstrated in sand/clay mix soils having hydraulic conductivities as low as 10^{-5} cm/s. Repeated results have shown that as soil densities increase and hydraulic conductivities decrease, the time required to flush the tracer fluid through the soil increases correspondingly (Quaranta, et al, 1996). The idea behind using the PVD system in a soil flushing scheme is that the drainage path of the groundwater is shortened which promotes subsurface liquid movement thus expediting the soil flushing process. The PVDs In this case are used for injection of a flushing solution into the *in situ* soil concurrent with an extraction process for removal of the contaminated solution. Balancing injection and extraction diminishes the potential for inducing a volumetric change in the soil thereby reducing the hydraulic conductivity and increasing the flushing time.

SUMMARY AND CONCLUSIONS

PVDs have been established as an efficient and cost effective method for improving soft soils with low hydraulic conductivities. Since 1940, researchers have continued to investigate and optimize the use of PVDs in soil improvement. Their efforts have resulted in development of design and analysis techniques that are shown to accurately predict consolidation effects. As the need continues to arise for the remediation of groundwater and contaminated low hydraulic conductive soils, PVDs are being turned to as a tool for embellishing existing remediation technologies. PVD-enhanced *in situ* soil flushing has been shown to be feasible, and on-going research continues to yield results indicating greater potential for this innovative technique.

ACKNOWLEDGMENT

This study was supported by the Department of Energy, Morgantown Energy Technology Center (METC). The authors would like to thank Dr. Rodney Geisbrecht, METC project manager, for his valuable input to this work.

REFERENCES

Atkinson, M.S. and Eldred, P.J.L. (1981) "Consolidation of Soil Using Vertical Drains," Geotechnique, Vol. 31, No. 1, pp. 33-43.

Barron, Reginald A. (1948) "Consolidation of Fine-Grained Soils by Drain Wells," Transactions, American Society of Civil Engineers, Vol. 113, pp. 718-742.

Bowders J. J. and M. A. Gabr "Strip-Drains for *In Situ* Clean Up of Contaminated Fine-Grained Soils," Geotechnical News, Volume 13 No.3, September 1995, pp. 21-25.

Casagrande, L. and Poulos, S. (1969) "On the Effectiveness of Sand Drains," Canadian Geotechnical Journal, Vol. 6, pp. 287-326.

Christiansen, J.A., C.R. Page, and J.P. Perez (1991) "Large-Scale Remediation Projects," In Proceedings: Hazardous Materials Control/Superfund Conference, pp. 127-130.

Davies, J.A. and Humpheson, C. (1981) "A Comparison Between the Performance of Two Types of Vertical Drain Beneath a Trial Embankment in Belfast," Geotechnique, Vol. 31, No. 1, pp. 19-31.

Cognon, J. M., Juran, I., and Thevanayagam S. (1994) "Vacuum Consolidation Technology- Principles and Field Experience," Special Publication Proceedings of the Conference on Vertical and Horizontal Deformations of Foundations and Embankments, Vol. 2, No. 40, pp. 1237-1248.

Koerner, R.M. (1993) Designing with Geosynthetics, 3rd Ed., Prentice-Hall, Englewood Cliffs, N.J., 652 pp.

Kjellman, Walter (1948) "Accelerating Consolidation of Fine-Grained Soils by Means of Card-Board Wicks," Proceedings, Second International Conference on Soil Mechanics and Foundation Engineering, Vol. 2, pp. 302-305.

Fellenius, B. H., and Castonguay, N. G. (1985) "The Efficiency of Band Shaped Drains," Report, National Research Council of Canada, Department of Civil Engineering, University of Ottawa, Canada, pp. 1-54.

Gabr M.A., R. Thomas, D. Szabo, and J. Quaranta (1995) "Effect of SDS Surfactant on the Hydraulic Conductivity of Sand/Kaolinite Soils," Proceedings of the Xth Pan American Conference on Soil Mechanics and Foundation Engineering, Vol II, Mexico City, October, pp. 662-673.

Hansbo, S. and Torstensson, B. A. (1977) "Geodrain and Other Vertical Drain Behavior," Proceedings of the 9th International Conference on Soil Mechanics and Foundation Engineering, Vol. 1, pp. 533-540.

Hansbo, Sven (1979) "Consolidation of Clay by Band-Shaped Prefabricated Drains," Ground Engineering, Vol. 12, No. 5, pp. 16-18 and 21-25.

Hansbo, S., Jamiolkowski, M. and Kok, L. (1981) "Consolidation by Vertical Drains," Geotechnique, pp. 45-66.

Hansbo, S. (1981) "Consolidation of Fine-Grained Soils by Prefabricated Drains," Proceedings Tenth International Conference on Soil Mechanics and Foundation Engineering, Rotterdam, Vol. 3, pp. 677-682.

Hansbo, S. (1983) "How to Evaluate the Properties of Prefabricated Drains", Proceedings of the Eighth European Conference on Soil Mechanics and Foundation Engineering, Helsinski, Vol. 2, pp. 621-626.

Holtz, R.D., (1987) "Preloading with Prefabricated Vertical Strip Drains", Proceeding 1st GRI Seminar, Very Soft Soil Stabilization Using High Strength Geosynthetics, Oct 22-23, Drexel University, Phil. PA. pp 104-129

Jacob, A.; Thevanayagam, S.; Kavazanjian, E. Jr.(1994) "Vacuum-Assisted Consolidation of a Hydraulic Landfill," Geotechnical Special Publication, Proceedings of the Conference on Vertical and Horizontal Deformations of Foundations and Embankments, ASCE, Vol. 2, No. 40 p 1249-1261.

Kjellman, Walter (1948) "Accelerating Consolidation of Fine-Grained Soils by Means of Card-Board Wicks," Proceedings, Second International Conference on Soil Mechanics and Foundation Engineering, Vol. 2, pp. 302-305.

Koerner R.M., Fowler, J., and Lawrence, C.A. (1986) "Soft Soil Stabilization Study for Wilmington Harbor South Dredge Disposal Area", Miscellaneous Paper GL-86-38, USAE Waterways Experiment Station, Vicksburg, MS. 81 pp.

Long, R. P., and Fontaine, L. L. (1992) "Performance of Wick Drains at Windsor, Connecticut," Transportation Research Record, No. 1369, Transportation Research Board, pp 1-7.

Quaranta J., M. A. Gabr, D. Szabo, and J. J. Bowders (1996) "Characteristics and Performance of Prefabricated Vertical Drains for Enhanced Soil Flushing," Paper submitted for Publication and Presentation in the 76th Transportation Research, Washington, D. C.

Richart, Jr., F.E. (1959) "Review of the Theories for Sand Drains," Transactions, American Society of Civil Engineers, Vol. 124, pp. 709-736, originally published, 1957, In the Journal of the Soil Mechanics and Foundations Division, ASCE as Proceedings Paper 1301.

Sarkar, Subal K., and Castelli, Raymond J. (1988) "Performance of Wick Drains in New Orleans Clays," National Research Council, Transportation Research Record No. 1159, pp 39-46.

Singh, G. and Hattab, T.N. (1979) " a Laboratory Study of Efficiency of Sand Drains in Relation to Methods of Installation and Spacing," Geotechnique, Vol. 29, No. 4, pp. 395-422.

Stanton, T.E. (1948) "Vertical Sand Drains as a Means of Foundation Consolidation and Accelerating Settlement of Embankments Over Marsh Land," Proceedings Second International Conference on Soil Mechanics and Foundation Engineering, Vol. V, Rotterdam, pp. 273-279.

Terzaghi, Karl (1943) Theoretical Soil Mechanics, John Wiley and Sons, NY, 510 pp.

Proceedings of the Seventh (1997) International Offshore and Polar Engineering Conference
Honolulu, USA, May 25-30, 1997

ISBN 1-880653-28-1 (Set); ISBN 1-880653-29-X (Vol. I)

Improvement of Marine Clay by Quick Lime Pile

B.S. Chun, Hanyang University, Seoul, Korea
S.S. Kim, Chung-Ang University, Seoul, Korea
H.K. Yoo, Hanyang University, Ansan, Korea
B.K. Kang and K.H. Koh, Samsung Heavy industry Co., Ltd., Seoul, Korea

ABSTRACT

This paper presents a case study involving the use of quick lime piles to improve the ground of marine clay for the construction of Shin-Wha industrial complex in Korea.

The shear tests were performed on the quick lime to analyze its strength and the result showed that the cohesion of the quick lime pile increased as the density of pile was increased.

Model tests were performed to investigate the effects of the quick lime pile on the surrounding soil. The strength of soil showed a tendency to increase with the increase of the area occupied by the quick lime pile. Test results were compared with the theoretical predictions based on the existing method applicable to the estimation of the shear-strength increase of soil by the quick lime pile. The prediction by the equation derived from Mohr-Coulomb failure criteria agreed reasonably well with the test results.

KEY WORDS : field test, improvement, marine clay, shear-strength, soil container, quick lime pile, water-content ratio.

INTRODUCTION

The quick lime pile has been widely used to improve the capacity of the foundation in embankment or excavation and to decrease an excessive settlement in Northern Europe and Japan (Broms, B.B. & Boman, P. 1979). The object soil can be improved by the expansion effect that compress the soil in the horizontal direction through the reaction between the quick lime pile and groundwater by dehydration. The quick lime pile method is mainly used to improve the very soft ground.

This is the study on the ground improvement by using the quick lime pile. We analyzed the characteristics of the quick lime produced in Korea, and studied shear-strength improvement and water-content ratio decrement to verify the improvement effect by the quick lime pile through both the laboratory test with the model box and the field test construction.

It was measured that the water-content ratio, void ratio, and shear-strength ralative to the curing time and area ratio of the quick lime pile in the laboratory test. We developed the theory of water-content ratio and shear-strength increment and then compared those with values by the field measurement.

After the field test construction of the quick lime pile, the increment of shear-strength and the decrement of water-content ratio of the ground ralative to the curing time of the quick lime pile were measured by the field vane test and the dutch cone penetration test.

CHARACTERISTICS OF QUICK LIME PRODUCED IN KOREA

The production area of the quick lime are widely distributed in Korea and it is mainly produced in Samchuk, Yongwal, Jungseon of Kangwon Province, Nonsan, Danyang, Jechun of Chungcheong Province, and Moongyung of Kyongsang Province of Korea.

In this study, the characteristics of the quick lime produced in Danyang and Nonsan were analyzed.

Specific Gravity of the Lime Produced in Korea

The specific gravities of the quick limes produced in Korea were compared with those of the limes produced in foreign countries as shown in Table 1. From these results, we can recognize that the specific gravity of the quick lime produced in Korea is in low level as compared with those of foreign countries.

Table 1 Specific gravity of the quick lime produced in Korea

Origination of the Quick Lime	Specific Gravity
Nonsan	3.04
Danyang	3.06
Foreign Countries Gada, 1983)	3.01~3.37

Chemical Composition of the Quick Lime

The chemical composition test of the quick lime presents the standards of weathering and quality of quick lime. Table 2 shows the results of the chemical composition tests of the quick limes produced in Korea.

Table 2 Chemical composition of the quick lime

Origination	SiO_2	Al_2O_3	Fe_2O_3	CaO	MgO	Na_2O	K_2O	Ig-Loss	Sum
Danyang	1.8	0.5	0.3	93.2	1.8	0.01	0.04	2.3	100
Nonsan	1.2	0.6	0.4	92.3	1.2	0.10	0.02	4.2	100

Slaking Test of the Quick Lime

The reaction state of the quick lime can be estimated by measuring the temperature which comes from the reaction between the quick lime and water. Table 3 shows the slaking-test result of the quick lime.

Table 3 Slaking-test result

Reaction Time (min)	1	2	4	5	6	8	10	15	20	25	30	35	45	55	65
Danyang	37.2	40.6	47.6	50.6	52.9	56.0	57.8	61.1	63.1	64.6	65.5	66.2	66.8	66.6	–
Nonsan	32.5	36.4	43.6	45.1	47.0	49.2	51.5	57.4	59.8	61.3	61.8	62.0	62.1	61.8	–

X-ray Diffraction Analysis and SEM Photographing

The results of X-ray diffraction analyses of the quick limes produced in Danyang and Nonsan are shown in Figs. 1~2.

We can hardly see small peaks in the quick lime produced in Danyang, so we can estimate that it is a pure quick lime with no impurities. But, as the quick lime produced in Nonsan shows some small peaks, its values of intensity at 37.4, 54.0, 32.3 of 2-theta are lower than those of Danyang.

By examining the quality and particle size of the quick lime from SEM photographing, we can affirm the reaction state in the ground. Fig. 3 shows the SEM photographing results of the quick limes produced in Nonsan and Danyang.

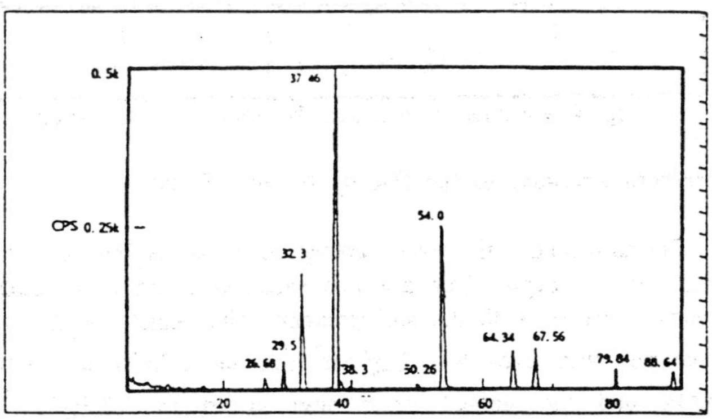

Fig. 1 X-ray diffraction analysis of the quick lime produced in Danyang

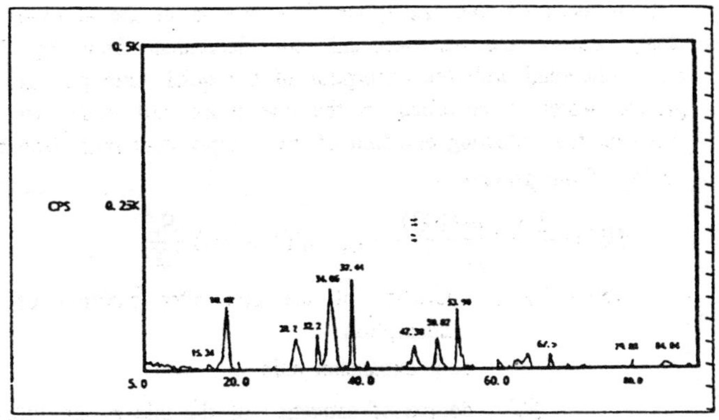

Fig. 2 X-ray diffraction analysis of the quick lime produced in Nonsan

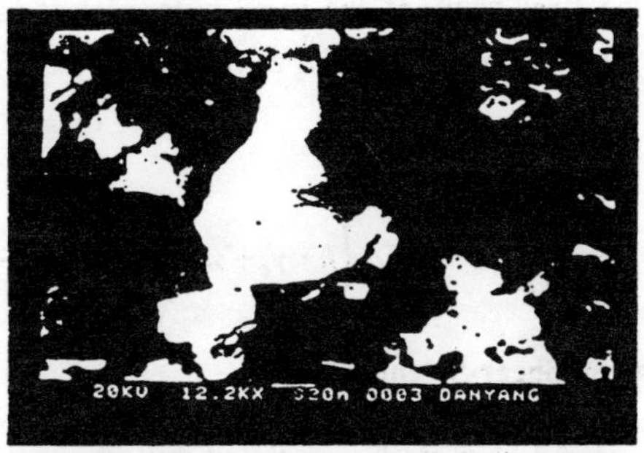

(a) The quick lime produced in Danyang

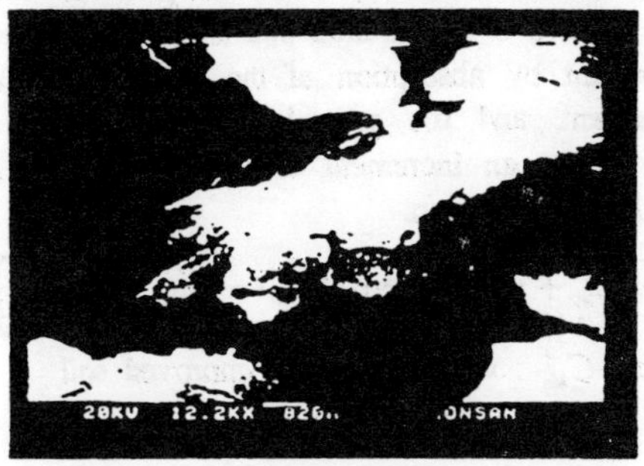

(b) The quick lime produced in Nonsan

Fig. 3 SEM photographing result of the quick lime

GROUND IMPROVEMENT THEORY OF THE QUICK LIME PILE

Decrement of Water-Content Ratio

As the quick lime pile is made in the weak ground, Ishida(1977) presented the following equation of the mean water-content ratio decrement (ΔW_h) relative to absorption.

$$\Delta W_h = \frac{1+(w_0/100)}{\gamma_t} \cdot a_s \cdot h \cdot \gamma_c \qquad (1)$$

where, w_0 : initial water-content ratio of the ground

a_s : area ratio of the quick lime pile

h : absorption coefficient of the quick lime pile

γ_t : unit weight of the ground

γ_c : unit weight of the quick lime pile

It is assumed that the quick lime pile is in the absolute dry state, because the water-content ratio decrement from Eq. 1 is only concerned with the absorption of the quick lime pile. As the ground water is absorbed to the pile more and more, we can represent the following equation of the water-content ratio decrement ΔW_s of the ground.

$$\Delta W_s = \frac{1+(w_0/100)}{\gamma_t}\cdot a_s \cdot n'(1+\varepsilon_v)\frac{S_r'}{100}\cdot \gamma_w \quad (2)$$

where, n' : porosity of the pile after reaction of the absorption

ε_v : bulk expansion ratio

S_r' : degree of saturation of the pile in equillibrium after absorption

γ_w : unit weight of the water

Now, the final value of the mean water-content ratio decrement $\Delta W(\%)$ of the ground can be calculated by using Eq. 3.

$$\begin{aligned}\Delta W &= \Delta W_h + \Delta W_s \\ &= \frac{1+(w_0/100)}{\gamma_t}\cdot a_s\left\{h\cdot\gamma_c + n'(1+\varepsilon_v)\frac{S_r'}{100}\cdot\gamma_w\right\}\end{aligned} \quad (3)$$

Increment of Shear-Strength

It is assumed that the ground is in the saturation or normal consolidation state with clay, and we explain the state where the absorption relative to quick lime pile is finished. The water-content ratio decrement by absorption of the pile is associated with the void decrement, and the ground consolidation due to the void decrement causes an increment of the cohesion. It is represented by the following equation:

$$C_u' = C_{u0} + m\cdot p_c\left[\exp\left(2.3\cdot\frac{G_s\times\Delta W}{100C_c}\right)-1\right] \quad (4)$$

where, C_u' : cohesion of the improved soil

C_{u0} : cohesion of the origin

m : strength-increment coefficient

p_c : consolidation yield stress of the origin

G_s : specific gravity of soil

ΔW : water-content ratio decrement of the ground

C_c : compression index of soil

Generally, the strength-increment coefficient m $(= C_u/p)$ by the consolidation has the value of 0.28~0.33 at any test, when the direct shear test is carried out with non-disturbed samples in the normal consolidation state under the undrained condition (Mikasa Masato, Nisigaki Yosihiko, Okazima Yowuichi, 1978).

In this study, we compared the caculated results from the failure theory by Mohr-Coulomb and the presented equations by Skempton(1957), Mayne(1985), Wroth(1984), Kalsson(1967), Bjerrum(1960) and Byung-Hee Kang(1977) with the measured results from the tests.

STRENGTH PROPERTIES OF THE QUICK LIME PILE

The quick lime tested in this study originated from Danyang in Korea, and it has the high quality in the contents of CaO and slaking temperature as shown in Tables 2 and 3.

Strength Property of the Quick Lime Pile by Density of the Pile

The direct shearing test on the quick lime pile was performed where density varies as 1.1 g/cm^3, 1.2 g/cm^3, 1.3 g/cm^3. The vertical load on the shear box was 0.3 kg/cm^2 and the salinity of the water reacted with the quick lime was 0.4%. The result is shown in Fig. 3.

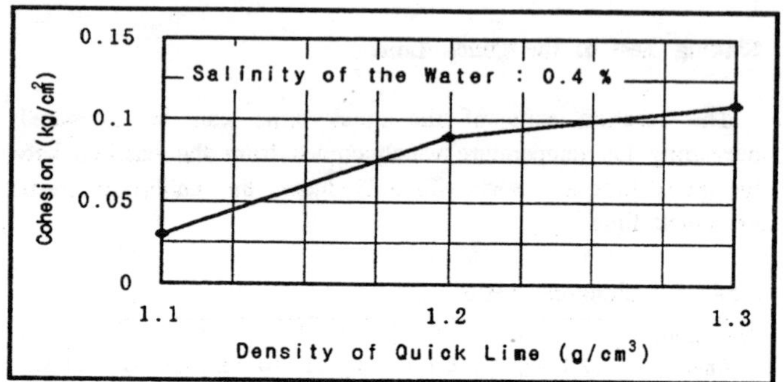

Fig. 3 Variation of chesion by the density of the pile

Strength Property by Salinity of the Water Reacted with the Pile

To investigate the influence of the salinity of the water on the cohesion of the pile, the density of the quick lime was fixed to 1.2 g/cm^3 and the vertical load to 0.3 kg/cm^2, but the salinity of the water was varied as 0%, 0.4%, 1.5%, 3%. Fig. 4 shows the test results.

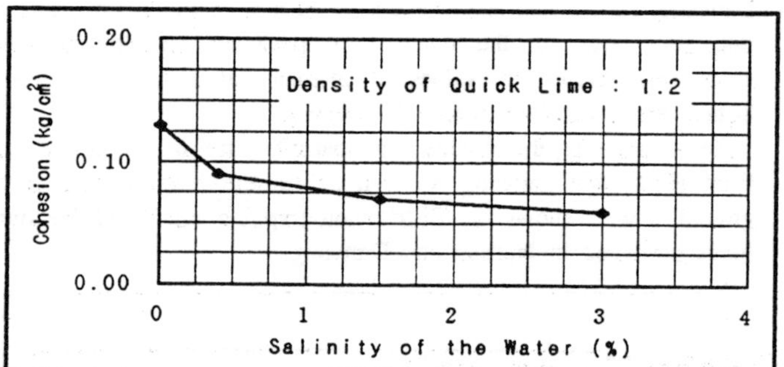

Fig. 4 Variation of cohesion by salinity of the water

Strength Property of the Pile by Normal Stress

The horizontal soil pressure acting on the quick lime pile varies with the soil depth. This test was executed to know the strength character varied with the soil pressure. The density of the quick lime pile was fixed to 1.2 g/cm^3 and the salinity of water to 0.4%, and the normal stress was varied as 0.3 kg/cm^2, 1 kg/cm^2, 2 kg/cm^2. Fig. 5 shows this test result.

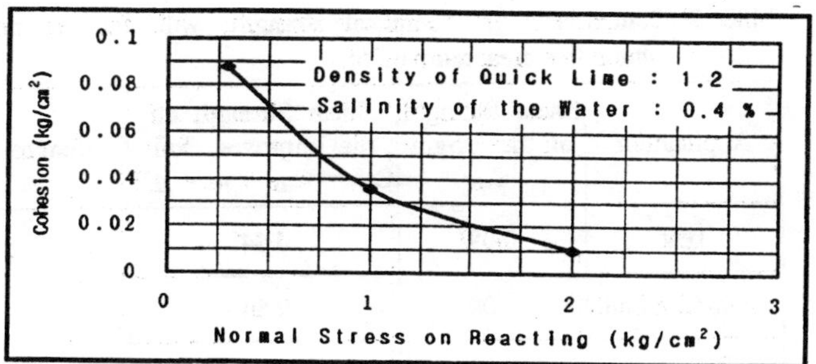

Fig. 5 Variation of cohesion of the pile by normal stress loaded when the lime reacts with water

LABORATORY TEST

Type of Soil

Physical-chemical properties of the ground soil are shown in Table 4. The soil sample was remolded in laboratory.

Table 4 Physical-chemical properties of the soil sample

Index	Test Results
W_n (%)	68.7
W_L (%)	32.2
W_P (%)	21.1
PI	11.1
Specific Gravity	2.68
O.M.C. (%)	15.2
γ_{dmax}	1.75
Unified Classification	CL
Content of NaCL (%)	0.4
pH	7.99
C_C	0.335
C_S	0.045

Test Equipment

Two types of soil containers(type of A, B) were manufactured to measure the shear-strength and water-content ratio of the ground surrounding the quick lime pile. (refer to Fig. 6, Fig. 7)

A-type container was manufactured to measure the variation of engineering properties with the spaces of the piles. And B-type container was manufactured to measure the variation of engineering properties with the area of the quick lime in the ground.

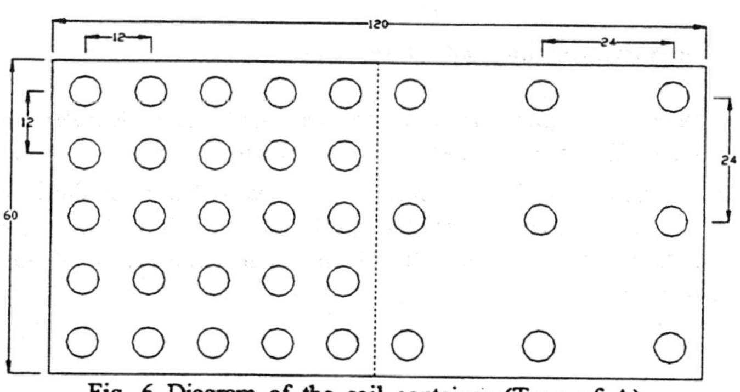

Fig. 6 Diagram of the soil container (Type of A)

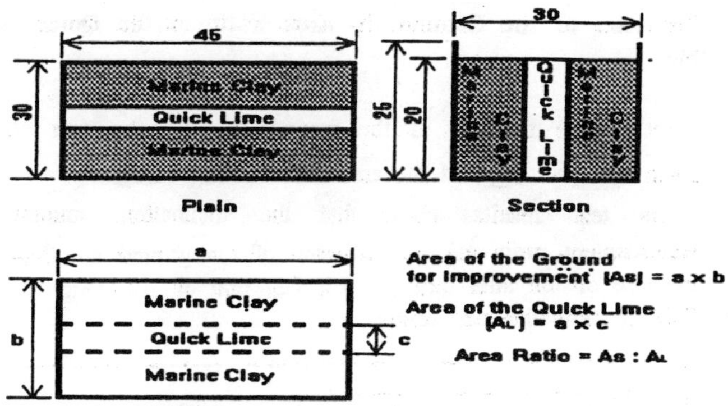

Fig. 7 Diagram of the soil container (Type of B)

ANALYSIS OF THE LABORATORY TEST RESULTS

Properties of the Ground by Varying the Spaces of the Quick Lime Pile

To investigate the variations of water-content and void ratio, the dense case was defined as the case of 12cm of the pile space, the loose case as the case of 24cm of the pile space in the A-type soil container. The test results are shown in Figs. 8~9.

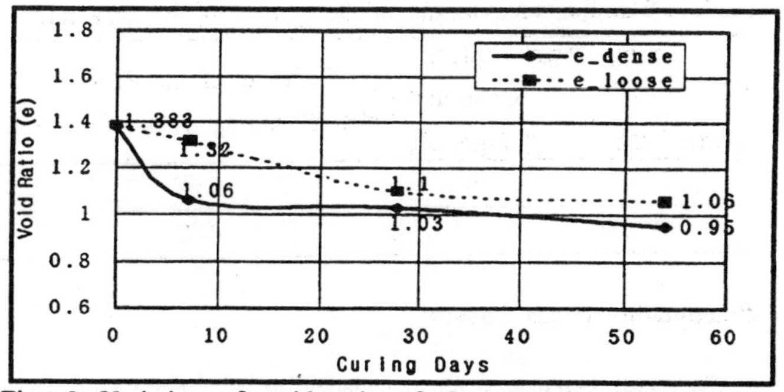

Fig. 8 Variation of void ratio of the ground with the curing days of the pile

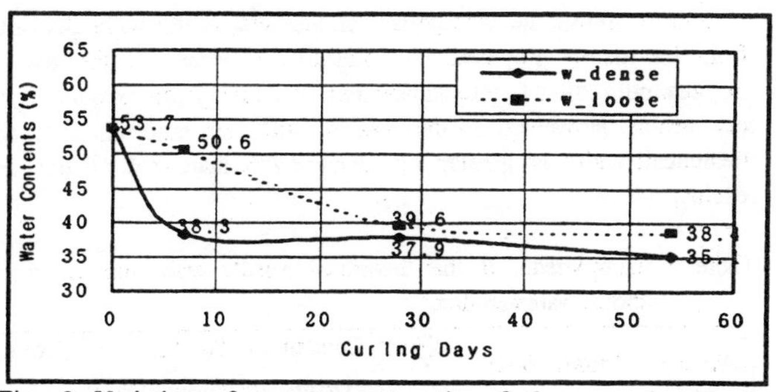

Fig. 9 Variation of water-content ratio of the ground with the curing days of the pile

From the results, we realized that the declination effects of water-content and void ratio are larger in case of the dense spacing than that of the loose spacing. Especially, in case of the dense space of piles, the inclination ratio is larger at the initial stage of curing than that in case of the loose space of piles, where the inclination ratio is linear.

Properties of the Ground by Area Ratio of the Quick Lime Pile

Let define the area of the ground as A_S, the area of the quick lime as A_L, and the area ratio as $A_S : A_L$.

The test results show that the inclination amount of water-content ratio and the cohesion of the ground are larger in the case of the area ratio of 5:1 than that of 10:1. Fig. 10 and Table 5 show the test results.

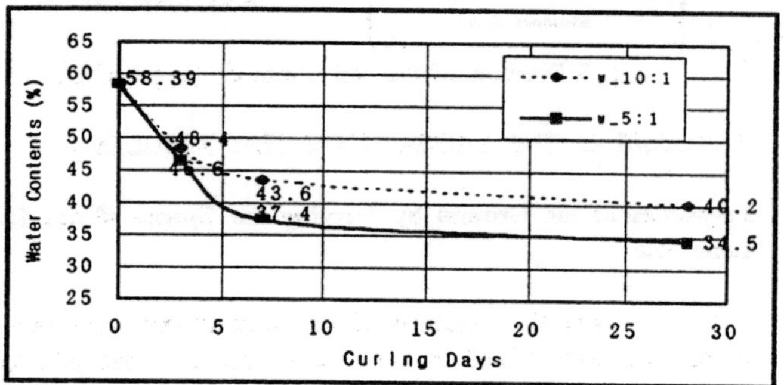

Fig. 10 Variation of water-content by the area ratio

Table 5 Variation of cohesion of the ground by the area ratio

Curing Days / Area Ratio	Initial Cohesion (kg/cm^2)	Cohesion of the Ground after 28 Curing Days of the Pile
5 : 1	0.1	0.64
10 : 1	0.1	0.20

Examination of the Theory

The used parameters for calculation of the inclination amount of water-content ratio was obtained from the test results or the reference(JSL, 1970).

Table 6 shows the comparison of the test results with the value from the theory proposed by Ishida(1977). Here, in the case of the area ratio of 5:1, the values from the theory are similar to the test results. However, in the case of the area ratio of 10:1, the theoretical value is greater by about 8.2% than that of the test results.

Table 6 Comparison of the theoritical results with the test result about water-content

Area Ratio	Initial Water Content Ratio (%)	Water Content of the Improved Ground (After 28 Curing Days)	
		Theoritical result	Test Result
5 : 1	58.4	37.4	34.5
10 : 1	58.4	48.4	40.2

After the test conducted with the soil container type of B, the theoretical value of $C_u/\Delta F$ was compared with the test results. ΔF for the prediction of C_u was derived from the consolidation curve of the soil.

Table 7 Comparison of theoretical strength with the test result about the area ratio of 5:1

Application	Shear Strength of the Origin C_{u0}	Shear Strength of the Improved Soil $C_{u1} = C_{u0} + m \cdot \Delta P$	Remark
Test	0.09	0.64	ΔP = 1.86
Mohr-Coulomb	0.09	0.56	
Bjerrum	0.09	0.56	
Karlsson	0.09	0.50	
Skempton	0.09	0.46	
Mayne	0.09	0.83	
Wroth	0.09	0.78	
Byung-Hee Kang	0.09	0.74	

From the results, it was proved that the values of $C_u/\Delta F$ calculated from the theories proposed by Mayne(1985), Wroth(1984) and Byung-Hee Kang(1977) were excessive, and these values calculated from the theories proposed by Mohr-Coulomb and Bjerrum(1960) were similar with the test result.

FIELD TEST ON THE QUICK LIME PILE

Condition of the Ground

The condition of the ground in which the quick lime pile was constructed is a shore reclamation zone for constructing Si-Hwa industrial zone in Korea. There is Si-Hwa adjacent to a retarding basin, refering to the geological survey report(The Bureau of Construction in Korea, 1986). There are silty sand layer from the surface to 2m in depth, silty clay from 2m to 6.6m, weathered soil from 9.5m to 18.7m, and weathered rock below 18.7m. The water-content is 22.18% through 42.89%. Non-Plastic zone is identified from 1.0m to 1.8m in depth. From 3.0m to 3.8m, the liquid limit is 34.1%, and the plastic limit is 14.13%.

Soils of 2m depth are sampled to execute the laboratory test. The result shows that the Atterberg limit is non-plastic and water-content is 42%.

Construction of the Quick Lime Pile

Nine lime piles were constructed, and their diameters are 400mm. The space between the piles is 1m, depth is 3m, density is 1.1 t/m^3 that is the density of a free falled quick lime. No casing is required to cast the quick lime in place. Earth auger was used for the boring. Fig. 11 shows the arrangement of quick lime piles.

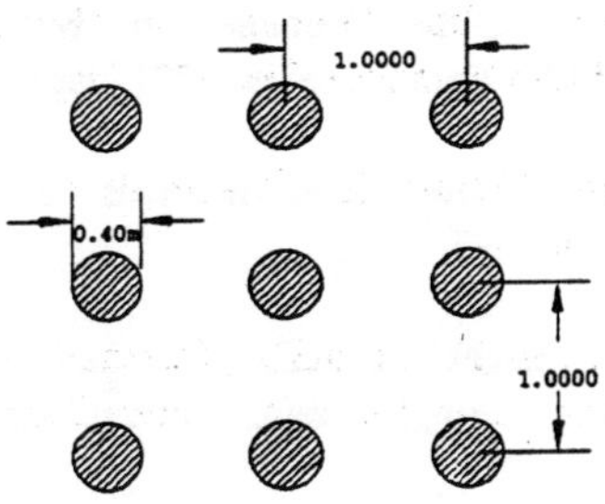

Fig. 11 Arrangement of the quick lime piles

ANALYSIS OF THE FIELD TEST RESULTS

Dutch cone penetration test and the vane test were executed every seven days to measure the ground improvement after the quick lime pile was constructed.

Dutch cone penetration test was executed at the depth of 2.5m and the vane test at the depth of 2m of the ground.

Effect of Decrease of Water-Content of the Ground

Comparing the theoretical values of the water-content decrease with the actual values measured in situ, we analyzed the actual water-content decrease effect in the ground around the quick lime pile and the amount of water-content decrease in this chapter.

To calculate the amount of water-contents decrease, major parameters acquired through the field and laboratory tests were used. The rest of parameters unable to be obtained from the field and laboratory tests was determined with the pre-existing reference(JSL, 1970).

Table 8 shows the amount of the theoretical water-content decrease and the measured water-content decrease.

Table 8 Comparison of the theoretical value with the test result about the decrease of the water-content ratio of the ground

Initial Water Content W_o	7 Days Curing	14 Days Curing	21 Days Curing	28 Days Curing	35 Days Curing	$\triangle W$ (Test)	$\triangle W$ (Theoretical)
42 %	36 %	33 %	31 %	31 %	30 %	12 %	12 %

Fig. 12 shows the relationship between the curing time and water-content decrease measured in situ.

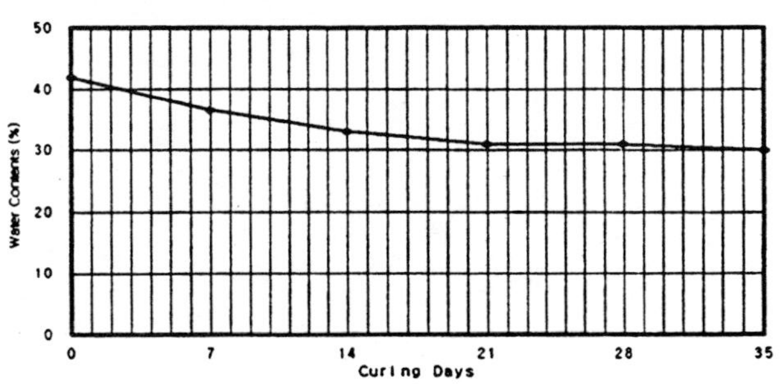

Fig. 12 Variation of water-content with the curing days of the pile

Increase of Shear-Strength of the Ground

Among the many parameters, the value of $C_u/\triangle F$ was adopted as 0.3 and the other properties for theoritical calculation was adopted from the test in laboratory and field.

Fig. 13 shows that the variation of cohesion measured by executing the vane shearing test. Fig. 14 shows the variation of cone index along the depth of the ground.

From the test results by the vane shearing test, the cohesion of the ground was found to be 1.78t/m^2 in the initial state and 5.29t/m^2 after 35 curing days. The theoretical value of shear-strength was 5.34 t/m^2.

The field test shows that the measured amount of decrease of the ground water-content is similar as the theoritical one but the measured shear-strength of the ground is different by about 0.05t/m^2 from the theoritical one.

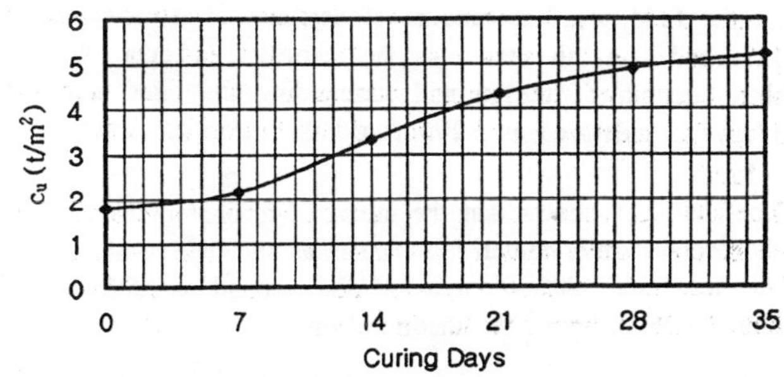

Fig. 13 Variation of cohesion of the ground with the curing days of the pile

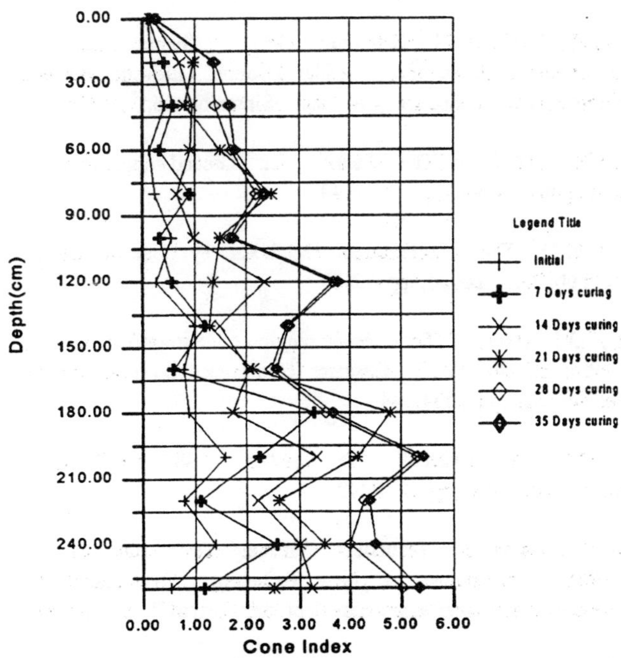

Fig. 14 Variation of cone index of the ground with the curing days of the pile

CONCLUSION

(1) The quality of Korean quick lime is excellent for a construction material because of its higher CaO content and specific gravity.
(2) The strength of the quick lime pile is characterized by the cohesion of the quick lime, which is $0.03kg/cm^2$ when the density of the pile is $1.1g/cm^3$ and $1.11kg/cm^2$ when the density is $1.3g/cm^3$.
(3) The shear-strength of the ground surrounding the quick lime pile is improved two times as large as the initial strength when the area ratio is 10:1, and six times as large as the initial value when the area ratio is 5:1.
(4) Among the theories of $C_u/\Delta F$, the ones proposed by Mohr-Coulomb and Bjerrum were similar to the test results.
(5) The amount of decrease of water-content of the ground is measured as 12% after inserting the quick lime pile cured for 35days, and the initial shear-strength of the ground increases to $5.29t/m^2$.
(6) In the field test, the amount of decrease of water-content is 12% and it is the same as the theoritical amount. And the shear-strength of the ground surrounding the piles cured for 35days is measured as $5.29t/m^2$ and the theoritical value is $5.34t/m^2$.
The field test shows that the measured amount of decrease of the ground water-contents is similar as the theoritical one but the measured shear-strength of the ground is different by about $0.05t/m^2$ from the theoritical one.

REFERENCES

Broms, B.B. & Boman, P. (1979). "Stabilization of Soil with Lime Columns", Ground Engineering, Vol. 12, No. 4, pp 23-32.

Chew, H.H., Takeda, T., Kichikawa (1993). "Chemico Lime Pile Soil Improvement Used for Soft Clay Ground", Eleventh Southeast Asian Geotechnical Conference, 4-8 May, Singapore, pp 319-324.

Chun, B.S. (1987). "The Method for Ground Improvement", Kunseulyungusa, Seoul, pp 145-148.

Gada (1983). "The Stabilization Methods by Use of Lime", Gazima Pub. Co., Tokyo, pp 12-38.

Henkel, D.J. (1960). "The shear-strength of Saturated Remolded Clay", Proc. of the ASCE Research Conference on shear-strength of Cohesive Soil, pp 533-544.

Ingles, O.G. & Metcalf, J.B. (1972). "Soil Stabilization", Butterworths, Australia, pp 127-143.

Institute of Construction Technology, Samsung Heavy Industries Co., Ltd. (1996). "Proceedings on New Technology Development for Seacoast-Reclamation and Improvement of Soft Ground", pp 6.1-6.81.

Ishida (1974). "The Improvement of Clay Loam by Use of Quick Lime", Soils and Foundations, Vol. 22, No. 5, pp 37-42.

Ishida (1977). "About the Decrement of Moisture Contents by Lime Pile". JSGE 12th Conference. No. 321, Japan, pp 1265-1268.

JSL (1970). "The Stabilization Methods by Use of Lime", Sangaidou, Japan, pp 6-34.

Mikasa Masato, Nisigaki Yosihiko, Okazima Yowuichi (1978). "The 13th Soil Engineering Research Lecture Magazine", Japan, pp 325-328.

Ming-jui Hung, Huei-wen Chang, Hin-chi Lei (1995). "Improvement Effects of Quicklime Pile on Soft Cohesive Soils", Proc. of the 5th Int. Offshore and Polar Eng. Conference, pp 555-559.

The Bureau of Construction in Korea (1986). "The Geological Survey Report of Si-Hwa Industrial Zone", pp 196-197.

Proceedings of the Seventh (1997) International Offshore and Polar Engineering Conference
Honolulu, USA, May 25-30, 1997

ISBN 1-880653-28-1 (Set); ISBN 1-880653-29-X (Vol. I)

Case Histories on Monitoring Vibro-Driven Piles in Reclaimed Grounds

H. Tsuboi
Fudo Construction, Co., Ltd., Osaka, Japan
T. Matsui
Osaka University, Osaka, Japan
Y. Ando
Fudo Construction Co., Ltd., Osaka, Japan

ABSTRACT

Vibration behavior of piles in reclaimed grounds has not fully understood so far. This paper concerns the monitoring results of vibro-driven piles at seven reclaimed lands. Monitoring items were the driving speed, vibratory acceleration, amplitude, electricity consumption and reaction force of the pile tip. The numerous time-series-monitoring data were then analyzed in connection with the soil profiles of reclaimed grounds. As a result, it was confirmed that there are significant correlations between such items as N-values, reaction forces and vibratory acceleration.

KEY WORDS: monitoring system, reaction force, reclaimed ground, sand compaction piles, vibro-driven piles

INTRODUCTION

Along the coastal lines of Japan, there are many reclaimed lands where a numerous number of sand compaction piles (SCP) have been installed using vibro-driven steel pipe casings (closed-end) for soil improvement. However, the vibro-driven casings are presently installed into the ground without adequate knowledge of their vibration behavior. In this study, closed-end casings used for soil improvement of SCP are regarded as virtual vibro-driven piles.

In order to establish a theoretical background for studying the vibration behavior of piles driven into the ground, this paper firstly describes the development of a monitoring system of vibro-driven piles. Monitoring items are the driving speed, vibratory acceleration, amplitude, electricity consumption and reaction force of the pile tip. Then, this monitoring system is applied to full-scale vibro-driven piles at seven reclaimed lands. The numerous time-series-monitoring data are analyzed in connection with the soil profiles of the reclaimed grounds.

DEVELOPMENT OF MONITORING SYSTEM FOR BEHAVIOR OF VIBRO-DRIVEN PILES

Vibro-driving force is used for placing piles, sheet piles, casings or cellular cofferdams into the ground. Using a thin hollow reaction force meter attached at around the tip of the casing of the same outer diameter and a vibratory acceleration meter, a monitoring system is developed for measuring such items as the trajectory of casing, the wave form of the reaction force at pile tip, vibratory acceleration, amplitude and electricity consumption (Tsuboi, H et al. 1996, Ogawa, M et al. 1997). In this chapter, the details of the monitoring system and the method of data processing are described below.

Monitoring Process and Processing Data

Fig. 1 shows the vibro-driven steel pipe casing and the thin hollow reaction force meter whose outer diameter is the same as that of pile. The vibration behavior of the pile is monitored by consecutively measuring such items as the reaction force, vibratory acceleration and driving time. Fig. 2 illustrates the outline of the monitoring system. Analog signals monitored by a ground level meter, level meter (sand level meter used for sand poured into the casing), inner barometer, reaction force meter, acceleration meter, and vibration pickup (amplitude meter) are converted into digital signals by the A-D converter as shown in Fig. 2. They are then transformed into physical quantities by calibration, and recorded on a hard disk in every 5 msec together with displaying them on the display screen. Because the mean values of the data obtained during the driving of every 10 cm are used in the ordinary analysis, the primary processing is carried out by a procedure described below. Then, through the secondary processing analytical program, the time-series records of a certain depth interval are obtained so that the correlation between monitoring items can be analyzed.

Primary Processing. All the data (factors) recorded by the monitoring system are processed by a certain depth interval (GL), and re-recorded on an external storage such as floppy disks. Fig. 3 shows the basic factors and processing pattern, together with the definition of the peak value in the primary processing. As seen in the figure, such basic factors as SL, *Fd*, etc. are processed in the processing pattern by their preset depth interval (10 cm). Then, the processing factors such as duration (dt), cumulative time (Σt) and average speed (v) in each preset depth interval are obtained and recorded. From these values obtained during the driving of every 10 cm, the peak and mean values are calculated and recorded as the primary processing data.

Secondary Processing. From the primary processing data, depth-distribution diagram of each vibration parameter and correlation diagram of each factor are produced. The driving speed is represented by gradient of depth-time curve obtained by level meter. Here, the time-driven depth ratio Tzi / Zi (sec/ cm) is defined as the ratio of the driven length Zi (cm) and the required time Tzi (sec). Therefore, the driving speed is the reciprocal number of the time-driven depth ratio. As the present depth interval is 10 cm in this study, the required time for Zi is represented by T_{10} (sec). The monitoring items and calculation methods are described below.

i) Driving speed (v):
Measured by correlating the depth with revolutions of a wire pulley synchronized with the vertical motion of vibro-driven pile. When obtaining Zi (cm) from the depth meter, the driving speed is represented as follows:

$$v = Zi / Tzi \text{ (cm/ sec)} \quad \text{Eq. 1}$$

Here, $Zi = 10$ cm.

Therefore, $$v = 10/ T_{10} \text{ (cm/ sec)} \quad \text{Eq. 2}$$

ii) Vibratory acceleration (α):
Digitally measured by the piezoelectric pickup.

iii) Amplitude (a):
Digitally measured by the piezoelectric pickup.

iv) Electricity consumption (E):
Using the measured values of electric current I (A) and voltage V (V), it is obtained by the following equation.

$$E = \sqrt{3} \cdot V \cdot I \quad \cos\phi \text{ (kW)} \quad \text{Eq. 3}$$

ϕ : Phase-contrast ($\cos\phi = 0.8$)

v) Reaction force at pile tip (Fd):
Measured by a hollow reaction force meter (capacity: 490 kN, maximum: 1470 kN) attached to the tip of vibro-driven pile.

The other data include those recorded by the sand level meter which follows the injection and discharge of the poured sand, and by the ground level meter which follows the tip trajectory of vibro-driven pile. Note that the vibratory acceleration takes the negative maximum value, while the reaction force takes the positive maximum peak value, because the measured values change with time according to the progress in the driving operation as in the wave form shown in Fig. 3.

SEVEN CASE HISTORIES

Fig. 4 shows the seven recently reclaimed lands chosen for monitoring. At all the sites, boring tests and the standard penetration test (SPT) were carried out for soil investigation. Table 1 gives the number of the vibro-driven piles that were monitored, the outline of soil properties, driving length and measuring items, which were the reaction force of the pile tip, vibratory acceleration, electricity consumption, driving time and amplitude. Major results of each site are described below. Note that monitoring data were taken in every 1 m of depth, in order to compare them with the N-value (SPT).

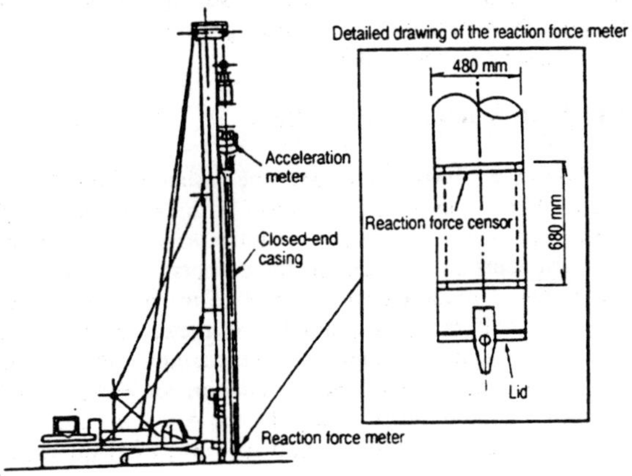

Fig. 1 Outline of pile driving and hollow reaction force meter at the tip of the vibro-driven pile (casing)

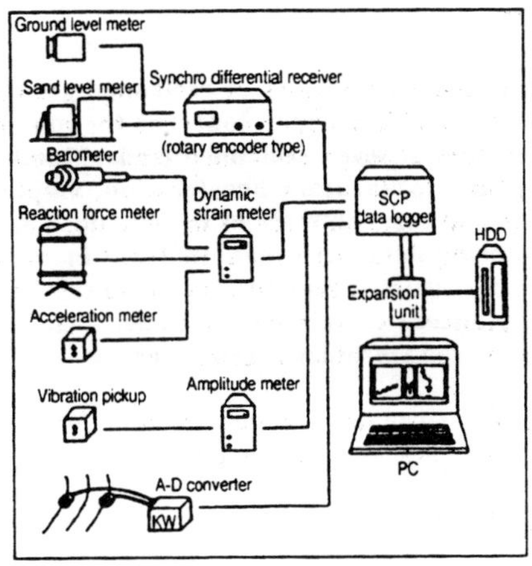

Fig. 2 Monitoring system and data processing system

Processing pattern	Basic factors: SL	Fd	α	KW	A	Processing factors: dt	Σt	v
Maximum positive peak value	O	O	O	O	O			
Maximum negative peak value	O	O	O	O	O			
Maximum positive negative values	O	O	O	O	O			
Mean positive peak value	O	O	O	O	O			
Mean negative peak value	O	O	O	O	O			
Integral of the interval	—	O	O	O	O			

GL : Depth
SL : Sand level
Fd : Reaction force
α : Acceleration
KW: Power
A : Amplitude
dt : Duration
Σt : Cumulative time
v : Average speed

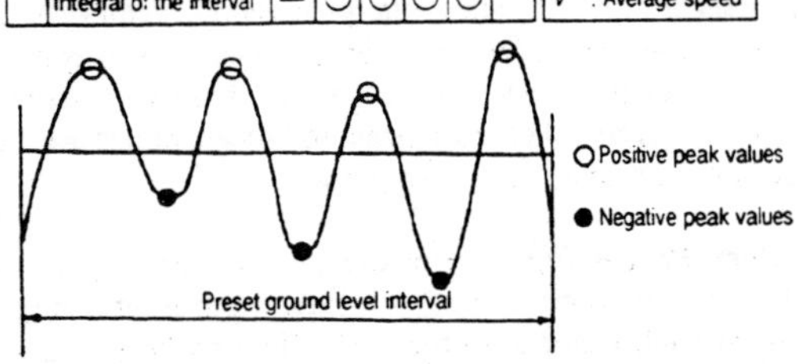

Fig. 3 Definition of the peak value of wave form and the processing patterns of each factor

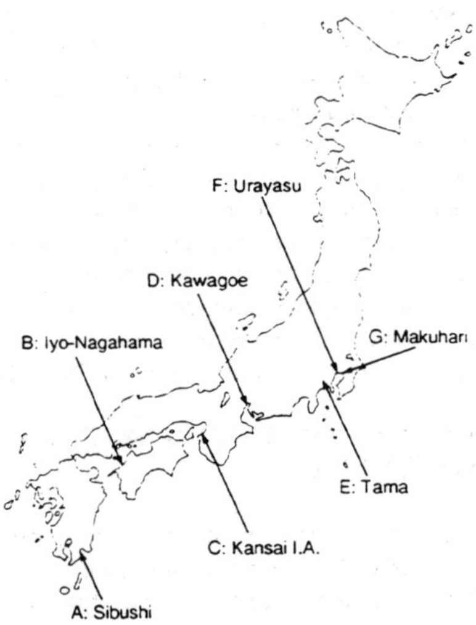

Fig. 4 Location of seven monitoring sites

Monitoring results in Shibushi site (Site - A)

This site was reclaimed with volcanic soil called Shirasu which is typical in Kyushu region. Shirasu is granular materials with large voids which are easily crushed by vibration. Though their bulk density is small, they can form a ground of greater bearing capacity when compacted. Fig. 5 shows the geophysical log and results of SPT, and Fig. 6 gives the monitoring results. As can be seen, the reaction force, vibratory acceleration and amplitude correspond to the N-values reasonably.

Monitoring results in Iyo-Nagahama site (Site - B)

The soil profile of this site consists mainly of fine sand. Fig. 7 gives the geophysical log and N-value of the monitoring site. This sandy ground was rather loose because of low N-values of around 5. Fig. 8 shows the monitoring results, which correspond to the N-values reasonably.

Monitoring results in Kansai International Airport site (Site - C)

The site is an offshore man-made island reclaimed on the soft seabed deposits of about 20 m thick and about 18 m deep. The soft seabed deposits were first improved by the sand drain method, on which a reclaimed island of about 30 m thick was constructed. The reclamation materials consist of crushed rocks smaller than 300 mm in diameter. Fig. 9 gives the geophysical log and the N-values of the ground after compacted with sand compaction piles (SCP). Greater N-values are obtained at several locations, because of the influence of the crushed rocks of larger size, but the N-values at this site might be around 20. Fig. 10 gives the monitoring results, showing that the reaction forces correspond well to the electricity consumption but not to the vibratory acceleration in this site.

Monitoring results in Kawagoe site (Site - D)

The site was reclaimed with dredged soil on the cohesive ground, and the reclamation soil consists of sandy soil and silty sand. Fig. 11 gives the geophysical log and the N-values, and Fig. 12 gives the monitoring results. The electricity consumption was not measured on this site. The overall results correspond to the N-values well.

Table 1 Details of case histories

Site symbol	Site (Prefectures)	Number of boring	Number of measured piles	Site geology	Driving length (m)	Measuring items				
						Reaction force	Vibratory acceleration	Electricity consumption	Driving time	Amplitude
A	Shibushi (Kagoshima Prefecture)	7	9	Sandy soil Volcanic soil (Shirasu)	14	○	○	○	○	
B	Iyo-Nagahama (Ehime Prefecture)	3	12	Sandy soil Mainly fine sand	10	○	○	○	○	
C	Kansai International airport (Osaka Prefecture)	6	6	Sand and gravels	15	○	○	○	○	
D	Kawagoe (Mie Prefecture)	2	12	Silty soil	10	○	○		○	
E	Tama (Tokyo Metropolitan)	3	3	Cohesive soil Including humus	8	○	○	○	○	
F	Urayasu (Chiba Prefecture)	1	1	Sandy soil	10	○	○	○	○	
G	Makuhari (Chiba Prefecture)	1	4	Silty sand	12	○	○	○	○	○

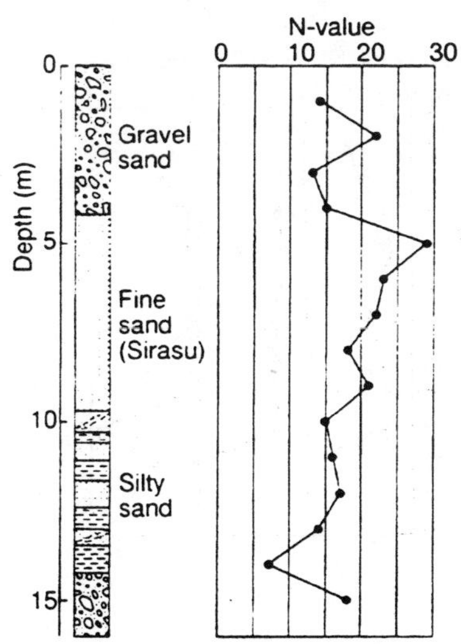

Fig. 5 Soil profile in Shibushi site (Site-A)

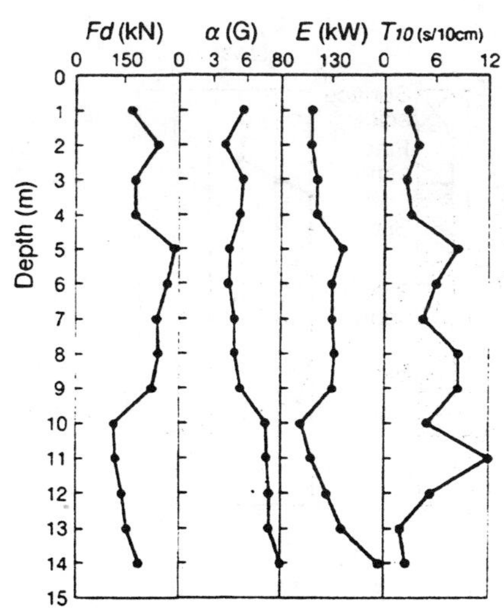

Fig. 6 Monitoring results of Shibushi site (Site-A)

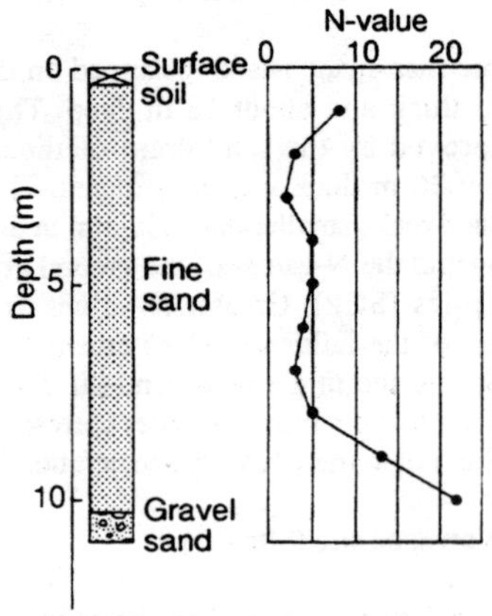

Fig. 7 Soil profile in Iyo-Nagahama site (Site-B)

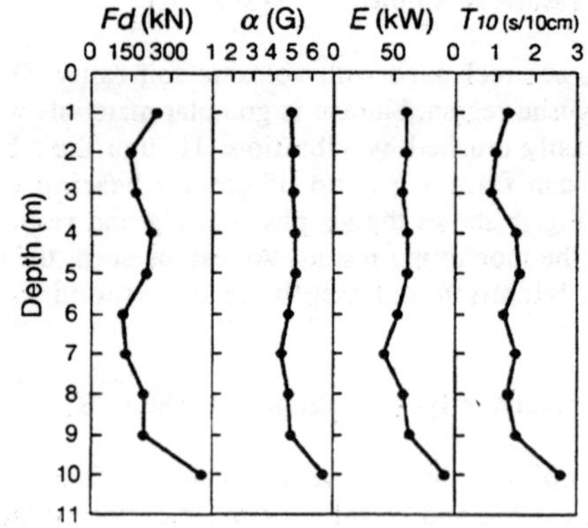

Fig. 8 Monitoring results of Iyo-Nagahama site (Site-B)

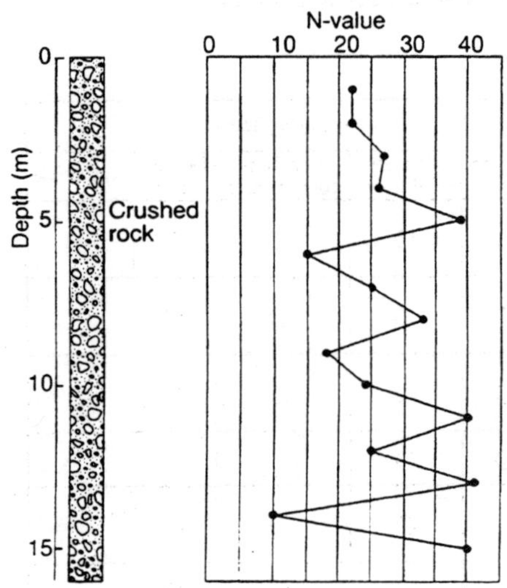

Fig. 9 Soil profile in Kansai International Airport site (Site-C)

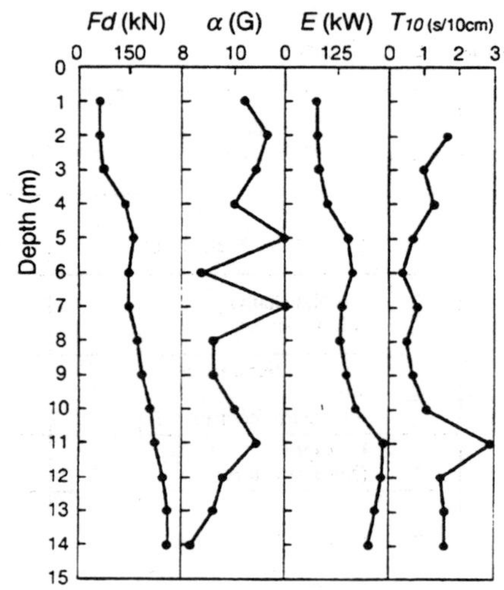

Fig. 10 Monitoring results of Kansai International Airport site (Site-C)

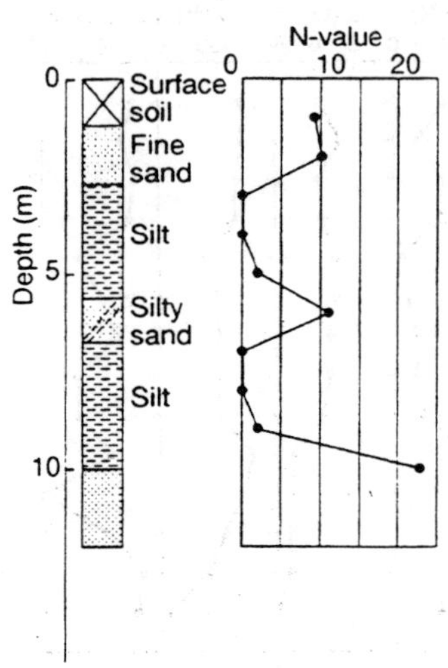

Fig. 11 Soil profile in Kawagoe site (Site-D)

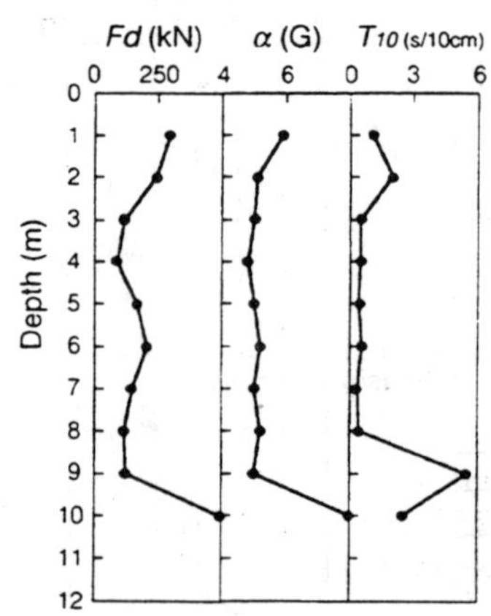

Fig. 12 Monitoring results of Kawagoe site (Site-D)

Monitoring results in Tama (Site - E)

The site is located close to the river and consists of cohesive soil and partly humus. Fig. 13 gives the geophysical log and the N-values, and Fig. 14 gives the monitoring results. Because the vibrator-load is larger than bearing capacity in the soft soil layer, only reaction forces under the self-weight of vibro-driven piles were measured. The driving operation was carried out by carefully hanging the pile from above. Therefore, T_{10} is about zero from −2 m to −6 m, showing that the monitoring results correspond to the N-values approximately.

Monitoring results of Urayasu site (Site-F)

The site is located facing Tokyo Bay, and was reclaimed in a large-scale with sandy soil as seen in Fig. 15. The monitoring results are shown in Fig. 16. Partly owing to the fact that the reclamation soil consists of uniform fine sand, the monitoring data correspond to each other very well. The correlation between N-values and reaction forces was excellent.

Monitoring results of Makuhari site (Site-G)

The site is located close to the Tokyo Metropolitan facing Tokyo Bay, and was reclaimed in a large-scale, designated as an efficient-land-utilization zone where many office buildings have been constructed. Fig. 17 gives the geophysical log and the N-values, and Fig. 18 shows the monitoring results including the reaction forces, vibratory acceleration, electricity consumption, time-driving ratio and amplitude of the depth pattern. All the items correspond to the N-values reasonably.

COMPREHENSIVE CONSIDERATION

The correlations between monitoring items are studied below based on the monitoring data obtained at the seven sites described above. Fig. 19 gives the relation between reaction force (*Fd*) and N-value, showing that there are not only a significant positive correlation as a whole except on Site-A and Site-C, but also more significant correlations on each site. Fig. 20 shows the relation between reaction force (*Fd*) and vibratory acceleration (α). A significant positive correlation is also seen as a whole except on Site-A and Site-C in which greater accelerations are observed. Fig. 21 gives the relation between reaction forces (*Fd*) and electricity consumption (*E*). Also, a significant correlation is recognized except on Site-A and Site-C. One of the reasons why Site-A and Site-C have different correlations from others in the above three figures might be that those soil profiles include a gravelly soil layer. Fig. 22 gives the relation between reaction forces (*Fd*) and time-driven depth ratio, which shows a positive correlation.

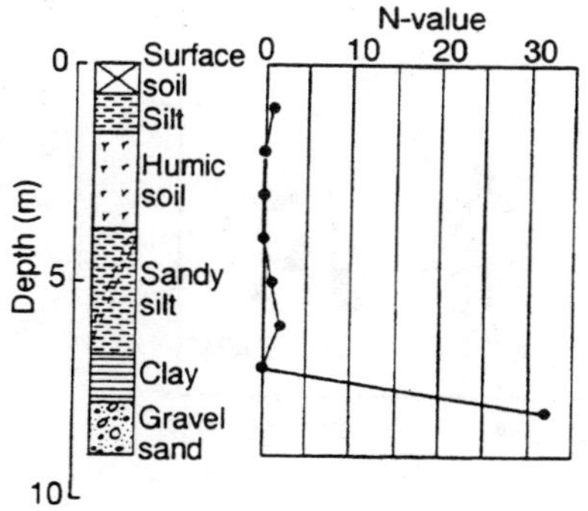

Fig. 13 Soil profile in Tama site (Site-E)

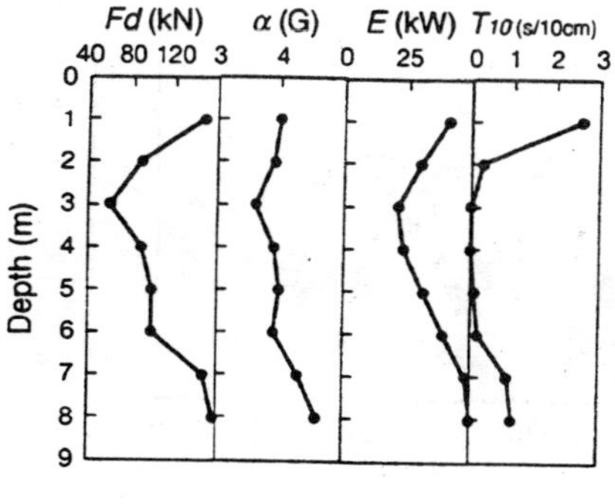

Fig. 14 Monitoring results of Tama site (Site-E)

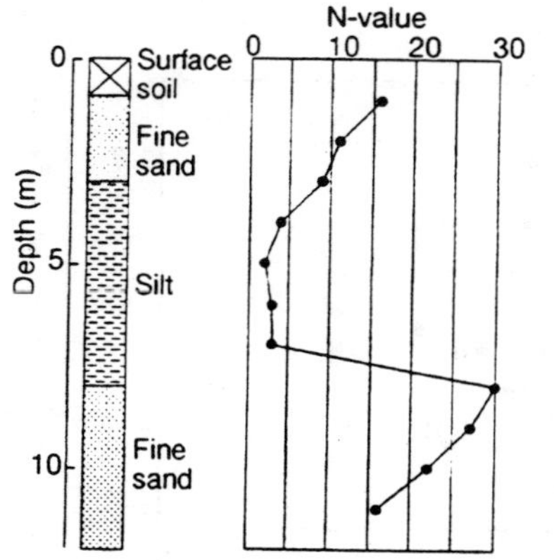

Fig. 15 Soil profile in Urayasu site (Site-F)

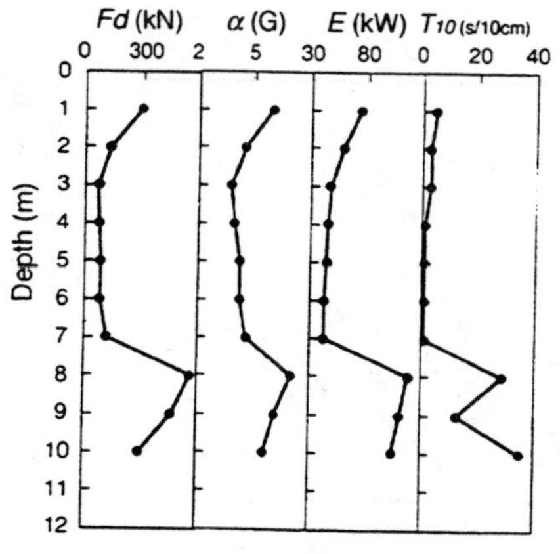

Fig. 16 Monitoring results of Urayasu site (Site-F)

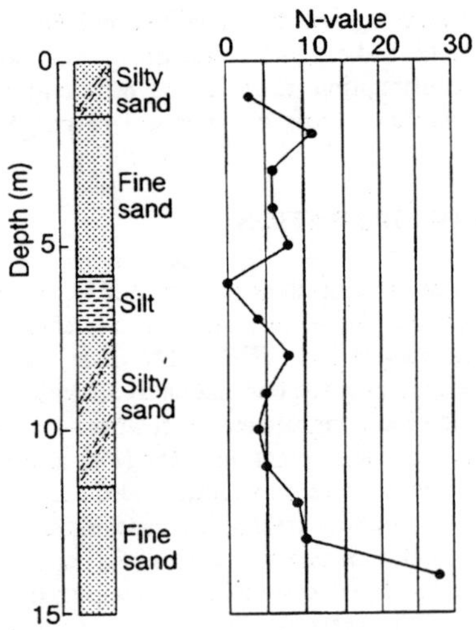

Fig. 17 Soil profile in Makuhari site (Site-G)

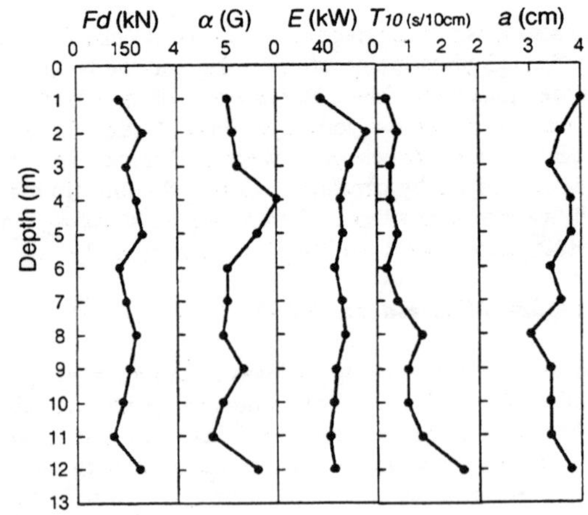

Fig. 18 Monitoring results of Makuhari site (Site-G)

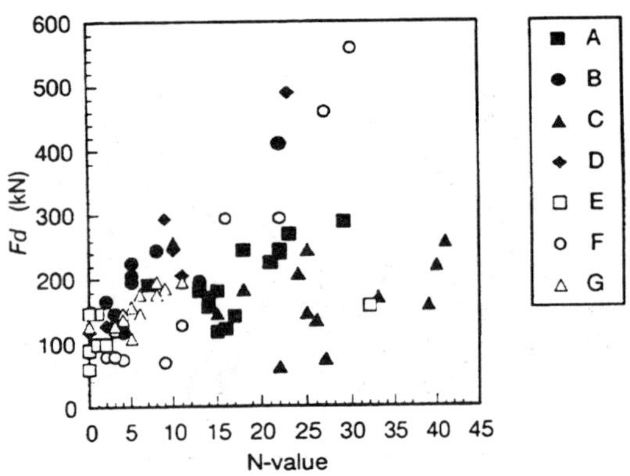

Fig. 19 Relation between Fd and N-value

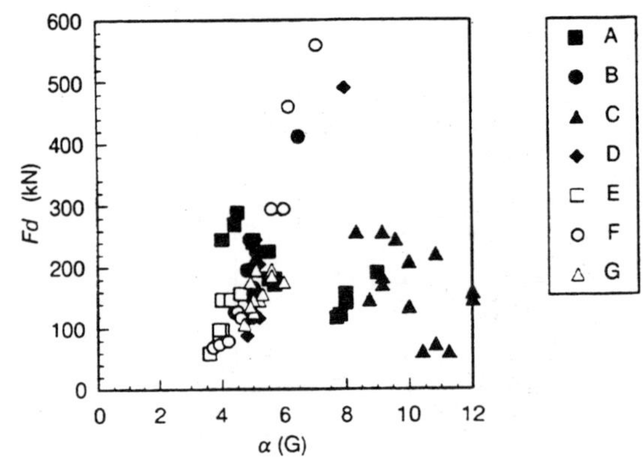

Fig. 20 Relation between Fd and α

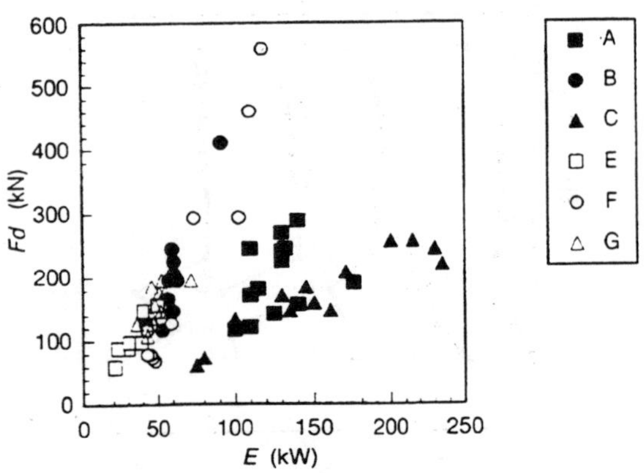

Fig. 21 Relation between Fd and E

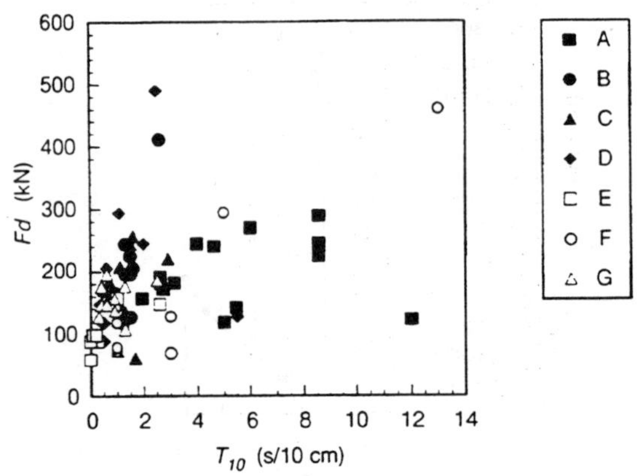

Fig. 22 Relation between Fd and T_{10}

It has thus been confirmed that there are significant correlations between the N-values, reaction forces, vibratory acceleration, electric power and time-driven depth ratio. However, further research works should be carried out for quantitative understanding of these correlations in the future.

CONCLUSION

In this paper, a monitoring system for behavior of vibro-driven piles was developed using a reaction force meter and vibratory acceleration meter attached at around the tip of piles. Using the system, seven reclamation sites in Japan were chosen for obtaining case histories data on monitoring vibro-driven piles. The monitoring results clearly showed that the N-values, reaction force, vibratory acceleration, electricity comsumption, and time-driven depth ratio significantly correspond to each other. Therefore, the development of this monitoring system has opened up the possibility of quantitative understanding of the correlation between vibro-driven piles and soils.

REFERENCES

Tsuboi, H, Ogawa, M, and Matsui, T (1996). "Monitorring for estimating vibration characteristics of penetration pile into ground," Proc. 50th Japan National Conf. on JSCE, pp70–71. (in Japanese)

Ogawa, M, Tsuboi, H, and Matsui, T (1997). "Classification of vibration performance of penetration piles into the ground," JSCE, No 560, III–38. (in Japanese)

Proceedings of the Seventh (1997) International Offshore and Polar Engineering Conference
Honolulu, USA, May 25-30, 1997

ISBN 1-880653-28-1 (Set); ISBN 1-880653-29-X (Vol. I)

Shape Effect of Mandrels on Smear and Well Resistance of Plastic Drain Boards

Soo-Sam Kim, Chung-Ang University
Yeon-Soo Jang, Dong-Guk University
Min-Soo Kang, Chung-Ang University
Kyung-Hwan Koh, Samsung Heavy Industry Co.

Seoul, Korea

ABSTRACT

In this study, the amount of smear zones around the PDB (prefabricated drain board) mandrels is compared with respect to the shape and the size of mandrels by penetrating model mandrels of various shapes into the reconstituted clay. The shapes of mandrels selected are circular, diamond, rectangular and cross shape. The cross shape is the one developed to reduce the smear effects and to increase the effective diameter of PDBs.

A comparative study is also performed to analyze the influence of the shape of mandrels on the consolidation time, relative distance of PDBs, and the drainage depth of weak clay using Hansbo solutions which can incorporate smear and well resistance effects.

KEY WORDS: Smear effect, prefabricated drain board, mandrel, well resistance

INTRODUCTION

Vertical drains are commonly used to accelerate consolidation of thick deposits of soft compressible clayey soils. Recently, the tendency of using prefabricated bandshaped drain boards (PDB) has increased compared to the use of sand drains.

Performance of PDBs is markedly influenced by their transverse and longitudinal permeability, as well as their mechanical properties. The mandrel which protects the drain during installation results in shear strains and disturbance to the soil surrounding the drain. The permeability of clay around the drain is highly influenced by disturbance caused by installation. This zone of disturbance, called the smear zone, depends mostly on the mandrel size and shape, the macroscopic soil fabric, and the installation procedure.

In this study, smear zone tests are carried out to investigate the amount of smear caused by the shape and size of the mandrel and the resulting stress conditions from soils remolding. Permeability tests for the horizontal flow were conducted to study the variation of horizontal permeability in the smear zone. A mandrel of cross shape is proposed to reduce the smear effect and the well resistance. The effectiveness of the cross shape mandrel is analyzed by using Hansbo's solution of consolidation with drains.

TEST EQUIPMENT AND PROCEDURES

Smear zone tests were carried out with the equipment which can give loads to model mandrels for penetrating into the reconstituted soil samples. These soil samples are contained in a plastic tube (Fig. 1).

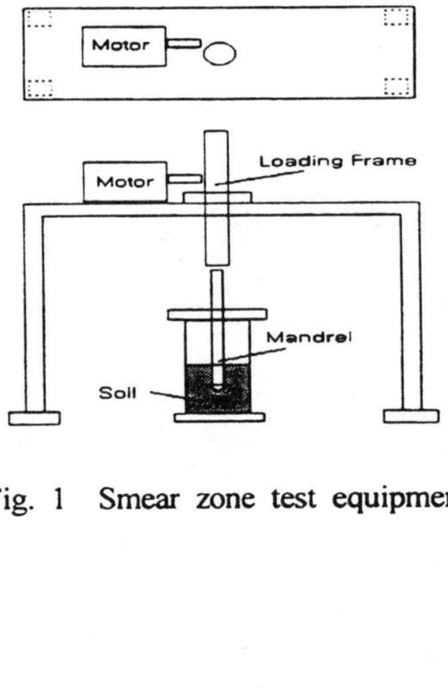

Fig. 1 Smear zone test equipment

Penetration rate of the mandrel can be controlled in the range of 1 mm/sec~6.5 mm/sec.
Fig. 2 shows the various types of mandrel used in this test.
The shapes of the model mandrel are made by resembling the mandrel used in the field. The model mandrels of diamond and 50mm rectangular shape have the same cross-sectional area as circular shape mandrel of 30mm diameter so that shape effect can be compared. The models of rectangular shape with different thickness are also made to estimate the effect of various thickness on the disturbance of clay.

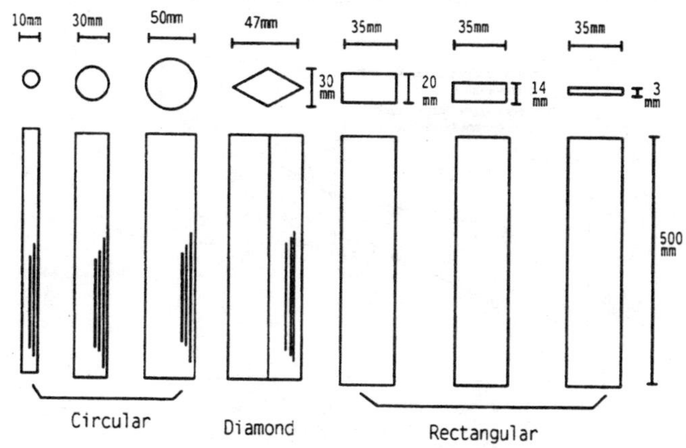

Fig. 2 Tested Mandrel Shapes

The clay samples used in the experiment are obtained from Busan in Korea, and reconstituted by the following procedure ;
It first passes through a 0.425mm sieve at a slurry state to get rid of remnants such as clamshells, gravels, etc. and remolded thoroughly to a slurry state. Second, it is poured into a cylindrical mold with inner diameter of 250mm and preconsolidated under a pressure of 49kPa. Drainage was allowed at the top and bottom of the tank through circular porous stone.
Third, preconsolidation is stopped after degree of consolidation, 90%, is achieved with $\sqrt{t}$ method. Then the clay samples were taken out and preserved.
Fourth, in order to observe the amount of smearing, soil sample is cut 30mm intervals and strips of soup noodles are laid horizontally on the surface of every layer shown in Fig. 3.
A model mandrel is driven into the prepared soil sample with the rate of 1.5 mm/sec. The soil sample is carefully cut vertically and the deformed shape of the lines of the noodles are sketched. The experiments are performed on the various conditions shown in Table 1.
The most important aspect of smeared zone is the decrease of permeability by the remolding of soil adjacent to the driven mandrel. For this reason, the permeability in smeared zone is measured after driving the mandrel into the soil sample. The specimens in the horizontal direction are taken from the smeared zone and tested with the permeability test.

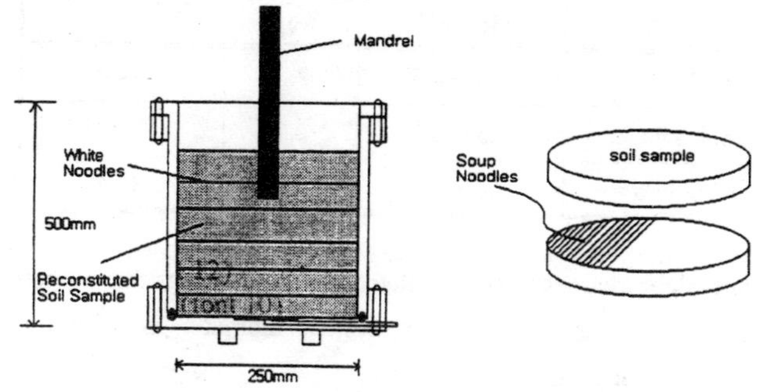

a) Vertical section b) Cut surface laid with noodles

Fig. 3 Preparation of Test Sample

TEST RESULTS AND ANALYSIS

Smear Zone Test

From the smear zone test results, the range of smeared zone with the size of circular mandrel is tabulated in Table 1.
This indicates that the smear zone increases with the increase of the size of mandrel. With the variation of preconsolidation pressure at the sample reconstitution stage, smearing effect increases with the increase of stiffness of soil (Table 2). The samples are penetrated by the circular 30mm diameter mandrel.
From the view point of mandrel shapes, diamond and rectangular shapes are more effective in reduction of smear zone than the circular shape.
Averaged water contents are shown in Table 1, 2 which are obtained from top, middle and bottom of the test samples after the penetration test was finished.
Fig. 4 shows the smear zone caused by model mandrels of diamond and rectangular shape. In order to compare the results with those of circular model mandrel, elliptical boundaries are converted to the equivalent circle.

Table 1 Test results of d_s/d_m with respect to the size of circular mandrel

Diameter	d_s/d_m*		Avg. water content (%)
	Range	Avg.	
10mm	2.1~2.5	2.3	40.3
30mm	2.3~3.3	2.7	39.1
50mm	3.7~4.0	3.9	39.0

* d_s: Diameter of smear zone d_m: Diameter of the mandrel

Table 2 Test results of d_s/d_m with variation of preconsolidation pressure of the clay sample

Preconsolidation Pressure	d_s/d_m		Avg. water content (%)
	Range	Avg.	
49 kPa	1.9~2.5	2.3	44.8
78.4 kPa	2.3~3.3	2.7	39.1
117.6 kPa	2.8~3.7	3.2	35.4

The equivalent diameter is calculated by converting the mandrel area to the circular area. The equivalent diameter is shown on horizontal axis in Fig. 5.

Their ranges of smear zone are arranged in Table 3. Smear zone ratio, d_s/d_m, between the diamond and the rectangular shape of mandrel show little difference and as the thickness of the rectangular mandrel decreases, the range of smear zone ratio decreases considerably.

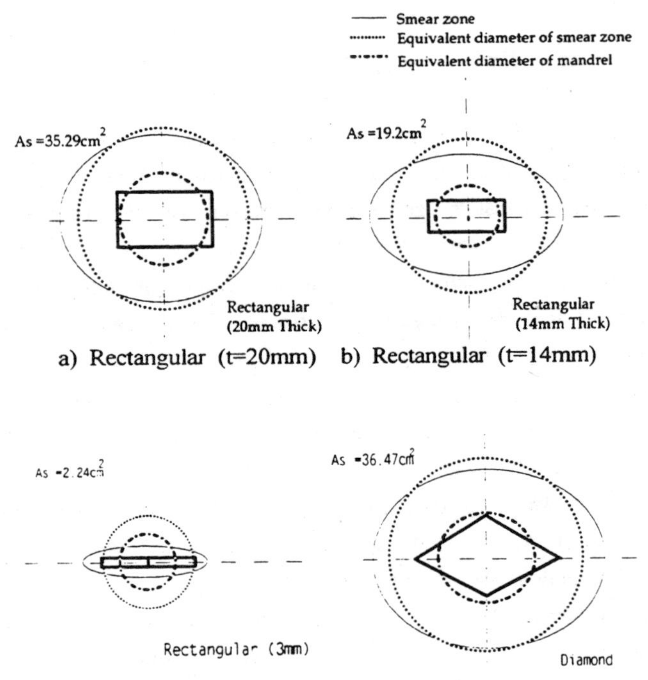

c) Rectangular (t=3mm) d) Diamond

Fig. 4 Comparison of smeared zones with various shapes and thickness of model mandrels

Fig. 5 shows d_s/d_m ratios from the test with circular mandrels of diameter, 10, 30, 50mm and rectangular mandrels with thickness of 3, 14, 20mm. It can be recognized that reducing the thickness of rectangular mandrels is more effective than reducing the size of mandrel for minimizing the smear zone around mandrels.

Table 3 Tested range of d_s/d_m with respect to the shape of mandrel

Shape of mandrel		d_s/d_m	$A_s\ (cm^2)$
Circular (30mm dia.)		2.7	51.5
Diamond		2.3	36.5
Thickness of Rectangular	20mm	2.2	35.3
	14mm	2.0	19.2
	3mm	1.5	2.2

* A_s : The equivalent area of smear zone

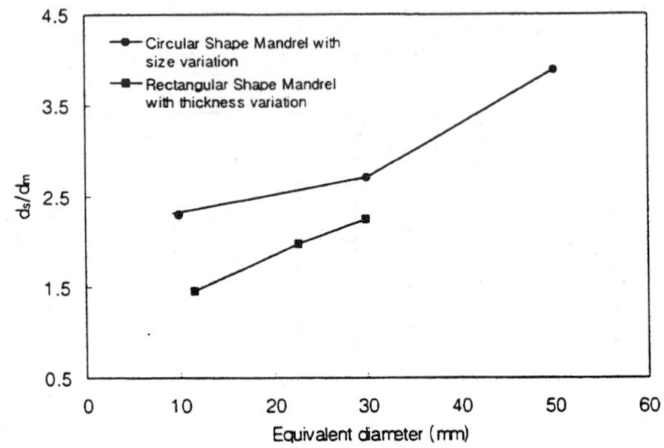

Fig. 5 Comparison of d_s/d_m with respect to the size and thickness of mandrel

Permeability Test

Three samples are cut vertically for the permeability test after driving the mandrel. The permeability samples are taken from the penetration test with the circular 30mm diameter mandrel. Specimen A is taken adjacent to the driven-mandrel which can be assumed to be fully disturbed (0~30mm apart from the mandrel). Specimen B is obtained 35~65mm distance from the mandrel and specimen C is from the part where it is believed not to be affected by smearing (70~110mm distance from the mandrel). Table 4 shows the results from the permeability test. The horizontal permeability decreases as it becomes closer to the disturbed zone.

Table 4 Test Results from Permeability tests for sampled specimens to radial direction

	Vertical permeability k_v	Horizontal permeability (mm/sec)		
		Vicinity A	Intermediate B	Undisturbed C
Test result	4.7×10^{-7}	4.9×10^{-7}	6.0×10^{-7}	9.3×10^{-7}

Smear Zone Test with Cross Shape Mandrel

The test results presented in the previous section showed the fact that the smear effect can be minimized by reducing the thickness of mandrel. However, simply reducing the thickness of mandrel may result in the torsional and buckling failure of mandrels during installation period. Hence, in this study, cross shape of mandrel is proposed as a the new type of mandrel which can reduce the smear effect and provide adequate stiffness to the mandrel. Additional smear zone tests were carried out for this new shape mandrels. Fig. 6 shows the cross-section of two types of the cross shape mandrel. Test results with those mandrels are shown in Fig. 7 and in Table 5.

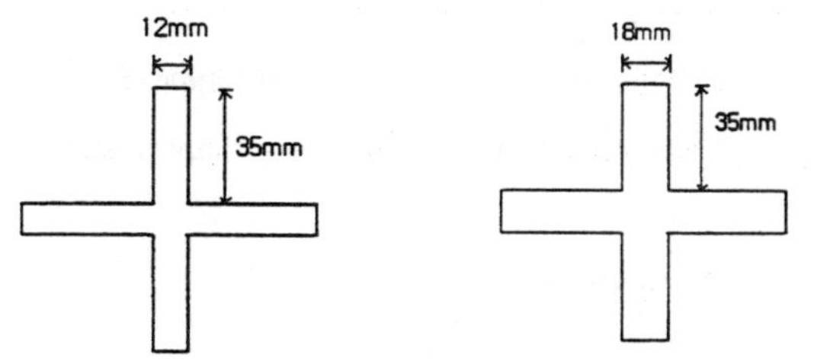

(a) Thickness 12mm (b) Thickness 18mm

Fig. 6 Cross shape mandrels

Table 5 Smear zone test results for cross shape mandrels

Mandrel thickness	d_s'/d_m'	A_s (cm^2)
12mm	2.13	81.0
18mm	2.42	164.0

Comparison of these results with those of diamond and the rectangular shape showed quite similar d_s/d_m ranges (Table 3). It can be recognized that if cross shape mandrel is used, not only the smear zone is reduced but also the drainage capacity is increased with improved drainage area of cross shape PDBs.

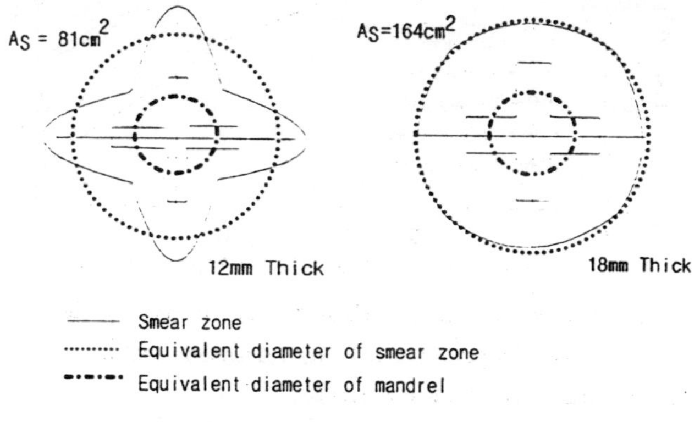

Fig. 7 Comparison of smeared zones with types of cross shape mandrels

ESTIMATION OF THE PERFORMANCE OF TESTED PDB MANDRELS

The performance of tested PDB mandrels with different shapes are compared using the theoretical solution of Hansbo for estimating consolidation effect (Hansbo, 1979).
This solution can include the smear and well resistance effect of plastic board drains. The modified general expression by Hansbo (1979) for average degree of consolidation is given as ;

$$U_h\ (z, T_h) = 1 - \exp\left(\frac{-8T_h}{F}\right) \quad (1)$$

and $F = F(n) + F_s + F_w$

where F is the factor which expresses the additive effect due to the spacing of the drains, $F(n)$; smear effect, F_s ; and well-resistance, F_w.
For typical values of the spacing ratio, n, of 20 or more, the spacing factor simplifies to :

$$F(n) = \ln\left(\frac{d_e}{d_w}\right) - \frac{3}{4} = \ln(n) - 0.75 \quad (2)$$

The smear effect and the well resistance factors are given as :

$$F_s = \left(\frac{k_h}{k_s} - 1\right)\ln\frac{d_s}{d_w} \quad (3)$$

$$F_w = \pi z(H - z)\frac{k_h}{q_w} \quad (4)$$

where

k_h/k_s : Ratio of the horizontal permeability to the permeability in the smear zone (= η)
q_w : Discharge capacity of the drain (= $\pi r_w^2 k_w$)
H : The length of the drain when drainage occurs at one end only (2/H for the drainage at both ends)

Soil Properties and Conditions of the Analysis

A domestic site in Korea where the PDB is actually installed is selected to estimate the performance of tested PDB Mandrels. Soil properties and the design considerations used for estimation of the performance are summarized in Table 6 and Table 7. The size of circular, diamond and rectangular shape of mandrel used in the analysis are shown in Fig. 8.
Fig. 9 shows the cross-section of cross shape mandrels.
Type A mandrel is for PDBs made by connecting four conventional PDBs and type B mandrel is for PDB made by connecting small size PDBs which have to be manufactured for use of this mandrel. Equivalent diameter of cross shape PDB (d_w') can be obtained by using equation (5).

$$d_w' = \frac{4(2a+b)}{\pi} \times \beta \qquad (5)$$

where β is the shape factor (0.75~0.78)
and a, b are long and short length of the PDB(Fig. 6).

Table 6 Soil properties used for the theoretical validation of mandrel performance

Classification	Properties
Soil type	CH, CL, MH
Water content	45 ~ 65%
Void ratio	1.2 ~ 1.8
Liquid limit	40 ~ 80%
Compressive index, C_c	0.34 ~ 0.83 (0.61)
Coefficient of consolidation, C_v	$1.4 \times 10^{-3} cm^2/sec$
Vertical permeability k_v	$5.7 \times 10^{-7} cm/sec$

Table 7 The site conditions of selected site

Site Condition	Design value
Height of surcharge (m)	3.8
Thickness of weak soil layer (m)	20
Drain spacing (m)	1.5
Estimated settlement (m)	1.683
Time for 90% consolidation (day)	210

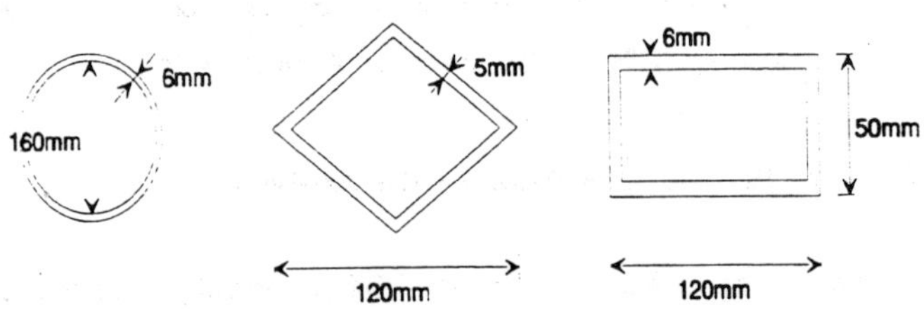

Fig. 8 The dimension of mandrels analyzed

The ratio of horizontal permeability to vertical permeability (k_h/k_v) and the ratio of horizontal permeability to smear zone permeability are both assumed 3.

For the calculation of required drainage capacity, equation (6) suggested by Kamon et al. (1992) is used.

$$q_{req} = \varepsilon_f \times U_h \times F_s \times H \times \pi \times C_h / 4T_h \qquad (6)$$

where $T_h = -F(n) \times \frac{\ln(1-U_h)}{8}$

ε_f : the limit of final settlement (25%)

F_s : Safety factor (=2)

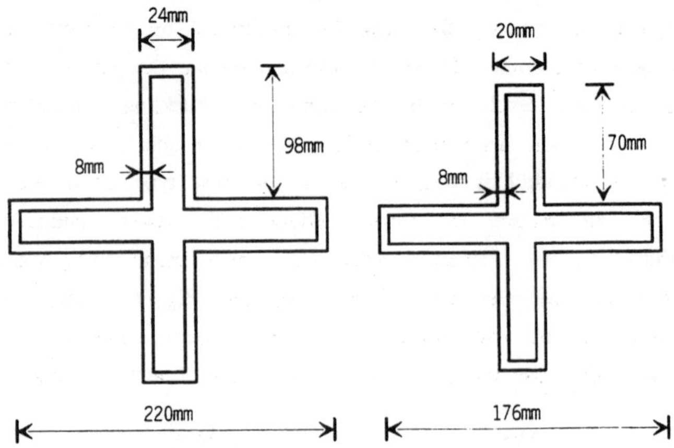

(a) type A (b) type B

Fig. 9 The dimension of two cross shape mandrels

The Results of Analysis

The results of consolidation analysis obtained from Hansbo's solution are shown in Fig. 10.

For the shape effect, the time for 90% consolidation of circular shape case is about 25% longer than for the diamond and the rectangular shape. The cross shape mandrels are much more effective than the diamond and rectangular shape in reduction of consolidation time. On the other hand, two types of cross shape mandrel show almost the same results.

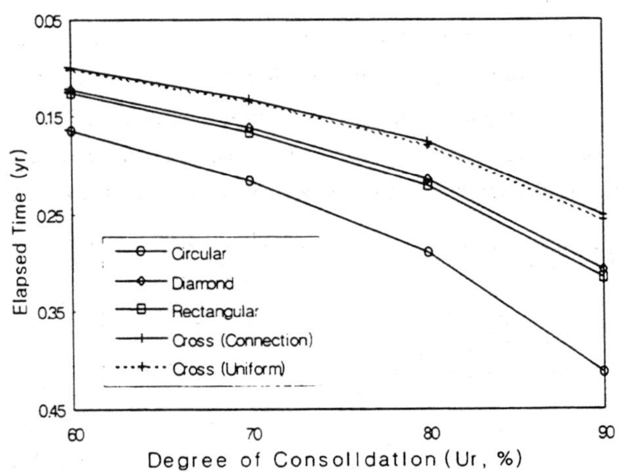

Fig. 10 Comparison of the elapsed time versus the degree of consolidation obtained from Hansbo's solution for the mandrels of various shapes

Sensitivity of drain installation spacings to the mandrel shape for obtaining 90% degree of consolidation for the duration of

0.5 year is also shown in Table 8.

Table 8 Sensitivity of drain spacing to shape of mandrels for obtaining same consolidation effects
(Ur = 90%, t = 0.5 yr, triangle installation)

Shape of Mandrel	Spacing (m)
Circular	1.64
Diamond	1.88
Rectangular	1.86
Cross (Connection)	2.05
Cross (Uniform)	2.03

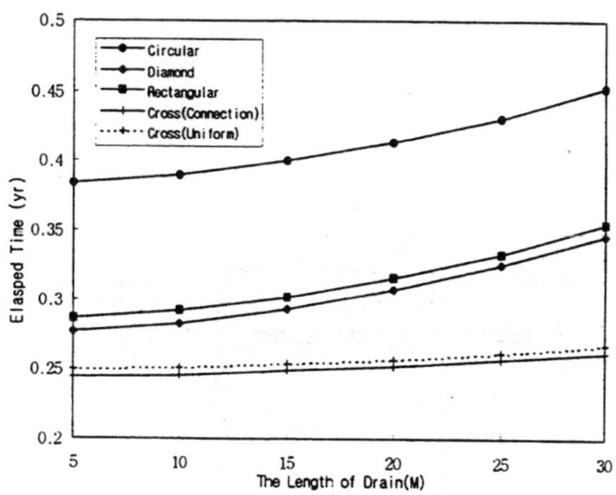

Fig. 11 Time required for the 90% consolidation with the various depth of installed drains

Most effective results are obtained using cross shape drains. Drain spacing using cross shape drain is about 25% larger than that using circular shape mandrel, and about 10% larger than that using diamond shape mandrel. But the difference between connected drain type and uniform type of cross shape mandrel is insignificant. The effectiveness of cross shape drain can also be found in Fig. 11 which shows the relation between the depth of installed drains and the time for obtaining 90% consolidation. Due to the fact that the equivalent diameter of cross shape drain is much larger than others, the time taken for 90% consolidation is not increased with the increase of the depth of installed drains.

CONCLUSIONS

In this study, the amount of smear zone around the driven mandrel is investigated in the laboratory with the smear zone model test equipment using various shape and size mandrels. From the test results, it can be concluded that smear effect is highly dependent on the shape and size of mandrel and stiffness of soil. Rectangular and diamond shape mandrels are more efficient for reducing the smear effect. Reducing the thickness of rectangular shape mandrels results in the reduction of smear zone around the rectangular shape mandrel.
Efficiency of cross shape mandrel and drain using Hansbo's solution of consolidation with drains is shown by the comparative analysis. Use of cross shape mandrel has similar smear zone ratio to those of rectangular and diamond shape. However, the drainage capacity of cross shape mandrel has improved greatly compared with that of other conventional plastic board drains.

ACKNOWLEDGEMENT

The authors wish to thank Sam-Sung Heavy Industry, Co., Ltd., in providing financial support of this research in the form of a research grant.

- Reference -

Akagi T. (1977), "Effect of Mandrel-Driven Sand Drains on Strength", Proc. 8th International Conference on Soil Mechanics and Foundation Engineering, Tokyo, Vol. 1, pp. 3~6.

Bergado D. T., Asakami Hiroshi, Alfaro M. C., and Balasubramaniam A. S. (1991), "Smear Effects of Vertical Drains on Soft Bangkok Clay", ASCE, Journal of Geotechnical Engineering, Vol. 117, No. 10.

Hansbo, S. (1979), "Consolidation of Clay by Band-Shaped Prefabricated Drains", Ground Engineering, Vol. 12, No. 5, pp.16~25.

Hansbo S., Jamiolkowski M. and Kok L. (1981), "Consolidation by Vertical Drains", *Géotechnique* 31, No 1.

Kamon, M., Pradhan T. B. S and Suwa S. (1991), "Evaluation of Design Factors of Prefabricated Band-Shaped Drains", GEO-COAST'91, pp.329~333.

Onoue Atsuo (1988), "Consolidation by Vertical Drains taking Well Resistance and Smear into Consideration", Soil and Foundations, Vol. 28, No. 4, pp.165~174.

Proceedings of the Seventh (1997) International Offshore and Polar Engineering Conference
Honolulu, USA, May 25-30, 1997

ISBN 1-880653-28-1 (Set); ISBN 1-880653-29-X (Vol. I)

Pullout Behavior of Model Suction Anchors in Soft Marine Clays

S. Narasimha Rao, R. Ravi and C. Ganapathy
Indian Institute of Technology
Madras, India

ABSTRACT

This paper reports the results of laboratory testing on the suction anchor under short-term monotonic and sustained loading. Model anchors of different L/D ratios were installed in a soft clay bed prepared in a test tank and the uplift load was applied through a wire rope and pulley arrangement. To measure the suction developed in the system, inductive pressure transducers of 100 kPa capacity were used. The sustained pullout load tests were conducted for a load level (F/P_u) of 30%, 50% and 75%. Test results reveals that in all the static tests, suction pressure is found to be a function of anchor displacement.
KEY WORDS: Anchor, Consistency Index, Reverse-end bearing, Soft clay, Suction, Sustained load level

INTRODUCTION

In the recent times, Suction anchor or suction caisson finds its cost-effective application in deeper water in resisting large uplift forces from floating offshore structures. Suction anchor is an inverted top capped hollow cylinder of a fairly large diameter with length to diameter ratio (L/D) of 1.0 to 2.0 embedded into the sea bed soil. In the installed position, when the anchor is subjected to any pullout movement, it mobilizes significant breakout resistance through the development of negative pore water pressure in the soil plug formed inside and at the anchor base. In a highly impermeable and saturated clay system, differential pressures due to negative pore water pressures and high ambient pressures generate a large suction called passive suction. The passive suction developed at the top of the soil plug tends to keep the soil plug inside the anchor. The passive suction generated at the anchor bottom caused a significant breakout resistance referred as suction induced reverse-end bearing. Fig.1 shows the suction anchor embedded in surface anchor condition with the various components contributing to the capacity. The literature shows that a few investigators have attempted to evaluate the behavior of suction anchors in clayey soil through laboratory and field model testing. **Finn and Byrne (1972)** carried out laboratory model studies to

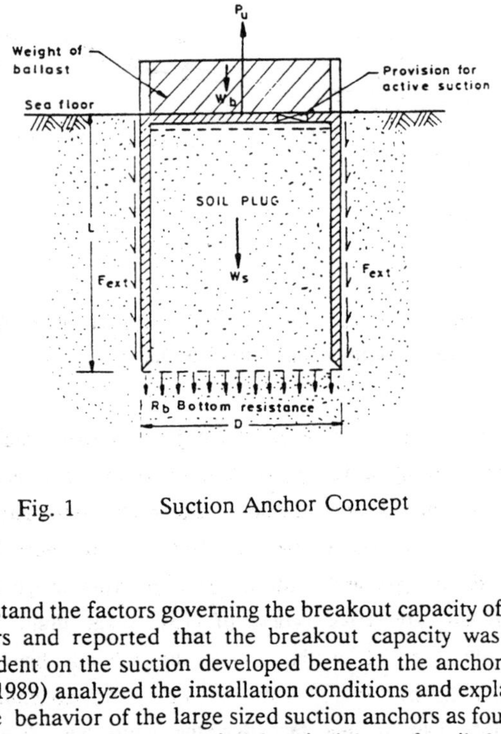

Fig. 1 Suction Anchor Concept

understand the factors governing the breakout capacity of suction anchors and reported that the breakout capacity was mainly dependent on the suction developed beneath the anchor. **Albert** et al (1989) analyzed the installation conditions and explained in service behavior of the large sized suction anchors as foundation for TLP through conventional principles of soil behavior. **Fuglsang** and **Steensen-Bach** (1991) made a parametric study on model suction anchors of embeddment ratios(L/D) 1.67, 2.0, and 3.33 in two type of soils. **Steensen-Bach** (1992) reported the modes of anchor failure under static loading and the suction pressure developed at the top of the soil plug. This paper provides the results of laboratory model study directed toward the evaluation of suction pressure variation under short-term monotonic loading and sustained loading conditions in a soft marine clay.

SCOPE

The main objectives of the present study are:

- To bring out the Load-Displacement behavior of the suction anchor under short-term loading conditions.
- To study the variation of passive suction (underpressure) at the top of the soil plug and anchor base during the breakout process under both short-term and sustained loadings.
- To study the failure pattern of the suction anchor system.

EXPERIMENTAL WORK

Model Anchors Fabricated

Laboratory tests were conducted on scaled (1 : 100) model anchors of 75 mm diameter and 3 mm skirt wall thickness. To study the influence of embeddment ratio (L/D) on the pullout behavior, three anchors of different ratios of length to diameter (L/D = 1.0, 1.5, 2.0) were fabricated. A steel hook rod welded to the centre of the anchor top cap was used to connect it to the loading system. On either side of the central hook, provisions were made in the form of threaded holes to fix the two screw types pressure transducers for measuring the suction pressure developed at the top of soil.

Soil Used

In the present investigation, a typical marine clay obtained from the coastal deposit of the east coast of India was used. This soil consists of 90 % silt and clay and 10 % sand. The liquid limit of the soil is 82 % and the plastic limit is 32 %.

Experimental Set up

The tests were conducted in a cylindrical test tank of size 750 mm diameter and 900 mm height. Based on the earlier testing carried out on a similar type of anchor in this laboratory (Narasimha Rao et al,1993), it was found out the boundary effect of the test tank on uplift resistance could be rendered insignificant by making its diameter about 8 times the anchor diameter. This is also in conformity with the investigations of Devi and Sutherland (1979). As the diameter of the test tank in the present work was kept 10 times that of the anchor diameter, it is believed that the boundary effect on the break out resistance of the soil are minimal. The schematic diagram of the test setup is shown in Fig. 2. To measure the suction pressure developed at the top of the soil plug and at anchor bottom inductive diaphragm type pressure transducers(P11) of HBM make of -100 kN/m^2 capacity were fixed on the anchor top and at the anchor base. Pullout load was applied in increments by placing weights on the load hanger attached to the anchor top through a wire rope and pulley arrangement. The outputs from the load cells, displacement transducer and pressure transducers were fed to a multichannel carrier frequency amplifier and the signals were amplified and recorded by means of a data acquisition system (Orion-3530 Data Logger).

Testing Procedure

The tests were conducted in a soil bed prepared at a soft consistency (I_c = LL-w_n/LL-PL)) of 0.40. During the formation of soil bed, at various depths undrained shear strength of soil bed was measured using vane shear apparatus and the average value reported is taken as C_u. Also a few random samples were taken out and checked for full saturation and homogeneity in a triaxial test where the pore pressure parameter-B obtained is nearer to unity. Fully saturated clay soil was placed in the test tank in layers of 50 mm thickness. After the clay bed was formed the anchor was installed by slowly pushing the skirt wall vertically into the soil bed up to the required position where the bottom pressure transducer had been already fixed at the anchor base inside the soil bed. Two days were allowed after the installation of the anchor for equilibration in the pore water pressure. This was based on the test carried out on the samples taken at intervals of 4 hours over a period of 2-3 days and pore water pressures were measured using the transducer system fitted to the triaxial testing assembly. In the first few trials made, it was found that a time interval of 20-24 hours was required for 95% equilibration of pore pressures.

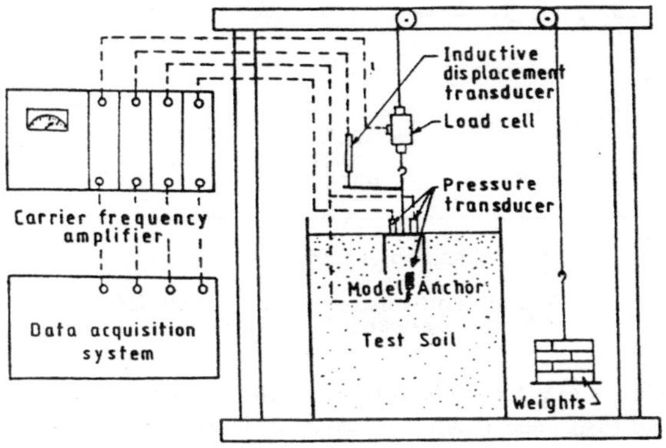

Fig. 2 Test set up

Testing program

The model suction anchors of different L/D (= 1.0, 1.5, 2.0) ratios were subjected to the following two phases of testing in a soil bed prepared at an I_c value of 0.4:

Short term pullout tests:

Under the short term monotonic loading, pullout tests were conducted to estimate the ultimate uplift capacity (P_u) of suction anchor which also brought out the variation of suction pressures with anchor movement at the top of the soil plug and at anchor bottom.

Sustained load tests:

In this phase of testing, model anchors were subjected to different sustained loads(F) which were less than the short term ultimate uplift capacities(P_u) as obtained from the above short term load tests. These load tests were conducted for a sustained load level of (F/P_u) of 25%, 50% and 75%. The variation of anchor displacement, suction pressure at the top of the soil plug and anchor base with time(t) for the given sustained load levels(F/P_u) were brought out in these testing.

RESULTS AND DISCUSSIONS

The performance of these anchors under various testing conditions is mainly brought through the measured pullout capacities and variations in the suction pressures and load displacement patterns observed.

a) Short term uplift load tests:

Load-Displacement behavior of suction anchor

The typical pullout load-displacement curves obtained from short-term pullout load tests on model suction anchors of different L/D ratios are shown in Fig.3. These are the curves obtained from tests on anchors embedded in clay bed formed at an I_c value of 0.4. From these curves, three distinct phases in the pullout behavior could be observed. In phase I, at the initial ranges of the deformation up to 2.0 mm (1% to 3% of the diameter), there is a steep rise in the pullout resistance and this is followed by phase II in which there is a gradual and slow increase in the pullout resistance with deformation. However In this phase(II), the deformation is quite large. In the last portion

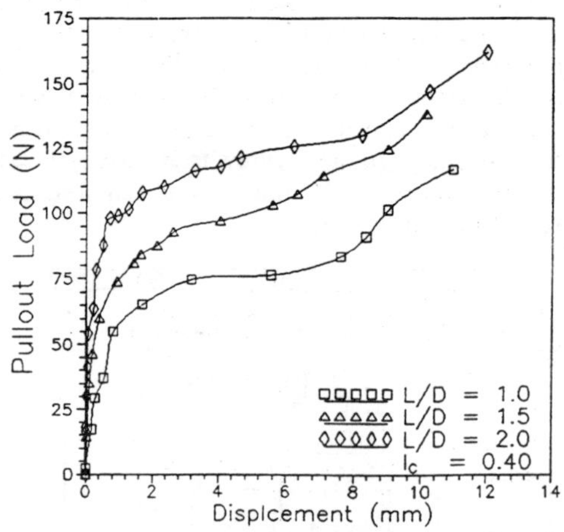

Fig. 3. Pullout load–displacement curves of Suction anchor (I_c=0.4)

(phase III), there is a reverse curve exhibiting sudden increase in the pullout resistance which is followed by sudden pullout of the anchor. Further, it is to be noted that in all these tests the soil remained intact inside the anchor confirming the formation of the soil plug which increased the pullout capacity.

For any load test conducted either on this type of cylindrical anchor or pile, the first component to be mobilized is skin the friction and this needs a very small deformation. From the load deformation presented in Fig.3, it is obvious that there is a steep rise in the capacity even at a deformation less than 1.5 mm (2% of the diameter). This indicates that the skin friction predominates in this phase. As the plug formation is developed, the skin friction mobilized is only the external skin friction. The phase I is followed by a large deformation process and this stage (phase II) can be called **suction induced reverse end bearing.**

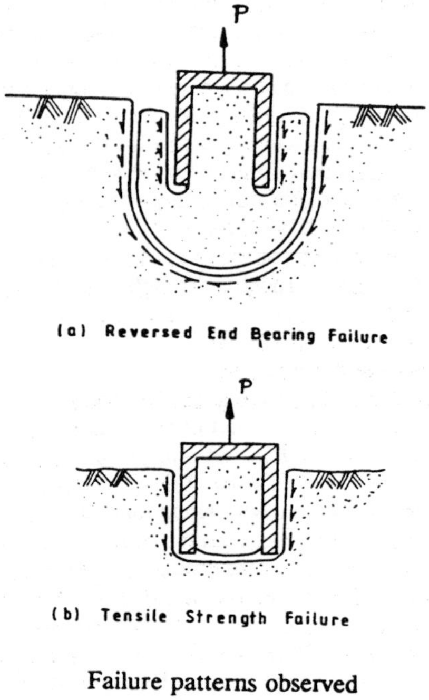

Fig.4. Failure patterns observed

This phase lasts to deformations of as much as 12 mm (16 % of the anchor diameter). Beyond this stage, the tensile strength of the soil at the anchor bottom mobilizes. When the pullout force exceeds the load capacity of these three components, there can be rupture in the soil mass at the base of the anchor and the anchor gets pulled out clear of the base. From the results presented in Fig.3, it is clear that with an increase in L/D, there is an improvement in the ultimate anchor capacity. Fig.4(a) shows the failure mode in which a shallow crater is formed and this is a circular slip surface of about 1.5 times the diameter from the anchor centre extending vertically up to the soil surface. However, at large deformation, the anchor is completely pulled out (Fig.4(b)) with the plug inside as if it is torn out at the anchor bottom (soil tensile strength is fully mobilized).

Variation in Suction pressure at the top of the soil plug

The variation in suction pressure at the top of the soil plug with anchor displacement is shown in Fig.5, for all the L/D ratios tested at a given consistency. These suction pressures are negative pressures and the transducers used in this setup are capable of measuring negative pressure, ie; pressure less than atmospheric up to -100 kN/m². These curves show that the suction pressure increases with an increase in the anchor displacement and does not exhibit a clear peak value even at the pullout stage. This confirms that suction remains developed during the entire pullout process and that the soil plug remains intact within the anchor. Also, the suction pressures measured varies as a function of anchor displacement.

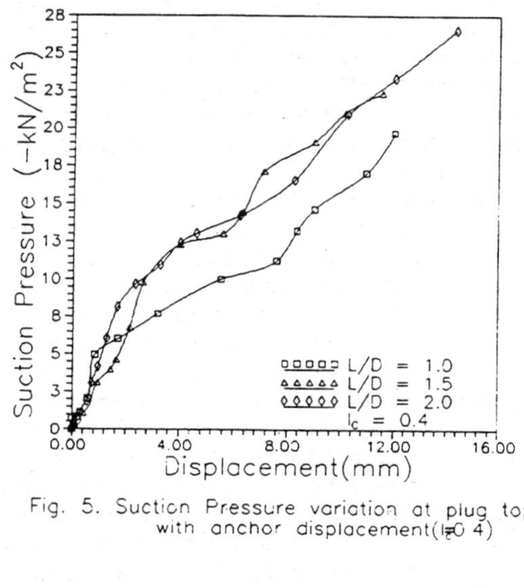

Fig. 5. Suction Pressure variation at plug top with anchor displacement(I_c=0.4)

Suction pressure variation at anchor bottom

Fig. 6. shows the typical plot of change in suction (-ve) pressure at the anchor base with vertical displacement for all the L/D ratios at an I_c value of 0.4. From these curves, it is observed that the suction pressure increases with displacement to reach a maximum value unlike the pressure variation at the top of the soil plug. The peak pressure value is mobilized at relatively low displacements and then it drops down to a constant value at larger displacements. This is attributed to the fact that the vertical anchor movement makes the location of the bottom pressure transducer well below the anchor base at higher displacements. Although it looks that the peak pressure value mobilized is higher for higher L/D ratios, it varies as a function

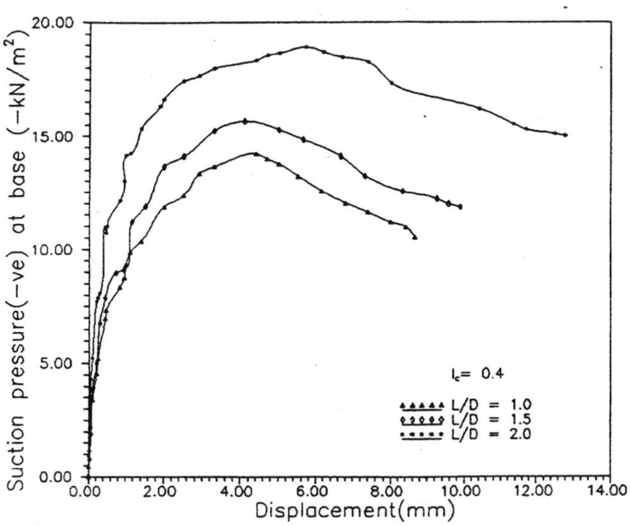

Fig.6 Suction pressure(−ve) variation at anchor base with upward displacement (I_c=0.4)

of anchor displacement only.

(b) Sustained load tests:

Sustained load tests were performed for all the given L/D ratios varying the sustained load levels (F/P_u) in the range of 25%, 50% and 75%. In these tests, by keeping the static load constant, time dependent observations of anchor displacement and pore pressure changes at the top of the plug and anchor base were made.

Fig. 7. shows the typical plot of anchor displacement with time for the different F/P_u values corresponding to L/D of 2.0 and I_c value of 0.4. From this plot, it can be observed that for any given sustained load levels(F/P_u), the anchor displacement increases initially with time and then it becomes more or less stable. Also, with the increase in sustained load levels, the corresponding displacement at which anchor gets stabilized also increases. This implies that for a given time(t), the magnitude of anchor displacement increases with the increase in F/P_u.

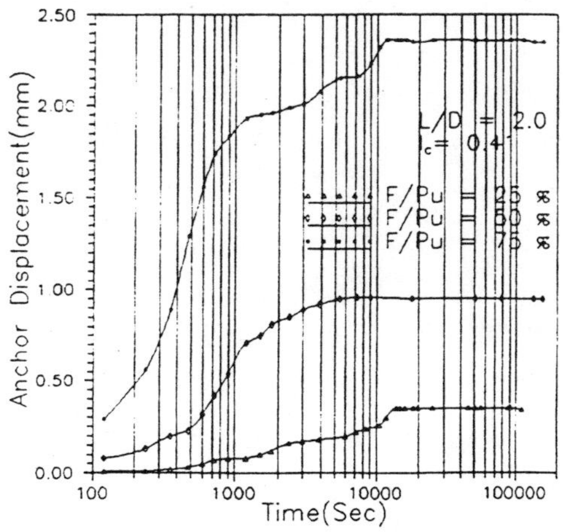

Fig.7. Anchor Displacement Vs Time for different sustained load level(L/D = 2.0).

Further, as the sustained load level(F/P_u) increases, the rate of strain(rate of change of vertical movement with time) increases significantly before the movement gets stabilized. Almost the same trend was observed for L/D of 2.0. In all the sustained load tests, it was observed that even for the maximum sustained load level of 75%, anchor failure in the form of complete pullout did not occur within the observed time frame.

The change in pore water pressure(-ve) at the top of the soil plug and at anchor bottom with time for different sustained load levels are shown in Fig. 8 and 9, respectively. These plots correspond to L/D of 1.0 and an I_c value of 0.4. From both the plots, one can note that the magnitude of suction pressures mobilized for a given F/P_u value stabilizes after a slight decrease with time.

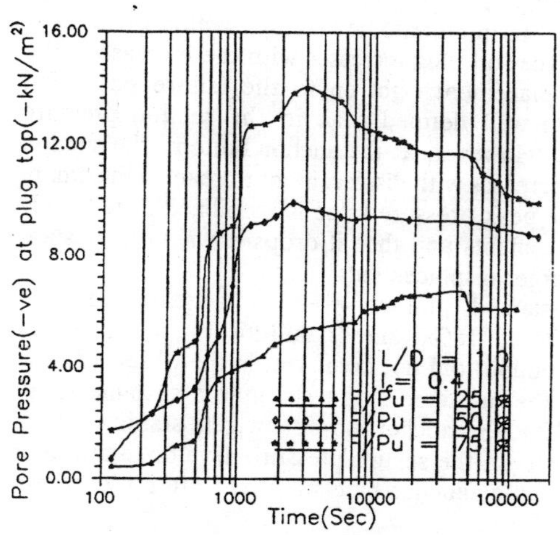

Fig. 8. Variation of suction pressure(-ve) at top of soil plug with time under different F/P_u

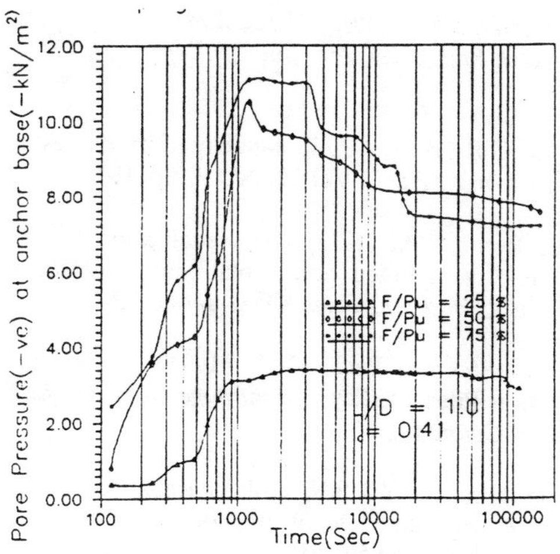

Fig. 9. Variation of suction pressure(-ve) at anchor base with time under different F/P_u

CONCLUSIONS

Based on the results of the present laboratory model testing program carried out the following important conclusions can be drawn:

(1) Displacements corresponding to 12 mm (16 % of anchor diameter) are necessary to mobilize the ultimate pullout capacities in suction anchors under short term loading. This large deformation enables the mobilization of skin friction, suction induced reversed end bearing and the tensile strength of the soil at the anchor base.

(2) In the specific conditions of the present testing program, the soil plug is moved with the anchor up to the failure point and contributes to the breakout capacity. Adequate suction pressure remains developed at the anchor top to hold the soil plug in position.

(3) Under short term monotonic loading, suction pressure at the top of the soil plug increases with the increase in the anchor displacement right up to the failure point and there is no well defined peak in the suction pressure mobilized. However, at the anchor bottom, the suction pressure increases with displacement to reach a maximum value. The peak pressure value is mobilized at relatively low displacements and then it drops down to a constant value at larger displacements.

(4) Under sustained pullout, even at a sustained load level(F/P_u) of 75%, anchor failure in the form of complete pullout did not occur within the observed time frame. In the sustained load tests, upto a sustained load level of 75%, the anchor movement gets stabilized with time. However, the suction pressure developed at top of soil plug and at anchor base stabilized after a decrease with time.

REFERENCES

Albert, L.F., Holtz,R.D and Magris, E (1989). "Super pile system, a feasible alternate foundation for TLP in deep water," *Marine Geotechnology,* Vol.8, pp 135-158.

Davie, J.R. & Sutherland, H.B (1978). "Modelling of clay uplift resistance," *J Geotech.Engng Div. Am. Soc.Civ.Engrs 104,* GT6, 755-760.

Das, B.M. & Seeley, G.R (1982). "Uplift capacity of pipe piles in saturated clay," *Soils and Fdns,* Vol 22, NO 2, pp 91-94.

Finn, W.D.L. and Byrne, P.M (1972). "The evaluation of the breakout force for a submerged ocean platform," *Offshore Technology Conference, Houston, Texas,* OTC paper 1604.

Fuglsang, L. and Steensen-Bach, J.O (1991). "Breakout Resistance of Suction Piles In Clay," *Centrifuge 91, Centrifuge Conference, Boulder, Colorado.*

Narasimha Rao, S., Prasad, Y.V,S.N, and Veeresh, C (1993). "Behaviour of embedded model screw anchors in soft clays," *Geotechnique,* Vol 43, No 4, pp 605-614.

Steensen Bach, J.O. (1992) Recent model tests with suction piles in clay and sand. *Offshore Technology Conference, 6844,* Vol.1, pp 323-330.

Proceedings of the Seventh (1997) International Offshore and Polar Engineering Conference
Honolulu, USA, May 25-30, 1997

ISBN 1-880653-28-1 (Set); ISBN 1-880653-29-X (Vol. I)

An Important Consideration for Seismic-Resistant Design of Offshore Piles

P.M. Rao
Fugro-McClelland Marine Geosciences, Inc.
Houston, TX, USA

M.W. O'Neill
University of Houston
Houston, TX, USA

ABSTRACT

Model-scale experiments were conducted in a vibrated calibration chamber to study the effect of horizontal ground shaking and the coupled effect of vertical motion of the seabed and the induced seaquake independently, on the performance of offshore tension piles driven into sand. Pile response to horizontal ground shaking was studied for various bias tension load levels using typical offshore California-type simulated seismic events of Richter magnitudes 7.0, 7.5 and 8.0. Conditions leading to stability, mobility, and the failure were established for the top 4.5 m of prototype piles based on the results. The seaquake effect was studied for typical M = 8.0 simulated seismic events in deep water (300 m). The results suggested that while horizontal ground shaking may not have a major impact on the degradation of soil strength in prototype piles in saturated sands, seaquakes in deep waters may affect the stability of offshore piles greatly during strong seismic events and therefore should be considered for designing piles.

KEY WORDS: tension, pile, calibration chamber, seaquake, ground shaking, liquefaction, offshore

INTRODUCTION

Vertical motion of the seabed during an earthquake produces waves that propagate through the seawater and impose large loads on structures in the sea and the seabed (Ambraseys, 1985). This phenomenon is referred to as a "seaquake." Seaquakes may induce changes in the pore water pressure in soils underlying the sea due to coupling of compression waves into the sea water and may consequently influence the stability of offshore piles.

A model-scale laboratory investigation was conducted to study the mechanism of the development of excess pore water pressure in the seabed due to simulated seaquake loading. Additionally, the relative effects of seaquake loading of the seabed and the horizontal component of seabed shaking on the ability of piles to sustain biased tension loads were investigated.

Model piles were driven into a calibration chamber containing saturated siliceous sand and loaded axially through a spring-mass system (representing the superstructure) to a known percentage of the static capacity of the pile. The horizontal ground shaking tests consisted of exciting the base of the soil column with the horizontal component of acceleration records of simulated California-type seismic events (scaled to simulate diffusion rates) to determine the effect of magnitude of biased load and earthquake magnitude (M = 7.0, 7.5 and 8.0) on pile response. The seaquake effect on piles was studied for a typical M = 8.0 simulated event in water depths of about 300 m. The hydrodynamic pressure exerted on the seabed for these conditions was computed analytically and applied on the surface of the soil column in the chamber by mechanical means.

PHYSICAL MODELING

Model-Prototype Scaling

In order to make physical interpretations of the test results and to scale the behavior of piles from model to prototype, static and dynamic scaling factors for various soil and pile parameters were established for the model testing conditions based on similitude criteria and dimensional analysis. The scaling rules that were employed in this study are given in Table 1. Scaling rules were similar to those for centrifuge testing. However, it is not possible to model the gradient in vertical stress with the pressure chamber, so the stresses that were modeled were those at a depth of 2/3 of the length of the pile in the layer of soil considered for the prototype. This stress, when mutiplied by the perimeter area of the penetrating portion of the pile, is proportional to the uplift resistance of both the prototype and model piles under the assumption that the vertical effective stress gradient in the seabed is linear and the resulting distribution of horizontal effective stress against the perimeter of the prototype pile is triangular. Ochoa (1990) provides an in-depth derivation of the scaling factors. In this paper the penetration of the prototype pile was taken as 4.5 m or , alternatively, the results may be viewed as applying to the top 4.5m of a pile penetrating deeper.

Dynamic scaling was employed to model the inertial effects and diffusion of excess pore water pressure. These scaling factors were derived using dimensional analysis assuming a stress scaling factor of 1, a scaling factor for density and viscosity of the soil and fluid of 1 and that the prototype velocity is preserved in the model. To scale length and thereby diffusion it was necessary to scale time and hence frequency. The time scaling factor for diffusion ($1/n_d$) is maintained equal to the time scaling factor for inertia if $k_p/k_m = n_d$. Ideally, the factor n_d should have been equal to the static length scaling factor obtained from the effective stress similitude described above. However, due to limitations of the servo-hydraulic testing machine, the highest possible value of n_d was 5.2, which was used. For the model study, it is also assumed that the permeability of the soil in the model, k_m (a very fine sand) is 1/5 the permeability of the prototype sand, k_p. Therefore, it was observed that n_d was 5.2 and n_s was 15 for the 4.5 m-long pile. This distortion was not considered serious enough to have a significant impact on the results.

For proper modeling of sea-seabed interaction phenomena (seaquake tests), both Froude and Mach (diffusion) similitude must be satisfied (Yamamoto, 1983). There exists an uncorrectable distortion between the frequency scaling factor for Mach and Froude similitude. A mean value of the two scaling factors equal to 4 was selected for the model seaquake study to scale the inertia and the diffusion effects for the 4.5-m-long pile. It must be understood that for modeling these effects for longer piles, the scaling factors in Table 1 stand to be modified.

Table 1: Model Test Scaling Factors [Ochoa, (1990)]

Parameter	Symbol	Dimension-less Group	Similarity Requirements = Model/Prototype
		STATIC	
Length	L		$N_L = 1/n_s$
		DYNAMIC	
Length	L		$N_L = 1/n_d$
Velocity	v		$N_v = 1$
Time (inertial terms)	t_1		$N_t = 1/n_d$
Frequency (Froude Similitude)	f	$4\pi^2 L/g$	$N_f = \sqrt{n_d}$
Frequency (Mach Similitude)	f	$4p^2\sigma k_s/L\eta g$	$N_f = n_d$
Acceleration	a		$N_a = n_d$
Soil Density [a]	ρ		$N_\rho = 1$
Liquid Density[a]	ρ_λ		$N\ \rho_\lambda = 1$
Particle Size	d	d/L	$N_d = 1/n_d$
Mass	m		$N_m = 1/n_d^3$
Force	F	F/ma	$N_F = N_m N_a = 1/n_d^2$
Viscosity	η		$N_\eta = 1$
Permeability	k	$k\eta/d^2\ \rho_\lambda a$	$N_k = N_d^2 N\ \rho_\lambda N_a N\eta^{-1} = 1/n_d$ $N_k = 1$ for 1-g cond.
Stress	σ	$\sigma L^2/F$	$N_\sigma = N_F N_L^{-2} = 1$
Strain	ε		$N_\varepsilon = 1$
Coefficient of Consolidation	c_v	$c_v \rho g m_v/k$	$Nc_v = N_k N_\rho^{-1} Nm_v^{-1} = 1/n_d$
Time (in diffusion terms)[a,b]	t_2	tc_v/L^2	$N_t = N_L^2 Nc_v^{-1} = 1/n_d$

(a) Model (m) amd Prototype (p) soil and pore fluid are the same

(b) From differential equation, $\partial u/\partial t = c_v(\partial^2 u/\partial L^2)$, $N_\sigma/N_t = Nc_v(N_\sigma/N_L^2) = N_t = N_L^2 Nc_v^{-1}$

(c) k_s = intrinsic permeability of the soil

Testing Assembly for Horizontal Shaking Tests

Experimental simulation was performed in a calibration chamber 0.53 m in height and 0.51 m in diameter. The steel chamber was equipped with lateral rubber membranes inside its sidewalls and a bladder beneath its top cap for applying the lateral and the vertical stress, respectively, to the soil deposited in the chamber. In addition, the top plate used to cover the chamber had ports for pile insertion and drainage during shaking. Drainage was permitted only at the top. The soil sample was prepared in the chamber by depositing very fine sand at an average dry density of 16.85 kN/m^3 using the pluvial (raining) technique. The properties of the sand used are tabulated in Table 2. The soil mass was then purged with carbon dioxide gas and saturated with deaired water.

Table 2: Properties of Sand

Parameter	Value
γ_{dry}	16.85 kN/m^3
$\gamma_{dry\ (max)}$	18.12 kN/m^3
$\gamma_{dry\ (min)}$	15.51 kN/m^3
D_{10}	0.107 mm
Permeability	3×10^{-2} cm/sec @ e = 0.54
Angle of Internal Friction (ϕ_d)	36.5 degs
Pile-Soil Interface Friction Angle (δ)	27 degs
Shear Modulus (G) @ 10^{-5} % strain at effective confining stress = 17.24 kPa	7×10^4 kPa

The model pile used for the study was a closed-ended steel pipe, 25.4 mm in diameter and 405 mm in length, with a wall thickness of 1.2 mm. The pile was driven by impact into the saturated soil mass, which had been consolidated under an isotropic effective stress of 17.24 kPa, representing the effective stress at the two-thirds depth around a 4.5-m-long by 380-mm-diameter prototype pile. The pile was driven offset from the center of the chamber with a steel cylinder placed in the center of the chamber so that near-uniform strain gradients were developed around the pile during the application of simulated ground motion through rotational motion at the base of the chamber. Teflon sheets were placed inside the membranes and against the internal cylinder to minimize shear at the boundaries of the sand column.

The pile instrumentation consisted of a single level of strain gages mounted on the exterior pile wall at its head to sense the axial load. The soil instrumentation consisted of miniature pore water pressure transducers located at the mid-depth of the pile, one in the "near-field," 1 diameter from the pile wall, and the other in the "far-field," 14 diameters from the pile wall. An LVDT was also mounted on the top plate of the chamber to monitor the relative movement between the pile-head and the chamber during the shaking event.

Biased tension load was applied on the pile-head by means of a dead weight and spring system during the seismic loading. A cable was attached to the pile-head that passed through frictionless pulleys, one supported by the frame of the testing machine and the other supported by an independent gantry. The other end of the loading cable was attached to springs and a container containing lead shot, which provided the biased load. The spring-mass system had a natural period of 2 sec, which was above the longest period in the wave train for the simulated seismic events selected for this study.

Accelerograms for two California seismic events, the Upland event of February 28, 1990, and the Oceanside event of July 13, 1986, were available at a deep sand site, offshore from Long Beach, California, maintained by the Minerals Management Service of the U.S. Department of the Interior. Both the earthquakes had a Richter magnitude of about M = 5.0 (Hauksson, 1988; Hauksson, 1991) and both were 75 km from the observation point. The Fourier amplitude spectra of the horizontal component of ground motion for both events were scaled to match a "target" spectrum that would represent a spectrum for a higher magnitude event for that region (e.g., M = 7.0, 7.5, and M = 8.0). Scaled spectra were generated for M = 7.0, 7.5 and 8.0 simulated events for each of the two earthquakes (Ochoa 1990). The scaled spectra were then transformed back to the time domain and integrated twice to obtain displacement-time histories for simulated M = 7.0, 7.5 and 8.0 events. The displacement and time axes of the simulated events were scaled by multiplying by a factor of 1/5.2 to model the diffusion characteristics of the pore water.

The seismic motion for the horizontal shaking tests was generated by applying these programmed time histories to the base of the chamber through the rotary motion of the actuator of a closed-loop servo hydraulic testing machine. The rotary motion was such that the offcentered pile experienced a tangential displacement-time history equal to the desired linear displacement-time history.

Hydrodynamic Seaquake Pressure

The hydrodynamic pressure, P_d induced on the seabed due to underwater propagation of compressional waves during vertical excitation of the seabed can be obtained as follows:

$$P_d = -\rho \frac{\partial \Phi}{\partial t} \quad (1)$$

where, ρ = density of seawater, and Φ is the seawater velocity potential evaluated on the seabed. The seawater velocity potential is obtained by solving the governing wave equation in the flow field for a known set of boundary conditions. The seawater is assumed to be an inviscid, compressible and irrotational fluid. The wave reflection or scattering due to the presence of structures in the sea is neglected. A one-dimensional physical representation of the analytical model is shown in Figure 1.

The simulated displacement excitation function (w) of the seabed is idealized as a harmonic function and is expressed as follows:

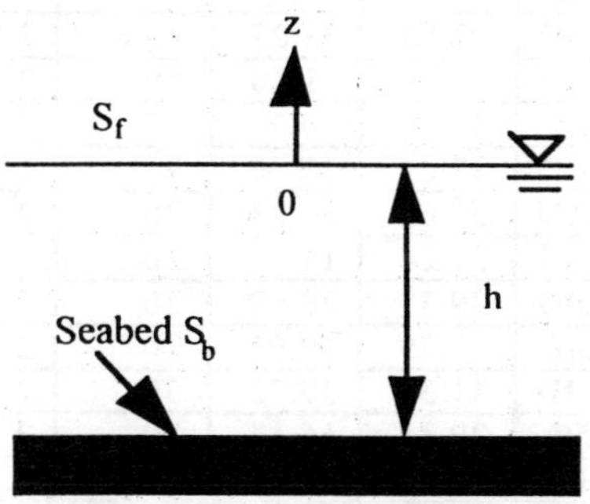

S_b - Seabed surface undergoing vertical excitation
S_f - Surface of the free water

Figure 1: Physical Representation of the Analytical Model

$$w = We^{i\Omega t}, \quad (2)$$

where, W is the displacement amplitude, Ω is the frequency of excitation and t is the time.

The seawater velocity potential is a function of both space and time. Therefore, it can take the form

$$\Phi(z,t) = \phi(z)e^{i\Omega t}, \quad (3)$$

where ϕ is the seawater velocity potential amplitude, Ω is the seafloor displacement excitation frequency, t is the time and e is the Naperian logarithm base.

The governing equation for the flow-field in one-dimensional form can be written as

$$\frac{\partial^2 \Phi}{\partial t^2} = c^2 \frac{\partial^2 \Phi}{\partial z^2}, \quad (4)$$

where c is the underwater acoustic speed or the wave velocity of the compressional wave in water. Alternatively, using equation 3, equation 4 reduces to the following form

$$\frac{\partial^2 \phi(z)}{\partial z^2} + \frac{\Omega^2}{c^2}\phi(z) = 0. \quad (5)$$

The boundary conditions on $\phi(z)$ are as follows:

$$\phi = 0, \text{ on } S_f \text{ (disregarding surface waves)} \quad (6)$$

$$\frac{\partial \phi}{\partial z} = V_b, \text{ on } S_b, \quad (7)$$

where V_b is the velocity amplitude of vertical seafloor vibration = $i\Omega W$.

A suitable form for ϕ that satisfies equations 6 and 7 has been given by Williams and Mau (1988) as

$$\phi(z) = \frac{i\Omega W}{k}\frac{\text{Sinkz}}{\text{Coskh}} \quad (8)$$

where, k is the wave number = Ω/c.

On the seabed, i.e., z = -h, the dynamic pressure generated, is given by

$$P_d = \rho\Omega^2 \frac{Wc}{\Omega}\tan\left(\frac{\Omega h}{c}\right)e^{i\Omega t} \quad (9)$$

Kobayashi et al. (1992) have provided the solution for velocity potential for a general two-dimensional problem, considering only a portion of the seabed undergoing vertical excitation (fixed length wave source). Idealized sinusoidal hydrodynamic pressures induced on the seabed due to seaquakes were computed for a typical M = 8.0 simulated seismic event, viz., the Oceanside event. Figure 2 shows the displacement-time history for the M = 8.0 simulated Oceanside event. The boxed portion of the time window was used to compute the equivalent harmonic displacement-time history to represent the irregular vertical displacement-time history shown in the figure. Two predominant frequencies can be observed:

(a) Low frequency,high amplitude signal: Ω = 0.12 Hz = 0.766 rad/sec; W = 18.5 mm.
(b) High frequency,low amplitude signal: Ω = 1 Hz = 6.28 rad/sec; W = 1.25 mm.

Letting c = 1430 m/sec and ρ = 1000 kg/m^3, the amplitudes of the hydrodynamic pressure induced on the seabed for typical water depths ranging from 200 to 300 m were computed and tabulated in Table 3 using the solutions provided by Williams and Mau (1988) and Kobayashi et al. (1992). The higher set of values was used.

Testing Mechanism for Seaquake Tests

The chamber configuration was modified in order to develop a mechanism for application of the hydrodynamic pressure on the surface of the sand confined in the calibration chamber. The sand was deposited at a controlled dry density into the chamber by raining and saturated with deaired water. Unlike the arrangement for the horizontal shaking tests, the sand was not deposited around the cylinder at the center of the chamber. A bed of coarse sand about 25 mm thick was placed on the top of the saturated bed of fine test sand so that the applied hydrodynamic pressure could be transmitted uniformly to the top surface of the test sand. The sand was then consolidated by applying the desired controlled stresses.

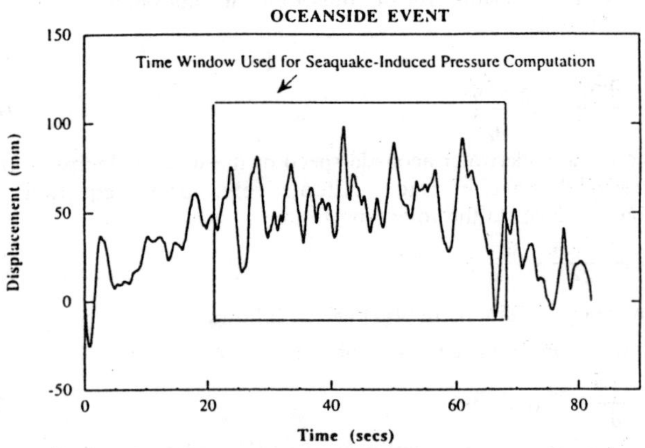

Figure 2: Displacement-Time Record of the Vertical Component of Scaled M = 8.0 Oceanside Event

A water compartment made up of two aluminum flanges with a rubber sleeve bonded to their ends was then fixed to the top plate of the calibration chamber. The top flange of this compartment had watertight openings through which cables for soil instrumentation were passed to the data acquisition system. The compartment was filled with glycerine over water. The procedure for the seaquake tests consisted of applying a prescribed hydrostatic pressure to the water confined in the compartment. This was accomplished by applying air pressure from the pressure panel to the glycerine-water column. The sinusoidal variation in the hydrodynamic pressure was superimposed on the hydrostatic pressure. It was developed by mechanically stroking water confined in the water compartment by cyclic motion of the calibration chamber mounted on the actuator of the servo-hydraulic testing machine, while the top flange of the water compartment was held in place by two reaction beams attached to the loading frame of the testing machine. The amplitude of the cyclic motion was 1-2 mm. depending upon the test condition, which was within the range of the scaled vertical displacements of the two components of the M = 8.0 simulated event. This ensured that the soil velocities developed in the model were similar to the seabed velocities during the earthquake.

The hydrodynamic pressures on the sand surface ranged from 14 kPa to 50 kPa, representing seaquakes during M = 8.0 seismic events in water depths ranging from 225 m to 300m. The soil instrumentation consisted of three miniature pore pressure transducers: two were placed along the center of the chamber in the soil at 1/3rd and 2/3rd depths in the chamber, i.e., 178 mm and 356 mm from the top of the chamber, to obtain the pore water pressure chamber distribution across the depth of the soil, and the other was mounted on the top plate of the calibration chamber within the water compartment to monitor the applied hydrodynamic pressure on the sand surface. Drainage is a critical issue in these tests. The mechanism of applying the hydrodynamic pressure being a closed system, drainage at the top was only possible during the unloading cycle of the applied pressure, which is a reasonable idealization of the prototype seabed condition during seaquakes.

The testing program is summarized in Table 4. Several tests were conducted in the absence of the pile to understand the mechanism of development of pore water pressure in the soil during seaquakes and the influence of parameters such as the predominant frequency of the vertical excitation, number of cycles of loading, and the initial effective stress in the soil. Two types of boundary conditions were used: (1) Tests with the application of only vertical overburden with rigid lateral boundaries simulated the one-dimensional, zero lateral strain conditions that exist in the field during seaquakes, and (2) Tests with equal vertical and lateral effective stresses at the boundary (identical to the ground shaking test). Model tests were conducted at scaled frequencies of 2.5, 5.0 and 10.0 Hz, which represented approximate prototype frequencies of about 0.6 Hz (mean value for M = 8.0 simulated event), 1.2 Hz, and 2.5 Hz (scaling factor = 4).

Table 3: Seaquake-Induced Pressure Amplitudes on the Seabed for Typical Water Depths

Water Depth, m	Low-Frequency High-Amplitude Signal Component Pressure Amplitude , kPa		High-Frequency Low-Amplitude Signal Component Pressure Amplitude, kPa	
	Williams and Mau (1988)	Kobayashi et al. (1992)	Williams and Mau (1988)	Kobayashi et al. (1992)
200	2.18	1.77	13.57	11.7
225	2.46	2.00	17.05	15.0
250	2.73	2.22	21.96	19.77
275	3.00	2.45	29.55	27.36
300	3.30	2.67	43.37	41.63

Pile response to the applied seaquake was studied for a mean effective stress of 17.24 kPa, and the frequency of the hydrodynamic pressure was 2.5 Hz. The details of model pile instrumentation, pile installation, and the mechanism of bias loading were similar to the horizontal ground shaking tests. The mechanism of the generation of the seaquake loading on the sand surface was similar to the sea-seabed interaction tests without the pile, and the duration of the tests was 16 sec, representing a duration of about 82 secs in the prototype. The testing arrangement is illustrated in Figure 3. The applied hydrodynamic pressure, pore water pressures in the soil mass, pile-head movement and pile-head load were recorded during the pile test.

Table 4: Summary of the Seaquake Tests

Test No.	σ'_v kPa	σ'_h kPa	p_s kPa	p_d kPa	t_d secs	f Hz
SS#1	0	RIGID	42.83	38.27	20	2.5
SS#2[a]	41.38	RIGID	43.00	41.30	20	2.5
SS#3	17.24	17.24	31.03	30.34	25	2.5
SS#4[b]	124.14	RIGID	37.24	33.52	9	2.5
SS#5[b]	124.14	RIGID	39.62	20.28	20	2.5
SS#6[b]	124.14	RIGID	37.24	14.23	20	10.0
SS#7[b]	124.14	RIGID	39.31	30.62	20	5.0
SS#8[b]	124.14	RIGID	53.79	49.65	20	2.5
SS#9	0	RIGID	41.38	39.72	50	2.5
SS#10	27.59	27.59	29.11	16.14	20	2.5
PS#2[a,c]	41.38	RIGID	31.03	28.97	20	2.5

σ'_v - initial vertical effective stress; σ'_h - initial lateral effective stress; p_s - applied bias hydrostatic pressure amplitude; p_d - applied hydrodynamic pressure amplitude; t_d - duration of seaquake; f - frequency of hydrodynamic pressure

a - Mean effective stress in the sample = 17.24 kPa

b - Mean effective stress in the sample = 51.72 kPa

c - Test conducted to study pile response

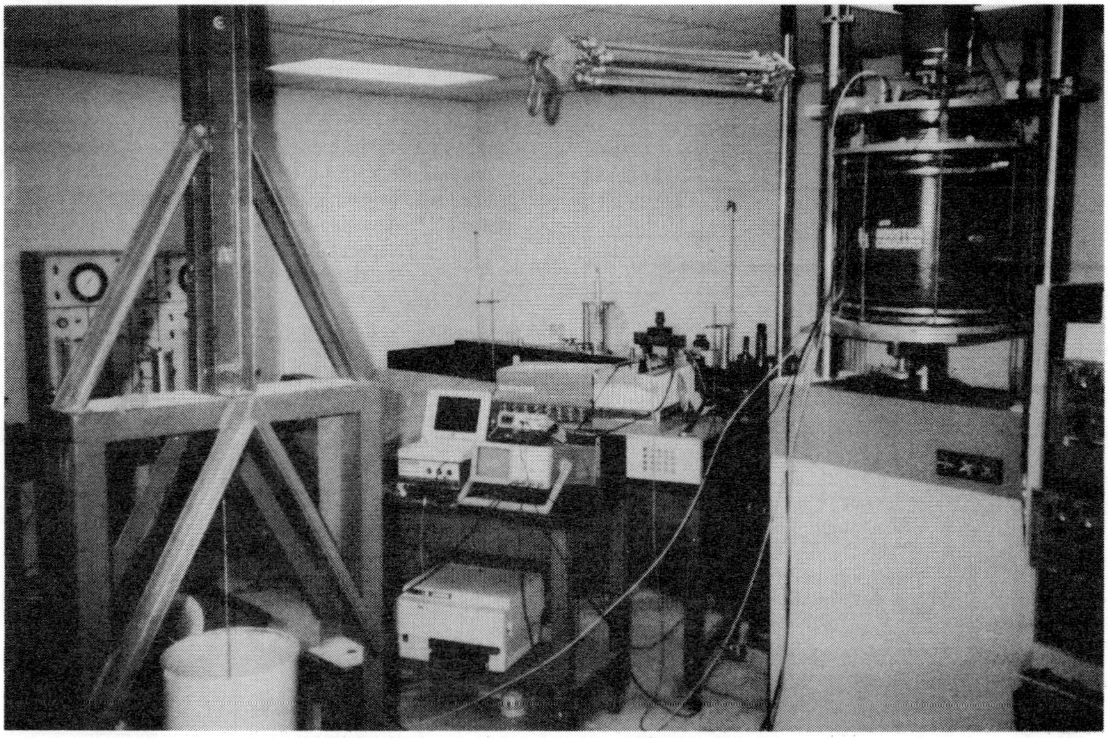

Figure 3: Testing Arrangement for the Seaquake Tests

PILE PERFORMANCE DURING HORIZONTAL GROUND SHAKING

Horizontal ground shaking tests were performed on the tensioned pile for various combinations of bias loading levels (45% to 90% of the static capacity) and magnitude of the simulated events (M = 7.0,7.5, and 8.0) (Ochoa, 1990; Rao et al., 1997). A summary of results of selected tests is presented in Table 5. Based on the experimental results, conditions were established for stability (sustained load and small pile movement), mobility (sustained load associated with substantial pile movement) and failure (extraction) for closed-ended driven piles in medium-dense saturated sand, loaded with a biased tension load. A movement of 0.15 mm (scaled movement = 2.3 mm) was established as an upper bound to pile stability. Piles that moved less than this amount vertically were considered stable, and those that moved in excess of 0.15 mm but were not completely extracted represented a "mobility" condition between stability and failure. There existed a threshold bias load below which the piles remained stable for each of the different magnitude events. Failure occurred at relatively lower levels of bias loading (about 75% of static capacity) during stronger events (M = 8.0) primarily as a result of degradation in soil strength due to the excess pore water pressure buildup (about 20% of initial effective stress) and the load due to the superstructure feedback on the pile during the seismic event (about 10% of the static uplift capacity). The pore pressures that developed in the soil mass during the M = 7.0 and M = 7.5 simulated events were not more than 4% of the ambient mean effective stress. As a result, extraction occurred only at 90% or higher bias loading for both events of these magnitudes.

SOIL AND PILE RESPONSE DURING SEAQUAKES

Mechanism of Pore Water Pressure Response

The seaquake tests consisted of imposing a periodic water pressure on the surface of the sand bed. Figure 4 shows a typical example of the seaquake-associated seabed pressure and the oscillatory pore pressures recorded at the 1/3rd and 2/3rd depths in the calibration chamber. The figure clearly shows that the amplitude of the seaquake-associated bottom pressure, p_d, attenuates from 49.65 kPa to about 36.55 kPa with depth. Additionally, the peak value of the pore water pressure generated in the soil mass shifts to the right, indicating a phase lag between the applied hydrodynamic pressure and the oscillatory pore water pressure.

The phenomenon of propagation of seaquake-induced pressure in the soil mass is governed by several factors, such as compressibility of water, compressibility of the soil skeleton, predominant frequency of the vertical excitation of the seabed and the diffusion characteristics of the sand. The influence of duration of the seaquake or the number of cycles of seaquake loading, predominant frequency of vibration, initial effective stress in the soil, and boundary condition of the imposed stresses on the phenomenon of propagation of seaquake-induced pore water pressure during simulated seismic events of Magnitude 8.0 was studied through parametric tests summarized in Table 4.

Figures 5,6,7, and 8 illustrate the attenuation more clearly. The oscillatory pore pressure, p_m, is normalized by the amplitude of the seafloor pressure, p_d. The effect of number of cycles of seaquake loading on the variation of pressure ratio is shown in Figure 5 for 2.5 Hz excitation frequency for the rigid lateral boundary tests. The number of cycles of loading had very little influence on the degree of attenuation. Figures 6 and 7 demonstrate the variation of pressure ratio as a function of initial vertical effective stress for a rigid lateral

Table 5: Summary of Results of Selected Horizontal Shaking Tests

No.	M	R.S.C. (kN)	B.L. (%)	P.S.C. (kN)	P.H.M (mm)	C.P.E.
UP#1	7.0	0.56	50	0.55	0.006	Stability
OS#1	7.0	0.49	46	0.47	0.06	Stability
OS#2	7.0	0.46	65	0.47	0.05	Stability
UP#4	7.0	0.62	75	0.59	0.17	Mobility
UP#7	7.0	0.52	85	0.48	0.30	Mobility
R2[a]	7.0	0.49	89	0.46	1.14	Mobility
UP#8	7.5	0.48	60	0.45	0.10	Stability
UP#11	7.5	0.45	80	0.39	0.53	Mobility
S4[a]	7.5	0.47	91	0.00	pull out	Failure
U1[a]	8.0	0.35	45	0.32	0.05	Stability
UP#16	8.0	0.64	61	0.56	0.99	Mobility
U2[a]	8.0	0.51	64	0.43	0.25	Mobility
UP#18	8.0	0.46	78	0.00	pull out	Failure

M = Event Magnitude; R.S.C. = Reference Static Capacity; B.L. = Applied Bias Loading as a Percentage of Static Capacity; P.S.C.= Post Shaking Capacity; P.H. M = Pile-Head Movement During Shaking (mm); C.P.E. = Condition of Pile Extraction; OS = Oceanside; UP = Upland; a = Tests conducted with the Oceanside event (Ochoa, 1990)

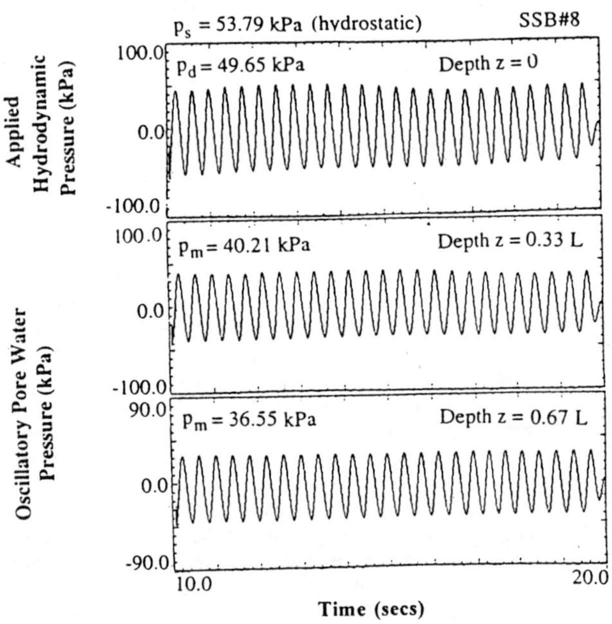

Figure 4: Pore Water Pressure Response

boundary condition and as a function of initial mean effective stress for a controlled stress lateral boundary, respectively. As is evident by comparing the figures, the initial effective stress in the soil did not have a major impact on the degree of attenuation for either type of boundary condition. Typically, the pressure at z/L = 0.67 ranged from 60% to 80% of the applied hydrodynamic pressure on the seabed for the rigid lateral boundary tests and was about 90% for the flexible lateral boundary. Therefore, it can be said that the compressibility of the soil skeleton may not have a significant influence on the degree of attenuation since the flexible boundary allows for more compression in the soil during application of hydrodynamic pressure than a rigid boundary. The predominant frequency of vertical excitation was the most obvious factor that correlated to attenuation and phase lag of the oscillatory pore pressure in the seabed (Figure 8). Doubling the frequency from 2.5 Hz to 5.0 Hz increased the attenuation by about 30%, and increasing the frequency to 10 Hz increased the attenuation further by about 40%.

No definitive correlation could be established between the phase lag and the initial effective stress condition of the sand. The variation of phase lag was within a narrow band ranging from 18 degrees to 24 degrees for tests conducted at an excitation frequency of 2.5 Hz with a rigid lateral boundary under varying initial vertical effective stress (See Figure 9).

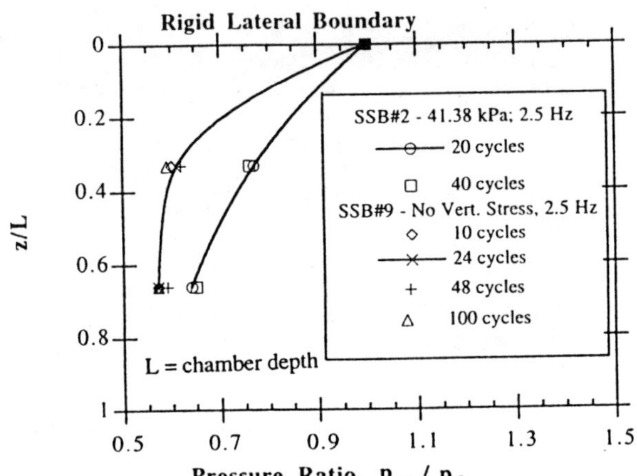

Figure 5: Variation of Pressure Ratio with Number of Cycles of Seaquake Loading for a Rigid Boundary Test Condition

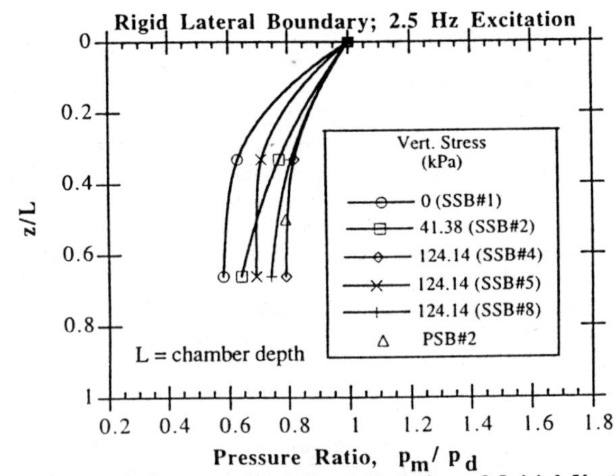

Figure 6: Pressure Ratio as a Function of Initial Vertical Effective Stress for a Rigid Boundary

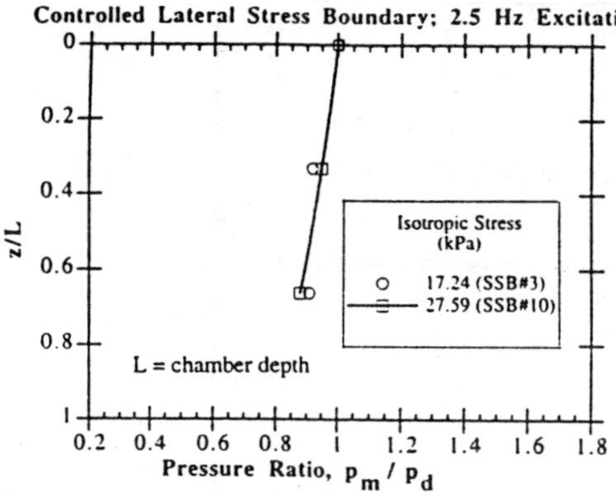

Figure 7: Pressure Ratio as a Function of Initial Mean Effective Stress for a Controlled Lateral Stress Boundary

Effective Stress Oscillation

The effective stress in underwater deposits of soil is held constant if the external water level, and the consequent total stress, is changed. However, as evident from the preceding section, this is not true when the boundary water pressure varies so quickly that the change of pressure on the surface of the deposit is not experienced uniformly throughout the deposit. This introduces a change in the effective stress in the soil during seaquake loading. At any given point in the soil mass, the change in effective stress must be equal to the difference between the applied hydrodynamic pressure, P_d and the measured pore water pressure, P_m in the soil mass, assuming no side wall shear effects. This difference represents the excess pore water pressure in the soil mass. Since the applied seaquake loading is periodic in nature, the induced excess pore water pressure component is also periodic. If the excess water pressure reaches or exceeds the ambient effective stress in the soil mass, the soil skeleton will instantaneously assume a liquefied state at that point. Conversely, if the excess pore pressures become less than the ambient effective stress in the soil mass, the vertical effective stress at that point attains values greater than on the surface. The increase in effective stress exerts compressive forces on the soil skeleton that can potentially densify the sand. The excess pore water component for a typical test conducted at a frequency of 2.5 Hz is shown in Figure 10. It is obvious that at times of 0.03 to 0.23 sec after the start of a given loading cycle, a decrease in effective stress can be expected and an increase in effective stress will occur in the time domain outside of the above range. The peak positive excess pore water pressure was about 85% of the initial mean effective stress of 17.24 kPa, therefore, a degradation in shear strength of about 85% could be expected instantaneously at the indicated depth.

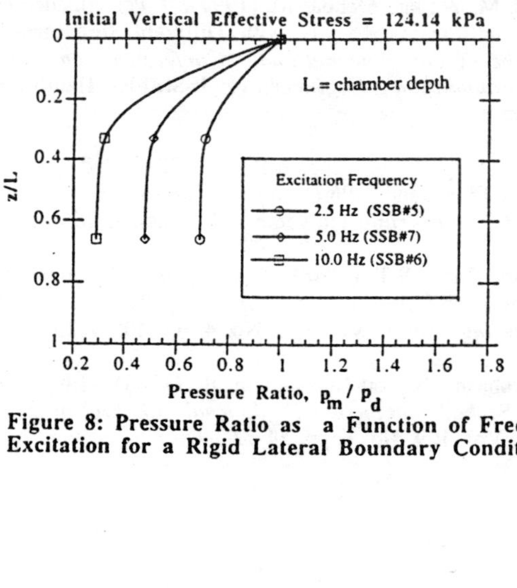

Figure 8: Pressure Ratio as a Function of Frequency of Excitation for a Rigid Lateral Boundary Condition

PILE BEHAVIOR DURING SEAQUAKE LOADING AND ITS COMPARISON TO HORIZONTAL GROUND SHAKING

Figure 11 shows the pile response during seaquake loading. The pile in this case was loaded in tension to about 45% of the static capacity. As expected from sea-seabed interaction tests discussed earlier, about 80% of the seaquake-associated pressure was transferred as pore water pressure in to the soil mass at about 50% of the pile depth. The pile failed within two cycles of seaquake loading (about 5 sec in the prototype). The instantaneous degradation in soil strength during the times in the loading cycles in which positive excess pore water pressure was established close to the pile caused the pile to ratchet upwards, consequently leading to complete extraction of the pile. For comparison, Figure 12 shows the result of a horizontal ground shaking test under similar conditions in which the pile was loaded to about 60% of the static capacity. The residual excess pore water pressure developed in the soil during horizontal ground shaking was much less than the peak value of excess pore water pressure developed during the seaquake loading. Based on the pile-head movement during the shaking, the pile response was classified as a mobility condition, which demonstrates that the seaquake loading produced more severe effects than the horizontal ground shaking, and therefore needs serious consideration for offshore pile design.

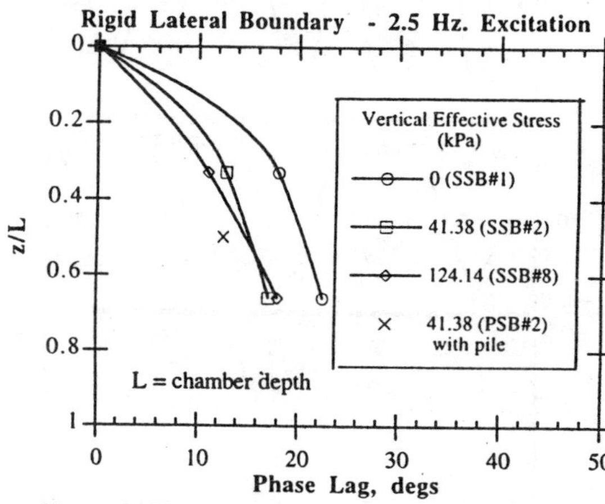

Figure 9: Phase Lag as a Function of Initial Vertical Effective Stress for a Rigid Lateral Boundary

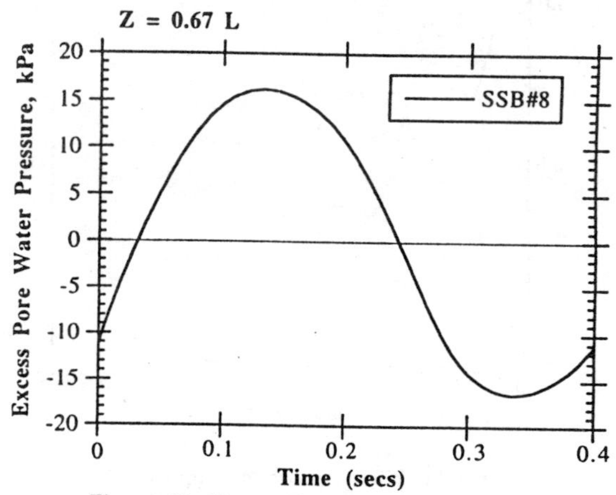

Figure 10: Excess Pore Water Variation During Seaquake Loading - SSB#8

CONCLUSIONS

(i) The seaquake induced in the overlying seawater during the vertical excitation of the seabed in an earthquake generates oscillatory excess pore pressure in the seabed, where the seaquake-associated seabed pressure is propagated into the seabed with attenuation and phase lag.

(ii) The degree of attenuation and the phase lag are greatly influenced by the predominant frequency of vibration of the seabed.

(iii) The mechanism of generation of excess pore pressures in the seabed during horizontal ground shaking is different from that during seaquake loading, with the latter being more severe.

(iv) While horizontal ground shaking may not have a major impact on the degradation of soil strength in prototype piles in saturated sands, seaquakes in deep water may affect the stability of offshore piles greatly and therefore need consideration in the design of offshore piles.

(v) Although the experiments were conducted on short piles, the results apply qualitatively to longer piles more typical of offshore structures.

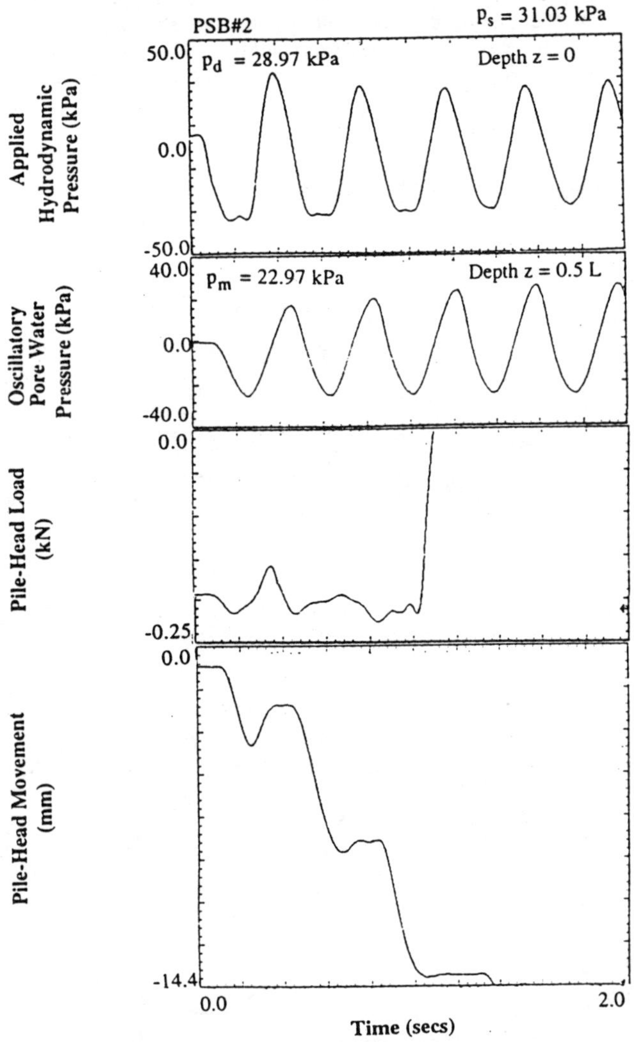

Figure 11: Pile Response to Seaquake - PSB#2

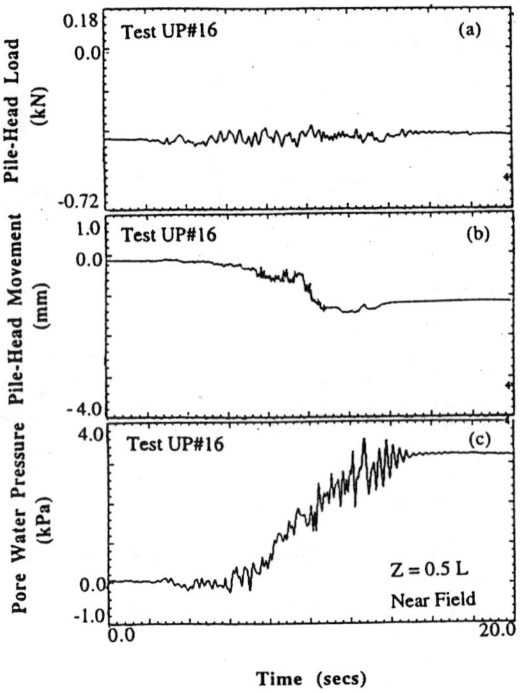

Figure 12: Pile Response to Horizontal Ground Shaking - UP#16

ACKNOWLEDGEMENT

This study was supported by a contract from the Minerals Management Service.

REFERENCES

Ambraseys, N. (1985). "A Damaging Seaquake," Short Communication, *Earthquake Engineering and Structural Dynamics*, Vol. 13, pp. 421-424.

Hauksson, E., and Jones, L. M., (1988). "The July 1986 Oceanside (M_L = 5.3) earthquake sequence in the continental borderland, Southern California," *Bulletin of the Seis. Soc. of America*, Vol. 78, No. 6, pp. 1885-1906.

Hauksson, E., and Jones,. L. M., (1991). "The 1988 and 1990 Upland Earthquakes: left-lateral faulting adjacent to the central transverse ranges," *Journal of Geophysical Research*, Vol. 96, No. B5, 1991, pp. 8143-8165.

Kobayashi, H., Sakuta, M., Okamoto, O., and Hirakawa, S. (1992). " Safety consideration of the TLP on earthquake wave propagated through seawater," *Proceedings of the 20th Annual OTC*, Paper 6888, Vol. 2, pp. 139-145.

Rao, P. M., O'Neill, M. W., and Ochoa, M. (1997). " Performance of Tension Piles During Seismic Events with Different Signatures," *Proceedings of the XIVth International Conference on Soil Mechanics and Foundation Engineering*, ISSMFE, Hamburg, Germany, September.

Ochoa, M., (1990). "Response of Tension Piles to Simulated Seismic Motion in Saturated Fine Sand," *Ph.D. thesis, Department of Civil and Environmental Engineering*, University of Houston, 286 pp.

Williams, A. N., and Mau, S-T, (1988). "Earthquake Response of Submerged Circular Arch," *Journal of Waterways, Ports, Coastal, and Ocean Engineering*, ASCE, Vol. 114, No. 4, pp. 405-422.

Yamamoto, T., Takahashi, S., and Schuckman, B., (1983) "Physical Modeling of Sea-Seabed Interactions," *Journal of Engineering Mechanics*, ASCE, Vol. 109, No. 1, pp. 54-72.

Proceedings of the Seventh (1997) International Offshore and Polar Engineering Conference
Honolulu, USA, May 25-30, 1997

ISBN 1-880653-28-1 (Set); ISBN 1-880653-29-X (Vol. I)

Horizontal Impact Loads on Large Bored Single Piles in Sand

Thomas Grundhoff
Ruhr-University Bochum
Germany

ABSTRACT: A series of centrifuge model tests are carried out to study the bearing behaviour of long flexible piles subjected to horizontal impact loads. The piles are modelled along with a fine grained sand. As model piles, aluminium pipes with an external diameter of 30 mm and a length of 695 mm are used. A free-head and a fixed-head supporting system for the pile head are considered. In these tests pile displacements at its top level, at the ground level and at its tip are measured. In addition, the applied load and the bending moment distribution along the embedded pile length as well as the wave propagation within the surrounding soil are recorded. In this paper the results of one free-head and one fixed-head pile test concerning the deflection and the bending moment distribution are presented. The time history response of both systems and the bearing behaviour of the pile-soil interface are found to be absolutely different.

KEYWORDS: Centrifuge, Collision, Dynamics, Impact, Lateral load, Piles

1 INTRODUCTION

The design and the performance of pile foundations under extreme dynamic loadings has a major influence on the feasibility, design, cost and reliability of a construction. Foundation performance consists of the ability of a system to sustain imposed loadings and maintain deformations within tolerable limits. Large bored piles constructed for structures such as, berthing structures, harbour dolphins and bridge piers in the environs of traffic- or waterways will serve not only as a structural element for the portion projected over the ground surface but also as a foundation to the structure. Concerning such pile constructions, the loading case due to horizontal impact has to be taken into account for the determination of the bearing behaviour of the pile-soil interface (Figure 1).

The current state-of-practice for the design of horizontal impact loaded piles is fundamentally empirical. The theoretical and experimental background of existing design methods does not allow to arrive at pile sizes such as, diameter and length with respect to surrounding soil conditions. In principle the design does not consider the resulting effects due to the high magnitude of impact loads.

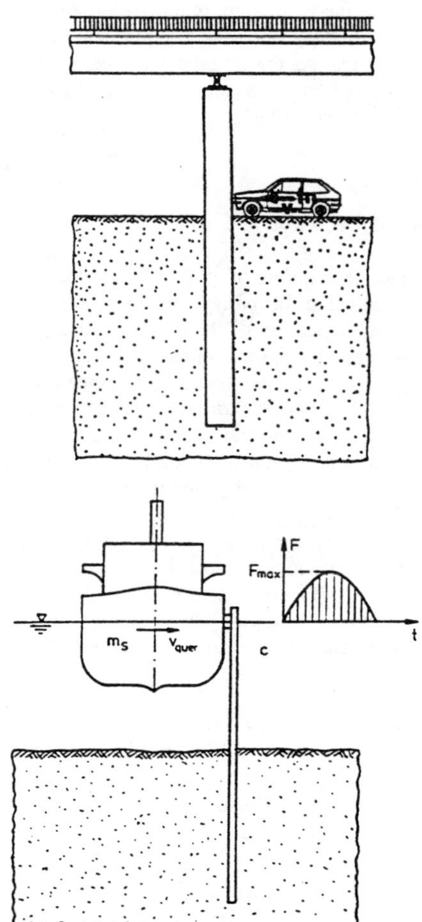

(a)

(b)

Fig. 1. Cases of horizontal impact: (a) motor vehicle impact and (b) ship impact.

Essentially there are two broad ways of design, which depend upon the assumption about the duration of the impact. The first method considers inertia forces of the system by a static design concept, in which the modulus of subgrade reaction will be set to an increased value (DIN 4014, 1990). In the second approach, inertia forces are not considered because of an assumed higher duration period of impact. Herewith the kinetic energy of the load mass has to be converted into a sufficient displacement work of the pile and its surrounding soil (EAU, 1990).

The response of a pile subjected to an impact load is complex and depends upon the mass, the stiffness and the ability of energy dissipation (damping) of the entire system. Under high levels of loading, inelasticity effects within the soil also exert an important influence on the dynamic characteristics of the system. The two principle sources of damping are hysteresis and radiation. Hysteretic damping (material damping) is the result of inelasticity in the system and represents heat and frictional losses as well as visco-elastic behaviour. Radiation damping is due to the propagation of energy in the soils away from the dynamically excited system (geometric damping). Thus, the regarded system could be represented by the elements pile, near-field soil and far-field soil. Each of these elements represents a separate damping component.

To get more information about the bearing behaviour of piles due to horizontal impact loads, model tests are executed in the Bochum geotechnical centrifuge ZI. The aim of these tests is to obtain further knowledge about the pile-soil interaction in the near-field zone and the soil behaviour in the far-field zone. In this paper, a particular emphasis is given to the time history response of a pile-soil system subjected to horizontal impact loads. As the investigations pertaining to this project are in progress, the system behaviour is evaluated based on the centrifuge model tests executed so far.

2 STATE-OF-PRACTICE

The mathematical description of physical systems involving impact loads is very complicated. In many cases this is possible only with an immense coarse simplification of the original pattern. BRUHNS (1996) pointed out: "Impact refers to the collision between bodies. It is characterized by the generation of relatively large contact forces which are exerted and removed in a very short interval of time and initiate stress waves which spread out from the region of contact. The complicated physical process occurring during impact conditions lead to serious difficulties in the mathematical analysis of this type of problems. Models representing the physical systems must be idealized to render them amenable to theoretical treatment, and the hypothesis which postulate dynamic behaviour of such bodies have to be verified by suitable experiments. As a consequence, complete solutions have been obtained only for simple configurations, utilizing the laws of conservation of mass, linear and angular impulse-momentum, and mechanical energy balance."

When dividing the system pile-soil into the three elements as mentioned above, an exemplary idealized system may consist of the elements mass, spring and viscous dashpot as shown in Figure 2 (ANTES ET AL., 1996). However, with specification of the stiffness and the damping, a mathematical approach would be possible.

Especially in case of horizontal impact loads on piles, only a very few investigators have attempted to analyse pile-soil system behaviour. The bearing behaviour of bored piles is investigated by POPP (1974) by conducting three full-scale tests. Three concrete piles 11 m in length with a diameter varying from 0.80 m to 1.20 m and with an embedded length of 6 m. The prefabricated piles are embedded in a 6 m deep circular test

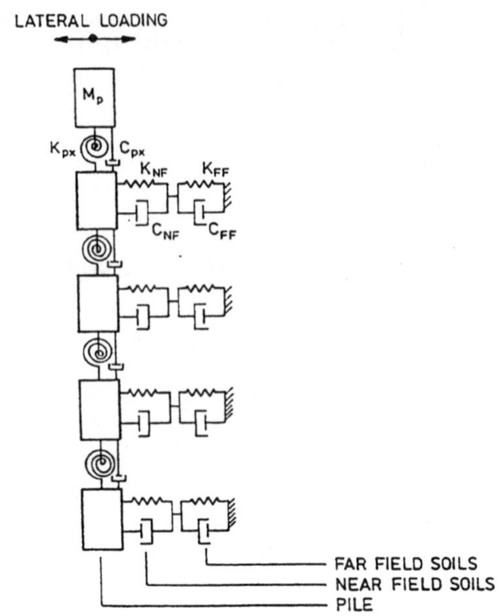

Fig. 2. Idealized discrete element system for dynamic pile response.

pit of 3.50 m diameter (with a concrete boundary). In load direction the pile heads are hinged with a non-adjustable support 4.75 m above the ground line. The impact was induced by special test vehicles. In these tests, the time history response of the loads at the pile head and at the location of the impact (1.20 m above GL) are measured. The bending moment distribution as well as the pile deflection variation is reported. In addition, the horizontal soil pressures in different depths are recorded. From this data, POPP derives two relations for the horizontal subgrade reaction due to lateral pile impact (Equation 1).

$$k_s\,(z) = k_{s,max} \cdot \left(1 - e^{-\chi \cdot z/l}\right) \qquad \text{for } l \le 6 \text{ m} \qquad (1a)$$

$$k_s\,(z) = k_{s,max} \qquad \text{for } l > 6 \text{ m} \qquad (1b)$$

with

$k_s\,(z) = k_{s,max} \cdot \left(1 - e^{-5.5 \cdot z/l}\right)$ for non-cohesive soil

$k_s\,(z) = k_{s,max} \cdot \left(1 - e^{-2.5 \cdot z/l}\right)$ for cohesive soils

$k_{s,max}$ - threshold value of modulus of horizontal subgrade reaction

z - ordinate of depth

l - embedded length.

POPP recommended that the relations given in Equation 1 are valid for the same soil conditions and for similar pile constructions. Nevertheless, his main statement is that a static design concept is sufficient in case of lateral impact endangered piles.

Accordingly, DIN 4014 (1990) recommended to calculate the bearing behaviour of piles subjected to horizontal impact loads by increasing the static modulus of subgrade reaction three times. For this, the maximum impact loads were replaced by lateral static loads 1.2 m above the ground surface.

By using such a static design concept, the effects influencing this problem are not covered in a realistic manner (FRANKE, 1992). Besides the magnitude and the duration of the impact, the kind of soil (cohesive, non-cohesive) as well as the soil condition (saturated, non-saturated) have to be considered. In connection with the above, FRANKE mentioned the kinematical degree of freedom of the diplaced soil. If a raft foundation on the soil surface will hinder the upward movement of the soil, an increased soil resistance has to be considered in the design.

WAAS (1986) justified the approach of a statical load and stiffness with the very short duration of impact loadings and reported damping do not occur and has no influence on the maximum demand of the system.

By formulating a design concept for this load case in a way, that the kinetic energy of an impact mass is replaced by a sufficient deformation work of the pile and the surrounding soil, a simple energy balance is not satisfying (KOLYMBAS, 1989). The equation of the kinetic impact energy with the deformation work of the embedded pile predict to large deformations and to small bending moments. This results of the fact, that wave energy spreading into the surrounding soil and viscous properties of the soil are not recognized with this kind of energy balancing.

For the design of dolphins in non-cohesive soils under ship impact loadings, the recommendations EAU (1990) ignore inertia forces because of the long time duration of this kind of impact (JENNE, 1958). Instead the suggested design concept determines a maximum impact load as well as a permissible deflection in dependence of the loaded system (dolphin and soil) and the load inducing system (type of ship) (COSTA, 1964). The calculation follows the limit analysis after BLUM (1932). However, large deformations as permissible for dolphins generally are above tolerable limits for building foundations and herewith this kind of design concept.

3 CENTRIFUGE MODEL TESTS

3.1 PRINCIPLES OF CENTRIFUGE MODEL TECHNIQUE

A model moving in a circular orbit with the radius r and the angular velocity ω is simultaniously subjected to both an acceleration g due to earth's gravity and a radial acceleration of an order expressed in Equation 2.

$$n \cdot g = \omega \cdot r^2 \qquad (2)$$

When the model is placed in a swinging basket (Figure 3) it is adjusted to the direction of the resultant acceleration, which results in a perpendicular to the model surface acting acceleration field. In practice the 1g component is neglected so that the resultant acceleration is set to ng. The value of factor n corresponds to the model geometrical scaling. Thus, weight stresses in the model are increased n times, this results in a geometrically identical stress condition for the model and the prototype (full scale). Because of the non-linear stress-strain behaviour of soil, this is important to get the same stress-strain relationship in model as in prototype. Each measured physical quantity must be converted from model scale to prototype scale by appropriate scaling factors. The most important scaling factors needed for the tests reported in this paper are shown in Table 1. For technical details of the Bochum geotechnical centrifuge ZI, which is used for the tests, the reader is refered to JESSBERGER ET AL. (1988).

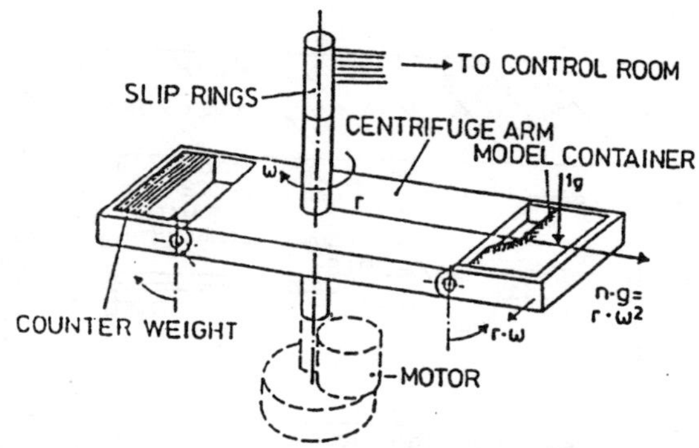

Fig. 3. Principle of geotechnical centrifuge (JESSBERGER ET AL., 1988)

Tab. 1. Main scaling rules for dynamic centrifuge model tests.

Parameter	Prototype at 1·g	Model at n·g
Length	1	1/n
Area	1	$1/n^2$
Volume	1	$1/n^3$
Mass	1	$1/n^3$
Stress	1	1
Strain	1	1
Force	1	$1/n^2$
Moment	1	$1/n^3$
Time	1	1/n
Frequency (acc.)	1	n

3.2 TEST DESIGN

The geometrical scaling factor for the tests reported here is selected as 1:50. So this leads to a necessary g-level of n = 50g, which is maintained during centrifuge tests. The model set-up is shown in Figure 4. The pile tests are run in a rectangular rigid model container (length/width/height = 850/425/815 mm). A sand-dummy (180 mm in thickness) is placed on the container bottom, in order to reduce the time during the test preperation. An absorbing material ("DUXSEAL") is used along the sides of the container, to reduce wave reflections.

The impact is generated through a newly developed mechanism in which a loading weight is shot on to the pile to induce impact load. The loading weight consists of a hard rubber cylinder and a special piezoelectric force transducer, which is used to measue the applied load H (t). The force transducer itself has got a small rubber tip. The total mass of the loading weight is 0.183 kg. The loading weight impact is realized with help of a steel compression spring, which is flexed by winding up a steel cable on to a lockable shaft. During flight, the shaft will be unlocked and the loading weight is shot on to the pile. With this arrangement the impact load is induced to the pile at e_M = 24 mm above the soil surface.

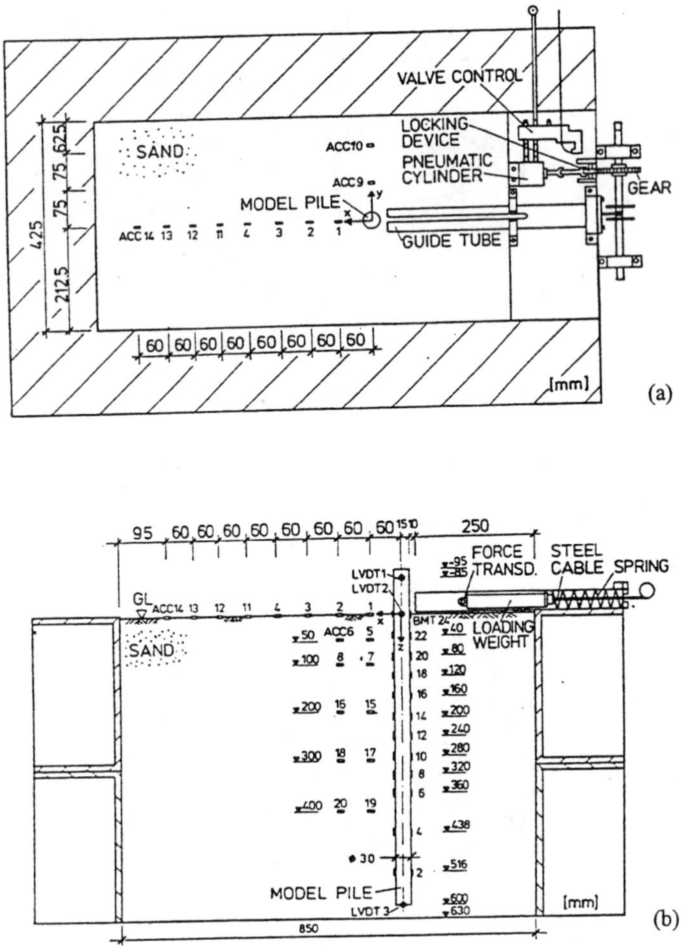

Fig. 4. Model set-up for single piles under horizontal impact load: (a) plan view, (b) cross section.

An aluminium pipe with an external diameter of D_M = 30 mm, an embedded length of l_{eM} = 600 mm and a free length of l_{fM} = 95 mm are used as model piles. The peripheral portion of the embedded length of the model pile is smeared with a thin layer of sand. The resulting weight of the model pile is m_M = 0.7 kg, which corresponds to a mass distribution of $\overline{m}_M = 1.007\ kg/m$. It therefore replicates a long flexible concrete pile behaviour in prototype. Table 2 shows the relations between model and prototype pile. Concerning the mass distribution it has to be remarked that there are two prototype values: The first one is converted from the model (scaled) and the second one is calculated by the unit weight of concrete (real). This fact has to be born in mind for the interpretation of the test results. Since this is known, it is optionally possible to fill the model piles with any additional mass. For measuring the bending moment distribution along the embedded pile length, it is equipped with 24 pairs of strain gauges called bending moment transducers (BMT). The pile was calibrated in such a way, that it is possible to measure the shear force Q (0,t) at the ground level by the first BMT (No. 24) as well. The degree of freedom (DOF) of the pile head is varied between free-head and fixed-head conditions (Figure 5).

Tab. 2. Model and prototype piles.

Parameter		Model	Prototype
External diameter D	[m]	0.030	1.50
Internal diameter d	[m]	0.026	-
Embedded length l_e	[m]	0.600	30.00
Free length l_f	[m]	0.095	4.75
Height of loading point e	[m]	0.024	1.20
Young's modulus E	[MN/m^2]	70000	30000
Moment of inertia I	[m^4]	1.7329E-8	0.252714
Flexual rigidity EI	[kNm^2]	1.21303	7581420
Mass distribution $\overline{m}$	[kg/m]	1.007	scaled: 2518 / real: 4503

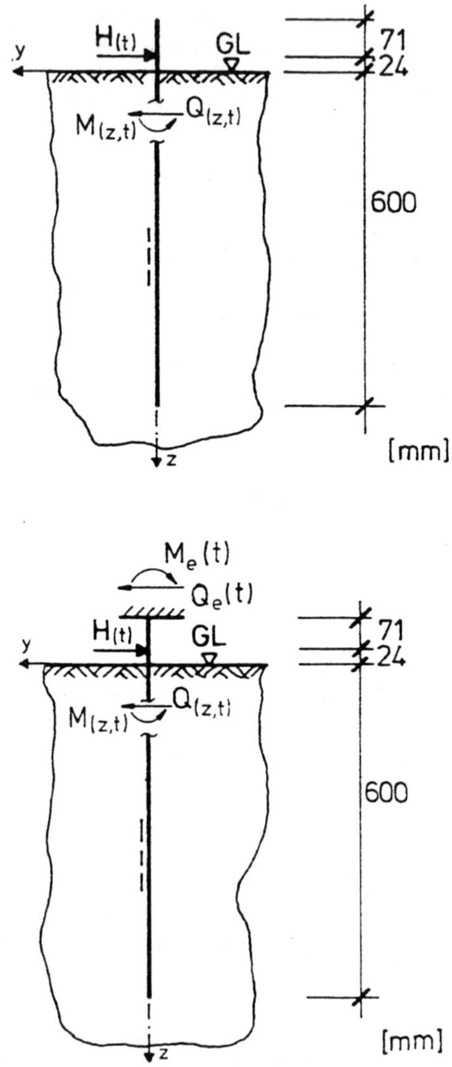

Fig.5. Static systems and definition of signs.

As a model soil the Bochum standard sand 942E is used (Table 3). With this kind of sand, the ratio of the average particle size to the external pile diameter amounts (d_{50} / D) = 1 / 143. The fine grained sand is introduced into the model container by pluviation around the already installed pile. By adopting this technique a consistent density within the sand at each level can be achieved (Table 4).

Tab. 3. Properties of the Bochum Standard Sand (942E).

Parameter		Value
Density of solid particles ρ_s	[g/cm³]	2.66
Compactness ρ_{min}/ρ_{max}	[g/cm³]	1.39 / 1.68
Porosity n_{min}/n_{max}	[-]	0.37 / 0.48
50 percent-diameter d_{50}	[mm]	0.21
10 percent-diameter d_{10}	[mm]	0.12
Uniformity coefficient $C_U = d_{60}/d_{10}$	[-]	1.92
Angle of internal friction ϕ (DIN 18137, 1990)	[°]	40.7

Extensive measurements are required in order to get the detailed knowledge about the pile-soil interaction behaviour in case of horizontal impact loads. In addition to the already mentioned force transducer and BMT's, 20 accelerometers (ACC's) and 3 LVDT's have to be recorded. The ACC's were used to measure the shear wave and pressure wave velocities within the sand. The LVDT's are needed to obtain the pile head displacement y (-0.085,t), the deflection at the ground level y (0,t) and the pile tip displacement y (0.6,t). The requirement to measure all transducer signals simultaneously with a sufficient scan rate, lead to serious problems in data acquisition. Therefore a special data acquisition equipment was installed, which made it possible to record each transducer signal with a scan rate of f_s = 100 kHz.

The ability of measuring shear wave and pressure wave velocities in small scale model tests in centrifuge was explained in detail by SIEMER (1996). Thereby, the derivation of significant dynamic soil properties such as the dynamic soil stiffness G_{dyn} and the dynamic Poisson ratio ν_{dyn} is possible.

3.3 TEST RESULTS

The results of one free-head and one fixed-head pile test, mainly concerning the time history responses of the pile loads, the bending moment distributions and the pile deflections are briefly presented. The measured values are presented in model dimensions. Table 4 presents the density index I_D of the model sand, maximum horizontal load, maximum shear force and the corresponding maximum displacements respectively.

Tab. 4. Soil density and maximum values of loads and deflections.

Test name		HIL2-1 (f)	HIL5-1 (f)
Pile head DOF		free	fixed
I_D	[-]	1.01	1.04
Max. H (t)	[kN]	1.208	1.622
Max. Q (0,t)	[kN]	1.729	- 1.897
Max. y (-0.085,t)	[mm]	- 3.410	- 0.128
Max. y (0,t)	[mm]	- 2.065	- 0.312
Max. y (0.6,t)	[mm]	- 0.017	- 0.003

Considering the value y (-0.085,t) = - 0.128 mm in Table 4, it has to be noted that the full restraint is achieved at z = - 0.095 m.
Figures 6 and 7 represent the time history responses of loads and deflections for both tests. The load-displacement relationships (Q (0,t) / y (0,t)) are shown in Figures 8 and 9. In Figure 6 (a) and 7 (a) I (t) and W (t) are the differences H (t)- Q (0,t) of the respective tests. For test HIL2-1 (free-head) it can be seen that up to $t_M \approx 0.0005$ sec no displacement at the ground line has occured. Until this time Q (0,t) is negative, which indicates the presence of large inertia forces. Immedeately after this time range the maximum amount of displacements are observed. This implies that the inertia forces were superseded and yielding occured within the near-field soil. Thereafter the displacements decrease to a constant value. This constant value of displacements are due to the occurence of permanent displacements within the soil. In addition, after this point of time the shear force has shown an oscillating tendency until t = 0.01 sec (Figure 6 (a)). This is due to the gradual diminishing of the pile oscillation. This damping process is also illustrated in Figure 8. In case of the fixed-head pile (HIL5-1) no significant oscillation is observed (Figure 9). Instead, the embedded part of the pile apparently experiences a small kinematic shift in horizontal direction.

The time histories of the bending moment distribution for both tests are shown in Figures 10 and 11. The sequence of time steps are marked as well in Figures 6 (a) and 7 (a). Below a depth of z_M = 0.35 m no more significant change in the flexural moments are recorded in both the cases. Obviously at this depth the point of restraint is reached. This would mean that the yielding and damping area of soil is restricted to this level, irrespective of the type of pile head support. The deviation of this influenced zone in horizontal direction depends of the degree of freedom of the pile head and as a result of the soil working as a support. In Figure 12 this influenced soil area is shown schematically. In addition to the above, Figure 13 presents a photograph of the pile in test HIL2-1.

4 CONCLUSIONS AND FURTHER WORK

Summarizing the observed results of two horizontal impact load tests, it is revealed, that the total behaviour of a pile itself depends of the type of pile head support. On the other hand it seems that the soil area which is involved on the pile bearing behaviour is restricted to a definite depth of embedment.

However, many more investigations have to be conducted, in order to evaluate the bearing behaviour of piles subjected to lateral impact loads. As a part of further investigations, the following points needed to be investigated:

(a) Variations of the pile mass, length and flexural ridigity;
(b) Variations of the type of impact (elastic, plastic or elasto-plastic) and impact energy;
(c) Analysis of acceleration measurements in time and frequency domain;
(d) Analysis of pile deflections and resulting soil resistence by back calculating the recordings of the bending moment distribution.

From these kinds of analysis, a detailed knowledge of the above mentioned problem will be obtained and herewith a mathematical approach would be possible.

ACKNOWLEDGEMENT

The work reported herein was carried out at the Ruhr-University Bochum and was made possible with financial support from the National German Science Foundation (DFG).

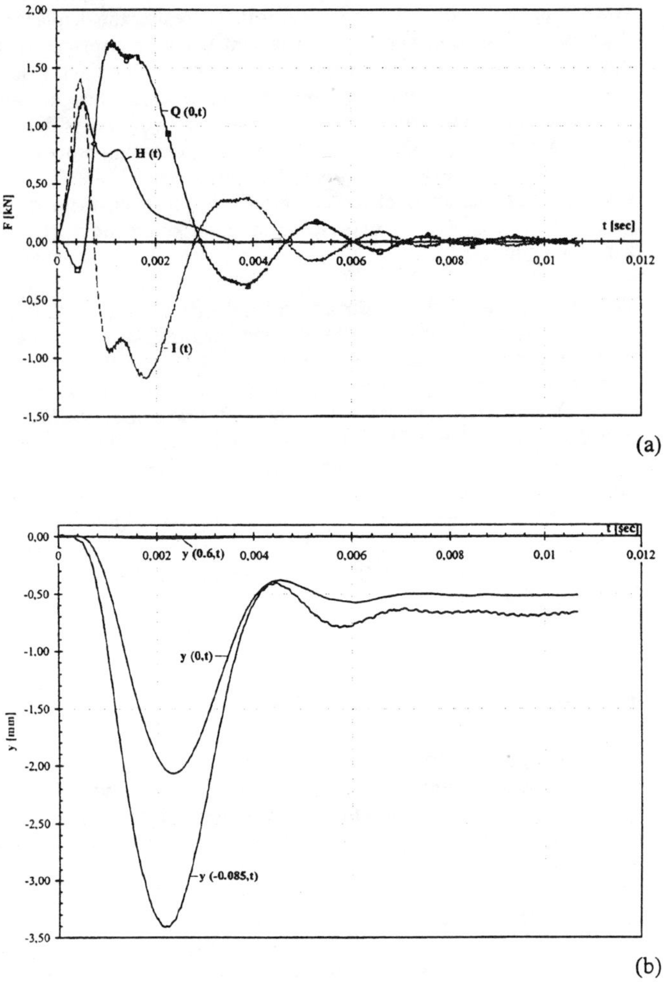

Fig. 6. Time history response curves for test HIL2-1 (free-head): (a) loads and (b) displacements.

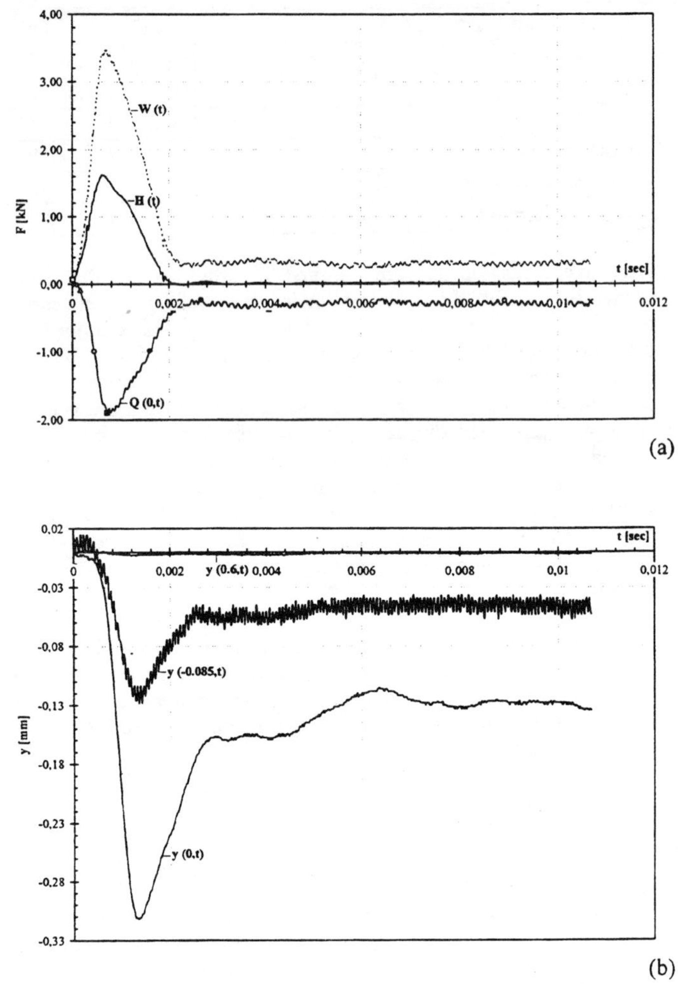

Fig. 7. Time history response curves for test HIL5-1 (fixed-head): (a) loads and (b) displacements.

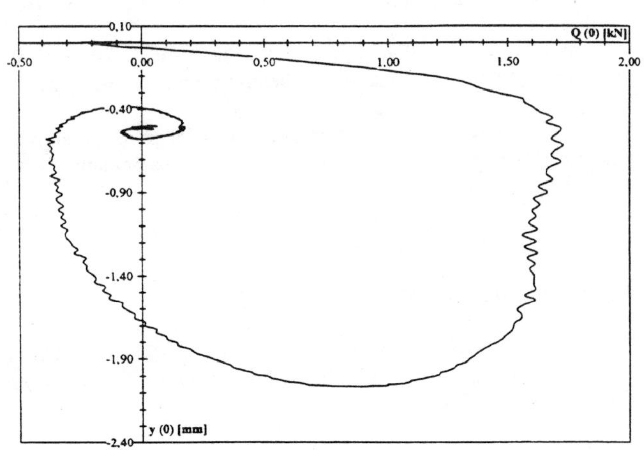

Fig. 8. Load-displacement relationships: HIL2-1 (free-head)

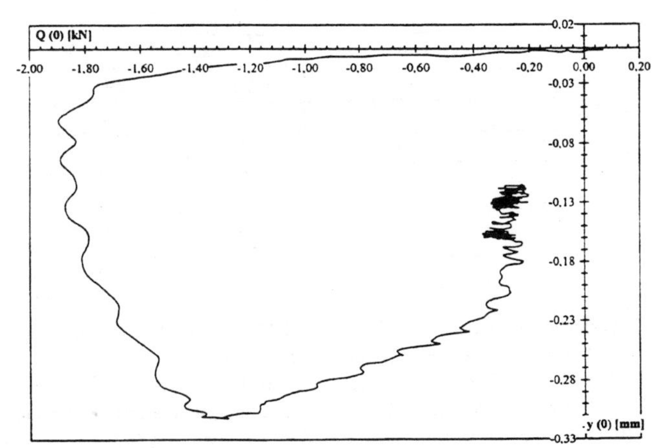

Fig. 9. Load-displacement relationships: HIL5-1 (fixed-head).

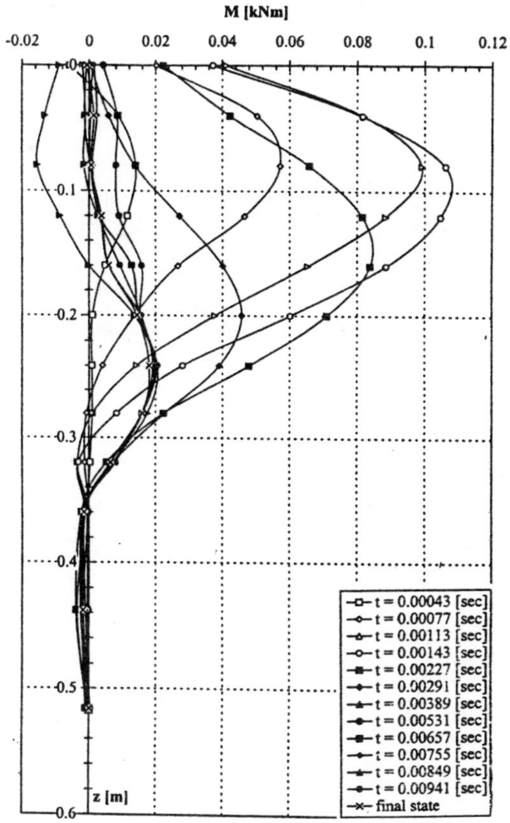

Fig. 10. Bending moment distributions: HIL2-1 (free-head).

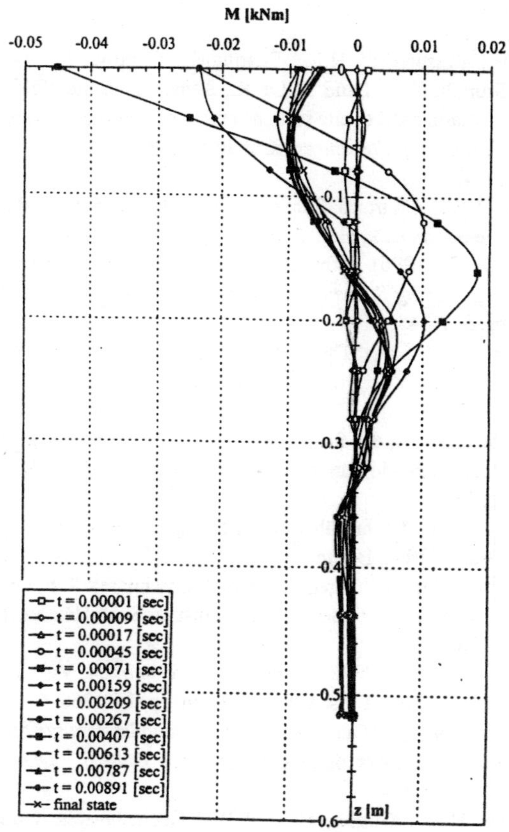

Fig. 11. Bending moment distributions: HIL5-1 (fixed-head).

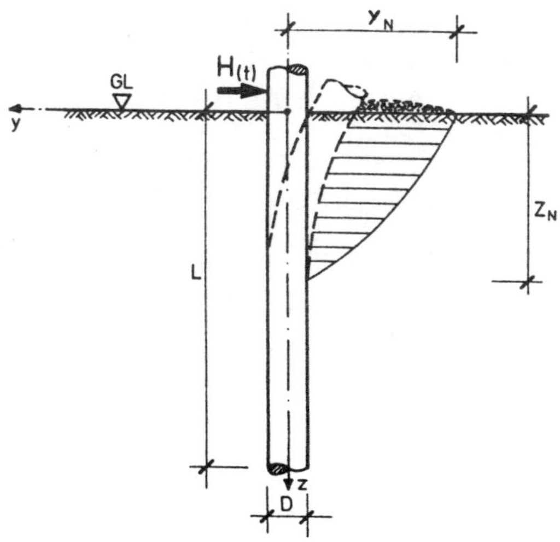

Fig. 12. Definition of yielded soil area.

Fig. 13. Impact of test HIL2-1.

REFERENCES

Antes, H., Jessberger, H.L., Schmid,G., Chouw, E., Forchap, E., Grundhoff, T., Laue, J., Le, R., Siemer, T. & Steinfeld, B. (1996). "Dynamic and cyclic soil-structure interaction," *Dynamics of Civil Engineering Structures*, Krätzig et al. (eds), Balkema, Rotterdam, pp. 451-531.

Blum, H. (1932). "Wirtschaftliche Dalbenformen und deren Berechnung," *Die Bautechnik*, Heft 5, p. 50.

Bruhns, O. T. (1996). "Impact loading problems," *Dynamics of Civil Engineering Structures*, Krätzig et al. (eds), Balkema, Rotterdam, pp. 321-345.

Costa, F. V. (1964). "The Berthing Ship," *The Dock and Harbour Authority*, Vol. XLV, pp. 523-525.

DIN 18137 (1990). "Baugrund, Versuche und Versuchsgeräte; Bestimmung der Scherfestigkeit; T1: Begriffe und grundsätzliche Versuchsbe-dingungen; T2: Triaxialversuch," Beuth Verlag, Berlin.

DIN 4014 (1990). "Bohrpfähle - Herstellung, Bemessung, Tragverhalten," Beuth Verlag, Berlin.

EAU (1990). "Empfehlungen des Arbeitsausschusses "Ufereinfassungen" - Häfen und Wasserstraßen," Ernst & Sohn, Berlin, E69, E111, E128.

Franke, E. (1992). "Pfähle," *Grundbau-Taschenbuch*, Vol. 3, Smoltczyk (ed), Wilhelm Ernst & Sohn, Berlin, pp. 173-268.

Jenne, L. (1958). "Die Größe der Erdwiderstandsziffer und des Raumgewichtes beim eingespannten Dalben," *Die Bautechnik*, Heft 4, p. 134.

Jessberger, H. L., Güttler, U. (1988). " Geotechnische Großzentrifuge Bochum - Modellversuche im erhöhten Schwerefeld," *Geotechnik*, Nr. 2, pp. 85-97.

Kolymbas, D. (1989). "Pfahlgründungen," Springer-Verlag, Berlin, pp. 58-59.

Popp, C. (1974). "Untersuchungen über den Querstoß beim Aufprall von Kraftfahrzeugen auf Gründungspfähle aus Stahlbeton und Stahl," *Deutscher Ausschuss für Stahlbeton (DAfStb)*, Heft 246, Wilhelm Ernst & Sohn, Berlin.

Siemer, T. (1996). "Zentrifugenmodellversuche zur dynamischen Wechselwirkung zwischen Bauwerk und Baugrund infolge stoßartiger Belastung," PhD. thesis, Ruhr-University Bochum, Heft 27.

Waas, G. (1986). "Pfahlgründungen unter dynamischer Belastung," *Bodendynamik - Grundlagen und Anwendung*, Haupt (ed), Friedr. Vieweg & Sohn, Braunschweig, pp. 189-224.

Proceedings of the Seventh (1997) International Offshore and Polar Engineering Conference
Honolulu, USA, May 25-30, 1997

ISBN 1-880653-28-1 (Set); ISBN 1-880653-29-X (Vol. I)

Simulation of Suction Pile Installation in Sand in a Geocentrifuge

H.G.B. Allersma, F.J.A. Plenevaux and J.-F.P.C.M.E. Wintgens
Delft University of Technology
Delft, The Netherlands

ABSTRACT

A test program has been carried out in the small geotechnical centrifuge of the University of Delft to investigate the installation of suction piles in sand. A test technique has been developed to simulate a pressure difference of more than 30 meters water. During installation, the displacements and pressures are measured. The influence of several parameters are tested and the mechanism which is active during installation can be examined in more detail. It appears that there is almost a linear relation between the pressure and the dimensions of the suction pile. Piles up to a ratio of H/D=4 could be penetrated without any problem. The penetration load during suction is approximately 8 times smaller than during mechanical penetration.

KEY WORDS: soil mechanics, suction pile, geotechnical centrifuge

INTRODUCTION

In recent years, suction piles have been applied increasingly often in offshore engineering (Wang et al., 1978; Senpere et al., 1982). Suction piles are attractive because of the convenient method of installation. A pile with a diameter of 9 m and a height of 10 m can be installed in 1 hour, by using a pump only. Several research programs have been carried out to investigate the bearing capacity of suction piles. However, little research has been performed to examine the behaviour of suction piles during installation. Some measuring data has been obtained from piles which are installed in practice. In these cases, however, the installation proceeds well, so that the limits of this foundation technique could not be examined. Since prototype tests are very expensive and time consuming, it is not so easy to investigate the process parameters during installation under different circumstances.
In small scale tests it is much easier to change parameters. The soil type can be varied as well as the dimensions of the suction pile and other process parameters. In a small scale test, however, problems arise concerning the stress dependent behaviour of sand. Furthermore, the required pressure gradients are so low that measurements are not accurate enough to visualize differences in design. These restrictions can be survived by performing the tests in a geotechnical centrifuge. Centrifuge test on suction piles are reported in literature by Fuglsang et al., 1991; Renzi et al., 1991 and Helfrig et al., 1976. The attention in these tests, however, were mainly focused on bearing capacity and the suction tests were performed on clay. At the University of Delft a test technique has been developed in the centrifuge in order to simulate the installation of piles in sand by suction. In a centrifuge the soil stresses over a similar depth is the same as in the prototype situation. At 150 times earth gravity a suction pile with a height of 10m and a diameter of 9m can be simulated with a can with a height 67mm and a diameter of 60mm. The effect of water depth has been simulated by performing the test in a container which can be pressurised, where the interior of the suction pile is connected with atmospheric pressure. During the test, penetration depth, pressure difference and flow rate of the water are measured.
More than 30 tests were carried out in the geotechnical centrifuge. Since the centrifuge is small the samples can be prepared rather quickly, so that several tests can be performed in a short period. Thanks to small-size computers and measuring techniques based on image processing, advanced tests can be performed, even in a small centrifuge.
The aim of this study was to examine the influence of several parameters on the installation of suction piles in dense sand.

PRINCIPAL OF SUCTION PILE

A suction pile is a large diameter steel cylinder which is closed at the top either by a dome formed section or by a flat stiffened plate. The pile is open at the bottom. Pump inlets and release valves are installed at the top as well as the lug to connect the pile chain.
The pile, which can be launched from an installation vessel or from the aft deck of a supply boat, must be landed softly on the sea floor. The cylinder is lowered to the sea floor with open valves so that the enclosed air can escape rapidly.
Once the pile has penetrated the sea floor by its own weight the release valves are closed. To avoid piping when pumps are started, additional weights are usually needed in cohesionless soils like sands to penetrate the skirt into the soil sufficiently.
Pumping the trapped water out of the pile by means of pumps mounted on the top causes a pressure difference between the external

Fig.1 The University of Delft geotechnical centrifuge.

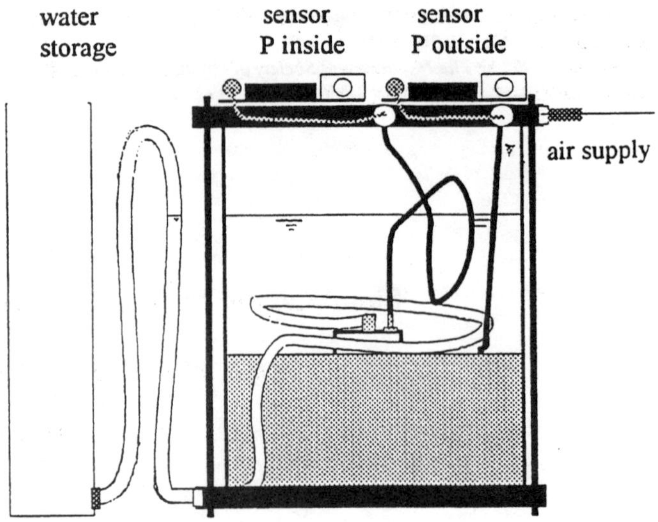

Fig.2 Diagram of the test container.

hydrostatic water pressure and the water pressure inside the pile. The pressure difference generates the driving force for soil penetration .
In permeable soils water will flow through the pores to the tip of the pile during pumping, so that the soil stress around the tip is reduced due to local fluidization. Due to this reduction, the driving force is sufficient to ensure penetration of the pile. Pumping is stopped after reaching the designed penetration depth. In general this depth is reached in about 20 min to 1 hour, depending on the soil configuration, the pile dimensions and the pump capacity.
After installation, pore pressures in the soil regain their initial values. The loads are therefore resisted by the soil in a similar way to a conventional driven or drilled pile with regard to friction. The pump, the air valves and possible measuring equipment mounted on the top of the pile, can be made retrievable for recovery after complete installation.

SIMULATION OF THE INSTALLATION IN A CENTRIFUGE

In order to model the stress dependent behaviour of sand in a correct way the test program was carried out in the geotechnical centrifuge (Fig.1) of the University of Delft (Allersma, 1994). This centrifuge has a diameter of 2.5 meter, which can accelerate models with a weight of 300N up to 300g. Through the use of several computer controlled devices, a large variety of geotechnical problems can be simulated in the centrifuge. In this test program the computer controlled air supply system was used to manipulate the suction pile. A powerful option of the centrifuge is that digital image processing (Allersma, 1990) can be used to perform measurements in flight. In these tests the displacement of the pile during installation was measured by this technique. An advanced computer controlled loading system was used to measure the load during mechanical installation of the suction pile.
To be able to simulate a large water depth the tests are carried out in a perspex container (Fig.2) which was placed in the centrifuge. The interior of the suction pile was connected with the atmosphere, so that the required pressure difference could be realised simply by increasing the pressure in the perspex container. This has the same effect as in the prototype situation where the ambient pressure is constant and the pressure inside the pile is reduced by suction. The top of the suction pile was connected with the space outside the container through a flexible plastic hose.

A second small hose was connected with a pressure transducer, in order to measure the pressure inside the pile. The pressure outside the suction pile was measured also, so that the pressure difference is known during installation. The flow rate of the water during installation was measured by elaborating video images showing the water table in a transparent cylindrical container. The limit of the pump capacity was simulated by a nozzle.
In all the tests performed, the sand inside the pile reached the top cover. The soil heave h inside the pile can therefore be calculated by the difference between the height of the pile and the final penetration depth H reached at the end of the suction test. A heave parameter a can be defined as

$$a = \frac{h}{H} \qquad 1)$$

All the tests are carried out in homogeneous sand layers. In order to reduce the particle size effect, a sand with a small grain size was used. The main characteristics of this sand are displayed in Table 1.

d_{90} (mm)	0.140	G_s	2.65
d_{60} (mm)	0.109	$\gamma_{d,min}$ (kN/m³)	12.32
d_{50} (mm)	0.102	$\gamma_{d,max}$ (kN/m³)	16.06
d_{10} (mm)	0.068	$CU=d_{60}/d_{10}$	1.62

	Triaxial Test	Cone Penetration Test	Shear Box
n %	45.1 - 41.3	41.6	41.8
φ	38° - 42°	45°	38°

	Permeability Test
n %	49.6 - 41.4
k (m/s)	$7.4\ 10^{-5}$ - $2.9\ 10^{-5}$

Table 1 Properties of the sand used.

The friction angle between the steel and the sand has been determined by a shear box test and was found to be 27°.
The sand samples were manually prepared. Dense samples were obtained by knocking and shaking the sample box. The box was filled by wet sand layer by layer. Best results are obtained at porosities close to the minimum porosity of the sand. The averaged porosity used was 41%, with a standard deviation of 1%.

TEST PROGRAM

In the first instance, the suction piles were penetrated into the sand soley by means of a computer controlled loading system. This data was used to examine the stress dependent behaviour of the sand and the data of the suction tests could then be compared with these results.
Several suction tests were performed under several different circumstances. The dimensions of the piles were varied, as were the density of the sand and the suction procedure.
Four different model piles were used in this test program. One of the piles was made from transparent material in order to observe the behaviour of soil inside the pile. The main characteristics of the piles, the height *H* and diameter *D* are given in Table 2.

pile type	model				prototype		
	H/D	H	D	wall	H	D	wall
	mm	mm	mm	mm	m	m	mm
medium	2	75	39	0.5	11.3	5.9	75
stubby	1	60	63.2	0.75	9.0	9.5	113
slender	4	108	25.2	0.27	16.2	3.8	41
plexiglas	1	61	69	2.0	9.2	10.4	300

Table 2 Main characteristics of the piles used

The dimension of the models has to be multiplied by 150 to get the dimensions of the prototype. The thickness of the wall was larger in comparison to most prototype situations. For technical reasons, thinner walls were hard to obtain. It is not impossible, however, to decrease the wall thickness.
The weight of the models was also larger compared with the weight of the prototypes. For the mechanical penetration tests the weight of the pile was compensated at the beginning of the test, so only the force of pushing and pulling the anchor was measured. For the suction test, the underwater weight was also taken into account.

DISCUSSION OF TEST RESULTS

In the mechanical penetration tests the pile was driven into the sand at a speed of approximately 2 mm per second. During displacement, the required load was recorded. An example of some tests with a suction pile with a diameter of 60 mm and a height of 67mm is shown in Fig. 3. Since the range of the loading system was at that moment 30 mm, the tests were performed in two steps. In the figure only the final 30 mm of penetration is shown. The tests were performed under different gravities and the load was divided by the actual gravity. Also, to enable a better comparison with other tests, the loads are converted to pressures at 150g conditions. As can be seen the sand behaves in a very stress dependent manner, in particular, at lower stress levels. The test at 150g was repeated in order to demonstrate that the experiments can be reproduced well.

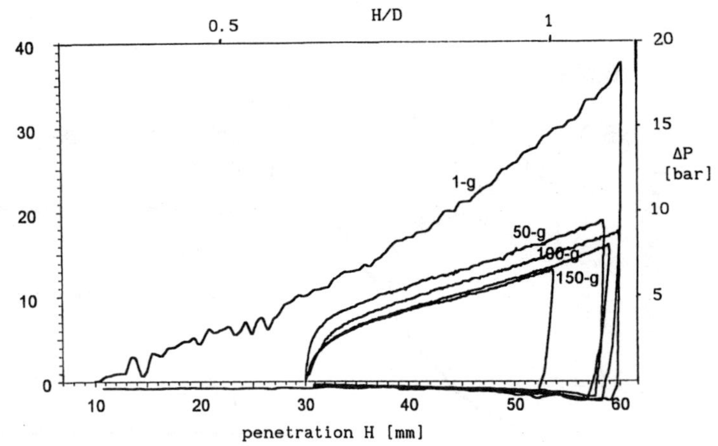

Fig.3 Mechanical penetration of suction piles at different g levels, the load is normalized.

Several suction tests were carried out in order to investigate the behaviour of suction piles with different geometries. Sufficient data was available to present the suction process in graphical form. A typical example of the course of a test is presented in Fig.4, where the pressures and the penetration depth are shown as a function of time. At t=0 the centrifuge was started, which is visible in the form of an increase in the hydrostatic pressure. At t=320s the air pressure in the test container was increased (starting pumps in prototype). A pressure difference was then visible between the inside and outside of the suction pile where a vertical displacement of the pile could also be detected. The pressure difference increases during penetration and stabilizes at the moment that penetration stops. At t=350s the air pressure in the container was decreased to zero. At about t=440s the air pressure was again increased to a higher value in order to examine whether more penetration could be achieved at a higher pressure difference. The pressure was decreased again and at t=500s the centrifuge was stopped. Also, the flow rate of the water during suction was available continuously, by using a calibrated relation between head loss in the tubes and flow rate.
From the suction tests, the maximum load required for a given final penetration can be derived. It appeared that the required load by suction is much lower than the load required for penetration of the pile by a mechanical loading system.
In all tests penetration proceeded gently. Even after stopping the water flow for a while the penetration continued at almost the same pressure difference when the water flow was restarted. It appeared that the axis of the suction pile tends to a vertical position even when the initial position is not vertical.
If the initial penetration depth of the suction pile is not sufficient, piping seems to occur. This phenomenon was observed with suction piles of Plexiglas, having a lower weight and a larger tip surface. In this test program no special attention has been paid to examine the relation between initial penetration, tip surface and piping.
If the initial pressure difference was sufficiently large to start penetration it could be measured how much extra water was required above the volume of the pile. It was found that this was only a few percent of the pile volume. Subsequent analysis of the sand plug in the suction pile has shown that only the sand close to the boundary was in a fluidized state. More to the centre the sand looks undisturbed. This phenomenon was also observed on line in a transparent pile which was penetrated into dry sand using a vacuum pump at 1g. However, if the diameter of the pile became too small, full fluidization could be observed, which results in a significant upheaval of the soil plug. Due to the upheaval it was not possible to penetrate the pile completely. In

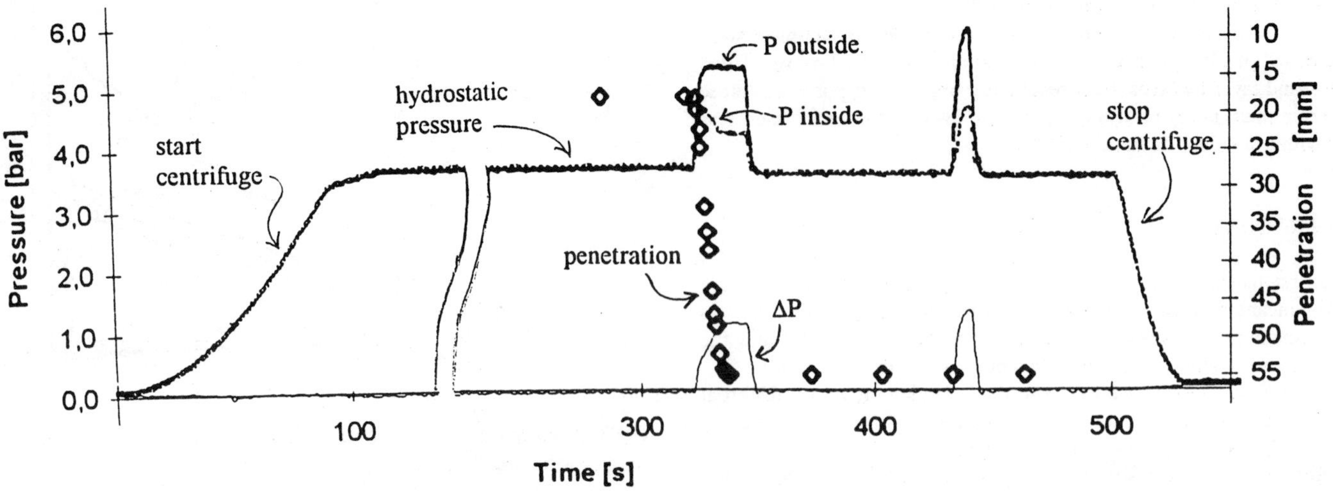

Fig.4 Graphical presentation of the course of the pressure and displacement in time.

the centrifuge tests model, piles up to a ratio H/D=4 penetrate without any problem. This ratio is more than the limit of about 3, which is usually taken in practice to avoid fluidization. A more specific test program is required to determine the critical H/D ratio.

The most important test results are summarized in Table 3. The total load F_{TOT} at maximum penetration is the sum of the underwater weight W' and the force derived from the maximum pressure difference $F_{\Delta P}$. For tests made in several steps (e.g. 19.1, 19.2, 19.3), the maximum load required to reach the maximum penetration after each step is given. The pressure difference ΔP also includes the own weight of the suction pile. Results obtained with the same anchor are very close to each other.

Test	Sample porosity	Model dimensions				Pressure difference	Prototype dimensions					
	n [%]	D [mm]	d [mm]	W' [g]	H [mm]	ΔP [bars]	D [m]	d [m]	W' [MN]	H [m]	$F_{\Delta P}$ [MN]	F_{TOT} [MN]
"Stubby" pile D												
12	46.2	62.5	0.75	85.2	54	1.15	9.4	0.11	2.8	8.05	5.0	7.8
13	46.2	62.5	0.75	85.2	58	1.43	9.4	0.11	2.8	8.73	7.0	9.8
15	41.6	62.5	0.75	85.2	55	1.50	9.4	0.11	2.8	8.25	7.4	10.2
23	39.8	62.5	0.75	85.2	52	1.58	9.4	0.11	2.8	7.87	8.0	10.8
"Medium" pile C												
19.1	39.2	38.5	0.50	68.8	53.3	1.84	5.8	0.08	2.3	7.99	2.5	4.8
19.2	39.2	38.5	0.50	68.8	61.2	1.96	5.8	0.08	2.3	9.18	2.8	5.1
19.3	39.2	38.5	0.50	68.8	66	2.17	5.8	0.08	2.3	9.96	3.4	5.6
20	38.9	38.5	0.50	68.8	68	2.27	5.8	0.08	2.3	10.19	3.7	6.0
"Slender" pile F												
16	41.9	25.3	0.25	36.1	98	2.44	3.8	0.04	1.2	14.72	1.5	2.7
17.1	41.9	25.3	0.25	36.1	84.3	2.34	3.8	0.04	1.2	12.65	1.4	2.6
17.2	41.9	25.3	0.25	36.1	94.5	2.82	3.8	0.04	1.2	14.17	2.0	3.2
17.3	41.9	25.3	0.25	36.1	100	2.98	3.8	0.04	1.2	14.97	2.1	3.3
22	40.9	25.3	0.25	36.1	101	2.53	3.8	0.04	1.2	15.08	1.6	2.8

Table 3 Summary of suction tests: total forces F_{TOT} for a final penetration H

In Fig. 5 and 6, it is shown how the pressure difference increases with the penetration depth. It appeared that a nearly linear relationship can be observed. Fig. 5 shows the relation with the density of the sand (test pile D 13, 15 and 23 of Table 3). As may be expected, the pressure

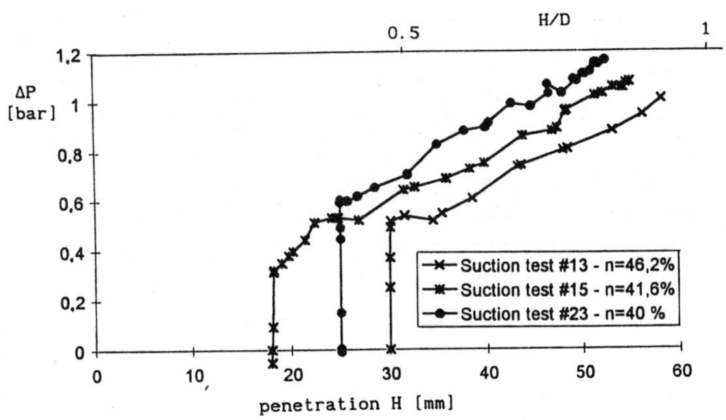

Fig.5 Graphical presentation of the influence of the density on the required pressure.

level increases if the density increases. In Fig. 6, it is shown with similar tests (test pile C 19 and 20 of Table 3) that there is not a large difference between suction at once or suction by steps. As can be seen, the pile starts penetrating if the same pressure difference is reached as the maximum value before the pressure is decreased to zero.

In order to examine the influence of the different parameters, the characteristics of the piles used were rather different. Therefore not all parameters could be compared in a quantitative way. If the pressure difference is considered in relation to the final penetration depth, it is clear that the pressure increases if the depth increases. In an attempt to normalize the penetration depth, the pressure difference ΔP is divided by the height H of the pile (a linear relation is assumed because of the results in Fig. 5 and 6). It appeared that the obtained relative values were rather close to each other (D:1.1; C:1.3; F:1.0). A relative number for the total load can now be obtained by multiplying the previous numbers by the pile surface. The following data is then obtained: D:7.1; C:3.8; F:1.0. A remaining difference between the piles is the circumference (linear with D) and the wall thickness (linear with d). Since the water flow is only effecting the soil close to the boundary, it

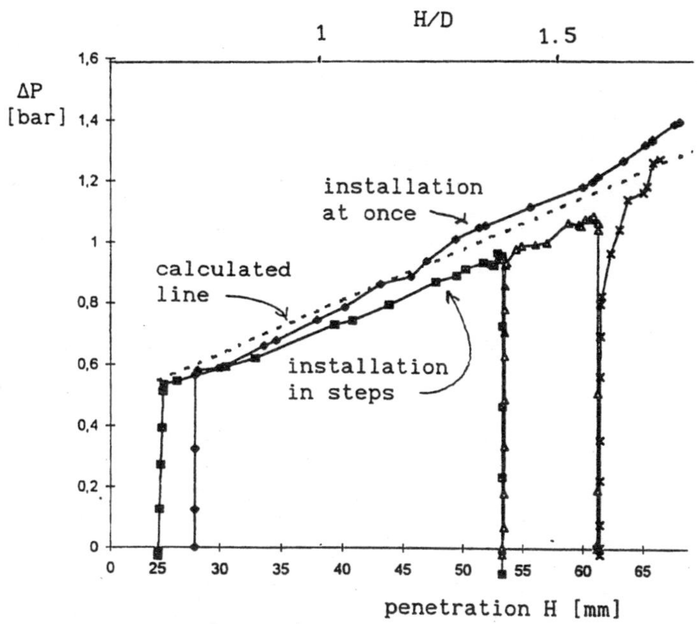

Fig.6 Suction test to demonstrate that a stop during the process has no significant influence. The dashed line represents the calculated pressure required to fluidize the sand at a similar depth.

can be assumed that the wall of the suction pile behaves as a long strip. Theoretically the length and wall thickness of a strip have a linear influence on the penetration resistance. If the previous values are also normalized by the wall thickness and circumference, the following ratios are found D:0.95; C:1.0; F:1.0. It appeared that the values are very close to each other which means that simple linear relations can be used to compare the results of the different suction tests. This also indicates that the dimensions of the tested piles (e.g. *H/D* ratio) have no special influence on the penetration load. The relation can be described by the following non-dimensional equation

$$\frac{\Delta PD}{\gamma Hd} = C \qquad 2)$$

where C is a constant and γ is the volume weight of the soil.
The roughness of the boundary of the suction pile can be assumed to be the same for all piles.

The penetration-pressure difference diagrams can be used to compare the driving load during suction with the mechanical driving load. A comparison between the results of penetration tests and suction test with the stubby pile D was made. In each test, the saturated sand samples have been used with porosity near the minimum porosity.
The mechanical penetration test required a load of 3.44 kN for a penetration of 50 mm. In prototype dimensions a force of 77 MN would be necessary for a final penetration of 7.5m. Suction tests performed in the same soil configuration required a total penetration load of about 10 MN. As expected the resistance during mechanical penetration is much higher than the force needed in the case of suction. Since the difference is much more than two times larger it can be concluded that the reduction of penetration load is mainly caused by reduction of the tip resistance.
For the soil heave analysis, only tests performed under the same conditions have been considered. These suction tests are performed in samples with almost the minimum porosity. By means of a nozzle, similar penetration speeds are obtained and a filter was used to prevent discharge of sand .

Anchor type	Mean diameter D [mm]	Wall thickness d [mm]	Ratio d/D [%]	Effective Height [mm]	Final penetration (average) [mm]	Soil heave (average) [mm]	a
Anchor D	62.5	0.75	1.2	60	55	5	.09
Anchor C	38.5	0.50	1.3	75	68.5	6.5	.095
Anchor F	25.3	0.25	1.0	108	100	8	.08

Table 4 Calculated results for heave factor *a*

As expected, the heave factor *a* tends to remain constant. The higher *a*, the greater the soil heave for the same penetration depth. A more accurate comparison between the different piles can be obtained if the volume of the sand heave is related to the volume taken by the wall of the pile in the soil. Pile D has comparatively a greater wall thickness than anchor F, as indicated by the *d/D* ratio in Table 4. At a same penetration depth, pile D will thus comparatively heave more sand because of the higher volume taken by the anchor wall. Therefore, the greater the *d/D* ratio, the greater the soil heave and the higher *a* becomes (see Table 4).
French experiments (Fau and Raynaud, 1986) carried out at 1g in saturated sand on 1/50 scale models have led to a heave factor of about 0.2. This large heave obtained at 1g can probably not occur with the prototype because a sand column with a low porosity can only be stable on a small height. In centrifuge modeling a more realistic behaviour of the soil is obtained. Therefore the sand heave probably reaches a lower value.

Numerical calculations with the FEM computer program Plaxis (1995) have shown that fluidization in the anchor will preferentially occur near the wall and the edges. In Fig. 6 the dashed line indicates the calculated pressure difference at which the sand starts to fluidize near the inner wall. This line, which is based on calculations with Plaxis, is in good agreement with the measured pressure difference. The finite element analysis can only be used to calculate the start of fluidization. At a later stage the sand properties change, so that it is not known what soil parameters have to be used in the calculation. The definition of the soil parameters is also a problem in calculating the driving load during suction. Therefore no further attempts were made in this stage to predict the driving load by FEM calculations.

CONCLUSION

The main objective of this research was to investigate to what extent suction pile installation can be simulated in a geotechnical centrifuge. The test technique developed appeared to be successful. It was possible to simulate the pump capacity by pressurizing the test cylinder and the relevant parameters could be measured sufficiently accurately.
This technique has been employed to simulate the installation of a suction pile in dense sand. The results of these suction tests are in good agreement with each other. As expected, the resistance during mechanical penetration is much higher than the force needed in the case of suction.
The analysis of results confirmed the following phenomena:

- Water flowing through the soil reduces the tip resistance and the boundary friction strongly (totally by a factor of 8).
  In dense sand, local fluidization seems to help in achieving complete penetration. Experiments have shown that in the pile the critical gradient is reached. Thus the sand along the inside wall of the pile must be fluidized. This involves a movement of the sand

grains under the tip of the pile, which reduces the tip resistance. In less dense sand the total force required is smaller.

- The inclination of a model pile decreases during penetration into the soil by suction.
- A pile with a ratio H/D=4 could be penetrated without any problem. However, more tetst are required to determine the relation between H/D ratio and initial penetration depth in more detail.
- The relation between the pressure difference ΔP, the pile diameter D, the height H and the wall thickness h appeared to be linear.

Adopting adequate scaling rules, test results have shown a good agreement with numerical calculations of the water flow. Theoretical calculation of the required pressure to reach the final penetration is quite delicate because the sand strength parameters in the fluidized area are not known. Experimental testing in the centrifuge seems therefore to be a very helpful tool for obtaining empirical data supporting the theoretical analyses.
A more intensive test program is required to validate the linear influence of the different pile dimensions. Also the influence of the surface roughness needs to be investigated in more detail.

REFERENCES

Allersma, H.G.B. (1990), "On line measurement of soil deformation in centrifuge tests by image processing", *9th Int. Conf. on Experimental Mechanics*, Vol.4, pp.1739-1748.

Allersma, H.G.B. (1994), "The University of Delft geotechnical centrifuge", *Int. Conf. Centrifuge94*, Balkema, Rotterdam, pp. 47-52.

Fau, D. and D. Raynaud (1986), "L'ancrage par succion. Etude sur modeles reduits." *Revue Française de Geotechnique*, No.36.

Fuglsang, L.D., J.O. Steensen-Bach (1991), "Breakout resistance of suction piles in clay", *Proc. Int. Conf. Centrifuge91*, pp.153-159.

Helfrig, S.C., R.L. Brazill, A.F. Richards, U. Lehigh (1976), "Pullout characteristics of a suction anchor in sand". *Offshore Technology Conf., OTC6029*, pp. 535-540.

Plaxis Manual (1995), Finite element code for soil and rock analysis. *Version 6, Balkema, Rotterdam.*

Renzi, R., W. Maggioni, F. Smits (1991), "A centrifugal study on the behaviour of suction piles." *Proc. Int. Conf. Centrifuge91*, pp.169-176.

Senpere, D. and G.A. Auvergne (1982), "Suction anchor piles- A proven alternative to driving or drilling. *Offshore Technology Conference, OTC4206*, pp. 483-494.

Wang, M.C., K.R. Demars and V.A. Nacci (1978), "Application of suction anchors in offshore technology."*Offshore Technology Conference, OTC 3203*, pp. 1311-1320.

Proceedings of the Seventh (1997) International Offshore and Polar Engineering Conference
Honolulu, USA, May 25-30, 1997

ISBN 1-880653-28-1 (Set); ISBN 1-880653-29-X (Vol. I)

Prediction of Lateral Capacity of Piles in Clays from Standard Penetration Tests

Yenumula V.S.N. Prasad
Newfoundland and Labrodor Hydro
St. John's, Canada

ABSTRACT

A method is suggested to calculate the ultimate load and ground-line displacement of laterally loaded piles in clays based on standard penetration resistance values. This method is based on the correlations between SPT-N values and pressuremeter limit pressures. The proposed method is verified with the published test data on prototype piles.

KEY WORDS: Clays, ground line displacement, lateral load, piles, standard penetration resistance, ultimate load

INTRODUCTION

In many Port and Harbour structures, pile foundations are usually subjected to lateral loads in addition to vertical loads. The bearing capacity and settlement of foundations under vertical loads in cohesionless soils can conveniently be estimated directly from the results of standard penetration tests for shallow and deep foundations (Terzaghi and Peck, 1967; Meyerhof, 1956; 1959; 1965; 1976). Very recently, Meyerhof (1995b) presented a method to estimate the behaviour of laterally loaded pile in cohesionless soils through SPT-N values. In that method, the existing correlations between the pressuremeter limit pressure p_l and standard penetration resistance N values are used.

In this paper, correlations between the results of static cone penetrometer, pressuremeter and standard penetration tests in cohesive soils are used to develop simple methods of analysis for preliminary estimates of ultimate lateral resistance and ground line displacements in rigid and flexible piles directly from standard penetration tests and pile geometry. The validity of the proposed method is examined by comparing with published field test data on prototype piles.

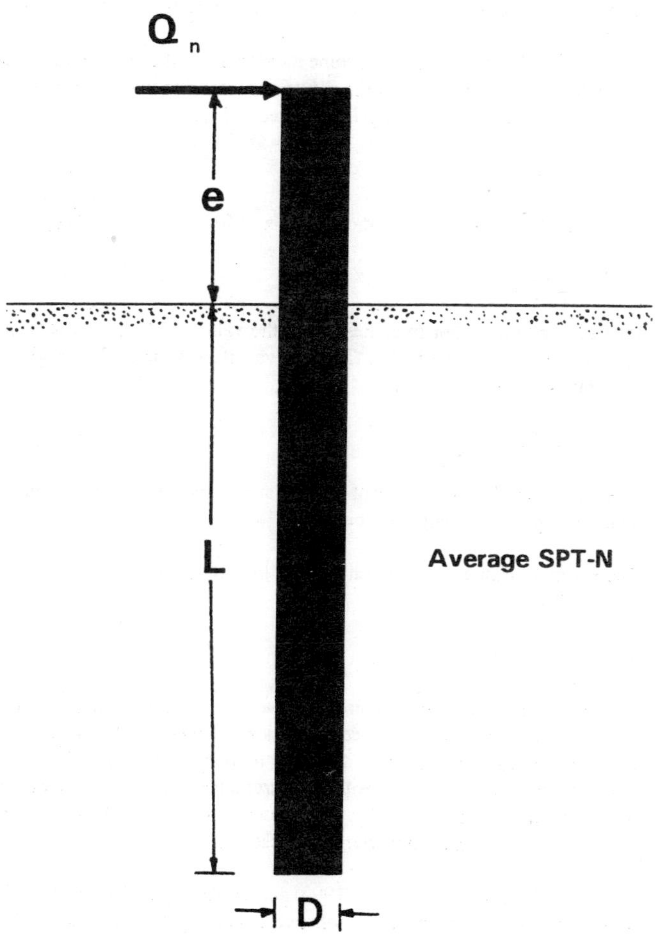

Fig. 1 Geometry of a pile under lateral load

THEORY

Lateral Resistance of Flexible Piles

If a vertical flexible pile of diameter D and depth L (Fig. 1) is driven to full embedment in cohesive soil of unit weight γ and standard penetration resistance N (blows/ 0.3 m) and is subjected at its free-head to an ultimate horizontal load Q_n applied at ground level, it has been shown (Meyerhof et al., 1981, Meyerhof and Sastry, 1987b) that, in the absence of structural failure,

$$Q_n = 0.4 \gamma D L^2_{eu} K_c \leq Q_{nl} \quad (1)$$

where

$$Q_{nl} = 0.4 \ p_l D L_{eu} \quad \text{for } L_{eu}/L_c \geq 1 \quad (2)$$

In Eq.2, L_c is the critical depth, which is about 5 times D in clays (Meyerhof, 1995a), L_{eu} is the ultimate effective depth (Meyerhof et al., 1988) and Q_{nl} is the ultimate horizontal load calculated based on limit pressure. In the above equations, K_c is the resultant net lateral soil pressure coefficient for the shaft of an equivalent rigid pile; and p_l is the limit pressure obtained from the pressuremeter tests (Baguelin et al., 1978).

If the load is acting at some eccentricity e, the above equations 1 and 2 are to be multiplied by the reduction factor due to the moment, r_c (Meyerhof et al., 1981) which is presented in Eq.3.

$$r_c = 1/(1+1.9e/L) \quad (3)$$

The ultimate effective depth ratio (L_{eu}/L) is (Meyerhof et al., 1988),

$$L_{eu}/L = 1.5 K_{rc}^{0.12} \leq 1 \quad (4)$$

for free-head piles. For fixed-head piles, the value of L'_{eu}/L is about one half the value for free-head piles. The relative stiffness factor (K_{rc}) is given by (Poulos and Davis, 1980)

$$K_{rc} = E_p I_p / E_s L^4 \quad (5)$$

where E_pI_p is the flexural rigidity of the pile; and E_s is average soil modulus along the embedded length of the pile.

The empirical relationship between p_l and N in cohesionless soils (Baguelin et al., 1978) is,

$$p_l = 50 \text{ N (kPa)} \quad (6)$$

But no such simple correlations were reported for cohesive soils. However, there are well established correlations between SPT-N value and cone penetration resistance, q_c. Similarly there are correlations between limit pressure, p_l measured with pressuremeter and cone penetration resistance, q_c. Hence, logically, correlations can be established between SPT-N values and pressuremeter limit pressure, p_l.

Kulhawy and Mayne (1990) quoted Szechy and Varga (1978) for identifying the consistency of cohesive soils using both SPT and CPT. The correlations are presented in Table 1. From the values presented, relationship between N and q_c can be represented by an Equation (7).

$$q_c = 200 \text{ N (kPa)} \quad (7)$$

The relationship between limit pressure, p_l obtained from pressuremeter and cone penetration resistance, q_c (Baguelin et al., 1978) is presented in Eq. (8).

$$q_c = p_l \text{ for soft clays} \quad (8a)$$

$$q_c = 2.5 \ p_l \text{ for firm to stiff clays} \quad (8b)$$

$$q_c = 4.0 \ p_l \text{ for very stiff to hard clays} \quad (8c)$$

Thus, for different consistencies of clays, direct relationship between p_l and N can be established using equations 7 and 8, as follows (in kPa):

$$p_l = 200 \text{ N for soft clays} \quad (9a)$$

$$p_l = 80 \text{ N for firm to stiff clays} \quad (9b)$$

$$p_l = 50 \text{ N for very stiff to hard clays} \quad (9c)$$

Substituting these average relationships (Eqs. 9) in Eq. 2, the ultimate lateral resistance of a free-head pile for $L_{eu}/L_c \geq 1$ is, approximately,

$$Q_{nl} = 80 \text{ N D } L_{eu} \text{ (kN) for very soft clays} \quad (10a)$$

$$Q_{nl} = 32 \text{ N D } L_{eu} \text{ (kN) for firm to stiff clays} \quad (10b)$$

$$Q_{nl} = 20 \text{ N D } L_{eu} \text{ (kN) for very stiff to hard clays} \quad (10c)$$

If the values of the $L_{eu}/L_c \leq 1$, the above capacities are to be reduced by a factor of L_{eu}/L_c.

Ultimate Lateral Resistance of Rigid Piles

For rigid piles, K_{rc} greater than about 0.01 (Poulos and Davis, 1980), Eq.10 simplifies to (for $L/L_c \geq 1$)

$$Q_{nl} = 80 \text{ N D L (kN) for very soft clays} \quad (11a)$$

$$Q_{nl} = 32 \text{ N D L (kN) for firm to stiff clays} \quad (11b)$$

$$Q_{nl} = 20 \text{ N D L (kN) for very stiff to hard clays} \quad (11c)$$

If the values of $L/L_c \leq 1$, the above capacities are to be reduced by a factor of L/L_c. Thus, Q_n can be directly calculated from the standard penetration resistance and pile geometry. For bored and non-displacement free-head piles the ultimate lateral resistance, Q_n is frequently about the same as those given above for displacement piles (Broms, 1964; Sastry and Meyerhof, 1987a). On the other hand, for fixed-head piles, the capacities are twice those of the corresponding free-head piles (Broms 1964).

Displacements of Piles under Working Loads

The lateral ground-line displacement, y_h of a vertical fully embedded flexible pile under horizontal load, Q (Q less than ultimate load) acting at ground level can be readily estimated from an equivalent rigid pile when (Poulos and Davis,1981; Meyerhof et al., 1988)

$$y_h = Q\ I_{yH} / E_s L_e \qquad (12)$$

where I_{yH} is elastic influence factor, which can be taken as 4 and E_s is the average soil modulus. The theoretical relationship between L_e/L and K_{rc} for free-head piles is, approximately, given by (Meyerhof et al., 1988)

$$L_e/L = 2.1\ K^{0.2}_{rc} \qquad (13)$$

and about one half this value for fixed piles.

The average value of E_s is given below from the established correlations between E_s and p_l shown (Baguelin et al., 1978) as follows:

$$E_s = 20\ p_l \text{ for soft and firm clays} \qquad (14a)$$

$$E_s = 30\ p_l \text{ for stiff to hard clays} \qquad (14b)$$

and, after substituting Eq. 9 into Eq. 14,

$$E_s = 4400\ N \text{ for soft clays} \qquad (15a)$$

$$E_s = 1600\ N \text{ for firm, stiff and hard clays} \qquad (15b)$$

Thus the ground line deflection of a pile subjected a load Q (kN) at ground level can be approximated using Eqs.12 and 15 as:

$$y_h = Q/\ 1000\ N\ L_e \text{ for soft clays} \qquad (16a)$$

$$y_h = Q/400\ N\ L_e \text{ for firm, stiff and hard clays} \qquad (16b)$$

For bored piles, the values of defections will be 1.5 to 3 times the deflections calculated for driven piles (Meyerhof and Sastry, 1987a).

FIELD CASES

For validation of the proposed method suggested above, a few test results from published field case were considered. Bhushan et al. (1979) reported the test results of concrete drilled piers in very stiff clays. Five results of straight piers 2, 4, 6, 7, 8 tested in sites A and B are considered in this analysis (see Table 2 for details). The pier 2, tested in the site A, the average undrained shear strength (c_u)of the soil was 260 kN/m². E_s value is calculated using the relation $E_s = 70\ c_u$ (Poulos and Davis 1980). The SPT-N values are 19, 34, 35, 44, 20 and 20 over a depth of 4.27m (14 ft).

GAI consultants (1982) have analysed the test results in site B and they suggested the value of E_s as 17225 kN/m². The SPT-N values are 19, 30, 32, 22, 43, 24, and 31 in the test site B over a depth of 4.27m (14 feet). For a rigid pile, N values are averaged along the pile embedment depth L. On the other end, for flexible pile, the field N values are averaged along an equivalent rigid pile of ultimate effective depth L_{eu}.

Using the Eq.11c for very stiff clays, capacities were calculated and are presented in Table.2. It can be seen that the capacities are very close to the predicted values. Ground line deflections are calculated using Eq. 16. As the drilled piers can be considered as bored piles, the calculated displacements are multiplied by a factor of 1.5. It can be seen that there is a good agreement between the predictions and actual observed ground line deflections (refer Table 3).

CONCLUSIONS

On the basis of correlations between the results of pressuremeter, static cone penetrometer and standard penetration resistance tests in clays, simple methods of analysis have been developed to estimate the ultimate lateral resistance and ground-level displacements directly from standard penetration tests and pile geometry. Comparisons of lateral load tests on several piles in the field show excellent agreement. The methods proposed in this paper are very useful for preliminary estimation of capacities in the field. However, the proposed method is to be verified with more field test results.

REFERENCES

Baguelin, F., Jezequel, J.F., and Shields, D.H. (1978). "The pressuremeter and foundation engineering.," Trans Tech Publications, Clausthal, Germany.

Bhushan, K., Haley, S.C., and Fong, P.T. (1979). "Lateral load tests on drilled piers in stiff clays," Journal of the Geotechnical Engineering." ASCE, Vol.105, GT8, pp.969-985.

Broms, B.B. (1964). "Lateral resistance of piles in cohesionless soils," Journal of Soil Mechanics and Foundation Engineering. ASCE, Vol.90, SM3, pp. 123:156.

GAI consultants, Inc., (1982). " Laterally loaded drilled pier research - Volume 2," EPRI report EL-2197, Project 1280-1, Palo Atlo, California.

Kulhawy, F.H. and Mayne, P.W. (1990). " Manual on estimating soil properties for foundation design," EPRI report, Palo Atlo, California.

Meyerhof, G.G. (1956). "Penetration tests and bearing capacity of cohesionless soils," Journal of Soil Mechanics and Foundation Engineering. ASCE, Vol.82, SM1, pp. 866-1-19.

Meyerhof, G.G. (1959). "Compaction of sands and bearing capacity of piles," Journal of Soil Mechanics and Foundation Engineering. ASCE, Vol.85, SM6, pp. 1-30.

Meyerhof, G.G. (1965). "Shallow foundations." Journal of Soil Mechanics and Foundation Engineering. ASCE, Vol.91, SM2, pp. 21-31.

Meyerhof, G.G. (1976). "Bearing capacity and settlement of pile foundations," 11th Terzaghi lecture. Journal of Soil Mechanics and Foundation Engineering. ASCE, Vol.102, GT3, pp.195-228.

Meyerhof, G.G. (1995a). "Behaviour of pile foundations under special loading conditions: 1994 R.M. Hardy keynote address," Canadian Geotechnical Journal, Vol.32, No.2, pp.204-222.

Meyerhof, G.G. (1995b). "Standard penetration tests and pile behaviour under lateral loads in cohesionless soils," Canadian Geotechnical Journal, Vol.32, No. 5, pp.913-916.

Meyerhof, G.G., and Sastry, V.V.R.N. (1985). "Bearing capacity of rigid piles under eccentric and inclined loads," Canadian Geotechnical Journal, Vol.22, No.3, pp. 267-276.

Meyerhof, G.G., and Sastry, V.V.R.N. (1987a). "Influence of installation method on the behaviour of rigid piles in clay subjected to moment and horizontal load." Canadian Geotechnical Journal, Vol.24, No.1, pp.145-148.

Meyerhof, G.G., and Sastry, V.V.R.N. (1987b). "Full-displacement pressuremeter method for rigid piles under lateral loads and moments," Canadian Geotechnical Journal, Vol.24, No.4, pp. 471-478.

Meyerhof, G.G., Mathur, S.K. and Valsangkar, A.J. (1981). "Lateral resistance and deflection of rigid walls and piles in layered soils," Canadian Geotechnical Journal, Vol.18, No.2, pp. 159-170.

Meyerhof, G.G., Sastry, V.V.R.N., and Yalcin, A.S. (1988). "Lateral resistance and deflection of flexible piles," Canadian Geotechnical Journal, Vol. 25, No. 3, pp. 511-522.

Poulos, H.G. and Davis, E.H. (1980)."Pile foundation analysis and design," John Wiley&sons,NY.

Szechy, K. and Varga, L. (1978). "Foundation Engineering-Soil Exploration and Spread Foundations," Akademial Kiado, Budapest, 508 p.

Terzaghi, K.,and Peck, R.B. (1967). "Soil mechanics in engineering practice," John Wiley & Sons, New York.

Table 1. Consistency of cohesive soils using SPT and CPT (After Szechy and Varga, 1978)

N value	Cone tip resistance q_c/p_a	Consistency	Consistency index
< 2	< 5	very soft	< 0.5
2 to 8	5 to 15	soft to medium	0.5 to 0.75
8 to 15	15 to 30	stiff	0.75 to 1.0
15 to 30	30 to 60	very stiff	1.0 to 1.5
> 30	> 60	hard	> 1.5

Note: p_a = atmospheric pressure

Table 2. Comparison of ultimate lateral capacities

Pile details	Pile geometry			e	r_c	N	L_c (m)	L_{eu} (m)	Ultimate load, Q_n (kN) actual	Ultimate load, Q_n (kN) predicted	Reference
	D (m)	L (m)	K_{rc}	(m)							
Pier No.2 Site A	1.22	4.58	0.0292	0.23	0.91	28	6.1	4.48	1850	2050	Bhushan et al. (1979)
Pier No.4 Site B	1.22	3.81	0.6406	0.23	0.89	28	6.1	3.81	1730	1450	Bhushan et al. (1979)
Pier No.6 Site B	1.22	4.73	0.2712	0.23	0.91	29	6.1	4.73	2050	2370	Bhushan et al. (1979)
Pier No.7 Site B	0.61	2.75	0.1490	0.23	0.86	29	3.1	2.75	715	750	Bhushan et al. (1979)
Pier No.8 Site B	0.61	4.73	0.0169	0.23	0.91	29	3.1	4.35	1420	1400	Bhushan et al. (1979)

Table 3 Comaprison of ground line deflections

Pile details	Pile geometry			N	L_e (m)	Load Q (kN)	y_h (mm) observed	y_h (mm) predicted	Reference
	D (m)	L (m)	K_{rc}						
Pier No.2 Site A	1.22	4.58	0.0292	28	4.58	1335	30	39	Bhushan et al. (1979)
Pier No.4 Site B	1.22	3.81	0.6406	28	3.81	1110	38	39	Bhushan et al. (1979)
Pier No.6 Site B	1.22	4.73	0.2712	29	4.73	890	19	24	Bhushan et al. (1979)
Pier No.7 Site B	0.61	2.75	0.1490	29	2.75	535	20	25	Bhushan et al. (1979)
Pier No.8 Site B	0.61	4.73	0.0169	29	4.73	890	30	25	Bhushan et al. (1979)

Proceedings of the Seventh (1997) International Offshore and Polar Engineering Conference
Honolulu, USA, May 25-30, 1997

ISBN 1-880653-28-1 (Set); ISBN 1-880653-29-X (Vol. I)

Response of Soil Plug to Horizontal and Vertical Excitation

Yongkyu Choi
Kyungsung University
Pusan, Korea

Michael W. O'Neill
University of Houston
Houston, TX, USA

ABSTRACT

The principal objective is to ascertain whether simulated high amplitude seismic motion (horizontal or vertical motion of the soil) degrades the soil plug in an open-ended pipe pile driven into submerged sand so significantly that soil plug resistance would have to be reduced or ignored after a seismic event. A model pipe pile was installed into loose sand in a pressure chamber by impact and loaded under static compression, and static compression with superimposed simulated horizontal and vertical seismic motion, while allowing the pile head to settle. After seismic motion, the static compression load was reapplied until the pile failed. Plug failure was not observed in any of the tests.

Key words: Earthquake, piles, soil plug, pore pressure, sand, calibration chamber test.

Introduction

During an earthquake, the three main components of excitation (horizontal excitation of the ground, the vertical excitation of the pile due to superstructure feedback produced by vertical excitation of the ground, and the seawater excitation induced by the vertical ground shaking, that is, the seaquake) could have effects on the soil plugs in open-ended pipe piles installed at offshore sites. While the ground and pile excitation could be modeled by exciting the soil and pile with simulated motions, seaquake excitation induced by the vertical ground shaking can be simulated by pulsing the water pressure at the soil surface.

The static behavior of soil plug in sand has studied by several researchers using model tests (Kishida et al., 1977; Klos et al., 1977; Paikowsky et al., 1990; O'Neill et al., 1991; Lee et al., 1993; Choi and O'Neill, 1997). The studies by ONeill et al., Lee et al., and Choi and ONeill were performed in calibration chambers. The effects of earthquake motion on soil plug stability have not been studied in detail, Choi and O'Neill (1997) showed that plug resistance relaxed during pile loading with constrained vertical movement during horizontal earthquake excitation of the soil. The principal objective of this paper is to extend the results of the previous study to conditions of horizontal seismic motion of the soil in which pile is embeded and vertical dynamic motion of the pile itself produced by vertical soil excitation. The specific issue is whether such action degrades the soil plug in an open-ended pipe pile driven into submerged sand so significantly that the plug resistance would have to be reduced or ignored after a seismic event. Seaquake effects, which should not be significant in shallow water, are not considered here, but will be reported elsewhere.

An open-ended model pile composed of separate inner and outer tubes was driven successively by impact driving into a fully saturated micro-fine sand mass that had been pluviated to a known relative density and saturated inside a pressure chamber. A static compression load test was performed to failure after each installation. Simulated horizontal or vertical earthquake motion was then applied to the soil with the pile initially loaded to its measured ultimate compressive resistance. This biased load was maintained during seismic excitation. Seismic loading was simulated using an earthquake scaling technique (O'Neill et al., 1990, 1992) from a real earthquake record at a deep sand site in which the seismic motion was applied to the soil through the base of the pressure chamber by a closed-loop hydraulic testing system.

(a) Photograph of excitation test equipment

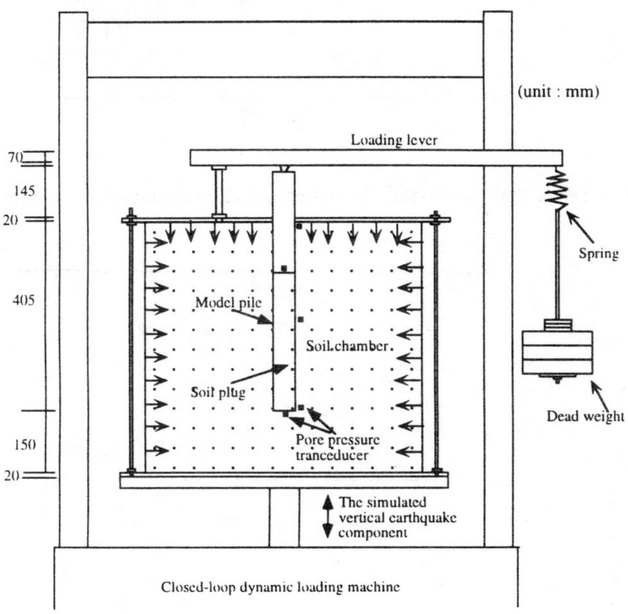

(b) Schmatic diagram of vertical excitation test

Fig. 1. The excitation test equipment

Experimental Arrangement

The testing system (Fig.1) consisted of a pressure chamber, the model pile, a simulated spring-mass system for maintenance of constant biased load on the pile head with a dynamic superposition that represented superstructure feedback, a closed-loop loading system, and data acquisition systems.

The assembled cylindrical test chamber was 564 mm in height and 610 mm in diameter. A flexible membrane of 3.2 mm thickness, which was placed against the inside of the cylinder wall and sealed to the outside, provided uniform total lateral confining pressure to the soil inside the chamber through regulated air pressure, while a separate flexible pressure membrane (pancake bladder), which was affixed to the underside of a rigid top plate, provided vertical confinement. Drains were provided on the top surface. Isotropic pressures were maintained. The horizontal seismic motion was applied to the base of the soil column by rotation of the test chamber and driving the pile off-centers. A central hollow cylinder was used in order to produce near-uniform strain fields in the vicinity of the test pile. Vertical seismic motion was applied to the base of pressure chamber by vertical excitation of the chamber and load was maintained on the pile, which was driven at the center of the chamber. Teflon sheets were placed on the insides of the membranes and on the outside face of the central hollow cylinder to reduce the friction between the sand and the membranes so as to produce predominantly vertically polarized shear waves. The test chamber had a diffuser (perforated tube) in the bottom of the chamber to saturate the soil, which also provided a rough surface for transfer of shear waves from the chamber into the sand column for the rotation tests. See Choi and O'Neill (1997) for futher details.

The sand was a very fine sand, classified as SM, with a coefficient of uniformity (Cu) of 1.66 and effective size (D_{10}) of 0.06 mm. The relative density of the pluviated sand was 50%, and its dry unit weight was 16.35 kN/m^3 . The coefficient of Darcian permeability was 1.21 x 10-3 cm/sec, and the effective-stress friction angle under submerged conditions between the pile surface and the sand was 22°.

A model pile, the components of which are shown in Fig. 2, consisted of separate outer and inner tubes. The pile was driven into the confined, saturated sand with an impact hammer of 49.8 N weight and 380 mm drop height. About one hour after driving, the pile was subjected to static compressive loading until the pile head settled 5-8% of the outside pile diameter. Failure loads were defined by the De Beer criterion (Choi and O'Neill, 1997). After the compression load applied to the pile head was removed, one blow with a hammer of 25 N weight and 300 mm drop height was applied to the pile head to recover the original stress conditions. A load equal to the load that produced about 95% of the static failure was reapplied. The soil in the chamber was then excited dynamically in a rotary or vertical mode, as explained further below. Total confining pressures of 20 kPa and 35 kPa were applied. After each dynamic test, the compression load was removed, and the compression load was reapplied to compare the ultimate pile capacity before shaking with that after shaking. Table 1 summarizes the testing plan.

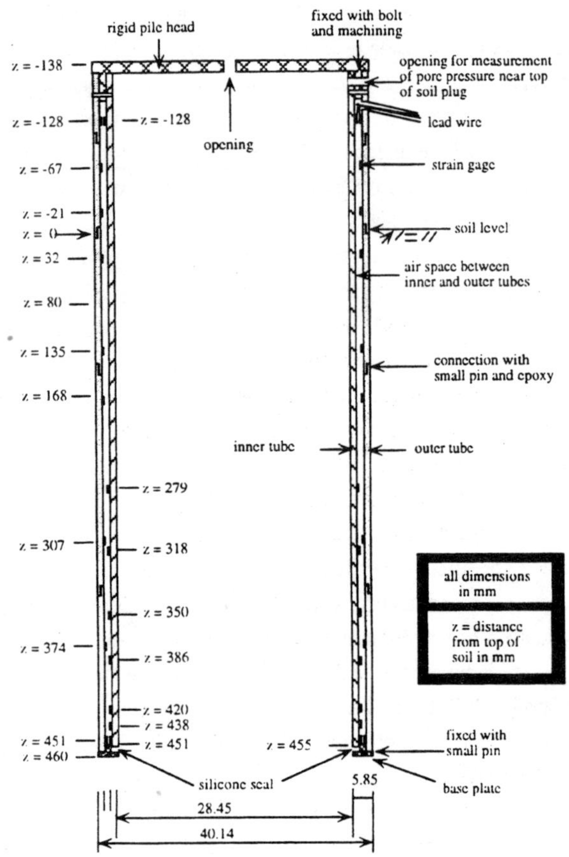

Fig. 2. Schematic drawing of specially designed model pile

Table 1. The detailed test plan

	Horizontal shaking test						Vertical shaking test					
Test type	h1-b	h1-e	a	h2-b	h2-e	h2-a	v1-b	v1-e	v1-b	v2-b	v2-e	v2-a
Relative density of sand (%)	50			50			50			50		
Confining pressure (kPa)	20			20			20			35		
Interaction between pile - structure	soft			stiff			soft			soft		
Static load test before shaking motion	X	X	X	X	X	X	X	X	X	X		X
Simuleted horizontal / vertical shaking motion		X	X		X		X	X		X		X
Static load test after shaking motion			X		X			X				X

The horizontal and vertical earthquake shakings were simulated using recorded data measured from SEMS, a 3-d strong motion seafloor accelerometer positioned just offshore of Long Beach, CA, at a distance of 74 km from the epicenter of an earthquake that occurred at Upland, CA, USA, in 1990. Simulated seismic motion was scaled to a Richter magnitude of 8.0 (Choi and O'Neill, 1997). Simulated seismic motion was applied by a closed-loop hydraulic testing system, and the resulting tangential and vertical displacement-time history of soil at the toe of the pile during simulated seismic shaking is shown in Fig. 3. Time and displacement scaling were employed to produce realistic diffusion behavior in the pore water. The pile head was allowed to deflect axially during seismic loading.

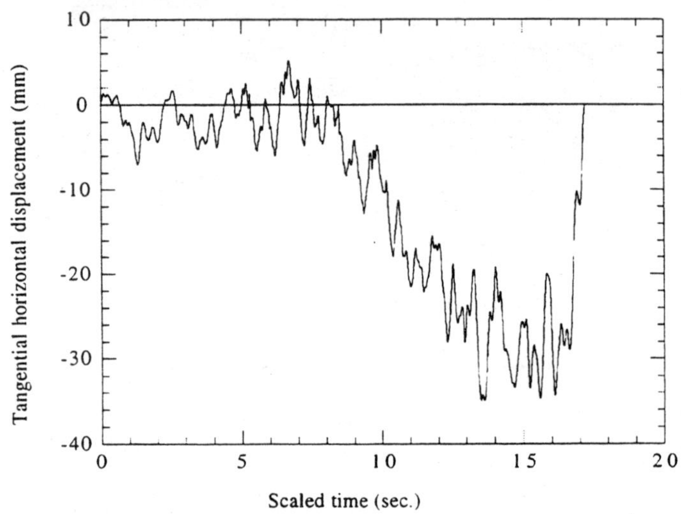

(a) Tangential horizontal displacement displacement-time history

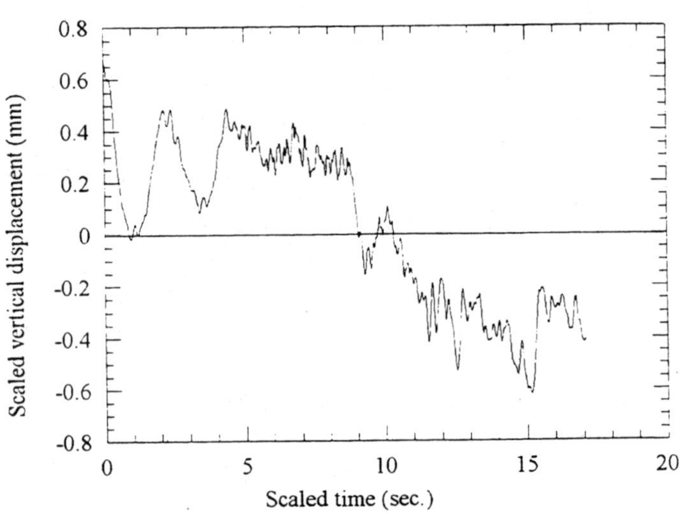

(b) Vertical displacement-time history

Fig. 3. Displacement-time history for the simulated earthquake excitation (M=8.0) applied to the test chamber

Model Pile, Instrumentation and Interpretation

The model pile was designed to measure the distribution of the soil plug resistance. The assembled model pile had a 40.14 mm outer diameter, 28.45 mm of inner diameter, and 5.85 mm composite wall thickness. On the outer tube, strain gage pairs were attached to the inside surface at levels near the toe, at 2.0, 4.0, 6.0, 8.0, 10.0, 12.0, and 13.5D_o from the toe, and at the soil surface, where D_o is the outside diameter of the pile. On the inner tube, they were attached to the outside surface, near the toe, at 0.5, 1.0, 2.0, 3.0, 4.0, and 16.5D_i from the toe, where D_i is the inside diameter of the pile. At the toe, a 5.85 mm-wide steel base plate was attached to the outer tube (only), so that both tubes deformed separately. At the toe, the gap between the base plate and the inner tube was sealed with silicone. The base-plate area was the same as the steel toe area of a solid-wall pile. A similar plate at the head of the pile was connected to both tubes. This arrangement allowed for direct measurement of plug resistance.

The strain gages were wired in full bridge configuration. Zero readings on all circuits were made before the pile was driven into the chamber so that the results presented later include the effects of residual stresses that developed during pile driving.

Results

Response of Soil Plug Horizontal Excitation

Fig. 4 shows the pile-head, load-settlement behavior obtained from the static load test before and after simulated earthquake motion and during the simulated earthquake load test, for an effective chamber pressure of 25 kPa. The applied compressive pile head load during excitation was greater than 95% of ultimate pile capacity before shaking, and the spring used in the spring-mass system was relatively soft. The pile head settlement during the applied seismic motion was 0.4 mm, but the pile did not approach failure. In the seismic motion test using a relatively stiff spring (Fig. 1) in the cantilever spring (h2 in Table 1)-mass system, the pile head settled during shaking by 3.2 mm (Fig. 6). Fig. 4 shows that the degradation of pile resistance during horizontal earthquake motion was 16% of ultimate pile capacity before shaking, which is obtained from the comparison of ultimate pile capacities before and after seismic motion was applied.

Fig. 5 shows pile-head displacement and pore pressure time histories developed during the simulated horizontal earthquake motion. Although relatively larger displacement (1.2 mm) occurred in the pile head at the moment of strong earthquake motion, only 0.4 mm of pile-head displacement remained after seismic motion ceased. Three of the bottom four traces were measured 0.5 D_o external to the outside face of the pile. The bottom trace is inside the plug. During earthquake motion, the pore pressure that developed under the toe of the soil plug and steel rim was relatively higher than elsewhere but much less than the confining pressure. This trend could have produced a gradient that caused upward flow in the plug and some degradation in capacity of the plug. Degradation upon reloading was observed in the soil plug and pile toe rim, and to some extent in the upper half of external pile skin. Comparision of pore pressure (u) near the top of soil plug with u beneath the toe suggests that the directions of the hydraulic gradients reversed during seismic motion.

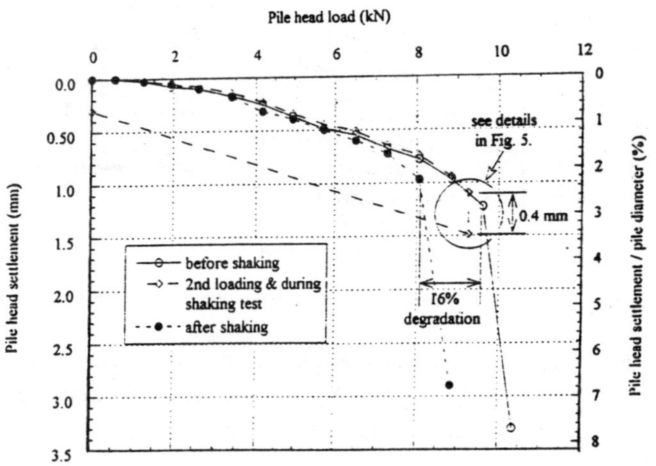

Fig. 4. Pile-head load vs. settlement in the simulated horizontal shaking test, Test h1 (confining pressure = 25 kPa, Q/Q_{ult} = 95%)

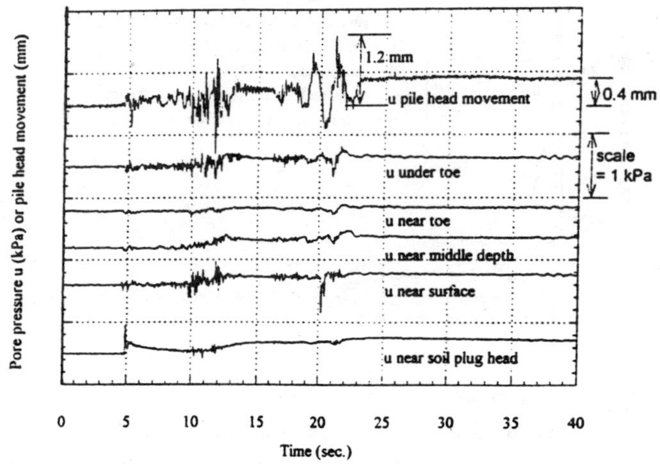

Fig. 5. Pile-head movement and pore pressure during the simulated horizontal motion, Test h1 (confining pressure + 25 kPa)

Fig. 6 compares the test results for tests h1 and h2

(non-failure conditions) during the simulated horizontal seismic motion, and results for a pile capacity relaxation test, which was performed with pile-head settlement held constant during the imposed seismic motion (Choi and O'Neill, 1997). In the former tests, 16% and 22% of ultimate pile capacity was lost, and, in the relaxation test, the ultimate pile load was reduced by 21%. In test h2, the pile settled abruptly during seismic excitation

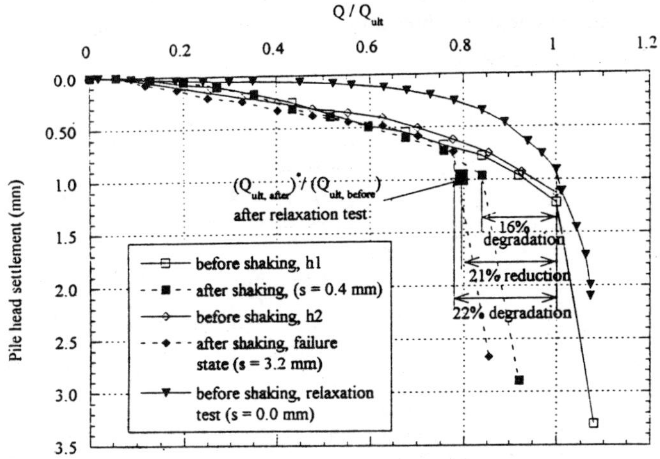

Fig. 6. Comparision of ultimate pile capacity in the static tests performed before/after the simulated horizontal earthquake motion (confining pressure = 20 kPa)

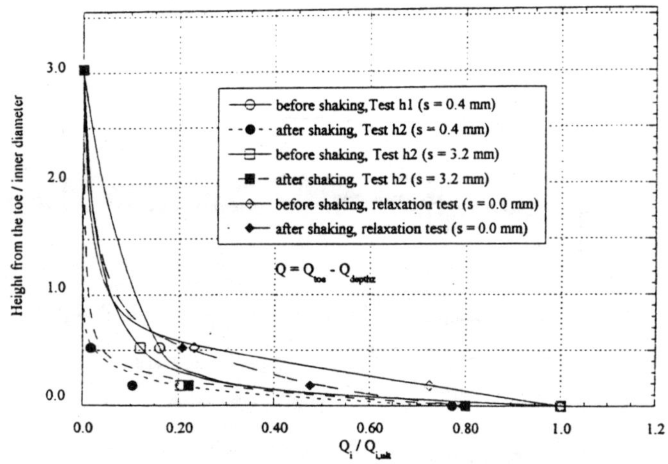

Fig. 7. Comparision of the normalized load in the inner tube, Q, developed during the static load test before/after the simulated horizontalearthquake motion (confining pressure = 20 kPa)

Fig. 7 shows the distribution and degradation of soil plugging resistance along the soil plug toe. Soil plugging was developed primarily in the bottom 2-3 D_i of the soil plug, and, during the seismic motion, degradation rates of tests h1 and h2 were 23% and 20% respectively, compared with 21% relaxation in the relaxation test. These degradations were only slightly greater than the overall degradation of the pile, but there was no evidence of plug slippage in the reloading test. This suggests that the pile degraded mostly by the reduction of ultimate bearing capacity of soil at the toe. The degradation of soil plugging resistance during horizontal excitation (which is not necessarily equal to capacity degradation) was about 20%, independent of the amount of pile settlement. While the pile could settle within the mass of surrounding external soil, the soil plug during shaking did not deform against the internal skin of the pile because the soil plug remained tight during seismic loading. Thus, seismic shaking did not likely degrade the soil plug so significantly that the plug resistance would have to be reduced after the simulated seismic event.

Response of Soil Plug to Vertical Excitation

Fig. 8 shows the pile-head, load-settlement relationships obtained from the static load tests before and after the simulated vertical excitation of the chamber, and during the simulated vertical shaking test. In this test (v2), the confining pressure was 35 kPa and the compressive pile head load was 95% of ultimate pile capacity. During the imposition of vertical earthquake motion, the pile head did not settle and the ultimate pile capacity did not degrade, suggesting no loss of plug resistance. Similar behavior was observed in test v1.

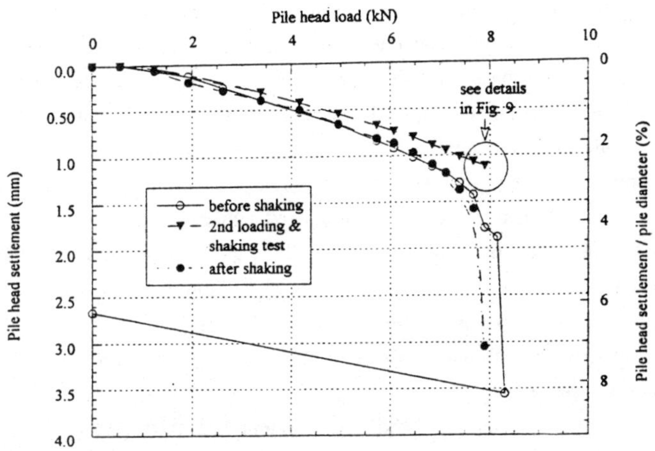

Fig. 8. Pile head load vs. settlement for the simulated vertical shaking test (confining pressure = 35 kPa)

Fig. 9 shows the pile head settlement, and pore pressure time histories during the simulated vertical excitation. The pore pressures were measured at the same locations as in the h series. The pore pressure response under the toe of soil plug was developed relatively higher than response the external to the pile surface. The response at the top of the plug was almost identical to that under the toe. Although pore pressures in the plug were observable, they were insufficient to cause plug failure. However, should conditions exist that produce higher pore pressures at the toe, such as a seaquake, plug failure may be possible.

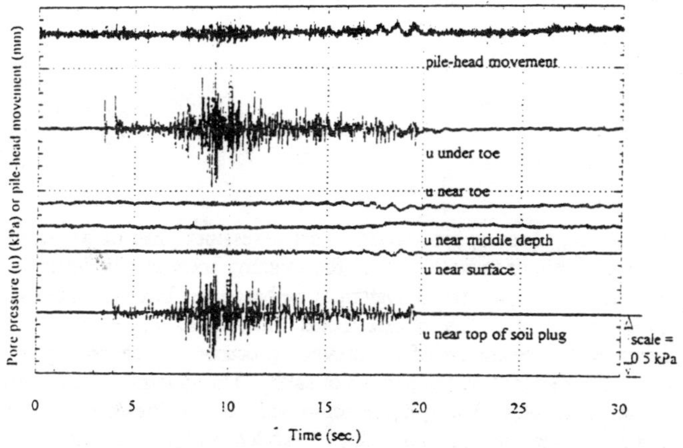

Fig. 9. Pore pressure response and pile head movement during simulated vertical shaking test v2 (confining pressure = 35kPa)

Conclusions and Comments

Soil plugging resistance and pore pressures were measured in and near an open-ended model pipe pile installed by impact driving into loose sand in a pressure chamber loaded under static compression superimposed with simulated horizontal or vertical seismic motion, where the pile head was free to settle. After seismic motion, the static compression load was reapplied until the pile failed. Some conclusions are as follows.

1. Horizontal motion applied to saturated sand in a pressure chamber representing the measured horizontal component of soil motion from a Richter magnitude 8.0 earthquake 74km from the pile location caused permanent reductions in axial compressive pile resistance and plug resistance of about 20% in an open-ended model steel pipe pile. The loss of capacity was primarily caused by a reduction of bearing capacity at the pile toe. There was no indication that the plug dislodged.
2. Vertical motion applied to the saturated sand representative of the measured vertical component from the same earthquake resulted in no loss of pile or plug resistance.
3. The loss in capacity during horizontal shaking was abrupt when the spring representing structral feedback was stiff, allowing for lager feedback loads at higher frequencies than occured with softer springs.
4. In both horizontal and vertical shaking tests, higher pore pressures occurred beneath the pile toe than elsewhere around the pile. Relatively higher pore pressure also occurred within the plug than exterior to the pile. This observation suggests that events that can produce higher pore pressures near the pile toe, such as seaquake, could produce pore pressures in the plug that could dislodge the plug before producing excessive loss of external skin friction.

References

Choi, Yongkyu, and ONeill, M.W. (1997). " Soil Plugging and Relaxation in Pipe Pile During Simulated Earthquake Motion," *J. of Geotechnical and Environmental Engineering*, ASCE, 1997. (Accepted for publication)

Kishida, H., and Isemoto, N., " Behavior of Sand Plugs in Open-ended Steel Pipe Piles," *Proc. of 9th ICSMFE*, Vol. 1, Tokyo, Japan, 1977, pp. 601-604.

Klos, J., and Tejchman, A., " Analysis of Behavior of Tubular Piles in Subsoil," *Proc. of 9th ICSMFE*, Vol. 1, Tokyo, Japan, 1977, pp. 605-608.

ONeill, M.W., and Raines, R.D., " Load Transfer for Pipe Piles in Highly Pressured Dense Sand," *J. of Geotechnical Engineering*, ASCE, Vol. 117, No. 8, 1991, pp. 1208-1226.

ONeill M.W., Vipulanandan, C., and Ochoa, M., "Response of Tension Piles to Simulated Seismic Motion in Saturated Fine Sand," *Report No. UHCEE 90-09*, Univ. of Houston, 1990.

ONeill, M.W., Vipulanandan, C., and Ochoa, M., "Laboratory Scale Modelling of Pullout Behavior of Bias-loaded Tension Piles During Earthquakes," *Proc. 24th Offshore Technology Conference*, Houston, Texas, USA, May 4-7, 1992, pp. 55-62.

Paik, K., Lee, J., and Lee, S.," Effects of Soil Conditions on the Behavior of Open-ended Steel Pipe pile," *J. of the Korean Geotechnical Society*, Vol. 9, No. 3, Sep., 1993, pp. 23-33.

Paikowsky, S.G., and Whiteman, R.V., "The Effects of Plugging on Pile Performance and Design, " *Can. Geotech. J.*, Vol. 27.3, 1990, pp. 429-440.

Proceedings of the Seventh (1997) International Offshore and Polar Engineering Conference
Honolulu, USA, May 25-30, 1997

ISBN 1-880653-28-1 (Set); ISBN 1-880653-29-X (Vol. I)

Dependence of Pile Reliability on Load Uncertainty

G. Singh and W.T. Lai
University of Leeds
Leeds, United Kingdom

ABSTRACT

So as to address the main challenges of pile design, rationally, various methods of reliability analysis are reviewed briefly and comprehensive Monte Carlo simulation recommended as the most dependable approach to meeting these challenges. It is argued that use of variance reduction or importance sampling should be avoided. A brief review of the sensitivity of a pile design in sand to various soil and size uncertainties as well as to model choice is followed by results, discussion and conclusions of a study into the design dependency on loading uncertainty. Increase in load uncertainty is shown to cause a large reduction in axial capacity, specially in the probability of failure range of 10^{-2} to 10^{-4}. Axial capacity becomes almost invariant at low probabilities of failure and it is recommended that the design probability of failure be restricted to $\leq 10^{-4}$.

KEY WORDS: Design, Models, Probability, Reliability, Sand, Simulation, Uncertainty.

INTRODUCTION

Consideration of load uncertainty is particularly important in the design of foundations for offshore structures and the axial reliability of a pile has usually the most dominant influence on the overall reliability of these structures. A very versatile, user-friendly, comprehensive and dependable Monte Carlo simulation methodology has been developed and used to quantify the influence of the mean and standard deviation of loading on a single pile in sand, using Terzaghi and Berezantsev models for pile-length/pile diameter (L/d) ratios of 20 and 40.

The problems of pile design manifest themselves primarily as three major challenges. The ***first*** challenge is presented by the need for improved understanding of the soil-pile interactions as well as for consideration of physical, statistical and human uncertainties. The proliferation of various formulae, leading to widely different answers (Lai et al, 1995), is a clear evidence of this problem. Along with the need for contributions to improved modelling it is vital to develop rational means of assessing these improvements. This assessment has to be based on rational comparison with and of the existing methods. In other words evolution of design methodology must be based on clearer objectives, and driven and monitored by reliable means. The ***second*** challenge is presented by the inertia of the prevailing practice of depending on the concept of factors (total or partial) of safety. This deterministic approach, whilst useful only in simple, common and non-significant situations, is irrational and discourages clearer understanding of the relative importance of the various influencing factors and models.

There has been a slow but steady, albeit controversial, move towards probability based approaches to engineering design. The ***third*** challenge is presented by the overwhelming inertia of these controversial approaches to the probabilistic design. This is discussed further.

At this stage it is important to point out that just as the reliability of a system cannot be obtained without dependable knowledge of the reliabilities of the elements in that system, the reliability calculations of a foundation of an offshore structure cannot be based on suspect assessment of the reliability of the individual piles.

This paper builds on the authors' previous presentations. These covered, respectively, the philosophical and pragmatic basis for the methodology used, the influence of the model choice on the calculated reliability, and sensitivity of design to various sand and pile related uncertainties (Singh et al., 1994 and 96; Lai et al., 95).

Various methods of reliability analysis are reviewed briefly so as to justify the chosen methodology. Particular attention is given to the numerous variations on the theme of Monte Carlo simulation put forward by various researchers. The paper will then present the results and discuss their significance from the value-engineering point of view.

METHODOLOGY FOR RELIABILITY ANALYSIS: REVIEW AND CHOICE

The commonly accepted classification of probabilistic methods, in the order of increasing levels of dependability (or reliability in the general sense of the word), is as follows:

LEVEL 0: Essentially deterministic methods that generally use one safety factor between characteristic strength and loading. The factor is obtained using higher level methods. For these factors to provide good value-engineering they must be derived through dependable/reliable higher level methods and not through those which are suspect themselves. Sadly in most current practices, these factors are calibrated with the latter (Level II).

LEVEL I: These are partly probabilistic using partial safety factors for the parameters. They have the same shortcomings as those of Level 0 methods. In any case, use of constant factors does not assure control of reliability level of a system. It also needs reliable calibration.

As pointed out earlier, whilst these two levels are useful in simple, common and non-significant situations, they are irrational and discourage clearer understanding of the relative importance of the various influencing factors and models.

LEVEL II: An overwhelming amount of literature on probabilistic methods belongs to this Level. Literature on this subject refers to First Order Reliability (FOR), First Order Second Moment (FOSM), First Order Mean Approach (FMA), First Order Design Point Approach (FDA), Approximate Full Distribution Approach (AFDA), Reliability Index Method (β method) etc. There is a proliferation of various versions and so called "improvements" of these that can be found in journals and speciality conference proceedings. A person new to this field may easily be influenced by this proliferation. These are essentially analytical approaches involving evaluation of the involved convolution integral. This is achieved, mostly, by either simplification/stylisation of: (a) the problem in hand or/and (b) the input distributions of the variables involved or/and (c) the output distribution of the objective function (reliability or its complement, probability of failure) (Singh and Chung, 1991). Sometimes the evaluation of the integral is performed by simplifying the process of integration itself through numerical (approximate) methods.

As pointed out by the authors in their previous presentation (Singh et al., 1994), not only are the results of these analytical approaches of suspect nature but the methods can be applied only to relatively simple problems. These methods may not be feasible for real engineering problems of even modest complexity. It is be useful to poin out that this level analysts often resort to Monte Carlo Simulation when they wish to validate their analytical approaches. Misuse of simulation can occur if the simplifications and stylisations of the problem are built into the simulation.

Pioneers of Systems Analysis (Operational Research) have long been cognisant of the fact that the simulation method can be used for realistic problems when all other methods fail. They rightly observed that simulation should be used as a last resort. This observation was made in the 50s, 60s and 70s. Whilst the employers of engineers have become more discerning in their objectives and the means available to the engineers have advanced dramatically, many reliability analysts are still caught in the inertia of the earlier level II work on reliability, done in the environment where there was (only) some justification for seeing simulation as a last resort. Computers are now fast and inexpensive.

While it is right and proper to acknowledge the valuable contribution of workers at this level - to the extent that they have succeeded in drawing attention to the need for taking a stochastic approach to decision making - it is high time that an unjustifiable fixation with this level is broken. Sinking of further efforts in this bottomless pit is unacceptable. There should be a move towards cutting opportunity losses. It is advicable that practitioners and educators work together to overcome this inertia by bringing about a change for dependable engineering as against the pursuit of theoretic elegance for its own sake. A huge 'superstructure' of *notional reliability* results has been and is being created using undependable 'foundation' of this level methods. This structure is bound to collapse, sooner or later. The earlier the profession builds a new superstructure on solid foundations of Monte Carlo Simulation the more *effective and efficient* it shall be.

LEVEL III: This is thought to be the most comprehensive and dependable level available to the profession at any particular stage of the evolution of the reliability art and science as well as of modelling and data developments. This level provides the benchmark against which results of the lower level methods are evaluated and calibrated. In the authors view (Singh et al., 1994) there is only one practical and dependable methodology available at present i.e. ***comprehensive*** Monte Carlo simulation (MCS) which meets the design needs from pragmatic and philosophical points of view. The authors listed the ideal features desired in any reliability methodology for the real life problems at ISOPE-94, and showed how these were provided by the authors' methodology of "VENTURER" (Singh, 1987). The methodology produces, effectively (dependably) and efficiently the desired results in graphical and tabular forms. The methodology avoids infliction of "paralysis of analysis" and of unhealthy degree of pursuit of theoretic elegance for its own sake.

There has been in very recent years some significant signs of acknowledgement and understanding of the power of MCS (Tichy, 1993) but this progress has been held back by the inertia of the Level II methods and the accompanying unjustifiable degree of preoccupation with speed.

Whilst desire to improve speed of analysis is understandable the sacrifice of effectiveness (dependability) for the sake of efficiency is unacceptable. Sadly, in recent years there have been a number of publications (Melchers, 1993) pursuing greater "efficiency" with MCS. These present various versions of variance reduction and importance sampling techniques which do produce results with lower number of simulation but the dependability of these results is suspect because of the assumptions and simplifications required for the sake of elegance and speed. As pointed out by the authors (Lai et al, 1996) these techniques are retrograde steps that should be avoided in a reliability based engineering design wherein probability distribution-tail and shape are sought. These "efficient" techniques may be used, if at all, only in studies wherein the decision maker is content with knowing only the mean and the variance.

CHOICE OF PARAMETERS

Soil (Sand): Detailed discussion of this has been presented by Singh et al. (1996). A summary is presented in Table 1. The chosen means and coefficients of variations (COVs) represent the middle of the ranges reported by various researchers. This should allow a fair assessment of the influence of load uncertainties. It is important to point out that the dependence of angle of internal friction, ϕ, on density, γ, has been recognised and incorporated with the relationship of $\phi = a*\gamma - b$. In this study a (= 6.5) and b (= 78.9) have been assumed to be deterministic although the methodology allows for them to be treated probabilistically.

Piles: These are also presented in Table 1.

Models: The authors reviewed a number of models (Lai et al., 1995) that are available to designers. These were assessed and compared with respect to the computed probabilities of failure for the same set of input data. That study facilitated model appraisal and selection in response not only to calculated risks but also to a designer's attitude to these risks. Of the various models studied those of Terzaghi and Berezantsev were found to be in the middle of a wide range observed. These two appear to represent reasonable compromise models in terms of deterministic design loads at failure probabilities of about 10^{-4} and they have reasonable sensitivity coefficients.

Loads: These have been treaded deterministically as well as probabilistically. Each load considered for the computation of probability of failure, P_f, has been treated as a mean value with COV of 0, 5, 10 and 20% with a normal distribution. The latter assumption has been made only to highlight the influence of COV. The methodology can handle any kind of input distribution and makes no assumptions about the output (objective) distribution.

SENSITIVITY OF PILE DESIGN TO MODELS AND VARIOUS UNCERTAINTIES: BRIEF REVIEW

Effects of model choice, sand density and other parameters, particle size of sand and pile dimensions have also been studied and reported in detail by the authors at the last two ISOPE conferences.

Models: Selection of bearing capacity factor N_q is not an easy task. All the four models studied (Berezantsev (1961), Coyle and Castello (1981), Meyerhof (1976) and Terzaghi (1943)) gave not only very different values of N_q but, as expected, very different probability of failure (P_f) values at a given value of pile capacity. At a failure probability of 10^{-4} Coyle and Castello model gave the lowest design axial load, followed respectively by Berezantsev, Terzaghi and Meyerhof models. One of the important observations was that in some cases a traditional factor safety greater than even three was not adequate for achieving a P_f of 10^{-4}. It was from these studies that Berezantsev and Terzaghi models were found to give middle range results.

Soil and Pile: As expected sand density has a very large influence on the computed capacity. Importance of paying greater attention to not only its mean but also its COV value has to be emphasised. Increase in COV value from 2 to 5% decreases the axial design load, at P_f of 10^{-4}, by almost threefold.

Although the COVs of coefficient of earth pressure, k_s, and angle of friction between sand and pile, δ, can be as high as 30% they are of minor practical significance.

The design axial loads for course sand are much higher than for medium and fine ones. However, the effects of variation of particle size are not as large as those of density.

Bearing in mind the fact that the COV of pile diameter is not likely to be more than 5%, it is safe to state that variation of its COV is also not of practical significance.

RESULTS OF THE STUDY OF DESIGN DEPENDENCY ON LOADING STOCHASTICITY

A summary of typical results is presented in Figure 1 and Table 2 for Berezantsev and Terzaghi models for L/d ratios of 20 and 40.

The figure shows that as the COV values of load increase the mean axial capacities decrease and the shape of the curves between P_f of 10^{-2} and 10^{-4} becomes more sensitive to the increases in COVs. For example, at COV of 0% the differences in the mean load at P_f of 10^{-3} and 10^{-4} are insignificant compared to those for COV of 20%. The differences become more pronounced as the L/d ratio increases. It is most interesting to note that at P_f of 10^{-5} and less the mean loads are almost constant for even the highest value of COV. This threshold P_f value rises to 10^{-3} at COV of 0%. Generally as the mean load value increases the various curves for the respective L/d ratios tend to converge to constant P_f values. The P_f value increases with decreasing L/D ratio.

The results show the importance of paying greater attention to load stochasticity. In the P_f range of 10^{-2} to 10^{-4} a large decrease is caused in the mean axial capacity as the load-COV increases from 0 to 20%. The designer should, therefore, select an appropriate value.

In Table 2 the third column shows the mean load capacity values calculated with the program VENTURER at P_f of 10^{-4}. The fourth column gives the results of the traditional deterministic methods. The total factor of safety shown in the fifth column was determined by dividing the fourth by the third column. The last column is obtained by dividing the deterministic capacities by the conventional factor of 3.

As can be seen in Table 2 (column 3) and the figure, almost for all values of load-COV and L/d the Terzaghi model gives higher mean loads than the Berezantsev model, albeit the difference becomes insignificant at L/d of 20. The Berezantsev model, in other words, is more conservative

It is useful to note that the findings of this study appear to support Meyerhof's recommendation that P_f values of 10^{-4} be used in foundation design. At lower loads the reliability of design rises sharply with even a very small reduction in the load. This is the first study of its kind which has produced this significant graph. The authors recommend that the P_f value used in design of piles for significant structures should be restricted to $\leq 10^{-4}$.

It is useful to see that increase in COV of load causes an increase in the value of total factor of safety required to ensure a P_f of $\leq 10^{-4}$. On the whole Terzaghi model requires lower total factors of safety. The question as to which is the best model (of the numerous available) remains unanswered.

The answer will *have* to depend on the designer's attitude to risk for any particular situation. A decision support system has been developed (Singh and Das 1995) to assist engineers and their employers to select the best from available alternatives.

CONCLUSIONS

For engineering design it is recommended that comprehensive Monte Carlo simulation be used without resorting to variance reduction or importance sampling devices. Effectiveness (dependability) of reliability analysis must not be compromised for the sake of efficiency (speed). From the study of the effects of load uncertainty, using Berezantsev and Terzaghi models, it was observed that:

- Berezantsev model is more conservative
- Increase in load-COV is accompanied by a large decrease in load capacity, specially in P_f range of 10^{-2} to 10^{-4}.

Table 1: Various soil parameters used in the analysis

Soil Parameter	Type of Distribution	Mean Value	COV (%)
Density of sand, γ	normal	18 kN/m^3	2%
coeff. of earth pressure, k_s	normal	1.5	10%
Angle of friction between pile and soil, δ *	normal	0.75ϕ	10%
Pile diameter, d	normal	0.5m	2%
L/d ratio (L - length of pile (m))	uniform (deterministic)	20 and 40	-

Table 2: Comparison Table for The Study of load uncertainty

L/d ratio	COV (%) of imposed load	Mean axial load at P_f of 10^{-4} (kN) **(a)**		Pile capacity based on deterministic method* (kN) **(b)**		Total factor of safety **(b/a)**		Deterministic capacity factored by 3 (kN)	
		Berezantsev	Terzaghi	Berezantsev	Terzaghi	Berezantsev	Terzaghi	Berezantsev	Terzaghi
20	0	1446	1467	4359	3360	**3.01**	2.29	1453	1120
	5	1300	1350	4359	3360	**3.35**	2.49	1453	1120
	10	1190	1200	4359	3360	**3.66**	2.80	1453	1120
	20	1000	1100	4359	3360	**4.36**	**3.05**	1453	1120
40	0	3177	3500	9400	7875	2.96	2.25	3133	2625
	5	2900	3160	9400	7875	**3.24**	2.49	3133	2625
	10	2650	2880	9400	7875	**3.55**	2.73	3133	2625
	20	2180	2360	9400	7875	**4.31**	**3.34**	3133	2625

* Using mean value of variables to compute pile capacity.

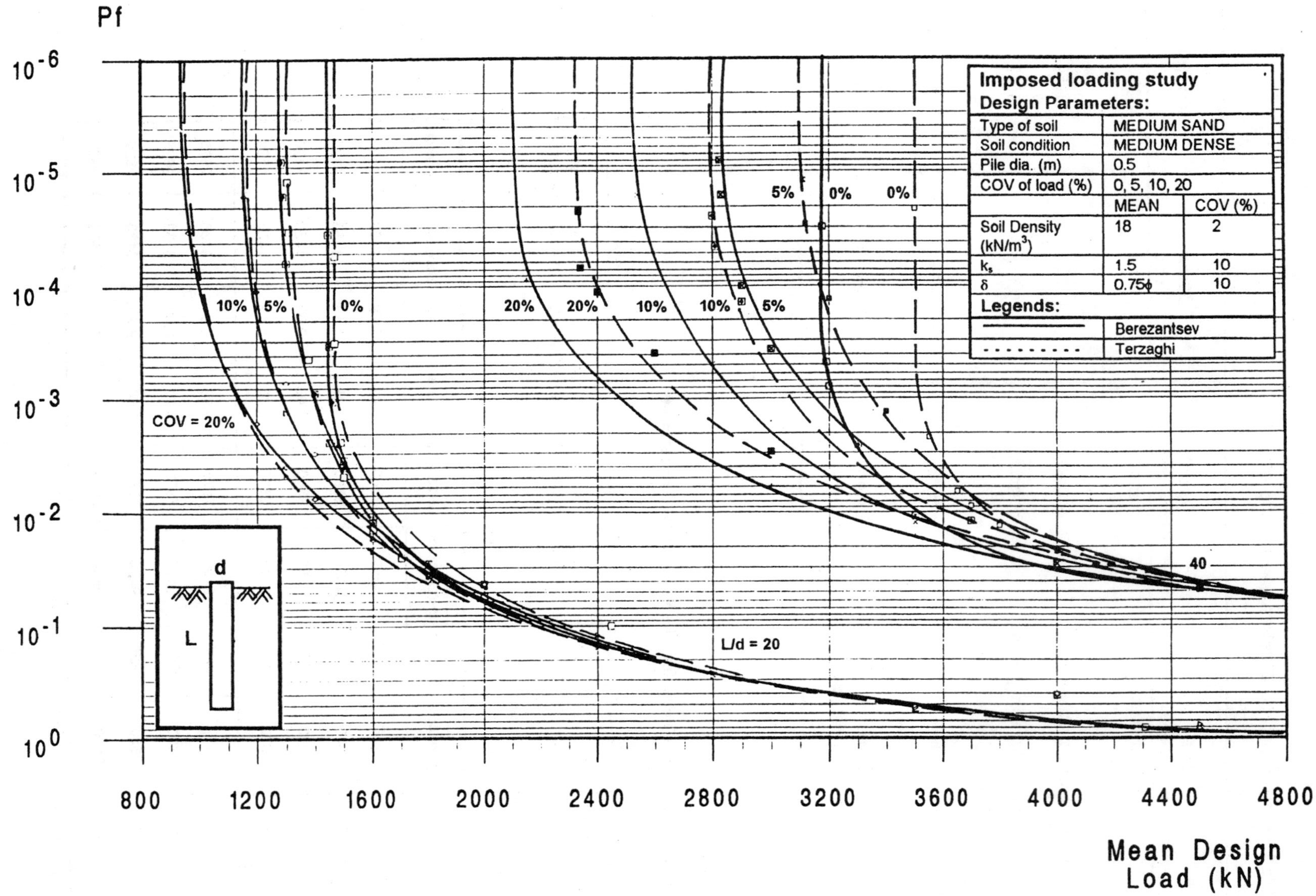

FIGURE 1: Influence of COV of Load on Probability of Failure.

- Influence of load-COV increases with L/d.
- The capacity becomes almost invariant at low values of P_f - this threshold P_f rises with decreasing COV.
- For significant structures, such as offshore, the design P_f value should be restricted to $\leq 10^{-4}$ - often requiring a factor of safety greater than 3.

REFERENCES

Berezantsev, V.G., Khristoforov, V.S., "Capacity and Deformation of Pile Foundations", Proceedings of the V International Conference on Soil Mechanics and Foundation Engineering, Vol.2, pp. 11-15.

Bowles, J.E. (1974a), "Analytic and Computer Methods in Foundation Engineering", McGraw-Hill Book Co., NY, 519 pp.

Coyle, H.M., and R.R. Castello (1981), "New Design Correlations for Piles in Sand", JGED, ASCE, Vol. 107, GT 7, July, pp. 965-986.

Meyerhof, G.G. (1976), "Bearing Capacityand Settlement of Pile Foundations", JGED, ASCE, Vol.102, GT 3, March, pp. 195-228 (Terzaghi Lecture).

Lai, W.T., Singh, G. and Das, B.M. (1995), "Influence of Model Choice on the Calculated Reliability of a Single Pile", Proc. Int. Conf., ISOPE-95, The Hague.

Lai, W.T., Singh, G. and Das, B.M. (1996), "Influence of Model Choice on the Calculated Reliability of a Single Pile", Int. Journal of Offshore and Polar Engineering, Vol. 6, No. 2, ISSN 1053-5381.

Melchers, R.E., (1993), "Modern Computational Techniques for reliability Estimation", Proc. Conf., Probabilistic Methods in Geotechnical Engineering, Australia, ISBN 90 54 10 3035.

Singh, G. (1987), "Risk and Reliability Appraisal on Microcomputer", Chartwell Bratt, *Studentlitteratur,* ISBN 91-44-27761-X.

Singh, G. and Chung, C. M. (1991) "Microcomputer-based Reliability Analysis in Geotechnics", Computers and Structures, Vol. 41 (6), PP. 1397-1402.

Singh, G., Das, B.M. and Lai, W.T. (1994), "Risk Assessment in Pile Design", Proc. Int. Conf., ISOPE-94, Japan, pp.492-497.

Singh, G. and Das, B.M. (1995), To engineer is to venture: be a frontier-venturer, Proceedings Second International Congress in Computing in Civil Engineering, ASCE, Atlanta.

Singh, G., Lai, W.T. and Das, B.M., (1996), "Sensitivity of Probabilistic Pile Design to Various Uncertainties", Proc. Int. Conf., ISOPE-96, USA.

Terzaghi, K. (1943), "Evaluation of Coefficient of Subgrade Reaction", Geotechnique, Vol. 5, No. 4, Dec., pp. 297-326.

Tichy, M., (!993), "Applied Methods of Structural Reliability", in "Topics in Safety, Reliability and Quality" Ed. Keuer, A.Z. Kluwer Academic Publishers

Proceedings of the Seventh (1997) International Offshore and Polar Engineering Conference
Honolulu, USA, May 25-30, 1997

ISBN 1-880653-28-1 (Set); ISBN 1-880653-29-X (Vol. I)

Vertical Capacity of Caisson Foundations in Calcareous Sediments

P.G. Watson and M.F. Randolph
University of Western Australia
Nedlands, Australia

ABSTRACT

This paper addresses the performance of caisson foundations in normally consolidated calcareous silt subjected to vertical loading. Results are presented from extensive centrifuge testing of a model caisson foundation, with particular focus on (a) the bearing capacity of caisson foundations compared to existing theory; (b) a comparison between drained and undrained bearing capacity; (c) the response to tensile loading; and (d) the effect of applying preload on the observed bearing capacity.

KEYWORDS : Caisson foundation, calcareous sediments, bearing capacity, preloading, tensile capacity, monotonic loading, cyclic loading

INTRODUCTION

In order to address recent problems associated with pile foundations, as experienced with the North Rankin A and Goodwyn A platforms, and in response to the discovery of extensive oil and gas fields in deep water off Australia's North-West Shelf, a joint industry project is being undertaken between the Special Research Centre for Offshore Foundation Systems at The University of Western Australia and various industry sponsors (see Acknowledgments). An integral part of this study is an examination of the performance of caisson foundations in calcareous sediments subjected to monotonic and cyclic loads characteristic of those developed throughout Australia's North-West Shelf. This paper presents the results of a series of centrifuge tests on a model caisson in calcareous silt, focusing in particular on the response to vertical loading.

As indicated, the current study is part of a larger investigation into the performance of caisson foundations in calcareous sediments. Unlike clays and sands, where abundant literature is available detailing the response of foundations to combinations of monotonic and cyclic loading, the susceptibility of calcareous sediments to large volume changes resulting from shear loading, and the risk of liquefaction associated with the build up of excess pore pressure during storm loading, means that extensive research is required to develop an understanding of caisson performance. Further, the consolidation characteristics of calcareous sediments are such that the overall time scale of consolidations events is within the lifetime of the structure. As such, preloading of the foundation either through the application of suction or by direct loading, and the dissipation of excess pore pressure after storm loading can result in beneficial increases in bearing resistance. The aspects of caisson performance examined in this paper include :

- the evaluation of bearing capacity from centrifuge test results, and a comparison with existing design approaches for caisson foundations,
- a comparison between partially drained and undrained monotonic loading,
- an examination of the undrained monotonic uplift (tensile) behaviour,
- an investigation into the effect of applied preload on the observed bearing capacity.

TEST PROGRAM

Geotechnical Centrifuge and Model Caisson

Testing has been conducted at the Geotechnical Centrifuge Facility at The University of Western Australia.

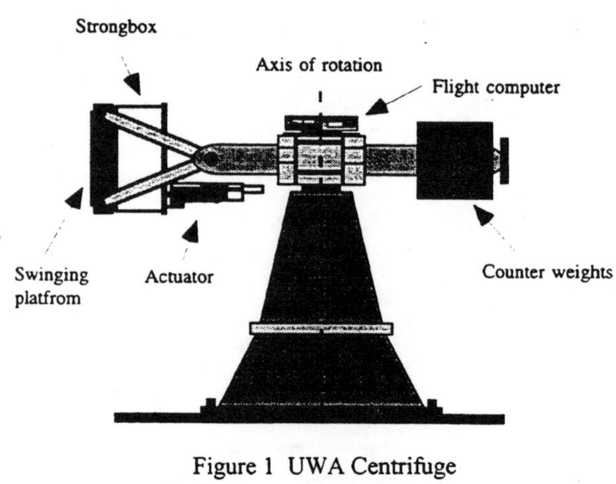

Figure 1 UWA Centrifuge

The UWA centrifuge, illustrated in Figure 1, is an Acutronic Model 661 geotechnical fixed beam centrifuge with a radius of 1.8 m and a maximum payload of 40 g-tonnes, and is capable of subjecting 200 kg sample packages to accelerated gravity fields up to 200 g. Control over actuators and other electronic equipment is achieved remotely from the main control room using power slip-rings, with instrumentation data signals and video signals being transferred from the test package to the control room via instrumentation slip-rings. Primary data acquisition and display of results is undertaken in the control room, with additional data being obtained from the in-flight computer which is capable of sampling selected channels at frequencies in excess of 50 Hz. A full description of the centrifuge and associated equipment is available in Randolph et al (1991).

In addition to the model caisson foundations designed and fabricated in-house, specialist equipment has been developed for the research project, including:

- A Stroud cell (Maddocks, 1981), purchased from Cambridge Insitu in 1995 and capable of recording vertical (V), horizontal (H) and moment (M) loads. This has been subjected to extensive calibration, and provides the principal means of determining loads applied to caisson foundations during testing.
- Unique actuators capable of providing high-speed, high-force actuation in both the vertical and horizontal directions, along with supporting electronic equipment and computer code. These enable the application of storm (cyclic) loading to caisson foundations at model frequencies up to 10-15 Hz, in addition to various combinations of monotonic vertical, horizontal and moment load.

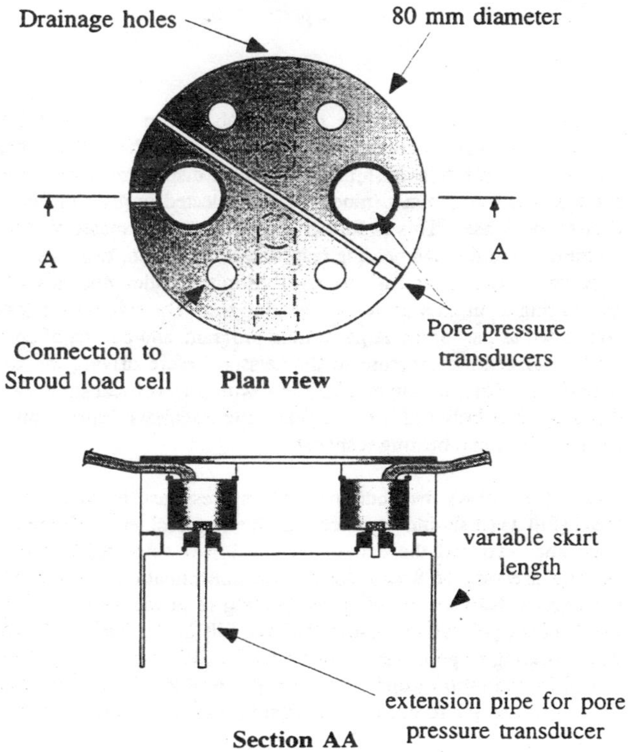

Figure 2 Model caisson foundation

The model caisson used for the test program, shown in Figure 2, consists of a circular caisson 80 mm diameter (D) with variable skirt lengths (d) ranging from 20% to 80% of the diameter. At 150 g this model represents a prototype caisson 12 m in diameter with skirts ranging from 2.4 m to 9.6 m. The design of the model ensures that deeper skirts can easily be fabricated to increase the d/D ratio, allowing investigation into foundations of variable geometry. For the purposes of this paper, the model caisson represents a prototype foundation 12 m in diameter with skirt length 4.8 m. Three miniature pore pressure transducers have been built into the model caisson enabling the measurement of excess pore pressure at skirt tip level, at the interface between the caisson top cap and the soil surface inside the caisson, and at the interface between the caisson top cap and the soil surface outside the caisson.

Soil Sample

The soil sample used throughout the caisson test program consists of calcareous silt obtained from the North-West Shelf. The soil sample has been based on the soil properties determined for a proposed offshore facility on the North-West Shelf, and is essentially a normally consolidated calcareous silt with undrained shear strength proportional to depth. To achieve the desired strength properties, the natural seabed material has been sieved to 300 μm and has been modified by the addition of ultra-fine commercially precipitated calcium carbonate powder (commercial name OMYACARB 1). In addition, silicon oil has been used as the pore fluid to provide appropriate consolidation parameters. Extensive material testing has been undertaken to characterise the centrifuge sample for comparison with the observed *in-situ* material properties, and includes:

- analysis of particle size distribution (performed by laser diffraction);
- specific gravity tests;
- Rowe cell (consolidation) and permeability tests;
- monotonic and cyclic simple shear tests;
- centrifuge testing including cone penetrometer, T-bar and pullout T-bar testing.

Results from two calcareous silt samples are presented in this paper, with the majority of caisson testing being conducted in the second sample. Table 1 presents the general soil properties of the calcareous silt obtained through laboratory testing (permeability and consolidation parameters determined from tests using silicon oil as the pore fluid and reported as the water equivalent value), while Figure 3 illustrates the results of centrifuge testing on the consolidated material in terms of deduced undrained shear strength *vs* depth. Future caisson testing will use calcareous silt 2 as the soil sample.

Table 1 : General soil properties, Calcareous silt 1 and 2

Parameter	
Average particle size, D_{50}	40 μm
Uniformity coefficient, D_{60}/D_{10}	55
Coefficient of curvature, $D_{30}^2/D_{60}D_{10}$	1.5
Specific gravity, G_s	2.76
Unit weight, γ'	7 - 8 kN/m^3
Voids ratio, e	1.16 - 1.44
Consolidation coefficient, c_v	800 - 1500 m^2/yr
Permeability, k	9×10^{-8} - 2×10^{-7} m/s
Average undrained strength gradient	Silt 1 : 2.1 kPa/m Silt 2 : 1.41 kPa/m

Each of the samples prepared for the centrifuge is prepared as a slurry and mixed under vacuum before being placed into the centrifuge strongbox. Plan dimensions of the strongbox are 650 x 390 mm, with a total depth of 325 mm. To avoid segregation and ensure a uniform normally consolidated sample, each package is subjected to a slow consolidation period of 2 - 3 days, with a gradual increase in gravity. The final model sample depth varied from 200 - 220 mm.

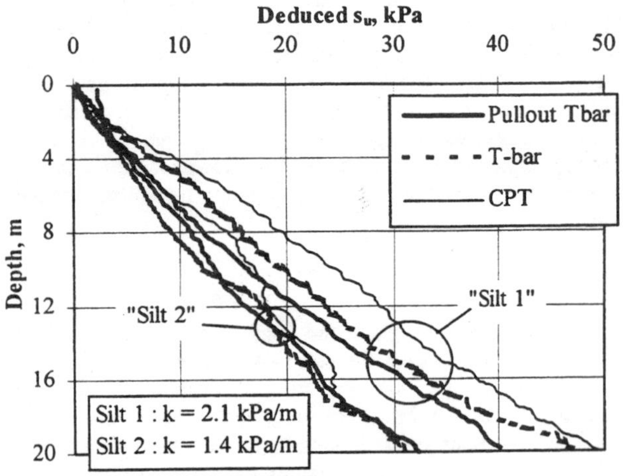

Figure 3 Results of "in-flight" centrifuge soil characterisation testing

General test procedure

The test procedure adopted for the foundation test program involved the following steps:

- Having consolidated the sample via a gradual rampup to 150 g, the soil characterisation testing (cone penetrometer, pullout T-bar and standard T-bar) was conducted to ensure the sample was uniform and normally consolidated, and to measure the soil strength profile.
- The actuators and model caisson required for the foundation test were then positioned on the centrifuge, and the sample reconsolidated at 150 g with the model caisson positioned out of the soil.
- The foundation was installed into the soil at 0.1 mm/s (model scale) to full touchdown, as determined by an increase in vertical load observed in the loadcell connected directly to the caisson top cap.
- The centrifuge was then ramped down to 1 g (maintaining vertical caisson position with minimal bearing pressure observed on the caisson), and the drainage holes at the top of the caisson, which were open during installation, were closed.
- The soil sample was then allowed to reconsolidate at 150 g with the actuator supporting the net weight of the model (allowing for buoyancy and gravity effects).
- Once the soil was consolidated, the specific foundation test was undertaken.

RESULTS

Derivation of net bearing pressure, q_{net}

As indicated previously, the Stroud cell attached directly to the caisson top cap provides a measure of the vertical, horizontal and moment load applied to the caisson. The observed vertical load is corrected for the changing weight of the foundation due to buoyancy effects and its displacement through a changing g-field, and divided by the area of the caisson foundation to derive the observed bearing pressure (q_m). In order to produce results in the traditional bearing capacity form, a correction for the self weight of the soil must be made (as noted by Tani and Craig, 1995). The net bearing pressure (q_{net}) can be deduced from the observed bearing pressure using

$$q_{net} + \sigma_v' = q_m + W_{sp} \quad (1)$$

where W_{sp} represents the weight of soil within the soil plug (equal to $\gamma'd$), and σ_v' is the effective stress at skirt tip depth (equal to $\gamma'z$, where z is the skirt tip depth). Figure 4 illustrates this relationship.

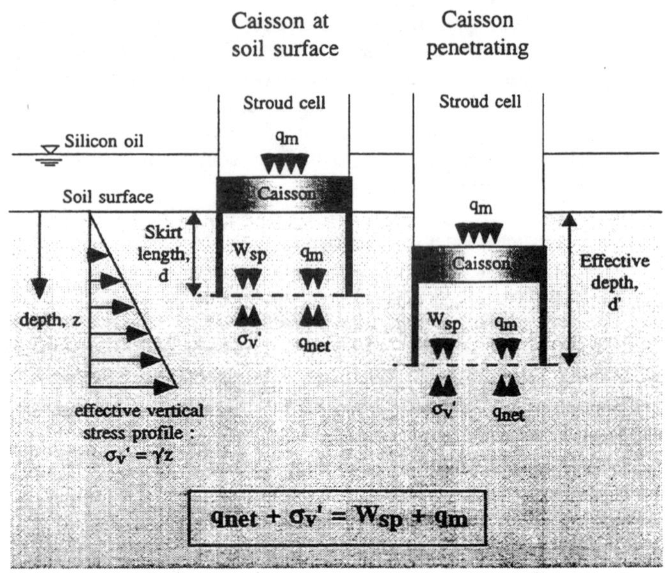

Figure 4 Derivation of net bearing pressure (q_{net}) from measured bearing pressure (q_m)

Bearing capacity

Following the test procedure outlined above, an undrained bearing capacity test was performed in silt 1 to derive the vertical bearing capacity *vs* depth relationship. The vertical displacement rate used for the test was 0.575 mm/s (model scale), selected to give an undrained bearing response. This is further discussed in a comparison between undrained and drained bearing response. In addition, two caisson tests were conducted in silt 2 which involved initial "sideswipe" tests before performing an undrained vertical bearing test. The sideswipe tests were conducted as part of the larger testing program, and consist of applying a nominal bearing pressure to the caisson before driving the caisson laterally at a constant lateral strain rate with zero vertical strain rate. At this stage it is believed that the sideswipe tests have little or no effect on the subsequent bearing response.

Figure 5(a) shows the undrained bearing response for silt 1, labelled UBC (Silt 1) (it should be noted that due to a problem with the control code and load cell the caisson in silt 1 was subjected to a vertical bearing pressure of 8 kPa during soil consolidation). Included is a comparison between the observed bearing response for silt 2 from the two "sideswipe" caisson tests (UBC/SS, Silt 2) and a modified version of the bearing response observed for silt 1 (UBC, Silt 2), proportionally scaled by the ratio of undrained strength profile. As can be seen, the undrained bearing response for both silt samples is highly comparable.

Figure 5(b) shows the undrained bearing response for the caisson foundation in silt 1 compared with published solutions for foundations in cohesive soil with strength proportional to depth. Several points need to be made in discussing this figure:

- An attempt has been made to correct the observed q_{net} response for side friction effects. This was performed assuming a coefficient of lateral earth pressure (K_o) of 0.4, estimated from triaxial tests

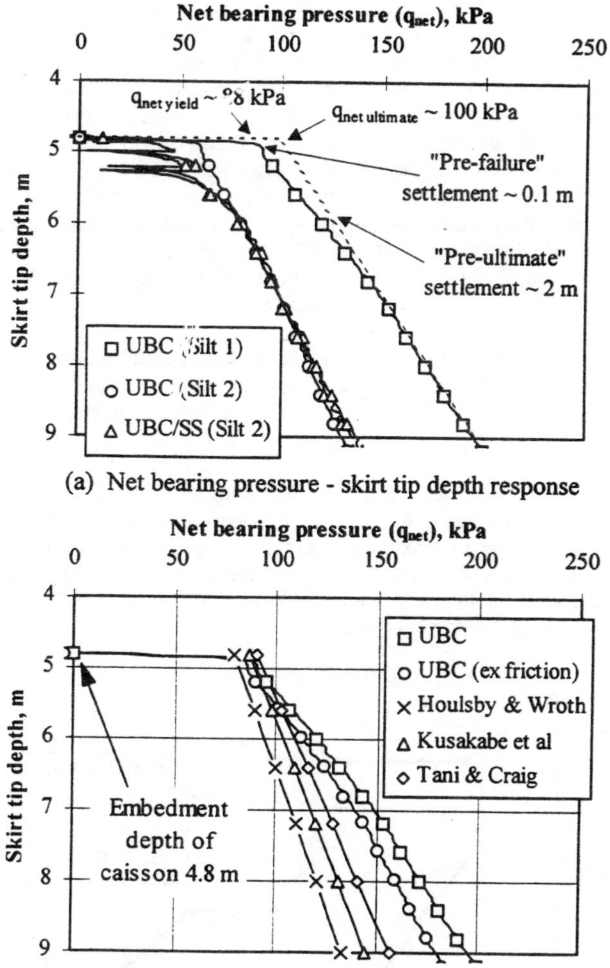

(a) Net bearing pressure - skirt tip depth response

(b) Comparison with existing theory

Figure 5 Undrained bearing response of caisson foundations in calcareous sediments

performed on the *in-situ* seabed material, and a constant wall friction angle (δ) of 20°. As the tests were conducted with silicon oil as the pore fluid, and the caisson skirts were smooth, this is expected to produce an overestimate of the maximum friction developed during the test. The results, labelled UBC (ex friction) in Figure 5(b), indicate that friction has only a moderate effect of the observed bearing capacity in the centrifuge caisson tests.

- Houlsby and Wroth (1983) used a slip line method (a lower bound solution) to examine the bearing capacity of large diameter circular gravity-base foundations in soils where cohesion increases linearly with depth using the relationship $s_u = s_{uo} + kz$, where s_{uo} is the undrained shear strength at the soil surface, k represents the linear strength gradient and z is the depth below the soil surface. The results published by Houlsby and Wroth can be compared with the results obtained by the current research by substituting for s_{uo} the undrained shear strength at the skirt tip depth, and including the observed shear strength gradient. As illustrated, the predicted bearing capacity is significantly lower than that observed. It should be noted that this method considers the caisson as a surface foundation acting on a soil with some surface shear strength, s_{uo}, and no allowance is made for depth effects due to the embedment of the skirts.
- Kusakabe et al (1986) looked at a similar problem to that analysed by Houlsby and Wroth, but used an upper bound collapse mechanism in deducing the relationship between undrained bearing capacity and soil strength. Again depth effects are ignored, and the deduced bearing capacity is below that observed in the undrained penetration.
- The most recent research into the bearing capacity of foundations in these soil types was conducted by Tani and Craig (1995), and included an examination into the effect of skirt tip depth on the bearing response. Tani and Craig examined foundations with a d/D ratio varying from 0 (directly comparable to Houlsby & Wroth) up to 0.3. In the current research, d/D is 0.4, and so an extrapolation of the results from Tani and Craig has been performed to estimate the undrained bearing capacity *vs* depth relationship for the current caisson foundation. Again it is illustrated that the predicted results significantly underestimate the observed undrained bearing capacity, although it is noted that the theory does predict the initial yield bearing capacity reasonably well. This result can be at least partly explained by a careful consideration of the geometry of the problem. As the caisson penetrates deeper into the sample, soil is restrained from flowing on top of the caisson (due to the thick top cap and the load cell of roughly equivalent diameter). This increases the skirt depth by increasing the effective depth of the foundation (as shown in Figure 4). The implication then is that as the caisson reaches maximum penetration (skirt tip depth = 0.8 D), the effective foundation depth (d') results in a d'/D ratio equivalent to 0.8. Given the results of Tani and Craig, it is likely that this increased ratio will significantly increase the theoretical bearing capacity. Further numerical research is required in this area to investigate this aspect of caisson behaviour.

Figure 5(a) also illustrates the "pre-failure" and "pre-ultimate" settlements, in this case 0.1 m (0.008 D) and 2 m (0.17 D) respectively, required to mobilise the yield and ultimate bearing capacity. For the undrained penetration in silt 1, it can be seen that $q_{net\ (yield)}$ is approximately 88 kPa, while extrapolation of ultimate response back to the original skirt tip depth produced a $q_{net\ (ultimate)}$ of 100 kPa.

At present numerical analysis and further centrifuge testing are being undertaken to examine the higher than predicted bearing capacity response. Early results indicate that the increased capacity during undrained penetration is likely to be related to depth effects, although attempts are being made to ensure the increased capacity is not due to the centrifuge test procedure, in particular the ramp down to 1 g to close the drainage holes. Centrifuge testing of gravity based structures (prototype diameter 40 m, d/D 0.1) for a related industry project (Randolph and Watson, 1996) have yielded results comparable with existing theory, indicating that the higher capacity may be induced by the high effective embedment of the caisson.

Drained vs undrained bearing capacity

In addition to the undrained bearing response examined above, a caisson penetration test was conducted in silt 1 under primarily drained conditions. The velocity (v) selected for the partially drained penetration was based on research done by Finnie (1993), where it was shown that the ratio vD/c_v can be used to predict the degree of consolidation. Finnie observed that for fully undrained tests $vD/c_v > 100$, while for fully drained conditions $vD/c_v < 0.01$. While it was possible to obtain fully undrained foundation tests, limitations over the minimum vertical displacement rate obtainable with the actuator control system meant that an achievable loading rate of 0.01 mm/s was selected ($vD/c_v = 2.5$), giving predominantly drained conditions. Figure 6 illustrates the significantly higher bearing capacity observed for the partially drained caisson test (PBC) compared with the undrained bearing capacity (UBC), as well as the ratio of partially drained to undrained bearing capacity (PBC/UBC).

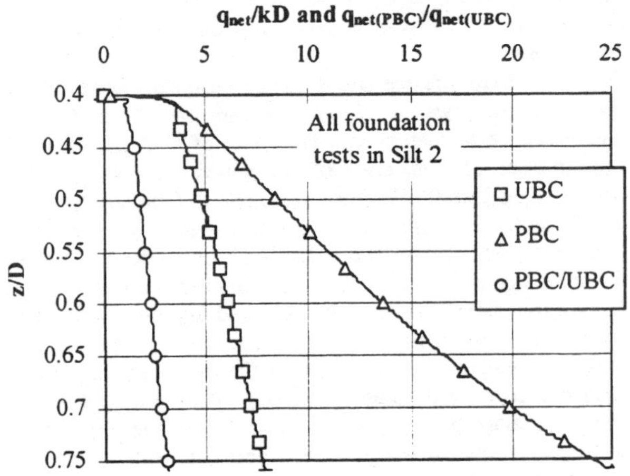

Figure 6 Effect of loading rate on bearing response

Tension capacity

Integral in the investigation of caisson foundations subjected to vertical loads is an examination of the tensile capacity. Figure 7 shows a comparison between the observed tension response for silt 2 (UTC, Silt 2) and the observed bearing response for both silt 1 and silt 2 (UBC, Silt 1 and UBC, Silt 2 respectively).

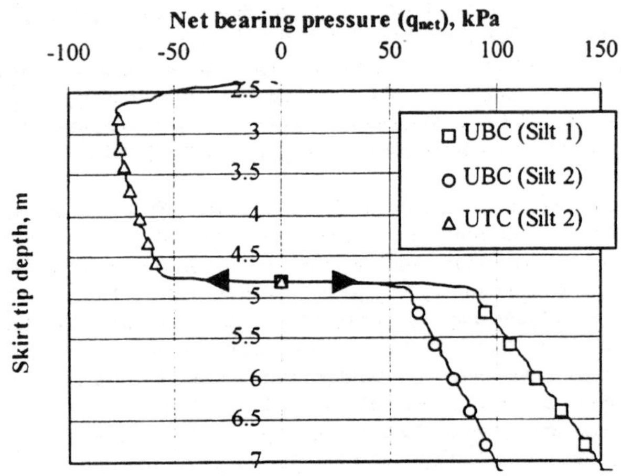

Figure 7 Comparison of undrained tensile and bearing capacity in calcareous sediments

As can be seen, the caisson subjected to tension responded in an initially stiff manner with a defined yield pressure ($q_{tension\ yield}$ ~ -55 kPa) before undergoing a gradual increase in tension pressure with further tensile strain. An ultimate tensile capacity was reached at $q_{tension\ ultimate}$ equal to -75 kPa, with the soil plug eventually falling out corresponding to a rapid drop in tension capacity. Similar to the bearing response, where it was noted that a compressive settlement of 0.17 D was required before the bearing capacity reached an ultimate value, a tensile displacement of 1.8 m (0.15 D) was required to mobilise the ultimate tensile capacity. This is similar to results published by Wang et al (1978) and Jones et al (1994), as follows:

- Wang et al investigated the tension capacity of suction piles in silt and clay using model piles 100 mm and 150 mm in diameter, with d/D ratios of 0.1 or 0.5. All tests were conducted at 1 g. Comparing the results from both models with a d/D ratio of 0.5, it was noted that the ultimate tensile pressure was only achieved after tensile displacements of between 0.1 D and 0.14 D were observed. However, no apparent yield tension was observed, and this is possibly due to the rate at which the test was conducted.
- Jones et al conducted similar tests in fine-grained sand using a model with D = 125 mm and d/D = 2.2, and under partially drained conditions observed that a pullout displacement of 0.2 D was required to mobilise the maximum tensile capacity.

From the results it can be seen that the tension capacity approaches that observed for the bearing capacity, giving credence to the belief that tension capacity can be thought of as a reverse bearing capacity problem.

An additional result obtained from the testing is illustrated in Figure 8, showing an example of the pullout behaviour of the caisson foundation after penetration to maximum depth. Also plotted is the total pore pressure (Figure 8a) and excess pore pressure (Figure 8b) response measured at the interface between the caisson top cap and soil surface (PT A) and at the skirt tip (PT B).

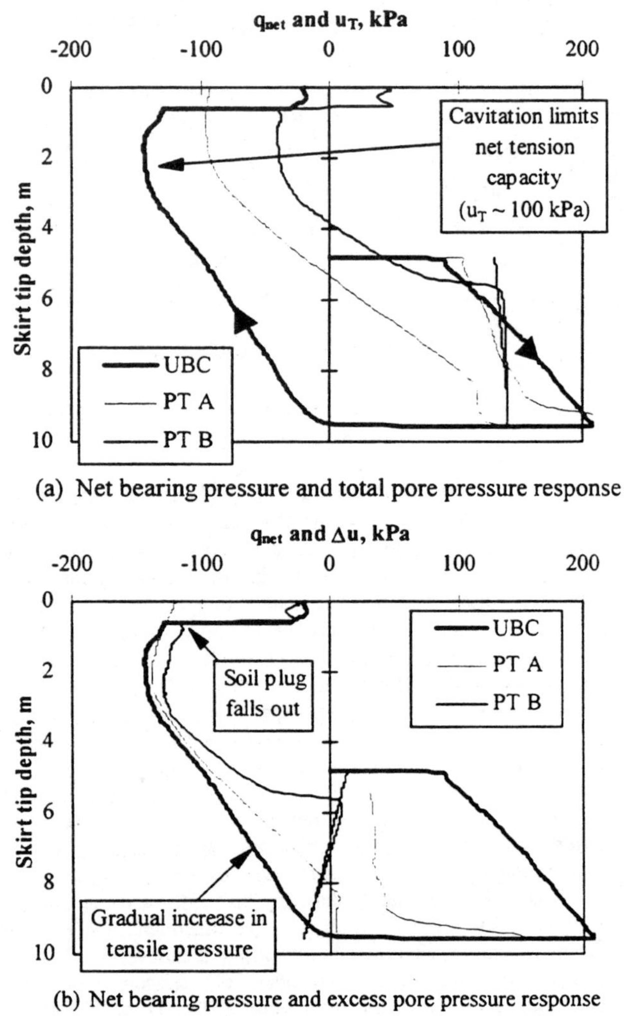

Figure 8 Pullout behaviour of caisson after penetration to depth

Several observations can be made from the foundation behaviour:

- The foundation appears to have no initial tension capacity ($q_{net} \approx 0$) when subjected to uplift at depth. Instead, the tension capacity displays a gradual increase with increasing tensile displacement. The reason for this behaviour is as yet unknown, but it is believed that the response may be related to the significant positive excess pore pressure developed during penetration. Further caisson testing to investigate the tensile capacity of caisson foundations in these soil types is currently underway, focusing on the pullout capacity of the foundations under vertical tensile pressure.
- Cavitation can be seen to limit the observed net tension capacity of the foundation, with the maximum tension capacity reached (and the soil plug falling out) at the same time as total pore pressure at the soil surface - caisson top cap interface reaches a minimum pressure of -100 kPa.
- It can be seen that during penetration the excess pore pressure recorded on the shear plane at skirt tip level reduces, suggesting that the soil is undergoing dilation. It is hoped to explore this observation further in later tests.

Effect of Preload

As indicated previously, the consolidation characteristics of calcareous sediments mean that reasonable attempts can be made to improve the capacity of the caisson foundations through the application of preload. This can be achieved either through the application of direct load, whereby a load greater than the design vertical load is applied to the foundation for a period of time to induce consolidation settlement, or through the application of suction (not included in this paper).

Two caisson tests were performed to examine the effect of direct preload on the bearing capacity of caisson foundations. This involved the application of a direct load to the caisson foundation equivalent to 40 - 50 % of the bearing capacity, and allowing either 100 % consolidation, or 50 % consolidation (as determined by the dissipation of excess pore pressure and the measurement of consolidation settlement). Figure 9(a) illustrates the measured bearing capacity of the two caisson tests compared with the original bearing capacity. In order to directly compare the measured bearing capacity from both silt samples, the observed net bearing pressure has been normalised by the soil strength gradient and the diameter of the foundation (q_{net}/kD), and has been plotted against depth normalised by foundation diameter (z/D).

Several observation can be made from the observed foundation behaviour:

- The bearing capacity response of the 50 % and 100 % consolidation preload caissons display an initial peak respectively 35 % and 45 % greater than the original bearing capacity. After the initial peak, the net bearing capacity of both caissons reduced to approximately 14% and 16% greater than the original bearing capacity respectively, before beginning to increase with further penetration. This is best illustrated in Figure 9(b), showing the relationship between the net bearing capacity of the preloaded caissons normalised by the original net bearing capacity.
- From the results it can be shown that applying preload can have a significant effect on the bearing capacity of caisson foundations, in particular the initial yield bearing capacity. Although not clear from the figures, the pre-failure settlements for the preloaded caissons (reflected in the stiffness of the foundation response) are significantly lower than those observed for the original caisson. It can also be seen that limiting the consolidation to 50 % had virtually the same effect on the observed net bearing response as 100 % consolidation. This is significant, as the prototype loading times for 50 % and 100 % are approximately 280 days and 7.7 years respectively. Given that bearing capacity will be predominantly affected by the strength of the soil at the skirt tips, and that excess pore pressure in this region will dissipate quickly due to high pore pressure gradients, this result is not surprising.
- Although it is initially surprising to note that with further penetration the net bearing pressure of the caisson does not reduce quickly to the original bearing capacity, it must be realised that penetration has only been conducted over a depth equivalent to 0.4 D below the skirt tips. As the consolidation will effect the soil for some depth beneath the caisson (in a "bulb" shape emanating from the caisson skirt tip), it will require significant penetration (possibly > 1 D) before the observed bearing capacity reduces to the original level. This also explains why the caisson subjected to 100 % consolidation (where the soil "bulb" was allowed to fully consolidate) displayed a higher bearing capacity at maximum penetration than the 50 % preloaded caisson.

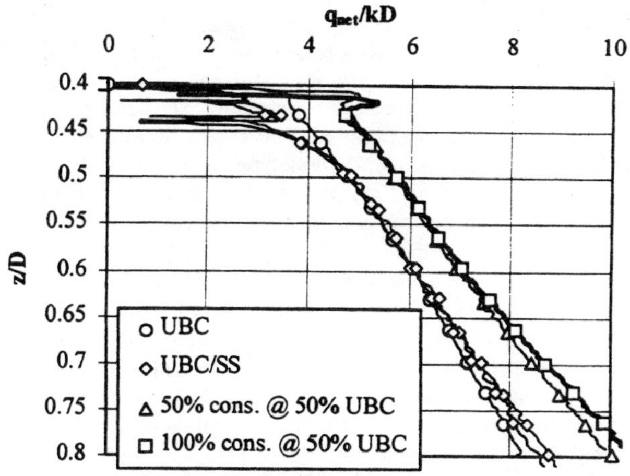

(a) Undrained bearing response of preloaded caissons

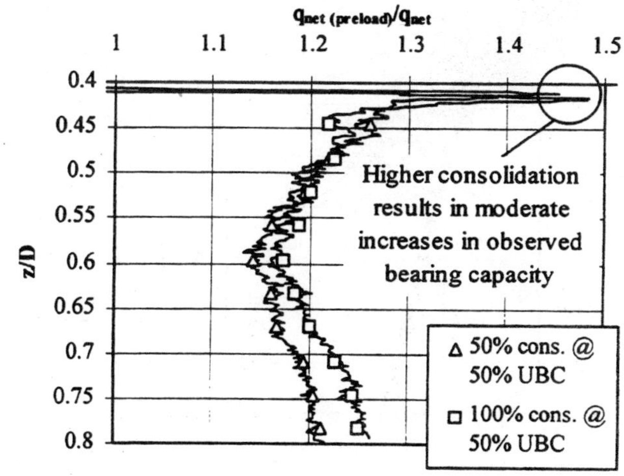

(b) Ratio of preloaded bearing capacity to original bearing capacity

Figure 9 Effect of direct preload on undrained bearing capacity

In addition to direct preload and suction preload, an increase in bearing capacity is likely to occur in situations were foundations are subjected to moderate amounts of cyclic loading (such as the application of wave and wind loading to offshore structures) and dissipation of pore pressure is allowed to occur. A preliminary investigation into this has been conducted by subjecting the model caisson to horizontal cyclic loading under a constant bearing pressure (at a level equivalent to that applied to the direct preload caissons), followed by full dissipation of excess pore pressure and vertical penetration. Six horizontal loading cyclic packages, each consisting of 1000 cycles, were conducted at a model frequency of 10 - 14 Hz (prototype frequency 0.04 - 0.06 Hz allowing for scaling effects and the use of silicon oil as pore fluid). The amplitude of the horizontal shear stress for the first three packages consisted of ±10% of the bearing pressure, while the second three packages consisted of horizontal shear stress at ±20% of the bearing pressure. Figure 10 illustrates the results obtained, where it can be seen that the combination of cyclic loading and dissipation of excess pore pressure has increased the bearing capacity by approximately 200%. Vertical penetration under the combination of horizontal cyclic loading and consolidation was observed to be approximately four times that observed for consolidation alone (Figure 9a). This aspect of caisson response is to be examined through an investigation into the performance of caisson foundations subjected to cyclic loads.

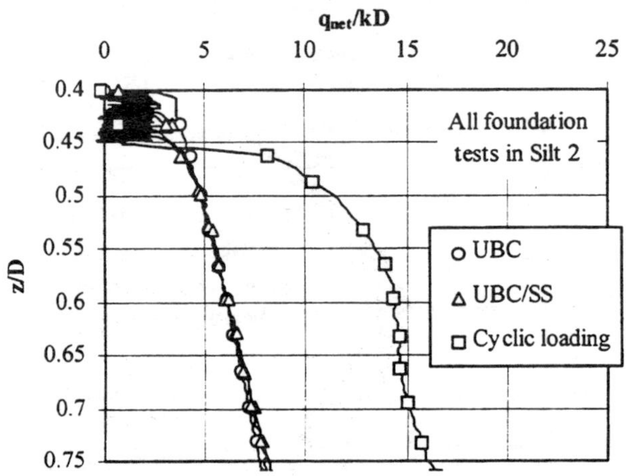

Figure 10 Effect of cyclic preload and consolidation on undrained bearing response

CONCLUSIONS

Centrifuge testing has been undertaken to examine the behaviour of caisson foundations subjected to vertical loads. The model foundation was an 80 mm caisson with skirt depth equal to 40 % of the diameter, tested at 150 g. The soil type used for the study was calcareous silt obtained from the North-West Shelf of Australia.

In order to address questions of reproducibility in physical model testing, each sample prepared for centrifuge testing was subjected to extensive "in-flight" soil characterisation to ensure continuity of foundation test results between samples. In addition, as the results outlined are part of a larger test program, published data are continually being supplemented by additional results, which show extremely consistent soil profiles and foundation response.

The results obtained from centrifuge testing indicate that the bearing capacity of caisson foundations penetrated to significant depth is underestimated by existing theory for shallow foundations on soils with strength proportional to depth. However, existing theory does predict the initial yield pressure accurately.

In addition, the tension capacity of these foundations appears to approach that of the bearing capacity, supporting the belief that caisson foundations subjected to tensile loads can be analysed as a reverse bearing capacity problem. However, in tension tests where the foundation was first penetrated to maximum depth before being immediately extracted, the caisson exhibited a gradual increase in tension capacity with increasing tensile displacement. This phenomenon is as yet unexplained, although testing is currently ongoing to explore this further.

Finally, there appear to be significant benefits to be derived from the application of preload to these foundations, either through the application of direct load or through dissipation of excess pore pressure after cyclic loading.

ACKNOWLEDGMENTS

The work described here has been funded through a joint industry project, funded by the Minerals and Energy Research Institute of Western Australia, the West Australian Petroleum Company and Woodside Offshore Petroleum; the project forms part of the activities of the Special Research Centre for Offshore Foundation Systems, funded through the Australian Research Council's Research Centres Program. The first author is supported by an Australian Postgraduate Award supplemented initially by a Samaha Research Scholarship, and more recently by an F S Shaw Memorial Postgraduate Scholarship at the University of Western Australia. The value of discussions with Dr Fraser Bransby during preparation of this paper is gratefully acknowledged. Also, special thanks are due to the workshop and centrifuge staff at UWA, without whose excellent technical support the experimental research would not have been possible.

REFERENCES

Finnie I.M.S. (1993) "*Performance of shallow foundations in calcareous soil*," Phd Thesis, The University of Western Australia.

Houlsby G.T. and Wroth C.P. (1983) "Calculation of stresses on shallow penetrometers and footings," *Proc. IUTAM/IUGG Symp. on Seabed Mechanics*, Newcastle, UK, 107-112.

Jones W.C., Iskander M.G., Olson R.E. and Goldberg A.D. (1994) "Axial capacity of suction piles in sand," *Seventh Int. Conf. on the Behaviour of Offshore Structures*, Vol 1 : Geotechnics, 62-76.

Kusakabe O., Hiroyuki S. and Nakase A. (1986) "An upper bound calculation on bearing capacity of a circular footing on a non-homogeneous clay," *Soils and Foundations*, 26(3), 143-148.

Maddocks D.V. (1981) "The development of a transducer to measure the effective soil stresses and the pore water pressure acting on the surface of a driven, steel pile," *CUED/D - Soils TR113*, Cambridge University Engineering Department.

Randolph M.F., Jewell R.J., Stone K.J.L. and Brown T.A. (1991) "Establishing a new centrifuge facility," *Proc. Int. Conf. on Centrifuge Modelling - Centrifuge 91*, Boulder, Colorado, 3-9.

Randolph M.F. and Watson P.G. (1996) *"Gorgon development : Centrifuge modelling of skirted raft foundations - Report to West Australian Petroleum Company,"* Geomechanics Group, The University of Western Australia, GEO:96185.

Tani K. and Craig W.H. (1995) "Bearing capacity of circular foundations on soft clay of strength increasing with depth," *Soils and Foundations*, 35(4), 21-35.

Wang M.C., Demars K.R. and Nacci V.A. (1978) "Applications of suction anchors in offshore technology," *Proc. of the 10th Annual Offshore Technology Conference*, Houston, OTC 3203, 1311-1320.

Proceedings of the Seventh (1997) International Offshore and Polar Engineering Conference
Honolulu, USA, May 25-30, 1997

ISBN 1-880653-28-1 (Set); ISBN 1-880653-29-X (Vol. I)

Finite Element Modelling of Skirted Strip Footings Subject to Combined Loadings

M.F. Bransby and M.F. Randolph
University of Western Australia
Nedlands, Australia

ABSTRACT

The response of skirted offshore foundations to combined vertical (V), moment (M) and horizontal (H) loading was studied using two dimensional finite element analysis. Despite using a simple constitutive model and geometry, new information was gained about the shape of the yield locus and the soil deformation mechanisms occurring at yield. The shape of the yield locus was similar to that predicted by previous workers in V-M and V-H space but differed significantly in M-H space. This behaviour was explained using upper bound plasticity mechanisms suggested by soil deformation mechanisms observed in the finite element analysis. This procedure was used to give a good approximation to the shape of the yield locus and should form the basis for future design methods.

KEY WORDS: Skirted shallow foundations, Undrained, Yield locus.

INTRODUCTION

Offshore structures may consist of multiple shallow foundations which act together in consort (eg. jack-up rigs, tension leg platforms) or single shallow foundations (eg gravity based platforms). Both of these shallow foundations undergo environmentally induced combined vertical, V, moment, M, and horizontal, H loading. In order to predict the behaviour of an entire offshore structure the response of these individual foundations to this type of loading must be understood

Recent research workers (eg. Tan, 1990; Murff, 1994; Butterfield and Gottardi, 1995; Houlsby and Martin, 1992; Dean *et al.*, 1992) have analysed the combined loading response of shallow foundations using plasticity methods. Yield loci in V-H-M space have been deduced for different foundation systems in different soil types. A yield surface is introduced whereby

$$f\left(\frac{V}{As},\frac{H}{As},\frac{M}{AsD}\right)=0 \qquad (1)$$

where A is the area of the foundation, D its diameter or breadth and s is a representative soil strength or effective stress. Associated plasticity is sometimes assumed so that this yield surface also describes a plastic potential defining the relative magnitudes of the incremental deformation during elasto-plastic yielding. Within the yield surface, behaviour is usually assumed to be linear elastic (Bell *et al.* 1991, Martin 1994). To use the normality relationships the load and displacement definitions must form work conjugate pairs so that normalised total system work, W is written as

$$\frac{W}{ADs}=\left(\frac{V}{As}\right)\frac{\delta v}{D}+\left(\frac{H}{As}\right)\frac{\delta h}{D}+\left(\frac{M}{ADs}\right)\delta\theta \qquad (2)$$

where $\delta v/D$ is the normalised vertical footing displacement, $\delta h/D$ is the normalised horizontal displacement and $\delta\theta$ is the footing rotation and these displacements are measured at the same point at which the loads are taken to act.

Previous workers have investigated either footing behaviour in drained or partially drained sand (eg. Butterfield and Gottardi, 1995; Tan, 1990) or shallow footing behaviour in undrained clay (Houlsby and Martin, 1992). In both cases there is no resistance to uplift loading and so the yield envelope appears as plotted in Figure 2 with M_{max} and

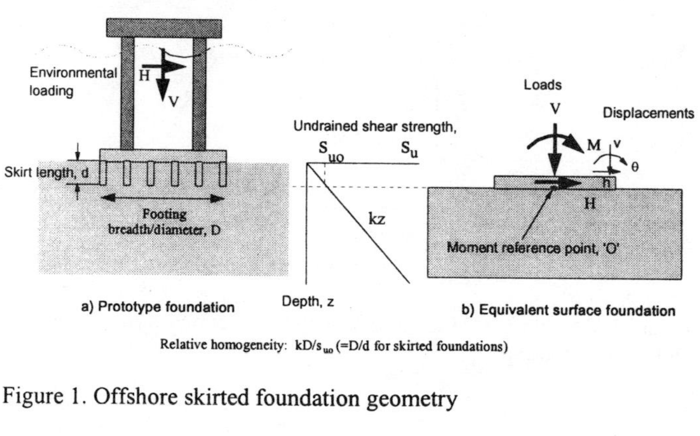

Figure 1. Offshore skirted foundation geometry

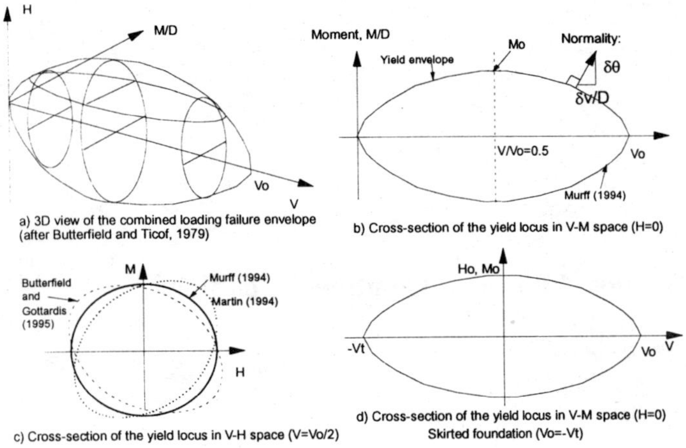

Figure 2. Yield envelope approach to combined loading response of footings

H_{max} occurring when $V/V_o \approx 0.4\text{-}0.5$

Murff (1994) suggested a form of yield surface of size given by its intercept on the M=0, H=0 axis designated as the ultimate vertical load V_o. The yield locus is given as

$$f = \left(\frac{V}{V_o}\right)^2 - \left(1 + \frac{V_t}{V_o}\right)\frac{V}{V_o} + \frac{V_t}{V_o} + \left[\left(\frac{H}{H_o}\right)^2 + \left(\frac{M}{M_o}\right)^2\right]^{0.5} = 0 \qquad (3)$$

where V_t is the normalised bearing capacity under purely vertical tension and H_o and M_o are the ultimate horizontal load and moment respectively. This is plotted in Figure 2 for V_t=0.

Skirted shallow foundations (or bucket foundations) are being used increasingly in areas such as the North Sea (Tjelta and Haaland, 1993) and there is increasing interest in their use in the North-West Shelf in Western Australia as they provide uplift resistance and transfer loads below the surface without the expense of pile driving. They consist of shallow foundations with thin skirts at the circumference which penetrate the soil vertically to a depth, d and therefore the foundation appears like an upturned bucket of diameter D and height d. They may form groups of buckets as in the corner of a TLP or a single large footing for a gravity base platform. The presence of the skirt allows suctions to develop when uplift loadings are applied so that there is an uplift capacity, and increased bearing capacity in compression by confining the upper soil within the skirts, forcing the failure down to the level of the skirt tips.

Because skirted foundations are able to resist uplift loading, they may resist moment loading even when V = 0. Initial centrifuge test results (Watson and Randolph, 1997) suggest that the ultimate uplift load, $V_t \approx -V_o$ and so Murff's (1994) gives a yield locus symmetrical about the V = 0 axis (Figure 2). However, as Murff's (1994) equation was developed for the behaviour of footings without skirts further investigation is required to find the shape of the yield locus and understand the system behaviour when skirts are present. Under particular investigation is the behaviour of these types of foundation in the normally consolidated calcareous sediments typical of the North-West Shelf in Western Australia.

GEOMETRY OF SKIRTED FOUNDATIONS

Investigation into seabed properties typical of the North-West Shelf suggests that the soil is normally consolidated, with undrained shear strength, s_u that varies proportionally with depth, z. This is expressed as

$$s_u = kz\,, \qquad (4)$$

where k is the undrained shear strength gradient.

A shallow foundation resting on the surface of this soil would have a small bearing capacity and large footing settlement would occur before significant bearing capacity could be achieved or alternatively deep foundations would be required. Tani and Craig (1995) and Watson and Randolph (1997) suggest that foundations with skirts of length, d transfer the footing load to depth d at which the soil undrained shear strength, s_{uo} = kd and so the bearing capacity will be related to s_{uo} (Figure 1). Tani and Craig (1995) proposed that the behaviour of the skirted foundation in undrained soil with shear strength proportional to depth can be approximated by analysing a foundation on the soil surface but with the soil shear strength profile described by $s_u = s_{uo}+kz$ (Figure 1). This finding was a result of comparison of different lower bound plasticity analyses of embedded foundations in soil with undrained shear strength proportional to depth. The conclusion suggests that soil above the level of the footing base does not increase the bearing capacity of the footing. Centrifuge test suggest that this is the case (Tani and Craig 1995, Watson and Randolph, 1997) but this requires further justification with upper bound plasticity analyses and further finite element analysis.

Beneath the skirt tips the undrained shear strength of the soil is increasing at k kPa/m and so this will affect the bearing capacity of the foundation. A dimensionless soil non-homogeneity factor kD/s_{uo} is introduced to describe this variation with depth. For a skirted foundation when s_{uo} = kd then $kD/s_{uo} \equiv D/d$.

Skirted foundations may be approximated as circular in plan. To understand the behaviour of a circular foundation undergoing vertical, moment and horizontal loading 3-d analysis is required. Initial finite element analysis of both circular footings and infinitely long strip footings suggest that their behaviour is similar (eg. compare yield loci shown in Figure 3 which were produced using 2D and 3D finite element analysis in elastic-plastic soil) although the actual V_u, H_u and M_u values will vary. As the main mechanistic behaviour of the circular footings may be observed by analysing a strip footing, detailed analysis of the strip footing will lead to insight applicable to the understanding of circular footing behaviour but requiring significantly less computing capacity and processing time. In addition, upper bound plasticity mechanisms are much simpler to develop for the strip footing case. Therefore, findings from a finite element study of a strip footing are reported in this paper.

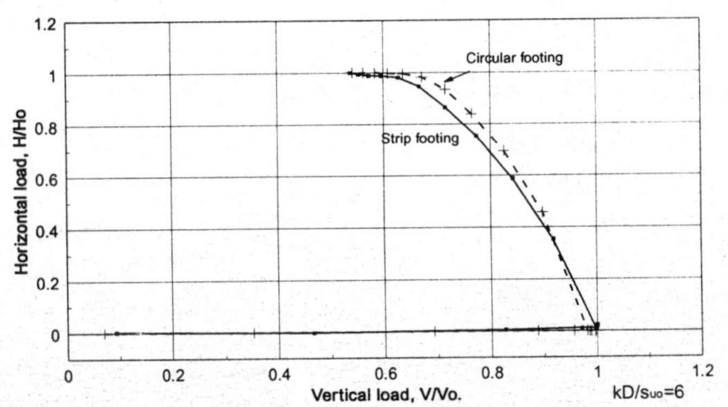

Figure 3. Combined V-H loading response of footings: strip footing and circular footing.

FINITE ELEMENT ANALYSIS

Geometry and soil model

The strip footing geometry is shown in Figure 1. The footing was rigid, rough and of breadth, D=12 m and rested on the soil surface. It was considered to be infinitely long to enable plane strain analysis. The soil was modelled as an elastic-plastic Tresca material with an undrained shear strength profile described using $s_u=s_{uo}+kz$, where $s_{uo}=10$ kPa and k was varied to investigate footing behaviour in soils with different undrained shear strength non-homogeneities (kD/s_{uo} values). However, most analyses were carried out with $kD/s_{uo}=6$ unless the effects of kD/s_{uo} variation was being studied specifically. Hence, unless otherwise stated in the text, $kD/s_{uo}=6$. The rigidity index for the soil was taken as, G/s_u = 200, but the findings of the analysis will be independent of soil stiffness as only ultimate behaviour is investigated. This simple soil model approximated the soil behaviour appropriately for investigation of the system failure (yield) locus but would have to be refined for investigation of pre-failure behaviour.

Footing loads and displacements are as defined as in Figure 1b. The reference point for moment (and vertical displacement) is in the centre of the footing at the soil-footing interface. The position of this reference point will affect the results (Martin, 1994; Butterfield and Gottardi, 1995).

Finite element analysis was carried out using CRISP90 (Britto and Gunn, 1990) on a 5DX66 Personal Computer and the finite element mesh consisted of 472 cubic strain triangles. A consolidation analysis was carried out with small time increments so that the soil behaviour was undrained.

Aims

The aim of the series of finite element analyses was to investigate the yield locus for the strip footing in V-H-M space. Of special interest was its shape, the soil deformation mechanisms occurring at yield and whether normality applied. The shape of the yield locus could also be compared with those presently recommended (eg. Murff, 1994; Martin, 1994; Tan, 1990).

Yield locus probes

Displacement controlled events were found to be most suitable for probing the yield locus as the exact shape of the yield locus could be ascertained and the ratio of plastic incremental displacements could be found at failure. Load controlled events were found to be less suitable because pre-failure yield meant that the point of global failure was difficult to ascertain as there was no discrete change in displacement increment vector direction or marked reduction in global system stiffness.

Two types of displacement probes were carried out. In the first a single probe was carried out at a constant ratio of $\delta v/D:\delta h/D:\delta\theta$ until the footing loading did not vary with increased displacement. By choosing a given ratio of $\delta v/D:\delta h/D:\delta\theta$ (eg. $(\delta h/D)/\delta v/D = 3$, $\delta\theta = 0$ as shown in Figure 4) both the shape of the yield locus and direction of plastic incremental displacement vector could be ascertained at failure. The second type of probe was used by Tan (1990) during his centrifuge tests and is known as a 'side-swipe' test. It consisted of two stages. In the first stage a given displacement (typically vertical) was prescribed to the footing until ultimate load was reached. Then a second displacement probe was applied during which the original vertical displacement increment was set to zero. The premise of the test is that because elastic stiffnesses are much greater than plastic stiffnesses, the stress path in the second stage will almost exactly follow the shape of the yield locus as negligible expansion of the yield locus will be required to balance the elastic deformation and so the shape of the yield locus may be determined in one test.

RESULTS

Vertical-horizontal (V-H) load space

The yield envelope was probed using both horizontal 'side-swipe' tests and constant displacement ratio tests (probes). In the 'side-swipe' test the initial probe was a vertical displacement probe with no rotation or horizontal displacement, $\delta v/D = 0.083$, $\delta h/D = 0$, $\delta\theta = 0$ (OA in Figure 4) which allowed calculation of V_o. The horizontal 'side-swipe' was carried out with , $\delta v/D = 0$, $\delta h/D = 0.083$, $\delta\theta = 0$ (AB in Figure 4) and the stress path followed the shape of the yield locus in V-H space (Figure 4).

The ultimate vertical load, V_o was compared with solutions from plasticity analyses. Lower bound analyses from Davis and Booker (1973) and Tani and Craig (1995) predict that V/Ds_{uo} = 10.49 when kD/s_{uo} = 6 whereas Houlsby and Wroth (1983) found V/Ds_{uo} = 10.37. These compare well with V/Ds_{uo}= 10.7 obtained from the finite element analysis. Equally, the measured ultimate horizontal load when H/As_{uo} = 1.034 compares well with the exact solution H/As_{uo} = 1.

Constant displacement ratio probes were carried out for a range of different $\delta h/\delta v$ values. The probe with $\delta h/\delta v = 3$ is shown on Figure 4. Initially the system behaves elastically when the stress state in within the yield locus (OC in Figure 4). The stress path reaches the yield locus near 'C' on Figure 4 and follows the yield locus until stresses no longer change at D. When stabilised (at 'D') on the yield locus there are no elastic displacements as there is no load change and because $\delta h/D = \delta h^e/D+\delta h^p/D$ and $\delta v/D = \delta v^e/D+\delta v^p/D$, where the superscripts 'e' and 'p' represent elastic and plastic deformations respectively, then $\delta h^p/\delta v^p \approx \delta h/\delta v$. Hence, the direction of the plastic displacement increment at the final failure point is known and normality can be investigated. Results suggest that normality holds since the estimated gradient of the yield locus correctly relates to the displacement increment vector.

As the elastic displacements become negligible at the stabilisation point on the yield locus, the incremental soil displacements will be those for the plastic failure mechanism and the soil deformation mechanism at that point on the yield locus can be found. Soil deformation mechanisms at different points on the yield locus are shown in Figure 4.

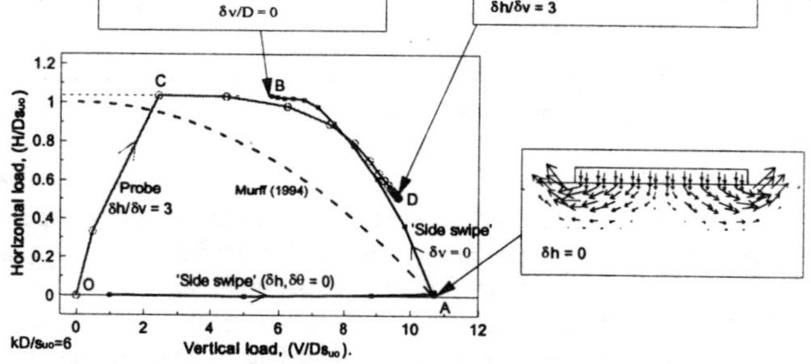

Figure 4. Yield locus and displacement mechanisms under V-H loading (M~0)

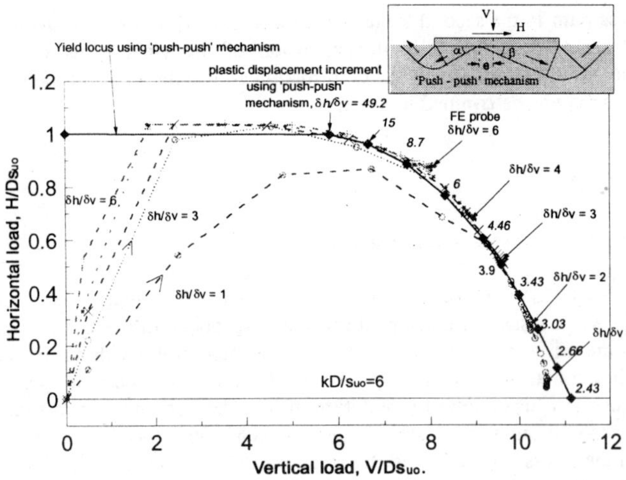

Figure 5. Yield locus and plastic displacement mechanisms under V-H loading (M ~ 0).

The shape of the yield locus is compared to the equation of Murff (1994). If $V_t = -V_o$ Murff's (1994) equation simplifies to

$$f = \left(\frac{V}{V_o}\right)^2 - 1 + \left[\left(\frac{H}{H_o}\right)^2 + \left(\frac{M}{M_o}\right)^2\right]^{0.5} = 0 \tag{5}$$

where H_o is the horizontal force when M = 0 and V = 0 and M_o is the moment when V = 0 and H = 0. When M=0

$$H = H_o\left(1 - \left(\frac{V}{V_o}\right)^2\right) \tag{6}$$

which is plotted on Figure 4 assuming that $H_o = s_{uo}A$ and using the V_o value from the finite element analysis. This gives a very conservative approximation to the yield locus obtained by finite element analysis.

The soil deformation mechanisms on the yield locus and the shape of the yield locus can be compared with those obtained using plasticity analysis. Upper bound plasticity analyses were carried out using the push-push mechanism suggested by the soil deformation mechanisms observed during finite element analysis and shown in Figure 5. Eccentricity e and angles α and β (defined in Figure 5) were varied to find the minimum moment for any given vertical loading and so tracing a theoretical upper-bound yield locus. This analysis predicts both the deformation mechanisms and the shape of the yield locus well (Figure 5).

Vertical-moment (V-M) load space

Similar probes were carried out in V-M load space as for the investigation of V-H space except that the moment 'side-swipe' tests consisted of a vertical penetration (δθ, δh/D = 0) followed by a rotation with δv/D = 0. The main findings are summarised on Figure 6 and show the deduced yield locus, displacement probes and soil displacement mechanisms on the yield surface. Again, normality holds on the yield locus. Murff's (1994) equation gives a good, though slightly conservative, fit to the shape of the yield envelope.

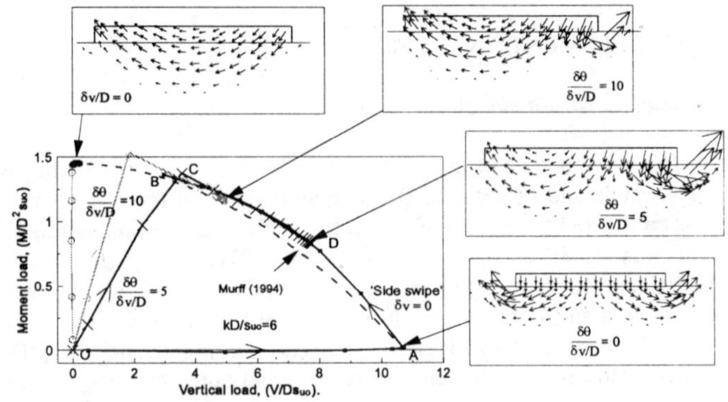

Figure 6. Yield locus and displacement mechanisms under V-M loading (H=0)

At the ultimate moment when H = 0, $M/D^2s_{uo} = 1.449$. This compares well with the value of $M/D^2s_{uo} = 1.491$ predicted using the plastic scoop mechanism shown in Figure 7 suggested by the soil deformation mechanism in Figure 6.

The soil deformation mechanism along the yield locus (Figure 6) suggests that plastic analysis of a combined scoop-wedge deformation mechanism (Figure 7) can be used to predict the shape of the yield locus in V-M space. Preliminary results of this analysis, carried out in the same way as the push-push mechanism are presented in Figure 7.

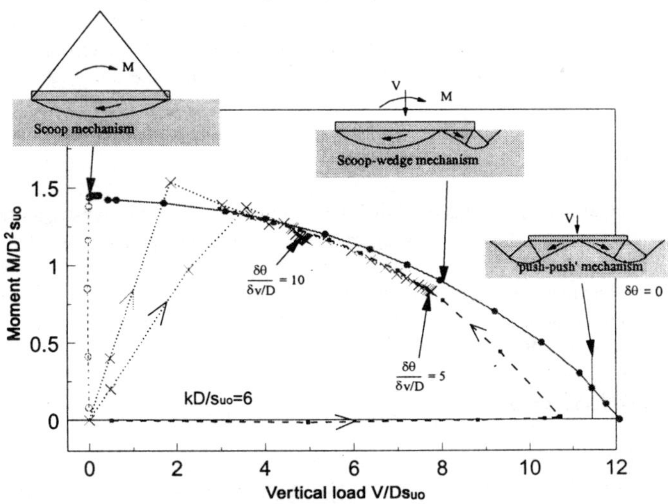

Figure 7. Yield locus and plastic displacement mechanisms under V-M loading (H ~ 0)

Horizontal-moment (H-M) load space

Initially, similar probes to those for the previous analysis were carried out under a combination of horizontal displacement and rotation. The probes revealed that the maximum moment that the footing could sustain did not occur when H = 0 but when there was a combination of positive horizontal load and moment. To reach the true ultimate moment a probe with δh = 0 was required (OAB in Figure 8) during which H increased. Once the yield locus was reached (at 'A') the stress path followed the yield locus (AB on Figure 8). At the true ultimate moment, M_u ('B' on Figure 8) normality applies as δh = 0 and the soil deformation mechanism is shown in Figure 8. Following this probe,

0 the side swipe probed the yield locus for increasing values of H (BC in Figure 8), whereas when $\delta h < 0$ the yield locus was probed with decreasing H (BD in Figure 8). The two side-swipe tests allowed determination of the entire yield locus in M-H space (with V = 0).

To confirm the shape of the yield locus and to investigate normality and the soil deformation mechanisms occurring on the yield surface, displacement probes were also carried out. Figure 8 shows the results from one probe with $\delta\theta/(\delta h/D) = 1$ which investigates a section of the yield locus with H > 0 and another probe with $\delta\theta/(\delta h/D) = -2$ which investigates the yield locus with H < 0. The results confirmed the shape of the yield locus found during the 'side swipe' tests and suggested that the assumption of normality was valid.

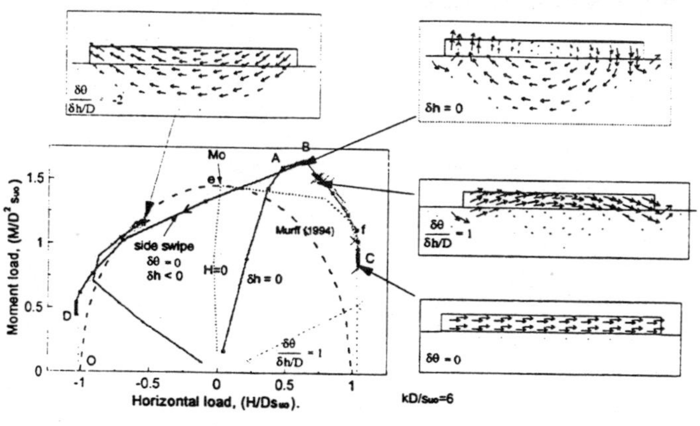

Figure 8. Yield locus and displacement mechanisms under H-M loading (V~0)

It is interesting to note that preliminary testing with H = 0, $\delta\theta > 0$ and then a side swipe with $\delta\theta = 0$, $\delta h/D > 0$ failed to show the existence of the true peak moment. Instead, a maximum moment, M_o with H = 0, V = 0 was reached in the initial probe ('e' on Figure 8) and the side swipe then missed the corner of the yield locus as elastic behaviour was seen with no change in M until the yield locus was reached (dotted line 'ef' in Figure 8). This suggests that this type of behaviour could be missed in an experimental investigation so that the yield locus would appear almost symmetrical.

The eccentricity of the yield locus in H-M space differs significantly from the shape of the yield locus suggested by Murff (1994) who assumed that maximum moment was sustained with no horizontal loading (Eq. 5, Figure 2c). Instead, here it is observed that maximum moment is sustained with a positive horizontal loading. Eccentricities of the yield locus in H-M space have been previously observed by Butterfield and Gottardi (1995) and Martin (1994). Martin analysed spud can foundations in soft clays and found an eccentricity in the same direction (Figure 2c) which he suggested was due to the kinematics of the system, shape of the footing and the normality property. Butterfield and Gottardi (1995) studied circular footings on drained sand and found that the eccentricity of the yield locus was different from that observed here with maximum moment sustained with small negative horizontal loads (Figure 2c). They also suggested redefining the position of the moment reference point to a point below the footing so that the yield locus was symmetrical about the H = 0 axis although no attempt was made to justify this transformation theoretically.

The soil deformation mechanisms observed on the yield locus in the finite element analyses suggested two different upper bound plasticity mechanisms which could be used in the same way as done for the push-push mechanism in V-H space. The first mechanism was the scoop mechanism (Figure 9) previously used to calculate ultimate moment. For any given horizontal load, L was varied to give the minimum moment and so a yield locus was obtained (Figure 9). The second mechanism consisted of a scoop and two wedges which was suggested by the soil deformation mechanisms observed close to ultimate moment during finite element analysis. This was used to calculate a second approximation to the yield locus (Figure 11). As a further cut off, when $|H/As_{uo}| \geq 0$ the simple slip mechanism (Figure 9) would prevail and this was used as an additional constraint on the yield locus. The use of these different plastic deformation mechanism allows the shape of the yield locus to be approximated very closely (Figure 9).

The scoop mechanism observed on the yield locus suggests that the type of simplifying transformation suggested by Butterfield and Gottardi (1993) would allow simpler expression of the yield locus. Indeed, if the reference point for moment loading is changed to the rotation point for the scoop (at a height L above the base of the footing) then the transormed yield locus is almost symmetrical. This allows the use of a simple mathematical form to describe the shape of the yield locus for different kD/s_{uo} values. This is beyond the scope of this paper and is reported separately.

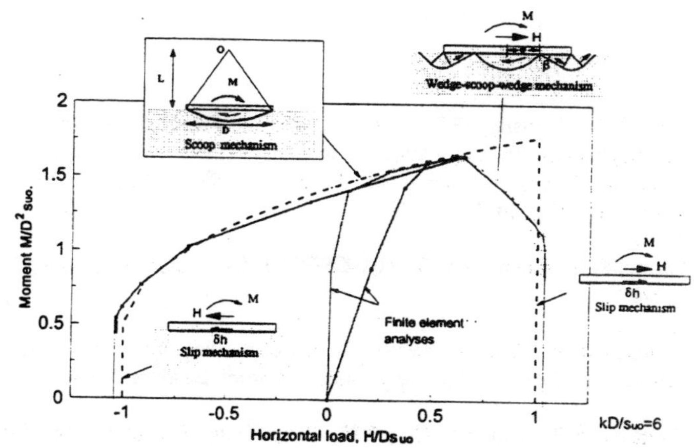

Figure 9. Plastic deformation mechanisms under H-M loading (V~0)

CONCLUSIONS

In an attempt to gain insight into the response of individual skirted foundations to combined vertical, horizontal and moment loadings, a finite element study was made of a strip footing in undrained soil with shear strength increasing linearly with depth. Displacement probes and 'side-swipe' tests similar to those conducted when model testing enabled investigation of the failure locus in V,H,M - space.

The shape of the yield locus was similar to those previously recommended in V-H and V-M space and plastic footing displacements obeyed normality. Soil deformation mechanisms on the yield locus were observed and upper-bound plasticity analysis using similar mechanisms gave good approximations to its shape. This will allow prediction of yield loci for different soil undrained shear strength profiles and suggests a similar procedure should be carried out with different geometries and extreme soil shear strength profiles.

The yield locus was found to be eccentric in H-M space with the maximum moment being sustained with a significant amount of horizontal load. Hence, previously recommended expressions for the yield loci and plastic potentials of shallow foundations may not be suitable for use with skirted foundations. The eccentricity of the yield locus was found to be due to the kinematics of the 'scoop' soil deformation mechanism when V, H = 0 and use of this and another plastic mechanism allowed close prediction of the shape of the yield locus in H-M space.

In conclusion, the study of this simple foundation led to knowledge about the yield locus and soil deformation mechanisms which suggest improved design methods for shallow foundations. Later, similar analyses of more realistic (3D) geometries with complex soil models and foundation loadings will allow further advances in the understanding of the combined loading response of skirted foundations to be made, but the behaviour is expected to be similar to that observed with the simple strip footing.

ACKNOWLEDGEMENTS

The work described here has been funded through a joint industry project, funded by the Minerals and Energy Institute of Western Australia, the West Australian Petroleum Company and Woodside Offshore Petroleum; the project forms part of the activities of the Special Research Centre for Offshore Foundation Systems, funded through the Australian Research Council's Research Centre's Program.

REFERENCES

Bell, R.W., Houslby, G.T., Burd, H.J., (1991). Finite element analysis of axisymmetric footings subjected to combined loads. *Proc of the Int. Conf. on Computer Methods and Advances in Geomechanics,* Cairns, Australia, Vol 3. pp. 1765-1770.

Britto, A.M., Gunn, M.J. (1990). *CRISP90. User's and programmer's guide.*

Butterfield, R., Ticof, J. (1979). The use of physical models in design. Discussion. *Proc. 7th Eur. Conf. Soil Mechanics.* Brighton Vol 4.

Butterfield, R., Gottardi, G. (1995). Simplifying Transformations for the Analysis of Shallow Foundations on Sand. *Proc. 5th Int. Offshore and Polar Eng Conf.*, The Hague, pp. 534-538.

Davies, E.H., Booker, J.R. (1973). The effect of increasing strength with depth on the bearing capacity of clays. *Geotechnique* 23, No. 4, pp. 551-563.

Dean, E.T.R., James, R.G., Schofield, A.N., Tan, F.S.C, Tsukamoto, Y. (1992). The bearing capacity of conical footings on sand in relation to the behaviour of spudcan footings of jack-ups. *Proc. Wroth Memorial Symp. "Predictive Soil Mechanics",* Oxford, pp 230-253.

Houlsby, G.T., Martin, C.M (1992). Modelling of the behaviour of foundations of jack-up units on clay. *Proc. Wroth Memorial Symp. "Predictive Soil Mechanics",* Oxford, pp 339-358.

Houlsby, G.T., Wroth, C.P., (1983). Calculation of stresses on shallow penetrometers and footings. *Proc. IUTAM/IUGG Symposium on Seabed Mechanics,* Newcastle,UK, September, pp. 107-112..

Martin, C.M., (1994). *Physical and numerical modelling of offshore foundations under combined loads.* D.Phil thesis, University of Oxford.

Murff, J.D., (1994). Limit analysis of multi-footing foundation systems.. *Proc of the 8th Int. Conf. on Computer Methods and Advances in Geomechanics,* Morgantown, 1, pp. 223-244

Tan, F.S., (1990). *Centrifuge and theoretical modelling of conical footings on sand.* Ph.D. thesis, University of Cambridge.

Tani, K., Craig, W.H., (1995). Bearing capacity of circular foundations on soft clay of strength increasing with depth. *Soils and Foundations.*Vol 35, No. 4, pp 21-35.

Tjelta, T.I., Haaland, G. (1993). Novel Foundation Concept for a Jacket Finding its Place. *Society for Underwater Technology. Vol. 28: Offshore Site Investigation and Foundation Behaviour.* pp. 717-728.

Watson, P.G., Randolph, M.F., (1997). Vertical Capacity of Caisson Foundations in Calcareous Sediments. *Proc. 7th Int. Offshore and Polar Eng. Conf.*, Honolulu.

Proceedings of the Seventh (1997) International Offshore and Polar Engineering Conference
Honolulu, USA, May 25-30, 1997

ISBN 1-880653-28-1 (Set); ISBN 1-880653-29-X (Vol. I)

Analysis of Anchor Mooring Lines in Sands

S. Bang
South Dakota School of Mines and Technology
Rapid City, SD, USA

R.J. Taylor
Naval Facilities Engineering Service Center
Port Hueneme, CA, USA

ABSTRACT

An analytical solution method capable of determining the geometric configuration and developed tensile forces of mooring lines associated with plate, pile, or drag anchors embedded in general seafloor soils has been developed. The solution method is capable of analyzing multi-segmented mooring lines with sinkers or floats embedded in layered seafloor soils, utilizing a systematic iterative solution search method based on specified boundary conditions. The results of the developed analytical solution method have been compared with field tests in sand, and found very effective. The effects of the soil friction angle and the anchor depth on the geometry and developed tension of the mooring line have been also briefly studied.

KEY WORDS: Anchor, mooring line, tension, sand, chain, cable

INTRODUCTION

The US Navy has recently developed an analytical solution method that can analyze offshore and deep water mooring lines associated with embedded drag and fixed anchors (Bang, 1996). Mooring lines may be comprised of multi-segments with different material and/or geometric properties and embedded in a general seafloor soil having either cohesion or friction or both. Any number of sinkers or floats can also be added to the suspended portion of the mooring line within the water. Solutions can be obtained with a fixed total length of the mooring line, a fixed horizontal length of the mooring line, or a fixed exit angle of the mooring line with the seafloor surface.

The validity of the developed method of analysis in analyzing mooring lines embedded in cohesive seafloor soils has been verified through comparisons with laboratory centrifuge model tests and full-scale field tests. The mooring lines in the centrifuge tests were single-segmented chains and wire cables, whereas the field tests utilized multi-segmented mooring lines consisting of stud-link chains and stranded wire ropes. Results of the tests conducted in cohesive soils and the resulting comparisons between the measured and calculated values have been reported (Bang, et al., 1996).

This paper describes the detailed algorithm of the analysis when it is used to analyze mooring lines embedded in general seafloor soils. The analysis considers the development of both adhesion and friction between the mooring line and the seafloor soil in determining the geometric configuration and the tension of the mooring line. The developed method of analysis was used to compare with the field test results of drag embedment anchors in sand. Finally, a simple analytical parametric study was conducted to identify the effects of the soil friction angle and the anchor depth on the anchor and mooring system.

MOORING LINE ANALYSIS

Fig - 1 shows a schematic diagram of a mooring line element embedded in a general seafloor soil. T and ϕ are the axial tensile force and the inclination angle at the beginning of the element. N, (f ds + N $\tan\delta$), and (w ds) are the normal force, the tangential force, and the buoyant weight of the mooring line, respectively.

From the equilibrium conditions of

$$\Sigma\ F_t = 0, \quad \Sigma\ F_n = 0, \quad \text{and} \quad \Sigma\ M = 0 \qquad (1)$$

one can solve for N, T and ϕ . Note that previously published solutions of the embedded mooring line analysis did not consider the moment equilibrium (Brian Watt Associates, 1983, Degenkamp and Dutta, 1989, Vivatrat, et al., 1982). The current solution method, which utilizes complete equilibrium conditions, therefore permits an additional degree of freedom in each chain/cable element. In the analysis, it is assumed that the soil tangential forces (f ds + N $\tan\delta$) remain at their limiting state at all time, since the dominant mode of the mooring line movement during deployment is sliding. The normal soil forces (N) however remain as unknowns because of the available additional degrees of freedom and therefore can be less than those defined by the limiting state, i.e., the soil bearing capacity. It is also noted that the assumption of the chain/cable element forming a circular arc as assumed in some of the previous solutions is no longer necessary in the current solution.

Eq (1) forms the basis of the recursion formula for the analysis of an embedded mooring line in seafloor, i.e.,

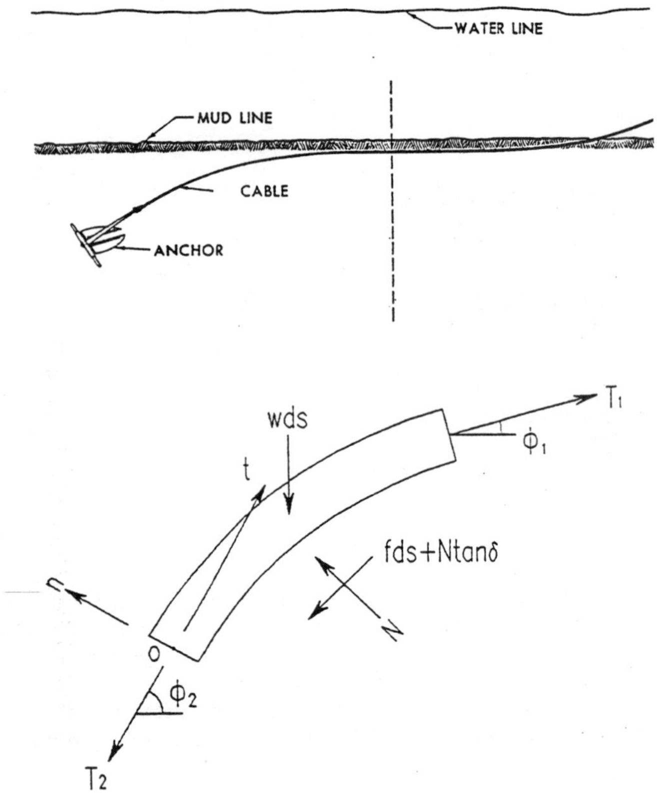

Fig - 1 Mooring Line Element

$$N = \frac{2T_1 - f\,ds}{\tan\phi_1 + \tan\delta}$$

$$T_2 = T_1 - (f + w \sin\phi_1)\,ds - N\tan\delta \qquad (2)$$

$$\phi_2 = \phi_1 + \frac{N - w\,ds\cos\phi_1}{T_2}$$

where

T_1 and T_2 : axial forces at the beginning and end of the element.

ϕ_1 and ϕ_2 : mooring line inclination angle at the beginning and end of the element.

δ : interface friction angle between the mooring line and the soil.

f:: tangential force per unit length.

w: buoyant weight of mooring line per unit length.

N: normal force.

ds: element length.

The solution process starts with a known mooring line inclination angle at the seafloor surface (ϕ_1). The catenary and embedded portions of the mooring line are then solved separately and added for the final solution. The following describes the solution details of the embedded portion of the mooring line.

At the seafloor surface, the inclination angle (ϕ_1) is given as part of the iteration process. From this known inclination angle and the known horizontal force at the water surface, the axial tension at the seafloor surface (T_1) is calculated. Using Eq (2), the axial tension and the inclination angle at the end of the element, ϕ_2 and T_2, are calculated. From compatibility, the values of the axial tension and the inclination angle at the end of the current element then become those at the beginning of the next element. Concurrently, the coordinates of the elements are calculated and recorded. This process continues until the depth to the element end reaches the specified anchor depth.

The tangential component, f, is estimated from (Bang, 1996)

$$f = E_s\, D\, \alpha\; \beta\; S_u \qquad (3)$$

where

E_s : equivalent diameter conversion factor for sliding force to convert mooring line diameter to circumferential area.

D: diameter of mooring line.

α : soil adhesion conversion factor.

β : contact area conversion factor.

S_u: undrained strength of soil

For stud link chains, the value of E_s is obtained by converting the chain link diameter to the circumferential area of a cylinder defined by a circle encompassing two perpendicular chain links, i.e., 3.6 times π . For cables, the value of E_s is simply π .

The soil adhesion conversion factor (α) is the ratio of the adhesion between the mooring line and the soil vs. the soil cohesion. The contact area conversion factor (β) is the ratio of the true contact area between the mooring line and the soil vs. the surface area of a cylinder defined by the mooring line. In case of stud link chains, the cylinder is defined by a circle encompassing two perpendicular chain links. For cables, the cylinder is simply defined by the cable diameter. The value of β may be less than 1.0, if a separation occurs on the backside of the mooring line while it is deployed.

The value of the normal force, N, is limited to the soil bearing capacity (Bang, 1996), i.e.,

$$N < N_{max} = Q\,ds$$
$$Q = E_b\, D\, q \qquad (4)$$

where

Q: bearing capacity of soil per unit length.

q: bearing capacity of soil per unit area.

E_b : equivalent diameter conversion factor for normal force.

D: diameter of mooring line.

For stud link chains, the value of E_b is obtained by converting the chain link diameter to the diametric projectional area of a cylinder defined by a circle encompassing two perpendicular chain links, i.e., 3.6. For cables, the value of E_b is 1.0.

The bearing capacity of mooring lines in sand is estimated from the lateral resistance of a rigid cylinder segment located at relatively deep depths. For cohesionless soils, Reese (1962) proposed the following bearing capacity equation, assuming that soil flow occurs around the segment cross-section.

$$q = K_a b \gamma z (\tan^8\beta - 1) + K_o b \gamma z \tan^4\beta \tan\phi \qquad (5)$$

where

b: width of the segment.

z: depth.

γ : soil unit weight.

K_o : at-rest lateral earth pressure coefficient.

K_a : active lateral earth pressure coefficient.

β : $45^o + \frac{\phi}{2}$.

ϕ : soil friction angle.

SOLUTIONS

The analysis of a mooring line is based on one of the three possible imposed boundary conditions; either the total length of the mooring line, the horizontal distance between the buoy and the anchor, or the mooring line angle at the seafloor surface is specified. The first two cases can be solved by utilizing an iterative solution technique, searching for the correct mooring line angle at the seafloor surface. In such cases the analysis includes the following steps.

1. Input necessary geometric, material, and loading properties.
2. Establish the solution upper and lower bounds of the mooring line angle at the seafloor surface.
3. Search for the solution iteratively.

The establishment of solution bounds and subsequent solution search are not needed if the mooring line angle at the seafloor surface is specified in input. The solution upper and lower bounds of the mooring line angle at the seafloor surface are those between which the true solution lies. These are obtained iteratively based on a specified margin of the angle at the seafloor surface.

Once the upper and lower bounds of the mooring line angle at the seafloor surface are obtained, the analysis method searches for the solution iteratively. Within the iterative solution search, the tension and angle of the mooring line are continuously monitored to identify whether any mismatch among the input parameters occurs or not, e.g., the adequacy of the specified mooring line size and the applied force at the buoy against the material and geometric properties.

The detailed solution of a multi-segmented mooring line embedded in a layered, seafloor soil is obtained from an additional iterative search. The initial material and geometric parameters to be used in the analysis are obtained from the assumed length of the buried mooring line. The results of the analysis with these assumed parameters are then compared with the true material properties of the multi-segmented mooring line, which in turn determines the necessity of additional iterations. For instance, at the first iteration, by comparing the solution of the buried length of the mooring line with the specified length of the first segment of the mooring line attached to the anchor, one can determine whether the entire first segment is completely buried or not. If it is determined that only part of the first segment of the mooring line is buried within the seafloor, Eq. (2) with an additional iterative search are used to obtain the complete solution of the buried portion of the mooring line. On the other hand, if it is determined that the entire first segment portion of the mooring line is buried, the buried portion of the mooring line is analyzed from Eq. (2) assuming that the remaining portion of the buried length of the mooring line beyond the first segment consists of second segment mooring line. Similar analysis as described above will reveal whether the entire second segment of the mooring line is buried within the seafloor or not, which in turn determines the necessity of further iterations. This process continues until the complete details of the mooring line composition within the seafloor are identified, which is followed by the analysis of the catenary portion of the mooring line.

COMPARISONS WITH FIELD TESTS IN SAND

The Naval Facilities Engineering Service Center, formerly the Naval Civil Engineering Laboratory, conducted a series of conventional drag-embedment anchors in sandy and cohesive seafloor soils (Taylor, 1982). The tests were performed to evaluate the performance and suitability of various Navy and commercial drag embedment anchors and to provide data to enable anchor selection for various applications.

In this paper, the results of selected field anchor drag tests conducted off Port Hueneme, California, in cohesionless soils are compared. The seafloor soil consisted of a poorly graded dense sand (SP based on the Unified Soil Classification System) with an approximate bulk wet density of 130 lb/ft^3.

Each anchor test was instrumented to determine the anchor depth, anchor shank pitch, anchor roll, and anchor load. A load cell was attached between the end of the shank and the mooring line. The mooring line load and mooring line angle at the barge and the barge displacement relative to a fixed buoy were also recorded. These data were needed to calculate the true anchor drag distance as well as to determine the contribution of the bottom resting mooring line to the total mooring capacity.

The developed solution method has been applied to compare with the results from three field tests. These three tests were selected because of their consistency in performance and data. Tables - 1 through 3 include the details of field tests, e.g., the test conditions, description of the mooring line and the anchor, and measurements of loads at anchor and deck. The details of the remaining field tests can be found in reference (Taylor, 1982).

Table - 1 shows the details of the comparison of the horizontal force at the barge deck between the predicted from the solution method and the measured from field test 303-3. Tables - 2 and 3 show similar comparisons for tests 304-7 and 304-8. It is noted that the relatively large differences in anchor force and deck tension are primarily due to the soil friction caused by the portion of the mooring line lying on the seafloor surface.

As can be seen from the tables, the predicted values of horizontal forces at the barge deck compare very well with the measured values.

Measured Anchor Force (kips)	10.6	12.9	16.8
Measured Anchor Shackle Depth (ft.)	0.7	0.9	1.2
Measured Deck Horizontal Tension (kips)	33.2	35.5	43.6
Calculated Buried Chain Length (ft.)	17.7	19.5	22.1
Calculated Force At Seafloor Surface (kips)	16.7	21.5	29.7
Calculated Deck Horizontal Tension (kips)	36.7	41.1	49.2

Water Depth: 42.2 ft.

Soil Buoyant Unit Weight: 67.6 pcf

Soil Friction Angle: 40 deg

Anchor: Moorfast 3,000 lbs

Mooring line: 62 ft. - 2.25 in. chain, 84 ft. - 2.185 in. chain, 170 ft. - 2.875 in. chain, 34 ft. - 2 in. wire

Frictional coefficient: 1.357

Table - 1 Field Test 303 - 3

Measured Anchor Force (kips)	13.0	14.0	27.6	29.2
Measured Anchor Shackle Depth (ft.)	0.4	1.1	1.5	1.9
Measured Deck Horizontal Tension (kips)	37.2	37.8	72.0	60.9
Calculated Buried Chain Length (ft.)	16.6	20.9	26.2	28.4
Calculated Force At Seafloor Surface (kips)	16.7	25.0	47.3	55.7
Calculated Deck Horizontal Tension (kips)	35.3	44.2	63.7	73.2

Water Depth: 41.0 ft.
Soil Buoyant Unit Weight: 67.6 pcf
Soil Friction Angle: 40 deg
Anchor: Stato 3,500 lbs
Mooring line: 62 ft. - 2.25 in. chain, 84 ft. - 2.185 in. chain
170 ft. - 2.875 in. chain, 34 ft. - 2 in. wire
Frictional coefficient: 1.357

Table - 2 Field Test 304 - 7

Measured Anchor Force (kips)	14.0	14.5	16.0	16.8	19.0	30.7
Measured Anchor Shackle Depth (ft.)	0.4	0.5	0.6	0.7	1.1	2.0
Measured Deck Horizontal Tension (kips)	35.5	39.2	39.5	40.8	47.4	59.1
Calculated Buried Chain Length (ft.)	17.0	17.7	18.8	19.3	22.2	29.2
Calculated Force At Seafloor Surface (kips)	17.8	19.3	22.0	23.0	31.2	59.4
Calculated Deck Horizontal Tension (kips)	37.6	38.8	41.7	42.2	49.8	76.3

Water Depth: 40.7 ft.
Soil Buoyant Unit Weight: 67.6 pcf
Soil Friction Angle: 40 deg
Anchor: Stato 3,500 lbs
Mooring line: 62 ft. - 2.25 in. chain, 84 ft. - 2.185 in. chain
170 ft. - 2.875 in. chain, 34 ft. - 2 in. wire
Frictional coefficient: 1.357

Table - 3 Field Test 304 - 8

The resulting errors vary from 10.5 to 15.7 % for test 303-3, from -11.6 to 20.2 % for test 304-7, and from -1.0 to 29.1 % for test 304-8. Considering the inconsistency among measured data and wide variations in field conditions, the results of the analytical solution compare very well with the field measurements.

ϕ (Deg)	H_a (ft.)	L_b (ft.)	T (lbs)	α_a (deg)	ΔT
30	0.5	54.20	48,481	2.7	3.0
	1.0	61.87	47,074	5.5	5.9
	1.5	66.27	45,696	8.4	8.6
	2.0	69.34	44,335	11.4	11.3
35	0.5	40.52	47,596	3.8	4.8
	1.0	46.10	45,310	7.7	9.4
	1.5	49.29	43,070	11.9	13.9
	2.0	51.51	40,865	16.2	18.3
40	0.5	30.28	46,092	5.4	7.8
	1.0	34.27	42,341	11.2	15.3
	1.5	36.53	38,689	17.5	22.6
	2.0	38.12	35,119	24.3	29.8
45	0.5	22.45	43,429	8.0	13.1
	1.0	25.23	37,165	17.2	25.7
	1.5	26.79	31,160	28.0	37.7
	2.0	27.85	25,384	41.0	49.2

ϕ: Soil Friction Angle
H_a: Anchor Depth
L_b Length of Buried Mooring Line
T: Tension at Anchor
α_a: Angle at Anchor
ΔT: % Capacity due to Mooring Line

Table - 4 Effects of Soil Friction Angle and Anchor Depth

EFFECTS OF SOIL FRICTION ANGLE AND ANCHOR DEPTH

A simple parametric study has been conducted to identify the effects of the soil friction angle and the anchor depth on the mooring line geometry and the distribution of tensile forces. The following properties were assumed to obtain the results included in Table - 4.

Water depth: 10 ft.
Anchor shackle depth: 2 ft.
Mooring line: 1 inch stud link chain
Effective soil unit weight: 50 pcf
Horizontal force at buoy: 50,000 lbs
Solution criterion: exit angle at seafloor surface of 0 degree
Soil friction angle: 30, 35, 40, and 45 deg.

As can be seen from Table - 4, the mooring line angle at the anchor (α_a) increases quickly as the soil friction angle (ϕ) increases. It increases from 2.7 degrees to 8.0 degrees at the anchor depth (H_a) of 0.5 ft. as ϕ changes from 30 to 45 degrees. The increase in α_a is more pronounced as the anchor depth increases. The length of the buried mooring line (L_b), as expected, decreases as ϕ increases. An approximately 20 % reduction in L_b results as the soil friction angle increase from 30 to 45 degrees for all values H_a. The percentage of mooring capacity attributed to the buried portion of the mooring line (ΔT) dramatically increases as ϕ increases. For instance, only 3.0 % of the total mooring capacity comes from the buried mooring line with ϕ of 30 degrees and H_a of 0.5 ft. This however increases to 49.2 % when ϕ is 45 degrees and H_a is 2.0 ft.

This parametric study indicates that the effect of the soil friction angle is very significant on the mooring line geometry and the load transfer mechanism, and more profound than that of the soil cohesion.

Fig - 2 Effect on Length of Buried Mooring Line

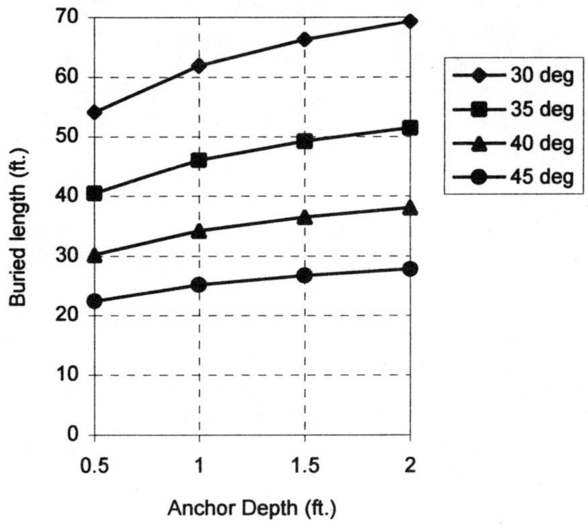

Fig - 3 Effect on Mooring line Angle at Anchor

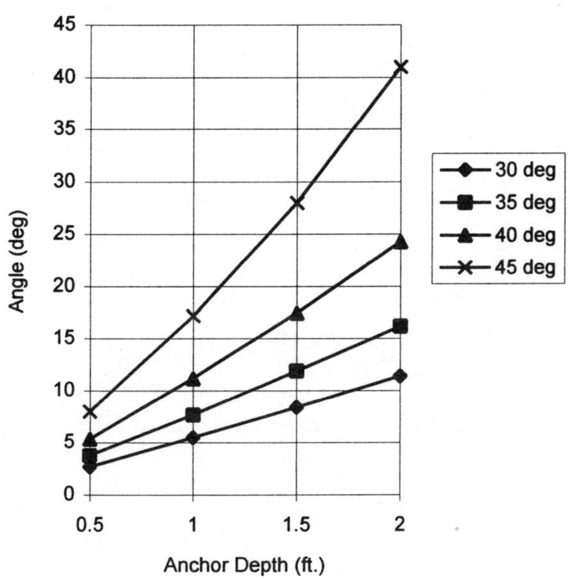

Fig - 4 Effect on Buried Mooring Line Capacity

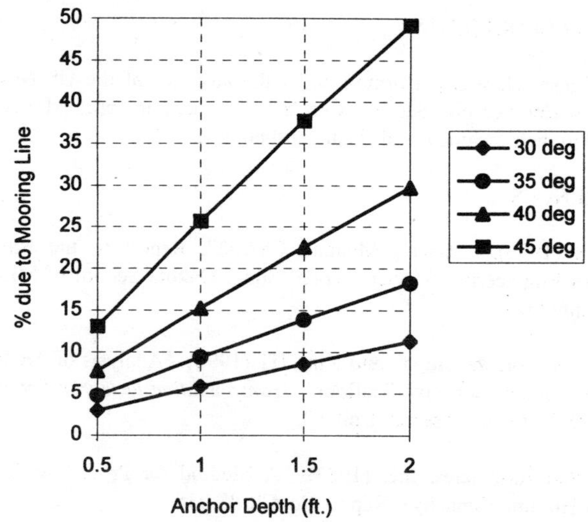

Results of the analysis are also plotted in Figs 2 - 4. The rate of increase in L_b with respect to ϕ decreases as can be seen from the decreasing spacing among the curves in Fig - 2. The rate of increase in α_a and ΔT on the other hand increases rapidly as ϕ increases as evidenced from the increasing spacing among the curves in Figs 2 and 3.

Figs 2 - 4 show that as H_a increases L_b, α_a, and ΔT increase. The rates of increase of these parameters however show distinctive individual effect. As H_a increases, L_b increases with decreasing rate as can be seen from the decreasing slopes of the curves in Fig - 2. On the other hand, as can be seen from Fig - 3, α_a increases with increasing rate as H_a increases for relatively dense sands, i.e., ϕ greater than 35 degrees, but with constant rate for relatively loose sands, i.e., ϕ less than 35 degrees. The rate of increase in ΔT with respect to H_a remains more less constant (Fig - 4).

CONCLUSION

An analytical solution method, satisfying the complete equilibrium requirements, has been developed to determine the total mooring line configuration and tensions when part of the mooring line is embedded in layered cohesive or cohesionless seafloor. The mooring line may consist of multiple segments of different material and/or geometric properties.

The results of field tests of drag embedment anchors in sand have been compared with the predictions made with the analytical solution method. It indicates that the developed solution method is capable of correctly analyzing mooring lines embedded in cohesionless seafloor soil deposits.

The effect of the soil friction angle was found to be very significant on the mooring line geometry and the tensile force distribution, and to be more profound than that of the soil cohesion. The anchor depth in sand also contributes significantly to the behavior of the mooring system and its load transfer mechanism.

Comparisons described in this paper have been made assuming typical values of material parameters associated with the mooring line without detailed measurements or calibration. Controlled laboratory tests, e.g., centrifuge model tests, need to be performed for this purpose. With well controlled laboratory tests results available, more meaningful comparisons with the field tests and parametric study may be attempted in the future.

ACKNOWLEDGMENTS

This research was performed under the auspices of the US Naval Facilities Engineering Service Center. The authors are extremely grateful to their technical and financial supports.

REFERENCES

Bang, S (1996). "User's Manual, CSAP2", Report to the Naval Facilities Engineering Service Center, South Dakota School of Mines and Technology.

Bang, S, Taylor, RJ, Jie, Y, and Kim, HT (1996). "Analysis of Anchor Mooring Line in Cohesive Seafloor", *Transportation Research Record*, No. 1526, National Research Council.

Brian Watt Associates, Inc. (1983). "A Method for Predicting Drag Anchor Holding Capacity", Report No. CR 83.036.

Degenkamp, G, and Dutta, A (1989). "Soil Resistance to Embedded Anchor Chain in Soft Clay", *Journal of Geotechnical Engineering*, Vol. 115, No. 19.

Reese, LC (1962). "Ultimate Resistance Against a Rigid Cylinder Moving Laterally in a Cohesionless Soil", *Society of Petroleum Engineers Journal.*

Taylor, RJ (1982). "Drag Embedded Anchor Tests in Sand and Mud", TN No. N-1635, Naval Civil Engineering Laboratory.

Vivatrat, V, Valent, PJ, and Ponterio, AA (1982). "The Influence of Chain Friction on Anchor Pile design", *Proc., 14th Annual Offshore Technology Conference*, Paper No. 4178.

Proceedings of the Seventh (1997) International Offshore and Polar Engineering Conference
Honolulu, USA, May 25-30, 1997

ISBN 1-880653-28-1 (Set); ISBN 1-880653-29-X (Vol. I)

Piles and Shallow Foundations Under Cyclic Loadings

Jan Laue, Thomas Grundhoff and Hans L. Jessberger
Ruhr-University Bochum
Germany

ABSTRACT

To describe the soil structure-interaction of cyclic loaded structures, centrifuge model tests are carried out on laterally loaded piles and vertically loaded shallow foundations. The principle of deformation behaviour of sand will be with briefly. In addition to the presentation of the test results, possibilities of research on cyclic problems in the Bochum geotechnical centrifuge are presented. The obtained data from centrifuge model tests under dry conditions is related to foundations of coastal and offshore structures. It is shown, that the observed mechanisms have to be regarded even in saturated conditions too. The evaluated data is compared with the results and equations obtained from cyclic and dynamic triaxial tests. Thus, the differences between element tests and the real soil-structure interaction behaviour are established.

KEY WORDS: Cyclic Loading, Shallow Foundations, Piles, Soil-Structure Interaction, Centrifuge Model Tests

1 INTRODUCTION

Foundations of coastal and offshore structures are quite often experienced with cyclic loadings. These loadings cause displacements and affects the structures resting on such foundations. In saturated conditions of sand, the developements of the displacements are treated as a problem of the combination of the influencing load and the development of pore pressures, which possibly leads to extreme liquifaction contitions. With the results of centrifuge model tests, presended in this paper, it will be shown that large deformations and changing of the boundary conditions also occur in dry soil. In addition to the influence of the pore pressure developement, the soil structure interaction under dry conditions can not be described sufficiently. Most existing fast calculation methods are based on ideal elastic or elastic perfect-plastic material models for the soil. They suffer under the validity only for small load amplitudes. Higher loads require different model laws, so that the description of the soil-structure interaction becomes much more complicated. Empirical models for the description of the displacement of structures have been formulated regarding the results of cyclic and dynamic tests, but they offen suffers under the boundary conditions of the test equipments or under the estimation of the influencing prameter. For several structures (e.g. BURLAND & BURBRIDGE 1985) the parameters can only be estimated for a still existent structure.

To achieve a better understanding of the displacements, in case of deep foundations, centrifuge model tests have been carried out on the behaviour of laterally loaded single piles regarding the subgrade reaction, the changing of the stiffness of the system and the development of the displacement in case of different load frequencies. In the case of circular shallow foundations, the frequency dependence and the changing of the stiffness of the system is considered by measuring the settlement and the accelerations on top of the foundation and in the surrounding halfspace. The influence of non harmonic loadings and preloadings are investigated and a procedure to calculate the settlement will be given.

2 BASICS OF SOIL BEHAVIOUR UNDER CYCLIC LOADINGS

The behaviour of soils under cyclic loadings have been studied under many different boundary conditions. Especially the development of strains and the related stiffness shall be taken into account in dependency on the frequency, the stress amplitudes and the number of load cycles. For example, the formulations of GÜTTLER (1984) and THIEL (1988) are presented hereafter.

The developement of strains during cyclic or dynamic loading can be described by logarithmic or exponential functions as shown in Figure 1 in principle. Güttler chose a logarithmic approach in form of:

$$\varepsilon = a_i + b_i \cdot \log N$$

with i = [el; pl]   pl := irreversible (plastic)
el := reversible (elastic)   N := Numbers of load cycles

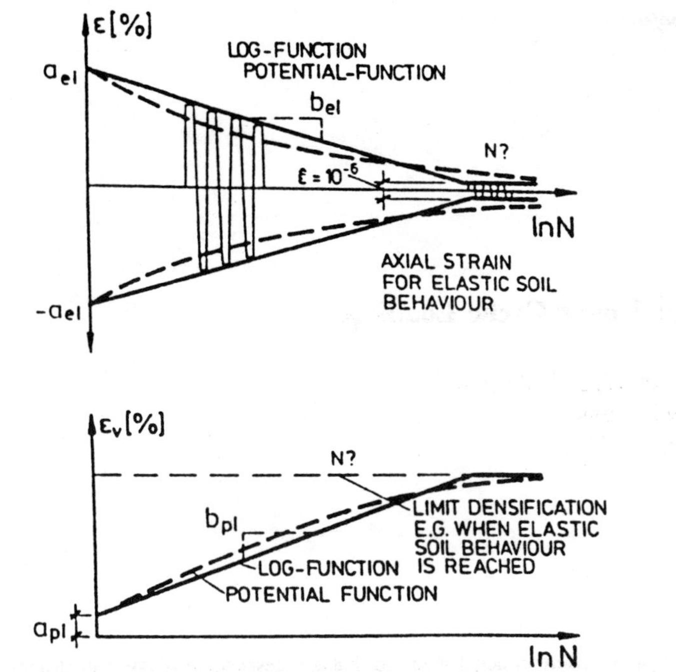

Fig. 1. Principle scheme of the measured and idealized reversible and irreversible strain development during cyclic loadings

The irreversible term represents the determined strain as an addition of all plastic portions. For the elastic formulation, a term of *cos (π*t)* has to be included. The parameters a_i (a_{el} and a_{pl}) corresponding to the strain portions of the first cycles and the b_i (b_{el} and b_{pl}) corresponding to the increasing or decreasing factor can directly be determined from the triaxial test results. They can be described additionally in terms of the load frequency and the load values (τ/τ_B or τ/σ_m). τ notifies the maximum shear amplitude of triaxial test using the Mohr-Coulomb plane. τ_A represents the maximum shear value for monotonic load conditions whereas σ_m represents the initial stress level of 0.5 x ($\sigma_1 + \sigma_3$). This equation follows the most used methods to estimate the final settlements, while the knowledge of the initial settlements is required. For several testing methods, equivalent equations can be find actually f.e. in Brown (1996).

For the expression of the dependence of stiffness in dependence on the number of load cycles, Thiel (1988) chose an exponential term including the influences of the initial stiffness during the first load cycle (A) and an index of hardening (B).:

$$E = A \cdot (T \cdot f)^{B} \qquad (1)$$

$$A = e^{s \cdot \ln(\tau/\sigma_m) + t} \cdot f^{n \cdot (\tau/\sigma_m) + \alpha_0} \qquad (2)$$

$$B = \Gamma \cdot (\tau/\sigma_m) \qquad (3)$$

with

with	T*f:= N	Number of load cycles
	α_0:=	initial stiffness coefficient
	n:=	rise of the initial stiffness coefficient dependent on the frequency
	Γ:=	gradient of hardening
	s,t:=	further material values
	σ,τ:=	values to describe the stiffness threshold value
	f:=	frequency
	s,n:=	material constants (comp. THIEL 1988)
	T:=	time for one load cycles

This description of the development of the stiffness depends on five parameters (α_o, n, Γ, s, t), which can easily be evaluated with three cyclic triaxial tests. More details about this equations and the derivation of the parameters can be found in THIEL (1988) and JESSBERGER ET AL. (1996). These equations have been prooven for several granular and artificial materials (LAUE & THIEL, 1992).

Both models, which are given here examplary define the soil behaviour with an increasing stiffness. After a certain number of load cycles, in both studies an elastic material behaviour has been found.

Comparing the findings in triaxial cells with test results gained in shear boxes, a different behaviour can be stated. Cyclic shear box tests using granular soil show under equivalent boundary conditions after an initial phase of hardening a softening up to a rupture. An example for a high load coefficient τ/τ_B is given in Figure 2 after KYSALA & FIRT (1974). This softening can even be found for smaller load coefficients increasing the numbers of load cycles.

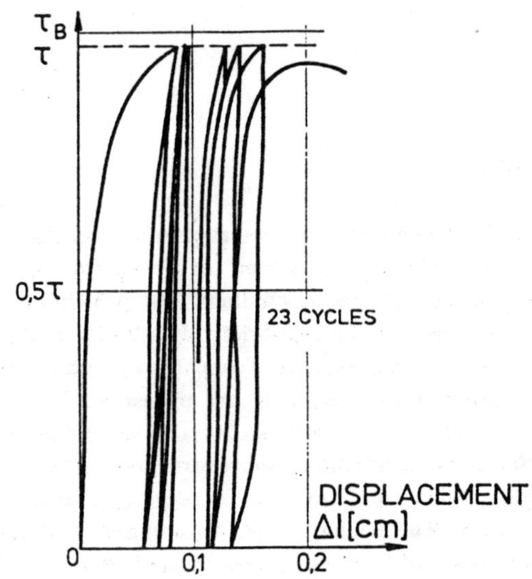

Fig. 2. Stress strain behaviour of sand under high loadings in a cyclic shear test.

Thus it is shown, that the test results and the interpretation of the behaviour of granular material are not only dependent on the boundary conditions as there are the initial stress state, the cyclic load value and the frequency but also on the test device. Using a test with ideal shear zones a softening of the regarded material with increasing numbers of load cycles is found more commonly than using a triaxial cell. Nevertheless, for weaker materials such as combustion ashes of municipal solid waste, a softening has also been found in triaxial tests for a high number of load cycles (N>50000).

These findings cannot easily be compared with the displacement behaviour of foundations. The observed behaviour of a complete system (foundation and soil) is the result of a combination of both mechanisms (hardening and softening), whereas usually the hardening effect for shallow foundations, the softening effect regarding zones of low density for deep foundations are taken into consideration. The influence of the softening is hardly depending on the system and the activation of shear zones, even in lower stress states.

3 PRINCIPLES OF CENTRIFUGE MODEL TESTING

Research in a geotechnical centrifuge has the advantage of executing tests on small scale models in a raised g-level, which represents the prototype stress field. Therefore, the feasibility of reproducing tests under defined boundary conditions on complete systems is given. Table 1 shows the scaling rules for the geotechnical centrifuge model technique with the scaling factor n as the factor for the raised g-level and the corresponding scaling factor of the reduction of the model scale. The scaling rules under wet and saturated conditions are not included here.

Tab. 1. Scaling rules for centrifuge model tests for cyclic loadings under dry conditions.

parameter	prototype at 1*g	model at n*g
length	1	1/n
area	1	$1/n^2$
volume	1	$1/n^3$
mass	1	$1/n^3$
stress	1	1
strain	1	1
force	1	$1/n^2$
frequency (acc)	1	n

In compensation to the reduction of the geometrical values, the scaling rule of frequency requires a load n-times faster than in the prototype scale. Cyclic tests in geotechnical centrifuges need special equipment for the data acquisition and the load device, which is especially restricted by the geometrical boundary conditions of the closed system during the centrifuge test.

The loads for the tests presented here are introduced to the system by specially developed hydraulic jacks with a maximum frequency of 100 Hz for small displacement values under a maximum g-level of n = 100. In Bochum, three dynamic hydraulic jacks (two for vertical loadings and one for horizontal loadings) are in use whereas at least two of them can be used and controlled independently. An external trigger starts simultaneously the loading and a fast data acquisition, which is described in more details in LAUE ET AL. (1996), and causes the further measurements of up to 64 independent channels at specified times. All tests have been conducted with the Bochum Standard Sand 942 D, a fine grained sand. This sand reproduces using the scaling law relationship a wider grained sand. The properties of this sand are given in Table 2. The sand is filled in the test container by pluviation. This methods guarantees a high reproducable density of I_D=98%. The density is controlled using small boxes placed inside the container at several depths and locations.

Tab. 2. Properties of the Bochum Standard Sand (942D).

Parameter		Value
Density of solid particles ρ_s	[t/m³]	2.66
Compactness ρ_{min}/ρ_{max}	[t/m³]	1.41 / 1.70
Porosity n_{min}/n_{max}	[-]	0.36 / 0.47
50 percent-diameter d_{50}	[mm]	0.23
10 percent-diameter d_{10}	[mm]	0.12
Uniformity coefficient $C_U = d_{60}/d_{10}$	[-]	2.08
Angle of internal friction ϕ	[°]	38

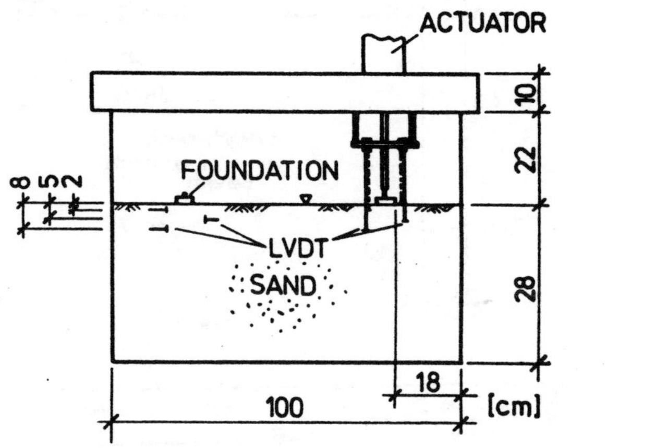

Fig. 3. Cross section of test setup for centrifuge model tests on cyclic loaded shallow foundations

4 VERTICALLY LOADED SHALLOW FOUNDATIONS

4.1 TEST DESIGN

Centrifuge model tests have been carried out on vertically loaded circular shallow foundations placed on top of a very dense sand, with the aim of finding a better solution for estimating settlements under cyclic and repeated loadings and of developing better understanding of the bearing mechanism. An exemplary setup of the tests is shown in Figure 3.

Different kinds of circular shallow foundations with various diameter and embedding depth between zero and one diameter, have been investigated in more than 40 test. The foundation load consists of a constant dead load as a variable percented of the rupture load superimposed by cyclic sinusoidal loadings with frequencies in model scale of 1 Hz, 10 Hz, 30 Hz und 60 Hz representing prototype frequencies between 0,03 and 2 Hz. One load period lasts 1000 loading cycles. After this, in several tests, the amplitude of the load has been increased and another 1000 cycles are applied to the observed foundation. The settlement is measured using the internal LVDT of the hydraulic jack. To control the load function, a force transducer is placed between the rod and the foundation. Additionaly, the bearing mechanism can be evaluated using measurements of displacement at specified points in the halfspace. These points have been defined in a distance of ½ D and D from the side of the foundation with D representing the diameter of the foundation. Measurements depths have been chosen at the surface, the bottom of the foundation and to 2, 5 and 8 cm (in model scale) below the base. The displacement is measured in vertical and tangential directions by the LVDT's. Additionally for the faster tests the acceleration due to the cyclic loading is observed inside the halfspace.

4.2 TEST RESULTS

A typical test result for the settlement of the foundation and the deformation behaviour of the surrounding subsoil is shown in Figure 4a for a foundation with a depth of embedment of t_p = 0,6 m (t_M = 20mm) a load frequency of 0,33 Hz, with load values of a statical dead load being 50% of the rupture load V_R and an amplitude V' of 50% of the dead load.

The index M indicates the model scale. After an initial settlement due to the dead load at every point in the halfspace, an increasing settlement of the foundation and heaving of the soil in the halfspace can be seen. The Figure 4b shows the settlement under the center of foundations with an embedding depth of t_P = 1,68 m for different load histories indicated by the corresponding numbers of the tests.

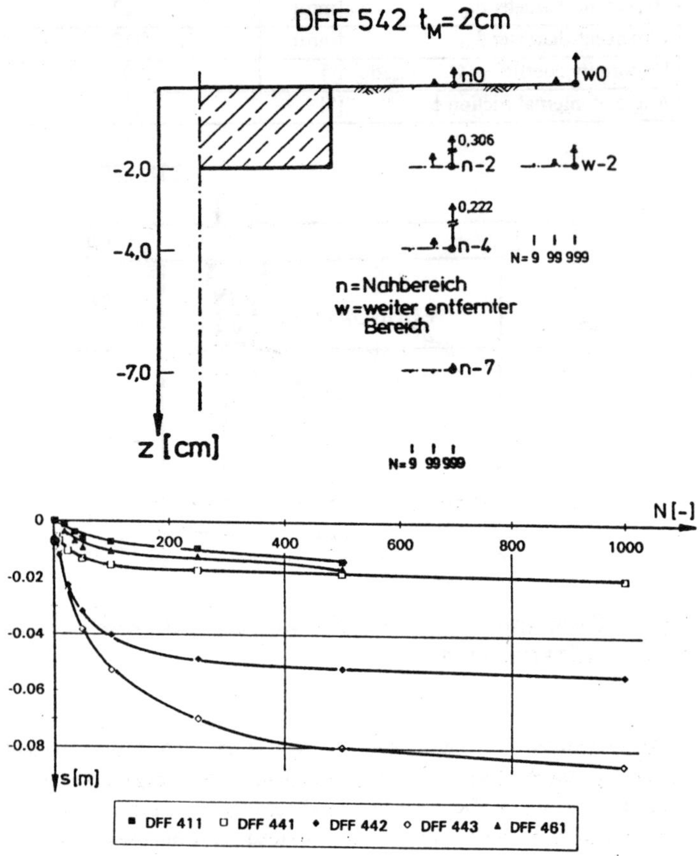

Fig 4. Typical results for one test with a load frequency f_P = 0,3 Hz in prototype scale for the vertical displacements (4a) and for different tests with t_P = 1.68 m.

The combination of the measurements of horizontal and vertical displacements in the surrounding of the foundation leads to an assumption of changing bearing zones below the foundation (Fig. 5).

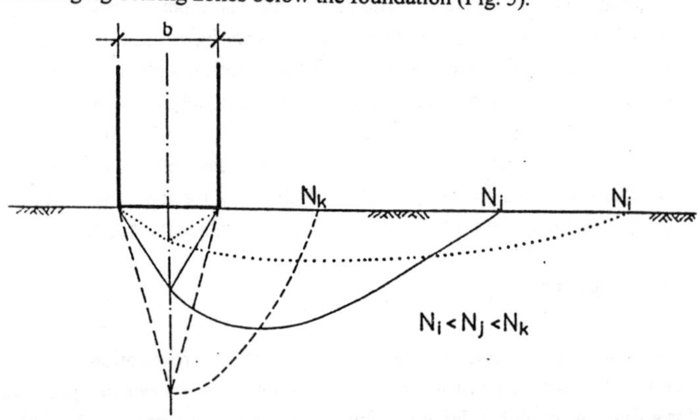

Fig. 5: Assumption on changing bearing zones with increasing number of load cycles

For low number of load cycles N a flat zone is influenced below the foundation where larger areas around the foundation deform. With an increasing number of cycles, the bearing mechanism change, to larger depths by reducing the horizontal influence zone.

An increase of settlements has been found for all of the tests being conducted. Even under high preloadings an increasing settlement after a certain numbers of load cycles has been observed. It has been found that the frequency is of less importance for the development of settlements. Figure 6 shows the settlement of shallow foundations under the applied load in principle. After applying the dead load regular or overloads forces the foundation. Even under overload conditions (line), the developement of settlements occurs after a certain number of load cycles (N_{crit}) in the same increasing value as for regular loads (dotted line). The increasing after N_{crit} is only dependent on geometric boundary conditions and the values of dead load and amplitude.

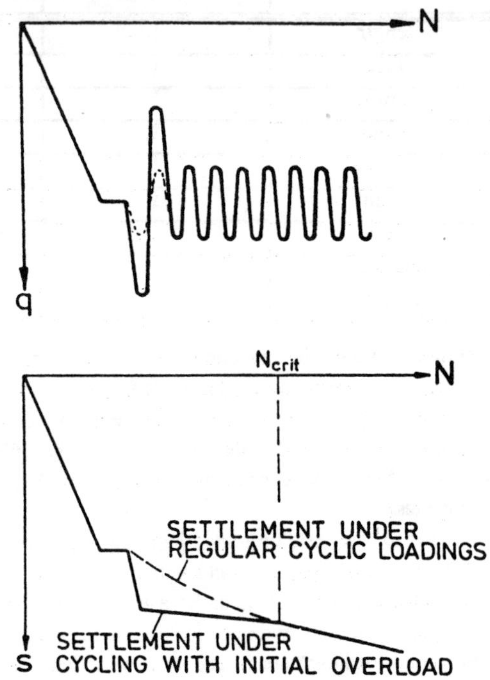

Fig. 6. Principle scheme of the interplay between load and settlement under regular and irregular loads

With this knowledge, a method to evaluate the settlement independent of the value of the initial settlement has been developed using a hyperbolic assumption and with this more realistic method of evaluating the settlement under static loadings and a logarithmic approach for the increasing:

$$s_N = \frac{1/k_s}{1/q_m - 1/q_R} + b \cdot \ln N \qquad (4)$$

with
k_S = elastical spring coefficient
q_m = initial load value of dead load
q_R = repture load
b = increasing factor
N = number of load cycles
S_N = after a given number of load cycles

Irregular overloads can be evaluated using the term independent of the number of load cycles and should be introduced as a negative starting value. Thus an estimation of N_{crit} is possible. The increasing factors varies between 0,003 for small amplitudes of 20% of the dead load up to 0,08 for high values of 80%. This method of predicting the settlement and the influence of preloading is discussed in detail in LAUE (1996 & 1997).The evaluation can be made using parameters of already known buildings or by means of centrifuge model tests, which are a useful tool especially for low cost experiments. The necessarity of further research on other geometries and soils is obvious. The validity of the assumption on the influenced halfspace will be proofed in further tests, increasing the number of measuring points in the halfspace and using two-dimensional tests.

5 LATERALLY LOADED SINGLE PILES

5.1 TEST DESIGN

The dynamic single pile tests are run in a rectangular rigid model container (length/width/height = 0.85/0.42/0.80 [m]). A hydraulic actuator, which is used for loading, is placed on the container rim. The pile is coupled with the actuator by a rigid bar, which is hinge connected to the pile head and the piston rod. A special force transducer within the bar is used for giving the actual load value to the hydraulic control system. The model setup of the cyclic single pile tests is shown in Figure 7.

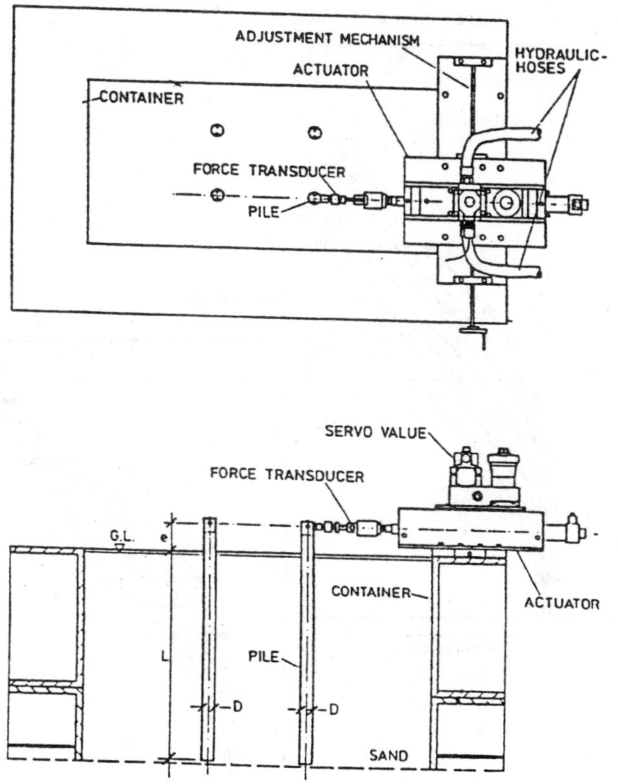

Fig. 7. Schematic model setup for cyclic single pile tests: (a) plan view, (b) cross section.

The model piles are made of aluminum tubes with an external diameter of D_M = 30 mm. and an embedded length of l_M = 60 cm. The index M represents the model scale. With this type of pile, long flexible concrete piles are modelled. That is to say, at an acceleration of 50g, which is used in the reported tests, concrete piles with an approximate diameter of D_P = 1.50 m and an embedded length of l_P= 30 m are modelled. Table 3 shows the charateristic values of model and prototype piles.

Tab. 3: Model and prototype piles.

Parameter	Model (M)	Prototype (P)
External diameter D [m]	0.030	1.50
Internal diameter d [m]	0.026	-
Embedded length l [m]	0.600	30.00
Young's modulus E [MN/m²]	70000	30000
Resisting moment I [m⁴]	1.7329E-8	0.252714
Bending resistence EI [kNm²]	1.21303	7581420

The piles are loaded at e_M = 8.5 cm above the ground line. They are load-controlled tested with a symmetrical, two-way sinusoidal load function. In model scale the load frequencies are chosen from f_M = 10 Hz up to f_M = 50 Hz. One test corresponds to 1000 load cycles, but only for specified cycles (1, 2, 4, 10, 40, 100, 400 and 1000) the pile displacement at the ground surface and the head, the pile load and the bending moment distribution along the pile are measured.

5.2 TEST RESULTS

Understanding dynamic and cyclic pile-soil interaction naturally requires detailed knowledge about the static pile-soil behaviour. Therefore, in the first instance static load tests on single piles and pile rows were run to get reference values (KOTTHAUS, 1992). At this point, the results of three typical cyclic loading tests are presented (Table 4).

Tab. 4: Presented cases of loading tests (model scale).

Test name	Frequency [Hz]	Cyclic load amplitude [kN]
DSP4-B	10	± 0.40
DSP4-D	50	± 0.10
DSP6-D	30	± 0.30

If a pile, embedded in dry sand, is loaded, the soil in front of the pile is compressed. A gap is formed behind the pile, and sand slides into this gap. Whenever the pile is cyclic loaded, this process will be repeated for each load direction. Near the ground surface it follows that a zone of compressed soil develops (Figure 8). Thus, the soil stiffness in this zone increases.

In order to investigate the effect of increasing soil stiffness, the ratios of maximum deflection at the ground line y(n) with respect to the deflection in the first cycle y(1) are studied. Figure 9 shows the relationship between y(n)/y(1) and the number of loading cycles N. Although there is a dispersion of data, the tendency of each curve is very similar. From cycle N = 400 onwards, the ratio y(n)/y(1) converged to approximately 0.3.

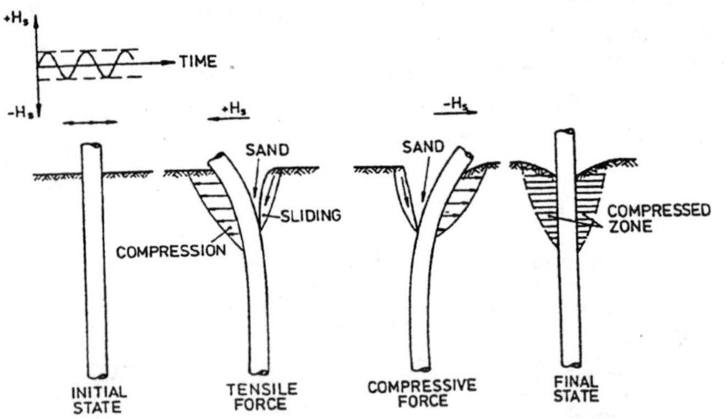

Fig. 8. Principle pile-soil interaction near the ground surface due to cyclic horizontal pile load (ANTES ET AL., 1996).

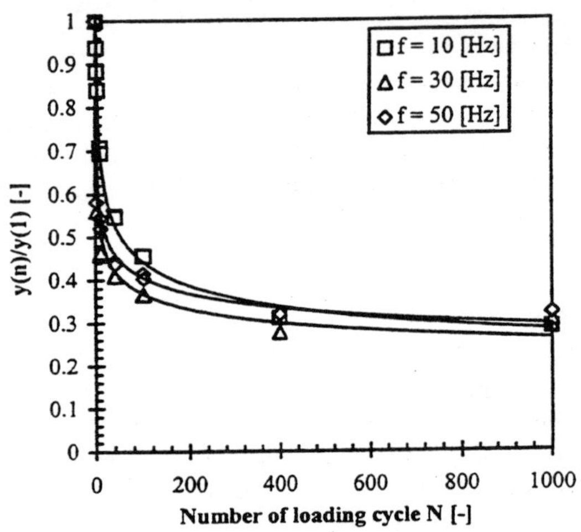

Fig. 9. Development of the ratio of deflection with the number of loading cycles.

To get a more detailed idea of the pile-soil interaction without any further analysis, the following relations have to be considered (Figure 10). The resulting bending moment M(z) of a laterally loaded pile is

$$M(z) = -\int_0^z\int_0^z p(\zeta)d\zeta^2 + H\cdot(e+z) \qquad [\text{kNm}] \qquad (5)$$

where $p(z) = -\dfrac{d^2M(z)}{dz^2}$ = soil resistance.

Rearranging equation (5) gives

$$\frac{M(z)}{H\cdot(e+z)} - 1 = -\left(\int_0^z\int_0^z p(\zeta)d\zeta^2\right)/(H\cdot(e+z)) \qquad [-]. \qquad (6)$$

Equation (6) is an expression for the ratio of the moment due to the soil resistance with respect to the moment due to the horizontal load. This enables viewing the change of the soil properties by simple calculations directly with the recorded data.

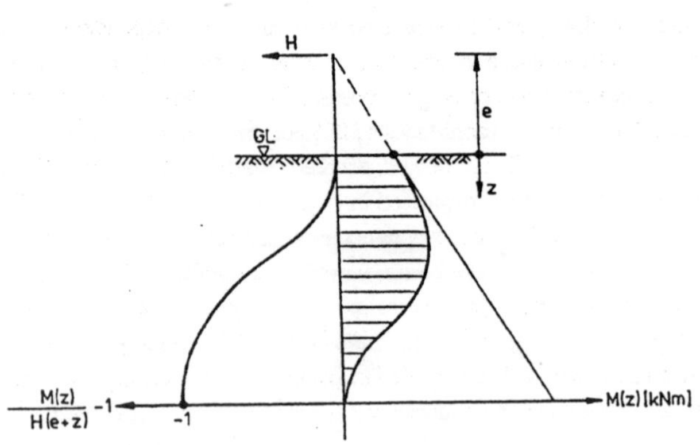

Fig. 10. Definition of the ratio of the moment due to soil resistance.

Figure 11 (a) shows the development of the bending moment of test DSP4-B versus the depth for different load cycles. Obviously, in depths larger than 3.5 cm (Model scale), the bending moment of each level decreases with the number of load cycles and the absolute maximum is moving upwards. This indicates the increasing soil stiffness. In depths smaller than 3.5 cm other effects are decisive. For instance, in the course of cyclic loading, a hollow arises around the pile. However, the resulting soil stiffness is low, which results in approximately constant bending moments in this area. Figure 11 (b) makes clear, that the point of clamping is moving upwards as well. When $M(z)/(H\cdot(e+z)) - 1 \approx 1$, no significant soil resistance will be caused.

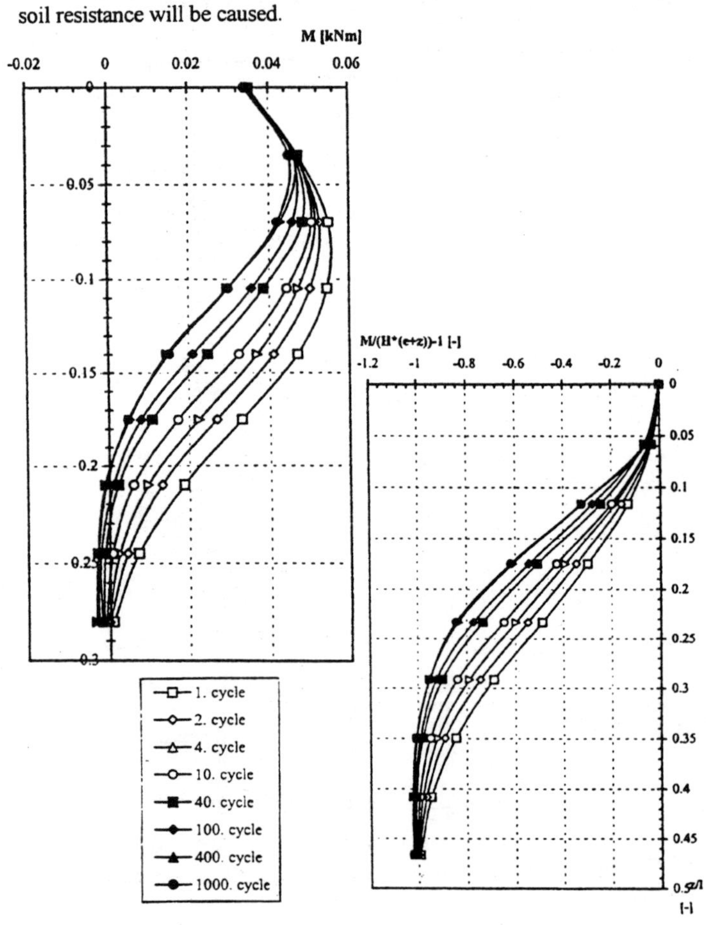

Fig. 11. Bending moment (a) and ratio of the moment due to the soil resistance (b) for test DSP4-B (model scale).

The ratio of the moment due to the soil resistance for different load frequencies is compared in Figure 12. There it is shown, that the point of clamping after thousand cycles appears to be independent from the load frequency. The clamping after thousand cycles is reached in a depth of z/l ≈ 0.35. A better understanding of this effect is given with Figure 13, where the development of M(z)/(H·(e+z))-1 with the number of loading cycles N is shown for two cases of frequency. Herein it can also be seen, that the final clamping depth is reached at loading cycle N = 400, independent of the frequency.

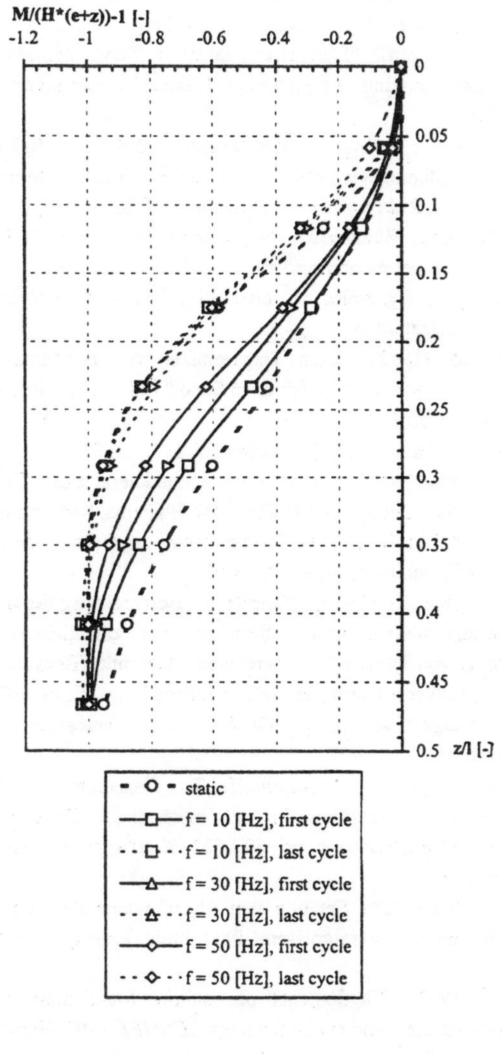

Fig. 12 Ratio of the moment due to the soil resistance for tests DSP4-B, DSP4-D and DSP6-D.

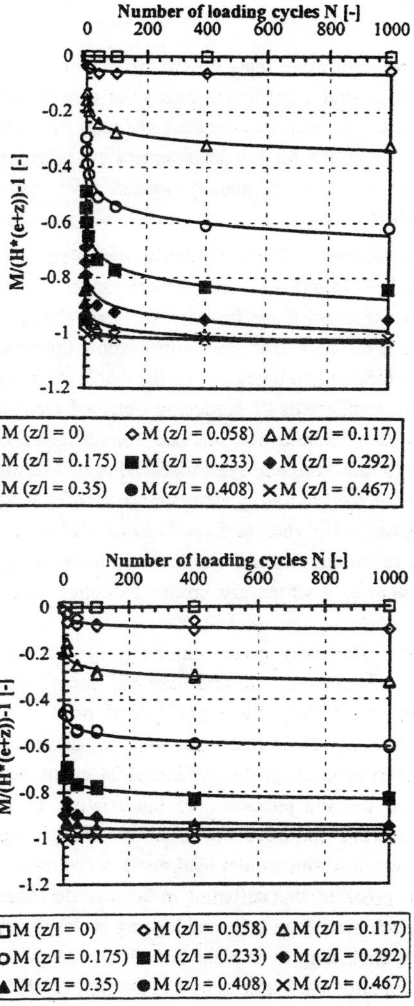

Fig. 13. Relationship between ratio of the moment due to soil resistance and the number of loading cycles: (a) DSP4-B and (b) DSP4-D.

6 CONCLUSIONS

The results of the centrifuge tests on the behaviour of cyclic loaded structures shows in general a hardening behaviour of both systems. This can be compared with the behaviour of the soil found by means of cyclic triaxial tests. The description for the development of settlements and also for the subgrade reaction of laterally loaded piles by exponential functions is possible.

Introducing the evaluation of THIEL (1988) in methods of calculating the settlement of a shallow foundation, an underestimation of the settlements in comparison to the test results can be obtained. The stiffness of the soil is overstated. Thus it is obvious, that some zones below the foundation also soften and that this mechanism takes part on the whole development of the displacements. The coefficients of hardening obtained from triaxial tests are smaller, than they should for realistic calculations. To gain the hardening factors, for the complete system instead of measuring at a site, it is possible to evaluate the hardening factors more exactly and more cheap with centrifuge model tests, whereas these factors differ in dependency of the geometry and of the load values. Additionaly it has been found, that preloadings lead only to a temporary limited reduction effect for cyclic loaded structures.

In contrast to the repeatedly given opinion (e. g. POULOS, 1988 and TAKAHASHI, 1991) of softening soil and therewith increasing deflections and bending moments of laterally cyclic loaded piles, increasing soil stiffness and decreasing deflections as well as decreasing bending moments were observed in all of the conducted 24 cyclic tests on single piles. Beyond that, the independency of the relative final state after thousand cycles from the load frequency is shown. This phenomenon may be caused by the two-way sinussoidal load, used in the reported tests. On the other hand it is possible, that softening in the regarded stress level will occur only after much more than thousand loading cycles. Further research on the mechanism shall help to explain this difference, which has been previously found for alternate loadings by e.g. DIETRICH (1974).

The test results on foundations in dry sand show different mechanism as reason for the incresing deformations, which can affect any building and which should be regarded also in saturated conditions for offshore structures especially for the estimation of the state of serviceability. To find more detailed methods, it is necessary to investigate the different bearing mechanism of cyclic and static loaded structures. This research can be made using combination of different laboratory test and the geotechnical centrifuge as a tool for low cost experiments on complete systems in the real prototype stress scale.

7 ACKNOWLEDGEMENT

The investigations presented in this paper are part of the investigations of the research project SFB151 "*Dynamics of Civil Engineering Structures*" sponsored by the National German Science Foundation (DFG) at the facility of civil engineering of the Ruhr-University Bochum.

8 REFERENCES

Antes, H., Jessberger, H.L., Schmid,G., Chouw, E., Forchap, E., Grundhoff, T., Laue, J., Le, R., Siemer, T. & Steinfeld, B. (1996). "Dynamic and cyclic soil-structure interaction," *Dynamics of Civil Engineering Structures*, pp. 451-531. Krätzig et al. (eds), Balkema, Rotterdam.

Brown, S.F. (1996). "Soil mechanics in pavement engineering" , 36th Rankine Lecture *Géotechnique* 46 No.3 pp. 383 - 426

Burland, J.B. & Burbidge, M.C. (1985). "Settlement of foundations on sand and gravel." Proc. ICE, Part 1 Vol 78.

Dietrich, T. (1974). "Biegsame Pfähle in Sand, beansprucht durch zyklische Neigung des Pfahlkopfs" *Annual report of the BAM*, 54 - 55, Berlin

Güttler, U. (1984). "Beurteilung des Steifigkeits- und Nachverdichtungsverhaltens von ungebundenen Mineralstoffen," PhD. thesis, Ruhr-University Bochum, Heft 8.

Jessberger, H.L., Wunderlich, W., Cramer, H., Laue, J. & Thiel, G. (1996). "Experimental and numerical simulation of soil behaviour" Chapter 4, Dynamics of civil engineering, Structures pp 267ff Balkema, Rotterdam

Kotthaus, M. (1992). "Zum Tragverhalten von horizontal belastetetn Pfahlreihen aus langen Pfählen in Sand," PhD. thesis, Ruhr-University Bochum, Heft 18.

Kotthaus, M., Grundhoff, T., Jessberger, H.L. (1994). "Single piles and pile rows subjected to static and dynamic lateral load," *Centrifuge 94*, pp. 497-502. Leung, Lee & Tan (eds), Balkema, Rotterdam.

Kysela, Z. & Firt, V. (1974). "Shear failure on slip surface by dynamic load." 4th Danube Conference, Bled.

Laue, J. & Thiel, G. (1992). "Beitrag zur Beschreibung des dynamischen Materialverhaltens von natürlichen und künstlichen Böden am Beispiel von Feinsand, aufbereitetem Bauschutt (Recycling-Material) und Müllverbrennungsasche". published in *Festschrift des 60. Geburtstages von Prof. Dr.-Ing. H.L. Jessberger*, Balkema, Rotterdam

Laue, J., Siemer, T., Grundhoff, T., Jessberger, H. L. (1996). "Experimental methods in soil mechanics," *Dynamics of Civil Engineering Structures*, pp. 597-622. Krätzig et. al. (eds), Balkema, Rotterdam.

Laue, J. (1996). "Zur Setzung von Flachfundamenten auf Sand unter wiederholten Lastereignissen," PhD thesis, Ruhr-University Bochum, Heft 25.

Laue, J. (1997). "Settlements of shallow foundations subjected to combined static and cyclic loadings" *ICSMFE 1997*, Hamburg

Laue, J. (1997). "On the stability of iron ore concentrate" *ISOPE* , Honolulu

Poulos, H. G. (1988). "Marine Geotechnics," Unwin Hyman, London.

Takahashi, K., Kikuchi, Y., Suzuki, M. (1991). "Lateral Resistance of Single Piles under Large Repeated Load," *GEO-COAST ' 91*, Yokohama, Vol 6, No 14, pp. 753-758.

Thiel, G. (1988). "Steifigkeit und Dämpfung von wassergsättigtem Feinsand unter Erdbebenbelastung," PhD. thesis, Ruhr-University Bochum, Heft 14.

Proceedings of the Seventh (1997) International Offshore and Polar Engineering Conference
Honolulu, USA, May 25-30, 1997

ISBN 1-880653-28-1 (Set); ISBN 1-880653-29-X (Vol. I)

Cyclic Load-Induced Settlement of Shallow Foundations on Geogrid-Reinforced Layered Soil

Braja M. Das
California State University
Sacramento, CA, USA

K.H. Khing
Southern Illinois University
Carbondale, IL, USA

ABSTRACT

Laboratory model test results for the cyclic load-induced settlement for a strip foundation supported by a stronger sand layer of limited thickness underlain by a weaker saturated clay with a geogrid reinforcement at the sand-clay interface are presented. In performing the tests, the foundation was initially subjected to an allowable static load, after which a cyclic load with a frequency of 1 cps was superimposed on it. The magnitudes of the static load and the amplitude of the cyclic load were varied. Based on the laboratory model test results, relationships for the foundation settlement and intensities of the static and cyclic loads are presented.

KEYWORDS: Cyclic load, geogrid, layered soil, settlement, strip foundation, ultimate bearing capacity

INTRODUCTION

Ultimate and allowable bearing capacities of shallow foundations on weak clay found in coastal areas can be improved to a great extent by placing a compact (or strong) layer of sand of limited thickness over the foundation. Limited theoretical and experimental studies are presently available in the literature which relate to the bearing capacity of shallow foundations on layered soil (e.g. Meyerhof, 1974; Meyerhof and Hanna, 1978). Of these studies, the theoretical development of Meyerhof and Hanna (1978) appears to be the most rational one.

More recently, a number of papers have been published which show that geogrids can be used for reinforcing soils supporting shallow foundations, thereby improving the load-bearing capacity. Khing et al. (1994) conducted a laboratory model study to determine the improvement of the bearing capacity of a surface strip foundation supported by a dense (stronger) sand layer of limited thickness underlain by a weak clay extending to a great depth with a single layer of geogrid at the sand-clay interface (Fig. 1). The results of that study can be summarized as follows:

1. The maximum benefit from inclusion of a geogrid to improve the ultimate bearing capacity occurred when the thickness of the sand layer, *H*, was about two-thirds the width of the foundation, *B*.
2. The optimum value of the width of the geogrid layer, *b*, required for deriving the maximum benefit is about 6*B*.
3. The maximum increase in the ultimate bearing capacity due to the inclusion of a single layer of geogrid was in the range of 20–30%.

Shallow foundations constructed in the manner shown in Fig. 1 (i.e., with geogrid inclusion at the sand-clay interface) may in some instances support vibrating machines. The cyclic loading of the machine will tend to induce additional permanent settlement of the foundation. The purpose of this paper is to present the results of a laboratory model study to evaluate the permanent settlement of a surface strip foundation on a layered soil (Fig. 1) reinforced with a layer of geogrid at the sand-clay interface while subjected to a cyclic loading of rectangular wave form with a frequency of 1 Hz, which is well below the resonant frequency of the foundation.

To obtain the maximum benefit from the geogrid reinforcement, for all tests reported here, H/B and b/B were kept at 2/3 and 6, respectively, as per the findings of Khing et al. (1994).

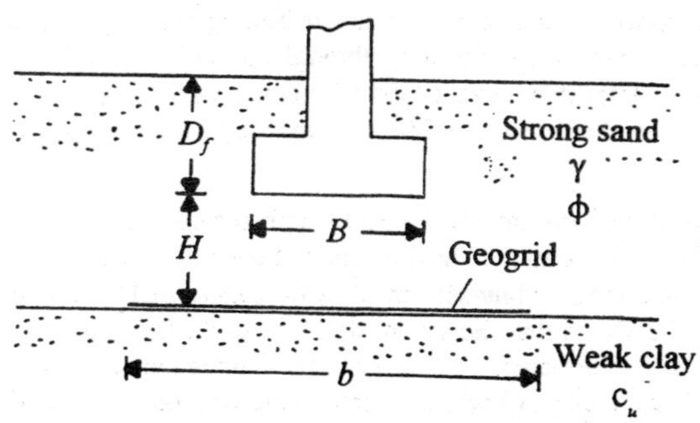

Fig. 1. Strip foundation on a strong sand layer underlain by a soft clay with geogrid reinforcement at sand–clay interface.

LABORATORY MODEL TESTS

Laboratory model tests were conducted with a model aluminum foundation measuring 76 mm (B) × 152 mm, the bottom of which was made rough by covering it with epoxy glue and rolling it in sand.

The sand used for the model tests had 100% passing through U.S. No. 20 sieve (0.85 mm opening) and 0% passing through U.S. No. 60 sieve (0.25 mm opening). The clayey soil used for the tests had 63% passing U.S. sieve No. 200 (0.075 mm opening). The liquid and plastic limits of the soil were 48% and 27%, respectively. TENSAR BX1000 geogrid was used as the reinforcing material. The aperture size of this geogrid was 25 mm in the machine direction and 33 mm in the cross-machine direction.

The test box used for the laboratory tests had inside dimensions of 152 mm (width) × 915 mm (length) × 915 mm (height). Since the width of the test box and the length of the model foundation were the same, a plane strain condition was generally established during the tests.

In conducting the model tests, the clay soil was pulverized in the laboratory and mixed with water. The wet soil was placed in several plastic bags and stored for about a week before testing. Moist clay was compacted in 25-mm thick layers in the test box to the desired height. The geogrid reinforcement ($b = 6B$) was centrally placed on top of the compacted clay. Sand of thickness H was poured into the box using a raining technique. The model foundation was then centrally placed on top of the sand layer.

The values of the dry unit weight, relative density of compaction, and angle of friction (direct shear test) for sand during the tests were 17.14 kN/m^3, 70%, and 40.3°, respectively. For the compacted clay, the average moisture content, moist unit weight, degree of saturation, and undrained shear strength (laboratory vane shear) were, respectively, 38.6%, 17.98 kN/m^3, 100%, and 14 kN/m^2.

Two types of tests were conducted, and they are described below.

Series A: This test was conducted to determine the static ultimate bearing capacity with reinforcement at the interface [$q_{u(R)}$] and the associated foundation settlement, s_u. Load to the foundation was applied by a hydraulic jack. A proving ring and a dial gauge were used to measure the load and corresponding settlement, respectively.

Series B: Cyclic load tests were conducted in this series. The tests were conducted by first applying a static load per unit area, $q = q_s$, on the foundation (Fig. 2a). Note that:

$$q_s = \frac{q_{u(R)}}{FS} \tag{1}$$

where FS is the factor of safety over the ultimate bearing capacity. Some time was allowed for the foundation to go through the settlement related to the static load. After that a cyclic load of intensity $q = q_d$ was superimposed on the static load (Fig. 2b). The resulting load pattern is shown in Fig. 2c.

A Universal testing machine was used to apply the static and cyclic load on the model foundation. The frequency of the cyclic load was 1 cps. The load and corresponding settlement(s) were measured by a load cell and an LVDT. Note that the amplitude of the cyclic load intensity was equal to $q_{d(max)}$. Permanent settlement of the foundation due to the cyclic load application only (s_d) was measured along with the number of load cycles using a data acquisition system. Table 1 gives the details of the parameters of the model tests conducted.

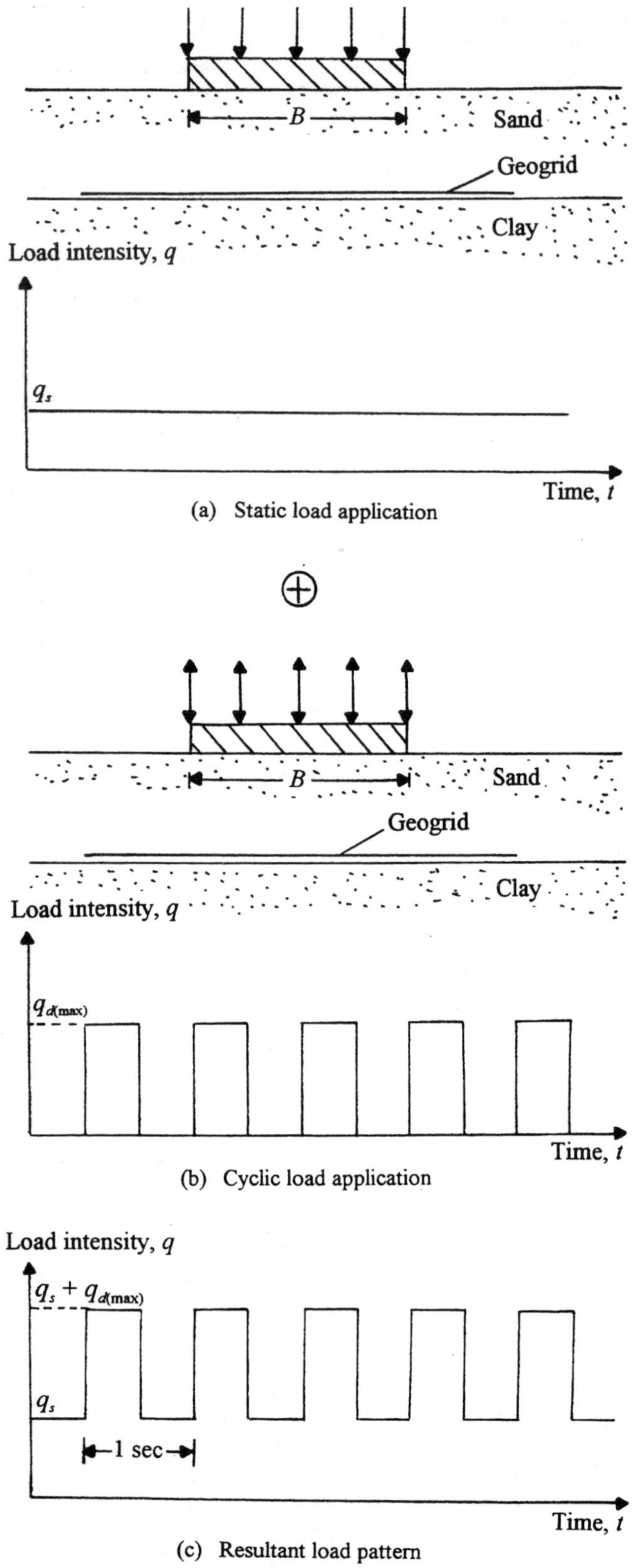

Fig. 2. Cyclic load tests — Series B

TABLE 1. Sequence of model tests of Series B ($H/B = 2/3$, $b/B = 6$)

Test No.	$FS = \dfrac{q_{u(R)}}{q_s}$	$\dfrac{q_{d(\max)}}{q_{u(R)}}$ (%)
B-1	7	14.35
B-2	7	19.91
B-3	7	33.33
B-4	5	14.35
B-5	5	19.91
B-6	5	33.33
B-7	3	14.35
B-8	3	19.91
B-9	3	33.33

MODEL TEST RESULTS

Series A: The load-displacement plot obtained from this test is shown in Fig. 3. Note that, for this test, H/B was kept at 2/3 and b/B was equal to 6. The ultimate bearing capacity with reinforcement, $q_{u(R)}$, is also shown in this figure. The magnitude of $q_{u(R)}$ for this case is about 70 kN/m² at a settlement level $s_u \approx 0.12B$.

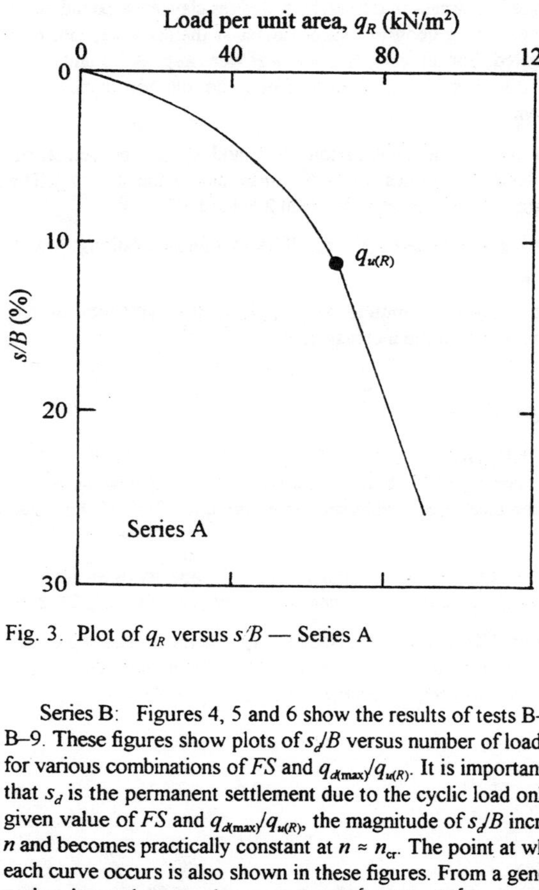

Fig. 3. Plot of q_R versus s/B — Series A

Series B: Figures 4, 5 and 6 show the results of tests B–1 through B–9. These figures show plots of s_d/B versus number of load cycles (n) for various combinations of FS and $q_{d(\max)}/q_{u(R)}$. It is important to realize that s_d is the permanent settlement due to the cyclic load only. For any given value of FS and $q_{d(\max)}/q_{u(R)}$, the magnitude of s_d/B increases with n and becomes practically constant at $n \approx n_{cr}$. The point at which n_{cr} for each curve occurs is also shown in these figures. From a general observation it can be seen that $n_{cr} \approx 2 \times 10^5$ to 3×10^5 irrespective of the values of $q_s/q_{u(R)}$ and $q_{d(\max)}/q_{u(R)}$. The magnitudes of the ultimate nondimensional settlement $s_d/B = s_{d(u)}/B$ which occurs at $n = n_{cr}$ have been

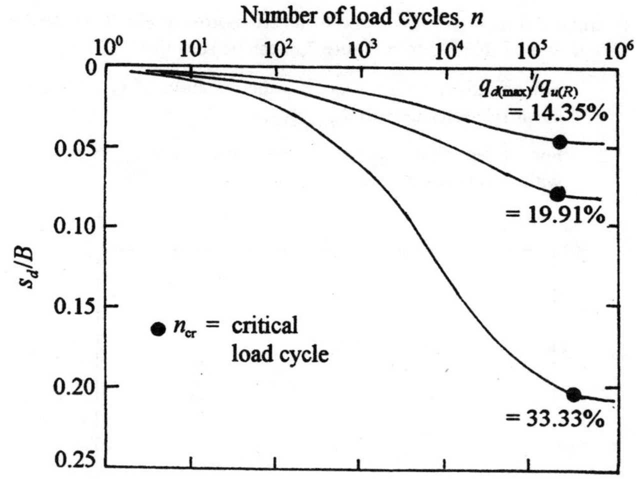

Fig. 4. Variation of s_d/B with n — $FS = 7$

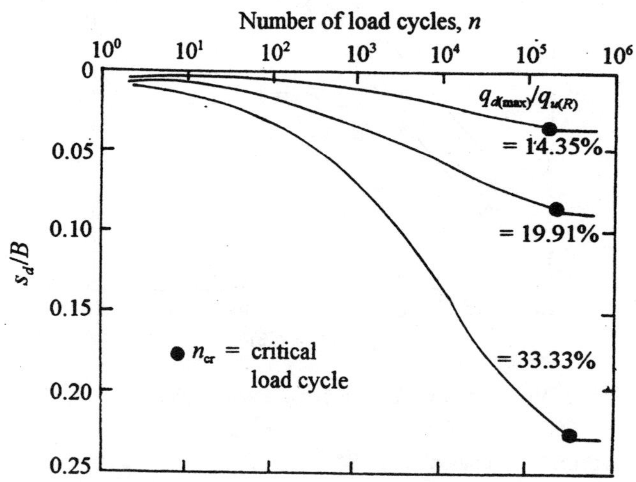

Fig. 5. Variation of s_d/B with n — $FS = 5$

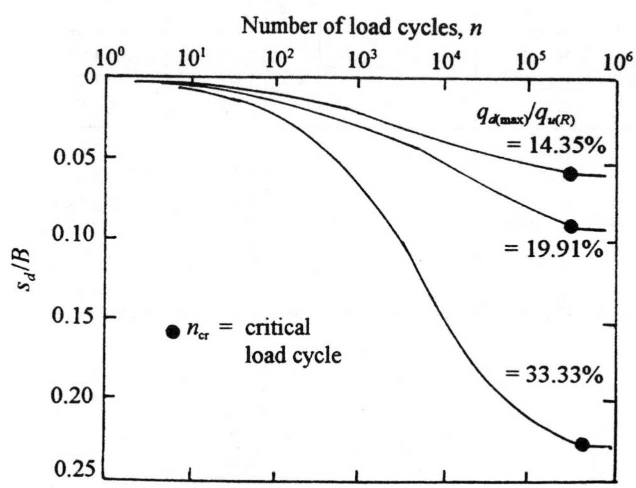

Fig. 6. Variation of s_d/B with n — $FS = 3$

determined from Figures 4, 5, and 6 and are plotted in Fig. 7. These are also given in Table 2. From Figure 7, it can be seen that:

1. For any given value of $q_s/q_{u(R)}$, the variation of $s_{d(u)}/B$ bears a linear relationship with $q_{d(max)}/q_{u(R)}$.
2. For a given value of $q_{d(max)}/q_{u(R)}$, the magnitude of $s_{d(u)}/B$ increases with the decrease in the static factor of safety, *FS*.

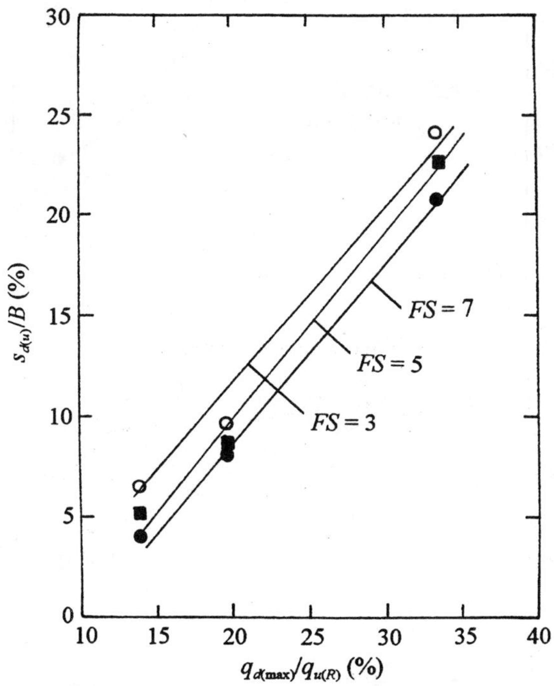

Fig. 7. Variation of $s_{d(u)}/B$ with $q_s/q_{u(R)}$ and $q_{d(max)}/q_{u(R)}$

TABLE 2. Variation of $s_{d(u)}/B$

Test No.	*FS*	$\frac{q_{d(max)}}{q_{u(R)}}$ (%)	$\frac{s_{d(u)}}{B}$ (%)
B-1	7	14.35	4.69
B-2		19.91	8.13
B-3		33.33	20.63
B-4	5	14.35	5.0
B-5		19.91	8.44
B-6		33.33	22.18
B-7	3	14.35	6.01
B-8		19.91	9.56
B-9		33.33	23.13

3. For the present cyclic load tests with interface reinforcement,

$$\frac{s_{d(u)}}{B} = 0.806\left[\frac{q_{d(max)}}{q_{u(R)}}\%\right] + C \tag{2}$$

(for $q_{d(max)}/q_{u(R)} >$ about 15%)

where C = a constant which is a function of $q_s/q_{u(R)}$.

The magnitudes of C for the tests conducted are shown in Table 3.

TABLE 3. Magnitude of *C*

FS	*C*
7	-8.06
5	-7.25
3	-5.24

CONCLUSIONS

Laboratory model test results for the cyclic load-induced permanent settlement of a surface strip foundation supported by a stronger sand layer of limited thickness underlain by a weaker clay layer extending to a great depth with a geogrid reinforcement at the sand-clay interface were presented. For all tests, *H/B* and *b/B* were kept at 2/3 and 6, respectively. Based on the model test results, the following conclusions can be drawn:

1. For any given combination of q_s and $q_{d(max)}$, the permanent settlement increases with n to a maximum value at $n = n_{cr}$. The magnitude of n_{cr} varies between 2×10^5 to 3×10^5.
2. For a given value of *FS*, $s_{d(u)}/B$ bears a linear relationship with $q_{d(max)}/q_{u(R)}$.
3. For a given magnitude of $q_{d(max)}/q_{u(R)}$, the magnitude of $s_{d(u)}$ increases with the increase in *FS*.

REFERENCES

1. Khing, KH, Das, BM, Puri, VK, Yen, SC, and Cook, EE (1994). "Foundation on Strong Sand Underlain by Weak Clay with Geogrid at the Interface," *Geotextiles and Geomembranes*, Vol. 15, pp. 199–206.
2. Meyerhof, GG (1974). "Ultimate Bearing Capacity of Sand Overlying Clay," *Canadian Geotechnical Journal*, Vol. 11, pp. 224–229.
3. Meyerhof, GG, and Hanna, AM (1978). "Ultimate Bearing Capacity of Foundations on Layered Soil Under Inclined Load," *Canadian Geotechnical Journal*, Vol. 15, pp. 656–572.

Proceedings of the Seventh (1997) International Offshore and Polar Engineering Conference
Honolulu, USA, May 25-30, 1997

ISBN 1-880653-28-1 (Set); ISBN 1-880653-29-X (Vol. I)

Dynamic Response of Square Foundation Vibration Using 3D Boundary Element Method

Sheng-Huoo Ni and Pay-Shiun Tsai
National Cheng-Kung University
Tainan, Taiwan, China

ABSTRACT

This paper investigates the response of square foundations subject to vertical and horizontal excitations using the frequency domain boundary element method. The vibration response of foundation is expressed in the form of dynamic flexibility. Key parameters of the vibration response are investigated. Including Poisson's ratio, material damping ratio, and foundation embeddment, the study shows that the effect of Poisson's ratio on the vertical mode of vibration is more significant than that on the horizontal mode of vibration. The effect of material damping ratio is insignificant on both modes of vibration. However, it can reduce the resonance amplitudes. The deeper the embeddment of foundation is the less dynamic flexibility of both vertical and horizontal modes of vibration will be.

KEY WORDS : Boundary element method, foundation vibration, dynamic flexibility

INTRODUCTION

The purpose of this study is to investigate the dynamic reponse of a three-dimensional square rigid foundation under vertical and horizontal harmonic forced vibration. Several factors affecting the dynamic response are investigated. Applications of BEM to solve various vibration problems have been done for many researches. Dominguez (1978) used frequency domain BEM to study the dynamic response of rigid foundation. Abascal and Dominguez (1986) used two dimensional frequency domain BEM and constant elements to obtain dynamic flexibility of a rigid strip foundation resting on a visco-elastic soil. In their study, the responses of the soil with a layer of bedrock and various stiffness of bedrock are investigated. Israil and Ahmad (1989) applied quadratic elements to analyze the dynamic flexibility of a rigid strip foundation resting on layered soils, in which the influence of Poisson's ratio, material damping ratio, embedded depth of foundation are surveyed, and the soil parameters of the Poisson's ratio of 0.33, material damping ratio of 0.05 and vertical vibration mode were used for computations. Wong and Luco (1976 solved the dynamic reponse of rigid foundations of arbitrary shapes The method used is based on dividing the contact area between the foundation and the soil into a number of subregions and in assuming that the contact stresses within each area to be uniform but of an unknown magnitude. Rizzo and Shippy (1985) also used BEM to obtain the dynamic flexibility of a rigid square foundation under a vertical vibration mode.

FREQUENCY DOMAIN BOUNDARY ELEMENT METHOD

Governing Equations

The soil medium is assumed to be a linear elastic homogeneous and isotropic half space in the study. The forced harmonic motion of a three-dimensional surface and embedded square foundation are considered. Under this condition, the governing equations of motion in the frequency domain take the form:

$$\mu\nabla^2\bar{u}(x,\omega)+(\lambda+\mu)\nabla\nabla\bar{u}(x,\omega)+\rho\bar{f}(x,\omega)=-\rho\omega^2\bar{u}(x,\omega) \qquad (1)$$

where λ and μ are Lame' constants, $\bar{u}(x,\omega)$ is the displacement vector, with x being the position vector and ω the excitation frequency, ρ and $\bar{f}(x,\omega)$ are the mass density and body force vector, respectively. The body force is neglected. The boundary conditions can be expressed as

$$\bar{u}(x,\omega)=\bar{u}_b(x,\omega),\ \bar{t}(x,\omega)=\bar{t}_b(x,\omega) \qquad (2)$$

x is on the boundary ∂D

in which $\bar{u}_b(x,\omega)$, $\bar{t}_b(x,\omega)$ are known displacements and known tractions in the frequency domain boundary ∂D, respectively.

Fundamental Solution

The fundamental solution satisfying three dimensional full infinite radiation condition(1965) can be expressed as

$$U_{ij} = \frac{1}{4\pi\mu}\left[L(k_T)\delta_{ij} + \frac{1}{k_T}M_{,ij}\right] \tag{3}$$

where $L(k_T) = \frac{e^{ik_T r}}{r}$; $M = L(k_T) - L(k_L)$; $M_{,ij} = \frac{\partial^2 M}{\partial x_i \partial x_j}$;

$r = |x - y|$; $k_T = \frac{\omega}{c_T}$; $k_L = \frac{\omega}{c_L}$; $c_T = \sqrt{\frac{\mu}{\rho}}$; $c_L = \sqrt{\frac{\lambda + 2\mu}{\rho}}$

δ_{ij} is Kronecker's delta, $\delta_{ij} = 1$ as $i = j$, and $\delta_{ij} = 0$ as $i \neq j$, c_T is the shear wave velocity, c_L is the dilatational wave velocity, k_T and k_L are the wave numbers of the shear wave and the dilatational wave, respectively. The traction can be obtained as

$$T_{ij}(\mathbf{x},\mathbf{y},\omega) = D_{ijk}(\mathbf{x},\mathbf{y},\omega)n_k \tag{4}$$

where n_k is outward normal vector at x, and

$$D_{ijk} = \lambda U_{il,l}\delta_{jk} + \mu(U_{ij,k} + U_{ik,j}) \tag{5}$$

To solve the singularity problem, the method suggested by Kitahara and Nakagawa (1985) is used in this study. In the method, the numerical error in the fundamental solution is reduced effectively when the field point approaches the loading point or in the condition of low frequency vibration.

On the basis of Betti's reciprocal theorem and the radiation condition apply to (3) and (4), the following boundary integral equation can be obtained.

$$\int_{\partial D}\left[t_i(x,\omega)U_{ij}(x,y,\omega) - T_{ij}(x,y,\omega)u_i(x,\omega)\right]d\Gamma = C_{ij}u_i(y,\omega) \tag{6}$$

where U_{ij} and T_{ij} are the infinite plane fundamental solutions, representing the displacement and traction at a point x in direction i due to a unit harmonic force applied at a point y in direction j, ∂D is the boundary surface of the domain of interest, Γ is the full domain boundary surface, C_{ij} is the discontinuity or jump term resulting from the singularity of the T_{ij} kernel. By using constant element, C_{ij} is

$$C_{ij} = \frac{1}{2}\delta_{ij} \tag{7}$$

If the boundary ∂D is discretized into constant elements, equation (6) can be expressed as

$$C_{ij}u_i + \sum_{j=1}^{n}\tilde{H}_{ij}u_j = \sum_{j=1}^{n}G_{ij}t_j \tag{8}$$

in which $\tilde{H}_{ij} = \int_{\Gamma_j} T_{ij}(x,y,\omega)d\Gamma_j$, $G_{ij} = \int_{\Gamma_j} U_{ij}(x,y,\omega)d\Gamma_j$ (9)

Equation (8) can be rewritten as

$$\sum_{j=1}^{n}H_{ij}u_j = \sum_{j=1}^{n}G_{ij}t_j \tag{10}$$

in which $H = C + \tilde{H}$, equation (10) can be expressed in a matrix form as

$$\mathbf{HU} = \mathbf{GT} \tag{11}$$

The contact condition at the soil-foundation interface is assumed to be perfectly bonded contact. In this bonded condition, the adjacent soil moves with the foundation such that the soil will not separate from the foundation during the foundation vibrating. All degrees of freedom are coupled with each other. For an embedded foundation the same bonded contact condition is also assumed. In this case, the bottom surface and four embedded sides of foundation will move with the soil without slipping. To confirm the foundation to be rigid and without torsional motion, a special input for the boundary is mode. In the vertical vibration mode, a unit vertical displacement only is imposed to each element of the foundation area. In the horizontal vibration made, a unit horizontal displacement in the x-direction is imposed to each element of the foundation area. The geometry and dimensions of the foundation for the surface and the embedded foundation are shown in Fig. 1.

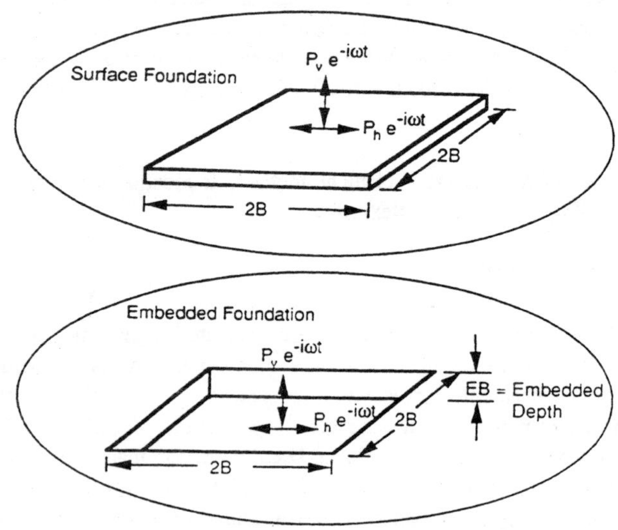

Fig. 1 Geometry and Dimension of Square Foundation

The boundary condition of the free surface is traction free. For example, if the element i is on the free surface, then

$$t_{ix} = t_{iy} = t_{iz} = 0 \tag{12}$$

Substituting the boundary condition into (11), are obtains

$$AX = F \tag{13}$$

Solving equation (13), one can obtain all the unknown function (x) on the boundary, then all the unknown nodal displacements and tractions can be solved.

The three-dimensional mesh, each node usually consists of many elements and has three components of traction and displacement. Therefore, it takes a lot of storage space and computation time to perform the numerical calculation. If the symmetry condition exists in a physical problem, then the degrees of freedom can be reduced to a great extent.

In the vertical vibration mode, the medium domain considered is symmetry with xz-plane, yz-plane and the diagonal plane of xz and yz-planes. Therefore, as shown in Fig.2 only one-eighth of domain need to be discretized for the analysis. In the horizontal vibration mode, the domain considered is symmetry with xz-plane, and is anti-symmetry with yz-plane. Therefore, as shown in Fig.3, one quarter of the full domain needs to be considered and discretized.

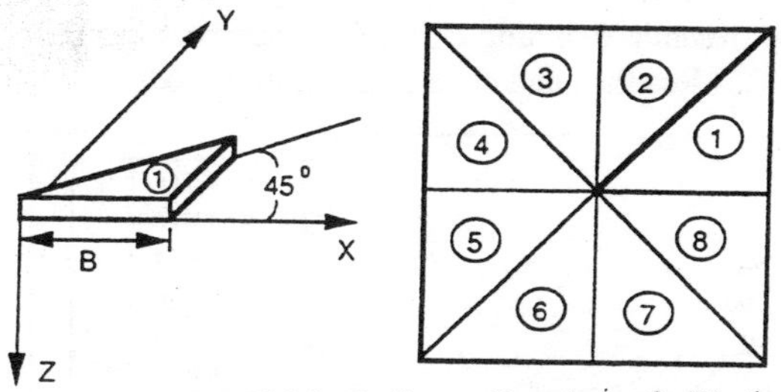

Fig. 2 One-Eighth of a Square Foundation for Vertical Vibration Mode

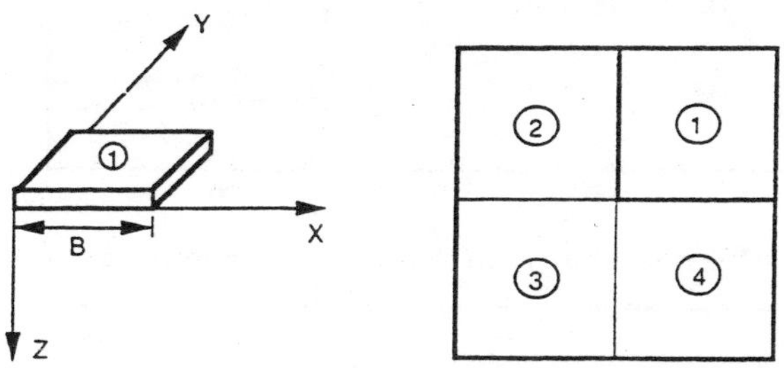

Fig. 3 One Quarter of a Square Foundation for Horizontal Vibration Mode

Material Damping

Both the geometrical damping and the material damping will attenuate the vibration energy of foundation. The geometrical damping have been included in the fundamental solution. The material damping can be incorporated by the correspondence principle (1980) into the governing equations. After a Fourier transformation, the governing equation of one-degree-of-freedom of vibration system can be written as

$$-m\omega^2\bar{u}(\omega) - i\omega c\bar{u}(\omega) + k\bar{u}(\omega) = \bar{f}(\omega) \tag{14}$$

The mechanism of the spring and dashpot can be expressed as a complex stiffness k^* by

$$k^* = k - i\omega c = k(1 - \frac{i\omega c}{k}) = k(1 - \frac{i\omega 2m\xi\omega_n}{m\omega_n^2}) = k(1 - 2i\xi\frac{\omega}{\omega_n}) = k(1 - 2i\beta) \tag{15}$$

in which β is the hysteric damping ratio.

As describe earlier, the fundamental solution has covered the geometric damping. The material damping of soil mass is taken into account by replacing the Lame' constants in a complex form

$$\mu^* = \mu(1 - 2\cdot i\cdot\beta) \quad , \quad \lambda^* = \lambda(1 - 2\cdot i\cdot\beta)$$

$$C_L^* = C_L\sqrt{(1 - 2\cdot i\cdot\beta)} \quad , \quad C_T^* = C_T\sqrt{(1 - 2\cdot i\cdot\beta)} \tag{16}$$

where β is material damping ratio of soil mass. To consider the material damping the wave number in the fundamental solution (in equation (3)) needs to be replaced by the complex wave number in the program code.

ANALYSIS METHOD

The traction and displacement of each element at the interface of soil and foundation can be solved from equation (3). In the vertical vibration case, the amplitude of harmonic force (P_v) is equal to the traction in the direction of z times the area of the particular element In a similar way, the amplitude of horizontal harmonic force (P_h) is equal to the traction in the direction of x times the area of the particular element considered in the horizontal vibration case.

The dynamic response of foundation vibration can be presented by the dynamic flexibility. For a harmonic force with a specified vibration frequency ω, the vertical ($F_{vv}(\omega)$) and horizontal ($F_{hh}(\omega)$) dynamic flexibility can be defined respectively as

$$F_{vv}(\omega) = \frac{u_v B\mu}{P_v} = \mathrm{Re}[F_{vv}(\omega)] + i \times \mathrm{Im}[F_{vv}(\omega)]$$

$$F_{hh}(\omega) = \frac{u_h B\mu}{P_h} = \mathrm{Re}[F_{hh}(\omega)] + i \times \mathrm{Im}[F_{hh}(\omega)] \tag{17}$$

where u_v and u_h are the vertical and horizontal displacement respectively. B is the half width of foundation, μ is the shear modules of soil, P_v and P_h are the amplitude of vertical and horizontal force, respectively. The real part of dynamic flexibility $\mathrm{Re}[F(\omega)]$ represents the displacement in phase with the foundation reaction force. The imaginary part of dynamic flexibility $\mathrm{Im}[F(\omega)]$ represents the displacement 90 degree out of phase with the reaction force.

The geometry and dimension of foundation considered is shown in Fig. 1 in which, B is the half-width of foundation. The vertical and horizontal harmonic forced vibration are considered separately. The foundation is assumed to be rigid and massless. The dynamic flexibility is plotted against dimensionless frequencies given by

$a_0 = \frac{\omega}{C_T}B$, where C_T is the shear wave velocity.

The effect of embedded depth of foundation on the response of dynamic flexibility is also investigated. For convenience, the embedded depth parameter E is defined as the embedded depth divided by the half-width of foundation. The dynamic flexibility has been normalized with dynamic shear modules of the soil. Therefore, the result obtained is considered to be independent of the dynamic shear modules of soil.

According to the study of Dominguez (1987), for the convergence in calculation of a three-dimensional square footing, the discretization for the interface in close proximity to the foundation should use element lengths about 1/4 to 1/6 of the Rayleigh wavelengths. The free surface discretization has to be extended at least to 7.5B for the surface foundation and 4 times embedded depth for the embedded foundation. In this study, the suggestion of Dominguez is followed in discretizing the boundaries. The discretization is selected at the interface of the foundation to a wavelength covering six elements in this study.

To establish the accuracy of the present program code based on the numerial model, a preliminary study is performed for a rigid square foundation of side of 2B, placed on the surface of soil mass with a Poisson's ratio $\nu = 1/3$, material damping ratio $\beta = 0.01$ % . The result obtained is shown in Fig. 4. in which the results obtained by Wong and Luco, and by Rizzo, are also illustrated. Good agreement between the results are obtained.

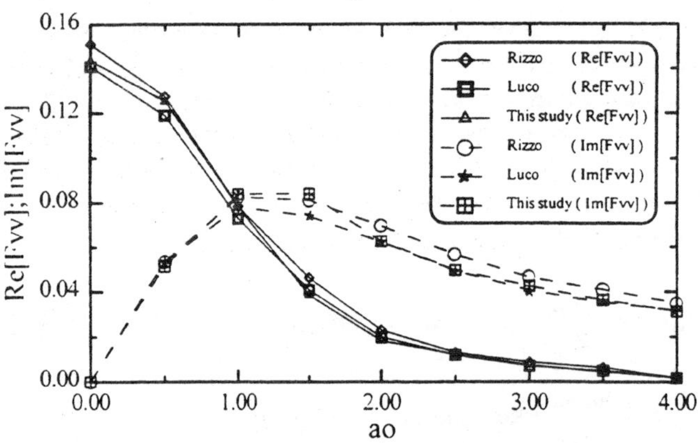

Fig. 4 Comparison of Accuracy of Vertical Dynamic Flexibility for Surface Foundation (β = 0.01%, υ = 0.33, E = 0)

PARAMETRIC STUDY

For the convenience of comparisons, the result is presented in the form of the dynamic flexibility. The factors affected the dynamic flexbility are studied, including Poisson's ratio of soil, material damping ratio, and embedded depth of the foundation. Poisson's ratio of 0.33 and 5 percent material damping are used throughout unless otherwise stated. The bonded contact is assumed along the soil-foundation interface.

Effect of Poisson's Ratio

In geneal, the effect of Poisson's ratio on P-wave is more significant than on S-wave. In the vertical vibration case, the relative contribution of the dilatational wave is higher than that of the other waves. Poisson's ratio significantly influences the dilatational wave velocity. The variation of Poisson's ratio affects the vertical reponse to a considerable extent. The result of the effect of Poisson's ratio is shown in Fig. 5, in which the effect of Poisson's ratio for a surface foundation is depicted. It can be seen that when soil is loaded vertically, with increasing Poisson's ratio there will be more lateral displacement. Thus, the lateral confinement of soil will offer more resistance, thereby softening the soil system. It can also be seen that for Poisson's ratio of 0.49, as a_0 greater than 2.5, the value of the dynamic flexibility becomes negative. This means that the direction of the displacement is opposed to the direction of the vertical loading In other words, there is a phase difference between the displacement and the force. Similar result, are obtained in plane strain case (1994).

The response of dynamic flexibility for the horizontal vibration is shown in Fig. 6 As shown in Fig. 6, increasing of the Poisson's ratio decreases the dynamic flexibility. But the decrement in horizontal vibration mode is less than that of vertical vibration mode.

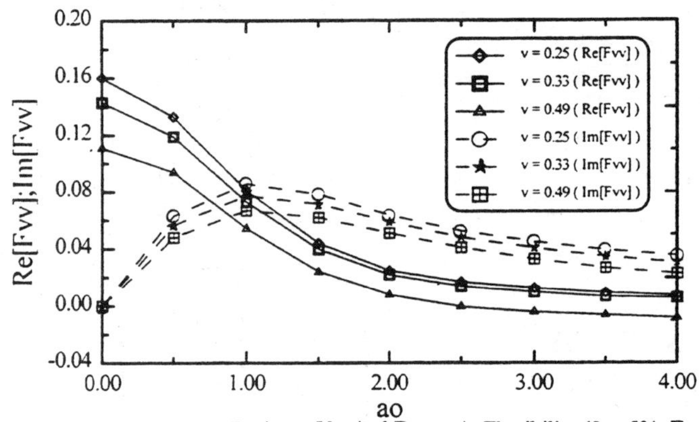

Fig. 5 Effect of Poisson's Ratio on Vertical Dynamic Flexibility (β = 5%, E = 0)

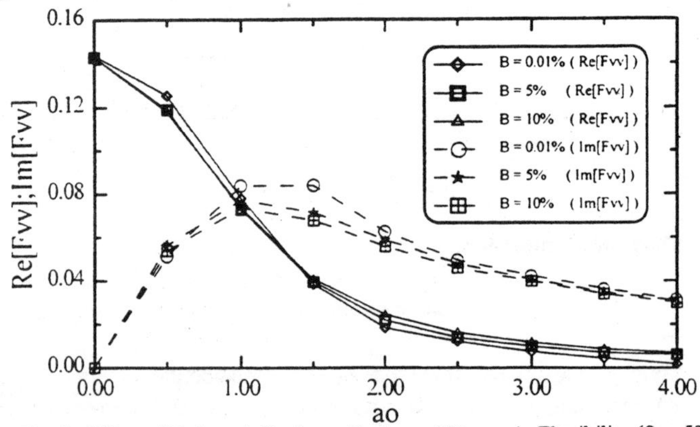

Fig. 6 Effect of Poisson's Ratio on Horizontal Dynamic Flexibility (β = 5%, E = 0)

Effect of Material Damping

The energy attenuation of body waves generated by the foundation vibration is due to the existence of the geometrical damping and internal material damping. The effect of material damping ratio on the dynamic flexibility is investigated. Results of computations are shown in Fig. 7 and Fig. 8 for the vertical vibration and horizontal vibration, respectively. As shown in both figures, increasing of the material damping ratio increases the values of the dynamic flexibility. The imaginary part of the dynamic flexibility is affected most, where the real part remains almost unchanged. However, the effect of material damping ratio on the dynamic flexlibility is insignificant when comparing with other factors. This may be due to the fact that the energy attenuation is mostly governed by the geometrical damping in the half space. As shown in the Fig.8, the larger material damping ratio, the less dynamic flexibility.

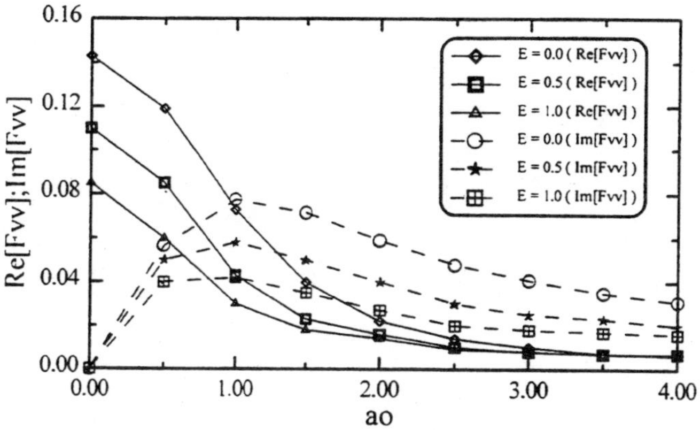

Fig. 7 Effect of Material Damping Ratio on Vertical Dynamic Flexibility ($\upsilon = 0.33\%$, E = 0)

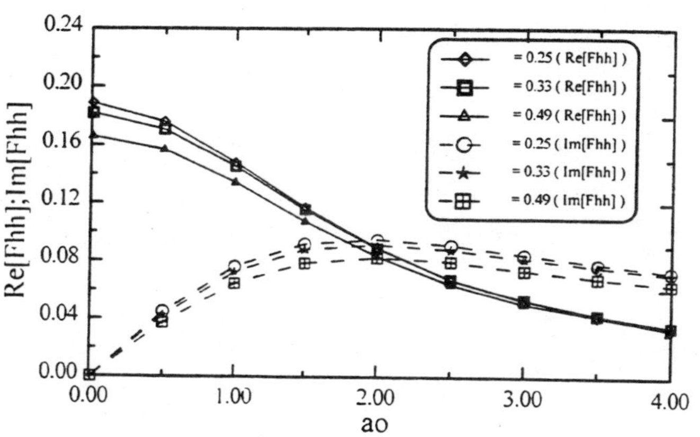

Fig. 8 Effect of Material Damping Ratio on Horizontal Dynamic Flexibility ($\upsilon = 0.33\%$, E = 0)

Effect of Embedded Depth of Foundation

The effect of the embedded depth of the foundation on the dynamic flexbility is shown in Fig. 9 and Fig. 10 for the vertical and horizontal vibration, respectively. The embedded depth of E = 0, E = 0.5 and E = 1.0 are considered. As shown in Fig. 9 for the vertical vibration, the dynamic flexbility decreases as the embedded depth parameter E increases. This may be because the contact area increases as the embedded depth increases.

For the effect of the embbed depth of foundation, it can be seen from Fig. 9 and Fig. 10 that the influence on the horziontal vibration mode is larger than that on the vertical vibration mode.

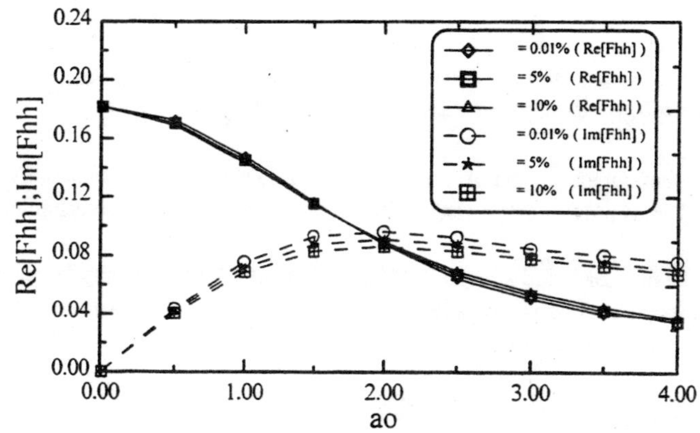

Fig. 9 Effect of Embedded Depth on Vertical Dynamic Flexibility ($\upsilon = 0.33\%$, $\beta = 5\%$)

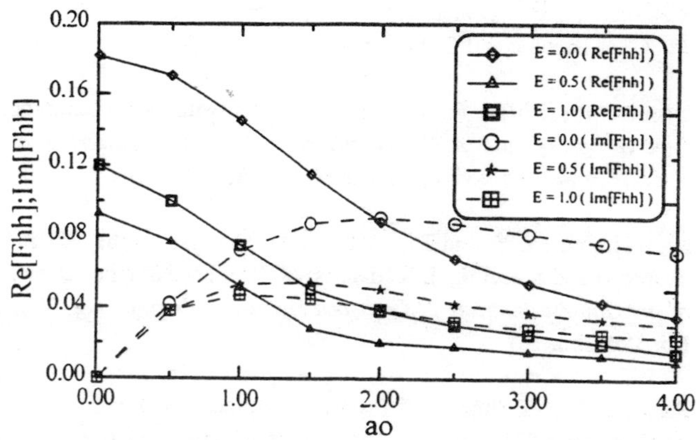

Fig. 10 Effect of Embedded Depth on Horizontal Dynamic Flexibility ($\upsilon = 0.33\%$, $\beta = 5\%$)

CONCLUSIONS

1. Due to eariness in discreting the boundary and involving the material damping ratio in the analysis, the frequency domain boundary element method is well suited for analyzing the response of thefoundation vibration of square footing.
2. The effect of Poisson's ratio on vertical vibration mode is more significant than on horizontal vibration mode. The dynamic flexibility decreases with increasing Poisson's ratio.

3. The effect of the material damping ratio on the foundation vibration is in significant. But it reduces the resonant amplitude significantly.
4. The dynamic flexibility decreases with increasing the embedded depth of the foundation for both the vertical and horizontal vibration mode. However, the effect of the embedded depth of the foundation on the horizontal vibration mode is more significant than on vertical vibration mode.

REFERENCES

Abascal, R., and Dominguez J, (1986). "Vibrations of Zoned Viscoelastic Soils, "*Journal of Engineering Mechanics*, ASCE, Vol. 112, No. 5.

Dominguez, J (1978). "Dynamic Stiffness of Rectangular Foundation, " Report No. R78-20, Department of Civil Engineering, MIT.

Dominguez, J, and Abascal R, (1987). "Dynamic of Foundations," Topics in Boundary Element Research Applications in Geomechanics.

Israil, ASM, and Ahmad S, (1989). "Dynamic Vertical Compliance of Strip Foundations in Layered Soils," *Earthquake Engineering and Structural Dynamics*, Vol. 18.

Kupradze, VD (1965). "Potential Methods in the Theory of Elasticity," Israel Program for Scientific Translations, Jerusalem.

Kitahara, M, and Nakagawa K, (1985). "Boundary Integral Equation Methods in Three Dimensional Elastodynamics," *Boundary Element IV*, C. A. Brebbia and G. Maier, ed..

Lysmer, J (1980). "Foundation Vibrations with Soil Damping," *Proceedings of 2nd ASCE Conference on Civil Engineering and Nuclear Power*, Knoxville, Tennessee, Vol. II, Paper 10-4.

Ni, S H, Feng, Z Y, and Tsai, P S (1994). "Analysis of the Vibration Response and Screening Effectiveness of Strip Foundations," *Journal of the Chinese Institute of Civil and Hydraulic Engineering*, Vol. 6, No. 3(in Chinese).

Rizzo, F J, and Shippy, D J (1985). "Boundary Integral Equation Analysis for a Class of Earth-Structure Interaction Problems," Final Report to NSF, Grant No. CEE80- 13461, Department of Engineering Mechanics, University of Kentucky, Lexington.

Wong, HL, and Luco, J E (1976). "Dynamic Response of Rigid Foundations of Arbitrary Shape," *Earthquake Engineering and Structural Dynamics*, Vol. 4.

Proceedings of the Seventh (1997) International Offshore and Polar Engineering Conference
Honolulu, USA, May 25-30, 1997

ISBN 1-880653-28-1 (Set); ISBN 1-880653-29-X (Vol. I)

The Study on the Electro Osmotic Consolidation of Marine Clayey Soil

Soo-Sam Kim
Chung-Ang University, Seoul, Korea
Kwang-Yeol Lee
Dong-Seo University, Pusan, Korea
Sang-Jae Han
Chung-Ang University, Seoul, Korea
Kyung-Hwan Koh
Samsung Heavy Industry Co., Seoul, Korea

ABSTRACT

As a part of supplementary ground improvement method, this research was focused on the fundamental study for the utilization of electro osmosis. In order to verify the efficiency of consolidation by applying electro osmosis mechanism, in this study, conventional theories of electro osmotic drainage were primarily examined by simple experimental studies. Based on the results of primary study, a governing equation for the electro osmotic behaviors in soils was established to conduct further study. A number of electro osmotic consolidation tests were conducted on newly modified apparatus. Two different soils that contains different concentration of sodium chloride were used for the tests to investigate the effects of concentration of electrolyte in soils on the electro osmotic drain consolidation. Consequently, the conventional studies were performed for verifying the effects of voltage gradients. Based on experimental study and theoretical analysis, it was found that the electro osmotic drain consolidation was highly affected by the concentration of electrolyte in soils and discovered that characteristics of electro osmotic drain consolidation for marine clay in Korea met good agreement with Esrig's theory.

KEY WORDS: Electro osmosis, consolidation time, voltage gradients, sodium chloride concentration

INTRODUCTION

Conventional ground improvement techniques, such as sand drain and plastic-board drain method, are currently used in combination with surcharge by the weight load. However, several major problems occur due to lateral displacement of the ground, clogging and smear zone.

As a part of supplementary ground improvement method, this research is focused on the fundamental study for the utilization of electro osmosis. This study, in order to verify the efficiency of consolidation and drainage by electro osmosis, conventional theories related to electro osmosis were primarily investigated to select and derive the governing equation for the appreciation of electro osmotic behavior of soil. The electro osmotic consolidation test was performed with a test apparatus which is modified from the conventional oedometer test device. .

During the test, the effect of electrolyte such as sodium chloride in soils was comparatively studied and also the effect of voltage gradients was investigated. Consequently those test results were compared with the well-known conventional theories.

EXPERIMENTAL APPARATUS

Electro osmotic consolidation tests were carried out with a system shown in Fig. 1 and 2. It consists of a testing cell, loading frame, a power supply, displacement / pore pressure measuring devices, water level observation panel and other devices. The testing cell is made of plexiglass in which electrodes are placed directly at the top and bottom of the soil sample. The electrodes used in this test are carbon plates to avoid some corrosive action where the electrode interfaces with soil sample. The soil sample has a diameter of 10cm and 5cm height.

Water squeezed or drained out of the soil sample during the consolidation process is collected to the water level measuring panel through the drainage valves attached to the top and bottom of the testing cell.

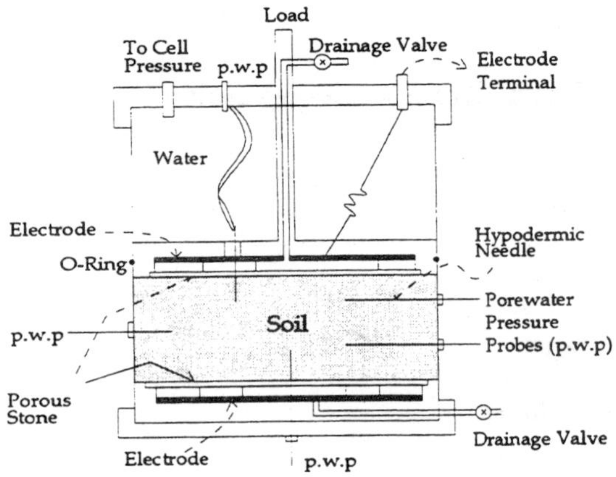

Fig. 1 Electro osmotic testing cell

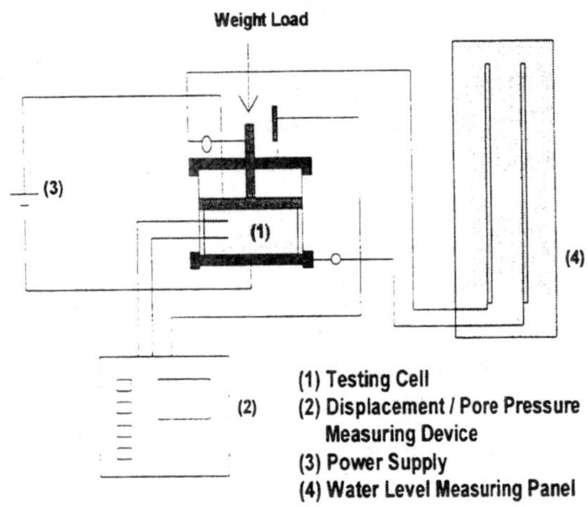

Fig. 2 Electro osmotic consolidation test apparatus

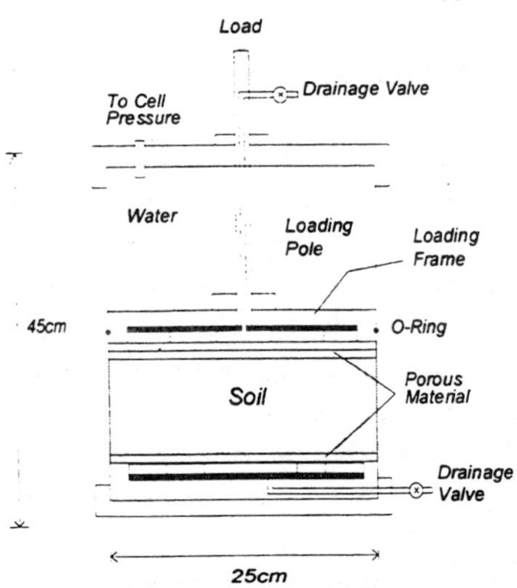

Fig. 3 Reconstituted sample equipment

SOIL SAMPLING

Soil samples used in this study are reconstituted with a device which is shown in Fig. 3. For the preparation of soil sample, distributed clayey soil which is collected from southern part of Korea and were passed No. 200 sieve(with passings of 0.074mm) in the form of solution. This soil was slurry state having more than 1000% of water content by weight. Then it was put into prepared cell of 25cm in diameter. Hydraulic pressure was applied up to 39.2 *kPa* to accelerate settlement of soil particles.

EXPERIMENTAL RESULT AND ANALYSIS

Soil properties

The soil samples used in this study were marine silty clay. The soil properties of samples are summarized in Table 2.

Table 1 Testing plan and sample condition

condition / type	VOLTAGE	SALINITY	SAMPLE CONDITION	INITIAL VOID RATIO
Non voltage Consol.(I)	0.0 v/cm	Natural	Reconstituted Sample (P_c=39.2 *kPa*)	1.30
E/O Consol.(I)	0.5 v/cm			
	1.0 v/cm			
	1.5 v/cm			
Non voltage Consol.(II)	0.0 v/cm	Washed	Reconstituted Sample (P_c=39.2 *kPa*)	1.32
E/O Consol.(II)	0.5 v/cm			
	1.0 v/cm			
	1.5 v/cm			
E/O STEP	3.0 v/cm	Washed	Reconstituted Sample (P_c=39.2 *kPa*)	1.32

Table 2 Soil properties of reconstituted soils

Specific Gravity	Plastic Limit (%)	Liquid Limit (%)	Classification by USCS
2.69~2.73	21.3~26.4	39.6~42.8	CL

To find the effect of electrolyte concentration, two different samples were prepared ; i.e. a natural state of concentration of sodium chloride and a reduced concentration of sodium chloride by washing in distilled water. The sodium chloride concentrations obtained from those soil samples are shown in Table 3 and 4.

Consolidation tests on natural state samples

When electro osmotic pressure is applied to the clayey soils in the field with surcharge of embankment load, excess porewater pressure by surcharge and negative porewater pressure by electro osmosis are developed concurrently in soils.

Therefore, it is necessary to analyze the compositive characteristics of consolidation both by loading and electro osmosis. In this study, consolidation test with both electro osmosis and non electro osmosis were carried out to investigate the effects of electro osmosis on the consolidation characteristic of clayey soils. A plot of void ratio versus logarithm of the pressure from conventional consolidation test is shown in Fig. 4. From the slopes of the straight line portion of the curve, the value of compression index, C_c, is calculated as 0.43. The range of hydraulic permeability was calculated from the consolidation tests. It was ranged from 0.42×10^{-7}cm/sec to 2.06×10^{-7}cm/sec.

Table 3 Concentrations of sodium chloride in soil samples at natural state

Concentration of Chloride Ion (%)	Concentration of Sodium Chloride (%)
1.83	3.02

Table 4 Concentration of sodium chloride in washed soil samples

Concentration of Chloride Ion (%)	Concentration of Sodium Chloride (%)
0.059	0.097

For the soils with salinity of 3.02%, electro osmotic consolidation tests were performed according to the variation of voltage gradients. The e-log p' curves obtained from those tests are shown in Fig. 5. In the beginning of consolidation, electro osmotic effects on consolidation tests were comparatively evident and the amount of settlement increases with the increase of voltage gradient.

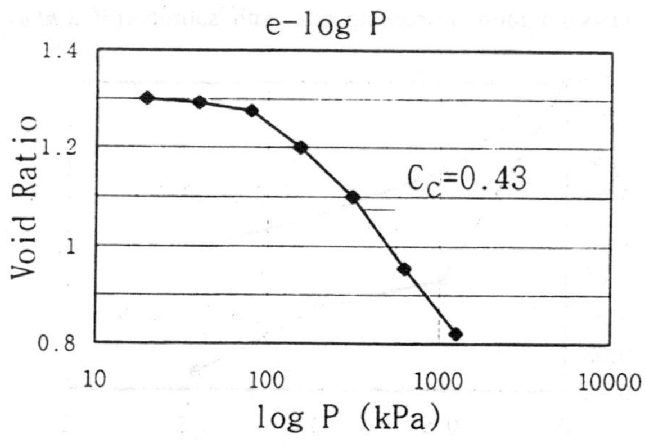

Fig. 4 e-logP from standard consolidation test on natural soils

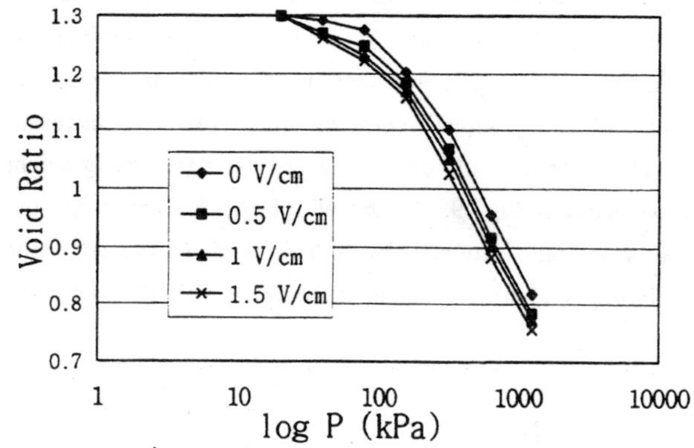

Fig. 5 e-logP from electro osmotic consolidation test for different voltage gradient on natural soils

In Fig. 6, the variation of void ratio according to the increase of voltage gradient is generalized with the initial value of void ratio, 1.3.

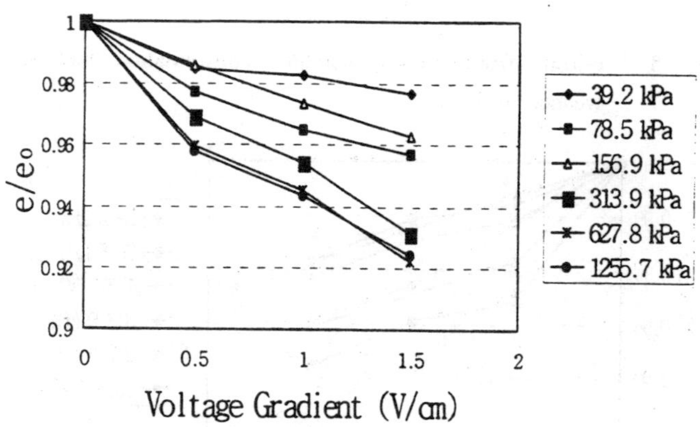

Fig. 6 Variation of voltage gradient depend on different loading steps

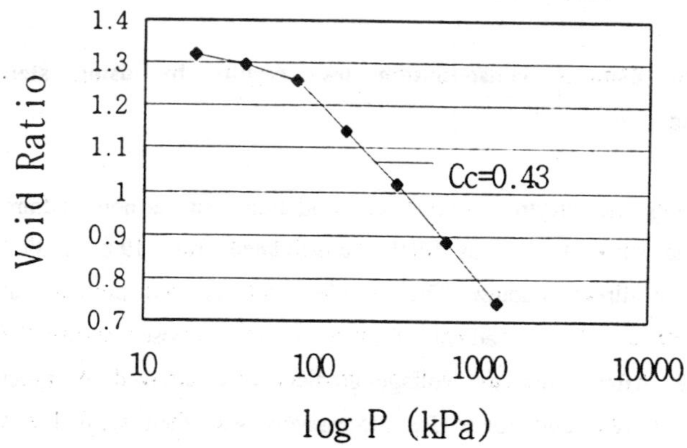

Fig. 7 e-logP from standard consolidation test on washed soils

In Fig. 8, the result curves of the electro osmotic consolidation test are shown by the relation of e-logp' by the variation of voltage gradient. As it can be seen in the diagram, with the increase of voltage gradient, compressibility of soils increases typically. The increase of compressibility by voltage gradient also can be seen in Fig. 9. At the all of the loading stage the change of void ratio divided by initial void ratio(e/e_0) value shows declining tendency with the increase of voltage gradient.

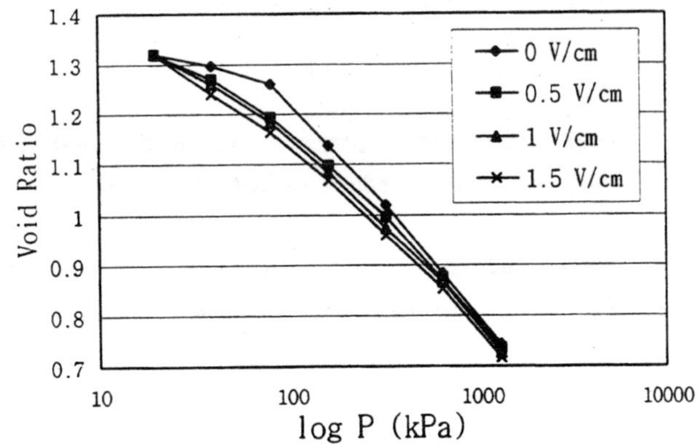

Fig. 8 e-logP from electro osmotic consolidation test for washed soils

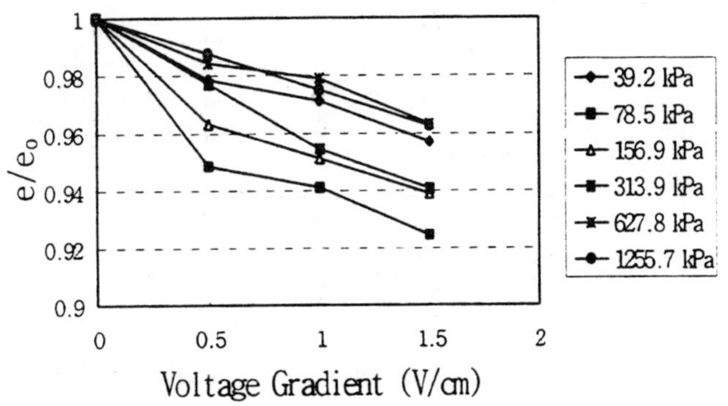

Fig. 9 Variation of voltage gradient depend on different loading steps

Electro osmotic consolidation test results by using step loading

During the electro osmotic consolidation test, a non sodium chloride soil sample was firstly consolidated from 19.6 to 39.2 *kPa* by direct loading. Then holding the applied pressure at 39.2 *kPa* a voltage gradient of 3volt/cm was imposed across the sample. After 24hrs, the voltage gradient was removed. A direct stress of 78.5 and 156.9 *kPa* successively was then applied and held constant again. After that voltage gradient of 3volt/cm was applied for the next 24hrs. Finally next step of loading was pressed to the sample.

This curve of direct loading stress increment following electro osmotic consolidation has a tendency to return to the virgin compression line corresponding to direct loading only(Fig. 10).

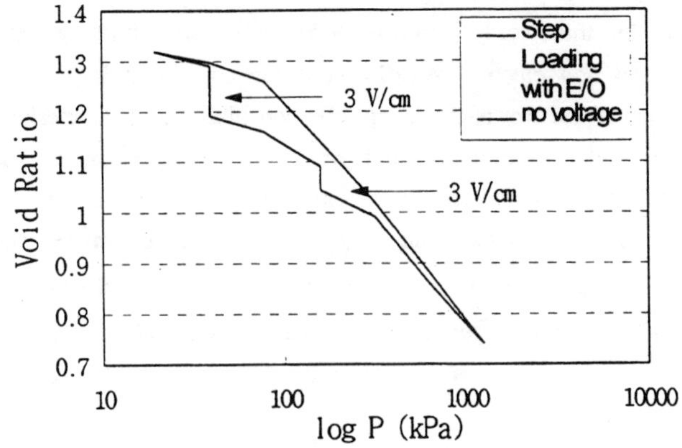

Fig. 10 e-logP relationships using step loading

EFFICIENCY OF ELECTRO OSMOTIC CONSOLIDATION

Variational analysis for normalized void ratio versus voltage gradient was conducted under a constant consolidation pressure i.e. 39.2, 78.5 and 156.9 *kPa*. The results of the analysis were presented in Fig. 11-Fig. 13

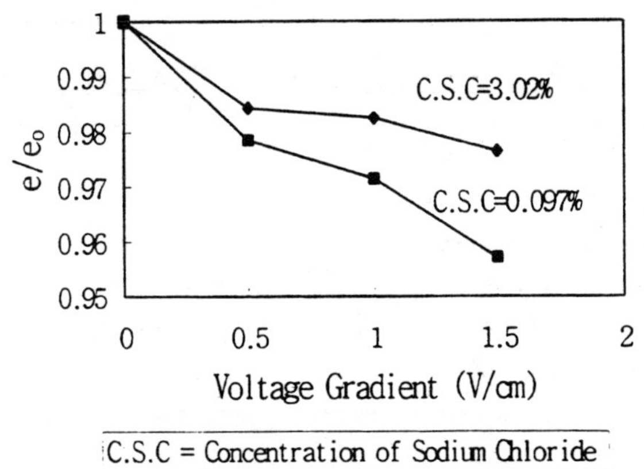

Fig. 11 Void ratio decreasing rate with salinity (39.2 *kPa*)

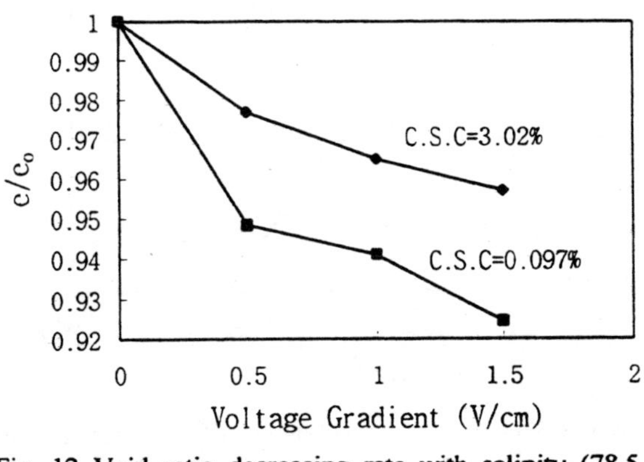

Fig. 12 Void ratio decreasing rate with salinity (78.5 *kPa*)

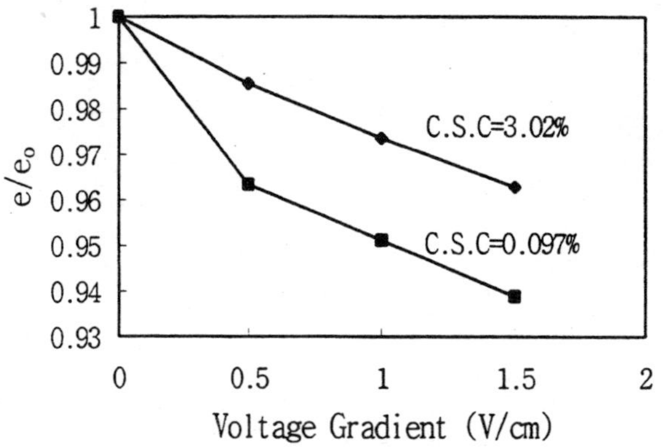

Fig. 13 Void ratio decreasing rate with salinity (156.9 *kPa*)

These figures indicated that the electro osmotic consolidation for soils having higher concentration of sodium chloride is less effective than the one for soils having lower concentration of sodium chloride.

Also, it was found that the consolidation settlement was highly affected by the concentration of electrolyte.

COMPARATIVE STUDY WITH CONVENTIONAL THEORIES

For the comparative study with conventional theories such as Esrig's and Johnston and Butterfield's, the profile of initial excess porewater pressure is assumed in two types. One is assumed that initial excess porewater pressure is constant with depth(A) and the other varies linearly with depth(B). With these assumption and together with test data in Fig 10, the relationship of average degree of consolidation and time factor is obtained. Test results obtained in this study are compared with Esrig's and Johnston and Butterfield's theories in Fig. 14. As a result, test data from this study is well consistent with Esrig's than Johnston and Butterfield's.

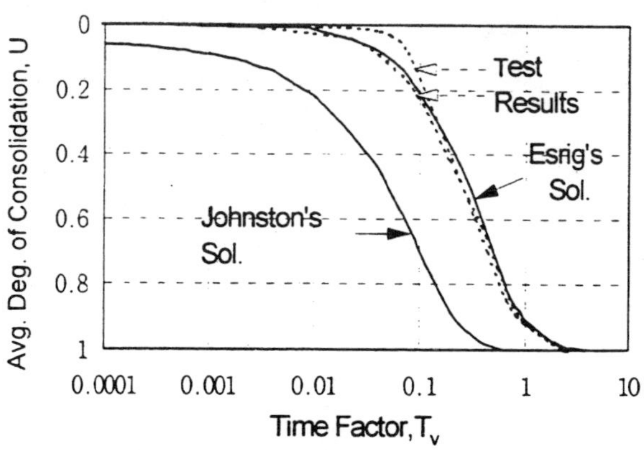

Fig. 14 The comparison of Esrig(1967), Johnston & Butterfield (1977) and this study (The relation of $U-Tv$)

The comparison of those curves with Esrig's solution for the 39.2 and 156.9 *kPa*(pre consolidation stress) can be seen in Fig. 15 and 16.

It is found to be more reasonable that the profile of initial excess porewater pressure is assumed to be linearly increased with depth in analysing the test data.

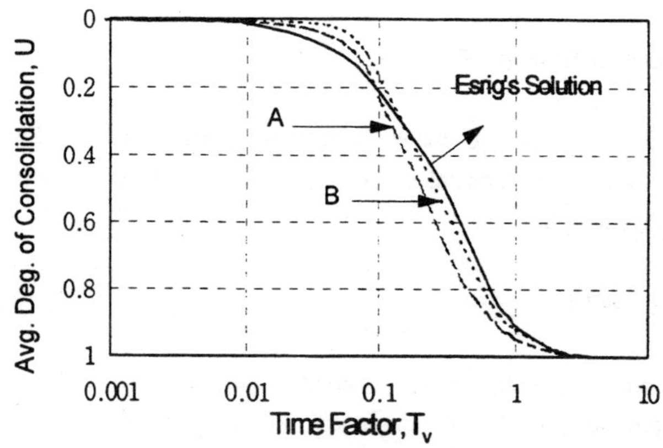

Fig. 15 The relation curves of $U-Tv$ according to assumed initial excess pore water pressure profile (P_C=39.2 *kPa*)

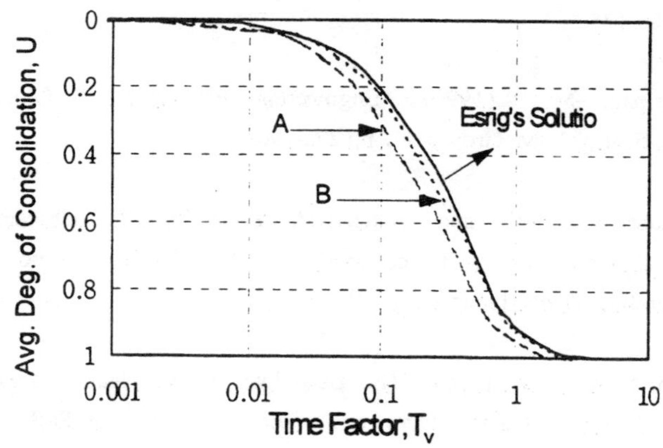

Fig. 16 The relation curves of $U-Tv$ according to assumed initial excess pore water pressure profile (P_C=156.9 *kPa*)

CONCLUSIONS

The electro osmotic consolidation tests was performed to soils with salinity of field condition and with reduced salinity. Based on the experimental results presented herein, the following conclusion can be made.

1) The void ratio of soils was decreased during the test with the increase of voltage gradient which also shows electro osmotic effect can accelerate the consolidation of soft soils.

2) With the increase of voltage gradient consolidation of soils

increases. And this behavior is highly affected by the concentration of electrolyte(sodium chloride) contained in soils.

3) The electro osmotic consolidation characteristics of this study show good agreement with Esrig's theory.

4) Here in this study, it is more reasonable that initial excess porewater pressure profile is assumed to vary linearly with depth.

ACKNOWLEDGEMENTS

The authors wish to thank Sam-Sung Heavy Industry, Co., Ltd., in providing financial support of this research in the form of a research grant.

REFERENCES

Casagrande, L.(1952), "Electro osmotic stabilization of soils", J. of Boston Society of Civil Engineers, Vol 39, pp.51-83.

Esrig, M. I. and Gemeinhardt, J. P.(1967), "Electrokinetic stabilization of Illitic clay", J. of the soil mechanics and foundations Div., Proceedings of the ASEC, Vol.93, No. SM3, pp.109-128.

Hausman, M. R.(1990), "Engineering Principles of Ground Modification", McGrow-Hill, pp.272-296.

Johnston, I. W. and Butterfield, R.(1977), "A laboratory investigation of soil consolidation by electro osmosis", Australian Geotechnics J., pp.21-32.

Kim, S. S, et al.(1996), "The Behaviour on the electro osmotic consolidation of soft clayey soil", 2nd Int. Conf. on Soft Soil Engineering, Nanjing, May 27-30, pp.960-965.

Wan, Tai-Yeu and Mitchell, J.K.(1975), "New apparatus for consolidation by electro osmosis", J. of the soil mechanics and foundations Div., Proceedings of the ASCE, Vol. 101, No. GT5, pp.503-507.

Proceedings of the Seventh (1997) International Offshore and Polar Engineering Conference
Honolulu, USA, May 25-30, 1997

ISBN 1-880653-28-1 (Set); ISBN 1-880653-29-X (Vol. I)

Consolidation Analysis of Natural Clay by Flow Surface History Variable Model

Nobuharu Abe
Osaka University
Osaka, Japan

ABSTRACT
In this paper, an elasto-viscoplastic constitutive model is proposed to predict the one-dimensional consolidation behavior of natural soft clay deposits. In order to describe the yielding and time dependent behavior of natural clays the proposed model is derived being based on the flow surface theory of viscoplastisity and the concept of internal state variable. The applicability of the model is examined by comparing calculated results with laboratory behavior of natural clays, and it is confirmed that dependence of both time-loding history and strain rate on the consolidation yielding behavior of natural clays can be expressed by the proposed flow surface history variable model.

KEY WORDS
Natural soft clay, Consolidation, Elasto-viscoplasticity, Flow surface theory, Internal state variable, Time-dependency, Time-loading history

INTRODUCTION

Clay behavior under compression has always been a topic in geotechnical engineering. Recently, in connection with coastal development, the consolidation yielding characteristics of pseudo-overconsolidated clays or lightly overconsolidated clays have become of major interest. On the other hand, the influence of time-loading history and strain rate on the compressibility of natural clays has been evidenced based on laboratory experiments by many researchers. Therefore, in order to predict the consolidation behavior of natural soft clay deposits realistically, has to be developed the elasto-viscoplastic constitutive model that can describe dependence of both time-loading history and strain rate.

In the following sections, first the elasto-viscoplastic model for natural clays is proposed, being based on the flow surface theory and the concept of internal state variable. And then the applicability of the proposed model is examined through numerical simulations for laboratory behavior of natural clays. It is also demonstrated that the proposed flow surface history variable model (FSHVmodel) can explain consistently the consolidation tests conducted by Aboshi (1973) . Finally, it is shown that the time-loading history and the strain rate are essential factors controlling the consolidation behavior of natural clays.

FLOW SURFACE HISTORY VARIABLE MODEL OF VISCOPLASTICITY

The flow surface history variable model (FSHV model) was derived, being based on the following basic assumptions (Matsui, Abe and Hayashi, 1989).

1) The deformation of clays is decomposed into the deformation at the reference state and the rate-dependent deformation.

2) The reference state assumed in the model is not the static equilibrium state, but the viscous state specified with a reference strain rate.

3) In order to introduce the logarithmic creep law, the creep equilibrium state and the time-loading history effect into the model, the following evolutional equation (creep characteristic equation) is assumed, shown as in Fig. 1:

$$\frac{\varepsilon_z^v + h_o}{\mu} = -\ln\left(\frac{\dot{\varepsilon}_z^v}{\dot{\varepsilon}_{zr}^v} + \delta\right) \qquad (1)$$

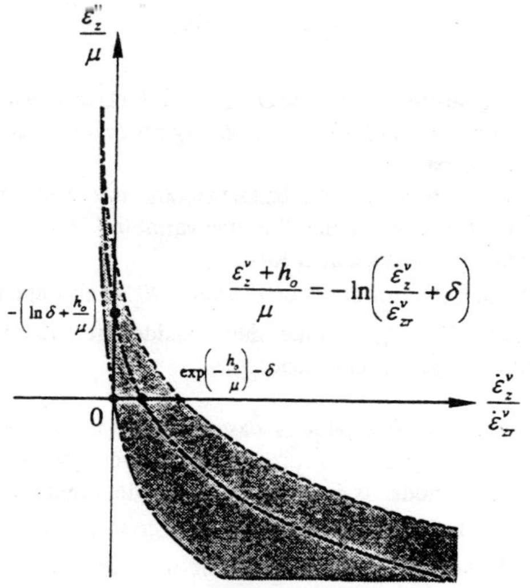

Fig. 1 Evolutional equation for FSHV model

where ε^v_z and $\dot{\varepsilon}^v_z$ are the viscous strain and its rate; $\dot{\varepsilon}^v_{zr}$ is the reference viscous strain rate which specifies the reference state of the model; μ is the coefficient of secondary compression and δ is the internal restraint strain rate which represents the internal restraint of clay structure; and h_o is the initial value of history variable.

4) The instantaneous viscoplastic response does not exist. Therefore, clays respond elastically at the moment of loading.

The flow function F of FSHV model is derived by time- integrating Eq. (1) under the initial condition that the instantaneous viscoplastic strain is zero, as follows:

$$F = \mu \ln\left[\frac{1}{\delta}\left[\left\{1-\exp\left(-\frac{\delta}{\mu}\dot{\varepsilon}^v_{zr}t\right)\right\}\exp\left(\frac{f-h_o}{\mu}\right)+\delta\exp\left(-\frac{\delta}{\mu}\dot{\varepsilon}^v_{zr}t\right)\right]\right] - \varepsilon^{vp}_z = 0 \quad (2)$$

where ε^{vp}_z is the viscoplastic strain, and t is time variable. The plastic strain f in Eq. (2) is given as follows:

$$f = \frac{\lambda-\kappa}{1+e_o}\ln\left(\frac{\sigma_z}{\sigma_{zo}}\right) \quad (3)$$

where σ_z and σ_{zo} are the vertical effective stress and its initial value; λ and κ are the compression and swelling indices, respectively; and e_o is the initial void ratio.

The history variable h is defined to be the viscous strain during loading and given by the following equation:

$$h = h_o + \varepsilon^v_z = \mu\ln\left[\frac{1}{\delta}\left[1-\left\{1-\delta\exp\left(\frac{-f+h_o}{\mu}\right)\right\}\exp\left(-\frac{\delta}{\mu}\dot{\varepsilon}^v_{zr}t\right)\right]\right] \quad (4)$$

The elasto-viscoplastic constitutive relation are obtained from the flow surface theory of viscoplasticity, together with the consistency condition as follows:

$$\dot{\sigma}_z = \left[\frac{\kappa}{1+e_o}\frac{1}{\sigma_z}+\frac{\partial F}{\partial \sigma_z}\right]^{-1}\left(\dot{\varepsilon}_z-\frac{\partial F}{\partial t}\right) = D^{ev}\left(\dot{\varepsilon}_z-\frac{\partial F}{\partial t}\right) \quad (5)$$

In order to analyze the behavior of natural clays when subsequent loading is applied, the following updating procedure is adopted at the moment of the loading.

1) Since it is assumed that the instantaneous viscoplastic strain is zero at the moment of any loading, the time variable t is reset to be zero at the moment of the subsequent loading.

2) At this time, the initial history variable h_o is also updated into h, by following Eq. (4), to take into consideration the time-loading history until the subsequent loading.

The proposed FSHV model is based on the assumption that both elastic and viscoplastic strains exit at every stage of loading. Therefore, this model is independent of loading-unloading criterion. However, when the history variable h is greater than $(-\mu\ln\delta)$, it is assumed that clays are in the elastic nucleus. In this zone clays behave elastically.

CONSOLIDATION ANALYSIS OF NATURAL CLAY

In this section, the applicability of the proposed FSHV model is examined through numerical simulations for consolidation behavior of natural clays by the finite element method.

Table 1 Material parameters and history variable

Material parameters	Batiscan clay	Keihinzima clay	Hiroshima clay
Compression index, λ	1.86	0.322	0.304
Swelling index, κ	0.0687	0.0324	0.033
Coefficient of secondary compression, μ	0.0316	0.028	0.00358
Reference strain rate, $\dot{\varepsilon}^v_{zr}$ (min^{-1})	7.38×10^{-5}	5.0×10^{-5}	5.0×10^{-5}
Internal restraint strain rate, δ	1.0×10^{-3}	1.0×10^{-3}	1.0×10^{-5}
Coefficient of permeability, k_{zo} (cm/min)	5.4×10^{-6}	3.4×10^{-5}	1.3×10^{-5}
Permeability change index, C_k	1.15	1.0	1.0
Initial value of history variable, h_o	0.28	0.27	0.067

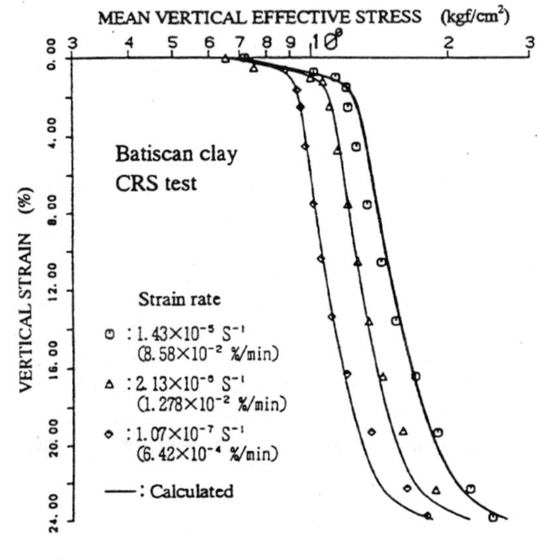

Fig. 2 Effective stress vs. strain relations for CRS tests on Batiscan clay (1kgf/cm^2 = 98 kPa)

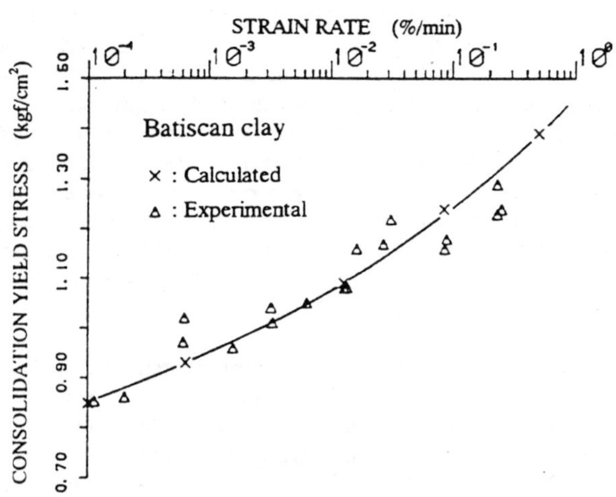

Fig. 3 Consolidation yield stress vs. strain rate relation on Batiscan clay (1kgf/cm^2 = 98 kPa)

Oedometer tests of Batiscan clay

The proposed FSHV model was applied, by Abe et al.(1994), to oedometer tests on Batiscan clay conducted by Leroueil, et al. (1985). This series of oedometer tests include eighteen CRS tests and nine long-term creep tests. After the specimens were preconsolidated under their in situ stress σ_{z0} = 65 kPa, they were strained with the strain rates varying between 1.7×10^{-8} s^{-1} and 4×10^{-5} s^{-1} in CRS tests and loaded to the final stresses σ_{zf} varying between 67 kPa and 151 kPa in creep tests. The consolidation yield stress σ_{zy} of Batiscan clay measured in conventional oedometer tests is 88 kPa. All tests were performed without back pressure on the specimen of the height of 19.0 mm and the diameter of 50.8 mm.

While material parameters listed in Table 1 were determined based on soil parameters in the reference (Kabbaj, et al., 1986), the initial history variable h_o was obtained by using the curve fitting method (calibration method) to the effective stress-strain relation of the CRS test at a constant strain rate of 1.07×10^{-7} s^{-1}. In the present case, h_o is estimated to be 0.28. The mean vertical effective stress of the specimen σ_z was calculated by the following equation in the analysis:

$$\sigma_z = \overline{\sigma}_z - 0.67u_b \qquad (6)$$

where $\overline{\sigma}_z$ is the total stress, and u_b is the pore water pressure at the bottom of specimen. The clay specimen was subdivided into 10 one-dimensional elements in the finite element analysis.

Fig. 2 shows the comparison between the calculated and the experimental results of CRS tests at three different values of the constant strain rate, 1.43×10^{-5}, 2.13×10^{-6} and 1.07×10^{-7} s^{-1}. Fig. 3 shows the consolidation yield stress vs. strain rate relations. It is found from both figures that the strain rate-dependence of yield stress of Batiscan clay can be simulated well by the proposed model. As seen from Fig. 2, the logarithm of effective stress vs. strain curves in the experiments tend to bend gradually even after yielding with increasing strain. In order to express such behavior, the strain-dependent compression index $\overline{\lambda}$ and coefficient of secondary compression $\overline{\mu}$ were employed in the present simulation.

$$\overline{\lambda} = \lambda_0 \exp\{-s(\varepsilon_z)^m\} \qquad (7)$$

$$\overline{\mu} = \mu_0 \exp\{-s(\varepsilon_z)^m\} \qquad (8)$$

where λ_o and μ_o are the reference values of compression index and coefficient of secondary compression, respectively; s and m are constants independent of ε_z.

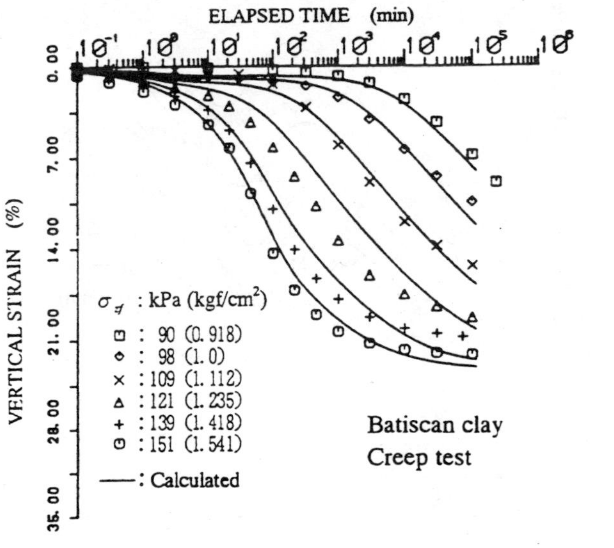

Fig. 4 Developments of strain with elapsed time for creep tests on Batiscan clay

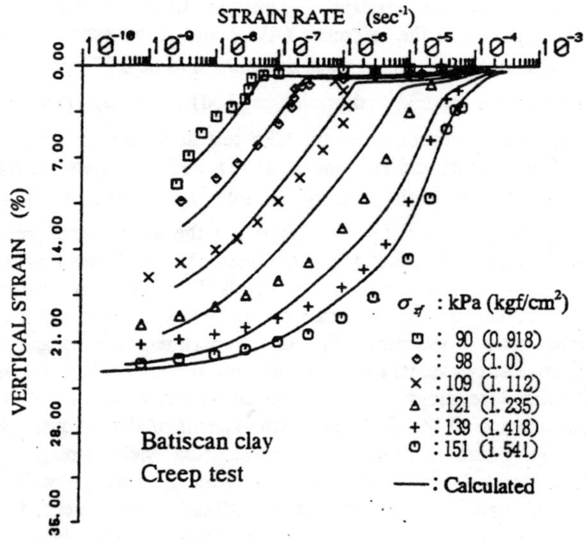

Fig. 5 Strain vs. strain rate relations for creep tests on Batiscan clay

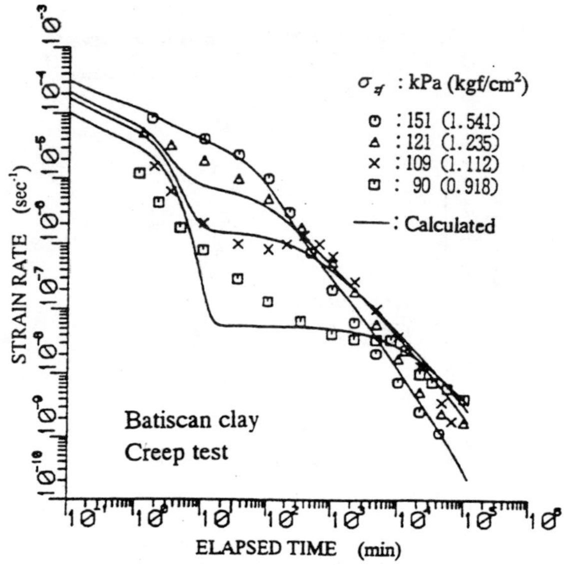

Fig. 6 Variations in strain rate with elapsed time for creep tests on Batiscan clay

Figs. (4), (5) and (6) show the comparison between the calculated and the experimental results of creep tests in which the final stress exceeds the consolidation yield stress. Experimental behavior of creep tests is expressed well by the calculated results using the initial history variable h_o obtained from the CRS tests.

CRS tests of Keihinzima clay

CRS tests on Keihinzima clay were conducted, by Akagi and Aida (1988a) at five different values of constant strain rate, 0.5, 0.2, 0.1, 0.05 and 0.02 %/min until the vertical effective stress reached to 12.8 kgf/cm^2. The specimens used in the experiment were the undisturbed clay taken by block sampling at Keihinzima in Tokyo. The specimens were preconsolidated under the pressure of 0.05 kgf/cm^2 for one to two hours, then they were strained with constant strain rates. The material parameters listed in Table 1 were determined from the reference (Akagi and Aida, 1988b). And the initial history variable h_o was determined to be 0.27 by applying the curve fitting method (calibration method) to the observed effective stress vs. strain relation of the CRS test at a constant strain rate of 0.1 %/min (Abe and Hashimoto, 1993). Fig. 7 shows the result of calibration for initial history variable h_o. Fig. 8 shows the comparison between the calculated and the experimental results of the CRS test. It is found from the figure that calculated results are in good agreement with experimental ones if we take into consideration the fact that the constant strain rate in the experiment was a slightly larger than 0.1%/min. On this experiment the mean vertical effective stress vs. its rate relations were reported. Fig. 9 shows the calculated relations of mean vertical effective stress and its rate relations for all the CRS tests with experimental relations. Arrows in the figure indicate the consolidation yield stresses. The calculated behavior explains the experimental behavior well though the calculated results express the slightly larger dependence of consolidation yield stresses on the strain rate than experiments.

Consolidation tests of Hiroshima clay

Aboshi (1973) conducted long-term consolidation tests on Hiroshima clay to make clear the change of settlement characteristics with the thickness of clay layers, including the effect of creep. The clay used in this experiment was not undisturbed natural clay, but the reconstituted clay that had been artificially consolidated for about 6 years. Thicknesses of the specimens were chosen as 2.0, 4.8, 20, 40 and 100 cm. After the specimens had been preconsolidated under 0.1 kgf/cm^2 and 0.2 kgf/cm^2 successively, they were loaded to a consolidation pressure of 0.6 kgf/cm^2. Since the information for material parameters and preconsolidation history was not given in the reference, material parameters and the initial history variable h_o were estimated by applying the curve fitting method to the settlement curve of the specimen with a thickness of 2.0 cm (Abe and Hashimoto, 1993).

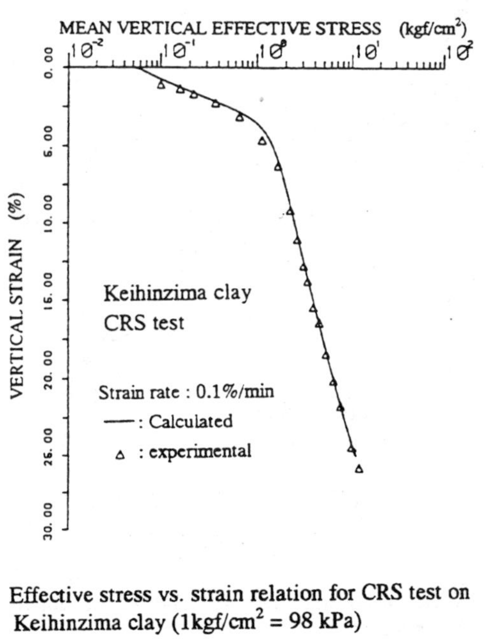

Fig. 7 Effective stress vs. strain relation for CRS test on Keihinzima clay (1kgf/cm^2 = 98 kPa)

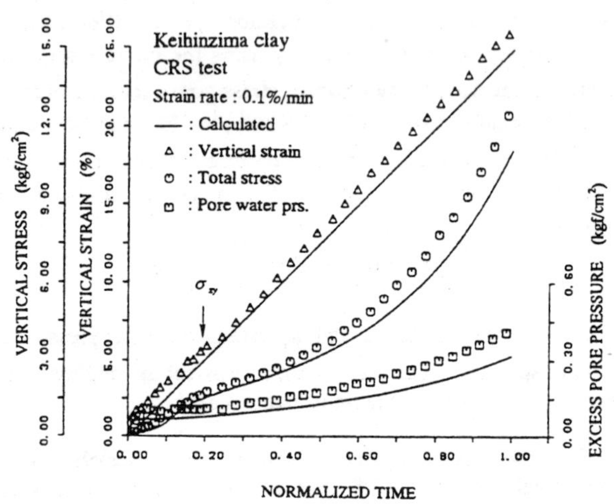

Fig. 8 Developments of total stress, strain and pore water pressure with elapsed time for CRS test on Keihinzima clay (1kgf/cm^2 = 98 kPa)

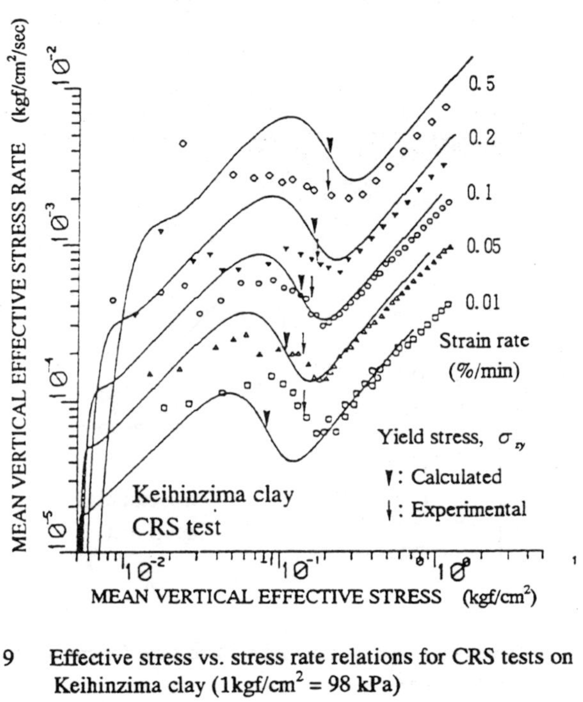

Fig. 9 Effective stress vs. stress rate relations for CRS tests on Keihinzima clay (1kgf/cm^2 = 98 kPa)

Fig. 10 shows simulation results with a constant initial history variable h_o of 0.049. The calculated results that express isotache behavior do not agree with experimental results. Therefore, the inverse analysis was performed to estimate the initial history variables of the specimens with different thicknesses. Fig. 11 shows the results of inverse analysis with respect to settlements, and Fig. 12 shows the initial history variable vs. drainage length of specimen relation. The longer the drainage length is, the larger the initial history variable becomes. These results are reasonable as the preconsolidation time becomes longer with the increase in the thickness of specimen. Aboshi's experiment makes us suppose that the change of long-term settlement characteristics is caused by the difference of time-loading history between specimens at preconsolidation .

CONCLUSIONS

In this paper, the elasto-viscoplastic flow surface history variable model for natural clays was proposed and its applicability was examined by laboratory experiments of natural clays. Furthermore, the dependence of both time-loading history and strain rate on the consolidation yielding behavior was considered.

The main conclusions can be summarized as follows:

(1) The proposed FSHV model can express the consolidation behavior of natural clays including time-loading history and strain rate dependent yielding, by estimating history variable h as the viscous strain accumulated until the current loading. The initial history variables h_o of natural clays can be obtained from the effective stress vs. strain relation of the CRS test.

(2) It is supposed from simulation results that the change of long-term settlement characteristics in the current consolidation is caused by the difference of time-loading history between specimens at preconsolidation .

REFERENCES

1)Abe, N. and Hashimoto, K. (1993). "One-dimensional consolidation analysis of natural clay deposits by history variable model," Proc. of 48th Annual Conf. of Japan Society of Civil Engrg, Vol.3, pp984-985 (in Japanese).

2)Abe, N., Hashimoto, K. and Sumizu, T. (1994). "One-dimensional consolidation analysis of natural clay deposits by history variable model (II)," Proc. of 49th Annual Conf. of Japan Society of Civil Engrg, Vol.3, pp260-261(in Japanese).

3)Aboshi, H. (1973). "An experimental investigation on similitude in the consolidation of a soft clay, including the secondary creep settlement," Proc. of 8th Int Conf. on Soil Mech. Found. Eng. Vol.4.3, p88.

4)Akagi, H. and Aida, H. (1988a). "One-dimensional consolidation behavior of undisturbed cohesive soils under rapid loading," Proc. of Symp. on the method of special consolidation test, pp127-134 (in Japanese).

5)Akagi, H. and Aida, H. (1988b). "Study on the one-dimensional consolidation characteristics of undisturbed cohesive soils under rapid loading condition," Proc. of 23rd Japan Nat. Conf. on Soil Mech. and Found. Eng. pp311-314.(in Japanese).

6)Kabbaj, M., Oka, F., Leroueil, S. and Tavenas, F. (1986). "Consolidation of natural clays and laboratory testing," Consolidation of Soils: Testing and Evaluations, ASTM STP 892, pp378-404.

7)Leroueil, S., Kabbaj, M., Tavenas, F. and Bouchard, R. (1985). "Stress-strain-strain rate relation for the compressibility of sensitive natural clays," Geotechnique 35 (2), pp159-180.

8)Matsui, T., Abe, N. and Hayashi, K. (1989). "Viscoplastic modelling of time-dependent behaviour clays," Proc. of Int. Symp. on Numerical Models in Geomech. Niagara Falls, pp100-107.

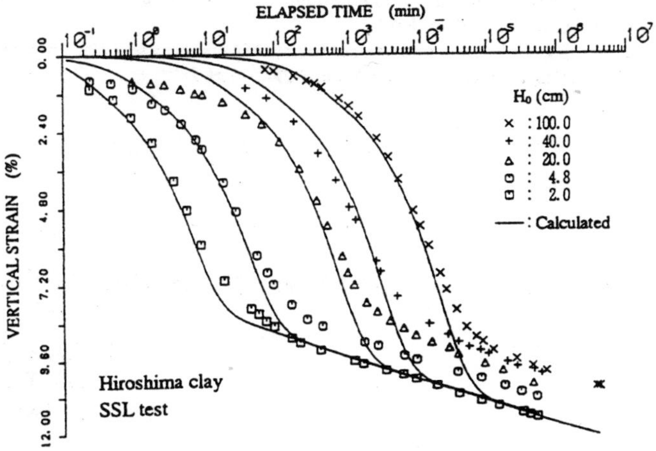

Fig. 10 Developments of settlement with elapsed time on Hiroshima clay

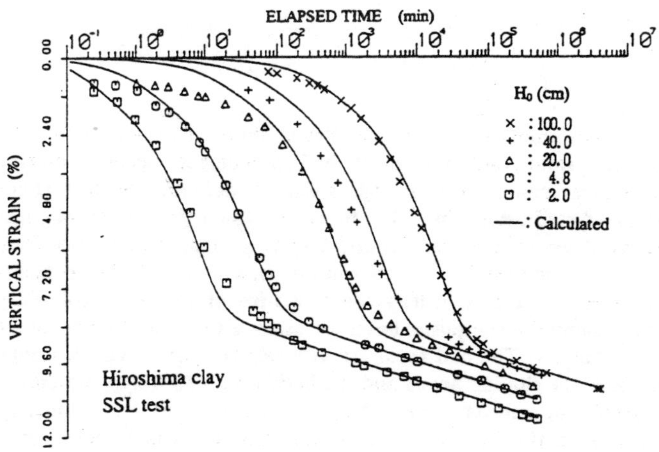

Fig. 11 Results of inverse analysis for initial history variables on Hiroshima clay

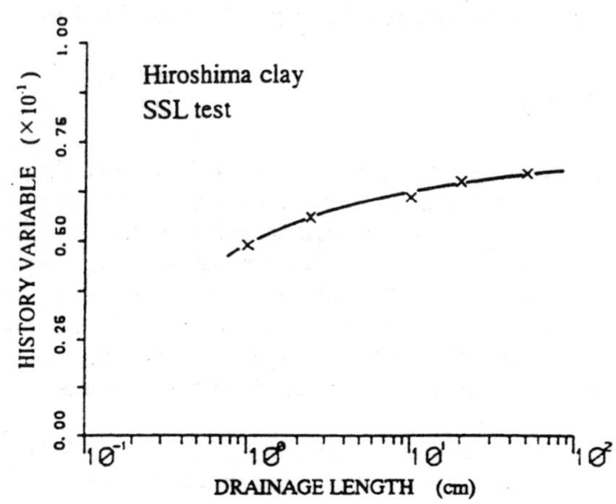

Fig. 12 History variable vs. drainage length of specimen on Hiroshima clay

Proceedings of the Seventh (1997) International Offshore and Polar Engineering Conference
Honolulu, USA, May 25-30, 1997

ISBN 1-880653-28-1 (Set); ISBN 1-880653-29-X (Vol. I)

Bed Shear Stress Distribution Over a Composite Beach Profile

E. Özkan Çevik and Y. Yüksel
Yildiz Technical University
Istanbul, Turkey

ABSTRACT

The mechanism of onshore-offshore movements of the bottom sand and the local bed movement around marine structures can be better understood by defining the bed shear stress distribution. In order to obtain the variation of the bed shear stress over horizontal and following sloping bottom ie from intermediate to shallow water conditions, a series of measurements have been made to clarify the structure of the wave boundary layer along the composite beach by using a Constant Temperature Anemometer. The bed shear stress measurements have been done on both the sloping beach and the horizontal bed which is offshore region. The results clearly show the effect of the wave shoaling process on the bed shear stress. The maximum bed shear stress value has been found at the plunging point due to strong vortex generation and the turbulent structure, however, the periodic component of the bed shear stress is still of the periodic character like the offshore region.

KEY WORDS: Bed shear stress, beach profiles, wave breaking.

INTRODUCTION

When the wave propagates into more shallow water, the process of shoaling leads to the increase of the wave height. But this process can continue until at a certain location and then the wave breaks. Although the shoaling process is characterized by small energy loss, the wave breaking is associated with a very large loss of energy. In the surf zone, the wave energy flux from offshore is dissipated to turbulence and heat. Due to this strong energy dissipation the wave height decreases during the propagation of the wave towards the shore in the surf zone.

Natural beaches tend to exhibit an equilibrium profile that may be planar and non-planar. During the last decade, the observations by a number of researchers have shown the time variation of the bed shear stress that has been measured in a wave flume under breaking and non-breaking wave conditions over a constant bed slope (Deigaard et al, 1991, Arnskov et al, 1993).

Deigaard et al (1991) indicated that the experimental results clearly show the effect of the turbulence generated by the wave breaking. The periodic component of the bed shear stress changes its sign twice for each wave period. But, their tests were only made with spilling breakers and weak plunging breakers.

Arnskov et al (1993) carried out their experiments under wave and current in order to measure the instantaneous bed shear stress over a hydraulic smooth bed. They used a dual-component hot-film probe which enables the determination of magnitude and direction of the instantaneous bed shear stress. They present the values of the maximum and the mean bed shear stress in the direction of the current for various wave-current combinations. Their results reveal that the r.m.s. value of the fluctuations in the bed shear stress in the case of current alone is approximately 40 % of the mean value. They also indicated that the fluctuations in the bed shear stress are significantly reduced with the superposition of waves, and this becomes more and more pronounced, as the wave height increases.

The bed shear stress variation is important for the bed load transport, for the near bed boundary condition, for the suspended sediment as well as for the local scouring process around marine structures. However, only a few investigations have been done for the bed shear stress variations under oscillating flow conditions so far.

The objective of this present work is to investigate the magnitude of bed shear stress under breaking and non-breaking waves over a composite beach profile.

EXPERIMENTAL SET-UP

The experiments were carried out in a 20 m long, 1 m high and 1 m wide wave channel in which periodic waves were produced by a paddle at one end. A smooth and impermeable beach was modeled by using steel plate at the other end, which was laid on a slope of 1:5. Horizontal bed conditions were made by using cement powder on sand bottom (Fig. 1).

In order to minimize the reflection, wave absorbers were installed at the both ends of the channel. The wave characteristics were recorded by a wave monitor using three wave gauges.

The bed shear stress was measured by a DANTEC 55R46 flush-mounted hot film probe as indicated by Deigaard et al (1991). The hot film probe was placed on the horizontal bottom and nine different positions on the sloping bottom. The probe itseif was placed in a plexiglas case, which is fitted into a hole. The transition between the probe, the plexiglas case and the bottom were made as smooth as possible using soft plastic. The data sampling was made at a frequency of 50 samples per second and data were evaluated by using a PC computer. Each record duration is about 3.3 minutes.

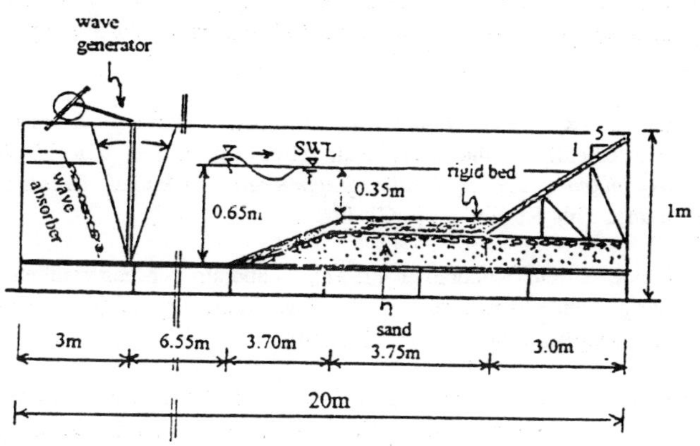

Fig. 1 Laboratory wave flume.

EXPERIMENTAL RESULTS

Ten different test conditions were considered (Table 1). For every test condition, typical recordings of wave profiles along the beach from deep water to shore line were made and also a video camera was used to ensure finding the locations of breaking and plunging points.

Table 1 Laboratory Test Conditions

Case	Wave Height H_o (cm)	Wave Length L_o (m)	Wave Period T (s)	Wave Steppness H_o/L_o	Breaker Type
1	8	4.8	2.635	0.016	P'
2	11	4.7	2.6	0.023	P
3	4.16	1.5	1.0	0.0277	P
4	13.12	4.7	2.578	0.0279	P
5	10.3	3.3	1.863	0.0312	P
6	7.4	2.1	1.3	0.0352	P
7	9.7	2.5	1.466	0.0388	P
8	10.6	1.7	1.12	0.062	P
9	10.15	1.3	0.9	0.078	S-P'
10	8.375	1.14	0.8	0.08375	S-P

P'- Plunging, S-P'- Spilling-Plunging

Fig. 2 shows that a typical time history of the bed shear stress in the surf zone. The bed shear stress is only recorded by its numerical value. As it is shown in Fig. 2 the sign shifts twice a wave period and the duration of the onshore bed shear stress is longer than the duration of the offshore bed shear stress. In the surf zone a significant turbulence contribution to the bed shear stress can be observed, especially for the onshore direction of the bed shear stresses the turbulence is large. However, the periodicity of the signal is still clear, this finding agrees well with that of Deigaard et al (1991). Because of the non-linearity of approaching waves, the direction and the magnitude of horizontal orbital velocity over the bed is bigger in the onshore direction. Consequently, the bed shear stress has shorter duration under the wave crests and also onshore flow is intensely turbulent. However the downrush is controlled by gravity and it has more smooth flow structure.

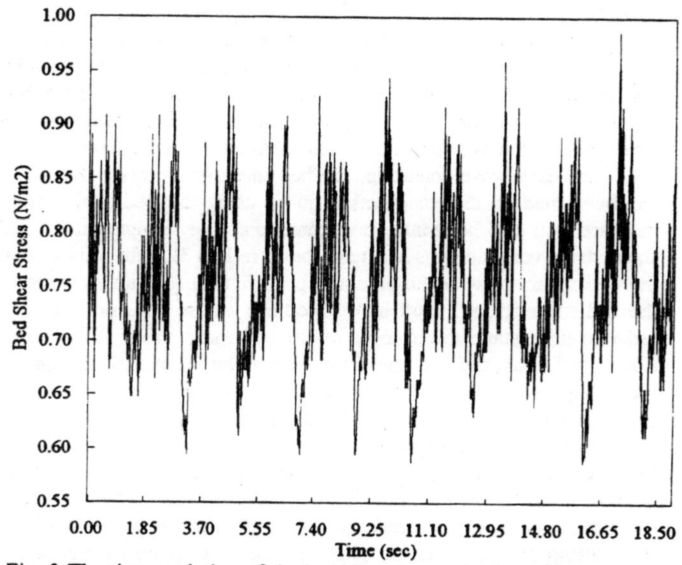

Fig. 2 The time variation of the bed shear stress in the surf zone.

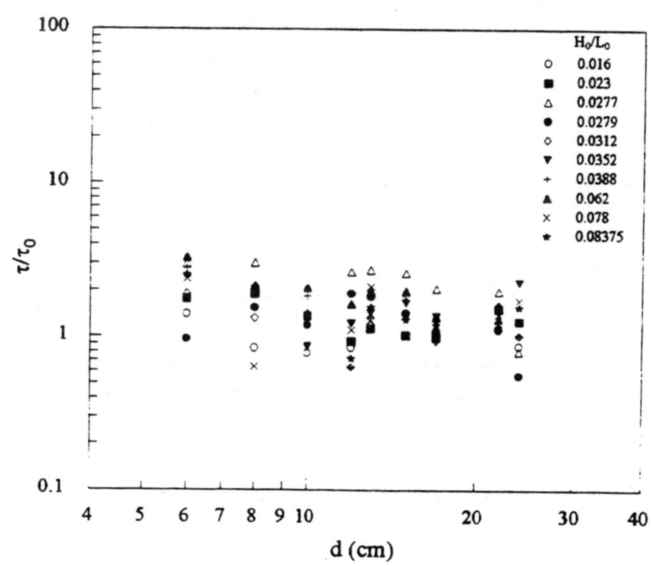

Fig. 3 The variation of the dimensionless bed shear stress with depth.

The bed shear stress measurements were carried out from the toe of the beach up to the plunging point for each wave condition. However, the signals from CTA gave uncomfortable results after plunging point. Due to the increase of the air concentration because of the wave plunging after this point; the plunging makes it impossible to obtain reliable measurements. The bed shear stress increases along the beach up to the plunging point, however, the bed shear stress gets bigger values between the breaking and the plunging points when they compared with the outside of surf zone and they reach the maximum values at the plunging points. The extreme values may thus be caused by the falling jet and strong vortex generation. Nadaoka et al (1988) indicated that the three dimensional large scale eddies referred to as "obliquely descending eddies" existed at the plunging point. The obliquely descending eddies bring highly intermittent intensive turbulence to the bottom with the large onshoreward momentum. Deigaard et al (1991) found that the level of the bed shear stress is not drastically increased in the surf zone compared to the situation immediately offshore the breaker point. However, the level of the bed shear stress has more increasing tendency in the surf zone than the outer zone. Same conclusion has been obtained in the current work.

For each wave condition, the bed shear stress along the beach has been made a dimensionless ratio by using the bed shear stress measured on the horizontal bed condition. The dimensionless bed shear stress versus depths has been shown in Fig. 3. Figure shows that the bed shear stress at the plunging point is almost twice as large as the bed shear stress for offshore conditions on the horizontal bed. In order to understand the changes of the bed shear stress for different wave conditions, the bed shear stress over the horizontal bottom for fixed depth were plotted against the various wave steepness in Fig.4. As it is indicated in Fig. 4 the intensity of shear stress increases with the decrease of offshore wave steepness H_0/L_0 on the horizontal bed. Nadaoka and Kondoh (1982) stated that the intensity of turbulence increase with the decrease of offshore wave steepness H_0/L_0.

No significant difference was observed between the results from the spilling breakers and the plunging breakers in the current work. This result agrees with Deigaard et al (1991).

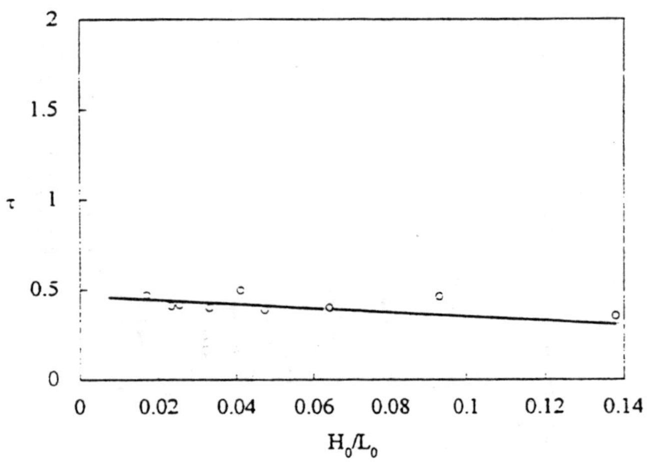

Fig.4 The variation of bed shear stress versus wave steepness.

CONCLUSION

In the surf zone there is a significant turbulence contribution to the bed shear stress, especially for the onshore direction. However, the magnitude of the bed shear stress is more intensive, the periodicity of the time variation of the bed shear stress still remains in the surf zone. The bed shear stress increases along the composite beach till the plunging point and reaches the maximum value at this point. Its value is almost twice that of the offshore horizontal bottom value.

ACKNOWLEDGMENT

This work was financially supported by Yıldız Technical University research fund. The authors sincerely wish to express their appreciation to the authorities of the fund.

REFERENCES

Arnskov, MM, Fredsøe, J and Sümer BM (1993). "Bed Shear Stress Measurements Over A Smooth Bed In Three-Dimensional Wave-Current Motion". *Journal of Coastal Engineering*, Vol 20, pp 277-316.

Deigaard, R, Bro Mikkelsen, M and Fredsøe, J (1991). "Measurements of the Bed Shear Stress in a Surf Zone," *Ints. Hydrodynamic and Hydraulic Engineering Technical University*, Denmark, Prog Re Vol 73, pp 21-30.

Nadaoka, K and Kondoh, T (1982). "Laboratory Measurements of Velocity Field Structure In The Surf Zone By LDV," *Coastal Engineering in Japan*, Vol 25, pp 125-145.

Nadaoka, K, Ueno, S and Igarashi, T (1988). "Sediment Suspension Due To Large Scale Eddies in The Surf Zone," *International Coastal Engineering Conference*, Malga, Spain, pp1646-1660.

Proceedings of the Seventh (1997) International Offshore and Polar Engineering Conference
Honolulu, USA, May 25-30, 1997

ISBN 1-880653-28-1 (Set); ISBN 1-880653-29-X (Vol. I)

Model Prediction of Stress-Strain Behavior on Cohesionless Soil

Robert Yun-Pin Chin
National Cheng Kung University
Tainan, Taiwan, China

ABSTRACT

This compares the observed stress-strain behavior of cohesionless soil with those predicted from constitute models, such as the Cam-clay model (Roscoe, 1963), the modified Cam-clay model (Roscoe, 1968), the Miura modified Cam-clay model (Miura, 1982), and the Cap model (Baladi, 1979). The prediction of the relationship between the stress ratio, $\eta(q'/p')$ and volumetric strain ratio ε_v / p_c' under cyclic loading for Cap model is more accurately than the other three constitute models. The proposed curve equation is expressed in this paper.

Key words: Stress-strain, Cam-clay model, Cap model

INTRODUCTION

This study compares the observed stress-strain behavior of cohesionless soils with those predicted from a number of stress-strain models developed at Cambridge University; Such as; (1) the incremental stress-strain Cam-clay model of Roscoe et al., (1963); (2) modified Cam-clay model of Roscoe et al., (1968) based on the concept of singular yield surface; (3) Miura modified Cam-clay model (1982) based on the concept of yield curve of sand in a particle-crushing region yield curves depicted by this model are well comparable with the experimental yield curve. It could also the experimental stress-strain curve in triaxial compression test; and (4) Cap model (Baladi, 1979) are also based on the concept of continuous yielding of soils, but it is expressed in terms of three dimensional state of stress and formulated on the basis of continuous mechanics principles. There are certain other differences between the Cap and Cam-clay models. In the Cap models, the portion of the cap above the fixed yield or failure surface is omitted, and only the yield surface is assumed. In the Cam-clay models, the moving cap plays the main role in defining yielding, and the fixed yield surface is used essentially to define the critical state. On the other hand, in the Cap model, both the fixed and moving surfaces are used to define the yielding process.

STRESS-STRAIN CONSTITUTE MODELS

The Critical State Concept

The critical state concept represents an idealization of the observed patterns of behavior of saturated remolded clay in triaxial compression tests but is assumed to apply also to sands. the critical state theory is a three-dimensional approach. The state of effective stress of a soil specimen can be expressed in terms of the stress invariants

$$p' = \frac{1}{3}\left(\sigma_1' + \sigma_2' + \sigma_3'\right), \quad \text{and}$$

$$q = \sqrt{\frac{1}{2}\left[\left(\sigma_1' - \sigma_2'\right)^2 + \left(\sigma_2' - \sigma_3'\right)^2 + \left(\sigma_3' - \sigma_1'\right)^2\right]} \qquad (1a)$$

where p' = the mean principal effective stress; q = the effective deviatoric stress; σ_1', σ_2', and σ_3' denote the axial and lateral stresses.

For the case of the triaxial compression test it is assumed that $\sigma_2' = \sigma_3'$ and these stress parameter reduce to

$$p' = \frac{(\sigma_1' + 2\sigma_3')}{3}, \quad \text{and} \quad q = (\sigma_1' - \sigma_3') \qquad (1b)$$

$$v = (\varepsilon_1 + 2\varepsilon_3), \quad \text{and} \quad \varepsilon = \frac{2(\varepsilon_1 - \varepsilon_3)}{3} \qquad (1c)$$

Equation of Cam-clay Model

The Cam-clay model is an elasto-plastic model in which increments of strain are divided into elastic and plastic components, that

$$d\varepsilon_v = d\varepsilon_v^e + d\varepsilon_v^p \qquad (2a)$$

$$d\varepsilon_s = d\varepsilon_s^e + d\varepsilon_s^p \qquad (2b)$$

The elastic strains are assumed to be small and reversible. However the plastic strains are irrecoverable and tend to be larger. As long as the stress point remains within the yield surface plotted in $p'-q$ space then the strains are purely elastic and $d\varepsilon_v^p = d\varepsilon_s^p = 0$ .suming that there is no recoverable component of shearing distortion and that energy dissipates according to

$$dw = Mpd\varepsilon_s, \quad \text{hence} \quad pd\varepsilon_v^p + qd\varepsilon_s^p = Mpd\varepsilon_s^p \qquad (2c)$$

In the above equation, M is the slope of at the critical state line in $p'-q$ diagram. The product of M and p is the deviatoric stress at the critical state corresponding to mean stress p (Fig.1). This product represents the shear stress acting on the micfractures within the sample due to particle slippage. The product of Mp and $d\varepsilon_s^p$ represents the total energy loss due to friction on all micfractures as a result of the shear strain increment It may ce contributed from both shear and volume strain as the equation in critical model (Roscoe and Burland, 1968). M is related to ϕ', the angle of friction obtained in triaxial compression tests, as

$$M = \frac{6\sin\phi'}{3-\sin\phi'} \qquad (2d)$$

Based upon these assumptions, The shape of the yield surface, η is given by

$$\eta = M \ln\frac{p_o}{p}, \qquad (2e)$$

The equation of state boundary surface for drained triaxial compression test is

$$\eta = \frac{\lambda M}{\lambda-\kappa} M \ln\frac{p_o}{p} \qquad (2f)$$

where λ is the gradient of the normal consolidation line in $e-\ln p'$ space. and κ is the gradient of the swelling and re-compression line in $e-\ln p'$ space (Fig. 2) .

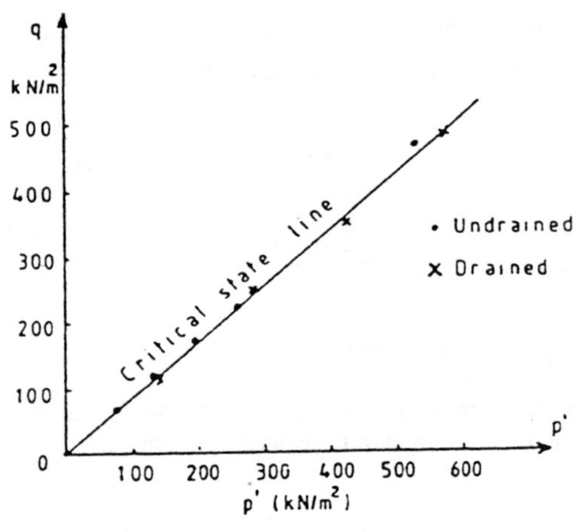

Fig. 1 Position of the critical line (after Smith, 1992)

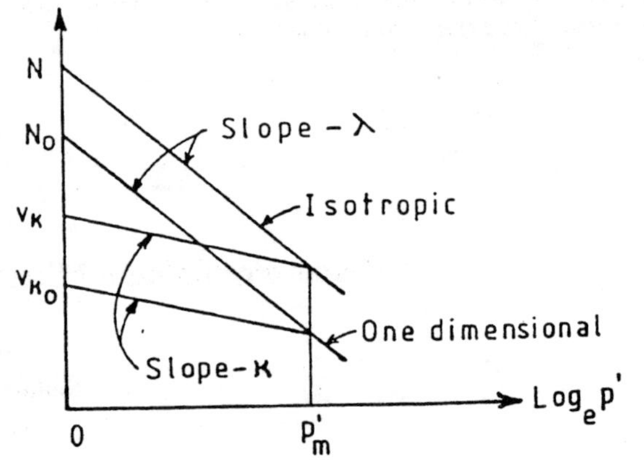

Fig. 2 Isotropic and one-dimensional consolidation (after Smith, 1992)

Equation of Modified Cam-clay Model

Modified Cam-clay model was developed by Roscoe (1968) as an extension of Cam-clay model to a three-dimensional stress state which is an isotropic, nonlinear elastic, elastic strain and hardening model. Only the volumetric strain is assumed to be partially recoverable.

In the modified Cam-clay model (Roscoe, 1968), the dissipated energy dW is assumed as

$$dW = p\sqrt{\left(d\varepsilon_v^p\right)^2 + M^2\left(d\varepsilon_s^p\right)^2} \qquad (3a)$$

The yield curves for the modified Cam-clay model can be written as

$$M^2p^2 - M^2p_o p + q^2 = 0 \qquad (3b)$$

The preceding equation of an equatio of an ellipse on $p-q$ plot. The equation of volumetric and deviatoric increments for drained triaxial compression test is expressed as,

$$d\varepsilon_v = \frac{1}{1+e}\left[\frac{\lambda dp}{p} + \frac{(\lambda-\kappa)2\eta d\eta}{M^2+\eta^2}\right] \qquad (3c)$$

The equation for undrained triaxial compression test is expressed as

$$d\varepsilon_s^p = \frac{-\kappa}{1+e}\left[\frac{dp}{p} + \frac{2\eta d\eta}{M^2+\eta^2}\right]\left[\frac{2\eta}{M^2-\eta^2}\right] \qquad (3d)$$

The equation of state boundary surface for drained triaxial test can be obtained by integrating the above equation. which is

$$\frac{p}{p_o} = \left[\frac{M^2}{M^2+\eta^2}\right]^{(1-\kappa/\lambda)} \qquad (3e)$$

Equation of Miura modified Cam-clay model

Miura (1982) found that the experimental yield curve lies within the Cam-clay and the modified equation curves in compression region, and it lies without the two curves in the extension region.Therefore, in order to obtain better relationship that agrees with the experimental yield curves in both compression and extension stress regions, Miura searched for a better expression for the energy dissipation per unit volume to give an experimental result for predicting the yield curve of sand in a particle-crushing region, that is

$$dW = p\sqrt{\left(d\varepsilon_v^p\right)^2 + \left(Md\varepsilon_s^p\right)^2 - \left(M\eta\right)^2 d\varepsilon_v^p\, d\varepsilon_s^p} \tag{4a}$$

The yield curve equation is derived as,

$$\frac{p}{p_o} = \exp\left[-\int_0^{\eta}\frac{\left(2+M^2\eta\right)\eta}{M^2\eta^3+\eta^2+M^2}d\eta\right] \tag{4b}$$

The equation for volumetric and deviatoric increment for drained triaxial compression test is

$$d\varepsilon_s^p = \frac{\lambda-\kappa}{1+e}\left[\frac{dp}{p}+\frac{\left(2+M^2\eta\right)\eta}{M^2\eta^3+\eta^2+M^2}d\eta\right] \text{x} \left[\frac{\left(2+M^2\eta\right)\eta}{M^2-\eta^2}\right]$$

$$d\varepsilon_v^p = \frac{1}{1+e}\left[\frac{\lambda dp}{p}+\frac{(\lambda-\kappa)\left(2+M^2\eta\right)\eta}{M^2\eta^3+\eta^2+M^2}d\eta\right] \tag{4c}$$

And the equation for undrained triaxial compression test is expressed as

$$d\varepsilon_s^p = -\frac{\kappa}{1+e}\left[\frac{dp}{p}+\frac{\left(2+M^2\eta\right)\eta}{M^2\eta^3+\eta^2+M^2}d\eta\right] \text{x} \left[\frac{\left(2+M^2\eta\right)\eta}{M^2-\eta^2}\right] \tag{4d}$$

The equation of state boundary surface for drained triaxial compression test can be obtained by integrating the above equation,

$$\frac{p}{p_o} = \exp\left[-\int_0^{\eta}\frac{dp}{p}+\frac{\left(2+M^2\eta\right)\eta}{M^2\eta^3+\eta^2+M^2}d\eta\right] \tag{4e}$$

Equation of Cap model

The Cap models (Baladi. 1979) are also based on the concept of continuous yielding of soils. but it is expressed in terms of three dimensional state of stress and formulated on the basics of continuous mechanics principles. There are certain other differences between the Cap Cam-clay models. In the Cap models. the portion of the cap above the fixed yield or failure surface (Fig. 3) is omitted. and only the yield surface marked ABC is considered. In the Cam-clay models, the moving cap plays the main role in defining yielding, and the fixed yield surface is used essentially to define the critical state. On the other hand. in the Cap models. both the fixed and moving surface are used to define the yielding process.

Elastic and plastic response functions:

For the elastic (recoverable) response of the material, it is assumed that the bulk modulus K is a linear function of pressure; thus

$$K = K_i K_1 \frac{J_1'}{3} = K_i K_1 p' \tag{5a}$$

in which K_i and K_1 are material constants that can be determined from the slope of the unloading curve from isotropic consolidation test (Baladi, 1979). The elastic shear modulus G is assumed to be constant. For the plastic behavior, the loading function is assumed to be isotropic and to consist of two parts: (1) an ultimate failure envelope that serves to limit the maximum shear strength of the material; and (2) a strain - hardening cap (Figs. 4 and 5).

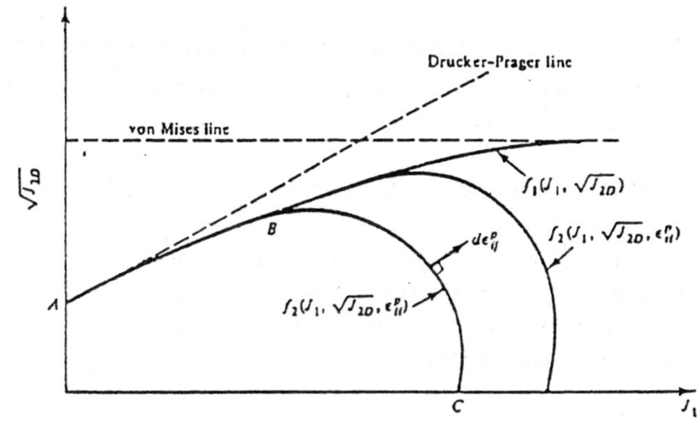

Fig. 3 Cap model (after Balidi, 1979)

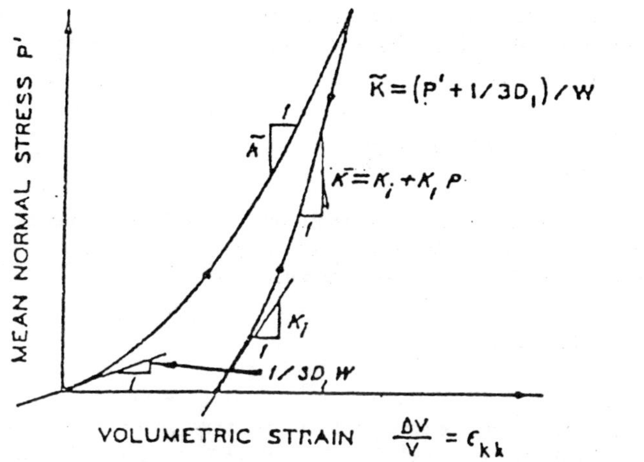

Fig. 4 Relationship for isotropic consolidation test (after Baladi, 1979)

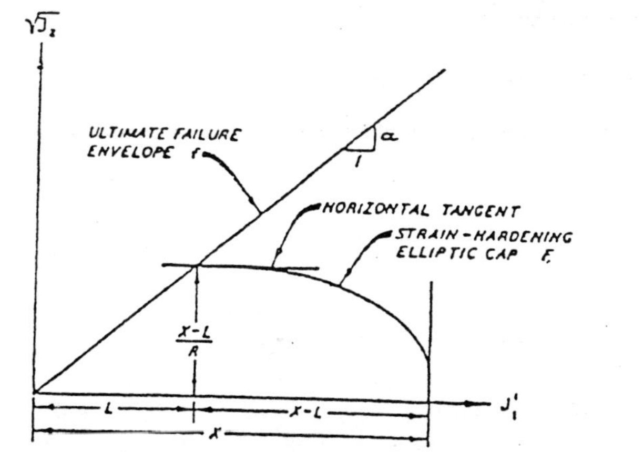

Fig. 5 Yield surface for elastic-plastic model (after Baladi, 1979)

The failure envelope portion of the loading function is denoted by

$$f\left(\sqrt{J_2}, J_1'\right) = \sqrt{J_2} - \alpha J_1' = 0 \qquad (5b)$$

where α is the slope of the failure envelope, and the strain - hardening yield surface by

$$F\left(\sqrt{J_2}, J_1', K\right) = (J_1' - L)^2 + R^2 J_2 - (X - L) = 0 \qquad (5c)$$

The hardening function that controls material compaction or dilation, or both, is assumed to be equal the plastic volumetric strain $\left(\varepsilon_{KK}^{p}\right)$ and to take the following form:

$$K = W\left[ln(D_1 X + 1) - D_2\, ln(D_3 X + 1)\right] \qquad (5d)$$

The intersection of the cap with the J_1' - axis is denoted by X. The ratio of the major to the minor axis of the elliptic cap is denoted by R. The value of J_1' at the center of the cap is denoted by L which is related to other parameters by

$$L = \frac{X}{1 + R\alpha} \qquad (5e)$$

The above equation indicates that the failure envelope (Eq. 5b) intersects each ellipse at the crown. The constants D_2 and D_3, however, are related to the material constants K_i, K_1 and W through the following equations:

$$D_2 = \frac{1}{WK_1}; \quad \text{and} \quad D_3 = \frac{1}{3K_i} \qquad (5f)$$

The parameter, α (Eq. 5b) is the frictional strength of the material and is related to the effective angle of internal friction, ϕ', through the following relation (for triaxial compression tests);

$$\alpha = \frac{2\sin\phi'}{\sqrt{3}(3 - \sin\phi')} \qquad (5g)$$

In summary, there are seven material constants that describe the behavior of the cap model, i.e., $K_i, K_1, W, D_1, G, \alpha$, and R. A more complicated model can be developed by replacing these constants with appropriate mathematical expressions that are functions of stress or strain invariant.

Isotropic consolidation phase

During the isotropic consolidation phase of the triaxial compression test.

$$\sigma_Z' = \sigma_r' = \frac{J_1'}{3} = p'; \qquad \varepsilon_Z = \varepsilon_r = \frac{\varepsilon_{KK}}{3} \qquad (5h)$$

The resulting expression provides the following relation between the elastic volumetric strain and effective pressure:

$$\varepsilon_{KK}^{e} = \frac{1}{K_1}\ln\left[\frac{K_1}{K_i}p' + 1\right] \qquad (5i)$$

The relation between the plastic volumetric strain, ε_{KK}^{p} and the effective pressure, p' is given in Eq. 5d. During isotropic consolidation, X in Eq. 5d should be replaced by $3p'$; thus

$$\varepsilon_{KK}^{e} = W\left[\ln(3D_1 p' + 1) - \ln(3D_3 p' + 1)\right] \qquad (5j)$$

The total volumetric strain takes the following form;

$$\varepsilon_{KK} = W\, ln(3D_1 p' + 1) \qquad (5k)$$

Shear phase

During the shear phase of a triaxial test, cell pressure is constant

$$\sigma_r = \text{constant} = p_c; \quad \text{and} \quad d\sigma_r = 0 \qquad (5l)$$

in which p_c is the confining pressure at the end of isotropic consolidation phase. For undrained test, the volumetric strain is constant during the shear phase and equals to the volumetric achieved at the end of the consolidation phase of the test. Accordingly, for an undrained triaxial compression test

$$\varepsilon_{KK} = W\ln(3D_1 p' + 1) = \varepsilon_{vc} \quad \text{and}$$

$$d\varepsilon_{KK} = d\varepsilon_{KK}^{e} + d\varepsilon_{KK}^{p} = 0 \qquad (5m)$$

The total volumetric strain becomes

$$\varepsilon_{KK} = \frac{1}{K_1}\ln\left(\frac{K_1}{K_i}p' + 1\right) + W\ln(D_1 X + 1) - WD_2\ln(D_3 X + 1)$$

$$= \text{constant} = \varepsilon_{vc} \qquad (5n)$$

The equation of the elliptic cap (Eq. 5c) for the triaxial test takes the following form:

$$X = L + 3\left[\left(p' - \frac{L}{3}\right) + \frac{R^2}{27}(\sigma_Z' - \sigma_r')\right] \qquad (5o)$$

EXPERIMENTAL PROGRAM

Test Material

Experimental works were performed on a local Li-Kang river sand, a natural sand deposit widely covering the South-Western region of the island of Taiwan. The sand specimens used in this investigation were obtained from a depth of about 5 m below ground surface. It is uniform, fine to medium, having a particle size range of 6.25 mm to 0.08 mm (i.e. No.4 to No.200 U.S. Sieves) with the grain shape varying sub-angular to angular composed mainly of quartz and shale. The properties of test sand are as follows; specific gravity of solid, G_s = 2.72, uniformity coefficient, C_u = 2.0, mean effective grain size D_{50} = 0.42mm, maximum dry density, $\gamma_{d\max} = 1.608\ \text{Mg/m}^3$, and minimum dry density, $\gamma_{d\max} = 1.288\ \text{Mg/m}^3$. All test specimens for this study were controlled at its initial relative density at 15 % with the initial void ratios, e = 1.046.

Test samples with the size of 100 mm in height and 50 mm in diameter were prepared directly on the triaxial cell base by pouring a pre-weighed amount of air-dried sand mixed with about 8% of water into a membrane lined mold in five equal layers with zero height of drop. All specimens were saturated to assure that pore water pressure parameter, B exceeded a value of 0.95. Consolidated drained and undrained triaxial compression tests were performed with the effective confining pressures varying in the range from 100 to 600 kPa.

Determination of material constant λ, κ and M

The value of λ and κ can be related to the commonly known quantities such as compression index and swelling index. λ is defined as the slope of virgin loading line on $e - \log p'$ plot while the swelling index, κ is defined as the slope of unloading curve on the same plot. An isotropic consolidation test was performed on the specimen at the initial relative density of 15% (e = 1.046). The specimen was loaded, doubling the load up to a high isotropic pressure of 800 kPa. Finally, the specimen was unloaded by the reverse step. Another constant, M with the frictional properties of sand, and is equal to the stress ratio, $\eta = q / p'$ at the critical state. The test results were presented in a previous paper (Chin, 1989), such as λ = 0.1505, κ = 0.00724 and M = 1.238.

The material constants of Cap model on Li-Kang sands were presented in the previous paper (Chin, 1989), which gives $D_1 = 34.2 Cm^2 / N$, W = 0.0491, K_1 = 160, $K_i = 500 N / Cm$, R = 0.943 and μ = 0.45.

PREDICTION OF THE STRESS-STRAIN RELATIONSHIP

Based on the drained tests, Fig. 6a shows that the axial strain of the computed results tend to underestimate the deviatoric stress q for the lower confined pressures of 100 kPa. For the results shown in Fig. 6(b) for the high confined pressures of 600 kPa, the Cap model analytical results seem to agree with the experimental curves due to the grain crushed under higher confined pressure. It is very difficult to predict the axial strain versus the volumetric strain relationship for the lower confined pressures of 100 kPa (Fig. 7a), and only the Cap model analytical results curve seem to agree with the experimental curves. In Fig. 7b, the Cap model analytical results are some what far apart from the experimental results, and the other three results are close to the experimental results for the higher confined pressures of 600 kPa.

Based on the undrained triaxial test, Fig. 8a shows that the experimental curve for the lower effective confined pressures of 100 kPa is agree with the modified Cam-clay model predicted curve. However, the Cap model analytical results seem to agree well with the experimental curve for the high effective confined pressures of 600 kPa (Fig. 8b), although it is still unable to fully match the observed curves.

Similarly, it is difficult to predict the axial strain versus the pore water pressure relationship for the lower effective confined pressures of 100 kPa in Fig. 9a, while the modified Cam-clay model predicted curves seem to agree with the experimental curves. On the other hand, in Fig. 9b for the higher effective confined pressures of 600 kPa, only the Cap model predicted curves are far apart from the observed curves while the other three predicted curves deviate only slightly from the experimental curves.

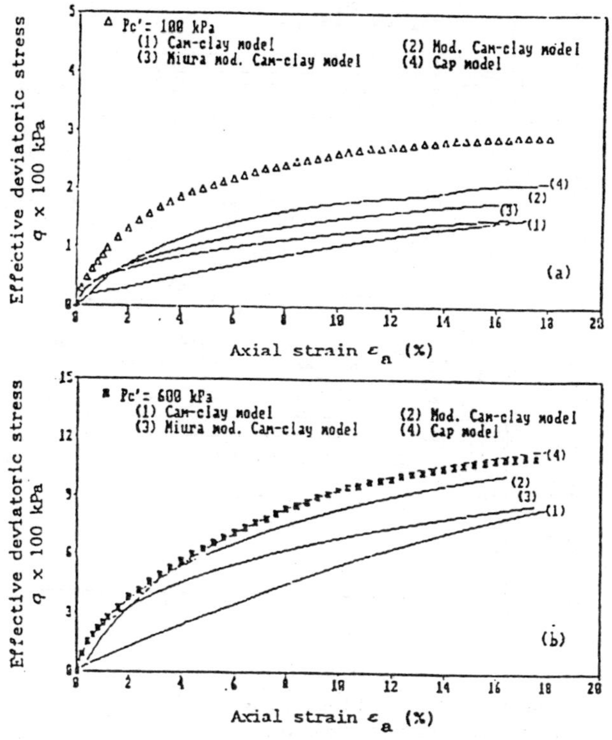

Fig. 6 Prediction of effective deviatoric stress and axial strain for drained test on Li-Kang sand, (e = 1.046)
(a) p_c' = 100 kPa (b) p_c' = 600 kPa

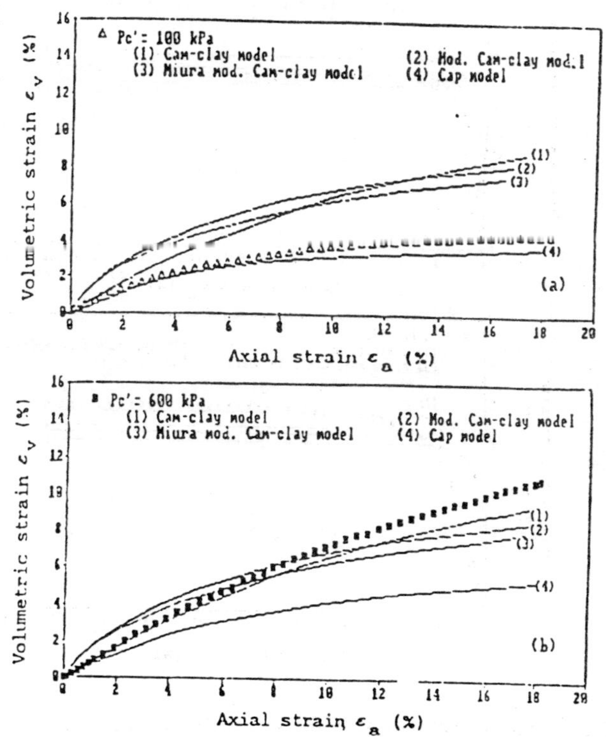

Fig. 7 Prediction of volumetric strain and axial strain for drained test on Li-Kang sand, (e = 1.046) (a) p_c' = 100 kPa (b) p_c' = 600 kPa

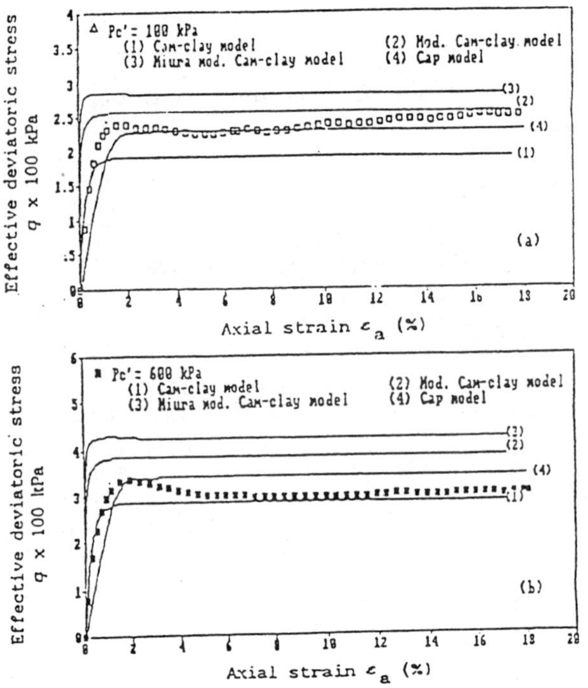

Fig. 8 Prediction of effective deviatoric stress and axial strain for undrained test on Li-Kang sand. (e = 1.046)
(a) p' = 100 kPa (b) p' = 600 kPa

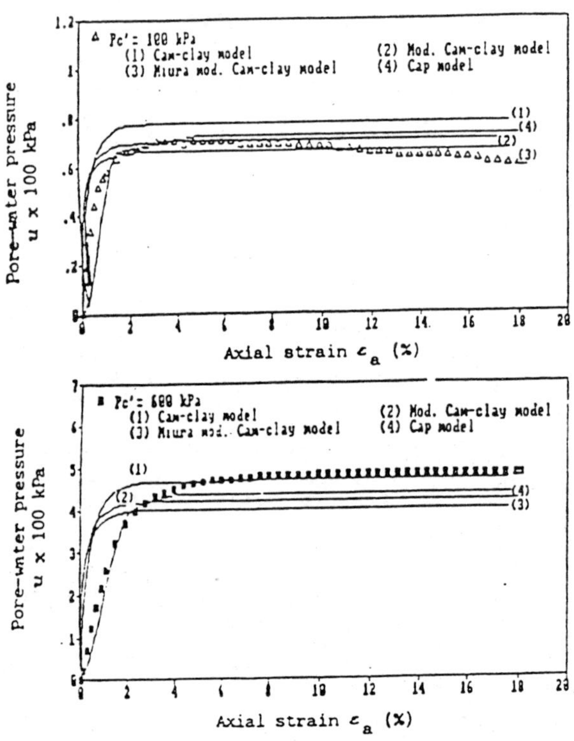

Fig. 9 Prediction of pore water pressure and axial strain for undrained triaxial test on Li-Kang sand, (e = 1.046)
(a) $p_c' = 100$ kPa (b) $p_c' = 600$ kPa

In the incremental stress-strain models: Cam-clay model, modified Cam-clay model, and Miura modified Cam-clay model, only the volumetric strain is assumed to be partially recoverable. The reason is that, elastic shearing strain is assumed to be identically zero. But the elastic volumetric strain is non-linearly dependent on the hydrostatic stress and is independent of the deviatoric (shear) stresses for certain stress histories, strain-oftening may occur. The Cap model (Baladi, 1979) is based on a loading function which serves as both a yield surface and plastic potential. Figure 10 shows the comparisons of experimental curve with predicted one from the effective stress on $q - p'$ plane for undrained triaxial test. These results show that only the modified Cam-clay model analytical results seems to agree with the experimental results, the other three models analytical results are far from the experimental results.

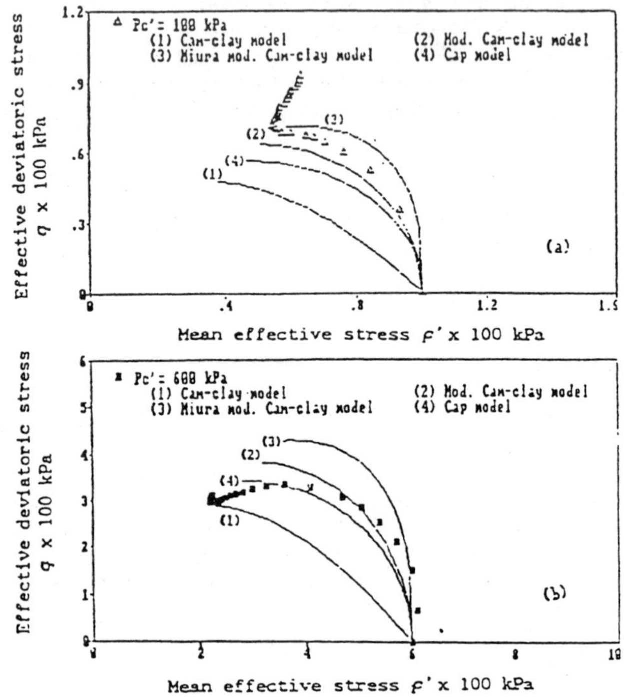

Fig. 10 Prediction of effective stress path on $q - p'$ plane for undrained triaxial on Li-Kang sand, (e = 1.046)
(a) $p_c' = 100$ kPa (b) $p_c' = 600$ kPa

PREDICTION OF STRESS RATIO VERSUS VOLUMETRIC STRAIN RATIO

Figure 11 shows the comparisons of experimental curve with the predicted the stress ratio versus volumetric strain ratio for drained triaxial tests. It is observed that the analytical curves proposed by the author seems to be able to predict this relationship more accurately than the other four analytical curves. The proposed curve equation may be expressed as follows:

$$\eta(q/p') = 1.52 \times \left[1 - \frac{\left(\frac{\varepsilon_v}{p_c'} - 5.14 \right)}{26.4} \right]^{(1/2)} \qquad (6)$$

where p_c' is the effective confined pressure and e is the volumetric strain ε_v in (%),

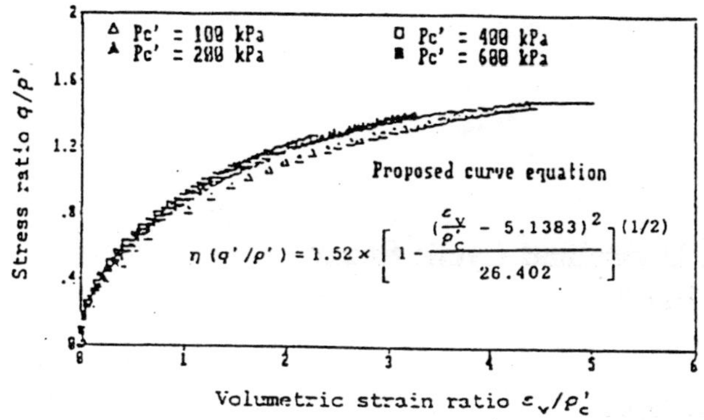

Fig. 11 Prediction of stress ratio, $\eta\ (q\ /\ p')$ - volumetric strain ratio, $\varepsilon_v\ /\ p_c'$ for drained triaxial test on Li-Kang sand (e = 1.046)

PREDICTION OF STRESS RATIO VERSUS PORE WATER RATIO

Fig. 12 shows the comparisons of experimental curve with the predicted the stress ratio versus pore water ratio for undrained triaxial tests. The analytical curves proposed by this study predicts the stress ratio, $\eta\ (q\ /\ p')$ versus pore water ratio, $U_w\ /\ p_c'$ relationship more accurately than do the other four analytical curves. The proposed curve equation is

$$\eta\ (q\ /\ p') = 1.707\ U_w\ /\ p_c' \qquad (7)$$

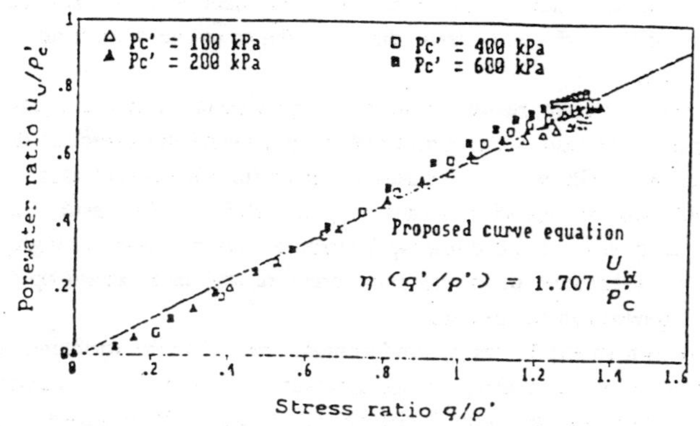

Fig. 12 Prediction of stress ratio, $\eta\ (q\ /\ p')$ versus pore water ratio, $U_w\ /\ p_c'$ for undrained triaxial test on Li-Kang sand (e = 1.046)

CONCLUSIONS

The main conclusions of this study are as follows:

(1) The model constant of λ value from isotropic consolidation test is an important influential factor to predict those behavior,

(2) The undrained stress-strain behavior of sand is similar to those of slightly over-consolidated clay. It is impossible to predict the stress path behavior by applying the boundary surface theory of normally consolidated clay,

(3) Drained and undrained triaxial tests of any type on sand and clay in loose or dense states at high or low pressure and with any required pre-compression all fall within the compass of predictions of strength, pore pressure and volume change. The predictions are found to be remarkably sound, but the actual numerical values often leave much to be desired.

(4) To predict the stress ratio, $\eta\ (q\ /\ p')$ versus the volumetric strain ratio, $\varepsilon_v\ /\ p_c'$ for drained triaxial test, the proposed curve equation is expressed as Eq. 6, and

(5) To predict the stress ratio, $\eta\ (q\ /\ p')$ versus the pore water pressure ratio, $U_w\ /\ p_c'$ for undrained triaxial test, the proposed curve equation is expressed as Eq. 7.

ACKNOWLEDGMENTS

This study was performed as part of research grant by the National Science Committee, R.O.C., NSC75-0410-E006-02. The support is gratefully acknowledged.

REFERENCES

Baladi, G.Y and Rohani, B (1979). "Elastic-Plastic Model for Saturated Sands, "*J. Of Geotechnical Engg., ASCE,* Vol. 105, GT4, pp. 465-480.

Chin, Y P (1989). "Evaluation of Cap Model on Saturated Sands, " *Proceeding, The Third Conference of Geotechnical Engg.*, Kenting, R.O.C., (in Chinese) pp. 363-377.

Miura, N and Yamamoto, N (1982) "On the Yield Curve of Sand in a Particle crushing Region, " *Proceeding J. SCE,* No.326, pp. 83-90.

Roscoe, K.H and Poorooshasb, H.B (1963), "A Theoretical and Experimental Study of Stress-Strain in The Triaxial Compression Tests on Normally Consolidated Clays," *Geotechnique*, Vol.8, pp. 12-28.

Roscoe, K.H. and Burland, J.B (1968), "On The Generalized Stress-Strain Behavior of Wet Clay," *Cambridge University Press.* Cambridge, UK.

Proceedings of the Seventh (1997) International Offshore and Polar Engineering Conference
Honolulu, USA, May 25-30, 1997

ISBN 1-880653-28-1 (Set); ISBN 1-880653-29-X (Vol. I)

Effect of Pre-Shearing with Drainage on Undrained Cyclic Shear Behaviour of Dense Sand

Ken-ichi Sato
Fukuoka University, Fukuoka, Japan
Kazuya Yasuhara
Ibaraki University, Hitachi, Japan
Nobuo Yoshida
Fukuoka University, Fukuoka, Japan

ABSTRACT

Wave-induced settlement and instability of seabed ground are recognized as important problems in the construction of nearshore structures such as breakwaters. The behaviour of seabed ground, in particularly when, subjected to wave loading, is closely associated with the undrained cyclic shear deformation and the generation of excess pore pressures due to stormy wave loading, leading to reduced effective stresses. During calmer periods between cyclic loading events, the generated pore pressure may get time to dissipate, and the effective stresses gradually increase. In this manner, the liquefaction and post-liquefaction recompression in seabed ground is usually repeated in every sequence of wave actions and calm periods.

In this paper, based on above-stated considerations, a series of cyclic torsional shear tests have been carried out to investigate the effect of drainage on cyclic and post-cyclic behaviour of pre-sheared dense sand. Results of experiments indicate that: (1) Dense sand subjected to drainage history with cyclic shearing stiffens or softens depending on the magnitude of the excess pore pressure and shear strain during undrained cycling following drainage. Behaviour during further sequence of undrained cyclic shearing is influenced by both the principal stress direction and the magnitude of cyclic stress ratio. (2) The recompression volumetric strain due to dissipation of cyclic-induced excess pore pressure can be related to the maximum shear strain observed during undrained cyclic shear, irrespective of the principal stress direction, cyclic stress ratio or previous drainage history.

KEYWORDS: Seabed, sand, wave loading, preshearing, drainage, principal stress direction, liquefaction, recompression

INTRODUCTION

The stress conditions for soil elements subjected to cyclic loading such as those beneath the foundations of breakwater and offshore structures are very complex. Based on this, the authors (Sato et al., 1993) performed an elastic stress analysis for the wave-induced stress distribution and stress states in the seabed underneath a breakwater. The results of this previous study indicated that: (1) the angle of maximum principal stress direction to the vertical axis is reversed by about 90 degrees to the direction of wave actions, (2) the stress path in $(\sigma_v - \sigma_h)/2p$ and τ_{vh}/p space traces a straight locus with a constant angle of principal stress direction from a pushing to a pulling wave or from a pulling to a pushing wave, depending on the location beneath the breakwater. Thus, it was concluded that there was virtually no rotation of the principal stress direction due to cyclic wave loads in soil deposits.

Following above results, a series of undrained static and cyclic torsional shear tests were carried out to investigate the effect of the inclination of the principal stress direction to the vertical axis on strength and deformation (Sato et al., 1996). The results of experiments show that the principal stress direction appears to highly affect the behaviour of excess pore pressure and shear strain up to initial liquefaction conditions.

As shown in Fig.1, the ground beneath the off-shore structures is subjected to cyclic loading by the interaction between wave actions and structures. In particular during stormy weather, the breakwater is subjected to continuous cyclic loading by extremely large waves which in turn induces cyclic stress under undrained conditions into the ground beneath the structure.

Since the excess pore pressure accumulates in the ground during storms, the effective stress decreases and may result into liquefaction especially in loose sand ground. On the other hand, when the waves become calm, the excess pore pressure rapidly dissipates because of the large coefficient of permeability of the sand ground, and densification occurs due to recompression of the ground (Zen and Yamasaki, 1990). The effect of drainage on this kind of ground is considered to be safer for stability of the structure than in the one before allowing drainage.

In this way, compression, resulting from decrease in effective stress and dissipation of excess pore pressure, often occurs in the ground subjected to wave induced cyclic loading. In order to clarify

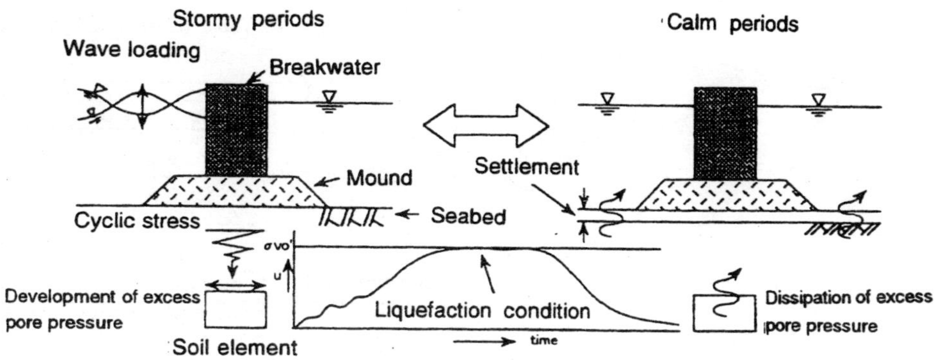

Fig.1 A schematic illustration of the ground beneath the breakwater

this behaviour more appropriately, it is very important for us to understand: (1) the cyclic shear behaviour during stormy waves, (2) compression characteristics caused by dissipation of generated excess pore pressure. As for the first, there are many studies on the effect of shear stress history that have shown that although an increase of density occurs due to recompression of liquefied specimen, a extremely low resistance behaviour is observed during the second sequence of cyclic shearing (Finn et al.,1970 ; Ishihara and Okada,1978,1982). As for the second, many studies have been carried out regarding settlement of sands following earthquakes (Lee and Albaisa,1974 ; Sasaki et al.,1982 ; Kokusho and Nagasaki,1984 ; Ishihara and Yoshimine, 1992 ; Shamoto et al.,1996).

Through another series of tests using the hollow cylinder equipment, this paper discusses the recompression behaviour after cyclic shearing and the behaviour of cyclic shearing after drainage history.

TEST PROCEDURES

The hollow cylinder apparatus and specimen

The tests were performed using a hollow cylinder apparatus. In this device, application of an axial load W, a torque T, and internal and external pressures p_i and p_o, respectively, generate an axial stress σ_v, torsional shear stress τ_{vh}, radial stress σ_r and circumferential stress σ_h in the wall of the specimen as illustrated in Fig.2. By controlling these stresses the magnitude of the three principal stresses σ_1, σ_2 and σ_3, can be independently controlled, together with the orientation of the major principal stress relative to the vertical axis, α, and the coefficient of intermediate principal stress, b. The hollow cylindrical specimen has a 30mm inner radius r_i, 50mm outer radius r_o and height h of 200mm. Toyoura standard sand (Gs= 2.64, e_{max}=0.991, e_{min}=0.623) which was formed into a hollow cylindrical specimen container by an air pluviation method was used in this study. The specimen was saturated by flooding with CO_2 and de-aired water in that order. All the tests in the present paper were conducted on dense specimens (Dr=80%), because the seabed ground consists of dense sand like the one in the North sea (Rahman et al., 1977 ; Eide and Andersen, 1984).

Stress and strain parameters

Table 1 shows a set of equations for calculating average stress and strain values on specimen elements (Hight 1983). In measuring and calculating results, correction was made for membrane tension. The amount of swelling of the tube used to connect the inside hollow part of the specimen with the burette was determined to make correction for the actual volume change of the inside hollow part of the specimen.

Description of the test

Considering stress conditions in seabed, a dense Toyoura sand having anisotropic particle structure was used. Samples were then consolidated isotropically to an effective stress 196 kPa using a back pressure of 98 kPa to ensure that a Skempton's B-value in excess of 0.98 could be obtained. Both inner and outer cell pressures were then increased to an isotropic stress condition with p_c'=196kPa. After consolidating for 2 hours, stress-controlled undrained tests were carried out. It should be noted that during all tests, the following two conditions were maintained in all steps. (i) The value of mean principal stress p' was held constant at 196 kPa and (ii) the value of the relative magnitude of the intermediate principal stress b ($=(\sigma_2-\sigma_3)/(\sigma_1-\sigma_3)$) was also held constant at 0.5. The undrained cyclic shear tests with fixed and reversed orientation of principal stress axes which were performed are schematically explained in Fig. 3. The tests were performed with the shear stress τ increased in increments of $\Delta\tau$=9.8kPa to failure, with the direction of the major principal stress maintained at 15°, 30°, 45°, 60°, and 75°. In order to specify the three dimensional stress condition, it is necessary to introduce other variables, α and τ, which are used to determine the direction of maximum principal stress and the shear stress on the stress plane as shown in Fig. 4. The angle α as defined in Fig. 4 represents the specimen inclination to the vertical axis at the initial time of sequence of cyclic loading. The number of cycles in undrained cyclic shear tests were five.

The test scheme adopted in this study is illustrated in Fig. 5. A specimen was first consolidated to confining pressure of 196 kPa along the path AB shown in Fig. 5(a). Cyclic stresses were then applied to the sample in undrained conditions until some amount of pore water pressure developed which is expressed by the distance between points B and C in Fig. 5(a). The pore water was drained while keeping the cell pressure unchanged in the triaxial chamber. The sample was then brought to the same consolidation pressure as

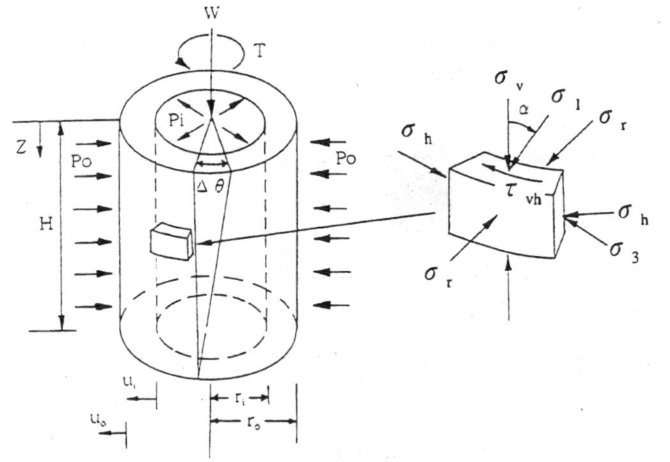

Fig.2 Hollow cylindrical specimen

Table-1 Stress and strain parameters

	Stress	Strain
Axial	$\sigma_z = \dfrac{W}{\pi(r_o^2 - r_i^2)} + \dfrac{(P_o \cdot r_o^2 - P_i \cdot r_i^2)}{(r_o^2 - r_i^2)}$	$\varepsilon_z = \dfrac{z}{H}$
Radial	$\sigma_r = \dfrac{P_o \cdot r_o + P_i \cdot r_i}{r_o + r_i}$	$\varepsilon_r = -\dfrac{u_o - u_i}{r_o - r_i}$
Circumferential	$\sigma_h = \dfrac{P_o \cdot r_o - P_i \cdot r_i}{r_o - r_i}$	$\varepsilon_h = -\dfrac{u_o + u_i}{r_o + r_i}$
Torsional shear	$\tau_{vh} = \dfrac{3 \cdot T}{2\pi \cdot (r_o^3 - r_i^3)}$	$\gamma_{vh} = \dfrac{\Delta\theta \cdot (r_o^3 - r_i^3)}{3H \cdot (r_o^2 - r_i^2)}$

r_o: Outer radial、r_i: Inner radial、H: Specimen height、z: Axial displacement、u_i: Inner radial displacement
u_o: Outer radial displacement、$\Delta\theta$: Torsioanl displacement

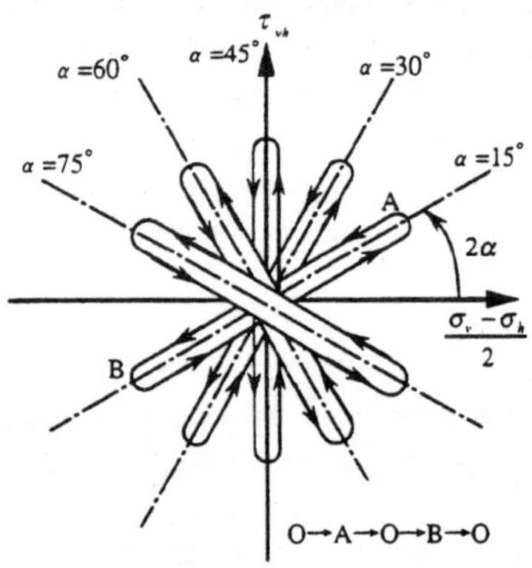

Fig.3 Loading schemes

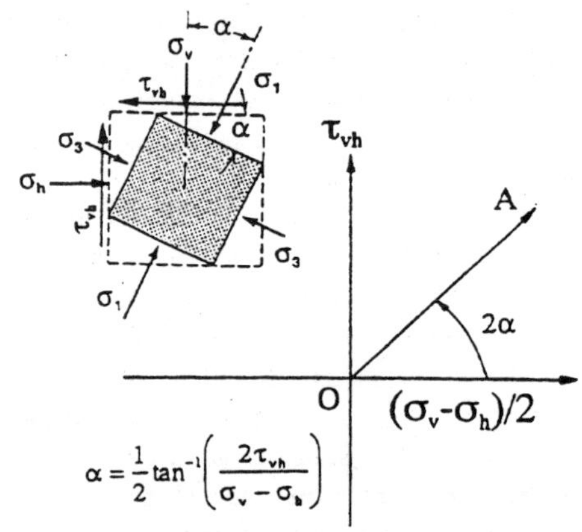

Fig.4 Definition of major principal stress direction α and shear stress τ

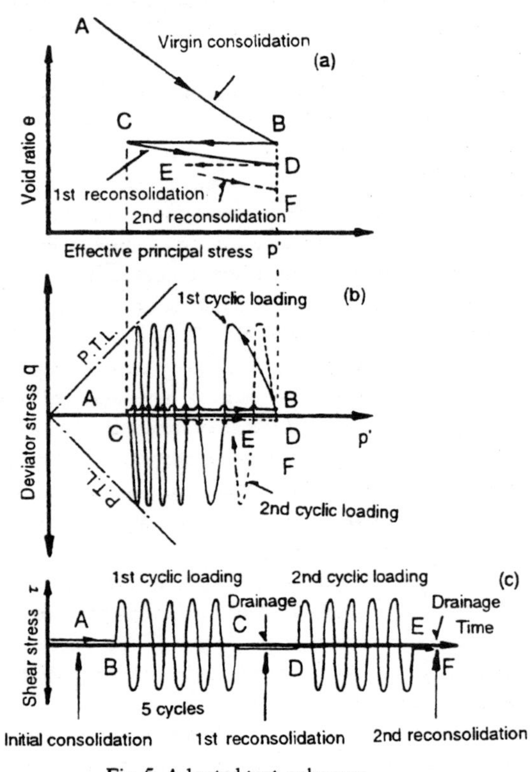

Fig.5 Adopted test schemes

Table-2 Initial condition of specimens

Test No.	Principal stress direction α(°)	Cyclic stress ratio τ/p$_c$'	Initial void ratio	
			1st loading	2nd loading
CD151	15	0.20	0.698	0.687
CD152	15	0.24	0.695	0.682
CD153	15	0.28	0.695	0.663
CD301	30	0.20	0.704	0.693
CD302	30	0.24	0.703	0.691
CD303	30	0.28	0.699	0.678
CD451	45	0.20	0.690	0.682
CD452	45	0.24	0.689	0.669
CD453	45	0.28	0.694	0.664
CD601	60	0.20	0.690	0.673
CD602	60	0.24	0.706	0.686
CD603	60	0.28	0.702	0.669
CD751	75	0.20	0.696	0.679
CD752	75	0.24	0.698	0.684
CD753	75	0.28	0.691	0.663

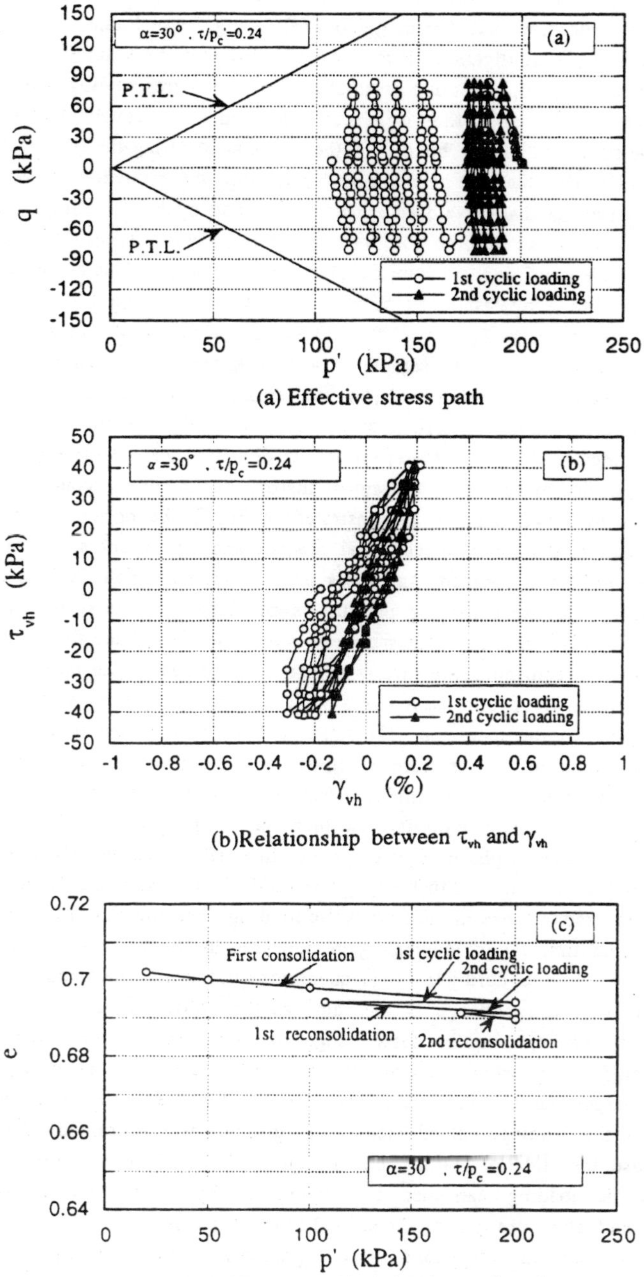

(a) Effective stress path

(b)Relationship between τ_{vh} and γ_{vh}

(c) Compression behaviour of sand

Fig.6 Results of undrained cyclic shear test following drainage history (α=30° , τ/p_c'=0.24)

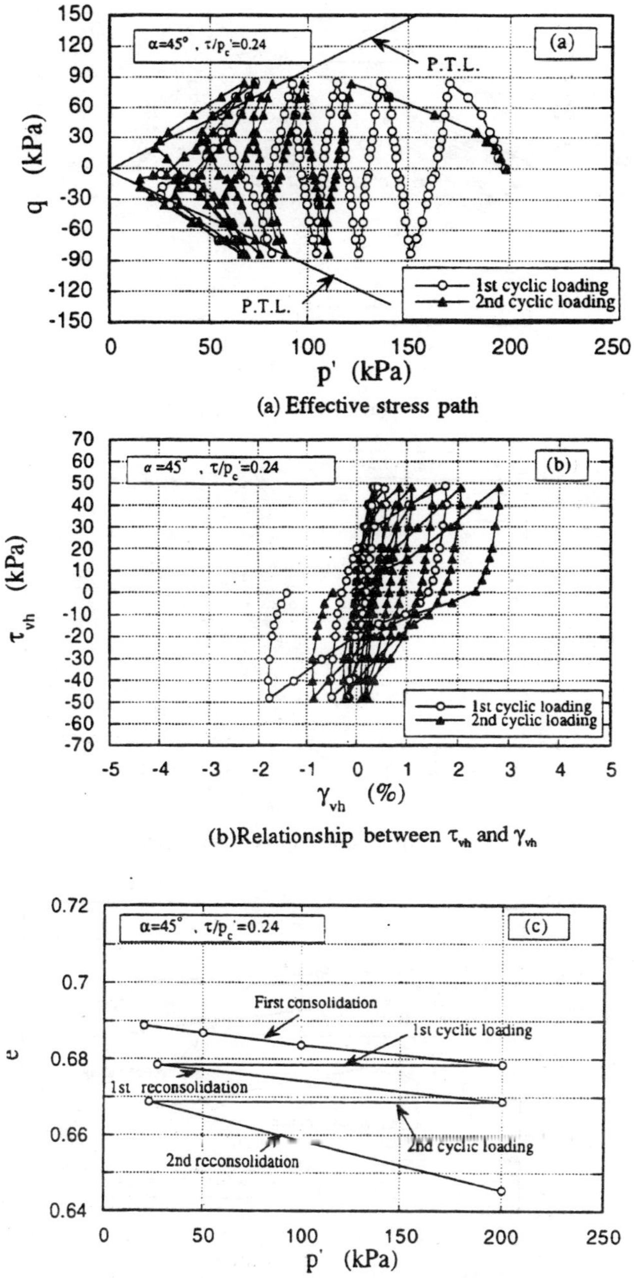

(a) Effective stress path

(b)Relationship between τ_{vh} and γ_{vh}

(c) Compression behaviour of sand

Fig.7 Results of undrained cyclic shear test following drainage history (α=45° , τ/p_c'=0.24)

designated by point D. The series of stress change above consisting of cyclic stresses and reconsolidation was considered to represent the stress history in the entire scheme of stress changes employed in this study. The specimens having experienced different pre-shearing were then subjected to cyclic shear stress in undrained conditions as illustrated by the stress path D to E in Fig. 5(a).

The pore water pressure and the shear strain that developed at this stage of loading were observed and the effect of the preshearing was evaluated. The initial conditions for the specimen are summarized in Table 2.

TEST RESULTS

Effect of drainage history on cyclic shear characteristics

Effect of principal stress direction on undrained shear characteristics following drainage history

The influence of the previous drainage history after liquefaction on the subsequent undrained cyclic shear behavior of dense sand is graphically illustrated in Figs. 6 through 8 with τ/p_c'=0.24 and α=30°, 45° and 60°. The positive and negative octahedral deviator stress q

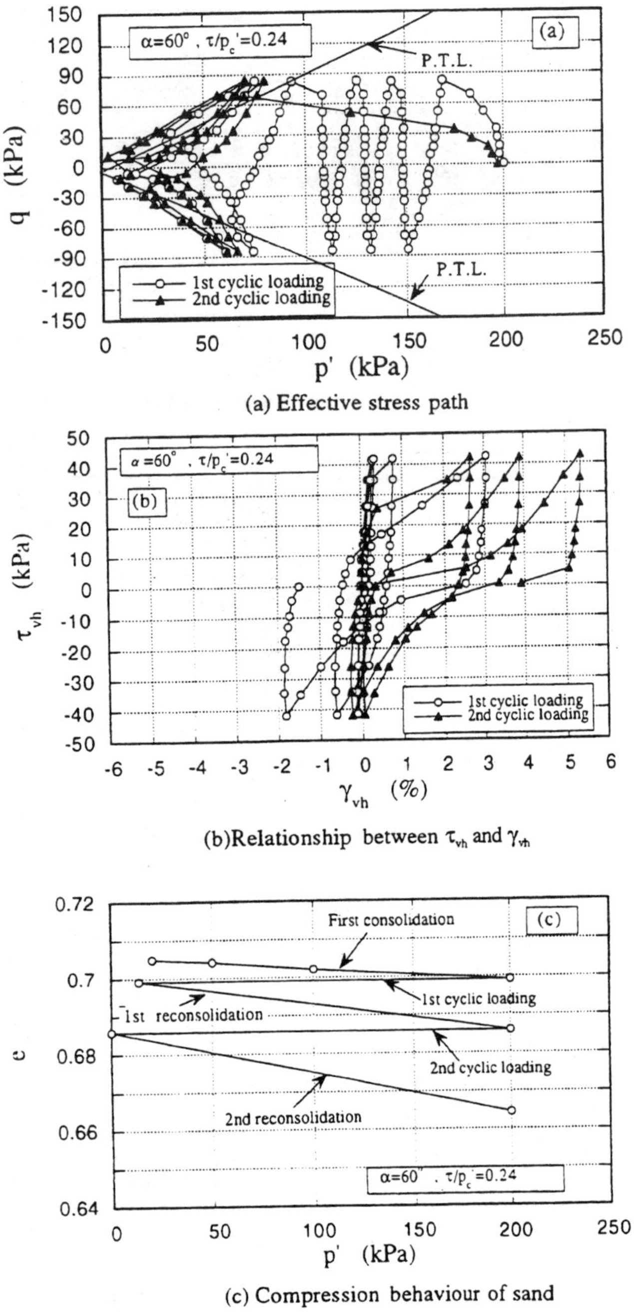

Fig.8 Results of undrained cyclic shear test following drainage history (α=60°, τ/p_c'=0.24)

were determined by the direction on which the torsional stress is applied.

Fig. 6 shows the results of a test in which the specimen was first subjected to the five times of cyclic loading (N=5) under α=30°, and was subsequently allowed to reconsolidate and was further subjected to the same amplitude of cyclic stress in the previous stage. It is noted in Fig. 6(a) and (b) that an excess pore pressure equal to 50% of the initial effective confining pressure develops in the first loading accompanied by a residual torsional shear strain -0.2%. Shown in Fig. 6(c) are the changes in void ratio versus effective confining pressure relations that occurred in the sample during the virgin isotropic consolidation before cyclic loading and also during the 2nd and 3rd reconsolidation that were carried out following the undrained cycles of shear stress application. It is also observed that the volume change during the reconsolidation is smaller than the corresponding volume change observed in the virgin consolidation before cyclic loading. The effect of drainage history on the dense sand during the first cyclic loading can also be recognized by noting the reduction in pore pressure that has developed during the second cycle of loading. Fig. 6(a) also shows that the pore water pressure of only 10% of the initial effective confining pressure developed in this stage. This decrease is marked in comparison with 50% of p_c' observed in the first loading. It is noted in Figs. 6(a) to (c) that a significant degree of workharding has taken place in the sample as a consequence of applying cyclic preshearing as exemplified by the small values of pore water pressures and torsional shear strains that have developed in the second loading. In this case, the specimen is densified and stabilization of the ground by the effect of drainage history has therefore occurred.

In the next example demonstrated in Fig.7, the principal stress direction was α=45° at the initial loading and , the cyclic stress ratio t/p_c'=0.24 was adopted. An excess pore pressure of about 90% of the initial effective confining pressure is developed and the stress paths show cyclic mobility. The sample was then reconsolidated to the same effective confining pressure as before by dissipating the pore water pressure. The cyclic shear stress with the same amplitude was again applied to the specimen. The stress paths demonstrated in Fig.7(a) show that the pore pressure which developed during the second cyclic loading was 50% of the initial effective confining pressure during the first cycle. Cyclic mobility can then be recognized when the stress path reaches a phase transformation line after a sudden movement towards the origin. Much larger torsional shear strains are developed compared to those in the first cyclic loading. The effect of preshearing on the first cyclic loading again appeared in the void ratio change characteristics during the subsequent consolidation. The magnitude of the void ratio change is the same as that observed in the virgin consolidation for the same range of increment in the effective confining pressure. However, the magnitude of the void ratio change is larger than that in the aforementioned test where α=30° was adopted and the development of excess pore pressure was restrained. In this case, the effect of drainage history on sand behaviour during the first cyclic loading can also be recognized by noting the reduced liquefaction strength on the second cyclic loading.

Recent investigations have gradually elucidated that the re-liquefaction is caused by the effect of structural anisotropy of liquefied sand (Finn et al.1970, Ishihara and Okada 1978,1982, Yoshimi et al.1992). In particular, Ishihara and Okada(1978) showed, through statically cyclic triaxial tests, that the decrease in effective stress on re-liquefaction highly depends on the direction of loading especially after passing the phase transformation line at the first cyclic loading. In other words, the effect on decrease in effective stress was largest when the last cycle in the first loading and the first cycle in the second cyclic loading are in opposite directions. This is because of the strong anisotopy created when the initial structure of the specimen was destroyed. In this paper, the coefficient of intermediate principal stress, b, was fixed at 0.5 during cyclic loading. Therefore, the results are compared with those of triaxial loading in which the value of b changes during cyclic loading. However, considering the principal stress direction from the

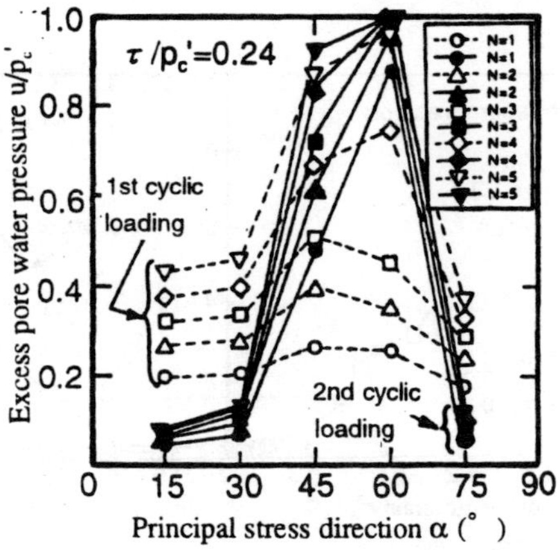

Fig.9 Relationship between principal stress direction α and excess pore water pressure u/p_c'

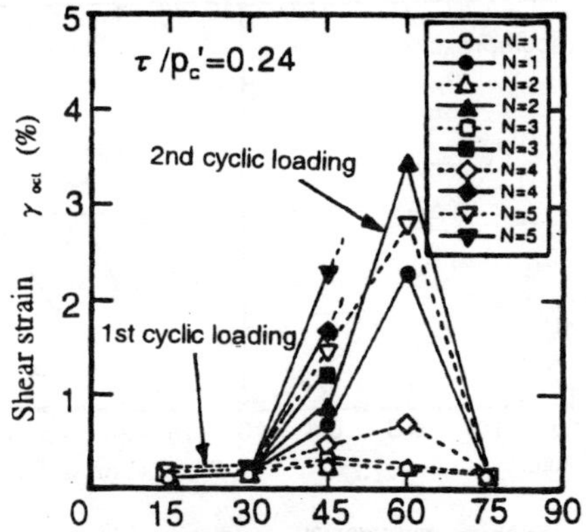

Fig.10 Relationship between principal stress direction α and deviator strain γ_{oct}

depositional surface of the specimen, the same condition was obtained as in the case of triaxial test by reversing the principal stress direction during loading and reloading by 90°. Because of the difference in the direction of cyclic loading between triaxial and torsional test, it was made clear that there is decrease in strength due to anisotropy effect.

Fig. 8 shows a results of the similar test with an increased principal stress direction α=60°. The pore pressure equal to the initial confining pressure was developed in the first loading and reached a liquefaction condition. The sample was then reconsolidated by allowing the dissipation of excess pore pressure and again subjected to the same magnitude of cyclic shear stress. At the first cycle of second cyclic loading, a pore pressure as much as 70% of the initial confining pressure developed and even though the specimen reached a liquefaction condition. This behaviour was more marked here than that in the case of α=45°. The double amplitude torsional shear strains in the case of increased principal stress direction up to α=60° during the first cyclic loading was larger than that in the aforementioned tests with α=30° and 45°. After drainage, much larger torsional shear strains were observed in the specimen. The decrease in void ratio due to drainage was larger than in the tests where α=30° and 45°. This resulted in the densification of the specimen. Accordingly, the effect of drainage history on recompression of sand could not be observed since there was no increase in liquefaction strength between the cases of α=45° and 60° in the second cyclic loading.

From the results of above three experiments, it is emphasized that the behaviour of cyclic shear characteristics following drainage history is influenced by the principal stress direction a in the first cyclic loading.

The effect of drainage history on the relationship between excess pore pressure and principal stress direction in each cycle with $\tau/p_c'=0.24$ is illustrated in Fig. 9. The magnitude of excess pore pressure following drainage history was different at each cycle and was influenced by principal stress direction α. This was also the case in the first cyclic loading. Especially, there was a marked development of excess pore pressure following drainage history with the increasing number of load cycles when the principal stress direction α was maintained between 45° and 60°. From these two results, the effective stress path reached the phase transformation line during the first cyclic loading and therefore the specimen exhibited cyclic mobility. Furthermore, in the results with other principal stress directions where the decrease in effective stress following drainage history was restrained, a little influence of the principal stress direction was also observed on the process of excess pore pressure development. The effect of drainage history on the relationship between shear strain and principal stress direction in each cycle with $\tau/p_c'=0.24$ is illustrated in Fig. 10. As in Fig. 9, Fig. 10 indicates that the behaviour of shear strain after drainage history is also influenced by principal stress direction, with the most influence occurring when the principal stress direction α is between 45° and 60°. The development process of excess pore pressure and shear strain following drainage history is clearly shown to be influenced by fabric anisotropy depending on the principal stress direction.

The behaviour of recompression following drainage history

Fig. 11 shows the relationships between the excess pore pressure ratio u/p_c' at the end of the first cyclic loading, the principal stress direction, and the volumetric strain followed by the dissipation of excess pore pressure. The results at the end of second cyclic loading are shown in Fig.12. From these results, the magnitude of excess pore pressure is different for each different principal stress direction. In particular, there is a marked development of excess pore pressure on the first and second cyclic loading when the principal stress direction α is between 45° and 60°. On the other hand, the volumetric strain increases with increasing excess pore pressure. There is a unique relationship between the volumetric shear strain and excess pore pressure regardless of the cyclic stress ratio or the principal stress direction. The relationship between the volumetric shear strain and excess pore pressure for recompression after sequences of cyclic loading is illustrated in Fig. 13. The results of the tests with different conditions conducted by Nagase and Ishihara(1989) are also shown in the figure. Volumetric strains on the first and second cyclic loadings show the same trend as the results obtained by Nagase and Ishihara. The volumetric strains on the second cyclic loadings tends to be smaller when compared to those in the first cyclic loading. This is considered to be due to the increase in specimen density as was also observed by Nagase and Ishihara. But, there is little scatter of volumetric strain data around $u/p_c'=1.0$ where the specimen is in liquefaction condition. This suggests that it is difficult to predict

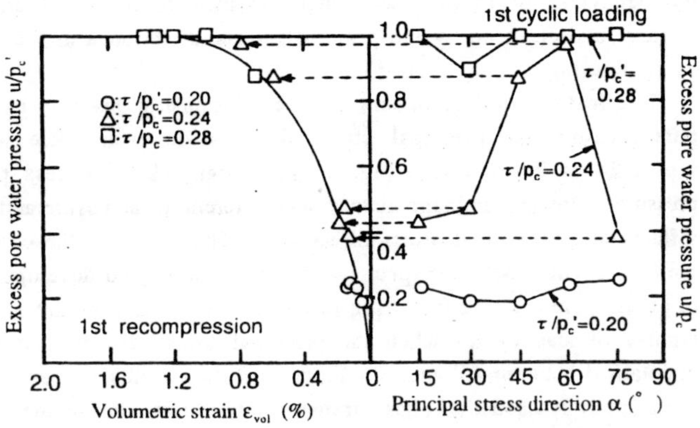

Fig.11 Relationships between u/p_c' and α, and the volumetric strain ε_{vol}

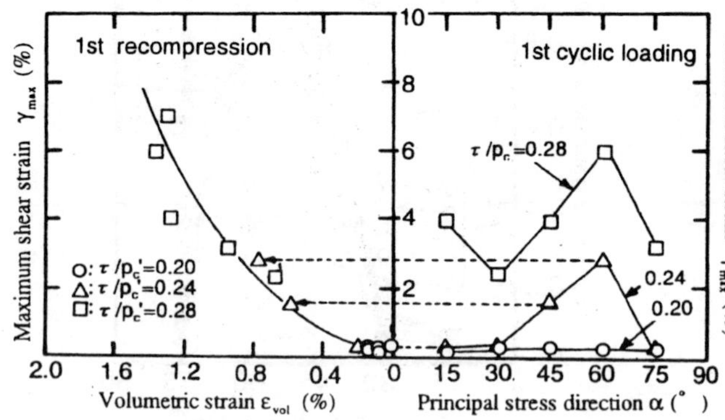

Fig.14 Relationships between γ_{max} and α, and the volumetric strain ε_{vol}

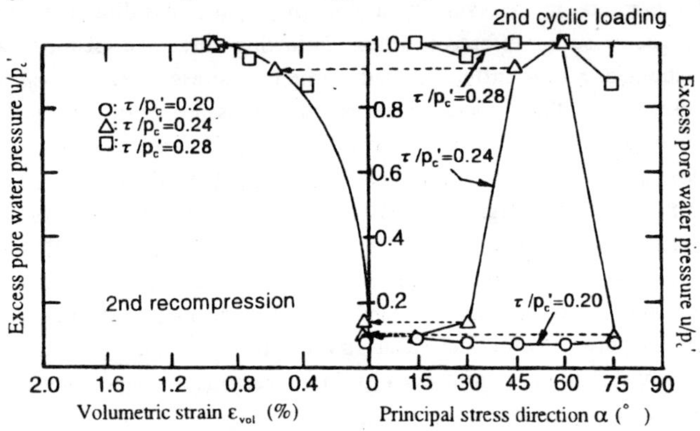

Fig.12 Relationships between u/p_c' and α, and the volumetric strain ε_{vol}

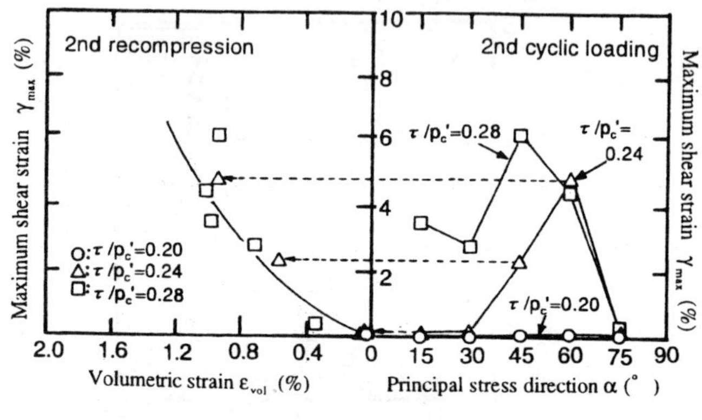

Fig.15 Relationships between γ_{max} and α, and the volumetric strain ε_{vol}

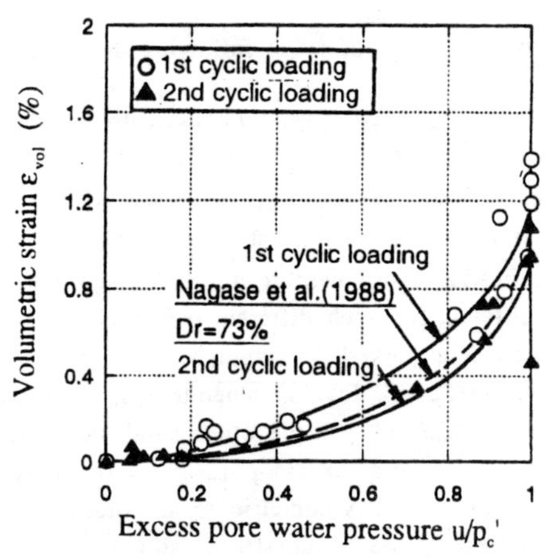

Fig.13 Relationships between u/p_c' and ε_{vol}

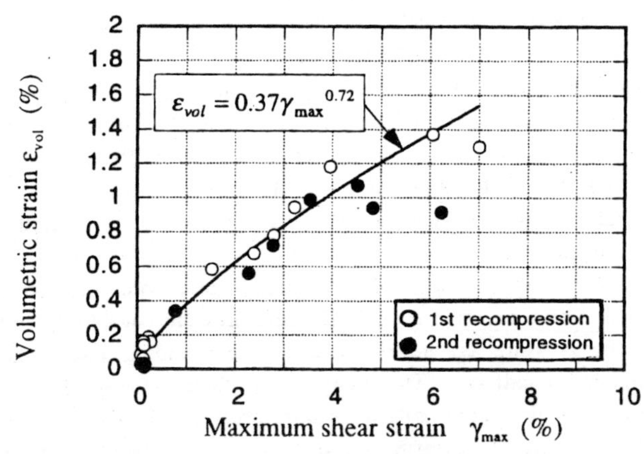

Fig.16 Relationships between γ_{max} and ε_{vol}

ground settlement by using the relationship between post-cyclic volumetric strain and excess pore pressure at liquefaction.

To overcome this difficulty, another way was attempted to correlate ε_{vol} to the cyclic-induced maximum shear strain γ_{max}. Fig. 14 shows the relationships between the maximum shear strain γ_{max} at the end of the first cyclic loading, the principal stress direction, and the volumetric strain which occurs following the dissipation of excess pore pressure. The results at the end of second cyclic loading are also shown in Fig. 15. From these results, the magnitude of maximum shear strain is influenced by the change in the principal stress direction. Also the magnitude of maximum shear strains at the first and second cyclic loading increases with cyclic stress ratio. On the other hand, the volumetric strain increases with maximum shear strain. It can be said from Fig.15 that there is a unique relationship between the volumetric shear strain and maximum shear strain regardless of the cyclic stress ratio or the principal stress direction. The relationship between the volumetric shear strain and maximum shear strain for the 1st. and 2nd recompression is summarized in Fig. 16. The relationship between maximum shear strain and volumetric strain as shown in Fig. 16 can be empirically expressed by the following equation:

$$\varepsilon_{vol} = 0.37\gamma_{max}^{0.72} \qquad (1)$$

Eq.1 provides a way of determining the volumetric strain from the maximum shear strain. Fig. 16 suggests that the volumetric strains after cyclic loading can well be correlated to the maximum shear strain, irrespective of cyclic stress ratio, or drainage history, or initial fabric anisotropy of specimens with the different principal stress directions.

CONCLUSIONS

A series of undrained cyclic torsional shear tests with different principal stress directions was carried out on saturated dense Toyoura sand to investigate drainage and pre-shearing effects. The following conclusions were obtained :

(1) The magnitude of excess pore pressure and shear strain after drainage history depends on the principal stress direction and the magnitude of cyclic stress ratio at cyclic loading. Dense sand undergoing drainage history after the first cycle and subsequent undrained cyclic loading stiffens or softens depending on the magnitude of the excess pore pressure and shear strain following drainage. Cyclic shearing behaviour after drainage is also influenced by both the principal stress direction and the magnitude of cyclic stress ratio.

(2) Despite the increase in density of specimen following drainage, the effect of drainage on decrease in effective stress is largest when the last cycle in the first sequence of cyclic loading and the first cycle in the second sequence of cyclic loading are in opposite directions. This is due to the strong anisotropy created by the destruction of the initial structure of the specimen, depending on the direction of loading between the last cycle of the 1st sequence and the 1st cycle of 2nd sequence of cyclic loading. The development process of excess pore pressure and shear strain following drainage history is clearly influenced by fabric anisotropy depending on the principal stress direction.

(3) There is scatter of post-cyclic recompression volumetric strain data around $u/p_c'=1.0$ due to dissipation of cyclic-induced excess pore pressures when the specimen is in liquefaction condition. On the contrary, the recompression volumetric strain can well be correlated to the maximum shear strain observed during cyclic shearing. This relationship holds irrespective of the angle of principal stress direction, cyclic stress ratio or drainage history.

REFERENCES

Eide, O. and Andersen K.H.(1984):Foundation engineering for Nothern North Sea, *NGI Publication*, No.154, pp.38-42.

Finn, W.D.L., Bransby, P.L. and Pickering, D.J. (1970) : Effect of strain history on liquefaction of sand, J.SMFD, *ASCE*, Vol.96, No.SM6, pp. 1917-1934.

Hight, D.W. (1983): Laboratory investigations of seabed clays, *Ph. D Thesis submitted to the University of London*.

Ishihara, K. and Okada, S. (1978):Effect of stress history on cyclic behavior of sand,*Soils and Foundations*, Vol.18, No.4, pp.31-45.

Ishihara,K. and Okada,S. (1982): Effect of large pre-shearing on cyclic behavior of sand, *Soils and Foundations*, Vol.22, No.3,pp.109-125.

Ishihara, K. and Yoshimine, M. (1992): Evaluation of settlements in sand deposits following liquefaction during earthquakes, *Soils and Foundations*, Vol.32, No.1, pp.173-188.

Kokusho, T. and Nagasaki, K. (1984) : The characteristics of settlement for dense sand after liquefaction, *Proc. of 19th annual Conf.* on JGS, pp.1661-1664, (In Japanese).

Lee,K.L. and Albaisa, A. (1974) : Earthquake induced settlements in saturated sands, J.GED, *ASCE*, Vol.100, No.GT.4, pp.81-92.

Nagase, H. and Ishihara, K. (1988) : Liquefaction-induced compaction and settlement of sand during earthquakes , *Soils and Foundations*, Vol.28, No.1, pp.83-94.

Rahman, M.S., Seed, H.B. and Booker J.R. (1977) : Pore pressure development under offshore gravity structures, J.GED, *ASCE*, Vol.103, No.GT.12, pp.1419-1436.

Sasaki, T., Tatsuoka, F. and Yamada, S. (1982) : Estimation of settlement during reconsolidation in sand ground, *Proc. of 17th annual Conf.* on JGS, pp.1661-1664, (In Japanese).

Sato, K., Takagi, M., Yasuhara, K. and Kenkyo K. (1993) : Behaviour of sand under wave-induced stress conditions beneath a breakwater, *Journal of the Faculty of Eng. Ibaraki Univ.*, Vol.41, pp.1-15 (In Japanese).

Sato, K., Yasuhara, K., Higuchi, T. and Yoshida, N. (1996) : Effect of principal stress direction on undrained cyclic shear behavior of dense sand, *Journal of Geotech. Eng.*, JSCE, No.541/III-35, pp.199-213, (In Japanese).

Shamoto, Y., Sato, M. and Zhang, J. (1996) : Simplified estimation of earthquake-induced settlements in saturated sand deposits, *Soils and Foundations*, Vol.36, No.1, pp.39-50.

Yoshimi,Y., Kuwabara, F. and Tokimatsu, K. (1992) : One-dimensional volume change characteristics of sands under very low confining stresses, *Soils and Foundations*, Vol.15, No.3, pp.51-60.

Zen, K. and Yamasaki, H. (1990) : Mechanism of wave-induced liquefaction and densification in seabed, *Soils and Foundations*, Vol.30, No.4, pp.90-104.

Proceedings of the Seventh (1997) International Offshore and Polar Engineering Conference
Honolulu, USA, May 25-30, 1997

ISBN 1-880653-28-1 (Set); ISBN 1-880653-29-X (Vol. I)

Effects of Compression Rate on Unconfined Compressive Strength of Cohesive Soils: Evaluation Based on the Effective Stress

Masayoshi Shimizu
Tottori University
Tottori, Japan

ABSTRACT

The effects of the compression rate on the unconfined compressive strength were investigated from the point of view that the strength must reflect the effective stresses of samples. Unconfined compression tests were carried out on two remolded and consolidated clays of high and low plasticity index values with the measurement of pore water pressure by means of the pressure plate technique. The effects of compression rate on unconfined compressive strengths (q_u) were quantitatively evaluated with the consideration of the initial suction (S_0) and the maximum preconsolidation pressure (p_0). Results show that the relationship between log $((q_u/2)/S_0)$ and $\log(p_0/S_0)$ can be linear for each rate of compression and that two coefficients specifying the linear relationship depend on the rate of compression. The conclusion was derived from the results that the increase in compression rate in the practical range causes the increase in compressive strength of only a few percents.

KEY WORDS: unconfined compressive strength; unconfined compression test; rate of compression; effective stress; suction

INTRODUCTION

The unconfined compression test is one of laboratory tests by which the undrained shear strength of saturated cohesive soils are estimated. The estimated strength is used for analyzing the stability of the ground in the undrained condition.

The unconfined compressive strength of a specimen is controlled by both the effective stress that the specimen possesses and the effective stress history that it had been subjected to (Shimizu, 1993). The specimen's effective stress can be evaluated by the measurement of the negative pore water pressure, i.e., the suction of the specimen; the effective stress history can be specified by the maximum and current effective vertical stresses that the sample had been applied in the ground. The difference between the specimen's effective stress and the current effective stress can be regarded as the disturbance that the specimen had been subjected to. The more the disturbance is, the less the estimated undrained strength is. The disturbance can not be avoided, and therefore the estimated strength is considered unreliable.

However, in many countries as in Japan, the design method based on the unconfined compressive strength has traditionally been employed in practical problems. The tradition will not be abandoned soon and the traditional design method will continue to be used. From this point of view, the efforts should be made to resolve factors that affect the unconfined compressive strength, remaining to be solved. One of such factors is the rate of compression. As stated above, the principle of effective stress is valid even for the behavior in unconfined compression tests. In this paper, the effects of the rate of compression on the unconfined compressive strength is investigated in terms of effective stresses.

The compressive strength determined from the unconfined compression test can be applied to the stability analysis so far as the following conditions are satisfied:

1) The problem is the so-called short-term stability problem;
2) The strength parameter ϕ_u is null;
3) The unconfined compression tests are conducted under the undrained condition; and
4) The sample is undisturbed.

This study is based on the hypothesis that these conditions are satisfied; and therefore, for example, such a case that the rate of shear in the ground is not the same as the rate of compression in laboratory tests is not discussed. In other words, the effect of the rate is discussed only in the context of laboratory tests.

SAMPLES AND METHOD

Samples

Two samples of different consistency were used for experiments (See Table 1). The sample of higher plasticity is a mixture of bentonite and a silty-clay called "Fujinomori Clay"; the ratio in mass of the former to the latter is 1:1.2. Particles larger than 74mm was sieved out from the original sample of Fujinomori Clay. The sample of lower plasticity is the fraction of Fujinomori Clay which passed a 420mm sieve.

Table 1: Consistency of samples

Sample	w_L (%)	w_P (%)	I_P
Higher plasticity	104.6	28.6	76.0
Lower plasticity	55.8	31.8	24.0

Each sample was remolded with distilled water at the water content twice as much as its liquid limit and consolidated in a one-dimensional consolidation mold, 25 cm in diameter and 20 cm in height. The maximum vertical pressure (p_0) in the consolidation is 49 kPa for the higher plasticity sample and 98 kPa for the lower plasticity sample.

Consolidated samples were divided to several small blocks which would be trimmed to be specimens for the unconfined compression test. A specimen is a cylinder of 3.5 cm in diameter and 8 cm in height.

Method

Unconfined compression tests with the measurement of suction were conducted. The difference of the pore air pressure (u_a) and the pore water pressure (u_w) is the suction (S). The suction was measured by the pressure plate technique, in which the pore water pressure (u_w), which is negative, is raised by applying a pressure in the pore air, so that u_w will be positive. The detail of the testing method was described in the previous paper (Shimizu and Tabuchi, 1993). Here the outline is presented.

Figure 1 shows the apparatus. The cell is a usual one used for the triaxial test. A ceramic disc plate is set onto the top surface of the pedestal. The cell pressure is applied by air pressure; it transmits, through an opening made in the loading cap, into the specimen's pore to be the pore air pressure (u_a). The pore air can not enter the ceramic plate as far as it is less than the air entry value (=200kPa) of the ceramic plate. Thus we can measure u_w separately from u_a.

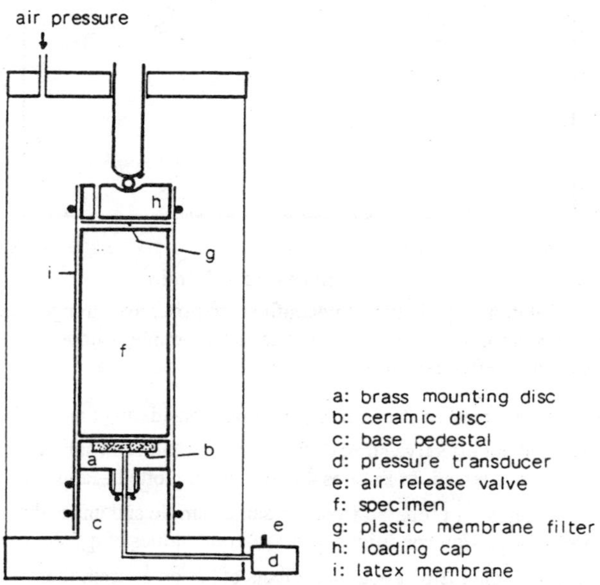

Figure 1: Apparatus used for unconfined compression tests with the measurement of suction.

The ceramic plate, the connection tube and the inside of the pressure transducer have to be saturated with water prior to tests. The saturation is very important to correctly measure the pore water pressure. The ceramic disc was de-aired by boiling; and the connection tube and the transducer were de-aired by flowing high pressure water.

We measured the initial suction (S_0) before vertical compression; S_0 is the suction that a specimen possesses just before the compression. We applied cell pressure, and therefore u_a, of a specified magnitude (=98kPa); then u_w increased with time. We determined S_0 as the difference between u_a and u_w when u_w becomes to be constant with time. The time necessary for u_w to be constant can be different from specimen to specimen; it is about two hours on average. When we apply the cell pressure, both u_a and u_w rise by the same amount equal to the applied cell pressure. This ensures that the initial suction of a specimen is not changed.

The vertical compression was done at a specified rate in the range of 0.01 to 1.0 (%/min). The measurement of u_w was done during the vertical compression, too.

For some specimens of the lower plasticity sample, the initial suction was intentionally lowered by letting them suck water before unconfined compression tests. This allows us to get results for a wide range of S_0.

RESULTS AND DISCUSSION

Effective stress behavior in unconfined compression tests

The definition of the effective stress for saturated soils was used, because all the specimens were almost fully saturated; the degree of saturation was more than 90% for all the specimens tested. Axial and radial effective stresses, σ'_1 and σ'_3, respectively, were determined by the following equations:

$$\sigma'_1 = q + u_a - u_w \quad (1)$$

$$\sigma'_3 = u_a - u_w \quad (2)$$

where q is the vertically applied load per unit area.

Figures 2(a), (b) and 3 show effective stress paths in unconfined compression tests.

Figure 2 (a) shows that, when the rate of compression is so low as 0.01%/min, the paths tend to end along a straight line. This implies that the definition of effective stress is valid.

On the other hand, Figure 2(b) shows that, when the rate is so high as 1.0%/min, we can not determine a straight line which may envelop all the paths. This does not deny the validity of the definition of the effective stress, but this implies that the pressure transducer incorrectly measure the pore water pressure. It will take finite time for the pore water pressure to transmit to the transducer, as it took rather long time to determine the initial suction, A possible reason why the transmission of the water pressure is not instantaneous is the insufficient degree of saturation in the ceramic plate itself and in the connection between the ceramic plate and the transducer. The method for raising the degree of saturation has to be improved.

Figure 3 shows the effective stress paths when the sample of lower plasticity was tested at a low rate of compression, 0.01%/min. The behavior is similar to that shown in Figure 2(a). We note here that the behavior of specimens that were let suck water prior to the compression is not peculiar.

We can observe, in Figures 2(a) and 3, that effective stress paths, including those at failure, depend on the values of effective stress at the beginning of shear, which are equal to the values of S_0. The Author showed in the previous study (Shimizu and Tabuchi, 1993) that effective stress paths for the specimens having the same value of the disturbance ratio r_d, defined as p_0/S_0, are similar even if they have different values of S_0 or p_0.

The disturbance ratio corresponds to the OCR, and therefore the finding above corresponds to the fact that effective stress paths in CU triaxial compression tests on saturated clay specimens having the same

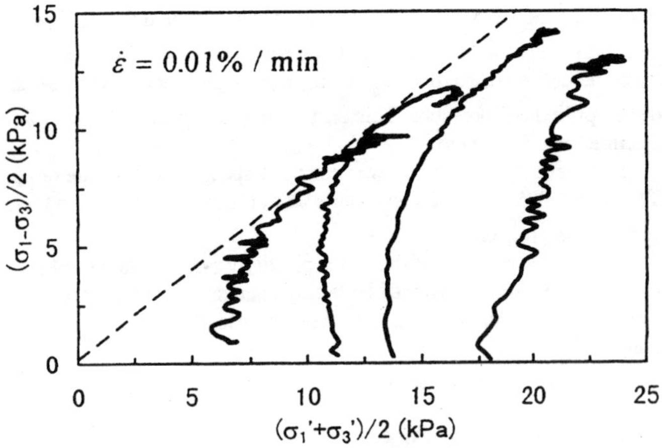

(a)

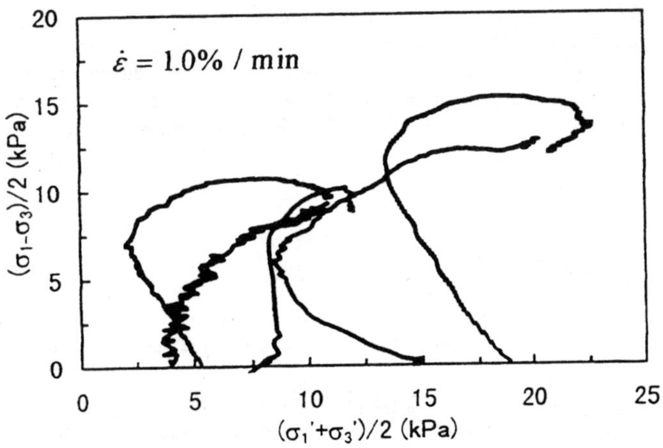

(b)

Figure 2: Effective stress paths during unconfined compression tests on the higher plasticity sample.

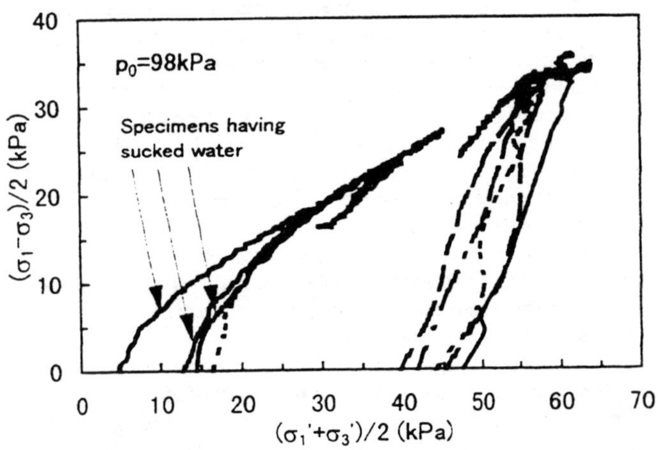

Figure 3: Effective stress paths during unconfined compression tests on the lower plasticity sample.

values of OCR are similar even if the maximum values of past consolidation pressure are different.

Effects of the rate of compression

At first, we examine the effects of the rate of compression ($\dot{\varepsilon}$) on the

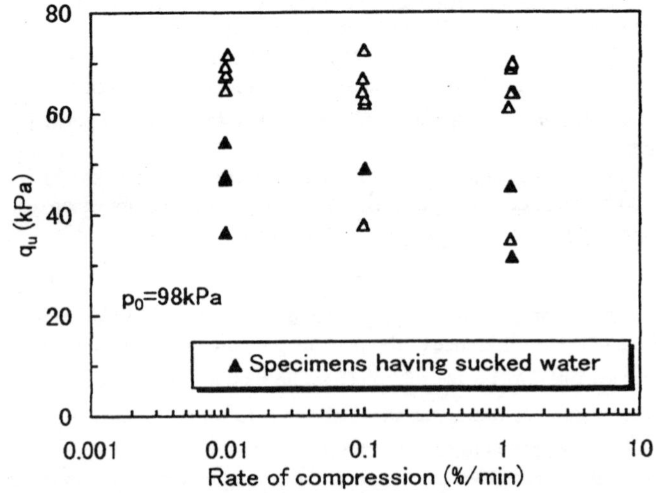

Figure 4: Relationships between unconfined compressive strength and the rate of compression for the lower plasticity sample; without consideration of effective stresses.

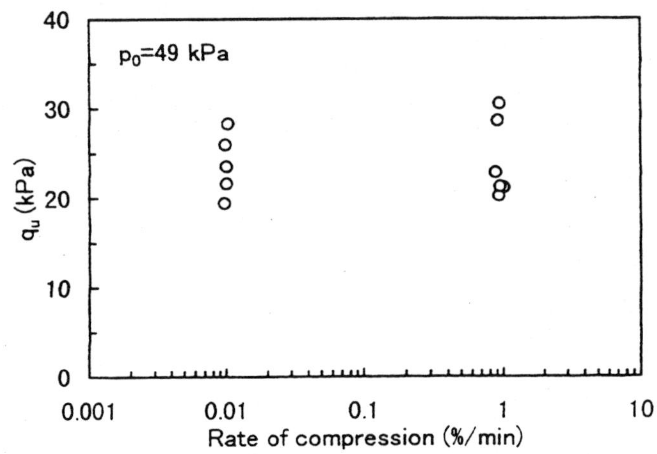

Figure 5: Relationships between unconfined compressive strength and the rate of compression for the higher plasticity sample; without consideration of effective stresses.

unconfined compressive strength (q_u) without considering the information about effective stresses that samples had been subjected to before they were tested. In Figures 4 and 5, q_u is plotted against $\dot{\varepsilon}$. Specimens that were prepared from the same sample and under the same preconsolidation conditions have different values of q_u. We can not derive any conclusion about the effects of $\dot{\varepsilon}$ on q_u from these figures, in which no information about effective stresses is considered.

As stated in the previous section, the behavior of effective stresses in unconfined compression tests on highly saturated clay samples is analogous to the behavior in triaxial compression tests on saturated clay samples that were overconsolidated. Furthermore, as known well, the mechanical behavior of overconsolidated saturated clays can be characterized with the OCR. These facts make us suppose that the effects of $\dot{\varepsilon}$ on q_u can be examined by taking account of the parameters p_0 and S_0, which express the history of effective stresses of specimens.

In figures 6 and 7, $(q_u/2)/S_0$ is plotted against $r_d(=p_0/S_0)$ in the logarithmic scale for the low and high plasticity samples, respectively.

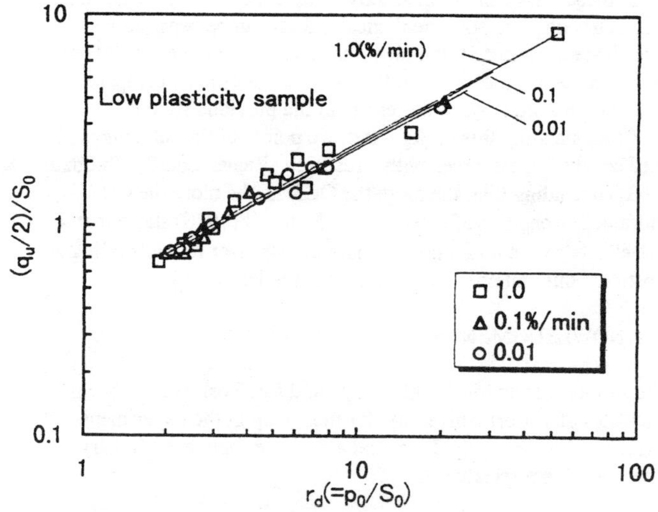

Figure 6: Unconfined compressive strength normalized by the initial suction versus disturbance ratio for the lower plasticity sample.

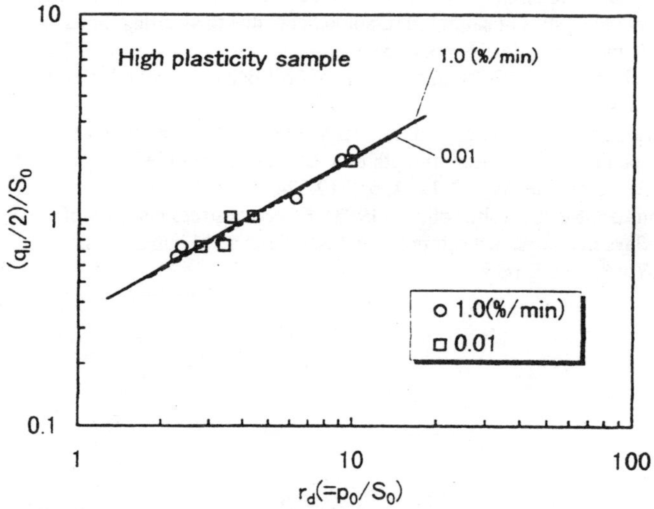

Figure 7: Unconfined compressive strength normalized by the initial suction versus disturbance ratio for the higher plasticity sample.

We can observe that a linear relationship for each value of $\dot{\varepsilon}$. These relationships correspond to the finding by Mitachi et al.(1976) that $\log(s_u/p_c)$ can be a linear function of log(OCR), where s_u is undrained strength and p_c is the consolidation effective stress in CU triaxial compression tests.

The straight lines drawn in the figures are regression lines determined using the least square method. They are expressed in the mathematical form as:

$$\frac{q_u/2}{S_0} = r_d{}^{\alpha} \cdot \beta , \tag{3}$$

where α and β are constants specifying the slope and the position of the lines, respectively. The values of α and β seem to vary with $\dot{\varepsilon}$ (Fig.8). For both samples, it seems that α depends on $\dot{\varepsilon}$ less than β does. And therefore, we assume that . α can be regarded as constant ~~but~~ β depends on $\dot{\varepsilon}$. The lines locate upwards with higher values of $\dot{\varepsilon}$.

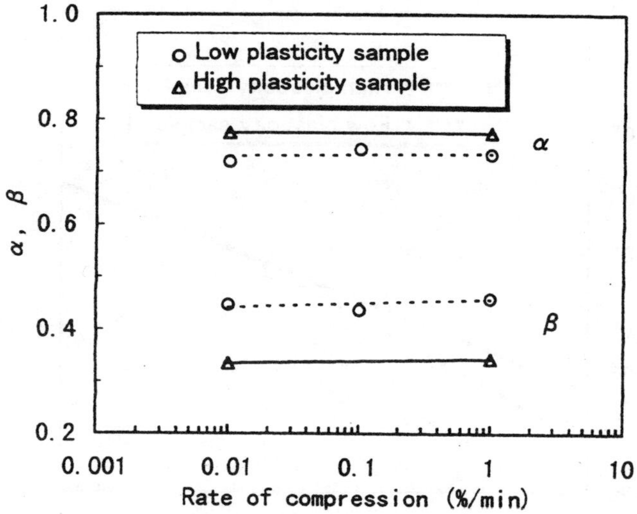

Figure 8: Relationships of coefficients, a and b, to the rate of compression.

From Equation (3), we can derive an equation expressing the effects of $\dot{\varepsilon}$ on q_u. For specimens having the same values of effective stress parameters, p_0 and S_0, Equation (3) can be modified to:

$$\frac{q_u}{q_{u0}} = r_d{}^{\alpha-\alpha_0} \cdot \left(\frac{\beta}{\beta_0}\right) , \tag{4}$$

where the quantities with the suffix "0" means those when $\dot{\varepsilon}$ =1.0%/min. As we can assume that $\alpha - \alpha_0 \cong 0$, Equation (4) reduces to

$$\frac{q_u}{q_{u0}} = \frac{\beta}{\beta_0} \tag{5}$$

We can determine in Figure 8 a regression line for the relationship between β and $\dot{\varepsilon}$. By substituting the expression for b in Equation (5), we can evaluate the relationship between q_u and $\dot{\varepsilon}$. The results are presented in Figure 9.

We can see in Figure 9 that q_u decreases with decreasing $\dot{\varepsilon}$, and the change in q_u is at most 10% for the low plasticity sample and 8% for the high plasticity sample in the examined range of $\dot{\varepsilon}$. For the practical range , say 0.1 to 2%/min, the change in q_u is only a few percents.

CONCLUSIONS

The findings in this study can be summarized as follows:
(1) Without the information about the effective stress, the unconfined compressive strength (q_u) scatters in relation to the rate of compression ($\dot{\varepsilon}$);
(2) With the consideration of the maximum preconsolidation pressure (p_0) and the initial suction (S_0), the effects of $\dot{\varepsilon}$ on q_u can be expressed by Equation (4) or Equation (5);
(3) For the samples used, the relationship between q_u and $\dot{\varepsilon}$ is independent of the disturbance ratio, $r_d(=p_0/S_0)$;
(4) For the samples having the same values of p_0 and S_0, q_u varies only

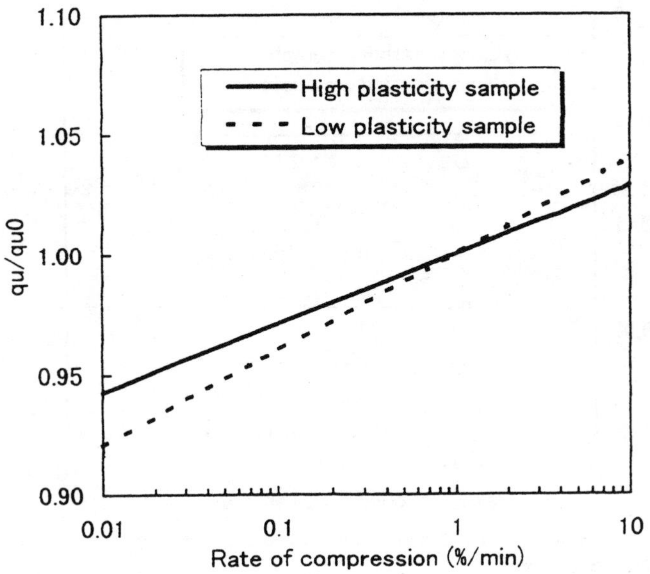

Figure 9: Relationships between the unconfined compressive strength and the rate of compression for specimens having the same effective stress conditions..

a few per cent with $\dot{\varepsilon}$ being changed in practically possible range, e.g., from 0.01 to 2%/min.
(5) The sample of lower plasticity tends to be affected more by the rate of compression than the sample of higher plasticity.

The findings above have to be interpreted by taking into account that we can not strictly control the drainage conditions in unconfined compression tests. Thus, the effects of $\dot{\varepsilon}$ on q_u reflect both the rheological effects and the effects of drainage conditions.

In those tests which can strictly control the drainage conditions, e.g., in triaxial tests, the rheological effects can be separately investigated. Almost all the previous researches on the rate effects in strengths use such tests. Therefore we can not directly compare the results in this study with the results in the previous studies.

Understanding this background, we point out the followings: the third conclusion disagrees with a research (Richardson & Whitman, 1963) concluding that, the more the OCR is, the more the CU undrained strength is affected by $\dot{\varepsilon}$. Mitchell (1976) states that the rate effects on the undrained strength are stronger for soils of higher plasticity; this statement disagrees with the last conclusion.

ACKNOWLEDGMENTS

The Author thanks Mr. Ryou Takada and Mr. Toshiyuki Tabuchi, graduates of Tottori University, for their help in the experiments conducted for this study. Financial aid by the Japanese Ministry of Education is appreciated.

REFERENCES

Mitachi, T. and Kitago, S. (1976): Change in undrained shear strength characteristics of saturated remolded clay due to swelling, Soils and Foundations, Vol.16, pp.45-58.
Mitchell, J. K. (1976):Fundamentals of Soil Behavior, John Wiley & Sons, Inc.
Richardson, A. M. and Whitman, R. V. (1963):Effect of strain rate upon undrained shear strength of a saturated remoulded fat clay, Geotechnique, Vol.13, No.4, pp.310-324.
Shimizu, M. and Tabucxhi, T. (1993): Effective stress behavior of clays in unconfined compression tests, Soils and Foundations, Vol.33, No.3, pp.28-39.

Proceedings of the Seventh (1997) International Offshore and Polar Engineering Conference
Honolulu, USA, May 25-30, 1997

ISBN 1-880653-28-1 (Set); ISBN 1-880653-29-X (Vol. I)

Estimation of *In-situ* Consolidation Parameters and Prediction of Settlement Behavior of Embankment

Takaharu Shogaki, Yoshikazu Maruyama and Keiji Kogure
National Defense Academy
Yokosuka, Japan

ABSTRACT

A method for estimating the preconsolidation pressure (σ'_p), the compression index (C_c), the e-log p curve, and the coefficient of consolidation (c_v) values for the *in-situ* soil was proposed using the plasticity index (I_p), the volumetric strain (ε_{vo}) at *in-situ* effective overburden pressure (σ'_{vo}) determined from the 24-hour e-log p curve, σ'_p, C_c and c_v for disturbed samples determined in the standard oedometer test (Shogaki, 1996).

In-situ consolidation parameters for Iwakuni clay in Japan were investigated in terms of the ε_{vo} value. The prediction of settlement behavior of the test embankment at the Iwakuni replacement airport was performed on the estimated *in-situ* consolidation parameters and the consolidation parameters obtained by the standard oedometer test.

KEY WORDS: consolidation parameter, holocene clay, sample disturbance, settlement analysis, standard oedometer test

INTRODUCTION

Various characterizations of sample disturbance in terms of undrained shear strength (s_u) have been performed by many researchers such as Skempton and Sowa (1963), Okumura (1974a), Nakase et al. (1985), Matsuo and Shogaki (1988), Shimizu and Tabuchi (1993) and Shogaki (1995). These studies focused on the direct application of the s_u value to the short term stability problem of a clay foundation under undrained conditions. When the preconsolidation pressure (σ'_p) is adopted for estimating the mobilized undrained shear strength (e.g., Ladd and Foott, 1974 ; Mesri, 1989), the effects of sample disturbance on both undrained shear strength and consolidation parameters should be established.

A guideline for the quality of specimen on sample disturbance based on the results of consolidation tests was proposed by Andresen and Kolstad (1977), and Lacasse and Berre (1988). The effects of sample disturbance on consolidation parameters, however, can not be determined based on these guidelines. A method for correcting consolidation parameters for sample disturbance using volumetric strain (ε_{vo}) measured in standard oedometer tests was proposed by Shogaki (1996).

The settlement behavior of the test embankment at the Iwakuni replacement airport has been measured for more than 8 years (Hiroshima Defense Administration Bureau, 1994). In this paper, the effect of sample disturbance on the consolidation parameters is investigated through a series of laboratory tests for the clays sampled from the Iwakuni sites, and the relationship between the ε_{vo} values and the corrected values of the *in-situ* consolidation parameters are discussed. The prediction of settlement behavior of the test embankment at the Iwakuni replacement airport is performed on the estimated *in-situ* consolidation parameters and the consolidation parameters obtained by standard oedometer test.

SOIL SAMPLES AND TEST PROCEDURES

The undisturbed soil samples used in this study were obtained from holocene marine clays located offshore at seven different sites at the Iwakuni replacement airport. Field sampling was performed with a fixed piston sampler with an inner diameter of 75 mm to enhance the quality of the samples. The index properties of these soils are shown in Table 1. The plasticity index (I_p) and the unconfined compressive strength (q_u) values are in the range of 30 to 72 and 89 kPa to 163 kPa, respectively. It was confirmed that the index properties of these soils are similar to those of the test embankment.

The equipment used to cause disturbance to the soil samples, together with the procedure, is shown in Fig.1 (Shogaki and Kaneko, 1994). As shown in Fig.1(a), the ratio (R_a) of the cross sectional area (A_E) of a device with an inner diameter of the sampling tube changed to values of 1.0, 0.95, 0.90, 0.80 and 0.70. These samples were designated S_1, S_2, S_3,

Table 1 Index properties of clays

Sites	w_n (%)	w_L (%)	I_p	σ'_{vo} (kPa)	σ'_p/σ'_{vo}	q_u (kPa)
R6-1	51	61	30	61	2.63	89
R6-2	82	92	58	134	1.12	151
R11	95	110	72	66	1.82	118
R16	80	88	52	140	1.21	86
R81	58	81	59	117	1.14	138
R6-3	63	68	37	165	1.02	143
R6-4	70	78	49	182	1.13	163

w_n : Natural water content

σ'_{vo} : Effective overburden pressure

σ'_p : Preconsolidation pressure

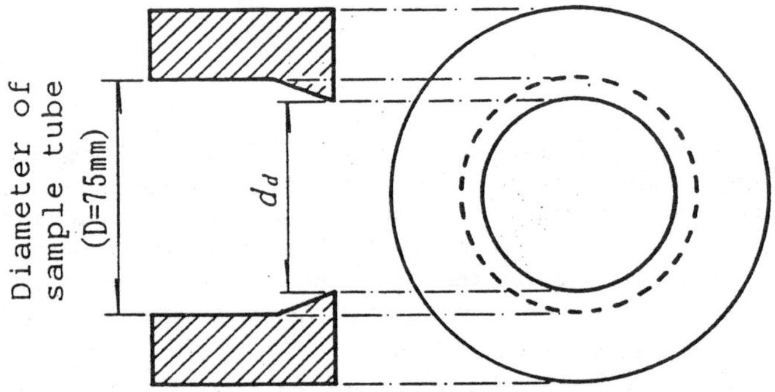

Sample	d_d (cm)	A_R (cm²)	R_a
S_1	7.50	44.18	1.00
S_2	7.31	41.95	0.95
S_3	7.12	39.86	0.90
S_4	6.72	35.47	0.80
S_5	6.30	31.44	0.70
S_6	Remolded soil		

a) Disturbing equipment

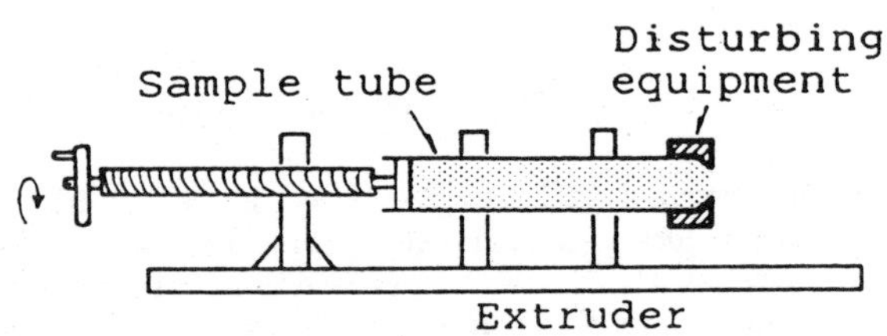

b) Procedure of disturbing

Fig.1 Equipment and procedure for disturbing the soil samples (after Shogaki and Kaneko, 1994)

S_4 and S_5, respectively. Sample S_1 is the original *in-situ* sample undisturbed in the laboratory and sample S_6 is completely remolded. Samples with a length of 80 mm were extruded from the end of edge of the sampling tube, in the order S_1, S_2 etc. It had been previously confirmed that the strength and deformation characteristics of the sample disturbed by this method are similar to those of a sample disturbed by *in-situ* sampling and preparation for laboratory tests (Shogaki et al., 1996).

The specimens used for the oedometer test were 60 mm in diameter and 20 mm in height. S-specimens for the unconfined compression test were 15 mm in diameter and 35 mm in height.

There are four to seven specimens for the unconfined compression test and one for the oedometer test. The q_u value is the mean value of q_u for the unconfined compression test specimens. The oedometer tests were performed using a load increment ratio of unity and the duration of loading for each load increment was one day. The values of c_v were determined according to Taylor's method (Taylor, 1948). The values of C_c and σ'_p were determined from the e-log p curves corresponding to 24-hour compression. Mikasa's method was used to identify preconsolidation pressure on the 24-hour e-log p curve. These methods have been employed in Japan as the Japanese industrial standard for determining one-dimensional consolidation properties of soils (JIS A 1217-1990). It was shown by Okumura (1974b) that there was no difference in the σ'_p values between the Mikasa method and the Casagrande method (1936).

In the unconfined compression tests, the specimens were sheared at a strain rate of 1 %/min. The value of q_u was determined to be the maximum stress corresponding to axial strain (ε_a) of less than 15 %.

EFFECTS OF SAMPLE DISTURBANCE ON STRENGTH AND CONSOLIDATION PROPERTIES

The stress-strain curves for the specimens of soil R16 are shown in Fig.2. The initial tangent modulus became small and the strain at failure increased as the degree of sample disturbance was increased. The void ratio of the soil R16 is plotted against the logarithm of consolidation pressure in Fig.3, and the σ'_p, C_c, q_u and ε_{vo} values of each specimen are also given in Table 2. When soils are greatly disturbed, the σ'_p, C_c, and q_u values become smaller, and the ε_{vo} value becomes larger. The ε_{vo} value is defined by Shogaki (1996) as

$$\varepsilon_{vo} = \frac{e_0 - e_1}{1 + e_0} \times 100 \quad (\%) \qquad (1)$$

where the e_0 is the initial void ratio of the sample. In the *in-situ* soil under the σ'_{vo} value, there was no sample disturbance, therefore, the ε_{vo} value is 0 since the e_1 value is equal to the e_0 value. The e_1 value decreases with sample disturbance because the e value decreases with sample disturbance under the same σ'_{vo} value (Schmertman, 1955; Shogaki and Kaneko, 1994), and the ε_{vo} value becomes larger. Therefore, the ε_{vo} value can be employed as an index to

The values of c_v and m_v of soil R16 are plotted against the logarithm of the mean values of consolidation pressure (p) in Figs.4 and 5, respectively. It is clear from the figures that when soils are disturbed, the c_v value is smaller, and the m_v value is larger. The effect of sample disturbance is typically positive in the overconsolidated region. These tendencies are similar to those of other clay (Shogaki and Kaneko, 1994).

The relationship between the value of ε_{vo} and the overconsolidation ratio (OCR) for S_1 samples is shown in Fig.6. The range of the two curves and other plots besides the Iwakuni clay in Fig.6 refer to the results obtained from the samples for other natural deposits (Shogaki, 1996). The results presented in this study have the same tendency as the range of the two curves shown in Fig.6. It can be judged from Fig.6 that the Iwakuni clays used in this study, therefore, have common quality for the sample disturbance.

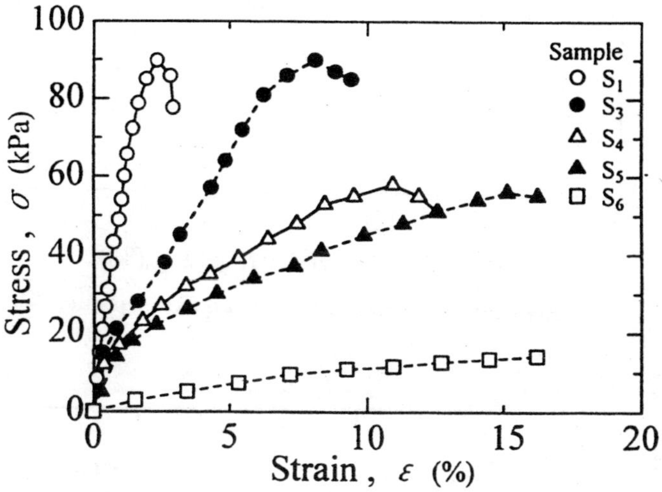

Fig.2 Stress-strain curves (R16 sample)

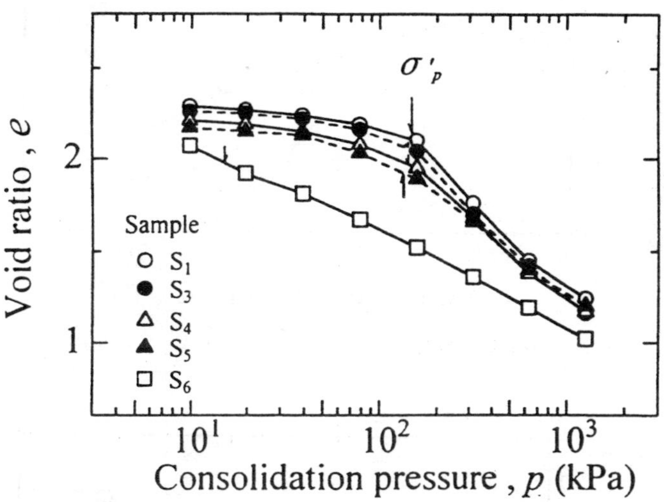

Fig.3 Relationship between e and p (R16 sample)

Table 2 Test results of each specimen (R16 sample)

Samples	σ'_p (kPa)	C_c	q_u (kPa)	ε_{vo} (%)
S_1	150	1.16	92	4.4
S_3	140	1.09	90	6.6
S_4	140	1.00	58	10.1
S_5	120	0.88	56	15.9
S_6	15	0.57	14	20.1

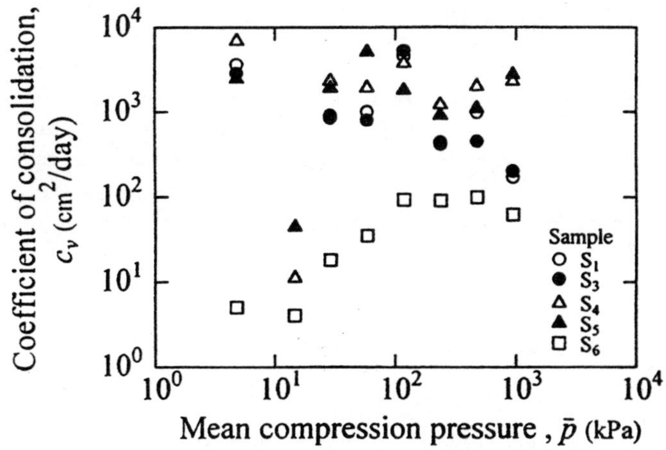

Fig.4 Relationship between c_v and $\bar{p}$ (R16 sample)

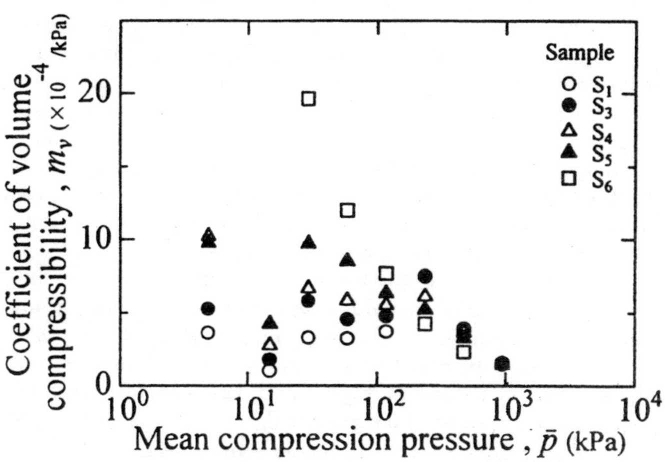

Fig.5 Relationship between m_v and $\bar{p}$ (R16 sample)

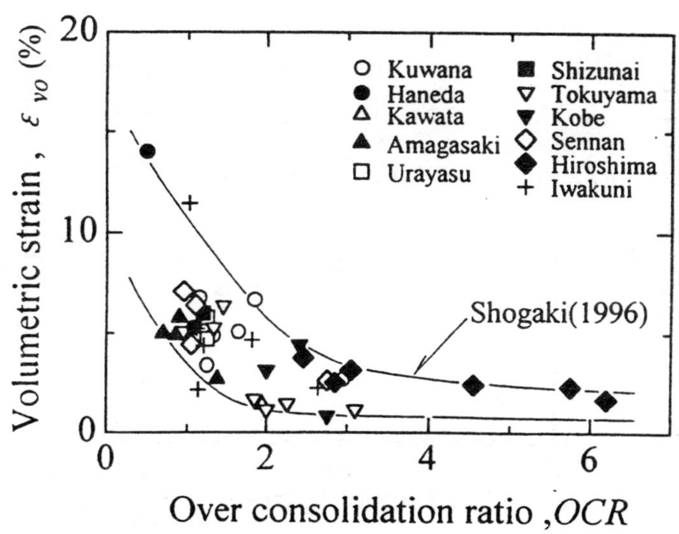

Fig.6

ESTIMATION OF IN-SITU CONSOLIDATION PARAMETERS

Figure 7 shows the relationship between the σ'_p values and ε_{vo} values for the R16 clay. When the ε_{vo} value increases with sample disturbance, the σ'_p value decreases because the degree of rigidity against compression decreases with sample disturbance. According to this extrapolation in Fig.7, the σ'_p value of *in-situ* S_0 soil ($\sigma'_{p(I)}$) for the R16 clay is 169 kPa since ε_{vo} value of the *in-situ* soil is 0 as described in the explanation of Eq.(1).

The relationships between the values of C_c, c_v, m_v, and the ε_{vo} value for the R16 clay are also shown in Figs.8, 9 and 10, respectively. These values of C_c, c_v and m_v are the mean value of the consolidation pressure range from σ'_p to $2\sigma'_p$. When the ε_{vo} value increases with sample disturbance, the C_c and c_v values decrease, and the m_v value increases. The *in-situ* C_c, c_v and m_v values can be decided as $C_{c(I)} = 1.28$, $c_{v(I)} = 1790$ cm^2/day and $m_{v(I)} = 3.21\times10^{-4}$ 1/kPa from Figs.8, 9 and 10, respectively, since ε_{vo} value of the *in-situ* soil is 0.

For the R16 sample, the void ratios are plotted against the logarithm of consolidation pressure in Fig.11. Figure 11 also shows the e-log p curve of the *in-situ* condition using the $\sigma'_{p(I)}$ and the $C_{c(I)}$ values given in Figs.7 and 8, respectively. The e_0 value of the *in-situ* condition was assumed to be that of the sample S_1. This assumption is generally accepted for saturated clays. It can be judged by comparison with the five different measured e-log p curves that the estimated e-log p curve of the *in-situ* soil is a reasonable curve.

Figures 12 and 13 show the relationships between $\eta_3 = c_{v(I)}/c_v$ and $\eta_4 = m_{v(I)}/m_v$ values, and the ε_{vo} values for all Iwakuni clay samples. It is well known that the c_v value increases with p in both the recompression and compression ranges. The c_v values in Figs.9 and 12, therefore, are the mean values of c_v to p values greater than σ'_p for each sample. The c_v value decreases with an increase in ε_{vo} value, since the c_v value also decreases with an increase in sample disturbance (Shogaki and Kaneko, 1994). The $c_{v(I)}$ and $m_{v(I)}$ values for the soil of the *in-situ* can be estimated from the ε_{vo}, c_v and m_v values of tested specimens.

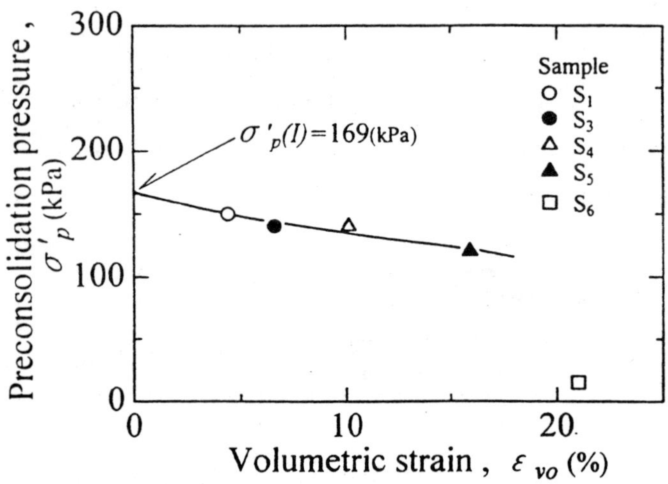

Fig.7 Relationship between σ'_p and ε_{vo} (R16 sample)

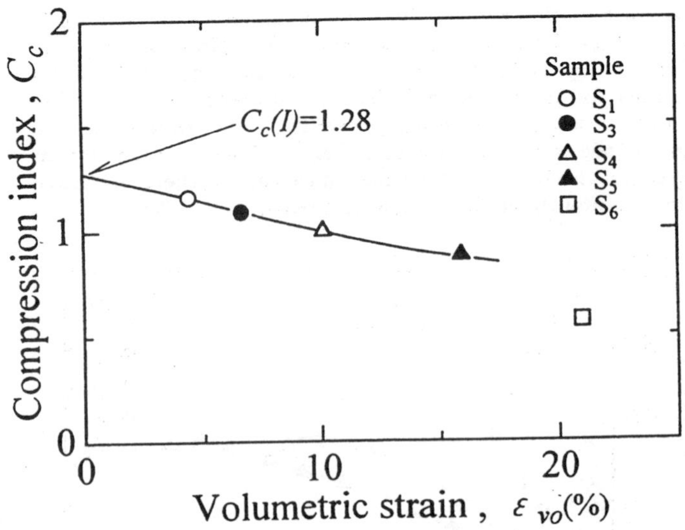

Fig.8 Relationship between C_c and ε_{vo} (R16 sample)

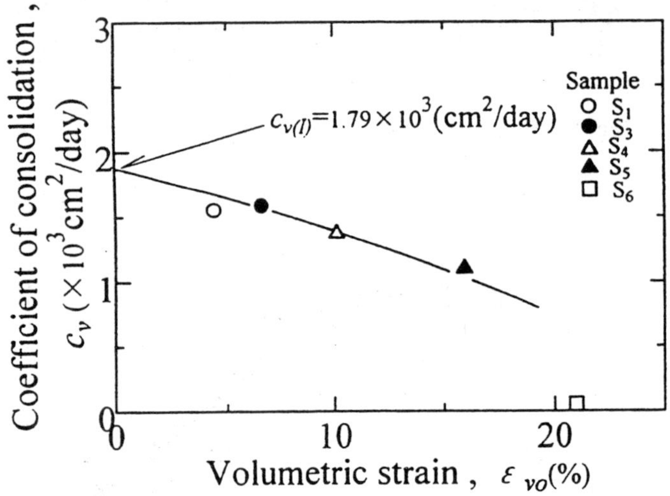

Fig.9 Relationship between c_v and ε_{vo} (R16 sample)

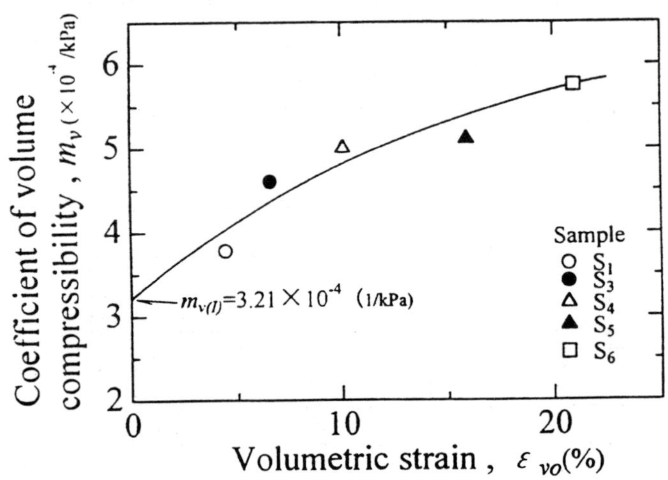

Fig.10 Relationship between m_v and ε_{vo} (R16 sample)

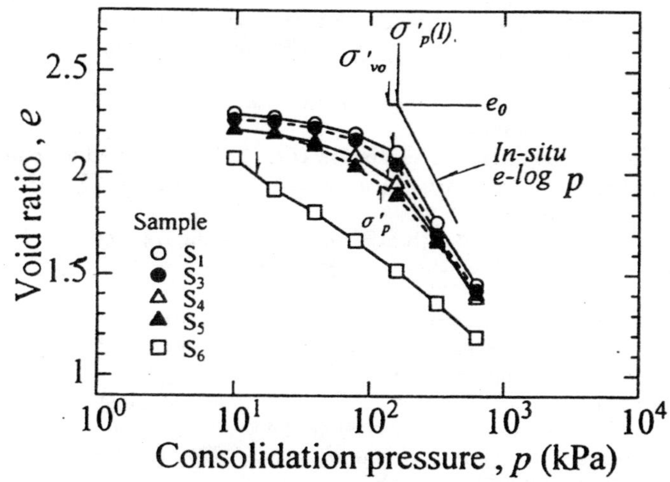

Fig.11 Relationship between e and p (R16 sample)

ESTIMATION OF $c_{v(I)}$ and $m_{v(I)}$, AND PREDICTION OF SETTLEMENT BEHAVIOR OF EMBANKMENT

The q_u, strain at failure (ε_f) of the unconfined compression test, and ε_{vo} values of the samples obtained from four different sites at test embankment are plotted against the depth (z) in Fig.14. The q_u value decreases and the ε_f value increases when the degree of sample disturbance is increased. In Fig.14, there is a tendency that the plots of the large ε_f value also have large ε_{vo} value, and this tendency is coincident with the ε_{vo} value increasing as the degree of sample disturbance was increased, as shown in Table 2.

Figure 15 shows the relationships between the values of m_v and c_v, and z for the E4 samples. The $c_{v(I)}$ and $m_{v(I)}$ values are obtained from the ε_{vo} values of each specimen in Figs.9 and 10, respectively. It can be seen in Fig.15 that the c_v value is about 30% less than the *in-situ* c_v value, and the m_v value is about 15% greater than the *in-situ* m_v value.

Figure 16 shows the relationships between the height (h)

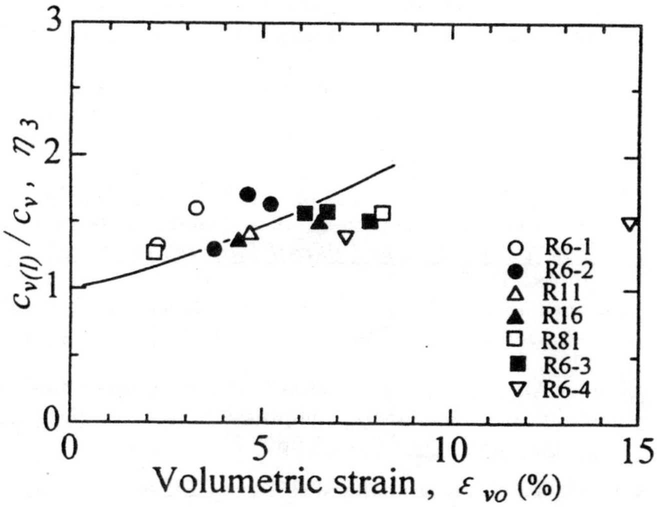

Fig.12 Relationship between η_3 and ε_{vo} (R16 sample)

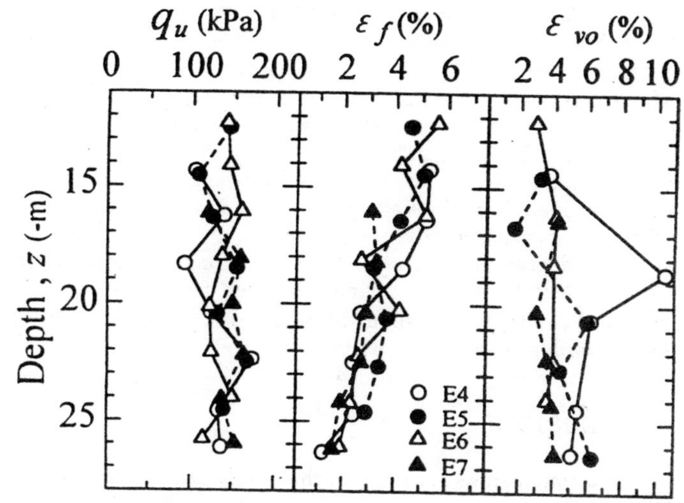

Fig.14 Relationships between z, and q_u, ε_f and ε_{vo} (Before fill)

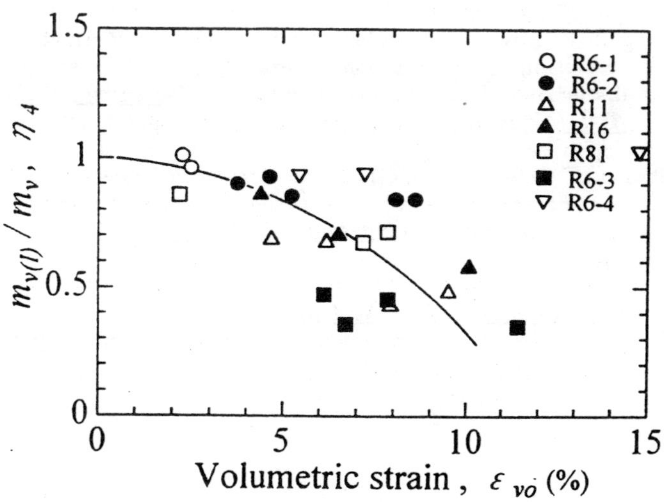

Fig.13 Relationship between η_4 and ε_{vo} (R16 sample)

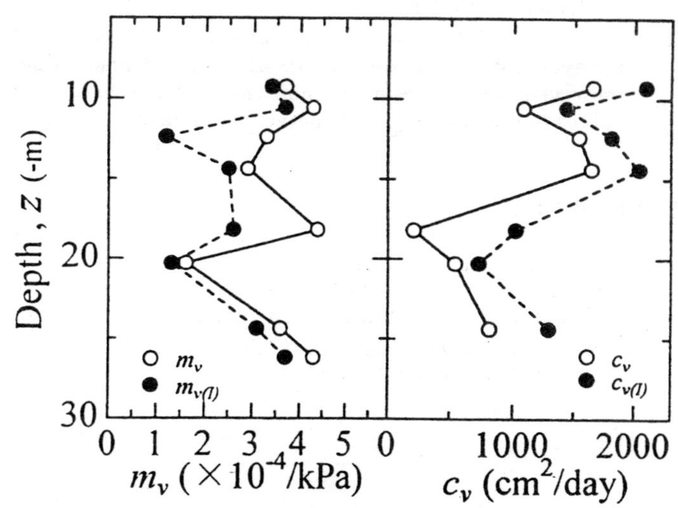

Fig.15 Relationships between z, and m_v and c_v (E4 site)

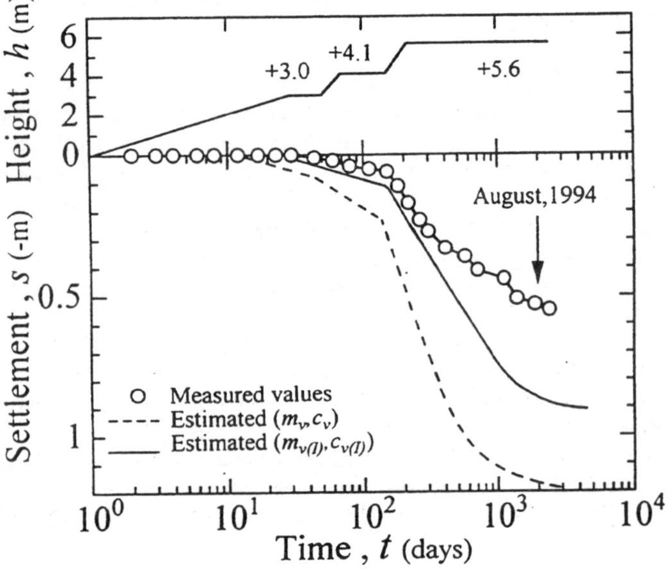

Fig.16 Relationships between h, s and t (E4 site)

of embankment, consolidation settlement (s) of ground under the embankment observed at the site of E4, and the logarithm of time (t). The solid and dotted lines in Fig.16 represent the relationships between the estimated settlement before fill and time obtained from the values of $m_{v(I)}$ and $c_{v(I)}$, and the values of m_v and c_v, respectively.

The degree of accuracy of the relationship between the estimated settlement and time obtained from the values of $m_{v(I)}$ and $c_{v(I)}$ improved about 50 % more than those of the m_v and c_v values of the test results.

CONCLUSIONS

The conclusions obtained in this study are summarized as follows:

1) The validity of the method for the estimation of the *in-situ* consolidation parameters is confirmed from the case history for the prediction of the actual consolidation settlement behavior.
2) The degree of accuracy of the consolidation settlement behavior estimated from the *in-situ* $c_{v(I)}$ and $m_{v(I)}$ values improved about 50% more than those of the m_v and c_v values of the standard oedometer test results.

REFERENCES

Andresen.A.AA.. and Kolstad.P.(1977): "The NGI 54-mm sampler for undisturbed sampling of clays and representative sampling of coarser materials", *Proceedings of the 1st International Symposium of Soil Sampling*. Singapore, pp.13-21.

Casagrande,A(1936): "The determination of the preconsolidation load and its practical significance", *Proceedings of the 1st ICSMFE*, Cambridge, Mass., Vol.1, pp.60-64.

Hiroshima Defense Administration Bureau (1994): "Report of Iwakuni air port."

Japanese Standards Association (1993): "Test method for one-dimensional consolidation properties of soils", *JIS A1217-1990*, pp.1-13.

Ladd,C.C. and Foott,R.(1974): "New design procedure for stability of soft clays", *Proceeding of ASCE. Vol.100. No.GT7*, pp.763-786.

Lacasse,S. and Berre,T.(1988): "Triaxial testing methods for soils, Advanced triaxial testing of soil and rock", *ASTM ATP 977*, pp.264-289.

Matsuo,M. and Shogaki,T.(1988): "Effects of plasticity and disturbance on statistical properties of undrained shear strength", *Soils and Foundations. Vol.28. No.2*, pp.14-24.

Mesri,G.(1989): "A reevaluation of s_u(mob) = 0.22 σ'_p using laboratory shear tests", *Canadian Geotechnical Journal. Vol.26. No.1*, pp.162-164.

Nakase,A., Kusakabe,O. and Nomura,H.(1985): "A method for correcting undrained shear strength for sample disturbance", *Soils and Foundations. Vol.25. No.1*, pp.52-66.

Okumura,T.(1974a): "The variation of mechanical properties of clay samples depending on its degree of disturbance", *Proceedings of Speciality Session. 4th Asian Conference. ISSMFE*, pp.73-81.

Okumura,T.(1974b): "Studies on the disturbance of clay soils and the improvement of their sampling techniques", *Report of the Port and Harbor Research Institute. Vol.10. No.193*, P.45 (in Japanese).

Schmertman,J.T.(1955): "The undisturbed consolidation behavior of clay", *Transactions. ASCE. Vol.120*, pp.1201-1233.

Shimizu,M. and Tabuchi,T.(1993): "Effective stress behavior of clays in unconfined compression tests", *Soils and Foundations. Vol.33. No.3*, pp28-39.

Shogaki,T. and Kaneko,M.(1994): "Effects of sample disturbance on strength and consolidation parameters of soft clay", *Soils and Foundations. Vol.34. No.3*, pp.1-10.

Shogaki,T.(1995): "Effective stress behavior of clays in unconfined compression tests", *Soils and Foundations. Vol.35. No.1*, pp.169-171.

Shogaki,T., Miwa,K. and Maruyama,Y.(1996): "A method for correcting consolidation parameters for sample disturbance using disturbed sample", *Proceedings of the 31st National Conference on JGS*, pp.645-646 (in Japanese).

Shogaki,T.(1996): "A method for correcting consolidation parameters for sample disturbance using volumetric strain", *Soils and Foundations. Vol.36. No.3*, pp.123-131.

Skempton,A.W. and Sowa,V.A.(1963): "The behavior of saturated clays during sampling and testing", *Geotechnique. Vol.13. No.4*, pp.269-290.

Taylor,D.W.(1948): "Foundamentals of soil mechanics. *John Wiley & Sons*, pp.208-249.

Proceedings of the Seventh (1997) International Offshore and Polar Engineering Conference
Honolulu, USA, May 25-30, 1997

ISBN 1-880653-28-1 (Set); ISBN 1-880653-29-X (Vol. I)

Dynamic Response of Caisson with Suction and Its Foundation Due to Wave

Tomiya Takatani
Maizuru National College of Technology, Kyoto, Japan
Yoshi-hiko Maeno
Nihon University, Tokyo, Japan
Shigeo Takahashi and Ken-ichiro Shimosako
Ministry of Transport, Yokosuka, Japan

ABSTRACT

To apply a steel caisson with skirt, which takes suction effect into the stability, to the coastal structure on soft seabed, dynamic response of the caisson structure with suction and its soft foundation under wave attack was simulated by an advanced caisson structure-seabed interaction analysis. Dynamic response of a caisson structure with suction and seabed around it was numerically evaluated for wave pressures acting on the front surface of the caisson, which were measured under the irregular wave in the experimental basin. The effects of the skirt length and suction on the dynamic response of caisson structure foundation were numerically investigated for some material properties of seabed foundation.

KEY WORDS : *Caisson-seabed interaction, Suction, Poroelastic seabed, Irregular wave, Pore water pressure*

INTRODUCTION

The offshore structure foundation plays a very important role for the stability of offshore structures installed in a great water depth. Judging from the fact that the reclamation works recently have developed from seacoast to offshore, the installation water depth of offshore structure becomes larger, and also it is not so easy to sufficiently secure the supporting force of seabed ground for the installation of offshore structure as before. While, severe sea conditions on site require the construction schedule to be shorten and easy. As described above, the construction of maritime structures has been developed from sites on rigid bed such as rock-base or sand layer in a shallow water to those on soft bed in a deep water. The demands mentioned above motivate to apply the suction effect to stabilize offshore structures, and actually it has been applied in the North Sea (Tsuzuki et al., 1995). In practice, the suction is effective not only for the increase of the penetration force in the installation of an offshore structure but also for the resistance to a pull-out force due to the interruption of atmospheric pressure after installation. Considering that the use of suction makes the construction of offshore structures easy, repeatedly available and effective on soft foundation, Maeno et al.(1993, 1994) have already investigated the feasibility of suction to a mooring system for floating structures in the ocean. A series of these researches reported that the suction effect is recognized in other types of mooring anchors such as a sinker, an end bearing anchor plate and so on.

To examine the feasibility of a steel caisson as the closing dike using for the reclamation works, Takatani et al.(1995a, 1996) have evaluated the dynamic response of a caisson structure with suction by using the research results obtained by Maeno et al.(1993, 1994). The dynamic response of a caisson structure with suction acting between the bottom of a caisson structure and seabed has been already discussed from the view points of the stiffness and the permeability of seabed, the skirt length of a caisson structure and suction ratio. Takatani et al.(1995b, 1996) obtained the following conclusions: (1)The larger suction acting on the edge of the base of caisson, the more stable against wave attack; (2)As the permeability of seabed decreases and its stiffness increases, the displacement response of a caisson structure decreases and its stability increases. In these researches mentioned above, the steel caisson is a structure with a skirt to get the suction effect, and the use of suction makes the installation of it easier on soft seabed and also has a feasibility of extension of its development toward offshore in a deep water.

In order to ensure the feasibility of a caisson structure with suction, the further examination on dynamic response of a caisson structure with suction and seabed around it under storm waves can be considered to be necessary. In this paper, dynamic response of a steel caisson with suction and the seabed is analyzed in the frequency domain by means of an advanced caisson structure-seabed interaction analysis, and also the effects of the permeability of seabed, skirt length and distribution shape of suction acting on compartments on the dynamic response of seabed around a steel caisson with suction are numerically investigated for some material properties of seabed foundation. Based on the frequency response, not only the displacement response of a caisson structure with suction and seabed but also the response of pore water pressure in the seabed is numerically evaluated in the time domain for the time series data of wave pressures acting on the front surface of a caisson, which were measured under the irregular wave with the conditions of the maximum wave height H_{max}=33.3cm, the significant wave height $H_{1/3}$=18.0cm and the period T=3.04s in the experimental basin in the Port and Harbour Research Institute, the Ministry of Transport, Japan.

ANALYTICAL METHOD ON FREQUENCY RESPONSES OF CAISSON AND SEABED

Based on a caisson structure-seabed interaction analysis proposed by Takatani et al.(1995a), an advanced analytical method of the displacement, stress, and pore water pressure in the seabed around a caisson structure with suction is mentioned in this section. The seabed is assumed to be a fluid-saturated poroelastic medium proposed by Biot(1956) which is consist of both soil particle skeleton and pore water. The stability of a caisson structure against wave attack is discussed from a view point of the response of seabed around it. In particular, the effects of the permeability of seabed, skirt length and suction acting on each compartment on the caisson structure base on the dynamic response of seabed around a caisson structure are numerically investigated by means of a caisson structure-seabed interaction analysis.

Figure 1 shows a sketch of a caisson structure with suction installed on seabed, which is assumed to be a rigid rectangular prism with W in both width and depth and H in height, and also has a skirt length of D. The rocking force at the center of the caisson structure base shown in this figure represents the wave attack acting on the front surface of a caisson structure. The displacement response of a caisson structure for the rocking vibration can be numerically evaluated by assuming a caisson structure to be a rigid structure in this analysis.

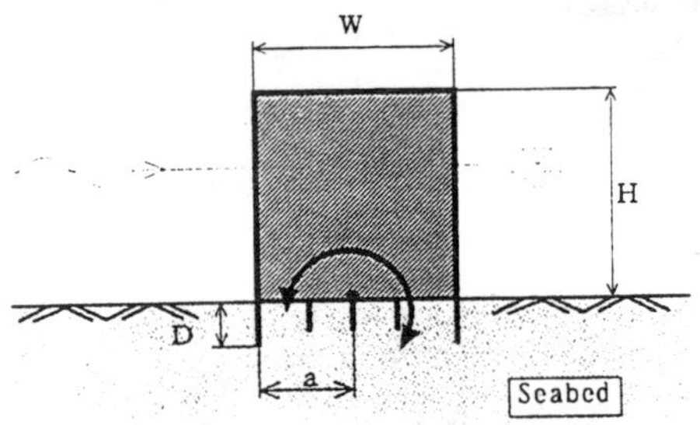

Figure 1 A caisson structure with suction on seabed

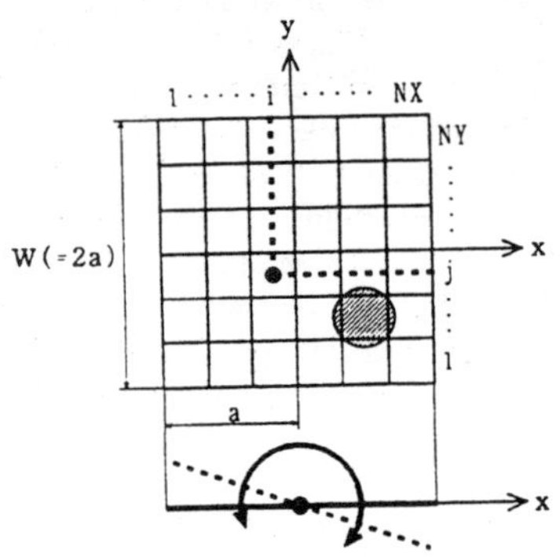

Figure 2 Compartments on the base of caisson structure

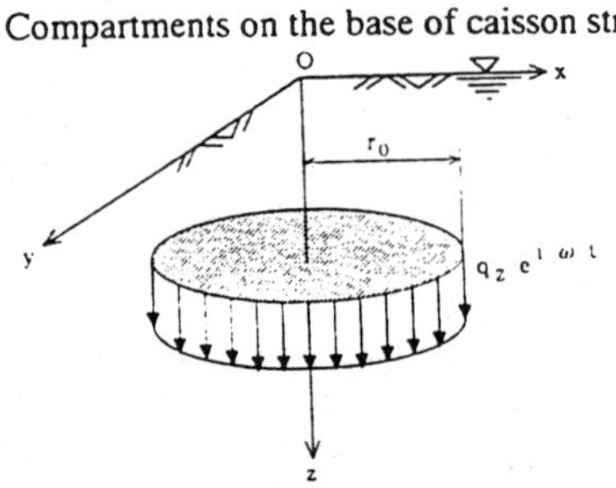

(a) A distributed load vertically acting over a circular plane

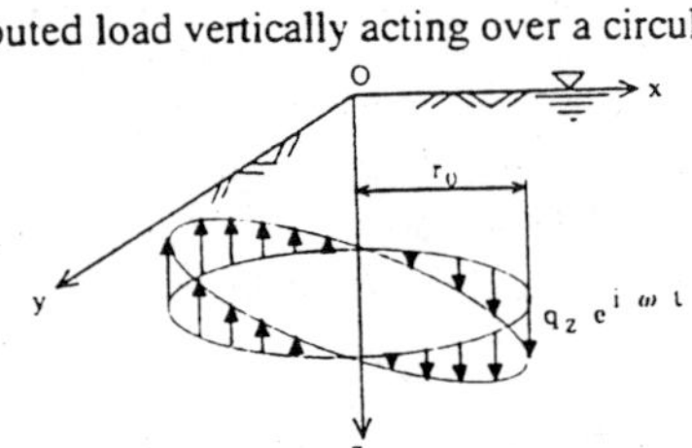

(b) A rocking load vertically acting on a circumference

Figure 3 Two load types using in analytical method

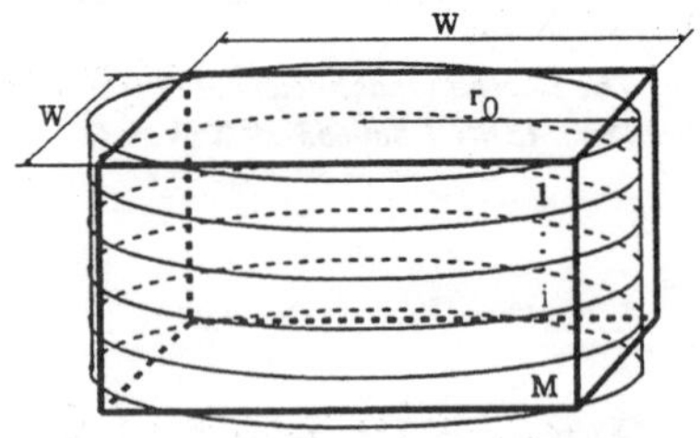

Figure 4 Division model of skirt part

In general, the different suction force is applied to each compartment on the base of a caisson structure in order to install it on-seabed without inclination under various sea conditions. Dividing the base of a caisson structure into square compartments as shown in Figure 2, the effect of suction for rocking vibration about y axis on the response of a caisson structure is numerically evaluated in this paper.

In the displacement response analysis of a caisson structure, a distributed load vertically acting over a circular plane shown in Figure 3(a) is used to express a suction effect acting on each compartment on the base of a caisson, and also a rocking load vertically acting on a circumference with radius r_0 shown in Figure 3(b) is done to express the frictional force acting on the skirt part of a caisson (Takatani et al.,1995a, 1995b), based on the same manner as the interaction analysis of a suction pile anchor embedded in a poroelastic seabed proposed by Maeno et al.(1994). The displacement solutions due to two loads in a poroelastic seabed mentioned above have been already derived by Takatani and Maeno (1995a).

The dynamic response analysis of the displacement, stress, and pore water pressure in seabed around a caisson structure with suction is mentioned as follows.

By dividing the base of a caisson structure into finite square elements as shown in Figure 2 and dividing the skirt part into finite ring elements, and also assuming the load acting on each square element to be constant as shown in Figure 4, the displacement at each square and ring element due to unit load acting on it can be superposed by a concept of the influence coefficient. The amplitude of an uniformly distributed load acting on each square element is calculated from a square element with the equivalent area to a circular plane with radius r_0 shown in Figure 3(a), and also that of a ring load acting on each ring element can be calculated from a circumference length of a square element with the equivalent length to a circle with radius r_0 shown in Figure 3(b). Simultaneous equations for both the unknown uniformly distributed load acting on each square element and the unknown ring load acting on each ring element can be formulated, provided that the displacement at each square and ring element is equal to the rotational displacement due to a rocking force acting to the caisson structure with suction. According to the condition that the total amount of unknown load acting to each element, which is obtained by solving simultaneous equations, is equal to a rocking force acting to the caisson structure, the displacement of a caisson structure with suction can be analytically evaluated.

On the other hand, the displacement, stress, and pore water pressure in the seabed around a caisson structure can be easily expressed by a superposition technique of the displacement and stress solutions due to both the uniformly distributed load $\bar{q}_z(i,j)$ acting on (i,j) square element and the ring load $\bar{q}_z(i)$ acting on i-th ring element as the following equation.

$$v = \sum_{i=1}^{NX}\sum_{j=1}^{NY} G(x,z|i,j)\bar{q}_z(i,j) + \sum_{i=1}^{M} G(x,z|i)\bar{q}_z(i) \quad (1)$$

where, v denotes the displacement, stress, and pore water pressure in the seabed around a caisson structure, and $G(x,z|i,j)$ and $G(x,z|i)$ are the influence coefficients on the displacement , stress, and pore water pressure at an estimating location due to the uniformly distributed load and the ring load, respectively. NX and NY are division numbers on the base of a caisson structure along x and y directions shown in Figure 2, respectively, and also M is a division number of the skirt part of a caisson structure. Further, the stress and pore water pressure solutions in a poroelastic seabed have been already reported in detail by Takatani et al.(1992).

FREQUENCY RESPONSE OF SEABED

In the numerical analysis in this paper, the material properties on a saturated coarse sand (Haplern and Christiano, 1986) shown in Table 1 are adopted for the soil conditions of a poroelastic seabed, and the dynamic response of seabed due to a rocking force acting to a caisson structure with suction is analyzed. Q shown in Table 1 is a coupling elastic coefficient between the volume change of solid and that of

Table 1 Material properties for a fluid-saturated poroelastic seabed

$\mu = 96\ MPa$, $\lambda = 432\ MPa$, $Q = 280\ MPa$,
$R = 274\ MPa$, $v = 0.3$ (Poisson's ratio),
$f = 0.48$ (porosity of poroelastic medium),
$\rho_f = 2.67\ g/cm^3$ (density of solid skeleton),
$\rho_s = 1.00\ g/cm^3$ (density of pore fluid)

fluid, and also R is related to the compressibility of fluid. The frequency response of the seabed around a caisson structure with suction as well as the caisson structure itself is numerically evaluated by using both parameters of non-dimen-sional frequency $\omega a/Vs$ and non-dimensional permeability $Ko\ (= kVs/ga)$, where a is a half width of caisson structure model, k is a permeability of seabed, g is a gravity of acceleration, ω is a circular frequency, and Vs is a shear wave velocity. The range of non-dimensional permeability Ko is adopted $10 \sim 10^{-3}$ judging from the permeability $k = 10^{-1} \sim 10^{-5}\ m/s$ of actual seabed in this analysis.

In addition to two parameters described above, the suction ratio Rc is used in this numerical analysis, which means a ratio of the amplitude of suction acting on outside regions at the base of a caisson to that on inside region with $2b$ in width drawn by slash lines as shown in Figure 2. As the ratio Rc increases, the amplitude of suction acting on inside region with $2b$ in width decreases as compared with those on both outside regions. Takatani et al.(1996) reported that the stability of a caisson structure with suction for wave attach increases with the increase of suction ratio Rc. In this analysis, suction ratio Rc is assumed to be 1.0. The division numbers NX and NY of the width of a caisson base as shown in Figure 2 are 20 respectively, whereas the number of compartment on the base of a caisson structure is 400.

Figure 5 shows the displacement variations at 9 locations in seabed under a caisson structure in the frequency domain for non-dimensional permeability $Ko = 0.1$, skirt length $D/a = 0.5$ and suction ratio $Rc = 1.0$. As is evident from this figure, the displacement in seabed under a caisson structure decreases with the depth z/a of an estimating location increasing. Judging from this result that the displacement responses at two estimating locations $z/a = 0.25$ and 0.5 are almost same regardless of non-dimensional frequency

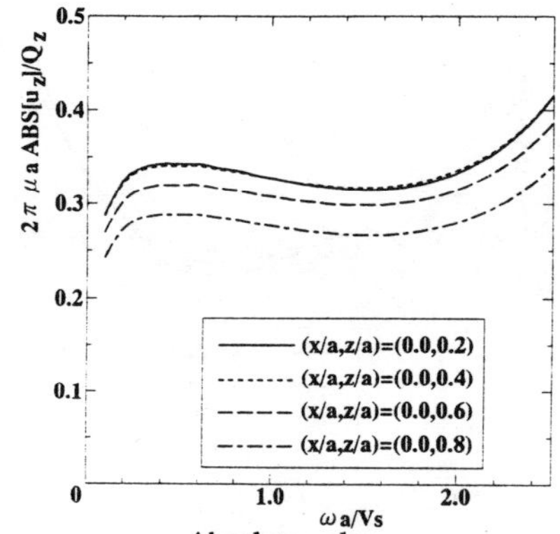

Absolute value

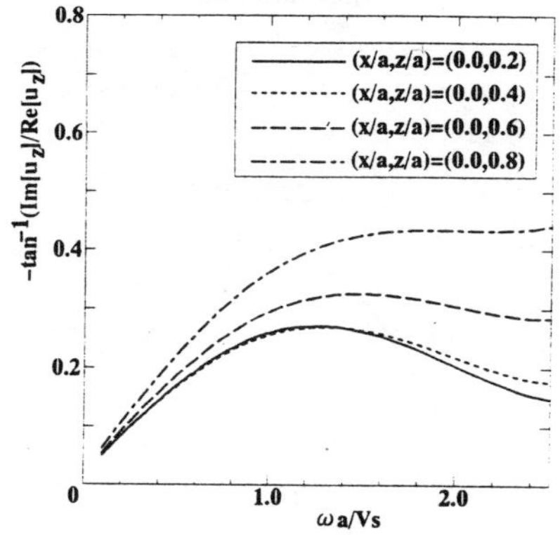

Phase angle

(a) $z/a = 0.2, 0.4, 0.6, 0.8$

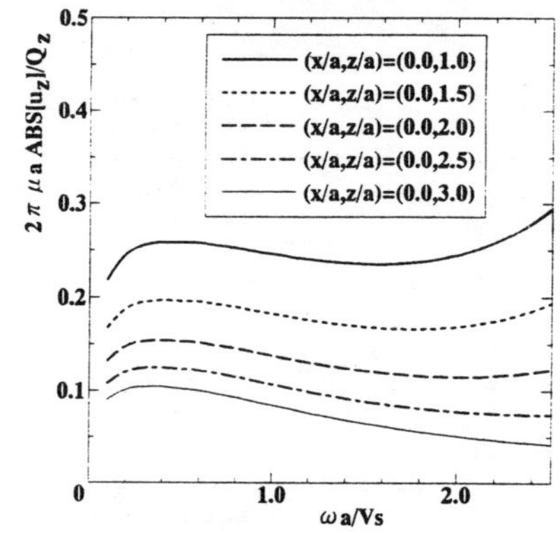

Absolute value

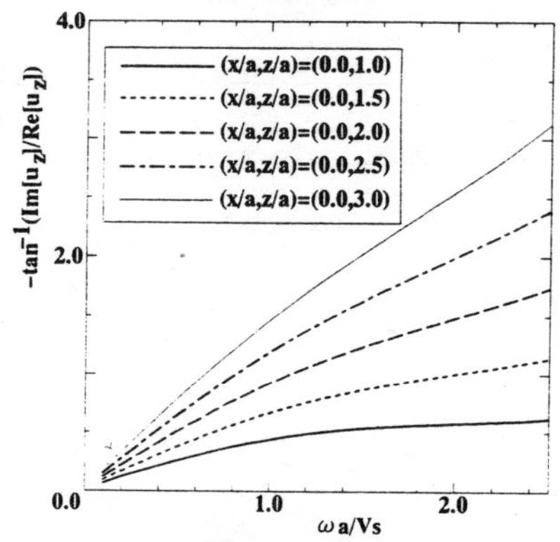

Phase angle

(b) $z/a = 1.0, 1.5, 2.0, 2.5, 3.0$

Figure 5 Displacement of seabed under a caisson structure ($Ko = 0.1, D/a = 0.5$)

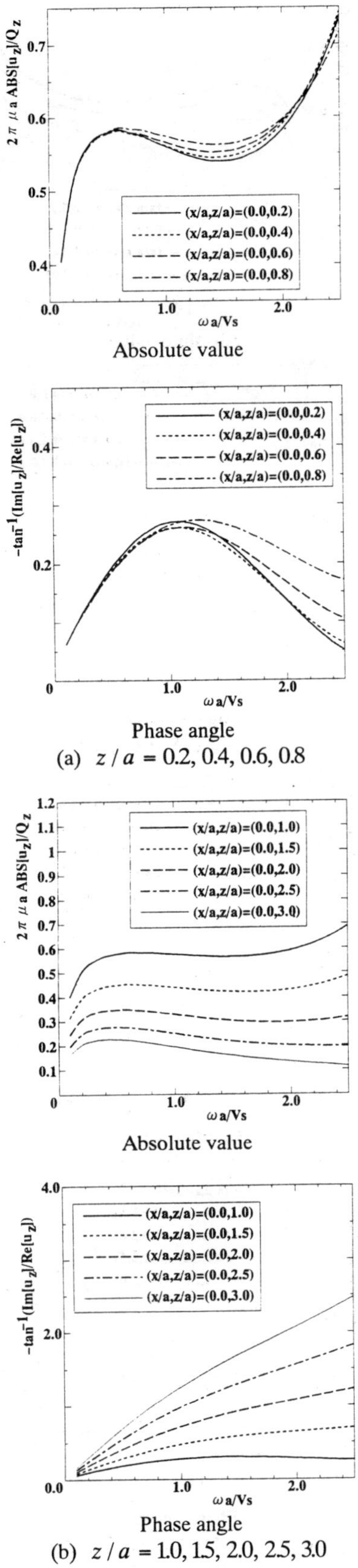

Figure 6 Displacement of seabed under a caisson structure ($Ko = 0.1, D / a = 1.0$)

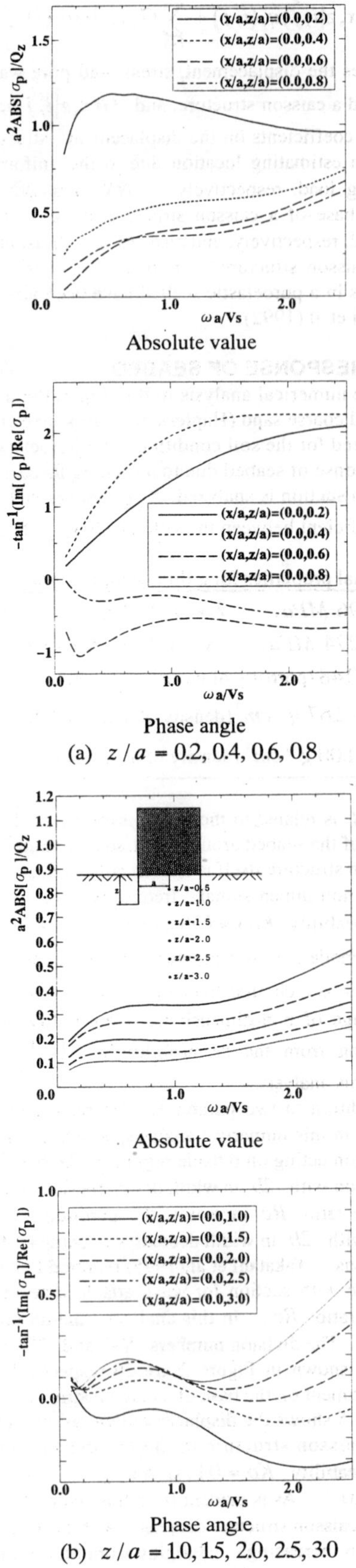

Figure 7 Pore water pressure of seabed under caisson structure ($Ko = 0.1, D / a = 1.0$)

$\omega a / Vs$, the displacement behavior of seabed surrounded with both the base and skirt part of a caisson is almost same. Also, the phase angle at these estimating locations $z / a = 0.25$ and 0.5 are almost same regardless of non-dimensional frequency $\omega a / Vs$. It is evident from this figure that the influence of both the suction and frictional force on the displacement at each location gradually disappears with the depth z / a of each location increasing. Consequently, the displacement response of seabed under a caisson structure greatly depends on the skirt length of a caisson structure.

Figure 6 shows the displacement variations at 9 locations in seabed under a caisson structure in the frequency domain for non-dimensional permeability $Ko = 0.1$, skirt length $D / a = 1.0$ and suction ratio $Rc = 1.0$. As is apparent from this figure, the displacement of seabed under a caisson structure decreases with the depth z / a of an estimating location increasing. The displacement variations at the range of $0.0 \le z / a \le 1.0$ are almost same regardless of non-dimensional frequency $\omega a / Vs$, and the movement of seabed surrounded with both the base and skirt part of a caisson is almost same. It is, however, apparent from Figure 6 that a remarkable variation according to non-dimensional frequency $\omega a / Vs$ is recognized within the range of $0.0 \le z / a \le 1.0$ in comparison with that in the skirt length $D / a = 0.5$ shown in Figure 5, and also that the response value at each location within $0.0 \le z / a \le 1.0$ is smaller than that in the skirt length $D / a = 0.5$. According to the frequency response results shown in Figures 5 and 6, it can be considered that the displacement response of seabed under a caisson structure with suction in the case of the skirt length $D / a = 1.0$ is smaller than that in the skirt length $D / a = 0.5$. On the other hand, the phase angle at the range of $\omega a / Vs < 1.0$ is almost same at each location, but the differences among the phase angles at locations are distinctly recognized with the increase of non-dimensional frequency $\omega a / Vs$. As the depth z / a of each location increases, the influence of both the suction and frictional force on the displacement at each location gradually disappears. It is noted that the displacement response at each location becomes small and the phase angle is large as the non-dimensional frequency $\omega a / Vs$ and the depth z / a of an estimating location increase. Consequently, the displacement response of seabed under a caisson structure greatly depends on the skirt length of a caisson structure as same as shown in the displacement response of a caisson itself (Takatani et al., 1996).

Figure 7 shows the pore water pressures at 9 locations in seabed under a caisson structure in the frequency domain for non-dimensional permeability $Ko = 0.1$, skirt length $D / a = 1.0$ and suction ratio $Rc = 1.0$. The response of pore water pressure at the location $z / a = 0.2$ in seabed under a caisson structure is a maximum value in comparison with other locations. The pore water pressure in seabed gradually decreases with the depth z / a of each location increasing, but the response value of pore water pressure becomes large with the increase of non-dimensional frequency $\omega a / Vs$. Accordingly, it is noted that the response of pore water pressure greatly depends on the non-dimensional frequency $\omega a / Vs$ as well as the depth z / a of an estimating location. On the other hand, it is apparent from this figure that a large difference between the phase angles at the locations $z / a = 0.4$ and 0.6 is recognized. For this reason, it can be considered that the influence of suction acting on each compartment appears distinctly in the response of pore water pressure. It should be noted that an influence range of suction acting on each compartment is almost same regardless of skirt length D / a , although the response results for skirt lengths $D / a = 0.25, 0.5$ and 0.75 are not shown in this paper on account of the limited space.

TIME-HISTORY RESPONSE OF SEABED

Based on the frequency response of a caisson structure with suction and the seabed around it as mentioned in previous section, the responses of both the caisson structure and seabed in the time domain are evaluated for the experimental data of the wave pressure acting on the front surface of a caisson structure with suction. The effects of the skirt length, permeability of seabed and suction ratio on the responses of both the caisson structure with suction and seabed are numerically investigated through some numerical examples. The outline of the wave pressure acting on the surface of caisson, which was measured in the wave basin in the Port and Harbour Research Institute, the Ministry of Transport, Japan, is explained as follows.

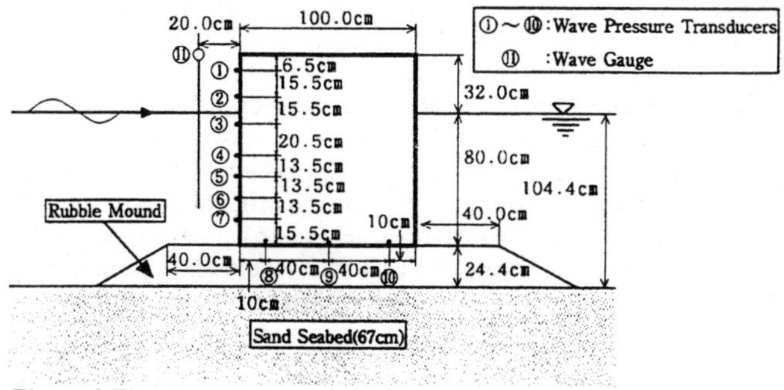

Figure 8 The cross section of a caisson model and the set-up of measurement instruments

Figure 8 shows a profile of the cross section of a caisson structure model and the arrangement of instruments for measuring the wave pressure on the caisson. The caisson structure in experiments is an acryl synthetic resin box with 15 mm in thickness, and is fixed on the mortar plate with 5 cm in thickness, and also the caisson structure model is adjusted with the lead weight inside of it. The wave pressure was measured by 10 wave pressure transducers installed on the surface of the

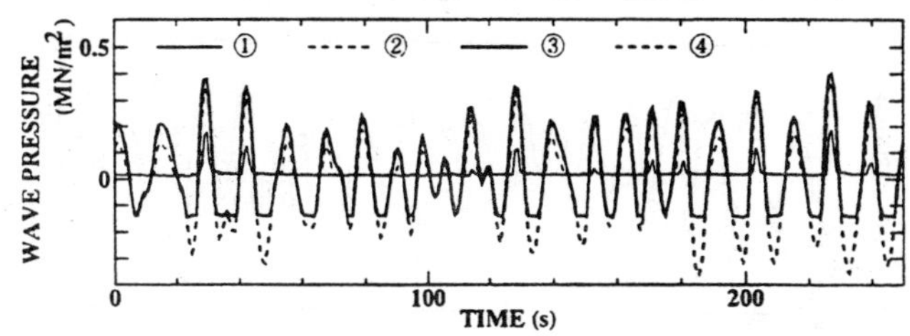

(a) Measuring points ① ~ ④

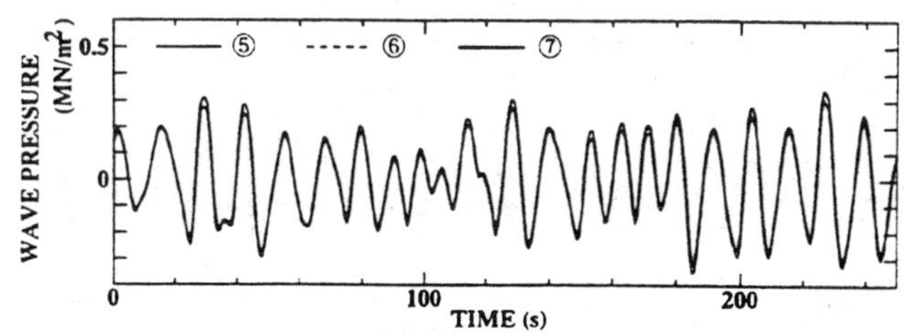

(b) Measuring points ⑤ ~ ⑦

Figure 9 The variations of wave pressure (Irregular wave with 3.04s in period, H1/3=18.0cm, Hmax=33.3cm)

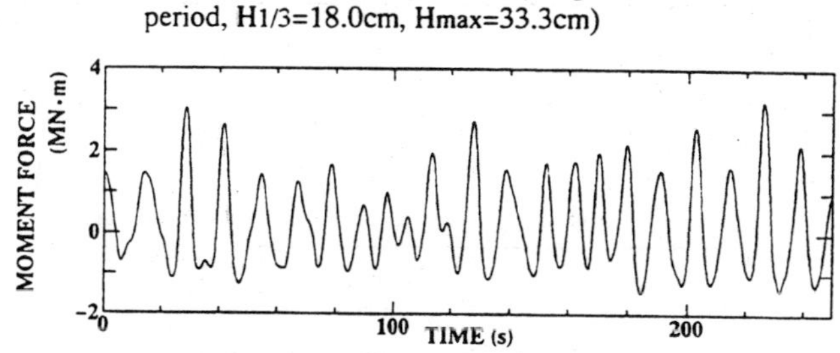

Figure 10 Moment force about y axis

caisson structure model, 7 of whose locations are on the front surfaceand 3 on the side surface of the caisson model, and the wave gauge is installed at 20 cm front of the caisson.

Figure 9 indicates the wave pressure data measured by 7 wave pressure transducers for the irregular wave with 3.04s in period, H1/3=18.0cm in significant wave height and Hmax=33.3cm in maximum wave height, which are transformed from the original data by considering the scale of a caisson structure model (=1/16). The size of a caisson structure model is assumed to be 16m in width and 17.9m in height in this paper. Each shape of wave pressure at the measuring points ① ~ ③ has flat parts in the trough, while that at points ④ ~⑦ is almost irregular without flat parts in the trough of wave pressure.

The moment force about y axis in the time history numerically calculated from 7 wave pressure data in Figure 9 is shown in Figure 10. Although the shapes of wave pressure data at points ① ~ ③ have flat parts in the trough of wave pressure, the calculated moment force shown in Figure 10 is almost irregular due to the contribution of measuring points ④ ~ ⑦.

Figure 11 shows the relationship between the displacement and stiffness of seabed around a caisson structure with suction. The displacement responses at 4 locations in seabed under a caisson structure in the time domain are numerically evaluated for non-dimensional permeability $Ko = 0.1$, skirt length $D/a = 1.0$ and suction ratio $Rc = 1.0$. As the stiffness of seabed increases with the shear wave velocity Vs increasing, the displacement response at each location in seabed becomes small. It is evident from the frequency response shown in Figure 6 that the phase in displacement response at each location under a caisson structure is gradually delayed as the depth z/a of each location increasing. Accordingly, it is necessary to give careful consideration to the behavior of both the displacement, and phase responses in case of installing a caisson structure with suction on the seabed, whose stiffness and shear wave velocity are small.

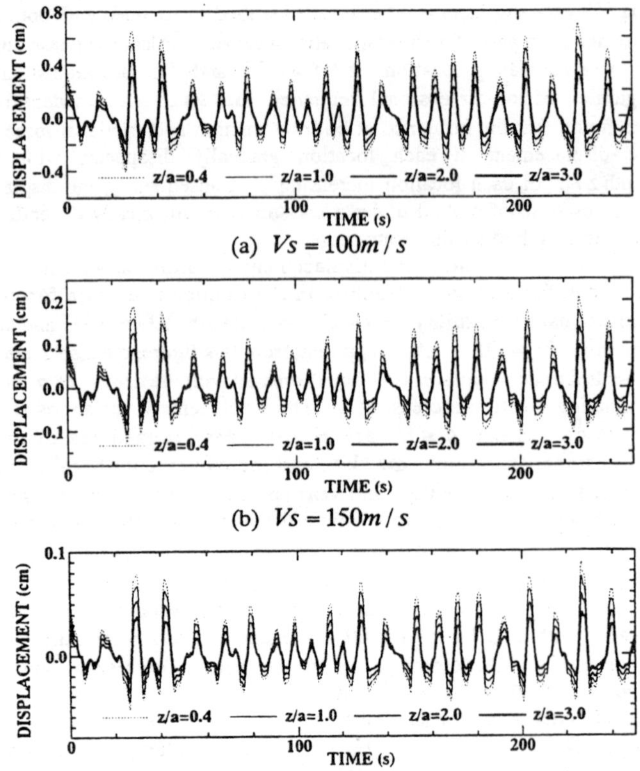

(a) $Vs = 100m/s$

(b) $Vs = 150m/s$

(c) $Vs = 200m/s$

Figure 11 Relationship between displacement response and stiffness of seabed ($Ko = 0.1, Rc = 1.0, D/a = 1.0$)

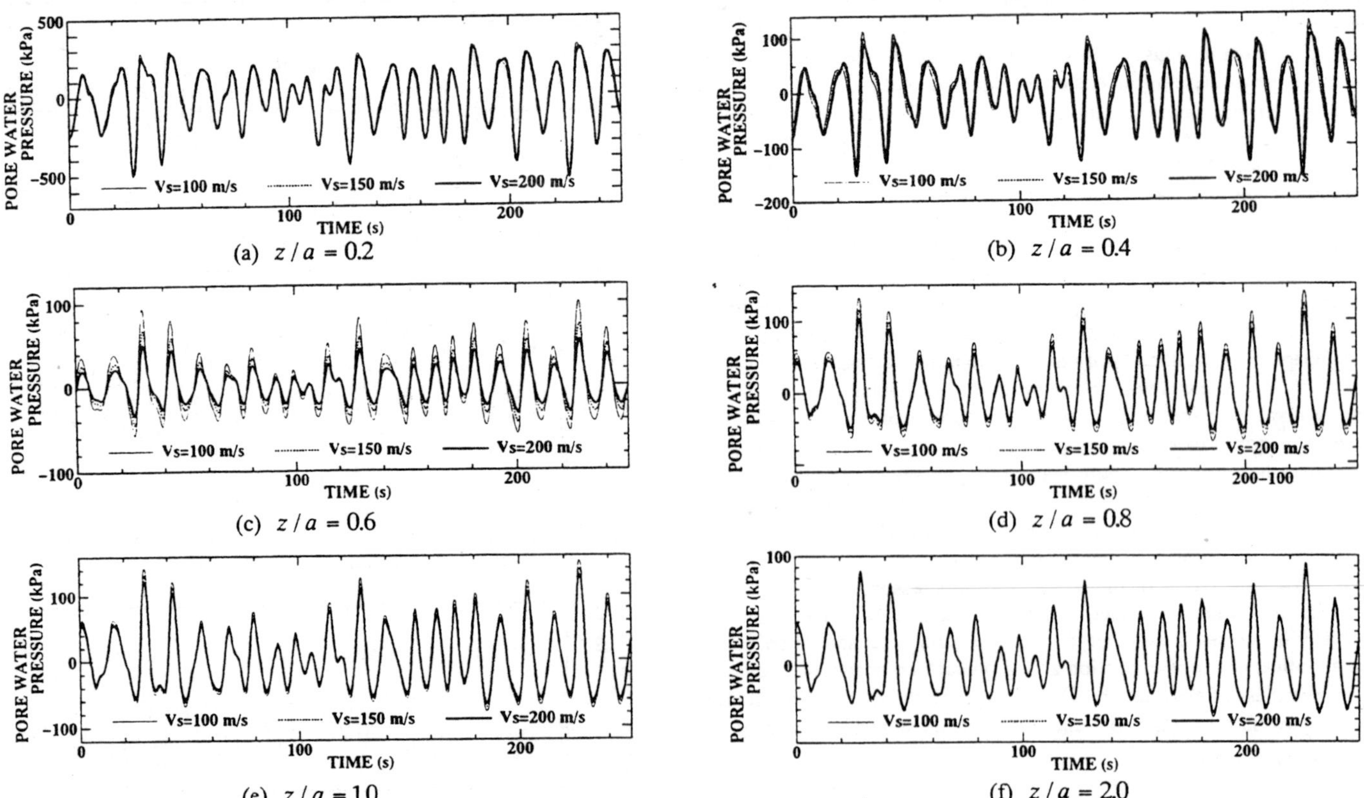

(a) $z/a = 0.2$

(b) $z/a = 0.4$

(c) $z/a = 0.6$

(d) $z/a = 0.8$

(e) $z/a = 1.0$

(f) $z/a = 2.0$

Figure 12 Relationship between pore water pressure response and distance from caisson base ($Ko = 0.1, Rc = 1.0, D/a = 1.0$)

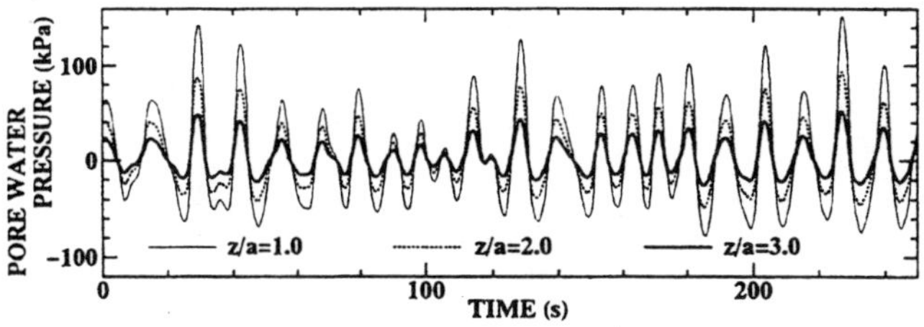

(a) $Vs = 100m/s$

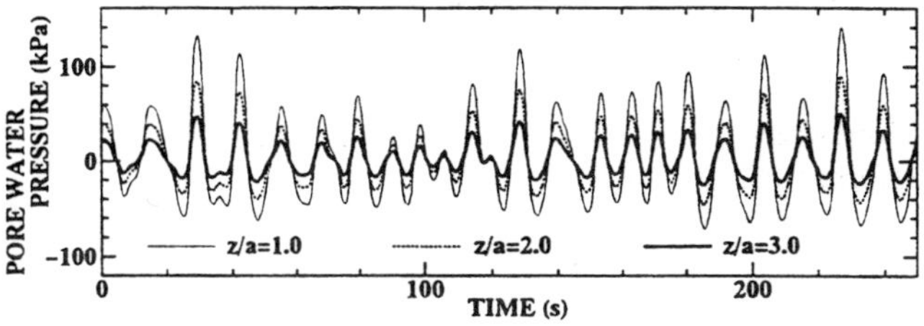

(b) $Vs = 150m/s$

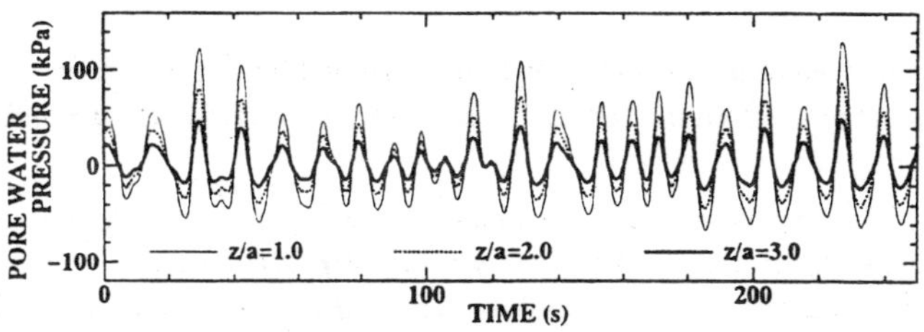

(c) $Vs = 200m/s$

Figure 13 Relationship between pore water pressure response and stiffness of seabed ($Ko = 0.1$, $Rc = 1.0$, $D/a = 1.0$)

Figure 12 indicates the relationship between the pore water pressure in seabed and the distance from caisson structure base. The pore water pressure responses at 6 locations in seabed under a caisson structure in the time domain are evaluated for non-dimensional permeability $Ko = 0.1$, skirt length $D/a = 1.0$ and suction ratio $Rc = 1.0$. As is evident from the frequency response shown in Figure 7, the response at location $z/a = 0.2$ is the maximum value in 6 locations. As the depth z/a of each location increases, the amplitude of pore water pressure become smaller and also the phase angle is different from the location $z/a = 0.2$ in the vicinity of the base of a caisson. While the influence of the shear wave velocity Vs on the pore water pressure response gradually disappears with the increase of the depth. It is, in particular, apparent from Figure 12(c) and (d) that a remarkable difference between pore water pressure responses at locations $z/a = 0.6$ and 0.8 is recognized.

Figure 13 shows the relationship between the pore waterpressure and stiffness of seabed around a caisson structure with suction. The pore water pressures at 3 locations in seabed under a caisson structure in the time domain are evaluated for non-dimensional permeability $Ko = 0.1$, skirt length $D/a = 1.0$ and suction ratio $Rc = 1.0$. As the stiffness of seabed increases with the shear wave velocity Vs increasing, the pore water pressure response at each location in seabed decreases. It is evident from the frequency response shown in Figure 7 that the pore water pressure response greatly depends on not only the suction and frictional force acting on the skirt part of caisson structure but also the stiffness of seabed around a caisson structure.

CONCLUSIONS

The dynamic response in a poroelastic seabed around a caisson structure with suction is numerically evaluated for an external force of wave pressures acting on the front surface of caisson, which were measured in the wave basin experiment. The stability of a caisson structure with suction and the response of seabed foundation against wave attack are investigated for some material properties of seabed foundation. The results obtained in this paper are summarized as follows:

1) The displacement response at an estimating location under a caisson structure becomes small and the phase angle is gradually delayed as the non-dimensional frequency $\omega a/Vs$ and the depth z/a of an estimating location increases. Also, the displacement response of seabed under a caisson structure greatly depends on the skirt length D/a of a caisson structure as seen in the displacement response of a caisson itself.
2) It is necessary to give careful consideration to the behavior of both the displacement and phase responses in case of installing a caisson structure with suction on the seabed, whose stiffness and shear wave velocity are small.
3) The dynamic response of pore water pressure under a caisson has a maximum value in the vicinity of compartments, and also gradually decreases with the distance from the base of a caisson increasing. An influence range of the suction acting on each compartment is almost same regardless of skirt length D/a.
4) The pore water pressure response in seabed foundation depends on not only the suction and frictional force acting on the skirt part of a caisson structure but also the stiffness of seabed around it.

ACKNOWLEDGMENT

This paper was sponsored by Grant-in Aid for Scientific Research (C) (No. 07650584) from the Ministry of Education, Science, Sports, and Culture in Japan. The support of the Ministry of Education, Science, Sports, and Culture for this research is gratefully acknowledged.

REFERENCES

Biot, M.A. (1956). "Theory of Propagation of Elastic Waves in a Fluid-Saturated Porous Solid", Journal of Acoustical Society of America, Vol.2, pp.168-178.

Halpern, M.R. and Christiano, P. (1986). "Response of Poroelastic Half-Space to Steady-State Harmonic Surface Traction", International Journal of Numerical and Analytical Methods in Geomechanics, Vol.10, pp.606-632.

Takatani, T., Maeno, Y., Takayama, T. and Hiraishi, T. (1992). "Frequency Response of Seabed around Anchor due to Steady-Sate Mooring Force", Proceedings of Civil Engineering in the Ocean, JSCE, Vol.8, pp.451-456. (in Japanese)

Takatani, T., Maeno, Y., Hiraishi, T. and Takayama, T. (1994). "Dynamic Response of Seabed around Suction Anchor", Proceedings of Civil Engineering in the Ocean, JSCE, Vol.10, pp.207-212. (in Japanese)

Takatani, T. and Maeno, Y. (1995a). "Dynamic Response of Caisson Structure with Suction Effect", Proceedings of Civil Engineering inthe Ocean, JSCE, Vol.11, pp.375-380. (in Japanese)

Takatani, T., Maeno, Y., Takahashi, S. and Shimosako, K. (1995b). "Dynamic Response of Caisson Structure with Suction on Poroelastic Seabed", Proceedings of Coastal Engineering in Japan, JSCE, Vol.42, pp.816-820.(in Japanese)

Takatani, T., Maeno, Y., Takahashi, S. and Shimosako, K. (1996). "Dynamic Response of Caisson with Suction on Soft Seabed", Proceedings of the 6th International Offshore and Polar Engineering Conference, Los Angels, Vol.1, pp.536-543.

Tsuzuki, M., Horsnall, M. R., Kikuchi, Y. and Zen, K. (1995). "Supporting Characteristics of Ocean Structure - 7. Foundation Structure with Suction Force", Tsuchi-To-Kiso, JSSMFE, Vol.43, No.3, pp.49-54. (in Japanese)

Maeno, Y., Takatani, T., Hiraishi, T. and Takayama, T. (1993). "On Application of Suction Anchor to Mooring System of Floating Structure", Proceedings of Coastal Engineering in Japan, JSCE, Vol.40, pp.906-910.(in Japanese)

Maeno, Y., Takatani, T., Hiraishi, T. and Takayama, T. (1994). "Dynamic Response of Suction Pile Anchor in Poroelastic Seabed due to Mooring Force", Proceedings of the 5th International Offshore and Polar Engineering Conference, The Hague, Vol.1, pp.420-425.

Proceedings of the Seventh (1997) International Offshore and Polar Engineering Conference
Honolulu, USA, May 25-30, 1997

ISBN 1-880653-28-1 (Set); ISBN 1-880653-29-X (Vol. I)

Discontinuous Deformation Analysis on Seafloor Sliding Simulation

L.K. Chien and Y.N. Oh
National Taiwan Ocean University
Keelung, Taiwan, China

ABSTRACT

The stability of seafloor is one of the geotechnical problems that need to be solved. A practical model for analysis is proposed in this paper. The model is based on the concept of Discontinuous Deformation Analysis (DDA) by Shi (1988). DDA is well known for its capability in analyzing large strain problems for discontinuous problems. However, DDA does not incorporate pore pressures and drained conditions between blocks. In this study, modifications were tried to improve this deficiency. Since DDA gives the progressive displacements and deformations of blocks, failure criteria can be observed. Thus, this paper demonstrates the applicability of DDA to analyze the stability of seafloor.

KEYWORDS: Seafloor, Discontinuous Deformation Analysis, failure mechanism.

INTRODUCTION

Exploration of the oceans has always been the attention of mankind. In the past few decades, there have been an increasing intensity of exploration as reclamation of land is conducted in the coastal area or in a deep-sea area. In the 1990, the Ministry of Economical Affairs in Taiwan declared a requirement of land reclamation including approximately 10,000 hectares in the West Taiwan Nearshore. The reclamation of land involves many geotechnical problems that must be solved. In the recent years, hydraulic sand fills method has been one of the most important reclamation method. In general, using this reclamation method, the filled soil pumped by machines would induce changes in the seafloor. Thus, the stability of seafloor is an important research topic.

Henkel (1970) was the first to take wave forces into account for submarine rotational slips. The effects of the passage of a sinusoidal transient wave pressure on the underwater slope were considered. Henkel (1970) demonstrated that the pressure changes on the seafloor associated with the passage of a wave are able to cause shear failure in soft sediments in water depth up to about 120m. The sediment was assumed to be rigid. Therefore, the analysis is considered appropriate for undrained loading.

Madsen(1978) and Yamamoto et. al.(1978) employed Biot's consolidation equations for the pore pressure and effective stresses in a pore-elastic horizontal seafloor. The seafloor was influenced by the passage of a sinusoidal wave. The analytical solution was obtained to discuss the instability of submarine sediments in horizontal seafloor.

Okusa(1985) obtained a similar analytical solution as Madsen and Yamamoto et al., but in terms of stress and pore pressure. The effects of harmonic ocean waves on the sediment in horizontal seafloor were discussed.

For the applications of Discontinuous Deformation Analysis (DDA), Huang and Ma, in 1992 have reported the computation of factors of safety using DDA. They obtained factors of safety using the conventional factors of safety definition. Yeung and Goodman (1992) found the safety margin for rock slopes using a different approach. They simply targeted a point in the slope and determined how low the strength reduction could be tolerated before that point moved. This provides a basis for the applications of DDA in slope stability assessment. However, DDA has not been applied to Marine Geoctechnical Engineering field.

In this paper, given the complexity of oceans it was decided that a numerical modeling was required to simulate the instability of seafloor. Therefore, a model based on Discontinuous Deformation Analysis (DDA) Model was proposed to simulate the instability of seafloor. In this study, a possible failure mechanism is assumed to be a circular slip surface failure. All possible failure mechanisms are found by trial using the slice methods. This paper provides a new simulation model to investigate the stability of seafloor based on Discontinuous Deformation Analysis (DDA).

DISCONTINUOUS DEFORMATION ANALYSIS (DDA)

The concept of discontinuous analysis was first defined by Goodman and Taylor in 1968, where they introduced the idea of "Joint Element". In 1971, Cundall revised the concept by introducing "Distinct Element". The discrete element method (DEM) has been used for over a decade for numerical modeling the mechanical behavior of discontinuous materials. Instead of a continuum, the DEM treats the discontinuous materials as an assemblage of distinct or discontinuous particles or blocks, each governed by physical laws. This numerical theory was first adapted to geomechanics by Cundall (1974), who studied the dynamic behavior of rock masses. As this numerical method has evolved, it has been widely applied to research in geotechnical engineering.

The concept of "Discontinuous Deformation Analysis" (DDA) was introduced by Shi in 1988, which enhanced the studies on discontinuity analysis. DDA is similar to Finite Element Method (FEM) where the equations are listed in the similar manners. However, DDA has a discontinuous boundaries among the elements. Therefore, based on the discontinuous concepts in DDA, the behaviors among the soil particles can be simulated.

The model in this study is based on the concept of Discontinuous Deformation Analysis (DDA) by Shi (1988) for rock slopes. In DDA, considering the equilibrium of forces, moment and kinematics for all blocks, a unique solution in displacement and deformation for each block is obtained. Also, DDA is well known for its capability in analyzing large strain problems for discontinuous problems.

The concept of DDA developed by Shi (1988) was mainly for two dimensional stability analysis of rocks. The rock medium is considered as an assembly of linear elastic blocks. The displacement and rotation at the center of gravity (u_o,v_o and r_o) as well as the strain tensors ε_x, ε_y and γ_{xy} for each blocks are treated as unknowns. For the *i*th block, these 6 unknowns are induced in the matrix $[D_i]$.

Individual blocks are connected and form a system by contacts between blocks and by displacement constraints on each block. For a system having *n* blocks, the simultaneous equilibrium equations have the form:

$$\begin{pmatrix} K_{11} & K_{12} & K_{13} & \cdots & K_{1n} \\ K_{21} & K_{12} & K_{23} & \cdots & K_{2n} \\ K_{31} & K_{32} & K_{33} & \cdots & K_{3n} \\ \vdots & \vdots & \vdots & \ddots & \vdots \\ K_{n1} & K_{n2} & K_{n3} & \cdots & K_{nn} \end{pmatrix} \begin{pmatrix} D_1 \\ D_2 \\ D_3 \\ \vdots \\ D_n \end{pmatrix} = \begin{pmatrix} F_1 \\ F_2 \\ F_3 \\ \vdots \\ F_n \end{pmatrix} \quad (1)$$

Fi is a 6× 1 submatrix which includes the loading condition applied on the *i*th block. It considers body forces as well as other applied forces. $[K_{ij}]$ is a 6×6 submatrix which defines the properties of block i. In this paper, the block are consider as linear elastic. The properties are defined by Young's modulus, E and Poison's ratio υ. Where $[K_{ij}]$ characterizes the contacts between blocks i and j. The contact properties include the friction angle and stiffness ratios which are related to the amount of shear and normal forces that could develop between blocks.

The simultaneous equilibrium equations are solved by minimizing the total energy associated with the forces and stresses. The result is a unique solution of $[D_i]$ which satisfies equilibrium of forces, moment, kinematics and stress for each block.

DDA detects contacts and interpenetrations between blocks. The condition of sliding between blocks is determined based on the Coulomb failure criteria. The normal and shear contact forces between blocks are computed based on the assigned contact stiffness and the amount of interpenetrations. For a given boundary condition, the computed of the corresponding rock mass movement is conducted in a series of time steps. In each time step, the $[D_i]$ is computed. Based on $[D_i]$ and the contact properties, the $[F_i]$ is updated and used in the next time step computation. The process repeated until the result shows no significant further displacement. For an unstable rock mass, the results from these time steps can be used to provide a graphic display of the block movement. The DDA has been applied successfully in describing the behavior of rock by Shi (1990).

In this paper, DDA is applied to seafloor stability analysis. DDA is well known for its capability in analyzing large strain problems for discontinuous problems. However, DDA does not incorporate pore pressures and drained conditions between blocks. In this study, modifications were tried to improve this deficiency.

METHOD AND STEPS OF ANALYSIS

Wave Theory

To apply the theory of wave motion, the following general assumptions are made for this study (as shown in Fig.1):

(1)the wave is a single component of the spectrum;

(2)the wave is part of a train that has constant height and period;

(3)the wave height is consider to be very small compared to the length;

(4)the wave form is sinusoidal, with crest and trough of equal vertical distances from the mean water level;

(5)the wave is propagating in a straight line;

(6)the depth of water is constant or changes very slowly;

(7)the water is considered non-viscous.

Based on small-amplitude water wave theory, the wave length can be expressed as follow(Poulos, 1988):

$$L = (gT^2/2\pi)\tanh 2\pi d/L \quad (2)$$

where, L is the wave length, T is the wave period, d is the water depth, g is the gravity force(m/s²).

While, the wave height(H) can be expressed as (Poulos, 1988):

$$H = H_0\left[\left(1+\frac{4\pi d/L}{\sinh(4\pi d/L)}\right)\tanh(2\pi d/L)\right]^{-1/2} \quad (3)$$

where, L is the wave length, T is the wave period, d is the water depth.

Wave Pressure

As the wave length and wave height are provided, analysis of pore pressures induced by wave loading start with the determination of wave pressure in on the seafloor. The wave pressure is determined by using small-amplitude water wave theory. The wave pressures on the seafloor are given by small-amplitude water wave theory as

$$p_b = p_o \sin(\lambda x - \omega t) \quad (4)$$

$$p_o = \frac{r_\omega H}{2\cosh \lambda d} \quad (5)$$

where, p_b is the wave pressure, p_o is the amplitude of wave pressure on the seafloor, γ_ω is the unit weight of water, λ is the wave number, H is the wave height, d is the depth of water and ω is the frequency (Finn et.al. 1983).

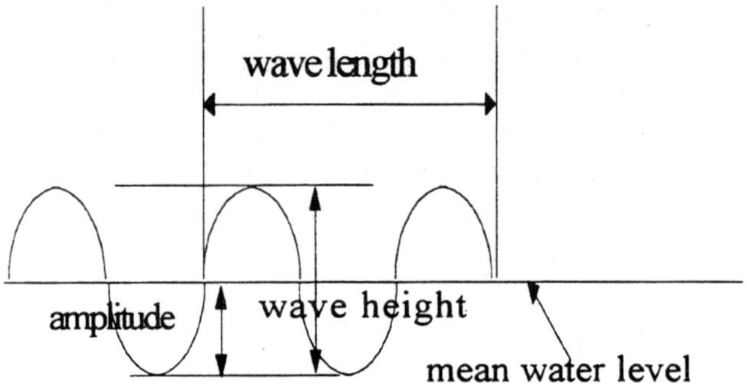

Fig. 1 Definition sketch of a wave.

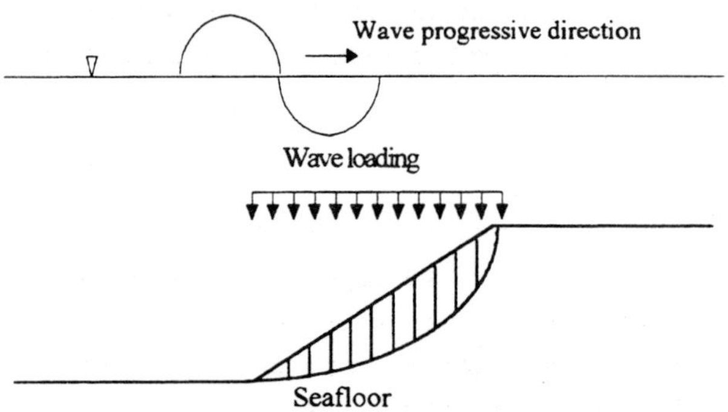

Fig.2 Wave loadings on seafloor.

In this study, the waves are represented as a series of harmonic waves based on simplifying assumptions. The wave pressures are assumed as loadings acting on the seafloor. Using slice method, the seafloor are treated as blocks (as shown in Fig.2).

The stability of seafloor, which is the topic of this paper, is a more complex problem requiring effective stress analysis for all the phenomena involved. The model in this study is a modification of DDA. This model assumes that pore pressures are generated by waves and can be determined from stress analysis. The model combines effective stress analysis, wave loadings, method of slices and DDA.

The seafloor is assumed to be rigid and the equilibrium is maintained. The pore pressures are generated in seafloor by wave loadings.

Analysis of loadings.

Discontinuous Deformation Analysis(DDA) has been widely used for several years to compute the response of rock slopes. In this paper, for the stability analysis of seafloor under wave loadings, the seabed or seafloor beneath the wave is characterized by elements or blocks (as shown in Fig.3).

The waves are assumed as a uniform harmonic wave based on the small-amplitude water wave theory. Lateral boundaries of the DDA model are assigned. The loads produced by the waves are assumed to be distributed as line loadings. The loadings are applied on the element or block as a straight line (as shown in Fig.3). The stresses and displacements within the blocks of the seafloor can be calculated.

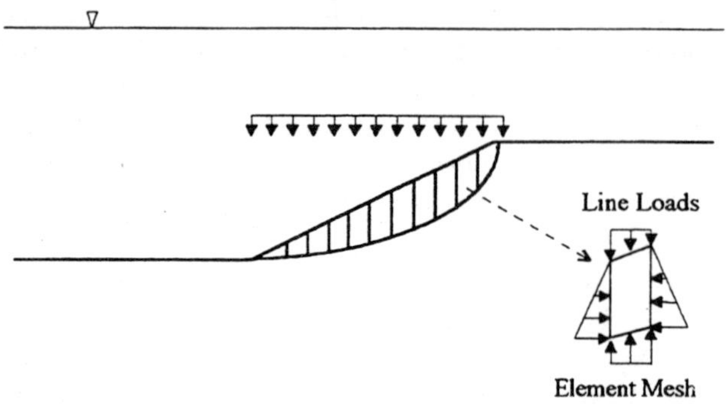

Fig. 3 Definition sketch of loadings on block.

Determination of the Slip Surface.

In this study, the slip surfaces are determined before the seafloor is generated. The soil conditions and required parameters are assumed. The slip surfaces (as shown in Fig.4) are generated by a program named STABL5. STABL5 is a program based on Janbu's method, and was developed by Purdue University for Indiana State Highway Commission. The program can generate circular slip surfaces, surfaces of sliding block character and more irregular surfaces of random shape. The original program was based on Janbu's Method, but Simplified Bishop's Method was later added for circular slip surfaces, and Spencer's Method was later added for both circular and general shape slip surfaces.

Soil Conditions

The soil parameters of the assumed seafloor used in this study are showed in Table 1. The soil characteristics adopted are those similar to the reclaimed soil in the west Taiwan Nearshore (Yeh, 1994).

The slope angle of the seafloor is assumed to be 25°. Those slope angles are adopted as calculation conditions. The water depth of 20 meters is adopted in this calculation example. The soil assumptions in this study are based on the environment properties in West Coast Taiwan. Further, a seafloor (as shown in Fig.5) with a sliding zone is studied. The soil is assumed cohesionless. The pore fluid density is assumed to be $\rho = 1$ t/m^3. The soil is assumed to be saturated.

Table 1 Soil Conditions.

Soil type	SP-SM
Max. dry density	1.62 kg/cm^3
Min. dry density	1.19 kg/cm^3
Poisson Ratio	0.2
Elastic Modulus, E	6.9 N/cm^3
Unit Weight	2.7 N/cm^3
Friction Angle	20°

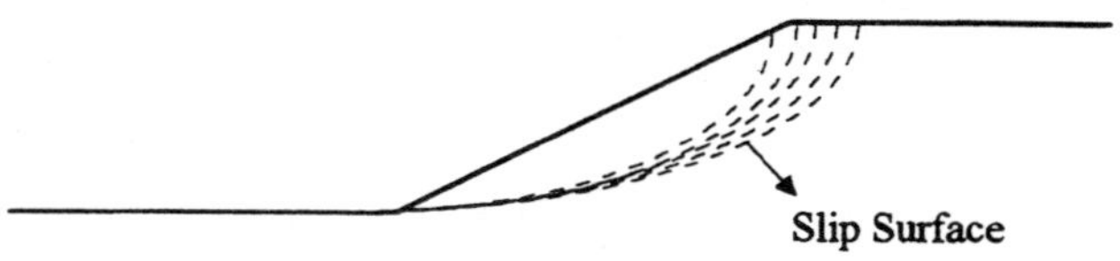

Fig. 4 Definition sketch of Slip Surfaces.

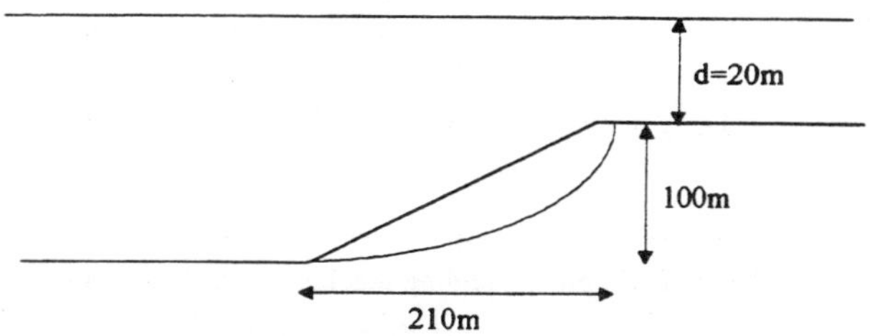

Fig. 5 Definition sketch of Seafloor.

Analysis Procedure

For wave load analysis, it was mentioned previously that small-amplitude water wave theory is used. In this section, the steps in a complete analysis are described. The goal here is to present an overall study of the flow of the analysis from step to step.

The flow chart for this study is shown in Fig.6. The steps are described as follow:

1. Input the wave parameters.
2. Input joints, fixed points and loading points.
3. Determine the slip surfaces of the seafloor.
4. Generate Block system from the input joints.
5. Input loading forces and material properties.
6. Identify all possible contacts between blocks.
7. Formulate the simultaneous equilibrium equations.
8. Add or Subtract and solve matrices.
9. Solve the system of equations.
10. Check for penetrations and tension at all contacts. Repeat step 6 through 9 if necessary.
11. Store new block geometry. Repeat step 6 through 10 if necessary.
12. Store analysis results.

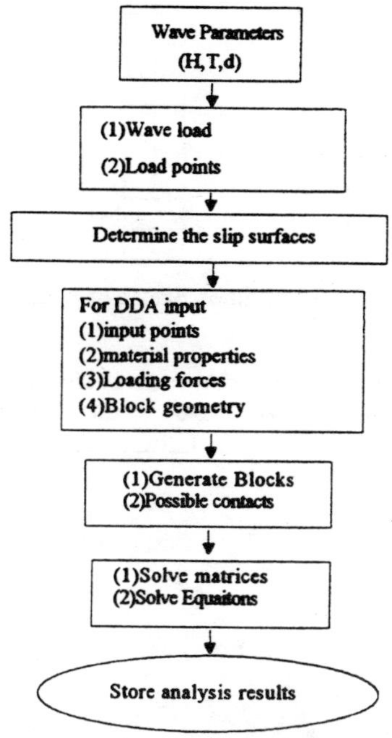

Fig.6 Flow Chart for DDA and Seafloor Analysis.

For analysis purposes, each time step for the block movements is recorded. The time step is not the "elapsed time" in physical manner. Time step is just number of repeated iterations in the model. Given the initial conditions, the displacement at the beginning of each time step is assumed to be zero.

VERIFICATIONS OF MODEL

Finn al.et.(1980) took wave forces into account for submarine rotational slips. Methods for estimating the stability of ocean floor under wave loading and determination of wave-induced pore-water pressure are considered by Finn al.et(1980).

Comparison between the model in this study and the results of Finn et. al.(1980) are made. Analyses are carried out for a seabed of 18.5m depth. The depth of water was 3.67m, and the wave characteristics were H=2.75m, T=7sec, and L=40m. Additional properties were assumed, Poisson Ratio=1/3, density of seawater=1025.6 kg/m^3.

A seabed is generated in blocks by DDA(as shown in Fig.7). The same parameters are adopted for analysis and comparisons. The comparison results are given in Fig.8. The results are similar tendency with soil depth ratio. The compared results are somewhat different. This occurs because for both analysis methods, analyses on wave-induced pore-water pressure are based on small-amplitude water wave theory. However, for DDA, only pore-pressure on the blocks are recorded. As for Finn et. al.(1980), the pore-pressure for varying depth is analyse.

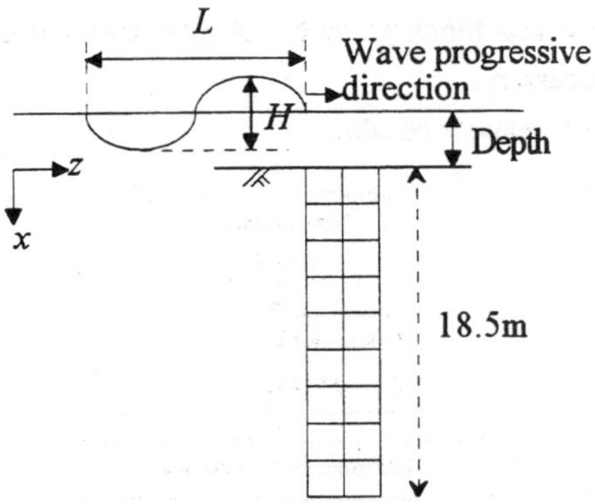

Fig.7 Definition Sketch of a Seabed.

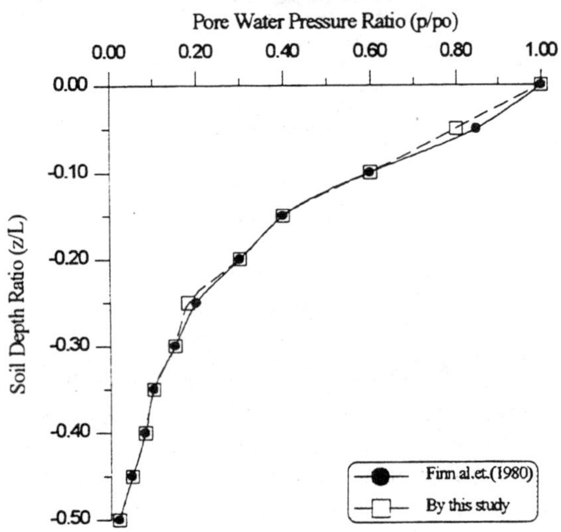

Fig.8 Comparison of Pore-Water Pressure by this study and Finn et. al.(1980).

DEMOSTRATIVE EXAMPLES

In this study, a factor of safety is defined as

$$FS = \frac{\tau}{\sigma} \qquad (6)$$

where *FS* is the factor of safety, τ is the shear stress along the failure plane, and σ is the total stress along the failure plane (Chen, 1993).

The factor of safety varies along the plane or slip surface. This definition is used as a mean for simplifying calculations.

Six Block Sliding

As shown in Fig.9, a slip surface in a seafloor with slope angle of 25°, slice cuts across the seafloor based on slice methods. The seafloor may failure in a sliding mechanism(as shown in Fig.10). In this mechanism, the block on the top position is the active block pushing the passive blocks on the bottom. The failure mechanism at final stage is shown by Fig.11.

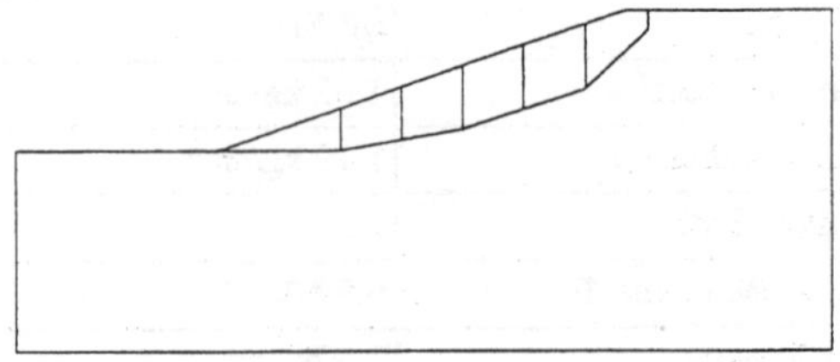

Fig.9 Six blocks sliding mechanism at initial stage (Step=1, FS=0.91).

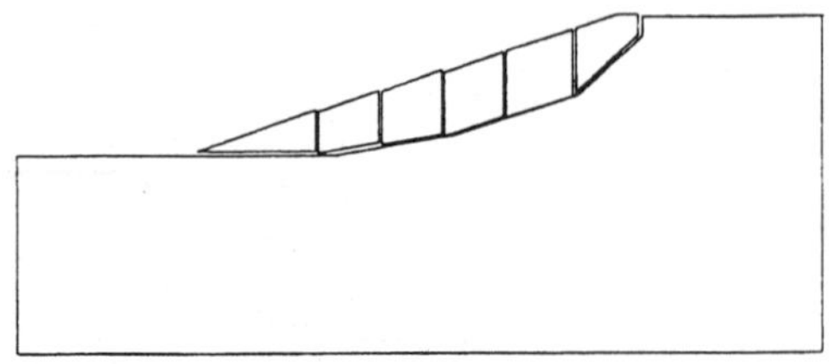

Fig.10 Six blocks sliding mechanism at sliding stage (Step=225, FS=0.91).

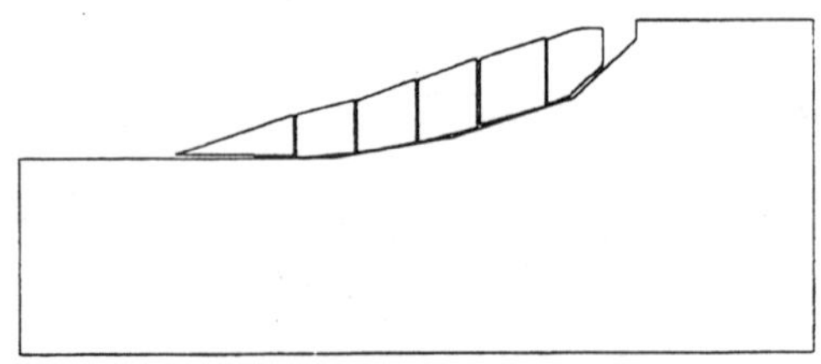

Fig.11 Six blocks sliding mechanism at final stage (Step=1920, FS=1.02).

Sliding on two Slip Surface

If there are two possible slip surfaces in a seafloor, the failure mechanism is controlled by the geometry of the slip surface. With the two slip surfaces shown in Fig.12, the sliding blocks break apart for movement to occur(as shown in Fig.13). The failure mechanism in this case involves the splitting up of the sliding blocks(as shown in Fig.14). This case shows the influence of slip surface on the failure mechanism of a seafloor.

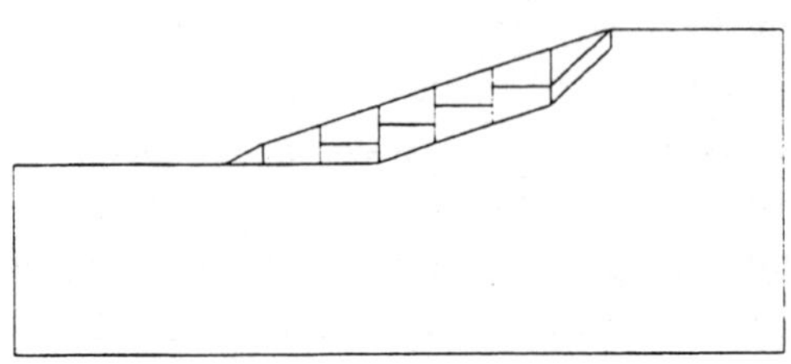

Fig.12 Sliding mechanism with two slip surfaces at initial stage (Step=1, FS=0.89).

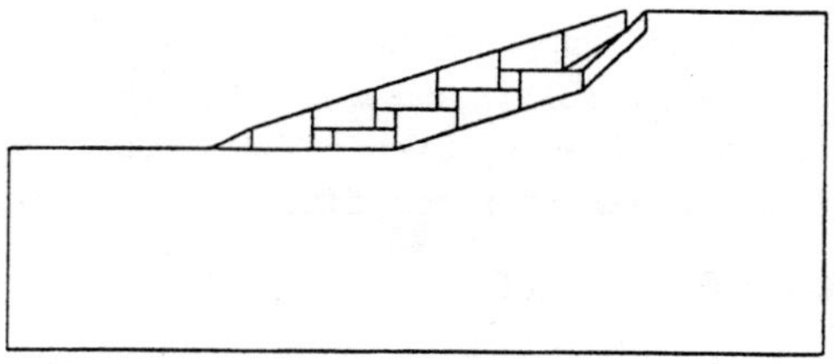

Fig.13 Sliding mechanism with two slip surfaces at sliding stage (Step=225, FS=0.90).

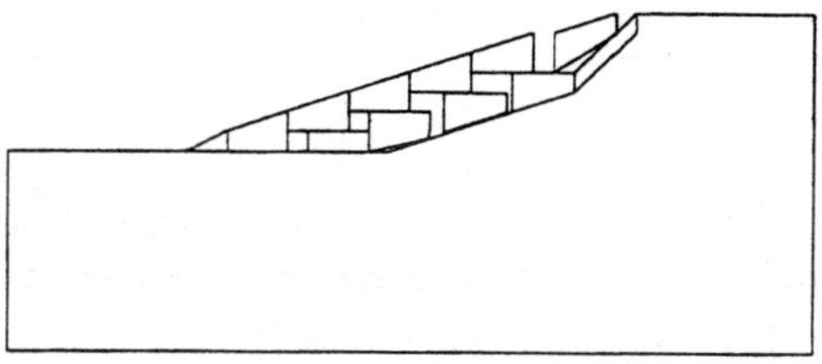

Fig.14 Sliding mechanism with two slip surfaces at final stage (Step=1955, FS=1.01).

Sliding with non-rectangular blocks

The sliding of block systems was studied by Goodman(1992), and most of the blocks are in a rectangular approach. In their analysis, the block movements are studied for rock slopes by using equilibrium approach. Analysis of a seabed with sandy soil typically gives a particle behavior. Therefore, in this study, the smaller circular blocks are also generated to simulate particle movements.

The circular blocks generated by DDA is shown in Fig.15, where 48 circular blocks are generated. The circular block system has two behaviors. A stable zone near the bottom, a sliding zone near the top of the seafloor and in between the seafloor(as shown in Fig.16 and Fig.17). The result seems similar to those obtained from "six blocks sliding". However, there are minor differences. For the "six blocks sliding", the results appear to indicate sliding and toppling movements in which the rectangular blocks slide and topple simultaneously. In this case, the blocks near the top of the seafloor topple.

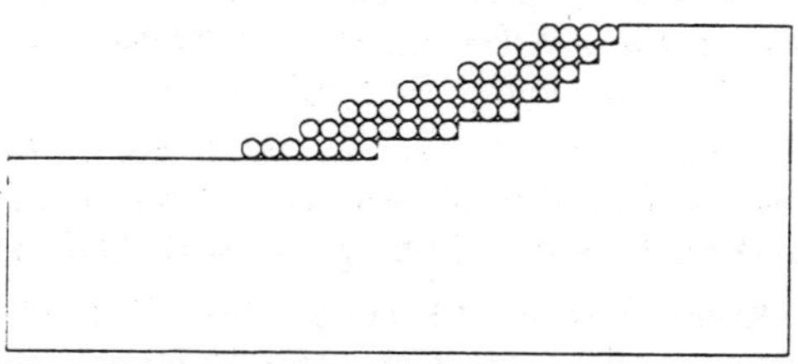

Fig.15 Failure mechanism with circular blocks at initial stage (Step=1, FS=0.88).

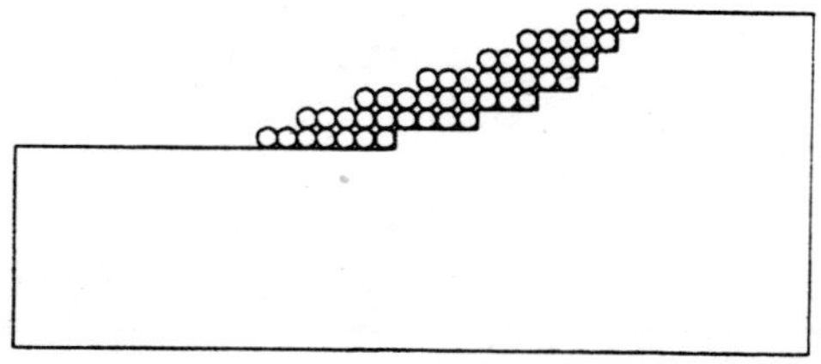

Fig.16 Failure mechanism with circular blocks at sliding stage (Step=225, FS=0.88).

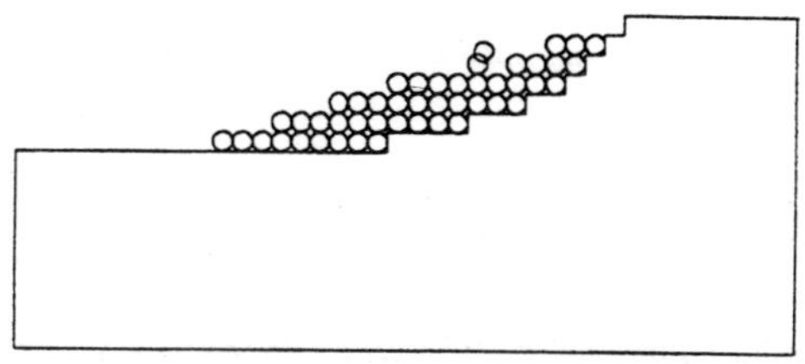

Fig.17 Failure mechanism with circular blocks at final stage (Step=2250, FS=1.02).

Sliding and Toppling with irregular seafloor surfaces

The sliding and toppling of blocks with varying seafloor surface are discussed. Here, rectangular blocks on a stepped base are studied(as shown in Fig.18). Results appear to indicate the blocks slide and topple simultaneously. As shown in Fig.19, the blocks slide and topple at the same time. This results demonstrate that failure mechanism of seafloor may involves slide and topple at the same time. This case typically shows that, sliding zone is near the seafloor bottom, and the toppling zone is in between and on the top of the seafloor(as shown in Fig.20).

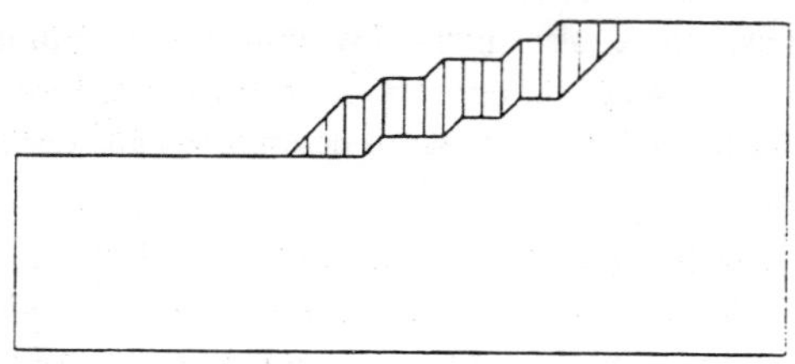

Fig.18 Failure mechanism with stepped base at initial stage (Step=1, FS=0.91).

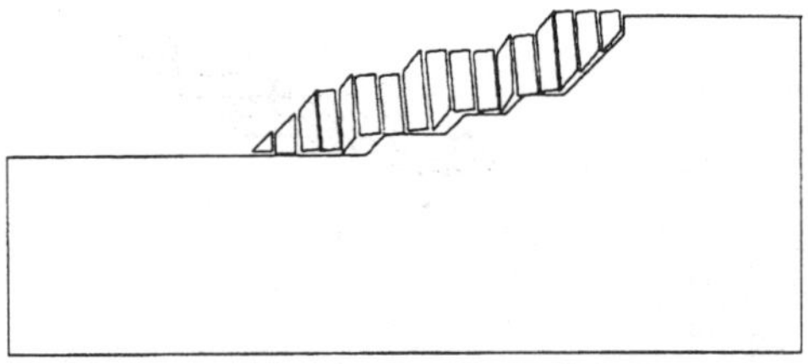

Fig.19 Failure mechanism with stepped base at sliding stage (Step=225, FS=0.91).

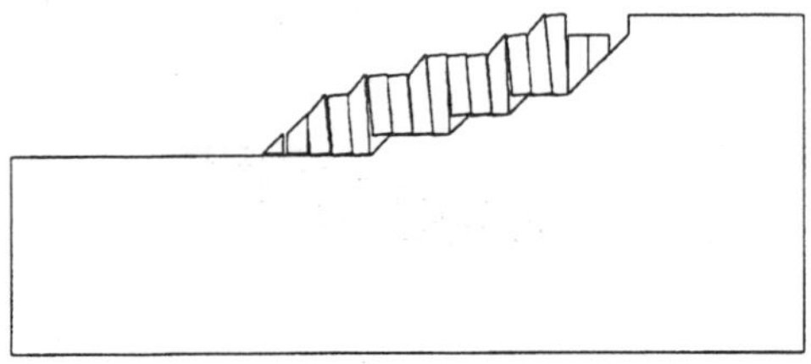

Fig.20 Failure mechanism with stepped base at final stage (Step=1525, FS=0.91).

Closing Remarks

As mentioned in previous discussion, by using the new model developed in this study, the progressive and deformations of blocks can be describe, and the failure mechanism can be demonstrated. Therefore, based on the model proposed in this study, the stability of the seafloor can be evaluated.

CONCLUSION

This study proposes a new method for the stability analysis of seafloor. The results prove that DDA is an applicable numerical method for seafloor stability analysis. This paper also demonstrates the applicability of DDA to seafloor sliding problems. DDA can be applied to demonstrate failure mechanisms of seafloor. It can also serve as a seafloor stability analysis.

The major advantage of DDA is the failure mechanisms need not to be assumed. Different patterns in the seafloor lead to different failure mechanisms. DDA also gives the progressive deformation and displacement of block movements. Thus, the failure mechanisms can be determined by analyzing rate of deformation and displacement of the blocks.

DDA is well known for its capability in analyzing discontinuous problems. However, DDA does not incorporate pore pressures and drained conditions between blocks. In this study, modifications were made to improve this deficiency. Comparisons were made with Fin al.et(1980) and show satisfactory results.

From a computational point of view, the proposed model still has many restrictions. Thus, more works need to be done. Further research will be carried out. First, a soil simulation tank model will be constructed to investigate the stability of slope and the failure mechanism. Second, more comparisons with other analysis methods will be carried out.

This study has improved the deficiencies of discontinuous deformation analysis (DDA), and is a starting way for the stability analysis of seafloor and prediction for unstable seafloor using discontinuous deformation method (DDA).

ACKNOWLEDGMENT

This research was supported by National Science Council of R.O.C., under Grant No. NSC-86-2611-E-019-030, and was greatly appreciated.

REFERENCES

Chen, S., 1993, *Slope Stability Assessment Based Upon Discontinuous Deformation Analysis*, Ph.D p+15XThesis, University of Pittsburg Pittsburgh.

Cundall, P. A., 1971, "A Computer Model for Simulating Progressive, Large Sacle Movements in Blocky Rock Systems", *Symposium of International Society of Rock Mechanics Nancy*, France, pp. 11-18.

Goodman, R. E., Taylor, R. and Brekke T.L., 1968, " A Model for the Mechanics Jointed Rock.":,Journal of the Soil and Foundation, ASCE, V.94, pp.637.

Henkel, P.J., 1970, "The Role of Waves in Causing Submarine Landslides", *Geotechnique*, Vol.20, No.1, pp.75~80.

Huang, An-Bin, Ma, Max Y. (1992), " Discontinuous Deformation Slope Stability Analysis", *Stability and Performance of Slopes and Embankments II*, Geotechnical Special Publication No.31, ASCE, pp.479-490.

Liam Finn, W.D., Siddharthan, R. and Martin G.R., 1983, "Response of Seafloor To Ocean Waves", *Journal of Geotechnical Engineering*, ASCE, Vol.109, No.4, pp.556-572.

Madsen, O.S., 1978, "Wave-induce Pore Pressure and Effective Stress in a Porous Bed", *Geotechnique*, Vol.28, No.4, pp377~393.

Okusa, S., 1985, "Wave-induced Stresses in Unsaturated Submarine Sediments", *Geotechnique*, Vol.35, No.4, pp517~532.

Poulos, H.G. (1988), "*Marine Geotechnics*", Prentice Hall, London, pp.344-411.

Shi, G.H., 1988, *Discontinuous Deformation Analysis: A New Numerical Model for the Statics and Dynamics of Block System*, Ph.D. Thesis, University of California, Berkeley.

Sinotech Engineering Consultants, Inc. 1990. *Yunlin offshore development project. Morphological baseline studies*, Taipei, Taiwan (in Chinese).

Yamamoto, T., Koning, H.L., Seumeiher, H. and Hijum, E.U., 1970, "On the response of A Poroelastic Bed to Water Waves", *Journal of Fluid Mechanics*, Vol.87, No.1, pp.193~206.

Yeh, Guo-Liang, 1994, *A study on the Dynamic Properties of Reclaimed soil in Taiwan Nearshore*, Master Thesis, National Taiwan Ocean University, Keelung, Taiwan (in Chinese).

Yeung, M.R. and Goodman, R.E. (1992), "Use of Shi's Discontinuous Deformation Analysis on Rock Slope Problems", *Stability and Performance of Slopes and Embankments II*, Geotechnical Special Publication No.31, ASCE, pp.461-478.

Proceedings of the Seventh (1997) International Offshore and Polar Engineering Conference
Honolulu, USA, May 25-30, 1997

ISBN 1-880653-28-1 (Set); ISBN 1-880653-29-X (Vol. I)

Improved Data Interpretation Method for SASW Tests at Complex Geotechnical Sites

Sung-Ho Joh, Brent L. Rosenblad and Kenneth H. Stokoe, II
University of Texas at Austin
Austin, TX, USA

ABSTRACT

The Spectral-Analysis-of-Surface-Waves (SASW) method is a nonintrusive seismic method for evaluation of shear stiffness profiles at geotechnical sites from surface wave measurements. Recent field experience has demonstrated the complexity of interpreting surface wave data at sites with large and abrupt stiffness changes. Geotechnical sites in polar environments as well as offshore sites containing gas hydrates have the potential to exhibit such strong stiffness contrasts which create multiple wave groups and result in complex phase spectra that can be easily misinterpreted. To aid in the correct interpretation of SASW data collected at such complex sites, a new phase interpretation scheme, called "Impulse Response Filtration" (IRF), is proposed. This filtered impulse response leads to a clearer interpretation of the phase differences between receivers. The IRF method is explained and demonstrated using experimental results from a pavement site which closely simulates frozen over unfrozen geotechnical sites representative of polar environments.

KEY WORDS: nonintrusive measurements, field seismic tests, shear stiffness, nondestructive testing, surface waves

INTRODUCTION

The Spectral-Analysis-of-Surface-Waves (SASW) method has been successfully employed in many geotechnical environments to non-intrusively determine the shear stiffness profile of these sites (Stokoe, et al., 1994). SASW testing involves measurement of the phase difference between vertical surface wave motions at two or more sensor locations. From the phase difference measurements, the phase velocity of the surface wave as a function of wavelength is determined. The phase velocity vs. wavelength relationship is referred to as a dispersion curve, which represents the "field signature" of the site under evaluation. The experimental dispersion curve is then matched with a theoretical dispersion curve in order to determine the shear stiffness profile at the site. Accurate interpretation of the phase velocity data measured in the field is, therefore, a critical step in developing a reliable shear wave velocity profile from the inversion procedure.

Field experience of the writers at pavement sites has demonstrated the complexity of interpreting near-surface data at sites with large and abrupt stiffness contrasts. At geotechnical sites, which often exhibit gradually increasing stiffness with depth, the data interpretation is quite straightforward, and robust shear stiffness profiles are obtained. However, at complex sites with abrupt changes in stiffness properties, multiple wave groups are present at some receiver spacings and the resulting phase difference plots are complex and difficult to interpret. Such phase plots can be easily misinterpreted, leading to erroneous dispersion data and eventually incorrect stiffness profiles. This behavior must be considered at geotechnical sites that have large and abrupt changes in stiffness properties such as polar environments, as well as some offshore sites containing gas hydrates, where frozen (stiffer) material is underlain by unfrozen (softer) material.

The SASW method has the potential to be a very valuable method for evaluating both the layering profile and stiffness properties at polar and offshore sites. Work has been underway at the Offshore Technology Research Center at Texas A&M University and the University of Texas at Austin for the past eight years to develop this method for these applications (Gauer, 1990; Lee, 1995; Luke, 1994; and Joh, 1996). To aid in the correct interpretation of data at complex geotechnical sites where stiffer material overlies softer soils, a new phase interpretation scheme, called "Impulse Response Filtration" (IRF), is proposed (Joh, 1996). IRF is based on the separation of wave groups by the filtration of the impulse response determined between two receivers. The separation of the wave groups is performed by applying an appropriate filter, determined from a Gabor spectrogram (Dziewonski et al., 1969), to the impulse response. This filtered impulse response leads to a clearer interpretation of the phase spectrum between receivers. The method is explained and demonstrated on experimental results from a pavement site with a large stiffness contrast. The pavement site can be considered to be a good representation of many of the difficulties that would occur in data interpretation at polar sites with frozen over unfrozen soils and at offshore sites with massive gas hydrate layers (Anderson et al., 1992).

SASW TESTING METHOD

The SASW method is a nonintrusive in-situ method of nondestructively measuring the shear stiffness properties of solid materials. The method has been used on many geotechnical sites including natural soil deposits, compacted earth materials, seabed materials, and solid waste landfills (Stokoe et al., 1988;

Kavazanjian et al., 1996; Luke, 1994); as well as concrete structures and highway and airport pavements (Kalinski et al., 1995; Nazarian and Stokoe, 1989). The test method involves actively exciting surface wave energy at one point and measuring the resulting vertical surface motions at various distances (receiver points) away from the source. Figure 1a shows the typical field testing arrangement. Measurements are performed along a linear array placed on the exposed surface. Fourier transforms are performed on the recorded time records (Fig. 1b) of two (or more) vertical receivers. The phase difference relationship between the receivers as a function of frequency (ϕ vs. f) is found from the cross power spectrum $G_{12}(f)$, defined by:

$$G_{12}(f)=S_1(f)\ S^*_2(f) \qquad (1)$$

where $S_1(f)$ is the Fourier transform of receiver 1 and $S^*_2(f)$ is the complex conjugate of the Fourier transform of receiver 2. A typical ϕ vs. f result is shown in Figure 1c. The ϕ vs. f plot in Fig. 1c is called a wrapped phase plot because of the "jumps" present in the plot. These "jumps" represent 360-degree phase shifts or full cycles of the wave. By properly counting these jumps, the phase plot can be unwrapped, as illustrated in Fig. 2a. From the unwrapped phase and frequency values, the phase velocity can be found from:

$$v = f * (360/\phi)*d \qquad (2)$$

where v is the phase velocity, f is the frequency, ϕ is the unwrapped phase angle and d is the receiver spacing. Therefore, a plot of phase velocity vs. wavelength can be determined as shown in Fig. 2b.

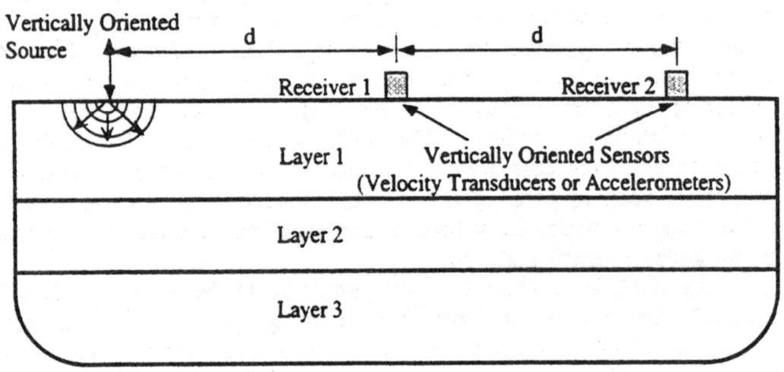

a. Field Arrangement of Source and Receivers

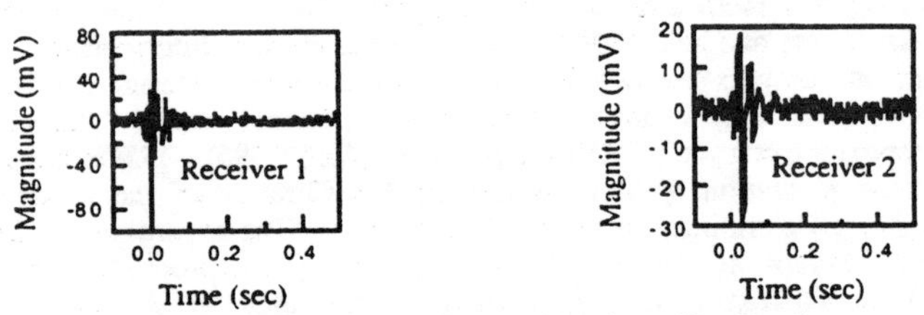

b. Time Records of Vertical Motion at Two Receivers

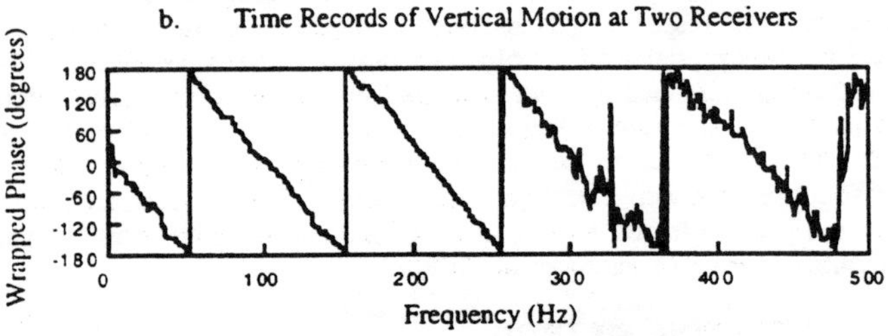

c. Wrapped Phase Spectrum Determined from Stress Waves Propagating between Receivers

Fig. 1 Typical SASW Field Test: Experimental Arrangement and Associated Spectral Measurement

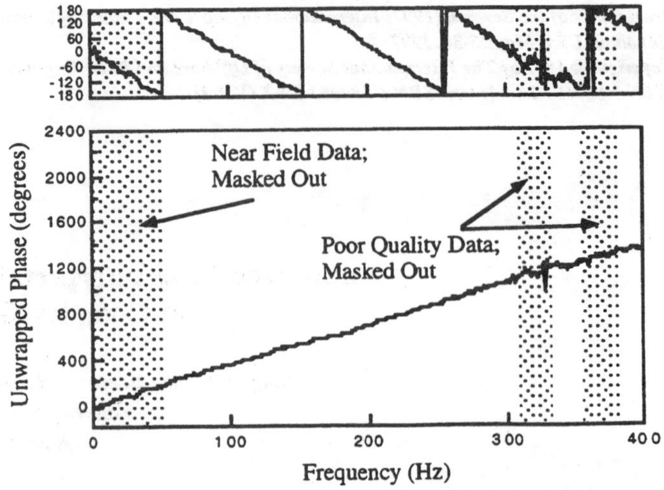

a. Unwrapped Phase Spectrum Derived from Fig. 1c

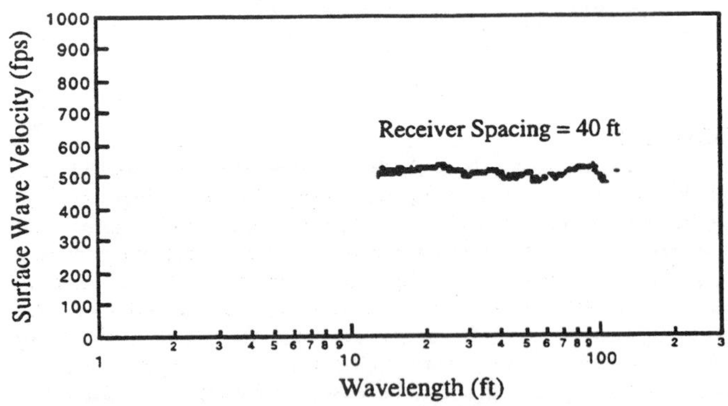

b. Phase Velocity Dispersion Curve Derived from Fig. 2a

Fig. 2 Unwrapped Phase Spectrum and Associated Dispersion Curve from Testing at One Receiver Spacing as Shown in Fig. 1

The SASW test procedure is repeated with many receiver spacings which cover a broad range of wavelengths as illustrated in Fig. 3. This process is followed so that wavelengths are measured which cover the shallow materials (high frequencies) to deep materials (low frequencies). A composite dispersion curve is created from measurements at all receiver spacings, as illustrated in Fig. 4a. Due to the large number of data points in the composite plot, an average dispersion curve with fewer points is calculated for the inversion process, as shown in Fig. 4b. Through an iterative inversion process of matching a theoretical dispersion curve with the average experimental dispersion curve, the shear stiffness profile can be evaluated (Joh, 1996). The stiffness profile is typically the final product of the SASW test. Therefore, the initial step of correctly interpreting the phase difference plots is vital to correctly determining the stiffness properties.

FIELD DATA FROM SASW TESTS PERFORMED AT A SITE WITH A STRONG STIFFNESS CONTRAST

The problems associated with testing geotechnical sites with a strong stiffness contrast are demonstrated in this paper by making use of results from tests performed on a runway at an airport in the United States. This pavement site clearly demonstrates the complexity of the measured phase spectrum. The site is also a good approximation of a frozen layer over unfrozen material. The site profile consists of a stiff upper layer (an asphalt layer over a

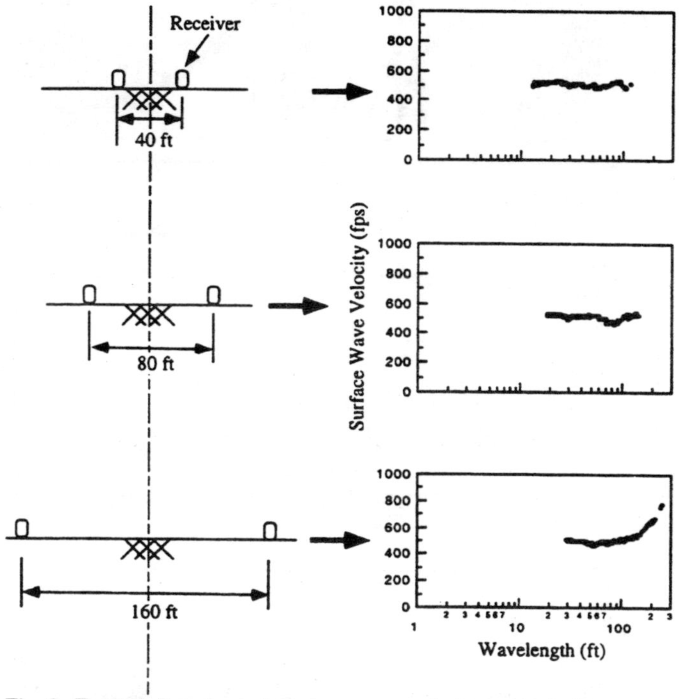

Fig. 3 Typical Receiver Arrangements and Associated Dispersion Curves

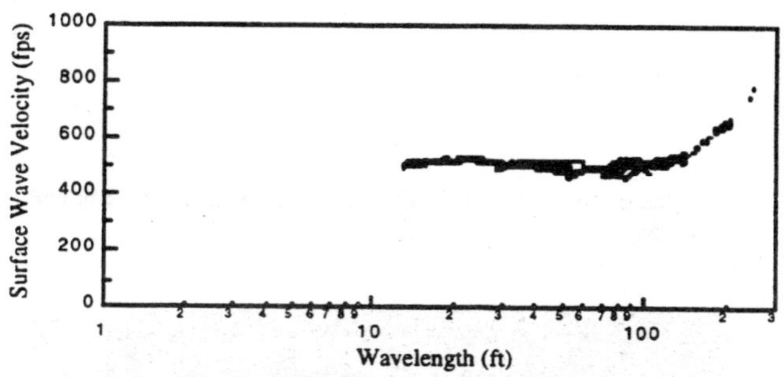

a. Composite Plot of Dispersion Curves from Three Receiver Spacings

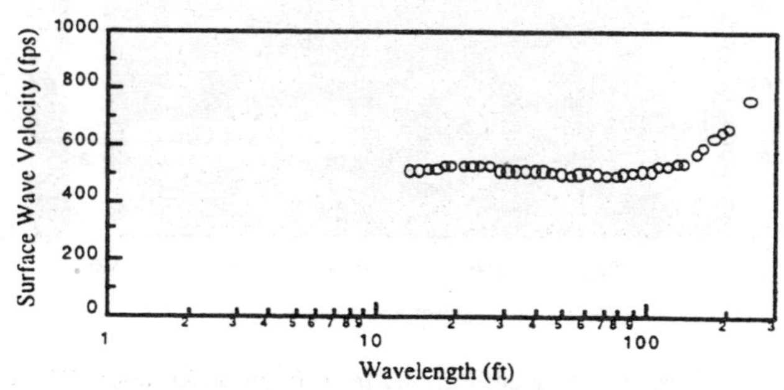

b. Average Dispersion Curve Developed from Dispersion Curves in Fig. 4a

Fig. 4 Combined Experimental Dispersion Curve from Three Spacings Shown in Fig. 3

concrete layer with a total thickness of about 2 ft (0.6 m)) underlain by a much softer subgrade. Figure 5a shows the time records recorded from one receiver pair at the site (d=32 ft (9.8 m)). The source in this case was a 10-lb sledge hammer. Figure 5b shows the corresponding phase difference plot in the frequency domain. As explained, correct interpretation of this plot is vital to determining the correct dispersion curve and, after inversion, the resulting stiffness profile at the site. A sample comparison of Fig. 5b with Fig. 1c (the results from a geotechnical site where the stiffness gradually increases with depth) demonstrates the complexity of the phase difference plot in two ways. First, unlike the ϕ vs. f plot shown in Fig. 1c, the number and location of 360-degree jumps in the phase plot in Fig. 5b can not be easily discerned. Secondly, and more importantly, even if the number of jumps could be identified, simply unwrapping the phase from these jumps would yield incorrect results. The complexity is due to the presence of multiple wave groups in the record, as discussed below. The filtration technique proposed below is a means of overcoming both of these problems by separating the record into individual wave groups and by determining the number of 360-degree jumps associated with each part of the record. The filtration technique is explained and demonstrated below using real data from the airport site.

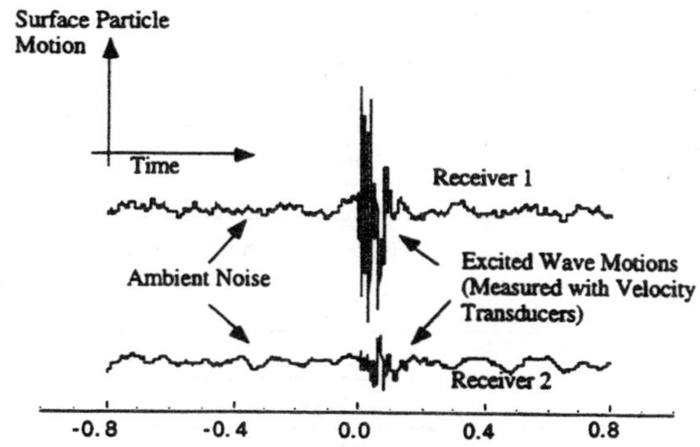

a. Measured Time Records

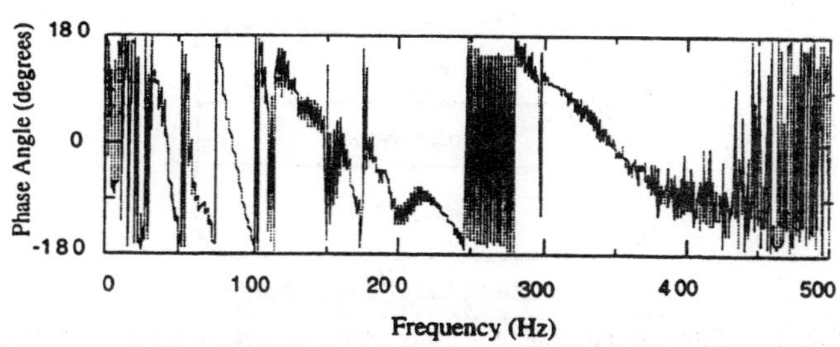

b. Phase Spectrum Derived from Time Records Shown in Fig. 5a

Fig. 5 Experimental Data from Stiff-over-Soft Site (Airport Pavement) with a Strong Stiffness Contrast (Concrete over Soil)

IMPULSE-RESPONSE FILTRATION TECHNIQUE (IRF)

In the case of layered systems with a strong stiffness contrast, multiple wave groups are observed due to the refraction and reflection of energy at the interface between the highly contrasting layers. Such strong contrasts are typically found in conventional seismic testing when the soil-rock interface is encountered. However, this condition may exist at many other sites where SASW testing can effectively be applied, as illustrated in Fig. 6. These multiple wave groups which travel at different velocities interfere with each other. As a result, the measured phase difference plot, which is the basis for SASW testing, is complicated. Without special analysis tools, unwrapping of the phase plot will likely be misinterpreted.

Figure 7 shows how the phase plot from Fig. 5b can be misinterpreted. In this case, the phase was unwrapped by assuming jumps at the numbered locations. Without further information, this unwrapping appears to be a good interpretation. However, it will be shown that the phase plot has been misinterpreted in two ways which will ultimately lead to the wrong dispersion curve, and hence an incorrect inverted profile. It is, therefore, important to investigate the wave groups present in the measured response to obtain a correct phase velocity dispersion curve. For this purpose, it is better to use the impulse response as a tool rather than individual time records so that the response of the material between receivers is investigated.

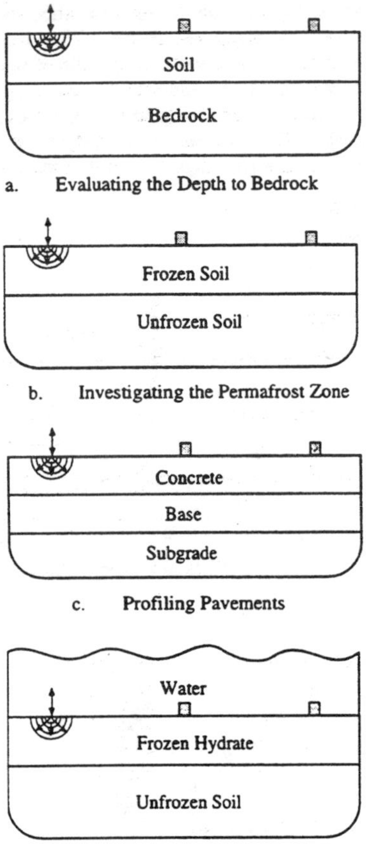

Fig. 6 Sites Exhibiting a Strong and Abrupt Stiffness Contrast where SASW Testing is Applicable

The impulse response can be calculated either by the deconvolution of the time signal from the second receiver out of the first receiver, or by the inverse Fourier transform of the transfer function. The impulse response, therefore, represents the response of the second receiver due to a unit impulse at the first receiver. The impulse response determined from the time records presented in Fig. 5a is shown in Fig. 8.

The wave groups present in the impulse response can be investigated using a Gabor spectrogram (Dziewonski et al., 1969 and Caiti et al., 1991). The Gabor spectrogram represents the original signal as a linear combination of time and frequency shifted Gaussian functions. Figure 9 shows the Gabor spectrogram determined from the impulse response shown in Fig. 8. From the Gabor plot, two distinct wave groups are observed. One is an earlier arrival in the higher frequency range (above about 105 Hz), while the other appears in the lower frequency range (below about 105 Hz). The complexity observed in the phase plot in Fig. 5b is due to

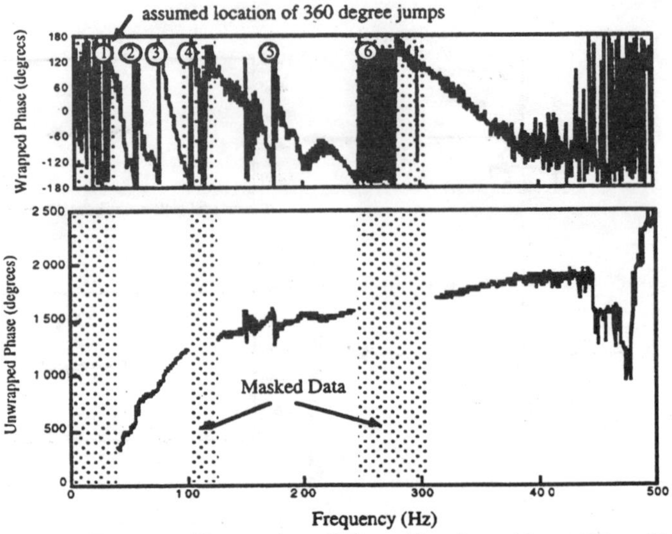

Fig. 7 Erroneous Unwrapping of Phase Plot from Airport Site with a Strong Stiffness Contrast

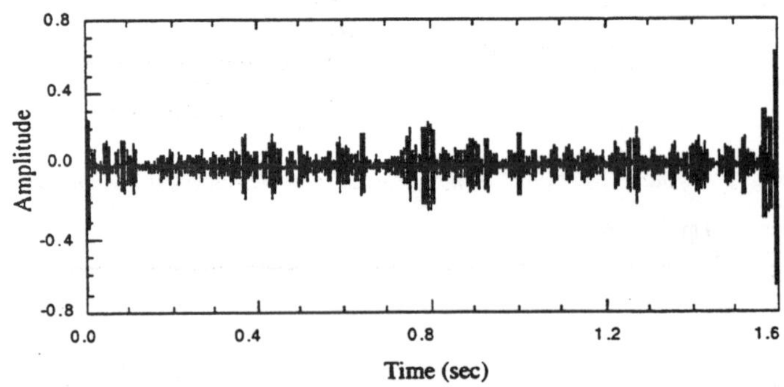

Fig. 8 Impulse Response of Data from Time Records Shown in Fig. 5a

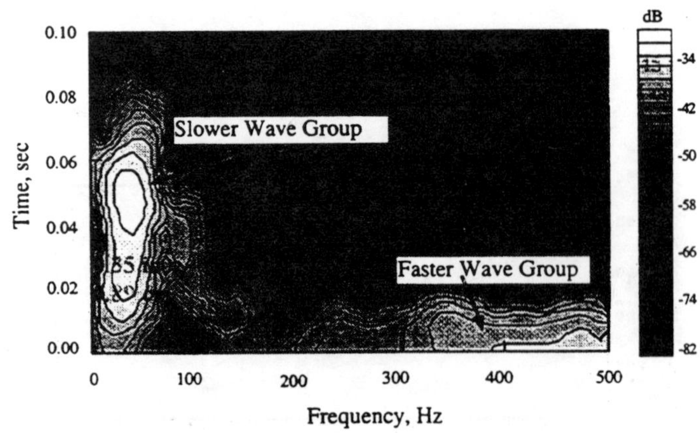

Fig. 9 Gabor Spectrogram Derived from Impulse Response Shown in Fig. 8

the presence of these two wave groups. The frequency value of 105 Hz is also observed to be the approximate boundary between the dissimilar portions of the phase plot in Fig. 5b. To correctly interpret this data, the contributions of each wave group must be separated. It is proposed that this can be done through a time domain filter, the attributes of which are determined from the Gabor spectrogram.

In the case of the Gabor spectrogram shown in Fig. 9, the boundary between wave groups is observed to be at approximately 0.02 seconds in the impulse response. Therefore, by applying an appropriate filter to the impulse response, the individual wave groups can be separated. The filters used to isolate the wave groups are shown with an expanded view of the impulse response in Figs. 10 a and b for the faster and slower wave groups, respectively. The window uses cosine taper functions leading into a region of unity where the signal is of interest.

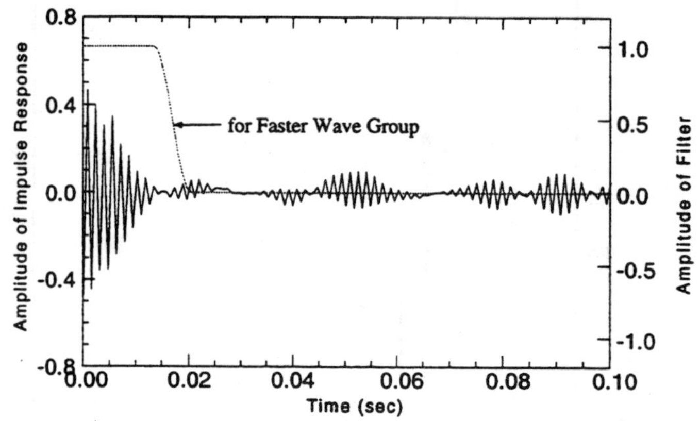

a. Filter Applied to Isolate Faster Wave Group

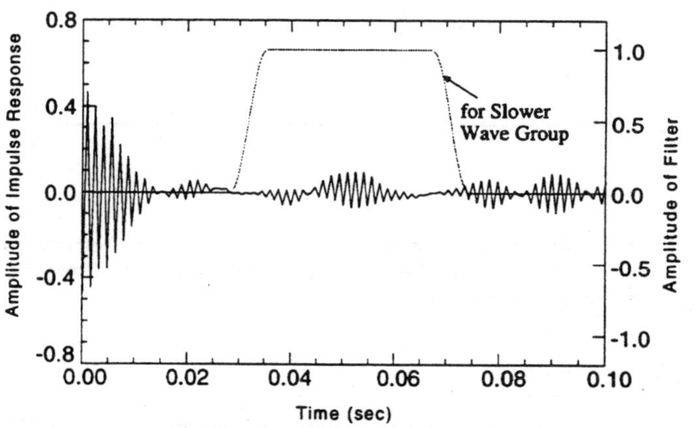

b. Filter Applied to Isolate Slower Wave Group

Fig. 10 Filters applied to Impulse Response to Isolate Wave Groups

The filters should be initially chosen to isolate as fully as possible the wave group of interest. This will produce a "sawtooth" pattern in a wrapped phase plot that, when superimposed on the measured phase plot, will indicate the locations of the jumps associated with that wave group. Figure 11a shows the original data with the isolated faster wave superimposed on top. Figure 11b shows the expanded portion of the wrapped phase plot with the slower wave group superimposed on the measured data. The original phase data can now be unwrapped based on the jumps indicated by the filtered phase data. The number and location of these jumps are shown in Figs. 11a and b.

The unwrapped filtered and original data as well as the misinterpreted data from Fig. 7 are plotted together in Fig. 12. The original interpretation does not match the IRF interpretation for two reasons. First, a jump in the phase that occurs at 9 Hz (as shown with IRF) was not taken into account in the original unwrapping. Second, in the first interpretation it was assumed that only one wave group existed, and the jumps were therefore counted consecutively.

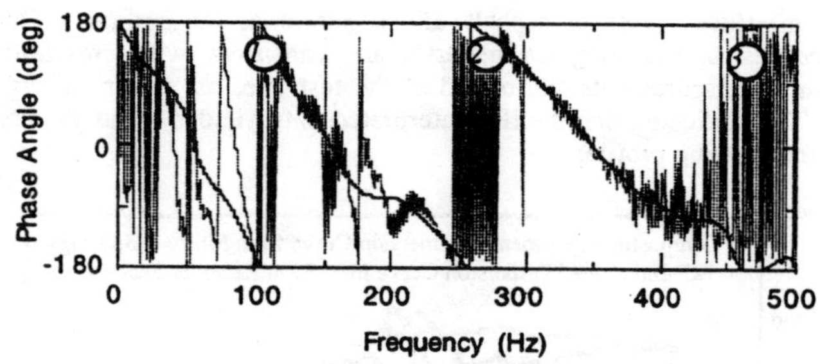

a. Unfiltered Phase Spectrum with Filtered Fast Wave Group Superimposed

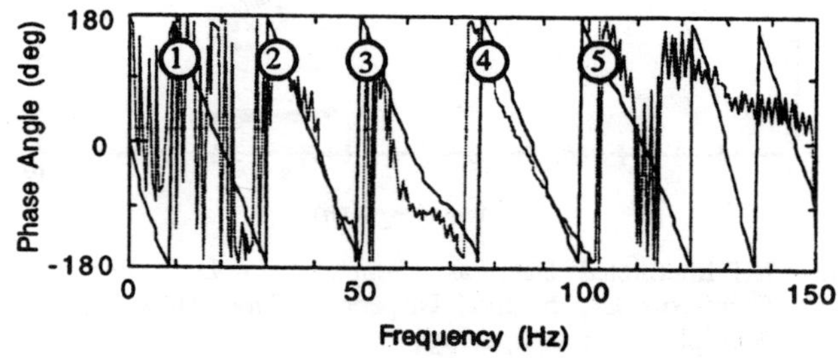

b. Expanded Unfiltered Phase Spectrum with Filtered Slow Wave Group Superimposed

Fig. 11 Comparison of Raw Phase Spectrum with Spectrum from Isolated Individual Wave Groups through Filtration of Impulse Response

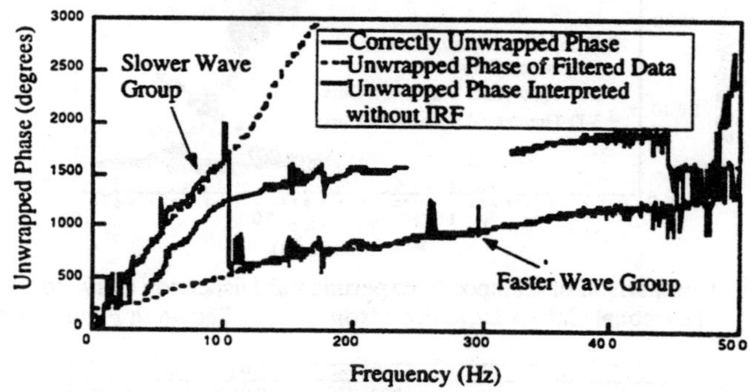

Fig. 12 Unwrapped Phase Spectrum based on Backbone Curves Shown in Fig. 11

The IRF interpretation demonstrates that two wave groups exist that must be isolated and dealt with separately.

To further illustrate the correctness of the IRF interpretation, the IRF-interpreted data from the 32-ft (9.8-m) receiver spacing are compared in Fig. 13 with other data obtained at other receiver spacings at this site. The IRF-interpreted data fit properly with the data from other receiver spacings. The composite experimental dispersion curve shown in Fig. 13 was used to extract the stiffness profile through iterative forward modeling. Figure 14a shows the match of two theoretical solutions to the experimental data. One theoretical solution, indicated as 2-D in Fig. 14a, considers the first mode of plane surface waves. The other solution, indicated as 3-D in Fig. 14a, includes all surface wave modes, as well as the contribution of body waves, from a point source excitation (Foinquinos, 1991). The 3-D solution was matched to the experimental data through an iterative forward modeling procedure. The final profile is shown in Fig. 14b. The thickness and velocity of the upper layers are consistent with construction plans indicating 1-ft

(0.3-m) thick layers of asphalt and concrete at the surface. The velocities of the subgrade material are consistent with crosshole seismic measurements performed at the test site, as shown in Fig. 15. This indicates that the IRF-interpreted data yielded valid results for the existing profile.

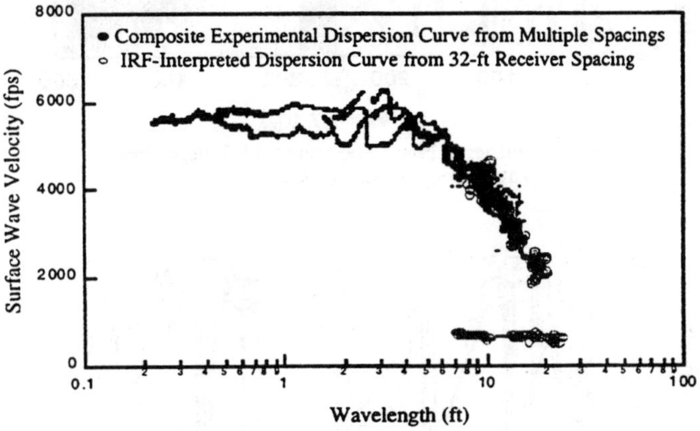

Fig. 13 IRF-Interpreted Data as Compared to Data from the Composite Experimental Dispersion Curve Obtained from Multiple Receiver Spacings

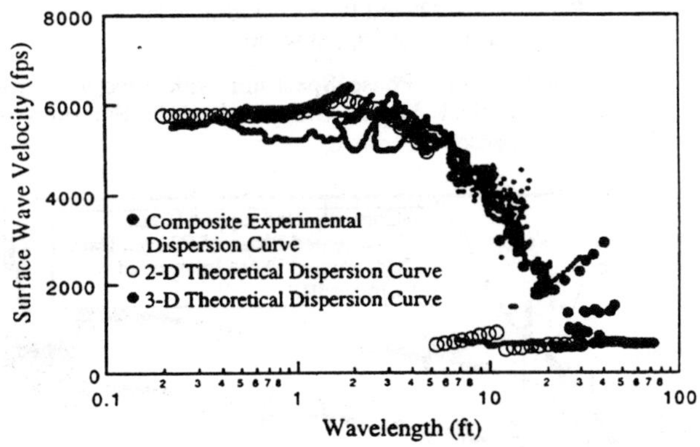

a. Comparison of Composite Experimental Dispersion Curve to Theoretical Curves Developed from Profile Shown in Figure 14 b

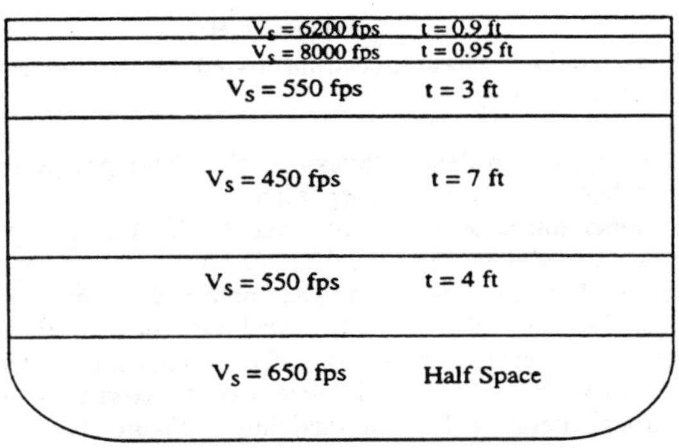

b. Profile Developed through Forward Modeling of Composite Experimental Dispersion Curve

Fig. 14 Comparion of Composite Experimental Dispersion Curve to Theoretical Dispersion Curve Obtained through Forward Modeling

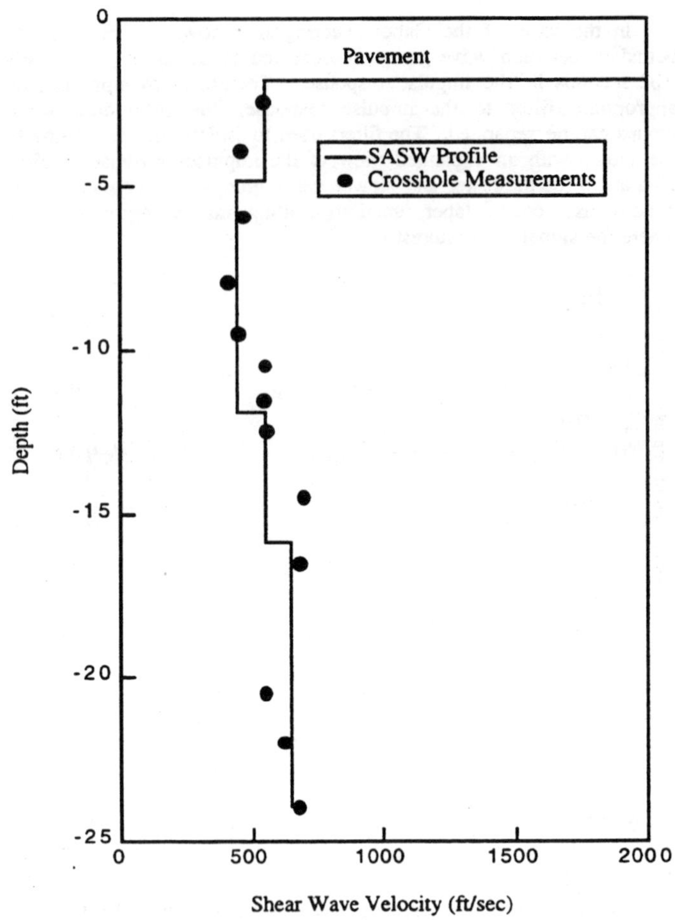

Fig. 15 Comparison of Shear Wave Velocity Profile Determined from the SASW Method to Nearby Crosshole Measurements of Shear Wave Velocity

CONCLUSION

In summary, the Spectral-Analysis-of-Surface-Waves (SASW) method holds great promise as a nondestructive and nonintrusive means of determining the stiffness properties and profiles at polar sites. However, testing at sites with strong stiffness contrasts creates special problems that are not encountered in conventional SASW testing. The strong presence of multiple wave groups and the associated complexity of the phase plot demand special tools. The Impulse Response Filtration (IRF) technique allows the wave groups of interest to be isolated such that the phase plot can be correctly interpreted. This will lead to reliable and robust stiffness profiles and properties at these sites.

ACKNOWLEDGMENTS

The authors gratefully acknowledge support from the Offshore Technology Research Center (OTRC). This Center is jointly administered by Texas A&M University and the University of Texas at Austin. Funding for the Center comes from the National Science Foundation (NSF) Engineering Research Center Program under Grant Number CDR8721512 and from the Offshore Industry.

REFERENCES

Anderson, A.L., E.D. Sloan, and J.M. Brooks, (1992). "Gas Hydrate Recoveries in the Gulf of Mexico: What is the Shallow Water Depth Limit for Hydrate Occurrence?," *Proceedings*, 24th Annual Offshore Technology Conference, Houston, Texas, pp. 381-385, May 4-7.

Caiti, A., T. Akal, and R.D. Stoll, (1991). "Determination of Shear Velocity Profiles by Inversion of Interface Wave Data," *Shear Waves in Marine Sediments*, J.M. Hovem et al., Eds, pp. 557-565.

Dziewonski, A. M., S. Bloch, and M. Landisman, (1969). "A Technique for the Analysis of Transient Seismic Signals," *Bull. Seismol. Soc. Am.*, Vol. 59, pp. 427-444.

Foinquinos, M. R., (1991). "Analytical Study and Inversion for the Spectral-Analysis-of-Surface-Waves Method," Master's Thesis, University of Texas at Austin.

Gauer, R.C., (1990). "Experimental Study of Applying the Spectral-Analysis-of-Surface-Waves Method Offshore," Master's Thesis, University of Texas at Austin.

Joh, S., (1996). "Advances in the Data Interpretation Technique for Spectral-Analysis-of-Surface-Waves (SASW) Measurements," Ph.D. Dissertation, University of Texas at Austin.

Kalinski, M.E., K.H. Stokoe II, J.O. Jirsa, and J.M. Roesset, (1995). "Nondestructive Identification of Internally Damaged Areas of Concrete Beam Using the SASW Method," Transportation Research Record 1458, pp. 14-19.

Kavazanjian, E. and N. Matasovic, K.H. Stokoe II, and J.D. Bray, (1996). "In Situ Shear Wave Velocity of Solid Waste from Surface Wave Measurements," 2nd International Conference on Environmental Geotechnics, Osaka, Japan, November 5-8.

Lee, B. S., (1995). "Analytical Study of Surface Wave Propagation Along the Seafloor and its Application to Underwater SASW Testing," Ph.D. Dissertation, University of Texas at Austin.

Luke, B.A., (1994). "In Situ Measurement of Stiffness Profiles in the Seafloor Using the Spectral-Analysis-of-Surface-Waves (SASW) Method," Ph.D. Dissertation, University of Texas at Austin.

Nazarian, S., and K.H. Stokoe II, (1989). "Nondestructive Evaluation of Pavements by Surface Wave Method," *Nondestructive Testing of Pavements and Backcalculation of Moduli, ASTM STP 1026*, A.J. Bush III and G.Y. Baladi, Eds., American Society for Testing and Materials, Philadelphia, pp. 119-137.

Stokoe, K.H. II, S. Nazarian, G.J. Rix, I. Sanchez-Salinero, J.C. Sheu, and Y.J. Mok, (1988). "In Situ Seismic Testing of Hard-to-Sample Soils by Surface Wave Method," *Proceedings*, Earthquake Engineering and Soil Dynamics II ASCE Specialty Conference, Park City, Utah, pp. 305-320, June.

Stokoe, K.H. II, S.G. Wright, J.A. Bay, and J.M. Roesset, (1994). "Characterization of Geotechnical Sites by the SASW Method," Technical Report: Geophysical Characterization of Sites, ISSMFE Technical Committee 10, edited by R.D. Woods, Oxford Publishers, Presented at the 13th International Conference on Soil Mechanics and Foundation Engineering, New Delhi, India, January.

Proceedings of the Seventh (1997) International Offshore and Polar Engineering Conference
Honolulu, USA, May 25-30, 1997

ISBN 1-880653-28-1 (Set); ISBN 1-880653-29-X (Vol. I)

Consolidation Characteristics of Pleistocene Clays Under Wide Range of Applied Pressures

Tamotsu Matsui, Kazuhiro Oda, Yasuyuki Nabeshima and Kwang Hoon Yang
Osaka University
Osaka, Japan

Abstract

A set of one-dimensional consolidation apparatuses is developed to investigate consolidation behavior of clays under a wide range of applied pressures. The performance of developed apparatuses is checked through a series of consolidation tests. Then, consolidation characteristics of undisturbed Pleistocene clays which were taken from Osaka Bay area are investigated. Especially, the effect of applied pressures on the consolidation characteristics is discussed comparing those of undisturbed Pleistocene clays and reconstituted Holocene clays. It is found that consolidation behavior of both undisturbed Pleistocene clays and reconstituted Holocene clays is similar at applied pressures of over 500 kgf/cm^2.

Key Words
compression curve, compression index, consolidation, Pleistocene clays

Introduction

Many large-scale reclamation projects such as Port Island, Phoenix project, Kansai International Airport, have been planned and carried out in Osaka Bay area. It is one of the most important problems in these projects to predict the consolidation settlement behavior of reclaimed lands during and after reclamation works. Therefore, it is essential to fully elucidate the consolidation characteristics of seabed deposits.

In Osaka Bay area, there are Holocene deposits, underlain by thick Pleistocene clay layers which alternately deposited with stiff sand and gravel layers. These Pleistocene clays usually shows stress overshooting in the one-dimensional compression curve and are considered to be in a quasi-overconsolidated state because of aging effect.

In this paper, the authors try to elucidate one-dimensional consolidation characteristics of undisturbed Pleistocene clays in a quasi-overconsolidated state under a wide range of applied pressures. Firstly, a set of one-dimensional consolidation apparatuses is developed, in which pressures from the minimum of 0.2 kgf/cm^2 to the maximum of over 1000 kgf/cm^2 can be applied. Secondly, the performance of developed apparatuses is checked through a series of consolidation tests for reconstituted kaolin clays. Finally, consolidation characteristics of undisturbed Pleistocene clays which were taken from Osaka Bay area are investigated under a wide range of applied pressures. Especially, the effect of applied pressures on such consolidation characteristics as one-dimensional compression curve and compression index is discussed.

Consolidation Apparatus

A set of consolidation apparatuses developed in this study consists of three major units, such as loading facilities, consolidation containers and measuring instruments. Notable points on these equipments are described below.

Loading facilities

Photographs 1 and 2 show the consolidation apparatuses for both low and high pressures, respectively. In this study, three loading facilities and pneumatic regulators are used, to conveniently control the applied pressure from 0.2 kgf/cm^2 to over 1000 kgf/cm^2. Table 1 shows the description of these loading facilities. For both low and middle pressures, conventional pneumatic cylinders are used, while for high pressure, the air-hydro method in which a lower pneumatic pressure can be mechanically converted into a higher oil pressure to generate high pressures. The generated high pressure can be consistently maintained by a magnet relay unit.

Consolidation containers

Figure 1 illustrates schematically consolidation container, which consists of consolidation ring, both upper and lower pistons and so on. Photograph 3 shows a consolidation ring and measuring instruments. Even at the maximum pressure of 1600 kgf/cm^2, the radial deformation of consolidation ring must be extremely small so as to satisfy the one-dimensional consolidation condition. In these apparatuses, the consolidation ring has the size of inner diameter of 60 mm, maximum outer diameter of 125 mm and height of 100 mm. Also, the inner side of consolidation ring is coated by teflon to reduce the friction between consolidation ring and specimens.

Photograph 1 Consolidation apparatus for low pressure

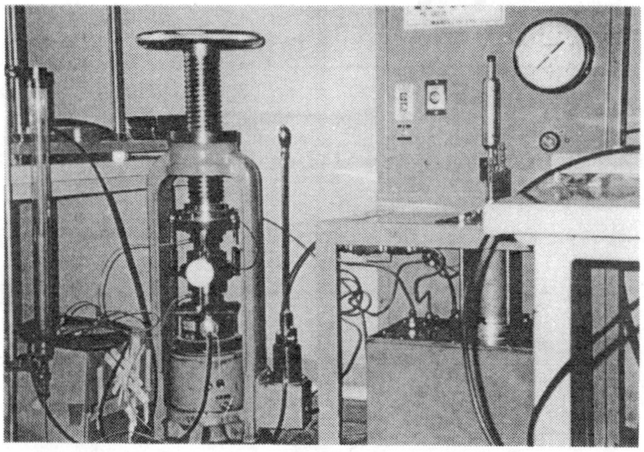

Photograph 2 Consolidation apparatus for high pressure

Table 1 Description of loading facilities

Applied pressure level	Device of loading	Range of generated pressure
low pressure	pneumatic cylinder	0.2-15 kgf/cm^2
middle pressure	pneumatic cylinder	15-300 kgf/cm^2
high pressure	Oil cylinder (Air-hydro method)	300-1600 kgf/cm^2

Measuring instruments

A load cell is set under lower pedestal to measure the load applied to specimens. The displacement of specimen is measured by a displacement gauge at the position of upper piston .

Checking of apparatus

A series of experiments is carried out to check the performance of the developed apparatuses. Kaolin clays are used as sample in these experiments. Table 2 shows physical properties of the kaolin clays. The kaolin clays are reconstituted at the water content of twice of liquid limit and consolidated at the pressure of 1.0 kgf/cm^2 under one-dimensional condition.

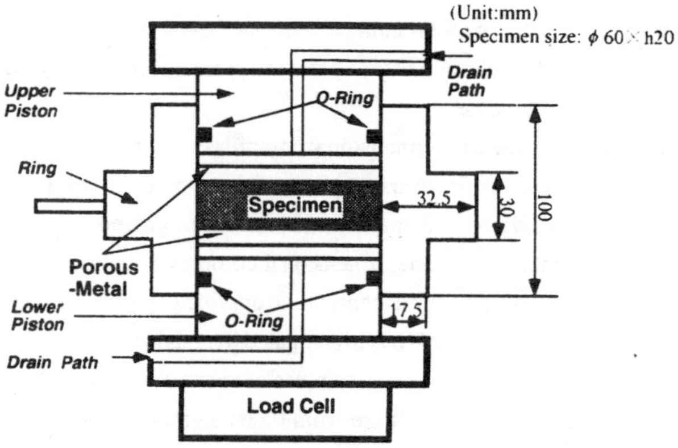

Figure 1 Consolidation container

Photograph 3 Consolidation ring and measuring instruments

Table 3 shows the loading stage together with applied pressures. The loading facilities for low, middle and high pressures are used for loading stages of 1 to 6, 7 to 9 and 10 to 12, respectively. The pressure at each loading stage is instantly applied to specimens and consistently maintained.

Figures 2 and 3 show the variations of settlement and average degree of consolidation, U, with the elapsed time, respectively. U is defined as

$$U = \frac{S_t}{S_f} \tag{1}$$

where S_t and S_f are settlements at an arbitrary time of t and at the end of consolidation, respectively. The consolidation settlements are affected by the applied pressure, so that time and settlement curves do not coincide each other. Most of time and U curves, however, are similar except for two cases where the applied pressures are 2.0 kgf/cm^2 and 3.9 kgf/cm^2. It is considered that the consolidation behavior in these two cases is affected by the yielding of the clays. It is found that the pressure level hardly affects the consolidation process.

Figure 4 shows the one-dimensional compression curve. The yielding of this kaolin clays occurs at the applied pressure of about 1.0 kgf/cm^2, which corresponds to the maximum pressure applied at the pre-consolidation process. The one-dimensional compression curve is almost straight in the semi-logarithmic chart at the applied pressures of over 1.0 kgf/cm^2. McConnachie (1974) has investigated the fabric change in kaolins due to consolidation. In his study, samples of reconstituted kaolins were consolidated at the pressure range from 0.001 kgf/cm^2 up to 1000 kgf/cm^2. It is shown that the observed compression curves are almost straight at the pressure range from 0.15 kgf/cm^2 to 1000 kgf/cm^2. This behavior is similar to that observed in this study. It is concluded that the one-dimensional consolidation behavior of clays under low to high pressures can be investigated by the developed apparatuses.

Table 2 Physical properties of kaolin clay

Specific gravity	2.62
Liquid Limit (%)	88.2
Plastic Limit (%)	28.3
Plastic Index	59.9

Table 3 Loading stage

Stage	Applied pressure (kgf/cm^2)	Incremental pressure (kgf/cm^2)	Incremental ratio for pressure	Loading facilities
1	0.4	-	-	low
2	0.9	0.5	1.25	low
3	2.0	1.1	1.22	low
4	3.9	1.9	0.95	low
5	8.0	4.1	1.05	low
6	16.2	8.2	1.03	low
7	36.1	19.9	1.23	middle
8	72.9	36.8	1.02	middle
9	148.9	76	1.04	middle
10	301.3	152.4	1.02	high
11	424.4	123.1	0.41	high
12	912.5	488.1	1.15	high

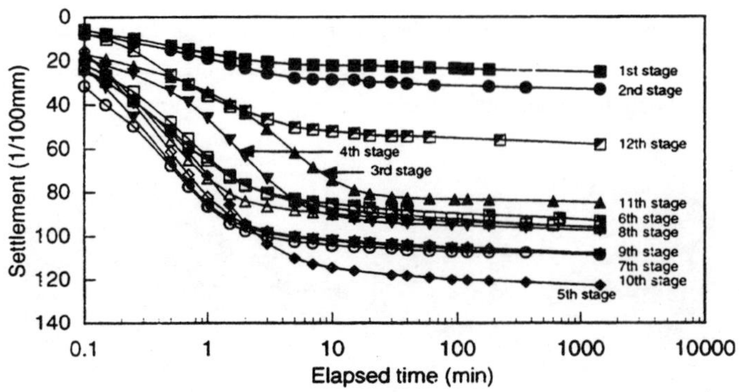

Figure 2 Relationship between elapsed time and settlement

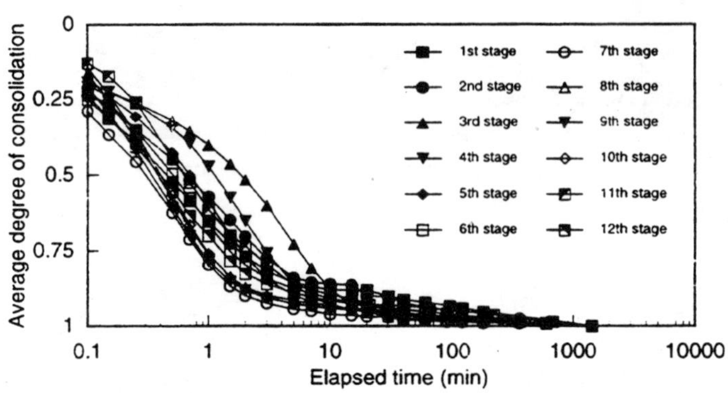

Figure 3 Relationship between elapsed time and average degree of consolidation

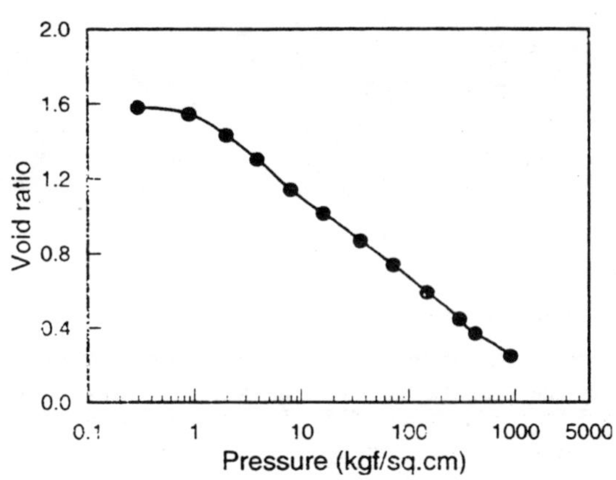

Figure 4 One-dimensional compression curve of Kaolin clays

Clay samples

Table 4 shows the physical properties of clay samples. Samples A and B were taken from Holocene clayey deposits and Upper Pleistocene clayey deposits, respectively. Sample A is reconstituted at the water content the twice of liquid limit and one-dimensionally consolidated at the applied pressure of 1.0 kgf/cm^2. Sample B is in undisturbed state. Atterberg limits of both samples lie the above A-line in the plastic chart, as shown in Figure 5. It is found that both clays have high plasticity.

Consolidation behavior of clays

One dimensional compression curve

Figure 6 shows the one-dimensional compression curves of Samples A and B. The yielding of Sample A occurs at the applied pressure of about 1.0 kgf/cm^2, which corresponds to the maximum pressure applied at the pre-consolidation process. The one-dimensional compression curve of Sample A is almost straight in the semi-logatithmic chart at the applied pressures from 1.0 kgf/cm^2 to about 50.0 kgf/cm^2. At the applied pressure of over 50.0 kgf/cm^2, the slope gradually decreases as increaseing the pressure.

The yielding of Sample B occurs at the applied pressure of about 8.0 kgf/cm^2, which is slightly greater than the estimated overburden pressure. The void ratio significantly decreases with the applied pressure just after the yielding, due to progressive collapse of natural soil structure. The slope of one-dimensional compression curve of Sample B gradually decreases as increaseing the applied pressure. Also, both of one-dimensional compression curves approximately coincide at the applied pressure of over 300 kgf/cm^2.

Burland (1990) suggested that the clays which are reconstituted at between 1.0 and 1.5 times the liquid limit and one-dimensionally consolidated have inherent mechanical properties which are independent of the natural state of soils. He termed the one-dimensional compression curve of these clays as 'intrinsic compression line (ICL)'. It was shown that the intrinsic compressive characteristics of ICL is dependent on the void ratio at the liquid limit, provided that the Atterberg limits of clays lie above the A line.

The Atterberg limits and specific gravity of Sample B are almost equal to those of Sample A, although the sedimentation age of Sample B is much older than that of Sample A. Also, it has been known that the mineral constituents of Holocene clays are similar to those of Upper Pleistocene clays in Osaka Bay area. Therefore, there may be few difference of 'intrinsic' compressive characteristics between Sample A and Sample B. The difference of consolidation behavior between Samples A and B at the applied pressure up to 500 kgf/cm^2 will occur because of the effect of soil structure. While, it is considered that the soil structure of Sample B becomes similar to that of Sample A at the applied pressure of over 500 kgf/cm^2.

Compression Index

Figure 7 shows the variation of compression index, C_c^*, with the applied pressure. C_c^* is defined as

$$C_c^* = -\frac{e_{i+1} - e_i}{p_{i+1} - p_i} \tag{2}$$

where e and p denote the void ratio and the applied pressure, respectively. Also, the subscript i denotes the loading stage.

C_c^* of Sample A monotonically increases as increasing the applied pressure within about 8.0 kgf/cm^2, and then gradually decreases at the

Table 4 Physical properties of natural clays

	Sample A (Holocene clays)	Sample B (Pleistocene Clays)
Specific gravity	2.69	2.65
Liquid Limit (%)	98.9	109.0
Plastic Limit (%)	38.1	35.4
Plastic Index	60.8	73.6

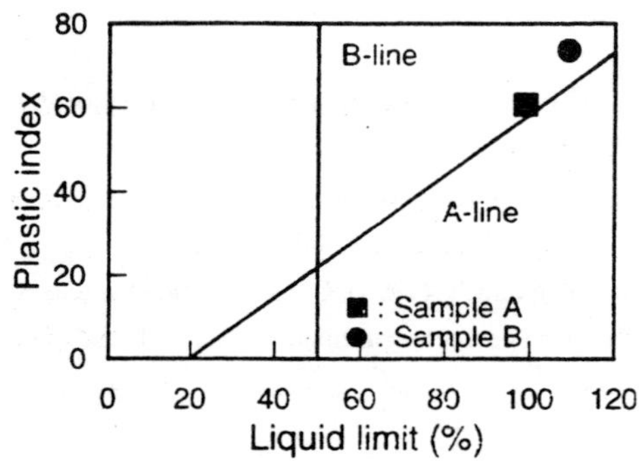

Figure 5 Plastic chart

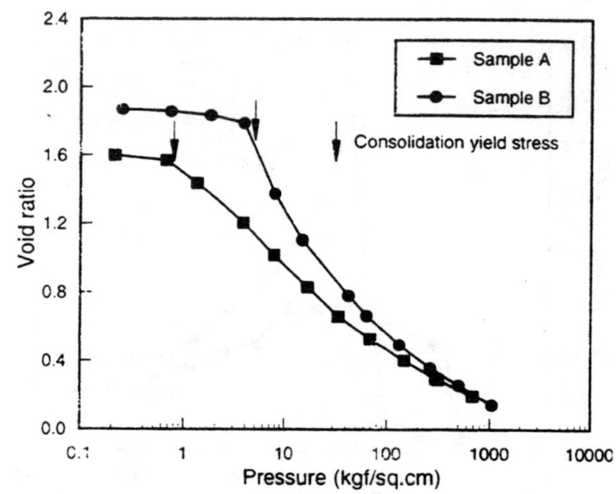

Figure 6 One dimensional compression curves of samples A and B

applied pressures of over 50 kgf/cm². While, C_c^* of Sample B dramatically increases as increasing the applied pressure at around the consolidation yield stress (pre-consolidation stress), and then significantly decreases as increasing the applied pressure from about 8.0 kgf/cm² up to about 500 kgf/cm². At the applied pressure of over 500 kgf/cm², C_c^* of Sample B is almost equal to that of Sample A.

Matsui and Sakagami (1994) investigated the consolidation behavior of Pleistocene clays in Osaka Bay area through several series of consolidation tests in which the maximum pressure up to 50 kgf/cm² was applied. They termed the stable part of compression curve at the pressure which is much higher than the consolidation yield stress as 'reference compression line'. They gave the relationship between liquid limit, w_L, and reference compression index, C_{cr}, which was defined as the slope of the reference compression line.

For undisturbed clays,

$$C_{cr} = 0.0080(w_L - 9.8) \tag{3}$$

For reconstituted clays,

$$C_{cr}' = 0.0066(w_L - 9.8) \tag{4}$$

The observed C_c^* at the normally consolidated state and the calculated C_{cr} and C_{cr}' are summarized in Table 5. For Sample A of reconstituted clays, C_{cr}' is almost equal to the maximum of C_c^*, which is measured at the applied pressure from 10kgf/cm² to 40kgf/cm². For Sample B of undisturbed clays, C_{cr} is in between the maximum and the minimum of C_c^* and almost equal to C_c^* at the applied pressure of about 40kgf/cm² as shown in Figure 7. Matsui and Sakagami (1994) estimated the compression curve at the applied pressure of about 40kgf/cm² as the reference compression line. Therefore, C_{cr} and C_{cr}' which are calculated by Eqs. 3

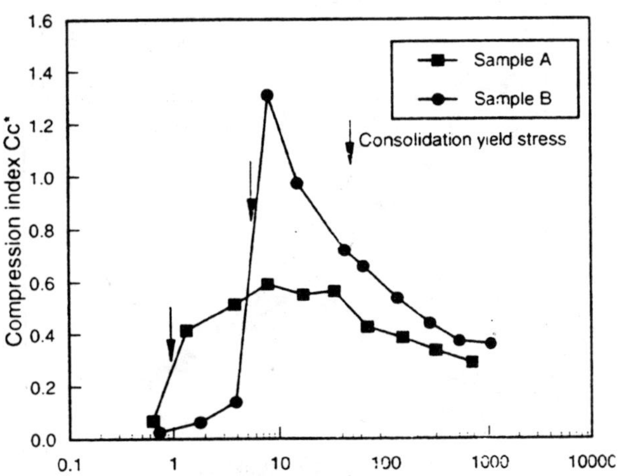

Figure 7 Variation of compression indices with applied pressure

Table 5 Observed and calculated compression indices

	Sample A	Sample B
Liquid limit (%)	98.9	109.0
Max. C_c^*	0.591	1.310
Min. C_c^*	0.288	0.360
C_{cr} (in Eq. 3)	-	0.794
C_{cr}' (in Eq. 4)	0.588	-

and 4, respectively, are almost equal to the C_c^* at the applied pressure of about 40kgf/cm².

Conclusions

In this paper, a set of one-dimensional consolidation apparatuses is developed to investigate the consolidation behavior of clays under a wide range of applied pressure. Main conclusions are summarized as follows:

1. A set of one-dimensional consolidation apparatuses is developed, in which pressures from the minimum of 0.2 kgf/cm² to the maximum of over 1000 kgf/cm² can be applied. Their performance was checked through a series of consolidation tests.

2. The one-dimensional compression curve of upper Pleistocene clays investigated in this paper matches that of the reconstituted Holocene clays at the applied pressure of over 500 kgf/cm². It is considered that the soil structure of undisturbed upper Pleistocene clays becomes similar to that of reconstituted Holocene clays at this significantly higher pressure level.

3. The compression indices of upper Pleistocene clays dramatically increase as increasing the applied pressure at around its consolidation yield stress, and then monotonically decrease. Both compression indices of upper Pleistocene clays and reconstituted Holocene clays approximately coincide at the applied pressure of over 500 kgf/cm².

References

1.McConnachie, I. (1974). Fabric changes in consolidated kaolin. Géotechnique 21, No. 2, 207-222.

2.Burland, J. B. (1990). On the compressibility and shear strength of natural clays. Géotechnique 40, No. 3, 329-378.

3.Matsui, T. & Sakagami, T. (1994). Consolidation characteristics of aged clays in relation to geotechnical history. Proc. 4th ISOPE Conference, vol. 1, 583-588.

Proceedings of the Seventh (1997) International Offshore and Polar Engineering Conference
Honolulu, USA, May 25-30, 1997

ISBN 1-880653-28-1 (Set); ISBN 1-880653-29-X (Vol. I)

Seepage Through an Embankment Computed by Grid Research Using FEM

Mahmudul Alam Chowdhury
Kagoshima University/ Miyazaki University

Masato Kunitake and Fumiyoshi Kondo
Miyazaki University

Miyazaki, Japan

ABSTRACT

A combination of saturated and unsaturated steady-state two-dimensional fluid flow analysis by the finite element method was applied for predicting the unknown soil parameters from the time-history of ground water level position of a superannuated earth dam. The saturated coefficients of permeability in the core wall and shell earthfill were estimated using finite element method, and grid points were arrayed along the assumed value of two concerned coefficients of permeability for applying a grid analysis procedure. The optimum values of saturated coefficient of permeability were confirmed with the minimization of a norm of the difference between the measured and calculated phreatic lines.
KEYWORDS: Phreatic line, superannuated earth dam, saturated coefficient of permeability, core wall, shell, inverse analysis.

I. INTRODUCTION

A study of the literature indicates that majority of investigators have been concerned with question of seepage problem of earth dam and earth structures. A relatively less elaborate study has been reported on inverse analysis method on the seepage and leakage of the saturated coefficient of permeability of old earth dams subjected to natural decay process. An experiment was conducted to study the effectiveness for predicting the soil parameters obtained from the monitored level of ground water movement of a superannuated earth dam by inverse analysis method.

Several researchers reported related information in this field during the past few decades. Fell *et al.* (1992) rated earth dam while subjected to natural degradation process cause damage and ultimately leads to structural failure. Leakage and piping through the dam embankment and from the foundation are the major causes of structural failure. The leakage problems in a superannuated Fukada earth dam has been selected as an experimental model. The aim of this study was to reduce the differences between the calculated and measured phreatic lines of the dam assuming two zones, core wall (impervious earthfill) and shell (pervious earthfill) for controlling seepage and predication of saturated coefficients of permeability by FEM. It involves estimation of seepage and unknown parameters by inverse analysis. The unknown parameters are the fluid head and the position of the free surface for which complex and nonlinear boundary conditions should be encountered. Information available in the literature indicates that the finite element method is found to be of general nature with respect to geometry, boundary conditions and soil properties, giving a powerful algorithm (programming sequence) for solving fluid flow problems in complex system. Several researchers have successfully applied this method for solving seepage flow problem (Desai, 1972), (Desai, 1973), (Freeze and Cherry, 1979), (Taylor and Brown, 1967) and (Zienkiewicz *et al.*, 1966).

The concept of seepage through the both saturated and unsaturated zones in an earth structure has been well studied by Neuman (1973a) using finite element method since the part of an earth dam above the phreatic line is generally unsaturated. Arai *et al.* (1985) report that the unknown permeability of porous medium can be determined from the monitored groundwater level by using a combination of saturated and unsaturated flow. Arai *et al.* (1991) and Chowdhury *et al.* (1996) further observed that inverse analysis enables to ascertain a set of unknown soil parameters by minimizing the sum of the squares of differences between the displacements observed in the field and the corresponding displacements calculated from values of soil parameters assumed in iterative calculation.

II. MATERIALS AND METHODS

2.1 Site Investigation and Data Collection

This experimental research was conducted to study the

seepage and leakage problems of a 200–year–old superannuated Fukada earth dam, exclusively used for irrigation purposes situated at Miyazaki, Japan. The dimension of this superannuated earth dam is 86 m long, 7 m high, with top width of the dam body is 1.86 m. The basin area is 86 km^2 having maximum water holding capacity of 50,000 m^3 in the reservoir. The dam irrigates about 150 km^2 of agricultural land. Leakage problem developed in the dam and the site investigation was conducted in 1983 (Hosoyamada *et al.*, 1987).

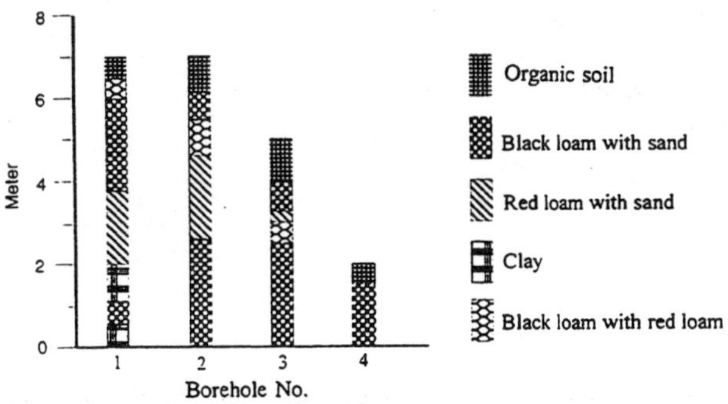

Fig. 1 Soil profile of the Fukada dam.

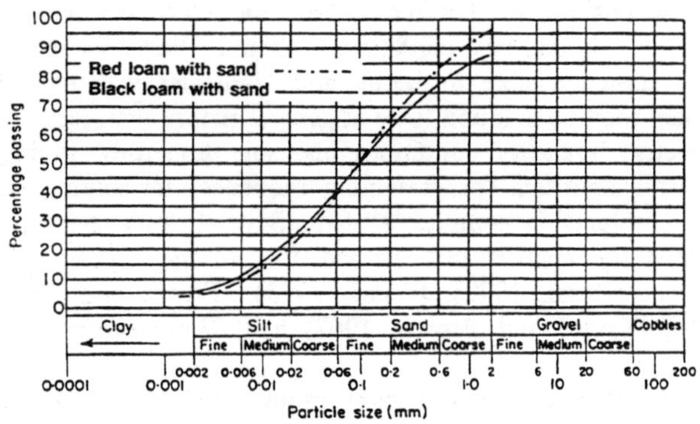

Fig. 2 Particle size distribution curves of embankment soils of the Fukada dam.

Four observation holes at some selected points from the top of the dam to the periphery were bored in 1984 at depth of 7 to 2 m for monitoring the ground water level in the dam. The soil profile collected from the borehole is depicted in Fig. 1. The profile consists of five main soil types which are organic, black loam with sand, red loam with sand, clay and black loam with red loam. Fig. 2 shows the particle size distribution curves and percentage passing through black loam with sand and red loam with sand. Hosoyamada *et al.*(1987) report the coefficient of permeability of organic soil as 1.01 x 10^{-4} m/s to 2.28 x 10^{-4} m/s and that of the paddy field soil as 6.67 x 10^{-4} m/s. The texture of the paddy field soil is black loam with sand and little organic matter.

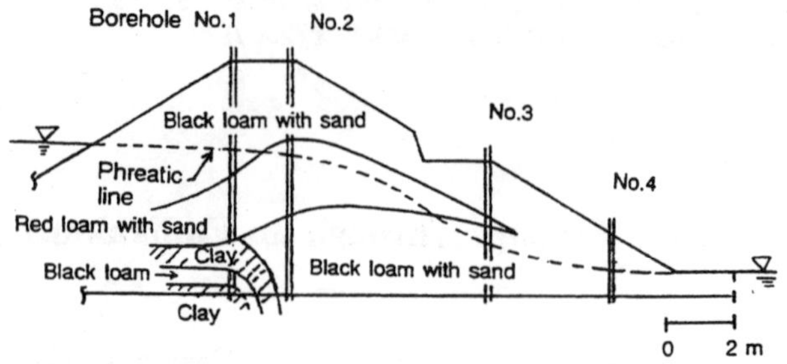

Fig. 3 Cross–section and phreatic line of the Fukada dam, 16 January 1984.

The particle size distribution curves indicates the soil profile contains a bit of clay (Fig. 2) and the saturated coefficient of permeability of a red loam with sand is lower permeability depends on the particle size distribution of the soil where the finer the particle size, the lower permeability. The variations in the water level, phreatic level, leakage through the dam body and rainfall from December, 1983 to December, 1985 were monitored (Hosoyamada *et al.*, 1987). From the measured data in water tables in the four boreholes, cross–section and phreatic line of the Fukada earth dam is drawn and shown in Fig. 3. It appears from the figure that the phreatic line runs over red loam soil having sand in the middle of the dam and then over black loam with sand at the periphery of the dam upto the red sandy loam layer at the edge.

2.2 Analytical Model

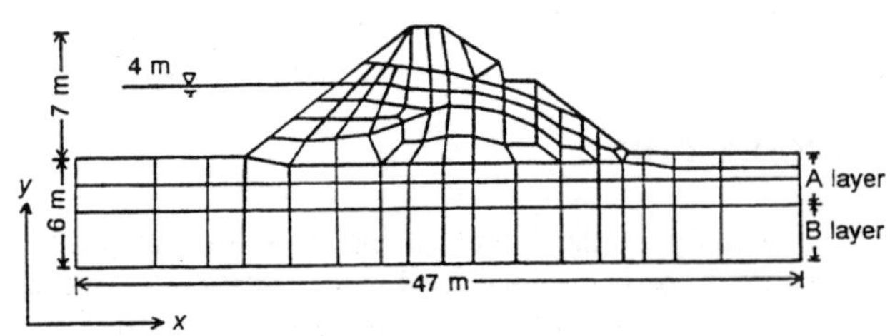

Fig. 4 Finite element–mesh used for the analytical model of the Fukada dam on double–layer foundation.

The dam is assumed two–dimensional and non–homogeneous and divided into three regions for the analytical purpose: the body of the dam, the upper ground foundation (A layer) and the lower ground foundation (B layer). Permeability under the foundation ground assumed to be constant along the width axis direction and the upstream water depth was 4 m. The soils of the dam are classified into five categories according to the saturated coefficient of permeability, k^s_1, k^s_2, k^s_3, k^s_4 and k^s_5. Applying the finite element method the body of the dam model was sectioned into 116 elements and 397 nodal points were used. Following the above assumptions, the analytical model of the Fukada dam is presented in Fig. 4.

2.3 Formulation of the Model

Neuman (1973a) concept of finite element method for steady-state fluid flow through the saturated-unsaturated porous media and Gauss-Newton's iteration algorithm were applied to arrive at an approximate solution of the problem. In case of an incompressible two-dimensional steady state fluid flow through a porous media of permeability, the pore water pressure p is defined as

$$p = \bar{\rho}_f \bar{g} (h - \bar{z}) \quad (1)$$

where $\bar{\rho}_f$ = fluid density, $\bar{g}$ = acceleration due to gravity, h = hydraulic head and $\bar{z}$ = elevation.

The governing equation for the porous medium is expressed as

$$\frac{\partial}{\partial x}\left(k^r k^s \frac{\partial h}{\partial x}\right) + \frac{\partial}{\partial y}\left(k^r k^s \frac{\partial h}{\partial y}\right) = 0 \quad (2)$$

where h is the hydraulic head or potential as the sum of the elevation head and pressure head, k^r = the relative hydraulic conductivity and k^s = saturated permeability tensor at saturation. $0 \le k^r < 1$ at unsaturation and $k^r = 1$ at saturation.

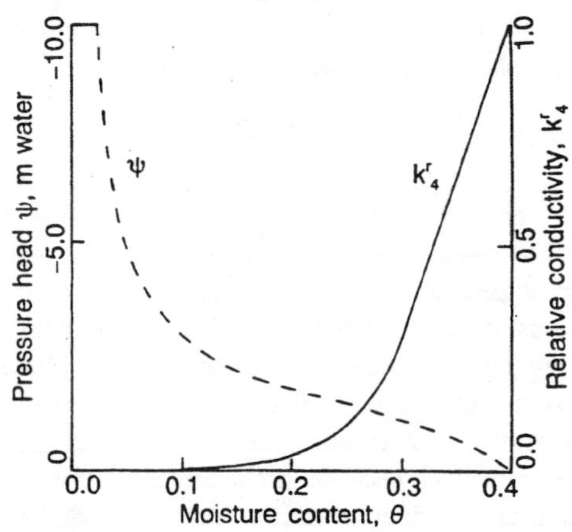

Fig. 5 ψ − θ and k^r_4 − θ function of a core for analyzing seepage through the Fukada dam (Neuman 1973a).

Hence, the hydraulic head $h = \psi + \bar{z}$, where the pressure head

$$\psi = \frac{p}{\bar{\rho}_f \bar{g}}.$$

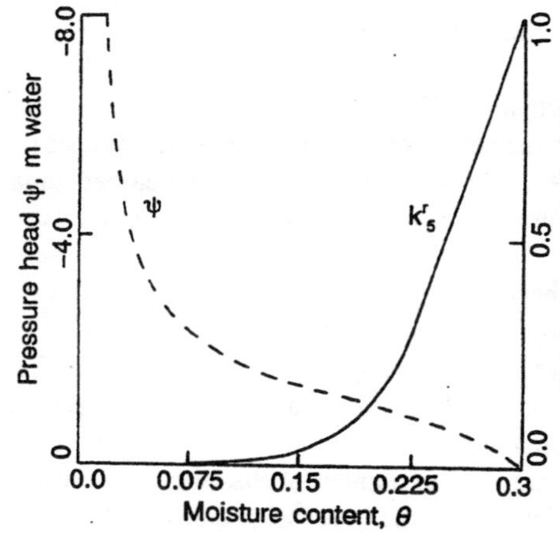

Fig. 6 ψ − θ and k^r_5 − θ function of a shell for analyzing seepage through the Fukada dam (Neuman 1973a).

The value of ψ is variable. It is greater than zero in the saturated zone below the phreatic line, equal to zero at the phreatic line and less than zero in the unsaturated zone. It appears that water is held in the soil pores under the force of surface tension called capillary suction or tension head in unsaturated zone. The pressure head ψ and the relative conductivity k^r are the functions of the moisture content θ in analyzing the unsaturated flow. Experimental results showed that ψ−θ and k^r−θ relationship significantly differed for the different types of soil. The relationship between the pressure head ψ, relative hydraulic conductivity k^r and moisture content θ is presented in Fig. 5 and Fig. 6 neglecting the hysteresis effect. ψ is a single-valued function of θ and k^r is also one of θ. The relative hydraulic conductivity function of the core wall and shell used in the analysis are shown in Figs. 5 and 6. The typical boundary conditions related with the free surface flow are

$$h = \bar{h} \quad on \quad S_1 \quad (3)$$

$$k^r k^s \frac{\partial h}{\partial x}\frac{\partial x}{\partial n} + k^r k^s \frac{\partial h}{\partial y}\frac{\partial y}{\partial n} = \bar{Q} \; on \; S_2 \quad (4)$$

here, S_1 is the part of the boundary on which h is prescribed; S_2 is the part on which the flow, $\bar{Q}$, is presented and n denotes the normal to the boundary S_2. The complied equations are obtained by adding element equations using the direct stiffness method. To consider the associated boundary conditions, the assembled equations can be written as

$$[k]\,\{h\} = \{F\} \quad (5)$$

where [k] is the permeability matrix, {h} is the nodal hydraulic head vector; and {F}= the nodal flux vector.

2.4 Inverse Analysis

Neuman (1973b) stated direct and indirect method to solve the parameters identification by inverse analysis procedure besides some methods those are already developed so far. In the indirect inverse analysis method the technique used is to decrease the actual difference between the observed and calculated head at some selected inspection points. The parameter optimization was obtained by examining the dependent variables at the spatial and time domains. The extent of optimization depends on the accuracy of the measured data collection. The indirect method has the advantage of solving the problem having limited number of observation points, requiring no differentiation of the measured data (Yeh, 1986). Thus present experiment involves indirect method to determine the unknown saturated coefficients of permeability to achieve better alignment of phreatic lines. The procedure was to start (a sequence of programmed equations) algorithm with the initial sets of estimates of parameters until the model approaches sufficiently closer to the observed parameters.

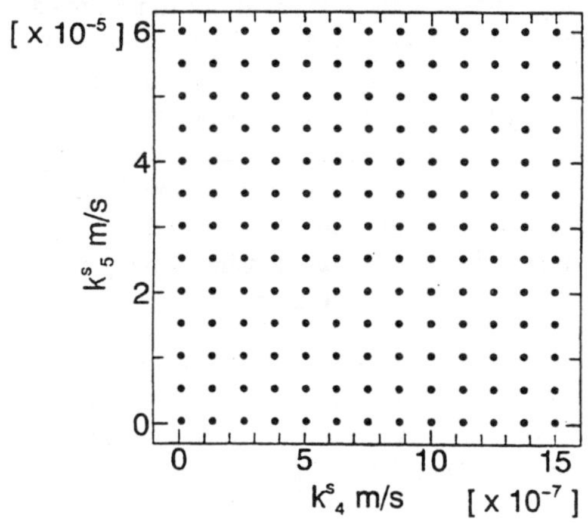

Fig. 7 Sequence of grid points for determining the values of k^s_4 and k^s_5 in case 3 of the Fukada dam.

A grid point analysis was used to evaluate the minimum difference between the measured and calculated phreatic lines, varying the values of k^s_4 and k^s_5. Assuming the two-dimensional model, finite element method analysis computer program was coded by Fortran Language, 13 grid points were arrayed along the values of k^s_4 and k^s_5 for estimating the saturated coefficients of permeability of k^s_4 and k^s_5. Initially two basic values one for k^s_4 and the another for k^s_5 were assumed stating the magnitude within their minimum and maximum range. The initial assumed basic values of k^s_4 and k^s_5 were plotted in the middle of a grid field having 13 points along the k^s_4 (X axis) and k^s_5 (Y axis). There were in all 169 grid points as shown in Fig. 7. With these basic values program was run to calculate for each and every grid point varying k^s_4 and k^s_5 values to reduce the difference of phreatic lines. Contour lines showing equal permeability were drawn using assumed values of k^s_4, k^s_5 and corresponding calculated sum of the squares of the residuals between the calculated and measured phreatic lines and this line was used in determining the optimum saturated coefficient of permeability values of k^s_4 and k^s_5. The calculation were repeated by changing the assumed values of k^s_4 and k^s_5 until a significant graphical contour line was obtained, Fig. 14. It was apparent from the graphical contour lines that the center of the contour lines indicated the unknown values of k^s_4 and k^s_5 for the minimum difference between the measured and calculated phreatic lines (Ogata *et al.*, 1995). Apply the inverse analysis procedure attempt was made to solve the problem by replacing the optimization problem to confirm the optimum saturated coefficient of permeability values of core wall and shell earthfill with the following objective function:

$$\text{minimize} \quad J = \frac{1}{M}\sum_{m=1}^{M}\left(h^c_m - h^b_m\right)^2$$

Where M = number of inspection positions including the observation boreholes, h^b_m = phreatic level observed at the m–th inspection position and h^c_m = the calculated phreatic level at the m–th inspection position, which minimized the difference between the measured and calculated phreatic lines by assuming the values of various soil parameters in iterative calculations (Arai *et al.*, 1985). To estimate the optimum saturated coefficient of permeability values, a minimum difference between the phreatic lines was considered.

III. RESULTS AND DISCUSSION

A homogeneous dam contains the same type soil material all over section and constant coefficient of permeability values are expected in the whole section. In case of non–homogeneous or the composite section, earth dams opposite situation prevails. The composite earth dams may contain two or more zones of different permeability. The coefficient of permeability of the materials in the shell is always higher than that in the core wall because shell contains coarse sand and gravel material and finer materials such as silt and clay in the core wall section.

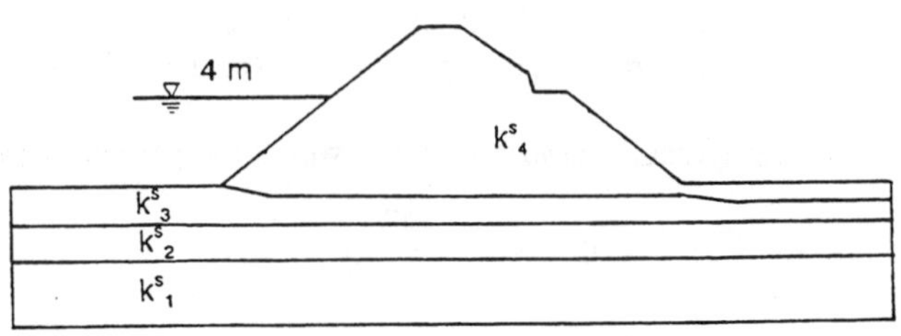

Fig. 8 Cross–section in case 1, assuming a homogeneous type dam model.

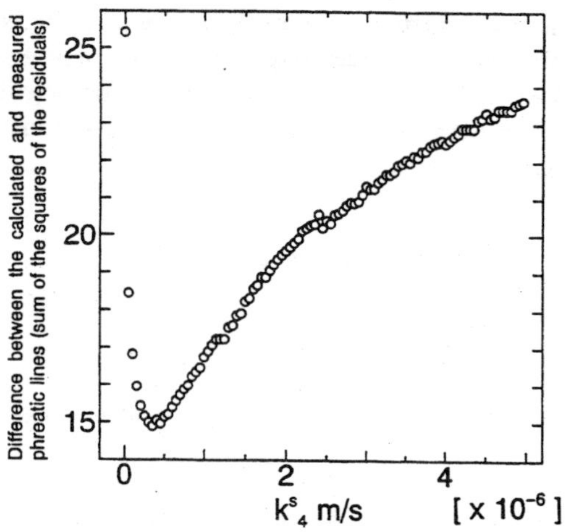

Fig. 9 Plotted result of grid research for determining the values of k^s_4 in case 1 of the Fukada dam.

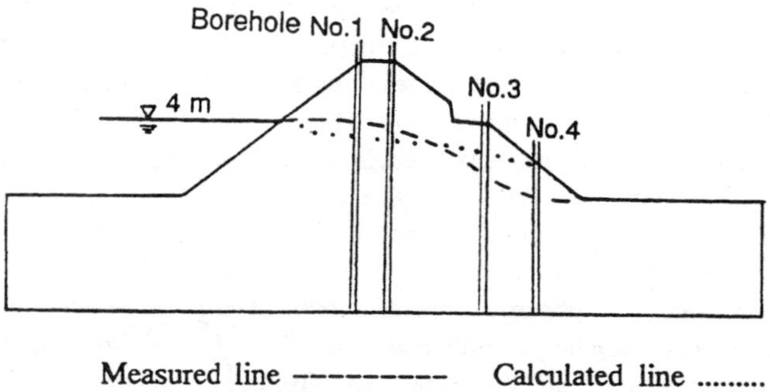

Fig. 10 Comparison of calculated and measured phreatic lines in case 1 of the Fukada dam.

Fig. 8 represents the body of the analytical dam model where it was considered to be of the homogeneous zone type. The saturated coefficients of permeability of three layers of foundation in Fig. 8, 11, and 12 were found to be 1 x 10^{-7} m/s of k^s_1, 5 x 10^{-7} m/s of k^s_2 and 2 x 10^{-6} m/s of k^s_3 respectively, and these values remained constant during computation of inverse problem. Set of analysis was done by varying the of value of k^s_4. The effort of this attempt was to carry out an acceptable alignment between the measured and calculated phreatic lines. During these iterative calculations the difference between the phreatic lines (sum of the squares of the residuals) varied with the changes of coefficient of permeability k^s_4 as shown in Fig. 9 some assumed coefficient of permeability of k^s_4 were responded better to carry out the reduction value in the difference of measured and calculated seepage lines by creating a narrower discrepancy between the phreatic lines. Here, Fig. 9 shows the optimum coefficient of permeability of k^s_4 (3.3505 x 10^{-7} m/s) for a minimum reduced value (sum of the squares of the residuals is 14.90595) in the difference of measured and calculated phreatic lines. And taking this estimated saturated coefficient of permeability it was possible to obtain a calculated phreatic line with the narrowest discrepancy as shown in Fig. 10. However, by considering the body of the analytical dam model as a homogeneous zone type, it was not successful to carry out a significant alignment of the measured and calculated phreatic lines to obtain a optimum reduced value in difference to solve the concerned seepage problem. It appears that while comparing measured phreatic line with the position of the free surface determined as a homogeneous type dam (Fig. 10) the change in hydraulic head mostly occurs in the region between borehole No.2 and No.4. Then it was assumed that there was a lower permeable zone which having a red loam with sand and it was assumed to be an impervious zone as shown in Fig. 3.

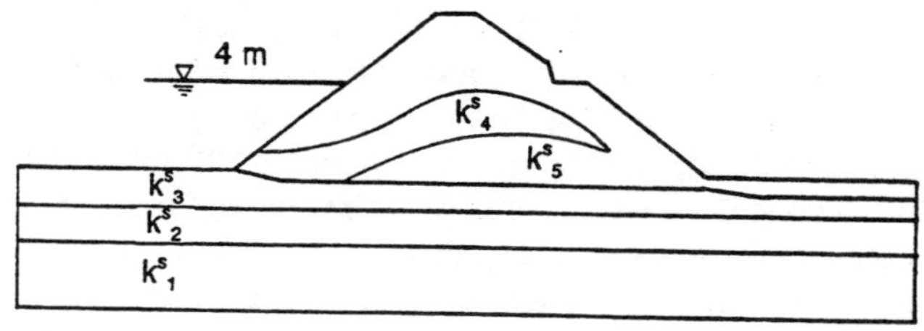

Fig. 11 Cross-section in case of impervious zoning 1 of the Fukada dam.

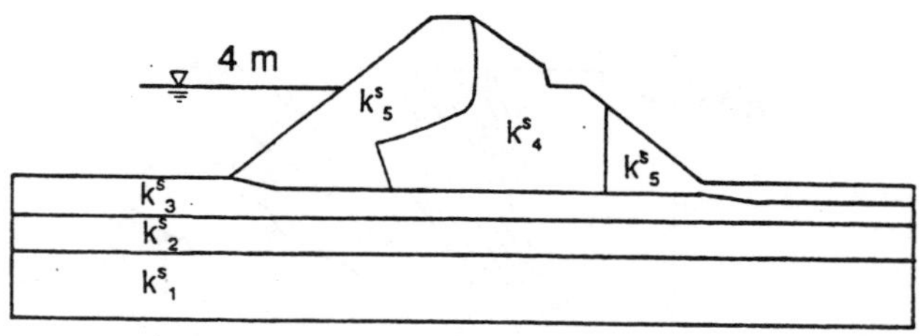

Fig. 12 Cross-section in case of impervious zoning 2 of the Fukada dam.

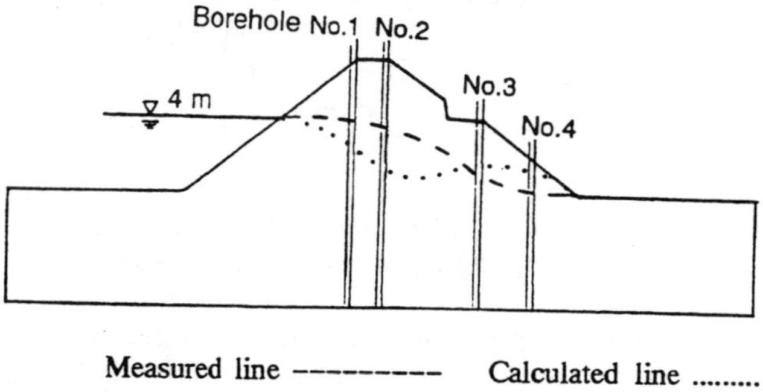

Fig. 13 Comparison of calculated and measured phreatic lines in case 2 of the Fukada dam.

In fact, there are no standard guideline to define zoning of embankment or earth dams has been reported so far. Fell (1992), Fredlund (1993) and Hosoyamada (1987) assumed impervious core wall earthfill from their result of site investigation. Accordingly two assumed zones are shown in Fig. 11 and Fig. 12 in order to solve this problem. From the site investigation report of Fukada earth dam, a soil material zone is expected near the location of boerhole No.2 which have a lower permeability observed from the significant downward change of the measured phreatic line. From the cross-section and soil profile three main soil zones may visualized in the body of the dam (Fig. 1 and Fig. 3) but zones have two different permeability values. The initial composite analytical earth dam model is established dividing the model into five soil zones, designated by k^s_1 to k^s_5 according to the coefficients of permeability of soil materials from the data on the initial soil parameters from the site investigation report of Fukada earth dam by Hosoyamada (1987). The soil material having core wall shows lower permeability k^s_4, and k^s_5 for shell are presented in Fig. 11. The saturated coefficients of permeability of core and shell earthfill, k^s_4 and k^s_5, are assumed as 1×10^{-7} m/s and 4×10^{-6} m/s and these were confirmed by inverse analysis method. While calculation was carried out using this initial assumed soil parameters, large deviation 1452×10^{-3} m was observed between the measured and calculated phreatic lines in case 2 as shown in Fig. 13. The measured discharge was 0.11181m^3/d/m and the calculated one was 0.17007m^3/d/m. Here, the calculated discharge was found to be 52% more than the measured discharge.

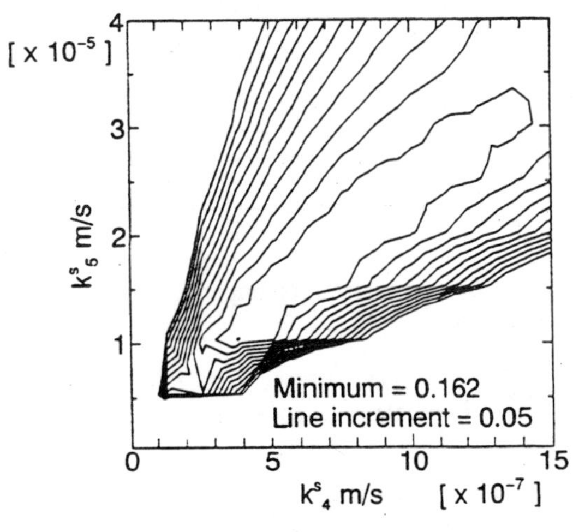

Fig. 14 Plotted result of grid research for determining the values of k^s_4 and k^s_5 in case 3 of the Fukada dam.

Then a wider impervious zone was assumed in the region between borehole No.2 and No.4, as shown in Fig. 12, because change in hydraulic head mostly occurs in the region around the impervious zone, as depicted by the concentrated distribution of equipotential lines in the impervious zone (Fredlund and Raharfjo, 1993). As the difference in the coefficients of permeability between the shell and the core wall increases, greater hydraulic head changes take place in the core wall. In this case 3, a wider impervious zone was assumed because the difference in the saturated coefficients of permeability between the shell and the core wall soil materials was one order.

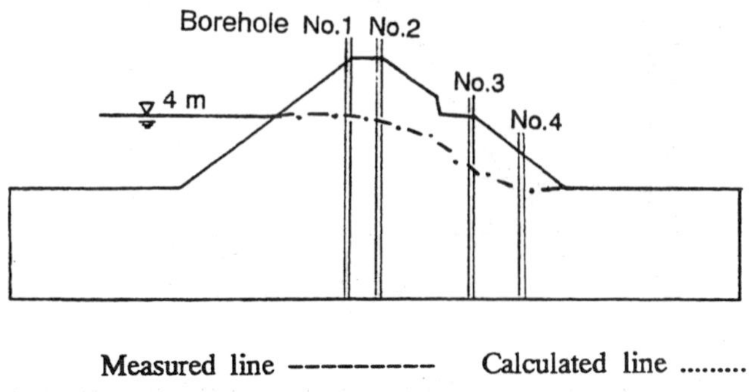

Fig. 15 Comparison of calculated and measured phreatic lines in case 3 of the Fukada dam.

A new assumption for the zone arrangement in the case 3 shown in Fig. 12. With these assumed soil parameters, calculations were carried out resulting in the calculated phreatic line closer to the measured one. Then in case 3, the optimum saturated coefficient of permeability values 3.38×10^{-7} m/s of k_4^s, and 0.898×10^{-5} m/s of k^s_5 were estimated from the obtained contour line as shown in Fig. 14. This resulted in a very negligible difference, about to nil, between the measured and calculated phreatic lines as shown in Fig. 15. Where, the calculated discharge obtained was 0.12847 m^3/d/m, being 14.9% more than the measured discharge. In case 3 the alteration of core wall and their estimated optimum permeability made it possible to carry out a precise alignment of the phreatic lines and discharges. The measured and calculated phreatic lines were well aligned, and the difference between them was almost negligible; thus, the calculated discharge was accepted as authentic.

IV. CONCLUSIONS

4.1 A significant reduction is attained between the calculated and measured phreatic lines but the calculated discharge greater. Because the measured volume is only the discharge through the body of the dam whereas the volume of discharge through the foundation of the embankment was not measured. But in this analysis both the discharge through the body of the dam and the foundation of the embankment were calculated. It appears that the calculated discharge was increased more than the measured one.

4.2 This study satisfied that observed groundwater and time-history of a superannuated earth dam allows to estimate its unknown soil parameters applying a combination of saturated and unsaturated flow in porous media with finite element method. Moreover it demonstrates the capability of FEM for solving a seepage problem comprising unknown parameter estimation by the inverse analysis approach.

4.3 The applied grid analysis approach was used to determine significant values of unknown saturated coefficient of permeability by calculating the differences between the deviation observed in the field and the corresponding deviation calculated from the involved saturated coefficients of permeability assumed in iterative calculation.

REFERENCES

Arai, K, Ohnishi, Y, and Kasagawa, M (1985). "Parameter estimation of unconfined aquifer by measuring change of groundwater level," *J Soils and Found*, Vol.25, No.4, pp 129–134.

Arai, K, Ohta, H, and Miyata, M (1991). "Comparison of static and statistical methods for back-analysis of elastic consolidation problems," *Proc of Computer Methods and Advances in Geomechanics*, pp 949–954.

Chowdhury, MA, Kunitake, M, Kondo, F, Inagaki, H, Ogata, H, Saito, M, and Kakuda, I (1996)."Estimation of the Parameters of Seepage Through an Earth Dam," *Proc. of the Int Offshore and Polar Eng Conf*, Los Angeles, USA, Vol. 1, pp 477–482.

Desai, CS (1972). "Seepage Analysis of Earth Banks Under Drawdown," *J Soil Mech and Found Div*, ASCE Vol. 98, No.Sm11, pp 1143–1162.

Desai, CS (1973). "*Seepage in Mississippi River Banks, Report I, Analysis of Transient Seepage using a Viscous-Flow Model and the Finite Difference and Finite Element Methods*," U.S. Army Engineer Waterways Experiment Station, Vicksburg, Mississippi.

Fredlund, DG, and Rahardjo, H (1993). "*Soil Mechanics for Unsaturated Soils*," John Wiley & Sons, Inc.

Freeze, RA, and Cherry, JA (1979). "*Groundwater*," Prentice-Hall, Inc., Englewood Cliffs, New Jersey.

Fell, R, MacGregor, P, and Stapledon, D (1992). "*Geotechnical Engineering of Embankment Dams*," A. A. Balkema, Rotterdam.

Hosoyamada, K, Kunitake, M, Kakita, K, and Higashi, T (1987). "*Technological development for safety of a fill dam used in agriculture. Investigation for leakage of superannuated fill dam* (in Japanese language)," ept of Agri Miyazaki Univ, pp 1–54.

Neuman, SP (1973a). "Saturated-Unsaturated Seepage by Finite Elements," *J Hydraul Div*, ASCE, Vol. 99, No.HY12, pp 2233–2250.

Neuman, SP (1973b). "Calibration of distributed parameter groundwater flow models viewed as a multiple-objective decision process under uncertainty," *Water Resour Res*, Vol. 9, No.4, pp 1006–1021.

Taylr, RL, and Brown, CB (1967). "Darcy Flow Solutions with a Free Surface," *J of the Hydraul Div*, ASCE, Vol. 93, No.HY2, pp 25–33.

Ogata, H, Kunitake, M, Kondo, F, and Nakazwa, T (1995). "Estimation of Parameters for the Temperature Analysis of Concrete using 3-D FEM," *Proc of the Int conf on Computional Eng Sci*, Hawaii, USA, Vol 1.

Yeh, WW-G (1986). "Review of Parameter Identification Procedures in Groundwater Hydrology: The Inverse Problem," *Water Resour Res*, Vol.22, No.2, pp 95–108.

Zienkiewicz, OC, Mayer, P, and Cheung, YK (1966). "Solution of anisotropic seepage by finite elements," *J Engr Mech Div*, ASCE, 92, EM1, 111.

Proceedings of the Seventh (1997) International Offshore and Polar Engineering Conference
Honolulu, USA, May 25-30, 1997

ISBN 1-880653-28-1 (Set); ISBN 1-880653-29-X (Vol. I)

Shear Behavior of Remolded Clay in Strain Path Controlled Mini-Triaxial Test

Tamotsu Matsui, Kazuhiro Oda and Yasuyuki Nabeshima
Osaka University
Osaka, Japan

ABSTRACT

For the purpose of sustaining stress and strain uniformity in a clay specimen and controlling arbitrary strain paths, the authors have developed a strain path controlled mini-triaxial testing apparatus, in which a small size specimen (22.5 mm in diameter and 45 mm in height) is used. In this paper, to investigate shear behavior of clay, the authors carried out a series of strain path controlled mini-triaxial tests using remolded kaolin clay. As a result, strain path dependent shear behavior is observed, which is quite different from that of drained or undrained triaxial test. That is, as increasing the strain increment ratio R between axial and volumetric strains, the maximum deviator stress increases while the maximum excess pore pressure decreases, and the effective stress path shifts off the failure line. Furthermore, the strain path dependent shear behavior of remolded clay is discussed, following that it is closely related to strain increment ratio and dilatancy.

KEY WORDS : dilatancy, mini-triaxial test, remolded clay, shear behavior, strain increment ratio, strain path controlled test, strain path dependency.

1. INTRODUCTION

The strain path controlled test has been developed by Chu and Lo (1991, 1994) to investigate asymptotic behavior of saturated sand. They emphasized the existence of a unique relationship between an asymptotic stress ratio and strain increment ratio in shear behavior of sand. In the strain path controlled test, volume change is directly controlled along the constant strain increment ratio R $(= d\varepsilon_v/d\varepsilon_1)$ which is a increment ratio between volumetric strain ε_v and axial strain ε_1, using digital pressure / volume control device (DPVC). This control system is also used by Uchida and Vaid (1994). They pointed out that volumetric strain distribution in a sand specimen is not uniform in a conventional drained test. The difference between strain path controlled triaxial test and conventional drained or undrained triaxial test is illustrated in Fig. 1. In a drained triaxial test, pore water is freely drained from drainage ends, although an excess pore pressure still remains even in thoroughly slow displacement rate at the center part of specimen where is the farthest part from drainage ends, as shown in Fig. 1(a). Also, in an undrained triaxial test, as pore water is not drained, an excess pore pressure appears along a shear band, as shown in Fig. 1(b). Therefore, in drained and undrained triaxial tests, the distribution of pore pressure is not uniform in the specimens. While, in a strain path controlled test, the distribution of pore pressure is sustained uniformly in the specimen using DPVC. This means that strain path controlled test meets the requirement as an element test, which is the stress and strain distributions uniformity in a specimen. Therefore, the shear behavior under the real element test condition can be investigated using strain path controlled test.

However, the strain path controlled tests have not been applied to saturated clay because of its low permeability. The authors try to carry out strain path controlled tests using mini-triaxial compression testing apparatus to

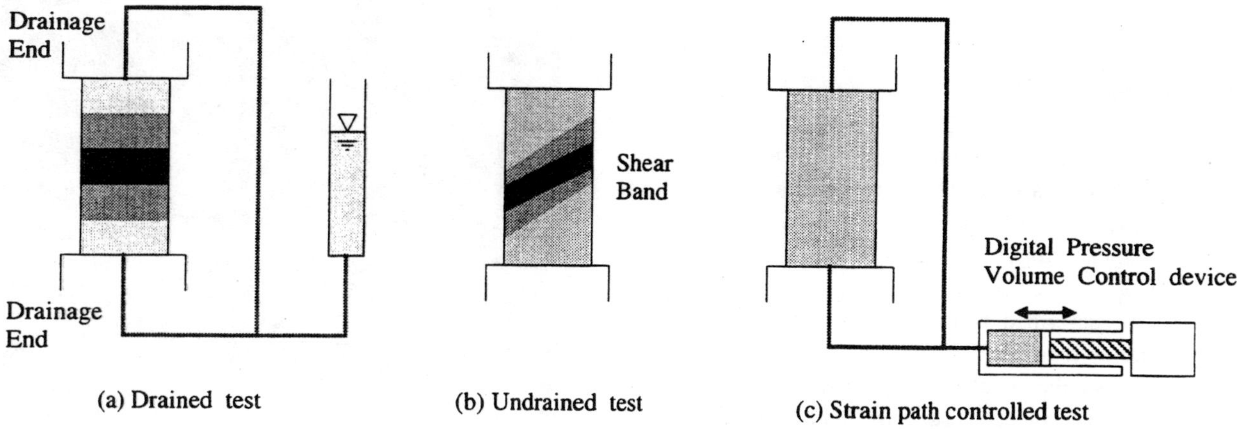

Fig. 1 Difference between drained triaxial test, undrained triaxial test and strain path controlled test

investigate shear behavior of clay. In this paper, the advantage and features of mini-triaxial test are described and the strain path dependent shear behavior of remolded clay is discussed under the real element test condition, focusing on strain increment ratio and dilatancy.

2. ADVANTAGE OF USING SMALL SIZE SPECIMEN

A specimen used in mini-triaxial compression testing apparatus is trimmed by 22.5 mm in diameter and 45 mm in height. This means that four specimens can be taken from a piece of clay sample (75 mm in diameter and 50 mm in height) taken by a thin-walled sampler, as shown in Fig. 2. Therefore, four kinds of triaxial tests can be carried out, using a clay sample of 50 mm long. If conventional triaxial compression testing apparatus is used, clay sample which is longer than 400 mm long must be required.

A small specimen has another advantage. Fig. 3 shows the experimental variation of volumetric strain with elapsed time during the consolidation process. The smaller the specimen size becomes, the faster the volumetric strain occurs because of the shorter drainage distance for the smaller specimen (Matsui et al., 1993). The consolidation time in conventional triaxial specimen, 50 mm in diameter and 100 mm in height, is about ten times that in small size specimen, 22.5 mm in diameter and 45 mm in height. This means that triaxial tests for smaller size specimen can be carried out more effectively from the viewpoint of saving time. Also, it is not easy to carry out drained shear tests for clay using a conventional specimen, because of low permeability of clays. It is suggested, however, that this difficulty can be overcome by using a small size specimen to shorten the drainage distance. Therefore, strain path controlled triaxial test can be easily carried out using mini-triaxial compression testing apparatus.

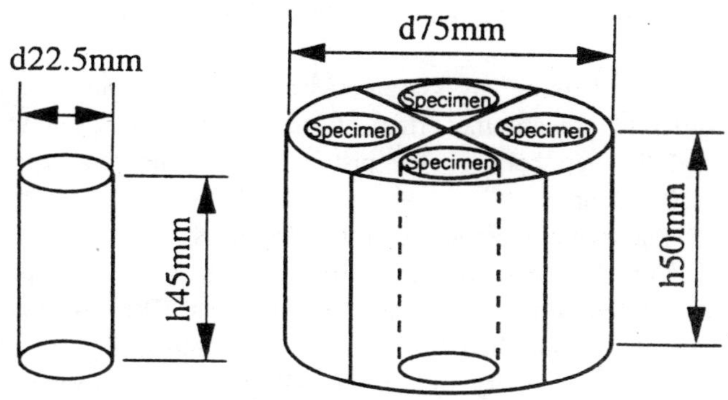

Fig. 2 Specimen for mini-triaxial test

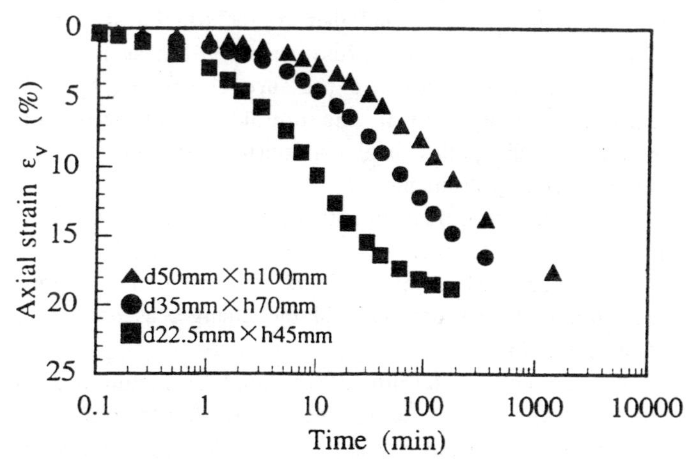

Fig. 3 Experimental variation of volumetric strain with elapsed time

3. STRAIN PATH CONTROLLED MINI-TRIAXIAL TEST

3.1 Testing Apparatus

Strain path controlled mini-triaxial testing apparatus consists of five main units. They are axial loading facilities, triaxial cell, pneumatic pressure control instruments, measuring instruments and digital pressure / volume control device (DPVC). Noticeable points on these equipments are described below.

(1) Axial loading facilities

Photo 1 shows the axial loading facilities, triaxial cell and measuring instruments. The axial loading facilities can be used both for displacement control or stress control. In the former, the displacement rate is controlled by DC motor. Maximum load is 490 kN and displacement rate can be changed in the range of 0.005 ~ 0.5 mm/min. In the latter, the axial load is applied by a double acting bellofram cylinder which can apply cyclic loading.

(2) Triaxial cell

The acrylic cell or aluminum cells is used according to cell pressure level, such that the acrylic cell is used for under 700 kPa and the aluminum cell for over 700 kPa.

The "leaking air system (Chan, 1975)" is adopted in order to prevent the effect of shaft friction on axial load and the gap between the piston rod and the rod support is 0.012 mm.

(3) Pneumatic pressure control instruments

Fig. 4 illustrates the line system of strain path controlled mini-triaxial testing apparatus. The pneumatic pressure is supplied by the air compressor which has the maximum capacity of 1 MPa. All pneumatic pressure is controlled by electric / pneumatic (E/P) transducers or regulators. The confined pressure is directly applied to triaxial cell. The back pressure is applied to the burette through drainage lines, and then to the triaxial specimen. The maximum applied pressure is 900 kPa in each instrument. The pneumatic pressure controlled by E/P transducer is applied to the double acting bellofram cylinder which applies axial load during consolidation.

(4) Measuring instruments

The measuring instruments in strain path controlled mini-triaxial testing apparatus are related to the measurements of axial load, axial displacement, pore pressure, cell pressure and volume change, as shown in Photo 1 and Fig. 1.

A load cell is set up at the head of piston rod to measure the axial load. The axial displacement is measured by a displacement gauge on the upper platen of triaxial cell. The cell pressure and the pore pressure are measured by electric pressure transducers. The volume change of the specimen is measured by a differential pressure transducer and a glass tube (6.0 mm in inner diameter), as shown in Fig. 4. In a consolidation process, the volume change and the pore pressure are measured. In a shear process, the amplifiers and personal computer with a 16-bit analog / digital (A/D) converter unit are used, to automatically measure axial load, axial displacement, pore pressure, cell pressure and volume change.

Photo 1 Strain path controlled mini-triaxial testing apparatus

(5) Digital pressure / volume control device (DPVC)

All strain path controlled tests are carried out by controlling the volume change of the specimen using the digital pressure / volume control device (DPVC), which is connected to the pore water through the top drainage line. The DPVC consists of stepping motor, screw gear and stainless steel tube piston, as shown in Photo 2 and Fig. 5. The stepping motor is controlled by personal computer through D/A board. The resolution of DPVC is 0.000750 cm^3 per 100 pulses.

3.2 Test Sample and Test Conditions

In the strain path controlled mini-triaxial test, remolded kaolin clay is used to investigate the mechanical behavior of clay. The physical properties is shown in Table 1. The kaolin clay is remolded for 6 hours with 260 % water content which is about three times of liquid limit. The kaolin slurry is consolidated under 98.1 kPa for a week. The strain path controlled test conditions are summarized in Table 2. The cell pressure, back pressure and axial strain rate are the same in each test case, while the strain increment ratio $d\varepsilon_v/d\varepsilon_1$ is different. The drained, undrained and mean effective stress constant triaxial tests are also carried out to compare with strain path controlled test

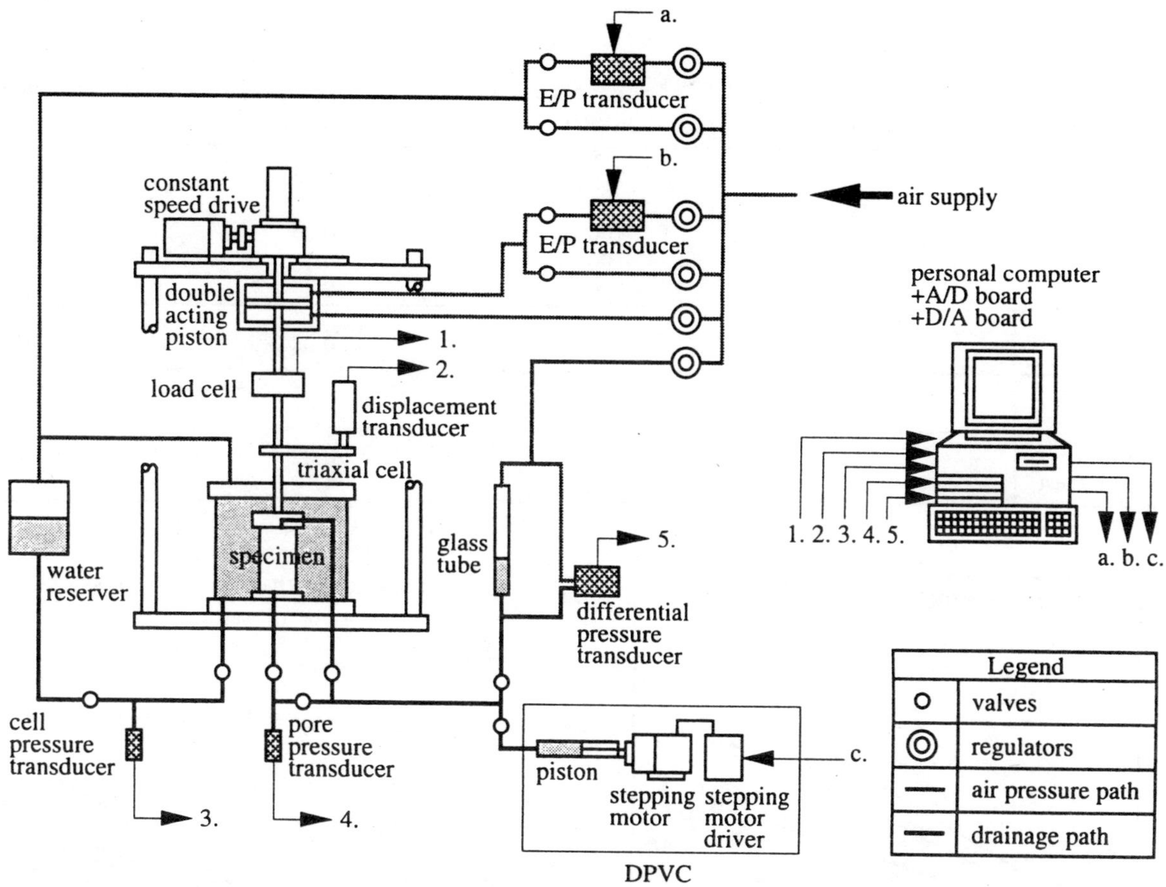

Fig. 4 Line system of strain path controlled mini-triaxial testing apparatus

Photo 2 Digital pressure / volume control device (DPVC)

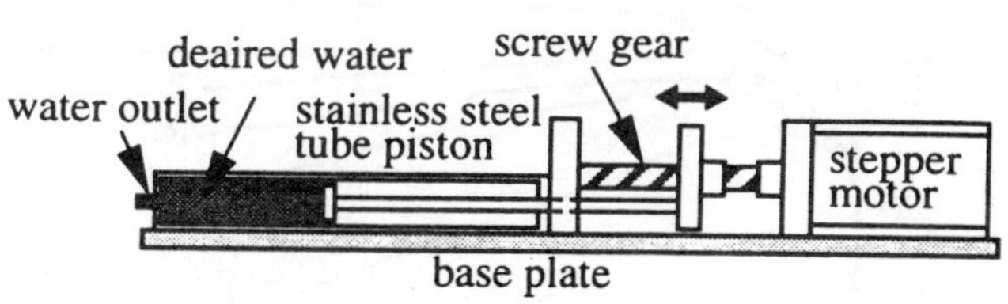

Fig. 5 Outline of Digital pressure / volume control device (DPVC)

results. The undrained test can be considered as a kind of strain controlled test at R=0.0, although the excess pore pressure distribution within the specimen is not necessarily uniform.

Table 1 Physical properties of kaolin clay

Specific gravity	Liquid limit (%)	Plastic limit (%)	Platicity index
2.62	87.5	28.3	59.2

4. STRAIN PATH CONTROLLED MINI-TRIAXIAL TEST RESULTS

4.1 Strain Path Dependent Shear Behavior

Fig. 6 shows the relationship between volumetric strain and axial strain. The strain increment ratio $d\varepsilon_v/d\varepsilon_1$ is the slope of the straight line. Therefore, the strain increment ratio of the undrained test is zero because no volumetric strain occurs. It is confirmed from this figure that control of volumetric strain is satisfactory for all tests. Fig. 7 shows the results of strain path controlled mini-triaxial tests. Fig. 7 (a) shows the relationship between deviator stress and axial strain. As increasing the strain increment ratio, the maximum deviator stress rapidly increases. Fig.

Table 2 Strain path controlled mini-triaxial test conditions

	Effective consolidation stress (kPa)	Back Pressure (kPa)	Axial strain rate (%/min)	Strain increment ratio
Strain path controlled tests	100	100	0.05	0.08
	100	100	0.05	0.17
	100	100	0.05	0.26
	100	100	0.05	0.33
	100	100	0.05	0.41
CU- test	100	100	0.04	0.0
CD - test	100	100	0.02	-----
Mean effective stress constant test	100	100	0.02	-----

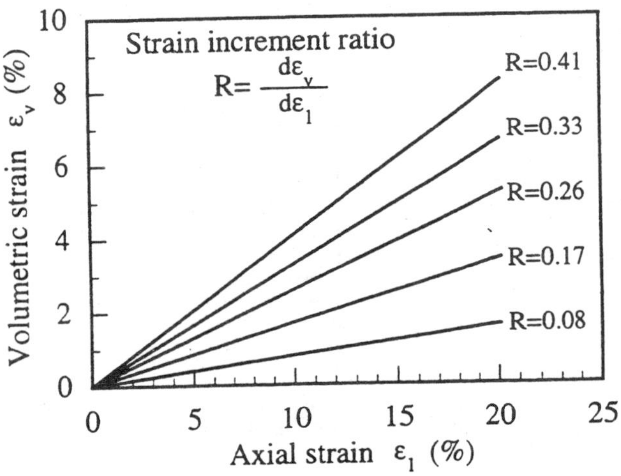

Fig. 6 Relationship between volumetric strain and axail strain

7 (b) shows the relationship between excess pore pressure and axial strain. This relationship also shows strain increment dependency. The maximum value of pore pressure decreases as increasing strain increment ratio. The excess pore pressure, in case where strain increment ratios are 0.33 and 0.41, eventually becomes negative value. Fig. 7 (c) shows effective stress paths, which are different from those of conventional drained and undrained triaxial tests. As increasing strain increment ratio, effective stress path shifts off the failure line which was decided through the undrained triaxial tests. The slopes of the effective stress paths, in case where strain increment ratios are from 0.17 to 0.41, seems to asymptotically approach a constant value. Fig. 7 (d) shows the relationship between effective stress ratio q/p' and axial strain. The effective stress ratios, in case where strain increment ratios are from 0.17 to 0.41, tend to approach an asymptotic effective stress ratio which is just a little bit smaller than critical stress ratio. The similar asymptotic behavior of saturated sand is reported by Chu and Lo (1994).

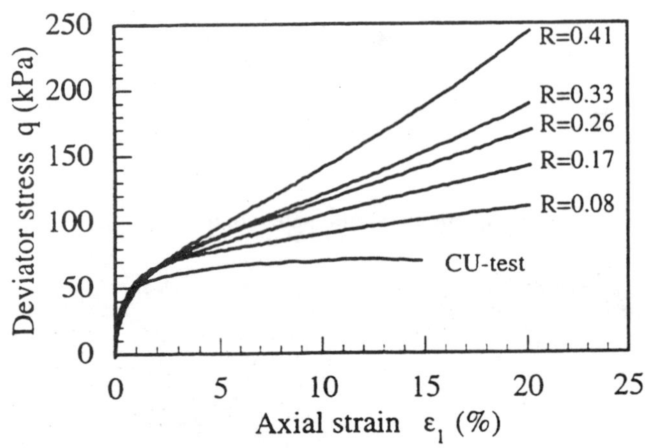

(a) Relationship between deviator stress and axail strain

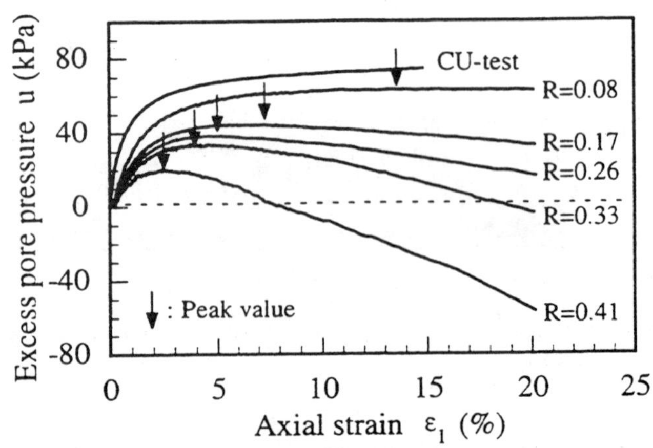

(b) Relationship between excess pore pressure and axail strain

Fig. 7 Strain path controlled mini-triaxial test results

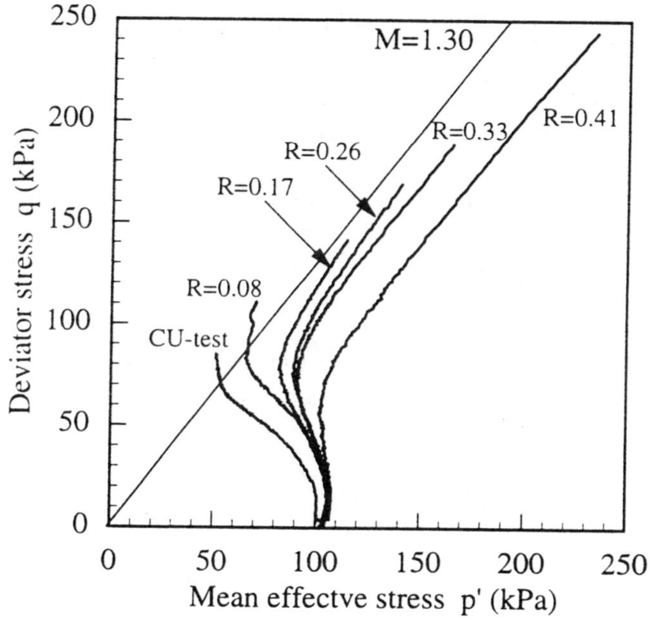

(c) Effective stress path

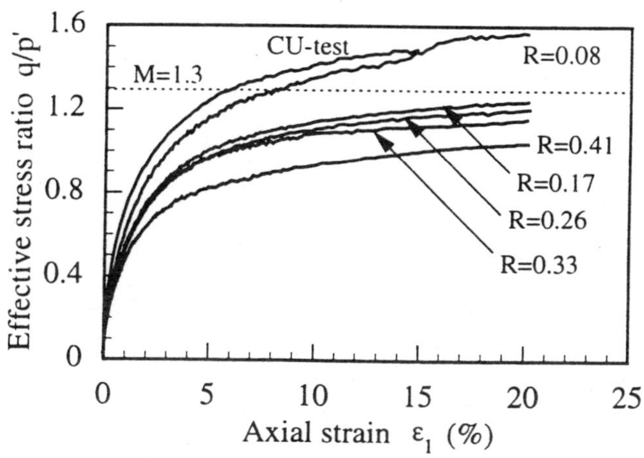

(d) Relationship between effective stress ratio and axail strain

Fig. 7 Strain path controlled mini-triaxial test results

Photo 3 shows the deformation of the specimens after undrained test and strain path controlled tests. It is confirmed that, as increasing strain increment ratio, the specimen is sheared keeping the strain uniformity more comparing to that of undrained test.

4.2 Strain Path Dependent Shear Behavior and Dilatancy

The relationship between strain path dependency and dilatancy is investigated. Fig. 8 shows relationship between

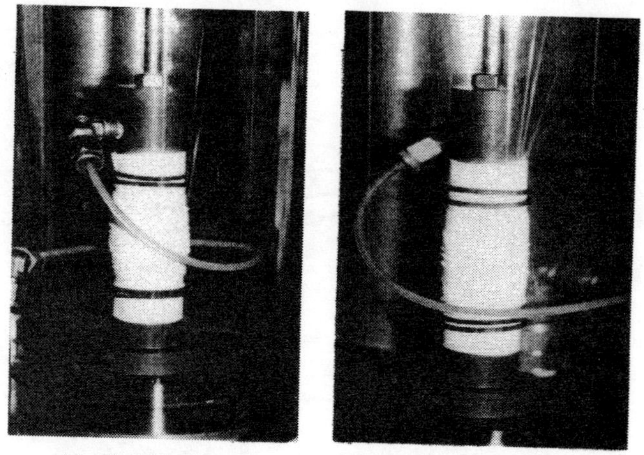

(a) Undrained test (R=0.0)

(b) Strain path controlled test (R=0.41)

Photo 3 Comparison of specimen deformation between undrained triaxial test and strain path controlled test

strain increment ratio and axial strain in the mean effective stress constant test which is performed at the same effective consolidation stress, as shown in Table 2. The data points in this figure are results of the mean effective stress constant test, and the solid line is their average line. The axial strains, where the average line goes across the broken lines which are strain increment ratios in the strain path controlled mini-triaxial tests, correspond approximately to the axial strains at the maximum excess pore pressure in Fig. 7 (b).

The difference between volumetric strain controlled by DPVC and dilatancy is shown with axial strain in Fig. 9. This figure shows similar tendency as Fig. 7 (b). This fact indicates that pore pressure behavior is closely related to strain increment ratio and dilatancy.

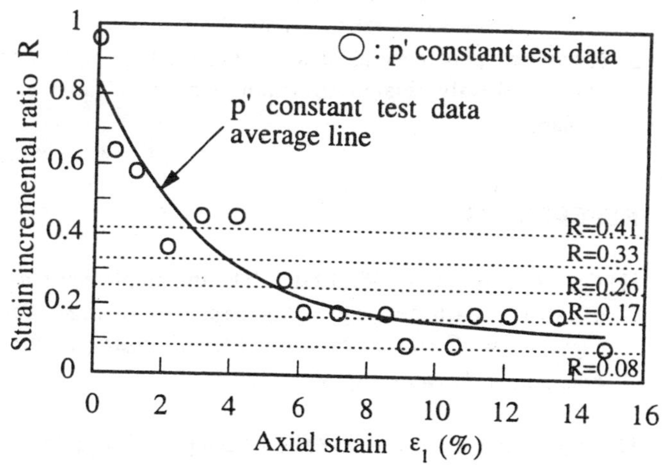

Fig. 8 Relationship between strain increment ratio and axail strain in mean effective stress test

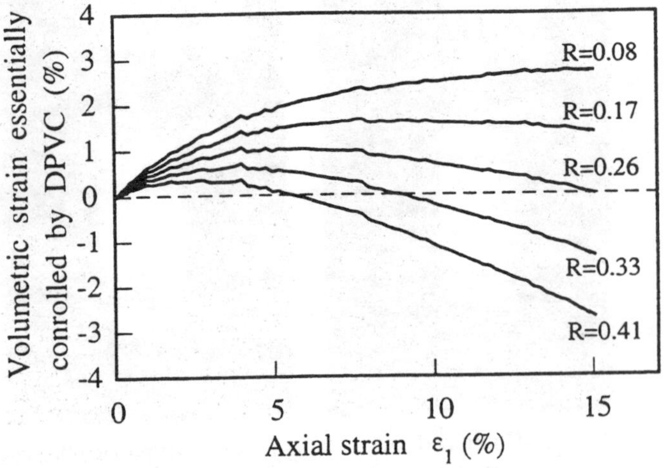

Fig. 9 Relationship between controlled volumetric strain and axial strain

5. CONCLUSIONS

The authors have investigated strain path dependent shear behavior of remolded clay by the strain path controlled mini-triaxial tests, which can keep stress and strain uniformity more in the specimen. The main results in this paper are summarized as follows:

1) The shear behavior of remolded clay depends on the strain path, which is different from that of conventional drained and undrained triaxial tests. That is, as increasing the strain increment ratio, the maximum deviator stress increases, while the excess pore pressure decreases, resulting that the effective stress path shifts off the failure line.
2) An asymptotic behavior is observed in both the effective stress path and the relationship between effective stress ratio and axial strain. The asymptotic value of effective stress ratio is just a little bit smaller than effective stress ratio of critical state.
3) The strain path dependent shear behavior of remolded clay is closely related to strain increment ratio and dilatancy.

REFERENCES

[1] Chu, J. and Lo, S.C.R. (1991). "On the implementation of strain path testing", *Proceedings of the 10th European Conference on Soil Mechanics and Foundation Engineering*, Vol.1, pp.53~56.

[2] Chu, J. and Lo, S.C.R. (1994). "Asymptotic behavior of a granular soil in strain path testing", *Geotechnique*, Vol.44, No.1, pp.65~82.

[3] Uchida, K. and Vaid, Y.P. (1994). "Sand behavior under strain path control", *Proceedings of 13th International Conference on Soil Mechanics and Foundation Engineering*, Vol.1, pp.17~20.

[4] Matsui, T., Oda, K. and Nabeshima, Y. (1993). "Development of mini-triaxial compression apparatus", *Proceedings of the 11th Southeast Asian Geotechnical Conference*, pp.171~176.

[5] Chan, C.K. (1975). "Low-friction seal system", *Journal of the Geotechnical Engineering Division*, A.S.C.E., Vol.101, No.GT9, pp.991~994.

Proceedings of the Seventh (1997) International Offshore and Polar Engineering Conference
Honolulu, USA, May 25-30, 1997

ISBN 1-880653-28-1 (Set); ISBN 1-880653-29-X (Vol. I)

Drained and Undrained Behavior of Sand Under Cyclic Loading

T. Nakai, T. Hoshikawa and M. Hinokio
Nagoya Institute of Technology
Nagoya, Japan

ABSTRACT

Various kinds of cyclic shear tests on sand under drained and underained conditions were performed to investigate stress-strain behavior (particularly soil dilatancy) under cyclic loading due to earthquake, ocean wave and others. Not only conventional triaxal tests but also true triaxial test in which three-principal stresses can be controlled independently were carried out. The stresses change their directions arbitrarily in the ground during earthquake or ocean wave loading. It is then important to investigate the stress-strain behavior including soil dilatancy in general stress condition. It is experimentally shown that the dilatancy characteristics of sand depends not only on the amplitude but also on the initial stress ratio and the direction of principal stresses.

KEY WORDS: cyclic loading, dilatancy, drained shear, sand, triaxial tests, true triaxial test, undrained shear

INTRODUCTION

Many elastoplastic models for sand under cyclic loading have been proposed (e.g. Hashiguchi,1980; Oka and Washizu,1981; Nakai, Fujii and Taki,1989), and many cyclic shear tests have been performed to investigate experimentally the cyclic shear behavior of sand and/or verify these models (e.g. Ishihara, Tatsuoka and Yasuda,1975; Wood,1982). The applicability of these models were confirmed within a limited stress conditions such as cyclic triaxial tests with equal amplitude after isotropic consolidation. However, it is necessary to check the validity of the models in general stress conditions with various amplitude to apply them to the analysis of practical dynamic problems of soil ground (e.g. liquefaction analysis). For this purpose, the authors performed several kinds of drained and undrained cyclic tests with different amplitudes and different shear stress histories, using conventional triaxial test apparatus and true triaxial test apparatus. All stresses in the present paper are effective stresses, unless otherwise stated.

BEHAVIOR OF SAND UNDER CYCLIC LOADING

To investigate stress-strain behavior of sand under cyclic loading, triaxial tests and true triaxial test on medium dense Toyoura sand (mean diameter D_{50}=0.2mm, uniformity coefficient U_c=1.3, specific gravity G_s=2.65, relative density D_r=73%) were performed in drained and undrained conditions. In every test, specimen was consolidated isotropically to 196kPa and then sheared along various stress paths under constant mean principal stress.

(1) triaxial tests

Triaxial tests were classified into two series (test Ⅰ and test Ⅱ). In test Ⅰ series, we discuss about the influence of drainage conditions (fully drained or fully undrained) on the stress-strain behavior under cyclic loading with a constant deviatoric stress ($q=\sigma_1-\sigma_3$) or principal stress ratio ($R=\sigma_1/\sigma_3$) amplitude. Table 1 shows the conditions of test Ⅰ series. All of the tests in test Ⅱ series were conducted under drained condition. Here, the influence of previous stress history on the successive cyclic stress-strain-dilatancy characteristics will be discussed. As shown in Fig.1, test Ⅱ-① is a cyclic shear test with constant magnitude of principal stress ratio $R(=\sigma_1/\sigma_3)$=4. On the other hand, in test Ⅱ-② and ③, the specimens are subjected to various cyclic shear stress indicated in the table of Fig.1 before cyclic shearing with constant magnitude of principal stress ratio R=4 in the same way as test Ⅱ-①.

Table 1 Conditions of test Ⅰ series

	test Ⅰ-①	test Ⅰ-②	test Ⅰ-③
Drainage Condition	Undrained	Undrained	Drained
Stress Amplitude	q=±294kPa	R=±3	R=±3

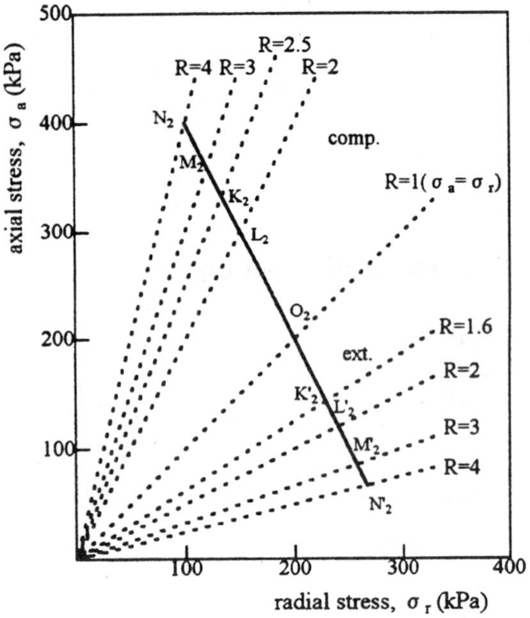

test Ⅱ-①	test Ⅱ-②	test Ⅱ-③
$N_2 \Leftrightarrow N_2' \times 10cyc.$	$L_2 \Leftrightarrow L_2' \times 5cyc.$ $M_2 \Leftrightarrow M_2' \times 5cyc.$ $N_2 \Leftrightarrow N_2' \times 10cyc.$	$N_2 \Leftrightarrow M_2 \Leftrightarrow L_2 \Leftrightarrow K_2 \times 5cyc.$ $N_2' \Leftrightarrow M_2' \Leftrightarrow L_2' \Leftrightarrow K_2' \times 5cyc.$ $N_2 \Leftrightarrow N_2' \times 10cyc.$

Fig. 1 Conditions of test Ⅱ

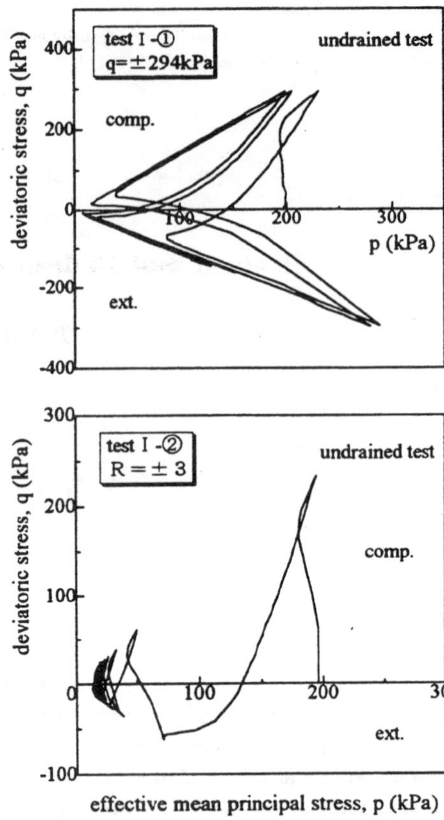

Fig. 2 Effective stress paths of test Ⅰ-①,②

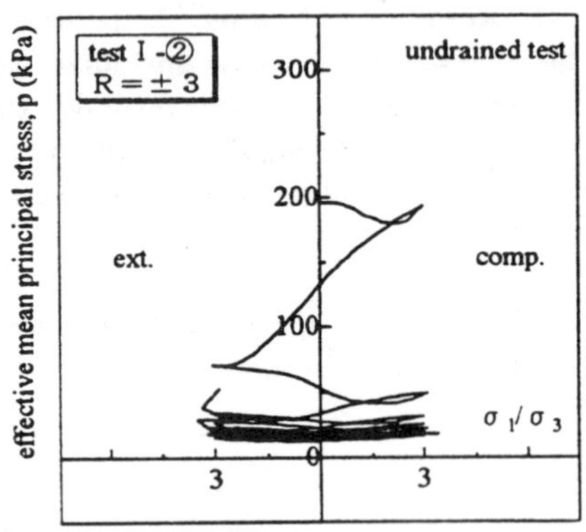

Fig. 3 Stress ratio vs. effective mean principal stress of test Ⅰ-②

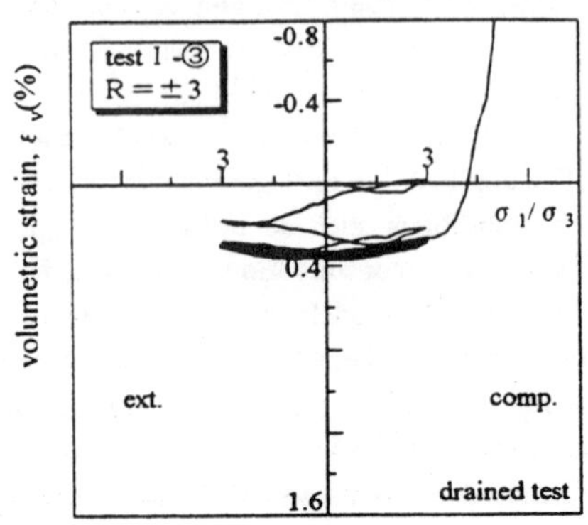

Fig. 4 Stress ratio vs. volumetric strain of test Ⅰ-③

Figure 2 shows the effective stress paths of tests Ⅰ-① and ② in p-q plane. Test Ⅰ-① with constant amplitude of deviatoric stress describes the liquefied state (p≒0kPa) and the successive cyclic mobility. On the other hand, sand in test Ⅰ-② with constant amplitude of principal stress ratio (σ_1/σ_3=3.0) does not liquefy and settled to a stable state eventually, though some decrease in mean stress is observed. Figure 3 rearranged the results of test Ⅰ-② in terms of principal stress ratio σ_1/σ_3 and effective mean principal stress p. The results of drained test Ⅰ-③ with constant amplitude of principal stress ratio (σ_1/σ_3=3.0) are shown in Fig.4, arranged using σ_1/σ_3 and volumetric strain ε_v. It can be seen that there is a close qualitative relation between Fig.3 and Fig. 4. This means that undrained cyclic behavior of sand can be properly analyzed by a constitutive model which describes the drained stress-strain behavior in general stress systems. The stress-strain curves of these tests are also shown in Fig.5.

Next, the dilatancy characteristics and the stable state under various cyclic loading will be investigated. Figures 6 and 7 show the principal stress ratio-volumetric strain relations and the principal stress ratio-axial strain relations of test Ⅱ series, respectively. We can see from Fig.6 that though the accumulated volumetric strain at stable state in test Ⅱ-② is almost the same as that in test Ⅱ-① without previous stress history, the volumetric strain at stable state in test Ⅱ-③ is much smaller. This difference of volumetric strain between test Ⅱ-② and test Ⅱ-③ depends on the dilatancy characteristics of previous

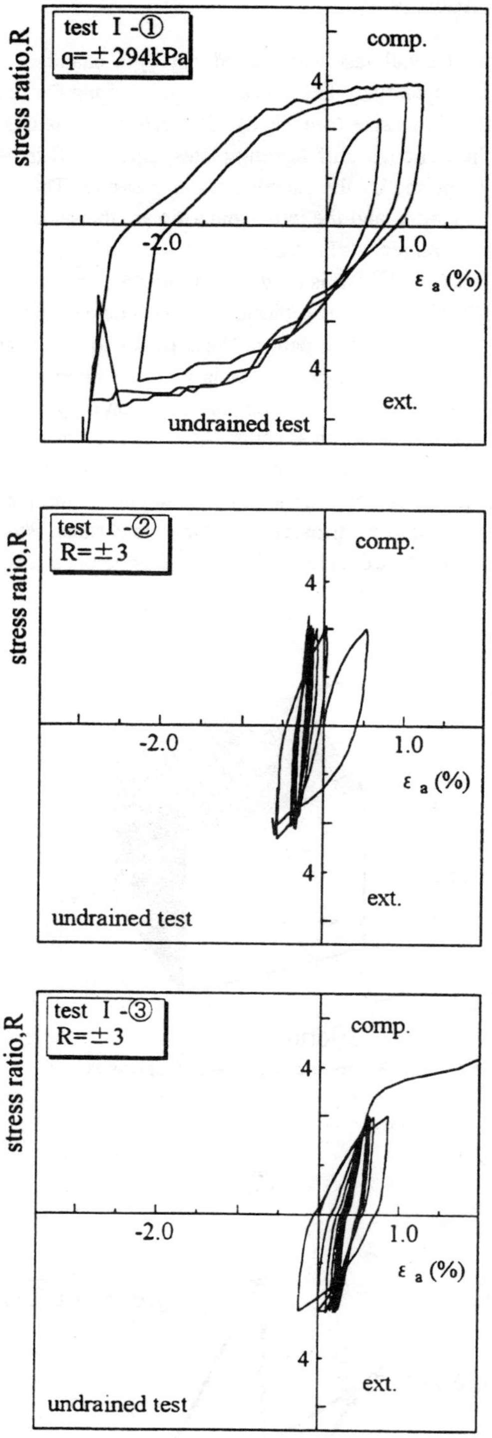

Fig. 5 Axial strain vs. stress ratio of test I -① to ③

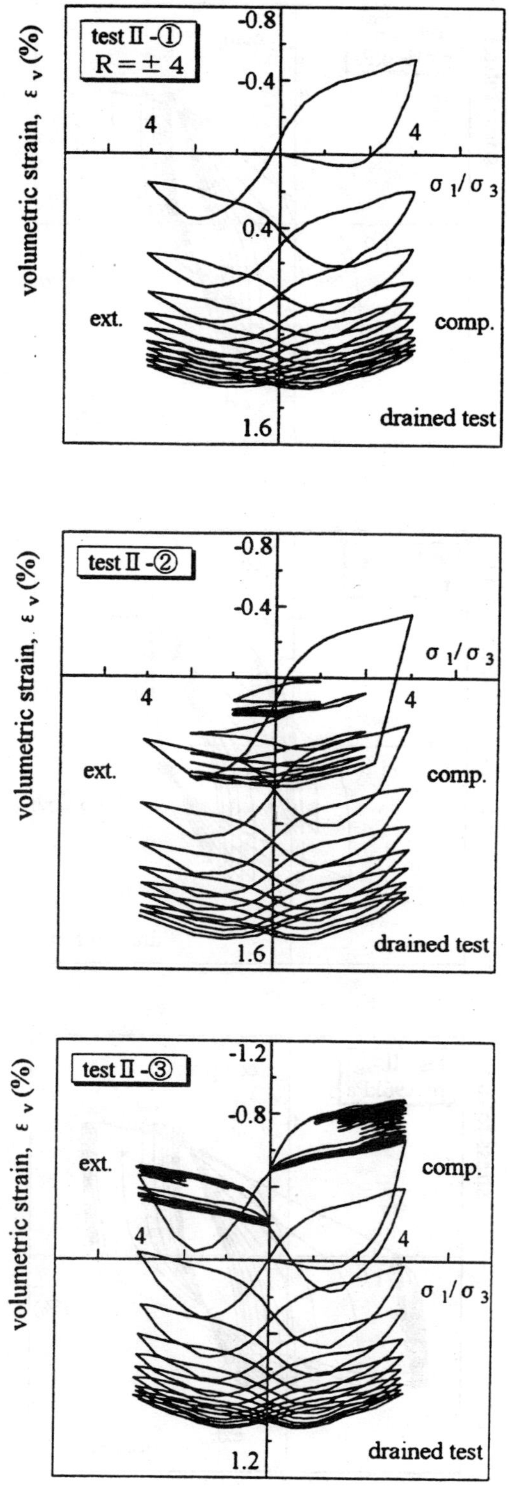

Fig. 6 Stress ratio vs. volumetric strain of test II -① to ③

shear histories. Namely, sand shows only volume contraction by cyclic loading symmetric with respect to isotropic stress, regardless of the amplitude (see test II -② in Fig.6). Contrary to this, eccentric cyclic loading causes volume expansion of sand as shown in test II -③ of Fig.6. It is also seen that this dilative characteristics of sand are more remarkable in triaxial compression than in triaxial extension.

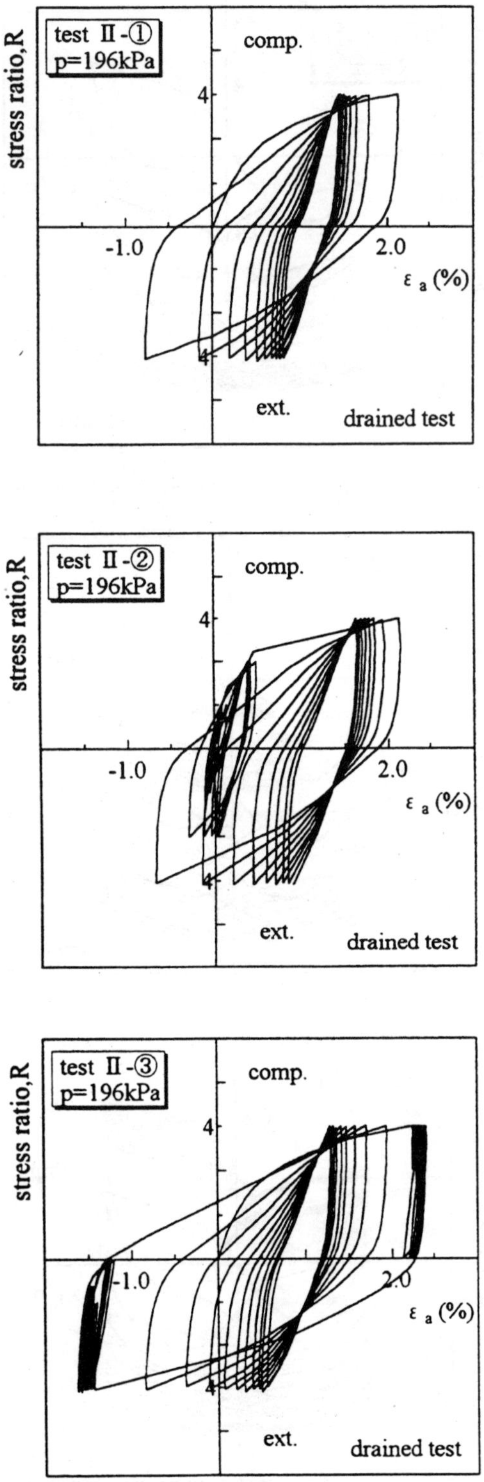

Fig. 7 Axial strain vs. stress ratio of test Ⅱ-① to ③

(2) true triaxial test

Cyclic true triaxial test was carried out by a chamber type true triaxial test apparatus. The specimen is rectangular prism of 10cm×10cm×7cm in size (see Fig.8). Two principal stresses (σ_a and σ_ℓ) are applied through stiff loading plates, and the other horizontal stress (σ_r) is applied by the chamber water pressure. The method of specimen preparation and the intial void ratio are the same as those in triaxial tests. Drained cyclic true triaxial test under constant mean principal stress (p=196kPa) is conducted along the stress paths in Fig.9. As shown in Table 2, sand is subjected to the cyclic deviatoric stresses consecutively from path-1 to path-6. Here, path-1 to path-4 are cyclic triaxial compression and extension tests with the same amplitude of stress ratio ($R=\sigma_1/\sigma_3=4.0$). However, the direction of principal stress axes are different in those stress paths. Now, stress paths in path-5 and path-6 cannot be produced by conventional triaxial test. Though every stress state at turning points of stress path in path-5 and path-6 is in triaxial compression or triaxial extension conditions, stress state between these points are under different three principal stresses.

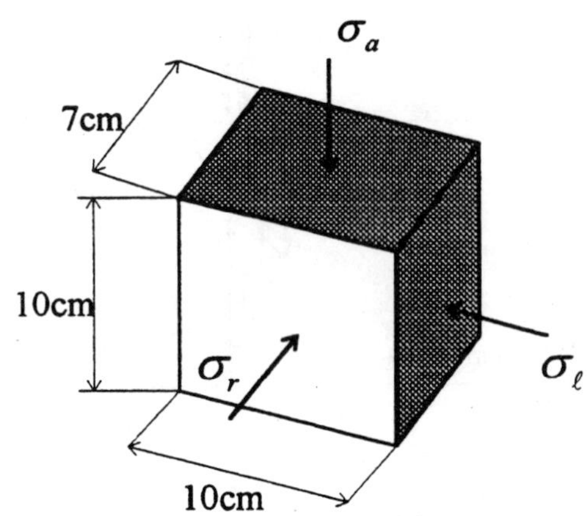

Fig. 8 Specimen and loading directions

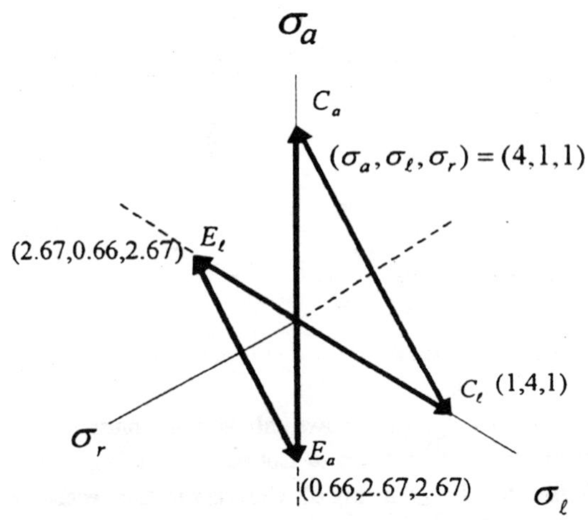

Fig. 9 Stress path on octahedral plane

Table 2 Stress paths and number of cycles of loading

Path No.	Stress Path	Num. of Cycle
Path-1	$C_a \Leftrightarrow E_a$	10
Path-2	$C_\ell \Leftrightarrow E_\ell$	10
Path-3	$C_a \Leftrightarrow E_a$	5
Path-4	$C_\ell \Leftrightarrow E_\ell$	5
Path-5	$C_a \Leftrightarrow C_\ell$	10
Path-6	$E_a \Leftrightarrow E_\ell$	10

Solid curves in Fig.10 to 13 show the test results of path-1 to path-4, respectively. Here, figure (a) is the relation between principal stress ratio ($R=\sigma_1/\sigma_3$) and volumetric strain ε_v, and figure (b) the relation between R and axial strain ε_a. Figure 10 (path-1) shows the same dilatancy characteristics of sand as those of test II-① in Fig.6, so that we can see that present true triaxial test is conducted with adequate accuracy. Now, it is seen from Fig.11 that subsequent volume contraction occurs after sand reaches a steady state under a cyclic loading between C_a and E_a in path-1, when sand is subjected to another cyclic load with same amplitude between C_ℓ and E_ℓ (only direction of cyclic stress path is changed). On the other hand, under cyclic loading in path-3 and path-4 following path-1 and path-2 (path-3 and path-4 are the same as path-1 and path-2 respectively), volume change hardly occurs as shown in Figs. 12 and 13. Figure 14 shows the observed relation between volumetric strain and principal stress ratio in path-5 and path-6. It is also seen from this figure that volume expansion occurs under cyclic loading in path-5 and path-6 in the same way as those under eccentric cyclic triaxial loading in Fig.6. Volume expansion in path-5 (cyclic loading between triaxial compression state and triaxial compression state) is more remarkable than those in path-6 (cyclic loading between triaxial extension and triaxial extension). This tendency also is the same as those in Fig.6.

These tests results indicate that the dilatancy characteristics of sand during cyclic loading depend on the initial stress ratio and the change in principal stress direction very much, even though its amplitude and frequency are the same. Further, this initial stress dependency of the dilatancy characteristics influences on the liquefaction phenomena of saturated sand deposit by earthquake and ocean wave particularly.

CONCLUSIONS

Stress-strain behavior of sand under various cyclic loadings are experimentally investigated under undrained and drained conditions. Closed qualitative correlation between effective mean principal stress in undrained test and volumetric strain in drained test is observed. Further, it is shown that the dilatancy characteristics depend on the initial stress ratio and the change in principal stress direction even if the amplitude and number of cycles are the same.

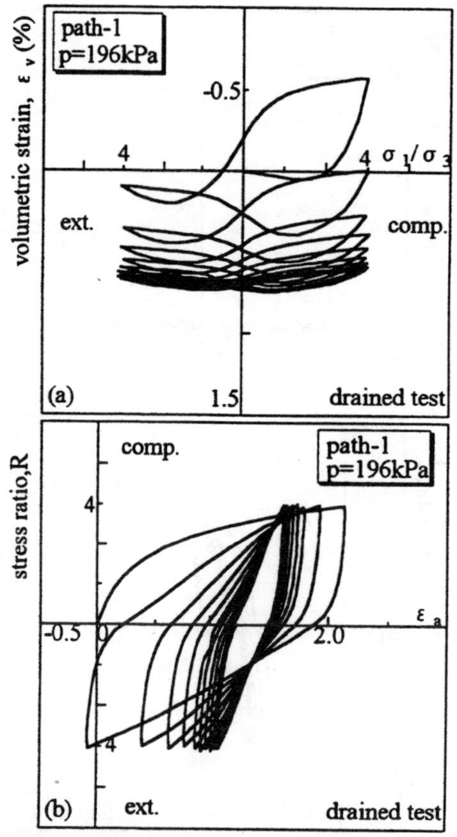

Fig. 10 Stress ratio vs. volumetric strain, axial strain vs. stress ratio of path-1

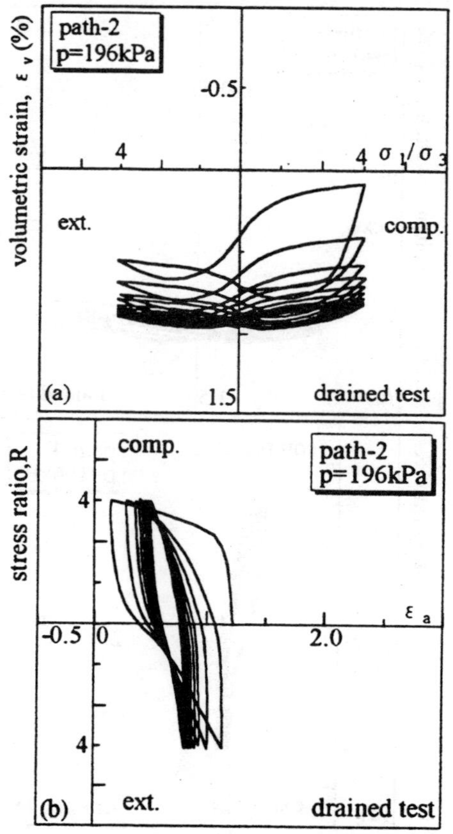

Fig. 11 Stress ratio vs. volumetric strain, axial strain vs. stress ratio of path-2

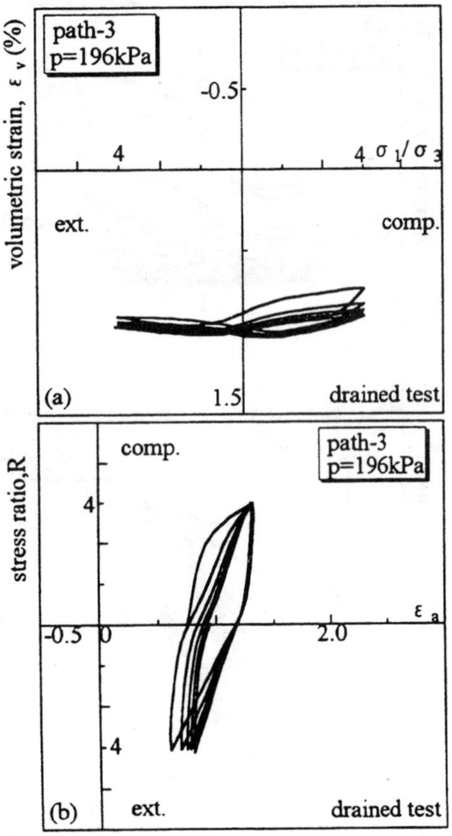

Fig. 12 Stress ratio vs. volumetric strain, axial strain vs. stress ratio of path-3

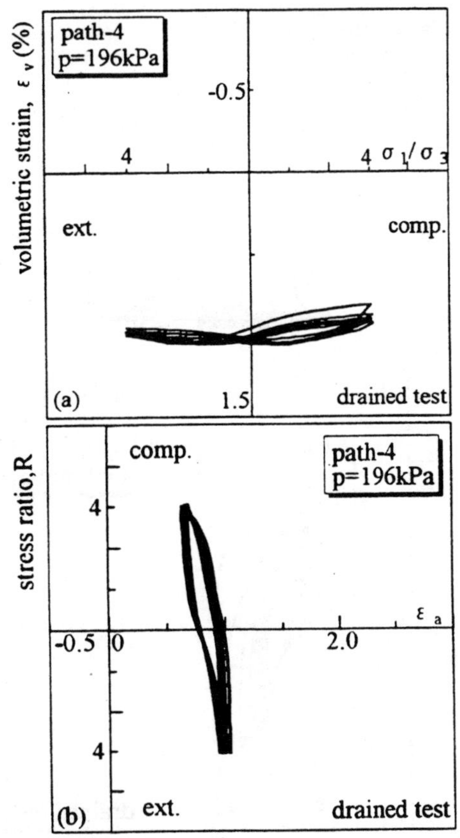

Fig. 13 Stress ratio vs. volumetric strain, axial strain vs. stress ratio of path-4

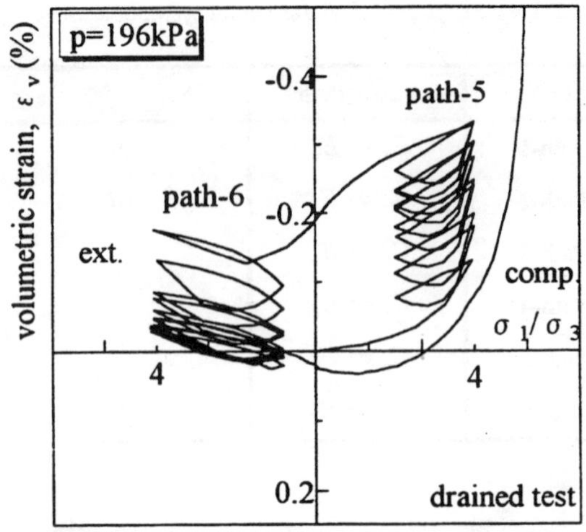

Fig. 14 Stress ratio vs. volumetric strain of path-5,6

REFERENCES

Hashiguchi, K (1980). "Constitutive equations of elastoplastic materials with elastic-plastic transition," J Appl. Mech, ASME, Vol 102, No 2, pp.266-272.

Ishihara, K, Tatsuoka, F, and Yasuda, S (1975). "Undrained deformation and liquefaction of sand under cyclic loading," Soils and Foundations, Vol 15, No 1, pp.29-44.

Nakai, T, Fujii, J, and Taki, H (1989). "Kinematic extension of an isotropic hardening model for sand," Proc. NUMOG Ⅲ, pp.36-45.

Oka, F, and Washizu, H (1981). "Constitutive equations for sands and over-consolidated clays under dynamic loads based on elastoplasticity," Proc Int Conf on Recent Advance in Geotechnical Earthquake Eng. and Soil Dynamics, Vol 1, pp.71-74.

Wood, DM (1982). "Laboratary investigation of the behavior of soils under cyclic loading," Soil Mechanic-Transient and cyclic load, Chap 20, John Wiley and Sons, pp.513-582.

Proceedings of the Seventh (1997) International Offshore and Polar Engineering Conference
Honolulu, USA, May 25-30, 1997

ISBN 1-880653-28-1 (Set); ISBN 1-880653-29-X (Vol. I)

Effect of Sample Disturbance on Consolidation Yield Stress of Ariake Clays

Zhenshun Hong and Katsutada Onitsuka
Saga University
Saga, Japan

ABSTRACT

Ariake clays have been widely deposited around Ariake Bay in Japan. The basic characteristics of Ariake clays are very soft yet sensitive. Quantitatively assessing the quality of soil samples and correcting the mechanical properties for sample disturbance is a well-known, but not well understood issue in geotechnical engineering. In this study, the undisturbed Ariake clays are deliberately disturbed to various level of disturbance. Then, a series of consolidation tests are performed to investigate the effect of sample disturbance on the consolidation yield stress. A new yet simple index for providing a quantitative measure of the degree of sample disturbance is proposed in this paper. Further, a new approach of correcting the consolidation yield stress for sample disturbance is also proposed based on the new index of sample disturbance.

Key Words: Ariake clay, compressibility, consolidation yield stress, sample disturbance, sensitivity

INTRODUCTION

Ariake clays have been widely deposited around Ariake Bay, located at the western part of Kyusyu Island, Japan. The basic characteristics of Ariake clays are very soft and sensitive. The natural water content is generally over 100%. The liquid limit is often greater than the natural water content. In addition, the thickness of the Ariake clay deposit varies generally 10-30m, with a maximum of about 40m. Hence, the settlement and the stability in the area of such a marine deposit are serious problems in the engineering practice. Accurately assessing the consolidation yield stress is indispensable in the settlement and the stability analyses. Stress release and mechanical disturbance during sampling and handling, however, may lead to a significant difference in the mechanical behavior. Quantitatively assessing the quality of soil samples and correcting the mechanical parameters for sample disturbance is a well-known, but not well understood issue in geotechnical engineering. In this study, a new yet simple index for providing a quantitative measure of the degree of sample disturbance is proposed, and a new approach of correcting the yield stress for sample disturbance is also proposed based on the new index of sample disturbance.

A NEW INDEX OF SAMPLE DISTURBANCE

In order to investigate the effect of sample disturbance on the compressive behavior for Ariake clays, a vibration equipment is used to deliberately disturb the undisturbed Ariake clays to various level. The vibration equipment is schematically shown in Fig.1. The soil specimens are put on the vibration table, and vibrated for a certain time. It is considered that the level of sample disturbance varies with the vibration time.

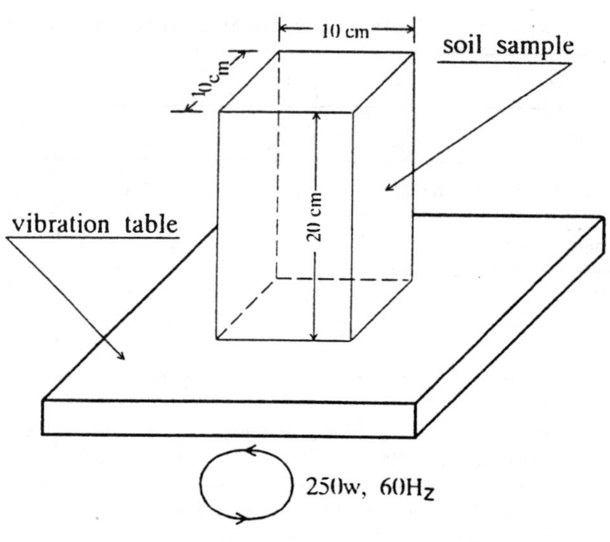

FIG.1 SCHEMATIC PLOT OF VIBRATION EQUIPMENT

The Ariake clays used herein were obtained in Ogi-gun, Saga Prefecture of Japan with a polyvinyl chloride pipe. The basic physical properties are shown in Table 1. A series of consolidation tests were performed on the samples with different level of sample disturbance. The samples are designated T-1 through T-12. The typical compression curves are shown in Fig.2. It can be seen that the consolidation test data can be well interpreted by two linear lines in the ln(1+e)-logp plot. The interpretation technique using a bilogarithmic plot was firstly proposed by Butterfield (1979). Onitsuka et al. (1995) have verified its validity for Ariake clays. The stress at the intersection point of the bilinear lines in the ln(1+e)-logp plot is the consolidation yield stress, p_{CL}. The bilinear lines bounded at the yield stress are termed as the preyield and the postyield lines,

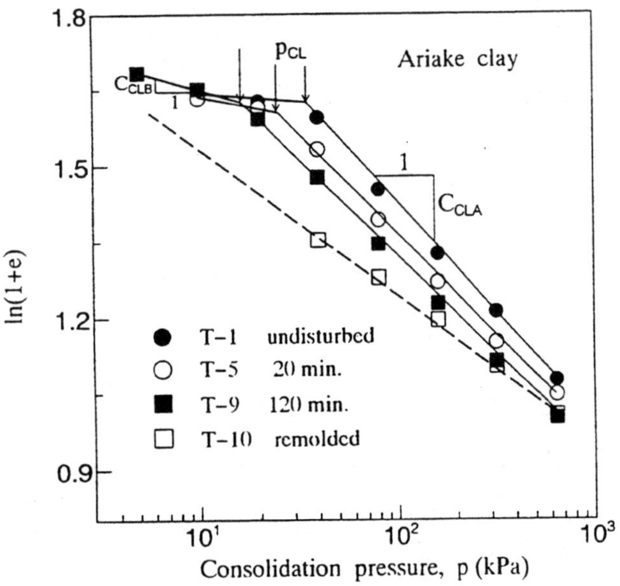

FIG.2 TYPICAL COMPRESSION CURVES FOR VARIOUS DISTURBANCE DEGREE

respectively. Their slopes are defined as the compression indices, designated C_{CLB} and C_{CLA} respectively. The consolidation test results are summarized in Table 2 for T-1 through T-12. It can be seen that the yield stress and the compression indices are significantly affected by the level of sample disturbance. The relationship between the compression index in the preyield state, C_{CLB}, and the vibration time, t, is shown in Fig.3. There is a strong tendency of the value of C_{CLB} increasing with the vibration time.

Schmertmann (1991) has illustrated that all natural soils possess a soil structure developed during their formation. Nagaraj et al. (1990) have indicated that the recompression path for sensitive clays is always much flatter than the rebound line from a stress level higher than the yield stress. A few in-situ load-settlement data (Flokes and Crooks, 1985; Pelletier et al., 1979) have indicated a negligible settlement up to the yield stress. Hence, the value of C_{CLB} can be negligible for the field compression path. On the other hand, the

Table 1. Basic physical properties

	Ariake clay
density of soil particles (g/cm^3)	2.661
natural water content (%)	154
liquid limit (%)	117
plastic limit (%)	52
sand content (%), d>0.075mm	0.9
silt content (%), 0.005mm<d≦0.075mm	38.6
clay content (%), d≦0.005mm	60.5

Table 2. Consolidation test results for an Ariake clay (T-1~T-12)

sample	T-1	T-2	T-3	T-4	T-5	T-6	T-7	T-8	T-9	T-10*	T-11*	T-12*
vibration time (min.)	0	0	10	10	20	40	60	40	120	remolded		
initial water content (%)	156.4	154.2	158.7	156.3	160.4	156.4	161.2	158.9	163.6	118.0	117.0	103.0
p_{CL} (kPa)	29.8	32.2	30.8	28.1	25.6	23.9	20.2	22.0	15.5	0	0	0
C_{CLB}	0.0516	0.0169	0.0288	0.0371	0.0596	0.0620	0.102	0.0868	0.118	—	—	—
C_{CLA}	0.410	0.429	0.380	0.380	0.417	0.403	0.422	0.402	0.395	—	—	—
C_{CLR}**	0.289									0.295	0.286	0.286
SD(%)=C_{CLB}/C_{CLR}×100%	17.85	5.85	9.97	12.84	20.62	21.45	35.29	30.03	40.83	100		

Note: *) T-10 through 12 are the artificially-consolidated samples under the maximum applied stresses 30 kPa, 30 kPa and 100 kPa, respectively.
**) The value of C_{CLR} is the average value of T-10 through 12.

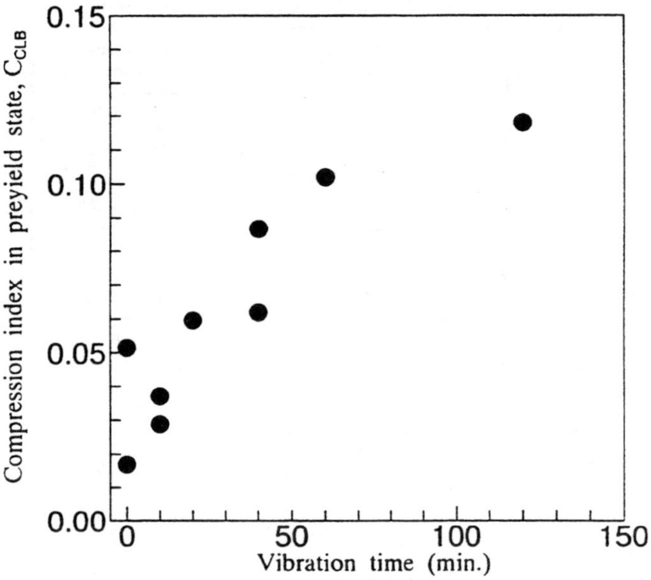

FIG.3 EFFECT OF SAMPLE DISTURBANCE ON COMPRESSION INDEX IN PREYIELD STATE

value of C_{CLB} increases with the increase in the level of sample disturbance. When the sample is completely disturbed, the value of C_{CLB} reaches the upmost, equal to that in the remolded state.

An e-logp compression curve for a remolded soil has often been assumed to be a straight line in the e-logp plot. But it is generally in a shape being slight concave upwards (Burland, 1990), as shown in Fig.4. These curves are alternatively plotted herein in the ln(1+e)-logp plot, as shown in Fig.5. It can be seen that they can be defined well by a straight line. The oedometer test data for 20 remolded soils with a wide spectrum of liquid limit values from various literature

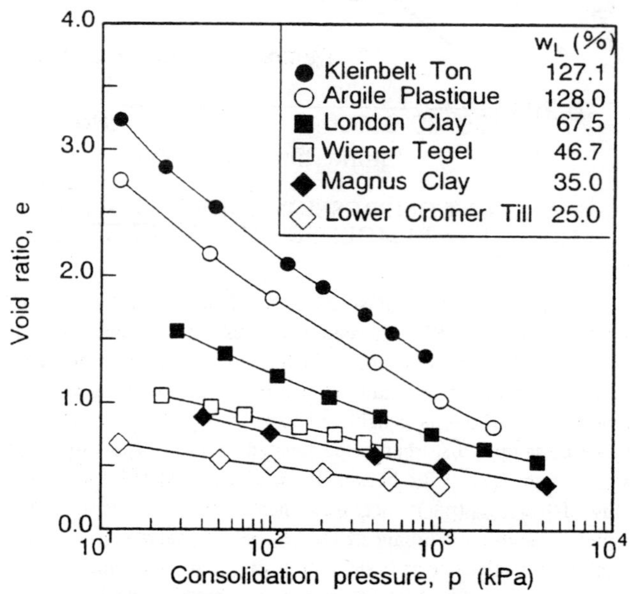

FIG.4 COMPRESSION CURVES OF REMOLDED SOILS IN e-logp PLOT (after Burland, 1990)

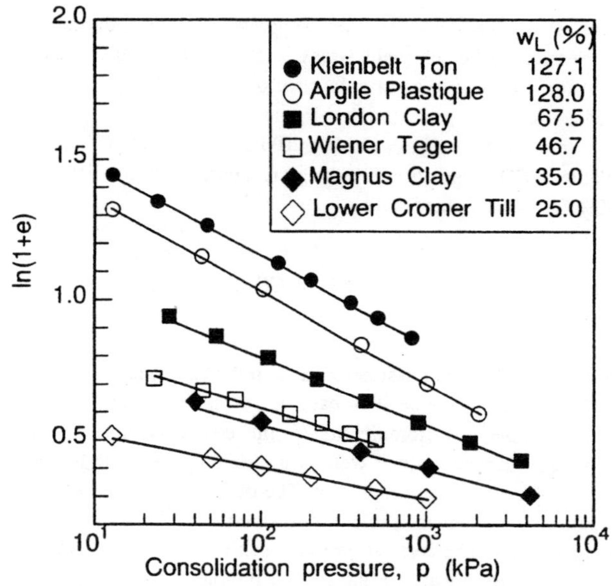

FIG.5 COMPRESSION CURVES OF REMOLDED SOILS IN ln(1+e)-logp PLOT

sources, shown in Table 3, are used to check the validity of the ln(1+e)-logp plot for a remolded soil. Regression analysis shows that the linear correlation coefficient of ln(1+e) versus logp for all these remolded soils is greater than 0.990. The slope of the compression path in the ln(1+e)-logp plot is defined as the compression index in the remolded state, designated C_{CLR}.

Table 3. Data for remolded natural soils

Soil	w_L(%)	e_L	Reference
Kleinbelt Ton	127.1	3.521	Burland, 1990
Argile Plastique	128.0	3.302	
London Clay	67.5	1.829	
Wiener Tegel	46.7	1.288	
Magnus Clay	35.0	0.956	
Lower Cromer Till	25.0	0.663	
Silty Clay	69.0	1.918	Nagaraj & Murthy, 1983
Boston Blue Clay	38.8	1.071	
Whangamarino Clay	136.0	3.740	
Lagunillas Clay	79.0	2.173	
Soft Clay	78.0	2.145	
SAIL Soil	158.3	4.474	
Black Cotton Soil	97.3	2.656	
Red Soil	45.3	1.208	
Vienna Clay	47.0	1.297	
Silty Clay	56.4	1.551	
Silty Sand	36.2	0.959	
Bedford Clay	87.0	2.384	
Residual Clay	59.0	1.617	Leonards & Ramiah, 1959
Glacial Silty Clay	28.0	0.762	

Since the magnitude of the compression index during the recompression up to the yield stress is directly affected by sample disturbance, the degree of sample disturbance can be quantitatively defined as Eq.1.

$$SD\,(\%) = C_{CLB}/C_{CLR} \times 100\% \quad \text{......(1)}$$

The value of SD ranges from 0% being responsible for the field state to 100% for the remolded state.

A GENERALIZED EQUATION FOR COMPRESSION INDEX OF REMOLDED SOILS

The value of the compression index for remolded soils, C_{CLR}, is not routinely measured though it is easy to do an oedometer test on a remolded sample. As aforementioned, the compression curve of a remolded soil can be defined well by a straight line in the ln(1+e)-logp plot, other than the e-logp plot. The oedometer test data for 20 remolded soils shown in Table 3, are interpreted herein using the ln(1+e)-logp plot. The correlation relationship between C_{CLR} and the liquid limit w_L is shown in Fig.6 for the remolded soils. Regression analysis gives the best fitted regression line as shown in Eq.2.

$$C_{CLR} = -0.390 + 0.332 \times \log(w_L) \quad \text{......(2)}$$

The coefficient of correlation, r, for above equation is 0.961. The consolidation tests are also performed on the remolded Ariake clays to check the validity of the above equation. The basic properties for the Ariake clays are shown in Table 4. The relationship between the compression index C_{CLR} and w_L for these remolded Ariake clays is shown in Fig.7. It can be seen that it is quite consistent with Eq.2. Hence, it can be easy to use Eq.2 to provide the C_{CLR} values for the soils with w_L values ranging from 25% to 175%. Accordingly, the degree of sample disturbance can be easily obtained from Eqs.1 and 2 for each sample only if the values of C_{CLB} and w_L are measured in laboratory.

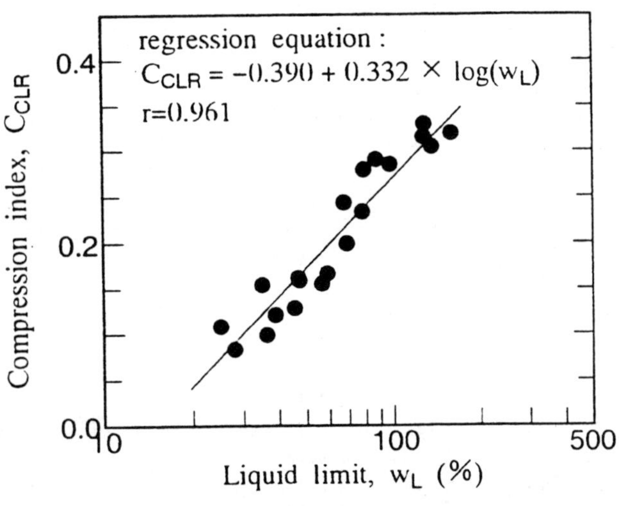

FIG.6 RELATIONSHIP BETWEEN C_{CLR} AND w_L

A METHOD OF CORRECTING YIELD STRESS FOR SAMPLE DISTURBANCE

As stated previously, for the samples obtained from the same depth

Table 4. Compression index of remolded Ariake clays

site	sample No.	initial water content, w_0(%)	liquid limit w_L (%)	compression index, C_{CLR}
Ashikari	RE-1	154.0	117.0	0.295
	RE-2			0.286
	RE-3			0.285
Higashiyoka	RE-4	230.0	175.2	0.378
	RE-5			0.342
	RE-6			0.356
Higashiyoka	RE-7	229.4	156.0	0.341
	RE-8			0.351
	RE-9			0.348
	RE-10			0.343
Ashikari	RE-11	148.7	120.0	0.332
	RE-12			0.313
Ashikari	RE-13	144.3	98.6	0.256

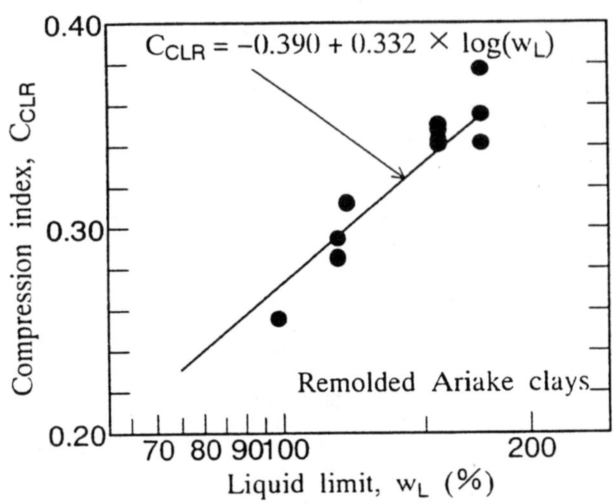

FIG.7 COMPRESSION INDEX C_{CLR} FOR REMOLDED ARIAKE CLAYS

at the same site, the yield stress is the function of the degree of sample disturbance, SD. The oedometer test data of an Ariake clay, shown in Table 2, on the samples disturbed to various level are used to investigate the effect of sample disturbance on the yield stress. The highest value of p_{CL} is considered to be responsible for the lowest degree of sample disturbance, designated $(p_{CL})_{best}$. The value of p_{CL} for T-2 is the best value herein. From Fig.8, it can be seen that the value of $[(p_{CL})_{best}-(p_{CL})]$ increases linearly with the increase in the degree of sample disturbance SD. Regression analysis gives a linear correlation coefficient of 0.982, indicating that the reduction of the yield stress measured in laboratory is almost caused by sample disturbance. The best fitted regression line is obtained as Eq.3.

$$[(p_{CL})_{best}-(p_{CL})] = -0.861+0.349 \times SD(\%) \quad \text{......(3)}$$

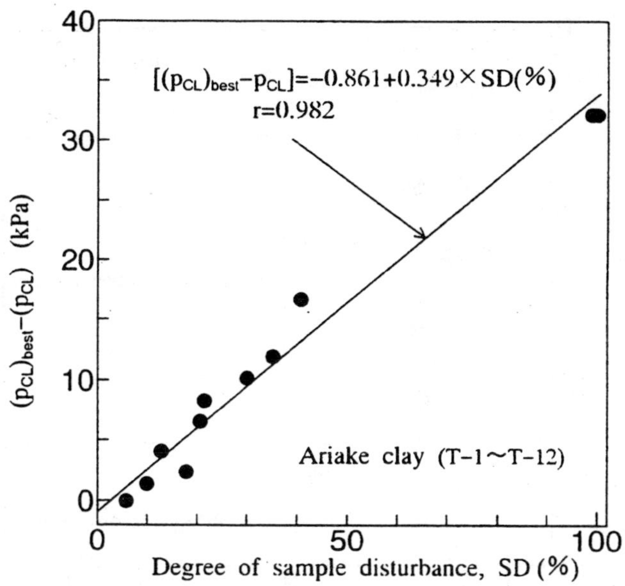

FIG.8 EFFECT OF SAMPLE DISTURBANCE ON YIELD STRESS FOR AN ARIAKE CLAY

The negative intercept indicates that even the best estimate value of p_{CL} is still responsible for some extent of sample disturbance. The corrected yield stress, p_{yf}, corresponds to that without disturbance (i.e., SD=0%). That is, the negative intercept is responsible for the difference of $[(p_{CL})_{best}-(p_{yf})]$. The value of p_{yf} for the samples T-1 through T-12 can be obtained by shifting the linear regression line

Table 5. Corrected yield stress for an Ariake clay

sample	p_{CL} (kPa)	SD(%)	R_p (kPa)	$p_{yf}=p_{CL}+R_p\times SD(\%)$ (kPa)
T-1	29.8	17.85	0.349	36.0
T-2	32.2	5.85		34.2
T-3	30.8	9.97		34.3
T-4	28.1	12.84		32.6
T-5	25.6	20.62		32.8
T-6	23.9	21.45		31.4
T-7	20.2	35.29		32.5
T-8	22.0	30.03		32.5
T-9	15.5	40.83		29.8
T-10	0	100		34.9
T-11	0	98.96		34.5
T-12	0	98.96		34.5

given in Eq.3 parallel to the intercept with the origin. The following equation therefore can be used to obtain the corrected yield stress p_{yf}.

$$p_{yf}=p_{CL}+R_p\times SD(\%) \quad ......(4)$$

The quantity R_p represents the correction coefficient for the yield stress. Its value is equal to the slope of the regression line of $[(p_{CL})_{best}-(p_{CL})]$ versus SD. The corrected yield stress p_{yf} for the samples T-1 through T-12 are summarized in Table 5. It can be seen that the corrected yield stresses are little changed among the samples. This result indicates that the defined SD can reflect the sample quality, and the effect of the scattered data on the consolidation yield stress can be almost eradicated with the method proposed herein.

APPLICATION OF PROPOSED METHOD

The above method is proposed based on the condition that the number of test data on the specimen from every depth at the same site is enough to do the regression analysis. The application of the proposed method on a natural soil deposit is discussed as follows. In the case of the natural water content w_n being not less than the liquid limit w_L, its compression path in the remolded state is a straight line in the ln(1+e)-logp plot. That is, the yield stress p_{CL} is zero when the degree of sample disturbance SD is equal to 100%. As stated previously, the yield stress decreases linearly with the increase in the degree of sample disturbance. The measured yield stress p_{CL} obtained from a so-called "undisturbed" sample is responsible a certain degree of sample disturbance SD. Hence, based on Eq.4 with consideration of the known condition of $p_{CL}=0$ when SD=100%, the corrected yield stress p_{yf} responsible for SD=0% can be calculated from the following equation.

$$p_{yf}=p_{CL}\times 100/(100-SD) \quad ......(5)$$

Two examples of the application using Eq.5 for correcting the yield stress are given for natural Ariake clay deposits as follows.

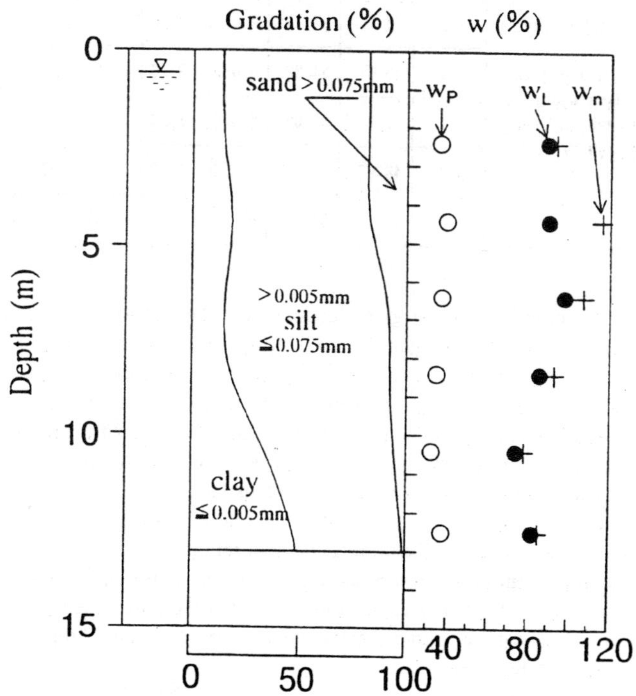

FIG.9 SOIL PROFILE FOR AN ARIAKE CLAY

Example 1 is given for a natural Ariake clay deposited in Hama-machi, Kashima City, Saga Prefecture. The undisturbed samples were obtained using a fixed piston thin-wall sampler. The sampled depth varies from 2.0m to 13.0m below the ground surface. The ground water level is 0.5m below the ground surface. The soil profile is shown in Fig.9. It can be seen that the natural water contents are larger than the liquid limit for all the samples. Using Eq.5, the corrected yield stress p_{yf} can be obtained. The results are shown in Table 6. In which the values of p_{cl} and C_{CLB} are interpreted from the oedometer tests data for the undisturbed samples. Since the oedometer test is not performed on the remolded samples, the values of C_{CLR} are calculated using the liquid limit w_L based on Eq.2: $C_{CLR}=-0.390+0.332\times\log(w_L)$. Then, the value of SD is calculated from the following equation: $SD(\%)=C_{CLB}/C_{CLR}\times 100\%$. Fig.10 shows the relationships of p_{yf} versus the overburden pressure p_{vo} and p_{CL} versus p_{vo} through the depth for the investigated Ariake clay deposit. It can be seen that the natural Ariake clay deposit can be divided into three layers: crust layer, upper Ariake clay and lower Ariake clay. The upper and lower Ariake clays are bounded at about 11-12m. This result is consistent with that reported by Ariake Bay Research Group (1965). Fossil assemblage analysis of the sediment have indicated that the upper layer is deposited under relatively stable marine conditions, whereas the lower layer is accumulated under rapid transgression (Ariake Bay Research Group, 1965; Ohtsubo et al., 1995). From Fig.10, it can be seen that the ratio of p_{yf} over p_{vo} can be approximately expressed by a constant through the upper Ariake layer.

Table 6. Corrected yield stress for an Ariake clay deposit (Example 1)

soil	depth (m)	p_{vo} (kPa)	p_{CL} (kPa)	C_{CLB}	w_L (%)	SD (%)	p_{yf} (kPa)
	2.4	15.75	40.43	0.0438	89.2	17.03	48.73
Ariake clay	4.4	22.75	27.75	0.0355	89.9	13.74	32.17
(obtained in Hama-machi,	6.4	30.67	34.22	0.0302	97.8	11.16	38.52
Kashima City, Saga Pref.)	8.4	39.77	50.05	0.0297	85.4	11.83	56.77
	10.4	48.97	60.07	0.0246	73.9	10.67	67.25
	12.4	57.75	93.13	0.0357	82.2	14.53	109.0

The other example is given for an Ariake clay deposit deposited in Mikazuki-machi, Ogi-gun, Saga Prefecture. The basic properties and the oedometer test results for the sampled undisturbed samples are shown in Table 7. It can be seen that the natural water content w_n is larger than the liquid limit w_L through the investigated depth. The corrected yield stress is obtained in the same way as that in Example 1. The corrected results are also shown in Table 7. The distributions of (p_{yf}-p_{vo}) and (p_{CL}-p_{vo}) through the depth are shown in Fig.11. As the same as that for Example 1, the natural Ariake clay deposit can be divided into three layers: crust layer, upper Ariake clay and lower Ariake clay. From Fig.11, it can be seen that the difference between p_{yf} and p_{vo} can be approximately expressed by a constant through the upper Ariake layer.

Above examples indicate that the relationships of p_{yf} versus p_{vo} can be expressed by their ratio or their difference through the depth for a natural soil deposit. In fact, a natural soil generally exhibits the value of overconsolidation ratio (OCR) in excess of unity though the

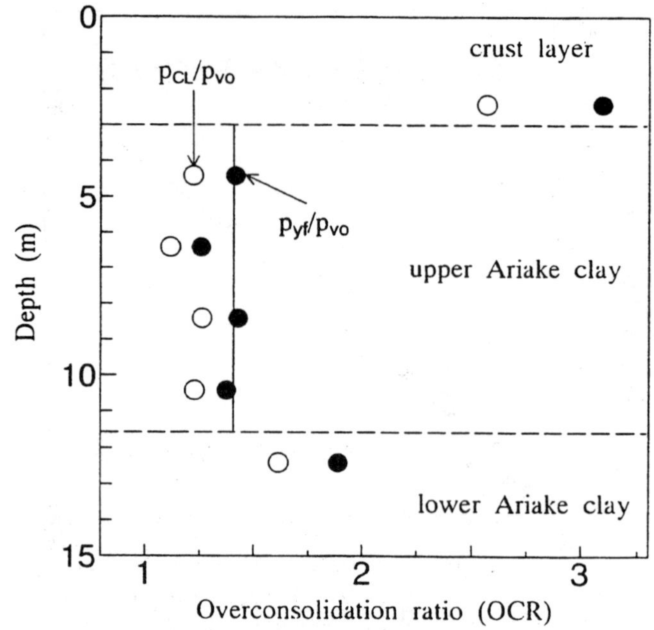

FIG.10 DISTRIBUTION OF OCR FOR AN ARIAKE CLAY (Example 1)

Table 7. Corrected yield stress for an Ariake clay deposit (Example 2)

soil	depth (m)	w_n (%)	w_L (%)	p_{vo} kPa	p_{CL} kPa	C_{CLB}	SD (%)	p_{yf} kPa
Ariake clay	4.4	121.2	105.0	28.6	53.5	0.0432	15.4	63.2
(obtained in Mikazuki-machi,	6.4	109.3	75.2	36.6	50.6	0.0543	23.3	66.0
Saga Prefecture)	8.4	96.0	72.6	45.7	64.9	0.0309	13.6	75.1
Boring No. 1	10.4	95.3	74.8	54.6	71.0	0.0274	11.8	80.5
	12.4	58.9	54.0	69.0	125	0.0407	22.0	159.7
Ariake clay	2.4	63.8	68.6	18.7	137	0.0338	15.4	162.3
(obtained in	4.4	110.2	69.1	26.9	47.6	0.0458	20.8	60.0
Mikazuki machi, Saga	6.4	103.8	64.3	35.1	52.2	0.0291	13.8	60.6
Prefecture)	8.4	114.3	67.3	43.4	49.6	0.0711	32.8	73.8
Boring No. 2	10.4	103.5	77.6	52.1	70.7	0.0407	17.1	85.3
	12.4	66.4	60.5	62.8	90.6	0.0454	22.5	117.0
Ariake clay	2.4	89.3	94.9	15.9	91.7	0.0256	9.6	101.5
(obtained in	4.4	114.9	99.9	23.6	63.2	0.0377	13.8	73.3
Mikazuki-machi, Saga	6.4	100.8	75.8	32.6	57.6	0.0336	14.3	67.2
Prefecture)	8.4	102.3	76.2	41.4	68.3	0.0380	16.2	81.5
Boring No. 3	10.4	90.2	68.2	51.1	70.2	0.0482	22.0	89.9
	12.4	80.2	65.2	62.4	78.5	0.0352	16.6	94.1

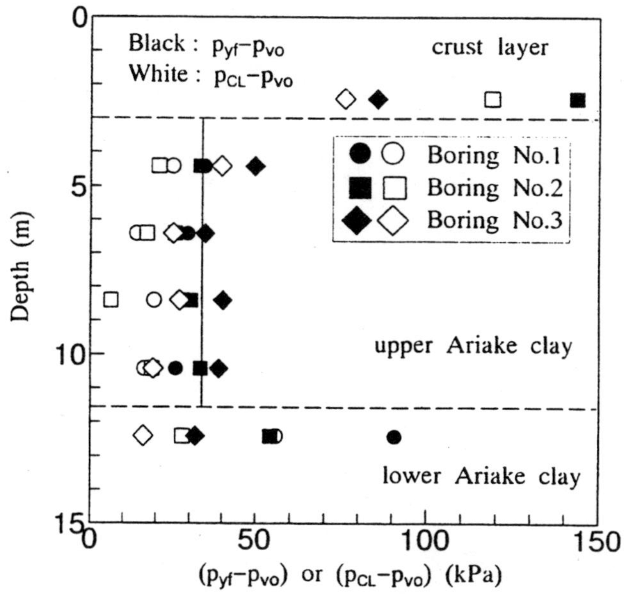

FIG.11 DISTRIBUTION OF YIELD STRESS FOR AN ARIAKE CLAY (Example 2)

site geology often indicates the normal consolidation (Mesri, 1975; Schmertmann, 1991). This phenomenon has been explained by the soil structure for natural soils, mainly caused by the secondary compression and the chemical bonding (Bjerrum, 1973; Hanzawa and Adachi, 1983). According to the concept proposed by Bjerrum (1973), the field yield stress, designated p_{yf} herein, developed by the secondary compression increases linearly with the increase in the overburden pressure, p_{vo}. The relationship between p_{yf} and p_{vo} responsible for a homogeneous layer of a natural soil subjected to the secondary compression, therefore, can be expressed in Eq.6 (Bjerrum, 1973; Hanzawa and Adachi, 1983).

$$p_{yf} = K_1 \times p_{vo} \quad ......(6)$$

The Ariake clay shown in Fig.10 is one example for this case.

In addition, for a homogeneous layer of a natural soil subjected to the chemical bonding, Hanzawa and Adachi (1983) have conceptually indicated that the field yield stress, designated p_{yf} herein, has a relationship with the overburden pressure, p_{vo}, as shown in Eq.7.

$$p_{yf} = p_{vo} + K_2 \quad ......(7)$$

The coefficients of K_j (j=1,2) are constants. The Ariake clay shown in Fig.11 is one example for such a case.

CONCLUSIONS

The conclusions obtained from this study are summarized as follows:

1) The specimens obtained from the same depth at the same site are deliberately disturbed by vibration to different level. The oedometer test results indicate that the compressibility during the recompression up to the yield stress increases with the increase in the level of sample disturbance. The compression index reaches the uppermost when the sample is completely remolded. For the field compression path, the compression can be negligible during the recompression.

2) A new index for sample disturbance is proposed as SD(%) = $C_{CLB} / C_{CLR} \times 100\%$. This index can reflect quantitatively the extent of sample disturbance.

3) The generalized relationship between the compression index in the ln(1+e)-logp plot, C_{CLR}, and the liquid limit, w_L, for the remolded soils can be expressed in the equation : $C_{CLR} = -0.390 + 0.332 \times \log(w_L)$. This equation can be used to provide the value of C_{CLR} when the oedometer test is not performed on the remolded soils.

4) The measured consolidation yield stress p_{CL} decreases linearly with the increase in the degree of sample disturbance.

5) A simple approach of correcting p_{CL} is proposed. The relationship between the measured p_{CL} and the degree of sample disturbance SD is investigated to obtain the correction coefficient R_p. Then, the consolidation yield stress corrected for sample disturbance p_{yf} can be calculated by following equation: $p_{yf} = p_{CL} + R_p \times$ SD(%).

6) In the case of the natural water content w_n being not less than the liquid limit w_L, the corrected yield stress p_{yf} responsible for SD=0% can be calculated from the following equation: $p_{yf} = p_{CL} \times 100 / (100\text{-SD})$. Further research is needed to discuss the application of the proposed method for the case of the natural water content w_n being less than the liquid limit w_L.

REFERENCES

Ariake Bay Research Group (1965). "Quaternary system of the Ariake and the Shiranui Bay areas with special reference to the Ariake clay," The Association for the Geological Collaboration in Japan, pp.1-86 (in Japanese).

Bjerrum, L. (1973). " Problems of soil mechanics and construction on soft clays and structurally unstable soils (Collapsible, expansive others)," State-of-the-Art Report, 8th. ICSMFE, Moscow, Vol.3, pp.109-159.

Burland, J.B. (1990). "On the compressibility and shear strength of natural clays," Geotechnique, Vol.40, pp.329-378.

Butterfield, R. (1979). "A natural compression law for soils," Geotechnique, Vol.29, No.4, pp.469-480.

Flokes, D.J. and Crooks, J.H.A. (1985). "Effective stress paths and yielding in soft clays below embankments," Canadian Geotechnical Jour., Vol.21, No.1, pp.21-35.

Hanzawa, H. and Adachi, K. (1983). "Overconsolidation of alluvial clays," Soils and Foundations, Vol.23, No.4, pp.106-118.

Leonards, G.A. and Ramiah, B.K. (1959). "Time effects in the consolidation of clays," ASTM, Special technical publication, No.254, pp.116-130.

Mesri, G. (1975). Discussion of "New design procedure for stability of soft clays," ASCE, Vol.101, No.GT4, pp.409-412.

Nagaraj, T.S. and Murthy, B.R.S. (1983). "Rationalization of Skempton's compressibility equation," Geotechnique, Vol.33, No.4, pp.433-443.

Nagaraj, T.S., Murthy, B.R.S., Vatsala, A. and Joshi, R.C. (1990). "Analysis of compressibility of sensitive soils," ASCE, Vol.116, No.GT1, pp.105-118.

Ohtsubo, M., Egashira, K. and Kashima K. (1995). "Depositional and post-depositional geochemistry, and its correlation with the geotechnical properties of marine clays in Ariake Bay, Japan," Geotechnique, Vol.45, No.3, pp.509-523.

Onitsuka, K., Hong, Z., Hara, Y. and Yoshitake, S. (1995). "Interpretation of oedometer test data for natural clays," Soils and Foundations, Vol.35, No.3, pp.61-70.

Pelletier, J.H., Oslon, R.E. and Rixner, J.J. (1979). "Estimation of consolidation properties of clay from field observation," Geotechnical Testing Journal, ASTM, Vol.2, No.1, pp.34-43.

Schmertmann, J.H. (1991) : "The mechanical aging of soils," ASCE, Vol.117, No.GT9, pp.1288-1330.

Proceedings of the Seventh (1997) International Offshore and Polar Engineering Conference
Honolulu, USA, May 25-30, 1997

ISBN 1-880653-28-1 (Set); ISBN 1-880653-29-X (Vol. I)

A Small Diameter Sampler and the Quality of Its Samples

Takaharu Shogaki
National Defense Academy
Yokosuka, Japan

Takashi Sudoh
Koa Kaihatsu Co. Ltd
Tokyo, Japan

ABSTRACT

A new small diameter sampler with two chamber hydraulic pistons (45 sampler) has been developed, and its applicability is examined though the unconfined compression tests and the standard oedometer tests. The quality of samples obtained from the 45 sampler is higher than those of the sampler normally employed in Japan for both Holocene and Pleistocene clay deposits.

KEY WORDS: clay, sample disturbance, shear strength, soil sampling, standard oedometer test, unconfined compression test

INTRODUCTION

A new small diameter sampler with two chamber hydraulic pistons (45 sampler) was developed, and its applicability is examined through the unconfined compression tests and the standard oedometer tests for samples obtained from Holocene and Pleistocene clay deposits.

In an engineering sense, there is no difference in shear strength and deformation characteristics between the S specimen (15 mm in diameter (d) and 35 mm in height (h)) and the O specimen (35 mm in d and 80 mm in h), which are examined for soils of the plasticity index (I_p) from 17% to 98%, unconfined compressive strength (q_u) from 20 kPa to 1000 kPa, overconsolidation ratio (OCR) from 0.7 to 78, and different sites (Shogaki, 1991; Shogaki et al., 1993). Ten S specimens can be taken from a sample, 75 mm in d and 50 mm in h, sampled from the sampler (75 sampler) normally employed in Japan as the Japanese Geotechnical Standard for the method for obtaining undisturbed soil sample using a thin-walled tube sampler with fixed piston (JGS 1221-1995). Of the ten specimens, the stress-strain curves of specimens located very close to the sampling tube wall are similar to those of specimens located in the central site of the plane of the sampling tube (Shogaki et al. 1995). This fact suggests the possibility of the development of a small diameter sampler and its applicability in practice. The new 45 sampler was developed from this information.

In this paper, an outline of the 45 sampler is shown, and its applicability is investigated through the unconfined compression tests and the standard oedometer tests for samples obtained from both 45 and 75 samplers. The quality of samples obtained from the 45 sampler is higher than those of the 75 sampler for both Holocene and Pleistocene clays.

OUTLINE AND CHARACTERISTICS OF THE 45 SAMPLER

The longitudinal section and the specifications of the 45 sampler are shown in Fig.1 and Table 1, respectively. The specifications of the 75 sampler are also shown in Table 1. The outer diameter of the 45 sampler takes into consideration the diameter, 66 mm, of a borehole for the standard

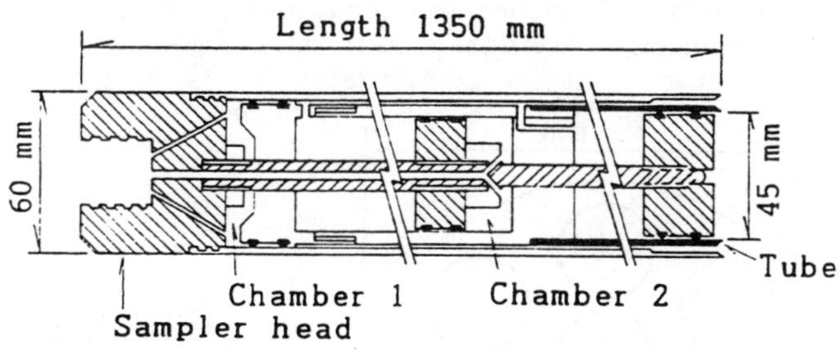

Fig.1 Longitudinal section of the 45 mm piston sampler

Table 1 Specifications of the 45-mm and 75-mm samplers

Sampler	45	75	Tube	45	75
Outer diameter (mm)	60	89	Inner diameter (mm)	45	75
Total length (mm)	1350	2165	Length (mm)	600	1000
Mass (kg)	11	24	Length of Sample (mm)	500	800

penetration test (SPT). The 45 sampler, which solves many of the problems of the 75 sampler used by the Japanese Geotechnical Society, has the following advantages.

1. It will contribute to the job of sampling by reducing the work load and increasing efficiency because the 45 sampler is 62% lighter in weight and 46% shorter in length than the 75 sampler.
2. The 45 sampler can take high quality samples for the sample disturbance, since the penetration of the sampling tube can be advanced smoothly because of the longer outer tube.
3. The penetration force of the 45 sampler is about twice as powerful as the 75 sampler, since the 45 sampler has two chamber hydraulic pistons as shown in Fig.1. Therefore, the 45 sampler is also better for hard soils than the 75 sampler.
4. When there is a change from sand layer to clay layer during the SPT, an undisturbed clay sample can be taken from the same borehole with the 45 sampler.

SOIL SAMPLES AND TEST PROCEDURES

The undisturbed soil samples used in this study were obtained from the Holocene and Pleistocene clay deposits located on the Holocene plain in Mito city, Japan. The sampling depths (z) of both samplers were minus 10 m below the ground surface for the Holocene clay deposit, and minus 42 m for the Pleistocene clay deposit. The sampling was performed at two sites about 3 m apart in horizontal distance at the same elevation.

The lengths of soil sampled from the 75 and the 45 samplers were 800 mm and 500 mm, respectively. In both clays, the ratio of the length of the sample before the withdrawal to that of penetration of the sampler, below the bottom of the borehole during sampling, was almost 1. After 10 mm of the soil at the edge of the sampler were struck off, samples with a length of 45 mm were extruded from the sampling tube. Figure 2 shows the sites of specimens for the plane of the sampling tube. To examine the effect of the site of the specimen for the plane of the sampling tube, ten S specimens from 75 mm in diameter and four S specimens from 45 mm in diameter, were taken as shown in Fig.2. The sites

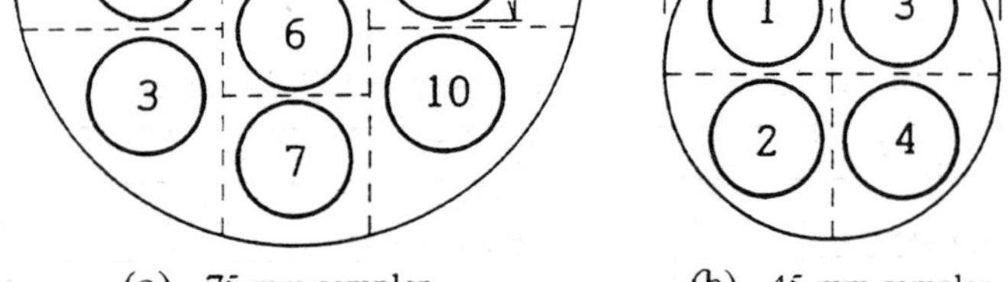

(a) 75-mm sampler (b) 45-mm sampler

Fig.2 Sites of specimens for the plane of the sampling tube

of each specimen as shown in Fig.2 are identical with each sample. The specimens used for the oedometer test were d= 60 mm and h= 20 mm, and d= 30 mm and h = 10 mm, for the sample taken from the 75 sampler, and d= 30 mm and h = 10 mm for the 45 sampler. In Japan, the standard size for the oedometer tests is d = 60 mm and h = 20 mm. There was no difference in consolidation characteristics beween the small and standard size specimens, which were tested from soils of the I_p = (27~81)%, q_u =(42~294) kPa and OCR =(0.8~2.6) (Shogaki and Miwa, 1996). The oedometer tests were performed using a load increment ratio of unity and the duration of loading for each load increment was one day. The values of the preconsolidation pressure (σ'_p) and the compression index (C_c) were determined from the void ratio (e) to consolidation pressure (p) curves corresponding to 24-hour compression. Mikasa's method was used to identify preconsolidation pressure on the 24-hour e-log$p$ curve. This method has been employed in Japan as the Japanese Industrial Standard for determining one-dimensional consolidation properties of soils (JIS A 1217-1990). It was shown by Okumura (1974) that there is no difference in the σ'_p values between the Mikasa method and casagrande method (1936).

In the unconfined compression tests, the specimens were sheared at a strain rate of 1%/min using portable unconfined compression apparatus (Shogaki, 1991). The value of q_u was determined to be the maximum stress corresponding to axial strain (ε_f) of less than 15 %. The secant modulus (E_{50}) is given by $q_u/2\ \varepsilon_{50}$, in which ε_{50} is the strain at the value of q_u /2. The $\bar{w}_n$, $\bar{\rho}_t$, $\bar{q}_u$, $\bar{E}_{50}$, and $\bar{\varepsilon}_f$ are the mean values of water content (w_n), wet density (ρ_t), q_u, E_{50} and ε_f.

EFFECT OF SAMPLER TYPES ON STRENGTH PROPERTIES

Effect of the site of the specimen for a longitude section of sampling tube on strength properties

The results of unconfined compression tests for the Holocene and the Pleistocene clays are shown in Fig.3. (a) and (b), respectively. The $\bar{w}_n$ and $\bar{\rho}_t$ values obtained from the 45 and the 75 samplers are similar for both clays. In the case of the Holocene clay as shown in Fig.3(a), it can be seen that the q_u and E_{50} values obtained from the 75 sampler are about 20 % and 70 % smaller, respectively, than those of the 45 sampler. On the other hand, in the case of the Pleistocene clay as shown in Fig.3(b), the q_u and E_{50} values obtained from the 75 sampler are about 30 % and 45 % smaller, respectively, than those of the 45 sampler caused by the sample disturbance of the 75 sampler.

The stress-strain curves by both samplers for the Holocene and the Pleistocene clays are shown in Fig.4. The initial tangent modulus and the q_u values obtained from the 45 sampler are larger than those of the 75 sampler.

The ratios of the $\bar{q}_u$ values of each sample to the largest value of those of $\bar{q}_u/q_{u(max)}$ are plotted against the sampling depths for the Holocene and the Pleistocene clays in Fig.5(a) and (b), respectively. The $\bar{q}_u/q_{u(max)}$ values in both parts of the cutting edge and its opposite side decrease in the case of the pleistocene clay, as shown in Fig.5(b). The small values of $\bar{q}_u/q_{u(max)}$ values are caused by sample disturbance. In these parts of the 45 sampler, the decline of the $\bar{q}_u / q_{u(max)}$ values, however, are little, and the samples in the part of 50 mm to 400 mm from the cutting edge of the sampling tube have a good quality.

Effect of the site of the specimen for plane of sampling tube on strength properties

$\bar{q}_u / \bar{q}_{u(5,6)}$ are the ratios of the q_u value, except for No.5

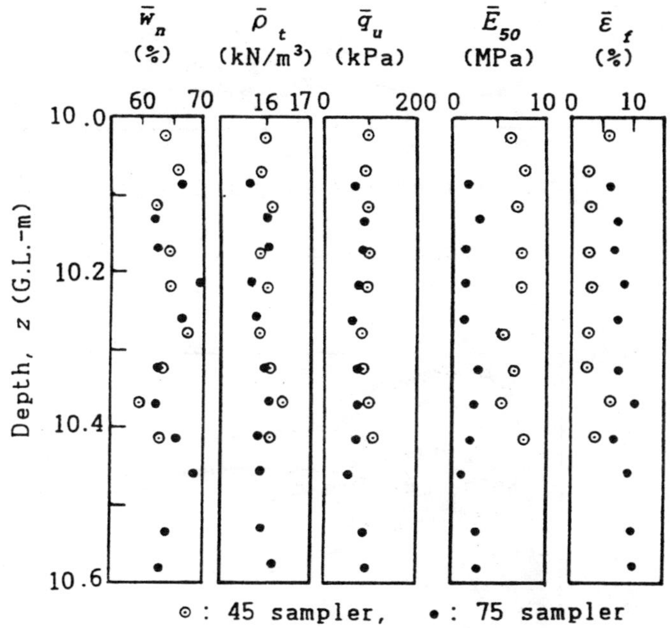

(a) Holocene clay (z = −10 m)

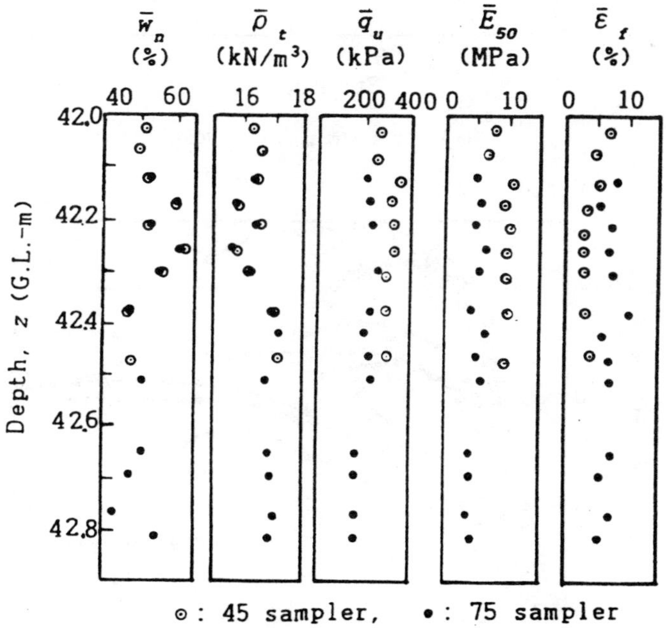

(b) Pleistocene clay (z = −42 m)

Fig.3 Results of the unconfined compression tests

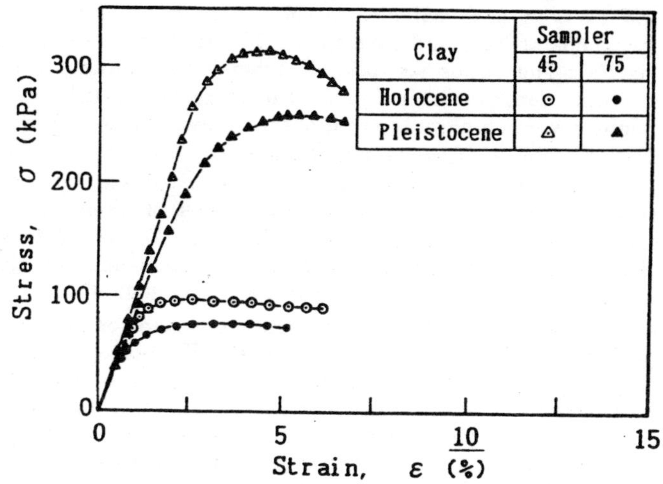

Fig.4 Stress-strain curves

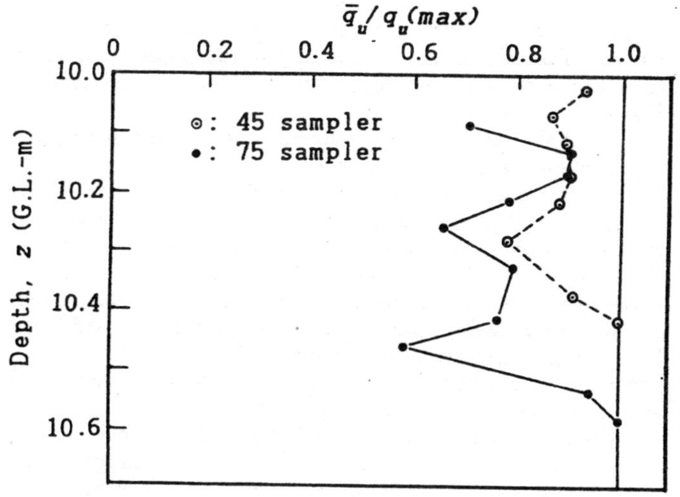

(a) Holocene clay (z = −10 m)

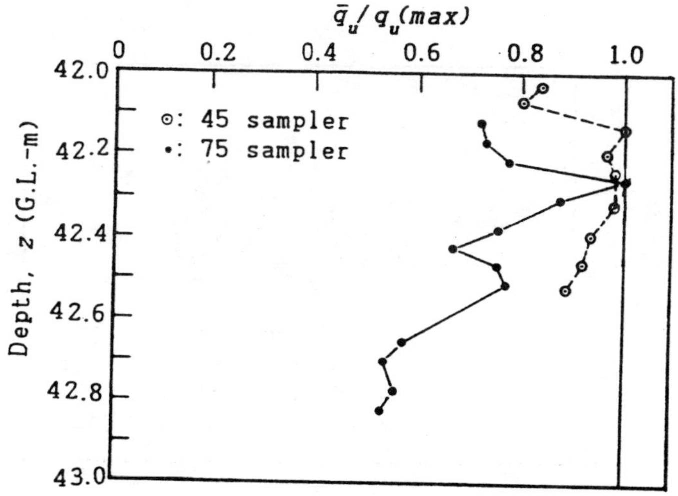

(b) Pleistocene clay (z = −42 m)

Fig.5 Relationship between $\bar{q}_u / q_{u(max)}$ and z

and No.6 in each sample, to the mean value of those of specimens of No.5 and No.6 in the 75 sampler, as shown in Fig.2(a). The ratios are plotted against the distance from sampling depths (*Ds*) in Fig.6. The $\bar{q}_{u(5,6)}$ values for the 45 sampler were employed to that of the 75 sampler at the same depths. The $\bar{q}_u/\bar{q}_{u(5,6)}$ values at $Ds \fallingdotseq 70 \sim 90$ mm are about 1.3 because the q_u value of specimens of No.5 and No.6 are small. This is caused by broken pieces of shell in the specimens. The $\bar{q}_u/\bar{q}_{u(5,6)}$ values from the 75 sampler are about 1 independent of the *Ds* values. This means that the q_u values of specimens closely located to the wall of the tube are similar to that of the central site (specimens of No.5 and No.6) of the plane of the sampling tube. Therefore, it can be judged from Fig.6 that the variation of the $\bar{q}_u/\bar{q}_{u(5,6)}$ values for each sampler reflect the variation of inherent q_u values without the sample disturbance caused by the penetration of the sampling tube and the extrusion of soil samples. The $\bar{q}_u/\bar{q}_{u(5,6)}$ values of the 45 sampler are larger than those of the 75 sampler for the reasons described in Fig.3.

$\bar{E}_{50}/\bar{E}_{50(5,6)}$ are the ratios of the E_{50} value, except for No.5 and No.6 in each sample, to the mean value of those of specimens of No.5 and No.6 in the 75 sampler, as shown in Fig.7. The $\bar{E}_{50}/\bar{E}_{50(5,6)}$ values of the 45 sampler are larger than those of the 75 sampler for the reasons described in Fig.3.

All of the fourteen stress-strain curves at the same depth of the Holocene and Pleistocene clays are shown in Figs.8 and 9, and the range of the q_u and E_{50} values of each sampler are also given in Tables 2 and 3, respectively. The stress-strain curves and the range of the q_u and E_{50} values of specimens obtained from the 45 sampler are larger than those of specimens from the 75 sampler regardless of the Holocene and Pleistocene clay deposits. It is clear from Figs.8 and 9 that the sample disturbance caused by friction between the sampling tube and the sample was not extended to the specimen sites, as shown in Fig.2.

Using the SEM (Scanning Electron Microscope), Shogaki and Matsuo (1985) examined the effect of microstructual disturbance caused by the penetration of the sampling tube and the extrusion of soil samples. For undisturbed Yokkaichi clay with I_p = 44 % obtained from the 75 sampler,

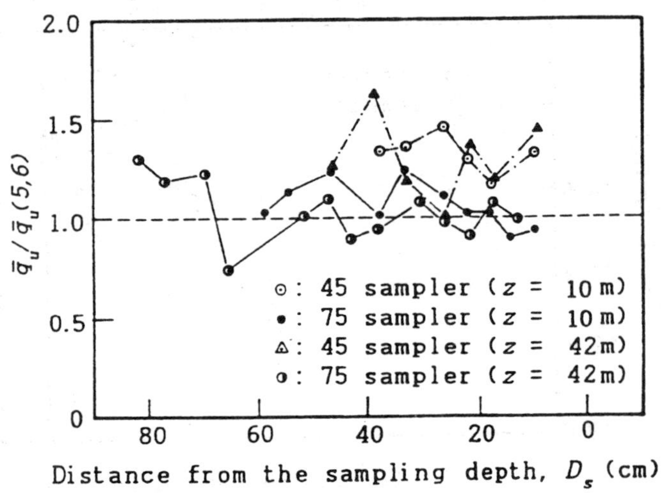

Fig.6 Relationship between $\bar{q}_u / \bar{q}_{u(5,6)}$ and D_s

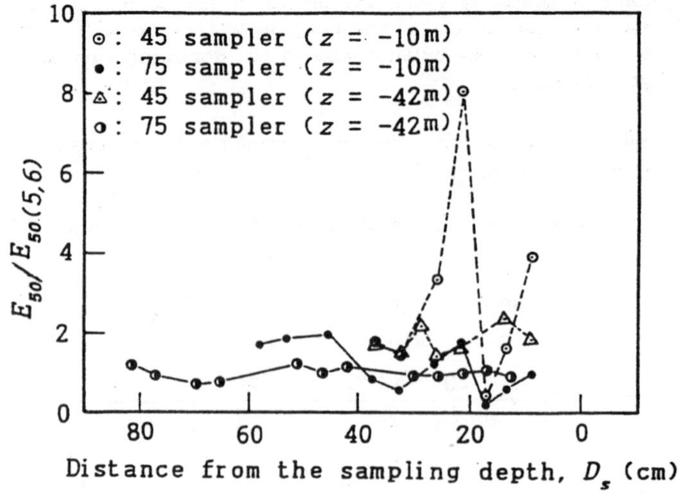

Fig.7 Relationship between $\bar{E}_{50} / E_{50(max)}$ and D_s

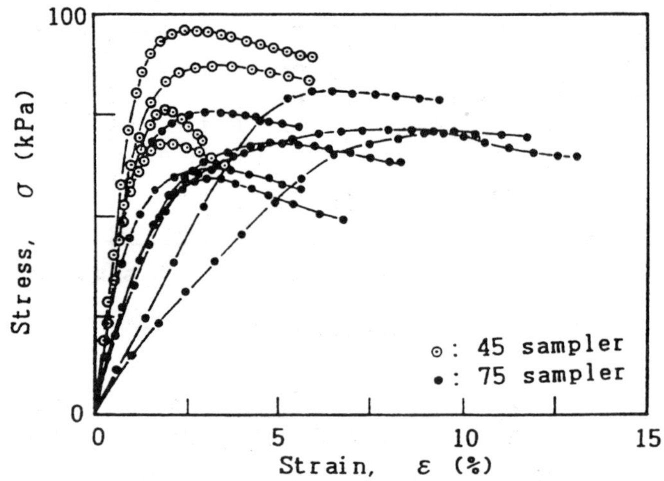

Fig.8 Stress-strain curves (Holocene clay)

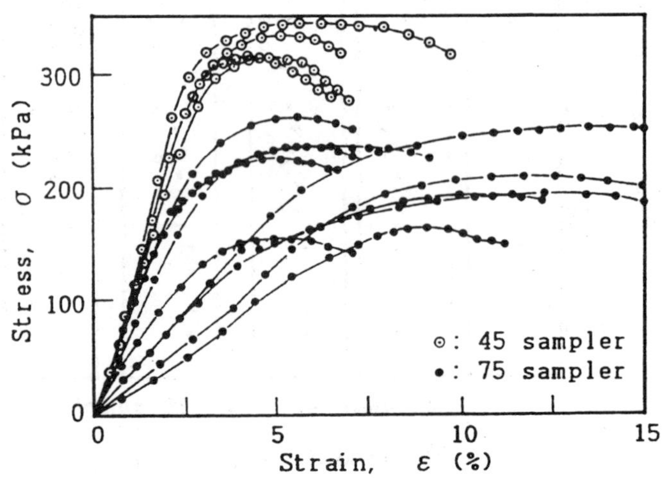

Fig.9 Stress-strain curves (Pleistocene clay)

Tabel 2 q_u and E_{50} values for both samplers (Holocene clay)

Sampler	q_u (kPa)	E_{50} (MPa)	$\overline{q}_u$ (kPa)	$\overline{E}_{50}$ (MPa)
45	68~96	5.4~8.4	82	6.8
75	61~82	1.2~6.7	73	2.9

Tabel 3 q_u and E_{50} values for both samplers (Pleistocene clay)

Sampler	q_u (kPa)	E_{50} (MPa)	$\overline{q}_u$ (kPa)	$\overline{E}_{50}$ (MPa)
45	311~346	9.6~11.6	326	10.7
75	153~260	2.1~8.8	212	5.3

there was complete remolding at the tube wall, but 2 mm from the wall, the microstructure was not greatly altered from that at the center of tube (Shogaki and Matsuo, 1985). It can be recognized that the results of the unconfined compression test in Figs.8 and 9 coincide well with the results of microstructural disturbance using the SEM obtained by Shogaki and Matsuo, (1985).

EFFECT OF SAMPLER TYPES ON CONSOLIDATION PROPERTIES

The symbols for each specimen plotted in Figs.10 to 12 are listed in Table 4. The relationships between the void ratio (e) and the logarithm of consolidation pressure (p) of the Holocene and Pleistocene clay deposits are shown in Fig.10, and the preconsolidation pressure (σ'_p), compression index (C_c) and swelling index (C_s) of each specimen are given in Table 5.

In Fig.10, the e values under the same p values differ in several specimens, which is caused by the different initial void ratios. Shogaki and Kaneko (1994) reported an experimental investigation that showed that σ'_p, C_c, the coefficient of

Tabel 4 Symbols for each specimen usedin Figs.10 to 12

Sample	Sampler	Specimen	Symbol
Holocene	45	1	○
		2	●
	75	1	□
		2	■
Pleistocene	45	1	△
		2	▲
	75	1	×
		2	◇

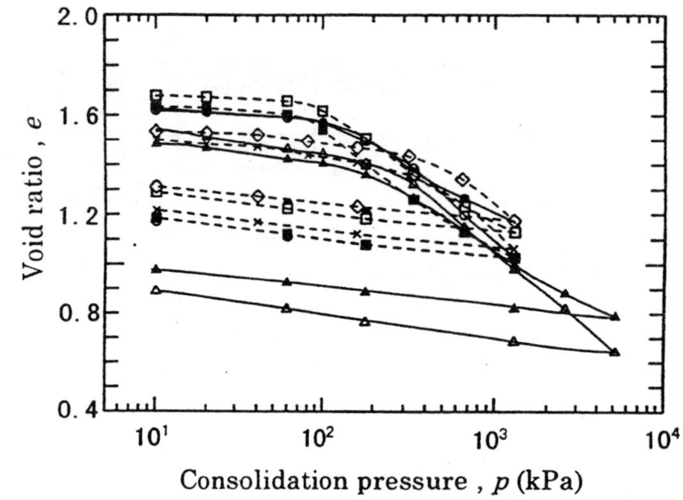

Fig.10 e -log p curves

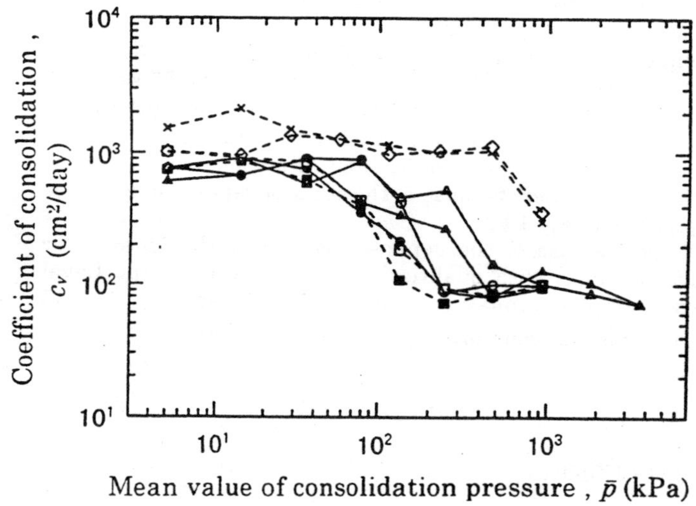

Fig.11 Relationship between c_v and $\overline{p}$

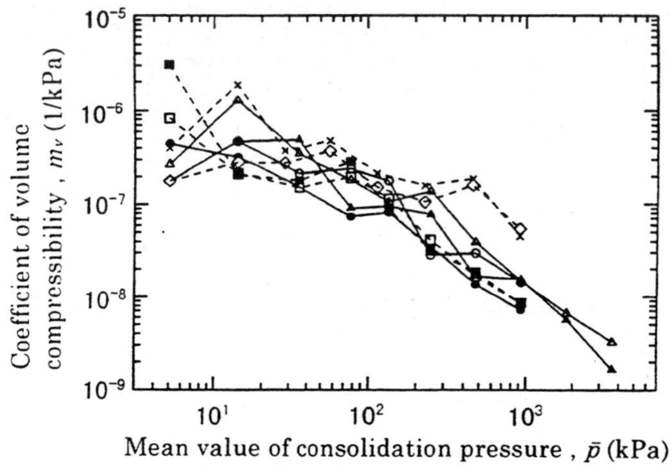

Fig.12 Relationship between m_v and $\overline{p}$

Tabel 5 Resuluts of standard oedometer tests

Sample	Sampler	Specimen	w_o (%)	ρ_t (kN/m^3)	I_p (%)	ε_{vo} (%)	σ_p (kPa)	C_c	C_a	q_u
Holocene	45	1	62.4	16.0	30	2.2	192.5	0.61	0.07	107
		2	63.6	16.0		1.9	153.3	0.48	0.06	100
	75	1	66.1	16.0	33	2.4	109.2	0.54	0.08	71
		2	68.4	16.1		6.0	85.0	0.54	0.09	68
Pleistocene	45	1	56.9	16.0	33	9.0	300.0	0.64	0.09	333
		2	54.4	16.1		8.9	232.9	0.48	0.07	282
	75	1	56.5	16.3	31	6.5	453.0	0.62	0.07	232
		2	60.8	15.9	42	4.3	462.5	0.57	0.06	216

consolidation (c_v) values become smaller, and the coefficient of volume compressibility (m_v) values become smaller when soils are greatly disturbed. In the case of the Holocene clay deposits, the σ'_p, C_c and q_u values obtained from the 75 sampler are smaller than those of the 45 sampler because of sample disturbance of sample obtained from the 75 sampler. The c_v and m_v values are also plotted against the logarithm of the mean values ($\bar{p}$) of p in Figs.11 and 12 , respectively. The c_v values obtained from the 75 sampler are smaller than those of the 45 sampler, because of the effect of sample disturbance described in Fig.10.

In the case of the Japanese soft clays, the 75 sampler gives samples with qualities similar to those of the Laval type sampler (Tanaka et al., 1995). Therefore, the 45 sampler is also comparable to the Laval type sampler.

CONCLUSIONS

A new small diameter sampler with two chamber hydraulic pistons (45 sampler) was developed. By using the 45 sampler, an undisturbed clay sample can be taken from the same borehole for the standard penetration test. The 45 sampler is better for hard soils than the sampler (75 sampler) normally employed in Japan. The quality of samples obtained from the 45 sampler is higher than those of the 75 sampler presently employed in Japan for both Holocene and Pleistocene clays.

REFERENCES

Casagrande.A. (1936): "The determination of the preconsolidation load and its practical significance", *Proc. 1st ICSMFE.* Cambridge, Mass. Vol.1, pp.60-64.

Japanese Standard Association, (1993): "Test method for One-dimensional consolidation properties of soils", JIS A1217-1990, pp.1-13.

Japanese Geotechnical Society, (1995): "The method for obtaining undisturbed soil sample using thin-walled tube sampler with fixed piston (JGS 1221-1995).", The method for geotechnical investigation, pp.152-158(in Japanese).

Okumula,T. (1974): "Studies on the disturbance of clay soils and the improvement of their sampling techniques", *Report of the Port and Harbor Research Institute*, Vol.10, No.193, pp.45 (in Japanese).

Shogaki,T., & M.Matsuo (1985): "Factor analysis approach to unconsolidated undrained shear strength of clay with some sonsideration on Microscopic structures", *Proc. of Symposium on Sampling*, pp.109-116 (in Japanese).

Shogaki,T (1991): "Strength properties of clay by portable unconfined compression apparatus", *Proc. Int. Conf. on Geotechnical Engineering for Coastal Development*, pp.85-88. Yokohama.

Shogaki,T., K.Watanabe & T. Sudo (1993): "Effects of size and shape of specimen on undrained shear strength of pleistocene clay", *Proc. Int. Symposium on Geotechnical Eng. of Hard Soils-Soft Rocks*, pp.803-807. Athens.

Shogaki,T & M.Kaneko (1994): "Effects of sample disturbance strength and consolidation parameters of soft clay", Soils and Foundations, Vol.34, No.3, pp.1-10.

Shogaki,T., H.Moro & K.Kogure (1995): "Statistical properties of soil data within thin-walled samplers", *Proc. 5th Int. Offshore and Polar Eng. Conf*, pp.406-413. Hague.

Shogaki,T. & K.Miwa (1996): " Effect of specimen size on consolidation parameters in standard oedometer tests", *Proc. Civil Eng. in the Ocean*, Vol.12, pp.225-229 (in Japanese).

Tanaka.H & M.Tanaka (1995): "Sample quality obtained by various types of samples", *Proc. 10th Asian Regional Conf.* ISSMFE, pp.83-86. Beijing.

Proceedings of the Seventh (1997) International Offshore and Polar Engineering Conference
Honolulu, USA, May 25-30, 1997

ISBN 1-880653-28-1 (Set); ISBN 1-880653-29-X (Vol. I)

Stability of Heaps of Iron Ore Concentrate

Jan Laue
Ruhr-University Bochum
Germany

ABSTRACT

Caused by the capsizing of a vessel loaded with heaps of wet iron ore concentrate investigations are carried out on the stability of such piles. Geotechnical centrifuge model tests have been conducted in the Bochum geotechnical centrifuge ZI to investigate the stability behaviour of wet iron ore concentrate heaps under static and cyclic loading conditions, representing the movements of the ship. The basic behaviour of the iron ore concentrate is described in comparison with sand in comparable graduation. The parameters for the used materials are evaluated with conventional soil mechanical laboratory tests and also with special tests for the transport of granular materials specified by the Internationale Maritime Organization. More than fourty tests were executed in the centrifuge with the help of a newly developed and constructed shaker combining the possibility of imposing vertical and swinging movements under cyclic loading conditions. An additionaly constructed swinging bracket allows large angles of rolling. Based on a serie of model the model tests, the validity of model laws for the investigation involving cyclic and dynamic problems of wet material are discussed. The findings on the behaviour of wet iron ore concentrate heaps under selfweight conditions, steadily cyclic movements (f.e. rolling of a ship) and sudden movements are shown. Variations of water content, temperature of the material, sample preparation and shape, are correlated to the installation and the behaviour of the heaps on board. With the help of this findings, requirements to reduce the destabilisation are formulated.

KEY WORDS: Centrifuge Model Tests, Iron Ore Concentrate, Transport Stability, Shaking Tests, Bulk Material

1. INTRODUCTION

The investigations presented in this paper have been intended by the capsizing of a vessel loaded with heaps of iron ore concentrate. Figure 1 shows the situation on bord in principle. Several overlapping heaps of iron ore concentrate have been placed in open air on a plane steely cargo area (ca. 100 m x 24 m) without shots. The heaps have a heigth of about 5m. Between the sides of the heap and the border of the cargo area, a gap is held free of around 1m at both sides. At both sides of the cargo space some outlets for water are installed.

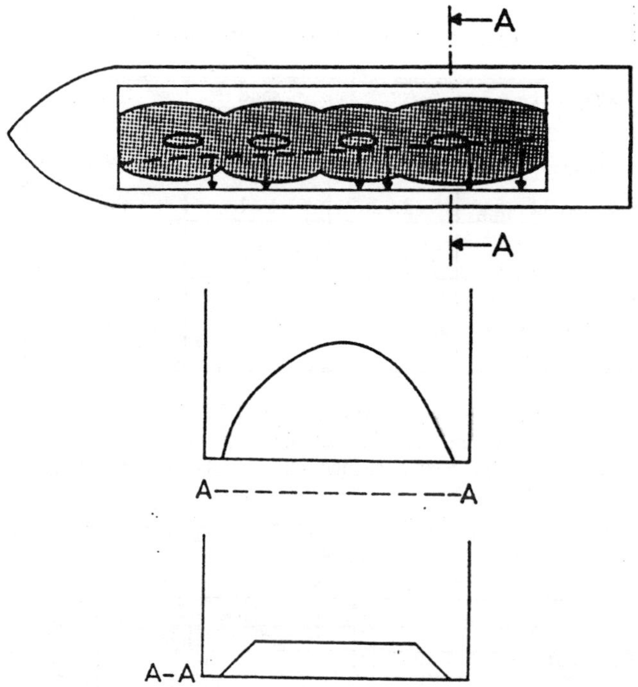

Fig. 1. Principle situation of cargo on board before and during the voyage.

The heaps are placed on the cargo space by free falling. In this case no trimming of the bulk was done. The iron ore had water contents of about 3.5% up to 6%. During the last voyage of the ship in winter 1990, the heaps break down, settled and changed their shape. This has been described by the survivors with the word liquifaction, which has to be discussed in detail. This "liquifaction" was also observed during several voyages in summertime too. At the end of the last voyage, changing the course under bad conditions (storm, height of waves), the vessel tilts stepwise from angles of 0° to 2°, 2° to 4° and 4° to 11°. After this the vessel turned around.

The task of the research work presented in this paper consists of finding solutions for the capsizing of the vessel and describing the transport behaviour of the wet iron ore concentrate in more detail from a geotechnical point of view. This includes a comparison of geotechnical laboratory investigations with the test methods claimed by the Internationale Maritime Organization (IMO) in the code for bulk cargoes (1985) to describe the transport stability and observations of the whole system by means of geotechnical centrifuge model tests under defined boundary conditions. To study the transport behaviour of the whole system, variations of the water content, climatic influences, shape of the heaps and different load conditions are regarded.

2. Laboratory Investigations

In respect to the well known behaviour of soils, the laboratory tests on iron ore concentrate are compared with the properties of a fine grained sand (Normsand 942 D) which is normally used representing non cohesive soils in the centrifuge model tests. The properties of the used iron ore concentrate are given with the equivalent parameters of the nearly same grain-sized sand in Table 1. The gradation curves of the original iron ore concentrate, the sand 942D and additionally a fine grained iron ore concentrate, which has been used conducting the "model the model" tests are given in Figure 2.

Tab. 1. Material properties of iron ore concentrate and sand 942 D.

parameter	ore	sand	units
weigth of grain ρs	5.17	2.66	t/m^3
minimal weight ρ_{min}	2.92	1.41	t/m^3
maximal weight ρ	3.57	1.70	t/m^3
angle of friction φ	40	38	°
average diameter of grain d_{50}	0.44	0.23	mm

The angle of friction given in Table 1 is valid for high densities. For lower densities reductions of this value have to be made. The minimal and maximal weight has been evaluated following the German rule DIN 18126 (1989).

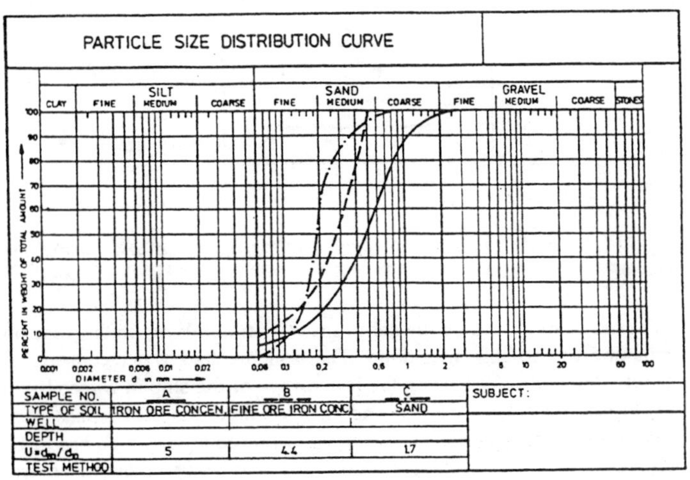

Fig. 2. Graduation curves for iron ore concentrate, fine iron ore concentrate and sand.

In addition to the physical and mechanical investigation of the material, tests requiered by the IMO to define the boundary conditions for transport are carried out to determine the maximum water content which can be allowed for the transport and a maximum angle of the bulk. The maximum allowable water content for the transport of this iron ore concentrate following the code for bulk cargoes of the IMO is w=6.8% in weight. The results of the tests on the angle of bulk are given in Fig. 3. The angle of the bulk was determined by creating heaps of 400 mm height with free falling iron ore from a defined height. The angle of the slope was measured at four locations, and the average value is given as the angle of bulk for different water content. For dry granular material without cohesion, the test result should give the angle of friction φ to be equal to the slope angle β. For moisture, β will be overestimated in correlation to larger heaps caused by capillary cohesion effects. The determination of a maximum possible slope angle using this methods overestimates the frictional capacity of granular materials.

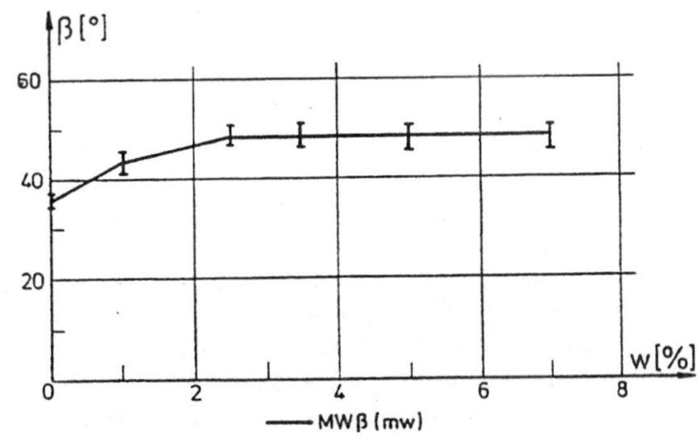

Fig. 3. Development of bulk angle in dependency of the water content.

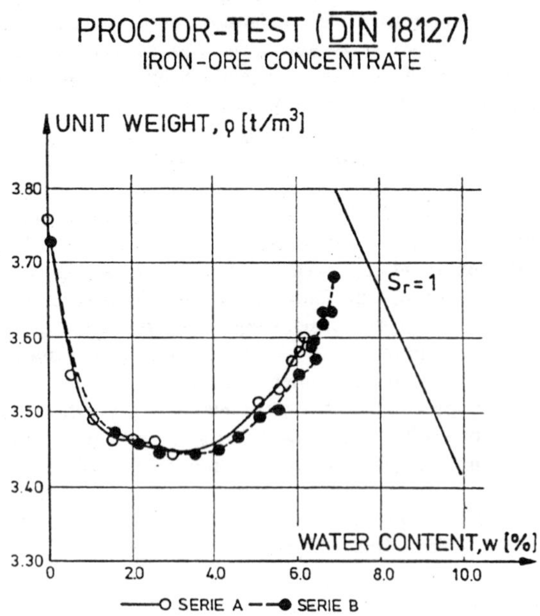

Fig. 4. Results of proctor tests on iron ore concentrate.

The material parameters of the iron ore concentrate can easily compared with sand. The solid weight of the iron ore concentrate is with ρ_s=5,17 t/m^3 nearly as double as high as the weight of the sand with ρ_s=2,63 t/m^3. The material properties like the friction angle and the friction properties between the soil and steel are nearly similar. At least the proctor tests (see Figure 4) shows unusual results. For sand a maximum can be determined easily. The outlook of the curve in case of iron ore looks different. The saturation line can be found too, so that it can be expected, that the upper part of the regular proctor line is shortcutted to one point. The conclusion has drawn in front of the last journey of the vessel, that a loose state with a medium water content should give the best safety for transport, which cannot be sufficient for safety reasons with the test results shown here. The influence of the density on the frictional parameters is obvious.

One important difference between sand and the used iron ore concentrate is the coefficient of permeability with $k=1.7*10^{-5}$ m/s. Thus both materials have nearly a same gradation curve, the permeability of the iron ore concentrate is two decades less than sand. Additionaly it shows a high potency of water capillarity. In air dryed conditions a water content of 2% weight is found.

3. CENTRIFUGE MODEL TESTS

To study the behaviour of the heaps on bord of the ship centrifuge model tests were conducted. The advantage of modeling a system in a geotechnical centrifuge is the possibility to apply real stress conditions on a scaled model. The principle and the basic of centrifuge modelling are given elsewhere e.g. LAUE ET AL. (1996).
At least, one problem of modelling in centrifuges have to be regarded in the case of simulating the behaviour of the wet moisture during cyclic loadings due to the different scaling of frequency and water velocity.

TAN & SCOTT (1985) define the permeability coefficient k with equation 1.

$$k_m = n * k_p \quad (1)$$

with k = coefficient of permeability
n = scaling factor and factor of raised g-level
Indices m = model scale
p = prototyp scale

Other authors (SCHOFIELD (1980), GOODINGS (1984) and TAN & SCOTT (1987)) define $k_M = k_P$ resulting in a n^2 faster laminar flow.

For dynamical processes, the scaling of the velocity has to be regarded additionally. The velocity of movements is scaled with 1. By this it has to be regarded, that the time for outflow phenomena is scaled by n due to the reduced length of the model. With the influence of frequency, which has to be increased by n, two possibilities of simulating flow of water and the deformation of the heaps are possible. By using the original fluid, the settlement and the change of shape of the heaps can be regarded in correct time scale. Due to the raised permeability a dynamically induced liquifaction will not or even less be find. Using a fluid with higher viscosity or changing the graduation curve, the permeabiliy coefficient can be reduced offending against the correct simulation of the time dependent settlement, but reproducing outflow phenomena or even liquifaction correctly. This is proofed by modelling of model tests presented in sation 5.3.

4. TEST DEVICES

For the research of the behaviour of the heaps a shaker has been developed allowing the overlay of vertical movements and tilting. A cut through the model container is given in Figure 5. A box with a basic plate of 300mm x 500mm (15m x 25m in prototyp scale) is placed in a larger circular container with a diameter of 1000 mm to use the given possibilities of the geotechnical centrifuge in Bochum. One side of the box, made of glass, allow the observation of the shape of the heaps by video. Pore pressure transducers are placed in the bottom plate at different locations. Displacement is controlled by LVDT's placed on the heaps. Additional LVDT's located on the side of the box control the load function. Sticks, which are placed inside the heaps give further possibilities to control movements and settlements in further tests. Two cameras control the test, one is placed on top of the outer container to observe the surface of the heaps, the other is placed on the same level as the box controling the shape and the water. A standing pipe allows the observation of water for saturation or outlet.

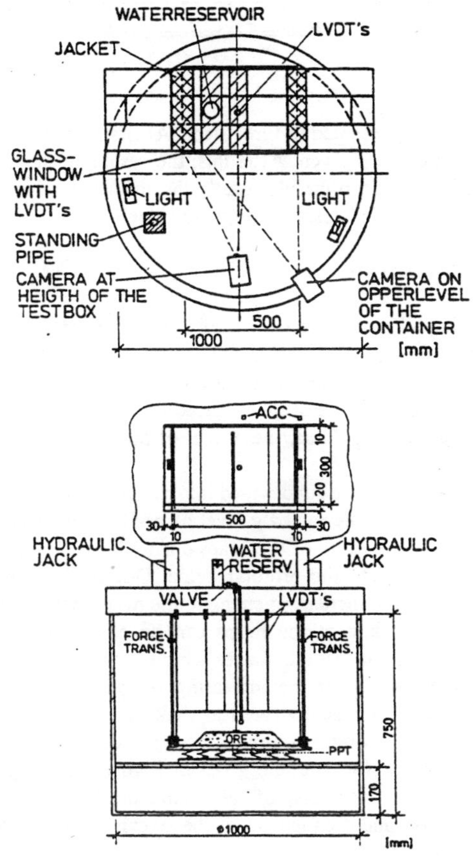

Fig. 5. Top view and cut through the model container.

The movement is applied by two hydraulic jacks which can be controlled independently. The hydraulic jacks allow maximum frequencies for low amplitude of 100 Hz. For lager amplitudes, the maximum possible frequencies reduces. To allow larger angles of tilting, the spring bed is replaced by a bracket. The bracket allow an overlay of a constant tilting with the rolling of the ship.
For simulating the influence of spread water and rain, a raining apparatus was placed on the beam in the middle of the jacks. All boundary conditions of the developed test set-up are given in Table 2, comparing the model scale with the simulated prototyp scale. The level of raised gravity is set to n = 50.

Tab. 2. Boundary conditions of the shaker and the swinging bracket.

	model scale	prototype scale
length [mm]	500	25000
width [mm]	300	15000
static tilting (shaker) [°]	5	5
dynamic tilting (shaker) [°]	4 (5Hz)	4 (0.1 Hz)
vertical movements (shaker) [mm]	18.4	920
maximal static tilting (bracket) [°]	13	13
dynamic tilting (shaker) [°]	4 (5Hz)	4 (0.1 Hz)
additional water [$1x10^{-3}$ m^3]	1.5	187500

5. Tests results

The test results are presented in three steps. At first, the findings of the tests using the shaker are given. After this, the results of the tests using the bracket are shown in principle. Lastly the influence of temperature and the "model the model" tests are discussed briefly.

5.1 Shaking table tests

Different load conditions have been studied using the shaker. The kind of preparation of the heaps, the influence of trimming and differences in the water content are the main topics of this research. To represent the loading procedure on site, the ore was placed in some in the box using a chute from a high of 5 m. Thus, the weight of the heaps is low (ρ= 3.1 t/m³).

Each centrifuge test can be subdevided into different steps. At first, the centrifuge swings up to the raised g-level of n=50. This raising up can be interpreted as the building of the heaps during loading, starting from the model scaled heap to the prototyp scale. Due to the reduction of capillary effects and the increasing strength, the heaps start to deform. A first change of the shape of the heaps can be observed whereas the resulting shape is simular to the shapes which have been found at the cargo space on board. After the increasing of gravity, different load functions, as there are sinusoidal vertical shakings, tiltings with angles of rolling up to 2° using maximum load frequencies f_M= 5Hz or overlayed functions as for example given in Figure 7 are applied. The overlayed movement consists of two sinusoidal functions using different amplitudes and frequencies at the two hydraulic jacks at the sides of the shaked box.

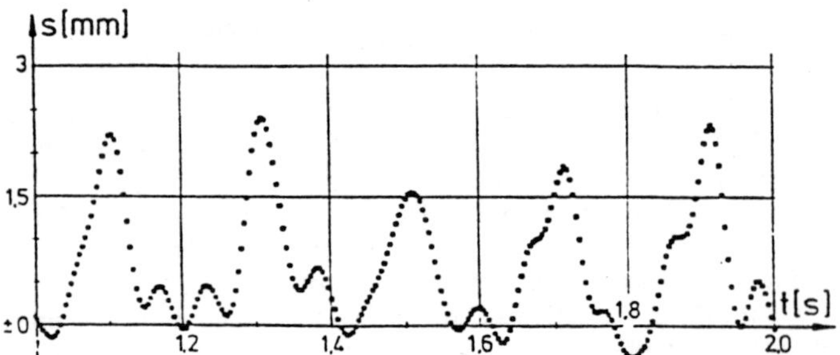

Fig. 6. Exemplary load function for overlayed movements at the center of the box.

Figure 7a and 7b show the development of the shape of heaps taken from two different tests, drawing the recorded outlook of the heap given by video of the camera placed inside the larger container.

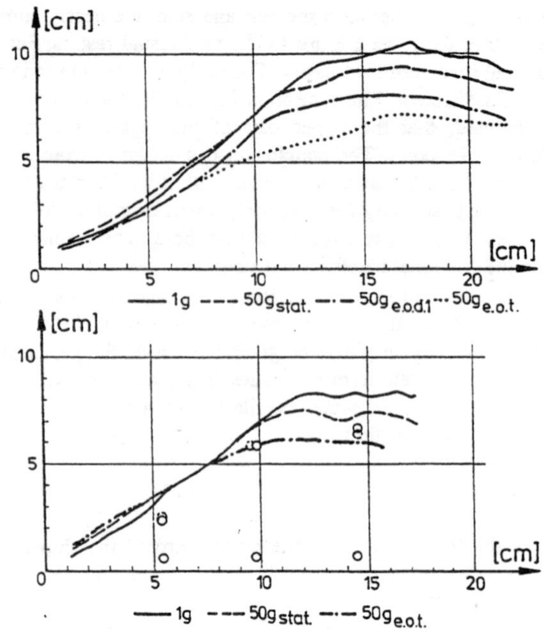

Fig. 7. Developement of the shape of heaps of two tests.

The given results have to be interpreted as follows. The line named "1g" is the viewable shape of the heaps at the beginning of the tests in the normal field of gravity. $50g_{stat}$ reprsesents the shape after increasing the gravity level to n = 50 without any additional dynamic loading, which is former stated as the initial settlement and deformation representing the deformation under selfweigth conditions. The influence of the dynamic loading is shown with the lines $50g_{e.o.d.1}$ (end of dynamic phase one) and $50g_{e.o.t.}$ (end of test). In the given case, the line $50g_{e.o.d.}$ is the shape of the heap after a load phase of 1000 cycles of sinusoidal vertical shaking. After the tests (e.o.t.) the heap is forced by several load phases with different load functions. A continuous deformation of the heaps had been regarded. Beneath a settlement of the heaps , in Figure 7b a horizontal movement of parts of the slope can be observed, following the path of circular markers, which have been used in some tests. It has to be mentioned, that the shapes presented in Figure 7 can only show the principle behaviour of the heaps. A further interpretation suffers under the influence of the boundary conditions as for example some parts of the three dimensional heaps are placed directly at the wall and also under the location of the camera. Additional measurements, as given in Figure 8 in example for one test, allow to interprete the measured settlement of the center of the heaps. The settlement is drawn after the initial settlement, which is not included in this figure.

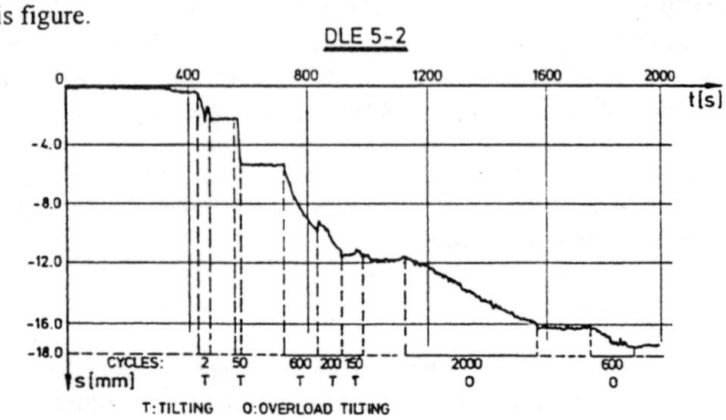

Fig. 8. Settlement of the center of the heaps for the test DLE 5.2.

The load history is included in principle in Figure 8 At the beginning of the test two cycles of tilting with an angle of 2° are applied slowly one after another. The next load phase consist of 50 cycles followed by 600, 200 and another 150 cycles. The last 150 cycles have a lower amplitude of movements. After this 1002 cycles of pure tilting with different amplitudes, a settlement of $s_M = 12$ mm in model scale representing s_P = 600 mm in prototype scale can be stated. Changing the load function to overlayed movements this settlement increases after another 2600 cycles up to $s_M = 18$ mm in model scale equivalent to $s_P = 900$ mm. Interpreting all tests fullfilled with the shaking table it is mentioned, that additional settlements and a deformation of the heaps occurs in dependency of the value and the kind of the load function with increasing numbers of cycles in nearly all tests. The test presented in Figure 8 shows a settlement of nearly 25% of the initial heigth after increasing the level of gravitiy. The worst influence of increasing deformation is the changing of the influencing load. This is obvious regarding Figure 9. The settlements of different tests are drawn in dependency of the logarithmic scaled number of load cycles. Four of the five tests presented in this figure are conducted with heaps. The test 6-1 is a comparative test with a trimmed, equivalent to a leveled or even, surface placed in the shaking box in comparable density.

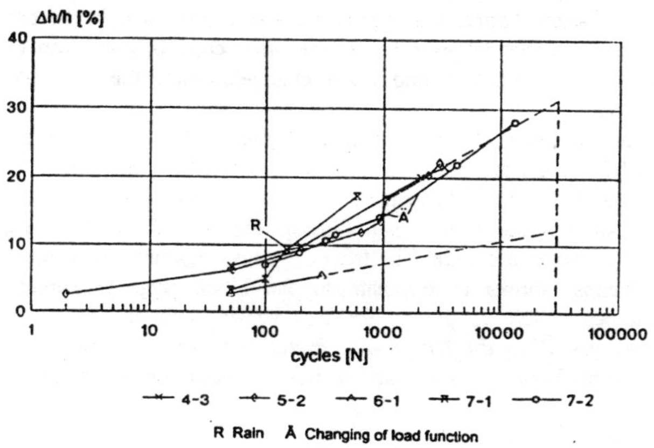

Fig. 9. Comparison of settlements of different tests.

Independent on the original shape, the settlements increases with increasing number of load cycles. The increasing is faster for heaps, than for trimmed conditions. A faster increasing of settlements can furtheron be recognized for a changing of load conditions (Ä) or an increasing of water content by rain or spread water (R). The dotted line shows the extrapolation of the measured values to the reestimated number of cycles N= 30000 for the regarded case of the capized vessel. Due to trimmed conditions, the settlements can be reduced from over 30% to 10% or less of the initial heigth depending on the density.

5.2 Swinging bracket tests

The bracket tests are conducted using only one of the hydraulic jacks. Small angles of rolling are applied with frequencies of f_m = 5Hz. After some cycles, a large angle of tilt is applied directly or stepwise. Figure 10 shows the resulting shape observed with the video camera. The test results presented here (DLE 9-3) are taken from a test with sand using a glycerin water mixture as pore fluid.

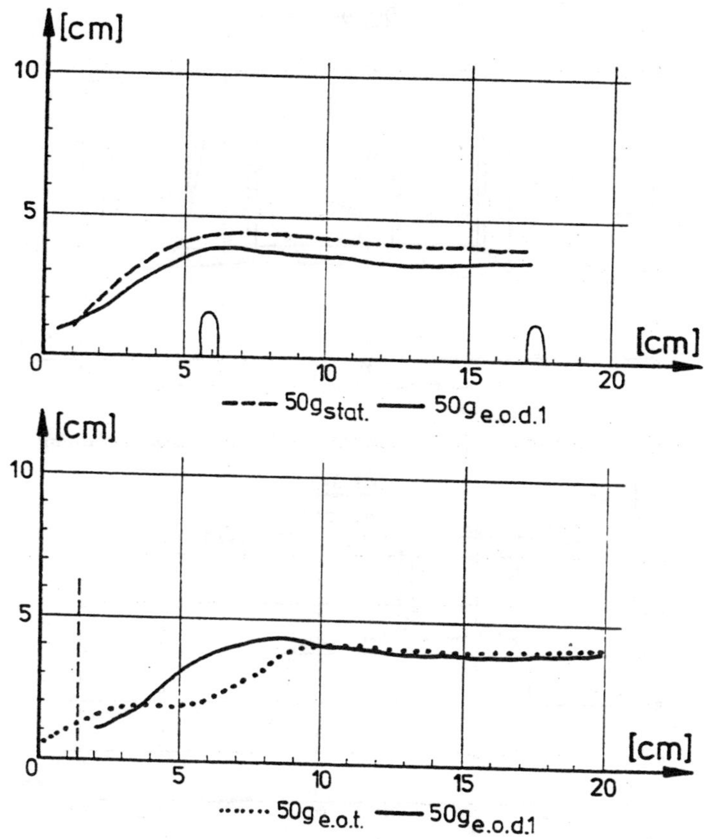

Fig. 10. Shape of the heap from the test DLE 9-3.

The upper figure compares the shape beyond the initial settlement with the shape after the first loading phase consisting of small angles of rolling. The lower figure relates this with the shape after tilting fastly to a maximum angle of around 11°. Mention, that the dotted line in the lower figure represents the zero axes of the upper figure of Figure 10. An additional settlement cannot be recognized in the center areas. A movement of the slope and a sliding of the material up to the border of the cargo space is obvious. Additionally a dewatering was observed in the upper part of the cargo space and in the top layers of the heaps. A lot of free water has been found in the lower parts of the cargo space in mostly every tests, including tests without saturating and less water contents.

To focuse on a better observation of the deformation and movements of the heaps, sticks have been placed inside the heaps up to the bottom plate in the middle axes. In this tests, the heaps are installed with a defined two dimensional shape over the whole width. The contour of the heaps is measured after the tests at defined locations. Figure 11 shows exemplary results from this measurements.

Before tilting, the heaps are saturated using a standing pipe. The free water is afterwards outletted to an initial heigth of 5mm representing a free water level in prototyp of 250 mm. After the tilting a complete changing of the heaps is observed. The sticks moved in orders of 3 cm representing movements in prototyp scale of 1.50m. The sticks tilt, so that a larger movements of the upper layers can be stated. Shear cracks can be observed at the higher side of the cargo space. At the upper parts of the heaps liquifaction can be observed. Additional informations towards this partly liquifaction can be gained by the measurements of the pore pressure transducers (Figure 12).

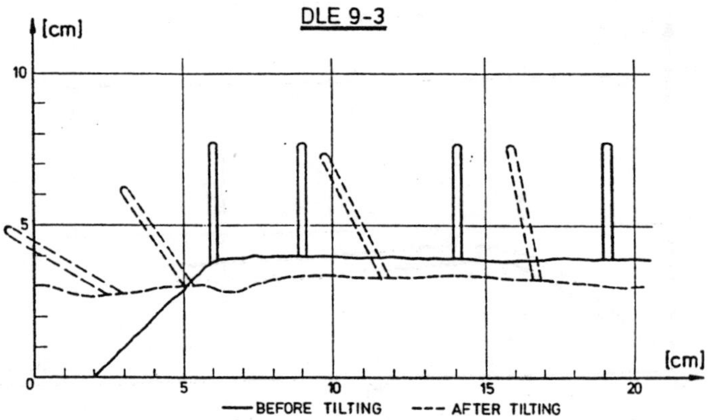

Fig. 11. Comparison of the measured contour before and after tilting.

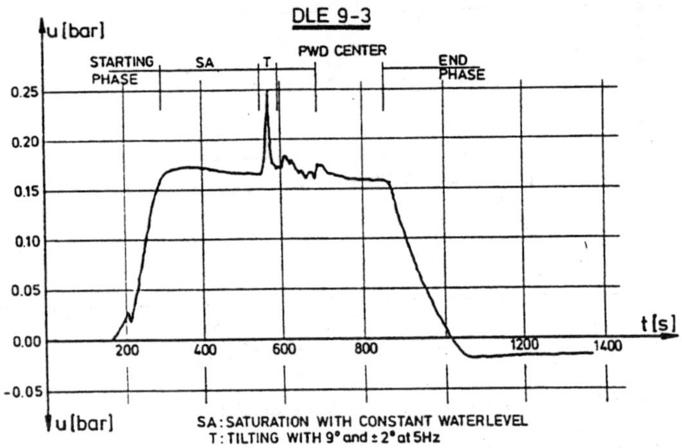

Fig. 12. Measurements of pore pressure in the bottom at the center of the cargo area.

An increase of pore pressure due to the tilting phase of about 70% can be observed. The measured value represents 60% of the overburden pressure, so that a total liquifaction as theoretically described for example in the German rules for the design of nuclear power stations (KTA) is not possible. In the upper layers and also due to changes in the water distribution and flow of water, partly liquifaction is to be notified. Additionally, the raise of water pressure reduces the friction capacity between the bottom of the cargo space and the heaps. At least by this test results, a movement of the mass point of large values is found.

5.3 Further influences and modelling of models

In order to simulate the pore pressure correctly, one of the test series concern the influence of the coefficient of permeability. Following the scaling laws for laminar flow, the velocity of water has to be reduced. This can be done decreasing the shape of the gradation curve (compare Figure 2), which directly reduces the value of permeability. Another possibility is the increasing of viscosity of the pore fluid as for example using a glycerin water mixture with a viscosity of n times the viscosity of water. The results of this tests are presented in Table 3 for the movements of the tip of the stick and the cut point of the sticks through the heap.

Tab. 3. Boundary conditions and movements of sticks.

Test	Stick 1	Stick 2	Stick 3	Stick 4	Kind
DLE 9-3	70/35	60/38	55/28	36/26	S - G
DLE 9-4	35/12	25/14	30/14	35/17	F - G
DLE 9-6	55/23	15/12	15/7	14/8	O - G
DLE 9-7	25/7	14/6	1/1	10/10	O - W
DLE 9-8	64/18	35	17/5	4/-3	F - W

S: sand
W: pore fluid water
F: fine grained ore
G: glycerin water mixture
O: original ore

The results show, that the two possibilities of correct reduction of velocity to the highest deformations. This is due to the measured and observed partly liquifaction. The test with fine iron ore and glycerin show larger movements than the test with the original iron ore concentrate and water. This is due to the doubled reduction of velocity while a larger statical water pressure acts horizontally in the centre of the heaps to the deeper side. In the other case, original iron ore and water, a fast overflow and water channels inside the cargo shall be mentioned.

The influence of cold water is discussed briefly. Three different possibilities are regarded. A complete frozen heap, the installing of the heaps with frozen grains and the use of cold water. The complete frozen block show no large movement during tilting. The friction at the surface of the cargo area is high. The frozen grained material which is placed in the heaps, shows in opposite to the block large movements and settlements. This is due to the less densitiy, which will be reached during installing the heaps with frozen material. Cold water is worth than normal water. It increases the possibilities of sliding of iron ore.

6. Conclusions

After this short description of the large research work undertaken to the transport stability of iron ore conmcentrate, different conclusions have to be drawn. They effect the test methods proposed by the IMO for the safe transport of granular cargo, the simulation of water flow phenomena in the centrifuge, observations to partly liquifaction during transport and recommendations for a safe transport. The capsizing of the vessel, as mentioned in the beginning, can be described incause of a worth by summation of factors of risk.

The tests required by the IMO for liquifiable cargos are misleading. The angle of the slope is only valid for small sizes of heaps. For material with high water capillarity, this definitioning dangerous, because the evaluated angle of slope will not be the angle for wet materials in larger scale. The value of the maximum allowable water content has to be discussed additionally for open storage conditions and unleveled transport. A separation of different zones, saturation and dry are found in any tests and also during several voyages of comparable freight, so that even in cases with water contents below the maximum allowable water content, saturated zones are existing. This leads to the observed partly liquifaction during the transport even in cases of larger movements. This phenomena increase with a reduction of weigth of the material as sand or grain. It shall be recommended additionally, that the unit of the water content should be given by volume percent in opposite to the percentage of weight. The value is hardly dependent on the self weight of the cargo and allow no direct feeling of the amount of water or fluid in the cargo space.

As recommendation for a safer transport, a trimming of the cargo is necessary. Also a better control of the water inside the cargo seems to be useful to minimize the risk of saturation and by this of liquifaction.

Deformations, flow phenomena and liquifactions for many materials can easily be studied using geotechnical centrifuges.In case of bulk cargoes further studies will help to identify the phenomena acting on board and to focuse of safe transport conditions.

7. References

DIN 18126 (1989): Soil, testing procedures and testing equipment, determination of denity of non-cohesive soils for maximum and minimum compaction, Boit Verlag Berlin

Goodings (1984): Relationsship for modeling water effects in geotechnical models. *Proc. of the appl. of Centrifuge modeling in Geotechnical design*, Craig ed. Manchester

IMO (1985): Codes for bulk cargoes, German edition; Bekanntmachung der Richtlinien für die sichere Behandlung von Schüttladungen bei der Beförderung mit Seeschiffen vom 7. November

Laue, J., Siemer, T., Grundhoff, T., Jessberger, H.L. (1996): Experimental methods in soil mechanics, *Dynamics of Civil Engineering*, pp 597 - 622, Krätzig et.al. (eds.), Balkema, Rotterdam

Schofield (1980): Cambridge Geotechnical Centrifuge Operations, 20th Rankine Lecture, *Geotechnique*, Vol 30, No.3

Tan & Scott (1985): Centrifuge scaling considerations for fluid particle systems, *Geotechnique,* Vol. 35, No. 4

Tan & Scott (1987): Discussion Geotechnique, Vol 37 No.1

Proceedings of the Seventh (1997) International Offshore and Polar Engineering Conference
Honolulu, USA, May 25-30, 1997

ISBN 1-880653-28-1 (Set); ISBN 1-880653-29-X (Vol. I)

A Feasibility Study of the Utilization of De-sulphurization Slag as Engineering Filling Materials

Der-Her Lee, Juu-En Chang and Yuin-Yao Jhi
National Cheng Kung University
Tainan, Taiwan, China

ABSTRACT

The mechanical and environmental characteristics of the mixture of de-sulphurization(DS) and blast furnace(BF) slags are investigated in this study. The experimental results show that the DS slag is a well-graded sandy soil. While achieving the maximum dry density, the required mixing ratio of the DS and BF slags is closed to 1. Similar results also existed in the case of shear strength, normal strength and CBR as compared with normal filling materials. In addition, the cured samples exhibit advantages of small void ratio, less compressibility, and higher CBR. Although pH value of slag is up to 11.3, it is reduced to 10 ~ 11, if mixed with BF slag by 1:1 ratio. The total quantity of heavy metals contained in the DS, BF, and mixing product are quite low. It is concluded that the mixture of DS and BF slag can be utilized as an alternative engineering filling material with negligible environmental impact.

KEY WORD: de-sulphurization, filling material, mixture, environment

INTRODUCTION

De-sulphurigation(DS) and blast furnace(BF) slags are the sub-products of a consistent process steelworks. The DS slag is solid material which is produced by adding DS agent into liquid-iron in blast furnace. And the BF slag is the compound of some impurities those are created in the steelmaking process, such as flux-supporting, iron ore, silicate, aluminum and other traces in coke. Lot of research reports indicated that BF slag is a widely useful resource. It can be used as cement additive, subgrade material, concrete aggregate, soil stabilizer, raw material of glass and kiln industry, flux-supporting of iron-smelting, and engineering filling material, etc.(Chiu et al. , 1993; Ou, 1992).

Furthermore the outward of DS slag is powder with particles. The major chemical contents of it are CaO, Fe, SiO_2 and some traces of Al_2O_3, MgO, S and P_2O_5, etc.. According to the investigation of CSC(Chiu et al. , 1993), the DS slag is being utilized as a resource of acid-soil improvement, cement material, and recycling material for ironwork or steelmaking furnace (Ou, 1992).

Nevertheless, the amount of DS slag used by the material ways is not so much that the recycling of DS slag is still a big problem of our country. Recently, reclaiming land from the area to establish nearshore industrial parks is a boom and a demand in Taiwan. The most popular filling material used in this country is offshore sand. However, the cost and quantity of reclamation work is not only influenced by the engineering properties of filling materials but also determined by the transportation distance from the filling sources to the field. Offshore sand can be pumped up to fill the lower land near the seashore or to form a offshore island by the hydraulic fill method. The hydraulic fill method is not only expensive but also always making a remarkable change in the offshore environment. The method can serve well only at the wide area near sea, i.e., it is not suitable for the filling work of small low land at inland. On the other hand, the DS slag is a compound of stable minerals of CaS, and it has many good engineering properties. It seems as a new and good material for filling works, especially for the inland and small area of filling engineering.

This paper tries to investigate the applicability of DS slag and BF slag as the new materials for filling and reclamation works. The focus of this paper is to determine the optimum mixing ratio of DS and BF slags. Following this ratio, the mixture of these slags can develop some good engineering characteristics which are needed by reclamation work.

PROCEDURES OF TESTING

There are two sets of laboratory tests had been carried out. The first one is a series of physical and mechanical tests. The purpose of this series is to fined out the optimum mixing ratio of these two slags, and then, to investigate the response of the optimum mixture under the tests of consolidation, swelling, direct shear and CBR, etc.. The detail of the first series of tests is as follows:

1. Physical properties tests: grain size analysis, specific gravity tests.

2. Compaction tests: in order to find out the maximum dry density and optimum moisture content for several kinds of mixture with different ratio of DS and BF slag. And then, determining the optimum mixing ratio.

The second set is environmental impact tests, such as the concentration of heavy metals in the leachate of DS , BF slag and their mixture at the optimum mixing ratio. From results of the tests, we can understand the properties of leached contamination from the mixture an0d evaluate the impact on the environment. Therefore, the contents of the second set of tests are:

1. Basically chemical properties, as shown in Table1.
2. The leaching tests of DS slag.
3. The leaching tests of the mixture of DS and BF slags with optimum mixing ratio.

RESULTS AND DISCUSSIONS

Mechanical Characteristics

The original grain size distribution curves of DS and BF slags are shown in the Fig.1 by solid lines with circle and triangle marks, respectively. The uniformity coefficient(Cu) of DS slags is 39, and BF slag is 65. On the other hand, the coefficient of curvature(Cd) obtained from these two curves are 0.059 and 3.46 for DS and BF slags. Therefore, according to the results of Cu and Cd as well as the grain size distribution curve, the DS slag can be classified into the group of well-graded sandy and silty sand(SW-SM) by using the Unified Classification System(ASTM). And, the BF slag belongs to the group of well-graded gravel(GW-GM). However, if the AASHTO Classification System is used, both DS and BF will be ranked as the same group of well-graded subgrade soil(A-1-a).

The largest particle of two slags is too large to satisfy the requirement of the standard of compaction test. Hence, in the preparation for testing specimens, we modified the grain distribution of these two slags and shifted the original curve to right hand side to form two broken lines presented in Fig.1. The uniformity coefficient and coefficient of curvature of the two modified distribution curves of DS and BF slags, are Cu=25.95, 17.46 and Cd=1.02, 1.77, respectively. Both DS and BF are classified into the same group not according to the ASTM but also AASHTO system. They belong to the group of SW-SM and A-1-b. The specific density of DS slag is 3.15, and BF slag is 2.82. Both of them are over the average value of soil, because about 20 % of iron particles are involved by them.

Compaction Tests

In the series of compaction tests, five mixtures of DS and BF slags with different mixing ratio(by weight) were selected to investigate the maximum dry density(γ_d) and optimum moisture content(O.M.C.). The ratios of DS slag to BF slag are 10:0, 7:3, 5:5, 3:7, and 10:0. Figure 2 shows the results of γ_d and O.M.C. of the five different mixtures, where, the alphabet B and C represent the DS and BF slag, respectively. Fig.2 also reveals that the compaction curve will move to left band side of the figure if the content of BF slag increased. However, the top value of the maximum γ_d of these five mixture is obtained from B5C5. It is a mixture of 50 % DS slag and 50 % of BF slag. From Fig.2, we know that the maximum γ_d of B5C5 is 2.14 gf/cm^3 and O.M.C. = 12.3 %, while, the greatest value of O.M.C. in the five mixture is 17.8 % which is gained from the pure DS slag .

Table1 Items and methods of the chemical properties

Items	Pre-digesting leaching solution	Measured method
Organically carbon%	Mebius methods	titrated by Fe_2^+(FAS)
EC	samples:water=1:1	EC meter
pH	samples:water=1:1	glass pH electrode
Basic	samples:water=1:1	titrated by acid
Cd,Cr,Cu Ni,Pb,Zn	extracted by 0.1N Hcl	ICP-MS analysis
Hg	dissolved by HNO_3+H_2SO_4	ICP-MS analysis
As	dissolved by 9.6N HCl	ICP analysis
C.E.C.	0.4N NaOAC+0.1N NaCl and 0.5N $Mg(NO_3)_2$	ICP analysis

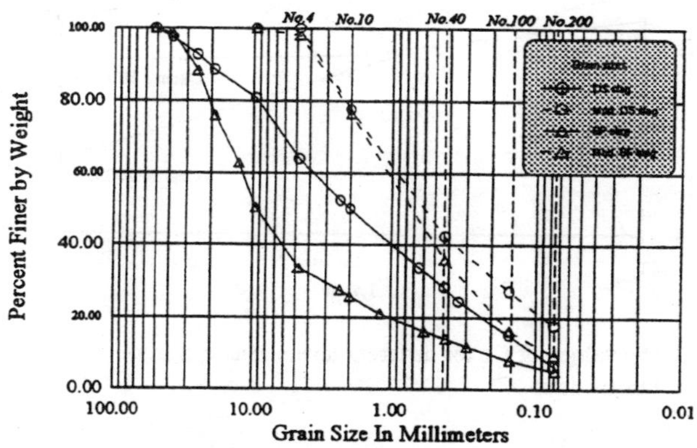

Fig.1 Grain sizes distribution curve of DS and BF slag

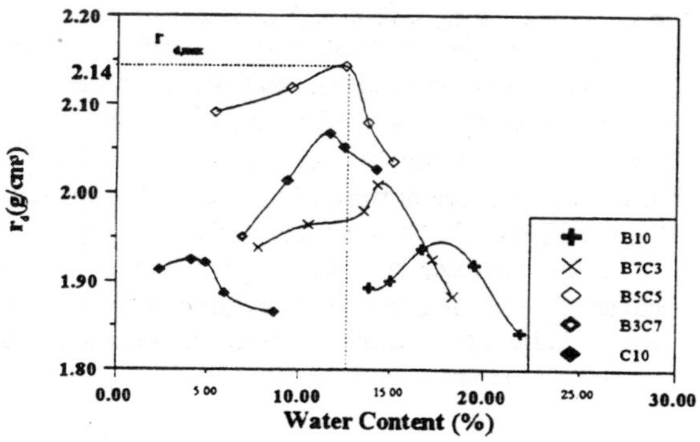

Fig.2 Compaction curves of DS mixed by BF in different ratio

Consolidation Tests

From the mixture B5C5, we can get the maximum γ_d ,therefore, it was selected to perform the consolidation, direct shear and CBR tests. Firstly, several compaction specimens of the mixture B5C5 with water content 12 ± 0.5 % were produced. Then, three consolidated specimens(named as L1, L2 and L3) were taken carefully from different layers of each compaction specimen. Some of these consolidation specimens were cured in water for 10 or 20 days and remarked by C10 or C20. Fig. 3 presents the compaction index(Cc) of the specimens under different curing time. This figure shows that the initial value of Cc was about 0.085 ~ 0.11, but it decreased 11 ~ 28 % and 31 ~ 35 % when the specimens were subjected to a water curing for 10 and 20 days, respectively. The phenomena may be brought about by the hydration created in the pozzolanic materials, such as CaO and SiO_2, of the DS and BF slags.

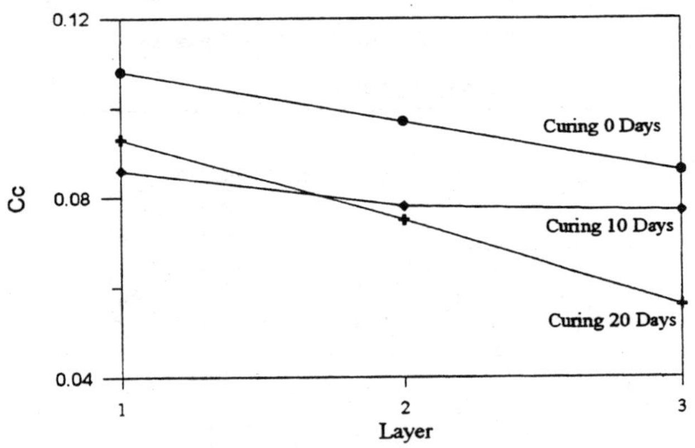

Fig.3 The values of Cc for different layers after different curing days

Direct Shear Tests

In direct shear tests, samples were taken from two layers(S1 and S2) of compaction cells. The curing sample was remarked by C10 same means as one of consolidation test. Table 2 shows results of direct shear tests. The value of peak and residual shear strength parameters of uncurled samples is about C = 0.7 ~ 1.9 kgf/cm^2 ,and ϕ = 35° ~ 36° ,and Cr = 0.59 ~ 0.86 kgf/cm^2, and ϕr = 32.4 ~ 35.6° , respectively. It is lager then one of typical filling soils. The value of curing samples, S1C10 and S2C10, is not clearly changed in quantity of C and Cr, but it was increased by 5° and 1.5° ~ 3° in parameter ϕ and ϕr, respectively. The curing effect caused by chemical reaction of DS and BF may be helpful for connection among soil particles, that make friction angle ϕ grows up.

Table2 Parameters of direct shear strength of mixture

Number	S1C0	S2C0	S1C10	S2C10
$C(kg/cm^2)$	0.7	1.86	0.7	1.86
ϕ	35.9	35.94	39.8	39.8
$C_r(kg/cm^2)$	0.59	0.86	0.76	0.49
- ϕ_r	32.42	35.6	34	38.6

California Bearing Ratio Tests (CBR)

In CBR tests, there were three sets of mixture, nine samples, to be cured individually in 0,10, and 20 days. We also measured the swelling potation of CBR samples under water. The value of CBR is percentage of standard punching pressure which penetrates the standard well-graded crushed stones through 2.0 ~ 5.0 centimeter depth.

Figure 4 shows the results of CBR for different curing conditions. The value of CBR is about 62.5% for the uncurled samples. The CBR grows up about 74% and 102% after 10 and 20 curing days, respectively. It is means that the more curing days the greater value of CBR can be obtained. It is good for subgrade materials. As comparison with normal filling material, the mixture used in this study is an excellent filling material insight of engineering. The swelling strain of CBR samples under water is about 0.8 ~ 1.8%. The swelling potation of the mixture is close to the value of middle and low plasticity clay(CL). It is greater than value of normal filling soils, so we must keep in mind for designing the filling procession using mixture of DS and BF slags.

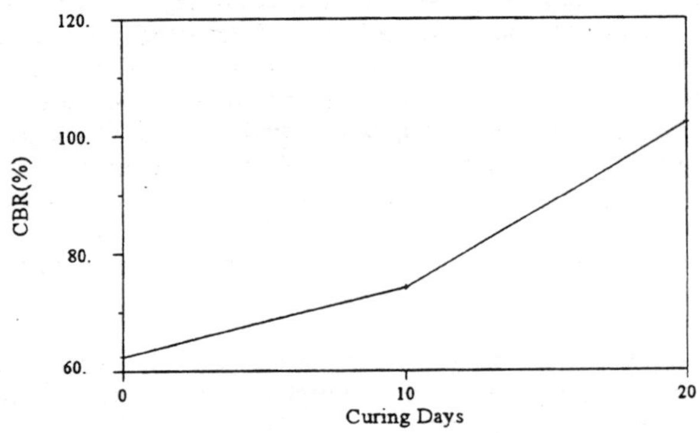

Fig. 4 Modified CBR value after different curing days

Environmental Characteristics of DS

Before the feasibility study of the utilization of DS and BF slags, it is necessary to develop the basic chemical characteristics of DS ,BF, and mixture. This study developed the value of pH, EC, AlK, organic substance , CEC, and heavy metals(Cu, Zn, Ni, Cd, Cr, Pb, As, Hg).

The results compared with the value of regional soil constituent(background values) shown in Table 3. Although pH value of DS slags is up to 11.3 trifle greater than the one of background values, it is reduced to 10 ~ 11, if it is mixed with BF slags by 1:1 ratio. The mixture can improve the acidity soils. Because of the small value of EC of DS and BF slags, it is no effect to the salt mine content of background soils.

The value of basically of DS slags is about 1120 mgHCO_3^-/kg greater then the value of background about 138 mgHCO_3^-/kg. It is illustrated that the filling material of mixture produces an acid-basic cushioning effect better than normal filling soils. Although the value of organic substance content of DS slags is 9.31%, the value is greater than background value of 3.1%, but the value of BF slags is reduced to 0.94%, so the value of the mixture of DS and BF slags is close to background value. Because the value of CEC of mixture is greater than background value, it is appear to develop stronger positive ion exchange potation. It means that the mixture is suitable to become an excellent filling material.

The contents of heavy metals in DS, BF, and mixture shown in Table 3 were developed by extracting from 0.1N HCl solution. Although the contents of Cd and Cr is greater than the background value, but it is less than the allowable value of standard heavy metal contents of Taiwan. Others of heavy metals were less than the background values. It is illustrated that the total quantity of heavy metals contained in the DS, BF, and mixing product are quit low ,and no effect of the one contained in regional soils.

Table3 Comparison of basic chemical characteristics of DS, BF, and mixture

Item sample	DS	BF	50%D+50%B	30%D+70%B	value 1*	value 2*	value 3*
pH	11.3	7.91	11.09	10.17	7.83	–	–
EC(μ mho/cm)	3620	349	2520	534	8710	–	–
Alk(mgHCO_3/kg)	1122	113.5	629.5	190.3	138.5	–	–
Organic (%)	9.31	0.94	5.66	2.35	3.10	–	–
CEC(meq/100g)	15.19	28.56	16.09	20.86	14.29	–	–
Cu (mg/kg)	0.40	3.86	1.33	0.9	0.92	3.88	5.8
Zn (mg/kg)	0.08	2.26	0.2	0.26	4.98	6.89	10.57
Ni (mg/kg)	0.97	0.75	2.14	2.38	1.31	1.94	1.88
Cd (mg/kg)	0.28	0.19	0.38	0.43	0.04	0.18	0.18
Cr (mg/kg)	0.49	1.92	1.33	1.47	0.52	0.9	0.97
Pb (mg/kg)	0.47	3.67	0.75	0.83	0.0.83	6.15	8.93
As (mg/kg)	7.11	7.68	7.23	7.38	8.80	7.39	7.55
Hg (mg/kg)	ND	ND	ND	ND	–	0.09	0.14

(*after Chang et al.,1995)

CONCLUSIONS

1. Based on the ASTM classification system, DS and BF slags are classified as SW-SM and GW-GM respectively. In the AASHTO classification system the DS and BF slags classificed as A-1-a. Both classification show that DS and BF slags are excellent subgrade material. When DS slags mixed with BF slags by 1:1 ratio, the maximum γ_d is equal to 2.14 g/cm^3 and O.M.C. is equal to 12.3 %.
2. The initial value of Cc of mixture is about 0.085 ~ 0.11, the more time as samples cured the more imcompression is reduced. The Cc value is declined by 11 ~ 28 % and 31 ~ 35 % of initial value after 10 and 20 curing days, respectively.
3. The parameters of peak and residual shear strength are of C = 0.7 ~ 1.9 kgf/cm^2, ϕ = 35 ~ 36 °,and Cr = 0.59 ~ 0.86 kgf/cm^2, ϕr = 32.4 ~ 35.6 ° , respectively. After 10 curing days cured, the values of C and Cr are not clearly changed, but value of ϕ and ϕr is increased by 5° and 1.5° ~ 3° individually.
4. The initial value of CBR of mixture is about 62.5 %. After 10 (or 20) curing days, the value of CBR is increased to value of 74 % (or 102 %).
5. The pH value of DS slags is 11.3 greater then the value of 7.8 of background soils. The pH value of BF slags is close to the one of background. The mixture of mixing ratio of 1:1 of DS and BF slags, pH value is reduced to 10 ~ 11, exhibits advantage of positive ion exchange potation, that is an excellent improvement of acid-soils.
6. The heavy metal contents of mixture of DS and BF slags are quit small, and they satisfy the allowable value of standard code of Taiwan.
7. The mixture of DS and BF slags with a ratio of 1:1 shows that it is a good engineering filling material in sight of mechanical and environmental engineering.

REFERENCES

Chiu, YF and Mou, JE, (1993) , "A Study on the Utilization of the Desulphurization Slags at China Steel Corp., " *3rd Conference on Industrial Waste Minimization*, pp.1 ~ 12, Taipei, Taiwan (in chinese) .

Ou, MC , (1992), "Introduction of Slag Utilization in CSC , " *Proceeding of the International Conference on Industrial Waste Minimization '92*, pp.427-433.Taipei, Taiwan.

Chang, JE , Chang, YK, Cheng, CH and Tsai, SH (1995) , "Monitoring and Management of the Water Quality Surrounding Landfill Sites," *Proceedings 8th Environmental Planning and Management Conference*, pp149-156, Tainan ,Taiwan (in chinese) .

Proceedings of the Seventh (1997) International Offshore and Polar Engineering Conference
Honolulu, USA, May 25-30, 1997

ISBN 1-880653-28-1 (Set); ISBN 1-880653-29-X (Vol. I)

Effect of Internal Structure on Mechanical Behavior of Osaka Pleistocene Marine Clay

A. Yashima, F. Oka and H. Shigematsu
Gifu University
Gifu, Japan

ABSTRACT

This paper examines an importance of yielding characteristic and an effect of the internal structure on mechanical behavior of Osaka Pleistocene marine clay. The first part considers yielding properties of the Upper-Most Pleistocene marine clay in Osaka Bay in a general stress space. An envelope of the yielding stress from an isotropic consolidation test, consolidated-undrained and -drained tests is obtained. The second part focuses on an effect of internal structure of clay on mechanical properties of the Pleistocene marine clay. A series of triaxial compression tests on the undisturbed and reconstituted samples are carried out.

KEY WORDS: Pleistocene marine clay, internal structure, cementation/bonding, yielding, triaxial compression test

INTRODUCTION

The Pleistocene marine clay is known to be a well-structured natural clay. It is therefore necessary to investigate an influence of particle structure (fabric) and bonding among particles on the mechanical behavior based on laboratory tests on the undisturbed clay samples. For this purpose, unconfined compression tests, oedometer tests, isotropic consolidation tests, consolidated-undrained compression tests and consolidated-drained compression tests have been carried out on the Osaka Upper-Most Pleistocene marine clay. Deposition of the Osaka Upper-Most Pleistocene marine clay (Ma12) is ended near the end of the last glacial period. It is from medium to stiff clay and the stratum is continuous over the entire Osaka basin with various thickness. Recent experiences in reclamation works in Osaka Bay Area have shown that the consolidation behavior of this Upper Pleistocene marine clay is very important in order to understand large settlements of the reclaimed islands. On the other hand, further research works are required to understand the mechanical behavior during the shearing process.

The undisturbed Osaka Upper-Most Pleistocene marine clay samples were taken with the block sampling technique from bottoms of the vertical shafts excavated at Tsurumi and Kyuhoji in the central part of Osaka Prefecture, Japan (Fig.1). High quality of the block samples were confirmed based on comparisons between unconfined compression strengths on the block samples and those in an extensive literature (Research Committee on Geotechnical Information of Osaka Bay Seabed deposit, 1995).

In the present study, the existence of a yield locus is firstly discussed. Yielding characteristic is a key information for constructing whole structure of the elasto-(visco)plastic model. It

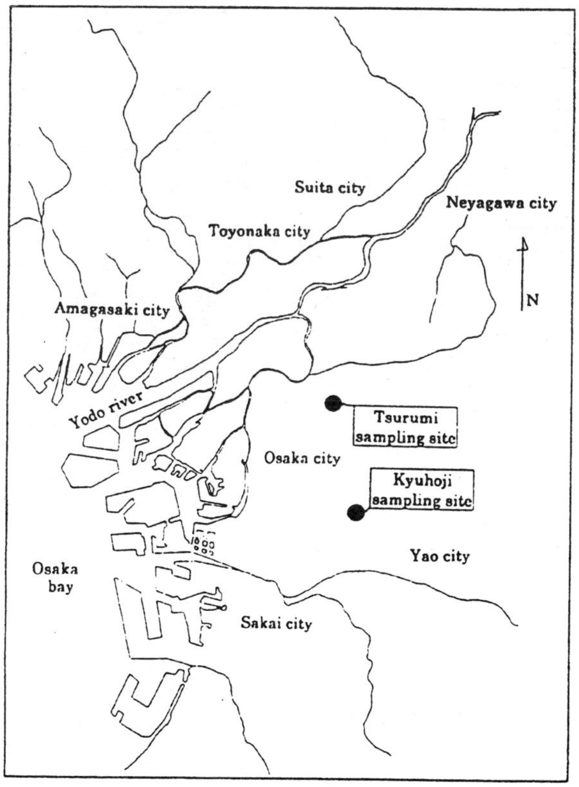

Fig.1 Sampling sites for the Upper-Most Pleistocene marine clay.

is, however, rather complicated to define the yield points in the stress space. In the present paper, the yield points are identified as the ones where the total strain develops extensively in a series of the laboratory tests.

Secondly, an effect of the internal structure on the deformation behavior is discussed. We conduct consolidated-undrained triaxial compression tests on the undisturbed samples, the reconstituted samples, and the samples to which larger stresses than the initial consolidation yield stress were applied before the shearing process. It is found that peak strength of the reconstituted clay is much smaller than that of the intact clay, whereas strength of the reconstituted clay at large strain is equal to residual strength of the intact one. It is also apparent that the strain-softening is less significant in the behavior of the reconstituted sample. A dilatancy characteristic on Osaka Upper-Most Pleistocene marine clay is then investigated. In the heavily overconsolidated region, the stress-strain relationship shows strain-hardening and strain-softening with a positive dilatancy. In the normally consolidated region beyond the initial consolidation yield stress, on the other hand, the strain-softening behavior is less significant with a negative dilatancy.

Many reclamation projects at water front areas as well as deep excavations through the Pleistocene marine clay layers are being planned in Japan. In such projects, large stresses beyond the consolidation yield stress are expected to the Pleistocene marine clay. It is therefore necessary to take a destructuration effect of the internal structure into account when a prediction of deformation behavior on the well-structured Pleistocene marine clay is conducted.

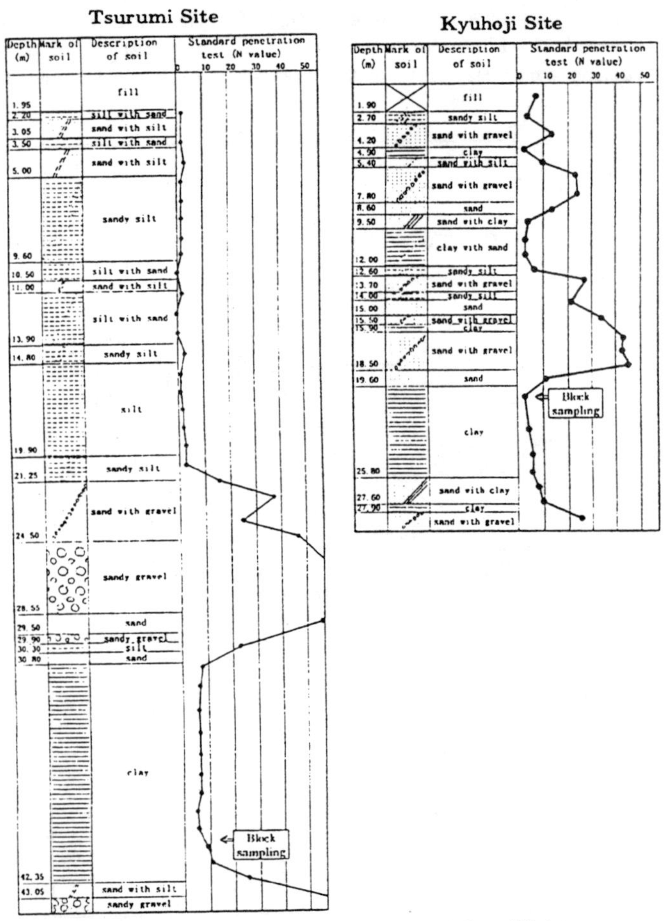

Fig.2 Geotechnical profiles of the Upper-Most Pleistocene marine clays from two sites.

BLOCK SAMPLING OF PLEISTOCENE MARINE CLAY

The Pleistocene clay has been usually sampled by the tube-sampling technique around Osaka Bay. In this study, however, to get high quality clay samples, the Upper-Most Pleistocene marine clays were block-sampled on the bottoms of the vertical shafts at Tsurumi and Kyuhoji in Osaka Prefecture, Japan. At Kyuhoji, a clay sample was taken at depth of 20 m below the ground surface. On the other hand, at Tsurumi, five blocks of clay were taken at depth of 40 m below the ground surface. Geotechnical profiles and N-values from the standard penetration tests at two sites are illustrated in Fig.2. Basic index and physical properties of the clays sampled at two sites are also summarized in Table 1.

To investigate quality of the samples obtained in this study, a series of the consolidated undrained compression tests are carried out on the overconsolidated and normally consolidated clays. Fig.3 summarized the rate of the strength increase defined by c_u/σ'_c in which c_u is the half of the compression strength and σ'_c is the consolidation pressure with respect to the normalized mean effective stress together with the data on tube-sampled specimen published by Tsuchida et al. (1984). The normalized mean effective stress less than 1 means that the clay specimen is overconsolidated. In the overconsolidated region, the effect of the sampling disturbance on the strength of clay is well known to be very significant (Oka et al., 1996). If the clay is disturbed during the sampling process, the strength of the specimen becomes lower and the strain at the failure becomes larger (La Rochelle et al., 1981; Hight et al., 1992; Oka et al., 1996). It is found from Fig.3 that the block-sampled specimen gives larger strength than those undertaken by tube-sampling technique. From this experimental evidence, the samples obtained by the block-sampling technique are considered to have higher quality than the many Osaka Pleistocene marine clay samples undertaken by the tube-sampling technique in an usual practice. In this study, extensive laboratory tests on Pleistocene marine clay samples obtained by the block-sampling technique are carried out.

Table 1 Index and physical properties of the Upper-Most Pleistocene marine clays from two sites.

	Kyuhoji clay	Tsurumi clay
sampling depth	G.L.-20 ~ -23(m)	G.L.-40 ~ -41(m)
effective overburden pressure	188(kPa)	284(kPa)
pre-consolidation pressure (isotropic consolidation test)	333 ~ 343(kPa)	580(kPa)
pre-consolidation pressure (oedometer test)	412(kPa)	725(kPa)
compression index (λ=C_c/2.303)	0.327	0.507
swelling index (κ=C_s/2.303)	0.028	0.026
unconfined compressive strength	160 ~ 240(kPa)	225 ~ 490(kPa)
density	2.693	2.722
natural water content	56.9(%)	61.6(%)
liquid limit	78.3(%)	94.5(%)
plasticity limit	28.2(%)	35.0(%)
liquidity index	0.574	0.45
plasticity index	50.1(%)	59.5(%)
activity	0.86	1.08
clay fraction	70(%)	67.9(%)
silt fraction	27(%)	30.7(%)
sand fraction	3(%)	1.4(%)

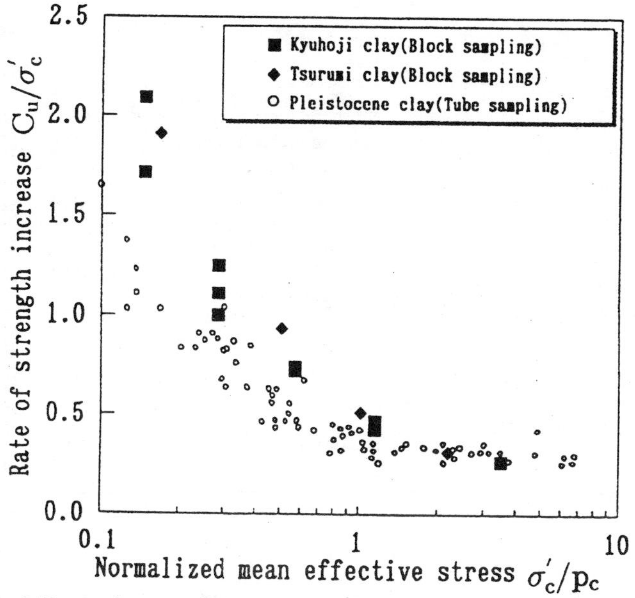

Fig.3 Rate of strength increase with respect to the normalized mean effective stress.

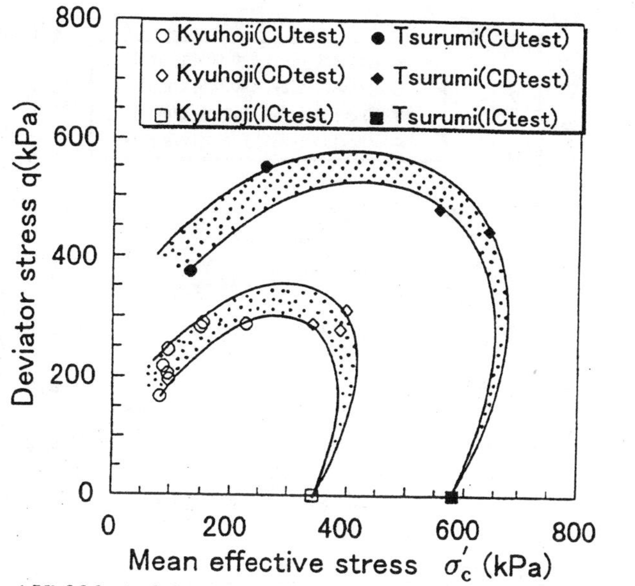

Fig.4 Yield loci of the Upper-Most Pleistocene marine clays from two sites.

YIELDING BEHAVIOR OF UPPER-MOST PLEISTOCENE CLAY

Concepts of yielding for isotropically-consolidated reconstituted clays have been investigated in many organizations. The Cam-clay type isotropic yield surface have been often obtained for isotropically-consolidated reconstituted clay. Natural clays are, however, generally anisotropic; most natural clays are structured and viscous. In a stress diagram, the yield locus delimits a zone in which soil behavior is mostly elastic from a zone in which it is mostly plastic or where there is failure. The yield points are identified here as the ones where total strain develops extensively in a series of the laboratory tests. In practice, the Casagrande method is used to define the yield points.

Isotropic consolidation tests, consolidated-undrained and -drained compression tests at different strain rates on the Upper-Most Pleistocene marine clay from Tsurumi and Kyuhoji were carried out to get the yield loci. Fig.4 illustrates the yield loci for the clays from the two sites. The yielding properties on the clays from the two sites are found to be anisotropic. Similar test programs have been already carried out on Japanese alluvial clays in our laboratories (Adachi et al., 1995; Oka et al., 1996). The alluvial clays tested were undertaken at Hachirogata, eastern Osaka, western Osaka and Ariake, Japan. Those alluvial clays are known as very sensitive clays. The normalized yield loci with respect to the isotropic preconsolidation pressure are summarized in Fig.5. The shapes of the yield loci for the Upper-Most Pleistocene marine clay are found to be very similar to those for the typical Japanese sensitive alluvial clays in the general stress space. This indicates that natural clay is initially anisotropic and that an anisotropic quasi-initial yield surface, not an isotropic one like that used in the original Cam-clay model, is necessary for a construction of an elasto-(visco)plastic constitutive model.

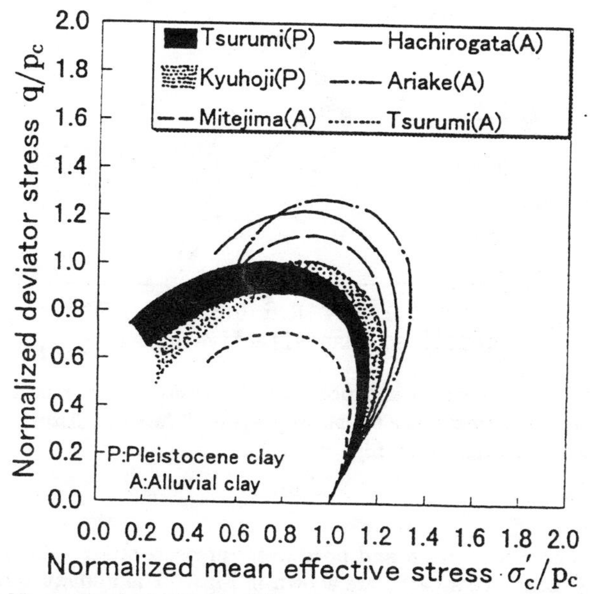

Fig.5 Normalized yield loci with respect to the isotropic preconsolidation pressure.

EFFECT OF INTERNAL STRUCTURE

When discussing the effect of internal structure of the Upper-Most Pleistocene marine clay on the mechanical properties, the extensive laboratory tests have to be carried out. We conduct the consolidated-undrained triaxial compression tests on the undisturbed samples, the reconstituted samples that were first remolded and then reconsolidated, and the samples to which larger stresses than the initial consolidation yield stress were applied before the shearing process. In this paper, tests results on the clay samples from Kyuhoji are summarized.

Firstly, an experimental result of the consolidated-undrained compression test on the normally consolidated undisturbed clay is compared with that of the normally consolidated clay with stress history in which the clay specimen was consolidated at 3 times larger isotropic stress than the preconsolidation stress. The de-

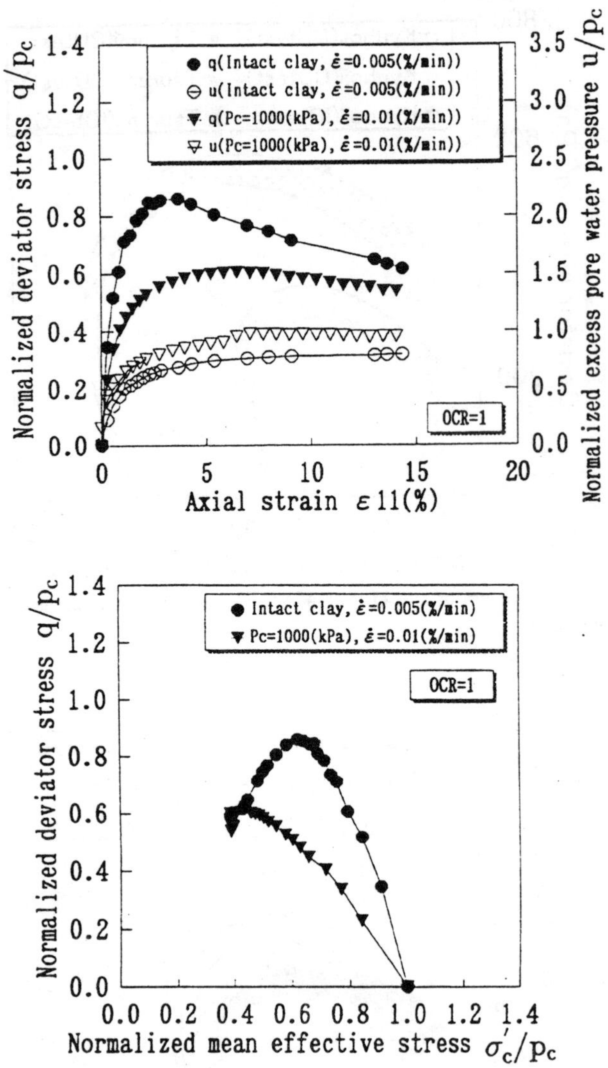

Fig.6 Stress-strain and pore water pressure-strain relations and effective stress paths for normally consolidated undisturbed sample and sample with high stress history.

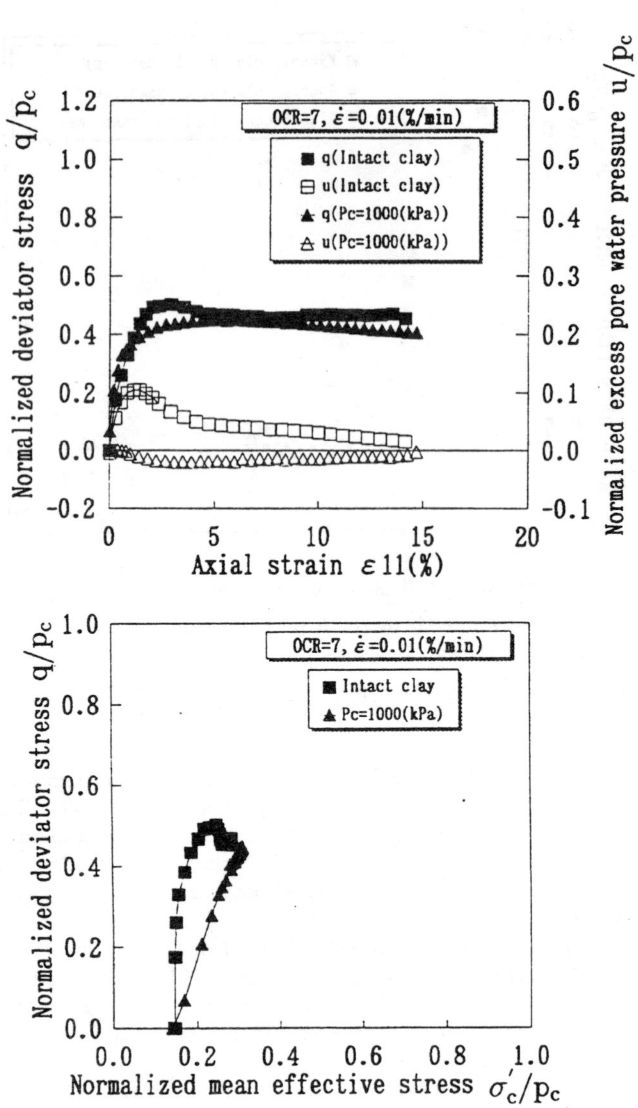

Fig.7 Stress-strain and pore water pressure-strain relations and effective stress paths for overconsolidated undisturbed sample and sample with high stress history.

viator stress-strain and pore water pressure-strain relations and effective stress paths are shown in Fig.6. The undisturbed specimen shows strain-hardening and -softening behavior. Once the clay sample is consolidated at larger stress beyond the preconsolidation pressure, a cementation/bonding between the clay particles is destroyed and the strain-softening becomes less significant even in the heavily overconsolidated region (OCR=7), when the clay is consolidated at larger stress beyond the preconsolidation pressure, the strain-softening behavior is not significant. On the other hand, the undisturbed clay shows strain-hardening and -softening behavior (Fig.7).

Secondly, to fully destroy the developed internal structure by aging of the Upper-Most Pleistocene marine clay, the reconstituted clay specimen was prepared in the laboratory. Figs.8 and 9 summarize the comparative results for the intact and reconstituted clay specimens. It is found that the peak strength of the reconstituted clay is much smaller than that of the intact clay, whereas the strength of the reconstituted clay at large strain is equal to the residual strength of the intact one. It is also apparent that the strain-softening is less significant in the behavior of the reconstituted sample. It is worth noting that the internal structure was destroyed by the remolding process. A dilatancy characteristic on Osaka Upper-Most Pleistocene marine clay is then investigated. In a heavily overconsolidated region, the stress-strain relationship shows strain-hardening and strain-softening with a positive dilatancy. In the normally consolidated region beyond the initial consolidation yield stress, on the other hand, the strain-hardening behavior is only observed with a negative dilatancy.

The consolidated-undrained compression tests performed on the intact, disturbed (with stress history beyond the preconsolidation pressure) and reconstituted specimens demonstrate that the Upper-Most Pleistocene marine clay has a significant internal structure which induces more brittle behavior, higher compressive strength and higher stiffness than the disturbed and reconstituted samples. This internal structure is easily destroyed by disturbance during sampling process, high stress history, inappropriate sample treatment, etc.

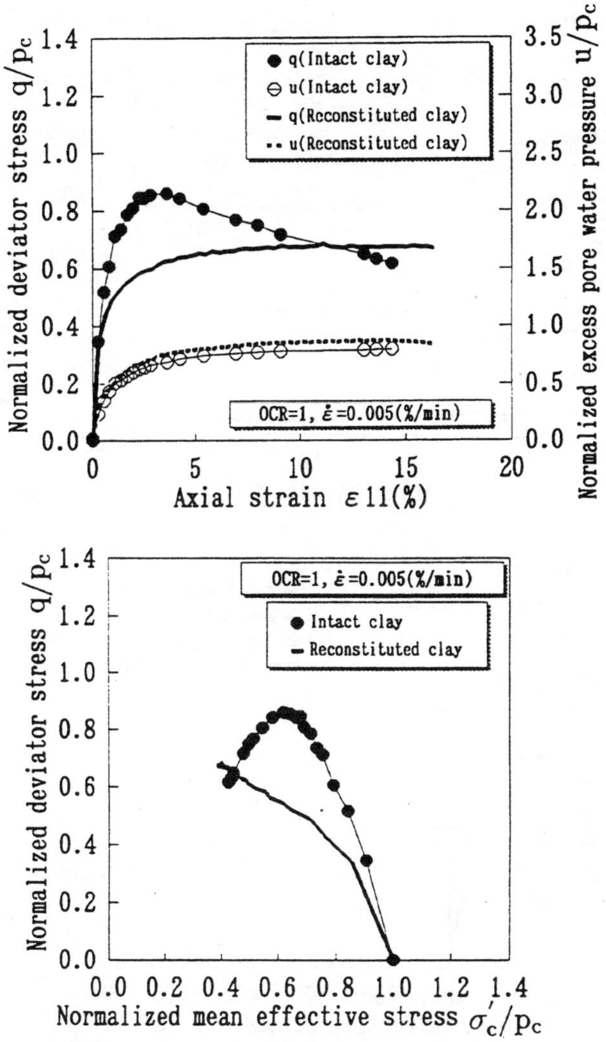

Fig.8 Stress-strain and pore water pressure-strain relations and effective stress paths for normally consolidated undisturbed and reconstituted sample.

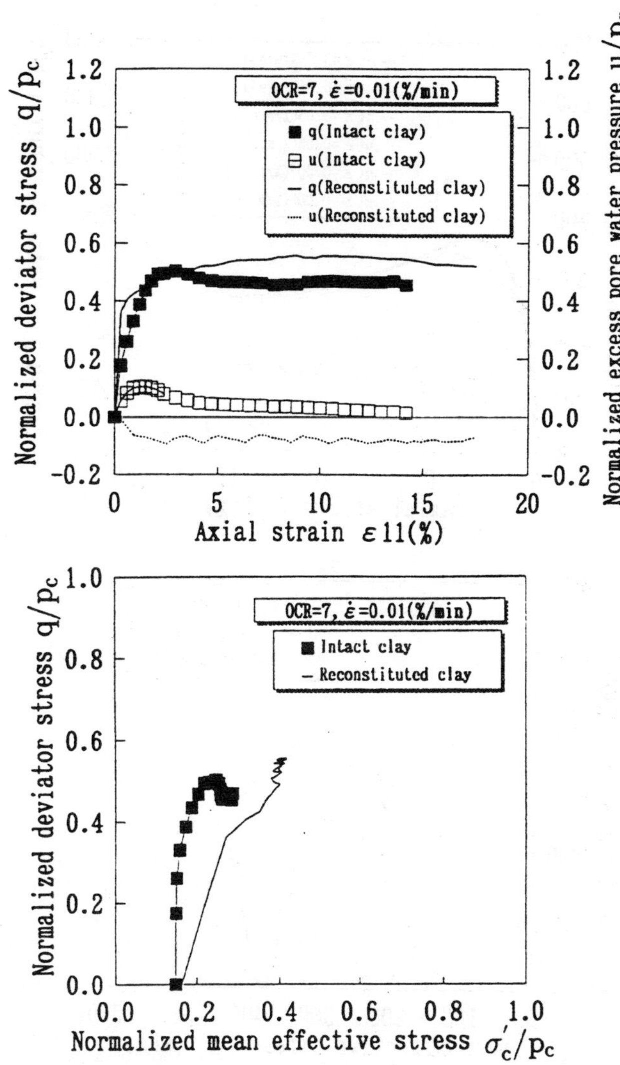

Fig.9 Stress-strain and pore water pressure-strain relations and effective stress paths for overconsolidated undisturbed and reconstituted sample.

STRAIN RATE EFFECT

Variation of the peak strength with strain rate is understood by carrying out the consolidated-undrained compression tests at different strain rates. Figs.10 and 11 show the experimental results on the undisturbed specimens for the normally and overconsolidated clays. For the normally consolidated clay, the tests performed at different strain rates showed the different stress-strain relations and effective stress paths. This strain rate effect has been often recognized on many reconstituted and undisturbed clay specimens in literature (see, for example Richardson and Whitman, 1963). For the overconsolidated clay with OCR of 3.5, on the other hand, this strain rate effect is not significant.

CONCLUSION

In the present paper, mechanical properties of the Upper-Most Pleistocene marine clay has been discussed based on extensive experimental results on the high quality clay samples. The following

- The Upper-Most Pleistocene marine clay has a significant internal structure which induces more brittle behavior, higher compressive strength and higher stiffness than the disturbed and reconstituted samples. This internal structure can be easily destroyed by disturbance during the sampling process, high stress history, inappropriate sample treatment, etc.
- For the normally consolidated Upper-Most Pleistocene marine clay, an effect of strain rate on the peak strength is significant. For the overconsolidated clay with OCR of 3.5, on the other hand, this strain rate effect is not significant.

conclusions are obtained from this study:

- Clay samples undertaken by the block-sampling technique on bottoms of the vertical shafts at Tsurumi and Kyuhoji in Osaka Prefecture, Japan are considered to have higher quality than many Pleistocene clay samples obtained by the tube-sampling technique.
- Yielding properties on the Upper-Most Pleistocene marine clays from the two sites are anisotropic. Shapes of the yield loci for the clays are very similar to those for typical sensitive alluvial clays in Japan.

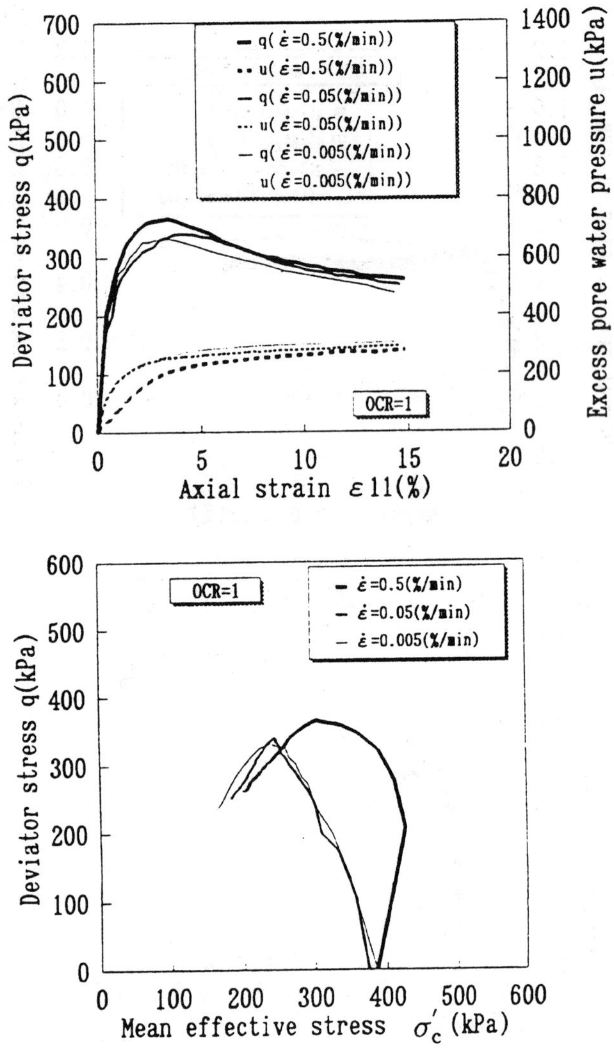

Fig.10 Stress-strain and pore water pressure-strain relations and effective stress paths for normally consolidated undisturbed samples at different strain rates.

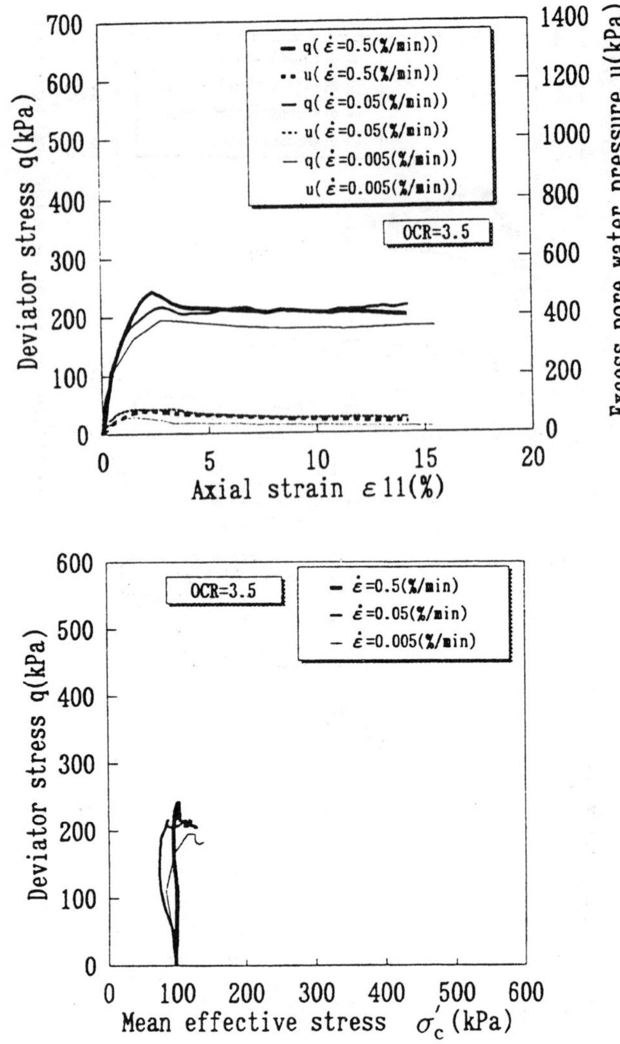

Fig.11 Stress-strain and pore water pressure-strain relations and effective stress paths for overconsolidated undisturbed samples at different strain rates.

REFERENCES

Adachi,T., Oka,F., Hirata,T., Hashimoto,T., Nagaya,J., Mimura,M. and Pradan,T.B.S. (1995), "Stress-strain behavior and yielding characteristics of eastern Osaka clay", *Soils and Foundations*, 35-3, pp.1-13.

Hight,D.W., Böese,R., Butcher,A.P., Clayton,C.R.I. and Smith,P.R. (1992), "Disturbance of the Bothkennar clay prior to laboratory testing", *Géotechnique*, 42-2, pp.199-217.

La Rochelle,P., Sarrailh,J., Tavenas,F., Roy,M. and Leroueil,S. (1981), "Causes of sampling disturbance and design of a new sampler for sensitive soils", *Canadian Geotechnical Journal*, 18-1, pp.52-66.

Oka,F., Yashima,A., Hashimoto,T. and Amemiya,M. (1996), "Application of Laval type large diameter sampler to soft clay in Japan", *Soils and Foundations*, 36-3, pp.99-111.

Research Committee on Geotechnical Information of Osaka Bay Seabed Deposit (1995), "Seabed deposits in Osaka Bay", *JSGE Kansai Branch* (in Japanese)

Richardson,A.M. and Whitman,R.V. (1963), "Effect of strain rate upon undrained shear strength of a saturated remoulded fat clay", *Géotechnique*, 13-4, pp.310-324.

Tsuchida,T., Kikuchi,Y., Nakajima,K. and Kobayashi,M. (1984), "Engineering characteristics of marine clay at Senshu-offshore in Osaka Bay, (No.3) static shear property", *Technical Note of the Port and Harbour Research Institute*, Ministry of Transport, Japan, 498, pp.87-114.

Proceedings of the Seventh (1997) International Offshore and Polar Engineering Conference
Honolulu, USA, May 25-30, 1997

ISBN 1-880653-28-1 (Set); ISBN 1-880653-29-X (Vol. I)

A Novel Procedure for Testing Model Drag Anchors

M.P. O'Neill and M.F. Randolph
University of Western Australia
Nedlands, Australia

S.R. Neubecker
Advanced Geomechanics
Australia

ABSTRACT

Physical modelling of drag anchor and chain systems plays a vital role in the development of new anchor designs. However, the large drag lengths required by these anchors to reach ultimate capacity, coupled with a natural scaling dissimilarity that exists between the anchor and the chain, make this modelling difficult to achieve in the laboratory or centrifuge. A new testing procedure has been developed that utilises a rigid loading arm and allows the model anchor to be tested without an anchor chain, thereby allowing a more detailed analysis of the geotechnical forces acting on the anchor. A series of tests have been conducted using this new procedure, and the results show that the anchor capacity and inclination can be used, in combination with the shear strength of the soil, to determine fundamental resistance parameters of the anchor. The results also validate the assumption that the anchor flukes approach an approximately horizontal orientation as the anchor reaches its ultimate capacity.

KEYWORDS: Anchors, drag anchor, model testing, scaling, ultimate holding capacity.

INTRODUCTION

The growing trend towards floating structures as the preferred method of offshore hydrocarbon extraction has focused greater attention on the anchoring systems used to moor such facilities. There are a variety of anchorage options available, including pile anchors and gravity anchors, but high capacity drag embedment anchors have proven to be the most economical and versatile mooring system.

Current methods for drag anchor design are based largely on the results of field tests that have been used to develop empirical charts to predict anchor holding capacity as a ratio of the dry anchor weight. These design charts place little importance on the soil type or properties, classifying it broadly as "sand" or "clay". However, work by Taylor (1980) and Dunnavant and Kwan (1993) indicated that anchor behaviour is strongly dependant on the detailed seabed soil properties. Hence, physical modelling of drag anchor and chain systems is essential in order to assess the performance of new anchor designs in varying soil conditions.

Drag embedment anchors require large drag lengths in order to mobilise their full holding capacity. This makes modelling under controlled conditions in the laboratory or centrifuge difficult to perform, since it requires large volumes of soil or extremely small-scale model anchors. Also, Neubecker and Randolph (1996) demonstrated that in an anchor and chain mooring system, a scaling dissimilarity exists between the anchor and the chain. It was found that larger anchors require disproportionately larger chains, making interpretation of small scale anchor/chain tests complicated.

A novel testing procedure for small scale drag anchors has been developed at the University of Western Australia that enables a more detailed analysis of the anchor-soil response to be conducted, and allows more accurate extrapolation to larger scales. This procedure involves dragging the model anchor with the anchor padeye at a fixed embedment depth and without a chain, thereby removing the need to consider the effect of the chain on the model and prototype performance.

A series of tests have been conducted in soft clay using this new procedure. Theory developed by Neubecker and Randolph (1996) was applied to the results, and it was found that the ultimate holding capacity of drag anchors can be described by two fundamental anchor parameters.

THEORY

It is generally accepted that a drag embedment anchor travels parallel to its flukes. This assumption was validated by Dunnavant and Kwan (1993), who conducted a number of centrifuge drag anchor tests and found that as the anchor reached its ultimate holding capacity, the anchor flukes became approximately horizontal. Apart from at very shallow embedments, experience shows that soil flows around the anchor, with no void forming behind the anchor.

Neubecker and Randolph (1996) simplified existing methods for calculating the holding capacity of anchors in cohesive soils, using simple bearing capacity and moment equilibrium calculations. They expressed the geotechnical resisting force, T_p, acting on the anchor parallel to the direction of the flukes as:

$$T_p = f\, A_p N_c s_u \qquad (1)$$

where A_p is the frontal projected anchor area in a plane perpendicular to the flukes, N_c is a bearing capacity factor taken as 9, s_u is the local undrained shear strength, and f is an anchor form factor.

There will also be geotechnical forces acting normal to the fluke, so that the resultant resistance force, T_w, acting on the anchor will make an angle, θ_w, with the fluke (Figure 1). It then follows that:

$$T_w = \frac{T_p}{\cos\theta_w} \qquad (2)$$

Neubecker and Randolph (1996) proposed that the form factor, f, and the angle of the resultant to the fluke, θ_w, may be taken as geometric properties of the anchor. They also noted that the value of θ_w may vary due to small differences in the strength of the soil over the vertical range occupied by the anchor, but that these variations were assumed to be minor and hence θ_w could be taken as constant.

It should be noted that the resultant geotechnical resistance, T_w, is independent of the weight of the anchor, but this weight must be taken into account when calculating the actual padeye force, T_a, as shown in Figure 1. As the anchor embeds, the upper fluke surface and resultant force, T_a, will be at angles of β and θ_a respectively to the horizontal. Allowing for the offset angle, ψ, between T_a and T_w (which will vary during embedment and with the relative magnitudes of T_w and the weight, W), the relationships between the angles may be expressed as:

$$\theta_a = \theta_w + \psi - \beta \qquad (3)$$

In the case of a model drag anchor test, conducted in the laboratory at 1 gravity, the anchor weight becomes insignificant in soil of moderate strength, and ψ may be taken as zero.

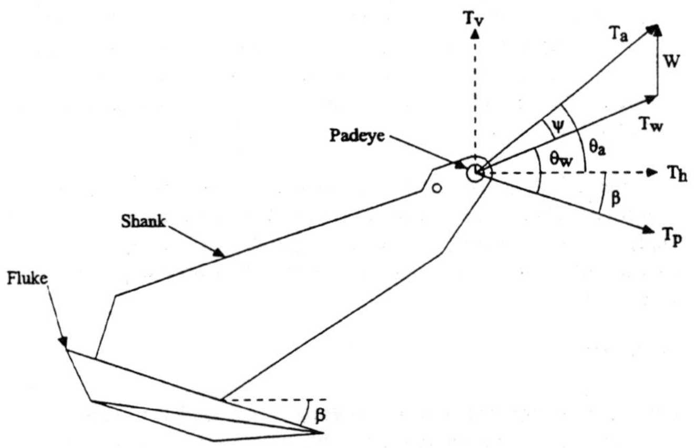

Figure 1: Force equilibrium of anchor

Every anchor type can be considered to have unique properties f and θ_w, which will be nearly independent of the anchor size and are capable of describing the behaviour of the anchor in any given cohesive soil strength profile. These properties can be determined by experimental modelling, comparison against published field data or by analysis using existing anchor simulation programs.

It is possible to implement the above equations into an incremental simulation of drag anchor behaviour that can be run using a simple spreadsheet. However, the kinematic behaviour of the anchor during embedment must first be identified. Neubecker (1995) outlined a method describing the motion of a drag anchor using polar coordinates (r_i, θ_i). As illustrated in Figure 2, and taking the anchor padeye as the origin, the displacements ΔX and ΔY of a point i on the anchor can be expressed by:

$$\Delta X = \Delta x + r_i \Delta\theta \sin\theta_i \qquad (4)$$

$$\Delta Y = \Delta y + r_i \Delta\theta \cos\theta_i \qquad (5)$$

where Δx and Δy are horizontal and vertical incremental displacements of the anchor padeye respectively and $\Delta\theta$ is the incremental rotation of the anchor. For the case where the embedment depth of the padeye is fixed, Δy is equal to zero. Taking $\Delta Y/\Delta X$ as equal to the tangent of the displacement vector angle to the horizontal, α, it can be shown that:

$$\Delta\theta = \frac{\Delta x}{\dfrac{r_i \cos\theta_i}{\tan\alpha} - r_i \sin\theta_i} \qquad (6)$$

It should be noted that the rotation of the anchor is primarily governed by kinematic constraints.

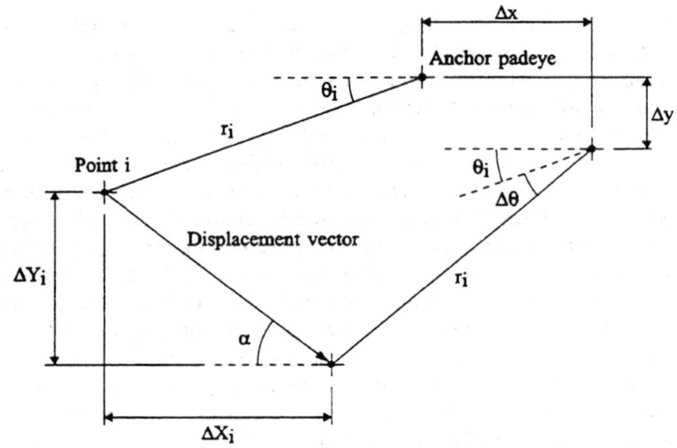

Figure 2: Absolute displacement of point i on anchor

Once an appropriate point (point i in Figure 2) on the anchor is identified and values of the coordinates r_i and θ_i are defined, the incremental simulation proceeds as follows:

1. Assume an anchor fluke orientation β and displace the anchor padeye horizontally an increment Δx.
2. Calculate the anchor resistance parallel to the direction of travel T_p using Equation (1).
3. Calculate the weightless anchor capacity T_w using Equation (2).
4. Determine $\Delta\theta$ using Equation (6).
5. Find $\theta_{i/new}$, where $\theta_{i/new} = \theta_i + \Delta\theta$
6. Find β_{new}, where $\beta_{new} = \beta - \Delta\theta$

7. Displace the anchor padeye a further horizontal increment Δx and repeat steps 2 through 6.

RESULTS

The model anchor testing apparatus, newly developed at the University of Western Australia, consists of a rigid loading arm connected via a pin joint to the padeye of the model drag anchor (Figure 3). The model anchor itself is a 1:80 scale model of a 32 tonne anchor with principal dimensions similar to a Vryhof Stevpris anchor (Vryhof, 1990) as shown in Figure 4.

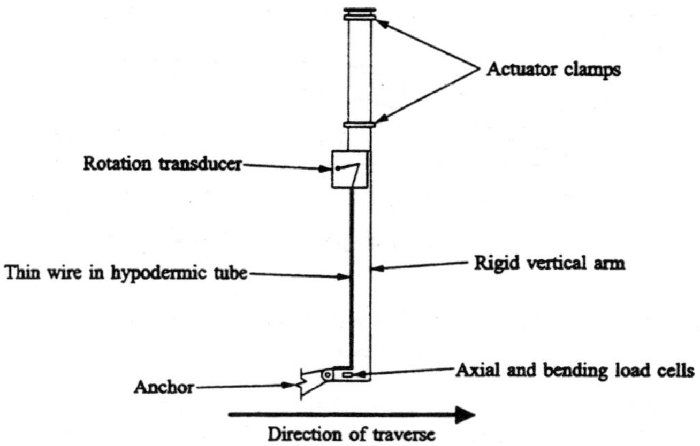

Figure 3: Anchor loading arm

Once fitted into the actuator, the loading arm and anchor are embedded vertically into the test sample to a predetermined padeye embedment depth with the upper edge of the anchor shank horizontal. The loading arm is then translated horizontally and the anchor allowed to rotate about its padeye in response to the geotechnical forces acting on it. Instrumentation fitted at the base of the loading arm next to the pin joint measures horizontal and vertical anchor loads, while a rotation transducer connected to the anchor via a thin wire measures the anchor orientation in the soil. This enables a more detailed analysis of the geotechnical forces acting on the anchor than is otherwise attainable from more conventional anchor tests performed with a chain.

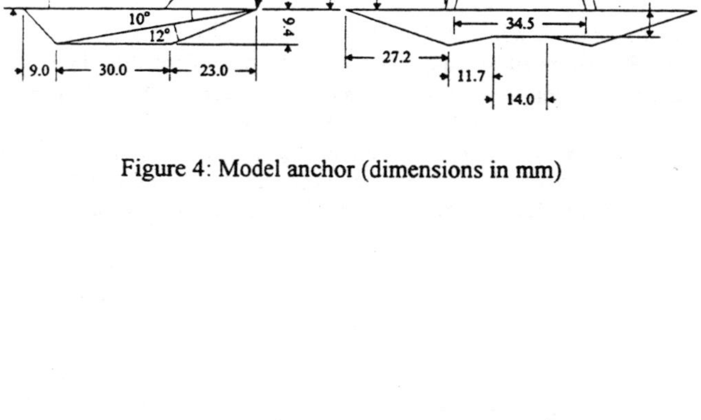

Figure 4: Model anchor (dimensions in mm)

Experimental Results

Three series of anchor drags, comprising eleven experimental drags in total, were conducted using the vertical loading arm and model anchor in soft kaolin clay. Tests 1 and 2 were conducted in samples with uniform undrained shear strengths of 19 kPa and 31 kPa respectively. Test 3 was conducted in a sample with an undrained shear strength profile that ranged between 5 kPa and 19 kPa over a depth of approximately 200 mm. Figure 5 presents the s_u profiles for each of the samples, determined using a T-bar apparatus (Stewart and Randolph, 1991). All tests were conducted in a strongbox of dimensions 650 mm long by 390 mm wide by 325 mm high.

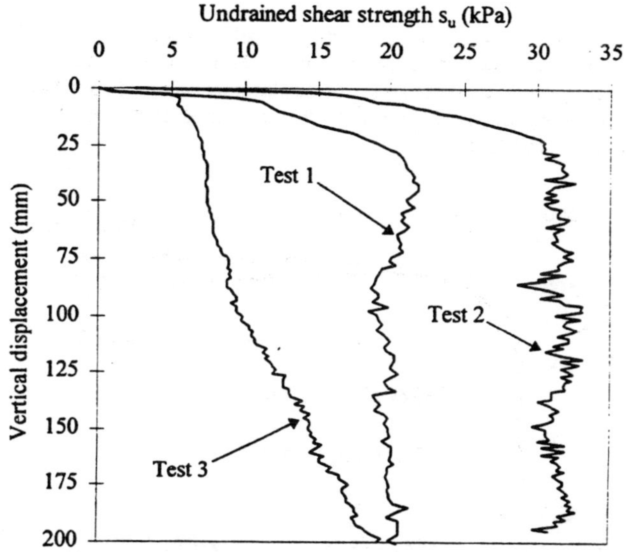

Figure 5: Undrained shear strength s_u profiles

Due to physical limitations associated with the actuator, the maximum attainable drag distance was 160 mm or 2.6 flukelengths, although most drags were conducted over somewhat shorter distances. Five of the drags were "one-stage" drags, in that the anchor was displaced horizontally 1.5 to 2.5 fluke lengths and then extracted from the test sample. The remaining six drags made up three "two-stage" drags, one in each test box. In these tests, once the anchor had been embedded to the required depth and dragged the maximum possible distance, the actuator was carefully removed, leaving the loading arm and model anchor relatively undisturbed within the test sample. The actuator was reversed to the starting position, reattached to the loading arm and displaced horizontally again to the limit of the actuator. This method enabled the behaviour of the model anchor to be analysed over a total drag distance of four to five fluke lengths.

Table 1 outlines the drag type and assumed undrained shear strength s_u for all experimental drags. Note that the undrained shear strength s_u assumed for each drag was calculated at the centroid of the projected area of the anchor flukes in a plane perpendicular to the flukes.

Figure 6(a) shows the plots of horizontal and vertical anchor loads against horizontal displacement for Test 1. Drags 1.1 and 1.2a, which both commenced with the flukes at a relatively high inclination (high β), showed an initial sharp rise in horizontal loads before reducing very slightly as the drags proceeded. Drag 1.2b, which was the second stage of a 2-stage drag, commenced at a much lower fluke inclination, and

Test Number	Drag Number	Drag Type	Average s_u (kPa)
Test 1	1.1	1-stage	19
	1.2 (a & b)	2-stage	19
Test 2	2.1 (a & b)	2-stage	31
	2.2	1-stage	31
	2.3	1-stage	31
Test 3	3.1 (a & b)	2-stage	13
	3.2	1-stage	14
	3.3	1-stage	15

Table 1: Anchor test summary

apart from the very start exhibited a nearly constant horizontal load. The vertical loads in Drags 1.1 and 1.2a increased steadily before levelling out at the later stages of displacement. Drag 1.2b exhibited a constant vertical load, similar to that measured at the conclusion of Drags 1.1 and 1.2a.

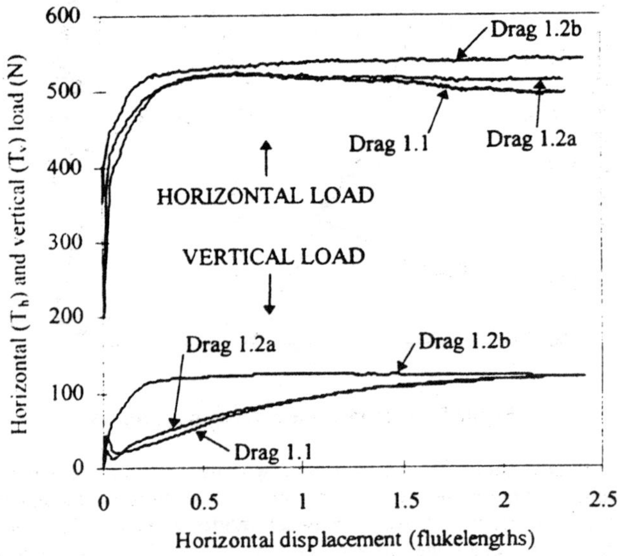

Figure 6(a): Horizontal (T_h) and vertical (T_v) load vs horizontal displacement - Test 1

The horizontal and vertical anchor loads for Test 3 (Figure 6(b)) displayed similar trends to those encountered in Test 1, although there was a slightly greater variation in the magnitude of the loads measured between drags. This was expected since Test 3 was performed in a soil with a non-uniform strength profile across the depth occupied by the model anchor.

Figure 7(a) shows the measured fluke angle β plotted against horizontal displacement for Test 1. The curves are somewhat "stepwise" in nature, due to slight sticking of the wire connecting the rotation transducer to the anchor. However, they give a good indication of the orientation of the anchor as it traversed through the soil. Drags 1.1 and 1.2a commenced at inclinations of 40 to 45 degrees, and exhibited a steady but decreasing rate of rotation. By the end of the drags, the fluke had rotated to a β value of approximately 15 degrees, and the curves appeared to be levelling out.

Drag 1.2b, which is a continuation of Drag 1.2a, showed only a few

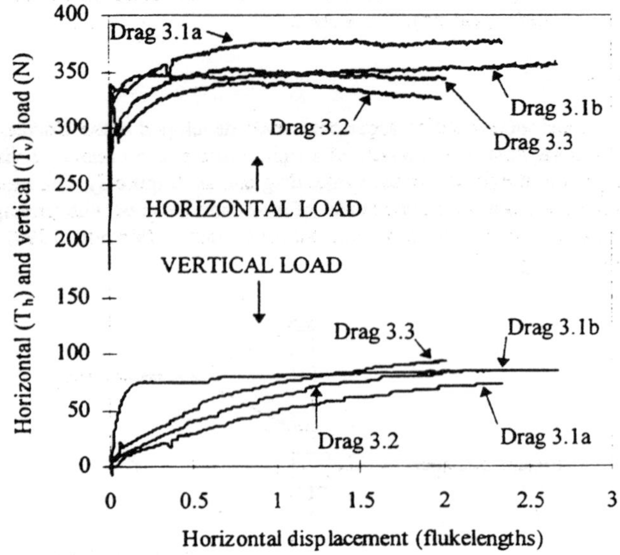

Figure 6(b): Horizontal (T_h) and vertical (T_v) load vs horizontal displacement - Test 3

degrees change in the orientation of the anchor over the entire 2.4 fluke lengths of displacement, suggesting that the anchor had reached a "steady state" at a fluke angle of approximately 12 degrees. The fact that the horizontal and vertical loads measured in Drag 1.2b (Figure 6(a)) also remained relatively constant further suggests that the anchor had attained its ultimate orientation and holding capacity.

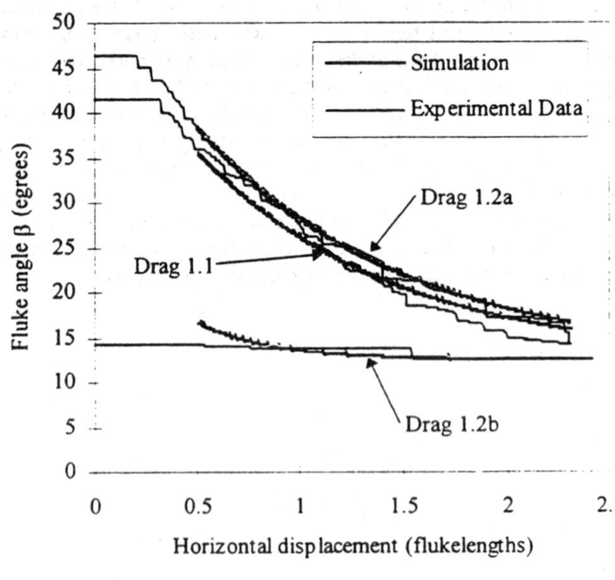

Figure 7(a): Fluke angle β vs horizontal displacement - Test 1

Test 3 again demonstrated a similar trend (Figure 7(b)). Drags 3.1a, 3.2 and 3.3 exhibited a steady but decreasing rate of rotation. Following a large reduction in β at approximately 0.6 flukelengths, Drag 3.1b (a continuation of Drag 3.1a) also ceased to rotate at β = 12 degrees. It should be noted that the final orientation of the anchor for all drags in Test 3 was verified via direct measurement of the embedment depth of the anchor padeye and back end of the anchor fluke at the conclusion of the drags. In the case of Drag 3.2, the final anchor orientation was further confirmed by direct observation of the anchor itself embedded in the sample.

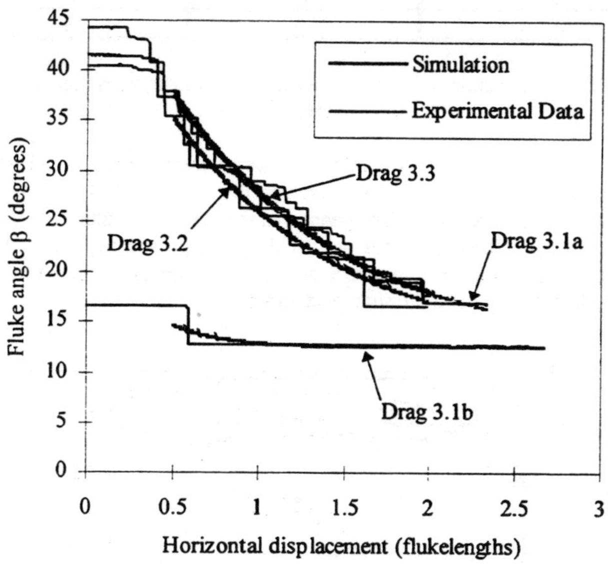

Figure 7(b): Fluke angle β vs horizontal displacement - Test 3

Hence, the experimental data suggest that the model anchor travels parallel to a plane lying approximately 12 degrees from the top face of the fluke. This conclusion seems reasonable, since the total angle of the leading fluke edge is 22 degrees (Figure 4), implying that the travel plane lies approximately midway between the top and bottom edges of the fluke.

Simulation Results

The incremental simulation procedure outlined previously was then implemented and compared with the experimental data using a least squares regression scheme to obtain a "best fit". Since the rotation of the anchor was independent of the strength or strength profile of the soil, simulation of the rotation was completed first by choosing an appropriate initial value of β. Simulation of the anchor loads to the experimental data was then performed by adjusting the values of f and θ_w.

An analysis of the rotation data suggested that a point on the very end of the fluke underneath the anchor padeye should be taken as the reference point for determination of r_i and θ_i, as illustrated in Figure 4. In the case of the simulated load curves, the projected anchor area A_p was taken in the plane normal to the top face of the anchor fluke, and included the area between the shanks. Also, common values of f and θ_w were adopted for all drags in each test.

Figures 7(a) and 7(b) include the simulated rotation curves which show excellent agreement with the measured rotation. Note that in each experimental drag the first half of a flukelength includes a region where little or no rotation occurs. This is primarily due to the drag commencing in soil disturbed during embedment of the anchor, and therefore the least squares regression scheme and simulated curves ignore this disturbed region by commencing at a horizontal displacement of 0.5 flukelengths.

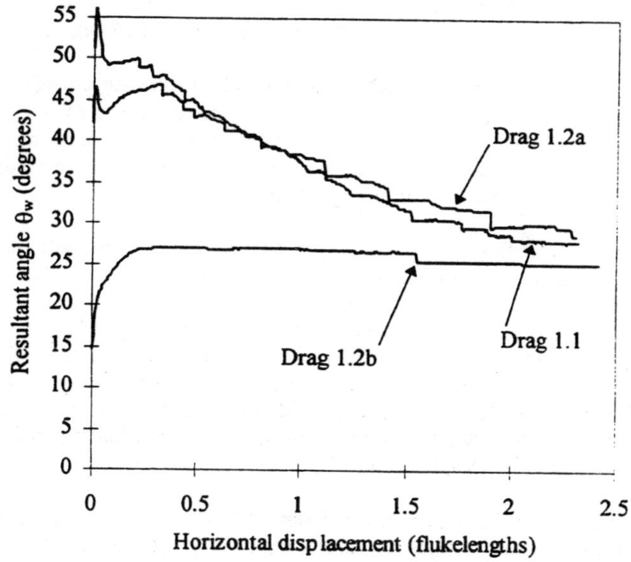

Figure 8(a): Resultant angle θ_w vs horizontal displacement - Test 1

At this point it should be restated that the theory developed by Neubecker and Randolph (1996) assumed that θ_w is constant. Figure 8(a) shows θ_w, derived from the experimental data, plotted against horizontal displacement for Test 1. Interestingly, for the two drags which commenced immediately after embedment of the anchor (Drags 1.1 and 1.2a), θ_w started at a value of approximately 50 degrees, and steadily decreased during displacement of the anchor before showing signs of levelling out towards the end of the drags at around 28 degrees. For Drag 1.2b (a continuation of Drag 1.2a), θ_w remained relatively constant at 27 degrees.

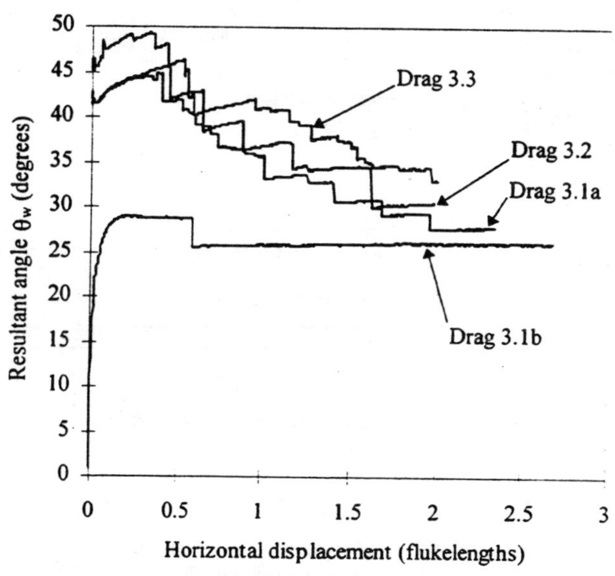

Figure 8(b): Resultant angle θ_w vs horizontal displacement - Test 3

Similarly for Test 3 (Figure 8(b)), in those drags that took place immediately after embedment (Drags 3.1a, 3.2 and 3.3), the value of θ_w dropped from approximately 50 degrees down to between 27 and 33 degrees, depending on the drag length of the test. Drag 3.1b, a continuation of Drag 3.1a, resulted in a constant θ_w of 26 degrees.

In each case where the anchor had reached its ultimate capacity, the value of θ_w remained steady. However, in those drags where the anchor had been vertically embedded prior to horizontal displacement, the resultant angle commenced at a much higher value before tending towards the ultimate value during displacement.

The cause of this discrepancy lies in the disturbance of the soil during embedment of the anchor. In most of the tests, the model anchor was embedded at a fluke angle of approximately 40 to 45 degrees (Figures 7(a) and 7(b)). At this orientation, the anchor must travel horizontally approximately 105 mm (1.7 flukelengths) before the back of the fluke has entered what would have been undisturbed soil prior to horizontal displacement. Hence, the model anchor does not have the opportunity to be fully subjected to soil undisturbed by the embedment procedure until after 2 flukelengths of drag.

A large initial θ_w implies that the vertical load response of the anchor during the early stages of displacement is greater than would be expected had the anchor been moving through undisturbed soil.

Hence, it was decided that in the case of all 1-stage drags and the first stage of all 2-stage drags, simulation of the load response of the anchor was valid only from a horizontal displacement of 2 flukelengths or greater. All load simulation curves for these tests remained unchanged by commencing at 0.5 flukelengths, but the least squares regression scheme employed to determine appropriate values of f and θ_w were adjusted to include experimental data only beyond 2 flukelengths.

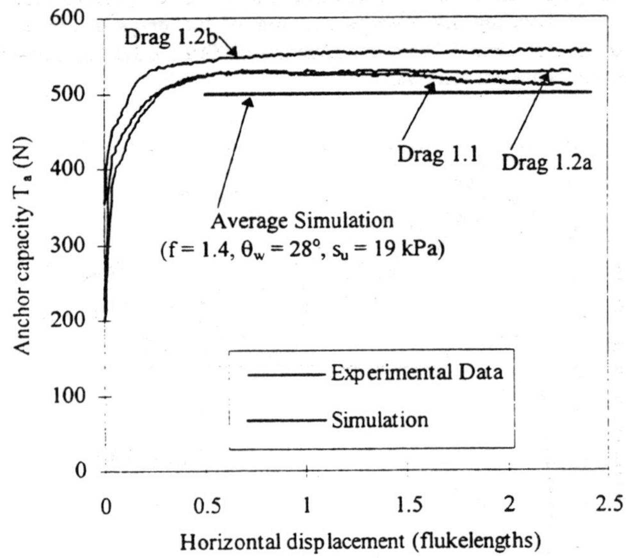

Figure 9(a): Anchor capacity T_a vs horizontal displacement - Test 1

Table 2 summarises the resulting f and θ_w values obtained from the least squares regression scheme, assuming that A_p = 1835 mm^2 (including the area between the shanks):

Experimental and simulated weightless anchor capacity curves for Test 1 are plotted against horizontal displacement in Figure 9(a). The simulated curves, based on the average values in Table 2, show good agreement with the experimental results.

Test Number	f	θ_w (Degrees)
Test 1	1.50	27.6
Test 2	1.21	30.6
Test 3	1.50	26.4
Average	1.4	28

Table 2: Summary of f and θ_w values from least squares regression schemes

Figure 9(b) shows the weightless anchor capacity curves for Test 3. Again, the simulated curves are based on the average f and θ_w values in Table 2. There are some discrepancies in the simulated values for Drags 3.2 and 3.3. However, since neither drag progressed beyond 2 fluke lengths, it is difficult to make any firm comparisons between the experimental data and the simulated values.

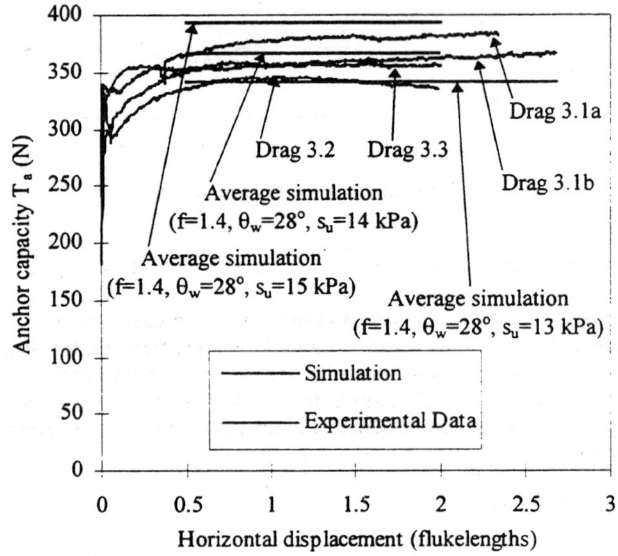

Figure 9(b): Anchor capacity T_a vs horizontal displacement - Test 3

CONCLUSIONS

A new procedure for testing model drag anchors has been successfully developed and tested. The procedure is aimed at quantifying the two key anchor parameters proposed by Neubecker and Randolph (1996), that determine the embedment response and ultimate holding capacity in non-dimensional terms. The particular advantage of the proposed approach is that the anchor parameters should be essentially identical for all anchors (of various weights) for a given design.

The new testing procedure involves dragging the anchors at a fixed embedment depth, measuring horizontal and vertical components of resistance, and also the current orientation of the anchor. The approach eliminates the complexity of modelling the anchor chain or cable, and significantly reduces the length of drag needed to reach steady-state conditions.

The test data indicate that, as the anchor reaches its ultimate holding capacity, the flukes become horizontal parallel to a plane lying approximately midway between the top and bottom leading edges of the fluke. For the anchor tested, the upper surface of the flukes is then about 12 degrees above horizontal. The resultant geotechnical resistance was found to act at 28 degrees to the upper surface of the flukes, and the form factor, f, that determines the component of resistance parallel to the upper surface of the flukes (see Equation (1)) was found to be 1.4.

The testing procedure is particularly effective at identifying anchor design parameters for different classes of anchor, and also as an aid to improving anchor designs to optimise the form factor and angle of resulting resistance, θ_w.

ACKNOWLEDGEMENTS

The work described here forms part of the activities of the Special Research Centre for Offshore Foundation Systems, funded through the Australian Research Council's Research Centre's Program. The first author is supported by a University Postgraduate Award. Special thanks are due to the workshop and centrifuge staff at UWA, without whose excellent technical support the experimental research would not have been possible.

REFERENCES

Dunnavant, T. W. and Kwan, C-T. T. (1993), Centrifuge modelling and parametric analyses of drag anchor behaviour, *Proc., 25th Annual Offshore Technology Conference*, Houston, Texas, Paper 7202.

Neubecker, S. R. (1995), The behaviour of drag anchor and chain systems, PhD Thesis, Department of Civil Engineering, The University of Western Australia.

Neubecker, S. R. and Randolph, M. F. (1996), The performance of drag anchor and chain systems in cohesive soil, *Marine Georesources and Geotechnology*, 14: 77-96.

Stewart, D. P. and Randolph, M. F. (1991), A new site investigation tool for the centrifuge, *Centrifuge 91*, Balkema, Rotterdam, pp 531-538.

Taylor, R. J. (1981), Performance of conventional anchors, . *Proc., 13th Annual Offshore Technology Conference*, Houston, Texas, Paper 4048.

Vryhof Anchors. 1990. *Anchor Manual.* Krimpen ad Yssel, The Netherlands.

Proceedings of the Seventh (1997) International Offshore and Polar Engineering Conference
Honolulu, USA, May 25-30, 1997

ISBN 1-880653-28-1 (Set); ISBN 1-880653-29-X (Vol. I)

Geotechnical Properties of Crude Oil-Contaminated Sand

E.C. Shin
University of Inchon, Inchon, Korea
J.B. Lee
Daewoo Engineering Company, Seoul, Korea
B.M. Das
California State University, Sacramento, CA, USA

ABSTRACT

Geotechnical properties of a crude oil-contaminated quartz sand were investigated. Laboratory tests were conducted to determine the compaction characteristics, angle of internal friction, and coefficients of permeability of natural sand and crude oil-contaminated sand. The effect of oil contamination on the ultimate bearing capacity of a surface strip foundation was also evaluated by laboratory model tests.

KEY WORDS: Bearing capacity, compaction, contamination, crude oil, permeability, sand, shear strength

INTRODUCTION

Every year, several oil spills occur around the world due to offshore operations. These oil spills ultimately cause contamination of the coastal areas. The physical properties of oil-contaminated soil may control the stability of slopes, bearing capacity of foundations of coastal structures and may cause other environmental damage to the coast lines. At the present time, only a few studies are available in the literature relating to the behavior of oil-contaminated sand (Cook et al., 1992; Evgin et al., 1989). The purpose of this paper is to present some geotechnical properties of crude oil-contaminated sand such as compaction, permeability, and shear strength determined from laboratory tests. These properties were compared with those of clean sand. For the present study, a poorly graded silica sand (Jumoon Jin sand of Korea) was used. Oman crude oil was used as the contaminant.

PROPERTIES OF SAND

As mentioned in the preceding section, a poorly-graded Jumoon Jin sand was used for the present study. The grain-size distribution of the sand is shown in Fig. 1. Other properties of the sand are given in Table 1.

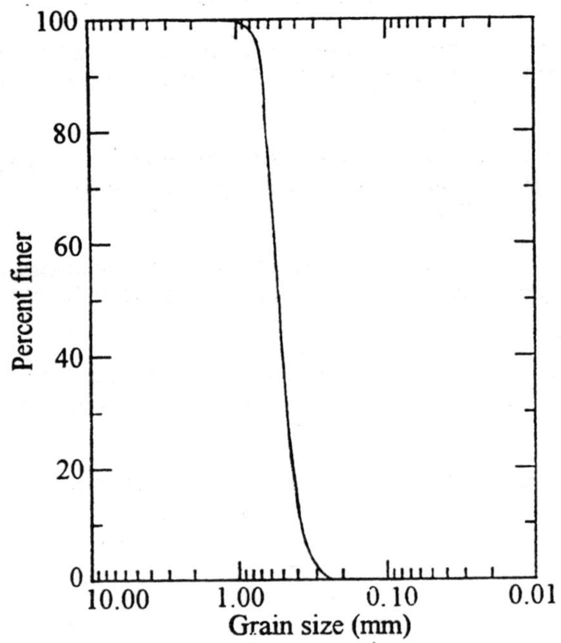

Fig. 1 Grain-size distribution of sand

TABLE 1. Properties of the Sand

Item	Quantity
Effective size, D_{10}	0.36 mm
Uniformity coefficient, C_u	1.61
Coefficient of gradation, C_c	1.10
Maximum dry unit weight, $\gamma_{d(max)}$	17.5 kN/m^3
Minimum dry unit weight, $\gamma_{d(min)}$	13.6 kN/m^3
Maximum void ratio, e_{max}	0.91
Minimum void ratio, e_{min}	0.53
Specific gravity of soil solids, G_s	2.65

PROPERTIES OF CRUDE OIL

The properties of Oman crude oil and grade 50 heavy-duty motor oil are given in Table 2.

TABLE 2. Properties of Oman Crude Oil and Grade 50 Heavy-Duty Motor Oil

Item	Quantity
Oman Crude Oil	
Specific gravity	0.854
Sulphur content	1.05%
Viscosity	424 Mpa.s at 24°C
Grade 50 Motor Oil	
Specific gravity	0.927
Viscosity	455 Mpa.s at 24°C

COMPACTION

Modified Proctor compaction tests (ASTM D-1557) on the sand were conducted with crude oil and water as lubricants. For the tests conducted with crude oil as lubricant

$$w_o = \frac{W_o}{W_s} \tag{1}$$

where w_o = oil content, W_o = weight of oil, and W_s = weight of soil solids.

Figure 2 shows the variation of the dry density of compaction with oil content (w_o) and moisture content (w). It may be seen from this figure that, for tests with crude oil as lubricant, $\gamma_{d(max)}$ was about 16.63 kN/m^3 at w_o = 9.4%. Similarly for tests with water as lubricant, $\gamma_{d(max)}$ was 16.21 kN/m^3 at w = 19.1%.

Since the specific gravities of oil and water are different, it appears that a better comparison can be made by the degree of saturation which corresponds to $\gamma_{d(max)}$. The degree of saturation, S, can be given as

$$S = \frac{wG_s}{e} \text{ (for water as lubricant)} \tag{2}$$

and

$$S = \frac{w_o G_s}{eG_o} \text{ (for crude oil as lubricant)} \tag{3}$$

where e = void ratio and G_o = specific gravity of crude oil

Using the preceding relationships, the degree of saturation at which maximum dry unit weights occur are as follows:

With water: S = 83.8%

With crude oil: S = 51.8%

Hence, a higher dry unit weight of compaction is obtained with crude oil as lubricant at a lower degree of saturation

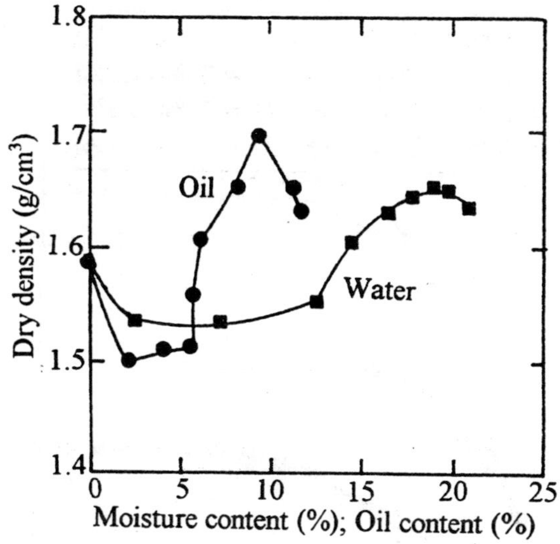

Fig. 2 Modified Proctor compaction test

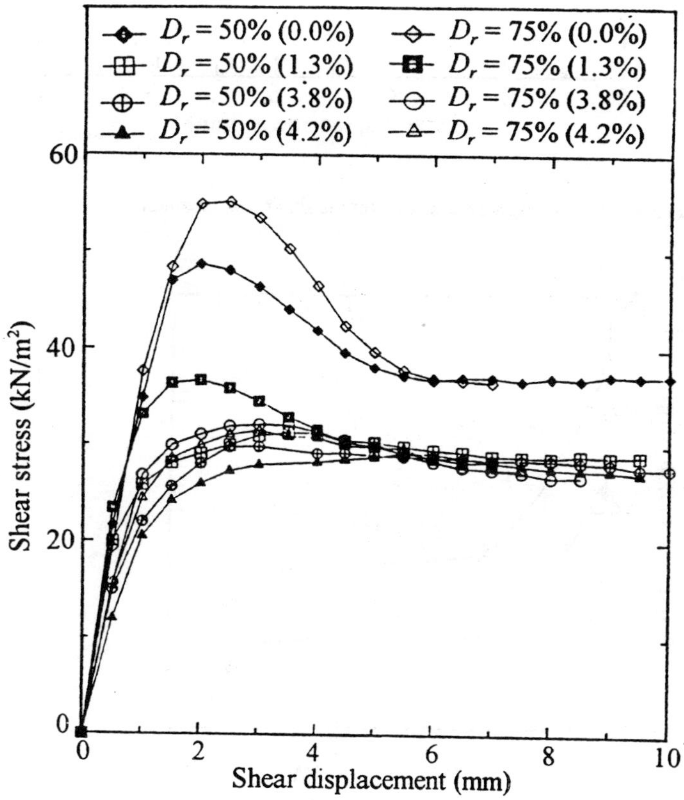

Fig. 3 Plot of shear stress versus shear failure

SHEAR STRENGTH

A number of direct shear tests were conducted with dry and crude oil-contaminated sand to evaluate the effect of crude oil on the angle of friction. These tests were conducted at relative densities of compaction (D_r) of 50% and 75% at oil contents (w_o) of 0%, 1.3%, 3.8%, and 4.2% and a normal stress of 60 kN/m^2. Figure 3 shows the plot of the variation of shear stress vs. shear displacement for all tests conducted. Similarly,

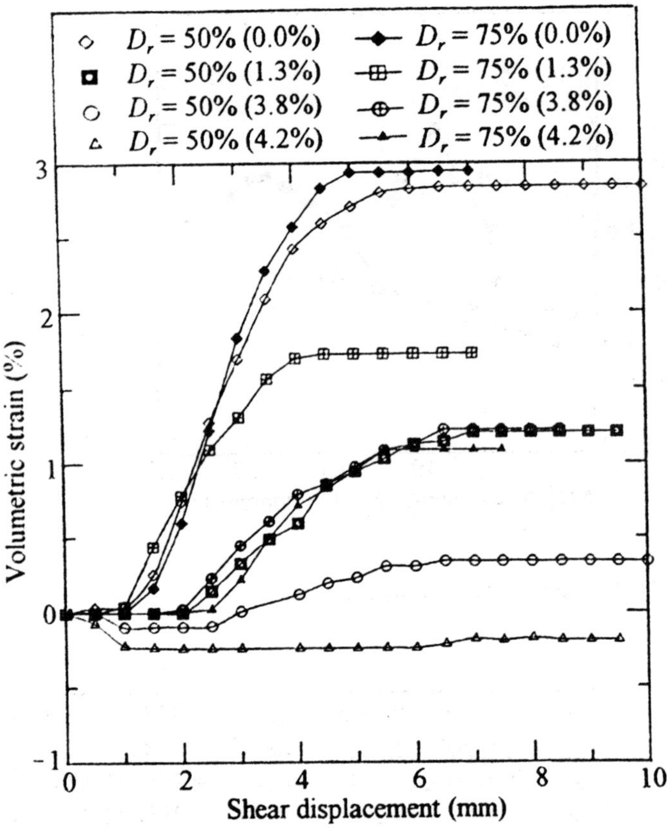

Fig. 4 Plot of volumetric strain versus shear displacement

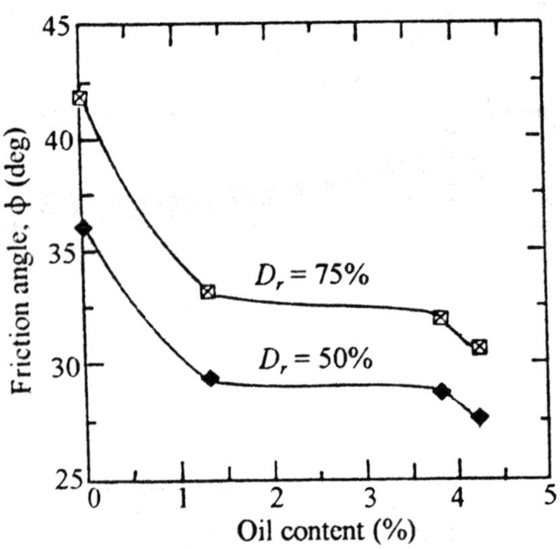

Fig. 5 Variation of soil friction angle with oil content

Fig. 4 shows the plot of volumetric strain versus shear displacement for these tests. Based on the experimental results shown in Fig. 3, the magnitudes of the peak friction angle ϕ were determined and are shown in Fig. 5. It may be seen from Fig. 5 that the peak friction angle decreases by about 25% when the oil content increases to about 4%. Evgin et al.

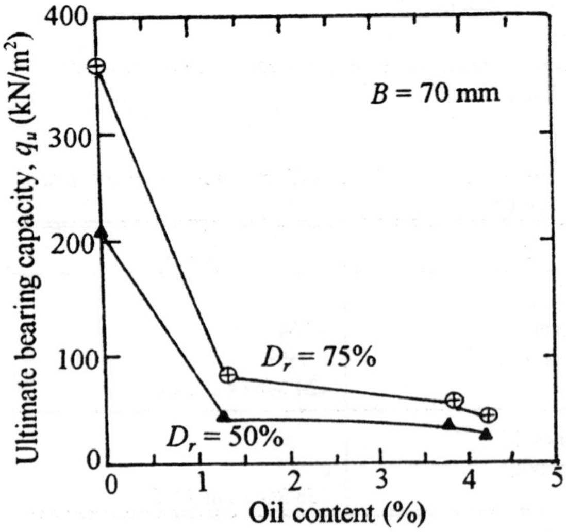

Fig. 6 Variation of ultimate bearing capacity of surface strip foundation with oil content

(1989) conducted triaxial tests on dense and loose crushed quartz sand contaminated with grade 50 heavy duty motor oil. The results obtained from their tests are shown in Table 3.

TABLE 3. Angle of Friction, ϕ, of Crushed Quartz Sand— Clean and Contaminated with Heavy Duty Motor Oil

Type of Sand	ϕ (deg)
Clean dense	41.5
Clean loose	32.2
Oil-contaminated dense	35.3
Oil-contaminated loose	26.2

The decrease in friction angle with oil contamination as shown in Fig. 5 and Table 3 is in the same order. The rapid decrease of ϕ with oil contamination may have a great influence in the stability of slopes, bearing capacity of foundations, and others of the like.

In order to evaluate the effect of oil contamination on the bearing capacity of foundations, a limited number of model bearing capacity tests were conducted on a ***rough surface strip*** foundation supported by clean and oil-contaminated sand. The width (B) of the model strip foundation was 70 mm. The rests of these tests are shown in Fig. 6. As can be seen from this figure, for any given degree of compaction, the ultimate bearing capacity decreased rapidly with the increase in oil content.

PERMEABILITY

A number of constant head permeability tests were conducted on compacted sand specimens. The details of the test conditions are given in Table 4. The tests for Series A and B were conducted with oil as a permeant and, in these cases, water was used to compact the soil. Similarly in Series C and D, the oil was used in compacting the soil and water was used as the permeant. The variations of the coefficient of permeability, k, obtained from these tests are shown in Figs. 7 and 8. For tests in Series A and B, the magnitude of k initially increased and then decreased with the increase in moisture content. The reason for this behavior is not known. However in Series C and D, the coefficient of permeability continuously decreased with the increase in the oil content.

TABLE 4. Details of the Permeability Tests

Series	Molding water content, w (%)	Molding oil content, w_o (%)	Relative density, D_r (%)	Type of permeant
A	0 4.2 8.5 16.9	-- -- -- --	45	Oil
B	0 3.9 7.8 15.6	-- -- -- --	85	Oil
C	-- -- -- -- --	0 0.74 1.6 3.1 5.6	45	Water
D	-- -- -- -- --	0 0.65 1.3 1.9 3.5	85	Water

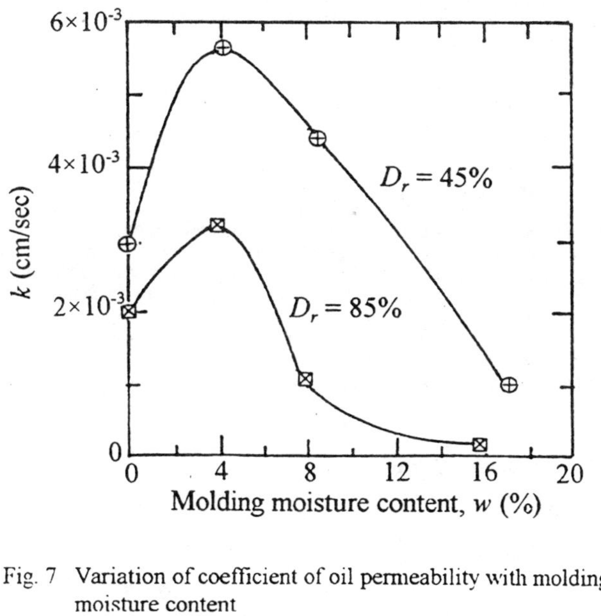

Fig. 7 Variation of coefficient of oil permeability with molding moisture content

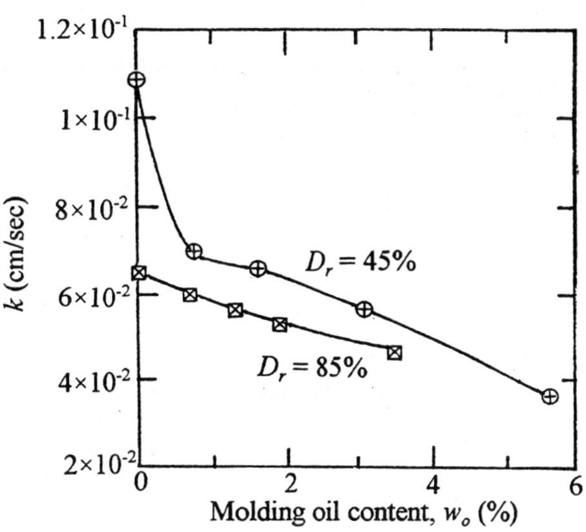

Fig.8 Variation of k with molding oil content

CONCLUSIONS

Laboratory test results for the effect of crude oil contamination on the geotechnical properties of sand have been presented. Based on the tests, the following conclusions can be drawn:

1. A higher maximum dry unit weight can be obtained from modified Proctor compaction tests when crude oil is used as a lubricant as compared to water.
2. The internal friction angle of the sand decreases rapidly with the increase in oil content.
3. The ultimate bearing capacity of shallow foundations decreases rapidly with the increase of oil content in the sand.
4. Water permeability through sand decreases with the increase of oil content in the sand.
5. Oil permeability through the sand also decreases with the increase of moisture content in the soil.

REFERENCES

Cook, EE, Puri, VK, and Shin, EC (1992). "Geotechnical Properties of Crude Oil-Contaminated Sand," *Proc., Second International Offshore and Polar Engineering Conference*, Vol. 1, pp. 384-387.

Evgin, E, Amor, FB, Altall, A, Lord, S and Konuk, I (1989). "Effect of Oil Spill on Soil Properties," *Proc., 8th International Conference on Offshore Mechanics and Arctic Engineering*, Vol. 1, pp. 715-720.

Proceedings of the Seventh (1997) International Offshore and Polar Engineering Conference
Honolulu, USA, May 25-30, 1997

ISBN 1-880653-28-1 (Set); ISBN 1-880653-29-X (Vol. I)

Stabilization of Petroleum Contaminated Drilling Wastes by Additives

M. Tuncan, A. Tuncan and H. Koyuncu
Anadolu University
Eskisehir, Turkey

ABSTRACT

Petroleum industry has resulted in the release of large quantities of waste containing a variety of organic and inorganic pollutants. During the petroleum production, petroleum contaminated drilling muds and cuttings were disposed of in open pits called "mud-pit". So, there has been large quantities of petroleum contaminated wastes which present high risks to the environment.

This study progressed in the field to demonstrate the construction feasibility and effectiveness in the site environment which was appropriate or not. The field study was conducted as a joint effort by researchers at Anadolu University and Turkish Petroleum Corporation (TPAO). Petroleum drilling wastes were collected from the mud-pit and mixed with pozzolanic materials such as lime, fly ash, bentonite clay, kaolinite clay, gypsum, klinker, portland cement and marble powder in order to stabilize the petroleum drilling wastes. The stabilized wastes were deposited into boxes constructed in the field. After stabilization, the end product was chemically and physically stable and durable and also easy to handle and work with. In this study, physical and chemical properties of stabilized mixtures were investigated.

The laboratory study showed that a mixture of petroleum drilling wastes + kaolinite clay (30%), petroleum drilling wastes + gypsum (10%) + bentonite clay (10%) and petroleum drilling wastes + klinker (15%) + lime (15%) + fly ash (15%) gave the best results with respect to oil content.

KEY WORDS: Petroleum Drilling Wastes, Stabilization, Oil Content, Electrical Conductivity, pH, Heavy Metals and Additives.

INTRODUCTION

Oil well drilling activities produce large quantities of wastes in the form of drilling muds and cuttings. These wastes were disposed of in open pits (mud-pit), and large quantities of unstabilized wastes remain at a number of sites. To reduce the high risks of environmental impacts, efficient and economical alternative is to stabilize the wastes and redeposit them into mud-pit. Oil well drilling activities consist of drilling muds and drilling cuttings. Drilling muds are used to cool, clean or lubricate bits, to carry cuttings up from the bottom and to control subsurface pressure (Rogers, 1963).

An economical and fast solution is to stabilize petroleum contaminated drilling wastes by mixing it in place with additives. Stabilization is the improvement of physical and mechanical properties of wastes, encapsulation of pollutants and reduction of solubility of toxic substances. The stabilization method was applied to petroleum drilling wastes to bind the oil and metals in a structure formed by cementing of pozzolanic materials to produce chemically and physically stable and mechanically handable mixtures.

Stabilization of petroleum drilling wastes is accomplished by pozzolanic reactions which include both chemical and physical interactions between the wastes and additives. Pozzolanic reaction between clay and lime involves reaction between calcium and silica or alumina to form cementing type mixtures such as calcium silicates or calcium aluminates. Stabilization with lime, fly ash, cement, bentonite clay, kaolinite clay, gypsum, klinker and marble powder is the simplest and inexpensive technique. These additives are used to increase pH, to contribute surface area and to retain hydrocarbons. Lime, fly ash, cement and clay stabilization are often used successfully to stabilize the petroleum wastes.

MATERIALS AND TESTING

Petroleum Drilling Wastes (PDW)

Petroleum drilling wastes contain drilling muds and cuttings which consist of both liquid and solid phase. Drilling muds are normally classified as oil-base and water-base mud (Rogers, 1963). Oil-base mud contains a continuous oil phase usually diesel oil, bentonite clay and chemicals. Water-base mud contains fresh or salt water phase, bentonite clay and chemicals. Water-base muds are preferred for their low cost. However, certain situations require oil-base muds for drilling operations in formations sensitive to water-base fluids. Drilling cuttings are the solid cuttings which are brought to the surface from the well during borings. In this study, mud consists of chloride, barite, bentonite clay, caustic, carboxy methly cellulose (CMC) and crude oil. Constituents of petroleum drilling wastes are given in Table 1.

Table 1. Some Properties of Petroleum Drilling Wastes.

Oil Content (%)	: 4.00
Water Content (%)	: 171
Organic Matter Content (%)	: 6.51
pH	: 8.74
Cation Exchange Capacity, CEC (meq/100g)	: 28.08
Electrical Conductivity (mS/cm)	: 35.8
Grain Size Analysis	
Sand (%)	: 7
Silt (%)	: 71
Clay (%)	: 22

Additives

One of the most popular pozzolans is fly ash. Most fly ash is a by-product of coal burning power plants. Fly ash act as a conditioner in the waste to absorb liquid and to coat waste globules.

Lime has been used as a practical soil stabilizing agent for a long time. The most common alkali used in the stabilization is lime. Hydrated form of lime is more common because it is easier and safer to store and handle. Lime neutralizes the waste and decreases the water content or plasticity and oil of wastes (Van Keuren et al.,1987). The solubility of hydrocarbons is also reduced by lime. If the lime is mixed with iron and aluminum bearing minerals, it will form a crude form of cement. Many types of coal power plant fly ash contain at least some amounts of iron and aluminum. If fly ash are mixed with the lime, then the mixture can react with the wastes in a manner analogous to portland cement. Lime and lime/fly ash mixtures tend to be less expensive than cement. They, however, tend to set up more slowly than the cement products and generally form a physically weaker solids. Lime can be used for the treatment of acidic wastes due to high acid buffering capacity. As a result, the pH of the stabilized product will remain high for a longer period of time minimizing the mobility of metals.

Bentonite clay and kaolinite clay were used as absorbent to improve the physical and chemical characteristics of wastes. They were selected as the soil medium since clays tend to retain hydrocarbons (Tuncan and Pamukcu, 1992). They are highly colloidal, plastic and inexpensive materials and feasible for metals and organics. The reaction of lime with clay produces a calcium silicate hydrate.

Portland cement was used to improve the interparticle binding and mechanical properties of wastes. It is a mixture of a number of minerals such as calcium silicates, calcium aluminates and iron oxides. The properties of additives are given in Tables 2 and 3.

Table 2. Some Properties of the Additives.

Additives	pH	CEC (meq/100gr)	Electrical Conductivity (mS/cm)	Water Content (%)	Organic Matter Content (%)
Fly Ash	12.34	9.46	10.38	0.3	1.1
Lime	12.60	16.43	6.29	0.5	24.4
Cement	12.63	21.74	4.97	1.9	3.0
Bentonit	9.50	90.22	2.69	12.7	6.8
Kaolinit	4.86	25.93	0.15	0.1	7.4
Gypsum	7.28	17.26	2.53	0.6	6.8
Klinker	12.67	8.98	11.24	0.7	1.4
Marble	9.69	7.57	0.24	4.1	1.5

Table 3. Total Metal Concentrations of Additives (mg/kg).

Mixture	Pb	Cr	Zn	Ni	Cu	Mn	Fe	Cd	As
Fly Ash	80	144	88	369	213	3318	49031	<0.5	<5
Lime	106	70	43	35	53	509	3024	<0.5	<5
Cement	120	111	77	24	473	2168	22229	<0.5	<5
Bentonit	92	10	71	8	81	2419	10724	<0.5	<5
Kaolinit	136	405	52	280	280	870	7550	<0.5	<5
Gypsum	77	55	79	9	40	99	663	<0.5	<5
Klinker	147	115	184	135	455	1595	20865	<0.5	<5
Marble	78	45	37	35	46	252	1288	<0.5	<5

Mixtures

Experiments were conducted in the field. A pilot field study was needed because of the determination of construction feasibility and the most appropriate mixing technique which is in the pit or out of the pit. One-additive, two-additive and three-additive mixtures are given in Table 4.

Testing

A testing program was conducted to investigate how much improvement could be accomplished in the physical and chemical properties of a petroleum drilling wastes by hydrated lime, fly ash, bentonite clay, kaolinite clay, gypsum, klinker, marble powder and portland cement treatment.

Petroleum drilling wastes were excavated with conventional construction equipments from the mud-pit. Specimens were prepared by mixing the petroleum drilling wastes with water at the predetermined optimum water content of the mixed system before adding additives. Additives were then mixed by workers to form a paste-like mixtures. The mixtures were divided into 25 sets. Each set was being treated with combination of additives at different percentages. The amount of the additive was measured as weight percentage of petroleum drilling wastes. The proportions of these additives were selected as 10%, 15%, 25% and 30% by weight of the petroleum drilling wastes. Final mixtures were then placed into the precast boxes. Boxes were 70 cm by 140 cm in cross section and 30 cm in height. One box was prepared for each mixture. Grass was planted in half of the box and the rest of the box would remain open to air.

After mixing petroleum drilling wastes with additives, water content and organic matter content tests were conducted on the untreated and treated fresh specimens. Oil content, pH, electrical conductivity and total metal content tests were conducted on the untreated and treated specimens after 30 days.

Table 4. One-Additive, Two-Additive and Three-Additive Mixtures.

Mixture	FA	L	PC	KC	BC	G	K	MP
Mixture 1:PDW	--	--	--	--	--	--	--	--
Mixture 2:PDW+	--	--	--	30	--	--	--	--
Mixture 3:PDW+	--	--	--	--	--	30	--	--
Mixture 4:PDW+	--	25	--	--	--	--	--	--
Mixture 5:PDW+	--	--	--	--	--	--	30	--
Mixture 6:PDW+	--	--	30	--	--	--	--	--
Mixture 7:PDW+	--	--	--	--	30	--	--	--
Mixture 8:PDW+	--	--	--	--	--	--	--	30
Mixture 9:PDW+	30	--	--	--	--	--	--	--
Mixture 10:PDW+	--	--	--	--	10	10	--	--
Mixture 11:PDW+	--	--	--	--	10	--	10	--
Mixture 12:PDW+	--	--	10	--	10	--	--	--
Mixture 13:PDW+	10	10	--	--	--	--	--	--
Mixture 14:PDW+	--	10	--	--	--	--	10	--
Mixture 15:PDW+	--	--	--	--	--	10	10	--
Mixture 16:PDW+	10	--	10	--	--	--	--	--
Mixture 17:PDW+	10	--	--	--	--	--	10	--
Mixture 18:PDW+	15	15	--	--	--	--	15	--
Mixture 19:PDW+	--	15	15	--	15	--	--	--
Mixture 20:PDW+	10	10	--	--	--	--	10	--
Mixture 21:PDW+	--	15	--	--	15	--	15	--
Mixture 22:PDW+	15	--	--	--	--	15	15	--
Mixture 23:PDW+	--	--	15	--	15	15	--	--
Mixture 24:PDW+	15	15	15	--	--	--	--	--
Mixture 25:PDW+	15	--	15	--	--	15	--	--

PDW=Petroleum Drilling Waste, FA=Fly Ash, L=Lime, PC=Portland Cement, KC=Kaolinit Clay, BC=Bentonit Clay, G=Gypsum, K=Klinker, MP=Marble Powder.

Determination of Oil Content

Specimens were dried at 40 ^oC for 24 hours. After drying, 20 grams of treated and untreated specimens were placed into the chamber of the soxhlet extractor and extracted with 175 ml of methylene chloride. The solvent reservoir was heated to 60 ^oC by electrical heater for 2 hours. Methylene chloride extracts oil efficiently and has a sufficiently low boiling point to be conveniently used in a soxhlet extraction apparatus (Mc Gill and Rowell, 1980).

Determination of Total Metals

Specimens were dried at 105 ^oC for 24 hours and grounded to pass through No. 100 sieve. Ten grams of treated and untreated specimens were then digested in 3 to one concentrated nitric acid and hydrochloric acid mixture for 2 hours at 80 ^oC. The diluted digests were then filtered through pore size of 0.2 mm cellulose nitrat filter. Copper, barium, manganese, zinc, lead, chromium, nickel, arsenic and cadmium were determined by atomic absorption spectrometer.

Determination of Cation Exchange Capacity (CEC)

CEC was determined by sodium saturation method (Chapman, 1965). First, specimens were saturated with sodium acetate to replace the cations held on the surface and in the interlayer positions. Specimens were washed with isoprophyl alcohol to remove the excess saturating salt. Specimens were washed with ammonium acetate, by which the adsorbed sodium was released into the solution. The amount of sodium was determined by atomic adsorption spectrometry.

Determination of Water Content and Ignitable Organic Matter Content

The water content was measured by placing the mixtures in an oven at 105 °C for 24 hours (ASTM D2216). The organic matter content of the mixtures was determined by the ignition loss method (ASTM D2974). Treated and untreated specimens were pre-dried at 105 °C for 24 hours. These specimens were then burned at 550 °C for 2 hours. The loss of weight of the specimen as the result of the second heating was assumed to be primarily due to organic matter ignition.

EXPERIMENTAL RESULTS

The variation of oil content, pH, electrical conductivity, water content and organic matter content of one-additive, two-additive and three-additive mixtures are given in Tables 5, 6 and 7, respectively. Addition of additives decreases oil content, electrical conductivity, water content and organic matter content of petroleum drilling wastes, in general. The most efficient mixtures are the petroleum drilling wastes + kaolinite clay (30%) with respect to oil content and petroleum drilling wastes + cement (30%) with respect to electrical conductivity for the one-additive mixtures.

The most effective mixtures are the petroleum drilling wastes + gypsum (10%) + bentonite clay (10%) with respect to oil and the petroleum drilling wastes + klinker (10%) + bentonite clay (10%) with respect to electrical conductivity for the two-additive mixtures. The most effective mixtures are the petroleum drilling wastes + klinker (15%) + lime (15%) + fly ash (15%) with respect to oil content and petroleum drilling wastes + cement (15%) + gypsum (15%) + bentonite clay (15%) with respect to electrical conductivity for the three-additive mixtures.

Addition of lime, fly ash, cement, gypsum, klinker, kaolinite clay and bentonite clay reduces oil content and electrical conductivity of treated specimens and increases pH values of wastes. pH is an important parameter which affect the immobilization of metals. With proper pH management which is provided in this case, metals are effectively immobilized. High pH is desirable because metal hydroxides have minimum solubility in the range of pH 7.5-11.

Table 5. Water Content, Organic Matter Content, Oil Content, pH and Electrical Conductivity Properties of One-Additive Mixtures.

Mixture	Oil Content (%)	pH	Electrical Conductivity (mS/cm)	Water Content (%)	Organic Matter Content (%)
PDW	4.00	8.74	35.8	171	6.51
Mixture:2	1.25	8.14	16.6	96	5.23
Mixture:3	1.43	8.11	23.4	99	5.49
Mixture:4	1.51	12.45	24.7	102	6.79
Mixture:5	1.64	12.05	15.2	93	4.34
Mixture:6	1.71	11.36	13.1	91	5.48
Mixture:7	1.94	8.20	18.5	102	4.86
Mixture:8	1.95	8.13	20.0	96	5.48
Mixture:9	2.06	9.26	20.7	93	3.93

Table 6. Water Content, Organic Matter Content, Oil Content, pH and Electrical Conductivity Properties of Two-Additive Mixtures.

Mixture	Oil Content (%)	pH	Electrical Conductivity (mS/cm)	Water Content (%)	Organic Matter Content (%)
PDW	4.00	8.74	35.8	171	6.51
Mixture:10	1.41	8.05	23.3	102	6.13
Mixture:11	1.45	12.66	11.6	100	6.07
Mixture:12	1.62	11.19	14.0	108	6.14
Mixture:13	1.77	13.55	17.4	96	5.49
Mixture:14	1.78	13.64	15.3	97	5.29
Mixture:15	1.78	12.97	14.3	101	5.34
Mixture:16	1.90	12.55	15.0	105	4.85
Mixture:17	2.32	12.76	13.8	102	4.66

Table 7. Water Content, Organic Matter Content, Oil Content, pH and Electrical Conductivity Properties of Three-Additive Mixtures.

Mixture	Oil Content (%)	pH	Electrical Conductivity (mS/cm)	Water Content (%)	Organic Matter Content (%)
PDW	4.00	8.74	35.8	171	6.51
Mixture:18	0.94	12.37	15.5	69	5.26
Mixture:19	1.17	12.18	11.3	79	6.30
Mixture:20	1.28	12.04	21.8	59	4.53
Mixture:21	1.31	12.39	13.4	75	5.99
Mixture:22	1.36	11.48	13.4	70	4.62
Mixture:23	1.44	9.45	9.0	76	5.98
Mixture:24	1.54	12.22	17.4	73	5.12
Mixture:25	1.72	10.83	13.7	93	6.06

Total metal concentrations of one-additive, two-additive and three-additive mixtures are shown in Tables 8, 9 and 10, respectively.

Lead occurs in soils, rocks and minerals. It is also found in substantial amounts in cement and lime. Petroleum industry wastes contain both inorganic and organic lead compounds. Lead concentration in the soil changes between 70-100 mg/kg (Altıntas et al., 1994). Allowable lead concentration in the petroleum drilling wastes is 100 mg/kg (Mucsy et al., 1984). If the pH of mixture increases, soluble lead concentration decreases. Lead Concentration can not be decreased under the allowable limits even after stabilization except mixture:3 and mixture:4.

Drilling muds contain both soluble and difficultly soluble chromates for corrosion control and chromium lignosulfonates. Chromium concentration in the soil is 300 mg/kg (Altıntas et al., 1994). Allowable chromium concentration in the petroleum drilling wastes is 100 mg/kg (Mucsy et al., 1984). Addition of additives to the petroleum drilling wastes decreased chromium concentration under the allowable limits.

Cadmium and its compounds are highly toxic. The primary treatment method is precipitation with alkali, usually lime. The pH of mixture is the most important factor controlling cadmium solubility. Solubility of cadmium decreases as pH increases. Between pH 8 and 11 about %99 of the cadmium is speciated as the solid carbonate (Conner, 1990). Cadmium concentration in the soil is 3 mg/kg (Altintas et al., 1994). Allowable cadmium concentration in the petroleum drilling wastes is 5 mg/kg (Mucsy, et al., 1984). Cadmium concentration in petroleum drilling wastes is under the allowable limits.

Zinc is an aquatic toxin according to Environmental Protection Agency (EPA). Zinc is found in all natural waters and soils, and is an important nutrient in plant and animal life. Zinc concentration in the soil is 300 mg/kg (Altıntas et al., 1994). Allowable zinc concentration in the petroleum drilling wastes is 300 mg/kg (Mucsy et al., 1984). Zinc concentration in petroleum drilling wastes is under the allowable limits.

Arsenic is classified as a nonmetal, although it is grouped with the metals for most environmental purposes. Most of arsenic compounds are highly toxic. Arsenic concentration in the soil is 5 ppm. Allowable arsenic concentration in the petroleum drilling wastes is 5 mg/l (Morgan and Novoa, 1984). Arsenic concentration in petroleum drilling wastes is under the allowable limits.

Copper is an aquatic toxin according to EPA. Copper is abundant in minerals and soils. Copper concentration in the soil is 100 mg/kg (Altintas et al., 1994). Allowable copper concentration in the petroleum drilling wastes is 100 mg/kg (Mucsy et al., 1984). Copper concentration in petroleum drilling wastes is under the allowable limits.

Nickel concentration in the soil is 50 mg/kg (Altıntas et al., 1994). Allowable nickel concentration in the petroleum drilling wastes is 100 mg/kg (Saarela, 1992). Nickel concentration in petroleum drilling wastes is under the allowable limits.

Manganese concentration in the soil changes between 20 and 800 mg/kg (Altıntas et al., 1994). Allowable manganese concentration in the petroleum drilling wastes is 50 mg/kg (Saarela, 1992). Manganese concentration can not be decreased under the allowable limits even after stabilization.

Barium is classified as an alkali earth metal. It is almost in the form of barite. Barite is used in oil well drilling as an additive to drilling mud. Allowable barium concentration in the petroleum drilling waste is 100 mg/l (Morgan and Novoa, 1984). Barium concentration can not be decreased under the allowable limits even after stabilization.

Table 8. Total Metal Concentrations of One-Additive Mixtures (mg/kg).

Mixture	Pb	Cr	Zn	Ni	Cu	Mn	Ba	Cd	As
PDW	345	110	257	60	43	469	2296	<0.5	<5
Mixture 2:	135	85	180	116	21	257	1500	<0.5	<5
Mixture 3:	95	48	124	37	20	232	----	<0.5	<5
Mixture 4:	105	88	140	33	21	265	850	<0.5	<5
Mixture 5:	126	93	160	31	68	418	1050	<0.5	<5
Mixture 6:	112	98	155	14	27	315	1700	<0.5	<5
Mixture 7:	134	80	181	94	25	432	----	<0.5	<5
Mixture 8:	132	92	148	76	40	416	----	<0.5	<5
Mixture 9:	173	84	160	147	39	488	480	<0.5	<5

Table 9. Total Metal Concentrations of Two-Additive Mixtures (mg/kg).

Mixture	Pb	Cr	Zn	Ni	Cu	Mn	Ba	Cd	As
PDW	345	110	257	60	43	469	2296	<0.5	<5
Mixture 10:	169	84	163	30	28	378	----	<0.5	<5
Mixture 11:	181	100	166	40	35	436	780	<0.5	<5
Mixture 12:	193	69	149	39	27	425	2142	<0.5	<5
Mixture 13:	180	94	173	39	31	462	1800	<0.5	<5
Mixture 14:	166	102	151	31	31	360	380	<0.5	<5
Mixture 15:	180	98	167	34	32	361	----	<0.5	<5
Mixture 16:	147	100	163	41	30	442	1850	<0.5	<5
Mixture 17:	177	116	187	43	34	459	1350	<0.5	<5

Table 10. Total Metal Concentrations of Three-Additive Mixtures (mg/kg).

Mixture	Pb	Cr	Zn	Ni	Cu	Mn	Ba	Cd	As
PDW	345	110	257	60	43	469	2296	<0.5	<5
Mixture 18:	138	94	151	38	36	342	1250	<0.5	<5
Mixture 19:	161	98	139	26	25	338	1300	<0.5	<5
Mixture 20:	142	96	200	33	56	389	----	<0.5	<5
Mixture 21:	171	58	145	33	61	328	130	<0.5	<5
Mixture 22:	172	92	166	35	72	371	----	<0.5	<5
Mixture 23:	165	69	175	36	33	349	----	<0.5	<5
Mixture 24:	173	59	163	31	44	364	370	<0.5	<5
Mixture 25:	160	69	164	32	60	366	----	<0.5	<5

CONCLUSIONS

The additive stabilization of petroleum contaminated wastes produce effective and economics over other remediation technologies. The treated specimens showed significant improvement in their physical and chemical properties with respect to those of untreated specimens.

The following conclusions have been drawn from this study.

1. Addition of additives decreases oil content, electrical conductivity, water content and organic matter content of petroleum drilling wastes, significantly.

2. Addition of additives increases pH of petroleum drilling wastes, as expected.

3. After stabilization, concentrations of chromium, zinc, nickel, copper were below the allowable limits. However, concentrations of lead, barium and manganese are over the limits.

The best results in all of the tests conducted were obtained with specimens treated with petroleum drilling wastes + kaolinite clay (30%), petroleum drilling wastes + gypsum (10%) + bentonite clay (10%) and petroleum drilling wastes + klinker (15%) + lime (15%)+fly ash (15%) with respect to oil content.

ACKNOWLEDGMENTS

This study was supported by the Scientific and Technical Council of Turkey (TUBITAK), Turkish Petroleum Corporation (TPAO) and Anadolu University, Eskisehir, Turkey.

REFERENCES

Altıntas, Ü., Anaç, D., Hakerlerler, H., Okur, B. and Tuncay, H. (1994). "Heavy Metal Contamination in Gediz Valley Irrigatable Land and Its Causes," *Research Project*, No. 91-ZRF-51, Ege University Research Found, Izmir, Turkey.

Chapman, H.D. (1965). "Cation Exchange Capacity", in Methods of Soil Analysis, by Black, C.A. Editor, Part 2 pp. 891-901, No. 9, in the Series Agronomy; *Am. Inst. Agronomy*, Madison, Wisconsin.

Conner, J.R., (1990). *Chemical Fixation and Solidification of Hazardous Wastes*, Van Nostrand Reinhold Publishing Company, New York.

McGill, W.B. and Rowell, M.J., (1980). "Determination of Oil Content of Oil Contaminated Soil," *The Science of the Total Environment*, Vol.14, pp. 245-253, Elsevier Scientific Publishing Company, Amsterdam, Netherland.

Morgan, D.S. and Novoa, J.I., (1984)."Oil Sludge Solidification Using Cement Kiln Dust," *Journal of Environmental Engineering*, Vol. 110, No.5, pp. 935-948.

Mucsy, G., Kranicz-Pap, E. and Urbanyi, G. (1984). "Disposal of Oil-Containing Sludges on Farmlands," *Hazardous and Industrial Waste Management and Testing*: Third Symposium, ASTM STP 851, pp. 135-151, Philadelphia, USA.

Rogers, F.W., (1963). *Composition and Properties of Oil Well Drilling Fluids*, Gulf Publishing Company, Houston, Texas.

Saarela, J., (1992). "Effects of Tailings Spread Around Three Mines in Finland," *Proceedings of the Mediterranean Conference on Environmental Geotechnology*, A.A. Balkema Publishing Company, pp. 483-489, Çesme, Turkey.

Tuncan, A. and Pamukcu, S., (1992). "Geotechnical Properties of Petroleum and Sludge Contaminated Marine Sediments," *Proceedings of the 2nd International Offshore and Polar Conference*, pp. 388-397, San Francisco, USA.

Van Keuren, E., Martin, J. and De Falco, A., (1987). "Pilot Field Study of Hydrocarbon Waste Stabilization," *Proceedings of the 19th Mid-Atlantic Industrial Waste Conference*, pp. 330-341, Technomic Publishing Company.

Proceedings of the Seventh (1997) International Offshore and Polar Engineering Conference
Honolulu, USA, May 25-30, 1997

ISBN 1-880653-28-1 (Set); ISBN 1-880653-29-X (Vol. I)

GIS Database of Irrigation Ponds Damaged by the 1995 Hyokoken-Nambu Earthquake

Kazunori Uchida
Kobe University
Kobe, Japan

ABSTRACT

Using the GIS database of irrigation ponds damaged by the 1995 Hyogoken-Nambu earthquake, the distribution characteristics and major causes of damage are examined. Furthermore, preliminary study of accelerograms for seismic response analysis is conducted with use of Fourier spectra of them measured at three locations. As a result of this investigation, it is concluded that the damaged ponds have a different distribution characteristic from the damage rate for wooden houses in urban areas, and that frequency and amplitude properties of the accelerograms acquired for seismic response analysis have been clarified to depend on the closest distance from the earthquake induced fault line.

KEY WORDS: GIS, database, earthquake disaster, irrigation pond, accelerogram, Fourier spectrum

INTRODUCTION

Over twelve hundreds of irrigation ponds were damaged by the 1995 Hyogoken-Nambu earthquake. Three quarters of these ponds were located in Awaji Island, but over one hundred ponds were also suffered in Ono and Kasai cities located with distances over 20 km from the fault line induced earthquake. The damage in rural areas has been little investigated, which should be compared with urban disasters such as buildings and waterfront facilities. In this study, a database of these damaged ponds using geographic information system, GIS, to investigate geotechnical causes of the damage. First, the GIS was introduced for the reconstruction and development of agriculture in Canada (Deuker, 1979). The GIS is a useful tool to easily recognize the disasters by superposing geographic and geological maps as well as geotechnical data (Matsui et al., 1995). Recently, GIS databases are often used for investigation of disaster (Iwasaki et al., 1996) and disaster prevention planning. Using the GIS database, distribution characteristics and major causes of damage in northern Awaji Island are evaluated in this paper. And also Fourier spectrum are analyzed for preliminary study of accelerograms acquired for earthquake response analysis of the damaged ponds.

GIS DATABASE OF DAMAGED PONDS

In Kobe University, we have been making efforts to complete GIS database of disaster by the 1995 Hyogoken-Nambu earthquake. The GIS database in this investigation is a part in agricultural field. The data is recorded in the database, *MS-ACCESS*, for location, capacity, dimension of embankment, geology of foundation, damage classification, photographs and figures of the damaged ponds in relation to GIS maps such as geographic and geological maps using the GIS software, *GeoBase*. This has some characteristics; it is easy to customize by using *Visual BASIC*; it has high capability to draw both vector-type and raster-type maps, and to draw maps of different scales without boundaries in high speed. The data can be searched from a spread sheet in the database and GIS maps vice versa. When a name of pond is clicked in a spread sheet, a geographic map near the pond is drawn quickly in the center of a window. On the other hand, a radio button beside a name of pond shown in a map is clicked, the data related to damage of the pond is shown in a table with photographs and figures of damage, as shown in later. It is very useful to superpose the damage data of ponds with geographic, geological maps, land use maps and geotechnical data as well as aerial photographs and figures.

Distribution Characteristics of Damaged Ponds

1 Distribution Characteristics from the Epicenter

First, the distribution characteristic is examined using distances between locations of damaged ponds and the epicenter. Figure 1 shows the locations of damaged ponds on a GIS map with distances from the epicenter. For superposing them on a GIS map, north latitudes and east longitudes of damaged ponds are converted to numerical coordinates using in the Japanese map system. The following results can be found from this figure and the GIS database:

1) About three quarters of all damaged ponds concentrate in northern part of Awaji Island.

2) About 10 % of damage distributes within 10 km, about 45 % within 20 km, and about 85 % within 30 km.

However, it is not so clear that this distribution directly relates to the distances from the epicenter. Because the 1995 Hyogoken-Nambu earthquake was a right-lateral strike-slip type earthquake caused by shallow active faults.

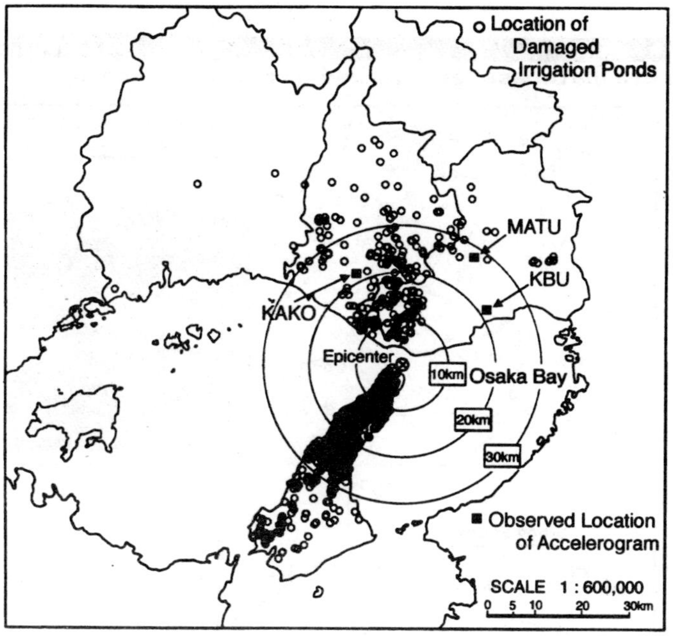

Fig. 1 Location of damaged ponds with distance from the epicenter

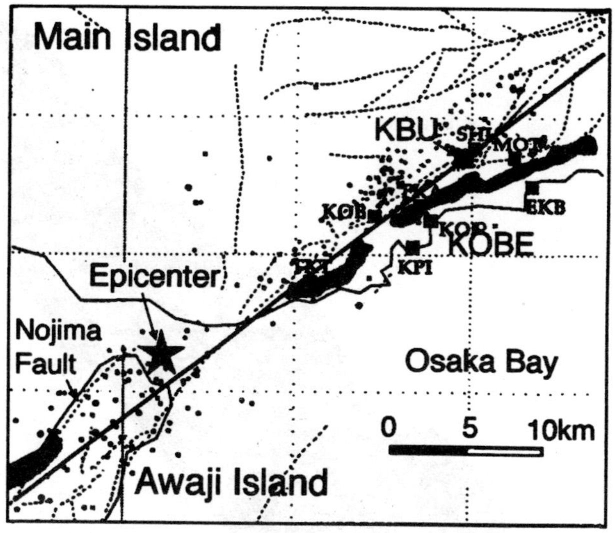

Fig. 2 Location of the fault line induced by the 1995 earthquake (Irikura,1996)

2 Distribution Characteristic from the Earthquake Induced Fault

Then, the distribution characteristic is discussed using closest distance between locations of the ponds and the fault line induced earthquake. In Figure 2, the fault line induced earthquake is modeled by a line passing two points; 34°30'19"N and 134°53'59"E; 34°46'02"N and 135°17'32"E. This line is evaluated from the result of strong earthquake analysis by Irikura (1996) and the distribution of after shocks occurred in January 17, 1995 (Katao and Ando, 1996). A relationship between numbers of damaged ponds and distance from the fault line induced earthquake is shown by bars in Figure 3(a), and by polygonal lines in Fig. 3(b), located in Awaji and Main Islands respectively. The following results can be obtained from these figures:

1) The distributions of the damaged ponds are fairly different in Awaji and Main Islands.

2) Almost of all damaged ponds in Awaji Island distribute within 10 km and their numbers decrease in logarithm with the distances from the fault line, which is the same trend as damping property of earthquake energy.

3) On the other hand, over one hundred of damaged ponds are located in 15 to 25 km distances from the fault in Main Island.

It is shown in Figs. 1 and 2 that major damaged urban areas distribute in southern east parts from the fault line, and that the damaged ponds distribute in northern west parts in Main Island. Thus, it is very important for investigation of the earthquake properties to compare the distribution characteristics between disasters in urban areas and the one of damaged ponds. Iwasaki et al. (1996) obtained the following conclusions for the relationship between wooden house damage and the distance from the fault line in urban areas:

1) Damaged wooden houses concentrate within about 10 km, and decrease to some percentages at further distances.

2) The damage-rate of damaged houses decrease linearly with distance from the fault line.

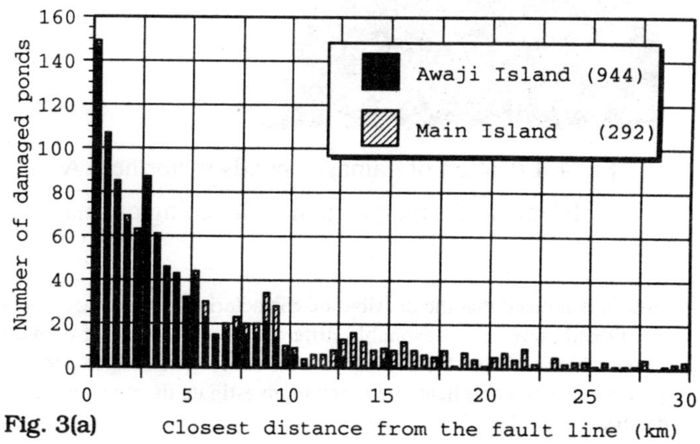

Fig. 3(a)

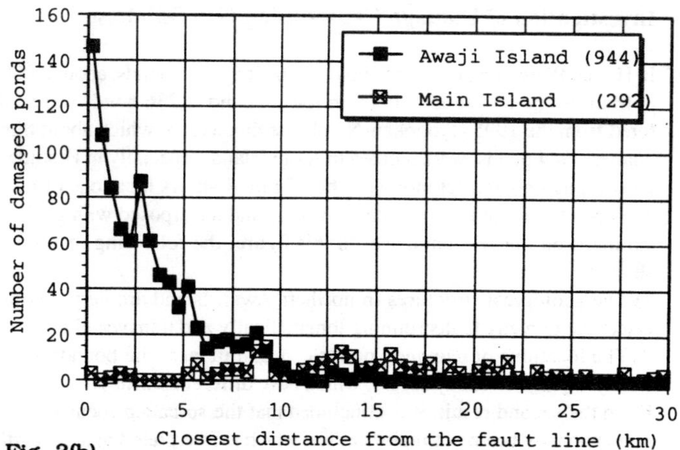

Fig. 3(b)

Fig.3 Distribution characteristic of closest distance from the fault line

Fig.4 Location of damaged ponds in northern Awaji Island and superposition with geological map

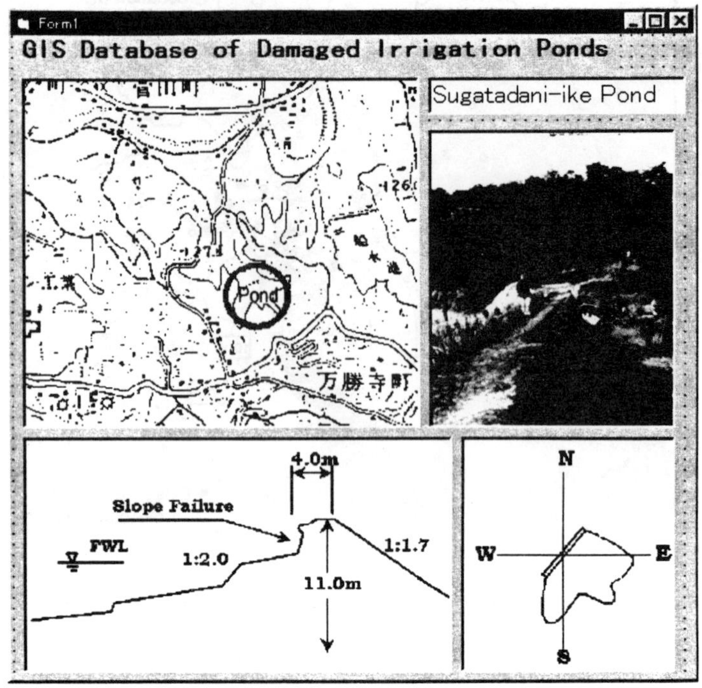

Fig.5 Display example of the GIS database

Now, it is noticed that the distribution characteristics of damaged ponds mentioned above are reasonably different from the ones of damaged wooden houses. From the comparison, the causes of damaged ponds are estimated to be complicated enough to investigate them using the GIS database.

Investigation of Damage Causes using GIS Database

In Hyogo Prefecture, there are about 51,000 irrigation ponds, about 27,000 in Main Island and about 24,000 in Awaji Island. 1,236 ponds were suffered from the 1995 Hyogoken-Nambu earthquake, of which about three quarters, 944 ponds, were located in Awaji Island, especially in Hokudan-cho, Tsuna-cho and Ichinomiya-cho. Figure 4 shows locations of damaged ponds and faults in northern Awaji Island superposed with geological maps on the GIS database. In this figure, the following results are shown:
1) The geological structures in northern Awaji Island are very complicated due to many faults running longitudinally and transversely.
2) The locations of damaged ponds concentrate near the boundaries of clayey layers and sandy layers, which have different stiffness.
From the second result, it is concluded that the so-called focusing phenomenon may be one of major causes, which is interpreted that the earthquake wave amplified in clayey layers focuses with the one refracted in stiff neighboring layers. The same cause may be appropriate for damaged ponds located in Main Island. The characteristic damage mode in Main Island is the failure of upper slope as shown in Figure 5. But to find causes of this failure, as well as analysis using the GIS database, numerical analysis should be conducted using geotechnical data of respective ponds and acceleration wave data.

PRELIMINARY STUDY OF ACCELEROGRAMS FOR SEISMIC RESPONSE ANALYSIS

Measuring stations of accelerograms are located in urban areas, but rarely located very close to the locations of damaged ponds. Therefore, accelerograms measured at the closest station are often used. Fortunately, accelerograms were obtained at the two stations close to the fault, MATU and KAKO, shown in Fig. 1. These two stations are located in about 11.46 and 25.14 km distant from the fault line induced earthquake. Figure 6 shows these accelerograms in three orthogonal directions to compare the ones measured at Kobe University, shown as KBU in Fig. 1, located almost on the fault line (0.65 km) as shown in Fig. 1. Comparing these accelerograms, it is shown that these waves have some similar shapes of pulse type waves and that acceleration amplitudes damp in logarithm with distance of propagation. Furthermore, Fourier spectrum analyses of the data are conducted to compare these frequency properties, as shown in Figure 7. From the figure, the followings can be found:
1) The frequency properties of the EW-component for KBU and MATU are almost the same in the range of 0.1 to 2 Hz, because the two are located in the N-S direction, as shown in Fig. 1.
2) The peak frequencies for the three components measured at the three locations are very similar with different amplitudes.
3) The predominant frequency of all three components is almost 0.8 Hz.
4) The KBU wave seems to have the inherent frequency of this fault earthquake, and the other waves include higher frequency components due to repeating reflection and interference during propagation.
For seismic response analysis of damaged ponds, accelerograms measured most close to each pond will be used with modified the maximum amplitude according to the distance from the fault line.

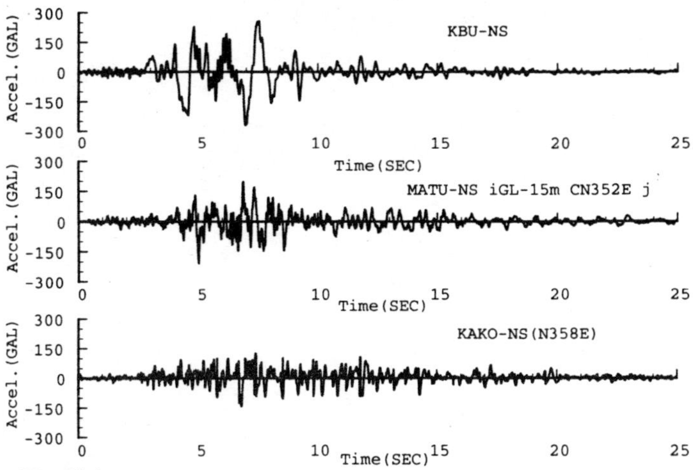

Fig.6(a)

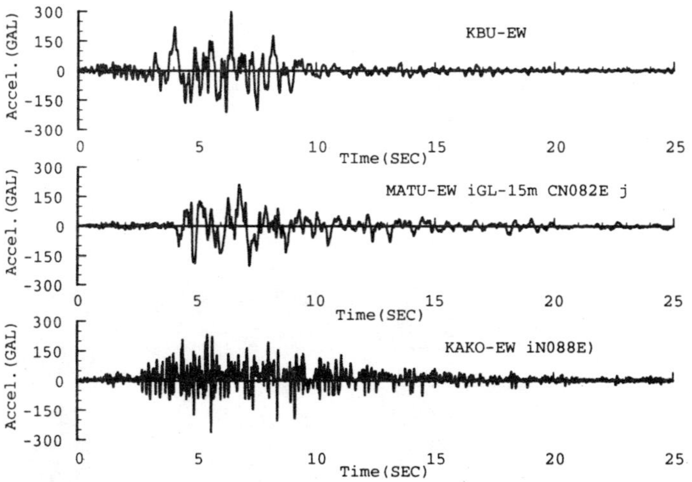

Fig.6(b)

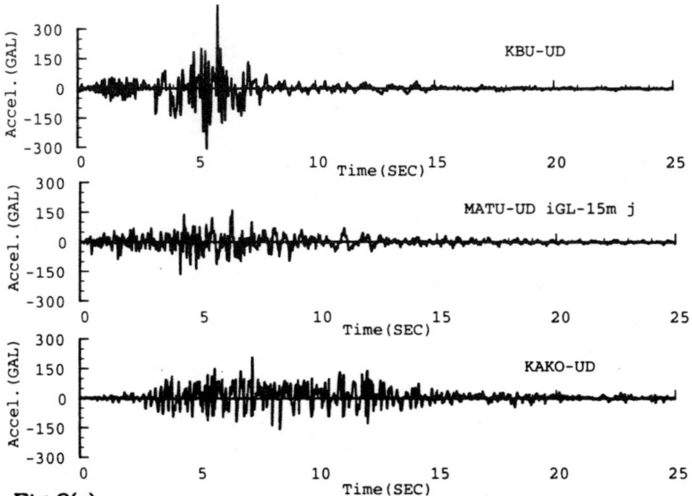

Fig.6(c)

Fig. 6 Accerelograms of NS, EW and UD components measured at KBU, MATU and KAKO

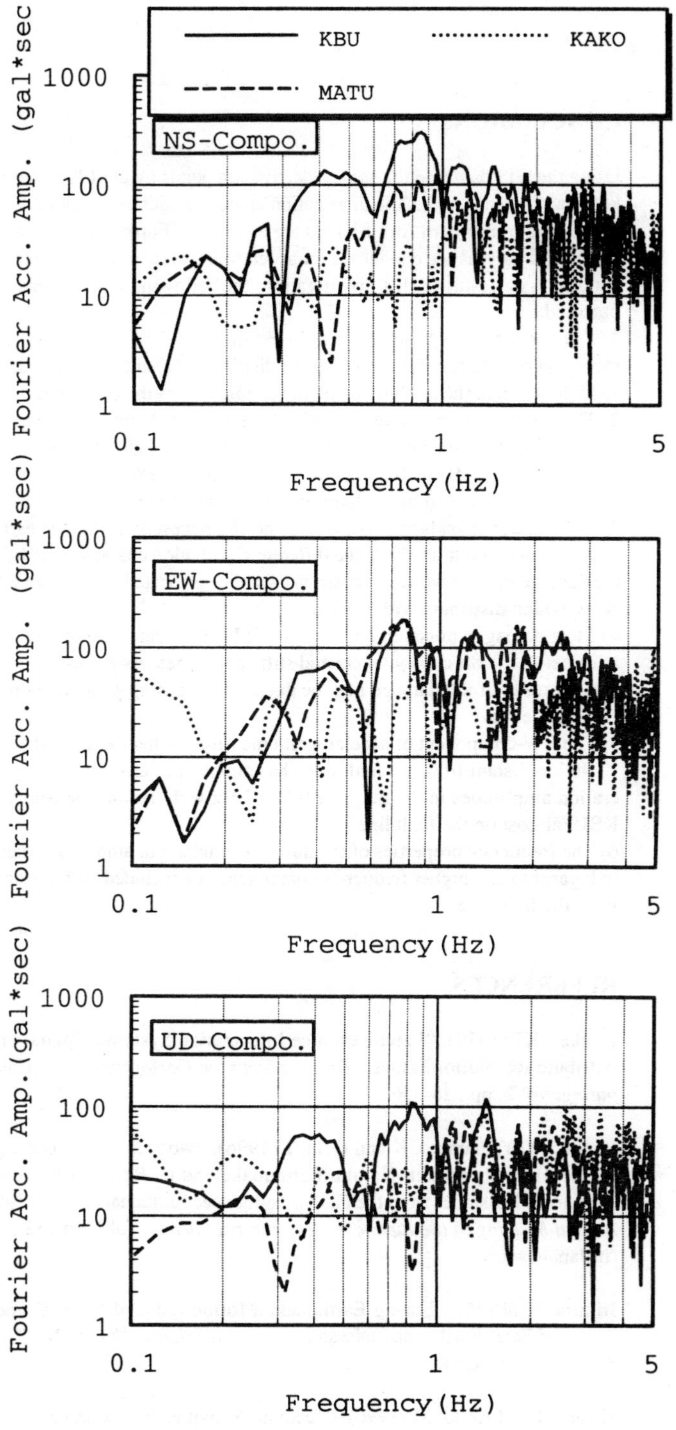

Fig. 7 Fourier spectrum of accelerograms measured at KBU, MATU and KAKO

CONCLUSIONS

Using the GIS database for damaged irrigation ponds by the 1995 Hyogo-ken-Nambu earthquake, the distribution characteristics and major causes of damage in northern Awaji Island are examined. Furthermoıe, Fourier spectrum analyses are conducted for accelerograms measured at three locations. As a result of this investigation, the following conclusions are obtained:

1) Damage analysis based on the closest distance from the fault line is more appropriate for understanding the distribution characteristic of damaged ponds than the analysis using the distance from the epicenter.

2) The distribution characteristic of damaged ponds located in Awaji Island is different from the one in Main Island. Almost of all damaged ponds distribute within 10 km, but over one hundred of the latter ponds locate in 15 to 25 km distant from the fault line, as shown in Fig. 3.

3) The damaged ponds mainly locate in northern west part from the earthquake induced fault line have the different distribution characteristic from the damage rate for wooden houses in urban areas located in southern east parts, which distribute within 10 km.

4) Most damaged ponds in northern Awaji Island locate near the boundaries between soft clayey layers and stiff sandy ones. As a result, one of major causes of the damage may be due to the focusing phenomenon of earthquake wave.

5) The EW-component of accelerogram measured at the location MATU, 11.46 km distant from the fault line, has almost the same Fourier acceleration amplitudes in the range of 0.1 to 2 Hz with the one measured at KBU, almost on the fault line.

6) The frequency properties of the three are similar, but amplitudes damp in logarithm and higher frequency components are included with distance from the fault line.

REFERENCES

Deuker, KJ (1979). "Land Resource Information Systems: Spatial and Attribute Resolution Issues," *Proc Int Symp on Cartography and Computing*, Vol 2, pp 328-336.

Iwasaki, Y, Yamamoto, K and Ochi, T (1996). "Wooden House Damage by the 1995 Hyogoken-Nambu Earthquake and its Relationship with Ground and Distance from the Fault Line Induced Earthquake," *Proc 1996 Annual Meeting of the Japanese Geotechnical Society*, Vol 1, pp 253-254 (in Japanese).

Irikura, K (1996). "Strong Earthquake Motion Induced "Belt-Shaped Zones of Fatal Earthquake Disaster," *Iwanami Science*, Vol 66, No 2, pp 86-92 (in Japanese).

Katao, H and Ando, M (1996). "Tectonic Activities before and after the 1995 Hyogoken-Nambu Earthquake," *Iwanami Science*, Vol 66, No 2, pp 78-85 (in Japanese).

Matsui, T et al (Research Committee on Geotechnical Information of Osaka Bay Seabed Deposits) (1995). "CD-ROM for Geotechnical Information of Seabed Deposits and its Interpretation," JSSMF Kansai Branch (in Japanese).

Proceedings of the Seventh (1997) International Offshore and Polar Engineering Conference
Honolulu, USA, May 25-30, 1997

ISBN 1-880653-28-1 (Set); ISBN 1-880653-29-X (Vol. I)

On the Role of Geologic Factors in Determining Peculiarities of Water Mass Structure in the Clarion-Clipperton Ore Field

G. Tkatchenko, R. Kotlinski and V. Stoyanova
Interoceanmetal Joint Organization
Szczecin, Poland

Dehua Lian, Guozhen Zhang and Yongyang Huang
COMRA
Beijing, China

ABSTRACT

Comprehensive geological, hydrophysical, and hydrochemical studies, carried out on reference transects in pioneer areas of IOM and COMRA have shown that the peculiarities of water mass structure and dynamics of its variability are substantially affected by geological factors such as deep-sea volcanism, hydrothermal activity, and seismicity. Those factors bring about episodic events of environmental pollution, comparable in scale with a potential anthropogenic pollution resulting from polymetallic nodule mining. This problem calls for a long-term monitoring of reference transects in pioneer areas, as a basis for investigating spatio-temporal variability of the marine environment, so that it can be protected from unacceptable changes brought about by the contractors' activities.

KEY WORDS: Interoceanmetal, COMRA, environmental reference transect, Benthic Impact Experiment

INTRODUCTION

Within 1993 - 1996, IOM and COMRA were involved in carrying out comprehensive geological and oceanographic surveys, according to the schedule agreed upon within the framework of a cooperative program, along their respective reference transects located within the pioneer areas of both organizations. Some results of those surveys were already discussed in earlier publications (Kotlinski et al., 1996; Tkatchenko et al., 1996) which dealt with variability of water masses in relation to geological processes reflecting interactions between the litho- and hydrosphere. Those interactions are particularly important in view of the fact that submarine volcanism and hydrothermal activity as well as contemporary tectonic processes related to movements of oceanic crust plates within the Clarion-Clipperton ore field and seismicity in the Clarion and Clipperton fractures underlie the unpredictable natural contamination of the oceanic water. The magnitude of key parameters involved, including heavy metals, nutrients, and suspended particulates, can be comparable to the anticipated anthropogenic pollution resulting from commercial polymetallic nodule mining.

The present work deals with this problem in view of the fact that IOM, while investigating its reference transect, obtained new data on distribution of heavy metals in the water column and thus gained an opportunity to compare those with similar data collected by COMRA from their reference transect. The IOM and COMRA transect coordinates are 9°45'N; 119°25'W to 12°26'N; 120°22'W and 7°22'N;144°22'W to 9°45'N;144°21'W, respectively.

RESULTS AND DISCUSSION

By the way of example, Figs 1 and 2 show distributions of some heavy metals in the water column along the IOM and COMRA transects. A comparison of the distributions with geological profiles reveals a clear spatial correlation of the anomalies found with certain geological objects, i.e., large zones of fracture faults and volcanic structures.

With respect to its geomorphology and geology, the IOM transect shows the presence of four large blocks differing one from the other in their hypsometry, sea floor relief morphology, and sedimentary cover thickness. The largest zone of fracture faults is observed around CTD stations 6 and 7 (fig.1) ; a large-scale discharge of mineralized water is observed along this zone, resulting in anomalous vertical distribution of nickel and zinc throughout the entire water column, from the bottom to the water surface; additi onally, an anomaly in vertical distribution of manganese and copper, extending almost throughout the water column, can be seen as well. When viewed horizontally, this zone of fracture faults seems to produce a front separating, within the transect studied, two water masses: northern and southern, differing one from the other in their physical features (Tkatchenko et al., 1996) and, as demonstrated by this study, also in their chemical characteristics.

The water mass in the northern part of the transect contains slightly elevated background concentrations of iron, copper, and zinc (Fig.1), while slightly elevated concentrations of manganese and nickel are observed in the southern part. In the latter, intensive discharges of mineralized water are recorded in the near-bottom water mass, accompanied by anomalies in distribution of manganese and nickel. As this zone also contains the commercially most productive planar deposits of polymetallic nodules, their presence there can be taken as an evidence of their hydrogenic origin.

The area of the largest volcanic structures (CTD stations 13 and 14) reveals substantial, in terms of the space they occupy, anomalies in distribution of nickel, zinc, and manganese, the latter being less extensive. The likely association of those anomalies with hydrothermal activity, including the post-volcanic phenomena, is quite clear. Unfortunately, distribution of nutrients along the IOM transect will be studied in 1997 at the earliest.

As shown in Fig. 1, distribution of suspended particulate matter reveals elevated particulate concentrations in the northern part of the transect where volcanic structures are located. The concentrations found are by an order of magnitude higher than those reported from the DOMES area (Ozturgut et al., 1978). During studies accompanying the IOM BIE involving the use of a benthic disturber (Tkatchenko, Radziejewska, 1996), resuspended

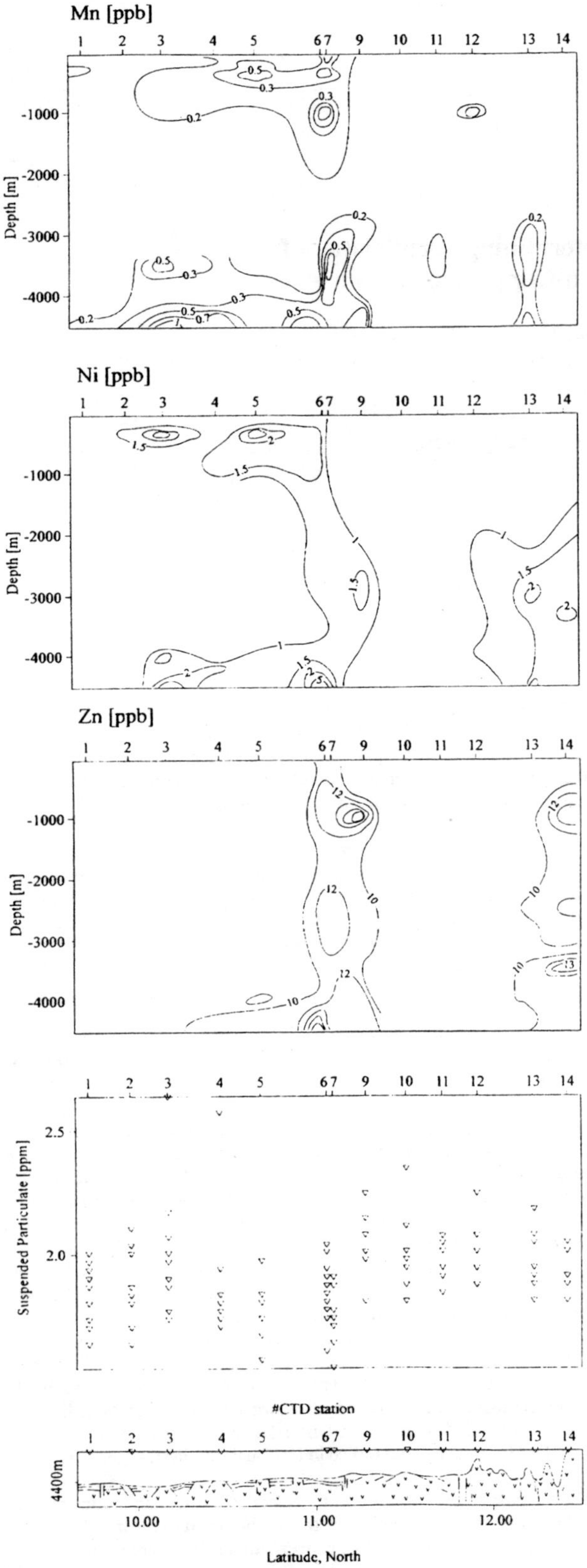

Fig.1 Distribution of Mn, Ni, Zn and suspended particulate matter as a function of depth and schematic cross-section through the IOM reference transect

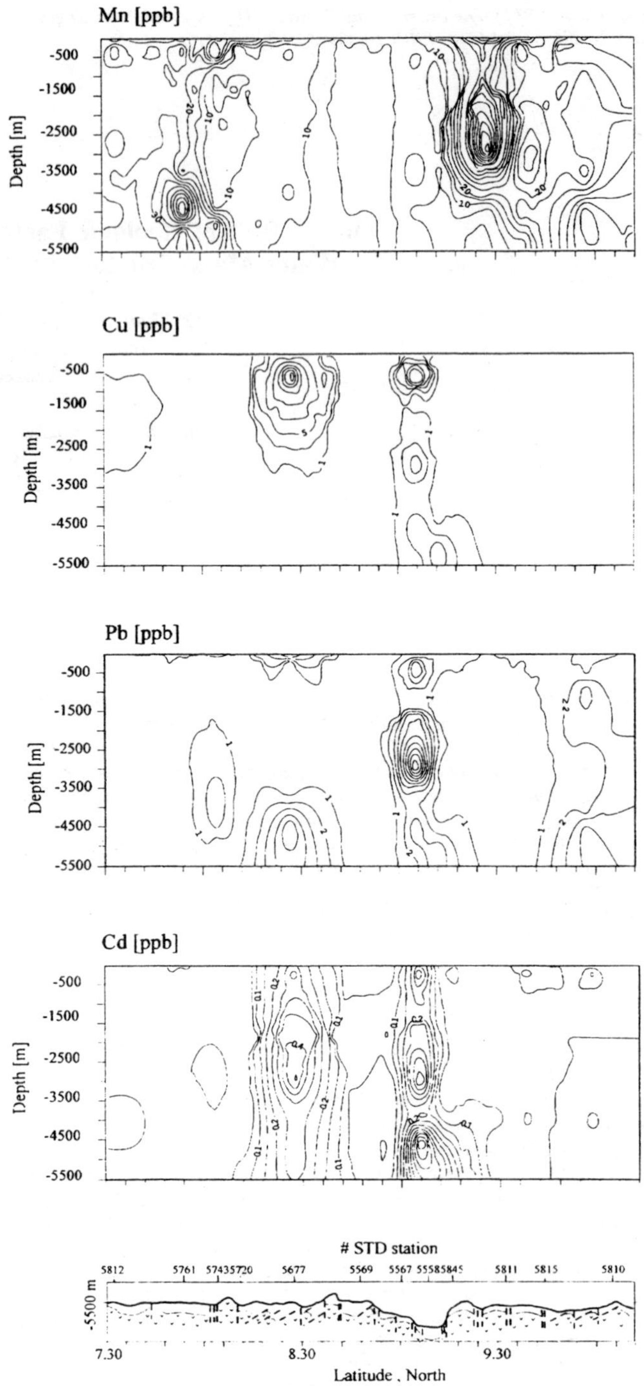

Fig.2 Distribution of Mn, Cu, Pb, Cd as a function of depth and schematic cross-section through the COMRA

sediment was collected with a sediment trap deployed in the vicinity of CTD station 6. Examination of the trap content with the electron microscope revealed a wide range of barite and aragonite grains as well as the presence of chalcopyrite (Photo), which confirms the presence of hydrothermal activity in the area.

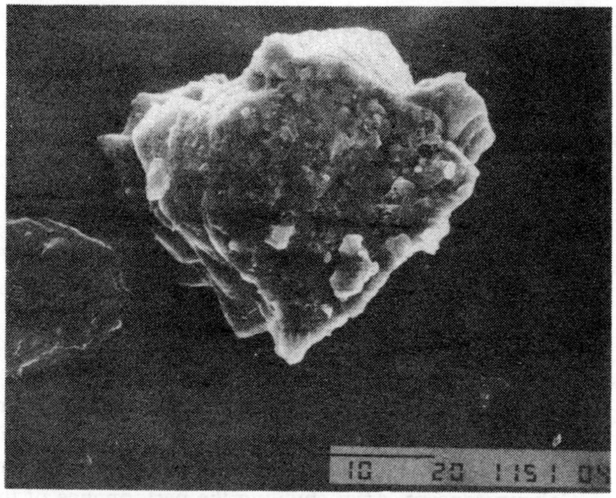

Photo. A chalcopyrite grain (x 2000) in resuspended sediment sample from a sediment trap deployed in the IOM BIE test site

Data on nutrients and heavy metals along the COMRA transect demonstrate a still more important role of active volcanism and fracture faults in controlling the distribution of hydrochemical anomalies (Tkatchenko et al., 1996). The volcanic activity around CTD stations 5743 and 5720 is corroborated by pronounced anomalies in distribution of phosphates and nitrates and manganese. The across-water-column anomalies in copper and lead (Fig. 2) as well as in cadmium are observed to be similar in their location to those recorded from the IOM transect, that is they are to be found above a zone of large fracture seen as a fault with graben. The bathymetric chart produced by the Seabeam shows that the section of the COMRA transect between CTD stations 5720 and 5569 coincides with a zone of submeridional fracture, reflected in the relief as a linear valley. This zone is associated (Fig. 2) with anomalies in distribution of lead, observed in the near-bottom, lower deep, and surface water masses; additionally, an anomaly in the distribution of copper can be seen in the upper deep, intermediate, and surface water masses. Moreover, a pronounced cadmium anomaly extending throughout the entire water column is present here as well. Discontinuities in the observed anomalies along the COMRA and IOM transect are an evidence of a cyclic nature of post-volcanic and hydrothermal activity and their different geochemical specialization.

To sum up, the authors feel obliged to stress the importance of international cooperation between all the pioneer investors in investigating the background and anomalous hydrochemical factors, whereby all sources of natural contamination of the water can be identified and described. Such studies should be aimed at obtaining data from which to derive a unified package of key oceanographic and ecological characteristics of the giant Clarion-Clipperton ore field.

ACKNOWLEDGMENTS

The authors thank Dr. Tereza Radziejewska for her help with the English translation and Mr A. Krawcewich for his assistance during the final computer preparation of the manuscript.

REFERENCES

Kotlinski, R, Stoyanova, V, Tkatchenko, G (1996) " Environmental Studies on a Reference Transect in th e IOM Pioneer Area", *Proc 6th Int Offshore Polar Engin Conf*, ISOPE, Los Angeles, Vol 1, pp 54-57.

Ozturgut, E, Anderson, GC, Burns, RE, Lavelle, JW, and Swift, SA (1978) "Deep Ocean Mining of Manganese Nodules in the North Pacific: Pre-mining Environmental Conditions and Anticipated Mining Effects", NOAA Tech. Memorandum ERL MESA-33, Environmental Research Laboratories, NOAA, Boulder, Co., 133 pp.

Tkatchenko, G , Dehua, L, Kotlinski, R, Guozhen, Z, Stoyanova, V, Yongyang, H (1996) "A Comparative Analysis of Results Provided by Comprehensive Studies on Reference Environmental Transects in Pioneer Areas of COMRA and IOM", *Proc Int Seminar on Deep* Sea-Bed Mining Technology, Beijing, Vol 2, pp C1-C11.

Tkatchenko, G , Radziejewska, T, Stoyanova, V, Modlitba, I, Parizek, A (1996) "Benthic Impact Experiment in the IOM Pioneer Area: Testing for Effects of Deep Seabed Disturbance", *Proc Int Seminar on Deep Sea-Bed Mining Technology*, Beijing, Vol 2, pp C55-C68.

Proceedings of the Seventh (1997) International Offshore and Polar Engineering Conference
Honolulu, USA, May 25-30, 1997

ISBN 1-880653-28-1 (Set); ISBN 1-880653-29-X (Vol. I)

Performance of a Wells Type Turbine Coupled to an Induction Generator for Wave Energy Applications

Ardhendu G. Pathak, V. Jayashankar and M. Ravindran
National Institute of Ocean Technology
Madras, India

ABSTRACT

This paper briefly discusses the recent progress related to power module performance in the Indian Wave Energy program. The prototype plant, located on the west coast of India, works on the principle Oscillating Water Column (OWC). A concrete caisson with two harbor walls in front serves as the OWC chamber. The power module consists of a self rectifying Wells type turbine connected to an induction generator. It is shown that the aerodynamic absorption characteristics of the turbine improves if the allowable slip of the induction generator is increased. This is demonstrated by computer simulations of the system for randomly varying aerodynamic input. Typical plots are presented and the physics behind the phenomenon is also discussed. The increase in slip in the prototype can be achieved by adding external rotor resistances in series with the rotor windings. The implications of this for the operation of the plant are also discussed. The conclusions are supported by analysis of the data obtained from the prototype wave power plant.

KEY WORDS
Wave Energy, Wave Power, Wells Turbine, Induction Generator, Simulations, Prototype, India

INTRODUCTION

The wave energy research in India started in 1983 with the formation of an inter-disciplinary Wave Energy Group at Indian Institute of Technology, Madras under the sponsorship of Department of Ocean Development, Government of India. After conducting detailed laboratory investigations, a prototype demonstration plant of peak capacity 150 kW was designed, constructed and commissioned at Vizhinjam, Trivandrum near the southern tip of Indian Coast in October 1991. The plant works on the Oscillating Water Column concept. After studying the performance of the plant, the power module was redesigned (Ravindran , 1995). The new power module of 55kW peak rating was commissioned in March 1996. The performance monitoring data of this module is being analyzed by the National Institute of Ocean Technology. The tests on the present power module will conclude by December 1996. The Prototype The power module is mounted atop a concrete caisson located at the end of a breakwater in the Vizhinjam harbor on West coast of India. The caisson sits on a prepared bed in 10m of water depth. The caisson top and the breakwater are connected by 45 m long bridge. The power module consists of a Wells type self rectifying air turbine with tapered chord blades, coupled to an induction generator. The generator is connected to the grid.

Specifications of the Operational Wave Energy Plant at Vizhinjam, India

a) OWC caisson dimensions
23.2m X 17m X 15.3 m

b) Wells Turbine
i)Configuration: Twin turbines coupled to a common electrical generator
ii)Axis
Horizontal
iii)Rating (Peak Power of each turbine)
38 kW (Total 76 kW)
iv)Speed at peak power
1800 RPM
v)Efficiency at peak power
0.6

c) Electrical Generator
i)Type
Slip-ring induction machine
ii) Rated speed
1483 RPM
iii) Power at rated speed
55 kW
iv) Voltage/ frequency
440 V three phase/ 50 Hz.

INSTRUMENTATION

Eleven parameters viz. wave height, stator current, electrical power, torque of two turbines, speed, differential pressure across the turbine, OWC pressure and slip of the generator are monitored through a PC based data acquisition system.

SIMULATIONS

The equation of motion for the turbine + generator system can be written as

$$I.\frac{\partial\omega}{\partial t} = T_t - T_g - T_l \qquad (1)$$

where

I	=	Moment of inertia
ω	=	Angular velocity of rotor
T_t	=	Torque produced by turbine
T_g	=	Torque imposed by generator
T_l	=	Torque equivalent of losses

The input to the turbine is the pulsating pressure drop across the turbine rotor due to the air flow from the OWC chamber. An equivalent circuit is used to model the generator. The unbalanced torque at any instant produces angular acceleration of the rotor. The details of calculations are contained in Koola (1995). The output of the model are instantaneous mechanical and electrical powers and angular velocity of the rotor.

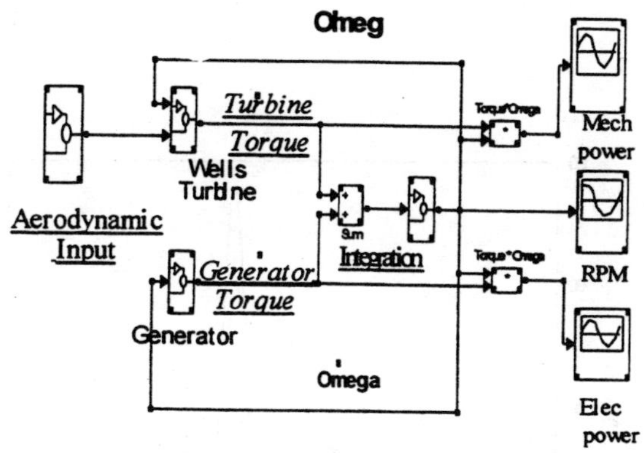

Fig. 1 SIMULINK model of Wells turbine and induction generator

Figure 1 shows a simplified SIMULINK™ model of the system. The blocks labelled Wells Turbine and Generator are subsystems comprising many sub-blocks which model turbine and generator behaviour. This model was used to check the effect of variation of various parameters on the system performance. Of particular interest is the effect of changing the allowable slip of the generator on the energy extraction by the turbine. To understand how the allowable slip of the generator affects the mechanical power delivered by the turbine it is necessary to consider the aerodynamics of the turbine.

The maximum power that can be extracted by the turbine is limited by its stalling behavior. A turbine blade stalls when the relative angle between the tangential speed and the axial velocity exceed about 14°. Thus as the axial velocity of the input air flow goes up, unless the angular speed of rotation of the turbine also goes up, the stalling angle is quickly reached. In other words, stalling can be avoided or delayed, if the turbine accelerates fast enough in response to the incoming air flow. Mechanically this can be achieved by having the inertia of the entire system as low as possible. The second alternative is to increase the allowable slip of the generator, allowing the system to reach higher speeds. The lower limit of the system moment of inertia is dictated by design consideration, and hence is considered fixed. The second, electrical, alternative is easy to implement. This can achieved by adding external resistances in series with the rotor windings of the generator. The upper limit on the allowable slip is determined by the maximum safe speed of operation. For the present module this limit is 2000 RPM.

RESULTS AND DISCUSSION

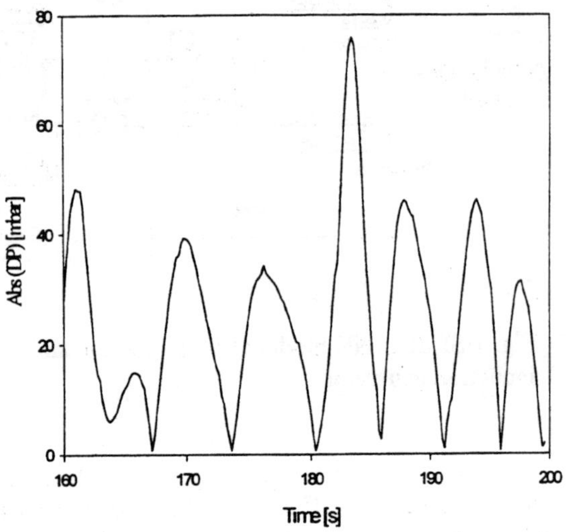

Fig. 2 Variation of absolute pressure drop across the turbine

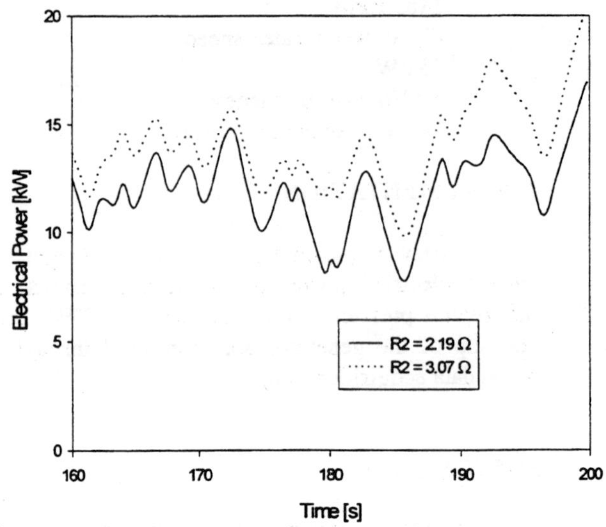

Fig. 3. Predicted variation of electrical power for two values of external rotor resistances

Figure 2 shows an example of a time series of the pressure drop across the turbine that is used as an input to the model. The model is first run assuming that a resistance R2 = 2.19 Ω is connected in series with the rotor resistances. Then with the same input, the model is run assuming that a resistance of value R2 = 3.07 Ω is connected.

Figure 3 shows the electrical output expected from the module and Figure 4 shows the instantaneous speed variation. It is clear from Figure 3 that added resistances in series with the rotor windings do in fact lead to increase in net output from the module. A number of simulations under varying levels of input confirmed this conclusion. However, this net increase in the power is obtained at the cost of increase in the shaft speed variation. Encouraged by these observations, the power plant was run with different external rotor resistances for varying input conditions. Figure 5 shows the average power of the power plant as a function of input pneumatic power to the turbine. Each dot on the graph represents an average over 15 minute period. The pneumatic input is inferred from the site measurements and the model characteristics of the turbine. The readings presented here cover a three month period from April to June 1996.

However, high value of external rotor resistances can cause the module to cross the safe speed limit, if there is sustained high wave activity. Also for low wave activity (off monsoon) period the I^2R losses in external resistances may compensate the higher mechanical power produced through improvement in the aerodynamic absorption. Thus it is desirable to have an intelligent controller for switching of the rotor resistances, to achieve optimum operation point at all conditions of input.

A controller that could maximize the power output from the plant based on switching of the rotor resistances was designed, fabricated and tested at the site. It requires an external resistance with a number of tappings. The controller action essentially involves the shorting of a portion of the externally connected rotor resistance depending on the speed. The resistance value is so chosen that the plant behaves as a high slip machine thus enabling the turbine to freely ride the wave. In case the speed exceeds a preset upper limit, the external resistance is reduced. The results are being analyzed and will be reported separately.

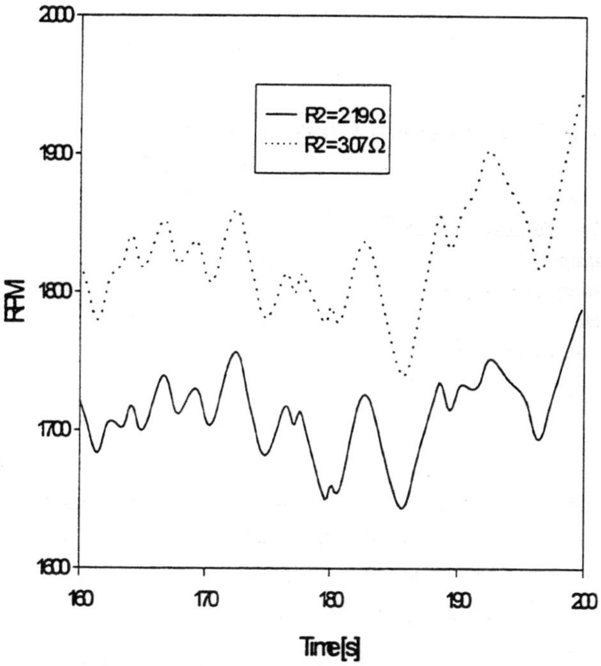

Fig. 4. Predicted variation of shaft speed for two values of external rotor resistances

SUMMARY

This paper presented a brief outline of the recent progress of the Indian Wave Energy program related to the performance of the power module. The performance of the Wells type self rectifying air turbine connected to an induction generator was discussed with the help of computer simulations. The simulations show that increase in allowable slip of the electrical generator by adding external rotor resistances leads to improvements in the aerodynamic absorption capabilities of the turbine. This conclusion is confirmed by analysis of the data collected from the prototype wave energy plant. Based on the experience gained from simulations and the operation of the plant, a rotor resistance controller is suggested to further improve the performance of the plant.

ACKNOWLEDGMENT

The wave energy program is funded entirely by the Department of Ocean Development, Govt. of India. Dr. V. Jagadeeshkumar and Dr. P.M. Koola of IIT Madras were responsible for instrumentation system of the wave energy plant. The authors also wish to thank the site staff of the wave energy project.

REFERENCES

[1] Koola P.M., Pathak A.G. and Ravindran M., " Turbo-Generator Simulation Studies for Enhanced Wave Power Conversion", Paper presented at the Second European Wave Power Conference, 8-10 Nov. 1995, Lisbon, Portugal

[2] Ravindran M., Pathak A.G., Koola P.M. and Latha G., " Indian Wave Energy Program : Progress and Future Plans", Paper presented at the Second European Wave Power Conference, 8-10 Nov. 1995, Lisbon, Portugal

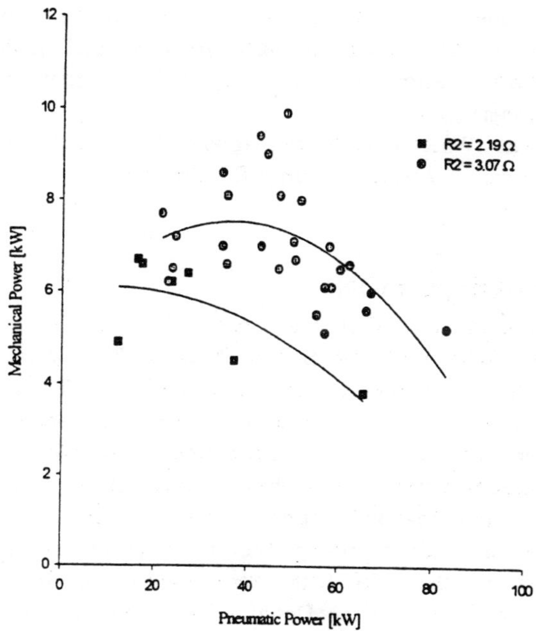

Fig. 5. Measured mean mechanical power output of the turbine as a function of pneumatic input to the turbines for two values of external rotor resistances

Proceedings of the Seventh (1997) International Offshore and Polar Engineering Conference
Honolulu, USA, May 25-30, 1997

ISBN 1-880653-28-1 (Set); ISBN 1-880653-29-X (Vol. I)

A Study of Sedimentation Processes in Seamangeum Coastal Area

Moon Seup Shin
Kunsan National University, Kunsan, Korea
Tetsuo Yanagi
Ehime University, Matsuyama, Japan
Sung Kun Hong
Kunsan National University, Kunsan, Korea

ABSTRACT

The purpose of this study is to find sedimentation patterns variation in Seamangeum coastal sea region. Water circulations are calculated diagnostically from the observed water temperature and salinity data and wind data and tidal residual current. Three dimensional movements of injected particles due to currents, turbulence and sinking velocity are tracked by the Euler-lagrange methoed.Calculated sedimentation patterns of riverine material are highly similar to the observed ones.

Key Words :Water circulations,Diagnostic,Water temperature, Salinity,Tidal residual current,Euler-lagrange methoed, Sedimentation

1. INTRODUCTION

The coastal region of Saemangeum(Kunsan) is located between 126°10′ E~126°50′ E and 35°35′ N~36°05′ N at the western coast of the Korean Peninsula(Fig.1). There are many small islands including extensive areas of semi-diurnally flooded and dewatered tidal flats. The Saemangeum coastal region has a range of 5.6m spring tide and the maximum tidal current speed is about 1.41m/sec in ordinary spring tide. Most of the sediments deposited on the tidal flats were transported from the Mankyung River and the Dongjin River. The soil in this area consists of silty sand with the depth of 10m to 30m. The wind in the winter season is strong from the direction of northwest.

The 33km sea dike and 40,100ha reclamation area is being constructed in the Saemangeum area. Such large-scale coastal land reclamation works will certainly influence the marine environment and ecosystem in this area. The purpose of this study is to find sedimentation patterns variation in Seamangeum coastal sea region. Three dimensional movements of injected particles due to currents,turbulence and sinking velocity are tracked by the Euler-lagrange methoed.

2. FIELD OBSERVATION

The field observation of the tide and tidal current in Saemangeum coastal region were carried out by Rural Development Corporation in Korea for about month (1994.8). Measured values of tides and currents in the field were used for the boundary conditions in the numerical calculation. The field observation points shows in Fig.1. The geometry of Saemangeum coastal region shows in Fig.2.

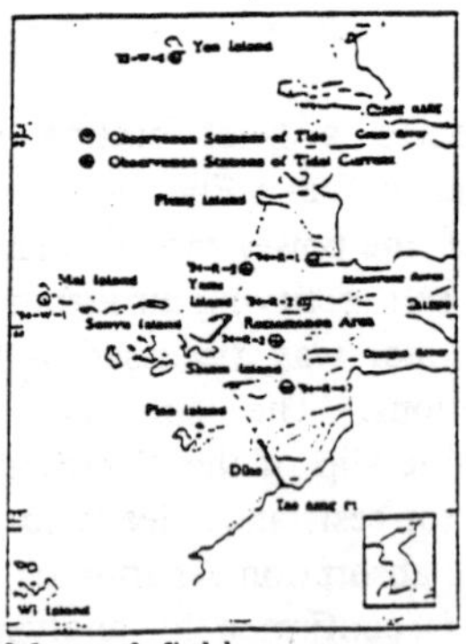

Fig.1 Map of field measurement

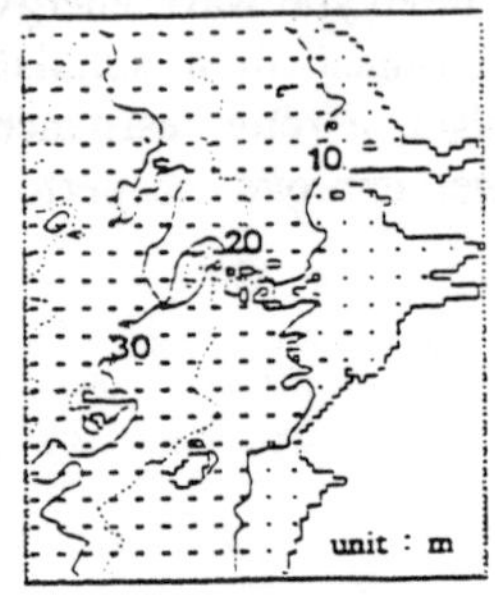

Fig.2 The geometry of Saemangeum coastal region

The tidal current ellipses of the field observation points(94-R-3,4) shows Fig.3. The tidal current stick diagrams shows Fig.4(a,b),5(a,b).The observed co-tidal (full line)and co-range(broken line) of M_2 tidal shows Fig.6.

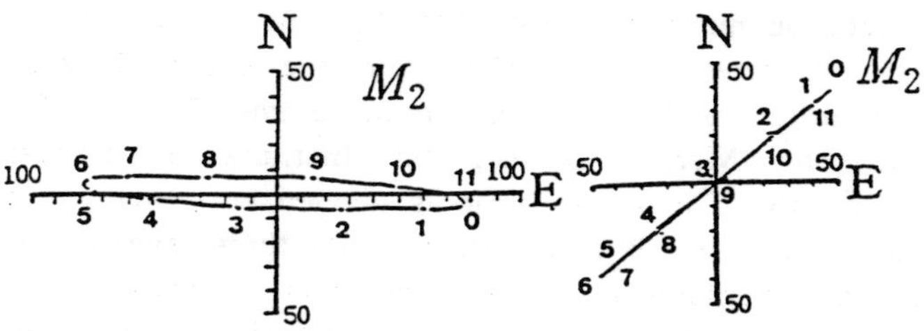

(94-R-3) (94-R-4)

Fig.3 Observed M_2 tidal current ellipse at saemangeum

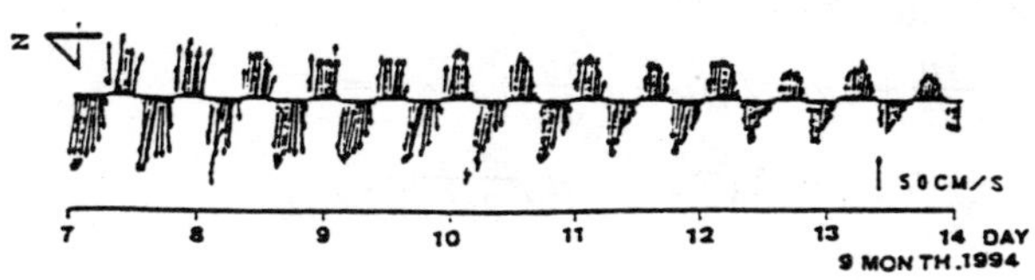

Fig.4(a) Tidal current stick diagrams at observation point(94-R-3)

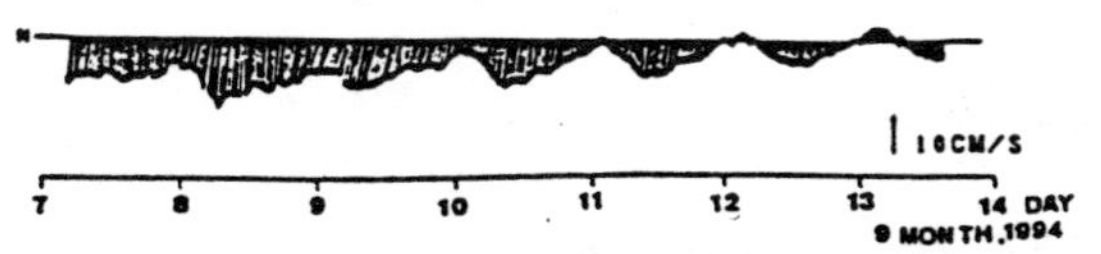

Fig.4(b) Tidal current stick diagrams of moving averaged at observation point(94-R-3)

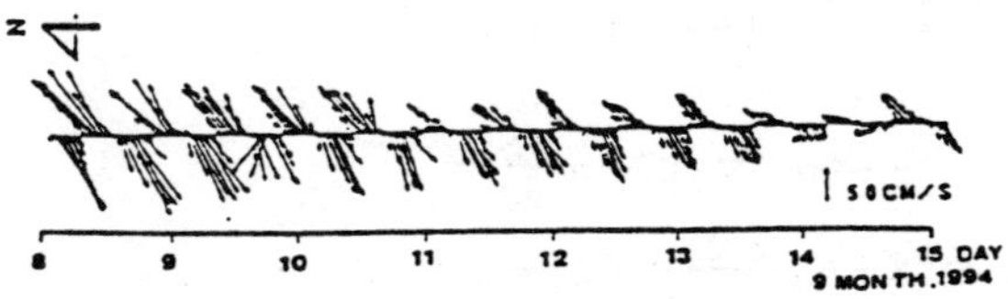

Fig.5(a) Tidal current stick diagrams at observation point(94-R-4)

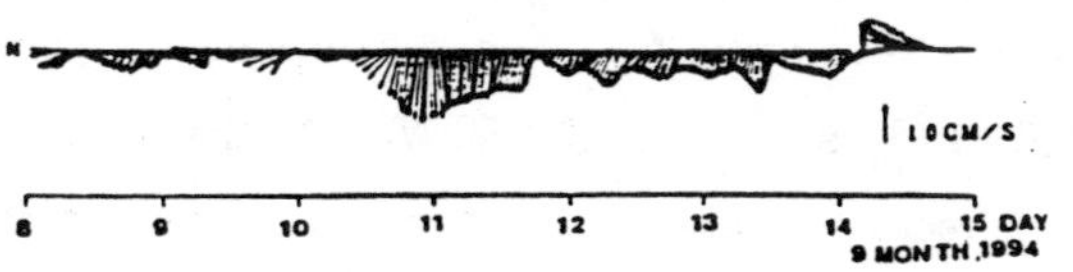

Fig.5(b) Tidal current stick diagrams of moving averaged at observation point(94-R-4)

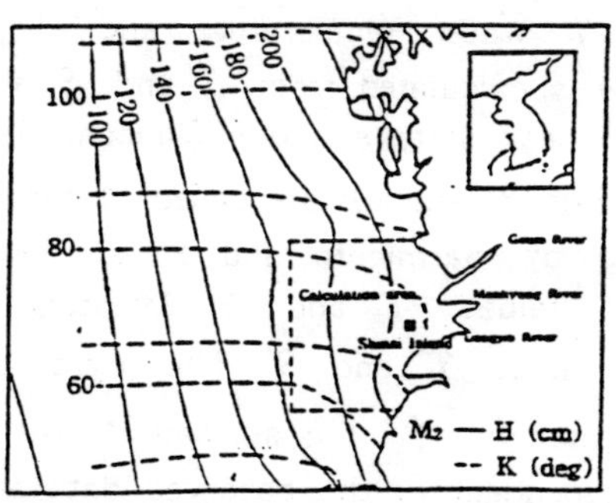

Fig.6 Observed co-tidal(full line)and co-range (broken line) of M_2 tidal.

3.NUMERICAL MODEL

3.1 Tidal current

A two-dimensional finite difference numerical model is used for the prediction of changes of tide and tidal current in the horizontal plane. The basic equations which govern the velocity field are the horizontal two-dimensional momentum equation of homogeneous fluid and continuity equation because the water is well mixed vertically in the Saemangeum area.

$$\frac{\partial u}{\partial t}+(u\cdot\nabla)u+fK\times u=-g\nabla\zeta-\frac{\gamma_b^2 u|u|}{H+\zeta}+\nu_h\nabla^2 u \qquad (1)$$

$$\frac{\partial\zeta}{\partial t}+\nabla\cdot\{(H+\zeta)u\}=0 \qquad (2)$$

where u is the vertically averaged velocity vector, ∇ is the horizontal differential operator, f is the Coriolis parameter($=8.466\times10^{-5}$ sec^{-1} *at* 35° 5′ *N*), K is the locally vertical unit vector, $g(=980cm\ s^{-2})$ is the gravitational acceralation, ζ is the sea surface elevation above the mean sea level, $\gamma_b^2(=2.6\times10^{-3})$ is the bottom frictional coefficient, H is the local depth, $\nu_h(=10^4\ cm^2\ s^{-1})$ is the coefficient of horizontal eddy viscosity. Only M_2 tide and tidal current is calculated here because it is the most dominant among the four main tidal constituents in the west sea of the Korea peninsula(Lee,1994). The CFL(Courant Friedrics Lewy) condition can be expressed as,

$$C=c\frac{\Delta t}{\Delta x}\leq 1 \qquad (3)$$

where C is Courant number, $\Delta x(=1km)$ the mesh size, Δt the time step, $c=\sqrt{gH}$ the phase velosity of long wave. Δt is decided to be 27sec from Eq.(3). The boundary condition along the open boundary is

$$\zeta=\zeta_o\sin(\frac{2\pi}{T}t-P) \qquad (4)$$

where, ζ_o, P are the amplitude and phase along the open boundary, respectively. $T(=12^h\ 25^m)$ M_2 the tidal period, t the time. The calculation area before and after the construction of dike, after the construction submerged breakwater is shown in Fig.3. respectively.

The observed harmonic constants of M_2 tide are given along the open boundary from Fig.6. The amplitudes range from 163cm to 204cm along the southern open boundary, from 167.5cm to 217.0cm along the northern open boundary, and from 163cm to 167.5cm along the western open boundary. The phases range from 54.25° to 65.45° along the southern open boundary, from 82.0° to 88.70° along the northern open boundary, and from 54.25° to 82.0° along the western open boundary as shown in Fig.6.

Eqs.(1,2) are approximated by the finite difference and are solved by the primitive method(Yanagi et al., 1983). The quasi-steady state was obtained at the fourth tidal cycle after the beginning of the calculation. The harmonic analysis of the sea surface elevation and the current field was carried out at the 5th tidal cycle.

The flow pattern of M_2 tidal current at the time of maximum flood and maximum ebb in the present Saemangeum coastal region are shown in Fig.7. The calculated co-range and co-tidal charts are shown in Fig.8. The calculated M_2 tidal current ellipses in observation stations(94-R-3,4) shows Fig.9. The calculated tide-induced residual current by M_2 tidal current shows Fig.10.

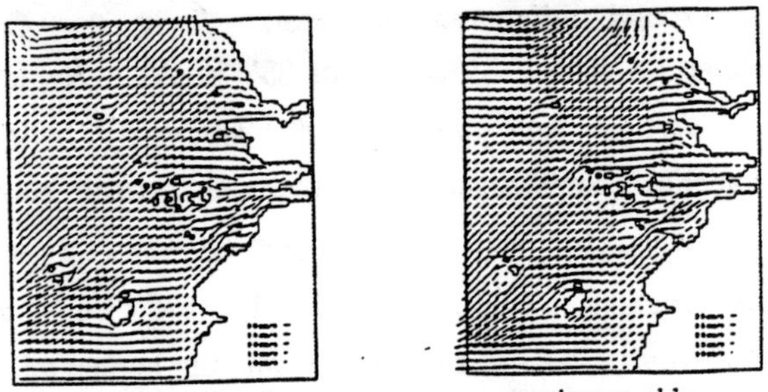

maximum flood maximum ebb

Fig.7. The flow pattern of M_2 tidal current at the time of maximum flood and maximum ebb in the present Saemangeum coastal region.

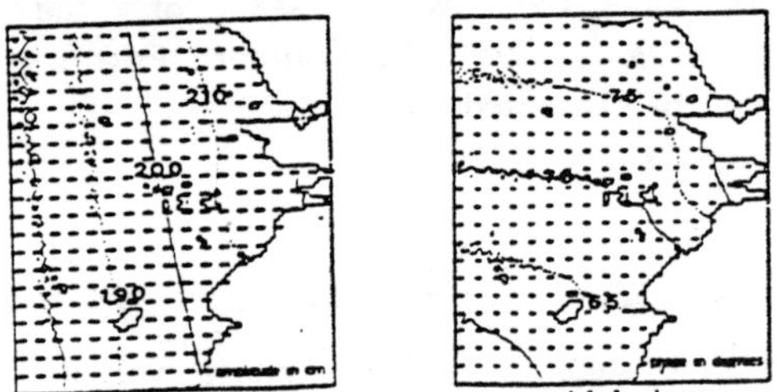

co-range charts co-tidal charts

Fig.8 Calculated co-range and co-tidal charts of M_2 tide

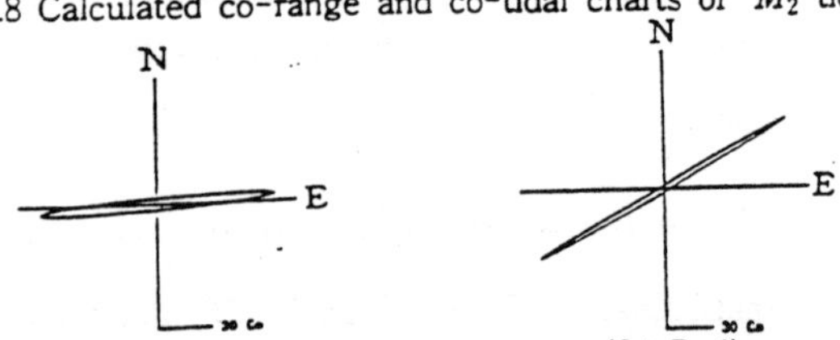

(94-R-3) (94-R-4)

Fig.9 Calculated M_2 tide current ellipse at Saemangeim

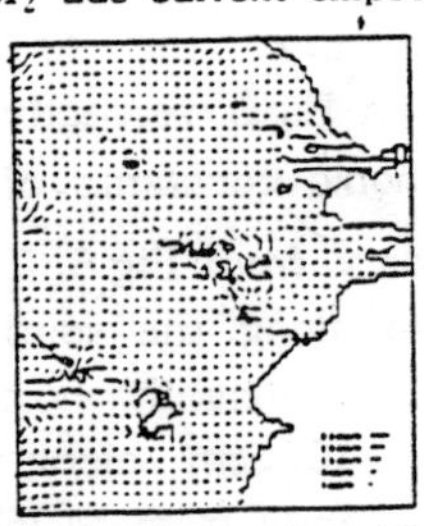

Fig.10. The calculated tide-induced residual current by M_2 tidal current.

4.VARIATION OF CIRCULATION

4.1 Numerical model

Variation of the water circulation in saemangeum coastal area is investigated with use of a robust diagnostic numerical model.water circulations in autumn are calculated diagnostically from the observed water temperature and salinity data from summer 1992 to May 1993 by Marine Development Institute in Kunsan University. The model basin is divided horizontally into 1km x 1km mesh and vertically into three layers(upper layer:0-5m,middle layer:5-10m,low layer:10-bottom).

Using conventional notation,the governing equations on the cartesian coodinate are as follows(Yanagi and Takahashi,1993):

$$\frac{\partial u}{\partial t}+(u\cdot\nabla_h)u+w\frac{\partial u}{\partial z}+fx\times u= -\frac{1}{\rho_o}\nabla_h p+A_h\nabla_h^2 u+A_v\frac{\partial^2 u}{\partial z^2}+T_s \quad (5)$$

$$\frac{\partial p}{\partial z}=-\rho g \quad (6)$$

$$\nabla_h u+\frac{\partial w}{\partial z}=0 \quad (7)$$

$$\frac{\partial T}{\partial t}+(u\cdot\nabla_h)T+w\frac{\partial T}{\partial z}= K_h\nabla_h^2 T+K_v\frac{\partial^2 T}{\partial z^2}+\gamma(T^*-T) \quad (8)$$

$$\frac{\partial S}{\partial t}+(u\cdot\nabla_h)S+w\frac{\partial S}{\partial z}= K_h\nabla_h^2 S+K_v\frac{\partial^2 S}{\partial z^2}+\gamma(S^*-S) \quad (9)$$

where, u is the horizontal velocity vector, w the upward velocity, f the Coriolis parameter($=8.466\times10^{-5}$ sec^{-1} at $35°.50\,N$), x the locally vertical unit vector, $g(=980cm$ sec$^{-2})$ the gravitational acceralation, ∇_h the horizontal gradient operator, t the time, p the pressure, ρ the density, ρ_o the reference density, T water temperature and S salinity. The density ρ is calculated from T and S with use of the usual nonlinear state equation(Wadachi, 1987). The last terms in Eqs.(8) and (9) are called γ terms which were introduced by Sarmiento and Bryan(1982) to prevent calculated values T and S deviating greatly from observed values T^* and S^*. The degree of modification is represented by γ. For a small γ, the model is near to be independent of the observed data and approaches a prognostic model. For a large γ, the model is restricted by the observed data and approaches a purely diagnostic model(Fujio and Imasato, 1991). A_v and K_v are the vertical eddy viscosity and diffusivity, respectively. A_h and K_h are the horizontal eddy viscosity and diffusivity, respectively.

The horizontal and vertical eddy viscosities and diffusivities are given as follow on the basis of the mixing length theory(Hino ,1989)

$$A_h = \frac{\alpha}{2\pi} U_{amp}^2 \times T_{tide}, \quad A_v = 10^{-5} \times A_h \tag{10}$$

$$K_h = \frac{\beta}{2\pi} U_{amp}^2 \times T_{tide}, \quad K_v = 10^{-5} \times K_h \tag{11}$$

where, U_{amp} and T_{tide} denote the amplitude and the period of M_2 tidal current, respectively. α and β the parameters used to adjust the horizontal eddy viscosity and diffusivity locating in a reasonable range. Considering that the amplitude of M_2 tidal current amplitude in Saemangeum Bay is almost in the range of 40 and 75 $cm/\sec$ by the calculation of tidal current before the dike construction, parameter α and β are equated to be 0.01. By this way, the horizontal eddy viscosity and diffusivity are almost in the range between 1.1×10^5 and $3.9\times10^5 cm^2/\sec$ before the dike construction. A_v and K_v range from $1.1\ cm^2\sec^{-1}$ to $3.9 cm^2\sec^{-1}$ in the calculation area.

The horizontal forces due to tidal stress $T_s=(T_{sx}, T_{sy})$ is calculated as follows(Yanagi & Yamamoto, 1993):

$$T_{sx} = -\overline{\left(u'\frac{\partial u'}{\partial x} + v'\frac{\partial u'}{\partial y}\right)} \tag{12}$$

$$T_{sy} = -\overline{\left(u'\frac{\partial v'}{\partial x} + v'\frac{\partial v'}{\partial y}\right)} \tag{13}$$

where, u' and v' are the x and y components of M_2 tidal current, respectively and the overbar denotes the average over one-tidal cycle. The boundary condition for momentum is no-slip condition at the lateral wall. Along the open boundaries, velocity is assumed to be zero. The bottom stress is given at the sea bottom as follows:

$$A_v\frac{\partial u}{\partial z} = \gamma_b \mid u \mid u \tag{14}$$

where $\gamma_b(=2.6\times10^{-3})$ is the bottom drag coefficient. The sea surface is assumed to be a free surface, and the momentum flux at the sea surface is given by

$$\rho_o A_v\frac{\partial u}{\partial z} = \rho_a C_d \mid W \mid W \tag{15}$$

where, $\rho_a(=0.0012 g\ cm^{-3})$ is the air density, $C_d\ (=0.0013\ )$ the sea surface drag coefficient, and $W\ (=4.10 m/\sec)$ the wind vector. The boundary conditions for water temperature and salinity are no-flux condition at lateral walls, at bottom, and at the sea surface. The leap-frog scheme with use of Dufort-Frankel method was adopted for the temporal acceleration term, viscosity term and diffusive term and the Euler-Backward scheme was inserted every eight time steps. The central difference scheme was adopted for the advection term and the semi-implicit scheme was used for the calculation of water elevation(Backhaus, 1983). The time step of the calculation is 30sec. The initial condition is at rest and the quasi-steady state is obtained about 6 days after the beginning of the calculation.

4.2 Observed Data

The field observation for the horizontal and vertical of water temperature ,salinity,density in Saemangeum coastal region was carried out from summer 1992 to May 1993 by Marine Development Institute in Kunsan University. Observation station of water temperature and salinity is shown in Fig.12. The model basin is divided horizontally into 1km x 1km mesh and vertically into three layers.0-5m,5-10m,10-bottom.The water temperature.salinity at every mesh are objectively interpolated from observed data with the use of a hyperbolic function(Yanagi and Igawa).The upper layer(0-5m) expresses the mixed layer above the seasonal pycnocline.Horizontal distributions of density in three layers.which are calculated from water temperature and salinity. Fig.14 shows the horizontal distributions of water temperature.salinity and density at three levels in autumn.

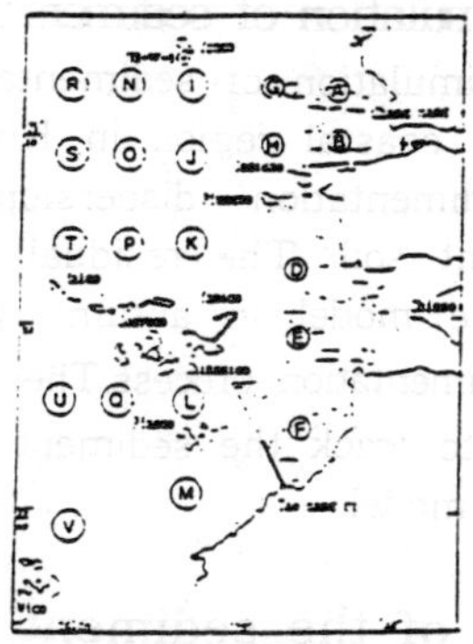

Fig.12 Observation station of water temperature and salinity

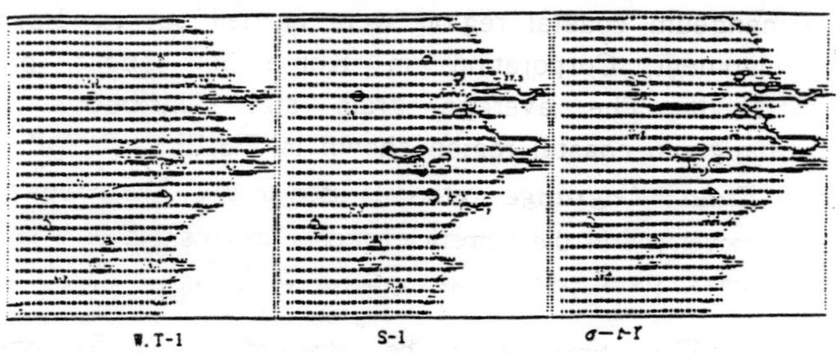

Fig.13 Distributions of salinity and density obtained in autumn.

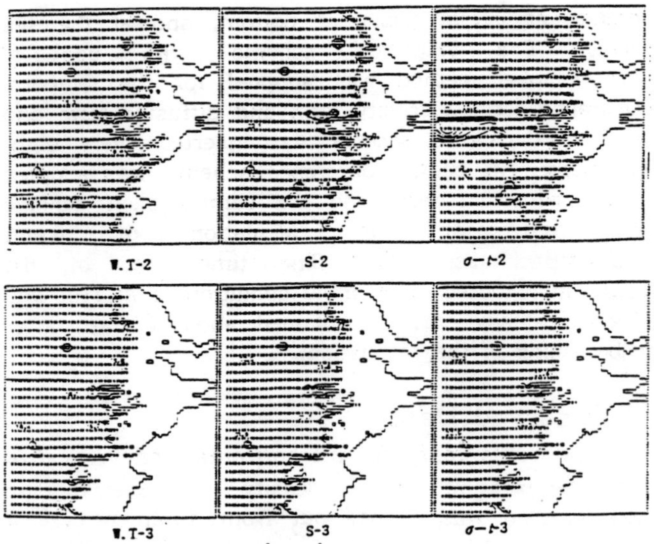

low layer
Fig.13 Distributions of salinity and density obtained in autumn.

Fig.14 The flow patterns at three layers of the autumn obtained by diagnostic model

5. Numerical simulation of sedimentation processes

A numerical simulation of sedimentation processes in the Saemangeum coastal region in Korea is tried. The degrees of sedimentation dispersion are compared observed sediment soil. The residual currents obtained by the diagnostic model in autumn were used in the calculation of sedimentation process.The *Euler–Lagrange* method is used to track the sediment soil movement in the numerical model.

5.1 Tracking of the sediment soil

The field observation of sea bed material in Saemangeum coastal region were carried out by Rural Development Corporation in Korea for about one year(1988). The average grain size distribution in Saemangeum coastal region shown in Fig.15.

The *Euler–Lagrange* method is used to track the sediment soil movement in the numerical model. The position of a soil grain($X^{n+1}, Y^{n+1}, Z^{n+1}$) at time $n+1$,which was at (X^n, Y^n, Z^n)at time n,can be calculated by the following equation utilizing the Taylor series(Yanagi et al.,1983).

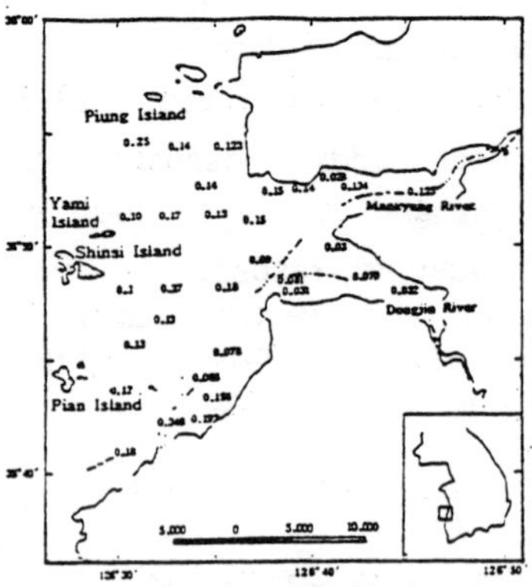

Fig.15 The average grain size distribution in Saemangeum coastal region.

$$X^{n+1}=X^n+u''\,\Delta t+(\nabla u'')u''\,t^2+R_x \qquad (16)$$

$$Y^{n+1}=Y^n+v''\,\Delta t+(\nabla v'')v''\,t^2+R_y \qquad (17)$$

$$Z^{n+1}=Z^n+w_s\Delta t+R_z \qquad (18)$$

where, u'' and v'' denote the calculated velocity of residual current in x and y directions,respectively in autumn. w_s means the sinking speed of soil grain and it is calculated from the Sotkes Law. R_x, R_y and R_z are the dispersion due to turbulence. The third terms in the right hand side of Eqs.(16)and (17) denote the effect of current shear. R_x, R_y and R_z are given by the following equation.

$$R_x \text{ and } R_y=\gamma\ \ (2\Delta t(D_{ht}+D_{hw}))^{1/2}$$
$$R_z=\gamma\ \ (2\Delta t(D_{vt}+D_{vw}))^{1/2} \qquad (19)$$

where, γ is the normal random number. The average value of γ is zero and its standard deviation is 1.0. The horizontal and vertical diffusivities due to tidal current D_{ht} and D_{vt} are given as follows on the basis of the mixing length theory,

$$D_{ht}=\frac{\beta}{2\pi}U_{amp}^2\times T_{tide},\quad D_{vt}=10^{-5}\times D_{ht} \qquad (20)$$

where, U_{amp} and T_{tide} denote the amplitude and the period of M_2 tidal current. respectively, β the parameter used to adjust the horizontal diffusivity locating in a reasonable range. The parameter β is decided to be 0.06. By this way,the horizontal diffusivity due to tidal current D_{ht} is almost in the range 1.5×10^4 and $1.7\times10^5 cm^2/\sec$ after the dike construction. D_{vt} ranges from $0.15\ cm^2\sec^{-1}$ to $1.7 cm^2\sec^{-1}$ in the calculation area. The horizontal and vertical diffusivities due to wave motion D_{hw} and D_{vw} are given on the basis of observed significant wave height and wave period. The field observations on significant wave height and wave period in the Saemangeum coastal area were carried out from December 1993 to October 1994 by Rural Development Corporation in Korea. The horizontal and vertical diffusivities by the significant wave motion D_{hw} and D_{vw} are given as follows,

$$D_{hw}(=3.9\times10^{4})=\alpha_{1}H^{2}_{1/3}\times T_{1/3},$$

$$D_{vw}(=3.9\times10^{-1})=10^{-5}\times D_{hw}$$

where, parameter α_1 and β_1 are 0.5 and 0.16, respectively. $H^2_{1/3}(=125cm)$ and $T_{1/3}(=5\,\text{sec})$ are the significant wave height and the significant wave period. The residual currents obtained by the diagnostic model in autumn were used in the calculation of soil grain dispersion.

5.2 Boundary condition and initial condition

Dispersion of soil grain has been calculated estuary of Dongjin and Mankyung,Geum River.In the numerical simulation, five hundred soil grain were released at the depth of 0.2m below the sea surface. O marks in Fig.(16~17) are the release positions of soil grain. When the soil grain reached at the sea bottom or at the coast, the position of the soil grain was brought back to the position before one time step, that is, the soil grain does not stop its movement at the coast or at the sea bottom. The dispersion of the soil grain is calculated one and three day after the release with residual currents obtained by the diagnostic model in autumn. Density and size of soil grain is 2.64 gr/cm^3 and 30,60,80,200,400㎛, respectively.

deposit suspend

Fig.16(a) The distribution deposit and suspend of suspend sediment in the Mankyung River(calculation time 24hr and size of soil grain:60㎛).

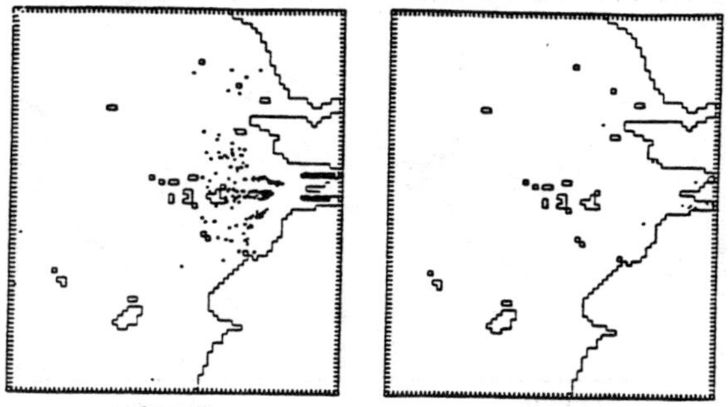

deposit suspend

Fig.16(b) The distribution deposit and suspend of suspend sediment in the Mankyung River(calculation time 72hr and size of soil grain:60㎛).

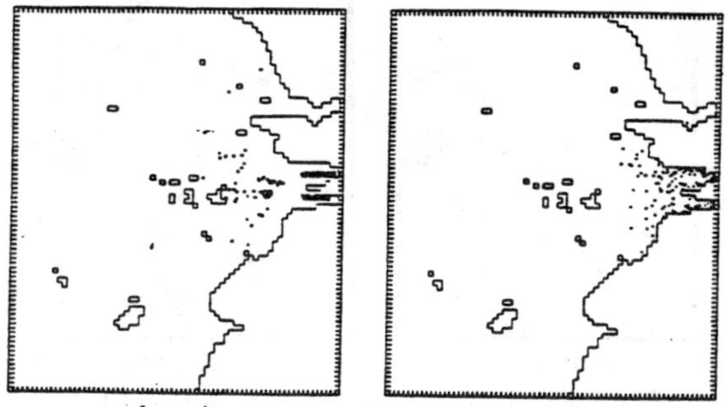

deposit suspend

Fig.17(a) The distribution deposit and suspend of suspend sediment in the Mankyung River(calculation time 24hr and size of soil grain:80㎛).

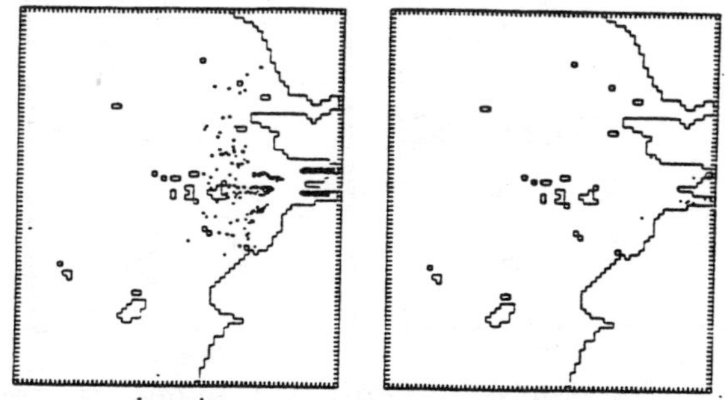

deposit suspend

Fig.17(b) The distribution deposit and suspend of suspend sediment in the Mankyung River(calculation time 72hr and size of soil grain:80㎛).

deposit suspend

Fig.18(a) The distribution deposit and suspend of suspend sediment in the Mankyung River(calculation time 24hr and size of soil grain:200㎛).

Fig.18(b) The distribution deposit and suspend of suspend sediment in the Mankyung River(calculation time 72hr and size of soil grain:200μm).

Fig.19(a) The distribution deposit and suspend of suspend sediment in the Mankyung River(calculation time 24hr and size of soil grain:400μm).

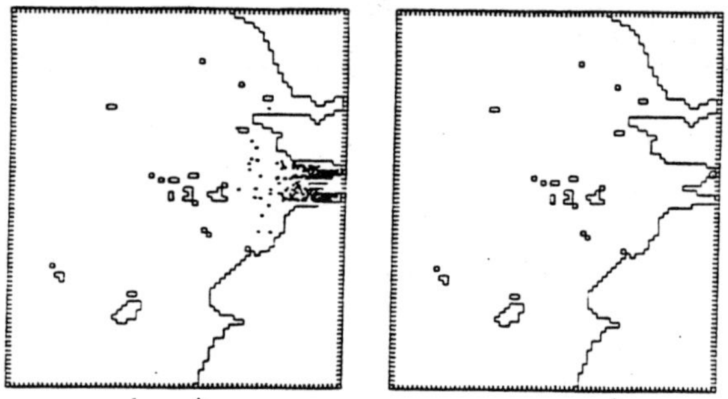

Fig.19(b) The distribution deposit and suspend of suspend sediment in the Mankyung River(calculation time 72hr and size of soil grain:400μm).

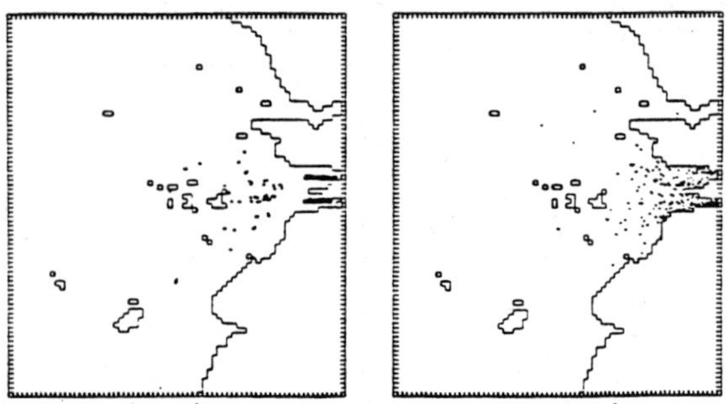

Fig.20(a) The distribution deposit and suspend of suspend sediment in the Dongjin River(calculation time 24hr and size of soil grain:30μm).

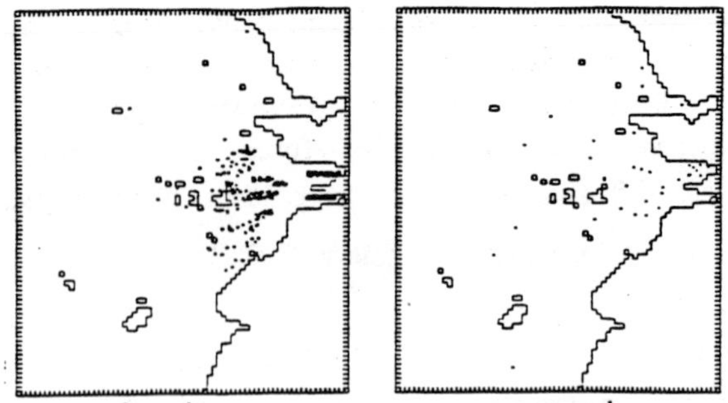

Fig.20(b) The distribution deposit and suspend of suspend sediment in the Dongjin River(calculation time 72hr and size of soil grain:30μm).

6.CONCULUSION

1.The calculated co-range and co-tidal charts of M_2 tide before the dike construction are similar to the observed ones.

2.Water temperature in autumn was the lowest at the mouths of Mankyung River and Dongjin River in the upper layer, but was nearly the same in the middle and lower layers. Salinity was the lowest at the northern area in the upper layer, but was uniform in the lower layer. Density was the lowest at the northern area in the upper layer, but was nearly uniform in the middle and lower layers.

3.About half of released soil grain move a quite distance from the release point three day after the releasing.

4.The dispersion range of soil grain for three day after releasing in the Mankyung River indicates that the soil grain was dispersed within the range of Mal and Yon Island when specific gravity of soil grain is 60μm. The soil grain moves a quite distance from the release point when size of soil grain is a small. These results indicate that size of soil grain and residual current is greatly influenced on the dispersion range of soil grain

REFERENCES

Shin Moon Seup, Jong Nam Lee, Sung Kun Hong : Development of Model by Soil Diffusion within Dam.Hydro-Port'94, October 19-21,1994.Yokosuka,Japan.

Yanagi T.:Fundamental study on the tidal residual circulation, J.Oceanogr.soc.Japan, 34(1978), 67-72.

Yanagi T.,Okada Shin-Ichi and Tsukamoto K.:Numerical Simulation of dispersal patterns of red sea bream juveniles, Pagrus major,in Nyuzu Bay,Japan,Journal of Marine System, 3(1992), 477-487.

Yanagi,T.,Okada S. and Tsukamoto K.:Numerical simulation of dispersal patterns of red sea bream juveniles, Pagrus major, in Nyuzu Bay, Japan,Journal of Marine System, 3(1992), 477-487.

Proceedings of the Seventh (1997) International Offshore and Polar Engineering Conference
Honolulu, USA, May 25-30, 1997

ISBN 1-880653-28-1 (Set); ISBN 1-880653-29-X (Vol. I)

Floating Wave-Energy Pump for Wide Wave-Range

H. Ueki, K. Kawaguty and S. Akamine
Nagasaki University
Nagasaki, Japan

ABSTRACT

The performance of reciprocating wave-energy pumps with floating caissons was analyzed for the conditions of various wave height and period. Wave-energy is converted to the up-and-down motion of the cylindrical float to which the piston is directly connected by a discharge pipe. As heaving amplitudes of the caisson and float increase with the wave height, the relative stroke of the piston to the cylinder also increases. In order to operate the pump safely the relative stroke should be controlled to be less than the length of the cylinder. A pump performance simulation was conducted for the case that a set of spring and damper was installed as one of shock absorber between the caisson and float to restrict the relative stroke of the piston and cylinder. It is clearly shown that the excessive energy can be dissipated by the shock absorber and that the wave-energy pump can be operated in wide wave-range.

KEY WORDS : Ocean Wave Energy Conversion, Pump, Double Action, Floating Caisson, Motion Control

NOMENCLATURE

- A : waterplane area
- a : cylinder cross sectional area
- D : diameter of float
- dx : clearance between tip of shock absorber and surface of float
- F_w : wave force
- F_f : kinetic friction force
- g : acceleration of gravity
- h : water depth
- H : pump head
- H_w : wave height
- k : wave number
- L : pipe length
- M : mass of float-piston
- M_z : added mass of float-piston
- N_z : damping coefficient
- P : in-cylinder pressure
- S : stroke
- t : time
- T_w : wave period
- V : flow velocity in pipe
- Z : displacement
- $\dot{Z}$: velocity
- $\ddot{Z}$: acceleration
- ρ : density of water
- ϕ : velocity potential

SUBSCRIPT

- 1 : upper chamber of cylinder
- 2 : lower chamber of cylinder
- d : pressure side
- s : suction side
- L : loss

INTRODUCTION

Many researchers have been making efforts to improve the performance of wave energy converters. McCormick (1982) reported specifications of OWC plant for various wave periods. Green et al (1983) suggested a practical OWC model. Richards et al (1984) reported about the R & D of pneumatic wave energy conversion system including the McCormick turbine. McCormick et al (1984) reviewed several types of wave energy converter and national or international projects for OWC. Assanis (1984) et al studied about wave energy extraction system for electric power generation. Kokkinowrachos et al (1987) showed hydrodynamic analysis methods for some wave-energy conversion systems. Szumko (1989) suggested mechanical models for oscillating water column (OWC). Recently, Setoguchi et al. (1996) investigated an air turbine for OWC power generation, and Curran and Gato (1996) reported about Wells turbine. Lee et al. (1996) reported about OWC power generation systems. Fukutomi and Nakase (1996) investigated a cross-flow water turbine. Most of above research works are concerned about the electric power generation by extracting wave energy. Eidsmoen (1996) investigated optimal control techniques for improving the energy absorption by a wave-energy converter (WEC). Clement (1996) used a two dimensional numerical wave tank to compute the non-linear radiation step response of oscillating water column wave-power device and examined the influence of several parameters of the device on the modes of the chamber free-surface.

A double-acting wave-energy pump with floating caissons has been suggested previously by authors (1993). The pump, which does not depend on the tidal amplitude or on the direction of the wave incidence, is planned to operate offshore. An analysis model developed by the authors (1994) for this floating pump could estimate the pump performance and also the motion of the caisson and float-piston. In the present study this analysis model was applied for a different type of caisson and various conditions of wave height and period. Further, the pump performance was investigated analytically in the case that a set of spring and damper was installed between the caisson and float as one of shock absorber for controlling the relative stroke of the piston and cylinder.

EXPERIMENTAL SETUP

Figure 1 shows the schematic of the tested floating wave-energy pump which was investigated experimentally and analytically by the authors (1993, 1994). The pressure piston is directly connected to the float by a discharge pipe. The cylinder and piston are set under the sea level in the middle of the caisson. The diameters of the piston and piston pipe are 100 mm and 45 mm respectively. The float is 730 mm in diameter and 200 mm in height. Four hollow columns act as the float of the caisson. The diameter and draft of the column is 0.216 m and 1.21 m respectively. The diameter of the cylinder base plate is 1.28 m. The stabilizer at the lowest part of the caisson is made up of a plate and pipe, and the diameter of the stabilizer plate is 0.8 m.

The pump performance test was conducted in a wave tank of 60 m in length, 4 m in width, and 2.3 m in depth. Sinusoidal waves with the wave height up to 0.22 m and the wave period up to 3.0 s could be generated in this wave tank. The flow-rate of the pump was measured from the discharge water volume per unit time at a constant pump head. The water surface displacement and the motions of both the caisson and float are measured by a magnetostrictive displacement sensor, or are recorded as video tape pictures.

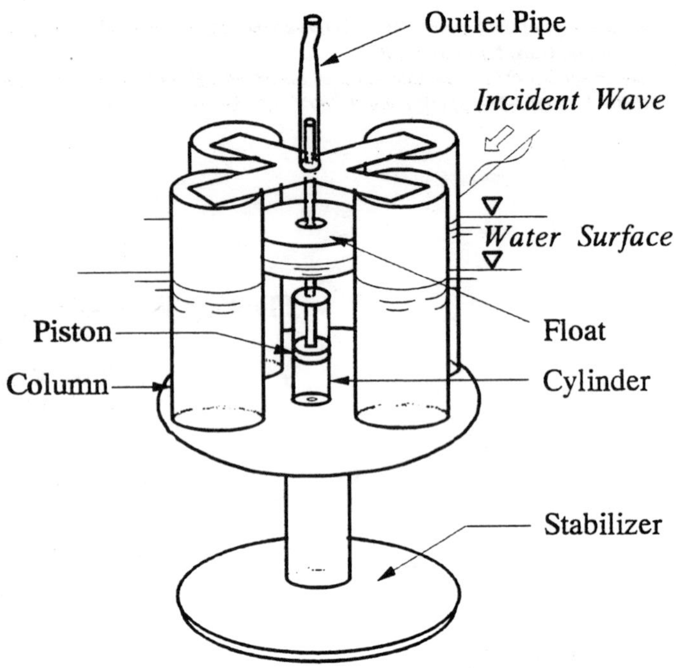

Fig.1 Floating Double-Acting Wave-Energy Pump

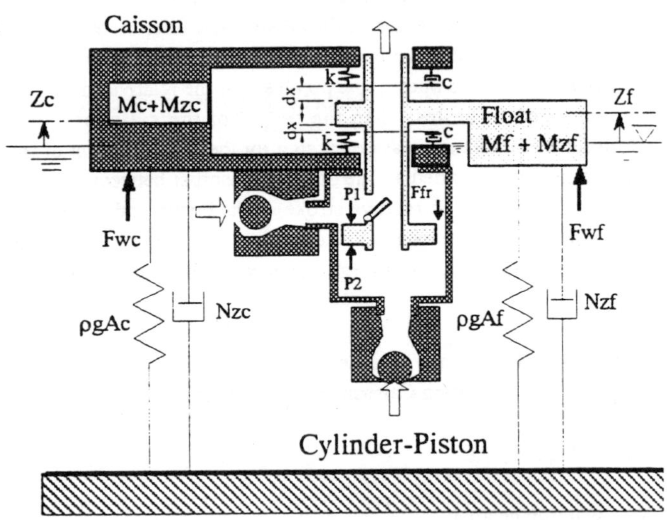

Fig.2 Oscillation Model for Caisson and Float-Piston

PUMP PERFORMANCE SIMULATION MODEL

Figure 2 shows the oscillation model for the float-piston and caisson. It is assumed that the float-piston and the caisson extract wave-energy independently, and they interact each other through both the in-cylinder pressure and the friction force between the piston and cylinder liner. A set of spring and damper was installed additionally between the caisson and the float as one of shock absorber in order to restrict the relative stroke of the piston and cylinder. The equations of motion for the float-piston and the caisson can be written as eqs. (1) and (2) respectively.

$$\left(M_f + M_{zf}\right)\ddot{Z}_f + N_{zf}\dot{Z}_f + \rho g A_f Z_f + \left(P_1 a_1 + P_2 a_2\right) + F_{fr}\mathrm{sgn}\left(\dot{Z}_f - \dot{Z}_c\right) + F_a = F_{wf} \tag{1}$$

$$\left(M_c + M_{zc}\right)\ddot{Z}_c + N_{zc}\dot{Z}_c + \rho g A_c Z_c + \left(P_1 a_1 + P_2 a_2\right) + F_{fr}\mathrm{sgn}\left(\dot{Z}_c - \dot{Z}_f\right) - F_a = F_{wc} \tag{2}$$

In the left hand side of above equations, the first term is the inertia force, the second is the damping force, and the third is the restoring force. The fourth term is the force acting to the piston and the fifth is the friction force between the piston and cylinder liner. The sixth term F_a is the force of the shock absorber consisted from a spring and damper. The right hand side indicates the external force of the waves. The equation was solved by Runge-Kutta-Gill method using a workstation.

F_a is written as follows,

$$F_a = k(Z_f - Z_c) - k \cdot dx + c(\dot{Z}_f - \dot{Z}_c) \quad ; Z_f - Z_c \le -dx$$
$$F_a = 0 \quad ; -dx \le Z_f - Z_c \le dx \qquad (3)$$
$$F_a = k(Z_f - Z_c) + k \cdot dx + c(\dot{Z}_f - \dot{Z}_c) \quad ; Z_f - Z_c \ge dx$$

where dx is the clearance between the tip of the shock absorber and float surface. The shock absorber is effective when the relative stroke of the float and cylinder increases more than dx.

The reciprocating mass of the float-piston M_f and the caisson M_c are 33.2 and 250 kg respectively. Added mass M_z, damping coefficient N_z and natural period T_n are estimated by a free oscillation test conducted in which the float-piston or caisson is caused to heave by release from an originally displaced position. For the float-piston, measured M_{zf}, N_{zf} and T_{nf} are 122.5 kg, 438.8 kg/s, and 1.22 s respectively. M_{zc}, N_{zc} and T_{nc} for the caisson are 1014.5 kg, 483.3 kg/s, and 5.89 s.

The wave force F_w acting to the float and the caisson can be calculated by equation (4),

$$F_w = \rho \frac{\partial \phi}{\partial t} A + M_z \left(-\frac{\partial^2 \phi}{\partial Z \partial t} \right) + N_z \left(-\frac{\partial \phi}{\partial z} \right) \qquad (4)$$

where ϕ indicates velocity potential of progressive wave. The first term in the right hand side of the equation is the Froude-Krylov force estimated by integrating the incident wave force on the float surface, and the second term is the diffraction force by incident wave scattering.

Figure 3 shows the pump simulation model. The stabilizer pipe is one of the water inlet with a ball valve (A). Another inlet (B) is at the top side of the cylinder. In the down-stroke of the piston, the water in the lower cylinder chamber is discharged through the flap valve (C) in the discharge pipe, and the water is sucked into the upper cylinder chamber through the valve (B) simultaneously. In the up-stroke of the piston, the water in the upper chamber is discharged through the flap valve and the water is sucked into the lower chamber simultaneously. P_1 indicates the pressure in the upper chamber of the cylinder and P_2 indicates the pressure in lower chamber. These in-cylinder pressures can be calculated by the Bernoulli's equation for unsteady flow,

$$P_i = \begin{cases} = \rho \left\{ \frac{1}{2} V_{si}^2 - \frac{1}{2} (\dot{Z}_f - \dot{Z}_c)^2 - (H_s + Z_f + H_{Lsi}) g - L_{si} \frac{dV_{si}}{dt} \right\} & (5) \\ = \rho \left\{ \frac{1}{2} V_{di}^2 - \frac{1}{2} (\dot{Z}_f - \dot{Z}_c)^2 + (H_d - Z_f + H_{Ldi}) g + L_{di} \frac{dV_{di}}{dt} \right\} & (6) \end{cases}$$

where i=1: upper chamber of cylinder,
i=2: lower chamber of cylinder.

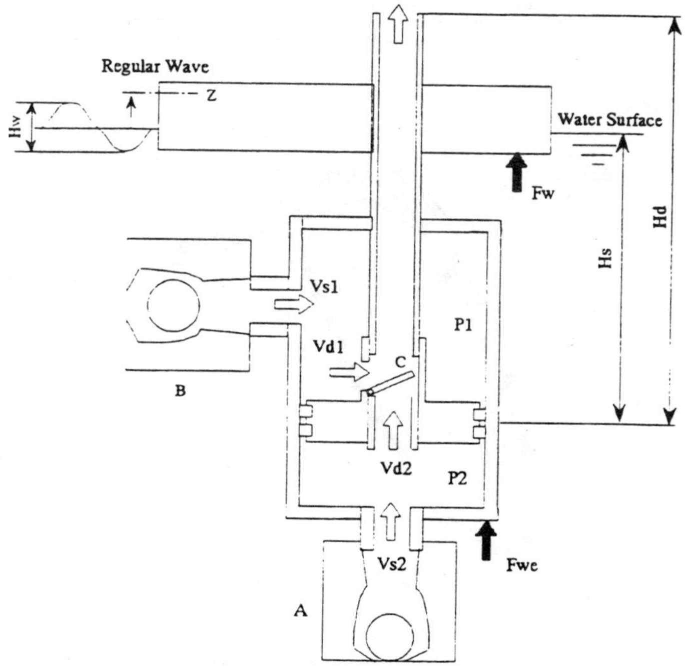

Fig.3 Pump Simulation Model

model	Hw(m)	Tw(s)	2.0	2.2	3.0
I	0.2	calc.	····	- - -	——
		exp.	△	◇	○
II	2.0	calc.	▪▪▪▪▪	▪ ▪ ▪	▬▬

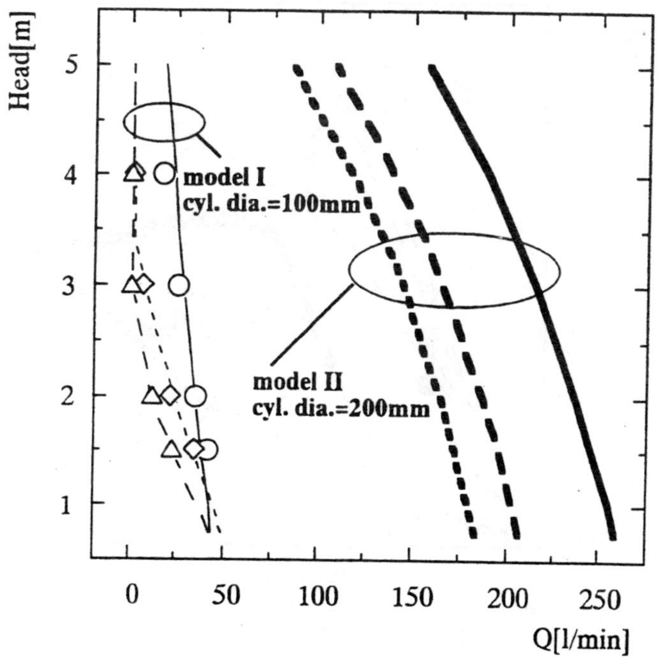

Fig.4 Performance of Floating Wave-Energy Pump

Equation (5) is applied for the suction mode and equation (6) is for the discharge mode.

Head loss H_L in the discharge pipe was estimated by Darcy-Weisbach equation. It was assumed that three non-return valves open or close simultaneously when a relative velocity of the float-piston and caisson changes its sign of direction. To make the calculation easy to converge, the in-cylinder pressure was smoothed in time domain. The kinematic friction force was estimated at 55 N constant.

In the present study, calculation was conducted in the wave height range from 1 to 6 m, and the wave period range from 2 to 8 s. A feasibility study was conducted for the model caisson that the cylinder diameter was 200 mm, the diameter and draft of the column were 0.326 m and 1.296 m, the diameter and draft of the stabilizer were 1.6 m and 3.0 m, the total mass was 500 kg, and the added mass was 4,000 kg. The estimated natural period and damping coefficient were 7.4 s and 1397 kg/s for the new model caisson. This new model is called type II while the original model is called as type I as shown in EXPERIMENTAL SETUP. The diameter of the float of these two models are the same value of 0.73 m.

PERFORMANCE OF WAVE-ENERGY PUMP

Figure 4 shows the flow-rate to head characteristics of the wave-energy pump without the shock absorber. Triangle, lozenge, and circle show the experimental data for the wave period of 2.0, 2.2, and 3.0 s respectively when the wave-energy pump with the original caisson (model I) was tested at the wave height H_w of 0.2 m in the wave tank. Thin solid, dashed, and dotted curves show the corresponding calculated results. The calculated results agreed well with the experimental ones as reported by the authors (1994). Thick three curves show the calculated results for the wave-energy pump with the new caisson (type II) in the condition of the wave height of 2.0 m and the water depth of 600 m. Even if the float size of these two types was the same, the flow rate of type II was much higher than that of type I. This comes from the fact that the wave height increased from 0.2 to 2.0 m and that the natural period of the new caisson (type II) is higher than the wave period.

MOTION OF CAISSON AND FLOAT-PISTON

Figure 5 shows the heaving amplitude of the caisson normalized by the wave height. Open square indicates the experimental result for the original caisson (type I) when the float-piston was removed. Thin dotted curve shows the corresponding calculated result. Other four marks show the cases when the pump head was varied from 1.5 to 4.0 m. The heaving amplitude of the caisson itself increases with wave period and shows the maximum in the wave period range from 4 to 5 s. The calculated and experimental results show that the caisson heaving increased remarkably during pumping operation. The caisson heaving increased with the pump

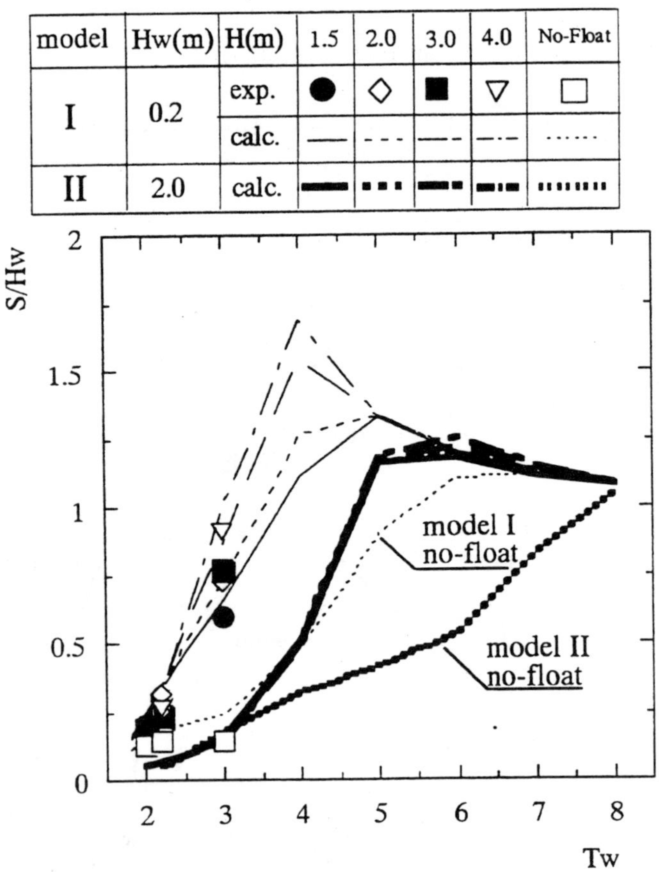

Fig.5 Caisson Heaving Amplitude

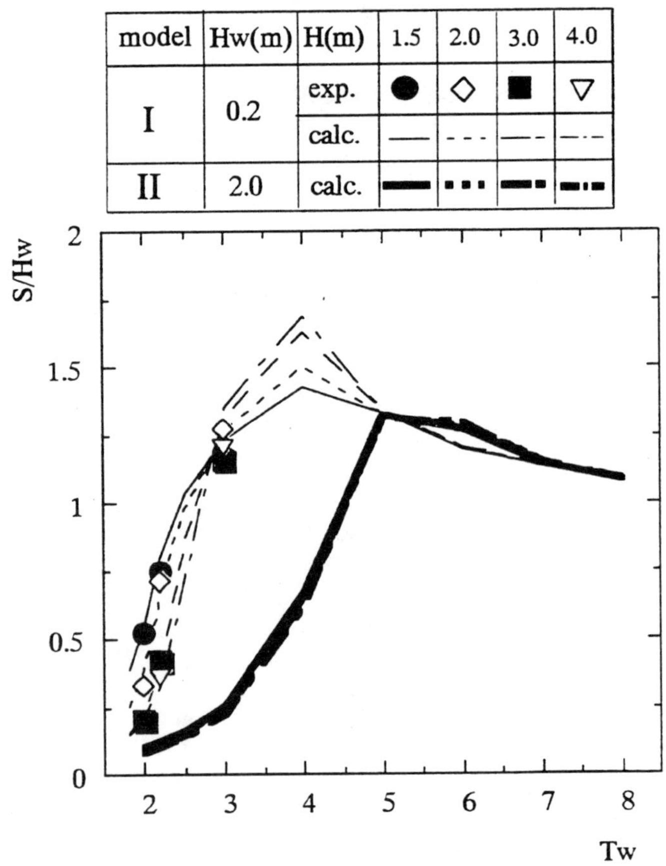

Fig.6 Float-Piston Heaving Amplitude

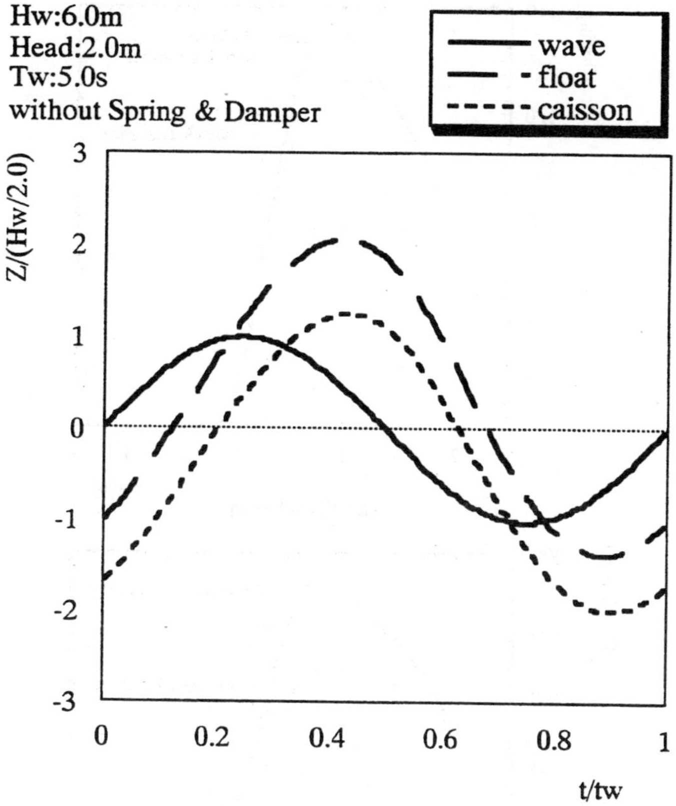

(a) Without shock absorber

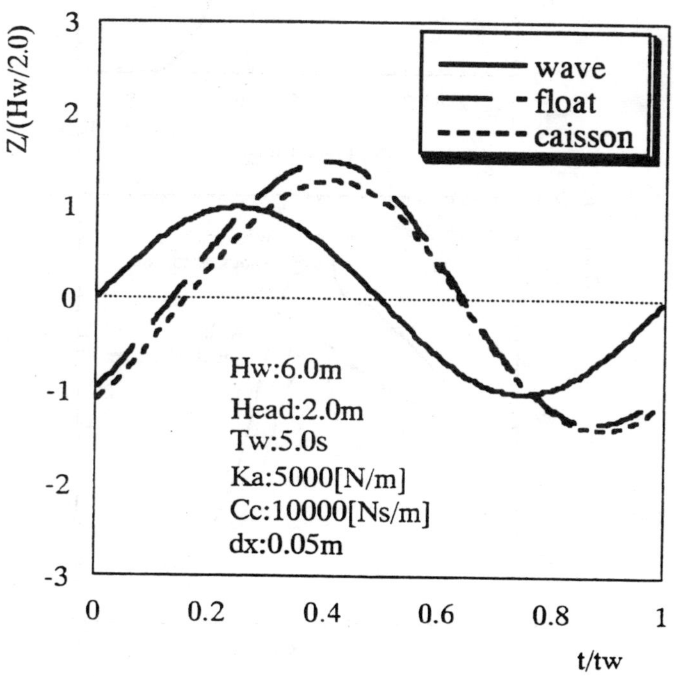

(b) With shock absorber

Fig.7 Time History of Caisson and Float Displacement (model II)

head when the wave period is shorter than 5 s. Thick curves show the cases of the type II when the wave height is 2.0 m. Although the heaving amplitude increases in case of pumping operation, the effect of the pump head is not remarkable in this case comparing to the original type I. The caisson heaving amplitude of type II is smaller than that of type I when the wave period is shorter than 5 s, but it agrees with that of type I when the wave period is longer than 5 s. The caisson heaving amplitude was found to get near to S/Hw=1 asymptotically.

Figure 6 shows the heaving amplitude of the float-piston normalized by the wave height. Marks and thin curves show the case of original model I and thick curves show the calculated results for the model II. The heaving amplitude of the float-piston of the model II is smaller than that of model I when the wave period is shorter than 5 s, but it agrees with that of model I when the wave period is longer than 5 s as seen in Fig.5. The normalized heaving amplitude of the float-piston is independent of the wave height and caisson type in the case of longer wave period. The heaving amplitude of the caisson and float showed the maximum value at the wave period less than the natural period of the caisson, i.e. 7.4 s.

EFFECT OF SHOCK ABSORBER ON PUMP PERFORMANCE

Figures 7 (a) and (b) show the time history of the vertical displacement of the caisson and float-piston for the cases of with and without shock absorber respectively. The wave period is 5.0 s and the pump head is 2.0 m. The clearance dx between the tip of the spring and damper on the float and caisson was 0.05 m. Spring constant Ka is 5,000 [N/m] and the damping coefficient Cc is 10,000 [Ns/m]. Figure 7 (a) shows the mean level of the float heaving is higher than that of the caisson heaving without the shock absorber. On the contrary, Figure 7 (b) shows the difference between the mean heaving levels of the float and caisson decreased remarkably with the shock absorber. The phase lag of the caisson displacement to the wave without shock absorber is comparable to that with shock absorber and that the float-piston with shock absorber performs a safe operation accompanied by the caisson at high wave levels.

Figures 8(a) , (b) and (c) show the variation of flow-rate due to wave period in the cases of the wave height of 1.0, 2.0, and 6.0 m respectively when the pump head is 2.0 m. Dashed and solid curves show the cases with and without shock absorber. It is clearly understood that the shock absorber is effective at the vicinity of the peak flow-rate curve in each wave height. This result can be applied to the case of finite amplitude waves or irregular waves.

Consequently, it is confirmed that the shock absorber is effectively applied to control the relative stroke of the piston and cylinder.

CONCLUDING REMARKS

The performance of the floating wave-energy pump in regular waves of various wave period and wave height was investigated analytically by solving the equation of motion of the caisson and float-piston. In order to control the relative motion of the caisson and float-piston, a shock absorber was installed between the caisson and float. The calculated results are summarized as follows.

1. The heaving amplitude of the caisson and float showed the

maximum value at the wave period less than the natural period of the caisson.

2. The heaving amplitude of the caisson and float got near asymptotically to the wave height in the case of longer wave period range.
3. The relative stroke of the cylinder and piston can be controlled by the shock absorber consisted of a spring and damper.

ACKNOWLEDGMENTS

The authors wish to thank Mr. T.Kawasaki of Nagasaki University for construction of the experimental apparatus and securing the experimental data.

REFERENCES

Assanis, D. N, Carmichael, A.D. (1984), "A Study of Wave Energy Conversion for An Offshore Structure", Proceedings of the 3rd International Offshore Mechanics Arctec Engineering Symposium, No.2, p.287

Clement, A. (1996), "Dynamic Non-Linear Response of OWC Wave Energy Devices", Proceedings of the Sixth (1996) International Offshore and Polar Engineering Conference, vol.1, p.91

Curran, R. and Gato, L.M.C.(1996), "Aerodynamic Losses Incurred by the Wells Turbine", Proceedings of the Sixth (1996) International Offshore and Polar Engineering Conference, vol.1, p.58

Eidsmoen, H. (1996), "Optimum control of a floating Wave-Energy Converter with restricted Amplitude", Journal of Offshore Mechanics and Arctic Engineering, Transactions of the ASME, vol.118, No.2, p.96

Fukutomi, J. and Nakase, Y. (1996), "Running Characteristics of a Cross-Flow Water Turbine in Oscillating Flow", Proceedings of the Sixth (1996) International Offshore and Polar Engineering Conference, vol.1, p.75

Green, W. L., Campo, J. J., Parker, J. E. Miller, J. A. (1983), "Wave Energy Conversion with an Oscillating Water Column on a Fixed Offshore Platform", Transactions of the ASME, Journal of Energy Resources Technology, Vol.105, No.4, p.487

Kawaguty, K. and Ueki, H. (1993), "Motion and Performance of Floating Double-Acting Wave-Energy Pump", Proceedings of the Third (199) International Offshore and Polar Engineering Conference, vol.1, p.190

Kokkinowrachos, K., Thanos, I., Zibell, H. G. (1987), "Hydrodynamic Analysis of Some Wave Energy Conversion Systems", Oceans, Vol.1987, No.2, p.566

Lee, C.H., Newman, J.N., and Nielsen, F.G.(1996), "Wave Interactions with an Oscillating Water Column", Proceedings of the Sixth

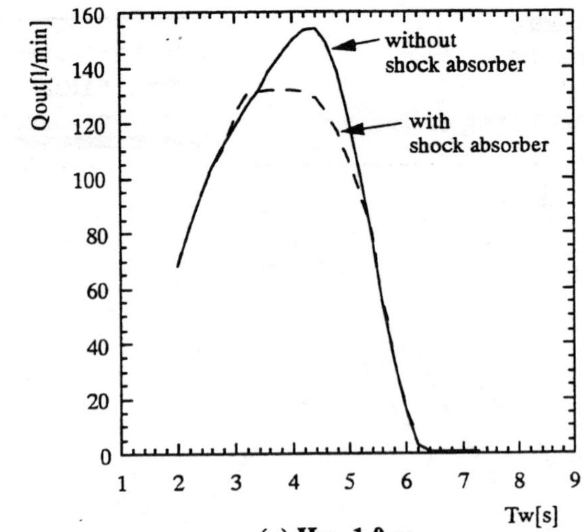

(a) Hw=1.0 m

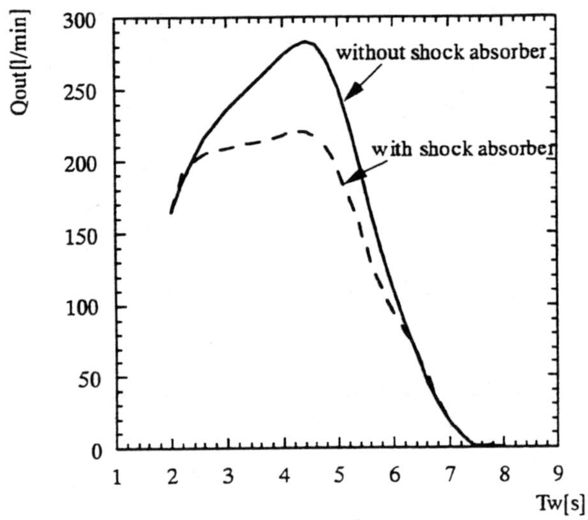

(b) Hw=2.0 m

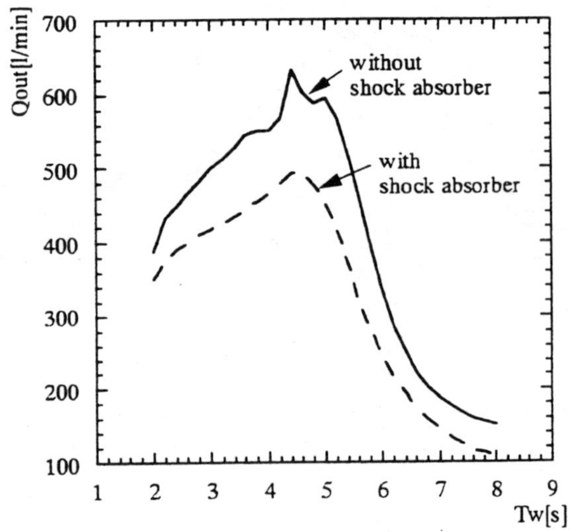

(c) Hw=6.0 m

Fig.8 Variation of flow rate due to wave period ;model II, H=2.0m, dx=0.05m, Ka=5000N/m, Cc=10000Ns/m

(1996) International Offshore and Polar Engineering Conference, vol.1, p.82

McCormick, M. E., Burcher, E. S. (1982), "Pneumatic Wave Energy Conversion", Energy Resource Environment, p.330

McCormick, M. E. (1984), "National and International Advances in Wave Energy Conversion", Proceedings of the 3rd International Offshore Mechanics Arctec Engineering Symposium, No.1, p.302

McCormick, M. E. (1991), "A Theoretical Analysis of a Self Propelled Backward Bent Duct Wave Energy Conversion System", Transactions of the ASME, Journal of Energy Resources Technology, vol.113, No.2, p.94

Richards, D., Weiskopf, F. B., Ritzcovan, P. J. (1984), "Pneumatic Wave Energy Conversion System Program Developments", Am Soc. Mech. Eng., No.84, p.8

Setoguchi, T. Kaneko, K., and Takao, M. (1996), "Air Turbine with Self-Pitch-Controlled Blades for Wave Energy Conversion", Proceedings of the Sixth (1996) International Offshore and Polar Engineering Conference, vol.1, p.70

Szumko, S. (1989), "Mechanical Model for Oscillating Water Column with Compressibility", Journal of Engineering Mechanics, vol.115, No.5, p.1851

Ueki, H., Kawaguty,K. and Kira., T. (1994), "Performance Analysis of Floating Double-Acting Wave-Energy Pump", Proceedings of the Fourth (1994) International Offshore and Polar Engineering Conference, vol.1, p.371

The Proceedings of the Seventh (1997) International OFFSHORE AND POLAR ENGINEERING CONFERENCE

VOLUME II, 1997

Offshore and Arctic Pipelines, Mechanics of Risers, Cables and Mooring, Underwater Vehicles, Control and Systems, Polar and Ice Engineering, JOIA Projects, Arctic and Antarctic Systems, Arctic Oil Pollution, Ice Remote Sensing

edited by:

Jin S. Chung, Colorado School of Mines, Golden, Colorado, USA
Robert M.W. Frederking, National Research Council Canada, Ottawa, Canada
Hiroshi Saeki, Hokkaido University, Sapporo, Japan
Alexander T. Bekker, Far Eastern State Technical University, Vladivostok, Russia

presented at:
The Seventh (1997) International Offshore and Polar Engineering Conference held in Honolulu, USA, May 25-30, 1997

organized by:
International Society of Offshore and Polar Engineers

sponsored by:
International Society of Offshore and Polar Engineers (ISOPE)
with cooperating societies and associations

International Standard Book Number: ISBN 1-880653-28-1 (Set)
ISBN 1-880653-30-3 (Vol. II)

Library of Congress Catalog Card Number: 97-70292

Cooperating Organizations:

Canadian Association of Petroleum Producers (CAPP)
American Society of Civil Engineers (ASCE) - Engineering Mechanics Division
Korea Committee for Ocean Resources and Engineering (KCORE)
Canadian Society of Civil Engineers (CSCE) - Engineering Mechanics Division
Chinese Society of Ocean Engineers (CSOE)
Chinese Society of Naval Architects and Marine Engineers (CSNAME)
Chinese Society of Theoretical and Applied Mechanics (CSTAM)
Russian Academy of Sciences
Singapore Structural Steel Society (SSSS)
Norwegian Petroleum Society (NPF)
The Institution of Engineers Australia (IE Australia)
Kansai Society of Naval Architects, Japan (KSNAJ)
IRO (The Netherlands)
Technical Research Centre of Finland (VTT)
The Society of Materials Science, Japan (JSMS)
The Offshore Engineering Society (OES), United Kingdom
Ukraine Society of Mechanical Engineers (USME)
IFREMER, France
Scott Polar Research Institute (SPRI), UK
The Institute of Engineers Indonesia (PII)

Printed and bound in USA

FOREWORD

We greatly appreciate the excellent responses and help we have received from colleagues around the world in the successful organization of the 7th International Offshore and Polar Engineering Conference (ISOPE-97), Honolulu, May 25 -30, 1997. The Conference features **91** sessions **of 517** refereed papers and **10** plenary sessions from 40 countries.

The purposes of the ISOPE conference are to:

- promote technological progress and activities, international technological transfer and cooperation, and opportunities for engineers to maintain and improve technical competence; and
- provide a timely international forum for technical activities, cooperation, opportunity and fellowship among researchers and engineers,

by developing focused session topics with papers of high quality in both originality and significance of the results which were accepted after rigorous review. The conference promotes interdisciplinary interaction between academia and industry.

The International Society of Offshore and Polar Engineers (ISOPE) has already held 12 successful international meetings:

- the 1st (1990) European Offshore Mechanics Symposium (ISOPE EUROMS-90) in Trondheim, 1990;
- the 1st (1990) Pacific/Asia Offshore Mechanics Symposium (ISOPE-PACOMS-90) Seoul; PACOMS-94 (SOSC-94) Beijing; and PACOMS-96 Pusan.
- the Annual ISOPE conferences, starting in Edinburgh, 1991 were held in San Francisco, Singapore, Osaka, The Hague and Los Angeles. Since 1992, the annual ISOPE Conference has been held with the world's largest technical program of its kind with refereed papers.
- the 1st (1995) ISOPE Ocean Mining Symposium (ISOPE OMS-95), Tsukuba; the 2nd ISOPE OMS-97 will be held in Seoul, November 24-25, 1997.
- the 1st (1996) International Deep-Ocean Technology (IDOT-96) Symposium and Workshop, Los Angeles.

This year the ISOPE-97 TPC expanded greatly the number of papers on recent worldwide developments in environment, advanced marine technology, coastal engineering, Sakhalin oil and gas. We shall continue to add other emerging topics to the future conferences.

On behalf of all the individual TPC members (listed in this volume), we would like to thank the reviewers (listed in this volume) for their help in maintaining the technical accuracy and quality of papers. We greatly appreciate the help from colleagues around the world, particularly from the authors and reviewers in meeting the deadlines. The efficient handling of the conference organization by the ISOPE staff is greatly appreciated. TPC received 517 mats, examined individual papers, page by page, paginated, and retyped all individual titles according to the ISOPE format, all in 3 weeks, ready for the printer. Only about 5% of the papers were redone by the authors, following the ISOPE mat format and font size. It was possible with the help of fax, international express mail and wordprocessor.

It is our pleasure to welcome participants from all over the world to the ISOPE-97 Conference in Honolulu. We hope that the participants will also join us at the ISOPE-98 Conference to be held in Montréal, Canada, May 24-29, 1998.

Co-chairmen of the ISOPE-97 Conference:

Jin Chung, Colorado School of Mines, USA
Bob Frederking, National Research Council, Canada
Ron Knapp, University of Hawaii, USA
Michel Olagnon, Ifremer, France
Yukio Ueda, Kinki University, Japan

Acknowledgement

The success of the 7th International Offshore and Polar Engineering Conference (ISOPE-97), USA, May 25-30, 1997, is credited to the teamwork of the members of the Technical Program Committee (TPC), the conference cosponsors, the cooperating organizations and the reviewers (see the list).

The TPC would like to express its sincere appreciation to the TPC members' organizations for their support of the TPC members, and University of Hawaii TPC colleagues for their support toward the successful conduct of the conference events, and to the ISOPE staff for the efficient handling of the conference organization and proceedings.

ISOPE-97 Technical Program Committee (TPC)

Prof. T. Yao, Hiroshima Univ., Higashi-Hiroshima City, Japan
Prof. R.W. Yeung, Univ. of California, Berkeley, CA, USA
Prof. S.C.S. Yim, Oregon State Univ., Corvallis, OR, USA

ISOPE-97 Executive Committee

Prof. Jin S. Chung (Chairman), Colorado School of Mines, USA
Dr. Robert M.W. Frederking, National Research Council, Canada
Prof. Ronald H. Knapp, University of Hawaii, USA
Prof.-Dr. Ram S. Puthli, Karlsruhe University, Germany

Conference Co-chairmen

Jin S. Chung, Colorado School of Mines, USA
Robert M.W. Frederking, National Research Council, Canada
Ron H. Knapp, University of Hawaii, Honolulu, USA
Michel Olagnon, Ifremer, France
Yukio Ueda, Kinki University, Wakayama, Japan

Hawaii Committee

Prof. R.H. Knapp (Chairman), University of Hawaii at Manoa, Honolulu, HI, USA
Prof. H.-J. Krock, University of Hawaii at Manoa, Honolulu, HI, USA
Prof. H.R. Riggs, University of Hawaii at Manoa, Honolulu, HI, USA

Partial List of Reviewers

Abe, N
Abel, A
Akyarli, A
Albert, D G
Alessandrini, B
Ali, P
Allyn, N
Andronicou, L
Aoki, S
Aoki, S I
Arai, M
Asai, K
Aso, K
Athanassios, T
Athanassoulis, G A
Bahlavouni, A
Bai, Y
Baier, V
Balasubramanian, S
Bangaru, N V
Barthelemy, J
Bea, R G
Bekker, A T
Berhault, C
Bertsimas, D J
Bitner-Gregersen, E
Bjørnsen, T
Blackmore, R
Boo, S Y
Borekci, O
Boswell, L F
Bouchard, G
Bowen, A
Brandes, H G
Bretz, G
Brown, A J
Bryden, I G
Burcharth, H F
Burgess, J J
Busch, W
Capanoglu, C
Caridis, P A
Carter, D J
Cavanié, A
Chan, D B
Chan, E S
Chandler, B
Chaplin, J R
Che, X L
Chen, H C
Cheung, K F
Chiang, E
Chien, L K
Chin, Y P
Cho, S R
Choi, H S
Choo, Y S
Chow, K W
Christensen, J
Chun, H H
Chung, J S
Chwang, A T
Clément, A
Cole, I S
Contento, G
Cornett, A M
Cox, D T
Croasdale, K R
Crookshank, N E
Dalane, J I
Damy, G
Das, B M
Deguchi, I
Demirbilek, Z
Den Haag, T
Dier, A
Dijkstra, G J
Dijkstra, O D
Dillon, W P
Doi, Y
Dos Santos, J F
Dursthoff, W
Dutta, P K
Duvat, G
Easson, W J
Engebretsen, T
Ertekin, R C
Fairchild, D P
Falzarano, J M
Farzaneh, M
Ferrant, P
Ferrante, A J
Fish, A M
Fithen, R
Fithen, R M
Ford, S F
Forristall, G Z
Francescutto, A
Frederking, R
Fujikubo, M
Fujino, M
Fujisaku, H
Fukumoto, M
Fuwa, T
Fylling, I
Garza-Rios, L
Goda, Y
Goel, A P
Gokce, T
Gokkus, U
Gouleilmos, A M
Grass, A J
Gresnigt, A M
Grilli, S
Grundy, P
Guerin, G
Gunbak, A R
Guo, T
Gutfraind, R
Handa, K
Hardy, C
Haritos, N
Harrison, G E
Haver, S
Herion, S
Hessler, R R
Higo, Y
Hiraishi, T
Hobbs, R E
Holder, G D
Hollebone, B
Hong, S W
Hong, S Y
Honji, H
Hopkins, M
Hosoyamada, T
Hover, F
Huang, T
Huang, T L
Huang, X L
Hyodo, M
Hyun, B S
Ifuku, M
Ikeda, Y
Isaacson, M
Iwata, K
Jacobsen, V
Jakobsen, J B
Jang, Y S
Jiang Q
Jin, D
Jo, C H
Johnson, B H
Jones, S J
Ju, F
Juhl, J
Kagawa, K
Kajiwara, H
Kaneko, K
Kao, C C
Kapdasli, S
Karahan, E
Karr, L A
Kashiwagi, M
Kawamura, Y

Kawano, K
Kershenbaum, N Y
Kielland, P
Kim, C H
Kim, D H
Kim, M H
Kim, S B
Kitamura, M
Kitamura, R
Kite-Powell, H L
Kiyokawa, T
Knapp, R H
Kioka, W
Kobayashi, A
Kobayashi, T
Kokawa, H
Kolle, J J
Koo, J Y
Koterayama, W
Krock, H J
Krogstad, H E
Kubo, M
Kujala, P
Kuriyama, Y
Kuroiwa, T
Kvamme, B
Kvenvolden, K A
Kyozuka, Y
Labeyrie, J
Laflamme, J
Laforte, J L
Lampela, K
Landrini, M
Langen, I
Le Boulluec, M
Le Goff, J P
Lee, S J
Lee, S K
Lee, Y G
Leira, B
Leonard, J
Li, J C
Li, Y C
Lian, D
Liu, Y H
Lin, G
Lin, M C
Lin, S S
Ling, S
Lipp, S
Liu, Y L
Liukkonen, S
Lozowski, E P
Luton, M J
Ma, Z L
Maeda, H
Magda, W
Makkonen, L
Manabe, Y
Mansard, E P D
Maple, J A
Marcus, H S
Martin, A J
Mase, H
Masson, D
Masuda, K
Masuda, Y
Matsui, T
McClimans, T
McClintock, F
McClure, G
McComber, P
Mebarkia, S A
Meon, G
Meylan, M H
Miao, G P
Mimura, N
Mitsuyasu, H
Miura, K
Mizutani, N
Modi, V J
Moe, G
Mogridge, G R
Mohr, W C
Molin, B
Moo-Young, H
Mori, K
Moshagen, H
Muggeridge, D B
Murakawa, H
Nabeshima, Y
Nadaoka, K
Naito, S
Nakai, T
Nakamura, M
Nam, D
Narayanan, R
Natvig, B J
Nerzic, R
Nestegard, A
Niedzwecki, J M
Niho, O
Nomura, S
Nwogu, N
O'Donoghue, T
Ogawa, K
Ogawa, Y
Oh, S K
Ohtaka, K
Olagnon, M
Onitsuka, K
Onoufriou, T
Oppenheim, B
Orisamolu, I R
Oshiwa, K
Otsuka, K
Oumeraci, H
Paik, J K
Pamukcu, S
Papanikolaou, A
Park, I K
Parsey, M R
Paté-Cornell, M E
Patrikalakis, N M
Pedersen, P T
Penna, R
Perkins, N C
Pervizpour, M
Pesce, C P
Powers, E J
Pradhan, B S
Prakash, S
Prasad, S
Prevosto, M
Puri, V K
Puthli, R S
Randall, R E
Raschke, S
Rigaud, V
Riggs, H R
Rijken, O
Riska, K
Roby, K
Rodd, J L
Roesset, J M
Ronold, K
Ryerson, C C
Saeki, H
Sakai, T
Sakakiyama, T
Sakamoto, Y
San, K
Sanni, S
Sarmento, A
Savenije, R Ph A C
Sayed, M
Sayer, P G
Scarsi, G
Schjølberg, I
Schriever, G
Senjanovic, I
Sergy, G
Setoguchi, T
Shen, F
Shen, H T
Shibayama, T
Shigematsu, T
Shin, J G
Shin, M S
Shinoda, T
Shogaki, T
Sibetheros, I
Sigbjörnsson, R
Silver, A L
Skjong, R
Sklenicka, V
Skop, R A
Skourup, J
Snijder, H H
Sodhi, D S
Sørensen, J D
Spencer, P
Squire, V A
Stansberg, C T
Stear, J
Strømsem, K
Su, M Y
Suga, Y
Sumer, B M
Sun, Z C
Suzuki, H
Suzuki, K
Szilder, K
Tabeta, S
Takagi, K
Takahashi, H
Takaki, M
Talley, W K
Tanaka, H
Tanaka, Y
Tanizawa, K
Tedeschi, R
Temperville, A
Teng, B
Thayamballi, A
Thiel, H
Timco, G W
Tomita, H
Tomita, Y
Toyosada, M
Triantafyllou, M S
Tsai, C P
Tsai, T K
Tsuboi, H
Tsunekawa, Y
Tsurusaki, K
Tuhkuri, J
Tulin, M
Uchida, K
Uchida, M
Ueda, Y
Utsunomiya, T
van der Vegte, G J
Vassalos, D
Venkataramana, K
Verley, R
Wadhams, P
Wang, B P
Wang, J
Wang, K H
Wang, Y S
Wang, Y X
Wang, Z Z
Watabe, T
Weave, T J
White, H B
Williams, A N
Williams, F M
Winterstein, S
Wright, B D
Wu, K Q
Xu, T
Yajima, H
Yamaguchi, S
Yamazaki, T
Yao, T
Yashima, A
Yasuzawa, Y
Yeung, R W
Yim, J Z
Yoneyama, H
Yoon, H K
Yoon, J H
Zdravkovich, M M
Zhang, D H
Zhang, H X
Zhang, J
Zhao, Y N
Zhou, L D

ISOPE Publication Order Information

International Journal of Offshore and Polar Engineers - Quarterly (March, June, September, December) (ISSN 1053-5381).

Editors: Jin S. Chung, J. Wardenier, R.M.W. Frederking, W. Koterayama. **Topics**: *offshore engineering, ocean engineering, environment, offshore mechanics, polar (Arctic and Antarctic) engineering, advanced marine technology, materials and technology and energy*: **Yearly subscription rates**: US$120. (Member, complimentary) and US$100 per a set of annual back issues.

Proceedings of the 1st (1990) European Offshore Mechanics Symposium (EUROMS-90), Trondheim, Norway, 20-22 August 1990. A single volume: (ISBN 0-9626104-4-5; LCCN 90-084691); approx. 590 pp., US$80. (Member, US$60.)

Proceedings of the 1st (1990) Pacific/Asia Offshore Mechanics Symposium (PACOMS '90), Seoul, Korea, 24-28 June 1990. For a set of 3 volumes, (ISBN 0-9626104-0- 2; LCCN 90-082039): approx. 1260 pp., US$150 per set and US$70 (Member, US$50) per volume.

Proceedings of Special Offshore Symposium China (SOSC-94) / 3rd (1994) Pacific/Asia Offshore Mechanics Symposium (PACOMS-94), Beijing, China, 16-18 April 1994. A single volume (ISBN 1-880653-15-X); 797 pp., US$100.

Proceedings of 4th (1996) Pacific/Asia Offshore Mechanics Symposium (PACOMS-96), Pusan, Korea, October 31-November 1, 1996. A single volume (ISBN 1-880653-27-3); 325 pp., US$60 (Member, US$50.)

Proceedings of 1st (1995) SOPE Ocean Mining Symposium, Tsukuba, Japan, November 21-22, 1995. A single volume (ISBN 1-880653-21-4, LCCN 95-77708): 238 pp., US$100 (Member, US$80.)

Proceedings of 2nd (1997) ISOPE Ocean Mining Symposium, Seoul, Korea, November 24-25, 1997. A single volume (ISBN 1-880653-33-8): 250 pp. est., US$100 (Member, US$80.)

Proceedings of the 1st (1991) International Offshore and Polar Engineering Conference (ISOPE-91), Edinburgh, U.K., 11-16 August 1991. For a set of 4 volumes, (ISBN 0-9626104-5-3; LCCN 91-071635): approx. 2190 pp., $320 (US$240; member) and $100 ($80; member) for a single volume.

Proceedings of the 2nd (1992) International Offshore and Polar Engineering Conference (ISOPE-92), San Francisco, USA, June 14-19, 1992. For a set of 4 volumes, (ISBN 1-880653-00-1; LCCN 91-78280): 2,899 pp. est., $340 (Member, $260) and $100 ($80; member) for a single volume. See below for postage and handling.

Proceedings of the 3rd (1993) International Offshore and Polar Engineering Conference (ISOPE-93), Singapore, 6-11 June 1993. For a set of 4 volumes, (ISBN 1-880653-05-2; LCCN 92-76219): 2,965 pp., $340 (ISOPE Member, $260) and $100 ($80; member) for a single volume. See below for postage and handling.

Proceedings of the 4th (1994) International Offshore and Polar Engineering Conference (ISOPE-94), Osaka, Japan, April 10-15, 1994. For a set of 4 volumes, (ISBN 1-880653-10-9; LCCN 93-80555): 2,700 pp. est., $340 (ISOPE Member, $260) and $100 ($80; member) for a single volume. See below for postage and handling.

Proceedings of the 5th (1995) International Offshore and Polar Engineering Conference (ISOPE-95), The Hague, June 11-16, 1995. For a set of 4 volumes, (ISBN 1-880653-16-8; LCCN 94-73796): 2,543 pp., $340 (ISOPE Member, $260). 2,589 pp., $340 (ISOPE Member, $260) and $100 ($80; member) for a single volume. See below for postage and handling.

Proceedings of the 6th (1996) International Offshore and Polar Engineering Conference (ISOPE-96), Los Angeles, May 26-31, 1996. For a set of 4 volumes, (ISBN 1-880653-22-2): 2,326 pp., $360 (ISOPE Member, $280). For a single volume, $110 (ISOPE Member, $80). See below for postage and handling.

Proceedings of the 7th (1997) International Offshore and Polar Engineering Conference (ISOPE-97), Honolulu, May 25-30, 1997. For a set of 4 volumes, (ISBN 1-880653-28-1): 3,548 pp., $380 (ISOPE Member, $300). For a single volume, $120 (ISOPE Member, $80). See below for postage and handling.

Postage (surface mail) and handling: add for each of ISOPE-92 through ISOPE-97 Conference Proceedings orders

A set	USA $15	Canada $20	Other countries $35
A volume	USA $8	Canada $10	Other countries $15

How To Order

To order, check or tick ☑ the appropriate boxes, fill in the information below, enclose the correct amount of payment or purchase order, and mail to ISOPE Order Department, P.O. Box 1107, Golden, Colorado 80402-1107 USA

❑ Member rate ISOPE Member No. _______ ❑ Non-member rate

❑ **International Journal of Offshore and Polar Engineering**
❑ **Proceedings of the 1st (1990) Pacific/Asia Offshore Mechanics Symposium**
❑ Vol. I ❑ Vol. II ❑ Vol. III ❑ A Set
❑ **Proceedings of the 3rd (1994) Pacific/Asia Offshore Mechanics Symposium (Special Offshore Symposium China)**
❑ **Proceedings of the 4th (1996) Pacific/Asia Offshore Mechanics Symposium**
❑ **Proceedings of the 1st (1990) European Offshore Mechanics Symposium**
❑ **Proceedings of the 1st (1995) ISOPE Ocean Mining Symposium**
❑ **Proceedings of the 2nd (1997) ISOPE Ocean Mining Symposium**
❑ **Proceedings of the 1st (1991) International Offshore and Polar Engineering Conference**
❑ **Proceedings of the 2nd (1992) International Offshore and Polar Engineering Conference**
❑ **Proceedings of the 3rd (1993) International Offshore and Polar Engineering Conference**
❑ **Proceedings of the 4th (1994) International Offshore and Polar Engineering Conference**
❑ **Proceedings of the 5th (1995) International Offshore and Polar Engineering Conference**
❑ **Proceedings of the 6th (1996) International Offshore and Polar Engineering Conference**
❑ **Proceedings of the 7th (1997) International Offshore and Polar Engineering Conference**

Proceedings of ISOPE-91~97 (*check*):
❑ Vol. I ❑ Vol. II ❑ Vol. III ❑ Vol. IV ❑ A Set

Method of Payment: Advance payment must accompany order (including the postage and handling charge above for 1992-97 Proceedings.) International money order in U.S. dollars, bank checks in U.S. dollars drawn on a U.S. bank, or credit card (Visa or MasterCard only) (*payable to ISOPE*) will be accepted. Remit with the mailing address including the postal (or zip) code for shipping.

❑ international money order in U.S. dollars is enclosed.
❑ bank check in U.S. dollars through a U.S. bank is enclosed.

❑ credit card no. ______________________________ *Circle one*: Visa MasterCard American Express

Exp. Date ________ Cardholder's name ____________________________

Signature ______________________ Date __________________

Ship To Name ____________________________

Company ____ Dept. ____________________________________

Address ______________________________________

City ____________________ State (Province) _______________

Country _______________________ Postal (or Zip) Code _____________

Prospective authors are invited to submit abstract to

ISOPE-98 Montréal, Canada, May 24-29

The 8th (1998) International

Offshore and Polar Engineering Conference

Montréal, Canada, May 24-29, 1998

First

Call For Papers

Abstract Deadline	**August 1, 1997**
Manuscript (for review) Deadline	**November 30, 1997**

The **International Society of Offshore and Polar Engineers (ISOPE)** is organizing about 60 sessions of refereed papers in cooperation with international technical program committee (TPC) members and cooperating organizations. Delegates from more than 35 countries are expected. All manuscripts will be peer-reviewed prior to the final acceptance for the conference proceedings. Papers of archival value may be further reviewed for the *International Journal of Offshore and Polar Engineering*.

Main fields of interest are:

Offshore Technology	**Environment**	**Materials & Welding**
Offshore Mechanics	**Geotechnics**	**Tubular Structures**
Ocean Engineering	**Energy, Resources**	**Risers, Pipelines, Cables**
Polar Engineering	**TLP, SPAR, FPS**	**Comp Methods/CAD**
Deep-Ocean Tech	**Adv Marine Tech**	**Safety, Risk , Reliability**
Underwater Systems	**Hydrodynamics**	**Coastal Engineering**
Measurements	**Underwater Eng**	**Emerging Topics**

The abstract should stress the originality and the significance of the results, and include the complete names, addresses and fax numbers of all authors.

Send 2 copies of your abstract in 300-400 words to: (1) One of the TPC members (session organizers) or (2) **ISOPE-97 TPC,** ISOPE, P.O. Box 1107, Golden, Colorado 80402-1107 USA: FAX 1-303-420-3760.

Sponsors: *ISOPE*. **Cooperating Organizations and Societies**: *ASCE (USA), CAPP (Canada), CSCE (Canada), CSOE (China), CSNAME (China), CSTAM (China), IE Australia, IFREMER (France), IRO (The Netherlands), JSMS (Japan), KCORE (Korea), KSNAJ (Japan), NPF (Norway), OES (UK), PII (Indonesia), RAS (Russia), Scott Polar Res Inst (UK), SSSS (Singapore), and VTT (Finland).*

The Proceedings of
The Seventh (1997) International
OFFSHORE AND POLAR ENGINEERING CONFERENCE

Honolulu, Hawaii, USA, May 25-30, 1997

VOLUME I

OFFSHORE RESOURCE & TECHNOLOGY
SAKHALIN OIL & GAS
RESOURCES & ENERGY (WAVE ENERGY, GAS HYDRATES, DEEP-OCEAN MINING)
TLP, VERY LARGE FLOATING PLATFORMS (VLFS) AND FPSO
OFFSHORE SYSTEMS
ENVIRONMENT (DEEP-SEA IMPACT, WASTE AND TRANSPORT, POLLUTANT, CONTAMINATION, ECOLOGY)
GEOTECHNICAL ENGINEERING

VOLUME II

OFFSHORE AND ARCTIC PIPELINES
MECHANICS OF RISERS, CABLES AND MOORING
UNDERWATER VEHICLES, CONTROL AND SYSTEMS
POLAR AND ICE ENGINEERING
JOIA PROJECTS
ARCTIC AND ANTARCTIC SYSTEMS
ARCTIC OIL POLLUTION
ICE REMOTE SENSING

VOLUME III

NUMERICAL WAVES
NONLINEAR WAVES
WAVE STATISTICS AND MEASUREMENTS
INTERNAL WAVES
HYDRODYNAMIC FORCES
DYNAMIC RESPONSES
AIRGAP AND METOCEAN
VORTEX SHEDDING AND FLOW-INDUCED VIBRATIONS
COASTAL HYDRODYNAMICS

VOLUME IV

TUBULAR STRUCTURES
WELDING AND ROBOTICS
FATIGUE AND FRACTURE
STEEL, CORROSION AND NDT
COMPOSITE MATERIALS
MECHANICS AND STRUCTURES
RELIABILITY, RISK AND SAFETY
COLLISION/IMPACT/DAMAGE
MARINE & UNDERWATER SYSTEMS
EARTHQUAKE ENGINEERING

CONTENT

POLAR AND ICE ENGINEERING

Proceedings of the Seventh (1997) International Offshore and Polar Engineering Conference
Honolulu, USA, May 25-30, 1997

ISBN 1-880653-28-1 (Set); ISBN 1-880653-30-3 (Vol. II)

The NorFra Pipeline Shore Approach: Engineering, Environmental and Construction Challenges

Bernt Thorbjørnsen, Henning Dale and Steinar Eldøy
Statoil
Stavanger, Norway

Massimo Mercanti
Snamprogetti
Italy

ABSTRACT

The NorFra pipeline is a 42 inch dry gas pipeline connecting Norwegian gas fields in the central area of the North Sea to the gas supply system in France. The pipeline enters the French coast at Dunkerque.

The present paper explains the NorFra pipeline project and put focus on the landfall selection process, environmental impacts and constraints, and the engineering and construction challenges. The paper emphasises the role played by coastal dynamics models established as part of the work. It will be explained how long and short term morphological processes i.e. local and global bank and dune erosion have been implemented in the engineering and construction solutions for the NorFra pipeline through the near shore area.

Problems regarding pipeline stability and pipeline installation will specifically be addressed. Landfall and shore approach construction work planning will be given attention. An environmental impact assessment (EIA) was prepared as a part of the application to the French authorities. The EIA played an active part in the final selection of the landfall site. The paper will present the technical concept finally selected and describe the engineering process leading to the final shore approach solution. The technical problems in the landfall will be focused and the practical use of the coastal dynamic model for the engineering and the construction work will be emphasised. The modelling work involved is described in a separate ISOPE'97 paper, ref. 1.

Keywords : NorFra pipeline, pipeline stability, pipelaying, coastal-modelling, construction, dredging, shore approach, environmental impact

INTRODUCTION

The NorFra pipeline connects central North Sea gas fields with France and enters the coastline at Dunkerque on the French coast of the English Channel. The coastal area is characterised by an irregular seabed dominated by a complex pattern of sand banks, strong tidal currents, severe wave attack from the North Sea and a high rate of natural sediment transport. Due to this the last 20 km section of the NorFra pipeline route has been a major engineering and construction challenge. The NorFra pipeline route is shown in figure 1.

The new landfall in France provides a third delivery point for gas supply to the Continental Europe, in addition to the existing ones in Germany and Belgium. The NorFra pipeline will allow an annual delivery of 14 billion m^3 of natural gas from 1998, representing about 35% of the French needs by the year 2005.

The main technical challenges in France were the pipeline shore approach, the location of the terminal, the route for the feeding pipeline to the existing gas grid, and the interfaces. Five landfall areas were spotted and technical, environmental and commercial criteria for the final selection were identified which concluded with a shore approach east of Dunkerque West Harbour. The area is a typical tidal dominated continental shelf where large sand banks, found in groups more or less parallel to the coastline, are the characteristic and dominant seabed feature. They are crossed almost perpendicularly by the NorFra pipeline and have a typical cross section of 1 - 3 km with a height of maximum 20 m. The water depth along the route varies from 5 m on the top of the sandbanks to 25 m in between the sandbanks. The technical challenge is due to the severe environmental conditions in combination with the irregular seabed topography. To understand the complex phenomena acting in this area coastal dynamic models have been established and used for engineering and construction purposes.

THE NORFRA PROJECT

The overall NorFra project consist of :

- modification of the existing North Sea Draupner platform
- construction of a pipeline with a nominal diameter of 42", a design pressure of 156.8 barg and a total length of 840 km of which 30 km is located in French coastal waters
- construction of a gas terminal located 4 km from the landfall point at the beach, with the main purpose of gas regulation, odorizing and fiscal metering.

LANDFALL SELECTION

During Conceptual Engineering six possible coastal locations in France were identified, see figure 2. After an in-depth evaluation the Dunkerque site was selected.

An environmental impact assessment (EIA) had to be prepared as a part of the application sent to the French authorities. This EIA had to include both a comparison and evaluation of impacts of different landfall alternatives, as well as an impact evaluation of the selected route. The EIA documented that the selected route with landfall in Dunkerque was also the preferred route from an environmental point of view, for the following reasons:

- this route avoids ecologically more sensitive areas further south along the coast, as well as highly urbanised areas onshore
- the route allows the onshore terminal to be located in an area causing minimal environmental impact
- landfall in Dunkerque gives the shortest offshore pipeline route in French waters
- the selected route reduces the dredging volumes, thus impact on the seabottom and marine life, and energy consumption and emissions during construction
- the selected offshore route avoids the more heterogeneous and sensitive areas for marine life located further to the Southwest
- the shorter offshore route causes less interference with fishery activities along the coast

The landfall and offshore routing is shown in fig. 3.

SEABED MORPHOLOGY

The region of Dunkerque is a part of the Flemish maritime plain. The plain is protected against the sea by coastal dunes. The coast is rather straight, low and sandy. The beach is in general fairly wide, up to more than 1500 m 20 km west of Dunkerque. In the vicinity of the landfall area, including the landfall area itself the coast has been modified by man. The "Braek" beach is made of an artificial sand embankment built since 1984.

The near shore area outside Dunkerque is a complex coastal area characterized by isolated sandbanks, separated by deep sandy throughs. The depth varies from 5 m on the top of the bank to 25 m in the throughs. A cross section profile is shown in figure 4.

OCEANOGRAPHIC CONDITIONS

The environmental regime is characterized by a strong tidal currents of more than 2 knots during spring tide. These are nearly parallel to the coast and to the banks close to the shore. Wave action is strong in the whole area with breaking on the beach and on the crest of the banks. A significant effect of the banks is an effective damping of the waves shorewards along the route, due essentially to the breaking and spreading effect of these quite shallow banks.

WAR REMNANTS

The route survey performed using sub-bottom profiler and magnetographical equipment identified several objects not otherwise known. Historically this part of the coast of France has been a highly exposed area for warfare. War remnants from both the 1. and the 2. World Wars have been and are still found.

The NorFra project had to consider this fact carefully in the planning of the pipeline route. In the latest edition of the British Admiralty map of the coastal area outside Dunkerque a former mined area was identified. A considerable effort was taken in investigating source material on mine-belt areas, removal programs and probability of detonation when finding. Both the English and the German naval departments supported the project by releasing essential data from their historical files for this purpose. The French navy also gave valuable support in performing a mine sweeping of the route.

This information was used in a probabilistic risk evaluation concluding that irrespective of position in the English Channel or the southern part of the North Sea the probability of detecting war remnants is low. The probability to trigger an unwanted detonation when remnants are found is also low. War remnants thus were concluded not to represent an essential constraint during the final routing process. Special precautions will, however, be taken during the dredging work in this area by installing a grid in front of the suction head.

CONCEPTUAL STUDIES

Seabed Morphology Criterion

A key premise was established at an early stage of the NorFra project: "The pipeline shall not be installed above the envelope of the lowest seabed level to be expected during its design life time".

Totally 7 historic bathymetric surveys in the actual nearshore area were identified and all the data was collected and used in the engineering of the near shore part of the pipeline route. The first bathymetric survey performed in this area dated back to 1861. The latest data was gathered by the NorFra project which performed a basic route survey in 1995.

Close examination of all the survey data including numerical comparison of all the variable soundings since 1861 together with study of relevant published papers on the local coastal morphology lead to the following main results:

- In the long term and at a regional scale, the Flemish Banks system appears as a rather stable system, with just a possible erosion of the crest of the banks, with a global long term tendency to migrate slowly shore-wards.
- On a shorter time scale some banks may experience fluctuating migration, in a range of some 100 to 200 m in a transverse direction, and with a period of a few decades.
- Moreover, in some specific areas, bed changes can be quite rapid and significant; bars in the channels, breach across the banks.

Sedimentation Study

A considerable amount of work was required to establish rates of sedimentation and scour of the predredged access channels for the lay barge. Local wave and current data along the nearshore route was necessary for this purpose. This work was performed through several steps. The most comprehensive one was to establish a computerised model of the phenomena and the boundary conditions for the model, ref. 1.

Pipeline Stability

The combination of rather deep (25-30m) and very shallow (5-10m) route sections together with a complex and rough environmental climate represented a considerable engineering challenge.

Long term stability

It was identified that long term stability could not be obtained without pipe burial through trenching and/or backfilling. The key premises were to identify pipe weight coating to satisfy both horizontal and vertical stability in a buried condition and to ensure layability. The analysis lead to a pipeline concrete thickness of 90 mm with density of 3400 kg/m^3 (submerged weight of 5500 N/m). The layability of this pipeline is ensured by means of a Flat Bottom Lay Barge with maximum operating draught of 5.5 m utilising additional buoyancy for laying in the sections deeper than 17-18 m only.

Short term stability

Results of short term pipeline stability calculations based on preliminary wave and current data gave excessive concrete thicknesses along parts of the nearshore route. The actual pipeline sections could not be laid by conventional lay barges based on calculated concrete thicknesses satisfying short term stability requirements. On the top of the sandbanks, concrete thickness of up to 19 cm was required to obtain short term stability, ref. fig. 5. Such concrete thicknesses is also on or beyond the limit for what is feasible from a production and logistic standpoint.

ENGINEERING APPROACH.

Installation Alternatives

It was decided at this stage that the project had to establish other solutions than the conventional one in order to enable pipeline installation along the nearshore route and at the same time ensuring short term stability of the pipeline. Two alternative installation concepts were considered, both concepts subject to the premise of not increasing the concrete thicknesses ensuring long term stability. The two installation concepts were:

- Progressive water flooding
- Progressive post trenching

A third fallback solution also based on flooding of the pipeline was considered for the event of an emergency situation during laying.

Progressive water flooding

A progressive flooding of the pipeline during laying can be utilised by means of flooding equipment connected to the pull-head on the beach and by introducing a proper tool (high friction plug) into the pipeline to control the length and sealing of the flooded section. The plug is controlled remotely or by use of friction and volume input control.

Progressive post trenching

Progressive post trenching means utilising trenching equipment working immediately behind the lay vessel in order to stabilise the pipeline immediately after laying.

Emergency flooding of the pipeline

Emergency flooding of the pipeline can be utilised through using a similar tool as for progressive flooding. The main difference is that the control requirements may be relaxed in the latter case. The water filling pump in this case needs only to be in an alert status to allow for short term water filling of the already laid pipeline section. It was established that the weather conditions requiring laydown of the pipeline were just severe enough to give an unstable pipeline situation. This means that if a summer storm which occurs during pipelay is strong enough to compromise the stability of the pipeline, this will also cause abandonment of the pipeline.

Environmental Modelling

In parallel with the engineering of the above installation concepts a study was performed with the objective to establish in detail and high quality wave and current data along the whole nearshore route. The approach was to establish coastal hydrodynamic models, in co-operation with coastal engineering specialists, capable of giving valid local design data along the whole nearshore route. Further details on this part of the work are presented in ref. 1.

EVALUATION OF ALTERNATIVES

Progressive Water Flooding

The alternative methods for nearshore installation of the pipeline described above were engineered and evaluated. Progressive water filling during laying was concluded to be a feasible solution for pipeline installation in the nearshore and landfall section. This was due to the fact that Statoil has considerable experience from development and use of this type of advanced plugs. This concept was thus selected as basis for the Installation Permit Application to French authorities.

A special task force was established for final engineering and testing of the high friction plug. The plug would have to hold typically 5 bar differential pressure during its travel through the complex sand bank area in order to provide control of the operation. A 5 bar differential pressure corresponds to a force of approximately 37 tons which has to be absorbed by the plug body.

Progressive Trenching

The progressive trenching concept was concluded to be less favorable. Progressive trenching requires the trenching spread to work in the tail of the laybarge. Both the laybarge and the barge towing the trenching equipment is moved forward by pulling anchor chains. The minimum distance between the two barges is derived from the requirement that no anchor chain interference is allowed. The minimum distance is 2000 meter. A normal lay speed of 0.5 km/day will give a pipeline exposure time of 4 days before burial which was not acceptable.

The inherent commercial and liability aspects in having two contractors working very close to each other were likewise considered non-attractive.

FINAL SELECTION OF THE NEAR SHORE SOLUTION

At the end of October last year a final selection of nearshore installation concept could be made. The following two facts were the basis for this selection:

- the final testing of the plug for progressive water flooding control was planned to take place during winter and spring 1997 and was consequently not completed

- the model studies were completed giving considerably improved data on wave action and current strength along the whole nearshore route.

Based on these data the short term pipeline stability was reevaluated. This work concluded that short term stability of the pipeline could be obtained along the whole nearshore section by a concrete coating of thickness less than 11.6 cm. For reference see fig. 6. Weight coating production feasibility and pipeline layability was investigated regarding the results from the revised stability analysis. It was concluded that it was possible to install and obtain short time stability of the pipeline by use of overweight concrete coating only.

The finally selected solution thus was the conventional concrete coated and airfilled laying of the pipeline for the whole nearshore and landfall section. The testing of the progressive flooding plug was decided to proceed as planned.

ENVIRONMENTAL IMPACT OF FINAL ROUTE AND INSTALLATION METHOD

Generally the environmental impact of the NorFra pipeline in the nearshore and landfall area are temporary and restricted basically to the construction period which includes dredging, trenching and backfilling work. Dredging works; which are required for different reasons, i.e. to satisfy pipeline stability, authority requirements, barge access and pipe on bottom comfort. Trenching and backfilling; to satisfy pipeline stability and authority requirements. The pipeline will thus be buried throughout the French waters, and only the landfall valve station and the receiving terminal will be visible in the landscape onshore.

The temporary impacts includes

- emission to air from the construction vessels (CO_2, NOx, SO_2)
- increased turbidity and change in water quality
- destruction and disturbance of marine life on the sea bottom
- disturbance on fishery activities during construction

Marine activities does normally give relatively high emissions of NOx and SO_2 resulting from fuel combustion (diesel engines). A solution aiming at reduced dredging activities does therefore reduce emissions to air. The construction activities are estimated to represent an increase in marine activities of 25 % compared to the generally high ship traffic in the area near Dunkerque, during the six months construction period. Existing onshore industry is however the major contributor to the air pollution in the area. In order to minimise the emissions from the construction of NorFra, it was decided to use low sulphur fuel on the vessels (0,2%).

Dredging and trenching activities generally causes increased turbidity. However, absence of fine sediments (mud and clay) in the sediments to be dredged and discharged reduces the potential for increased turbidity. Furthermore, experience from another pipeline landfall (Europipe in Germany) indicate that the turbidity increase will be only very local. It is thus expected that the ecological impact from increased turbidity will be insignificant.

The marine life at the sea bottom will be influenced by dredging activities. However, as the sea bottom along the selected route is relatively homogenous (consisting of sand) the area is not particularly rich from a biological point of view. Experience from the Europipe project in Germany indicate a regeneration time of 1-2 years for the marine life at the sea bottom.

There are significant fishery activities along the pipeline route near the coast. Different fishing gears are used, with fixed nets, beam trawls and shrimp trawls being the most important. As the pipeline will be buried or trenched in the French waters, no interaction between fishery and the pipeline is expected during the whole operational life of the pipeline. The most important fishery season is winter and spring. Construction activities are planned for avoiding construction works in the most intensive fishery season. Fishing activities will only be restricted in the vicinity of the construction vessels during pipelaying and trenching.

The environmental impact assessment bridged the technical solutions and was a valuable support to the technical work as well as a requirement from the French authorities.

CONSTRUCTION CHALLENGES

The construction work will have started when this paper is presented. This chapter will therefore only summarise our knowledge at present when writing.

Vessel administration

The most challenging part performing activities such as civil work, dredging, laying, trenching, dumping and backfilling in a very limited area with considerable traffic to a main industrial harbour is the logistics and vessel co-ordination. This was also a key issue raised by the French authorities, and we were given requirements regarding crossing of the main shipping channel. The work should be limited to a total closure of less than 5-7 days over two periods, the periods for laying and trenching. The dredging work should only give partial closure. A safety study indicated requirements for pipeline burial by backfilling in the shipping channel and it was decided to also utilise long time stability through backfilling, hence trenching was not required. The planning process lead to the conciliation that by marginally increasing the dredging work in the shipping channel as compensation for lowering by trenching, full closure could be limited to one period and to 3-4 days only. The total number of vessel required for the landfall is estimated to slightly less than 50 during the construction period.

Access trench sedimentation

When planning the construction work in the landfall uncertainties were related to the required maintenance to keep the dredged channel (for barge access) and trench (for pipeline) open between finish dredging and start pipelaying. This is related to sedimentation rate, time span and weather conditions. The hydrodynamic model simulations also provided improved estimates of the sedimentation rates.

Initially a maintenance dredging volume of 30% where anticipated included. However, the new results showed low sedimentation rate during good weather condition, i. e. approximately 5-10% cross section area reduction over a 6 weeks period. In storm condition this was raised to maximum 20%. This represented a considerable improvement of the basis for construction work planning.

CONCLUSIONS AND FINAL REMARKS

The shore approach section of the NorFra pipeline represented a significant engineering, environmental and construction challenge. The environmental impact assessment (EIA) proved to be an important tool in the landfall selection and gave valuable support in the planning of the construction work. The wave and current model development and consequently the results applied for the final engineering and the construction planning must be considered as the key to the success in establishing a rational an well founded near shore pipeline stability design and installation solution.

It should be pointed out that the knowledge of the environmental processes along the nearshore route regarding wave and current climate as well as sedimentation rates will be of further use during the construction work in June and July 1997.

ACKNOWLEDGEMENT

The authors are highly indebted to Statoil and partners of the NorFra project for admission to publish this paper.

REFERENCES

1. Moshagen H., Thorbjørnsen B., Strass P., Latteux B., and Venturi M.: *"The NorFra Pipeline. Challenging Shore Approach Handled by Coastal Dynamics Modelling"*, Proc. ISOPE, Honolulu, 1997.

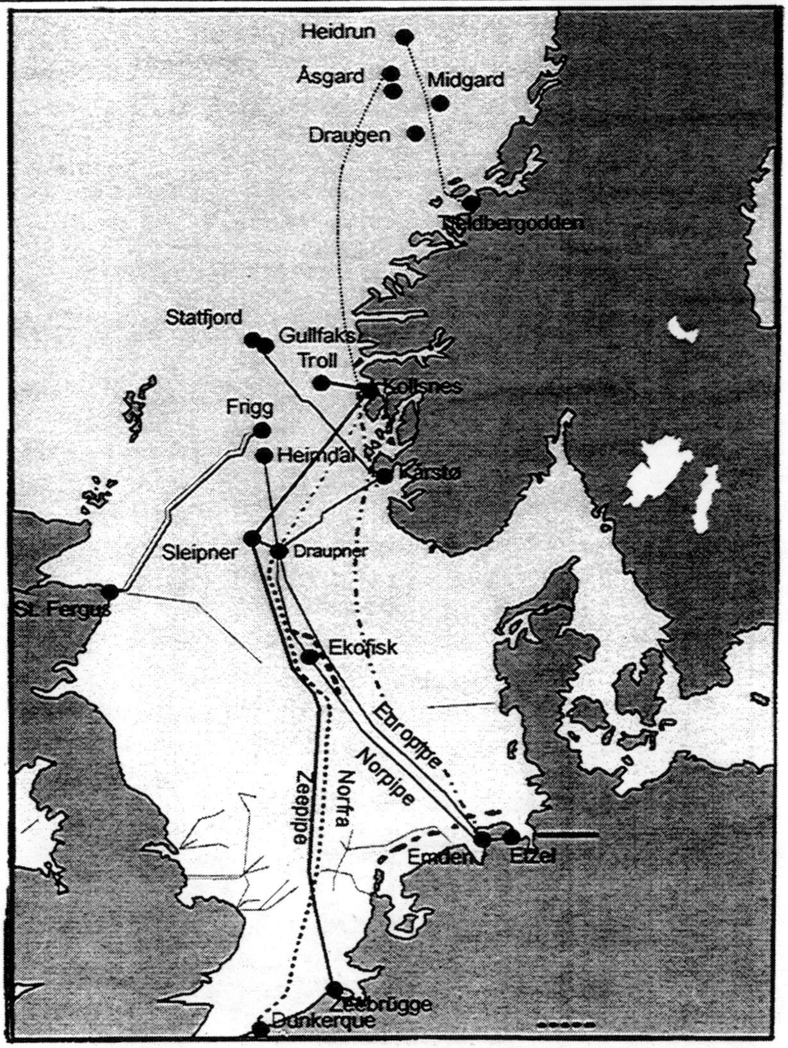

Figure 1. Norwegian Gas Pipeline Infrastructure

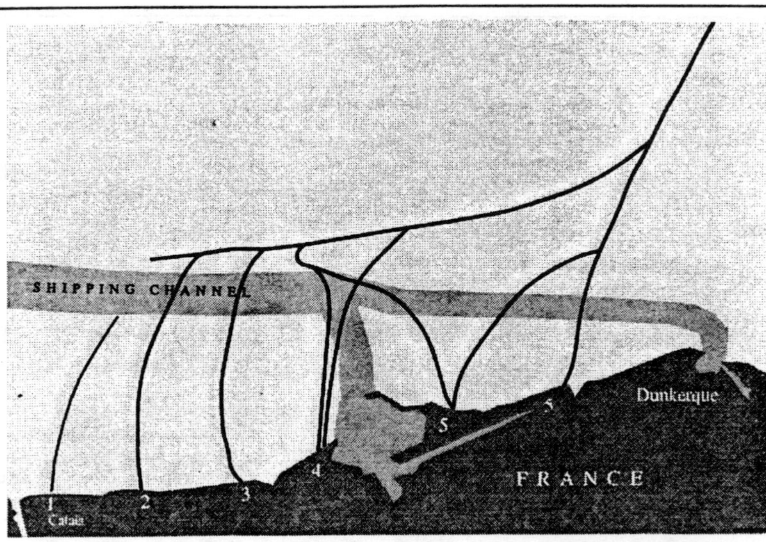

Figure 2. Alternative coastal locations for landfall in France.

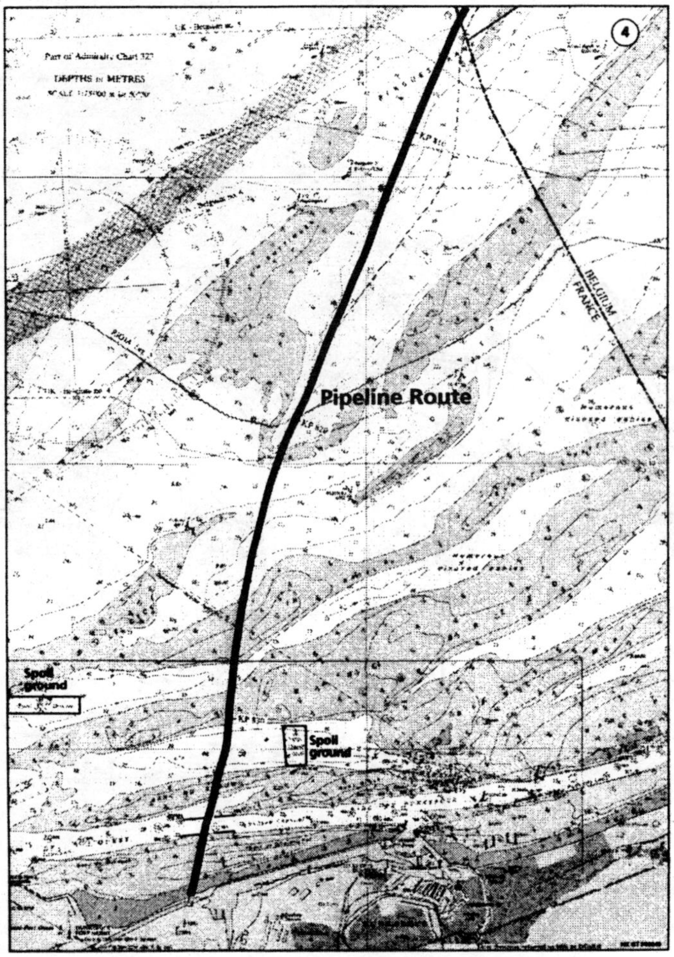

Figure 3. NorFra pipeline route in France.

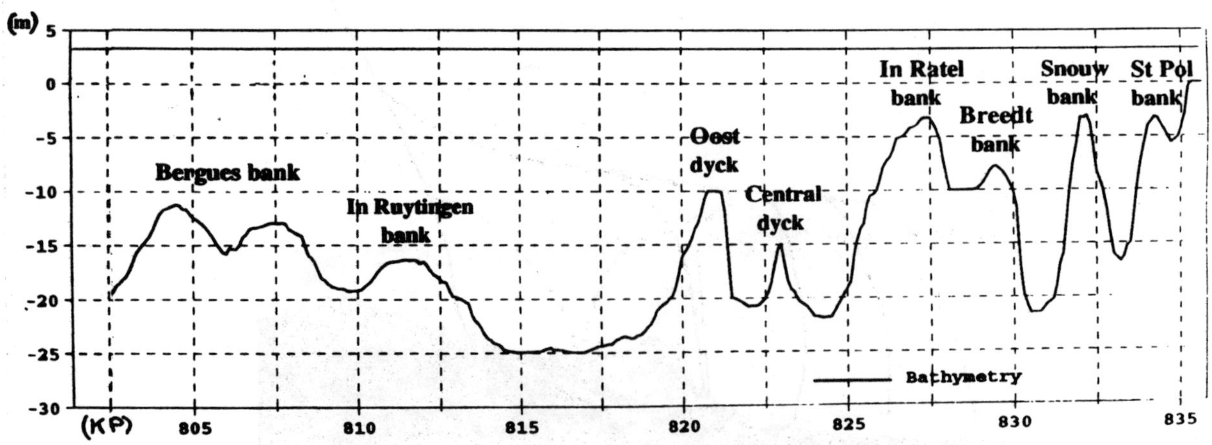

Figure 4. Cross section profile, pipeline route in France.

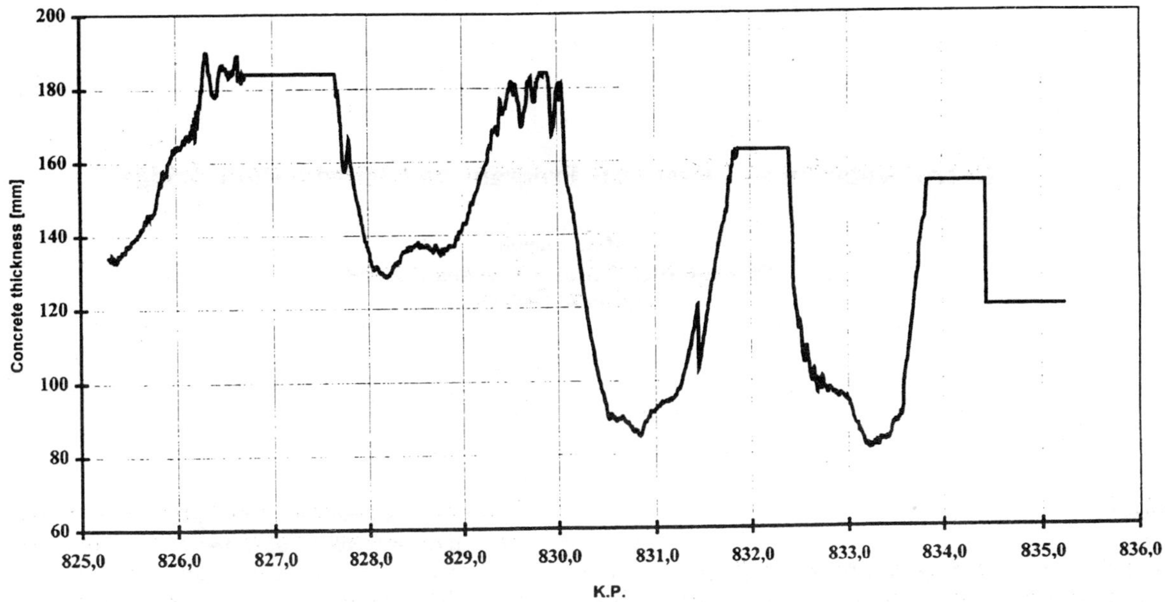

Figure 5. Minimum required concrete thickness, temporary on bottom stability.

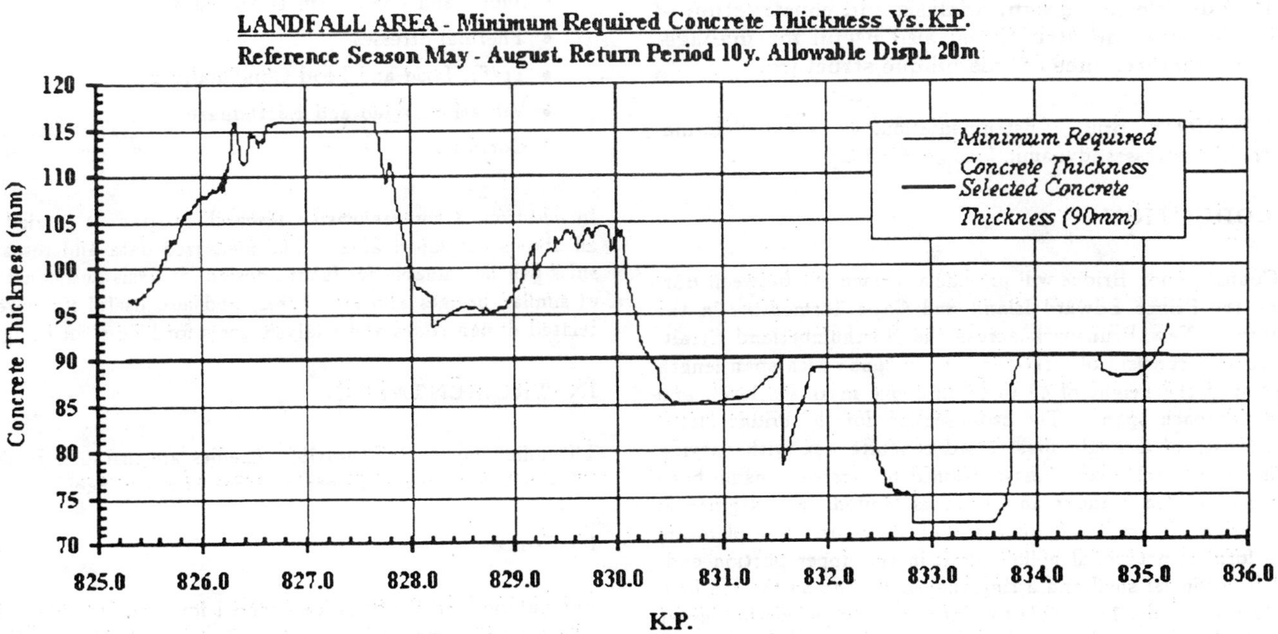

Figure 6. Minimum required concrete thickness based on numerical model study.

Proceedings of the Seventh (1997) International Offshore and Polar Engineering Conference
Honolulu, USA, May 25-30, 1997

ISBN 1-880653-28-1 (Set); ISBN 1-880653-30-3 (Vol. II)

Instrumentation and Research Program on Confederation Bridge

M.S. Cheung
Ottawa-Carleton Bridge Research Institute
Ottawa, Canada

ABSTRACT

The 12,880 m long Confederation Bridge will be the world longest spanned box girder bridge over salt water. Several design criteria for this bridge, with 250 m span length, 100 year design life and 4.00 to 4.25 safety index, are not covered by any code or standard. A comprehensive monitoring and research program is being carried out to monitor and study the bridge behavior under ice forces, short- and long-term deformations, thermal stresses, traffic load and load combinations, dynamics, wind and earthquake, and corrosion. The measured data and research results will be valuable for design, analysis and construction of similar bridges and structures, also useful for management and maintenance of this unique structure.

KEY WORDS: Bridge, instrumentation, ice forces, deformation, thermal stresses, dynamic load, traffic load.

INTRODUCTION

The Confederation Bridge will provide a surface link between Borden on the Prince Edward Island and Cape Tormentine in the Province of New Brunswick across the Northumberland Strait. The crossing consists of forty-three main spans with span length of 250 m at the height of 40 to 60 m above mean sea level, and eleven approach spans. The cross-section for the bridge structure consists of a single prestressed concrete box with varying depths within each span. Post-tensioned prestressing, using both internally bonded tendons and external tendons, and segmental construction are employed for the superstructure. The piers are of the form of octagonal hollow shaft in the upper portion ending with a conical shell and a ring-shaped footing on the sea bed. This 12,880 m long, 11 m wide two lane bridge is the world longest spanned, box girder bridge over salt water. The design life of the bridge is 100 years, and its design procedure is not covered by any code or standard worldwide.

This bridge is constructed and prestressed in segments in numerous stages, and will be operated under severe environment conditions. Consequently, accurate prediction of the behavior of such complicated bridge is a very difficult undertaking. In order to verify the design loads and load effects, two consecutive spans, one rigid frame span and one drop-in span, and four piers of this structure are to be instrumented to monitor its short-term and long-term performance under the following environmental loads and effects:

- Ice forces
- Short- and Long-Term Deformations
- Thermal Stresses
- Traffic Load and Load Combinations
- Vibration, Wind and Earthquake
- Corrosion.

In addition, a comprehensive research program is carried out in all above-mentioned areas. The measured data and research results will be valuable for future design, analysis and construction of similar bridges and structures, and also useful for developing related design codes and analysis tools for long-span bridges.

INSTRUMENTATION

The following instrumentation schemes are installed to monitor the short- and long-term performance of the bridge.

Ice Forces

Northumberland Strait is ice covered for 4 months every winter. The ice cover may be termed as first-year pack ice (Williams, 1996), consisting of discrete floes which may contain rafted and ridged ice. Floe size can reach 3 to 4 km in diameter, with level ice thicknesses to 1 meter. The embedded ridges can have keel depths as high as 18 meters. The ice cover is very dynamic under the action of wind and currents, which at the bridge, are largely tidal. When the ice floe encounters the conical ice shield of the bridge pier, the motion of the ice causes the leading edge to slide up the slope of the structure, inducing a bending moment within the ice sheet, which causes the ice to fail in flexure (Fig. 1) (Croasdale and Cammaert, 1996). The resulting ice pieces are pushed up and over the conical surfaces by the continuing advance of the interacting ice sheet.

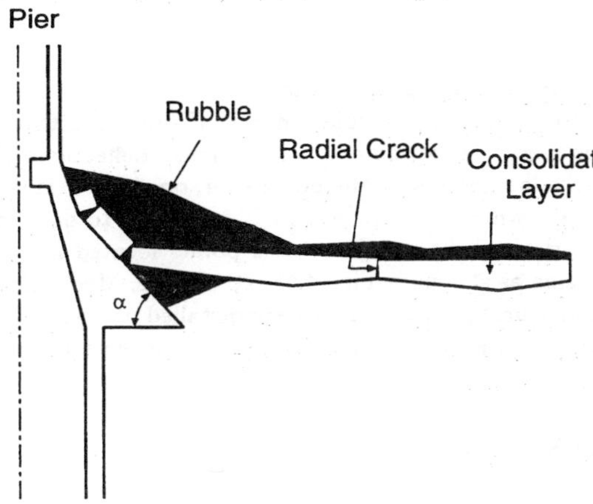

Fig. 1 Ice Interacting with Cone

When one of the embedded ridges encounters the pier, the failure of the ice floe is complicated by the presence of the broken ice rubble in the keel and sail, which must also clear round the pier, and which add to the overall load. Because of the complexity of the mechanics of the interaction, there are many different theories for the determination of the ice load leading to wide variance of the resulting design loads. For this reason, the monitoring program includes a scheme to measure global and local ice forces, and to observe the behavior of the ice interacting with the bridge piers and the prevailing ice conditions.

The ice force monitoring program encompasses three components: the indirect measurement of the global ice force, direct measurement of local ice pressure, and observation of the ice interaction kinematics, and prevailing ice conditions. The global ice force can only be measured indirectly as there is no technology currently available to measure the total load which would be compatible with the pier design. The load will be determined by measuring the structural response of the two piers which are part of the instrumented frame. The response will be measured by very sensitive biaxial tiltmeters placed at two locations in the pier shafts of Pier 31 and Pier 32 (Fig. 2). These have the capability of measuring tilt as low as 0.1 μ rad, which corresponds to an ice load of about 5 kN. In order to provide better dynamic resolution of the response, these are coupled with highly sensitive accelerometers measuring the corresponding lateral accelerations. These accelerometer outputs will also be used for the dynamic monitoring of the bridge, as explained later. The combination of instruments will be capable of measuring ice forces as low as 5 kN with frequencies to 3 Hz.

To measure local ice pressures, a total of 56 m^2 of ice load panels have been installed on pier P31 (Fig. 3). These consist of thin double-walled panels with internally strain-gauged load sensors, measuring loads on area of 0.25 m^2. These panels are installed on the conical ice shield (40 m^2) and on the pier barrel below the cone (16 m^2). The lower panels are designed to measure loads from interactions with ridge keels, while the cone panels are designed to measure the loads from the sheet ice. The panel results will be used to determine the distribution of ice loads, and will be valuable in the validation of ice load algorithms and the development of new algorithms.

The final component of the ice force monitoring project is the observation system. This system consists of four time lapse video cameras, two sonars, and a laser. Three of the cameras have been deployed to monitor the kinematics of the ice interacting with the two instrumented piers, P31 and P32. The fourth camera will be used to monitor the prevailing ice conditions in the strait. The cameras will be linked by a digital multiplexer to a 24 hour time lapse recorder. Two upward-looking sonars have been placed on the seabed immediately (40 meters) to the west of P31. These will be used to measure the keel depth of the ridges which are interacting with the instrumented pier. The laser will be arranged to measure the freeboard of the ice at the same location as the sonars are measuring the draft.

Short- and Long-Term Deformation

In design and earlier analyses, all factors which influence the time-dependent deformation of the structure have been accounted for. These include: variation of the modulus of elasticity of concrete with its age, creep and shrinkage of concrete, relaxation of prestressed steel, foundation movements, etc. However, only assumed values were used in design. Therefore, these values should be verified.

1. Tests for the modulus of elasticity and strength of concrete:
For concrete, both the modulus of elasticity E_c and the strength f_c' vary with time. The development of $E_c(t)$ with aging will be determined for three arbitrarily chosen concrete casts. For each of these casts 3 × 6 cylinders are stored and tested for E_c and f_c' values at 6 selected ages. The results of these stages are to be used to determine expressions to give the value of $E_c(t)$ and $f_c'(t)$ for any value of t, within the range of interest for this bridge. The results will be used to calibrate the expressions assumed in design.

2. Tests for creep and shrinkage of concrete:
The creep and shrinkage properties will be determined for three arbitrarily chosen concrete casts. All concrete specimens will be subjected to the same weather and curing as the prefabricated bridge elements. Five pairs of concrete cylinders are loaded at five different stages of the construction period to determine creep plus shrinkage values. Companion non-loaded specimens of the same number will be monitored to measure shrinkage values. The test devices and the companion specimens are kept inside the box girders.

3. Tests fot steel relaxation:
Three samples of steel used at each of three different prestressing stages subjected to three levels of initial stress, will be used for the long-term relaxation tests under temperature-controlled environment. Four tests will be conducted to evaluate the prestress

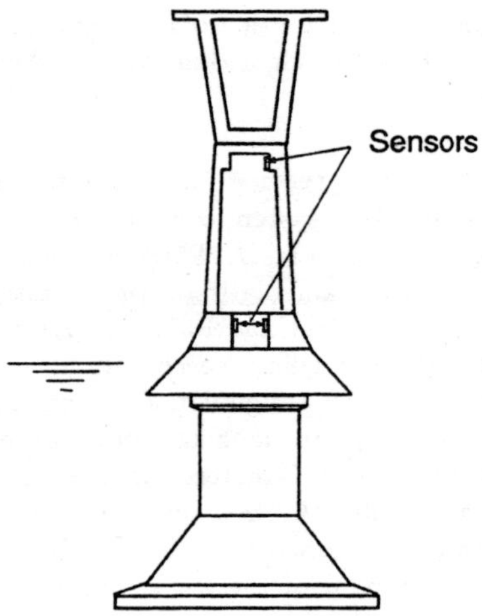

Fig. 2 Location of Tiltmeters

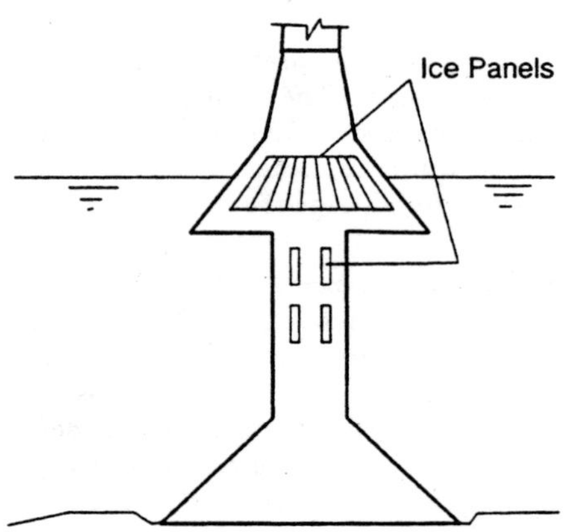

Fig. 3 Location of Ice Force Panels

loss on externally prestressed cables.

4. Pier footing movements:
In order to determine the foundation settlements, the displacements at the junction of the pier base with pier shaft (4 m above mean sea level) will be measured. The vertical translation with respect to stable onshore benchmarks will be monitored by precise leveling technique, while pier rotations about two horizontal axes will be determined using auto reflection measurements (Tesky, 1988).

5. Displacements of superstructure:
In order to monitor the displacements of superstructure, precise leveling is used to determine relative vertical deflections at the 17 benchmarks located inside the box girder, on the centre line of the bottom slab, whilst auto reflection is employed to determine the relative horizontal displacement at 4 points located in each of 11 cross sections as shown in Fig. 4. Furthermore, Mechanical strain gauges and vibrating wire gauges are installed at selected sections and locations of girders and piers (Fig. 5) to measure strains in the superstructure.

Thermal Stress

Thermal effects of a typical 250 m span were monitored during construction and their measurement will continue after completion of structure. Using thermocouples, vibrating wire strain gauges, temperature and strain are measured at selected sections. Pyranometers will be installed to record the intensities of solar radiation.

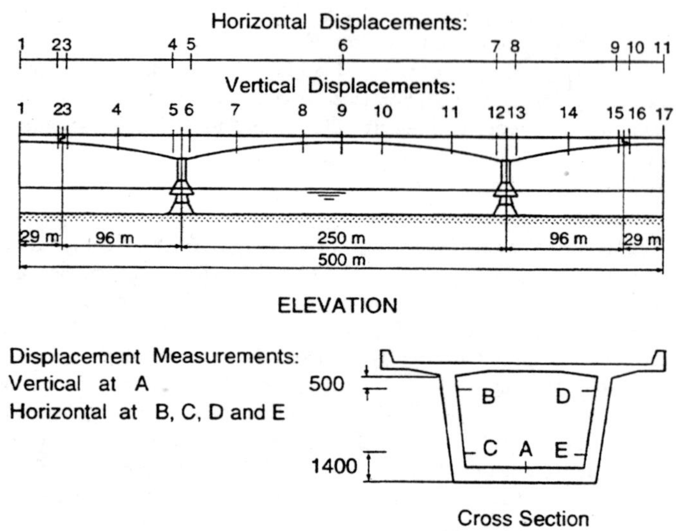

Fig. 4 Location of Displacement Measurements

During the construction stage, the main source of thermal stress is the heat of hydration caused by the chemical reaction of the cement with water. The resulting temperature rise and subsequent restrained non-uniform cooling may harm concrete strength and develop significant thermal stresses and potential cracks, especially in thick components built with high strength and high performance concrete. Both the maximum temperature and maximum temperature gradient have significant influence on the future concrete strength and potential crack formation.

The temperature will be recorded by thermocouples at seven sections (four along the deck, two in piers and one in ice shield) and in one ring footing as shown in Fig. 6. In each section a number of thermocouples are installed along the thickness direction at selected positions. In addition to temperature reading, strain measurements are taken using vibrating wire gauges at these sections to check the strain distribution. The typical arrangements of the thermocouples and strain gauges are shown in Fig. 7 and 8 respectively for section 1 and 2 in the box girder and section 7 in the ice shield. Section 1 has the largest differences in member thickness of the box girder and the large temperature differences are expected within this section. Section 7 is the most massive component and a high temperature was anticipated during construction.

For theoretical temperature studies on bridges the environmental data are necessary. To collect the information three pyranometers will be installed, one above the top surface of the bridge and one on each web of the box girder, for measuring the intensity of solar radiation. In addition, several thermocouples will be installed to record the ambient temperature as well as the temperature inside the box girder (Fig. 9).

After completion of the instrumented part of the bridge, all the strain measurements and most of the temperature readings at the seven selected sections will continue. Additional measurements will be made to record the deformation of the bridge. These measurements will provide information about the axial, vertical, horizontal and torsional deformation of the structure, including the movement at the expansion joints. The overall deformations of the superstructure and horizontal movements of piers will be measured using high precision survey equipment. Cable-tension linear transducer will be used to measure movement at the two expansion joints (Fig. 10).

In order to be able to determine thermal stresses from measured values of temperature and strain, the thermal properties of the concrete used for the bridge will be tested in laboratory. Relaxation type tests will be performed, where the stresses are recorded in axially restrained concrete prisms subjected to temperature changes developing at different rates. A test set-up developed at The University of Calgary (Chui and Dilger, 1993) allows any time-temperature history to be generated and the resulting stress history to be recorded automatically. This temperature loading process will be done repeatedly at different concrete ages until this monitoring program terminates.

Traffic Load and Load Combinations

Monitoring of the traffic loads includes:
1. Measurement of vehicle configurations and loads,
2. Multiple presence of vehicles,
3. Static and dynamic load effects on the structure

Vehicle configurations and loads will be monitored using recording video camera and existing weigh station near the approach span.

Statistic analysis of data collected through video systems will be

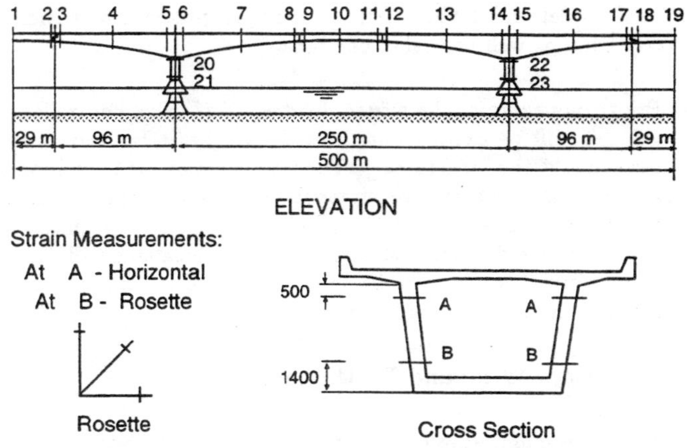

Fig. 5 Location of Strain Measurements

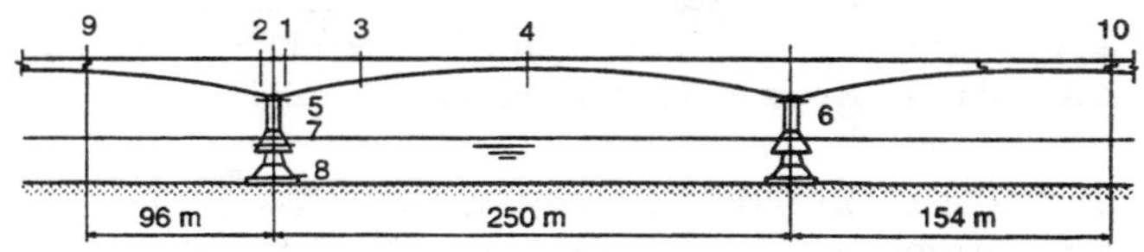

Fig. 6 Sections for Thermal Effect Monitoring

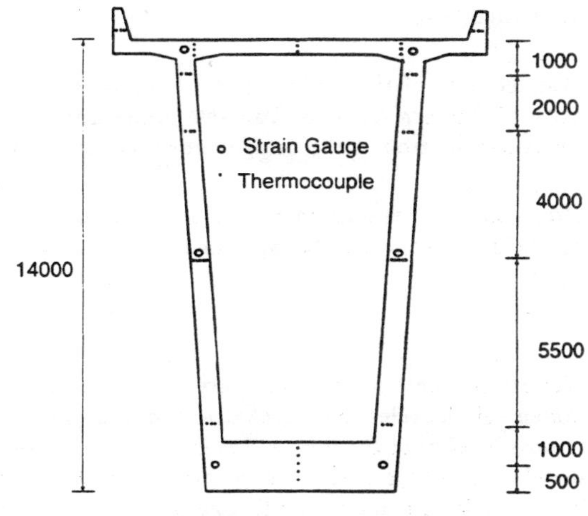

Fig. 7 Thermal Effect Monitoring at Sections 1 & 2

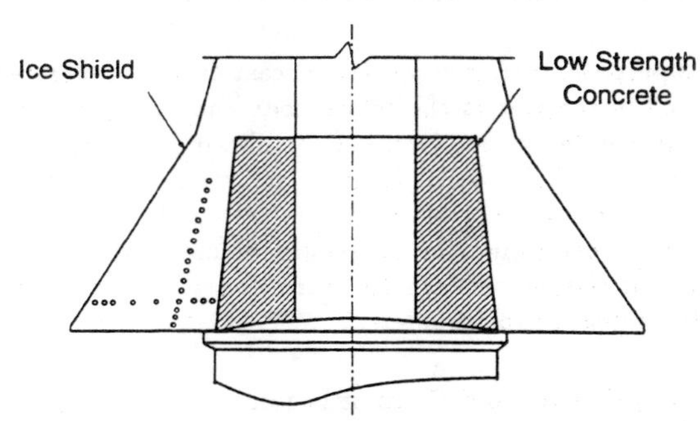

Fig. 8 Thermocouples in Ice Shield

used to determine the effects of multiple presence of vehicles in two lanes as well as within one lane.

Prior to opening of the bridge, loading tests using controlled test vehicles will be carried out to establish the static and dynamic behavior of the structures. The strain gauges used for monitoring thermal effects will be employed to measure longitudinal and transverse stresses due to these vehicles. After these initial load tests, periodic surveys of strains at selected locations will continue under normal traffic to verify the influence of multiple presence of vehicles on the static and dynamic loads.

Vibration, Wind and Earthquake

long term monitoring of dynamic response of the bridge to wind, earthquake and traffics will be conducted using 74 accelerometers at selected sections on the instrumented box girder, drop-in span, pier shafts and pier bases (Fig. 11). At each section, the acceleration will be measured in the longitudinal, lateral and vertical directions to the bridge span. The results are also used to determine ice forces on piers as mentioned earlier.

To determine the horizontal and vertical components of wind load, three components of wind speed (longitudinal, lateral and vertical to the bridge span) will be measured at a location on the instrumented span. In addition, wind speed and wind direction will be measured at one navigation span, which is the highest elevation above the water and will, therefore, be subjected to the highest horizontal wind speed.

Wind tunnel tests of the bridge model have been conducted at University of Western Ontario. Valuable information about wind speed distribution around the bridge girder has been obtained.

Two strong motion sensors have been installed in pier bases to monitor the ground motion during possible earthquake.

Corrosion

The concrete elements of the bridge are exposed to an aggressive environment of chlorides and sulphates. Consequently, there exists the possibility that the ions will migrate through the concrete and corrode the reinforcements with catastrophic results. Therefore, it is very important to continuously and non-destructively monitor the penetration of these ions into the concrete and to measure the corrosion rate of the steel reinforcement.

For this purpose, corrosion probes are cast into the concrete structure, both in the pier at the splash zone and on the bridge decking at varying depths and locations, to obtain long-term corrosion measurements. Laboratory studies using these corrosion probes and their associated corrosion meters are conducted prior to installation to determine the frequency of measurement required and the correlation between the corrosion rate of steel and the chloride content of the concrete.

Data Acquisition and Transmission

A data acquisition system is being installed for acquiring, pro-

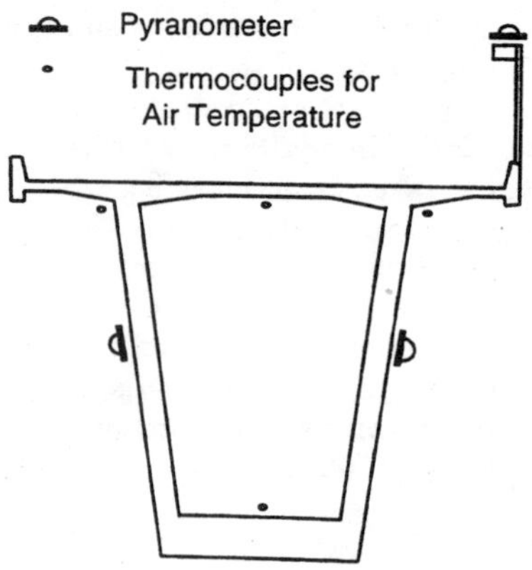

Fig. 9 Air Temperature Monitoring

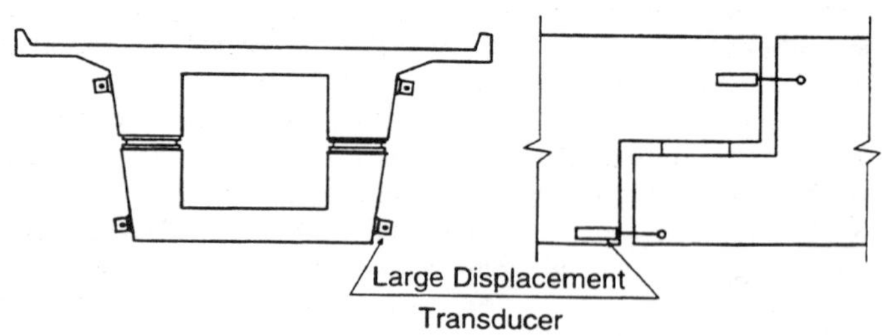

Fig. 10 Displacement Monitoring at Expansion Joint

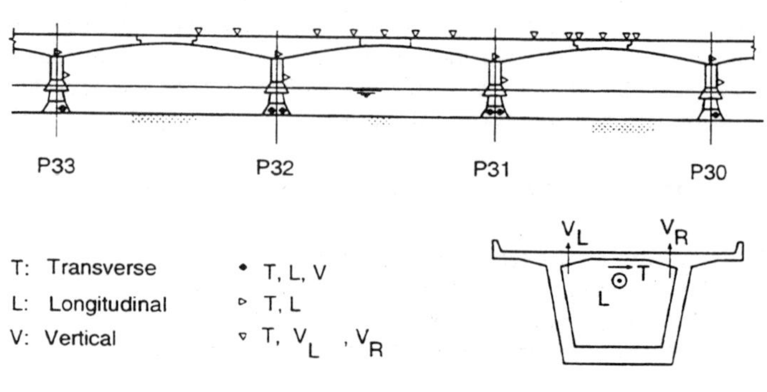

Fig. 11 Locations of Accelerometers

cessing and storing data from all the monitoring components. The system consists of an industrial personal computer and several low speed and high speed data loggers with total about 1000 channels. Fiber optic cables are used to connect the data loggers to the computer. The sensors for ice forces, traffic loads and dynamic loads are led to high speed data loggers whilst those for deformation and thermal stresses are led to low speed ones. The data loggers sample the measured data, convert the data to engineering units, carry out basic data processing and automatically transmit the data to the computer. The data loggers operate in two modes of acquisition: time-averaged and time-series. In time-averaged mode, the data will be stored only for time-averaged statistical values (such as mean, variance, and minima/maxima) over a fixed length of time. If any threshold value (load, strain, acceleration or wind speed etc.) is reached, the time-series mode will be initiated, and the system will store time history data for detailed analysis of bridge loads and responses during an extreme event such as a storm, a heavy vehicle pass-over or an earthquake.

The data files will be transmitted to central computers located at Carleton University and Public Works Canada in Ottawa using the high speed, high volume data transmission network newly developed by National Test Network.

RESEARCH PROGRAM

A comprehensive research program on the aforementioned loads, load effects and bridge responses has been carried out since the beginning of design and continuing throughout the construction process. Many tests and researches have played important role on decision making and parameter selection. The major subjects are presented as follows:

Ice Forces

1. Derivation of transfer functions for determination of global ice forces, extraction of effects of wind, and relation between ice force time history and structural response.

2. Statistical analysis of ice load and the local ice pressures to develop ice load spectra, and comparison with the ice load spectra specified in design codes and in the bridge design criteria.

3. Analysis of observations of kinematics of ice interactions with conical ice shield, and relation to nature of original ice feature.

4. Comparison of the measured loads with those obtained using generally accepted algorithms for ice loads on conical structures, and ice loads for unconsolidated ice ridge keels. Assessment of these algorithms with respect to the observed behavior and load, for given feature.

5. Modification and/or creation of ice load algorithms for sheet ice and ridged ice interactions with conical structures.

Short- and Long-term Deformation

During the construction stages the instrumentation will give the changes in displacements and strains at each important event. After completion of the bridge, the measurements will be taken regularly. The displacements at different stages will be predicted by step-by-step analysis using computers and the measured material properties. The computed displacements will be compared to the actually recorded displacements, using the most up-to-date technique of analysis. Comparison of the predicted and measured data will allow calibration of the applied technique and assumptions.

Thermal Stresses

The collected temperature data will be used for the following purposes:

1. To establish extreme values of temperatures and temperature differences at various cross sections;

2. To calibrate and improve the temperature prediction model;

3. To establish extreme values of stress in the structure on the basis of the measured temperature distribution;

4. To correlate the temperature measurement with measurements of diurnal and seasonal deformation of the structure.

5. To establish the number of freeze-thaw cycles in various components of the bridge.

The research program will include extensive statistical evaluation of the measured data obtained from the field and laboratory testing. The evaluation will establish the extreme values for temperature differences across individual components and cross-sections, and the overall response of the structure. The results will be used to calibrate and develop new computer models for predicting temperature effects in concrete bridges, and to provide a data base for calibration of design code.

Traffic loads and load combinations

The traffic survey data will be analyzed, first, to determine the statistical distributions of vehicle axle weight, axle spacing, as well as vehicle speed. The forces and load effects calculated from the traffic data will then be compared with the corresponding design values. Using the traffic survey data and historical account on the trends of legal axle and gross vehicle weight in this region and other parts of Canada, the growth of traffic on the bridge (load growth factor) can be determined. The resultant rate of growth and projected traffic volume and load pattern in the next 100 years will be calculated. A multi-lane load distribution factor based on the traffic survey data will be established by employing a suitable probabilistic model to be developed in the analysis.

Using the obtained statistical information, live load models for long-span bridges, including truck and lane loads, will be developed to represent the load effects at various limit states of the bridge. Statistical data on the traffic load distribution required by the industry for future planning and code development, and methods of their simulation, will be studied.

The developed live load models will consist of simulated vehicle loads, and combinations of concentrated and distributed loads. A computer program will be developed to simulate a traffic jam on the bridge by a randomly generated traffic load distribution, and further to determine the adequacy of the live loads and load factors used in the design. Thus, a reliable traffic load model for long-span bridges will be created.

The live load factors will be determined using the second-moment probabilistic equation. Load factors for various limit states will be established by using different limit state functions. Based on the statistical characteristics of the survey data, the reliability and safety index β of the bridge can be evaluated.

Dynamics, Wind and Earthquake

The numerical modeling techniques for long-span bridges, such as equivalent beam model, thin-walled beam model (Razaqpur and Li, 1991), Finite strip method (Cheung, Li & Chidiac, 1996) and finite element method (Jategaonkar, Jaeger & Cheung, 1985), will be investigated. The applicability of these techniques will be assessed using the field data under different loading conditions and excitation levels. Based on this information, more accurate and more efficient numerical methods for simulating the nonlinear dynamic behavior of long-span bridges will be developed.

Vibration behavior of the bridge will be recorded, and spectral analyses of the measured data will be carried out to determine the mode shapes, frequencies and damping characteristics of the bridge. The vibration can be induced by controlled impact, pull and release, and ambient excitations such as wind, traffic and other transient loading. The long-term environmental effects on the dynamic behavior and performance of the bridge will be investigated. Moreover, the dynamic response to ice impact will also be studied. The results will be used to develop and calibrate numerical models for dynamic analysis of long-span bridges.

The measured wind data and observed structural behavior under wind load will be compared with the results of wind tunnel test. A reliability based approach to determine the critical wind speed distribution function for long-span bridges from the results of scaled model wind tunnel test will be developed. The measured wind data will also be used to establish the dynamic wind load model for long-span bridges using equivalent static loads. Furthermore, advanced numerical modeling techniques will be developed to accurately model the wind induced nonlinear vibrational behavior of long-span bridge structures.

Because very few earthquake records are available for the eastern North America, the ground motion during an earthquake will be simulated based on seismological analysis of this region, and the statistical characteristics of the seismic activities around the bridge will be determined. The earthquake response of the bridge will be studied, including the spatial variation effect of the ground motion due to propagation of the seismic waves, using linear modal and linear and nonlinear direct time integration time history analysis techniques. As all the bridge piers are placed directly on bedrock, propagation of earthquake wave will be simulated by taking the three orthogonal components of the ground motion as the input at one end of the bridge, and adding the appropriate time delays to the other support inputs based on the traveling distance and the propagation speed of seismic waves. The traveling wave effect is significant for long-span bridges because of the large size of the structures. Direct measurement of the near-field spatial variation of seismic wave propagation and its effect on large structures is rare. The Confederation Bridge offers an unique opportunity to obtain information and to conduct research in this area.

CONCLUSION

A comprehensive research program has been carried out to monitor and study responses of Confederation Bridge to ice forces, short- and long-term deformations, thermal stresses, traffic load and load combinations, dynamics, wind and earthquake, and corrosion. This program will provide important information for management and maintenance of the bridge. The measured data and research results will be valuable for design and construction of similar bridges and structures, and also useful for the development of computer simulation techniques and design codes for long-span bridges.

ACKNOWLEDGEMENT

This instrumentation and research project is funded jointly by Natural Sciences and Engineering Research Council of Canada, and a consortium of industrial partners led bu Strait Crossing Development Inc. The project team includes Dr. T. Brown, Dr. W. Dilger, Dr. A. Ghali, Dr. D. Lau and Dr. G. Tadros.

REFERENCES

Cheung, MS, Li, W, and Chidiac, SE (1996). *Finite Strip Analysis of Bridges*, E & FN SPON, London, UK.

Chui, JJ, and Dilger, WH (1993). "Temperature Stresses and Cracking due to Hydration Heat," *Symposium on Creep and Shrinkage of Concrete, Proceedings of the fifth International RILEM Symposium*, Barcelona.

Croasdale, KR, and Cammaert, AB (1996). "An Improved Method for the Calculation of Ice Loads on Sloping Structures in First-Year Ice," *Proceedings of First International Conference on Exploration of the Russian Arctic Offshore*, St Petersburg, Russia.

Jategaonkar, R, Jaeger, LG, and Cheung, MS (1985). *Bridge Analysis Using Finite Elements*, CSCE, Montreal, Canada.

Razaqpur, AG, & Li, HG (1991). "Thin-Walled Multicell Box-Girder Finite Element," *J Struct Div ASCE*, Vol 117, No 10.

Tesky, WF (1988). "Special Survey Instrumentation for Deformation Measurements," *Proc. ASCE, Journal of Surveying Engineering*, Vol 114, No 1.

Williams, FM (1996). "Ice Features and Ice Mechanical Properties in Northumberland Strait," *Proceedings of the 6th International Offshore and Polar Engineering Conference*, Vol II, Los Angelos.

Proceedings of the Seventh (1997) International Offshore and Polar Engineering Conference
Honolulu, USA, May 25-30, 1997

ISBN 1-880653-28-1 (Set); ISBN 1-880653-30-3 (Vol. II)

Experimental Study of the Dynamic Behaviour of a Towed System

G. Duvat
DCE/Bassin d'Essais des Carènes
Paris, France

C. Large
DCN INGENIERIE CENTRD SUD
Toulon, France

ABSTRACT

The purpose of this paper is to describe the methods used to perform a set of sea trials of an instrumented system and to present some of the results obtained. The aims of the tests were to improve the knowledge of the vibratory behaviour of towed cables, the cable shapes and the towed body motions during different kind of maneuvers in order to validate software.

The system consisted of a ship towing a submerged body and a cable. All parts of the system were instrumented in order to measure movements and forces during the maneuvers.

In this paper we describe, in the first part, the means of measurement and the difficulties encountered during the sea trials. In the second part we present some of the results (cable shapes during straight lines maneuvers at constant speed, motions and angular displacements of the towed body, local accelerations and cable slopes).

KEY WORDS

Cable - Towed system - Strumming

NOMENCLATURE

C_X, C_Z, C_M	=	hydrodynamic coefficients of the towed body in the vertical plane
D_C	=	diameter of the cable (m)
C_N	=	normal drag coefficient of the cable
C_T	=	tangential drag coefficient of the cable
F_N	=	normal drag force of the cable per unit length (N/m)
F_T	=	tangential drag force of the cable per unit length (N/m)
L_{TB}	=	length of the towed body
S_{TB}	=	reference surface of the towed body
X, Y, Z	=	Cartesian coordinates (m).
		The origin of the right handed coordinate system is at the aft lower edge of the towed body on the side of the attachment point (see figure 2a) the X-axis is directed forward the Z-axis is upwards the Y-axis is in the transverse direction
W_{TB}	=	weight of the towed body in air (N)
V_S	=	free running speed of the ship (m/s)
u, w	=	translation components of the body speed in the X and the Z directions (non-dimensionalised by V_S)
q	=	rotating components of the body speed around the Y-axis (non-dimensionalised by $L_{TB}V_S$)
ρ_C	=	cable mass per unit length in air (kg/km)
ρ_W	=	water density (kg/m^3)
θ_i	=	local slope angle between the cable and the horizontal (°)

1. INTRODUCTION

In the paper we describe, in the first part, the means of measurement and the difficulties encountered during sea trials of a towed system and in the second part the results obtained.

The sea trials presented in this paper had two main aims : first, using the global results (tensions at the cable extremities, body immersion, cable slopes, angular displacements and rates of the immersed body) in order to validate software of simulation of the three-dimensionnal unsteady behaviour of a towed system composed of a cable towing an immersed body. The second objective was to use the acceleration measurements in order to improve the knowledge of the vibratory behaviour of the cable and to validate or improve the existing models of strumming at the Bassin d'essais des carènes.

In the paper, we will limit ourselves to the measurements performed during straight lines maneuvers at constant speed. We will present rough results because the fine analysis is still not made. However, some reliable input data will be given andthe coherence of the measurements be verified.

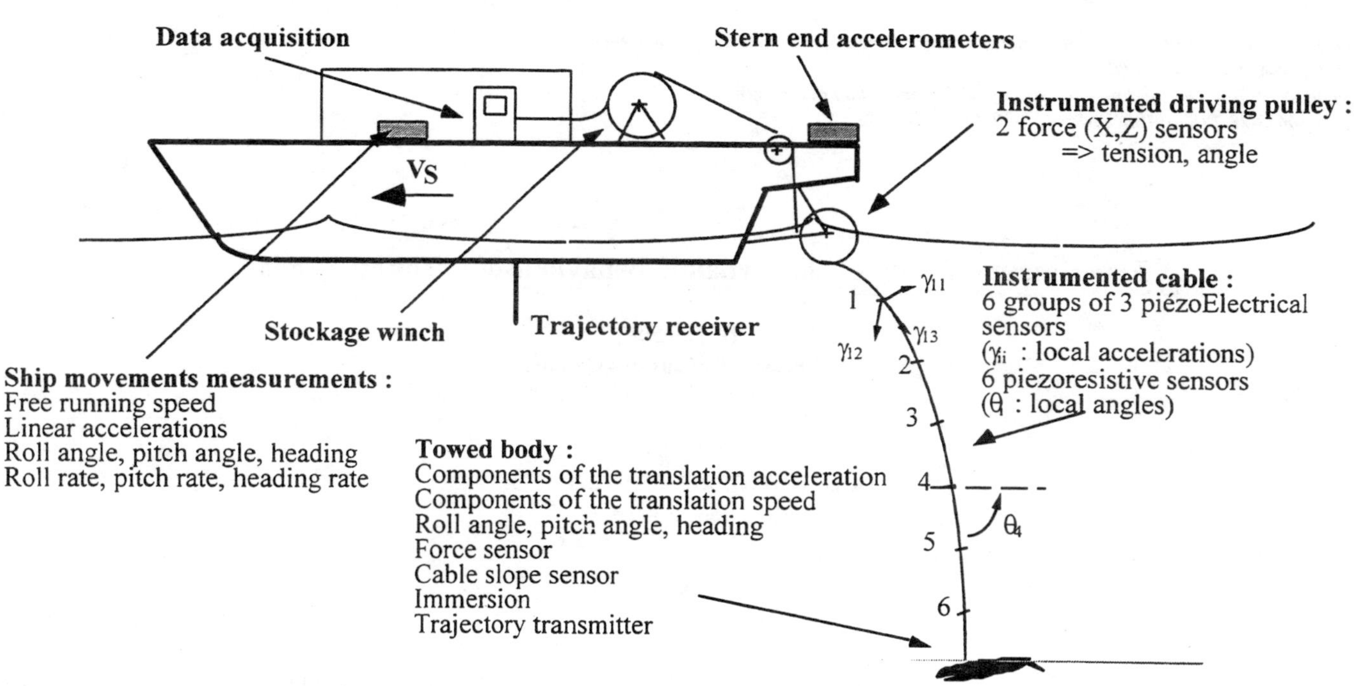

- Fig 1 : Towed system configuration -

Some global and local results on straight lines maneuvers with constant acceleration followed by speed stabilization and unsteady turns with measurements at the beginning and after stabilization will be shown during the presentation of the paper at the congress.

2. TOWED SYSTEM CONFIGURATION

We already had some experience and some data about towing heavy systems from a fixed underwater point. We therefore decided to study configuration with both extremities submerged. In order to respect project costs, we often use ready-made equipement :

- the towing ship is the French Navy's BEGM "THETIS", already equipped with handling, recording and measurement systems.
- the towed body was a submarine immergence system we used many times in the past : we have changed its buoyancy, and put some existing sensors inside.
- the Instrumented Cable N°2, called "IC2", has been entirely designed and manufactured. A previous experiment [1] helped us in the designing process.
- Last, a rigid chassis, supporting the fixed submerged pulley, and also used for launching and recovering the towed body, has been designed to run down the ship stern (see appendix I).

3. MEASUREMENT METHODS AND MATERIALS

The towed system configuration is shown on figure 1.

3.1. Recordings and readings

The ship was equipped with a reading/recording system called ACTIS of analogic and numeric signals. The signals issued from the sensors came into the system, which allowed the user to check instantaneously each data, on the time or the frequency domains. After each trial, data were recorded on a magnetic disc or tape.

3.2. Measurement of the ship parameters

The knowledge of the ship movements was necessary to calculate the input excitation of the towing point on the pulley. The chassis was supposed rigid and to have the same movements as the ship. The parameters listed on fig. 1 were measured with THETIS instrumentation. Local vibrations of the stern were recorded with accelerometers fixed under the quarter deck.

3.3. Measurement of the towed body parameters

All the sensors supply power came from the ship to the towed body through the instrumented cable.

3.3.1. Hydrodynamical data

The numerical data are presented in the paragraph 5.

The towed body has a NACA profile (see figure 2a), with the interior modified to accept different measurement devices, and to change its buoyancy to an acceptable value.

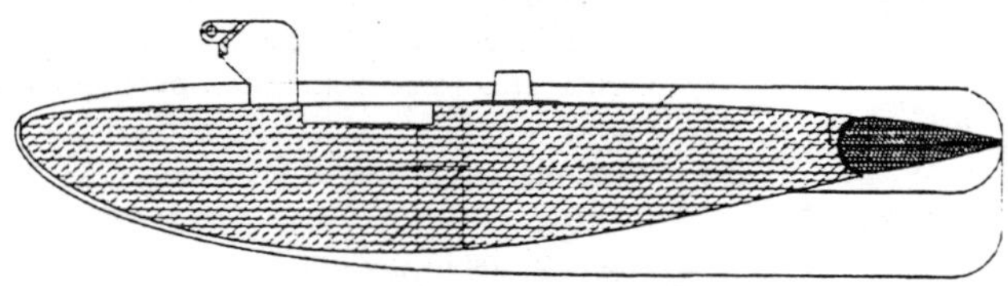

- Fig 2a : Side view of the towed body -

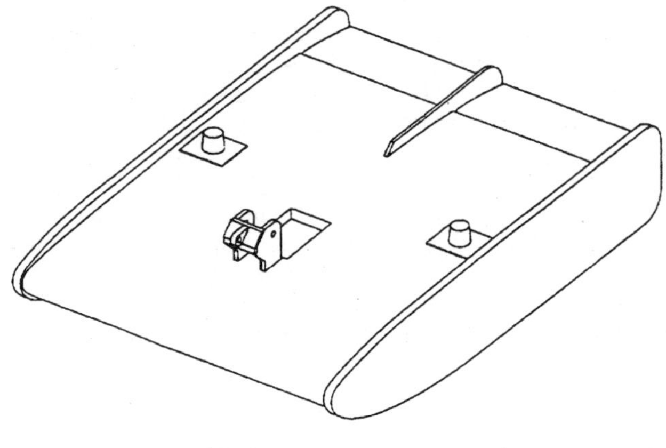

- Fig 2b : CAD representations of the body in towing position -

Hydrodynamic coefficients were obtained both by calculation and measurements on a smaller-scaled body. The position of the center of gravity and the buoyancy were obtained by calculation and measurement in air and in water, and the center of volume is obtained in the same way in water using the previous result.

3.3.2. Immersion

It was measured with a pressure transducer SENSOTEC TJE/AP122. The sensor is watertight and is fixed along one side of the body. The choice of the full scale of the sensor (300 m) is the result of a compromise between a reasonable figure of maximum immersion obtained by a simple calculation and the use of a security coefficient. Before its mounting in the body, the transducer was tested in a hydrostatic tank from 0 to 30 bars. The calibration curve obtained was used during the trials.

3.3.3. Yawing, rolling, pitching of the body

The values were measured with an gyroscope referenced SFENA B487AAM. The sensors were mounted into a watertight tank fixed perpendicularly to the main axis of the body, nearly opposite the attachment points.

3.3.4. Position of the body with respect to the ship

The position was measured with a emitter referenced ORE 4326A. The receiver was fixed on a pole under the ship hull. The emitter was fixed vertically in the body : its upper side extended out of the profile. To avoid inertial and hydrodynamical troubles, a false beacon of same shape and size was mounted symetrically to the real beacon on the other side of the system (see figure 2b).

3.3.5. Towing point of the body

A towing point was mounted on the forward upper part of the body. This point was fitted with a pin to allow vertical plane rotations of the body. The longitudinal position of the pin (which is a kind of two-teeth fork) was adjustable, as well as the lateral position of the cable termination with respect to the teeth of the fork. These adjustements were necessary before conducting the trials, to control the hydrodynamic stability of the body.

3.4. IC2 parameters Measurement

3.4.1. Tension and cable slope on the ship towing point

The horizontal and vertical components T_X and T_Z of the forces on the pulley axis are measured when the cable is under the tension T and makes the angle β with respect to the horizontal axis (see fig. 3). Tx and Tz are measured by a bar instrumented with resistive gauges, referenced SIXAXES FA2.2/03-50 kN.

$$tg\alpha = \frac{Tx}{Tz} \qquad (1)$$

$$\beta = 2\alpha - \sigma \qquad (2)$$

$$R = \sqrt{T_x^2 + T_z^2} \qquad (3)$$

$$T = \frac{R\sin\alpha}{\cos\beta - \cos\sigma} = \frac{R}{2\sin(\sigma - \alpha)} \qquad (4)$$

$$\sigma = 67° \qquad (5)$$

Eq. (1) to (5) allow the calculation of T and β.

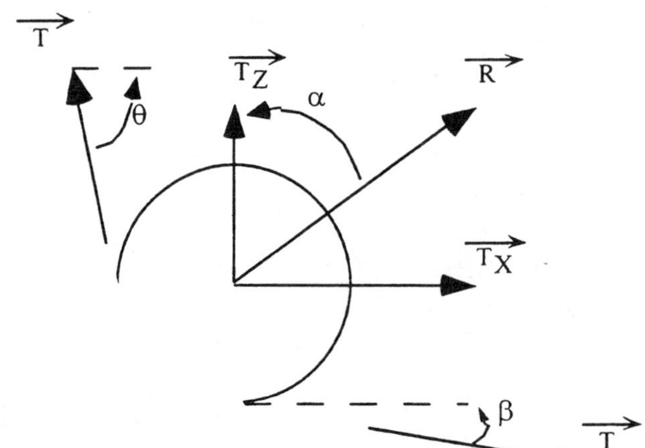

- Figure 3 -

R and α were calculated and the results sent to the ACTIS system.

3.4.2. Cable tension at the attachment point

The cable end which formed a pin with the towed body, referenced SEFAC 942.22.967, included inside a load sensor with resistive gauges, referenced FGP/FN6094/01, that gives the value of the tension in the direction of the cable.

3.4.3. Cable slope at the attachment point

A watertight rotating potentiometer was mounted in the bar which connected the fork on the body the cable end. Its resistance variation was proportional to the cable with respect to the vertical axis of the body. The reference of the potentiometer is MCBH19/08.

3.4.4. Mounting of the accelerometers inside the cable

Mounting accelerometers outside the IC2 had many drawbacks [2] :

- the hydrodynamic flow is locally modified,
- the inertial effects are modified as well,

- a rigid connection between the cable and the accelerometer is difficult to obtain,
- conditions of storage, handling, and protection of the cable are more demanding,
- some electrical conductors must be used with the IC2 during the trials ; it's more dangerous than the use of one only cable.

For all these reasons, and because we had an experience with the "inside technology" [1], we decided to mount the accelerometers inside the cable. The sensors were glued inside cylindrical boxes made of 316L steel; these boxes shelter the sensors from high mechanical pressures applied by the armor wires of the cable when the IC2 was bent or under tension. The electrical conductors of the core of the cable were welded to the sensors inside the boxes (see figure 4). The boxes were mechanically connected each other by a small wire rope along the axwas of the cable. There were 6 groups of 2 boxes. The distance between two boxes of the same group was 1m. The boxes and the conductors were now covered by a watertight jacket made of polyurethane. Over thiss jacket, the counter-wound armor of galvanized steel wires was layed. The armor was covered with a jacket of HYTREL (Polyester) to obtain a smooth surface in contact with water. In consequence, the movements of the boxes were the same as those of the whole cable.

See appendix 2 for the detailed description of the IC2.

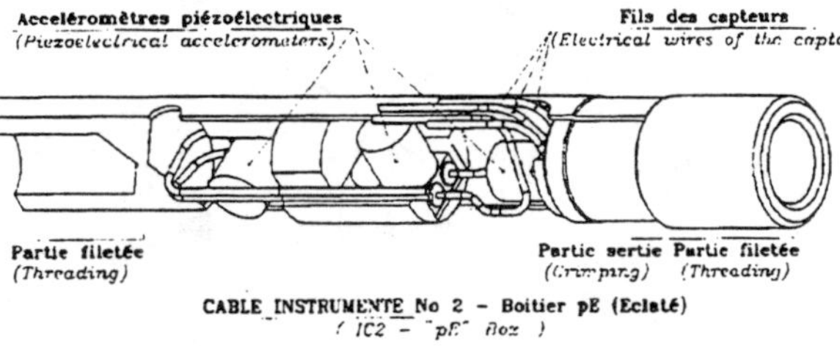

- Fig. 4 : Accelerometer boxes -

The IC2, manufactured by the DE REGT Special Cable BV society (Nederland), had the following external characteristics :

. Outer diameter : 26 mm
. Breaking strength (calculated) : 132 kN
. Mass in air : 1335 kg/km
. Mass in seawater (1.026 density) : 790 kg/km
. Overall length : 300 m

3.4.5. Acceleration measurements along the cable

For each of the 6 pairs of boxes pR/pE, each pE box measured the acceleration of the IC2 along 3 axes. The longitudinal of the cable X_C, and two transverse axes Y_C and Z_C. This method did not follow the variation of the position of the transverse axes, because of the self-rotating of the cable during the trials, despite its antigiration construction. Because of its construction the cable did not twist. However, it was still impossible to know the transverse orientations.
Our previous experiment [1] help us in the choice of the full scale of the accelerometers of the boxes because they were hidden inside the cable. The measurements were obtained with 6 x 3 monoaxial piezoelectrical accelerometers, mounted on the axis referenced KISTLER 8614A1000 (Y_C ,Z_C directions) and KISTLER 8614A500 (X_C direction). The sensors were tested before and after their mounting into the boxes.

3.4.6. Estimation of the cable shape

Because of the self rotation problem (see § 3.4.5), we decided to measure only the slope between the cable and the horizontal, so we supposed that the cable shape will stay in a vertical plane containing the axis of symetry of the towing ship.

Calculation of the σ angle is obtain by eq (6) : $\sin\sigma = \frac{g1}{g}$, with g1 = projection of gravity g along the longitudinal axis of the cable.

Measurement of g1 is realized with 6 x 1 monoaxial piezoresistive accelerometer (one per box pR) referenced ENTRAN EGA125-10D. These sensors were sensitive to shocks, so we glue them with a special damping resin. The sensor position into the box was not modified by the use of this glue.

4. SOLVING OPERATIONNAL TROUBLES

4.1. Separating the 0Hz from the rest of the signal

Only the 0-Hz signal was useful in the measurement of the cable shape. We used standard high-frequency filters which were tested in our laboratory. We measured different σ with a pR box excited at different frequencies. The test was successful, and therefore, we decided to keep that technology.

4.2. Handling of the system

4.2.1. Storage of the cable

A minimum cable bending diameter of 1000mm was required to protect the accelerometer of the IC2. 1000mm was the diameter of the pulley, and the cable was stored in one layer on the of more than 2000mm diameter winch. When the towed body staysed on the quarter deck, waiting for a trial, we kept the minimum curvature of the cable lying on the deck, far away from the critical 1000mm diameter.

4.2.2. Lauching and recovering of the towed body during trials

The chassis was just wide enough to allow the towed body to pass through it. The ship was also very sensitive to the sea state, and it was very difficult to set the body into the water or to keep it back to the ship if its movements exceeded a certain amplitude. The towed body was not stowed along the longitudinal ship axis, so the cable needs to be led back through the pulley each time.

We solve the two problems by the following ways :

1. Starting or finishing a trial in small sea states or in sheltered water as soon as possible.
2. Launching or recovering the towed body onlywhen the system was stabilized along the vertical axis of the chassis.
3. Using a cable-lead, which consisted in a kind of very heavily reinforced shovel supported by a hydraulically manoeuvered pinned arm. During a trial, the shovel stayed near the submerged pulley, in order to maintain the cable into the groove of the pulley if the cable accidentaly lose contact with it.

5. INPUT DATA OF THE SYSTEM

Under the hypothesis of no interaction between the towing ship and the body, during straight lines maneuvers at constant speed, it is not necessary to describe the ship motions because the towing speed is sufficient. However, the simulation need accurate data on the cable and the towed body. The maneuvers described in this paper being straight lines at constant speed, we can limit the data expressions to the vertical plane.

5.1. The cable

The main characteristics of the instrumented cable are given at the paragraphe 4.4.

The figure 5 below presents the deployed cable lengths and give the positions of the differents groups of sensors placed inside the cable.

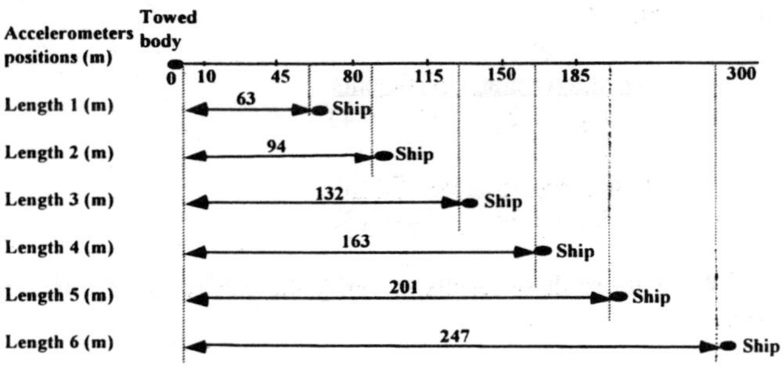

- fig 5 : Sensors positions with respect to the cable deployed lengths -

5.2. The towed body

The main characteristics of the towed body are the followings :

- $L_{TB} = 1.784m$,
- $S_{TB} = 2.14m^2$
- $W_{TB} = 4840N$
- Position of the center of volume :
  $X_C=0.98m$ - $Z_C=-0.1m$
- Position of the center of gravity :
  $X_G=1.04m$ - $Z_G=-0.12m$
- Formula for the C_X coefficient:
  $$C_X = -0.037u^2 + 0.021uw + 0.317(w^2+w|w|) - 0.05qu - 0.188qw^2 - 0.36w^2(q+|q|)$$
- Formula for the C_Z coefficient:
  $$C_Z = -0.105u^2 - 1.6uw - 1.4w^3 - 0.35qu - 2.5qw^2 - 0.8w^2(q+|q|)$$
- Formula for the C_M coefficient measured at the center of volume :
  $$C_M = -0.018u^2 + 0.385uw + 0.1065w^2 - 0.099w|w| - 0.118qu - 0.172qw - 0.0375q|q| - 0.548qw^2 + 0.348wq|q|$$

All the hydrodynamic results presented were obtained during tests in towing tank.

6. GLOBAL RESULTS

We will present in this paragraph some of the first results and will verify their coherence.

Knowing the values of the towed body incidence allows to calculate the tension at the lower extremity and the slope at this point.

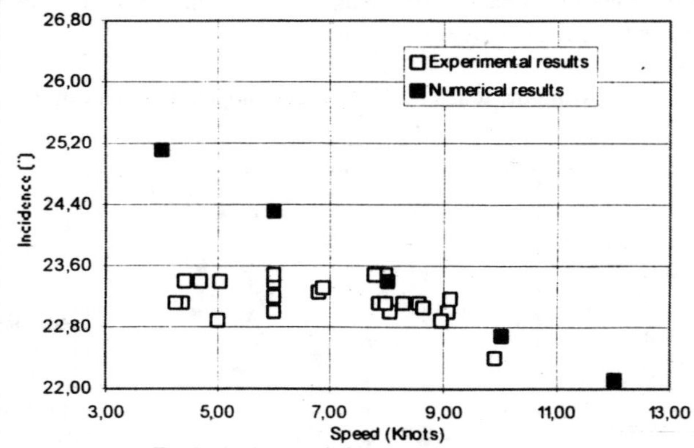

- Fig 6 : Incidence of the towed body vs speed -

The experimental values on the figure 6 confirm the right choice of the attachment point which was choosen to give a constant incidence in the range of speed.

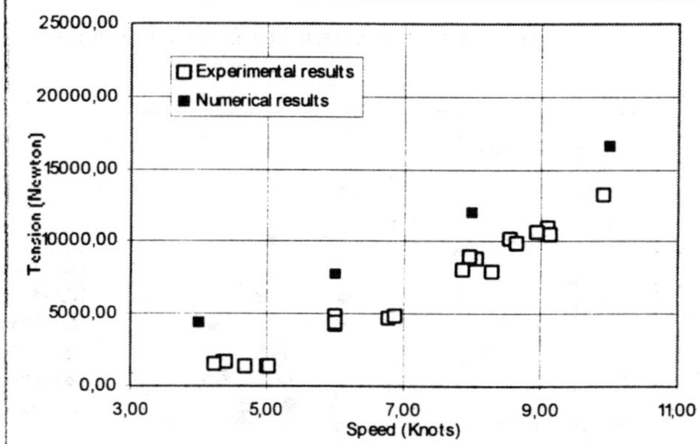

- Fig 7 : Tension at the lower extremity vs speed -

With our equilibrium model, we have recalculated the body incidence, the tension and the slope angle at the attachment point as you can see on the figures 6, 7 and 8.

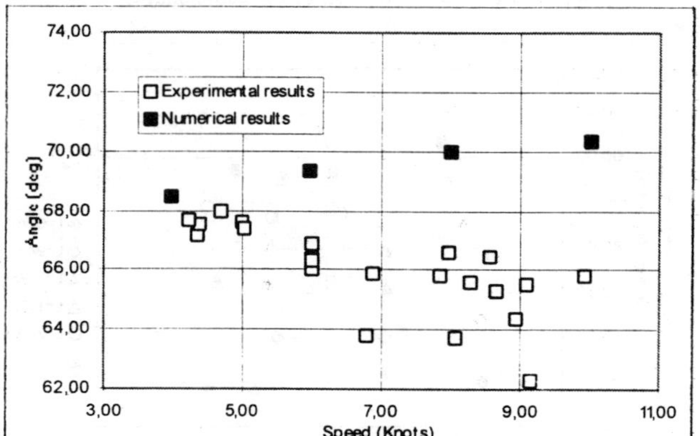

- Fig 8 : Angle at the lower extremity vs speed -

The simulation and experimental results are quiet different. The differences can be explained by the non taking account of the uncertainties on positions of the center of volume, center of gravity and hydrodynamic coefficients due to the C_X, C_Z and C_M measurements on model in towing tanks.

Knowing the tension and the cable slopes between cable

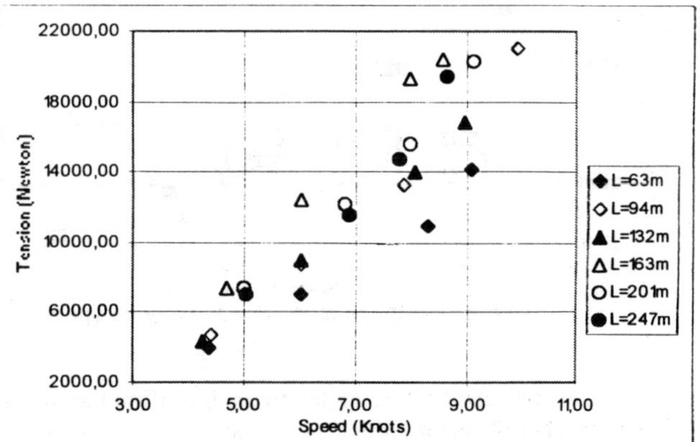

- fig 9 : Tension at the upper extremity vs speed -

and the horizontal at both extremities and the immersion

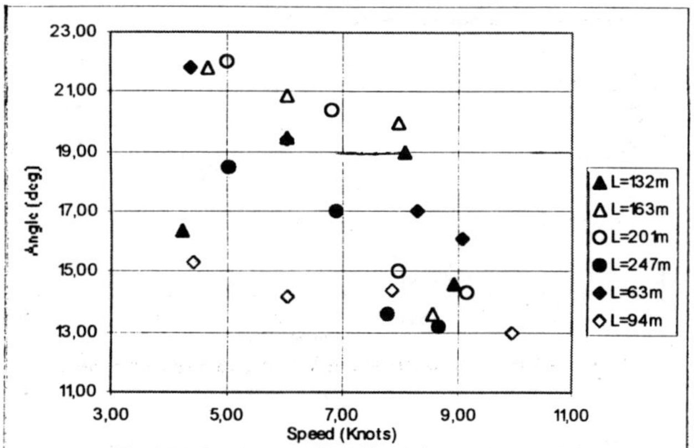

- Fig 10 : Angle at the upper extremity vs speed -

allows accurate determination of the cable shape.

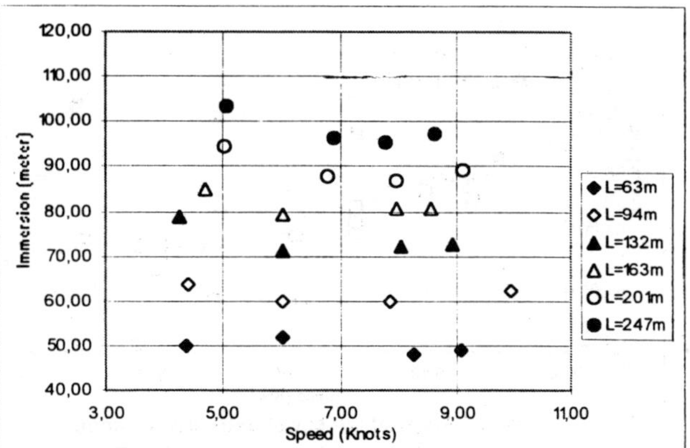

- Fig 11 : Immersion of the towed body vs speed -

With that kind of results, it is also possible to check the available strumming models (for instance [3] and [4]) or to guide development of a new one. This is our aim but it is still not made.

It is also possible to identify drag cable coefficients choosen for that kind of cable. For instance, it is possible to identify drag coefficients for two classical drag models (for instance Richtmyer and Landweber):

- Richtmyer drag forces modelisation :

$$F_N = 0.5C_N\rho_W D V_S^2 \sin^2\theta_t \quad (7a)$$
$$F_T = 0.5C_T\rho_W D V_S^2 \cos^2\theta_t \quad (7b)$$

- Landweber drag forces modelisation :

$$F_N = 0.5C_N\rho_W D V_S^2 \sin^2\theta_t \quad (8a)$$
$$F_T = 0.5C_T\rho_W D V_S^2 \quad (8b)$$

After indentification using both tension and immersion step (complete set of results presented in this paper) with less than 15% error, we obtain :

- Richtmyer drag coefficients :
  $C_N = 1.29 \qquad C_T = 0.43$

- Landweber drag coefficients :
  $C_N = 1.34 \qquad C_T = 0.32$

We will use these results to verify the coherence of some local results.

7. LOCAL RESULTS

As we said in paragraph 3.4.6, it is possible to measure the local slope with the signal of the piezoresistive sensors of the cable. An exemple is given on the figure 12.

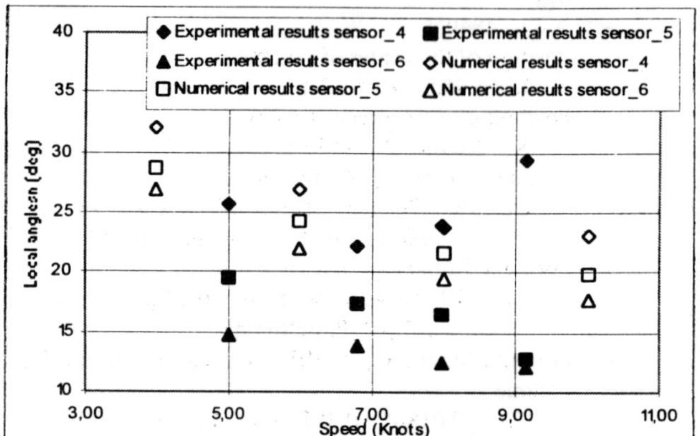

- Fig 12 : Local angles (L=201m) vs speed -

Using Landweber drag forces modelisation with the identified coefficients and with the experimental values of tension and slopes at the lower extremity, we have recalculated the slopes at the positions of the accelerometers in order to evaluate the reliability of the measurements (see figure 12). Experimental and simulation results are quiet similars and allow us to trust in the first ones.

On the figure 13, we present the first vibratory results. The DSP (spectral density of power) for the length 201m of the transverse piezoelectrical #3 signal show a subordination to the free running speed : there is an energy incrase with its.

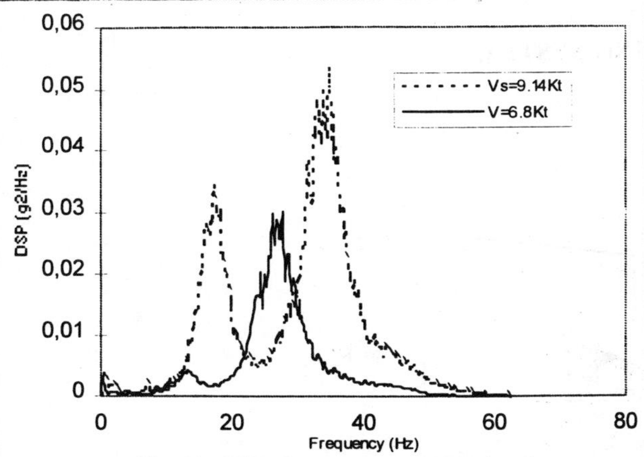

- Fig. 13 : DSP of transverse pE #3 signal -

CONCLUSIONS

In this article, we have described how to make use of instrumented cable and towed body for a sea trials campaign, presented the problems encountered and shown that these sea trials allowed to obtain reliable measurements.

Only a little set of our results has been given and we have particularly insisted on global results which the first ones we analyzed. However a little part of the local measurements has been presented in order to show we obtain interesting and reliable results. We are now sure that we can improve the knowledge of vibratory behaviour of cable but after a fine analysis still not made at this time.

ACKNOWLEDGEMENT

This work was supported by the Direction des Reherches et Etudes Techniques (DRET) and sea trials performed in cooperation with the GESMA, DCN.

REFERENCES

[1] - « Specifications of sea trials of instrumented cable and towed body »
André CASTEX
First towed system conference - Val de Reuil (France) - 1992

[2] - « Dynamic behaviour of a submarine cable in a uniform sea flow »
Ryoto KISHIMOTO - Iwao KITAZAWA
Bulletin of JSME n° 200 - February 1982

[3] - « Strumming predictions for the SEACON II experimental mooring »
Richard A. SKOP - Owen M. GRIFFIN - Steven E. Ramberg
Offshore Technology Laboratory n°2884 - May 1977

[4] - « Method for calculation of vortex induced vibration of flexible lines and dynamic of towed systems »
V. ZAITSEV - S.I. DEVNIN
Krylov Shipbuilding Research Institute report - 1994

APPENDIX I
TOWED BODY LAUCHING SYSTEM

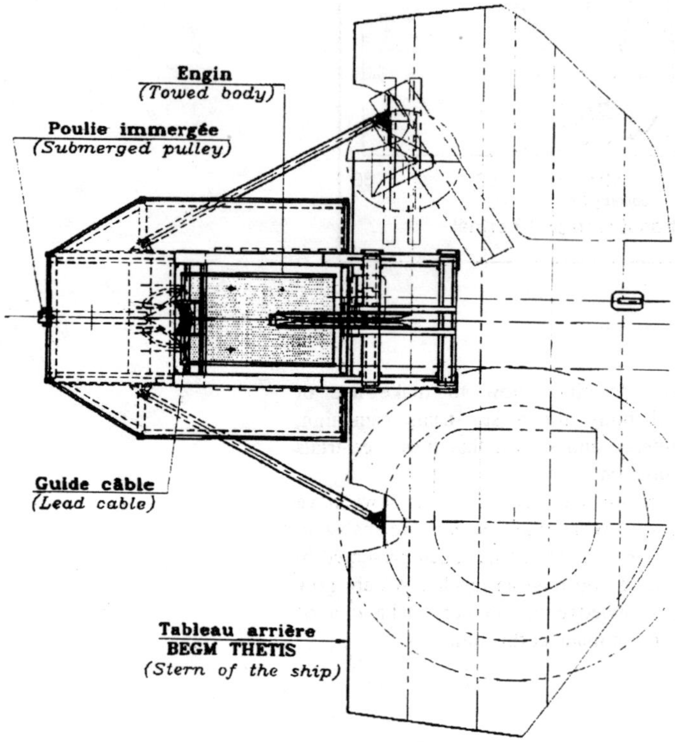

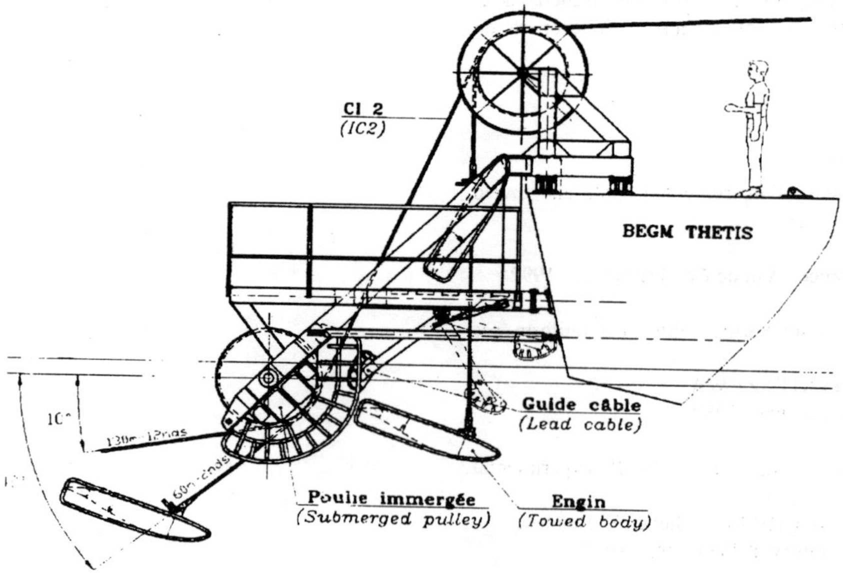

APPENDIX II
IC2 STRUCTURE

DE REGT
SPECIAL CABLE

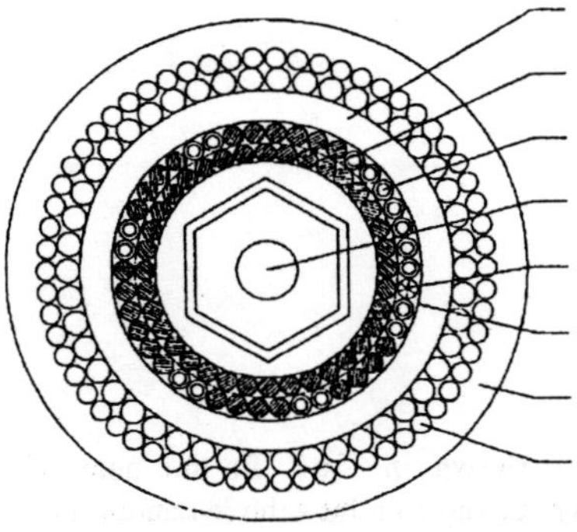

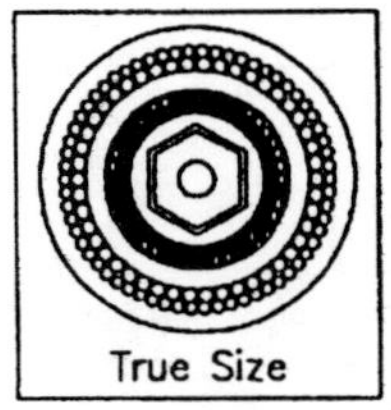

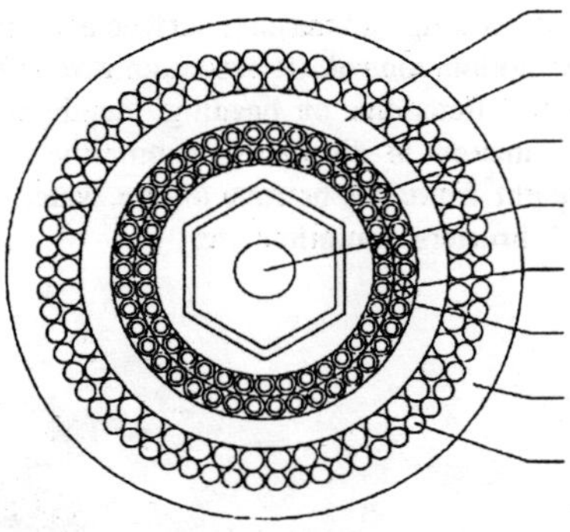

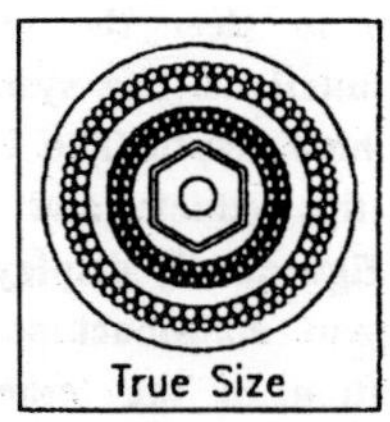

"Interstices not filled."

"The weight in seawater applies to an assumed seawater density of 1026 kg/m3."

*) Note : The values are corrected for the overlength of the elements.

Mechanical Properties :

Breaking strength	Bending radius		Weight	
calculated	*recommended min.*		*calculated*	
	static	dynamic	air	seawater
kN	mm	mm	kg/km	kg/km
132	-	–	1335	790

Optical Fibre Properties :

Component	Mode	Numerical Aparture	Attenuation at 850 nm dB/km
50/125	Multi	0.2	1.9

Electrical Properties :

Components	Rcond	Rinsul 500 V DC	Volt. Rating	High Volt. test at 5 min.
	Ohm/km	Mohm*km	V	kV
0.29 m2	64.6	500	60	–

'roceedings of the Seventh (1997) International Offshore and Polar Engineering Conference
Honolulu, USA, May 25-30, 1997

'SBN 1-880653-28-1 (Set); ISBN 1-880653-30-3 (Vol. II)

Advanced Tethered/Non-Tethered Control System for a Large Underwater Vehicle

Xiaocheng Shi and Xinqian Bian
Harbin Engineering University
Harbin, China

BSTRACT

This paper introduces a kind of tethered/non-tethered ɔntrol system for a large search and salvage underwater ehicle. In the non-tethered condition (manned condition), perator in the vehicle can directly use an integrated control nd display system to steer the vehicle with a joystick and ıtomatic control functions. The system can realize hold speed, old heading and hold depth. The information of the heading, epth, position of the vehicle and status of the thrusters is ıtegratively and figuratively displayed on a screen. Thus the 'stem creates a good man-machine interface, which is extreme eneficial to operation. In the tethered condition (non-manned ɔndition), operator in the ROV support vessel can remotely ɔntrol the vehicle to complete the search and salvage works ith a multi-core optical cable connecting an operator console ı board to the vehicle underwater. All of information about otions and other status of the vehicle, including the images ' TV and sonar, is communicated to the console from ıderwater. And all of information about control commands to ırusters and manipulators is communicated to the vehicle om the console on board. The system has obtained successful a test and has delivered to the user.

EY WORDS: tethered/non-tethered control system, integrated ›ntrol and display, large submarine

INTRODUCTION

The tethered/non-tethered control system introduced in this paper the key equipment of a large submarine made in China. The function the submarine is to accomplish missions of searching, torpedo lvage and lifesaving in deep water region. The submarine has 7 ·usters, 3 main thrusters, 2 horizontal and 2 vertical slot thrusters. ıd its air weight is 16 tonnes, its length is 8 meters. The submarine's outlook is shown in fig.1. In the non-tethered condition (manned condition), operator in the vehicle can directly use an integrated control and display system to complete underwater maneuver and jobs. In the tethered condition (non-manned condition), the difficult remote underwater jobs are completed by a advanced computer-optical communication system. Integrated control and display system is the kernel equipment for submarine's motion control. The control system's function is ensuring submarine's surface and underwater motion control and other motion control. It has computed aided manual manipulation, auto control functions of heading, depth and speed. The display system's function is displaying submarine's information of heading, depth, height from the bottom of the water, position, pitch and roll attitude, 7 thrusters' condition, etc.

Fig. 1 Submarine's outlook

Fig. 2 Tethered control operation

II. SYSTEM COMPOSITION

The advanced tethered control system of the submarine is composed by the tethered control device on the ship, a multi-core optical cable, the tethered control device in the submarine, and the integrated control and display system. The situation that the operator manipulates the vehicle to complete the remote underwater jobs on the ship, is shown in Fig. 2.

The tethered control equipment on the ship include three parts, a display console on the ship, a optical adapter and a optical winch.. The display console on the ship concentrates all of needed state information by the submarine and manipulative function and can display underwater circumstance of submarine surrounding via TV picture. The manipulative control command on the ship is passed the submarine underwater, at same time, and four pictures of underwater TV system, a picture of color image sonar and a picture of BOS sonar may be synthetically transferred to the console on the ship form submarine by the multi-core optical cable. The optical adapter accomplishes photoelectric signal change and optical cable multiplex. The winch achieves normal accept and release of optical cable and cable signal deliver is accomplished by optical slide.

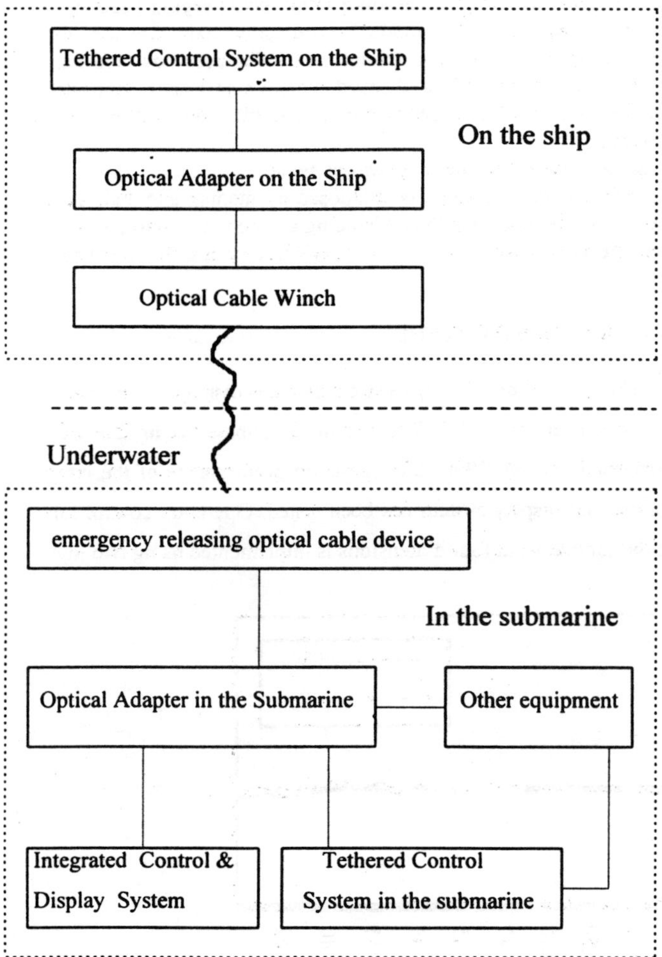

Fig. 3 System composition diagram

The cable control equipment underwater includes emergency releasing optical cable device, optical adapter and tethered control device in the submarine. The former is key device which ensure 6mm cable is not break in the tethered condition. The optical adapter accomplishes change of photoelectric signal and multiplex of optical cable. The tethered control device accomplishes collecting signals and controlling other equipment and then achieves submarine movement control with the integrated control and display system. The tethered control system's composition diagram is shown in fig.3 .

The integrated control and display system is composed by manipulation case, main chassis, CRT and interrelated outer equipments.

There are two control joysticks, many function keys, numerical tube displaying submarine's heading and alarm system on the manipulating case's face. The case is managed by 8098 CPU. There is belt on the manipulating case, it can be drawn out from the fixed place and placed on one's knee near to the viewing window for operating.

The main chassis is designed using hierarchical structure, divided into two parts. Upper part is host computer. Lower layer is auxiliary power source. The two parts all can be drawn out from the main frame along the guide, which is convenient for maintenance and changing circuit board. The 486 CPU board is used as main CPU board.

Integrated control and display system's composition diagram is shown in fig. 4.

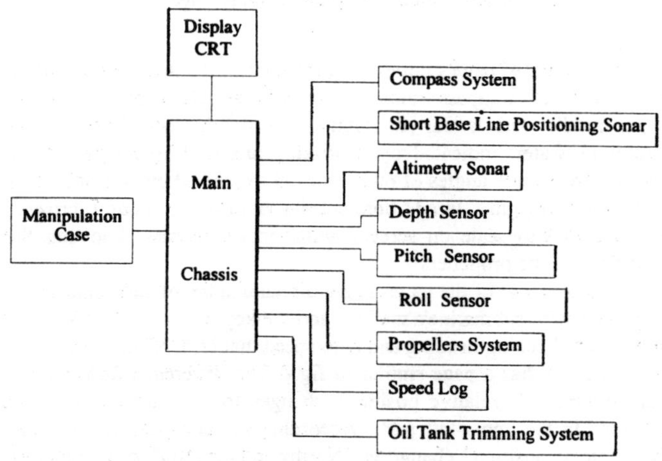

Fig.4 Integrated control and display system composition diagram

III. INTEGRATED CONTROL AND DISPLAY SYSTEM

Submarine's motion control is accomplished by the coordinating of manipulation case and comprehensive display pages.

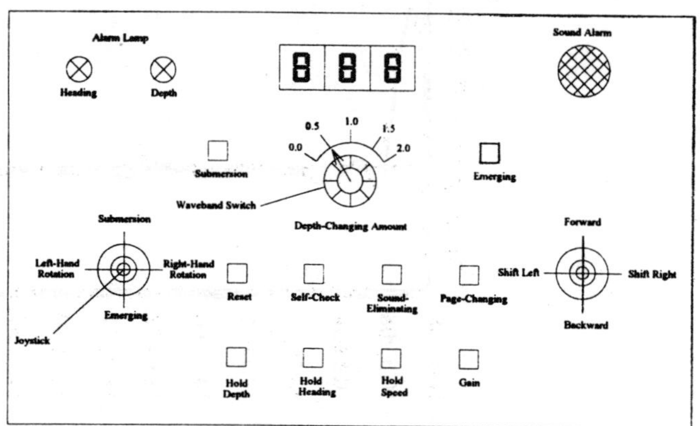

Fig.5 Face figure of manipulation case

A. Manipulation case's face

Face figure of manipulation case is shown in fig.5.

B. Integrated display pages

One of the System's Integrated display pages is shown in fig.6.

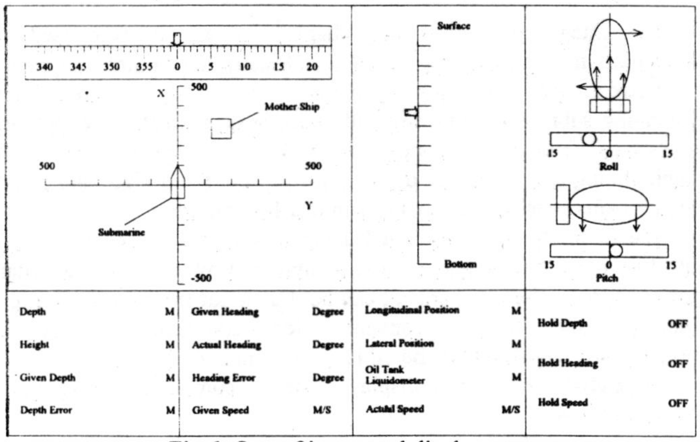

Fig.6 One of integrated display page

The content of page's upper-left part is current submarine's heading in the way of compass-imitator, and submarine's relative position. Upper-middle part is current submarine's depth and distance from bottom of water, indicated by a striking arrow. Upper-right part is current submarine's angles of pitch and roll expressed by two little balls shifting left or right. In addition, current rotating speeds of 7 thrusters are displayed to scale on screen, which make operator knowing the conditions of the propellers.

The lower part of the pages is the digital display of state parameters associated with submarine's motion, and the key states of Auto keeping depth, Auto keeping heading and auto speed that is "ON" or "OFF".

There still has a page similar to fig.4. The difference from fig.4 is that "submarine's relative position" changes to "submarine's absolute position" in the upper-left part. Accordingly, "Longitudinal position" and "Lateral position" change to "Northward position" and "Eastward position". These two vivid pages integrate all of state information required by submarine's motion controls. Operator can clearly know submarine's current position and status through the pages displayed on CRT screen.

C. Manual operation mode

At manual operation mode, the two joysticks in manipulation control submarine's 7 thrusters. Right joystick controls submai forward, backward, shift left and shift right. Pushing this joysti slanting direction makes submarine move slantwise. Left joy controls submarine's emerging, submersion, left-hand rotating and r hand rotating. The system can realize submarine's vector coi according to manual operation command and current speed, and i reasonable selection and thrust distribution to thruster's system using as far as possible energy to be required thrust.

D. Auto operation mode

1) Auto keeping depth

Pressing down "Hold depth" key, submarine will automatically on the given depth with Auto keeping depth mode. At Auto kee depth mode, controller uses bow and rear vertical propellers to co submarine's depth in the submarine's basic balanced state. Accordi rotational speed and direction of vertical thrust, the system determine dimension and direction of submarine's unbalanced f and make fine balanced tuning by using trim adjustment elect valve to control the oil to float bag and taking oil out of float bag.

2) Auto keeping heading

Pressing down "Hold heading" key, submarine will automati keep heading in the given heading with Auto keeping heading n At Auto keeping heading mode, according to main thrusters' rot speed, if speed is less than 1 knot, controller uses bow and rear thrusters to control heading; if speed is more than 1 knot, controller rotation speed difference between right and left main thrusters to co heading.

3) Mixed manual and auto operation mode

Mixed operating mode is composed by manual and auto oper mode. At this mode, depth and heading can be under auto control r other freedom motion control can be realized by manual operation.

IV. EXPERIMENT RESULT

The installation of integrated control and display system have performed on Oct. 1995. The test of the submarine in lake has performed on Feb. 1996. The operation performance of the integ control and display system has been tested. Operation control fun of the submarine at four dimensions is implemented using two joys

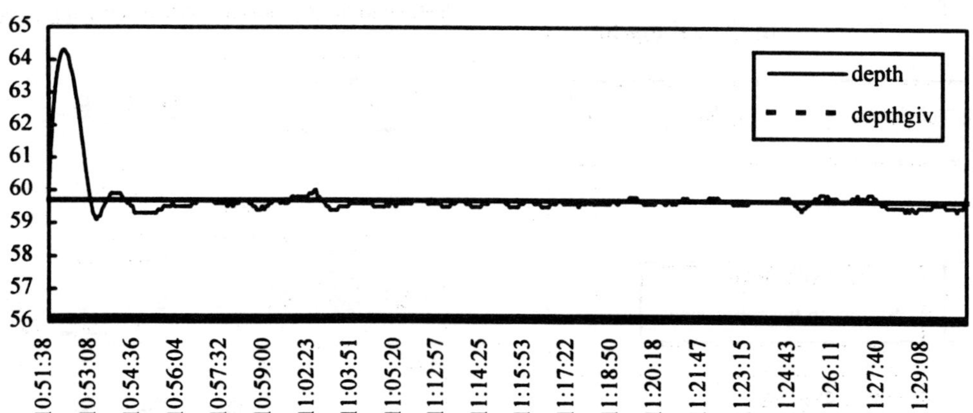

Figure 7. A typical curve of auto keeping depth

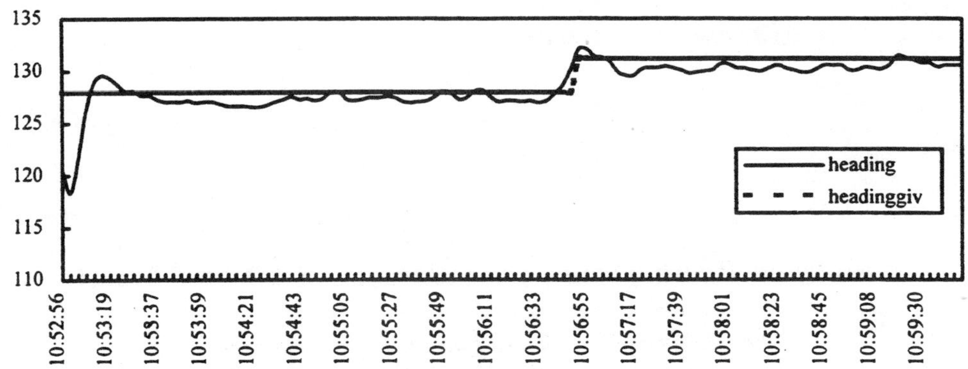

Figure 8. A typical curve of auto keeping heading

The auto keeping depth function makes that the submarine can hold in giving depth with precision of 0.2m at 0-2 knot speed. According to the submarine buoyancy state, the system can automatically make fine balanced tuning with left over buoyancy less than 10kg for the vehicle of 16 tonnes. A typical curve of auto keeping depth is plotted in figure 7.

The auto keeping heading function makes that submarine can hold heading in giving heading with precision of 2 degree at 0-2 knot speed. Figure 8 plots a typical curve of auto keeping heading.

The auto keeping depth, heading and speed functions are very convenient for underwater search and operation of submarine.

V. CONCLUSIONS

This paper introduces that the integrated control and display system is first one for the manned submarine in China. It optimizes the submarine operating performance and improves operation efficiency.

VI. REFERENCES

[1] X. Bian, and X. Shi etc. , "Design and Experimental Evaluation of the Dynamic Positioning Control System for Undersea Vehicles", in Proceedings of the INTERVENTION/ROV'92 Conference & Exposition, pp.273-276

[2] S. Lafferty, "Data Acquisition and Multi-Media Management for ROV's", in Proceedings of the INTERVENTION/ROV'92 Conference & Exposition, pp.286-290

[3] X. Bian etc. ,"3-D Trajectory Control for an Autonomous Underwater Vehicle HSEI-I", in Proceedings of the Third (1993) International Offshore and Polar Engineering Conference, pp.502-506.

[4] X. Shi etc. , "Control System Architecture for an Autonomous Underwater Vehicle HSEI-I", in Proceedings of the Third (1993) International Offshore and Polar Engineering Conference, pp.507-511.

Proceedings of the Seventh (1997) International Offshore and Polar Engineering Conference
Honolulu, USA, May 25-30, 1997

ISBN 1-880653-28-1 (Set); ISBN 1-880653-30-3 (Vol. II)

A Study on Hydrodynamic Characteristics of a Towed Vehicle

Satoru Yamaguchi, Wataru Koterayama and Eiji Sasaki
Kyushu University
Kasuga, Japan

ABSTRACT

A depth, pitch and roll controllable towed vehicle "Flying Fish" is being developed to measure the ocean current, temperature, salinity, dissolved oxygen, pH, turbidity, chlorophyll and total inorganic hydrocarbon. Field experiments on the performance of Flying Fish began in 1994. The control system used in these field experiments is designed based on the PID controller theory because the parameters in the PID controller are easily modified to correspond with conditions of the hardware. The hardware of Flying Fish was decided on field experiments conducted in 1996, and an even more advanced control scheme will be designed to obtain more accurate motion control beginning in 1997. For this we need a more precise model for the dynamics of Flying Fish. A series of model tests were carried out in an experimental tank using a 1/5 scale model to obtain hydrodynamic coefficients for the numerical model. The hydrodynamic coefficients were estimated based on the wing theory and the slender body theory and a regression analysis was made using the data from field experiments. Numerical simulations using hydrodynamic coefficients obtained from these methods were compared with the results of field experiments.

KEY WORDS

Towed Vehicle, Model Experiments, Field Experiments, Hydrodynamic Coefficients, Numerical Simulation

1 INTRODUCTION

The towed vehicle "Flying Fish" is a depth, pitch and roll controllable observation system; the principal features are shown in Table 1 (Koterayama, et al., 1994). The Flying Fish is equipped with many sensors to obtain physical and chemical data in the ocean. An acoustic Doppler current profiler instantaneously measures currents to a depth of 350m, and a total carbon analyzer utilizes many kind of reagents in glass bottles. A motion control system is indispensable for measurements using these sensors, because the motion of the vehicle affects the accuracy of any measurement. A profile of the Flying Fish is shown in Fig. 1. Vehicle depth is controlled by a main wing, and pitching and rolling motions by horizontal tail wings. These wings are actuated by the three AC servo motors shown. Although, a PID control system is currently used to maintain the depth, pitch and roll angle, a more advanced system is needed to improve constancy of the attitude of the motion because the system is a multi-input-output system which has three actuators and three desired values. The purpose of this paper is to report accurate hydrodynamic coefficients of the vehicle and to examine the characteristics of these coefficients. Three types of approach were used to acquire the coefficients: model experiments, a method based on the wing theory and the slender body theory, and a regression analysis using data from field experiments. The hydrodynamic coefficients obtained by these techniques are compared and evaluated by numerical simulation using these coefficients.

2 HYDRODYNAMIC COEFFICIENTS OF THE VEHICLE

The motions of the vehicle are described on the basis of a coordinate system fixed on the body as shown in Fig. 2. Positive direction of x, y and z axes are forward of the body, to the right of the x axis and downward, respectively. Origin o is center of the main wing. The numerical model of longitudinal motion of the vehicle is expressed by the following equations (Newman, 1977, Ohkusu, et al., 1987).

$$\begin{aligned}&(m+A_{11})\dot{u}+(mz_G+A_{15})\dot{q}\\ =\;&-(m+A_{33})qw+(mx_G-A_{35})q^2\\ &-(m-\rho\nabla)g\sin\theta\\ &+X_{uu}u^2+X_{ww}w^2+X_{\delta\delta}\delta^2+X_{\gamma\gamma}\gamma^2+X_T \qquad (1)\end{aligned}$$

$$(m+A_{33})\dot{w}-(mx_G-A_{35})\dot{q}$$
$$= (mz_G+A_{15})q^2+(m-\rho\nabla)g\cos\theta$$
$$+Z_w w+(Z_q+mu)q+Z_\delta\delta+Z_\gamma\gamma+Z_T \quad (2)$$

$$(I_{yy}+A_{55})\dot{q}-(mx_G-A_{53})\dot{w}+(mz_G+A_{51})\dot{u}$$
$$= -(mz_G+A_{15})qw-(mz_G-\rho\nabla z_B)g\sin\theta$$
$$-(mx_G-\rho\nabla x_B)g\cos\theta$$
$$+M_w w+(M_q-mx_G u)q+M_\delta\delta+M_\gamma\gamma$$
$$+l_b X_{uu}u^2+l_b X_{\gamma\gamma}\gamma^2-x_T Z_T+z_T X_T \quad (3)$$

where

(u, w)	:	velocity vector based on body fixed coordinates system
q	:	angular velocity
(xG, zG)	:	coordinates of center of gravity
(xB, zB)	:	coordinates of center of buoyancy
(xT, zT)	:	coordinates of a towing point
(X, Z)	:	external force
M	:	external moment
m	:	mass of the vehicle
∇	:	displacement of the vehicle
I_{yy}	:	moment of inertia
A_{ij}	:	added mass
ρ	:	density of water
g	:	acceleration of gravity
l_b	:	z coordinate of center of the body

Three types of approach were used to estimate the hydrodynamic coefficients. First, the hydrodynamic coefficients are estimated based on the wing theory and the slender body theory. Some small parts and interference of the body and wings are neglected in this method. Secondly, model experiments using a 1/5 model were carried out in a circulating water tank. A model experiment is the usual method for design of a vehicle (Koterayama, et al., 1988). Thirdly, the hydrodynamic coefficients are estimated by the regression analysis using data from field experiments.

2.1 A simple estimation method

On the basis of the wing theory and the slender body theory, the hydrodynamic coefficients of the vehicle are estimated

Table 1 Principal features of Flying Fish

Operating depth	0~200m
Dimension(L B H)	3.84m, 2.26m, 1.4m
Weight in air	1300kg
Weight in water	0kg
Instrumentation	ADCP, CTD, DO, Turbidity, pH, Chlorophyll, CO2
Towing velocity	0~12knots
Motion control	Heave, Pitch, Roll

(Ohkusu, et al., 1987). Those regarding the main and horizontal tail wings are estimated using the wing theory, while those related to the body are calculated based on the slender body theory and formulas about a spheroid. x axis component of added mass coefficient in x axis motion A_{11} is estimated as

$$A_{11}=A_{B11}+A_{M11}+2(A_{H11}+A_{V11})+2(A_{F11}+A_{S11}) \quad (4)$$

where subscripts B, M, H, V, F and S mean body, a main wing, horizontal tail wing, vertical tail wing, after float and front float, respectively. The body and floats are assumed to be spheroids and calculated by the following equation:

$$A_{B11}=k\frac{\pi\rho}{6}{d_{max}}^2 L \quad (5)$$

where k is a coefficient from the formula, d_{max} is the maximum diameter of the body, and L is length of the body. The wings are assumed to be plates and calculated as

$$A_{M11}=\frac{\pi\rho}{4}t^2 b \quad (6)$$

where t, b are thickness and span of the wing, respectively. Added mass for z axis is estimated as follows:

$$A_{33}=A_{B33}+C_{aw}(A_{M33}+2A_{H33})+2(A_{F33}+A_{S33}) \quad (7)$$

where C_{aw} is a corrective coefficient of wings. Added mass of the body and floats is calculated as

$$A_{B33}=C_a\rho\nabla_B \quad (8)$$

where C_a and ∇_B are added mass coefficient and displacement of the body, respectively. Added mass of wings is calculated as

$$A_{M33}=\frac{\pi\rho}{4}c^2 b \quad (9)$$

Added moment of inertia about the y axis is estimated as follows:

$$A_{55}=A_{B55}+A_{M55}+2A_{H55}+2(A_{F55}+A_{S55}) \quad (10)$$

Added moment of inertia of the body is calculated using strip of a two-dimensional cylinder. Added moment of inertia of wings are calculated using the formula for a long flat plane.

$$A_{B55}=\int_{x_t}^{x_e} x^2 C_{aB}\rho S(x)dx+l_b^2 A_{B11} \quad (11)$$

$$A_{M55}=\frac{1}{128}\pi\rho c^4 b+l_m^2 C_{aw}A_{M33} \quad (12)$$

where x_t and x_e are coordinates of tip points of the body, C_{aB} is an added mass coefficient of a cylinder and $S(x)$ is a sectional area of the body. l_m is a x coordinate of center of the main wing.

Lift, induced drag and moment for the main wing and the horizontal wings are estimated on the basis of three dimensional wing theory as follows:

$$X'_{\delta\delta} = -\frac{4\pi\lambda}{e(\lambda+2)^2}\frac{S_w}{Ld_{max}} \quad (13)$$

$$Z'_\delta = k_M\frac{2\pi\lambda}{\lambda+2}\frac{S_w}{Ld_{max}} \quad (14)$$

$$M'_\delta = -k_M\frac{2\pi\lambda}{\lambda+2}\frac{S_w}{Ld_{max}}\frac{l_p}{L} \quad (15)$$

where k_M is a corrective coefficient and S_w is area of the main wing. λ is an aspect ratio of the wing. e is Oswald's airplane efficiency, and lp is a x coordinate of center of pressure. These values for the tail wings can be calculated in the same way.

Damping coefficients X_{ww}, Z_w, M_w, Z_q and M_q are estimated based on the wing theory, the slender body theory and experimental results as follows:

$$X'_{ww} = Z_\delta - X_{\delta\delta} + (1-\varepsilon)(Z_\gamma - X_{\gamma\gamma}) + \pi\frac{d_{max}}{L} \quad (16)$$

$$Z'_w = -(Z_\delta + (1-\varepsilon)Z_\gamma) - \pi\frac{d_{max}}{L} \quad (17)$$

$$M'_w = Z_\delta\frac{l_p}{L} - (1-\varepsilon)Z_\gamma\frac{l_H}{L} + 0.4\pi\frac{d_{max}}{L} \quad (18)$$

$$Z'_q = -Z_\gamma\frac{l_H}{L} - \pi\frac{d_{max}}{L} \quad (19)$$

$$M'_q = -Z_\gamma\left(\frac{l_H}{L}\right)^2 - 0.6\pi\frac{d_{max}}{L} \quad (20)$$

where $l_H \approx l_h + c_H/4$, l_h is a x coordinate of center of a horizontal tail wing, and c_H is chord length of a horizontal tail wing. $1-\varepsilon$ is a coefficient about down wash.

The hydrodynamic coefficients are shown in no-dimension by dividing forces by $\frac{1}{2}\rho U_0^2 L d_{max}$.

2.2 Model Experiments

Model experiments were carried out to estimate the hydrodynamic coefficients in a circulating water tank. The length of measuring region of the tank is 4.4m and its breadth and height are 1.5m and 1.3m, respectively. The 1/5 model used in the experiments is shown in Fig. 3. The measuring region of the tank was sufficiently large for the model. The effects of the tank wall is about 2% for velocity of the flow so it is negligible. Tests on resistance, wing angle changing and attack angle changing were carried out for stated conditions; forced surge, forced heave and forced pitch tests were practiced for dynamic conditions. In these experiments, the constant flow velocities were form 0.4(m/sec) to 1.1(m/sec) for stated conditions and were 0(m/sec) and 0.6(m/sec) for dynamic conditions. The experimental results are shown in Fig. 4~Fig. 17. The resistance for x axis is shown in Fig. 4 in no-dimension. The horizontal axis shows velocity and the vertical axis shows x'_{uu}. A line in this figure is shown calculated by the least squares method. Experimental results for a main wing are indicated in Fig. 5 ~ Fig. 7. The horizontal axis shows the main wing angle and the vertical axis shows F'_x, F'_z and M'_y, respectively. Fig. 8 ~ Fig. 10 show results of attack angle changing tests. The horizontal axis shows the attack angle. F'_x is not symmetric for zero of the attack angle because the form of the vehicle is not symmetric for xy plane. Additionally, change of x component of the force is very small, so that there may be a measurement error in the results. There is a difference between the gradient of pitching moment in the positive region and the negative region. This difference is assumed to be caused by shading of the main wing by the body. The results of forced surge tests are shown in Fig. 11. A'_{11} is added mass coefficient for x axis in surge motion. The horizontal axis is circular frequency in no-dimension. A'_{11} has a tendency to increase with increase of circular frequency. The results of forced heave tests are shown in Fig. 12 ~ Fig. 14, where A'_{33} is additional mass coefficient for z axis in heave motion. These values are almost constant for the circular frequency of the motion. Z'_w and M'_w have a tendency to decrease with increase of circular frequency. This is assumed to be caused by shed vortex generated by oscillation of the wing. Results of forced pitch tests are shown in Fig. 15 ~ Fig. 17. A'_{55} is an added moment of inertia in pitch motion. There is a slight decrease in the value of Z'_q and M'_q in the high frequency region.

We use the mean value of results obtained by the above tests for numerical simulations. The obtained hydrodynamic coefficients suffer from the effects of Reynolds number, so accuracy of the coefficients should be examined by comparison with the results of regression analysis using field experiment data.

2.3 Parameter estimations using field experiment data

In earlier field experiments, much data about motion of the vehicle was acquired including pitch angle, roll angle, depth, main wing angle and horizontal tail wing angle of the vehicle (Koterayama, et al., 1995). We attempted to estimate hydrodynamic coefficients of the vehicle using these data from field experiments. The regression analysis based on the least squares method was used; that is, the equations of motion were rewritten in the form that has known inertia, gravity, buoyancy and tension term on the left side and unknown hydrodynamic force term on the right side. In the estimation, the lift coefficients of wings are treated to be known term and tension of the towing cable is calculated on a condition of equilibrium because tension of the cable and x directional movements were not measured in the field experiments. The lift coefficients of the wings acquired by model experiments can be used for the calculation as effects of Reynolds number are small for the lift of wings. The models for the regression analysis are shown as follows:

$$s_z = -A_{33}\dot{w} + Z_w w + Z_q q + Z_\delta \delta + Z_\gamma \gamma \quad (21)$$

$$s_{M_y} = -A_{55}\dot{q} + M_w w + M_q q + M_\delta \delta + M_\gamma \gamma \quad (22)$$

Each unknown term on the right side is treated as an explanation parameter. For example, the explanation parameters are $\dot{w}$, w, q, δ and γ in the first equation. Here, the term related to the cross-coupling coefficients and cross terms are neglected and M_γ is treated as known term on the estimation because there is a strong correlation between γ and δ. Results of fittings using the regression analysis for the equation in the z direction are shown in Fig. 18. The solid line is the standard parameter and symbols are results of the fitting. For the standard parameter vectors and the explanation parameter vectors, the data acquired in a field experiment are used. Interval of the data is 0.2(sec) and number of the data used for this regression analysis is 2000 points. Results of fittings for the equation of pitching motion are shown in Fig. 19 in the same form.

Estimated values of the hydrodynamic coefficients are shown in Table 2 compared with these from model experiments and simple estimation method. The results by simple estimation method show good agreement with the data from the field experiment in spite of simplicity of former. The estimated values by the regression analysis are different from the other results, especially for the values about pitching moments. As the pitching motion is very sensitive about ordinates of the center of gravity and the center of buoyancy, the differences assumed to be caused by errors of these factors. And the pitching motion was controlled to be very small in the field experiments so it is difficult to get data good for the estimation. On the other hand Z'_w and A'_{33} show good agreements with values by model experiment as the heaving motion was excited sufficiently in the field experiments. It is assumed that the effects of difference of Reynolds number between the model test and the field experiment are small for these coefficients.

Table 2 Hydrodynamic coefficients

	Model experiment	Simple method	Regression analysis
X'_{uu}	-0.2324	—	—
X'_{ww}	-0.11995±0.5859	0.84297	—
$X'_{\delta\delta}$	-0.9350	-1.9513	—
$X'_{\gamma\gamma}$	-0.4944	-0.6927	—
Z'_w	-3.2327	-3.3484	-3.2273
Z'_q	-1.1955	-1.2296	-0.1021
Z'_δ	1.3217	1.8620	—
Z'_γ	0.7916	1.0910	—
M'_w	-0.4329±0.296	-0.0737	-0.2017
M'_q	-0.7055	-0.7159	-0.1371
M'_δ	-0.1137	-0.1394	-0.2581
M'_γ	0.5392	0.6159	—
A'_{11}	0.0415±0.005	0.01990	—
A'_{33}	0.7099±0.0205	0.7239	0.7084
A'_{55}	0.1024±0.0228	0.07679	1.0867

3 NUMERICAL SIMULATIONS AND A FIELD EXPERIMENT

Numerical simulations were carried out to evaluate the hydrodynamic coefficients obtained by the above described studies. The motion of the vehicle is calculated on the basis of equations (1) ~ (3) and motion of a towing cable is calculated by the lumped mass method. Ship motions are given as top end motion of the towing cable. In the calculations, length of the cable is 300m and the ship speed is about 6 knots. Time history of depth of the vehicle is shown in Fig. 20 compared with the results of numerical simulation using the hydrodynamic coefficients obtained by model experiment with the data from the field experiment. In this figure, the vehicle is ascending to a depth of 5m. The depth has difference caused by error on the calculation of condition of equilibrium by the lumped mass method. But the difference is sufficiently small and the transient response shows good agreement. Comparison between the numerical simulation and the field experiment is shown in the same form in Fig. 21 ~ Fig. 23. Amplitude of the oscillation differs between the simulation and the field experiment in pitch angle and tail wing angle. These differences are assumed to result from error in the hydrodynamic coefficients obtained by the model experiment. The sensitivity of the coefficients for the maximum pitching angle was investigated by a series of numerical simulations as shown in Fig. 24. According to the results, M'_γ, Z'_w, M'_w and Z'_δ affect the maximum pitching angle strongly so the measuring of these coefficients should be practiced carefully to get more accurate mathematical model.

4 CONCLUSIONS

The hydrodynamic coefficients of the towed vehicle are estimated by three methods and the results compared. The set of coefficients is evaluated by comparison between the field experiment and numerical simulations using these hydrodynamic coefficients. These results lead to the following conclusion. Added mass coefficients and damping forces depend slightly on the oscillation frequency. The coefficients acquired by simple estimation method show good agreement with results from model experiments except for lift and induced drag of wings and damping force M'_w. These differences cause errors in the numerical simulation especially for calculation of the pitching motion. The estimated values obtained by the regression analysis show agreement with model experiments about Z'_w and A'_{33}. The results of pitching motion are believed to include large errors because that motion used for the analysis is minimal. In field experiments, the pitching motion of the vehicle must be controlled for accurate measurement and safety, which makes it difficult to obtain data good for the regression analysis. Amplitude of the pitching motion calculated using the coefficients by model experiments is smaller than that of the field experiment. It is assumed that the damping coefficients about pitching motion acquired by the model experiments are too large comparing with value in the field experiments.

REFERENCES

Koterayama, W, Nakamura, M, Kashiwagi, M, Kyozuka, Y and Ohkusu, M (1988). "A Preliminary Design of a Depth- and Roll-Controllable Towed Vehicle for Ocean Measurements," J The Society of Naval Architects of Japan, Vol 163, pp130–140 (in Japanese).

Koterayama, W, Yamaguchi, S, Nakamura, M and Moriyama, A (1994). "A Numerical Study for Design of Depth, Pitch and Roll Control System of a Towed Vehicle" Proc Fourth Int Offshore and Polar Eng Conference, Vol 2, pp337–344.

Koterayama, W, Yamaguchi, S and Nakamura, M (1995). "Field Experiments on an Intelligent Towed Vehicle "Flying Fish"" Proc Fifth Int Offshore and Polar Eng Conference, Vol 2, pp287–291.

Newman, J. N. (1977). "Marine Hydrodynamics," MIT press.

Ohkusu, M, Kashiwagi, M and Koterayama, W (1987). "Hydrodynamics of a Depth Controlled Towed Vehicle," J The Society of Naval Architects of Japan, Vol 162, pp99–109 (in Japanese).

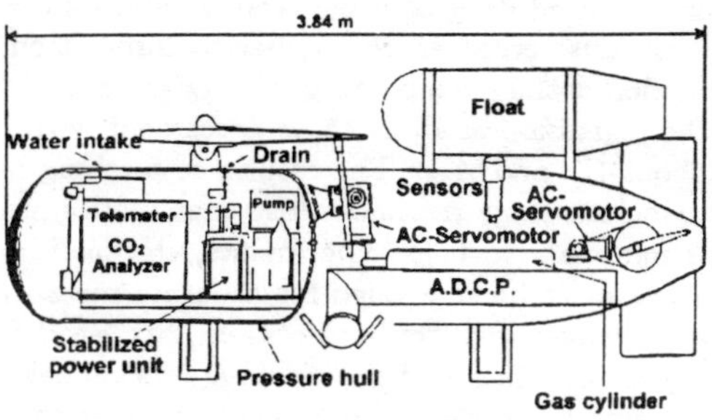

Fig. 1 Profile of Flying Fish

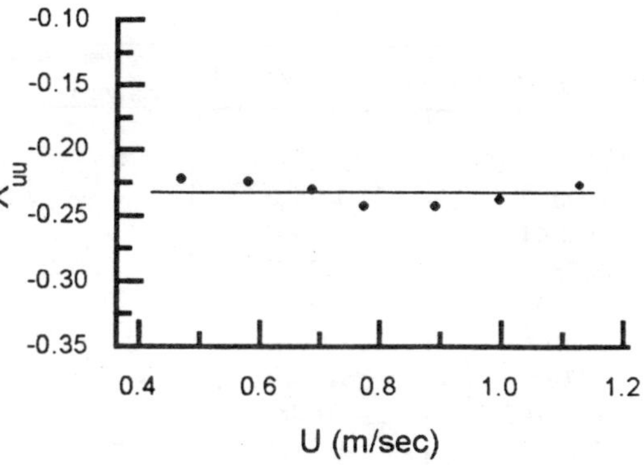

Fig. 4 Resistant coefficient

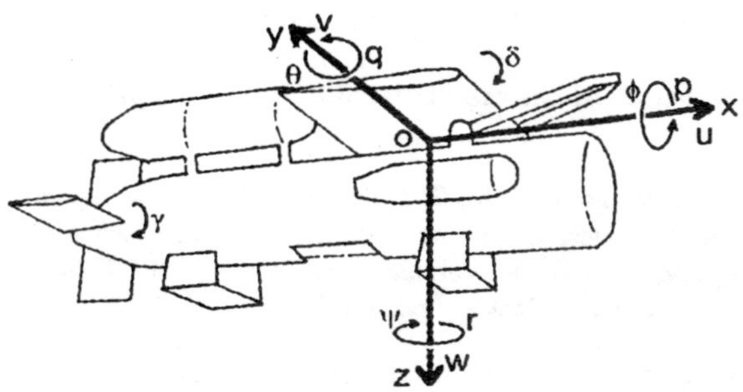

Fig. 2 Coordinate system

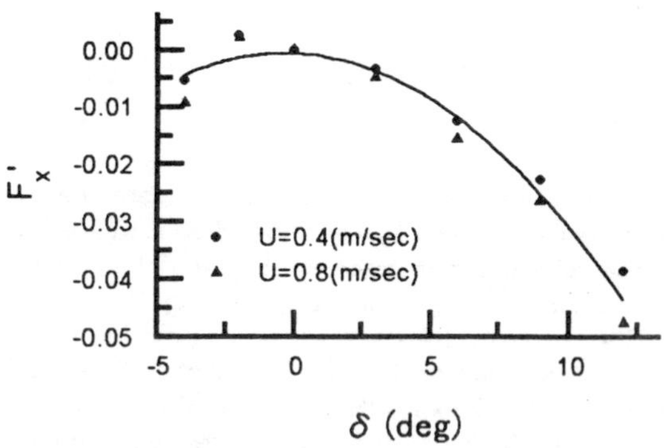

Fig. 5 Drag coefficient of a main wing

Fig. 3 Model experiment

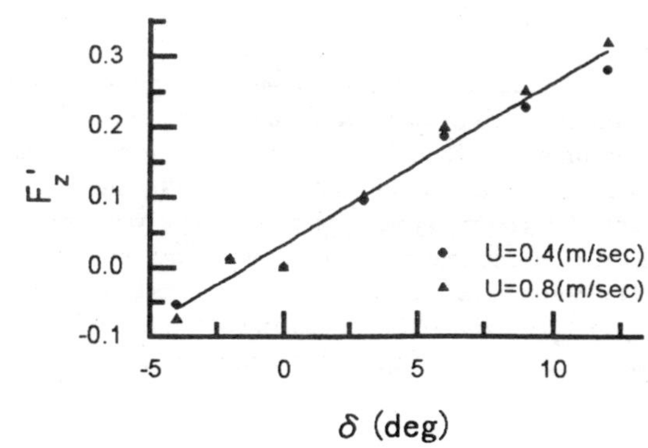

Fig. 6 Lift coefficient of a main wing

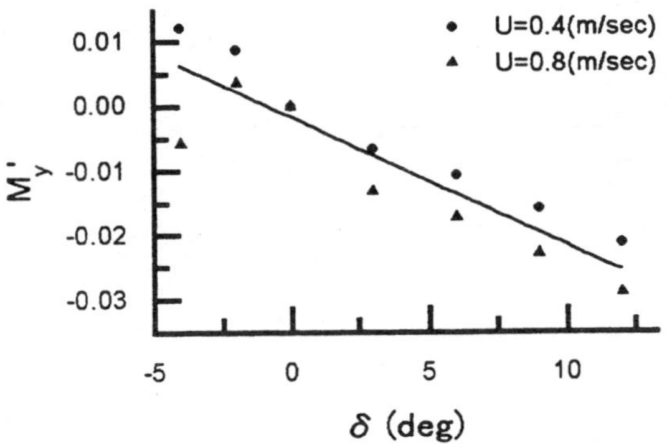

Fig. 7 Moment coefficient of a main wing

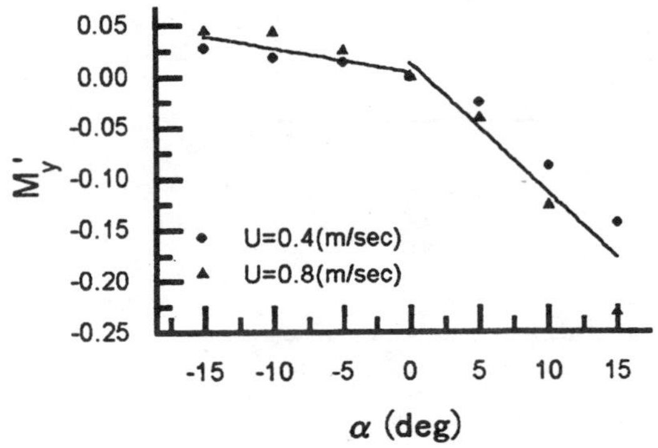

Fig. 10 Moment induced by attack angle

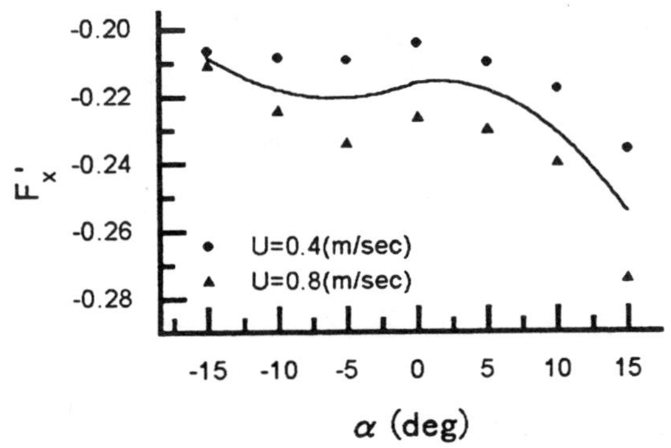

Fig. 8 X force induced by attack angle

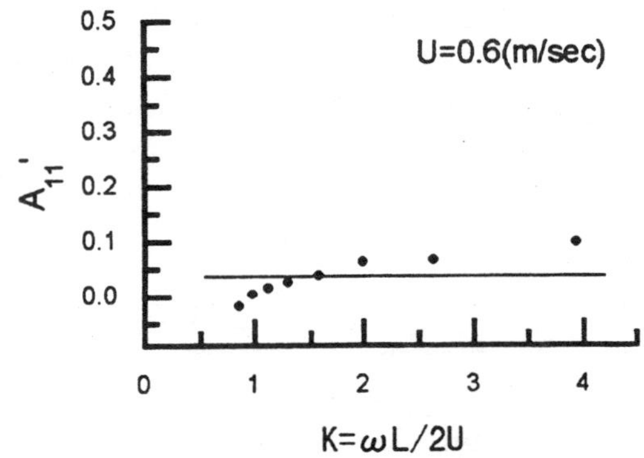

Fig. 11 Added mass coefficient in surge motion

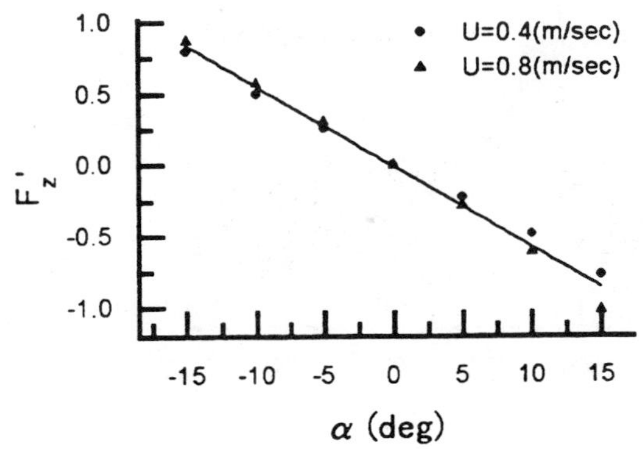

Fig. 9 Z force induced by attack angle

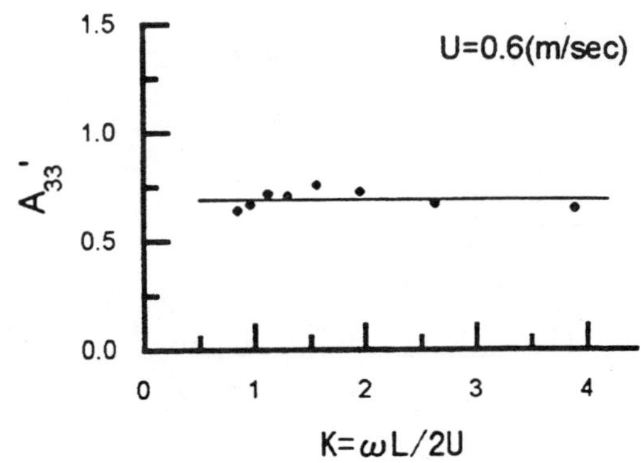

Fig. 12 Added mass coefficient in heave motion

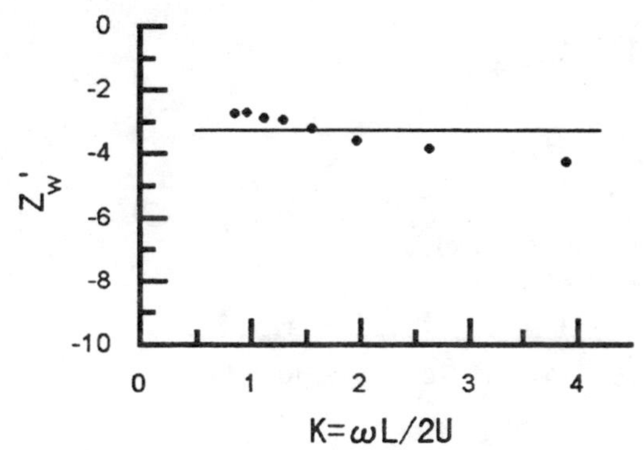

Fig. 13 Damping coefficient in heave motion

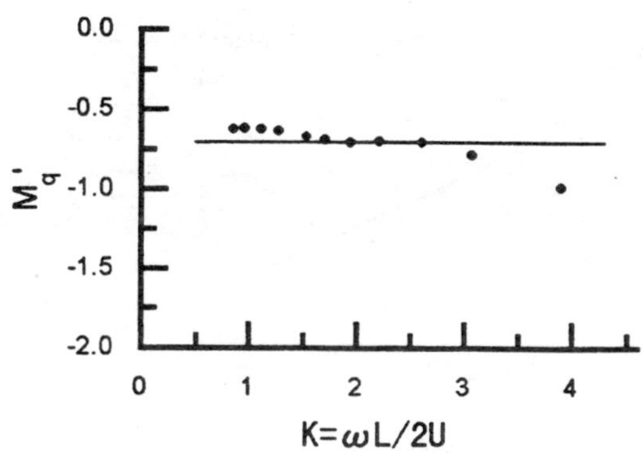

Fig. 16 Damping coefficient in pitch motion

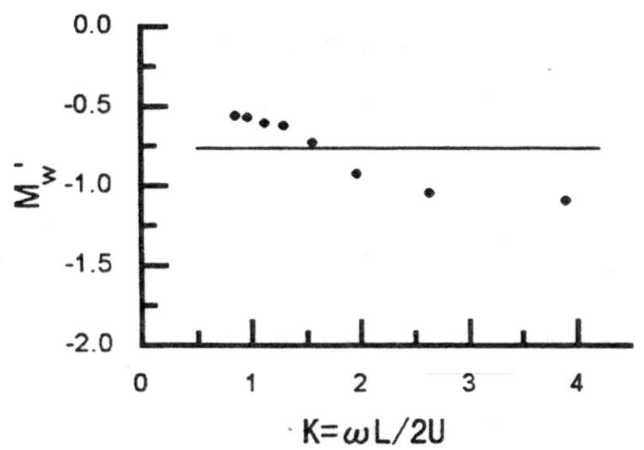

Fig. 14 Coupled damping coefficient in heave motion

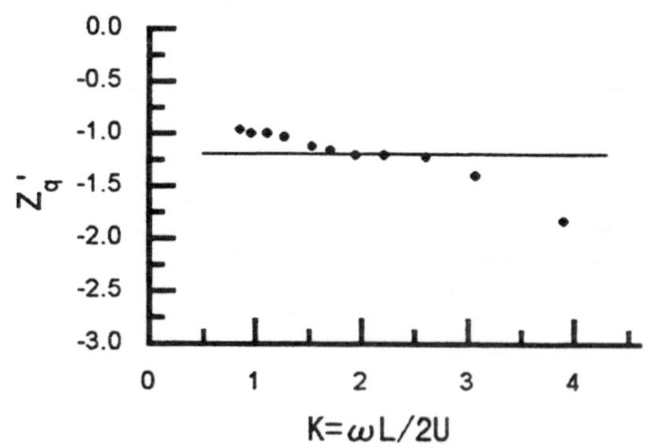

Fig. 17 Coupled damping coefficient in pitch motion

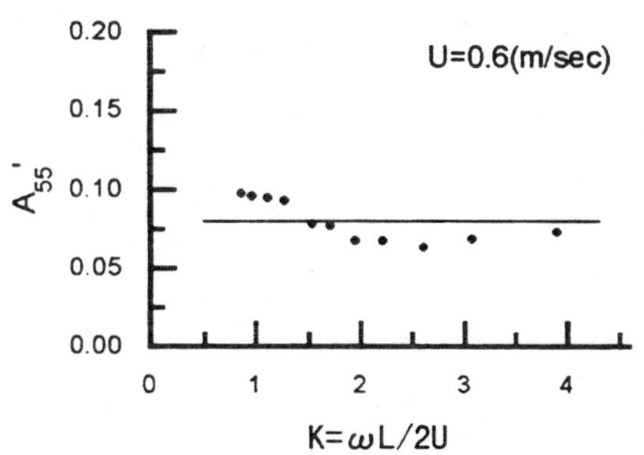

Fig. 15 Added moment of inertia in pitch motion

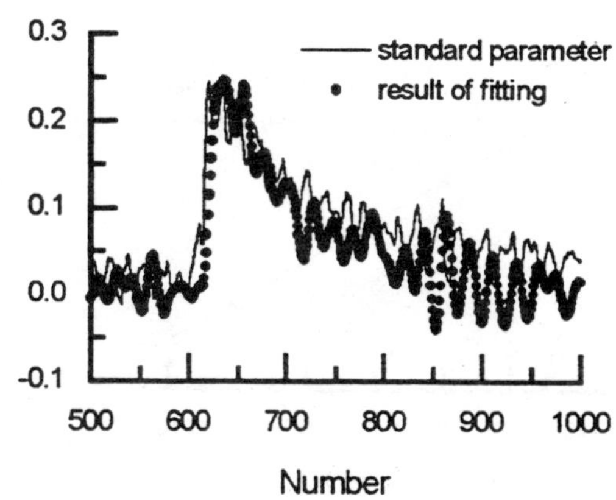

Fig. 18 Estimation about z force

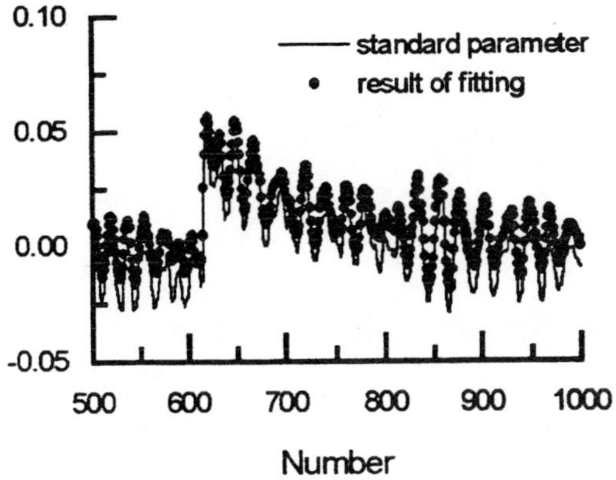

Fig. 19 Estimation about pitching moment

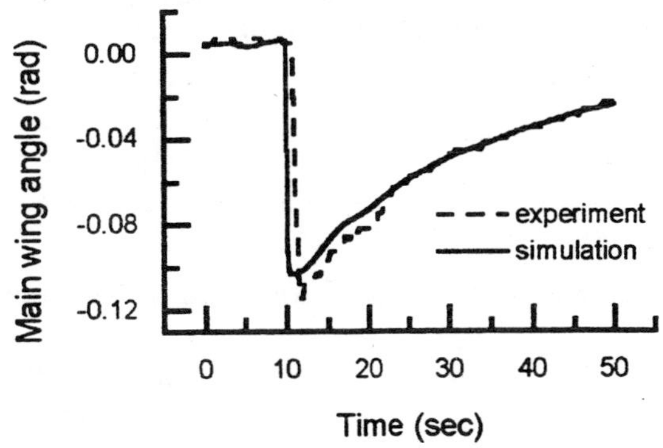

Fig. 22 Time histories of main wing angle

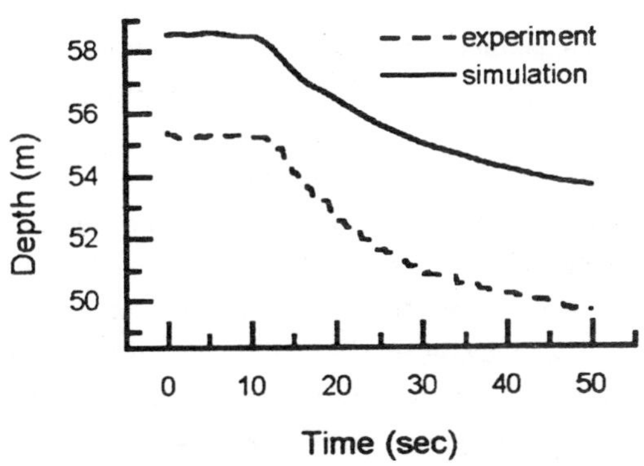

Fig. 20 Time histories of depth

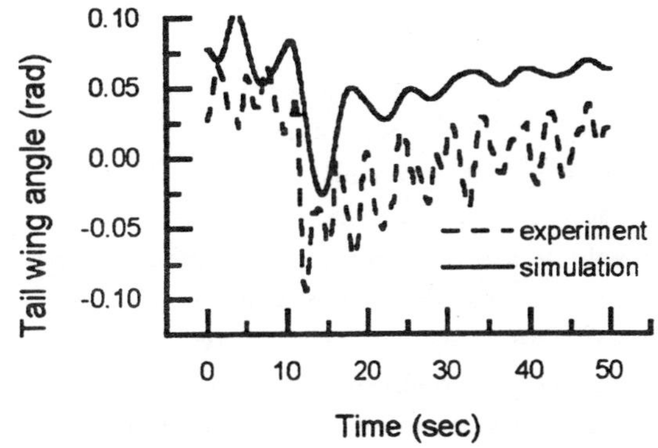

Fig. 23 Time histories of tail wing angle

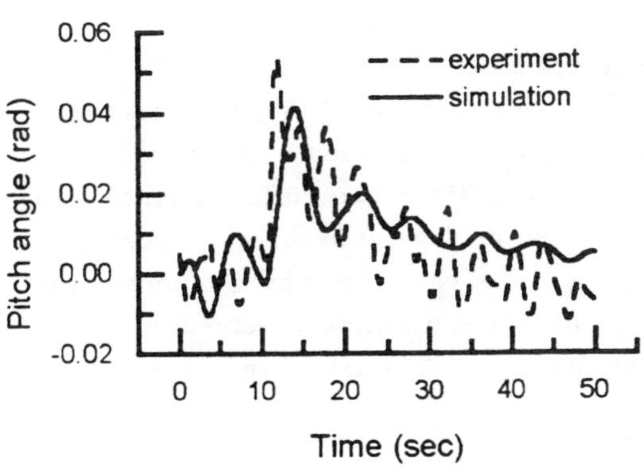

Fig. 21 Time histories of pitch angle

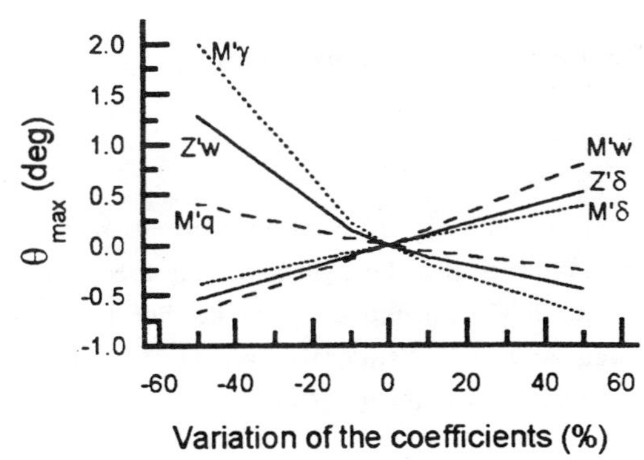

Fig. 24 Sensitivity analysis on the coefficients

Proceedings of the Seventh (1997) International Offshore and Polar Engineering Conference
Honolulu, USA, May 25-30, 1997

ISBN 1-880653-28-1 (Set); ISBN 1-880653-30-3 (Vol. II)

Coupled Motion Simulation of Underwater Towed and Self-Propulsive Vehicle

Ke-Qiang Zhu and Wei-Yang Li
Zhenjiang Shipbuilding Institute
Jiangsu, China

ABSTRACT:

This paper presents a dynamic model to simulate the motion of a Ship – Cable – Underwater vehicle system. The nonlinear ship maneuvering equation, Lumped parameter model of cable and six degree of freedom equations of vehicle are all included in the model. The model can simulate the three – dimensional transient coupled motion of the complex multi – body system in typical ship manevering conditions and can be used in either a towing problem or a tethered underwater vehicle problem.

The paper also provides examples illustrating the validity of the model. one of them is compared with a laboratory test. The results are shown to be in fairly good agreement with the test.

KEY WORDS: Cable, ROV, Hydodynamics

INTRODUCTION

The dynamics of TROV are nonlinear in nature due to rigid body coupling and the hydrodynamic forces on the vehicle. The ROV configuration is neither simple nor streamlined, and in addition, a number of thrusters are installed in close proximity to give it the required maneuverability. The ROV usually have comparable velocities along all three axes. Therefore the technique that depend on linearization of the equations of motion about a single forward operating speed cannot be used as effectively as they can with aircraft and submarines. The full, six – degree – of – freedom equations of motion of a slowmoving ROV are potentially very complex and satisfactory performance of the vehicle cannot be obtaned without consideration of cross – coupling between the different degrees of freedoms.

This paper presents a dynamic model to simulate the motion of a ship – cable – ROV system. The nonlinear ship maneuvering equation, lumped paramter model of cable and six – degree – of – freedom equations of ROV are all included. The model can simulate the three – dimensional transient coupled motion of the complex multi – body system in typical ship maneuvering conditions and can be used in either a towing or a tethered underwater vehicle problems.

TROV SIX – DEGREE – OF – FREEDOM EQUATIONS OF MOTION

Coordinate Systems

The dynamic model uses two orthogonal coordinate systems: (Fig (2.1)) global coordiuate system, (E,ξ,η,ζ), which remains fixed at the ocean surface mother ship with origin E, ζ pointing down into the water normal to the surface and ξ and η chosen in any two convenient mutually perpendicular horizontal directions with the only restriction being that the axes form a right – handed system; and a local coordinate system, (o, x, y, z), which is fixed on the vehicle with origin at 0, x pointing through the nose of the vehicle, z pointing through the belly of the vehicle and y completing the right – handed system. The position and orientation of the vehicle in global coordinates can be specified by r_{bo}, the vector from E to o, and the Euler angles ψ, θ, φ. Transformation of forces and motions from global to local coordinates can be accomplished by using the transformation of forces and motions from global to local coordinates can be accomplished by using the transformation matrix $[T_B]$ and from local to global by using its inverse which is just $[T_B]^T$ since $[T_B]$ is an orthogonal matrix,

$$[T_B]=\begin{bmatrix} X_1 & X_2 & X_3 \\ Y_1 & Y_2 & Y_3 \\ Z_1 & Z_2 & Z_3 \end{bmatrix} \tag{2.1.1}$$

where

$$X_1=\cos\theta\cos\psi$$
$$X_2=\cos\theta\sin\psi$$
$$X_3=-\sin\theta$$
$$Y_1=\sin\varphi\sin\theta\cos\psi-\cos\varphi\sin\psi$$
$$Y_2=\sin\varphi\sin\theta\sin\psi+\cos\varphi\cos\psi$$
$$Y_3=\sin\varphi\cos\theta$$
$$Z_1=\cos\varphi\sin\theta\cos\psi+\sin\varphi\sin\psi$$
$$Z_2=\cos\varphi\sin\theta\sin\psi-\sin\varphi\cos\psi$$
$$Z_3=\cos\varphi\cos\theta$$

The development of the dynamic model is carried out in the local coordinate system since the motion of the vehicle is usually described in reference to this system.

Inertia Force on ROV

Since(Prctically) all vehicles have a longitudinal symmetry plane xoz, the added mass coefficients A_{ij} with $(i+j)=$odd are zeros. The mass matrix[M] consisting of all the coefficients of inertia and added mass terms corresponding to vehicle acceleration components is defined as

$$M_{ii}=m+A_{ii} \quad (i=1,2,3)$$
$$M_{jj}=I_{jj}+A_{jj} \quad (j=4,5,6)$$
$$M_{24}=M_{42}=A_{24}-mz_G$$
$$M_{51}=M_{15}=A_{35}-mx_G$$
$$M_{26}=M_{62}=A_{26}+mx_G$$
$$\text{the other } M_{ij}=0$$

All remaining added mass and inertial terms are combined into a dynamics vector $\{F_I\}_{6\times1}$

$$\{F_I\}_{6\times1}=\begin{cases} m[qw-vr-x_G(q^2+r^2)+z_Gpr]+\sum_{j=1}^{6}(qA_{3j}-rA_{2j})V_{Rj} \\ m[ru-wp+z_Gqr+x_Gpq]+\sum_{j=1}^{6}(rA_{1j}-pA_{3j})V_{Rj} \\ m[pv-uq-z_G(p^2+q^2)+x_Grp]+\sum_{j=1}^{6}(pA_{2j}-qA_{1j})V_{Rj} \\ (I_{zz}-I_{yy})qr+mz_G(pw-ru)+(r^2-q^2)J_{yz}-J_{zx}qp+ \\ \quad J_{yx}pr+\sum_{j=1}^{6}(qA_{6j}-rA_{5j}+V_{Ry}A_{3j}-V_{Rz}A_{2j})V_{Rj} \\ (I_{xx}-I_{zz})rp+m[z_G(qw-rv)-x_G(pv-qu)]+(p^2- \\ \quad r^2)J_{xz}-J_{xy}rq+J_{zy}qp+ \\ \quad \sum_{j=1}^{6}(rA_{4j}-pA_{6j}+V_{Rz}A_{1j}-V_{Rx}A_{3j})V_{Rj} \\ (I_{yy}-I_{xx})pq+mx_G(ru-pw)+(p^2-q^2)J_{xy}- \\ \quad J_{yz}pr+J_{xz}rq+ \\ \quad \sum_{j=1}^{6}(pA_{5j}-qA_{4j}+V_{Rx}A_{2j}-V_{Ry}A_{1j})V_{Rj} \end{cases} \tag{2.2.2}$$

where V_R is the velocity of the vehicle relative to the fluid.

Weight and Buoyancy and Thrusters

The gravitational force and buoyant force are defined in terms of the global coordinate system so they must be transformed to the local coordinate systems:

$$\vec{F}_G+\vec{F}_B=(m-\rho\nabla_B)g[-\sin\theta \quad \cos\theta\sin\varphi \quad \cos\theta\cos\varphi]^T \tag{2.3.1}$$

$$(\vec{R}_G\times\vec{F}_G+\vec{R}_B\times\vec{F}_B=-g(mz_G-\rho\nabla_Bz_B)[\cos\theta\sin\varphi \quad \sin\theta \quad 0]^T \tag{2.3.2}$$

We combined the two forces into a dynamics vector $\{F_{GB}\}$

The resultant force and moment of a thruster configuration consisting of N thrusters can be expressed as the vector sum of the force and moment from each individual thruster:

$$\vec{F}_T=\sum_{i=1}^{N}\vec{F}_{Ti} \tag{2.3.3}$$

$$\vec{M}_T=\sum_{i=1}^{N}\vec{M}_{Ti}+\sum_{i=1}^{N}\vec{R}_{Ti}\times\vec{F}_{Ti} \tag{2.3.4}$$

Magnitudes of thrust and moment due to the ith thruster are written as (MA SAO NOMOTO et al 1986)

$$|\vec{F}_T|=K_{Ti}\rho n_i^2 D_i^4 \tag{2.3.5}$$

$$|\vec{M}_{Ti}|=K_{Qi}\rho n_i^2 D_i^5 \tag{2.3.6}$$

Where D_i is the diameter of the thruster,

n_i is the angular speed of the thruster shaft

K_{Ti}, K_{Qi} are the thruster and torque coefficients of the thruster.

The thruster and torque coefficients must be expresed as functions of the advanced coefficient

$$J=V_R/nD$$

Where V_R is the axial speed of the thruster relative to the fluid.

We combined F_T and M_T into a dynamics vector $\{F_T\}_{6\times1}$.

Non-Inertia Hydrodynamic Forces

Generally the configuration of the vehicle is of an open-fram type with buoyant material on its top. The flow around a vehicle of this type is complex, and it is difficult to estimate its hydrodynamic characteristics by calculations. So, the coefficients of hydrodynamic forces which are necessary for the evaluations of the maneuverability were obtained by hydrodynamic tests. The hydrodynamic coefficients due to the translational motions at constant velocities such as drag coefficient, moment coefficient, etc. can be obtained by a series of static hydrodynamic tests. The hydrodynamic coefficieuts due to the accelerations and angular accelerations of the vehicle should be obtained by means of Planar Motion Mechanism. The damping coefficients of the rotational motions can also be obtained in this series of tests.

The non-inertia hydrodynamic forces can be expressed as:

$$\left\{F_F\right\}_{X,Y,Z}=-\frac{\rho}{2}V_R^2\nabla_R^{2/3}C_{Fx,y,z} \tag{2.4.1}$$

$$\left\{F_F\right\}_{K,M,N}=-\frac{\rho}{2}V_R^2\nabla_R C_{Nx,y,z}$$

$$-\frac{\rho}{2}\nabla_R^{5/3}\{p|p|,q|q|,r|r|\}^T\cdot C_{Np,q,r} \tag{2.4.2}$$

Where $C_{Fx}=C_{Fx}(\alpha,\beta)$ $\quad C_{Nx}=C_{Nx}(\gamma)$

$C_{Fy}=C_{Fy}(\beta,\gamma)$ $\quad C_{Ny}=C_{Ny}(\alpha)$

$C_{Fz}=C_{Fz}(\alpha,\gamma)$ $\quad C_{Nz}=C_{Nz}(\beta)$

and

$$\alpha=\mathrm{tg}^{-1}(V_{Rz}/V_{Rx})$$
$$\beta=\mathrm{tg}^{-1}(V_{Ry}/V_{Rx})$$
$$\gamma=\mathrm{tg}^{-1}(V_{Rz}/V_{Ry})$$

We combined these forces into a dynamics vector $\left\{F_F\right\}_{6\times 1}$

Umbilical Cable Force

Umbilical cable force is one of the most important nonliear disturbance force on ROV when the current is strong and cable length is long enough. The movement of a flexible cable moving in fluid with changed velocity can be described by following governing vectorial equation:

$$(\lambda_c+\lambda_a)\partial\vec{v}/\partial t=g[\lambda_c-\rho A(1+\varepsilon)]\vec{\zeta^0}+\frac{1}{2}\rho(1+\varepsilon)d$$

$$(C_{DN}|\vec{U}_{RN}|\vec{U}_{RN}+C_{DT}|\vec{U}_{RT}|\vec{U}_{RT})+\lambda_a\frac{\partial\vec{U}c}{\partial t}+$$

$$\frac{\partial}{\partial S}\left(\frac{\vec{T}c}{1+\varepsilon}\frac{\partial\vec{r}}{\partial S}\right) \tag{2.5.1}$$

Where, λ_c λ_a the mass and added mass per unit length respectively.

ε, the strain or elongation per unit length.

$\vec{\zeta^0}$, the unit vector of ζ axis.

$\vec{U}_{RN}$, $\vec{U}_{RT}$ the components of the cable element velocity normal and tangential to the cable respectively.

C_{DN}, C_{DT}, the cable coefficients of normal and tangential drag.

d the cable effective diamter.

U_c the vector of fluid velocity.

The governing equation (2.5.1) can be cast into a discrete form by various methods such as the spatial finite difference method(A.P.K. DE ZOYSA, 1979, C.M. ABLOW, 1983), weighted residual method etc. The advantages of semi-analytical techniques (S. Huang, 1993) are dependent on the specific problem and the orientation of the analyst.

In contrast, the Lumped Parameter Method (LPM) handles special cases, such as embeded instrument, floats, weights etc. quite naturally by distributing their properties to the nearest nodes (H. Wang YL 1986). There are no complicated matrices to construct and any arbitrarilly complex heterogeneous structure can be effectively treated. Many of the limitations of the semi-analytical methods could be avoided. In particular the governing partial differential equations for a complex system are never be derived because the form of he spatial variations are implied by the collection of discrete elements. The continuous nature of the time coordinate produces a couple set of ordinary differential equations which can be integrated as an initial value problem. The most common approach used in LPM assumes the continuous cable system is modeled as a set of discrete massless springs (H. T. Wang, 1982). A given node typically receives half of the attributes (mass drag, weight, etc) of each connected element. Cable element tensions are determined from the instantaneous position of the nodes using a $\nabla l/l$ strain approximation and a given stress strain relation. The spatial coordinates of the lumped masses are the unknowns.

The lumped parameter equations we used have the following form :

$$\frac{\lambda_i\, d^2\vec{r_i}(t)}{dt^2}=\vec{T}_i-\vec{T}_{i-1}+\vec{W}_i+\vec{D}_i \tag{2.5.2}$$

Where:

$$\vec{T}_i=\frac{\pi}{4}d_i^2E_i(1/l_i-1/|\vec{r}_{i+1}-\vec{r}_i|)(\vec{r}_{i+1}-\vec{r}_i)$$

$$\vec{D}_i=\frac{\rho}{4}(l_{i-1}d_{i-1}+l_i\ d_i)(C_{ti}|\vec{U}_{Rti}|\vec{U}_{Rti}+C_{ni}|\vec{U}_{Rni}|\vec{U}_{Rni})$$
$$\vec{U}_{Rti}=[(d\vec{r}_i/dt-\vec{U}_c)\cdot\vec{\tau}_i]\vec{\tau}_i$$
$$\vec{U}_{Rni}=d\vec{r}_i/dt-\vec{U}_{Rti}$$

$$\vec{\tau}_i=\begin{cases}\vec{r}_2-\vec{r}_1 & (i=1)\\ (\vec{r}_{i+1}-\vec{r}_{i-1})/|\vec{r}_{i+1}-\vec{r}_{i-1}| & (i=2,N-1)\\ \vec{r}_N-\vec{r}_{N-1} & (i=N)\end{cases}$$

Where:

$\vec{T}_i$ = magnitude of the force exerted on point mass i by the spring connecting point masses i and i+1

d_i = diameter of the link (segment)

E_i = modulus of elasticity

l_i = length of the i'th segment

$\vec{r}_i$ =position vector of the i'th lumped mass with respect to the fixed inertial reference frame.

$\vec{W}_i$ =the gravitational force of ith lumped mass

$\vec{D}_i$ =the drag force lumped into the i'th node

λ_i =mass per unit length of cable

C_{ti}, C_{ni}=hydrodynamic force coefficients for the i'th link

$\vec{U}_{Rti}$, $\vec{U}_{Rni}$=tangential and normal velocity of the i'th node respectively

$\vec{\tau}_i$ =unit tangential vector at the i'th node

The components of tension on ROV can be obtained by using the transformation matrix $[T_B]$ from the global coordinate system to the body coordinate system at the tether point.

$$\vec{F}_c=[T_B]\vec{T}_N=\{X_c \quad Y_c \quad Z_c\}^T \tag{2.5.3}$$

The moment on ROV is

$$\vec{R}_c\times\vec{F}_c=\{R_{cz}Y_c, R_{cz}Z_c-R_{cz}X_c, -R_{cz}Y_c\}^T$$

Where, $\vec{R}_c=\{R_{cx} \quad R_{cy} \quad R_{cz}\}^T$ is the components of tether point vector at the body coordinate system.

We combined these 6 components into a dynamics vector $\{F_c\}_{6\times1}$ Finally all terms are combined to produce the final form of the dynamic model of TROV

$$[M]\{Q\}_{6\times1}=-\{F_I\}_{6\times1}+\{F_{GB}\}_{6\times1}+\{F_T\}_{6\times1}+\{F_C\}_{6\times1}+\{F_F\}_{6\times1} \tag{2.5.4}$$

where $\{Q\}=\{u \quad v \quad w \quad p \quad q \quad r\}^T$

BOUNDARY CONDITIONS

The tether cable must have the same motion at its attaching point respectively as surface ship and the lower ROV. This constraint provides the upper and lower boundary conditions for the cable. In our programe a set of maneuvering equations for the surface ship has been included. It will be useful in determining the effect of the cable on the dynamic stability and manuvering response of the towing vehicle when the drag of the cable is an appreciable fraction of the drag of the towing vehicle.

We did not assume that the point coincide with the center of mass of the vehicle. So, the effect of the rotational modes of the ROV on the motion of the tether cable could be included. The equations that form the lower boundary conditions are the six-Degree-of-F-reedom equations of motion of ROV. In order to make use of these boundary conditions in the numerical solution, a transformation of cable force from the spatial coordinate system to the system aligned with the ROV is used. The position and velocity of the lower end of the cable is given as(Naomi Kato, 1987):

$$\vec{r}_N(t)=\vec{r}_b(t)+[T_B]^T\vec{R}_c \tag{3.1}$$

$$d\vec{r}_N(r)/dt=[T_B]^T(\vec{U}+\vec{\Omega}\times\vec{R}_c+d\vec{R}_c/dt) \tag{3.2}$$

where $\vec{R}_c$ is the vector of cable termination point in the body system.

$\vec{r}_b$, is the vector of local coordinate original in the global system.

Since the interest in the present paper is on motions with long time scales, those factors which are perturbations to the overall mean long-time motion such as ocean waves are not considered.

NUMERICAL METHOD OF SOLUTION

A set of coupled, nonlinear, first-order differential equations of a ship-cable-ROV multi-body system can be solved numerically from an initial condition. A static equilibrium cable configuration can be assumed to provide the initial conditions for the problem. A computer program has been developed that integrates these governing equations using the Runge-Kutta method. The step size of numerical integration can be changed. If the two sets of values do not all agree within the specified accuracy, the step size is automatically reduced until the specified accuracy is met. In order to save the CPU time, the Fouth Runge-Kutta method is used for cable dynamics and a simple Euler integration method is used for surface ship and underwater ROV. Execution time depends largely on the value of the Young's modulus of the cable, the number of cable nodes, the amount of time over which dynamic motions are desired, and the type of problem under consideration.

Extensibility is needed to prevent infinite wave speeds and hence very small time intervals in the integration. To account for this, the Young's modulus used in the model will usually chose less than the physical one (H. T. Wang, 1982).

COMPARISON WITH EXPERIMENT

Comparison of the simulated with the measured results can give an idea of the accuracy of the simulation method, One experiment called "Anchor-last deployment". (Meggitt et al 1978) is chosen for comparison. In this experiment, a sphere of 50.8mm diameter weighed 1.09 N in air was attached to a rubber tether 4.14mm in diameter and 1.83m long, the tether were fully immersed and allowed to reach an initial static configuration catenary. The sphere located horizontally 1.676m from the fixed end at t=0. and were relesed then. Measurements were made for the position of specified points on the tether and the tension at the fixed end of the tether as a function of time. (Fig(5.1))

The comparison indicates that a different choice of the drag coefficients could bring the simulated and measured time histories of positions and tension into better agreement in spite of some transient behavior still differing from the measured values. For example the modes of oscillation that present in the measurements are not present in the simulation. This discrepancy may be due to inaccurate drag coefficients or certain unmodeled effects. such as rotation of the sphere or bending stiffness of the cable. Besides, since the tension is not a state variable in the integration, it is calculated from the positions of the nodes, so the integrator does not directly control its accuracy. It seems that the drag coefficients commonly used in the cable simulation not considering the internal damping of the cable are likely too low.(Fig(5.2))

SAMPLES

A computer simulation was performed for the three - dimensional coupled motion of a TROV system. It is an improvement to the traditional quasi - static method. The model is adaptable to many kinds of complex spatial movement of ROV. Numerica values of the parameters in the mathematical models are assumed as follows; the length of cable is 300 meters, the diamter of cable is 3.2cm, the weight in water is 2.25N/m, Young's modulus is $2\times10^{11}N/m^2$. The parameters of a ROV is after the Dolphin 3K (MASAO NOMOTO et al, 1986).

Fig(6.1) shows the configuration of a three - dimensional submersible cable. The surface exitation end is taken to be fixed. The lower ROV is moving in a given circular turn speed. Constant current speed is 0.5 knots. The figure shows the mannar in which the upper cable nodes lag the motion of the lower ROV end and also shows how current affects the configuration obviously. The biggest tension would occur during the state against flow. It is clear that the current plays a significant role.

Fig(6.2) shows the static equilibrium configuration of cable at different speed of current, which can be used as initial values in dynamic simulation to save computation time.

Fig(6.3) shows the fluctuations of cable tensions at the upper and lower end due to pure heaving or/and surging motions of the surface ship. The position of the upper end of the cable was input as sinusoidal functions of time, and the ROV keeps its position and attitude. The calculation was done for the following severe oceanic conditions: the amplitude of the ship heaving and surging are all taken as 5 meters. Period is 6 second. The calculation showed that the surface ship's excitation that has minium angle between excitation and tangent of cable of upper end would induce maximum amplitude of peak tension of cable. Therefore, the cable force is sensitive to surging and heaving motion of surface ship when the ship is located in the upperstream and downstreem of ROV respectively. A comparison of this model's predictions to experimental data is intended for the future.

SUMMARY

Nonlinear mathematical model of space motion is setup for a tethered remotely operated vehicle. The coupled equations of motion for the vehicle under the nonlinear force due to the umbilical cord are solved. The traditional calculation method of the quasi - static additional cable force is improved. The simulation results of the dynamic cable force agree well with the known experiments. The influence of the parameters, such as: how the result is affected by changes in the number of nodes and drag coefficient of cable, how the current affects the configuration of cable obviously, and the surface excitation on the system performance is also discussed.

References

A. P. K. DE ZOYSA, (1979). "Steady - state Analysis of Undersea Cables". Ocean Engng. Vol.5. PP209 - 223

C. M. ABlow and Schechter. (1983). "Numerical Simulation Undersea Cable Dynamics". Ocean Engng. Vol. 10. No. 6

H. T. Wang (1982). "Three - Deimensional Long - time Motion Behavior of Ocean Cable Systems", Preceedings of the Second International Offshore Mechanics and Arctic Engineering Symposium, Houston Texas, PP266 - 274.

H. Wang YL (1986). "Nonlinear analysis of mooring lines", OMAE, Vol.3.

MASAO NOMOTO and MUTSUO HATTORI (1986). "A Deep ROV Dolphin 3k Design and Performance Analysis." IEEE, J. Oceanic Engineering. Vol OE - 11. No3. PP373 - 391.

Meggitt D. J. et al (1978). "Small - size Laboratory Experiments on Displacement Dynamics of Cable Systems". Vol.1. TM - 44 - 78 - 11. Naval Civil Engineering Laboratory Port Hucneme. CA.

Naomi kato, (1987). "Vibration Analysis of Underwater Towed System". J. The Society of Naval Architects of Japan. Vol. 162.

S. Huang and D. Vassalos (1993). "A Semi - analytic Treatment of three Dimensional Statics of Marine Cables". Ocean Engng. Vol. 20.

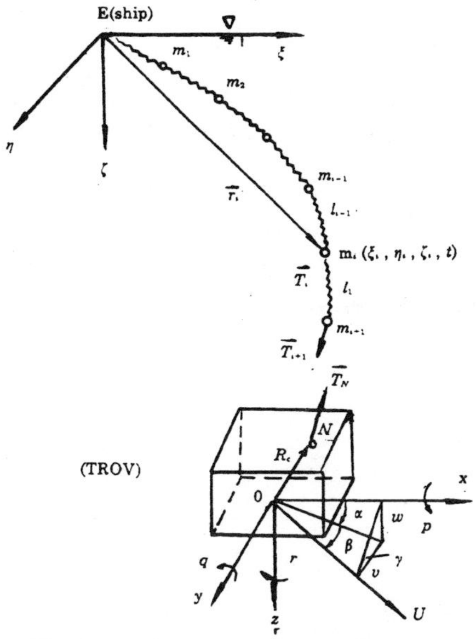

Fig (2.1) ROV, cable coordinate systems

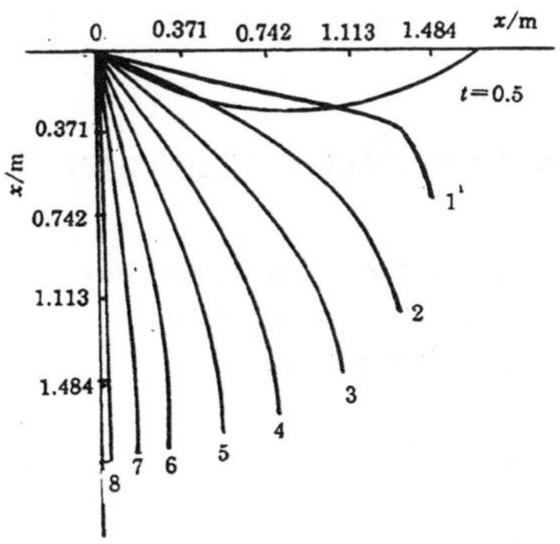

Fig(5.1) Simulated Configuration for Anchor Drop

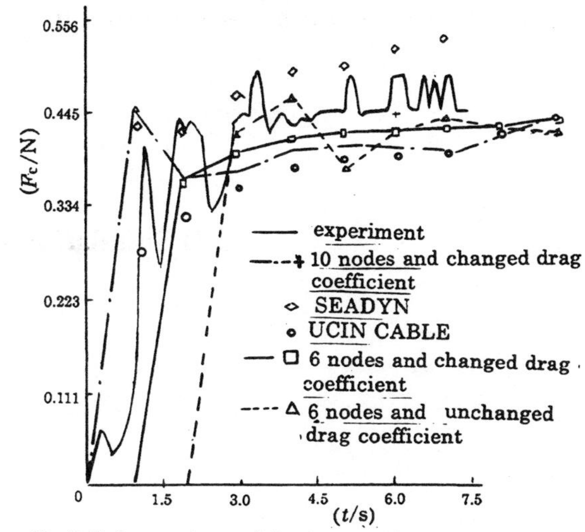

Fig(5.2) Comparison of the present theory with the test

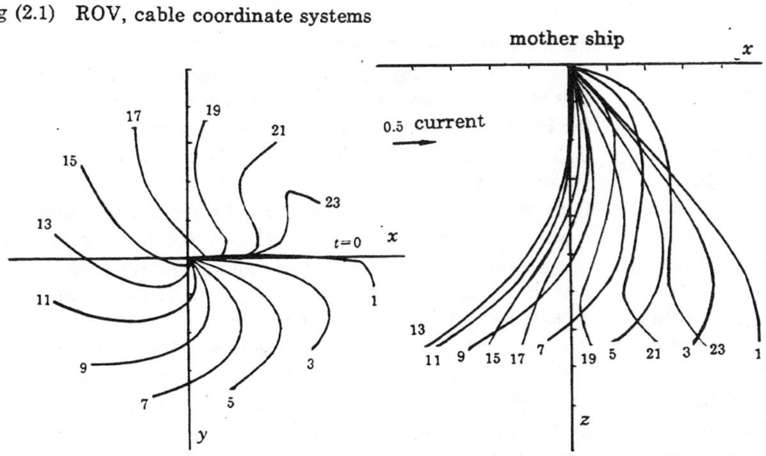

(a)Projection of Cable configuration in Vetical Plane

(b)Projection of Cable configuration in Horizontal plane

Fig(6.1) Three dimensional dynamic configuration

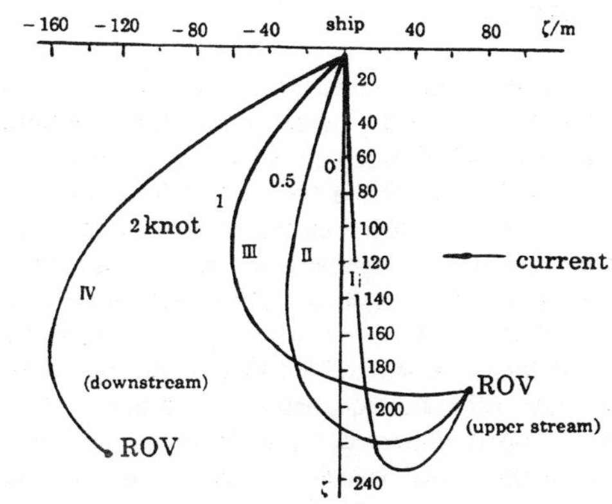

Fig(6.2) Static equilibrium cable configuration

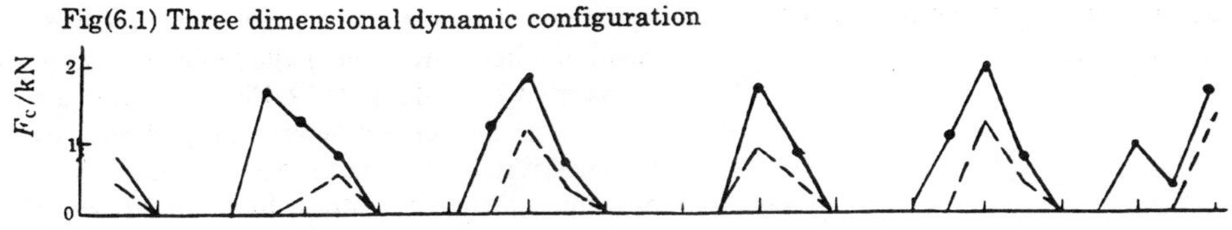

(a) surging motion of the surface ship

—— ship side

– – – ROV side

(b) Heaving motion of the surface ship

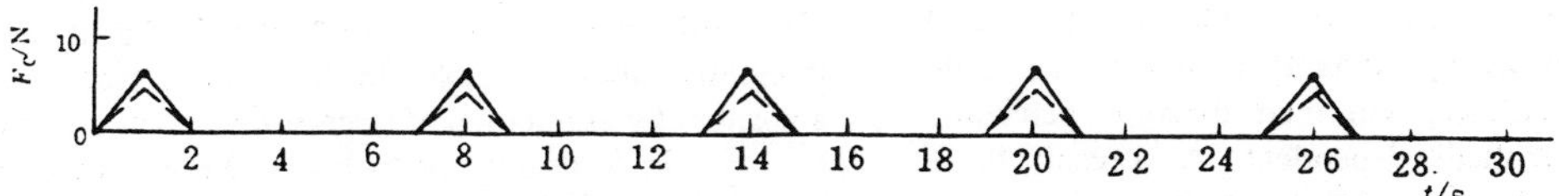

(c) Surging and heaving at the same time

Fig(6.3) Fluctuations of cable tension at the upper and lower ends

Proceedings of the Seventh (1997) International Offshore and Polar Engineering Conference
Honolulu, USA, May 25-30, 1997

ISBN 1-880653-28-1 (Set); ISBN 1-880653-30-3 (Vol. II)

3-D Simulation of a Two-Part Underwater Towed System

Jiaming Wu and Allen T. Chwang
University of Hong Kong
Hong Kong

ABSTRACT

A fully three dimensional model of a two-part underwater towed system is studied. In the model, the governing equations of cables are established based on Ablow and Schechter's method. The boundary conditions for the two-part underwater towed system are derived. The six-degree-of-freedom equations of motion for submarine simulations are adopted to estimate the hydrodynamic performance of a towed vehicle. The established governing equations for the system are then solved using a centered finite difference method. In this paper several algorithms are used to solve this special form of finite difference equations. The results in this paper indicate that the two-part underwater towed system improves the dynamic behavior of the towed vehicle and is an easy way to decouple the towing ship motion from the towed vehicle. Because the model uses an implicit time integration, it is stable for large time steps and is an effective algorithm for simulation of a large-scale towed system.

KEY WORDS: two-part tow, underwater, towed system, numerical simulation, cable, three dimension

1. INTRODUCTION

Underwater towed systems are fundamental tools for many marine applications which include naval defense, seabed mapping and ocean environmental measurements. These systems can be as simple as a single cable with its towed vehicle, or they may be composed of multiple towed cables and multiple towed bodies.

In simulating the hydrodynamic performance of an underwater towed system, the coupling effect between the towing ship and the towed system is generally neglected, and the hydrodynamic model is usually composed of two parts: the cable and the towed vehicle. It is well known that the equations of motion for the cable and the towed vehicle are nonlinear and their dynamic behaviors during various operations are mutually dependent. As a result, these equations are strongly coupled. In order to study the complete problem, they must be solved simultaneously as a whole. For such a complicated problem, it is not easy to solve analytically and numerical methods are usually employed.

The most prevalent approaches used nowadays in determining the hydrodynamic performance of an underwater cable are the lumped mass method (Walton and Polachech 1960) and the finite difference method (Sanders 1982, Ablow and Schetchter 1983). However, Howell (1991) stated that the explicit time domain integration scheme used in the lumped mass method made the method conditionally stable. Burgess (1991) pointed out that the time integration used in this algorithm requires the time step be so chosen that the Courant-Friedrichs-Levy wave condition is satisfied for the highest natural frequency of the lumped mass system. This restricts the use of very small time steps. However, Hearn and Thomas (1991) believed that the collapse of numerical procedure at large time steps in the method is not due to the instability of the numerical scheme, but is caused by the failure of the Newton-Raphson iterative procedure adopted to determine the correct tension levels when solving the nonlinear equations of motion. The reason for the collapse of the numerical procedure in the lumped mass method may not be clear, however it is true that time steps in this method must be chosen very small in order to avoid the failure in numerical procedure on the basis of experiences (Burgess 1989, Hearn and Thomas 1991).

In the finite difference method, the governing equations for the underwater cable are derived from the balance of forces at a point of cable. Among various finite difference methods, the model developed by Ablow and Schechter (1983) is worthy of note. In this model, the cable is treated as a long thin flexible circular cylinder in arbitrary motion. It is assumed that the dynamics of cable are determined by gravity, hydrodynamic loading and inertial forces. The governing equations are formulated in a local tangential-normal coordinate frame which has the unstretched distance along the cable. The differential equations are then approximated by finite difference equations centered in time and in space. By solving the equations, the motion of underwater cable can be determined in the time domain. The principal advantage of this method is that it uses an implicit time integration and is stable for large time step sizes. It is a good algorithm for simulation of large-scale underwater cable motion.

In describing the hydrodynamic behavior of a towed vehicle, the six-degree-of-freedom equations of motion are usually adopted. In resent years, several hydrodynamic models of underwater towed systems have been developed by various authors. Chapman (1984)

proposed a model to describe the dynamic behavior of an underwater towed system. The model is mainly used to simulate the behavior when the towing ship travels in a circular trajectory and is not fully three dimensional. Koterayama *et al.* (1988) developed a three-dimensional model for a single cable towed system, in the model the behavior of the towed cable is determined by the lumped mass method.

The research on hydrodynamics of an underwater towed system at present is mainly concentrated on a single cable. Owing to the demand of towed vehicle operations, there is a need for these systems to have ability to move with stable attitude in ocean environments such as wave actions, and the towed vehicle generally requires that the perturbations from the wave induced motion of a tow ship are minimized. One easy method is to use a two-part underwater towed system to decouple the surface vessel motion from the towed vehicle. Ranmuthugala and Gottschalk (1993) proposed a two-dimensional hydrodynamic model of a two-part underwater towed system. In the model the kinematic properties of the towed vehicle are not discussed.

In this paper a three-dimensional hydrodynamic model to simulate a two-part underwater towed system is proposed. In the model the governing equations of cables are established based on Ablow and Schechter's method. The six-degree-of-freedom equations of motion for submarine simulations are adopted to estimate the hydrodynamic performance of a towed vehicle. The established governing equations are then solved using a centered finite difference method. In this paper several algorithms are used to solve this special form of finite difference equations.

The numerical results indicate that the two-part underwater towed system improves the dynamic behavior of the towed vehicle. Because the model uses an implicit time integration, it is stable for large time steps. It gives more flexibility in choosing different time steps for different maneuvering problems. It is an effective algorithm for the simulation of a large-scale towed system. The model presented in this paper can easily be extended to simulate multiple towed cables and towed bodies.

2. GOVERNING EQUATIONS AND BOUNDARY CONDITIONS

The system described in this paper consists of primary, secondary and depressor cables, and the towed vehicle and the depressor are attached to the end of secondary and depressor cables respectively (Fig. 1). The primary cable and depressor cable are negative buoyant, while the towed vehicle and the secondary cable are neutrally buoyant.

Three different coordinate systems are used in the derivation of equations, i.e. the fixed inertial coordinate system (X,Y,Z) and local coordinate systems for the cables (t,n,b) and the towed vehicle (x,y,z) as shown in Fig. 1. The governing equations for cables are given first, then three cables are modeled separately and interfaced dynamically at the conjunction. The speed of the tow ship at water surface and dynamic equations of the towed vehicle and the depressor are taken as boundary conditions for governing equations of cables. For results presented in this paper, computations are started from a steady state solution which is taken as the initial condition for the whole system.

2.1 Cable modeling

The cable is treated as a long, thin, flexible circular cylinder. It is assumed that the dynamics of cable are determined by gravity, hydrodynamic loading and inertial forces, no bending or torsional stiffness is considered.

The equations are written in orthogonal coordinates (**t, n, b**) local to each point of the cable. The orthogonal coordinates (**i, j, k**) represent the fixed inertial frame, **i, j** being in the horizontal plane and **k** pointing downward. The origins of systems (**i, j, k**) and (**t, n, b**) are coincident. The orientation of the local frame is so chosen that **t** is tangent to the cable in the direction of increasing s, **b** is in the plane of **i** and **j**, **t** and **n** lie in a vertical plane. At any point of cables the two frames (**i, j, k**) and (**t, n, b**) are related by

$$(\mathbf{t}, \mathbf{n}, \mathbf{b}) = (\mathbf{i}, \mathbf{j}, \mathbf{k})[\mathrm{D}], \tag{1}$$

where

$$[D] = \begin{bmatrix} \cos\vartheta\cos\varphi & -\cos\vartheta\sin\varphi & \sin\vartheta \\ \sin\vartheta\cos\varphi & -\sin\vartheta\sin\varphi & -\cos\vartheta \\ \sin\varphi & \cos\varphi & 0 \end{bmatrix}.$$

Equation (1) represents a rotation through angle ϑ about the **k** axis to bring **i** axis into the plane of **t** and **n**, rotation about the new **i** axis through $\pi/2$ to bring **k** into coincidence with **b**, and rotation about **b** through φ to bring **i** and **j** into coincidence with **t** and **n**. The relative position of two coordinate systems is shown in Fig. 2.

Similar to Ablow and Schechter (1983), at any point of cables, the governing equations can be written in the form of

$$MY' = N\dot{Y} + Q, \tag{2}$$

$$Y = (T, v_t, v_n, v_b, \vartheta, \varphi)^T, \qquad Y' = \frac{\partial Y}{\partial s}, \qquad \dot{Y} = \frac{\partial Y}{\partial t},$$

$$M = \begin{bmatrix} 1 & 0 & 0 & 0 & 0 & 0 \\ 0 & 1 & 0 & 0 & v_b\cos\varphi & -v_n \\ 0 & 0 & 1 & 0 & -v_b\sin\varphi & v_t \\ 0 & 0 & 0 & 1 & -v_t\cos\varphi + v_n\sin\varphi & 0 \\ 0 & 0 & 0 & 0 & -T\cos\varphi & 0 \\ 0 & 0 & 0 & 0 & 0 & T \end{bmatrix},$$

$$N = \begin{bmatrix} \dfrac{-mev_t}{1+eT} & m & 0 & 0 & m_1 v_b\cos\varphi & -m_1 v_n \\ e & 0 & 0 & 0 & 0 & 0 \\ 0 & 0 & 0 & 0 & 0 & 1+eT \\ 0 & 0 & 0 & 0 & -(1+eT)\cos\varphi & 0 \\ -\dfrac{em_1 v_b}{1+eT} & 0 & 0 & m_1 & m_1 v_n\sin\varphi - mv_t\cos\varphi & 0 \\ -\dfrac{em_1 v_n}{1+eT} & 0 & m_1 & 0 & -m_1 v_b\sin\varphi & mv_t \end{bmatrix},$$

$$Q = \begin{bmatrix} -w\sin\varphi + (1+eT)^{1/2}\frac{1}{2}\rho\pi dC_t v_t|v_t| \\ 0 \\ 0 \\ 0 \\ \frac{1}{2}\rho dC_n (v_n^2 + v_b^2)^{1/2} v_b (1+eT)^{1/2} \\ -w\cos\varphi + \frac{1}{2}\rho dC_n (1+eT)^{1/2} v_n (v_n^2 + v_b^2)^{1/2} \end{bmatrix},$$

where T denotes the tension force of the cable, v_t, v_n and v_b are the velocity components along the **t, n,** and **b** directions respectively, m is the mass per unit length of cable, m_1 is equal to $m+\rho A$ (ρ being the density of water and A the cross-sectional area of the unstretched cable), e is equal to $1/EA$ (E being the Young's modulus of cable), w is the submerged weight per unit length of cable, and d is the cable diameter.

2.2 Boundary and initial conditions for the system

The velocity components of three cables at conjunction point must be identical and resultant forces of three cables at that point must be zero, i.e.

$$\mathbf{V}_{PN} = \mathbf{V}_{D0} = \mathbf{V}_{S0}, \tag{3a}$$

$$\sum \mathbf{T} = 0, \tag{4a}$$

or

$$[D]_{PN}\begin{bmatrix} v_t \\ v_b \\ v_n \end{bmatrix}_{PN} = [D]_{D0}\begin{bmatrix} v_t \\ v_b \\ v_n \end{bmatrix}_{D0} = [D]_{S0}\begin{bmatrix} v_t \\ v_b \\ v_n \end{bmatrix}_{S0}, \tag{3b}$$

$$[D]_{PN}\begin{bmatrix} T_t \\ 0 \\ 0 \end{bmatrix}_{PN} + [D]_{D0}\begin{bmatrix} T_t \\ 0 \\ 0 \end{bmatrix}_{D0} + [D]_{S0}\begin{bmatrix} T_t \\ 0 \\ 0 \end{bmatrix}_{S0} = 0, \tag{4b}$$

where subscripts P, D, and S denote the primary, depressor and secondary cables respectively, and subscripts 0 and N denote the first and last node.

The relation between the tow ship velocity and the towing point velocity of the primary cable is

$$[v_t, v_n, v_b]_{P0} = [S_x, S_y, S_z][D]_{P0}, \tag{5}$$

where S_x, S_y and S_z are the velocity components of the tow ship in inertial coordinates.

In this paper the depressor is assumed to be a weighted ball. Therefore, at the end of the depressor cable (node DN) the boundary condition is

$$\mathbf{T}_{DN} = -\tfrac{1}{2}\rho C_{DD} S_D |\mathbf{V}_{DN}|\mathbf{V}_{DN} - (C_I\rho\Delta + m_0)\dot{\mathbf{V}}_{DN} + w_0\mathbf{k}, \tag{6}$$

where C_{DD} is the drag coefficient of depressor, S_D the projected area of depressor, C_I the added-mass coefficient of depressor, Δ the volume of depressor, m_0 the mass of depressor and w_0 the submerged weight of depressor.

The velocity coupling relation between the end of the secondary cable (node SN) and the towing point of a towed vehicle is

$$[\mathbf{V}_V + \mathbf{w}\times\mathbf{r}_T] = [E][D]_{SN}\mathbf{V}_{SV}, \tag{7}$$

where $\mathbf{V}_V = (u,v,w)^T$ and $\mathbf{w}$ = $(p,\ q,\ r)$ are translational and angular velocities of the towed vehicle in the vehicle-fixed coordinate, $\mathbf{r}_T = (x_T, y_T, z_T)$ is the towing point coordinates in the vehicle-fixed frame, $\mathbf{V}_{SV}$ is expressed in local coordinates of cable and

$$[E] = \begin{bmatrix} \cos\theta\cos\psi & \cos\theta\sin\psi & -\sin\theta \\ -\sin\psi\cos\phi + \sin\phi\sin\theta\cos\psi & \cos\phi\cos\psi + \sin\phi\sin\theta\sin\psi & \sin\phi\cos\theta \\ \sin\phi\sin\psi + \cos\phi\cos\psi\sin\theta & -\sin\phi\cos\psi + \cos\phi\sin\theta\sin\psi & \cos\phi\cos\theta \end{bmatrix},$$

where θ, ϕ and ψ are the pitch, roll and yaw angles of the towed vehicle.

The six-degree-of-freedom equations of motion for a towed vehicle in surge, sway, heave, roll, pitch, and yaw can be written as (Abkowitz 1967)

$$\begin{aligned} &m\left[\dot u - vr + wq - x_G(q^2 + r^2) + y_G(pq - \dot r) + z_G(pr + \dot q)\right] \\ &\quad = X_H + X_W + X_C + X_T, \end{aligned} \tag{8}$$

$$\begin{aligned} &m\left[\dot v + ur - wp + x_G(pq + \dot r) - y_G(p^2 + r^2) + z_G(qr - \dot p)\right] \\ &\quad = Y_H + Y_W + Y_C + Y_T, \end{aligned} \tag{9}$$

$$\begin{aligned} &m\left[\dot w - uq + vp + x_G(pr - \dot q) + y_G(qr + \dot p) - z_G(p^2 + q^2)\right] \\ &\quad = Z_H + Z_W + Z_C + Z_T, \end{aligned} \tag{10}$$

$$\begin{aligned} &I_x\dot p + (I_z - I_y)qr + I_{xy}(pr - \dot q) - I_{yz}(q^2 - r^2) - I_{xz}(pq + \dot r) \\ &\quad + m\left[y_G(\dot w - uq + vp) - z_G(\dot v + ur - wp)\right] = K_H + K_W + K_C + K_T, \end{aligned} \tag{11}$$

$$\begin{aligned} &I_y\dot q + (I_x - I_z)pr - I_{xy}(qr + \dot p) + I_{yz}(pq - \dot r) + I_{xz}(p^2 - r^2) \\ &\quad - m\left[x_G(\dot w - uq + vp) - z_G(\dot u - vr + wq)\right] = M_H + M_W + M_C + M_T, \end{aligned} \tag{12}$$

$$\begin{aligned} &I_z\dot r + (I_y - I_x)pq - I_{xy}(p^2 - q^2) - I_{yz}(pr + \dot q) + I_{xz}(qr - \dot p) \\ &\quad + m\left[x_G(\dot v + ur - wp) - y_G(\dot u - vr + wq)\right] = N_H + N_W + N_C + N_T, \end{aligned} \tag{13}$$

where the left-hand sides represent inertial forces and moments and the right-hand sides denote external forces on the towed vehicle. Subscript H reflects hydrodynamic contributions, W buoyant and weight effects, C forces arising from control surface (control wing, flap *etc.*) and T towing forces. The symbols in the equations are based on the standard notation.

The towed vehicle being investigated in this paper is assumed to be a circular cylinder with symmetry about its center of uniform mass located at the mid point of the vehicle. There is no control surface in such a type of towed vehicle, thus control forces do not exist. The hydrodynamic forces acting on the towed vehicle are given by

$$X_H = X_{\dot u}\dot u - X_{uu}u|u|, \tag{14}$$

$$Y_H = Y_{\dot v}\dot v - \tfrac{1}{2}\rho C_{Dv}D_v \int \left[(v + xr)^2 + (w - xq)^2\right]\frac{(v + xr)}{U_{cf}(x)}dx, \tag{15}$$

$$Z_H = Z_{\dot w}\dot w - \tfrac{1}{2}\rho C_{Dv}D_v \int \left[(v + xr)^2 + (w - xq)^2\right]\frac{(w - xq)}{U_{cf}(x)}dx, \tag{16}$$

$$K_H = K_{\dot p}\dot p, \tag{17}$$

$$M_H = M_{\dot q}\dot q + \tfrac{1}{2}\rho C_{Dv}D_v \int \left[(v + xr)^2 + (w - xq)^2\right]\frac{(w - xq)x}{U_{cf}(x)}dx, \tag{18}$$

$$N_H = N_{\dot r}\dot r - \tfrac{1}{2}\rho C_{Dv}D_v \int \left[(v + xr)^2 + (w - xq)^2\right]\frac{(v + xr)x}{U_{cf}(x)}dx, \tag{19}$$

where C_{Dv} is the drag coefficient of the towed vehicle in sway and heave, D_v the diameter of the towed vehicle, and

$$U_{cf} = \left[(v + xr)^2 + (w - xq)^2\right]^{1/2}. \tag{20}$$

The integration in the above equations is over the length of the towed vehicle.

Hydrodynamic restoring forces and moments due to vehicle weight W and buoyancy B are given by

$$X_W = -(W - B)\sin\theta, \tag{21}$$
$$Y_W = (W - B)\cos\theta\sin\phi, \tag{22}$$
$$Z_W = (W - B)\cos\theta\cos\phi, \tag{23}$$
$$K_W = (y_G W - y_B B)\cos\theta\cos\phi - (z_G W - z_B B)\cos\theta\sin\phi, \tag{24}$$
$$M_W = -(x_G W - x_B B)\cos\theta\cos\phi - (z_G W - z_B B)\sin\theta, \tag{25}$$
$$N_W = (x_G W - x_B B)\cos\theta\sin\phi + (y_G W - y_B B)\sin\theta. \tag{26}$$

The towing forces and moments on the towed vehicle in local coordinates of the vehicle are

$$\mathbf{F}_T = \begin{bmatrix} X_T \\ Y_T \\ Z_T \end{bmatrix} = -[\mathrm{E}][\mathrm{D}]_{SN}\mathbf{T}_{SN}, \tag{27}$$

$$[K_T, M_T, N_T] = \mathbf{r}_T \times \mathbf{F}_T. \tag{28}$$

To complete the hydrodynamic model of the towed system, the expression for the rates of change of Euler angles for the towed vehicle is

$$\dot{\phi} = p + q\sin\phi\tan\theta + r\cos\phi\tan\theta, \tag{29}$$
$$\dot{\theta} = q\cos\phi - r\sin\phi, \tag{30}$$
$$\dot{\psi} = q\frac{\sin\phi}{\cos\theta} + r\frac{\cos\phi}{\cos\theta}. \tag{31}$$

For results presented in this paper, constant velocity of the tow ship $\mathbf{V}_{TS}$ is assumed to be the initial state, that is, $\mathbf{V}_{TS} = -V_0\mathbf{i}$ and

$$v_t = -V_0\cos\varphi, \quad v_n = V_0\sin\varphi, \quad v_b = 0, \quad \vartheta = 0, \quad \frac{\partial}{\partial t} = 0. \tag{32}$$

3. FINITE DIFFERENCE APPROXIMATION

The governing equation of cables (2) with the conditions described in Section 2.2 determines the towed system parameters completely. In this research the partial differential equation (2) is approximated by a finite difference equation centered in time and in space. Three cables are divided into a series of segments of length ΔS_j and time is divided into a series of time steps Δt. The node numbers in three cables are $0 \sim N_P$, $0 \sim N_D$, $0 \sim N_S$ respectively, 6 node variables $Y_j = (T, v_t, v_n, v_b, \vartheta, \varphi)_j^T$ $(j = 0 \sim N,\ N = N_P, N_D$ or $N_S)$ are included at each node of cables. The finite difference approximation of equation (2) is written at mid point of each cable segment (Burgess 1991),

$$\begin{aligned}
&\left[M_{j+1}^{i+1} + M_j^{i+1}\right]\frac{Y_{j+1}^{i+1} - Y_j^{i+1}}{\Delta S_j} + \left[M_{j+1}^{i} + M_j^{i}\right]\frac{Y_{j+1}^{i} - Y_j^{i}}{\Delta S_j} = \\
&\left[N_{j+1}^{i+1} + N_{j+1}^{i}\right]\frac{Y_{j+1}^{i+1} - Y_{j+1}^{i}}{\Delta t} + \left[N_j^{i+1} + N_j^{i}\right]\frac{Y_j^{i+1} - Y_j^{i}}{\Delta t} \\
&+ Q_{j+1}^{i+1} + Q_j^{i+1} + Q_{j+1}^{i} + Q_j^{i}
\end{aligned} \tag{33}$$

The finite difference equations include all node variables in three cables, there are $6(N_P + N_D + N_S + 3)$ unknown variables Y_j of all three cables with $6(N_P + N_D + N_S)$ finite difference equations in Eq. (33). The rest 18 scalar equations are provided by equations (3) to (7) in finite difference form. However, Eq. (7) introduces 9 new variables, that is, 3 translational velocity components (u, v, w), 3 angular velocity components (p, q, r) and 3 azimuth angles (θ, ϕ, ψ) of the towed vehicle. Therefore, 9 more equations are needed to solve the problem. These 9 equations are given by equations (8) to (13) and (29) to (31).

4. NUMERICAL SIMULATIONS AND RESULTS

The computer program based on the above established model gives the three dimensional shape of the system, the tension and velocity components along the three cable nodes and also the velocities and azimuth angle of the towed vehicle at each time step. The direction of towing path and the towing point speed may be changed at any time. Thus the model can easily be used to predict the system response during different maneuvers. In this section numerical simulations are given. The particulars of cables, depressor and the towed vehicle used in the investigations are

Depressor:	diameter = 0.85 m,	weight in water = 2000 kg,	
Towed vehicle:	$D_v = 0.3$ m,	$L = 2.0$ m, $y_T = z_T = 0$,	$x_T = -L/2$,
Cable:	$d = 0.02$ m,	$C_t = 0.02$,	$C_n = 2.0$,
	$S_P = 300$ m,	$m_P = 2.7$ kg/m,	
	$S_D = 6$ m,	$m_D = 2.7$ kg/m,	
	$S_S = 20$ m,	$m_s = \rho A$ kg/m,	

where L is the length of towed vehicle, C_t and C_n are tangential and normal drag coefficients of cables and S_P, S_D and S_S are the lengths of three cables.

For demonstration of the validity of the model on describing 3-D hydrodynamic performances of the system, an example of a tow ship traveling in wave fields is presented. The incident wave propagates in the X direction with amplitude η_0 and period T_0. The values of tow ship response functions for heave and surge are r_h and r_s. The tow ship travels sinusoidally on the horizontal plane along the path $-V_0 t\,\mathbf{i} + \eta_y\cos(2\pi t/T_y)\,\mathbf{j}$. The encountering frequency for the tow ship heave and surge is $\omega_e = \left(kV_0 + 2\pi/T_0\right)$. With the wave effects considered, the trajectory of the towing point can be written as

$$\mathbf{r}_{TS} = \left\{-V_0 t + \eta_0 r_s \cos\left[\left(kV_0 + \frac{2\pi}{T_0}\right)t + \varphi_s\right]\right\}\mathbf{i} + \eta_y \cos\left(\frac{2\pi}{T_y}t\right)\mathbf{j}$$
$$+ \eta_0 r_h \cos\left[\left(kV_0 + \frac{2\pi}{T_0}\right)t + \varphi_h\right]\mathbf{k}, \qquad (34)$$

where k is the wave number, φ_s and φ_h are phase differences of surge and heave, the values used in Eq. (34) are given in Table 1.

Table 1. The values used in the numerical simulation

$V_0(m/s)$	$\eta_0(m)$	$T_0(s)$	$\eta_y(m)$	$T_y(s)$	r_s	r_h
6.28	1.0	6.0	8.0	150.0	0.1	0.9

Fig. 3 shows time histories of vertical displacements of the depressor and the towed vehicle. It can be found that the amplitude of the towed vehicle is about one sixth of that of the depressor, which indicates that the two-part tow method does reduce the heaving response of the towed vehicle. According to the results of our numerical simulation, for the two-part underwater towed system, the heaving ratio between the towed vehicle and the depressor depends on the encountering frequency of the tow ship in waves and the length of the secondary cable *etc*. The higher the encountering frequency, the smaller the ratio (Fig. 4); the longer the secondary cable, the smaller the ratio (Fig. 5). In Fig. 3 the low frequency oscillations of the depressor and the towed vehicle are caused by the sinusoidal movement of the tow ship in the horizontal plane.

The two-part tow method basically has no influence on the sway or surge behavior of the towed vehicle. The time histories of the depressor and the towed vehicle are similar in surge or sway (Fig. 6 and Fig. 7). In Fig. 8 the roll, pitch and yaw angles of a towed vehicle during the above maneuver are shown. It can be seen from Fig. 8 that even though there is no other control measure, the pitching angle of the towed vehicle remains very small ($|\vartheta| \le 1.0^\circ$), which is within the permitted limit of most towed vehicle's requirements. Fig. 9 shows time histories of cable tensions at three cable ends (node P0, DN and SN). As it is obvious, the maximum tension occurs at node P0. The amplitudes of tension vary with different cable ends, the greater the mean value of tension, the greater the amplitude.

In the above numerical simulation, the nodes on the cables are equally spaced. The discrete segment numbers of the primary, depressor and secondary cables are 30, 3 and 5 respectively. The algorithms adopted in this paper guarantee that the choice of the node number on cable does not constitute a substantial obstacle on the calculation speed or cause the numerical calculation unstable. The time steps are held fixed at 0.3 *sec* in the simulation and the number of Newton iterations needed for convergence at each time step is about 4. The selection of the time step in the example is based on the nature of the high frequency oscillation of the towing point. For low frequency maneuvers such as a tow ship traveling in still water plane, the time step can be taken as large as 20 to 30 *sec* with no numerical instability.

5. CONCLUSIONS

A fully three dimensional model of a two-part underwater towed system is proposed. The computer program based on the model can give three dimensional shape of the system, the dynamic properties of cables and the towed vehicle. The results of numerical simulations indicate that the two-part tow method improves the behavior of the towed vehicle in vertical plane motions (heave and pitch), but no great difference is observed between the horizontal movements (surge and sway) of the towed vehicle and those of the depressor. Because of the finite difference method and other algorithms adopted in the model, more freedom is given to determine the time step and the node number distributed along cables. The model provides a good algorithm for simulation of the large scale motion of an underwater towed system. The model presented in this paper can easily be extended to simulate multiple towed cables and towed bodies .

REFERENCES

Abkowitz, MA (1969). *Stability and Motion Control of Ocean Vehicles,* MIT Press, pp 32-50.

Ablow, CM, and Schechter, S (1983). "Numerical Simulation of Undersea Cable Dynamics," *Journal of Ocean Engng,* Vol. 10, No. 6, pp 443-457.

Burgess, JJ (1989). "Imprementation of a Lumped Mass Model with Explicit Time Integration for the Simulation of Undersea Cable Mechanics," *AT&T Bell Labs Technical Memorandum, 46238-890628-01 TM.*

Burgess, JJ (1991). "Modeling of Undersea Cable Installation with a Finite Difference Method," *Proc. of the First Int. Offshore and Polar Eng. Conf.,* Edinburgh, United Kingdom, Vol. 2, pp 222-227.

Chapman, DA (1984). "Towed Cable Behaviour during Ship Turning Manoeuvres," *Journal of Ocean Engng,* Vol. 1, No. 4, pp 327-361.

Hearn, GH, and Thomas, DO (1991). "The Influence of Practical Time Integration Schemes on Dynamic Mooring Line Analysis," *Proc. of the 23th Offshore Technology Conference* OTC Paper 6604, pp 397-409.

Howell, CT, *et al.* (1991). "Numerical Analysis of 2-D Nonlinear Cable Equations with Applications to Low-Tension Problems," *Proc. of the First Int. Offshore and Polar Eng. Conf.,*San Francisco, USA, Vol. 2, pp321-326.

Koterayama, W, *et al.* (1988). "The Motion of a Depth Controllable Towed Vehicle," *Proc. of the Seventh International Conference on Offshore Mechanics and Arctic Engineering,* Houston, USA,Vol. 1, pp 423-430.

Ranmuthugala S.D. and Gottschalk S.A.(1993). "Dynamic Simulation of a Two-Part Underwater Tow," *Offshore Australia Conference.*

Sanders, JV (1982). "A Three Dimensional Dynamic Analysis of a Towed System," *Journal of Ocean Engng,* Vol. 9, No. 5, pp 483-499.

Walton, TS and Polachech, H (1960). "Calculation of Transient Motion of Submerged Cables," *Mathematics of Computation* Vol. 14, pp 27-46.

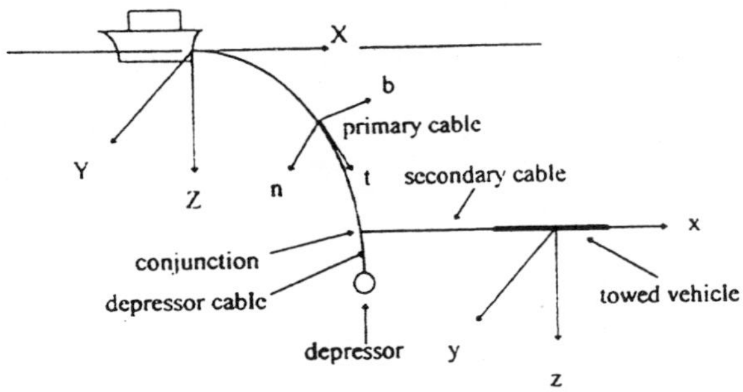

Fig. 1 A two-part underwater towed system

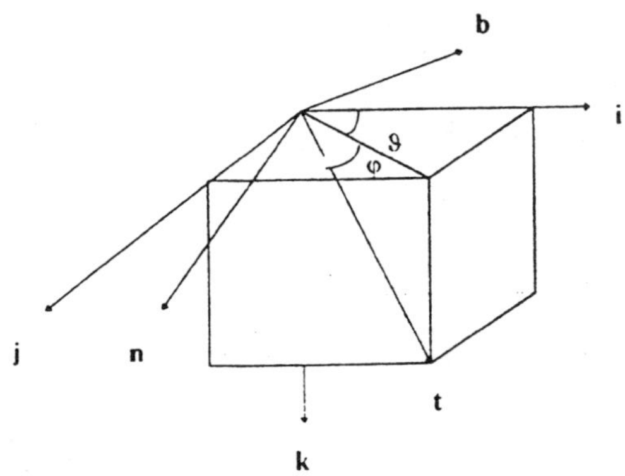

Fig. 2 The relative position of two coordinate systems

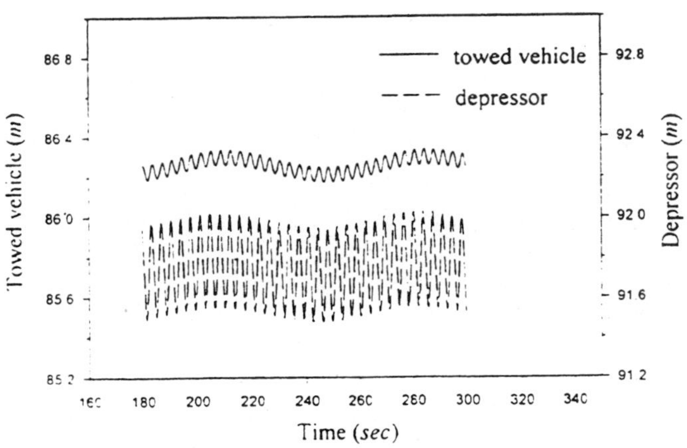

Fig. 3 Time histories of vertical displacements of the depressor and the towed vehicle

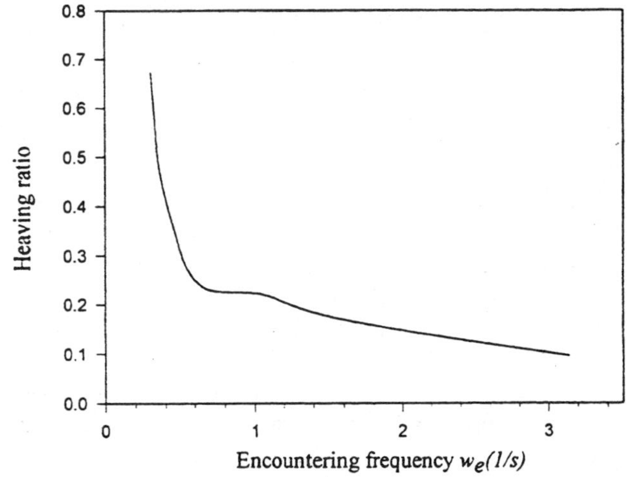

Fig. 4 Heaving ratio between the towed vehicle and the depressor *versus* the encountering frequency

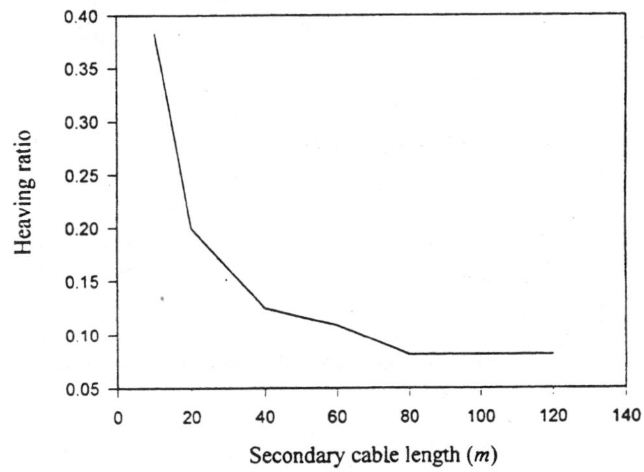

Fig. 5 Heaving ratio between the towed vehicle and the depressor *versus* the secondary cable length

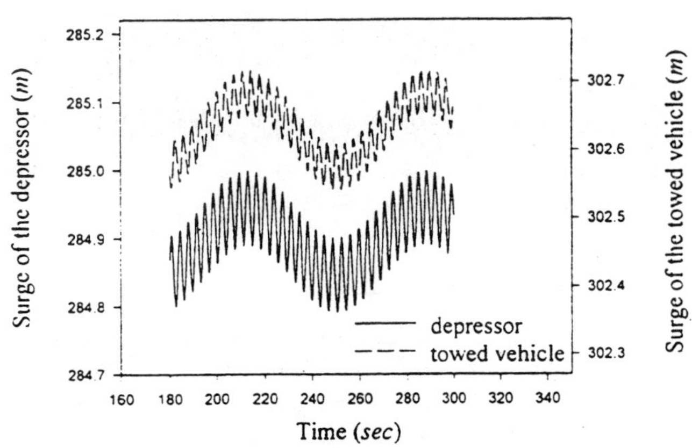

Fig. 6 Surge movements of the towed vehicle and the depressor

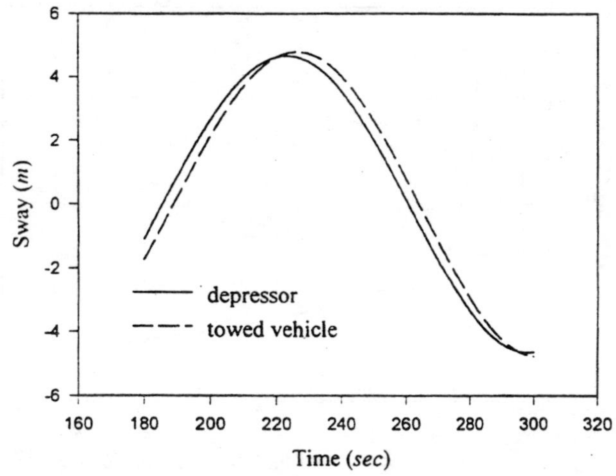

Fig. 7 Time histories of the dpressor and the towed vehicle in sway

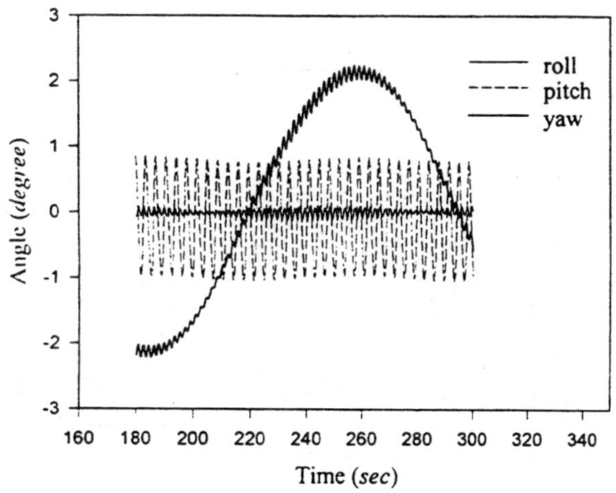

Fig. 8 The roll, pitch and yaw angles of the towed vehicle

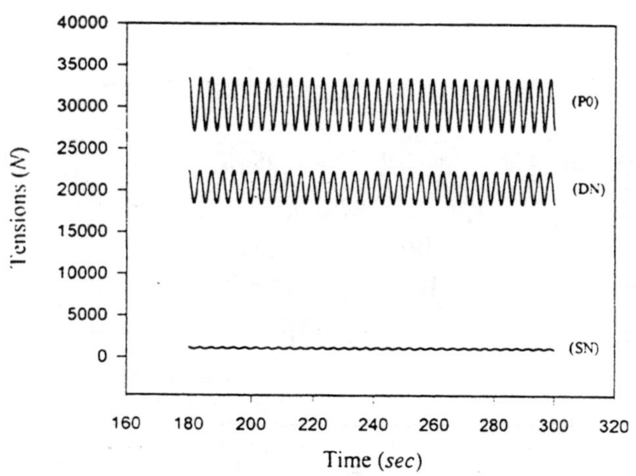

Fig. 9 Cable tensions at the cable ends

Proceedings of the Seventh (1997) International Offshore and Polar Engineering Conference
Honolulu, USA, May 25-30, 1997

ISBN 1-880653-28-1 (Set); ISBN 1-880653-30-3 (Vol. II)

LMI-Based Design of Robust Controllers for an Underwater Vehicle

Hiroyuki Kajiwara
Kyushu Institute of Technology
Iizuka, Japan

Wataru Koterayama, Masahiko Nakamura and Shigeru Yugawa
Kyushu University
Kasuga, Japan

ABSTRACT

The paper is concerned with a robust control problem for an underwater vehicle with uncertainties of hydrodynamic coefficients. Based on sensitivity analysis of hydrodynamic coefficients, a linear model with 4 vertices is introduced. The problem is formulated in the framework of multi-model based design of H_∞ control law with pole region constraints, and solved by an interior-point method numerically. Some successful simulation and experimental results are shown.

KEYWORDS: Underwater Vehicle, Hydrodynamic Coefficients, Multi-Model Based Design, Robust Control

INTRODUCTION

Underwater vehicles have been widely used in the ocean engineering. The vehicles could be classified into two groups based on the propulsive ways, that is, vehicles towed by cable and vehicles propelled by thrusters. In the case of towed vehicle, the body shape should be streamlined in order to move fast for wide area survey of sea bottom. On the other hand, in the case of self-propulsive vehicle, the blunt shape is generally adopted to make it unstable from the viewpoint of hydrodynamics, and it is supposed to be stabilized in order to realize good maneuverability as a *control system*. There is also another possibility to reduce a number of thrusters using multivariable control theory. Therefore, CACSD (Computer-Aided Control System Design) technology is expected to play a rule in balancing between function of body (hardware) and function of controller (software), in order to achieve the desired performance of the control system.

The Photo.1 shows an underwater vehicle treated in the research. In CACSD, a nonlinear simulator is necessary to evaluate the control system designed. So it is inevitable to model the controlled object mathematically. In particular, we must estimate hydrodynamic coefficients by swinging the vehicle in a water tank. Here a problem occurs on the accuracy we should achieve in the experiment. That is, we should answer whether or not the uncertainties of hydrodynamic coefficients could be allowed from the viewpoint of CACSD based on robust control theory. If the answer is "no", we will be required to achieve heigher accuracy in the estimation of hydrodynamic coefficients.

The paper is organized as follows. In the section 2, 4-vertex linear model is introduced based on sensitivity analysis of hydrodynamic coefficients. In the section 3, we formulate the problem in the framework of multi-model based design of H_∞ control law with pole region constraints. The problem is solved by an interior-point method numerically, and some successful simulation and experimental results are shown in the section 4

Photo.1 An Underwater Vehicle

UNCERTAINTIES OF HYDRODYNAMIC COEFFICIENTS

The motion equations of the underwater vehicle shown in Photo.1 are given as

$$\dot{\eta} = J(\eta)\nu \tag{1}$$

$$M\dot{\nu} + C(\nu)\nu + D(\nu)\nu + G(\eta) = \tau \tag{2}$$

$$\eta = \begin{bmatrix} X \\ Y \\ Z \\ \phi \\ \theta \\ \psi \end{bmatrix}, \quad \nu = \begin{bmatrix} u \\ v \\ Z \\ p \\ q \\ r \end{bmatrix}, \quad \tau = \begin{bmatrix} F_x \\ F_y \\ F_z \\ M_x \\ M_y \\ M_z \end{bmatrix}$$

where η consists of positions in the surge, sway, heave and Euler angles about the roll, pitch, yaw in the earth-fixed frame, ν consists of linear velocities in the surge, sway, heave and angular velocities around the roll, pitch, yaw in the body-fixed frame, τ consists of surge, sway, heave forces and roll, pitch, yaw moments. J, M, C, D, G, τ are velocity transformation matrix, an inertia matrix, a matrix of Coriolis and centripetal terms, a damping matrix, a vector of gravitational forces and moments, vector of thruster forces respectively.

The motion equation (2) has 11 longitudinal physical parameters, and 14 lateral-directional ones. Intervals including these parameters are estimated by the watertank experiments. For example, the estimation results on hydrodynamic coefficients M_q, Z_q are given in Fig.1. We have parameter intervals of M_q, Z_q as shown in Table 1.

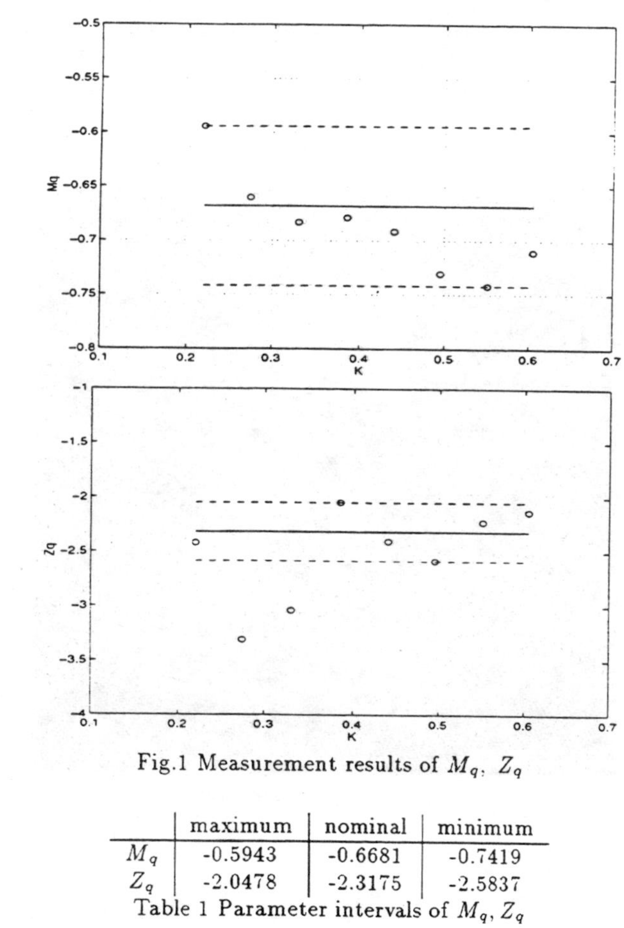

Fig.1 Measurement results of M_q, Z_q

	maximum	nominal	minimum
M_q	-0.5943	-0.6681	-0.7419
Z_q	-2.0478	-2.3175	-2.5837

Table 1 Parameter intervals of M_q, Z_q

Our concern is to study whether or not these uncertainties of hydrodynamic coefficients could be allowed from the viewpoint of robust control theory. For the purpose, we will investigate parameter sensitivities and concentrate on a few parameters. In Fig.2, nonlinear simulations are shown under the three values as in Table 1 for each longitudinal hydrodynamic coefficient. We can see which parameters are sensitive on their parameter interval.

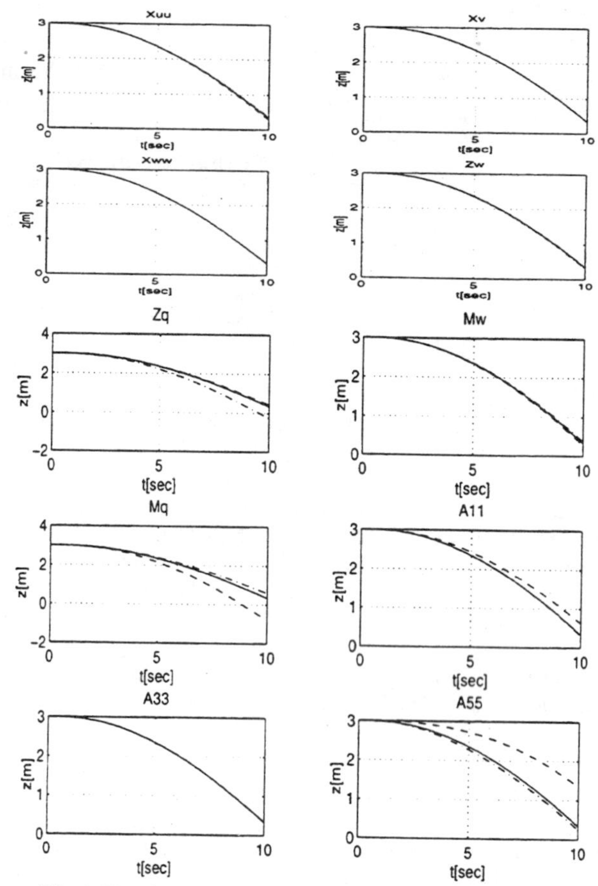

Fig.2 Simulations to check parameter sensitivities

Fig.2 indicates that added masses A_{11}, A_{55} and hydrodynamic coefficients M_q, Z_q have high sensitivities. In the following, for simplicity (because added masses don't depend on velocity variation), we will concentrate on the two parameters M_q, Z_q.

These dimensionless values are converted to dimensional ones in the nonlinear simulator as follows.

$$M_q - 0.5\rho U L^3 d \times M_q, \; Z_q - 0.5\rho U L^2 d \times Z_q \tag{3}$$

where U is the speed and ρ, L, d are constant. These show that the hydrodynamic coefficients are variable depending on the speed. So in estimating the parameter intervals for robust control, we should take account of such variations. Note that some unstability may occur when the vehicle is going down. Fig.3 shows a simulation result on the speed variation in the case of the vehicle going down at -1[m] via an LQ controller.

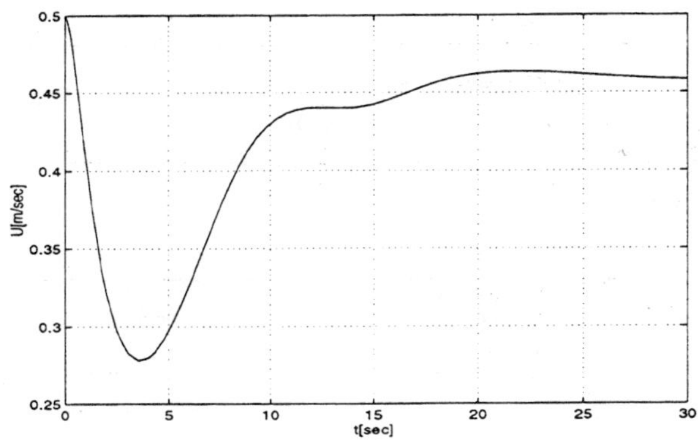

Fig.3 Speed variation in going down 1[m]

The initial speed 0.5[m/s] is reduced to 0.2780[m/s]. This means the parameter box may be shifted as shown in Fig.4.

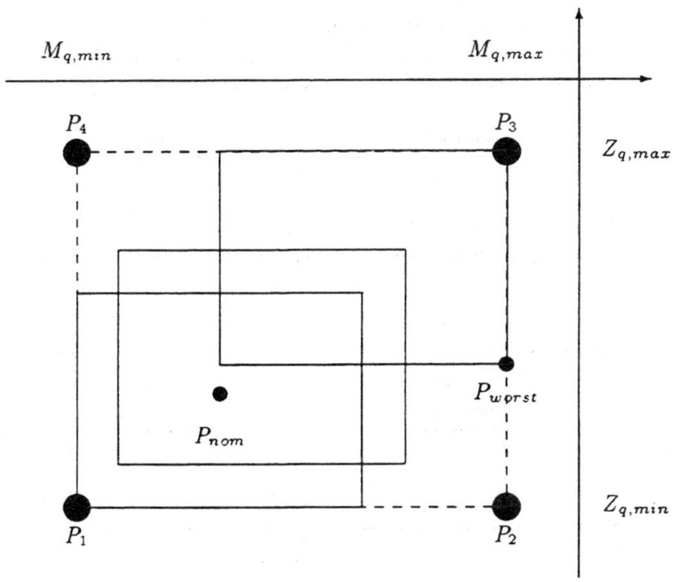

Fig.4 Parameter box of M_q, Z_q

LMI-BASED DESIGN OF CONTROL LAW

Linearizing (1) and (2) around an equilibrium state of running with 0.5[m/s], the following model is obtained in which M_q, Z_q enter in an affine manner. See Appendix.

$$\dot{x} = \overbrace{(A_0 + M_q A_{M_q} + Z_q A_{Z_q})}^{A(M_q, Z_q)} x + Bu \tag{3a}$$

$$y = Cx \tag{3b}$$

Here u consists of thruster inputs, and y consists of the depth, roll, pitch, yaw angles. Now for simplicity, consider the parameter box $P_1P_2P_3P_4$ in Fig.4. The vertices are

$$P_1(M_{q,min}, Z_{q,min}), P_2(M_{q,max}, Z_{q,min})$$

$$P_3(M_{q,min}, Z_{q,max}), P_4(M_{q,max}, Z_{q,max})$$

and a nominal point is assumed to be $P_{nom}(M_{q,nom}, Z_{q,nom})$ Then a polytopic representation of the matrix $A(M_q, Z_q)$ is as

$$A(M_q, Z_q) = p_1A_1 + p_2A_2 + p_3A_3 + p_4A_4 \tag{4}$$

$$p_1 = \frac{(M_{q,nom} - M_{q,min})(Z_{q,max} - Z_{q,nom})}{(M_{q,max} - M_{q,min})(Z_{q,max} - Z_{q,min})} \tag{5a}$$

$$p_2 = \frac{(M_{q,max} - M_{q,nom})(Z_{q,max} - Z_{q,nom})}{(M_{q,max} - M_{q,min})(Z_{q,max} - Z_{q,min})} \tag{5b}$$

$$p_3 = \frac{(M_{q,nom} - M_{q,min})(Z_{q,nom} - Z_{q,min})}{(M_{q,max} - M_{q,min})(Z_{q,max} - Z_{q,min})} \tag{5c}$$

$$p_4 = \frac{(M_{q,max} - M_{q,nom})(Z_{q,nom} - Z_{q,min})}{(M_{q,max} - M_{q,min})(Z_{q,max} - Z_{q,min})} \tag{5d}$$

$$p_1 + p_2 + p_3 + p_4 = 1 \tag{5e}$$

where A_1, A_2, A_3, A_4 are the values of $A(M_q, Z_q)$ at M_q, Z_q corresponding to P_1, P_2, P_3, P_4. The model (3) at any point in the parameter box is given as

$$\begin{bmatrix} A & B \\ C & 0 \end{bmatrix} = p_1S_1 + p_2S_2 + p_3S_3 + p_4S_4 \tag{6}$$

$$S_i = \begin{bmatrix} A_i & B \\ C & 0 \end{bmatrix} \quad (i = 1, \cdots, 4)$$

Our robust control problem is to stabilize a number of linear models described by (6) via employing a fixed controller

$$\dot{x}_K = A_K x_K + B_K y' \tag{7a}$$

$$u = C_K x_K + D_K y' \tag{7b}$$

If this problem could not solved satisfactorily, we should improve the accuracy of hydrodynamic coefficients M_q, Z_q. Therefore we tried to apply the LMI (Linear Matrix Inequality) based methodology to the robust control problem which can treat multi-model and multi-objective specifications. This means that the controller designed might stabilize all points in the parameter box $P_1P_2P_3P_4$. The controller should have the following form.

$$\begin{bmatrix} A_K & B_K \\ C_K & D_K \end{bmatrix} = p_1S_{K1} + p_2S_{K2} + p_3S_{K3} + p_4S_{K4} \tag{8}$$

$$S_{Ki} = \begin{bmatrix} A_{Ki} & B_K \\ C_K & D_K \end{bmatrix} \quad (i = 1, \cdots, 4)$$

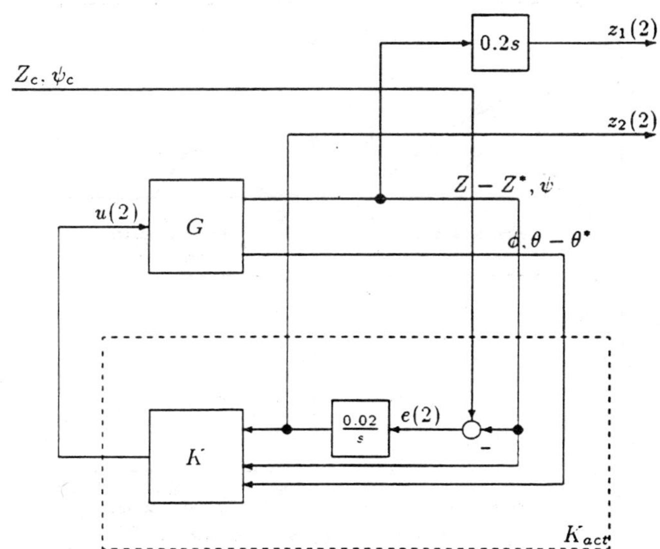

Fig.5: An Interconnection Structure

We give the following specifications for each vertex closed-loop system.

- H_∞-norm criterion: In the interconnection shown in Fig.5, satisfy

$$\left\| \begin{bmatrix} Z_c \\ \psi_c \end{bmatrix} \mapsto \begin{bmatrix} z_1 \\ z_2 \end{bmatrix} \right\|_\infty < \gamma \tag{9}$$

for $\gamma > 0$ as small as possible.

- Pole-region constraints: The closed-loop poles must stay in a left half circle with radius 50 and centered at the origin.

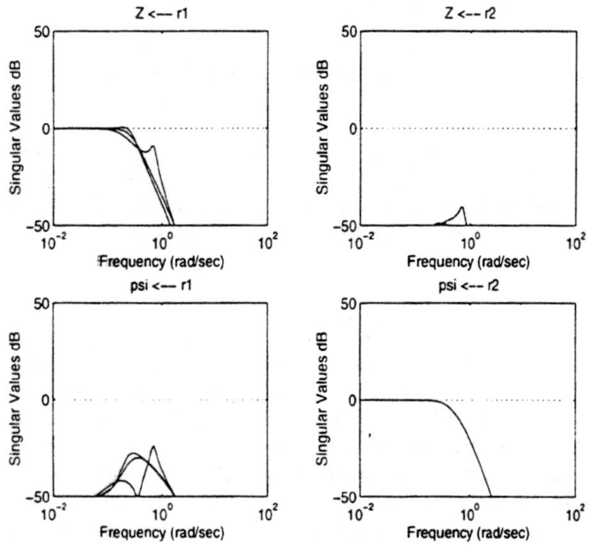

Fig.6a Closed-loop gains at 4 vertices
(1.1): From $r_1 = Z_c$ to $Z - Z^*$, (1,2): From $r_2 = \psi_c$ to $Z - Z^*$
(2,1): From $r_1 = Z_c$ to ψ, (2,2): From $r_2 = \psi$ to ψ

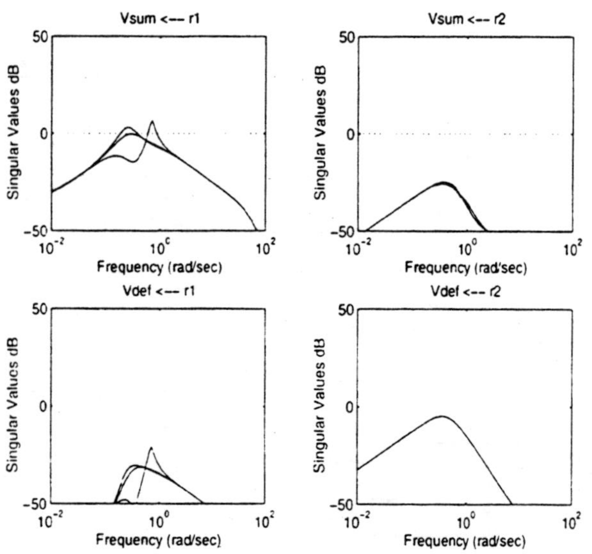

Fig.6b Closed-loop gains at 4 vertices
(1,1): From $r_1 = Z_c$ to $V_{sum} - V^*_{sum}$,
(1,2): From $r_2 = \psi_c$ to $V_{sum} - V^*_{sum}$
(2,1): From $r_1 = Z_c$ to $V_{def} - V^*_{def}$,
(2,2): From $r_2 = \psi_c$ to $V_{def} - V^*_{def}$

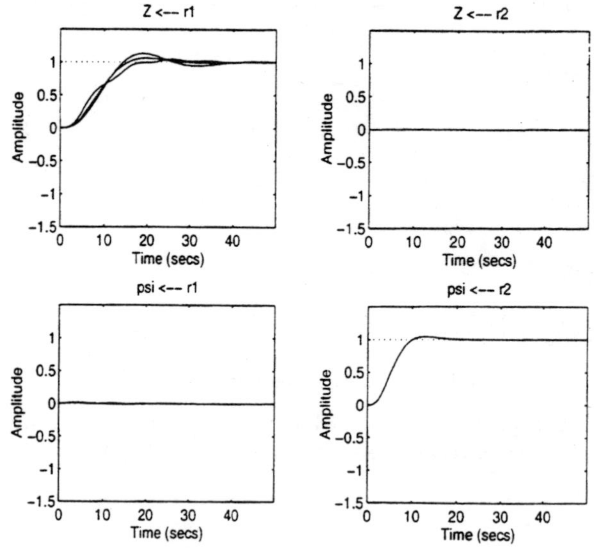

Fig.7a Step responses at vertices
(1,1): From $r_1 = Z_c$ to $Z - Z^*$, (1,2): From $r_2 = \psi_c$ to $Z - Z^*$
(2,1): From $r_1 = Z_c$ to ψ, (2,2): From $r_2 = \psi$ to ψ

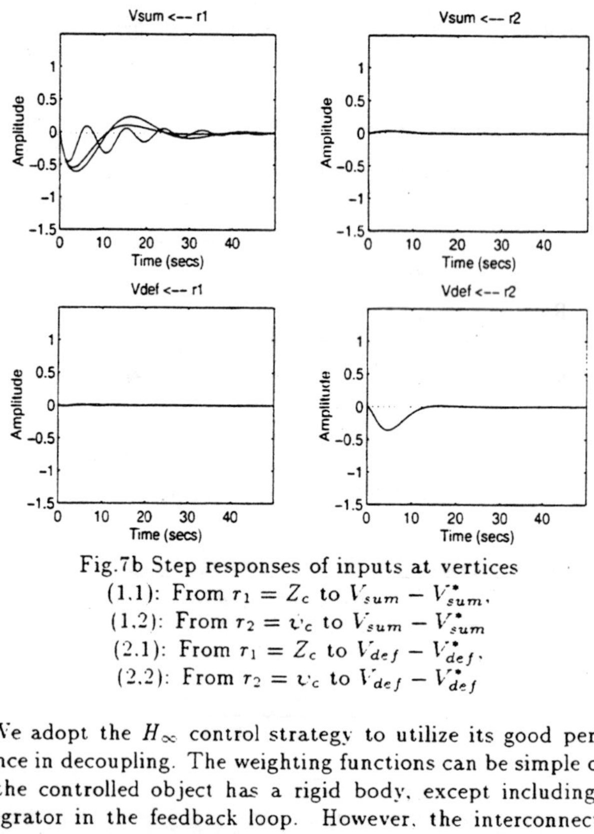

Fig.7b Step responses of inputs at vertices
(1.1): From $r_1 = Z_c$ to $V_{sum} - V^*_{sum}$,
(1.2): From $r_2 = \psi_c$ to $V_{sum} - V^*_{sum}$
(2.1): From $r_1 = Z_c$ to $V_{def} - V^*_{def}$,
(2.2): From $r_2 = \psi_c$ to $V_{def} - V^*_{def}$

We adopt the H_∞ control strategy to utilize its good performance in decoupling. The weighting functions can be simple ones as the controlled object has a rigid body, except including an integrator in the feedback loop. However, the interconnection often results in a high-gain controller. Thus we impose the second specification of pole-region constraint. After representing the two specifications as LMIs for 4-vertex linear models , we solve a system of LMIs for γ as small as possible by our original M-file **hinfgs2** using the LMI-CT on Matlab. There are 14 LMIs and 1419 decision variables. The computation time was 12385 seconds by a SUN-WS. Fig.6 and 7 shows the closed-loop gains and step responses in the case of the fixed controller (7). We can observe the good decoupling property.

EXPERIMENTAL RESULTS

Based a physical consideration, the worst case on the stability is corresponding to the point where $|M_q|$ is smaller and $|Z_q|$ is larger as seen from Fig.2. For example, the candidate is the point P_{worst} in Fig.4. We made experiments with the multi-model based controller discussed in the previous section, and with LQ and H_∞ controllers designed using the single model corresponding to P_{worst}. Fig.8a,b,c show the experimental results. We can see some dominance of the multi-model based controller.

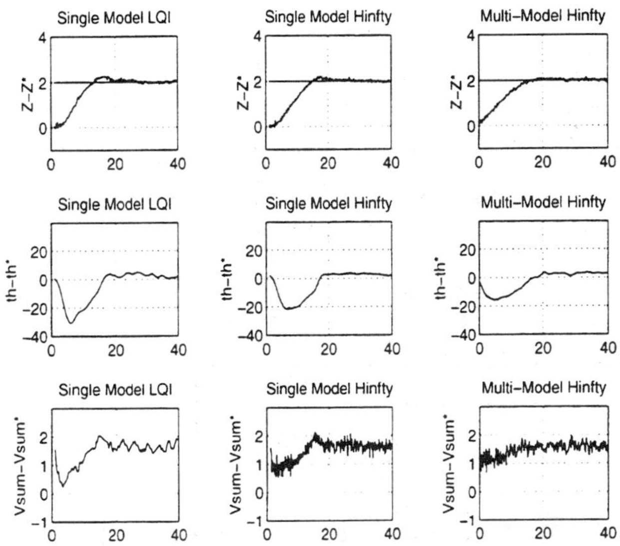

Fig.8a Experiments in going down from 2[m] to 4[m]
Left column: Single-Model Based LQI Controller
Middle column: Single-Model Based H_∞ Controller
Right column: Muti-Model Based H_∞ Controller

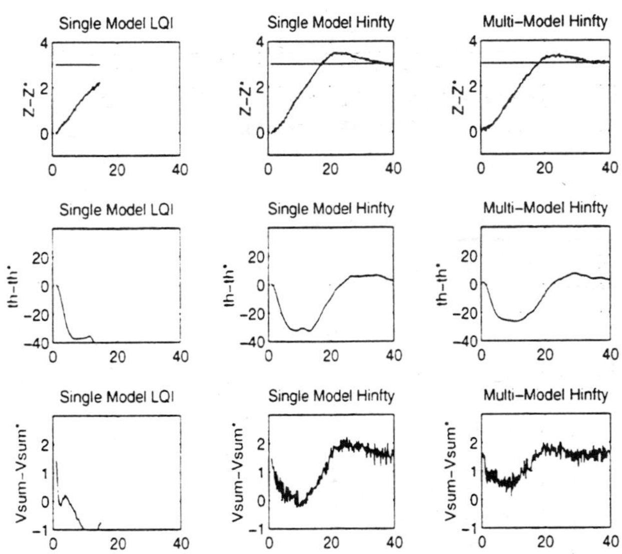

Fig.8b Experiments in going down from 1[m] to 4[m]
Left column: Single-Model Based LQI Controller
Middle column: Single-Model Based H_∞ Controller
Right column: Muti-Model Based H_∞ Controller

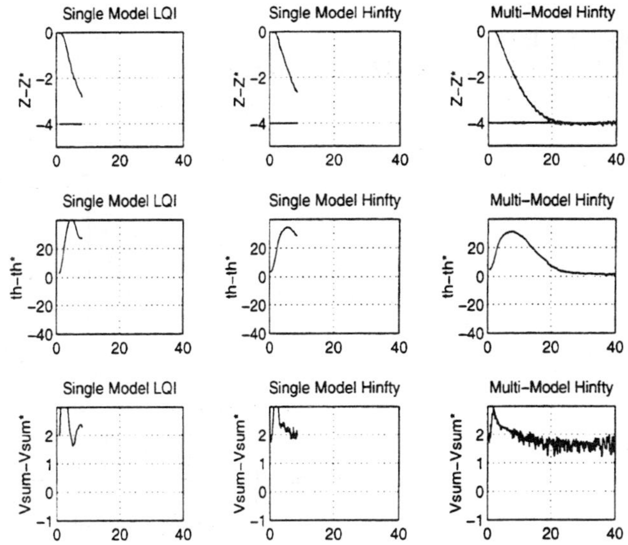

Fig.8c Experiments in going up from 5[m] to 1[m]
Left column: Single-Model Based LQI Controller
Middle column: Single-Model Based H_∞ Controller
Right column: Muti-Model Based H_∞ Controller

CONCLUDING REMARKS

It is shown that the current uncertainty of hydrodynamic coefficients are covered by the robust control strategy discussed in the paper. The further research will be concerned with applications of a gain-scheduled control strategy against the nonlinearity of the underwater vehicle.

REFERRENCES

- T.I.Fossen (1994). "Guidance and Control of Ocean Vehicles", John Wiley & Sons
- P.Gahinet,A.Nemirovski,A.J.Laub,M.Chilali (1995). LMI Control Toolbox, The MathWorks
- W.Koterayama, M.Nakamura and O.Kishimoto (1993). Development of an ROV for Sea Bottom Investigations Over a Wide Area, Proceedings of ISOPE'93, pp.455-462
- H.Kajiwara, W.Koterayama, M.Nakamura, H.Terada, T.Morita (1993). Control System Design of an ROV Operated Both as Towed and Self-Propulsive Vehicle, Proc. of ISOPE'93, pp.451-454
- H.Kajiwara, W.Koterayama, M.Nakamura, I.Sato (1994). H_∞ Synthesis of Depth Control Law for a Towed ROV with a Weight Shift Mechanism, Proc. of ISOPE'94, pp.358-361
- H.Kajiwara, W.Koterayama, M.Nakamura (1994). On Liner Parameter-Varying Models of a Remotely Operated Under Water Vehicle, Proc. of Techno-Ocean'94 International Symposium, vol.I, pp.415-420
- H.Kajiwara, P.Gahinet, W.Koterayama, M.Nakamura (1996). Robust Control for an Underwater Vehicle with Uncertainties of Hydrodynamic Coefficients, MOVIC'96, vol.3, pp.366-369

APPENDIX

1) Linear M·del of DELTA (in part)

$$E\dot{x} = A'x + B'u$$

where

$$x = \begin{bmatrix} u - u^* \\ v \\ w - w^* \\ p \\ q \\ r \end{bmatrix}, \quad u = \begin{bmatrix} x_{dm} - x_{dm}^* \\ V_R - V_R^* \\ V_L - V_L^* \end{bmatrix}$$

and

$$E = \left[\begin{array}{ccc|ccc} m + A_{11} & 0 & 0 & 0 & mz_G & 0 \\ 0 & m + A_{22} & 0 & -mz_G & 0 & mx_G \\ 0 & 0 & m + A_{33} & 0 & -mz_G & 0 \\ \hline 0 & -mz_G & 0 & I_{xx} + A_{44} & 0 & -I_{xz} \\ mz_G & 0 & -mx_G & 0 & I_{yy} + A_{55} & 0 \\ 0 & mx_G & 0 & -I_{zx} & 0 & I_{zz} + A_{66} \end{array}\right]$$

$$A' = -\left[\begin{array}{ccc|ccc} 0 & 0 & 0 & 0 & (m + A_{33})w^* & 0 \\ 0 & 0 & 0 & -(m + A_{33})w^* & 0 & (m + A_{11})u^* \\ 0 & 0 & 0 & 0 & -(m + A_{11})u^* & 0 \\ \hline 0 & -(A_{22} - A_{33})w^* & 0 & mz_G w^* & 0 & -mz_G u^* \\ 0 & 0 & 0 & 0 & m(x_G u^* + z_G w^*) & 0 \\ 0 & 0 & 0 & -mx_G w^* & 0 & mx_G u^* \end{array}\right]$$

$$+ \left[\begin{array}{ccc|ccc} 2(X_{uu} + X_{uuk}|Z|)u^* + b_t V_R & \pm X_v & 2X_{ww}w^* & 0 & 0 & 0 \\ 0 & Y_v \pm Y_{vv}v^* & Y_p & 0 & 0 & Y_r \pm Y_{rr}r^* \\ e_t V_R & 0 & Z_w & 0 & \mathbf{Zq} & 0 \\ \hline 0 & K_v & 0 & K_p & 0 & K_r \\ 2(l_z X_{uu} + l_k X_{uuk}|Z|)u^* + (z_{TH}b_t - l_{pTHx}e_t)V_R & 0 & M_w^* & 0 & \mathbf{Mq} & 0 \\ 0 & N_v^* \pm N_{vv}v^* & 0 & 0 & 0 & N_r \end{array}\right]$$

$$B' = \left[\begin{array}{ccc} 0 & \frac{1}{2}(b_t u^* + 2c_t V_R^*) & \frac{1}{2}(b_t u^* + 2c_t V_L^*) \\ 0 & 0 & 0 \\ 0 & \frac{1}{2}(e_t u^* + 2f_t V_R^*) & \frac{1}{2}(e_t u^* + 2f_t V_L^*) \\ \hline 0 & \frac{1}{2}l_{pTHy}(e_t u^* + 2f_t V_R^*) & -\frac{1}{2}l_{pTHy}(e_t u^* + 2f_t V_L^*) \\ dmg & -\frac{1}{2}l_{pTHx}(e_t u^* + 2f_t V_R^*) + \frac{1}{2}z_{TH}(b_t u^* + 2c_t V_R^*) & -\frac{1}{2}l_{pTHx}(e_t u^* + 2f_t V_L^*) + \frac{1}{2}z_{TH}(b_t u^* + 2c_t V_L^*) \\ 0 & -\frac{1}{2}y_{TH}(b_t u^* + 2c_t V_R^*) & \frac{1}{2}y_{TH}(b_t u^* + 2c_t V_L^*) \end{array}\right]$$

2) LPV Model of DELTA

$$\dot{x} = \overbrace{E^{-1}\underbrace{(A_0' + M_q A_{M_q}' + Z_q A_{Z_q}')}_{A'(M_q, Z_q)}}^{A(M_q, Z_q)} x + \overbrace{E^{-1}B'}^{B} u$$

where

$$A_0' = A'(M_q, Z_q) - (M_q A_{M_q}' + Z_q A_{Z_q}')$$

and

$$A_{M_q}' = \begin{bmatrix} 0 & 0 & 0 & 0 & 0 & 0 \\ 0 & 0 & 0 & 0 & 0 & 0 \\ 0 & 0 & 0 & 0 & 0 & 0 \\ 0 & 0 & 0 & 0 & 0 & 0 \\ 0 & 0 & 0 & 0 & 1 & 0 \\ 0 & 0 & 0 & 0 & 0 & 0 \end{bmatrix}, \quad A_{Z_q}' = \begin{bmatrix} 0 & 0 & 0 & 0 & 0 & 0 \\ 0 & 0 & 0 & 0 & 0 & 0 \\ 0 & 0 & 0 & 0 & 1 & 0 \\ 0 & 0 & 0 & 0 & 0 & 0 \\ 0 & 0 & 0 & 0 & 0 & 0 \\ 0 & 0 & 0 & 0 & 0 & 0 \end{bmatrix}$$

Proceedings of the Seventh (1997) International Offshore and Polar Engineering Conference
Honolulu, USA, May 25-30, 1997

ISBN 1-880653-28-1 (Set); ISBN 1-880653-30-3 (Vol. II)

The Problem of Synthesis of the Most Effective System for Supporting a Heat Balance in a Diving Chamber of the Hyperbaric Lifeboats

O.A. Kozyrko, Y.V. Zacharov, E.I. Trushlyakov and P.A. Chegrintsev
Ukrainian State Maritime Technical University
Nikolaev, Ukraine

ABSTRACT

Problems of the synthesis of the most effective system of supporting a heat balance of the divers in a pressure chamber of hyperbaric lifeboats and the concept of its efficiency are discussed in the paper. The complex criterion of efficiency developed to be used at the phases of the synthesis, complex design, research of the effect of the individual parameters of the system on its efficiency and optimization of its parameters is presented in the paper. The method of estimation of reliability on the basis of Kvirk's index is modified so that it could be applied for composite units (parallel and series combinations).

KEY WORDS

Diver, efficiency, hyperbaric lifeboat, life-supporting system, reliability, synthesis.

INTRODUCTION

Continued concern is being shown about the problems concerning deep-sea divers in compression, if the vessel they are operating from has to be abandoned. The hyperbaric lifeboats (HL) are widely used in the world practice to rescue the deepsea divers from an emergency vessel - carrier of a deep-sea diving complex.

One of the principal problems facing the forecasting research and design of life-supporting systems (LSS) of HL is its modular and parametrical synthesis. A problem is in choosing the most effective structure of the LSS meeting a proposal requirement. In other words the problem of the synthesis of the LSS is the problem of researching its efficiency.

In some cases when the system is designed on the basis of existing technical means the process of choosing the most effective modular combination is not difficult. However, for a lot of problems the solution of the problem is not obvious. In this case the success of the synthesis of the most effective variant of the LSS is based to a great extent on the choice of criterion of efficiency. It raises the problem of developing the complex criterion of efficiency of the LSS of HL.

The object of the paper is the system of supporting the heat balance of the divers in a pressure chamber of the HL and its efficiency.

THE PRESENT STATE OF THE PROBLEM OF SUPPORTING A HEAT BALANCE OF DIVERS IN A PRESSURE CHAMBER OF THE HYPERBARIC LIFEBOATS.

The system of supporting the heat balance of the divers in a pressure chamber of the HL is incorporated into the LSS of HL. It is intended to heat and/or cool the divers in the chamber.

In the world practice the heating of the chamber atmosphere is based on the use of heat of the lifeboat's propulsive engine exhaust gases or heat of water from the cooling system of the same engine.

In such kind of heating the heat power of LSS being had at its disposal depends on the engine operation mode, in other words on HL speed. It means that there are some conditions under which the recovery of the engine's waste heat will not be sufficient to heat the divers in the pressure chamber. It causes a necessity to compare the required heat power of LSS with the engine's waste heat energy in the range of the operational pressure of the chamber atmosphere, outside air temperature.

Moreover, the reliability analysis of LSS with such kind of heating showed that a HL propulsive engine is the least reliable element of LSS and in a case of the engine failure, the LSS has no sources of heat and electrical energy. No electrical energy at the HL means no cool energy either. Therefore, the LSS of HL should include an emergency sources of heat and electrical energy.

Possible ways of supporting the heat balance of divers in a pressure chamber of HL are:

1) Heating:

- Normal heating:
  - recovery of waste heat of HL's engine;
  - use of an auxiliary heat source in combination with the recovery of waste heat of HL's engine:

    Possible auxiliary sources of the heat energy:
    - waterheater operating on the liquid fuel;
    - electroheater powered by the diesel electrogenerator of HL;
    - electroheater powered by electrical accumulators taking a charge from the diesel electrogenerator of HL;

• cooling units operating as heating pumps;
• heat storage device of different types;
• chemical heat sources.

1.2) Emergency heat sources:
• waterheater operating on the liquid fuel;
• electroheater powered by electrical accumulators;
• cooling units operating as heating pumps;
• heat storage device of different types;
• chemical heat sources.

2) Cooling:
• use of cooling units of different types;

Weight of the sources of heat and electrical energy of required energy capacity becomes a factor that limits the sea endurance of the LSS of HL, cause increase in the HL displacement and complicate the procedure of the HL launch from an emergency vessel. Moreover, the possibility of using the HL under unfavorable weather and other conditions imposes the range of requirements upon the LSS of HL among which there are operationability under any space orientation, quickness of putting the system into operation, high level of operation reliability and minimal weight-dimension parameters.

All this raises the problem of developing the methodical base on the synthesis of the LSS of HL and researching its efficiency.

The synthesis of the system supporting a heat balance of divers in a pressure chamber of the HL is suggested to carry out in accordance with the block-scheme shown at the Fig. 1.

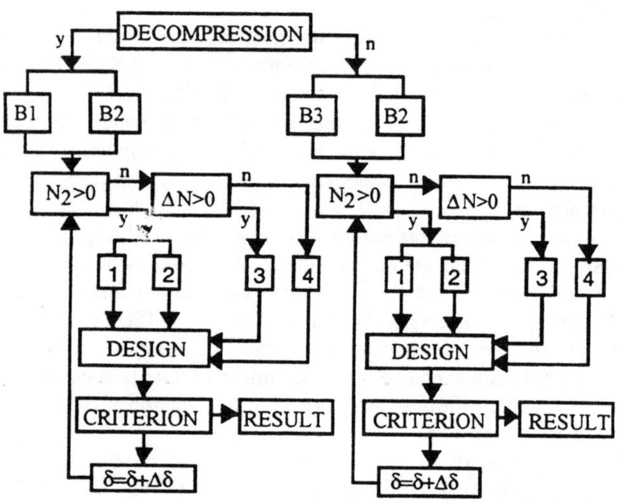

B1, **B2** and **B3** are initial data which specify the most severe presupposed operational conditions of life supporting system;
1, **2**, **3** and **4** are possible results of synthesis of system supporting heat balance of divers.

Fig. 1. Block-scheme of synthesis of system supporting heat balance.

The circuit of supporting the heat balance in the diving chamber must be designed to meet the proposal requirements even under the most severe presupposed conditions. For the heating system such conditions are when the heat losses are maximal and for the cooling system - when the heat inflow is maximal.

The most severe presupposed conditions for the cooling system are specified with the initial data **B1** and **B3** including the minimal presupposed operational pressure inside the diving chamber, maximal presupposed outside air temperature, temperature inside the diving chamber and thickness of heat insulation. The initial data **B1** and **B3** have a difference in the minimal presupposed operational pressure depending on the existence of the decompression opportunity. The most severe presupposed conditions for the heating system is specified with the initial data **B2** including the maximal design operational pressure inside the diving chamber, minimal presupposed outside air temperature, temperature inside the diving chamber and thickness (δ) of heat insulation and not depending on the decompression opportunity.

The criterion of sufficiency of the engine's waste heat energy recovery for heating the chamber atmosphere under any conditions of rescue operation is $\Delta N=N_1+1.1N_2>0$ for the engine propulsive power equaled to zero (N_1 - heat power of the engine being had at its disposal to be recovered, kW; $N_2<0$ - heat losses of the pressure chamber, kW). It means that if $\Delta N>0$ then the recovered heat is sufficient for the heating even in the case when the engine drives only HL's systems and equipment and does not drive the propeller.

The heat power N_1 of HL engine being had at its disposal to be recovered for a certain engine's operation mode can be calculated as $N_1=Ne(1-\eta)/\eta$, where Ne -effective engine power for a certain operation mode, kW; η-engine efficiency.

The effective engine power can be calculated as $Ne=Ne_1+Ne_2$, where Ne_1-effective propulsive power of HL engine, kW; Ne_2-total power of HL's equipment, kW.

Calculation of heat losses N of a HL's pressure chamber is carried out in accordance with original methods.

Thus, the structure of the heat-supporting circuit of HL's LSS is specified by the values ΔN and N_2 under presupposed conditions of HL use.

There are four possible results of synthesis of the heat-supporting system. If $N_2<0$ and $\Delta N<0$ for both initial data then the system must include a normal heating circuit with an auxiliary heat source **3** and emergency heating circuit **4**. If $N_2<0$ and $\Delta N>0$ for both initial data then the system must include a normal heating circuit without an auxiliary heat source **2** and emergency heating circuit **4**. If $N_2>0$ for the initial data **B3** (**B1**), $N_2<0$ and $\Delta N<0$ for the initial data **B2** then the system must include a cooling unit **1**, normal heating circuit with an auxiliary heat source **3** and emergency heating circuit **4**. If $N_2>0$ for the initial data **B3** (**B1**), $N_2<0$ and $\Delta N>0$ for the initial data **B2** then the system must include a cooling unit **1**, normal heating circuit without an auxiliary heat source **2** and emergency heating circuit **4**. If $N_2<0$ for the both initial data then the system must include a cooling unit **1** only.

The deep-sea diving requires the use of respiratory gas mixtures with different gas compound. The necessity in the gas mixtures is caused with change of biological activity of the gas' components and its effect on a human organism at a high pressure.

The principal types of the gas mixtures used in the diving technology at present are the mixtures on the basis of He and H_2 in which the content of a principal component is up to 98 %.

The thermodynamical features of the gas mixtures (heat conductivity, density, heat capacity) directly affect the coefficient of the heat evolution of the mixture, heat loss/inflow of the diving chamber and load (fluid and mixture flow rate and heat load) of the system's elements. The weight of some elements is in direct proportion to its load only. The weight of other elements (heat exchangers) is in direct proportion to its load and in inverse proportion to the coefficient of heat evolution of the mixtures.

Therefore, the choosing of the mixtures' compound should be carried out on the basis of the complex estimation of efficiency of using the gas mixture for every system's variant with taking into account the medical-biological requirements and technical task (first of all the mixture pressure).

The complex estimation of the efficiency is suggested to carry out on the basis of the following complex criterion of efficiency.

THE COMPLEX CRITERION OF EFFICIENCY OF THE LIFE-SUPPORTING SYSTEMS FOR THE HYPERBARIC LIFEBOATS.

At present there is no a mathematical apparatus for universal, exact and detailed estimation of the efficiency. The mathematical expressions suggested by different authors do not enable to solve the set problem completely because the expressions do not take into account all factors determing the system efficiency.

Do not separating the efficiency into economical and technical we can say that the criterion of efficiency should reflect the ratio between the result (effect) and outlay for it obtaining. The weight, fuel consumption and other can be used as the outlay.

The confusing between the concepts of efficiency and execution of functional missions arises when estimation of the life-supporting system efficiency. The functional mission of the life-supporting systems is supporting a specified compound and parameters of atmosphere inside a pressure chamber. The physical meaning of the problem solution is in bringing a specified heat quantity for compensation of heat losses, supplying a specified oxygen quantity and disposal of CO_2 for supporting a specified compound of the atmosphere. If we rate the heat producing, oxygen flow rate and other to outlays we will limit the frames of accepting the solution. However, in this case only the economical analysis will be carried out.

Therefore the concept of efficiency should be extended with a range of parameters: reliability, comfort level, ecological purity etc.

In accordance with the fundamental principles of the efficiency quantitative estimation the efficiency is examined as a some hierarchical aggregate of features which are interesting in the examined conditions of the object use.

At present there is no a common point of view and method for building a hierarchical tree of the life-supporting systems features and different approaches can be used for the tree building.

It should be born in mind that the more features are taken into account the higher accuracy of the quantitative efficiency estimation is achieved. In addition to the problem of features number taken into account for the efficiency estimation the problem what kind of features should be taken into account is very important.

In accordance with the method of qualimetria concerning the problem of quantitative estimation of quality any feature reflecting what ever aspect of the efficiency can be determined with two numerical parameters: ponderability β (importance) and estimation X. It should be noted that in accordance with the principles of qualimetria the using of β and X for the efficiency estimation enables to multiply them and to obtain the complex estimation of the efficiency because the problem is not in calculation of the absolute efficiency value but in numerical estimation of the efficiency for comparison of alternatives.

On the basis of carried out analysis the parameters of destination K_1 and reliability K_2 are accepted as the efficiency parameters for the systems of life-supporting divers in a pressure chamber of the hyperbaric lifeboats. The parameters of a comfort level and ecological cleanliness should not be considered as the parameters concerning the efficiency of HL' LSS because they are outside the main problem - the problem of rescue of a human life.

The potential productivity X_1 of 1 kg of LSS reduced weight, potential productivity X_2 of 1 kg of reduced fuel weight, potential productivity X_3 of 1 kW of reduced engine power and relative increase of HL displacement X_4 are accepted as the destination parameters.

The potential productivity A is the quasi-work performed over the divers when rescue operation.

$$A=A_1+A_2,$$

where A_1-potential energy of the divers at the depth H, kJ; A_2-work performed over the divers when they are transported with HL speed V (m/s) for duration τ (s), kJ.

$$A_1=\frac{m\cdot N\cdot g\cdot H}{1000},$$

where m-weight of a diver, kg; N-number of divers; g=9.81 m/s^2-acceleration of free falling; $H=P_{rgm}/0.0101$-depth corresponding to respiratory gas mixture pressure P_{rgm}, MPa.

$$A_2=\frac{\alpha\cdot m\cdot N\cdot\eta_1}{V}\cdot V\cdot\tau\cdot 3600=\frac{V\cdot F\cdot m\cdot N\cdot\eta_1}{D\cdot 1000}\cdot\tau\cdot 3600$$

,

where α-power-to-weight ratio of HL, kW/kg; η_1 - efficiency of the propeller-propulsive complex of HL; V-speed of HL, m/s; τ-sea endurance of LSS, h; $F=C_x\cdot\rho\cdot V^2\cdot\Omega/2$-hydrodynamic resistance force of HL, N (C_x -total drag resistance; ρ-sea water density, kg/m ; Ω-wet surface, m^2); D-HL displacement, kg.

$$X_1=\frac{A_1+A_2}{\Delta M_{LSS}},$$

where ΔM_{LSS}-reduced weight of LSS, kg.

$$\Delta M_{LSS}=M_{LSS}+\Delta M_1+\Delta M_2+\Delta M_3+\Delta M_4,$$

where M_{LSS}-weight of LSS, kg; ΔM_1 -additional weight of the engine of HL driving LSS, kg; ΔM_2-additional fuel weight, kg; ΔM_3 -additional weight of the electrogenerator of HL, kg; ΔM_4-weight of heat insulation of a diving c/chamber, kg.

ΔM_1, ΔM_2, ΔM_3, ΔM_4 are calculated using the known formulae:

$$\Delta M_1=q_1\cdot\Delta N,$$

where q_1-engine weight per one kW of power, kg/kW; ΔN-additional necessary engine power caused with installation of LSS, kW.

$$\Delta N=\frac{\Delta N_g}{\eta_g}+\Delta N_\alpha,$$

where ΔN_g -additional necessary electrogenerator power caused with installation of LSS, kW; η_g-electrogenerator's efficiency; ΔN_α - additional necessary engine power to hold the power-to-weight ratio of LSS for its increased displacement caused with installation of LSS, kW, $\Delta N_\alpha=\alpha\cdot\Delta M_{LSS}$.

$$\Delta M_2=b_d\cdot\tau\cdot\Delta N,$$

where b_d -engine's fuel consumption per one kW of power for one hour, kg/(kW·h).

$$\Delta M_3=q_3\cdot\Delta N_g,$$

where q_3-electrogenerator weight per one kW of power, kg/kW.

$$X_2 = \frac{A_1 + A_2}{\Delta M_2}, \quad X_3 = \frac{A_1 + A_2}{\Delta N},$$

The complex criteria have a disadvantage in its structure causing the unlimited intersubstitution of reaching one object with another. So, for an example, increasing the sea endurance of LSS for HL we can get the best variant of LSS with its excessively large weight. This disadvantage can be eliminated in a sufficient extent by using the parameter of relative increase of HL displacement X_4.

$$X_4 = \frac{\Delta M_{LSS}}{D_{HLB}},$$

The suggested method of complex estimation of HL's LSS efficiency is based on the following points:

1) the goal of the method is the numerical comparison of efficiency of alternative variants of HL's LSS and not the obtaining of an absolute efficiency value;

2) the design of the LSS is a trade-off engineering solution in which some features are given more preference to other.

For the comparison analysis the special parameters of efficiency should be presented in a dimensionless view by rating them to the parameters accepted as the base ones (parameters of an existing sample, specified mode, desired combination etc.).

With this in mind, in accordance with the recommendations (Brahman, 1984) the destination parameter has the following structure:

$$K_1 = \frac{\left(\frac{X_1}{X_{1B}}\right)^{0.23} \cdot \left(\frac{X_2}{X_{2B}}\right)^{0.31} \cdot \left(\frac{X_3}{X_{3B}}\right)^{0.15}}{\left(\frac{X_4}{X_{4B}}\right)^{0.31}},$$

The weight coefficients were calculated using the method of ranging the particular parameters on the basis of its importance (Jones, 1972).

The analysis of the scheme of divers rescue showed that the reliability parameters should reflect the system readiness to be emergency put into operation and the system operation reliability in any possible conditions at the divers rescue.

The coefficient of operational readiness suits for estimation of emergency duration for putting the system into operation K_r:

$$K_r = \frac{\tau_s}{\tau_{min}} < 1$$

where τ_s -time of emergency putting the system into operation, s; τ_{min}-minimal time of divers' entering a pressure chamber of HL and its unjoining from a vessel deep-diving complex.

There are the problems with exact calculation of K_r. Therefore this parameter is used as a limiting parameter.

The operational reliability is characterized with the probability P of nonfailure operation (Chordas, 1983):

$$P = e^{-\lambda\tau},$$

where λ-system's failure intensity, 1/h; τ-LSS's sea endurance, h.

Taking into account the difficulties of exact calculation of λ the Kvirk's nonreliability index I_k is suggested to be used as the reliability parameter at early design phases.

The method of Kvirk is not acceptable for design of composite units, it is acceptable for design of individual elements only (Jones, 1972). This method does not enable to estimate the reliability level of the composite units (parallel and series combinations). Therefore the index of Kvirk is suggested to be modified in the following way:

$I_k = 0.999$ - for the case of the least nonreliability level,
$I_k = 0.990$ - for the case of the medium nonreliability level,
$I_k = 0.900$ - for the case of the greatest nonreliability level.

In a series system all subsystems must operate successfully if the system is to function. In accordance with the reliability theory the reliability level is calculated as follows:

$$K_2 = \prod_{i=1}^{n} \left(I_{k(i)}\right) \text{ - for series system combinations.}$$

A parallel system is a system that is not considered to have failed unless all components have failed. In accordance with the reliability theory the reliability level is calculated as follows:

$$K_2 = 1 - \prod_{i=1}^{n} \left(1 - I_{k(i)}\right) \text{ - for parallel system combinations,}$$

where *i*- system element *i*, *n*-number of system elements,

With this in mind, the criterion of the system efficiency has the following view:

$$\begin{cases} K_r < 1 \\ K = (K_1)^{0.4} \cdot \left(\frac{K_2}{K_{2B}}\right)^{0.6} \end{cases}$$

One variant is more efficient than another one if K>1 otherwise the variant is less efficient.

The complex criterion of efficiency can be also used for researching the effect of individual parameters of LSS of HL on its efficiency.

The parametrical compound of the criterion of efficiency shows that the factors determining the efficiency of the LSS of HL are the reliability and weight of LSS's equipment, power consumpted by the equipment of LSS, physical and thermodynamical parameters of respiratory gas mixture, thermodinamical parameters of the processes in the equipment of LSS, region of use and parameters of HL.

CONCLUSIONS

The following conclusions can be made on the basis of the paper:

1. The engine of the HL is the least reliable element of the LSS. Therefore the LSS of HL should include emergency sources of heat and electrical energy the weight of which limits the sea endurance of the system, causes the increase of HL displacement and complicates the procedure of HL launch from an emergency vessel. It arises the problem of developing the methodical basis on researching the efficiency of the LSS of HL with the goal of developing the most effective systems.
2. The efficiency of the system of supporting the heat balance of the divers in a pressure chamber of the HL depends in a great extent on its modular structure which is determined by the range of operational pressure inside the diving chamber, outside and inside atmosphere temperature, compound of respiratory gas mixture and parameters of heat insulation.
3. The efficiency of the LSS of HL as a whole should be determined by the parameters of destination and reliability. The parameters concerning a comfort level and ecological cleanliness should not be considered as the efficiency parameters of the LSS of HL because they are outside the main problem - the problem of rescue of a human life.
4. The suggested modification of the method of Kvirk eliminates its disadvantage in impossibility to estimate the reliability level of composite units (parallel and series combinations).
5. The factors determining the efficiency of the LSS of HL are the reliability and weight of LSS's equipment, power consumpted by the

equipment of LSS, physical and thermodynamical parameters of respiratory gas mixture, thermodinamical parameters of the processes in the equipment of LSS, region of use and parameters of HL.

6. The developed complex criterion of efficiency can be not only used for the system synthesis but also for researching the effect of individual parameters of LSS of HL on its efficiency.

For further information contact Mr.O.Kozyrko by the following address:

P.O.Box 115, Nikolaev 327040, Ukraine
Tel.: (380-512)-34-83-24.

LITERATURE

Brahman TR (1984). "The choosing of alternatives in engineering.", *M.: Radio and Communication.*

Chordas GS (1983). "The design of Ship Systems" - Handbook, *Leningrad "Shipbuilding".*

Jones JC (1972). "Design Methods. ", *Wiley - Interscience, a division of John Wiley & Sons Ltd, London.*

Proceedings of the Seventh (1997) International Offshore and Polar Engineering Conference
Honolulu, USA, May 25-30, 1997

ISBN 1-880653-28-1 (Set); ISBN 1-880653-30-3 (Vol. II)

Bifurcation and Normal Forms of Dive Plane Reversal of Submersible Vehicles

Wei Kang and Fotis A. Papoulias
Naval Postgraduate School
Monterey, CA, USA

ABSTRACT

The problem of dive plane reversal of submersible vehicles at low speeds is analyzed using bifurcation theory. Simulation and numerical results are supported by a formal analysis procedure which includes calculation of normal forms and invariant resonant terms. It is shown that the primary loss of stability occurs in the form of a pitchfork bifurcation. A feedback control strategy is proposed which guarantees local stability of the nominal equilibrium state across the bifurcation point. Other applications, currently under consideration, for this general approach of bifurcation control are outlined.

INTRODUCTION

One of the most critical functions of a submersible vehicle is accurate depth keeping at the commanded depth. Such a function can be carried out either manually or automatically, especially in cases where human intervention is impossible or undesirable. Due to the technological significance of the problem and the numerous scientific applications of submersible vehicle systems, design of appropriate depth keeping control laws has received wide attention in the past. Response accuracy and stability are the primary considerations in designing a depth keeping control law. Of paramount importance here are the robustness properties of the particular design; i.e., its ability to maintain accuracy and stability in the presence of incomplete sensor and environmental information, as well as actual/mathematical model mismatch. The scope of the work reported in this paper is to demonstrate a potential loss of stability that may occur when a submersible is operating at low speeds, and to provide a general analysis technique for its avoidance. This is a static loss of stability and should not be confused with the dynamic loss of stability that is usually associated with higher speeds. It is shown that such a loss of stability is accompanied by a slow divergence of trajectories away from the commanded path. Solution branching occurs in the form of generic pitchfork bifurcations [Golubitsky & Schaeffer (1985)]. The use of bifurcation theory allows us to determine the crucial vehicle parameters that govern the problem of solution branching, and to develop guidelines to prevent its occurrence.

It is proved in this paper that, at the point of bifurcation, the system of the submersible vehicle is not controllable. Therefore, the bifurcation is unavoidable no matter what kind of feedback is used. Due to the qualitative change of the system performance as the parameter passing across the bifurcation point, a single smooth feedback can not stabilize the system on both sides of the critical parameter value. Piecewise smooth feedback is introduced in this paper to achieve more robust stability control near the bifurcation point. The design technique follows the ideas in [Kang (1997)], in which the quadratic invariants from the resonant terms in the uncontrollable dynamics are employed to characterize the stability of the system.

Finally, we present a new look at the problem of dive plane reversal [Clayton & Bishop (1982)] based on solution branching results. The term dive plane reversal refers to a well-known phenomenon in submarine operations where during low speed depth keeping there is a need for reversing the direction of dive plane deflection in order to execute a given change in depth. Physically, this can be explained by considering the relative magnitude of the hydrodynamic and hydrostatic forces. At moderate and high speeds, the normal force on the submarine's hull due to the angle of attack exceeds the normal dive plane force and the boat responds to ordered dive plane angles as expected. The phenomenon of dive plane reversal occurs at speeds below a certain critical speed in which the normal hull force is less than the normal dive plane force and the response of the boat is reversed.

Vehicle modeling in this work follows standard notation [Gertler & Hagen (1967)] and numerical results are presented for the DARPA SUBOFF model [Roddy (1990)] for which a set of hydrodynamic coefficients and geometric properties is available. Special emphasis is given in identifying the proper non-dimensional parameters in the problem, so that extension of the results to full scale models and other designs is possible using a minimal set of experimental data and/or analytical results. Unless otherwise mentioned, all results are presented in standard nondimensional forms with the exception of angular deflections which are shown in degrees.

PROBLEM FORMULATION

Equations of Motion

Assuming that vehicle motion is restricted in the vertical plane, the mathematical model consists of the coupled nonlinear heave and pitch equations of motion. In a moving coordinate frame fixed at the vehicle's geometrical center, Newton's equations of motion for a port/starboard symmetric vehicle are expressed as follows,

$$\begin{aligned}
&m(\dot{w} - Uq - z_G q^2 - x_G \dot{q}) = \\
&\quad Z_{\dot{q}}\dot{q} + Z_{\dot{w}}\dot{w} + Z_q U q + Z_w U w \\
&\quad - C_D \int_{\text{tail}}^{\text{nose}} b(x) \frac{(w - xq)^3}{|w - xq|}\, dx \\
&\quad + (W - B)\cos\theta + U^2 (Z_{\delta_s}\delta_s + Z_{\delta_b}\delta_b)\,, \qquad (1) \\
&I_y \dot{q} + m z_G w q - m x_G(\dot{w} - Uq) = \\
&\quad M_{\dot{q}}\dot{q} + M_{\dot{w}}\dot{w} + M_q U q + M_w U w \\
&\quad - C_D \int_{\text{tail}}^{\text{nose}} b(x) \frac{(w - xq)^3}{|w - xq|}\, x\, dx \\
&\quad - (x_G W - x_B B)\cos\theta - (z_G W - z_B B)\sin\theta \\
&\quad + U^2 (M_{\delta_s}\delta_s + M_{\delta_b}\delta_b)\,. \qquad (2)
\end{aligned}$$

In equations (1) and (2), W is the vehicle weight, B the buoyancy, (x_G, z_G) the coordinates of the center of gravity, (x_B, z_B) the coordinates of the center of buoyancy, δ_s the stern plane angle, and δ_b the bow plane angle. The cross flow integral terms in the above equations become important during hovering operations or low speed maneuvering, whereas at high speeds U (and consequently low angles of attack with respect to the water) their effect is minimal. The drag coefficient, C_D, is assumed to be constant throughout the vehicle length for simplicity. This does not affect significantly the results that follow. The remaining of the symbols in (1) and (2) follow standard notation and are explained in the Nomenclature. We use U for the vehicle forward speed instead of the standard symbol u [Gertler & Hagen (1967)] in order to emphasize the fact that propulsion dynamics are assumed decoupled from the basic model and the forward speed remains constant in the equations of motion. All changes in the forward speed are assumed to be quasistatic, which is consistent with standard bifurcation theory assumptions.

The vehicle pitch rate is

$$\dot{\theta} = q\,, \qquad (3)$$

and the rate of change of depth,

$$\dot{z} = -U\sin\theta + w\cos\theta\,, \qquad (4)$$

where θ is the pitch angle with respect to the nominal horizontal direction. The vehicle geometry and definitions for most of the above symbols are shown in Figure 1. The forward velocity U is assumed to be kept constant by the propulsion control during depth keeping. Any changes in U are assumed to take place in a quasi–steady way; i.e., at a rate much slower than vehicle motions in the dive plane.

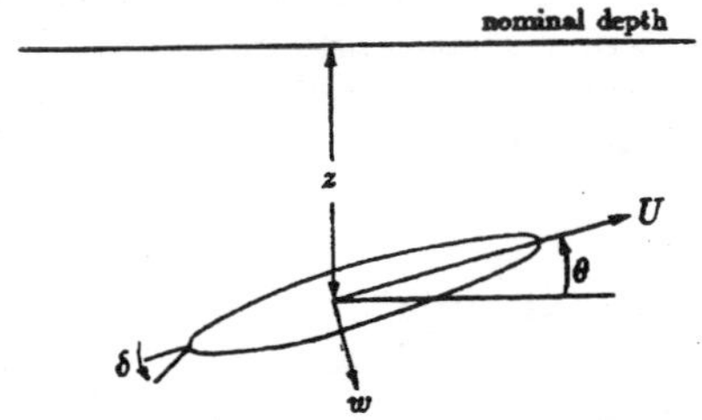

Figure 1: Vehicle geometry and definitions of symbols

Control Law

Equations (1) through (4) can be written as a set of four nonlinear coupled differential equations in the form,

$$\begin{aligned}
\dot{\theta} &= q\,, && (5) \\
\dot{w} &= a_{11}Uw + a_{12}Uq + a_{13}z_{GB}\sin\theta \\
&\quad + b_{11}U^2\delta_s + b_{12}U^2\delta_b \\
&\quad + d_w(w,q) + c_1(w,q)\,, && (6) \\
\dot{q} &= a_{21}Uw + a_{22}Uq + a_{23}z_{GB}\sin\theta \\
&\quad + b_{21}U^2\delta_s + b_{22}U^2\delta_b \\
&\quad + d_q(w,q) + c_2(w,q)\,, && (7) \\
\dot{z} &= -U\sin\theta + w\cos\theta\,, && (8)
\end{aligned}$$

where the coefficients a_{ij} are functions of the vehicle geometric properties and hydrodynamic coefficients [Papoulias & Riedel (1994)]. In equations (5) through (8), the vehicle is assumed to be neutrally buoyant ($W = B$), level ($x_G = x_B$), and statically stable ($z_G > z_B$). The terms I_w and I_q represent the cross flow drag integrals in (1) and (2), and $z_{GB} = z_G - z_B$ is the metacentric height. Without loss in generality we can assume z_B to be zero, so that $z_{GB} = z_G$.

During most cruising operations, effective depth control can be achieved by using the linearized version of equations (5) through (8), where the linearization is performed around level flight path at the commanded depth. The linear system of equations that can then be utilized for depth control law design is,

$$\begin{aligned}
\dot{\theta} &= q, && (9) \\
\dot{w} &= a_{11}U_0 w + a_{12}U_0 q + a_{13}z_{GB}\theta + b_1 U_0^2 \delta, && (10) \\
\dot{q} &= a_{21}U_0 w + a_{22}U_0 q + a_{23}z_{GB}\theta + b_2 U_0^2 \delta, && (11) \\
\dot{z} &= -U_0\theta + w, && (12)
\end{aligned}$$

where U_0 is the nominal speed for gain selection, α is defined as the bow plane to dive plane deflection ratio, and we have denoted

$$\begin{aligned}
&\delta_s = \delta\,, \quad \delta_b = \alpha\delta\,, \\
&b_1 = b_{11} + \alpha b_{12}\,, \quad b_2 = b_{21} + \alpha b_{22}\,. \qquad (13)
\end{aligned}$$

A linear full state feedback control law has the form [Friedland (1986)],

$$\delta = k_1\theta + k_2 w + k_3 q + k_4 z\,, \qquad (14)$$

where the gains k_1, k_2, k_3, k_4 are computed such that the closed loop system (9) through (14) has the desired dynamics. If the desired characteristic equation has the general form,

$$\lambda^4 + \alpha_3\lambda^3 + \alpha_2\lambda^2 + \alpha_1\lambda + \alpha_0 = 0\,, \qquad (15)$$

the controller gains can be computed by equating coefficients of the actual and desired characteristic equations. This procedure yields a linear system of four equations which can be solved for the four gains k_i in (14).

SOLUTION BRANCHING

First Observations

For demonstration purposes, assume that gain selection in (14) is based on a nominal speed $U_0 = 5$ ft/sec and a control time constant 3 dimensionless seconds. All initial conditions in the simulations are zero with the exception of depth which is given an initial offset. The above set of nominal speed and gains selection generates a fast response and rejection of the initial disturbance as was observed in various numerical simulations. Reduction in the forward speed U, beyond a certain level, resulted in marginal convergence or even divergence away from the commanded level flight path. It became apparent from extensive simulation studies that below a certain critical speed, U_c, the vehicle stabilized to a nonzero steady state off the commanded set point. Preliminary studies of the problem in this section are conducted through steady state, eigenvalue, and controllability analyses. A more rigorous analysis based on a general theory and normal form calculation is presented in the next section.

Steady State Solutions

Since we maintain the assumption $x_G = x_B$, it is expected that at steady state the heave velocity w will be small and, consequently, the cross flow integral drag terms I_w, I_q can be neglected. Steady state solutions are computed from $\dot{\theta} = \dot{w} = \dot{q} = \dot{z} = 0$, and using equations (5) through (8) we get

$$q = 0 ,$$
$$a_{11}Uw + a_{13}z_{GB}\sin\theta + b_1U^2\delta = 0 ,$$
$$a_{21}Uw + a_{23}z_{GB}\sin\theta + b_2U^2\delta = 0 ,$$
$$w\cos\theta - U\sin\theta = 0 ,$$

or,

$$\begin{aligned} w &= U\tan\theta , \\ \delta &= \frac{(a_{13}a_{21} - a_{23}a_{11})z_{GB}}{(b_2a_{11} - b_1a_{21})U^2}\sin\theta , \end{aligned} \tag{16}$$

where θ is a solution to,

$$\begin{aligned} &(a_{11}b_2 - a_{21}b_1)U^2\tan\theta + \\ &(a_{13}b_2 - a_{23}b_1)z_{GB}\sin\theta = 0 . \end{aligned} \tag{17}$$

Equation (17) admits the trivial solution $\theta = 0$ always, which results in $w = \delta = z = 0$; i.e., level steady state motion at the commanded depth. Besides $\theta = 0$, (17) may produce two more symmetric solutions in θ given by,

$$\cos\theta = \frac{(a_{11}b_2 - a_{21}b_1)U^2}{(a_{23}b_1 - a_{13}b_2)z_{GB}} . \tag{18}$$

Equation (18) is meaningful if $|\cos\theta| \leq 1$ which yields,

$$U^2 \leq \frac{a_{23}b_1 - a_{13}b_2}{a_{11}b_2 - a_{21}b_1}z_{GB} . \tag{19}$$

Substituting the expressions for a_{ij}, b_i in (19) we can find the critical value of the forward speed,

$$U_c^2 = \frac{Z_\delta W}{M_w Z_\delta - Z_w M_\delta} z_{GB} , \tag{20}$$

where we have denoted,

$$Z_\delta = Z_{\delta_s} + \alpha Z_{\delta_b} , \quad M_\delta = M_{\delta_s} + \alpha M_{\delta_b} .$$

For values of $U \leq U_c$, three steady state solutions exist, the trivial solution at the commanded path and two solutions symmetrically located with respect to the trivial solution.

A Zero Eigenvalue

The closed loop linearized system at any forward speed U is given by,

$$\begin{aligned} \dot{\theta} &= q , \\ \dot{w} &= (a_{11}U + b_1U^2k_2)w + (a_{12}U + b_1U^2k_3)q \\ &\quad +(a_{13}z_{GB} + b_1U^2k_1)\theta + b_1U^2k_4z , \\ \dot{q} &= (a_{21}U + b_2U^2k_2)w + (a_{22}U + b_2U^2k_3)q \\ &\quad +(a_{23}z_{GB} + b_2U^2k_1)\theta + b_2U^2k_4z , \\ \dot{z} &= -U\theta + w , \end{aligned}$$

with characteristic equation of the form,

$$A\lambda^4 + B\lambda^3 + C\lambda^2 + D\lambda + E = 0 . \tag{21}$$

Comparing equation (21) with the system of linear equations that was used for the determination of the four gains k_i, we can find that the coefficient E is given by,

$$\begin{aligned} E \;=\; &[(b_1a_{21} - b_2a_{11})U^4 \\ &+(b_1a_{23} - b_2a_{13})z_{GB}U^2]k_4 . \end{aligned} \tag{22}$$

A real eigenvalue of (21) crosses zero when $E = 0$, which yields the same critical value U_c as (20). This is a typical pitchfork bifurcation which generically occurs in certain dynamical systems when a real eigenvalue of the linearized system crosses zero [Guckenheimer & Holmes (1983)]. At the pitchfork bifurcation point, the trivial equilibrium solution becomes unstable and two symmetric equilibrium positions appear. These pitchfork bifurcations are classified as supercritical if the two symmetric solutions appear after the trivial solution loses its stability, and in such a case they are asymptotically stable. In case that the opposite is true, the new solutions are unstable and they co-exist with the stable trivial equilibrium solution, the pitchfork is said to be subcritical.

The Critical Speed Coefficient

Equation (20) suggests that the critical parameter that governs loss of stability and solution branching is a nondimensional speed coefficient based on the vehicle metacentric height.

$$U' = \frac{U}{\sqrt{g z_{GB}}} . \tag{23}$$

The critical speed coefficient is then,

$$U_c' = \left[\frac{Z_\delta W}{(M_w Z_\delta - Z_w M_\delta)g}\right]^{1/2} . \tag{24}$$

This critical speed coefficient is increasing with increasing bow plane to dive plane deflection ratio, α. The nonzero solutions in θ can be computed from (18), the steady state dive plane angle from (16), and the depth deviation from the control law (14). These computations are valid if the value of δ is less than its saturation limit δ_{sat}, which is typically set at ± 0.4 radians. In case where the computed δ exceeds δ_{sat}, the nonzero equilibrium pitch angle cannot be determined from (18). This means that $\dot{z} \neq 0$ at steady state, and the simplest condition is then $\dot{z}$ =const. [Papoulias (1993)]. This allows for a semi–equilibrium state in which $\ddot{z} = \dot{\theta} = \dot{q} = \dot{w} = 0$. The steady state solutions can then be computed from,

$$a_{11}Uw + a_{13}z_{GB}\sin\theta + b_1U^2\delta_{sat} = 0 ,$$
$$a_{21}Uw + a_{23}z_{GB}\sin\theta + b_2U^2\delta_{sat} = 0 ,$$

or,

$$\sin\theta = \frac{(b_2a_{11} - b_1a_{21})U^2\delta_{sat}}{(a_{13}a_{21} - a_{23}a_{11})z_{GB}} . \quad (25)$$

Controllability

The controllability matrix [Friedland (1986)] of the linear, time invariant, single–input system,

$$\dot{x} = Ax + b\delta , \quad (26)$$

is the square matrix,

$$\mathcal{C} = \left[b, Ab, A^2b, A^3b\right] . \quad (27)$$

As long as $\mathcal{C}$ is a nonsingular matrix, system (26) is state controllable, which means that an arbitrary change in the state vector x in finite time is possible through the use of a certain amount of control effort δ. In our case we have,

$$A = \begin{bmatrix} 0 & 0 & 1 & 0 \\ a_{13}z_{GB} & a_{11}U & a_{12}U & 0 \\ a_{23}z_{GB} & a_{21}U & a_{22}U & 0 \\ -U & 1 & 0 & 0 \end{bmatrix}, b = \begin{bmatrix} 0 \\ b_1U^2 \\ b_2U^2 \\ 0 \end{bmatrix} .$$

Numerical calculations show that the determinant of $\mathcal{C}$ becomes zero at precisely the critical speed, which could also have been defined as the speed at which the closed loop system (26) becomes uncontrollable. Therefore, we can summarize our findings by stating that there exists a vehicle critical speed for which (i) the straight line level flight path becomes unstable, (ii) additional nonzero symmetric stable equilibrium positions appear, and (iii) the closed loop depth control system is uncontrollable.

THEORETICAL ANALYSIS

Invariants and Stability

Before a stabilizing feedback is designed for the system, we introduce the quadratic invariants and their relation with the stability of nonlinear control systems. Quadratic invariants of nonlinear control systems were first found for linearly controllable systems in the development of the normal forms (see [Kang & Krener (1992)], [Kang (1996)], [Kang (1997a)]). Further investigation of the invariants for systems with uncontrollable linearization reveals that the invariants characterize the stability of nonlinear control systems near bifurcation point, [Kang (1997b)]. In this section, we focus on a control system in the following form

$$\begin{aligned} \dot{x}_1 &= x_2 + O(x,\mu)^2 \\ \dot{x}_2 &= x_3 + O(x,\mu)^2 \\ &\cdots\cdots \\ \dot{x}_{n-1} &= u + O(x,\mu)^2. \\ \dot{x}_n &= \gamma_{x_n\mu}x_n\mu + \gamma_{x_1\mu}x_1\mu + |J(x)|J(x) \\ &\quad + h(x,\mu) + O(x,\mu)^3. \end{aligned} \quad (28)$$

The state variable is x. The control input is u. In this paper we use the following linear state feedback

$$u = a_1x_1 + \cdots + a_nx_n. \quad (29)$$

There is a parameter in the system denoted by μ. The equilibrium point of our interest is $(x,\mu) = (0,0)$. The function $J(x)$ is C^2. However, $|J(x)|J(x)$ is C^1. Assume that $J(0) = 0$. The linearization of $J(x)$ is denoted by

$$J_1x_1 + J_2x_2 + \cdots + J_nx_n.$$

The function $h(x,\mu)$ is C^1. It is of $O(x,\mu)^2$ and it satisfies

$$\begin{aligned} \frac{\partial h}{\partial(x,\mu)} &= 0, \quad at \ (0,0,0), \\ h(x,\mu) &= 0, \quad if \ x_2 = x_3 = \cdots = x_{n-1} = 0. \end{aligned} \quad (30)$$

It is obvious that the linearization of (28) is not controllable at the origin. The uncontrollable dynamic system is given by the last equation in (28) (the equation of $\dot{x}_n$). The first equation in (30) implies that the system has a zero uncontrollable mode. It was proved that the coefficients $\gamma_{x_n\mu}$ and $\gamma_{x_1\mu}$ are invariant under certain change of coordinates and state feedback ([Kang (1997b)]). We are more interested in their relation with the stability of the system. It was proved in [Kang (1997b)] that the stability of the system in the presence of bifurcation can be determined through the values of the invariants. In [Kang (1997b)], the result was proved for systems which are at least C^2. However, $|J(x)|J(x)$ and $h(x,\mu)$ in (28) are C^1 functions. Therefore, the results in [Kang (1997b)] are not applicable. Nevertheless, the idea in that paper can be generalized to derive some useful results for systems in the form of (28). The main result in this subsection is

Theorem 1 *Given a system (28) with a state feedback (29). Assume that all the eigenvalues of the matrix*

$$\begin{bmatrix} 0 & 1 & 0 & \cdots & 0 \\ 0 & 0 & 1 & \cdots & 0 \\ & & \cdots & & \\ 0 & 0 & 0 & \cdots & 1 \\ a_1 & a_2 & a_3 & \cdots & a_{n-1} \end{bmatrix} \quad (31)$$

are on the left half plane. Then, the closed-loop system has two branches of equilibrium points near the origin which satisfy

$$\begin{aligned} x_1 &= -\frac{a_n}{a_1}x_n + O(x_n,\mu)^2, \\ x_2 &= O(x_n,\mu)^2, \cdots, x_{n-1} = O(x_n,\mu)^2, \\ \mu &= -\frac{\left|\frac{a_n}{a_1}J_1 - J_n\right|\left(\frac{a_n}{a_1}J_1 - J_n\right)}{\gamma_{x_n\mu} - \frac{a_n}{a_1}\gamma_{x_1\mu}}|x_n| + O(x_n)^2. \end{aligned} \quad (32)$$

or

$$x_1 = -\frac{a_n}{a_1}x_n + O(x_n,\mu)^2, \\ x_2 = O(x_n,\mu)^2, \cdots, x_{n-1} = O(x_n,\mu)^2, \\ x_n = O(\mu)^2 \tag{33}$$

The system is locally asymptotically stable at equilibrium points satisfying (33) if

$$\left(\gamma_{x_n\mu} - \frac{a_n}{a_1}\gamma_{x_1\mu}\right)\mu < 0. \tag{34}$$

If

$$\left(\gamma_{x_n\mu} - \frac{a_n}{a_1}\gamma_{x_1\mu}\right)\mu > 0,$$

the system is unstable at points on (33).

Remark. The assumption on the eigenvalues of (31) implies that the controllable part of the system is stabilized by the feedback. This is always achievable by pole placement or LQR. ◁

Proof. To find the equilibrium points of the system near the origin, set the right side of the system be to zero. From the first $n-1$ equations, we have

$$x_1 = -\frac{a_n}{a_1}x_n + O(x_n,\mu)^2, \\ x_2 = O(x_n,\mu)^2, \cdots, x_{n-1} = O(x_n,\mu)^2. \tag{35}$$

Substituting this into the last equation, we get

$$\left|\left(-\frac{a_n}{a_1}J_1 + J_n\right)x_n\right|\left(-\frac{a_n}{a_1}J_1 + J_n\right)x_n \\ + \left(\gamma_{x_n\mu} - \frac{a_n}{a_1}\gamma_{x_1\mu}\right)x_n\mu + O(x_n,\mu)^3 = 0$$

Since this equation contains terms of $O(x_n,\mu)^3$, it has two solutions satisfying

$$\mu = \frac{\left|\frac{a_n}{a_1}J_1 - J_n\right|\left(\frac{a_n}{a_1}J_1 - J_n\right)}{\gamma_{x_n\mu} - \frac{a_n}{a_1}\gamma_{x_1\mu}}|x_n| + O(x_n)^2. \tag{36}$$

or

$$x_n = O(\mu)^2. \tag{37}$$

The relations (32) and (33) are proved by (35), (36) and (37). To prove the stability of the closed-loop system, consider the matrix in the linearization of the system (only the lowest degree terms)

$$\begin{bmatrix} 0 & 1 & 0 & \cdots & 0 & 0 \\ 0 & 0 & 1 & \cdots & 0 & 0 \\ & & & \cdots & & \\ 0 & 0 & 0 & \cdots & 1 & 0 \\ a_1 & a_2 & a_3 & \cdots & a_{n-1} & a_n \\ \frac{\partial f_n}{\partial x_1} & \frac{\partial f_n}{\partial x_2} & \frac{\partial f_n}{\partial x_3} & \cdots & \frac{\partial f_n}{\partial x_{n-1}} & \frac{\partial f_n}{\partial x_n} \end{bmatrix}$$

where f_n represents the right side of the last equation in (28). It is equivalent to

$$\begin{bmatrix} 0 & 1 & 0 & \cdots & 0 & 0 \\ 0 & 0 & 1 & \cdots & 0 & 0 \\ & & & \cdots & & \\ 0 & 0 & 0 & \cdots & 1 & 0 \\ a_1 & a_2 & a_3 & \cdots & a_{n-1} & 0 \\ \frac{\partial f_n}{\partial x_1} & \frac{\partial f_n}{\partial x_2} + \frac{a_n}{a_1} & \frac{\partial f_n}{\partial x_3} & \cdots & \frac{\partial f_n}{\partial x_{n-1}} & \frac{\partial f_n}{\partial x_n} - \frac{a_n}{a_1}\frac{\partial f_n}{\partial x_1} \end{bmatrix} \tag{38}$$

The eigenvalues of (38) includes the eigenvalues of the matrix (31) and the value of

$$\frac{\partial f_n}{\partial x_n} - \frac{a_n}{a_1}\frac{\partial f_n}{\partial x_1} \\ = \frac{\partial}{\partial x_n}f_n(-\frac{a_n}{a_1}x_n, x_2, \cdots, x_n, \mu) \\ = (\gamma_{x_n\mu} - \frac{a_n}{a_1}\gamma_{x_1\mu})\mu \\ + \frac{\partial}{\partial x_n}\left(\left|J(-\frac{a_n}{a_1}x_n, x_2, \cdots, x_n)\right| J(-\frac{a_n}{a_1}x_n, x_2, \cdots, x_n)\right) \\ + \frac{\partial}{\partial x_n}\left(h(-\frac{a_n}{a_1}x_n, x_2, \cdots, x_n, \mu)\right) \tag{39}$$

At the equilibrium points given by (33), the value of (39) is

$$(\gamma_{x_n\mu} - \frac{a_n}{a_1}\gamma_{x_1\mu})\mu + O(\mu)^2$$

Therefore, the sign of the eigenvalue is determined (locally) by the sign of

$$(\gamma_{x_n\mu} - \frac{a_n}{a_1}\gamma_{x_1\mu})\mu$$

This proves the stability test condition (34). ◁

Normal Form Calculations

To stabilize the system (5)-(8) using Theorem 1, it is necessary to find the invariants $\gamma_{z\mu}$ and $\gamma_{x_1\mu}$. For this reason, a linear change of coordinates and a linear feedback are found which transform the system into the form of (28). In fact, the linearization of (28) is in the normal form obtained in [Kang (1997a)]. The following notations will be frequently used in this section.

$$N = a_{23}b_1 - a_{13}b_2, \\ D = a_{11}b_2 - a_{21}b_1, \\ P = a_{22}b_1 - a_{12}b_2 + b_2. \tag{40}$$

The critical value of the forward speed satisfies

$$U_c^2 = \frac{N}{D}z_{GB}$$

Since the stability around the bifurcaton point $U = U_c$ is our focal problem, it is assumed in this section that the nominal speed of the vehicle is $U_0 = U_c$. The actual speed is $U = U_c + \mu$, where μ is treated as a parameter of the system. The linear change of coordinates is given by

$$x_1 = b_1\theta - b_2 z \\ x_2 = b_1 q + b_2 U_0\theta - b_2 w \\ x_3 = -DU_0 w + PU_0 q + Nz_{GB}\theta \\ x_4 = b_1 q - b_2 w - (b_1 a_{22} - b_2 a_{12})U_0\theta + DU_0 z \tag{41}$$

This transformation satisfies the following equations.

$$\begin{bmatrix} \theta \\ z \end{bmatrix} = \frac{1}{det(T)}\begin{bmatrix} -DU_0x_1 + b_2x_2 - b_2x_4 \\ -PU_0x_1 + b_1x_2 - b_1x_4 \end{bmatrix} \\ \begin{bmatrix} q \\ w \end{bmatrix} = T^{-1}\begin{bmatrix} x_2 \\ x_3 \end{bmatrix} + \begin{bmatrix} 0 \\ U_0 \end{bmatrix}\theta \tag{42}$$

where T is the matrix

$$T = \begin{bmatrix} b_1 & -b_2 \\ PU_0 & -DU_0 \end{bmatrix} \tag{43}$$

Throughout this paper, we assume that $det(T) \neq 0$ which guarantees that (41) is invertible. Under the new coordinates (41), the system (5)-(8) is transformed into the following form

$$\begin{aligned}\dot{x}_1 &= x_2 + O(x,\mu)^2 \\ \dot{x}_2 &= x_3 + O(x,\mu)^2 \\ \dot{x}_3 &= F(x,\mu) + G(x,\mu)\delta \\ \dot{x}_4 &= \frac{2D^2U_0^2}{det(T)}\mu x_1 + \frac{2DU_0b_2}{det(T)}\mu x_4 \\ &+|J(x)|J(x) + h(x,\mu) + O(x,\mu)^3\end{aligned} \tag{44}$$

where $h(x,\mu)$ is a C^1 function and the first derivatives of h equal zero at $(x,\mu) = (0,0)$. Furthermore, $h(x,\mu) = 0$ if $x_2 = x_3 = 0$. The functions F and G have the following terms with lowest degree in (θ, w, q, z)

$$\begin{aligned}F_1 &= U_0 z_{GB}(Pa_{23} - Da_{13})\theta + U_0^2(Pa_{21} - Da_{11})w \\ &+ \left(U_0^2(Pa_{22} - Da_{12}) + Nz_{GB}\right) q + O(\cdot)^2 \\ G_0 &= U_0^3(Pb_2 - Db_1) + O(\cdot)^2.\end{aligned} \tag{45}$$

The proof of (44) involves complicated algebraic computation. However, the idea is straightforward. After taking the derivatives of the functions in (41), substitute the variables in the original coordinates by the new coordinates (x,μ). In the equation of x_4, the quadratic terms involving the variables x_2 and x_3 are absorbed in the function $h(x,\mu)$. The functions d_w and d_q has the decomposition

$$\begin{aligned}d_w(w,q) &= d_w(w,0) + (d_w(w,q) - d_w(w,0)) \\ d_q(w,q) &= d_q(w,0) + (d_q(w,q) - d_q(w,0))\end{aligned}$$

The functions $d_w(w,0)$ and $d_q(w,0)$ have the form $|J(x)|J(x)$ and the other parts are absorbed in $h(x,\mu)$. Since $h(x,\mu)$ includes the cross flow drag integrals, it is not C^2. From the linear normal form (44), it is obvious that the system is not linearly controllable at $\mu = 0$ (or equivalently $U = U_c$). This is a theoretical proof of the fact observed in the numerical experiment, as pointed out in the section of controllability.

DIVE PLANE REVERSAL

The Physical Problem

A brief description of the need for dive plane reversal was given in the Introduction in terms of the relative magnitudes of control surface and hull generated hydrodynamic forces. The phenomenon is usually described in terms of the neutral angles, as well. The neutral angles are represented by a plot of the pitch angle and stern plane deflection that is required for straight and level flight versus the ship speed, and for a given loading condition, x_{GB}. At high speeds these neutral angles remain essentially constant. Both the pitch angle, θ, and the dive plane angle, δ_s, have the same sign and they can generate enough lift to counter the hydrostatic imbalance moment. As the speed is decreased, it becomes increasingly difficult to maintain the pitch angle on the hull and counteract the metacentric moment. Below the critical speed both θ and δ_s must reverse sign to maintain neutral trim. The results that were obtained through the use of bifurcation theory in this work formalized the above qualitative conclusions and

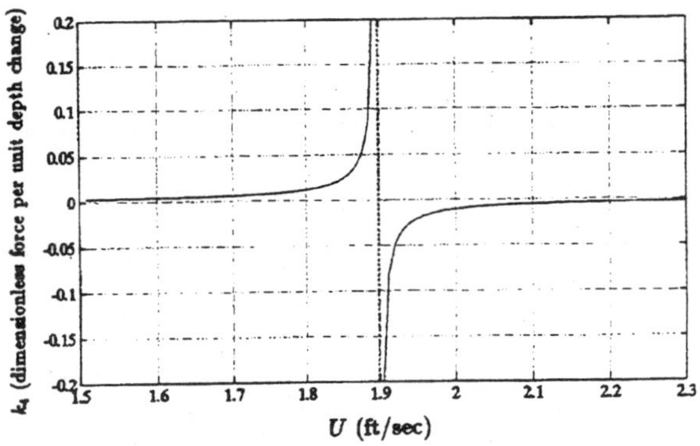

Figure 2: Normal force required per unit change in depth

identified the important nondimensional parameters that govern dive plane reversal.

Another way of viewing the problem is through the term k_4 in the control law (14). This represents a certain measure of the normal force that is required in order to respond to a unit change in depth. In terms of the forward speed U_0, this term can be directly computed from the equations that determine the four gains k_i. The corresponding plot is shown in Figure 2, in dimensionless values. It can be seen that k_4 approaches infinity at the critical speed U_c. This agrees with our previous conclusion that the system is uncontrollable at that precise speed. Furthermore, the sign of k_4 is reversed as the critical speed is crossed, which results in the need for dive plane reversal in order to execute the same ordered depth change.

Bifurcation Control

It is proved that the system (5)-(8) is equivalent to (44) at the bifurcation point $U = U_c$. Notice that (44) is in the form of (28). Therefore, Theorem 1 is applicable. Based on this theorem, it can be proved that the following design synthesis leads to a stabilizing state feedback.

Step 1. Using linear feedback design method such as pole placement or LQR find three numbers a_1, a_2 and a_3 such that all the eigenvalues of

$$\begin{bmatrix} 0 & 1 & 0 \\ 0 & 0 & 1 \\ a_1 & a_2 & a_3 \end{bmatrix}$$

are on the left half plane.

Step 2. Choose a number $\bar{a}_4$ so that

$$\bar{a}_4 > \left|\frac{a_1 b_2}{DU_0}\right|. \tag{46}$$

Step 3. The state feedback is a piecewise smooth function of (θ, w, q, z) given in the following formula.

$$\delta = \frac{1}{G_0}\left(\sum_{i=1}^{4} a_i x_i - F_1\right), \quad a_4 = \begin{cases} -\bar{a}_4 & if \ \frac{\mu}{det(T)} > 0 \\ \bar{a}_4 & if \ \frac{\mu}{det(T)} < 0 \end{cases} \tag{47}$$

The variables x_i, $i = 1, \cdots, 4$ are functions defined by (41). The functions F_1 and G_0 are defined by (45). The matrix T is given by (43).

The system (44) is in the linear normal form (28) if we set

$$F_1 + G_0\delta = a_1x_1 + a_2x_2 + a_3x_3 + a_4x_4.$$

The invariants from resonant terms are

$$\gamma_{x_1\mu} = \frac{2D^2U_0^2}{det(T)}, \quad \gamma_{x_4\mu} = \frac{2DU_0b_2}{det(T)}. \tag{48}$$

From Theorem 1, we know that the system has two branches of equilibrium points. One of them satisfies $x_n = O(\mu)^2$. Because of (33) and (42), $\theta = O(\mu)^2$ on this branch of equilibrium points. We know that $(\theta, w, q, z) = (0, 0, 0, 0)$ is the only branch on which $\theta = O(\mu)^2$ (see [Papoulias & Riedel (1994)]). Therefore, the branch of equilibrium points satisfying $x_n = O(\mu)^2$ is actually the branch with

$$\theta = 0. \tag{49}$$

To stabilize the system at an equilibrium satisfying (49), the coefficients a_1 and a_4 should satisfy (34). From (48) and the fact that $a_1 < 0$, this inequality is equivalent to

$$(a_1b_2DU_0 - a_4D^2U_0^2)\frac{\mu}{det(T)} > 0. \tag{50}$$

the value of a_4 determined by (46)-(47) guarantees the inequality (50). Therefore, we have the following theorem.

Theorem 2 *The feedback designed following the three steps stabilizes the system (5)-(8) at the equilibrium points with* $\theta = 0$.

CONCLUDING REMARKS

The main conclusions of this work can be summarized as follows:

- There exists a critical vehicle speed at which one real eigenvalue of the closed loop depth keeping linearized system matrix, crosses zero. The nominal equilibrium state becomes unstable in a divergent way for speeds less than critical.
- Below the critical speed two symmetric stable equilibrium positions appear which are characterized by nonzero pitch angle and depth.
- The closed loop depth control system becomes uncontrollable at precisely the critical speed. The depth deviation error gain approaches infinity at that speed and it changes its sign.
- The primary bifurcation parameter is a nondimensional speed coefficient which is based on vehicle speed and metacentric height. This can help transition experimental results from models to full scale.

Acknowledgments

This work was in part supported by the Naval Surface Warfare Center, Carderock Division, and by the Naval Postgraduate School. The authors would like to thank the principal reviewer of the paper for his careful reading of the manuscript and useful suggestions.

REFERENCES

Clayton, B. R. and Bishop, R. E. D. (1982) *Mechanics of Marine Vehicles.* Gulf Publishing Company, Houston.

Friedland, B. (1986) *Control System Design: An Introduction to State-Space Methods.* McGraw Hill, New York.

Gertler, M. and Hagen, G. R. (1967) Standard equations of motion for submarine simulation. David Taylor Research Center, Report 2510.

Golubitsky, M. and Schaeffer, D. G. (1985) *Singularities and Groups in Bifurcation Theory I.* Applied Mathematical Sciences 51, Springer–Verlag, New York.

Guckenheimer, J. and Holmes P. (1983) *Nonlinear Oscillations, Dynamical Systems, and Bifurcations of Vector Fields.* Applied Mathematical Sciences 42, Springer–Verlag, New York.

Kang, K. and Krener, A. J. (1992) Extended quadratic controller normal form and dynamic feedback linearization of nonlinear systems. *SIAM J. Control and Optimization,* **30**, No. 6.

Kang, K. (1996) Extended controller form and invariants of nonlinear control systems with a single input. *J. Mathematical Systems, Estimation, and Control,* **6**, No. 1.

Kang, K. (1997a) Bifurcation and normal form of nonlinear control systems – part I/part II. *SIAM J. Control and Optimization.* In press.

Kang, K. (1997b) Bifurcation control via state feedback for single input nonlinear systems – partI/ part II. *IEEE Trans. Automat. Contr.* Submitted for publication.

Papoulias, F. A. (1993) Dynamics and bifurcations of pursuit guidance for vehicle path keeping in the dive plane. *J. of Ship Research,* **37**, 2.

Papoulias, F. A. and Riedel, J. S. (1994) Solution branching and dive plane reversal of submarines at low speeds. *J. of Ship Research,* **38**, 3.

Roddy, R. F. (1990) Investigation of the stability and control characteristics of several configurations of the DARPA SUBOFF model (DTRC model 5470) from captive–model experiments. David Taylor Research Center, Report DTRC/SHD–1298–08.

Proceedings of the Seventh (1997) International Offshore and Polar Engineering Conference
Honolulu, USA, May 25-30, 1997

ISBN 1-880653-28-1 (Set); ISBN 1-880653-30-3 (Vol. II)

New Experimental Results on Shallow-Water Autonomous Underwater Vehicle Motion in Controlled Environments

P.E. An, S.M. Smith, S.E. Dunn, X. Chen and L. Mu
Florida Atlantic University
Boca Raton, FL, USA

ABSTRACT

Ocean Explorer (OEX) series autonomous underwater vehicles (AUV) which were developed at Florida Atlantic University provide a low cost, long range platform on which to carry out oceanographic, military and commercial missions in shallow water. A high performance controller is thus needed for an AUV as its motion can be significantly affected by sea state conditions because of its small size and its shallow water operations. The main focus of this paper is to present part of the vehicle self-motion data that were collected in a controlled environment where waves were generated in a well-defined manner, and the environment was chosen to be the MASK (Maneuvering And Sea-Keeping) facilities at the David Taylor Model Basin Research Center. This study is expected to provide a better understanding of the wazzu vehicle dynamic characteristics with known disturbances, which can then be used to validate mathematical models by means of semi-empirical methods. effort It should be noted that the self-motion study carried out in a controlled environment using a small AUV is believed to be the first ever undertaking.

KEY WORDS: Autonomous underwater vehicles, self-motion, sea-keeping, regular waves, fuzzy logic control.

1. INTRODUCTION

With matured sensor and electronics technologies, the new-generation autonomous underwater vehicle designs are increasingly focused on reduction of vehicle size and modularity in payload configuration [Smith, 1995]. These focuses naturally lead to desirable deployment features such as portability in terms of reduced ship time/cost and field reconfigurability in terms of down-time cost. These small and highly maneuverable underwater vehicles results in a cost-effective mobile sensing platform on which to carry out oceanographic, military and commercial missions, such as search/survey and target detection / classification based on sonar imaging techniques. However, as the vehicle size is reduced, its motion to time-varying sea-state conditions is increasingly affected especially in shallow-water regimes [An, 1996b], [Peterson, 1994]. Analyzing the relationship between the AUV motion and the seaway characteristic thus enables us to design a stable, efficient controller which compensates wave motion.

The objective of this paper is to characterize the shallow water motion of small autonomous underwater vehicles with respect to different sea states generated at the David Taylor Model Basin where regular or random waves can be generated accurately for a range of frequencies and heights. This paper is mainly focused on the closed-loop motion characteristics of the FAU Ocean Explorer vehicle in response to wave disturbances. The analysis presented in this paper includes power spectrum density estimation (PSD) of wave and vehicle pitch and roll motion and cross-variance function between the pitch and stern plane whereas those of other motion variables are referred to [An, 1996a].

2. BACKGROUND

2.1 Hull Configuration of Ocean Explorer

The Ocean Explorer is basically 7 feet long and 21 maximum inches in diameter and is designed with

- a tear-drop shaped fiber-glass hull based on a modified version of the Gertler Series 58 Model 4154 body shape.
- aft-mounted cruciform control surfaces.
- a 3-bladed propeller 18 inches in diameter .
- intelligent Ni-Cd battery packs. The battery packs can supply up to 12 hours of continuous missions at 3-knot cruising speed.
- main computer and electronics board (MC68030 at 50 MHz on the VME bus). Each of the components is embedded with a LonWorks Neuron node, and the control communication is achieved via LonTalk protocol.
- sensors include Watson AHRS-C302RS (3-axis acceleration, angles and rates), SIMRAD mesotech 809 (altitude), Druck PTX 1649 (water depth) , Sonic Speed (water speed), DGPS, LBL and USBL positioning system.

A TSK fuzzy controller was used to control vehicle motion such as heading, pitching and depth [Tagaki, 1985]. For an example, the pitch controller takes in pitch and pitch rate as inputs and produces a desired stern plane angle which was then fed to the vehicle. The detailed description of the fuzzy controller can be found in [An, 1996a].

2.2 Vehicle Encountering Frequency

Encountering frequencies for different wave frequencies and headings are listed in Table 1 to which the readers can be referred in further analysis [Kinsman, 1970].

heading / wave freq.	head Sea	beam (90° [1])	follow sea (Hz)	wave length (m)
0.3 Hz	0.36	0.3	0.24	16.92
0.7 Hz	1.01	0.7	0.39	3.18

Table 1 Encountering Frequency

3. EXPERIMENT

3.1 Mission description

There are totally 36 valid experiments which have been carried out at the David Taylor Model Basin. A simplified sketch of the Maneuvering And Sea-Keeping basin (MASK) is shown in Figure 1, where waves are generated either by the long wave maker or short wave maker, but not both simultaneously. The corresponding long and short beaches which minimize the reflected waves are also shown in the figure. Since the facilities are extremely effective at shielding the earth's magnetic field, the heading reference from the Watson sensor produces information which is close to a random signal. In order to facilitate closed-loop self-propulsion capabilitythe yaw rate-gyro output was numerically integrated to produce a heading estimate, which was found to be accurate based on at-sea data if maneuvering was kept to a minimum. For all the missions, the north direction was artificially defined as that shown , in the figure. There is an underpass

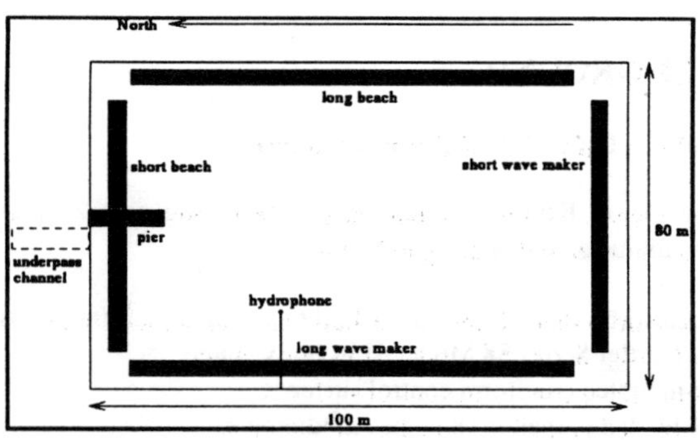

Figure 1 David Taylor Model Basin

which allows the AUV to be launched and recovered using a overhang crane located at the end of the channel.In addition, there is a pier on the short beach side which serves as a dock for the AUV data upload between missions. Along the long wave maker, there is a hydrophone which records the distance to the waves, and can be used to calibrate the wave frequency and height.

[1] Vehicle heading is defined with respect to the direction of wave propagation.

The experiments were designed for AUV moving at various sea condition such as different wave frequencies (0.3 Hz , 0.4 Hz, 0.6 Hz, 0.7 Hz , random wave (sea-state 2)), different depths (1.5m , 2m, 3m), different headings (0° , 90° , 180° , 200° , 240° , 270° , 340°), different wave heights (0.2m, 0.4m for the regular waves and 0.64m for the random waves) and different control condition (open-loop, closed-loop). From a typical wazzu time history (data sampling rate was 8 Hz), the vehicle's motion can be separated into 3 stages (see Figure 2). The vehicle first dived to a desired depth; maneuvered by tracking the commanded heading at the wazzu specified depth; and finally emerged to the surface. The PSD of vehicle motion was estimated based on time history data starting from 30 through 62 seconds (256 samples) within each mission such that the vehicle is ensured to be operated in a steady state condition. The Welch's averaged periodogram method was used in obtaining the PSD channel estimates and the window function was based on Hanning type. The window size was chosen to be 128 which results in resolution of approximately 0.06 Hz. This choice was based on the constraints of the tank dimension, vehicle nominal speed and the sampling rate. The wave PSD was independently estimated using the hydrophone data collected at a rate of 20Hz. The window size was chosen to be 512, and the PSD resolution was 0.04 Hz.

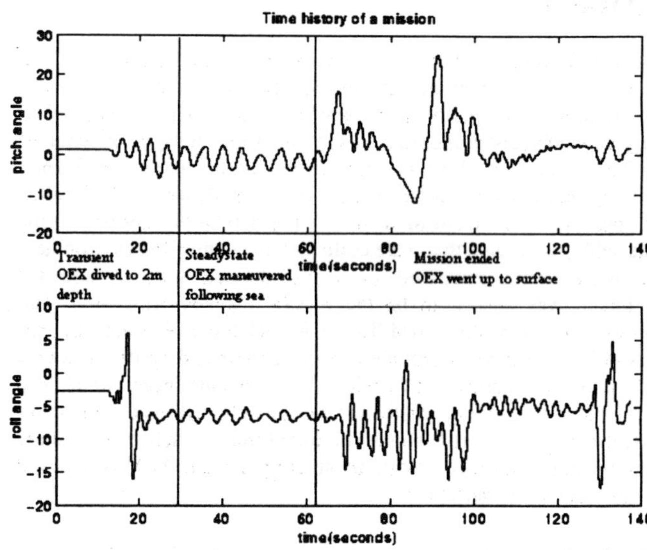

Figure 2 Typical motion time history

3.2 Wave frequency

The wave frequency of interest ranged from 0.3 Hz to 0.7 Hz[2]. After analysis of the recorded wave data, some periodic wave frequencies were observed to be close to their specified values (see Figure 3). However in cases of 0.7 Hz, a multitude of undesired peaks of small magnitude was found in the estimated PSD (see Figure 4: small peak at 0.05 Hz). The situation was worse for random waves. Typical PSDs of random waves generated near 0.4 Hz reveal undesired components which can be significant (see Figure 10).

[2] Frequency less than 0.3 Hz was not attempted because of the reflected wave disturbances.

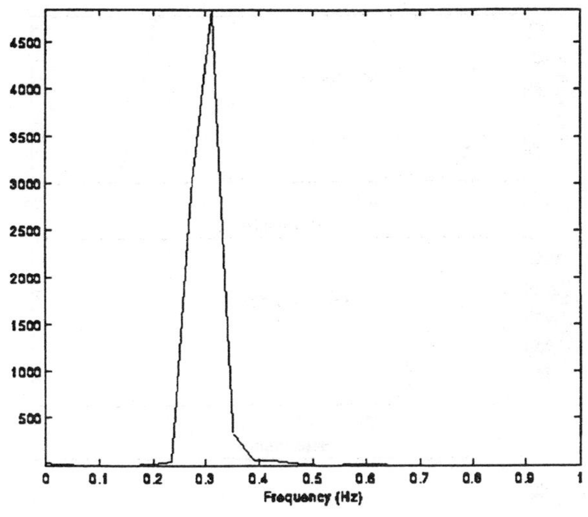

Figure 1 PSD of a 0.3 Hz wave

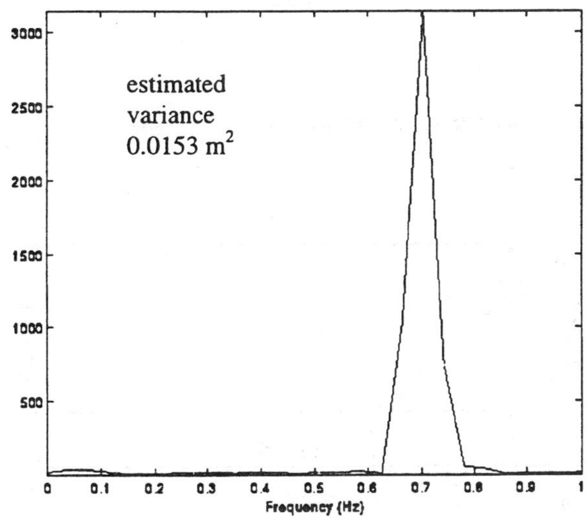

Figure 2 PSD of 0.7 Hz wave

sea condition	head sea		following sea		beam sea	
	freq.[1]	peak val.[2]	freq	peak val.	freq wazz u	peak val.
heading	0.2 0.4	18 8	0.2	80	0.2 0.3	55 25
stern plane	0.4	1900	0.2	29000	0.2	6500
rudder	0.2 0.4	290 75	0.2	1300	0.2 0.3	950 400

Table 1 Peak frequencies and variance estimates

wave freq. 0.3 Hz

sea condition	head sea		following sea		beam sea	
	freq	peak val.	freq	peak val.	freq	peak val.
heading	0.2 to 0.4 1.1	2.2 0.9 1.7	0.1 0.3	45 48	0.2 0.3 0.7 0.9	30 45 20 5
stern plane	1.1	22000	0.3 0.9	50000 5000	0.3	7500
rudder	0.2 0.4 1 1.1	85 50 15 50	0.2[3] 0.45	900 100	0.3 0.7 0.9	1600 1900 600

Table 2 Peak frequencies and variance estimates
wave freq. 0.7 Hz

3.3 Periodic waves

Figure 5 and 6 show the PSDs of vehicle pitch and roll motion at 2m operating depth given the wave height of 0.2m and wave frequency of 0.3 Hz. In these figures, the vehicle maneuvered in head sea, following sea and beam sea conditions. Figure 7 and 8 show similar results except that the wave frequency was changed to 0.7 Hz. To provide some ability for cross-referencing, the peaks and variances of relevant AUV motion variables are included in Table 2 and 3.

Several comments can be made from observing the listed figures and data. Firstly, the dominant peaks in the pitch and roll PSD in both head and following sea conditions were found to be highly correlated withtheir corresponding encountering wave frequencies (refer Table1). As the controller attempts to compensate the wave disturbance, largest peaks of stern plane and rudder were also found near their encountering frequencies. Secondly, small peaks in the pitch and roll PSDs were observed at lower frequencies in head sea condition which appeared to be caused by AUV motion. Whereas in both following sea and beam sea conditions the same small peaks appeared to merge with the downward shifting encountering frequencies given the low resolution of PSD estimation. Thirdly, the effects of pitching on rolling are much larger in both head and following sea conditions whereas in the beam sea condition the correlation was noticeably reduced. Lastly, the pitch and stern plane motion exhibited significantly larger variances as compared to those in the head and beam sea conditions. This suggests that much control effort is often wasted in the following sea maneuvering when the controller is improperly designed by ignoring the wave disturbances.

[1] Unit: Hz
[2] Unit: deg^2/Hz
[3] broad band

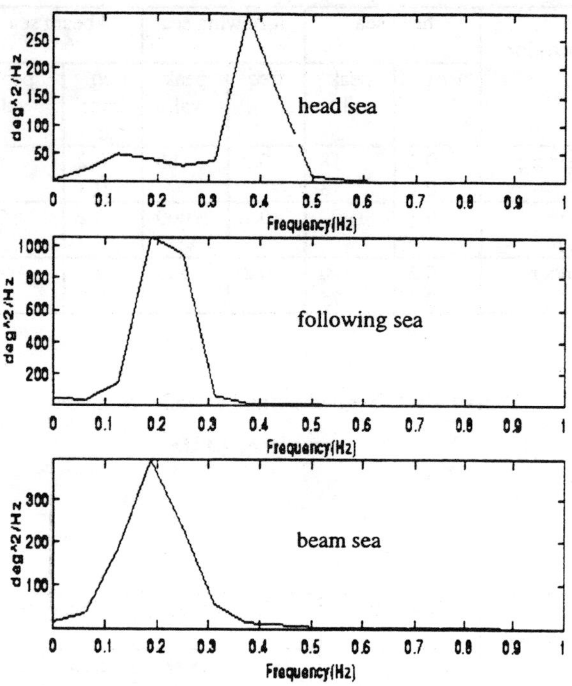

Figure 5 PSD of pitch motion (Wave frequency 0.3 Hz)

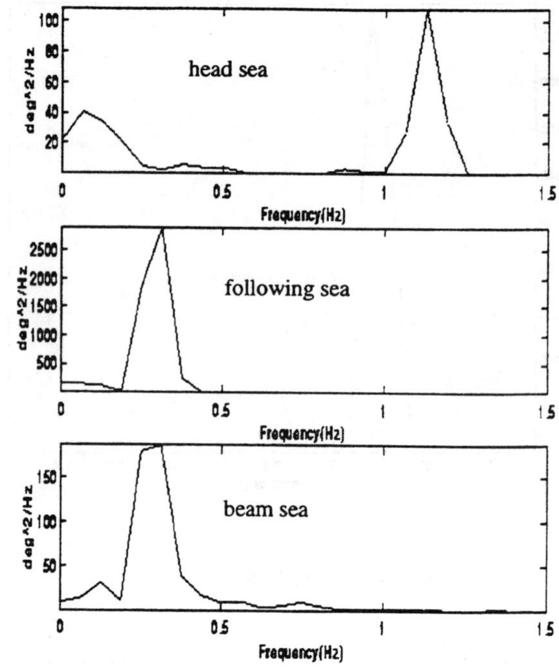

Figure 7 PSD of pitch motion (Wave frequency 0.7 Hz)

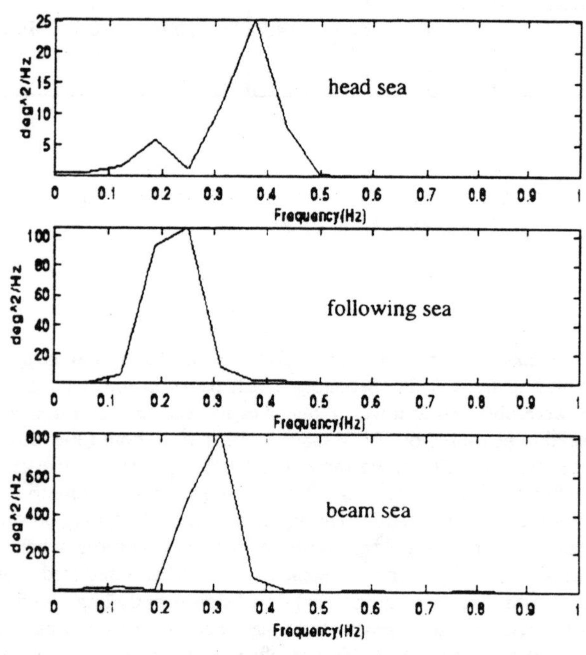

Figure 6 PSD of roll motion (Wave frequency 0.3 Hz)

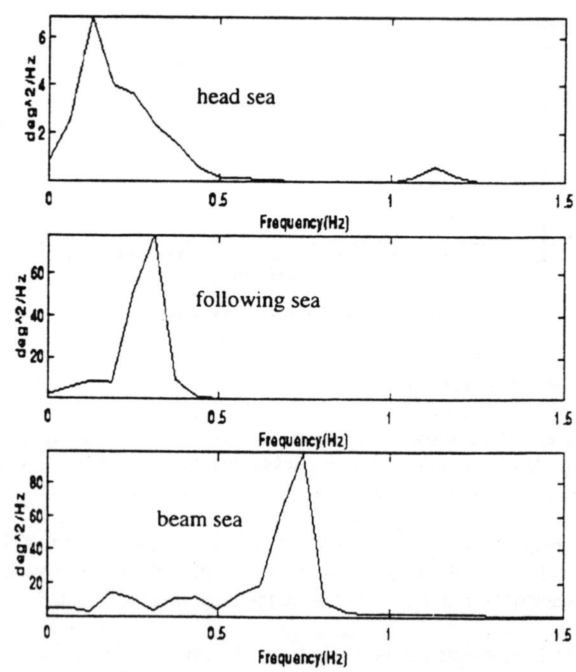

Figure 8 PSD of roll motion (Wave frequency 0.7 Hz)

3.4 Random waves

Figure 9 and 11 show the time histories of wave, vehicle pitch and roll motion in head and following sea condition respectively given significant wave height of 0.64m and 1.5m operating depth. Figure 10 and 12 show the corresponding estimated PSDs, and the peaks and variances of relevant AUV motion variables are listed in Table 4. In the wave PSD, two significant peaks were found to be near 0.25 and 0.42 Hz. Notice that there was a small peak near 0.5 Hz in both pitch and roll PSD during the head sea mission, and a similar peak near 0.25 Hz during the following sea mission. These peaks appeared to be generated by the encountering waves. It should also be noted that the vehicle surfaced during both the head and following sea missions, causing a large component near 0.05 Hz to appear in both the pitch and roll PSDs.

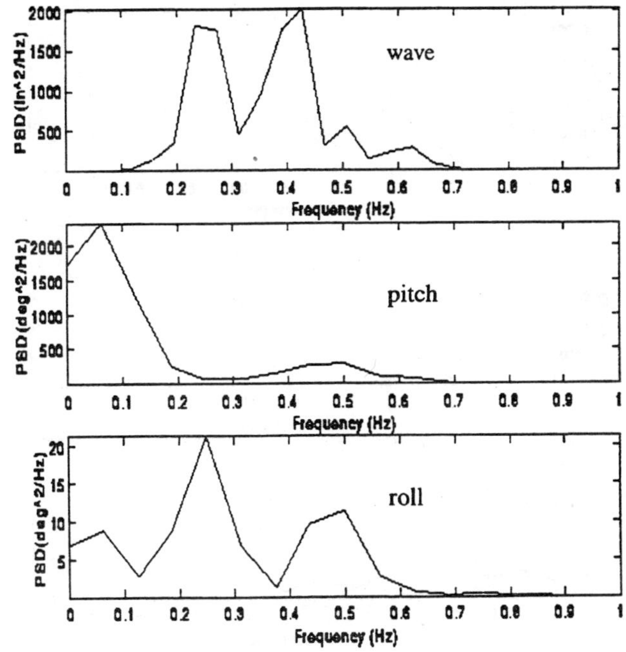

Figure 10 PSDs of wave and pitch and roll

(random wave , head sea)

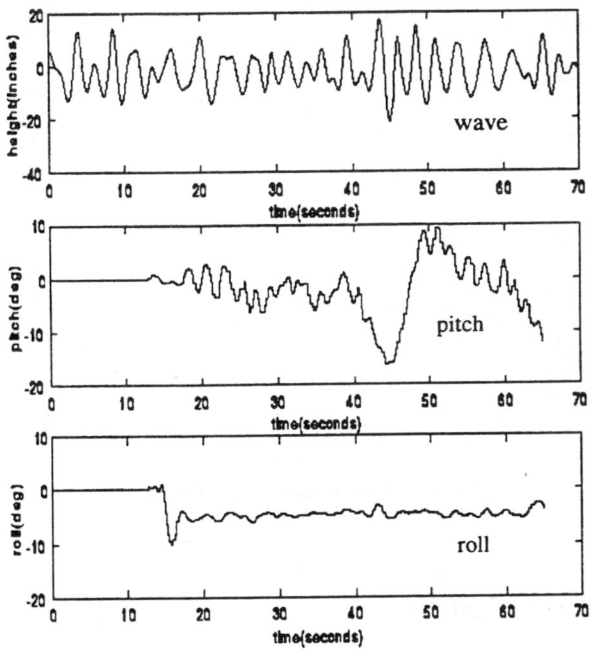

Figure 9 Time history for wave , pitch and roll

(random wave ,head sea)

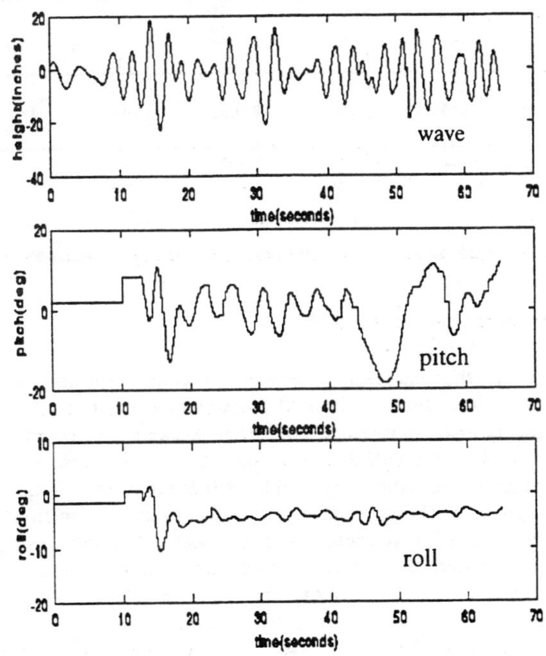

Figure 11 Time histories of wave ,pitch and roll

(random wave , following sea)

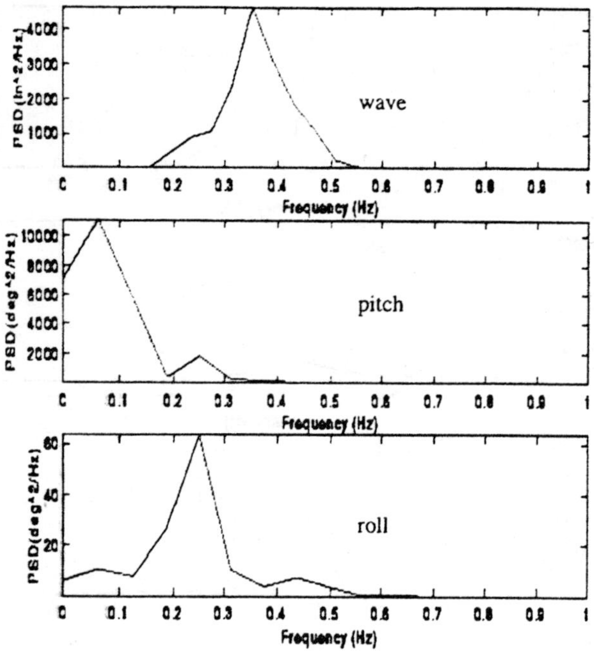

Figure 12 PSDs of wave, pitch and roll

(random wave, following sea)

sea condition	head sea		following sea	
	freq.	peak val.	freq.	peak val.
heading	0.05 0.2 0.4	14 13 7	0.05 0.25 0.5	10 80 15
stern plane	0.1 0.5	3100 20000	0.05 0.25 0.5	1000 3000 400
rudder	0.2 0.4 0.6	450 150 50	0.05 0.25 0.5	100 3500 950

Table 4 Peak frequencies and variance estimates random wave

3.5 Cross-variance Function

To study the relationship between the control surface and AUV motion one can estimate a desired cross-variance function. Figure 13 shows the computed variance coefficient between pitch and stern plane when OEX traveled 2m below the water surface in following sea condition given periodic wave height of 0.2m and wave frequency of 0.3 Hz. Figure 14 shows the PSD of the cross-variance function. In the figure, one can observe there was a big peak at 0.24 Hz which was caused by the stern plane motion in compensating the wave effect. By looking at the time history, one can observe that the AUV began surfacing after 60 seconds. The stern plane was commanded for the climbing operation. This can be seen in the cross-variance coefficient which reaches its maximum negative value[6] around 60 seconds (exclude the wave effect) and gradually tends to zero as time shift increases further.

[6] The signs for pitch and stern plane angles are based on the SNAME nomenclature.

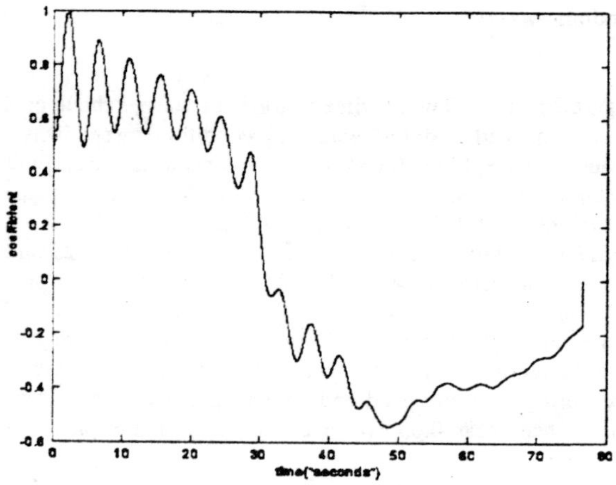

Figure 13 Cross-variance Function
Pitch vs. Stern plane

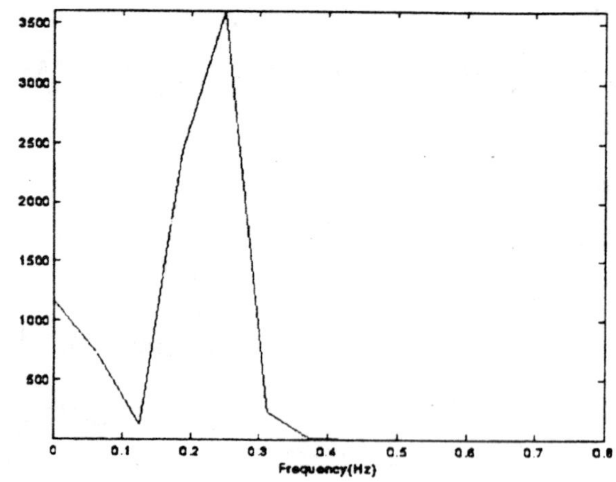

Figure 14 PSD of cross-variance function

4. CONCLUSION

One important goal in this controlled study is to build up a database for our OEX vehicle motion in shallow water with known disturbances which in turn can provide important insights into the vehicle self-motion for more complex ocean environments. From the results presented the dominant vehicle motion frequency appeared to be near the encountering frequency. It is important to note that the ability to collect and analyze the wave data at the MASK facilities enables us to properly evaluate the self- motion characteristics. One should bear in mind that the vehicle is less likely to respond to any controller as the wave frequency is increased, and a proper controller design must incorporate knowledge of both rejectable and non-rejectable disturbances. Our future work includes validating the vehicle's hydrodynamic coefficients with the self-motion data. Another piece of work including estimating the encountering frequency (AUV as wave buoy).

ACKNOWLEDGMENT

The authors would like to thank Grant Rossignol at the Carderock Division at the David Taylor Model Basin for his logistical support. This work is one of our current projects sponsored by the Office of Naval Research.

REFERENCES

An, P.E., Smith, S. M. (1996a). "An Experimental Study on Small AUV Self-Motion ", Internal Report, Dept. of Ocean Engineering, Florida Atlantic University.

An, P. E. *et al* (1996b). " A Quantitative Measure of Sea-State Effect on Small Autonomous Underwater Vehicle Motion in Shallow Water" , Oceanology International 96, Brighton, UK, pp.211-233.

Smith ,S. M. *et al* "The Ocean Explorer AUV: A Modular Platform For Coastal Oceanography" , UUST, Durham, New Hampshire, September.

Fossen, T. I. (1994). Guidance and Control of Ocean Vehicles, John Wiley Publishing.

Kinsman B. (1970). Wind Waves. Dover Publishing Press.

Peterson R.S., Nguyen T.C., Rodriguez R.R. (1994). "Motion Minimization of AUVs for Improved Imaging Sensor Performance Beneath a Seaway", AUV Proceedings, Boston, MA.

Tagaki T., Sugeno M. (1985). "Fuzzy Identification and its Applications to Modeling and Control", IEEE Trans. on SMC, Vol. 15, No. 1, pp.116-132.

Proceedings of the Seventh (1997) International Offshore and Polar Engineering Conference
Honolulu, USA, May 25-30, 1997

ISBN 1-880653-28-1 (Set); ISBN 1-880653-30-3 (Vol. II)

Exploring the Possibility of Placing Traditional Marine Vessels Under Oscillating Foil Propulsion

J. Czarnowski, R. Cleary and B. Kreamer
Massachusetts Institute of Technology
Cambridge, MA, USA

ABSTRACT

One hundred and fifty million years of evolution has enabled fish to develop extremely efficient technologies for aquatic propulsion and maneuvering. Recent experimental, MIT Robotuna (Barrett, 1996), and computational studies confirmed this. However, because these studies were based on models with undulating bodies, direct application to ships and submarines is not possible. Our goal was to apply the promising "fish" technologies to conventional marine vessels pragmatically. Success could mean the savings of millions of marine transport dollars and subsequent environmental damage, autonomous ocean research vessels covering more territory, and military submarines evading their foe because their "fish wake" is difficult to detect. An extensive literature review and discussions with marine hydrodynamics experts showed that efficient propulsion may be obtained by the oscillation of a foil without an undulating body. For example, large whales and other oscillating foil propelled animals requiring high efficiency employ the thunniform mode of swimming, which consists primarily of tail oscillation. Other animals, such as marine turtles or penguins, obtain efficient propulsion by pectoral flipper oscillation alone. In this spirit, a boat foil propulsion design was developed with two foils moving out of phase so that thrust would be the only net force and vertical foil orientation allowing for the possibility of thrust maneuvering with thrust vectoring. The challenge of actuating the foils with the appropriate "fish" motion was first met with a prototype mechanical device capable of oscillating two foils with limited independent adjustment of the foil's heave, pitch, frequency and phase between the motions. First-ever autonomous operation and wake visualization experiments were conducted and yielded useful results. The shortcomings of the first mechanism where met with a remote controlled four-degree of freedom robotic mechanism. Advanced mechanical design and state of the art electronics allow this 3.4 meter long oscillating foil propelled boat to produce accurate efficiency measurements over a wide range of operating parameters. Results from preliminary experiments are very encouraging.

KEYWORDS: Oscillating foil propulsion. swimming boats. biological propulsion.

INTRODUCTION

Evolutionary forces of survival combined with the boundless creativity provided by genetic mutation over geological time scales have resulted in swimming creatures developing a rich store of marine propulsion and maneuvering technologies. Primarily in this century, much work has been done to understand and apply some of these natural marine technologies. This quest for understanding was begun most notably by a biologist Gray whose experiments and computations led to the long standing Gray's Paradox (Gray. 1936). Claiming as much as seven fold disagreement between available power and required power for fish swimming. Gray's Paradox remained essentially unresolved for sixty years. Recent power and efficiency research performed with a robotic tuna (Barrett. 1996) has given new understanding to Gray's Paradox by providing concrete experimental results verifying the hypothesis that some forms of fish propulsion are highly efficient due to a combination of an undulating body and an oscillating caudal foil. However, because these studies were based on models with undulated bodies, direct application to ships and submarines is not currently feasible. Therefore, there exists a need to develop pragmatic applications of the promising "fish" technologies to conventional marine vessels. For example, converting only 3% of the United States shipping fleet to a propulsor with 10% higher efficiency would mean a monetary savings of $15.000.000 and 120 million fewer liters of petroleum fuel being burned (1992 U.S. Fuel and Shipping Statistics). Additionally, surface and submarine autonomous ocean research vessels whose current ranges are limited by available power supplies could go further and deeper than was previously possible. Finally, with a quieter wake, military submarines may go undetected longer. The biggest problem with applying fish swimming technology to a vessel or submarine is the need for a flexible, undulating body. Fish flesh, skin and bones are naturally flexible but the level of difficulty associated with engineering a flexible-hulled vessel capable of carrying cargo safely or people comfortably using current technologies is not economically feasible. If only the undulating body requirement could be removed without significantly compromising the efficiency, pragmatic application may be feasible.

Analytical and numerical techniques (Lighthill, 1975; Wu. 1961; Longvinovich. 1971; Cheng and Murillo. 1984; Karpouzian et al., 1990; McCune and Tavares. 1993), and experimental (Scherer.

1968; Delaurier and Harris, 1982; Lai et al., 1993; Triantafyllou and Barrett and Streitlien and Anderson, 1996) studies have shown that an oscillating foil will operate at overall propulsive efficiencies as high as eighty-five percent. Computational studies of oscillating foil propelled vessels have been made, (Faulkner, 1990; Yamaguchi and Bose, 1994) as well as some experimental attempts (Bowlus and Kelly and Siekmann, 1965; Tsutahara and Kimura, 1987). Computation by Yamaguchi and Bose reported 85% propulsive efficiency for a 200,000DWT Tanker under oscillating foil propulsion. Tsutahara and T. Kimura's experiment on a one meter long catamaran yielded a maximum of 75% efficiency, but employed the Weis-Fogh Mechanism which exploits the propulsive advantages of instantaneous fluid circulation about two foils that are forced together and rapidly pulled apart, instead of undergoing pure oscillation. Patents have been issued for oscillating foil propulsion devices (Gongwer's "Aqueon" Human Swimming Aid, 1965; Warner's Hand Powered Small Boat Propulsor, 1962), but no substantiated evaluation of their performance has been made. Widely used, but without patent, is the centuries old practice common in Asia of propelling small craft by hand oscillation of an oar extended aft of the stern. The thrust of the study described in this paper is to carry out a conclusive experimental investigation into the possibility of placing traditional marine vessels under oscillating foil propulsion, because, to date, one has not been performed.

PROBLEM DEFINITION

The following three objectives define the problem of exploring the possibility of placing traditional marine vessels under oscillating foil propulsion: (1) choose the most appropriate foil arrangement with respect to the vessel, (2) design, construct and test a proof of concept prototype capable of providing qualitative efficiency and wake mechanics information, (2) design, construct and test an advanced foil propulsion mechanism capable of yielding accurate propulsive efficiency and corresponding wake mechanics information over the relevant experimental parameter ranges.

As seen in Figure 1, a design of two vertically oriented foils placed aft of the transom was chosen. Two foils were chosen to eliminate the adverse yawing or pitching which occur in a one foil design. Vertical orientation was chosen to allow for the possibility of maneuvering with the propulsor. The foils were placed aft of the transom to eliminate complications in extending shafts through the underside of the vessel.

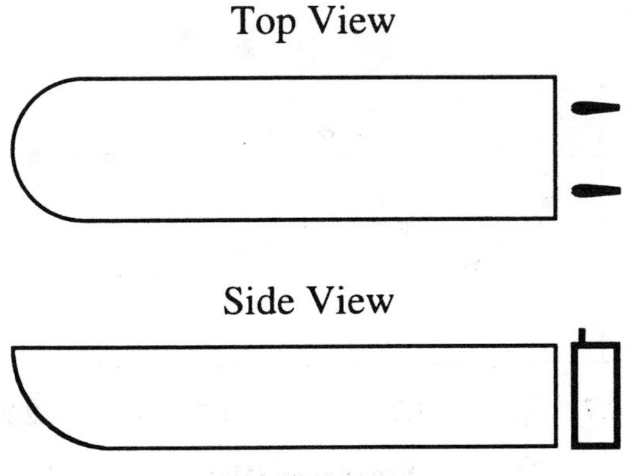

Figure 1 Chosen foil arrangement and orientation.

Figure 2 illustrates the parameters of the foil kinematics. The foils, of a particular cross-sectional shape, chord, c, aspect ratio, AR, and pitching location, b, undergo harmonic heaving, $h(t)$, and pitching, $\theta(t)$, with maximum values, h_o, and θ_o.

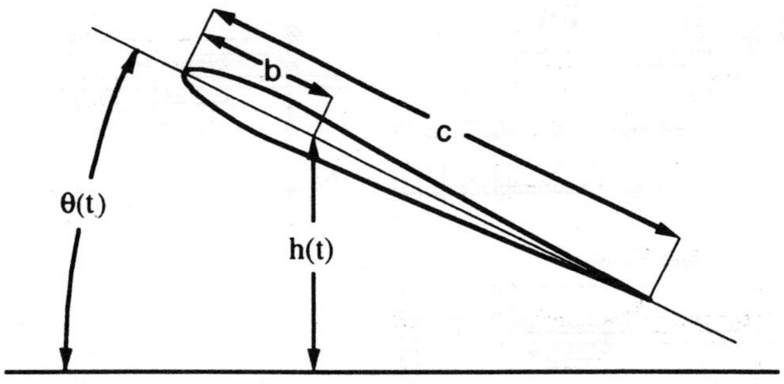

Figure 2 Foil kinematics parameters.

Both foils oscillate at frequency, f, and relative phase between heaving and pitching, φ. An average power, P, is required to produce harmonic heaving and pitching and subsequent average forward thrust, T. At a forward velocity, V, the boat will have hydrodynamic resistance, R. Operating free of tether, the resistance will equal the average thrust. Three important relations between the above parameters are:

- the heave amplitude to chord ratio, $h*$: $h* = h_o / c$
- the Strouhal Number, St: $St = f h_o / V$
- the propulsive efficiency, η_p: $\eta_p = VR / P$

The problem investigated experimentally was to determine how variations in the foil oscillation parameters affect the propulsive efficiency. The following sections describe the design, construction and testing of mechanisms capable of producing the required foil kinematics and of measuring the resulting forces and power consumption.

PROTOTYPE DESIGN AND EXPERIMENTS CONDUCTED

A model aircraft carrier of 2.2 meter length, 7.6 cm draft, 29 cm beam and rectangular transom was used as the test platform for the prototype oscillating foil mechanism. The mechanical design of the prototype, Figure 3, and final form, Figure 4, utilized a combination of four-bar linkages, slave arms, gears, pulleys and tendons to achieve harmonic pitching and heaving of the two NACA 0012, 7.6 cm chord, $AR = 1$, foils. The mechanism was constructed of aluminum and powered by one 12 Watt Pittman DC motor operating from a small lead acid battery. The mechanism allows $h*$ to be varied between zero and 0.5, θ_o from zero to 60 degrees and φ from zero to 90 degrees. f may be varied from zero to 2 Hz. The foils are constructed of painted Fiberglas laminated on a foam and wood substrate and are each equipped with three dye injection ports located along the leading edge. Do to the significant mechanical energy losses in the mechanism, a power measurement system on the prototype mechanism was not devised.

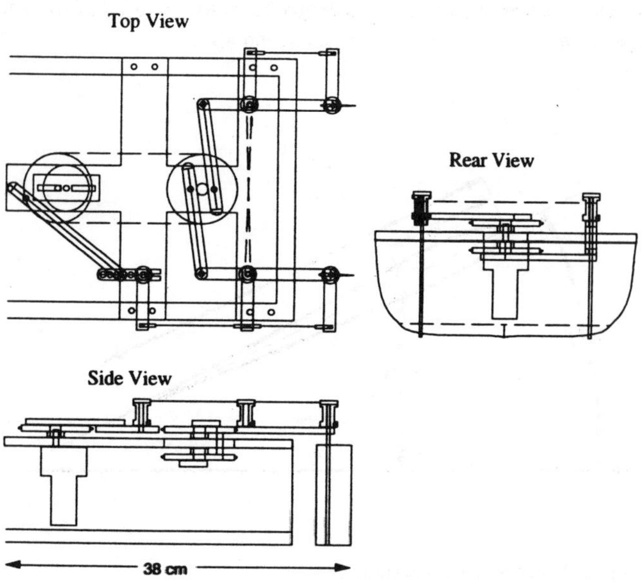

Figure 3 Prototype mechanical design CAD drawings.

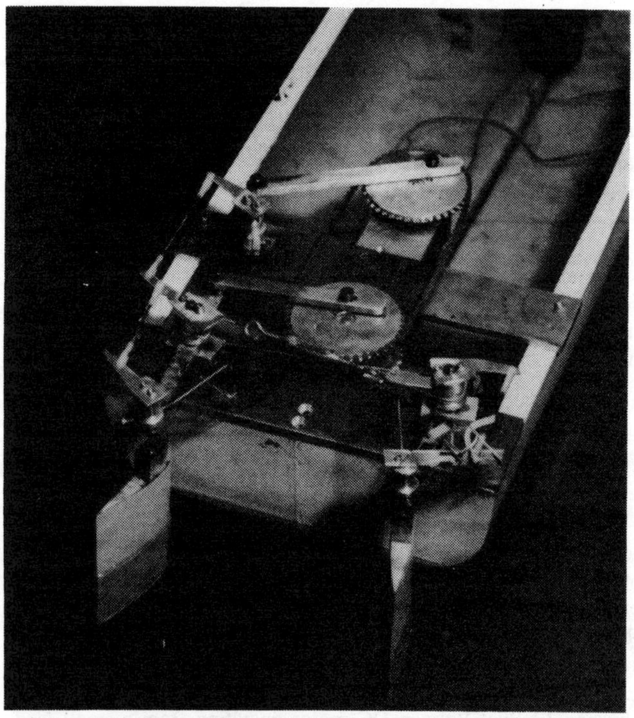
Figure 4 Photograph of prototype being tested on a model aircraft carrier.

Without being able to measure P, only qualitative efficiency experiments could be performed. Assuming that over a small range of frequencies (1 - 2 Hz) the mechanical power required to actuate the foils through one cycle is roughly constant, it is possible to gain an initial estimate of how the efficiencies of certain operating parameter combinations collate to each other by comparing the total number of oscillations it takes to move the model a certain distance. Experiments were conducted where $h*$ and θ_0 were held constant while f and subsequently St were varied. The model was un-tethered, set on a straight course and was allowed to reach equilibrium speed. It was timed and the number of oscillations were counted as the model traveled a known distance. The un-tethered model was kept on a straight course in a fresh water tank, (1.4 meter deep, 30 meter long, 2.6 meter wide), by running a thin line stretched tightly along the length of the tank through low friction eyelets attached at the bow and stern of the model, Figure 5.

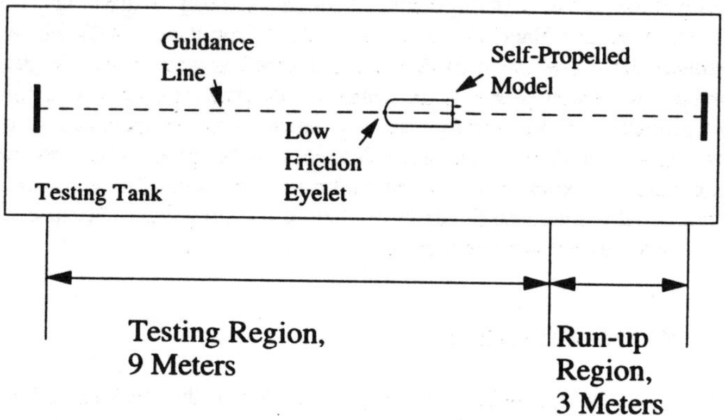

Figure 5 Prototype relative efficiency experimental setup.

While operating with $h* = 0.4$ and $\theta_0 = 25$ degrees, Figure 6 shows how f affects the relative efficiency (average number of cycles for all runs to cover the distance divided by the number of cycles to traverse the distance). The maximum relative efficiency occurred during the two runs with $f \sim 1.7$ Hz. Figure 7 illustrates the role of St in the relative efficiency. It is seen that the two runs of highest relative efficiency, ($f \sim 1.7$ Hz) occurred at $St \sim 0.23$, which is in agreement with previous work showing that high efficiency at similar $h*$ and θ_0 occur around $St = 0.25$ (Triantafyllou and Barrett and Streitlien and Anderson, 1996).

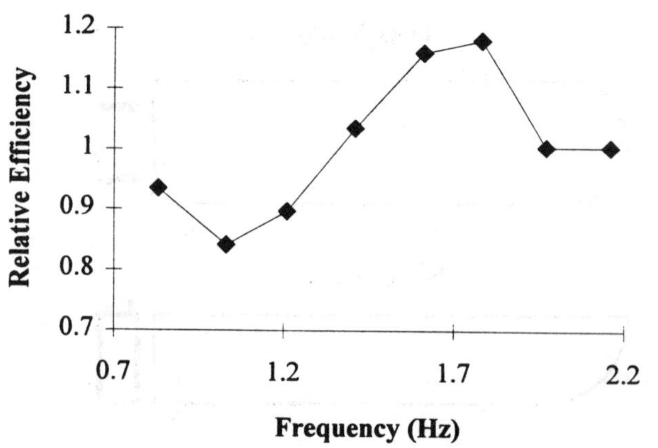

Figure 6 Dependence of relative efficiency on frequency.

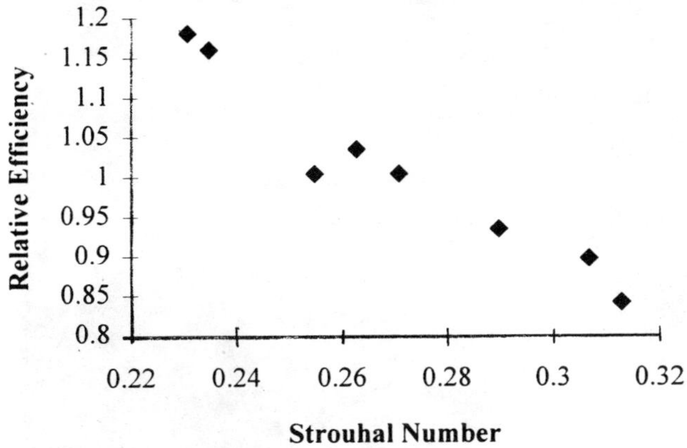

Figure 7 Dependence of relative efficiency on Strouhal Number.

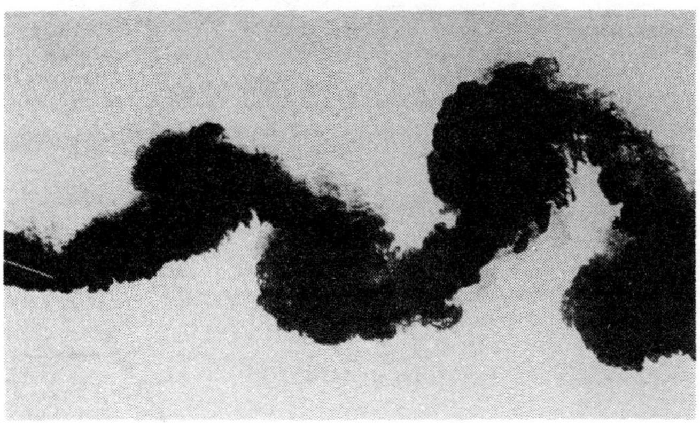

Robotic tuna wake.

Boat wake.

Figure 8 Comparison of dye visualization of the robotic tuna wake and the oscillating foil propelled prototype boat wake.

Using the same parameters as those initial experiments that yielded high relative efficiencies, wake visualizations were performed. In order to recreate the original un-tethered experiments while supplying the foils with pressurized dye and obtaining video recordings of the wake structure, the model was attached to a towing carriage. The carriage held a color CCD video camera viewing the wake region and supplied the foils with red food coloring under pressure. The experiments were conducted by setting $h*$, θ_0 and f to the original un-tethered values and then towing the model at the previously recorded speed. An interesting comparison is that of the boat wake to that of a robotic tuna, (Barrett, 1996), operating at a comparable values of *St*, Figure 8. It was encouraging to see the boat wake closely resembling its biological analogy. Another notable aspect of the wake structure is the interaction of the two parallel reverse Karman Streets. This interaction could not be observed in any previous single foil experiments.

The prototype mechanism fulfilled its role as a proof of concept model and provided some useful qualitative experimental results. However, due to the amplitude limits on $h*$, θ_0 and f, and, especially the lack of accurate power measurement, an advanced mechanism was needed.

ADVANCED MECHANISM DESIGN AND EXPERIMENTS CONDUCTED

The advanced mechanism assumed the same foil arrangement and orientation as the prototype, but had the requirements of accurately measuring its power consumption, allowing full adjustment of the operating parameters over a larger range, and being capable operating un-tethered under supervisory control. Refer to figures 9-11. A new, larger model, 1/42 scale Destroyer DDG-51 *Arleigh-Burke* of 3.4 meter length, 48.3 cm beam, 15.2 cm draft and nearly square transom was chosen for testing the advanced mechanism. The mechanism uses four independent computer-controlled motors to provide heave and pitch actuation for the two foils. Pulleys attached to the motors pull tendons which move linear slides supporting the foam/Fiberglas/wood NACA 0012, 7.6 cm chord, $AR = 2$, foils. Heave motion for each foil spans half the beam and θ_0 can vary between zero and 270 degrees. Two 250 Watt BEI brushless DC motors power the heave motion and allow for oscillations up to 4 Hz at maximum $h*$. The pitch motion is uncoupled from the heaving with telescoping universal joints and is powered by two 30 Watt Pittman brush DC motors. Copley amplifiers provide power to the pitch motors and BEI Controllers power the heave motors. HP optical encoders give motor positions to an MEI digital signal processor motion controller. A C program running on an embedded personal computer, Dell 486, operates the MEI card and a DAS8-JR data acquisition card. Electrical power into the motors is measured by multiplying the input voltage and current with analog multipliers. A Futaba radio control unit coupled with the PC-104 provides remote data acquisition and foil actuation control. Two 12 V lead acid automotive batteries and several DC-DC converters provide power for all components.

The propulsive efficiency can be determined in un-tethered experiments by measuring only the speed of the boat and its power consumption and having a priori knowledge of the boats resistance vs. velocity characteristics because it is known that the resistance equals the thrust at the testing velocity. Thus, the power being used by the mechanism multiplied by the overall propulsive efficiency will equal the resistance at that velocity multiplied by that velocity. The resistance versus velocity characteristics for the test model were obtained at the MIT Ship Model Resistance Towing Tank. Figure 12.

Figure 9 Photo of the DDG-51 destroyer model being tested in the MIT Towing Tank.

Figure 11 Advanced mechanism being tested on the model destroyer.

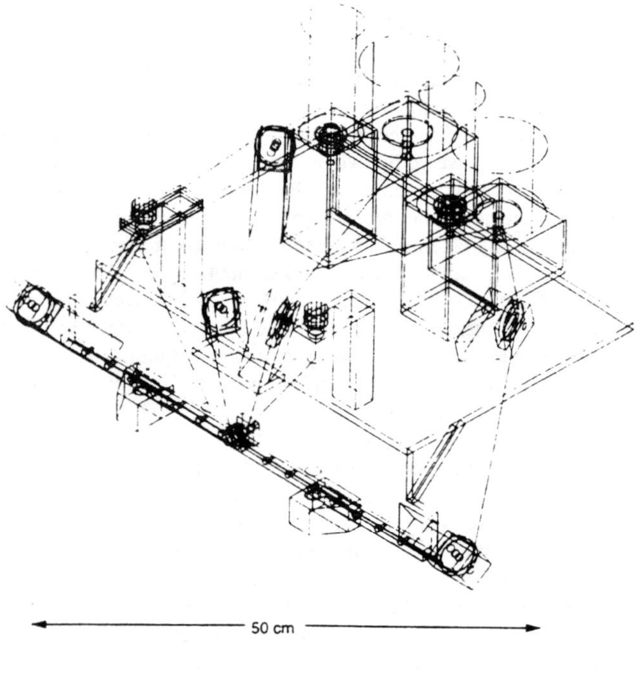

Figure 10 Advanced mechanism design CAD drawings.

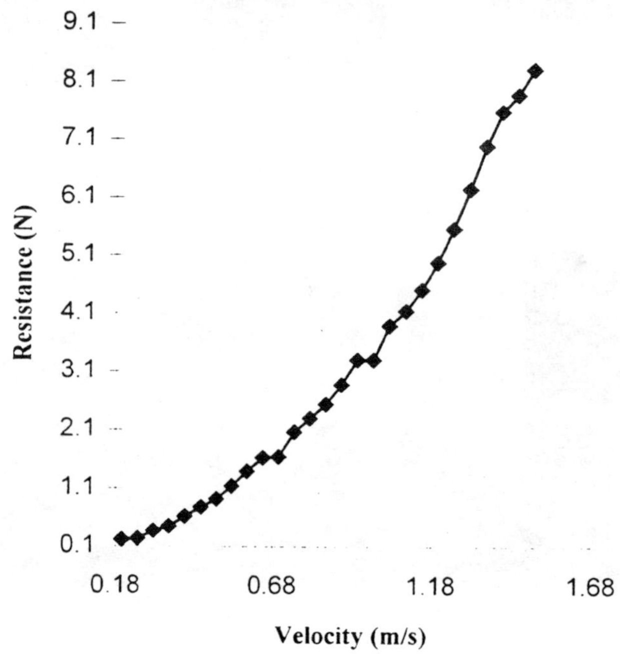

Figure 12 Drag vs. Velocity results for the destroyer model.

At this time, a few preliminary experiments have been conducted with the advanced mechanism. The model was placed in the large tank, the foils were actuated, the model speed was allowed to stabilize, and then power measurement was taken as the model was timed between two points. Preliminary tests at $h* = 0.75$, $\theta_o = 20$ degrees and $f = 2$ Hz gave and encouraging $\eta_p = 59\%$. Significant efficiency increase is expected as optimal operating parameters are determined through more experiments. Also, experimental speed and length of the preliminary runs were limited by the tank size.

CONCLUSIONS AND FUTURE WORK

The prototype mechanism produced high relative efficiencies at the appropriate Strouhal range (0.2-0.3) and possibly more significantly, is the high correlation observed between the boat wake and that of the robotic tuna. Experimentation has begun with the advanced mechanism and the preliminary results are encouraging with overall efficiency measurements around 60%. From here, the advanced mechanism will be tested extensively over a wide range of operating parameters in a large body of water. The wake of high efficiency runs will be visualized and studied. Finally, work to separate hydrodynamic inefficiencies from electrical and mechanical losses will be attempted to determine an upper bound on the overall hydrodynamic propulsive efficiency.

References

Barrett, DS (1996). "Propulsive Efficiency of a Flexible Hull Underwater Vehicle," *Ph.D. Thesis*, Massachusetts Institute of Technology, 1996.

Bose, N, and Yamaguchi, H (1994). "Oscillating foils for marine propulsion," *Proceedings of the Fourth International Offshore and Polar Engineering Conference*, pp 539.

Bose, N, Lien, J, and Ahia, J (1990). "Measurements of the bodies and flukes of several cetacean species," *Proc. R. Soc. Lond*, pp 163.

Bowlus, GH, Kelly, HR, and Siekmann, J (1965). "On the swimming of hinged hydrofoils," *Sonderabdruck aus - "Ingenieur-Archiv"*, pp 34.

Cheng, HK, and Murillo, LE (1984). "Lunate-tail swimming propulsion as a problem of curved lifting line in unsteady flow," *Journal of Fluid Mechanics*, pp 143.

Delaurier, JD and Harris, JM (1980). "Experimental Study of oscillating-wing propulsion," *American Institute of Aeronautics and Astronautics*, pp 368.

Faulkner, D (1990). "Ecological marine propulsion devices," *Proceedings of the International Symposium on Ecological Design*, pp 1.

Freymuth, P (1990). "Thrust generation of an airfoil in hover modes," *Experiments in Fluids*, pp 17.
Gongwer, CA (1965). "Oscillating foil swimming aid," *United States Patent.*.

Gray, J (1936). "Studies in Animal Locomotion," *Journal of Experimental Biology*, pp 192-199.

Hoppe, KG (1989). "The dynamo-elastic oscillation foil propeller," *Engineering*, pp 57.

Karpouzian, G, Spedding, G, and Cheng, HK (1990). "Lunate-tail swimming propulsion," *Journal of Fluid Mechanics*, pp 329-351.

Katz, J, and Weihs, D (1978). "Hydrodynamic propulsion by large amplitude oscillation of an airfoil with chordwise flexibility," *Journal of Fluid Mechanics*, pp 485.

Kelly, HR, Rentz, AW, and Seikmann, J (1963). "Experimental studies on the motion of a flexible hydrofoil," *Fluid Mechanics*, pp 30.

Kimura and Tsutahara, M (1987). "An application of the Weis-Fogh mechanism to ship propulsion," *Journal of Fluids Engineering*, pp 109.

Lai, PSK, Bose, N, and McGregor, RC (1993). "Wave propulsion from a flexible-armed, rigid-foil propulsor," *Marine Technology*, pp 30.

Lang, TG (1965). "Speed, power, and drag measurements of dolphins and porpoises," *San Diego Naval Undersea Center Report*, pp 553.
Lighthill, J (1975). *Mathematical Biofluiddynamics*, SIAM.

Liu, P, and Bose, N (1993). "Propulsive performance of three naturally occurring oscillating propeller planforms," *Ocean Engineering*, pp 57.

Longvinovich, GV (1971). "Hydrodynamics of a thin, flexible body," *Hydrodynamic problems in bionics*, pp. 4.

McCune, JE, and Tavares, TS (1993). "Perspective: Unsteady wing theory - The Karman/Sears legacy," *Journal of Fluids Engineering*, pp 115.

Oritsland, TA (1991). "Foil-propelled boat," *Human Power*. pp 3.

Scherer, JO (1968). "Experimental and theoretical investigation of large amplitude oscillating foil propulsion systems," *United States Army Engineering research and Development Laboratories*.

Streitlien, K, and Triantafyllou, MS (1995). "Force and moment on a joukowski profile in the presence of point vortices," *American Institute of Aeronautics and Astronautics*, pp 603.

Triantafyllou, MS, and Barrett, DS, Streitlien, K, and Anderson, JM (1996). "Flapping foils of High Propulsive Efficiency,".

Triantafyllou, MS, Triantafyllou, GS, and Gopalkrishnan, R (1991). "Wake mechanics for thrust generation in oscillation foils," *Phys. Fluids A*, pp 2835.

Uldrick, JP (1968). "On the propulsion efficiency of swimming flexible hydrofoils of finite thickness," *Journal of Fluid Mechanics*, pp 29.

Warner, NS (1963). "Boat Propulsion and Steering Device," *United States Patent*.

Wu, TY (1961). "Hydromechanics of swimming propulsion," *Journal of Fluid Mechanics*, pp 10.

Proceedings of the Seventh (1997) International Offshore and Polar Engineering Conference
Honolulu, USA, May 25-30, 1997

ISBN 1-880653-28-1 (Set); ISBN 1-880653-30-3 (Vol. II)

Concept Design of a Flexible-Hull Unmanned Undersea Vehicle

Jamie M. Anderson
Charles Stark Draper Laboratory
Michael S. Triantafyllou
Massachusetts Institute of Technology
Peter A. Kerrebrock
Charles Stark Draper Laboratory

Cambridge, MA, USA

ABSTRACT

In recent years, research in the propulsion and maneuvering mechanisms used by fish has demonstrated the utility of bio-propulsion for use on undersea vehicles. Despite recent advances in unmanned undersea vehicle (UUV) technology, little progress has been made in improving propulsive efficiency and maneuverability. Most underwater vehicle designs employ a conventional propeller as the main propulsor and shrouded thrusters and/or control fins for maneuvering. Two types of vehicle designs are prevalent: torpedo shaped bodies streamlined for speed and range, or box-shaped bodies designed for maneuvering and station keeping. Unfortunately, most future UUV missions require all of these capabilities: high transit speed, long range/duration, maneuverability and station keeping ability. Thus we look to fish as a potentially optimal UUV design in that they are able to cruise great distances at significant speed, maneuver in tight spaces and accelerate and decelerate quickly.

This paper summarizes the relevant design issues and current work in the development of a flexible-hull UUV which propels and maneuvers like a fish. Following the morphology and kinematics of a yellowfin tuna, the Charles Stark Draper Laboratory Vorticity Control UUV (VCUUV) will be the first demonstration of a freely swimming Thunniform (tuna-like motion) robotic vehicle. Simulation of the required kinematics and loads indicate that Thunniform motion can be actuated with a rigid forebody comprising 60% of the total vehicle length and four rigid links actuating the tail section and caudal fin. Three different actuation concepts were compared by analyzing possible components and arrangements to attain the required loads and mission duration. A recirculating hydraulics concept was chosen for prototyping due to the versatility of the design for study of a variety of swimming speeds and maneuvers.

KEY WORDS

Fish swimming, vorticity control, tuna, unmanned undersea vehicle (UUV).

INTRODUCTION

Improvements in propulsive efficiency and maneuverability of unmanned undersea vehicles (UUV's) are necessary to achieve the challenging new missions in littoral zone warfare, mine reconnaissance and scientific missions in cluttered environments. Increasingly sophisticated (and power intensive) mission payloads, and the need for greater range and endurance compete for finite energy supplies on board UUV's. Improved propulsive efficiency and energy capacity will directly increase mission time and reduce the launch and recovery overhead. More work may be done in a vehicle deployment: a larger area may be surveyed, or perhaps more time may be spent in one location studying a site of interest. Even modest increases in mission duration will directly improve mission performance.

Maneuverability has been a particularly vexing problem for underwater vehicle designers. There are two common actuators for

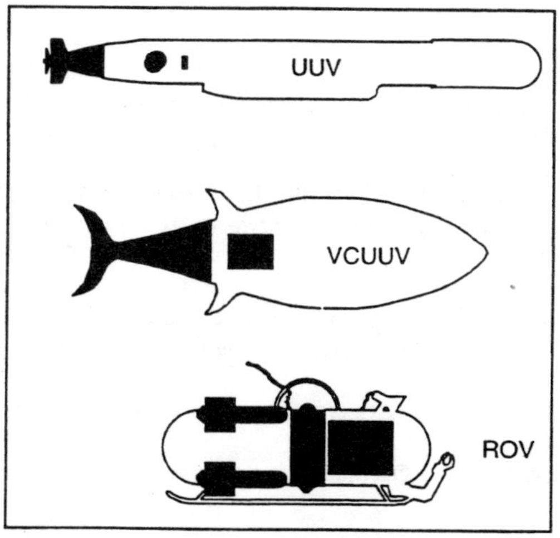

Figure 1: Underwater vehicle comparison. The shaded portion of each vehicle indicates the propulsion and maneuvering plant displacement: UUV (without cross axis thrusters) 4.5%, VCUUV 20%, ROV 50%.

vehicle control: thrusters (either shrouded or tunnel) usually implemented on remotely operated vehicles (ROVs), and lifting surfaces (tail fins or canards) used on torpedoes and similarly shaped vehicles. Thrusters operate well under static conditions (bollard thrust), but degrade rapidly in performance as vehicle speed increases. Flow across the axis of a thruster significantly reduces the magnitude, and can drastically change (even reverse) the direction of thrust. Additionally, tunnel thrusters exhibit significant time lag when reversed due to entrained water in the tunnel itself.

Lifting surfaces fail at low speed due to the dependence of lift force on the square of the vehicle speed. At low speeds, little lift force is generated and there is a time lag in force generation due to the Wagner effect. Below a critical speed control authority is lost; thus, in an inspection or mapping mission a vehicle using control surfaces would not be able to slow down and closely inspect a target. For instance, in a mine reconnaissance mission, an important capability is close inspection without fear of collision or surface capture due to second order wave effects. A common approach to designing a hydrodynamically efficient autonomous vehicle that can also maneuver at low speed is to have a torpedo shaped body with lifting surfaces, and cross axis thrusters. However, it has been shown that this approach has a detrimental effect on vehicle range, since the weight and lost displacement attributed to the thrusters can be a significant fraction of the vehicle energy supply. In some design studies where vehicle envelope displacement is highly constrained (such as will fit in a submarine torpedo tube), the addition of cross axis thrusters can only be achieved if the energy supply is removed entirely.

Figure 1 shows the vehicle displacement dedicated to the propulsion and maneuvering plant for conventional UUV's, ROV's and our new concept, the Vorticity Control UUV (VCUUV), an autonomous underwater vehicle that has the form of a tuna and propels with tuna-like swimming motion. The VCUUV will be capable of long transits like a UUV and high maneuverability like an ROV, with modest displacement cost to the vehicle (approximately 20-30% of the envelope displacement).

Fish have optimized their interaction with water through millions of years of evolution. Although there are morphological and "mission" differences between species, many are optimized for swimming steadily. For instance, many tunas, sharks and mammals cruise continuously, performing remarkably long transits across oceans. Although these animals have distinctly different evolutionary histories, they swim notably similarly. Triantafyllou and Triantafyllou (1995) observed that when the swimming speed, amplitude of tail motion and frequency of tail beat are considered together in the nondimensional Strouhal number, most fish swim within the same optimal range, $0.25 < St < 0.35$. The Strouhal number is defined analogously to the case of vortex shedding behind a bluff body:

$$St = \frac{fA}{U} \qquad \text{Equation (1)}$$

where f is the frequency of the tail beat, A is the tip to tip excursion of the tail fin and U is the swimming speed.

In recent years, researchers at the Massachusetts Institute of Technology have investigated rigid flapping foils as a means of generating large propulsion forces at very high efficiencies. This research has culminated in the development of *Robotuna* a full scale model of one of the most energy efficient marine distance swimmers in the ocean (Triantafyllou and Triantafyllou, 1995). The *Robotuna* has successfully demonstrated very high propulsive efficiency, the ability to generate large propulsive forces and apparent drag reduction (Barrett, 1996). The MIT research has shown that the propulsion and maneuvering performance of fishes is related to their ability to control their wake vorticity. Contrary to drag producing bodies, fish generate an oscillatory wake comprised of alternating vortices arranged in a jet pattern. Manipulation of wake vorticity appears to be a dominant factor affecting the propulsive performance of fish (Anderson, 1996).

At the Charles Stark Draper Laboratory, the next generation of flexible hull robots is in development. The Vorticity Control Unmanned Undersea Vehicle (VCUUV) will be a freely swimming, tuna-shaped vehicle which demonstrates tuna-like swimming and maneuvering performance. The vehicle is intended as a technology demonstration of the propulsion and maneuvering system, not as a fully autonomous ocean going vehicle. Thus, the vehicle sensor suite and control algorithms will be simplified such than the vehicle can perform basic maneuvers required to demonstrate the technology. The vehicle will be operated in shallow, controlled environments where the propulsion system performance can be quantified.

BACKGROUND

Much of our background knowledge and confidence that a tuna shaped UUV will perform well stems from the MIT *Robotuna* project. *Robotuna* is a 49 inch long biologically inspired animated tow tank model built to study swimming efficiency and how it relates to body kinematics (Barrett, 1996). The body shape and flexibility mimics the yellowfin tuna, a fast swimming pelagic fish similar in shape and kinematics to the renowned bluefin tuna. *Robotuna* was meant to be flexible in that the design allows extreme body motion combinations well beyond the capabilities of a real tuna, enabling complete access to the swimming parameter space. The robot is comprised of six cable driven links which can be independently actuated. Each link is driven by a cable which runs via pulleys through the body and mast to a servomotor mounted out of the water on the tow tank carriage. The linkage mechanism was designed with high attention to detail to reduce friction and backlash. The outer skin of the tuna is a composite assembly of lycra-spandex, latex and foam which acts as an elastic spring during oscillation.

The *Robotuna* was exercised in the MIT Ocean Engineering Testing Tank by prescribing a set of kinematic parameters (angular deflections of each joint, tow speed, phase relationships) and measuring the net power transmitted to the linkages and the reaction force between the tuna and carriage. Optimal motion was defined as that set of motion parameters that produced no net force on the carriage (self propelled status) with minimum energy input. Because of the large parameter space, a genetic algorithm was employed to assist with the search for optimal motion. "Genes" were applied to each independent kinematic variable and generations of fish containing these "genes" were tested and their "fitness" evaluated. Then, the best performers were rewarded with "offspring" which further evolved (procreation and mutation), eventually producing the self propelled optimum motion (SPOM). Interestingly, the SPOM closely matches the observations of actual yellowfin tuna motions by Dewar and Graham (1994).

The *Robotuna* project has demonstrated that significant drag reduction can be achieved even without careful tuning of the drive mechanism. A measure which quantifies the drag reduction during swimming is the thrust power ratio:

$$TPR = \frac{DU}{P_{in}} \qquad \text{Equation (2)}$$

where D is the drag of the "dead fish" (while held straight, not swimming), U is the speed of the towing sled and P_{in} is the mean power to oscillate the mechanism. If the thrust power ratio is greater than one, the "live fish" drag is less than the "dead fish" drag. In the self-propelled *Robotuna* experiments, the thrust power ratio achieved values in excess of one (1.2-1.4) with apparent drag reductions near

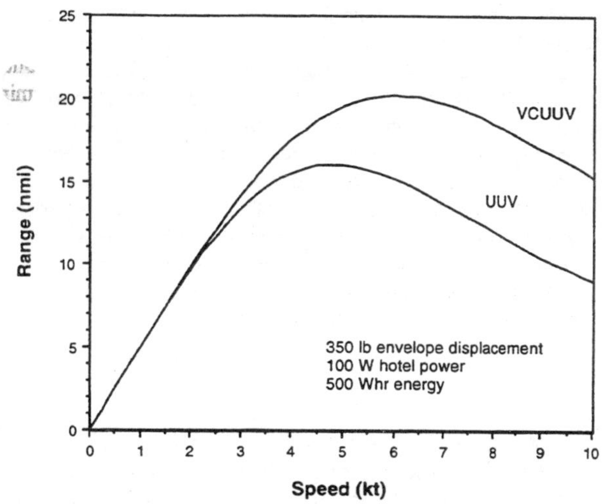

Figure 2: Speed (knots) and range (nautical miles) comparison for conventional UUV and VCUUV

50%. For comparison, the thrust power ratio for the Draper built DARPA UUV is between 0.7 and 0.85.

The pronounced effect of drag reduction on vehicle mission is illustrated in Figure 2. For a given vehicle displacement, hotel power and battery capacity, the range is plotted as a function of speed for a conventional UUV and the VCUUV which is assumed to have half the drag of the former. The maximum range and optimal speed both increases by 25% for a 50% reduction in the drag. Although this is a simplistic model, the benefits of reduced drag are readily apparent.

KINEMATICS OF THUNNIFORM SWIMMING

Self propelled optimal motion (SPOM) for the *Robotuna* is a good starting point for the VCUUV. The genetically evolved SPOM closely matches biological observations for live, naturally swimming tuna. The key parameters are the vehicle speed, oscillation frequency and amplitude, propulsive wavelength, shape parameters which describe the envelope of transverse motion, tail phasing and angle of attack.

The necessary kinematics for straight swimming are described by the following equations:

$$y(x,t) = a(x)\sin(kx - \omega t) \qquad \text{Equation (3)}$$

$$a(x) = c_1 x + c_2 x^2 \qquad \text{Equation (4)}$$

$$\theta_{tail}(t) = \theta_0 \sin(\omega t + \varphi) \qquad \text{Equation (5)}$$

where $y(x,t)$ is the lateral body deflection at a given x location along the body. The lateral deflection is a traveling wave which grows in amplitude from head to tail according to the shape envelope $a(x)$. The tail is articulated independently of the body motion with unique pitch amplitude θ_{tail} and phase relationship φ to the body motion. The independent kinematic variables are listed in Table 1 along with their values for self propelled optimal motion (Barrett, 1996).

Figure 4(a) shows the backbone envelope for the parameters listed in Table 1. A vertical line drawn at $\frac{x}{L} = 0.65$ shows one proposed cut off point for tail oscillation for the VCUUV where the upstream portion of the body remains laterally rigid. The effect that this has on propulsive efficiency is unknown at this time but is expected to be minimal. Preliminary sensitivity experiments with the *Robotuna* have indicated that swimming efficiency is relatively insensitive to parameter variation near the SPOM. The sensitivity of the kinematics will be further investigated during experiments in the Winter of 1996-7. Figure 4(b) shows the actual body shapes and movement for the backbone envelope of Figure 4(a). A small amount of yawing motion at the nose shows the rigid body recoil due to the tail motions and is a consequence of the tuna apparatus which has a fixed pivot point. Real tunas yaw slightly (snout tip ± 2 - 5% of body length) but the effect is only significant at low speeds (Dewar and Graham, 1994).

PRELIMINARY VEHICLE DESIGN

Morphology

Nature has demonstrated that a vehicle shaped like a tuna can swim efficiently at significant speed and maneuver well. It is our task to recreate this ability in a robotic vehicle. In building a UUV that is propelled and maneuvered by tuna-like motion it is important to keep in mind that a tuna and the VCUUV will ultimately have far more differences than similarities. This is inevitable, since the VCUUV is a mechanism that cannot possibly replicate all characteristics of a fish, and the VCUUV will be burdened with a mission and equipment that are uniquely human in purpose and design. Designing a successful VCUUV requires application of the same process of system optimization as for conventional UUV's. The challenge comes in gleaning the essential characteristics of fish propulsion and maneuvering from the natural model, without allowing them to dominate the design of the UUV to the detriment of mission performance.

Our primary design concern is the shape of the hull envelope. Although it may be possible to propel a torpedo shaped body with a tuna tail, we do not know how the hull shape affects the stability of the vehicle in different operating conditions. Thus, in this first attempt at an autonomous tuna robot, we will adhere to the shape of an actual tuna. Tunas are very streamlined with a nearly elliptical fusiform hull

Table 1: Swimming kinematic variables for Robotuna self propelled optimum motion from Barrett (1996)

Variable	Description	Value
U	Swimming speed in body lengths per second (held constant for *Robotuna*)	0.65 L/s †
St	Strouhal number = $\frac{fA_{TIP}}{U}$	0.156
A_{TIP}	Peak to peak amplitude of caudal fin tip in body lengths	0.115 L
α	Maximum angle of attack of caudal fin during swimming cycle	16.2°
φ	Phase lead of caudal fin motion with respect to lateral body motion	95°
λ	Propulsive wavelength in body lengths	1.27 L
c_1	Coefficient of linear term in Equation (4)	0.00372
$c_2 L$	Coefficient of quadratic term in Equation (4)	0.203

† The body length L is the precaudal length (length from the nose to the point where the tail fin begins).

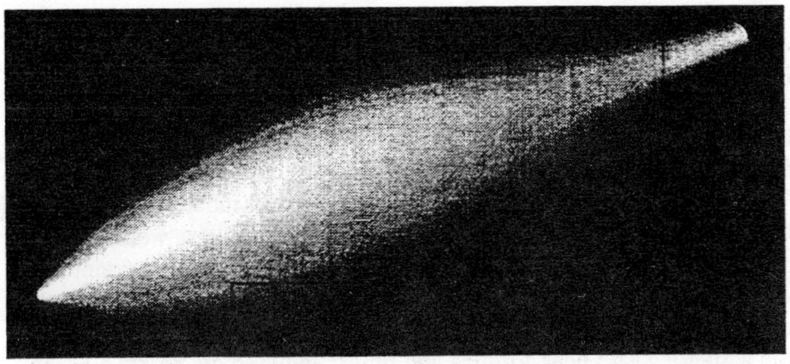

Figure 3: Tuna body envelope after casting and measurement.

shape well suited for packaging vehicle components. The *Robotuna* hull closely follows lines obtained from a taxidermist's model of a yellowfin tuna, with an elliptical curve fit on the individual cross sections. The VCUUV outer envelope will follow the shape of an actual yellowfin tuna measured at Draper Laboratory.

We obtained a freshly caught 34 lb (151 N), 40 inch (1 m) long yellowfin tuna that was shipped on ice directly after capture with minimal damage and the gut cavity intact. It was immediately frozen while hanging to straighten the backbone. The fish was then thawed, cast in plaster, dissected and disposed of (except for the fins which were retained to be measured). The plaster mold was used to create the urethane cast which was measured at the CSDL machine shop using a

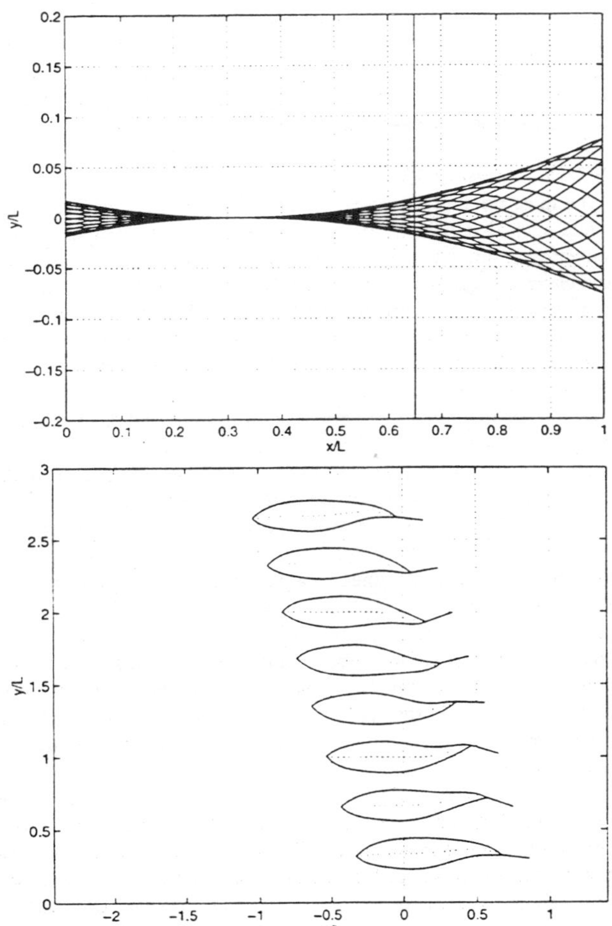

Figure 4: (a) Backbone envelope for self propelled optimal motion, (b) Body shape and movement for parameters in (a).

numerically controlled milling machine. The surface data were corrected, smoothed, scaled and input to ProEngineer™, a solid modeling computer aided design program. Figure 3 shows the tuna solid model which is used to estimate true envelope displacement and centers of gravity and buoyancy while the vehicle arrangements evolve.

It is important to note that as a general rule, the VCUUV vehicle design will adhere to the measured tuna morphology. Several finer features of the tuna will not be duplicated in the initial design. They are considered refinements that may be added in the future to enhance performance. They are: caudal fin flexibility, fully articulated (sweep and pitch controlled) pectoral fins, retractable first dorsal fin, finlets near the caudal fin and the fleshy penducular keel. However, we will include primitive (pitch only) pectoral fins, stiff second dorsal and anal fin and a stiff caudal fin.

Vehicle Size

Overall vehicle size is an important design constraint which was addressed early in the project. Issues relevant to vehicle sizing are: ease of packaging (enough buoyancy to support required actuators, batteries, *etc.*), size of the testing tank, material costs and ease of handling (movement within CSDL, fitting into elevators, fitting into a van or truck, *etc.*). Our primary concern was that the vehicle must be large enough to accommodate commercial off-the-shelf (COTS) actuators and electronics. Although COTS products tend to be costly, they have recently become "plug and play" components that can be installed quickly with minimal customization. The time constraint of our project to two years requires that we proceed with readily available items. Thus, we placed an upper limit of eight feet (2.4 m) overall length which corresponds to approximately 450 lb (2000 N) envelope displacement. For comparison, the MIT *Robotuna* is 4.1 feet (1.2 m) in length and approximately 60 lb (267 N) displacement. Preliminary VCUUV arrangements estimates indicate that the actual buoyant displacement will be approximately 300 lb (1334 N). At present, the preliminary design carries a 29 % weight margin which can be used for mission payload or energy supply.

Number of Articulated Links

The VCUUV design follows the same methodology used in *Robotuna*: a planar kinematic chain of independently operated rigid links. However, unlike the *Robotuna*, the VCUUV will have much of the forward body (60 % of the total body length) as a rigid pressure hull which will house dry electronics, batteries and some propulsion system components. The important issue here is the number and location of the link pivot points. *Robotuna* has six actively operated links and two passive links arranged in cosine spacing with increased resolution towards the tail. This design is quite conservative in that body movements beyond those of a typical tuna are possible. The VCUUV will be able to flex aft of the rigid pressure hull only, limiting the kinematics to Thunniform (localized small tail movements as in tunas and mackerels) motion only.

As a design tool in our linkage selection process, we developed a simulation of the traveling wave equations for steady swimming and body shape envelope which allows for arbitrary linkage number and locations. The number and lengths of links are input to the program and the link motions are computed to match the desired kinematics of Equations (3-5) with the *Robotuna* SPOM parameters. Figure 5 shows the simulation for three evenly spaced links plus the independently articulated tail. The solid line shows the vehicle hull which follows the linkage movement and the dashed line is the desired hull motion. Three links and the caudal fin appear to be sufficient for steady swimming. Despite the low number of independent links, very smooth hull motions are easily attained when the articulated region also includes a flexible spline element.

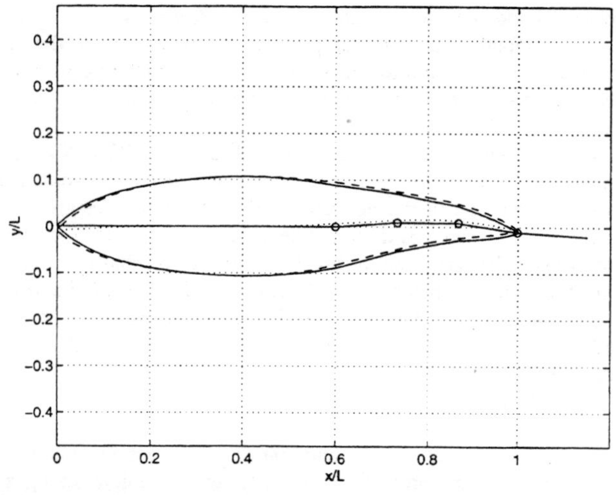

Figure 5: VCUUV linkage with three evenly spaced links plus the caudal fin. The solid hull outline represents the hull following the links. The dashed outline is the desired hull motion. Circles represent linkage pivot points.

Estimated Input Torque and Power

The linkage simulation was also used to estimate the maximum joint torques and power that are required to swim steadily. A strip theory model was developed which accounts for each link's inertia, the added inertia due to entrained water outside of the vehicle envelope and viscous drag force on each differential segment. The strip theory model provides a conservative estimate of the input torques and power. Several effects are unaccounted for such as the effect of the vehicle forward speed on the local velocity at each link, and the effect of imposed movement of forward links on any given link. Figure 6 shows the torque and power for each joint and the total instantaneous power for the parameters given in Table 1. The total power requirements are quite modest with net peak to peak power of ± 10 Watts. This power value is consistent with the predicted surface drag for a rigid tuna, 8 feet (2.4 m) overall, swimming steadily at 2.6 knots

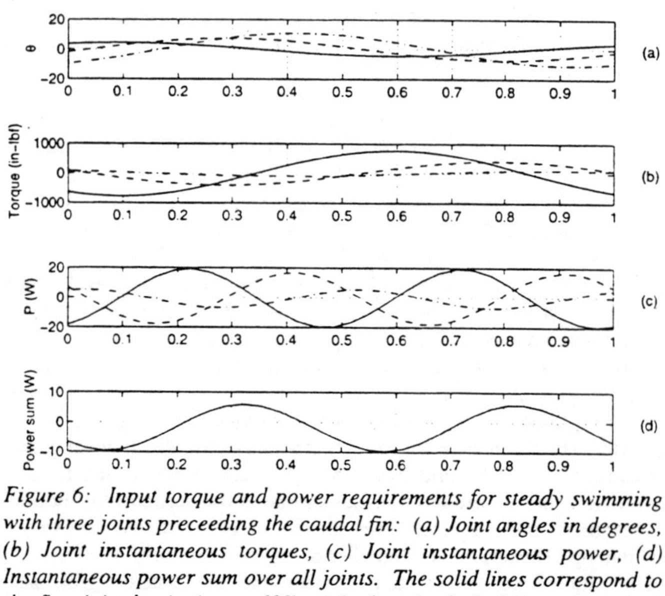

Figure 6: Input torque and power requirements for steady swimming with three joints preceeding the caudal fin: (a) Joint angles in degrees, (b) Joint instantaneous torques, (c) Joint instantaneous power, (d) Instantaneous power sum over all joints. The solid lines correspond to the first joint beginning at 60% of the length, dashed lines the second joint and dot-dash lines the last joint just before the tail.

(1.3 m/s). Input power to the caudal fin is estimated to be approximately 16 W using experimental data for 2-D flapping foils (Triantafyllou *et al.*, 1997).

For the most conservative sizing, each link actuator must be able to move all sections aft of the link pivot at the displacement and frequency for the design speed. This loading corresponds to a typical "power up" test where each link would be tested independently. This case is extremely conservative as the "power up" test would not be done at the swimming frequency; however, this case is important as it would help us size the actuators for sudden failure. For this case, the torques required are much higher due to the increased inertial load on the actuator. The peak torques and power required are nearly 5000 in-lb and 150 W for the first joint, 1000 in-lb and 50 W for the second joint and 175 in-lb and 15 W for the last joint (not including the torque required by the tail). Although these values are much higher than the steady swimming loads, they do illustrate the need for failure prevention/safety systems as very large loads may occur during rapid starting or failure of the mechanism.

Buoyancy and Stability

In order that the vehicle remain stable during steady swimming and turning, it is important that the articulated caudal region be nearly neutrally buoyant. This is to ensure that the movement of caudal region does not shift excess buoyancy or weight off the center axis of the vehicle, generating a rolling moment. Thus, we are striving to keep the caudal region free flooded and as near neutrally buoyant as possible. In the interests of time and cost, the vehicle will not include a sophisticated variable buoyancy control system. Instead, the buoyancy will be adjusted manually to near neutral and then adjusted while swimming with a bias angle applied to the pectoral fins. Depth control will be managed with pectoral fin pitch which requires that the center of gravity and buoyancy lie near the pectoral fin axis.

DESIGN CANDIDATES

The most notable characteristic of a fish-like UUV is that it must articulate such that fish motion can be replicated and propulsion and maneuvering achieved. This difference drives the propulsion and maneuvering mechanism of the VCUUV to be a larger subsystem than its counterparts in a conventional UUV but less than for a conventional ROV as shown in Figure 1. Such an allocation of vehicle weight and volume must be more than offset by gains in propulsion and maneuvering efficiency to benefit the vehicle overall.

We have considered three articulation design candidates in detail: a tendon/servomotor drive similar to the *Robotuna*, a piezo/beam resonant drive and a recirculating hydraulics drive. Resonance of the mechanical structure is an important issue that depends on the intended use of the vehicle. One concern with the selection of actuators that are sized for a resonant mechanical linkage (that is, one with elastic storage and mass elements tuned to a particular oscillation frequency), is that it may prove impossible to generate enough force to turn. Turning requires a bias curvature of the entire aft linkage which necessitates actuation authority beyond that required to get the vehicle swimming in a straight line. Since our primary objectives of this work are to create a vehicle that can maneuver as well as swim, resonant structures are not appealing unless there is a secondary system to actuate the turning curvature.

One significant benefit of a resonant mechanical linkage is the high efficiency between the output work and the input power. Such a system would be able to essentially close the loop around the mechanical system power and the output work. On the *Robotuna*, only the actual mechanical power input to the vehicle was measured, not the power input to the servomotors. We will employ a similar methodology in measuring system efficiency; that is, the actual power input to the vehicle (link force times velocity) will be quantified and compared to the output work (steady drag times swimming speed).

Our design employs a conservative approach with an actuator system that can apply more force than predicted so that we may best demonstrate the technology with combined swimming and turning maneuvers.

The first concept considered was a tendon/servomotor drive for which the major design issue is how to efficiently convert rotary motion (assuming electric motors are used) to oscillatory motion. The approach was demonstrated to work well in Robotuna, however, here a fundamental limitation to efficiency can be found. Electric motors peak at only one point in the speed-torque curve. Unfortunately, as shown in Figure 6, peak torque is required at the moment in time when a link is at the limit of its travel, and the motor is stalled. A short time later when the link is traveling at maximum velocity the applied torque is zero, and the motor is running at no-load speed. At both times the motor efficiency has reached zero, having passed through maximum efficiency in between. The motor will cycle from zero to maximum efficiency and back to zero four times each tail beat, resulting in a low average motor efficiency. Additional losses occur downstream of the motor. Since it was deemed impractical to have tendons penetrating the hull, the motor/gearboxes were located in the free flooded tail section, placing sever constraints on their overall size. The result was small high speed motors driving multistage gearboxes, with high losses. Additional concerns, including the need for a dedicated controller for each motor, an unfavorable weight distribution, and cost lead us to seek alternative approaches.

The second design concept uses piezo polymers or ceramics to effect resonant bending of a beam that makes up the spine of the articulated segments. An array of piezo elements attached to the surface of a beam would generate a strain wave to bend the beam to match the kinematic requirements. This concept uses no mechanical components and has no rotating parts. Issues include the power density and strain of piezo materials, high voltages required to drive them, and efficiency. Although this design scheme (beam) differs greatly from the other designs employing rigid links, it could prove useful as the vehicle could potentially be much smaller since the actuators are small and lightweight. Another advantage is that the piezo elements can be used in reverse as displacement sensors. However, there are serious issues that must be resolved including the high voltage requirement, the possibility of damage due to overstrain, how the caudal fin would be articulated, and how turning would be effected (it is likely that the piezos do not have enough energy density to apply bias curvature to the beam). The piezo-beam design concept was abandoned because of the risks associated with implementing a relatively new technology.

The third concept uses one motor/pump assembly to distribute power through an hydraulic system which replaces the sheaves and tendons of the previous concept with hydraulic cylinders driving each segment. Hoses connect each cylinder to a servovalve fed from a high pressure accumulator and hydraulic pump in the pressure hull (Figure 7). This approach eliminates the problems associated with cable tendons, but comes with its own maintenance problems, such as leakage through dynamic seals and bending fatigue of the hoses. The hydraulic design has the advantage that the motor runs more nearly at a constant load (assuming fish swim with a somewhat constant expenditure of power) but has the disadvantage of requiring a pump and proportional servo valves which have inherent losses that decrease overall system efficiency.

Figure 8 shows preliminary wet arrangements in the tail section of the vehicle. The design is very compact along the centerline of the tail structure which allows the tail skin support structure to be easily removed for mechanism maintenance. Required maneuvering curvatures are easily attained with the configuration as shown.

The recirculating hydraulic system has the benefit that it is the only design capable of operating off the design point (estimated loads)

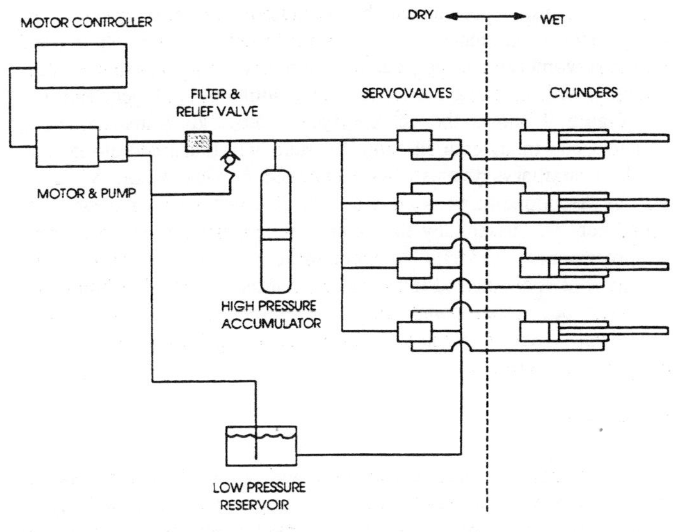

Figure 7: Recirculating hydraulic system block diagram

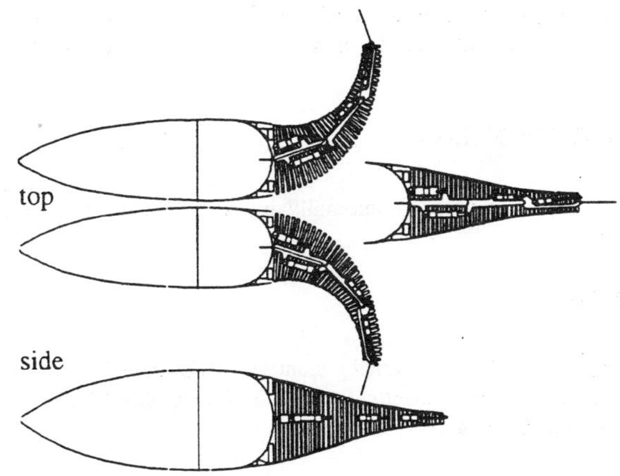

Figure 8: Hydraulic cylinder arrangements

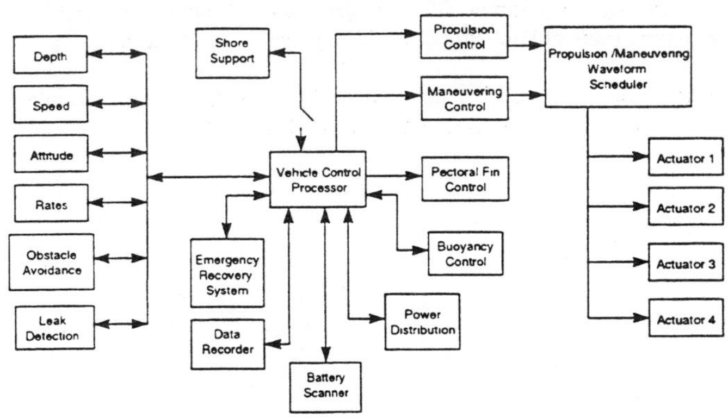

Figure 9: VCUUV system block diagram

because the system pressure can be raised to overdrive the system. The low system efficiency is not significantly lower than the tendon/servomotor concept, yet the mechanical system is much less complex, with significant reduction in the number of hull penetrations.

Figure 9 shows the vehicle system block diagram. Sensors, actuators and their required control systems are illustrated along with standard auxiliary systems. As mentioned previously, the VCUUV will be a rudimentary autonomous vehicle without much onboard "intelligence". Essentially the vehicle will operate in an open loop, pre-programmed capacity to demonstrate simple swimming and turning, not sophisticated intelligent autonomous systems technology. However, our system components (processor, guidance sensors, *etc.*) have been selected with an eye towards future capabilities in complex missions and control.

SUMMARY

In this paper we have presented design issues and current work in the development of a flexible-hull UUV at the Charles Stark Draper Laboratory. Although we have considered several elements to the design, several issues remain to be resolved such as power distribution, control system design (especially the combined control of swimming straight and executing a maneuver) and whether or not the hydrodynamic damping is sufficient to prevent recoil of the forebody. The objectives that we have set for the VCUUV project are ambitious but the potential payoffs in efficiency and maneuverability justify the risk.

ACKNOWLEDGMENTS

This project is funded internally by the Charles Stark Draper Laboratory.

REFERENCES

Anderson, JM (1996), "Vorticity Control for Efficient Propulsion", *Ph.D. Thesis, Massachusetts Institute of Technology/Woods Hole Oceanographic Institution Joint Program.*

Barrett, DS (1996), "Propulsive efficiency of a flexible hull underwater vehicle", *Ph.D. Thesis, Massachusetts Institute of Technology.*

Dewar, H and Graham, JB (1994), "Studies of tropical tuna swimming performance in a large water tunnel, Part III: Kinematics", *Journal of Experimental Biology,* Vol 192, pp 45-59.

Triantafyllou, MS, Barrett, DS, Streitlien, K and Anderson, JM (1997), "Flapping Foils of High Propulsive Efficiency", in preparation.

Triantafyllou, MS and Triantafyllou, GS (1995), "An Efficient Swimming Machine", *Scientific American,* March, 1995.

Proceedings of the Seventh (1997) International Offshore and Polar Engineering Conference
Honolulu, USA, May 25-30, 1997

ISBN 1-880653-28-1 (Set); ISBN 1-880653-30-3 (Vol. II)

AUV Platform Requirements for Ambient Turbulence Measurement

Manhar R. Dhanak and Ken Holappa
Florida Atlantic University
Boca Raton, FL, USA

ABSTRACT

An Autonomous Underwater Vehicle (AUV) provides a suitable platform for making in–situ flow measurements in the ocean environment. It allows surveys of the water column over a substantial range, potentially during a storm and has an advantage over towed bodies in that its self motion is decoupled from the low frequency heaving motion of mother ships. Further it is more versatile than bottom–mounted or suspended probes. Here we examine the potential of an AUV as a platform for making oceanic turbulence measurement, with particular reference to Florida Atlantic University's 2.2m long autonomous underwater vehicle (AUV). Measures necessary to make the platform sufficiently quiet for turbulence measurements are discussed.

KEY WORDS: AUV, small–scale ocean turbulence, shear probes, pressure probe, autonomous measurement platform

INTRODUCTION

Environmental sensing missions typically involve measuring fluid and flow properties of the water column. Traditionally, such measurements are carried out with probes mounted on tethered vehicles (Lueck, 1987) or on fixed platforms on the sea–bed or suspended from buoys (Gregg, 1977). Measurements have also been made using full–scale submarines and ships. The development of small AUVs now offers the possibility of making four–dimensional space–time measurements in an affordable way, potentially during a storm when processes such as sediment transport and turbulent mixing occur with great intensity and induce major changes to the bed form and the water column.

In this paper we are concerned with using an AUV to measure ambient small–scale turbulence in the water column. The measured signals lie in the range 0–100 Hz and therefore require stringent constraints on the low frequency self motion and vibrational noise associated with the platform. The principal advantage of an AUV over towed probes is that it is free of low frequency vibrations associated with towing lines and with the unsteady motion of towing ships, especially in rough seas. Steep–line towed probes[1] suffer from the broad–band high–frequency vibrations of the towing lines and the low–frequency noise associated with the heaving and rolling motion of towing ships. Although careful long–line towing[2] eliminates some of these problems, the probe remains coupled with the surface sea–state, especially with the response of the towing ships to surging waves. The noise associated with towed probes lie within the basic energy range of small scale turbulence and is clearly undesirable. A self propelled vehicle, however, induces vibrational noise which can contaminated the measured signals. Such noise is typically expected to be narrow–banded, offering the possibility of subtracting it, during post processing, from the measured broad–band turbulence signals. The motion of the AUV induced by eddies of size comparable to that of the vehicle is a further source of contamination.

In this paper we describe our experience in dealing with the problem on the *Ocean Explorer*, an autonomous underwater vehicle built at Florida Atlantic University. We examine the characteristics of the noise transmitted to the turbulence measurement package and the nature of the contamination of the measured flow signals. The AUV and the turbulence package is described in §2 and the measurements are described in §3.

THE AUV AND THE TURBULENCE PACKAGE

These are described in detail in Holappa and Dhanak (1996). Briefly, the Ocean Explorer is a 2.4m long automated underwater vehicle (figure 1) designed and built at Florida Atlantic University. Its hull is a Gertler body of 0.53m maximum diameter. Apart from the turbulence package, the only appendages on the body are two stern planes, the rudder and a three–bladed propeller and a thin circular propeller guard. The motor and the controls are located in the rear half of the submarine and the front half, which can be custom designed, contains the pay–load (the turbulence package in the present case). The latter section can easily be mounted on or off the submarine. The automated submarine carries out a pre–programmed mission, which includes turning the turbulence data acquisition computer on and off during the mission. This minimizes the use of the turbulence package's battery. The submarine has several safety features which allow it to surface in case of difficulty.

The turbulence package includes two shear probes, a dynamic Pitot tube, a micro–structure thermistor and three locally mounted accelerometers housed in a single pressure case, located about 0.3m ahead of the bow of the vehicle to minimize flow disturbances associated with the motion of the latter. Mean current speed is measured with a Marsh–McBirney electromagnetic current meter. The turbulence package also utilizes the CTD and self–motion packages in the rear section of the submarine to get estimates of the local density depth and vehicle attitude.

The pressure vessel also contains electronic conditioning packages for the probes, a small on–board PC with 540Mbyte hard–disk capacity used for data acquisition and storage, with ethernet capability, and a battery power supply for the computer.

FIRST MEASUREMENTS

First measurements made in Boca Lake, on the east coast of Florida at about 2m depth of water are shown in figure 2. The vehicle was operating on a depth–following mission; the depth, measured by an on–board sensor, is maintained constant by the controller through appropriate adjustment of the stern plane angles. The figure shows the power spectra of the gradient of the z component of velocity as measured with a shear probe together with the corresponding spectra of the signal measured by the z–accelerometer. The accelerometer gives an indication of the vehicle self–noise transmitted to the turbulence package. The peak at 20Hz can be identified with the motors of the AUV while the peak at around 80Hz is believed to be associated with the spinning hard disk of the on–board data–logging computer. It is evident that there is significant coherence between the two signals shown in the figure and that the flow measurements are substantially contaminated. In essence the signal to noise ratio is rather high. The level of the contamination can be examined if we filter the accelerometer signal from the shear probe measurement. The coherence function $\gamma^2(f)$ is shown in figure 3. We filter out the vibrational noise using $R_{sp\_clean}(f) = (1 - \gamma^2(f))R_{sp\_meas}(f)$ where $R_{sp\_clean}(f)$ is the cleaned signal and $R_{sp\_meas}(f)$ is the spectrum, shown in figure(2), of the measured shear probe signal. $R_{sp\_clean}(f)$ is shown in figure 4 and is compared with the Nasmyth universal spectrum for isotropic turbulence corresponding to a dissipation rate of 5 x 10^{-8} w/kg. The filtration substantially removes the peaks in the mid–frequency range, but does not remove the significant hump in the spectrum in the low frequency range. This is associated with the pitching motion of the AUV. Further, contamination at higher frequencies (> 100 Hz) is not removed by the process. From the results, it is clear that in order to measure such low levels of turbulence, the measurement package needs to be well isolated from the vibrational noise sources, if filtration techniques are not to be used. Further, a better pitch control is essential. In a region of higher level of turbulent activity, where the signal to noise ratio is larger, better results can be expected. Figure 5 shows ocean measurements made off the Florida coast (Holappa and Dhanak, 1996) at a depth of 8m during significant activity in the water column. The measurements suggest dissipation rates of $O(10^{-7}$w/kg), which is about 1/2 order larger than that corresponding to the data in Boca Lake. Rubber isolation mounts were used to isolate the mechanical tail section of the AUV from the rest of the vehicle; the tail section houses the motors and the gears. Further, rubber mounts were used to isolate the turbulence package from the body of the vehicle. Consequently, the high–frequency vibrational noise transmitted to the pressure vessel containing the turbulence package is significantly reduced, as can be seen from the z– accelerometer spectrum; the noise level from the motors is of the same order, but appears to have shifted a little towards higher frequency. Clearly there is room for improvement in our isolation mounting. The measured shear probe signal is not significantly contaminated by the vibrational noise, partly as a result of the isolation measures introduced and partly due to the larger signal to noise ratio. It may be noted that the low frequency hump is smaller and occurs at a lower frequency than in the Boca Lake data. This is because the vehicle was operating with the controller shut off, stern plane fixed at $+2^0$, rudder fixed at 0^0 and the thruster fixed for an approximate speed of 1 m/s. This was done to minimize the influence of the controller and resulting vehicle motion on the turbulence measurement.. The hump at lower frequencies (below 1 Hz) is associated with the vehicle motion. This motion was due to the wave forcing of the vehicle and is broad banded in nature. In the middle range of frequencies the shear spectra are well approximated by the Nasmyth universal spectrum.

Further improvements are being made, including fine tuning the controller software to allow turbulence measurement while actively controlling the vehicle, and use of better isolation mounting to further reduce vibrational levels. Details of the measures being pursued will be discussed. These measures will allow measurement of lower levels of turbulence and associated lower rates of dissipation.

CONCLUSION

For an AUV to be a useful platform for measurement of turbulence, a good pitching control is necessary and the turbulence package needs to be well isolated from the motors of the self– propelled vehicle. Further, any other noise, such as that associated with the spinning hard disk of a data logging computer, which lies in the measurement range of frequencies, needs to be eliminated. On Florida Atlantic University's AUV, such noise–reduction measures currently allow measurement of turbulent energy dissipation rates of $O(10^{-7}$ w/kg). Further measures are being taken to enable measurement of lower rates of dissipation.

ACKNOWLEDGEMENTS

This work is supported by the Office of Naval Research. The authors have benefitted significantly from discussions with Dr Rolf Lueck.

REFERENCES

Holappa, K and M. Dhanak. 1996. Proceedings of Oceans 96 MTS/IEEE Meeting. Ft. Lauderdale, Florida.

Gregg, M. C., 1977. *J. Phys. Oceanogr.*, 7, 436–454.

Lueck R G, 1987. *Deep–sea Res.*, **34**, 1677–1688.

Figure 1. The Ocean Explorer showing the pressure vessel containing the turbulence probes.

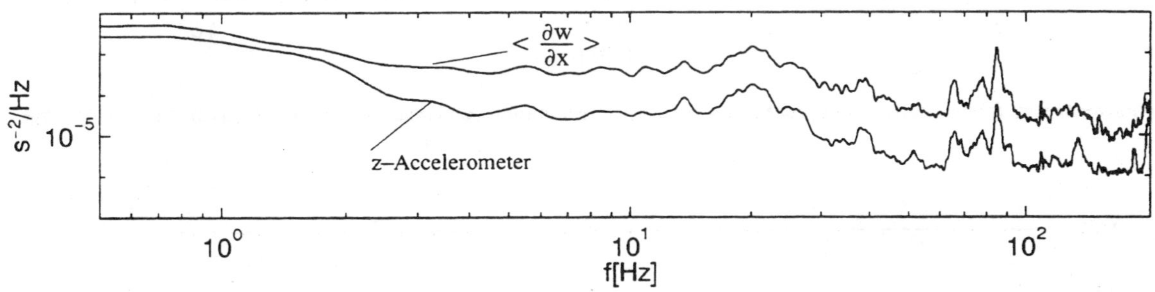

Figure 2. Power spectra for the cross–plane accelerometer and shear probe measurements made in Boca Lake, on the east coast of Florida at a depth of -2m with the AUV moving at a speed of 1m/s.

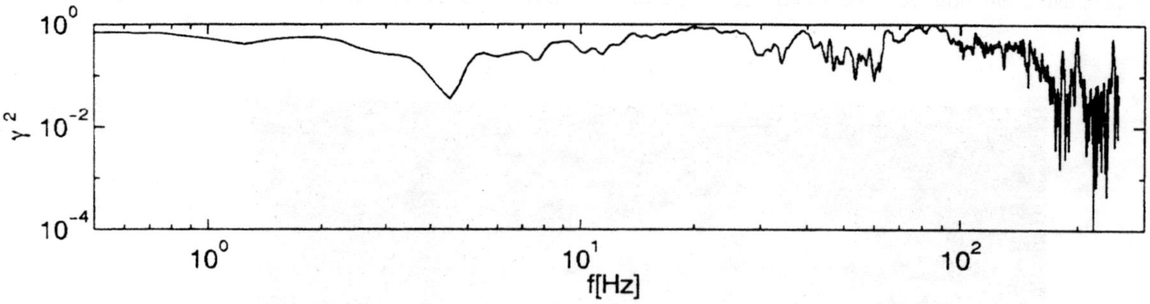

Figure 3. Coherence function for the two spectra shown in figure 2.

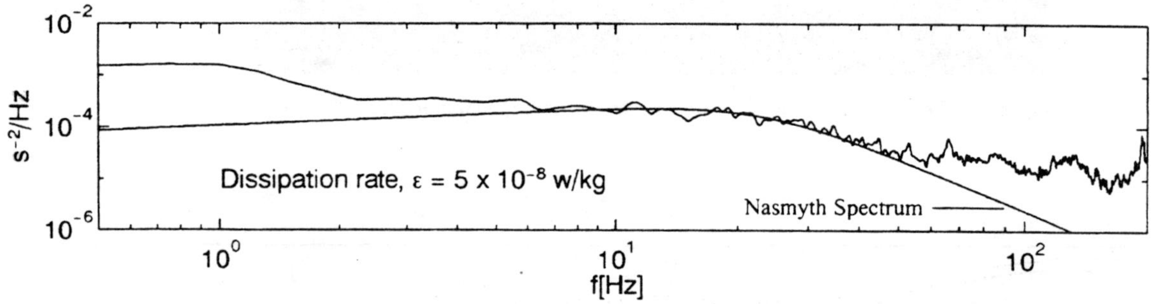

Figure 4. Power spectra for the cleaned shear probe measurements shown in figure 2 compared with the Nasmyth universal spectrum.

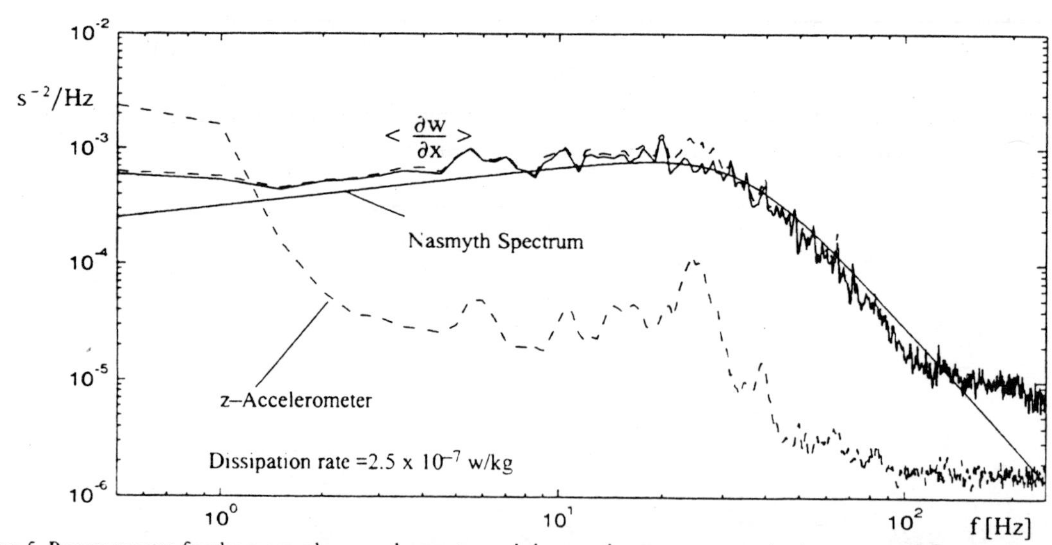

Figure 5. Power spectra for the cross–plane accelerometer and shear probe measurements in the ocean, off the east coast of Florida near Boca Raton, at a depth of −8m with the stern plane and rudder fixed and using a noise isolation package, at a speed of 1m/s.

Proceedings of the Seventh (1997) International Offshore and Polar Engineering Conference
Honolulu, USA, May 25-30, 1997

ISBN 1-880653-28-1 (Set); ISBN 1-880653-30-3 (Vol. II)

A Hybrid Navigation System for Underwater Robotic Vehicles

Pan-Mook Lee and Chong-Moo Lee
Korea Research Institute of Ships and Ocean Engineering, KIMM, Taejon, Korea
Seong-Wook Cheong
Samsung Heavy Industry Co, Taejon, Korea
Jae-Seok Oh and Jun-Ho Oh
Korea Advanced Institute of Science and Technology, Taejon, Korea

ABSTRACT

This paper presents a hybrid navigation system for underwater robotic vehicles to track precisely in rough sea environment conditions. The tracking system is composed of multi-sensor systems such as an inclinometer, a tri-axis magnetometer, a flowmeter, and an super short base line(SSBL) acoustic navigation system. Due to the inaccuracy of the attitude sensors, the heading sensor with measurement errors, and the flowmeter, the predicted position slowly drifts and the estimation error of position becomes larger. On the other hand, the measured position is liable to change abruptly due to the corrupted data of the SSBL system in the case of low signal to noise ratio or large ship motions. By introducing a sensor fusion technique with the position data of the SSBL system and those of the attitude heading flowmeter reference system (AHFRS), the hybrid navigation system updates the three-dimensional position robustly. A Kalman filter algorithm is derived on the basis of the error models for the flowmeter dynamics with the use of the external measurement of the SSBL. A failure detection algorithm decides the confidence degree of external measurement signals by using a fuzzy inference. Simulation is included to demonstrate the validity of the hybrid navigation system.

KEY WORDS : Sensor Fusion, AHFRS(Attitude, Heading and Flow Reference System), SSBL(Super Short Baseline), Failure Detection, Fuzzy Decision Logic, Kalman Filter

INTRODUCTION

The current approach to cost-effective achievement of high-accuracy navigation in underwater vehicle is a system that include one or more inertial navigation systems(INS), and one or more navigation reference sensors such as a global positioning system(GPS) receiver, flowmeter, depthmeter, heading sensor, and positioning sonar system. Functional integration of INS and GPS can improve the performance of vehicles and have actively processed during last decade in aerospace engineering fields (Brumback and Srinath, 1987, Chen and Chui, 1990 and Da, 1994).

In underwater vehicle, however, the integrated INS/GPS tracking system requires better qualities of INS system than the system of an airplane because of the slow motion characteristics of the underwater vehicle and long working intervals in deep sea. The GPS signal cannot penetrates seawater media to reach on the vehicle undersea but only works at sea surface. Kwak and *et al.* (Kwak, 1993) have proposed an experimental investigation of GPS/INS integration for small autonomous underwater vehicle(AUV) navigation, which is not a INS system in real application but a motion reference system. Bennamoun and *et al.* (Bennamoun, 1996) have proposed an integrated GPS/INS/SONAR navigation system for AUV navigation. The system is navigated using GPS and INS at surface and navigated using INS and SONAR at submerged condition. They showed that the integrated system yields better results and less error and improves tracking performance.

This paper presents a hybrid navigation system for underwater robotic vehicles to track precisely in rough sea environment conditions. The objective of this paper is to develope an estimation algorithm of the position of an underwater vehicle, which is based on sensor fusion technique and failure detection method with the use of a low quality attitude sensors, a flow sensor and a low quality acoustic tracking systems.

The tracking system is composed of multi-sensor systems such as an inertia measurement unit, a tri-axis magnetometer, a flowmeter, and a super short base line(SSBL) acoustic navigation system. The first device, no longer an inertial navigation system since the drift effects of the inertial sensors are dominant due to the low quality of this system, provides 6 degree of freedom reference motions and attitude of the vehicle in body fixed coordinates. Although the SSBL can track the position of the vehicle, it is easily contaminated by the ship motions and the ocean environmental noises. The flowmeter detects only forward directional velocity of the vehicle and it acquires the inaccurate data due to the relative current velocities and the errors of measuring equipments.

The hybrid navigation system predicts the position of the vehicles with the use of the reference motion data, the forward speed data, the attitude and the heading of the vehicle by the integration and the filtering method. The predicted position slowly drifts and the estimation error of position becomes larger due to the inaccuracy of the inertial sensors, the heading sensors and the flowmeter. On the other hand, the measured position of the SSBL is liable to change abruptly due to the corrupted data of the SSBL system in the case of low signal-to noise ratio or large ship motions.

By introducing a sensor fusion technique with the position data from the SSBL system and those of the attitude heading flowmeter reference system (AHFRS), the hybrid navigation system updates the three-dimensional position. A Kalman filter algorithm is derived on the basis of the error models for flowmeter dynamics with the use of the external measurement of the SSBL. It is assumed that the measurement signal in the attitude and the heading of the vehicle have no failure and have only white Gaussian noises and static offset errors. When failure occurs in the SSBL tracking system, the filter have to ignore the position signal from the SSBL to avoid the contamination of the integrated tracking system. A failure detection algorithm with the use of a fuzzy inference is arranged in the fusion algorithm to evaluate the confidence degree of external measurement signals. Simulation on the tracking problem of an underwater robotic vehicle is included to demonstrate the validity of the hybrid navigation system.

SENSOR CHARACTERISTICS AND MODELING

Attitude Heading Flow Reference System(AHFRS)

In general, a strap-down inertial measuring unit(IMU) which usually have three rate gyros and three accelerometers measures the attitude of a vehicle in body coordinate. If an IMU is equipped with high precision sensors, it could track the position of a moving vehicle in open space by dead reckoning navigation and correct the position error intermittently with the assist of external measuring devices. In this paper, however, we consider a low quality IMU sensors and additive assistant sensors for motion of an underwater vehicle, and fuse the several signals to estimate the position of the vehicle precisely. The AHFRS is a motion reference sensor unit composed of three parts: a inertial measuring unit, a magnetic compass unit, and a flowmeter. Since the gyros in the IMU is a vibrating type and have a large drift characteristics, the IMU diverges to wrong direction while integrating the angular velocity to get the attitude of the vehicle even in short time duration. Therefore, it is no longer a navigation system stand-alone but only reference sensors for the relative motion of the vehicle. Roll and pitch angle can be calculated from the accelerometers and the gyros. The gyros detect the high frequency motion of the AUV and the accelerometers work as an inclinometer in low frequency motion reciprocally. A magnetic compass is combined with the rate gyro to measure the yaw angle.

It is assumed that the measurement signal in the attitude and the heading of the vehicle have no failure and have only white Gaussian noises and static offset errors. The modelling of the IMU is given as follows:

$$\phi_{attitude} = K_{\phi}\phi_{real} + \Delta\phi_{offset} + \Delta\phi_{noise} \quad (1)$$

where $\phi_{attitude}$ is three dimensional rotation angle in each axis in body coordinate and ϕ_{real} is the real attitude of the vehicle. K_{ϕ}, $\Delta\phi_{offset}$ and $\Delta\phi_{noise}$ are respectively gain, offset error and measurement noise of the IMU sensor unit, which depend on misalignments and drifts of the sensors.

Autonomous underwater vehicles generally move forward direction. Since the forward speed of an AUV is dominant and can be measured by a simple flowmeter, it is natural to measure the forward velocity of the vehicle with a measurement device such as a mechanical type flow sensor or a acoustic doppler sensor. A flowmeter measures not absolute speed of the vehicle but relative speed with respect to ocean current. In addition, it cannot measure rapid change of speed of the vehicle and have a low and narrow frequency bandwidth characteristics. The influence of ocean current can be modelled as a variable gain and the sensor dynamics can be considered as a first order lag filter. Thus, the flowmeter is modelled like as

$$V_{fm} = K_V V_{real} + \Delta V_{offset} + \Delta V_{noise} \quad (2)$$

where V_{fm} is the measured forward velocity by flowmeter and V_{real} is its real value. K_V is sensor gain and ΔV_{offset} is the effect of sea current and mechanical friction of the measuring device. When the sea current is severe, the effect of the current should be compensated to obtain more precise velocity of the vehicle with a flowmeter. An acoustic sonar tracking system is cooperated with the AHFRS to provide the external measurement of an optimal filter algorithm.

Super Short Baseline(SSBL) System

A super short baseline(SSBL) system is adopted to track the underwater vehicle as shown in Fig. 1. A transponder module on the mother ship sends a ping and receives the interrogated signal from the underwater vehicle, while the transponder on AUV detects the ping and replies to the mother ship. The SSBL is an acoustic device that measures the relative position of the vehicle by processing the phase difference of the received acoustic signal. In this way SSBL system measures the direction angle, the slant angle and range of the vehicle from the mother ship. Since the SSBL is sensitive to the phase difference of the received signal, it shows often wrong tracked position data depend on environmental noises, self-generated noises, the motion of mother ship, and the multi-path propagation of the acoustic signal. An ill-estimated position data, that is, a failure should be removed through a well designed filter to obtain optimal estimates for a tracking system. It is a difficult problem to decide that the measured position data is right or wrong in the middle of estimation process. The failure mode is divided into a soft and a hard failure depending on the size of the deviation from the real position. Soft failure occurs in rough sea conditions or noisy environment conditions, while hard failure occurs in the conditions of low signal-to-noise ratio, the multi-pathed acoustic signal, ill-compensation of mother ship motion or critical failure in the signal processing system. The measurement model of the SSBL system is needed to decide that the measured states are successful or erroneous at current time. It can be modelled such as

$$P_{ssbl} = P_{real} + \Delta P_{ssbl} + P_{\delta}\delta(t-\tau) \quad (3)$$

where P_{ssbl} is the relative position vector of the underwater vehicle in navigational coordinate and P_{real} is its true position and ΔP_{ssbl} is the component of measurement noises or the soft failure. $P_{\delta}\delta(t-\tau)$ is the hard failure signal at instance $t = \tau$. The slant angle of the SSBL is very sensitive to the range and operating angle of the vehicle. If a depth sensor is available, this problem is easily overcome by compensating the slant angle with the depth information.

NAVIGATION FILTER ALGORITHM FOR UNDERWATER VEHICLES

Kalman Filter for AUV with Fuzzy Failure Detection and Recovery System

In multiple navigation sensor system, it is significant how to combine the navigation data of each sensor. Kalman filter is commonly used to fuse the navigation informations of sensors. In general, one navigation sensor is used as a major sensor and the others are redundant systems or minor sensors to compensate major sensor error (Maybeck, 1979). In a Kalman filter for navigation system, the state estimates are propagated with a system model and updated by external measurements. If one makes an appropriate model for the major sensor and knows the noise levels of multiple feasible measurements, one could obtain the optimal estimation of the states with the use of a Kalman filter and the multiple navigation sensor system. However, when external measurements contain large noises such as a hard failure, the Kalman filter produces the contaminated estimates of the states. A special filter or a monitoring system is required to remove the failed signal of the multi sensor system. Da (Da, 1994) have proposed a detection algorithm of a sensor failure with the use of chi-square test for the INS navigation system of planes. He have also proposed a switchable monitoring algorithm to enhance the sensitivity of failure detection. Napolitano *et al.* (Napolitano, 1995) have suggested a failure detection method with the use of neural network based scheme. On the other hand, this paper presents a failure detection method based on the fuzzy inference logic.

The navigation system of the underwater vehicle in this paper is composed of a dead reckoning system named AHFRS and an external measurement SSBL system. The AHFRS is a measurement sensors for dead reckoning system to evaluate position and velocity information of the vehicle. The attitude of the underwater vehicle changes slowly and the velocity of the vehicle is nearly constant in short duration. In the proposed dead reckoning system, the estimates of position is obtained by integrating the forward speed of the vehicle according to the attitude angle and the heading angle in navigation coordinate. The integration of the relative velocity of the vehicle, however, results in the accumulation of errors due to directional sea currents, measurement errors and the unmodelled lateral motions of the vehicle. The flowmeter can measure only forward speed excluding the speed of side direction, which induces speed error according to the change of heading and attitude of the vehicle. The dead reckoning system based on the AHFRS is a useless thing by itself, thus, it is required to use the external measurements in the navigation filter. The SSBL system is adopted as the externals in this paper.

The navigation filter should have a sensor failure detection algorithm and recovery function, since the SSBL system is liable to produce wrong navigation data for some operating conditions. Considering the factors: observability of failure from measurements, elapse time to detect failure, sensitivity to soft failure and hard failure, and ability to remove fault signals and recovery from failure, we have developed a failure detection algorithm based on a fuzzy inference logic to detect and recover the failure in the SSBL system. The fuzzy detection scheme does not depend on system order, non-linearity, time variance feature of the system and can produce a fuzzy decision that gives information about the grade of failure. When the hard failure occurs in the SSBL system, the Kalman filter does not take the external measurements and propagate the state estimates with the use of the predicted motion from the AHFRS. In the case of the soft failure, the Kalman filter fuses the external measurements and the predicted motion of the vehicle with weighting factors, which are decided by the fuzzy logic depending on the grade of failure. With this scheme, the Kalman filter algorithm continuously propagates the state estimates where the dead reckoning system AHFRS is active in failure mode and the SSBL system plays the role of the external measurement in normal mode.

Structure of Fuzzy Kalman Filter Algorithm

The indirect feedforward Kalman filter algorithm is implemented for the navigation of an underwater robotic vehicle. The measurement models (1) and (2) on the estimates of position and the motion of the vehicle reduce to

$$\begin{aligned}
\{x_{ahf},\ y_{ahf},\ z_{ahf}\}^T &= \{x_{true}+\delta x,\ y_{true}+\delta y,\ z_{true}+\delta z\}^T \\
V_{fm} &= V_{true}+\delta V \\
\psi_{ahf} &= \psi_{true}+\delta\psi \\
\gamma_{ahf} &= \gamma_{true}+\delta\gamma
\end{aligned} \tag{4}$$

where, ψ_{ahf} and the γ_{ahf} are the measured heading angle and the measured pitch angle of the vehicle, respectively. By considering the error models (4), the true velocity of the vehicle in navigation coordinate can be obtained approximately from the velocity of the flowmeter and the attitude and heading angle of the vehicle as follows

$$\begin{aligned}
\dot{x}_{true} &= V_{true}\sin\psi_{true}\cos\gamma_{true} \\
&\cong V_{fm}\sin\psi_{ahf}\cos\gamma_{ahf} - V_{fm}\delta\psi\cos\psi_{ahf}\cos\gamma_{ahf} \\
&\quad -\delta V\sin\psi_{ahf}\cos\gamma_{ahf} + V_{fm}\delta\gamma\cos\psi_{ahf}\sin\gamma_{ahf} \quad (5)\\
\dot{y}_{true} &= V_{true}\sin\psi_{true}\cos\gamma_{true} \\
&\cong V_{fm}\cos\psi_{ahf}\cos\gamma_{ahf} + V_{fm}\delta\psi\sin\psi_{ahf}\cos\gamma_{ahf} \\
&\quad -\delta V\cos\psi_{ahf}\cos\gamma_{ahf} + V_{fm}\delta\gamma\cos\psi_{ahf}\sin\gamma_{ahf} \quad (6)\\
\dot{z}_{true} &= V_{true}\sin\gamma_{true} \\
&\cong V_{fm}\sin\gamma_{ahf} - \delta V\sin\gamma_{ahf} - V_{fm}\delta\gamma\cos\gamma_{ahf}. \quad (7)
\end{aligned}$$

Each directional component of the measured velocity in navigation coordinate can be represented by $u_{fm} = V_{fm}\cos\psi_{ahf}\cos\gamma_{ahf}$, $v_{fm} = V_{fm}\sin\psi_{ahf}\cos\gamma_{ahf}$ and $w_{fm} = V_{fm}\sin\gamma_{ahf}$. In addition, by defining $\delta\dot{x}$, $\delta\dot{y}$ and $\delta\dot{z}$ as $u_{true}-u_{fm}$, $v_{true}-v_{fm}$ and $w_{true}-w_{fm}$, respectively, the equation (4) reduces to

$$\begin{aligned}
\delta\dot{x} &= -v_{fm}\delta\psi - \delta V\sin\psi_{ahf}\cos\gamma_{ahf} + w_{fm}\sin\psi_{ahf}\delta\gamma \quad (8)\\
\delta\dot{y} &= u_{fm}\delta\psi - \delta V\cos\psi_{ahf}\cos\gamma_{ahf} + w_{fm}\cos\psi_{ahf}\delta\gamma \quad (9)\\
\delta\dot{z} &= -\delta V\sin\gamma_{ahf} - \delta\gamma\sqrt{u_{fm}^2+v_{fm}^2} \quad (10)\\
\delta\dot{V} &= -K_V\delta V \quad (11)\\
\delta\dot{\psi} &= -K_\psi\delta\psi \quad (12)\\
\delta\dot{\gamma} &= -K_\gamma\delta\gamma. \quad (13)
\end{aligned}$$

We can define the following error equations of the AHFRS for the design of a Kalman filter, which is derived by considering the measurement error of the AHFRS as system disturbances and replacing the difference between the SSBL signal and the estimated position as the external measurement:

$$\dot{X} = FX+\xi \tag{14}$$
$$Z = HX+\eta \tag{15}$$

where,

$$F = \begin{bmatrix} 0 & 0 & 0 & -\sin\psi_{ahf}\cos\gamma_{ahf} & -v_{fm} & w_{fm}\sin\psi_{ahf} \\ 0 & 0 & 0 & -\cos\psi_{ahf}\cos\gamma_{ahf} & u_{fm} & w_{fm}\cos\psi_{ahf} \\ 0 & 0 & 0 & -\sin\gamma_{ahf} & 0 & -\sqrt{u_{fm}^2+v_{fm}^2} \\ 0 & 0 & 0 & -K_V & 0 & 0 \\ 0 & 0 & 0 & 0 & -K_\psi & 0 \\ 0 & 0 & 0 & 0 & 0 & -K_\gamma \end{bmatrix}$$

$$H = \begin{bmatrix} 1 & 0 & 0 & 0 & 0 & 0 \\ 0 & 1 & 0 & 0 & 0 & 0 \\ 0 & 0 & 1 & 0 & 0 & 0 \end{bmatrix}$$

$$X = \{\delta x\ \delta y\ \delta z\ \delta V\ \delta\psi\ \delta\gamma\}^T$$

$$\xi = \{\xi_{\delta x}\ \xi_{\delta y}\ \xi_{\delta z}\ \xi_{\delta V}\ \xi_{\delta\psi}\ \xi_{\delta\gamma}\}^T$$

and δx, δy, and δz are the position error of the vehicle in navigation coordinate, ξ is the disturbance of the AHFRS for each component. We assume that the system disturbances and the measurement noises are white Gaussian and the covariance matrix are defined as follow:

$$E[\xi(t)] = 0 \quad , \quad E[\xi(t)\xi(t+\tau)] = Q(t)\delta(t-\tau)$$
$$E[\eta(t)] = 0 \quad , \quad E[\eta(t)\eta(t+\tau)] = R(t)\delta(t-\tau), \qquad (16)$$

$\xi(t)$ and $\eta(t)$ are assumed to be statistically independent, $Q(t)$ is positive-semidefinite and $R(t)$ is positive-definite. When no failure occurs, the general Kalman filter algorithm reduces to the following equations for the discrete optimal estimation of states:

$$\bar{x}(t_{k+1}) = \Phi(t_{k+1}, t_k)\hat{x}(t_k) \qquad (17)$$
$$\bar{P}(t_{k+1}) = \Phi(t_{k+1}, t_k)\hat{P}(t_k)\Phi^T(t_{k+1}, t_k) + Q(t_k) \qquad (18)$$
$$K(t_{k+1}) = \bar{P}(t_{k+1})H^T(t_{k+1})[H(t_{k+1})\bar{P}(t_{k+1})H^T(t_{k+1}) + R(t_{k+1})]^{-1} \qquad (19)$$
$$\hat{x}(t_{k+1}) = \bar{x}(t_{k+1}) + K(t_{k+1})[z(t_{k+1}) - H(t_{k+1})\bar{x}(t_{k+1})] \qquad (20)$$
$$\hat{P}(t_{k+1}) = [I - K(t_{k+1})H(t_{k+1})]\bar{P}(t_{k+1}) \qquad (21)$$

where,

$\bar{x}(t_{k+1})$: priori state estimation
$\hat{x}(t_{k+1})$: posterior state estimation
$\bar{P}(t_{k+1})$: priori variance estimation
$\hat{P}(t_{k+1})$: posterior variance estimation
$\Phi(t_{k+1}, t_k)$: transient matrix ($= \exp^{F(t_k)(t_{k+1}-t_k)}$)
$\hat{x}(0) = x(0)$ and $\hat{P}(0) = P(O)$.

In the case of the soft failure, the Kalman filter fuses the external measurements and the predicted motion of the vehicle by regulating the weighting factors. The schematic diagram of the fuzzy Kalman filter with failure detection is shown in Fig. 2. We define two variables $\delta\bar{V}$ and α as fuzzy input sets to detect sensor failure.

$$\delta\bar{V} = |\bar{V}_{fm}(t+T) - \bar{V}_{ssbl}(t+T)| \qquad (22)$$
$$\alpha = \sqrt{(x_{ssbl}(t+T) - V_{fuse}(t)T\cos\psi(t)\cos\gamma(t))^2 + (y_{ssbl}(t+T) - V_{fuse}(t)T\sin\psi(t)\cos\gamma(t))^2 + (z_{ssbl}(t+T) - V_{fuse}(t)T\sin\gamma(t))^2} \qquad (23)$$

where,

T : SSBL sampling interval
$\bar{V}_{fm}(t+T)$: the measured mean forward speed at $[t_k, t_{k+1}]$
$\bar{V}_{ssbl}(t+T)$: the evaluated forward speed from SSBL data
$V_{fuse}(t)$: optimal filtered forward speed

The tendency of the sensor failure can be decided with the aids of the fuzzy inference. Because the behavior of the underwater vehicle changes slowly as mentioned in previous section, it is reasonable to predict the motion of the vehicle with the signal from the AHFRS when failure occurs. As shown in Fig. 2, each position data of AHFRS system and SSBL system are summed with weighting factors where the weighting factor is calculated from the fuzzy logic corresponding to the tendency of failure. As the grade of failure increases, the fuzzy logic decides that the confidence of the external measurements goes down and decreases the weighting factor of the SSBL system. The proposed failure detection and Kalman filter algorithm detects the failure in the middle of real-time processing and estimates the states stably whether the external measurements are fail or not.

TRACKING SIMULATION AND RESULTS

The performance of the proposed failure detection and Kalman filter method is illustrated by an underwater robotic vehicle. Simulations are carried out for the 2-dimensional position tracking of the vehicle under sea current conditions.

The motions of the vehicle are generated with second order differential equations for forward, lateral and yaw direction. The damping coefficients generally depend on the velocity of the vehicle and are coupled with the motions (Nahon, 1996, Healey *et al.*, 1996). In order to simulate the motion of the vehicle, we simply select the coefficients as constant values at normal operating speed and assume that the motions are independent with one another, since the purpose of this paper is to demonstrate the effectiveness of the proposed method. The values used in the simulation is shown in Table 1. The sea current flows with constant velocity 0.5 m/esc from 150 degree to -30 degree from the y-axis in navigation coordinate. The vehicle initially move toward the direction of the y-axis with velocity 2.0 m/sec. The constant force and moment as shown in Table 1 are given to the vehicle after $t = 0$. The flowmeter of the AHFRS modelled as a first order lag filter, of which the time constant is 1 second.

Figure 3 shows the result for the simulated tracks of the vehicle during 250 seconds, where the real line is the simulated track of the vehicle by solving the differential equation and the dotted line is the evaluated track with the use of the signals from the AHFRS system. The mark 'o' and 'x' represent the start point and the end point respectively. The position error of the AHFRS enlarges as the time increases due to the effects of the error of the flowmeter and measurement errors of the AHFRS. This shows that the AHFRS can not be applicable to the navigation system of the underwater vehicle without the aids of external measurements.

The measured positions are generated including two kinds of the failures: hard failures occur at every 20 seconds and soft failures exist at the time intervals [50 70], [130 150] and [200 220]. The hard failures are chosen as arbitrary large value 50 meters. On the other hand, the soft failures are randomly generated on the basis of the deviations of the slant angle of a SSBL system, the LXT tracking system (ORE, 1995), where the errors of the slant angle is selected to 2 or 5 degrees about the standard deviation.

When the estimates of the position are obtained by processing the signals of the SSBL system alone, we have to choose a threshold level to decide the occurrence of failure in the SSBL system. Fig. 4 and Fig.5 show the position estimates of the vehicle with the use of the Kalman filter and the SSBL system including failures, and Fig. 6 represents the errors about the true x and y positions. We could improve the performance of the tracking system by ignoring the failed signals. However, it is hard to decide the threshold level, since it depends on the operating conditions of tracking system, its range, and sea states. Furthermore, the estimates of the position might have a bad influence from the soft failure which exists under the threshold level. This paper replaces the threshold level with the confidence level of the failure.

The fuzzy rules for the detection of failure are given at Table 2. Fig. 7 represents the membership functions for the inputs $\delta\bar{V}$ and α and the output 'status'. Bisector method is selected as the defuzzification and the simulation carried out with the fuzzy toolbox in the Matlab.

Fig. 8, Fig. 9 and Fig. 10 respectively represent the simulated planar motions of the vehicle, x and y components of the position data, and the errors about the true tracks of the vehicle through the proposed failure detection and Kalman filter algorithm about the fusion of the multi sensor system. The estimates of track of the vehicle are improved remarkably for both the hard failure and the soft failure. Fig. 11 shows the weighting factors at each time step. If the value is zero, only the signal from the SSBL is considered as the external measurement. If the value is unit, the signal form the SSBL is ignored completely and the filter system estimates the position with the dead reckoning system AHFRS. In the case of the soft failure mode, the value exists in the range of [0 1] and the fused signal is adopted as the external measurement according to the status from the fuzzy inference logic. From Fig. 10, we can track the vehicle with the accuracy less than 5 meters, where the errors are caused by the components of the ill-measurement of the AHFRS. If the hard failure occurs continuously in a certain long interval, then the errors of the tracking system would increase proportionally to the errors of the flowmeter and the duration of the hard failure.

Through the tracking simulation for the measurement system including the hard failures and the soft failures, the effectiveness of the failure detection and Kalman filter algorithm is illustrated. The proposed method improves the performance of the position estimates of the underwater robotic vehicle in rough sea conditions and is actually applicable to the tracking system of the vehicle equipped with the AHFRS and the SSBL system.

CONCULSION

This paper suggests a hybrid navigation system with the attitude heading flowmeter reference system(AHFRS) and the super short baseline(SSBL) system for underwater robotic vehicles to track precisely in rough sea conditions. A Kalman filter algorithm is derived on the basis of the error models for the flowmeter dynamics with the use of the external measurement of the SSBL. A failure detection algorithm decides the confidence degree of external measurement signals by using a fuzzy inference logic. The estimates of track of the vehicle are remarkably improved with the use of the sensor fusion technique for both the hard failure and the soft failure. The proposed method is applicable to the tracking system of the vehicle equipped with the AHFRS and the SSBL system. The effectiveness has been verified through simulation.

REFERENCES

Brumback, BD, and Srinath, MD (1987). "A Fault-Tolerant Multisensor Navigation System Design," *IEEE Trans. on Aerospace and Electronic Systems*, Vol. 23, No. 6, Nov., pp 738–755.

Chen, G, and Chui, CR (1990). "A Modified Adaptive Kalman Filter for Real-Time Applications," *IEEE Trans. on Aerospace and Electronic Systems*, Vol. 27, No. 1, Jan., pp 149–154.

Da, R (1994). "Failure Detection of Dynamical Systems with the State Chi-Square Test," *Journal of Guidance, Control, and Dynamics*, Vol. 17, No. 2, March-April, pp 271–277.

Bennamoun, M, Boashash, F, and Faruqi, F (1996). "The Development of an Integrated GPS/INS/SONAR Navigation System for Autonomous Underwater Vehicle Navigation," *IEEE Symposium on Autonomous Underwater Vehicle Technology*, pp 256–261.

Kwak, SH, Stevens, CD, Clynch, JR, McGhee, RB, and Whalen, RH (1993). "An Experimental Investigation of GPS/INS Integration for Small AUV Navigation," *Proc. of the 8-th Int. Symp. on Unmanned Untethered Submissible Technology*, Sep. 27–29, pp 239–251.

White, KA, Smith, SM, Ganesan, K, Kronen, D, Rae, GJS, and Langenbach, RM (1996). "Performance Results of a Fuzzy Behavioral Altitude Flight Controller and Rendezvous & Docking of an Autonomous Underwater Vehicles with Fuzzy Control," *IEEE Symposium on Autonomous Underwater Vehicle Technology*, pp 117–124.

Napolitano, MR, Neppach, C, Casdorph, V, and Naylor, S (1995). "Neural-Network-Based Scheme for Sensor Failure Detection, Identification, and Accommodation," *Journal of Guidance, Control, and Dynamics*, Vol. 18, No. 6, Nov.-Dec., pp 1280–1286.

Maybeck, PS (1979). "Stochastic Models, Estimation, and Control : Vol.1," Academic Press Inc..

Nahon, M (1996). "A Simplified Dynamics Model for Autonomous Underwater Vehicles," *IEEE Symposium on Autonomous Underwater Vehicle Technology*, pp 373–379.

Healey, AJ, Marco, DB, and McGhee, RB (1996). "Autonomous Underwater Vehicle Control Coordination Using A Tri-Level Hybrid Software Architecture," *Proc. of Int. Conf. on Robotics and Automation*, Minneapolis, April, pp 2149–2158.

O.R.E. Inc (1995). "Operation and Maintenance Manual for the ORE International, Inc. LXT Tracking System," Falmouth, Massachusettes.

Table 1 Vehicle model and test conditions for simulation

Item	Value	Unit
Mass	300	Kg
Added mass coefficient	1.0	
Yaw moment of inertia	390.63	$Kg \cdot m^2$
Damping coefficient (Forward direction)	26.7	$N \cdot sec/m$
Damping coefficient (Lateral direction)	133.5	$N \cdot sec/m$
Damping coefficient (Yaw Direction)	38.2	$N \cdot sec$
Thrust force	40.0	N
Yaw moment	1.92	$N \cdot m$
Current speed	0.5	m/sec
Direction of current	150	degree

Table 2 Fuzzy inference rules for the detection of failure

Rule No.	δV	α	Status
1	–	big	Failure
2	big	mid2	Failure
3	medium	mid2	Swfail
4	small	mid2	Noisy
5	big	mid1	Noisy
6	medium	mid1	Noisy
7	small	mid1	Valid
8	–	small	Valid

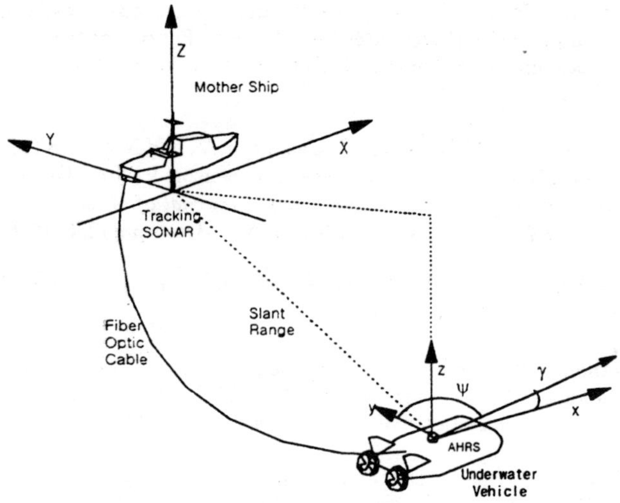

Fig. 1. The hybrid navigation system for an underwater vehicle equipped with the SSBL and AHFRS.

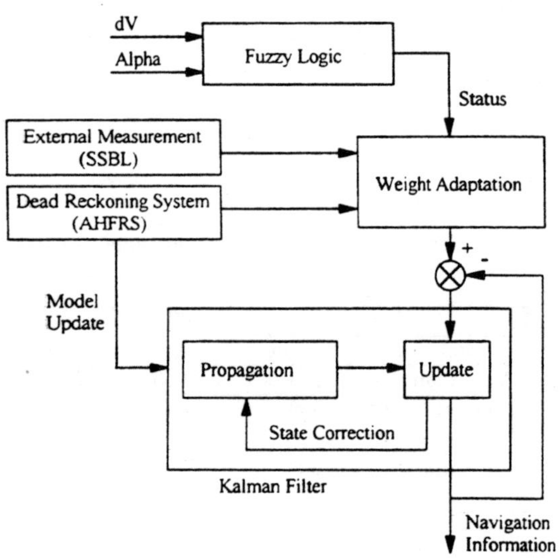

Fig. 2. Schematic diagram of the hybrid navigation system.

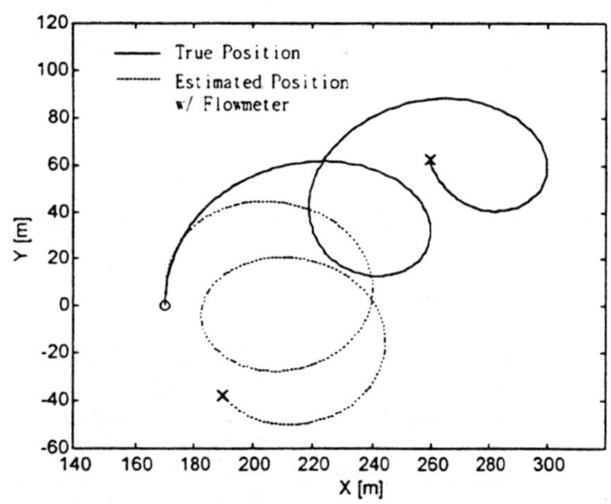

Fig. 3 The simulated tracks of the vehicle and the estimated tracks with the use of AHFRS.

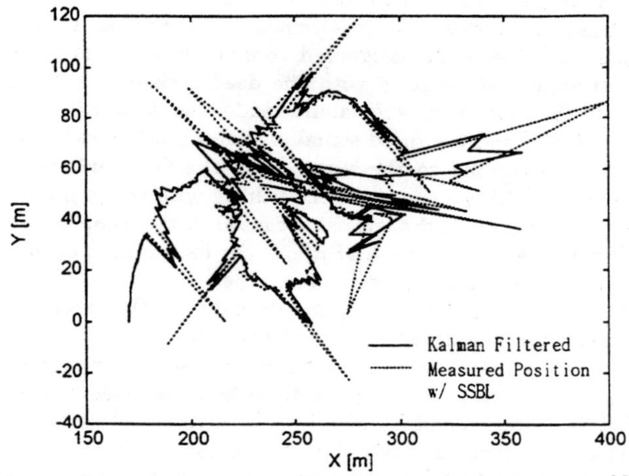

Fig. 4. The measured and the filtered tracks of the vehicle with SSBL.

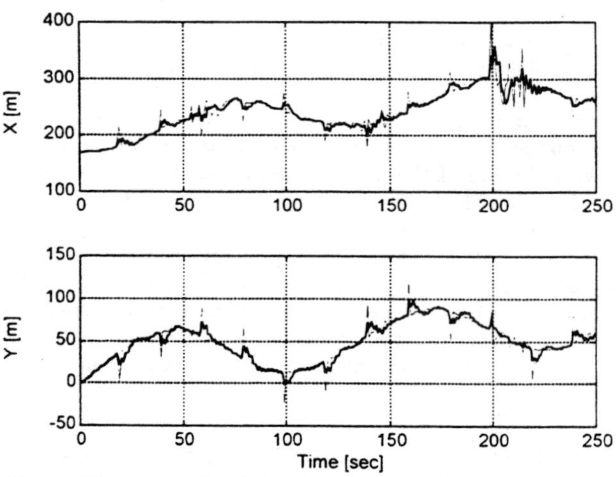

Fig. 5. The measured and the filtered tracks of the vehicle with SSBL.

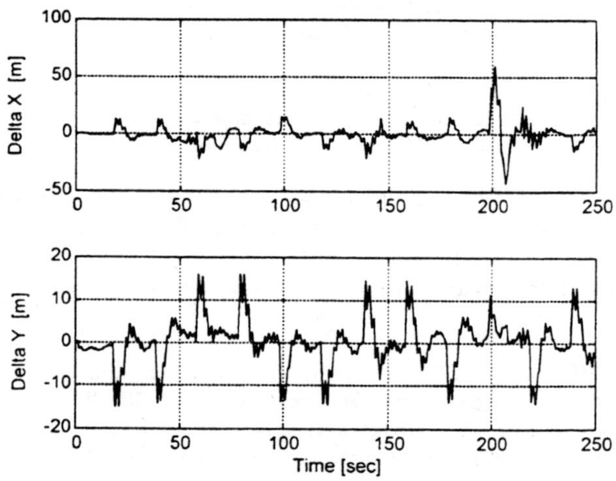

Fig. 6. The errors of the estimated position with SSBL.

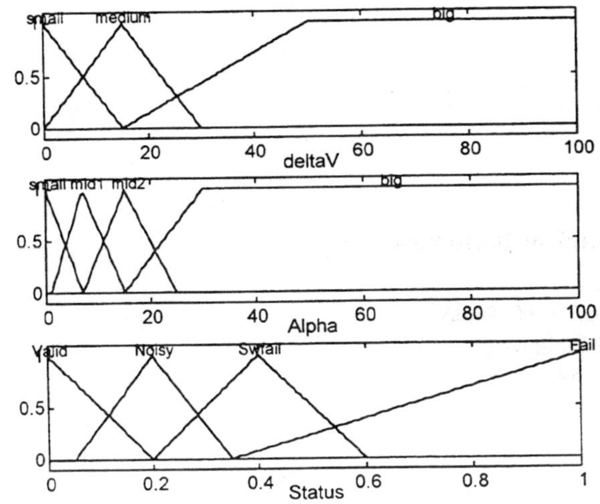

Fig. 7. The membership functions of the inputs δV and α and the output of the fuzzy inference.

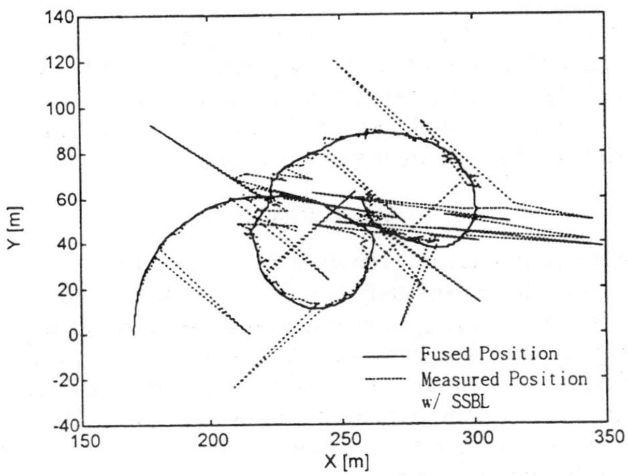

Fig. 8. The measured and the fused tracks of the vehicle by the proposed navigation system.

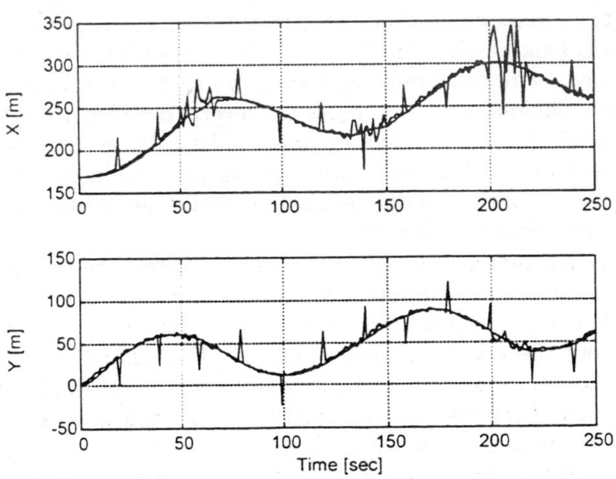

Fig. 9. The measured and the fused tracks of the vehicle by the proposed navigation system.

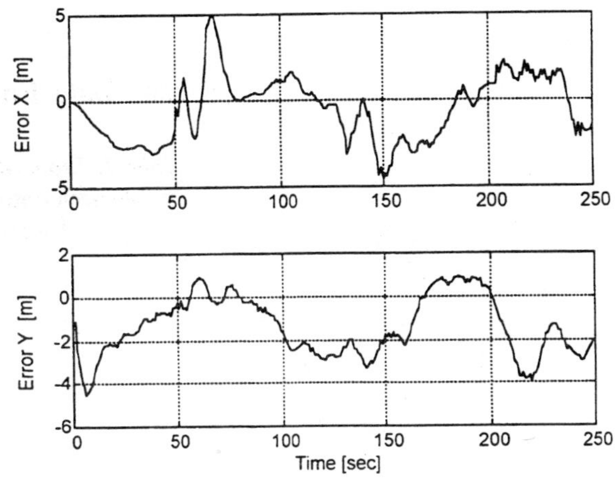

Fig. 10. The errors of the estimated position by the proposed navigation system.

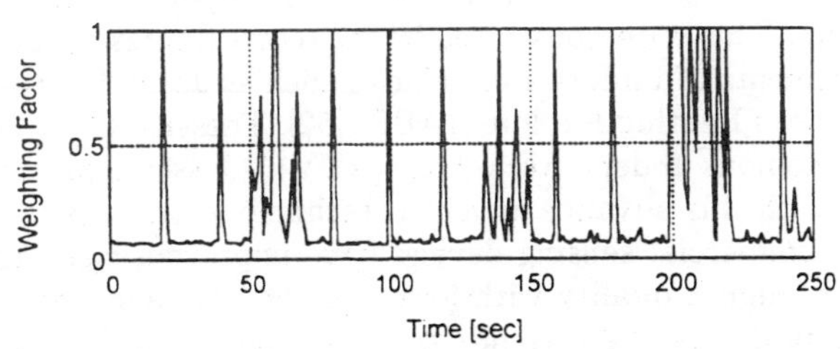

Fig. 11. The weighting factors in the process of the position estimates for the failure detection algorithm.

Proceedings of the Seventh (1997) International Offshore and Polar Engineering Conference
Honolulu, USA, May 25-30, 1997

ISBN 1-880653-28-1 (Set); ISBN 1-880653-30-3 (Vol. II)

Sensor Data Fusion in Marine Robotics

John J. Leonard and Christopher M. Smith
Massachusetts Institute of Technology
Cambridge, MA, USA

Abstract - Sensor data fusion is an important and difficult requirement in marine robotics. This paper examines the role of sensor fusion in three different application areas: navigation of autonomous underwater vehicles, acoustic scene reconstruction, and ocean data assimilation. The research issues encountered in these problems include management of uncertainty, modeling of sensor physics, vehicle and environment dynamics, feature extraction, data association, state estimation, and sensor management. Previous research in these areas is reviewed and suggestions for future research are presented.

Keywords:
underwater robotics; data fusion; sonar; navigation.

INTRODUCTION

Improved methods for sensor data interpretation are critical for realization of the full potential of today's state-of-the-art marine robotics technology. Marine robotics technology has undergone a phase of dramatic increase in capability in recent years. This is demonstrated by the recent impressive accomplishments of underwater vehicles such as Jason [2], the Autonomous Benthic Explorer (ABE) [52], Theseus [11], and the autonomous underwater vehicle (AUV) Odyssey II [6]. In concert with this advance in vehicle technology, the capabilities of commercial sensing devices have improved, yielding devices of higher quality with lower weight, cost and power consumption. As these trends continue, the primary challenges in subsea robotics are shifting from development of system and sensor technology to problems of *information processing* and *sensor-based control*. State-of-the-art marine robot systems can produce vast amounts of data, but our ability to use marine robot systems to their full potential is limited by the difficulty of processing this information in autonomous fashion.

This paper examines the role of sensor fusion in three different application areas: navigation of autonomous underwater vehicles, acoustic scene reconstruction, and ocean data assimilation. The next three sections present a description of each of these problems. Subsequently, some of the fundamental research issues shared by these application areas are described and related research in these topics is reviewed. Finally, suggestions are made for future research.

AUV NAVIGATION

Navigation is a critical requirement for any type of mobile robot, but this is especially true for autonomous underwater vehicles. Good navigation information is essential for safe operation and recovery of an AUV, especially for under-ice deployments [4] or in regions of high currents [51]. In addition, often for the data to be gathered by an AUV to be of military, commercial, or scientific value, the location from which the data has been gathered must be precisely known.

Because the ocean is impenetrable to electromagnetic energy except at very low frequencies, navigation systems such as LORAN and GPS are unavailable, and instead acoustic systems must be used [5]. One method for acoustic navigation is long baseline (LBL) navigation, in which the vehicle operates within a pre-calibrated array of acoustic beacons. This technique has been in use for several decades [27], [41], primarily for tethered and manned underwater systems. Because multipath acoustic propagation can result in a high degree of spurious data, reliable operation for autonomous systems remains a challenge.

For an autonomous LBL navigation system, rejection of outliers (data association) is an important function [62]. The AUV must make decisions regarding the quality of the data without the aid of a human operator. The reliability of the solution must be very high, because a survey AUV like the Odyssey II cannot "stop and think" when confronted with a difficult situation; the vehicle must be continually in motion to maintain control authority. If the navigation algorithm were to go awry during sea-trials, sending the vehicle hundreds of meters off-course, loss of the vehicle is a very real possibility. Additionally, the navigation algorithm must execute in real-time on a vehicle with limited computational resources.

A significant difficulty in acoustic navigation can be caused by an error in the assumed sound speed profile. Even if the sound speed profile is known at the start of an AUV mission, the acoustic propagation environment can change during the mission. To address navigation in complex and dynamic acoustic environments, Deffenbaugh has developed a technique for very long baseline navigation [17, 18]. The approach uses the extra information provided by the multipath arrivals to invert for sound speed profile variations in space and time, and in the process provide a more accurate position estimate.

For some applications of AUVs, the use of acoustic beacons is undesirable or impractical. If an accurate *a priori* map of the environment is available, one approach to globally-referenced position estimation is to use measurements of geophysical parameters, such as bathymetry, magnetic field, or gravitational anomaly [58, 59, 40, 23]. These approaches are based on matching sensor data with an *a priori* environment map, under the assumption that there is sufficient spatial variation in the parameter(s) being measured to permit accurate localization.

In practice, an up-to-date, high-quality map may be unavailable in the operating area of interest. This motivates research into the problem of concurrent mapping and localization. The goal of concurrent mapping and localization is for the AUV to build a map of its environment and to use that map to navigate in real time. While this has not to our knowledge been attempted with underwater vehicles, the problem has seen some attention in the land robotics community [34]. However, with a few exceptions [44, 49], implementations of concurrent mapping and localization have been restricted to simulation.

A seminal technique for concurrent mapping and localization, called the stochastic map, was published by Smith, Self, and Cheeseman [53]. The stochastic map consists of a single state vector that represents the estimates of the vehicle and feature locations and an associated covariance matrix. As the vehicle moves around its environment, taking measurements of environmental features, the stochastic map is updated using an extended Kalman filter. Moutarlier and Chatila [44] and Rencken [49] have implemented suboptimal versions of the stochastic map using land robots.

The major problems encountered by the stochastic map are the failure of the extended Kalman filter to properly track the highly nonlinear transformations involved in geometric estimation and the fact that the technique scales (at best) quadratically with the number of features present. Also, Smith, Self, and Cheeseman ignored errors which may arise from ambiguity in the source of sensor data. Some of these drawbacks have recently been considered by Uhlmann [61] in a theoretical investigation of concurrent mapping and localization. However, there are many important issues to address in future research to realize this capability on-board an AUV.

ACOUSTIC SCENE RECONSTRUCTION

Sensing the environment is a critical requirement in marine robotics. While a variety of optical sensors are available for underwater use [56, 46], sonar is a natural choice for investigating the turbid marine environment due to its superior propagation characteristics. The aim of acoustic scene reconstruction is to produce a model of an unknown underwater environment based on sonar data obtained from multiple vantage points.

Scene reconstruction is related to the problem of acoustic object recognition, which attempts to identify objects of interest based on known models. Both signal processing and image processing approaches have been pursued. Signal processing techniques attempt to recognize objects by matching templates or resonant signatures of a sonar waveform [16, 30]. Image processing approaches aim to imitate the methods by which skilled human operators interpret visual sonar displays [57].

There are two different approaches to 3-D acoustic scene reconstruction. Stochastic backprojection builds a grid-based, volumetric map of the environment by probabilistically estimating measurement sources [55]. Feature-based reconstruction attempts to associate sonar returns with discrete features in the environment, building a model of the environment composed of geometric primitives [42]. In both approaches, computational complexity is a serious concern.

Stewart pioneered the stochastic backprojection approach and has applied it to a variety of data sets from real ocean settings, such as profile data from the USS *Monitor* [55]. The representation employed is a grid-based, volumetric representation, similar to the occupancy grid developed by Moravec and Elfes for land robot navigation [21]. Singh has extended this work through introduction of a measure of entropy to monitor the amount of information conveyed by new sensor readings in construction of a map [52].

In feature-based scene reconstruction, the first step in interpretation is the extraction of geometric features from raw sensor data, based on a model of sensor physics [35]. After features have been extracted from sensor data, the next step is data association: grouping measurements that share a common origin and rejecting spurious measurements [14]. Once measurements have been partitioned into sets that originate from common environmental features, the shape model (a map of geometric primitives) can be updated [42]. Results to date have been limited to effectively two-dimensional scenes.

OCEAN DATA ASSIMILATION

Ocean data assimilation is the process of integrating real data with a model of an ocean process [38, 50]. One of the largest problems in accurately applying models of ocean flow (or meteorological models, for that matter) is that, because of their dynamic three-dimensional nature and the fact that energy is present at multiple (non-separable) scales, any specification of initial conditions for a forecast will ultimately become uncorrelated with the state of the system [26]. This is due to unmodeled dynamics (such as those at scales which are not modeled for reasons of computation efficiency) and the spread of errors in the initial state. Ocean data assimilation attempts to rectify this inadequacy by providing additional information as the system evolves.

Two different scenarios for ocean data assimilation are *model validation* and *model registration.* In model validation, the objective is to test the validity of a proposed model of an ocean process by comparison with real data. An example of model validation is provided by the proposed study of convective overturning in the Labrador Sea [28, 63]. In model registration, there is a model that is believed to be accurate and the objective is to integrate real data into the model to come up with a specific state estimation for the process of interest. Examples of this are found in numerical weather prediction [26] and oceanographic process field estimation [50].

Ocean data assimilation is a daunting task for two reasons. First, the theoretical and computational challenges of modeling the ocean are tremendous [66]. In addition, our opportunity to observe the ocean is extremely limited. To address the latter issue, the autonomous ocean sampling network (AOSN) project is an ambitious effort that attempts to create a fundamentally new capability to observe the oceans [15]. The idea behind AOSN is to establish a network of small, low cost AUVs supported by acoustic and satellite communication links and power recharging stations. Development of adaptive sampling strategies for real-time ocean data assimilation is an important component of this effort [64].

Munk, Worcester, and Wunsch have pioneered the method of ocean acoustic tomography, which employs acoustic arrays to make precise travel time measurements of a parcel of ocean [45]. In comparison to the point measurements provided by conventional ocean sensing technologies, tomographic measurements are spatially integrating [45]. Recently, the Haro Strait experiment combined moving source tomography, AUV sampling, acoustic propagation modeling, and ocean field prediction [51]. In the future, there promises to be a number of exciting field deployments that will combine the latest technologies for modeling and observing the ocean.

RESEARCH ISSUES

The three applications described above are quite distinct. We believe, however, that these problems have an underlying structural commonality; navigation of AUVs, acoustic scene reconstruction, and ocean data assimilation are all challenging problems in sensor data fusion. The difficulties shared by these problems include the interpretation of uncertain and ambiguous data, modeling of sensor physics, representation of information, and management of computational complexity.

Sensor data fusion can be defined as the purposeful combination of measurements with a priori models, knowledge, and information. Some of the key issues in sensor data fusion are [19, 24, 37]:

- management of uncertainty
- representing, detecting, and tracking features
- modeling of sensor physics
- modeling of platform and environment dynamics
- data association
- state estimation
- sensor management

This section examines these issues from the perspective of marine robotics.

Management of uncertainty

Uncertainty is ubiquitous in marine robotic sensing. In order to properly carry out sensor data fusion, the relative uncertainties of the quantities to be fused must be known or estimated. Uncertainty in marine robotics tasks arises primarily from three sources:

1. uncertainty in the *values* of measurements (noise and biases)
2. uncertainty in the *origins* of measurements (data association error)
3. uncertainty in the *motion* of the sensor or vehicle (navigation error).

The interplay between these different sources of error is what makes sensor data fusion difficult [14]. For example, coupling between navigation and sensing error in concurrent mapping and localization leads to correlations between vehicle and feature position estimates [54, 25, 61].

Since the vehicle must combine uncertain quantities to carry out sensor data fusion, the question of uncertainty representation is an important one. Most data fusion architectures use a Kalman filtering framework for combining measurements and state estimates. This technique is based on a Bayesian representation of uncertainty. Although in a strictly Bayesian paradigm, first-order probabilities are the only meaningful quantities, it is sometimes useful to consider higher-order uncertainty, or the amount of uncertainty in the estimate of the uncertainty of the state estimates [33]. This line of reasoning is followed in the Dempster-Shafer theory of evidence [67] and in other work on higher order probability [13].

In addition to properly combining uncertain data, the vehicle must manage the strategies for addressing these errors so that it can make real-time decisions based on current estimates of its state. Consideration of all three sources of error introduces major computational complexity concerns [14]. Additionally, there may be problems in determining the degree of cross-correlation between uncertain estimates [61].

Representing, detecting, and tracking features

In order to extract useful information from sensor data, the vehicle needs to make some assumptions about its environment. These assumptions form the basis for feature representation. The major dichotomy in feature representation methods is the question of whether to consider the environmental context of the vehicle as one continuous feature (or field), which may be broken down by a grid or pixels, or as a set of discrete features (or objects), which may be independently modeled.

The question of whether a continuous or a discrete geometric representation is more appropriate depends directly on the requirements of the task. For a human-in-the-loop application such as piloting a remotely-operated vehicle (ROV), an occupancy grid offers a method for displaying a large amount of data to the human operator. For object grasping and autonomous recognition of man-made objects, however, an occupancy grid would entail an unduly small cell resolution size, resulting in prohibitive storage and processing requirements. A feature-based geometric approach offers the potential of a compact representation capable of efficiently characterizing a scene and providing direct input to higher level reasoning schemes. This can only be possible, however, if the noise and ambiguity of the input data can be handled in a computationally efficient manner.

Selection of the appropriate features for a given task will depend on a good model of sensor physics as well as environment and platform dynamics. Detection of features based on a sensor model is essentially a problem in data association and track initiation. Tracking features is the subject of state estimation.

Modeling of sensor physics

The key to using any sensor is to have a good sensor model [20]. The objective in developing a sensor model is to support two capabilities:

- Prediction: what data should the sensor produce when observing a known scene from a given position?
- Explanation: given observed sensor data, what is the geometry of the scene that produced the data?

In feature-based acoustic scene reconstruction, a specular sonar model has proven useful in developing geometric constraints for shape estimation. In a specular wavelength regime, isolated features will show up in sonar scans as circular arcs [31]. Circular arcs extracted from sonar data can be used as features for prediction and explanation of sonar data [35].

In acoustic navigation of AUVs, an understanding of acoustic propagation effects is essential, and can be incorporated into the navigation processing to provide robust performance in dynamic environments [17].

Platform and environment dynamics

Dynamic models of the vehicle or sensor platform and the environmental processes being observed are essential in properly representing the dynamic characteristics of the states the vehicle is estimating. Dynamic models also allow the vehicle to make predictions about what it should be sensing, and what is happening in other regions.

One can distinguish between simulation models of vehicle and environment dynamics, which must faithfully reproduce behavior, and models for vehicle use, which must be manageable by the vehicle in real-time. There is the problem of how to most accurately represent these dynamic processes. But there is the further problem of the appropriate model to be used by the AUV; this choice will also depend on computational complexity and speed, fidelity and robustness, operating regime, and expected environmental context.

Providing a model for the vehicle will depend heavily upon the expected range of contexts the vehicle will face. The model should explicitly capture those dynamic phenomena which are of importance or primary interest, while hiding the complexity of the remaining dynamic context.

One example of the integration of environment dynamics with an underwater vehicle simulation occurred in the Haro Strait experiment, in which the AUV Odyssey simulator [7] was coupled with an ocean current model [22] to provide predictions of vehicle performance in high current environments [51].

Data association

Data association is the task of resolving uncertainty in the origins of measurements. The objective is to determine the correspondence between measured data and features of the environment/scene/object being observed, while rejecting spurious measurements. When a continuous feature representation is used, data association is addressed as an integral part of state estimation. For discrete feature models, ambiguity in measurement source must be resolved prior to updating state estimates.

The field of multiple target tracking has provided a number of advanced data association techniques. These can be classified in two categories: target-based and measurement-based. In either approach, the first step is validation gating, a preliminary processing stage in which many infeasible matches between measurements and targets are rejected [3, 61].

An example of a target-based approach is the joint probabilistic data association filter [3], which provides state estimates based on the weighted sum of all the measurements that are in the validation region of a target. These techniques have fixed computational requirements, but produce estimates that average good and bad data. In addition, track initiation for new targets is not explicitly incorporated.

Multiple hypothesis tracking is a measurement-based approach to grouping measurements that originate from the same geometric feature, while rejecting spurious measurements [48]. Each new observation initializes a tree of possible interpretation hypotheses that classify the measurements according to different target models. The tree is grown as new observations are validated with hypothesized targets, and is subsequently pruned to choose the single best interpretation of all past measurements [43, 32]. An important feature of the approach is that ambiguous assignment decisions can be deferred until more data is acquired and a better decision can be made. Cox and Leonard applied multiple hypothesis tracking to land robot map building in a dynamic environment [14]. Moran and Leonard extended the approach to apply to underwater acoustic shape reconstruction of curved objects in two dimensions [42]. Because of the multiple state estimates involved in multiple hypothesis tracking, computation complexity and scaling are significant problems. Current topics of research include pruning strategies and track initialization.

State estimation

Once any necessary data association decisions have been made, state estimation techniques provide the mechanism of integrating sensor measurements obtained from multiple sensors and/or sensing locations. Techniques for robust and efficient state estimation are a central part of research in sensor data fusion [24]. Many approaches are based on the Kalman filtering [3] and Bayesian decision theory [9].

Two important issues are the inherently nonlinear relationships between the parameters being estimated, and correlations between state estimates. Correlations may be due to *a priori* assumptions, correlated measurements, and correlated noise [65]. These correlations may be explicitly modeled, but it should be pointed out that some correlations may be unobservable (and hence unmodelable). For situations when unmodeled correlations are present, Uhlmann has developed an alternative technique for estimation updating called Covariance Intersection [61]. This technique has important implications for concurrent localization and mapping.

Sensor management

Sensor management is the problem of deciding what to sense and from where to sense. Manyika writes "when presented with several sensing options or configurations, the option making the best use of sensor resources to achieve sensing goals must be chosen. [39]" For example, an underwater robot trying to build a map of an unknown environment needs to decide where to move in order to best improve the map it is creating.

In some cases, for example when a simple uniform survey is adequate, sensor management can be performed prior to the mission. However, when the environment is changing or discrete features are being sought (e.g., a search and mapping mission), an adaptive approach to sensor management is needed.

Adaptive sensor management has been studied under many guises, such as action selection [60], active perception [1], and adaptive sampling [8, 64]. A general formulation of adaptive sensor management is hard to obtain because the metrics by which one would evaluate different sensing strategies are highly dependent on the sensor employed and the task at hand.

Manyika has investigated sensor management in the context of decentralized sensor fusion architectures [39]. This work relates to marine data fusion in the context of using multiple AUVs for cooperative mapping. The use of multiple AUVs raises the issue of communication management, because of the bandwidth limitations, time delays, and unreliability of underwater communication [12, 29].

FUTURE RESEARCH

We believe that the problem of enabling an AUV to navigation in, and build a map of, an unknown environment provides a great challenge which can focus future research efforts. Realization of this capability — concurrent mapping and localization — would draw on recent advances in all the research issues discussed above. The reliable extraction of feature location estimates from sonar data in natural environments presents a formidable challenge. However, experiments can begin in environments with man-made objects or distinctive landmarks.

Data association remains a key source of computational complexity. To this point, *ad hoc* approaches to data association have not proven to be robust. It may be possible that multiple hypothesis tracking can offer a foundation on which to develop a comprehensive framework for management of uncertainty in robotic sensing. Attaining this goal will require substantial innovations in the approach, incorporating new methods for non-linear representation and track initiation. A fundamental issue is the initialization of 3D geometric primitives when individual measurements are sparse and provide weak geometric constraints. Because multiple hypothesis tracking is exponentially complex, improved computational efficiency will be essential.

Advanced simulation environments can be a useful aid in developing new methods for data fusion. Simulation has been vital to developers of AUVs because it offers a cost effective alternative to expensive and hazardous field testing [47, 10]. A virtual ocean environment goes beyond the simulation of the AUV itself to incorporate simulations of sensor physics and ocean environment dynamics, with realistic error models. While simulation is no substitute for real data from a deployment at-sea, components of a virtual environment can serve as a human-computer interface for a real experiment [36].

SUMMARY

This paper has surveyed some of the problems of sensor data fusion in marine robotics. Advances in this field have the potential to play a key role in marine applications such as climate change assessment, marine habitat monitoring, oil exploration, and underwater inspection and repair. The high costs and complexity of ocean operation result in a high potential payoff when robotic technology can be successfully applied.

ACKNOWLEDGEMENTS

This work was supported in part by the Naval Undersea Warfare Center, Newport, Rhode Island, under contract N66604-94-C-D057.

REFERENCES

[1] R.K. Bajcsy. Active perception. *Proceedings of the IEEE*, 76(8):996–1005, August 1988.

[2] R. Ballard. Soundings: The Newsletter of the JASON Project for Education, Waltham, MA, http://seawifs.gsfc.nasa.gov/scripts/JASON.html, 1996.

[3] Y. Bar-Shalom and T. E. Fortmann. *Tracking and Data Association*. Academic Press, 1988.

[4] J. G. Bellingham, C. Chryssostomidis, M. Deffenbaugh, J. J. Leonard, and H. Schmidt. Arctic under-ice survey operations. In *Proc. Int. Symp. on Unmanned Untethered Submersible Technology*, pages 50–59, 1993.

[5] J. G. Bellingham, T. R. Consi, U. Tedrow, and D. Di Massa. Hyperbolic acoustic navigation for underwater vehicles: Implementation and demonstration. In *Proceedings AUV '92*, pages 304–309, 1992.

[6] J. G. Bellingham, C. A. Goudey, T. R. Consi, J. W. Bales, D. K. Atwood, J. J. Leonard, and C. Chryssostomidis. A second generation survey AUV. In *IEEE Conference on Autonomous Underwater Vehicles*, Cambridge, MA, 1994.

[7] J. G. Bellingham and J. J. Leonard. Task configuration with layered control. In *Proceedings of the IARP 2nd Workshop on Mobile Robots for Subsea Environments*, pages 193–302, Monterey, CA, USA, May 1994.

[8] J. G. Bellingham and J. S. Willcox. Optimizing AUV oceanographic surveys. In *IEEE Conference on Autonomous Underwater Vehicles*, Monterey, CA, 1996.

[9] James O. Berger. *Statistical Decision Theory and Bayesian Analysis*. Springer-Verlag, 1980.

[10] D. P. Brutzman. NPS AUV integrated simulator. Master's thesis, Naval Postgraduate School, Cambridge, MA, March 1992.

[11] B. Butler. Field trials of the Theseus AUV. In *Proc. Int. Symp. on Unmanned Untethered Submersible Technology*, pages 6–15, September 1995.

[12] J. Catipovic. Performance limitations in underwater acoustic telemetry. *IEEE J. Ocean Engineering*, 15(3):205–216, July 1990.

[13] T. Chavez. Modeling and measuring the effects of vagueness in decision models. *IEEE Trans. Systems, Man, and Cybernetics*, 26(3):311–323, May 1996.

[14] I. J. Cox and J. J. Leonard. Modeling a dynamic environment using a Bayesian multiple hypothesis approach. *Artificial Intelligence*, 66(2):311–344, April 1994.

[15] T. Curtin, J. G. Bellingham, J. Catipovic, and D. Webb. Autonomous ocean sampling networks. *Oceanography*, 6(3):86–94, 1993.

[16] C. Dean and M. Werby. Target shape and material composition from resonance echoes of submerged elongated elastic targets. In *Proc. SPIE Conference on Automatic Object Recognition*, volume 1471, pages 54–65, Orlando, FL, 1991.

[17] M. Deffenbaugh. A matched field processing approach to long range acoustic navigation. Master's thesis, Massachusetts Institute of Technology, 1994.

[18] M. Deffenbaugh, H. Schmidt, and J. Bellingham. Acoustic positioning in a fading multipath environment. In *IEEE Oceans*, pages 596–599, 1996.

[19] H. F. Durrant-Whyte. *Integration, coordination and control of multi-sensor robot systems*. Boston: Kluwer Academic Publishers, 1987.

[20] H. F. Durrant-Whyte. Sensor models and multi-sensor integration. *Int. J. Robotics Research*, 7(6):97–113, 1988.

[21] A. Elfes. Sonar-based real-world mapping and navigation. *IEEE Journal of Robotics and Automation*, RA-3(3):249–265, June 1987.

[22] M.G.G. Foreman, R.A. Walters, and R.F. Henry. A model for simulating currents in the eastern juan de fuca and the southern strait of georgia. In *IEEE Oceans*, pages 335–340, 1993.

[23] E. Geyer, P. Creamer, J. D'Appolito, and R. Gains. Characteristics and capabilities of navigation systems for unmanned untethered submersibles. In *Proc. Int. Symp. on Unmanned Untethered Submersible Technology*, pages 320–347, 1987.

[24] G. Hager. *Task-directed Sensor Fusion and Planning: A Computational Approach*. Boston: Kluwer Academic Publishers, 1990.

[25] P. Hebert, S. Betge-Brezetz, and R. Chatila. Decoupling odometry and exteroceptive perception in building a global world map of a mobile robot: The use of local maps. In *Proc. IEEE Int. Conf. Robotics and Automation*, pages 757–764, April 1996.

[26] J. R. Holton. *An Introduction to Dynamic Meteorology*. San Diego, CA: Academic Press, 1992.

[27] M. Hunt, W. Marquet, D. Moller, K. Peal, W. Smith, and R. Spindel. An acoustic navigation system. Technical Report WHOI-74-6, Woods Hole Oceanographic Institution, 1974.

[28] H. Jones and J. Marshall. Convection with rotation in a neutral ocean: a study of open ocean deep convection. *Journal of Physical Oceanography*, 23:1009–1039, June 1993.

[29] J. H. Kim. Remote monitoring and control of autonomous underwater vehicles. Master's thesis, MIT, 1997.

[30] R. Kuc. Fusing binaural sonar information for object recognition. In *IEEE/SICE/RSJ International Conference on Multisensor Fusion and Integration for Intelligent Systems*, pages 727–735, 1996.

[31] R. Kuc and M. W. Siegel. Physically based simulation model for acoustic sensor robot navigation. *IEEE Transactions on Pattern Analysis and Machine Intelligence*, PAMI-9(6):766–778, November 1987.

[32] T. Kurien. Issues in the design of practical multitarget tracking algorithms. In Y. Bar-Shalom, editor, *Multitarget-Multisensor Tracking: Advanced Applications*, pages 43–83. Boston: Artech House, 1990.

[33] P. Lehner, K. Blackmond Laskey, and D. Dubois. An introduction to issues in higher order uncertainty. *IEEE Trans. Systems, Man, and Cybernetics*, 26(3):289–293, May 1996.

[34] J. J. Leonard and H. F. Durrant-Whyte. Simultaneous map building and localization for an autonomous mobile robot. In *Proc. IEEE Int. Workshop on Intelligent Robots and Systems*, Osaka, Japan, 1991.

[35] J. J. Leonard and H. F. Durrant-Whyte. *Directed Sonar Sensing for Mobile Robot Navigation*. Boston: Kluwer Academic Publishers, 1992.

[36] J. J. Leonard, S. T. Tuohy, J. G. Bellingham, B. A. Moran, J. H. Kim, H. Schmidt, N. M. Patrikalakis, and C. Chryssostomidis. Virtual environments for AUV development and ocean exploration. In *Proc. Int. Symp. on Unmanned Untethered Submersible Technology*, pages 436–443, September 1995.

[37] Zhi-Quan Luo and John N. Tsitsiklis. Data fusion with minimal communication. Technical Report LIDS-P-2065, MIT Laboratory for Information and Decision Systems, 1991.

[38] P. Malanotte-Rizzoli and A. R. Robinson, editors. *Ocean processes in climate dynamics: global and mediterranean examples*. Kluwer Academic Publishers, 1994.

[39] J. S. Manyika and H. F. Durrant-Whyte. *Data Fusion and Sensor Management: A decentralized information-theoretic approach.* New York: Ellis Horwood, 1994.

[40] M. B. May. Gravity navigation. In *Record of the 1978 Position Location and Navigation Symposium*, pages 212–218, San Diego, CA, USA, November 1978.

[41] P. H. Milne. *Underwater Acoustic Positioning Systems.* London: E. F. N. Spon, 1983.

[42] B. A. Moran, J. J. Leonard, and C. Chryssostomidis. Curved shape reconstruction using multiple hypothesis tracking. *IEEE J. Ocean Engineering*, 1996. To Appear.

[43] S. Mori, C. Chong, E. Tse, and R. Wishner. Tracking and classifying multiple targets without a priori identification. *IEEE Transactions on Automatic Control*, AC-31(5), May 1986.

[44] P. Moutarlier and R. Chatila. An experimental system for incremental environment modeling by an autonomous mobile robot. In *1st International Symposium on Experimental Robotics*, Montreal, June 1989.

[45] W. Munk, P. Worcester, and C. Wunsch. *Ocean Acoustic Tomography.* Cambridge University Press, 1995.

[46] S. Nagahdaripour and S. Zhang. Determining the size and position of cylindrical objects from optical images in underwater environments. In *Proc. Int. Symp. on Unmanned Untethered Submersible Technology*, pages 232–242, September 1995.

[47] G. Pappas, W. Shotts, M. O'Brien, and W. Wyman. The DARPA/NAVY unmanned undersea vehicle program. *Unmanned Systems*, 9(2):24–30, 1991.

[48] D. B. Reid. An algorithm for tracking multiple targets. *IEEE Transactions on Automatic Control*, AC-24(6), December 1979.

[49] W. D. Rencken. Concurrent localisation and map building for mobile robots using ultrasonic sensors. In *Proc. IEEE Int. Workshop on Intelligent Robots and Systems*, pages 2192–2197, Yokohama, Japan, 1993.

[50] A. R. Robinson. Physical processes, field estimation, and an approach to interdisciplinary ocean modeling. Technical report, Department of Earth and Planetary Science, Harvard University, October 1994.

[51] H. Schmidt, J. Bellingham, M. Johnson, D. Herold, D. Farmer, and R. Pawlowcisz. Real-time frontal mapping with AUVs in a coastal environment. In *IEEE Oceans*, pages 1094–1098, 1996.

[52] H. Singh, D. Yoerger, R. Bachmayer, A. Bradley, and W. Stewart. Sonar mapping with the autonomous benthic explorer (ABE). In *Proc. Int. Symp. on Unmanned Untethered Submersible Technology*, pages 367–375, September 1995.

[53] R. Smith and P. Cheeseman. On the representation and estimation of spatial uncertainty. *International Journal of Robotics Research*, 5(4):56, 1987.

[54] R. Smith, M. Self, and P. Cheeseman. Estimating uncertain spatial relationships in robotics. In I. Cox and G. Wilfong, editors, *Autonomous Robot Vehicles.* Springer-Verlag, 1990.

[55] W. K. Stewart. *Multisensor Modeling Underwater with Uncertain Information.* PhD thesis, Massachusetts Institute of Technology, 1988.

[56] W. K. Stewart. Remote sensing issues for intelligent underwater systems. In *International Conference on Computer Vision*, pages 230–235, 1991.

[57] G. Trimble, J. Vilaro, D. Okamura, R. Lum, and K. Dutta. Underwater object recognition and automatic positioning to support dynamic positioning. In *Proc. Int. Symp. on Unmanned Untethered Submersible Technology*, pages 273–279, 1991.

[58] S. T. Tuohy, J. J. Leonard, J. G. Bellingham, N. M. Patrikalakis, and C. Chryssostomidis. Map based navigation for autonomous underwater vehicles. *International Journal of Offshore and Polar Engineering*, 6(1):9–18, March 1996.

[59] C. Tyren. Magnetic anomalies as a reference for ground-speed and map-matching navigation. *The Journal of Navigation*, 35(2):242–254, May 1982.

[60] Toby Tyrrell. Defining the action selection problem. In *Proceedings of the Fourteenth Annual Conference of the Cognitive Science Society*, pages 1152–1157, 1992.

[61] J. Uhlmann. *Dynamic Map Building and Localization: New Theoretical Foundations.* PhD thesis, University of Oxford, 1995.

[62] J. Vaganay and V. Rigaud. Supervised navigation: optimal algorithm example and needs for autonomous supervision. In *Proc. IEEE Int. Conf. Robotics and Automation*, May 1995.

[63] M. Visbeck. Labrador sea experiment home page, http://www.ldeo.columbia.edu/~visbeck/labsea/, 1997.

[64] J. Scott Willcox, Yanwu Zhang, James G. Bellingham, and John Marshall. Auv survey design applied to oceanic deep convection. In *IEEE Oceans '96*, 1996.

[65] Alan S. Willsky, Martin G. Bello, David A. Castanon, Bernard C. Levy, and George C. Verghese. Combining and updating of local estimates and regional maps along sets of one-dimensional tracks. *IEEE Transactions on Automatic Control*, AC-27(4):799–813, August 1982.

[66] C. Wunsch. *The Ocean Circulation Inverse Problem.* Cambridge University Press, 1996.

[67] Ronald R. Yager, Mario Fedrizzi, and Janusz Kacprzyk, editors. *Advances in the Dempster-Shafer Theory of Evidence.* John Wiley and Sons, Inc., 1994.

Proceedings of the Seventh (1997) International Offshore and Polar Engineering Conference
Honolulu, USA, May 25-30, 1997

ISBN 1-880653-28-1 (Set); ISBN 1-880653-30-3 (Vol. II)

Reentry of Deepsea Riser by Adaptive Control

Dongho Nam
Junior College of Inchon
Inchon, Korea

Hideyuki Suzuki and Koichiro Yoshida
University of Tokyo
Tokyo, Japan

ABSTRACT

This paper presents reentry of a deepsea riser by an adaptive control. Reentry operation is one of the major problems regarding a deepsea riser, as the lower end of riser must be accurately positioned over the tarket point on the seabed. But the deepsea riser shows complex 3-D elastic response due to flexibility, complex hydrodynamic forces and nonlinearity of the riser dynamics. The required positioning accuracy of the riser bottom end is high. Moreover, elastic deformation must be controlled for securing structural integrity.

In water, estimation error of system parameter and nonlinearity of hydrodynamic force may make the result of conventional optimal control inaccurate. In adaptive control, hydrodynamic terms like added mass and drag coefficient in the riser dynamics are identified and control forces on the floating body and riser are calculated simultaneously. An adaptive algorithm for MIMO linear discrete time system without requiring a persistent excitation is adopted in this study. The effectiveness of adaptive control logic is tested by numerical simulations and model experiment. The present control system shows good overall performances and can be applied to the control of the deepsea riser in terms of 2-dimensional responses.

Key Words : Adaptive control, Riser, Reentry, Persistently exciting input, Unmodeled dynamics, Parameter identification

INTRODUCTION

The riser reentry in deepsea drilling is the operation to make the bottom end of the riser connect with BOP (blowout preventer). It is one of the major critical operations regarding deepsea riser. The very long riser shows complex 3-dimensional coupled nonlinear response and this response makes reentry operation of the riser increasingly difficult and the guideline reentry usually performed in shallow water is not possible anymore. This paper concerns automatic reentry of the riser which is essential for deepsea riser as shown in Fig.1. Position control of the lower end of a 6000-m vertical pipe had been discussed for the manganese nodule mining and the elastic response of the 3-D responses are considered in Brink and Chung (1981). But the required accuracy can be higher for the reentry of the deepsea riser.

Previously, Suzuki, Yoshida and Nam (1994) presented a basic research on automatic reentry of the riser by an optimal control. The coupled equations of motion of the floating body and the riser were formulated first and then effectiveness of optimal control system was demonstrated by computer simulation and a basin experiment in terms of 2-dimensional responses.

However, estimation error of system parameter and nonlinearity of hydrodynamic force may make the result of conventional optimal control inaccurate in water. The nonlinear hydrodynamic term like drag coefficient is easily varying according to operation condition and environment. In this study, adaptive control is adopted in order to deal with hydrodynamic force terms for the sake of better 2-D dynamic response.

The majority of adaptive controllers can be classified as

either direct or indirect. In indirect adaptive control, the system parameters are identified first, then control law is updated using the most recent parameter estimates. A well-known problem with this scheme is, however, that the estimated system model may not be controllable or stabilizable for singular points. To avoid having the parameter estimates converge to a singular point, one can use a persistently exciting input to force the parameter estimates to converge to the true parameter values (Astrom and Wittenmark, 1989). The idea is to use an external input with many different frequencies which enrich the input/output information allowing perfect identification of the system parameters.

But the problem of choosing a persistently exciting input in the presence of unmodeled plant dynamics is a possibility that any uncertainty in the plant could counteract the exciting input. Moreover, to use a persistently exciting input is not practical and it may do damage to a very flexible structure like a riser. Therefore, an adaptive algorithm, which does not require a pesistently exciting input derived for MIMO(multi-input multi-output) linear discrete-time system by Ossman and Kamen (1987), is selected as an adaptive controller in this study. In the present paper, adaptive control system is formulated and tested by computer simulation and basin experiment.

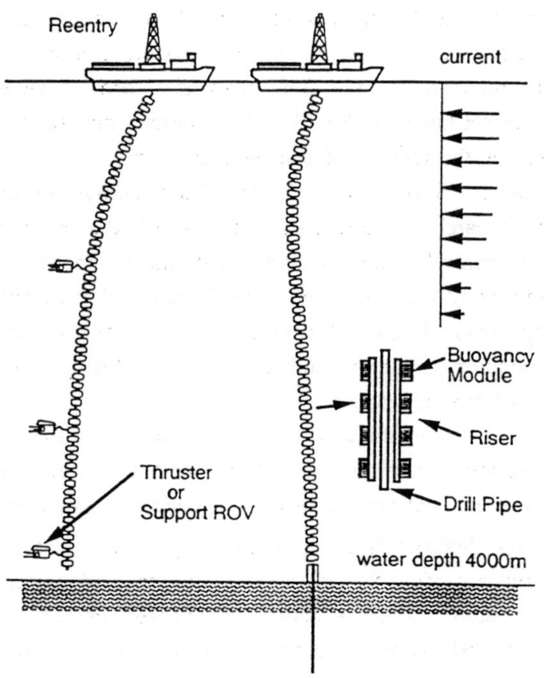

Fig.1 Schematic diagram of riser reentry.

ADAPTIVE CONTROL ALGORITHM

Ossman and Kamen(1987) developed a recursive estimation algorithm that forces the estimates of the unknown system parameters to converge to the known intervals. It is similar to the least square estimation algorithm, except that a correction term, which is introduced in design of an adaptive observer for SISO continuous time system by Kreisselmeier (1986) at first, has been added to ensure that estimates do converge to the known intervals. They combined the parameter estimation algorithm with a MIMO extension of the adaptive regulator proposed by Samson and Fuchs (1981). The result is an adaptive regulator that can be applied to both nonminimum and minimum phase discrete time systems and which does not require a persistently exciting input. The closed loop system is globally stable in the sense that the inputs and outputs converge to zero for any initial conditions in the plant.

System Definitions and Assumptions

The assumptions made on the unknown plant are: 1) an upper bound on the system order is known, 2) the system parameters belong to known bounded inervals, and 3) the plant is stabilizable for all possible values of the unknown system parameters ranging over the known intervals.

The system to be regulated is the MIMO linear discrete time system described by

$$y(k) = -\sum_{j=1}^{n} A_j y(k-j) + \sum_{j=1}^{n} B_j u(k-j) \qquad (1)$$

In eqn.(1), y(k) is output vector and u(k) is control input vector. The sysem described by eqn.(1) can rewritten in the following form convenient for parameter estimation:

$$y(k) = D^T \phi(k-1) \qquad (2)$$

$$\textit{where},\ D^T = [-A_1 \cdots -A_n\ B_1 \cdots B_n]$$
$$\phi^T(k-1) = [y(k-1) \cdots y(k-n)\ u(k-1) \cdots u(k-n)]$$

In certain applications, some of the entries in the system matrix D will be known a priori. Since it is not necessary to estimate known parameters, a scheme for separating the known parameters from the unknown parameters is advantageous. This is accomplished by rewriting eqn.(2) in the following form:

$$y(k) = \theta^T \phi_a(k-1) + \psi^T \phi_b(k-1) \qquad (3)$$

The matrix θ contains all of the unknown entries in D while the matrix ψ contains only those entries of D which are known a priori. The vectors $\phi_a(k-1)$ and $\phi_b(k-1)$ are regression vectors whose components come from

$\phi(k-1)$. The vector norm used throughout this paper is the Euclidean norm.

Parameter Estimation

Consider the linear discrete-time system described by ARMA(Auto-Regressive Moving Average) model like eqn.(3) and the parameter estimation algorithm is given by

$$\begin{aligned}\theta(k) &= \theta(k-1) - P(k-1)f(\theta(k-1)) \\ &+ \frac{P(k-1)\phi_a(k-1)}{\eta_{k-1}^2 + \phi_a^T(k-1)P(k-1)\phi_a(k-1)} \\ &\times[y^T(k) - \phi_b^T(k-1)\psi - \phi_a^T(k-1)\theta(k-1)]\end{aligned} \tag{4}$$

$$P(k) = P(k-1) - \frac{P(k-1)\phi_a(k-1)\phi_a^T(k-1)P(k-1)}{\eta_{k-1}^2 + \phi_a^T(k-1)P(k-1)\phi_a(k-1)} \tag{5}$$
$$0 < P(0) = P^T(0) < 2I$$

$$f_{ij}(k-1) = \begin{cases} \theta_{ij}(k-1) - \theta_{ij}^{max}, & when \ \theta_{ij}(k-1) > \theta_{ij}^{max} \\ \theta_{ij}(k-1) - \theta_{ij}^{min}, & when \ \theta_{ij}(k-1) < \theta_{ij}^{min} \\ 0, & when \ \theta_{ij}(k-1) \in [\theta_{ij}^{min}, \theta_{ii}^{max}] \end{cases} \tag{6}$$

$$\eta_{k-1} = \begin{cases} 1, & when\ the\ determinant\ of\ P(k) > \varepsilon, \\ \max(1, \|\phi_a(k-1)\|), & otherwise. \end{cases} \tag{7}$$

where ε is any small positive number. It is assumed that each component θ_{ij} of the unknown sysem parameter matrix θ belongs to known bounded interval $[\theta_{ij}^{min}, \theta_{ij}^{max}]$.

This parameter estimation algorithm differs from the recursive least-squares estimator in the following three ways: 1) the initial covariance matrix P(0) must be less than 2I, whereas the least-squares algorithm allows for any positive definite initial covariance matrix, 2) the data normalization introduced through the term η_{k-1} which is simply equal to one in the least-squares algorithm, 3) the major difference is the addition of a correction term $P(k-1)f(\theta(k-1))$. This term forces the estimates θ_{ij} of the components of system parameter matrix to converge to the sets $[\theta_{ij}^{min}, \theta_{ij}^{max}]$. This property eliminates the need for a persistently exciting input. The estimation algorithm eqn.(4)-(7) has the following properties (Ossman and Kamen, 1987).

1. P(k) converges to a positive semidefinite matrix $P_\infty < 2I$.
2. $|\theta_{ij}(k)| < \infty$ for all k.
3. $\dfrac{\|e(k)\|^2}{\eta_{k-1}^2 + \phi_a^T(k-1)P(k-1)\phi_a(k-1)} \rightarrow 0 \quad as \quad k \rightarrow \infty$

The prediction error e(k) is defined as

$$e(k) = y(k) - \theta^T(k-1)\phi_a(k-1) - \psi^T\phi_b(k-1)$$

4. $\|e(k)\| \le a(k)\|\phi(k-1)\| + b(k)$

where a(k) and b(k) are positive scalar sequences which converge to zero.

5. $f(\theta(k)) \rightarrow 0$ as $k \rightarrow \infty$, which implies θ_{ij} converges to $[\theta_{ij}^{min}, \theta_{ij}^{max}]$.
6. $|\theta_{ij}(k) - \theta_{ij}(k-p)| \rightarrow 0$ *as* $k \rightarrow \infty$ *for any integer* p

Adaptive Regulator

The state-space observer form realization for the system described by eqn.(1) is given by

$$x(k+1) = Fx(k) + Gu(k), \quad y(k) = Hx(k) \tag{8}$$

where

$$F = \begin{bmatrix} -A_1 & I & & \\ -A_2 & & I & \\ \vdots & & & \ddots \\ -A_{n-1} & & & I \\ -A_n & & & 0 \end{bmatrix}, G = \begin{bmatrix} B_1 \\ B_2 \\ \vdots \\ B_{n-1} \\ B_n \end{bmatrix}, H = [I \ 0 \ \dots \ 0] \tag{9}$$

The control force is given by

$$u(k) = -L(k)x(k) \tag{10}$$

$$L(k) = [G^T(k)R_kG(k) + I]^{-1}G^T(k)R_kF(k) \tag{11}$$

$$\begin{aligned} R_{k+1} &= Q + L^T(k)L(k) \\ &+ (F(k) - G(k)L(k))^TR_k(F(k) - G(k)L(k)) \end{aligned} \tag{12}$$

where L is feedback gain, Q is weighting matrix, R is the solution to the Riccati difference equation. This asymptotic approach requires the solution of only one iteration of the Riccati difference equation at each point in time, so it offers considerable savings in on-line calculations.

Since knowledge of an upper bound on system order is one of the assumptions made on most adaptive controllers, it may be possible that this controller also goes unstable for system with unmodeled dynamics. Moreover, transient response including convergence rate and overshoot is not considered, so the effectiveness of this adaptive algorithm must be verified by computer simulation and basin experiment.

CONTROL SIMULATION

We refer to the research results by optimal control (Suzuki, Yoshida and Nam, 1994) for the formulation of adaptive control system for the reentry of the riser. Simulation calculations are carried out for the experimental model and real system. The experimental model has two thrusters, one at midpoint of the riser and the other at lower end of the riser. Four thrusters are assumed to be installed on the riser for real system. Each of the experimental model and real system have one thruster at the floating body also. The characteristic values of the experimental model and real system is shown in Table 1. Throughout this paper, F0 indicates the thruster on the floating body, F1 and F2 are those installed at lower end of the riser and at midpoint of the riser. In the real system, thrusters F3 and F4 are installed further on the riser at one fourth and three fourths of the riser length from the riser top.

The inertia coefficient C_m and drag coefficient C_d in the Morison equation to evaluate hydrodynamic forces, depend on KC number and Reynolds number (Faltinsen, 1990). In the simulation, it is assumed to be $C_m=2.0$ and $C_d=1.0$ as the true values of uncertain hydrodynamic parameters.

Simulation Result for the Experimental Model

Using the central difference method (Bathe and Wilson, 1976), discretize the experimental model system by time intervals $\Delta t=0.005$sec and the following ARMA model is obtained.

$$y(t)=Cy(t-1)+Ky(t-2)+Pu(t-1) \quad (13)$$

$$y(t)=[Y(t),\ \theta(t),\ Z_1(t)]^T,\quad u(t)=[u_0(t),\ u_1(t),\ u_2(t)]^T \quad (14)$$

$$C=\begin{bmatrix} c_{11} & c_{12} & c_{13} \\ c_{21} & c_{22} & c_{23} \\ c_{31} & c_{32} & c_{33} \end{bmatrix},K=\begin{bmatrix} k_{11} & k_{12} & k_{13} \\ k_{21} & k_{22} & k_{23} \\ k_{31} & k_{32} & k_{33} \end{bmatrix},P=\begin{bmatrix} p_{11} & p_{12} & p_{13} \\ p_{21} & p_{22} & p_{23} \\ p_{31} & p_{32} & p_{33} \end{bmatrix} \quad (15)$$

where Y is sway displacement of the floating body and θ is rigid body rotation angle of the riser and Z_1 is the first deflection mode of the riser. Coordinate system and definition were demonstrated in the previous paper (Suzuki, Yoshida and Nam, 1994). u_0, u_1, and u_2 are control forces required at the floating body(F0), at the lower end of the riser(F1) and at midpoint of the riser(F2), respectively.

Fig.2 shows the simlation result with initial true value ($C_m=2.0$, $C_d=1.0$). Initial displacement of the floating body and inclination of the riser are 1.0m and 0.1rad. The arrows in the figure indicate the positions of installed thrusters. The approximate value of current velocity measured in the experiment is used in the simulation. Up to 10th deflection modes of the riser are considered as the influence of unmodeled dynamics in the calculation. We set the weighting matrix Q=I(unit marix) and the initial Riccati solution R(0)=I in eqn.(12).

Despite the presence of unmodeled dynamics, the coupled responses of the floating body and riser are successfully controlled in the current and the lower end of the riser is brought to the target and kept there. The function of control is confirmed. Average control forces required are 9.1gf at the floating body and 0.93gf at the riser. Positioning error for the lower end of the riser is 2mm which is 0.1% of the water depth 2m.

Table 1. Principal data of model and real system.

Floating body	Model	Real system
Mass(kg)	14.083	1.5×10^7

Riser	Model	Real system
Length(m)	1.777	4000
Outer diameter(m)	0.064	0.406
Young's modulus(N/m^2)	6.3×10^9	2.1×10^{11}
Moment of inertia of section(m^4)	4.667×10^{-12}	5.541×10^{-4}
Weight in air(N)	22.54	3.0×10^7
Weight in water(N)	1.13	1.5×10^6

Simulation Result for the Real System

Using the central difference method, discretize the real system by time intervals $\Delta t=0.1$sec and the following ARMA model is obtained.

$$y(t)=Cy(t-1)+Ky(t-2)+Pu(t-1) \quad (16)$$

$$y(t)=[Y(t),\ \theta(t),\ Z_1(t),Z_2(t),Z_3(t)]^T,\quad u(t)=[u_0(t),u_1(t),u_2(t),u_3(t),u_4(t)]^T \quad (17)$$

where system parameters C, K, P to estimate are 5×5 matrix similar to 3×3 matrix in eqn.(15). Z_1, Z_2, Z_3 are the first, second, third deflection mode of the riser. u_0, u_1, u_2, u_3, u_4 are control forces required at the floating body(F0),

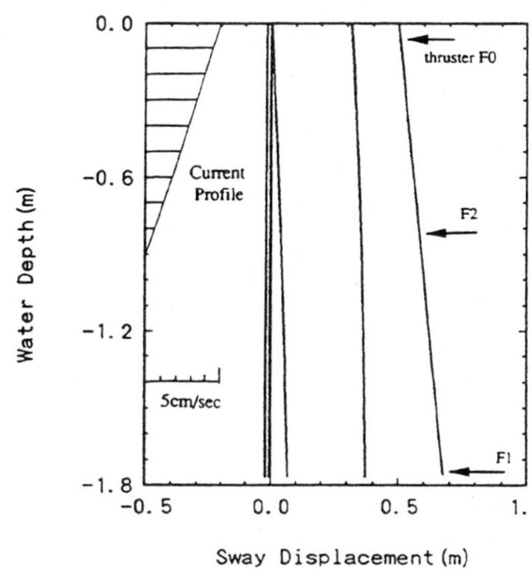

Fig.2 Simulation result of model with true initial value. (Time interval of displayed lines = 3.0sec)

at lower end of the riser(F1), at midpoint of the riser(F2), at one fourth(F3) and three fourths(F4) of the riser length from the riser top, respectively.

Fig.3 shows the simulation result for the real system with 4000m riser. The simulation is computed with the initial system parameters deviated from true values by 20% ($C_m=1.6, C_d=0.8$). Initial displacement of the floating body and inclination of the riser are 1000m and 0.1rad. Current velocity is 0.25m/sec at the surface of the sea and zero at water depth 1000m. Up to 12th deflection modes of the riser are considered as the influence of unmodeled dynamics in the calculation. We set the weighting matrix $Q=10^4I$ and the initial Riccati solution R(0)=I in eqn.(12).

Effectiveness of the control to keep the lower end of the riser right over the target suppressing the deflection of the riser, is verified despite presence of unmodeled dynamics. The positioning error of the lower end of the riser is 2m. The estimated system parameters not presented here also rapidly converge to constant values in the simulation. Average control force required at the floating body(F0) is 1.4tonf and that required at the riser(F1+F2+F3+F4) is about 700kgf.

EXPERIMENT

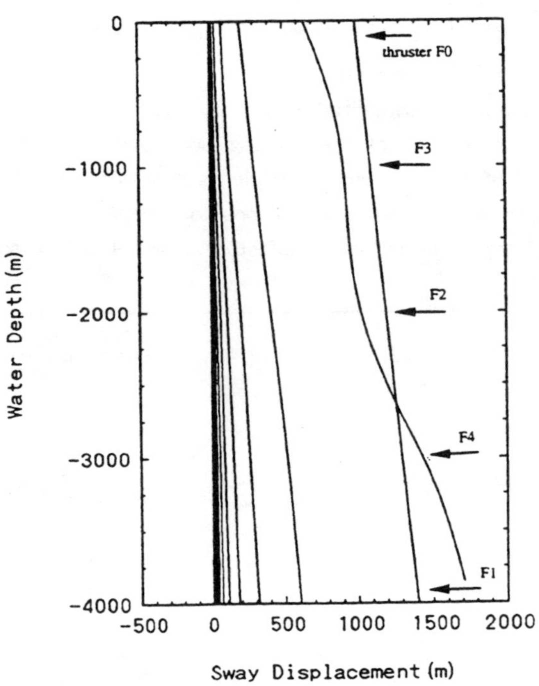

Fig.3 Simulation result of real system with 20% initial error. (Time interval of displayed lines = 10min)

Experimental Setup and Model

Simulation results for the experimental model and real system are very satisfactory. However, to carry out a basin experiment is necessary in order to confirm the function of adaptive control system because the equations of motion used in the simulation are linearized ones and the effect of unmodeled dynamics which cannot be considered in the simulation also exists.

To utilize the experimental results for the real system, the experimental model must satisfy some similarity law. But it is very difficult to make the experimental model which can satisfy geometrical similarity in case of a very long and slender structure like a riser (Hartnup and Airey, 1982). Therefore, considering the principal parameters of dynamical similarity, the experimental model is made in the present study.

Ratio of the mass of the floating body to that of the riser for real system (about 5:1 in the preliminary design) is very important factor for coupled responses between the floating body and the riser and this value is kept to be same in the experimental model. The ratio of weight in water to weight in air (1/20 in the real system) is kept same in the riser model and on the other hand the bending rigidity is made as small as possible.

The maximum thrust of a propeller type thruster used in the experiment is 120gf. Displacement of the floating body and deformation of the riser are measured by ultrasonic ranging sensors and triangular surveying technique.

Experimental setup is not presented here, as it was demonstrated in the previous paper (Suzuki, Yoshida and Nam, 1994), to which details are referred. Deflection of the riser is measured up to the 3rd mode by modal filter (Meirovitch and Baruh, 1985) and the control is run at 5Hz. Block diagram of control is shown in Fig.4.

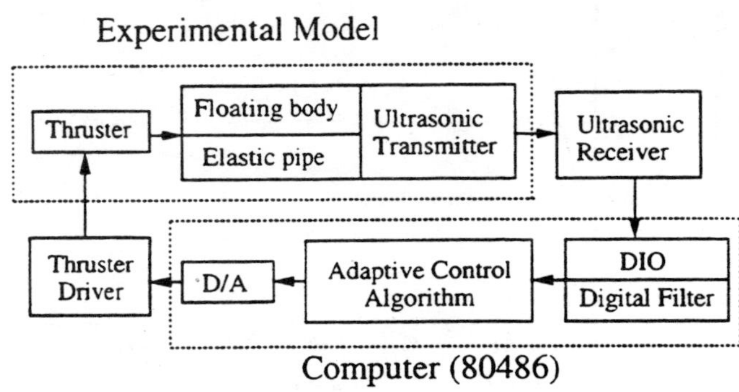

Fig.4 Block diagram of control system.

Basin Experimental Result and Discussion

In the reentry experiment, the control starts in the current with displaced position of the floating body and initial inclination of the riser. Deformation and inclination of the riser are suppressed and the lower end of the riser is brought right over the target on the basin floor and then the position is kept.

Sway displacement of the floating body is shown in Fig.5. Control accuracy is less than 1cm which is 0.5% of the water depth 2m. Surge(out of plane) motion of the floating body which is not discussed in the present study is independently controlled by conventional PID control and it makes the experiment purely 2-dimensional problem.

The result of surge control is shown in Fig.6 and the performance of yaw control not presented here is also satisfactory.

Fig.7 shows the sway displacement of the lower end of the riser from the target. Solid line shows the case in which both the floating body and the riser are controlled and the lower end of the riser is kept right under the floating body. Broken line shows the case in which only the floating body is controlled. By control of the riser, displacement of the lower end of the riser from the target, which is induced by current, is decreased from 6cm to 3cm, that is, reduced by 50% quantitatively. Average control force required for position keeping of the floating body is 20gf. This value corresponds to about 20tf in the real system. The control forces required at midpoint and lower end of the riser are 3.0gf corresponding to about 3.0tf in the real system.

System parameters are estimated from the experiment. Fig.8 and Fig.9 show only the dominant diagonal elements

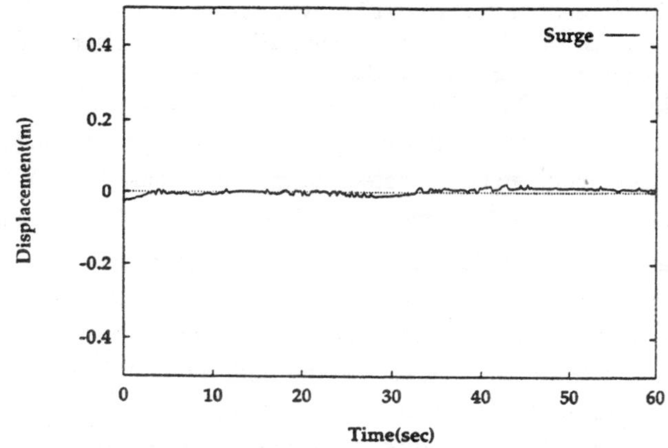

Fig.6 Surge displacement of floating body in the experiment.

of the matrix C, P in eqn.(15). The estimated parameters also rapidly converge to constant values. The estimation algorithm used in this study has no need to calculate inverse matrix and the regulator requires the solution of only one iteration of the Riccati difference equation at each point in time. This adaptive controller has the great practical merit of computational simplicity.

But the estimated parameters do not converge to the true values because we do not use a persistently exciting input to force the parameter estimates to converge to the true values. We set the weighting matrix Q=I and the initial Riccati solution R(0)=I in eqn.(12) and $C_m=2.0, C_d=1.0$ as initial system parameters.

The basin experimental results by adaptive control show that sway motion of the floating body and the rigid body rotation of the riser are controlled suppressing deflection of the riser. But the accuracy of position keeping of the lower end of the riser is not satisfactory. One of the reasons is

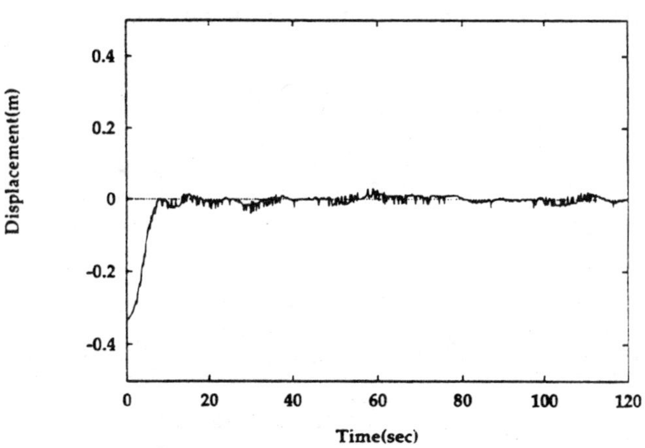

Fig.5 Sway displacement of floating body in the experiment.

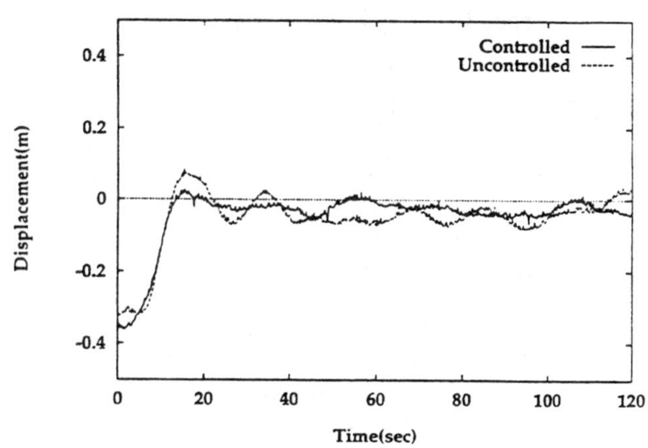

Fig.7 Displacement of lower end of riser in the experiment.

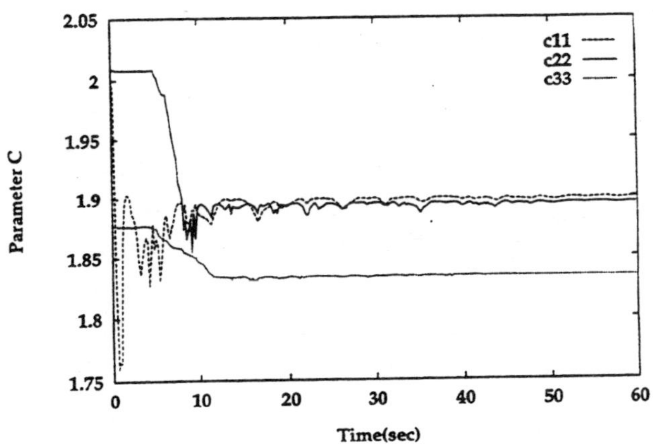

Fig.8 Estimated value of parameter C in the experiment.

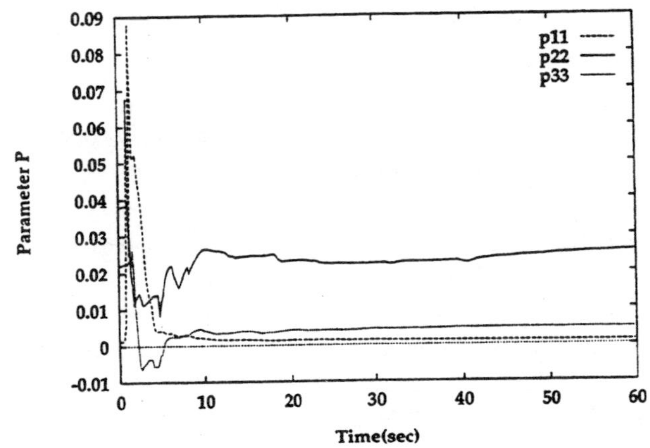

Fig.9 Estimated value of parameter P in the experiment.

thought to be modeling too simple control system like eqn.(13)-(15). We designed the adaptive control system considering the capability of control hardware used in the experiment for real-time control.

Moreover, the effect of electrical noise of the thruster is observed in the experiment. Fluctuation of control force due to the noise may make the control system unstable, so we set the weighting matrix Q to calculate small control forces at the riser. This is another reason why the control effect is not so significant for the lower end of the riser.

CONCLUSION

Basic investigations are carried out for automatic reentry system of a deepsea riser by an adaptive control. It is adopted in order to deal with the uncertain hydrodynamic terms like added mass and drag coefficient. Selecting an adaptive algorithm for the MIMO linear discrete time system, which does not require a persistent excitation, the main conclusions can be drawn as follows:

(1) The simulation result with 20% initial error in the presence of unmodeled dynamics shows that suppressing the deformation of the riser, the position of the lower end of the riser is kept right over the target on the sea floor and the estimated system parameters also rapidly converge to constant values. The robustness of the adaptive control system designed in this study is confirmed.

(2) The basin experiment is performed by adaptive control which is well known to be very difficult to implement because of on-line parameter estimation. Control accuracy for displacement of the floating body is less than 1cm, 0.5% of the water depth. By the control of the riser, the displacement of the lower end of the riser from the target is reduced by 50% compared with uncontrolled case within the range of realistic value of the thrusters. The estimated parameters also rapidly converge to constant values as demonstrated by the simulation result. Effectiveness and stability of the present adaptive system are confirmed and the possibility for realization of the automatic reentry system by adaptive control is obtained.

(3) With the research results by optimal control (Suzuki, Yoshida and Nam, 1994), promising results are obtained for the synthesis of a 2-dimensional control system for position and elastic response of a very long and flexible offshore structural system in the present study. The present result can be extended to a 3-dimensional real system through further study.

REFERENCES

Astrom, KJ. and Wittenmark, B (1989), *Adaptive Control*, Addison Wesley.

Bathe, KJ. and Wilson, EL (1976), *Numerical Methods in Finite Element Analysis*, Prentice-Hall.

Brink, AW, and Chung, JS (1981), "Automatic Position Control of a 30,000 Tons Ship during Ocean Mining Operations," *OTC4091*, pp.205-224.

Faltinsen, OM (1990), *Sea Loads on Ships and Offshore Structures*, Cambridge University Press.

Hartnup, G, and Airey, R (1982), "Hydrodynamic Tests on Marine Risers," *OTC4319*, pp.223-233.

Kreisselmeier, G (1985), "An Approach to Stable Indirect Adaptive Control," *Automatica*, Vol.21, pp.425-431.

Meirovitch, L, and Baruh, H (1985), "The Implementation of Modal Filters for Control of Structures," *J. Guidance & Control*, Vol.8 No.6 pp707-716.

Ossman, KA, and Kamen, EW (1987), "Adaptive Regulation of MIMO Linear Discrete Time Systems Without Requiring a Persistent Excitation," *IEEE Trans. AC* Vol.AC-32, No.5, pp.881-889.

Samson, C, and Fuchs, JJ (1981), "Discrete Adaptive Regulation of Not-Necessarily Minimum-Phase Systems," *Proc. IEEE* Vol.128, Pt.D, No.3 pp.102-108.

Suzuki, H, Yoshida, K, Ishida, S, and Nam, DH (1994), "Active Control of Riser Deformation and Vessel Motion for Entry/Reentry System," *Proc. of the 4th Int. Conf. on Offshore and Polar Engineering*, Vol.2, pp.216-223.

Proceedings of the Seventh (1997) International Offshore and Polar Engineering Conference
Honolulu, USA, May 25-30, 1997

ISBN 1-880653-28-1 (Set); ISBN 1-880653-30-3 (Vol. II)

Application of Thrusts to Elastic Joints of Long Vertical Pipe in 3-D Nonlinear Motions: Part I. MSE and FEM Modeling

Jin S. Chung
Colorado School of Mines
Golden, CO, USA

Baorong Cheng and Zhao-Chang Zheng
Tsinghua University
Beijing, China

ABSTRACT

In order to be able to apply external (thrust control) forces on elastic joints along a very long vertical pipe in deep ocean, the joints are modeled by finite elements (FEM), as well as a point mass-spring elements (MSE) consisting of a point mass and connecting springs. The elastic joint is modeled by 2 computational methods: mass-spring elements (MSE) and finite elements (FEM). An MSE is less complex than FEM in modeling a point mass, to which the external force (e.g., thrust) vector is applied. The MSE treats the mass-spring only and does not include the elastic properties. The flexible joint was successfully modeled previously by FEM. The previous FEM model is modified in order to apply the external force to the joint or joint's mass. When the FEM is used, the external force is applied to nodes of the pipe element adjacent to the joint. With the MSE, the external force vector can be applied directly to the point mass of the elastic joint. Preliminary results are discussed. The accuracy and advantages of the two methods are compared in detail with numerical examples in Part II of the present paper.

† Visiting research scholar at Colorado School of Mines, Golden, Colorado, USA.

KEY WORDS: Modeling, mass-spring element, finite element, thrust vector, vertical pipe, elastic joint, three dimension, nonlinear, motions, deep-ocean.

INTRODUCTION

It is now a general consensus that most new researchers involved in developing recent deep-ocean mining pipe systems prefer a self-propelled, seafloor miner (or collector) to operate at 4,000 - 6000-m depth, which is similar to those of Chung and Chung et al. (1980-1995). Further research of such a self-propelled miner will likely lead to a change in the connecting systems of the lift (or vertical flow) pipe to the miner, similar to that of Chung, Whitney and Loden (1980 and 1981).

For such mining systems (Chung and Tsurusaki, 1994), it is desirable for the designers and operators to explore ways to reduce the lateral deflection and axial motion and stress of a vertical pipe, while keeping minimal interruption of vertical slurry transport or lift. Still, the real values of the hydrodynamic damping along the pipe during the at-sea operation are not clearly definable in the prediction of lateral motion amplitudes and static deflection of the pipe bottom end. The current flow vector in the water column is not accurately known during the at-sea operations and varies as a function of time and vertical distance from the free surface. Moreover, the hydrodynamic drag coefficients can fall into a critical Reynolds number region (Chung and Felippa, 1981). This question of actual hydrodynamic damping values may remain unanswered and may have to be dealt with by the control system design for the field operations of a deep-ocean pipe system (Chung, Whitney and Loden, 1980; Chung and Felippa, 1981).

For a new entire pipe system, the pin joins a horizontal or nearly horizontal rigid pipe to a long near-vertical lift (or hoist) pipe system and connects to a self-propelling seafloor miner on the seafloor (Chung, 1997). This pipe system was studied in the late 70's by an international consortium. Its primary purpose was not to interrupt solid (e.g. manganese nodules) transport by the 2-phase (solid-liquid) flows from the miner to the pipe, while allowing a greater freedom for a miner to maneuver under the least influence of the pipe motion. This concept can apply to both manganese nodule production from 4,000-6,000-m water depth and cobalt-rich manganese crust production from 800-2,500-m depth.

For this new concept, one of the most crucial operational parameters is static, as well as dynamic, interactions of the pipe bottom end with the miner maneuvering on the seafloor. One of the initial concepts was to install elastic joints along the pipe, and was successfully modeled with the elastic joints in the updated 3DNLPIPE code. It was previously demonstrated (Chung and Cheng, 1995; Cheng and Chung, 1996) that the elastic (flexible) joints can be made very effective in changing eigenfrequencies and axial, as well as bending, characteristics, for designers, when a certain number of joints with proper stiffness are installed at proper locations along the vertical pipe. However, this modeling did not include the application of the external lateral forces or thrust vectors on the joints, and the mass of the joints was ignored.

Part I of the present investigation emphasizes the modeling of the elastic joint (Fig. 1) by 2 computational methods: mass-spring element (MSE) and finite elements (FEM). MSE modeling applies the thrust (external) forces directly to the elastic joint, while the FEM modeling applies to the pipe elements (FEM). Part II (Cheng and Chung, 1998) emphasizes the accuracy and advantages of the 2 methods.

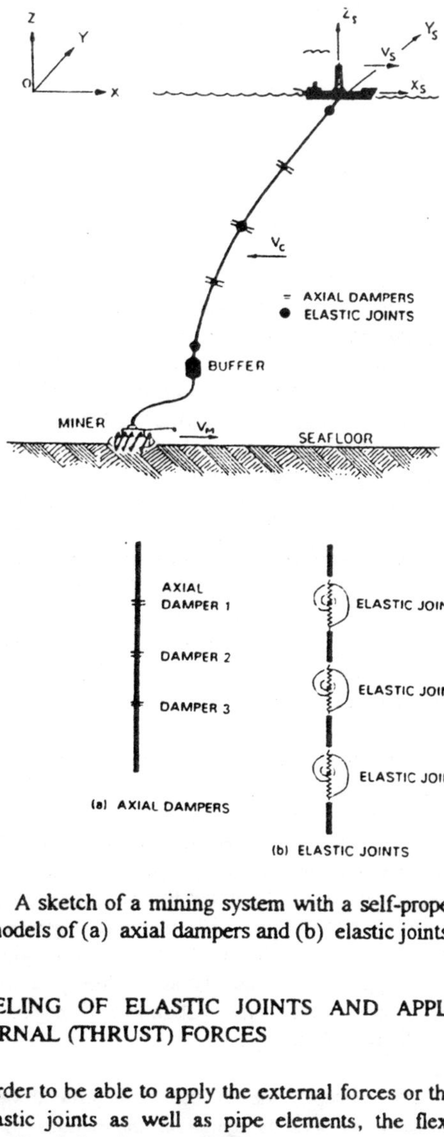

Fig. 1 A sketch of a mining system with a self-propelled miner and models of (a) axial dampers and (b) elastic joints

MODELING OF ELASTIC JOINTS AND APPLICATION OF EXTERNAL (THRUST) FORCES

In order to be able to apply the external forces or thrust vectors on the elastic joints as well as pipe elements, the flexible joints are modeled by finite elements (FEM), as well as a mass-spring elements (MSE) consisting of a point mass and connecting springs (Fig. 2). The point mass is a mass-spring with its mass, but with its size being ignored.

Elastic joints, to which the external forces are applied, are modeled by MSE, as well as FEM. An MSE is simpler than FEM to model an elastic joint. Although the joint was successfully modeled previously by FEM (3DNLPIPE code, Cheng and Chung, 1994), the present code needs to be modified in order to apply the external force to the joint or joint's mass. When the FEM is used, the external forces are applied to the nodes of the beam elements adjacent the joint: 3 displacement and 3 rotational springs (Fig. 2). The MSE allows the external force to be applied directly to the point mass of the joint. Note that MSE may be classified as a type of DEM (discrete element methods), but the MSE treats the mass-spring only and does not include the elastic properties.

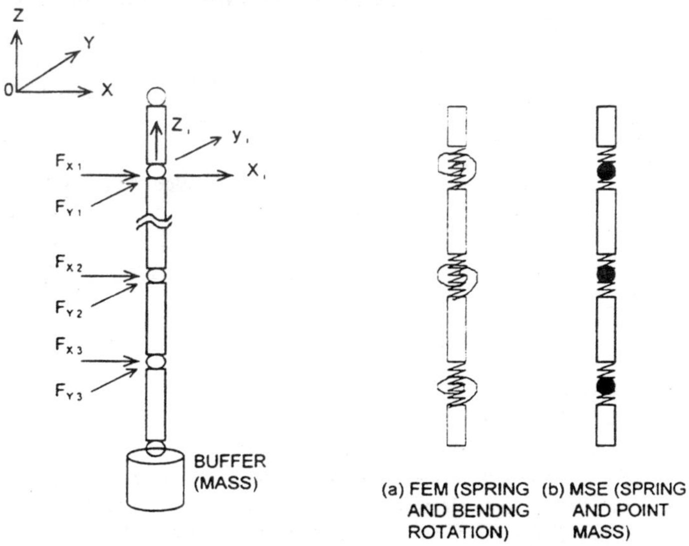

Fig. 2 Modeling of elastic joint by (a) FEM (spring and bending rotation) and (b) MSE (spring and point mass) and application of thrust vector: Joint positions at z_J = –1,000, –9,000 and –17,000 ft

The Entire Pipe System

The entire pipe system consists of a very long pipe with elastic joints, as was previously modeled using FEM by Cheng and Chung (1995). The pipe is modeled by finite elements, and the elastic joints are modeled by 2 different methods: mass-spring elements (MSE) and finite elements (FEM).

Mass-spring Element (MSE) Modeling

Each mass-spring element (MSE) is modeled by 1 rigid body element and 2 spring elements connecting the adjacent pipe elements under the following assumptions:

- The size of the elastic joint is very small, compared to the pipe element. This allows the mass-spring element in modeling the elastic joint to be considered as a point mass for the response computation. In dynamic analysis, the mass and moment of inertia of the elastic joint can be lumped to the point mass. For the calculation of the hydrodynamic forces and the application of the thrust vectors to the MSE, however, the size of the elastic joint is accounted for.
- The elastic joint has the same coordinate position as that of the nodes of the adjacent pipe elements.

- The spring element consists of 3 displacement springs and 3 rotational springs represented by a diagonal stiffness matrix. Their stiffness is considered linear elastic.

Similar approaches for the mass-spring element were previously reported in Duncan (1952), Huang (1958) and Qian and Zhang (1991).

EIGENVALUE ANALYSIS

Let X be an independent displacement vector in the generalized coordinates of the entire pipe system. This vector can consist of 2 displacement subvectors relating to the pipe node and the elastic joint:

$$X^T = (X^T_{pipe}, X^T_{ej})^T \tag{1}$$

where X_{pipe} = generalized displacement vector of the pipe node and X_{ej} = generalized displacement vector of the elastic joint.

For the e-th pipe's finite element, the kinetic energy, T^e_{pipe}, may be expressed as:

$$T^e_{pipe} = \frac{1}{2}(\dot{X}^e_{pipe})^T m^e (\dot{X}^e_{pipe}) \tag{2}$$

When the axial force is included, the potential energy of the e-th pipe's finite element may be expressed as:

$$U^e_{pipe} = \frac{1}{2}(X^e_{pipe})^T (k^e_s + k^e_g)(X^e_{pipe}) \tag{3}$$

where X^e_{pipe} and $\dot{X}^e_{pipe}$ = displacement and velocity vectors, respectively, of the node of the e-th pipe element, m^e = element mass matrix, k^e_s, k^e_g = structural stiffness matrix and geometric stiffness matrix, respectively, and can easily be obtained by FEM.

For the r-th elastic joint, the potential energy, U^r_{ej}, of its spring element joined to the adjacent pipe elements is expressed as:

$$U^r_{ej} = \frac{1}{2}(X^{PT}, X^{rT}, X^{QT}) k^r_j \begin{pmatrix} X^P \\ X^r \\ X^Q \end{pmatrix} \tag{4}$$

where X^r = displacement vector of the r-th MSE, X^P and X^Q = displacement vectors of the pipe nodes adjacent to the r-th MSE, respectively, and k^r_j = the stiffness matrix of the r-th elastic joint. k^r_j can be expressed as:

$$k^r_j = \begin{pmatrix} k_s & -k_s & 0 \\ -k_s & 2k_s & -k_s \\ 0 & -k_s & k_s \end{pmatrix} \tag{5}$$

where k_s is a diagonal matrix consisting of 3 displacement stiffness and 3 rotational stiffness. These stiffness can be determined from the linear elastic material of the joint.

Although the mass moment of inertia of the elastic joint is not assigned to the pipe nodes adjacent to the joint, it can be modeled by MSE directly for the joint to have the kinetic energy, T^r_{ej}. The kinetic energy modeled by MSE can be expressed as:

$$T^r_{ej} = \frac{1}{2}\dot{X}^{rT} m^r_{ej} \dot{X}^r \tag{6}$$

where T^r_{ej} = kinetic energy of the r-th elastic joint, $\dot{X}^r$ = velocity vector of the r-th MSE, and m^r = mass matrix of the r-th elastic joint, which can be determined using geometric and material properties of the joint.

The total potential energy (U) and kinetic energy (T) of the entire pipe system are the sum of those of the pipe elements and the elastic joints:

$$U = \sum_e U^e_{pipe} + \sum_r U^r_{ej} = \frac{1}{2}X^T K X \tag{7}$$

$$T = \sum m^e_{pipe} + \sum_r m^r_{ej} = \frac{1}{2}\dot{X}^T M \dot{X} \tag{8}$$

where K and M = stiffness matrix and mass matrix, respectively, of the entire pipe system, which can be assembled by every pipe element modeled by FEM and the elastic joint modeled by MSE as:

$$K = \sum_e L^{eT}(k^e_s + k^e_g)L^e + \sum_r L^{rT} k^r_j L^r \tag{9}$$

$$M = \sum_e L^{eT} m^e L^e + \sum_r L^{rT} m^r L^r \tag{10}$$

where L = index matrix, k^e_s, k^e_g = the structural stiffness matrix and geometric stiffness matrix of the e-th pipe element, respectively, k^r_j = the stiffness matrix of the spring element of the r-th elastic joint, and m^e, m^r = the mass of the e-th pipe element and the mass of the r-th elastic joint, respectively.

Thus, the governing equation of eigenvalue may be written as:

$$KA = \lambda MA \tag{11}$$

where λ = eigenvalue and A = eigenvector matrix.

STATIC EQUILIBRIUM EQUATION

The static equilibrium equation for the e-th pipe element is modeled by FEM (Chung, Cheng and Huttelmaier, 1994a and b) as:

$$(k^e_L + k^e_{NL})\Delta u^e = (F^e_{pipe})_{t+\Delta t} + (R^e_I)_t \tag{12}$$

where

$$k^e_L = \int B^T_L c B_L dV = \text{linear stiffness matrix of e-th element}$$

$$k^e_{NL} = \int B^T_{NL} \sigma B_{NL} dV = \text{nonlinear stiffness matrix of e-th element} \tag{13}$$

$$(R^e_I)_t = \int B^T_L \sigma dV$$

and Δu^e = incremental displacement vector of the e-th pipe element, $(F^e_{pipe})_{t+\Delta t}$ = external (fluid) force vector applied to the e-th pipe element at time step, t+Δt, $(R^e_1)_t$ = associated internal force vector of the e-th pipe element at Δt, B_L and B_{NL} = linear and nonlinear strain displacement matrices, respectively, c = matrix representing the material law for 3-D isotropic elasticity, and σ = updated Cauchy stress matrix (Bathe, 1982).

As to the r-th elastic joint modeled by MSE, its static equilibrium equation at t+Δt is expressed as:

$$k^r_j \begin{pmatrix} u^i \\ u^r \\ u^j \end{pmatrix}_{t+\Delta t} = (F^r_{ej})_{t+\Delta t} \qquad (14)$$

where $(F^r_{ej})_{t+\Delta t}$ = the external force vector (e.g., thrust force) applied to the r-th elastic joint at t+Δt, u^i, u^j = displacement vectors at the elastic joint and the 2 adjacent FEM-modeled pipe nodes, i and j, u^r = displacement vector at r-th MSE, and k^r_j = stiffness matrix of the r-th spring element, as defined in Eq. 5.

Eq. 14 can be rewritten in the form of an incremental equation as:

$$k^r_j \begin{pmatrix} \Delta u^i \\ \Delta u^r \\ \Delta u^j \end{pmatrix} = (F^r_{ej})_{t+\Delta t} - k^r_j \begin{pmatrix} u^i \\ u^r \\ u^j \end{pmatrix}_t = (F^r_{ej})_{t+\Delta t} + (R^r_2)_t \qquad (15)$$

where $(R^r_2)_t$ = the internal force vector of the r-th elastic joint associated with $(F^r_{ej})_{t+\Delta t}$.

Assembling the pipe element modeled by FEM and the elastic joint modeled by MSE, the static equilibrium equation for the entire pipe system can be expressed as:

$$(K_S + K_L + K_{NL})\Delta u = (F_{pipe})_{t+\Delta t} + (F_{ej})_{t+\Delta t} + (R_1)_t + (R_2)_t \qquad (16)$$

where

Δu = incremental displacement vector of the entire pipe system

K_S = stiffness of the elastic joint assembled with k^r modeled by MSE

K_L, K_{NL} = linear and nonlinear stiffness, respectively, assembled with k^e_L and k^e_{NL} by FEM

$(R_1)_t, (R_2)_t$ = internal force vectors assembled with $(R^e_1)_t, (R^r_2)_t$ at time step t by MSE

$(F_{pipe})_{t+\Delta t}, (F_{ej})_{t+\Delta t}$ = external force vectors at t+Δt applied to the pipe node and the elastic joint, respectively. Buoyancy force of the pipe is included (Chung and Cheng, 1995).

FEM Modeling and Thrust Vector Applications. Previously, the elastic joint was modeled by FEM for the cases where the mass and moment of inertia of the elastic joints are not included and no external forces (or thrust forces) are applied to the joints (Cheng, Chung and Zheng, 1995).

The elastic joint can be modeled using FEM for the cases where the mass and moment of inertia of the elastic joints are included and the external forces (or thrust forces) are applied to the joints. The mass and external force of a joint need to be divided into 2 equal halves, and each half is added to the nodes of the adjacent pipe elements. In terms of the pipe responses to the external forces applied to the joints, the accuracy of the results from the FEM depends on the configuration and size of the joints and the magnitude of the joint mass.

DYNAMIC ANALYSIS

Governing Equation of Motion

The governing equation of motion of the entire pipe system at time step t + Δt can be written as:

$$M\ddot{u}_{t+\Delta t} + C\dot{u}_{t+\Delta t} + (K_S + K_L + K_{NL})\Delta u = (F_{pipe})_{t+\Delta t} + (F_{ej})_{t+\Delta t} + (R_1)_t + (R_2)_t \qquad (17)$$

where M = mass matrix of the entire pipe system, which includes the mass of the elastic joint as defined in Eq. 10, C = the damping matrix = $C_s + C_h$, C_s = structural damping and C_h = hydrodynamic damping.

Incremental updated Lagrangian formulation, implicit time integration and Newton-Raphson iteration technique are adopted. Among several kinds of integration schemes which may be used in solving Eq. 17, the Newmark scheme is effective and used in achieving numerical stability, as it was demonstrated in the previous papers (e.g., Chung, Cheng and Huttelmaier, 1994), to which details are referred.

External Forces and Excitation

Dynamic excitation considered in this paper includes the external hydrodynamic forces, which consist of the wave forces on the pipe and the forces induced by the pipe-top motion, which is associated with the wave-induced ship motion. The hydrodynamic force vector consists of its normal component, F_N, and the tangential component, F_T, as follows (Chung, Whitney and Loden, 1980 and 1981; Chung and Tsurusaki, 1994):

$$\mathbf{F}_N = (\ /4)\, \rho_W d^2 C_M \dot{\mathbf{v}}_{RN} + \rho_W d^2 \dot{\mathbf{v}}_{WN}$$
$$(\rho_W/2) d C_D |\mathbf{v}_{RN}| \mathbf{v}_{RN}$$

$$\mathbf{F}_T = (\pi \rho_W/2) d C_f |\mathbf{v}_{RT}| \mathbf{v}_{RT} \qquad (18)$$

where d = the outer diameter of pipe, ρ_w = mass density of water, and C_M, C_D and C_f = hydrodynamic inertia, drag and friction coefficients, respectively. The relative velocity vector, $\mathbf{v}_R$, is defined by:

$$\mathbf{v}_R = \mathbf{v}_P - \mathbf{v}_C - \mathbf{v}_W \qquad (19)$$

where $\mathbf{v}_P$ = pipe velocity, $\mathbf{v}_C$ = steady current velocity, $\mathbf{v}_W$ = water particle velocity due to wave, $\mathbf{v}_{RN}$ = the normal component of $\mathbf{v}_R$, $\dot{\mathbf{v}}_{RN}$ = the normal component of the relative acceleration, $\dot{\mathbf{v}}_R$, and $\dot{\mathbf{v}}_{WN}$ = the normal component of the water particle acceleration.

The excitation motion of the pipe top is caused by the ship motion, which, in turn, is induced by waves. For numerical examples, it is represented for the horizontal (x-) and axial or vertical (z-) motions of the pipe top, respectively, as follows:

$$x_t = x_0 \sin \quad t \quad \text{and} \quad z_t = z_0 \sin \quad t \tag{20}$$

where x_0 and z_0 = the amplitudes of the excitation motion of the pipe top in the x- and z-directions, respectively, and ω = the corresponding excitation frequency.

	Pipe 1	Pipe 2
Length (ft)	18,000	18,000
Outer Diameter (ft)	1.6667	1.6667
Inner Diameter (ft)	1.5608	1.5608
Young's Modulus (lb/ft^2)	4.32x10^9	4.32x1 0^9
Pipe Weight in Air (lb/ft)	126.4	126.4
Pipe Weight in Water (lb/ft)	114.3	114.3
Buffer Weight (L.T.)	200	200
Number of Elements	18	18
Joint Properties: Axial (z) Stiffness, K_a (lb/ft) Transverse (x, y) Stiffness, K_x (lb/ft) Rotational Bending Stiffness, K_b (ft-lb/rad) Rotational Torsional Stiffness, K_θ(ft-lb/rad) Mass and Mass Moment of Inertia Positions of 3 Joints at z = −1,000, −9,000, −17,000 ft	 ∞ ∞ 10^6 ∞ 0	 ∞ 2,000 10^6 ∞ 0
Pipe Top: pinned (no rotation about pipe axis allowed: $\theta_z = 0$)		
Pipe Bottom End: free but can be controlled by thrusters in the x- and y-directions.		

Table 1 Principal dimension of pipe and properties of pipe and elastic joints.

Initial Condition and Static Equilibrium State

For the linear system, the zero position of the generalized coordinate is usually placed at the static equilibrium position, as the static stress has no effect on the amplitude of the dynamic response. However, when the dynamic response is calculated at the first time step for the deep-ocean vertical pipe system involving geometrical nonlinearity, the effect of the static axial stress must be considered. This is because it produces terms in the nonlinear stiffness matrix and in the equivalent force vector.

The static load includes self-weight and buoyancy of the pipe, external normal drag due to steady current and steady flow-induced torsional moment caused by asymmetric arrangement of cable around the pipe and buffer orientation relative to the flow. The static equilibrium configuration of the pipe and the element static stress are obtained by solving the static equation, Eq. 16, without the mass and damping terms.

In order to link the start of the dynamic response with the static equilibrium, the undeflected or vertical pipe configuration is taken as zero position of the generalized coordinates.

PRELIMINARY EXAMPLES

Preliminary results are presented here, and more detail investigation will be reported in Part II of this paper. Computations of pipe responses are made for 2 different transverse joint stiffness: *Pipes 1 and 2* (Table 1).

	MSE	FEM
Bending (x) Deflection at Bottom (ft) at Joint, z = −17,000 (ft)	 −173.3 −170.6	 −173.3 −170.6
Biaxial Bending (y) Deflection at Bottom (ft) at Joint, z = −17,000 (ft)	 −0.289x10^{-4} −0.3158x10^{-3}	 −0.289x10^{-4} −0.3158x10^{-3}
Axial Displacement at Bottom (ft)	−21.98	−21.98
Twist Angle, θ_z (rad) at Bottom at Joint, z = −17,000 ft	 0.6613 0.6538	 0.6613 0.6538
Max. Internal Axial Force (lb)	2.45x10^6	2.45x10^6
Max. Internal Bend. Moment (ft-lb) at Joint, z = −1,000 ft at Joint, z = −17,000 ft	 1660 413	 1660 413

Table 2 Comparison of static responses of *Pipe 1* with 3 elastic joints without mass by FEM and MSE models for joint properties: $k_a = \infty, k_b = 10^6 \text{ft}-\text{lb}/\text{rad}, k_x = \infty, k_\theta = \infty$

	MSE	FEM
Bending (x) Deflection at Bottom (ft) at Joint, z = −17,000 (ft)	 −181.7 −178.4	 −181.7 −177.9
Biaxial Bending (y) Deflection at Bottom (ft) at Joint, z = −17,000 (ft)	 −0.289x10^{-4} −0.3158x10^{-3}	 −0.289x10^{-4} −0.3158x10^{-3}
Axial Displacement at Bottom (ft)	−21.98	−21.98
Twist Angle, θ (rad) at Bottom at Joint, z = −17,000 ft	 0.6613 0.6538	 0.6613 0.6538
Max. Internal Axial Force (lb)	2.45x10^6	2.45x10^6
Max. Internal Bend. Moment (ft-lb) at Joint, z = −1,000 ft at Joint, z = −17,000 ft	 1660 409	 1660 409

Table 3 Comparison of static responses of *Pipe 2* with 3 elastic joints without mass by FEM and MSE models for joint properties: $k_a = \infty, k_b = 10^6 \text{ft}-\text{lb}/\text{rad}, k_x = 2{,}000 \text{ lb}/\text{ft}, k_\theta = \infty$

Static Responses

The properties and effect of the elastic joints are embodied in the solution for the static equilibrium and in the internal force and moments of the joints along the pipe. The stiffness, number and position of the joints were extensively analyzed previously (Cheng, Chung and Zheng, 1995; Chung and Cheng, 1996).

In establishing a static equilibrium state, the flow-induced torsional moment along the pipe induced by the cable-pipe arrangement asymmetric to the current flow or nodal torsional, $(M_T)_N$, and buffer torsional moment at the pipe's bottom-end, $(M_T)_B$, are accounted for (Chung and Whitney, 1993).

The static pipe responses are computed with both the FEM and MSE models for the 2 pipe systems. The results from the present 2 numerical models are compared (Tables 2~3).

Vibration Characteristics of *Pipe 1*. The thrust (force) vector is designed such that the elastic joints and the buffer reach a steady state near the vertical or desired pipe position. Pipe 1 has rigid transverse (x, y) stiffness, K_x (lb/ft) = ∞, and zero mass (Table 1). For *Pipe 1* ($k_x = \infty$), both models predict identical dynamic responses.

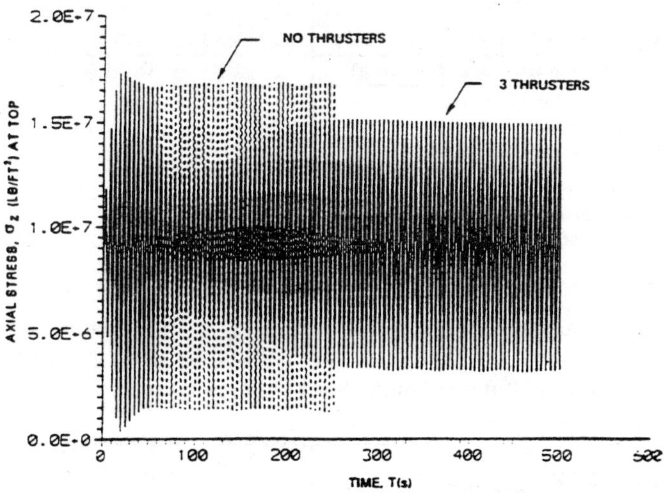

Fig. 3 Comparison of axial stresses σ_Z at top of Pipe 1 with 3 elastic joints without mass by FEM and MSE models for joint properties: $k_a = \infty, k_b = 10^6 \text{ft} - \text{lb} / \text{rad}, k_x = \infty, k_\theta = \infty$

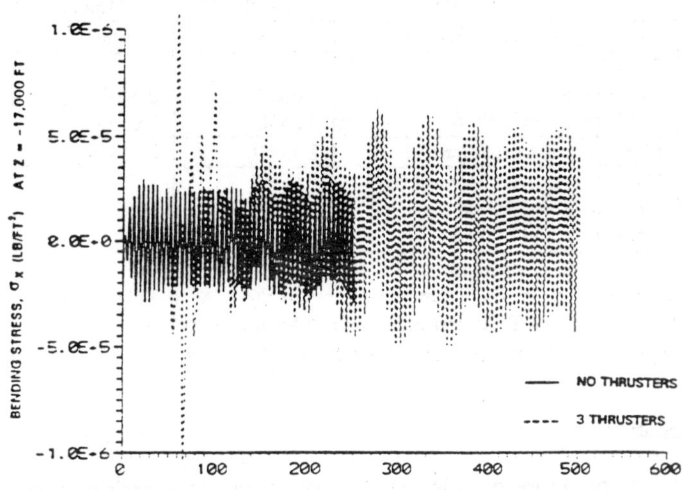

Fig. 4 Comparison of bending stresses σ_X of Pipe 1 with 3 elastic joints without mass by FEM and MSE models for joint properties: $k_a = \infty, k_b = 10^6 \text{ft} - \text{lb} / \text{rad}, k_x = \infty, k_\theta = \infty$

Thrusters are modeled to control the pipe motion displacements in the x- and y-directions. The applications of the 2 types of ramp thrust loading show that the thrust activation at different time, t_s = 5 s and 50 s, during transient pipe motion does not influence the final steady-state position of the pipe in the (x, y) plane.

Application of the thrust vector of a ramp loading reduces the vibration amplitudes of the axial motion and stress by about 16% (Fig. 3). The thrust activation at time, t_s, during transient pipe motion, does not influence the final steady-state axial vibrations of the pipe. The amplitude of bending stress is slightly increased by the thrust vector (Fig. 4). However, it is noted that the maximum stress level is orders-of-magnitude larger for axial stress than for bending stress.

Ramp Loading of Thrust Force. A ramp loading function with the thrust force vector, $\mathbf{F_t}$ (lb) (Fig. 2) in the x-direction in this investigation is applied to the joints of the pipe system, which is already subject to the hydrodynamic forces. The thrust is activated at t_s and increased linearly to its full power in t_f, and then is kept constant for t_f: $t_f - t_s = 5$ s. Usually, it takes about 500 s for both pipe systems to reach a steady state. The structural damping, C_s, is not included in the present numerical examples, and it helps in achieving further numerical stability, when included.

Two types of ramp thrust loading are applied (Fig. 5) in order to study effects of time delay in reaching steady state. Both pipes are identical in their principal dimensions and have 3 identical elastic joints, and the properties of the joints are described in Table 1.

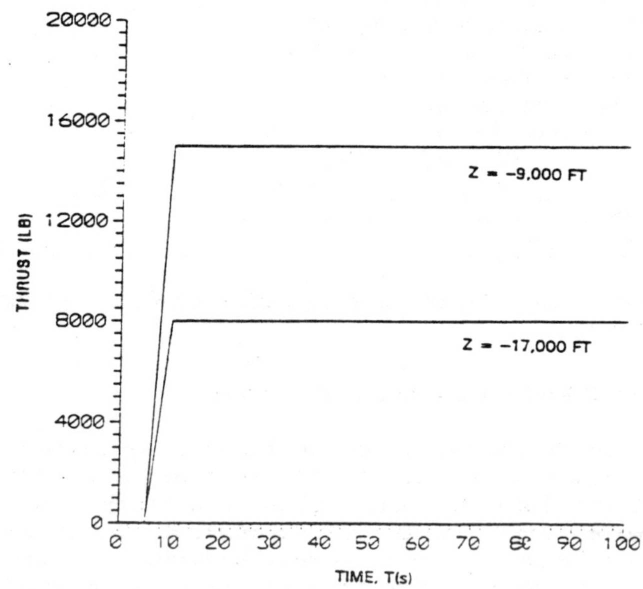

Fig. 5 Effects of ramp thrust loading on joints of Pipe 1.

MSE and FEM Predictions: Thrust Effect on Pipe Vibrations

As the joint of *Pipe 1* has no mass and its transverse (x- and y-) stiffness is infinitely rigid, both MSE and FEM predictions of all the responses are identical, when the thrusts (e.g., a thrust ramp loading; Fig. 5) are applied to the joints. However, responses of *Pipe 2*

$(k_x = 2{,}000 lb / ft)$ show difference in results between predictions by 2 models. Details of computations by FEM and MSE models and comparisons of the axial (z-), bending (x-, y-) and torsional responses for *Pipe 2* and another pipe with a mass or buffer mass at the bottom end of pipe will be discussed in Part II.

CONCLUSIONS

In order to apply the external (thrust control) forces on the elastic joints along a vertical deep ocean pipe, the elastic joints along a very long vertical pipe are modeled by 2 computational methods: mass-spring elements (MSE) and finite elements (FEM). The MSE is special *combined element,* and consists of a point mass and connecting springs where the point mass is a mass-spring with its mass, but with its size being ignored. The FEM model, which previously used to model the elastic joint, is modified in order to apply the external force to the joint. Using the FEM, the external force is applied to a point of the beam very close to the joint. Using the MSE, the external force can be applied directly to the point mass of the elastic joint. The MSE is simpler than FEM in modeling the elastic joint.

The accuracy and advantages of the two methods are compared with numerical examples in Part II of this paper (Cheng and Chung, 1998), and control strategy for pipe behavior with thrusters will follow..

ACKNOWLEDGEMENT

The first two authors gratefully acknowledge the support of the National Science Foundation under Research Grants BCS-9207967 and BES-9531182. The last two authors acknowledge partial support of National Natural Science Foundation of China under the Research Grant No 59335101 while they were in China.

REFERENCES

Bathe, KJ (1982). *Finite Element Procedure in Engineering Analysis*, Prentice Hall.

Brink, AW and Chung, JS (1980, 1982). "Automatic Position Control of a 300,000-Ton Ship Mining System," *J Energy Resources Tech*, ASME, 1982, Vol 104, pp 285-293, and also *Proc Offshore Tech Conf*, 1980.

Cheng, B-R and Chung, JS (1995). "Effects of Flexible Joints Along a Long Pipe on the 3-D Nonlinear Coupled Pipe Responses," *Proc 5th Int Offshore and Polar Engineering Conf*, ISOPE, The Hague, The Netherlands, Vol. 2, pp 236-243.

Cheng, B-R, and Chung, JS (1996). "Effects of Axial Dampers and Elastic Joints on the Three-Dimensional Responses of a Deep-Ocean Pipe Coupled with Torsion," *Proc 6th Int Offshore and Polar Eng Conf*, ISOPE, Los Angeles, USA, Vol 1, pp 37-45.

Cheng, B-R, and Chung, JS (1997). "Effects of Axial Dampers and Elastic Joints on 3-D Dynamic Responses of a Deep-Ocean Pipe with Torsional Coupling," *Int J Offshore and Polar Eng*, ISOPE, Vol 7, No. 1, pp 36-43.

Cheng, B, and Chung, JS (1998). "Modeling of Joint Between Long Vertical Pipe and Horizontal Pipe," to be submitted for publication.

Chung, JS, and Cheng, B-R (1995). "Effects of Elastic Joints on the 3-D Nonlinear Responses of a Deep-Ocean Pipe: Modeling and Boundary Conditions," *Int J Offshore and Polar Eng*, ISOPE, Vol 6, No 3.

Chung, JS (1978). "A Motion Analysis of a Riser Upper Joint Interacting With a Floating Vessel," *J Pressure Vessel Tech*, ASME, Vol 100, pp. 91-97.

Chung, JS (1994). "Added Mass and Damping on an Oscillation Surface-Piercing Circular Column With a Circular Footing," *Int J Offshore and Polar Eng*, Vol 4, No 1, pp 11-17.

Chung, JS (1994). "Deep-Ocean Cobalt-Rich Crust Mining System Concepts," *Proc MTS-94 Conf*, Marine Tech Soc, Washington, DC, pp 98-104.

Chung, JS (1995). "Assessment of Options in Design of Deep-Ocean Mining Pipe Systems," *Proc 1st (1995) ISOPE - Ocean Mining Symp*, Tsukuba, Japan, ISOPE, pp 23-30.

Chung, JS (1997). "A New Concept of Pipe System: Seafloor Miner to Lift Pipe," *Proc 2nd (1997) ISOPE - Ocean Mining Symp*, Seoul, Korea, ISOPE.

Chung, JS and Felippa, C (1981). "Nonlinear Static Analysis of Deep-Ocean Mining Pipe - Part II: Numerical Studies." *J Energy Resources Tech*, ASME, Vol 103, pp 16-25.

Chung, JS, Whitney, AK and Loden, WA (1980, 1981). "Nonlinear Transient Motion of Deep-Ocean Mining Pipe." *Proc Offshore Tech Conf*, Houston, 1980, and *J Energy Resources Tech*, Vol 103, 1981, pp 2-10.

Chung, JS, and Whitney, AK (1981). "Dynamic Vertical Stretching Oscillation of a Deep-Ocean Mining Pipe," *Proc Offshore Tech Conf*, Houston.

Chung, JS and Whitney, AK (1993). "Flow-Induced Moment and Lift for a Circular Cylinder with Attached Cable," *Int J Offshore Mech & Polar Eng*, ISOPE, Vol 3, No 4, pp 280-287.

Chung, JS and Tsurusaki, K (1994). "Advance in Deep-Ocean Mining Systems Research," an invited plenary speech, *Proc 4th Int Offshore and Polar Eng Conf*, ISOPE, Vol 1, Osaka, pp 18-31.

Chung, JS, Cheng, B-R, and Huttelmaier, H-P (1994a). "Three-Dimensional Coupled Responses of Deep-Ocean Pipe: Part I. Excitation at Pipe Ends and External Torsion," *Int J Offshore and Polar Eng*, ISOPE, Vol 4, No 4, pp 320-330.

Chung, JS, Cheng, B-R, and Huttelmaier, H-P (1994b). "Three-Dimensional Coupled Responses of Deep-Ocean Pipe: Part II. Excitation at Pipe Top and External Torsion," *Int J Offshore and Polar Eng*, ISOPE, Vol 4, No 4, pp 331-339.

Chung, JS, Whitney, AK, Lezius, D, and Conti, R (1994). "Flow-Induced Torsional Moment and Vortex Suppression for a Circular Cylinder with Attached Cable," *Proc 4th Int Offshore and Polar Eng Conf*, ISOPE, Vol 3, Osaka, Japan, pp 447-459.

Chung, JS, and Cheng, B-R (1995). "Eigenvalues for a Long Vertical Deep-Ocean Pipe With Elastic Joints," *Proc. Flow-Induced Vibrations Symp*, ASME, Honolulu, July 24-28, PVP-Vol 298, pp 153-160.

Chung, JS and Cheng, B-R (1998). "Application of Thrusts to Elastic Joints on Long Vertical Pipe for the 3-D Nonlinear Motions: Part II. Numerical Examples of MSE and FEM Results," to be submitted to publication.

Duncan, WJ (1952). "A Critical Examination of the Representation of Massive and Elastic Bodies by Rigid Mass Elastically Connected," *Quart J Mechanics and Appl Math*, pp 97-108.

Felippa, C and Chung, JS (1981). "Nonlinear Static Analysis of Deep Ocean Mining Pipe. - Part I: Modeling and Formulation," *J Energy Resources Tech*, ASME, Vol 103, pp 11-15.

Huang, T (1958). "On Natural Frequencies of Uniform Beam Analyzed on Discrete Systems," Highway Bridge Impact Investigation: 8th Prog Rept, Univ of Illinois.

Qian, LX, and Zhang, X (1991). "Rigid Finite Element Method in Structural Analysis," *Comput Mech*, Balkama, Hong Kong, pp 1165-1171.

Whitney, AK, Chung, JS, and Yu, BK (1981). "Vibrations of a Long Marine Pipe Due to Vortex Shedding," *J Energy Resources Tech*, ASME, Vol 103, pp 231-236.

Proceedings of the Seventh (1997) International Offshore and Polar Engineering Conference
Honolulu, USA, May 25-30, 1997

ISBN 1-880653-28-1 (Set); ISBN 1-880653-30-3 (Vol. II)

Dynamics of Deep Water Marine Risers - Asymptotic Solutions

Geir Moe and Bjørn Larsen
Norwegian University of Science and Technology
Trondheim, Norway

ABSTRACT

An asymptotic solution of the differential equation describing the motions of marine risers in more than 1000 m waterdepth is presented. It is focused on the case in which the riser is subjected to wave forms travelling away from the excited zone. Four cases are defined, and analyzed by this method. Case 1 for top tension 3000 kN is compared to solutions using the FEM program Risana. Unexpected large discrepancies were found. A discussion of possible causes is included, but no certain explanation was found.

INTRODUCTION

Drilling in 1500 m plus waterdepth is now passing from a very unusual event to something relatively commonplace. Production at the same depths must be expected to become similarly commonplace in the near future. Rigid metal risers ('marine risers') which are a practical necessity for drilling activities, and an attractive alternative for production risers, will be focused on herein.

Structural analyses of marine risers are routinely done by numerical tools, and the better of these numerical programs are now reported to work well, even though they may be somewhat cumbersome for very long risers. Some impressive results have also been obtained by very simple analytical approximations. Thus static solutions, and solutions for the dynamic case in which inertia and tension forces dominate, have been determined quite accurately by analytical methods, see e.g. Sparks (1980) and Brouwers (1982). The underlying assumption in both cases is that bending stiffness is only of importance near the bottom and in the wave active zone, where the tension can be approximated as constant. In the remaining range which Brouwers call 'the riser main section', the decrease of the tension with increasing depth is modelled, but bending is neglected. Finally the total solution is put together by matching of the solutions from the three zones (Bottom, main zone and top).

As water depth increases the typical situation will be that the riser is excited at the top end by vessel motions and by waves which typically penetrate to about 100 m depth. Below, between 100 m and the bottom, at say 1500 m, there will normally be no dynamic loading in the in-line direction, apart from motions induced in essentially steady currents, typically Vortex Induced Vibrations (VIV) or flow interaction phenomena between neighbouring members. These phenomena will be disregarded in this paper(, apart from a possible adjustment of the mean drag coefficient). For a complete analysis VIV and interaction effects should be considered afterwards, but that is outside the scope of the present paper. Internal waves may occur due to stratification of the water masses. These will typical have so high period that they can be treated as a slowly varying currents.

This paper will present approximate solutions for the bending free, unloaded riser which is excited harmonically at the upper end. It will be shown under which conditions damping dominates in the solution. Then an approximate, analytic solution for the damping dominated case will be presented. The damping will depend on the current profile on the actual location, and certain simple profiles will be treated herein. The authors are not aware of any other similar treatment of this problem.

A traditional solution based on the modal technique relies on the assumption that excitation and damping forces are small, while inertia and tension stiffness forces are large, and at special frequencies ('resonance frequencies') the two balance each other, permitting very large motion-to-load ratios. The shapes of the vibrations at these frequencies, so-called 'modal shapes', may then be predicted on the basis of these two terms alone. Since both of them are conservative, neither will add or subtract energy from the system. The motion amplitudes will therefore usually be determined by a balance between the work done by the relatively small damping and the excitation forces which will not however change the shape at which the structure vibrates. In contrast: for the present case the damping forces will also have a strong influence on the shape of the vibrations. Damping forces may even dominate over inertia forces, at least on part of the riser, since the excitation is assumed to act only near the top. The solution will show that the envelope of the motion will decay with depth and that there will be a wave travelling downwards from the top end. The present solutions will be compared to available numerical solutions for a suitable case.

THE DIFFERENTIAL EQUATION

For a nearly vertical geometry, i.e. neglecting curvature effects and also neglecting extensibility effects, the riser differential equation may be formulated as follows, see e.g. Sparks (1980) or Brouwers (1982)

$$EI\,x'''' - (T(z)x')' + c_s\,\dot{x} + m_0\ddot{x} = F \qquad (1)$$

Here EI is bending stiffness, z is the vertical co-ordinate, $x = x(z,t)$ is the horizontal riser displacement, T(z) is the axial tension, apostrophe (') represents differentiation with respect to the vertical co-ordinate z, and dot ($\dot{}$) represents differentiation with respect to time t. m_0 is the mass per unit length of the riser, appendages and content. A linear structural damping term ($c_s\dot{x}$) has also been included. F is the external hydrodynamic force which is usually expressed as

$$\begin{aligned} F &= K_m\dot{u}_w - m_a\ddot{x} + K_d(u_w + u_c - \dot{x})|u_w + u_c - \dot{x}| \\ K_m &= \rho\pi R^2 C_m \\ m_a &= \rho\pi R^2(C_m - 1) \\ K_d &= \rho R C_d \end{aligned} \qquad (2)$$

see e.g. Browers (1982). Here ρ is water density, R is the external riser radius, u_c is the current velocity, u_w horizontal wave particle velocities, and C_m and C_d are the inertia and drag coefficients in the Morison formula. Bending may be important close to the bottom and top, and possibly in the wave zone or if concentrated loads are applied. For the present case the bending term in (1) is tentatively neglected. The purpose is either to cover conditions in 'the riser main zone' which by definition is below the wave active zone, or to model the effect of riser top motions in the absence of waves. Then also $u_w=0$, and (1) simplifies to

$$\begin{aligned} -(T(z)x')' + c_s\,\dot{x} + m_{tot}\,\ddot{x} &= F_2 = K_d(u_c - \dot{x})|u_c - \dot{x}| \\ m_{tot} &= m_0 + m_a \end{aligned} \qquad (3)$$

The right hand side in (3a) depends on the velocity in a complicated manner, but if the current velocity u_c is always larger than the riser velocity $\dot{x}$, then the situation simplifies drastically:

$$F_2 = K_d(u_c^2 + \dot{x}^2 - 2u_c\dot{x}) \qquad (4)$$

The first term in the bracket represents a static load and produces a steady offset that does not affect the dynamic analysis. The second term is of second order in amplitude and may be split into a static term plus one term at the double frequency of $\dot{x}$. The latter is to be neglected in the first order solution, but could be dealt with as a second order contribution in a perturbation scheme. The last term in (4) is a first order, linear damping term which should be transferred to the left hand side and added to the structural damping term. Furthermore the riser tension may be written as follows

$$\begin{aligned} T(z) &= T_b + wz \\ w &= g[(\rho_s - \rho_w)A_s + (\rho_i - \rho_w)A_i] \end{aligned} \qquad (5)$$

in which w is the wet weight of the riser, i.e. weight per unit length of riser and content, minus the product of the unit weight of water and the gross cross sectional area of the riser. A_s and A_i are cross sectional area of riser wall and internal area respectively, while ρ_s, ρ_i, ρ_w are the densities of riser wall, riser fluid and sea water, respectively. It is now useful to introduce a shifted vertical co-ordinate S having its origin at a point where $T(z)$ according to (5) becomes zero, see below. Then the dynamic equation (3) changes to (6b)

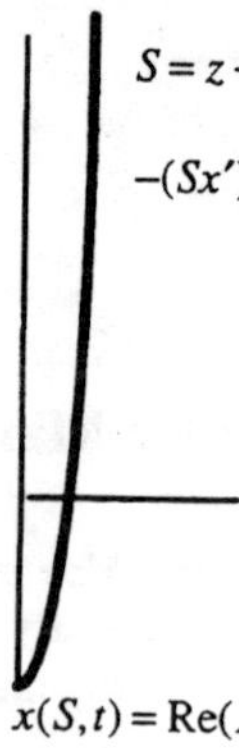

$$\begin{aligned} &S = z + T_b / w \\ &-(Sx')' + \frac{c_s + 2K_d u_c(S)}{w}\,\dot{x} + \frac{m_{tot}}{w}\,\ddot{x} = 0 \end{aligned} \qquad (6)$$

Apostrophe now denotes differentiation with respect to S, and the current $u_c(S)$ must now be given in terms of the new co-ordinate. In addition to (6) boundary conditions are needed. Consider for the time being the riser to be fixed at the bottom end and undergoing harmonic motion at the top end, viz.

$$\begin{aligned} &x(S,t) = \mathrm{Re}(X_0 e^{i\omega t}), \quad at\, S = L + T_b / w \quad (z = L) \\ &x(S,t) = 0, \qquad\qquad at\, S = T_b / w \qquad (z = 0) \end{aligned} \qquad (7)$$

It is assumed that the solution is harmonic in time

$$x(S,t) = \mathrm{Re}(U(S)e^{i\omega t}) \qquad (8)$$

Substitution into (6b) yields

$$SU'' + U' + \{m_{tot}\omega^2 / w - i\omega[c_s + 2K_d u_c(S)] / w\}U = 0 \qquad (9)$$

This equation may now be made nondimensional by introduction of a nondimensional vertical co-ordinate $s=S/L$ and a horizontal co-ordinate $u=U/L$, viz.

$$\begin{aligned} &s = S / L = z / L + T_b / (wL) \\ &u = U / L \\ &c(s) = \omega L[c_s + 2K_d u_c(s)] / w \\ &m = \omega^2 m_{tot} L / w \\ &su'' + u' + (m - ic(s))u = 0 \end{aligned} \qquad (10)$$

Expressing the boundary conditions in the same variables

$$\begin{aligned} &s_b = T_b / (wL) \\ &s_t = T_t / (wL) = s_b + 1 \\ &u(s_t) = X_0 / L \\ &u(s_b) = 0 \end{aligned} \qquad (11)$$

This is a linear, ordinary differential equation in the independent variable s with an irregular singular point at infinity and an ordinary singular point at $s=0$ (Bender & Orzag, 1984, Sec. 3.1). For these cases asymptotic solutions may be found by postulating a solution of the form

$$\begin{aligned} &u(s) = e^{y(s)} \\ &u' = y'e^{y(s)} \\ &u'' \cong (y')^2 e^{y(s)} \end{aligned} \qquad (12)$$

In the latter expression the second derivative of $y(s)$ has been neglected, assuming that the square of the first derivative dominates. This is usually true in these situations, it may be verified when the solutions for $y'(s)$ is available. Substitution of (12) in the differential equation (10e) results in a quadratic equation for $y'(s)$. Its roots are

$$y' = -\frac{1}{2}s^{-1} \pm [1/(4s) - m + ic(s)]^{1/2} s^{-1/2} \tag{13}$$

Consider the terms in the square bracket in (13) It will be argued that the first term, *1/(4s)*, can be neglected relative to the second one, m. From (10) is seen that s which is between s_b and s_t, is equal to the ratio between the local riser tension and the wet weight of the riser. This ratio of the bottom riser tension and the riser wet weight is normally at least equal to 1/4, in which case the first term in the square bracket varies between 1 and 1/5. The nondimensional mass may be written as

$$m = \frac{L\omega^2}{g} \frac{m_{tot}}{m_{wet}} \tag{14}$$

The latter of the ratios in (14) will typically be 2-4, while in 1000 m waterdepth and for waves at 15 sec period the first ratio in (14) is about 17 and the nondimensional m is hence about 35-70. *1/(4s)* may therefore be neglected relative to m in (13).

SOLUTION FOR THE CONSTANT CURRENT CASE

Consider initially the case where the current u_c is constant, so that $c(s)$ in (10c) becomes a constant, c_0. Then (13) may be written as

$$\begin{aligned} y' &= -\frac{1}{2}s^{-1} \pm (A + iB)s^{-1/2} \\ A + iB &= (-m + ic_0)^{1/2} \end{aligned} \tag{15}$$

in which in (15b) the principal root is used. That means that A and B both will be positive, real numbers. Integrating to find y and thence from (12a) $u(s)$

$$\begin{aligned} y &= -\frac{1}{2}\ln s \pm 2(A + iB)s^{1/2} \\ u &\propto s^{-1/2} e^{\pm(2As^{1/2} + 2iBs^{1/2})} \end{aligned} \tag{16}$$

The solution for u(z) is thus of the form

$$\begin{aligned} u = s^{-1/2}\{&\tilde{C}_1 \exp(2As^{1/2} + 2iBs^{1/2}) \\ &+ \tilde{C}_2 \exp(-2As^{1/2} - 2iBs^{1/2})\} \end{aligned} \tag{17}$$

The constants above may be complex and may be written in polar form as $\tilde{C}_i = C_i \exp(i\theta_i)$ in which of course C_i and θ_i are real. Substituting into (8)

$$\begin{aligned} x = X/L = C_1 s^{-1/2} e^{2As^{1/2}} &\cos(\omega t + 2Bs^{1/2} + \theta_1) \\ + C_2 s^{-1/2} e^{-2As^{1/2}} &\cos(\omega t - 2Bs^{1/2} + \theta_2)\} \end{aligned} \tag{18}$$

Here the first term represents a downward travelling wave inside a diminishing envelope, and the second term an upward travelling wave inside an envelope that diminishes when travelling upwards. (The former because the exponential will dominate in the expression for the envelope.)

Using the two boundary conditions in (11) C_1 and C_2 may now be determined.

SPECIAL SOLUTIONS

Case I: No damping, homogenous boundary conditions

For the undamped case there must be no energy input to the system, hence the top boundary condition must be $u(s_t)=0$, which is a special case of (11c). From (15b) substituted in (17)

$$\begin{aligned} A &= 0 \\ B &= \sqrt{m} \\ u &= s^{-1/2}\{C_1 e^{2i\sqrt{ms}} + C_2 e^{-2i\sqrt{ms}}\} = C s^{-1/2} \sin(2\sqrt{ms} + \theta) \end{aligned} \tag{19}$$

It is here made use of the fact that vibrations of undamped systems will take place without any phase differences along the vibrating member, so that in (8) the real part can be taken of $U(S)$ and $\exp(\omega t)$ independently. Please note that $2(ms)^{1/2}$ in the expression for u above is real, so that θ in the last expression must be real. The top and bottom boundary conditions now require that the argument of the above sine function, viz. $[2(ms)^{1/2} + \theta]$, is equal to π times some integer number for $s = s_b$ and s_t. Subtracting the expression for the bottom boundary condition from the expression for the top one, the unknown phase angle θ will cancel, yielding

$$2\sqrt{ms_t} - 2\sqrt{ms_b} = n\pi, \quad (n = 1, 2, 3, \ldots) \tag{20}$$

According to (14) the nondimensional mass m is a function of ω^2, and hence the frequencies ω_n corresponding to n in (20) can be determined, yielding

$$\omega_n = \frac{n\pi w}{2\sqrt{m_{tot}}(\sqrt{T_t} - \sqrt{T_b})} = \frac{n\pi(\sqrt{T_t} + \sqrt{T_b})}{2L\sqrt{m_{tot}}} \tag{21}$$

The latter expression is obtained by multiplying numerator and denumerator by the parenthesis in the denumerator in the last expression. This is identical to the result in Moe and Chucheepsakul (1988). It may be worth mention that the above simplifies to the well known expression for natural frequencies of a string with constant tension if $T_t = T_b$.

Case II: The constant damping case

It is convenient to start with the long string case, which is also the case expected to occur in practice for deep water risers. Using the solution (18) for this case the first term which represents a wave travelling from the top and downward will produce a negligible motion at the bottom support, satisfying the bottom boundary condition automatically. Hence the second term in (18) will vanish, i.e. $C_2 = 0$. Then the phase angle and the amplitude follows directly from (18), (8) and (11c).

$$X = X(s) = X_0\left\{\frac{s_t}{s}\right\}^{1/2} e^{-2A(\sqrt{s_t} - \sqrt{s})} \cos\{\omega t + 2B(\sqrt{s} - \sqrt{s_t})\} \tag{22}$$

It is seen that this represents a wave that travels downwards inside a diminishing envelope dominated by an exponential.

As an illustrative example consider the situation in which damping dominates, and set $m = 0$. Then from (15b)

$$A = B = c_0/\sqrt{2} \tag{23}$$

This should be inserted in (22). It is seen that in this case the envelope diminishes by a factor $\{535\ \sqrt{s_t/s}\}$ over one wavelength along the riser, and since the square root is close to unity, so for $m = 0$ it follows that the motion dies out over one wavelength of the riser.

* * *

If the riser is not so long that the motion at the bottom support is negligible, (22) can be used to find $X(s_b)$, the harmonic motion at the bottom end, as a function of $\tilde{X}_0$, an initially unknown amplitude at the top end. Then the second term in (18) is activated, giving an upwards travelling wave of the form

$$X_2 = X_2(s) = \tilde{X}_0\sqrt{\frac{s_t}{s}}e^{-2A(\sqrt{s_t}-2\sqrt{s_b}+\sqrt{s})} \cos\{\omega t - 2B(\sqrt{s_t} - 2\sqrt{s_b} + \sqrt{s})\} \quad (24)$$

At the top end this wave will have the amplitude $\tilde{X}_2 = Z\ \tilde{X}_0$, in which Z is a complex ratio

$$Z = \exp\{-4(A+iB)(\sqrt{s_t} - \sqrt{s_b})\} \quad (25)$$

The sum of these two complex displacement amplitudes $\tilde{X}_0 + \tilde{X}_2$ should be equal to the prescribed top displacement X_o, thus the complex amplitude of the downward wave is

$$\tilde{X}_0 = |\tilde{X}_0| e^{i\theta_x} = X_0/(1+Z) \quad (26)$$

Using this amplitude in both terms of the solution, (22) and (24)

$$\tilde{t} = t + \theta_x/\omega$$

$$X = |\tilde{X}_0|\sqrt{s_t/s}\ \{e^{-2A(\sqrt{s_t}-\sqrt{s})}\cos[\omega\tilde{t} - 2B(\sqrt{s_t} - \sqrt{s})] \\ -e^{-2A(\sqrt{s_t}-2\sqrt{s_b}+\sqrt{s})}\cos[\omega\tilde{t} - 2B(\sqrt{s_t} - 2\sqrt{s_b} + \sqrt{s})]\} \quad (27)$$

To simplify this expression the constant θ_x from (26) has been incorporated into the time variable.

As a check consider (27) for the undamped case, i.e. set $c_0=0$, so that $A=0$ and $B=\sqrt{m}$, as for Case I. Then the exponential functions in (27) become unity, and the motion amplitude at the top will be zero, provided the two cosine functions above are equal at position $s=s_t$, at all times t. This will happen if the arguments of the cosine functions differ by $(2n\pi)$, where n is an integer. Then (20) follows. It is seen that the boundary condition at the top must be homogenous, i.e. $X_o = 0$ in (7a), since X is identically zero here. Then the motion amplitude in (27) is indeterminable, we are dealing with an 'eigenvalue problem'. The argument of the second cosine may be written

$$\omega\tilde{t} - 2\sqrt{m}(\sqrt{s_t} - 2\sqrt{s_b} + \sqrt{s}) = \\ \omega\tilde{t} + 2\sqrt{m}(\sqrt{s_t} - \sqrt{s}) + 2n\pi = \\ \omega\tilde{t} + \alpha(s) + 2n\pi \quad \alpha(s) = 2\sqrt{m}(\sqrt{s_t} - \sqrt{s}) \quad (28)$$

Subtracting the two cosine terms above and dropping the quantity $(2n\pi)$ in the argument of the cosine, one finds

$$\cos[\omega\tilde{t} - \alpha(s)] - \cos[\omega\tilde{t} + \alpha(s)] = 2\sin(\omega\tilde{t})\sin[\alpha(s)]$$

$$X(s) = A_x s^{-1/2}\sin(\omega\tilde{t})\sin[\alpha(s)] = \\ A_x s^{-1/2}\sin(\omega\tilde{t})\sin[2\sqrt{m}(\sqrt{s_t} - \sqrt{s})] \quad (29)$$

Here it is seen quite clearly that the solution will be in accordance with modal theory, i.e. there will be no phase differences in the motions for different points along the riser.

Case III: Linear damping variation along the depth

Returning to (13), the starting point for the asymptotic expansion and again neglecting the term 1/(4s), but now assuming that the nondimentional damping is linear in the variable s, viz.

$$c(s) = c_1 s + c_0$$

$$D_1 = (A+iB)^2 = -m + ic_0 \quad (30)$$

$$y' \equiv -\tfrac{1}{2}s^{-1} \pm \sqrt{\{D_1 + ic_1 s\}/s}$$

Please note that c(s) should always be positive for $s>s_b$. (30) may be integrated to yield

$$y = -\tfrac{1}{2}\ln s \pm \sqrt{\{D_1 s + ic_1 s^2\}} \pm \frac{D_1}{\sqrt{ic_1}}\ln[\sqrt{ic_1 s} + \sqrt{D_1 + ic_1 s}\,] \quad (31)$$

Finally one finds

$$u\sqrt{s} \propto \{\sqrt{D_1 + ic_1 s} + \sqrt{ic_1 s}\}^{\pm\{D_1/\sqrt{ic_1}\}}\exp\{\pm\sqrt{D_1 s + ic_1 s^2}\} \quad (32)$$

Each of the two solution terms in (32) must be multiplied by an arbitrary constant and added to give the general solution. That means that it is permissible to multiply each of the above terms by any nonzero constant, in this case it is advantageous to multiply (32) by the factor

$$\{\sqrt{D_1}\}^{\mp D_1/\sqrt{ic_1}} \quad (33)$$

Introducing also a new variable, D_2, one obtains

$$D_2 = ic_1/D_1$$

$$u\sqrt{s} \propto \{\sqrt{1+D_2 s} + \sqrt{D_2 s}\}^{\pm\{\sqrt{D_1}/\sqrt{D_2}\}}\exp\{\pm\sqrt{D_1 s(1+D_2 s)}\} = \\ \{\sqrt{1+D_2 s} \mp \sqrt{D_2 s}\}^{\{\sqrt{D_1}/\sqrt{D_2}\}}\exp\{\pm\sqrt{D_1 s(1+D_2 s)}\} \quad (34)$$

The latter equality for the case with a minus in the exponent follows directly by using the definition of negative powers and then multiplying the nominator and denominator by the differences between the two square roots.

As a control one may let c_1 go to zero to return to the constant damping case. Noticing that $\sqrt{D_2 s}$ is the leading order term, so that the other root in the equations (34) approaches 1, it can be seen that both the first «power» term and the second «exponential» term in the last expression above approaches $\exp(\pm\sqrt{D_1 s})$, thus (16) is recovered.

The solution will now depend on the boundary conditions and may be obtained in the same manner as for Case II above, i.e. starting with the dampened, downward wave as given by (34) above, in which the plus sign is chosen, and the top motion is specified as $\tilde{X}_0$. If this solution does not die out at the lower riser end, then an upward travelling wave must also be considered in the same manner as before.

Case IV Damping of the form $c(s)=c_j s^j$, and negligible mass

Returning to (13) and again neglecting the term 1/(4s) one has

$$y' = \frac{1}{2}s^{-1} \pm [ic_j s^{j-1}]^{1/2}$$

$$y = -\frac{1}{2}\ln s \pm \sqrt{ic_j}\,\frac{2}{j+1}s^{\frac{j+1}{2}} \qquad (35)$$

$$u \propto s^{-\frac{1}{2}}\exp[\pm\sqrt{ic_j}\,s^{\frac{j+1}{2}}]$$

This solution can be manipulated in the same manner as for Case II to incorporate the boundary conditions.

As a control (32) may be Specialised for the case $m=c_0=0$. Then the exponential and $s^{-1/2}$ appears directly, while the sum of the two roots is positive, and when raised to the zeroth power, the result becomes 1. Thus the same result is obtained as from (35) with $j=1$, viz.

$$u \propto s^{-\frac{1}{2}}\exp[\pm\sqrt{ic_1}\,s] \qquad (36)$$

TEST CASE AND NUMERIC SOLUTION

Specifications for riser cross sections in steel and titanium respectively are given in Table 1. These risers are intended for drilling from a MODU (Mobile Offshore Drilling Unit) in deep water at the Vøring platau off the coast of Nordland in Northern Norway. In Table 2 further data for three steel risers cases and three titanium riser cases are indicated.

These risers were subjected to

1. a static loading due to a current of 1.65 m/s at the surface, decreasing linearly to 0.5 m/s at the seafloor.
2. a dynamic loading consisting of a top riser motion amplitude of 0.5 m and a period equal to 17.5 sec in the above mentioned current.

The riser top motion amplitude under extreme conditions is probably in the order of 6 m, but for that situation the damping would be nonlinear and the present analytic damping model would not be directly applicable.

Numeric analyses were performed by means of the FEM computer program Risana, developed by professor Larsen of the Department of marine structures, NTNU, Larsen (1996a), (1996b)

For the above riser configurations under static loading numerical results were compared to an asymptotic solution in which the main section of the riser was modelled as bending free, but for which bending was considered locally near the riser top and the sea floor where the riser axial tension could be considered as constant. Only minor differences was found between the two solutions, for deflections typically 2-3 % and for stresses up to about 6 %. The theory for the asymptotic solution has been developed earlier, Moe (1988), and is essentially the same as the solution presented in Sparks (1980).

The dynamic case was analyzed by the presented asymptotic solution and numerically by Risana. An example of an asymptotic solution is given in Figure 2 in which snapshots of the riser shape at different points in time are given. Also in the same figure are indicated two characteristic quantities of this solution: The nondimensional top quarter wave length, θ , and the nondimensional distance from the riser top to the point at which the envelope of the solution is reduced to 10 % of the top amplitude, $A_{0.1}$.

The general character of the two solution were quite similar, but the asymptotic solution decayed faster along the riser than the numeric solution. In some cases the increase is very large, up to 100 %, usually it is in the order of 10 to 50 percent, see Table 3 for some situations under Case 1. These results were disappointing, and efforts have been spent to find the reason for the differences. The reason does not appear to be discretisation problems in the FEM analysis, since runs with an increased number of elements produced practically speaking the same answers. Input errors or misunderstanding of the Risana input format can not be ruled out, but appears unlikely, and since Risana is a well tested program, it is unlikely that program errors will manifest itself when used on such relatively simple cases as the present ones. The most likely source of the discrepancies thus appears to be the asymptotic solution. The method as such usually produce quite good results, Bender & Orszag (1984), but the application of the method for the present case may be flawed.

The neglected term y'' in the development of (13), and 1/(4s) in (13) are indeed small for the present case. Thus with the data of test Case 1 from formula (14) the nondimensional mass becomes $m=22.5$, while 1/(4s) near the top end of the riser is less than 0.25 and can be neglected relative to m. Thus the square root term dominates in (13). The leading term of $(y')^2$ is then

$$(y')^2 \cong [-m + i\,c(s)]\,s^{-1} \qquad (37)$$

which for $m >> 1$ and $s > 1$ clearly dominates over

$$y'' \approx \frac{1}{2}s^{-2} \mp \frac{1}{2}[-m + ic(s)]^{1/2}s^{-3/2} \qquad (38)$$

Here the variation in the square root as a function of s is neglected relative to the dominating influence of $s^{-1/2}$. Thus from (37) and (38) is seen that $(y')^2$ is at least 10 times larger than y'', which therefore ought to be negligible.

Finally the importance of the bending stiffness relative to the tension stiffness will be assessed. Assuming that bending and tension are the dominvative terms, as will be the case in a "boundary zone" close to the ends or a concentrated load, (or every when, for very high modes of vibrations), upon introduction of $M = EIx''$, (1) simplifies to

$$M'' - \frac{T}{EI}M = 0 \qquad (39)$$

which has the solution

$$M = M_o \exp\left[\left(\frac{T}{EI}\right)^{1/2}(x - x_t)\right] \qquad (40)$$

for a long riser under constant tension and subjected to a moment M_o at the upper end $x = x_t$. The moment will be $(M_o / 10)$ at a distance $\Delta x = x_t - x = 2{,}3(EI/T)^{1/2}$ which for the present case 1 with T = 3000 kN, yields $\Delta x = 38m$. This is about 20 % of the comparable envelope reduction length $A_{0.1}$ for the case in which bending is neglected and indicates that bending is indeed of minor influence for this case. However more detailed investigations are called for, before a firm conclusions can be drawn on this point.

CONCLUSIONS

Asymptotic solutions have been presented for the behaviour of long risers whose top point undergoes harmonic motions while the damping depends linearly on the riser velocity. The damping coefficient can be described in three ways: constant, varying linearly with depth or being proportional to s^j in which j is an integer, and s has its origin at a point where the effective riser tension, if extrapolated vertically, would become zero.

The solution have been compared to numeric analyses with the FEM program Risana, and considerable discrepancies were seen. At the present time no explanation can be given for these discrepancies, however some possibilities have been discussed.

REFERENCES

Bender,C.M, Orszag,S.A.: Advanced mathematical methods for scientists and engineers, McGraw-Hill, Singapore, 1984

Brouwers,J.J.H.: «Analytical methods for predicting the response of marine risers», Proc. of the Koninklijke Nederlanse Akademie van Wetenschappen, Series B, Vol 85 (4), 1982

Larsen, C.M.: «Risana input description», Report from Department of Marine structures, NTNU, Trondheim, 1996a

Larsen, C.M.: «Design analyses for deep water», (in Norwegian), Paper presented at the course «Riser technology. The key to success for floating production», arranged by NIF, Trondheim, 6 feb 1996, 1996b

Moe,G.: «Response analysis of nearly vertical risers», Report K-3-88, Dep. of structural engineering, NTNU, Trondheim, 1988

Moe,G., Chucheepsakul, S.: «The effect of internal flow on marine risers», Proceeding of ISOPE 88, Houston, Tx, 1988

Sparks,C.P.: «Mechanical Behaviour of Marine Risers», Mode of influence of principal parameters», J. Energy Res. Techn., Dec 1980, Vol.102

Table 1 Riser dimensions and properties

	STEEL X65	TITANIUM gr.5 EL
Riser OD	0.519 m	0.519 m
Riser WT	0.037 m	0.020 m
Design pressure	345 Bar	345 Bar
Density	7.85 T/m^3	4.44 T/m^3
Coupling mass pr. joint	734.7 kg	734.7 kg
Kill/choke lines OD	0.127 m	-
Kill/choke lines WT	0.051 m	-
Density kill/choke lines	1.6 T/m^3	1.6 T/m^3
Drill string OD	0.168 m	-
Drill string WT	0.018 m	-
Structural damping	2%	2%
Allowable yield stress	448 Mpa	760 Mpa
Modul of elasticity	207000 Mpa	114000 MPa
Density mud	1.2 T/m^3	1.2 T/m^3
Flex. joint stuffiness	1547 kNm/rad	1547 kNm/rad

Table 2 Specifications of the parameters defining the investigated cases

Key parameters	Case 1	Case 2	Case 3	Case 4	Case 5	Case 6
Material	Steel	Steel	Steel	Titanium	Titanium	Titanium
Water deep	1000	1500	2000	1000	1500	2000
Buoyancy area	1.155	1.155	1.155	0	0.452	0.452
Top tension	4414.5	4414.5	4414.5	4414.5	4414.5	4414.5
Static offset	90	120	175	50	113	120

Table 3 Analytic solution. Percentage difference from numeric results.

Case 1	Sub. weight 0.304 T/m		Sub. Weight 0.268 T/m		Sub. weight 0.162 T/m		Sub. Weight 0.0675 T/m	
	θ	$A_{0.1}$	θ	$A_{0.1}$	θ	$A_{0.1}$	θ	$A_{0.1}$
T=4414.5 kN	-	-	18.5 %	46.9 %	15.7 %	59.4 %	100 %	74.0 %
T=3500 kN	-	-	47.3 %	33.4 %	3.87 %	50.8 %	-	-
T=3000 kN	84.5 %	11.6 %	50.3 %	28.8 %	-	-	38.3 %	67.5 %

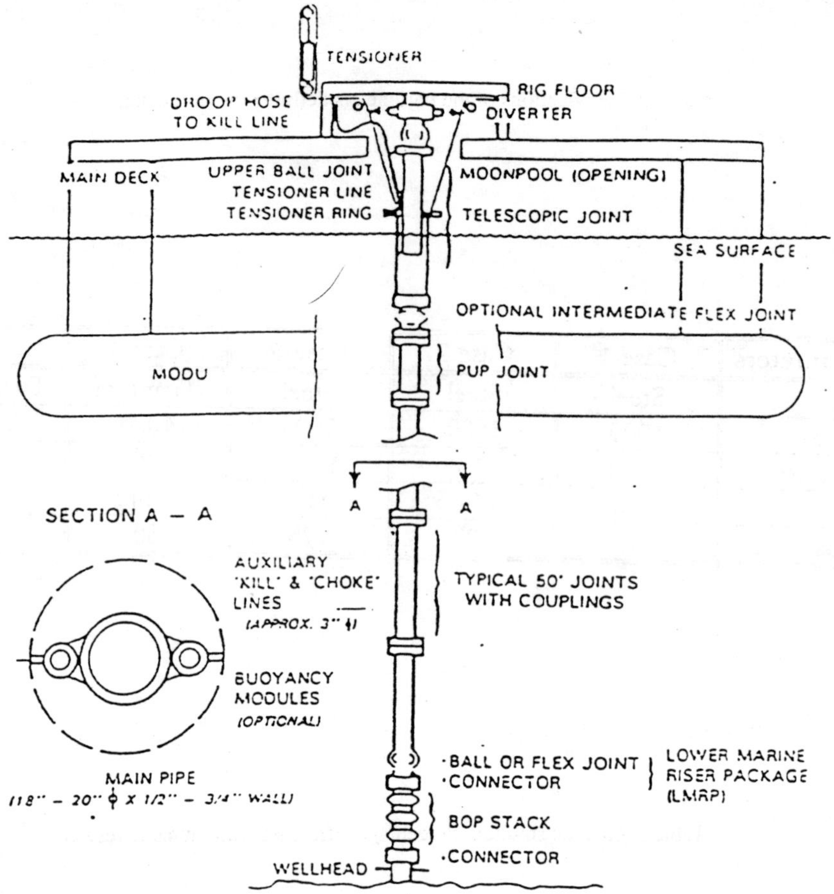

Figure 1 Steel drilling riser for a MODU

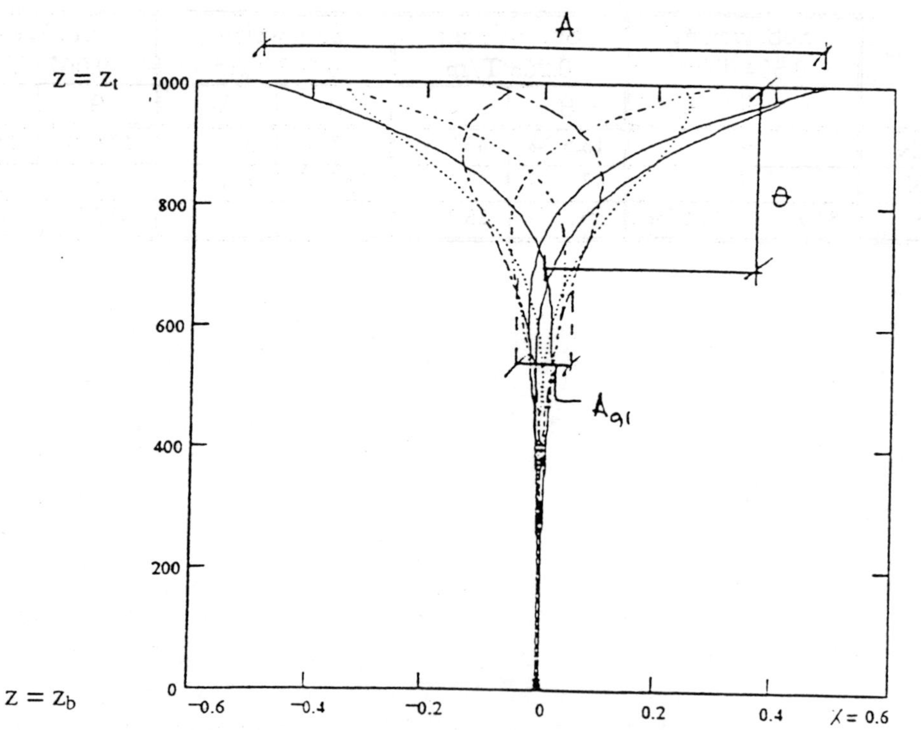

Figure 2 Riser shape at different time instances. x in m, z in m (different scales). Also the parameters A, $A_{0.1}$ and θ

Proceedings of the Seventh (1997) International Offshore and Polar Engineering Conference
Honolulu, USA, May 25-30, 1997

ISBN 1-880653-28-1 (Set); ISBN 1-880653-30-3 (Vol. II)

Effect of Axial Deformation on Natural Frequencies of Marine Cables

Somchai Chucheepsakul
King Mongkut's Institute of Technology
Bangkok, Thailand

Tseng Huang
University of Texas at Arlington
Arlington, TX, USA

ABSTRACT

In this paper, the effect of axial deformation on the natural frequencies for the in-plane vibration of marine cables is presented in which the emphasis is placed on the derivation of equations of motion with a few simplified numerical examples. The model formulation developed in this study is based on the virtual work-energy functional of marine cables which involves strain energy due to axial stretching, and virtual work done by hydrostatic pressure, and other external forces. The difference of Euler's equations between equilibrium and motion states is considered which yields the equations of motion of marine cables. The numerical examples for the simple case of specified top tension are given.

KEY WORDS : Dynamics of marine cables, axial deformation, natural frequencies.

INTRODUCTION

Marine cables employed in offshore engineering applications may experience not only to large displacements or sags but also to large axial deformation. These effects have influences on the cable behaviors, thus it is of essential for engineers designing marine cables to have a good understanding of these effects. A large number of research papers dealing with mechanics of cables are found in literature, our intension is not to cite all of them herein but some of these related to cable dynamics are mentioned. Literature review of the past and recent developments in the field of cable dynamics can be found in Irvine and Caughey (1974), Irvine (1981), and Triantafyllou (1991). Other research papers dealing with dynamics of marine cables are, for examples, Goodman and Breslin (1976), Triantafyllou (1982), Larsen and Fylling (1982), Burgress (1993), and Vassalos and Huang (1996). Most of the previous research work have considered the axial extensibility in the formulation, however this effect has been received a little attention and has not been thoroughly examined, especially its influence on the lateral vibration. To the authors' knowledge the effect of axial extensibility to the lateral vibration is not reported elsewhere. Therefore, it is the purpose of this paper at the present development to demonstrate this effect on the natural frequencies of vibration of marine cables.

The main objective of this paper is to derive the equations of motion for in-plane vibration of marine cables. The formulation based on work-energy principle is a direct extension of the previous work by Huang (1992) in which the cable in motion state is included. Work-energy functional consists of strain energy due to axial stretching, virtual work due to hydrostatic pressure, and work done by the external forces. Poisson's effect on the cross sectional area and volume changes is not considered. The fluid drag and inertia terms are included in the model formulation, however for ease of numerical computation these effects are neglected. The equations of motion are obtained by considering the difference of work-energy functional between the two states; equilibrium and motion states. As a result, two linearized undamped free vibration equations of motion are obtained. One equation represents the motion in the horizontal direction and the other represents the vertical motion, these two equations are coupled. The method of Galerkin finite element is used to obtain the mass and stiffness matrices and then the eigenvalue problem is solved. It should be noted that the reference static equilibrium state of marine cables has to be determined first in which the detailed procedure for the case of specified top tension was given in Chucheepsakul et al. (1995). Preliminary numerical examples to demonstrate the effect of axial deformation on the natural frequencies are given.

DERIVATION OF THE EQUATION OF MOTION

Variational Formulation

Figure 1 shows the equilibrium configuration and a displaced configuration of the marine cable. Let α be a convected parameter to define cable curves. A material point at the equilibrium position is defined as $x_o = x_o(\alpha)$ and $y_o = y_o(\alpha)$. Figure 2 shows components of forces acting on an infinitesimal element. Figure 3 shows the same element at the displaced position in which u and v are the horizontal and vertical components of displacement of a material point at (x_o, y_o) measured from the equilibrium position.

Consider an element having an unstretched length dS. At the equilibrium position, its length ds_o is

$$ds_o = \sqrt{x_o'^2 + y_o'^2} \cdot d\alpha \quad (1)$$

in which the superscript prime denotes a differentiation with respect to α. In terms of strain ε_o and unstretched arc length dS, it is

$$ds_o = (1+\varepsilon_o)dS \quad (2a)$$

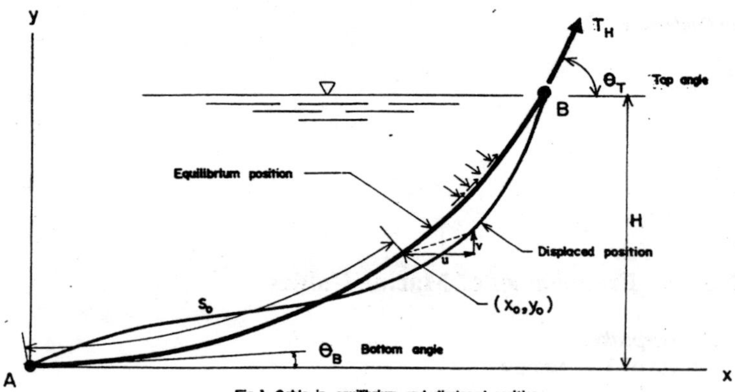

Fig. 1 Cable in equilibrium and displaced positions

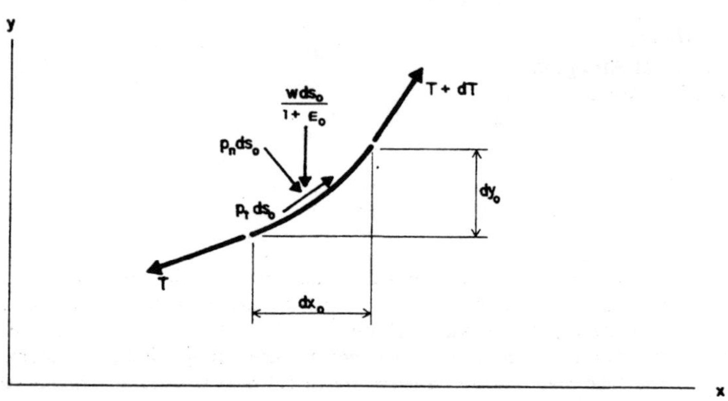

Fig. 2 Infinitesimal element

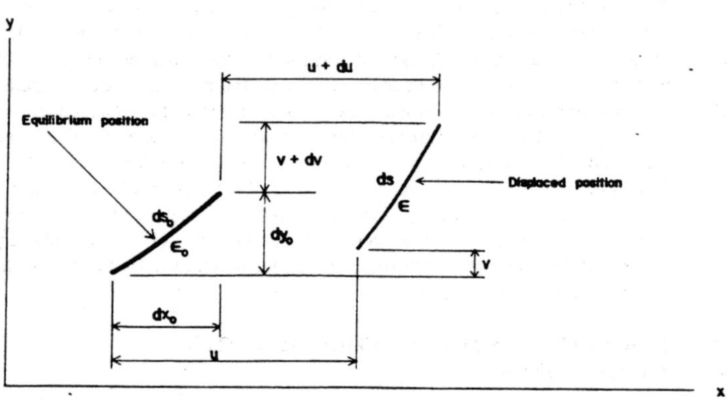

Fig. 3 Displaced position

Thus

$$dS = \frac{\sqrt{x_o'^2 + y_o'^2}}{1+\varepsilon_o} \cdot d\alpha \tag{2b}$$

Likewise at the displaced position, the arc length ds, the strain ε and the displacement u and v are related by,

$$ds = \sqrt{(x_o' + u')^2 + (y_o' + v')^2} \cdot d\alpha \tag{3}$$

$$ds = (1+\varepsilon)\, dS \tag{4}$$

From Eqs.(1) to (4), one can write the strain at the displaced position as

$$\varepsilon = \frac{ds - dS}{dS} = \frac{1+\varepsilon_o}{\sqrt{x_o'^2 + y_o'^2}} \sqrt{\left(x_o'^2 + u'\right)^2 + \left(y_o'^2 + v'\right)^2} - 1 \tag{5}$$

and its variation can be written as

$$\delta\varepsilon = \frac{1+\varepsilon_o}{\sqrt{x_o'^2 + y_o'^2}} \frac{(x_o' + u')\delta u' + (y_o' + v')\delta v'}{\sqrt{(x_o' + u')^2 + (y_o' + v')^2}} \tag{6}$$

Strain Energy of Cable due to Axial Deformation

The first variation of strain energy due to axial deformation δU in the cable at the displaced position is written as

$$\delta U = \int \frac{1}{2} EA \varepsilon \delta \varepsilon dS \tag{7}$$

where E is Young's modulus and A is the undeformed cable cross sectional area. Using Eqs.(2), (5) and (6), Eq.(7) becomes

$$\delta U = \int EA \left[1 + \varepsilon_o - \frac{1}{\sqrt{1+2\gamma}}\right] \left[\frac{(x_o' + u')\delta u' + (y_o' + v')\delta v'}{\sqrt{x_o'^2 + y_o'^2}}\right] \tag{8}$$

in which

$$\gamma = \frac{x_o' u' + y_o' v' + \frac{1}{2} u'^2 + \frac{1}{2} v'^2}{x_o'^2 + y_o'^2} \tag{9}$$

By noting that

$$\frac{1}{\sqrt{1+2\gamma}} \approx 1 - \frac{1}{2}(2\gamma) + \frac{1}{2!}\left(-\frac{1}{2}\right)\left(-\frac{3}{2}\right)(2\gamma)^2 + \cdots$$

$$\approx 1 - \gamma + \frac{3}{2}\gamma^2 + \cdots \tag{10}$$

Thus Eq.(8) becomes

$$\delta U \cong \int EA \left[\varepsilon_o + \frac{x_o' u' + y_o' v'}{x_o'^2 + y_o'^2}\right] \frac{(x_o' + u')\delta u' + (y_o' + v')\delta v'}{\sqrt{x_o'^2 + y_o'^2}} \cdot d\alpha \tag{11}$$

After neglecting the quadratic terms of the derivative of u and v, the linearized integrand of δU is

$$\delta U = \int \frac{EA}{\sqrt{x_o'^2 + y_o'^2}} \left\{ \left[\varepsilon_o (x_o' + u') + \frac{x_o'^2 u' + x_o' y_o' v'}{x_o'^2 + y_o'^2}\right] \delta u' \right.$$

$$\left. + \left[\varepsilon_o (y_o' + v') + \frac{x_o' y_o' u' + y_o'^2 v'}{x_o'^2 + y_o'^2}\right] \delta v' \right\} d\alpha \tag{12}$$

Virtual Work Done by the Hydrostatic Pressure

As commonly accepted, the hydrostatic pressure acting on the element at the displaced position is replaced by a pseudo−buoyancy force and a pseudo−tension force which are incorporated into the effective weight and effective tension respectively. However, due to the Poisson's effect the cross sectional area becomes $A(1-\mu\varepsilon_o)^2$ and the volume become $A(1-\mu\varepsilon_o)^2(1+\varepsilon_o)\, dS$. Here μ is the Poisson's ratio. In this study both the quantities $(1-\mu\varepsilon_o)^2$ and $(1-\mu\varepsilon_o)^2(1+\varepsilon_o)$ are assumed to be unity. The virtual work done by the pseudo−axial tension is written as

$$\delta U_b = \int \rho_w g A (H - y_o)\, \delta(ds) \tag{13}$$

in which ρ_w is the density of sea water and H is the water depth.

Since $\quad \delta(ds) = (\delta\varepsilon)\, dS \quad$ (14)

Then from Eqs.(6) and (2) give

$$\delta(ds) = \frac{(x_o'+u')\delta u'+(y_o'+v')\delta v'}{\sqrt{(x_o'+u')^2+(y_o'+v')^2}}\cdot d\alpha \tag{15}$$

Equation (13) becomes

$$\delta U_b = \int \rho_w g A(H-y_o)\left[\frac{(x_o'+u')\,\delta u'+(y_o'+v')}{\sqrt{x_o'^2+y_o'^2}\cdot\sqrt{1+2\gamma}}\right]d\alpha \tag{16}$$

$$= \int \rho_w \frac{gA(H-y_o)}{\sqrt{x_o'^2+y_o'^2}}\left[(x_o'+u')\,\delta u'+(y_o'+v')\,dv'\right]\left[1-\frac{x_o'u'+y_o'v'}{x_o'^2+y_o'^2}\right] \tag{17}$$

After linearization, one has

$$\delta U_b = \int \rho_w \frac{gA(H-y_o)}{\sqrt{x_o'^2+y_o'^2}}\left\{\left(x_o'+u'-\frac{x_o'^2u'+x_o'y_o'v'}{x_o'^2+y_o'^2}\right)\delta u'\right.$$

$$\left.+\left(y_o'+v'-\frac{x_o'y_o'u'+y_o'^2v'}{x_o'^2+y_o'^2}\right)\delta v'\right\}d\alpha \tag{18}$$

Work Done by External Forces

The virtual work done, δW, by the external forces, including the inertia terms and others, due to the virtual displacements δu and δv from the displaced position is

$$\delta W = \int\left\{\left[P_n y_o'+P_t x_o'-\left(\frac{w}{g}\right)\frac{\sqrt{x_o'^2+y_o'^2}}{(1+\varepsilon_o)}\cdot\ddot{u}\right]\delta u\right.$$

$$\left.+\left[-P_n x_o'+P_t y_o'-\frac{w_e\sqrt{x_o'^2+y_o'^2}}{1+\varepsilon_o}-\left(\frac{w}{g}\right)\frac{\sqrt{x_o'^2+y_o'^2}}{(1+\varepsilon_o)}\cdot\ddot{v}\right]\delta v\right\}d\alpha \tag{19}$$

in which P_n and P_t are the normal and tangential components of the Morison drag and inertia terms, w_e is the effective weight of cable, w is the weight of the cable per unit unstretched length, and dot denotes the derivative with respect to time. It should be mentioned that under stretching the cable mass in the formulation is conserved as it should be. This can be reasoned as follows : In Eq. (19), w/g is the mass per unit unstretched length and $(w/g)/(1+\varepsilon_o)$ is the mass per unit stretched length at the equilibrium position, and $\sqrt{x_o'^2+y_o'^2}\,d\alpha$ is the stretched arc length within the infinitesimal region $d\alpha$ at the equilibrium position (corresponding to unstretched region dS). Thus $\left(\frac{w}{g}\right)\frac{\sqrt{x_o'^2+y_o'^2}}{1+\varepsilon_o}\cdot d\alpha = \frac{w}{g}dS$, which can also be noted by Eq.(2b). Furthermore the mass is also conserved as the cable moves from the Equilibrium Position to the Displaced Position (Fig.3), since the convected parameter α is used. The normal and tangential of the Morison drag and inertia terms per unit arc length can be expressed as

$$P_n = \frac{1}{2}\rho_w DC_{Dn}V_n|\,V_n\,|+\rho_w AC_M a_n \tag{20}$$

$$P_t = \frac{1}{2}\rho_w \pi DC_{Dt}V_t|\,V_t\,|+\rho_w AC_M a_t \tag{21}$$

in which

$$V_n = (V_c+V_w-\dot{u})\cos\phi-\dot{v}\sin\phi \tag{22}$$

$$V_t = (V_c+V_w-\dot{v})\sin\phi+\dot{v}\cos\phi \tag{23}$$

$$a_n = (a_w-\ddot{u})\cos\phi \tag{24}$$

$$a_t = (a_w-\ddot{v})\sin\phi \tag{25}$$

where V_c is the current velocity, V_w and a_w is the horizontal component of wave particle velocity and acceleration respectively, C_{Dn} and C_{Dt} are the coefficient of normal drag and tangential drag respectively, and C_M is the inertia coefficient. At the present study it is appropriate to express P_n and P_t in the following forms for simplicity.

$$P_n = \frac{1}{2}\rho_w DC_{Dn}V_c^2\cos^2\phi+\frac{1}{2}\rho_w DC_{Dn}f_1\left(v_w,\dot{u},\dot{v},\dot{u}\dot{v},\dot{u}^2,\dot{v}^2\right)$$

$$+\rho_w AC_M f_2(a_w,\ddot{u}) \tag{26}$$

$$P_t = \frac{1}{2}\rho_w \pi DC_{Dt}V_c^2\sin^2\phi+\frac{1}{2}\rho_w \pi DC_{Dt}g_1\left(v_w,\dot{u},\dot{v},\dot{u}\dot{v},\dot{u}^2,\dot{v}^2\right)$$

$$+\rho_w AC_M g_2(a_w,\ddot{u}) \tag{27}$$

where f_1 and g_1 are the functions of wave particle velocity, components of cable velocity and their nonlinear terms, f_2 and g_2 are the functions related to wave particle acceleration and cable acceleration, and $\sin\phi$ and $\cos\phi$ terms are $x_o'/(1+x_o'^2)^{1/2}$ and $1/(1+x_o'^2)^{1/2}$ respectively.

Euler's Equations

The virtual work of the marine cable is written as

$$\delta\pi = \delta U+\delta U_b-\delta W_{ext} \tag{28}$$

Let consider the integrands of $\delta U+\delta U_b$

$$\delta U+\delta U_b = \int\left[\left\{\left[\frac{EA\varepsilon_o+\rho_w gA(H-y_o)}{\sqrt{x_o'^2+y_o'^2}}\right](x_o'+u')\right.\right.$$

$$\left.+\left[\frac{EA-\rho_w gA(H-y_o)}{\left(x_o'^2+y_o'^2\right)^{3/2}}\right]\left(x_o'^2u'+x_o'y_o'v'\right)\right\}\delta u'$$

$$+\left\{\left[\frac{EA\varepsilon_o+\rho_w gA(H-y_o)}{\sqrt{x_o'^2+y_o'^2}}\right](y_o'+v')\right.$$

$$\left.\left.+\left[\frac{EA-\rho_w gA(H-y_o)}{\left(x_o'^2+y_o'^2\right)^{3/2}}\right]\left(x_o'y_o'u'+y_o'^2v'\right)\right\}\delta v'\right]d\alpha \tag{29}$$

If y_o is used as the parameter α, then $x_o' = dx_o/dy_o$ and $y_o' = 1$, and let T be the effective tension at the equilibrium position which is defined as $EA\varepsilon_o+\rho_w gA(H-y_o)$, then Eq.(29) becomes

$$\delta U+\delta U_b = \int_o^H\left\{\left[\frac{T}{\sqrt{1+x_o'^2}}(x_o'+u')+\left[\frac{EA-\rho_w gA(H-y_o)}{\left(1+x_o'^2\right)^{3/2}}\right]\left(x_o'^2u'+x_o'v'\right)\right]\delta u'\right.$$

$$\left.+\left[\frac{T}{\sqrt{1+x_o'^2}}(1+v')+\left[\frac{EA-\rho_w gA(H-y_o)}{\left(1+x_o'^2\right)^{3/2}}\right]\left(x_o'u'+v'\right)\right]\delta v'\right\}dy_o$$

$$\tag{30}$$

For cable in static equilibrium, $\delta\pi = 0$. Thus, integration by parts of Eq.(30) is performed and afterward it is evaluated at $u = v = u' = v' = 0$ and the dot terms of u and v are not considered. Then

$$\delta\pi = \frac{Tx_o'}{\sqrt{1+x_o'^2}}\delta u\Big|_o^H - \int_o^H\left[\left(\frac{Tx_o'}{\sqrt{1+x_o'^2}}\right)' + (P_n + P_t x_o')\right]\delta u dy_o$$

$$+\frac{T}{\sqrt{1+x_o'^2}}\delta v\Big|_o^H - \int_o^H\left[\left(\frac{T}{\sqrt{1+x_o'^2}}\right)' + \left(-P_n x_o' + P_t - \frac{w_e\sqrt{1+x_o'^2}}{1+\varepsilon_o}\right)\right]\delta v dy_o$$

$$= 0 \tag{31}$$

Equation (31) yields two differential equations for static equilibrium corresponding to the virtual displacements δu and δv.

$$\left(\frac{Tx_o'}{\sqrt{1+x_o'^2}}\right)' + P_n + P_t x_o' = 0 \tag{32}$$

$$\left(\frac{T}{\sqrt{1+x_o'^2}}\right)' + \left(-P_n x_o' + P_t - \frac{w_e\sqrt{1+x_o'^2}}{1+\varepsilon_o}\right) = 0 \tag{33}$$

Here P_n and P_t in Eqs.(26) and (27) are reduced to

$$P_n = \frac{1}{2}\rho_w DC_{Dn}\frac{V_c^2}{1+x_o'^2} \tag{34}$$

$$P_t = \frac{1}{2}\rho_w \pi DC_{Dt}\frac{V_c^2\cdot x_o'^2}{1+x_o'^2} \tag{35}$$

For cable in motion, integration by parts of Eq.(30) is performed and it is evaluated at $u \neq 0,\ v \neq 0,\ u' \neq 0$, and $v' \neq 0$.

Then

$$\delta\pi = \frac{T(x_o'+u')}{\sqrt{1+x_o'^2}}\cdot\delta u\Big|_o^H + \left[\frac{EA-\rho_w gA(H-y_o)}{(1+x_o'^2)}\right](x_o'^2u' + x_o'v')\delta u\Big|_o^H$$

$$-\int_o^H\left[\left(\frac{T(x_o'+u')}{\sqrt{1+x_o'^2}}\right)' + (P_n + P_t x_o') + \left(\frac{w}{g}\frac{\sqrt{1+x_o'^2}}{(1+\varepsilon_o)}\right)\ddot{u}\right]\delta u dy_o$$

$$-\int_o^H\left[\frac{EA-\rho_w gA(H-y_o)}{(1+x_o'^2)^{3/2}}\right](x_o'^2u'' + x_o'v'')\,\delta u dy_o$$

$$+\frac{T}{\sqrt{1+x_o'^2}}(1+v')\delta v\Big|_o^H + \left[\frac{EA-\rho_w gA(H-y_o)}{(1+x_o'^2)^{3/2}}\right](x_o'u' + v')\delta v\Big|_o^H$$

$$-\int_o^H\left[\left(\frac{T(1+v')}{\sqrt{1+x_o'^2}}\right)' + \left(-P_n x_o' + P_t - \frac{w_e}{1+\varepsilon_o}\sqrt{1+x_o'^2}\right) + \frac{w\sqrt{1+x_o'^2}}{g(1+\varepsilon_o)}\cdot\ddot{v}\right]\delta v dy_o$$

$$-\int_o^H\left[\frac{EA-\rho_w gA(H-y_o)}{(1+x_o'^2)^{3/2}}\right](x_o'u'' + v'')\,\delta v dy_o \tag{36}$$

Thus, Euler's equations associated with the cable motion δu and δv are written respectively as

$$\left(\frac{T(x_o'+u')}{\sqrt{1+x_o'^2}}\right)' + (P_n + P_t x_o') + \frac{EA-\rho_w gA(H-y_o)}{(1+x_o'^2)^{3/2}}(x_o'^2u'' + x_o'v'') + \frac{w}{g}\frac{\sqrt{1+x_o'^2}}{(1+\varepsilon_o)}\ddot{u}$$

$$= 0 \tag{37}$$

$$\left(\frac{T(1+v'')}{\sqrt{1+x_o'^2}}\right)' + \left(-P_n x_o' + P_t - \frac{w_e\sqrt{1+x_o'^2}}{1+\varepsilon_o}\right) + \frac{EA-\rho_w gA(H-y_o)}{(1+x_o'^2)^{3/2}}(x_o'u'' + v'')$$

$$+\frac{w}{g}\frac{\sqrt{1+x_o'^2}}{(1+\varepsilon_o)}\cdot\ddot{v} = 0 \tag{38}$$

Subtracting Eq.(32) from Eq.(37) yields the equation of motion in u direction and subtracting Eq.(33) from Eq.(38) yields the equation of motion in v direction which they can be written as

$$m_u\ddot{u} + \rho_w AC_M f_2(a_w,\ddot{u}) + \frac{1}{2}\rho_w DC_{Dn} f_1(v_w,\dot{u},\dot{v},\dot{u}\dot{v},\dot{u}^2,\dot{v}^2)$$
$$+ f_3(u'',v'') = 0 \tag{39}$$

and

$$m_v\ddot{v} + \rho_w AC_M g_2(a_w,\ddot{v}) + \frac{1}{2}\rho_w DC_{Dt} g_1(v_w,\dot{u},\dot{v},\dot{u}\dot{v},\dot{u}^2,\dot{v}^2)$$
$$+ g_3(u'',v'') = 0 \tag{40}$$

where

$$m_u = \frac{w}{g}\frac{\sqrt{1+x_o'^2}}{(1+\varepsilon_o)} \tag{41}$$

$$m_v = \frac{w}{g}\frac{\sqrt{1+x_o'^2}}{(1+\varepsilon_o)} \tag{42}$$

$$f_3(u'',v'') = \frac{T\cdot u''}{\sqrt{1+x_o'^2}} + \left[\frac{EA-\rho_w gA(H-y_o)}{(1+x_o'^2)^{3/2}}\right](x_o'^2u'' + x_o'v'') \tag{43}$$

$$g_3(u'',v'') = \frac{T\cdot v''}{\sqrt{1+x_o'^2}} + \left[\frac{EA-\rho_w gA(H-y_o)}{(1+x_o'^2)^{3/2}}\right](x_o'u'' + v'') \tag{44}$$

At the present study, the effect of drag and inertia terms are neglected. Thus, Eqs.(39) and (40) yield the equations of motion for undamped free vibration of marine cables.

$$\begin{bmatrix} m_u & 0 \\ 0 & m_v \end{bmatrix}\begin{Bmatrix}\ddot{u}\\ \ddot{v}\end{Bmatrix} + \begin{bmatrix} k_{uu} & k_{uv} \\ k_{vu} & k_{vv}\end{bmatrix}\begin{Bmatrix}u''\\ v''\end{Bmatrix} = \{0\} \tag{45}$$

where

$$m_u = m_v = \frac{w}{g}\frac{\sqrt{1+x_o'^2}}{(1+\varepsilon_o)} \tag{46}$$

$$k_{uu} = \frac{T}{\sqrt{1+x_o'^2}} + \left[\frac{EA-\rho_w gA(H-y_o)}{(1+x_o'^2)^{3/2}}\right]x_o'^2 \tag{47}$$

$$k_{uv} = k_{vu} = \left[\frac{EA-\rho_w gA(H-y_o)}{(1+x_o'^2)^{3/2}}\right]x_o' \tag{48}$$

$$k_{vv} = \frac{T}{\sqrt{1+x_o'^2}} + \left[\frac{EA-\rho_w g A(H-y_o)}{(1+x_o'^2)^{3/2}}\right] \quad (49)$$

METHOD OF SOLUTION

The reference static equilibrium configuration is determined by using the numerical procedure of the authors' previous work (Chucheepsakul et al., 1995) in which (x_o, y_o) and its derivatives as well as ε_o are obtained. In the procedure, the range of independent variable y_o, which is the total water depth, is divided into finite elements. To solve the equations of motion, Eq.(45), the method of Galerkin finite element (Cook et al., 1989) is used in which the assumed displacement fields u and v are approximated by

$$\{u\} = \begin{Bmatrix} u \\ v \end{Bmatrix} = [N]\{d\} \quad (50)$$

where $$\{d\} = \lfloor u_1 \;\; u_1' \;\; v_1 \;\; v_1' \;\; u_2 \;\; u_2' \;\; v_2 \;\; v_2' \rfloor \quad (51)$$

and $$[N] = \begin{bmatrix} N_1 & N_2 & 0 & 0 & N_3 & N_4 & 0 & 0 \\ 0 & 0 & N_1 & N_2 & 0 & 0 & N_3 & N_4 \end{bmatrix} \quad (52)$$

Here N_i, i = 1, 2, 3, 4 represent the weight functions and are the cubic polynomial shape functions, which are the same ones used in static analysis.

Substitution of Eq.(48) into (45) and following the Galerkin's method yields

$$\sum_{j=1}^{\text{Nel}}\left(\int_o^h [N]^T\begin{bmatrix} m_u & 0 \\ 0 & m_v \end{bmatrix}[N]dy_o\{\ddot{d}\} + \int_o^h [N']^T\begin{bmatrix} k_{uu} & k_{uv} \\ k_{vu} & k_{vv} \end{bmatrix}[N']dy_o\{d\}\right) = \{0\} \quad (53)$$

which is the global equation of free vibration

$$[M]\{\ddot{D}\} + [K]\{D\} = \{0\} \quad (54)$$

where $[M]$ is the total mass matrix defined as

$$[M] = \sum_{j=1}^{\text{Nel}}\int_o^h [N]^T\begin{bmatrix} m_u & 0 \\ 0 & m_v \end{bmatrix}[N]dy_o, \quad (55)$$

$[K]$ is the total cable stiffness defined as

$$[K] = \sum_{j=1}^{\text{Nel}}\int_o^h [N']^T\begin{bmatrix} k_{uu} & k_{uv} \\ k_{vu} & k_{vv} \end{bmatrix}[N']dy_o, \quad (56)$$

$\{\ddot{D}\}$ and $\{D\}$ are the global acceleration and displacement respectively. For free vibration, Eq.(52) leads to the following eigenvalue problem

$$([K]-\omega^2[M])\{\overline{D}\} = \{0\} \quad (57)$$

where ω_i is the natural frequency of vibration and $\{\overline{D}\}$ is the corresponding mode shape. The boundary conditions for this problem are as follows : $u = v = 0$ at $y_o = 0$ and H.

NUMERICAL EXAMPLES AND RESULTS

To verify the model formulation, a simple case of horizontal cable suspended by its own weight is considered. The input data of cable configuration given by West et al. (1975) is used in the calculation. The program input parameters are accordingly modified as follows : $\rho_w = V_C = x_H = 0$, T_H = 13,000 kN, H = 854 m, D = 0.023 m, E = 1.794 kN/m^2. In addition, the distributed external load of $w\sqrt{1+x_o'^2}/(1+\varepsilon_o)$ representing the uniform weight of cable per unit stretched length which is equal to 9.48 kN/m is applied. Furthermore, the tension in cable is modified in order to handle the corresponding static configuration. Results of static analysis yield the equilibrium configuration of cable suspended horizontally by its own weight with sag of 70.7 m. The in-plane natural frequencies for the first four modes are favourably compared with those found in the literature and given in Table 1. The effect of top tension and elastic modulus for this example problem is also investigated and results are given in Table 2. It is seen that for a fixed value of top tension, the natural frequencies increase with the increase of elastic modulus and for a fixed value of elastic modulus, the frequencies increase as top tension increases.

Table 1. Comparison of the natural frequencies (rad/sec) for in-plane modes of horizontally suspended cables

ω rad/sec	This study	Gambhir and Batcheler (1978)	West et al. (1975)	Irvine (1981)
ω_1	0.809	0.795	0.811	0.831
ω_2	1.174	1.155	1.175	1.247
ω_3	1.654	1.627	1.653	1.662
ω_4	2.004	1.998	2.027	2.078

Table 2. Results of the natural frequencies for a fixed value of top tension with three different values of elastic modulus and for a fixed value of elastic modulus with three different values of top tension.

ω	T_H = 13,000 kN			$E = 1.794\times10^8$ kN/m^2		
rad/sec	$E = 1.794\times10^6$ kN/m^2	$E = 1.794\times10^7$ kN/m^2	$E = 1.794\times10^8$ kN/m^2	T_H = 13,000 kN	T_H = 15,000 kN	T_H = 17,000 kN
ω_1	0.526	0.806	0.809	0.809	0.878	0.945
ω_2	0.829	0.993	1.174	1.174	1.264	1.344
ω_3	1.273	1.327	1.654	1.654	1.782	1.908

In order to demonstrate the effect of extensibility on the natural vibration of marine cables, preliminary numerical investigation was carried out using the input data given in Table 3. To simplify the numerical computation as mentioned earlier, the drag and inertia terms in the motion state are neglected. However, in the static equilibrium state, the drag terms resulted from the steady current velocity are included in the calculation. In the present study, numerical results are given for the different values of specified top tension and elastic modulus.

Table 3. Program parameters

Top tension	T_H	= 27.5, 30, and 35 kN
Total water depth	H	= 300 m
Top end location	x_H	= 150 m
Effective weight of cable	w_e	= 27.83 N/m
Cable diameter	D	= 0.023 m
Elastic modulus	E	= $2\times10^6, 2\times10^7, 2\times10^8$ kN/m^2
Density of sea water	ρ_w	= 1021 kg/m^3
Uniform current velocity	V_c	= 1.0 m/sec
Normal drag coefficient	C_{Dn}	= 1.0
Tangential drag coefficient	C_{Dt}	= 0.03
Mass coefficient	C_M	= 0.0

Table 4 shows numerical results of the first three natural frequencies of the marine cables with three different values of top tension which are 27.5, 30.0, and 35 kN. As expected, the natural frequencies increase with the increase in top tension. Plots of equilibrium configuration and the first three coupled mode shapes for T_H = 27.5 kN are shown in Fig. 4.

Table 4. Natural frequencies of vibration versus top tension for $E = 2\times10^8$ kN/m^2

ω rad/sec	T_H = 27.5 kN	T_H = 30 kN	T_H = 35 kN
ω_1	1.561	1.916	2.202
ω_2	2.218	2.320	2.500
ω_3	3.134	3.306	3.557

To demonstrate the effect of elastic modulus, three different values of elastic modulus of 2×10^6, 2×10^7, and 2×10^8 kN/m^2 are used for the specified of top tension of 27.5 kN. Numerical results of the first three natural frequencies are given in Table 5. It can be seen that the frequencies increase as E increase.

Table 5. Natural frequencies of vibration versus elastic modulus for top tension = 27.5 kN

ω rad/sec	$E = 2\times10^6$ kN/m^2	$E = 2\times10^7$ kN/m^2	$E = 2\times10^8$ kN/m^2
ω_1	0.906	1.500	1.561
ω_2	1.585	1.615	2.218
ω_3	2.393	2.426	3.134

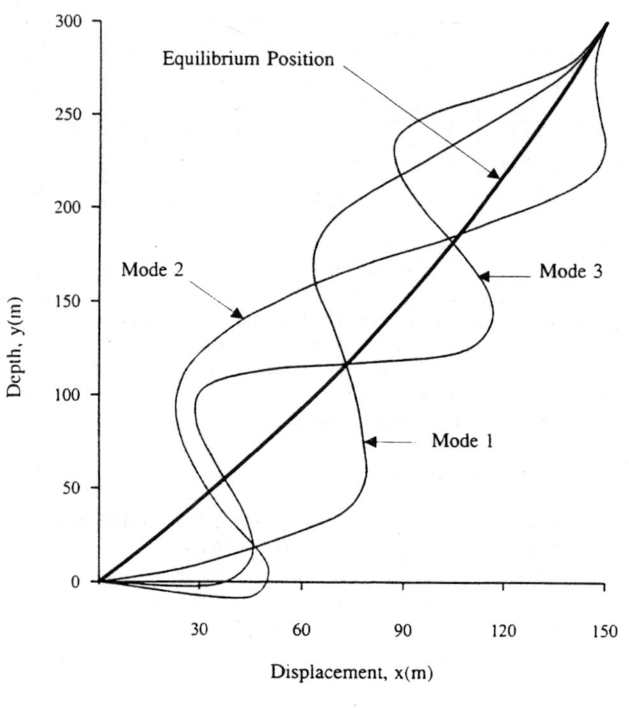

Figure 4 In-plane coupled mode shapes of u and v

CONCLUDING REMARKS

The derivation of equations of motion for in–plane vibration of marine cables with axial deformation is presented. In the formulation, work–energy functional consisted of strain energy due to axial deformation, additional virtual work due to hydrostatic pressure and work done by the other external forces in the equilibrium state and motion state is considered. The Poison's ratio effect on the cross sectional area and volume changes and the fluid drag and inertia terms are neglected in the present study. Numerical results are given for the case of specified top tension. It is found that, for the examples considered herein, the natural frequencies increase with the increase in top tension as well as the increase in elastic modulus.

Although at the present study a limited amount of numerical results have been presented, an extensive numerical study needed to be done and the results will be reported in the near future.

REFERENCES

Burgress, J (1993). "Bending Stiffness in a Simulation of Undersea Cable Deployment," *Int J Offshore and Polar Eng*, Vol 3, No 3, pp 197–204.

Chucheepsakul, S, Huang, T, and Laohapotjanart, P (1995). "Effect of Axial Deformation on the Equilibrium Configurations of Marine Cables," *Proc 5th Int Offshore and Polar Eng*, ISOPE, Vol 2, pp 244–248.

Cook, RD, Malkus, DS, and Plesha, ME (1989). *Concepts and Applications of Finite Element Analysis*, Third edition, John Wiley & Sons, pp 461–465.

Gambhir, ML and Batchelor BD (1978). "Parametric Study of Free Vibration of Sagged Cables," *Computer and Structures*, Vol 8, pp 641–648.

Goodman, TR and Breslin, JP (1976). "Statics and Dynamics of Anchoring Cables in Waves," *J Hydronautics*, Vol 10, No 4, pp 113–120.

Huang, T (1992). "A Static Equilibrium Formulation including Axial Deformation for Marine Cables," *Proc 2nd Int Offshore and Polar Eng*, ISOPE, Vol 2, pp 252–255.

Irvine, HM (1981). *Cable Structures*, The MIT Press, Cambridge, Mass.

Irvine, HM and Caughey, TK (1974). "The Linear Theory of Free Vibration of a Suspended Cable," *Proc Royal Society (London) Series A*, Vol 341, pp 299–315.

Larsen, CM and Fylling, IJ (1982). "Dynamic Behavior of Anchor Lines," *Norwegian Maritime Res. 3*

Triantafyllou, MS (1982). "Preliminary Design of Mooring Systems," *J Ship Res*, Vol 28, No 1, March, pp 25–35.

Triantafyllou, MS (1991). "Dynamics of Cables, Towing Cables and Mooring System," *Shock and Vibration Digest*, Vol 23, No 7, pp 3–8.

Vassalos, D and Huang S (1996). "Dynamics of Small–Sagged Taut–Slack Marine Cables," *Computers and Structures*, Vol 58, No 3, pp 557–562.

West, HH, Geschwindner, LF, and Suhoshi, JE (1975). "Natural Vibration of Suspension Cables," *ASCE Journal of the Structural Division*, Vol 101, No ST11, pp 2277–2291.

Proceedings of the Seventh (1997) International Offshore and Polar Engineering Conference
Honolulu, USA, May 25-30, 1997

ISBN 1-880653-28-1 (Set); ISBN 1-880653-30-3 (Vol. II)

Structural Analysis of Flexible Pipe Using Finite Element Method

Flavio Torres Lopez da Cruz
PETROBRAS-CENPES
Rio de Janeiro, Brazil

Carlos Alberto Nunes Dias
São Paulo University
São Paulo, Brazil

ABSTRACT

The principles of three dimensional modeling of a flexible pipe are here determined. Each layer is modeled using the elements which most adequately represent its structural behavior, within the limitations of the Finite Elements Method. The interaction of two steel layers are modeled by solid elements, when there is a plastic layer between them, or by beam elements in the case of direct contact. The tension layers are represented by helicoidal springs made of beam elements and the pressure layers are represented by cylindrical thin shells consisting of orthotropic plates. In this case , due to theoretical manipulation, it is possible to represent the helicoidal shaped layers as a trivial cylinder, making it feasible to model complex pipe structures .

KEY WORDS : *F.E.M. , FLEXIBLE , PIPE.*

1.0 INTRODUCTION

Flexible pipes are composite products made up of several steel and plastic concentric layers which, working together, give strength and leakproofness, maintaining the necessary flexibility to conduct the several fluids of the oil industry, mainly in the oil production offshore. This composite material, in a concentric layered structure, shall be able to resist severe loads, such as tension, internal and external pressure, torsion and bending which act in the usual operating conditions, of an offshore production platform, in the particular case of Brazil , for very deep waters.

The Flexible Pipes were employed for the first time in offshore oil production by the end of the 70s. After the oil discovery at Campos Basin, and due to the oil price increase since 1973, PETROBRAS began to use the so called E.P.S. (Early Production System) which consists of a conventional floating platform used to store, in advance, the oil production at a specific location. In this situation, for the production flow it was decided to use the flexible pipe due to its ease of installation as well as the possibility of reutilization.

From the increasing use of flexible pipe, arises the need to better model its mechanical structural behavior. Above all, the tendency of oil production in increasing water depth is to reach technological and economical barriers. In the case of PETROBRAS , this is more relevant due to its pioneering and leadership in ultra deep water (over 1000 meters depth).

The evaluation of the strength of a flexible pipe in the early phase of the production system design lies on the analysis of the structural interaction between its several layers, in order to determine the correct dimensions and types of the layers, taking into consideration all the applied loads, in particular those related to the installation phase.

The mechanical loads acting on the pipe during the installation bring additional difficulties for both the interaction of the layers and the theoretical evaluation of its consequences. The non-axisymmetric loads due to the installation apparatus may seriously reduce the strength of the pipe and are, in most cases, the dimensioning load condition for the flexible pipe.

The nature of mechanical loads due to pipe installation procedures makes impractical the analytic assessment of the stress field on its layers. If we take into account that many times a significant portion of the pipe structure remains under loaded in the normal operation, the importance of a good stress evaluation for the installation conditions, which is the limiting situation, is clearly of great relevance for the product feasibility.

The most usual installation analysis method for flexible pipe structures[Goto et al, 1986] considers a two dimensional transverse section and axisymmetric loads (tension and pressure) and so cannot take into consideration the three dimensional state of strains that arises from the linear tractioners during the laying operations. By using FEM (Finite Element Method), this work proposes a methodology for stress analysis of the flexible pipe layers useful for all possible of loading conditions, from installation to the normal operation.

For the data preparation tasks, a pre-processor program (called GERFLEX)was developed to make easier the tiresome activities of FEM modeling Several analysis cases are needed to assess reliability, qualifying the proposed methodology as a tool for flexible pipes structural analysis.

2.0 PROBLEM DESCRIPTION

A flexible pipe must be able to satisfy the basic strength requirements, like a usual steel pipe, but with the further requirement to have very small bending rigidity. For this, each layer is conceived to fulfill specific functions which impose constraints on dimensions, material and geometry.

During the pipe installation, the inner layers feel the contact pressure caused by the diameter reduction of the double cross wound armors (figure 1) which is the mechanism to resist the external pipe traction. To grip the pipe by friction, the launching equipment applies radial forces whose effects are added to those of the contact pressure. These are the main actions on flexible pipes that deserve attention during installation (figure 3).

Even considering the manufacturing diversity, it is possible to classify the flexible pipe, used by petroleum industry, as composed by the following generic layers with its corresponding structural functions.

(1) **Interlocked Steel Carcass.** It is manufactured by cold forming strips (one or two) by means of a spiraling machine. To assure the required bending flexibility the strips are loose interlocked in a small spiral pitch.

The Interlocked Steel Carcass does not perform any sealing function and is therefore only submitted to the contact pressure that comes from the internal thermoplastic leak proof sheath, giving mechanical support to it.

(2) **Spiral Pressure Layer.** Depending on the pipe loading severity, this optional spiral layer is included having a basic function to improve the resistance to the radial effects of pipe internal and external pressures as well as to give additional resistance to the shrinking of the armors.

Of course the Spiral Pressure Layer also contributes to the resistance against the external mechanical loads of the laying equipment. As for the carcass, its small spiral pitch does not affect the bending flexibility of the pipe while it still allows the contribution to the radial loads resistance.

(3) **Double Cross wound Traction Armors.** The armors are composed by two layers of flat steel wire in a helicoidal shape around the inner layers of the flexible pipe.

These internal and external armors are wound in big spiral pitches so that they can directly resist the axial load applied to the pipe itself, by converting this traction into squeeze contact pressure over the inner layers [Patel,Witz and Tan, 1993]. A further function is to provide additional internal pressure resistance to the pipe, keeping it balanced against torsion.

(4) **Thermoplastic Sheaths.** The main function of the plastic sheaths is to ensure the pipe is leak proof and thermally insulated, both internally and externally.

A further function may be to provide, in dynamic applications, smooth friction between the layers as well as, from the structural point of view, due to the small rigidity of the plastic material, to work like mattress spreading the internal loads over the inner layers.

3.0 USE OF THE FINITE ELEMENT METHOD (F.E.M.)

In the case of stress analysis for the pipe launching, the analytical models [McNamara and Harte, 1990] are not quite suitable because the mechanical loads are not axisymmetric. On the other hand, we believe that, in general, two dimensional modeling is not quite acceptable basically for the following reasons:

(a) a model for the cross section cannot be able to detect stress changes in the axial pipe direction and the same may be said in respect to a longitudinal section model that cannot detect changes of the tangential stress.

(b) a model for the cross section needs the help of an analytical expression to compute the squeeze during the pipe traction. In spite of its extended use, this kind of approach is doubtless a source of undesirable uncertainty.

(c) the helicoidal geometry of all the spirals (wires of the armors and strips of carcass and pressure layer) is in fact three dimensional.

So, to obtain the stress field in both axial and radial directions, avoiding the restrictions of an analytical approach, we decided to use 3D modeling which requires numerical solution of the structural equilibrium equations.

For the 3D FEM modeling, the main difficulties arise from the complex spatial geometry of the pipe layers of different types and materials, resulting in a model with a large number of nodes, elements and equations to be solved. Further, the existence of non linear behavior due to material, damping, contact and sliding between layers as well as due to big displacements, can introduce serious obstacles. Some possible 3D models would have no chance to be solved in a simple computer with the usual FEM software. Hence, for the purposes of this work, we had to restrain the study scope to the range where the structural pipe behavior remains linear, with no intention to get ultimate states.

Thus it is obvious that this 3D modeling has no direct application on new product development, but it can be applied without apprehension to evaluate the pipe capacity up to the linear bound of its structural behavior, a situation that comprises the laying and the normal operations of the flexible pipe, even in deep water.

4.0 MODELLING DESCRIPTION

Following the parametric modeling dictates, the perimeter of the pipe is divided into a polygon with N sides. This same subdivision was applied on the longitudinal direction over a length corresponding to a complete armor spiral pitch, in order to offer a regular shape to the finite elements (figure 3).

The number N and the total length of pipe to be analyzed (defined in terms of complete pitches of the armors) controls the size of the model that must conform to the computer capacity. For this, the following conflicting aspects must be taken into account :

(a) hardware and software resources that, in practice, limit the maximum number of degrees of freedom of the model.

(b) the greater N the better the polygonal contour of the cross section, improving the model behavior.

(c) the local loads, induced by the mechanical action of the traction mechanism during launching, require mesh refinement in order to assure a correct reproduction of its effects. Thus the model length must be longer than the extension of the linear tractioner feet, avoiding the effects of the boundary conditions at the pipe extremities.

In the modeling, the contact between layers can use two different approaches depending on the structures' features. For a plastic sheath between two metallic layers brick elements are directly attached to the nodes of the elements at these layers. On the other hand, for the cases where a direct contact between metallic layers is present, small beam elements in the radial direction are used to connect the nodes between layers.

In this context, we avoid the use of the so called 'contact elements' in order to simplify the modeling. There is no strong reason to perform a non linear analysis as the one due to the contact elements since, any way, it is always possible to employ an alternative strategy of element deletion in case traction occurs .

Specially for the case of the contact between the external and the internal armors, the torsional rigidity of the beams controls the pantographic effect of these two layers. The best results were found neglecting the torsional constant of these beams.

The element types for each layer (figure 3), observing the most plausible structural behavior, can be summarized as:

(1)Thermoplastic Sheaths. Brick-8 Isoparametric Elements, with isotropic linear material, describing the entire volume of the layer.

(2)Traction Armors. 3D Beam Elements wounded in N helicoids through the mean surface of the layer, each helicoid representing a set of corresponding wires so that the effect of all the wires is computed.

(3)Interlocked Steel Carcass and Spiral Pressure Layer. Thin Rectangular Plate Element, with orthotropic linear material, forming a cylindrical shell through the mean surface of the layer.

The cylindrical shell must have very small stiffness in the axial direction of the flexible pipe and a high stiffness in the radial direction, in order to represent the mentioned steel spiral layers. This directional dependency, imposed by the helicoidal geometry of the layers, may be simulated by use of fictitious orthotropic material, as will be seen later.

The model construction for a pipe, with different materials, complex geometry of distinct pitch spiral layers, and so on, is a task that requires long time of execution, is not free of mistakes and needs help. For this reason we developed a computer program that, from the general properties of the flexible pipe, generates automatically the complete model with nodes, elements, loads and boundary conditions.

The program GERFLEX only asks for the geometric and material properties of the pipe and allows some control on mesh refinement, so that the user can have a coarse mesh, for quick but inaccurate analysis, or a very fine one for precise assessment of the stress field. With fine meshes it is possible to get models with up to 100,000 equations, which means that the user must be careful about his computer limitations. For the present work we have prepared models with up to 35,000 degrees of freedom which in a 90Mhz PENTIUM microcomputer require about six hours of processing.

It must be pointed out that for the case of mechanical loads (figure 3), the program automatically calculates all the external forces on the pipe, considering the traction acting on it and the dry friction over the external layer. This means that all local pressure and shear external forces, applied by traction mechanisms during installation, may be correctly computed.

5.0 BASIC FORMULATION OF THE ORTHOTROPIC SHELL

Essentially, the only innovation introduced in the proposed modeling lies in the concept of an orthotropic material shell to represent the spiral layers that resist the radial pressures : **Interlocked Steel Carcass and Spiral Pressure Layer.** Considering that, for small displacements and strains, these layers behave globally like a helicoidal linear spring and, locally like, a linear shell, we have to determine first the Elasticity Matrix .

ELASTICITY MATRIX

Taking an orthotropic material with helicoidal orientation, the Elasticity Matrix ,for each plate element, is [Timoshenko and Woinowsky-Krieger, 1976]

$$C = \begin{bmatrix} E'_x & E'' & 0 \\ E'' & E'_y & 0 \\ 0 & 0 & G \end{bmatrix} \qquad (1.a)$$

where

$$E'_x = E_x / (1 - \nu_x \nu_y) \qquad (1.b)$$
$$E'_y = E_y / (1 - \nu_x \nu_y) \qquad (1.c)$$
$$E'' = \nu_x E_y / (1 - \nu_x \nu_y) = \nu_y E_x / (1 - \nu_x \nu_y) \qquad (1.d)$$

and

C = Elasticity Matrix ,

E_x ,E_y = Young Modulus to be used as fictitious material properties in both directions over the helicoid,

ν_x ,ν_y = Poisson Coefficients, idem.

In this context, the angle of the helicoid with the longitudinal axis of the pipe is given as material property orientation for all the plate elements that constitute the orthotropic shell. Since there is no structural interaction in the normal tangential direction of the helicoid, we may take,

$$E'' = 0.0$$
$$E'_y = E'_x / 10^6$$
$$\nu_x = \nu_y = 0.0$$

Thus, it remains find $\mathbf{E'_x}$ and $\mathbf{G}$.

If we were actually treating a tubular structure with orthotropic material - as a fiber glass pipe - the determination of the Elasticity Matrix would be complete. However, the material of our steel pipe is not, in fact, orthotropic but the pipe geometry has itself an orthotropic behavior. So, in order to get the unknowns $\mathbf{E'_x}$ and **G**, we have to proceed by considering the structural arrangement, with different rigidities in orthogonal directions, like a grid.

For this, we need first to take the relations for stiffness and stress in an orthotropic thin rectangular plate [Timoshenko and Woinowsky-Krieger, 1976] of thickness t :

(1)Bending Stiffness

$$D_x = (t^3 / 12) E'_x \qquad (2.a)$$
$$D_y = (t^3 / 12) E'_y \qquad (2.b)$$

(2)Torsion Stiffness

$$D_{xy} = (t^3 / 12) G \qquad (2.c)$$

(3)Membrane Stiffness

$$B_x = t E'_x \qquad (3.a)$$
$$B_y = t E'_y \qquad (3.b)$$
$$B_{xy} = t G \qquad (3.c)$$

(4)Bending Stresses

$$\sigma^f_x = M_x / (6 t^2) \qquad (4.a)$$
$$\sigma^f_y = M_y / (6 t^2) \qquad (4.b)$$
$$\sigma^f_{xy} = M_{xy} / (6 t^2) \qquad (4.c)$$

where $\mathbf{M_x}$,$\mathbf{M_y}$ are the bending moments and $\mathbf{M_{xy}}$ the torsional moment,(all per unit length).

(5)Membrane Stresses

$\sigma m_x = N_x / t$ (5.a)
$\sigma m_y = N_y / t$ (5.b)
$\sigma m_{xy} = N_{xy} / t$ (5.c)

where N_x, N_y are the in plane normal forces and N_{xy} the shear force,(all per unit length).

GRID ANALOGY.

Taking our strip spiral tube as a grid with distinct stiffness in two orthogonal directions, we may write for the rigidities :

$D_x = (E_{mat} \; I_{xx}) (n / p)$ (6.a)
$D_y = 0$ (6.b)
$D_{xy} = (G_{mat} \; J_t) (n / p)$ (6.c)
$B_x = (E_{mat} \; A_s) (n / p)$ (6.d)
$B_y = 0$ (6.e)
$B_{xy} = (G_{mat} \; A_s) (n / p)$ (6.f)

where

E_{mat} = Young Modulus of the material,
G_{mat} = $E_{mat} / [2 (1 + \nu_{mat})]$
ν_{mat} = Poisson coefficient of the material,
n = number of strips per helicoid,
p = spiral pitch,
I_{xx} = bending inertia of the strip,
J_t = torsion constant of the strip,
A_s = cross section area of the strip,

and

$D_y = B_y = 0$, because there is no wire in y direction (the normal tangential direction of the strip).

Now, equating the corresponding equations for the rigidities, in order to assure for the shell the same rigidities of the spiral layer, we have :

$D_x = (E_{mat} \; I_{xx}) (n / p) = (t^3 / 12) E'_x$ (8.a)
$D_{xy} = (G_{mat} \; J_t) (n / p) = (t^3 / 12) G$ (8.b)
$B_x = (E_{mat} \; A_s) (n / p) = t E'_x$ (8.c)
$B_{xy} = (G_{mat} \; A_s) (n / p) = t G$ (8.d)

from where we can extract the unknowns $\mathbf{E'_x}$ and **G**, as well as the thickness (**t**) of the orthotropic shell that is not the same as the layer thickness (**h**). Depending on the strip cross section, **t** is in general smaller than the layer thickness **h**.

An attempt to solve equations (8) shows, however, that we cannot satisfy all of them. Hence, a better choice is to neglect the less important shear membrane stiffness ($\mathbf{B_{xy}}$) of equation (8.d). Diving equations (8.a) and (8.c), we obtain

$$t = (12 D_x / B_x)^{1/2} = (12 I_{xx} / A_s)^{1/2} \quad (9.a)$$

and, once the thickness **t** is known, from (8.a) or (8.c) we have

$$E'_x = (E_{mat} \; I_{xx}) (n / p) (12 / t^3) = (E_{mat} \; A_s) (n / p) (1 / t) \quad (9.b)$$

and from (8.d)

$$G = (G_{mat} \; A_s) (n / p) (1 / t) \quad (9.c)$$

As can be seen, the thickness **t** is a fictitious value that must be used as the plate element thickness in order to assure that the orthotropic shell of the model has the rigidities of the corresponding strip spiral layer.

If the relative rigidities between the several layers are imposed, as we intend to do, the internal force distribution may be obtained with good precision. However, as will be seen later, in our approach there is a need of apply corrections for the stresses.

STRESS CORRECTION.

At first, let the membrane stress (σm_x) on the actual strip be

$$\sigma m_x = (p N_x) / (n A_s) \quad (10.a)$$

and, on the other hand, on the plate of shell model

$$\sigma m_x = N_x / t \quad (10.b)$$

so that, if ,as we want, $\mathbf{N_x}$ (normal internal force per unit length at x direction) is the same for both cases, the correction α_m is determined by dividing (10.a) by (10.b), from where

$$\alpha_m = (p t) / (n A_s) \quad (10.c)$$

This means that the membrane stress given by the Finite Element Analysis must be multiplied by α_m to be correct. If , for example, the strip cross section is rectangular (b x h) , with no empty parts (p=b) and the helicoid is made of only one strip (n=1), we would have

$$t = (12 I_{xx} / A_s)^{1/2} = [12 (b h^3 / 12) / (bh)]^{1/2} = h \quad (11.a)$$

$$\alpha_m = (p t) / (n A_s) = (b h) / (1 . b h) = 1.0 \quad (11.b)$$

from where one can realize that the correction factor is due to the empty parts in the strip cross section.
Doing a similar derivation for the bending , we finally find :

$$\alpha_f = (p h) / (n A_s) \quad (12)$$

as the correction factor that must be applied to the bending stress obtained on the shell model, where **h** is the actual thickness of the layer or, what is the same, the web height of the strip cross section.

6.0 CONVERGENCE TEST

It is a good practice[Bathe, 1982] in FEM applications to check the convergence of the model by successive mesh refinements. Among the several distinct cases of risers and flowlines for which we did so, for the sake of brevity, we selected here the one seen in figure 4.
Considering the evident lack of compatibility between the elements used (mixing solid and plate elements) , the observed convergence is quite acceptable.
The basic measure for the tests was the global stiffness of the flexible pipe for which it is possible to find good manufacturer information. In this context, the models present some inconsistency only for the global

bending rigidity because of their incapacity to properly compute the effects of the large displacements that occur due to the high bending flexibility of the pipe.

7.0 EXAMPLES OF STRESS ANALYSIS.

In accordance with the main objective of this work, examples of stress analysis are summarized in tables 1 and 2 for two different types of flexible pipe for the launching condition.

The tables, for a 2.5" flowline and a 4" riser, show the maximum stress in the layers, allowing comparisons with the materials' U.T.S. values. As can be seen (Table 1), for the design limit of 300m water depth the launching does not require so much of the 2.5" flowline carcass. On the other hand, the calculated safety factor for the armors no longer differ from those recommended by the manufacturers (between 1.5 and 2.0) but still suggest some over dimensioning. For the 4" riser (Table 2) we can see that the carcass and pressure layer have excessive resistance while the safety factors for the armors are very closed to those used by the industry.

Some additional information about the modeling features and behavior can be extracted from figure 5 where two different sets of results for an 11" riser are presented. The first set obtained by the linear approach explained here, with the program GERFLEX as pre-processor for the FEM analysis using ALGOR. The second set refers to a non linear solution obtained with model using the FEM analysis program ABAQUS. For the level of applied loads, no meaningful differences exist.

For an 4" flowline model, figures 6 show deformed shapes and stresses to illustrate the armor squeezing and how the mechanical loads act on the flexible pipes during installation.

8.0 CONCLUSIONS.

A complete procedure for three dimensional stress analysis of flexible pipes was developed, computing the layer structural interactions in a small specimen, just as could be done in a test machine .The procedure, able to simulate the launching operations, provides satisfactory results within the limits imposed by the assumption of linear behavior.

layer → stress(Mpa) ↓	interlocked carcass	internal armor	external armor
maximum stress	50.4	348.2	317.0
allowable stress	486	1,368	1,368
safety factor	9.64	3.93	4.32

Table 1 - Maximum Von Mises Stress for a 2.5" FLOWLINE Laying Operation with Three Feet Tractioners .

Layer → Stress (Mpa) ↓	interlocked carcass	pressure layer	internal armor	external armor
maximum stress	24.9	154.2	563.6	557.9
allowable stress	495	900	1,260	1,260
safety factor	19.88	5.84	2.24	2.26

Table 2 - Maximum Von Mises Stress for a 4.0" RISER Laying Operation with Three Feet Tractioners .

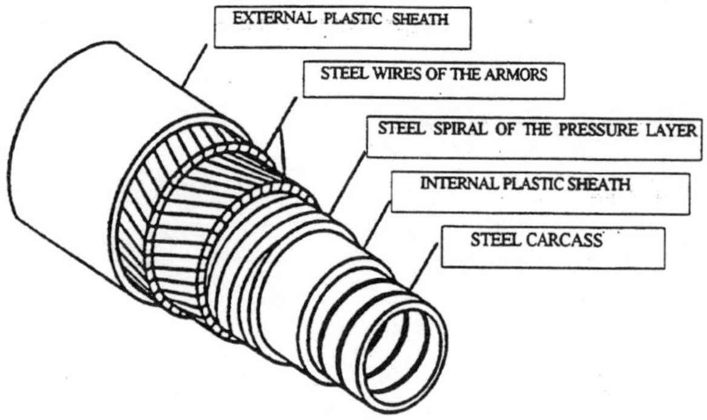

FIGURE 1.0 Typical Layered Structure of a Flexible Pipe

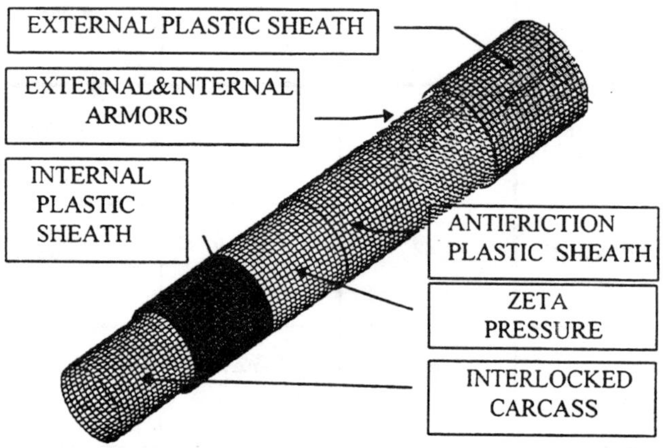

FIGURE 2.0 Unflaked View of a Typical Finite Element Model

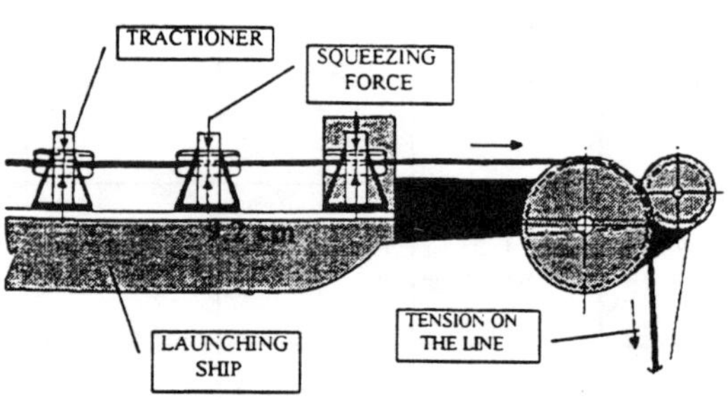

FIGURE 3.0 The Launching Operation

FIGURE 4.0. Convergence Test for 9.5" RISER, based on stiffness properties

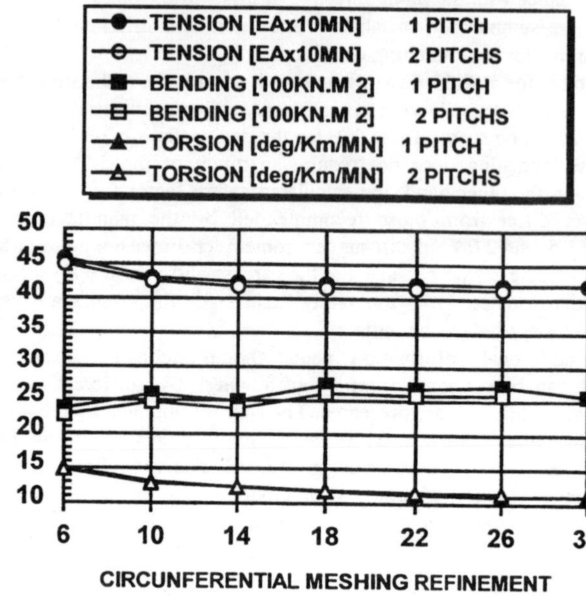

FIGURE 5.0. Von Mises Stress on the Carcass for 11 " RISER under Launching Loads

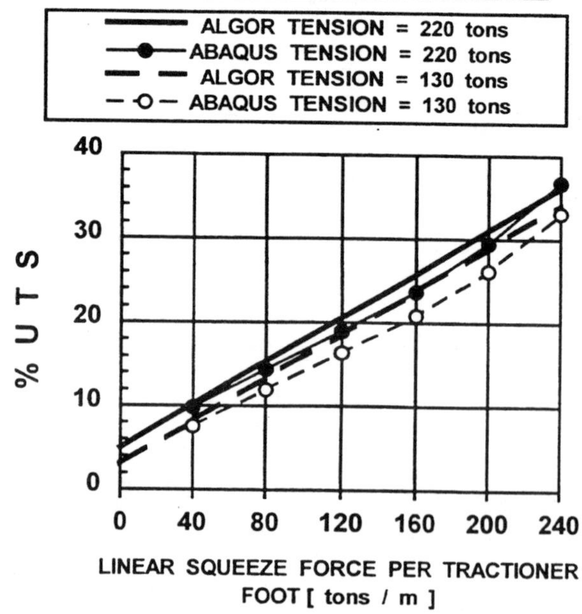

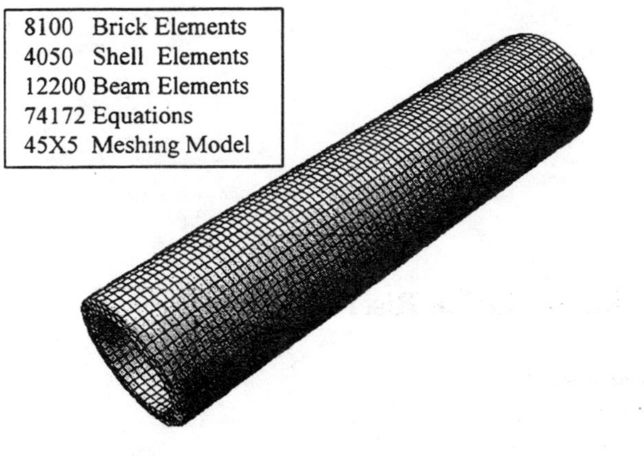

FIGURE 6.1. Overall View of a 4" FLOWLINE Model Two Pitches Long

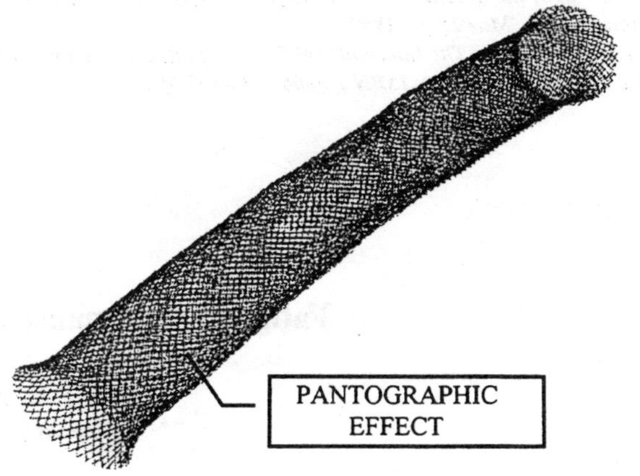

FIGURE 6.2. Deformed Shape of the Armors, Squeezing the Internal Layers

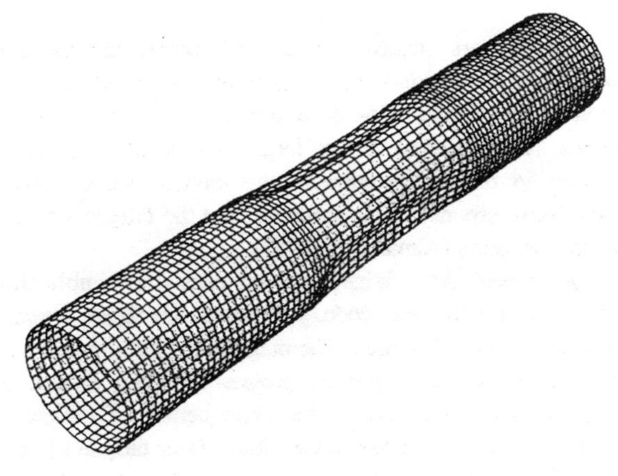

FIGURE 6.3. Deformed Shape of the Carcass due to Tractioner Forces

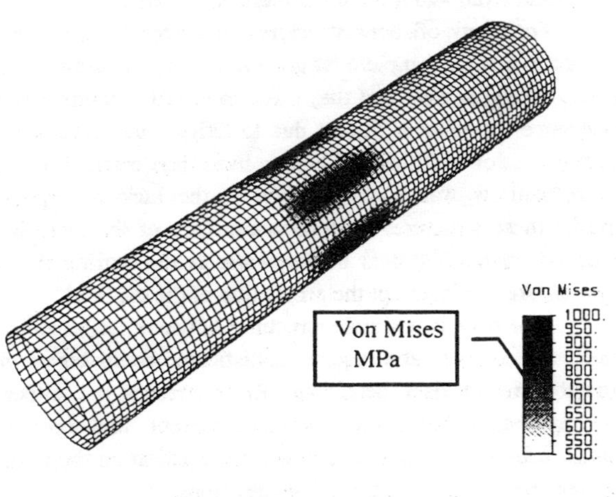

FIGURE 6.4. Stress on the Carcass due to Tractioner Forces

REFERENCES

ALGOR Finite Elements Analysis System. *Algor Interactive System , Inc.*

ABAQUS Standard Version 5.5. *Hibbit , Karlsson , Sorensen , Inc.*

Bathe K . J . (1982). " Finite Element Procedures in Engineering Analysis". *Prentice-Hall, Inc. , Englewood , New Jersey*

Chen B and Nielsen R (1991)."Investigation on Flexbody™ Collapse of the Flexible Pipe".Marine *Structure Journal Wellstream Corporation*

Goto Y et al (1986)."Analytical Study on Mechanical Strength of Flexible Pipes".*9th Annual ASME Energy Sources Technology Conference & 1st OMAE Offshore & Artic Frontier Specialist Symposium*, pg. 337-341

McNamara JF and Harte AM (1990)."Verification of Analytical Modelling Procedures for the Stresses Analysis of Flexible Pipe Cross Sections".*9th ASME et Al Offshore Mechanics & Artic Engineering Conference Proceedings*

Patel MH , Witz JÁ and Tan Z (1993)."A Flexible Riser Design Manual".*University College London BPP Offshore Technology Series*

Timoshenko SP and Woinowsky-Krieger S (1976)."Theory of Plates and Shells".*McGraw-Hill Book Company, Inc., 2nd Edition*

Proceedings of the Seventh (1997) International Offshore and Polar Engineering Conference
Honolulu, USA, May 25-30, 1997

ISBN 1-880653-28-1 (Set); ISBN 1-880653-30-3 (Vol. II)

Fatigue Performance of Deep Water Rigid Marine Risers

G.F.M. de Souza and E. Gonçalves
Polytechnic School - São Paulo University
São Paulo, Brazil

ABSTRACT

The inherent statistics aspects related to loads due to lateral moored vessel movement and steel mechanical properties are investigated in order to obtain reliable parameters to evaluate the long-term non-linear dynamics response of rigid risers, as for the fatigue life analysis.

KEY WORDS:Fatigue, Reliability, Risers

INTRODUCTION

Generally,offshore structures are designed to perform specific functions, with an adequate safety and a good economy standard.

As part of the design process, there are the requirements of structural strength based on criteria refering to failure modes, such as, rupture by overloading, fatigue failure of individual components, buckling or even an unstable fracture,induced by overloads or corrosion cracking. Design criteria are stated in terms of stress caused by external loads, such as, functional loads, due to steel weight, ballast, etc, which may be considered quasi-static, and environmental loads due to waves, current and wind, which have a dynamic behaviour.

The prevision of the static behaviour is used on the structural synthesis process, when the basic dimensions of the individual components are established, through the employment of a design loading, safety coefficients and a group of specific expressions that efficiently represent the most probable collapse modes of the structural parts. On the point of view of static behaviour, it is always employed deterministic calculation models, (Hughes, 1983).

The dynamic behaviour of the structure is verified in an analysis, when it is evaluated the effect of the dynamic loading over the structural parts. Usually, during this design stage, the fatigue phenomena is analysed, aiming the definition of the possibility of crack propagation through the structure life.

For offshore structures it is well-known that the dynamic external loading due to wind, wave and current is very complex, so the evaluation of the fatigue life of a tubular joint, or other welded connections, is subjected to a large uncertainty, arising from uncertainties in the computation of the environmental forces, the global and local response of the structure, and the fatigue strength and damage accumulation,(Almar-Ness,1985).

At present, most fatigue analyses consist of simple checking of safety according to some code. Computed life times are checked against an antecipated life time. The magnitude of computed life time is also used to define inspection programs. When fatigue is the governing ultimate limit state or the consequences of fatigue failure can be dramatic, more confident information about fatigue life must be produced. This more detailed analysis must be based on physical and mechanical models of the loading, the fatigue strength and the damage accumulation model, considering uncertainties involving these variables. Reliability methods deal with the uncertain nature of loads, resistance, etc, and lead to assessment of the failure probability of the structure and a rational measure of the safety coefficient.

For many offshore structures operating in deepwater, it is important to make a complete fatigue evaluation, including reliability technics, taking in view that they often have little redundancy and the consequences of element failure due to fatigue are more severe with respect to loss of facilities and human lives than traditional structures in more shallow waters, in addition to the lack of experience in operating these structures in deep water. One of the most important part of a production system in deepwater is the marine riser, which fatigue failure can interrupt the production of the oil field.

The riser is a simple structure; however, its mechanics and loadings are complex and require sophisticated computer programs to accurately predict riser behaviour. Riser stresses are either static, caused by static offset of the top vessel, current and top tension, or dynamic, caused by vessel motion and wave action on the riser, which level must be limited by fatigue considerations.

The purpose of this paper is to present the application of structural concepts for the evaluation of the fatigue resistance of a rigid marine riser operating in deep water, including reliability technics. The analysis is based on modelling the dynamic stress ranges acting on the structure as a wide band process, in order to account for the non-linearities in riser dynamic response. The riser fatigue analysis is based on the application of design codes, for deterministic approach, and statistical modelling of dynamic loads due to lateral moored vessel movement,induced by wave action, and steel mechanical properties, in order to obtain reliable parameters to evaluate the long-term non-linear dynamic response of rigid risers, as for reliability analysis.

THEORETICAL REVIEW

Considering known the dynamic stress acting on the structure, it is presented a calculation procedure that determinates the riser operational life, based on deterministic and probabilistic design criteria.

Riser Equation of Motion

The riser is modelled as a tubular "beam-column" subjected to inner and outer pressure as well as tension and lateral loads, which governing differential motion equation is:

$$q_y = m_y \frac{\partial^2 y}{\partial t^2} + \frac{\partial^2}{\partial z^2}\left[EI(Z)\frac{\partial^2 y}{\partial z^2}\right] - \left[\frac{\partial T(Z,t)}{\partial Z} + \gamma_i A_i - \gamma_e A_e\right]\frac{\partial Y}{\partial Z} - \left[T(z,t) + p_e A_e - p_i A_i\right]\frac{\partial^2 y}{\partial z^2} \quad \text{(EQ.1)}$$

where q_y ... load for unit length, resulting from hydrodynamic drag induced by currents, waves and vessel motion, calculated by Morrison` s equation.

Z ... coordinate along riser.

y ... lateral displacement of riser

EI .. flexural rigity of riser

$m_y \frac{\partial^2 y}{\partial t^2}$ ... (mass/unit length) times acceleration

$\frac{\partial^2}{\partial z^2}\left(E I \frac{\partial^2 y}{\partial z^2}\right)$ ... change in shear load in riser

$\frac{\partial}{\partial z}\left(T \frac{\partial y}{\partial z}\right)$ ... lateral forces due to change in tension T, and angle of tension, along the length.

$(\gamma_i A_i - \gamma_e A_e)\frac{dy}{dz} - (T(z,t) + p_e A_e - p_i A_i)$ lateral forces per unit length due to internal and external pressure on a curved pipe.

γ_i, γ_e ... specific weight of fluids inside and out.

A_i, A_e ... inner and outer cross sectional area of pipe.

p_i, p_e ... inner and outer pressures.

The objective of riser analysis is to solve equation 1, for the unknown riser displacements y. The water loads (q_y) depend on the velocity and acceleration of the riser as well as the velocity and accelaration of the water from waves and currents; Thus the equation is non-linear and normally solved in time domain. So, the riser lateral displacement time response is wide band, taking in view the non-linearities in equation of motion. Dynamic stresses are caused by riser displacements, induced by vessel motion and wave and current action, and suffer influence from the riser top tension.

For long risers, operating in deep water, the lateral stiffness of the riser is due to tension, similar to a taut string, and the top of the riser becomes more flexible and may be subjected to greatly increased wave and vessel induced dynamic bending stress, which might affect the fatigue resistance of the structure.

The stress fluctuation history acting on the riser is calculated based on the bending-moment, axial and shear force time history. The last ones were determined employing a computer that solves the non-linear dynamic equilibrium equation of the riser (Fowler, 1986), considering as external dynamic loading the first order lateral movement of the vessel, where the riser is top-connected, due to wave loading. As static loading on risers is used the static offset of the vessel, also considering the maximum second order lateral movement evaluated for the vessel, riser top tension and marine current.

Since long suspended riser, as found in deep water production systems, presents a non-linear dynamic response, particularly for high sea states, the stress spectrum, based on the stress history, should be considered to be wide band. For such cases, when using a probability distribution to determine the stress fluctuation probability density distribution, it is recommended the application of distribution appropriate to wide band spectra, such as the generalized gamma distribution (Ochi, 1978), which is applyed in this study. This distribution is written as:

$$f_X(x) = \frac{c}{\Gamma(m)} \lambda^{cm} x^{cm-1} e^{-(\lambda x)^c}, \quad 0 \le x < \infty \quad \text{(EQ.2)}$$

Even in case that the stress sprectrum is narrow band, this probability density function can be applied, because the Rayleigh distribution, usually employed in this type of random process, is a special case of generalized gamma distribution, namely m=1 and c=2.

The stress fluctuation histogram, which will be the base for the definition of the stress fluctuation probability density function, is determined from a stress time-history record using the "Rainflow" Method, which is considered the most accurate method for counting stress cycles when the mean stress is not zero, (Fuchs et al, 1980).

Fatigue Damage Assessment

The assessment of fatigue capacity and fatigue failure of structures is in general carried out by applying two different methodologies:

i) Crack growth rate curves and the fracture mechanics model.

ii) S-N curves and the Palmgren-Miner damage model.

For design purpose, it is assumed that there is no initial crack in the riser. Therefore the crack propagation model can be discarded and the SN curves with Palmgren-Miner rule is used to estimate the fatigue life. This procedure is specified in most design codes for offshore structures, e.g., API-RP2A (API, 1987) and AWS D1-1, (AWS, 1990).

A fundamental assumption is that the curve characterizing fatigue behaviour under constant amplitude loading can be written as:

$$NS^{b} = K_s \qquad \text{(EQ.3)}$$

in which b and K_S are empirical constants. EQ.3 defines the number of cycles to failure, N, as a function of stress range, S. It is assumed that EQ.3 applies for all S> S_F considering the existance of the endurance limit in the model.

A second fundamental assumption is that Palmgren-Miner's rule applies,neglecting load sequence effects over fatigue process. In this model, it is assumed that the damage on the structure per load cycle is constant at a given stress range and equal to:

$$\delta = \frac{1}{K_s} \cdot S_a^b \, S_{a_i} \qquad \text{(EQ.4)}$$

The total damage accumulated during n cycles of amplitude S_{a_i} is given by:

$$d\delta = \sum_{i=0}^{n-1} \delta_i \text{ where } \delta_i = \frac{1}{K_s} \cdot S_{a_i}^b$$

or

$$d\delta = \frac{1}{K_s} \sum_{i=0}^{n-1} S_{a_i}^b \qquad \text{(EQ.5)}$$

In this formulation it is assumed that the accumulated damage $d\delta$ is independent of the sequence in which stress will occur.

Taking in view that the stress range is usually modelled as a random variable, and based on the hypothesis that the number of stress ranges acting on the structure is sufficiently large, the accumulated damage is written as:

$$d\delta = \frac{1}{K_s} E(n)\, E(S_{a_i}^b) \qquad \text{(EQ.6)}$$

where E(n) is the expected value of number of stress cycles acting on the structure and $E(S_{a_i}^b)$ is the expected value of $S_{a_i}^b$.

Considering that the long-term stresses acting on the riser are induced by environmental loadings, that can be described as a set of stationary short-term sea states, the total fatigue damage can be obtained by summing up the accumulate damage over all sea states, applying the "Lifetime Weighted Sea Method", (Hughes, 1983). The total damage D may be written as:

$$D = \frac{1}{K_s} \cdot \sum_{j=1}^{k} p_j \, [E(n)]_j \cdot [E(S_{a_i}^b)]_j \qquad \text{(EQ.7)}$$

where:

p_j ... probability of occurrence of an environmental loading during the structure life;

$[E(n)]_j$ ... expected value of number of stress cycles acting on the structure, for environmental condition j;

$[E(S_{a_i}^b)]_j$ ... expected value of $S_{a_i}^b$, for environmental condition j.

Equation EQ.7 can be written as:

$$D = \frac{1}{K_s} \cdot T \cdot \sum_{j=1}^{k} p_j \, f_j [E(S_{a_i}^b)]_j \qquad \text{(EQ.8)}$$

where T is the riser life and f_j is the stress range frequency in the jth sea state.

According to Miner's rule, fatigue failure occur if D>Δ, where Δ is the critical cumulative damage, which is often taken as 1. Letting D=Δ, the basic damage expression of EQ.8 can be expressed in terms of time to failure,

$$T = \frac{\Delta \cdot K_s}{\sum_{j=1}^{k} p_j \cdot f_j \cdot [E(S_{a_i}^b]_j} \qquad \text{(EQ.9)}$$

which would represent the predicted riser operational life.

For design purpose, in order to account for scatter in SN curve experimental data, several current design codes employ an special SN curve, named "Design SN-Curve", that represents a lower limit for the fatigue resistance of welded steel test specimens, reducing the uncertainties in material fatigue resistance. Furthermore, the hypothesis on cumulative damage adds to a number of uncertainties in fatigue life prediction, because experimental studies of variable amplitude loading indicate that fatigue failure occurs for damage ratios below 1,0, (Almar-Ness, 1985). Many design building codes limit the critical damage ratio, in order to account for uncertainties in Miner's rule, such as the AWS D1.1 Code that set limit to the critical damage to 1/3 for critical members whose sole failure mode would be catastrophic or the APIRP 2A Code, that prescribes a maximum value of 0,5 for the critical damage.

Taking in view the fatigue failure assessment of the rigid marine riser, the operational life can, preliminary, be defined by the application of EQ.9, considering the SN curve and critical damage defined by a structural code. As the experience in operating rigid risers in deep water is very little, with a few flowlines being used at Campos Basin, Brazil, it is hard to choose a deterministic procedure in order to define the structure fatigue life. The design codes may be used as a guide for synthesis purpose, but for structural analysis, as for fatigue assessment, more complex calculation procedure could be employed, in order to define reliable parameters for the structure, including its operational life, based on modelling the uncertainties involving the physical and mechanical models of the loading, the fatigue strength and the damage accumulation model, leading to the assessment of the structure fatigue failure probability, and a rational measure of the safety coefficient.

The initial approach for fatigue reliability analysis may be the statistical modelling of the critical cumulative damage defined by Miner's rule. According to many investigations, (ASCE, 1982), the critical cumulative damage follows a lognormal distribution.

In fatigue design, according to structural codes, there is a limit fatigue cumulative damage factor admissible for the structure, named D. The probability of fatigue failure (p_F) can be expressed as:

$$p_F = \Phi\left(\frac{\ell nD - \lambda_\Delta}{\xi_\Delta}\right) \quad \text{(EQ.10)}$$

where

$\Phi(.)$ ... standard normal definition function

λ_Δ ... mean value of $\ell n\, \Delta$.

ξ_Δ ... standard derivation of $\ell n\, \Delta$

So, the failure probability computed by EQ.10, could be the fatigue failure probability of the riser associated with the operational life defined by EQ.9. It can be realized that this probability depends on the scatter of the critical cumulative damage distribution, being greater with an increase in critical damage coefficient of variation.

A more complex approach for fatigue failure probabilistic analysis was developed by Wirsching, 1984, where not only the critical cumulative damage is statistically modelled, but also the material fatigue strength is a random variable, in order to account the scatter in S-N data. Furthermore, inaccuracies in fatigue stress estimation are described by a random variable, so the basic operational life expression can be expressed in terms of time to failure,

$$T = \frac{\Delta K_s}{B^b \cdot \Omega} \quad \text{(EQ.11)}$$

where

$$\Omega = \sum_{j=1}^{k} p_j \cdot f_j \cdot [E(S_{a_i}^b)]_j$$

Considering that Δ, K_S and B are random variables, the time to failure (T) is also a random variable. Being Ts the desired life of the structure, the probability of fatigue failure is given by:

$$p_F = P(T \le Ts) = \Phi(-\beta) \quad \text{(EQ.12)}$$

where β ... reliability index.

Wirsching, 1984, suggested that each random variable has a lognormal distribution, and the reliability index is written as:

$$\beta = \frac{\ell n \tilde{T} - \ell n\, T_s}{\tau_{\ell nT}} \quad \text{(EQ.13)}$$

where

$$\tau_{\ell nT} = \{\ell n[(1+C_\Delta^2)(1+C_{Ks}^2)(1+C_B^2)^{b^2}]\}^{1/2}$$

$\tilde{T}$ is the median value of T and equal to

$$\tilde{T} = \frac{K_s\, \Delta}{B^b\, \Omega}$$

and "~" expresses median values and the C's denote the coefficients of variation of each variable.

This approach employes the median value of material fatigue strength and critical cumulative damage to define the fatigue failure probability, related to a desired life for the structure. Furthermore, the stress ranges acting on the structure are calculated considering a narrow band process, with a correction factor related to the scatter in stress range power spectrum.

Finally, a more complex approach can be applied for fatigue evaluation, based on probabilistic theory, where a limit state function related to fatigue failure is formulated, written as (Souza, 1994):

$$g_s(X) = \Delta - \frac{1}{K_s} \cdot \sum_{j=1}^{h} p_j \cdot (\sum S_{a_i}^b)_j \quad \text{(EQ.14)}$$

Taking in view equation (EQ.11), for a given operational life, the fatigue collapse will occur when $g_s(x) \le 0$, and the vector of random variables has the elements:

Δ ... critical cumulative damage, for which fatigue failure will occur, in order to quantify modelling error associated with Palmgren-Miner rule;

K_S .. constant parameter of SN curve, to account uncertainties in fatigue strength;

$[\sum_i S_{a_i}^b]_j$.....sum of stress amplitudes acting on the structure, for each load condition j, during riser's life.

Since the number of stress cycles for a sea state is generally large, the probability density function of $\left[\sum_i S_{a_i}^b\right]_j$ can be well approximated by a normal distribution for a fixed value of b, (Ricles and Léger, 1990).

For a generalized gama stress fluctuation probability density function, the mean of $\left[\sum_i S_{a_i}^b\right]_j$ is equal to:

$$E\left[\sum S_{a_i}^b\right]_j = E\left[S_a^b\right]_j \cdot T \cdot \left(\frac{n_s}{t_s}\right) \quad \text{(EQ.15)}$$

where

T ... life estimated for the structure

n_s ... number of stress fluctuation observed during simulation time.

t_s ... simulation time

$E\left[S_{a_i}^b\right]_j$ ... mean value of stress fluctuation, and

for a generalized gama stress fluctuation probability density function:

$$E\left[S_{a_i}^b\right]_j = \frac{1}{\lambda^b} \cdot \frac{\Gamma\left(m + \frac{b}{c}\right)}{\Gamma(m)}$$

where λ, m and c are generalized gama parameters and Γ (.) is the gama function.

The standard deviation $\sigma\left(\sum S_{a_i}^b\right)_j$ was approximated by the following relation, (Crandall, 1958):

$$\frac{\sigma\left(\sum S_{a_i}^b\right)_j}{E\left(\sum S_{a_i}^b\right)_j} = \left[\frac{f_1(b)}{\xi\left(\frac{n_s}{t_s}\right)T}\right]^{1/2} \quad \text{(EQ.16)}$$

where f_1 (b) ... function tabulated by (Crandall, 1958).

ξ ... damping coeficient.

The SN curve coefficient and critical cumulative damage are modelled as lognormal distributions.

Considering that the random variables are statistically independents the fatigue failure probability is computed using the Advanced First-Order Second-Moment (FOSM) reliability method, (Ciria, 1977). This method threats the fatigue failure assessment on its fundamental essence, without excessive simplifications, including the representation of $\sum S_{a_i}^{b}$ as a random variable, and the wide band stress range distribution.

All these calculation procedure can be employed in the fatigue failure analysis of rigid marine risers, although the accuracy of the results will vary, depending on the hypothesis related to each formulation.

FATIGUE ANALYSIS OF RIGID RISERS

The fatigue analysis is applied on the evaluation of the fatigue failure probability of a rigid riser operating at Campos Basin, in Brazil, linked to a GVA 4500 semi-submersible platform. Figure 1 shows the configuration studied and the main dimensions and properties of the member. The riser is considered to be flooded with petroleum.

The sea state data used for the analysis is summarized in Table 1, given the significant wave height (Hs) and the average wave period (Ts), specifically for the application of ITTC spectrum, and the probability of ocurrence for each sea state.

The moored vessel motion analysis was determined through the evaluation of the vessel frequency transfer function and the wave spectrum:

$$Svessel(\omega)=|H(\omega)|^2 \cdot Swave(\omega) \qquad \text{(EQ.17)}$$

where

H(ω) ... movement transfer function;

Svessel(ω) ... vessel movement spectrum;

Swave(ω) ... wave spectrum.

Taking in view the necessity of time domain definition of the riser movement, the vessel movement spectrum was divided in a serie of 75 or more regular movement, that were composed in order to define the irregular lateral vessel movement in time domain.

The vessel movement was assumed to be plane, alligned with the current profile, considered linear, from the bottom of the sea until the water line. The maximum current velocity is 0,943 m/s. For the marine riser it was considered an added mass coefficient $C_m = 2{,}0$ and a drag coefficient $C_d = 0{,}7$. In Figure 2 it is shown the efforts acting on the riser for sea state 3, and in Figure 3 it is shown the stress fluctuation probability density function for the same sea state, giving confidence in the application of generalized gamma function for stress range representation.

For design purpose, in order to define the riser operational life, it is applied the AWS D1.1 Code, considering the AWS-X SN curve, and the critical cumulative damage equal to 0,3.

For reliability analysis, five different cases are considered, and described in Table 2. In case (1), only the critical cumulative damage is considered a random variable, modelled as a lognormal distribution, with median and coefficient of variation recommended by (ASCE, 1982). In case (2), applying the methodology proposed by (Wirsching, 1984), the critical damage, the AWS-X SN curve coefficient, Ks, and the coefficient B are considered as random variables, modelled as lognormal distributions. In case (3), applying the methodology proposed by (Souza, 1994), the critical damage and the AWS-X SN curve coefficient are modelled as lognormal distributions, and the stress sum is modelled as a normal distribution, derived from a generalized gamma distribution. Cases (4) and (5) are small modifications of case (3), the first one considering an increase in SN curve scatter and the second considering an increase in critical damage scatter, in order to examine the sensitivity of the reliability index to a greater uncertainty in data related to fatigue phenomena.

For all cases, the Goodman Criterion is used to consider the effect of mean stress over the material fatigue resistance.

Analysis of Results

Considering the "wide-band" stress distribution, and applying the Design Criterion proposed by AWS Code D.1.1, the operational life recommended for the rigid riser is:

$$T = \frac{\Delta \cdot Ks}{\Omega} \cdot 0{,}3 = \frac{1 \cdot 1{,}6391x10^{14}}{7{,}074x10^{12}} \cong 7\,\text{years}$$

The operational life for the structure could be extended to 12 years, if the safety factor of 0,5 is applied on Palmgren-Miner Rule, as recommended by API-RP 2A (API, 1987). So, there is a great variation in the operational life, taking in view the use of a safety factor for modelling the uncertainties of Miner`s rule.

Furthermore, the SN design curve recommended by AWS consider a 95% survival probability for the material when cyclic loaded at a given stress level, so the safety factor employed for the riser operational life definition is greater than the value 0,3 recommended by the Design Code. So, it is necessary a verification of this Design Criterion, taking in view the probabilistic nature of the variables involved in fatigue analysis.

Taking in view the dispersion in critical cumulative damage, and according to ASCE Advisory Committee (ASCE, 1982), it may be modelled with a lognormal distribution, with median 1,0 and coefficient of variation equal to 0,53. Based on equation EQ.10, the safety factor employed on riser design may be translated into a failure probability. The safety factor employed by AWS corresponds to a 0,8% failure probability, while the factor 0,5, advised by API, corresponds to 8% failure probability. It must be realized that a small change in this safety factor cause a great variation in the failure probability. Although this procedure still not consider the statistical distribution of SN curve, it gives some information about the fatigue failure probability associated with the safety factor recommended by the Design Criterion, being useful for the design process initial stages.

According to the data presented in Table 2, a more complex reliability analysis could be done, applying the calculation procedure proposed by Wirsching, (1984), summarized in equation EQ.13. The stress range acting on the riser are defined according to narrow band theory, and the mean value of $S^b_{a_i}$ for each sea state is modelled by a Rayleigh distribution,

$$\bar{S}^b = (2\sqrt{2}\sigma_0)^b \Gamma(\frac{b}{2}+1),$$

where σ_0 is the root mean square of stress history process.

The damage associated to each sea state is corrected, in order to account for narrow band hypothesis, by a factor proposed by Wirsching, 1984, being a function of the spectral width parameter.

In Figure 4 is presented a graphic that relates the reliability index and the riser operational life. These results were obtained by using a range of values for T, and applying equation EQ.13 for reliability index calculation.

It is possible to notice that there is a decrease in reliability index as T becomes large. For the design operational life, according to AWS, T=7 years, the reliability index is aproximatily 2,5, corresponding to a failure probability of 0,62%, which is almost the same value obtained by the application of equation EQ.10, considering only the critical damage as a random variable. Moreover, the fatigue failure probability for an operational life of 25 years is 6,65%, very close to the value 5,6% obtained by Wirsching (1979) for a jacket structural joint operating in North Sea.

Finally, the fatigue failure probability is calculated according to the principles of First Order Second Moment Reliability Method, with the limit state function presented in equation EQ.14, according to the data presented in Table 2.

According to the data presented in Figure 4 ,the reliability index presents the same behaviour defined according to Wirsching methodology, decreasing with an increase in riser operational life, although the reliability index obtained by FOSM application, for the usual life defined for offshore structures, is smaller than that obtained by the lognormal format defined by Wirsching, being more conservative than this criterion. The reliability index corresponding to the operational life defined by AWS, is 1,90, giving a fatigue failure probability of 2,87%, higher than the defined by Wirsching methodology. Moreover the fatigue failure probability for an operational life of 25 year is 5,05%, close to the value 5,6% obtained by Wirsching (1979) for an offshore structure. The effect of uncertainty and scatter in the SN data on the reliability index is also shown in Figure 4, and for the conditions analysed, considering the variables as uncorreleted, there is no effect over the reliability index defined in base case. For the case (5), where there is an increase in the scatter of the critical damage, there is a decrease in the reliability index obtained for each riser age T, when compared to the results defined in base case, being maintained the linear relation between reliability index and riser age.

It is interesting to notice that, for uncorreleted variables, the reliability index was found to be invariant to the scatter in SN data, at least for the values adopted in this study.This fact is a consequence of the minimization procedure proposed by Ciria, 1977, and is also related to large scatter of the critical damage distribution.

The procedure proposed by Souza, 1994 can be considered more complex and complete than the others because it considers the stress distribution as wide band, modelling the riser non-linear dynamic behaviour, and also the sum of stress amplitude is modelled as a random variable, increasing the complexity of the reliability index calculation, giving more conservative results about the fatigue failure probability.

It could be assessed that the ability to establish a β-T relationship is important form the standpoint that it gives an indication of the reliability decay with respect to time and thus serve as a guideline as to when member inspection for fatigue damage is necessary.

Finally, it is also important to notice that there is not an international rule that defines the minimum requirement for the reliability index for offshore structures, taking in view the fatigue analysis, so is is difficult to define if the results obtained for the riser reliability index can be considered satisfatory, according to an specific design requirement. Taking in view studies from (Wirsching, 1979), analysing the fatigue failure of an offshore structure tubular joint in North Sea, where a reliability index of 1,59 was found, it is possible to say that riser would operate for a 25 years period, although the AWS design code recommend only a 7 years operational life.

CONCLUSIONS

A method based on structural component reliability procedures has been presented to evaluate the fatigue failure of marine rigid risers operating in deep water. The fatigue life for this type of structure, taking in view the long operational life, can not be predicted only by deterministic procedures, due to uncertainties arising from loading , fatigue strength and in the damage accumulation models. All these uncertainties were threated considering the variables as probabilistic, and the first-order reliability method was applied to determine a nominal reliability level.

Numerical examples, considering a rigid marine riser operating in deep water, demonstrated that reliability can be used as a guideline for the fatigue process of the structure, taking in view the operational life and variations in data, and that the design codes for usual offshore structures consider large safety factors, that must be carefully analysed before being applied for deep water structures, due to lack of experience in operating these structures for long time periods.

The methodology presented in this study allows the estabilishment of parameters for fatigue design of riser structures less conservative than those presented in offshore structures design codes.

For the case analysed, it was found a great dependence between reliability index and the scatter of the fatigue damage model, so there is a necessity to have more statistical data about the damage for offshore structures, in order to have reasonable statistical properties for the critical damage.

The results for reliability index have not shown a dependence on the scatter of the SN curve, at least for the cases analysed in this paper.

Specifically for the nominal reliability level computed for the structure, although it can not be compared with values suggested by Design Criterion, in the initial design stages, when there are different design concepts, the fatigue failure probability of the solutions can be evaluated and compared to each other, giving more information to the designer about the solutions behaviour, taking in view the future decisions involved in the structure design.

REFERENCES

Alman-Ness, A. (ed), (1985). "Fatigue Handbook: Offshore Steel Structures", Tapir Publishers, Trondheim, Norway.

API-American Petroleum Institute, (1987). "API-RP-2A Recommended Practice for Planning, Designing and Constructing Fixed Offshore Plataforms", API, Washington, DC, USA.

ASCE-American Soceity of Civil Engineers, (1982). "Fatigue Reliability Under Variable Amplitude Loading", Journal of the Structural Division of the American Society of Civil Engineers, vol. 108, pp 47-69.

AWS-American Welding Society, (1990). "AWS D1.1 Structural Welding Code: Steel", AWS, Miami, USA.

CIRIA-Construction Research and Information Association, (1977). "Report 63. Rationalisation of Safety and Serviciability Factors on Structural", CIRIA, London, U.K.

Crandall, S.H., (1958). "Random Vibration". John Wiley, New York.

Fowler, J.R., Young, R.D., Long, J.R., (1983). "Marine Riser Design is Simplified". Oil & Gas Journal, May, pp. 129-145.

Fuchs, H.O., Stephens, R.I., (1980). "Metal Fatigue in Engineering", John Wiley & Sons, New York, USA.

Hughes, O.F., (1983). "Ship Structural Design: A Rationally-Based Computer-Aided Optimization Approach", John Wiley & Sons, New York, USA.

Ochi, M.M., (1978). "Generalization of Rayleigh Probability Distribution and its Application". Journal of Ship Research, v.22, n. 4, pp. 259-265.

Ricles, J.M., Léger, P., (1990). "Fatigue Reliability of Offshore Structures", Proceedings of the 7th International Symposium of Offshore Engineering, pp. 529-543, Rio de Janeiro, Brasil.

Souza, G.F.M., (1994). "Fatigue Reliability Analysis of Rigid Marine Risers", Doctoral Thesis, São Paulo University, São Paulo, SP, Brazil.

Wirsching, P.H., (1979). "Fatigue Reliability in Welded Joints of Offshore Structures", Proceedings of the 11th Offshore Technology Conference, pp. 197-206, Houston, USA.

Wirsching, P.H., (1984). "Fatigue Reliability for Offshore Structures", Journal of Structural Engineering, vol. 110, pp. 2340-2356.

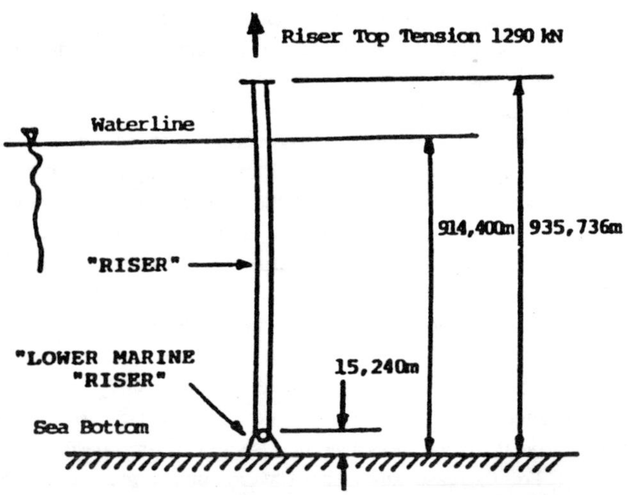

GEOMETRIC PROPERTIES	
Riser Length (m)	920,496
Distance From Bottom to Lower Marine Riser (m)	15,240
Distance From Waterline to Riser Top (m)	9,144
Riser External Diameter (m)	0,406
Riser Internal Diameter (m)	0,374
Choke Line External Diameter (m)	0,102
Choke Line Internal Diameter (m)	0,069
Kill Line External Diameter (m)	0,102
Kill Line Internal Diameter (m)	0,069
Buoyance Units External Diameter (m)	0,610
PHYSICAL PROPERTIES	
WEIGHT (KN/m)	
- AIR	2,745
- WATER	0,889

Figure 1 - Riser Structural Concept

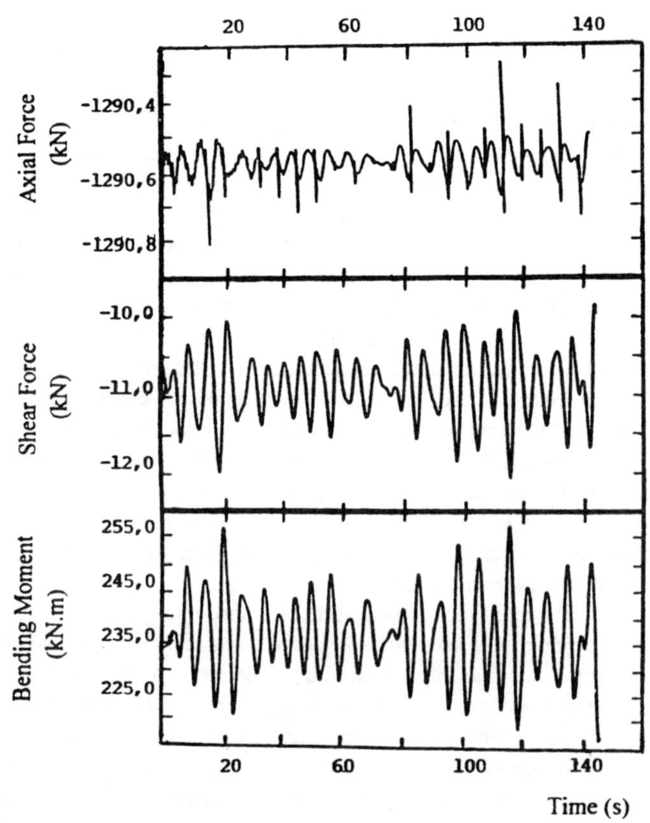

Figure 2 - Internal Efforts Time-History.Sea State 3

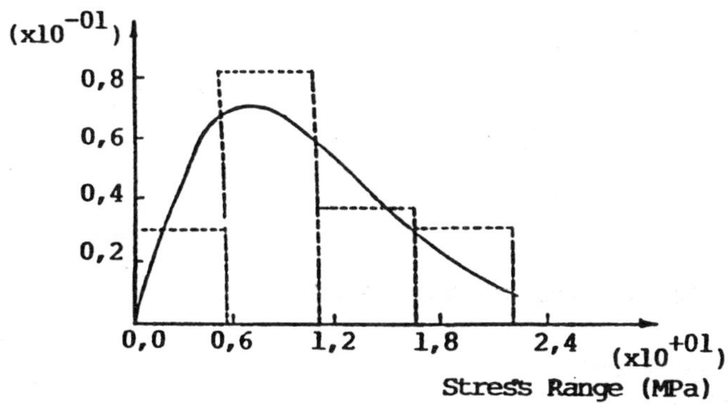

Figure 3 - Stress Range Density Probability Function.Sea State 3

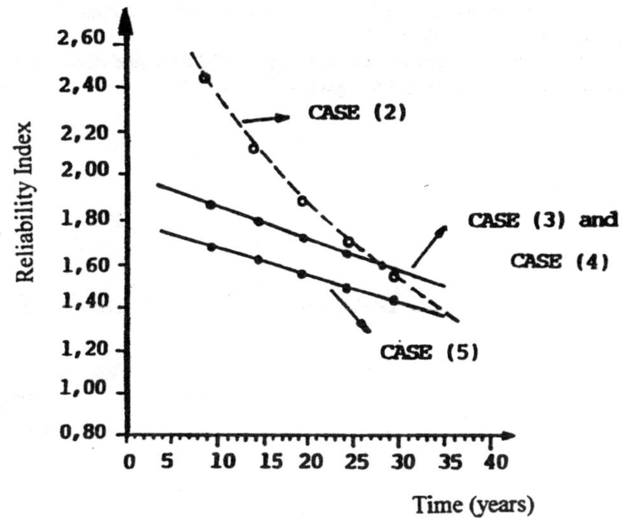

Figure 4 - Riser Reliability Index as a Function of Operational Life.

TABLE 1 - SEA STATES APPLIED DURING THE ANALYSIS

SEA STATES	SIGNIFICANT WAVE HEIGHT	AVERAGE WAVE PERIOD	PROBABILITY OF OCCURRENCE
1	0,78	5,24	0,0229
2	1,25	5,27	0,2561
3	1,75	5,77	0,3852
4	2,25	6,26	0,1962
5	2,75	6,89	0,0880
6	3,25	7,72	0,0328
7	3,75	7,89	0,0100
8	4,25	8,20	0,0068
9	4,75	9,00	0,0020
			1,0000

TABLE 2 - STATISTICAL PROPERTIES OF RANDOM VARIABLES

PARAMETERS	CASES	CASE 1	CASE 2	CASE 3	CASE 4	CASE 5
SN CURVE	K_s	-	$9{,}6762 \times 10^{14}$	$9{,}6762 \times 10^{14}$	$9{,}6762 \times 10^{14}$	$9{,}6762 \times 10^{14}$
	σ_{K_s}	-	1,0233	1,0233	2,0233	1,0233
	b	4,38	4,38	4,38	4,38	4,38
CRITICAL DAMAGE	Δ	1,0000	1,0000	1,0000	1,0000	1,0000
	σ_Δ	0,4976	0,4976	0,4976	0,4976	0,5500
INNACU-RACIES IN STRESS	B	-	1,0000	-	-	-
	σ_B	-	0,1000	-	-	-

Where

K_s ... Mean Value

σ_{K_s} ... Standard Deviation of K_s

b ... Slope

Δ ... Mean Value

σ_Δ ... Standard Deviation of Δ

B ... Mean Value

σ_B ... Standard Derivation of B.

'oceedings of the Seventh (1997) International Offshore and Polar Engineering Conference
onolulu, USA, May 25-30, 1997

'BN 1-880653-28-1 (Set); ISBN 1-880653-30-3 (Vol. II)

Dynamical Measurements of the Bending Stiffness of Taut Cables

Peter Hagedorn and Ulrich Gutzer
TH Darmstadt
Darmstadt, Germany

ABSTRACT

The variation of the bending stiffness of stranded cables with the normal force and the curvature is of interest in a variety of technical problems. Due to the complex geometry of stranded cables analytical predictions are difficult to obtain. Moreover, even numerical solutions depend on a number of parameters, e.g. friction coefficients, which again have to be determined experimentally, see Raoof (1996) and Knapp (1995). The variable bending stiffness is closely related to damping effects, where the stick-slip behavior leads to energy dissipation. On the other hand, the slip between the layers of a stranded cable reduces its bending stiffness. Static measurements show a large range for the bending stiffness in dependence of the curvature as presented by Papailiou (1995).

In a recent publication, Hagedorn and Gutzer (1996), we discussed self damping measurements of cables in a short test rig. A local mechanical model for the cable was formed by a number of springs in series with COULOMB friction elements, acting in parallel and describing the local energy dissipation. The same test rig is also used for the dynamic determination of the bending stiffness from the measured nonlinear oscillations.

In this paper, we first derive a global measure of the cable's bending stiffness from the measured free vibrations, leading to ABEL's integral equation. In the next step, we solve the nonlinear differential equation of the taut cable — with the bending stiffness as a function of curvature — in order to derive the same globel stiffness measure. Comparing the two results, the bending stiffness can be identified from the measurements. The dynamically determined bending stiffness is similar to the outcome of the static tests. The results can be used for improving cable calculations in many engineering applications, since they are of importance for the curvature distribution, i.e. the shape of a taut cable and the stresses.

INTRODUCTION

For many technical applications, the knowledge of the stiffness of stranded cables is of importance. Due to the *stick-slip* behavior between the strands, there is — besides the energy dissipation due to dry friction — a change of the bending stiffness with the curvature. The corresponding parameters are difficult to obtain analytically and therefore have to be determined via experiments. Moreover, the bending stiffness depends on the normal force in the cable.

In this contribution we present results based on experiments carried out on a short test span, where we measure the circular frequency of an oscillating rigid clamp. The frequencies are very low, the cable acts therefore mainly as a rotational spring with some additional hysteretic damping and its inertia can be neglected.

Mathematical transformations are then used to compute a measure for the rotational stiffness from the measured frequencies, obtaining it as a function of the rotation angle of the clamp. Alternatively, the rotational stiffness is derived from the solution of the nonlinear differential equation of the taut cable with an approximation for the *bending stiffness varying with curvature*. Comparing the two results, we are able to identify the missing parameters, giving the bending stiffness as a function of the curvature.

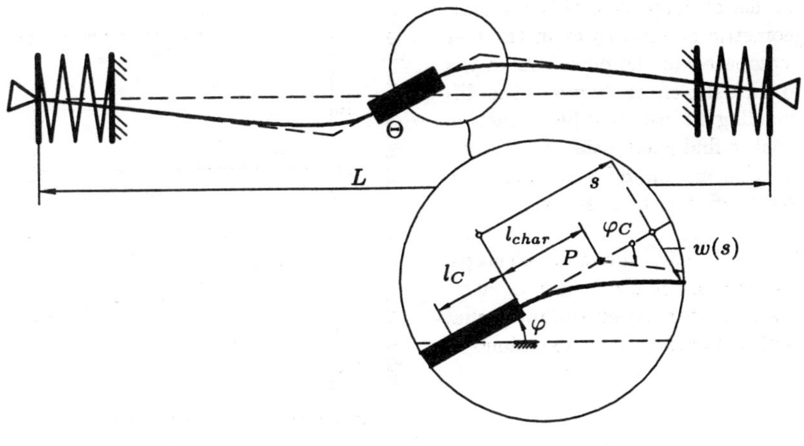

Figure 1: Test rig and detail of the bending boundary layer as seen from above

In recent papers by the authors in (1995) and (1996), our test rig was described together with some results for the damping properties of the investigated cable. Here, we discuss the analysis of the experimental data with regard to the evaluation of the cable's stiffness properties. Although all the data refer to the bending stiffness of conductors of overhead transmission lines, the technique is applicable to cables of different types and designs.

THE TEST RIG

The test rig consists of a rotational oscillator, restoring torque and damping being provided by the cable whose characteristics are to be measured, see figure 1. The rotor is a rigid body with rotational inertia Θ, connected to the middle of the conductor by a 'rigid' clamp of length $2l_C = 0.3$ m. The cable tension T (normal force) is kept approximately constant during the small amplitude angular oscillation of the rigid body. This is achieved by using helical springs at the supports. The total length of the conductor in the test rig is $L \approx 4$ m.

The frequency being low compared to the first eigenfrequency of the two cable segments, the dynamics of the cable can be neglected and the main inertia contribution is due to the rotational inertia Θ of the rigid body. Bending of the cable takes place mainly in relatively small regions on either side of the clamp, in the so called *bending boundary layers*.

Free vibrations with angular amplitude $\hat{\varphi}$ and angular frequency $\omega(\hat{\varphi})$ are measured. The differential equation of the rotational oscillator is given by

$$\Theta\ddot{\varphi} + c_{rot}(\varphi)\,\varphi = 0\,. \tag{1}$$

In a first approximation, we assume the rotational stiffness c_{rot} to remain constant, i.e. not to depend on the angle φ. For this linear system, the rotational stiffness c_{rot} can be calculated, given the rotational inertia and a measured value of the circular frequency $\omega = \sqrt{c_{rot}/\Theta}$. On the other hand, c_{rot} is related to the cable's bending stiffness EI and to other parameters, so that EI can also be found from the measured value of ω

We therefore first examine the relation between c_{rot} and the bending stiffness. In Anderson and Hagedorn (1994) it was shown, that the asymptotic line to a clamped cable with some bending axis always intersects the axis of the clamp at the point P, which does not depend on the angle φ_C, as long as EI remains constant, see the detail in figure 1. It was also shown, that the distance between the point P and the end of the clamp is always equal to the characteristic length

$$l_{char} = \sqrt{EI/T}\,, \tag{2}$$

for any cable tension T. The characteristic length is therefore a measure of the bending stiffness for a given tension T.

The restoring torque acting on the rotor is

$$c_{rot}\varphi = 2T(l_C + l_{char})\sin\varphi_C\,, \tag{3}$$

where $2l_C$ is the clamp length. By linearizing and simplifying the geometry of the cable in the test rig for small angles φ we obtain

$$c_{rot} = 2TL\,\varphi\,\frac{l_C + l_{char}}{L - 2(l_C + l_{char})}\,. \tag{4}$$

Since c_{rot} can be calculated from the measured $\omega = \sqrt{c_{rot}/\Theta}$, we can use (4) to find the characteristic length l_{char} and the bending stiffness EI, if all the other parameters are known. Solving (4) for l_{char} leads to

$$l_{char} = \frac{c_{rot}L}{2(TL + c_{rot})} - l_C\,, \tag{5}$$

which gives the characteristic length for $c_{rot} = \Theta\omega^2$. The bending stiffness is then found from the relation $EI = Tl_{char}^2$.

We should remember, that the relations (5) and (4) were derived assuming a constant bending stiffness and negligible geometric nonlinearities in the test rig, see Gutzer and Hagedorn (1995). This leads to a constant angular frequency ω, independent on the amplitude $\hat{\varphi}$. Figure 2 shows the relation between the vibration amplitudes $\hat{\varphi}$ and the angular frequency ω found for a typical conductor in our test rig, drawn on linear and semilogarithmic scales. This clearly indicates, that the system is nonlinear in the restoring torque, so that the idea of a constant c_{rot} has to be given up.

A careful discussion of the order of magnitude of the nonlinear terms shows, that the geometric nonlinearities in the test rig are still negligible when compared to the deviation in the behavior of the cable. Due to stick-slip effects between the different layers of strands, the bending stiffness will be a function of the curvature κ. It is our goal to find good estimates for this function $EI(\kappa)$. Of course, a non-constant EI will lead to an elastic line in the boundary layer different from an exponential curve.

The determination of $EI(\kappa)$ will be carried out in two steps: first the function $c_{rot}(\varphi)$ will be determined from the frequency-amplitude relation and next, a relation between the rotational stiffness $c_{rot}(\varphi)$ and $EI(\kappa)$ will be found. This is explained in the following two chapters.

DETERMINATION OF c_{rot} FROM THE FREQUENCY-AMPLITUDE RELATION

In this chapter, the measured angular frequencies depicted in figure 1 will be used to derive the rotational stiffness as a function of the rotational angle φ.

The differential equation (1) of the rotational oscillator has, if we neglect damping, the first integral

$$\frac{1}{2}\Theta\,\dot{\varphi}^2 + U(\varphi) = E = U(\hat{\varphi}), \tag{6}$$

with the potential energy

$$U(\varphi) = \int c_{rot}(\varphi)\varphi\, d\varphi \tag{7}$$

and the total energy E of the system. This yields the equation of the phase curves, $\dot{\varphi}(\varphi)$. Integration with respect to time leads then to

$$t - t_0 = \int_{\varphi_0}^{\varphi} \frac{d\bar{\varphi}}{\dot{\varphi}(\bar{\varphi})}, \qquad \varphi(t_0) = \varphi_0 \tag{8}$$

and from integration over one quarter of a period, we obtain

$$\frac{\omega(\hat{\varphi})}{2\pi} = \left[4\int_0^{\hat{\varphi}} \frac{d\varphi}{\sqrt{2[U(\hat{\varphi}) - U(\varphi)]/\Theta}}\right]^{-1}. \tag{9}$$

This is a closed form relationship between $c_{rot}(\varphi)$, contained in $U(\varphi)$, and the measured frequency-amplitude relation $\omega(\hat{\varphi})$.

The integral in (9) can be evaluated in closed form for very simple functions $c_{rot}(\varphi)$ only. On the other hand, equation (9) can be considered as an *integral equation* for the unknown monotonic function $U(\varphi)$. With the substitutions

$$x = U(\varphi), \quad y = U(\hat{\varphi}), \quad \zeta(x) = d\varphi/dx \tag{10}$$

and the normalized period, defined by

$$\tau(\hat{\varphi}) = \frac{\pi}{\omega(\hat{\varphi})\sqrt{2\Theta}}, \tag{11}$$

we recognize ABEL's well known integral equation

$$\bar{\tau}(y) = \int_0^y \frac{\zeta(x)}{\sqrt{y-x}}\, dx. \tag{12}$$

Here the normalized period as a function of the maximum potential energy y was introduced according to (10):

$$\hat{\varphi} = U^{-1}(y) = \varphi(y) \quad \Rightarrow \quad \bar{\tau}(y) = \tau(\varphi(y)). \tag{13}$$

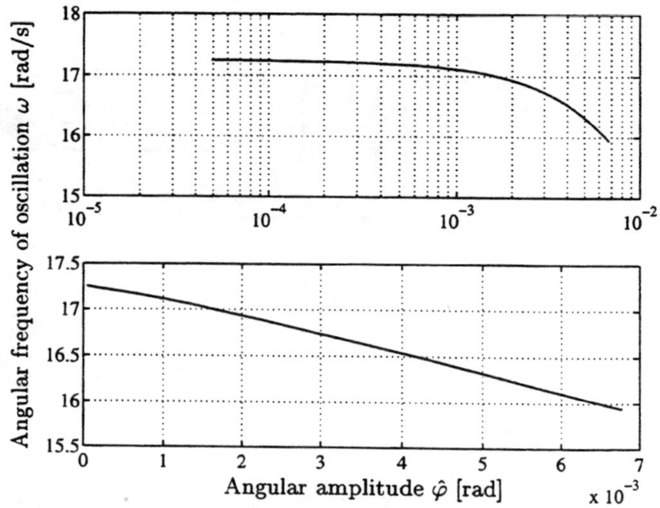

Figure 2: Measured frequency-amplitude relation $\omega(\hat{\varphi})$

The closed form solution of (12) for $\zeta(x)$ is given by

$$\zeta(x) = \frac{1}{\pi}\frac{d}{dx}\int_0^x \frac{\bar{\tau}(y)}{\sqrt{x-y}}\, dy \tag{14}$$

and integration with respect to x yields after resubstitution,

$$\varphi(U) = \frac{1}{\pi}\int_0^U \frac{\tau(\varphi(x))}{\sqrt{U-x}}\, dx. \tag{15}$$

This formula gives an implicit relation between the angle φ and the potential energy U derived from the measured frequency-amplitude relation. It can be solved by iteration, starting with $\varphi_{(0)}(U)$ taken from the linear case, $c_{rot} = const$, where we have

$$\varphi(t) = \hat{\varphi}\cos(\omega t + \gamma) \quad \Rightarrow \quad U_{(0)}(\hat{\varphi}) = \frac{1}{2}\Theta\,\omega^2\hat{\varphi}^2. \tag{16}$$

The monotonic relation $\varphi(U)$ found allows the numerical calculation of the inverse function $U(\varphi)$. Finally, the rotational stiffness $c_{rot}(\varphi)$ is calculated from U by

$$c_{rot}(\varphi) = \frac{1}{\varphi}\frac{dU}{d\varphi}, \tag{17}$$

according to (7). Thus we have obtained the function $c_{rot}(\varphi)$ from the frequency-amplitude relation.

The numerical computation of the integral in (15) from the measured τ leads to problems due to the singularity at the upper integration limit. Using the approximation

$$\tau(\varphi(x)) = \bar{\tau}(x) = \sum_{j=0}^{J} \bar{\tau}_j x^j, \tag{18}$$

it can be calculated analytically, with $z = x/U$, according to

$$\varphi(U) = \frac{1}{\pi}\sum_{j=0}^{J} \bar{\tau}_j U^{j+1/2}\int_0^1 z^j\,(1-z)^{-1/2}\, dz. \tag{19}$$

The integral represents the beta function $\mathrm{B}(j+1, -\frac{1}{2}+1)$, converging for $x > 0$, $y > 0$ to $\mathrm{B}(x,y) = \Gamma(x)\Gamma(y)/\Gamma(x+y)$.

Using (5) and with c_{rot} taken from (17), we now define a characteristic length depending on the angle φ:

$$l_{char}(\varphi) = \frac{c_{rot}(\varphi)\, L}{2[TL + c_{rot}(\varphi)]} - l_C \,. \tag{20}$$

This characteristic length, being a global stiffness measure, will be compared in the next chapter to the characteristic length, derived from the elastic line of the taut cable, which is connected to the bending stiffness $EI(\kappa)$.

THE ELASTIC LINE OF A CABLE WITH VARYING BENDING STIFFNESS

Now, we establish a connection between the bending stiffness of the cable as a function of curvature and the elastic line resulting. With the bending stiffness $EI(\kappa)$ varying with the curvature, the elastic line of the taut cable under the normal force T follows from the nonlinear differential equation

$$EI(\kappa)\,\kappa'' - T\kappa = 0\,. \tag{21}$$

Primes denote derivations with respect to the spatial coordinate s and we assume $EI'\kappa'$, $EI''\kappa \ll EI\kappa''$. Due to slip effects — reducing the stiffness with increasing curvature — the function $EI(\kappa)$ can be assumed to be monotonic. The curvature has its maximum at the clamp so that

$$d\kappa/ds < 0 \tag{22}$$

for $s > 0$. Multiplication of (21) with κ' and integration yields

$$\kappa'^2/2 - \Xi(\kappa) = const\,, \quad \text{with} \quad \Xi(\kappa) = \int \frac{T\kappa\, d\kappa}{EI(\kappa)} \tag{23}$$

and, defining the curvature amplitude $K = \kappa(s_0)$, another integration leads to

$$s(\kappa) = \int_K^{\kappa} \frac{d\bar{\kappa}}{\sqrt{2\,\Xi(\bar{\kappa})}} + s_0\,. \tag{24}$$

Hence, the solution of the differential equation (21) is reduced to an integration and the construction of the inverse function $\kappa(s)$, which is always possible due to (22).

Using the simple *ansatz*

$$EI(\kappa) = C\kappa^D \tag{25}$$

and applying the boundary conditions

$$\kappa(s_0) = K\,, \quad \kappa'(s \to \infty) = 0 \tag{26}$$

we find the solution

$$\kappa(s,K) = \left(\frac{s - s_0}{\alpha} + K^{D/2}\right)^{2/D}, \quad \alpha = \sqrt{\frac{2C(2-D)}{TD^2}}\,. \tag{27}$$

The elastic line $w(s,K)$ is then calculated by integration,

$$w(s,K) = \iint \kappa(s,K)\, ds\, ds + C_3\, s + C_4 \tag{28}$$

and with $w(s_0,K) = w_0$, $w'(s_0,K) = w_0'$ and the abbreviations

$$\beta = \frac{\alpha D}{2+D}\,, \quad \gamma = \beta^2\,\frac{2+D}{2+2D} \tag{29}$$

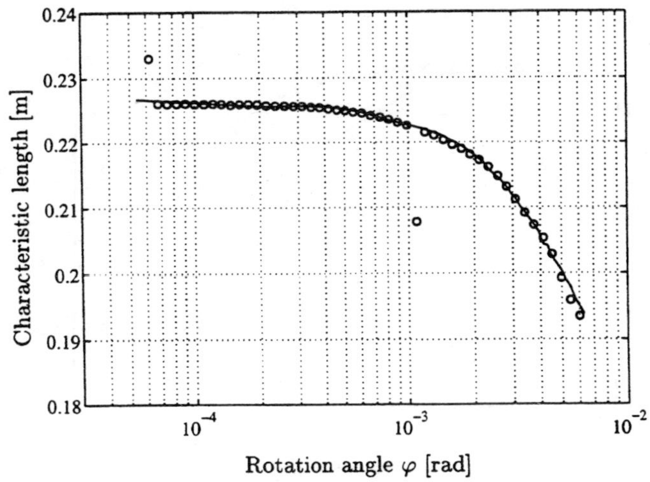

Figure 3: Characteristic length $\bar{l}_{char}$ from the elastic line ('—'), fitted to l_{char} from the frequency-amplitude relation ('o')

we obtain

$$\begin{aligned} w'(s,K) &= w_0' + \beta\,[\,\kappa(s,K)^{1+D/2} - K^{1+D/2}\,]\,, \\ w(s,K) &= w_0 + \gamma\,[\,\kappa(s,K)^{1+D} - K^{1+D}\,] \\ &\quad + [\,w_0' - \beta K^{1+D/2}\,]\, s\,. \end{aligned} \tag{30}$$

The ansatz (25) implies together with the condition (22), that D has a negative value and therefore $d^2EI/d\kappa^2 > 0$. If we want to model *arbitrary* functions $EI(\kappa)$, we can define them piecewise:

$$EI(\kappa) = C_j \kappa^{D_j} \qquad \text{for} \qquad \kappa_{j-1} \le \kappa < \kappa_j\,, \tag{31}$$

with $\kappa_j > 0$ being the largest curvature in the κ-range j. The equations (30) are valid for each range j, if we substitute K by κ_j. The values of the κ_j, $j = 1, \ldots, N$ follow from the $2N$ parameters C_j, D_j due to the continuity of $EI(\kappa)$.

Starting with $\kappa_k = K$ at $s = 0$, it is possible to compute the loci $s_j(\kappa_j, \kappa_{j+1})$ of the boundaries between the κ-ranges, since we know $s(\kappa)$ for the corresponding κ-range j from (24). The constants α_j, β_j and γ_j can be defined for each j, finally leading to complicated summation formulae for curvature and elastic line not reproduced here.

EVALUATION OF THE BENDING STIFFNESS $EI(\kappa)$ VIA PARAMETER IDENTIFICATION

As the final step, the parameters C_j, D_j will now be derived from the measurements in our test rig. For that purpose, we define a characteristic length $\bar{l}_{char}$ for the cable with variable bending stiffness $EI(\kappa)$. It is defined in complete analogy to the case of constant EI depicted in the detail of figure 1: $\bar{l}_{char}$ is the distance from the end of the clamp to the point P of the intersection of the clamp axis with the tangent on the cable at the end of the free cable length l. The elastic line and its slope are given according to the chapter before and a short calculation leads to

$$\bar{l}_{char}(K) = l - w(l,K)/\tan w'(l,K)\,, \tag{32}$$

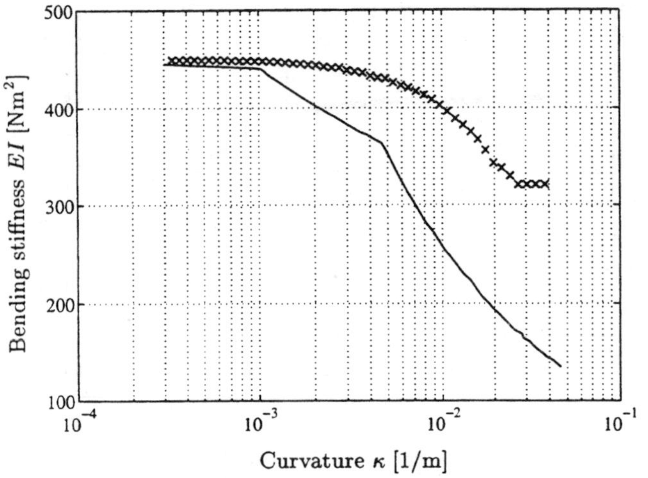

Figure 4: Approximated bending stiffness $EI(\kappa)$ for $N = 3$ and identification starting curve ('x')

with $l = L/2 - l_C$. It is a function of the curvature amplitude K and the unknown parameters $C_j, D_j, j = 1, \ldots, N$, see (31). The geometric condition

$$w(l, K) = \frac{1}{2} L \sin \varphi(K) \quad (33)$$

relates K and φ in the test rig. The largest curvature K for a certain φ, found at the clamp, will be significantly higher with the stiffness decreasing with curvature as if modeled with $EI = const$.

From these two equations, the characteristic length can be computed as a function of φ, enabling the identification of the $2N$ parameters by a comparison with the characteristic length $l_{char}(\varphi)$ from equation (20):

$$\bar{l}_{char}(\varphi, C_j, D_j) \stackrel{!}{\equiv} l_{char}(\varphi) \,. \quad (34)$$

The parameters C_j and D_j may be identified minimizing the sum of error squares $[\bar{l}_{char}(\varphi_i) - l_{char}(\varphi_i)]^2$ for a number of values φ_i. These parameters therefore define the bending stiffness $EI(\kappa)$, which allows, for given N, the best possible approximation of the measured relation $\omega(\hat{\varphi})$.

A comparison for $N = 3$ is shown in figure 3 and the resulting bending stiffness - curvature relation can be seen in figure 4, together with the curve produced by the parameters used to start the identification algorithm. The curves show good agreement with those obtained from static tests, see Papailiou (1995). The theoretical bounds for the bending stiffness are given by complete sticking between the strands, EI_{max}, and slip without any friction, EI_{min}. The values for the cable measured are 577 Nm^2 and 16 Nm^2, respectively. With the curvature exceeding approximately 10^{-3} 1/m the intensity of stick-slip transitions increases rapidly, indicating, that the outermost layer of spiral strands starts slipping relatively to the next layer of the cable.

CONCLUSIONS

In the paper presented, we described a method to obtain an approximation for the stiffness of cables as a function of the curvature. Using a relatively inexpensive, small test rig we can compute this quantity from the nonlinearity of the measured free oscillations.

The stiffness measure derived dynamically as described is, naturally, useful for calculations in cable vibrations. It should be noted, that the stick-slip behavior between the layers of the stranded cable leads to history effects due to the dry friction involved. These effects may be of interest for static cable devices, but they are usually difficult to reproduce. Therefore another advantage lies in the fact, that predeformation influences only yield an offset in the oscillations measured, which can be eliminated easily.

The elastic line of a cable with the bending stiffness decreasing with curvature can differ significantly from the elastic line computed for $EI = const$, depending on the cables properties and the load applied. Generally, the maximum curvature will increase. The results may therefore be useful for offshore and polar engineering applications, if one is interested in the shape of a taut cable with variable bending stiffness.

ACKNOWLEDGEMENT

The support of *Richard Bergner GmbH + Co., Schwabach, Germany* as well as of NATO through the *NATO Grant for Collaborative Research No. 961143* is gratefully acknowledged.

REFERENCES

[1] Anderson, K. and Hagedorn, P. *On the Energy Dissipation in Spacer-Dampers in Bundled Conductors of Overhead Transmission Lines.* Journal of Sound and Vibration, 180(4) 3, 529-538, 1995.

[2] Gutzer, U. and Hagedorn, P. *On Laboratory Measurements of Damping and Bending Stiffness of the Conductors of Overhead Transmission Lines.* Proceedings of the International Symposium on Cable Dynamics, Liege, Belgium, October 19-21, 529-536, 1995.

[3] Hagedorn, P. and Gutzer, U. *On the Mechanical Hysteresis and Bending Stiffness of Conductors and Cables.* Proceedings (II) of the Sixth International Offshore and Polar Engineering Conference, Los Angeles, USA, May 26-31, 175-181, 1996.

[4] Knapp, R. H. and Le, T. T. *KNAPP-SAC: Stress Analysis of Cables.* Knapp Engineering Inc., 1984-1995.

[5] K. Papailiou. *Die Seilbiegung mit einer durch innere Reibung, die Zugkraft und die Seilkrümmung veränderlichen Biegesteifigkeit (= The Bending of Cables with a Bending Stiffness Varying with Internal Friction, Normal Force and Curvature).* Doctoral Dissertation, ETH Zürich, 1995.

[6] M. Raoof. *Effect of Lay Angle on Various Characteristics of Spiral Strands.* Proceedings (II) of the Sixth International Offshore and Polar Engineering Conference, Los Angeles, USA, May 26-31, 189-196, 1996.

Proceedings of the Seventh (1997) International Offshore and Polar Engineering Conference
Honolulu, USA, May 25-30, 1997

ISBN 1-880653-28-1 (Set); ISBN 1-880653-30-3 (Vol. II)

Analysis of Resonant Tangential Response in Submerged Cables Resulting from 1-to-1 Internal Resonance

B.L. Newberry and N.C. Perkins
University of Michigan
Ann Arbor, MI, USA

ABSTRACT

Even in very dissipative fluid environments, cables can exhibit resonant response due to comparatively light damping along the cable tangent. Past studies by various researchers have demonstrated such resonant tensioning in suspensions subjected to dominantly tangential excitation. Recent studies have also confirmed that resonant cable motions exist even when the excitation in directed purely normal to the cable tangent. These recent studies propose one such resonant mechanism describing a nonlinear interaction between a normal dominant cable mode and a tangential dominant cable mode having natural frequencies in a 1-to-3 ratio. Herein, an alternative mechanism is proposed for two such modes having natural frequencies in a 1-to-1 ratio. An asymptotic analysis is used to determine the cable/fluid parameters that control this resonant response and conclusions are drawn comparing the 1-to-3 and 1-to-1 resonance mechanisms.

KEY WORDS

Cable Dynamics, Mooring Lines, Fluid/Structure Interaction, Nonlinear Dynamics

1. INTRODUCTION

Cables are used in a variety of engineering applications and are often selected based on their inherent flexibility. Ocean engineering systems utilize cable elements for the towing and station keeping of vessels, for the towing and positioning of instrumentation, for tethers and umbilicals, and for electrical and optical signal transmission. In-air cable applications range from structural elements in bridges and guy towers to transmission lines. Compared to most structural elements, cables possess properties which exhibit strong directional dependence. First, cables are very stiff in the tangential direction compared to any lateral direction. For cables deployed underwater, this stiffness directionality is augmented by drag directionality; large drag in the lateral directions compared to very small drag in the tangential direction. Moreover, the lateral and tangential deformations are frequently coupled.

Large and/or sustained dynamic tension may degrade the performance and positioning of instruments attached to a cable, may adversely affect the station keeping of ocean structures and vessels, and may promote cable fatigue, fretting, and wear. Dynamic tension phenomena have been reported in both laboratory (Papazoglou et al., 1990; Welch and Tulin, 1993) and full scale (Berteaux et al., 1993; Grosenbaugh, 1995) experiments on submerged cables conducted under qualitatively different excitation regimes. The tensioning phenomena reported in the work of Papazoglou et al., Berteaux et al., and Grosenbaugh result from external excitation directly applied to the tangential direction. However, indirect dynamic tensioning mechanisms also exist, as a consequence of the coupling of lateral and tangential deformations. A resonance mechanism recently examined by the authors (Newberry and Perkins, 1996a; Newberry and Perkins, 1996b) derives from the internally resonant interaction of a normal dominant cable mode and a tangential dominant cable mode having commensurable natural frequencies near a 1-to-3 ratio. This 1-to-3 internal resonance is one of a family of resonance mechanisms associated with modal coupling. Common to all these resonance mechanisms in submerged cables is the direct or indirect excitation of lightly damped tangential dominant cable modes.

The present paper focuses on a second resonance mechanism that develops from the interaction of a normal dominant mode and a tangential dominant mode having nearly equal natural frequencies (1-to-1). A two degree-of-freedom model is constructed that captures the 1-to-1 resonance. An analytical solution for resonant response is derived using asymptotics which reveals the key cable/fluid parameters that control the resonant response and the associated dynamic tension. Comparisons are drawn between this 1-to-1 mechanism and the afore mentioned 1-to-3 mechanism.

2. SUBMERGED CABLE MODEL

Consider a cable extending between two level supports and forming a shallow-sag equilibrium configuration as shown in the schematic of Figure 1. H and V are the span and sag of the equilibrium suspension which has length L. For a shallow-sag suspension $V/H < 1/8$ (Irvine and Caughey, 1974). Deformation from equilibrium U is decomposed into the components U_1, U_2 and U_3 aligned with the equilibrium tangential, normal, and binormal directions respectively. The cable is assumed to be a one-dimensional elastic continuum with negligible flexural, torsional, and shear rigidities. This study will focus on the dynamics of the cable within the equilibrium plane of the suspension (i.e. U_1 and U_2).

Fluid structure interaction effects are incorporated into the present model of the submerged cable. These effects are decomposed into the hydrostatic pressure on the equilibrium configuration, the hydrodynamic added inertia in the lateral directions (an added mass coefficient of one is assumed), and the hydrodynamic drag in the lateral directions. The hydrodynamic drag is modeled using Morison's relation and assumes that the surrounding fluid is quiescent (i.e. no current).

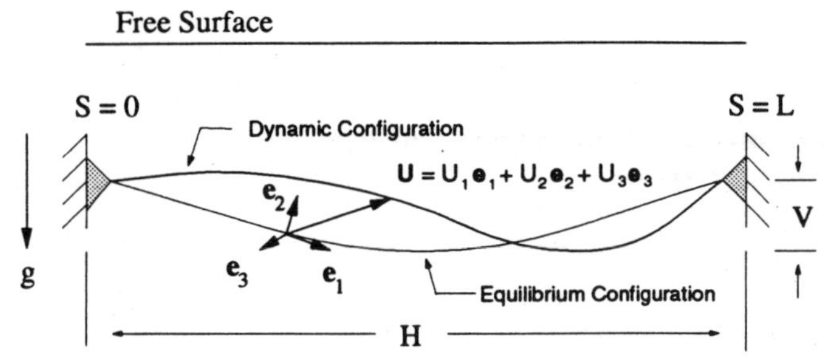

Fig. 1 Shallow-sag cable suspension schematic. Deformation is defined with respect to the Frenet triad defined by the equilibrium configuration.

The cable is assumed to be subject to domain excitation in the normal ($\mathbf{e}_2$) direction. This excitation, which may ultimately derive from vortex shedding, wave motion, or from the motion of attached bodies, is herein prescribed as a spatially and temporally harmonic field. This idealized harmonic excitation serves to identify possible resonance mechanisms that may also exist for more complicated (broader band) excitation sources. A summary of the derivation of the three-dimensional equations of cable motion is provided in (Perkins, 1986; Newberry and Perkins, 1996a) and the equations governing in-plane response are

$$\rho A U_{1,tt} = (T_{d,s})(1+z_1) + (T_o + T_d)(z_{1,s} - K z_2) - C_s U_{1,t} \quad (1)$$

$$(\rho + \rho_w) A U_{2,tt} = (T_{d,s}) z_2 + (T_o + T_d)(z_{2,s} + K z_1) + K T_d + F_2(s,t) - \Upsilon(U_{2,t}) \quad (2)$$

where

$$\begin{aligned} z_1 &= U_{1,s} - K U_2 \\ z_2 &= U_{2,s} + K U_1 \\ \epsilon_d &= z_1 + \frac{1}{2}\left(z_1{}^2 + z_2{}^2\right) \\ T_d &= E A \epsilon_d \\ \Upsilon(U_{2,t}) &= \frac{1}{2}(\rho_w C_d d) U_{2,t} |U_{2,t}| \end{aligned} \quad (3)$$

The parameters appearing in (1) - (3) include the equilibrium tension T_o (spatially uniform for a shallow-sag cable), the nominal Young's modulus E, the cable cross-sectional area A, the equilibrium cable curvature K (spatially uniform for a shallow-sag cable), the structural damping coefficient for the tangential direction C_s, the fluid drag coefficient C_d, and the water ρ_w and cable ρ densities respectively. The function ϵ_d is the dynamic component of the cable strain (tension), $F_2(s,t)$ is the domain excitation in the normal direction, and $\Upsilon(U_{2,t})$ is the hydrodynamic drag. The notation $()_{,s}$ and $()_{,t}$ denotes partial derivatives with respect to the independent spatial (arc length) and time variables, respectively. Note that structural damping has been neglected in the lateral direction as it is negligible compared to fluid drag. This continuum model is presently reduced to a (discrete) two degree-of-freedom model that captures a 1-to-1 resonance of two cable modes. The accuracy of the discrete model is assessed in Section 2.2.

2.1 Reduced-Order Discrete Model

Prior to discretization, the continuous model is nondimensionalized using the characteristic length scale L and time scale $\sqrt{L/g}$. Furthermore, the nonlinear Morison's relation (3) utilized in the continuous model for the lateral fluid drag is replaced with a two-term Fourier approximation assuming monochromatic normal response:

$$\Upsilon(u_{2,t}, s) = C_L u_{2,t} + C_{NL} u_{2,t}^3 \quad (4)$$

The accuracy of this approximation is discussed in (Newberry and Perkins, 1996a). This Fourier approximation is analytic in contrast to the nonanalytic form apparent in (3) and greatly facilitates the subsequent analysis. Letting U denote the amplitude of the normal response having frequency Ω and (normalized) mode shape $\theta(s)$, the coefficients in (4) are

$$C_L = \left(\frac{16\Omega U|\theta(s)|}{15\pi}\right)\left(\frac{\rho_w C_d d}{2\rho_c g A}\right)(gL) \quad (5)$$

$$C_{NL} = \left(\frac{32}{15\pi\Omega U|\theta(s)|}\right)\left(\frac{\rho_w C_d d}{2\rho_c g A}\right)(gL) \quad (6)$$

and are amplitude and frequency dependent.

The reduced-order model is formulated using global Galerkin discretization based on two linear modes of the cable suspension. The selected modes have nearly equal natural frequencies ($\omega_1 \approx \omega_2$) and are of like symmetry (either both even or both odd) with respect to the cable midspan in keeping with the principal observations of (Newberry and Perkins, 1996a). Each mode describes specific contributions of normal and tangential response which remain coupled due to the equilibrium curvature. The first mode $\mathbf{\Phi}_1(s)$ is dominated by normal response while the second mode $\mathbf{\Phi}_2(s)$ is dominated by tangential response. The discretization procedure detailed in (Newberry and Perkins, 1996b) leads to the ordinary differential equations governing the modal coordinates α_1 and α_2:

$$\begin{aligned} &\ddot{\alpha_1} + \hat{A}_1\dot{\alpha_1} + \hat{A}_2\dot{\alpha_2} + \hat{A}_3\dot{\alpha_1}^3 + \hat{A}_4\dot{\alpha_2}^3 + \omega_1^2\alpha_1 + \\ &\hat{A}_5\alpha_1^2 + \hat{A}_6\alpha_2^2 + \hat{A}_7\alpha_1\alpha_2 + \hat{A}_8\alpha_1^3 + \hat{A}_9\alpha_2^3 + \hat{A}_{10}\alpha_1^2\alpha_2 + \\ &\hat{A}_{11}\alpha_1\alpha_2^2 + \hat{A}_{12}\dot{\alpha_1}^2\dot{\alpha_2} + \hat{A}_{13}\dot{\alpha_1}\dot{\alpha_2}^2 + \hat{A}_{14}\,sin(\Omega t) = 0 \end{aligned} \quad (7)$$

$$\begin{aligned} &\ddot{\alpha_2} + \hat{B}_1\dot{\alpha_1} + \hat{B}_2\dot{\alpha_2} + \hat{B}_3\dot{\alpha_1}^3 + \hat{B}_4\dot{\alpha_2}^3 + \omega_2^2\alpha_2 + \\ &\hat{B}_5\alpha_1^2 + \hat{B}_6\alpha_2^2 + \hat{B}_7\alpha_1\alpha_2 + \hat{B}_8\alpha_1^3 + \hat{B}_9\alpha_2^3 + \hat{B}_{10}\alpha_1^2\alpha_2 + \\ &\hat{B}_{11}\alpha_1\alpha_2^2 + \hat{B}_{12}\dot{\alpha_1}^2\dot{\alpha_2} + \hat{B}_{13}\dot{\alpha_1}\dot{\alpha_2}^2 + \hat{B}_{14}\,sin(\Omega t) = 0 \end{aligned} \quad (8)$$

where ω_i, i=1,2, denote the eigenfrequencies of the two modes and $\hat{A}_j$ and $\hat{B}_j$, j = 1 - 10, are the coefficients obtained through the Galerkin discretization.

2.2 Accuracy of Discrete Model

The accuracy of the discrete model is now assessed for the case of a 1-to-1 resonance between a normal dominant mode and a tangential dominant mode. To this end, a sample suspension is selected which is representative of a synthetic, shallow-sag marine cable in water (McKenna and Wong, 1979; Dexter, 1979). The properties of the suspension, which are listed in Table 1, have been selected to achieve significant coupling between the 15th normal dominant mode (symmetric) and the 2nd tangential dominant mode (symmetric). The natural period of either mode is approximately 1.6 seconds. The resulting discrete model coefficients are shown in Table 2 of Appendix A.

Table 1 Physical Properties of the Example Synthetic Marine Cable Suspension.

Parameter	Value
Cable Density (ρ_c)	1150 kg/m^3
Water Density (ρ_w)	1020 kg/m^3
Cable Span (H)	1000 m
Cable Sag (V)	10.4 m
Cable Length (L)	1000.87 m
Cable Diameter (d)	0.05 m
Equilibrium Tension (T_e)	30000 N
Elastic Modulus (E)	440 MPa

Figure 2 presents steady-state results obtained both by direct finite difference simulation of the continuous model and by Runga-Kutta simulation of the reduced-order model above. The finite difference algorithm follows that detailed in (Newberry and Perkins, 1996a). The simulation results are presented as spatial envelopes of the tangential (U_1) and normal (U_2) displacements. The excellent agreement between the two solutions demonstrate that the normal response is controlled by the 15th normal dominant mode while the tangential response is controlled by the 2nd tangential dominant mode.

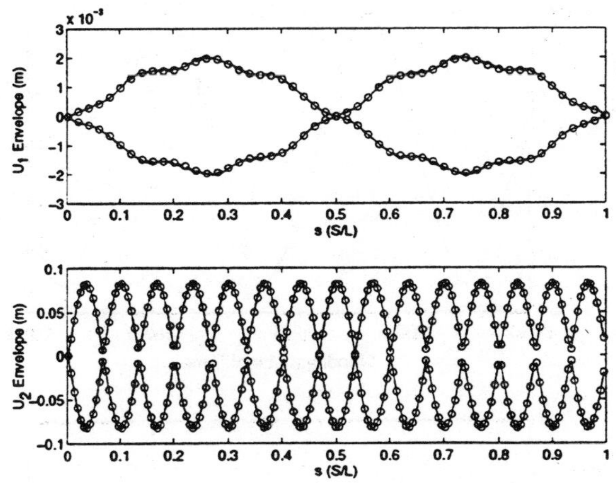

Fig. 2 Comparison of solutions obtained from the discrete model and the continuous model. Simulations performed for $C_d = 2$ and 1% critical structural damping. Continuous model results ——; discrete model results o o oo.

3. ASYMPTOTIC ANALYSIS

An analytical solution to the resonant response is presently pursued and used to identify the principal cable/fluid parameters controlling this mechanism. To this end, the suspension is assumed to be forced near primary resonance

$$\Omega = \omega_1 + \epsilon\sigma_1 \tag{9}$$

where Ω represents the excitation frequency, ϵ is a small parameter and σ_1 represents an external detuning parameter. Furthermore, consider the case of near 1-to-1 tuning between the normal and tangential dominant modes controlled by the internal detuning parameter σ_2:

$$\omega_2 = \omega_1 + \epsilon\sigma_2 \tag{10}$$

Ordering assumptions consistent with the physics of the resonant motion are now imposed. Assuming weakly nonlinear response, it is reasonable to assume that the amplitude of the tangential dominant mode will be much less than the amplitude of the normal dominant mode. This assumption is based on the facts that 1) the normal response is directly excited, 2) the cable is comparatively stiff in the tangential direction, and 3) that any energy delivered to the tangential direction derives from coupling terms in (1) which are all proportional to the small equilibrium curvature. Thus

$$\begin{aligned} \mathcal{O}(\alpha_1) &= \epsilon \\ \mathcal{O}(\alpha_2) &= \epsilon^2 \\ \mathcal{O}(k) &= \epsilon \end{aligned} \tag{11}$$

The coefficients in the discrete model associated with curvature coupling and dissipation are now scaled recognizing that 1) the coupling terms are proportional to the equilibrium curvature, 2) the drag coefficients are amplitude dependent, and 3) the structural damping is assumed small ($\mathcal{O}(C_S) = \epsilon$). These observations lead to scaling relations for the following coefficients:

$$\begin{aligned} \hat{A}_1 &\longmapsto A_1\epsilon & \hat{B}_1 &\longmapsto B_1\epsilon^2 \\ \hat{A}_2 &\longmapsto A_2\epsilon^2 & \hat{B}_2 &\longmapsto B_2\epsilon \\ \hat{A}_3 &\longmapsto A_3/\epsilon & \hat{B}_4 &\longmapsto B_4/\epsilon \\ \hat{A}_5 &\longmapsto A_5\epsilon & \hat{B}_5 &\longmapsto B_5\epsilon \\ \hat{A}_7 &\longmapsto A_7\epsilon & \hat{B}_7 &\longmapsto B_7\epsilon \\ \hat{A}_{14} &\longmapsto A_{14}\epsilon^2 & \hat{B}_{14} &\longmapsto B_{14}\epsilon^3 \end{aligned} \tag{12}$$

The hat notation will now be dropped from all the coefficients for convenience.

The following expansion is proposed using the method of multiple-scales with two time scales ($T_0 = t$ and $T_1 = \epsilon t$):

$$\begin{aligned} \alpha_1 &= \epsilon\alpha_{11}(T_0, T_1) + \epsilon^2\alpha_{12}(T_0, T_1) \\ \alpha_2 &= \epsilon^2\alpha_{21}(T_0, T_1) + \epsilon^3\alpha_{22}(T_0, T_1) \end{aligned} \tag{13}$$

Substituting this expansion into (7) and (8) and collecting terms of like power in ϵ leads to the following sequence of equations.

$\mathcal{O}(\epsilon)$ equations:

$$\frac{\partial^2\alpha_{11}}{\partial T_0{}^2} + \omega_1^2\alpha_{11} = 0 \tag{14}$$

$\mathcal{O}(\epsilon^2)$ equations:

$$\begin{aligned} \frac{\partial^2\alpha_{12}}{\partial T_0{}^2} + \omega_1^2\alpha_{12} &= -2\frac{\partial^2\alpha_{11}}{\partial T_0 T_1} - A_1\frac{\partial\alpha_{11}}{\partial T_0} - A_3\frac{\partial\alpha_{11}}{\partial T_0}^3 - \\ &\quad A_{14}\, sin(\Omega T_0) \\ \frac{\partial^2\alpha_{21}}{\partial T_0{}^2} + \omega_2^2\alpha_{21} &= 0 \end{aligned} \tag{15}$$

$\mathcal{O}(\epsilon^3)$ equations:

$$\begin{aligned} \frac{\partial^2\alpha_{22}}{\partial T_0{}^2} + \omega_2^2\alpha_{22} &= -2\frac{\partial^2\alpha_{21}}{\partial T_0 T_1} - B_1\frac{\partial\alpha_{11}}{\partial T_0} - B_2\frac{\partial\alpha_{21}}{\partial T_0} - \\ &\quad B_3\frac{\partial\alpha_{11}}{\partial T_0}^3 - B_5\alpha_{11}^2 - B_{14}\, sin(\Omega T_0) \end{aligned} \tag{16}$$

Beginning at ϵ order, the generator solution for the coordinate of the normal dominant mode is

$$\alpha_{11}(T_0, T_1) = \delta_{11}(T_1)\, sin(\omega_1 T_0 + \phi_{11}(T_1)) \tag{17}$$

where the amplitude and phase are dependent on the slow time scale. Proceeding to the next $\mathcal{O}(\epsilon^2)$ equations and removing secular terms provides the following evolution equations governing the slowly varying amplitude and phase:

$$\begin{aligned} 2\omega_1 \frac{\partial \delta_{11}(T_1)}{\partial T_1} + \frac{3A_3\omega_1^3\delta_{11}(T_1)^3}{4} + \\ A_1\omega_1\delta_{11}(T_1) = A_{14}\, sin[\theta_{11}(T_1)] \\ 2\omega_1\delta_{11}(T_1)\frac{\partial \theta_{11}(T_1)}{\partial T_1} + \\ 2\sigma_1\omega_1\delta_{11}(T_1) = A_{14}\, cos[\theta_{11}(T_1)] \end{aligned} \tag{18}$$

where

$$\theta_{11}(T_1) = \phi_{11}(T_1) - \sigma_1 T_1 \tag{19}$$

Periodic solutions for α_{11} are obtained from the fixed points of the evolution equations (18), as determined by

$$A_3^2\delta_{11}^{*\,6} + \left[\frac{8A_1A_3}{3\omega_1^2}\right]\delta_{11}^{*\,4} + \left[\frac{16A_1^2 + 64\sigma_1^2}{9\omega_1^4}\right]\delta_{11}^{*\,2} = \frac{16A_{14}^2}{9\omega_1^6} \tag{20}$$

$$\theta_{11}^* = tan^{-1}\left(\frac{3A_3\omega_1^2\delta_{11}^{*\,2} + 4A_1}{8\sigma_1}\right) \tag{21}$$

where δ_{11}^* and θ_{11}^* denote the constant amplitude and phase defining the periodic solution. The amplitude δ_{11}^* may first be determined from (20) which is cubic in $\delta_{11}^{*\,2}$. This cubic equation admits a unique real-valued solution for $\delta_{11}^{*\,2}$ as A_1 and A_3 are positive.

Attention now focuses on determining the coordinate of the tangential dominant mode α_2. Beginning at $\mathcal{O}(\epsilon^2)$, the generator solution is

$$\alpha_{21}(T_0, T_1) = \delta_{21}(T_1)\, sin(\omega_2 T_0 + \phi_{21}(T_1)) \tag{22}$$

where $\delta_{21}(T_1)$ and $\phi_{21}(T_1)$ again denote slowly varying amplitude and phase variables, respectively. Removing secular terms at $\mathcal{O}(\epsilon^3)$ leads to the evolution equations

$$\begin{aligned} 2\omega_2\frac{\partial \delta_{21}(T_1)}{\partial T_1} + B_2\omega_2\delta_{21}(T_1) + \tau_1\, cos(\theta_{21}(T_1)) + \\ \tau_2\, sin(\theta_{21}(T_1)) = 0 \\ 2\omega_2\delta_{21}(T_1)\frac{\partial \theta_{21}(T_1)}{\partial T_1} + 2\sigma_3\omega_2\delta_{21}(T_1) - \\ \tau_1\, sin(\theta_{21}(T_1)) + \tau_2\, cos(\theta_{21}(T_1)) = 0 \end{aligned} \tag{23}$$

where

$$\begin{aligned} \theta_{21}(T_1) &= \phi_{21}(T_1) - \sigma_3 T_1 \\ \sigma_3 &= \sigma_1 - \sigma_2 \\ \tau_1 &= \frac{3B_3\, cos(\theta_{11}^*)\omega_1^3\delta_{11}^{*\,3}}{4} + B_1\, cos(\theta_{11}^*)\omega_1\delta_{11}^* \\ \tau_2 &= \frac{3B_3\, sin(\theta_{11}^*)\omega_1^3\delta_{11}^{*\,3}}{4} + B_1\, sin(\theta_{11}^*)\omega_1\delta_{11}^* - B_{14} \end{aligned} \tag{24}$$

Fixed points of (23) provide the following amplitude and phase for periodic tangential response

$$\delta_{21}^* = \sqrt{\frac{\tau_1^2 + \tau_2^2}{\omega_2^2(B_2^2 + 4\sigma_3^2)}} \tag{25}$$

$$\theta_{21}^* = tan^{-1}\left[\frac{2\sigma_3}{-B_2}\right] + tan^{-1}\left[\frac{\tau_2}{\tau_1}\right] \tag{26}$$

with uniqueness following that for δ_{11}^*.

The periodic solution to leading-order for each coordinate is

$$\begin{aligned} \alpha_1 &= \epsilon\delta_{11}^*\, sin(\omega_1 T_0 + \phi_{11}(T_1)) = \epsilon\delta_{11}^*\, sin(\Omega T_0 + \theta_{11}^*) \\ \alpha_2 &= \epsilon^2\delta_{21}^*\, sin(\omega_2 T_0 + \phi_{21}(T_1)) = \epsilon^2\delta_{21}^*\, sin(\Omega T_0 + \theta_{21}^*) \end{aligned} \tag{27}$$

Note that the normal and tangential coordinates respond at the excitation frequency. Moreover, the stability of this periodic motion is guaranteed and can be readily confirmed through stability analysis employing the evolution equations (18), (19), (23), and (24).

3.1 Accuracy of Asymptotics

The accuracy of the asymptotic solution (27) is presently assessed through comparison with numerical solutions obtained from the discrete model (7) and (8). Figure 3 provides such a comparison for the example suspension of Table 1 subject to excitation sufficient to induce an amplitude in the normal direction of ten diameters. Note the excellent agreement (both qualitatively and quantitatively) between the asymptotic and numerical solutions. Moreover, this agreement improves as the excitation amplitude decreases (as expected). The nonlinearities associated with fluid drag represent rather small corrections to the underlying linear equations in this instance. In other instances, however, the nonlinear corrections are indeed larger and their inclusion here poses no complications.

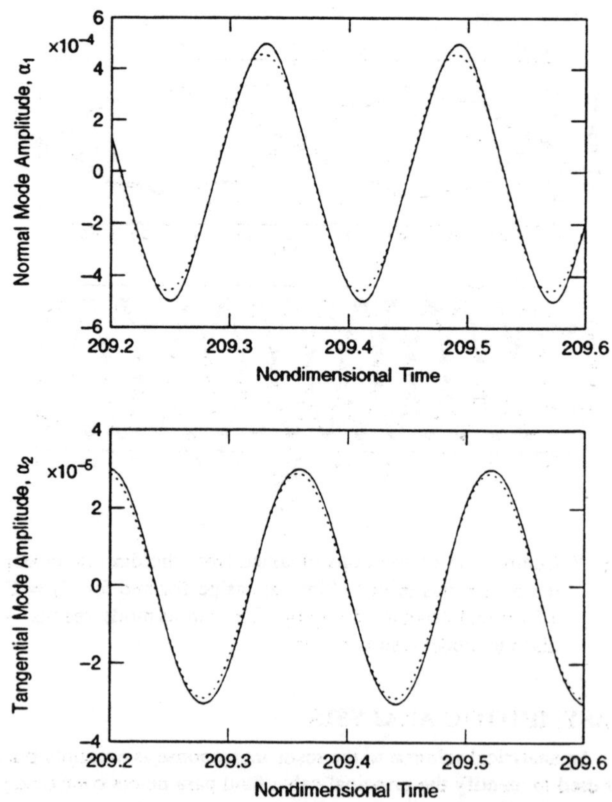

Fig. 3 Comparison of the asymptotic solution with numerical solution of the discrete model. Results based on $C_d = 2$, 1% critical structural damping, and excitation amplitude sufficient to produce a normal response amplitude of ten cable diameters. Numerical solution = ——, asymptotic solution = – – – –.

4. RESONANCE MECHANISM

The asymptotic solution provides considerable insight into which suspension/design parameters control the resonance mechanism. These parameters include measures of 1) the external detuning, 2) the excitation level, 3) the structural damping, and 4) the fluid drag. To begin, one needs to understand how the coefficients in the discrete model (7) and (8) are related to the four effects above. These relationships, which are developed in Appendix B; refer to (B1) - (B4), are utilized here. Use of (B2) and (B4) in (25) leads to

$$\delta_{21}^{*} \approx \left| \underbrace{\frac{[4B_{1b}\omega_1 C_S]\delta_{11}^{*}}{4\omega_2\sqrt{B_2^2C_S^2+4\sigma_3^2}}}_{LinearComponent} + \underbrace{\frac{[4B_{1a}C_d\omega_1 + 4B_{14} + 3B_3C_d\omega_1^3]\delta_{11}^{*\,2}}{4\omega_2\sqrt{B_2^2C_S^2+4\sigma_3^2}}}_{NonlinearComponent} \right| \qquad (28)$$

which shows how the linear and nonlinear components of δ_{21}^{*} depend on the four effects listed above.

Consider first the influence of external detuning σ_1, as captured by the parameter $\sigma_3 = \sigma_1 - \sigma_2$ in (28). Figure 4 illustrates how the asymptotic solution varies with external detuning σ_1 for two cases of internal detuning σ_2. The frequency response calculations were generated holding the amplitude of the normal dominant mode constant (one-half diameter). Thus the excitation amplitude varies as the external detuning is swept. This variation is slight, however, since the normal response is over-damped as evident by the illustrated results for θ_{11}^{*}. By contrast, the tangential response is lightly damped and possesses a strong, narrow-band resonance centered about $\sigma_1 = \sigma_2$. This light damping is verified by the sudden phase inversion depicted in the results for θ_{21}^{*}.

For a given value of internal detuning, the tangential resonance reaches an extremum when the external detuning matches the internal detuning ($\sigma_3 = \sigma_1 - \sigma_2 = 0$); refer to (28). Numerical simulations confirm that the largest resonant response is obtained when the internal and external detunings vanish. Note however that this limiting case can not be analyzed using the present asymptotic analysis without reordering α_2. In the limit $\sigma_2 \to 0$, the eigenvalues ω_1^2 and ω_2^2 exhibit the curve-veering phenomenon (Perkins and Mote, 1986; Burgess and Triantafyllou, 1988) and the associated mode shapes $\Phi_1(s)$ and $\Phi_2(s)$ are extremely sensitive to small changes in σ_2. In particular, these mode shapes are no longer dominated by the normal or tangential components and therefore the assumption $\mathcal{O}(\alpha_2) = \epsilon^2$ is no longer valid.

The dependence on the excitation level can be investigated by noting how the amplitude of the tangential dominant mode δ_{21}^{*} depends on that of the normal dominant mode δ_{11}^{*}. It is evident from (28) that δ_{21}^{*} contains components that vary both linearly and quadratically with the normal response amplitude δ_{11}^{*}. For the present suspension ($C_d = 2$, 1% critical damping), computations using (28) reveal that the nonlinear contribution is a small fraction of the response; refer to Figure 5. The relative importance of the nonlinear component, however, depends strongly on the suspension parameters. For greatly decreased dissipation, the nonlinear portion becomes dominant as will be demonstrated in the following.

Dependence on the structural damping and fluid drag are now assessed. Beginning with structural damping, consider the case where $\sigma_3 = \sigma_1 - \sigma_2$ is small. For this case, the nonlinear component of (28) is (approximately) inversely proportional to the structural damping coefficient (C_S), while the linear component is unaffected by the structural

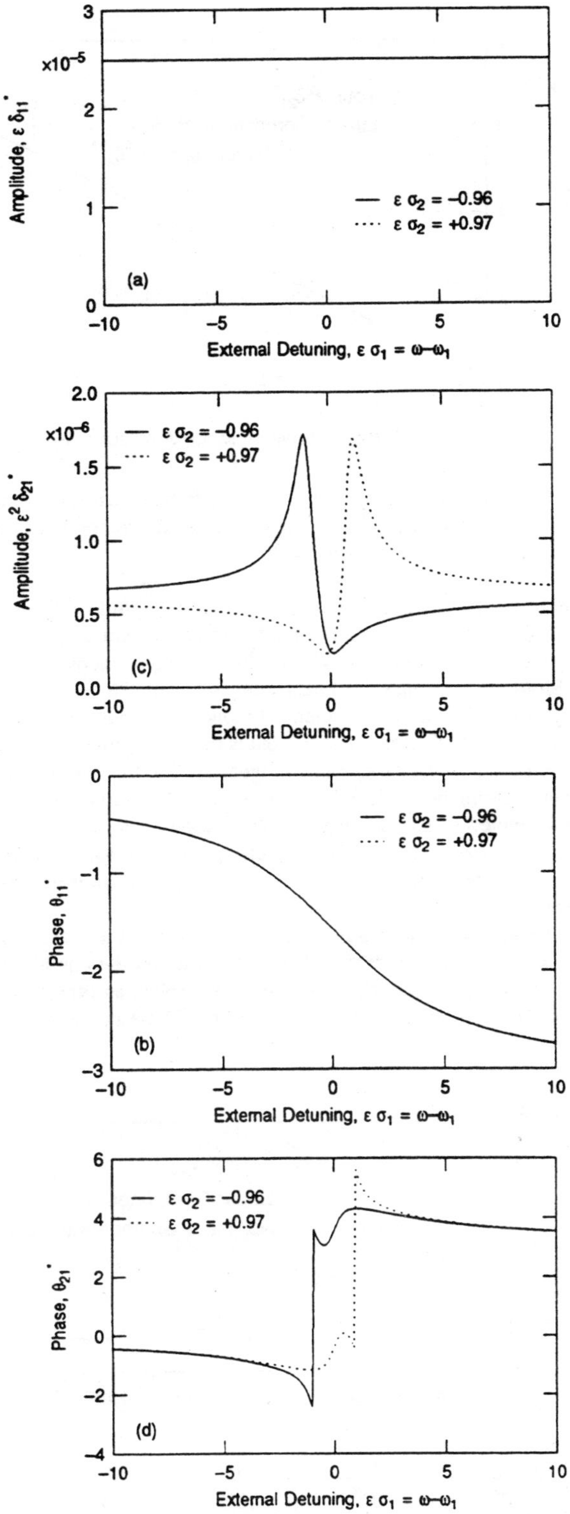

Fig. 4 Dependence of the resonance mechanism on external detuning (σ_3). Frequency and phase response for normal (a,b) and tangential (c,d) dominant modes. Results based on the perturbation solution using a peak-to-peak normal response amplitude of one cable diameter, $C_d = 2$ and 1% critical structural damping.

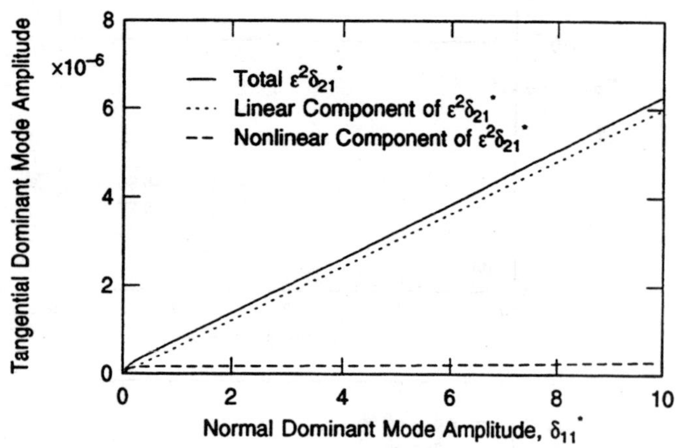

Fig. 5 Dependence of δ_{21}^* on δ_{11}^*. Results based on the perturbation solutions using $C_d = 2$ and 1% critical structural damping, for the case $\sigma_3 = \sigma_1 - \sigma_2 = 0$.

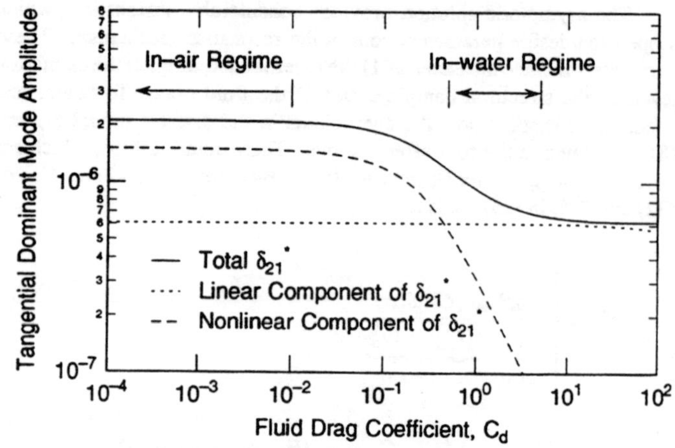

Fig. 7 Dependence of δ_{21}^* on fluid drag. Results based on the perturbation solutions using a normal response peak-to-peak amplitude of one cable diameter and 1% critical structural damping, for the case $\sigma_3 = \sigma_1 - \sigma_2 = 0$.

damping as illustrated in Figure 6. The total amplitude $\epsilon^2\delta_{21}^*$ is asymptotic to that associated with the nonlinear component as $\zeta \to 0$ and is asymptotic to that associated with the linear component as $\zeta \to 1$.

Similarly note that the linear component of (28) is unaffected by changes in the fluid drag coefficient. By contrast, the nonlinear component is independent of C_d for $C_d < 0.1$ but is inversely proportional to C_d in the regime $C_d > 0.1$ as shown in Figure 7. The total amplitude of the tangential dominant mode exhibits both a low-C_d asymptote and a high-C_d asymptote. Ocean engineering applications (i.e. cables submerged in-water) lie closest to the high-C_d asymptote. The low-C_d asymptote, approximate for suspensions in-air, leads to increased tangential response, as expected.

Further parametric studies are both possible and warranted. Among them include the dependence of the resonance on the equilibrium curvature (tension), the elastic properties of the cable, and the order of the modes that might participate in the resonance. These studies are presently underway.

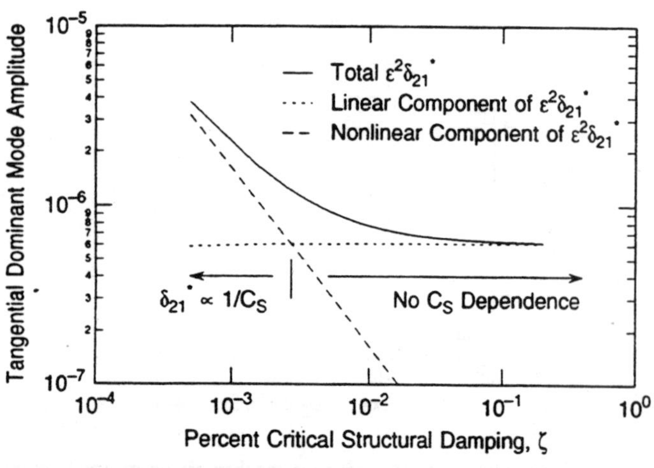

Fig. 6 Dependence of δ_{21}^* on structural damping. Results based on the perturbation solutions using a normal response peak-to-peak amplitude of one cable diameter and $C_d = 2$, for the case $\sigma_3 = \sigma_1 - \sigma_2 = 0$.

5. CONCLUSIONS AND COMPARISONS

As previously mentioned, alternative resonance mechanisms exist for fluid-loaded cable suspensions. One such mechanism is nonlinear and results in 1-to-3 internally resonant response of a normal dominant mode and a tangential dominant mode (Newberry and Perkins, 1996b). The similarities and differences between these two resonance mechanisms are summarized below.

1) For drag coefficients consistent with ocean engineering applications (for example $1 < C_d < 3$), the 1-to-1 resonance mechanism herein induces a tangential response that scales linearly with the amplitude of the normal response. By contrast this scaling is quadratic for the 1-to-3 resonance mechanism. Thus, for increasing lateral response levels, the 1-to-3 mechanism ultimately leads to large response.
2) The 1-to-3 mechanism develops as a consequence of *nonlinear* fluid/structure interaction and the amplitude of the tangential dominant mode is proportional to the fluid drag coefficient. By contrast, the 1-to-1 mechanism is generated by linear drag. Moreover, for drag coefficients consistent with ocean engineering applications (for example $1 < C_d < 3$), the response of the tangential dominant mode (1-to-1 mechanism) is inversely proportional to the fluid drag coefficient.
3) Structural damping has a consistent impact on both mechanisms; the amplitude of the tangential response is reduced with increasing structural damping.
4) The 1-to-1 mechanism leads to much larger tangential response amplitudes (one order of magnitude) than the 1-to-3 mechanism for equivalent lateral response amplitudes.

Both mechanisms demonstrate that resonant response may be realized in submerged cable suspensions. These resonances develop for a lightly damped tangential dominant mode. Thus, cable theories that ignore these modes (such as theories based on quasi-static cable stretching) cannot predict these resonances.

6. ACKNOWLEDGEMENTS

The authors gratefully acknowledge the support of the United States Office of Naval Research through grant no. N00014-96-1-0446.

7. REFERENCES

Berteaux, HO; Bocconcelli, A; Eck, CF and Kery, S (1993). "The Dynamic Mooring Experiment," *Sea Technology*, Vol. 34, No. 3, March, pp. 30-36.

Burgess, JJ; Triantafyllou, MS (1988). "The Elastic Frequencies of Cables," *Journal of Sound and Vibrations*, Vol 120(1), pp. 153-165.

Dexter, SC (1979). *Handbook of Oceanographic Engineering Materials*, John Wiley & Sons, Inc., New York, 1979.

Grosenbaugh, MA (1995). "The Effects of Hydrodynamic Damping on the Tension of Oceanographic Surface Moorings", Proceedings of the *International Symposium on Cable Dynamics*, 1995, pp. 85-92.

Irvine, HM, and Caughey, TK (1974). "The Linear Theory of Free Vibrations of a Suspended Cable," *Proc. of the Royal Soc. London*, Series A, Vol. 341, pp. 299-315.

McKenna, HA, and Wong, RK (1979). "Synthetic Fiber Rope, Properties and Calculations Relating to Mooring Systems," *Deepwater Mooring and Drilling*, ASME, OED, Vol. 7, pp. 189-203.

B. L. Newberry and N. C. Perkins (1996a). "Investigation of resonant tensioning in submerged cables subjected to lateral excitation." In *Proc. 6th International Offshore and Polar Engineering Conference* (Los Angeles, CA), Vol. II, pp. 210-216, 1996.

B. L. Newberry and N. C. Perkins (1996b). "Asymptotic Analysis of a Tension Resonance Mechanism in Submerged Cables," In *Proc. of the 1996 International Mechanical Engineering Congress and Exposition of the ASME*, Nonlinear Dynamics and Controls Symposium (Atlanta, GA), November 17-22, to appear.

Papazoglou, VJ; Mavrakos, SA and Triantafyllou, MS (1990). "Nonlinear Cable Response and Modal Testing in Water," *Journal of Sound and Vibrations*, Vol 140(1), pp. 103-115.

Perkins, NC (1986). "Three Dimensional Theory and Analysis of Travelling Sagged Cable Dynamics," Ph.D. Dissertation, University of California, Berkeley, CA.

Perkins, NC and Mote Jr., CD (1986). "Comments on Curve Veering in Eigenvalue Problems," *Journal of Sound and Vibrations*, Vol 106(3), pp. 451-463.

Welch, SM, and Tulin, MP (1993). "An Experimental Investigation of the Mean and Dynamic Tensions in Strumming Cables," *International Journal of Offshore and Polar Engineering*, Vol. 3, No. 3, Sept., pp. 205-212.

8. APPENDIX A: Discrete Model Coefficients

Table 2 Discrete Model Coefficients for the Test Suspension (Table 1).

Parameter	Value	Parameter	Value
$\hat{A}_1$	3.36e+00	$\hat{B}_1$	6.15e-02
$\hat{A}_2$	3.26e-02	$\hat{B}_2$	7.80e-01
$\hat{A}_3$	7.10e+06	$\hat{B}_3$	1.70e+05
$\hat{A}_4$	2.83e+01	$\hat{B}_4$	3.98e+00
ω_1	3.99e+01	ω_2	3.90e+01
$\hat{A}_5$	-3.00e+02	$\hat{B}_5$	9.91e+02
$\hat{A}_6$	-1.69e+02	$\hat{B}_6$	-3.40e+00
$\hat{A}_7$	8.41e+01	$\hat{B}_7$	-6.27e+02
$\hat{A}_8$	2.42e+07	$\hat{B}_8$	5.74e+05
$\hat{A}_9$	-2.92e+02	$\hat{B}_9$	1.44e+04
$\hat{A}_{10}$	1.20e+06	$\hat{B}_{10}$	2.92e+04
$\hat{A}_{11}$	1.55e+04	$\hat{B}_{11}$	1.92e+04
$\hat{A}_{12}$	2.69e+05	$\hat{B}_{12}$	8.51e+03
$\hat{A}_{13}$	4.51e+03	$\hat{B}_{13}$	1.60e+02
$\hat{A}_{14}$	-8.34e-03	$\hat{B}_{14}$	-1.99e-04

9. APPENDIX B: Proportionality Relations

Recognize that the normal dominant mode is over-damped, $\theta^*_{11} \approx -\pi/2$; refer to Figure 4. As such, the amplitude of the tangential response can be simplified:

$$\delta^*_{21} = \sqrt{\frac{\tau_1^2 + \tau_2^2}{\omega_2^2(B_2^2 + 4\sigma_3^2)}} \approx \frac{|4B_1\omega_1\delta^*_{11} + 4B_{14} + 3B_3\omega_1^3{\delta^*_{11}}^3|}{4\omega_2\sqrt{B_2^2 + 4\sigma_3^2}} \qquad (B1)$$

Upon completion of Galerkin discretization, the linear drag terms (denoted A_1, B_1, and B_2) are each linear combinations of two components: one proportional to linear fluid drag and one proportional to the structural damping. Note also that the fluid drag directly impacts the normal response, but only affects the tangential response via coupling in proportion to the equilibrium curvature. By contrast, the structural damping directly impacts the tangential response, but only affects the normal response through (curvature) coupling. As such, B_2 is dominated by the structural damping, A_1 is dominated by the fluid drag, and B_1 receives contributions from both dissipation mechanisms in proportion to the equilibrium curvature. Recalling the amplitude dependent nature of C_L and C_{NL}, refer to (5) and (6), one can conclude that

$$\begin{aligned} A_1 &\propto \mathcal{A}_1 C_d \delta^*_{11} \\ B_1 &\propto \mathcal{B}_{1a} C_d \delta^*_{11} + \mathcal{B}_{1b} C_S \\ B_2 &\propto \mathcal{B}_2 C_S \\ B_3 &\propto \mathcal{B}_3 \frac{C_d}{\delta^*_{11}} \end{aligned} \qquad (B2)$$

The forcing terms A_{14} and B_{14} scale linearly with the excitation amplitude; a result that is evident from the Galerkin discretization. Neglecting small corrections due to nonlinear drag, the asymptotic solution for δ^*_{11} reduces to

$$\delta^*_{11} = \frac{A_{14}/\omega_1}{\sqrt{A_1^2 + 4\sigma_1^2}} \qquad (B3)$$

Recognizing A_1 depends on δ^*_{11} and that σ_1 is small leads to

$$\begin{aligned} A_{14} &\propto \mathcal{A}_{14} {\delta^*_{11}}^2 \\ B_{14} &\propto \mathcal{B}_{14} {\delta^*_{11}}^2 \end{aligned} \qquad (B4)$$

Proceedings of the Seventh (1997) International Offshore and Polar Engineering Conference
Honolulu, USA, May 25-30, 1997

ISBN 1-880653-28-1 (Set); ISBN 1-880653-30-3 (Vol. II)

Installation Options for Deployment of Light-Weight Telecommunications Cable

James J. Burgess
Lucent Technologies
Whippany, NJ, USA

Abstract

The mechanics of cable when deployed over up-slopes and down-slopes is studied for installations in which there is no bathymetric survey. Two installation methods are considered; 1) cable is deployed from the surface, and 2) the installation platform follows the bottom features at low altitude. The bottom following platform is shown to be the best installation method for this case, because when deploying from the surface advance knowledge of sea-floor up-slopes is required. The limitation of bottom following is in undulating terrain where there is a limit on the rate of descent the vehicle can attain without creating the condition of zero cable tension at the vehicle.

Introduction

Successful cable installations are typically preceded by careful survey of the installation route. This is done to prevent the formation of cable suspensions over sea-floor terrain. Suspensions reduce the life of cable systems because they increase the threat of fouling by fishing gear and anchors, and they expose the cable to currents that induce motion that can abrade the cable where it contacts the sea-floor. Surveying is a substantial part of cable installation expense. A deployment method that eliminates it would reduce the cost of undersea systems. Eliminating route survey would also be an advantage in tactical military applications where there may be no opportunity for route survey.

The long standing method for cable installation is deployment from surface ships. Routes are carefully surveyed, and the ship speed, and cable payout rate are carefully controlled to prevent suspensions. This requires a cable strong enough to withstand the deployment tensions which includes the cable wet weight times deployment depth plus the dynamic tension due to installation platform sea-keeping motions, and some unpredictable component due to equipment failures.

An alternative to surface deployment is deployment from an underwater vehicle. Two types of deployment can be envisioned; 1) from a vehicle in straight and level flight, 2) from a bottom following vehicle. The first alternative is not considered here. It does reduce the required cable strength, but it has the same difficulties in filling bottom terrain as surface deployment, that is, advance knowledge of sea-floor features is necessary. In theory, deployment from a bottom following vehicle, eliminates the need for prior knowledge, because by following close to the bottom most large suspensions can be eliminated.

The dynamics of cable due to sea-state are not considered in this paper. For surface deployment, sea-state induced dynamics are reasonably well understood. For sea-state less than 4-5 dynamic tension are typically about 20% of the static load. Surface deployed cables are designed with factors of safety of 3 or more above the maximum predicted static load to prevent cable breakage. In the case of undersea vehicle deployment sea-state induced dynamics are likely to have most importance in shallow water. If the vehicle were to operate in regions (and/or sea-states) in which wave induced water motion reached the bottom, then the results of this paper would require modification to account for the oscillation wave forces that would act on the cable.

Cable Suspensions

A cable suspension can be created where the water depth increases or decreases along the cable track. An increase in depth is referred to as a down-slope, and a decrease in depth is referred to as an up-slope. A suspension will form on a down-slope when the cable payout rate is not high enough to cover the increased track length down the slope. A suspension will form on an up-slope when the angle of the slope (measured downward from the horizontal plane) is greater than the cable deployment angle. In deployments that are preceded by bathymetric survey, suspensions are prevented by establishing a platform velocity and payout rate schedule such that the cable angle is always greater than up-slope angles, and the payout rate is always high enough to "fill" down-slopes. If survey data is not available then:

1. the installation platform must be capable of varying its speed, course, and payout rate in reactions to real-time depth measurements, and
2. filling of sea-floor topography will be limited by the mechanics of the cable in the water column.

Down-Slope Suspensions

Down-slope suspensions can always be prevented if the deployment platform;

1. has a depth sounder,
2. maintains a constant depth while passing over the down-slope,
3. can vary forward speed, and
4. can vary cable payout rate.

The capability to fill down-slopes can be tested by comparing the payout rate needed to fill a down-slope with the rate at which the cable tension becomes zero. A down-slope may be filled as long as the required fill rate does not cause tension to become zero. A cable without tension at the installation vehicle can not be controlled, and may foul the vessel propulsion unit.

Cable payout rate is described in terms of cable slack, which is defined by the relation:

$$\varepsilon = \frac{V_p - V_s}{V_s} \qquad 1)$$

where V_p is the cable payout rate, and V_s is the installation platform speed. The slack necessary to prevent a down-slope suspension is known as the required "fill" and is computed with the formula(Roden, 1974)

$$f = \frac{\sin\alpha + \sin\beta}{\sin(\alpha + \beta)} \qquad 2)$$

where; f is the required fill, α is the descent angle of the cable measured down from the horizontal, and β is the down-slope angle measured up from the horizontal. The angles are defined in Figure 1.

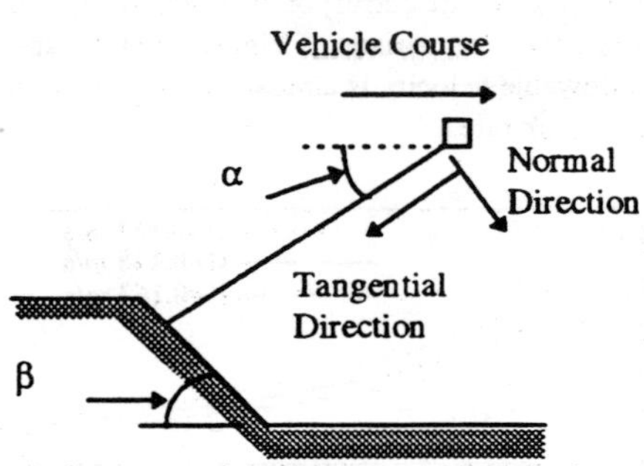

Figure 1: Definition of the cable and bottom slope angles.

Equation 2 is derived by comparing the distance traveled by the ship as it traverses the down-slope to the length of cable necessary to follow the contour of the sea-floor. As long as the cable slack, ε, is greater than the required fill, f, the cable will lie on the sea-floor without suspensions.

The angle alpha is the angle at which the cable wet weight projected normal to the cable balances with the normal component of the hydrodynamic drag. If there is no current, it may be determined by solving the equation:

$$H^2 \cos\alpha = V_n|V_n| \qquad 3)$$

where $V_n = V_s \sin\alpha$, is the normal component of the cable velocity relative to the water, H is the sink rate of a unit length of cable falling perpendicular to its length, ignoring end effects, and V_s is the platform forward speed. When there is no current, this equation may be reduced to

$$\alpha = \cos^{-1}\left\{0.5\left[\left(\frac{H}{V_s}\right)^2 + \sqrt{\left(\frac{H}{V_s}\right)^4 + 4}\right]\right\}, \qquad 4)$$

by taking advantage of a trigonometric identity.

The quantity H, when expressed in the units of degree-knot, is also referred to as the "hydrodynamic constant." It is computed with

$$H = \sqrt{\frac{w}{0.5 C_n \rho d}} \qquad 5)$$

where w is the cable wet weight, C_n is the normal drag coefficient, ρ is water density, and d is cable diameter.

To retain cable control it is necessary to keep the cable under tension. The tension in a cable will be positive as long as the cable weight projected tangent to the cable is larger than the tangential component of the hydrodynamic drag acting on the cable. If there is no current, the payout rate at which zero cable tension occurs can be determined substituting the value of cable angle determined in equation 4 into the tangential force balance,

$$w\sin\alpha = \frac{1}{2} C_t \rho \pi d V_t |V_t| \qquad 6)$$

where $V_t = V_s(1+\varepsilon-\cos\alpha)$, is the tangential velocity of the cable, and C_t is the tangential drag coefficient. With the positive direction defined as pointing away from the platform in the direction of previously deployed cable.

Assuming that ε is always positive equation 6 reduces to

$$V_s = \frac{w\sin\alpha}{(1 + \varepsilon - \cos\alpha)0.5 C_t \rho \pi d} \qquad 7)$$

Determination of the correct drag coefficient is very difficult in the analysis of undersea cable mechanics. The drag coefficients are affected by vortex-induced vibrations of the cable, and very little data has been gathered in controlled settings to establish a reliable relationship between cable properties, deployment speed and drag coefficients. Data for normal drag coefficient have typically been obtained during sea-trials, and the values reported range from 1.0 to 3.0 (Burgess, 1994, Kipp, 1989, Shashaty, 1988, Caron et. al., 1963). Data for the tangential drag coefficient have only been obtained in test tanks, and no data has been gathered for cables during deployment. The data that most closely matches the conditions of cable deployment (order of magnitude Reynolds number from 100 to 1000, with small angle of incidence to the flow) were reported by Pao et. al. who performed tests in a flume with a fixed, inclined cable a low Reynolds numbers and incidence angles between 15 and 60 degrees. They report that the tangential drag coefficient may be estimated with the relation

$$C_t = 0.720 R_t^{-0.5} \cos^{-1.12} \alpha \qquad 8)$$

where $R_t = V_t d/\nu$, the tangential Reynolds number, α is the incidence angle to the flow, and ν is the fluid kinematic viscosity.

It must be noted that during deployment the cable is not fixed, and the incidence angle to the flow is often less than 15 degrees. The influence of the cable payout, and small incidence angle on the drag has not been investigated.

If it is assumed that the deployment vessel remains at a constant depth when passing over a down-slope then equations 2, 4, and 6 can be solved for the maximum vehicle velocity at which cable can be deployed with sufficient slack to fill the bathymetry. Consider two cables: 1) a 0.01334m diameter surface deployed cable with wet weight equal to 2.3519 N/m, and 2) a 0.00127 m diameter proposed underwater vehicle cable with wet weight equal to 0.02846 N/m. In the remainder of this paper these cables will be referred to as the SD (Surface Deployed cable), and UV (Underwater Vehicle cable) respectively.

On a down slope of 90 degrees the maximum deployment speeds of the UV and SD cables are 1.2, and 21.8 m/sec. respectively. To compute these numbers, the normal drag coefficient was set to 2.0, and the tangential drag coefficient was computed with equation 8. The SD cable can be deployed at higher speeds because it has a higher sink rate than the UV cable (0.414 m/s and 0.138 m/s respectively). At any speed, it deploys at a larger angle than the UV cable, and has a larger fraction of its weight projected along the cable, so it can be deployed with more slack. The ability to deploy greater slack at a speed means that the deployment platform can fill steeper down-slopes at that speed, or it can move faster over that slope.

Larger cable sink rate does not necessarily translate to a practical improvement in the ability to fill down-slopes. The maximum deployment speeds for the SD cable are much greater than the deployment speed of a typical surface platform (3.09 m/sec. or less), so the additional sink rate of the SD cable is not essential for filling down-slopes. The UV cable must be deployed at lower speed over down-slopes but the speed is compatible with typical speeds of surface platforms. For down-slopes of less than 60 degrees the vehicle can travel at over 3.09 m/sec., while over a vertical down-slope the vehicle must slow to about 1.2 m/sec. Down-slopes can be filled by an underwater vehicle, if the vehicle speed is properly controlled. The higher sink rate of the surface cable is not essential for filling down-slopes.

The sensitivity of UV cable maximum down-slope deployment speed to variations in the sink rate, and tangential drag coefficient can be computed with equation 7. Results are shown in Figures 2 and 3. Figure 2 shows the maximum velocity that cable can be deployed over bottom slopes ranging from 30 to 90 degrees, for three values of the coefficient in Equation 8. The coefficient in Equation 8 is raised and lowered by 10 %. Reducing the tangential drag coefficient increases the speed at which cable can be deployed, more so at low slope angles than at high. At low angles the speed increases 2 m/s across the range, and at high angles the speed increases about 0.5 m/s.

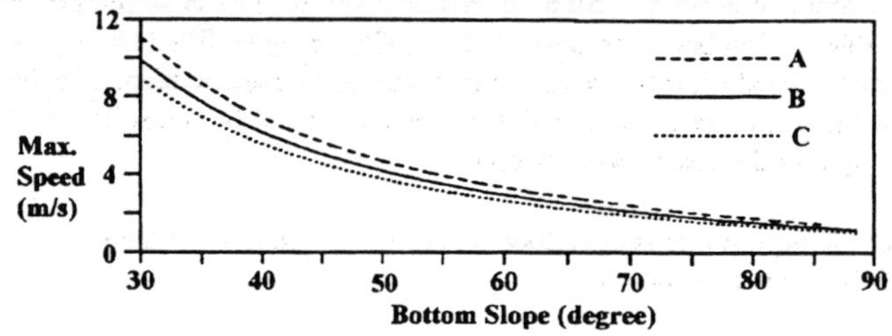

A: $Ct = 0.684 * R_t^{-0.5} \cos^{-1.12}(\alpha)$
B: $Ct = 0.720 * R_t^{-0.5} \cos^{-1.12}(\alpha)$
C: $Ct = 0.792 * R_t^{-0.5} \cos^{-1.12}(\alpha)$

Figure 2. Sensitivity to tangential drag coefficient of the maximum platform speed at which downslopes can be filled with the UV cable: C_n=2.0, w = 0.285 N/meter, d=0.127 cm.

The sensitivity of the maximum velocity to the sink rate is shown in Figure 3. The sink rate is varied by plus and minus 10% of the sink rate of the UV cable. Reducing the sink rate decreases the speed at which cable can be deployed. The decrease in speed is larger at low slope angles than high slope angles (about 3 m/sec. at 30 degrees, and about 0.75 m/sec. and 85 degrees). The sink rate can be changed by altering cable wet weight, diameter, or normal drag coefficient. The 10% variation in sink rate corresponds to variation in these parameters of 21%. Comparison of Figures 2 and 3 show that the maximum allowable velocity is almost as sensitive to the tangential drag as it is to sink rate.

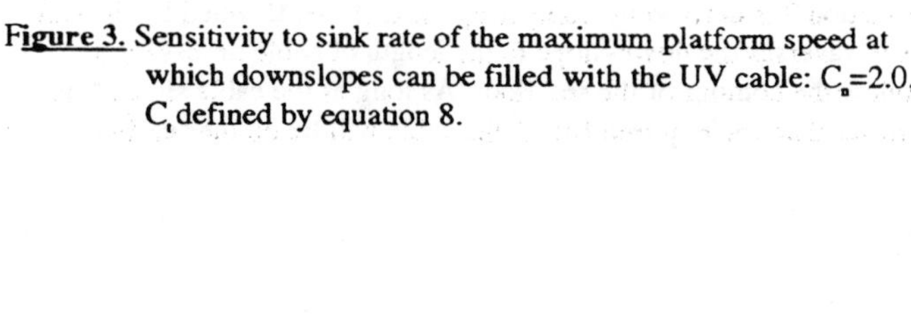

Figure 3. Sensitivity to sink rate of the maximum platform speed at which downslopes can be filled with the UV cable: C_n=2.0, C_t defined by equation 8.

An underwater deployment platform should have bottom following capability. Bottom following would allow the filling of up-slopes (as discussed below), but will introduce an additional complication on the down slope. The descent of the vehicle increase the normal component of cable drag. The cable will deploy at smaller angles to the horizontal, and cable tension will go to zero at lower values of the cable slack, than in straight and level flight.

When the vehicle descends at an angle θ measured down from the horizontal plane, the tangential and normal components of velocity relative to the cable are given by

$$V_t = V(\sin\theta\sin\alpha - \cos\theta\cos\alpha) \qquad 9)$$

$$V_n = V(\cos\theta\sin\alpha + \sin\theta\cos\alpha) \qquad 10)$$

Assume that the vehicle deploys cable with no slack during a descent, and that the cable is not restrained by friction on the sea-floor. Equations 7 and 8 may be inserted in Equations 3 and 6, and the maximum descent angle for a vehicle at a given speed, with a specific cable can be determined. The descent angle so derived represents the maximum steady state descent angle. Figure 4 shows a plot of the maximum descent angle possible with the UV cable over a range of vehicle speed from 0.5 to 3.75 m/sec.. It is seen that as the vehicle speed is increased the allowable descent angle decreases. The descent angle increases rapidly to a value of 90 degrees as the vehicle speed approaches the cable sink rate. The limitation on vehicle descent angle does not effect the ability to fill down-slopes (by slowing the vehicle, and reducing descent angle, any down-slope can be filled), but it will effect the ability to fill up-slopes in undulating terrain, because the vehicle may not reach the base of the up-slope before it must climb. For example, assume the UV cable is deployed from a vehicle that has a minimum speed of 0.5 m/sec (because of a need to generate hydrodynamic force on control surfaces), and series of sinusoidal shaped sea-floor hills of height 10 meters (amplitude 5 meters) are encountered.

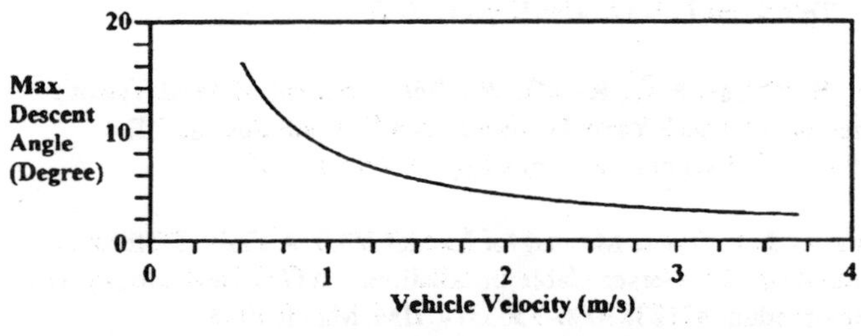

Figure 4: Maximum vehicle descent angle that maintains positive cable tension as a function of vehicle speed.

The maximum vehicle descent angle from figure 4 is about 16 degrees. If the vehicle is maintained at an altitude of 10 meters above the sea-floor, then the vehicle can fill the up-slopes of the hills only if the half wave length of the hills is greater than about 73 meters. This is illustrated in Figure 5.

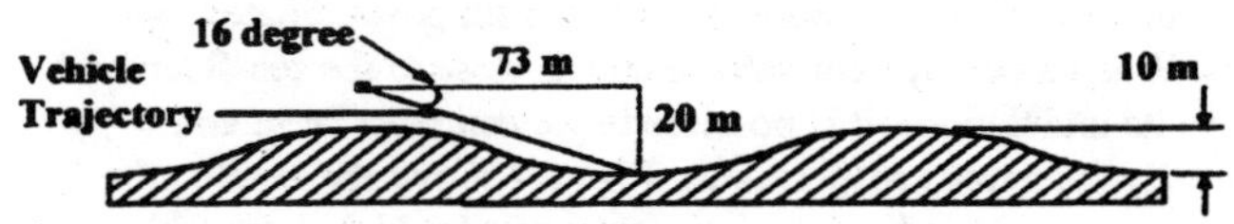

Figure 5: Illustration of cable deployment limitation in undulating terrain due to limitation on maximum vehilce descent angle.

Up-Slope Suspensions

When no preliminary survey is conducted, and the deployment platform maintains a constant depth, it is not possible to prevent up-slope suspensions over unexpected obstacles, without stopping and recovering cable. To prevent an up-slope suspension the platform speed must be set so that the cable angle is greater than the up-slope angle. If the vessel is slowed (to increase the angle) upon detecting an up-slope in excess of the cable angle, the new cable shape is transmitted down the cable at a speed roughly equal to the cable sink-rate. If the deployment vessel slows at the base of an up-slope, and resumes speed at the top, a suspension will be created. This is illustrated in Figure 6 that shows a simulation (Burgess,1993) of surface deployment of the SD cable in a expected water depth of 100 meters. The deployment vessel encounters a very large obstacle and slows down to change the cable angle. The obstacle is a ridge that has 45 degree up-slope, and rises from a depth of 100 meters to 50 meters, and has a ridge top that is 150 meters wide. The figure shows "snap-shots" of the cable profile every 20 seconds. In the figure the deployment vessel slows at the base of the obstacle to a speed that will match the cable angle to the up-slope , at the top of the slope the vessel resumes its original speed. It is seen that the segment of the cable with the correct slope is too short, and arrives at the sea-floor too late (too far down track) to prevent a suspension. The resulting suspension spans about 375 meters. Any surface deployment that is not preceded by accurate bathymetric survey is likely to create long and undesirable suspensions.

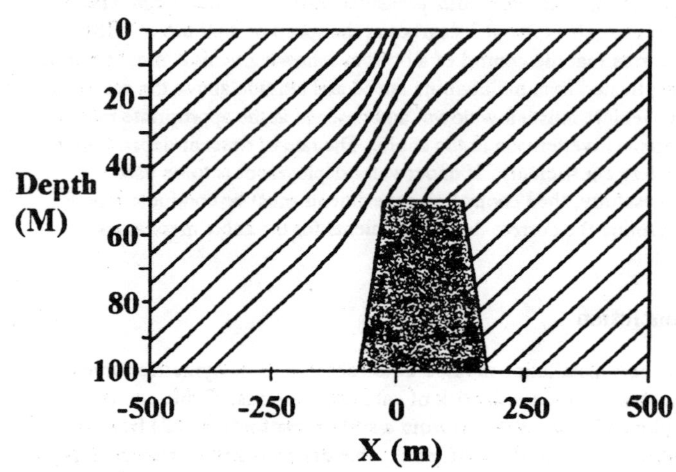

Figure 6: Deployment of SD cable pver an unexpected ridge at a speed of 3.09 m/sec, with 1% cable slack.

A bottom following underwater vehicle can fill unanticipated up-slopes. When a deployment vehicle travels close to the sea-floor, and follows its undulations, it is possible to prevent most large and potentially dangerous suspensions. Figure 7 shows the results of a simulation in which a hypothetical deployment vehicle lays cable over the same obstacle used above, at a speed of 3.09 m/sec., and 1% slack. Each line in the Figure is a "snap-shot" of the cable geometry, the snapshots are separated by a time of 20 seconds. The vehicle maintains an altitude of 10 meters above the se-floor, and it has a vertical turn radius of 30 meters. It also has a forward looking sonar with a range of 20 meter. It can be seen from the figure that no suspensions are created over this obstacle.

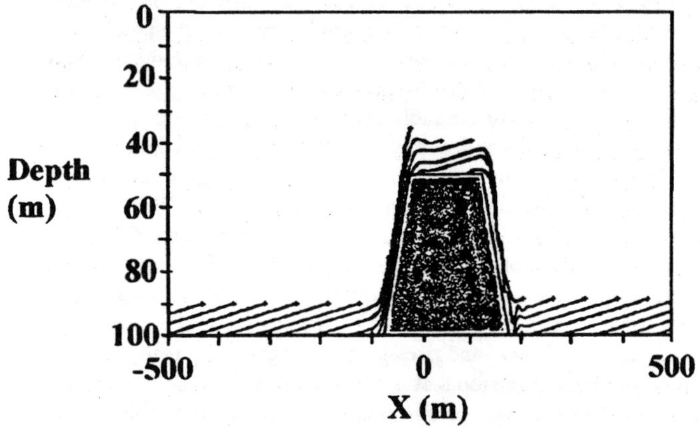

Figure 7: Deployment of UV cable from an ideal bottom following vehicle over an unexpected ridge at a speed of 3.09 m/sec., with 1% cable slack.

Close inspection of Figure 7 reveals a region in which the cable has "buckled" near the base of the down-slope. The "buckling of the cable indicates that the tension has gone to zero. In this simulation the zero tension occurs because the vehicle is descending at 45 degrees, while it is known from the results presented in figure 4 that the maximum descent angle is about 3 degrees at the speed of 3.09 m/sec.. This confirms that the control of a bottom following vehicle must be more complicated than maintaining speed and altitude above the sea-floor. The vehicle must slow down, and descend at the appropriate rate, to keep positive tension in the cable. The rate of descent depend upon the drag coefficients. If underwater deployment is to be used successfully, then controlled experiments must be conducted to determine the correct drag coefficients for the cable mechanics models.

Conclusion

The best installation method in the absence of bathymetric survey is one that minimizes the risk of cable suspensions. Cable can be deployed in two ways; 1) from a surface platform, and 2) from an underwater vehicle. A different cable design is need for each of these methods. Surface deployment requires a cable strong enough to withstand the deployment loads of the static cable weight, plus sea-state induced dynamic tensions. Underwater deployment requires a small diameter, light weight cable that can easily be stowed on an underwater vehicle.

The higher sink rate of a surface deployed cable (due to the need for strength) makes it better for filling down-slopes, but the low sink rate of an underwater vehicle cable does not restrict the ability to fill down-slopes. The vehicle speeds and allowable payout rates are consistent with typical cable deployment speeds.

When an up-slope is encountered the underwater vehicle has a distinct advantage over surface deployment. From the surface, up-slope suspensions can only be prevented by changing platform speed well in advance of the up-slope. Prior knowledge of the up-slope is required to prevent a suspension. An underwater vehicle that follows bottom contours can fill the same features.

The results demonstrate that deployment from an underwater vehicle is the best method for deploying cable when there is not route bathymetric survey. The limitation of this method will be in undulating terrain where the need to keep cable under tension at the platform will limit the platform descent angle, and therefore its ability to fill the undulation up-slope.

References

Roden, C. E. Submarine Cable Mechanics and Recommended Laying Procedures, AT&T Bell Labs Technical Publications Department, Homdel, N.J. 1974

Burgess, J. J., "Simulations of the Array Sea-Trial Cable Installations," AT&T Bell Laboratories Memorandum #ITD-94022276M, May 1994

Kipp, C. R., "ULSS Deep Water Trunk Cable Hydrodynamic Constant Determination," AT&T Bell Laboratories Memorandum # 46234-890320-01IM, March 1989

Shashaty, A. J., "Hydrodynamic Constant of 0.450 Cable," AT&T Bell Laboratories Memorandum # 46234-880602-01IM, June 1988

Caron, C. S., Hale, A. L., Roden, C. E., "SD Hydrodynamic Trial," Bell Telephone Laboratories Report, 1963

Pao, H. P., Ling S. C., Kao, T. W., "Measurement of Axial Viscous Force on a Smooth Yawed Cylinder in a Uniform Stream," The Catholic University of America Report, March 1990

Burgess, J. J., "Users Manual for Imodel-V3.0, A Finite Difference Simulation of Undersea Cable Installation,", AT&T Bell Laboratories Memorandum #712383000-930331-01IM, March 1993

Proceedings of the Seventh (1997) International Offshore and Polar Engineering Conference
Honolulu, USA, May 25-30, 1997

ISBN 1-880653-28-1 (Set); ISBN 1-880653-30-3 (Vol. II)

Axial Fatigue of Spiral Strands in Offshore Platform Applications

M. Raoof
Loughborough University
Loughborough, United Kingdom

M. Alani
University of Derby
Derby, United Kingdom

ABSTRACT

Steel cables are used for mooring certain types of offshore platforms such as semi-submersibles, guyed towers, etc, and are proposed as elements of the mooring systems for wave energy devices. Their various characteristics (especially under cyclic loading) are, therefore, of prime importance as regards a safe, economical, and functional performance of such structures.

In a series of previous publications, a theoretical model capable of predicting axial fatigue life of multi-layered spiral strands from first principles, is reported. The analytical model is an extended version of a previously reported orthotropic sheet model. The theoretical predictions for some substantial spiral strands were backed by large scale experimental data.

Recently, some very carefully documented axial fatigue test data on 25 and 45mm.O.D. spiral strands were reported by the team in University of Alberta. The important point about this data is the extensive reporting of the number and position of individual wire fractures along the test specimens, and (unlike the vast majority of other published data in the literature) the full reporting of changes in a number of other properties of the cables (such as their axial stiffness) during the axial cyclic tests, with a careful description of the mode(s) of failure. In the present paper, the encouraging support for the theoretical predictions by this recently reported set of axial fatigue test data is presented.

The effect of mean axial load on the axial fatigue life is subsequently examined, using an extensive series of theoretical parametric studies. The parametric studies were carried out using 3 types of realistic 127mm.O.D.. spiral strand constructions with lay angles of 12, 18.01 and 24 degrees (which cover the current manufacturing limits). Using the results based on the parametric studies, the plausibility of using a simple Goodman (or Gerber) transformation for steel cables, as originally advocated by others, is critically examined. It is demonstrated that contrary to previous claims, presenting the axial fatigue test data in terms of an equivalent load range (with a minimum load of zero), rather than the traditional use of stress range, does not lead to any improvements in the fit of the data to a straight line. Indeed, it is theoretically shown that for a wide range of mean axial loads, the plots based on the stress range suffer from less scatter of data about the fitted line, and despite the extra efforts involved, the concept of equivalent load range does not offer any practical advantages.

KEY WORDS: Spiral strand, wire rope, fatigue, axial load, offshore platforms, bridges

INTRODUCTION

Steel cables are used for mooring certain types of offshore platforms such as semi-submersibles, guyed towers, etc., and are proposed as elements of the mooring systems for wave energy devices. The loading spectrum of the individual mooring lines is obviously very complex: it depends mainly upon the type of structure, its location, and the type of mooring system adopted. The most significant aspects of the service conditions are long lives (in excess of maybe twenty years), and the random nature of the imposed loading. Permanent immersion of most of the cable in sea water is another important consideration.

A catenary moored platform, for example, will typically have a pair of mooring lines at each corner of the structure. Such platforms maintain position and derive stability through their mooring lines, which are pretensioned to restrict motion amplitudes. Platform motions will vary in response to wave, wind, and current loadings. Such movements will result in tension fluctuations (and, hence, axial fatigue failures) in the mooring lines whose catenary shape can change significantly resulting in also restrained bending fatigue failures around the fairlead pulleys and other potential sources of restrained bending.

In the past, the only reliable source for obtaining information on steel cable (wire rope and/or spiral strand) fatigue characteristics has been in the form of laboratory tests on cable specimens which can be very costly and time consuming. The results of such tests often exhibit considerable scatter and the number of variables involved makes the interpretation of the extensive work by manufacturers and technical societies rather difficult. Divergence of opinion on various aspects of cable behaviour under even closely controlled laboratory conditions is not uncommon. Even for the same diameter and loading, different cable constructions may exhibit significantly different fatigue lives. In published work, there seems to be an undue amount of repetition and imprecise conclusion. The conflicting desires for publicity and commercial secrecy are also evident.

The growing need for progressively larger diameter spiral strands (and wire ropes) has necessitated a re-examination of the traditional methods of analysis and design, which are largely based on commercial experience relating to significantly smaller diameter cables. It is, then, obvious that extrapolation of test data relating to much smaller diameter cables to predict the performance of much larger diameter specimens is a very risky process.

In a series of publications, the first author and his associates have put a substantial analytical and large scale experimental effort into assessing contact forces and the associated relative displacements between wires, taking full account of frictional effects, in multilayered large diameter spiral strands and wire ropes, where a spiral strand is composed of a group of helically laid wires with a common axis. The term rope, on the other hand, is applied to an assembly of a number of helical strands in one or more helical layers over a central core which may in itself be made up of a fibre or an independent wire rope, with the choice being largely dependent on the use for which the rope is intended. The theoretical basis of the work has been fully discussed elsewhere, in the context of the work of others, in particular Costello and his associates (Costello, 1990). For the present purposes, much attention will be devoted to the axial fatigue life prediction of multi-layered spiral strands with particular attention paid to the influence of mean axial load on spiral strand axial fatigue life.

BACKGROUND

Extensive surveys of the literature relating to the axial fatigue of steel cables have already been reported by Raoof (1991a,b, 1992) and Chaplin and Potts (1991), and there is little point in repeating such a survey here. For the present purposes, however, it is felt appropriate to merely mention a number of important factors relating to axial fatigue life prediction of steel cables which should prove of value in terms of explaining the main features of the theoretical work to be discussed later in this paper.

The criterion of fatigue failure for steel cables is of necessity more complex than that applied to continuous structures, where crack length measurements or a simple observation of loss of integrity may suffice. Discard criteria are based on a mixture of past experience, personal preferences, and national or international standards for each particular type of rope application. The occurrence of an unacceptable number of wire breaks is, by far, the most common measure adopted for wire rope fatigue damage assessment. This, however, is not free from serious pitfalls. Wire breakages may be internal and/or external, and reliable methods for detecting internal wire failures (under service conditions, especially in underwater applications) have still to be developed, although some progress has been made in fairly recent years. The situation is complicated further because multiple breaks may occur along any individual wire as a result of significant levels of internal friction.

End terminations often concentrate very significant numbers of wire fatigue breakages in their vicinity. As a result of this, dramatic reductions occur in the fatigue life of the cable. In published work, the length of the test specimen is often too short so that the sample is dominated by end effects. Even for the same type of end termination, cable fatigue life is dependent on the skill of the craftsman making the termination.

In published work, a number of axial fatigue tests on a wide variety of cable constructions have been reported. However, much of the test data is of limited practical value for correlation purposes. Indeed, some is even misleading. The primary reason for this unsatisfactory state of affairs is that there has been little or no uniformity in test parameters between the reported test programmes. Samples of different length with a variety of end terminations from different manufacturers who use different cable constructions with widely varying wire material properties often exhibit large degrees of scatter.

Finally, the degree of scatter even for nominally identical tests carried out on specimens from the same batch is often quite significant (Raoof, 1991a).

Experimental constant axial fatigue loading on cables is generally in the form of a nominally sinusoidal stress perturbation with an amplitude $S'_a = \frac{1}{2}(S_{max} - S_{min})$, superimposed on a steady mean axial stress S_m. The results relating to the number of fatigue cycles to failure, N, are usually presented in terms of two variables, for example, S_{max} - N (or S_a - N, where the stress range $S_a = S_{max} - S_{min}$) curves for various values of the ratio R, which is defined as $R = \frac{S_{min}}{S_{max}}$. To take the effect of mean axial load on axial fatigue life into account, Yeung and Walton (1985) followed by Chaplin and Potts (1991) have, in the absence of any other more reliable technique, used equations of the form

$$\frac{S_{eq}}{S_u} = \frac{\frac{S_a}{S_u}}{1-\left(\frac{S_m}{S_u}\right)^n} \qquad (1)$$

in replotting some of the limited available data on spiral strand axial fatigue life, with a view to reducing some of the scatter in composite plots of the experimental results. In Equation (1), S_a is the range of applied axial stress (ie. twice the amplitude) and S_m is the mean. S_{eq} is an equivalent stress analogous to those defined by Goodman and Gerber who originally proposed a similar technique for plain specimens (as opposed to helically laid cables): in Equation (1), Goodman used n=1 and Gerber took n=2. Obviously, Equation (1) attempts to obtain a single curve for all possible combinations of S_m and S_a, for each given strand construction with the provision that the transformation is done to a range with a minimum load of zero (Yeung and Walton, 1985), Fig.1. Figs 2a and 2b (after Chaplin and Potts, 1991), are intended to illustrate the effect of plotting in terms of an Equivalent Load Range to reduce the spread caused by mean load.

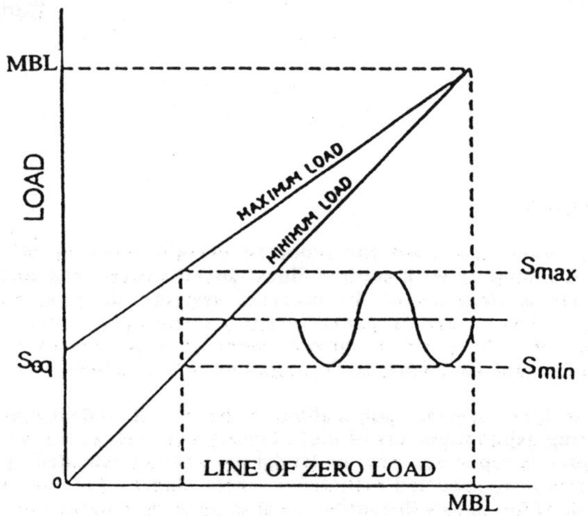

Fig. 1 Concept of 'Equivalent Load Range' applied to fatigue assessment of wire ropes - after Yeung and Walton (1985).

However, it is interesting to note that in Fig. 2b (which is based on n=1 in Equation (1)), the scatter in the fatigue results is found to increase with decreasing levels of the Equivalent Load Range (ie. increasing fatigue life). Chaplin and Potts offer no explanation as to why this happens, and one cannot (in the absence of a sound theoretical model) ascertain whether this finding is due to the peculiar nature of the plotting technique or whether experimental uncertainties due to factor(s) such as definition of fatigue life, length of the specimens and termination effects, etc, may be responsible for such a scatter.

Among a number of detailed theoretical and experimental studies of various characteristics of spiral strands, Raoof (1990,1991a) has addressed the axial fatigue behaviour of large diameter multi-layered spiral strands where, using results from a previously reported orthotropic sheet concept (Hobbs and Raoof, 1982, Raoof and Hobbs, 1988), an insight is given into the pattern of normal (Hertzian) stresses throughout realistic strand constructions. Based on such information in conjunction with available axial fatigue data on single component wires, an analytical model is developed which is capable of predicting strand axial fatigue life from first principles. The so-developed model is capable of throwing some light into various cable characteristics of practical importance. The match between experimental data on some substantial spiral strands with diameters ranging from 39-127mm and theoretical predictions has been shown to be very encouraging (Raoof, 1990, 1991a, Alani and Raoof, 1995, Raoof, 1996a). The theory enables one to predict both internal and/or external first wire fractures which may either happen at the end terminations or in the free field (ie. sufficiently away from the ends so that the end effects are minimal).

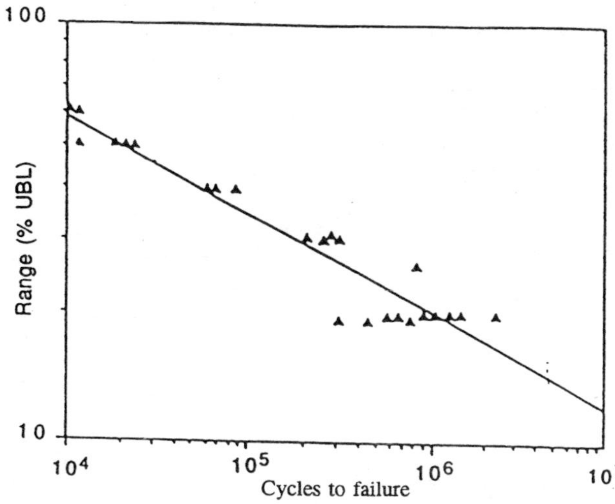

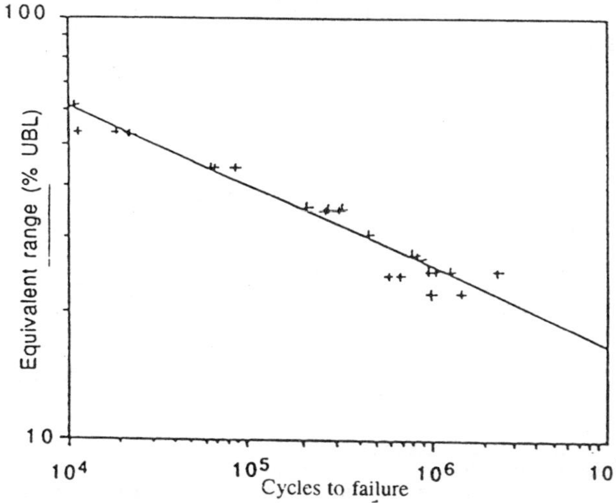

Fig. 2 Tension-tension results for 19 mm six strand rope illustrating the effect of plotting in terms of an Equivalent Load Range to reduce spread caused by mean load - after Chaplin and Potts (1991) : (a) life plotted against actual load range; (b) life plotted against Equivalent Load Range.

PURPOSE OF PRESENT PAPER

Recently, some very carefully documented axial fatigue data on 25 and 45mm diameter spiral strands were reported by Papanikolas (1995). Among other things, the number and position of individual wire fractures along the test specimens and the changes in a number of other cable characteristics (such as their axial stiffness) during uniform axial cyclic tests, with a careful description of the assumed mode(s) of failure, were all reported. There is little point in repeating such findings here, as they are all fully reported by Papanikolas (1995). For the present purposes, the encouraging support for the axial fatigue theory of the first author provided by all the axial fatigue data of Papanikolas will be reported: such test data will provide even further support for the theoretical model, the predictions of which have already been shown to be supported by a wide range of test data obtained on spiral strands with widely different constructions and wire material properties, produced by different manufacturers, and tested at a number of different institutions.

The effect of mean axial load on the axial fatigue life of spiral strands is subsequently examined, using an extensive series of parametric studies, based on the proposed theoretical model. For this purpose, a cable manufacturer has provided the first author with construction details as reported elsewhere (Raoof, 1996b) of 3 types of realistic multi-layered 127mm O.D. spiral strands with lay angles of 12, 18.01 and 24 degrees (which cover the current manufacturing limits), with the lay angle kept constant for all the individual layers throughout each spiral strand construction. Their other constructional details such as number and diameter of wires in various layers, etc, have intentionally been kept nominally the same: following previous findings of the first author (Raoof, 1991c, 1996b), lay angle is the primary geometrical parameter controlling various characteristics of spiral strands.

Using such theoretical parametric studies, the plausibility of using a simple Goodman (or Gerber) transformation for spiral strands, as originally advocated by others, is critically examined.

As a prerequisite to the above, however, it is perhaps useful to describe the salient features of the theoretical model as previously reported by Raoof (1990, 1991a). This will provide the reader with a fairly sound understanding of the basis for the final numerical results.

THEORY

Using the orthotropic sheet model (Hobbs and Raoof, 1982, Raoof and Hobbs, 1988), it is now possible to obtain reliable estimates of interwire contact forces (and stresses) throughout multi-layered structural stands. Experimental observations suggest that individual wire failures are largely located over the trellis points of interlayer contact and it is now believed that this is as a result of high stress concentration factors in these locations.

Once the maximum effective Von-Mises stress $\bar{\sigma}'_{max}$, over trellis points of contact for a given strand mean axial load is calculated, the stress concentration factor, K_S, is defined as (Knapp and Chiu, 1988).

$$K_s = \frac{\bar{\sigma}'_{max}}{\bar{\sigma}'} \qquad (2)$$

where, $\bar{\sigma}'$ is the nominal axial stress in the wire (corresponding to the strand mean axial load), which may be calculated using the method developed by Raoof and Hobbs (1988).

Using the so-obtained values of K_S in conjunction with axial fatigue data on single wires, a theory has been developed which predicts the axial fatigue life of strands (under constant amplitude cyclic loading) from first principles.

The model is based on the observation that for carbon steel wires, the fatigue stress-number of cycles plot (S-N curve) possesses an endurance limit, S', below which no damage occurs. Traditionally the magnitude of S' is compared to the ultimate wire strength S_{ult}: tests on single galvanised wires suggest an approximate value of $S' = 0.27\ S_{ult}$. The reduced magnitude of the endurance limit, S_e, which takes interwire contact and fretting plus surface conditions and size effects etc. into account, may be defined as

$$S_e = K_a K_b S' \qquad (3)$$

where, $K_b = \dfrac{1}{K_s}$ and K_a is constant.

The so-obtained values of the parameter S_e, then, are used to produce the S-N curves for fatigue life to first outer (or inner) wire fractures in spiral strands using the S-N curves available in the literature for axial fatigue life of individual wires of a given grade (Raoof, 1990, 1991a).

Fig. 3 presents the correlations between theoretical predications and axial fatigue test data on a 51 mm 1X139 strand as obtained from Bridon Ropes. It should be noted that with epoxy resin end terminations, the majority of initial wire fractures were found to occur away from the ends, and the scatter in the test data is found to be covered by a K_a factor within the range $0.5 \le K_a \le 1.0$, with fatigue life defined as the number of cycles to first outer layer wire fracture.

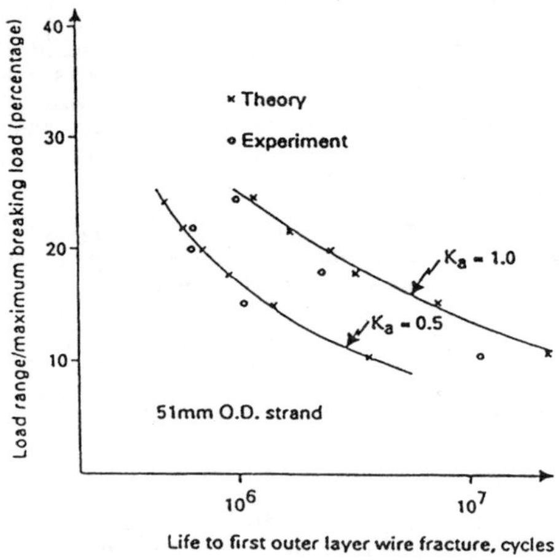

Fig. 3 Comparison of theory and test data for axial fatigue life of a 51 mm diameter spiral strand - after Raoof (1990).

As discussed at some length elsewhere (Raoof and Hobbs, 1994), for the end terminations to have no effect on the wire fractures, the minimum length of a test specimen should be around 10 lay lengths, with the wire fractures happening within the central region which extends by 2.5 lay lengths on either side of the middle section of the test specimen. In order words, wire fractures within a distance of 2.5 lay lengths from either end are influenced by the end effects. For the test results relating to the 51mm diameter strands, because of the short length of test specimens (much less than 10 lay lengths), certain wire fractures (although happening away from either end) have obviously been influenced by the detrimental end effects, suggesting $K_a = 0.5$ as an appropriate factor in the presence of end effects. Otherwise, for wire breakages which are found to happen away from either end, and, in addition, are not influenced by the end effects, one may assume $K_a = 1.0$.

RESULTS

Further Experimental Support for the Theoretical Model

An investigation of the axial fatigue strength of multi-layered spiral strands was recently reported by Papanikolas (1995). In his work, the effects of stress range, cable construction, and length to diameter ratio were experimentally investigated using sixteen full-size multi-layered wire strands. The strands tested were of 45 or 25mm diameter, had a length-to-diameter ratio of 75 to 133, and were subjected to a stress range between 15% and 30% of the ultimate tensile strength of the strands. Tables 1 and 2 present the construction details for the 45 and 25mm spiral strands, respectively. Table 3 presents the testing parameters for the axial fatigue test program of Papanikolas (1995).

Table 1 Strand make-up for the 45 mm diameter cable

Layer	No of Wires	Lay Direction (1)	Wire Diameter (mm)	Lay Length (mm)	Strand O.D. (mm)	Lay Angle (Degree)
1	30	LL	4.034	506	44.37	14.06
2	24	RL	4.034	395	36.27	14.38
3	18	LL	4.034	301	28.29	14.21
4	12	RL	4.034	206	20.16	13.82
5	6	RL	4.034	100	12.29	14.54
6	1	-	4.315	-	4.315	-

(1) LL = Left lay RL = Right lay

Table 2 Strand make-up for the 25 mm diameter cable

Layer	No of Wires	Lay Direction (1)	Wire Diameter (mm)	Lay Length (mm)	Strand O.D. (mm)	Lay Angle (Degree)
1	12	LL	5.056	263	25.34	13.62
2	6	RL	5.056	161	15.43	11.44
3	1	-	5.260	-	5.260	-

(1) LL = Left lay RL= Right lay

Table 3 Testing parameters for the axial fatigue test program

Specimen	Diameter (mm)	Length (1) (mm)	Max Stress (MPa)	Stress Range (MPa)
SP1	44.37	3315	505	303
SP2	44.37	3323	540	371
SP3	44.37	3336	540	371
SP4	44.37	3320	540	371
SP5	44.37	3323	505	303
SP6	44.37	3310	522	337
SP7	44.37	3328	522	337
SP8	44.37	3326	556	404
SL1	25.34	3357	517	310
SL2	25.34	3376	517	310
SL3	25.34	3379	465	207
SL4	25.34	-	517	310
SL5	25.34	3387	542	362
SS1	25.34	2153	542	362
SS2	25.34	2160	491	258
SS3	25.34	2152	517	310

(1) The length is measured from inside-to-inside of the sockets

In his test program, Papanikolas used various criteria for terminating the axial fatigue tests: these included deterioration of modulus of elasticity, complete fracture of the test specimen, and, in some cases, occurrence of three broken wires. For the 45mm diameter strand, two specimens were tested to complete destruction. For six other specimens, testing was stopped after there was a large number of wire breaks and there was a significant deterioration of the modulus of elasticity. For the eight 25mm diameter strands tested, six were tested until a minimum of three wires (out of 19) were broken. In the other two cases, the tests were terminated before any wire breakages. Location of wire breaks were all carefully reported with the internal wire breakages recorded after disassembling the specimens. Space limitations do not allow presenting such data here. Indeed, there is little point in doing so, particularly because these are all fully reported by Papanikolas (1995). It will, perhaps, suffice to say that, for example, in the case of the 45mm diameter specimens, the average number (with a rather high standard deviation) of wire breakages in specimens Sp2, Sp4, and Sp7 were 22.5%, 12.3%, 13.6%, 41.1%, 9.3%, and 1.3% in layers 1, 2, 3, 4, 5, and 6, respectively, where, outer layer is assumed to be layer 1.

For the 25mm diameter strand which had 63.2% of the wires located in the external layer, examination conducted on five specimens (that were later disassembled) showed that 89% of the wire breaks occurred in the outer layer. Finally, based on examination of the disassembled specimens, it was observed that wire breaks were distributed more or less randomly along the length of the cable.

Figs. 4a,b present the correlations between theoretical predictions and axial fatigue data of Papanikolas (1995) for the 25mm.O.D. spiral strand. The ultimate tensile strength for the wire material $S_{ult} = 1631$ N/mm^2, and the ultimate breaking load of the strand is 596 kN. The theoretical model predicts the fatigue life to first wire fracture in the outer layer and/or the second (penultimate) layer (ie. layer 2), with the fatigue life for the experimental data relating to the 5% wire breakage, which in this case, also relates to the first wire breakage. In Fig. 4a, $K_a = 1.0$, and $K_a = 0.5$ is assumed for the theoretical plots in Fig. 4b.

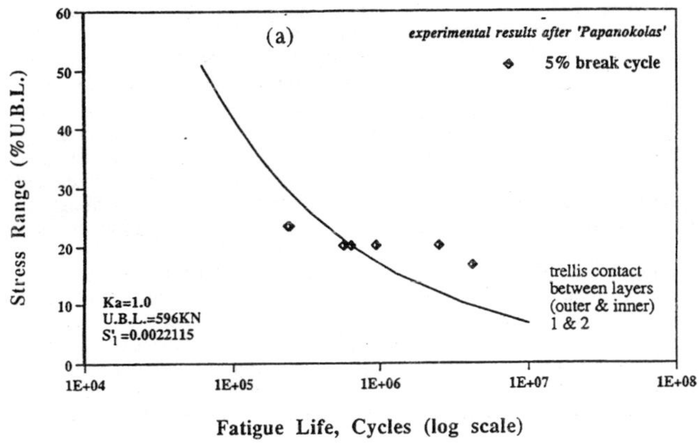

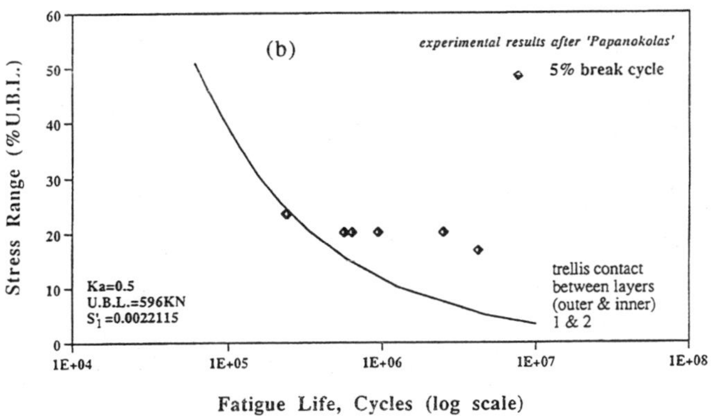

Fig. 4 Correlations between the theoretical predictions and axial fatigue test data of Papanikolas (1995) for the 25 mm diameter spiral strand : (a) K_a = 1.0; (b) K_a = 0.5.

The rather wide scatter of test data which relate to specimens tested in the same test rig, with identical construction and end termination details, using the same grade of wire, and fairly similar lengths, is noteworthy. All specimens were subjected to the same mean axial load of 362 kN, with the tests covering a rather narrow range of stress range (expressed as a percentage of Ultimate Breaking Load). Admittedly, the presently adopted failure criterion for the test data of 5% wire fracture (which, as mentioned before, is in this case equivalent to the occurrence of first wire fracture) is by no means a universally accepted failure criteria. For such case of a small diameter cable, 5% (ie. first) wire fractures may suffer from secondary effects such as presence of, for example, accidental damage in the individual wires which can lead to premature failures. Nevertheless, it is encouraging to find that the theoretical plot using K_a = 0.5, Fig. 4b, provides a very reasonable lower bound solution to all the experimental results, as it takes the influence of end effects, and other uncertainties as regards the surface conditions, size effects, etc., at the trellis points of interlayer contact, into account.

Figs. 5a,b present the correlations between theoretical predictions and axial fatigue test data relating to the 45 mm diameter strand. The ultimate tensile strength of wire material S_{ult} = 1667 N/mm^2, and in Fig. 5a, the assumed Ka = 1.0, while K_a = 0.5 is used for producing the theoretical data in Fig. 5b. The ultimate breaking load of the cable is 1842 kN, and all the cables were tested under a mean axial load of 354 kN. The theoretical S-N curves for the trellis contact points between all the layers for this spiral strand are presented in Figs. 5a,b, with the theoretical predictions relating to the axial fatigue life to first wire fracture. As regards the experimental data, the number of cycles to, both, the first wire fracture and also 5% wire fracture, are presented, where an examination of the individual test data suggests that the number of fatigue cycles between the first and 5% wire fractures can vary quite considerably from one test to another. Yet again, the range of stress range (expressed as a percentage of cable Ultimate Breaking Load) adopted in the test program, is rather narrow. The theoretical curve relating to the first wire fracture in the innermost layer is found to provide an accurate lower bound to all the test data relating to 5% wire fracture. Only two individual experimental results relating to the first wire fracture are found to give number of axial fatigue cycles lower than this theoretical solution. It is, perhaps, worth mentioning that compared to the test data for the 25 mm diameter strand, the experimental data relating to the 45 mm diameter cable are found to exhibit relatively less scatter as regards fatigue life to first wire (and also 5% wire) fracture. This observation, however, may just be coincidental.

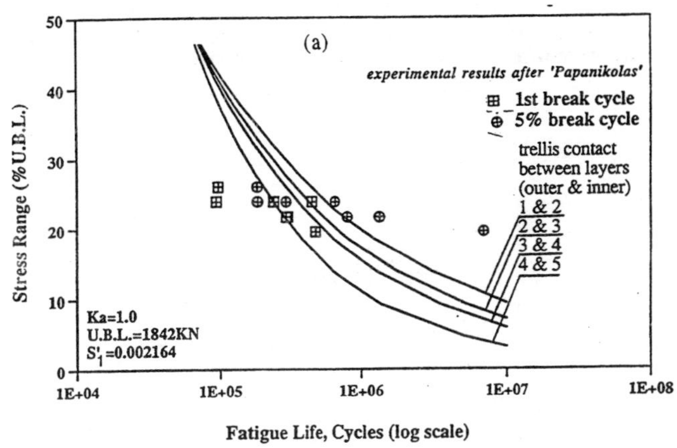

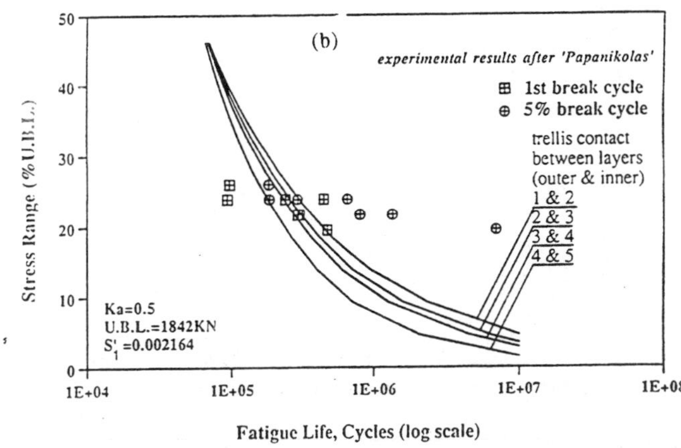

Fig. 5 Correlations between the theoretical predictions and axial fatigue test data of Papanikolas (1995) for the 45 mm diameter spiral strand : (a) K_a = 1.0; (b) K_a = 0.5.

Coupled with the previously reported encouraging support provided by large scale experimental data for the predictions of the theoretical model (Raoof, 1990, 1991a, Alani and Raoof, 1995, Raoof, 1996a), it may reasonably be concluded that the theoretical model is a reliable one, and may be used for predicting the axial fatigue life to first outer (or inner) wire fracture of any spiral strand construction.

Effect of Mean Axial Load

In what follows, results based on an extensive series of theoretical parametric studies using the previously mentioned three construction details for 127 mm diameter spiral strands with lay angles of 12, 18.01, and 24 degrees are reported: using such results, then, the potential shortcomings of using Goodman and/or Gerber transformations to cater for the variations in the mean axial load under axial fatigue conditions, will be critically examined. Full details of the parametric studies are already reported elsewhere (Alani and Raoof, 1997), and for the present purposes, only the salient features of the work will be reported.

Figs. 6a-c present theoretical plots of Equivalent Load Range (expressed as a percentage of Ultimate Breaking Load, U.B.L.) against axial fatigue life to first outer layer wire fracture, for all the three 127 mm O.D. spiral strand constructions with lay angles of $\alpha = 12$, 18.01, and 24 degrees. The values of Equivalent Load Range in these plots are calculated by using Eq. (1) with an assumed value of n = 1. The assumed value of the parameter $K_a = 0.5$, with the data presented for four different values of the strand mean axial strain $S_1' = 0.001$, 0.002, 0.002867, and 0.004. An examination of the plots in Figs. 6 clearly suggests that for decreasing levels of the strand mean axial strains, S_1', the individual plots of Equivalent Load Range versus axial fatigue life (corresponding to a given strand lay angle) undergo a clockwise rotation with the combined plots covering a wide range of S_1' values (for given values of α and K_a) exhibiting increasingly wider scatter as the magnitude of Equivalent Load Range is decreased.

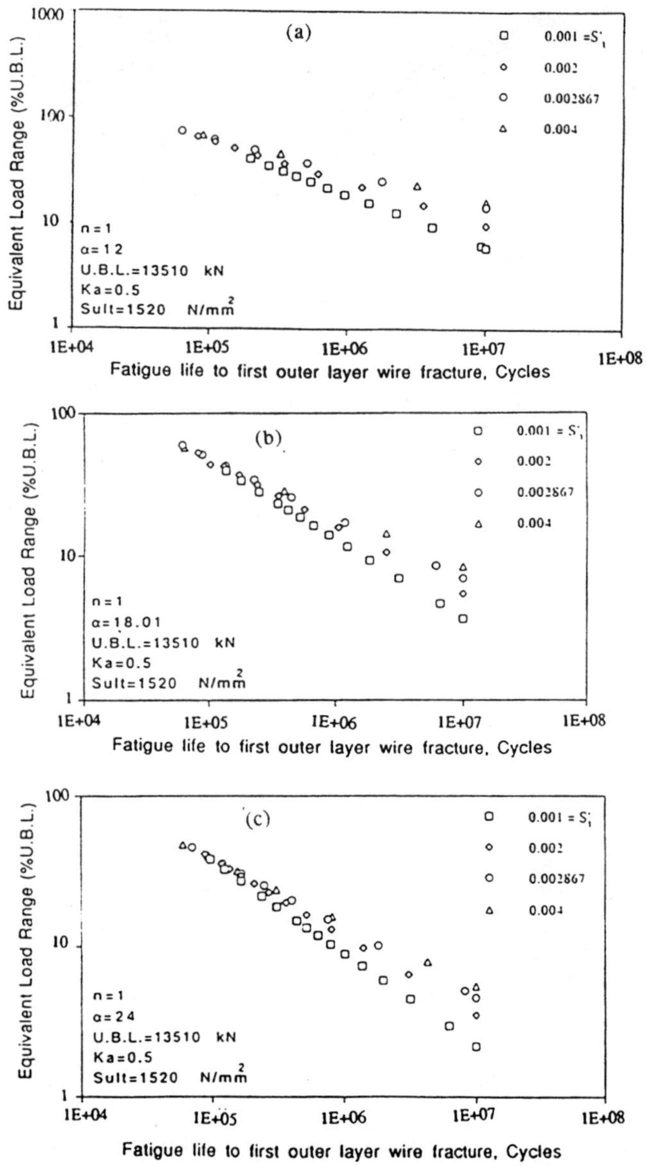

Fig. 6 Theoretical plots of Equivalent Load Range versus axial fatigue life to first outer layer wire fracture with $K_a = 0.5$ and n = 1, for the three spiral strand constructions used in the theoretical parametric studies, with $0.004 \leq S_1' \leq 0.001$: (a) $\alpha = 12$ degrees; (b) $\alpha = 18.01$ degrees; and (c) $\alpha = 24$ degrees.

The plots in Figs. 6a-c and others reported elsewhere (Alani and Raoof, 1997) suggest that the endurance limit (for a given strand construction and grade of wire) increases with increasing values of the strand mean axial strain, S_1' (for an explanation of this, perhaps, initially puzzling conclusion, reference should be made to Alani and Raoof, 1997). Moreover, the endurance limit decreases with increasing lay angle. These conclusions are of particular practical importance in offshore platform applications, where the small amplitude forces are the ones with the highest number of occurrences, and a knowledge of small amplitude/long life behaviour is necessary for these applications.

A comparison of the plots in Figs. 6 and the experimental results of Chaplin and Potts (1991) as reproduced in Fig. 2b (which relate to a 19 mm six strand rope) is instructive: as previously mentioned, in Fig. 2b the test data presented in terms of an Equivalent Load Range (calculated on the basis of a simple Goodman transformation) also show significantly more scatter for decreasing values of the Equivalent Load Range, which is in line with the present theoretical predictions.

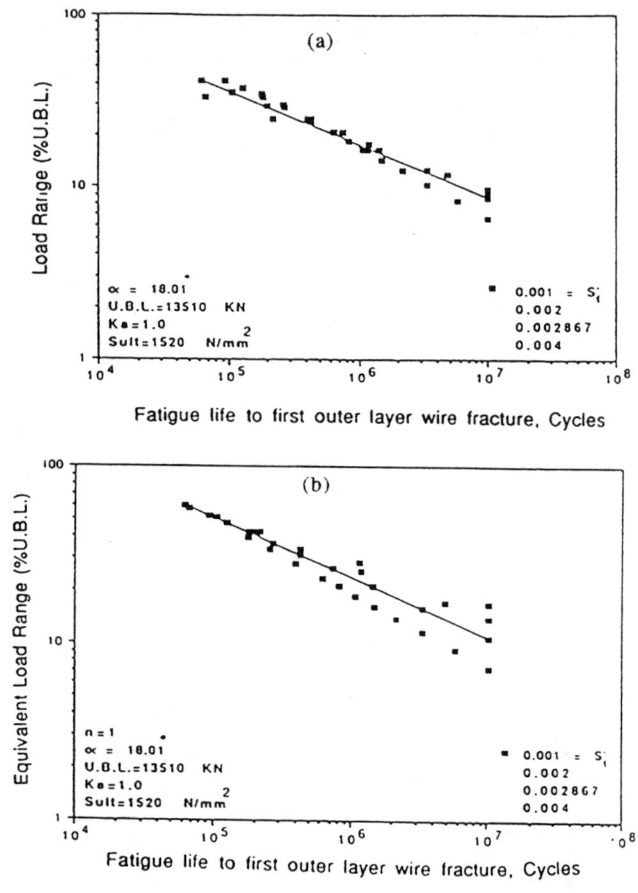

Fig. 7 Theoretical predictions of stress range versus axial fatigue life to first outer layer wire fracture for the strand construction with a lay angle of 18.01 degrees, $0.001 \leq S_1' \leq 0.004$, and $K_a = 1.0$, using different concepts: (a) stress range; (b) Equivalent Load Range with n = 1.

The question, however, is raised as to whether the use of a Goodman and/or Gerber transformation does indeed lead to a significant improvement in the fit of the fatigue data to a straight line as previously claimed by Chaplin and Potts (1991). Fig. 7a presents the theoretical predictions of stress range versus fatigue life to first outer layer wire fracture for the strand construction with a lay angle of 18.01 degrees for a range of assumed mean axial strains $0.001 \leq S_1' \leq 0.004$ and a K_a value of 1.0, with a least square line fitted to the theoretical data points. Fig. 7b, on the other hand, presents similar plots of Equivalent Load Range versus axial fatigue life, using K_a = 1.0. The assumed value of n = 1 in Equation (1) is used for producing the theoretical plots in Fig. 7b. Comparison of the results in Figs. 7a,b strongly suggest that the use of a Goodman transformation has not resulted in any decreases in the scatter of the theoretical data points about the fitted straight line. Table 4 presents a much more comprehensive set of theoretical results based on K_a = 1.0 relating to all the three types of 127 mm O.D. spiral strand constructions with lay angles of 12, 18.01, and 24 degrees. Of particular interest are the correlation factors R for the fitted lines for each case: unlike previous claims, the values of R in Table 4 suggest less scatter (ie. a better correlation) for the plots based on the axial load range (as opposed to those associated with the concept of Equivalent Load Range based on either n = 1 or 2). Even when the failure criterion is changed to that of first wire fracture in the innermost layer (assuming K_a = 1.0), the results in Table 4 (which assume n = 2) again show no reductions in the scatter of the theoretical results using the concept of Equivalent Load Range. Indeed, the correlation factors, R, for the load range versus fatigue life plots are found to be higher in magnitude than those of the associated Equivalent Load Range versus axial fatigue life plots. Finally, Table 5 presents similar results to those in Table 4, but with K_a = 0.5. In Table 5, fatigue failure is defined as the number of axial cycles to first outer layer wire breakage. Yet again, the correlation factors, R, relating to the load range versus fatigue life plots are found to be as good as those based on the alternative approach of Equivalent Load Range with either n = 1 or 2.

Table 4 Details of theoretical fatigue data for all three spiral strand constructions based on K_a = 10

Lay angle α (degrees)	K_a	Type of assumed cyclic axial load (% U.B.L.)	n	Location of first wire fracture used in the definition for axial fatigue life	a	b	Correlation factor R
12.00 18.01 24.00	1.0	Load range	- - -	outer layer	117.24 108.32 95.081	-15.170 -14.691 -13.147	0.955 0.962 0.964
12.00 18.01 24.00	1.0	Equivalent Load Range	1.0	outer layer	172.61 149.66 120.28	-22.559 -20.519 -16.747	0.888 0.941 0.941
12.00 18.01 24.00	1.0	Equivalent Load Range	2.0	outer layer	131.54 117.85 99.803	-17.084 -16.046 -13.830	0.948 0.965 0.960
12.00 18.01 24.00	1.0	Load range	- - -	innermost layer	124.01 112.46 102.47	-17.445 -16.152 -14.901	0.947 0.943 0.941
12.00 18.01 24.00	1.0	Equivalent Load Range	2.0	innermost layer	139.91 121.66 106.50	-19.738 -17.511 -15.500	0.947 0.945 0.940

Table 5 Details of theoretical fatigue data for all three spiral strand constructions based on Ka = 0.5

Lay angle α (degrees)	K_a	Type of assumed cyclic axial load (% U.B.L.)	n	Location of first wire fracture used in the definition for axial fatigue life	a	b	Correlation factor R
12.00 18.01 24.00	0.50	Load range	- - -	outer layer	122.59 111.83 101.15	-17.097 -16.005 -14.710	0.953 0.948 0.940
12.00 18.01 24.00	0.50	Equivalent Load Range	2	outer layer	138.31 121.07 105.16	-19.349 -17.368 -15.306	0.952 0.950 0.940
12.00 18.01 24.00	0.50	Equivalent Load Range	1	outer layer	135.497 124.539 128.070	-19.008 -18.006 -18.136	0.964 0.946 0.963

Finally, the fitted straight lines to all the theoretical data points are expressed as

$$y = a x + b \tag{4}$$

The above equation corresponds to an individual strand construction, with each line covering the full assumed range of S_1, with the values of the constants a and b for each individual fitted line also given in Tables 4 and 5.

CONCLUSIONS

Practical implications of the available literature on axial fatigue life of steel cables, and their shortcomings have been critically addressed. It is concluded that a number of unresolved problems still remain in the field.

The salient features of a theoretical model for predicting the axial fatigue life of spiral strands to first outer (or inner) wire fractures are presented. Based on the encouraging support for the theoretical predictions provided by a set of carefully conducted large scale test data (reported by others) as presented in this paper, and some extensive previously reported encouraging correlations between theoretical results and test data on spiral strands with diameters ranging from 39 to 127 mm, with widely different construction details and grade of wire, from different manufacturers, and tested by a number of different institutions, it is suggested that the theoretical model is a reliable one, and may safely be used for predicting the axial fatigue life of any spiral strand construction.

Using an extensive series of theoretical parametric studies on 3 different 127 mm spiral strand constructions with lay angles of 12, 18.01, and 24 degrees (which cover the full range of current manufacturing limits), certain new findings of practical importance are reported. For example, the endurance limit for spiral strands is found to increase in magnitude, with increasing values of the strand mean axial strain (or load). Moreover, the endurance limit is theoretically found to decrease with increasing lay angle. Such conclusions are of practical importance in offshore platform applications, where the small amplitude forces are the ones with the highest number of occurrences, and a knowledge of small amplitude/long life behaviour is necessary.

Using the data based on theoretical parametric studies, the plausibility of using a simple Goodman (or Gerber) transformation for steel cables (as opposed to plain specimens), as originally advocated by others, is critically examined. It is demonstrated that contrary to previous claims, presenting the axial fatigue data in terms of an Equivalent Load Range (with a minimum load of zero), rather than the traditional use of stress range, does not lead to any improvements in the fit of the data to a straight line. Indeed, it is theoretically shown that for a wide range of spiral stand mean axial strains (loads), the plots based on the stress range suffer from less scatter of data about the fitted line (ie. there exists a higher degree of correlation), and despite the extra efforts involved, the concept of Equivalent Load Range does not offer any practical advantages.

ACKNOWLEDGEMENTS

The long standing co-operation of Bridon Ropes' Personnel, Doncaster, UK, is gratefully acknowledged.

REFERENCES

Alani, M., and Raoof, M. (1995), "Axial Fatigue Characteristics of Large Diameter Spiral Strands", Proceedings of 5th International Offshore and Polar Engineering Conference, J.S. Chung et al. (eds), The Hague, The Netherlands, Vol.II, June, 260-265.

Alani, M., and Raoof. M. (1997), "Effect of Mean Axial Load on Axial Fatigue Life of Spiral Strands", International Journal of Fatigue, Vol.19, No.1, Jan. 1-11.

Chaplin, C.R., and Potts, A.E. (1991), "Wire Rope Offshore - A Critical Review of Wire Rope Endurance Research Affecting Offshore Applications". University of Reading Research Report for Department of Energy, HMSO Publication OTH91341.

Costello, G.A. (1990), "Theory of Wire Rope", Springer-Verlag, New York Inc.

Hobbs, R.E., and Raoof, M. (1982), "Interwire Slippage and Fatigue Prediction in Stranded Cables for TLP Tethers", in 'Behaviour of Offshore Structures'. Chryssostomidis, C., and Connor, J.J. (eds), Hemisphere Publishing/McGraw-Hill, New York, Vol.2, 77-99. (Proceedings, 3rd International Conference on Behaviour of Offshore Structures, 1982, held at M.I.T., Cambridge, MA, U.S.A).

Knapp, R.H., and Chiu, E.Y.C. (1988), "Tension Fatigue for Helically Armored Cables", Journal of Energy Resources Technology, ASME, Vol.110, March, 12-18.

Papanikolas, P.K. (1995), Axial Fatigue of Multi-Layered Wire Strands. PhD thesis submitted to the University of Alberta.

Raoof, M. (1990), "Axial Fatigue of Multi-layered Strands", Journal of Engineering Mechanics, ASCE, Vol.116, No.10, 2083-2099.

Raoof, M. (1991a), "Axial Fatigue Life Prediction of Structural Cables from First Principles", Proceedings of The Institution of Civil Engineers, Part II, Vol. 89, March, 19-38.

Raoof, M. (1991b), "Cable Fatigue Prediction in Offshore Applications", Proceedings of 10th International Conference on Offshore Mechanics and Arctic Engineering, ASME, Stavanger, Norway, Vol.IIIB, June, 403-411.

Raoof, M. (1991c), "Methods for Analysing Large Spiral Strands", Journal of Strain Analysis, Institution of Mechanical Engineers, Vol.26, No.3, 165-174.

Raoof, M. (1992), "A Critical Review of Draft API Recommended Practice 2FP1 Regarding Fatigue Life Estimation of Moorings", Proceedings of 11th International Conference on Offshore Mechanics and Arctic Engineering, ASME, Calgary, Canada, Vol.III, Part B, June, 521-532.

Raoof, M. (1996a), "Axial Fatigue Life Prediction of Bridge Cables", Proceedings of Third International Conference on Bridge Management, Inspection, Assessment, and Repair, Harding, J.E., et al (eds), E & F N Spon, April, 523-531.

Raoof, M. (1996b), "Effect of Lay Angle on Various Characteristics of Spiral Strands", Proceedings of 6th International Offshore and Polar Engineering Conference, Chung, J.S., et al (eds), Los Angeles, California, U.S.A., Vol.II, May, 189-196.

Raoof, M., and Hobbs, R.E. (1988), "Analysis of Multi-layered Structural Stands", Journal of Engineering Mechanics, ASCE, Vol.114, July, 1166-1182.

Raoof, M., and Hobbs. R.E. (1994), "Analysis of Axial Fatigue Data for Wire Ropes", International Journal of Fatigue, Vol.16, October, 493-501.

Yeung, L.C.T., and Walton, J.M. (1985), "Accelerated Block Tension Fatigue Testing of Wire Ropes for Offshore Use", Organisation Internationale pour l'Etude de l'Endurance des Cables (OIPEEC), Round Table Conference, National Engineering Laboratory, Glasgow, June, 2.5.1-2.5.14.

Proceedings of the Seventh (1997) International Offshore and Polar Engineering Conference
Honolulu, USA, May 25-30, 1997

ISBN 1-880653-28-1 (Set); ISBN 1-880653-30-3 (Vol. II)

Transition in the Wake of Flexible Cables and Beams

C. Evangelinos and G.E. Karniadakis
Brown University
Providence, RI, USA

Abstract

We investigated the effect of flow state on the amplitude of vibration in three-dimensional flow-structure interactions at low Reynolds number. In particular, we used a parallel spectral element code to study transition in the wake of flexible cables and beams. We found that cables tend to enhance transition whereas beams tend to delay it. Consequently, at Reynolds number around 200 the amplitude of crossflow vibration of beams is approximately half diameter, which is about half of the amplitude of vibration of cables in similar conditions.

Key Words: Transition, flexible structures, flow-induced vibrations

1 Introduction

Fluid flows over flexible cylinders arise in many engineering situations, such as marine cables towing instruments, flexible risers used in petroleum production and mooring lines, and other marine applications (see [3], [12], [11], [5], [13]). It is therefore important to understand and be able to predict the hydrodynamic forces and motion of cables caused by flow-induced vibration. In previous work, we have used direct numerical simulation to investigate flow-structure interactions in two- and three-dimensions (see [2], [7], [8], [9]). We have used a simple string equation to model the motion of the structure thereby neglecting the effect of bending stiffness. In addition, due to great computational expense, previous simulations were limited to low Reynolds number.

In the current work, we extend some of the preliminary findings of [9] for the effect of flow on the amplitude of vibration as the flow undergoes transition from laminar to early turbulent state. In particular, it was found in [9] that for cable-flow interactions, the amplitude of vibration increases from about half a diameter to one diameter, as the Reynolds number changes from $Re = 100$ to 200. As a reference point, we note that the wake of a stationary cylinder becomes three-dimensional and turbulent in the regime $Re = 200 - 300$ (see [6], [1] and references therein). The question therefore arises as to what is the transition regime for a freely moving flexible cylinder, and what is the corresponding effect on the amplitude of vibration. For a rigid cylinder, crossflow forced vibrations tend to make the wake more "two-dimensional" and therefore they tend to delay transition, [10]. However, for flexible freely moving cylinders no such studies exist nor it is known how the type of structure (i.e. its bending stiffness) influences the state of the wake.

To simplify the problem we consider long flexible cylinders, which are modeled as linear cables and beams. The complete formulation is given in section 2 along with a review of the numerical method. In section 3, we present a few representative simulations of flow past cables and beams at lockin as well as non-lockin states, and we conclude in section 4 with a brief summary of this ongoing investigation.

2 Mathematical Formulation

2.1 Governing Equations

We consider the interaction of incompressible flow past a long *flexible cylinder*. The equations that describe this problem are the coupled Navier-Stokes and the cable or beam equations. In a stationary, Cartesian coordinate system (x', y', z') these equations are:

$$\frac{\partial \boldsymbol{u}'}{\partial t} + (\boldsymbol{u}' \cdot \nabla)\, \boldsymbol{u}' \quad = \quad -\nabla p + Re^{-1}\nabla^2 \boldsymbol{u}',$$

$$\nabla \cdot \boldsymbol{u}' = 0,$$

where $Re = Ud/\nu$ is the Reynolds number based on the free-stream velocity U and the cylinder diameter d; ν is the kinematic viscosity.

In the case of flow-induced vibration, the equation of motion of the cable for its two directions of motion (i.e. in the x and y-directions) is given by a slightly modified forced wave equation:

$$\frac{\partial^2 \boldsymbol{\xi}}{\partial t^2} = c^2 \frac{\partial^2 \boldsymbol{\xi}}{\partial z^2} - \omega_n^2 \boldsymbol{\xi} + \frac{1}{m}\boldsymbol{F}(z,t), \qquad (1)$$

where $\boldsymbol{\xi}(z,t) = (\zeta(z,t), \eta(z,t))$ gives the cable displacement in the streamwise and crossflow directions and $c = \sqrt{T/m}$ gives the phase speed of waves in the cable. The cable has mass per unit length m and tension T. To maintain a mean displacement, the cable is lightly elastically supported by linear springs with spring constant k, giving a natural frequency of $\omega_n = \sqrt{k/m}$. The spring constant is selected sufficiently small to have negligible effect on the cable response. The fluid force on the cable is denoted by $\boldsymbol{F}(z,t)$. The components of $\boldsymbol{F}(z,t)$ in the streamwise and crossflow directions are the drag and lift force on the cable. Internal damping is neglected here as we are interested in the maximum possible response of the system.

Similarly, the equation of motion for a beam is

$$\frac{\partial^2 \boldsymbol{\xi}}{\partial t^2} = -a^2 \frac{\partial^4 \boldsymbol{\xi}}{\partial z^4} - \omega_n^2 \boldsymbol{\xi} + \frac{1}{m}\boldsymbol{F}(z,t), \qquad (2)$$

where $a^2 = EI/m$ with EI the bending stiffness, and $\omega_n = \sqrt{k/m}$ as before.

To simplify the solution of the fluid equations we use a coordinate system attached to the moving cylinder. This maps the time-dependent and deforming problem domain to a stationary and non-deforming one as shown in figure 1. This mapping is described by the following transformation:

$$\begin{aligned} x &= x' - \zeta(z',t'), \\ y &= y' - \eta(z',t'), \\ z &= z'. \end{aligned}$$

Accordingly, the velocity components and pressure are transformed as follows:

$$\begin{aligned} u &= u' - \frac{\partial \zeta}{\partial t} - w'\frac{\partial \zeta}{\partial z}, \\ v &= v' - \frac{\partial \eta}{\partial t} - w'\frac{\partial \eta}{\partial z}, \\ w &= w', \\ p &= p'. \end{aligned}$$

The Navier-Stokes equation and continuity equation are transformed to

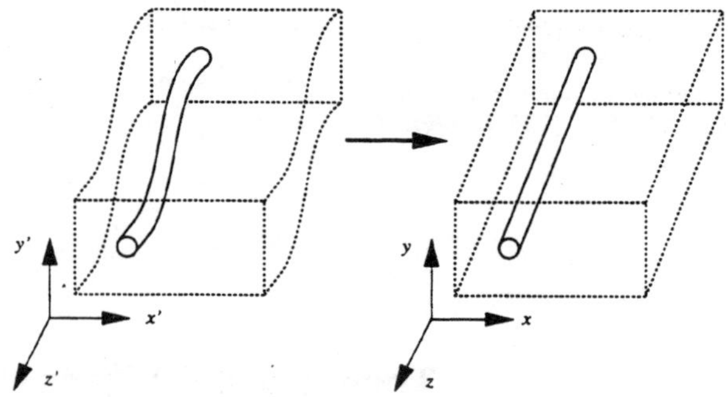

Figure 1: The coordinate system is attached to the moving flexible cylinder, producing an undeformed, stationary computational domain.

$$\begin{aligned} \frac{\partial \boldsymbol{u}}{\partial t} + (\boldsymbol{u}\cdot\nabla)\boldsymbol{u} &= -\nabla p + Re^{-1}\nabla^2 \boldsymbol{u} + \mathbf{A}(\boldsymbol{u}, p, \boldsymbol{\xi}), \\ \nabla \cdot \boldsymbol{u} &= 0, \end{aligned}$$

where the forcing term $\mathbf{A}(\boldsymbol{u}, p, \boldsymbol{\xi})$ is the extra acceleration introduced by the transformation, consisting of both inviscid and viscous contributions.

2.2 Numerical Method

To solve the three-dimensional Navier-Stokes equations, we use a parallel spectral element/Fourier method [4]. Spectral elements are used to discretize the x-y planes, while a Fourier expansion is used in the z-direction (i.e. along the flexible cylinder). Consequently, all flow and cylinder variables are assumed to be periodic in the spanwise direction. The computational domain extends $35d$ (cylinder diameters) downstream, and $15d$ above and below the cylinder. The "structure span", i.e. wavelength of vibrations in the flexible cylinder was $L/d = 4\pi$ for all simulations. Each x-y plane is discretized by a mesh consisting of $K = 158$ elements, with each element having $N = 9$ or 11 points per direction corresponding to 8th or 10th polynomial interpolation order. Typically, 64 z-planes (32 Fourier modes) are used giving a total of approximately 800,000 to 1,200,000 mesh points depending on the value of N (see figure 2). A non-dimensional time step of $\Delta t U/d = 0.001$ is used giving over 5,000 time steps per shedding cycle. For each simulation, the cylinder initial position and velocity are set and the simulations are run for at least 25 shedding cycles, or until the statistics are relatively stationary.

3 Simulation Results

The simplest periodic solutions to the vibrating homogeneous string equation (1) are standing waves and traveling waves. Both a standing wave response,

$$y(z,t) = A\cos(\omega t)\cos(2\pi z/L),$$

and a traveling wave response,

$$y(z,t) = A\cos(\omega t \pm 2\pi z/L),$$

satisfy the vibrating string equation, $y_{tt} = c^2 y_{zz}$, when $\omega = 2\pi c/L$, where c is the phase speed. In previous work [9], we found that in flow past flexible cables the traveling wave response is the preferred response in the absence of boundary conditions. However, there are cases where standing wave responses are stable, for example in the case of very small spanwise wavelength.

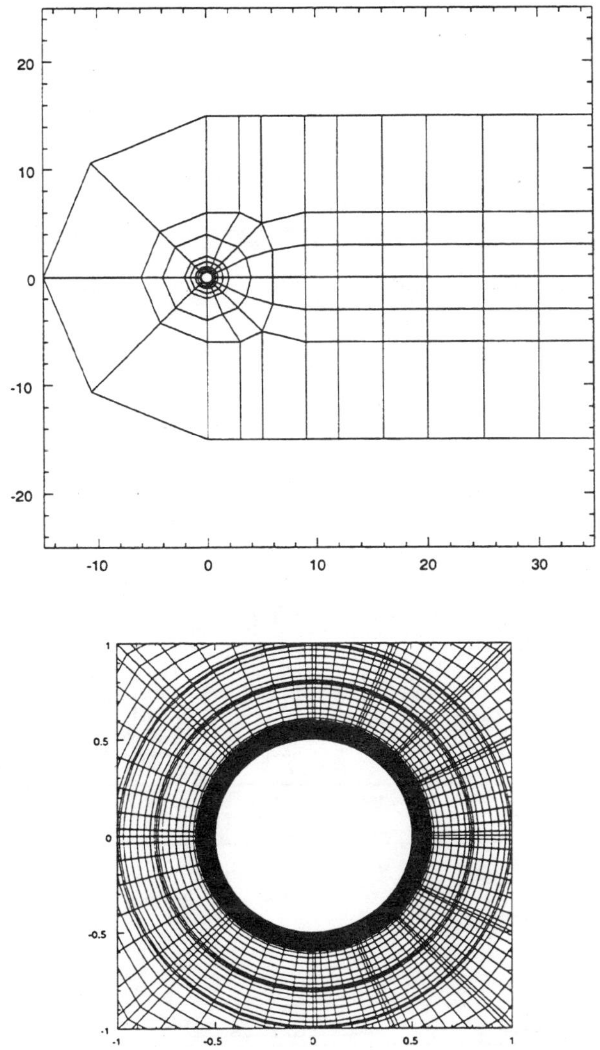

Figure 2: Spectral element mesh in the $x-y$ plane; Fourier expansions are used in the periodic spanwise direction. Close up (bottom) of the discretization around the cylinder for polynomial interpolation order 8th.

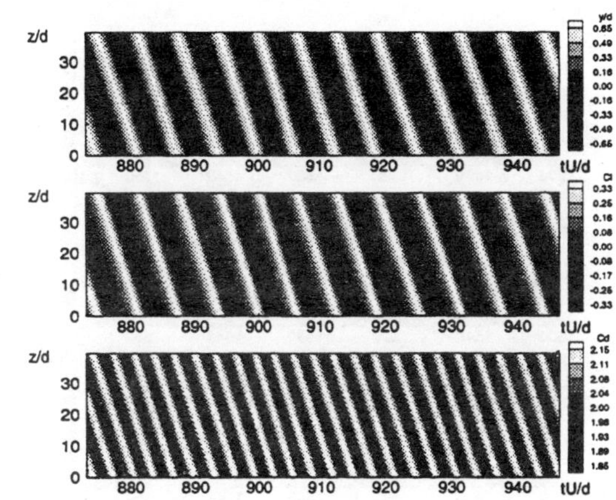

Figure 3: Cable crossflow displacement (y/d), lift coefficient (C_l) and drag coefficient (C_d) versus non-dimensional time and spanwise distance (z/d) along cable for L/d=12.6 and Re = 100.

In figure 3 we plot the response of a cable for laminar flow at $Re = 100$. In particular, we plot the displacement y/d, lift coefficient C_l and drag coefficient C_d, versus time tU/d, and position along cable z/d. At this Reynolds number an identical traveling wave response is obtained, irrespective if the structure motion is described by equation (1) or equation (2). This is the case for a stable, single-mode periodic spanwise motion and phase speed $c^2 = n^2\beta^2 EI/m$, where $\beta = 2\pi/L$ and n is the mode number. Consequently, a beam will respond in an identical manner to a cable for single mode vibrations. This was confirmed directly for both standing and traveling wave responses.

In general, however, the coupled response of a cable is different than the response of a beam. These differences are more pronounced in the transition regime as we demonstrate next. In particular, we consider simulations of flow past cables and beams at $Re = 200$ and $Re = 300$. Both lock-in and non-lock-in cases are investigated.

In figure 4 we plot the response of a cable (left) and of a beam (right) at $Re = 200$ at lockin. The reduced velocity $V = U/fd$ is $V_c = 5.41$ for the cable and $V_b = 5.08$ for the beam. In both cases we see that a traveling wave response is the stable solution. This structure motion causes an oblique shedding at about the same angle with respect to the cylinder axis. There are two important differences between the cable and the beam response. First, for the beam we obtain a single-mode periodic response whereas for the cable we observe an aperiodic response even at synchronization state. Second, the amplitude of oscillation for the beam is approximately $0.5d$ whereas for the cable is $1d$. Correspondingly, there are differences in lift and drag forces as shown in figure 4.

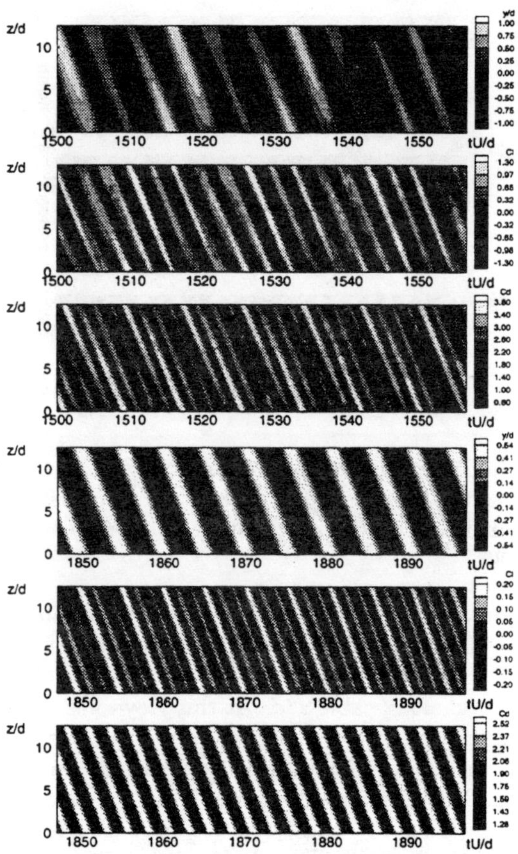

Figure 4: Crossflow displacement (y/d), lift coefficient (C_l) and drag coefficient (C_d) versus non-dimensional time and spanwise distance (z/d) at lockin. Cable (3 top plots) and Beam (3 lower plots). $Re = 200$.

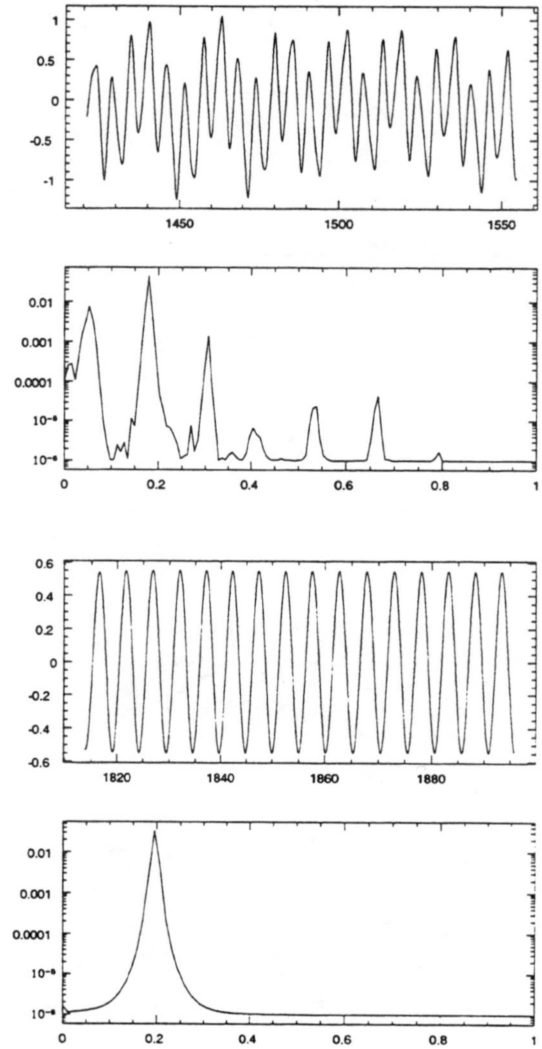

Figure 5: Time history and corresponding spectrum of the cable (2 top plots) and beam motion (2 lower plots) at lockin. $Re = 200$

These results are typical for all lockin states for the beam; however, no clear lockin state for the cable was obtained presumably because of the disorganization of the flow in the wake. In figure 5 we plot the spectrum of the structure motion at a fixed spanwise location. We see that unlike the beam a subharmonic appears in the frequency spectrum of the cable, responsible for the modulated traveling wave response plotted in figure 4.

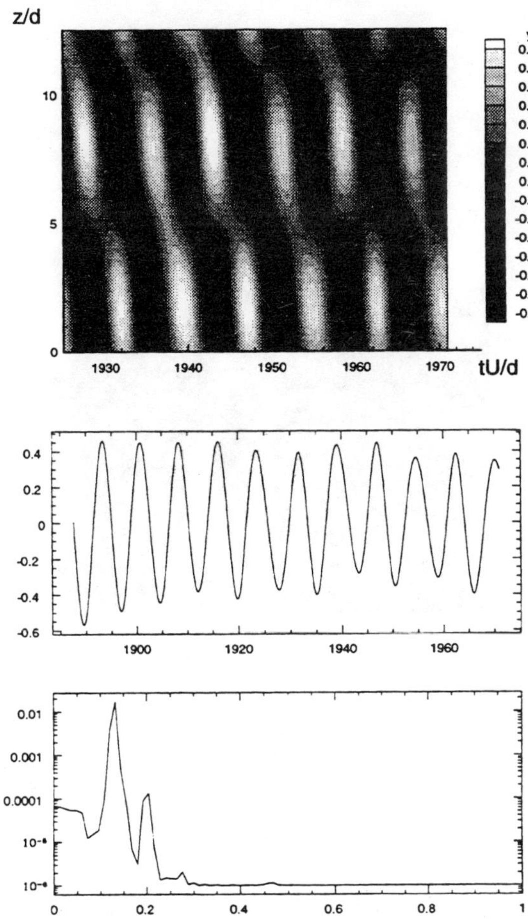

Figure 6: Crossflow displacement (top) and corresponding spectrum (bottom) for a vibrating beam left of lockin. $Re = 200$.

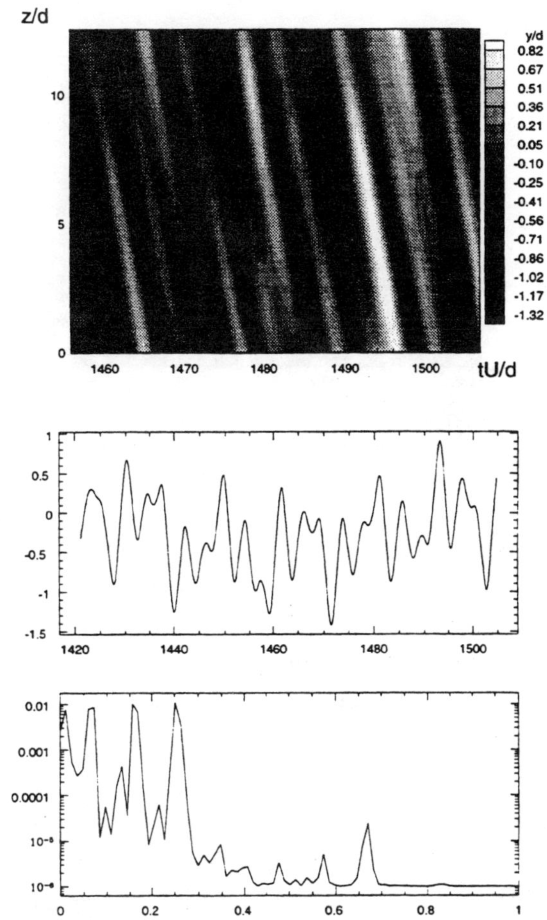

Figure 7: Crossflow displacement (top) and corresponding spectrum (bottom) for a vibrating cable left of lockin. $Re = 200$.

Table 1: Summary of crossflow displacements at $Re = 200$.

Cable	V_c	4.93		5.41	5.95	6.58
	y/d	0.75		1.00	0.97	0.95
Beam	V_b	4.98	5.05	5.95	6.62	7.41
	y/d	0.54	0.55	0.53	0.53	0.45

For larger values of the reduced velocity (left of lockin) quasi-periodic responses are also possible for the beam motion. In figure 6 we plot the evolution of the amplitude of the motion and corresponding spectrum as before for $V_b = 7.47$. We observe that a modulated traveling wave is obtained in this case due to the two incommensurate frequencies as shown in the corresponding spectrum. For such a detuned case, the response of the cable is very irregular as shown in figure 7, where the amplitude evolution and corresponding amplitude are plotted for a reduced velocity

$V_c = 6.65$. On the right of lockin a weaker modulation of the motion is caused both for the beam and the cable. In table 1 we summarize the values of crossflow amplitude obtained for the simulations we performed at $Re = 200$. The striking difference is that the amplitude of the beam motion is approximately half of the amplitude of the cable motion. As shown in figure 3 the amplitude of the motion at lower Reynolds number $Re = 100$ is approximately $0.5d$, i.e. close to the amplitude of a beam at $Re = 200$.

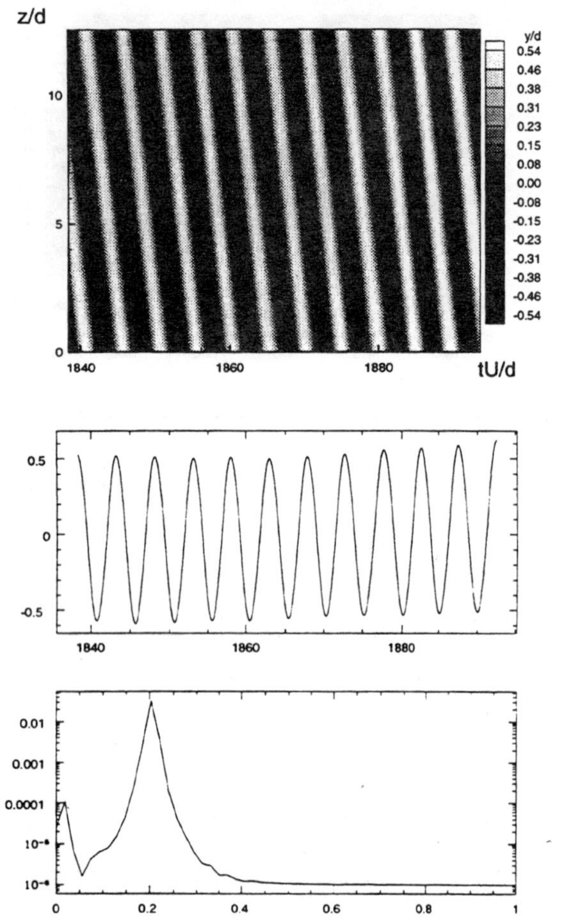

Figure 8: Crossflow displacement (top) and corresponding spectrum (bottom) for a vibrating beam at lockin. $Re = 300$.

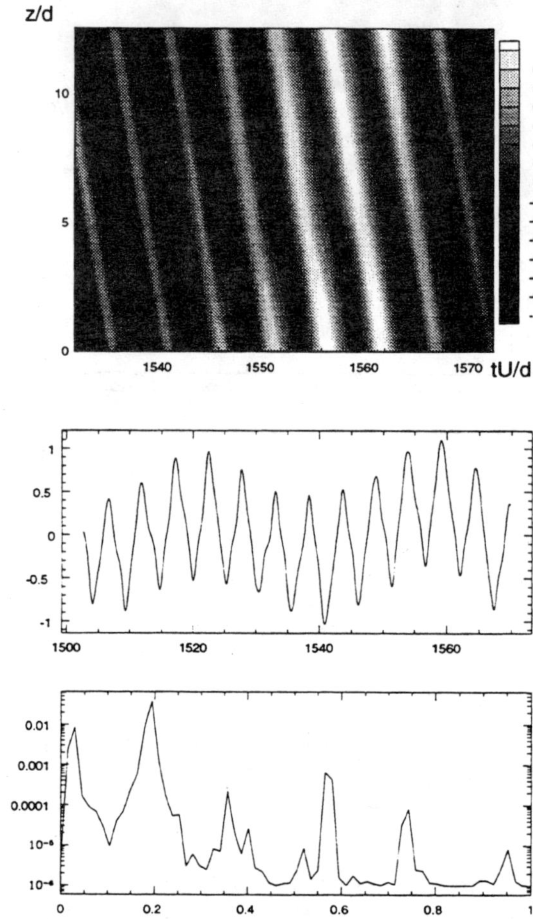

Figure 9: Crossflow displacement (top) and corresponding spectrum (bottom) for a vibrating cable at lockin. $Re = 300$.

At higher Reynolds number the same picture emerges, i.e. the lockin state for the beam is a single mode response whereas for the cable there are other frequencies involved. In figure 8 we plot the motion of a beam at $Re = 300$ and corresponding frequency spectrum as before at reduced velocity $V_b = 4.93$. A very weak modulation is present but otherwise a clear lockin state emerges. In figure 9 we plot the evolution of the crossflow motion of the cable at $Re = 300$ and at reduced velocity $V_c = 4.93$. Here again we see that an aperiodic motion is obtained with amplitude approximately twice as large as the amplitude of the motion of the beam.

3.1 Flow Structures

The differences in crossflow displacement between the cable and the beam at Reynolds number 200 and 300 are dictated by the differences in the induced flow structures in the near wake. To appreciate this we consider plots of instantaneous vorticity in x-y planes at different locations along the span. In all cases oblique shedding of vortices takes place induced by the traveling wave response of the structure.

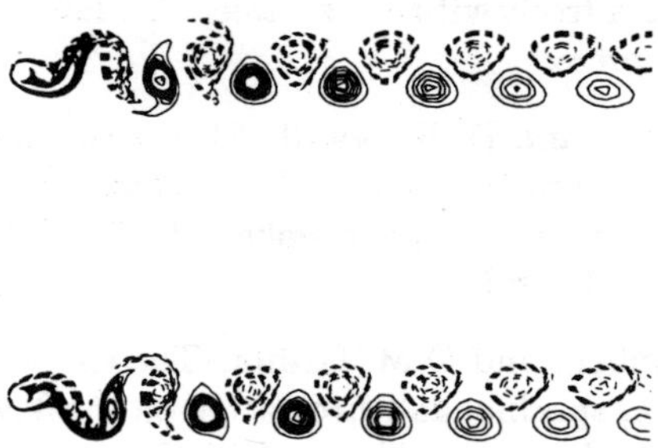

Figure 10: Slices in the x-y plane showing spanwise vorticity at $Re = 100$.

We first plot vorticity contours at $Re = 100$ in figure 10. For a traveling wave response the vorticity pattern is similar to a two-dimensional von Karman street formed behind a stationary cylinder. The different slices in the span show phase difference at the same instance but otherwise the same picture emerges at all z-locations. As mentioned earlier at this Reynolds number there is no difference between the cable or the beam response.

As the Reynolds number increases, the flow in the wake of the cable becomes very disorganized. In figure 11 we plot vorticity contours for the cable at $Re = 200$ at the lockin state described in figure 4. We see that a non-coherent vortex street appears, characteristic of a three-dimensional vortex street in the transition and early turbulent regime, [6]. In contrast, corresponding plots of vorticity for the beam at $Re = 200$ at lockin (see figure 12) show a vortex street quite similar to the laminar vortex street shown in figure 10.

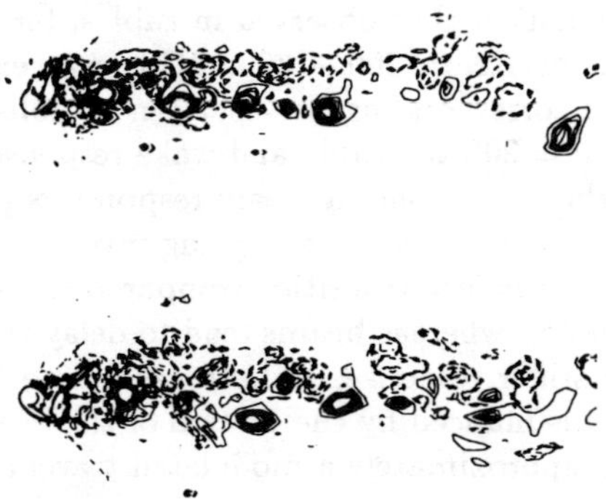

Figure 11: Cable: Slices in the x-y plane showing spanwise vorticity at $Re = 200$.

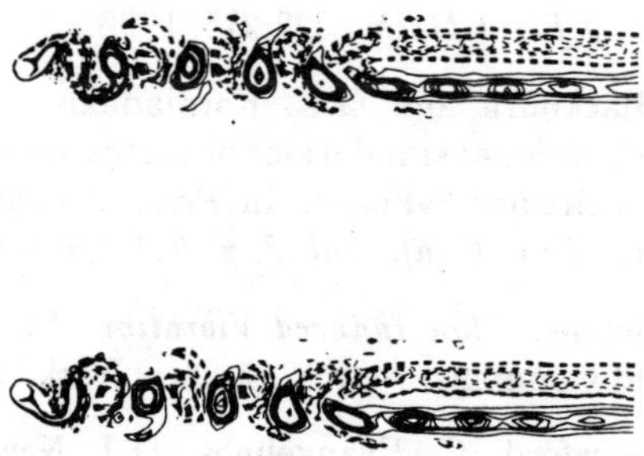

Figure 12: Beam: Slices in the x-y plane showing spanwise vorticity at $Re = 200$.

4 Summary

We have used direct numerical simulation to investigate differences in the coupled response of beam-flow and of cable-flow interaction. We have considered here the transition from laminar to turbulent regime that takes place between $Re = 200$ and $Re = 300$. We have employed linear models to describe the motion of a cable and of a beam, and have assumed that the effects from the supporting ends are neglected for sufficiently long structures.

The main finding of this work is that the crossflow amplitude observed for flow-induced vibrations of beams is approximately half of that observed in cables, for cases at or near lockin for $Re = 200 - 300$. This was explained in terms of the corresponding flow pattern established in the wake. At $Re = 200$ the cable and wake responses are aperiodic but the corresponding beam response is periodic. A tentative conclusion from this ongoing investigation is that cables tend to enhance transition compared to stationary or moving cylinder, whereas beams tend to delay transition by providing a stiffer response. In both cases an oblique shedding process is induced by the motion of the beam or cable that follows approximately a modulated traveling wave.

5 Acknowledgments

This work was supported by the Office of Naval Research, under the supervision of Dr. T. F. Swean. Computations were performed on the T3D and T3E at the Pittsburgh Supercomputing Center, and on the IBM SP2 at Cornell Theory Center and the Center for Fluid Mechanics at Brown University.

Bibliography

[1] D. Barkley and R.D. Henderson. Three-dimensional Floquet stability analysis of the wake of a circular cylinder. *J.Fluid Mech.*, 322:215, 1996.

[2] H.M. Blackburn and G.E. Karniadakis. Two- and three-dimensional simulations of vortex-induced vibration of a circular cylinder. In *Proc. 3rd Int. Offshore and Polar Eng. Conf., vol. 3, p. 715, Singapore*, 1993.

[3] R.D. Blevins. *Flow Induced Vibration.* Van Nostrand Reinhold Company, New York, New York, 1977.

[4] C.H. Crawford, C. Evangelinos, D.J. Newman, and G.E. Karniadakis. Parallel benchmarks of turbulence in complex geometries. *Computers and Fluids*, 25:677, 1996.

[5] F.S. Hover, M.A. Grosenbaugh, and M.S. Triantafyllou. Calculation of dynamic motions and tensions in towed underwater cables. *IEEE J. Oceanic Engineering*, 19:449, 1994.

[6] G.E. Karniadakis and G.S. Triantafyllou. Three-dimensional dynamics and transition to turbulence in the wake of bluff objects. *J. Fluid Mech.*, 238:1–30, 1992.

[7] D.J. Newman and G.E. Karniadakis. Direct numerical simulations of flow over a flexible cable. In *Sixth Int. Conf. on Flow-Induced Vibrations, Imperial College, April 10-12, 1995, editor P.W. Bearman, Balkema*, 1995.

[8] D.J. Newman and G.E. Karniadakis. Simulations of flow over a flexible cable: A comparison of forced and flow-induced vibration. *J. Fluids and Structures*, 10:439, 1996.

[9] D.J. Newman and G.E. Karniadakis. Simulations of flow past a freely vibrating cable. *J. Fluid Mechanics*, 1997. in press.

[10] A. Ongoręn and D. Rockwell. Flow structure from an oscillating cylinder – part 1: Mechanisms of phase shift and recovery in the near wake. *J. Fluid Mechanics*, 191:197–223, 1988.

[11] S.E. Ramberg and O.M. Griffin. The effects of vortex coherence, spacing, and circulation on the flow-induced forces on vibrating cables and bluff structures. *Naval Research Laboratory Report 7945*, 1976.

[12] J.K. Vandiver. Dimensionless parameters important to the prediction of vortex-induced vibrations of long, flexible cylinders in ocean currents. *MIT Sea Grant Report, MITSG 91-93*, 1991.

[13] D.R. Yoerger, M.A. Grosenbaugh, M.S. Triantafyllou, and J.J. Burgess. Drag forces and flow-induced vibrations of a long vertical tow cable – part 1: Steady-state towing conditions. *J. Offshore Mechanics and Arctic Engineering*, 113:117, 1991.

Proceedings of the Seventh (1997) International Offshore and Polar Engineering Conference
Honolulu, USA, May 25-30, 1997

ISBN 1-880653-28-1 (Set); ISBN 1-880653-30-3 (Vol. II)

Simplified Catenary Force-Deflection Analysis

John F. Flory
Tension Technology International
Morristown, NJ, USA

ABSTRACT

This paper presents a new method of performing force-deflection analysis on catenary mooring leg systems.

The traditional catenary equations are cumbersome for performing such analyses, principally because they reference deflection to the catenary generation point. The generation point is the origin used in the equations which describe, or mathematically "generate", the catenary curve shape. The catenary generation point is not stationary; its horizontal and vertical positions change as the catenary deflects.

The new form of catenary equation references the catenary deflection to a stationary point corresponding to the non-deflected position. It features only one principal equation, which directly relates applied force to deflection. It avoids the difficulties of calculating the scope of the catenary and the position of the catenary generation point. It avoids the use of a hyperbolic function. Thus it greatly simplifies the solution of catenary mooring leg systems.

KEY WORDS

Catenary, Mooring, Analysis, Equations

NOMENCLATURE

The nomenclature is summarized in Table 1. The first time a term is used, it appears in *italics*. Some of these terms differ from those used in the traditional method of catenary analysis.

INTRODUCTION

The catenary is the shape of a weighted flexible line suspended between two points under the influence of gravity. This paper introduces a new method of calculating the catenary top deflection caused by an applied force, by using an untraditional coordinate system based on the undeflected position of the catenary.

The physical catenary material is called *chain*, although the catenary may be comprised of any other flexible tension member, for example rope or cable. The chain has a *unit weight* (weight per unit length) $\boldsymbol{w}$.

A *horizontal force* is applied to the *top* of the catenary. The subscript $\boldsymbol{t}$ designates variables associated with the top of the catenary. Thus the horizontal force at the catenary top is designated as $\boldsymbol{H_t}$.

The *ground* is the surface on which the unlifted portion of chain rests. The *scope* is the distance along the catenary between any two points. The length of chain which is lifted off the ground is designated by $\boldsymbol{S}$.

New Catenary Coordinate System

Here we address only the case in which the catenary is tangent to the ground, such that no uplift is applied at the bottom of the chain, for example by an anchor. The unlifted portion of the catenary rests on a horizontal surface called the *ground plane*. The point at which the catenary touches the ground plane is the *ground point*.

The *base point* is the lowest point on the catenary curve, designated here by the subscript $\boldsymbol{b}$. In this ground tangent catenary case, the base point is on the ground plane and is coincident with the ground point.

Elevation refers to a vertical distance above the base point and is designated by $\boldsymbol{E}$. The top of the catenary is assumed to move only horizontally. Thus the elevation of the top above the base point $\boldsymbol{E_t}$ is a constant. For the ground tangent catenary case in a typical marine mooring analysis, this is the water depth.

In the *undeflected position*, with no horizontal force, the catenary hangs straight down and touches the ground plane at a point directly beneath the catenary top. This undeflected position is illustrated in Figure 1. The principal advantage of the method of catenary analysis presented here is that the defection of the top of the catenary is referenced directly to this undeflected catenary position.

Deflection refers to a horizontal distance from the undeflected catenary position. When horizontal force is applied to the top of the catenary, it deflects from the undeflected position by a distance $\boldsymbol{D_t}$ in the direction of that force. This deflected catenary is shown in Figure 2.

As the top deflects, additional catenary chain is lifted off the ground plane, and the base point deflects in the direction opposite the applied force. This *base deflection* is designated by $\boldsymbol{D_b}$.

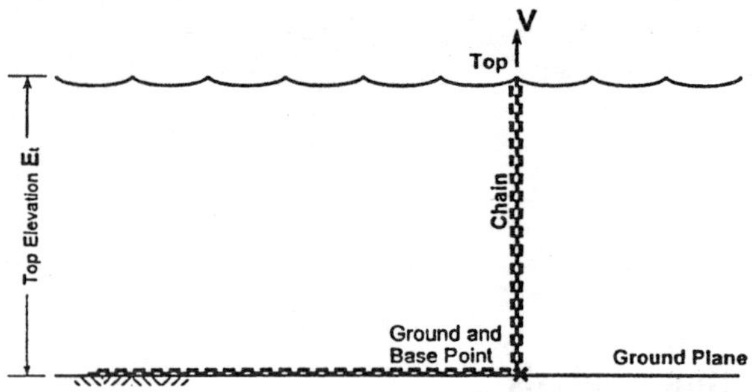

Figure 1 Catenary in Undeflected Position

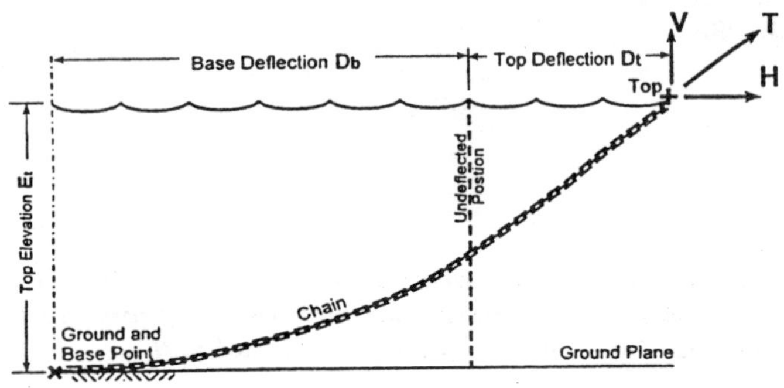

Figure 2 Catenary in Deflected Position with New Terminology

Traditional Catenary Coordinate System

The mathematic equations traditionally used to describe the shape of the catenary curve are based on a different coordinate system than that used here. The origin of that coordinate system is here called the catenary *generation point*. This traditional catenary coordinate system is shown in Figure 3.

The generation point is located at a distance C directly below the base point. This distance C is here called the *catenary parameter*. It appears in several important catenary equations, but otherwise the catenary parameter has no physical significance.

In the undeflected catenary position the generation point corresponds to both the base point and the ground point. Thus $C = 0$ in the undeflected position.

The traditional catenary coordinate system, with its origin at the generation point, has the disadvantage that the location of this generation point changes as force is applied. The generation point depresses below the ground plane, and it also moves away from the undeflected catenary position. Calculation of the actual deflection of the top of the catenary are greatly complicated by these movements of the generation point in the traditional catenary coordinate system.

Relationships Between Traditional and New Coordinate Systems

In the traditional coordinate system, the vertical distance of a point on the catenary above the generation point is designated by Y. Here that vertical dimension property is called *height* to distinguish it from elevation, which is measured from the ground plane. The height of the catenary top above the generation point is then Y_t. It generally has no direct physical significance.

Since the catenary top is at an elevation E_t above the base point, and the catenary generation point is at a distance C below that base point :

$$Y_t = C + E_t \tag{1}$$

Here we will use the term ***distance*** to refer to horizontal measurements from the catenary generation point. In the traditional catenary coordinate system, the horizontal distance from the catenary generation point to a point on the catenary is designated by X. In this new catenary coordinate system, that distance of the catenary top from the generation point is the sum of top deflection and base deflection :

$$X_t = D_t + D_b \tag{2}$$

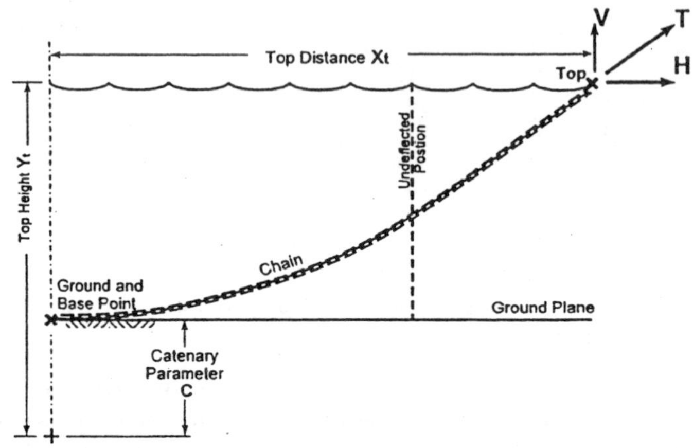

Figure 3 Catenary In Deflected Position with Traditional Catenary Coordinate System Terminology

The Traditional Catenary Equations

The following traditional catenary equations are taken from the Navy Fleet Moorings Design Manual DM 26.5.[1] These equations have been adapted to the catenary coordinate system and terminology introduced here. For general reference, they are presented in several different but equivalent forms.

Catenary Force Relationships

Using the nomenclature introduced above, the forces on and within the catenary can be calculated by the following equations.

One of the important properties of the catenary is that the horizontal force is related to the catenary parameter :

$$H = wC$$
$$= T\cos\theta$$

$$C = \frac{H}{w}$$
$$= \frac{T}{w}\cos\theta \tag{3}$$

The horizontal force applied to the catenary top is H. Because no other horizontal forces are applied to the catenary, the horizontal force H is the same at any point along the catenary. The horizontal force at the top is thus the same as that at the base point, $H_t = H_b$.

Another important property of the catenary is that the tangential force can be related to the height of the catenary :

$$\begin{aligned} T &= wY \\ &= w(C+E_t) \\ &= w\left(\frac{H}{w}+E_t\right) \\ &= H+E_t w \end{aligned}$$

$$Y = \frac{T}{w} \quad (4)$$

Since the unit weight of the catenary line is w, the total weight of the catenary is wS. Because the sum of vertical forces must balance, the vertical force at the top of the catenary is :

$$\begin{aligned} V &= wS \\ &= T\sin\theta \\ &= (H+E_t w)\sin\theta \end{aligned}$$

$$\begin{aligned} S &= \frac{V}{w} \\ &= \frac{T}{w}\sin\theta \\ &= \left(\frac{H}{w}+E_t\right)\sin\theta \end{aligned} \quad (5)$$

The total or tangential force T at any point in the catenary can be determined by vector addition of the horizontal and vertical force components :

$$T = \sqrt{H^2+V^2} \quad (6)$$

Traditional Catenary Shape Equations

Using the nomenclature introduced above, the catenary shape can be calculated with the following equations :

$$Y^2 = S^2 + C^2$$

$$\begin{aligned} S &= \sqrt{Y^2-C^2} \\ &= \sqrt{(C+E_t)^2-C^2} \\ &= \sqrt{2CE_t+E_t^2} \end{aligned} \quad (7)$$

$$\begin{aligned} Y &= C\cosh\frac{X}{C} \\ &= C\cosh\left(\frac{D_t+D_b}{C}\right) \end{aligned} \quad (8)$$

$$\begin{aligned} S &= C\sinh\frac{X}{C} \\ &= C\sinh\left(\frac{D_t+D_b}{C}\right) \end{aligned} \quad (9)$$

Catenary Angle Equations

The angle from horizontal of the slope of the catenary at any point is designated as Θ. It can be found from any of the following equations:

$$\begin{aligned} \theta &= \arctan\frac{V}{H} \\ &= \arctan\left(w\frac{S}{H}\right) \end{aligned} \quad (10)$$

$$\begin{aligned} \theta &= arcos\frac{H}{T} \\ &= arcos\left(\frac{H}{H+E_t w}\right) \end{aligned} \quad (11)$$

$$\begin{aligned} \theta &= \arcsin\frac{V}{T} \\ &= \arcsin\left(\frac{wS}{H+E_t w}\right) \end{aligned} \quad (12)$$

Traditional Force-Deflection Solution of the Catenary

The available catenary equations, given above, do not provide a direct relationship between the force H on the catenary and the resulting horizontal deflection D_t of the top from the catenary undeflected position.

An equation of the form $H = f(X)$ can not be derived, because equations (8) and (9) are transcendental; that is, the dependent parameter C appears on both sides of the hyperbolic trigonometric function. This type of equation can not be solved directly but can be solved by trial-and-error methods.

The relationship $X = f(H)$ can be solved through the use of a series of equations, as will be demonstrated at the end of this paper. That calculation procedure is cumbersome, because X relates to the catenary generation point which is not in a fixed position. Like the base point, it moves from the undeflected position when force is applied. This complicates the analysis of catenary mooring systems.

NEW CATENARY FORCE–DEFLECTION SOLUTION

By defining the catenary origin at the undeflected position of the catenary, a new set of catenary equations can be developed which greatly simplify the analysis of catenary mooring systems.

The catenary undeflected origin position is shown in Figure 1. With no horizontal force applied to the chain, it hangs straight down and touches the ground at a point immediately beneath its top. This undeflected ground point is used as the origin for the following catenary equations.

The vertical force at the top in this undeflected position is equal to the unit weight of chain times the top elevation or water depth :

$$V_b = E_t w \tag{13}$$

Figure 2 shows the catenary shape after application of a horizontal force. When the top of the catenary deflects, the chain scope increases as chain is lifted off the ground plane. The change in scope is equal to this additional length of lifted chain. As a result, the base point deflects along the ground by D_b . (It is assumed that there is no slack in the chain which lies on the ground.)

Thus for any given horizontal deflection D_t of the catenary top from the origin, the scope of chain between the base point and the catenary top is :

$$S = D_b + E_t \tag{14}$$

Non-Dimensionalized Equations

For convenience, the catenary dimensions and forces are non-dimensionalized. All dimensions are divided by the catenary top elevation E_t , which is the water depth.

All forces are divided by $E_t w$, the weight of a length of chain corresponding to the water depth or catenary top elevation in the undeflected position. Thus for example, the horizontal force parameter becomes $H / E_t w$.

With knowledge of the horizontal force H on the catenary, the tangent force at the top of the catenary can be directly determined by the following adaptation of Equation (4) :

$$\frac{T_t}{E_t w} = \frac{H}{E_t w} + 1 \tag{15}$$

From this equation and Equation (11), the slope at the top of the catenary is found :

$$\begin{aligned} \theta_t &= arcos \frac{H}{T_t} \\ &= arcos \left(\frac{H}{H + E_t w} \right) \end{aligned} \tag{16}$$

The vertical force V_t at the top of the catenary can then be obtained from the preceding and Equation (5) :

$$\begin{aligned} \frac{V_t}{E_t w} &= \frac{H + E_t w}{E_t w} \sin \theta_t \\ &= \left(\frac{H}{E_t w} + 1 \right) \sin \theta_t \\ &= \frac{H}{E_t w} \left(\frac{1}{\dfrac{H}{H + E_t w}} \right) \sin \theta_t \\ &= \frac{H}{E_t w} \frac{\sin \theta_t}{\cos \Theta_t} \\ &= \frac{H}{E_t w} \frac{1}{\tan \theta_t} \end{aligned} \tag{17}$$

When the vertical force is known, the catenary scope can be determined by Equation (5). Adapting that equation to the non-dimensional equation form, it becomes :

$$\frac{S}{E_t} = \frac{V_t}{E_t w} \tag{18}$$

Using Equations (2) and (14), the horizontal deflection of the top from the zero-deflection position can then be expressed as :

$$\frac{D_t}{E_t} = 1 + \frac{X_t}{E_t} - \frac{S}{E_t} \tag{19}$$

A New Form of Catenary Equations

The above set of equations could be used to solve for the relationship $D_t = f(H)$. But the relationship can be conveniently expressed by a single equation as follows.

The inverse hyperbolic sine can be expressed as a natural log function:

$$\text{arcsinh } k = \ln \left(k + \sqrt{k^2 + 1}\right) \tag{20}$$

Using this equation to transform Equation (9), it becomes :

$$\begin{aligned} D_t + D_b &= C \, arcsinh \frac{S}{C} \\ &= C \ln \left\{ \frac{S}{C} + \sqrt{\left(\frac{S}{C}\right)^2 + 1} \right\} \end{aligned} \tag{21}$$

Substituting Equations (14) for D_b and (3) for C in the above equation and then rearranging, it becomes :

$$D_t = E_t - S + \frac{H}{w} \ln\left\{\frac{Sw}{H} + \sqrt{\left(\frac{Sw}{H}\right)^2 + 1}\right\} \quad (22)$$

The equation is made non-dimensional by dividing all terms by E_t :

$$\frac{D_t}{E_t} = 1 - \frac{S}{E_t}$$

$$+ \frac{H}{wE_t} \ln\left\{\frac{Sw}{H} + \sqrt{\left(\frac{Sw}{H}\right)^2 + 1}\right\} \quad (23)$$

A new variable Q is defined as the ratio of the applied horizontal force to the initial vertical force in the undeflected position (water depth times unit weight of chain).

$$Q = \frac{H}{wE_t} \quad (24)$$

Note that Q is the ratio of the applied horizontal force to the weight of suspended chain in the particular water depth (vertical tension) at the undeflected catenary position. Thus Q has a simple physical significance.

Equation (7) is rearranged, substituting Equation (4) for C, and introducing the term Q, to achieve :

$$S^2 = 2\,C E_t + E_t^2$$

$$\frac{S^2}{E_t^2} = \frac{2\,H}{w E_t} + 1$$

$$= 2\,Q + 1$$

$$\frac{S}{E_t} = \sqrt{2\,Q + 1} \quad (25)$$

Substituting these into Equation (23), it can then expressed as :

$$\frac{D_t}{E_t} = 1 - \sqrt{2\,Q+1}$$

$$+ Q \ln\left\{1 + \frac{1}{Q} + \frac{\sqrt{2\,Q+1}}{Q}\right\} \quad (26)$$

This equation can readily be used in a spreadsheet or a computer program and can be preprogrammed into some pocket calculators.

Because the variable H appears both outside and inside the **ln** term, it is not possible to invert Equation (25) to directly solve for $H = f(D_t)$. However, it is relatively convenient to use the above equation to solve for that relationship by trial and error.

Example Calculations

The use of this new form of catenary equation is demonstrated in Table 2. Here the new method is compared with the traditional method.

The catenary anchor leg consists of a 100 mm stud link chain with a unit weight of 219 kg/m in a water depth of 183 m. A horizontal force of 200000 kg is applied to this anchor leg. What is the horizontal deflection of the catenary top due to this horizontal force?

Solving this problem using the above traditional catenary equations involves seven calculation steps. It requires calculating the catenary parameter C, the vertical and tangent forces at the top of the catenary, the scope of the catenary S, and the distance D_b through which the base point deflects from the undeflected catenary in the process of calculating the top horizontal deflection D_t. The traditional method also involves the inverse hyperbolic sine (arcsinh) function, which is not available on many calculators and spread sheets.

Solving this catenary problem by the new method introduced here involves only two steps (only one step if equation (23) is used). It employees the natural log (ln) function, which is much more accessible than the inverse hyperbolic sine function. The second step involves a long equation, but this can readily be programmed on a spread sheet or in a computer program.

The principal advantage of this new method is that it directly relates the applied force to the deflection at the top of the catenary. It does not require calculating the parameters C, S and D_b , and the algebraic "bookkeeping" involved in their use.

Conclusions

A new form of catenary equation has been developed, in which the horizontal deflection of the top of the catenary is directly related to a fixed point, the undeflected position of the catenary with no applied horizontal force.

Calculating the force– deflection relationship of a catenary is greatly simplified through the use of this new form of catenary equation. It avoids the many intermediate steps required in solving this relationship with the traditional catenary equations. The traditional catenary solution involves calculating the catenary scope and determining the point at which the deflected catenary touches the base plane, a point which changes position as the catenary shape changes.

The new method involves only two (or alternatively one) equations. It can readily be programmed in a spread sheet or a computer program.

The new method is particularly well suited for determining the force–deflection characteristics of a catenary anchor leg mooring system, in which the water depth is a fixed elevation. The only parameters involved are the chain weight, the water depth, and the applied horizontal force. The method then directly yields the resulting horizontal deflection of the catenary top.

Reference:

1. Naval Facilities Engineering Command, *Fleet Moorings, Basic Criteria and Planning Guidelines*, Design Manual 26.5, June, 1985, page 135 +.

TABLE 1 DEFINITIONS OF TERMINOLOGY AND VARIABLES

TERM		DEFINITION
base deflection	D_b	the deflection of the base point from the undeflected position
base point	b	the lowest point on the complete catenary curve (subscript)
catenary parameter	C	catenary parameter, the vertical distance from the generation point to the base point
catenary system top elevation	E_t	elevation of top of upper segment above ground plane of a multi-segment catenary system
chain		the physical catenary material, may be chain, rope, cable, or other similar material
deflection	D	horizontal distance of a point on the catenary from the origin
distance	X	horizontal distance of a point on the catenary from the generation point
elevation	E	vertical distance of a plane or point from the ground plane
generation point		the point from which the catenary shape is mathematically "generated"
ground	g	the surface upon which the unlifted chain rests (subscript)
ground deflection	D_g	the horizontal measurement from the undeflected position to the ground point
ground plane		the horizontal plane through the point at which the catenary touches the ground
ground point		the point at which the catenary touches the ground
ground tangent case		the case in which the catenary is tangent to the ground
height	Y	the vertical distance of a point on the catenary above the generation point
horizontal force	H	horizontal force on the catenary (same at all points)
origin		the position of the undeflected catenary
scope	S	distance along the catenary between two points
slope	θ	angle from the horizontal of the catenary at a point
tangential force	T	horizontal force on the catenary at a point
top	t	the upper end of the chain (subscript)
top deflection	D_t	the horizontal distance from the origin to the top
top elevation	E_t	the vertical distance of the top above the ground plane
top height	Y_t	the vertical distance of the top above the generation point
undeflected position		the position of the catenary with no applied horizontal force
vertical force	V	vertical force on the catenary at a point
weight	w	unit weight of chain

Table 2 EXAMPLE CATENARY FORCE–DEFLECTION CALCULATION

DEFINITION OF PROBLEM

Water Depth	E_t	= 183 m
Chain Weight	w	= 219 kg/m
Horizontal Force	H	= 200,000 kg

Find Horizontal Deflection of Top, D_t

SOLUTION BY TRADITIONAL METHOD

Step	Description	Calculation	Result
1	Determine Catenary Parameter by Eq (3)	$C = H / w = 200000 / 183$	= 913.2 m
2	Determine Tangential Force by Eq. (4)	$T = H + E_t\, w = 200000 + 219 * 183$	= 240077 kg
3	Determine Vertical Force by Eq. (6)	$V = \sqrt{(T^2 - H^2)} = \sqrt{(240077^2 - 200000^2)}$	= 132804 kg
4	Determine Catenary Scope by Eq. (5)	$S = V / w = 132804 / 183$	= 606.4 m
5	Determine Horizontal Distance by Eq. (9)	$X = C * \text{arcsinh}\,(S / C)$ $= 913.2 * \text{arcsinh}\,(606.4 / 913.2)$	= 568.9 m
6	Determine Deflection of Catenary Base by Eq. (14)	$D_b = S - E_t = 606.4 - 183$	= 423.4 m
7	Determine Deflection of Catenary Top by Eq. (2)	$D_t = X - D_b = 568.9 - 423.4$	= **145.5 m**

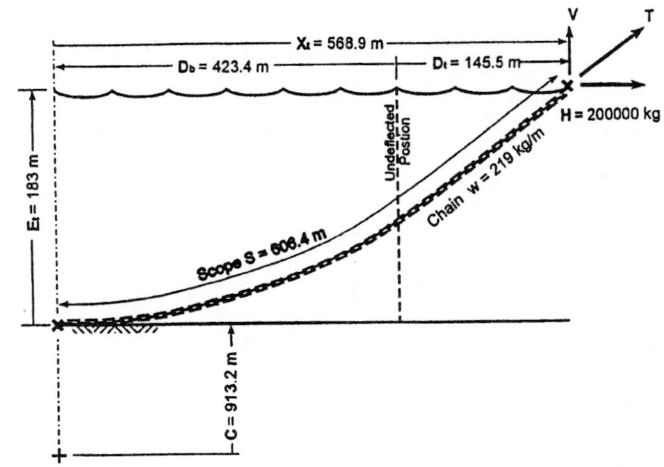

SOLUTION BY NEW METHOD

Step	Description	Calculation	Result
1	Determine Parameter Q by Eq. (24)	$Q = H / w\, E_t = 200000 / (219 * 183)$	= 5
2	Determine Deflection of Catenary Top by Eq. (26)	$D_t = E_t * [\, 1 - \sqrt{(2 * Q + 1)} + Q * \ln\{1 + 1/Q + Q * \sqrt{(2 * Q + 1)} / Q\}]$ $= 183 * [\, 1 - \sqrt{(2 * 5 + 1)} + 5 * \ln\{1 + 1/5 + 5 * \sqrt{(2 * 5 + 1)} / 5\}]$	= **145.5 m**

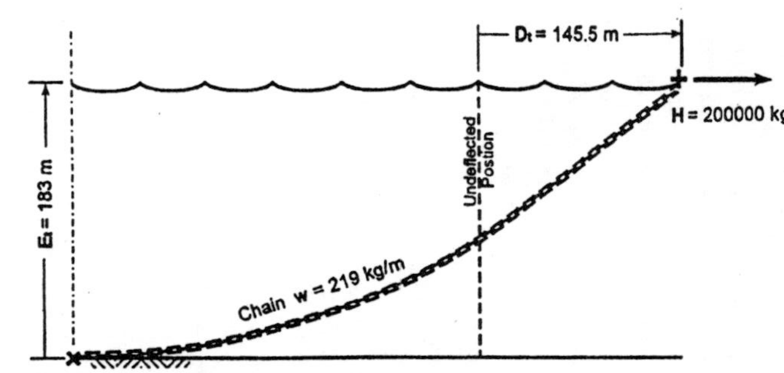

Proceedings of the Seventh (1997) International Offshore and Polar Engineering Conference
Honolulu, USA, May 25-30, 1997

ISBN 1-880653-28-1 (Set); ISBN 1-880653-30-3 (Vol. II)

Nonlinear Contributions in Predicting Dynamic Tensions on Mooring Lines for High and Low Frequencies of Excitation

J.K. Chatjigeorgiou and S.A. Mavrakos
National Technical University of Athens
Athens, Greece

ABSTRACT

This paper presents an efficient numerical method for the solution of the nonlinear dynamic problem of a mooring line in 3-D space. The method is based on the spectral description of the unknown parameters and uses a higher order implicit numerical scheme for the integration in the time domain. The mathematical model accounts for bending effects, linear and nonlinear constitutive relation and hysteresis phenomena. Extensive numerical results are given and compared with experimental data and other numerical predictions based on methods developed in the past. In all cases the dynamic behavior of the mooring line is calculated for several frequencies of excitation but special attention has been paid for the high and low values of them.

KEY WORDS

Mooring lines, Collocation method, Higher order implicit scheme, Synthetic cables.

INTRODUCTION

The evaluation of the dynamic behavior of catenary mooring lines has extensively been studied by several investigators in the past (Triantafyllou et al, 1985; Fylling et al, 1987; Kwan and Bruen, 1991) and became an indispensable part of the design procedure required by several regulatory organizations (Det Norske Veritas, 1989; American Petroleum Institute, 1991, and others).

Nonlinear phenomena and interactions among slowly varying large amplitude drift motions, fast varying wave induced motions and vortex-induced vibrations make a comprehensive dynamic analysis of mooring lines a very difficult task (Triantafyllou, 1994). As a result, many of the developed solution schemes are based on simplifying assumptions which allow approximate treatment of the system of the nonlinear governing equations and are adequate for particular range of the problem's parameters. For example, neglecting some (Chatjigeorgiou, 1996) or all (Shin & Triantafyllou, 1989) higher order nonlinear terms from the differential equations describing the dynamic behavior of high-tension steel catenary mooring lines, or, assuming additionally, that the dynamic tension along their length can be considered as constant for excitation frequencies well below their first elastic ones (Papazoglou et al., 1990; Mavrakos & Chatjigeorgiou, 1996), a significant reduction of the computational burden can be achieved. Although these kinds of assumptions give acceptable results for a specific range of excitation frequencies (Mavrakos et al, 1991; Mavrakos et al, 1996), there seems to be a lack of corresponding information for the range of the low and high frequencies.

In this paper an efficient algorithm is briefly presented dealing with the solution of the fully nonlinear system of differential equations of dynamic equilibrium, including bending effects and time variation of the strain along the mooring line. The algorithm makes use of the so-called collocation method for the space discretization of the mooring line, with the aid of Chebyshev polynomials (Fletcher, 1984). In this way the parabolic system of differential equations is reduced to a system with only one independent variable, that is the time, and integrated using higher order implicit techniques.

Numerical results concerning the most important parameter of a mooring line, i.e., the tension, at the top of several types of mooring lines are presented. They have been obtained using both approximate numerical schemes and the fully nonlinear solution developed here. The results of the various methods are compared to each other and to existing experimental data, thus assessing critically the importance of the various nonlinear contributions. Special attention has been paid to highlight the importance that the nonlinear phenomena have in the high and low frequency range of excitations.

DYNAMIC EQUILIBRIUM IN 3-D SPACE

The cable is modeled as a slender rod with bending stiffness. Let s denote the unstretched Lagragian co-ordinate measured from the lower cable end up to a material point of the cable, **v** (u, v, w) its velocity vector, m the mass per unit unstretched length, **F** (H, S_n, S_b)

the force vector along the cable, **R** (R_{D1}, R_{D2}, R_{D3}), the vector of drag forces and ε the local strain. Then the equations of motion of a cable in 3D-space can be expressed along the local tangential, **t**, normal, **n**, and binormal, **b**, directions of its moving configuration (moving dynamic reference) as (Howell, 1992; Burgess, 1992)

$$m\left(\frac{\partial u}{\partial t}+w\frac{\partial \theta}{\partial t}-v\frac{\partial \phi}{\partial t}\cos\theta\right)=\frac{\partial H}{\partial s}+S_b\Omega_2-S_n\Omega_3 \qquad (1)$$

$$-w_0\cos\phi\cos\theta+R_{D1}$$

$$m\left(\frac{\partial v}{\partial t}+\frac{\partial \phi}{\partial t}\left(u\cos\theta+w\sin\theta\right)\right)+m_a\frac{\partial v_{2r}}{\partial t}=\frac{\partial S_n}{\partial s} \qquad (2)$$

$$+\Omega_3\left(H+S_b\tan\theta\right)+w_0\sin\phi+R_{D2}$$

$$m\left(\frac{\partial w}{\partial t}-v\frac{\partial \phi}{\partial t}\sin\theta-u\frac{\partial \theta}{\partial t}\right)+m_a\frac{\partial v_{3r}}{\partial t}=\frac{\partial S_b}{\partial s} \qquad (3)$$

$$-S_n\Omega_3\tan\theta-H\Omega_2-w_0\cos\phi\sin\theta+R_{D3}$$

$$EI\frac{\partial \Omega_2}{\partial s}=EI\Omega_3^2\tan\theta+S_b\left(1+\frac{H}{EA}\right)^3 \qquad (4)$$

$$EI\frac{\partial \Omega_3}{\partial s}=EI\Omega_2\Omega_3\tan\theta-S_n\left(1+\frac{H}{EA}\right)^3 \qquad (5)$$

$$\frac{\partial u}{\partial s}+\Omega_2 w-\Omega_3 v=\frac{1}{EA}\frac{\partial H}{\partial t} \qquad (6)$$

$$\frac{\partial v}{\partial s}+\Omega_3\left(u+w\tan\theta\right)=\left(1+\frac{H}{EA}\right)\frac{\partial \phi}{\partial t}\cos\theta \qquad (7)$$

$$\frac{\partial w}{\partial s}-\Omega_3 v\tan\theta-\Omega_2 u=-\left(1+\frac{H}{EA}\right)\frac{\partial \theta}{\partial t} \qquad (8)$$

$$\Omega_2=\frac{\partial \theta}{\partial s} \qquad (9)$$

$$\Omega_3=\frac{\partial \phi}{\partial s}\cos\theta \qquad (10)$$

where m_a and w_0 are the added mass and the submerged weight of the cable per unit unstretched length, E is the Young's modulus of elasticity, A is the cable's cross sectional area and I its moment of inertia. Moreover ϕ,θ are the Euler angles and Ω_2,Ω_3 the curvatures of the line. At this point we must note that the stress-strain relation it was already assumed linear i.e.,

$$\varepsilon=\frac{H}{EA} \qquad (11)$$

The drag forces R are evaluated using the following equations

$$R_{D1}=-\frac{1}{2}\pi\rho_w D_0 C_{D1}v_{1r}\left|v_{1r}\right|\left(1+\varepsilon\right)^{\frac{1}{2}} \qquad (12)$$

$$R_{D2}=-\frac{1}{2}\rho_w D_0 C_{D2}v_{2r}\left|v_{2r}^2+v_{3r}^2\right|^{\frac{1}{2}}\left(1+\varepsilon\right)^{\frac{1}{2}} \qquad (13)$$

$$R_{D3}=-\frac{1}{2}\rho_w D_0 C_{D3}v_{3r}\left|v_{2r}^2+v_{3r}^2\right|^{\frac{1}{2}}\left(1+\varepsilon\right)^{\frac{1}{2}} \qquad (14)$$

where C_{D1},C_{D2},C_{D3} are the drag coefficients in the three directions of the moving coordinate system, ρ_w is the water density and D_0 is the unstreched cable's diameter. Finally $\mathbf{V_r}$ (v_{1r}, v_{2r}, v_{3r}) is the vector of the relative velocity between the cable and the current, the latter one being denoted by **J** (J_1, J_2, J_3).

SOLVING THE DYNAMIC SYSTEM

The solution of the governing parabolic system of nonlinear differential equations in the time domain, demands the evaluation of the ten unknown parameters of the problem at any time step and at any point along the mooring line. These unknown parameters are functions of two independent variables; the time, denoted by the variable t, and the space described though the Lagragian coordinate s. Thus the vector of unknowns can be written as follows:

$$\bar{Y}=\bar{Y}(s,t),\quad \bar{Y}=\left(u,v,w,S_b,S_n,H,\phi,\theta,\Omega_2,\Omega_3\right)^T \qquad (15)$$

The application of spectral or pseudospectral methods (Fletcher, 1984) for the solution of parabolic systems of nonlinear differential equations, allows the removal of the spatial dependence of the involved quantities reducing the initial system to a final one with exclusive time dependent unknowns.

In the present contribution, the spatial dependence of the relevant dynamic quantities in [0, L], L being the suspended cable's length, is accounted for with the aid of Chebyshev polynomials (Gottlieb & Orszag, 1977). In doing so, we introduce a new non-dimensional space variable x, which is defined in the interval [-1,1] by the relation:

$$x=\frac{s}{L/2}-1 \qquad (16)$$

and we invoke series expansions for any arbitrary unknown function of the problem in terms of Chebyshev polynomials, i.e.

$$a(x;t)=\sum_{i=1}^{N}A_i(t)T_{i-1}(x) \qquad (17)$$

In this way, the dynamic behavior of the mooring line is fully described, with the knowledge at any time step of the vector

$$\bar{Y}_i=\bar{Y}_i(t),\quad \bar{Y}_i=\left(U_i,V_i,W_i,S_{bi},S_{ni},H_i,\Phi_i,\Theta_i,\Omega_{2i},\Omega_{3i}\right)^T \qquad (18)$$

with i=1,...,N.

COLLOCATION METHOD - BOUNDARY CONDITIONS

By substituting (17) into the equations (1) - (10), a new system of differential equations can be obtained

$$F_j(x;t)=0, \quad j=1,\ldots,10 \tag{19}$$

which contains the 10N unknown elements of the vector (18). These represent the actual unknowns of the problem. Equations (19) should be supplemented by appropriate boundary conditions imposed at the cable's boundary points. For the present contribution, the following conditions have been taken into account:

$$\begin{aligned} &u(1,t)=h_u(t), \quad v(1,t)=h_v(t), \quad w(1,t)=h_w(t) \\ &u(-1,t)=v(-1,t)=w(-1,t)=0 \\ &EI\Omega_2(1,t)=EI\Omega_3(1,t)=EI\Omega_2(-1,t)=EI\Omega_3(-1,t)=0 \end{aligned} \tag{20}$$

That is, the velocities in the three directions of the moving coordinate system are assumed to be known functions of time at the top and zero at the touchdown point of the mooring line at the bottom. Moreover, the moments at the two boundary points are assumed to be vanished.

For the solution of the system (19) and the calculation of the 10N unknowns of the dynamic problem, N-2 collocation points are chosen along the mooring line dividing it into N-1 segments of equal length. The total number of points equals to N adding the two boundary points of the mooring line. In all these N-2 points, the equations (19) must be satisfied, i.e.

$$\begin{aligned} &F_j(x_k;t)=0 \\ &j=1,\ldots,10 \\ &k=2,\ldots,N-1 \end{aligned} \tag{21}$$

In this way, a system of 10*(N-2) equations are formed, which should be supplemented by the boundary conditions, see Eqs. (20), and the relations

$$F_j(-1,t)=F_j(1,t)=0, \quad j=4,5,6,7,8 \tag{22}$$

to create a system of 10*N differential equations with the same number of unknowns. The nonlinear system, is numerically integrated with respect to the time, using the method of finite differences. The used arithmetic scheme is an implicit one, whereas the Newton-Raphson method is applied to solve the system of the 10*N nonlinear algebraic equations. By the way of example, an arbitrary of these equations, including time derivatives, can be expressed as follows:

$$\sum_{i=1}^{N} \frac{y_i^{n+1}-y_i^n}{\Delta t} T_{i-1}(x_k)+\ldots-\Re^{n+1}(x_k)=0 \tag{23}$$

where $\Re^{n+1}$ denotes the summation of all the terms of the equation which do not include differentiation with respect to time.

Stability problems have been avoided through the use of higher order implicit schemes (Hoffman, 1993). For this purpose, Equation (23) has been written in the following form:

$$\sum_{i=1}^{N} \frac{y_i^{n+1}-\gamma\left(\alpha_0 y_i^n+\alpha_{-1}y_i^{n-1}+\alpha_{-2}y_i^{n-2}+\ldots\right)}{\Delta t} T_{i-1}(x_k)+\ldots \tag{24}$$

$$-\gamma\beta\Re^{n+1}(x_k)=0$$

Here, the parameters α,β,γ are constant numbers which are depended on the order of implicitness (Gear, 1971; Hoffman, 1993).

INITIAL CONDITIONS

For the solution of the above mentioned nonlinear system, initial conditions of the unknown functions at t=0, should be imposed. In the present contribution, the static counterparts of these dynamic quantities have been considered as initial guesses. The system of equations involves time dependent coefficients of Chebyshev polynomials and not the physical parameters of the dynamic problem. As a result, the initial values of the physical variables should be transformed into initial values of the corresponding time dependent coefficients. The latter can be obtained by invoking the orthogonality relation of the Chebyshev polynomials and the fact that any initial guess, $a_0(x)$, of an unknown variable at t = 0, can be expressed in terms of these polynomials as:

$$a_0(x)=\sum_{i=1}^{N} A_i T_{i-1}(x) \tag{25}$$

where A_i is a constant number.

Next, by multiplying both sides of (25) with $T_{j-1}(x)$ and applying the discrete orthogonality condition which the Chebyshev polynomials satisfy (Press et al, 1986):

$$\sum_{k=1}^{m} T_i(x_k)T_j(x_k)=\begin{cases} 0 & i\neq j \\ \dfrac{m}{2} & i=j\neq 0 \\ m & i=j=0 \end{cases} \tag{26}$$

with x_k (k=1,...,m) being the k-th root of $T_m(x)$

$$x_k=\cos\left(\frac{\pi\left(k-\frac{1}{2}\right)}{m}\right), \quad k=1,\ldots,m \tag{27}$$

the coefficients A_i can be obtained as:

$$A_j=\begin{cases} \dfrac{1}{N_k}\displaystyle\sum_{k=1}^{N_k} a_0(x_k)T_{j-1}(x_k) & j=1 \\ \dfrac{2}{N_k}\displaystyle\sum_{k=1}^{N_k} a_0(x_k)T_{j-1}(x_k) & j>1 \end{cases} \tag{28}$$

In evaluating the summations in (28), linear interpolation has been used to obtain the values of function $a_0(x)$ at the discrete points x_k.

NONLINEAR STRESS-STRAIN RELATION

The system of the governing differential equations (1)-(10) has been obtained under the assumption of a linear stress-strain relation. This is a reasonable assumption for steel cables until their yield point, but it is not applicable to the case of synthetic lines. In such cases, non-linear stress-strain relation should be taken into account in the problem's formulation. The latter is implemented by replacing the tension H and its spatial derivatives with strain associated quantities. It is:

$$H = f(\varepsilon) \tag{29}$$

$$\frac{\partial H}{\partial s} = \frac{\partial f(\varepsilon)}{\partial \varepsilon}\frac{\partial \varepsilon}{\partial s} = f'(\varepsilon)\frac{\partial \varepsilon}{\partial s} \tag{30}$$

In this way, the vector of the unknowns becomes

$$\vec{Y}_i = \vec{Y}_i(t), \quad \vec{Y}_i = \left(U_i, V_i, W_i, S_{bi}, S_{ni}, E_i, \Phi_i, \Theta_i, \Omega_{2i}, \Omega_{3i}\right)^T \tag{31}$$

with

$$\varepsilon(x;t) = \sum_{i=1}^{N} E_i(t) T_{i-1}(x) \tag{32}$$

RESULTS AND DISCUSSION

For the subsequent developments the tension along the mooring line will be denoted by T. Both frequency and time domain techniques have been used to solve the dynamic problem and to obtain the extensive numerical results. The frequency domain solutions allow treatment of the linearized dynamic equilibrium equations and take into account the effects of the quadratic drag only trough their equivalent linearization. As far as the time domain analysis is concerned, three different types of solutions have been used to obtain the results presented in the following. The first, denoted as "First Nonlinear Solution", solves a simplified second-order dynamic equation in which only one non-linear term is retained and the dynamic tension is assumed to be constant along the line (Papazoglou et al, 1990; Mavrakos & Chatjigeorgiou, 1996). The second, called "Second Nonlinear Solution", extends the previous one by including the dynamic tension variation along the line and by accounting for an additional second-order term, which is proportional to the spatial derivative of the dynamic tension (Chatjigeorgiou, 1996). Finally, the results referred as "Fully Nonlinear Solution" in the sequel have been obtained through the method outlined in the preceding sections which solves the fully non-linear dynamic equations including bending effects.

The various numerical predictions are first compared with existing experimental data. The latter have been obtained in a series of experiments conducted at the Towing Tank facilities of the Department of Naval Architecture and Marine Engineering, National Technical University of Athens, and they have been reported in previous work (Papazoglou et al, 1990). The geometric and inertia characteristics of the tested cable are given in Table 1.

Here, T_{br} denotes the breaking strength of the cable's material. The water depth was 2.97 m, with the upper end of the cable fixed on an oscillator. Moreover, a special anchor was constructed that allowed the insertion, in series, of springs to reduce the equivalent elastic stiffness of the line. The experimental results and the associated numerical predictions are shown in the Figures 1 - 3 together with the corresponding spring constants, the pretension level and the amplitude of excitation.

Table 1. Cable properties

D_0 (mm)	3
L (m)	60.273
E (N/m^2)	7.8×10^{10}
w_0 (N/m)	0.364
m (Kgr/m)	0.044
T_{br} (N)	8240
Material	Stainless steel AISI 316

It is obvious that the numerical results of the "Fully Nonlinear solution" compare very well with the experimental data. This confirms the ability of the specific method to predict the actual dynamic behavior of a mooring line. Moreover, as the excitation frequencies of the wire cable are well below its first elastic frequency, the dynamic tension variation along the line can be considered as negligible, a fact which enlightens the very good agreement between the first and the second nonlinear solution. Although there are some differences between the experimental data and the numerical predictions obtained through the "First-" and "Second-nonlinear-solution", the latter exhibit almost the same trend. Finally, the fact that the "Fully nonlinear solution" approximate better the experimental measurements than the rest numerical schemes (Figure 1), indicates the importance of nonlinear phenomena for large amplitudes of excitation, as the present case is (33 times the cable's diameter). It seems that for large excitation amplitudes, such phenomena can not be properly modeled through the simplified analyses.

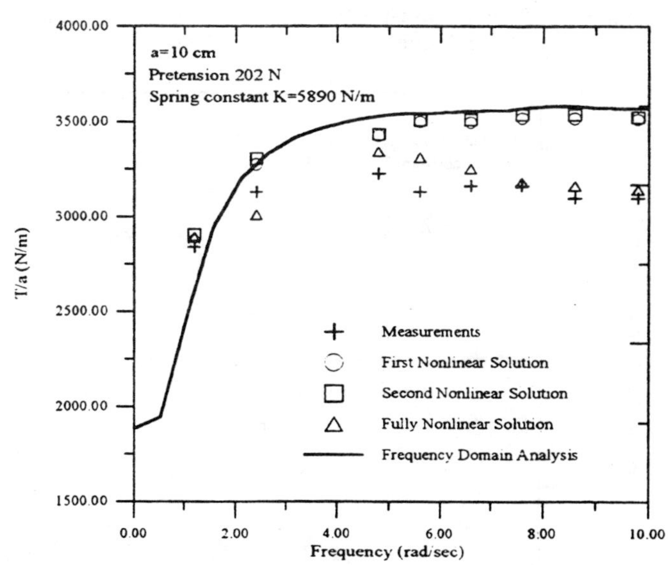

Figure 1. Dynamic tension at the top a steel mooring line anchored with a hard spring.

In Figure 2 comparisons between the "First-" and the "Fully-nonlinear" solutions are given together with the linearized predictions obtained from the frequency domain analysis. The same as previous cable is treated which is now equipped with a softer spring at its connection to the anchor and excited with a smaller amplitude (ratio of the top amplitude to cable's diameter equals to 13).

A favorable agreement between the "Fully nonlinear" and the linearized frequency domain analysis can be established, even for high frequencies of excitation. At the same time, the results obtained through the "First nonlinear solution" exhibit a remarkable deviation from the rest numerical predictions for frequencies exceeding $\omega \cong 5.0$ rad/sec where the maximum attainable value of the dynamic tension amplification is obtained. The differences can be traced back to the fact that the underlying assumption of zero variation of the dynamic tension along the cable is not valid any more for excitation frequencies beyond the first transverse eigenfrequency of the cable.

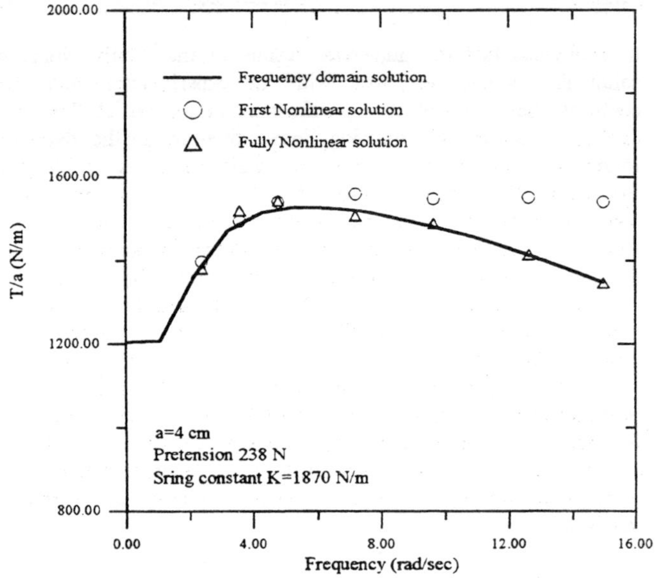

Figure 2. Dynamic tension at the top a steel mooring line anchored with a soft spring.

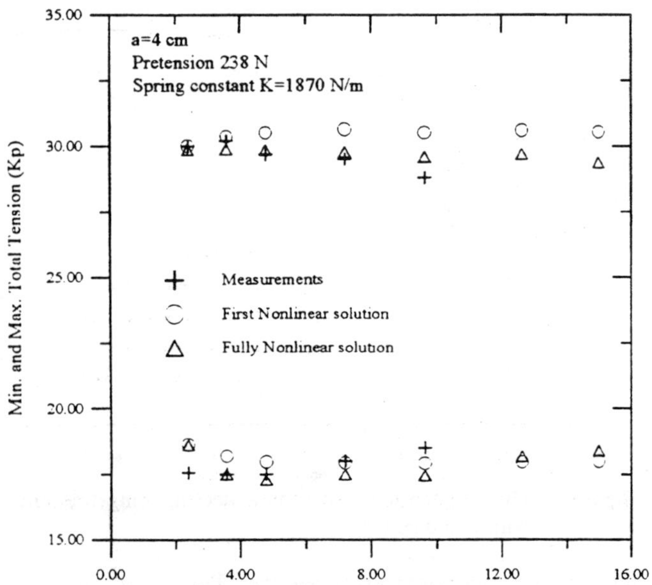

Figure 3. Total tension at the top of a steel mooring line anchored with a soft spring.

Summarizing the findings from Figures 1-3, we conclude that, for large excitation amplitudes, nonlinear phenomena seem to play an important role in evaluating the dynamic behavior of steel cables in the whole frequency range of interest (even for low frequencies). On the other hand, for small excitation amplitudes, nonlinear effects beyond those originated from the quadratic drag which are usually accommodated through the linearized frequency domain analysis, seem to be not important for the dynamic behavior of steel marine cables for low and high frequencies.

Next, we are dealing with a cable from synthetic material and we compare its dynamic behavior with the corresponding one of a wire mooring line. The synthetic and wire cables are chosen to have the same weight per unit length in air. Their properties are given in the following Table 2.

Table 2. Properties of steel and synthetic cable.

	Synthetic Cable	Steel Wire Cable
D_0 (m)	0.08	0.062
L (m)	1000	1000
E (N/m^2)	$0.8 * 10^9$	$0.7 * 10^{11}$
w_0 in air (N/m)	200	200
w_0 submerged (N/m)	149.4	169.61
m (Kg/m)	20.38	20.38
m_a (Kg/m)	5.15	3.09
Pretension (N)	$4 * 10^5$	$4 * 10^5$
Depth (m)	500	500
C_{D1}	0.05	0.05
C_{D2}	1.2	1.2

The two mooring lines are subjected to harmonic excitation in surge direction at their top end with an amplitude equal to 1 m. In cases where hysteresis phenomena are included in the study, the nonlinear stress-strain relation of the synthetic line has to be obtained from experiments. In the present case, the stress-strain relation of the investigated mooring line is given in Figure 4.

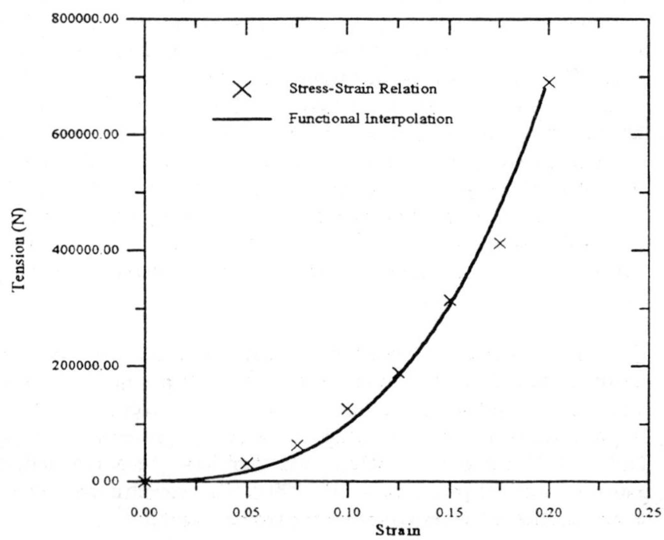

Figure 4.Nonlinear Stress-Strain relation of a synthetic line.

The experimental data have functionally been approximated with the following equation:

$$T = a_1\varepsilon^3 + a_2\varepsilon^2 + a_3\varepsilon$$

where

$a_1=8 * 10^7$; $a_2=1 * 10^6$; $a3=9 * 10^4$

Using this functional approximation of the actual nonlinear stress-strain relation, the equations (29) and (30) can be evaluated and introduced into the initial dynamic system, see Eqs. (1)-(10).

In Fig.5 the dynamic tension amplification for the synthetic and the wire line are plotted against the frequency of excitation. The results have been obtained using frequency domain analysis techniques. It is evident that the first elastic frequency of the synthetic cable is shifted within the wave frequency range of interest.

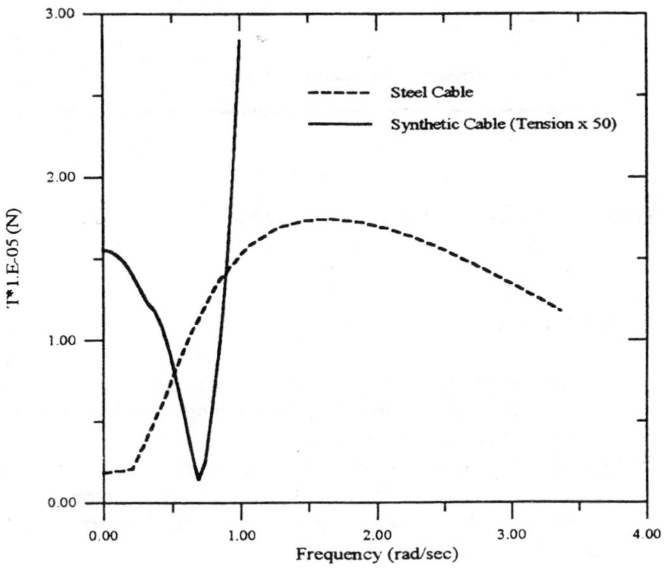

Figure 5. Dynamic tension at the top of a synthetic and a steel wire mooring line

The same synthetic cable has additionally been treated by the various time domain simulation techniques, mentioned above, using both linear and non-linear stress-strain relations. The corresponding results together with their frequency domain counterparts are plotted in Fig. 6. It can be seen that the various numerical predictions based on the linear constitutive relation compare satisfactory between each other in the low excitation frequency range, i.e. for $\omega \leq 0.2$ rad/sec, thus indicating the insignificance of the associated nonlinear effects. Beyond this value the "First nonlinear solution" is unable to accurately predict the dynamic behavior of the synthetic line, because the underlying assumption of zero variation of the dynamic tension is not valid any more. At the same time, the "Second-" and the "Fully-nonlinear" solutions give comparable results for excitation frequencies up to $\omega = 0.5$ rad/sec. For higher frequencies, however, the two methodologies seem to give divergent results. This can be traced back to the fact that the fully nonlinear solution accounts for additional contributions arising from the time differentiation of the dynamic quantities in the compatibility relations, see Eq. (6).

In Figure 6 numerical results based on a nonlinear stress-strain relation are also given. It can be seen that the associated dynamic tension amplification is greater than the corresponding one obtained through the linear constitutive relation, see Eq. (11). Moreover, it should be pointed out that the adoption of a nonlinear stress-strain relation influences not only the dynamic- but also the static-equilibrium of the mooring line. This can be seen in the Figure 7.

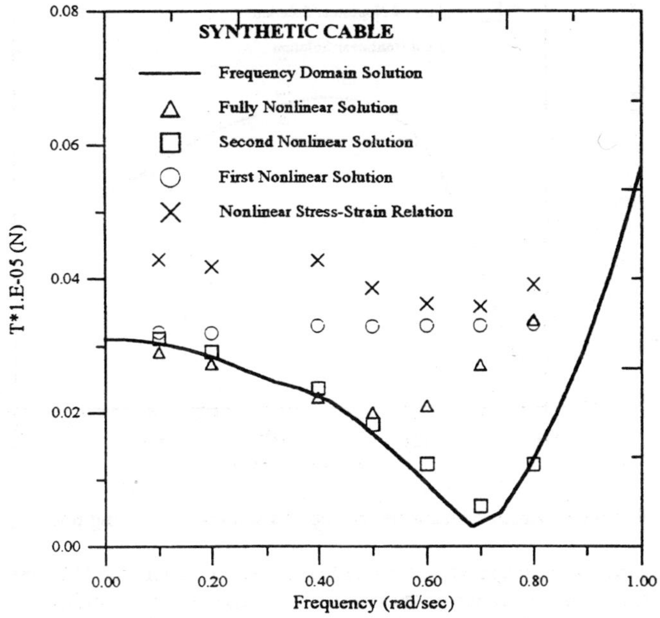

Figure 6. Dynamic tension at the top of a synthetic mooring line.

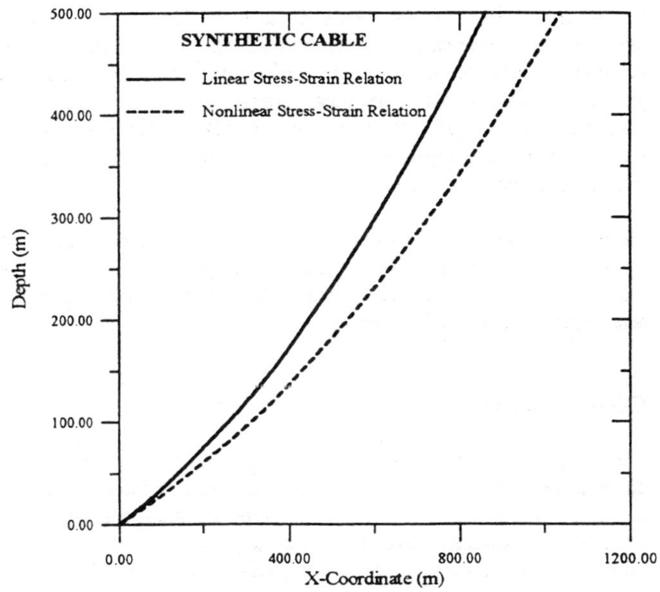

Figure 7. Static shape of a synthetic mooring line.

Next, we are concerned with the dynamic behavior of the wire mooring line, the characteristics of which are given in Table 2. All remarks made previously in relation with the wire cable examined in Figs. 1-3, remain valid for the present case too. In Fig. 8, the dynamic tension at the top of the steel cable is plotted against the frequency of excitation. All the already mentioned numerical schemes have been used, assuming a linear stress-strain relation.

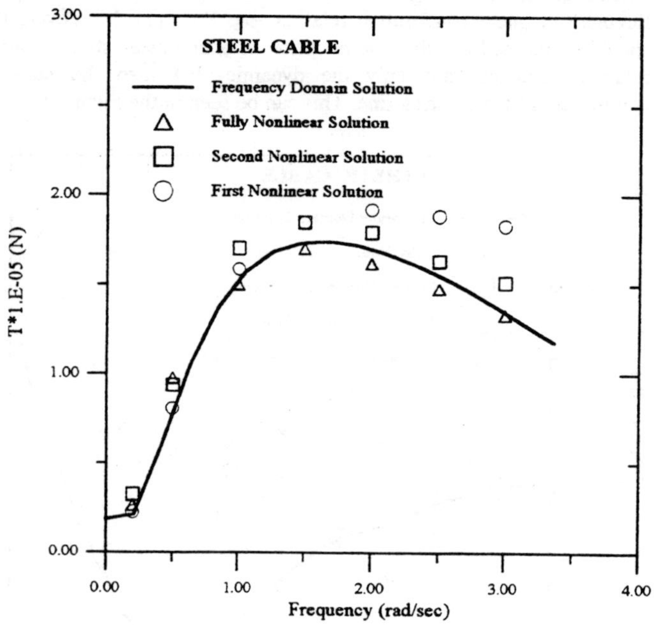

Figure 8. Dynamic tension at the top of a steel wire mooring line

It can be seen again that for small amplitudes of excitation (16 times the cable's diameter in the present case), the dominant nonlinearity arises from the quadratic drag and this is well predicted through the frequency domain analysis. Additional nonlinear phenomena, which the time domain simulation schemes account for, seem to be of minor importance. This is also reflected to the fact that the two dominating nonlinear solutions, i.e. the "Second-" and the "Fully - nonlinear", although based on principally different methods*, give results in good correlation between each other and to the frequency domain predictions for the whole frequency range of interest (the "first nonlinear solution" give reasonable results only for frequencies up to the first transverse eigenfrequency of the cable, i.e. for $\omega \leq 1.5$ rad/sec in the particular case).

Finally, for reasons of completeness, representative time histories of the total tensions at the cable's upper end are given in Fig. 9, for $\omega = 0.2$ rad/sec. They are calculated using various time domain techniques with comparable results.

CONCLUSIONS

The influence of several nonlinear contributions on the prediction of the dynamic behavior of a mooring line has been investigated, using different solution methods. Time variation of the strain and nonlinear stress-strain relations have been incorporated in the numerical schemes. It was shown that nonlinear phenomena are very important for increasing excitation amplitudes, thus the fully nonlinear mathematical model should be used for their accurate evaluation. It was found additionally, that contributions arising from time differentiation of the dynamic quantities in the compatibility relations may be significant in predicting the dynamic behavior of an elastic mooring line even for small excitation amplitudes.

* *Centered differences in frequency domain, Spectral Galerkin method and time integration using Newmark's method for the second nonlinear solution and finally collocation method and time integration using finite differences for the fully nonlinear solution.*

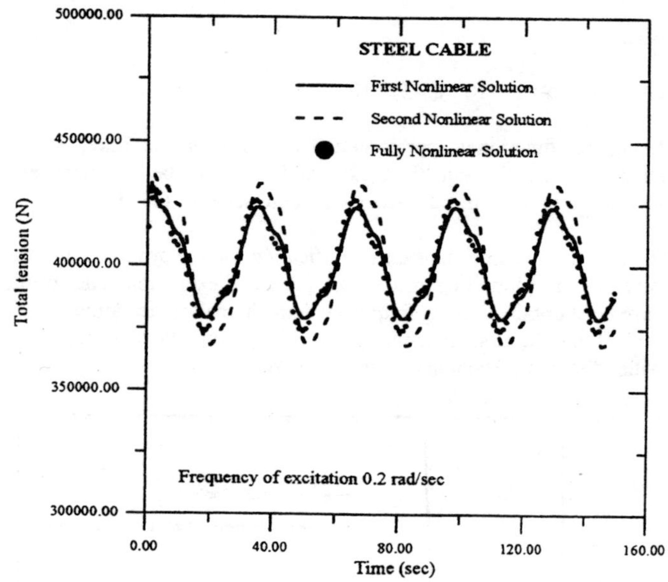

Figure 9. Tension's time history at the top of a steel wire mooring line.

REFERENCES

A.P.I., *Recommended practice for the design, analysis and maintenance of mooring for floating production systems* (RP 2FP1). American Petroleum Institute, 1991.

Burgess, J.J. Equations of Motion of a Submerged Cable with Bending Stiffness. Proceedings of the *11th International Conference on Offshore Mechanics and Arctic Engineering*, pp. 283-289, ASME 1992.

Chatjigeorgiou, J.K. Nonlinear Dynamics of Steel Mooring Lines (in Greek). Ph. D. Progress report, Dept. Nav. Arch. and Mar. Engng., National Technical University of Athens, Report No FLS96-2, Febr. 1996.

DnV, *Rules for Classification of Mobile Offshore Units. Position Mooring* (POSTMOOR), Part 6, Chap. 2. Det Norske Veritas, Oslo, 1989.

Fletcher, C.A.J. *Computational Galerkin Methods*. Springer-Verlag, New York Ing., 1984

Fylling, I.J., Ottera, G.O. & Gottliebsen, F. Optimisation and Safety Considerations in the Design of Stationkeeping Systems. Proc. *PRADS '87 Symp*., Trondheim, Norway, 1987, p.445.

Gear, C.W. *Numerical Initial Value Problems in Ordinary Differential Equations*, Prentice-Hall, Englewood Cliffs, New Jersey, 1971.

Gottlieb, D., Orszag, S.A. *Numerical Analysis of Spectral Methods: Theory and Applications*, SIAM Philadelphia, 1977.

Hoffman, J. D. *Numerical Methods for Engineers and Scientists*. McGraw-Hill, 1993.

Howell, C.T. Investigation of the Dynamics of Low-Tension Cables. Ph. D. Thesis, M.I.T., Cambridge, Massachusetts, USA, June 1992.

Kwan, C.T., Bruen, F.J. Mooring Line Dynamics: Comparison of Time Domain, Frequency Domain and Quasi - Static Analyses.

Proceedings, *Offshore Technology Conference (OTC'91),* OTC 6657, Houston, Texas, 1991.

Mavrakos, S.A, Papazoglou, V.J., Triantafyllou, M.S., Brando, P. Experimental and Numerical Study on the effect of Buoys on Deep Water Mooring Dynamics. Proceedings *1st International Offshore and Polar Engineering Conference (ISOPE'91),* Edinburgh, U.K., 1991, Vol. II, 243-251.

Mavrakos, S.A., Papazoglou, V.J., Triantafyllou, M.S., Chatjigeorgiou, J.K. Deep Water Mooring Dynamics. *Marine Structures*, 1996, **9**, 181-209.

Mavrakos, S.A., Chatjigeorgiou, J.K. Dynamic behavior of Deep Water Mooring Lines with Submerged Buoys. *Computers & Structures* (in print, 1996).

Papazoglou V.J., Mavrakos, S.A., Triantafyllou, M.S. Nonlinear Cable Responce and Model Testing in Water. *J. Sound and Vibration*, 1990, **140**, 103-115.

Press, W.H., Flannery, B.P., Teukolsky, S.A., Vetterling, W.T. *Numerical Recipes*. Cambridge University Press, Cambridge, England, 1986.

Shin, H., Triantafyllou, M.S. Dynamic Analysis of a Cable with an Intermediate Submerged Buoy for Offshore Applications. Proceedings, *8th International Conference on Offshore Mechanics and Arctic Engineering (OMAE'89),* The Hague, The Netherlands, 1989, Vol. II, 675-682.

Triantafyllou, M.S., Bliek, A., Shin, H. Dynamic Analysis as a Tool for Open Sea Mooring System Design. *Transactions SNAME*, 1985, **93**, 303-324.

Triantafyllou, M.S. Cable Mechanics for Moored Floating Systems. Proceedings, *7th International Conference on the Behavior of Offshore Structures (BOSS'94),* M.I.T., Cambridge, Massachusetts, 1994, **2**, 57-77.

Proceedings of the Seventh (1997) International Offshore and Polar Engineering Conference
Honolulu, USA, May 25-30, 1997

ISBN 1-880653-28-1 (Set); ISBN 1-880653-30-3 (Vol. II)

A Synthesis Procedure of Mooring System to Reduce Costs and Simplify the Underwater Layout

B.L.R. Andrade, M.R. Martins and H.L. Brinati
University of São Paulo
São Paulo, Brazil

ABSTRACT

This paper presents the formulation and application of a methodology to select the lines characteristics to be used in the early stage of the mooring system design. It is essentially a synthesis procedure that makes use of a simplified model, in order to reduce the computation time. The results obtained with this model are corrected by a quasi-static analysis of the actual system, which is incorporated in the procedure.

The synthesis procedure is applied in the design of a mooring system for a semisubmersible platform, considering the design constraints and different performance indexes. The analysis of the results illustrates how the choice of the performance index affect the selection of the lines configuration.

KEY WORDS: Mooring system, underwater layout

INTRODUCTION

As the oil production goes further into deep waters and there is an increase in the number of rigs operating in a given area, more severe requirements are specified in the design of the mooring systems for the floating units, in order to minimize the acquisition and installation costs, and to simplify the underwater layout. New anchoring technologies have been proposed top meet these requirements; an attractive solution is the taut-leg mooring systems which make use of lines of synthetic materials, like the polyester, and special anchors that are able to withstand vertical loads.

It does not matter whether a conventional or a taut-leg mooring system will be used, the question is how to determine the best combination of the mooring line variables in order to optimize a given performance index.

The design of a mooring system, today, is performed in accordance to renowned methodologies, like those suggested in the API recommendations or in the DNV rules, and make use of different codes like DMOOR (1994) and PROSAM (1995). The common feature of these methodologies and codes is that they are analysis tools; they should be applied to a mooring system with previously selected characteristics: number and pattern of lines; diameter and mechanical properties of each line segment; total line length and length of each segment; initial tension or horizontal distance to anchor at neutral position for each line. With these informations, as well as the data related to the environmental loads and hydrodynamic characteristics of the rig, the above methodologies and codes perform the calculation of lines tension and rig motion, which should then be checked against the design criteria.

However, since these methodologies or codes are not a synthesis procedure, they do not provide any help to the designer in the initial choice of the system parameters, not even in the adjustment of an already analyzed system, in order to meet the design requirements or to improve the performance index.

In these conditions the search of the optimum solution among the possible ones becomes an exhaustive trial and error process. Furthermore there is no guaranty that the final configuration obtained by this procedure is the best solution in terms of the adopted performance index.

In order to solve this problem, the authors developed a synthesis procedure that provides the best configuration of the mooring lines in accordance to the specified performance index, satisfying a given set of geometric and tension constraints. The utilization of an analysis code is required as a final checking of the design criteria.

The object of this paper is to present the description of the proposed methodology and illustrate its application in the design of conventional and taut-leg mooring systems. It is presented, in the sequence, the assumptions and the adopted model, as well as the calculation steps of the procedure. Finally, the synthesis procedure is applied in the design of mooring systems for PETROBRÁS semisubmersible platform SS-20, in a water depth equal to 1000 meters.

THE SYNTHESIS PROCEDURE

General description

The synthesis procedure was conceived to be used in the initial cycle of the mooring system design. The procedure purpose is to identify optimum set of mooring lines characteristics in such way that the designer could select, for further analysis, the particular set which optimizes a specified performance index like, for instance, the minimum weight, minimum cost, minimum line length, minimum horizontal distance to anchor, etc..

The determination of the best solutions is performed using the two opposed legs representation of the mooring system as proposed by D'Souza (1993), in order to accelerate the search of the solution. It was not used non-linear programming methods to obtain the optimum solution because the great number of design variables would make the computation time too long, which is not convenient for a calculation tool intended to be used in the early stage of the mooring system design.

The synthesis procedure is carried out by a computer program (PROTAN) which requires the following input data: environmental conditions (wave, wind and current), hydrodynamic parameters of the rig (drag coefficients, drift coefficients and RAOs); number and pattern of lines, design criteria and restrictions like maximum and minimum allowable line tensions, maximum and minimum line lengths, maximum initial tension, maximum total offset, maximum horizontal distance to anchor, or any other mechanical or geometrical restrictions.

The designer should also provide the following information: type of mooring system (conventional or taut-leg one), number of different segments, material for each line segment (chain, wire rope or polyester). It was assumed, for simplification, that all the lines have the same geometry and composition. The synthesis procedure makes use of the input data and a database of physical and mechanical properties of different line materials in order to : *i)* check whether there are mooring configurations, for each existing line diameter within a specific range, which satisfy the design requirements; *ii)* find for each line diameter, among the existing solutions, the optimum one, that means the line configuration for which the material mechanical strength is best used.

The synthesis procedure provides the following results for the optimum mooring configurations: total line length, length of each segment, initial tension and the horizontal distance to anchor at neutral position. The designer may use the results to select the line diameter that best meets the performance index adopted. The selected solution should then be submitted to an analysis code for a final checking of the design criteria.

The two opposed lines model

The two opposed lines model, as shown in Figure 1, is a representation of the actual mooring system obtained by linearizing the mooring lines response to displacements in the horizontal plane. The main reason for using this model is that it allows a quick evaluation of the maximum restoring force that a given mooring system could provide without violating the design criteria and restrictions.

The basic ideas of this model can be explained considering the simple case of a symmetrical mooring system with 8 equal lines; Figure 2a illustrates a floating unit subjected to a mean environmental load Fa in the direction of the x axis. As a result of the load application, the unit would be displaced by an amount δ, in the horizontal plane, and the mooring restoring force would balance the environmental load.

Assuming that the displacement δ is small when compared to the water depth and linearizing the mooring lines response, the force balance gives:

$$Fa = 2 \cdot (2 \cdot K_l) \cdot \delta \qquad (1)$$

where K_l is the line horizontal stiffness for a horizontal displacement in the plane of the line.

Let's consider a mooring system with two opposed lines; let's assume further that these lines have the same geometry and material of the actual system lines and are located in the vertical plane that contains the direction of the environmental load. Figure 2b shows a parallel spring model, with equivalent stiffness $2K_l$, representing this two lines system.

It may be seen from equation (1) that the horizontal stiffness of the whole system is twice the stiffness of the two opposed lines model. It is also clear that the restoring force of the whole system would be twice that of the model when both are displaced by the same amount δ. So, when the two lines model is displaced by any amount in the horizontal plane, the restoring force that would arise would be half the restoring force of the actual system. Furthermore, as the displacement and the lines are equal in the model and in the system, the tension in the most loaded line of the system will be equal the tension of the most loaded line of the model.

The factor 2, that appears in this example, is called here "responsibility coefficient", and is used in the synthesis procedure to evaluate the maximum restoring force and the stiffness of the mooring system, for a given line length and amplitude of dynamic motion (1st plus 2nd order motions). Similar results, with different value for the responsibility coefficient, could be obtained for non symmetrical patterns and for different numbers of mooring lines.

For taut-leg mooring systems the results concerning the lines tension and restoring force, obtained by the two opposed lines model, are, in general, very close to the actual system ones. For systems with conventional catenary lines significant differences between these values may appear and iterative corrections must be made during the synthesis procedure to reduce these differences to acceptable levels.

In order to illustrate the application of this model in the synthesis procedure, let's consider the problem of selecting a symmetrical mooring system with 8 equal lines to keep positioned a unit subjected to a mean environmental load (F_a) acting in the X axis direction, as shown in Figure 2a. Let's assume that the tension in the mooring lines and the rig displacement may be evaluated by the quasi static method proposed by API (1987) and that, for simplicity, the design criteria and restrictions are just the following ones: (a) the maximum allowable tension on the lines is equal to 50% of the material breaking load; (b) the minimum allowable tension in the lines is equal to 5% of the material breaking load.

Suppose now that the lines are to be composed of only one kind of material. Once chosen this material, the problem is to find the best combination of the lines' characteristics (material diameter, line length, initial tension, etc.) in such way that the mooring system complies with the design criteria and restrictions and be the best according to some performance index.

In order to obtain the optimum solution, let's assume the same problem applied to a equivalent mooring system composed by two equal and opposed lines, as shown in Figure 1.

Given a line diameter and a total line length it is possible, using the catenary equations, to obtain the lines configuration for which one line is subjected to a tension equal to 50% of the material breaking load and the other is subjected to a tension equal to 5% of the breaking load. These are the limiting configurations for the lines in order that the design criteria and restrictions are still met. In accordance to API's quasi static method, this could only occur when the unit would present the maximum offset, which is given by the sum of the mean offset with the dynamic motion.

This situation is represented by the points labeled with the number 2 in Figure 3, assuming that the two lines reach these limiting configurations at the same time. This figure shows the curve of tension at the top of the line against horizontal distance to anchor, calculated by catenary equations.

From the position that corresponds to the maximum offset, it is possible to evaluate the horizontal span of the lines at the neutral position, which is given by the mean value of the horizontal spans of the two lines at the maximum offset position. Using the catenary equations it is also possible to evaluate the initial tension. The neutral position is represented by the point 0 in Figure 3. In this figure, the difference between the horizontal spans at points 0 and 2 represents the maximum offset which the unit can exhibit without violating any design criteria.

For a given dynamic motion δ_d, it is possible to evaluate, from the position of the maximum offset, the position corresponding to the mean offset, as it is shown in Figure 3 by the points 1. In a similar way, the horizontal spans of the lines may be obtained at the mean offset position and, then, the tension on the lines could be evaluated at this position.

At the mean offset position, the difference between the horizontal components of the lines tensions gives the maximum restoring force (*FR*) that the two lines system could provide.

From the above analysis, it may be concluded that there is a correspondence between dynamic motion and maximum restoring force of the model. By repeating this procedure for others values of total line length and dynamic motion it is possible to draw curves relating the maximum restoring force to the total line length for different values of dynamic motion (δ_{di}), as it is shown in Figure 4. Each point in these curves indicates the maximum restoring force that a system with two opposed lines could provide while complying with the design criteria. It is important to point out that these curves would change whenever any design criteria or restriction is modified.

The results obtained for the two lines model can be extended to the actual mooring system, provided that they have lines with the same materials and the same diameter. According to equation 1, if the model is subjected to a force equal to half of the environmental load, the model and the system would have equal mean offsets.

Taking this result into account, it is possible to use Figure 4 to select the lines configuration of the actual system. If a restoring force equal to half of the mean environmental load is taken in this figure, we may select the line length for each value of dynamic motion.

It must be pointed out that the values of dynamic motion were arbitrarily selected and the actual dynamic motion of the unit is not known yet, since it depends on the mooring system stiffness which is a function of line length. Therefore it must be found the line length for which the dynamic motion will be equal to the one considered in the determination of the restoring force for this length, as it will be explained later.

Points 1 and 2 in Figure 3 were obtained assuming that there are only constraints for the maximum and minimum tension at the top of the line. If, additionally, others constraints are imposed in the design, like maximum and minimum tension at the anchor, maximum and minimum horizontal span, etc., the determination of points 1 and 2 should take this into account.

It is important to point out that the line lengths selected with this procedure give rise to mooring systems that best use the available mechanical strength of the lines material.

Procedure Outline

The synthesis procedure rely basically on the evaluation of the maximum restoring force that a mooring system, with a given line material and diameter, can provide while satisfying the design criteria and restrictions. Once this force is determined for different line lengths, a comparison is made with the design environmental load, and the line length, for which the material mechanical strength is best used, is then selected.

The main assumptions adopted in the synthesis procedure are:

- the environmental loads acting on the unit are collinear;
- the mooring system does not affect the first order motion of the unit;
- the rig motion in the horizontal plane is determined by a combination of the 1st and 2nd order motions in the direction of the environmental load;
- the lines tensions due to the rig dynamic motions can be evaluated by the quasi-static method suggested by API (1987);
- all the lines of the mooring system are made from the same material and have the same diameter and total length.

The basic ideas of the synthesis procedure are summarized in the graphics shown in Figure 5.

- Graphic 1 - Restoring Force (*FR*) x Line Length;
- Graphic 2 - System Stiffness (*S*) x Line Length;
- Graphic 3 - System Stiffness x Total dynamic motion.

Once adopted a line material and diameter, the first step consists in the calculation of the maximum restoring force that the mooring system can provide for a previously specified range of line lengths and dynamic motion; it is used the two opposed lines model in this evaluation. The results of this step are the curves shown in graphic 1 of Figure 5.

In the second step it is evaluated the mooring system stiffness at the mean offset, associated with each point on the curves of graphic 1. Again, by using the two lines model and the pre-established values of dynamic motion, the system stiffness can be calculated for each line length, leading to the results shown in graphic 2.

Now, knowing the actual environmental load (*Fa*) acting on the unit, it is possible to obtain from graphic 1 the line lengths for which the mooring system will provide the same value of restoring force, for each value of dynamic motion. For each line length the plane stiffness is obtained from graphic 2, and a third graphic can be plotted showing the system stiffness against the corresponding dynamic motion, as shown by curve 1 of graphic 3.

Assuming a linear spring-mass-damper model to simulate the 2nd order motion of the unit, a second curve can be plotted in graphic 3 (curve 2) showing the relationship between the system stiffness and the dynamic motion. The point where curves 1 and 2 match indicates the solution of the synthesis procedure. With the values of the system stiffness and dynamic motion of the matching point, it can be obtained from graphic 2 the line length which corresponds to the optimum solution for the pair of line material and diameter initially settled.

To enhance the pointed solution, a final analysis with the whole mooring system is made to check the errors introduced by the two opposed lines model. If these errors exceed some specified value, the model is corrected and the process repeated until the differences are negligible.

The process can be repeated for other diameter values of the same material or for different materials. As final results the designer can have the optimum line geometry for different diameters and then select the one that best satisfies the desired performance index.

The synthesis procedure described here in, including the correction of the model, is carried out by a computer program (PROTAN). This program can handle conventional and taut-leg mooring systems and accepts multicomponent mooring lines.

AN ILLUSTRATIVE EXAMPLE

The synthesis procedure was applied to the design of the mooring system for PETROBRÁS semisubmersible platform SS-20 in a water depth equal to 1000 meters.

Platform and environmental data

The platform main characteristics are:

- displacement: 22642 tons;
- number of columns: 6;
- columns diameter: 12m;
- pontoons length: 80m;

The platform is headed north and the environmental condition considered is a concurrent collinear combination of a 100 years wave and wind with a 10 years current. In this condition the maximum steady force acting on the platform is 3805 kN in the 315 degrees off the bow counterclockwise direction. The damping coefficient in this condition is 2470 tons/sec (excluded wave drift damping).

Mooring system characteristics

The mooring lines in both configurations considered in the analysis (conventional and taut-leg) have three different segments, with the below described characteristics:

a) taut leg mooring line
   - 1st segment (near the anchor): grade 4 mooring chain, with 50 meters;
   - 2nd segment: polyester, with length selected by the procedure;
   - 3rd segment (near the fairlead): 150m length of grade 4 mooring chain;

b) conventional mooring line
   - 1st segment (near the anchor): grade 4 mooring chain, with length defined by the procedure;
   - 2nd segment: wire rope, with length selected by the procedure;
   - 3rd segment (near the fairlead): 150m length of grade 4 mooring chain.

The length of the 3rd segment in both mooring systems is fixed in accordance with the operating requirements. The length of the 2nd segment is determined in both cases by the synthesis procedure for each material diameter.

The length of the 1st segment, near the anchor, is also determined by the synthesis procedure in the case of conventional mooring systems. For taut-leg mooring systems the length of the 1st segment is fixed according to the requirements of anchor deployment operation.

The study has been performed considering 10 different lines for taut-leg and conventional systems, each one corresponding to a different level of line break strength or diameter.

Design restrictions

The following general design restrictions have been imposed in the synthesis of each mooring system:

- maximum initial line tension: 1500 kN;
- maximum vertical tension on the line: 3000 kN;
- maximum offset: 10% of water depth;
- maximum horizontal distance to anchor: 3000m.

The specified constraints imposed to each mooring system were the following ones:

a) taut-leg mooring systems
   - maximum line tension: 60% of the line break load;
   - maximum line length on the sea bottom: 50m;
   - minimum line tension on the polyester segment: 5% of the line break load;

b) conventional mooring systems
   - maximum line tension: 50% of the line break load;
   - minimum line length on the sea bottom: 100m.

The difference in the maximum line tension allowed for the two kinds of mooring systems is due to the difference expected in the dynamic tension caused by the 1st order motion of the rig. In the case of taut-leg systems this dynamic tension will be of the same order of the quasi-static one and a grater value of maximum tension could be allowed in a quasi-static evaluation. In the case of conventional systems the dynamic tension may be much greater than the quasi-static value.

NUMERICAL RESULTS AND DISCUSSION

Figures 6 to 12 show the main characteristics of the optimum solutions determined by the synthesis procedure for the taut-leg and conventional mooring system of the SS-20 platform. The results are presented as a function of the material's breaking load.

The values of the lines total length and initial tension that define the optimum configuration for each level of breaking load are shown in Figures 6 and 7. The other figures present the values of lines horizontal span, maximum offset, maximum line tension, minimum line tension and maximum vertical force on the lines.

The optimum solutions pointed by the synthesis procedure are not the only ones that comply with the assumed design criteria, but are those which take the best advantage of the mechanical strength of the material and restoring capability of the mooring system, that means, those which comply at the limits with the design constraints.

As it is shown in Figure 6, for the taut-leg system, the synthesis procedure found optimum solutions for all the 10 breaking load levels (or for all 10 different diameters) considered in the analysis. For the conventional mooring system the procedure have pointed solutions only for the 7 largest values of breaking load. For the three smallest values there were no solutions that complied with the design criteria and constraints.

It may be seen from Figure 6 that for taut-leg systems the total line length of the optimum solutions decreases as the breaking load increases until a minimum value is reached for a 5 MN breaking load. For conventional systems the total length is practically the same for all the whole breaking load range.

It may be seen from Figure 7 that for taut-leg systems the initial tension of the optimum solutions increases as the breaking load increases until a maximum value equal to 1500 kN, which correspond to the maximum allowable value of the initial tension, is obtained for 5MN breaking load. For a further increases in the breaking load, the initial tension remains equal to the limiting value. For conventional systems, the initial tension of the optimum solutions is equal to the limiting value for the whole breaking load range.

The change of tendency in the curves of total line length and initial tension, observed in figures 6 and 7, is due to a change in the prevailing constraints of the optimization procedure. Generally speaking, there are, at least, two prevailing constraints in the search procedure, that means, for the optimum solutions two constraints will be satisfied at their limiting values.

In the case of the conventional mooring system, the optimum solutions were determined, for all the breaking load levels, by the same pair of constraints. The first one is the maximum offset, which was assumed to be equal 10 percent of the water depth. Figure 9 shows that all the optimum solutions present this offset. The second constraint is the initial tension on the lines which for all the solutions reach the maximum value, as it may be seen in Figure 7.

For the taut-leg systems, the prevailing constraints change as a function of the breaking load level. For the three smallest values considered the optimum solutions were determined by the maximum and minimum allowable tension on the lines, as it is shown in Figures 10 and 11. For braking loads greater than 5 MN, the initial tension becomes a prevailing constraint instead of the maximum allowable tension, as it may be seen in Figures 7 and 10. The minimum allowable tension on the lines remains as the other prevailing constraint in this range of breaking load. An interesting conclusion is that taut-leg system with breaking load greater or equal 5 MN will be subjected to maximum tensions lower than those permitted by the design criterium.

The results shown in Figures 6 to 12 may be used by the designer to select the system configuration which best satisfy a given performance index, like the minimum costs system, the minimum line length system, the minimum horizontal distance to anchor, minimum total offset system or even a configuration which best meets a combination of these indexes.

For instance, in the case of taut-leg mooring systems, if the performance index assumed is the minimum line length or minimum horizontal distance to anchor, which would contribute to simplify the underwater layout, it would be selected a system composed by lines with breaking load equal to 5 MN, as it may be noticed from Figures 6 and 8. If the performance index is the smallest total offset, it would be selected a mooring system with the greatest line breaking load.

The results shown in the figures may be used to compare the performance of taut-leg and conventional mooring systems. It may be seen from figures 6 and 8 that taut-leg systems present line lengths and horizontal distance to anchor much smaller (1/2 to 1/3) than the conventional ones, which demonstrate the great advantages of the taut-leg systems in terms of underwater layout. A more complete comparison between taut-leg and conventional mooring system is presented in Martins et all (1997).

The optimum solutions given by the synthesis procedure were subjected to a quasi-static analysis using DMOOR code. There was a good concordance between the values given by the two methods; the horizontal distance to anchor is exactly the same and there is a very slight difference in the maximum offset. Table 1 presents a comparison of the results obtained by the two procedures for taut-leg and conventional mooring systems with breaking loads equal to 5 MN and 6 MN, respectively.

The next step of the development of the synthesis procedure is the introduction of a simplified dynamic analysis of the line, like that performed by PROSAM, Aranha et all (1995). The synthesis procedure with this inclusion will become a selfcontaint design tool.

		Synthesis Procedure	DMOOR
Taut-Leg System	Horizontal Distance to Anchor (m)	881.21	881.32
	Maximum Offset (m)	24.57	25.12
Conventional System	Horizontal Distance to Anchor (m)	2272.97	2272.85
	Maximum Offset (m)	99.98	98.58

Table 1 - Validation of synthesis procedure

CONCLUSION

The paper described a synthesis procedure that is a very useful tool to design taut-leg mooring systems as well as conventional ones. Differently from most existing softwares, the proposed procedure, besides performing a mooring system analysis, provides the optimum line configuration for a given set of design constraints and a specified performance index.

A good agreement with the results obtained with DMOOR code was shown in the paper. It should be mentioned that in a previous study, Martins (1996), the solutions obtained by the procedure were analyzed using renowned softwares like DMOOR, version 4, and the results obtained indicate that the simplifications adopted do not introduce significant errors in the determination of the most efficient solution.

The analysis carried out with the semisubmersible platform SS-20 showed the feasibility of taut-leg mooring systems. Although they present some operational problems, these disadvantages are surpassed by the favorable characteristics that the taut-leg system present in terms of reduction of rig offsets and horizontal spans.

The results of the analysis carried out with the semisubmersible platforms could be extended to ships, as it may be seen in Andrade and Martins (1995).

ACKNOWLEDGMENTS

The results presented in this paper were obtained from a joint research project developed with PETROBRAS and the authors express their gratitude to the company for providing the opportunity to test the methods presented here.

The gratitude is extended to Mr. Arthur Eugênio Ribeiro, undergraduate student of USP, for his co-operation in software development.

REFERENCES

American Petroleum Institute (may 1987). "API Recommended Practice 2P, Analysis of Spread Mooring for Floating Drilling Units", second edition;

Andrade, BLR, and Martins, MR (may 1995). "Evaluation of the Taut-Leg Mooring Configuration for a Production Platform. Project report", (in Portuguese), PETROBRÁS/USP agreement;

Aranha, JAP, Martins, CA, and Martins, MR (1995). "PROSAM - Design of Mooring Systems", (in Portuguese), User's Guide, version 2;

DNV Position Mooring (POSMOR), July 1989;

D'Souza, RB, Dove, PGS, and Kelly, PJ (1993). "Taut-Leg Spread Moorings: A Cost-Effective Station Keeping Alternative for Deep-Water Platforms", 25th Annual OTC, Texas, USA, paper number OTC 7203;

Martins, MR, Brinati, HL, and Andrade, BLR (1997). "A Comparative Study of the Conventional and Taut-Leg Mooring Systems", 16th International Conference on Offshore Mechanics and Arctic Engineering, Yokohama, Japan;

Martins, MR (1996). "A Synthesis Procedure for Mooring Systems" (in Portuguese), M.Sc. Thesis, Dept. of Naval Architecture and Ocean Engineering, USP, São Paulo, Brazil;

Noble Denton & Associates (1994). "DMOOR - Deterministic Mooring Analysis - User's Guide", version 4.0.

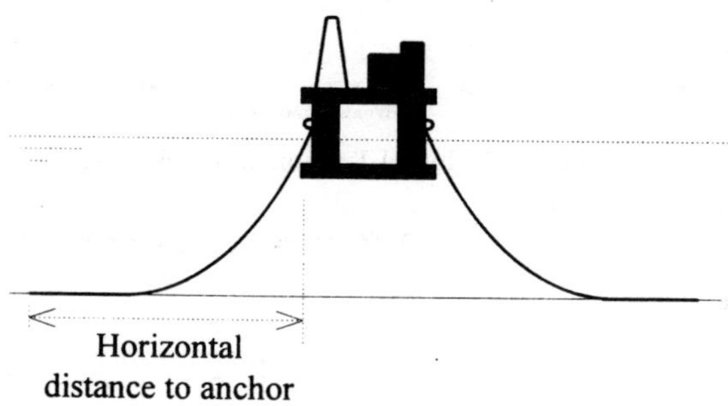

Figure 1 - Mooring systems composed by two equal opposed lines

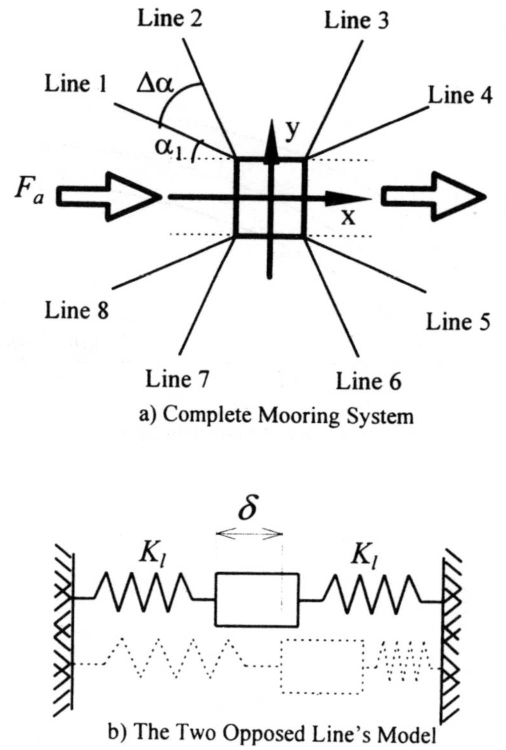

Figure 2 - Mooring system and the model.

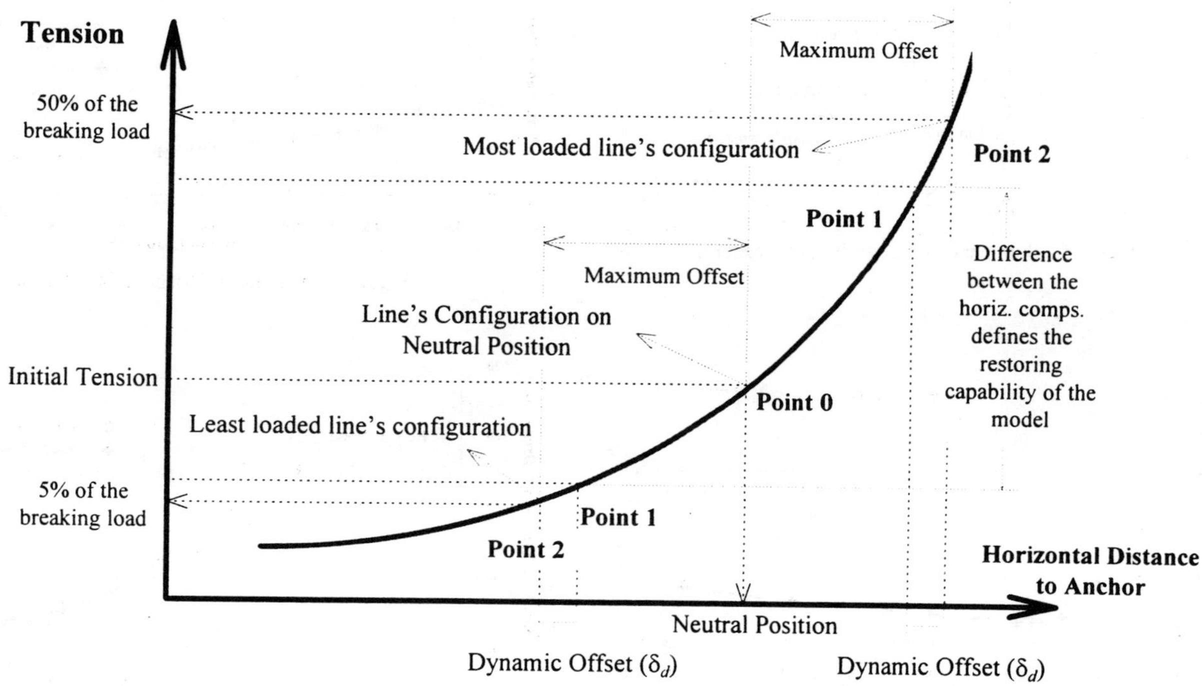

Figure 3 - Tension x Horizontal Distance to Anchor

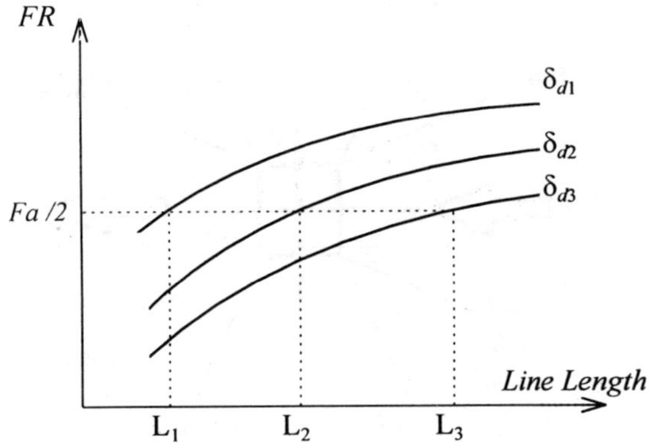

Figure 4 - Maximum restoring force

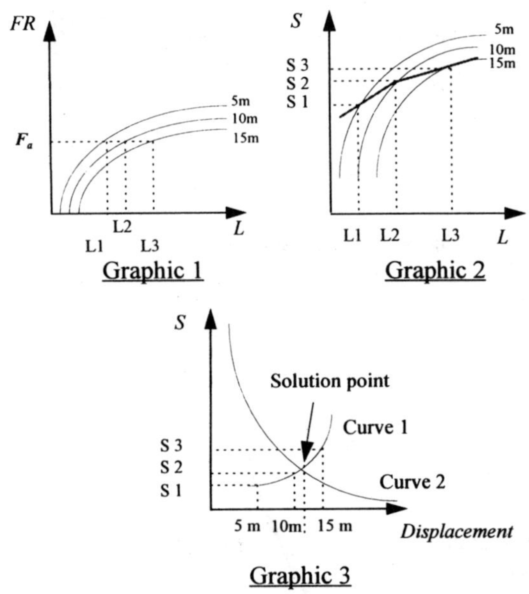

Figure 5 - Summary of the synthesis procedure.

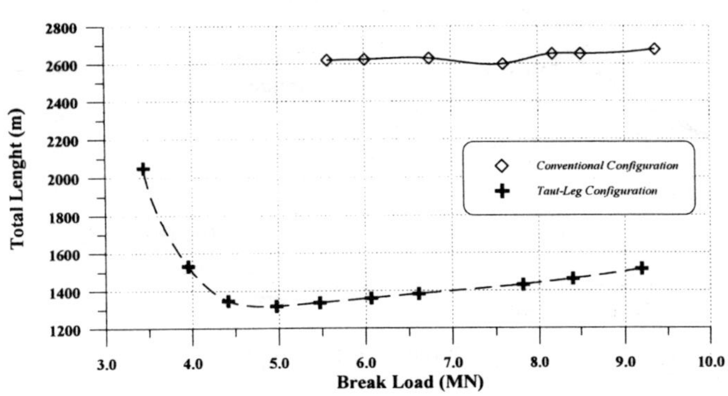

Figure 6 - Maximum Length x Break Load

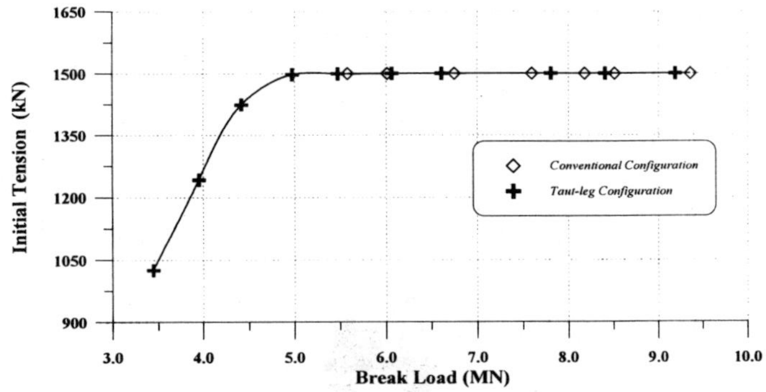

Figure 7 - Initial Tension x Break Load

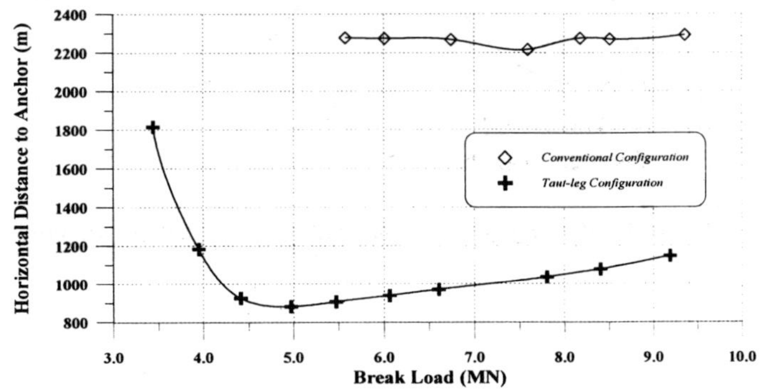

Figure 8 - Horizontal Distance to Anchor x Break Load

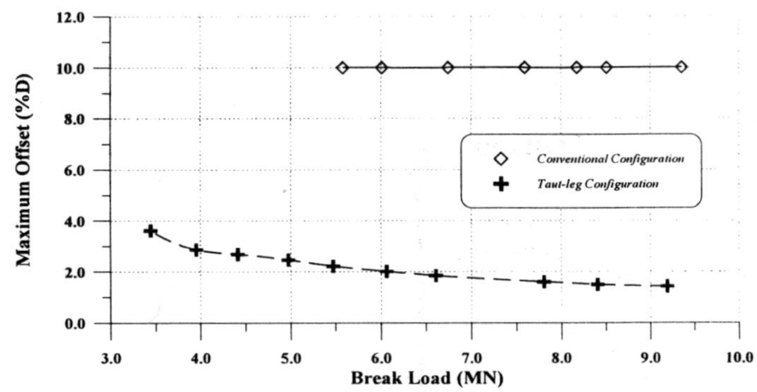

Figure 9 - Maximum Offset x Break Load

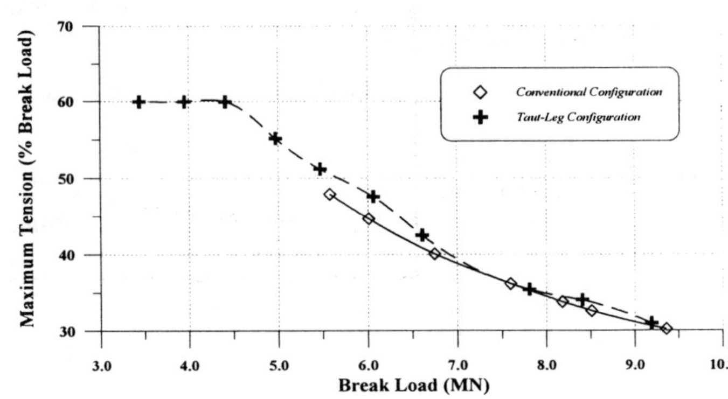

Figure 10 - Maximum Tension x Break Load

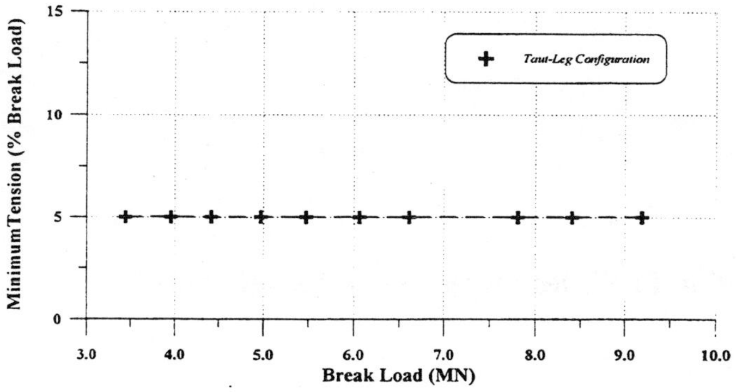

Figure 11 - Minimum Polyester Tension x Break Load

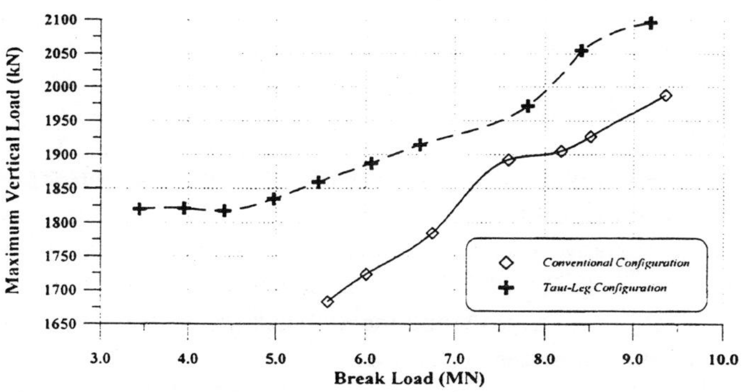

Figure 12 - Maximum Vertical Load x Break Load

Proceedings of the Seventh (1997) International Offshore and Polar Engineering Conference
Honolulu, USA, May 25-30, 1997

ISBN 1-880653-28-1 (Set); ISBN 1-880653-30-3 (Vol. II)

Transport Support for Oil and Gas Production Facilities on the Sakhalin Offshore

A.M. Polomoshnov and S.V. Astafiev
Oil and Gas Institute
Okha, Russia

ABSTRACT

In the next few years, marine vessels are expected to considerably increase their activities due to the beginning of offshore hydrocarbon field construction under 'Sakhalin-1" and "Sakhalin-2" Projects.

Oil production complexes will be designed for each offshore field. They will include several basic facilities such as:
- conventional ice-resistant platforms
- subsea pipelines
- onshore facilities
- subsea loading terminals
- subsea wellhead templates

Basically, problems of securing continuous operation of oil and gas production complexes will be associated with the existence of ice cover.

DESCRIPTION OF ICE COVER

Depending on winter severity, ice appears on the north-eastern Sakhalin offshore in mid-November or late November and disappears in mid-June. Ice concentration in this region reaches 9-10 points in late January. Persistent hummocking and grounding of hummocks is observed in February on the background of level ice thickness growth. Ice cover parameters reach their maximum values in March-April (Table 1).

Heavy ice belt exists in this time period within the close floating ice zone. This heavy ice is formed in western Okhotsk Sea (The Shantar Islands), upon compacting and consolidation in Sakhalin Bay it is driven to the area of offshore oil and gas fields. The heavy ice belt is 15-40 km in width and does not include giant and vast ice floes. Ice massif itself is noted for large vertical thickness and hummocking (Polomoshnov et al., 1990).

A flaw polynya 20-40 km in width is observed in this region annually in April-May.

Comparative analysis of ice cover parameters for the Arctic seas and Sakhalin offshore oil and gas field area indicates that they differ only slightly and the Okhotsk Sea may be positively classified among the Arctic seas.

Table 1

Values for ice cover parameters in March-April for the north-eastern Sakhalin offshore

Value ice cover	Parameter	
	average	maximum
Thickness of level ice of thermal origin	0.85 m	1.6 m
Ice floe dimensions	0.6 x 1.1 km	25 x 41 km
Thickness of rafted ice	1.5 m	2.7 m
Thickness of hummocked ice	7.0 m	35 m
Hummocked ice	3 point	5 point
Ice drift velocity	47 cm/sec	201 cm/sec
Floating ice concentration	9 - 10 point	10 point
Fast ice width	0.5 -1.0 km	7 km

VESSELS REQUIRED FOR OPERATIONS IN ICE CONDITIONS

Vessels capable of providing support for the operation of offshore platforms in the Arctic seas should be classified based on their functions:

- survey vessels (acquisition of meteocean data, ecological surveys, etc.);
- support vessels (delivery of drilling and production equipment);

- ice-breakers (escort of support vessels, safety provision for production platforms);

- tankers (delivery of produced hydrocarbons to consumers);

- rescue boats (rescue of personnel from operation platforms in case of accident, elimination of pollution effects);

In Russia, ice-breakers are used which are conventionally grouped into four classes: ЛЛ1, ЛЛ2, ЛЛ3, ЛЛ4 (Classification rules and construction codes for marine vessels, 1986).

Class ЛЛ1 ice-breakers perform all types of ice-breaking the whole year round in the Arctic seas along coastal and high latitude routes. They can move almost unlimitedly in compact ice and hummocked ice. Total propeller shaft power is 47807 kwt (65000 h.p.) and higher (Ice-breakers "Arctic" and "Russia").

Class ЛЛ2 ice-breakers perform independently all types of ice-breaking in the Arctic seas in summer season. In winter season they operate along coastal routes. They can move in compact ice below 2,0 m in thickness. Total propeller shaft power is 22065 to 47807 kwt (30000 to 65000 h.p.). "Makarov" and "Ermak" belong to ice-breakers of this ice class.

Class ЛЛ3 ice-breakers perform independently all types of ice-breaking in winter season in non-Arctic freezing seas, in shallow water zones and estuaries of rivers flowing into the Arctic seas. They can operate the whole year round along coastal routes in the Arctic seas together with higher ice class ice-breakers. They can move in compact ice 1,5 m in thickness (Ice-breakers "Moscow", "Vladivostok", etc.). Total propeller shaft power is 11032 to 22065 kwt (15000-30000 h.p.).

Class ЛЛ4 ice-breakers perform independently all types of ice-breaking in port and near-port water areas all the year round. They can move in compact ice 1,0 m in thickness. Total propeller shaft power is below 11032 kwt (15000 h.p.). "Khariton Laptyev" and "Fyodor Litke" belong to ice-breakers of this ice class.

In addition to ice-breakers, some ships can move independently in ice conditions which are classified as ice ships and are assigned the following symbols: УЛА, УЛ, Л1 , Л2 and Л3.

Table 2 gives correlation between ice classes according to the Russian Register of Shipping and similar ice classes of foreign ships.

OPERATION OF VESSELS IN THE ARCTIC SEAS OFFSHORE FIELDS

The demand for active development of Prudho-Bay fields discovered in 1968/69 offshore of Alaska caused development of efficient oil transport system.

Based on preliminary studies, oil transport along north-western passage was selected as a preferable option.

To verify this option, some oil companies purchased and updated the tanker "Manhattan" (150000 t water displacement) which was equipped with 43000 h.p. power system. Beginning August 1969, the tanker escorted by two icebreakers was involved in various transport activities, its mean velocity in ice-covered waters being 3-6 knots. During its way back the tanker got two holes as a result of riding up an iceberg in Lancaster Straight. In winter-spring, the second trip of this tanker was undertaken under the escort of the ice-breaker "Sent-Loran". Several times the tanker was blocked in the Beaufort Sea hummocked ice. In Lancaster Straight the tanker and the ice-breaker were ice nipped in the zone of compact ice and stayed there for 6 days essentially motionless.

On July, 13 "Manhattan" returned to the eastern USA coast with damages received in ice conditions (one bow fracture and two stern fractures).

These trips indicated that "Manhattan" could move independently in compact ice 135 cm in thickness only by hollowing out (back-and-forth movement mode). Based on calculations, 12500 kwt engines would be required for "Manhattan" to move uninterruptedly at a speed of 4 knots against 250 cm thick compact winter ice. The experiment revealed insufficient maneuverability of the tanker. Tanker circulation diameter in compact ice was 23 km, which is 13 times higher than that in clear water (Arikainen,1985).

Two powerful ice-breakers "Polar Star" and "Polar Sea" were constructed by the end of 70-ies. Water displacement was 13000 t and power-to-weight ratio about 4,5 h.p./ t. Field tests did not turn to be a success (screw and engine failure), but extremely unsuccessful was operation of "Polar Star" in winter of 1981.

In February, the ice-breaker was in heavy multi-year ice. The efforts to free the ice-breaker by using blasting charges did not lead to the expected result. Moreover, screw propeller shaft was broken. The ice-breaker entered the Bering Straight only by mid-May upon termination of ice compression and appearance of fractures in ice (Arikainen, 1989).

An expedition on the ice-breaker "Ermak" (class ЛЛ2 or CAC-5) was undertaken offshore of Sakhalin in March 1982. This ice-breaker had been working for two days in water areas of Odoptu and Piltun-Astokh fields. Ice conditions during this time period were close to average multi-year ice conditions. The expedition resulted in the following conclusions:

- The ice-breaker "Ermak" could move only along fractures and in thin ice zones, passing around ice breccia and hummocked ice.

- The ice-breaker could not move ahead in compression zones and was blocked by ice several times.

- The efforts to move in hummocked ice and ridges resulted in ice-breaker blocking in ice.

- Ice-breaker escort of УЛА and УЛ class transport vessels is possible in such season only along fractures and thin ice zones by passing around heavy ice zones.

- Precise ice-breaker positioning (holding it at a fixed point) during a number of days is not possible in this season due to high ice drift velocities (Report on ice-breaker "Ermak" expedition, 1982).

- Fig. 1 gives working parameters for the ice-breakers "Ermak" and "Moscow" in ice fields in the Far East seas.

SCENARIOS FOR TRANSPORT SUPPORT OF OIL AND GAS PRODUCTION FACILITIES ON THE SAKHALIN OFFSHORE

Transport activities to be carried out on the Sakhalin offshore during ice seasons may be conventionally classified as regular as non-regular.

Non-regular activities (metocean and ecological surveys, etc.) are not strictly time-dependent and may be carried out upon waiting for ice cover weakening or along arbitrary routes with ice-breaker escort, while regular activities (delivery of equipment to platforms or tanker loading with produced hydrocarbons) must be performed in strict compliance with the schedules.

Tanker loading from the platform or subsea loading terminal is the most complicated transport operation. The maim problem in case of tanker at an offshore loading terminal is the required tanker positioning (holding it at a fixed point) during its loading.

On average, this operation will take 1,5 - 2 days for 40000 t deadweight tanker. Note should be taken that daily drift of ice massif is 15-20 km (Polomoshnov et al., 1990). Tanker loading should be in strict compliance with the schedule. Failure to comply with the schedule may cause decrease in the amount of produced hydrocarbons and even termination of production flow in the fields. Such situation is probable for the fields on the north-eastern Sakhalin offshore in March-April in case of unavailable alternative option for tanker transport of hydrocarbons.

To select the most efficient and ecologically safe option for hydrocarbon loading under the Projects, several scenarios are considered for oil and gas transport from offshore fields by subsea pipelines to the coast and further by onshore pipelines to one of the ports (De-Kastri, Prigorodnoye and Khoe).

In our opinion, a preferable option is the whole year round tanker loading at an offshore loading terminal in De-Kastri (Chikhachiov Bay). For many years, timber has been loaded and carried by ships the whole year round in this port. The experience indicates that class Л1 and УЛ ships navigate independently in early and late winter along this route. In January-March, ships going from ice edge to Chikhachiov Bay are escorted by an ice-breaker (Class ЛЛ3 in severe winters and Class ЛЛ4 in mild and moderate winters). Average maximum ice thickness varies from 38 cm in severe winters to 20-30 cm in mild and moderate winters. Generally, 40-50 ships are received in De-Kastri during winter season.

Ice conditions in the port of Khoe are more complicated as compared to those in Chikhachiov Bay. Annually, fast ice 20-30 km in width is formed here, its thickness reaching 100-105 cm by March. High compression is observed in this water area upon fast ice breaking off. Such compression results in forming hummock ridges and grounded hummocks.

Ice conditions in the port of Prigorodnoye are not very complicated. However, a pipeline along the entire Sakhalin Island need to be constructed to provide for loading at an offshore loading terminal under this option.

CONCLUSIONS

1. Ice conditions in the area of the north-eastern Sakhalin offshore fields are complicated enough for the operation of vessels.

2. Experience in using special vessels of various ice classes to support oil and gas facilities operation in winter navigation indicates that offshore transport operation in strict compliance with the schedules as required for continuous operation of production platforms is not possible at present.

3. Equipment delivery to drilling platforms by support vessels and tanker loading are the most difficult operations in winter seasons.

4. During tanker loading in severe ice conditions the probability of tanker damage exists, which poses higher environmental pollution threats .

5. The alternative option for oil and gas transport from offshore platforms to consumers by subsea pipelines is more reliable and ecologically safe.

1.6. Based on the available oil and gas transport system in Sakhalin, the most economically attractive option is oil and gas delivery by pipelines to De-Kastri and further transport by tankers.

REFERENCES

Polomoshnov, AM, Truskov, PA., and Yakunin, LP (1990). "Influence of Grounded Ice Hummocks on Pipelines," PACOMS - 90, Seoul, pp. 401 - 407.

Classification rules and construction codes for marine vessels. Leningrad, Transport, 1986 (in Russian).

Arikainen AI (1989). "In the Ice of North-American Arctic," Leningrad, Gidrometizdat (in Russian).

Report on ice-breaker "Ermak" expedition. Okha, 1982 (in Russian).

Polomoshnov AM, Astafiev VN, and Truskov PA (1990). "Ice Investigations on Sakhalin Offshore," *V Intern. Symposium on Okhotsk Sea Ice*. Mombetsu, Hokkaido, Japan, 4 - 6 February, pp. 147 - 156.

Table 2

Correlation between ice classes for ships in various countries

Ice class according to Russian Register of Shipping	Ice class according to Finnish-Sweden classification (Lloyd's Register of Shipping, Bureau Veritas, Norwegian Veritas, NKK)	Ice class according to Canadian classification	German Lloyd's	ABC
ЛЛ1	-	CAC-1A	-	-
ЛЛ2	-	CAC-2	-	-
ЛЛ3	-	CAC-2	-	-
ЛЛ4	-	CAC-3	-	-
УЛА	-	CAC-5	-	F2
УЛ	1A Super	CAC-7	E4	A1
Л1	1A	-	E3	AO
Л2	1B	-	E2	BO
Л3	1C	-	E1	CO
Л4	-	-	-	-

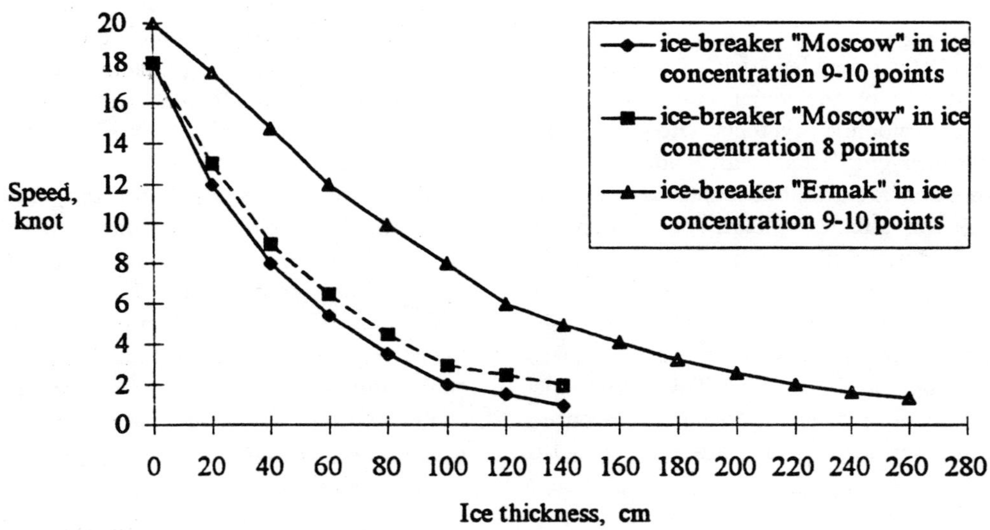

Fig. 1. Ice-breaker speed as a function of compact level ice thickness

Proceedings of the Seventh (1997) International Offshore and Polar Engineering Conference
Honolulu, USA, May 25-30, 1997

ISBN 1-880653-28-1 (Set); ISBN 1-880653-30-3 (Vol. II)

A Numerical Simulation to Determine Ice Scour and Pipeline Burial Depth

Ki-Young Yoon
Hyundai Heavy Industries Co.
Ulsan, Korea

Kyungsik Choi and Han-Il Park
Korea Maritime University
Pusan, Korea

ABSTRACT

The interaction of grounded ice features with underlying seabed is one of the major considerations in the design of Arctic pipeline systems. Regarding the development of offshore gas fields near Sakhalin Island, which is an ice-infested area, in this paper an ice scour model to determine the subsea pipeline burial depth is studied. Using a simplified ice-seabed interaction process, ice scour depth is easily calculated. This numerical model can simulate the scouring process for various environmental parameters such as ice mass, incoming velocity, soil strength etc. The paper also deals with various interacting forces during the scouring process and discusses ice loads which are transmitted to seabed.

KEY WORDS: Ice scour model, Ice-seabed interaction, Pipeline burial depth, Parametric study

1. INTRODUCTION

Ice scour or ice gouging phenonenon is often found in Arctic seas, especially in the shallow coastal regions of water depth less than 30 m. During spring seasons, massive ice features such as icebergs or pressure ridges drift and gouge the shallow seabed. The interaction of grounded ice with underlying seabed is a very important design factor for the Arctic pipeline systems. During the ice scour process the force exerted by the ice features on seabed is much greater than the strength of pipelines, hence severe damage may occur to pipelines. In the sense of pipeline's safety, the best way to avoid direct contact with ice is to bury pipelines inside seabed trenches. Pipelines should be buried beneath the maximum scour depth, but the cost for trenching goes higher with increasing pipeline burial depths. Hence it is necessary to estimate the ice scour depth correctly prior to the construction of Arctic offshore pipelines.

Ice scour is typical in the Arctic environments. According to field observation data, ice scour is found in water depth of 20-30 m in the Canadian Beaufort Sea (Palmer et al., 1989). Scour depth is 0.6-2.1 m and does not exceed maximum 5 m. For the Alaskan Beaufort Sea average scour depth is 0.58 m and maximum 4 m with standard deviation of 0.59 m in water depth of 30-35 m. For water depth shallower than 6 m or deeper than 58 m, scour depth does not exceed 1 m (Machemehl and Jo, 1989). At Sakhalin offshore, ice scour is usually found with scour depth of 0.5-1.0 m and with maximum of 2.13 m in water depth of 5-10 m (Astafiev et al., 1991).

In order to estimate the burial depth for Arctic offshore pipelines, the mechanism of ice scour must be fully understood. So far much research has been focused on the ice scour mechanism and its modelling. They fall into the following three categories.

First, there are several research activities to investigate ice scour geometry from field observation data. These statistical methods estimate scour depth from annual average frequency of ice scour and probability density function for ice scour depth. Usually this requires time and effort in gathering field data and neglects the ice scour mechanism itself. Especially the ice scour data from photo scanning or sonar surveys may include very old relics scour and moreover the sediment transportation can refill new scour trench. Hence it is quite difficult to figure out exact scouring date and field observation data does not necessarily represent original scour depth.

Secondly, experimental techniques were utilized for ice scour study. From model tests, the correlations among major parameters which influence ice-seabed interaction process and scour depth and ice forces exerted on the underlying seabed are estimated. However in those experiments ice model is usually so simple that the model has a limited capability to represent real environmental conditions at specific site.

Thirdly, theoretical or numerical methods can be used in parallel with experimental techniques. Several ice scour models have been introduced so far. Among them two scour models, that is, Chari's model (Chari, 1980) and Kioka's model (Kioka and Saeki, 1995) are most significant. In Chari's scour model it was assumed that when the original kinetic energy of ice is all dissipated through plastic deformation and friction of seabed soil, ice scour process stops. In this model vertical motion or rotation of ice was not considered. Kioka and Saeki performed a series of ice model gouging tests under constant speed and they proposed a scour model to calculate scour depth and also driving force on ice model. Kioka's model considered vertical movement of ice assuming a simple scour curve.

In this paper an ice scour model to determine the subsea pipeline burial depth is studied by combining Chari's and Kioka's models, which include the horizontal and vertical motion of ice. Using a simplified ice-seabed interaction process, ice scour depth and interaction forces during the scouring process are calculated for various environmental parameters such as ice mass, velocity, soil strength etc. A sample calculation is carried out for Sakhalin offshore site which is an ice-infested sea and influences of the environmental parameters on ice scour depth and interaction forces are studied using a parametric study.

2. ICE SCOUR MODEL

An iceberg or ice ridge drifts to shore with a certain velocity by wind and current until it contacts with seabed. During ice scour process, the ice feature gouges seabed soil and then it moves in vertical and horizontal directions depending on soil strength and resistance forces. In development of ice scour model in this study it is assumed that seabed soil has enough strength to support ice motion upward and scoured soil by the ice is completely deposited in front and sides of the ice feature. During scour process volume change of soil and change in soil strength are not considered. Also rotational motion of ice is neglected.

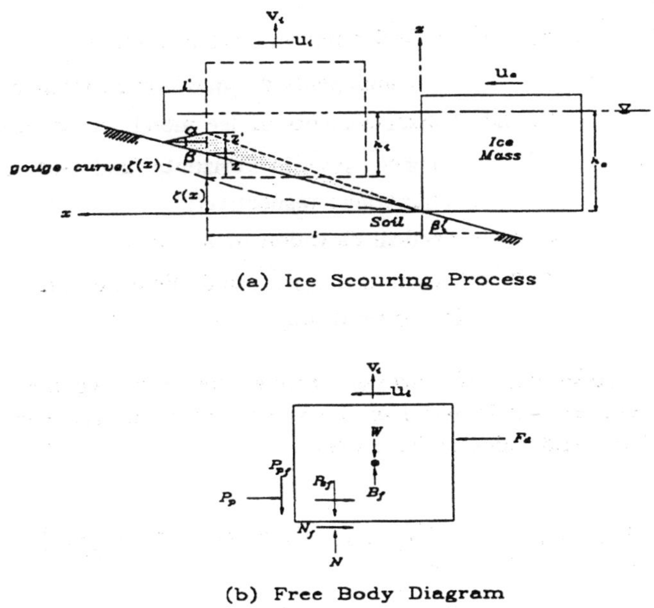

Fig.1 Schematic diagram of ice scouring process

In Fig.1 an ice model idealized as a rectangular shape is moving toward a sloping shoreline and then scour begins. During the ice-seabed interaction process various types of resistances and interaction forces act on the ice model and they increase as the scour proceeds. The resistances and interaction forces include frontal and side passive pressure of soil, vertical reaction and friction forces, weight and buoyancy of the ice feature.

At an instance of $x=l$ energy conservation gives the following equation.

$$\frac{1}{2} M u_o^2 = \frac{1}{2} M u_i^2 + \frac{1}{2} M v_i^2 - \int_0^l F_d \, dx + \int_0^l P_h \, dx + \int_0^{\zeta(l)} P_v \, dz + \int_0^{\zeta(l)} W' dz \tag{1}$$

where M = ice mass

u_o = initial velocity of ice

u_i, v_i = horizontal and vertical velocities of ice

F_d = horizontal driving forces on ice (due to current)

W' = effective weight of ice (= $W-B_f$)

W, B_f = weight and buoyancy force of ice

P_h, P_v = horizontal and vertical force components in total soil resistance

dx, dz = displacements of ice along the scour curve $\zeta(x)$ in horizontal and vertical directions respectively

$\zeta(l)$ = vertical elevation of ice at $x=l$

x, z = instantaneous ice scour distance and ice scour depth

Assuming no volume change, the total volume of scoured soil is equal to the volume of surcharge soil in front and sides of ice model.

$$\frac{1}{2}Bx^2\tan\beta - B\int_0^l \zeta(x)dx = \frac{l'z'B}{2} + \frac{z'^2(l+l')^2}{3l}\cot\alpha \quad (2)$$

where B = width of ice model

l, z = ice scour length and scour depth

l', z' = length and height of front surcharge soil

α, β = slopes of front surcharge soil and seabed

Ice scour depth z and frontal passive pressure of soil P_p are expressed as in eq.(3) and eq.(4). For the seabed soil characteristics, a Rankine passive pressure concept is adopted.

$$z = x\tan\beta - \zeta(x) \quad (3)$$

$$P_p = \frac{1}{2}\gamma h_s^2 B K_p \quad (4)$$

where $K_p = \dfrac{\cos\beta + \sqrt{\cos^2\beta - \cos^2\phi}}{\cos\beta - \sqrt{\cos^2\beta - \cos^2\phi}}$

ϕ = internal friction angle of soil

$h_s = z + z'$

Side passive pressure of soil is expressed depending on the contact geometry of ice model and side surcharge soil.

$$P_s = \int_0^{h_s\cot\beta'}\int_0^{\varepsilon\tan\beta'} \gamma'\ \eta\ K_{ps}\ d\eta\ d\varepsilon \quad \text{for } h_s \le L\tan\beta'$$

$$P_s = \int_{h_s\cot\beta'-L}^{h_s\cot\beta'}\int_0^{\varepsilon\tan\beta'} \gamma'\ \eta\ K_{ps}\ d\eta\ d\varepsilon$$

$$\text{for } h_s > L\tan\beta' \quad (5)$$

where $\tan\beta' = \dfrac{[\zeta(x) + h_s]}{x}$

$$K_{ps} = \tan^2\left(\frac{\pi}{4} + \frac{\phi}{2}\right)$$

Therefore frontal and side friction forces on ice model are:

$$P_{pf} = \mu_k P_p$$
$$P_{sf} = \mu_k P_s \quad (6)$$

where μ_k is a dynamic friction coefficient between ice and soil.

Equation of motion in vertical direction is as follows:

$$(M + m_a)\frac{d^2\zeta(x)}{dt^2} = -W' + N \mp P_{pf} \mp P_{sf} \quad (7)$$

where M = mass of ice

m_a = added mass of ice

N = vertical reaction force of seabed soil

In eq.(7), signs of P_{pf} and P_{sf} vary according to the sign of $\frac{d\zeta}{dx}$ when the ice moves along scour curve $\zeta(x)$.

In this study ice scour depth and interaction forces are calculated by solving above nonlinear equations with a Newton-Raphson numerical technique. After the first contact between ice feature and seabed, the ice transverses into soil. Due to the slope of seabed a horizontal movement causes vertical elevation and then scour depth changes. Because of the reaction force between ice and seabed, ice moves in vertical direction and settles in equilibrium position. The trajectory of ice movement has a close relationship with the size and mass of ice, initial incoming velocity, types and strength of soil etc. Kioka and Saeki (1995) assumed a simple parabolic function for the scour curve. Matching with the field observation data, a linear function with 10% vertical elevation of total scour depth is assumed in this study.

Load transmission to the buried pipelines through seabed soil is also calculated. Ice load (horizontal and vertical) acting directly upon seabed can be calculated from present scour model and then pressure distribution on seabed surface is estimated by assuming a contact area. Then the stresses at the position where a pipeline is buried are calculated by assuming the soil foundation as a semi-infinite, isotropic and uniform elastic foundation. Actual behavior of soil foundation at the pipeline position may be regarded as pure plastic or elasto-plastic. There may be a gap between the pipeline and surrounding soil. Because the material strength of pipeline is much higher than soil strength, the deformation of pipeline itself is negligible comparing to soil displacements. In this study displacement of buried pipeline is assumed as the displacement of soil at the same location.

3. CALCULATION RESULTS

3.1 SAMPLE CALCULATION

Information for present ice scour model to analyze ice-seabed interaction is size and mass of ice, initial incoming velocity of ice, initial draft of ice, unit weight and internal friction angle of soil, slopes of surcharge soil and seabed. Input data for Sakhalin offshore site for a sample calculation is shown in Table 1 and calculated results are given in Table 2.

Fig.2 shows changes in horizontal velocity of ice model from initial speed of 1.5 m/s to the final position. Results using Chari's model are also shown in this figure to compare each other. There is a slight difference in velocities for two models at the same

Table 1 Input data for Sakhalin offshore

Input	Descriptions	Sakhalin Offshore	
Ice Properties	Weight (ton)	100,000	
	B x L x H (m^3)	20 x 200 x 27.8	
	Velocity (m/s)	1.5	
	Initial Draft (m)	24.39	
	Unit Weight (ton/m^3)	0.9	
	Drag Coefficient	1.0	
	Added Mass Coefficient	1.0	
Soil Properties	Effective Unit Weight (ton/m^3)	1.0	
	Shear Strength (ton/m^2)	5.0	
	Internal Friction Angle (°)	30	
	Dynamic Friction Coefficient	0.1	
	Surcharge Soil Slope	1/10	
	Seabed Slope	1/100	1/1000
Sea Water Condition	Unit Weight (ton/m^3)	1.025	
	Current Velocity (m/s)	1.5	
Scour Curve		$\zeta(x) = c_1 x$	

c_1 = 9.0909E-04

Table 2 Maximum forces and ice scour depth for Sakhalin offshore

Seabed Slope	Description	Chari's Model	Present Model $\zeta(x) = c_1 x$	Remarks
1/100	Scour Depth	1.0	1.0	$M = 10^5$ ton u_o=1.5 m/s γ_s =1.87 ton/m^3 Z=0.5-1.0 m (max. 2.13 m)
	Hor. Force	243	177	
	Ver. Force	125	521	
	Side Friction	-	61	
	Drag Force	54	55	
	Hor. Stress	12.1	3.65	
	Ver. Stress	5.7	0.13	
1/1000	Scour Depth	0.35	0.29	
	Hor. Force	89	44	
	Ver. Force	37	181	
	Side Friction	-	44	
	Drag Force	57	57	
	Hor. Stress	11.7	1.82	
	Ver. Stress	4.96	0.05	

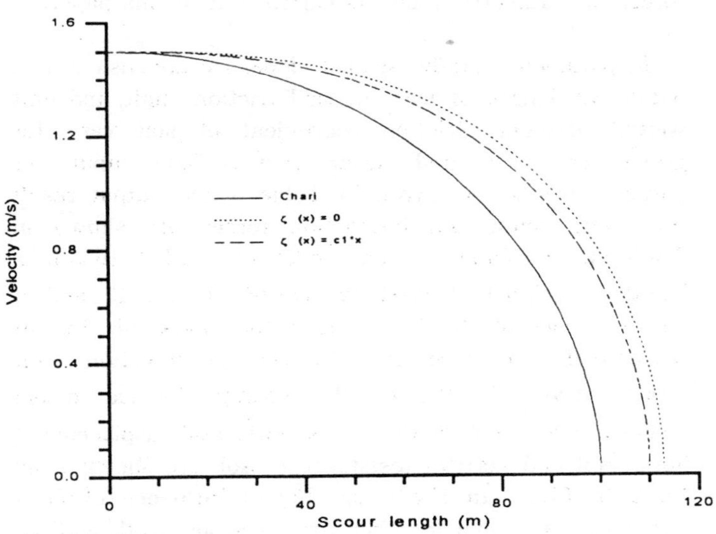

Fig.2 Ice model velocity during ice scour

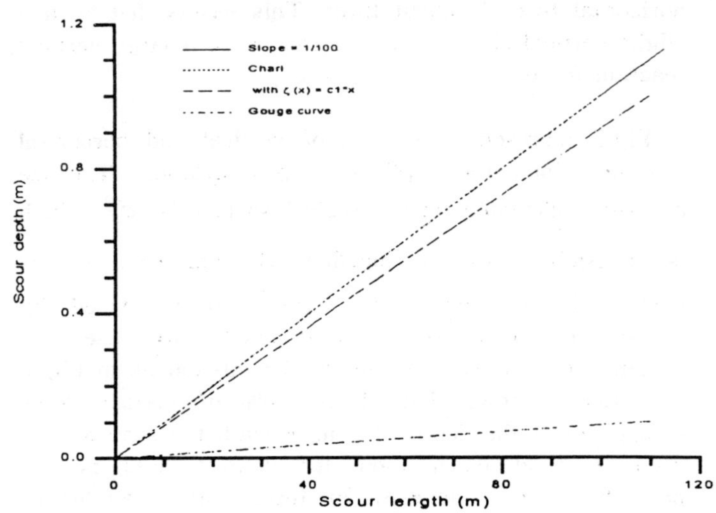

Fig.3 Ice scour depth versus scour length

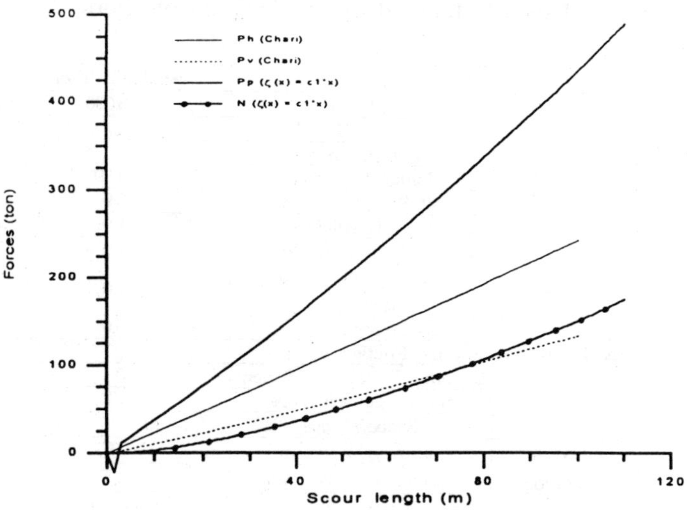

Fig.4 Applied forces and soil resistances

location and total scour lengths also do not differ much. Ice scour depths are, as shown in Fig.3, 1.1 m without assuming vertical elevation ($\zeta(x)=0$) and 1.0 m for vertical elevation ($\zeta(x)=c_1x$) respectively. These values are not far from field observation data, 1.2 m for Sakhalin offshore (Gulati, 1993). Fig.4 shows the horizontal resistance and vertical reaction force acting on seabed by the ice. Present model gives 208 ton for horizontal force and 99 ton for vertical force in case of zero vertical elevation (not shown in the figure). These are smaller than the result of Chari's model by 14-21%. In case of 10% vertical elevation, present model gives 177 ton for horizontal force and 521 ton for vertical force. These are about 3 times greater than Chari's model in vertical force. But this result is almost equivalent to the calculation of finite element method by Yang et al. (1993) where the ratio of vertical force to horizontal force is about three. This means that even a slight vertical elevation of ice can cause a large vertical reaction force.

Fig.5 represents variation of vertical and horizontal stresses along the depth of soil foundation. The ice pressure calculated on the seabed surface is q_o, which is transmitted to the pipelines through soil. q_o is calculated from maximum vertical force divided by ice-seabed contact area. For simplicity two cases are assumed for contact area in this calculation as in Fig.5. The result shows that almost 70% of vertical force disappears at the depth of 2 m beneath the seabed. The accurate field observation for contact area is not possible at the moment. In future study an actual contact area from observed data should be used.

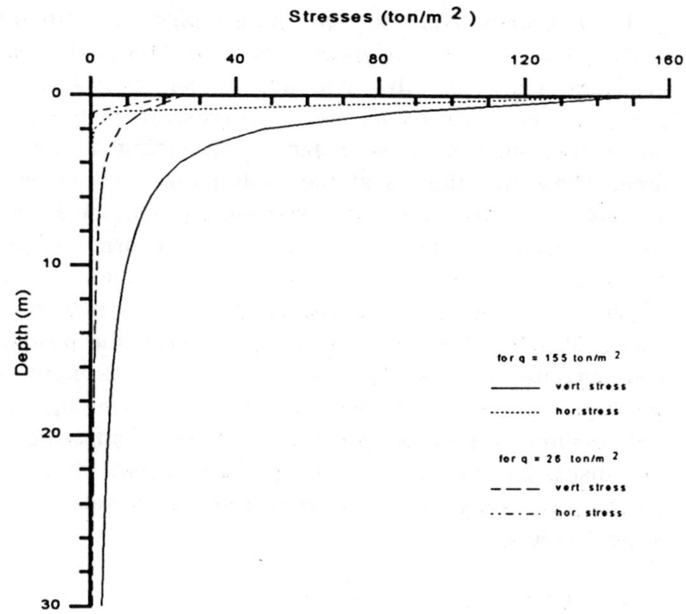

Fig.5 Stresses transmitted through soil depth due to ice-soil interaction

3.2 PARAMETRIC STUDY

Environmental parameters which influence the ice scour process are ice properties (size and mass, initial incoming velocity, initial draft etc.), soil properties (seabed slope, unit weight of soil, internal friction angle etc.) and ocean environments (current velocity, density of sea water, water depth etc.). In order to determine ice scour depth and interaction forces, field data for these environmental parameters should be collected prior to construction of pipelines. To study major environmental parameters affecting scour depths and interaction forces, a parametric study is carried out in this paper.

In parametric study, seabed slope, ice dimension (i.e., width and length of ice), internal friction angle and unit weight, dynamic friction coefficient of soil vary for given ice mass and scour curve. Data input for parametric study is given in Table 3 and output result for scour depth and interaction forces are shown in Table 4. In Table 3, ice model of 1:2:0.8 (width x length x height) is used for cases 11, 12, 13 and a longer model of 1:5:0.8 is used for cases 14, 15, 16 respectively. In case of 10% vertical elevation with scour curve $\zeta(x)=c_1x$, the change in ice model velocity, ice scour depth versus horizontal displacement, horizontal and vertical resistance of soil are shown from Fig.6 to Fig.8. In Fig.9 and Fig.10 influence of each parameter is shown together for ice scour depth and ice interaction forces.

Changes in seabed slope from 1/400 to 1/50 affect scour depth from 0.27 to 1.27 m. Fig.9 shows that increment in seabed slope gives shorter scour length and deeper scour depth. The influence of seabed slope is stronger than other parameters. Frontal resistance of soil varies with range of 89.4-401.1 ton as shown in Fig.10. It is expected that ice scour depth increases as ice mass increases. Scour depth varies from 0.49 m to 1.09 m according to ice mass variation from 25,000 ton to 200,000 ton. Frontal resistance of soil varies with range of 114.2-399.1 ton. As ice size increases, the contact area between ice and soil increases. Scour depth decreases from 1.17 m to 0.66 m according to surcharge soil slope variation from 3.75° to 30°. Frontal resistance of soil varies from 165.3 ton to 354.2 ton. As the unit weight of soil increases from 1.31 ton/m^3 to 2.63 ton/m^3 ice scour depth decreases from 1.14 m to 0.63 m. Frontal resistance of soil varies from 167.3 ton to 380.4 ton. As the internal friction angle and dynamic friction coefficient increase, ice scour depth and frontal resistance of soil decrease as shown in Fig.9 and Fig.10.

4. CONCLUSIONS

In order to understand ice scour mechanism and to estimate safe burial depth for pipelines which is one of the most important design considerations for the Arctic pipeline systems, in this paper a numerical ice scour model is proposed. Using a simplified ice-seabed interaction process, ice scour depth is easily calculated. This numerical model can simulate the scouring process for various environmental parameters such as ice mass, incoming velocity, soil strength etc. The paper also deals with various interaction forces during the ice scour process in sloping seabed conditions.

The proposed scour model is used for Sakhalin offshore site by comparison with the corresponding Chari's scour model. The calculated results fall within the acceptable range of observed field data. Present model gives smaller horizontal forces than Chari's model by 14-21% in case of zero vertical elevation. In case of vertical elevation the present model gives about 3 times greater vertical force than Chari's model. This means that a slight elevation can cause a significant increase in vertical reaction force. Therefore scour curves should be verified from the actual field observation data. In this paper a parametric study was carried out to understand the influence of environmental parameters on ice scour process. It is found that the unit weight of soil, seabed slope and ice mass are the major factors.

For future study a more realistic ice shape, i.e., iceberg or ice ridge shape will be considered. Damage of ice itself and deformation of soil and pipelines should be included in scour modelling.

REFERENCES

Astafiev, V.N., Polomoshnov, A.M. and Truskov, P.A., (1991) "Stamukhi on the Northern Sakhalin Offshore", Proceedings of the First International Offshore and Polar Engineering Conference, Edinburgh, The United Kingdom, Vol.II, pp.462-466.

Been, K., Kosar, K., Hachey, J., Rogers, B.T. and Palmer, A.C., (1990) "Ice Scour Models", Proceedings of the 9th International Conference on Offshore Mechanics and Arctic Engineering, Vol.5, pp.179-188.

Chari, T.R., (1980) "A Model Study of Iceberg Scouring in the Northern Atlantic", Journal of Petroleum Technology, Vol.32, pp.2247-2252.

Clark, J.I. et al., (1994) "Development of a Design Methodology for Pipelines in Ice Scoured Seabeds", ASME Proceedings of the 13th International Conference on Offshore Mechanics and Arctic Engineering, Houston, USA, Vol.V, pp.107-125.

Green, H.P., Reddy, A.S. and Chari, T.R., (1983) "Iceberg Scouring and Pipeline Burial Depths", Proceedings of the 7th International Conference on Port and Ocean Engineering under Arctic Conditions, Helsinki, Finland, Vol.I, pp.280-288.

Gulati, K.C., (1993) "Design Concepts for Sakhalin Offshore Production Platforms", Proceedings of the 12th International Conference on Port and Ocean Engineering under Arctic Conditions, Hamburg, Germany, Vol.2, pp.487-496.

Kioka, S. and Saeki, H., (1995) "Mechanisms of Ice Gouging", Proceedings of the Fifth International Offshore and Polar Engineering Conference, The Hague, The Netherlands, Vol.II, pp.398-402.

Lanan, G.A., Niedoroda, A.W. and Weeks, W.F., (1986) "Ice Gouge Hazard Analysis", Offshore Technology Conference, Houston, Texas, OTC-5298, pp.57-66.

Machemehl, J.L. and Jo, C.H., (1989) "Ice Gouge Study in the Alaskan Beaufort Sea", Journal of Korean Committee for Ocean Resources and Engineering, Vol.3, No.2, pp.545-550.

Palmer, A.C., Konuk, I., Love, J., Been, K. and Comfort, G., (1989) "Ice Scour Mechanisms", Proceedings of the 10th International Conference on Port and Ocean Engineering under Arctic Conditions, Lulea, Sweden, Vol.I, pp.123-132.

Pilkington, G.R. and Marcellus, R.W., (1981) "Methods of Determining Pipeline Trench Depths in the Canadian Beaufort Sea", Proceedings of the 6th International Conference on Port and Ocean Engineering under Arctic Conditions, Quebec, Canada, Vol.II, pp.674-687.

Yang, Q.S. et al., (1993) "Analysis of Subscour Deformation by Finite Element Method", Proceedings of the Fourth Canadian Conference on Marine Geotechnical Engineering, St. John's, Newfoundland, Canada, pp.739-754.

Table 3 Input data for parametric study

Case No.	Case-11	Case-12	Case-13	Case-14	Case-15	Case-16
Weight	10,000	100,000	1,000,000	100,00	100,000	1,000,000
Ice Breadth	19.08	41.10	88.55	14.06	30.29	65.25
Ice Length	38.16	82.21	177.11	70.29	151.43	326.24
Ice Height	15.26	32.88	70.84	11.25	24.23	52.20
Initial Draft	13.40	28.87	62.20	9.87	52.20	45.83
Initial Velocity	1.5 m/s					
Unit Weight	0.9 ton/m^3					
Drag Coefficient	0.4					
Added Mass Coefficient	0.6					
Soil Properties						
Unit Weight	1.75 ton/m^3					
Internal Friction Angle	10°					
Kinetic Friction Angle	0.1 or 0					
Surcharge Soil Angle	15°					
Water Unit Weight	1.025 ton/m^3					
Seabed Slope	1/100					
Current Velocity	1.5 m/s					
Scour Curve	$\zeta(x) = c_1 x$					

Table 4 Maximum forces and ice scour depth for parametric study

Case No.	Case-11	Case-12	Case-13	Case-14	Case-15	Case-16
Scour Depth, Z	0.52	0.83	1.33	0.58	0.93	1.48
Scour Length, X	57.2	91.4	146.3	64.1	102.2	162.3
Drag Force, F_d	11.2	53.4	249.2	6.0	28.7	136.4
Frontal Resistance, P_p	41.7	264.9	1670.5	31.0	198.5	1256.3
Side Friction, P_{sf}	7.9	38.4	180.0	9.2	43.5	203.4
Bottom Friction, N_f	5.1	35.3	248.3	7.1	50.0	354.7
$P_p + P_{sf} + N_f$	54.7	338.6	2098.8	47.3	292.0	1814.4
Effective Weight, W'	38.8	287.6	2137.5	59.1	436.6	3219.2
Bottom Reaction, N	50.8	352.5	2482.7	71.3	500.0	3546.8
Frontal Vert. Friction, P_{pf}	4.2	26.5	167.1	3.1	19.9	125.6
Side Vert. Friction, P_{sf}	7.9	38.4	180.0	9.2	43.5	203.4
$N - P_{pf} - P_{sf}$	38.7	287.6	2135.6	59.0	436.6	3217.8
Applied Area (m^2)						
$A_{F_d} = B \times h$	254.7	1183.4	5496.7	138.0	641.5	2980.8
$A_{P_s} = B \times H_s$	39.3	145.4	536.0	29.1	108.1	399.0
A_{P_d}	51.7	157.0	433.1	64.6	177.6	484.6

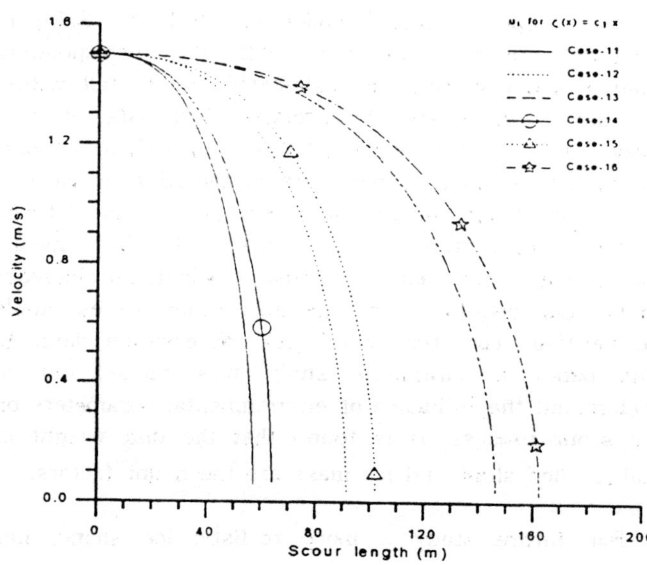

Fig.6 Ice velocity versus scour length for $\zeta(x) = c_1 x$

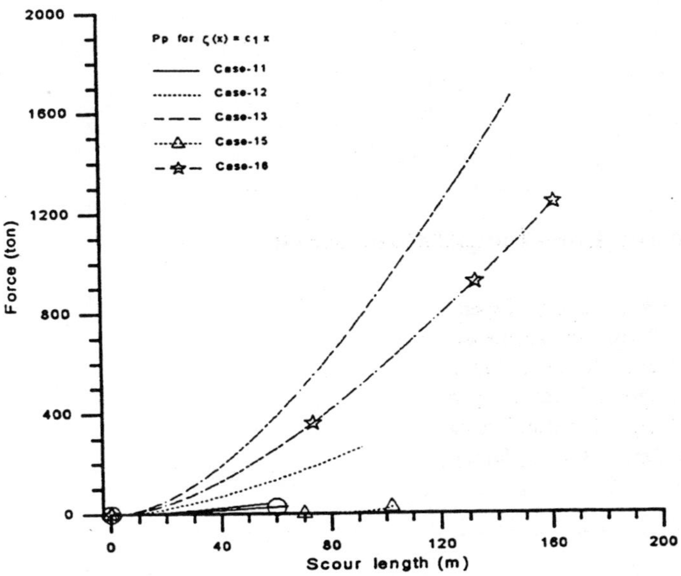

Fig.7 Frontal passive pressure versus scour length for $\zeta(x)=c_1x$

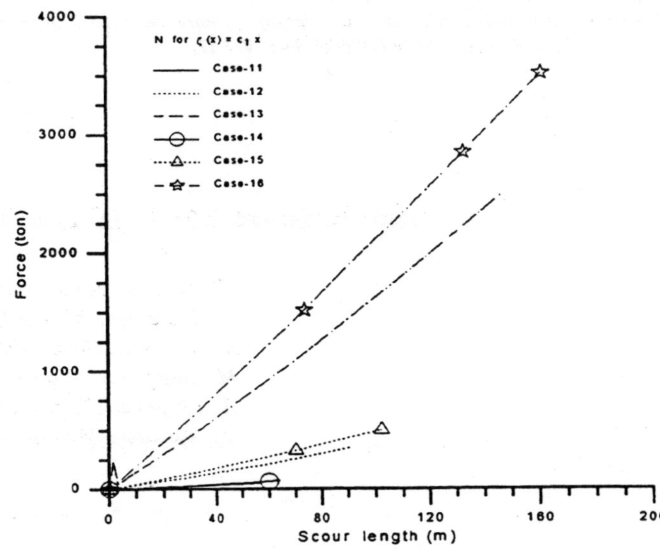

Fig.8 Vertical reaction force versus scour length for $\zeta(x)=c_1x$

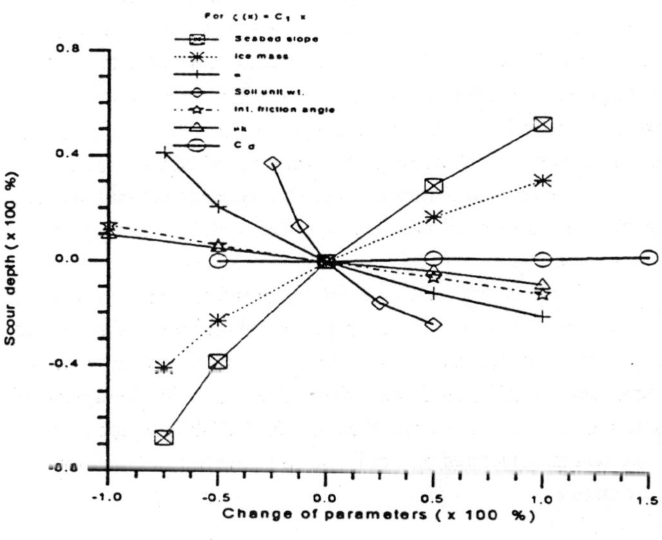

Fig.9 Effect of parameters on ice scour depth

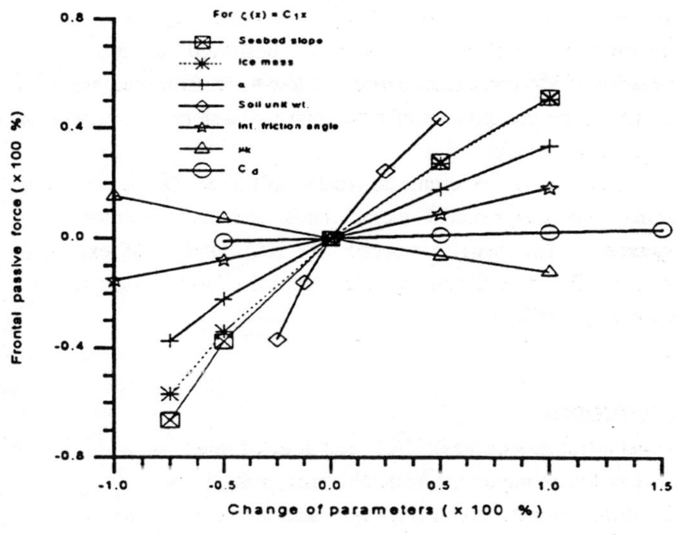

Fig.10 Effect of parameters on the frontal resistance of soil

Proceedings of the Seventh (1997) International Offshore and Polar Engineering Conference
Honolulu, USA, May 25-30, 1997

ISBN 1-880653-28-1 (Set); ISBN 1-880653-30-3 (Vol. II)

Sour Resistant X65 UOE Line Pipe for Low-Temperature Service

Y. Terada, Nippon Steel Corp., Kimitsu, Japan
H. Tamehiro, Nippon Steel Corp., Futtsu, Japan
H. Ishikawa, Nippon Steel Corp., Kimitsu, Japan
M. Sugiyama, Nippon Steel Corp., Futtsu, Japan
R. Chijiiwa, Nippon Steel Corp., Kimitsu, Japan
N. Ayukawa, Nippon Steel Corp., Tokyo, Japan

ABSTRACT

With the aim of developing sour resistant API X65 UOE line pipe having good low-temperature toughness, the effects of aluminum content on the simulated HAZ toughness and the formation of MA-constituent were examined. It was found that HAZ toughness improves remarkably when aluminum content is below 0.01%. Reduction in the aluminum content is considered to suppress the formation of MA-constituent harmful to low-temperature toughness by facilitating the precipitation of Fe_3C from untransformed austenite on cooling after welding.

Based on the experimental results, API X65 UOE pipe for sour service was commercially manufactured and its properties were examined. The newly-developed API X65 UOE pipe exhibited excellent HIC and SSC resistance, as well as excellent low-temperature toughness at -46℃.

KEY WORDS

low-temperature toughness, HAZ (heat-affected zone),
MA-constituent(martensite-austenite constituent),
HIC(hydrogen-induced cracking), SSC(sulfide stress cracking)

1. INTRODUCTION

As energy resources are exploited in the Arctic and offshore regions under severe environments, the property requirements for line pipe for transporting oil and gas have become diversified and sophisticated. Heavy wall API X65 pipe with excellent low-temperature toughness at -30℃ and excellent HIC and SSC resistance is commercially used for sour applications in the North Sea[1)2)]. Recently, sour resistant API X65 pipe has been required even for low-temperature service at -46℃. It is, however, difficult to satisfy both excellent HAZ toughness and HIC resistance. In order to improve HAZ toughness, it is important to refine the grain size and suppress the formation of MA-constituent harmful to toughness. Therefore, titanium-oxide bearing steel (Ti-O steel), which has improved HAZ toughness by utilizing intragranular ferrite (IGF), has been developed[3)4)5)]. IGF nucleates during austenite to ferrite transformation after welding from fine precipitates consisting of Ti_2O_3, TiN and MnS dispersed in steel. It is, however, difficult to attain both excellent HAZ toughness and HIC resistance because MnS, which facilitates the nucleation of IGF, deteriorates HIC resistance. On the other hand, it is known that a reduction in silicon content improves HAZ toughness by decreasing the amount of MA-constituent[6)]. However, when silicon content is lowered, it becomes difficult to attain high strength for sour resistant line pipe steel with low carbon and low manganese contents.

Therefore, a study was made on the manufacturing technology of sour resistant API X65 UOE line pipe for low-temperature service at -46℃. In this paper, the effect of aluminum content on the formation of MA-constituent in HAZ was investigated from the viewpoint of improving HAZ toughness, and then API X65 UOE line pipe for sour service was manufactured on a mill and its properties and performance were examined.

2. EFFECT OF ALUMINUM CONTENT ON HAZ TOUGHNESS

2.1 Experimental Procedure

Table 1 shows the chemical compositions of the steels used. They are calcium-treated API X65 class steels with low C-low S-Ni-Mo-Nb system. Steels A to D with different aluminum contents were vacuum-melted and cast into 16kg ingots. The ingots were hot-rolled into 15mm thick plates after reheating at 1100℃ for 1 hour. Simulated weld thermal cycle test specimens of 12mm in thickness, 12mm in width and 120mm in length were machined from the mid-thickness of rolled plates. The peak temperature and cooling time between 800 ℃ and 500℃ (Δ t8/5) of the simulated thermal cycle were set to 1400 ℃ and 54s, respectively. Charpy impact tests of simulated HAZ with steels A to C were carried out and the formation of MA-constituent was observed by optical microscope after etching the surface of specimens using LePera's reagent[7]. Furthermore, the austenite to ferrite transformation temperatures of steels A to C were measured by dilatometer using cylindrical specimens of 3mm in diameter and 10mm in length.

Table 1 Chemical compositions of steels used.

(mass%, * ppm)

Steel	C	Si	Mn	P	S*	Nb	Ni	Mo	Ti	Al	Ca*
A	0.066	0.18	1.46	0.006	3	0.034	0.26	0.26	0.013	0.004	21
B	0.061	0.15	1.45	0.006	2	0.034	0.25	0.26	0.013	0.014	20
C	0.052	0.18	1.46	0.006	3	0.033	0.26	0.26	0.013	0.060	19
D	0.077	0.15	1.29	0.006	2	0.046	0.36	0.26	0.017	0.331	20
E	0.047	0.23	1.25	0.004	3	0.038	0.37	0.27	0.012	0.004	20
F	0.045	0.23	1.25	0.004	3	0.038	0.36	0.27	0.013	0.025	20

Steels E and F with different aluminum contents were vacuum-melted and cast into 150kg ingots. The ingots were hot-rolled into 25mm thick plates after reheating at 1200℃ for 1 hour. The simulated thermal cycle test specimens of 12mm in thickness, 12mm in width and 120mm in length were machined from the mid-thickness of rolled plates. The peak temperature and Δ t8/5 of the simulated thermal cycle were set to 1400℃ and 21s, respectively. The HAZ microstructures of steels E, F and D were observed by transmission electron microscope (TEM).

2.2 Experimental Results

Figure 1 shows the effect of aluminum content on the simulated HAZ toughness and the area fraction of MA-constituent. As the aluminum content is reduced, the simulated HAZ toughness is improved and the area fraction and size of MA-constituent is decreased. When the aluminum content is higher than 0.01%, a large amount of upper bainite microstructure was formed in the simulated HAZ. In the steel A of 0.004%Al, the formation of MA-constituent is suppressed remarkably, resulting in the improvement of simulated HAZ toughness. The reduction in the amount of MA-constituent is especially important to improve the HAZ toughness in API X65 class steels with predominantly upper bainite microstructure.

Figure 2 shows the TEM micrographs of the simulated HAZ of steels E (0.004%Al) and F (0.025%Al). The HAZ microstructures were predominantly composed of upper bainite and the formation of IGF was not observed in both steels. Under the microscope, however, an amount of MA-constituents in the microstructure of the steel F is larger than that of the steel E. A typical MA-constituents observed among bainite laths in the steel F is shown in Fig.2.(a). An internal twins was often introduced into the martensite phase. Judging from the low carbon content of the steel F, it is found that the carbon is highly concentrated to the martensite phase. It was also confirmed by a dark field imaging technique and a micro diffraction method that a residual austenite phase was sometimes observed near the martensite phase. On the other hand, a precipitation of Fe_3C was observed at most bainite lath boundaries instead of the formation of MA-constituent in the steel E. A typical TEM micrograph is shown in Fig.2.(b). From these observations, it is considered that the formation of MA-constituent is related to the carbon concentration in the untransformed austenite and the precipitation of Fe_3C during the growth of bainite laths.

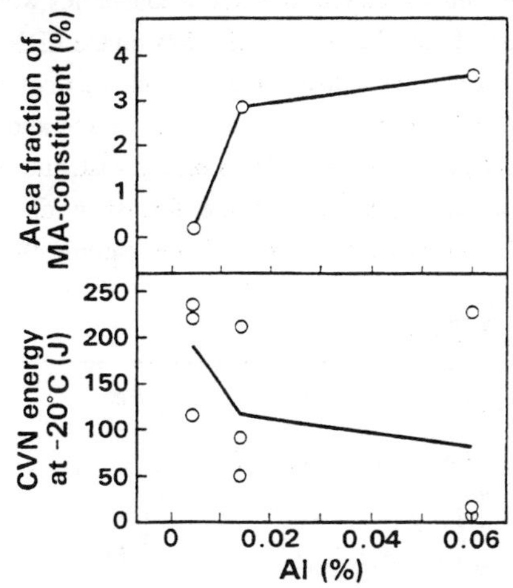

Fig.1 Effect of aluminum content on Charpy V-notch energy and area fraction of MA-constituent in simulated HAZ.

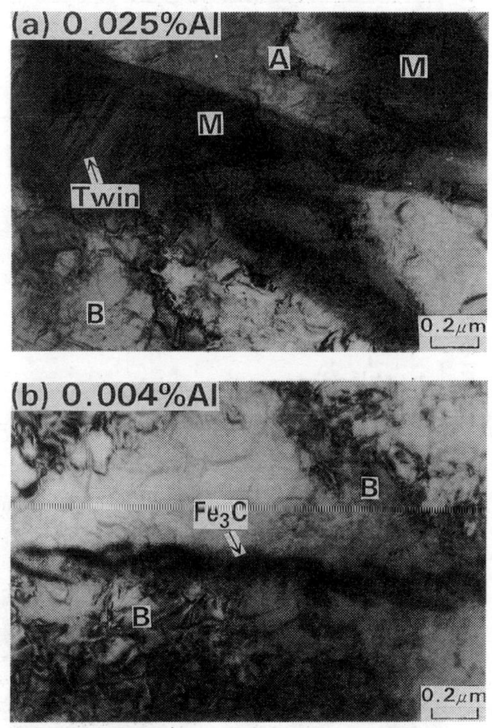

Fig.2 Transmission electron micrographs showing a typical MA-constituent (a) and Fe_3C (b) in the microstructure of simulated HAZ of steel F and E. In the figure, M, A and B indicate martensite, austenite and bainite.

2.3 Suppression Mechanism of MA-constituent by Reducing Aluminum Content

The effect of aluminum content on the formation of MA-constituent was studied from the viewpoint of (1) ease of the precipitation of Fe_3C from the untransformed austenite and (2) difference in the carbon concentration at the interface between growing bainite lath and austenite.

Figure 3 shows the TEM micrograph of the MA-constituent including thin Fe_3C along the bainite lath boundaries. The changes of carbon and aluminum contents across the boundaries were examined by EDX analysis. Since the peak level is low on the EDX spectra, it was clear by the measurement that the aluminum was not detected in the Fe_3C phase. In the case of high aluminum content above 0.010%, the precipitation of Fe_3C from the untransformed austenite is considered to be strongly delayed, resulting in the formation of MA-constituent because aluminum is hardly dissolved into carbide the same as with silicon[8].

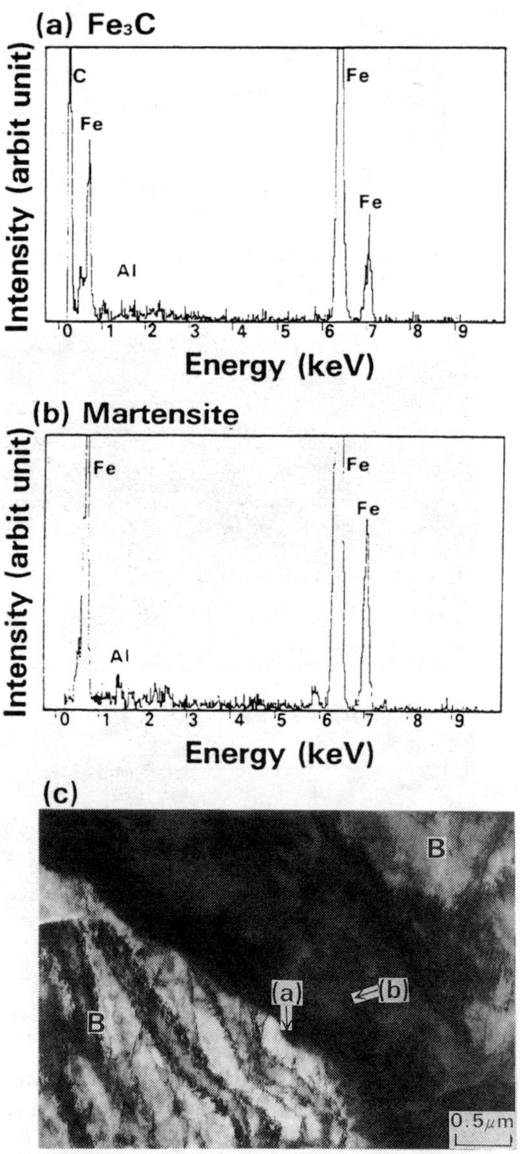

Fig.3 Transmission electron micrograph showing MA-constituent including thin Fe_3C along bainite lath boundary. Each EDX spectrum Fig.3.(a) and (b) are taken from the area indicated in the Fig.3.(c). In the TEM micrograph, B indicates the bainite phase.

Figure 4 shows the effect of aluminum content on the austenite to ferrite transformation temperature. Though the transformation temperature slightly drops with reducing aluminum content, a remarkable difference in transformation temperature between specimens was not recognized. In addition, the austenite to ferrite equilibrium transformation temperature (Ae_3) of Fe-0.05%C-Al ternary system was calculated by Thermo-Calc. The Ae_3 temperatures of 0.001%Al, 0.03%Al and 0.3%Al steels were 890℃, 894℃ and 927℃, respectively. The difference in Ae_3 temperature was as small as 4℃ between 0.001%Al and 0.03%Al steels.

The effect of the austenite to ferrite transformation temperature on the formation of MA-constituent is considered from the viewpoint of the carbon concentration at the interface between bainite lath and austenite. On the assumption that Fe_3C precipitates when the carbon concentration at the interface of growing of bainite lath exceeds the solubility of carbon on cooling after welding, the driving force of precipitation of Fe_3C becomes bigger where the carbon concentration at the interface is higher.

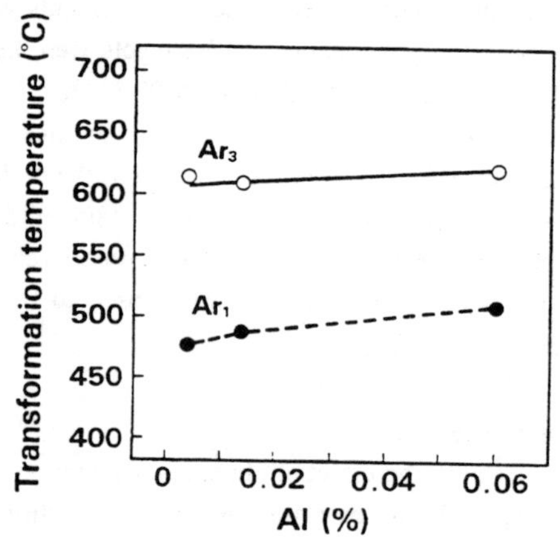

Fig.4 Effect of aluminum content on austenite to ferrite transformation temperature.

Figure 5 shows the difference in carbon concentration at the interface between bainite lath and austenite with austenite to ferrite transformation temperature. As the austenite to ferrite transformation temperature in a low aluminum steel is low, it is easy for carbon to concentrate at the interface between bainite lath and austenite, and consequently the driving force of Fe_3C precipitation in the low aluminum steel becomes bigger than that in a high aluminum steel. However, since the difference in the austenite to ferrite transformation temperature was small between 0.001% Al and 0.02% Al steels, the effect of the difference in carbon concentration at the interface between bainite lath and austenite on the formation of MA-constituent is considered to be negligible.

Therefore, the suppression of MA-constituent in the HAZ resulting from reducing the aluminum content is considered attributable mainly to the ease of the precipitation of Fe_3C from the untransformed austenite on cooling after welding.

The suppression mechanism of MA-constituent in the HAZ by reducing aluminum content is shown schematically in Fig.6. At the early stage of austenite to ferrite transformation, a carbon atom concentrates into the untransformed austenite as the bainite lath grows during cooling after welding irrespective of the aluminum content. In the case of low aluminum steel, the Fe_3C easily precipitates along the interface between bainite lath and untransformed austenite when the carbon concentration at the interface exceeds the solubility of carbon. The growth of bainite lath and precipitation of Fe_3C progresses alternately. In the case of high aluminum steel, however, the untransformed austenite is stabilized due to the difficulty of Fe_3C precipitation because aluminum atoms are hardly dissolved into carbides. When the temperature of untransformed austenite region decreases below Ms temperature through the carbon concentration and the temperature drop, a part of the untransformed austenite transforms to martensite and the remainder was retained as austenite, and finally MA-constituent is formed.

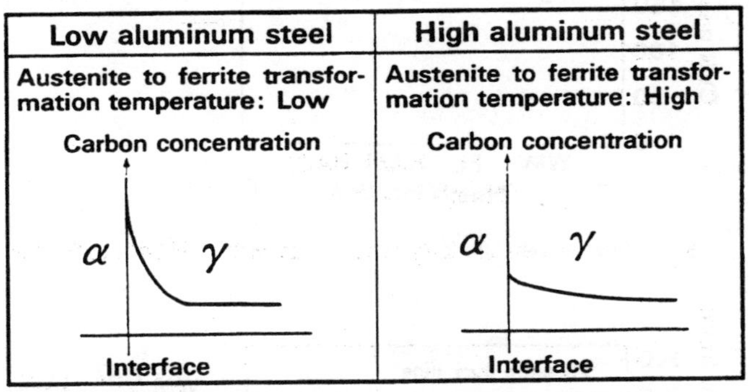

Fig.5 Difference in carbon concentration at interface between bainite lath and austenite with austenite to ferrite transformation temperature.

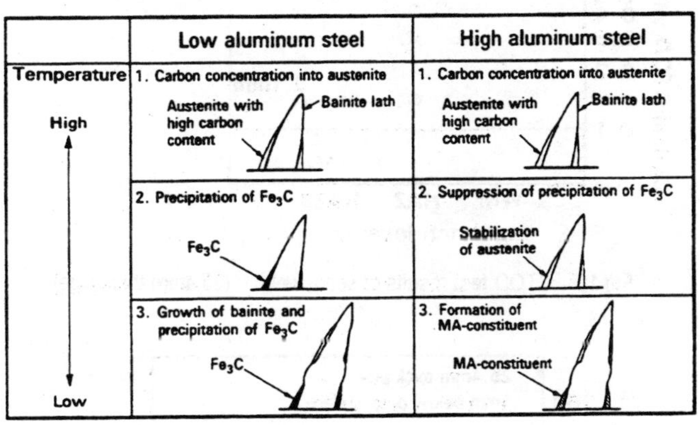

Fig.6 Suppression mechanism of MA-constituent by reducing aluminum content.

3. PROPERTIES OF SOUR RESISTANT X65 UOE LINE PIPE FOR LOW-TEMPERATURE SERVICE

Based on the results of the basic study described above, sour resistant X65 line pipe for low-temperature service was experimentally manufactured in the mill.

Figure 7 shows the manufacturing process of the sour resistant X65 UOE pipe. In the pretreatment of the hot metal, the hot metal was subjected first to desiliconization, and then to dephosphorization and desulphurization. In the refining process, decarburization, desulphurization, degassing and calcium treatment were applied to make the steel pure and clean. In order to reduce the center segregation of continuously cast slabs, soft reduction technology was applied during casting. After reheating, 240mm thick slabs were controlled-rolled into 20.0, 25.4 and 36.0mm thick plates, and then subjected to interrupted accelerated cooling in the heavy plate mill. UOE pipes were manufactured using these plates.

Table 2 shows the chemical composition of the UOE pipe body. The steel is a calcium-treated, clean steel of low C-Ni-Mo-Nb-Ti-low Al base. In order to decrease the detrimental effect of the center segregation of continuously cast slabs on the hydrogen-induced cracking (HIC) resistance, the manganese content was reduced to a low level of 1.24%. Sulfide shape was perfectly controlled by adding calcium. Table 3 shows the UOE pipe size and the mechanical properties of the base material. The strength satisfies the requirements of the API grade X65 specification. Low-temperature toughness and Battelle drop weight tear test (BDWTT) properties are also satisfactory. The high Charpy impact energy suggests that the steel has high ability to arrest unstable ductile fractures.

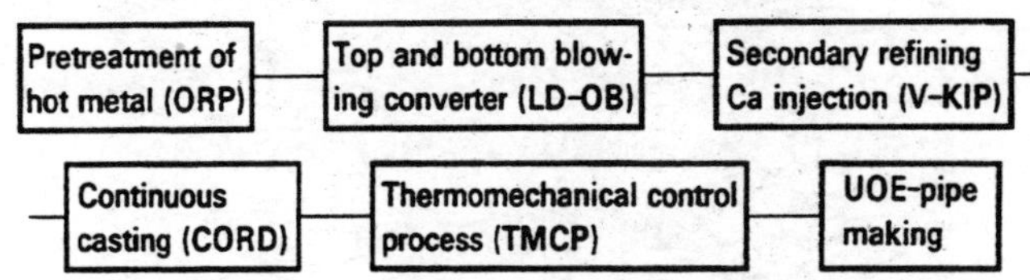

Fig.7 Manufacturing process of X65 UOE line pipe.

Table 2 Chemical composition of steel.

(mass%, * ppm)

C	Si	Mn	P*	S*	Ni	Mo	Nb	Ti	Al	Ca*	Ceq[1]	P_{CM}[2]
0.03	0.24	1.24	40	2	0.35	0.27	0.036	0.011	0.005	22	0.314	0.124

1) Ceq = C+Mn/6+(Ni+Cu)/15+(Cr+Mo+V)/5.
2) P_{CM} = C+Si/30+(Mn+Cu+Cr)/20+Ni/60+Mo/15+V/10+5B.

Table 3 Pipe size and mechanical properties of UOE pipe body.

Pipe size		Direction	Tensile properties[1]			CVN impact properties[2]			BDWTT	
OD (mm)	WT (mm)	of test	YS (MPa)	TS (MPa)	El (%)	vE_{-40} (J)	vE_{-60} (J)	vT_{rs} (°C)	SA_{-30} (%)	85% shear FATT (°C)
(API specification)			≥448	≥530						
1066.8 (42in)	20.0	L	486	569	48	–	–	–	–	–
		T	486	550	48	288	293	<-80	SA_{-40} 88	<-40
914.4 (36in)	25.4	L	487	547	56	–	–	–	–	–
		T	472	538	56	vE_{-46} 277	265	<-80	100	<-40
762.0 (30in)	36.0	L	572	616	59	–	–	–	–	–
		T	545	596	60	283	276	<-80	100	<-32

1) API rectangular specimen.
2) 2/3 subsize specimen.

Figure 8 shows the optical microstructure at the mid-thickness of 36.0mm thick pipe. The microstructure is composed of uniform and fine-grained ferrite-bainite through the optimization of the chemical composition and TMCP (thermo-mechanical control process) conditions.

Table 4 shows the mechanical properties of the seam weld, which was made by double submerged-arc welding. The transverse weld tensile strength was higher than that of the base material and the specimens fractured in the base material. At a test temperature of -46℃, this pipe shows high Charpy impact energy in both the weld metal and HAZ.

Figure 9 shows the Charpy V-notch impact energy at different notch locations of the seam weld of 36.0mm thick pipe. Cross section notches were made at the WM (weld metal centerline), FL (fusion line), HAZ1 (1mm from fusion line) and HAZ2 (2mm from fusion line). The Charpy impact energy was very high in all locations and the pipe has been proved to possess superior HAZ toughness for low-temperature service.

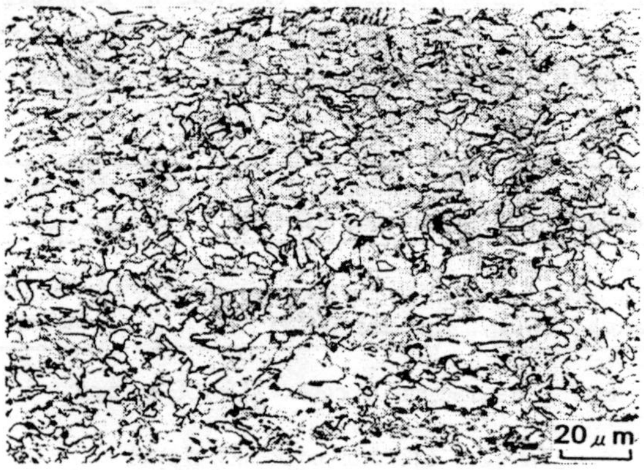

Fig.8 Optical microstructure of base material at mid-thickness. (36.0mm thick pipe)

Table 4 Mechanical properties of seam weld.

Pipe size		Direction of test	Transverse weld tensile properties[1]		CVN impact properties[2]		
OD (mm)	WT (mm)		TS (MPa)	Location of fracture	Test temp. (°C)	Energy in WM (J)	Energy in HAZ (J)
1066.8 (42in)	20.0	T	586	BM	−46	227	297
914.4 (36in)	25.4	T	566	BM	−46	172	293
762.0 (30in)	36.0	T	623	BM	−46	175	282

1) With reinforcement.
2) 2/3 subsize specimen.
Average value, notch location.
WM : Weld metal centerline
HAZ : 1mm from fusion line

Figure 10 shows the CTOD test result of the seam weld of 25.4mm thick pipe. The CTOD test was carried out according to BS5762. Cross section fatigue notches were made at the WM, HAZ (50%HAZ-50%WM) and HAZ3 (3mm from fusion line). All critical CTOD values were more than 0.1mm at - 46℃.

Figure 11 shows the hardness distribution across the seam welded joint of 25.4mm thick pipe. The hardness of the weld metal was higher than that of the base material. The hardness of the weld metal and HAZ were as low as below 250Hv.

Figure 12 shows the HIC test results of 20.0mm and 36.0mm thick UOE pipes. HIC test specimens were cut from every 90° circumferential location except the seam weld. Ultrasonic testing was employed over the entire specimen to detect HIC and HIC resistance was evaluated in terms of crack area ratio (CAR). In NACE solution,

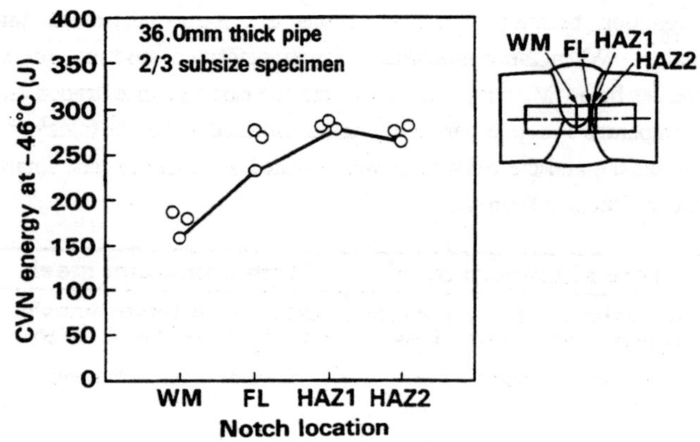

Fig.9 Low-temperature toughness of seam weld. (36.0mm thick pipe)

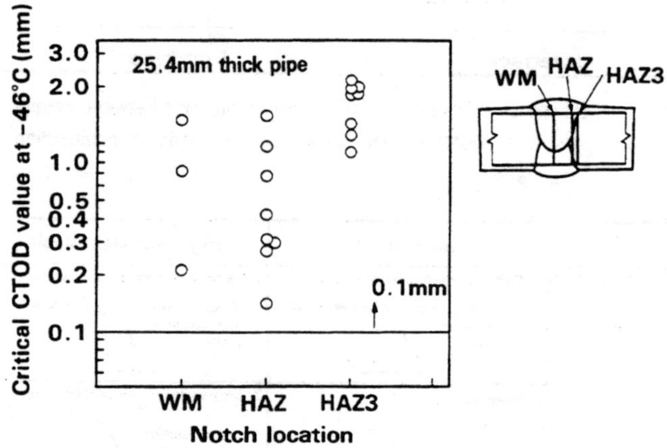

Fig.10 CTOD test results of seam weld. (25.4mm thick pipe)

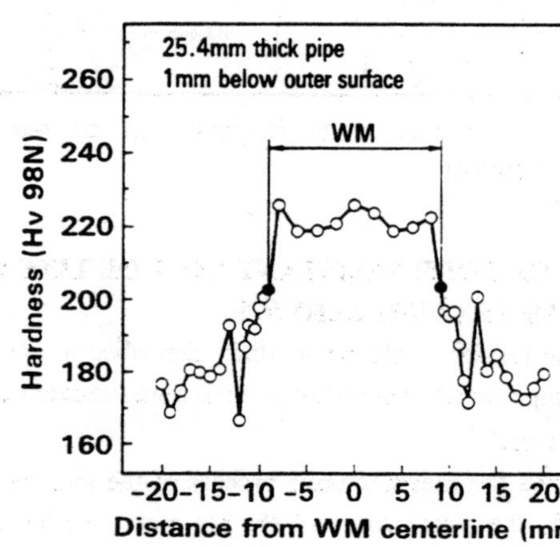

Fig.11 Hardness distribution across seam weld.

all test specimens of both pipes were free of HIC, that is to say, this X65 line pipe has excellent HIC resistance.

Table 5 shows the results of SSC test. The test specimens were machined from the base material and seam weld. The SSC test was carried out according to the method-A specified in NACE TM-01-77. The test duration was 720 hours. No cracking occurred at an applied stress of actual yield stress. It was confirmed that the X65 grade UOE pipe shows excellent SSC resistance in the base material and seam weld.

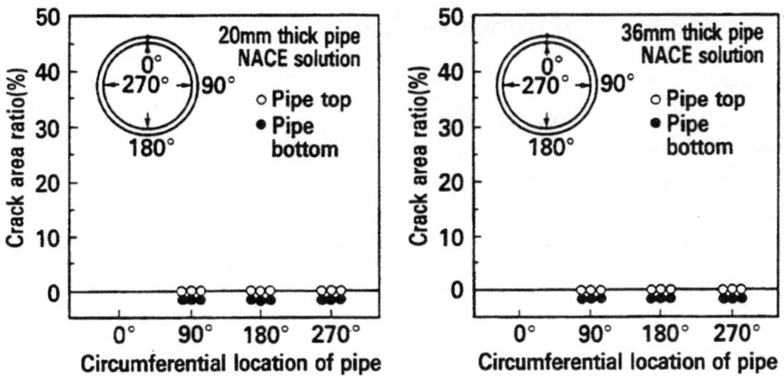

Fig.12 HIC test results of X65 UOE pipe.

Table 5 SSC test results according to NACE TM-01-77. (36.0mm thick pipe)

Solution	Location	No.	Method	Applied stress	Test result	Duration(hr)
NACE	BM	BM-1	A	0.8 actual YS	No failure	720
		-2	A	1.0 actual YS	No failure	720
	Seam weld	SW-1	A	0.8 actual YS	No failure	720
		-2	A	1.0 actual YS	No failure	720

4. CONCLUSION

A study was made on the manufacturing technology of sour resistant API X65 UOE line pipe for low-temperature service at - 46℃. It was found that simulated HAZ toughness improves remarkably through the suppression of the formation of MA-constituents when aluminum content is below 0.01%. Reducing aluminum content is thought to suppress the formation of MA-constituent by facilitating the precipitation of Fe_3C from untransformed austenite on cooling after welding. Based on the experimental results, X65 UOE pipe for sour service was commercially manufactured and its properties were examined. The newly-developed API X65 UOE pipe exhibited excellent HIC and SSC resistance, as well as excellent low-temperature toughness at - 46℃.

ACKNOWLEDGMENTS

The authors express their appreciation to Mr.H.Miyazaki, Mr.M.Fujioka, Mr.K.Nishioka and Dr.M.Abe of Nippon Steel Corporation for their helpful comments in this study.

REFERENCES

1) H.Tamehiro, T,Yoshino, N.Yurioka and M.Abe, "High-Strength Sour Service Line Pipe for Offshore Use", *Offshore Mechanics and Arctic Engineering (OMAE)*, The Hague, Netherlands, 1989, Volume V, 339

2) H.Tamehiro, Y.Horii, M.Yamaguchi, Y.Ogata, H.Akazaki, Y.Kawada, H.Miyazaki, K.Shinada and H.Ogawa, "Sour Service Large-Diameter Line Pipe Having Good Field Weldability and Sulfide Stress Corrosion Cracking Resistance", *Offshore Mechanics and Arctic Engineering (OMAE)*, Stavanger, Norway, 1991, Volume V, 85

3) R.Chijiiwa, H.Tamehiro, M.Hirai, H.Matsuda, H.Mimura, "Extra High Toughness Titanium-Oxide Steel Plates for Offshore Structures and Line Pipe", *Offshore Mechanics and Arctic Engineering (OMAE)*, Houston, Texas, U.S.A., 1988, Volume V, 165

4) K.Nishioka and H.Tamehiro " High-Strength Titanium-Oxide Bearing Line Pipe Steel for Low-Temperature service", *MICROALLOYING - 88* , Chicago, Illinois, U.S.A., 1988, 597

5) Y,Yoshida, H.Tamehiro, R.Chijiiwa, K.Funato, N.Doi, K.Tanaka and M.Kibe, "Production of High-Strength TMCP Steel Plate for Offshore Structures", *Offshore Mechanics and Arctic Engineering (OMAE)*, Glasgow, Scotland, UK, 1993, Volume IIIA, 207

6) Y.Ohtani, S.Watanabe, Y.Kawaguchi and Y.Yamaguchi, "Improvement in Toughness of Large Heat-Input Weld Bond of 60kg/mm^2 Grade High Tensile Strength Steel Plates due to Boron Addition and Low Silicon Content", *Tetsu to Hagane*, 1978, vol.64-14, 2205-2214 (In Japanese)

7) F.S.LePera, "Improved Etching Technique to Emphasize Martensite and Bainite in High-Strength Dual-Phase Steel", *Journal of Metals*, March (1980), 38-39

8) X.Xue and T.N.Baker, "Influence of aluminium on carbide precipitation in low carbon microalloyed steels", *Materials Science and Technology*, May (1993), vol.9, 424

Proceedings of the Seventh (1997) International Offshore and Polar Engineering Conference
Honolulu, USA, May 25-30, 1997

ISBN 1-880653-28-1 (Set); ISBN 1-880653-30-3 (Vol. II)

Technological Regimes of the Natural Gas Transport by Marine Pipelines with Schtokmanovskoe Gas Condensate Field, Barentsevo Sea

N.G. Figarov
Russian Research Institute for Pipeline Construction, Moscow, Russia
V.V. Rusakova
Management of Perspective Development of GAZPROM, Moscow, Russia
A.S. Tsourikov
Russian Research Institute of Natural Gases, Moscow, Russia

ABSTRACT

In the present report the algorithm of realization on the computer of the solution of a conjugated problem of thermal interaction of the subsea pipeline with environment is formulated and results of the prediction of change of conditions of operation of projected pipeline system during the natural gas transport from Schtokmanovskoe gas condensate field to the shoreline zone are considered. The offered method of the analysis is based on use of set of analytical and numerical mathematical models, describing processes thermogasdynamics and heat mass transfer at the transport of natural gas. The peculiarities of natural gas transport from Schtokmanovskoe gas condensate field are stipulated by there are not compressor stations throughout the marine route (560 km up to coast) and complicated geotechnical and climatic conditions (large depths of Barentsevo sea, large elevation differences of a relief seabed, low temperatures of a sea water and others).

KEY WORDS - Mathematical model, prediction, temperature, pressure, gas pipeline, Freeze Bulb, heaving, ice formation.

NOMENCLATURE

A, dimentionless parameter in Eqs. 20 and 21;

a, thermal diffusivity;

c_p, specific heat of gas;

d_o, inside diameter of steel pipe;

d, outside diameter of steel pipe;

d_c, outside diameter of subsea pipeline including a thickness of an arctic cement coat;

g, gravitational constant;

h, depth of the underground mode of installing the subsea pipeline;

L, total length of the subsea pipeline;

n, cartesian surface normal;

P, gas pressure;

Q, latent heat per unite volume;

q, environmental heat flux at the outside surface of subsea pipeline in radial direction;

R_{air}, Renoir air constant;

r, radial distance from the center of subsea pipeline;

S, moving boundary of phase change;

T, gas temperature;

t, time;

W, velosity of gas flow;

V, volume flow rate of gas;

v_b, sea current near the bottom;

w_{gr}, water content of the seabed grounds;

x, y, z, cartesian coordinate system;

z, distance along the subsea gas pipeline;

z_{gas}, gas compressibility factor.

Greek symbols

$\alpha(\rho(t))$, local heat transfer coefficient from the outside surface of subsea pipeline to the brine;

α_{gas}, local heat transfer coefficient from the gas flow to the internal steel wall of subsea pipeline;

γ, mass density;

Δ, gas gravity;

δ, wall thickness of steel pipeline;

δ_c, thickness of the weight cement coat outside of subsea pipeline;

ζ, given accuracy of calculations;

λ, thermal conductivity;

λ_c, thermal conductivity of the arctic cement coat outside of subsea pipeline;

λ_s, thermal conductivity of steel pipe;

ξ, pipeline hydraulic factor;

ρ, radius of Freeze Bulb or ice formation outside pipeline;

ρ_v, ρ_{v-1}, average radiuses of Freeze Bulb or ice formation outside pipeline for (v) and (v−1) - time steps, respectively;

τ, dimentionless parameter in Eq. 26.

χ, heaving factor.

Subscripts

a, average value gas state parameters;

b, brine;

c, critical gas state parameters;

f, initial freezing point of brine;

gr, seabed grounds;

gas, natural gas parameters;

i, j, number zone of phase change (frozen or thawed);

k, space interval (length or radius);

L, final (outlet) gas parameters;

M, frozen zone;

o, initial (inlet) gas parameters;

s, ambient (brine or seabed grounds);

T, thawed or unfrozen zone;

v, time step;

и, at all insulation, including entire materials are composed a pipeline construction (steel of pipe, inner and outside corrosion protected coating, weight cement coat and other);

л, ice formation.

INTRODUCTION

In the present report actual problems of the natural gas transport from Schtokmanovskoe gas condensate field, located in Barentsevo sea on distance about 560 km from coast are considered as well as a technique of calculation of change of pressure and temperature of gas along the pipeline from their initial values at the deposit (T_0 = 50 °C ; P_0 = 15 MPa, P_0 = 16 MPa and P_0 = 17 MPa) up to a shoreline is summarized. The examples are given for two constructive decisions, when string of the pipeline is welded from pipes by a diameter of 48 in. and 40 in. The transport of natural gas throughout the pipeline is accompanied by change of initial pressure and temperature of gas due to Joule-Tomson effect of gas along the pipeline, and also intensive heat transfer with an environment, so the temperature of gas can decreases below 0 °C.

The whole route of the subsea pipeline from Schtokmanovskoe gas condensate field up to a shoreline is conditionally subdivided into sites: conductive environmental heat transfer (shallow-water - on which pipeline is buried in seabed) and convective environmental heat transfer (deep-water, where the pipeline is laid on seabed). Shallow-water sites reach up to depth of the sea no more than 110 m and just for them freezing processes in seabed grounds around the gas pipeline are predicted which should be taken into consideration during detailed engineering.

In order to predict of quantitative data on change of temperature of gas along the pipeline and dynamics of Freeze Bulb a mathematical model describing real processes of heat mass transfer of a product with an environment was accepted (P-609-86, 1987) and its numerical implementation is carried out. With the purpose of simplification a complex problem, and taking into account, that such approach will a little bit have an effect on final practiced results, a three dimentional problem is submitted in two conjugated problems: one-dimentional - along an axis of the pipeline and two-dimentional - in a plane of an orthogonal axis. One-dimentional problem is solved in view of influence to temperature and pressure of the moving gas of Joule-Tomson effect, and two-dimentional problem takes into account an opportunity of environmental phase change phenomena (ambient is brine or seabed grounds) in accordance with Stefan equation. Developed algorithm of the calculation, which implements analytical and numerical methods, allows to receive the engineering solution of this three dimentional problem and to give the prediction of formation of a temperature field around the pipeline and distribution of temperature of gas along the pipeline on any period of operation of gas pipeline system with sufficient for practical purposes by accuracy (Dubikov,Figarov,Tsourikov,1994).

Results of calculation have shown, that temperature of the gas pipeline approaches temperature of an environment already through 85 - 125 km. Further decreasing gas temperature along the pipeline causes an ice formation, the magnitude of which depends on intensity convective heat transfer (sea bottom current around the pipe). Presence even of small underwater currents stipulates in sharp reduction of ice formation thickness or to its absence. On shallow-water sites, where there is underground mode of installing the pipeline, occurs frost penetration in seabed grounds. The depth of freezing beneath outside of pipeline during the term of its operation (33 years) reaches 6.6 m. If the pipeline is buried in a coarse-grain almost slightly-heave marine soil, the moving its axis, caused of Freeze Bulb, will be small (no more than 35 mm for entire term operation), that will insignificantly affect a stress-deformed condition of a gas pipeline, but in a silty soil the magnitude of vertical heaving of an axis of pipeline can reach 72 cm and should be taken into account during detailed engineering.

MATHEMATICAL MODEL

For approached calculations of gas flow throughout pipelines in many cases it is possible to be content with the simplified stationary scheme, when is accepted, that the vector of velocity flow in each section of the pipeline is directed along an axis, and magnitudes of velocity (W), pressure (P), density (γ_{gas}) and the temperature (T) are considered constant on section of a pipe. Then the problem of predicting of change of pressure and temperature of gas along the pipeline in view of dynamics of phase changes freezing processes around the pipe is described by system of the equations (Loytsansky 1973):

$$\frac{1}{\gamma_{gas}}\frac{\partial P}{\partial z} + \frac{\partial}{\partial z}\left(\frac{W^2}{2}\right) + \xi\frac{W^2}{2d_0} \pm g = 0,$$

$$c_p\frac{\partial T}{\partial z} = -\frac{\pi d_0}{\gamma_{gas} V} q + T\frac{\partial \gamma_{gas}^{-1}}{\partial T}\frac{\partial P}{\partial z} + \xi\frac{W^2}{2d_0},$$

$$W = \frac{4V}{\pi d_0^2} \quad (1)$$

By attaching to these equations the real gas equation of state

$$P = z_{gas}\,\gamma_{gas}\,g\,\frac{R_{air}}{\Delta}\,T \quad (2)$$

and by expressing density of a thermal flux on an outside surface of a gas pipeline by means of a heat flux boundary conditions for conduction mechanism (for shallow-water sites, case of buried pipeline in seabed) and for convection mechanism (for deep-water sites, where the pipeline is laid on seabed), respectively :

$$q = \oint_{(l)} q_r\,dl, \qquad q_r = \begin{cases} \lambda_i \left.\dfrac{\partial T_i}{\partial n}\right|_{d_c/2} \\ \alpha(\rho(t)) \cdot (T_s(t) - T_f) \end{cases} \quad (3)$$

We shall receive closed system for definition of four unknown functions W, γ_{gas}, P, T, which are describing a distribution of velocity, density, pressure and temperature of gas flow along the pipeline.

Natural boundary conditions, which correspond to considered processes of heat mass transfer in operated gas pipeline, are:

$$z=0 \quad P(0) = P_0;\ T(0) = T_0(t);\ W(0) = W_0;\ \gamma_{gas}(0) = \gamma_{0gas};\ q(0) = q_0$$

$$z=L \quad W(L) = W_L;\ \gamma_{gas}(L) = \gamma_{Lgas};\ \gamma_{0gas}W_0 = \gamma_{Lgas}W_L;\ q(L) = q_L \qquad (4)$$

For definition of environmental heat flux at the outside surface of subsea pipeline in radial direction and prediction of phase change processes into environment (seabed grounds, brine) was solved a problem, which is described by the following heat conduction equation:

$$\nabla^2 T_i = \frac{1}{a_i}\frac{\partial Ti}{\partial t}, \qquad i = 1, 2 \qquad (5)$$

For a semi-infinite medium $\{-\infty < y < +\infty;\ 0 < x < +\infty\}$ with circular cutout $\{y^2 + (x-h)^2 \le (d/2)^2\}$ with the following initial and boundary conditions:

$$T=0 \qquad T_i(x, y, 0) = T_s(0) \qquad (6)$$

$t > 0$

On a surface of seabed

$$x = 0 \qquad T_i(0, y, t) = \varphi(t) \qquad (7)$$

On a wall of the pipeline

$$x,y \in \{y^2 + (x-h)^2 = (d/2)^2\} \qquad T_i(x,y,t) = \psi(T(z),t)$$

$$x,y \in \{y^2+(x-h)^2=(d_k/2)^2\} \qquad T_k\big|_{d_k} = T_i\big|_{d_k}, \quad \lambda_k \frac{\partial T_k}{\partial n}\Big|_{d_k} = \lambda_i \frac{\partial T_i}{\partial n}\Big|_{d_k} \qquad (8)$$

On the moving boundary of phase change

$$x,y \in \{S_j(x,y,t)=T_f\} \qquad T_i(x,y,t)=T_T = T_M = T_f = \text{const}$$

$$x,y \in \{S_j(x,y,t)=T_f\} \qquad \pm \lambda_M \,\text{grad}T_M + \lambda_T \,\text{grad}T_T = Q_i \frac{dS_j}{dt} \qquad (9)$$

Outside of a zone of thermal influence of the pipeline

$$x \to \infty \qquad T_i(\infty, y, t) = \text{const} \qquad (10)$$

(All variables and parameters, used in the equations (1) - (10) and in the subsequent equations are defined in Nomenclature).

On a deep-water sites there are a convective environmental heat transfer, which is provided to Newton's law. Then the heat transfer process between the pipeline and brine can be expressed in the following heat convection equation(P-609-86, 1987):

$$q = \oint_{(l)} \alpha(\rho(t)) \cdot (T_s(t) - T_f)\, dl; \quad \alpha(\rho(t)) = 476.5 \cdot (v_w/\rho(t))^{-0.5} \qquad (11)$$

The system of the equations (5) - (10) for a case, when intensity of heat convection is insignificant and it is possible an ice formation outside of the pipeline, can be reduced to the following equation and boundary conditions:

$$\frac{1}{a_i}\frac{\partial}{\partial r}\left[r \frac{\partial T_i}{\partial r}\right] = \frac{\partial Ti}{\partial t}, \qquad i = 1, 2 \qquad d_k/2 \le r \le \rho(t) \qquad (12)$$

$$t = 0 \qquad T_i = T_{\text{н}} = T_{\text{л}} = T_s(0), \qquad \rho(0) = d_k/2 = d_c/2 \qquad (13)$$

$t > 0$

$$r = d/2 \qquad T_i = T_{\text{н}}^1 = f(t) \qquad (14)$$

$$r = d_{k-1}/2 \qquad T_i = T_{\text{н}}^{k-1} = T_{\text{н}}^k, \quad \lambda_{k-1}\frac{\partial T_{\text{н}}^{k-1}}{\partial r}\Big|_{d_{k-1}/2} = \lambda_k \frac{\partial T_{\text{н}}^{k}}{\partial r}\Big|_{d_{k-1}/2} \qquad (15)$$

$$r = d_k/2 \qquad T_i = T_{\text{н}}^k = T_{\text{л}}, \quad \lambda_k \frac{\partial T_{\text{н}}^{k-1}}{\partial r}\Big|_{d_k/2} = \lambda_k \frac{\partial T_{\text{л}}}{\partial r}\Big|_{d_{k-1}/2} \qquad (16)$$

$$r = \rho(t) \qquad \lambda_{\text{л}} \frac{\partial T_{\text{л}}}{\partial r}\Big|_{r=\rho(t)} - \alpha(\rho(t)) \cdot (T_s(t) - T_f) = Q_b \frac{d\rho}{dt} \qquad (17)$$

$$\rho(t) < r < \infty \qquad T_{\text{л}} = T_s(t) \qquad (18)$$

The equations (1) - (18) describe interaction of a non-isothermal flow of a compressed fluid, thus simultaneously in the above-stated problem were used following approved in practice of the assumption: the flow of gas in the pipeline is one-dimentional and single-phase; thermobarographic parameters of a flow satisfy to a condition of local thermodynamic balance; the influence of gas viscosity and kinetic effects are insignificant; the influence of deformations of walls of the pipeline to process of transport of gas is not taken into account; processes of hydraulic resistance on an internal wall of the pipeline and external heat transfer are quasi-steady-state; compressed fluid - the gas, satisfies to the thermodynamic calorimetric equation of a real gas equation of state (in polynomial form); the convective heat transfer and migration heat transfer in frozen and thawed zones of a seabed grounds is not taken into account; thermophysical properties of a medium and materials of projected pipeline are considered as constants, not dependent on temperature.

At the solution of system of the equations (1) - (18), the pipeline is discreted by means of piecewise sections $L_k = z_k - z_{k-1}$ in limits of which the gas pipeline provides to one of constructive modes of installing the pipe (buried in seabed or simply laid on seabed). The length of each section of subsea gas pipeline is limited by elevation difference of edges (z_k and z_{k-1}). This elevation difference should not be exceeded a magnitude of one hundred meters. On edges of each section of a gas pipeline additional conditions are carrying out, according to which the final parameters of a gas flow of each previous section or its element, are initial for the subsequent section:

$$z = z_k \qquad \begin{cases} T(z_k+0) = T(z_k - 0) \\ P(z_k+0) = P(z_k - 0) \\ W(z_k+0) = W(z_k - 0) \\ V(z_k+0) = V(z_k - 0) \end{cases} \qquad (19)$$

In limits of each section or its element the environmental heat flux (q) is averaged throughout the length section (L_k) and is considered independent from coordinate systems. Above-stated approach to the solution of the given class nonlinear three dimentional problems allows, keeping accuracy necessary for practice of final results, essentially to simplify a procedure of the calculation. It is reached by use on characteristic sites of the route of the known approached analytical solutions. For multifront phase change problems, the analytical solutions of which are accompanied by significant difficulties, systems of the transcendental equations are used.

The depth of freezing seabed grounds or ice formation outside of gas pipeline is predicted from the following equations, which are derived from in view of made assumptions from the solutions of the equations (5) and (12), respectively:

$$t_v = \lambda_M^{-1} \cdot Q \cdot (T_f - T_0)^{-1} \cdot \rho_v \cdot (\rho_v - \rho_{v-1}) [\ln(2 \cdot \rho_v / d_c) + A_{gr}] + 0.5 \cdot a_0^{-1} \times$$

$$\times (\rho_v - \rho_{v-1}) [\rho_v - 0.5 \cdot \rho_v^{-1} \cdot (\ln(2 \cdot \rho_v / d_c) + A_{gr})^{-1} \cdot (\rho_v^2 - 0.25 \cdot d_c^2 \times$$

$$\times (1 - 2 \cdot A_g))] ; \quad A_{gr} = \lambda_M \cdot (\alpha_{gas}^{-1} + \lambda_s^{-1} \cdot \ln(d/d_0) + \lambda_c^{-1} \cdot \ln(d_c/d)) \qquad (20)$$

$$\rho_v^{0.5} \cdot [\ln(2 \cdot \rho_v / d_c) + A_b] = 4 \cdot 10^{-3} \cdot (T_0 - T_f) \cdot (T_f - T_s) \cdot v_b^{-0.5}$$

$$A_b = \lambda_л (\alpha_{gas}^{-1} + \lambda_s^{-1} \cdot \ln(d/d_0) + \lambda_c^{-1} \cdot \ln(d_c/d)) \qquad (21)$$

The computed implementation of mathematical model (1) - (21) has used to carry out calculating of technological regimes of the natural gas transport by subsea pipelines from Schtokmanovskoe gas condensate field. The calculation thermal and hydraulic regimes of the gas transport is carried out by means of numerical solution the system of equations (1) - (3) for each section with use of analytical expressions (20) - (21). Above all should be determine average parameters of temperature ($T_a^{(v)}$) and pressure of gas ($P_a^{(v)}$) in the beginning of each section for some constant time step (t_v) in particular equal to one month (t_v = 730 hours). Then parities of signs of temperatures of gas (T_0) and environment (T_s) on the Centigrade scale should be established, for example let there will be the case of equality of signs of compared temperatures:

$$\text{sign}(T_A^{(v)} - 273) = \text{sign}(T_s - 273) \qquad (22)$$

As the first approximation values of pressure and temperature for section of pipeline without the account of environmental phase change processes under condition of $\rho_{v-1} = \rho_v = d_c/2$ are accepted. If predicted temperature T (z) throughout the length of section (L_k) does not change the sign and has the same sign with temperature of an environment in accordance with equation (22), the calculation comes to an end on this sole iteration for the given time step (t_v). In case calculated temperature in the first approximation T (z) throughout the length of the section will change the sign, the length of the initial section should be decreased up to length, in limits of which a gas temperature does not change the sign and subsequent subiterations (T_μ) for the given time step (t_v) are carrying out.

The second (T_μ) and the subsequent ($T_{\mu+1}$) subiterations for the given time step (t_v) are carrying out, if calculated temperature $T_\mu(z)$ after the first subiteration throughout the length of the section (L_k) will change its sign or initial temperature $T_a^{(v)}$ has a sign, opposite to a sign of temperature of environment. In this case in second and subsequent subiterations for the given time step (t_v) the average temperature of gas from the previous subiteration is calculated, on magnitude of which by means of the equations (20) - (21) will be defined Freeze Bulb or ice formation, heat flux (q) from the equation (3) and then again a distribution of temperature T (z) throughout the length of the section of a gas pipeline is computed. For each time step (t_v) all subsequent subiterations similar two above described are carried out until fulfillment of a condition of convergence of computing process. Criterion of convergence of subiterative process for each time step (t_v) is comparison of two subsequent values of considered final temperatures $T_\mu(z)$, $T_{\mu+1}(z)$ of the section of a gas pipeline:

$$|(T_{\mu+1}(z) - T_\mu(z)) / T_\mu(z)| \leq \zeta \qquad (23)$$

The practice of computation has shown, if an accuracy of calculations equals $\zeta = 10^{-2}$ then quite enough two subiterations.

The computations of technological regimes of the gas transport throughout the total length of marine pipeline (change of pressure P (z) and temperature T (z) of gas along the pipeline) are made for all term of operated pipeline system by means of impose on each section a condition (19) in order to pass to the subsequent section for each time step (t_v) under a satisfaction of given limiting conditions:

$$\begin{cases} P_{z=L} \leq P_{min}, \ P_{min} = \text{const} \\ T_{z=L} \leq T_{sup}, \ T_{sup} = \text{const} \end{cases} \qquad (24)$$

The total calculations accuracy of pressure $P_{z=L}$ and temperature $T_{z=L}$ at the end of 200 km - pipeline in accordance with given technique constitutes no less than $5 \cdot 10^{-2}$ MPa and 0.2 °C, respectively. Numerical examples carried out on the above-stated technique and their comparisons to field data have shown, that the considered mathematical model of interaction of a gas pipeline with an environment allows with sufficient for practical purposes by accuracy to analyse changes of temperature and pressure of gas along the pipeline, as well as a change environmental temperature regime.

GEOTECHNICAL CONDITIONS & DETAILED ENGINEERING

The marine route of a gas pipeline is about 560 km crosses various geological structures, forming the diverse relief forms are represented on Fig. 1. About 85 % of the route (470 km) is placed on the territory with depths of the sea more than 110 m and only on two sites (North - Kaninskoe Plateau - 90 km and Shoreline district is about 0,8 km) the depth of the sea less than 110 m.

On North - Kaninskoe Plateau, where supposing a gas pipeline is buried in seabed, grounds are composed by marine Quartenary silty sands and sandy loam by thickness more than 8 m with the increased contents of clay particles (up to 30 - 40 %).

On Shoreline district, where also supposing a gas pipeline is buried in seabed, grounds are composed everywhere laying marine modern sands by thickness of about 7 - 12 m and include entire grain-size spectrum from large (1-2 mm grains) up to fine particles(0.5-0.25 mm).

In spite of everywhere spreading ambient temperatures below 0°C a permafrost is absent beneath a seabed throughout entire the route. A such phenomenon is consequence due to more low freezing temperature of brine as compared with to ambient temperatures.

The economically quantities of gas recovery of Schtokmanovskoe gas condensate field is supposed to be carried out In volume $50 \cdot 10^9$ m^3 per year or $70 \cdot 10^9$ m^3 per year for that a Construction three strings of a gas pipeline by a diameter 40 inches or 48 inches is required, respectively. As In any pipeline Construction, the recommended inner corrosion protected coat as well as external coating system including also outside concreting of a pipes by means of superheavy arctic cement by thickness no less than 80 - 100 mm. Natural gas without preliminary compression should provide a subsea gas pipeline with pressure 15 - 17 MPa at temperature + 50 °C. The principal data for investigations of natural gas flow transport of the Schtokmanovskoe gas condensate field are summarized below.

Model water parameters of Barentsevo sea (Dobrovolsky, 1977) **consists of**

$T_b = -0.9$ °C; $T_f = -1.0$ °C; $\gamma_b = 1.022$ g/cm^3; $\lambda_л = 2.21$ W/m K; salinity is about 24.7 ‰; $v_b = 0.22$ m/s.

Model seabed ground parameters of North - Kaninskoe Plateau consists of

$T_{gr} = -0.9$ °C; $T_f = -1.0$ °C; $\gamma_{gr} = 1.89$ g/cm^3; $w_{gr} = 0.37$; $\lambda_T = 1.35$ W/m K; $\lambda_M = 1.45$ W/m K; $a_T = 7.1 \cdot 10^{-7}$ m^2/s; $a_M = 8.9 \cdot 10^{-7}$ m^2/s; $\chi = 0.12$.

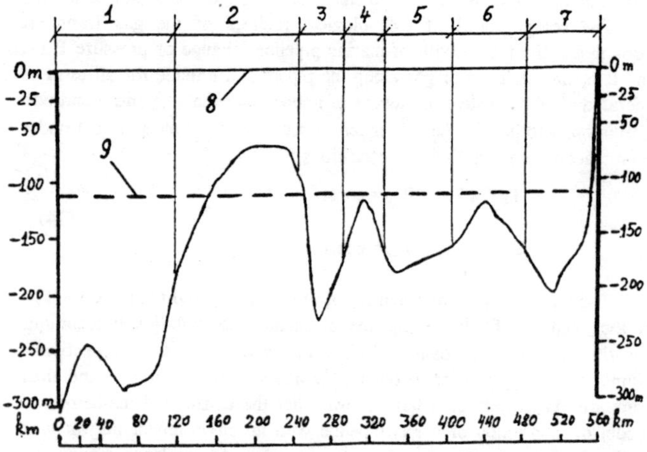

Fig.1 Longitudinal topographical section of Barentsevo Subsea Crossing from North (Schtokmanovskoe gas field - 0 km) to South (Shoreline - 560 km).

1 - Central Depression; 2 - North - Kaninskoe Plateau; 3 - Kaninsky Trench; 4 - South - Kaninskoe Plateau; 5 - North - Murmanskay Low-Lands; 6 - Murmanskay Hills; 7 - Kolsky Trough; 8 - Sea Level; 9 - Conventional Depth of one - hundred and ten metres beneath it the under-ground mode of installing the pipe is not erected.

Model seabed ground parameters of Shoreline district consists of T_{gr} = -0.9 °C; T_f = - 1.0 °C; γ_{gr} = 1.95 g/cm^3; w_{gr} = 0.32; λ_T = 1.35 W/m K ; λ_M = 1.45 W/m K; a_T = 6.3 · 10^{-7} m^2/s; a_M = 9.7 · 10^{-7} m^2/s ; χ = 0.05 .

Gas properties and pipeline configuration data consists of respectively:

Δ = 0.6 ; P_0 = 15.0 MPa, P_0 = 16.0 MPa and P_0 = 17.0 MPa;

T_0 = 323 K; P_c = 4.62 MPa ; T_c = 199.3 K.

$$c_p = 1.695 + 1.638 \cdot 10^{-3} \cdot T_a + 1.96 \cdot 10^6 \cdot T_a^{-3} \cdot (P_a - 0.1) \qquad (25)$$

$$z_{gas} = 1 - 0.0241 \cdot P_a \cdot P_c^{-1} \cdot \tau^{-1} \qquad (26)$$

$$P_a = 2/3\,(P_{0k} + P^2\,(P_{0k} + P)^{-1}) \qquad (27)$$

$$\tau = 1 - 1.68 \cdot T_a \cdot T_{ck}^{-1} + 0.78 \cdot (T_a \cdot T_{ck}^{-1})^2 + 0.0107 \cdot (T_a \cdot T_{ck}^{-1})^3 \qquad (28)$$

$$T_a = L_k^{-1} \cdot \int_{z_{k-1}}^{z_k} T(z)\, dz \qquad (29)$$

$$\alpha\,(\rho(t)) = 476.5 \cdot (v_b / \rho(t))^{-0.5}; \qquad (30)$$

if there is not ice formation $\rho(t) = d_c/2$.

$\{ d_0$ = 956 mm; α_{gas} = 174.5 W/m^{2}K; ξ = 1.00124 · 10^{-2} ;
$\{ \delta$ = 30 mm; δ_c =80 mm; h = 2.0 m
$\{ \lambda_s$ = 46.52 W/m K ; λ_c = 1.55 W/m K

$\{ d_0$ = 1147 mm; α_{gas} = 264.9 W/m^{2}K; ξ = 9.64405 · 10^{-3};
$\{ \delta$ = 36 mm; δ_c =100 mm; h = 2.0 m
$\{ \lambda_s$ = 46.52 W/m K ; λ_c = 1.55 W/m K

RESULTS & DISCUSSION

To determine the best possible regimes of natural gas transport by subsea pipelines, the model was applied to a very complicated conditions of the Schtokmanovskoe gas condensate field. Profiles of pressure and gas temperature, ice formation, frost penetration in seabed and heaving values of Freeze Bulb were computed for entire term of operated submarine pipelines.

Predicted profiles of gas pressure along the pipeline. Fig. 2 shows predicted profiles of gas along the route for various regimes of gas transportation (P_0 = 15.0 MPa, P_0 = 16.0 MPa and P_0 = 17.0 MPa at the T_0 = 323 K) for two different diameters of pipeline 40 in. and 48 in, respectively. Fig. 2 shows that initial pressure and volume flow rate strongly influences onto the profile of gas along the pipeline. Results indicate that the outlet pressures at the inlet pressure 17 MPa about 1.5 times are greater then the outlet pressures at the inlet pressure 15 MPa. On Fig. 2 also distinctly can see some change of curvature computed structures of pressure, which are caused by elevations difference along the route.

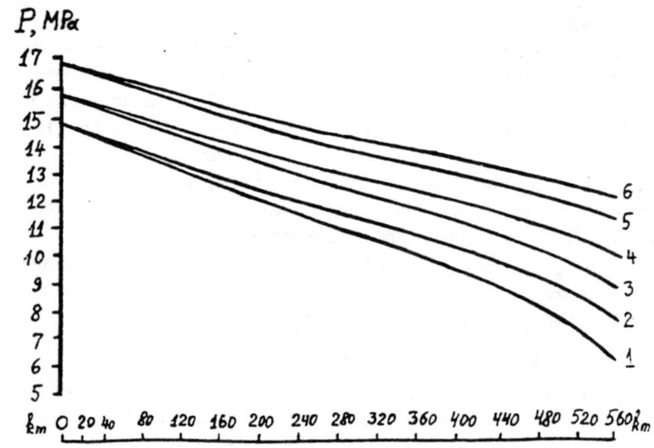

Fig. 2 Predicted profiles of pressure of gas along the subsea pipeline shut-in Barentsevo Sea condition.

1, 3, 5 - profiles of pressure of gas for 40 in. pipeline, volume flow rate of gas 5 · 10^7 m^3 per day, initially temperature of gas at inlet + 50 °C & pressures of gas at inlet - 15 MPa, 16 MPa and 17 MPa, respectively.
2, 4, 6 - profiles of pressure of gas for 48 in. pipeline, volume flow rate of gas 7 · 10^7 m^3 per day, initially temperature of gas at inlet + 50 °C & pressures of gas at inlet - 15 MPa, 16 MPa and 17 MPa, respectively.

Predicted profiles of gas temperature along the route and ice formation outside the subsea pipeline. Fig. 3 shows predicted profiles of gas temperature along the pipeline for two technological regimes. On Fig. 3 as well as on Fig. 2 distinctly can see some change of curvature computed temperature profiles, which are caused by difference in intensity convective and conductive heat transfer of a gas pipeline with an environment. More intensive convective heat transfer of a gas pipeline with brine on two deep-water sites of a route (0 - 120 km and 245 - 559 kms) appears in more sharp dropping a gas temperature, which already on 85 km and 125 km (depending on volume flow rate of gas) approaches an environmental temperature.

At downtrend of gas temperature up to magnitude - 5.7 °C or - 4.8 °C (critical gas temperature of initial ice formation outside a pipeline by a diameter of 40 in. and 48 in., respectively), an ice formation outside subsea pipeline begins In accordance with a condition of thermal equilibrium, which is derived from the equation (17) at $\rho(t) = d_c/2$ = const.

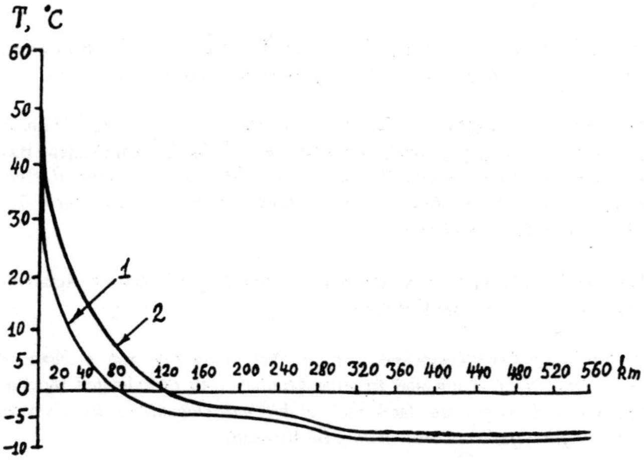

Fig. 3 Predicted profiles of temperature of gas along the subsea pipeline shut-in Barentsevo Sea condition.

1 - profile of temperature of gas for 40 in. pipeline, volume flow rate of gas $5 \cdot 10^7$ m^3 per day, initially temperature of gas at inlet +50 °C & pressure of gas at inlet - 15 MPa, respectively.
2 - profile of temperature of gas for 48 in. pipeline, volume flow rate of gas $7 \cdot 10^7$ m^3 per day, initially temperature of gas at inlet + 50 °C & pressure of gas at inlet - 17 MPa, respectively.

On first frontier between two sites (convective and conductive heat transfer) at a distance 121 km computed gas temperatures reaches -0.1 °C and -3.1 °C for the pipeline by a diameter of 48 in. and 40 in., respectively, therefore on the given zone there is not ice formation outside pipeline. On second frontier between two sites (convective and conductive heat transfer) at a distance 559 km computed gas temperatures reaches -6.7 °C and -8.2 °C for the pipeline by a diameter of 48 in. and 40 in., respectively. Therefore on the given shoreline zone outside a gas pipeline is subjected to ice formation the maximum thickness of which reaches 42 mm and 15 mm, accordingly for the pipeline by a diameter of 40 in. and 48 in. Ice formation outside a gas pipeline occurs very quickly (not duration than first hours) and an increment ice at once ceases at achievement of a condition of thermal equilibrium, which is established on boundary ice - brine. The maximum ice formation, as have shown results of computations does not exceed 5 cm, that will not practically affect on ballasting of a gas pipeline because of change of buoyancy force, arising in result of ice formation outside a subsea pipeline.

Predicted depths and heaving values of freeze bulb beneath of underground mode of installing the subsea pipeline. On Fig. 4, 5 the results of predicting of dynamics of freeze bulb formation around the gas pipeline crossing North - Kaninskoe Plateau (at the distance of 240 km) and Shoreline district (at the distance of 560 km) are presented. Fig. 4, 5 shows that the depths of seabed freezing it is more around the pipeline with a smaller diameter (40 in.), than with a large diameter (48 in.), at the practically identical values of physical and thermophysical soil properties. It is explained by designed features of the pipeline of a smaller diameter (40 in.), possessing a smaller thickness of the weight cement coat considered as insulation coating, together with lower gas temperatures in comparison with gas temperatures of the greater pipeline diameter (48 in.). Gas temperatures accordingly are equal on North - Kaninskoe Plateau - 3.8 °C and - 4.9 °C, and on Shoreline district -6.7 °C and -8.2 °C. Parameters of freeze Bulbs and heaving values of the pipeline by a diameter of 40 in. and 48in. accordingly are equal on North - Kaninskoe Plateau 5.84 м, 72 mm ; 5.65 м, 70 mm and on Shoreline district are equal 6.56 м, 35 mm; 6.20 м, 35 mm, respectively (Fig. 4,5).

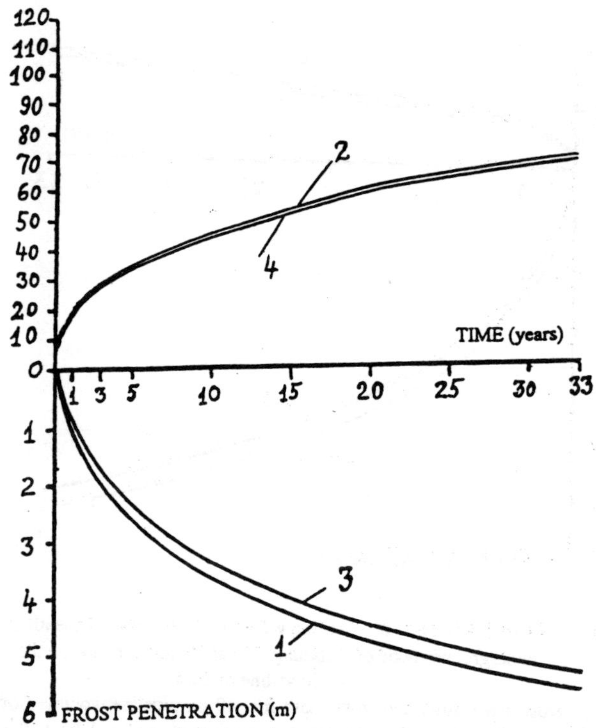

Fig. 4 Predicted Depths & Heaving values of Freeze Bulb beneath of underground mode of installing the subsea gas pipeline on the North - Kaninskoe Plateau.

1, 2 - frost penetration and frost heave for 40 in. pipeline, volume flow rate of gas $7 \cdot 10^7$ m^3 per day, initially temperature of gas at inlet +50 °C & pressure of gas at inlet - 17 MPa, respectively.
3,4 - frost penetration and frost heave for 48 in. pipeline, volume flow rate of gas $5 \cdot 10^7$ m^3 per day, initially temperature of gas at inlet +50 °C & pressure of gas at inlet - 15 MPa, respectively.

A large difference in heaving values is caused by various geotechnical properties seabed grounds of North - Kaninskoe Plateau (very high heaving silty sands and sandy loam) in comparison with coarse-grain slightly-heaving sands of Shoreline zone (Smith, Dallimore and Kettle, 1985). The intensive heaving process which is in progress in the seabed grounds of North - Kaninskoe Plateau can by an essential image affect onto change the projected underground mode of installing the subsea gas pipeline, so it should be taken into account during detailed engineering of marine pipeline systems.

CONCLUSIONS

In the report the basic aspects of a developed technique of predicting of interaction transported product by marine pipelines with an environment are performed. Constructed on stationary and quasi- stationary approximations of considered process, it represents the computed technique, allowing with sufficient for practical implementation an accuracy to receive the answers on arise the questions during detailed engineering of the subsea pipeline systems:

- the developed mathematical model allows to give the prediction of change of gas temperature and gas pressure along the pipeline concerning a thermal interaction of gas flows with environment;
- a transport of gas by marine pipelines on large distances in subarctic regions without erection of intermediate compressor stations causes drop a gas temperature below 0 °C;

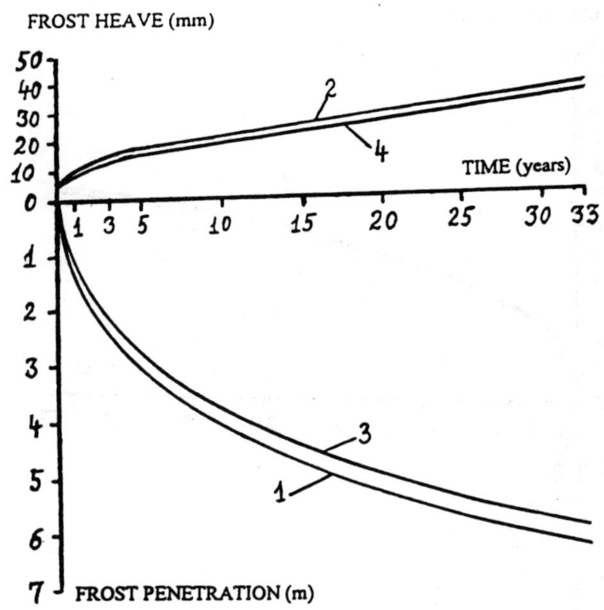

Fig. 5 Predicted Depths & Heaving values of Freeze Bulb beneath of underground mode of installing the subsea gas pipeline on the Shoreline district.

1, 2 - frost penetration and frost heave for 40 in. pipeline, volume flow rate of gas $7 \cdot 10^7$ m^3 per day, initially temperature of gas at inlet +50 °C & pressure of gas at inlet - 17 MPa, respectively.

3,4 - frost penetration and frost heave for 48 in. pipeline, volume flow rate of gas $5 \cdot 10^7$ m^3 per day, initially temperature of gas at inlet +50 °C & pressure of gas at inlet - 15 MPa, respectively.

- on an example of transport of gas with Schtokmanovskoe gas condensate field up to a shoreline zone (a distance of 560 km) has shown, that during detailed engineering should be taken into account results of the prediction change of temperature of gas along the pipeline;
- on sites, where not apply an under-ground mode of installing the pipe , the ice formation outside a subsea pipeline depends on intensity of environmental heat transfer. At absence intensive heat transfer from pipeline to environment occurrence ice formation outside the pipeline and in consequence it causes change the buoyancy of subsea gas pipeline. At presence intensive heat transfer from pipeline to environment (as for example on Schtokmanovskoe gas condensate field) there is no ice formation outside the subsea pipeline (beginning North - Kaninskoe Plateau) or its occurrence very minute (beginning Shoreline zone);
- on sites, where apply an under-ground mode of installing the pipe, is occurred a Freeze Bulb around the pipeline. At presence very high heaving seabed grounds there is the displacement of the pipeline from its initial position (as for example on Schtokmanovskoe gas condensate field a magnitude of Freeze Bulb reaches 6.6 м and heaving of a pipe can reach on separate sites more than 70 cm).

Thus, the theoretical analysis presented here indicates that there is a good potential in the prediction technique applying to detailed engineering of the natural gas pipeline system of the Schtokmanovskoe gas condensate field.

REFERENCES

Dobrovolsky A.D. (1977): "Convective Circulation in the Barentsevo Sea," *Moscow State University, Geographical Faculty.* (in Russian)

Dubikov G.I., Figarov N.G. and Tsourikov A.S. (1994): "Subsea permafrost and gas pipeline construction of the Baydaratskaya Bay crossing in Siberia, Russia," *Proc. of the Seventh International Cold Regions Engineering Specialty Conference, Edmonton, Alberta , 7-9 March,* (1994), pp.843-864.

Loytsansky L.G. (1973):"Mechanics of liquid & gas", *Russian Academy of Science, Moscow*. (in Russian).

P-609-86 (1987): "Guidelines for prediction thermal and mechanical interaction of freezing and thawing grounds with complicated pipeline systems and associated facilities," *Published Documents by Russian Institute for Pipeline Construction.* (in Russian)

Smith M.W. , Dallimore S.R. and Kettle R.J. (1985): "Observations and prediction of frost heave of an experimental pipeline," *Proc. of the Fourth International Symposium on Ground Freezing, Sapporo, 5-7 August,* (1985), pp. 297-304.

Proceedings of the Seventh (1997) International Offshore and Polar Engineering Conference
Honolulu, USA, May 25-30, 1997

ISBN 1-880653-28-1 (Set); ISBN 1-880653-30-3 (Vol. II)

An Introduction to the DNV 1996 Rules for Submarine Pipeline Systems

Leif Collberg, Tommy Bjørnsen and Kim J. Mørk
Det Norske Veritas
Høvik, Norway

ABSTRACT

An introduction is given to the new revision of Det Norske Veritas (DNV) "Rules for Submarine Pipeline Systems", denoted DNV'96 in the following.

The new Rules constitute a complete revision of the previous Rules from 1981 and provide a significant step forward for pipeline design and operation. The Rules adopt the latest research achievements and practical experience gained through some of the most challenging pipeline projects recently executed.

The major changes compared to the old DNV'81 Rules are:

- **introduction of Limit State based design format with reliability based partial safety factors**
- **new materials**
- **new design scenarios**
- **new installation methods**
- **implementation of results from major R&D projects**
- **provide a recipe for reliability based design methods**
- **give credit for improved materials and quality**
- **requirements to condition assessment and re qualification**

A brief introduction to the major changes is given in the following. The paper concludes with a short example, illustrating the use of the Limit Stated concept.

INTRODUCTION

The pipeline industry has traditionally been rather conservative. The main design requirements in codes for both land based and offshore pipelines have remained more or less as they were at the introduction of the ASME codes several decades ago. Significant technical progress has been achieved in pipeline design, materials, fabrication, installation and operation of offshore pipelines during the last ten to fifteen years without necessarily being implemented in codes. The major changes in the offshore oil and gas industry taken place recently have forced the pipeline industry to focus on cost reduction through innovative design approaches, new materials, standardisation and optimised construction and operational procedures. The industry has also been forced to implement new concepts such as safety and risk.

On this basis, there has been a need for update of recognised codes to implement the latest developments and knowledge within the pipeline industry.

DNV has recently completed a major update of the well known DNV Rules for Submarine Pipeline Systems from 1981 referred to as DNV'96. DNV'96 are based upon the principle of Limit State philosophy adopting a Load and Resistance Factor Design (LRFD) format (or partial safety factor format), and include the latest research achievements and practical experience gained within pipeline technology.

The results from the SUPERB Joint Industry Project, see e.g. Sotberg *et. al.* (1996), performed by SINTEF, Snamprogetti and DNV have formed the basis for the capacity formulations within the Rules.

DEVELOPMENT AND LAYOUT

The objective updating the Rules was to:

- provide an international acceptable standard of safety with respect to strength and performance
- serve as a technical reference document and guideline for designers, clients and contractors
- specific minimum requirements in relation to DNV certification word wide

DNV therefore in 1993 initiated a revision of the '81 Rules with the aim to fulfil the above targets.

There were two important factors in such a revision. One was to include the latest research achievements especially from the SUPERB project. SUPERB's application of limit stated based design with calibrated partial safety factors constituted, hence, an important input.

The second important factor was to establish close co-operation with members of the NORSOK initiative; Statoil, Norsk Hydro, Saga Petroleum and Phillips Petroleum Company Norway. By this co-operation the combined experience from pipeline operators would be adopted in line with the NORSOK spirit.

Four working groups with representatives from the different oil companies were established:

- Administration and co-ordination
- Design
- Material

- Installation

The working groups have, based upon a very early draft, contributed to the different sections in the Rules. The Rules were in February -96 issued for internal hearing (including the participating oil companies). An external hearing revision, including the received comments were issued in September 1996. This revision was also sent for comments to all sponsors of the SUPERB project (Conoco, EMC, Exxon, Shell, Statoil, NPD, HSE). The final Rules were closed and approved in December 12th 1996.

The intention of the Rules is to provide the user with specific guidance in areas with well proven technology and functional requirements otherwise. Even though this may seem restrictive for development of new design concepts, the adopted limit state philosophy gives extensive flexibility for new design scenarios.

The Rules are in principle self contained but are supported by different guidelines covering specific aspects, usually as result from different research projects, ref. Figure 1.

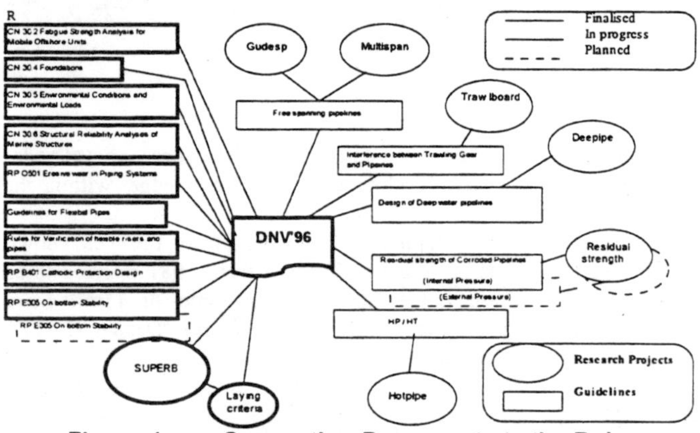

Figure 1 Supporting Documents to the Rules

DEFINITIONS

Some important definitions have been introduced or clarified, in DNV'96 and are explained in the following:

Design Pressure is the maximum internal pressure during normal operation, referred to a specified reference height, to which the pipeline or pipeline section is to be designed. The Design Pressure must take account of steady flow conditions over the full range of flow rates as well as possible packing and shut-in conditions over the whole length of the pipeline or pipeline section which is to have a constant Design Pressure.

Incidental Pressure is the maximum permitted internal pressure due to incidental operation, such as surges due to sudden closure of valves or failure of the Pressure Regulation System, of the pipeline or pipeline section.

Local Design Pressure and *Local Incidental Pressure* is the internal pressure at any point in the pipeline system or pipeline section for the corresponding Design Pressure or Incidental Pressure. This is equal to the Design/Incidental Pressure at the reference height plus the static head of the transported medium due to the difference between the reference height and the height of the section being considered.

Figure 2 illustrates the relations between the different systems and pressures.

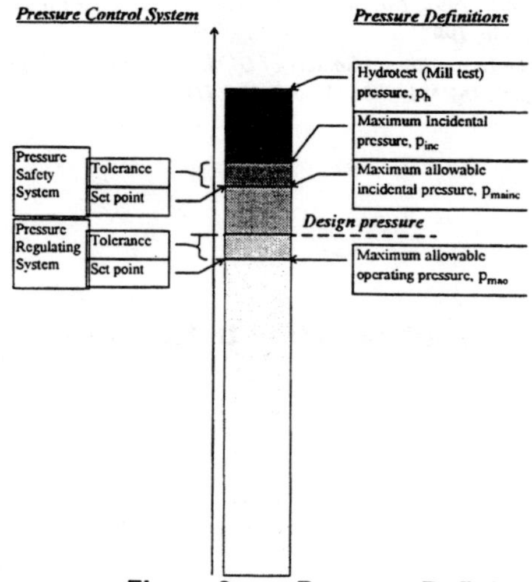

Figure 2 Pressure Definitions

The term *Buckling* is in the Rules used in a broad sense describing general instability of the pipe. Distinction is made between *Global Buckling*, usually referred to as column buckling (or bar/Euler buckling) and *Local Buckling*. Local buckling denotes buckling where collapse implies deformation of the cross section of the pipe and includes the commonly used terms; collapse, buckling, wrinkling etc.

The definitions of *hydrostatic test, pressure test* and *mill test* have, in practice, been re-defined to represent different tests for different disciplines. Therefore, the following clarification is made, note that this is in line with current codes.

Hydrostatic test (hydrotest), often referred to as mill test, denotes the structural test at the steel mill applied to each single pipe joint while the *Pressure Test* is a test performed on the complete pipeline section after installation and is succeeded by a leakage test. Hydrostatic (at the mill) test is considered being the strength test, which always shall be performed, while pressure test is a quality and leak test which normally shall be performed.

A *Limit State* is the limit between an acceptable and unacceptable condition expressed in mathematical terms for a given failure mode. The limit states are grouped into different *Limit State Categories*.

Load and Resistance Factor Design (LRFD) - to verify that factored design loads do not exceed factored design capacities for any of the complete set of limit states to be considered in design. This is one deterministic solution from a reliability calibration adopting the limit state concept.

SAFETY PHILOSOPHY

General

In compliance with modern design codes the following safety hierarchy is adopted:

- Overall safety objective - to be specified by the operator
- Systematic review - to identify and evaluate the consequences of single failures and series of failures (typically risk analyses)
- Quality Assurance - providing requirements to work, production and verification. This is often required to be in compliance with the ISO 9000 system.

- Limit state design - to verify that factored design loads do not exceed factored design capacities for any of the complete pre-defined set of limit states to be considered in design
- Probabilistic design- as an alternative to the basic design criteria in the Rules

The safety philosophy adopted implies that gross errors (human errors) are controlled by requirements to the organisation of the work, competence of persons performing the work, verification of the design and Quality Assurance during all relevant phases. This is in accordance with DNV'96 design philosophy.

Reliability-based Calibration

Any design code provides a certain safety margin against failure in design. This inherent safety margin is mainly related to the choice of safety factors sometimes selected on a more or less arbitrary basis. This has caused different (often unknown) safety levels for different design checks.

In DNV'96 the limit state design principle is adopted. This implies that the performance of the pipelines is described in terms of a set of limit states for which adequate safety margins are quantified as described in next section. For all limit states a set of safety factors are calibrated for each safety class using a structural reliability approach.

It introduces flexibility in specific conditions and provides design with a consistent safety level without compromising the safety objective.

However, iIn a sound calibration process a varying degree of conservatism need to be introduced for individual design scenarios depending on the knowledge of the prevailing loads, pipe capacities, etc. Thus, the calibrated design criteria being generally applicable may be expected to be conservative on average.

DNV'96 allows, as an alternative, probabilistic design provided an acceptable reliability method is applied by competent personnel. The acceptable target failure probabilities should be in compliance with the implied safety in the Rules. By performing a reliability analysis for a specific design case or for a more restrictive scope of scenarios the inherent conservatism may be reduced. For an introduction to reliability based design procedures, see e.g. DNV CN 30.6 and Jiao *et. al.* (1997).

Target Safety Level

Determination of appropriate target safety levels is fundamental to the process of developing new design criteria through the application of reliability methods. A target safety level is defined as the maximum acceptable failure probability level for a particular limit state design to be accepted.

The evaluation of the target safety levels for pipelines should primarily be based on the implied safety in currently accepted design practice, using uncertainty measures representative at the time when the code was made. Further, the nature of failure and the actual consequence potential in terms of hazard to human health and safety, damage to the environment, economic losses, and the amount of expense and effort required to reduce such hazard potential should be taken into account.

The consequence of failure will normally depend on;

- type of fluid
- location of pipeline
- limit state category
- phase

The categorisation of fluids is in accordance with the ISO regulative (ISO 1996) and is defined in Table 1. Further, location class definitions is given in Table 2.

In the Rules, a novel safety class concept is introduced. Based on the fluid category, location class and phase, the pipeline is classified into a Safety Class. See Table 3. However, items altering this normal classification may be consequence of pollution (e.g. in arctic areas) or extraordinary economical failure consequences.

Table 1 Categorisation of Fluids

Category	Description
A.	Typical non-flammable water-based fluids.
B.	Flammable and/or toxic substances which are liquids at ambient temperature and atmospheric pressure conditions. Typical examples would be oil, petroleum products, toxic liquids and other liquids which could have an adverse effect on the environment if released.
C.	Non-flammable substances which are gases at ambient temperature and atmospheric pressure conditions. Typical examples would be nitrogen, carbon dioxide, argon and air.
D.	Non-toxic, single-phase gas which is mainly methane.
E.	Flammable and toxic substances which are gases at ambient temperature and atmospheric pressure conditions and which are conveyed as gases or liquids. Typical examples would be hydrogen, methane (not otherwise covered under category D), ethane, ethylene, propane, butane, liquefied petroleum gas, natural gas liquids, ammonia, and chlorine.

Table 2 Definitions of Location Classes

Location Class	Description
1	The zone where no frequent human activity is anticipated along the Pipeline route
2	The part of the Pipeline/Riser in the near platform (manned) zone or in areas with frequent human activity. The extent of zone 2 should be based on appropriate risk analyses. If no such analyses are performed a minimum distance of 500 m could be adopted.

Table 3 Normal Classification of Safety Classes

Phase	Fluid Category A and C		Fluid Category B, D and E	
	Location Class		Location Class	
	1	2	1	2
Temporary	Low	Low	Low	Low
Operational	Low	Low	Normal	High

With no implicit safety level available, the Rules provides recommendation on target failure probabilities versus safety class and limit state category as given in Table 4.

The basis for the values in Table 4 rely on a conservative assessment of implied safety in current accepted design practice guided by accident statistics and engineering judgement. For details, see Sotberg *et. al.* (1997) and Mørk *et. al.* (1997).

Table 4 Recommended Target Safety Levels

Limit State Category	Probability Bases	Safety Classes		
		Low	Normal	High
Serviceability	Annual per Pipeline[1)]	10^{-2}	10^{-3}	10^{-3}
Ultimate	Annual per Pipeline[1)]	10^{-3}	10^{-4}	10^{-5}
Fatigue	Lifetime probability per Pipeline[2)]	10^{-3}	10^{-4}	10^{-5}
Accidental	Annual per km[3)]	10^{-4}	10^{-5}	10^{-6}

1 Or the length of the period in the temporary phase
2 No inspection and repair is assumed, temporary and in-service conditions considered together.
3 Refers to the overall allowable probability of severe consequences.

LIMIT STATE DESIGN

Introduction

Limit state design implies that the pipeline capacity is characterised by the actual capacity for each individual failure mode (limit state). The adopted limit state format applies explicit design checks for physical limit states rather than implicit design check presumably covering a number of failure modes. The first step is therefore to identify all relevant limit states and secondly to group these limit states into limit state categories due to consequence of a failure. As an example, a brittle fracture will usually be more serious than local yielding. This is reflected in the target safety level and subsequently in the set of safety factors for the different safety classes. Typical limit states and corresponding limit state categories for a pipeline may be:

Serviceability Limit State (SLS) Category

- Ovality / Ratcheting Limit State
- Accumulated plastic strain Limit State
- Damage due to or loss of weight coating
- Yielding

Ultimate Limit State (ULS) Category

- Bursting Limit State
- Local Buckling Limit State (pipe wall buckling Limit State)
- Global Buckling Limit State (normally for load-controlled condition)
- Unstable Fracture and Plastic Collapse Limit State

Fatigue Limit State (FLS) Category

- Fatigue due to cyclic loading

Accidental Limit State (ALS) Category

- Dropped objects
- Trawl gear hooking
- Earthquake

Advantages with limit state methodology

The most important advantage with the limit state methodology in the new Rules, is that it gives a large flexibility with respect to different design scenarios and allows calibration of the partial safety factors to give a consistent level of safety for a broad scope of design .

An example of the opposite may be the hoop stress formulation in the old codes. In order to increase the pipeline safety close to the platform a reduced usage factor has been prescribed. The hoop stress usage factor will, however, results in higher safety for internal pressure and only implicitly accommodate for third party activities and dropped objects etc.

This example illustrates clearly, that a limit state based code, will give the following advantages:

- larger flexibility with consistent safety level
- not restraint to existing design concepts
- ease in extension of codes to allow for new design scenarios

LRFD format

The objective of limit state design in a LRFD format is to verify that factored design loads do not exceed factored designresistance for any of the considered failure modes. The level of safety is considered to be satisfactory if the design load effect (S_d) does not exceed the design resistance (R_d)

$$S_d(\gamma_F \cdot S_{F,c}, \gamma_E \cdot S_{E,c}, \gamma_c) \leq R_d(R_c / \gamma_R) \qquad (1)$$

where:

γ_F = Basic Load Factor for functional loads
$S_{F,c}$ = Characteristic load effect from functional loads
γ_E = Basic Load Factor for environmental loads
$S_{E,c}$ = Characteristic load effect from environmental loads
γ_c = Condition specific Load Factor
R_c = Characteristic resistance
γ_R = Resistance Factor (=$1/\eta_R$ in SUPERB)

The different γ-factors are referred to as partial safety factors. In the LRFD-format, it is the combination of the different partial safety factors and characteristic values, that gives the safety level. In the calibration, it was decided to keep the load factors independent on safety class whereas the resistance factors may vary.

Characteristic values for load effects and resistance are usually given as percentile values of the respective probability distributions. E.g. the 100 year wave may be selected as a characteristic load, having a small probability to be exceeded while the SMYS may be selected as a characteristic resistance value with high exceedence probability. The characteristic values are to be based on reliable data using recognised statistical techniques.

The "conservative" characteristic values are then further modified by means of partial safety factors, usually larger than unity.

Note, in LRFD-format an *increase* in; safety factor, partial safety factors, load factors and resistance factors gives *increased* safety level, whereas in the allowable stress format, *decrease* of usage factor gives *increased* safety level. In order to avoid confusion, the term *usage factor* should be used in connection with allowable stress format only.

The same limit states, but with different (resistance) factors apply to all phases. Hence, there are no different limit states for installations etc. (installation is normally classified as safety class low, allowing higher utilisation than during operation.)

LOAD EFFECTS

General

The design load effect, left hand side of Eq. 1, is based on (or is a function of) a characteristic load multiplied by load factors, ref. Table 5. The load factors are independent of safety class but are to be adjusted by the condition specific load factor (γ_c), where appropriate, ref. Table 6. The condition factor adjust for difference in load uncertainty for different specific conditions. The factored loads are combined according to the Limit State function for the particular failure mode.

An important issue of the load combinations is the implicit handling of system effects. A pipeline is, from a probabilistic point of view, a series system, i.e. the pipeline fails if the weakest link fail. Where some parts, or complete sections of a pipeline are exposed to identical loads, e.g. as for an empty pipeline resting on a flat sea bottom, a system effect exists (weakest link). Note that environmental induced load effect normally imply that a system check can be excluded, e.g. for a pipeline during installation where it is less likely that the weakest part of the pipeline is located at the most exposed position when the largest wave hit the vessel.

The system effect has been included in the local buckling check as a separate load combination, load combinations a) ref. Table 5, where environmental loads are irrelevant and a system factor of 1.1, (Mørk*et. al.* 1997) has been applied on the functional load, giving a load factor of ≈1.2.

Table 5 Load Combinations Specified by the Rules

Limit State Category	Load Comb. id	Functional load γ_F	Environmental load γ_E
SLS & ULS	a	1.2	-
	b	1.1	1.3

Table 6 Load condition factors

Condition	γ_c
Pipeline resting on un-even seabed	1.07
Continuously stiff supported pipe	0.82
Pressure Test	0.91
Otherwise	1.00

Pressure

The design pressure for a pipeline section is related to a reference height. Local design pressure is equal to the design pressure plus the static head of the transported medium due to the difference between the reference height and the height of the part of the pipeline considered. The wall thickness design shall be based on the local design pressure, i.e. the steady flow pressure profile is not used for the design of the pipeline.

For long export lines, this may be unrealistically conservative and the Rules allows for "sectioning", i.e. division of the pipeline into different parts with different design pressures. The design pressure for the different sections shall be determined in such a way that the local design pressure is not exceeded during normal operation. This implies specific requirements to the pressure control system.

During incidental operations, i.e. operations which not is a part of the ordinary operation, the Rules allows a incidental pressure of 10% above the design pressure. Deviations from this may be accounted for by applying an auxiliary LRFD pressure containment criteria.

The hydrostatic test is considered as a strength test verifying the lowest yield strength of the linepipe. The hydrostatic pressure is determined by giving a hoop stress of 0.96 of SMYS. This test is compulsory unless equal check can be documented by other methods.

The pressure test is primary a leak test and shall normally be performed. Since this test not is a strength test, the test pressure is related to the local incidental pressure, hence, not utilising the corrosion allowance. Pressure test shall normally be performed. The pressure test shall always be performed when installation operations may have jeopardised the pipe integrity, i.e. burial operations etc.

DESIGN CRITERIA

Introduction

The design resistance, the right hand side of Eq. 1, is based on a characteristic Resistance divided by a Resistance Factor. The resistance factors are dependent of safety class.

The given limit states in the Rules have no limitations to water depth. The Rules state that "Applied in deep water where experience is limited special considerations are to be given to:

- new failure mechanisms
- validity of parameter range
- new characteristic loads and load combinations
- dynamic effects"

Further experience, e.g. from the ongoing JIP Deepipe, may be incorporated by adjusting the partial safety factors, typically the condition factor.

The set of partial safety factors are explicit in the different limit states. All limit states are given on LRFD format except for the pressure containment formula which is, as a base case, given as an allowable stress design format (with load factor equal to unity). As an alternative, an ASD format is also given for the buckling criteria where there is internal over pressure.

Simplicity in design formats have been aimed for and implemented where possible.

Pressure Containment

The hoop stress formula or wall thickness design formula is referred to as pressure containment formula. The hoop stress is compared to both the yield strength (serviceability limit state category) as well as the tensile strength (ultimate limit state category) with corresponding usage factors.

The hoop stress formula is based upon the mean diameter and the minimum wall thickness and is outlined in Jiao *et. al.* (1996). For operational condition, the corroded cross section shall be considered.

Both the usage factors and the hoop stress formula are in agreement with the usage factors proposed for the ISO-standard. For material fulfilling additional requirements, a separate set of usage factors are given allowing higher utilisation. The two criteria reads:

$$(p_{ld} - p_e) \cdot \frac{D-t}{2 \cdot t} \leq \eta_S \cdot SMYS \tag{2}$$

$$(p_{ld} - p_e) \cdot \frac{D-t}{2 \cdot t} \leq \eta_U \cdot SMTS \tag{3}$$

or alternatively on the LRFD format

$$(p_{li} - p_e) \cdot \frac{D-t}{2 \cdot t} \leq \eta_S \cdot 1.1 \cdot SMYS \qquad (4)$$

$$(p_{li} - p_e) \cdot \frac{D-t}{2 \cdot t} \leq \eta_U \cdot 1.1 \cdot SMTS \qquad (5)$$

where:

p_{ld} = Local design pressure
p_{li} = Local incidental pressure
p_e = External pressure
D = Nominal outside diameter
t = pipe wall thickness
t_{nom}-t_{fab} Hydrostatic and pressure test
t_{nom}-t_{fab}-t_{corr} Operation
t_{nom} = Nominal wall thickness, i.e. the minimum wall thickness plus the negative fabrication tolerance
t_{fab} = Fabrication negative tolerance
t_{corr} = Corrosion allowance
η_S, η_U Usage factors given by Table 7

Table 7 Pressure containment usage factors ***(material fulfilling additional requirements)***

	Safety Class			Pressure
	Low	Normal	High	Test
η_s	0.85	0.80	0.80	0.96
η_u	0.74	0.70	0.67	0.84

Buckling

Buckling, in the in a broad sense, include the following limit states:

- Local buckling
    - Interaction (moment and external pressure)
    - External pressure system check
    - Propagating buckling
- Global buckling

Local buckling interaction check. Two formulas are given for the bucking capacity: load controlled and displacement controlled. A short introduction will be given here. A more detailed background presentation is given in Mørk *et. al.* (1997).

A "load-displacement" (or moment-curvature) capacity curve for a pipe is generally independent on whether the applied load is load controlled or displacement controlled until the characteristic (i.e. maximum capacity) has been reached, ref. Figure 3.

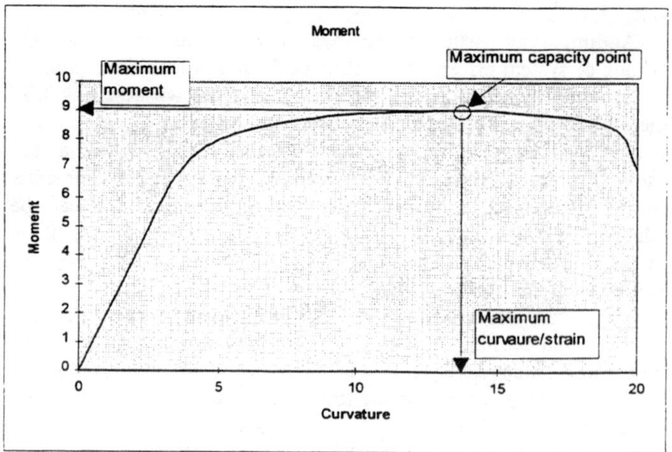

Figure 3 Moment-curvature curve

The essential difference is the consequence of exceeding the characteristic capacity point which is considered critical for a load controlled condition but not (necessarily) for a displacement controlled condition. This must be reflected in the limit state if the safety level shall be consistent.

A classification into load controlled or displacement controlled condition is not obvious. Unless some conservative assumptions can be made, like calculating the resulting curvature when the restraint loads has vanished, the classification may be based on a sensitivity study.

Note also that the maximum moment on a moment-curvature curve usually is well defined, whereas the corresponding strain is not. This may be one argument why the displacement controlled test data points vary significantly more than for load controlled condition.

A common misunderstanding of displacement controlled moment capacity is that a pipe in displacement controlled condition is curvature controlled. This is not the case (for such case the Rules specifies a condition factor of 0.82 to be applied). Bending moment tests are usually carried out as "4-point bend" test, ref. Figure 4. This gives no curvature control of the tested section.

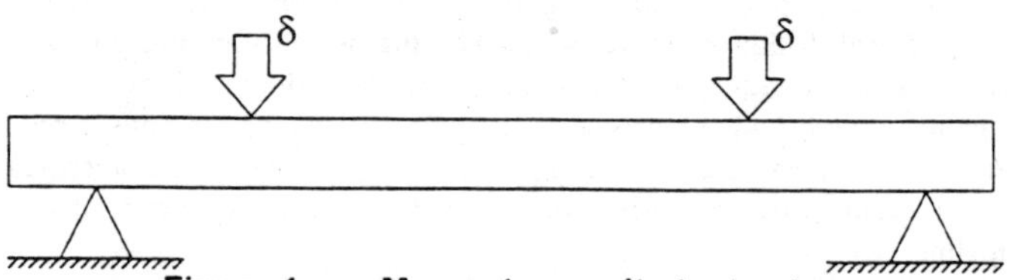

Figure 4 Moment capacity test set-up

The test set-up also illustrates that if the point of maximum capacity is exceeded in displacement controlled condition, a localised buckling failure can appear (since it is not curvature controlled). Hence, the Rules do not allow exceeding the point of maximum capacity.

The interaction load controlled buckling check describes a quadratic relations ship between the moment and external pressure capacity (which obviously always is load controlled):

Load controlled:

$$\left(\frac{\gamma_F \cdot \gamma_c \cdot M_{F,c} + \gamma_E \cdot M_{E,c}}{M_c/\gamma_R}\right)^2 + \left(\frac{p_e}{p_c/\gamma_R}\right)^2 \leq 1 \qquad (6)$$

A similar formulation for displacement controlled buckling is given in the Rules.

The external pressure capacity, p_c, is given by the 3rd order polynomial (identical to BS8010):

$$(p_c - p_{el})(p_c^2 - p_p^2) = p_c \cdot p_{el} \cdot p_p \cdot f_0 \cdot \frac{D}{t} \qquad (8)$$

where

$$p_{el} = \frac{2 \cdot E\left(\frac{t}{D}\right)^3}{1-\nu^2}$$

$$p_p = 2 \cdot SMYS \cdot \left(\frac{t}{D}\right)$$

$$M_c = M_p \cdot f_M$$

$$M_p = SMYS \cdot D^2 \cdot t$$

$$N_p = SMYS \cdot D \cdot t \cdot \pi$$

γ_i = different partial safety factors, defined in, Table 5, Table 6 and Table 8

The last term, f_M, adjust the moment capacity to compensate for internal pressure and axial load. This implies that this equation is valid only in case of moment or pressure dominated scenarios. In case of axial force dominated scenarios an additional design check may be required.

Table 8 Buckling resistance factors
(material fulfilling additional requirements)

	Safety Class		
	Low	Normal	High
γ_R	1.19	1.32	1.52

The buckling and collapse criteria are compared with "current" design practice reflected by DNV'81 and BS8010 inFigure 5 for safety class Low and Normal.

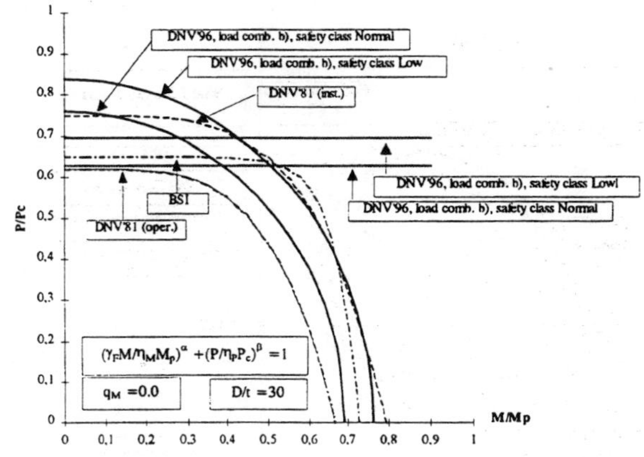

Figure 5 Comparison of buckling interaction in different codes

According to BS8010 buckling may be avoided if the bending moment, axial force and pressure satisfy the allowable combination given in Annex C. Further, it is specified that the initial ovalisation of the pipe cross section is not to be taken less than 2.5%. This introduces an implicit "average" usage factor in the order of 0.6-0.7.

An implicit usage factor in BSI 8010 limiting the bending stress to 0.96 (1.0) of the yield stress is applied. Further, P_{ext}/P_{crit}=0.3 is assumed in DNV'81.

The comparison is for illustration only. However, the following comments apply:

- The overall performance of DNV'96 complies well with DNV'81 (installation) and BS 8010 for safety class LOW.
- The allowable bending moment utilisation complies very well with existing practice
- The collapse system check for safety class Low and Normal corresponds closely to BS 8010 and DNV (operational)

A simplified buckling check for pipelines with internal over pressure, based on the the Von Mises equivalent stress is given as an alternative to the limit state formulation in the Rules.

The Von Mises equivalent stress check versus the original buckling check is illustrated in Figure 6

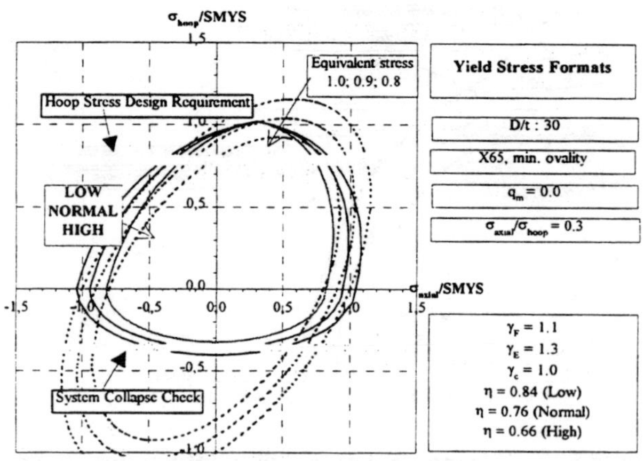

Figure 6 Comparison between simplified buckling criteria (von Mises) and buckling limit states

The following assumptions are used:

- X65 steel, minimum ovality (=0.25%)
- D/t=30 is applied to illustrate an "average" performance.
- $\sigma_{axial}/\sigma_{hoop}$=0.3 is assumed reflecting an upper-bound value of σ_{hoop}/SMYS close to 0.8.
- A typical "cut-off" effect from system collapse check (load combination a)) and hoop stress requirements is included for illustration.

An equivalent stress criterion using η_{eq}=1.0 and 0.9 in compliance with the new draft ISO-code (ISO, 1996) is shown corresponding to safety class Low and Normal, respectively. The case η_{eq}=0.8 corresponding to safety class High is shown for comparison only. The following comments apply:

- An equivalent stress criterion is not appropriate in general apart from the internal over-pressure case. The sensitivity to D/t is rather small in this case.
- An overall fit to the external over-pressure case is not feasible since the buckling format is normally controlled by the collapse capacity rather than the actual compressive hoop stress for D/t larger than ≈15. This is acknowledged in the ISO code by requiring an additional buckling and collapse check.

Ovality

The ovality definition in DNV is given in

$$f_o = 2 \cdot \frac{D_{max} - D_{min}}{D_{max} + D_{min}} = \frac{D_{max} - D_{min}}{D_{nom}} \qquad (9)$$

The ovality shall normally be less than 3%. The ovalisation is included explicitly in the external pressure capacity while not in the moment capacity. Exceeding 3% ovality, the reduction in moment capacity as well as other operating constraints must be evaluated.

Accumulated Plastic Strain and Fracture Mechanics

Accumulated plastic strain is not a failure but a measure degradation for base material properties as well as for defects and is, as such, used both for base material degradation requirements (accumulated plastic strain) and for fracture.

The rules specifies the strain limits or requirements given in Table 9.

Table 9 Testing requirements given for accumulated plastic strain ranges

Strain range	Base Material	Fracture Mechanics
ε_c<0.5%	None	None
0.5%<ε_c<2.0%	None	ECA
2%<ε	Testing	ECA

Note that any strain concentration factor, SNCF, shall be included in the accumulated plastic strain. Engineering criticality assessment (ECA) is, hence, required for all conditions were plastification is experienced.

The above ranges implies that additional testing of fracture mechanic evaluation is not required for load controlled conditions where all strains are expected to be less than 0.5% (except for the pressure containment limit state where the longitudinal weld CTOD requirements shall be checked if the higher utilisation factors are used)

Fatigue

The fatigue limit state has not been changed compared to the previous revision of the Rules. It has been emphasised, however, that use of the Minor sum equal to 0.3, compared to the usual 0.1, shall be accompanied with an inspection scheme capable of detecting fatigue cracks.

The fatigue limit state applies the ASD formulation in the sense that a single usage factor, Minor sum, is used instead of load factors and resistance factors. In the formal guideline on free spanning pipelines (Mørk *et. al.* 1997) ref. Figure 1, a LRFD format will be introduced.

Other criteria

The presented limit states present some of the limit states to be checked during design. Some additional important criteria are also stated:

- Free spanning pipelines
- Stability
- Trawl board interference
- Insulation
- Dropped objects

The planned guideline on free spanning pipelines will, in addition to the fatigue calculations also include load effect calculation, partial safety factors, interpretation of measurements etc.

The pipeline stability may either be calculated by simplified formulas or the DNV RP E 305 (1988).

Trawl board interference criteria is supported by the new DNV Guideline on Interference between trawl gear and pipelines (1997) where calculations methodology is outlined in details, both for a simplified calculation and for a simplified case.

Dropped objects and other accidental scenarios may be identified by a risk analyses, and shall be designed for

LINEPIPE MATERIAL

The material section has been extended to include some new materials. The linepipe requirements are classified into quality level I,

II or III, where level I and II are required for applying displacement controlled criteria. Further, depending on the use of pipeline, supplementary requirements are given for:

- sour service (S)
- fracture arrest properties (F)
- plastic deformation (P)
- dimensions (D)

An introduction to the material requirements and description of modification compared to the DNV-81 is given in Lohne *et. al.* (1997).

INSTALLATION

Installation requirements to structural properties, is covered by the limit state formulation, while requirements to documentation, specifications, testing etc. are given in this section for installation. The following types of installations are covered:

- S-lay
- J-lay
- Towing
- Installation methods introducing plastic deformations

REQUIREMENTS TO RE-QUALIFICATION

The safety of the pipeline will vary during the lifetime of the pipeline. This variation in safety will normally be due to :

- Variations in loads (e.g. decrease of pressure)
- Accumulation of damage (e.g. fatigue)
- Deterioration mechanisms (e.g. corrosion)

The acceptable (target) safety levels specified by the Rules shall be met during the entire life time (normally implying that the safety level decrease during operation down to the minimum at the end of operation lifetime). Re-qualification is required when the design basis for the design has changed, jeopardising the overall safety of the system. Typical examples requiring a re-qualification may be:

- change in environmental loads
- change in functional loads
- deterioration mechanisms exceeding values specified in the design basis
- lifetime extension

The Rules specify that the standard to which the pipeline was built shall be used in the re-qualification unless the lifetime extension exceeds the smaller of 15% of the original design life time or 5 years. A more detailed discussion regarding re-qualification may be found in Collberg *et. al.* (1996.)

A significant part of a typical re-qualification is to determine the residual strength of a corroded pipe. DNV (Bjørnøy *et. al.* 1997) is currently conducting a research projects related to this topic. DNV plan to issue a guideline on residual strength of corroded pipes during 1997, ref. Figure 1.

DESIGN EXAMPLE

Design Data

The example is taken from Appendix A of the Rules. Further details is given in the appendix, including the choice of:

- location class
- fluid category
- safety class

Further, illustration of pressure containment and buckling limit states are given in the appendix.

The example pipeline is an oil flow line from a sub-sea installation at 1100 meter water depth to a fixed structure at 300 m water depth. The total length of the pipeline from the sub-sea structure to the jacket is 20 km routed of areas with uneven seabed leading to pipeline spanning. The basic data are given in Table 10. For this example only the pipeline located close to the platform at 300 meter water depth has been considered.

Table 10 Design basis data for examle

Pipe Data			
Outside Diameter	D	500	mm
Nominal thickness	t_{nom}	20	mm
Fabrication tolerances	t_{fab}	5%	
Corrosion allowance	t_{corr}	2	mm
Ovality (DNV definition)	f_0	0.5%	
Material Data			
Material Grade	X65		
Young's Modulus	E	210.000	MPa
Poisson Ratio	v	0.3	
Specified Min. Yield Stress	SMYS	448	MPa
Spec. Min. Tensile Stress	SMTS	551	MPa
Specified Min. Yield Stress	SMYS(T)	403	MPa at 150°C
Spec. Min. Tensile Stress	SMTS(T)	496	MPa at 150°C
Additional criteria fulfilled			
Process Data - Oil			
Design pressure	p_d	200	barg @ -1100 m
External pressure	p_e	111	barg
Loads			
Installation			
Functional Moment	M_F	1000	kNm
Environmental Moment	M_E	300	kNm
Functional Axial Load	N_F	200	kN
Environmental Axial Load	N_E	20	kN
Operation			
Functional Moment	M_F	500	kNm
Environmental Moment	M_E	1	kNm
Functional Axial Load	N_F	10	kN
Environmental Axial Load	N_E	1	kN

Results

Only two limit states have been checked in the example, pressure conatianment and local buckling. These have been checked for:

- installation
- pressure test
- operation

The results are presented in form of minimum required wall thickness in Figure 7.

The following comments are made to the results:

- For the pressure containment, operation is the governing condition. However, during some circumstances, pressure test may be governing. However, the following conditions will make the pressure test less restricting:
  - Corrosion allowance

- De-rating of yield and tensile strength due to higher temperature during operation
- Buckling during operation tend to be dimesioning (provided that the pipeline is exposed to buckling e.g. due to HP/HT pipelines or freee spanning pipelines) compared to installation when:
  - Corrosion allowance is applied
  - Yield and tensile strength is derated due to higher temperature during operation
  - Water depth is less deep
  - higher safety class than during installation

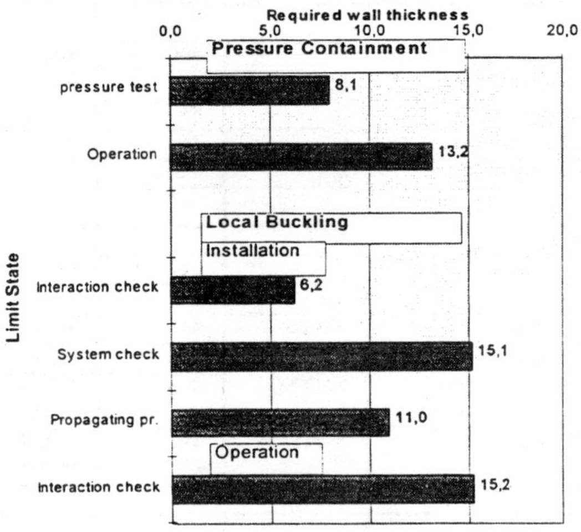

Figure 7 Required wall thickness per limit state

The resulting required wall thickness is 15.2 mm given by buckling during operation.

CONCLUSION

The new 1996 DNV Rules for Submarine Pipeline Systems have been introduced. The Rules are based upon the limit state methodology, using a load and resistance factor design format (LRFD) with calibrated partial safety factors. A modern safety philosophy has been incorporated in the Rules by introduction of the "safety class" concept.

The two major advantages with the new rules and the LRFD format are:

- Flexibility, not restricting the user to traditional design
- Consistent safety level

Further, recent research results and project experience have been included in the Rules, also allowing for:

- New installation methods
- New materials

Through the structure of the Rules; limit states, load factors, condition factors and resistance factors, further improvement and extension gained through new research and experience is made possible.

DNV encourage the industry companies to contact DNV in order to make new knowledge and design experience accessible in future revisions of the DNV Rules. This will enable more companies to compete in providing cost effective solutions to the oil industry.

REFERENCES

Bjørnøy O. H., Cramer E. H., Sigurdsson G., "Probabilistic calibrated design equation for burst strength assessment of corroded pipes," ISOPE'97 Honolulu

BS 8010 1993"Code of practice for Pipelines; Part 3. Pipelines subsea: design, construction and installation", BSi

Collberg L., Cramer E., Bjørnøy O.H., 1996 "Re-qualification of pipeline Systems" ISOPE'96 Los Angeles

DNV 1988, "On-bottom Stability Design of Submarine Pipelines", Det Norske Veritas Recommended Practice E305

DNV 1992, "Structural Reliability Analysis of Marine Structures", Classification Note 30.6, Det Norske Veritas

DNV 1997, "Guideline: Interference between Trawling Gear and Pipelines", Det Norske Veritas

ISO 1996, "Pipeline Transportation Systems for the Petroleum and Natural Gas Industries" Committee Draft 13623, Rev. B4 9/7/96

Jiao G., Sotberg T., Bruschi R., Verley R., Mørk K., 1996, "The SUPERB Project: Wall Thickness Design Guideline for Pressure Containment of Offshore Pipelines", OMAE'96, Los Angeles

Jiao G., Mørk K., Bruschi R., Sotberg T., 1996, "The SUPERB Project: Reliability Based Design Procedures and Limit State Based Design Criteria for Offshore Pipeines", OMAE'97, Yokohama

Lohne P., Sæther T., 1997, "Requirement to Linepipe Material and Manufacture", OMAE'97 Yokohama

Mørk K., Spiten J., Torseletti E., Ness O. B., Verley R., 1997, "The SUPERB Project & DNV'96: Buckling and Collapse Limit State", OMAE'97 Yokohama

Mørk K., Vitali L., Verley R., 1997, "Design Guideline for Free Spanning Pipelines Subjected to Vortex Induced Vibrations", OMAE'97 Yokohama

Sotberg T., Bruschi R., Jiao G., Mørk K., 1997 "The SUPERB Project: Recommended Target Safety Levels for Limit State Based Design of Offshore Pipelines". OMAE'97 Yokohama

Sotberg T., Bruschi R., Mørk K., Jiao G., 1996 "The SUPERB Project: Reliability Based Design Guideline for Submarine Pipelines" OTC 8220, Houston, 1996

Proceedings of the Seventh (1997) International Offshore and Polar Engineering Conference
Honolulu, USA, May 25-30, 1997

ISBN 1-880653-28-1 (Set); ISBN 1-880653-30-3 (Vol. II)

Stability of Pipeline in Curved Route During Offshore Pipeline Installation

H. Shin
University of Ulsan, Ulsan, Korea
J.B. Kim
Hyundai Heavy Industries Co., Ulsan, Korea
Chul H. Jo
Inha University, Incheon, Korea

ABSTRACT

The pipeline stability during laying for a curved route has been investigated with various offshore pipeline projects which were successfully performd recently. During a curved laying operation, it is very difficult to keep pipeline in the designed route if pipeline being in contact with seabed doesn't create enough resistant force against the external force. To maintain the designed configuration in a curved route, a sufficient lateral force created by the pipeline/soil interaction should be provided. The lateral force is a function of submerged pipeweight, contact length, soil property, embedment, size of pipe, etc. For a decided pipe for the project, the laterral resistance required to maintain the route is dependent on the external force, contact length, soil property and embeded depth. This paper introduces how to control the laybarge to keep the curved route and shows how to estimate the allowabel stable radius for variouse pipelines during laying in a curved route.

KEY WORDS

pipelay, curved pipelay, offshore pipeline installation

INTRODUCTION

The main pipe properties (diameter, wall thickness, grade, type, etc.) are decided during a basic design stage considering internal/external pressures, buckle propegation, flow rate, corrosion, pipe staility, hydrotests, etc. in design life. The weight coating is usually applied to make pipes heavier which increases dry and submerged weights. Heavier pipes have more resistance force against external force created by mainly wave and current. The thickness of concrete coating effects greatly on lay conditions. The stress during installation should not be exceeded the allowable stress as defined in the international codes and standards. The weight coating also affects on the allowable pipeline radius for curved route. Even with the bare pipes satisfying the stress limit, the concrete coating to them limits the radius of pipeline during installation.

This paper illustrates the laybarge path to maintain the designed pipeline route in a curved area. With the experience from variouse projects, the restrictions and conditions for this type of operation are explained in a practical point of view. The mimimum allowable bending radius of coated pipeline for installation is also investigated and explained. The consideration factors for pipeline stability are also mentioned. The example of calculation to come up with the minimum radius of curved pipeline which ensures the design route after laying is introduced in the paper. For characterized site conditions and restrications, an additional weight coating to obtain the minimum feasible radius for a curved route is also investigated.

DESIGN PARAMETERS

1. Pipeline and Laybarge

The case pipeline outer diameter was 323.9 mm, wall thickness 9.5 mm and concrete coating thickness 30 mm. The detail design parameters for case pipeline are specificed in Table 1. For the installation simulation, the pipelay barge, HD-2500, and the associated stingers were applied. For the water depth for case study of 12 meter and with the pipe parameters specificed in Table 1, the normal lay analysis was performed. The optimized result indicated the required tension of 12 tons as shown in Table 2. This data was used in the pipelay simulation.

2. Soil Condition

The soil condition is one of the important factors deciding the lateral force of pipeline in contact with seabed since soil type characterizes the shear strength and its properties. Each project site has different soil type and even along the route, the soil properties change in many cases. Before the project execution, it is very important to conduct a site survey to make sure of the exact subsea conditions. From the pre-construction survey, the required data such as water depth, tidal range, soil properties, seabed profile, any obstacles along the reoute, etc. are to be obtained. For this case study, actual site conditions in India were applied. The soil was mostly clay with undrained shear strengh of 5 kPa.

PIPELINE STABILITY STUDY

During the design period, the pipeline is designed to be stable against external forces, mainly current and wave. However, some locaitons require additional protection to anchoring, shipping, etc. The laying of pipeline in trenched area is not considered in the study since the curved pipeline is usually trenched after laying even if required. Only the stability duing installation was applied in the investigation.

1. Stability Check at Design Stage

1) Installation Stability

For the installation analysis, it is generally accepted that 1 year return period environmental ocnditions are applied. However, 4 month return period environmental conditions were utilized for the study since those were demanded conditions for the actual project.

2) Post-Installation Stability

The pipelines shall be designed to be stable in one year return period environmental criteria if they are flooded with sea water. This situation represents the pipeline abandonment.

3) In-Place Stability

The stability of pipelines filled with production fluid shall be checked for critical environmental conditions during whole design life, i.e. which was twenty years for this study. But the in-place stability calculations for the condition where self burial has yet to occur is based on five year environmental criteria considering a self burial after a certain period. The stability criteria used is defined in DnV RP E305. Procedures shall be developed for stabilization or trenching of the pipelines if they have failed on self buried after five years. Table 3 shows wave and current data for 4 month, 1 year and 5 year return period. From the stability analysis for the above three cases, the concrete weight coating thickness was decided as 30 mm with density of 2800 kg/m^3.

2. Stability during Laying in Curved Route

1) Laybarge path to keep the curved pipeline route

For a curved route, the path of laybarge does not follow the pipeline route due to the configuration of pipeline sagbend as shown in Figure 1. Therefore the actual laybarge paths and headings can be calculated by Equations 1 to 7.

$$a = \frac{i}{n}c \tag{1}$$

$$x_i = R \sin a \tag{2}$$

$$y_i = R\,(1-\cos a) \tag{3}$$

$$x_p = x_i + DT \cos a \tag{4}$$

$$y_p = y_i + DT \sin a \tag{5}$$

$$X_p = X_1 + x_p \sin b + y_p \cos b \tag{6}$$

$$Y_p = Y_1 - x_p \cos b + y_p \sin b \tag{7}$$

DT in the above Equations 4 and 5 is the horizontal touchdown distance of pipeline from the barge stern. DT varies with respect to actual laying tension. Various DTs can be calculated from the tension deadband analysis as shown in Figure 2. The DT value for the maximum tension during pipeline installation in curved route will be applied in actual laying.

2) Pipeline Stability

The curved pipeline routes are specified as allowable constant radius between both tangent points (TP1 and TP2 as shown in Figures 1 and 3) from one straight route to another. Pure deformation can be applied providing that the as-laid pipeline follows the design route on the flat seabed and the residual moment can be calculated for pure bending using Equation 8.

$$M = \frac{EI}{R} \tag{8}$$

where, EI is flexural rigidity of pipe and R radius of the curved pipeline route which is larger than the allowable radius from Equation 9.

$$R_{all} = \frac{ED}{2\sigma_{all}} \tag{9}$$

where σ_{all} is allowable stress of pipe material which can be assumed as 85 % of SMYS (Specified Minimum Yield Strength) neglecting a small quantity of residual tension.

However, the actual curved pipeline route can be established with a stepwise turning by keeping the laybarge heading with one or two joints advance as per the radius as shown in Figure 3. The incremental change of heading can be also utilized instead of the defined radius.

$$\beta = \frac{l}{R} * \frac{180}{\pi} \tag{10}$$

where l is the distance which the laybarge keeps the same heading (normally one pipe joint, 12.2 meter). Therefore the laybarge heading can be increased by β degree for every l lenght to keep the radius R.

Similarly the bending moment can be defined as the following Equation for small l.

$$M = EI\frac{d\theta}{ds} = EI\frac{\beta}{l} \tag{11}$$

As lateral force between pipe and soil is larger than the restoring force caused from bending, the curved pipeline will be stable (refer to Figure 3). Lateral force from pipe-soil interaction is composed of friction, cohesion and soil shear force but the soil shear force will not be considered since the pipeline sinkage during pipelay is negligible.

For unit length of the pipe at seabed, the friction force, f and the cohesion C_u are given by the follwoing Equations 12 and 13:

$$f = \mu w \tag{12}$$

where μ is lateral friction conefficient and w is unit submerged weight

$$Cu = \Delta A\ \alpha\ S_u \tag{13}$$

where, α is adhesive coefficient (0.5-0.7), S_u is undrained shear strenght and ΔA is effective unit area, i.e. arc length as written in the following Equation;

$$\Delta A = \pi D\ \gamma / 360 \tag{14}$$

where D is the outer diamter of pipe including the concrete coating and γ is cross angle for the effective zone in degree.

The moment due to lateral force from pipe-soil interaction is givein in Equation 15 and the specified curved pipeline shape can be kept under the condition in Equation 16.

$$M_s = \frac{1}{2}(f + C_u)l^2 \tag{15}$$

$$M_s \geq M \tag{16}$$

In addition to the moment from the friction force and cohesion in Equation 15, another moment from the axial tension can be also considered;

$$M_t = T\ Sin\ \beta\ l \tag{17}$$

where T is th axial tension at the end of pipe section lenght l.

The axial tension will decrease as laying farther and it will become finally the residual tension. Therefore, the moment from the axial tension can be neglected for conservative approach.

Even the excessive curvature, the heading change can be applied during pipelaying in the curved route. The pipeline curve can be established based on the moment as shown in Equation 15. The final radius can be estimated by the following relationship;

$$R_s = \frac{EI}{M_s} \tag{18}$$

3) Application to Practical Laying

Based on the pipeline characteristics listed in Table 1, the allowable bending radius and the allowable stable radius can be calculated.

- Allowable Bending Radius

Applying the allowable stress of 50 % of SMYS (52 ksi) to Equation 9 for a conservative result, the allowable bending radius was estimated as 374 m.

- Allowable Stable Radius

With the soil conditions and pipeline configuration applied in the case study as defined in (19), the allowable stable radius was calculated as 644 meter.

$$\mu = 0.5,\ \alpha = 0.5,\ \gamma = 30° \tag{19}$$

The stable radius can be changed according to parameters shown in (19). Therefore, this parameters shall be selected properly for acurate estimation of the stable radius for the case.

The above result shows that the feasible curved radius of the case pipeline is governed by the allowable stable radius instead of the allowable bending radius. During actual laying of the case pipeline, 450 meter radius was applied but during survey it was found that that radisu was quite large which almost ran cross over the adjacent 6-inch gas line.

The case pipeline was designed based on the minimum submerged weight suitable for dynamic stability but this weight was not enough to achieve 450 meter radius. The required unit submerged weight for the 450 meter radius can be calculated as 95 kg/m instead of 50.88 kg/m in consideration of no further sinkage of pipeline. Therefore, the concrete coating thickness can be increased accordingly.

CONCLUSION

From the application and comparison of both stability cases, the conclusion reaches as follows:

- The stability of the pipeline for the curved route shall be checked additionally in basic design stage so that the weight coating thickness shall be decided properly.

- During a practical laying, the minimum stable radius can be calculated based on this study and can be used further as a guideline for the pipeline installation in a curved route.

- This study can be used as a simple calculation tool at site for the estimation of the conservative stable radius. A further study will be necessary to incorporate all relating constraints and factors.

REFERENCES

- JP Kenny Design Report 031041-01-P-3-08, Stability Analysis and Weight Coat Design, Kuala Lumpur, Malaysia, 1995

- DnV RP E305, On bottom Stability Design of Submarine Pipelines, Oct. 1988

- J. P. Rammant and E. Backx, Offshore Pipeline Installation Senstivity Analysis for a Conventional Lay-barge, Appl. Ocean Res. 1980, Vol. 2, No.1, pp 13-21.

Table 1. Ppeline Specification

DESCRIPTION	VALUE
Outside Diameter (mm)	323.9
Wall Thickness (mm)	9.5
External Corrosion Coating (mm)	4 (FEB coating)
Concrete Coating Thickness (mm)	30
Line Type	Water Injection
Elastic Modulus (Mpa)	2.07 x 105
Moment of Inertia (m4)	1.16 x 10-4
Steel Grade(MPa)	358
Dry Unit Weight (kg/m)	174.52
Empty Submerged Weight (kg/m)	50.88
Flooded Submerged Weight (kg/m)	125.75
Contents Flooded Submerged Weight (kg/m)	125.35

Table 2. Design Laying Parameters

DESCRIPTION	VALUE
Lay Tension (ton)	12
Water Depth (m)	12
Touchdown X-Distance (m)	89.3
Minimum Design Radius (m)	500

Table 3. Wave and Current Data

	RETURN PERIOD		
DATA	**4 MONTH**	**1 YEAR**	**5 YEAR**
H_s	3.0 m	4.9 m	6.9 m
T_{Hs}	n/a	11.9 sec	11.9 sec
H_{max}	n/z	9.9 m	10.2 m
Current	0.6 m/sec	0.86 m/sec	0.86 m/sec

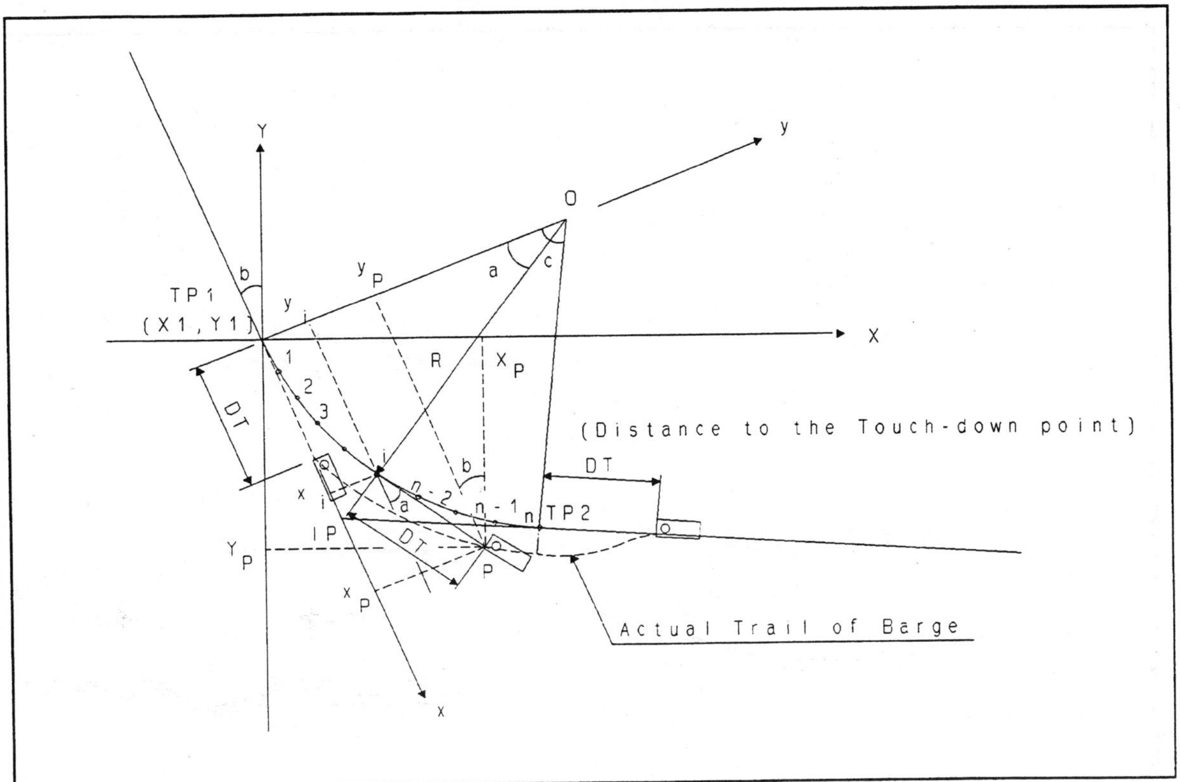

Figure 1, Laybarge Path for the Curved Pipeline Route

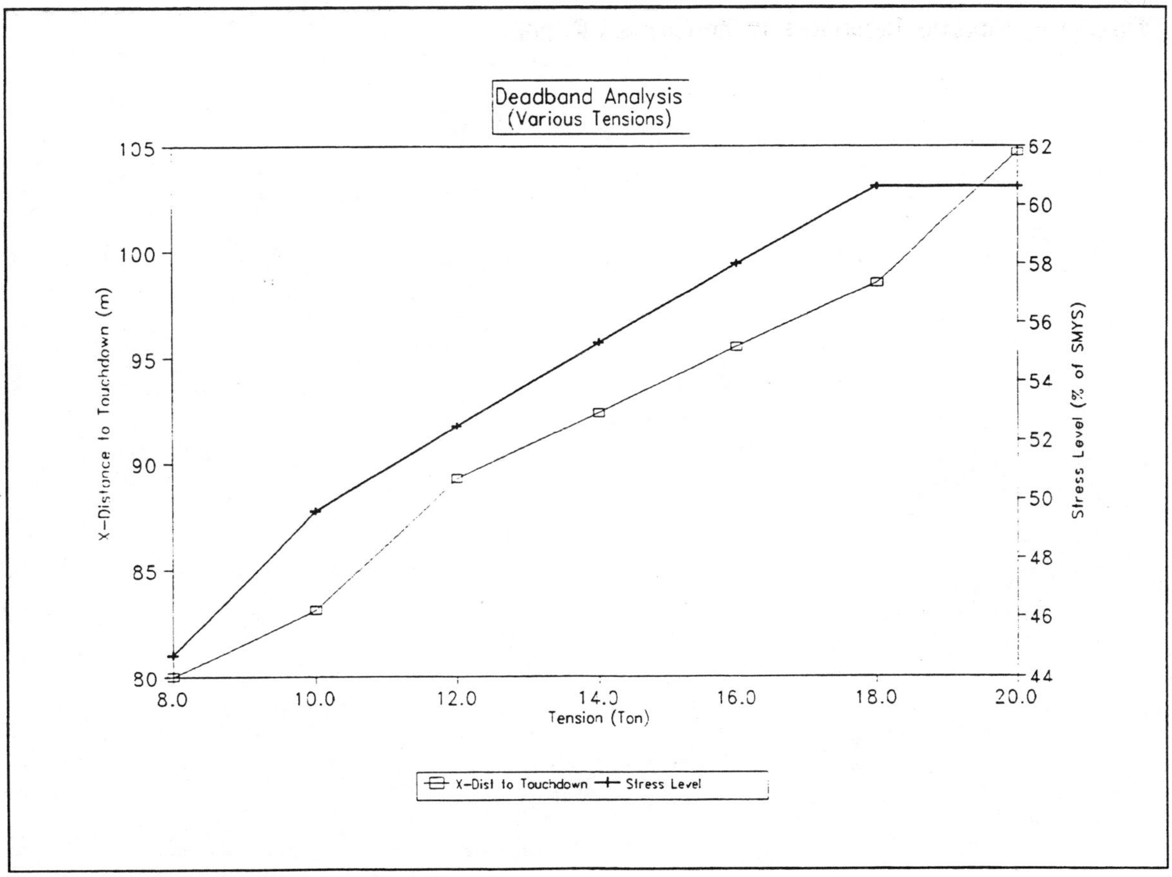

Figure 2, DTs for Tension Deadband

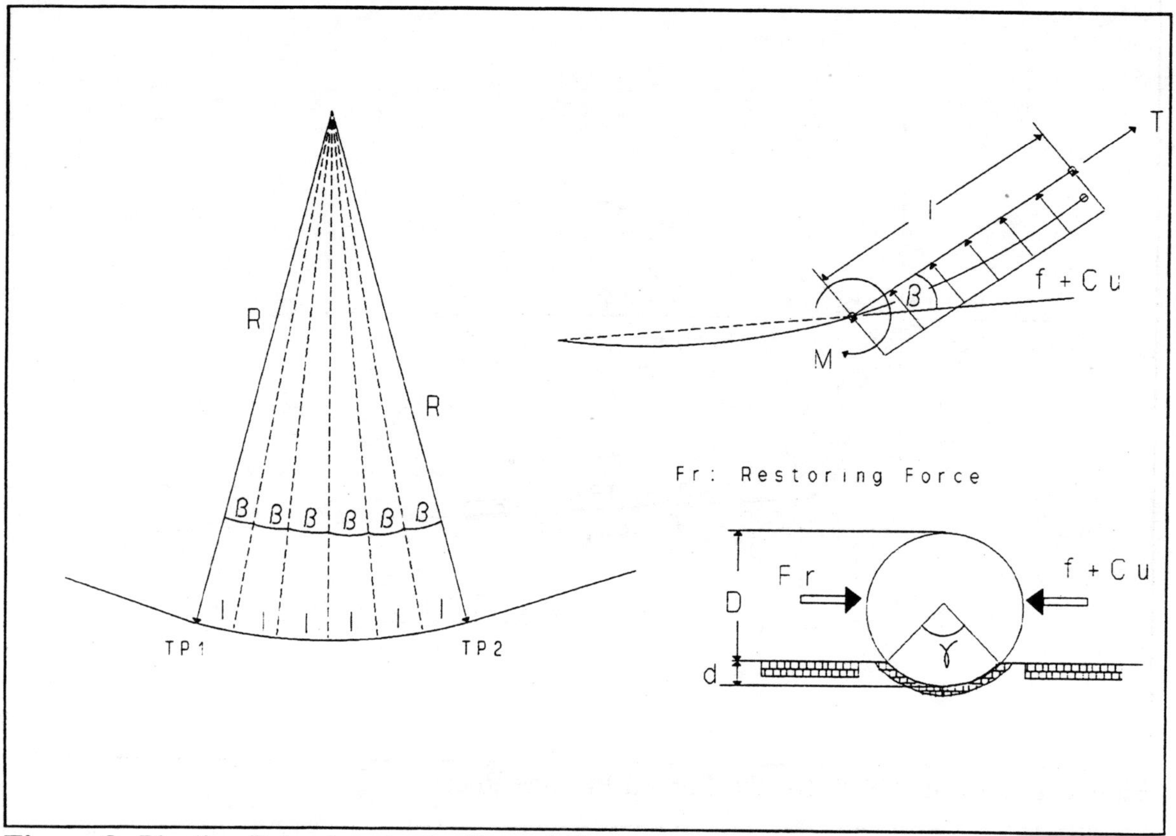

Figure 3, Pipeline Behaviors in the Curved Route

Proceedings of the Seventh (1997) International Offshore and Polar Engineering Conference
Honolulu, USA, May 25-30, 1997

ISBN 1-880653-28-1 (Set); ISBN 1-880653-30-3 (Vol. II)

Prevention of Hydrate Formation in Pipelines by Electrical Methods

Jens Kristian Lervik and Harald Kulbotten
Norwegian Electric Power Research Institute (EFI)
Trondheim, Norway

Gunnar Klevjer
Norwegian University of Science and Technology
Trondheim, Norway

ABSTRACT

Norwegian oil companies, cable manufacturers and pipeline installation companies are presently conducting studies on electrical heating of multiphase subsea pipelines to prevent hydrate formation (plugs). The evaluation of technical feasibility and cost estimates have been completed both for a 50 Hz direct resistive heating system and for a system based on electromagnetic induction. The electrical rating of the systems depends on the heat requirement, pipe material and the pipe length. The feasibility of the concepts has been verified through full scale subsea tests. Results from the measurements are used to determine the characteristic parameters of the two systems on fields in the North Sea. The study includes both carbon steel, duplex steel and martensitic steel (steel of 13% Cr content) pipes, with diameters from 8" to 20". The heat requirement mainly depends on the thermal conductivity of the pipe insulation and seabed (gravel), and in the case of melting plugs, the heat capacity of the pipe, thermal insulation and the hydrate must also be taken into consideration.

A temperature rise of approx. 20 deg. C above sea temperature is necessary to prevent plugs in pipelines during planned shut-downs or reduced flow rate. Calculations show that acceptable temperature requirements are achieved for thermally insulated pipes, both on the seabed and for buried pipes. With buried pipes, acceptable temperature conditions may be obtained without thermal insulation. A typical value of the corresponding heat requirement in the pipe is approximately 100 W/m related to a heat transmission coefficient of 5W/m^2 °C for a 10" pipe.

Due to the electrical heating, possible corrosion problems caused by alternating currents (AC) are investigated. The coherence between corrosion and current level has been evaluated. The tests indicate that the induction heating system implies no modifications of the conventional anode system. However, the direct heating concept requires additional anodes at the ends of the heated section of the pipeline. The number of anodes is correlated to the operational time of the electrical heating system. If this time is estimated to 2 years during the lifetime of the pipeline, approximately 20 sacrificial anodes at each pipe end of the heated section is sufficient as corrosion protection.

KEY WORDS

Electric heating, pipeline, hydrate prevention, corrosion protection, thermal rating, subsea.

1. INTRODUCTION

Electrical heating is attractive to prevent plugs in pipelines during planned shut-downs or reduced flow rate. The heating system must give a temperature rise of approx. 20 deg. C above sea temperature. Feasibility studies and cost estimates have been carried out both for a 50 Hz direct resistive heating system and for a system based on electromagnetic induction.

1.1 General Requirements for Electric Heating Systems

The basic design parameters for heating are as follows:

- Seawater temperature.
- Required steady state pipe temperature.
- Operation time during the lifetime of the pipeline to determine the size of the sacrificial anodes at both ends of the pipeline .
- Temperature profile of the pipe at normal production rates for the rating of the supply cable.
- Electric and magnetic characteristics of the pipe.
- Power frequency.
- Supply cable current.
- Geometry, including pipe and cable dimension and location of feeder cables.
- Thermal properties of the pipe insulation and surroundings.
- Resistivity of seawater, seabed and water depth.

1.2. Specific Requirements and Advantages of the Different Heating Concepts

1.2.1 Direct Heating System

In the direct heating system the power cables are connected to each pipe end, and the feeder cable to the well stream end of the pipe is located close to and parallel to the pipe, see **Figure 1**. The supply current will be divided between seawater and pipe, and in the middle part of the pipeline this current distribution is constant. The current will be transferred between seawater and pipe at the cable connections at the ends (current transfer zones) as indicated in **Figure 1**. An important design basis concerning the current transfer zones is the dimension and location of sacrificial anodes.

The system is applicable for conventional pipes of carbon steel, duplex steel and martensitic steel and requires no additional measures on the pipeline except extra anodes at the pipe ends.

The heat development and the efficiency of the system depends on the current in the pipe, the distance between pipe and cable, the impedance and other system parameters.

The concept is based on a single phase 50 Hz system. It is applicable for different power frequencies, but since 50/60 Hz is the relevant power system frequency in offshore installations, tests and calculations are based on 50 Hz. Use of 50 Hz will also have the advantage of commercially available components.

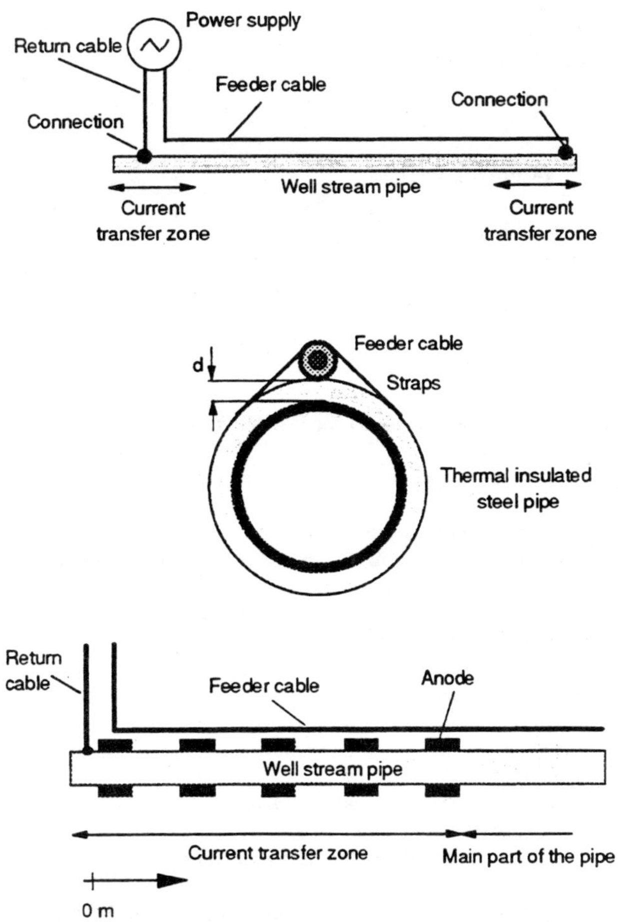

Fig. 1 Illustration of a system for direct heating of pipelines. The thermally insulated pipe is connected to power (feeder) cables at both ends and the return current is divided between pipe and seawater. d-distance between cable and steel pipe surface. The sketch at the bottom indicates location of anodes at the transfer zone.

1.2.2 Induction Heating System

In the induction heating system [3], see Figure 2, the heating of the pipe is based on electromagnetic induction by a single phase or a three phase system for a power frequency in the range from 50 - 400 Hz with unscreened supply cables. The effect of the system is strongly dependent on the distance between the pipe and induction cables. These cables are therefore recommended to be located in grooves cut in the thermal insulation and covered by protective strips. Due to the cut in the insulation this will cause an increase of the thermal conductance compared to an insulation without grooves. The maximum long term operating temperature of the cable is limited by 90°C in the cable insulation. As the cables are located close to the pipe, the well stream temperature should be considered (concerning the thermal rating of the cables).
Due to the wave behaviour, the cable current will vary along the pipeline. This will affect the generated heat in the pipe and the system impedance, dependent on the power frequency, and may introduce restrictions to the maximum length of the heated pipeline.

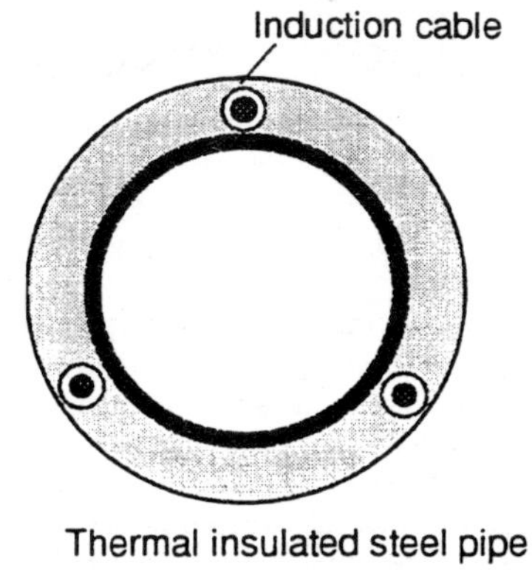

Fig. 2 Sketch of an induction heating system for pipelines. The induction cables are located in grooves in the thermal insulation, and the heating of the pipe depends on material parameters, distance cable-pipe and cable current.

2. QUALIFICATION TESTS

Both heating concepts are tested on scaled models in the laboratory. These tests are, together with calculations, used as a basis for the full scale test programs which are carried out to qualify the methods.

2.1 Direct Heating System

The main parameters in the direct heating system are to be verified through full scale tests. These tests have consequently included the following measurements:

- the length of current transfer zones;
- steady state current distribution between pipe and seawater;
- pipe heat development, as a function of the spacing between pipe and feeder cable and cable current;
- impedance of the direct heating system;
- DC potential difference between seawater and pipe (from a corrosion point of view).

EFI has earlier made a preliminary test on a scaled model of the direct heating system, and this work confirms the feasibility of the concept, [2]. This report also includes calculations of electrical and thermal parameters of a 27 km 10" pipeline, and gives general information about current transfer at the cable connections (see Figure 1) and methods for calculation of generated heat in the pipe.

The full scale tests were carried out on a carbon steel pipe with an O.D. of

219,1 mm and a w.t. = 7 mm. The pipe length was 208 m and was PP(polypropylene)-coated. The water depth at the test site was approx. 20 m. The feeder and supply cables (unscreened XLPE-insulated 1200 mm^2 aluminium conductors) were connected to each end of the pipe by cadweld bolts connected to angle iron pieces which were welded to the pipe. Two locations of the feeder cable are tested: close to the pipe (piggy-back, 50 mm from the pipe surface) and 500 mm from the pipe. It should be noticed that the project work is based on a system for use on thermally insulated pipes.

13 bracelet (aluminium) anodes were connected to the pipe at the ends, equally spaced over a length of approx. 50 m from the cable/pipe connection. The thickness and length of each anode was 50 mm and 1 m respectively.

The current in the pipe was measured by several Rogowski coils, which were located close to each anode to determine the transfer current between seawater and anodes.

At 5 locations the pipe coating was removed at a size of 100 x 100 mm to simulate possible damages of the coating in a real installation. At these "damaged areas" the DC potential referred to a reference zinc (Zn) electrode was measured.

The power supply is connected to the public 400 V network. The main components of the supply system were a circuit breaker and a step transformer (150 kVA), and a capacitor bank (15 mF) for reactive compensation.

2.1.1 Results from Measurements and Calculations

The test period lasted for 10 weeks, and several measurements of the current distribution were conducted during the tests. The duration of the individual tests was from 7 hours to 120 hours within a current range of 760 to 1450 A at 50 Hz.

Computer calculations of system parameters for direct heating, i.e. the steady state current distribution, generated heat and impedances, are performed by the finite element method.

The measured values were used to evaluate the results from the calculations, and are necessary when estimating the accuracy of the calculation models. Prior to and during the test the conductivity of the seawater was measured, and proved to be constant at 3,2 [1/Ωm].
The laboratory tests were mainly to determine the electric and magnetic characteristics of the pipe material, and were used as data inputs in the calculations. Current distribution around the circumference, and generated heat were measured and compared with the calculated values.

The inaccuracy of the measurements of 50 Hz currents is less than 1 % of the measured value, 5 % for the generated heat at 50 Hz, and 0,2 % for the DC potential.

Measurements of the current in the pipe were carried out at for a supply current in the range 760 - 1450 A at 50 Hz. The results are shown in the diagram in Figure 3. The current is given proportionally to the supply current for distances between pipe and cable of 50 mm and 500 mm respectively. There were no significant difference in the relative current distribution between pipe and seawater for the current range 760-1450 A.

At a distance between pipe and cable of 50 mm the stationary current in the pipe was approx. 54 % of the supply current and approx. 43 % at 500 mm distance.

The transfer zone length is defined as the length between the cable/pipe connection to the position where the current in the seawater is 90% of the final (steady-state) value. As seen in Fig. 3 the current transfer zone between pipe and seawater was approx. 40 m.

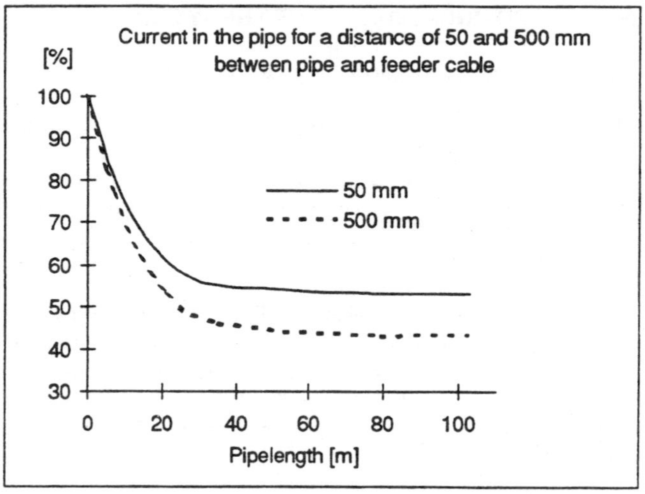

Fig. 3 Current distribution in the pipe at a distance of 50 mm (piggy-back) and 500 mm between pipe and cable. The cable is connected at the pipe end (position 0 m, see Fig. 1). 8" carbon steel pipe.

The maximum anode current occurs at the anode close to the cable/pipe connection. For the anode configuration in the test installation (13 anodes) a design system current of 2 kA gives a maximum anode current of 110 A. This value is reduced if the number of anodes in the current transfer zone is increased. This leads to a maximum current density of approx. 100 A/mm^2 for the anode surfaces. This value was compared with the results from corrosion tests in the laboratory, where weight loss measurements were carried out on a model involving the actual materials (carbon steel and aluminium alloy). These tests indicate that for current (AC) densities lower than 240 A/mm^2 there is no significant influence on the corrosion rate on steel when the steel is coupled to sacrificial anodes of aluminium. Also the influence on the corrosion rate on the sacrificial anode coupled to the steel is moderate for the same current densities.

2.2 Induction Heating System

The qualification work on this system was reported to ISOPE-92, [7], and the main objectives of the full scale tests were:

- heat generation by induction,
- current distribution between steel pipe, sea bed and surrounding water,
- wave phenomena causing changes in the magnitude of the cable current (AC) along the pipeline,
- to study impact of induction heating on corrosion of steel pipe.

The results from the measurements were in good agreement with the theoretical calculations, and proved the feasibility of the system. This confirms that calculation methods have sufficient accuracy for practical ratings of induction heating systems.

Corrosion aspects were studied on a pipe with a conventional number of anodes as for a non-heated pipeline. The results from the study indicated no significant effect of alternating currents on sacrificial aluminium anodes, carbon steel and duplex steel.

3. ELECTRICAL RATING OF THE SYSTEMS

3.1 Electric and Magnetic Properties of Pipes

The electrical parameters for the heating concepts (impedance, voltage drop, current distribution, generated heat) are calculated by a computer program based on the finite element method. It is necessary to determine the following material data of the pipe at the actual operating temperatures to calculate these parameters:

- the electric resistivity,
- the relative magnetic permeability.

These parameters are not available in the literature and have to be measured. Laboratory methods are established to measure the material data. It is important to be aware of variation in these data even in pipes from the same batches. Experiences from laboratory measurements on a steel quality (Duplex) indicate that 10 % variation of the generated heat may be expected.

Table 1 shows measured values for different steel materials. The relative permeability reflects the magnitude of the magnetic flux density in relation to the magnetic field. This characteristic is a non linear correlation which must be considered by use of the finite element method calculations.

Table 1 Input material data for calculations by the finite element method of electrical parameters of the heating systems.

Steel material	Resistivity at 20°C [$\Omega mm^2/m$]	Relative permeability
13% Cr	0,70	50-80
Duplex	0,78	40-60
Carbon	0,21	500-1500

3.2 Basic Design Criteria

Some basic design criteria should be considered when rating the heating systems:

- The maximum operating temperature of the cable is 90 °C.
- The required (minimum) pipe temperature is approx. 20 °C above seawater temperature.

Additional requirements as the heating time from seawater temperature to the required pipe temperature may be specified.

For rating of the concepts the following items must be concidered:

- the required generated heat is calculated when the pipe is laying on the seabed, i.e. surrounded by seawater;
- the cable conductor cross section is determined when the pipe is buried at the maximum trenching depth;

Well stream temperatures above 100 °C may restrict the use of the heating system because of the max. operation temperature of the cable.

3.3 Rating Cases for the Heating Concepts

3.3.1 Design Data

The calculations have been performed on 10" pipelines, 10 and 30 km length, Carbon steel and 13% Chromium steel (13% Cr).
Seawater temperature: 6 °C.
Required stationary steel pipe temperature: 30 °C.
The heat transfer coefficient of the pipe: 5W/m^2 °C.

3.3.2 Rating of Direct Heating

The results are shown in Table 2 and 3. "Transfer current" in the tables means the current transferred between the pipe and seawater and corresponds to the total current in seawater in the main part of the transmission length. A cable copper conductor cross section of 630 mm^2 was used in the calculations. The calculated voltage, "U", is the voltage to ground.

Table 2 Electrical parameters for piggy-back configuration of direct heating at 50 Hz of a 10 km pipeline.
10" Carbon steel and 13% Cr steel pipe. Seawater temperature 6 °C. Required stationary pipe temperature is 30 °C.
Heat transfer coefficient for the pipe: 5W/m^2 °C.

Steel material	Supply current [A]	Transfer current [A]	P [MW]	U [kV]
Carbon	1118	650	1,5	3,4
13% Cr	1282	296	1,5	3,6

Table 3 Electrical parameters for piggy-back configuration of direct heating at 50 Hz of a 30 km pipeline.
10" Carbon steel and 13% Cr steel pipe. Seawater temperature 6 °C. Required stationary pipe temperature is 30 °C.
Heat transfer coefficient for the pipe: 5W/m^2 °C.

Steel material	Supply current [A]	Transfer current [A]	P [MW]	U [kV]
Carbon	1152	670	4,7	10,5
13% Cr	1320	305	4,7	11,1

The current in the Carbon steel pipe is approx. 42% of the supply current, and 77% in the 13% Cr steel pipe respectively.

The calculations (based on the measured material data of the pipe) prove a significant difference in the value of the transfer current for the two steel qualities. For Carbon steel the anodes must be rated for a transfer current of approx. 700 A and for 13 % Cr steel approx. 300 A. A transfer current of 700 A requires 10-15 additional bracelet anodes (1 m length of each anode) at the transfer zones.

The number of anodes are dependent on the operational time of the heating system. If this time is estimated to 2 years during the lifetime of the pipeline, approx. 20 anodes at each pipe end of the heated section is sufficient as corrosion protection. This is related to a carbon steel pipe and a transfer current of 1000 A.

The generated heat in the pipe decreases with increasing distance between cable and pipe. Calculations have shown that for a distance of 0,5 m the generated heat in the pipe is approx. 75 % of the value for the piggy-back configuration.

The generated heat will vary along the pipeline. For a pipe length of 20 km there is no significant variation in the heat generation along the pipeline. At a pipe length of 30 km the variation of the heat development along the pipeline is max. 8 % and should be considered. A heat variation of 8 % implies an increase of approx. 4 % of the current in the feeder cable at the end.

For all rating cases the cable temperature is within the maximum operating temperature even for buried pipes (1,2 m clay coverage).

3.3.3 Rating of Induction Heating

Calculations have been performed on the two cases for a three-phase induction heating system.

The results are shown in Table 4. The calculated voltage, "U", is the voltage to ground. A cable copper conductor cross section of 150 mm^2 was used in the calculations.

Table 4 Electrical parameters for a three-phase induction heating system. Pipe length 10 km at 400 Hz and 30 km at 200 Hz. Carbon steel and 13% Cr steel pipe. Seawater temperature 6 °C. Required stationary pipe temperature is 30 °C. Heat transfer coefficient for the pipe: 5W/m^2 °C.

Pipe length [km]	Steel material	Supply current [A]	Frequency [Hz]	P [MW]	U [kV]
10	Carbon	365	400	1,8	8,7
	13% Cr	320	400	1,6	7,0
30	Carbon	577	200	8,9	24
	13% Cr	470	200	6,9	16

The variation of the heat development along the pipeline for pipe lengths of 10 and 30 km is 13 % and 45 % respectively for Carbon steel pipe. Corresponding values for 13% Cr steel are 10 % and 36 %.

For a pipe length of 30 km the calculated supply current implies that the temperature limit of the cable is exceeded if the pipeline is buried.

4. COSTS OF THE IMPLEMENTATION OF ELECTRICAL HEATING SYSTEMS

4.1 Illustrative Examples

Cost estimates are obtained for the heating concepts on two different examples of pipelines. Both pipelines are thermally insulated and laying on the seabed (not buried). Required temperature rise is 25 °C. The examples are based on the following data:

- 10 km length, 20" pipe of Duplex steel, thermal insulation thickness 30 mm and resistivity of 5 m °C/W.
- 30 km length, 10" pipe of Carbon steel, thermal insulation thickness 60 mm and resistivity of 5 m °C/W.

The direct heating system in this study is based on the piggy-back concept. The induction heating system is based on a three phase power supply. The power frequency is 200 and 400 Hz for the 10 km and 30 km pipeline respectively. The cables are located in grooves cut in the thermal insulation and covered by protective strips.

4.2 Estimated costs

The cost studies include estimates of the different aspects of supplementary costs caused by the implementation of the heating systems. This implies both manufacture, fabrication and installation schedules for the pipeline:

- modification of thermal insulation and weight coating (when induction heating is used),
- additional laying costs,
- cables and accessories for the heating system, including risers, splices and terminations,
- anodes,
- topside electrical equipment and installation,
- engineering.

The results are presented in Table 5.

Table 5 Cost estimates for the investments and installation of the heating concepts.

Case	Total costs [MNOK*]	
	Direct heating	Induction heating
10 km/20" pipe	30	60
30 km/10" pipe	50	110

*1 MNOK ≈ $150.000

There are several aspects of the heating concepts that make the cost estimates uncertain. Details for risers and splicing of cables have not been available at this stage, and an allowance is made for the cost estimates for this study. Nor are details of the repair methods for the heating concepts available at present, which implies that this study has to be continued concerning maintenance costs of the heating systems.

5. CONCLUSION

The technical solutions of the two heating concepts are qualified through full scale subsea tests. The methods are applicable for both carbon steel, duplex steel and martensitic steel (steel of 13% Cr content) pipes, with diameters from 8" to 20".

The heating systems may be rated to keep the temperature of the pipe above the hydrate formation temperature.

The heat development in the pipe is dependent on the dimensions, the electric and magnetic characteristics of the pipe. The resistivity and relative permeability are the input parameters for calculating the electrical parameters by the finite element method. The theoretical calculations proved to be in good agreement with the results from the full scale tests. This confirms that calculation methods have sufficient accuracy for practical ratings of the heating systems.

When rating the heating system, the required heat development in the pipe is determined for the case with the pipe laying on the sea bed. The cable conductor cross section is determined by the thermal conditions for buried pipe. The concepts include non-screened cables, limited to 24 kV at the time being. The maximum operating temperature of the cable is 90 °C.

The direct heating concept implies some modifications of the conventional corrosion protection system. An increase in the number of sacrificial anodes located at the both ends of the heated pipeline is recommended. The additional number of anodes is correlated to the rating current, steel quality, pipe dimensions and operational time of the electrical heating system.

The full scale test has proved that a number of 20 additional anodes is sufficient at the current transfer zones (at the pipe ends) of the direct heating system, if the operational time is estimated to 2 years during the lifetime of the pipeline. The tests have indicated that there is no significant influence on the corrosion rate on steel when the steel is coupled to sacrificial anodes of aluminium.

Tests on the induction heating system indicated no significant effect of alternating currents on sacrificial aluminium anodes, carbon steel and duplex steel. Accordingly this concept requires no additional measures concerning corrosion protection.

The cost study on the different heating concepts has indicated that the direct heating system will have economic advantages compared to the use of induction heating. However, there are several aspects concerning technical details of the heating concepts that make the cost estimates uncertain.

ACKNOWLEDGEMENTS

This paper presents results from joint projects funded by Statoil and Saga Petroleum. Conceptual studies, laboratory and full scale tests are carried out in collaboration with Coflexip Stena Offshore Norway, Alcatel Contracting Norway Aker Engineering and Sintef Materials Technology. The authors wish to acknowledge representatives from these companies for their technical support during the work.

REFERENCES

[1] G. Klevjer, T. Verlo. Electrical heating of subsea pipelines, EFI TR 3761, ISBN 82-594-0210-6, 1990-18-12.

[2] J. K. Lervik, G. Klevjer. Scaled model test of a direct heating system, EFI TR F3946, ISBN 82-594-0410-9, 1992-11-04.

[3] J. K. Lervik, H. Kulbotten, G. Klevjer. Rating of Combipipe for the Sleipner Vest Project, EFI TR F4354, ISBN 82-595-0869-4,1995-10-27.

[4] U. Steinsmo, J. M. Drugli. A study of the influence of alternating currents on the corrosion of pipelines, STF34 F91183, 1991-09-27.

[5] J. M. Drugli, U. Steinsmo. AC Influenced Corrosion, STF24 F97206, 1997-01-09.

[6] J. K. Lervik, H. Kulbotten, G. Klevjer, T. Lauvdal, Direct Electrical Heating of Subsea Pipelines, ISOPE -93.

[7] C. H. Ahlén, G. Klevjer, T. Lauvdal, J. K. Lervik, K. Rønningen. Electrical Induction Heating System of Subsea Pipelines, ISOPE-92.

Proceedings of the Seventh (1997) International Offshore and Polar Engineering Conference
Honolulu, USA, May 25-30, 1997

ISBN 1-880653-28-1 (Set); ISBN 1-880653-30-3 (Vol. II)

Force and Vortex Shedding Characteristics of a Circular Cylinder Near a Plane Boundary

C. Lei, K. Kavanagh and L. Cheng
University of Western Australia
Nedlands, Australia

ABSTRACT

The effect of a plane boundary and the velocity gradient on force and vortex shedding characteristics of a smooth circular cylinder is investigated experimentally at Reynolds numbers ranging from 1.0×10^4 to 2.2×10^4. It is observed that the drag coefficient and the Strouhal number are affected by both the gap ratio and the velocity gradient, while the lift coefficient is dominated by the gap ratio only. A new method for identifying the vortex shedding suppression point is proposed.

KEY WORDS: Vortex shedding, drag, lift, Strouhal number, boundary layer

INTRODUCTION

Flow around a circular cylinder near a plane boundary has been the subject of much research due to its engineering relevance. Although a great deal of work has been published on this subject, there remain some confusing, or even controversial aspects of that research.

When fluid flows around a circular cylinder above a critical Reynolds number, vortex shedding occurs. The vortex shedding changes the local pressure each time a vortex is shed from the cylinder. Therefore, the cylinder is subjected to a fluctuating drag and lift in addition to the mean drag and lift. The drag, lift and vortex shedding frequency are normally expressed in non-dimensional forms as the drag coefficient C_D, lift coefficient C_L, and Strouhal number St.

These three non-dimensional parameters depend on the nature of both the approach flow (such as Reynolds number, flow pattern and freestream turbulence etc.) and the cylinder (such as aspect ratio and surface roughness etc.). When the cylinder is located near a plane boundary, the bed proximity and the velocity profile in the boundary layer also have some influence on these parameters. The effect of a plane boundary on force and vortex shedding characteristics has been investigated by many researchers.

The experiments conducted by Bearman and Zdravkovich (1978) in a wind tunnel revealed that for gap ratios greater than 0.3 the Strouhal number was remarkably constant. However, Angrilli et al.(1982) observed that the vortex shedding frequency increased as the gap ratio decreased. The maximum increase was about 10% at the gap ratio of 0.5. Grass et al.(1984) investigated the influence of the boundary layer velocity gradient and the bed proximity on vortex shedding. The result showed that in both shear and uniform approach flows, with and without velocity gradients respectively, the Strouhal number progressively increased as the gap ratio was reduced below 2. The maximum increase of the Strouhal number, recorded in an approach flow with a large velocity gradient, was about 25 percent. Buresti and Lanciotti (1992) also measured the Strouhal number in different boundary layers and at different gap ratios. They found that in a very thin boundary layer, the Strouhal number increased at small gap ratios, while with relatively thick boundary layers, the Strouhal number decreased at very small gap ratios. The maximum variation of the Strouhal number was only about 5%.

Zdravkovich (1985) examined the drag coefficient with various boundary layers and with a broad range of Reynolds number. The results indicated that the drag coefficient was a constant value when the cylinder was out of the boundary layer. As the cylinder moved into the boundary layer, the drag coefficient decreased with the ratio of gap to thickness of boundary layer(G/δ). This study concluded that C_D was dominated by G/δ. Buresti and Lanciotti (1992) also measured the drag coefficient in different boundary layers, but the results indicated no correlation between C_D and G/δ.

There is good agreement among previous studies that the vortex shedding is suppressed at gap ratios less than 0.3. Some of the studies, such as that of Bearman and Zdravkovich (1978) and Grass et al. (1984), reached this conclusion by analyzing the power spectra of hot-wire signals at discrete locations in the wake of the cylinder. Normally there is a well-defined peak in the spectrum of hot-wire signals if vortex shedding does exist. The frequency corresponding to this peak is referred as the vortex shedding frequency. The gap ratio, at which the well-defined peak disappears, is defined as the vortex shedding suppression gap ratio. With this method, Bearman and Zdravkovich

(1978), and Grass et al. (1984) identified that vortex shedding was suppressed at gap ratios less than about 0.3 from independent studies.

Generally the hot-wire measurements perform well in identifying the suppression of vortex shedding. However, it is not a systematic method in that, first of all, the location of the hot-wire anemometer can be chosen randomly, and secondly, the identification of the disappearance of the peak is not straightforward. Sometimes hot-wire signals at different locations give completely different spectra when the gap ratio is close to the vortex shedding suppression point. For example, the spectral analysis results, given by Bearman and Zdravkovich (1978), showed that at gap ratio of 0.20 the spectrum measured in the upper wake indicated the vortex shedding, while the spectrum measured in the lower wake indicated no regularity. The same phenomenon can be found from the results presented by Grass et al. (1984) where the regularity was found in the lower wake instead at the gap ratio of 0.25. Therefore, there is a need for a more systematic method in identifying the vortex shedding suppression point.

From the above review, it is clear that there still remain some uncertainties regarding the influence of the bed proximity on the vortex shedding frequency (or Strouhal number) and the drag coefficient. Thus, the study of flow around a circular cylinder is far from complete.

The purpose of this research is to investigate the variations of the Strouhal number, the drag coefficient and the lift coefficient in different boundary layers of steady flows. The effects of the gap to diameter ratio (G/D), the velocity gradient (ΔU/U), and the ratio of gap to thickness of boundary layer (G/δ) are investigated. A new method for identifying the vortex shedding suppression point is proposed.

EXPERIMENTAL DETAILS

Experiments were conducted in the boundary layer wind tunnel at the Department of Civil Engineering in The University of Western Australia. The wind tunnel has two test sections, a high speed section and a low speed section. The cylinder was located at the mid-height in the low speed section of the wind tunnel. This section is 7.5 m long, 2.8

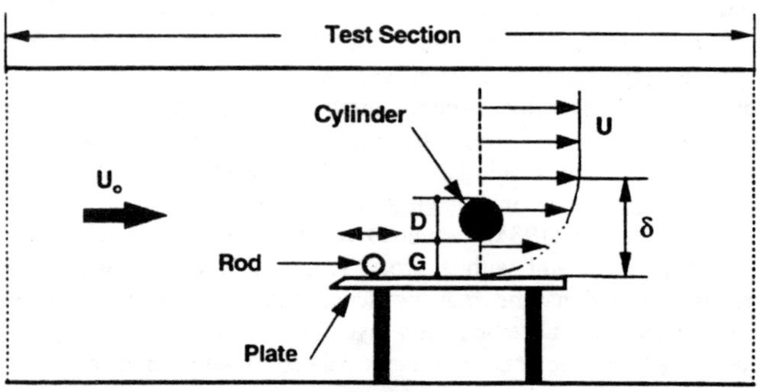

Fig. 1 Layout of the model set-up in the wind tunnel

m wide, and 2.2 m high. The maximum wind speed in this section is about 8 m/s. The turbulence level in the wind tunnel depends on the locations and experimental conditions. The freestream turbulence intensity in our experiments varied from 5.5% to 7.5%. Such a high turbulence intensity was caused by the fluctuation of the driving motor at a very low frequency (≤ 1 Hz). The measured integral length scale of the turbulence was very large (>>10D). This turbulence did not have much effect on our tests within the Reynolds numbers of our experiments. For details, the readers are advised to refer to Surry 1972, Farell 1981, Bearman and Morel 1983, and Basu 1986.

The layout of the model set-up is shown in Fig. 1. The cylinder was an aluminium tube with a diameter of 50 mm and an aspect ratio of 26. The plane boundary was modeled by a smooth wooden plate with a leading edge. The distance from the leading edge to the cylinder center was about 14 diameters. A steel rod of diameter of 20 mm was located between the leading edge and the cylinder. The velocity profile at the cylinder position was adjusted by putting the rod at different distances from the leading edge. The thickness of the boundary layer was determined by measuring the velocity profile of the boundary layer at cylinder location. When measuring the velocity profile, the cylinder was removed from the wind tunnel. The freestream velocity was measured at a position of about 26 diameters upstream of the cylinder.

The cylinder was pressure tapped at the mid-span. There were 20 tap holes on the surface of the cylinder at 18° intervals and arranged in two lines. The tap holes were connected to 20 pressure transducers (SenSym model SX01 with a full range of 0-1 psi) through plastic tubes with a internal diameter of 1.25 mm. Data were sampled and processed by National Instruments LabVIEW system.

The test set-up is shown in Table 1. Three boundary layers were generated. Under each boundary layer condition, the force and vortex shedding characteristics were examined for a wide range of gap ratios,

Table 1 Experimental arrangement

Code	D (mm)	δ/D	G/D	Re×10^{-4}
D050-BL1		0.5	0.0 ~ 3.0	
D050-BL2	50	1.5	0.0 ~ 3.0	1.0 ~ 2.2
D050-BL3	(L/D=26)	4.0	0.0 ~ 5.0	

and at each gap ratio, different Reynolds numbers were tested. The Reynolds number, in terms of the cylinder diameter and the velocity of the approach flow at the cylinder center, ranged from 1.0x10^4 to 2.2x10^4.

In this study, the pressure distributions on the surface of the cylinder were measured at the mid-span, and the drag and lift were calculated by integrating the pressures around the circumference of the cylinder. The vortex shedding frequency was obtained by spectral analysis of the lift.

RESULTS AND DISCUSSIONS

In this paper, the drag and lift are normalized by the freestream velocity U_o as follows:

$$C_D = \frac{F_D}{(1/2\rho U_o^2)\times D\times L}, \quad \text{and}$$

$$C_L = \frac{F_L}{(1/2\rho U_o^2)\times D\times L},$$

where F_D and F_L are the drag and the lift, respectively, ρ is the density of the air, and L is the length of the cylinder.

The effect of the Reynolds number on the mean drag coefficient, the mean lift coefficient, and the Strouhal number is investigated at all gap ratios. The results exhibit that the Reynolds number within our experimental range does not have significant effect on these parameters. Therefore, these parameters are averaged at each gap ratio for the Reynolds numbers in the experimental range.

1. Mean drag coefficient, C_D

The variation of the mean drag coefficient with gap ratio is shown in Fig. 2. It is obvious that the drag coefficient in the thinner boundary

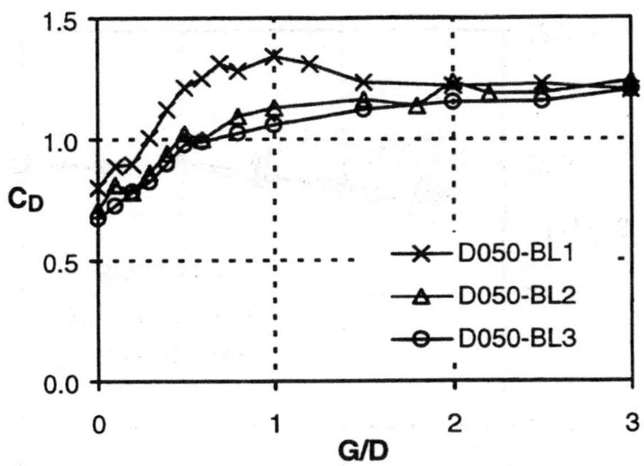

Fig. 2 Mean drag coefficient versus gap ratio

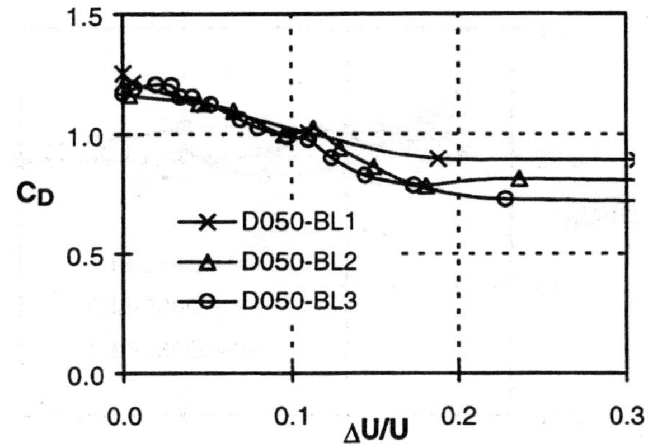

Fig. 3 Mean drag coefficient versus velocity gradient

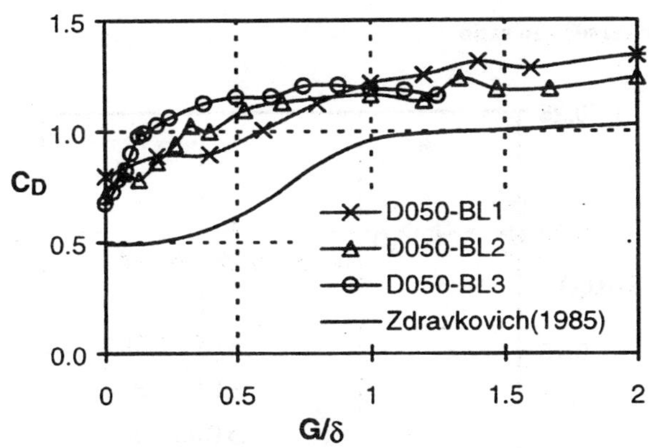

Fig. 4 Mean drag coefficient versus ratio of gap to thickness of boundary layer

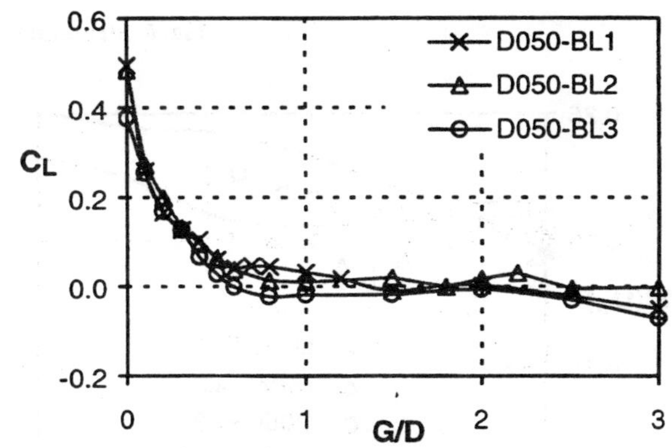

Fig. 5 Mean lift coefficient versus gap ratio

layer is greater than that in the thicker boundary layers. In relatively thick boundary layers, the drag coefficient reduces with the gap ratio, while in the thinner boundary layer, such as test "D050-BL1", the drag coefficient increases at the gap ratio of 1.5, reaches its peak value at G/D=1.0, and decreases thereafter. Generally speaking, the variation of the drag coefficient with the gap ratio is non-monotonic.

Fig. 3 shows the variation of the mean drag coefficient with the velocity gradient. The velocity gradient is calculated by the difference of velocities between the top and the bottom of the cylinder divided by the velocity at the cylinder center. It is seen that the mean drag coefficient is very sensitive to the velocity gradient when the gradient is small. In the range of velocity gradient from 0 to 0.1, the three curves are nearly equal. The drag coefficient decreases rapidly as the velocity gradient increases in this range. However, for velocity gradients greater than 0.2, the variation of the drag coefficient with velocity gradient is very small.

To examine the effect of the thickness of the boundary layer on the drag coefficient, the results are also plotted with G/δ as the horizontal axis and C_D as the vertical axis (Fig. 4). For the purpose of comparison, the experimental result by Zdravkovich (1985) is also plotted in the same figure. The monotonic trend of C_D with G/δ, as suggested by Zdravkovich, was not found in this study.

2. Mean lift coefficient, C_L

The variation of the mean lift coefficient with the gap ratio is presented in Fig. 5. Generally speaking, all three curves are very similar. Therefore, it can be concluded that the lift coefficient is dominated by the gap ratio.

Fig. 5 also indicates that when the gap ratio is greater than 1.0, the mean lift coefficient is close to zero and does not change much with the gap ratio. This reveals that the effect of the solid boundary on the mean lift is very limited for gap ratios greater than 1.0. Usually for gap ratios greater than 2.0, the cylinder can be treated as an isolated cylinder.

3. Strouhal number, St

In the literature, two different definitions of the Strouhal number have been used for the vortex shedding from a circular cylinder near a plane boundary. These are:

$St(U_o) = f\,D/U_o$ and

$St(U_c) = f\,D/U_c$,

where f is the vortex shedding frequency, U_o and U_c are the freestream velocity and the velocity of approach flow at the cylinder center, respectively. To investigate the effect of the definitions, the experimental results are analyzed in both ways in the present study.

The variation of the Strouhal number with the gap ratio is presented in Fig. 6. In Fig. 6a the Strouhal number is calculated in terms of the

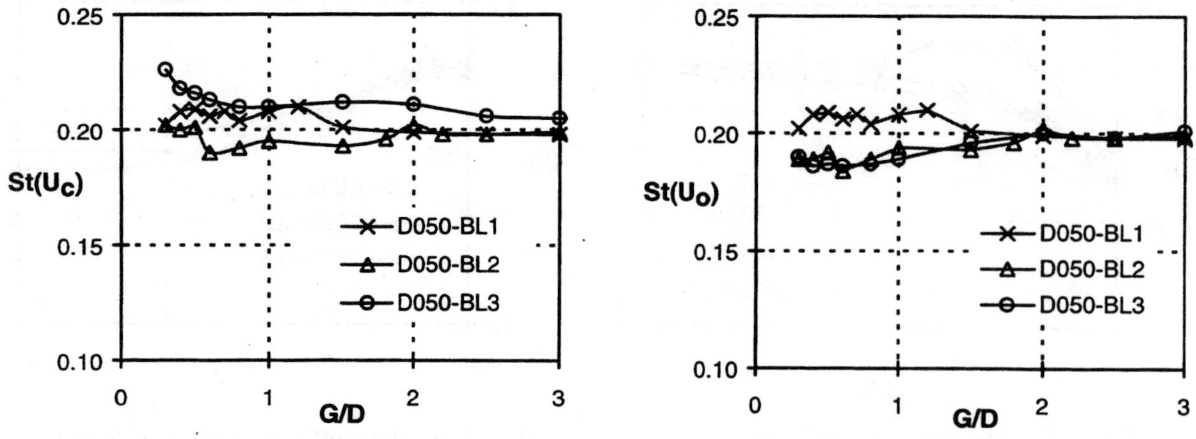

(a) Strouhal number in terms of U_c

(b) Strouhal number in terms of U_o

Fig. 6 Strouhal number versus gap ratio

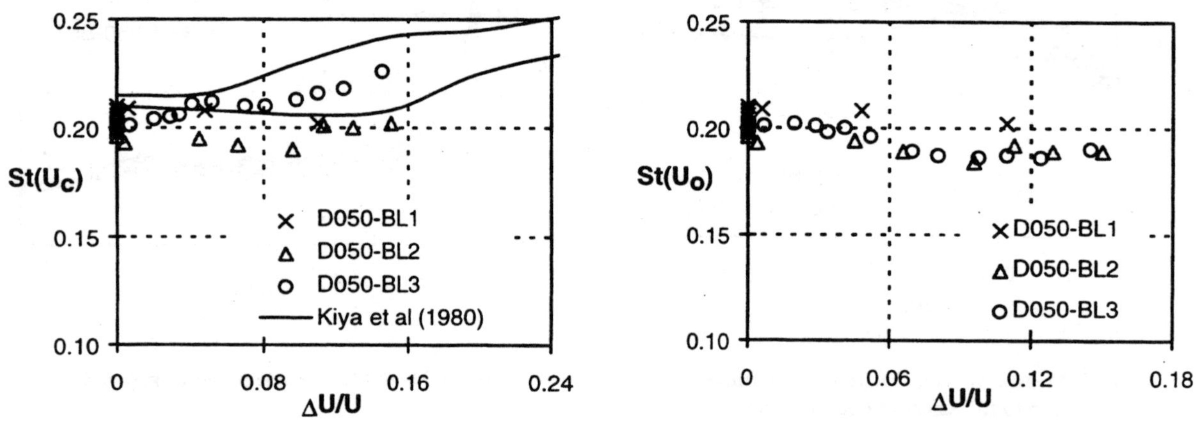

(a) Strouhal number in terms of U_c

(b) Strouhal number in terms of U_o

Fig. 7 Strouhal number versus velocity gradient

velocity of the approach flow at the cylinder center. It is quite apparent that the Strouhal number increases as the gap ratio decreases in test "D050-BL3", which has a very thick boundary layer. There is a peak Strouhal number at the gap ratio of 0.3 with a maximum increase of about 9%. In the other two tests with thinner boundary layers, the variation of the Strouhal number is not significant.

When the Strouhal number is calculated in terms of the freestream velocity, its behavior is different (see Fig. 6b). It is seen in Fig. 6b that for gap ratios greater than 2.0, the Strouhal number is constant for all boundary layer conditions. As the gap ratio is reduced from 2.0 to 0.3, the Strouhal number increases in a very thin boundary layer (see curve "D050-BL1"), and decreases in thick boundary layers (see curves "D050-BL2" and "D050-BL3"). The maximum variation of the Strouhal number in this study is about 7%. This observation agrees with Buresti and Lanciotti (1992) to a certain degree.

Fig. 7 demonstrates the variation of two Strouhal numbers with the velocity gradient. Fig. 7a is the Strouhal number in terms of the velocity at the cylinder center. The result of Kiya et al.(1980) is also plotted in the same figure (see the two solid envelope lines). It can be seen that the Strouhal number increases slightly with the velocity gradient in test "D050-BL3". This increase agrees with the result of Kiya et al. In the other two tests, the Strouhal number almost remains constant. The difference between these two studies is distinct, and is due to different experimental conditions. Kiya et al's results were obtained from experiments with shear flows, while Fig. 7a was obtained from boundary layer flows. With boundary layer flow, the effects of both the gap ratio and the velocity gradient are incorporated in the measurement. In addition, the Reynolds number in our experiments is much higher than the Reynolds number used by Kiya et al..

The variation of the Strouhal number in terms of the freestream velocity with the velocity gradient is plotted in Fig. 7b. It is obvious that the velocity gradient has little influence on the Strouhal number in test "D050-BL1". The Strouhal number in the other two tests slightly decreases with increasing velocity gradient. The maximum decrease of the Strouhal number is about 7%.

4. Vortex shedding suppression gap ratio

The onset and suppression of vortex shedding can be observed with different methods. As mentioned above, some scholars identified the vortex shedding suppression point by spectral analysis of hot-wire signals in the wake of the cylinder. This method is found not to be

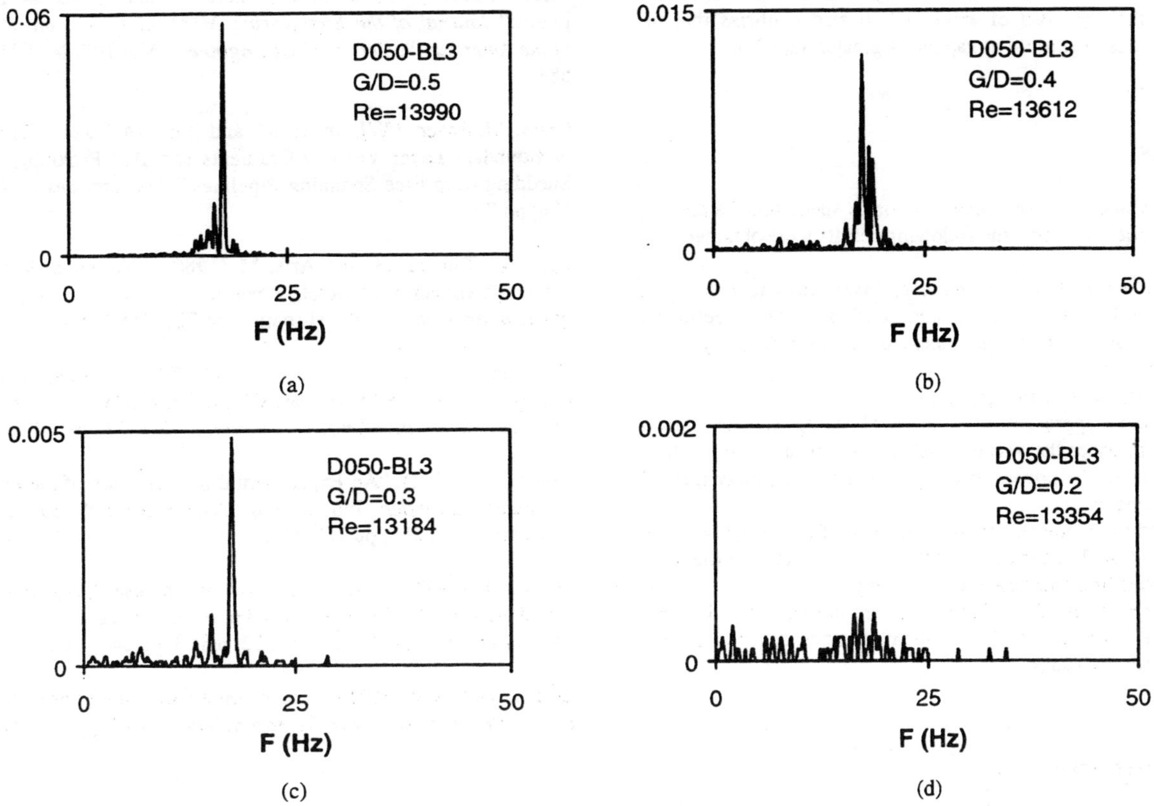

Fig. 8 Power spectra of lift at different gap ratios

systematic. An alternative method, which has been used by some other scholars, is based on the observation of the spectra of the lift.

Fig. 8 gives four typical power spectra of the lift at different gap ratios in test "D050-BL3". The graphs are plotted with different scale so that the peaks can be seen clearly. At G/D=0.5, the peak is sharp and the peak intensity is high (about 0.06). The frequency at this peak point is the same as the vortex shedding frequency. As the gap ratio reduces to 0.4 and 0.3, the spectra develop many small peaks, and the peak intensities decrease to about 0.012 and 0.005. At G/D=0.2, there is still a peak with the frequency of the vortex shedding, but the intensity of this peak is very low, and the peak is not distinguishable from background noises.

Similar characteristics of the spectra of the lift were observed in other two tests with different boundary layers. These observations indicate that the strength of vortex shedding becomes weaker as the gap ratio decreases. At the gap ratio of about 0.2, the vortex shedding frequency becomes indistinguishable. Therefore, it is proved that the vortex shedding is suppressed at a gap ratio of about 0.2.

Since both the regularity and the amplitude of the fluctuating lift are strongly influenced by vortex shedding, a more systematic method, which is based on the observation of the variation of root-mean-square (RMS) lift coefficient, is straightforward. Fig. 9 shows the variation of RMS lift coefficient with gap ratio.

In Fig. 9, it is apparent that the curves can be divided into three intervals based on the gap ratios. The first interval is from the gap ratio of 0 to 0.2, in which the fluctuation of the lift is very small, and the variation of RMS lift coefficient with gap ratio is also small. This indicates that no regular vortices are shed from the cylinder. The second interval is between the gap ratio of 0.2 to about 2.0. In this interval, the vortex shedding occurs, and the amplitude of the fluctuating lift increases with the gap ratio. That is to say, the effect of the boundary on

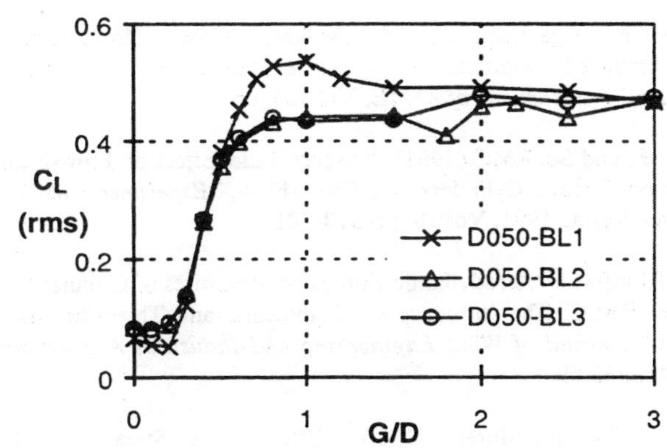

Fig. 9 RMS lift coefficient versus gap ratio

vortex shedding becomes weaker and weaker as the gap ratio increases. For gap ratios between 1.0 to 2.0, the influence of the boundary is limited. The third interval covers the range with gap ratios greater than 2.0. In this interval, regular vortices are shed from the cylinder, and the RMS lift coefficient is almost constant for all boundary layers. This indicates that the boundary has no effect on vortex shedding.

This observation confirms, in a different way, that the vortex shedding is suppressed at the gap ratio of about 0.2. It also confirms that the boundary influence is negligible at gap ratios greater than 2.0.

CONCLUSIONS

In this experimental study on force and vortex shedding of a circular cylinder near a plane boundary, the following results were obtained:

1. The velocity gradient in the boundary layer tends to reduce the mean drag coefficient. No monotonic trend of the drag coefficient with the gap ratio and the thickness of the boundary layer was found in this study.
2. The lift is dominated by the gap ratio.
3. The onset and suppression of vortex shedding can be clearly observed from either the variation of the root-mean-square (RMS) lift coefficient, or the power spectra of the lift. The former method is more systematic.
4. Vortex shedding is suppressed at a gap ratio of about 0.2. For gap ratios greater than 2.0, the effect of the solid boundary on the force and vortex shedding characteristics is negligible.
5. Both the gap ratio and the velocity gradient influence the Strouhal number. The behavior of the Strouhal number depends on the criterion of normalization.

ACKNOWLEDGMENTS

The authors wish to acknowledge the support of the Australian Research Council (Grand No.0415412001) for this project. Thanks are also given to Mr. Garry Smith and Mr. Wladek Bzdyl for their patient help with the experiments, and to Mr. Simon Petersen for suggestions given to the authors.

REFERENCES

Angrilli, F, Bergamaschi, S, and Cossalter, V (1982). "Investigation of Wall Induced Modifications to Vortex Shedding from a Circular Cylinder," *Transactions of ASME*, Vol 104, pp 518-522.

Baban, F, and So, RMC (1991). "Aspect Ratio Effect on Flow-Induced Forces on Circular Cylinders in a Cross-Flow," *Experiments in Fluids*, Springer-Verlag 1991, Vol 10, pp 313-321.

Basu, RI (1986). "Aerodynamic Forces on Structures of Circular Cross-section. Part 2. The Influence of Turbulence and Three-dimensional Effects," *Journal of Wind Engineering and Industrial Aerodynamics*, Vol 24, pp33-59.

Bearman, PW, and Morel T (1983). "Effect of Free Stream Turbulence on the Flow Around Bluff Bodies," *Progress in Aerospace Sciences*, Vol 20, pp 97-123.

Bearman, PW, and Zdravkovich, MM (1978). "Flow around a Circular Cylinder near a Plane Boundary," *Journal of Fluid Mechanics*, Vol 89, part 1, pp 33-47.

Buresti, G, and Lanciotti, A (1992). "Mean and Fluctuating Forces on a Circular Cylinder in Cross-Flow near a Plane Surface," *Journal of Wind Engineering and Industrial Aerodynamics*, Vol 41, No 1-3, pp 639-650.

Farell, C (1981). "Flow Around Fixed Circular Cylinders: Fluctuating Loads," *Journal of the Engineering Mechanics Division, Proceedings of the American Society of Civil Engineers*, Vol 107, No EM3, pp 565-588.

Grass, AJ, Raven, PWJ, Stuart, RJ, and Bray, JA (1984). "The Influence of Boundary Layer Velocity Gradients and Bed Proximity on Vortex Shedding from Free Spanning Pipelines," *Transactions of ASME*, Vol 106, pp 70-78.

Kiya, M, Tamura, H, and Arie, M (1980). "Vortex Shedding from a Circular Cylinder in Moderate-Reynolds-Number Shear Flow," *Journal of Fluid Mechanics*, Vol 141, part 4, pp 721-735.

Niemann, H-J, and Holscher, N (1990). "A Review of Recent Experiments on the Flow past Circular Cylinders," *Journal of Wind Engineering and Industrial Aerodynamics*, Vol 33, pp 197-209.

Norberg, C (1994). "An Experimental Investigation of the Flow around a Circular Cylinder: Influence of Aspect Ratio," *Journal of Fluid Mechanics*, Vol 258, pp 287-316.

Surry, D (1972). "Some Effects of Intense Turbulence on the Aerodynamics of a Circular Cylinder at Subcritical Reynolds number," *Journal of Fluid Mechanics*, Vol 52, part 3, pp 543-563.

Zdravkovich, MM (1985). "Forces on a Circular Cylinder near a Plane Boundary," *Applied Ocean Research*, Vol.7, No 4, pp 197-201.

Proceedings of the Seventh (1997) International Offshore and Polar Engineering Conference
Honolulu, USA, May 25-30, 1997

ISBN 1-880653-28-1 (Set); ISBN 1-880653-30-3 (Vol. II)

Wave Induced Forces on a Submarine Pipeline

F. Raichlen and A. Watanabe
California Institute of Technology
Pasadena, CA, USA

ABSTRACT

The wave-induced forces on an underwater sewage outfall have been investigated in a laboratory model with the pipeline supported off the bottom on a pedestal faced with rock layers. Velocities were measured around the structure and upstream and downstream. From measurements of the horizontal force the virtual mass coefficient was found to be approximately 30% to 40% larger for the cylinder with the support structure than for the cylinder alone mounted the same distance off the bottom. The lift coefficient decreased significantly with increasing clearance between the cylinder and the still water surface. The porosity of the rock support structure does not significantly affect either the virtual mass or the lift.

KEY WORDS

added mass, lift, pipelines

INTRODUCTION

The failure of a large sewage outfall which occurred several years ago when sections of the pipe were displaced raised the question as to whether storm waves could have contributed to the damage. The outfall was a 3.26 m diameter reinforced concrete pipe supported on rock bedding with its centerline 3.13 m above the bottom. Large rock ballast extended from the springline of the pipe to the bottom intended as armor for the bedding. Failure occurred in a depth of about 10.7 m; thus, a clearance of about 6 m existed between the crown of the pipe and the still water level. It was expected that the structure could be exposed to large waves with periods from 8 sec (or shorter) to 22 sec throughout its lifetime. This paper addresses this engineering problem in a somewhat generalized manner, and experimental results of force and velocity measurements are presented for a modeled outfall pipe and its rock ballast protection for a range of depths and incident wave conditions.

The forces on undersea pipelines resting on the bottom and partially buried have been investigated by a large number of researchers; only the results of several of these studies will be mentioned. In one of the earliest studies dealing with this problem, Wilson and Reid (1963) presented virtual mass (inertial) and lift coefficients for a bottom-sited pipe from potential flow considerations. Among others, Sarpkaya and Rajabi (1979) investigated experimentally the forces on bottom-mounted smooth and rough cylinders in a periodic flow and determined the variation of the drag, lift, and inertial coefficients. These parameters were shown by them to vary primarily with the Keulegan-Carpenter number and to some extent with the ratio of the Reynolds number to the Keulegan-Carpenter number. The experiments were conducted in a large U-shaped water tunnel. Such an enclosed facility has the advantage of creating a time-varying velocity which is essentially independent of distance, but by its very nature it precludes the investigation of free surface effects on the imposed forces. Nath et al. (1976) have also investigated the forces acting on a cylinder located near the bottom. In their experiments the cylinder was oscillated back-and-forth in a free surface tank; free surface effects were not investigated.

A study was conducted by Cheong et al. (1989) of the inertial dominated forces on a cylinder located near the bottom in a wave tank so the influence of the proximity to the bottom and to the free surface could be investigated. As found by others, as the cylinder approached the bed large negative lift forces, i.e., forces directed toward the bed, were measured. Any effect of the free surface for the experiments conducted appeared to be within the range of scatter of the experimental data. In a comparison of their data with those of others there appeared to be some agreement with potential flow theory, at least within the limits of the variation in experimental data (as much as 30%).

The subject of pipelines placed on the ocean bottom and sheltered by a trench has been investigated extensively by Jacobsen (1988). The experiments with sheltered pipes were conducted in a tank by oscillating the pipe and its simulated "seabed" with a periodic motion and measuring the imposed forces. The cylinder and bed were arranged at one wall of the tank so that the effects of the free surface were minimized. It was found that by placing a pipeline in a trench the horizontal and vertical forces induced by waves were reduced considerably.

The problem which was investigated experimentally and is reported herein is somewhat different from these. In this model a cylinder is located in a limited water depth supported with its centerline about one diameter above the sea bed and with the support structure armored with rock. The armor rock extends up to the springline of the cylinder. (In the prototype armor rock is used to protect the pipe from excessive forces, to reduce the effects of localized bottom erosion, and to protect the rock bedding material placed between the bottom of the pipe and the sea bed.) Thus, the objective was to determine the influence of a supporting structure on the forces acting on a cylinder and to compare these to the forces acting on the cylinder alone located at the same distance above the bottom. The virtual mass and lift coefficients for the cylinder, with and without rock protection, were determined experimentally for the structure exposed to solitary waves and to periodic waves with $0.05 < h/L < 0.32$, where h = depth and L = wave length, ($0.3 < kh < 2.0$, where $k = 2\pi/L$). Velocity measurements also were made upstream, downstream, around the cylinder, and near the face of the supporting rock structure for various wave conditions to assist in understanding differences in the added hydrodynamic mass observed for the structures.

ANALYTICAL CONSIDERATIONS

The wave-induced forces on a cylinder were proposed first by Morison et al. (1950) to consist of a viscous drag force component and an inertia force component which could be linearly superimposed to give the total force. Using this approach, the horizontal force per unit length was expressed as:

$$dF_H = dF_{D_{Horiz.}} + dF_{I_{Horiz.}} \tag{1}$$

where $dF_{D_{Horiz.}}$ represents the drag force acting on a section of the cylinder of length dl and diameter d. The horizontal inertial force acting on the same section is denoted as $dF_{I_{Horiz.}}$. The individual components are defined as follows:

Drag Force:

$$dF_{D_{Horiz.}} = \frac{\rho}{2} C_D A_p u|u| \tag{2}$$

Inertia Force:

$$dF_{I_{Horiz.}} = \rho C_M \, d(Vol) \frac{du}{dt} \tag{3}$$

where C_D is a drag coefficient, ρ is the fluid density, and u is the water particle velocity. Since the forces were measured only on the cylinder for various support configurations, A_p is the projected area of the cylinder, i.e., A_p is the product of the diameter and the elemental length: $A_p = (d \cdot dl)$. In Eq. 3 the displaced volume of the fluid is denoted as d(Vol). For the same reason it is defined as the product of the cross-section area of the cylinder and the elemental length: $d(Vol) = (\pi d^2/4) \cdot dl$. The virtual mass coefficient associated with this volume is defined as $C_M = 1 + M'/M$, where M' is the added hydrodynamic mass and M is the mass of the fluid displaced by the cylinder. The acceleration of the fluid particles is denoted as du/dt.

For the application of Eqs. 2 and 3 to the present problem the velocities and the accelerations are evaluated at the elevation of the centerline of the cylinder and, as is traditionally done, upstream where the wave is unaffected by the structure, i.e., the incident wave. Although water particle velocities were measured, the velocities and accelerations used in Eqs. 2 and 3 were predicted, where appropriate, from linear or nonlinear theories.

The lift force can be expressed as:

$$dF_L = \frac{\rho}{2} C_L A_p u^2 \tag{4}$$

where the lift coefficient is C_L. Since lift acts upward or downward independent of the direction of the horizontal velocity, the velocity in Eq. 4 is squared rather than expressed as the product of the velocity and the absolute value of the velocity as in Eq. 2. The projected area, A_p, is defined in the same way as in Eq. 2.

Both solitary and periodic waves were used in these experiments, and for the horizontal forces, the importance of the drag force relative to the inertia force was determined experimentally by comparing the time of occurrence of the maximum force to the water-surface time history recorded at the centerline of the cylinder; this will be discussed later. For all wave conditions tested it was found that the maximum horizontal force occurred nearly simultaneously with the time when the maximum water particle acceleration would be expected to occur. This indicated that the drag force, at most, was small compared to the inertia force.

The lift coefficient was evaluated only using solitary waves where the horizontal velocities were unidirectional and relatively constant with depth. As will be shown, generally, the maximum vertical force occurred simultaneously with the arrival of the wave crest at the centerline of the cylinder. Although the effects of vertical accelerations could not be eliminated from the analysis of the vertical force data using this method, it can be shown that the contribution of the vertical inertia forces to the total vertical force acting on the cylinder would be less than about 10% to 15% of that due to the lift forces for the solitary waves which were used.

EXPERIMENTAL EQUIPMENT

A glass-walled wave tank approximately 15.2 m long, 39.4 cm wide, and 61 cm deep with a hydraulically actuated programmable piston-type wave generator was used for the experiments. A rubble covered beach was located at the end of the tank opposite the wave generator. Resistance wave gages measured the water-surface time history; these were mounted to carriages which could be moved on rails which are horizontal to within ± 0.3 mm. Generally the water-surface time histories were obtained at two locations: midway between the structure and the wave generator and at the centerline of the cylinder.

A horizontal cylinder 8.90 cm in diameter and 38.73 cm long composed of a sealed PVC pipe was used to represent the undersea pipeline. (Considering the failed ocean outfall mentioned earlier, this cylinder represented a model with a scale of $L_r = 1/36.6$.) The cylinder was connected to a force cell located above the water-surface which consisted of two portal gages connected so that vertical and horizontal forces could be measured independently. The portal gauges were composed of leaf springs with displacements measured by linear variable differential transformers. A stainless steel rod 1.27 cm in diameter connected the cylinder to the force cell. It was sheltered from the flow by an independently supported hollow sheet metal "airfoil" shape. Each force cell was calibrated statically by applying known weights in a horizontal or vertical direction for the appropriate transducer using a pulley-line system.

The cylinder was mounted to the force cell with its centerline 8.56 cm above the bottom of the wave tank. A section representing the rock bedding and armor rock was placed beneath it, and independent of the cylinder. It was composed of two parts: a wooden section representing the small diameter bedding rock underlying the outfall pipe and about two layers of rock with a nominal size of about 2.5 cm cemented to the wooden section to represent the outer armor rock. The rock pieces were individually placed and cemented at their contact points to adjacent pieces. In this way a porous rock matrix was constructed which was stable against wave attack but allowed flow through the layer similar to that which could occur if loose rock had been used. The rock layers and the wood core were designed so that a gap of approximately 3 mm existed between the cylinder and the underlying material to allow the cylinder-force transducer combination to deflect without the cylinder coming into contact with the rock protection. The arrangement is shown schematically in Fig. 1 along with the nomenclature used. For these experiments the following dimensions were used: d = 8.90 cm, a = 8.56 cm, b = 24.5 cm, and c ≈ 6 cm.

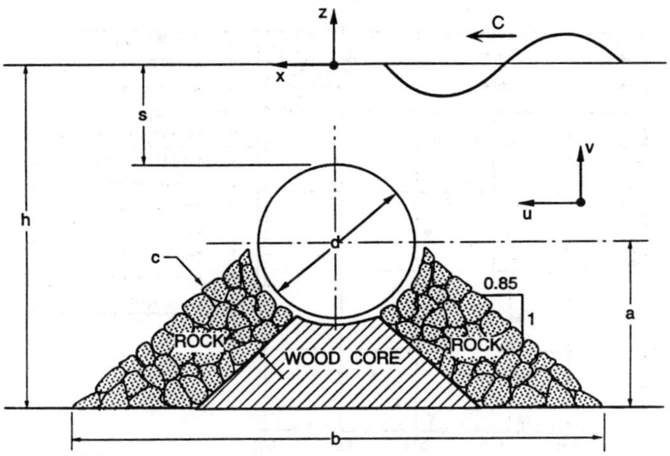

Figure 1 Schematic drawing of the experimental model of the cylinder with its rock protection.

Water particle velocities were measured using a portable four-beam, two-component, fiber optic laser Doppler velocimeter (LDV) manufactured by Thermo Systems, Inc. The instrument was used in the "back scatter" mode thereby providing flexibility in the location of the velocity measurements. The measurements were made on a vertical plane approximately 10 cm in from the glass wall of the wave tank, i.e., at a location about one-quarter of the tank width from the tank wall.

Several water depths were used for the investigation. Since only one cylinder-support geometry was tested, the depths are defined by the ratio of s/h. The clearance to depth ratios used were: s/h = 0.41, 0.57, and 0.71 corresponding to depths of: h = 22.05 cm, 30.26 cm, and 44.86 cm, respectively.

For these experiments the following range of non-dimensional parameters was used: (1) Keulegan-Carpenter parameter ($\mathbf{K} = u_m T/d$): $0.3 < \mathbf{K} < 5.6$, (2) the product of wave number and depth: $0 < kh < 2.0$ ($0 < h/L < 0.32$), and (3) Reynolds number ($\mathbf{R} = u_m d/\upsilon$): $1.3\times10^4 < \mathbf{R} < 1.9\times10^4$.

RESULTS AND DISCUSSION OF RESULTS

As mentioned earlier, two types of waves were used: translatory waves (solitary waves) and oscillatory waves. Solitary waves form a reasonable model of a nearshore breaking wave, and they have the advantage that the horizontal velocity is approximately constant with depth. A typical measured water-surface time history midway between the structure and the wave generator for the solitary wave is presented in Fig. 2 along with the results of a second order wave theory; the agreement between experiment and theory is good. In Fig. 3 the water-surface time history is presented for a typical periodic wave at a similar location. Even though the height-to-depth of this wave is small, i.e. H/h ≈ 0.1, the water-surface time history appears to agree better with the second order theory than it does with small amplitude wave theory.

Velocities were measured at the centerline of the cylinder and at locations upstream and downstream for both solitary and periodic waves. For both wave conditions the water depth was

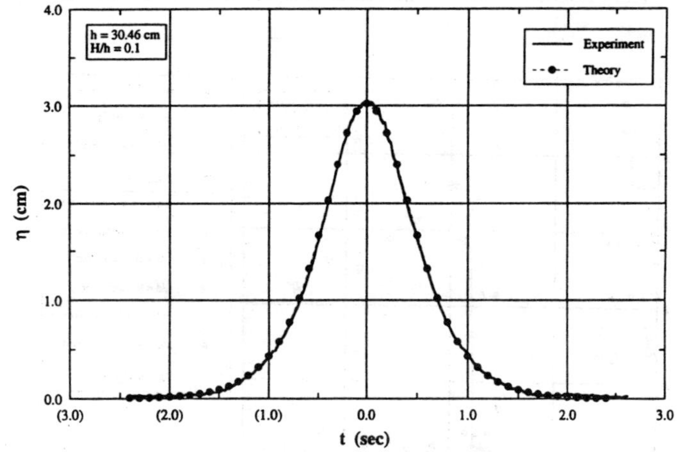

Figure 2 Water surface-time history for solitary wave at x/d = - 38.5: theory compared to experiment.

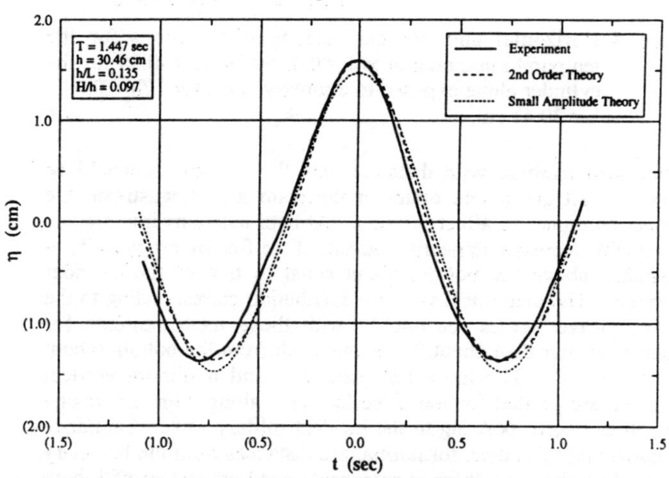

Figure 3 Water surface-time history for periodic wave at x/d = - 38.5: theory compared to experiment.

30.41 cm and the relative wave height was about H/h ≈ 0.1. The purpose of these measurements was to demonstrate the essentially undisturbed upstream and downstream conditions for reasonable distances from the cylinder and the relatively localized effect of the cylinder alone and the cylinder-support structure combination on the velocities.

For a solitary wave, the variation of the temporal maximum horizontal and vertical velocities with distance from the bottom at x/d = 0, +0.94, and –0.94 is presented in Fig. 4 for the cylinder alone along with the theoretical velocity distributions for the incident wave corresponding to second order solitary wave theory. (As indicated in Fig. 1, positive distances are measured downstream from the cylinder centerline and negative distances are upstream.) The abscissa is the ratio of the velocity to the small amplitude shallow water wave celerity, $u/\sqrt{gh}$ for the horizontal component and $v/\sqrt{gh}$ for the vertical component. The ordinate is the non-dimensional distance measured from the bottom. As expected, at x/d = 0 the strong influence of cylinder blocking can be seen. Near the free surface, at the three locations, the horizontal velocities are approximately the same and within 10% of the theoretical value for the incident wave. The velocity increases monotonically as the crown of the cylinder is approached becoming nearly twice that near the free surface. (For comparison, the ratio of the velocity at the crown of a bottom-mounted cylinder to the free stream velocity from steady potential flow is about 2.2.) The velocities at x/d = 0 measured between the cylinder and the

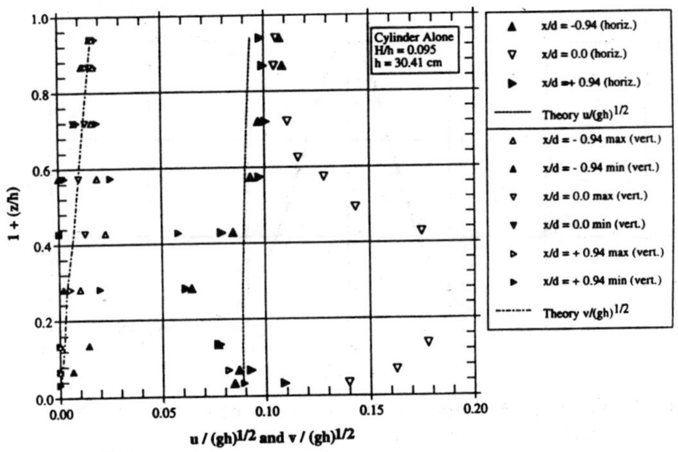

Figure 4 Horizontal and vertical velocity distributions for the temporal maximum at x/d = 0.0, +0.94, and -0.94 for the cylinder alone exposed to a solitary wave for H/h = 0.095, h = 30.41 cm.

bottom also increase with distance from the bottom, as would be expected. At about one diameter upstream and downstream the cylinder continues to affect the flow. At both locations investigated the velocity decreases to nearly one-half of the free stream velocity at a distance above the bottom about equal to that of the cylinder centerline. The theoretical velocity distribution corresponding to the incident wave agrees reasonably with these measurements for distances greater than about 0.6 of the depth from the bottom (about two diameters). The measured maximum and minimum vertical velocities are plotted for the three locations along with the second order theory corresponding to the incident solitary wave. Similar to the horizontal velocities, for normalized distances from the boundary greater than about 0.6 there is reasonably good agreement with both the measured maximum and minimum vertical velocities at all locations with the second order theory, and as the cylinder is approached the measurements diverge significantly from the theory.

The variation of the horizontal and vertical velocities with relative depth are presented in Fig. 5 for the cylinder with the rock protection exposed to a solitary wave. Measurements are presented at about thirty-eight cylinder diameters upstream, about four cylinder diameters downstream, and at the cylinder centerline. Both the horizontal and vertical velocities measured upstream and downstream of the cylinder agree reasonably well with the second order theory. For comparison, the velocities measured at x/d = 0 corresponding to the cylinder alone are presented. These results show that the effect of blockage is greater with the rock protection than with the cylinder alone positioned off the bottom. The variation of the vertical velocity with depth shows a smaller effect of the cylinder and its rock protection on the flow then for the cylinder alone. The second order solitary wave theory corresponding to the flow without the cylinder in place agrees fairly well with the experimental data away from the structure.

The velocities at locations upstream and downstream of the cylinder alone and the cylinder and its rock protection were measured with the structures also exposed to periodic waves. The variation of the temporal maximum normalized horizontal velocities with depth for the two conditions and the same wave are presented in Figs. 6 and 7 for upstream and downstream locations, respectively. For comparison the theoretical temporal maximum velocities for the incident wave are presented. The relative wave height was: H/h = 0.92, and the wave number times depth was: kh = 0.502. The velocities are normalized by the small amplitude shallow water celerity. Giving attention first to the maximum horizontal velocities at the location of the centerline of the cylinder (x/d = 0) it is seen that, similar to the case with solitary waves, the cylinder with the rock protection results in larger velocities

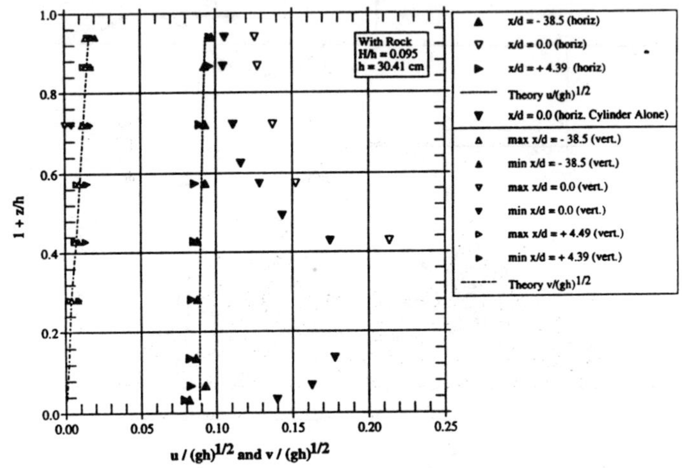

Figure 5 Horizontal and vertical velocity distributions for the temporal maximum at x/d = 0.0, -38.5, and +4.39 for the cylinder with rock protection and the cylinder alone exposed to a solitary wave for H/h = 0.095, h = 30.41 cm.

over the cylinder than for the case with the cylinder alone. At x/d = 0 the effect of the structure on the velocities is about the same for periodic waves as for the solitary wave of about the same relative height. Downstream of the cylinder, Fig. 7, the effect of the structure, with or without rock protection, on the velocity is small below the crown of the cylinder. This is similar to that seen at the upstream locations shown in Fig. 6. In summary, it appears that the effect of the structure on the flow is similar for solitary waves and for the periodic waves. This suggests the effect of the rock protection on the virtual mass would be about the same for solitary waves and for periodic waves; this will be discussed in more detail later.

The effect of the porosity of the rock protection on the flow near the cylinder-rock structure was investigated using the temporal maximum velocities. Since the rock protection was constructed with individual rock pieces glued to each other at contact points, the porosity of the overlayer of armor rock which would be realized in nature was reasonably modeled in the laboratory. Velocity measurements made near the rock-face are presented in Figs. 8 and Fig. 9 for solitary and periodic waves, respectively. Shown as solid lines and dashed are the profiles of the rock-face at locations where measurements were made. The arrows indicate the direction and the magnitude of the temporal maximum velocities. (Scales are shown both for distance and velocity.) The existence of temporal maximum velocities in two directions behind the structure for solitary waves is probably evidence of circulation in the wake of the structure. For solitary and periodic waves at nearly all of the measurement locations the vectors corresponding to the temporal maximum velocities are nearly parallel to the rock face indicating, at least for the temporal maximum, little flow into the structure. These results show that the armor rock layer apparently behaves more as an impermeable matrix than a permeable one at these locations. However, this conclusion

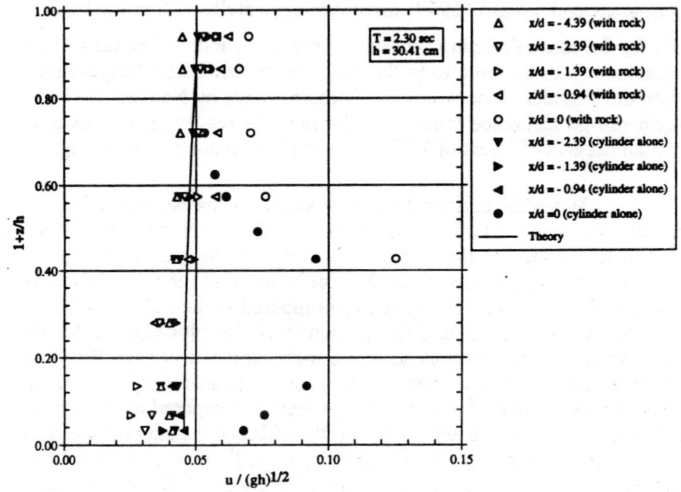

Figure 6 Horizontal velocity distributions for the temporal maximum at x/d = -4.39, -2.39, -1.39, -0.94, and 0 for the cylinder with rock protection and the cylinder alone for a periodic wave: H/h = 0.092, h = 30.41 cm, T = 2.3 sec, kh = 0.505.

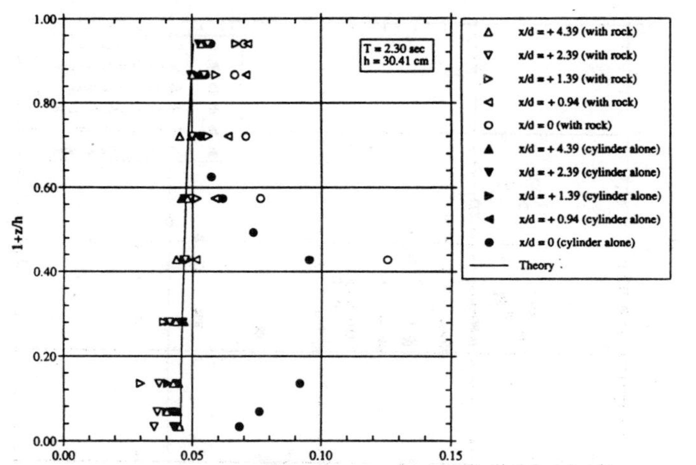

Figure 7 Horizontal velocity distributions for the temporal maximum at x/d = +4.39, +2.39, +1.39, +0.94 and 0 for the cylinder with rock protection and the cylinder alone for a periodic wave: H/h = 0.092, h = 30.41 cm, T = 2.3 sec, kh = 0.502.

must be applied with caution, since the velocities at only seven or eight locations were measured compared to approximately 200 to 300 pieces of rock on the two faces.

The virtual mass coefficient, C_M, was evaluated for three configurations: the cylinder alone, the cross-section described by Fig. 1 consisting of an armor rock outer layer with a wooden core, and a cross-section similar to that shown in Fig. 1 but with the supporting structure constructed completely of wood. (The reason for the latter condition was to explore the influence of the porosity of the outer rock layer on the hydrodynamic added mass and the lift coefficients.) The displaced volume per unit length for the cylinder was 62.22 cm^2. (The ratio of the cross-section area of the cylinder to the cylinder plus the support structure was 0.353.)

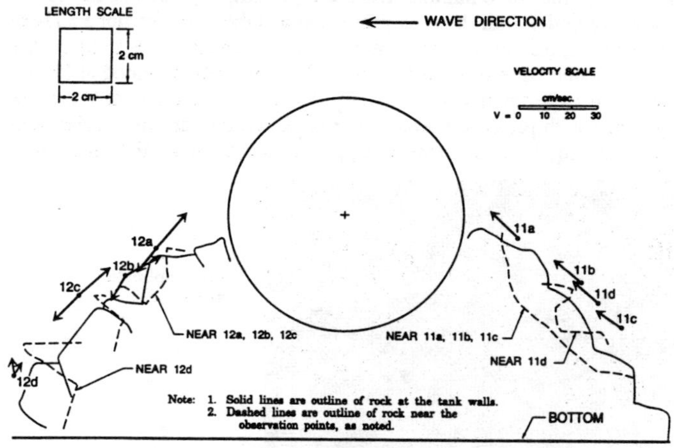

Figure 8 Temporal maximum velocities near the surface of the rock face for a solitary wave: H/h = 0.095, h = 30.41 cm.

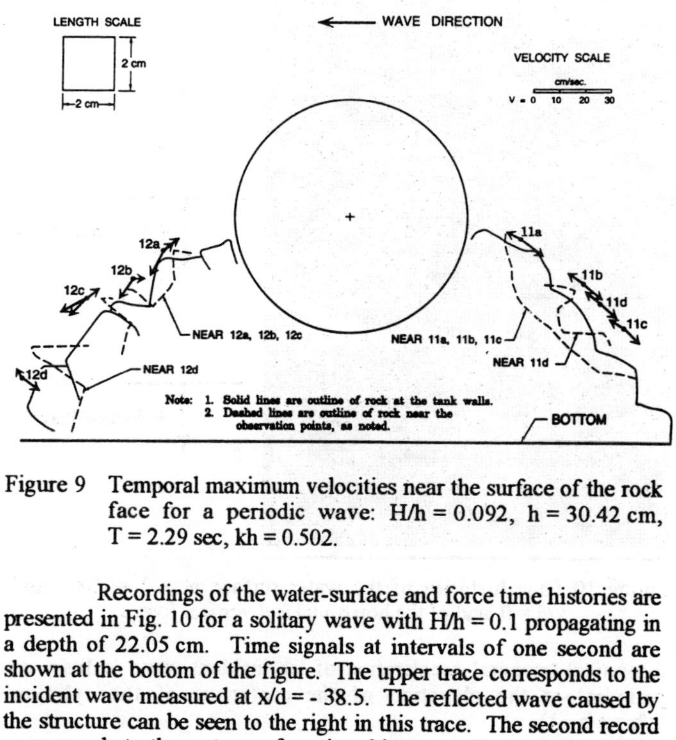

Figure 9 Temporal maximum velocities near the surface of the rock face for a periodic wave: H/h = 0.092, h = 30.42 cm, T = 2.29 sec, kh = 0.502.

Recordings of the water-surface and force time histories are presented in Fig. 10 for a solitary wave with H/h = 0.1 propagating in a depth of 22.05 cm. Time signals at intervals of one second are shown at the bottom of the figure. The upper trace corresponds to the incident wave measured at x/d = - 38.5. The reflected wave caused by the structure can be seen to the right in this trace. The second record corresponds to the water-surface time history at the centerline of the cylinder (x/h = 0). The variation of the horizontal force with time is presented in the third part of Fig. 10, and the variation of the vertical force with time is shown in the fourth part.

The times of the maximum and the minimum horizontal force shown in Fig. 10 are reasonably close to the occurrence of the maximum slopes of the wave measured at the centerline of the cylinder suggesting that the inertia force controls the horizontal force. The general shape of the horizontal force-time history which is similar to the acceleration-time history for a solitary wave supports this assumption. It is seen that the maximum vertical force occurs approximately at the time of the maximum wave amplitude at x/d = 0. Records for other cases were similar.

The variation of the virtual mass coefficient, C_M, with the product of the wave number and the depth, kh, is presented in Fig. 11 for the cylinder with the rock protection and the cylinder alone. Three relative depths above the crown of the cylinder were used in the experiments: s/h = 0.41, 0.57, and 0.71. The data obtained at kh = 0 are from experiments with solitary waves; the data for kh > 0 were obtained with periodic waves. As mentioned earlier, the acceleration used in Eq. 3 to determine C_M corresponded to the maximum

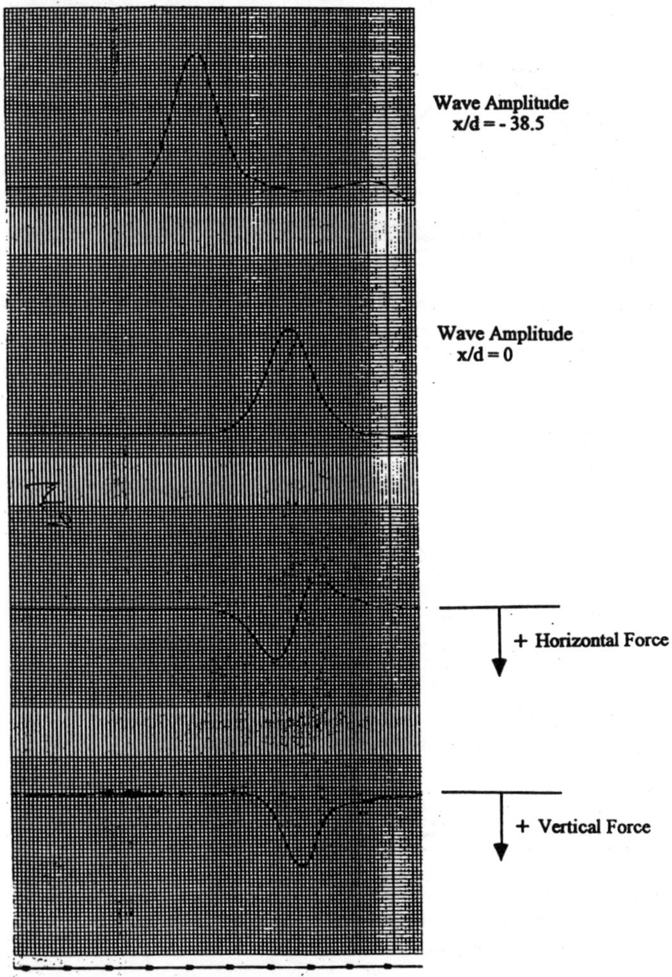

Figure 10 Time histories of the water surface at x/d = -38.5 and x/d = 0 and of the horizontal and vertical forces.

theoretical temporal acceleration of the incident wave (solitary or periodic) at the elevation of the centerline of the cylinder (z/h = a/h - 1).

Referring first in Fig. 11 to the data corresponding to the cylinder alone, it is seen that the virtual mass coefficient decreases slightly with kh and is approximately 2.3 for most of the range of the relative wave lengths which were investigated. This is reasonably close to the theoretical value of $C_M = 2$ for a cylinder in an infinite fluid indicating the effects of the bottom and the free surface on the hydrodynamic added mass are probably small. Indeed, for the depths shown there does not appear to be an orderly variation in the virtual mass coefficient with depth, or more importantly, considering the effect of the proximity of the structure to the water surface, with the relative depth of water above the cylinder, i.e. s/h.

Wilson and Reid (1963) propose a value of $C_M = 3.29$ for a cylinder located on the bottom from potential flow considerations. Sarpkaya and Rajabi (1979) have experimentally determined $C_M \approx 3.4$ for the case of a bottom-mounted smooth cylinder. The ratio of the clearance from the bottom to the diameter for the current experiments with the cylinder alone was 0.46. Measurements by Nath et. al. (1976) for a similar clearance ratio of 0.5 for periodic waves gave an average virtual mass coefficient of 2.27 which is close to that shown in Fig. 11.

When the rock protection is added, as can be seen in Fig. 11, the coefficient of virtual mass increases by about 30% to 40%. This is due to the change in flow around the structure as can be seen in the increased velocities (and, thus, the accelerations) over the cylinder for the cylinder with the rock protection compared to the cylinder alone as was shown in Figs. 5 and 6. In addition, the blockage under the cylinder due to the support structure must radically change the wake contributing also to the observed increase in the added hydrodynamic mass for the cylinder with the rock protection compared to that for the cylinder alone. Figs. 6 and 7 showed the effect of the support structure on the velocities was about the same for periodic waves as for solitary waves. This is consistent with the differences of the virtual mass coefficients with kh observed for the cylinder with and without the support structure. Whereas there does not appear to be an effect on C_M of the proximity to the water surface for the case of the cylinder alone, there is a small effect which can be seen in Fig. 11 for the rock protected cylinder. The closer the crown of the cylinder is to the free surface, the larger C_M. There is a tendency for C_M to decrease with increasing kh for both cases. This could be due to a decrease in wake

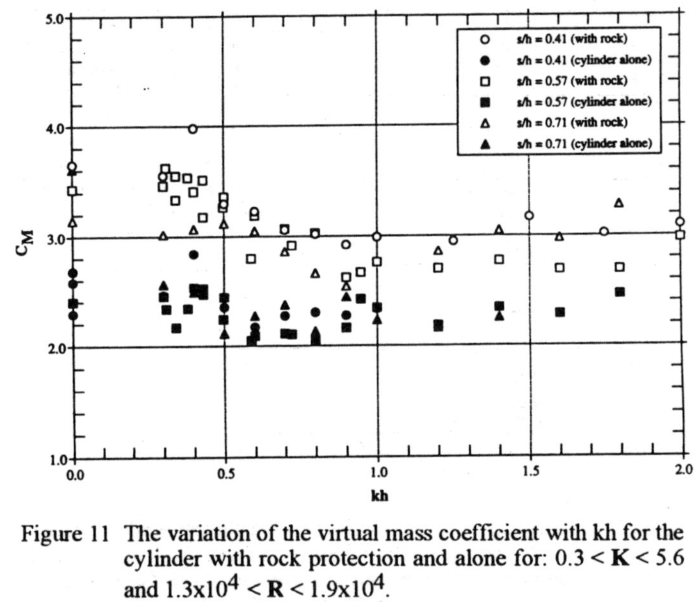

Figure 11 The variation of the virtual mass coefficient with kh for the cylinder with rock protection and alone for: $0.3 < \mathbf{K} < 5.6$ and $1.3\text{x}10^4 < \mathbf{R} < 1.9\text{x}10^4$.

size associated with shorter wavelengths for a given depth (or s/h). The change appears to be larger for the cylinder with the rock support structure possibly due to a correspondingly larger wake region.

Additional experiments were conducted with an impermeable wooden shape replacing the rock, and the virtual mass coefficients for this case are compared to those obtained with the rock layered cross-section in Fig. 12. In general, the impermeable supporting structure leads to somewhat larger virtual mass coefficients than the arrangement with the rock outer layer. Therefore, there must be some effect, although apparently relatively small, of the permeability of the rock layer on the added hydrodynamic mass. A trend toward decreasing C_M with increasing kh is also seen as was seen in Fig. 11.

The variation of the virtual mass coefficient with the Keulegan–Carpenter number, $\mathbf{K} = u_m T/d$, is shown in Fig. 13; u_m is the maximum water particle velocity and T is the wave period. The Keulegan-Carpenter number is proportional to the ratio of the excursion of the water particles to the diameter of the cylinder. The two cases shown in Fig. 13 are for the cylinder alone and the cylinder with the rock protection. For the limited range of the Keulegan-Carpenter parameter realized in these experiments, there does not

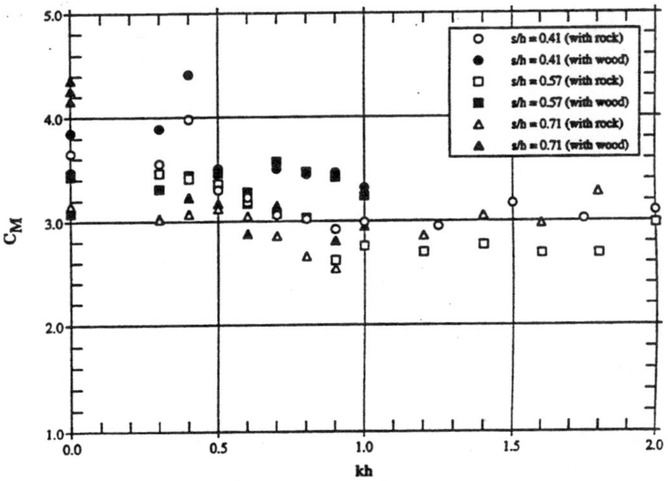

Figure 12 The variation of the virtual mass coefficient with kh for the cylinder with rock protection and with wood for: $0.3 < \mathbf{K} < 5.6$ and $1.3\mathrm{x}10^4 < \mathbf{R} < 1.9\mathrm{x}10^4$.

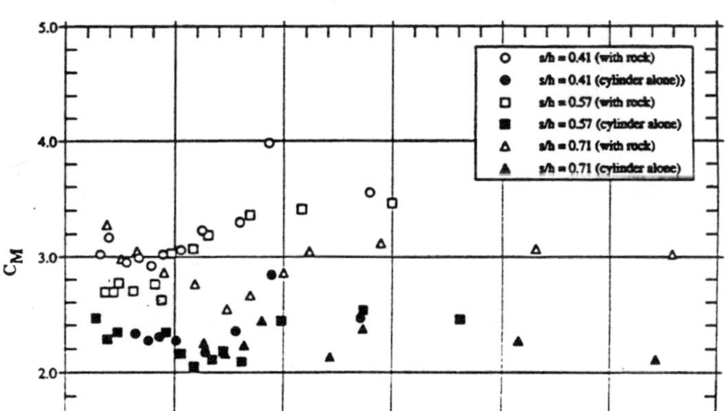

Figure 13 The variation of the virtual mass coefficient with the Keulegan-Carpenter number for the cylinder with rock protection and alone for: $0 < \mathrm{kh} < 2$ and $1.3\mathrm{x}10^4 < \mathbf{R} < 1.9\mathrm{x}10^4$.

appear to be an orderly variation of the virtual mass coefficient with **K** for either case. The small increase in C_M for $1 < \mathbf{K} < 3$ may be related more to variations in the local acceleration than to trends associated with the Keulegan–Carpenter number. On the other hand, the small increase of C_M with **K** seen may be associated with wake effects, and it is consistent with the increasing C_M with decreasing kh shown in Fig. 11. It should be realized that the ratio of the depth-to-wavelength, expressed as the product of the wave number and the depth, kh, and the Keulegan-Carpenter number are directly related through the relation of the water particle excursion to kh for the fixed size of the structure in these experiments. Although the Keulegan–Carpenter number relates important length scales of the problem, for engineering purposes in connection with the analysis of the failure of a section of the ocean outfall modeled in these experiments, in the nearshore region the more relevant parameter is the relative wavelength, h/L (or kh).

The variation of the lift coefficient, C_L, with the proximity to the free surface, as described by the ratio s/h, is shown in Fig. 14. These experiments were conducted with solitary waves and the data are presented for the condition of the cylinder with the rock support structure and with the impermeable wooden section for $H/h \approx 0.1$. There was no evidence of a lift force for the experiments with the cylinder alone. The velocity used in Eq. 4 corresponds to the maximum theoretical velocity of the incident wave at the elevation of the centerline of the cylinder. As the depth of water above the crown of the cylinder decreases, the lift coefficient increases significantly demonstrating the important effect of the free surface on lift. For the same incident wave, as the clearance, s/h, decreases the local velocity near the crown of the cylinder must increase thereby increasing the lift. Since the lift coefficient is defined using the temporal maximum velocity of the incident wave, an increase in the lift coefficient would be expected. The lift coefficient is most important in the prototype when considering the possibility of the pipe being moved out of its rock matrix by waves from extreme storm events. The experiments conducted with the impermeable base result in lift coefficients which are consistently greater (but not significantly so) than those associated with the armor rock suggesting an effect of the porosity of the

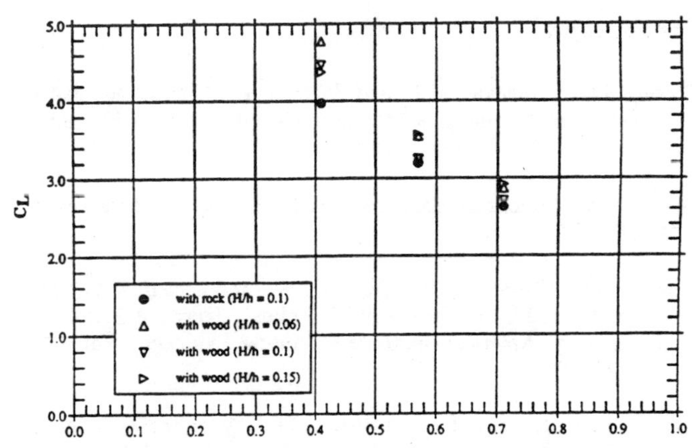

Figure 14 The variation of the lift coefficient with relative clearance, s/h, for the cylinder with rock protection and with wood for: $0.3 < \mathbf{K} < 5.6$ and $1.3\mathrm{x}10^4 < \mathbf{R} < 1.9\mathrm{x}10^4$

supporting base on the local velocities near the crown of the cylinder. This was not clearly evident in the velocity measurements reported earlier.

Sarpkaya and Rajabi (1979) have determined the lift coefficient for a circular cylinder resting on the bottom in a flow without a free surface is about $C_L \approx 6$. The lift force decreases rapidly with clearance from the bottom, indeed potential flow yields negative lift coefficients (C_L = -0.65) for a relative clearance of 0.5. Nath et.al. (1976) show a range of experimental data for a relative clearance of 0.5 of $-0.65 < C_L < +0.65$. The effect of the rock or wood protection in closing the region under the cylinder and providing a somewhat streamlined shape compared to that for a cylinder alone resting on the bottom may be one reason for the smaller lift coefficients for these experiments compared to those of Sarpkaya and Rajabi (1979).

CONCLUSIONS

There are several major conclusions which can be drawn from this study. As would be expected, the velocities are strongly affected by the presence of the structure, and this effect is about the same for solitary waves and for periodic waves of the same relative height. The virtual mass coefficient for a cylinder supported off the bottom by a closed structure with sloping sides is larger than that for the cylinder without the support structure located at the same distance above the bottom. For the cylinder alone and for the cylinder-rock structure tested there is a discernible, but small, influence of the ratio of the depth-to-wavelength (or the Keulegan-Carpenter number) on the virtual mass coefficient. Although there appears to be an effect of the permeability of the rock armor layer on the virtual mass and lift coefficients, the effect also is minor. The effect of the proximity of the structure to the free surface on the virtual mass coefficient is small, but the effect on the lift coefficient is significant.

ACKNOWLEDGMENTS

The experiments were conducted while AW was a Summer Undergraduate Research Fellow at Caltech. Dr. Fei Zhuang of the University of Southern California conducted the velocity measurements with the assistance of AW, and his help is gratefully acknowledged. The comments made by anonymous reviewers were very helpful and are appreciated.

REFERENCES

Cheong, H.-F., Shankar, N.J., and Subbiah, K., "Inertia dominated forces on submarine pipelines near seabed", Jour. of Hydraulic Res., IAHR, Vol. 29, No. 1, 1989.

Jacobson, V., "Forces on Sheltered Pipelines, Proc. Offshore Techn. Conf., 1988.

Morison, J.R., O'Brien, M.P., Johnson, J.W., and Schaaf, S.A., "The forces exerted by surface waves on piles", Journ. of Petroleum Technology, American Institute of Mining Engineers, Vol. 189, TP2846, 1950

Nath, J.H., Yamamoto, T., and Wright, J.C., Wave forces on pipes near the ocean bottom", Proc. Offshore Techn. Conf., 1976.

Sarpkaya, T. and Rajabi, F., "Hydrodynamic drag on bottom-mounted smooth and rough cylinders in periodic flow", Proc. Offshore Techn. Conf., 1979.

Wilson, B.W. and Reid, R.O., (discussion), "Wave force coefficients for offshore pipelines", JWWH, ASCE, Vol. 89, No. WW1, 1963.

Proceedings of the Seventh (1997) International Offshore and Polar Engineering Conference
Honolulu, USA, May 25-30, 1997

ISBN 1-880653-28-1 (Set); ISBN 1-880653-30-3 (Vol. II)

Local Scour Around Submarine Pipelines Under Wave Conditions

E. Özkan Çevik and Y. Yüksel
Yildiz Technical University
Istanbul, Turkey

ABSTRACT

Experiments have been carried out to determine the scour depth below submarine pipelines that are placed rigid position on bed under only wave conditions. The relative scour depth depends on Keulegan-Carpenter number (KC) and frequency parameter (β) and their ranges are 1.31 and 35, $3.5x10^2$ and $1.3x10^4$ respectively. Experimental results agree with the previous works with variation KC number of the relative scour depth. It was found that the relative scour depth is also proportional to the frequency parameter and the grain size gradation doesn't affect the scour depth below the submarine pipelines.

KEY WORDS :Local scour, submarine pipelines, seabed interaction.

INTRODUCTION

Scour below submarine pipelines plays very important role in the construction process and the management of submarine pipelines. However, rapid development of offshore oil fields has increased the construction of submarine pipelines for transport of crude oil to onshore refineries, and also disposal of waste water into the seas has been increasingly used. Interaction between the pipelines and an erodible bed under current and/or wave conditions tend to cause to scour around the pipelines. Scour under the submarine pipeline may expose part of the pipe and causes it to suspend in water. If the free span of the pipe is long enough, the pipe may experience resonant flow-induced oscillation leading to structural failure.

Design of a submarine pipeline is a rather complicated problem due to complex flow field and sea bed feature. Numerous researches have been carried out on the scouring process by many investigators. However, little has been published regarding studies of the structural stability of the submarine pipeline as affected by only wave condition.

Scour depth has been extensively studied in the steady current case (Kjeldsen et al, 1973, Bijker and Leeuweistein, 1984, Mao, 1986, Hansen et al, 1986). Scour occurring in waves has been investigated by Leeuwestein et al (1985), Ibrahim and Nalluri (1986). The most important work of these investigations under wave conditions was made by Sümer and Fredsøe (1990).

Sümer and Fredsøe (1990) carried out an experimental investigation on scour below pipelines exposed to waves. They concluded that the ratio of the scour depth to the pipe diameter depends on only the Keulegan-Carpenter number.

Scouring process depends on local flow structures, that means formation of vortices around submarine pipeline resting on a plane bed. The onset of scour is due to the combined action of the vortices and under flow, which leads to the formation of a small opening under the pipe as more and more sand particles are carried away. However, in the course of this process, vortex mechanism is going to be changed and finally local transport capacity of the flow around the pipeline will not be enough for carrying sands away because of the scour volume will be sufficiently large, and than scour process will reach a dynamic equilibrium condition. Flow structure for small structures such as submarine pipelines depend on the Reynolds number, the Keulegan-Carpenter number and also the gap ratio G/D (G is the gap between the cylinder and the sea bed).

The aim of this paper is to give the investigations of the scour below pipelines exposed to waves for different experimental conditions. Thus, presented results have been compared to Sümer and Fredsøe (1990)'s work and then it has been tried to get practical result in engineering point of view.

EXPERIMENTAL SET-UP AND PROCEDURES

The experiments were carried out in a 20 m long, 1m high and 1 m wide wave channel of in which periodic waves were produced by a paddle at one end. In order to minimize the reflection, wave absorbers were installed at both ends of the channel. The wave characteristics were recorded by a wave monitor using three wave gauges.

Four model pipes were used with diameters D=32.3 mm, 49 mm, 77 mm, 114 mm. The pipe surface in the tests acted as a hydraulically smooth surface. The mean water depth was held constant at 35cm over the sand bottom. The sand grain size was d_{50}=1.28 mm which is different from Sümer and Fredsøe (1990) and Lucassen (1984)'s works. All the tests were carried out in the initial condition of the cylinders locating on bed (G/D=0). Local scours under the fixed submarine pipes exposed to waves were measured when the dynamic equilibrium conditions were reached.

Ten different wave conditions were considered as shown in Table 1 where H_0/L_0 indicates deep water wave steepness, H_0 is deep water wave height and L_0 is deep water wave length.

Table 1 Experimental Conditions.

Case	Wave Height H_0 (cm)	Wave Length L_0 (m)	Wave Period T (s)	Wave Steepness H_0/L_0	Keulegan-Carpenter Number	Pipe Re Number	Pipe Diameter D (mm)
1	8.0	4.8	2.635	0.016	8.5	3.7×10^4	114.4
2	11.0	4.7	2.6	0.023	7.2	3.2×10^4	"
3	4.16	1.5	1	0.0277	1.31	1.5×10^4	"
4	13.12	4.7	2.578	0.0279	10	4.4×10^4	"
5	10.30	3.3	1.863	0.0312	5.4	3.3×10^4	"
6	7.40	2.1	1.3	0.0352	3.6	3.2×10^4	"
7	9.70	2.5	1.466	0.0388	3.8	3.0×10^4	"
8	10.60	1.7	1.12	0.062	2.5	2.6×10^4	"
9	10.15	1.3	0.9	0.078	1.7	2.2×10^4	"
10	8.38	1.0	0.8	0.0875	1.9	2.7×10^4	"
11	8.0	4.8	2.635	0.016	12.7	2.5×10^4	77.0
12	11.0	4.7	2.6	0.023	10.7	2.2×10^4	"
13	4.16	1.5	1	0.0277	2	1.0×10^4	"
14	13.12	4.7	2.578	0.0279	14.9	3.0×10^4	"
15	10.30	3.3	1.863	0.0312	8	2.2×10^4	"
16	7.40	2.1	1.3	0.0352	5.4	2.2×10^4	"
17	9.70	2.5	1.466	0.0388	5.6	1.7×10^4	"
18	10.60	1.7	1.12	0.062	3.7	1.5×10^4	"
19	10.15	1.3	0.9	0.078	2.6	1.8×10^4	"
20	8.38	1.0	0.8	0.0875	2.8	1.6×10^4	"
21	8.0	4.8	2.635	0.016	19.9	1.4×10^4	49
22	11.0	4.7	2.6	0.023	16.9	6.4×10^3	"
23	4.16	1.5	1	0.0277	3.1	1.9×10^4	"
24	13.12	4.7	2.578	0.0279	23.4	1.4×10^4	"
25	10.30	3.3	1.863	0.0312	12.5	1.4×10^4	"
26	7.40	2.1	1.3	0.0352	8.5	1.3×10^4	"
27	9.70	2.5	1.466	0.0388	8.9	1.3×10^4	"
28	10.60	1.7	1.12	0.062	5.9	1.1×10^4	"
29	10.15	1.3	0.9	0.078	4.0	9.4×10^3	"
30	8.38	1.0	0.8	0.0875	4.5	1.2×10^4	"
31	8.0	4.8	2.635	0.016	30.2	1.1×10^4	32.3
32	11.0	4.7	2.6	0.023	25.6	9.0×10^3	"
33	4.16	1.5	1	0.0277	4.6	4.3×10^3	"
34	13.12	4.7	2.578	0.0279	35.4	1.3×10^4	"
35	10.30	3.3	1.863	0.0312	19.0	9.3×10^3	"
36	7.40	2.1	1.3	0.0352	12.9	9.1×10^3	"
37	9.70	2.5	1.466	0.0388	13.4	8.4×10^3	"
38	10.60	1.7	1.12	0.062	8.9	7.3×10^3	"
39	10.15	1.3	0.9	0.078	6.1	6.2×10^3	"
40	8.38	1.0	0.8	0.0875	6.7	7.7×10^3	"

Even in the simple case of local scour around cylinders in unidirectional flow, the number of parameters involved is fairly large and their relationships can be complex. In this work, when previous works have considered, following dimensionless groups might be found for relative scour depth (S/D).

$$S/D = f(KC, Re, H/d, \theta, H/L) \qquad (1)$$

where S, D, H, d, L are scour depth, pipe diameter, wave height, water depth, wave length respectively and dimensionless parameters are called

$KC = U_m T/D$; Keulegan-Carpenter number,

in which U_m is maximum orbital velocity above the bed, T is wave period,

S/D; relative scour depth,

H/L; wave steepness,

$Re = U_m D/\nu$; pipe Reynolds number,

in which ν is kinematic viscosity of water,

H/d; depth parameter,

$$\theta = \frac{U_*^2}{(s-1)gd}; \text{ Shields parameter,}$$

in which U_* is undisturbed bed shear velocity, s is relative density of sediment and g is the acceleration of gravity.

As it is known from previous works, flow structure around cylinders depend on Reynolds number, Keulegan-Carpenter number, physical characteristics of the wave, object shape and orientation, and the proximity of other boundary due to the complex mechanism of flow field around the objects (Sarpkaya, 1976, Yüksel and Narayanan, 1994). However local scouring process has been formed by changing of flow field due to the presence of obstacles even under clear water conditions. The formation of vortex is the main cause of scour hole under the pipelines.

Now if Eq.1 is again considered, the ratio of Keulegan-Carpenter number to Reynolds number will give frequency parameter, β, which is known as $\beta = Re/KC = D^2 T/\nu$. Relative scour depth can also be found to depend on the following parameters.

$$S/D = f(KC, \beta, H/d, \theta, H/L) \qquad (2)$$

Although in the clear-water case, the variation of scour depth with θ is pronounced, in the live-bed case there is no any effect of θ on the variation of scour depth. Sümer and Fredsøe (1990) indicated that S/D increases from 0 at very small θ values up to values of 0.4-1.0 when the θ-value approaches the live-bed case. Hence Shields parameter can be eliminated in Eq.2 because all the experiments were carried out under live bed conditions (larger than 0.03) in this present work.

$$S/D = f(KC, \beta, H/d, H/L) \qquad (3)$$

If water depth is constant, the dimensionless wave height (H/d) can also be eliminated and vortex structure around the cylinders under wave conditions depend on mostly wave frequency which is involved in frequency parameter, β, finally Eq.4 can be obtained as follows;

$$S/D = f(KC, \beta) \qquad (4)$$

As it is shown from Eq.4, relative scour depth depends on frequency parameter. Especially in shoaling zone where any wave theories are available and it is quite difficult to estimate wave kinematics. So in engineering applications usage of frequency parameter is more practical than the Keulegan-Carpenter number.

ANALYSIS OF RESULTS

During the experiments the near bed orbital velocity was measured by a DANTEC hot film probe. The data sampling was made at a frequency of 50 samples per second and data were evaluated by using a PC computer. Each record duration is about 3.3 minutes. The maximum velocities were considered because U_m was needed in order to evaluate the KC number and Re number. More over the orbital velocities were obtained using Linear theory and they were found well agreement between measured and computed values.

Fig.1 shows the variation of relative scour depth, S/D, with the KC number. The figure clearly demonstrates that the relative equilibrium scour depth under pipes correlates remarkably well with Keulegan-Carpenter number. The scour variation with the KC-number from this presented experimental work can be represented by the following least-square fit for d_{50}=1.28 mm.

$$S/D = 0.045\,(KC)^{0.76} \tag{5}$$

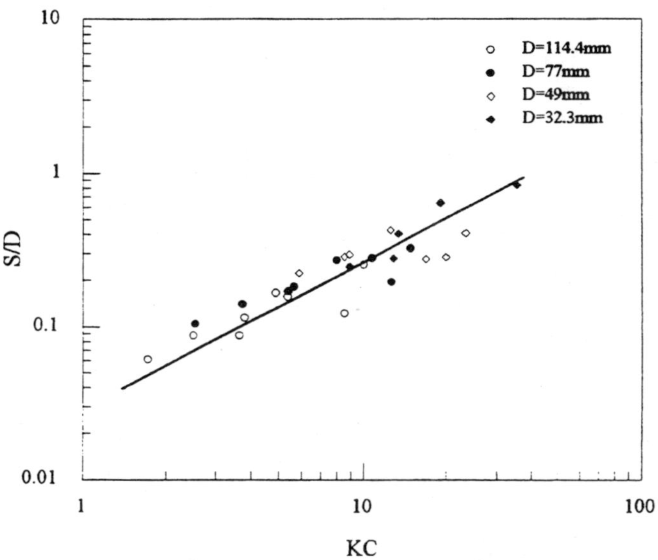

Fig. 1 Relative equilibrium scour depth versus Keulegan-Carpenter number. Correlation number, r = 0.90.

In order to compare the presented results with Sümer and Fredsøe (1990) and Lucassen (1984), the data were plotted on Fig.2 with the results given by these researchers. It has been shown that the presented data has well tendency with their data. The new correlation between relative scour depth and KC number is quite high. This result also indicates grain size distributions have not any effect on the scour depths under the submarine pipelines. The results show that the scour depth increases with increasing KC-number as explained from the flow pattern described in the previous works. The new relation between relative scour depth and Keulegan-Carpenter number was found as follows:

$$S/D = 0.05KC^{0.68} \tag{6}$$

Eq.6 shows that there is not any distinct difference with Sümer and Fredsoe (1990)'s relation. They gave following equation

$$S/D = 0.10KC^{0.5} \tag{7}$$

which includes also tidal conditions.

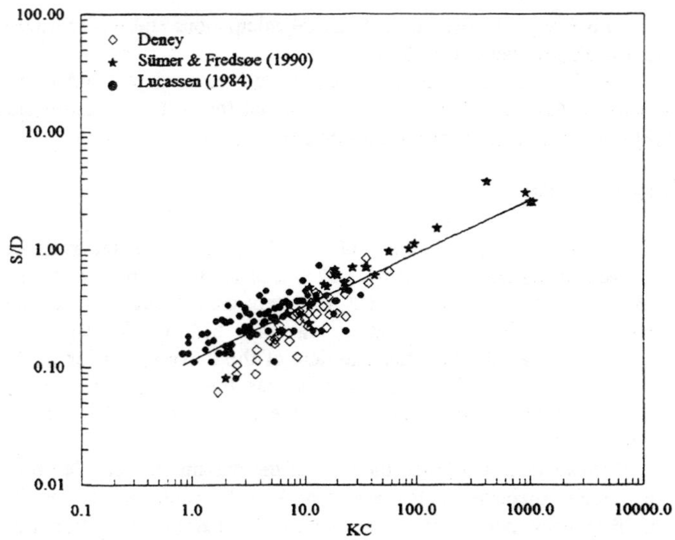

Fig. 2 Presented data with Lucassen (1984) and Sümer and Fredsøe (1990).

Since Eq.4 reveals dependence on KC as well as β, it is important to examine the effect of the parameter in further details. It has been shown in Fig.3 that, the correlation between the relative scour depth with frequency parameter, β, was found quite high. That means, wave period has significant effect on the scouring process. This result has also been observed from previous works about vortecies around the cylinders (Yüksel et al, 1995).

In addition, it is very easy to use this graph for engineering applications because there is no need for any calculation of wave kinematics, especially for shallow water regions, since the theories give insufficient results near breaking conditions.

The scour depth variation in Fig.3 can be represented by the following least square fit:

$$S/D = 15.79(\beta)^{-0.55} \tag{8}$$

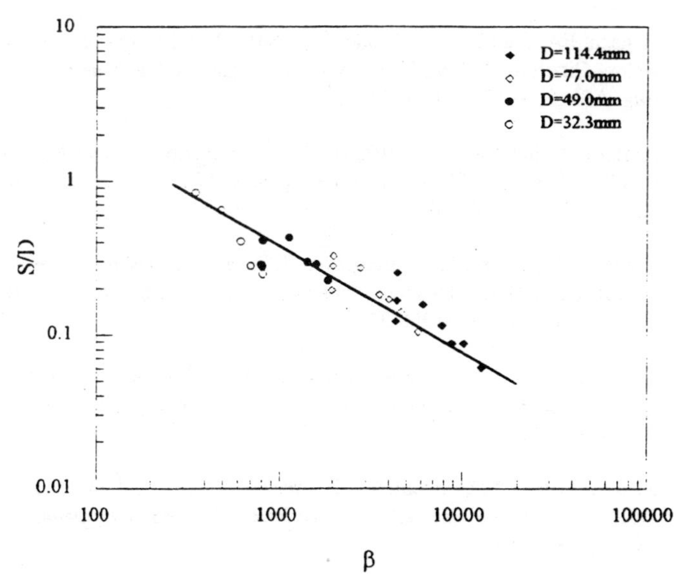

Fig. 3 Relative equilibrium scour depth versus frequency parameter. Correlation number, r=0.92.

Eq.8 was obtained after statistical calculations and power linear regression gave better correlation.

Eq.8 gives the maximum equilibrium scour depth under the pipelines. Also this figure clearly shows that the relative equilibrium scour depth decreases with increasing β parameter.

CONCLUSION

In this work the range of the Keulegan-Carpenter number achieved in the tests was KC=1.3-57.0. The Reynolds number range was Re= $4.2x10^3$-$4.4x10^4$. All the tests were carried out in the initial condition of cylinders locating on the bed (G/D=0.0). The experimental results show that the ratio of the maximum scour depth under the pipe to cylinder diameter versus the Keulegan-Carpenter number are in-line with Sümer and Fredsøe (1990) and Lucassen (1984).

However it was found the ratio of the maximum scour depth to the cylinder diameter is also proportional to the minus square root of the frequency parameter, β, in which the parameter is between $3.5x10^2$ and $1.3x10^4$. The correlation between the relative scour depth, S/D, and frequency parameter, β, were found well. It shows that the scour process depends on the cylinder diameter and the wave period. This result also gives more practical meaning because β parameter doesn't depend on the orbital velocity at the given location on the sea bed. However it is known that the wave height has also effect on scour depth, so frequency parameter must be analyzed more systematically by taking into consideration of wave heights.

ACKNOWLEDGMENTS

The authors wish to thank The Research Fund of Yıldız Technical University for the financial support.

REFERENCES

Bijker, EW, and Leeuwestein, N (1984). "Interaction Between Pipelines and the Seabed Under the Influence of Waves and Currents," *Seabed Mechanics*, B Denness,ed., Graham and Trotman, Gettysburg, Md., pp 235-242.

Hansen, EA, Fredsøe, J, and Mao, Y (1986). "Two-Dimensional Scour Below Pipelines," *Proc 5th Int Symp on Offshore Mech and Arctic Eng*, ASME, Vol 3, pp 670-678.

Ibrahim, A and Nalluri, C (1985). "Scour Prediction Around Marine Pipelines," *Proc Fifth Int Symp on Offshore Mech and Arctic Eng*, ASME, Vol 3, pp 697-684.

Kjeldsen,SP, Gjørsvik, o, Bringaker, KG, and Jacobsen, J (1973). "Local Scour Near Offshore Pipelines," *Second Int Conf. POAC Conference*, Iceland, pp 308-331.

Leeuwestein, W, Bijker, EW, Peerbolte, EB, and Wind, HG (1985). "The natural Self-Burial of Submarine pipelines," *Proc 5th Int Conf on Behaviour of Offshore Struct*, pp 717-728.

Mao, Y (1986). "The Interaction Between a Pipeline and an Erodible Bed," *Phd Thesis Presented to the Technical University of Denmark*, Lyngby, Denmark.

Sarpkaya, T (1976). "In-line and Transverse Forces on Smooth and Sand Roughened Cylinders in Oscillatory Flow at High Reynolds Numbers," Naval Post Graduate School, Tech Report, No NPS-69 SL 76062, Monterey, CA.

Sümer, BM, and Fredsøe, J (1990). "Scour Below Pipelines in Waves," *J of Waterway, Port, Coastal and Ocean Eng*, ASCE, Vol 116, No 3, pp 307-323.

Yüksel, Y and Narayanan, R (1994). "Wave Forces on Horizontal Cylinder Resting on Sloping Bottom," *Proc 4th Int Offshore and Polar Eng Conf*, Osaka, Vol II, pp 85-92.

Yüksel, Y, Atlı, V, and Özkan Çevik , E (1995). "Flow Field Along a Flat Surface With a Parallel Placed Cylinder," *Proc 5th Int Offshore and Polar Eng Conf*, The Hague, Vol 2, pp 157-161.

Proceedings of the Seventh (1997) International Offshore and Polar Engineering Conference
Honolulu, USA, May 25-30, 1997

ISBN 1-880653-28-1 (Set); ISBN 1-880653-30-3 (Vol. II)

The Numerical Simulation of Wave Forces on Seabed Pipeline by Three-Step Finite Element Method and Large Eddy Simulation

Yucheng Li, Bing Chen and Guozhang Lai
Dalian University of Technology
Dalian, China

ABSTRACT

A three-step finite element method (FEM) and Large Eddy Simulation (LES) was applied to incompressible turbulent flow around seabed pipeline in relative high Reynolds number. Both two dimensional and three dimensional numerical simulation was carried on in this paper to determine the 3-D effect. The results of numerical simulation agree quite well with the wave forces acting on pipeline model measured in physical model test.

INTRODUCTION

Analysis of the hydrodynamic forces acting on circular cylinder is very important in the design of pipelines and cables on seabed. Both experimental investigate such as P.W. Bearman et al. (1977) and numerical simulation has been done in the past. In the past two years, Y.C. Li and B. Chen (1995) and Y.C. Li and C. Dai (1996) had studied the wave forces on pipeline and flow field around pipeline in physical model test. Y.G. Lee and S.W. Hong et al (1994) used a finite difference method (FDM) with SGS turbulent model to simulated the vortex motion behind a circular cylinder above a horizontal plane boundary in a mean flow. Ge Wang and Guozhang Lai et al (1995) used FDM-FEM with LES method to study the oscillating flow around circular cylinder near a horizontal plane boundary. To make the solution stable in high Reynolds number, both of them employed high-order upwind scheme in space, and a special process or bodyfitted coordinate system had been introduced in numerical method to deal with the curve boundary of circular cylinder.

On the other hand, the finite element method appears to be very convenient to problems in complex geometry and a three-step algorithm based on a Taylor series expansion in time had been derived (Chun and Kawahara, 1993) to ensure the solution stability in high Reynolds number. To compare with upwind finite element methods (such as Petrov-Galerkin FEM), the three-step FEM can reduce false diffusion. To compare with two-step Taylor-Galerkin FEM, the three-step FEM can be used for longer time step with little loss of accuracy. For the purpose of developing a powerful numerical method for simulating the wave forces acting on seabed pipeline, three-step finite element method coupled with LES turbulent model were developed by authors in this study.

Furthermore, a 3-dimensional numerical simulation are also carried on in this study to determine the 3-D effect on pipeline problem. The results of both 2-D and 3-D numerical simulation are compared with physical test data.

GOVERNING EQUATIONS

The governing Equations are the N-S Equations and the continuity Equation of incompressible viscous fluid. The dimensionless parameters as follows were used.

$$\vec{V} = \vec{V'}/U_m \tag{1}$$

$$P = P'/\rho U_m^2 \tag{2}$$

$$X = X'/D \tag{3}$$

$$t = D/U_m \tag{4}$$

$$Re = U_m D/\nu \tag{5}$$

in which V', P', X' are real value of velocity, pressure and length, respectively. D is the diameter of pipeline. U_m is the maximum velocity of oscillating flow. ν is kinematics viscosity of fluid. ρ is the density of fluid. V, P, X, t and Re are dimensionless velocity, pressure, length, time and Reynolds number, respectively. To apply LES turbulent model, the dimensionless governing Equations are as follows:

$$\frac{\partial u_i}{\partial t} + u_j \frac{\partial u_i}{\partial x_j} = -\frac{\partial P}{\partial x_i} + \frac{1}{Re}\frac{\partial}{\partial x_j}\left(\frac{\partial u_i}{\partial x_j}\right) + \frac{\partial}{\partial x_j}\left(\tau_{ij}\right) \tag{6}$$

$$\frac{\partial u_i}{\partial x_i} = 0 \tag{7}$$

where $$\tau_{ij} = \nu_t\left(\frac{\partial u_i}{\partial x_j} + \frac{\partial u_j}{\partial x_i}\right) \tag{8}$$

$$\nu_t = (c\Delta)^2\left[\frac{\partial u_i}{\partial x_j}\left(\frac{\partial u_i}{\partial x_j} + \frac{\partial u_j}{\partial x_i}\right)\right]^{\frac{1}{2}} \tag{9}$$

in which c is Smagorinsky constant, it is taken as 0.1 in this paper, Δ is the characteristic length of mesh element. For 2-D case $\Delta = (S_{ij})^{1/2}$, S_{ij} is the area of mesh element (i, j=1, 2). For 3-D case $\Delta = [\Delta x_i \, .\Delta x_j]^{1/2}$ (i=1, 2,

3),x_i is the characteristic length of 3-D element in i direction.

NUMERICAL COMPUTATION METHOD

The three-step solution algorithm is as follows:

$$\frac{u_i^{n+1/3}-u_i^n}{\Delta t/3}=-u_j^n\frac{\partial u_i^n}{\partial x_j}-\frac{\partial P^n}{\partial x_i}+\frac{1}{Re}\frac{\partial}{\partial x_j}\left(\frac{\partial u_i^n}{\partial x_j}\right)+\frac{\partial}{\partial x_j}\left(\tau_{ij}^n\right) \quad (10)$$

$$\frac{u_i^{n+1/2}-u_i^n}{\Delta t/2}=-u_j^{n+1/3}\frac{\partial u_i^{n+1/3}}{\partial x_j}-\frac{\partial P^n}{\partial x_i}+\frac{1}{Re}\frac{\partial}{\partial x_j}\left(\frac{\partial u_i^{n+1/3}}{\partial x_j}\right)+\frac{\partial}{\partial x_j}\left(\tau_{ij}^{n+1/3}\right) \quad (11)$$

$$\frac{u_i^{n+1}-u_i^n}{\Delta t}=-u_j^{n+1/2}\frac{\partial u_i^{n+1/2}}{\partial x_j}-\frac{\partial P^{n+1}}{\partial x_i}+\frac{1}{Re}\frac{\partial}{\partial x_j}\left(\frac{\partial u_i^{n+1/2}}{\partial x_j}\right)+\frac{\partial}{\partial x_j}\left(\tau_{ij}^{n+1/2}\right) \quad (12)$$

Spatial discretization of Equations (10)~(12) can be performed using the standard Galerkin method. Before calculating the velocity u_i^{n+1} from Equation (12), the pressure P^{n+1} has to be solved. By taking the divergence on both sides of Equation (12), and apply incompressible constraint $\nabla\bullet V^{n+1}=0$, the following Poison Equation of pressure is derived:

$$\frac{\partial}{\partial x_i}\left(\frac{\partial P^{n+1}}{\partial x_i}\right)=\frac{1}{\Delta t}\frac{\partial u_i^n}{\partial x_i}-\frac{\partial}{\partial x_i}\left(u_j^{n+1/2}\frac{\partial u_i^{n+1/2}}{\partial x_j}\right)+\frac{1}{Re}\frac{\partial}{\partial x_i}\left[\frac{\partial}{\partial x_j}\left(\frac{\partial u_i^{n+1/2}}{\partial x_j}\right)\right]+\frac{\partial}{\partial x_i}\left[\frac{\partial}{\partial x_j}\left(\tau_{ij}^{n+1/2}\right)\right] \quad (13)$$

By using Galerkin finite element method and introduce the normal component of Equation (12) on boundary

$$\left(\frac{\partial P^{n+1}}{\partial x_i}\right)\cdot n_i=-\frac{u_i^{n+1}-u_i^n}{\Delta t}\cdot n_i-u_j^{n+1/2}\frac{\partial u_i^{n+1/2}}{\partial x_j}\cdot n_i+\frac{1}{Re}\frac{\partial}{\partial x_j}\left(\frac{\partial u_i^{n+1/2}}{\partial x_j}\right)\cdot n_i+\frac{\partial}{\partial x_j}\left(\tau_{ij}^{n+1/2}\right)\cdot n_i \quad (14)$$

into the weak formulation of Equation (13) results in

$$\int_V\frac{\partial\phi}{\partial x_i}\frac{\partial P^{n+1}}{\partial x_i}dV=-\frac{1}{\Delta t}\int_V\phi\frac{\partial u_i^n}{\partial x_i}dV-\int_V\frac{\partial\phi}{\partial x_i}u_j^{n+1/2}\frac{\partial u_i^{n+1/2}}{\partial x_j}dV-\int_S\phi\frac{u_i^{n+1}-u_i^n}{\Delta t}dS \quad (15)$$

where Φ is the weighting function of pressure. Detail procedure had been derived by Chun and Kawahara (1993). For 2-D case, four nodes linear element is used and the Poisson Equation of pressure is solved by direct LU matrix decomposition method. For 3-D case, eight nodes linear elements are used and the Poisson Equation of pressure is solved by BI-CGSTAB iterative method (H. A. Van Der Vorst, 1992).

MESH AND BOUNDARY AND INITIAL CONDITIONS

The finite element mesh of 2-D case is shown in Fig. 1. For 3-D case, the mesh system is composed of some vertical layer , in each layer the shape of elemental grid is the same as 2-D case.

At the outer boundary of calculation domain the velocity are set with predetermined values (in oscillating flow the values vary with time). At the solid boundary of pipeline surface and seabed the velocity are set to zero (nonslip condition). For 3-D case, on both vertical side boundary (vertical layer at $z=0$ and $z=z_{max}$) only the normal velocity component are set to zero (slip condition). The pressure on the top of outer boundary is set to a reference value.

At the beginning of calculation (t=0) the velocity and pressure in whole calculation domain are set to zero.

COMPUTATION RESULTS AND DISCUSSION

To verify the three-step finite element method and LES turbulent model, authors firstly carry on a numerical simulation of a circular cylinder above a horizontal plane boundary under mean flow as described by Bearman (1978) and Lee (1996). The gaps between the cylinder and the plate are 0.1 and 0.4 cylinder diameter. Based on cylinder diameter and mean flow velocity, the Reynolds number is 4.8×10^4. The distance from the outer boundary of calculation domain to the cylinder center is 55 cylinder diameter. A mesh system of 120×120 and dimensionless $\Delta t=0.003$ are used.

The velocity vector field in the vicinity of the cylinder are shown in Fig. 2. The time-averaged pressure distribution around the circular cylinder and along the flat plate are shown in Fig. 3. The comparison of pressure distribution with results of Bearman (1978) and Lee (1996) are also include in Fig. 3. Based on the comparison, the pressure distribution around the circular cylinder get from authors' results is similar with that of Lee (1996). However, the pressure distribution on flat plate behind the cylinder of authors' result is more closer to measured data of Bearman (1978).

COMPARISON WITH MEASURED WAVE FORCES ACTING ON PIPELINE MODEL

For the efficiency of three-step FEM and LES model had been verified in last section, authors further to calculated the wave forces acting on pipeline model and to compare calculated results with measured results of physical model test (Y.C. Li, 1996). It had been pointed out (Ge Wang, 1995) that wave flow around pipeline can be simulated by oscillating flow because of the ignoble particle vertical velocity near the seabed. In this study authors also use oscillating flow to simulate wave flow. The velocity of oscillating flow is determined by Stokes wave theory at the position of pipeline center but with the pipeline removed from flow field, and the velocity varies in the same period of wave.

2-D case:

The wave parameters are as follows: wave height H=11.2cm, wave period T=1.04s, water depth d=30cm, pipeline diameter D=4cm. As defined in Eq. (5), the Reynolds number is 8320. The dimensionless $\Delta t=5\times10^{-3}$. A mesh system of 100×100 are used. The distance between outer boundary of calculation domain and pipeline center is 100 pipeline diameter. The gap between pipeline model and seabed boundary is 0.25 pipeline diameter.

The comparison of calculated results and measured results are shown in Fig. 4.

It is indicated by Fig. 4: The time history of calculated dimensionless horizontal force and vertical force agree quite well with the dimensionless measured data especially when $t/T<3.5$. When $t/T>3.5$, the difference between calculated peak value of vertical force and measured data become little bigger, this is perhaps due to the influence of reflection from outer boundary.

3-D case:

The wave parameters are as follows: wave height H=9.12cm, wave period T=1.01s, water depth d=30cm, pipeline diameter D=4cm, length of calculation section of pipeline L=12cm. The Reynolds number is 6570. The dimensionless $\Delta t=5\times10^{-3}$. To reduce the total number of nodes the outer boundary of calculation domain extends to 8 pipeline diameter and a coarse mesh system of 65×58×31 is used. Only three wave periods was reached in time to avoid the influence of outer boundary. The mesh system in one vertical layer is shown in Fig. 1, and a 2-D calculation is done use this mesh for comparison.

The comparison of results of 3-D and 2-D dimensionless calculated results and dimensionless measured data is shown in Fig. 5. The results of 3-D calculation is taken from the calculated forces acting on whole length of pipeline and be divided by the length of pipeline L. It is indicated by Fig. 5: The shape of calculated time history of 3-D and 2-D horizontal force are very similar, and both of them agree quite well with measured data in amplitude but with a small shift in phase. The calculated time history of 3-D and 2-D vertical force are also agree well in phase, but the absolute peak value of 3-D results are often small than that of 2-D case. This is because the vortex in 3-D flow field can be diffused easier than in 2-D case. Fig. 5 shows the results of 3-D calculation agree a little better with measured data than 2-D results.

Fig. 6 shows the comparison of 3-D and 2-D velocity vector field in the vicinity of pipeline. It is indicated by Fig. 6: The flow fields at different vertical layer of z=0.75D, 1.5D and 2.25D are almost same. So the flow field around pipeline can be regarded as 2-D flow field. However, there is a little difference between the flow field around pipeline in 2-D case and that in 3-D case. For example, there is a apparent vortex in front of the pipeline in 2-D case but in 3-D case the shape of this vortex is not apparent. This is due to the 3-D effect.

CONCLUSION

The 2-D and 3-D numerical simulation method by three-step FEM and LES turbulent model is developed successfully to calculated the wave forces acting on seabed pipeline.

Both 2-D and 3-D numerical simulation results agree well with measured data and 3-D results seems a little better than 2-D results. However, considering the long time that 3-D calculation consumed, the 2-D numerical simulation is efficient enough in pipeline problem.

ACKNOWLEDGMENT

The measured data of wave forces acting on pipeline model is provided by Mr. C. Dai. His help on the arrangement of measured data is greatly appreciated.

REFERENCES

Chun B.J. and Mutsuto Kawahara (1993), The Analysis of Unsteady Incompressible Flows by a Three-step Finite Element Method, International Journal for Numerical Methods in Fluid, Vol. 16, pp. 793-811, 1993.

Ge Wang, Guozhang Lai and Yucheng Li (1995), A Large Eddy Simulation Method of Vortex Motion Behind a Circular Above a Horizontal Plane Boundary Under a High Reynolds Number, Proc. of National Conference on Hydrodynamics, Dalian, China. (in Chinese)

H.A.Van Der Vorst(1992), BI-CGSTAB: A Fast Smoothly Converging Variant of BI-CG for the Solution of Non symmetric Linear Systems. SIAM J. Sci. Stat. Comput. Vol. 13, No. 2, March 1992

James W. Deardorff (1970) A Numerical Study of Three-dimensional Turbulent Flow at Large Reynolds Numbers, J. Fluid Mech., Vol. 41, part 2, pp. 453-480, 1970.

P.W. Bearman and M.M. Zdravkovich (1978), Flow Around a Circular Cylinder Near a Plane Boundary, J. Fluid Mech., Vol. 89, part 1, pp. 33-47, 1978.

Young-Gill Lee, Sung-Wan Hong, Kuk-Jin Kang (1994), A Numerical Simulation of Vortex Motion Behind a Circular Cylinder above a Horizontal Plane Boundary, Proceedings of the Fourth (1994) International Offshore and Polar Engineering Conference, Osaka, Japan, April 10-15, 1994.

Yucheng Li, Bing Chen et al.(1996), Physical Model Test of Pipelines under Wave Action, Marine Science Bulletin, Vol. 15, No. 5 pp. 68-73, 1996. (in Chinese)

Yucheng Li, Haigui Kang, Chin Dai and Guangwei Li (1996), The Flow Visualization and the Hydrodynamic Characteristic of a Circular Cylinder near a Plane Boundary in Wave-Current Coexisting Field. (to be published on China Coastal Engineering, in Chinese)

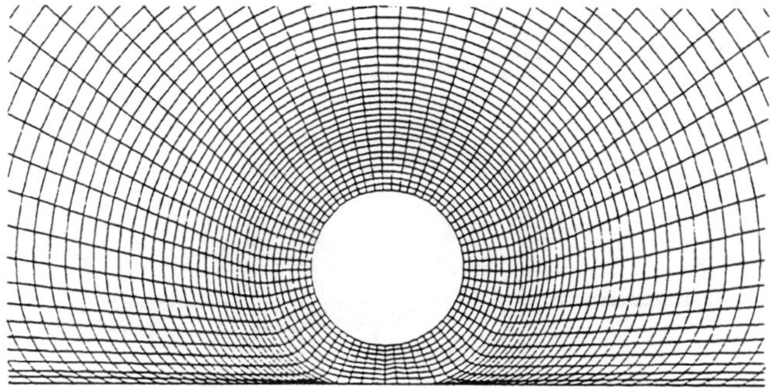

Fig. 1 Mesh system in the vicinity of the pipeline and seabed at G/D=0.25

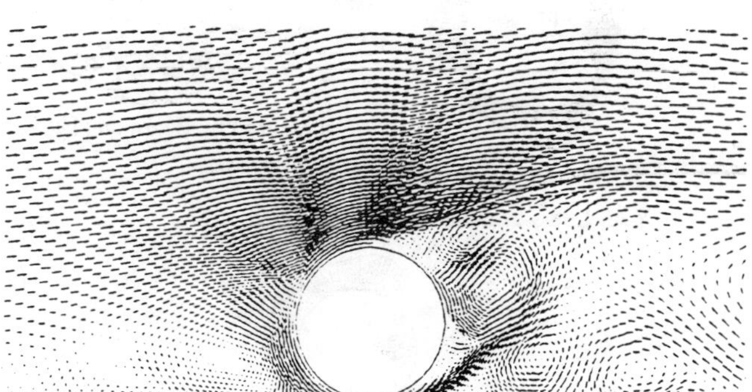

Re=48000. $\Delta t=0.003$ T=45

(a) G/D=0.1

Fig. 2 Velocity vectors in the vicinity of the circular cylinder.

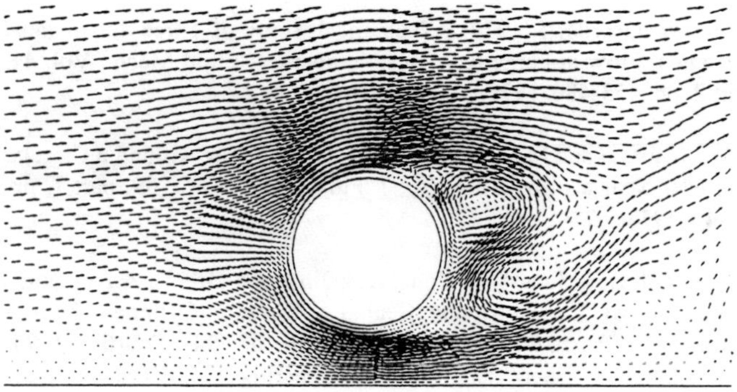

Re=48000. Δt=0.003 T=45

(b) G/D=0.4

Fig. 2 (continue)

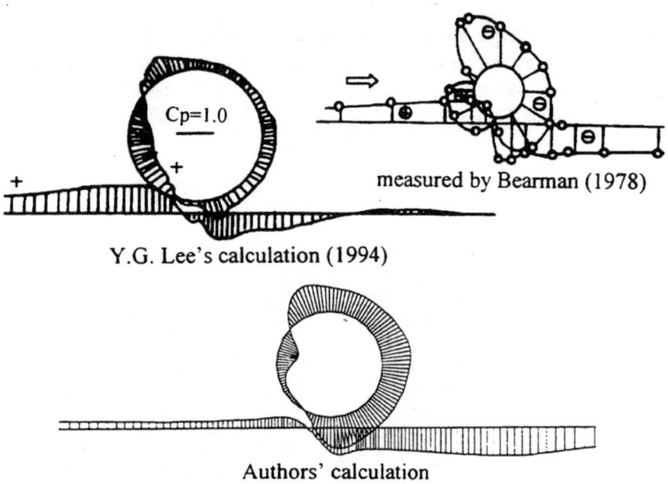

(a) G/D=0.1

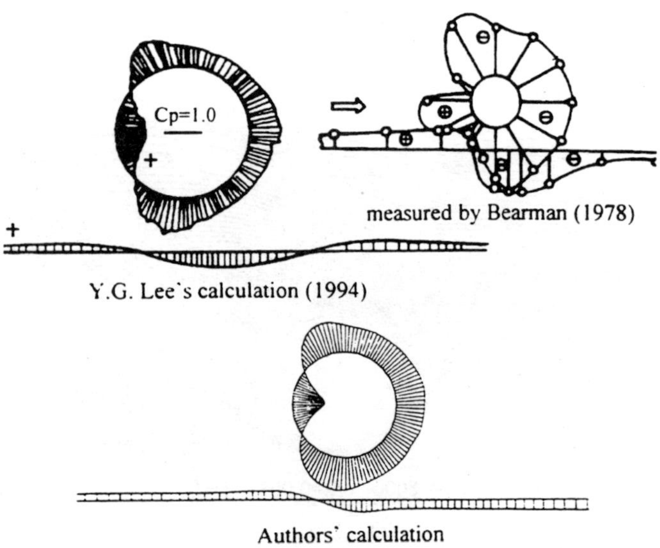

(b) G/D=0.4

Fig. 3 Time-averaged pressure distributions around the circular cylinder and along the flat plate.

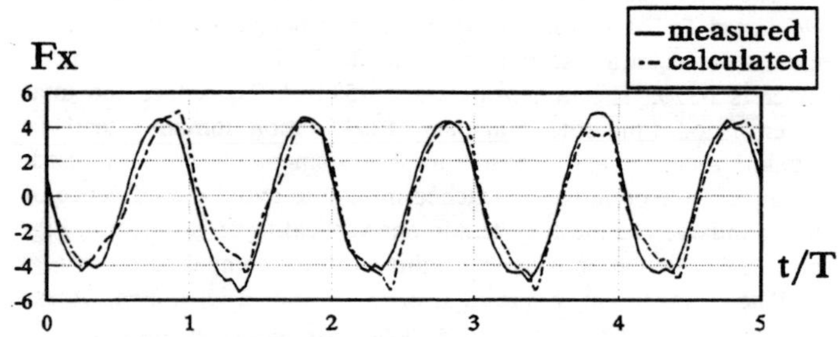

(a) Time history of dimensionless horizontal force

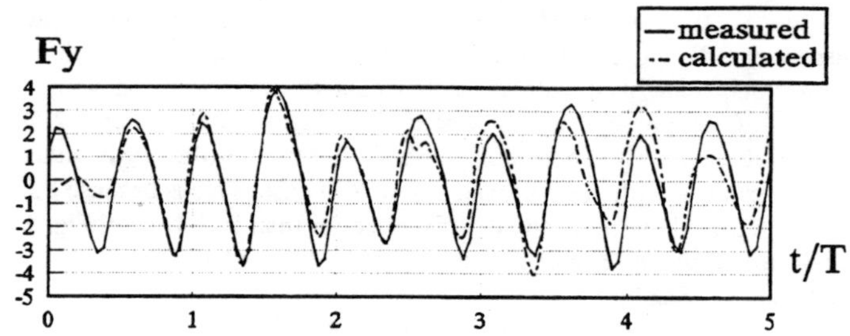

(b) Time history of dimensionless vertical force

Fig. 4 Comparison of 2-D calculated results and measured results

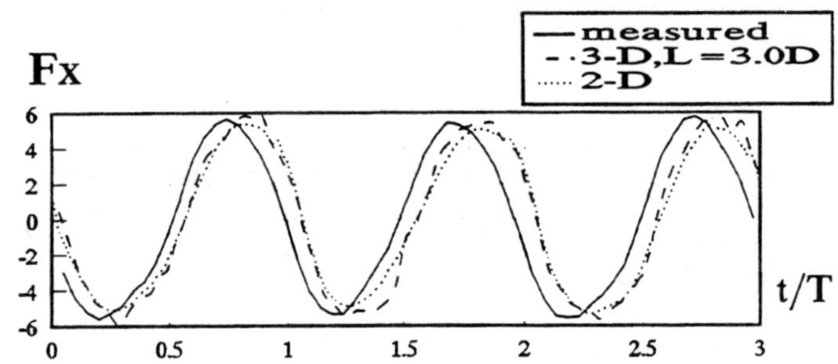

(a) Time history of dimensionless horizontal force

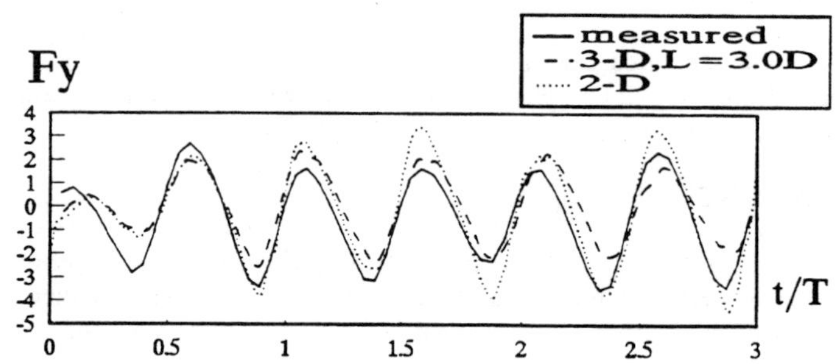

(b) Time history of dimensionless vertical force

Fig. 5 Comparison of 3-D and 2-D calculated results and measured results

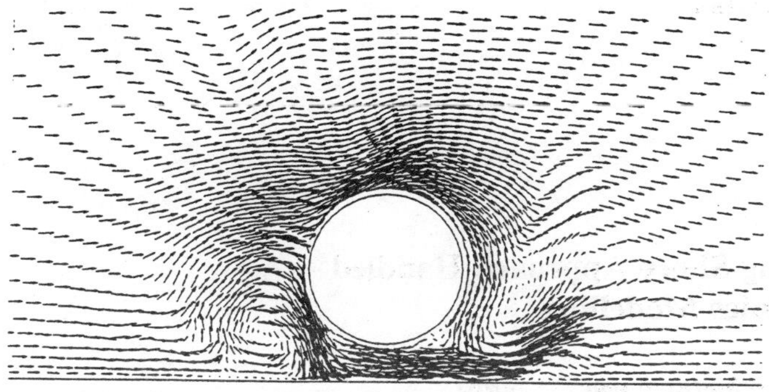

(a) 2-D , t/T=3.0

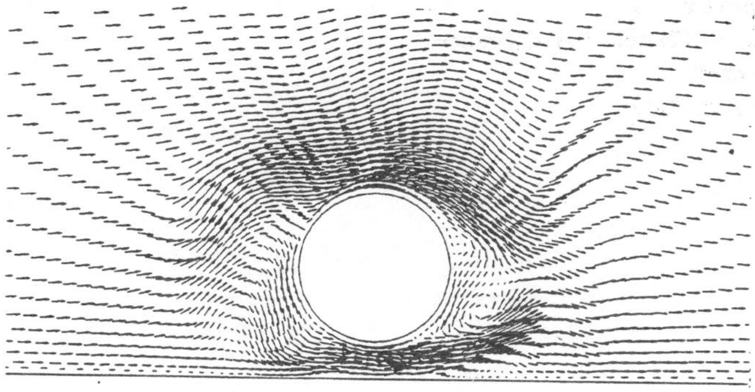

(b) 3-D , t/T=3.0 , z=0.75D

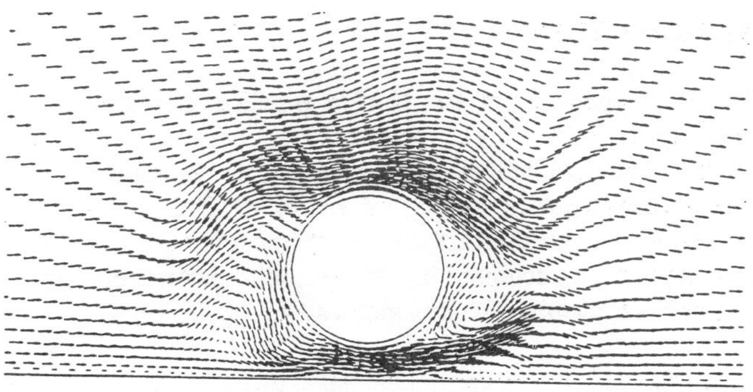

(c) 3-D , t/T=3.0 , z=1.5D

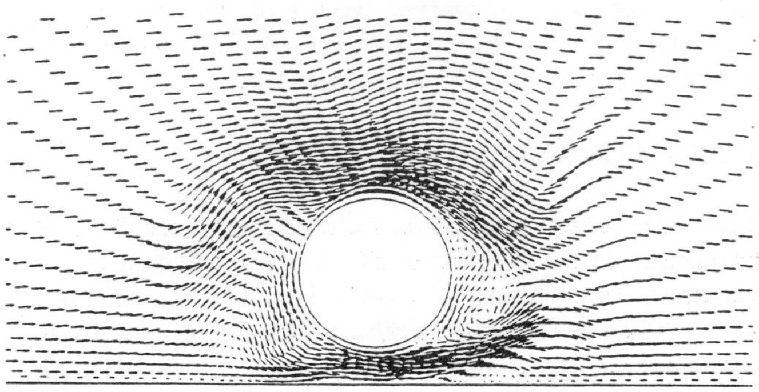

(d) 3-D , t/T=3.0 , z=2.25D

Fig. 6 Comparison of 3-D and 2-D velocity vector field in the vicinity of pipeline.

Proceedings of the Seventh (1997) International Offshore and Polar Engineering Conference
Honolulu, USA, May 25-30, 1997

ISBN 1-880653-28-1 (Set); ISBN 1-880653-30-3 (Vol. II)

The Norfra Pipeline: Challenging Shore Approach Handled by Coastal Dynamics Modelling

Hermann Moshagen, Bernt Thorbjørnsen and Per Strass
Statoil, Stavanger, Noway
Bernard Latteux
Laboratoire National d'Hydraulique, Chatou, Paris, France
Marco Venturi
Snamprogetti, Fano, Italy

ABSTRACT

The NorFra pipeline is a 42 inch dry gas pipeline connecting Norwegian gas fields in the central area of the North Sea to the gas supply system of France. The pipeline enters the French coast at Dunkerque. The nearshore area is characterised by an irregular seabed morphology dominated by a complex pattern of sand banks, strong tidal currents, direct exposure to waves coming in from the North Sea and a high rate of natural sediment transport. Due to these factors the last 20 kilometers of the NorFra offshore pipeline route was a major engineering and construction challenge.

The sand banks are crossed more or less perpendicularly by the NorFra pipeline and are typically 1 - 3 km wide and 15-20 m high. Tidal currents vary considerably within the area and may locally exceed 1.5 m/s. The predominant waves in the area come from the North Sea. Heavy waves may reach the area of the pipeline route also from the western part of the English Channel.

The present paper describes the wave, tide/current and sediment transport models set up as part of the NorFra shore approach engineering and construction planning work and demonstrates how the application of these models were essential in establishing the final design and installation solution for the NorFra pipeline through the nearshore area.

KEYWORDS: Pipeline, shore approach, installation, coastal dynamics, modelling, waves, tides, currents.

INTRODUCTION

The shore approach of the NorFra pipeline is planned at Dunkerque, France. The engineering, construction and environmental challenges met in crossing the nearshore area outside Dunkerque are described in ref. 1. This nearshore area extends over the last 20 km of the offshore pipeline route and is characterised by an irregular pattern of large sand banks typically 1 - 3 km wide and 15 - 20 m high.

Due to the very rough environmental climate in this area the wave and current loads were found to be very high. The resulting necessary submerged weight of the pipeline for horizontal stability during the construction phase was excessive and called for alternative solutions.

ENGINEERING APPROACH

The NorFra project selected the following approach for final engineering of the pipeline stability and installation in the nearshore area. Firstly a modified pipeline installation concept based on the existing pipeline stability design was to be engineered. In parallel a study was started to establish wave and tide/current simulation models for the actual nearshore area. The over all objective of the modelling work was to reanalyse the pipeline stability design based on qualified in detail wave and current data produced by the above hydrodynamic models.

MODELLING PHILOSOPHY

The merits of this type of dynamic coastal modelling are basically twofold:

- wave, tide and current data can be given in detail along the entire shore approach route including the effects of a complex seabed bathymetry

- significant and interesting parts of the technical solution can be made part of the model and the resulting hydrodynamic effects on the environment or on the technical installation itself may be studied directly in the model.

Thus the planned dredged channels for layvessel access through the shallowest sand banks along the route could easily be included in the simulation models and a later reanalysis of the stability of the pipeline - to be installed in the middle of the channel - would benefit from this.

Additionally the scope of the modelling work also included establishing of a sediment transport model working on the basis of the environmental data produced by the wave, tide and current models. The objective here was to simulate the side slope scouring and rate of sediment backfilling into the dredged access channels and thereby assist in planning of the construction work. Eventually some beneficial effect also on pipeline stability from a possible high initial sediment backfilling rate into the dredged channels could be envisaged

MODELLING TECHNIQUES

Wave modelling

The nearshore area that was modelled for the above purpose is shown in figure 1. The model was a 2D so called "third generation" spectral wave code named TOMAWAC developed at the Laboratoire National d'Hydraulique (LNH) at Chatou, France, ref. 8. The coastal area covered by the model was resolved numerically in 13 592 nodes corresponding to 26 861 area elements, see figure 2. The mesh size of the computational grid varied over the area according to the variations of the seabed bathymetry or the anticipated importance of local variations of the hydrodynamic parameters.

Boundary and wind conditions of the wave model were provided by the NESS hindcast database. Representative grid points of the database were selected for this purpose.The effect of wave braking was developed on the basis of the work by Battjes and Janssen, ref. 9, and implemented in the wave model.

Validation of the wave model was an important part of the work. The model was run for a typical summer storm selected from the NESS database. Supplementary wave measurements were also available for this storm event within the actual area. Figure 3 shows the comparison of calculated significant wave heights to the measured values at this location. The wave model was concluded to work satisfactorily for the need of the pipeline project.

Current modelling

The same nearshore area as for the wave model was covered by the current model. The model is a 2D - i.e. vertically integrated - model developed by LNH named TELEMAC 2D, ref. 10. The model solves the shallow water equations in two horizontal dimensions and is formulated numerically by a finite element scheme.

To run the model the water level was prescribed along the entire open boundary. Boundary data were derived from existing global data i.e. a larger model covering the southern part of the North Sea and the English Channel.

The validation of the current model was based on comparisons of computed and measured values of the current velocity for a situation dominated by tidal flow. The comparison was made at various locations within the model area. Figure 4 gives an impression of this comparison for a location in the vicinity of the pipeline route. The comparison shows a reasonably good over all agreement between calculated and measured current. However, a slight time shift and an approximately 20% underestimation of the maximum flow velocity can be seen. For the needs of the NorFra project a 20% increase of the calculated current values was recommended.

Sediment transport modelling

A sediment transport model was established on the basis of the wave and current models described above. The objective of the sediment transport model was to establish a computational tool to estimate the actual scouring of the dredged access channel side slopes as well as backfilling into the dredged access channels. Dredged channels are needed for shallow water laybarge access at three of the shallowest sand banks in the area.

The sediment transport model was developed on the basis of internationally recognized formulas and theories, thus the works of Bijker (1971), ref. 2, Fredsøe and Deigaard (1985), ref. 3, Lean (1980), ref. 4, Major-Mora et al. (1976), ref. 5, Van Rijn (1989), ref. 6 and Bonefille et al. (1969), ref. 7.

The sediment model was calibrated against sedimentation measurements for a prototype experimental trench which was located in the western part of Dunkerque harbour. The comparison with prototype data showed reasonably good agreement between measured and computed values.

However, the comparison also clearly demonstrated the importance of having high quality wave and current data for scouring and sedimentation calculations.

MODELLING RESULTS

Waves

Significant effects of the sand banks on the wave conditions along the route are clearly demonstrated in figure 5. Boundary and wind conditions were again deduced from the NESS data basis. The TOMAWAC model was then run for a summer storm situation of July 1970 and the results compared for the situations with and without access channels.

In order to see the effect of the access channel on the wave action in the channel itself the wave model was modified to include also the laybarge access channels. In order to represent properly the shape of the channels in the model the number of computational nodes was increased to 17 459 with a corresponding number of area elements of 34 581. The comparison of significant wave heights over the Snouw Bank channel cross section are shown in figure 6. Comparison of mean wave periods for the same section showed a similar change over the same cross section. It can be seen that a 10 - 12 % reduction of the significant wave height is found in the center of the channel while the reduction in mean wave period was found to be about 3%. Significant changes of the wave direction was also found.

Currents

A considerable variation in maximum current velocity along the shore approach route is demonstrated by figure 7. The results are given for a spring tide situation.

The current model was modified to study the effect of the access channels on the flow velocity in the channels themselves. Then the model was run for the same spring tide situation and the results compared. A cross section at the Snouw Bank access channel is again chosen for this comparison.

Figure 8 shows the result of the computations for a selected section of this channel. It can be seen that a flow velocity reduction of more than 20% is found in the center of the channel. Significant changes in flow direction was also found.

Application in design and construction planning

The wave and current models were finally run for a number of cases defined for the purpose of reanalysing pipeline stability along the nearshore route. This application of the above wave and current models gave as a result a significant improvement of the pipeline stability. And this again provided the basis for selecting a conventional pipelay concept for the nearshore route.

Finally the sedimentation model was applied for estimation of the rate of sediment backfill into the dredged access channels for actual working scenarios. Figure 9 shows a typical result for a selected cross section of the Snouw Bank for a period of severe summer season working conditions. Further details on the design and installation solutions are found in ref.1.

CONCLUSIONS

The NorFra pipeline project was challenged on pipeline stability and installation in the nearshore area outside Dunkerque due to the crossing of large sand banks and severe environmental conditions. However, significant improvements of the engineered solutions were obtained through in detail simulation of the wave and current environment in the area by use of dynamic coastal modelling.

The modelling approach may be used successfully in cases where the pipeline has to cross difficult areas nearshore. Some basic premises for application of the approach should, however, always be observed:

- establishing qualified models of this type requires sufficient and high quality seabed and environmental data in order to produce reliable results

- operation of this type of models requires highly specialised personnel in order to guarantee qualified model operation as well as interpretation of the results

- the work needed from start of establishing models through calibration and validation work until final results may be timeconsuming. The need for this type of modelling effort should therefore be clarified at an early stage of the pipeline project.

AKNOWLEDGEMENT

The authors are highly indebted to Statoil and partners of the NorFra project for admission to publish this paper.

REFERENCES

1. Thorbjørnsen B., Dale H., Eldøy S. and Mercanti M.: *"The N2 orFra Pipeline. Shore Approach Engineering, Environmental and Construction Challenges".* Proceedings ISOPE, Honolulu, 1997.

2. Bijker E.W.: *"Longshore Transport Computations"*. Journal of the Waterways, Habours and Coastal Engineering Division, ASCE, Vol. 97, no. WW4, 1971.

3. Fredsøe J. and Deigaard R.: *"The State of the Art of predicting natural Backfilling of Sand in Navigation Channels".* XIth World Dredging Conference, Brighton, UK, 1985.

4. Lean G. H.: *"Estimation of Maintenance Dredging for Navigation Channels".* H.R.S., Wallingford, England, 1980.

5. Major-Mora R., Mortensen P. and Fredsøe J.: *"Sedimentation Studies on the Niger River Delta".* Proceedings Coastal Engineering Conference, Honolulu, Hawai, 1976.

6. Van Rijn L.: *"Handbook. Sediment Transport by Currents and Waves".* Delft Hydraulics, Report no. H 461, 1989.

7. Bonefille R., Lepetit J.P. and Leroy J.: *"Avan-Port Ouest de Dunkerque".* Report no. 6, EDF LNH T 684 DHM, 1969.

8. Benoit M., Marcos F. and Becq F.: *"Development of a third generation shallow-water wave model with unstructured spacial meshing".* Proc. ICCE, Orlando, 1996.

9. Battjes J.A. and Janssen J.P.F.M.: *"Energy loss and set-up due to breaking of random waves"*, Proc. ICCE, Hamburg, 1978.

10. Hervouet J.M., Hubert J.L. and Janin J.M.: *"The computation of free surface flows with TELEMAC. An example of evolution towards hydroinformatics".* Journal IAHR, special issue, no. 32, 1994.

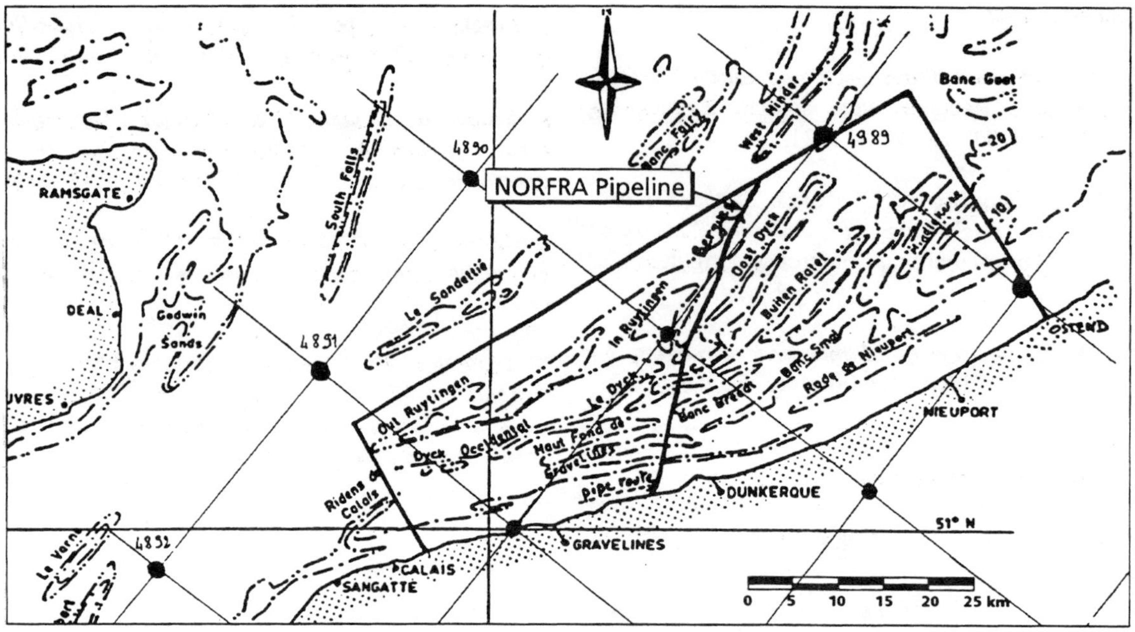

Figure 1. Nearshore area covered by model

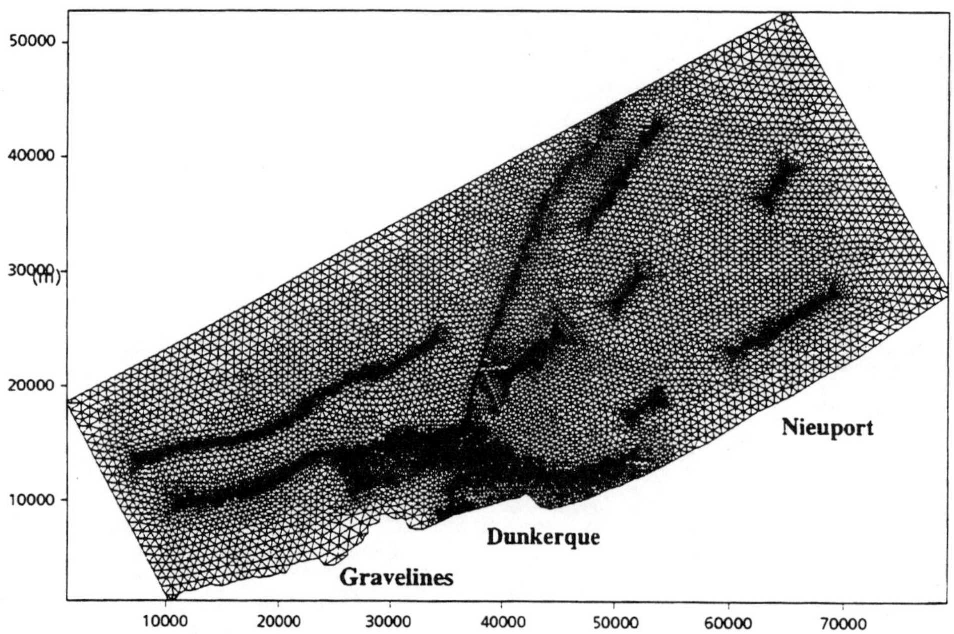

Figure 2. Discrete representation of model area

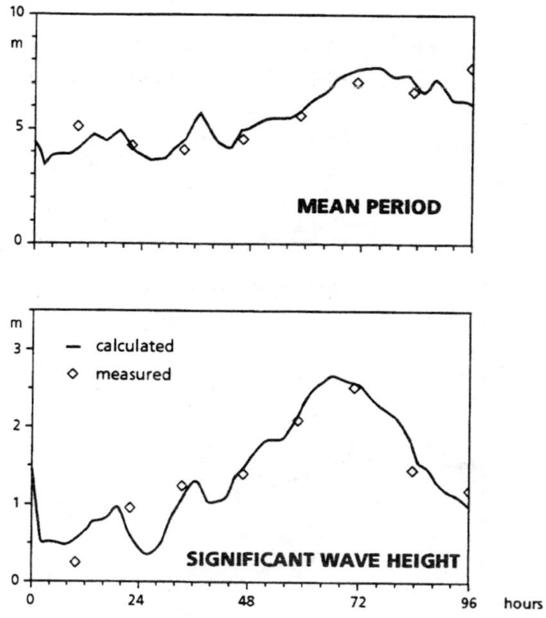

Figure 3. Wave model validation

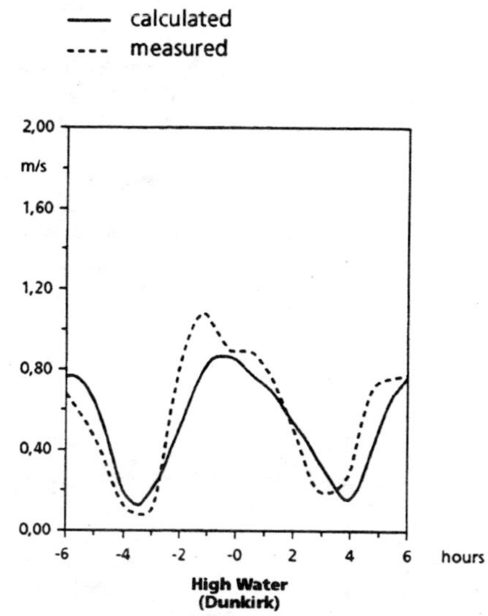

Figure 4. Current model validation

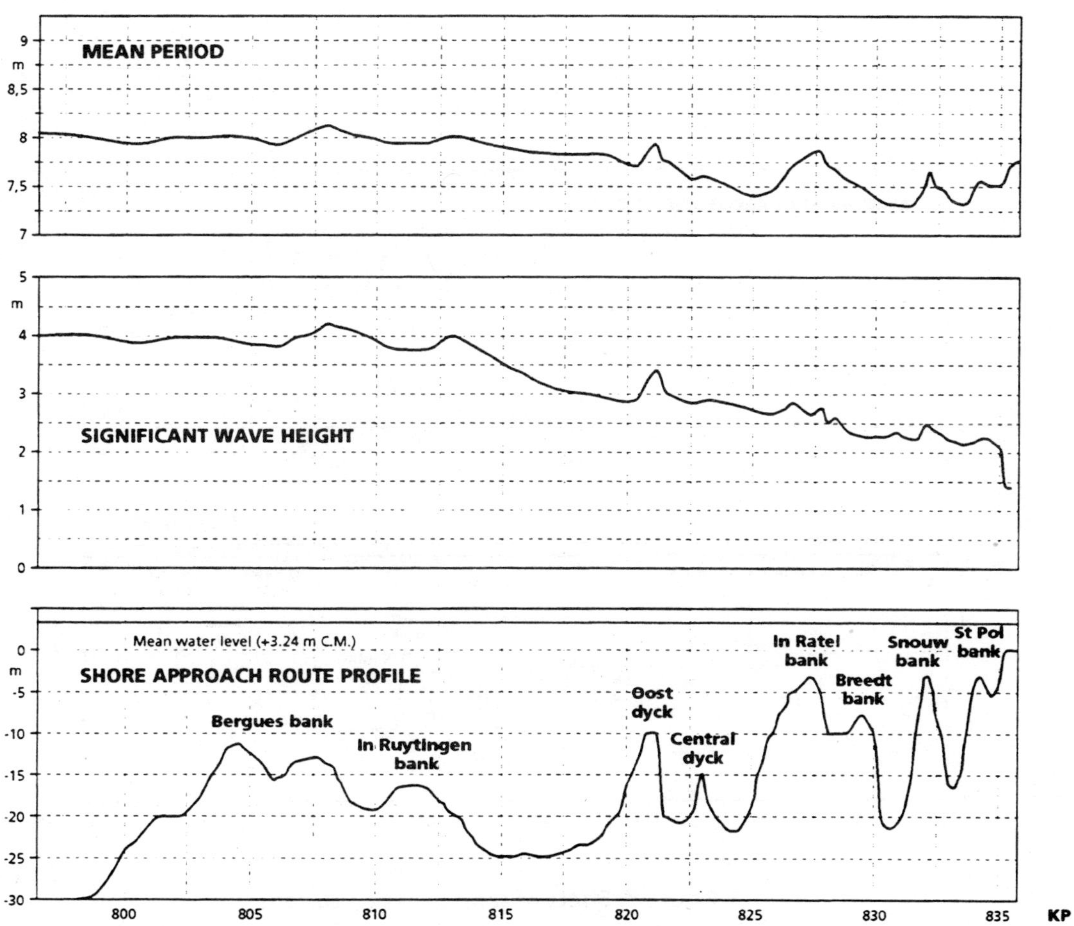

Figure 5. Effect of sand banks on waves

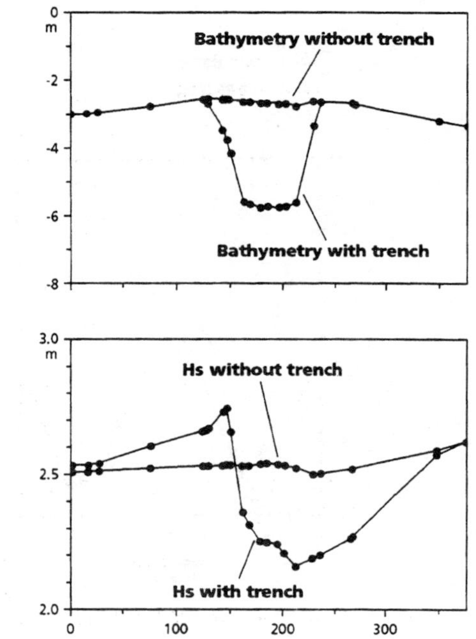

Figure 6. Effect of access channel on wave height

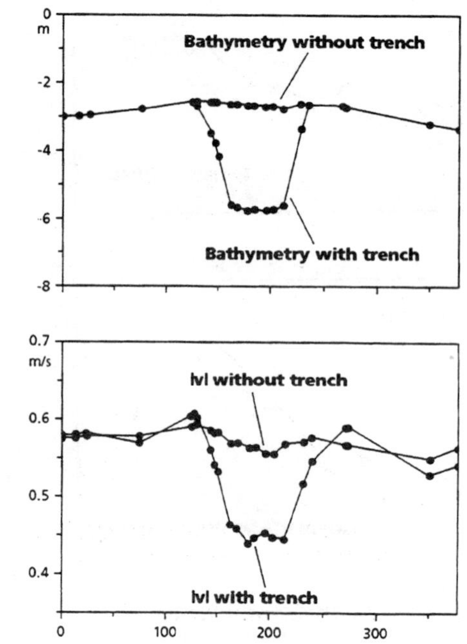

Figure 8. Effect of access channel on current

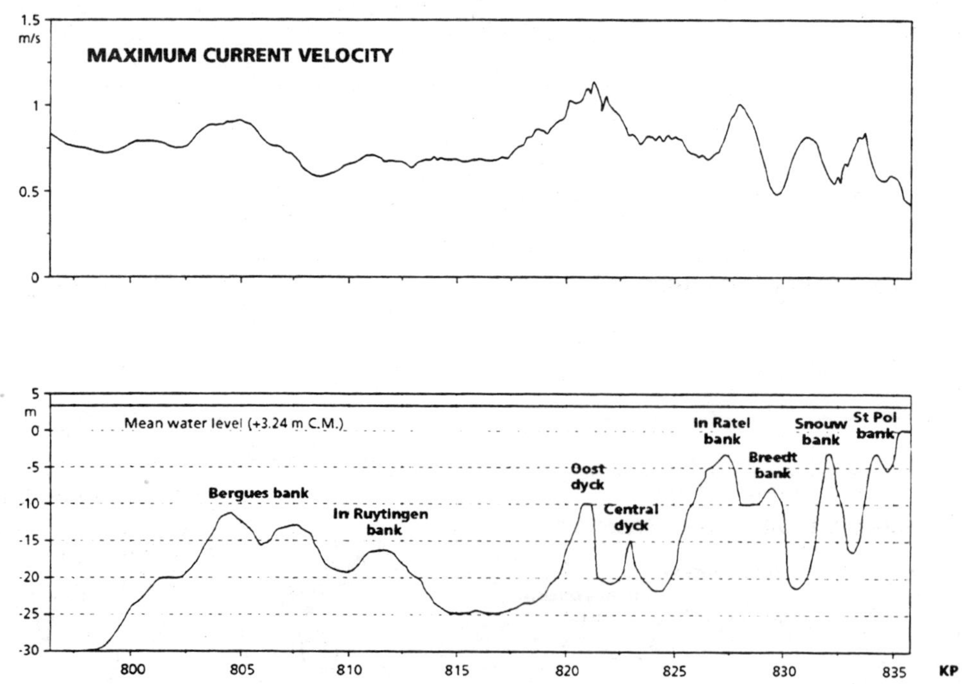

Figure 7. Effect of sand banks on current

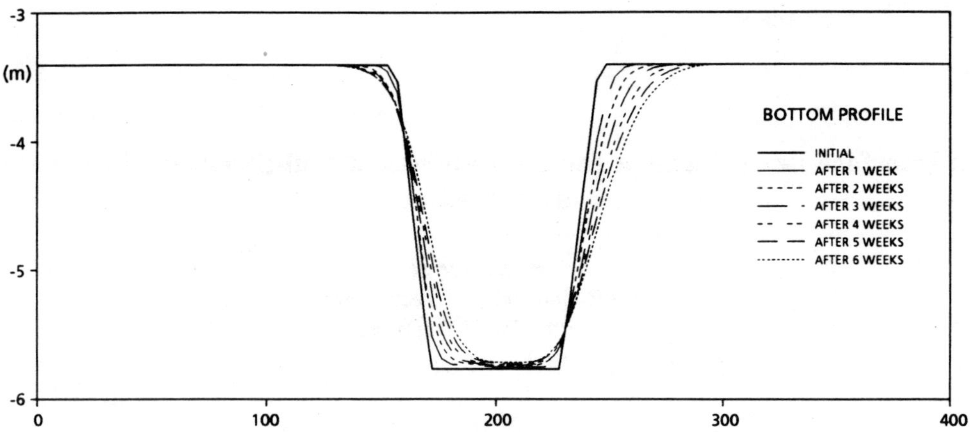

Figure 9. Time evolution of access channel cross section

Proceedings of the Seventh (1997) International Offshore and Polar Engineering Conference
Honolulu, USA, May 25-30, 1997

ISBN 1-880653-28-1 (Set); ISBN 1-880653-30-3 (Vol. II)

An Interface Beam Element for the Analyses of Soil-Structure Interactions and Pipelines

H. Karadeniz
Delft University of Technology
Delft, The Netherlands

ABSTRACT

In this paper, an inerface 3D beam element is introduced for the analysis of framed structures which interact with an elastic medium. The formulations of the element are based on the assumption that the elastic medium can be represented by a two parameters model of the Winkler and the shear type (Pasternak model). It is assumed, in general, that the properties of the elastic medium can be defined in any coordinate directions which may be different from the member coordinate directions. In the element formulation, the effects of shear deformations and a constant axial force of the beam are taken into account. The stiffness matrices of the beam and the elastic medium are explicitly derived and presented. Numerical examples are provided to demonstrate the element introduced.

INTRODUCTION

The finite element formulation of beams, in general, has been a subject for a long time. In the literature, numerous works about beam elements can be found for various analysis purposes. For rigidly connected members, formulations of a three dimensional (3D) beam element were presented by Tezcan (1966) and Oran (1973) including the derivation of the tangent stiffness matrix. In these works, the Euler-Bernouilli beam theory was used. However, in some works, see e.g. Karadeniz (1994) and Suares et al. (1995), joint flexibilities have also been considered in the beam element formulation. Most of beam elements which can be found in the literature are mainly for the analysis of super-structures. In practice, the effects of underlying, or surrounding, soil medium are represented by massless spring systems at the member ends as presented in Bowles (1974). Formulation of a 2D beam which may be finite or infinite resting on an elastic foundation has been well established so far, see e.g. Miranda and Nair (1966), Selvadurai (1979) and Ting (1982). By using the exact lateral displacement which is obtained from the solution of the differential equation, the stiffness matrix and consistent force vectors of a 2D beam on a Winkler type foundation are presented by Ting and Mockry (1984). In their work, effects of the shear deformation is not considered whereas they are taken into account in Aydogan (1995) in which an exact displacement field is used. A finite element formulation of a 2D Timoshenko beam resting on a two parameters (Winkler & shear types) foundation has been presented by Yokoyama (1991). However, all these works have a limited number of applications and they may not correctly represent the soil-structure interface characteristics. For an accurate analysis of the soil-structure interaction phenomenon a continuum finite, or boundary, element modelling is required as it is described in Boulon et al. (1995), Faruque (1986) and Selvadurai (1995). In reality, since stress-strain relations of the soil are not linear a non-linear analysis procedure, see e.g. Crisfield (1991), must be used to obtain correct results. However, for small deformations, linear uncoupled constitutive relations for the soil may reasonably be assumed. For space framed structures, formulation of an interface element is the main objective of this paper since it can find a large number of applications in practice. Before such an element is introduced, some preliminary formulations of a space beam subjected to external loadings and an axial force are presented first.

DIFFERENTIAL EQUATIONS OF A 3D BEAM

In this section, differential equations of internal forces and displacements of a 3D beam element are presented. It is assumed that the beam is subjected to distributed external loadings, q_y and q_z, in the local y and z coordinate directions. After the deformation state of an infinitesimal member, the internal force components of the beam are as defined in Fig.1. The displacement components in the local y and z coordinate directions (x coordinate direction is defined along the member axis) are respectively u_y and u_z. For a Timoshenko beam, the derivatives of these displacements with respect to x can be stated as,

$$\frac{du_y}{dx} = \gamma_y + \theta_z \qquad \frac{du_z}{dx} = \gamma_z - \theta_y \tag{1}$$

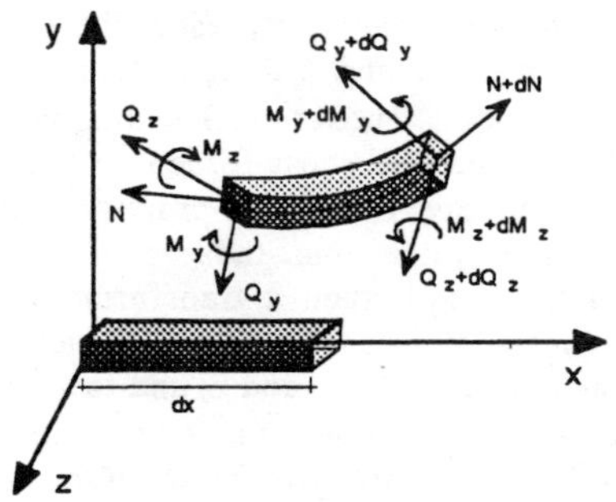

Fig. 1. Force components of an infinitesimal member after the deformation state.

where θ_y and θ_z are the rotations about the local y and z axes respectively, γ_y and γ_z are the shear deformations defined by,

$$\gamma_y = \frac{Q_y}{GA_y} \qquad \gamma_z = \frac{Q_z}{GA_z} \tag{2}$$

in which A_y and A_z are the effective shear cross-sectional areas in the local y and z directions respectively, and G denotes the Young modulus. For small rotations, and having neglected second order and higher terms, the differential equations of the shear forces and the bending moments can be obtained as written by,

$$\frac{dQ_y}{dx} = -q_y - N\frac{d\theta_z}{dx} \qquad \frac{dM_z}{dx} = -Q_y \tag{3}$$

$$\frac{dQ_z}{dx} = -q_z + N\frac{d\theta_y}{dx} \qquad \frac{dM_y}{dx} = Q_z \tag{4}$$

On the other hand, the bending moments can be written in terms of curvatures as,

$$M_y = EI_y\frac{d\theta_y}{dx} \qquad M_z = EI_z\frac{d\theta_z}{dx} \tag{5}$$

in which E denotes the elasticity modulus, I_y and I_z are respectively inertia moments about the local y and z axes. Throughout Eqs.(1-5), differential equations of the displacement components can be obtained as,

$$EI_z\frac{d^4u_y}{dx^4} = q_y - \frac{EI_z}{GA_y}\frac{d^2q_y}{dx^2} + N\frac{d^2u_y}{dx^2} \qquad EI_y\frac{d^4u_z}{dx^4} = q_z - \frac{EI_y}{GA_z}\frac{d^2q_z}{dx^2} + N\frac{d^2u_z}{dx^2} \tag{6}$$

When this equation is compared with the classical bending theory of a space beam, the right-hand sides can be considered as pseudo-distributed loading terms (updated distributed loads) which will be indicated by q_y^* and q_z^* respectively so that Eq.(6) can be rewritten as,

$$EI_z\frac{d^4u_y}{dx^4} = q_y^* \qquad EI_y\frac{d^4u_z}{dx^4} = q_z^* \tag{7}$$

where the updated loading terms are,

$$q_y^* = q_y - \frac{EI_z}{GA_y}\frac{d^2q_y}{dx^2} + N\frac{d^2u_y}{dx^2} \qquad q_z^* = q_z - \frac{EI_y}{GA_z}\frac{d^2q_z}{dx^2} + N\frac{d^2u_z}{dx^2} \tag{8}$$

Eqs.(7) and (8) will be used for the finite element formulation of a space beam resting on or embedded in an elastic medium in general.

The shape functions of a space beam element in lateral directions are obtained from the solution of the differential equations of the displacements, u_y and u_z, under a zero loading condition whereas linear shape functions are used for the axial and torsional deformation states. The deformation vector at a point between the member ends, {u}, is defined by,

$$\{u\} = [N]\{d\} \tag{9}$$

where {d} is the deformation vector at the member ends. The vectors, {u} and {d}, are defined by,

$$\{u\}^T = [u_x, u_y, u_z, \theta_x, \theta_y, \theta_z] \tag{10}$$

$$\{d\} = \begin{bmatrix} \{u\}_1 \\ \{u\}_2 \end{bmatrix} \tag{11}$$

in which $\{u\}_1$ and $\{u\}_2$ are the values of {u} at the member ends (1) and (2) respectively. θ_x, θ_y and θ_z are respectively rotations about x, y and z coordinates. The shape function matrix, [N], in Eq.(9) is presented in the Appendix.

MODELLING OF SOIL MEDIUM AND CALCULATION OF INTERFACE LOADING

The soil is a complex medium that its properties show large discrepancies from location to location and also in time. For a soil-structure interaction, the soil medium can be idealised in different ways as to be elastic, or elastic-plastic and time dependent, models among which Winkler-, elastic continuum- and two parameter elastic models, see e.g. Selvadurai (1979), may be largely used in practice. The Winkler model assumes that the soil reaction is related to the surface deflection only whereas the two-parameter soil model takes into account not only the surface deflection but also the shear deformation of the soil. In this paper, the two-parameter soil model is adopted since it represents the actual soil behavior and the corresponding reaction. The two-parameter soil model can be considered in different ways, see Selvadurai (1979), among which the Pasternak model is used here. This model assumes that the soil is represented by two independent models, one is for purely shear deformation (Shear layer) and the other one is for the deflection (Winkler model) as it is shown in Fig.2 in two dimensions on the x-z plane. The shear force Q_{sz} due to the shear deformation of the soil is defined by,

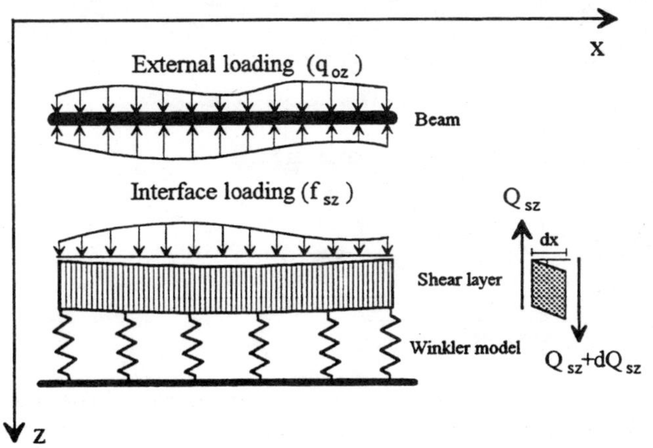

Fig. 2. Modelling of beam-soil interactions.

$$Q_{sz} = G_{sz}\frac{du_z}{dx} \tag{12}$$

where G_{sz} denotes an elastic shear constant of the soil in the z direction. From the equilibrium of an infinitesimal soil element, the interface loading, f_{sz}, can be obtained as written by,

$$f_{sz} = k_{sz}\,u_z - G_{sz}\frac{d^2u_z}{dx^2} \tag{13}$$

in which the first term on the right hand side is due to the Winkler model with a subgrade modulus of k_{sz}. This simple model of the interface loading in the z coordinate direction given by Eq.(13) can be extended in other (x and y) coordinate directions including the rotational deformations as stated in the vectorial form by,

$$\{f_s\}' = [k_s]'\ \{u\} - [G_s]'\ \frac{d^2\{u\}}{dx^2} \tag{14}$$

where the vector $\{u\}$ is as defined by Eq.(10), $[k_s]'$ and $[G_s]'$ are diagonal matrices which contain respectively subgrade and shear moduli of the soil. These matrices are defined as,

$$[k_s]' = \begin{bmatrix} k_{sx} & 0 & 0 & 0 & 0 & 0 \\ 0 & k_{sy} & 0 & 0 & 0 & 0 \\ 0 & 0 & k_{sz} & 0 & 0 & 0 \\ 0 & 0 & 0 & r_{sx} & 0 & 0 \\ 0 & 0 & 0 & 0 & r_{sy} & 0 \\ 0 & 0 & 0 & 0 & 0 & r_{sz} \end{bmatrix}, \quad [G_s]' = \begin{bmatrix} G_{sx} & 0 & 0 & 0 & 0 & 0 \\ 0 & G_{sy} & 0 & 0 & 0 & 0 \\ 0 & 0 & G_{sz} & 0 & 0 & 0 \\ 0 & 0 & 0 & 0 & 0 & 0 \\ 0 & 0 & 0 & 0 & 0 & 0 \\ 0 & 0 & 0 & 0 & 0 & 0 \end{bmatrix} \tag{15}$$

The subgrade modulus of the soil for the axial direction, k_{sx}, can be considered to represent an axial skin friction, see e.g. Chang et al. (1995), and the end bearing resistance of the member if the member end is free in the soil such as a pile end. The subgrade moduli, k_{sy} and k_{sz}, are due to deflections (u_y and u_z) in the y and z directions whereas r_{sy} and r_{sz} represent the subgrade modulli due to rotations in the y and z directions and the modulus r_{sx} represents the torsional skin friction. The diagonal terms of $[G_s]'$, G_{sx}, G_{sy} and G_{sz}, are the soil shear-force-constants in the x, y and z directions respectively. If the soil properties are defined in a different coordinate system than those of member as shown in Fig. 3 for a layered soil, then a transformation is essentially required. In general, it is assumed that a member has a local coordinate system given by x, y and z, and the soil properties are defined in a different coordinate system given by x', y' and z' as shown in Fig.4. The transformation between these two coordinate systems is written by,

$$\{x\}' = [T]'\ \{x\} \tag{16}$$

where $\{x\}$ and $\{x\}'$ denote respectively the (x, y, z) and (x', y', z') coordinates and $[T]'$ is the tranformation matrix. Having transformed the interface loading vector, $\{fs\}'$ given by Eq.(14), to the member coordinate directions (x, y, z) it can be stated that,

$$\{f_s\} = [k_s]\ \{u\} - [G_s]\ \frac{d^2\{u\}}{dx^2} \tag{17}$$

in which $[k_s]$ and $[G_s]$ are determined from,

$$[k_s] = [T_s]^T[k_s]'\ [T_s]$$
$$[G_s] = [T_s]^T[G_s]'\ [T_s] \tag{18}$$

where the transformation matrix $[T_s]$ is defined by,

$$[T_s] = \begin{bmatrix} [T]' & 0 \\ 0 & [T]' \end{bmatrix} \tag{19}$$

Eq.(17) is the statement of the beam-soil interface loading which will be used in this paper. As it can be realised from Eq.(18), the matrices of the subgrade moduli and the shear constants, $[k_s]$ and $[G_s]$, are not any more diagonal which means that interface loading components are coupled.

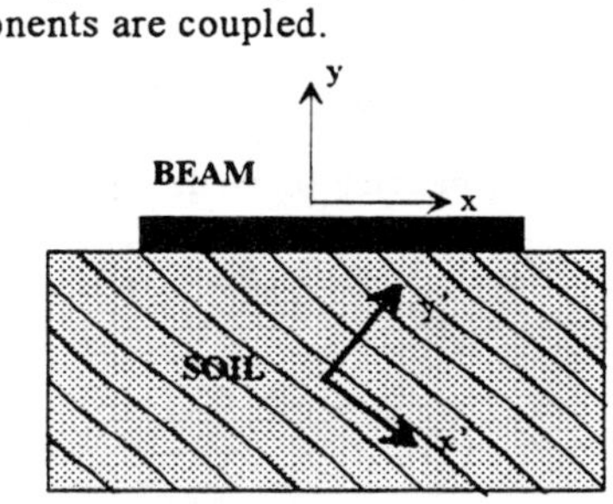

Fig. 3. A beam on a layered soil medium.

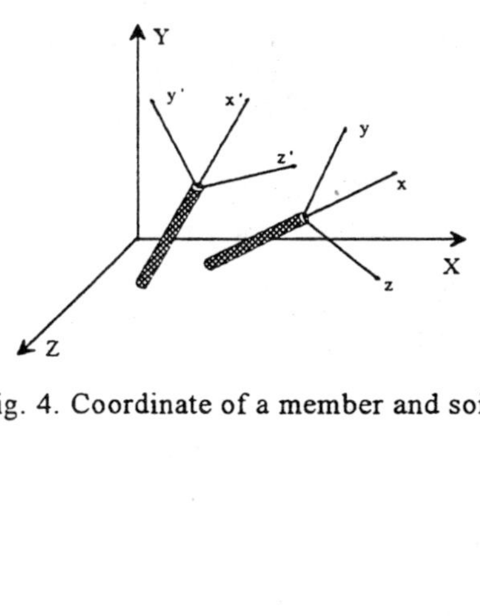

Fig. 4. Coordinate of a member and soil.

DERIVATION OF AN INTERFACE ELEMENT FOR THE BEAM-SOIL INTERACTION

Having presented the preliminary information, an interface element can be formulated easily by using the principle of the stationary total potential energy. It is assumed here that the classical beam theory is used with the pseudo-distributed loading terms. In general, these loading terms can be stated vectorially as written by,

$$\{q\}^* = \{q\} - [L_q]\frac{d^2\{q\}}{dx^2} + [L_n]\frac{d^2\{u\}}{dx^2} \tag{20}$$

where $\{q\}$ is the applied load vector and the matrices $[L_q]$ and $[L_n]$ are as defined by,

$$[L_q] = \begin{bmatrix} 0 & 0 & 0 & 0 & 0 & 0 \\ 0 & \alpha_y & 0 & 0 & 0 & 0 \\ 0 & 0 & \alpha_z & 0 & 0 & 0 \\ 0 & 0 & 0 & 0 & 0 & 0 \\ 0 & 0 & 0 & 0 & 0 & 0 \\ 0 & 0 & 0 & 0 & 0 & 0 \end{bmatrix} \quad \text{and} \quad [L_n] = \begin{bmatrix} 0 & 0 & 0 & 0 & 0 & 0 \\ 0 & N & 0 & 0 & 0 & 0 \\ 0 & 0 & N & 0 & 0 & 0 \\ 0 & 0 & 0 & 0 & 0 & 0 \\ 0 & 0 & 0 & 0 & 0 & 0 \\ 0 & 0 & 0 & 0 & 0 & 0 \end{bmatrix} \tag{21}$$

where $\alpha_y = \frac{EI_z}{GA_y}$ and $\alpha_z = \frac{EI_y}{GA_z}$. In order to formulate an interface element the effective applied load vector is expressed as, see Fig.2,

$$\{q\} = \{q_o\} - \{f_s\} \tag{22}$$

where $\{q_o\}$ denotes the external applied load vector. Having introduced Eqs.(16) and (22) into Eq.(20) and neglected comparatively small quantities, the pseudo-load vector $\{q\}^*$ can be stated as written by,

$$\{q\}^* = \{q_o\} - [k_s]\{u\} + [L]\frac{d^2\{u\}}{dx^2} \tag{23}$$

where the matrix $[L]$ is defined by,

$$[L] = [G_s] + [L_q][k_s] + [L_n] \tag{24}$$

Next, the member internal force vector, $\{F\}$, and the member strain vector, $\{\delta\}$, are defined as written by,

$$\{F\}^T = [\,Q_x, Q_y, Q_z, M_x, M_y, M_z\,]$$
$$\{\delta\}^T = [\,\frac{du_x}{dx}, \gamma_y, \gamma_z, \frac{d\theta_x}{dx}, \frac{d\theta_y}{dx}, \frac{d\theta_z}{dx}\,] \tag{25}$$

in which Q_x is the axial force, M_x is the torsional moment, γ_y and γ_z are as defined by Eq.(2). The constitutive relation between these vectors is stated by,

$$\{F\} = [E]\{\delta\} \tag{26}$$

where the matrix $[E]$ is defined to be,

$$[E] = \begin{bmatrix} EA & 0 & 0 & 0 & 0 & 0 \\ 0 & GA_y & 0 & 0 & 0 & 0 \\ 0 & 0 & GA_z & 0 & 0 & 0 \\ 0 & 0 & 0 & GI_x & 0 & 0 \\ 0 & 0 & 0 & 0 & EI_y & 0 \\ 0 & 0 & 0 & 0 & 0 & EI_z \end{bmatrix} \tag{27}$$

in which A denotes the cross-sectional area of the element and GI_x denotes the torsional rigidity of the beam. By using Eqs.(1) and (9) the member strain vector $\{\delta\}$ can be stated in terms of the member-end deformation vector $\{d\}$ as written by,

$$\{\delta\} = [B]\{d\} \tag{28}$$

where the matrix [B] can be obtained in terms of the shape functions and their derivatives. The strain energy of the element can now be stated as written by,

$$U = \frac{1}{2}\int \{\delta\}^T[E]\{\delta\}\,dx \tag{29}$$

The work done by external applied loads (here, the pseudo-loads) can be stated as written by,

$$W_q = \int \{u\}^T\{q_o\}\,dx - \frac{1}{2}\int \{u\}^T[k_s]\{u\}\,dx - \frac{1}{2}\int \frac{d\{u\}^T}{dx}[L]\frac{d\{u\}}{dx}\,dx \tag{30}$$

Having minimised the total potential energy of the element, $\Pi = U - W_q$, with respect to the member-end deformations, the following equilibrium equation will be obtained :

$$([K_b] + [K_w] + [K_s] + [K_n])\,\{d\} = \{P\} \tag{31}$$

In this equation, $[K_b]$ denotes the linear member stiffness matrix and $\{P\}$ denotes the consistent force vector due to applied external loads. This stiffness matrix is calculated from,

$$[K_b] = \int [B]^T[E][B]\,dx \tag{32}$$

and the consistent force vector is calculated from,

$$\{P\} = \int [N]^T\{q_o\}\,dx \tag{33}$$

The matrix $[K_n]$ in Eq.(31) denotes the contribution of the axial force (N), which is assumed to be constant, to the total stiffness matrix. This contribution will be calculated from,

$$[K_n] = \frac{1}{2}\int \frac{d[N]^T}{dx}[L_n]\frac{d[N]}{dx}\,dx \tag{34}$$

The matrices, $[K_w]$ and $[K_s]$, in Eq.(31) denote respectively contributions of the Winkler- and the shear type soil-models to the total stiffness matrix. The contribution of the Winkler type soil-model will be calculated from,

$$[K_w] = \int [N]^T[k_s][N]\,dx + \frac{1}{2}\int \frac{d[N]^T}{dx}([L_q][k_s] + [k_s]^T[L_q]^T)\frac{d[N]}{dx}\,dx \tag{35}$$

in which the second term on the right hand side is due to shear deformations of the beam. Finally, the contribution of the shear type soil-model will be calculated from,

$$[K_s] = \int \frac{d[N]^T}{dx}[G_s]\frac{d[N]}{dx}\,dx \tag{36}$$

The components of the total member stiffness matrix ($[K_b]$, $[K_n]$, $[K_w]$ and $[K_s]$) can be explicitly obtained from the integrations given above by using the shape functions provided in the Appendix.

NUMERICAL EXAMPLES

The first example is a plane frame taken from Aydogan (1995) who used a 2D interface beam element based on the exact solution of the differential equation. This example is used to validate the formulation of an interface element presented herein. The topology, dimensions and the loading are as shown in Fig.5a. In order to compare the results with those in Aydogan (1995), axial deformations of members are ignored in the analysis. It is also assumed that the frame is fixed in the horizontal direction at the bottom to prevent a possible displacement, and it is free for vertical deformations. Only a Winkler type foundation with a subgrade constant, $k_s = 4.5x10^8$ N/m^2 , is assumed. The elasticity modulus of the frame material is assumed to be, E = 25x10^9 N/m^2 and the member properties are, A_y=0.0500 m^2 and I=13.3333E-4 m^4 for the 2nd. story columns (A_y is the shear cross-section and I is the inertia moment), A_y=0.0625 m^2 and I=26.0417E-4 m^4 for the 1st. story columns, A_y=0.1080 m^2 and I=67.872E-4 m^4 for the story beams and A_y=0.7125 m^2 and I=2865.822E-4 m^4 for the bottom beams. Two cases of the analysis are carried out as,

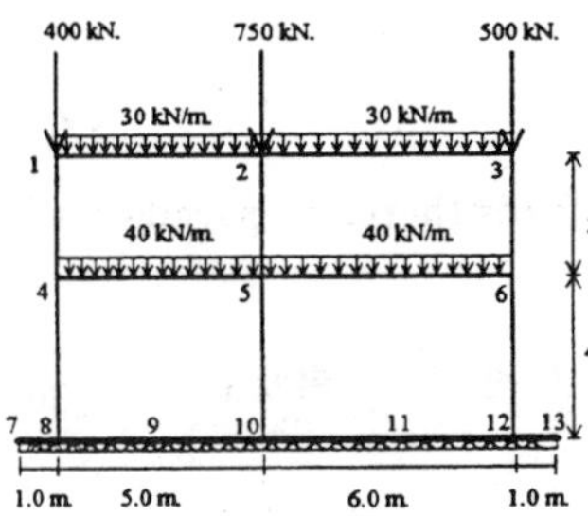

Fig. 5a. A plane frame on the ground.

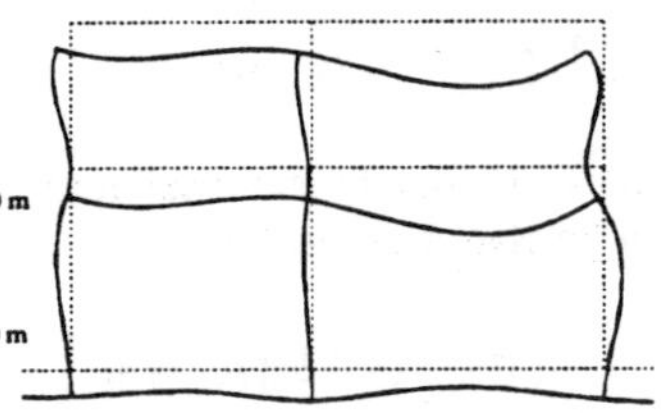

Fig. 5b. The deformation of the frame.

a) Shear deformation effects of beam are not considered, and
b) Shear deformation effects of beam are considered.

The deformation pattern of the frame is as shown in Fig.5b. In Aydogan (1995), numerical results for the deflections and the bending moments are available at the joint number 10 of the bottom beams. These results which are written in (.) are compared with those obtained herein. The deflections are 0.5164mm. (0.5125) with the shear effects and 0.4601mm. (0.4470) without the shear effects. The maxima of absolute moments are 540.07kN.m (578.6) with the shear effects and 618.04kN.m (606.3) without the shear effects, from which the influence of the shear deformation of the beam is clearly seen.

The second example is an eigenvalue calculation of an example jacket structure supported on piles as shown in Fig.6. The topological data are as given in Fig.6a. The underlying soil is assumed to have four layers which are equally spaced, and the subgrade modulli of each layer are the same in all directions with the values of k_s=0.5x10^7 N/m^2 for the first layer, k_s=1.0x10^7 N/m^2 for the second and third layers and k_s=2.0x10^7 N/m^2 for the fourth layer, see also Fig.6a. The member dimensions are, D=2.0m. and h=50mm. for the deck legs (D is diameter and h is thickness), D=1.2m. and h=16mm. for the jacket legs and vertical diagonals, D=0.8m. and h=8mm. for the

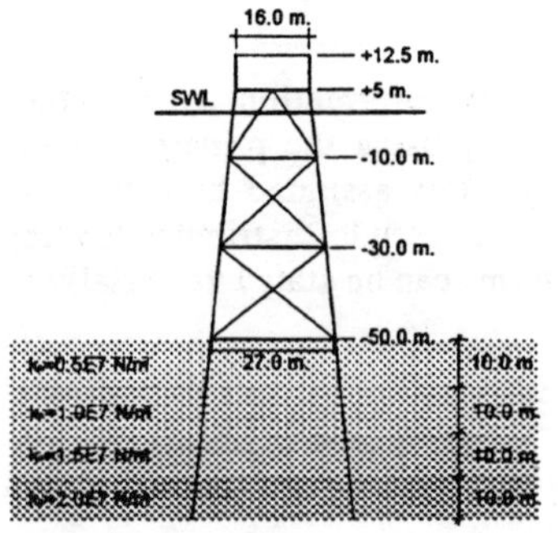

a) A side view, topological and soil data.

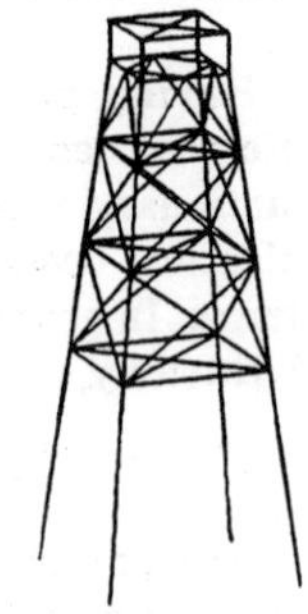

b) A 3D view

Fig. 6. An example jacket structure, topological and soil data, and a 3D view.

bracings and diagonals at the (+5m.) level, D=1.2m. and h=14mm. for the bracings and diagonals at the (-10m.) level and for the bracings at the (-30m.) level, D=1.2m. and h=16mm. for the diagonals at the (-30m.) level, and D=1.14m. and h=50mm. for the piles. It is assumed that the mass of the deck is M=3000ton, the legs are full of water and added masses of surrounding water are also taken into account. The four lowest natural frequencies and modes shapes of the structure are calculated by using the soil-structure interface elements for the piles. The first and second mode shapes are obtained to be flexural in orthogonal directions with the same natural frequency of $\omega_1 = \omega_2 = 1.631$ rad/sec. as shown in Fig.7. The third natural mode shape is torsional with the frequency of $\omega_3 = 3.475$ rad/sec., and the fourth natural mode shape is extensional in the vertical direction with the frequency of $\omega_4 = 7.025$ rad/sec. The effect of soil-structure interaction on the first natural mode shape is clearly seen from Fig.7. These eigenvalue results are approximately equal to those obtained from a structural model having support conditions of the piles at 10.0m. below the sea bottom with zero horizontal displacements and rotations in all directions, and an elastic support condition for vertical displacements with k_z=4.3E8 N/m. The natural frequencies obtained from this model are

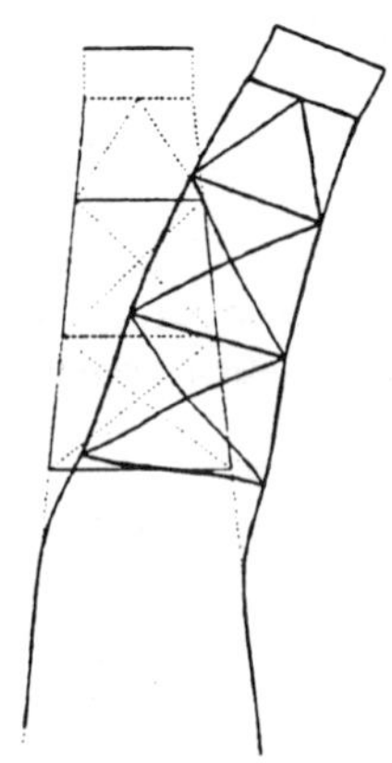

a) A side view

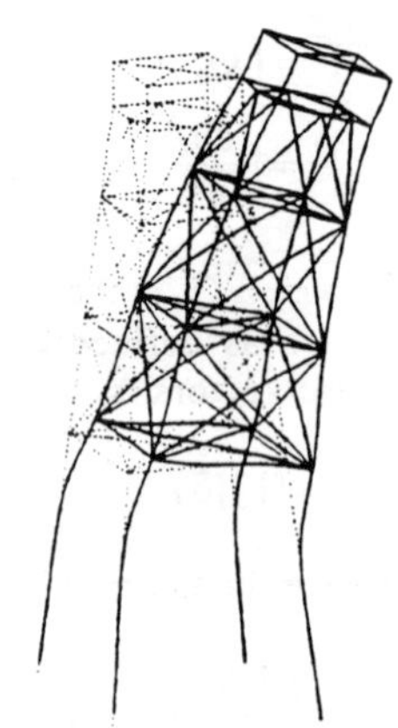

b) A 3D view

Fig.7. The first natural mode shape of the example jacket structure

$\omega_1=\omega_2=1.633$ rad/sec., $\omega_3=3.639$ rad/sec. and $\omega_4=6.988$ rad/sec. which are close to those of the soil-structure interaction model.

The third example is a fictitious pipeline passing over two gaps (e.g. canals) and an obstacle (e.g. a dike). This example is included only for demonstration purposes of a pipeline application to calculate internal member forces under the dead loads. The pipeline topology and the dimensions are as shown in Fig.8.The pipe is assumed to be made of steel and full of water.

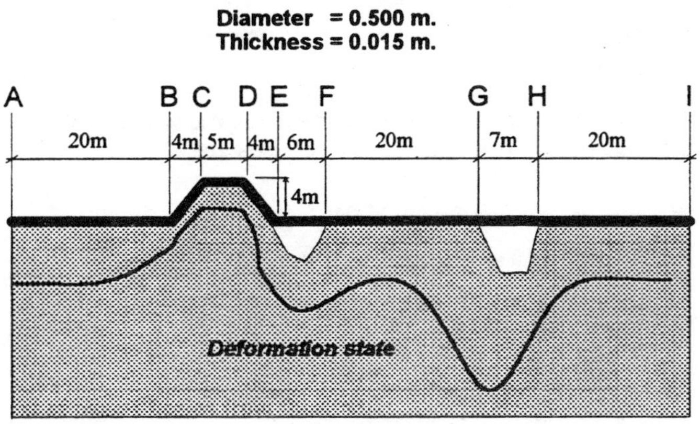

Fig.8. A fictitious pipeline and the static deformation state.

The calculated dead load which is uniformly distributed along the pipeline for all sections is equal to q=3.45 kN/m. It is also assumed that the pipeline is subjected to an axial force at the end (I) with the value of Fx = 5.0 kN. A zero rotation state and a settlement of 2.2 mm. at the both ends (A and I) are assumed. The horizontal displacement at the end A is constrained to be zero. The underlying soil is modelled by two parameters (Winkler and shear types) with the values of $k_{sx}= 0.15x10^6$ N/m^2 (axial skin friction), $k_{sy}= 1.5x10^6$ N/m^2 (subgrade modulus) and $G_{sy}= 0.5x10^6$ N/m (soil shear constant) for the regions of (A-B), (F-G) and (H-I) while for the obstacle (B-E) these values are assumed to $k_{sx}= 0.45x10^6$ N/m^2, $k_{sy}= 4.5x10^6$ N/m^2 and $G_{sy}= 1.5x10^6$ N/m. Under these conditions the pipeline is analyzed by the FEM with 4 equally spaced elements in the regions of (A-B), (F-G) and (H-I) and 2 equally spaced elements in the free spans, (E-F) and (G-H) while the obstacle region is represented by 3 elements. In the analysis, shear deformations of the pipe are taken into account. The deformation pattern of the pipeline is as shown in Fig.8 while the values at some critical points are presented in Table 1. The corresponding member internal forces are also presented in Table 2. The maximum deflection and bending moment are obtained at the center of the free span, (G-H), as they are expected.

Table 1. Displacements of the example pipeline

Location	Displacements (mm.)		Rotation (radian)
	Horizontal	Vertical	
A	0.00000	-2.2000	0.0
B	0.03737	-0.9683	$1.348x10^{-2}$
C	0.04386	-0.9581	$-2.865x10^{-3}$
D	0.05326	-1.0270	$-2.912x10^{-3}$
E	-0.6536	-1.7520	$-4.140x10^{-2}$
(E-F) center	-0.6482	-3.0060	$-3.050x10^{-2}$
F	-0.6428	-3.3540	$5.744x10^{-3}$
G	-0.6107	-5.2780	$-5.507x10^{-2}$
(G-H) center	-0.6058	-6.4200	$-1.995x10^{-4}$
H	-0.6009	-5.2920	$5.459x10^{-2}$
I	-0.5764	-2.2000	0.0

Table 2. Member internal forces of the example pipeline.

Member	Member End	Normal (kN)	Shear (kN)	Moment (kNxm
(A-B)	A	8.9501	0.8023	2.0617
	B	9.0061	6.9620	-10.6077
(B-C)	B	11.2912	1.4453	10.6077
	C	8.7096	-3.9770	4.4524
(C-D)	C	8.9707	-3.3464	-4.4524
	D	9.0800	-1.7334	-0.2478
(D-E)	D	7.6462	-5.1948	0.2478
	E	16.2413	-4.0188	-11.1778
(E-F)	F	8.6425	6.3739	12.6785
(G-H)	G	6.7641	12.0655	-8.0756
	Center	6.7641	0.0095	29.1738
	H	6.7641	12.0844	8.0094
(H-I)	I	5.0000	0.5012	-0.2844

CONCLUSIONS

A 3D beam interface element is presented in this paper. This element can be successfully used in the analysis of space frame structures which may be, partly or completely, buried in- or resting on the ground such as piles, offshore structural systems and pipeline systems, etc. Since all components of the element stiffness matrix are explicitly provided, a numerical integration procedure is avoided when evaluating the element stiffness matrix. As it is demonstrated by the examples, the element is very efficient in modelling structural systems and the surrounding media which may display different characteristics (material properties) from location to location. This interface element can also be used successfully in the non-linear analysis due to large deformations, such as calculation of the buckling load. Finally, it can be said that this element may have a permanent interest in practice.

REFERENCES

Aydogan, M., "Stiffness-matrix formulation of beams with shear effect on elastic foundation", *Journal of the Structural Division, ASCE*, Vol.121, No.9, 1265-1270, 1995.

Boulon,M., Garnica, P. and Vermeer, P.A., "Soil-structure interaction: FEM computations" in "Mechanics of Geomaterial Interfaces", Selvadurai, A.P.S. and Boulon, M.J. (Editors), Elsevier, 147-171, 1995.

Bowles, E., "Analytical and Computer Methods in Foundation Engineering", McGraw-Hill, 1974.

Chang, H., Liao, H. and Lu, Y., "Estimation of skin friction of pile shaft in remolted clay", *International Journal of Offshore and Polar Engineering, IJOPE*, Vol.5, No.3, 235-239, 1995.

Crisfield, M.A., "Non-linear Finite Element Analysis of Solids and Structures, Volume 1 : Essentials", John Wiley & Sons, 1991.

Faruque, M.O., "An axisymmetric interface element for soil-structure interaction problems" in "Mechanics of Material Interfaces", Selvadurai, A.P.S. and Voyiadjis, G.Z. (Editors), Elsevier, 189-201, 1986.

Karadeniz, H., "An algorithm for member releases and partly connected members in offshore structural analysis", Proceedings of Offshore Mechanics and Arctic Engineering (OMAE), Vol.1, 471-476, 1994.

Miranda, C. and Nair, K., "Finite beams on elastic foundation", *Journal of the Structural Division, ASCE*, Vol.92, No.ST2, 131-141, 1966.

Oran, C., "Tangent stiffness in space frames", *Journal of the Structural Division, ASCE*, Vol.99, No.ST6, 987-1001, 1973.

Selvadurai, A.P.S., "Elastic Analysis of Soil-Foundation Interaction", Elsevier, Amsterdam, 1979.

Selvadurai, A.P.S., "Boundary element modelling of geomaterial interfaces" in "Mechanics of Geomaterial Interfaces", Selvadurai, A.P.S. and Boulon, M.J. (Editors), Elsevier, 173-197, 1995.

Suarez, L.E., Singh, M.P. and Matheu, E.E., "Seismic response of structural frameworks with flexible connections", *Computers & Structures*, Vol.58, No.1, 27-41, 1995.

Tezcan, S.S., "Computer analysis of plane and space structures", *Journal of the Structural Division, ASCE*, Vol.92, No.ST2, 143-173, 1966.

Ting, B.Y., "Finite beams on elastic foundation with restrains", *Journal of the Structural Division, ASCE*, Vol.108, No.ST3, 611-621, 1982.

Ting, B.Y. and Mockry, E.F., "Beam on elastic foundation finite element", *Journal of the Structural Division, ASCE*, Vol.110, No.10, 2324-2339, 1984.

Yokoyama, T., "Vibrations of Timoshenko beam-columns on two-parameter elastic foundations", *Earthquake Engineering and Structural Dynamics*, Vol.20, 353-370, 1991.

APPENDIX

Shape functions of the member displacements :

$$[N]=\begin{bmatrix} 1-\xi & 0 & 0 & 0 & 0 & 0 & \xi & 0 & 0 & 0 & 0 & 0 \\ 0 & N_{y1} & 0 & 0 & 0 & N_{y2} & 0 & N_{y3} & 0 & 0 & 0 & N_{y4} \\ 0 & 0 & N_{z1} & 0 & N_{z2} & 0 & 0 & 0 & N_{z3} & 0 & N_{z4} & 0 \\ 0 & 0 & 0 & 1-\xi & 0 & 0 & 0 & 0 & 0 & \xi & 0 & 0 \\ 0 & 0 & N_{\theta_{y1}} & 0 & N_{\theta_{y2}} & 0 & 0 & 0 & N_{\theta_{y3}} & 0 & N_{\theta_{y4}} & 0 \\ 0 & N_{\theta_{z1}} & 0 & 0 & 0 & N_{\theta_{z2}} & 0 & N_{\theta_{z3}} & 0 & 0 & 0 & N_{\theta_{z4}} \end{bmatrix} \quad \text{(A.1)}$$

in which,

$$\xi = \frac{x}{L} \quad \text{(A.2)}$$

The elements of this shape function matrix which contain the effects of shear deformations in the beam are,

$$\begin{aligned} N_{y1} &= 1 - \mu_y (\Phi_y \xi + 3\xi^2 - 2\xi^3) \\ N_{y2} &= L\,\mu_y [\tfrac{1}{2}(2+\Phi_y)\xi - \tfrac{1}{2}(4+\Phi_y)\xi^2 + \xi^3] \\ N_{y3} &= \mu_y (\Phi_y \xi + 3\xi^2 - 2\xi^3) \\ N_{y4} &= L\,\mu_y [-\tfrac{1}{2}\Phi_y \xi - \tfrac{1}{2}(2-\Phi_y)\xi^2 + \xi^3] \end{aligned} \quad \text{(A.3a)}$$

$$\begin{aligned} N_{z1} &= 1 - \mu_z (\Phi_z \xi + 3\xi^2 - 2\xi^3) \\ N_{z2} &= L\,\mu_z [-\tfrac{1}{2}(2+\Phi_z)\xi + \tfrac{1}{2}(4+\Phi_z)\xi^2 - \xi^3] \\ N_{z3} &= \mu_z (\Phi_z \xi + 3\xi^2 - 2\xi^3) \\ N_{z4} &= L\,\mu_z [\tfrac{1}{2}\Phi_z \xi + \tfrac{1}{2}(2-\Phi_z)\xi^2 - \xi^3] \end{aligned} \quad \text{(A.3b)}$$

$$\begin{aligned} N_{\theta_{y1}} &= -\frac{6\mu_z}{L}(\xi^2 - \xi) \\ N_{\theta_{y2}} &= 1 - \mu_z [(4+\Phi_z)\xi - 3\xi^2] \\ N_{\theta_{y3}} &= \frac{6\mu_z}{L}(\xi^2 - \xi) \\ N_{\theta_{y4}} &= -\mu_z [(2-\Phi_z)\xi - 3\xi^2] \end{aligned} \quad \text{(A.3c)}$$

$$\begin{aligned} N_{\theta_{z1}} &= \frac{6\mu_y}{L}(\xi^2 - \xi) \\ N_{\theta_{z2}} &= 1 - \mu_y [(4+\Phi_y)\xi - 3\xi^2] \\ N_{\theta_{z3}} &= -\frac{6\mu_y}{L}(\xi^2 - \xi) \\ N_{\theta_{z4}} &= -\mu_y [(2-\Phi_y)\xi - 3\xi^2] \end{aligned} \quad \text{(A.3d)}$$

in which the constants Φ_y and Φ_z are the shear deformation parameters which are defined as written by,

$$\Phi_y = \frac{12}{L^2}\frac{EI_z}{GA_y} \quad \text{and} \quad \Phi_z = \frac{12}{L^2}\frac{EI_y}{GA_z} \quad \text{(A.4a)}$$

The parameters μ_y and μ_z in the Eqs.(A.3) are defined by,

$$\mu_y = \frac{1}{1+\Phi_y} \quad \text{and} \quad \mu_z = \frac{1}{1+\Phi_z} \quad \text{(A.4b)}$$

Proceedings of the Seventh (1997) International Offshore and Polar Engineering Conference
Honolulu, USA, May 25-30, 1997

ISBN 1-880653-28-1 (Set); ISBN 1-880653-30-3 (Vol. II)

Expansion Analysis of Subsea Pipe-in-Pipe Flowline

Gary E. Harrison, Naum Y. Kershenbaum and Han S. Choi
McDermott Engineering Houston
Houston, TX, USA

ABSTRACT

This paper presents an analytical method, investigation results, and applications of thermal expansion phenomena of subsea, insulated, pipe-in-pipe systems. Thermal expansion of a subsea, insulated, pipe-in-pipe arises from transmission of a hot product. To maintain the desired temperature of the hot product in the carrier pipe, thermal insulation is applied between the carrier pipe and the external jacket pipe to achieve the desired pipe-in-pipe design arrangement. A new analysis method has been developed to investigate thermal expansion of an insulated pipe-in-pipe system. Temperature gradient, pressure, soil resistance, and interaction forces between the carrier pipe and jacket pipe are considered in the analysis. Lateral deviation of the pipe-in-pipe system is also considered. A spreadsheet or computer program is described to calculate expansion, strains/stresses, and other results of the pipe-in-pipe system. The new analysis method has been applied in subsea pipeline design.

KEY WORDS : Pipe-In-pipe, carrier pipe, jacket pipe, pipe expansion, lateral deviation

INTRODUCTION

Subsea pipelines, such as flowlines from the subsea wells, are often operated with an elevated temperature product substantially warmer than the ambient seawater temperature. A flowline typically connects a subsea well to a subsea manifold or a platform where reservoir temperature and pressure is high (i.e., HP/HT flowline). To avoid wax or hydrate problems, the fluid temperature in a flowline must frequently be maintained within a desired temperature range. Heat loss from the flowline can be prevented by applying insulation material between the carrier pipe and the external jacket pipe using the typical pipe-in-pipe design concept. A new analysis method has been developed to investigate thermal expansion of an insulated pipe-in-pipe system. Additionally, the oil or gas production may be associated with high internal pressure. Both temperature and pressure result in pipeline expansion defined by temperature gradient, thermal expansion coefficient, internal pressure, Poisson's ratio, the interaction forces between the carrier pipe and jacket pipe, and the pipeline restraint conditions on the seabed.

A pipeline, buried or unburied, has both an active portion and a fully-restrained portion. A buried subsea pipeline has a much larger contact area with the seabed and, consequently, most of the pipeline is restrained by the combined effect of soil pressure and friction. In other words, a buried subsea pipeline, even under high temperature and internal pressure, has minimum mobility and consequently less possibility of expansion. However, the difficulty and expense of pipeline burial operations, especially in deep water, limit use of pipeline burial and motivate designers to develop unburied pipeline designs, thus providing reduced construction cost. An unburied pipeline's relatively low contact area with the seabed increases its mobility on the sea bottom, when operating with an elevated temperature and pressure loading. Thus an unburied pipeline may allow substantial longitudinal elongation and lateral deviation. In many cases, pipeline expansion results in an expansion loop or expansion offset end arrangement to accommodate the pipeline end movement.

Expansion analyses of a single wall pipeline were presented by AGA (1987), Choi (1995), Hobbs et al (1989), and Palmer et al (1981). Choi introduced an iterative method to solve the interaction of offshore pipeline expansion and restraint at the end of the pipeline and studied the effects of thermal gradients and restraining forces to provide accurate pipeline design criteria. Lateral deviation and self-limiting stabilization of a single wall pipeline under axial compressive load on a resistive soil medium was introduced by Kershenbaum et al (1996). Kershenbaum et al utilized an energy variational method to investigate subsea pipeline lateral deviation with plastic and elastic soil resistance. By incorporating the results of lateral deviation phenomenon in the design process, offshore pipeline construction costs may be reduced significantly; without pipeline burial or an expansion loop.

A primary objective of this paper is development of a new analytical method of thermal expansion for offshore pipe-in-pipe systems using the conventional approach with a structurally-coupled carrier and jacket pipe. Lateral deviation of the pipe-in-pipe system is also considered. By implementing a more realistic analytical model into the design, installation, and operation of offshore pipe-in-pipe flowlines, pipeline expansion loops or expansion joints may be avoided. The authors believe this paper may be the first to establish a simple analytical method of a pipe-in-pipe system.

PIPE-IN-PIPE SYSTEM

A typical pipe-in-pipe arrangement is presented in Figure 1. Hot and high pressure product is flowing in the carrier pipe whose wall thickness and material grade are determined by the maximum operating pressure or wellhead shut-in pressure. Insulation material is applied to the carrier pipe to prevent the heat loss. The outer jacket pipe protects the insulation and carrier pipe. The jacket pipe's wall thickness and material grade are determined by either buckle propagation or collapse criteria due to installation bending and external seawater pressure. The carrier and jacket pipe have structural connections through the bulkheads. In addition, centralizers are used to prevent contact of the carrier and jacket pipe. Pipe-in-pipe systems are used for many high temperature flowlines from subsea wells and for very cold product pipelines such as loading and unloading of Liquefied Natural Gas (LNG).

ELEMENTS OF PIPE EXPANSION

The elements of single wall pipeline expansion are well described by Choi (1995). Pipeline expansions are due to temperature, pressure, friction, and reactions from the soil and adjacent tie-in structures. Pipeline elongation due to thermal expansion can be obtained by integrating temperature gradient over the active or moving length of the pipeline. The detailed derivation of temperature gradient is presented in Appendix A. Fundamental equations of subsea pipeline expansion developed over the last two decades are summarized as follow.

Expansion of a pipeline due to a thermal loading is;

$$\Delta L_t = \alpha \cdot \int_0^{L_a} \Delta T_x \cdot dx = \frac{\alpha \cdot (T_i - T_a)}{K_2} \cdot (1 - e^{-K_2 \cdot L_a}) \qquad (1)$$

where, α = pipe material thermal expansion coefficient,
L_a = active length of a pipeline (moving or friction length),
x = longitudinal direction of pipeline,
$\Delta T_x = (T_i - T_a) \cdot e^{-K_2 \cdot x}$
= temperature difference at x from inlet,
T_i = inlet temperature,
T_a = ambient temperature, and
K_2 = index of temperature decay.

Pipeline elongation due to pressure loading is;

$$\Delta L_p = \left(\frac{p \cdot A_i}{A_S \cdot E} - \nu \frac{p \cdot D}{2 \cdot t \cdot E}\right) \cdot L_a \qquad (2)$$

where, p = internal - external pressure,
A_i = pipe flow section area,
A_S = pipe steel section area,
E = pipe modulus of elasticity,
ν = Poisson's ratio of pipe, and
D, t = pipe external diameter, and wall thickness.

Pipeline elongation due to frictional resistance is;

$$\Delta L_f = -\int \varepsilon_f \cdot dx = -\frac{f \cdot W \cdot L_a^2}{2 \cdot A_s \cdot E} \qquad (3)$$

where, ε_f = frictional strain,
f = friction factor, and
W = submerged weight of the pipe.

Pipeline elongation due to external compressive force, F, is;

$$\Delta L_F = -\frac{F \cdot L_a}{A_s \cdot E} \qquad (4)$$

From the force balance, a moving length can be calculated. However, if a pipeline has a temperature gradient, the thermal expansion term is a function of moving length and a closed form solution is impossible. An iterative numerical method is required to calculate moving length.

$$F_t + F_p - F_f - F = 0 \qquad (5)$$

where, $F_t = \alpha \cdot A_s \cdot E \cdot (T_i - T_a) \cdot e^{-K_2 \cdot L_a}$
= longitudinal thermal expansion force,
$F_p = P \cdot A_i - \nu \cdot A_s \cdot \sigma_h$ = longitudinal pressure force,
$F_f = f \cdot W \cdot L_a$ = frictional force, and
F = external longitudinal force.

EXPANSION OF PIPE-IN-PIPE

The carrier pipe expands due to temperature and pressure, and is restrained by Poisson's shrinkage, friction force from the centralizers, and reactions from the jacket pipe at bulkheads/spacers. It is assumed that bulkhead/spacers are not fully restricted between the carrier and jacket pipe. The total expansion length of the carrier pipe can be expressed as follow;

$$\Delta L_c = \Delta L_{tc} + (\varepsilon_{pc} + \varepsilon_{fc} + \varepsilon_{Fc}) \cdot L_{ac} \qquad (6)$$

where, L_{ac} = active length of carrier pipe,
$\Delta L_{tc} = \frac{\alpha \cdot (T_i - T_a)}{K_{2c}} \cdot (1 - e^{-K_{2c} L_{ac}})$ = thermal expansion,
$\varepsilon_{pc} = \left(\frac{p_i \cdot A_{ic}}{A_{sc} \cdot E} - \nu \frac{p_i \cdot D_c}{2 \cdot t_c \cdot E}\right)$ = pressure strain,
$\varepsilon_{fc} = -\frac{f_c \cdot W_c \cdot L_{ac}}{2 \cdot A_{sc} \cdot E}$ = frictional shrinkage between carrier pipe and bulkheads,
$\varepsilon_{Fc} = -\frac{F}{A_{sc} \cdot E}$ = shrinkage due to reaction between carrier pipe and jacket pipe.

It should be noted that W_c is a weight of the carrier pipe in air, p_i is internal pressure only, and the subscript, $_c$, refers to the carrier pipe. Interaction force, F, is compressive in the carrier pipe and tensile in the jacket pipe.

The jacket pipe expands due to frictional forces from centralizers and reaction forces from the carrier pipe, and is restrained by external pressure and friction with the soil on the seabed. Due to the on-bottom stability and insulation in the pipe-in-pipe system, the temperature in the jacket pipe is same as ambient temperature. The total expansion length of the jacket pipe can be expressed as follow;

$$\Delta L_j = \varepsilon_{fj} \cdot L_{ac} + (\varepsilon_{Fj} + \varepsilon_{pj} + \varepsilon_{sj}) \cdot L_{aj} \tag{7}$$

where, L_{aj} = active length of jacket pipe,

$\varepsilon_{fj} = \dfrac{f_c \cdot W_c \cdot L_{ac}}{2 \cdot A_{sj} \cdot E}$ = frictional strain from the centralizers,

$\varepsilon_{Fj} = \dfrac{F}{A_{sj} \cdot E}$ = strain from reactions of carrier pipe,

$\varepsilon_{pj} = (\dfrac{p_j \cdot A_{ij}}{A_{sj} \cdot E} - \nu \dfrac{p_j \cdot D_j}{2 \cdot t_j \cdot E})$ = strain from external pressure,

$\varepsilon_{sj} = -\dfrac{f_j \cdot W_j \cdot L_{aj}}{2 \cdot A_{sj} \cdot E}$ = frictional strain with the soil on the seabed.

It should be noted that W_j is the total submerged weight of the pipe-in-pipe system (submerged weight of jacket pipe + weight of carrier pipe in air), $p_j = -\gamma \cdot h$ is an external hydrostatic pressure on the jacket pipe, and the subscript, $_j$, indicates the jacket pipe.

The total elongation of the carrier pipe and jacket pipe must be identical, i.e.

$$\Delta L_c = \Delta L_j \tag{8}$$

The equilibrium force in the carrier pipe is balanced at the moving length, L_{ac}, by the friction forces from centralizers and reaction forces from the jacket pipe.

$$\alpha \cdot E \cdot A_{sc} \cdot (T_i - T_a) \cdot e^{-K_2 \cdot L_{ac}} + p_i \cdot A_{ic} - \nu \cdot A_{sc} \cdot \sigma_{hc} = f_c \cdot W_c \cdot L_{ac} + F \tag{9}$$

where, $\sigma_{hc} = \dfrac{p_i \cdot D_c}{2 \cdot t_c \cdot E}$

The equilibrium force in the jacket pipe is balanced at the moving length, L_{aj}, by the external pressure and soil friction force.

$$f_c \cdot W_c \cdot L_{ac} + F = f_j \cdot W_j \cdot L_{aj} - p_j \cdot A_{ij} + \nu \cdot A_{sc} \cdot \sigma_{hj} \tag{10}$$

The pipe-in-pipe expansion can be calculated from equations (8), (9), and (10). The unknowns are moving lengths of the carrier and jacket pipe and interaction force between the pipes (L_{ac}, L_{aj}, F). The thermal expansion terms are functions of moving lengths, thus closed form solutions can not be obtained explicitly. The above equations are readily solved by implementing an iterative numerical method using a corrective predicator function of moving length until force equilibrium is achieved.

LATERAL STABILITY

Lateral deviation of a single wall pipeline was studied by Kershenbaum et al (1996). Lateral deviation occurs rapidly and establishes an equilibrium configuration when a pipeline is under compression. In the case of pipe-in-pipe, the carrier pipe will be in compression from the expansion caused by the high temperature and pressure, and restraint from the bulkhead reactions of the jacket pipe. The amplitude and mode number of the lateral deviation are controlled by the diameter of jacket pipe and the interval of centralizers. The critical compressive force in a carrier pipe due to Euler's stability was derived from Timoshenko (1936);

$$F_c = \frac{\pi^2 \cdot EI}{l^2} \tag{11}$$

where, l is a half wave length of the lateral deviation and controlled by the distance between the centralizers. If the critical Euler buckling force in equation (11) is greater than reaction force, F, between the carrier pipe and jacket pipe, the carrier pipe will remain straight and lateral deviation will not occur. If the Euler buckling force is less than reaction force, the carrier pipe will deviate laterally within the jacket pipe. Equation (11) indicates the Euler buckling force is great when the centralizers are spaced closely enough to prevent sinusoidal buckling. The amplitude of lateral deviation is restrained by the gap between the two pipes and as follow;

$$A_c = \frac{2}{n \cdot \pi} \cdot \sqrt{(\Delta L_c - \Delta L_j) \cdot L} \tag{12}$$

where, $n = L/l$ is a mode number and L is the total pipeline length. Once the carrier pipe deviates laterally and establishes a self-limiting stabilization, the expansion length of carrier pipe is greater than that of jacket pipe, i.e. $\Delta L_c \succ \Delta L_j$.

The maximum curvature and bending stress due to the lateral deviation of the carrier pipe are;

$$K_c = \frac{A_c \cdot \pi^2}{l^2} \quad \text{and} \quad \sigma_{bc} = \frac{E \cdot D_c \cdot K_c}{2} \tag{13}$$

The jacket pipe is under tension and lateral deviation is not expected in most cases. If a jacket pipe and a carrier pipe are structurally connected by frequent bulkheads, then lateral deviation may occur in the jacket pipe. However, continuous or near-continuous connection of a carrier pipe and a jacket pipe is an impractical design concept. In most cases of pipe-in-pipe system, the bulkheads will bend to allow flexibility when the carrier pipe expands. The jacket pipe will be under tension or a small amount of compression, substantially less than the critical buckling force.

STRESS/STRAIN CALCULATIONS

The longitudinal expansion stresses depend on the restraint conditions. The stress in a carrier pipe, without bending stress, can be expressed as follow;

$$\sigma_{lc} = \frac{p_i \cdot A_{ic}}{A_{sc}} - \frac{f_c \cdot W_c \cdot x}{A_{sc}} - \frac{F}{A_{sc}} \quad \text{if } x \prec L_{ac}$$
$$\sigma_{lc} = -\alpha \cdot E \cdot \Delta T_x + \nu \frac{p_i \cdot D_c}{2 \cdot t_c} \quad \text{if } x \geq L_{ac} \tag{14}$$

The longitudinal expansion stress in the jacket pipe is as follow;

$$\sigma_{lj} = \frac{F}{A_{sj}} + \frac{f_c \cdot W_c \cdot x}{A_{sj}} + \frac{p_j \cdot A_{ij}}{A_{sj}} - \frac{f_j \cdot W_j \cdot x}{A_{sj}} \quad \text{if } x \prec L_{aj}$$
$$\sigma_{lj} = \nu \frac{p_j \cdot D_j}{2 \cdot t_j} \quad \text{if } x \geq L_{aj} \tag{15}$$

At the moving lengths of the carrier and jacket pipe, $x = L_{ac}$ or L_{aj}, the stresses in moving part and restrained part become identical and equations (14) and (15) will lead to equations (9) and (10).

The hoop stress in the carrier pipe is due to internal pressure only and hoop stress in the jacket pipe is due to the external pressure only.

$$\sigma_{hc} = \frac{p_i \cdot D_c}{2 \cdot t_c} \quad \text{and} \quad \sigma_{hj} = \frac{p_j \cdot D_j}{2 \cdot t_j} \tag{16}$$

For the calculation of combined stresses, the stress expressions in equations (13) through (16) must be converted into strain expressions. The Needleman-Hutchinson relation between the stress and strain is as follow;

$$\varepsilon = \frac{\sigma}{E} \quad \text{if } \sigma \prec \sigma_p$$
$$\varepsilon = \frac{\varepsilon_p}{h} \cdot \left[\left(\frac{\sigma}{\sigma_p} \right)^h + h - 1 \right] \quad \text{if } \sigma \geq \sigma_p \tag{17}$$

where, h is Hutchinson's exponent, ε_p, σ_p are proportional limits of strain and stress. The combined longitudinal strains, considering lateral deviation of the carrier pipe only, are as follow;

$$\varepsilon_{clc} = \varepsilon_{bc} + \varepsilon_{lc}$$
$$\varepsilon_{clj} = \varepsilon_{lj} \tag{18}$$

Torsional and shear strains are usually negligible in the pipe expansion problem. Then combined von Mises strains are as follow;

$$\varepsilon_{cvc} = (\varepsilon_{hc}^2 - \varepsilon_{hc} \cdot \varepsilon_{clc} + \varepsilon_{clc}^2)^{0.5}$$
$$\varepsilon_{cvj} = (\varepsilon_{hj}^2 - \varepsilon_{hj} \cdot \varepsilon_{clj} + \varepsilon_{clj}^2)^{0.5} \tag{19}$$

To calculate the combined von Mises stresses, the inverse relationship of equation (17) are applied to equations (18) and (19).

EXAMPLE CALCULATIONS

Example calculations were performed for the following pipe-in-pipe system.

Carrier Pipe :	Outer Diameter	= 6.625-inch
	Wall Thickness	= 0.688-inch
	Pipe Grade	= API 5L X-70
	Flowrate	= 3000 BOPD
	Max. Pressure	= 7000 psi
Jacket Pipe :	Outer Diameter	= 10.75-inch
	Pipe Grade	= API 5L X-52
	Wall Thickness	= 0.279-inch at 1000 ft water
		= 0.312-inch at 2000 ft water
		= 0.360-inch at 3000 ft water
		= 0.400-inch at 4000 ft water
		= 0.500-inch at 5000 ft water
Temperature:	Inlet	= 160 °F
	Ambient Water	= 40 °F

The wall thicknesses of the jacket pipe were selected for various water depths to allow an installation bending strain of 0.25 % with consideration of an approximate strain concentration factor of 1.5. A detail of sample input and output of the computer calculation is attached in Appendix B.

Figure 3 depicts pipe expansion and the effect of the friction factor of carrier pipe with respect to centralizers at 5000 ft water depth. A constant soil friction factor of 0.5 of the jacket pipe is used in Figure 3. As the friction factor in the carrier pipe increases, the reaction in the jacket pipe decreases, while the total expansion length changes with a peak value. Using many different combinations of friction factors, a worst case can be identified. Figure 4 shows the effects of the soil friction factors of the jacket pipe at 5000 ft water depth. A constant friction factor of 0.5 of the carrier pipe centralizers to jacket pipe is used in Figure 4. As the soil friction increases, the reaction force increases and the expansion length decreases. Figure 5 depicts the pipe expansion length vs. water depth. As water depth increases, the external seawater pressure on the jacket pipe restrains more of the expansion of the carrier pipe. Thus interaction force between carrier and jacket pipe increases and total expansion length decreases. Consequently, the expansion problem of a pipe-in-pipe system in ultra deep water may not be significant. Figure 6 depicts a typical expansion stresses in a pipe-in-pipe system without bending stress.

CONCLUSIONS

1. The simple analytical method can be used to estimate the expansion lengths of pipe-in-pipe systems and can be applied to design of subsea pipeline tie-in systems.
2. Pipe-in-pipe systems yield less longitudinal expansion compare to single wall pipe systems and may be able to avoid tie-in spool-piece offsets or "dog-legs".
3. As water depth increases, expansion length of the pipe-in-pipe system decreases.
4. Carrier pipe allowed much less lateral deviation due to inherent restraints in the pipe-in-pipe system.
5. Jacket pipe will not deviate laterally except fully restraint, HP/HT pipe-in-pipe system.

REFERENCES

AGA (1987). *Pipeline Riser System Design and Application Guide*, American Gas Association Project PR-178-622 reported by Brown and Root.

Choi, HS (1995). "Expansion Analysis of Offshore Pipelines Close to Restraints", *Proc of 5th Int Offshore and Polar Engineering Conf*, Vol 2, pp 81-88.

Hobbs, RE, and Liang, F (1989). "Thermal Buckling of Pipelines Close to Restraints", *Proc of 8th Int Conf on Offshore Mechanics and Arctic Engineering*, Vol 4, pp 121-127.

Kershenbaum, NY, Harrison, GE, and Choi, HS (1996). "Subsea Pipeline Lateral Deviation Due To High Temperature Product", *Proc of 6th Int Offshore and Polar Engineering Conf*, Vol 2, pp 74-79.

Palmer, A, and Ling, MTS (1981). "Movement of Submarine Pipelines Close to Platforms", *Proc 13th Offshore Tech Conf*, OTC 4067.

Timoshenko, SP (1936). *"Theory of Elastic Stability"*, by the United Engineering Trustees, Inc.

APPENDIX A : THERMAL GRADIENT CALCULATION

Temperature distribution in the pipe at x distance from the inlet can be expressed;

$$\Delta T_x = T_x - T_a = (T_i - T_a)\cdot e^{-K_2 \cdot x}$$

where $K_2 = \dfrac{U}{c_p \cdot \rho \cdot Q}$

Q is flowrate,
c_p is specific heat capacity of product,
ρ is mass density of product.
U is overall heat transfer coefficient defined by

$$U = \left[\frac{1}{K_p} + \frac{1}{K_s} + \frac{1}{K_w} + \frac{1}{K_i}\right]^{-1} \qquad \text{(BTU/hr-ft-°F)}$$

where, K_p, K_s, K_w, K_i are heat transfer coefficient of product, steel pipe, seawater, and insulation coating.

$$K_p = 0.023 \cdot k_p \cdot \mathrm{Re}^{0.8} \cdot \mathrm{Pr}^{0.3} \cdot \pi$$

$$K_s = \frac{2 \cdot \pi \cdot k_s}{Ln(D_o / D_i)}$$

$K_w = 2033$ with 1.5 ft / sec current if pipe is not buried

$$K_w = \frac{2 \cdot \pi \cdot k_{soil}}{Ln[\{2H + (4H^2 - D_o{}^2)^{0.5}\} / D_o]} \qquad \text{if pipe is buried}$$

$$K_i = \frac{2 \cdot \pi \cdot k_i}{Ln\{(D_o + 2t_i) / D_o\}}$$

where, k_p, k_s, k_{soil}, k_i are thermal conductivity of product, steel pipe, soil , and insulation coating. Re is Reynolds number and Pr is Prandtle number.

APPENDIX B : SAMPLE CALCULATION

Input Information	Flowline	Jacket
Pipe Properties		
Pipe Outside Diameter (in)	6.625	10.750
Pipe Wall Thickness (in)	0.688	0.500
Corrosion Allowance (in)	0.000	0.000
P/L Submerged Weight (klf)	0.1135	0.0036
Void Submerged Weight (klf)	-	0.0036
Prdct.Filled Air Wei./PIP.Sub.Wei.(klf)	0.1135	0.1171
P/L Original Length (ft)	20000	20000
Hutchinson Proportional Limit (kips)	56	36.1
Hutchinson Exponent	17	10.8
Thermal Insulation Thickness (in)	1.81	0.00
Insulation Thermal Condivity (BTU/hr-sft-F/ft)	0.0160	26.00
Poisson's Ratio	0.3	0.3
Pipe Steel Cros.Sect.Area, (in^2)	12.83	16.10
Pipe Section Mom.Inertia,in^4	57.30	211.95
Steel Module Young, (ksi)	29000	29000
Collar Spacing (ft)	20	
Flowline Deviation Amplitude,inches	0	
Steel Therm.Expansion Coeff.	6.50E-06	6.50E-06
Environmental Conditions		
Water Depth, feet	5000	5000
P/L Initial Temperature (°F)	160.00	160.00
Ambient Water Temperature (°F)	40.00	40.00
Water Specific Weight (#/cft)	62.40	62.40
Steel Conductivity (BTU/hr-sft-F/ft)	26.00	26.00
Collar / Soil Friction factor	0.22	0.50
Design Internal Pressure, (ksi)	7.00	-2.22
Product Properties		
Product Flowrate (BOPD)	3000	1
Product Specific Gravity	0.865	0.001
Product Heat Capacity (BTU/#-F)	0.562	0.500
Product Conductivity (BTU/hr-sft-F/ft)	0.030	0.018
Product Viscosity (centipose)	0.380	0.020
FORCE OF INTERACTION,kips	194.500	-194.500
Calculated Data		
BUCKLING COMPR.FORCE, kips	284.720047	-284.720047
Flowline Max.Curvature,inch^-1	0	
Coefficient "K2" Flowline	1.078E-05	
Coefficient "K2" Jacket Pipe	1.969E-01	
Coefficient "K oil / gas"	5.881E+01	1.421E-03
Coefficient "K st"	7.017E+02	1.673E+03
Coefficient "K sw"	2.033E+03	2.033E+03
Coefficient "K ins"	2.306E-01	8.781E+07
Reynold's Number	1.197E+05	1.000E+00
Prandtl Number	1.723E+01	1.344E+00
Over.Heat Trans.Coeff.Flowline/Jacket	2.296E-01	1.421E-03
Active Length (ft)	4160.100	4160.100
P/L Total Elong.(feet)	0.661	0.660
Pipe Hoop Stress.(ksi)	23.003	-23.889
Pipe Hoop Strain (ksi)	7.932E-04	-8.238E-04
Total Longitudinal Stress (ksi)	-11.508	1.775
Total Longitudinal Strain	-3.968E-04	6.122E-05
Combined Stress (ksi)	30.435	24.824
Combined Strain	1.049E-03	8.560E-04
Bending Stress (ksi)	0.000	
Bending Strain	0.000E+00	

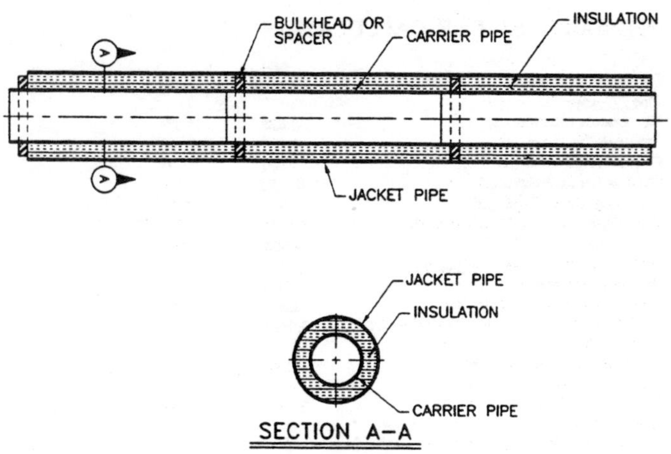

Fig. 1 Typical Pipe-In-Pipe Design Arrangement

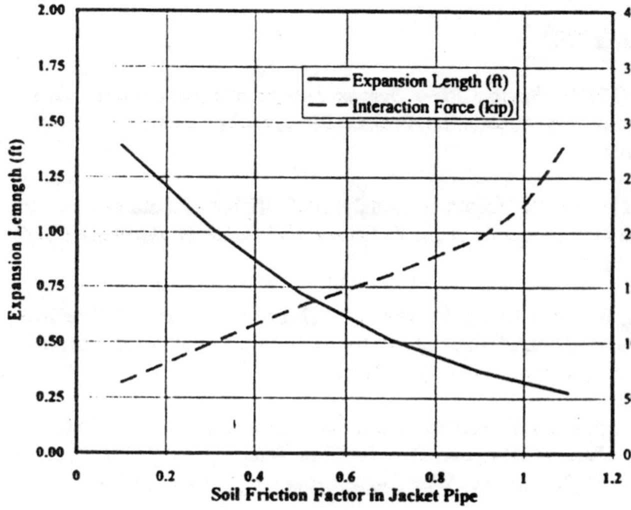

Fig. 4 Effect of Soil Friction Factor in Jacket Pipe

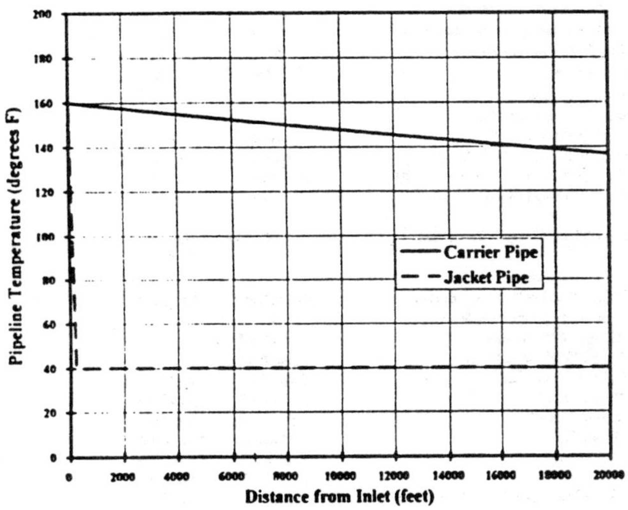

Fig. 2 Temperature Distribution of Pipe-In-Pipe System

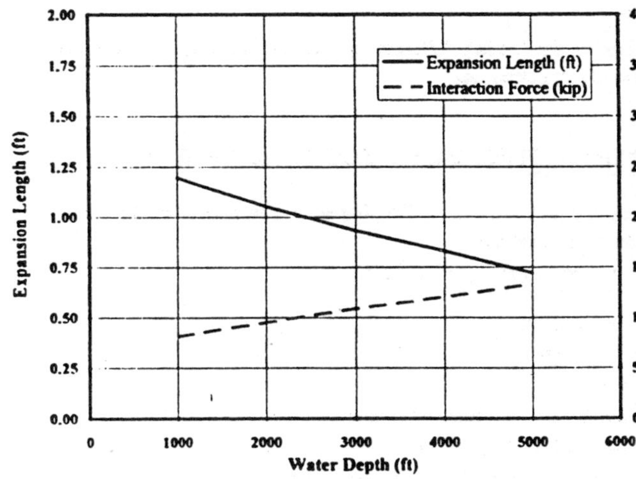

Fig. 5 Pipe Expansion vs. Water Depth

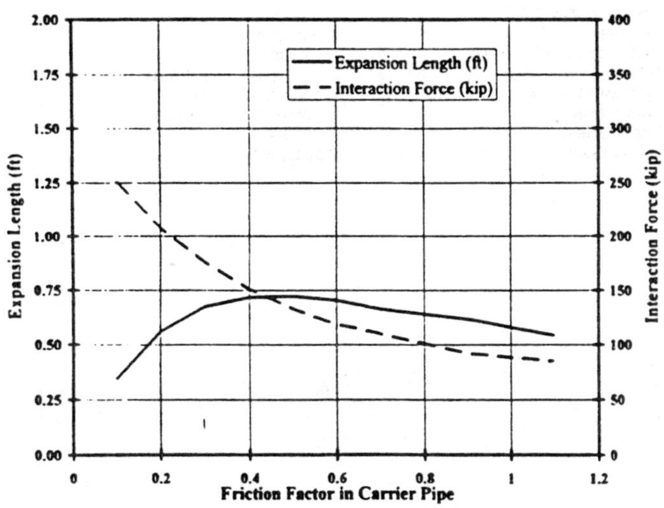

Fig 3. Effect of Friction Factor in Carrier Pipe

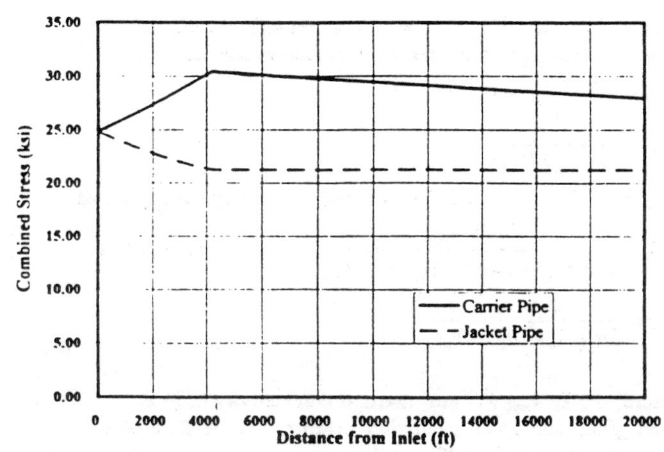

Fig 6. Combined Stress due to Pipe-In-Pipe Expansion (without Bending Stress)

Proceedings of the Seventh (1997) International Offshore and Polar Engineering Conference
Honolulu, USA, May 25-30, 1997

ISBN 1-880653-28-1 (Set); ISBN 1-880653-30-3 (Vol. II)

3-D Dynamic Buckling and Cyclic Behaviour of HP/HT Flowlines

Per R. Nystrøm, Knut Tørnes, Yong Bai and Per Damsleth
J P Kenny A/S
Stavanger, Norway

ABSTRACT

This paper discusses the 3-D buckling behaviour of High Pressure - High Temperature (HP/HT) flowlines on a very uneven seabed. Large thermal forces, coupled with large vertical imperfections, tend to buckle the flowline at many places along the route. Previously, the industry has primarily focused on pipeline buckling as a 2-D problem, i.e. either as vertical buckling or horizontal buckling. As a framework for limit state design, J P Kenny A/S investigated the mechanism behind the 3D interaction between vertical and horizontal buckling behaviour. A series of 2-D and 3-D FEM simulations demonstrate the effects of horizontal and vertical imperfections . A transient analysis of dynamic 'snap' buckling in terms of time-varying stresses, strains and displacements reveals the flowline response as the structural system becomes axially unstable. Finally, the cyclic shutdown/restart behaviour of the buckled flowline is demonstrated.

Keywords: Pipeline, flowline, HP/HT, buckling, limit state design, transient behaviour, cyclic, uneven seabed, shakedown, 3D FEM.

INTRODUCTION

HP/HT flowlines exposed on a very uneven seabed will tend to undergo global buckling because seabed friction increases the effective axial force as the pipeline expands. The combination of large vertical imperfections with high axial loads require realistic assessment of the flowline behaviour with respect to free spanning, expansion, buckling etc., to optimise routing and seabed intervention work.

In addition, it is necessary to demonstrate that the buckled flowline is structurally sound within the allowable stress/strain limits. The conventional industry design approach would require a large amount of seabed intervention in the form of burial and rockdumping to completely prevent the pipe from buckling. The estimated cost of such operations can, in some cases, threaten the viability of the field development.

A more optimum strategy may be to use a limit state design approach (Bai and Damsleth, 1997) and demonstrate the fitness for purpose of the HP/HT flowline through 3D FEM simulation of its global behaviour.

In order to apply a limit state design approach it is necessary to identify all the potential flowline failure modes and quantify and justify the allowable stress/strain limits. The possible failure modes for the considered flowline include:

- Ovalisation for serviceability limit state
- Bursting
- Buckling/collapse due to bending/pressure
- Fracture of welds due to bending/tension
- Low cycle fatigue due to shutdowns
- High cycle fatigue due to vortex-shedding induced vibrations
- Ratcheting

A series of 2D and 3D FEM simulations determined the global buckling behaviour of HP/HT flowlines and established the associated stress/strain levels for input to the design. The work included:

- Flowline 3D buckling behaviour on an uneven seabed
- Transient effects as the flowline 'snaps' into the buckled state
- Cyclic shutdown/restart behaviour of the buckled flowline .

PIPELINE BUCKLING

A pipeline may buckle as seabed friction builds up to resist axial expansion caused by temperature and pressure loads (Hobbs 1984, Pedersen and Jensen 1988). The compressive forces set up by the seabed friction, in addition to any other external forces such as restraint forces at tie-in points, is commonly referred to as the **effective axial force**.

Other important parameters that govern the buckling behaviour is the size and shape of any out-of-straightness (OOS), the structural stiffness and the seabed friction coefficients. In theory, a perfectly straight pipeline will not buckle, i.e. the pipe will remain in equilibrium for all values of axial load in the undeformed configuration. If a small imperfection is introduced into the system, the pipeline will buckle when the effective axial compressive force reaches a certain limit, i.e. the Euler buckling load. At this point, the pipe is said to have reached its ultimate capacity, which is the maximum axial load the pipeline can sustain.

When initial imperfections are small, buckling is expected to occur as a dynamic 'snap', i.e. a sudden uncontrolled buckle as the axial equilibrium becomes unstable. However, as the initial imperfections become larger, the buckling is less sudden and eventually for a large initial imperfection, the snap is eliminated and replaced by a gradual magnification of the initial out-of-straightness. A typical schematic of possible equilibrium paths for different initial imperfections is presented in Figure 1, (Hobbs 1984). Consider the case where there is a small initial imperfection, a . If the pipeline temperature is increased, the effective compressive force will initially increase from 0 to A. On reaching the critical load at A, the system becomes axially unstable, i.e. the imperfection is small and the force is large enough to cause a dynamic 'snap' between the two equilibrium states A and D. As the pipeline deflects to its new equilibrium configuration at D, the effective force in the section is relieved. An example of this behaviour could be a pipeline operating fully exposed on a flat seabed with relatively small initial OOS created during pipelaying. The pipeline may reach its ultimate axial capacity and 'jump' to another equilibrium configuration by buckling horizontally.

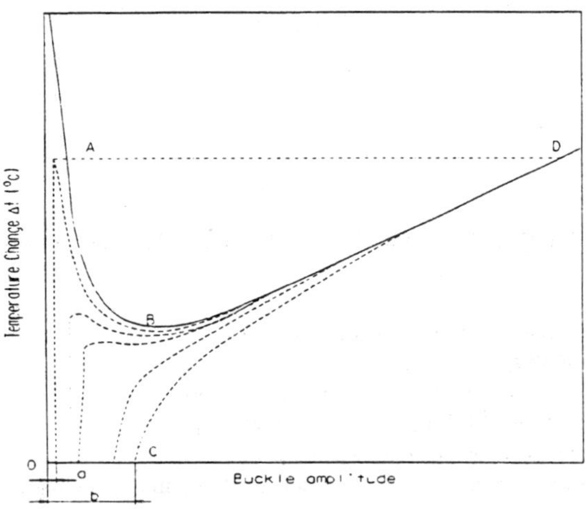

Figure 1: Buckling Equilibrium Paths

At slightly larger initial imperfections the buckle is less 'dramatic', i.e. the force required to buckle the pipe is less and hence also the resulting 'snap' movement. Eventually, at an even larger imperfection, b, the snap is eliminated and a gradual magnification of the initial imperfection takes place (C to D). An example of this behaviour is an exposed pipeline crossing over a large vertical imperfection such as a freespan shoulder. As the effective axial force increases, the pipeline will undergo a gradual vertical deflection, but no snap will occur.

The general effect of seabed friction resistance on horizontal buckling behaviour is shown in Figure 2 (Hobbs 1984). With decreasing seabed friction for a given temperature change, the resulting horizontal buckling force also reduces. A lower friction is further associated with larger buckle amplitude and wave length. This means that for a pipeline resting on a flat seabed, higher friction will result in a higher critical buckling force.

A pipeline exposed on a flat seabed surface will tend to buckle in the horizontal plane rather than vertically since this will be the path of least resistance, i.e. since the friction restraining force is less than the submerged weight. A pipeline which is placed in a trench or which is covered by rock will tend to buckle vertically due to the greater lateral resistance. Upheaval buckling takes place as the pipeline reaches its ultimate axial capacity or as a result of cyclic vertical creep (Pedersen and Jensen 1988).

Analytical methods for solving lateral and upheaval buckling problems have been well developed by the industry during the last two decades. Hobbs (1984) proposed solutions for both upheaval and lateral buckling. Pedersen and Jensen (1988) present an analytical solution for upheaval creep considering the effect of initial imperfections and Taylor and Gan (1986) proposed solutions for both lateral and vertical modes of buckling.

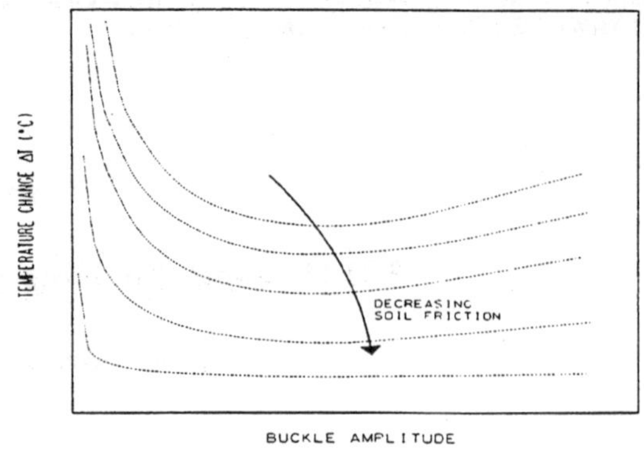

Figure 2: Buckling Equilibrium Paths - Effect of friction

Hobbs and Liang (1989) and Ellinas et al (1990) proposed to use intermittent rock dumping as a means of controlling buckling. More recently, Sævik and Levold (1995) and Kershenbaum et al (1996) proposed methods for lateral snaking of pipelines. The interaction between vertical and horizontal buckling modes was discussed by Nes et al (1996). These methods give valuable results for many buckling scenarios. However, pipeline buckling is primarily treated as 2D problems (either as upheaval or horizontal buckling).

FLOWLINE BUCKLING ON AN UNEVEN SEABED

The background for the detailed investigation of the buckling behaviour presented in this paper, is an ongoing detailed design project performed by JP Kenny A/S comprising a number of small diameter HP/HT flowlines laid on a very uneven seabed. In order to realistically assess the expansion and buckling behaviour, both 2D and 3D FEM models have been developed in which the seabed is generated from detailed survey data (Tørnes and Nystrøm 1997).

A typical operating configuration resulting from a static 2D analysis of a 10" oil flowline with a design temperature of 120 °C and an operating pressure of 170 bar, is presented in Figure 3. The flowline appears to have buckled vertically at two locations, namely at KP 1.4 and KP 1.8 with approximate amplitudes of 1.4 and 1.1 meters while the maximum equivalent stresses at the two buckles were found to be 350 MPa and 210 MPa, respectively.

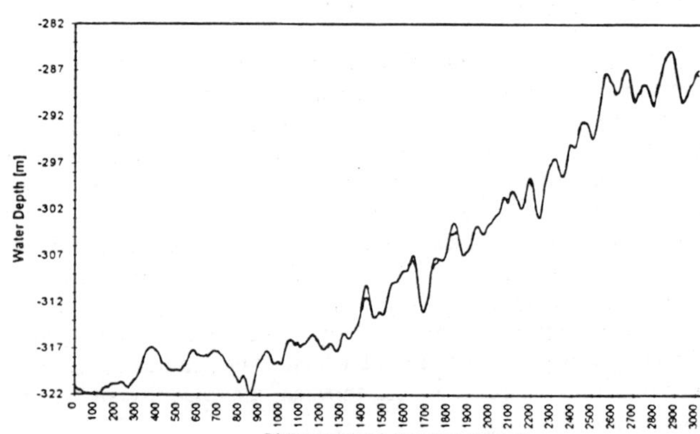

Fig. 3: 2D analysis results: Operating configuration

A better picture of the buckling mechanism can be achieved by tracing the deflection and effective force in the buckle as the flowline is elevated to its maximum design temperature.

First consider the vertical "buckle" that appeared at KP 1.8 in the 2D model. Figure 4 shows how the vertical displacement of the buckle develops as the temperature is increased. This is a typical deflection plot associated with a large initial imperfection, i.e. like the equilibrium path curves shown on the right hand side of Figure 1. From this it can be concluded that a vertical dynamic snap buckle has not occurred in this location. Figure 4 also shows how the effective force is developed with increasing temperature. Although the rate of increase in the effective force has been reduced at higher temperature, it is clear that the section has not yet lost its axial capacity to carry further load. If the latter was the case, the effective force would have stabilised at a constant value or, alternatively, be reduced as the temperature is increased. This means that the section has not undergone vertical buckling but is still in the pre-buckling, large deformation phase.

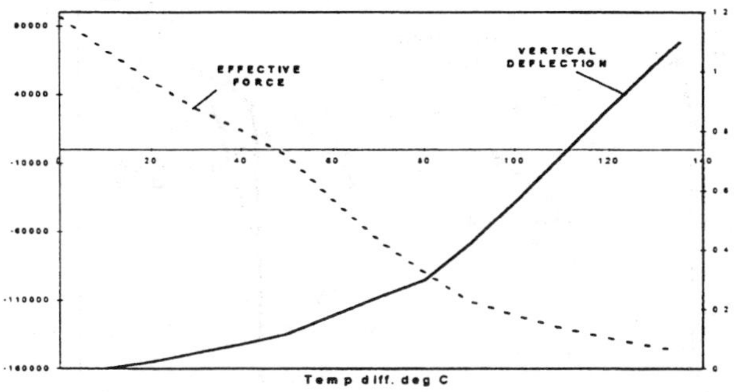

Fig. 4: 2D results for buckle at KP 1.8 : Buckle amplitude and effective force plot

A similar plot for the vertical buckle at KP 1.4, resulting from the 2D analysis, is presented in Figure 5. The deflection curve is associated with a smaller initial imperfection than the buckle at KP 1.8, i.e. moving further to the left in Figure 1. Since the vertical amplitude increases linearly for temperatures above 70°C, i.e. there is no sudden increase in the vertical deflection at a certain temperature, it can be concluded that no dynamic snap has occurred for this buckle either. However, the effective force plot shows that a maximum of approximately 170 kN compression is reached at 80°C, beyond which the effective force reduces slowly. In other words the ultimate axial capacity of the pipe section has been reached and the flowline has undergone large deflection vertical buckling at this point.

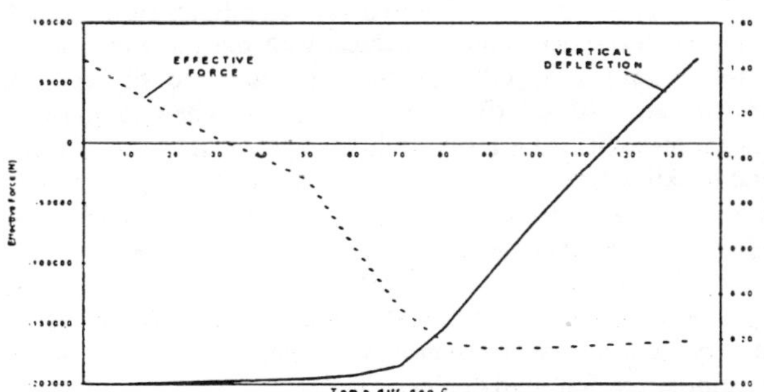

Fig. 5: 2D results for buckle at KP 1.4 : Buckle amplitude and effective force plot

For a pipeline resting on a very uneven however, the seabed vertical imperfections induce a more abrupt curvature than the horizontal imperfections created during laying. As a result, the critical buckling load required to lift the pipeline vertically is lower than the corresponding force needed to buckle horizontally. Although the initial movement occurs in the vertical plane, it was expected that as the pipe's contact force with the seabed diminishes the critical lateral buckling force would decrease and a lateral buckle could be initiated.

In order to investigate the pipeline behaviour on an irregular seabed as realistically as possible, a 3D pipeline - seabed FE model was developed where the 3D seabed bathymetry is imported as X,Y,Z data directly from a Digital Terrain Model (DTM) program. The 3D pipe-seabed model can then be presented as shown in Figure 6.

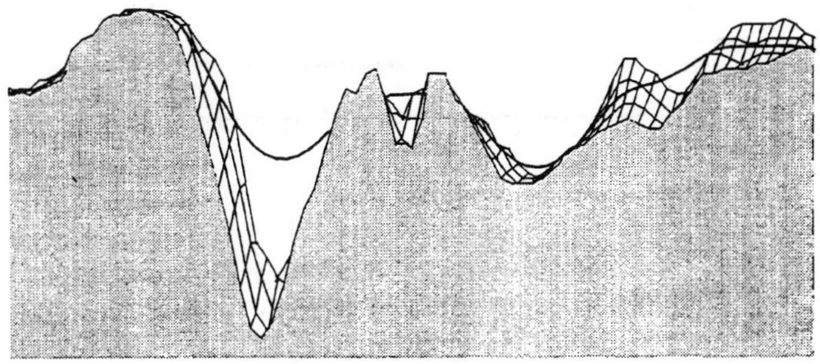

Fig. 6: 3D FEM In-place analysis model - Typical model details from an as-laid flowline simulation.

The vertical configuration of the flowline at full operating temperature resulting from the 3D analysis (not presented here) showed that the vertical buckle at KP 1.4 had disappeared in the 3D analysis. The upwards deflection at KP 1.8 was, on the other hand, also present in the 3D analyses albeit the height of the buckle was reduced.

Figure 7 shows a plan view of the deflected shape. At KP 1.4, i.e. at the location with the vertical buckle in the 2D analysis, a large horizontal buckle with an amplitude of 3.8 m has occurred in the 3D model. The buckle shape is close to the mode 3 described by Hobbs (1984). It can be seen that in addition to the buckle itself, smaller lateral deviations from the centerline are also found along the route with a maximum of approximately 0.5 m. These are due to a combination the effective axial force, weight of the flowline, friction and local seabed features.

The resulting maximum equivalent stress in the 3D horizontal buckle was found to be 302 MPa which is significantly less than what was found in the corresponding 2D analysis (350 Mpa).

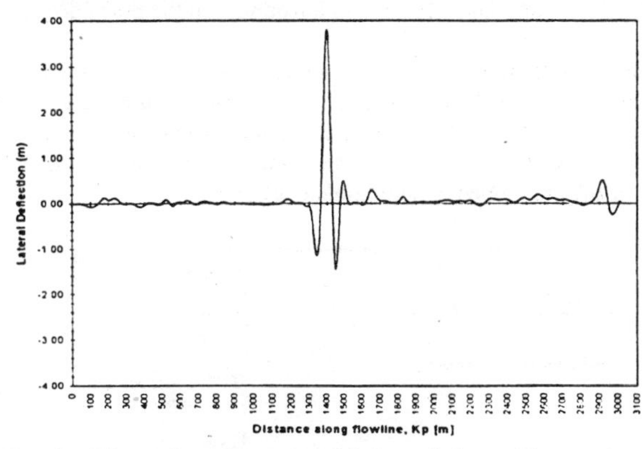

Fig. 7: 3D results: Plan view of deflected shape (Free ends)

The corresponding displacement and effective force results from the 3D analysis is presented in Figure 8. At about 80°C the flowline starts to lift vertically off the seabed. When the temperature is further increased to approximately 90°C, the flowline drops down to the seabed at the same time as it undergoes a sudden lateral displacement of 1.75 m, i.e. an interaction between the vertical and the horizontal mode has taken place. Beyond this temperature, the flowline is further deflected laterally as the temperature is increased until it reaches a final maximum horizontal deflection of 3.8 m at the design temperature. The plot of the effective force shows that it drops suddenly when the section deflects sideways which is typical for a dynamic 'snap' type of buckle.

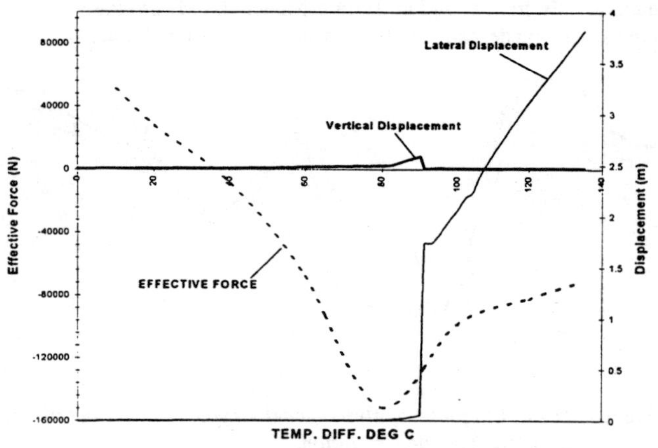

Fig. 8: 3D results for buckle at KP 1.8 : Buckle amplitude and effective force plot

INTERACTION BETWEEN VERTICAL AND HORIZONTAL MODES OF BUCKLING

More detailed studies were performed in order to gain further understanding of the complex mechanism between the vertical and horizontal mode of buckling. This included a number of 2D and 3D FE simulations using a typical 3D seabed model.

3D buckling model

In the first examples, a flat seabed with one realistic vertical seabed imperfection was used (1.8 m high and 50 m long, see Figure 11). A 10" oil flowline with a residual lay tension of 50 kN and no internal pressure is assumed. The seabed is assumed to be flat in the lateral plane, however a pipe imperfection of W_{om}/L_o=0.001 according to studies by Taylor and Gan (1986) was introduced in the pipe horizontal plane.

In order to determine the effect of the restraining force, two friction coefficients were checked (0.3 and 1.0). Temperature loads were applied to the model, in addition to the residual lay tension.

Figure 9 shows that, for a friction coefficient of 0.3, the flowline starts to lift off the seabed at approximately 13°C.

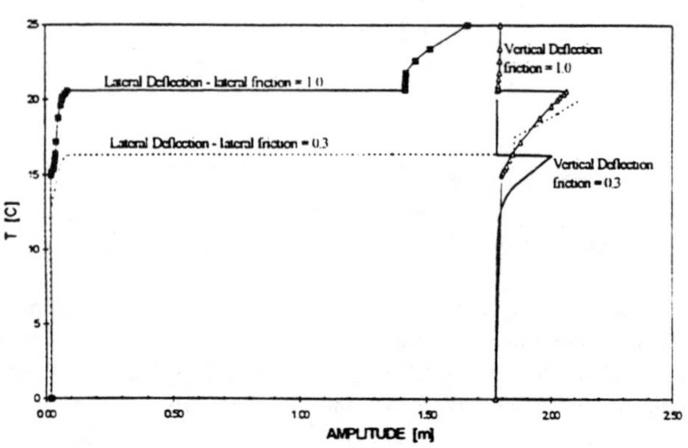

Figure 9: Static 3D analysis, Temperature vs. Deflection.

Little horizontal movement has yet occurred. At 16°C, the vertical displacement reaches 0.25 m but, suddenly drops down again to zero. In the same instant the pipe 'snaps' laterally by 1.85 m to a new equilibrium configuration. As stated previously the pipe is now fallen back onto the seabed again.

The same effect can be seen for the higher friction case (μ=1.0), the differences being that the vertical movement starts at 15°C, the pipe lifts slightly higher, the lateral buckling temperature is approx. 21°C and the resulting lateral displacement is 1.4 m.

Even though the higher friction case causes the pipeline to buckle laterally at a higher temperature, Figure 10 shows that the flowline buckles at approximately the same effective force for both friction cases. This is an important finding for the understanding of the 3D buckling behaviour.

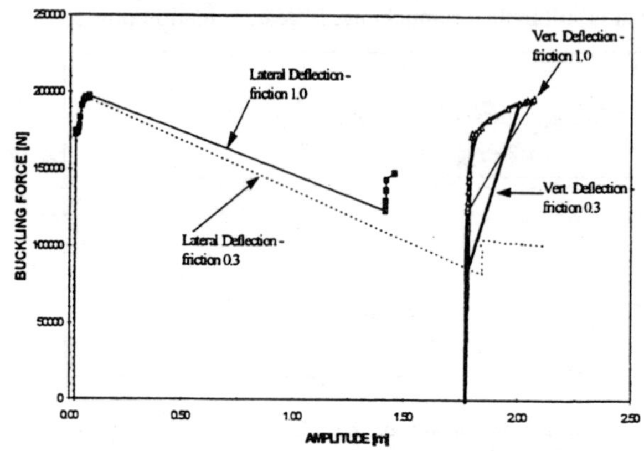

Figure 10: Static 3D analysis, Buck. Force vs deflection

It means that when the pipe lifts above the seabed, it will not experience the peak break out resistance caused by the embedment over the whole lift off length. As it buckles laterally it falls back on the vertical imperfection. However, since it has moved 1.8 m laterally it will only experience the sliding friction resistance of 0.3.

On an uneven seabed, therefore, it appears that the critical lateral buckling force is little affected by the lateral friction resistance.

This phenomenon can be explained by the fact that for a pipeline resting on an uneven seabed, the vertical imperfections can be significantly larger than the horizontal imperfections. As the flowline initially 'buckles' vertically, it looses lateral restraint (less contact, less lateral restraint) whereupon it interacts with the horizontal mode and buckles laterally. Specifically, the pipe starts to lift vertically at approximately 170 kN (Ref. Figure 10) and starts to loose lateral frictional restraint. As a result of this reduction in lateral restraint, the lateral buckling force also decreases. The critical lateral buckling force is reached at an effective axial force of approx. 200 kN. In this condition, an infinitely small increase in the force gives the pipe two options: (1) continue vertical lift off, or (2) lateral buckling. Because the pipe now has reached the point of axial instability in the horizontal plane, the path of least resistance is chosen, and a lateral 'snap' to a new equilibrium configuration occurs.

In order to demonstrate this 3D buckling behaviour further, several 2D buckling analyses were undertaken. One 2D vertical buckling analysis was performed applying the same seabed imperfection as in 3D, but restrained from any lateral movement. In addition, 2D lateral buckling simulations were carried out with the same horizontal OOS as the previously discussed 3D analyses but, on a flat seabed surface. These 2D lateral buckling cases were analysed by modelling a length from the buckle center and outwards with zero friction to simulate the lift off length in 3D. Figure 11 shows the vertical deflected shape from the 3D buckling analysis (μ=0.3) just as the pipe interacts with the horizontal plane.

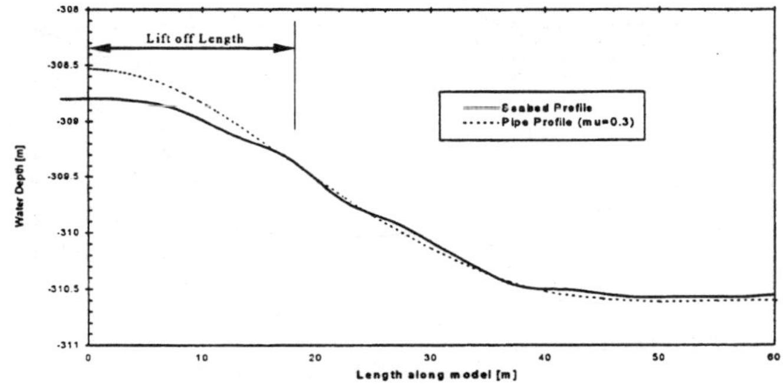

Figure 11: Vertical Pipe - Seabed Profile (Half model)

Figure 12 presents the lateral buckling force for various vertical lift off lengths. It demonstrates that the critical lateral buckling force reduces as the length of pipe lifting off the seabed increases.

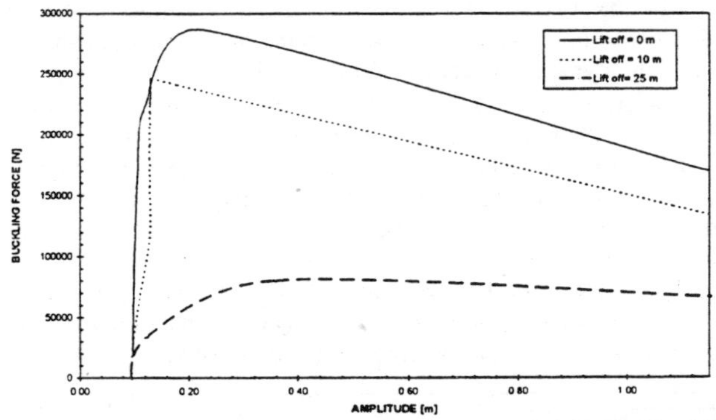

Figure 12: Lateral Buckling Force vs. Amplitude for different lift off lengths

By plotting the 2D lateral and vertical analysis results against each other it can be demonstrated that at a certain force and the corresponding lift off length, the lateral buckling force becomes smaller than the force required to lift the pipe further vertically. At this point the pipe interacts with the horizontal mode and buckles laterally, Ref. point A in Figure 13. The buckling force and lift off length corresponds well with the results from the 3D analyses, see Figures 9 and 10.

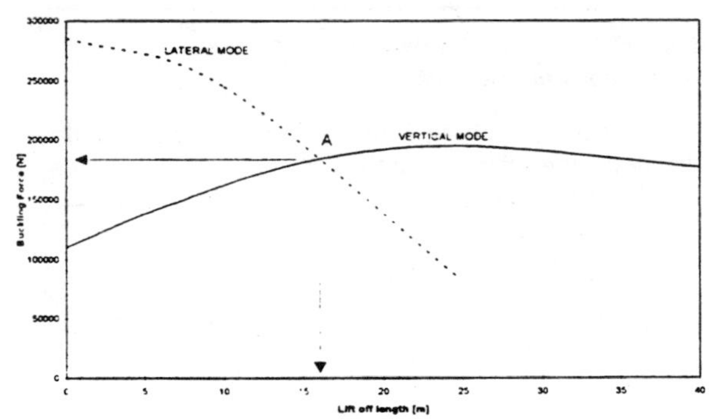

Figure 13: 3D buckling interaction

It should be noted that, in theory, not all pipes which lift off vertically will buckle laterally. A situation could occur where the flowline operational loads only are capable of lifting a section of the pipeline in Figure 13 for a length of 10 m. In this case the vertical mode would still be governing.

In design work, if the analyses results shows the pipe exposed vertically above the seabed, it will be necessary to investigate what other effects may apply, e.g. current, wave and fishing gear loads.

DYNAMIC BUCKLING ANALYSIS

In contrast to the gradual vertical uplift discussed, the lateral buckle is a sudden loss of axial stability which results in a dynamic 'snap' movement. An important objective of the work performed has been to investigate what actually happens with the pipe as it buckles in terms of stresses/strains, displacements, effective axial force etc. In the static buckling analysis when the pipe reaches the ultimate sustainable axial load, then the program has to find the new equilibrium shape by trial and error until an acceptable solution is found within given tolerances. Unfortunately, the static analysis does not report any results during this dynamic 'snap' period. The static Arc-length method may be applied to simulate snap-through response (Hinton 1992). However, during the snap-through process, the flowline experiences acceleration and velocity leading to dynamic effects which may be significant. Therefore, a transient analysis of the dynamic snap behaviour will produce a different equilibrium shape than the static simulation. The transient flowline buckling behaviour was simulated using the general equilibrium equation for a dynamic system:

$$[M]\{\ddot{u}\} + [C]\{\dot{u}\} + [K]\{u\} = \{F(t)\}$$

where

$[M]$ = structural mass matrix

$[C]$ = structural damping matrix

$[K]$ = structural stiffness matrix

$\{u\}$ = nodal displacement vector

$\{\dot{u}\}$ = nodal velocity vector

$\{\ddot{u}\}$ = nodal acceleration vector

$\{F(t)\}$ = time-dependent nodal force vector

The dynamic response is caused by a change from potential energy to kinetic energy. Nodal accelerations are integrated in time steps to satisfy the equilibrium equations until the pipe comes to rest. To account for inertia effects for displacing the surrounding water, mass elements were coupled to each pipe node. For comparison reasons, only temperature load is applied, i.e. the resulting strains will be lower than that of a flowline with full operational loads. The pipe-seabed initial geometry is the same as previously discussed for the 3D static analysis cases.

In ANSYS, stress/strain results can be obtained from 8 positions around the pipe circumference as presented in Figure 14 (Kohnke 1996).

Figure 15 shows the total axial strain from the 3D transient dynamic analysis (μ=0.3) at the buckle point vs. time. At first, during heat up, the highest strains occur at the top of the pipe (Position 90°). However, as the pipeline has lifted off the seabed for a certain distance, the horizontal plane interacts and a lateral 'snap' buckle is initiated. The strain at Position 0° shows that at time 32 s a gradual change in the curvature is started and peaks at 37 s. The maximum total axial strain of 0.09 % occurs at Position 45° as a resultant of the bending at Positions 0°and 90°.

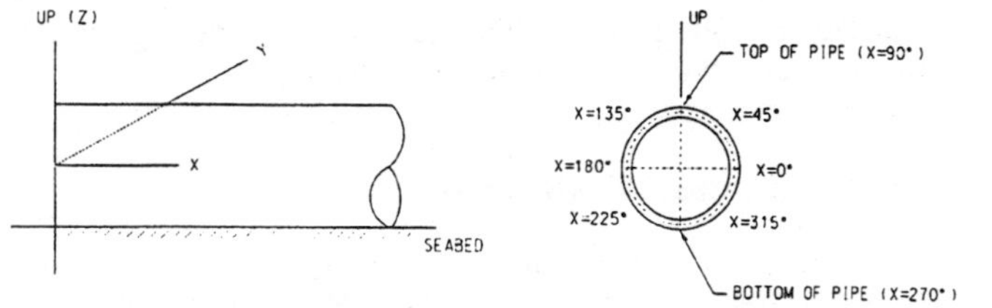

Figure 14: Definition of Position x at pipe circumference

As the snap movement continues the strains decrease and eventually as the 'snap' movement terminates at approx. 39 s the maximum strain at Position 45° stabilises around 0.065 % strain. Hence, the flowline has experienced a 35 % higher total axial strain during the 'snap' movement than in the post buckled configuration.

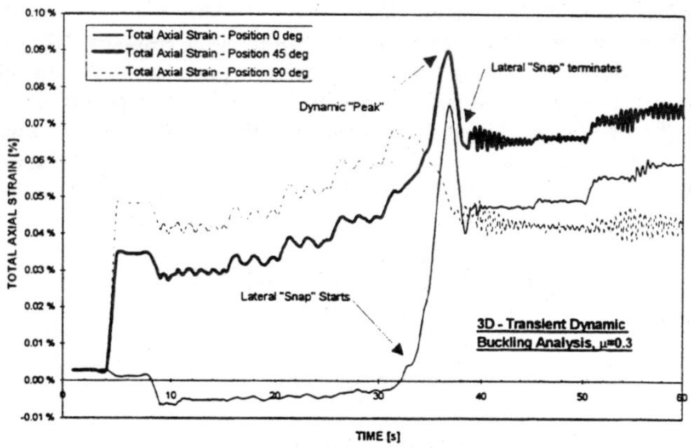

Figure 15: Total Axial Strain vs. Time

A similar effect has been seen in a trawl board pull-over response analysis (by the authors) simulating the transient response of a pipe under high axial compressive load as it buckles when subject to pull-over loads. It was demonstrated that the pipe arrived at a maximum strain after a couple of seconds from impact. After that, the energy absorbed by the pipe started to dissipate away from the buckle center, thereby reducing the curvature and relieving stresses and strains at the impact point.

Even more interesting is the comparison of the dynamic results with the static results for the same case as presented in Figure 16. The static case reveals a maximum total axial strain after buckling of 0.07% which is 30 % less than the maximum strain found during the 'snap' period in the dynamic analysis. In other words, **the strain experienced during the dynamic 'snap' may be significantly higher than the post buckled static result**. Other friction cases show a similar difference between the static and dynamic results. The post buckled strains compare well with only 0.005 % difference in the total axial strain. Furthermore, as temperature increases, dynamic and static strains grow with the same rate.

As the dynamic analysis results above show, the pipe buckles laterally in a continuous smooth manner which is due to the inertia effects. When the buckled pipe reaches its maximum curvature, the kinetic energy continues to displace the pipe laterally away from the buckle point in a wave form. Although this energy wave dissipates relatively quickly, it causes some relaxation of the curvature at the buckle center. As a result of this , the strains reduce as the pipe approaches a stationary condition again.

A curiosity is the speed at which the 'snap' buckle takes place. Figure 15 shows that the dynamic buckle takes approximately 7 - 8 seconds with an approximate displacement of 1.8 m (from Fig. 18). Thus, the 'snap' is in reality a relatively slow movement with a mean velocity of approximately 0.25 m/s.

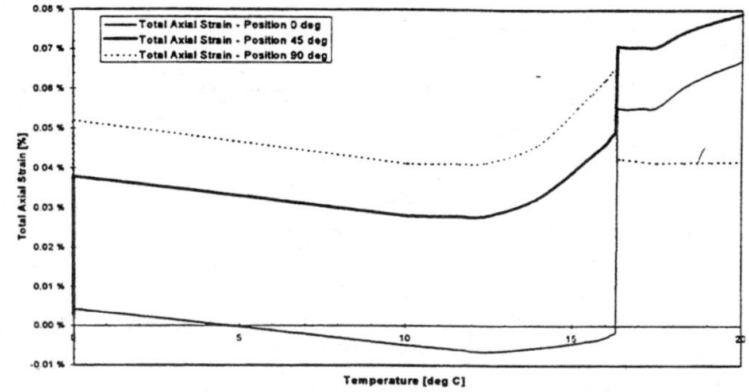

Fig.16: Total Axial Strain vs Temp.(Static case, μ=0.3))

Figure 17 presents the displacements vs. time from the dynamic analysis. The vertical lift off is approximately the same as the static analysis result, see Figures 9 and 10. The step wise lift off movement is caused by the way ANSYS applies the transient temperature load, which is stepped not gradual. The static lateral displacement is slightly larger than the dynamic movement. This may be because the pipeline mass is accelerated as it reaches axial structural instability. In a finite element system, the total energy (=deformation energy + initial energy) is a constant during the time (deformation) history. In a static analysis, the potential energy converted into kinetic energy is neglected and, therefore, the displacement from the static analysis may become larger than that from the more correct dynamic analysis. Another, difference could be that the pipe deflects laterally 1.8 m in the dynamic analysis before establishing contact with the seabed again compared to the static results which indicate contact as the pipe snaps laterally.

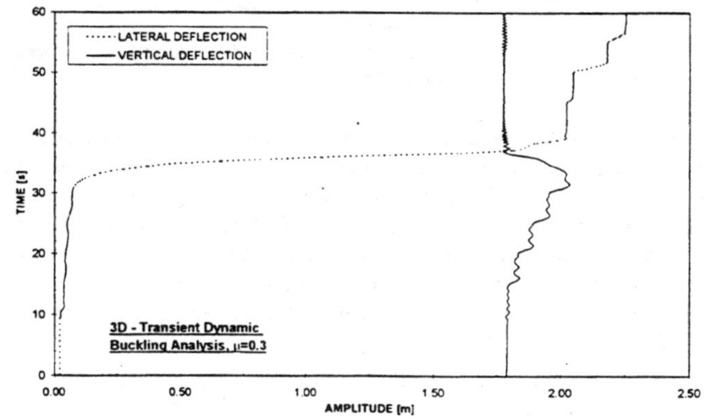

Figure 17: Amplitude vs. Time

The axial effective force results in Figure 18 correlate well with the static results for the same friction coefficient.

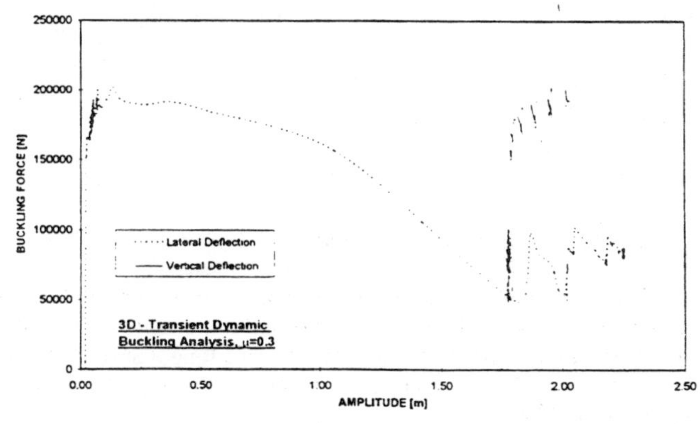

Figure 18: Buckling Force vs. Amplitude

CYCLIC IN-PLACE BEHAVIOUR DURING SHUTDOWN OPERATIONS

The aim of the pipeline structural design and in-service assessment is to ensure that the pipeline adequately fulfills its requirements with respect to serviceability and safety. Safety criteria are set such that the possibility of loss of life, pollution or property losses are acceptably small.

Fortunately, the understanding of the true collapse load has improved over the last decades, aided primarily by the development of non-linear finite element methods that take proper account of plasticity and large deflections. Efficient computational tools are now available for accurately evaluating the collapse behaviour of pipeline systems. In addition, it is also possible to account for cyclic effects. In this work the general F.E.M. package, ANSYS, has been used to establish the cyclic in-place behaviour of the flowline during shutdown and re-start of production.

The considered flowline is subjected to cyclic temperature and internal pressure due to shutdown events. In the limit-state design, it is necessary to ensure that the cyclic behaviour of the flowline is acceptable. This is achieved by the following tasks:

1. Apply cyclic load to the 3D static analysis model
2. Simulate cyclic behaviour of the steel pipe cross-section
3. Check the low-cycle fatigue strength of the girth welds

This paper discusses only the first two tasks which are the global cyclic response to shutdown operations.

Two measures of the pipeline strength may be defined. The first is the static strength in which the load is scaled up until collapse occurs. The second is the cyclic strength in which repeated start-up and shutdowns are taken into account. In the latter case, the limit state may be reached after several cycles and be caused by various effects. The cyclic capacity is defined as the maximum differential loads (temperature and pressure) between start-up and shutdown at which the structure "shakes down" to an elastic state. Some concepts can be mentioned:

a) Cycling leads to ductility exhaustion, causing fracture.
b) The pipeline remains intact, but the strains increase in each cycle until they are no longer acceptable.
c) The strains in each cycle decrease until, eventually the structure behaves purely elastic. This is called "shakedown".

However, if the load intensity were to be increased for the pipeline of case c), further cycling would have to take place to achieve a new shakedown configuration. The maximum load intensity at which the structure shakes down to an elastic state is the cyclic capacity. It is not the intention of this paper to calculate the cyclic capacity, but merely to check the in-place pipeline behaviour under the expected cyclic loads against the allowable strain range.

The analyses used for the in-place cyclic analysis were the same ones used for the 3D static and dynamic buckling analyses discussed previously. The analysis employed is static but includes the following non-linearities:

- Large deflection including stress stiffening effects (geometric non-linearity);
- Elasto-plastic material characteristic (including temperature derating effects and kinematic-hardening);
- General contact elements;
- Element 'birth and death'.

ANSYS provides several options for rate independent plasticity that occurs in a material when a certain level of stress is reached. The multi-linear kinematic strain-hardening option has been used in the cyclic analysis since this will simulate the cyclic behaviour better than, for instance, the isotropic hardening rule.

The investigated flowline is assumed to be subjected to the following cyclic strains:

- 1 strain cycles during reeling-on and off the lay vessel's drum.
- 1 strain cycle during bending over the stinger and bending in the sag bend.
- 100 cycles of planned and unplanned shutdowns.

The planned cycles of shutdowns are estimated to 40-50. However, there are some unplanned shutdowns and 100 is conservatively considered to be the upper limit of the total cycles of shutdowns during the entire operation period.

Only the 100 cycles of planned and unplanned shutdowns are considered in the analysis. It is assumed that the shutdowns are constant, i.e. the same load range is used for all the cycles. These shutdowns are conservatively assumed to be:

- content temperature goes from 130°C to ambient, i.e. $\Delta T=125°C$
- internal pressure goes from 370 bar to 0 bar

Due to the uneven seabed terrain, the various buckles occurring along the route (as shown in Figure 3) will experience different levels of axial feed in from adjacent areas. This variation may for example be caused by spans absorbing expansion, refer Figure 3. As a result of such local effects along the route, it must be considered whether all buckles must be investigated in detail or a general approach can be adapted considering only the "worst" buckle.

In order to demonstrate some of the governing parameters w.r.t. cyclic buckling behaviour, the following cases were analysed by F.E.M.

A) Lateral buckles occur approximately every 700 m, giving a 3D half symmetry model of 350 m length. Friction Coefficient, $\mu=0.3$.
B) Same as A) but with a friction coefficient, $\mu = 1.0$.
C) Lateral buckles occur approximately every 1400 m, giving a 3D half symmetry model of 700 m length. Friction Coefficient, $\mu=0.3$.

The general characteristic of the analyses is as follows: Analysing a pipeline on a 3D seabed surface having both a vertical and horizontal out-of-straightness. Applying internal pressure and temperature loads up to full operational load. Thereafter, reducing the internal pressure to 0 and gradually lowering the pipe wall temperature to ambient. For all cases 21 shutdowns were simulated.

Figure 19 presents Case A) the lateral displacement of the buckle center point vs. No. cycles. From these results it can be seen that after 6 - 7 shutdowns the lateral displacement converges towards approximately 7.1 m.

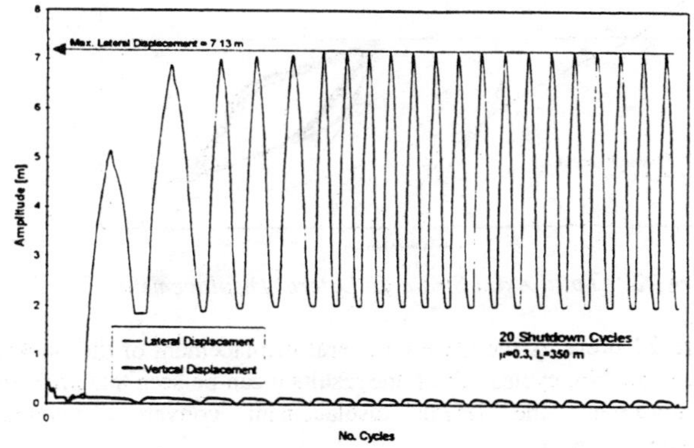

Figure 19: Displacement vs. No. Shutdown cycles

The corresponding total axial strains are shown in Figure 20 as a function of lateral displacement. The maximum tensile and compressive strains occur during the first cycle and shake down to +0.14 % and -0.11 % respectively.

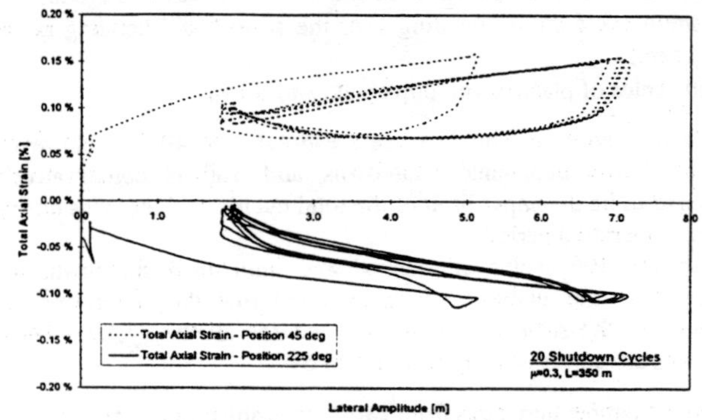

Figure 20: Total Axial Strain vs. Lateral Displacement

For Case 2) with a higher friction coefficient (μ = 1.0), the lateral displacement converges towards a maximum level of approximately 9.0 m after approximately 10 cycles, as shown in Figure 21.

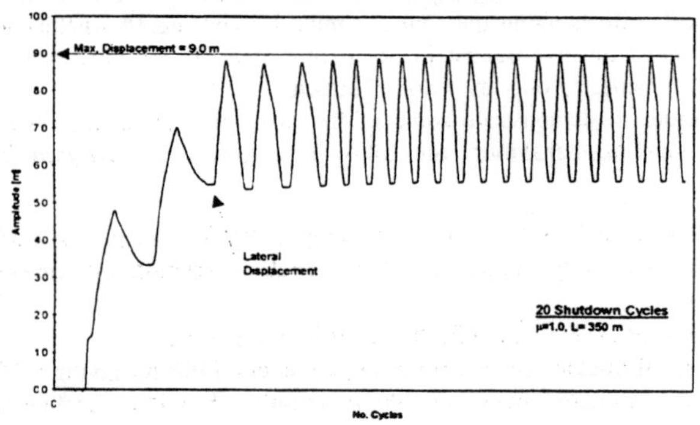

Figure 21: Displacement vs. No. Shutdown cycles

The maximum strains occur during the first cycle (see Figure 22), and shakes down to a lower level in the subsequent cycles. A shakedown level of +0.22 % and -0.19 % are found for the tensile and compressive strains respectively.

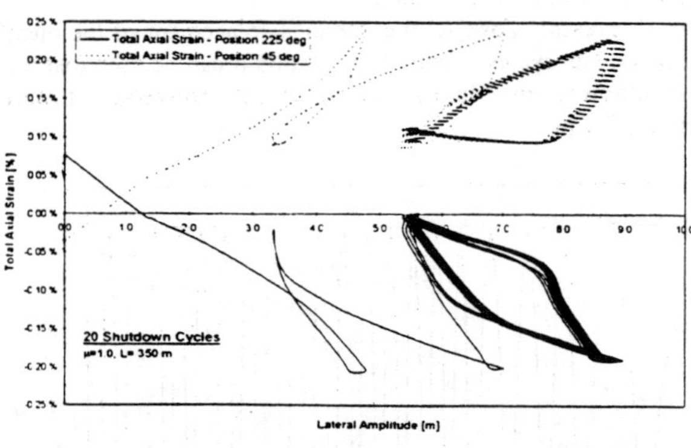

Figure 22: Total Axial Strain vs. Lateral Displacement

Figure 23 presents the Case C) lateral displacement of the buckle center point vs. No. cycles. From the results it can be seen that after 16 - 17 shutdowns the lateral displacement converges towards approximately 11.8 m.

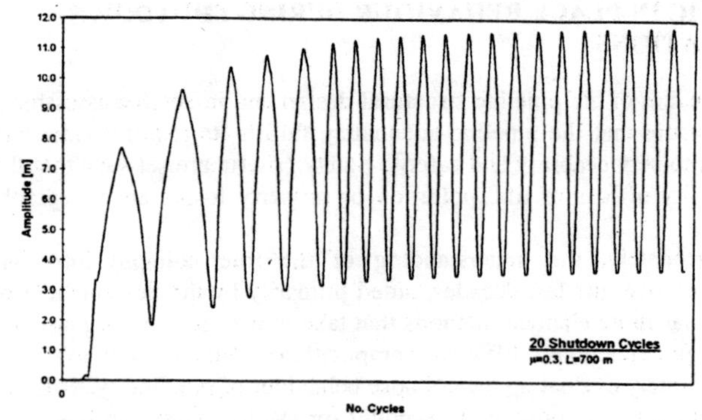

Figure 23: Lateral Displacement vs. No. Cycles

The corresponding total axial strains are shown in Figure 24 as a function of lateral displacement. The maximum tensile and compressive strains occur during the first cycle and shake down to +0.17 % and -0.13 % respectively.

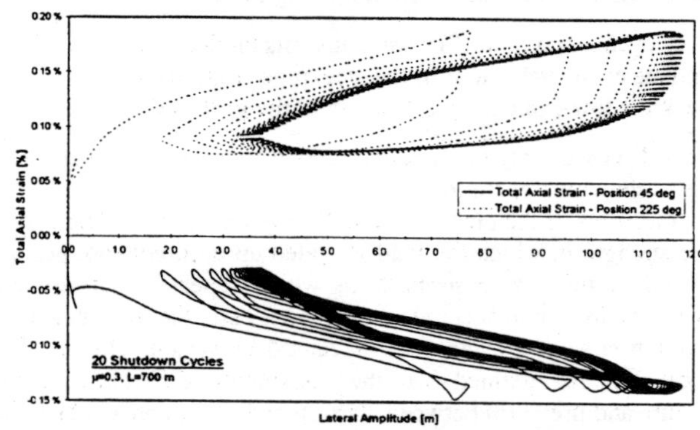

Figure 24: Total Axial Strain vs. Lateral Displacement

A very small increase in the lateral displacement of approximately 15 mm was found between the last two cycles, however, no increase in the total axial strain occurs between the last two cycles. The results generally show that the strains are reduced after subsequent shutdown cycles. Plastic axial strains are not increasing. For each start-up cycle, the growth of the total axial strain and displacement becomes smaller and smaller and eventually tends to zero, indicating that a shakedown situation exists.

It should be noted that the shutdown load assumptions are very conservative, since it is assumed that the pressure and temperature goes all the way to ambient for each shutdown. In reality, the majority of the shutdown incidents will be of short duration, causing only small reductions in temperature and pressure.

CONCLUSIONS

- The work performed shows that representing the pipe-seabed system in a realistical 3D pipe-seabed analysis model can eliminate the conservatism due to the limitations of a traditional 2D analysis approach. The global behaviour of the HP/HT flowline is used to define the expected stress/strain levels as a framework for establishing the governing failure mode of the limit state design.
- For flowlines resting on a very uneven seabed terrain, the vertical seabed imperfections result in more abrupt curvature than the horizontal imperfections created during laying. As a result, the axial load that initially forces the pipe vertically is found to be lower than the corresponding load needed to buckle the pipe laterally. It is observed that the pipe initially moves in the vertical

plane as a gradual magnification of the initial imperfection due to its large initial OOS. As the pipe lifts off the seabed, a length of the pipe becomes laterally unrestrained and thus the critical axial force needed for lateral buckling decreases. Eventually, if sufficient pipe has lifted off the seabed, the lateral buckling force becomes less than the force required to lift the pipeline further in the vertical plane. At that moment, an interaction with the horizontal mode occurs and the pipe goes through dynamic 'snap' movement.

- On an uneven seabed, it appears that the critical lateral buckling force is little affected by the lateral friction coefficient.

- The 3D lateral buckling results show significantly lower stresses and strains in the buckles compared to 2D vertical buckling results for the same seabed and loads.

- The transient dynamic buckling analyses show a local stress/strain peak occurring during the dynamic 'snap' period. These stresses/strains may be considerably larger than what the post-buckled static results show.

- The 3D transient dynamic buckling results demonstrate approximately the same buckling forces and resulting displacements as the static analyses.

- The cyclic shutdown/startup analyses show that the flowline bending tends to 'shake down' to an elastic state after a few cycles. For all cases, the largest stresses/strains are found to occur during the first operation start up. The deflected shape of the flowline migrates asymptotically towards a limiting value after few cycles.

REFERENCES

Bai, Y and Damsleth, P (1997): "Limit State Design of Pipelines", *Proc. of the In. Conf.on Offshore Mechanics and Arctic Eng.*

Ellinas, CP, Supple, WJ and Vastenholt, H (1990): "Prevention of Upheaval Buckling of Hot Submarine Pipelines by Means of Intermittent Rock Dumping", *Proceedings 22nd Offshore Technology Conference, pp 519-528.*

Hinton, E (1992): "Introduction to Nonlinear Finite Element Analysis", *NAFEMS, Glasgow.*

Hobbs, RE (1984): "In-Service Buckling of Heated Pipelines", *Journal of Transport Engineering, pp 178-179.*

Hobbs, RE and Liang, F (1989): "Thermal Buckling of Pipelines Close to Restraints", *Proceedings 8th Int. Conf on Offshore Mechanics and Arctic Eng. pp 121-127.*

Kershenbaum, NY, Harrison, GE and Choi, HS (1996): "Subsea Pipeline Lateral Deviation Due to High Temperature Product", *Proceedings 6th Int. Offshore and Polar Eng Conf, Vol 2, pp 74-79.*

Kohnke, P (1996): "ANSYS v. 5.3 Theory Manual", *ANSYS Inc.*

Nes, H, Sævik, S, Levold, E and Johannesen, A (1996): "Expansion Control Design of Large Diameter Pipelines", *Proceedings 15th Int. Conf on Offshore Mechanics and Arctic Eng, Vol 5,pp 279-285.*

Pedersen, PT and Jensen JJ (1988): "Upheaval Creep of Buried Heated Pipelines with Initial Imperfections", *Journal of Marine Structures.*

Sævik, S and Levold, E (1995): "High Temperature Snaking Behaviour of Pipelines", *Proceedings 5th Int. Offshore and Polar Eng. Conf., Vol. II, pp 63-71.*

Taylor, N and Gan, AB (1986): "Submarine Pipeline Buckling - Imperfection Studies", *Thin Walled Structures, Vol 4, pp 295-323.*

Tørnes, K, Nystrøm, PR, Damsleth, PA and Sortland, LH (1997): "The Behaviour of High Pressure, High Temperature Flowlines on Very Uneven Seabed", *Proceedings 7th Int. Offshore and Polar Eng. Conf.*

Proceedings of the Seventh (1997) International Offshore and Polar Engineering Conference
Honolulu, USA, May 25-30, 1997

ISBN 1-880653-28-1 (Set); ISBN 1-880653-30-3 (Vol. II)

Elastic-Plastic Analysis of a Thick-Walled Submarine Pipeline Considering Plastic Compressibility

C.H. Shin
Korean Register of Shipping
Taejon, Korea

ABSTRACT

Some thick-walled submarine pipelines in operation are subjected to combined loading of high internal pressure and temperature. In this study, an elastic-plastic analysis of a thick-walled pipe under combined internal pressure and radial temperature gradient is carried out considering the plastic compressibility. The Drucker-Prager plastic potential is used to describe the plastic compressible yielding and the strength-differential effect. For an elastic-plastic analysis of a thick-walled pipe, an elastic-viscoplastic model is used and a general elastic-viscoplastic constitutive equation is formulated. The influence of the plastic compressibility and temperature is investigated for a thick-walled pipe, using the finite element method.

KEY WORDS: Thick-walled pipeline, elastic-plastic analysis, plastic compressibility, strength-differential effect, viscoplastic model, finite element method

INTRODUCTION

In order to avoid possible failures in an operating submarine pipeline, the strength analysis is performed. Specifically, an elastic-plastic analysis is performed for a thick-walled pipeline under combined loading of high pressure and temperature. In the elastic-plastic analysis, plastic incompressibility is generally assumed. However, it is known that plastic compressibility should be considered for high strength steels or sintered powder materials since their plastic deformations are dependent on hydrostatic stresses (Betten and Borrmann, 1984; Spitzig *et al.*, 1976).

In Fig. 1, the effect of hydrostatic pressure on the yield stress of the HY-80 submarine-hull steel is shown. The flow stress increases with hydrostatic pressure. Specifically, the flow stress is greater in uniaxial compression than in uniaxial tension. This phenomenon is known as the strength-differential or S-D effect, and can be explained by employing the Drucker-Prager (1952) plastic potential.

For elastic-plastic analysis, an elastic-viscoplastic model is employed and a general elastic-viscoplastic constitutive equation is formulated. As is well known, a viscoplastic model for the plastic shear flow was proposed for the first time by Bingham (1922). In 1932, Hohenemser and Prager constructed a viscoplastic constitutive equation. using the relation of stress-strain rate derived from experiments. Based on their results, Perzyna (1966) proposed a generalized elastic-viscoplastic model, which considered the material stability criterion postulated by Drucker (1959).

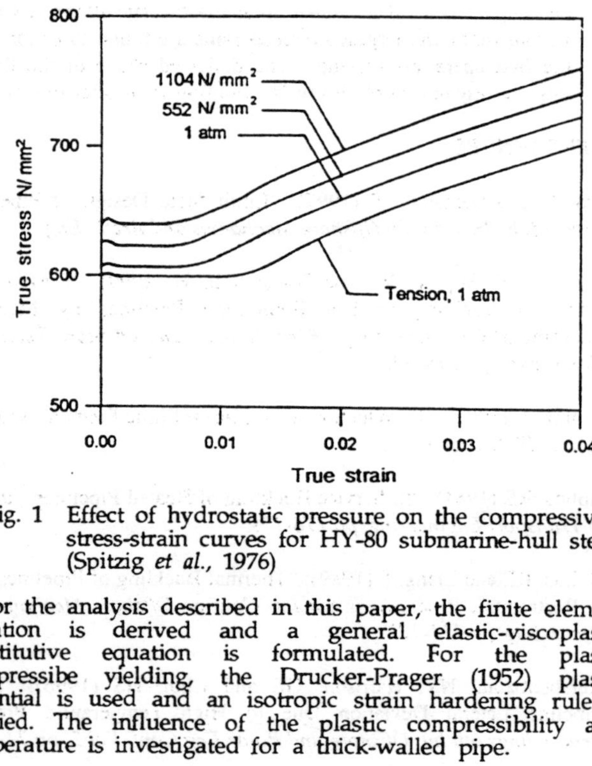

Fig. 1 Effect of hydrostatic pressure on the compressive stress-strain curves for HY-80 submarine-hull steel (Spitzig *et al.*, 1976)

For the analysis described in this paper, the finite element equation is derived and a general elastic-viscoplastic constitutive equation is formulated. For the plastic compressibe yielding, the Drucker-Prager (1952) plastic potential is used and an isotropic strain hardening rule is applied. The influence of the plastic compressibility and temperature is investigated for a thick-walled pipe.

FINITE ELEMENT EQUATION

Since the viscoplastic model is given as a function of time, strain rate instead of strain is used in the constitutive equation. At time $t=t_n$, the strain rate ($\dot{\varepsilon}^n_{ij}$) can be expressed as the sum of an elastic (${}^e\dot{\varepsilon}^n_{ij}$) and inelastic (${}^{vp}\dot{\varepsilon}^n_{ij}$) component, i.e.

$$\dot{\varepsilon}_{ij}^{n} = {}^{e}\dot{\varepsilon}_{ij}^{n} + {}^{vp}\dot{\varepsilon}_{ij}^{n}. \quad (1)$$

The stress increment for the corresponding time increment $\Delta t_n = t_{n+1} - t_n$ can be written as

$$\Delta\sigma_{ij}^{n} = D_{ijkl}\,\Delta^{e}\varepsilon_{kl}^{n}, \quad (2)$$

where the elastic stress-strain transformation tensor D_{ijkl} for isotropic materials is given by

$$D_{ijkl} = \mu(\delta_{ik}\delta_{jl} + \delta_{il}\delta_{jk}) + \lambda\delta_{ij}\delta_{kl} \quad (3)$$

with μ and λ being the Lamé constants.

Using an implicit scheme, a viscoplastic strain increment $\Delta^{vp}\varepsilon_{ij}^{n}$ can be defined as

$$\Delta^{vp}\varepsilon_{ij}^{n} = [(1-\theta)\,{}^{vp}\dot{\varepsilon}_{ij}^{n} + \theta\,{}^{vp}\dot{\varepsilon}_{ij}^{n+1}]\,\Delta t_n, \quad (4)$$

where θ is a parameter to perform time integration.

Introducing the pseudoviscoplastic tensor

$$H_{ijkl}^{n} = \frac{\partial\,{}^{vp}\dot{\varepsilon}_{ij}^{n}}{\partial\sigma_{kl}^{n}}, \quad (5)$$

equation (4) can be expressed as follows:

$$\Delta^{vp}\varepsilon_{ij}^{n} = \Delta t_n\,{}^{vp}\dot{\varepsilon}_{ij}^{n} + C_{ijkl}^{n}\,\Delta\sigma_{kl}^{n}, \quad (6)$$

where

$$C_{ijkl}^{n} = \Delta t_n\,\theta\,H_{ijkl}^{n}. \quad (7)$$

Then, the stress increment can be calculated from equations (2) and (6) as follows:

$$\Delta\sigma_{ij}^{n} = \hat{D}_{ijkl}^{n}\,(\Delta\varepsilon_{kl}^{n} - \Delta t_n\,{}^{vp}\dot{\varepsilon}_{kl}^{n}), \quad (8)$$

where

$$\hat{D}_{ijkl}^{n} = [D_{ijkl}^{(-1)} + C_{ijkl}^{n}]^{(-1)}. \quad (9)$$

The matrix form of equation (8) is

$$\Delta\boldsymbol{\sigma}^{n} = \hat{\mathbf{D}}^{n}(\Delta\boldsymbol{\varepsilon}^{n} - \Delta t_n\,{}^{vp}\dot{\boldsymbol{\varepsilon}}^{n}). \quad (10)$$

It is assumed that the yield criterion and the stress-strain relations are unaffected by the variations in temperature. Then, thermal effects can be included in the analysis of an expanding pipe by modifying the stress-strain equations. Subtracting the stress increment due to initial thermal strain $\boldsymbol{\varepsilon}_0$ from the above stress increment (10), the following stress-strain relationship can be obtained:

$$\Delta\boldsymbol{\sigma}^{n} = \hat{\mathbf{D}}^{n}(\Delta\boldsymbol{\varepsilon}^{n} - \Delta t_n\,{}^{vp}\dot{\boldsymbol{\varepsilon}}^{n}) - \mathbf{D}\Delta\boldsymbol{\varepsilon}_0. \quad (11)$$

The equilibrium equation at any instant of time $t = t_n$ can be written as

$$\int_V \mathbf{B}^{T}\boldsymbol{\sigma}^{n}\,dV - \boldsymbol{R}^{n} = 0, \quad (12)$$

where V is the volume, $\boldsymbol{R}^{n}$ denotes the vector of equivalent nodal forces, and $\mathbf{B}$ is the strain-displacement matrix satisfying the following relation:

$$\Delta\boldsymbol{\varepsilon}^{n} = \mathbf{B}\,\Delta\hat{\boldsymbol{u}}^{n}, \quad (13)$$

in which $\Delta\hat{\boldsymbol{u}}^{n}$ represents the increment of nodal displacements. Using equations (11) and (13), the incremental form of equation (12) can be expressed as

$$\mathbf{K}_t^{n}\,\Delta\hat{\boldsymbol{u}}^{n} = \Delta\boldsymbol{V}^{n}, \quad (14)$$

where $\mathbf{K}_t^{n}$ is the tangential stiffness matrix and $\Delta\boldsymbol{V}^{n}$ is the vector of incremental pseudoforces:

$$\mathbf{K}_t^{n} = \int_V \mathbf{B}^{T}\,\hat{\mathbf{D}}^{n}\,\mathbf{B}\,dV \quad (15)$$

$$\Delta\boldsymbol{V}^{n} = \Delta\boldsymbol{R}^{n} + \Delta t_n\int_V \mathbf{B}^{T}\,\hat{\mathbf{D}}^{n}\,{}^{vp}\dot{\boldsymbol{\varepsilon}}^{n}\,dV. \quad (16)$$

In the above equation (16), the vector of incremental nodal forces $\Delta\boldsymbol{R}^{n}$ can be expressed as

$$\Delta\boldsymbol{R}^{n} = \int_A \mathbf{N}^{T}\Delta\boldsymbol{p}\,dA + \int_V \mathbf{B}^{T}\,\mathbf{D}\Delta\boldsymbol{\varepsilon}_0\,dV, \quad (17)$$

where $\mathbf{N}$ is the displacement interpolation matrix, $\boldsymbol{p}$ is the vector of boundary tractions, and A is the surface area (Owen and Hinton, 1980).

By using the nodal displacement increments from equation (14), the stress at time $t = t_{n+1}$ can be evaluated from equations (11) and (13):

$$\boldsymbol{\sigma}^{n+1} = \boldsymbol{\sigma}^{n} + \Delta\boldsymbol{\sigma}^{n}. \quad (18)$$

However, since the stress in equation (18) is calculated from the linearized equation (14), the actual equilibrium equation (12) is not satisfied exactly. To reduce this error, unbalanced forces expressed as

$$\boldsymbol{\psi}^{n+1} = \boldsymbol{R}^{n+1} - \int_V \mathbf{B}^{T}\boldsymbol{\sigma}^{n+1}\,dV \quad (19)$$

are added to the force increment at the next time $t = t_{n+1}$.

CONSTITUTIVE EQUATION

It is assumed in viscoplasticity theory that a quasistatic yield surface exists and that the material is viscoplastic when the stress point lies outside the yield surface. Thus, the viscoplastic strain rate is postulated as a function of the excess stress above the static yield criterion. The viscoplastic constitutive equation is set up outside the yield surface with the following assumptions:

- The material behavior within the yield surface is elastic and reversible
- The material behavior outside the yield surface depends on time and is irreversible
- The material shows elastic-plastic behavior at an infinitesimally small deformation rate.

In this study, the Perzyna (1966) model is modified to find the elastic-viscoplastic compressible behavior of materials. For simplicity, the superscript n denoting time is omitted hereafter.

$$\begin{aligned}\dot{\varepsilon}_{ij} &= {}^{e}\dot{\varepsilon}_{ij} + {}^{vp}\dot{\varepsilon}_{ij} \\ &= \frac{1}{2G}\dot{\sigma}_{ij}' + \frac{1-2\nu}{E}\frac{1}{3}\dot{\sigma}_{kk}\delta_{ij} + \gamma\langle\phi(f)\rangle\frac{\partial\sigma_e}{\partial\sigma_{ij}},\end{aligned} \quad (20)$$

where f is a yield function in the form

$$f = \frac{\sigma_e(\sigma_{ij}')}{\sigma_y} - 1, \quad (21)$$

G is the shear modulus, E is the elastic modulus, ν is the Poisson ratio, γ is a fluidity parameter, δ_{ij} is the Kronecker delta, σ_e is the equivalent stress, σ_{ij}' is the deviatoric stress, and σ_y is the uniaxial yield stress.

The notation $\langle\ \rangle$ in equation (20) implies

$$\langle\phi(f)\rangle = 0 \quad \text{for } f \leqq 0 \quad (22a)$$

$$\langle\phi(f)\rangle = \phi(f) \quad \text{for } f > 0. \quad (22b)$$

For simplicity, the notation $\langle\ \rangle$ is omitted hereafter and only the viscoplastic component is considered:

$${}^{vp}\dot{\varepsilon}_{ij} = \gamma\,\phi(f)\frac{\partial\sigma_e}{\partial\sigma_{ij}}, \quad (23)$$

where $\phi(f)$ can be expressed by the power law:

$$\phi(f) = f^{m} = \left(\frac{\sigma_e}{\sigma_y} - 1\right)^{m}. \quad (24)$$

In the above equation, the equivalent stress σ_e can be calculated according to the theory of the maximum strain energy for the plastic potential $F(J_1, J_2')$ in which plastic compressibility is considered:

$$\sigma_e = \frac{J_1 \frac{\partial F}{\partial J_1} + 2J_2' \frac{\partial F}{\partial J_2'}}{\left(\frac{\partial F}{\partial J_1}\right)_e + \frac{2}{3}\sigma_e\left(\frac{\partial F}{\partial J_2'}\right)_e}, \tag{25}$$

where

$$J_1 = \sigma_{ii} \tag{26a}$$

$$J_2' = \frac{1}{2}\sigma_{ij}'\sigma_{ji}'. \tag{26b}$$

The subscript e to the parentheses in equation (25) represents the uniaxial equivalent stress state. Therefore, the viscoplastic constitutive equation (23) can be expressed as follows:

$$^{vp}\dot{\varepsilon}_{ij} = \gamma\phi(f)\left(\frac{\partial\sigma_e}{\partial J_1}\delta_{ij} + \frac{\partial\sigma_e}{\partial J_2'}\sigma_{ij}'\right). \tag{27}$$

Denoting the second invariant of the viscoplastic strain rate tensor by

$$I_2' = \frac{1}{2}\,{}^{vp}\dot{\varepsilon}_{ij}\,{}^{vp}\dot{\varepsilon}_{ji}, \tag{28}$$

the equivalent stress can be expressed from equations (21) and (23) as

$$\sigma_e = \sigma_y\left[1 + \phi^{-1}\left\{\frac{\sqrt{I_2'}}{\gamma}\left(\frac{1}{2}\frac{\partial\sigma_e}{\partial\sigma_{ij}}\frac{\partial\sigma_e}{\partial\sigma_{ji}}\right)^{-\frac{1}{2}}\right\}\right], \tag{29}$$

where ϕ^{-1} is the inverse function of ϕ. The above equation represents the dynamic yield condition for materials and can be expressed for the uniaxial stress state as

$$\sigma_e = \sigma_y\left[1 + \phi^{-1}\left(\frac{^{vp}\dot{\varepsilon}_e}{\gamma}\right)\right], \tag{30}$$

where $^{vp}\dot{\varepsilon}_e$ denotes the equivalent viscoplastic .strain rate. From this equation, the dynamic strength is higher than the static strength in the viscoplastic range of materials.

Since the elastic-plastic behavior ($\sigma_e = \sigma_y$) of materials can be obtained from the limiting case $^{vp}\dot{\varepsilon}_e \to 0$ or $\gamma \to \infty$ in equation (30), the viscoplastic model can be employed for the analysis of elastic-plastic problems. Specifically, with a high yield stress, the inelastic component in equation (20) vanishes ($^{vp}\dot{\varepsilon}_{ij} = 0_{ij}$) so that the elastic behavior of materials can be obtained.

In this study, the Drucker-Prager (1952) potential

$$F = \alpha J_1 + \sqrt{J_2'}, \tag{31}$$

which can handle the plastic compressibility and S-D effect, is employed. The equivalent stress for this potential can be calculated from equation (25):

$$\sigma_e = \frac{\alpha J_1 + \sqrt{J_2'}}{\alpha + 1/\sqrt{3}}. \tag{32}$$

The material constant α is obtained by the following equation from the uniaxial test results:

$$\alpha = \frac{1}{\sqrt{3}}\frac{\sigma_{yc} - \sigma_{yt}}{\sigma_{yc} + \sigma_{yt}}, \tag{33}$$

where σ_{yc} and σ_{yt} are the compressive and tensile yield stress, respectively. Therefore, the viscoplastic strain rate can be written as

$$^{vp}\dot{\varepsilon}_{ij} = \gamma\,\phi(f)\,(c_0\,\delta_{ij} + c_1\,\sigma_{ij}'), \tag{34}$$

where

$$c_0 = \frac{\alpha}{\alpha + 1/\sqrt{3}} \tag{35a}$$

$$c_1 = \frac{1}{\alpha + 1/\sqrt{3}}\frac{1}{2\sqrt{J_2'}}. \tag{35b}$$

Using the formula given by Betten *et al.* (1982), the equivalent viscoplastic strain rate can be calculated as

$$^{vp}\dot{\varepsilon}_e = \frac{\sqrt{3}\alpha + 1}{\sqrt{6\alpha^2 + 1}}\sqrt{\frac{2}{3}\,{}^{vp}\dot{\varepsilon}_{ij}\,{}^{vp}\dot{\varepsilon}_{ji}}. \tag{36}$$

The plastic Poisson ratio can be expressed from equation (34) as

$$^{p}\nu = \frac{1}{2}\frac{1 - 2\sqrt{3}\alpha}{1 + \sqrt{3}\alpha}. \tag{37}$$

The effective range of the plastic Poisson ratio is given by

$$0 \leqq {}^{p}\nu \leqq 1/2 \tag{38}$$

so that the following restrictions are obtained:

$$0 \leqq \alpha \leqq 1/\sqrt{12} \quad \text{or} \quad 1 \leqq \sigma_{yc}/\sigma_{yt} \leqq 3. \tag{39a,b}$$

Thus, in this study, the Drucker-Prager (1952) potential appears to be valid only for materials satisfying the above range (39a,b). Specifically, the Mises yield condition is given for $\alpha = 0$ ($\sigma_{yc}/\sigma_{yt} = 1$).

NUMERICAL CALCULATIONS

The elastic-plastic analysis of a thick-walled pipe under combined loading of internal pressure and temperature is performed using the modified Perzyna (1966) viscoplastic model. For the analysis, the following characteristic properties are used and the plane strain condition is assumed:

inner radius	$a = 180\ mm$
outer radius	$b = 200\ mm$
elastic modulus	$E = 2.1 \times 10^5\ N/mm^2$
tangent modulus	$E_t = 2.1 \times 10^4\ N/mm^2$
Poisson ratio	$\nu = 0.3$
yield stress	$\sigma_y = 240\ N/mm^2$
outer temperature	$T_b = 0$ ℃
thermal expansion coefficient	$\alpha_t = 1.2 \times 10^{-5}$ /℃

If a steady state distribution of temperature is assumed, temperature T satisfies the Laplace equation $\nabla^2 T = 0$, the solution of which may be written as

$$T = T_a\frac{\ln(b/r)}{\ln(b/a)}, \tag{40}$$

where T_a is the temperature at the inner surface of the pipe.

To solve the elastic-plastic problem, the parameters γ and m are chosen as $\gamma = 0.001$ and $m = 1$, and the calculations are repeated until the viscoplastic strain rate is equal to zero.

In Figs. 2 and 3, the distributions of circumferential stress (σ_θ) and equivalent stress (σ_e) are respectively shown for $\alpha = 0$ and $T_a = 0$ ℃. From the comparison of these figures, it can be seen that the position of the greatest circumferential stress in Fig. 2 indicates the elastic-plastic boundary. The circumferential stress in the plastic region increases with internal pressure since the material strain-hardens ($E_T/E = 0.1$). Chakrabarty (1988), however, showed that the circumferential stress in the plastic region decreases with internal pressure for an elastic-perfectly plastic material.

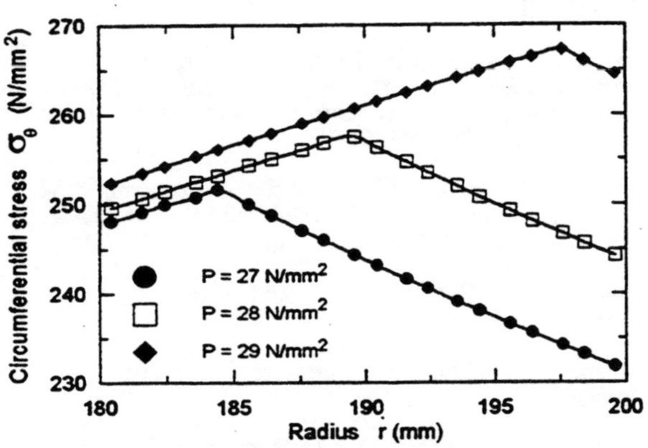

Fig. 2 Circumferential stress distribution in a thick-walled pipe for $\alpha=0$ and $T_a=0$ ℃

In Fig. 3, it can be seen that yielding begins at the inner surface of the pipe and a plastic zone spreads outward with increasing internal pressure. The equivalent stress in the plastic region is higher than the yield stress due to the material hardening.

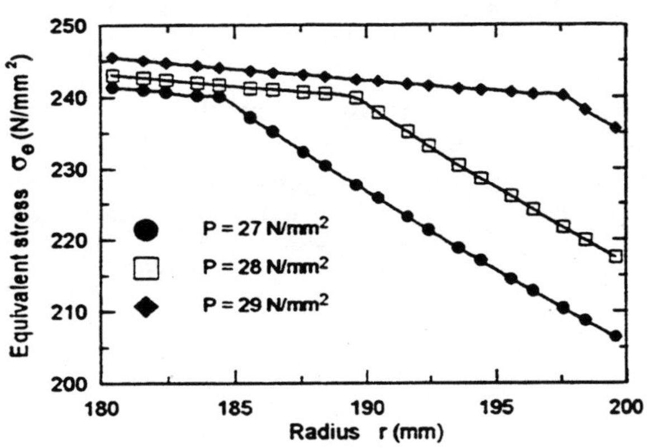

Fig. 3 Equivalent stress distribution in a thick-walled pipe for $\alpha=0$ and $T_a=0$ ℃

Fig. 4 shows the equivalent stress distribution for $\alpha=0$ and $T_a=50$ ℃. Under combined loading of internal pressure and temperature, the initial yielding can occur anywhere in the pipe. In Fig. 4, yielding begins at the outer surface of the pipe and the plastic zone spreads inward with increasing internal pressure. It is also seen that the equivalent stress increases with the radius of the pipe in the elastic region.

Fig. 5 shows the equivalent stress distribution for $\alpha=0.1155$ $(\sigma_{yc}/\sigma_{yt}=1.5)$ and $T_a=50$ ℃. Considering the plastic compressibility under the combined loading, yielding begins at the inner surface of the pipe and the plastic zone spreads outward with increasing internal pressure as in Fig. 3.

In Fig. 6, the equivalent stress distribution for $p=13\ N/mm^2$ and $T_a=50$ ℃ is shown for four different values of α, respectively: $\alpha=0$ $(\sigma_{yc}/\sigma_{yt}=1)$, $\alpha=0.1155$ $(\sigma_{yc}/\sigma_{yt}=1.5)$, $\alpha=0.1925$ $(\sigma_{yc}/\sigma_{yt}=2)$, $\alpha=0.2887$ $(\sigma_{yc}/\sigma_{yt}=3)$. When the material is assumed to be plastic incompressible ($\alpha=0$), the pipe is entirely elastic. The equivalent stress has its greatest value at the outer surface of the pipe. However, the stress distribution is changed with increasing α and the equivalent stress of the inner surface becomes higher than that of the outer surface in case of $\alpha=0.1155$. Therefore, it is clear that yielding begins at the inner surface and the plastic zone spreads outward with increasing α. For $\alpha=0.2887$, the pipe becomes fully plastic. It can be seen from this figure that the plastic compressibility of the material has an effect on the stress distribution of the pipe.

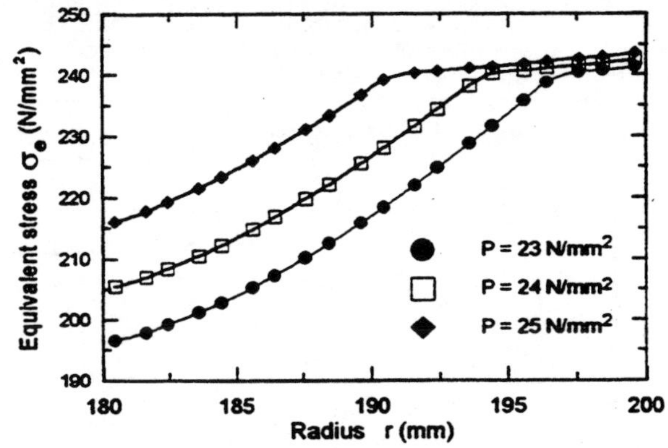

Fig. 4 Equivalent stress distribution in a thick-walled pipe for $\alpha=0$ and $T_a=50$ ℃

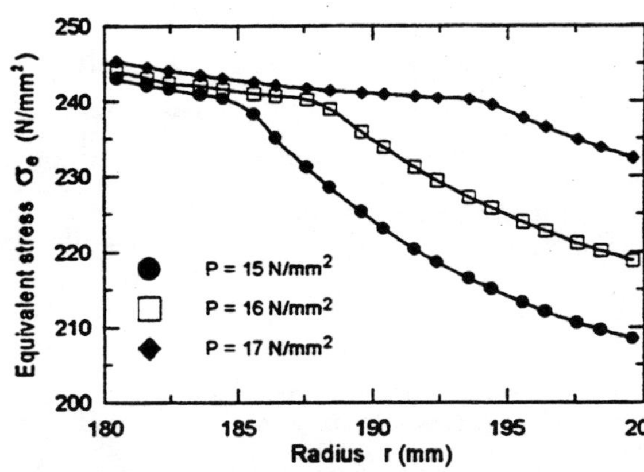

Fig. 5 Equivalent stress distribution in a thick-walled pipe for $\alpha=0.1155$ and $T_a=50$ ℃

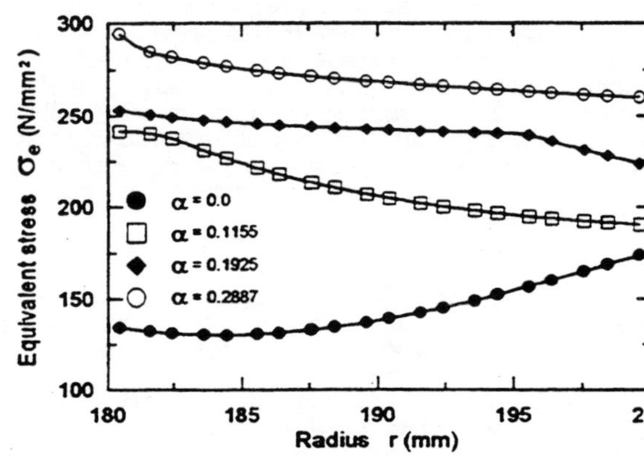

Fig. 6 Equivalent stress distribution in a thick-walled pipe for $p=13\ N/mm^2$ and $T_a=50$ ℃

In Fig. 7, the internal pressure (p_f), just when the pipe becomes fully plastic, is presented for different values of temperature. For $T_a = 0$ ℃, the plastic compressibility α has a small effect on the fully plastic pressure. However, the effect of the plastic compressibility increases with increasing temperature. The plastic compressibility of the material has a significant effect on the fully plastic pressure for $T_a = 50$ ℃. It can also be found that the pipe of a plastic compressible material sustains lower internal pressure than that of a plastic incompressible material. Therefore, the strength-differential effect should be taken into account for the elastic-plastic analysis of the pipe under combined internal pressure and temperature.

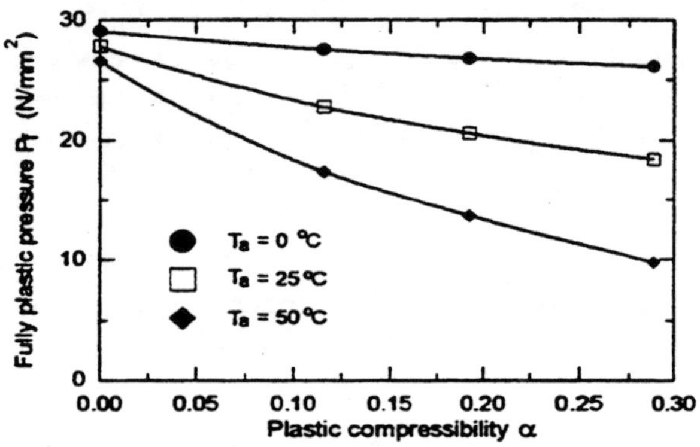

Fig. 7 Strength-differential effect on the fully plastic pressure

When the material strain-hardens to a sufficient degree, the critical internal pressure is reached after some fully plastic expansion has taken place. If the internal pressure exceeds this critical value, the expansion of the pipe is unstable and leads, eventually, to bursting.

CONCLUSIONS

In this study, an elastic-plastic analysis of a thick-walled pipe under combined loading of internal pressure and temperature was performed, using the modified elastic-viscoplastic model. For the elastic-plastic analysis, the finite element program has been developed, which can deal with the problem of plastic compressibility. To take into account the plastic compressible yielding and the strength-differential effect which describes the difference between the compressive and tensile yield stress, the Drucker-Prager (1952) plastic potential was used.

From the calculation results, it was shown that the plastic compressibility of the material has an effect on the stress distribution of the pipe. Specifically, the plastic compressibility has a significant effect on the fully plastic pressure of the pipe under temperature loading. It was also found that the pipe of a plastic incompressible material sustains higher internal pressure than that of a plastic compressible material.

Therefore, for the elastic-plastic analysis of a thick-walled submarine pipeline which has the strength-differential effect in its material properties, the plastic compressibility should be taken into account. The results from the plastic incompressible analysis would be on the unsafe side and the small increase in the internal pressure under temperature loading could lead to the catastrophic material failure.

It is noted that the proposed viscoplastic model can also be utilized for the analysis of creep problems, and the creep-strength-differential effect can be handled using the Drucker-Prager (1952) potential for creep potential.

REFERENCES

Betten, J, and Borrmann, M (1984). "Einfluss der plastischen Kompressibilität und des Strength-Differential-Effektes auf das Fliessverhalten von Sinter- und Polymerwerkstoffes," Rheol. Acta, Vol 23, pp 109-116.

Betten, J, Frosch, HG, and Borrmann, M (1982). "Pressure-dependent Yield Behaviour of Metals and Polymers," Mat. Sci. and Eng., Vol 56, pp 233-246.

Bingham, EC (1922). "Fluidity and Plasticity," McGraw-Hill, New York.

Chakrabarty, J (1988). "Theory of Plasticity," McGraw-Hill, Singapore.

Drucker, DC (1959). "A Definition of Stable Inelastic Material," J. Appl. Mech., Vol 26, pp 101-106.

Drucker, DC, and Prager, W (1952). "Soil Mechanics and Plastic Analysis or Limit Design," Quart. Appl. Mech., Vol 10, pp 157-165.

Hohenemser, K, and Prager, W (1932). "Über die Ansätze der Mechanik isotroper Kontinua," Zeit. für angew. Math. und Mech., Vol 12, pp 216-226.

Owen, DRJ, and Hinton, E (1980). "Finite Elements in Plasticity: Theory and Practice," Pineridge Press Limited, Swansea.

Perzyna, P (1966). "Fundamental Problems in Viscoplasticity," In Advances in Appl. Mech., Academic Press, New York.

Spitzig, WA, Sober, RJ, and Richmond, O (1976). "The Effect of Hydrostatic Pressure on the Deformation Behavior of Maraging and HY-80 Steels and its Implications for Plasticity Theory," Met. Trans., Vol 7A, pp 1703-1710.

Proceedings of the Seventh (1997) International Offshore and Polar Engineering Conference
Honolulu, USA, May 25-30, 1997

ISBN 1-880653-28-1 (Set); ISBN 1-880653-30-3 (Vol. II)

Intrinsic Coordinate Elements for Large Deflection of Offshore Pipelines

Poh C. Andrew Ngiam
Kvaerner R J Brown Pte Ltd
Singapore

ABSTRACT

This paper describes a numerical method for solving large deflections of elasticas and offshore pipelines. The method is based on finite element analysis using intrinsic coordinates, namely the rotations at nodal points, in which the element stiffness is independent of element orientation and the displacement vector is expressed in terms of nodal values of cross-sectional rotation. The generation of the vector of integrating coefficients for numerical integration adopts quadrature method, which is subsequently assembled into quadrature matrix. The element lengths are embodied in the vector of integrating coefficients and such lengths do not need to be uniform, This feature fits in very well with the intrinsic coordinate elements. The transformations that are required from intrinsic to Cartesian coordinates affect only the load vector in the equilibrium equation. Consequently computational advantages can be expected from the intrinsic coordinates formulation, particularly the large deflection problems.

KEY WORDS

Intrinsic coordinate elements, offshore pipelines, large deflection, quadrature matrix.

INTRODUCTION

During installation an offshore pipeline can undergo large deflections with small strain, i.e. the classical elastica problem where axial and shear deformations are neglected. The equations for large deflection are nonlinear and the length of the suspended pipeline during installation is not known a priori.

Numerical methods, particularly the finite element method, are often used to solve the problems. The finite element formulation usually refers the load-displacement interaction to a Cartesian coordinate system. A review of finite element analysis of offshore pipelines can be found in Ngiam (1990). In our present finite element formulation, we look at the displacement field which is specified in terms of cross sectional rotation $\psi(s)$, where s is measured along the deformed axis of an element.

This method is explored for a variety of elastica examples, including pipeline problems.

THE FINITE ELEMENT FORMULATION AND EQUILIBRIUM EQUATION

The present development is based on the assumption that the element is elastic in bending and can undergo large displacements with small strain, i.e. the classical elastica problem with axial and shear deformation neglected (Frisch-Fay, 1962). The displacement field of an element is specified in terms of the cross sectional rotation $\psi(s)$, where s is measured along the deformed axis of the element. The coordinates (s,ψ) are referred herein as the intrinsic coordinates (Wang, 1986)

The strain energy due to bending, U^e of an element of length and flexural rigidity, EI is:

$$U^e = \int_0^l \frac{EI}{2}\left[\frac{d\psi}{ds}\right]^2 ds \qquad (1)$$

The displacement field $\psi(s)$ is defined by a continuous polynomial, so that the rotations within any element may be interpolated from the nodal values on that element. Nodes are positioned at each end of an element so that quality of nodal values will satisfy interelement compatibility. A two-node element is therefore the minimum configuration to satisfy compatibility. That element may be called a linear element, since the rotation varies linearly. Additional internal nodes of rotation can also be generated within the element for more accurate representation of the displacement field. In this paper, a three-node element is called as a quadratic element and a four-node element is a cubic element.

The polynomial $\psi(s)$ may be expressed as a Lagrangian interpolation polynomial $\psi(s)=[\beta]\{\bar{\theta}\}$, where $[\beta]$ is the coefficient function and $\{\bar{\theta}\}$ the vendor of nodal rotations. Substituting the Lagrangian interpolation polynomial into (1) gives

$$U^e \approx \frac{1}{2}\{\bar{\theta}\}^T\left[\int_0^l EI[\beta]^T[\beta]\,ds\right]\bar{\theta} \approx \frac{1}{2}\{\bar{\theta}\}^T[\bar{K}]\{\bar{\theta}\} \qquad (2)$$

where the element matrix $[\bar{K}]$ represents the integral term. The $[\bar{K}]$ matrices, for the linear, quadratic and cubic elements, respectively, are:

$$\frac{EI}{l}\begin{bmatrix}1 & -1\\ -1 & 1\end{bmatrix}$$

$$\frac{EI}{3l}\begin{bmatrix}7 & -8 & 1\\ & 16 & -8\\ sym & & 7\end{bmatrix}$$

$$\frac{EI}{l}\begin{bmatrix}3.7 & -4.725 & 1.35 & -0.325\\ & 10.8 & -7.425 & 1.35\\ & & 10.8 & -4.725\\ sym & & & 3.7\end{bmatrix} \qquad (3)$$

The strain energy of an elastica is the sum of the strain energies of the component elements, i.e. $U=\Sigma U^e=\frac{1}{2}\{\theta\}^T[K]\{\theta\}$, where $\{\theta\}$ and $[K]$ are assembled from the element displacement vectors and $[\bar{K}]$ matrices respectively. The work product, W, due to external point loads, P_i, uniformly distributed load, w and a moment, M_r, may be expressed as:

$$W=\Sigma P_i d_i+\int w\,v\,ds+M_r\theta_r \qquad (4)$$

where, d_i is the conjugate displacement and v the vertical displacement. If the load P_i is resolved horizontally and vertically, with ϕ the angle of inclination of P_i, equation (4) can be expressed as,

$$W=\Sigma P_i\cos\phi\, u_i+\Sigma P_i\sin\phi\, v_i+w\int v\,ds+M_r\theta_r \qquad (5)$$

where u_i is the conjugate horizontal displacement. Now,

$$u_i=\int_0^{s_i}(\cos\psi-1)\,ds=\int_0^{s_i}\cos\psi\,ds-s_i \qquad (6)$$

and

$$v_i=\int_0^{s_i}\sin\psi\,ds \qquad (7)$$

Such integrals can be evaluated numerically, using nodal values of the integrand.

$$\int_0^{s_i} f(s)\,ds \approx \lfloor C(0,i)\rfloor\{f\} \qquad (8)$$

where $\lfloor C(0,i)\rfloor$ is a vector of coefficient, depending on i, and $\{f\}$ is the vector of nodal values of $f(s)$. The calculation of the vector $\lfloor C(0,i)\rfloor$ for the quadratic element is discussed as follows:

For the case of quadratic elements discussed in this paper, consider nodes O to n where n is even. Let

$$s_{2j-1}-s_{2j-2}=s_{2j}-s_{2j-1}=\frac{l_j}{2} \qquad (9)$$

Based on a parabolic approximation of the integral f(s),

$$\int_{s_{2j-2}}^{s_{2j-1}} f(s)\,ds=I(2j-2,2j-1)\approx\frac{l_j}{24}\left[5f_{2j-2}+8f_{2j-1}-f_{2j}\right]$$

$$\int_{s_{2j-2}}^{s_{2j}} f(s)\,ds=I(2j-2,2j)\approx\frac{l_j}{6}\left[f_{2j-2}+4f_{2j-1}+f_{2j}\right] \qquad (10)$$

If $I(O,k)=\int_{s_o}^{s_k} f(s)\,ds$, then the integral approximations are,

$$I(O,2j-1)=I(O,2j-2)+(2j-2,2j-1)\approx\lfloor C(0,2j-1\rfloor\{f\}$$

$$I(O,2j)=I(O,2j-2)+I(2j-2,2j)\approx\lfloor C(0,2j)\rfloor\{f\} \qquad (11)$$

Details on generating the integrating coefficients for linear and cubic elements can be found in Ngiam (1990).

Using these integrating vectors and letting $\lfloor D_i\rfloor=\lfloor C(0,i)\rfloor$, equations (6) and (7) become:

$$u_i=\lfloor D_i\rfloor\{\cos\theta\}-s_i$$
$$v_i=\lfloor D_i\rfloor\{\sin\theta\} \qquad (12)$$

The term $\int v\,ds$ in (5) can also be evaluated numerically,

$$\int_0^{s_N} v\,ds=\lfloor C(0,N)\rfloor\{v_i\}$$
$$=\lfloor C(0,N)\rfloor\{\lfloor D_i\rfloor\{\sin\theta\}\}$$
$$=\lfloor E\rfloor\{\sin\theta\} \qquad (13)$$

Substituting (12) and (13) into (5),

$$W=\Sigma P_i\cos\phi_i\left(\lfloor D_i\rfloor\{\cos\theta\}-s_i\right)+\Sigma P_i\sin\phi_i\lfloor D_i\rfloor\{\sin\theta\}$$
$$+w\lfloor E\rfloor\{\sin\theta\}+M_r\theta_r$$
$$=\lfloor G\rfloor\{\cos\theta\}+\lfloor H\rfloor\{\sin\theta\}+M_r\theta_r+constant \qquad (14)$$

where $\lfloor G\rfloor$ and $\lfloor H\rfloor$ are coefficient vectors. It will be noted that W is a linear function of $\{\sin\theta\}$ and $\{\cos\theta\}$ for any loading.

The total potential energy of the system is $\Pi = U - W$, Considering a variation $\delta\Pi$, it can be shown that for an elastic system, the equilibrium state corresponds to a minimum in the total potential energy (Bathe et al. 1976). Hence, after partial differentiation the equilibrium equation is:

$$[K]\{\theta\} - \{Q\} = 0 \qquad (15)$$

where {Q} is the load vector which is obtained by partial differentiation of (14) with respect to θ_i.

$$Q_i = G_i \sin\theta_i + H_i \cos\theta_i + M_r \delta_{ri} \qquad (16)$$

For a structural system where lengths of members are usually given and fixed, the [K] in (15) is constant and symmetric, and only the load vector is a function of the nodal rotations, $\{\theta\}$. Consequently, this will offer significant reduction of computational effort in the repeated solutions required for non-linear problems.

NUMERICAL EXAMPLES

Cantilever Loaded By Concentrated End Load

Consider a horizontal cantilever loaded by a vertical concentrated load, P at its free end and the self weight of the cantilever is ignored. Using the fixed end as the origin, the total potential energy is:

$$\Pi = \int_o^L \frac{EI}{2}\left(\frac{d\psi}{ds}\right)^2 ds - P\int_o^L \sin\psi ds \qquad (17)$$

The equilibrium equation is obtained after partial differentiation of (17), with respect to nodal rotations.

$$\sum_j K_{ij}\theta_j - PC_i \cos\theta_i = 0 \qquad (18)$$

where i,j=0,.......N, and C_i is an integrating coefficient in C (0, N).

For small deflection problem, $\cos\theta_i \rightarrow 1.0$ in (18) and with the boundary condition of θ_o=0, the nodal rotations θ_j can be obtained after solving (18). Using linear, quadratic and cubic elements, the results of nodal rotations match the analytical solutions.

For a large deflection problem, the error in an approximate solution to the equilibrium equation can be defined by the residuals, R_i

$$R_i = K_{ij}\theta_i - PC_i \cos\theta_i \qquad (19)$$

The [K] matrix in (19) is constant, and only the load vector is a function of nodal rotations.

The Newton-Raphson correction is used for the approximate solution of (19).

$$\{d\theta\} = -[J]^{-1}\{R\} \qquad (20)$$

where [J] is the Jacobian matrix, which is the partial differentiation of R_i with respect to θ_r.

$$J_{ir} = K_{ir} + PC_i\delta_{ir}\sin\theta_i \qquad (21)$$

where the second term on the right hand side of (21) contributes only to the diagonal of the J matrix. The J matrix is banded and symmetric. Several Newton-Raphson corrections may be necessary to achieved required solution accuracy and some cases may require the solution to proceed through several load increments.

Consider a cantilever with a non-dimensional applied load, $PL^2 / EI = 10$. Using seven equally spaced nodal points (0,.....6) and four incremental steps, with Newton-Raphson iteration for each step, iterations were terminated when correction to the end rotation was less than 10^{-6}. It was found that fewer than four iterations were required for each load step. Results for the end rotation and end deflection for linear, quadratic and cubic elements against the elliptic integral solutions (Mattiasson, 1980) and linear theory solutions are summarized in Table 1. The values of x and y in the table were recovered using integrating vectors as $x_i = \lfloor D_i \rfloor\{\cos\theta\}$ *and* $y_i = \lfloor D_i \rfloor\{\sin\theta\}$.

TABLE 1

Type of Element	End Rotation θ_N (rad.)	End Deflection y_N / L	Horizontal Projection x_N / L
Linear	1.42652	0.79940	0.44773
Quadratic	1.43079	0.81119	0.44385
Cubic	1.43080	0.81056	0.44302
Mattiasson	1.43029	0.81061	0.44500
Linear theory	5.00000	3.33333	1.00000

When comparing the end rotation against Mattiasson's solution, the linear elements under-estimate the end rotation and both the quadratic and cubic elements over-estimate it, with quadratic element giving a better solution.

Cantilever Loaded By Uniformly Distributed Load

Consider a cantilever under uniformly distributed load w. The total potential energy is

$$\Pi = \int_o^L \frac{EI}{2}\left(\frac{d\psi}{ds}\right)^2 ds - w\int_o^L v\, ds \qquad (22)$$

where v is the vertical displacement. The second integral, as it has been shown in (13), can be expressed to be the product of a coefficient vector, $\lfloor E \rfloor$ and a vector $\{\sin\theta\}$.

The equilibrium equations are obtained after partial differentiation of (22) with respect to nodal rotations.

$$\sum_j K_{ij}\theta_i - wE_i \cos\theta_i = 0, \quad i,j = 0,.....N \qquad (23)$$

In equation (23), $\cos\theta_i \rightarrow 1.0$ for small deflection solutions, and by substituting the boundary condition of $\theta_o=0$ into (23), the nodal rotations are obtained. It is observed that the solution of the nodal rotations based on quadratic elements agree with the analytical solution. But the solutions based on linear and cubic elements show some variations. Taking seven nodal points, 0,.....6 as an example, the errors of the nodal rotations are summarized in Table 2.

TABLE 2

Nodal Point	Linear Element	Cubic Element
1	0.000386	0.000086
2	0.000772	-0.000086
3	0.001158	0.0
4	0.001543	0.000086
5	0.001929	-0.000086
6	0.002314	0.0

The solution based on linear element have the largest errors, which increase linearly; and the nodal rotations are over-estimated. The solutions based on cubic element have very small errors and with nodes 3 and 6 having the exact solutions. The negative sign denotes the nodal rotation is under-estimated.

For a large deflection problem, consider the non-dimensional weight wL^3 / EI to be 6. Using seven nodal points and four load steps, with Newton-Raphson corrections in each step, iterations were terminated when correction to end rotation was less than 10^{-6}. Only three iterations were required for each load step. Results for the end rotation and end deflection are summarized in Table 3.

TABLE 3
CANTILEVER UNDER UNIFORMLY DISTRIBUTED LOAD OF $wL^3 / EI = 6$

Type of Element	End Rotation θ_N (rad.)	End Deflection y_N / L	Horizontal Projection x_N / L
Linear	0.7951	0.5488	0.8029
Quadratic	0.7906	0.5541	0.8034
Cubic	0.7908	0.5543	0.8030
Linear theory	1.0000	0.7500	1.0000

For a range of nodal point values, end rotations based on linear, quadratic and cubic elements are compared against a solution of 0.7904, which is obtained using power series expansion (Appendix A).

Single Point Lift of a Continuous Pipeline

Lifting of a continuous pipeline is of practical significance since, for example, an engineer may want to know how high a pipeline can be lifted safely in order to place supports beneath it. In this respect consider a diver wanting to lift a pipeline to provide sufficient clearance for pipeline welding.

In the analysis consider that a pipeline resting on a level and rigid sea bed is lifted by a vertical force, 2P (Figure 1). It is assumed that the pipeline will slide without frictional resistance and the curvature of the pipeline at the sea bed lift-off point equals zero. For small deflections, the pipeline lifted length, 2L, the lifted height at the lift point, d and the corresponding moment, M_n can be shown to be as follows:

$$L = \frac{3P}{2w} \; ; \; d = \frac{wL^4}{72EI} \; ; \; M_n = \frac{wL^2}{6} \tag{24}$$

For large deflections, the numerical solution based on intrinsic coordinate elements is as follows:

Because of symmetry, only half of the lifted pipeline section need be considered. The boundary conditions of $\psi_0 = \psi_n = 0$ can be used as the constraint equations and the corresponding Lagrangian multipliers are M_0 and M_n. Using 0 (lift off point at the sea bed) as the origin, the equilibrium equations are obtained after partial differentiation of the total potential energy function with respect to nodal rotations as,

$$\sum_j K_{ij}\psi_i - (w\,E_i + P\,C_i)\cos\psi_i + M_0\,\delta_{oi} + M_n\,\delta_{ni} = 0 \tag{25}$$

$$\psi_0 = 0 \; ; \quad \psi_n = 0 \tag{26}$$

where i = j = 0, 1, 2, ..., n. The equilibrium equations plus the boundary condition of $M_0 = 0$ constitute a set of n+4 equations, represented by

$$R_k(\psi, M_0, M_n, L) = 0 \; ; \quad k = 0,1,2,\ldots,n+3 \tag{27}$$

For an approximate solution, the values of R_k are known as residuals. The Newton-Raphson procedure is used for the solution of the residual equations and requires starting values of the variables. Since the pipeline is lifted gradually from the flat sea bed, the starting values for nodal rotations may be assumed zero. The starting values of the lifted length, L and the moment, M_n can be established based on the small deflection solutions in Eq. (24).

The dimensionless terms of characteristic length, lifted length, lift force, moment and lift height are considered in the solution,

$$\hat{L} = \left(\frac{EI}{W}\right)^{1/3} ; \; \bar{L} = \frac{L}{\hat{L}} \; ; \; \bar{P} = \frac{P}{w\hat{L}} \; ;$$

$$\bar{M} = \frac{M}{w\hat{L}^2} \; ; \; \bar{Y} = \frac{y}{\hat{L}} \tag{28}$$

The large deflection problem has been previously solved by Kooi and Kuipers (1984) using a finite difference method. Some of the solutions using quadratic elements are presented in Table 4 and compared with theirs. In the table the results presented are based on a dimensionless lift force $2\bar{P}=20$, which is the greatest load adopted by Kooi and Kuipers. The deflection due to this load is extremely large. The results are obtained using quadratic elements of equal length and four incremental load steps.

TABLE 4
SINGLE POINT LIFT OF A PIPELINE
DIMENSIONLESS LIFT FORCE EQUALS TO 20

Number of Elements	Half Length $\bar{L}$	Lifted Height $\bar{Y}$	Bending Moment $\bar{M}_n$
4	11.258732	9.763271	5.914088
8	11.199737	9.733902	4.491175
16	11.202644	9.806287	4.346398
32	11.202700	9.801202	4.416090
Kooi/Kuipers	11.202701	9.811415	4.431179
Linear theory	15	703.125	37.5

Kooi and Kuipers obtained the values quoted above by Richardson extrapolation, from solutions using 333 and 653 unknowns. Very similar results have been obtained here using only 32 elements or 68 equations. Their values of 11.20300 and 11.202701 for the dimensionless half length using 333 and 653 equations respectively, a difference of 0.0003. It can be noted that the difference is less than 0.0001 for the solutions using 16 and 32 elements (see Table 4) or 36 and 68 unknowns respectively. It was also reported by Ngiam (1990) that for the dimensionless lift forces smaller than 4.0, the solutions are reasonably consistent with those based on the linear theory equations.

The deflected profiles of the pipeline under the dimensionless lift force of 5, 10, 15 and 20 are presented in Figure 2. Plots of the dimensionless lift force against the dimensionless bending moment at the lift point and against the dimensionless lifted length are shown in Figures 3 and 4 respectively.

Pipeline Lifted by Cables

A multi-point lift is executed by having several cables tied at predetermined spacings along a pipeline while the other ends of the cables are hooked up to davits on board a barge. The pipeline is lifted from the seabed to its required position when the cables are winched by the davits. The lifting process can be performed in shallow water or relatively deep water.

Consider a pipe lift problem where the pipeline is to be lifted from a level seabed as shown in Fig. 5. Cables are attached to points along the pipeline such that each point defines a boundary between two adjacent elements. The pipeline will be lifted from the seabed to its required position, with the dotted lines in the figure showing the movements of the lift points and the pipe tip. It can be seen that the cable angles vary in the lifting process.

Let V_r and H_r be the vertical and horizontal force components of the cable loads, P_r , where r = 1, 2 ..... m. The symbol m corresponds to the total number of lift points.

It is assumed that the seabed is horizontal and the foundation is rigid so that the curvature and hence the moment of M_o of the pipe are zero at the lift off point, 0.

With the pipe lift off point as the origin, the potential energy of the system is

$$V^T = \int_o^L \frac{EI}{2}\left(\frac{d\psi}{ds}\right)^2 ds - w \,.\int_o^L \upsilon(s)\, ds - V_r .. \int_o^{L_r} \sin\psi\, ds$$
$$- H_r \,. \int_o^{L_r} (\cos\psi - 1)\, ds + M_o\, \psi_o \qquad (29)$$

Where L is the entire pipe suspended length which consists of the sagbend length and the lengths of pipe between the lift points. The latter, in practice are the predetermined spacings of pipe. The symbol L_r in Eq. (29) is the length of the pipe measured from 0 to the respective lift point, and w the pipe unit weight which is assumed to be uniform. The boundary conditions are

$$\psi_o = 0 \quad ; \quad M_o = 0 \qquad (30)$$

The sagbend length, L_1 increases as the pipeline is lifted progressively. As mentioned earlier, the directions of the cable loads are likely to change as the pipe is lifted. Therefore, V_r and H_r in Eq. (30) are unknowns, which require a further 2m equations, where m is the number of lift points. The ratio of the cable component loads, V_r and H_r is equal to the ratio of the vertical and horizontal projections of the length of the winched cable as shown in Fig. 5. Hence

$$V_r \,.\, \bar{x}_r - H_r \,.\, \bar{Y}_r \;=\; 0 \qquad (31)$$

where $\bar{x}_r$ and $\bar{y}_r$ are the horizontal and vertical projections of the winched cables, respectively. The projections may be obtained based on the cable geometry in Fig. 6,

$$\bar{x}_r = L_r + a_r - \int_o^{L_r} \cos\psi\, ds \quad ; \quad \bar{y}_r = y_r^F - \int_o^{L_r} \sin\psi\, ds \qquad (32)$$

Eq. (31) provides only m equations; another m equations are still needed.

The pipe may be lifted in a controlled manner by either:

(a) prescribing cable loads, P_r ,
(b) or by prescribing displacements at lift points, y_r ,
(c) or by prescribing cable shortened length, l_r^f .

The equilibrium equations of each of the three options, are, respectively

$$P_r^2 - V_r^2 - H_f^2 \;=\; 0 \qquad (33)$$
$$y_r - \int_o^{L_r} \sin\psi\, ds \;=\; 0 \qquad (34)$$
$$\left(l_r^f\right)^2 - \bar{x}_r^2 - \bar{y}_r^2 \;=\; 0 \qquad (35)$$

Each option in Eqs. (33), (34) and (35) provides m more equations. The equilibrium equations of Eqs. (29), (30), (31), (32) and/or (33), or (34), or (35) constitute a set of (n+3+2m) equations, which can be represented by

$$R_k\,(\psi, L_1, M_o, V_r, H_r) \;=\; 0, \qquad k = 0,...((n+2+2m) \qquad (36)$$

For an approximate solution, the values of R_k are known as residuals. The Newton-Raphson solution of the residual equations requires starting values of the unknown variables in the iterative process. Initially the nodal rotations may be set to zero. The sagbend length, L_1 is recommended to be in the range of two to four times the water depth. The value chosen is not found to be critical in the iterative solution. The procedures discussed in this section handle a variety of lifting problems, from single to multi-point lift problems.

The example considered is a two-point lift problem where the pipeline is lifted from the seabed. The pipe has an outside diameter of 273 mm, 12.7 mm of wall thickness, and 50.8 mm of concrete coating. The air and submerged unit weights of the pipeline are 2.252 and 1.082 kN/m, respectively.

Let the height of each davit, y_1^F and y_2^F , both be 7 m above water level, and the davit spacing be 44.5 m. Using the option of prescribed displacement, the first and second lift points are lifted 12 m and 24 m above seabed, with 25 m of water depth.

In this particular example, the entire pipeline is divided into three spans; namely, the sagbend, the mid-span and the free span (see Fig. 5).

The sagbend increases as the pipe is lifted progressively, but the mid-span and the free span do not change throughout the lifting exercise.

The values of a_r for cables 1 and 2 are 5 m and 3 m respectively, the mid-span and free span lengths are 46.5 and 13 m, respectively.

Results for a range of discretisations are shown in Table 5 wherein the no. of elements for the sagbend, mid-span and the free span are 8, 8 and 2 respectively. Only one displacement step is used for all cases, and six iterations are required. Iterations were terminated when the root mean square sum of the relative incremental change was less than 10e-5.

The output cable lengths at the first and second lift points are 21.056m and 9.519m respectively. These lengths are used as input to a commercial finite element software, OFFPIPE. Using quadratic elements, the results are compared with OFFPIPE outputs favourably.

TABLE 5

Description	Quadratic Element	OFFPIPE Output
No. of Elements	8,8,2	32, 1, 2
Sagbend Length (m)	64.506	64.0
Max. Sagbend Moment (kN-m)	155.159	156.133
Cable Forces (kN)		
P_1	62.998	64.108
P_2	101.656	99.370
Pipe Tension at O (kN)	88.586	86.910

FUTURE WORK

The formulation should be extended to include pipeline spanning along an uneven seafloor wherein effects due to temperature, and pressure, are included. Future work may also include 3-D model, torsion-coupled problems, pipe-soil interaction, riser and drill string problems.

ACKNOWLEDGEMENT

The author wishes to thank KRJB for allowing to use the in-house facilities and Ms Jasmine Quek for typing the manuscript.

CONCLUSIONS

A new finite element procedure has been developed for large deflections of pipeline problems, with the deformation defined in terms of nodal rotations. The element stiffness matrix is independent of element orientation. Computational advantage is obtained because only load vector in the equilibrium equation is affected in the transformation from Cartesian to intrinsic coordinates. Results for large deflection of elastica problems compared favorably with established solutions.

REFERENCES

Bathe, K.J. and Wilson, E.L. (1976). Numerical Methods in Finite Element Analysis. Prentice Hall.

Frisch-Fay, R. (1962). Flexible Bars. London, Butterworths.

Kooi, B.W. and Kuipers, M. (1984). A unilateral contact problem with the heavy elastica, International Journal of Non-Linear Mechanics, Vol. 19, No. 4, pp 309-321.

Mattiasson, K. (1981). Numerical Results From Large Deflection Beam and Frame Problems Analysed by Means of Elliptic Integrals. International Journal For Numerical Methods In Engineering. Vol. 17, No. 1, 145-153.

Ngiam, P.C.A. (1990). PhD Dissertation "Finite Element Analysis Using Intrinsic Coordinate Elements", University of New South Wales, Australia.

Rohde, F.V. (1952). Large Deflections of A Cantilever Beam With Uniformly Distributed Load. Quarterly of Applied Mathematics. Vol. 11, No. 3, 337-338.

Schmidt, R and Dadeppo, D.A. (1970). Large Deflections of Heavy Cantilever Beams and Columns. Quarterly of Applied Mathematics. Vol. 28, 441-444.

Wang, C.Y. (1986). A Critical Review of the Heavy Elastica. International Journal of Mechanical Science. Vol. 28, No. 8, 549-559.

APPENDIX A

For large deflection solutions to a cantilever under distributed load, the power series method may be used and the series expanded in order to meet any desired degree of accuracy. Early work in this area was carried out by Rohde, 1952. Interested readers can find much of the work done in this field summarized in Frisch-Fay, 1962, and by Schmidt and Dadeppo, 1970. All early authors in this field expressed their results in the form of graphs, which may not be of adequate precision. The current work briefly explained here is in fact the series expansion where precision is compared as the number of terms are expanded.

The basic differential equation to this problem is,

$$\frac{d^2\psi}{ds^2} - \frac{ws}{EI}\cos\psi = 0 \qquad (A.1)$$

$$Let,\ \ \psi(s) = \sum_{i=0}^{oo} a_i s^i \qquad (A.2)$$

and the boundary conditions at s=0 (the free end of the cantilever) are $\psi = \psi_o$ and $d\psi / ds = 0$, we get $a_o = \psi_o$ and $a_1 = 0$. Equation (A.2) can now be simplified as $\psi = \psi_o + \alpha$, where $\alpha = \sum_{i=2}^{oo} a_i s_{i.}$

The second order differentiation of (A.2) gives

$$\frac{d^2\psi}{ds^2} = \sum_{i=2}^{oo} i(i-1)a_i s^{i-2} \qquad (A.3)$$

and the cosine term in (A.1) can be expressed as

$$\cos\psi = \cos(\psi_o + \alpha) = \cos\psi_o \cos\alpha - \sin\psi_o \sin\alpha \qquad (A.4)$$

Expanding the α terms in (A.4) and substituting (A.3) and (A.4) into (A.1). After equating the like terms, the coefficients, a_i are obtained. The non-zero coefficients are a_3, a_6, a_9,............

For s = L, ψ = 0, (A.2) becomes

$$0 = \psi_o + \sum_{i=1}^{oo} a_{3i} L^{3i} \qquad (A.5)$$

The series in (A.5) can be expanded as required until the desired precision is obtained.

Table A.1 is a brief summary of the end rotations of various values of wL^3 / EI when the power series terms are expanded from a_9 to a_{21}.

TABLE A.1

wL^3 / EI	a_9	a_{15}	a_{21}
2	0.3216	0.3216	0.3216
4	0.5894	0.5886	0.5886
6	0.7931	0.7903	0.7904
8	0.9451	0.9400	0.9401
10	1.0592	1.0526	1.0525

FIGURES:

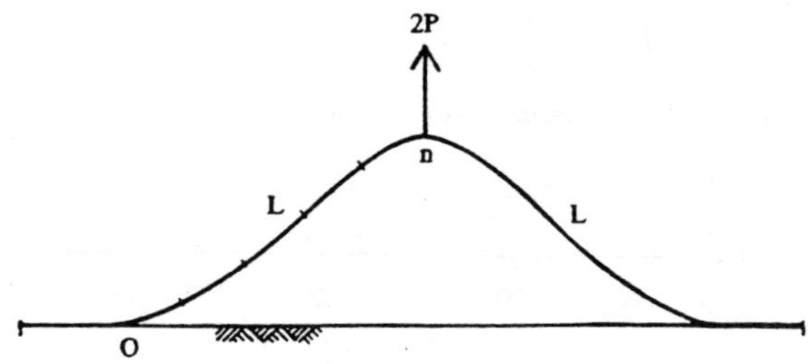

Fig. 1 Single Point Lift of a Continuous Pipeline (Zero Friction)

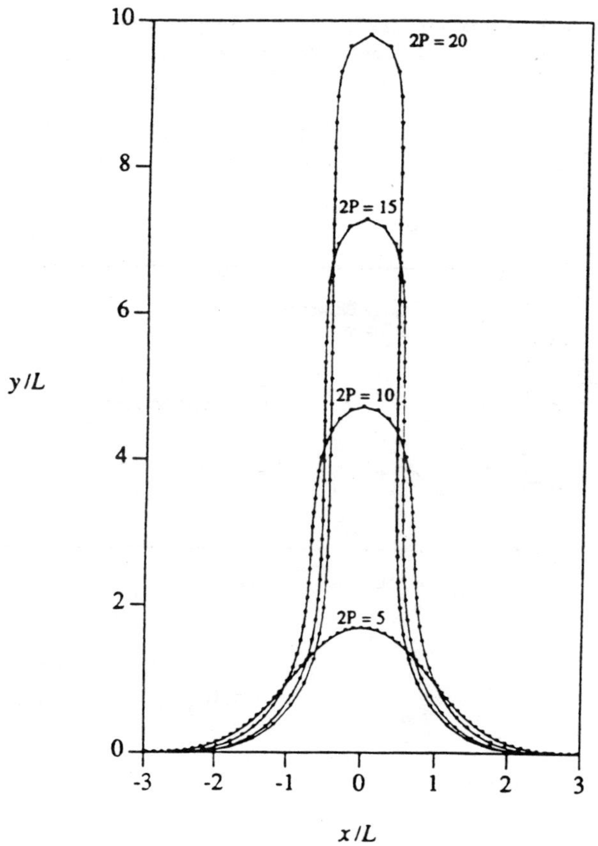

Fig. 2 Deflected Geometry of a Pipeline

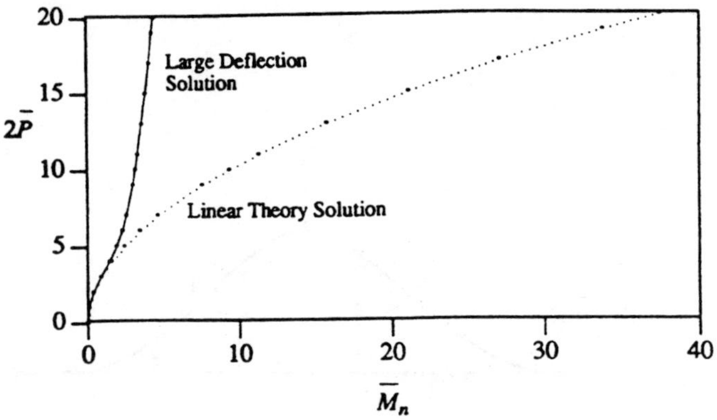

Fig. 3 Dimensionless Lift Force Vs. Bending Moment at the Lift Point (Zero Friction)

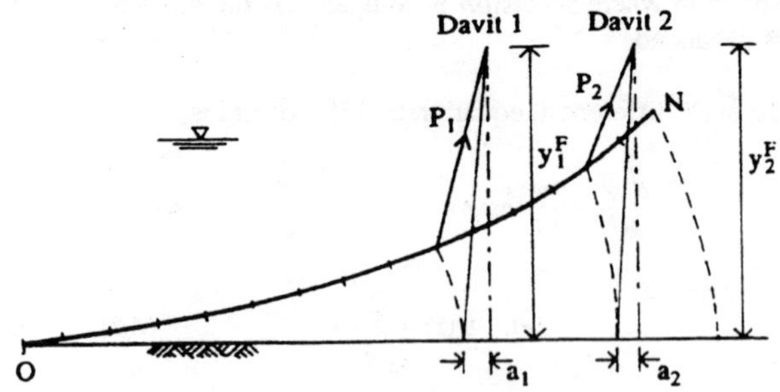

Fig. 5 Two-Point Lift Problem

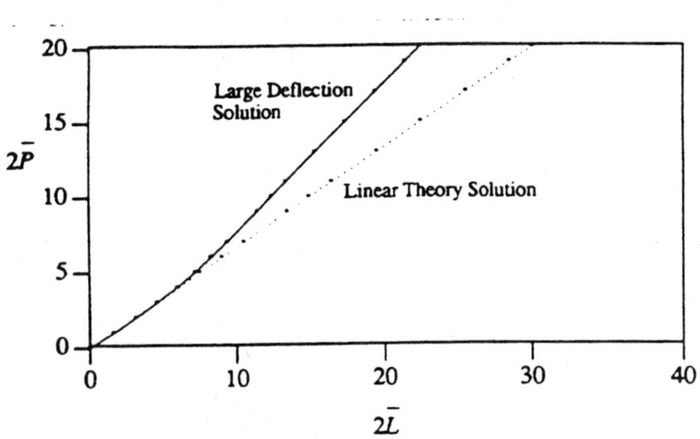

Fig. 4 Dimensionless Lift Force Vs. Lifted Length (Zero Friction)

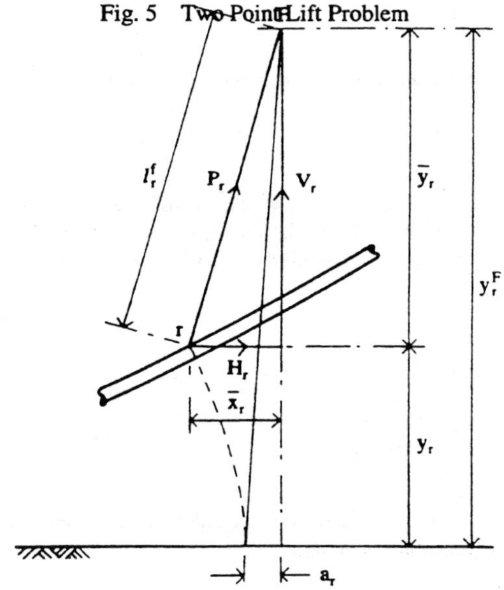

Fig. 6 Cable Component Loads and Cable Projections

Proceedings of the Seventh (1997) International Offshore and Polar Engineering Conference
Honolulu, USA, May 25-30, 1997

ISBN 1-880653-28-1 (Set); ISBN 1-880653-30-3 (Vol. II)

The Behaviour of High Pressure, High Temperature Flowlines on Very Uneven Seabed

Knut Tørnes, Per R. Nystrøm, Per Damsleth and Lars H. Sortland
J P Kenny A/S
Stavanger, Norway

ABSTRACT

The design of High Pressure - High Temperature (HP/HT) flowlines on very uneven seabed has become an important issue in recent years. The behaviour of such flowlines is largely governed by the tendency to buckling, either vertically if trenched or covered, or laterally if the flowline is left fully exposed on the seabed. The flowline behaviour is further complicated when thermal expansion is allowed to feed into freespans which influences the buckling driving forces. This paper will present recent work performed by JP Kenny A/S to gain further insight into how expansion, seabed friction and freespans influence each other as the flowline heats up. Elasto-Plastic 2-D and 3-D FE models demonstrate the general expansion behaviour of the flowline to study the interaction between seabed features and flowline buckling. A realistic Digital Terrain Model (DTM) using 3-D survey data is incorporated in the FE models. The analyses demonstrate the effects of different methods for controlling the global buckling. A minimum cost approach for seabed intervention is proposed.

KEY WORDS: Pipeline, flowline, HP/HT, buckling, expansion, in-place analysis, uneven seabed, 3D FEM

INTRODUCTION

The main concern in the design of small diameter flowlines operating at high pressure and temperature is the tendency of the flowline to undergo global buckling at some critical value of the effective axial force (Euler buckling load). The resulting large vertical or horizontal deformations may give rise to unacceptably high stresses and strains in the pipe wall. The conventional design strategy in such cases is to prevent the buckling by applying some form of cover, i.e. by continuous trenching and backfilling and/or rock dumping. However, the protection used to cover the pipe will restrain it which, in turn, increases the axial effective force to the point where upheaval buckling may occur. A large amount of cover may then be required to prevent upward movement of the pipe.

Several alternative methods have been proposed to control pipeline expansion and buckling. Ellinas and Supple (1990) proposed intermittent rock dumping as a means of controlling buckling by placing discrete rock berms along the top of the flowline at certain intervals to allow limited buckling to occur between each berm. When a section undergoes buckling, the rock berms on either side must be sufficiently large to prevent thermal expansion to feed through from adjacent exposed sections that have not yet undergone buckling. For a high temperature flowline on a relatively flat seabed, this method may require a large number of berms with corresponding volume of rock.

Vermeulen and Williams (1995) presented pre-snaking as a method for preventing buckling. Each section of the flowline is bent to a predefined angle while installing it from a lay vessel. The bends acts as a series of expansion spools to reduce the effective axial force and the tendency to global buckling. The main drawback of the method is that there is a limit on pipe size where this is practicable and that additional bending equipment will be required on the lay barge. Available space on the lay ramp may exclude this method for certain lay vessels.

Sævik and Levold (1995) and Fyrileiv and Aamlid (1995) proposed to install the pipeline in horizontal curves on the seabed to initiate lateral deflections at pre-defined intervals in order to release the effective forces in a controlled manner. The curved lay away method has been used as an alternative to tie-in spools (e.g. Troll Phase I). The drawback with the method is that the number of curves and the lay radii required may not be easy to achieve for small diameter flowlines on a seabed with relatively low friction.

Several analytical solutions have been proposed to solve pipeline buckling problems: Hobbs (1984), Pedersen and Jensen (1988) and Taylor and Gan (1986) proposed analytical solutions for both lateral and vertical modes of buckling. Although these methods give many valuable results for various buckling scenarios, they are not easily adaptable for modelling realistic initial conditions arising from the flowline load history and irregular seabed, nor do they consider the interaction between vertical and horizontal buckling modes. It is also difficult for a strictly analytical method to take into account non-linear material properties at high strain levels.

This paper discusses the expansion and buckling of a HP/HT flowline laid on a very irregular seabed and proposes some novel design strategies that may be used to control its behaviour.

Global flowline buckling tends to be initiated at vertical seabed imperfections. On an uneven seabed, the effective compressive forces will not build up sufficiently until the flowline fills in the bottom of freespans. Therefore, a route with a large number of spans would appear to be beneficial with respect to buckling provided the spans do not require intervention to mitigate Vortex Shedding Induced Vibrations.

However, it is apparent that the behaviour of HP/HT flowlines resting on an irregular seabed is extremely complex. It is important to fully understand the relationship between the axial force, displacement and seabed imperfections as the flowline pressure and temperature are increased to full operating conditions.

In the following, a description of 2-D and 3-D FE models to study this behaviour will be presented. The results from the analyses demonstrate the effects of possible intervention strategies for controlling the expansion and buckling behaviour.

HEATED FLOWLINE ON A PERFECTLY FLAT SEABED

First consider an ideal flowline with free ends laid on a perfectly flat seabed as illustrated in Figure 1 a.

The effective axial force in the as-laid empty condition will be equal to the residual lay tension. As the ends are released the flowline will tend to contract. Frictional resistance will develop which will oppose the movement until equilibrium is achieved. The effective axial force distribution in the as-laid condition is shown in Fig. 1 b.

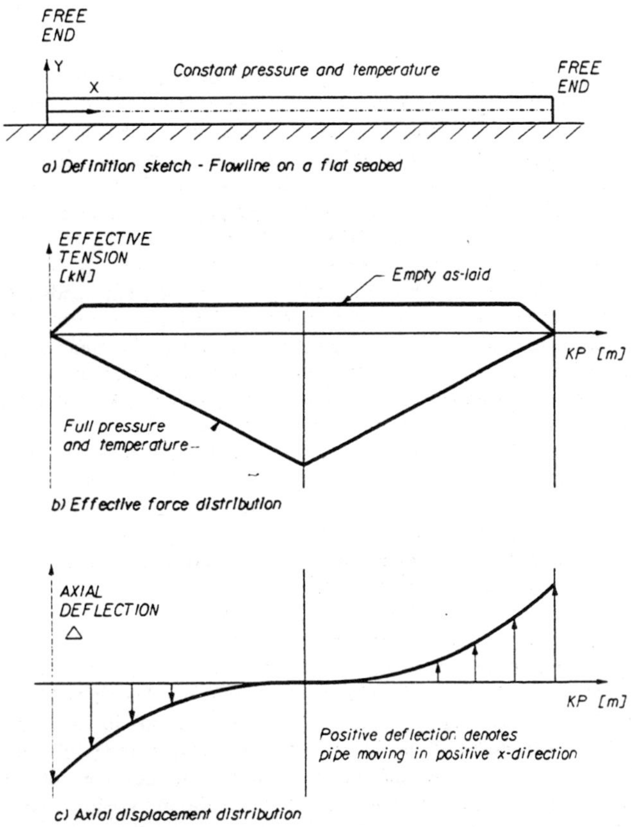

Fig. 1 Expansion behaviour of flowline on a flat seabed with free ends

When pressure and temperature is applied to the flowline, this will cause it to expand and the frictional resistance of the seabed soil will be mobilised and thus set up an effective compressive force opposing the expansion. The diagram in Fig. 1 b shows how the compressive effective force reaches a maximum value at midline and reduces linearly to zero towards each end of the flowline. A state of force equilibrium exists at the midpoint where the flowline is stationary. The corresponding axial displacement distribution along the line, which is shown in Fig. 1 c, can be obtained by integrating the axial strain from this point.

This example assumes that the flowline is too short for the frictional resistance to restrain the flowline completely. For longer lines, a point is eventually reached where the strain caused by the frictional force exactly counter balances the sum of the pressure and thermal strain. Beyond this point further expansion movement is prevented and thus the flowline is fully restrained.

If the effective compressive force reaches a critical level, i.e. the Euler buckling load, the flowline will undergo global buckling. On a flat seabed the pipeline is expected to buckle laterally rather than vertically because the restoring force created by the friction is less than the pipe submerged weight. The buckle produces a local reduction in the effective force as shown in Fig. 2 as well as feed-in of axial expansion from either side of the buckled section.

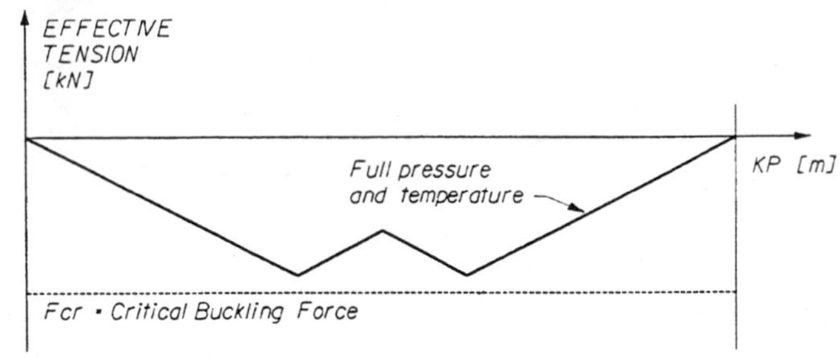

Fig. 2 Effective force distribution of a buckled flowline with free ends

The effective force distribution and expansion behaviour of a HP/HT flowline resting on an irregular seabed is more complex than the behaviour of a flowline laid on a perfectly flat seabed.

FINITE ELEMENT MODEL DESCRIPTION

The general purpose FEM package ANSYS was used to develop both 2D and 3D computer models to simulate flowline in-place behaviour. In general, the complete load path response is simulated from pipelay and precommissioning to production start-up and shut down cycles. A realistic model of the seabed is defined by importing survey data directly into the model.

The Static Large Deflection Analysis includes the following non-linearities:

- Large deflection including stress stiffening effects (geometric non-linearity);
- Elasto-Plastic Material (including temperature derating);
- General contact elements;
- Element 'birth and death'.

In addition to the static pipeline in-place analyses that will be presented in the following, ANSYS has also been used extensively to perform dynamic analysis of both the 2D and 3D as-laid and operational flowline configurations: 2D modal analyses have been performed to estimate multiple span natural frequencies in order to determine their susceptibility to Vortex shedding Induced Vibrations (VIV). In addition, 3D time-history analyses were used to simulate fishing gear pullover load effects. Since the emphasis in this paper is concentrated on the static flowline behaviour, these topics will not be discussed here.

GENERAL BEHAVIOUR ON AN UNEVEN SEABED

In order to study the behaviour of HP/HT flowlines placed on a very irregular seabed, a number of sensitivity analyses had to be performed. An important objective was to gain a good understanding of the influence of various parameters on the expansion and buckling behaviour. The main parameters investigated are:

- Seabed profile (incl. influence of trenching/presweeping);
- Friction;
- Residual lay tension;
- End constraints;
- Effect of intermittent rock dumping.

The various material and operating parameters used in the analyses are tabulated in Table 1 in the Appendix.

Case 1 is a 3.1 km long, 10" flowline with a design temperature of 135 °C and a design pressure of 170 bar. A seabed friction factor of 0.3 has been assumed.

Figure 3 shows the as-laid empty configuration of this flowline resulting from the 2D ANSYS analysis where spans are formed, the longest of them between 50 m and 100 m long. The maximum bending stress occurs in the deep span near KP 1.100 and was found to be less than 200 MPa.

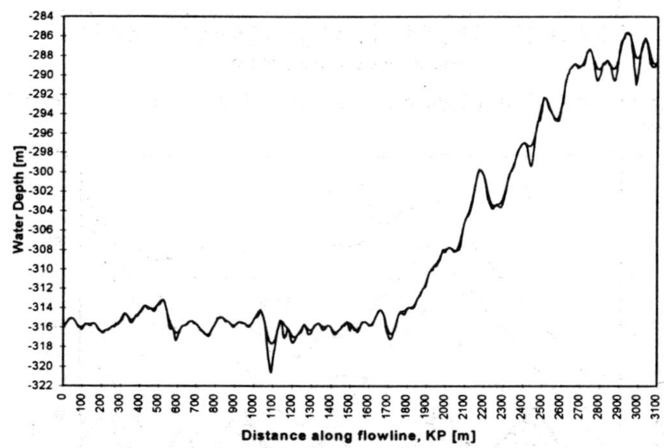

Fig. 3 Case 1: Empty as-laid configuration (Free ends)

Figure 4 shows the same flowline at full design temperature and pressure. Two buckles have formed at KP 1.65 and KP 2.15 with approximate heights of 2.8 m and 3.1 m, respectively. In addition, two smaller buckles have formed at the shoulders of the span at KP 1.1. A maximum equivalent stress of approximately 340 MPa occurs at the bottom of the span at KP 1.100. Similar stress levels occur in the two upward buckles.

Figure 4 also shows the net axial displacement of the flowline between start up and full operating condition. In this figure, positive displacement denotes movement to the right, i.e. towards KP 3.1. Similarly, negative displacement means the flowline is moving to the left. The figure demonstrates that, contrary to an ideal flowline that expands from the midpoint towards the ends, a small diameter flowline on an irregular seabed will initially expand towards free spans. Expansion is fed towards the spans from either side and thus the spans act as expansion loops able to absorb a limited amount of axial movement. At several points along the line, typically on either side of large spans or imperfections, the flowline remains stationary as the temperature increases. The net axial displacements at the flowline ends are 1.5 m at KP 0.0 and 0.8 m at KP 3.1, which are considerably less than the theoretical displacement of 2.9 m for the same flowline placed on a perfectly flat seabed (see below).

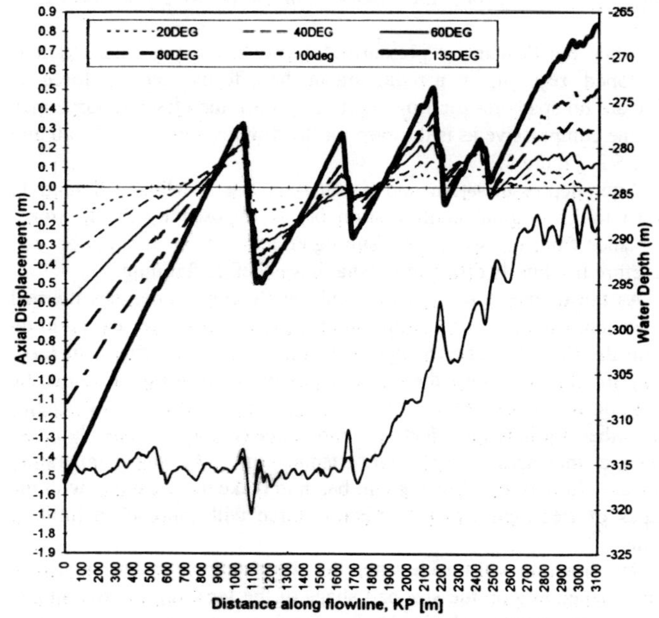

Fig 4 Case 1: Axial displacement. and operating configuration (Free ends)

Figure 5 shows the development of the effective axial force from the empty as-laid condition to full operating pressure and temperature. The only remaining effective force in the flowline after the ends are released is due to the residual lay tension of 5 tonnes.

After pipelay and prior to hydrotest the flowline will be flooded with water. The added weight of the water will tend to cause the flowline to sag more into the spans, while it is restrained by the increased frictional resistance at the shoulders.

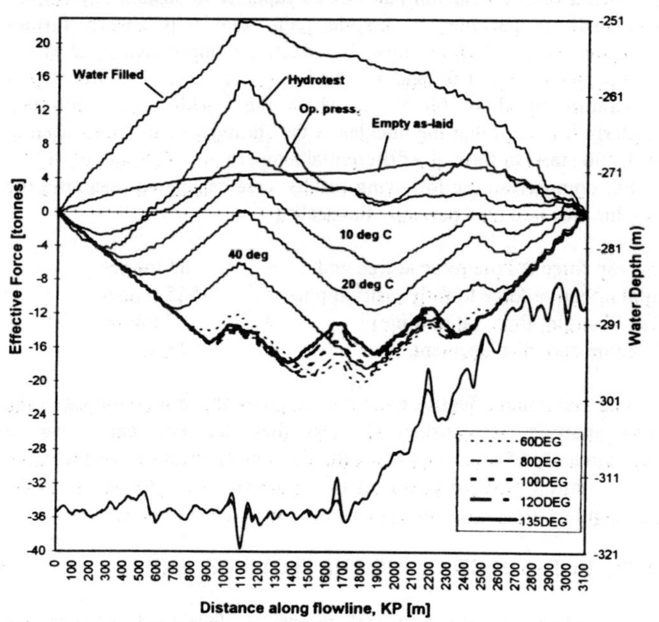

Fig 5 Case 1: Effective force distribution and operating configuration (Free ends)

The axial force restraining the flowline from feeding into the span is commonly referred to as sag tension and will add to the total effective force. Figure 5 shows that when the flowline is filled with water, the effective force increases to a maximum of approximately 23 tonnes.

When the flowline is pressurised during hydrotest, axial strain is developed, resulting in a reduction in the effective tension force. If there are no spans for the flowline to feed into, the effective force may become compressive as is the case for the flowline between KP 0.0 and KP 0.6 in Figure 5.

After depressurisation and dewatering, the flowline will tend to return to its original configuration but is opposed by the frictional resistance from the soil. The resulting effective force distribution will therefore be slightly different to what it was before flooding.

As the temperature increases and the flowline undergoes thermal expansion, frictional forces opposing the expansion movement build up along the flowline. The change in the direction of the frictional force along the line, wherever there is a stationary point in Fig. 4, limits the net build up of the effective force as opposing friction forces balance each other. Initially the effective compressive (negative) force along the flowline increases linearly with temperature. As the temperature increases further, the flowline will begin to make more contact with the slopes of freespans and the effective force will approach a limiting value.

The flowline will buckle when the effective force reaches a critical value, depending on the size and shape of the local out-of-straightness (OOS) imperfection. The buckled region looses its capacity to sustain any higher axial load and the pipe section is said to have reached its ultimate capacity. The post buckling behaviour is dependent on the initial OOS. For a large initial imperfection, the buckling is associated with a gradual magnification of the initial OOS as the temperature is increased while the effective force remains constant. For a smaller OOS, the effective force decreases while the deflection increases linearly with increased temperature. At very small imperfections, the buckling will take place as a dynamic "snap" when the ultimate capacity is reached, resulting in an instant drop in the axial effective force together with an instantaneous transverse deflection.

Once a buckled section has lost its capacity to sustain any further axial load, it becomes a flexible point that will absorb further expansion. Figure 5 shows how the effective compressive force at the two buckles (at KP 1.65 and KP 2.15) starts to drop at a differential temperature of about 60 °C at which the buckles were initiated. Similarly it is seen that the buckles at the shoulder of the deep span at KP 1.100 starts to form at a differential temperature of about 100 °C.

For comparison the following results were calculated assuming the flowline was laid on a perfectly flat seabed:

End cap force ("True force at free end")	=	65 tonnes
Total effective force to fully restrain pipe	=	-345 tonnes
Max. restraint force (at midline)	=	-27 tonnes
Flowline end displacement	=	2.9 m

The maximum effective compressive force of about 20 tonnes in the FEM analysis, is considerably less than the end cap force of approximately 65 tonnes and thus the flowline is always in true tension.

The initial positive peaks on the effective force plot at the lower temperatures are a result of sag tension due to the content weight.

EFFECT OF CONSTRAINING THE FLOWLINE ENDS

In addition to the frictional resistance developed between the flowline and the seabed, external forces imposed by end constraints will add to the total effective compressive force. For instance if the flowline were fully fixed at the ends, e.g. due to rock dumping to protect the line from dropped objects, one could expect that the axial expansion and effective force distribution along the line would change dramatically.

Case 2 analysed the flowline behaviour with the ends fully fixed. The resulting axial displacement and effective force distribution are presented in Figures 6 and 7, respectively.

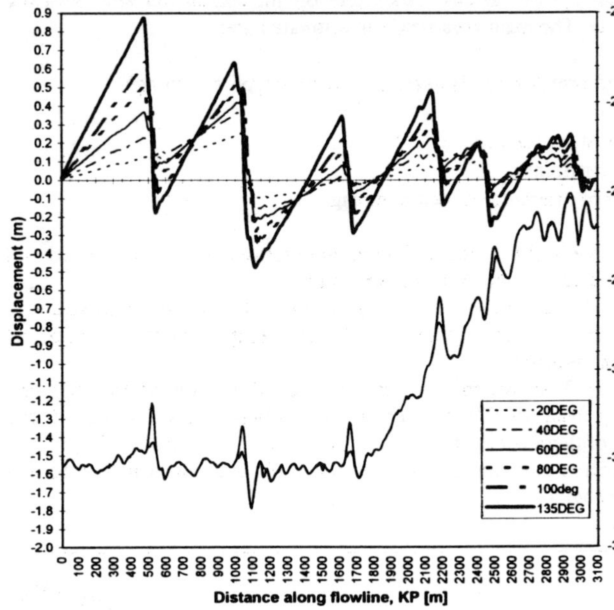

Fig 6 Case 2: Axial displacement and operating configuration (Fixed ends)

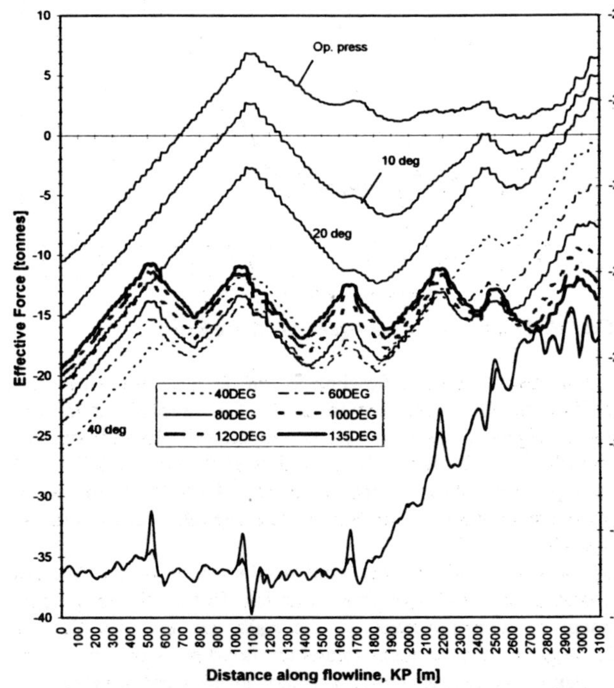

Fig 7 Case 2: Effective force distribution and operating configuration (Fixed ends)

Compared to Case 1, a new buckle has now formed near each of the flowline ends while the buckle at the shoulder of the deep span at KP 1.100 has increased in height. However, the size and location of the two buckles at KP 1.65 and KP 2.15 remain almost unchanged. The axial displacement and the effective force are only affected near the flowline ends by the new boundary conditions. Further away, at midsection, both the effective force and the axial displacement distribution are almost identical to that found for the flowline with free ends, showing the buckling to be mainly governed by local seabed features.

EFFECT OF PRESWEEPING OR TRENCHING

The susceptibility of a flowline to undergo buckling is dependent on both the effective force and the size and shape of any seabed imperfections. A possible way to suppress a particular buckle could be to remove or smooth out the imperfection that triggers the buckle by presweeping or trenching. In order to investigate this strategy, several analyses of Case 1 with modified seabed profiles were performed. The results of these analyses are not shown here, but it was found that reducing the imperfection height at KP 2.15 to suppress the buckle, caused new buckles to be triggered at the imperfections at KP 2.40 and KP 2.50. Removing these imperfections as well caused the buckle at KP 1.100 to grow considerably in size. The conclusion of these analyses is that suppressing buckles in one area will lead to new buckles in other areas. Although this method may be useful to remove a buckle at a particular location, it is not considered to be viable for the entire flowline.

This is confirmed in Case 3, where a 9.1 km flowline with a design temperature of 125 °C and a design pressure of 170 bar is considered. The flowline is modelled with free ends, trenched to a depth of 1-1.5 m, with a seabed friction factor of 0.3. For the untrenched case, approximately twelve buckles formed along the flowline with similar maximum stress levels to those found in Case 1. Figure 8 shows the operating configuration for the trenched flowline. One should keep in mind that this flowline is three times longer than the one in Figure 5. The trenched flowline was found to buckle in only five areas. In general, the buckle amplitude and corresponding stresses were found to be far higher than in the untrenched case; The maximum amplitude and equivalent stress were found to be 10 m and 470 MPa respectively. This contrasts with the untrenched case where the maximum buckle height and equivalent stress were found to be 5 m and 400 MPa respectively.

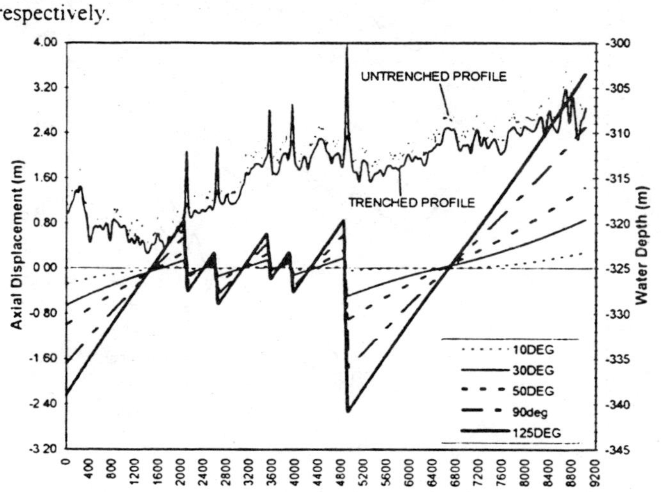

Fig 8 Case 3: Axial displacement and operating configuration (free ends)

The resulting effective force distribution is shown in Figure 9. As for the previous cases, this plot also shows the characteristic peaks indicating where buckling has occurred. It is further apparent that the buckling took place at a differential temperature of less than 10 °C and that a dynamic "snap" occurred since the effective force is not linearly reduced until after the temperature has reached about 30 °C. It should further be noted that due to the absence of sag tension, unlike the flowline in Case 1, this flowline is in effective compression during hydrotest and before applying any temperature increase.

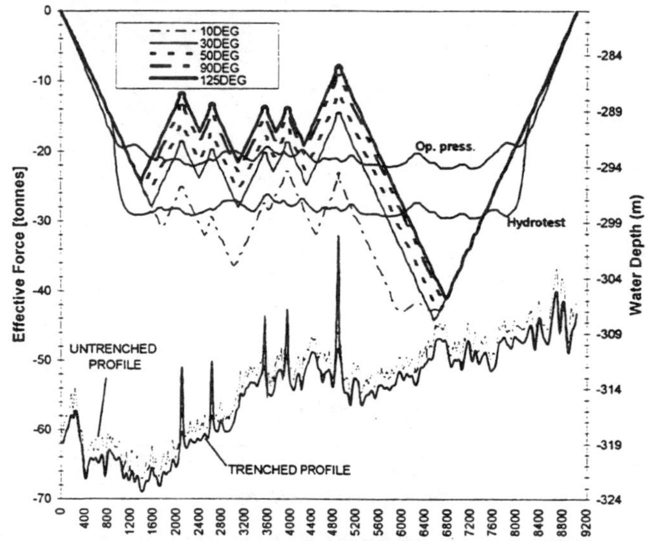

Fig 9 Case 3: Effective force distribution and operating configuration (Fixed ends)

CONTROLLING BUCKLING WITH INTERMITTENT ROCK DUMPING

Another method of controlling the flowline displacement and buckling behaviour is to use intermittent rock dumping. In order to ensure the preferred flowline behaviour is achieved, selective sections of the flowline may be covered with rock in order to locally increase the frictional resistance. However if the rock is dumped at arbitrary locations this may dramatically change the effective force distribution in a way that is not desired. An example of this was intermittent rock dumping applied on a trenched profile which resulted in a maximum effective force far exceeding the maximum effective force without rock berms (results not shown here). This caused new buckles to be initiated that otherwise would not have been triggered.

In Case 1 it was shown that a flowline laid on an irregular seabed will expand in the direction of least resistance. At several points along the flowline, typically on either side of large spans or imperfections, the flowline will be stationary as the temperature increases. In order not to disturb the natural displacement distribution, intermittent rock berms should preferably isolate the stationary points.

Case 4 considers the same flowline analysed in Case 1, but now with two 0.5 m high and 50 m long rock berms placed on either side of the stationary point at KP 1.85. The resulting effective force distribution is shown in Figure 10. It shows that the effective force between the rock berms has almost doubled compared to the case where no berms are included, although the effective force distribution and axial displacement distribution is almost unchanged in the remaining flowline. Since the flowline will tend to expand in the direction of least resistance, the axial displacement distribution can be ensured by increasing the effective force locally. On the other hand, the actual expansion feed-in into the buckled regions were almost unchanged by the placement of the rock berms. This is expected since the thermal expansion of a section can be obtained by the integral of axial thermal strain over the section in question. For example if a 300 meter section of the flowline was to be fully restrained with a vast rock berm, the expansion would only be reduced by a maximum of 0.5 m.

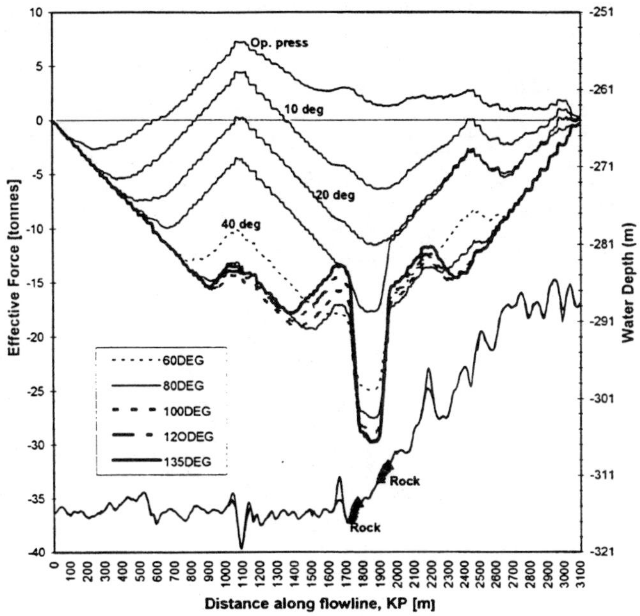

Fig 10 Case 4: Effective force distribution and operating configuration (Free ends)

SENSITIVITY TO SEABED FRICTION

In order to investigate the effect of varying the longitudinal seabed friction, an 11 km, 10" flowline was analysed for a wide range of friction factors (results not shown here). The results showed the high sensitivity friction has on the number and amplitudes of buckles at full operating conditions; The analyses resulted in 4, 12 and 28 buckles for friction factors of 0.1, 0.3 and 1.0 respectively.

It is apparent that, for a given imperfection, the effective force required to buckle the flowline is constant and independent of the level of the axial seabed friction. Yet, the accumulation of the effective force away from the buckled section is directly related to the build up of seabed friction. At low friction values, the effective force may therefore not build up sufficiently to trigger a buckle at an adjacent imperfection which would have triggered a buckle for a higher friction factor.

The net result is that a very low friction factor will make the flowline behave very similar to a trenched flowline, i.e. few buckles with large feed-in of thermal expansion into each buckle. As expected, the buckles triggered at lower friction factors also occurred when the friction was increased but with several new buckles. Similarly, the stationary points that were present for the low friction case were also present for the higher friction in addition to several more stationary points, i.e. at each side of the new buckles.

3D IN-PLACE ANALYSES

The analyses presented in the preceding sections are based on 2D FE models of the flowline. It was demonstrated that buckling is of major importance for a HP/HT flowline laid on an uneven seabed. In the 2D analyses, vertical buckles occurred in locations with large seabed imperfections. However, it is envisaged that as soon as the flowline lifts off the seabed, an interaction between the vertical and horizontal mode of buckling will occur, causing the flowline to buckle laterally.

In order to study the lateral buckling behaviour and to enable a direct comparison between the 2D and 3D results, Case 5 considers a 3.1 km, 10" flowline which are analysed in both 2D and 3D. This example is similar to Case 1 except that the longitudinal profile has changed as well as wall thickness and some material and operating parameters (see Table 1).

2D Results

Figure 11 shows the operating configuration resulting from the 2D analysis. The flowline appears to have buckled in two locations, namely at KP 1.4 and KP 1.85. The maximum equivalent stress of 365 MPa occurred at the deep span at KP 2.25.

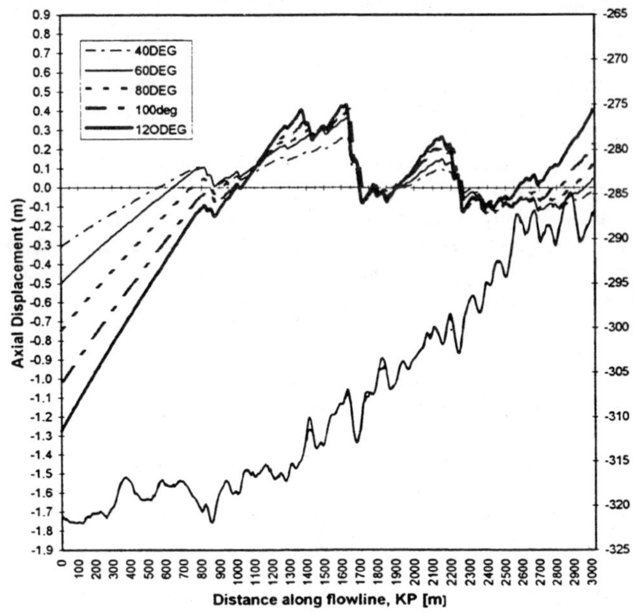

Fig 11 Case 5 - 2D results: Axial displacement and operating configuration (free ends)

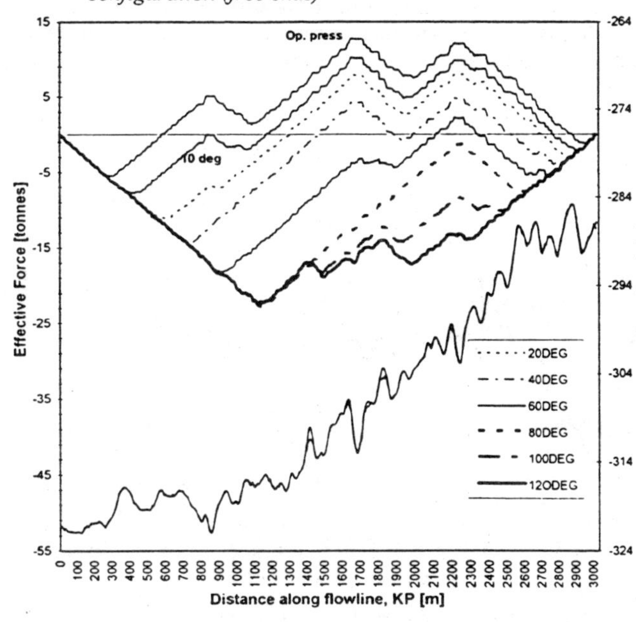

Fig 12: Case 5 - 2D results: Effective force distribution and operating configuration (free ends)

The maximum equivalent stresses in the two buckles were found to be 350 MPa and 210 MPa respectively. The axial displacement distribution is also shown in Fig. 11. Figure 12 shows the effective force distribution along the flowline as the temperature is increased. It shows the characteristic peak at KP 1.4 indicating that the flowline has buckled. At KP 1.85 no such peak exists, i.e. the effective force was not reduced prior to reaching the maximum temperature. This means that the flowline has not yet reached the level at which it can no longer sustain the axial load and thus the section has not buckled.

3D Results

The seabed elements are generated from the available survey data and are represented by a 3D grid as shown in Figure 13.

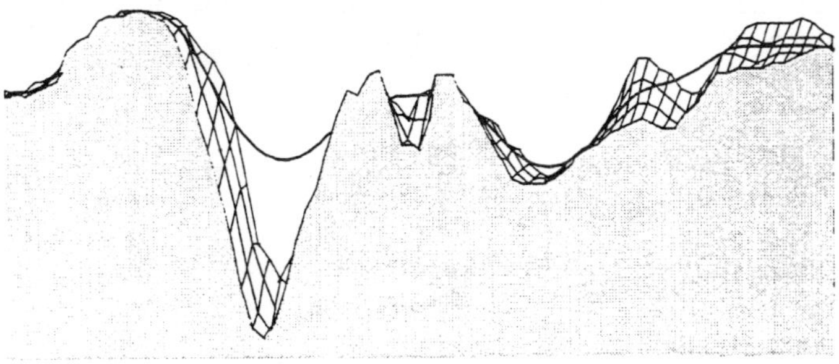

Fig 13: 3D FEM In-place analysis model - Typical model details from an as-laid flowline simulation.

The axial displacement plot (Figure 14) shows very similar results to that obtained from the 2D analysis (Figure 11).

The axial distribution of the effective force is presented in Figure 15. It shows almost identical results to the 2D simulation (Figure 12) up to and including a temperature of 80°C. Beyond this point there is a considerable reduction in the effective force at KP 1.4, indicating that buckling has taken place.

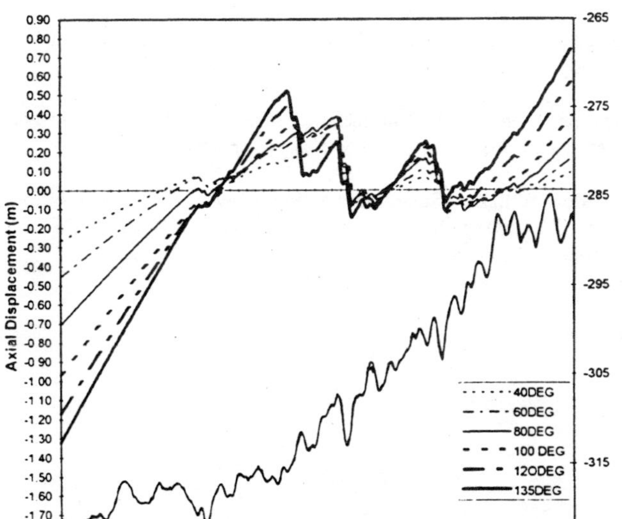

Fig 14: Case 5 - 3D results: Axial displacement and operating configuration (free ends)

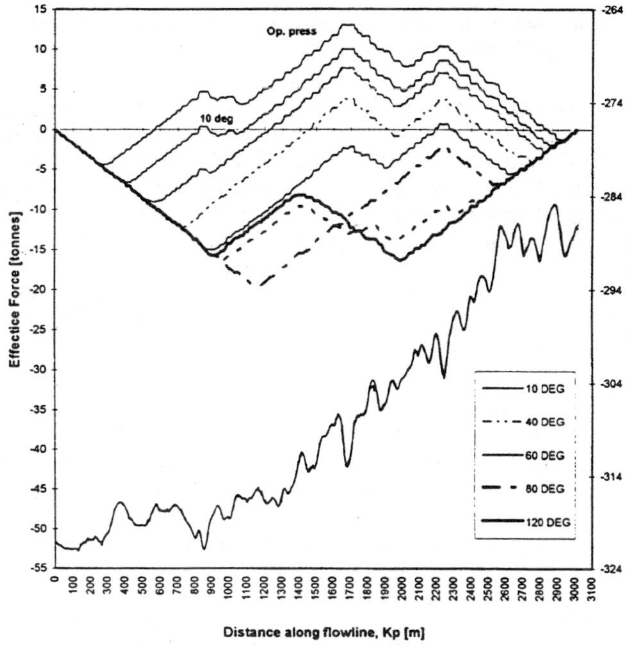

Fig 15 Case 5 - 3D results: Effective force distribution and operating configuration (free ends)

Figure 14 shows the vertical configuration of the flowline at full operating temperature resulting from the 3D analysis. By comparison to the 2D results, it is apparent that the vertical deflection at KP 1.4 has disappeared in the 3D analysis. The upwards deflection at KP 1.85 is also present in the 3D analyses, albeit the height of the buckle is reduced.

Figure 16 shows a plan view of the deflected shape. At KP 1.4, i.e. at the location with the vertical buckle in the 2D analysis, a large horizontal buckle with an amplitude of 3.8 m has occurred in the 3D model.

The resulting maximum equivalent stress in the horizontal buckle was found to be 302 MPa, which is significantly less than in the corresponding 2D analysis (350 MPa).

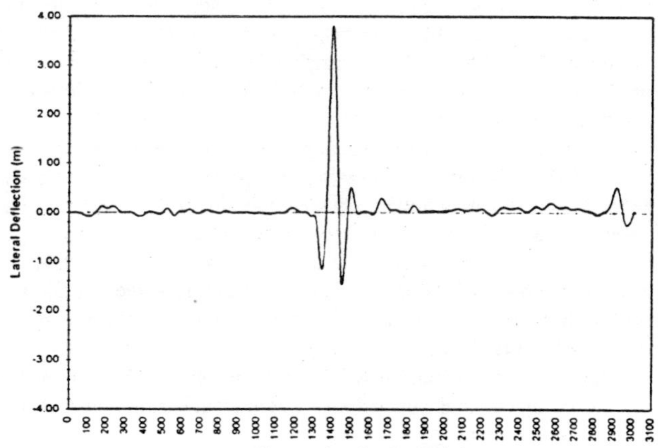

Fig 16 Case 5 - 3D results: Plan view (Free ends)

Other 3D buckling effects such as vertical and horizontal interaction, dynamic snap and cyclic behaviour are presented in the accompanying paper by Nystrøm and Tørnes (1997).

The following summarises the main observations from a number of similar 2D and 3D analyses,:

- In most cases where the 2D analysis predicted a vertical buckle to occur, it is likely that the flowline will buckle horizontally at the same location. At large vertical imperfections the flowline is likely to lift off the seabed initially as the temperature increases. The lateral restraint (i.e. friction) is reduced and the flowline will reach its critical lateral buckling load at a lower temperature than otherwise would have triggered a horizontal buckle.
- There is a significant reduction in the maximum stress level in the lateral 3D buckle compared to the corresponding vertical 2D buckle.
- In all cases investigated, the lateral buckle is associated with a dynamic snap with a corresponding larger feed-in of expansion to the buckled region than is the case for the vertical 2D buckle.
- The drop in the axial effective force after the ultimate capacity is reached affects a longer section of the flowline away from the buckled region compared to the corresponding vertical 2D buckle. This reduction in the effective force at a longer distance from the buckle may suppress an adjacent buckle that is indicated in the corresponding 2D analysis. The expansion that would otherwise have fed into the suppressed buckle would then be redistributed to the remaining buckles.

PROPOSED INTERVENTION STRATEGIES

The preceding discussion of HP/HT flowlines shows that the behaviour of flowline on a very uneven seabed is more complex than an ideal flowline on a perfectly flat seabed. If the expansion is allowed to feed into spans, the overall build up of effective force and buckling is delayed. Compared to a smooth seabed profile, more buckles occur with less total feed in to those buckles which are initiated.

Based on these studies, it is proposed that the following intervention strategies are considered for an HP/HT flowline on a very uneven seabed:

- If the friction for the flowline route is well defined and the in-place analyses show that stress levels are acceptably low, one option may be to leave the flowline as it is on the undisturbed seabed. Intermittent rock dumping may be used to improve the predictability of the flowline behaviour by ensuring that the axial deflection and effective force distribution will be as shown in the analysis. The rock berms should be placed at locations characterised by low axial movement.
- The stresses in the buckled sections can be controlled by limiting the feed-in to the buckle. Allowing more buckles to occur with less feed-in will result in lower stresses overall for the pipeline. The locations of rock berms will be determined by the lengths of flowline on the seabed building up critical axial effective force at either side of a buckling point. The amount of axial feed-in allowed will depend on the pipe, its bending stiffness, weight and the acceptable stress level.
- Because the seabed friction will often be uncertain, the rock berms should be placed to ensure adequate behaviour for the lower bound of the friction range.

A minimum intervention approach as outlined in the above, will require a thorough assessment using an adequate FEM package. Sufficient correlation of 2D and 3D results will permit use of mainly 2D simulations of flowline in service behaviour to save analysis time.

CONCLUSIONS

FEM simulations of global HP/HT flowline behaviour has led to an unconventional design approach to ensure the safety of the pipeline. The unevenness of the seabed is used to absorb expansion into freespans and initiate buckling. The build up of axial force is limited while the buckled shape absorbs further thermal and pressure expansion of the flowline. Correlation of 2D with 3D simulations allow 2D analyses to be used to develop a seabed intervention strategy which can save the operator seabed intervention costs. The methodology used demonstrates the advantages of FEM simulation to exploit the unique features of a project in order to optimise the pipeline design.

REFERENCES

Hobbs (1984). " In-Service Buckling of Heated Pipelines", *Jour. of Transp Eng , pp 178-189..*

Ellinas, Supple and Vastenholt (1990). "Prevention of Upheaval Buckling of Hot Submarine Pipelines by Means of Intermittent Rock Dumping", *Proc 22nd Offshore Tech Conf, pp 519-528.*

Fyrileiv, Aamlid and Venås (1995). "Analysis of Expansion Curves for Subsea Pipelines", *Proc 6th Int. Offshore and Polar Eng. Conf., Vol. II, pp. 66-73..*

Nystrøm, Tørnes, Bai and Damsleth (1997). "3D Dynamic Buckling and Cyclic Behaviour of HP/HT Flowlines", *Proc 7th Int. Offshore and Polar Eng. Conf.*

Sævik and Levold (1995). "High Temperature Snaking Behaviour of Pipelines", *Proc 5th Int. Offshore and Polar Eng. Conf., Vol. II, pp. 63-71.*

Pedersen and Jensen (1988). "Upheaval Creep of Heated Pipelines with Initial Imperfections", *Marine Structures, Vol. I, pp. 11-22.*

Taylor and Gan (1986). "Submarine Pipeline Buckling - Imperfection Studies.", *Thin Walled Structures, Vol. IV, pp. 295-323.*

Vermeulen, Williams, Salami and Ellinas (1994). "Pipeline Installation for High Pressure/High Temperature Applications" *Proc ASPECT '94 Conference, Aberdeen.*

APPENDIX:

The various parameters considered in the analyses cases discussed in the foregoing are tabulated in Table 1 below:

Parameter	CASE 1, CASE 2 and CASE 4	CASE 3	CASE 5
Inside diameter (mm)	228.6	228.6	228.6
Wall thickness (mm)	11.5 (0-2.6 km) 17.7 (2.6-3.1 km)	11.5 (0-8.5 km) 13.8 (8.5-9 km)	11.5 (0-2.6 km) 17.7 (2.6-3 km)
Submerged Op. Weight (N/m)	544 (0-2.6 km) 875 (2.6-3.1 km)	544 (0-8.5 km) 665 (8.5-9 km)	544 (0-2.6 km) 875 (2.6-3 km)
Max. temp diff. (°C)	135	125	120
Design pressure (bar)	170	170	170
Seabed friction factor	0.3	0.3	0.3
Residual lay tension (kN)	50000	5000	50000
Material	Duplex	X65	-
Term. exp. coeff. , α	1.35E-5 1/K	1.17E-5	1.35E-5
Youngs Modulus, E (MPa)	200,000 at 20 °C 190,000 at 100 °C 185,000 at 100 °C	207,000 (No derating)	207,000 (No derating)
Yield stress (MPa)	450 at 20 °C 370 at 100 °C 345 at 150 °C	448 (No derating)	448 (No derating)
Ramberg-Osgood parameter, $\sigma_{0.7}$	427.55 at 20 °C 343.36 at 100 °C 322.65 at 150 °C	427 (No derating)	427 (No derating)
Ramberg-Osgood parameter, n	15.27 at 20 °C 12.7 at 100 °C 14.68 at 150 °C	24.3 (No derating)	24.3 (No derating)

Table 1 Definition of study cases

Proceedings of the Seventh (1997) International Offshore and Polar Engineering Conference
Honolulu, USA, May 25-30, 1997

ISBN 1-880653-28-1 (Set); ISBN 1-880653-30-3 (Vol. II)

Bends in Steel Pipelines: A New Analytical Model to Calculate Stresses, Strains and Deformations

J. Mennink
Eindhoven University of Technology, Eindhoven/TNO-Building and Construction Research, Delft
A.M. Gresnigt and J. Blaauwendraad
Delft University of Technology, Delft
P.Ph.C. Coors
N.V. Nederlandse Gasunie, Groningen

The Netherlands

ABSTRACT

The structural behaviour of bends loaded by combinations of bending moment and internal or external pressure is determined using the principle of virtual work. A new analytical model to describe the strains, stresses and deformations in both the elastic and elastoplastic region has been developed.

The new model will enable further research on the analysis of the stress-strain variations that occur in the low cycle fatigue range of bends loaded by combinations of bending moment and internal or external pressure. Both elastic and elastoplastic geometrical non-linear results are presented.

Keywords: Pipeline, bend, fatigue, analysis, non-linear.

1. INTRODUCTION

Bending moments on bends in pipelines, may cause severe ovalization, strains and stresses. Compared with straight pipelines, bends show a reduced bending stiffness, much greater ovalization and a complex pattern of strains and stresses in both longitudinal and in circumferential direction.

The operational fluctuations in internal pressure in an ovalized bend will give variations in bending moment, curvature and ovalization. The resulting variations in stresses and strains may lead to alternating yielding and eventually lead to failure due to low-cycle fatigue.

Present analytical models can describe accurately the strains and stresses in the elastic region, but after yielding has occurred, these models give less accurate results, in particular for the plastic strains. For static loading, this is not a problem, but for cyclic loading a more accurate knowledge of the occurring strain variations is necessary to evaluate the danger (limit state) of low cycle fatigue.

There are several situations where low cycle fatigue has been or will be considered as a limit state. Examples are bends in buried pipelines that experience differential settlements causing high bending moments, offshore pipelines where risers are subjected to severe bending moments due to temperature differences and finally also pipelines in district heating.

The research therefore was concentrated on obtaining a model that can describe the stresses and strains in both the elastic- and elastoplastic regions due to a bending moment and internal pressure.

Much research has been carried out on the behaviour of bends and elbows loaded by combinations of internal pressure and bending. Von Karman (1911) presented the first theoretical solution for smooth bends under in-plane bending moment loading. Numerous publications followed, which extended his work to include other forms of loading and geometry's. Well known are for example the publications of Vigness (1943), Kafka and Dunn (1956), Rodabaugh and George (1957), Findlay and Spence (1979), Thomson and Spence (1983), Whatham (1986).

It appears that the existing analytical models as initially developed by e.g. Von Karman (1911) and further by Rodabaugh and George (1957) and Gresnigt (1986, 1995), which are based on potential energy, could not be used for an accurate description of the strains and stresses in the elastoplastic region. It is known that the latter behaviour can be described by models based on virtual work.

Baumjohann (1994) developed an interesting model, based on virtual work to describe the elastoplastic region. This model, however, does not take into account an extension of the pipe. It also regards the yield stress in lateral direction to be independent of the stresses in circumferential direction and vice versa. These two assumptions do not hold if internal pressure is applied. Therefore, it was decided to develop new models for the analyses of bends, based on virtual work and including an extension of the pipe.

Before introducing elastoplastic material behaviour, the principle of virtual work was used to set up analytical models for the elastic region. From here, the model can be extrapolated in the elastoplastic region.

In this paper, a description of the new analytical model is given. For a quick and easy application, a special computer programme NLAB (Non Linear Analysis of Bends) was developed. The results of this analysis were compared to:

- FEM calculations.
- Existing theories based on potential energy.
- Experimental results.

2. MODEL

2.1 General

Most previous analytic models regard bends under curvature. They do not account for the extension of the pipe due to internal pressure. To be able to describe the strains due to both bending and internal pressure, the developed model also regards extension of the pipe.

The bend will be calculated non-linear, both physically and geometrically. Physically non-linear by applying plastic theorems after yielding. Geometrically non-linear by using the deformed bend characteristics in the calculation.

The following assumptions are made:

- The material is isotropic.
- There are no internal stresses as a result of the manufacturing process.

2.2 Geometry

We will only consider the bend itself. The influence of the connecting straight pipes will be a subject of later studies. In previous research (Gresnigt,1986), it is demonstrated that if a bend with a large opening angle ($\alpha_0 \approx 60° - 90°$) is regarded, the influence of the connecting pipes on the stresses and strains in the middle of the bend will be relatively small.

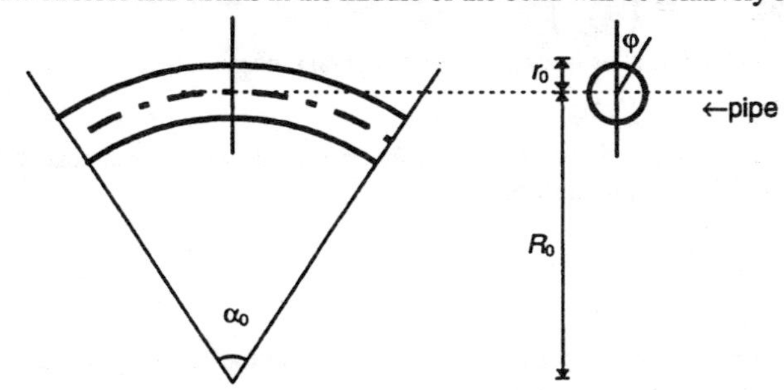

Figure 1. Model geometry.

The bend has the following characteristics, see figure 1:

α_0	=	the opening angle of the bend.
φ	=	the angle to describe the position on the pipe wall.
r_0	=	the radius of the pipe.
R_0	=	the radius of the bend.
$t(\varphi)$	=	the wall thickness of the pipe, this as a function of φ.

From these characteristics follow the following variables:

$R(\varphi)$	=	the bend radius of a fibre at angle φ.
	=	$R_0 + r_0 \cos(\varphi)$
$L(\varphi)$	=	length of a longitudinal fibre at angle φ.
	=	$\alpha_0 R(\varphi)$

2.3 Displacements

2.3.1 Longitudinal deformation

We assume a constant curvature over the bend and an elongation of the centre line of the bend. Considering the fact that a circle under a constant curvature again forms a circle, we increase the radius of the bend R_0 with a displacement w_R and the opening angle α_0 by γ, see figure 2.

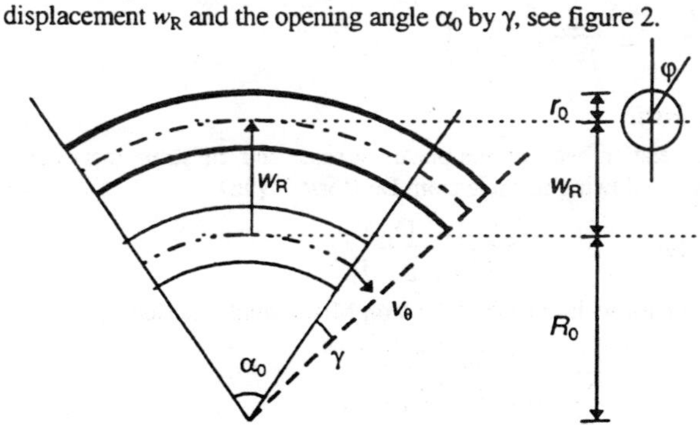

Figure 2. Longitudinal displacements.

Because a constant deformed state is assumed, the deformations can be described by constants:

w_R	=	A_1
v_θ	=	A_2
γ	=	$A_2 R_0$

2.3.2 Circumferential deformation

Because the deformation of the cross-section causes the large stiffness decrease of the bend, it has to be described properly.

We will use the traditional description going back to Von Karman (1911). He describes both the radial and tangential displacements as a function of n parameters B_j and the position on the wall described by the angle φ. The advantage of this description is that in circumferential direction the centre of the wall is strain-less.

To be able to describe an increase of the diameter and therefore a constant strain due to the internal pressure, an additional constant displacement B_0 is introduced.

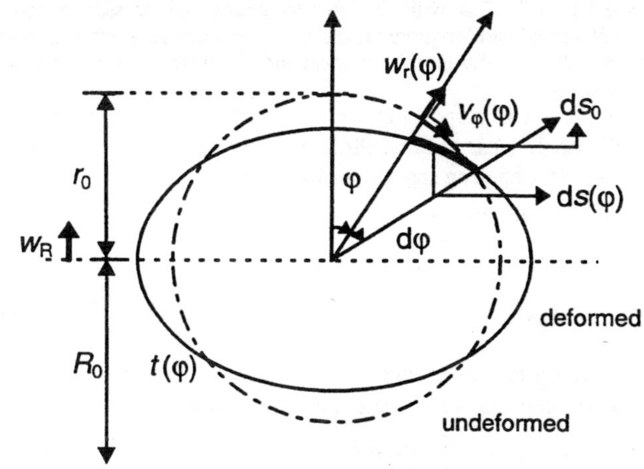

Figure 3. Circumferential deformation.

The variables in figure 3 have the following meaning:

$w_r(\varphi)$	=	radial displacement of the pipe wall
	=	$B_0 + 2\sum_{j=1}^{n} (j\, B_j \cos(2\, j\, \varphi))$
$v_\varphi(\varphi)$	=	tangential displacement of the pipe wall
	=	$-\sum_{j=1}^{n} (B_j \sin(2\, j\, \varphi))$
φ	=	circumferential positioning angle.
B_j	=	unknown displacement parameter.
ds_0	=	undeformed circumferential length.
	=	$r_0\, d\varphi$
$ds(\varphi)$	=	increase in circumferential length at an angle φ.
	=	$r^{**}(\varphi)\, d\varphi$
$r^{**}(\varphi)$	=	parameter to describe the deformed pipe radius.

2.3.2.1 Displacement vector

For the displacements we will use the following parameters:

- For the longitudinal displacements: A_1 and A_2.
- For the circumferential displacements: $B_0, B_1, B_2, .., B_n$

We can express these in one displacement vector:

$$\underline{u} = \begin{pmatrix} A_1 \\ A_2 \\ B_0 \\ B_1 \\ .. \\ .. \\ B_n \end{pmatrix}$$

This is a vector with n + 3 variables.

2.4 Loads

To be able to derive the complete moment-curvature diagram, including the behaviour after the maximum moment has been reached, we will use displacement-control in the calculation. Because we can link the curvature directly to the change in opening angle γ, we will control the calculation on the displacement A_2.

The bend will be loaded by internal pressure p and a forced deformation γ, see figure 2. This will give rise to strains and stresses in both the circumferential and longitudinal direction, resulting in a bending moment M. The relation between this moment and the curvature will be presented in chapter 3.

For calculation purposes in the elastoplastic range, the increase of the opening angle γ has to be applied incrementally in steps $\Delta\gamma_i$. The total increase of the opening angle in loadstep i equals:

$$\gamma_i = \Delta\gamma_i + \gamma_{i-1}$$

2.5 Strains

2.5.1 Longitudinal strains

For an increment of the longitudinal strain we obtain:

$$\Delta\varepsilon_\theta(\varphi) = \frac{L^{**}(\varphi) - L^{*}(\varphi)}{L^{*}(\varphi)}$$

$L^*(\varphi)$ = longitudinal length of a fibre at step i-1

$$= \alpha_0^{*} R^{*}(\varphi)$$

where:

$$\alpha_0^{*} = \alpha_0 + \frac{\Delta A_2}{R_0}$$

$$R^{*}(\varphi) = R(\varphi) + w(\varphi)$$

$L^{**}(\varphi)$ = longitudinal length of a fibre at step i

$$= \alpha_0^{**} R^{**}(\varphi)$$

$$w(\varphi) = w_R + w_r(\varphi)\cos(\varphi) - v_\varphi(\varphi)\sin(\varphi)$$

$$= \left(\alpha_0^{*} + \frac{\Delta A_2}{R_0}\right)\left(R^{*}(\varphi) + \Delta w(\varphi)\right)$$

$$\approx L^{*}(\varphi) + \alpha_0^{*}\Delta w(\varphi) + \frac{R^{*}(\varphi)}{R_0}\Delta A_2$$

$$\Delta\varepsilon_\theta(\varphi) = \frac{1}{R^{*}(\varphi)}\left(\Delta w(\varphi) + \frac{R^{*}(\varphi)}{\alpha_0^{*} R_0}\Delta A_2\right)$$

2.5.2 Circumferential strains

The increment in circumferential strain $\Delta\varepsilon_\varphi(\varphi)$ can be divided into 2 parts; an increase in diameter due to ΔB_0 and a curvature of the pipe wall.

Increase of the diameter:

$$\Delta\varepsilon_{\varphi 1}(\varphi) = \frac{\Delta B_0}{r_0 + B_0}$$

Curvature of the pipe wall:

$$\kappa = \frac{1}{r_0^2}\left(\frac{d\Delta v_\varphi(\varphi)}{d\varphi} - \frac{d^2\Delta w_r(\varphi)}{d\varphi}\right)$$

$$\Delta\varepsilon_{\varphi 2}(\varphi) = \frac{1}{r_0^2}\left(\frac{d\Delta v_\varphi(\varphi)}{d\varphi} - \frac{d^2\Delta w_r(\varphi)}{d\varphi}\right)$$

$$= \frac{z}{r_0^2}\sum_{j=1}^{n}(8j^3 - 2j)\cos(2j\varphi)$$

Total circumferential strain:

$$\Delta\varepsilon_\varphi(\varphi) = \frac{\Delta B_0}{r_0 + B_0} + \frac{z}{r_0^2}\sum_{j=1}^{n}(8j^3 - 2j)\cos(2j\varphi)$$

2.5.3 Total strains

In vector notation we can rewrite the increment of strains:

$$\underline{\Delta\varepsilon} = \begin{pmatrix}\Delta\varepsilon_\theta \\ \Delta\varepsilon_\varphi\end{pmatrix}$$

or, using the formulas from the previous paragraphs:

$$\underline{\Delta\varepsilon} = \underline{f}^T \underline{\Delta u}$$

where the transposed vector $\underline{f}^T$ gives the relation between the strains ad the increase of the displacements $\underline{\Delta u}$.

2.6 Stresses

2.6.1 General

We assume a bi-linear stress-strain relation, both for the longitudinal- and for the circumferential direction

Figure 4. Stress-strain relation.

Because we have strains in both directions, the yield stress and therefore the increase in stresses and strains is not uniquely defined. The relation between the increment of the stress- and strain vector equals:

$$\underline{\Delta\sigma} = \underline{D}\,\underline{\Delta\varepsilon}$$

2.6.2 Elastic

With the assumption of Hooke's Law we have a direct relation between the stresses and strains, we will neglect the wall throat contraction but we do include the lateral contraction ν.

$$\underline{\Delta\sigma}_{el} = \underline{D}_{el}\,\underline{\Delta\varepsilon}$$

with:

$$\underline{D}_{el} = \frac{E}{1-\nu^2}\begin{pmatrix}1 & -\nu \\ -\nu & 1\end{pmatrix}$$

2.6.3 Plastic

For the relation between the plastic stresses and the strain increment we will use the following formula from De Borst (1996).

$$\underline{\Delta\sigma}_{pl}(\varphi,z) = -\left(\frac{\underline{D}_{el}\,\underline{n}\,\underline{n}^T\,\underline{D}_{el}}{\underline{n}^T\,\underline{D}_{el}\,\underline{n}}\right)\underline{\Delta\varepsilon}$$

where $\underline{n}$ equals, with the use of the Von Mises yield criterion:

$$\underline{n} = \frac{1}{2\sigma_{yield}}\begin{pmatrix}2\sigma_\theta & -\sigma_\varphi \\ -\sigma_\theta & 2\sigma_\varphi\end{pmatrix}$$

This can be rewritten to:

$$\underline{\Delta\sigma}_{pl} = \underline{D}_{pl}\,\underline{\Delta\varepsilon}$$

where $\underline{D}_{pl}$ equals $\underline{0}$, if no yield occurs.

2.6.4 Total stresses

The increment of the total stresses equals:

$$\underline{\Delta\sigma} = (\underline{D}_{el} + \underline{D}_{pl})\,\underline{\Delta\varepsilon}$$

$$= \underline{D}\,\underline{\Delta\varepsilon}$$

with:

$\Delta\underline{\varepsilon} \quad = \underline{f}^T \Delta\underline{u}$ (see paragraph 2.5.3)

$\Delta\underline{\sigma} \quad = \underline{D}\,\underline{f}^T \Delta\underline{u}$

The total stresses after load increment *i* equals:

$\underline{\sigma}_i \quad = \Delta\underline{\sigma}_i + \underline{\sigma}_{i-1}$

2.7 Virtual work

2.7.1 General

The virtual work equation equals the sum of the absorbed strain energy due tot the deformation and the work done by the internal pressure. For a virtual small variation of the displacements $\delta\underline{u}$, it follows:

$$\mathrm{VW} = \int_V \delta\underline{\varepsilon}^T \underline{\sigma}\, dV - \int_A \delta v\, p\, dA = 0$$

In this equation the following parameters have to be explained:

$\underline{\sigma}$ = total stress vector
$\delta\underline{\varepsilon}^T$ = variation of the transposed strain vector
V = total volume of steel of the bend
p = internal pressure
A = total surface on which the internal pressure is acting
δv = variation of the displacement in the direction of the internal pressure: $\delta v = \underline{g}\,\delta\underline{u}$
$\delta\underline{u}$ = the virtual displacement field
$\underline{g}$ = vector to relate specific virtual displacements δv to $\delta\underline{u}$.

For the virtual work equation we have to integrate over the cross-section angle φ. Deformation parameters will deform the left and right side of a cross-section symmetrically. Therefore, instead of integrating from $\varphi = 0$ to $\varphi = 2\pi$, we will integrate twice from $\varphi = 0$ to $\varphi = \pi$.

2.7.2 Equation system

By using:

$\Delta\underline{\varepsilon} \quad = \underline{f}^T \Delta\underline{u},$ from paragraph 2.5.3

$\underline{\sigma}_i \quad = \Delta\underline{\sigma}_i + \underline{\sigma}_{i-1}$ from paragraph 2.6.4

$\quad = \underline{D}\,\underline{f}^T \Delta\underline{u} + \underline{\sigma}_{i-1}$

and:

$\delta v \quad = \underline{g}\,\delta\underline{u},$ from paragraph 2.7.1

we can rearrange the virtual work equation to:

$$\mathrm{VW} = \int_V \delta\underline{u}^T \underline{f}\,(\underline{D}\,\underline{f}^T \Delta\underline{u} + \underline{\sigma}_{i-1})\, dV - \int_A \delta\underline{u}^T \underline{g}^T p\, dA = 0$$

Because the virtual displacements $\Delta\underline{u}$ are independent of *V* and *A*, we get:

$$\mathrm{VW} = \delta\underline{u}^T \left(\int_V \underline{f}\,(\underline{D}\,\underline{f}^T \Delta\underline{u} + \underline{\sigma}_{i-1})\, dV - \int_A \underline{g}^T p\, dA \right) = 0$$

or simplified:

$$\mathrm{VW} = \delta\underline{u}^T (\underline{K}\,\Delta\underline{u} + \underline{f}_{in} - \underline{f}_{ex}) = 0$$

This must hold for <u>every</u> variation of the displacement field $\delta\underline{u}$. This leads to the eigenvalue problem:

$$\underline{K}\,\Delta\underline{u} = \underline{f}_{ex} - \underline{f}_{in}$$

This is an equation system with $n + 3$ equations and $n + 3$ variables. Due to the geometric and physical non-linearity's, this system has to be solved numerically.

2.7.3 Stiffness matrix $\underline{K}$ and internal vector $\underline{f}_{in}$

The stiffness matrix $\underline{K}$ is described by:

$$\underline{K} = \int_V \underline{f}\,\underline{D}\,\underline{f}^T dV$$

The internal vector $\underline{f}_{in}$ equals:

$$\underline{f}_{in} = \int_V \underline{f}\,\underline{\sigma}_{i-1}\, dV$$

Integration over the volume means:

- Integration over the pipe length, equalling to multiplying with:
  $L^*(\varphi) = \alpha_0^* R^*(\varphi)$
- Integration over the pipe wall, from:
  $z = \frac{-1}{2} t(\varphi)$ to $z = \frac{1}{2} t(\varphi)$
- Integration over the pipe circumference $ds(\varphi)$:
  $ds(\varphi) = r^{**}(\varphi)\, d\varphi$

Where we integrate from $\varphi = 0$ to $\varphi = \pi$ and multiply this by two.

2.7.4 External vector $\underline{f}_{ex}$, due to internal pressure

The virtual work done by the internal pressure equals the displacement of a surface of the pipe in the direction of the internal pressure.

$$\underline{f}_{ex} = \int_A \underline{g}^T p\, dA$$

where:

A = the surface of the end plane.
p = internal pressure.

The virtual work vector $\underline{f}_{ex}$ is due to:

1. the displacement of the end plane of the bend: $\underline{f}_{ex1}$
2. the change in cross-section surface: $\underline{f}_{ex2}$
3. the outward displacement of the entire bend: $\underline{f}_{ex3}$

*1. **Displacement of the end plane (figure 5)***

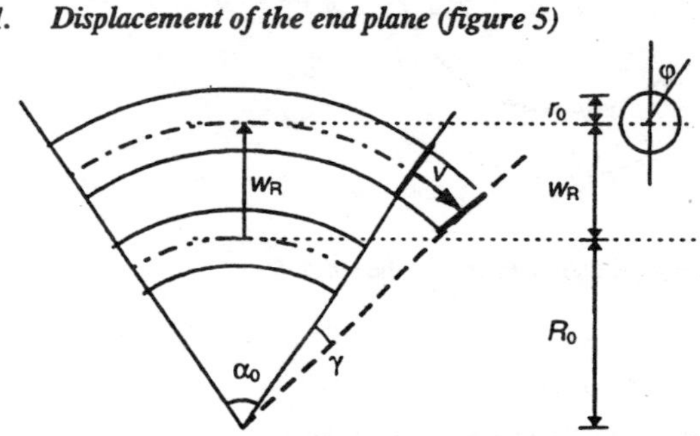

Figure 5. Displacement of the end plane.

The end plane will move over a distance v due to the displacement γ:

$v \quad = R^*(\varphi)\,\gamma$

The virtual work equation uses the variation of this displacement $\delta\gamma$

$\delta v \quad = R^*(\varphi)\,\delta\gamma$

Because we control the increase in opening angle γ, the variation of the increase $\delta\gamma$ equals to zero. Therefore, no virtual work is done.

$\underline{f}_{ex1} \quad = \underline{0}$

2. Deformation of the cross-section (figure 6)

This work equals the change in surface of the cross section, as a function of φ, integrated over the length $L^*(\varphi)$.

The deformation of the cross section results in :

$$\delta\underline{u}^T\underline{f}_{ex2} = 2p \int_0^{\pi} L^*(\varphi) \left(\left(r_0 + w_r + \frac{dv_\varphi}{d\varphi} \right) \delta w_r + \left(\frac{1}{2} r_0 + w_r \right) \frac{d\delta v_\varphi}{d\varphi} \right) d\varphi$$

$$= \delta\underline{u}^T \int_{A_2} \underline{g}_2^T p\, dA_2$$

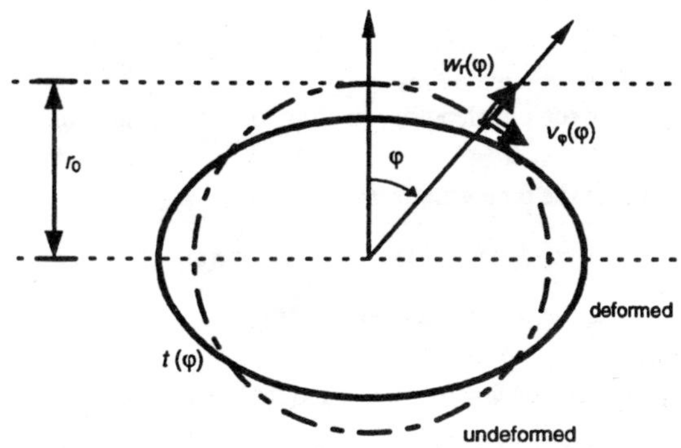

Figure 6. Deformation of the cross-section.

3. Outward displacement of the bend (figure 7)

Due to the modelling, the total bend can move outwards (or inwards).

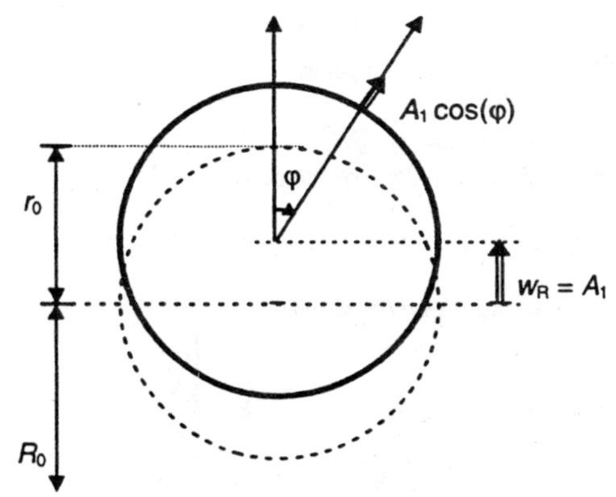

Figure 7. Outward displacement of the bend.

In this case:

$$\delta v \quad = \delta A_1 \cos(\varphi)$$

where the surface A equals the total bend surface.

$$\delta \underline{u}^T \underline{f}_{ex3} = -2 \int_0^{\pi} \frac{R^*(\varphi)}{R_0} p\, r^{**}(\varphi)\, \delta A_1 \cos(\varphi) d\varphi$$

$$= \delta \underline{u}^T \int_{A_3} \underline{g}_3^T p\, dA_3$$

3. EVALUATION

3.1 General

For a quick and easy application of the analytical model, a special computer programme NLAB (Non Linear Analysis of Bends) has been developed. The results obtained with NLAB have been compared to:

- FEM calculations.
- Existing theories based on potential energy.
- Experimental results.

Because in the new analytical model, the influence of the connected straight pipes is not yet included, it is not possible to directly compare the OPL test results (Gresnigt, 1986) with NLAB. Therefore, in this paper only "long" bends are considered (no influence of connected straight pipes).

In section 3.3, NLAB is compared with the analytical models of Rodabaugh and George (1957) and of Gresnigt (1986). In section 3.2 also a comparison is made with the results of a finite element calculation with the DIANA programme. In these comparisons a bend with R_0 = 480 mm is taken. The comparison is carried out for the moment-curvature and moment-ovalization diagrams.

In section 3.3, NLAB is compared with the same analytical models and with test results on a bend with R_0 = 1143 mm, that was tested by Rodabaugh and George (1957).

In section 3.4, the strains in the elastoplastic range calculated with NLAB are presented.

3.2 Results for a bend with R_0 = 480 mm

In the OPL research programme, (Gresnigt 1986), several tests were carried out to validate the analytical models that were developed in that programme. Also some finite element calculations were performed.

In this section we consider one of the tests (no. 72) with the following geometry:

R_0	=	480	mm
r_0	=	78.5	mm
t_0	=	2.9	mm

Results from NLAB are compared with the model of Rodabaugh and George (1957) and the model that was developed in the OPL programme: "BOCHT" (Bocht is the Dutch word for bend). For this case it gives the same results as the model of Rodabaugh and George.

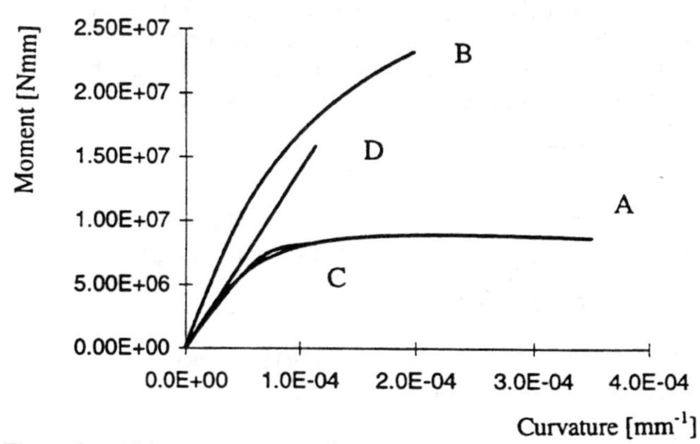

Figure 8. Moment-curvature diagram.

A	=	NLAB calculation,	$p = 0$
B	=	NLAB calculation,	$p = 29$ bar
C	=	DIANA finite element calculation,	$p = 0$
D	=	k_p	$p = 0$

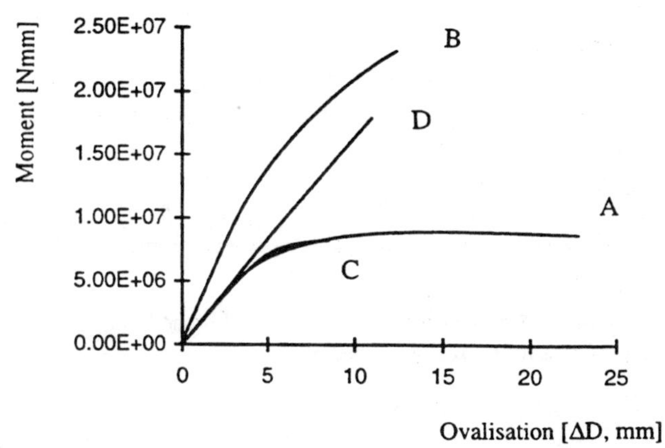

Figure 9. Moment-ovalization diagram. For an explanation of the lines see figure 8.

In figure 8 and figure 9 the calculated moment-curvature and moment-ovalization diagrams are presented. In these figures also a line k_p is given. The factor k_p, relates the flexibility of the bend to that of a straight pipe.

$$k_p = \frac{EI_{\text{straight pipe}}}{EI_{\text{bend}}}$$

The results of the analytical models agree well with the k_p values calculated with the well known formulations in the American Codes.

From figure 8 we can see the stiffening effect due to the internal pressure. There is a good agreement between DIANA and NLAB and between the analytical models (elastic range).

From figure 9 we again see the stiffening effect due to the internal pressure. Note also the good agreement between NLAB, k_p and the numerically obtained results from DIANA.

3.3 Elastic results for a bend with R_0 = 1143 mm

3.3.1 General

In this section the elastic results for the stresses with NLAB are compared with the same analytical models and with test results on a bend with R_0 = 1143 mm.

Rodabaugh and George have presented test results and results of their analytical model for a bend with the following geometry:

R_0 = 1143 mm
r_0 = 374.1 mm
t_0 = 13.08 mm

The stresses are presented using the stress-intensification factor i_{sf} on the vertical axis. This factor was proposed by Rodabaugh and George (1957) and equals the actual stresses in the bend due to bending, divided by the maximum stress in a straight pipe with the same diameter and wall thickness due to bending.

$$i_{sf} = \frac{\sigma_{max,bend}}{\sigma_{max,straight}} \quad \text{where:} \quad \sigma_{max,\,straight} = \frac{M}{I} r$$

On the horizontal axis, the angle φ is presented, where φ = 0° is the outside of the bend; φ = 90° is the side and φ = 180° is the inside of the bend, see figure 1.

3.3.2 Comparison of the models and test results

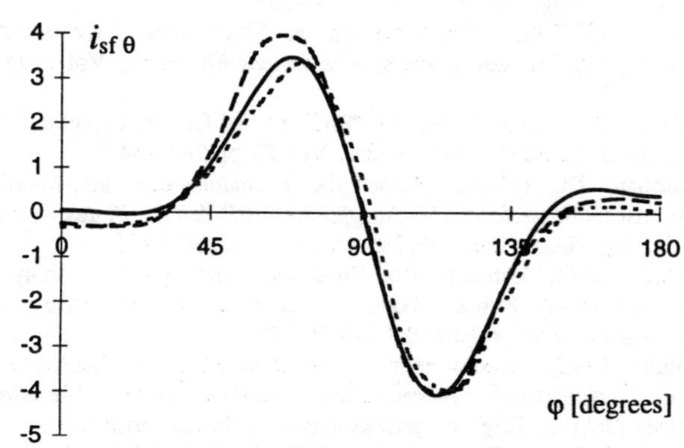

Figure 10. Comparison elastic longitudinal stresses for pressure p = 0.
– – = $i_{sf,\theta}$ for the middle of the pipe wall, model Rodabaugh and George.
- - - - = $i_{sf,\theta}$ average of the measurements inside and outside of the pipe, Rodabaugh and George.
—— = $i_{sf,\theta}$ for the middle of the pipe wall, NLAB.

In figure 10 and figure 11, the calculated and measured stresses for the longitudinal and for the circumferential direction are presented using the notations $i_{sf,\theta}$ and $i_{sf,\varphi}$ for the stress intensity factor in longitudinal direction and circumferential direction, respectively.

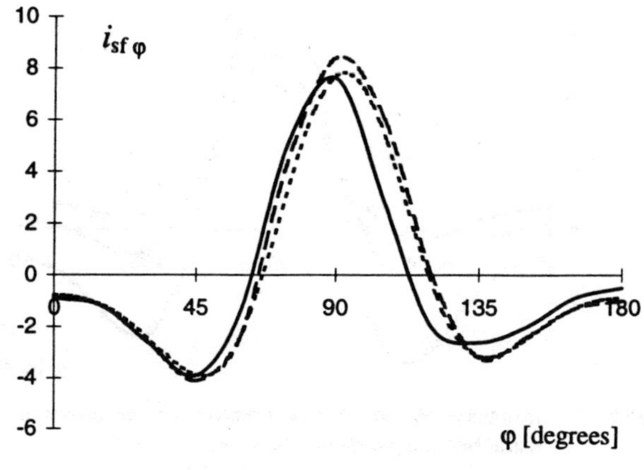

Figure 11. Comparison of elastic circumferential stresses for the outside of the pipe wall.
– – = $i_{sf,\varphi}$ model Rodabaugh and George.
—— = $i_{sf,\varphi}$ average of the measurements.
- - - - = $i_{sf,\varphi}$ NLAB.

There is a good similarity between all results of the reviewed models in the figure and also with BOCHT (not presented here).

3.3.3 Effect of internal pressuure

In figure 12 and figure 13 the influence of the internal pressure on the bending stresses is given, as calculated with NLAB, again using the notation $i_{sf,\theta}$ and $i_{sf,\varphi}$ for the stress intensity factor in longitudinal direction and circumferential direction, respectively.

Only the influence of the internal pressure on the bending stresses is presented. The constant stress due to expanding of the pipe (hoop stress and axial stress due to internal pressure only) is not shown in the diagrams.

Note the large decrease of the maximum bending stresses due to an increase in internal pressure. The stiffness of the bend increases with increasing internal pressure.

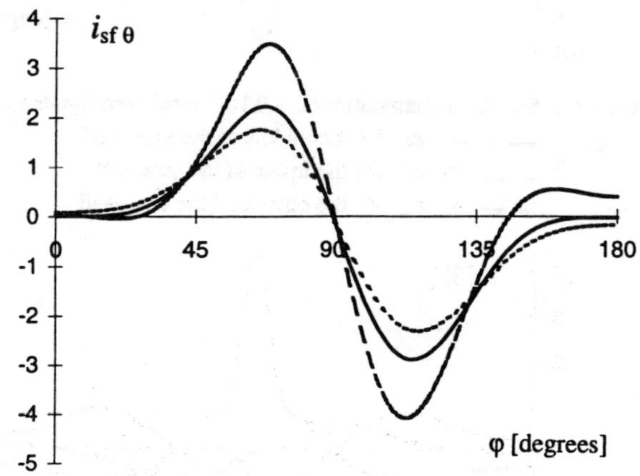

Figure 12. Influence of internal pressure on the elastic longitudinal stresses, results from NLAB.
– – = $i_{sf,\theta}$ for the middle of the pipe wall, p = 0
—— = $i_{sf,\theta}$ for the middle of the pipe wall, p = 29 bar
- - - - = $i_{sf,\theta}$ for the middle of the pipe wall, p = 58 bar

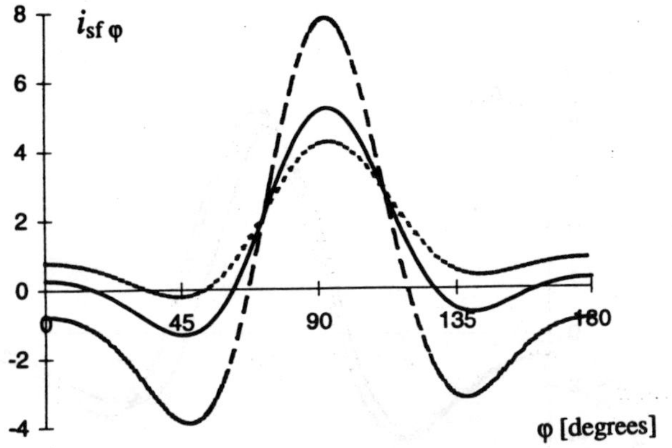

Figure 13. Influence of the internal pressure on the elastic circumferential bending stresses, results from NLAB.

— — = $i_{sf,\varphi}$ for the outside of the pipe wall, p = 0

——— = $i_{sf,\varphi}$ for the outside of the pipe wall, p = 29 bar

- - - - = $i_{sf,\varphi}$ for the outside of the pipe wall, p = 58 bar

3.4 Elastoplastic strains

In figure 14 and figure 15 the elastoplastic strains obtained with NLAB for pressure $p = 0$ are given for two curvatures. The calculations are carried out for the bend with $R_0 = 480$ mm as described earlier in section 3.2. The strain concentration at the middle is due to the formation of a plastic hinge. The notations ε_θ and ε_φ are used for the strains in longitudinal and circumferential direction, respectively.

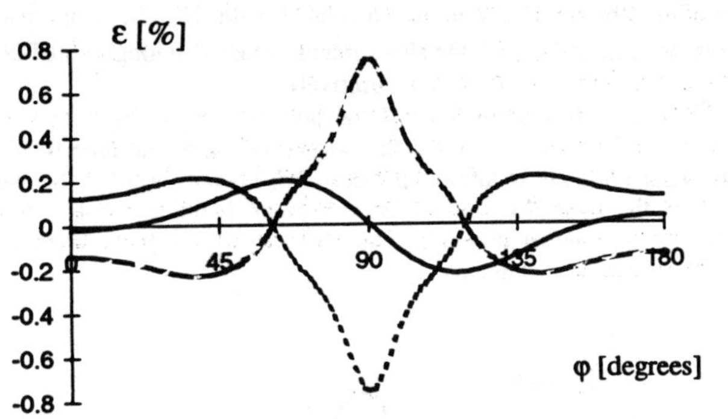

Figure 14. Strains at a curvature of 1.0 10^{-4} mm^{-1} (see figure 8).

——— = ε_θ for the middle of the pipe wall

- - - - = ε_φ for the inside of the pipe wall

— — = ε_φ for the outside of the pipe wall

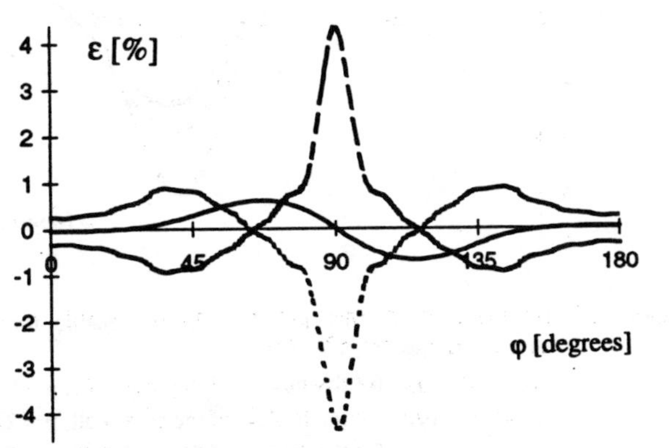

Figure 15. Strains at a curvature of 3.5 10^{-4} mm^{-1} (see figure 8). For an explanation of the lines see figure 14.

4. CONCLUSIONS

The conclusions from the results of the new analytical model, are:

a. The results calculated with the new model show a good agreement with existing theories and tests in the elastic area.

b. In the elastoplastic area the results indicate reasonable agreement with other theories and finite-element calculations.

c. Further refinement of the model is necessary to make it more accurate for the elastoplastic area. This is necessary to be able to extend it for the description of strain variations due to variations in internal pressure and/or curvature into the elastoplastic area.

d. From the results we see the importance of the internal pressure. Good implementation of the internal pressure is therefore essential for the extension into the elastoplastic area.

5. ACKNOWLEDGEMENT

The authors wish to thank the NV Nederlandse Gasunie for their support in the research.

6. REFERENCES

Baumjohann, F. (1994). "Zum Kollaps von Rohren und Rohrbögen unter Biegebeanspruchung," *Der Stahlbau 63*, Heft 1.

Blaauwendraad, J. (1996). "Elasticiteitstheorie; Energieprincipes en Variatiemethoden (Elasticity Theory; Energy Principles and Variation Methods)," *Lecture notes*, TU-Delft.

De Borst, R. (1996). "Computational Methods in Non-linear Solid Mechanics," *Lecture notes*, TU-Delft.

Dhalla, A.K. (1987), "Collapse Characteristics of a Thin-walled Elbow: Validation of an Analytical Procedure," *Journal of Pressure Vessel Technology*, November 1987, Vol 109, pp 394-401.

Findlay, G.E. and Spence J. (1979). "Stress Analysis of Smooth Curved Tubes with Flanged End Constraints," *Int J Pressure Vessels and Piping*, Vol 7, pp 83-103.

Gresnigt, A.M. (1986), "Plastic Design of Buried Steel Pipelines in Settlement Areas," *Heron*, Vol 31, 1986, no.4.

Gresnigt, A.M. and Van Foeken, R.J. (1995), "Strength and Deformation Capacity of Bends in Pipelines," *International Journal of Offshore and Polar Engineering*, Vol. 5. no. 4, pp 294-307.

Gross, N. (1952-1953), "Experiments on Short-radius Pipe Bends," *Proceedings of The Institution of Mechanical Engineers*, Vol 1, series B, pp. 77-87.

Kafka P.G. and Dunn M.B. (1956). "Stiffnesss of Curved Circular Tubes with Internal Pressure," *Trans ASME*, Vol 78, pp 247-254.

Von Karman, Th.. (1911), "Ueber die Formänderung dünnwandiger Rohre, insbesondere federnder Ausgleichrohre," *Zeitschrift des Vereines deutcher Ingenieure*, Band 55, No. 45, 11 November 1911.

Mennink, J. (1996), "Bends in Steel Pipelines; Deriving a New Analytical Model Based on Virtual Work to Calculate Stresses, Strains and Deformations," *TNO*, report 96-CON-R0830.

NEN 3650 (1992), "Requirements for Steel Pipeline Transportation Systems" *Nederlands Normalisatie Instituut (Dutch Standards Institute)*, Delft (an English translation is available on request).

Rodabaugh, E.C. and George, H.H. (1957), "Effect of Internal Pressure on Flexibility and Stress Intensification Factors of Curved Pipe or Welding Elbows," *Trans. ASME.*

Thomson, G. and Spence J. (1983). "Maximum Stresses and Flexibility Factors of Smooth Pipe Bends with Tangent Pipe Terminations under In-Plane Bending," *J Pressure Vessel Tech*, Vol 105, pp 329-335

Vigness, I., (1943), "Elastic Properties of Curved Tubes. *Trans. ASME*, pp 105-120.

Whatham, J.F. (1986). "Pipe Bend Analysis by thin Shell Theory," *J Applied Mech*, Vol 53, pp 153-180.

Westgård, I..T., (1994), "Fatigue of Bends in Steel Pipelines; Determination of Strain Reversals due to Variations in Internal Pressure," *TNO*, Report 94-CON-R1356.

Proceedings of the Seventh (1997) International Offshore and Polar Engineering Conference
Honolulu, USA, May 25-30, 1997

ISBN 1-880653-28-1 (Set); ISBN 1-880653-30-3 (Vol. II)

Analytical Methods for the Determination of Allowable Free Span Lengths of Subsea Pipelines

H.I. Park and C.H. Kim
Korea Maritime University
Pusan, Korea

ABSTRACT

Subsea pipelines are exposed to high potential risks of damage. One of main risk factors for the safety of a subsea pipeline is its free spanning. In this paper, static and dynamic free spans of pipelines are analysed and the allowable length of free spans are determined. The variation of the allowable lengths is examined for specialized boundary conditions. The free span is modelled as a beam on elastic foundations and the boundary condition is replaced by linear and rotational springs at each end. Non-dimensional curves are obtained to determine easily exact allowable lengths of free spans for subsea pipelines and are thus useful for the design of the subsea pipelines with free spans.

KEY WARDS : Subsea pipeline, Allowable span length, Static and dynamic analysis, Specialized boundary conditions, Non-dimensional curve.

INTRODUCTION

Subsea pipelines play an important role in the overall tasks of offshore oil and gas production. However, when subsea pipelines are installed and operated, they are exposed to harsh marine environmental conditions and thus the potential risks of damages are high. The pipelines are possibly damaged by direct factors of waves and currents and also lose their stability by indirect factors such as seabed scouring. The economic loss and recovery cost due to the pipeline failure are enormous. In addition to that, in the case of an oil pipeline failure, the bad effect on the marine environment is uncountable. Thus the unstable factors should be carefully investigated in order to prevent such a failure.

One of the important factors related to the pipeline unstability is seabed scouring induced by waves and currents. If the scouring happens around pipelines, a certain length of pipeline is exposed and there exist free spans on pipelines. The free span also occurs when pipelines are laid on an irregular seabed. If free spans happen to subsea pipelines, the pipelines can be buckled due to wave and current loads and their self-weights. The free spanned pipelines are also subjected to oscillatory forces created by vortex shedding in presence of currents and waves and are under a great number of stress cycles resulting in the fatigue failure.

The free span of pipeline should be analysed from the static and dynamic points of view. The static analysis of the free span is to find the allowable length to prevent the bending failure due to the distribution loads of self-weight or fluid loading. The dynamic analysis is to decide the allowable length of free span of which natural frequency is far away from the frequency of vortex shedding not to happen the resonance on the free span. In the design of subsea pipelines, the shorter length between the static and dynamic free spans is selected (DnV, 1981).

The research on the pipeline free span should be made in several points but the investigation into vortex shedding takes a main stream (Tsahalis, 1983, 1984; Jacobsen et al, 1984). These studies are mainly focused on the dynamic behaviour of fluid around the pipeline without considering the boundary conditions of the free span ends. Most investigations into pipeline spans consider the boundary conditions of the free spans as simply supported or fixed ones. Hobbs (1986) investigated the natural frequency of the free span for different boundary conditions based on Hetenyi's results (1946).

In this study, the allowable lengths of static and dynamic free spans are studied for the boundary conditions and non-dimensional curves are plotted to obtain easily the allowable free span lengths.

GOVERNING EQUATION

The model of the free span of pipeline adopted in this paper is shown in Fig. 1. Generally the boundary conditions of free span is supported by an elastic foundation and can be modelled as linear and rotational springs as shown in Fig. 2. If the elastic modulus of the foundation is k (N/m^2) and the pipe bending stiffness is EI, the equivalent spring stiffness, λ_H, is expressed as follows (Hetenyi, 1946) :

$$\lambda_H = \left(\frac{k}{4EI}\right)^{0.25} \tag{1}$$

The linear spring stiffness, k_1, and the rotational spring stiffness, k_2, are represented by using λ_H and k.

$$k_1 = k/(2\lambda_H) \tag{2}$$

$$k_2 = k/(4\lambda_H^3) \tag{3}$$

As can be seen from Fig. 2, if the beam is constrained by linear and rotational springs at the both ends and a distribution load is applied, the static governing equation of a free span can be written as follows [Dawson, 1983].

$$EI\frac{d^4y}{dx^4} = w \tag{4}$$

here, EI = the bending stiffness of pipe, N·m^2

w = the distribution load, N/m

On the other hand, the undamped dynamic equation of a free span is

$$\frac{d^4y}{dx^4} - \lambda^4 y = 0 \tag{5}$$

$$\text{here, } \lambda = \left[\frac{m\,\omega_n^2}{EI}\right]^{0.25} \tag{6}$$

m = the effective mass plus added mass of pipe per unit length, kg/m

ω_n = the angular frequency of the free span pipe, rad/sec

Boundary conditions of the model are as follows :

$$\text{At } x = 0\ ;\ EI\frac{d^3y}{dx^3} = -k_1 y_0,$$

$$EI\frac{d^2y}{dx^2} = k_2\theta_0 \tag{7}$$

$$\text{At } x = \frac{L}{2}\ ;\ EI\frac{d^3y}{dx^3} = 0,\ \frac{dy}{dx} = 0 \tag{8}$$

STATIC FREE SPAN LENGTH

Equation (4) is integrated to yield the moment equation (Park and Kim, 1996).

$$EI\frac{d^2y}{dy^2} = \frac{w}{2}x^2 - \frac{wL}{2}x + \frac{k_2wL^3}{12[2EI + k_2L]} \tag{9}$$

By using the relation of $-EI\frac{d^2y}{dx^2} = M_x$, a bending moment equation at the arbitrary point of the free spanned pipe can be obtained as follows :

$$M_x = -\frac{w}{2}x^2 + \frac{wL}{2}x - \frac{k_2wL^3}{12[2EI + k_2L]} \tag{10}$$

For the given pipeline characteristic values of L, EI and w, the bending moment varies within the two extreme ranges of $k_2 = \infty$ (fixed-fixed conditions : F-F condition) and $k_2 = 0$ (simply supported conditions : S-S condition) which are illustrated in Fig. 3. The maximum bending moment depends on k_2 and occurs at $x = 0$ or $x = \frac{L}{2}$. The bending moments are as follows :

$$\text{At } x = 0\ ;\ M_0 = -\frac{k_2wL^3}{12[2EI + k_2L]} \tag{11}$$

$$\text{At } x = \frac{L}{2}\ ;\ M_{L/2} = \frac{wL^2[6EI + k_2L]}{24[2EI + k_2L]} \tag{12}$$

When $k_2 = \frac{6EI}{L}$, M_0 is equal to $M_{L/2}$ and takes the following value.

$$M_0 = M_{L/2} = \frac{wL^2}{16} \tag{13}$$

Thus, when k_2 is less than $\frac{6EI}{L}$, M_0 is less than $M_{L/2}$ and the maximum bending moment occurs at $x = \frac{L}{2}$. When k_2 is greater than $\frac{6EI}{L}$, M_0 is greater than $M_{L/2}$ and the maximum bending moment occurs at $x = 0$.

The maximum bending moment must be less than the allowable bending moment of the pipe. From this fact, the allowable static free span length can be obtained.

The allowable bending moment of the pipe is as follows:

$$M = \frac{2\sigma_B I}{D} \tag{14}$$

here, σ_B = the allowable bending stress of free span, N/m^2.

I = the bending moment of inertia of pipe, m^4.

D = the outer diameter of pipe, m.

From equation (11) and (14), the following equation can be derived.

$$1 - \frac{24\sigma_B I}{DwL^2} - \frac{48\sigma_B EI^2}{k_2 DwL^3} = 0 \tag{15}$$

From equations (12) and (14)

$$1 + \frac{6EI}{k_2 L} - \frac{48\sigma_B I}{DwL^2} - \frac{96\sigma_B EI^2}{k_2 DwL^3} = 0 \tag{16}$$

In order to make the relation of k_2 and L from equation (15) and (16), the following nondimensional parameters are introduced.

$$\alpha = \frac{DwL^2}{\sigma_B I} \tag{17}$$

$$\beta = \frac{k_2^2 \sigma_B}{DwE^2 I} \quad \frac{\sigma_B}{2Dw}\sqrt{I/kE} \tag{18}$$

Thus, equations (15) and (16), can be respectively expressed as follows:

$$\beta = \left[\frac{48}{\alpha\sqrt{\alpha} - 24\sqrt{\alpha}}\right]^2 \tag{19}$$

$$\beta = \left[\frac{96 - 6\alpha}{\alpha\sqrt{\alpha} - 48\sqrt{\alpha}}\right]^2 \tag{20}$$

The equations of (19) and (20) are plotted in Fig. 4 where equation (19) corresponds to the line A and equation (20) does to the line B. Curves A and B respectively imply the tendency of fixed end condition and simply supported condition.

The crossing point of the two curves indicates the boundary condition of span ends where the bending moments at x = 0 and x = L/2 are equal and correspond to β = 9/8 ($k_2 = \frac{6EI}{L}$) and α = 32. Thus, when $0 \leq \beta \leq 9/8$, the curve B in Fig. 4 should be used and when $9/8 \leq \beta < \infty$, the curve A should be used to obtain α for different β.

If the value of k_2 which depends on the boundary conditions is known, β can be calculated from equation (18). Then, α can be obtained from Fig. 4. Finally by using equation (17), the allowable static free span length of a pipe can be obtained as follows:

$$L = \sqrt{\frac{\alpha\sigma_B I}{Dw}} \tag{21}$$

DYNAMIC FREE SPAN LENGTH

Dynamic free span analysis is to determine the maximum length of a free span to prevent the fatigue failure due to a vortex induced vibration. In order to minimize the vibration of a free span, the vortex shedding frequency should be away from the natural frequency of the free span. The vortex shedding frequency is as follows :

$$f_s = \frac{SV}{D} \tag{22}$$

here, S : Strouhal number

V : flow velocity

The Strouhal number is a function of Reynolds number or drag coefficient but is usually taken as 0.2 for the most practical pipeline design (Mousselli, 1981).

On the other hand, the vortex-induced vibrations of free spans are a function of the reduced velocity as given in the following.

$$V_r = \frac{V}{f_n D} \tag{23}$$

here, f_n : the natural frequency of free span

Pipeline vibrations may occur in the cross-flow and in-line flow directions. By far the more serious vibration are those which occur in the cross-flow direction. As can be seen from Fig. 5, if the reduced velocity approaches 5 with the increase of flow velocity, the cross-flow vibration starts. In order to minimize the cross-flow vibration, the reduced velocity is usually maintained to be less than 3.5 (DnV, 1981). Applying this condition to equation (22) and using equation (23) yields to the relation between two frequencies.

$$f_n \geq \frac{f_s}{3.5\,S} \tag{24}$$

Equation (24) indicates the minimum criterion of the

natural frequency of a free span not to fail due to a vibration-induced fatigue.

In reality there are infinite modes of vibration in a free span but if the natural frequency of the first mode satisfies the above condition, the others naturally satisfies the above condition. Thus if we consider only the first vibration-mode of a free span, we can obtain the required angular frequency of the free span from equations (22) and (24) as follows :

$$\omega_1 \geq \frac{2\pi V}{3.5D} \tag{25}$$

Next, the governing equation of dynamic motion of a free span is to be solved. The general solution of dynamic equation (5) is represented as follows:

$$y = A\sin\lambda x + B\cos\lambda x + C\sinh\lambda x + D\cosh\lambda x \tag{26}$$

Here, the constant A, B, C and D are to be obtained by using boundary conditions.

By using the boundary conditions of equations (7) and (8), the following equations are obtained.

$$B[\tan(\frac{\lambda L}{2}) - \frac{k_1}{EI\lambda^3}] + D[\tanh(\frac{\lambda L}{2}) - \frac{k_1}{EI\lambda^3}] = 0 \tag{27}$$

$$B[1 - \frac{k_2}{EI\lambda}\tan(\frac{\lambda L}{2})] - D[1 + \frac{k_2}{EI\lambda}\tanh(\frac{\lambda L}{2})] = 0 \tag{28}$$

Non-dimensional parameters, ψ and θ are introduced for convenience.

$$\psi = \frac{k}{m\,\omega_n^2} \tag{29}$$

$$\theta = \lambda L \tag{30}$$

The constants B and D are eliminated to yield the following equation using the parameters ψ and θ.

$$[\sqrt{2}S - \psi^{3/4}C]\cdot[\sqrt{2}CH) + \psi^{1/4}SH] + [\sqrt{2}SH - \psi^{3/4}CH]\cdot[\sqrt{2}C + \psi^{1/4}S] = 0 \tag{31}$$

here. $S = \sin(\frac{\theta}{2})$, $C = \cos(\frac{\theta}{2})$

$CH = \cosh(\frac{\theta}{2})$, $SH = \sinh(\frac{\theta}{2})$

If the boundary of a free span is the simply supported or fixed condition, θ can be easily obtained by using the simple boundary condition equations. The value of θ for fixed end condition and simply supported condition are 4.712 and π respectively. The relation between ψ and θ is plotted in Fig. 6 and the cases of simply supported and fixed condition are also given. From Fig. 6, the dynamic free span of pipe can be obtained for different boundary conditions and pipe characteristics. That is to say, if the boundary conditions and pipe characteristics are given, ψ can be calculated from equation (29) and the corresponding θ can be obtained from Fig. 6. The value of λ also can be obtained from equation (6). It is noted that in calculation of θ and λ, the ω_n in equation (6) and (29) should be replaced by ω_1 in equation (25). These values yield the allowable dynamic free span length of pipe by using the following equation.

$$L = \frac{\theta}{\lambda} \tag{32}$$

It can be seen from Fig. 6 that the allowable free span length is proportional to the seabed soil strength.

CONCLUSION

In this paper, the static and dynamic free spans of pipelines are analysed and the allowable length of free spans are determined. By introducing dimensionless parameters, the variation of the allowable lengths is examined for specialized different boundary conditions. Non-dimensional curves are obtained to determine easily the exact allowable lengths of free spans for subsea pipelines and are thus useful for the design of the subsea pipelines with free spans. Further studies need to be carried out for more generalized boundary conditions which are different at right and left ends of a free span.

REFERENCES

Dawson, TH (1983). *Offshore Structural Engineering*, Prentice-Hall, Inc., New Jersey.

Det norske Veritas (1981). *Rules for Submarine Pipeline Systems*.

Hetenyi, M (1946). *Beams on Elastic Foundations*, University of Michigan Press, Ann Arbor.

Hobbs, RE (1986). "Influence of Structural Boundary Conditions on Pipeline Free Span Dynamics", *Proceedings of the 5th Offshore Mechanics and Arctic Engineering Symposium, ASME*, Vol. 3, Tokyo.

Jacobsen, V et al. (1984). "Vibrations of Offshore Pipelines Exposed to Current and Wave Action", *Proceedings of the 3rd Offshore Mechanics and Arctic Engineering Symposium, ASME*, New York.

Mousselli, AH (1981). *Offshore Pipeline Design, Analysis, and Methods*, Penn Well Publishing Company, Oklahoma.

Park, HI and Kim CH (1996). "Design of Static Free Span for the Safety of Subsea Pipelines," *J of the Society of Maritime Safety*, Vol. 2.
Tsahalis, DT (1983). "The Effect of Seabottom Proximity on the Vortex-Induced Vibrations and Fatigue Life of Offshore Pipelines", *ASME Journal of Energy Resources Technology*, Vol. 105.
Tsahalis, DT (1984). "Vortex-Induced Vibrations of a Flexible Cylinder near a Plane Boundary Exposed to Steady and Wave-Induced Currents," *ASME Journal of Energy Resources Technology*, Vol. 106.

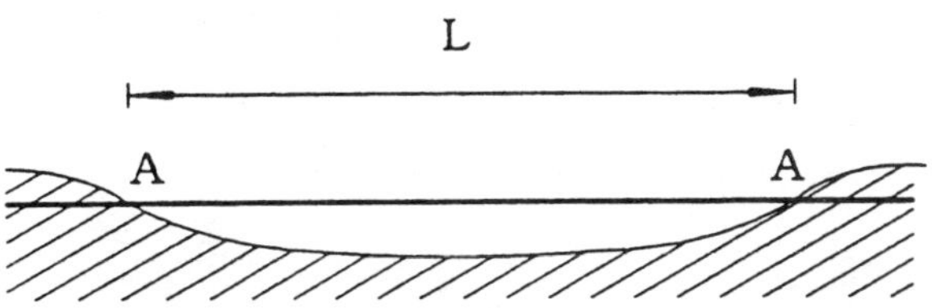

Fig. 1 Configuration of a free span

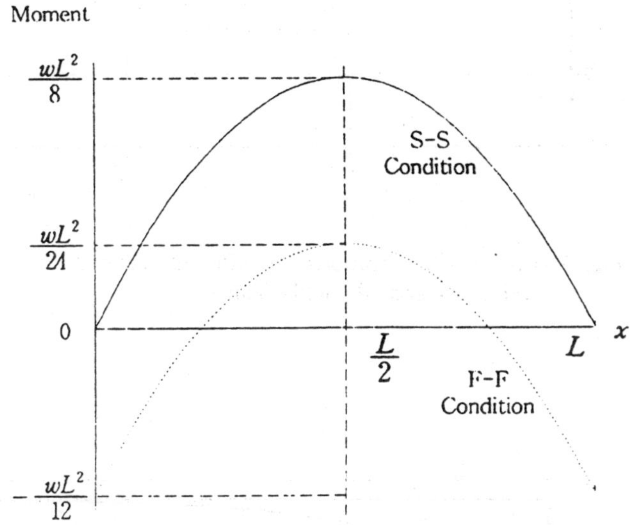

Fig. 3 Bending moment distribution along free span length for simply supported (S-S) and fixed-fixed (F-F) end conditions

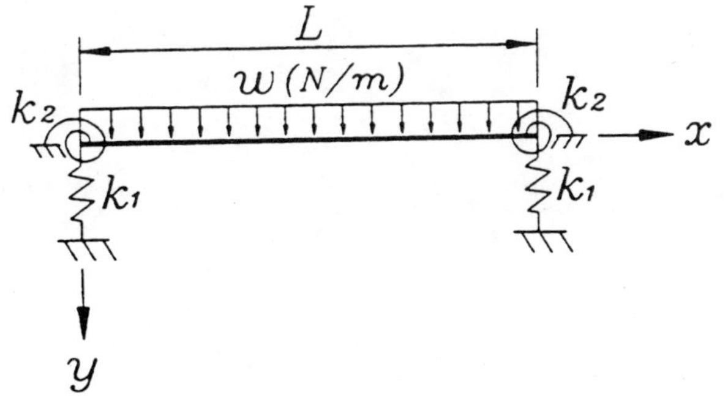

Fig. 2 Simplified beam model on elastic foundation

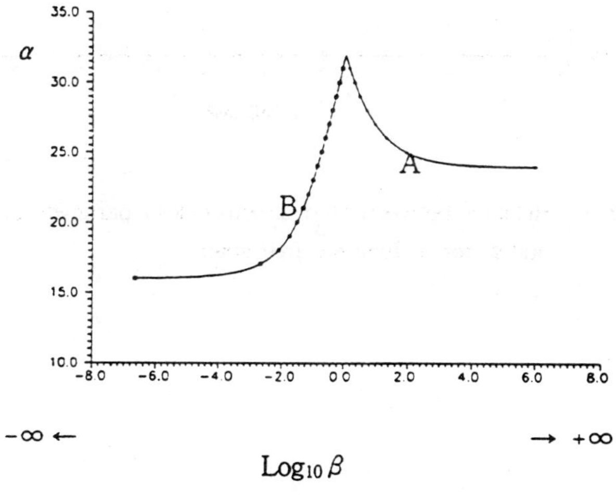

Fig. 4 Relation between two dimensionless parameters, α and β for a static free span

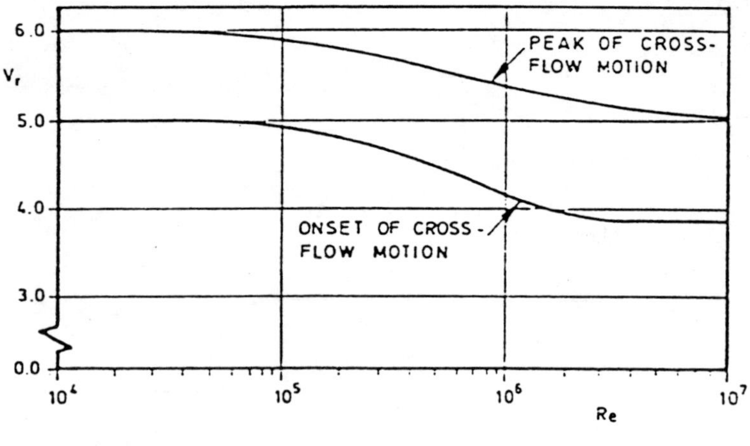

Fig. 5 Cross-flow vibrations for different reduced velocities and Reynolds number

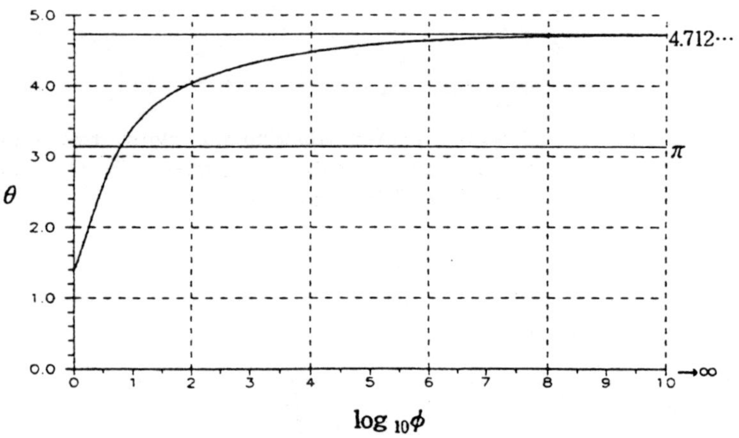

Fig. 6 Relation between two dimensionless parameters, θ and ϕ for a dynamic free span

Proceedings of the Seventh (1997) International Offshore and Polar Engineering Conference
Honolulu, USA, May 25-30, 1997

ISBN 1-880653-28-1 (Set); ISBN 1-880653-30-3 (Vol. II)

Non-Linear Finite Element Prediction of Wrinkling in Corroded Pipe

Daniel P. Nicolella and Marina Q. Smith
Southwest Research Institute
San Antonio, TX, USA

ABSTRACT

This paper describes a full scale testing and finite element investigation of the wrinkling behavior of corroded pipe. It is applicable to corroded pipe subjected to multi-axial loading such as longitudinal bending, internal pressure, and axial compression.

The finite element procedure incorporates the non-linear material behavior of the pipe steel by the use of a multilinear kinematic hardening plasticity model. Corrosion is simulated numerically by a reduced wall thickness section. Simulated pipe structural behavior is critically compared to a full scale test.

During the testing program, a full scale, 48-inch diameter section of the Trans-Alaska pipeline is subjected to multi-axial loading. Corrosion in the pipe is physically simulated using a reduced wall thickness section of the pipe. Deflections, curvature, and loading are measured for comparison to the finite element simulations.

The results indicate a good correlation between the numerical model and full scale test in predicting the, deflection, curvature, and moment capacity of the pipe at wrinkling.

KEY WORDS: Corroded pipe, , combined loading, non-linear finite element analysis, wrinkling.

INTRODUCTION

The integrity analysis of degraded pipe is attracting an increasing amount of attention from pipeline operators faced with making a repair/replace/ignore decision with regard to maintenance and operation of aging pipelines. While pressure based guidelines exists which allow pipeline operators to define the operational margins of safety against rupture (Battelle, 1989), reliable procedures for the prediction of wrinkling in degraded pipes subjected to combined loading are virtually non-existent. After nearly 20 years of service, some areas of the Trans-Alaska Pipeline have sustained normal settlement and/or corrosion inherent to aging pipelines. The settled areas can be identified reliably using a "smart pig" which measures curvature of the pipeline based on changes in pig orientation. A smart pig can also be used to measure changes in pipe wall thickness, and hence, identify areas with external corrosion without prior excavation of the pipe. Some of the settled areas may need to be excavated and re-leveled to guard against wrinkling of the pipe wall. Some corroded areas may need to be excavated and repaired to guard against rupture. Since excavation involves substantial cost and environmental risk, rational procedures are needed for deciding whether or not to excavate.

Several studies have been conducted which attempt the use of finite element techniques for the simulation of rupture or buckling in pipelines subjected to a variety of combined loads (Zhou *et al.*, 1993; Cronin *et al.*, 1996). Analytical models include both geometric discontinuities representing general or pitting corrosion and weld mismatches, which serve to initiate rupture and buckling, respectively. In general, criteria used to define the rupture point in the numerical models varies between the studies depending on the definition of the failure strength for the pipe material. However, the wrinkling or buckling criteria for each study reviewed incorporates a limit-load state, wherein the resisting moment decreases with increasing deformations. Evidence of the occurrence of a maximum moment is provided through the softening behavior of the moment-curvature curve beyond the peak moment.

In order to provide operators with guidelines that are both reliable and representative of realistic loading conditions, SwRI has worked with Alyeska Pipeline Service Company (APSC) to develop a theoretically based combined loading rupture criterion, and to include a less empirical wrinkling prediction procedure in the PC assessment tool SAFE (Shell Analysis Failure Envelope) (SwRI, 1992; SwRI, 1994). In the research program performed for APSC during which SAFE was developed, advanced three-dimensional, large deformation, elastic-plastic finite element (FE) analyses are used to simulated full-scale physical tests of artificially corroded pipe (SwRI, 1996). Upon completion of the simulations, the FE procedure is used in "surrogate experiments" for a parametric study where key factors affecting pipe capacity for rupture (e.g., corrosion dimensions, internal pressure, axial load magnitudes, and bending moment) are investigated. Results of the parametric study in conjunction with the full-scale FE simulations and physical test results, provide the database of failure conditions from which the rupture and global buckling, or "bending collapse", failure criteria embodied in SAFE are evolved. The objective of the research presented herein is therefor to develop a finite element procedure that will accurately predict the wrinkling behavior of corroded pipe subjected to combined axial, bending, and internal pressure loading.

MATERIALS AND METHODS

The analysis model used for this work employed the ABAQUS computer program with the nonlinear kinematic hardening procedure for modeling plasticity in metals proposed by Mroz (Mroz,1967). This theory has been encoded in a series of subroutines supplied to SwRI by

SSD Engineering Consultants (SSD, 1996) for use in the finite element computations. Development of the database for calibration of the wrinkling procedure incorporated in SAFE is accomplished through a combined testing and analytical program; wherein the testing program is used to validate the numerical analytical analyses. The criteria used to determine success of the procedure is the ability of the model to accurately define the wrinkling moment and curvature for the physical test. In addition, the trend of the moment-deflection and moment-curvature curves should accurately represent that of the test.

Full Scale Wrinkling Test

A baseline test representing the most typical set of corrosion geometry and loading conditions found in-service was physically tested providing data that was used to verify the FE procedure. Detailed deflection, rotation, and strain information, in conjunction with applied load magnitudes were recorded and critically compared with the predictions obtained from the simulation.

The testing facility utilized for the combined loading corrosion tests for rupture is shown in Figure 1. The vertically oriented lateral loading rams are 10-inch diameter hydraulic actuators with a maximum load capacity of 390 kips per ram (with 5000 psi hydraulics), each with a maximum stroke length of 8 inches. Axial loads are applied to the specimen using twin 14-inch diameter hydraulic actuators with a maximum load capacity of about 770 kips per ram (with 5000 psi hydraulics). By using the dual rams and a lever with a 3-to-1 mechanical advantage, the applied axial load capacity is boosted to 4,620 kips. Lateral loads can be applied using either load control or displacement control. Internal pressure is applied to the specimen filled with water using a water pump.

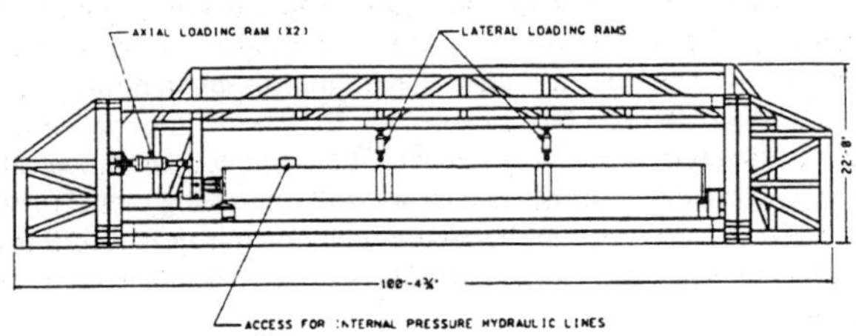

FIGURE 1. Combined loading test frame with pipe specimen installed.

The test specimen is comprised of a 17'-4" long section X65 pipe steel with an outside diameter of 48 inches and nominal wall thickness of 0.462 inch placed between the two 20-ft, 11-3/8-inch long end sections. (Figure 2a shows the complete pipe specimen.) These sections are fabricated from X65 steel, and have an outside diameter of 48 inches and nominal wall thickness of 0.562 inch. At the vertically oriented lateral load points, the specimen is internally and externally stiffened to prevent any local deformation of the pipe.

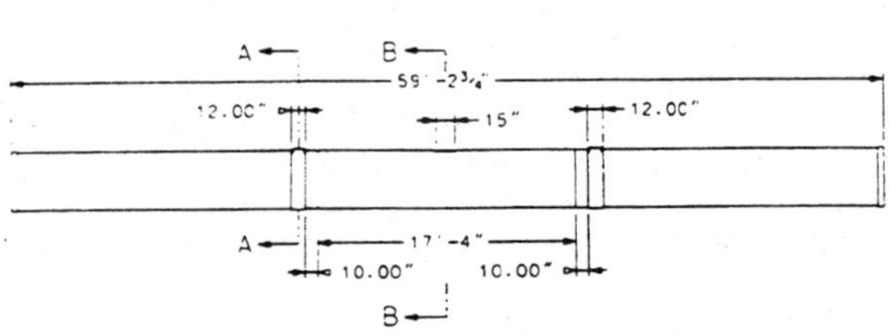

FIGURE 2a. Actual pipe specimen dimensions.

The artificial corrosion patch is ground into the pipe at the center of the test section. For the nominal test case evaluated, the 15-inch, by 15-inch, by 15 percent deep corrosion patch was located at the 12:00 position, or compression side of the pipe with regard to axial bending as in Figure 2b.

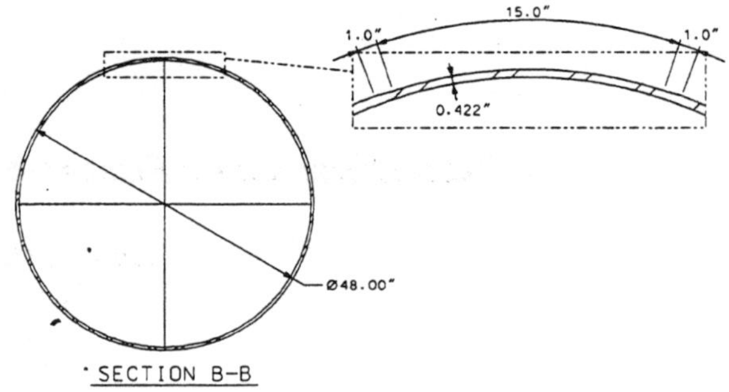

FIGURE 2b. Circumferential view of thinned region.

To emulate the sequence of load application experienced by the pipe in-situ, the first load step, bending due to static weight of the specimen, is imposed by filling the pipe with water. Axial loading of the pipe, simulating the restraint on thermal expansion that exists when the temperature of the transported fluid differs from the ambient temperature at installation, is applied second (DT = 75°F). The magnitude of this force is determined by defining the temperature differential present in the pipe section under consideration and calculating the compressive force in the nominal wall of the central test piece required to restrain the thermal expansion of the specimen in the field. The third step in the loading sequence simulates the stresses that arise when the pipe is pressurized, while the axial force from step 2 remains constant. For an in-service pipe, internal pressurization of the fluid generates hoop tension stresses in the pipe wall, and an axial tension stress due to Poisson's ratio. A temperature differential (ΔT) between the temperatures at installation and operation generates a compressive stress (for positive ΔT's) on the pipe section since axial movement is restrained. However, for the closed test specimen, an additional tensile axial stress is developed from the pressure acting on each of the end plates. To accurately characterize the in-service stresses with the closed specimen, a compensating axial force equal to the product of the internal end plate area multiplied by the pressure must be applied. Represented mathematically, the total axial force applied to the specimen, Fapplied, is

$$F_{applied} = F_{\Delta T} + F_{IP} - F_v$$

where FΔT is the axial force resulting from the fluid temperature difference, FIP is the force applied to compensate pressure acting on the pipe end plates, and Fv is the force resulting from the hoop expansion and Poisson's ratio. Note that the axial force FIP increases as the internal pressure increases.

While maintaining the axial load and internal pressure (load steps 2 and 3) on the specimen, lateral deflection at the load points is increased incrementally in load step 4 to just beyond the peak in the moment versus deflection curve to insure that the wrinkling point has been achieved. Wrinkling of the pipe specimen is indicated by a reduction in the applied bending moment with increasing lateral displacement. In load step 5, the test specimen is pressurized while maintaining constant lateral deflection until burst. Load step 5 is not simulated in the finite element analysis. Table 1 outlines the test specimen parameters and Table 2 outlines the loading sequence used in the test.

TABLE 1. Parameters used in nominal test.

Test Parameter	Value
Location of thinned region	12:00 (compression side)
Axial length of thinned region	15 inches
Circumferential length of thinned region (arc length)	15 inches
Depth of thinned region	15% of 0.496 inch[a] or 0.074 inches

TABLE 2. Full-Scale Test loading sequence.

Load Step	Description
1	Static weight of the specimen (steel plus water)
2	Axial load representing applied fluid temperature differential
3	Internal pressurization with compensating axial load
4	Axial bending to wrinkling
5	Internal pressurization with compensating axial load to rupture.

Finite Element Model

Material Model

The baseline test was simulated using the ABAQUS/Mroz combined finite element procedure. The plasticity model proposed by Mroz (SwRI, 1992) is used to model the pipe material post yield behavior. Computation of the material constitutive behavior performed by ABAQUS is accomplished by invoking a user developed material model subroutine, or UMAT. This UMAT has been designed to interface with the ABAQUS S4R four-noded, large deformation shell elements selected for the simulation. Because the material properties definition for the shell elements is provided via the UMAT, the transverse shear stiffness of the element is calculated externally from ABAQUS. Using the ABAQUS recommendation, the transverse shear stiffness for the shell element is calculated in full consideration of the shell element thickness. Thus, for all verification analyses, the transverse shear stiffness for a given element was calculated in accordance with the procedures prescribed by the ABAQUS user guide (Hibbitt, Karlsson & Sorensen, Inc., 1995).

Input of the material properties in true-stress strain space is required for all finite strain elements in ABAQUS, including the S4R four noded shell elements used in the quarter pipe model. The required input hoop shifts (applied to simulate the pre-straining induced during manufacture), stress, and strain were provided by SSD, who developed these parameters from tensile tests of the pipe steel used in the test. The internal ring stiffener at the lateral load point and the end plate of the quarter pipe model did not employ the Mroz material model, and were represented as linear elastic material since yielding in these regions is not allowed in the test. Thus, modeling of the additional stiffeners which span from 3:00 to 9:00 and 12:00 to 6:00 in Figure 2b were not required. The Mroz model input parameters for the steel model are given in Table 3.

TABLE 3. Mroz parameters for steel model.

F_v	ε_y	H_{SHIFT}	E
0	0	0	
45.6499	0.1503	14.8661	303.8
53.4973	0.1881	9.7377	207.2682
61.5266	0.2937	4.3522	16.0353
66.3323	0.4681	1.0823	27.5517
67.8	0.6449	0.078	8.3055
76.3915	2.4937	1.4375	4.647

For > 2.5%, E = 2.498 ksi / %

Single element analysis

Use of the Mroz material model with ABAQUS through the UMAT discussed above was verified in the single element uniaxial static and cyclic loading analyses. One ABAQUS S4R shell element was used for each of the single element verification analyses. The element was modeled as a square 5 inches by 5 inches by 0.5-inch thick. Each single element analysis was performed using one shell element to represent the an alternate TAPS(SwRI, 1995) pipe steel. Behavior of the single element was evaluated under static and cyclic tension and compression loading conditions. Loads were applied through fixed displacements representing 5.0% strain to simulate longitudinal tension and compression and hoop tension and compression loading. Biaxial cyclic stress behavior was simulated by first applying a hoop tensile stress of 45.0 ksi. Next, an axial compressive displacement of 1.0% was applied followed by an axial tensile displacement to 1.0% tensile strain. A compressive axial displacement of 1.0% was applied to produce a net axial strain of 0.0%

Full Scale Test Simulation

Using the Patran graphics package to generate the pipe model, the geometric outside diameter of the pipe specimen is exactly modeled. A quarter symmetry finite element model was developed for the pipe specimen installed in the four-point bending frame by claiming symmetry of the loads and specimen geometry about the pipe mid-span (the X-Z plane in Figure 3) and the axial mid-plane (the Y-Z plane in Figure 3). Changes in the pipe wall thickness through the stiffened and thinned regions are represented by defining the wall thickness relative to the outside diameter of the pipe.

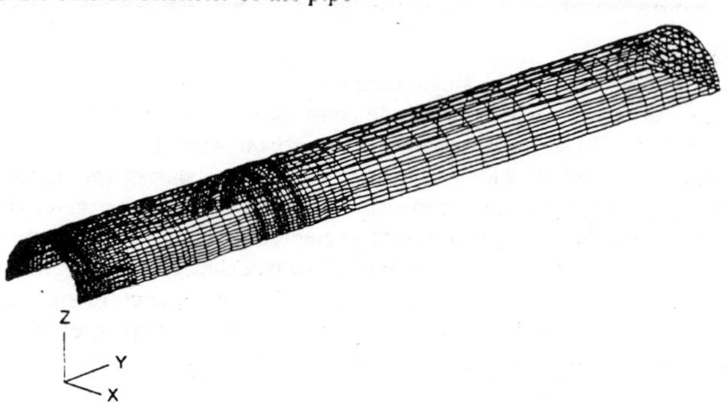

Figure 3. Quarter symmetry finite element mesh of simulated corroded pipe.

The depth of the corrosion patch in the model (12:00 position in the test), is 0.422-inch, equal to a 15% reduction in the actual wall thickness of 0.496-inches, observed at the 12:00 mid-span location in the test specimen. The average measured wall thickness of 0.493 inch was used to represent the central 0.462-inch nominal wall section of the pipe specimen used in the full-scale test. The wall thickness of the end section was modeled as a nominal 0.562 inch, which was verified by minimal wall thickness measurements taken for each actual end section of the pipe specimen.

Loading was applied to the model in four steps, corresponding to the actual test loading protocol. Step 1 modeled the weight of the steel and the water with the use of the body force load boundary condition option with ABAQUS. In loading step 2, an axial load was applied to the model end plate to represent the axial compression load applied to the test article to simulate stresses imposed by a temperature differential between the pipe at installation and the fluid being transported through the pipe. Internal pressure equal to 500 psi was applied to the interior surface of the shell elements in Step 3. During this phase of loading, the axial load was also increased to a maximum of 1,430,000 lb. to counter act the internal pressure. While maintaining the axial load and internal pressure on the specimen, lateral deflection at the load points in increased incrementally in load step 4 to just beyond the peak in the moment versus deflection curve to insure that the wrinkling point has been achieved. Bending loads were applied under displacement control in the test and simulation until wrinkling of the pipe occurred. Within each displacement increment, the modified Riks algorithm was used to ensure that the solution successfully converged on the load-displacement curve. The simulation was terminated when the pipe wrinkled, identified by a reduction in total moment capacity. Table 4 defines the load magnitudes applied to the finite element model.

TABLE 4. Load steps applied in finite element quarter pipe model.

Load Step	Description	Load Magnitude(s)
1	Weight of water plus weight of steel.	65,416 lb
2	Axial load simulating temperature differential (75°F)	1,088,290 lb
3	Internal pressure and compensating axial load	500 psi 1,430,000 lb
4	Lateral (bending) deflection to wrinkling	0 to wrinkling inches

RESULTS

The purpose of the single element finite element analyses was to verify the proper operation of the constitutive model encoded in the ABAQUS/UMAT when used in conjunction with the linear, large displacement shell element (S4R). Figure 4 shows the typical ABAQUS element stress-strain response transformed to the engineering stress-strain space. The element engineering stress-strain behavior agrees well with the curve developed in the mechanical tests, indicating that the UMAT is functioning properly. The discrepancy in the knee region is due to the difference in the strain increment steps used in the analysis, and overall, this procedure accurately predicts the material tensile stress-strain behavior.

Cyclic behavior was also verified by comparing the ABAQUS/UMAT behavior with a previously developed analytical procedure (SSD, 1996). Figure 5 shows the behavior of the finite element procedure during a cyclic loading sequence, indicating the proper response of the procedure in capturing the loading and unloading behavior of the material model.

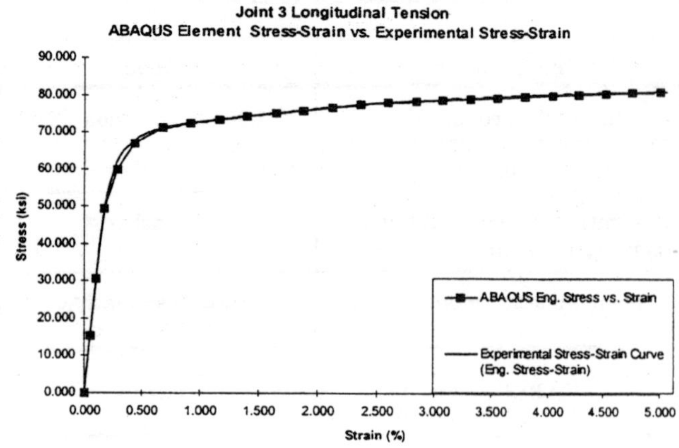

FIGURE 4. ABAQUS single element longitudinal tension engineering stress-strain compared to experimental stress-strain curve.

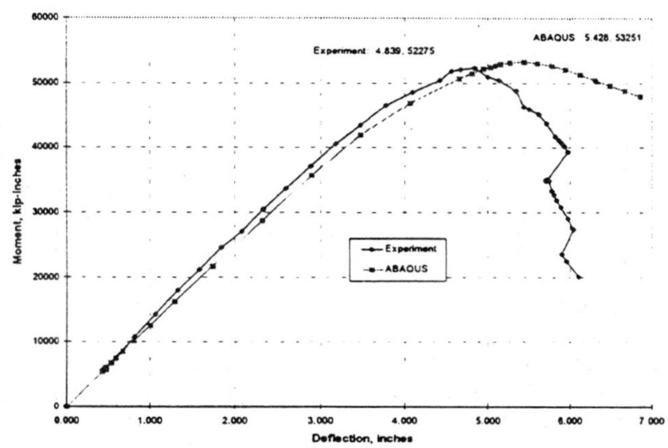

FIGURE 5. Comparison of ABAQUS and experimental moment-deflection curves at pipe mid-span.

A results summary for the finite element simulation of the baseline full-scale test are provided in Table 5. Moment-displacement, moment-curvature and load-displacement results for each load step are given. The pipe incipient wrinkling point is identified as the point of maximum moment capacity on the moment versus deflection curve, and the displacements, rotations, and curvatures at wrinkling are identified as corresponding to this point of maximum moment. Results given in the table are "total" quantities which include static or gravity results.

Figures 5 and 6 directly compare the combined test results for mid-span and load point moment-deflection, and those produced by the finite element analysis. As shown, the portion of the ABAQUS curve representing lateral displacement (load step 4) compares well with that observed in the full-scale test. Behavior past the peak point in the moment-deflection curve should not be directly compared with the physical test behavior since the analysis continues to impose lateral deflection beyond wrinkling, while the full-scale test maintains load point deflection while internal pressure and axial loading is increased to produce rupture (load step 5 in Table 2). The moment versus curvature (based on rotation predictions) at the 5.5-ft and 10-ft gage lengths for the combined result are compared with the finite element simulation in Figures 7 and 8. Again, the trend of the curve in the lateral displacement region is well correlated.

TABLE 5. Finite element analysis results summary.

State	Static Weight	Axial Load	Internal Pressure	Wrinkle
Mid Span Moment (kip-in.)	5,401	5,914	5,993	53,251
Mid Span Deflection (in.)	0.432	0.473	0.476	5.428
Mid Span Curvature (1/ft)(10^{-6})	-103	-114	-115	-1550
Load Point Deflection (in.)	0.370	0.405	0.407	4.538
Load Point Curvature (1.ft)(10^{-6})	-103	-108	-115	-1550

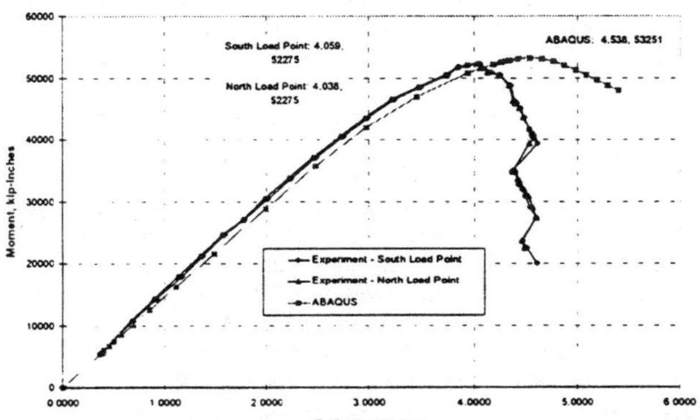

FIGURE 6. Comparison of ABAQUS and full-scale test moment-deflection curves at lateral load points. (Due to symmetry, only a single curve is predicted by ABAQUS.)

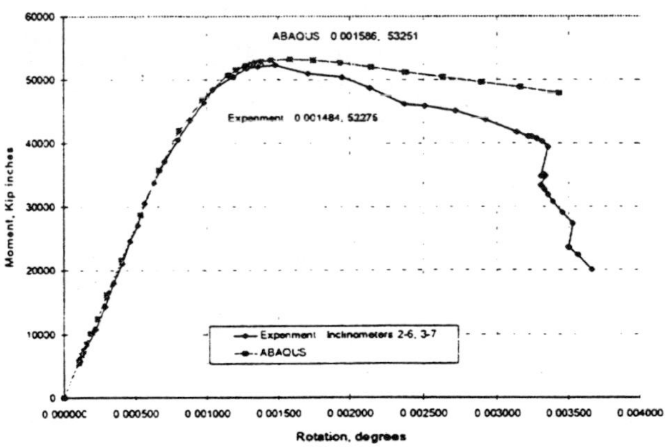

FIGURE 7. Comparison of ABAQUS and full-scale test moment-curvature curves at 5.5-ft gage length.

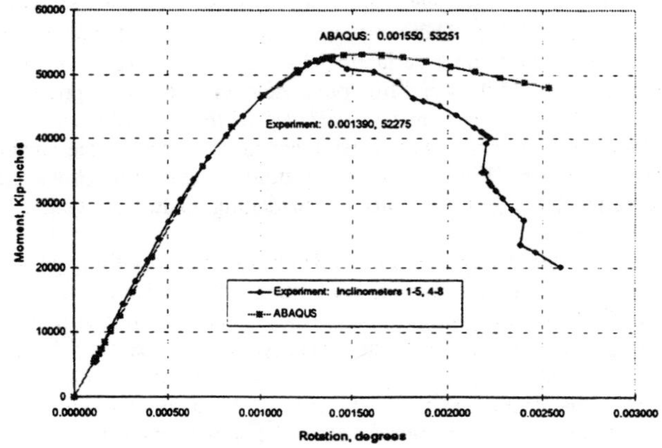

FIGURE 8. Comparison of ABAQUS and full-scale test moment-curvature curves at 10-ft gage length.

DISCUSSION AND CONCLULSIONS

The results of this research indicate that the combined ABAQUS/Mroz procedure is well suited for the wrinkling analysis of corroded pipes. The single element verifications of the procedure show that the combination of the ABAQUS S4R elements and Mroz constitutive model with true stress-strain material definitions accurately simulates the material response achieved in the tension and compression testing of pipe material coupons. The loading and unloading characteristics of the material at strains from -1% to 1% are also well represented when compared with cyclic loading analyses performed by SSD.

The global and local response of a degraded pipe section physically tested under combined loads provides information essential to the ultimate verification of the FE procedure to be applied in subsequent "surrogate experiments". When compared with the full-scale test data, the displacement, rotation, and strain response predicted up to and at wrinkling by the analysis are well correlated with those observed in the test. Table 6 gives a comparison of the physical test measurements versus the numerical predictions. The moment predictions are well correlated with differences between the measured

TABLE 6. Comparison of full-scale test results vs. finite element model predictions.

Measure	Test	FE Model	% Difference
Midspan Moment (kip-in.)	52,275	53,251	1.8
Midspan Deflection (in.)	4.839	5.428	10.8
Load Point Deflection (in.)	4.048	4.538	10.8
5.5 ft Rotation (degrees)	0.001484	0.001586	6.4
10 ft Rotation (degrees)	0.001390	0.001550	10.3

and predicted values being less than 2%. The midspan deflection and pipe curvatures however show differences on the order of approximately 11%. These discrepancies may be explained by the application of the Mroz material model and the local material post yield behavior. Beyond the wrinkling point, however, the model predicts a much stiffer response than that achieved in the test, which may be attributed to differences which were not investigated by the current research program. First, the Mroz constitutive model utilized was not designed to allow for the definition of hardening modulii above the 3.5 percent strain range.

The material is assumed to harden beyond the wrinkling point at the same final end slope through completion of the analysis. Moreover, the loading and unloading characteristics of the element and material model could not be fully evaluated since cyclic analytical (or full-scale test) data at strains approaching wrinkling and beyond is non-existent. To improve accuracy of the post-wrinkling behavior, hardening (and softening modulii) representative of the actual pipe material should be incorporated, and cyclic data at higher strain values for comparison with the FE procedure should be obtained. Other possible causes of the discrepancy between measured and predicted values could be slight imperfections in the geometry of the test specimen and possible non-symmetries in the test loading configuration. A more accurate model would be one such that the test specimen and the test loading and boundary conditions are modeled exactly without taking advantage of idealized system symmetries. This would dictate the use of a fully detailed finite element model that exactly models the geometry of the test specimen, taking into consideration ovalities, and "non-straightness" or camber in pipe geometry. In addition, without the assumption of symmetric load application, the model could be designed to accurately represent the exact magnitude of individual loads applied to the test article. These improvements are beyond the scope of the current study, and will be investigated at a later date.

ACKNOWLEDGMENTS

Funding for this work was provided by the Southwest Research Internal Research Program and Alyeska Pipeline Service Company. We would like to thank Mark Anderson, Dave Hackney, and Bill Flanders of APSC; and James Maple, consultant to APSC for their insight and guidance and for participating in numerous discussions regarding the objectives of the overall program. In Addition, we thank Graham Powell, Jim Hart, and Nasir Zulfigar of SSD Engineering Consultants, who provided single element finite element verification data, provided the Mroz constitutive model parameters, and performed parallel analytical pipe wrinkling analysis in addition to many hours of knowledgeable and important technical discussion.

REFERENCES

Battelle (1989). "Project PR 3-805: A Modified Criterion for Evaluating the Remaining Strength of Corroded Pipe," Final Report to the American Gas Association, December 22, 1989.

Cronin, DS, Roberts, A, Pick, R (1996). *"Assessment of Long Corrosion Grooves in Line Pipe,"* Proceedings of the 1st International Pipeline Conference–Volume 1, p. 401.

Hibbitt, Karlsson & Sorensen, Inc. (1995). ABAQUS/Standard User's Manual Version 5.5.

Mroz, Z (1967). "On the Description of Anisotropic Work-Hardening," *J. of Mech. Physics and Solics*, Vol. 15, pp 163-175.

SSD Engineering Consultants, Inc. (1996). "1996 Summary Report: SSD Work in Support of the Extended SAFE Program," Draft Report to Alyeska Pipeline Service Company, October 1996.

Southwest Research Institute (SwRI) (1992). "Development of Methodologies for Determining the Integrity of Corroded Pipelines, Phase II Final Report," Report to Alyeska Pipeline Service Company, April 27, 1992.

SwRI (1994). "Development of Methodologies for Determining the Integrity of Corroded Pipelines, Phase III Final Report," and "SAFE: Shell Analysis Failure Envelope, Program and Theoretical Manual," Reports to Alyeska Pipeline Service Company, February 1994.

SwRI (1995). "Mechanical Testing of TAPS Steel," Report to Alyeska Pipeline Service Company and SSD, Inc., November 1995.

SwRI (1996). "Full Scale Testing and Finite Element Simulation of a Thinned Pipe–Part A: Representative Test Case," Final Report to Alyeska Pipeline Service Company, November 1996.

Zhou, Z and Murray DW (1993). "Numerical Structural Analysis of Buried Pipelines," Structural Engineering Report No. 181, Department of Civil Engineering, University of Alberta, Canada, February 1993.

Proceedings of the Seventh (1997) International Offshore and Polar Engineering Conference
Honolulu, USA, May 25-30, 1997

ISBN 1-880653-28-1 (Set); ISBN 1-880653-30-3 (Vol. II)

JOIA Project of Study on Ice Load

Hiroshi Saeki, Hokkaido University, Japan
Ken-ichi Hiyayama, Iwate University, Japan
Tetsuro Kawasaki, Mitsubishi Heavy Industries, Japan
Satoshi Akagawa, Shimizu Corporation, Japan
Kazuyuki Kato, Ishikawajima-Harima Heavy Industries Co., Japan
Kazuhiko Kamesaki, NKK Corporation, Japan
Koji Saka, Saka Consulting, Japan
Akira Kurokawa, Japan Ocean Industries Association, Japan

ABSTRACT

It is very important to secure steady and abundant energy resources for a long time, since Japan is poor in natural resources. The oil exploration in cold regions and ice - infested waters in the Arctic Circle will become more significant to Japan. In these circumstances, Japan Ocean Industries Association (JOIA) planned the research project, "Study on Ice Load Acting on Marine Structures", which is funded by the Ministry of International Trade and Industry. The research project was started in 1993 in the scheme of 5 years. The main purpose of this research project is the development of ice load prediction methods and numerical simulation techniques for ice - structure interaction. Sub tasks of this study are as follows :

(1) Observation of failure mode of ice and understanding of the relation between global ice load and nonsimultaneous local ice pressure through the ice tank tests and the field tests.
(2) Development of analytical simulation methods for ice-structure interactions concerning vertical and inclined structures.

This report presents the outline of this research project and the summary of results in the early stage.

INTRODUCTION

Structures used for developing oil fields under ice - infested water are classified into a variety of types according to the sea region, object, and scale. The structures may be roughly divided into vertical sided structures and sloping structures in a view point of design concept for ice load and stability of structures.

JOIA have planned the research project of fundamental research on ice load from 1993 to 1997 considering the status of the ice load prediction method and technology.

The research project has been carried out by the three groups, namely, technical survey and evaluation (TSAE) working group (W/G), ice tank labolatory tests (ITLT) W/G and medium scale field indentation tests (MSFIT) W/G.

The study on the ice load acting on vertical sided structures is being carried out by ITLT - Vertical Structure W/G and MSFIT W/G, and the ice load on sloping structures is being conducted by ITLT - Sloping Structure W/G.

In the following sections, we introduce outlines of these research project and the summary of the results from 1993 to 1995.

The details of experimental study will be presented in different reports.

TECHNICAL SURVEY AND EVALUATION OF TECHNOLOGY

The survey and evaluation working group (W/G) is responsible for the following three major tasks. The W/G conducted technical surveys and evaluations on the current status and trend of ice load researches in the world in order to provide the results of the survey into the on-going research and to meet the requirements of the needs for research. The W/G collected, analyzed and evaluated the information and data in collaboration with National Research Council (NRC) and Center for Cold Ocean Resources Engineering Research (C-CORE) in Canada.

The outline of the major projects on ice load researches in the world have been obtained through the surveys to this date, however, the details of the projects have not yet been obtained. We will continue further work to get more detailed information on the specific projects of our interest.

Collecting the information on the current status and trend of ice load research

1) Overview of the current status and trend of ice load researches in the world
   a) Overview of the current status and trend of ice load researches

(by country, research institute, research area, methodology, problem to be broken through)

b) Overview of the current status of typical ice load researches (theme, outline of the research, research institute, principal investigator, budget, period, research result, problems to be broken through)

c) Problems to be broken through by research area (In particular, laboratory tests, outdoor experiments, and methods to estimate the ice loads exerted on the offshore structures)

2) The current status of exploration and development of the oil and gas fields in ice infested waters in the world

a) Overview of the current status of exploration and development of the oil and gas fields in ice infested waters

b) Technical problems to be broken through

3) Availability of data on natural conditions such as ice conditions, weather conditions, sea conditions and sea bottom conditions in the Sea of Okhotsk around Sakhalin Island

Publication of the research results and the international collaboration on the issues of ice load research

1) Publication of the research results in the domestic and international conferences
2) Exchange of information with overseas research institutes
3) International collaboration

Problems to be broken through in Japan's ice load research and future plan

1) Discussion on the future plan after the current research program is finished in 1998
2) Discussion on the feasibility of ice experiments in Sakhalin

ICE LOAD ON VERTICAL SIDED STRUCTURES

It is well recognized that dynamic response caused by ice interaction is the key issue for even designing a massive wide structure operating in an ice-infested waters after the lesson of Molikpaq event in 1986. Due to complex mechanism involved in crushing failure, many problems associated with ice structure interaction is not resolved yet. One of the open issues is to clarify the mechanism of simultaneous failure while ice sheets move against vertical sided structures. It is deduced that ice failure mechanisms and structure dynamic response characteristics is deeply involved. Once simultaneous failure takes place, largely fluctuating foundation load might significantly affect structure foundation. In these situations, it is still not known from the design view point what local ice pressure and global load are to be selected in the function of compliance of structures, ice strength and indentation velocity etc. Many questions still arise when we proceed a practical design. To eliminate these design issues as much as possible, a five year research program was designed. The program objectives are ;

1) to develop computer code capable of predicting local, global and foundation load and dynamic response of structure for ice sheet and ridge ice.
2) to perform intensive ice tank tests to validate the developed computer code. Also to develop a local ice structure interaction model by experimental results.
3) to evaluate seabed strength and deformation subject to dynamic response of structure
4) to validate the computer code developed by medium scale field indentation test results

Experimental study

From 1993 to 1995, the following tests have been carried out using two ice tanks in Japan.

1) Mechanical properties confirmation tests of model ice (1993)
The compressive strength ,bending strength and other mechanical properties of granular and columnar model ice used for the following tests were confirmed.
2) Small indentor test (1994)
To clarify the ice failure mechanism and the relation between indentation velocity and crushing depth, small indentor tests were conducted. In the test, the indentor was pushed against the vertical ice sheet in air to facilitate observation.
3) Indentation tests by stiff rectangular and circular models (1993)
Indentation tests were performed using four different size rectangular models of 100mm, 400mm, 800mm and 1,500mm in width and one circular model with a diameter of 800mm. Segmented local ice pressure panels were installed to measure local ice pressure characteristics.
4) Compliant rectangular model indentation tests (1994)
To investigate the influence of structure stiffness, indentation tests were performed using compliant 800mm wide rectangular model. The model has segmented local ice panels in water line and disk springs at bottom to vary stiffness in three levels. Fig.1 shows the schematic of a test arrangement. Ice thickness ranged from 60 to 94mm and the compressive strength varied from 115 to 300 KPa. The tests were performed in five speed ranging from 6 to 105mm/sec.

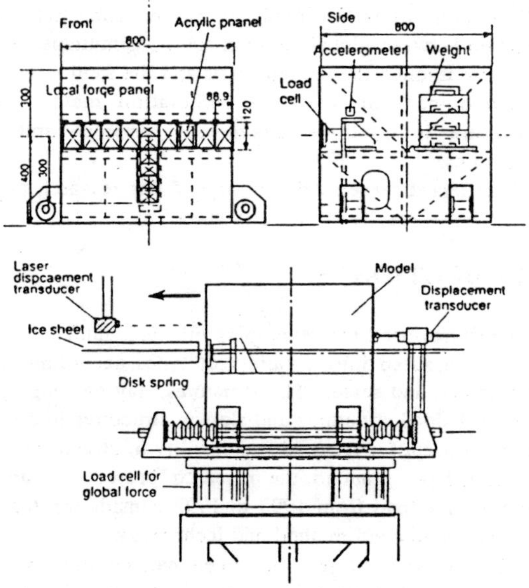

Fig.1 Schematic of a test arrangement

5) Indentation tests in ridge ice (1995)
To investigate the characteristics of ice load in ridge ice, indentation tests were conducted with three different stiffness.

Ice load prediction model

The model is being developed based on the model proposed by Kärnä (1992). The image of the model is shown in Fig.2. The programs main features are outlined below.

1) Ice load will be obtained to solve
   - the equation of motion for rigid ice floe
   - elastic response equation of structure
   - elastic deformation equation of ice floe .

   By solving these equations simultaneously, the relative displacement between ice floe and structure is obtained.
2) As a function of this relative displacement and relative velocity, local ice loads for each segmented contact area is obtained. A summation of local loads is obtained as global load.
3) Obtained loads are fed back to the above three equations as external loads and a new relative displacement is calculated .

By this method, time histories of local ice load, global ice load and structure response (displacement, velocity, acceleration) can be predicted. In the calculation program of ice load, it is tried to reflect some important test results such as crushing pressure distribution in order to simulate realistic ice failure phenomenon.

Through the study, it was confirmed that the developed code can simulate characteristics of both simultaneous and non-simultaneous failures. Some improvements are being planned to incorporate damaged area propagation and etc. to represent simultaneous failure more rationally. Fig. 3 shows an example of the comparison between ice tank indentation test and the simulation.

In addition to the above simulation, a study of ice load influence on soil foundation is being planned. It is estimated that the duration of ice load acting on structures is longer than that of an earth - quake load. Therefore the evaluation of ground strength under such cyclic load is a very important subject for the design of offshore structure in ice - infested waters.

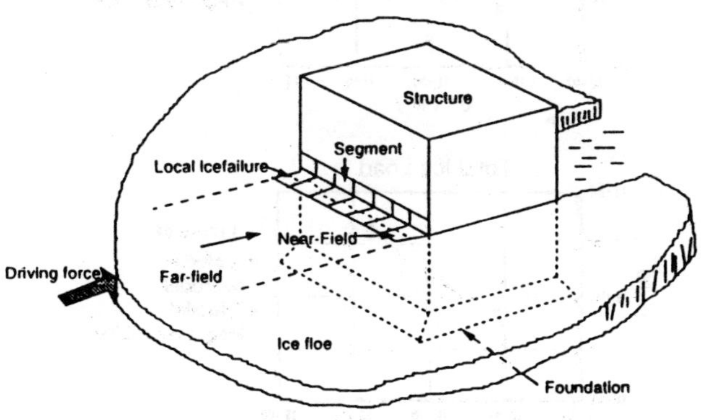

Fig.2 Image of a simulation model

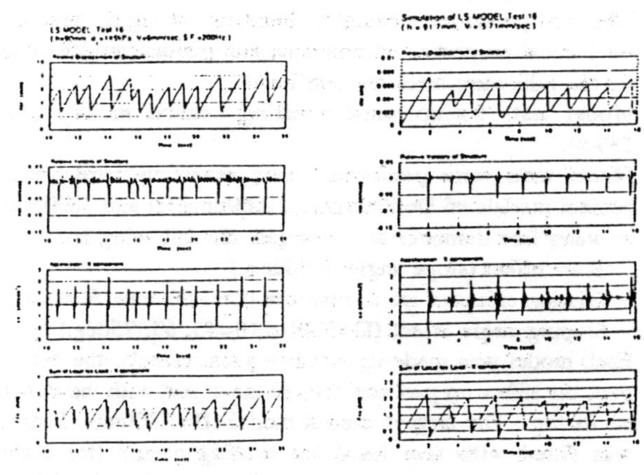

Fig. 3 Comparison of indentation test and simulation

Test results

Several results are already published. Major results are summarized as follows.

- The width of independent failure area is approximately the same as ice thickness, and it is more rational to represent indentation rate by the ratio of velocity to ice thickness, V/h rather than velocity to structure width , V/4D or V/2D. (Kamesaki et al., 1995a)
- Experimental results show that simultaneous failure is likely to occur when compliance of structure increases and indentation velocity decreases.
- Local ice pressure increases when its area decreases, however its characteristic is complex and can be a function of indentation velocity, ice thickness and compressive strength etc. (Kamesaki & Yamauchi, 1996a)
- Global load is a function of structural stiffness. When compliance of structures increases, global load tends to increase. (Kamesaki & Yamauchi, 1996b)

ICE LOAD ON SLOPING STRUCTURES

Ice load acting on sloping structures would be less than that on vertical sided structures, because the ice failure would be in bending and not in crushing mode. Thus, many kind of sloping structures have been considered in ice - infested waters in offshore oil fields, among them conical shaped structures is to be investigated in the view point of the simplicity of structural shape.

In this research project, according to the observation of failure mode of ice and understanding of the relation between global ice load and nonsimultaneous local ice load which were obtained in the experimental studies mentioned later, Distinct Element Method (DEM) was applied to simulate the non - continuous failure mode of ice and ice load.

Experimental study

From 1994 to 1995, the following tests have been carried out using three ice tanks in Japan.

1) Mechanical properties confirmation tests of model ice (1994).
The compressive strength, bending strength and other mechanical properties of columnar and granular model ice used for the following tests are confirmed.

2) Model tests by up ward breaking conical models (1994, 1995).
Model tests were performed using seven up ward breaking conical models of D=500mm(S), 1000mm(M) and 1800mm(L) in water line diameter to investigate the following issues :
- Scale effect (slope angle θ=60deg.)
- Friction effect of ice to structure (D=1,000mm, θ=60deg.)
- Sloping angle effect (D=500mm, θ=45, 60, 75deg.)

Each model was made up of three parts, namely, the breaking part, the ride - up part and the columnar part with the deflector on its top. The largest conical model (D=1,800mm, L model) was fitted with five local ice loading panels (for 30degs. curvature each) all around its breaking part. Fig.4 and Table 1 show the shape of conical model and their dimensions.

3) Model tests by down ward breaking conical models (1995).
Model tests were performed using six down ward breaking conical models using the former up ward breaking conical models to investigate the following issues :
- Scale effect (D=1,000mm, 1,800mm, θ= 60deg.)
- Sloping angle effect (D=500mm, θ=45, 60deg.)

4) Indentation tests by ship bow type model (1995).
Indentation test were performed using one wedge type bow (Lpp=3000mm, B=700mm, D=500mm) for the comparison with DEM simulation.

5) Indentation tests in ridge ice (1994, 1995).
To investigate the characteristics of ice load in ridge ice, indentation test were performed using up ward and down ward breaking conical models.

Table 1 Principal dimension of model

	Representative Width D (mm)
S model	500
M model	1,000
L model	1,800

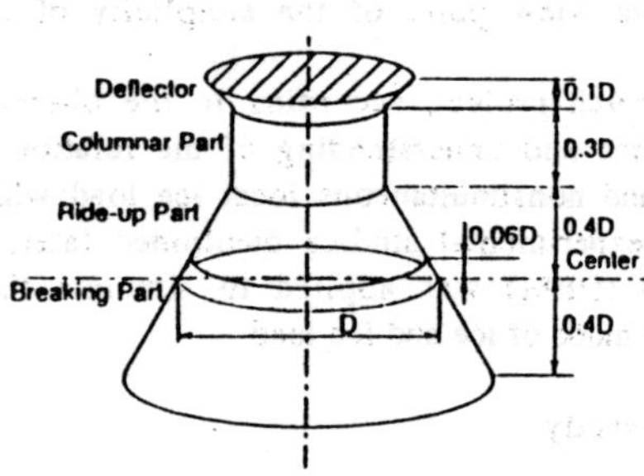

Fig. 4 Conical model

Ice - structure interaction simulation

Firstly, the current state of the art of numerical model for simulation of ice - structures interaction was reviewed and PFC (Particle Flow Code) - 2D and 3D - computer code was selected to apply mentioned simulation.

The computer code have the following capabilities

1) Input :
- ability to vary structure geometry
- ability to vary ice parameters including strength, internal friction angle, shape, velocity, density of the ice.

2) Output :
- global ice load and local ice load distribution
- tracking the ice fragments before and after contact with the structure
- modeling the ice failure and crack patterns.

Summary of the study

1) The relation between the ice load acting on each of the S, M and L models and model representative width cubed (D^3) is shown in Fig.5. Since the straight line represents the regression line with inclination of nearly 1 in each diagram, the ice loads can be considered to increase proportionally to be cube of model size.

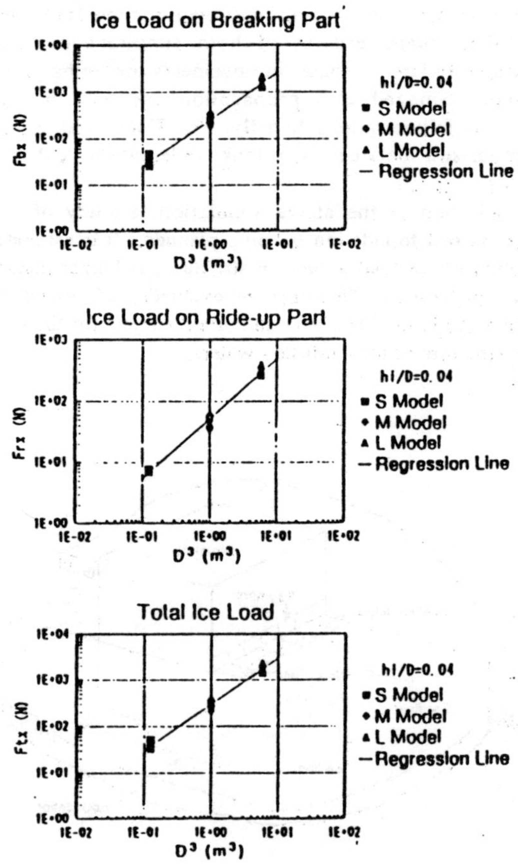

Fig.5 Scale effect of ice load

2) The result of ice to structure interaction simulation using PFC - 3D Code is shown in Fig.6 and Fig.7. It is to be mentioned that the behavior of ice ride - up and the tracking the ice fragments after bending failure of ice sheet were in good agreement with the experimental observation.

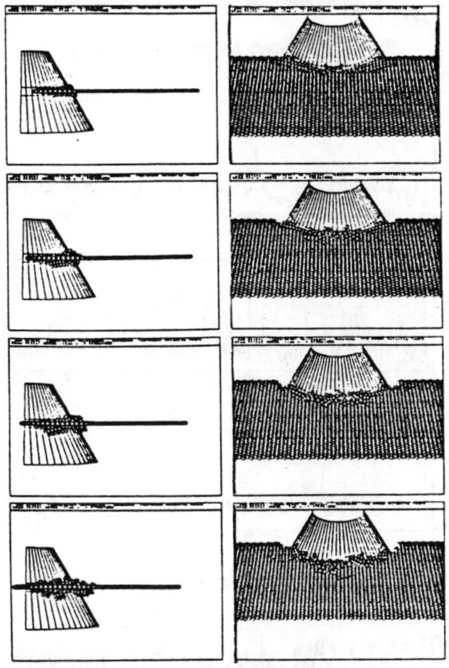

Fig.6 Failure and rubble behavior of ice sheet

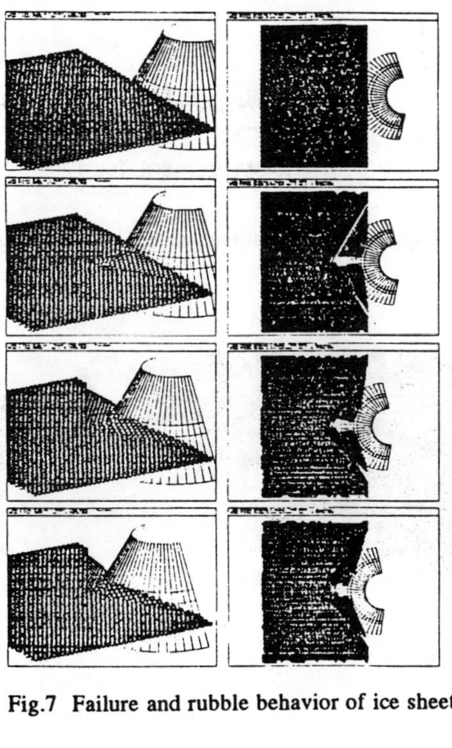

Fig.7 Failure and rubble behavior of ice sheet

MEDIUM SCALE FIELD INDENTATION TESTS (MSFIT)

JOIA project is also carring out ice indentation field tests. The main purpose of MSFIT is to obtain an understanding of the pressure-area curve in ice/vertical structure interaction. The maximum ice contact area with structure is planned to be 1.44 (m^2). The test site is located at Futamigaoka harbor on Lake Notoro connected to the Sea of Okhotsk, as shown in Fig.8. The MSFIT is planning to investigate the following ;

1) the ice failure mode,
2) the area of an independent ice failure zone in non-simultaneous failure,
3) mechanical properties of the test ice sheet,
4) the strained area of the test ice sheet by image sensing, and
5) the characteristics of the total ice load and local ice pressure.

In MSFIT, field tests are planned for two years as shown in Table 2.

Table 2 Test conditions (Planned)

No.	Width of model structure W (m)	Ice thickness h (cm)	Aspect Ratio (W/h)	Indentation speed V(cm/sec)	Fiscal year
1	1.5	6~12	12.5~25	0.3	1995/4~1996/3
2	3.0	12~24	12.5~25	0.6	1995/4~1996/3
3	4.5	12~18	25~37.5	0.9	1996/4~1997/3
4	6.0	12~24	25~50	1.2	1996/4~1997/3

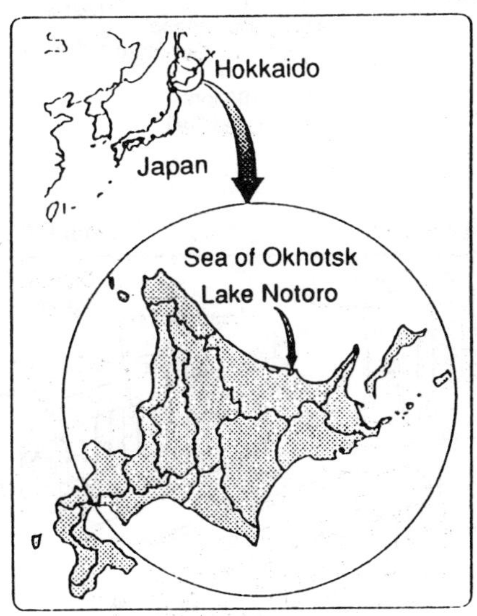

Fig.8 Location of the test field

Fig.9 shows the indentation test apparatus schematically. Load cells as well as a 100tonf load cell for total ice load measurement are mounted to measure local ice pressures. The maximum indentation depth is 100cm. The hydraulic servo-system used controls the indentation speed (V). The strain rate is planned to be 10^{-3} sec^{-1} as V/(2W). Fig.10 shows examples of time series of total load and local ice pressures. In the initial stage of ice/structure interaction, all time series give a prominent first peak, and then, random peaks in subsequent load ranges. Fig.11 shows the photo of the ice indentation test. The width of the model structure (W) and ice thickness (h) are 3m and 12cm, respectively. The model structure breaks an ice volume of h*W* 1m (stroke), and stable crushing failure mode was obtained. The ice failure mode observed is crushing and spalling (flaking, wedge, etc.) with horizontal cleavage cracks. A series of MSFIT test results will enhance the reliability of the design ice load on a structure. Detailed results will be presented in different papers.

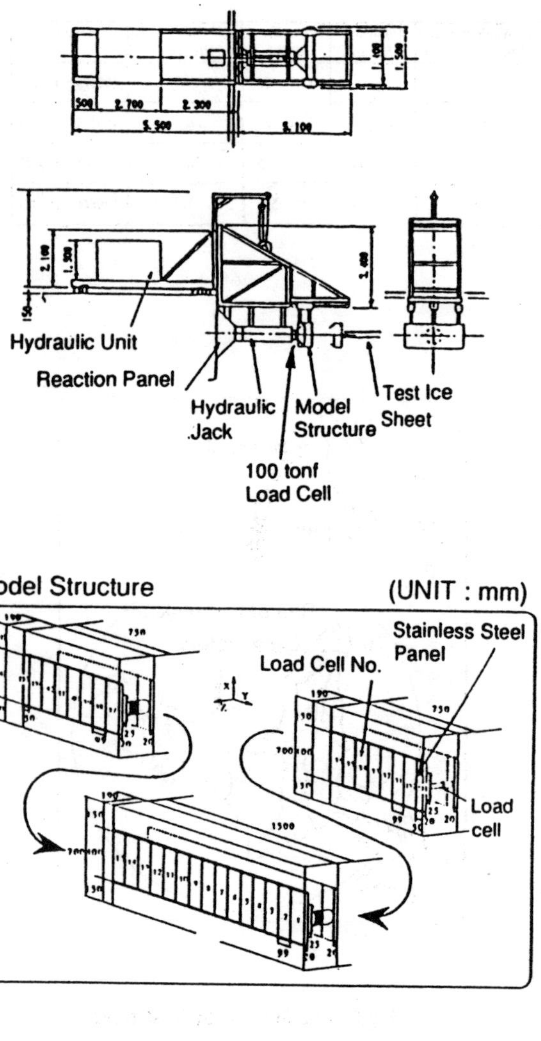

Fig.9 Indentation test apparatus

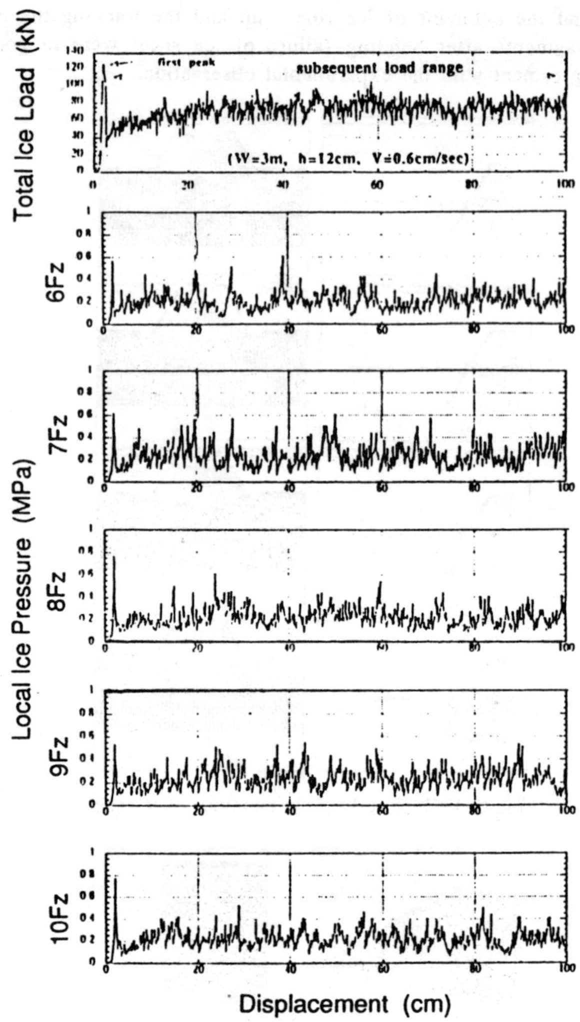

Fig.10 Total load and local ice pressure vs. displacement

Fig.11 Photo of ice indentation test

CONCLUSIONS

This report presents the summary of JOIA research project in the early stage from 1993 to 1995. The detail of each study will be presented in other reports.

JOIA project will be continued in the time schedule from 1996 to 1997, the results of study will be reported in the near future.

At the end of this report, authors would like to express their gratitude to Mr.Hiroyasu, Kawaguchi, the manager of Research and Development Dept. of JOIA for this publication.

REFERENCES

JOIA Committee of the Fundamental Research on Ice Load (1993). JOIA Study on Ice Load, Cold region technology conference '93, Sapporo

Kamesaki K., et al.(1995a). Experimental study on independent failure zone of ice-crushing The 10th international symposium on Okhotsk sea, sea ice & peoples, Monbetsu

Kamesaki K., Yamauchi Y. (1995b). Stiffness variation indentation test with segmented pressure panels and its response Cold region technology conference '95, Sapporo

Kamesaki K., Yamauchi Y.(1996a). Local pressure characteristics by indenter with segmented local ice pressure panels The 11th international symposium on Okhotsk sea, sea ice & peoples, Monbetsu

Kamesaki K., Yamauchi Y., Kärnä T.(1996b). Ice load as a function of structural compliance, 13th International Symposium on Ice, Beijing, China

Kärnä T.(1992). A procedure soil - structure - ice interaction, Technical Research Center of Finland, Seminar Series, Finland

Kato K., et al.(1995). Effects of Cone Angle and Infusion Velocity on Ice Forces on Conical Structures, Cold region technology conference '95, Sapporo

Kato K., et al.(1996a). Experimental Studies on Ice Forces on Conical Structures Using Segmented Model, The 11th international symposium on Okhotsk sea, sea ice & peoples, Monbetsu

Kato K., et al.(1996b). Experimental Study on Ice Forces on a Conical Structure Using Segmented Model, 13th International Symposium on Ice, Beijing, China

Ohta M., Ishikawa S., Kawasaki T.(1995). Experimental Study of Ice Loads on Slope - Sided Structures, Cold region technology conference '95, Sapporo

Ohta M., Ishikawa S., Kawasaki T.(1996a). Experimental Study of Ice Loads on Conical Structures, The 11th international symposium on Okhotsk sea, sea ice & peoples, Monbetsu

Ohta M., Ishikawa S., Kawasaki T.(1996b). Experimental Study of Ice Loads on Conical Structures, 13th International Symposium on Ice, Beijing, China

Proceedings of the Seventh (1997) International Offshore and Polar Engineering Conference
Honolulu, USA, May 25-30, 1997

ISBN 1-880653-28-1 (Set); ISBN 1-880653-30-3 (Vol. II)

Distinct Element Simulation of Ice Sheet Failure Against Offshore Structures

K. Katsuragi and M. Ochi
Mitsubishi Heavy Industries, Ltd.
Kobe, Japan

H. Seto and T. Kawasaki
Mitsubishi Heavy Industries, Ltd.
Nagasaki, Japan

ABSTRACT

Appropriate estimation of ice–structure interaction is of great importance in the design of offshore structures in ice-infested regions. This paper describes numerical simulations of two-dimensional ice–structure models performed under the research project, "Study on Ice Loads acting on Marine Structures", planned by the Japan Ocean Industries Association (JOIA) and funded by the Ministry of International Trade and Industry.

To trace the process behind successive ice-failure against a structure, distinct element methods (DEM) have been adopted taking advantage of their applicability to non-continuous media compared to standard finite element methods. In addition, two codes, PADIC developed by MHI and PFC-3D developed by ITASCA, have been investigated.

The conclusion of this study is that these distinct element codes show promise as powerful working tools for investigating ice–structure interaction, such as in prediction of ice sheet ride-up to and ice load against artificial islands.

KEY WORDS: Ice–structure interaction, Offshore structure, Distinct Element Method, PADIC, PFC-3D

INTRODUCTION

Sloping-wall structures have been considered effective in ships and other offshore structures that operate in ice-infested regions because of their ability to decrease ice load by bending and breaking ice sheets. To estimate the ice load interacting on such sloping-wall structures, a number of conventional methods can be used that are based on elasticity analysis in mechanics for continuous media. The main output of these methods is total ice load, which is analyzed in a quasi-static manner by separating it into load needed to break up the ice sheet and load needed to eliminate the sheet. This approach, however, is difficult to apply to rubbly ice. There has therefore been a need for the development of an estimation program that takes into account continuous failure of an ice sheet of any shape and the dynamic influence of rubbly ice, as an ice-load estimation model for sloping-wall structures.

In this research, we are therefore focusing our attention on distinct element methods (DEM) that can quantitatively estimate the dynamic behavior of non-continuous media, and are proceeding with the development of a new ice-load estimation method that can take into account the physical properties of ice and the flowability of rubbly ice.

This paper reports on the theory behind the developed DEM program and on simulation results for the mutual interaction between ice sheets and marine structures.

DISTINCT ELEMENT METHOS THEORY

GOVERNING EQUATIONS

Distinct element methods are used to analyze non-continuous media such as the ground, which here would be treated as an aggregate of non-continuous elements like sand and rocks. In particular, these methods consider equations of motion for these individual elements and trace the behavior of each element in terms of the traslation of force via the contact points between elements. Here, the contact force between elements is expressed as an elastic spring (spring constant K) and viscous dashpot (viscosity coefficient η). If we take the mass of an element to be m, radius of an element to be r, and moment of inertia to be I, then the equations of motion for translation (u) and rotation (ϕ) can be expressed by Eq.(1) as follows.

$$m\ddot{u} + \eta\dot{u} + Ku = F \tag{1a}$$

$$I\ddot{\phi} + \eta r^2\dot{\phi} + Kr^2\phi = G \tag{1b}$$

Here, F and G are external forces with respect to translation and rotation.

Cundall has developed a DEM that performs difference approximations according to time increment Δt and calculates acceleration $\ddot{u}$ from the contact force between elements. According to Cundall, the contact force in the normal direction at the time of contact can be given by the following equations.

$$\Delta e = K \cdot \Delta u \tag{2a}$$

$$\Delta d = \eta \cdot \frac{\Delta u}{\Delta t} \tag{2b}$$

Here, Δu, Δe and Δd represent increments in displacement, elastic resistance and viscous resistance in time interval Δt. Consequently, the elastic resistance and viscous resistance at time n

can be expressed by the following equations.

$$e^n = e^{n-1} + \Delta e \quad (3a)$$

$$d^n = \Delta d \quad (3b)$$

However, because Cundall's DEM does not include tensile force between elements, the following condition must be added to Eq.(3).

$$\text{when } e^n < 0 \qquad e^n = d^n = 0 \quad (4)$$

Accordingly, compressive force f in the normal direction between elements at time n can be calculated as follows.

$$f^n = e^n + d^n \quad (5)$$

Acceleration $\ddot{u}$ when compressive force f is applied can be calculated by the following equation.

$$\ddot{u} = \frac{f^n}{m} \quad (6)$$

Because recoil due to collisions between elements quickly dampens out, viscosity coefficient η is often given by the following equation.

$$\eta = 2\sqrt{mK} \quad (7)$$

POROUS SUBSTANCE MODELING BY DEM

In Cundall's DEM, elements are treated as being independent and transmitting force between each other only when coming into contact. It is therefore difficult to apply this DEM to media like ice sheets in which tensile force plays a role.

Hakuno et al. have therefore proposed a pore–spring model for the porous substance between elements. This model considers that there is also tensile force at contact between elements as in a normal elemental spring, and as such can be thought of as a natural extension of Cundall's model. In addition, two conditions can be considered as failure criteria in the pore–spring model: that due to tensile force in the normal direction as represented by Eq.(8), and that due to shear deformation in the tangential direction as represented by Eq.(9).

$$R \geq \beta \times L \quad (8)$$

Here, R is the distance between elements, L is the natural length of the pore–spring between elements, and β is the failure threshold value (1.0).

$$f_s = C + \sigma \cdot \tan\theta \quad (9)$$

Here, f_s is the shear force between elements, C is the Coulomb adhesive strength, σ is the true stress on the contact surface, and θ is the friction angle.

DEALING WITH THE BUOYANCY EFFECT

To evaluate the behavior of ice after major failure in the mutual interaction between ice sheets and structures, the buoyancy effect must be taken into account. To therefore treat this effect as an extension of the DEM model, we have developed a buoyancy spring model in which buoyancy acts on each individual element and the magnitude of buoyancy is proportional to the area of the element submerged under water.

If we let S be this area submerged under water, $\rho_{(water)}$ the water density, and g the acceleration of gravity, then the magnitude of buoyancy F_g is defined by the following equation.

$$F_g = S \times \rho_{(water)} \times g \quad (10)$$

Comparing Eq.(10) with Eq.(2a) of a normal elemental spring, we add viscous resistance Δd_g proportional to element speed as a damping effect in the buoyancy spring model.

$$\Delta d_g = \eta' \times \dot{v} \quad (11)$$

Here, $\eta' = 2\sqrt{m \times \rho_{(water)} \times g}$ and $\dot{v}$ is element speed.

FAILURE MODEL EVALUATION BY THE 3–POINT BENDING PROBLEM

To evaluate the applicability of DEM to ice–sheet failure behavior analysis, we carried out verification calculations for the 3–point bending problem. Calculations conditions are given in Table.1 and typical calculation results are shown in Figs.2 and 3. The following was found on the basis of this analysis.

- If the number of element pile–ups is small, unnatural break–up easily occurs in the horizontal direction. In addition, break–up can be eased by increasing the elements affected by the load and distributing the load.
- Break–up in the horizontal direction can be suppressed by failure in the normal direction by tensile force and by increasing the failure threshold value.
- Break–up easily occurs if there is absolutely no failure in the tangential direction by shear force.
- Failure mode in a test object is heavily influenced by the element layout pattern.

Based on the above, it is necessary to investigate the quantitative relationship between failure criteria and the pore–spring's anisotropy, spring constant, etc.

ICE–STRUCTURE INTERACTION SIMULATION

The current state of the art of numerical model for simulation of ice–structure interaction was reviewed and PFC (Particle Flow Code) - 2D and 3D - computer code was selcted to apply mentioned simulation.

The computer code have the following capabilities

1) Input:
   - ability to vary structure geometry
   - ability to vary ice parameters including strength, internal friction angle, shape, velocity, density of the ice
2) Output:
   - global ice load and local load distribution
   - tracking the ice fragments before and after contact with the structure
   - modeling the ice failure and crack patterns

We demonstrate the ice ride–up behavior on the artificial beach using PFC–2D in Fig.4. And we present the time history of horizontal ice load in Fig.5. In Fig.5, we have good agreement between our numerical results and Croasdale's model. The result of ice to structure interaction simulation using PFC–3D Code is shown in Fig.6 and in Fig.7. It is to be mentioned that the behavior of ice ride–up and the tracking the ice fragments after bending failure of ice sheet were in good agreement with the experimental observation.

CONCLUSIONS

This research has developed an analysis technique according to a discrete element method (DEM) to estimate ice load on marine structures. The following results have been obtained.

1) With respect to a model for continuous media, we have incorporated the Hakuno model that proposes a porous substance connecting separated elements, and have verified the model's effectiveness through calculations for the 3-point bending problem.
2) With respect to the buoyancy effect, we have developed a new buoyancy model analogous to a normal elastic spring.
3) Finally, in the analysis of how ice sheets floating on the water ride up to sloping walls, we have simulated the pattern of ice sheet bending and failure against time and have shown the applicability of analysis codes to the estimation of dynamic ice load.

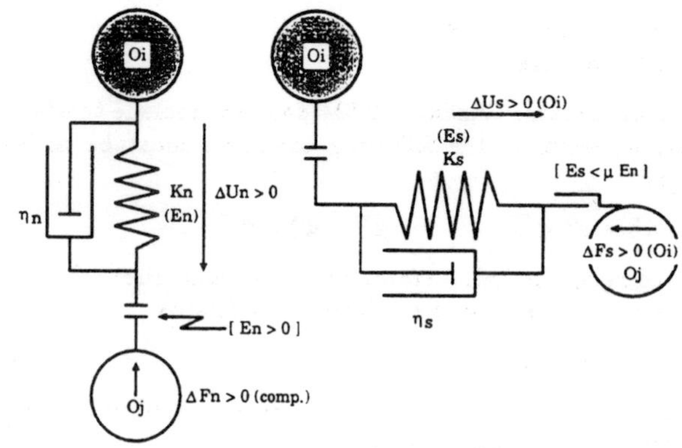

Fig. 1: Conventional DEM model

ACKNOWLEDGEMENT

This research has been conducted under the research project, "Study on Ice Loads acting on Marine Structures", planned by the Japan Ocean Industries Association (JOIA), to which the authors express their gratitude.

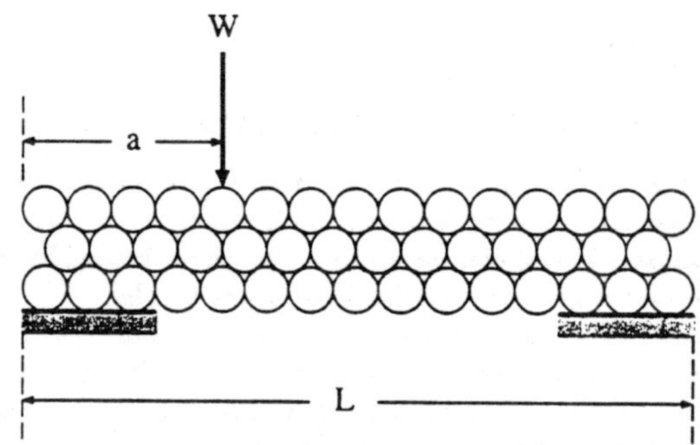

Fig. 2: 3-point bending problem

REFERENCES

[Cundall *et al.* 1979] Cundall, P. A.; Strack, O. D. L. 1979."A discrete numerical model for granular assemblies." *Géotechnique* 29, No.1: 47-65.

[Hakuno *et al.* 1989] Meguro, K.; Hakuno. M. 1989. "Fracture Analyses of Concrete Structures by the Modified Distinct Element Method." *Structural Eng./ Earthquake Eng. Japan Society of Civil Engineers*, Vol.6 No.2: 283-294.

Tab. 1: 3-point bending problem

element radius	$r = 0.5[\text{cm}]$
time step	$\Delta t = 10^{-5}[\text{sec}]$
load weight	$(Wg/\rho g) = 1000[\text{cm}^3]$
failure threshold	$\beta = 1.001$
Coulomb adhesive strength	$C = 2 \times 10^4[\text{g/cm}^2]$

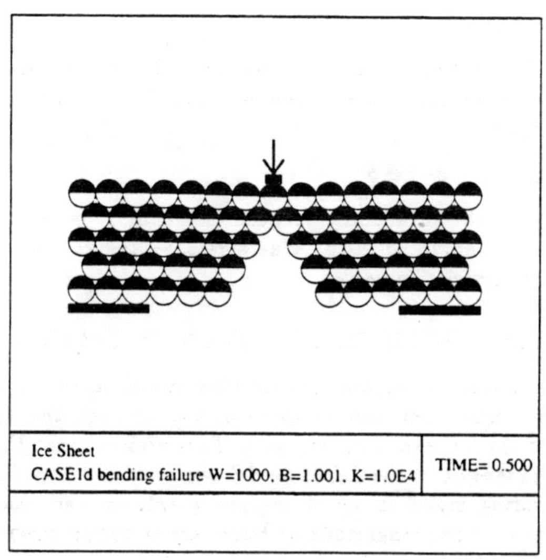

Fig. 3: Analysis of a bend test under vertical force

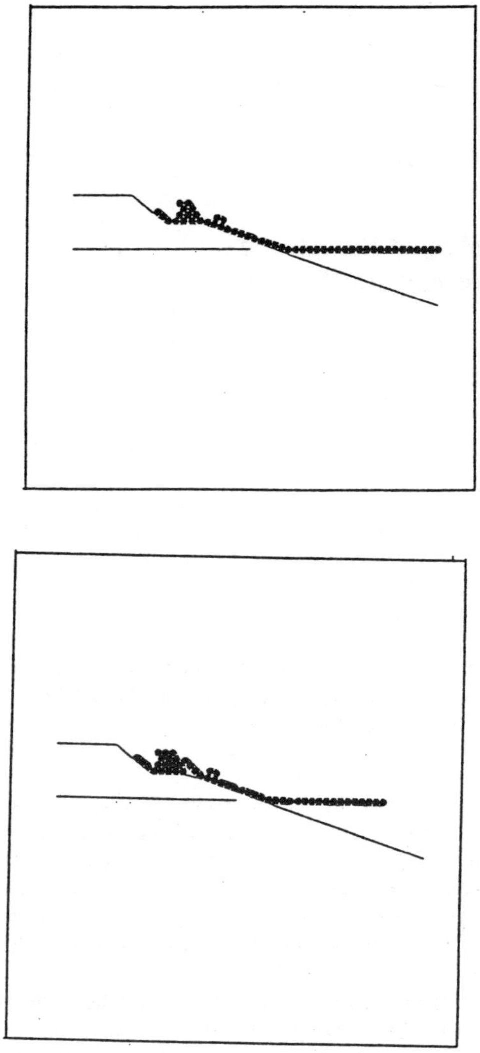

Fig. 4: Simulation of ice ride-up on the artificial island

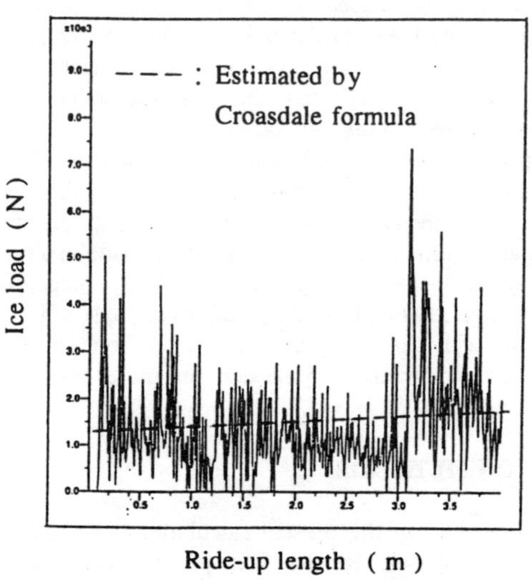

Fig. 5: Time history of horizontal ice load

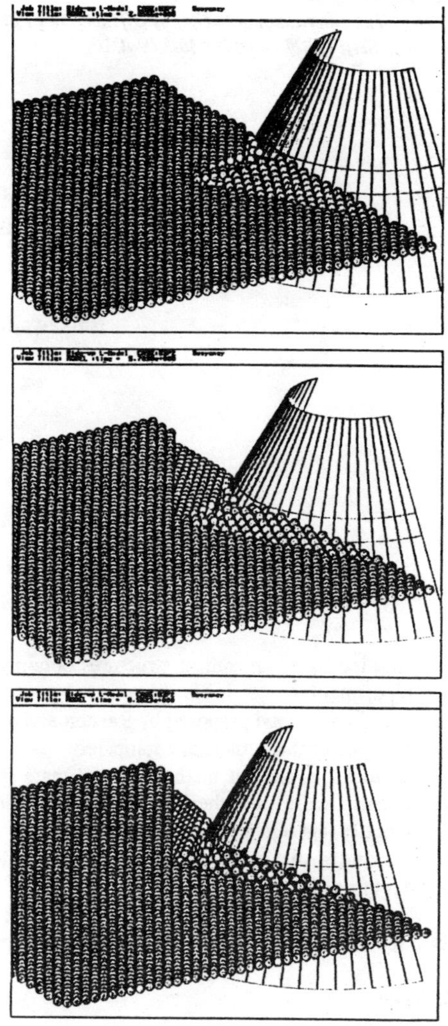

Fig. 6: Failure and rubble behavior of ice–sheet(a)

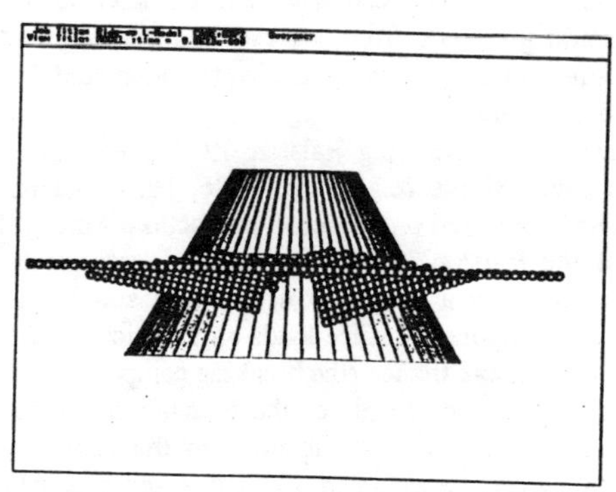

Fig. 7: Failure and rubble behavior of ice–sheet(b)

Proceedings of the Seventh (1997) International Offshore and Polar Engineering Conference
Honolulu, USA, May 25-30, 1997

ISBN 1-880653-28-1 (Set); ISBN 1-880653-30-3 (Vol. II)

Simulation of Ice Loads on a Conical Shaped Structure: Comparison with Experimental Results

Kazuyuki Kato
Ishikawajima-Harima Heavy Ind. Co., Ltd.
Yokohama, Japan

Koh Izumiyama
Ship Research Institute
Mitaka, Japan

ABSTRACT

In this study, the ice loads on conical structures obtained by three methods, the experimental ice loads obtained by an ice tank experiment, a theoretical method proposed by Ralston and a numerical method proposed by one of the authors are compared.

It is concluded that the present method can estimate the ride-up component more realistically than the Ralston's equation does. And this results in a better agreement with the experimental ice loads. Above all, it is concluded that the proposed method can give a better simulation of the total ice loads on a conical structure.

KEY WORDS: Ice loads, conical structure, simulation, ice tank

INTRODUCTION

A conical shaped arctic structure is seemed to be one of the most basic shapes for arctic structures. The conical shaped structure is considered to be a typical inclined structure which breaks an oncoming ice sheet by flexure. And failing the ice sheet by flexure needs less load than failing it by crushing. However, it is not expected that any perfect conical structure would be used in reality. Still, it is important to develop reliable methods to estimate ice loads on a conical structure. Because it should give us a first look estimation of ice loads on any inclined structures, and may give us an idea to understand the ice loads on an inclined structure.

Several theoretical works (e.g. Ralston, 1977, Croasdale ,1980) and many experimental works (e.g. Kato, 1986, Hirayama and Obara, 1986) have been conducted on ice - cone interaction. Among them, it is no doubt that the Ralston's theoretical work is superb in terms of estimating ice loads on a conical structure in a simple manner. He considered three components of ice loads on a conical structure. They are a component to break the ice (the breaking component), component due to rotation of broken ice piece (the foundation component) and component due to ice pieces riding up onto the cone (the ride-up component). However, he assumed a constant and probably extreme ride-up component in his theoretical work. Thus ice loads estimated by his equation would tend to overestimate the ice loads.

One of the authors has proposed a simulation method to simulate the ice loads on a conical structure (Izumiyama, 1992). With basic information such as measures of a structure, ice thickness and flexural strength and size of failed ice pieces, etc., his method can produce time series for the breaking, foundation and ride-up components provided an ice sheet progresses in a constant speed. Brief description of each component will be given in the next chapter.

Experimental work in an ice tank has been performed using segmented models (Kato et al., 1995, Kato et al., 1996). In this experimental work, the effort was toward to separate the breaking and ride-up components experimentally. Since it was impossible to separate the breaking and foundation component in the experiment, the foundation component was considered to be included in the breaking component.

The calculated results based on the said simulation method are compared with the experimental results in terms of characteristic values of ice loads which correspond to the maximum level of ice loads. Thus the characteristic values can be referred to as maximum ice loads. The comparisons are made in terms of the maximum total ice loads, the maximum breaking components, the mean ride-up components and the distributions of angle of interaction.

The calculated results based on the simulation method are also compared with the ice loads calculated from the Ralston's equation. The comparisons are performed on the same quantities as the experimental ice loads except the angle of interaction which the Ralston's equation does not take into account.

In the paper, the calculation method to simulate ice loads on a conical structure which has been proposed one of the authors is briefly outlined. The experimental study results of which are used for comparison is also outlined. The comparisons are made in terms of the maximum total ice loads, the maximum breaking components, the mean ride-up components and the distributions of angle of interaction. The comparison with the ice loads calculated from the Ralston's equation are also conducted. Discussion on those comparisons is described.

CALCULATION METHODS

Fig.1 is a flow chart of the present calculation. The calculation is composed of two phases. In the first phase, occurrence of the circumferential crack is simulated. This phase gives information on the location of the contact point between cone and ice and the size of the crack. In the second phase ice force is calculated based on information of the contact point and the crack size given by the first phase of thc calculation. Details of the calculation follow.

Method to simulate crack pattern

In the first phase of the calculation crack formation is simulated. Bending is the dominant failure mode of ice sheet interacting with a conical structure. There are two types of cracks observed in the ice. They are radial cracks and circumferential crack. In the first phase, formation of circumferential cracks is simulated. Fig.2 shows the method to simulate the crack formation. The semicircle and a curve around it in the figure show the water line of a cone and ice edge, respectively. The ice edge advances toward the cone at a given step. When the edge contacts the water line a circumferential crack is formed to give new ice edge. The crack is an arc with its center at the contact point.

The radius of the circumferential crack, R_c, is given by the following equation.

$$R_c = \alpha l_{BR} \tag{1}$$

$$l_{BR} = \sqrt{\frac{\sigma_U h}{\gamma}} \tag{2}$$

where σ_U, h and γ are flexural strength of ice in upward bending, ice thickness and density of water, respectively. α is a random variable with the normal distribution of a mean, μ, and a standard deviation, σ.

Equations (1) and (2) are assumed based on experimental results by Izumiyama et al. (1992). They made a detailed analysis of crack size observed in a model experiment to find a good correlation between crack radius and l_{BR}. Their analysis gave the values of μ=0.937 and σ=0.268. In the present simulation values of μ=0.642 and σ=0.150 are used based on the analysis of the present model test at IHI ice tank briefly described in the next section. The difference in the parameters can attribute to different model ice. Izumiyama et al. used model ice doped with ethylene glycol, while it is urea-doped model ice in the present test.

Method to calculate ice force

Fig.3 schematically shows interaction between a conical structure and ice sheet. It is well recognized that the ice force on a conical structure interacting with a sheet ice comprises three components. They are the breaking, foundation and ride-up components. After the contact with a cone ice deforms and radial and circumferential cracks gather. This event is associated with the breaking component, F_{BR}, a component to break ice. A fan-shaped ice piece then rotates as the ice sheet advances. This is associated with the foundation component, F_{FD}, a component due to weight, buoyancy and friction of the rotating ice piece. The ice piece then rides up onto the cone. This gives the ride-up component, F_{RU}, a component due to weight and friction of ice piece riding-up onto the cone surface.

There can be several ways to calculate the breaking component. In the present simulation the plastic limit approach is used. The present approach is similar to that by Ralston (1977), but different in an assumption concerning ice - cone contacts. Ralston assumed a full contact between the ice sheet and cone (Fig.4-a), while deformation of ice for each crack is considered in the present calculation (Fig.4-b). It is assumed that ice sheet inside the circumferential crack deforms in a way that it forms a conical surface with its apex at the contact point. The conical deformation gives an uniform curvature in the circumferential direction and no curvature in the radial direction. For this deformation, the plastic limit approach gives the following equation.

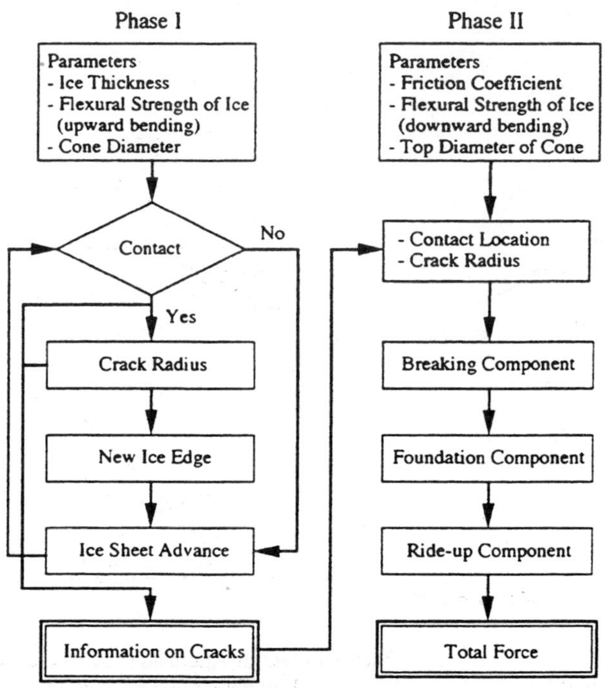

Figure 1 Flow chart of the calculation

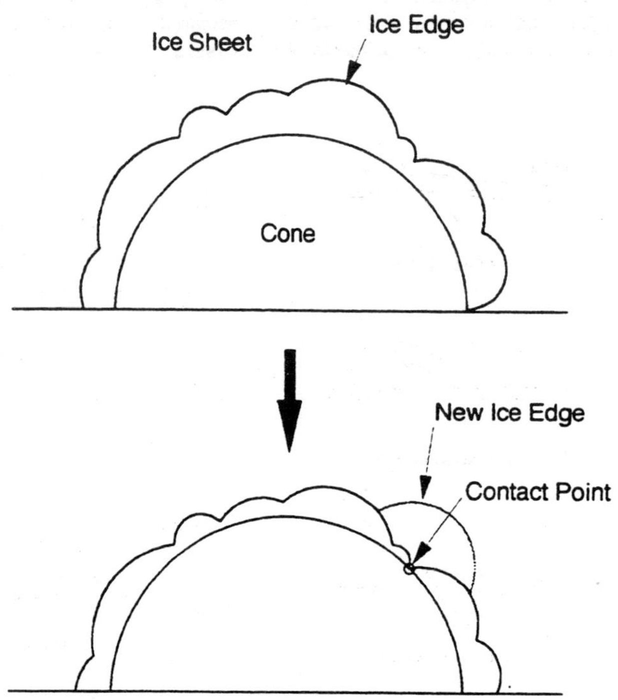

Figure 2 Method to Simulate the Formation of Circumferential Cracks

$$D_B = \int_r \int_\theta \frac{\sigma_D h^2}{6} \frac{1}{r} \frac{\partial \omega}{\partial r} r \, d\theta dr + \int_\theta \frac{\sigma_D h^2}{6} \frac{\partial \omega}{\partial r}\bigg|_{r=R} R d\theta \quad (3)$$

where D_B, ω, R, r and θ are the rate of energy dissipation due to deformation of ice, rate of vertical deformation of ice sheet, crack radius, radial distance and circumferential angle, respectively. σ_D and σ_U are flexural strength of ice in downward and upward bending, respectively. The first term of the right hand side of the equation (3) is a component due to deformation of ice sheet inside the deforming region and the second is a component due to the moment of resistance along the crack. The energy dissipation rate given by the equation (3) is equal to the product of the vertical component of the force to break ice sheet, F_{BZ}, and rate of vertical displacement of ice at the contact point, ω_0. This gives

$$F_{BZ} = \frac{h^2}{6}(\sigma_D + \sigma_U)\Phi \quad (4)$$

where Φ is an angle at the circumference of the crack. The horizontal and vertical components of the force are calculated by considering the equilibrium of the force components and friction force. Coulomb's law is assumed to calculate the friction force.

Calculations of the foundation and ride-up components are relatively straightforward compared with that of the breaking component. These components can be calculated by considering the weight, buoyancy and friction force of the broken ice piece. In calculating these components, it is assumed that the point of application of the force moves along the line AB shown in Fig.5. Ride-up of broken ice piece is assumed to terminate when the ice piece reaches the top level or the back side of the cone.

The three components of ice force are superpositioned to calculate the force during an unit event which is composed of the breaking of ice, rotation and ride-up of the broken ice piece (Fig.6-a). The same calculation is made for other events. Calculated ice forces are then superpositioned in time for all the events (Fig.6-b).

EXPERIMENT

Six conical structure models were used in the experiment. All the models had the water line diameter of 500 mm. The height above water line to the end of slope was all the same being 200 mm. The surface of the models were coated by the same paint and the frictional coefficient was estimated to be 0.1 from previous experience.

All the models consisted of three segments. They were named the breaking, ride-up segment and cylindrical segments. The breaking and ride-up segments formed sloping part of the model structures. The cylindrical segment was the vertical cylinder above the sloping part. The height from the water line to the separation between the breaking and ride-up segments was termed as the separation height and was denoted as Hs. The gaps between the neighboring segments were set to be 1 mm. Sketch of a model is seen in Fig.7 and the descriptions of models are summarized in Table 1. Each segment was connected to the center bar via triaxial load cell. The center bar was connected to the test carriage via triaxial load cell, which is hereafter named the total force load cell. Conceptual figure showing the alignment of segments is seen in Fig.8. By this alignment, twelve ice force components were measured.

The experiment was performed in the IHI Ice Model Tank (Tsutsumi and others, 1991). The model ice sheets were grown by seeding and freezing a 0.75% urea-in-water solution at a predetermined ambient temperature. The target ice thicknesses were 20 mm and 30 mm and the target flexural strengths were 10 kPa, 15 kPa and 20 kPa. The flexural strength was measured by the upward bending cantilever beam bending test. The towing velocities employed in the experiment were 7 mm/s, 70 mm/s and 140 mm/s. Thus total of at least 18 test runs were conducted for one model.

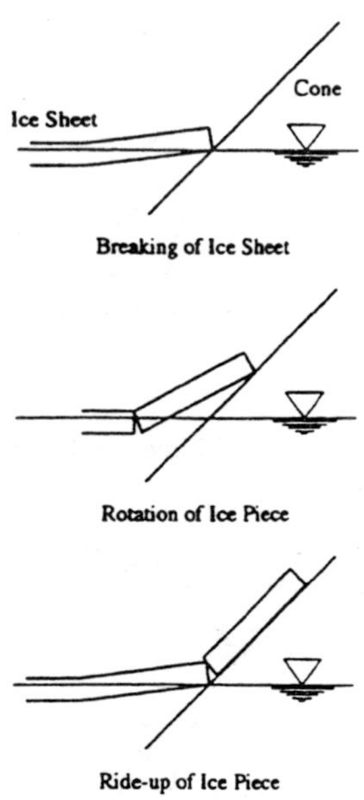

Figure 3 Schematic of Ice-Cone Interaction

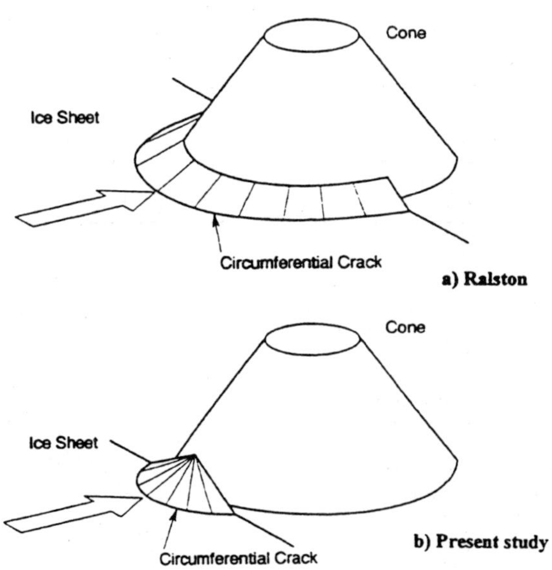

Figure 4 Deformation of Ice

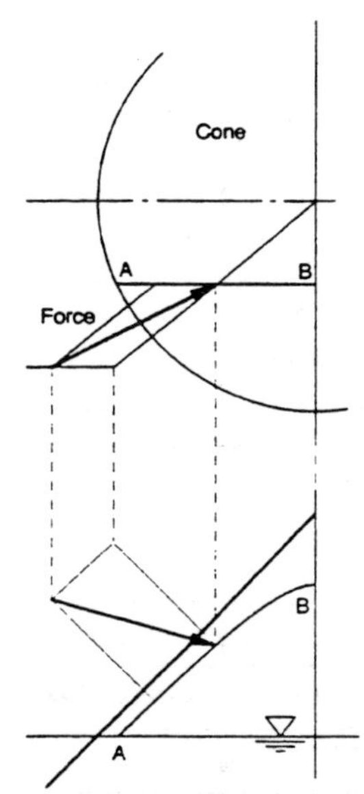

Figure 5 Point of Application of Force during Foundation and Ride-up Phases

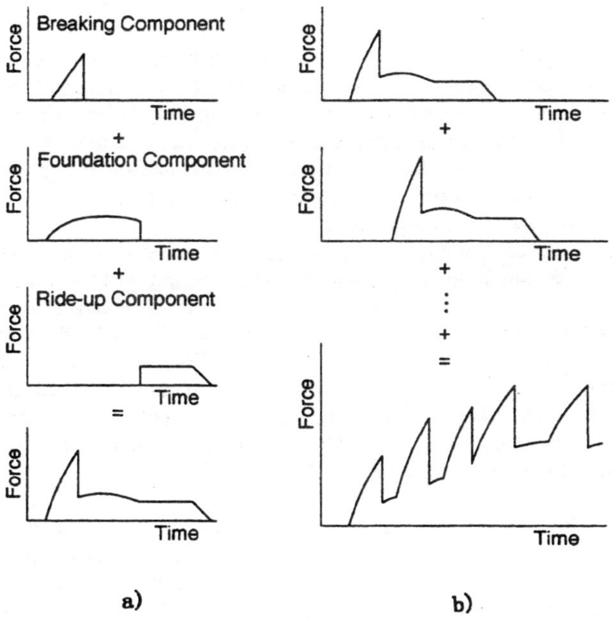

Figure 6 Superposition of Ice Forces

The experimental procedure and results were fully described in the previous reports by one of the authors (Kato et al., 1995, Kato et al. 1996). The main results obtained from the experimental study can be summarized as;

1. The ice loads measured at both the breaking and ride-up segments were affected by the separation height qualitatively and also quantitatively. It was concluded that the measured ice loads at the ride-up segment for the models with Hs=10 mm (UBC45-10, UBC60-10) were affected by the loads associated with ice sheet failure. Thus the measured ride-up components for these models were questionable.
2. The measured ice loads at the ride-up segment for the models with Hs=30 mm (UBC45-30, UBC60-30) could correspond to the ride-up component with a slightly underestimated manner. Thus the measured ice loads at the breaking segment could correspond to the breaking component with a slight overestimation.
3. The velocity dependency, the higher ice loads incorporated with the higher velocity, was observed. It was pronounced for the breaking components. Therefore the total ice loads also showed the remarkable velocity dependency. It was partly due to an inertia effect on ice sheet failure as pointed out by Sorensen (1979). And a dynamic amplification due to weak system in the experimental set up may be partly responsible for this large velocity dependency.

The foundation component should contribute both the breaking and ride-up components in the experiment. It is impossible to separate its contribution to the breaking components from that to the ride-up components. Since it is supposed that the contribution to the ride-up component is less than the contribution to the breaking component. Then we assume that the ice loads measured at the breaking segment contains the foundation component. Therefore, comparisons of the breaking component in the next section are made between the ice loads measured at the breaking segment and the sum of the breaking and foundation components calculated by the present method.

Table 1 Summary of the Models Used

Model ID	Direction	α	Hs
UBC45-30	upward	45 deg.	30 mm
UBC45-10	upward	45 deg.	10 mm
UBC60-30	upward	60 deg.	30 mm
UBC60-10	upward	60 deg.	10 mm
DBC60-30	downward	60 deg.	30 mm
UBC75-30	upward	75 deg.	30 mm

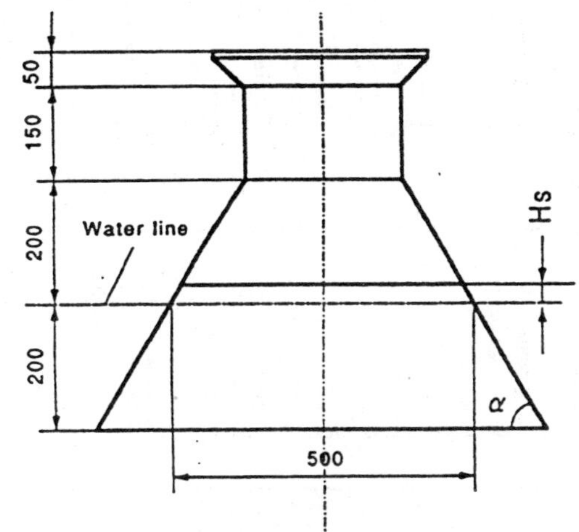

Figure 7 Sketch of a Model

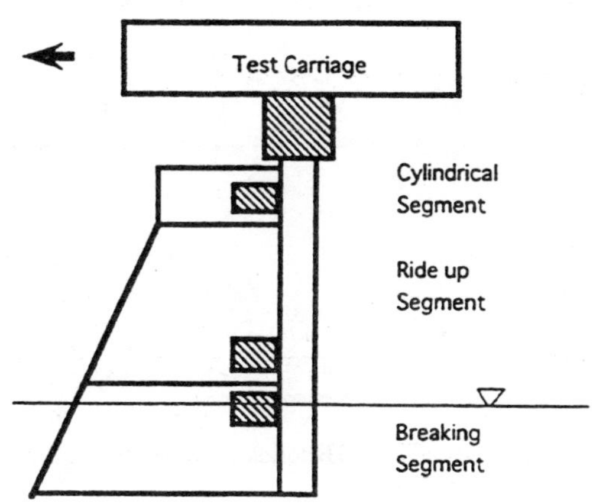

Figure 8 Conceptual sketch of segments alignment

COMPARISON

In this section, ice loads obtained from the simulation which employed the approach outlined in the previous section and ice loads obtained from the experiment are compared. With basic information such as measures of a structure, ice thickness and flexural strength, and size and distribution of failed ice pieces, etc., the proposed method can produce time series for the breaking, foundation and ride-up components provided an ice sheet progresses in a constant speed.

An example of time series of ice load obtained through the simulation is shown in Fig.9. From these time series, the characteristic quantities listed below are reduced and those characteristic quantities are compared with the same characteristic quantities reduced from the experimental results.

1. maximum total ice loads
2. maximum breaking components
3. mean ride-up components
4. distributions of angle of interaction

It should be noted that the comparisons are only made on the experimental results with v=7 mm/s. Because the simulation does not take any velocity dependence into account.

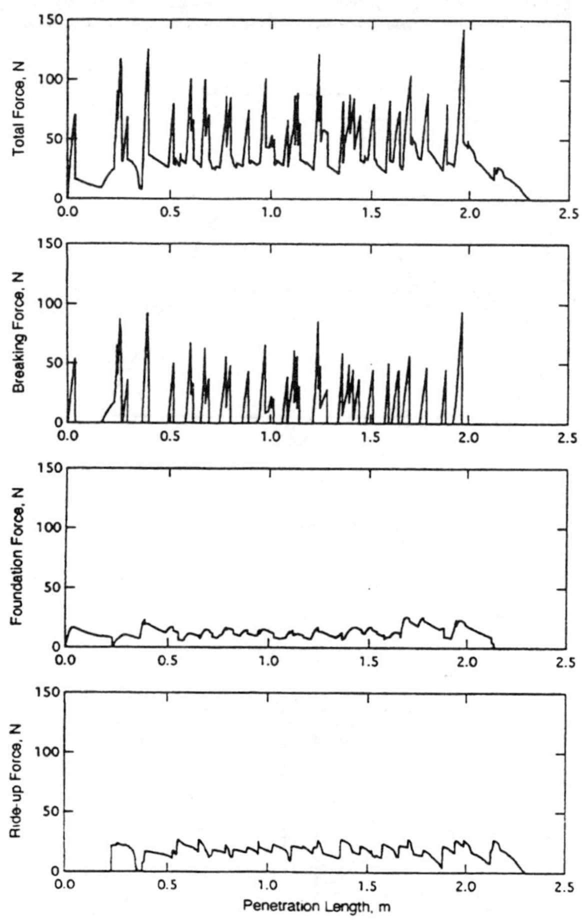

Figure 9 Examples of Time Histories Obtained by the Simulation

Maximum total ice loads

The maximum total ice loads are reduced by procedure described below. Every peak which is larger than 70% of the maximum load in a time series is picked up. The maximum ice load is defined as the mean of such peaks plus three times the standard deviation of these peaks. This procedure adopted for both the time series for horizontal and vertical ice loads. It is noted here that the horizontal ice load is a vector sum of the x- and y-direction loads, and the vertical ice load is the same as the z-direction load. The x-direction is defined as a direction of carriage movement, and the y-direction is perpendicular to the x-direction. The z-direction is defined as the vertical direction to x-y plane.

The procedure described above is adopted to both the simulated and experimental time series. For the simulated time series, a last 3/4 part is used to reduce the maximum total ice load. For the experimental one, the part where phenomena is steadily progressing is used.

The comparison for the horizontal ice load is shown in Fig. 10-a. And that for the vertical ice load is shown in Fig.10-b.

Maximum breaking components

The same comparisons are performed on the breaking components. The comparison for the x-direction load is shown in Fig. 11-a. That for the z-direction load is shown in Fig.11-b. It is noted that the comparisons are made only on the experimental results from the models identified as UBC45-30 and UBC60-30. Because the experimental results from the models with Hs=10 mm are questionable in terms of the separation of ice load components (Kato et al., 1996). And it is noted again that the simulated breaking component is the sum of the breaking and foundation components.

Mean ride-up components

For the ride-up component, a mean ice load is chosen to characterize the time series. The mean values are calculated in the same part in the time series as above two ice loads are calculated. The comparison for the horizontal ice load is shown in Fig. 12-a. And that for the vertical ice load is shown in Fig.12-b. Again the comparisons are made only on the experimental results from the models with Hs=30mm.

Distributions of angle of interaction

The angle of interaction is defined as;

$$\theta = \tan^{-1}(Fy/Fx) \,, \tag{5}$$

where Fx is a peak load in a time series of breaking component in x-direction, Fy is a corresponding load in y-direction.

Every angle of interaction corresponding to every peak in a time series is obtained from both the simulated and experimental time series. The frequency is counted 5 degree pitch and summed over all the tests having the same cone angle. The histogram of relative frequency are shown in Figs. 13-a and 13-b for α=45 degrees and α=60 degrees, respectively.

Discussion

The comparisons of total ice loads indicate that the simulated ice loads correspond well to the experimental ones. Of course there is some scatter, especially for α=60 degree, still it can be concluded that the proposed approach simulates the total ice loads on a conical structure.

The comparisons of the breaking components indicate that a good agreement is seen in the plots for smaller ice loads where ice thickness and flexural strength are supposed to be small. And relatively large scatter is revealed for larger ice loads, especially in the z-direction loads. It can be concluded that agreement between the simulated ice loads and experimental ones is tenable. It could be concluded that the proposed approach can simulate the breaking component of ice loads on a conical structure.

The comparisons of the ride-up components are scattered. And the points are scattered in the either sides of the 1:1 fit line. The range of loads associated with the present calculation is quite narrow, whereas that for the experimental ones is relatively wide. This results in such a scatter. Although it is not certain that taking an averaged quantity is a

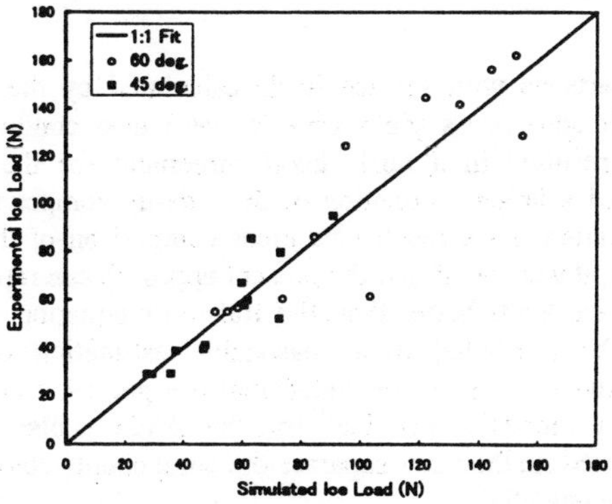

Figure 10-a Comparison of the measured and simulated total ice loads - horizontal ice load

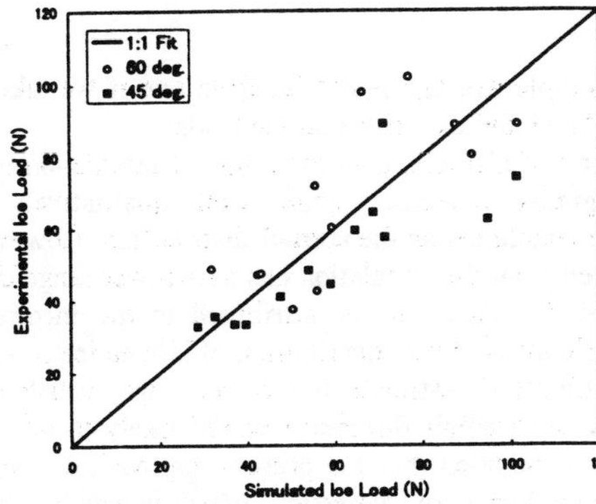

Figure 10-b Comparison of the measured and simulated total ice loads -vertical ice load

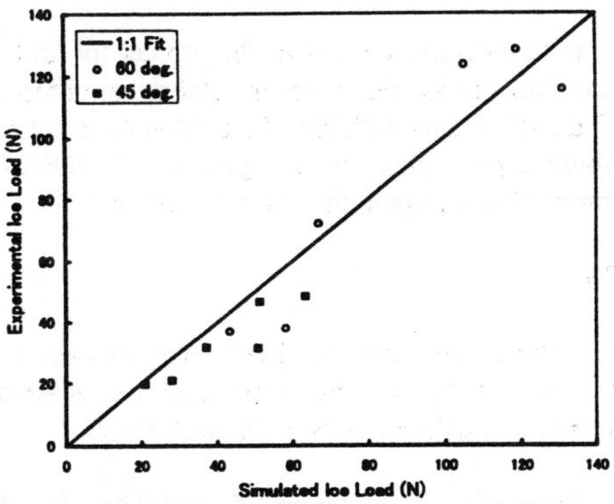

Figure 11-a Comparison of the measured and simulated breaking components - X-direction

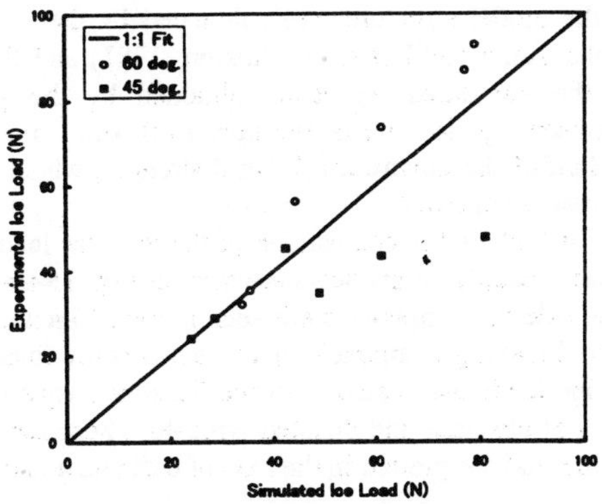

Figure 11-b Comparison of the measured and simulated breaking components - Z-direction

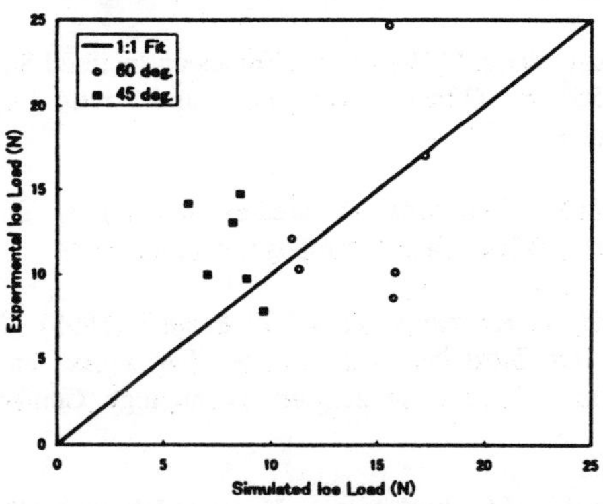

Figure 12-a Comparison of the measured and simulated ride up components - horizontal ice load

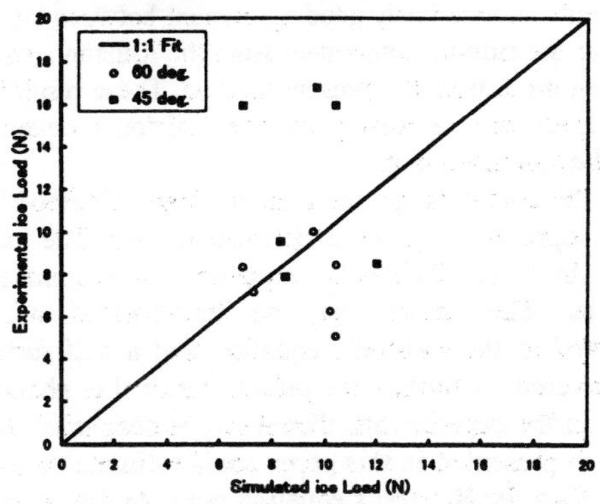

Figure 12-b Comparison of the measured and simulated ride up components - vertical ice load

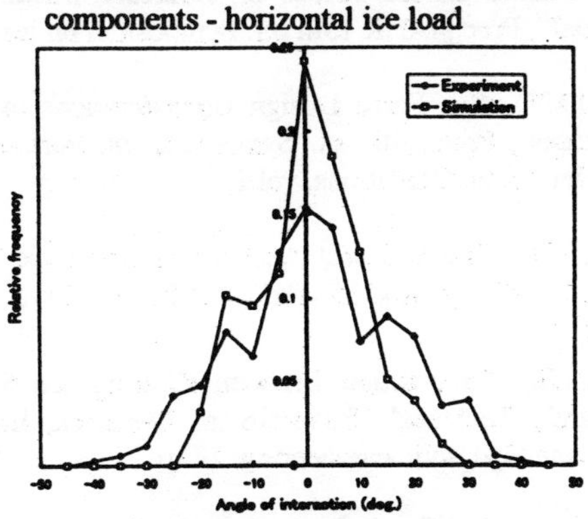

Figure 13-a Comparison of the distribution of angle of interaction - α=45 degrees

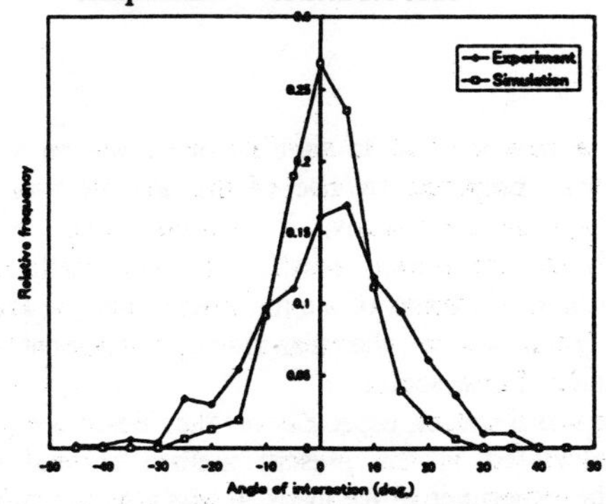

Figure 13-b Comparison of the distribution of angle of interaction - α=60 degrees

suitable way. A simple average is not far from one if we take ratios between the simulated and experimental ice loads.

The comparisons of histograms of the angle of interaction indicate that both histograms resemble quite well qualitatively. Both histograms can be concluded as the normal distribution. However, the histograms obtained from the simulation has a narrower range than the experimental ones do. This may be attributed to the incorrectness involved in the estimation of size distribution of broken ice pieces.

Since it is difficult to estimate the correct size distribution of broken ice pieces, such small difference would likely to take place. Thus it should be concluded that the present approach can simulate contact point change during an interaction, which is usually observed in a tank experiment and in a field observation.

Comparison with Ice Loads Calculated by the Ralston's Equation

Ice loads on the model structures are calculated by the Ralston's equation with Johansen's yield criteria (Ralston, 1981) and they are compared with the simulated ice loads obtained by the present approach. In Johansen's yield criteria, the upward flexural strength is assumed to be a half of the downward flexural strength, which is also assumed in the present approach.

Figs.14-a and 14-b show the comparison of the total ice loads. The comparisons of the breaking components are seen in Figs.15-a and 15-b and those of the ride-up components are seen in Figs.16-a and 16-b. It is noted that the breaking component again includes the foundation component. The ice loads associated with the Ralston's equation are plotted in the axis of abscissa and denoted with -R. Those associated with the present method are plotted in the axis of ordinate and denoted with -I.

From Figs.14-a and 14-b it is seen that the Ralston's equation gives higher total ice loads compared with the present calculation. The breaking components show a fairly good agreement between these two ice loads. However the ride-up component from the Ralston's equation are well larger than those from the present method. These result in that larger total ice loads are calculated by the Ralston's equation in comparing with the simulated ones.

As a result of the comparisons, the total ice loads obtained by the present approach represent well the experimental ones. The total ice loads obtained by the Ralston's equation overestimate the experimental ones. The reason may be incorporated with the assumption involved in the Ralston's equation that a half surface of structure is all covered by broken ice pieces. Since that phenomena was not observed in the experiments, then it can be concluded that the numerical approach presented in this paper could estimate the ride-up component better than the Ralston's equation does. And as a result, a better estimation of the total ice loads can be achieved by the present method.

SUMMARY

In this paper, a new method to simulate ice loads on a conical structure which was proposed by one of the authors was briefly outlined. The experimental study, its results were used for comparisons with the simulation results was also outlined. The comparisons were made in terms of the maximum total ice loads, the maximum breaking components, the mean ride-up components and the distributions of angle of interaction

Although there was a scatter, especially for the ride-up components, the ice loads calculated by the present method showed a good agreement with the experimental ice loads. It was also concluded that the contact points variation observed in the experiment could be expressed by the present method

The comparisons with the ice loads calculated by the Ralston's equation with Johansen's yield criteria were also conducted. The comparisons resulted in a fairly good agreement for the breaking component and a larger estimation of the ride-up component by the Ralston's equation. As a result of a cross comparison of those three total ice loads, it was found that the present approach can represent the experimental ice loads better than the Ralston's equation does. The reason may be attributed to a reasonable estimation of ride-up component. Above all, it is concluded that the proposed method can simulate total ice loads on a conical structure. And it is also concluded that the proposed method can separate the components consisting ice loads more realistically.

ACKNOWLEDGMENTS

This study was conducted as a part of the project for understanding the basic characteristics of ice forces on an offshore structure, which is organized and funded by Japan Offshore Industries Association (JOIA). The authors should express their sincere gratitude to JOIA for giving them an opportunity to participate this valuable project.

REFERENCES

Croasdale, K.R (1980). "Ice forces on fixed, rigid structures, In IAHR Working Group on ice Forces on Structures", a State-of-the-Art Report, T. Carstens, ed., CRREL Special Report 80-26.

Izumiyama, K, Kitagawa, H, Koyama, K and Uto, S (1992). "A Numerical Simulation of Ice-Cone Interaction", Proceedings, 11th IAHR Symposium on Ice, Vol. 1.

Hirayama, K and Obara, I (1986). "Ice Forces on Inclined Structures", Proceedings, 5th Int. Offshore Mechanics and Arctic Engineering Symposium, vol.4.

Kato, K (1986). "Experimental studies ice forces on conical structures", Proc., IAHR 8th Int. Symposium on Ice, vol.1.

Kato, K, Adachi, M, Kishimoto, H and Ichikawa T (1995). "Effects of Cone Angle and Intrusion Velocity on Ice Force on Conical Structures", Proc., 11th Cold Region Technology Conference (in Japanese).

Kato, K, Adachi, M, Kishimoto, H and Ichikawa T (1996). "Experimental Studies on Ice Forces on Conical Structures Using Segmented Model", Proc., IAHR 13th Int. Symposium on Ice, vol.1.

Ralston, T.D (1977). "Ice Force Design Considerations for Conical Offshore Structures", Proc., 4th Int. Symposium on Port and Ocean Engineering under Arctic Conditions, vol.2.

Ralston, T.D (1979) "Plastic Limit Analysis of Sheet Ice Loads on Conical Structures", Physics and Mechanics of Ice, Per Tryde ed.

Sorensen, C (1978). "Interaction between Floating Ice Sheet and Sloping Structure", Technical University of Denmark, Institute of Hydrodynamics and Hydraulic engineering, 175p

Tsutsumi, T, Ogiwara, S, Mizoguchi, S, Kato, K, Yokota, K, Ito, A, Akiyama, A and Ohgane, E (1991) "On IHI's New Experimental

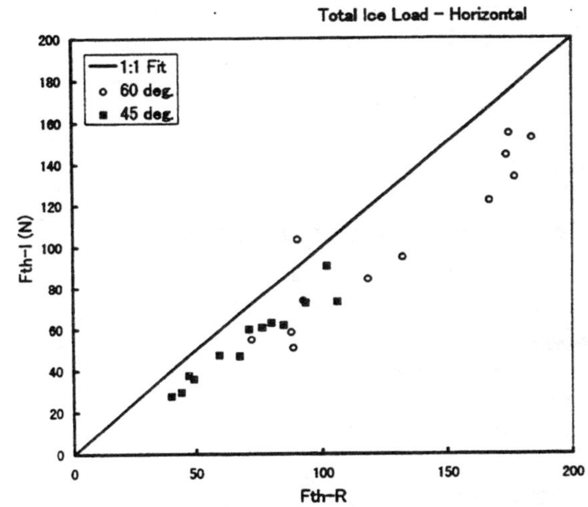

Figure 14-a Comparison of the Ralston's and simulated total ice loads - horizontal ice load

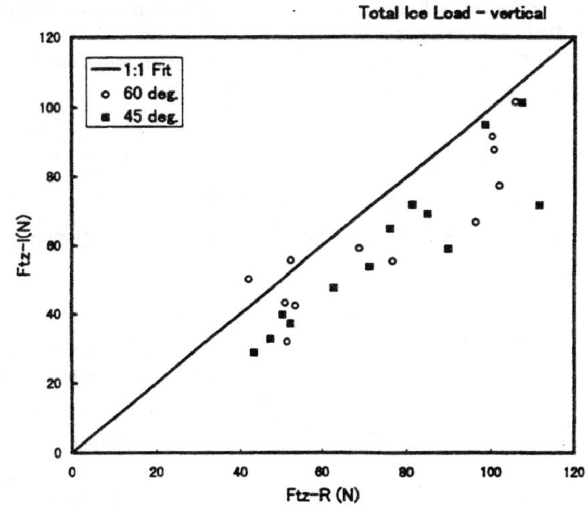

Figure 14-b Comparison of the Ralston's and simulated total ice loads - vertical ice load

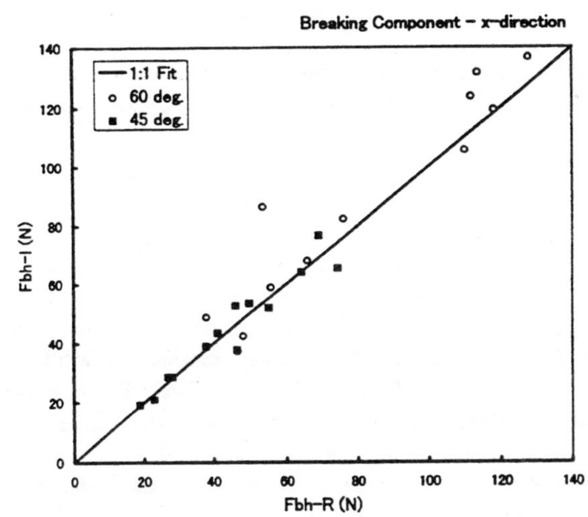

Figure 15-a Comparison of the Ralston's and simulated breaking components - X-direction

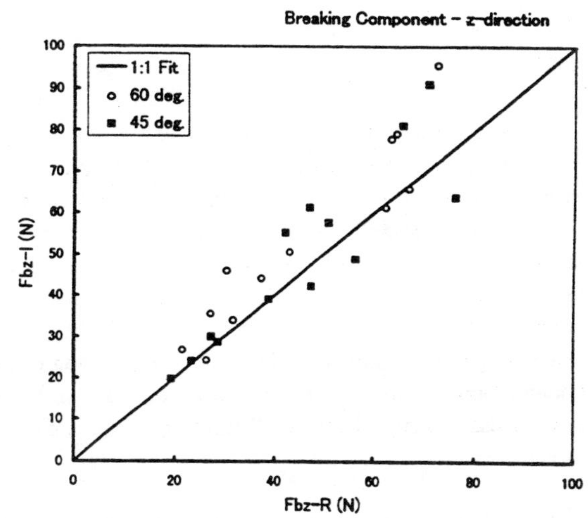

Figure 15-b Comparison of the Ralston's and simulated breaking components - Z-direction

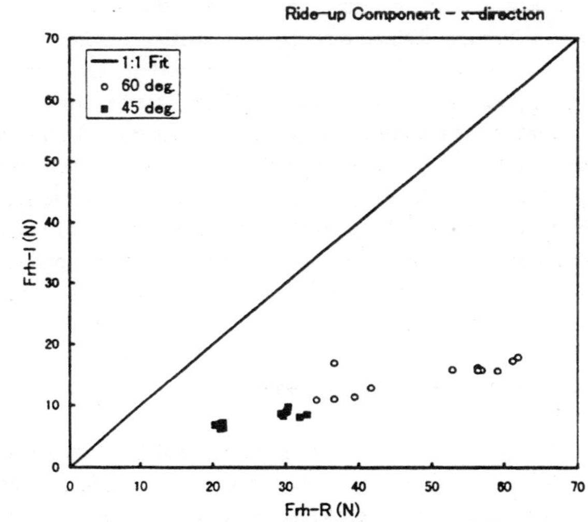

Figure 16-a Comparison of the Ralston's and simulated ride-up components - X-direction

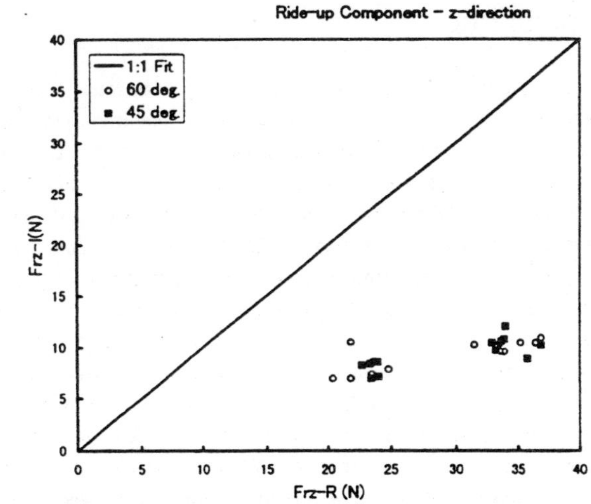

Figure 16-b Comparison of the Ralston's and simulated ride-up components - Z-direction

Proceedings of the Seventh (1997) International Offshore and Polar Engineering Conference
Honolulu, USA, May 25-30, 1997

ISBN 1-880653-28-1 (Set); ISBN 1-880653-30-3 (Vol. II)

Medium Scale Field Indentation Tests - Physical Properties and Strength of the Ice Sheet

H. Matsushita and T. Takawaki
Mitsui Engineering and Shipbuilding Co., Ichihara, Japan
K. Masaki, M. Hanada, H. Honda, A. Nishihata and H. Saeki
Hokkaido University, Sapporo, Japan
K. Hirayama
Iwate University, Morioka, Japan

ABSTRACT

Mechanical tests were conducted under around -2 ℃ with small size specimens prepared to obtain compressive and bending strength of the ice sheet facing the indentor, which is one year sea ice. Physical characteristics of the sea ice were summarized using various relationships as follows : density vs. salinity, compressive strength vs. density or salinity, compressive vs. bending strength, compressive strength vs. grain structure and inhomogeneity in compressive strength in the horizontal direction. In addition to these, the correlation of the results between indentation tests and mechanical tests is also discussed. And probability distribution of the compressive strength is concidered.

KEY WORDS: ice sheet, one year ice, physical properties, grain structure, mechanical strength, indentation test, probability distribution

1. INTRODUCTION

In establishing a model to estimate sea ice loads acting on structures, strength characteristics of ice sheets are important basic parameters. We found the strength characteristics of the ice sheets (compressive strength and bending strength) by conducting compression and bending tests using small specimens of test sea ice formed over a short period of time (3 - 25 days) to use in indentation tests at Notoro Lagoon. Both compression and bending tests were made at a seawater temperature of about -2 ℃ because the test ice was under water at the time of the indentation test. Basic physical properties of the sea ice (density, salinity and grain structure) and ice temperature were also measured. On the basis of these experimental results, we have clarified relationships between physical properties of sea ice and its strength characteristics. We paid particular attention to the grain structure of sea ice, as well as to relationships between the horizontal distribution of compressive strengths of sea ice and the distribution of local ice pressures obtained from the indentation test.

2. TEST METHOD

2.1. Strength test

Sixteen compressive specimens (ϕ 45 × L90 or ϕ 75 × L150 mm) and six three-point bending specimens (W45 × B45 × SL450 mm or W75 × B75 × SL750 mm) were taken from the fore part of the test sea ice (1.5 m wide or 3 m wide) for each test case of the indentation tests (Figure 2.1). To maintain ice temperatures (seawater temperature) at the time of indentation tests, the specimens were kept in a seawater bath after being processed, so that they could be taken out and used promptly in the strength tests. Each strength test was carried out on the same day as the indentation test or within 3 days at the longest.

An experimental device (Figure 2.2) was used to make the compression and bending tests. Increasing loads were applied to specimens at constant displacement rates by starting a motor. In the compression test, a displacement gauge attached to the load axis measured displacement of the specimens, while in the bending test, the gauge was set on the bottom of the specimens and measured deflection. Moreover, the loads were measured with a load cell, and relationships between loads and displacement and between loads and deflection were determined on an X-Y recorder. The compression and bending tests were carried out at a strain rate of 10-3/sec., and the maximum strength and Young's modulus of each specimen were obtained by determining relationships between loads and displacements.

2.2. Measurements of physical properties

The salinity, density, temperature, and grain structure of the test sea ice were determined. Samples for investigating the grain structure were taken from representative parts of the sea ice (upper, middle, and lower parts, in both horizontal and vertical directions). Proportions of columnar and granular crystals were determined from values of samples measured in the middle part of each test ice sheet used to investigate the grain structure.

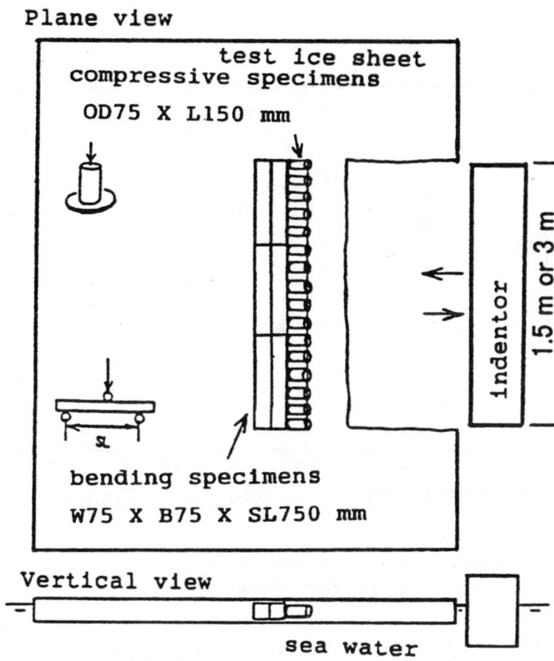

Figure 2.1 Location of specimens

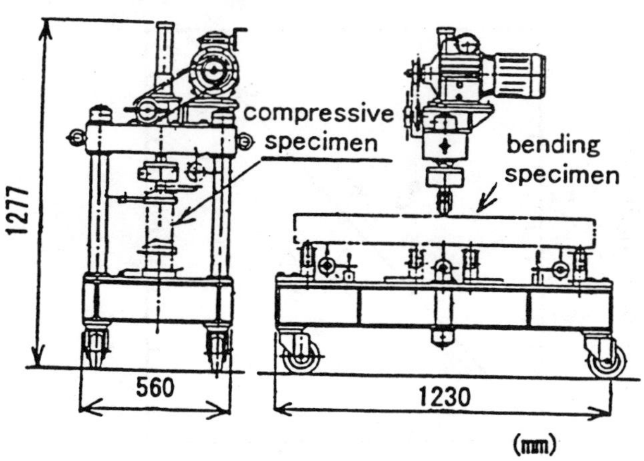

Figure 2.2 Experimental device

3. EXPERIMENTAL RESULTS AND DISCUSSION

3.1. Relationship between ice salinity and density

As shown in Figure 3.1, ice density and salinity have great dispersion, but a proportional relationship between them exists. Moreover, the figure shows that density and salinity are not related to the proportion of columnar crystals.

Proportion of Columnar crystals	100 %	50	20	0	–
mark	●	◒	⊖	○	△

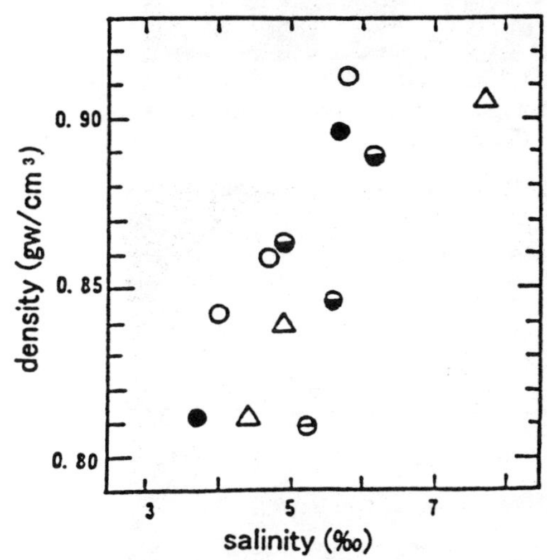

Figure 3.1 Relationship between sea ice density and salinity

3.2. Grain structure

Photographs of typical grain structure are shown in Figure 3.2 (a) to (c). The test sea ice consists of granular grain structure in the upper part and of columnar grain structure in the lower part.

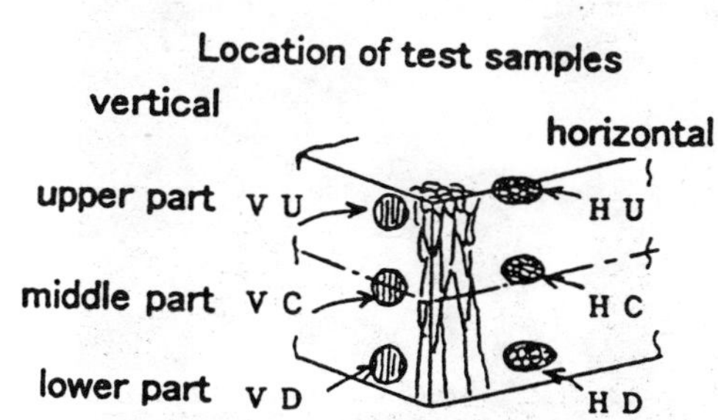

Figure 3.2 (a) Photograph of ice grain structure (T2-S3, VC)

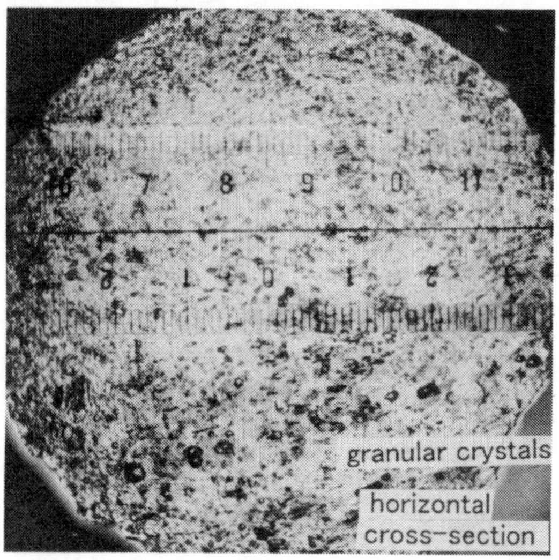

Figure 3.2 (b) Photograph of ice grain structure (T2-S3, HU)

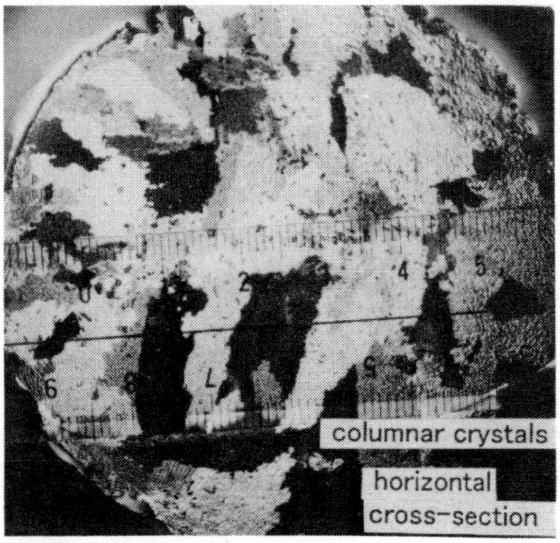

Figure 3.2 (c) Photograph of ice grain structure (T2-S3, HD)

3.3. Relationship between compressive strength, salinity and density

As shown in the study by Masaki et al.1), the compressive strength of drift ice and natural sea ice in Saroma Lagoon increases with ice density. In our experiment, however, the compressive strength of the test ice tended not to depend on salinity and density but to depend on the grain structure (the proportion of columnar crystals), as shown in Figure 3.3. This suggests that the grain structure of the ice was not consolidated because the ice was formed over a short period of time.

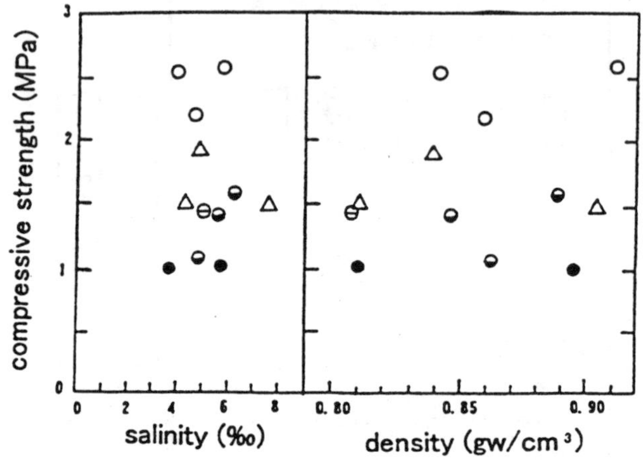

Figure 3.3 Relationships between ice compressive strength and salinity and density

3.4. Relationship between compressive strength and bending strength

Compressive strength is proportionately related to bending strength although the dispersion is large (Figure 3.4). The average compressive strength is about four times greater than the average bending strength.

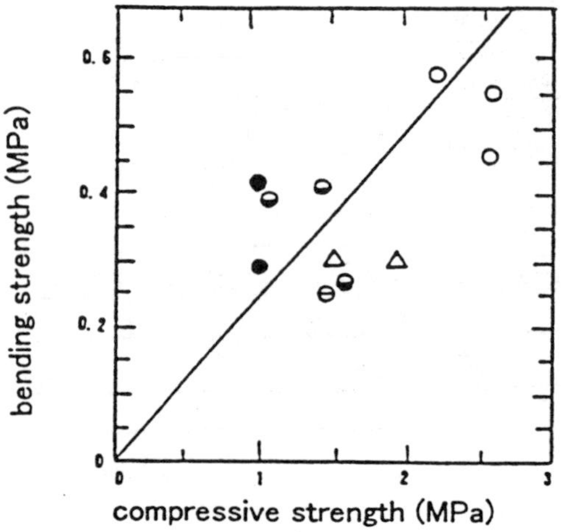

Figure 3.4 Relationship between ice bending strength and compressive strength

3.5. Young's modulus

A proportional relationship exists between Young's moduli of bending and those of compression although the dispersion is large (Figure 3.5). Furthermore, Young's moduli of bending have a larger dispersion than those of compression. Young's moduli were between 100 and 400 MPa at the time of compression, and between 100 and 600 MPa at the time of bending. As shown in Figure 3.6, ice grain boundaries are pushed in the compression test but are pulled in the bending test. Thus the deformability of the grain boundary directly influences the bending deformation. Because the test sea ice used in the experiment was formed over a short period of time, grain boundary strengths could be dispersed. As a result, Young's moduli of bending probably varied widely.

While Young's moduli of compression have a proportional relationship to the compressive strengths (Figure 3.7 (a)), there is not such a relationship between Young's moduli of bending and the bending strengths (Figure 3.7 (b)).

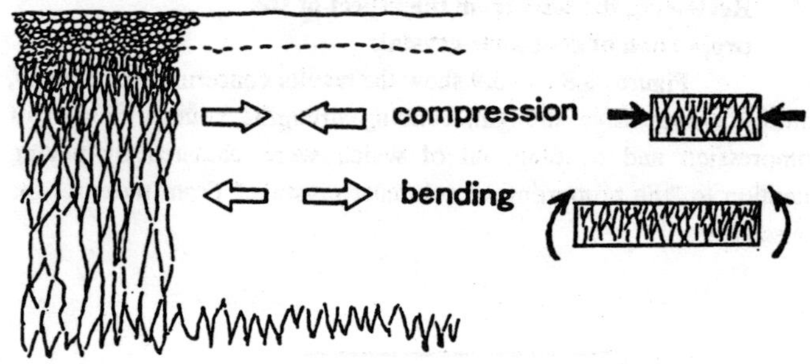

Figure 3.6 Relationship between load direction in compression and bending tests and ice grain structure

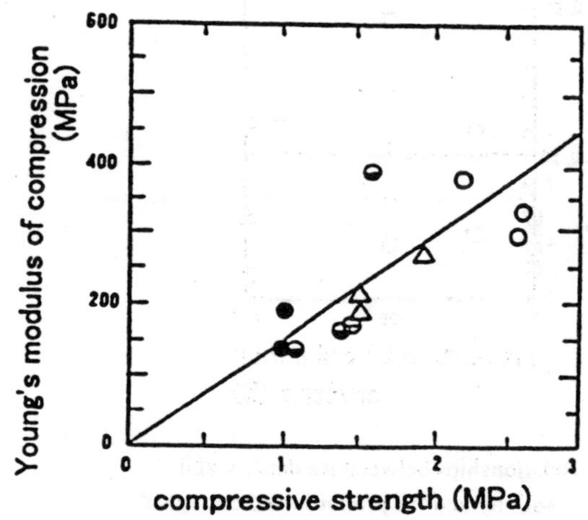

Figure 3.7 (a) Relationship between Young's modulus of compression and ice compressive strength

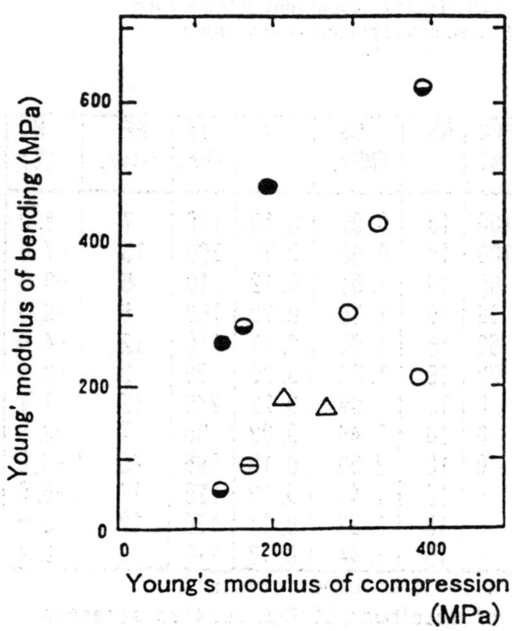

Figure 3.5 Relationship between Young's moduli of bending and of compression

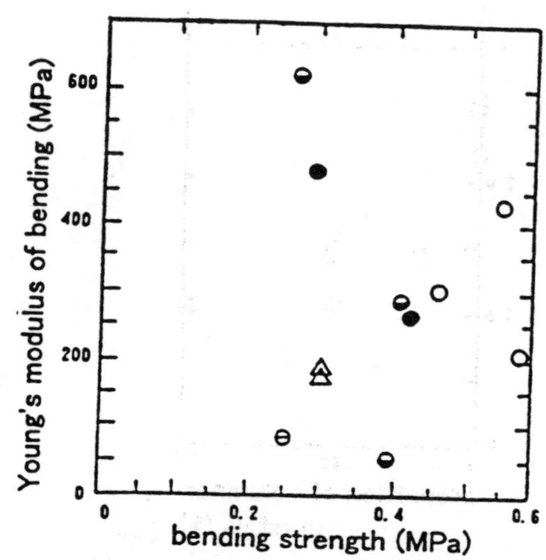

Figure 3.7 (b) Relatiionship between Young's modulus of bending and ice bending strength

3.6. Reviewing the tests from the aspect of the proportion of columnar crystals

Figures 3.8 and 3.9 show the results concerning ice density, salinity, compressive strength, bending strength, Young's moduli of compression and bending, all of which were obtained by paying attention to "the proportion of columnar crystals." From these results we see that:

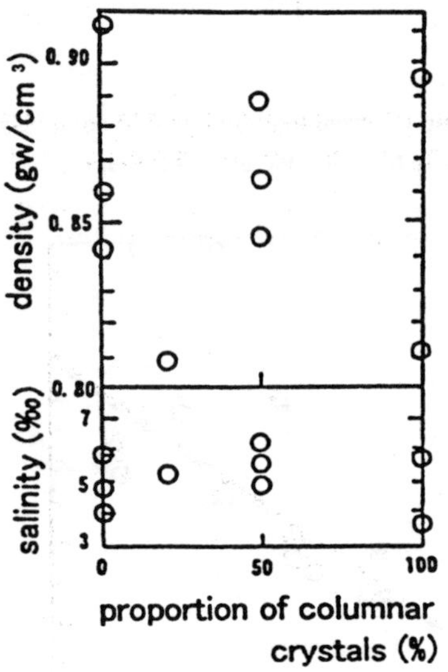

Figure 3.8 Relationships between ice density and salinity and the proportion of columnar crystals

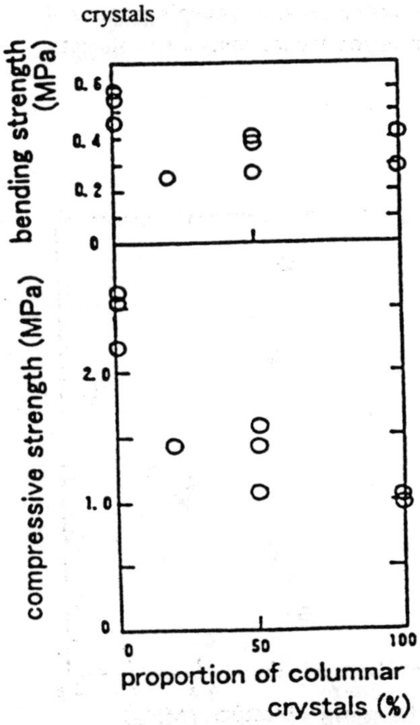

Figure 3.9 Relationships between ice bending strength and compressive strength and the proportion of columnar crystals

(1) Ice density and salinity are not related to the proportion of columnar crystals (Figure 3.8).

(2) Ice compressive strength decreases with increases in the proportion of columnar crystals (with decreases in the proportion of granular crystals) (Figure 3.9). This indicates that the compressive strength depends on crystal grain size, and that the compressive strength increases with decreases in grain size.

(3) In cases where the proportion of columnar crystals exceeds 0 % (100 % granular crystals), the bending strength is less than those found at 0 % columnar crystals and tends to be almost constant (Figure 3.9). This shows that the bending strength depends on the grain boundary strength of columnar crystals.

Here, the test results can be explained well by assuming that both compressive strength and bending strength are proportional to D-1/2 (D: crystal grain size). As the ratio of grain size of granular crystals (about 2mm) to that of columnar crystals (10 - 20 mm) is 1/5 - 1/10, the ratio of D-1/2 of granular crystals to that of columnar crystals is 2 - 3. This agrees well with the test results (ratio of compressive strength: 2.1 - 2.5; ratio of bending strength: about 2.0).

3.7. Dispersion of ice compressive strengths

Ice compressive strengths were dispersed, depending on the sampling point (Figure 3.10). As shown in Table 3.1 and Figure 3.11, coefficients of variation increase with an increase in the proportion of columnar crystals. The dispersion is thought to be due to differences in the crystal grain sizes, in the interfacial strengths of grain boundaries and in the volume of voids in the ice, all of which are caused by differences in mean temperature affecting the growth rate of ice. As shown in Figure 3.12, the proportion of columnar crystals

Table 3.1 Statistical data on conditions of test sea ice formation and compressive strength

TEST SEA ICE	PCC (%)	NS	CS (MPa)	CV	IT (mm)	PF (day)	MT (℃)
T2-S4	100	16	0.97	0.19	120	7	-8.0
T4-S8	100	16	0.96	0.33	200	13	-7.0
T2-S3	50	16	1.51	0.12	110	5	-6.0
T3-S5	50	16	1.35	0.22	150	8	-6.5
T2-S3-2	50	16	1.02	0.11	110	12	-4.0
T1-S1	20	16	1.39	0.29	65	3	-8.4
T4-S7	0	15	2.09	0.22	240	12	-7.1
T2-S30	0	16	2.46	0.08	95	7	-3.5
T5-S9	0	16	2.50	0.13	195	25	-6.6
T2-S4-2	-	16	1.43	0.12	120	12	-5.0
T3-S6	-	16	1.83	0.11	120	12	-5.5
T3-S30	-	15	1.44	0.17	115	6	-8.4

PCC:Proportion of columnar crystals,
NS:Number of specimens, CS:Compressive strength,
CV:Coefficient of variation, IT:Ice thickness
PF:Period of formation, MT:Mean temperature

increases with decreases in mean temperature. Moreover, we see that test ice sheets which were formed more rapidly at low mean temperatures have a higher coefficient of variation (Figure 3.13).

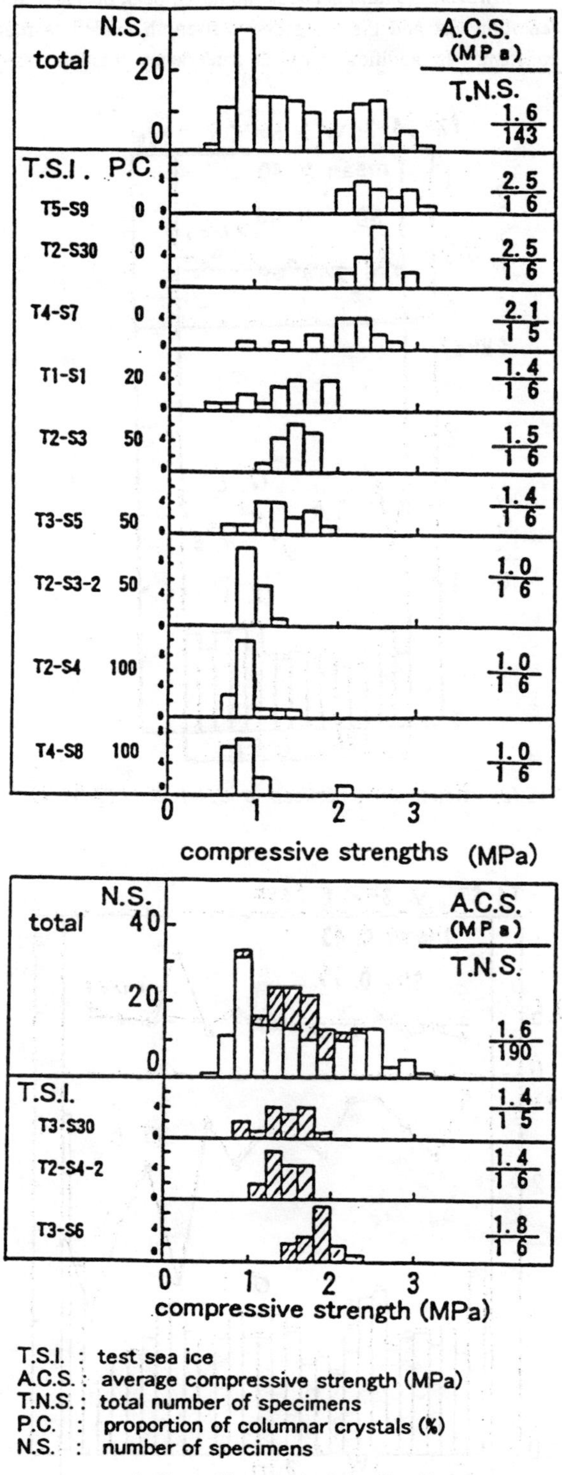

Figure 3.10 Dispersion of compressive strengths number of specimens

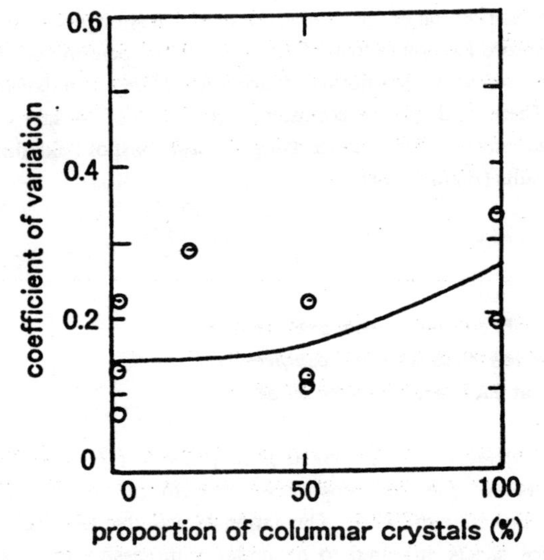

Figure 3.11 Changes in coefficient of variation due to the proportion of columnar crystals

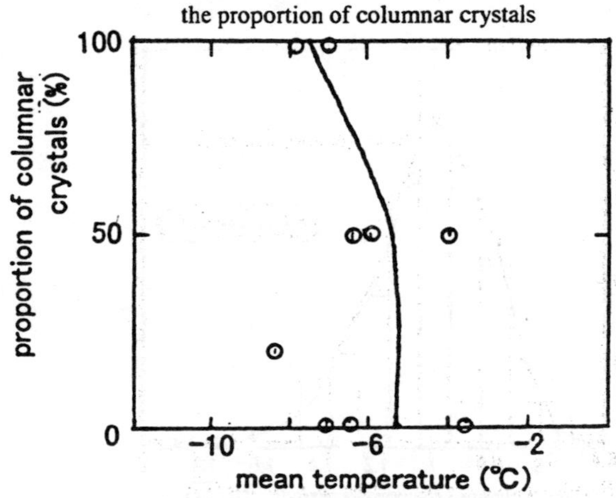

Figure 3.12 Changes in the proportion of columnar crystals due to changes in mean temperature

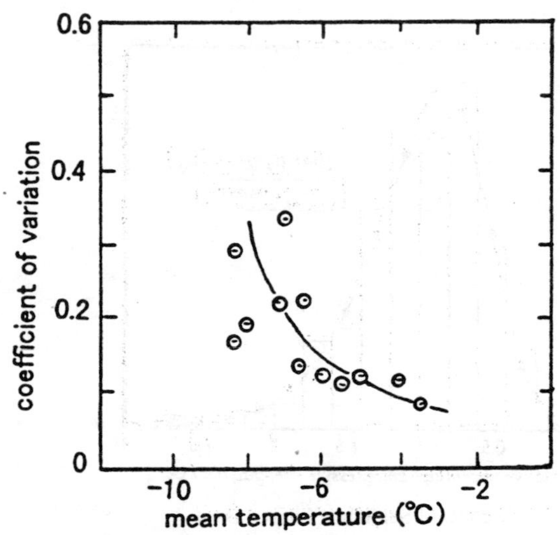

Figure 3.13 Changes in coefficient of variation due to changes in mean temperature

The distribution of the relative compressive strengths of the test sea ice in Notoro Lagoon (a total of 190 specimens), determined in the experiment, follow a log-normal distribution (three population parameters). These findings are consistent with those of Masaki et al.1). Investigations of natural sea ice along the east coast of Sakhalin show similar results (Figure 3.14).

3.8. Relationship between initial peak pressure during indentation tests and compressive strength at each measurement point

We believed that the initial peak pressure (σ 1) during indentation tests and the ice compressive strength (σ c) at each measurement point are correlated. Our rationale was that the middle parts of test ice sheets appeared to be under compressive pressure during indentation tests similar to that caused by compression tests

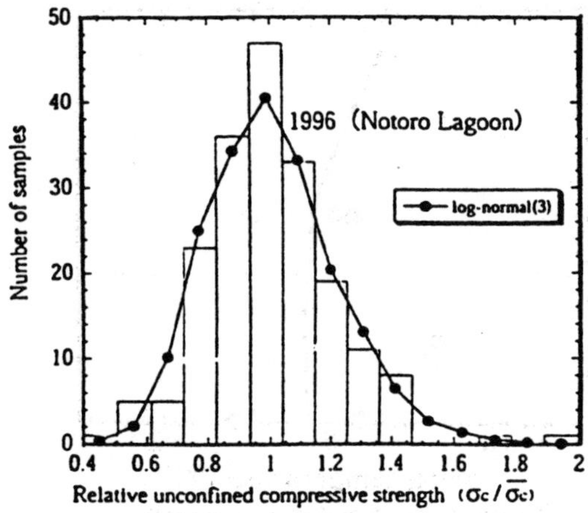

(a) Okhotsk Region (Notoro Lagoon)

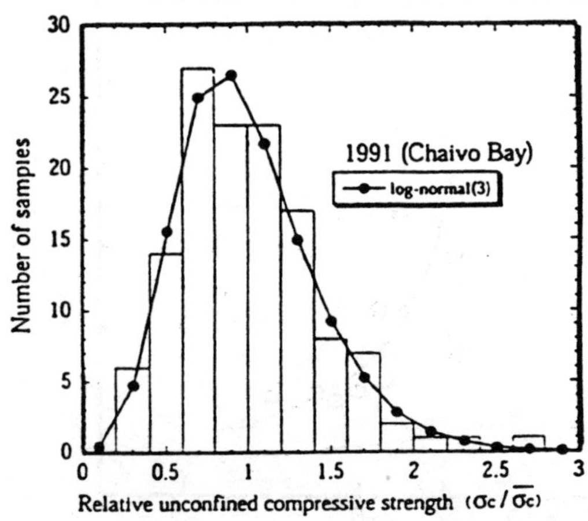

(b) Northeastern coast of Sakhalin (Chaivo Bay)

Figure 3.14 Frequency distribution of unconfined compressive strengths 1)

To confirm this assumption, we plotted σ c and σ 1 at each measurement point (Figure 3.15 (a) to (c)). The resulting correlation coefficients in each test case are shown in Table 3.2. From these figures, a clear correlation between σ c and σ 1 at each measurement point is not observed.

However, the ratio between the initial peak pressure during each indentation test and the compressive strength tended to remain almost constant. In addition, there is a tendency for this ratio to increase

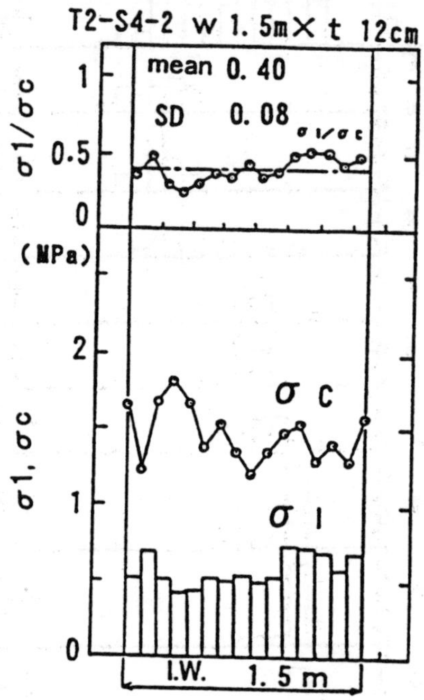

Figure 3.15 (a) Relationship between σ c and σ 1 (T2-S4-2)

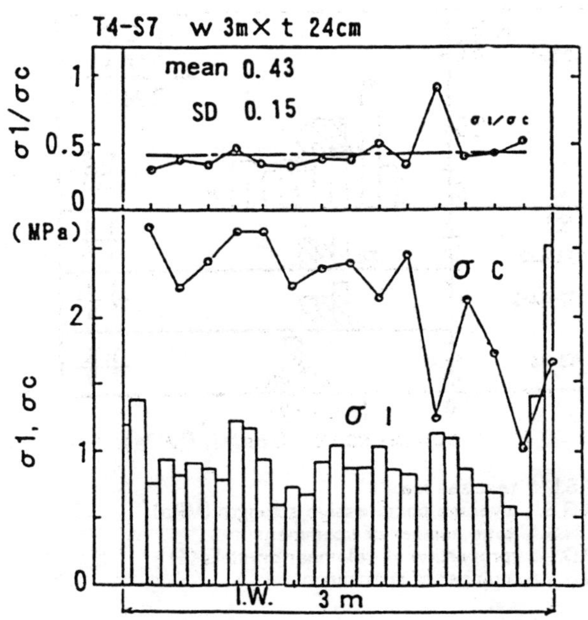

Figure 3.15 (b) Relationship between σ c and σ 1 (T4-S7)

with an increase in ice thickness (Figure 3.16). This pattern seems to indicate that compressive components of the pressure during indentation tests increase with an increase in ice thickness.

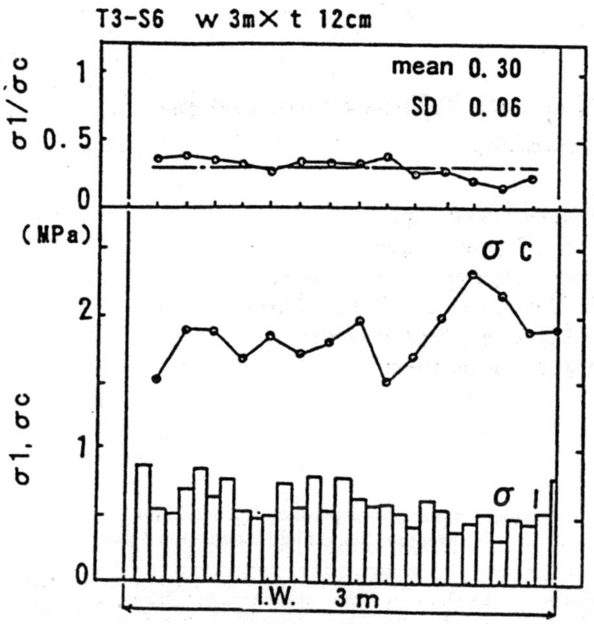

Figure 3.15 (c) Relationship between σ_c and σ_1 (T3-S6)

Table 3.2 Correlation coefficients of initial peak pressure (σ_1) during indentation tests and compressive strength (σ_c)

Test case	Correlation coefficient
T2-S4-2	0.377
T1-S1	0.247
T4-S7	0.304
T3-S6	0.195

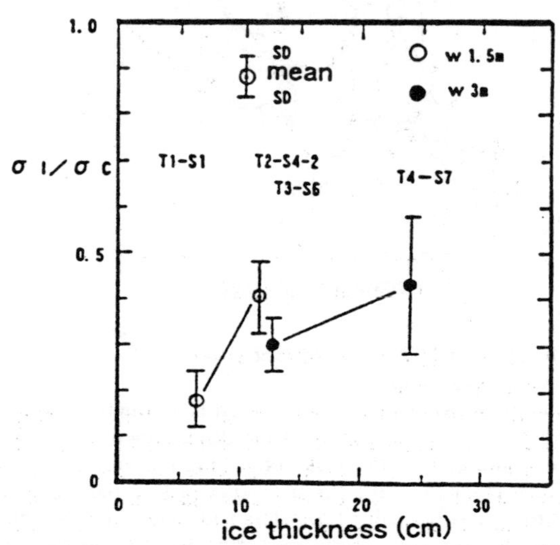

Figure 3.16 Relationship between σ_1/σ_c and ice thickness

4. CONCLUSIONS

An experiment to estimate ice loads acting on structures was carried out, using test sea ice formed in Notoro Lagoon over a short period of time (3 - 25 days) for indentation tests. Strength characteristics (compressive strength, bending strength, etc.) and physical properties (density, salinity and grain structure) of the ice were investigated at an ice temperature of about -2 ℃, and we found correlations between them. The main results are as follows:

(1) Ice density and salinity do not greatly influence ice compressive strength at about -2 ℃.

(2) The compressive strength is about four times greater than the bending strength.

(3) Compressive strength decreases with increases in the proportion of columnar crystals (with decreases in the proportion of granular crystals). This indicates that compressive strength depends on crystal grain size.

(4) In cases where the proportion of columnar crystals exceeds 0 % (100 % granular crystals), the bending strength is less than those found at 0 % columnar crystals and tends to be almost constant. This indicates that the bending strength depends on the grain boundary strength of columnar crystals.

(5) Dispersion in compressive strength occurs due to differences in sampling points in the horizontal direction.

(6) Sea ice that formed more rapidly at low mean temperatures has a higher coefficient of variation of compressive strength.

(7) A clear correlation does not exist between the initial peak pressure during ice indentation tests and the compressive strength at each measurement point. However, the ratio between initial peak pressure and compressive strength is almost constant for each test ice sheet. Also, this ratio tends to increase with an increase in ice thickness.

This study was conducted in 1996 as part of a larger "study of ice loads acting on offshore structures" (Chairman: Prof. Hiroshi Saeki at Hokkaido University) funded by the Japan Ocean Industries Association (JOIA). We will continue to conduct experiments in 1997 to clarify the strength characteristics of sea ice in Notoro Lagoon, and to compare results with those of the indentation test.

* Reference

1) Masaki, T. Okubo, Y. Honda, H Otsuka, N. and Saeki, H. (1996) "Characteristics of Unconfined Compressive Strength of Sea Ice in the Sea of Okhotsk", *Proceedings of The 13th International Symposium on Ice,* IAHR 96, Beijing, China, Volume 1, pp 72-79.

Proceedings of the Seventh (1997) International Offshore and Polar Engineering Conference
Honolulu, USA, May 25-30, 1997

ISBN 1-880653-28-1 (Set); ISBN 1-880653-30-3 (Vol. II)

Medium-Scale Field Indentation Tests (MSFIT) - Ice Failure Characteristics in Ice/Structure Interactions

Takahiro Takeuchi, Shimizu Corporation, Tokyo, Japan
Takaharu Masaki, Hokkaido University, Sapporo, Japan
Satoshi Akagawa and Muneo Kawamura, Shimizu Corporation, Tokyo, Japan
Naoki Nakazawa and Takashi Terashima, Pacific Consultants Co., Tokyo, Japan
Hideki Honda and Hiroshi Saeki, Hokkaido University, Sapporo, Japan
Ken-ichi Hirayama, Iwate University, Morioka, Japan

ABSTRACT

Medium-Scale Field Indentation Tests (MSFIT) including ice indentation tests, tests on physical properties of ice sheets and image sensing on ice deformation have been conducted since Jan. 1996 as part of a JOIA project. The purpose of these tests is to understand the changes in ice load on structure with respects to indentation area between ice and structure. This paper reportes the plan of this project and the results of ice indentation tests conducted in the winter of 1996.

Keywords: Non-simultaneous failure, ice indentation test, ice load

1. INTRODUCTION

The Japan Ocean Industries Association (JOIA) started a five-year program, "A Study of Ice Loads Acting on Offshore Structures" (heads of the committee: Prof. Hiroshi Saeki of Hokkaido University and Prof. Ken-ichi Hirayama of Iwate University) in 1993 as an enterprise consigned by the Ministry of International Trade and Industry (M.I.T.I.). The project attempts to understand domestic and foreign research on frozen sea technology through a survey on trends of this technology. It also attempts to establish a new method of estimating ice loads through various systematic test results of interactions between offshore structures and ice sheets by indoor water tank tests and field experiments. The purpose of MSFIT is to investigate the relationship between indentation ice pressure and indentation area, and ice failure characteristics in indentation tests(Table 1). The tests consisted mainly of indentation tests, as well as ice physical property tests and image processing. In the indentation tests with a maximum planned ice-structure contact area of 1.44 m^2, we observed the ice loads during interaction between an ice sheet and a structure and the failure characteristics of ice sheets. The physical tests clarified the strengths of ice sheet and its crystal structures. The image processing identified the failure modes of ice sheets and the distorted area within them observed in the indentation test. The tests were conducted inside the harbor of Notoro fishing port, Notoro Lake, Hokkaido, connected to the sea of Okhotsk (Fig.1). The salinity of the lake is almost the same as that in the sea of Okhotsk. This paper reports the ice failure characteristics and ice load from the results of indentation tests between Jan. and early March 1996.

Table 1 Summary of Planned test series for Medium-Scale Field Indentation Tests

Year conducted (Planned)	Indentor width W(m)	Ice thickness h(cm)	Aspect ratio W/h	Indentor velocity V(cm/sec)	Indentation depth (cm)
Jan.-Mar.1996	1.5	6-12	12.5-25.0	0.3	100
Jan.-Mar.1996	3.0	12-24	12.5-25.0	0.6	100
Jan.-Mar.1997	4.5	12-18	25.0-37.5	0.9	100
Jan.-Mar.1997	6.0	12-24	25.0-50.0	1.2	100

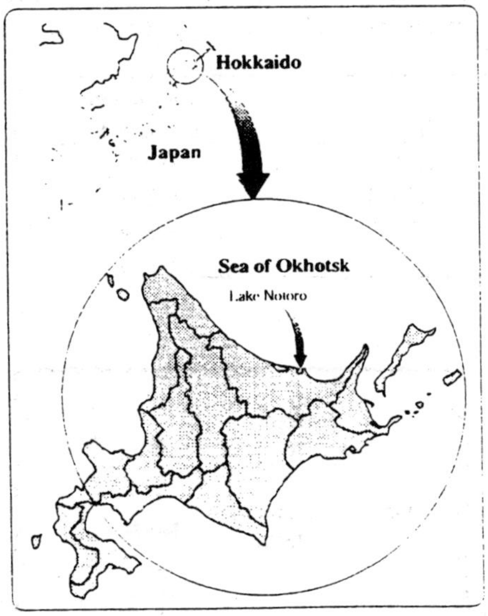

Fig.1 Location of test sites

2. EXPERIMENTAL SETUP AND PROCEDURES

2.1 Experimental Apparatus

A schematic drawing of the indentation test structure setup is shown in Fig.2. This apparatus is of the displacement controlling type and has a hydraulic servo-jack controlling the velocity of the model structure attached to the tip of a servo-jack in the ice sheet. The model structure is basically 1.5 meters wide and can be widened to a maximum of 3.0 meters. Total weight of the test apparatus for jack is around 18tonf. The maximum force capability of the servo-

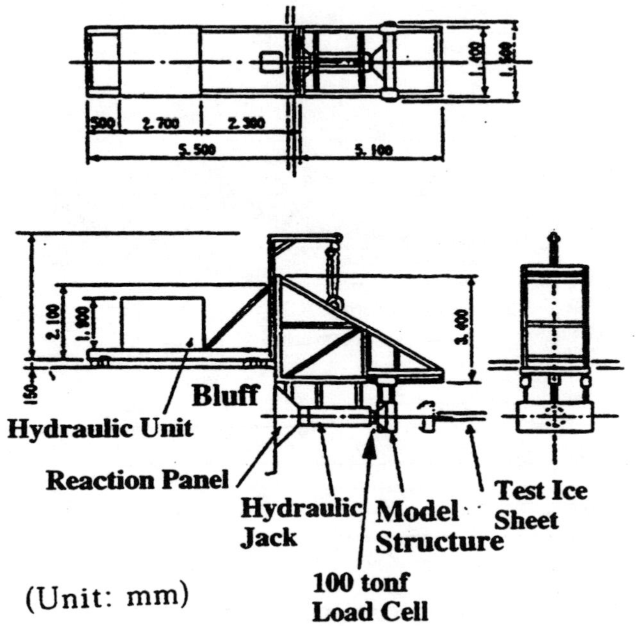

Fig.2 Schematic drawing of indentation test structure setup.

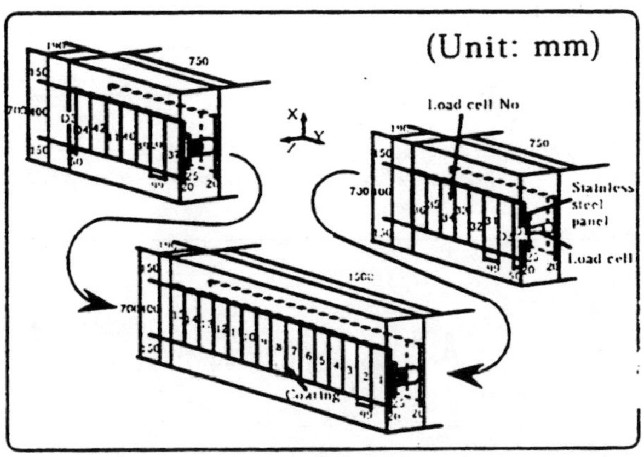

Fig.3 Shematic diagram of model structure.

jack was 100 tonf. The capability of the indentation velocity ranged from 0.1 cm/sec to 3.0 cm/sec, and the maximum stroke of the model structure is 120cm. Since the inside of the harbor where the tests were conducted is influenced by tides, which change the level of ice sheets, the vertical location of the servo-jack in the apparatus is adjustable. A schematic diagram of the model structure is shown in Fig.3. The model structure had load cells for local load measurements attached behind surface-treated panels made of stainless steel. The panels are completely separated from other. The surface roughness of the panels was within 1mm.

2.2 Testing Procedures

Tests were conducted along several tracks in the field ice sheet as shown in Fig.4. Numbers 1 through 8 in the figure show the positions of newly made ice sheets used for the tests. The ice sheets had a plastic rod with a thermocouple attached to measure the ice temperature in the direction of ice thickness. The ice sheets were prepared of removing snow, until the ice sheet reached the desired thickness. Then, a heat-insulating material was put on the surface to prevent the growth of the skeleton layer of the ice sheet. The indentation tests were conducted after moving the apparatus for the indentation tests, a mobile measurement house on a truck and a

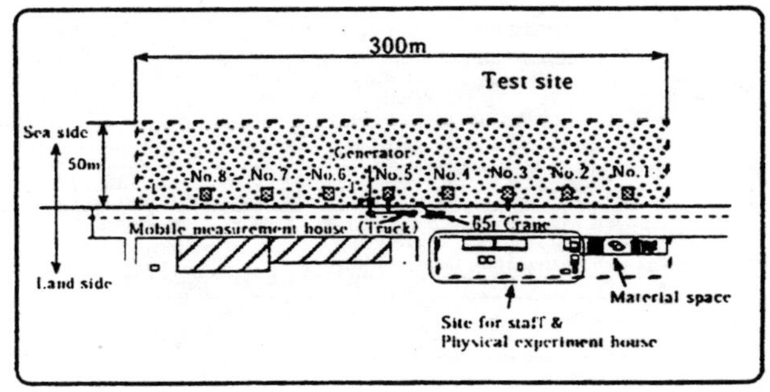

Fig.4 Locations of test facilities and newly formed ice sheets.

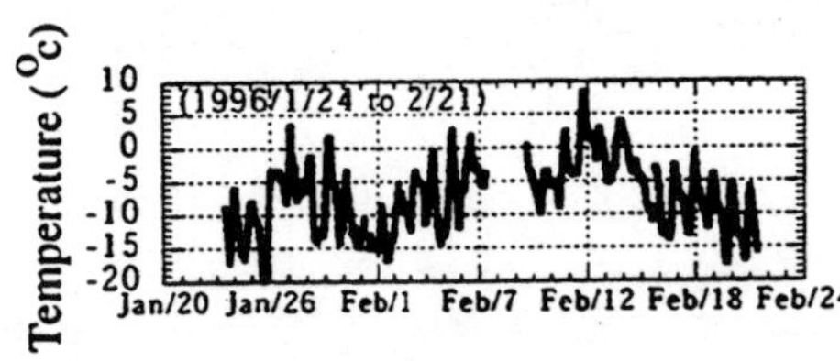

Fig.5 Air temperatures during formation of test ice sheets

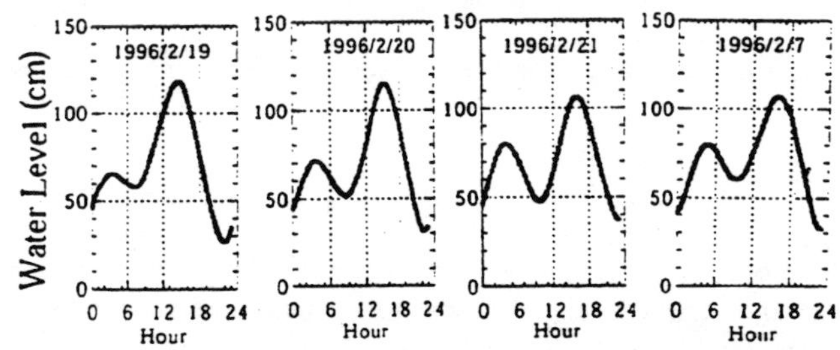

Fig.6 Tidal conditions of test sites on indentation test date.

generator, to the predetermined test site by the day before the tests. Immediately before the tests, we adjusted the positions of the model structure and the leading edge of the ice sheet so that they were parallel to each other, prepared the video taping equipment (including image processing apparatus), with connected the displacement meter to the model structure by a wire and started to measure the ice temperature in the ice thickness direction. The indentation tests were conducted as much as possible during the mornings when the temperature was low. The measurement devices for the physical tests were housed in a physical test measurement room. After the indentation tests, ice specimens were cut out from the test ice sheet at the tested position, and cylindrical shapes of ice specimens (height 15cm, diameter 7.5cm) were formed on land for physical tests. After the indentation tests, the flakes of destroyed ice and the patterns of ice sheet failure were observed. The shapes of the leading edge of ice sheets were measured to evaluate the roughness and phase of local ice failure of the test ice sheets. Sieves were also used to measure the size of ice fragments for later analysis. These procedures were repeated for each test.

2.3 Environmental and Test Conditions

The test period was from January through March 1996. The test velocity V was determined as being almost the same as the indentation rate $V/2W = 10^{-3}$ sec^{-1} (where W = indentation width). The measured items were total ice load, local ice load, moment with respect to y-direction (Fig.3), velocity, acceleration and inclination of model structure, failure made of ice sheet, ice thickness and ice temperature. Air temperatures under the process of forming test ice sheets are shown in Fig.5. Changes in water levels at the test sites due to tidal changes are aloso shown in Fig.6 for the test dates. Cracks usually occurred near the bluff. The locations of these cracks were considered to ensure that they did not to occur inside test ice sheets.

Table 2 Summary of test results of Jan.~Mar. 1996.

Test no.	Widh of model structure W(m)	Ice Thickness h(cm)	Indentation Velocity V(mm/sec)	First peak Ice Force (kN)	Subsequent Ice Force(kN)		
					Max	Mean	Standard Deviation
T-1	1.5	6.5	3.03	17.10	7.35	1.97	1.45
T-2	1.5	12.0	3.04	44.58	40.43	25.35	4.84
T-3	3.0	12.0	6.08	126.35	105.55	71.86	9.60
T-4	3.0	24.0	6.11	573.75	303.17	181.38	50.12

Test no.	Compressive Strength (MPa)	Contained ratio of columnar ice (percent)	Salinity (ppt)	Density (g/cm^3)	Date of test	Date of start of forming ice sheet
T-1	1.39	20	5.2	0.81	1996/2/19	1996/2/16
T-2	1.43	--	4.4	0.81	1996/2/20	1996/2/7
T-3	1.83	--	4.9	0.84	1996/2/21	1996/2/8
T-4	2.10	0	4.7	0.86	1996/2/7	1996/1/24

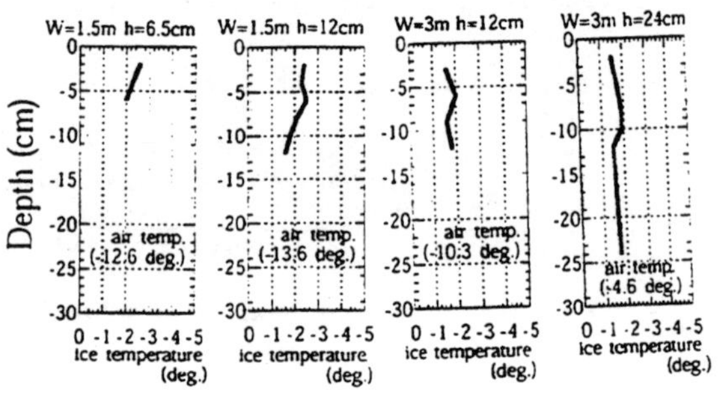

Fig.7 Ice temperatures in the direction of ice growth for each test.

3. TEST RESULTS

Four test results are reported as shown in Table 2. Ice temperatures in the direction of ice thickness were measured as described in Fig. 7. The ice temperature in each test was between -1.5 and -2.6(°C) , and its gradient was small. It is close to the sea water temperature (-1.7~-1.8 °C).

3.1 Ice Load

(1)Total ice load

The results of total ice load, model structure displacement, acceleration and inclination with respect to elapsed time are shown in Figs 8, 9, 10, and 11. All data were acquired at sampling frequencies of 50kHz for planned V=3mm/sec, and 100kHz for planned V=6mm/sec. The total ice loads acting on the model structure are evaluated as a total of the values measured with a load cell for a local load measurement. The time series of the total ice load shows an increase at the start of indentation with a high peak value, called the first peak force, which depends on the contact condition between the leading edge of ice sheet and the model structure at the start of the motion. After the first peak force occurred, a force less than the first peak force with small variation also occurred, called the subsequent force, which is related to the local failure of the ice sheet. The actual displacement with respect to elapsed time was close to the values previously controlled by a hydraulic servo-jack. The inclination of the model structure and acceleration was small, except those of the test of 7th of Feb., due to the smaller ice load.

(2) Local Ice Load

Among the load cells for local load measurement, cells No.1 to No.15 (Fig.3) measured three components (X, Y, Z-dir., Moment on Y), and cells No.36 to No.42, D1 to D4 measured only one vectorial component. Fig.12 shows an example of the local ice loads measured in the Z direction for the test case of Feb. 20th (W=1.5m, h=12cm). In the figures, the first peak force of the total ice load (Fig.9) coincided with the peak of the local ice pressure. After this peak, more than one peak (forces) acted randomly on each panel. These peaks did not occur simultaneously among those panels. This implies that ice failure mode changes from simultaneous to non-simultaneous near the occurrence of the first peak force.

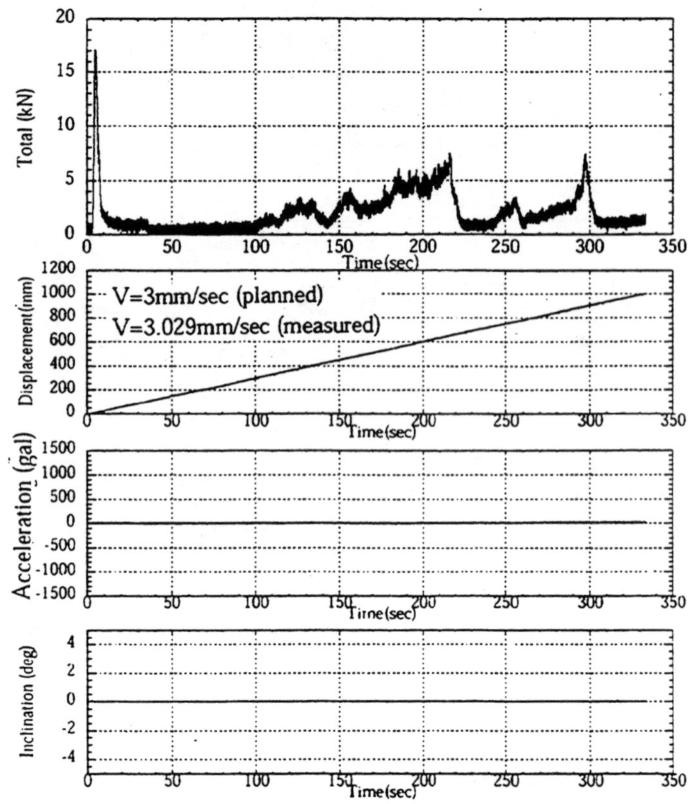

Fig.8 Plots of total ice force, displacement, acceleration and inclination with respect to time.(W=1.5m, h=6.5cm)

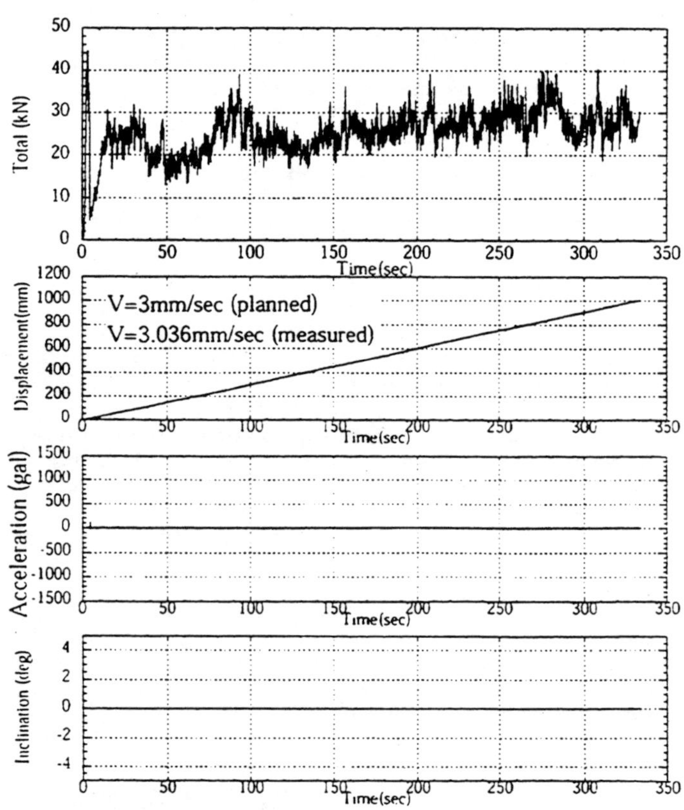

Fig.9 Plots of total ice force, displacement, acceleration and inclination with respect to time.(W=1.5m, h=12cm)

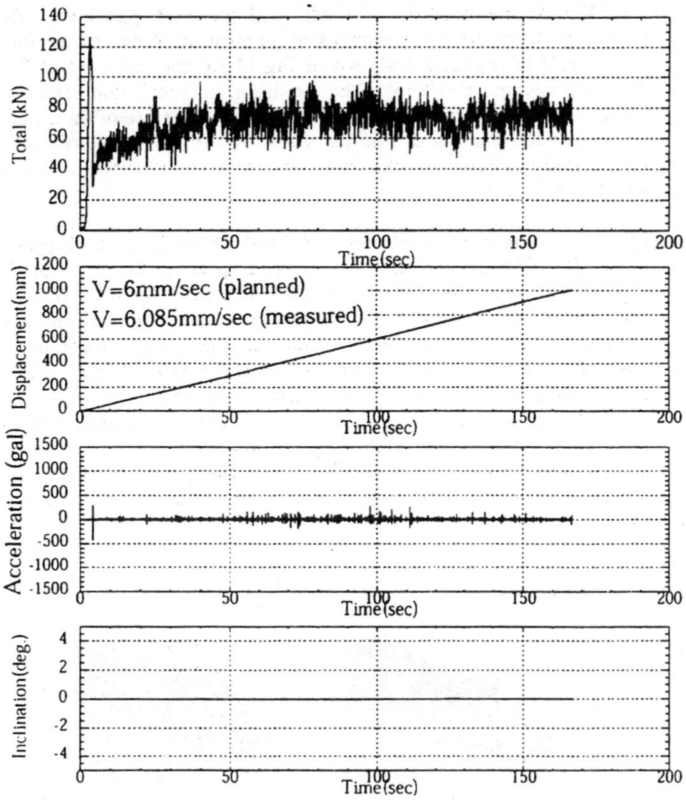

Fig.10 Plots of total ice force, displacement, acceleration and inclination with respect to time.(W=3m, h=12cm)

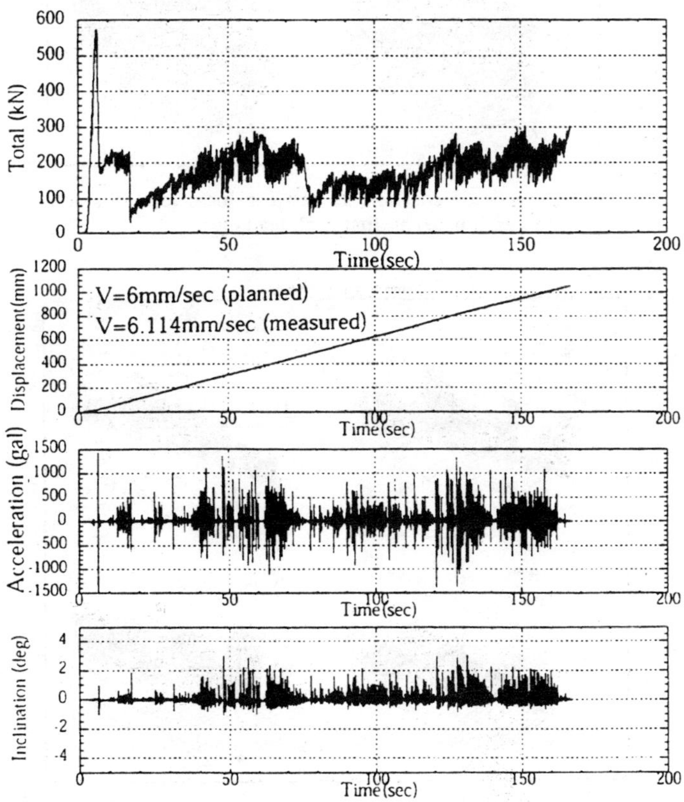

Fig.11 Plots of total ice force, displacement, acceleration and inclination with respect to time.(W=3m, h=24cm)

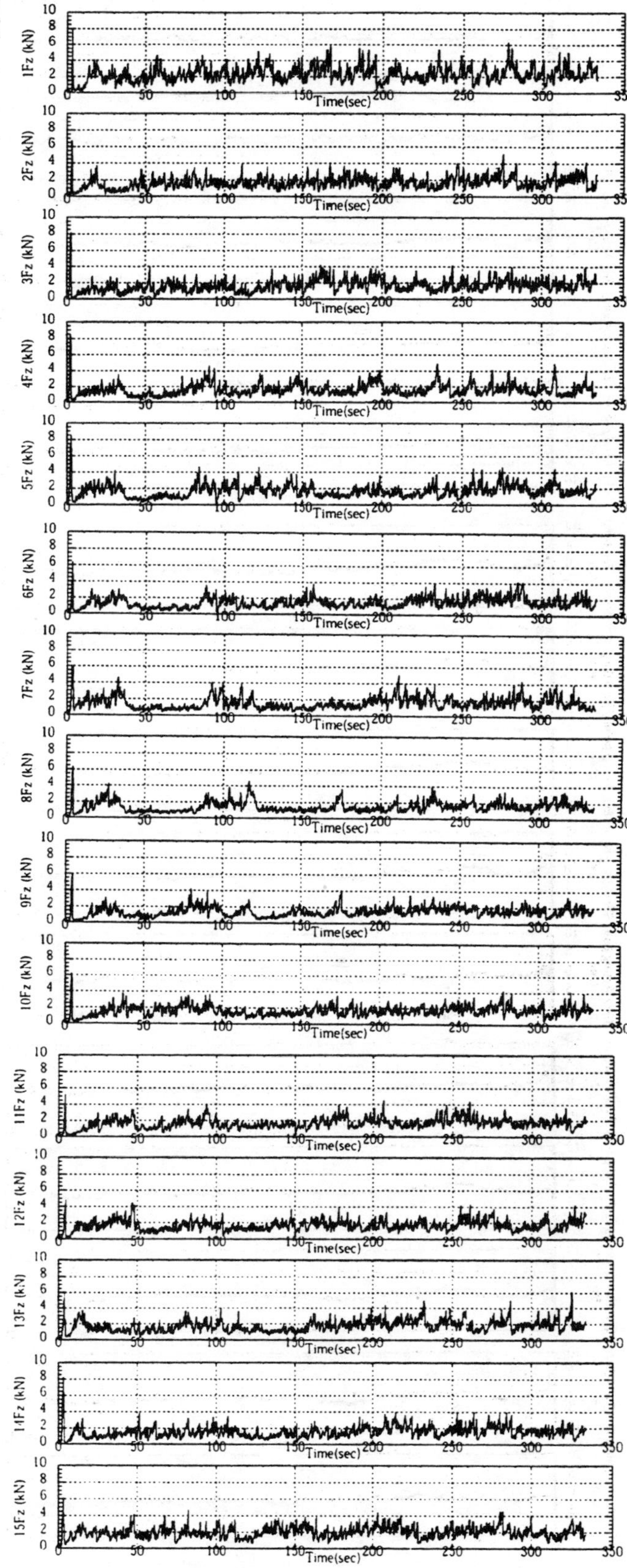

Fig.12 Plots of local ice loads in the Z direction (for W=1.5m, h=12cm).

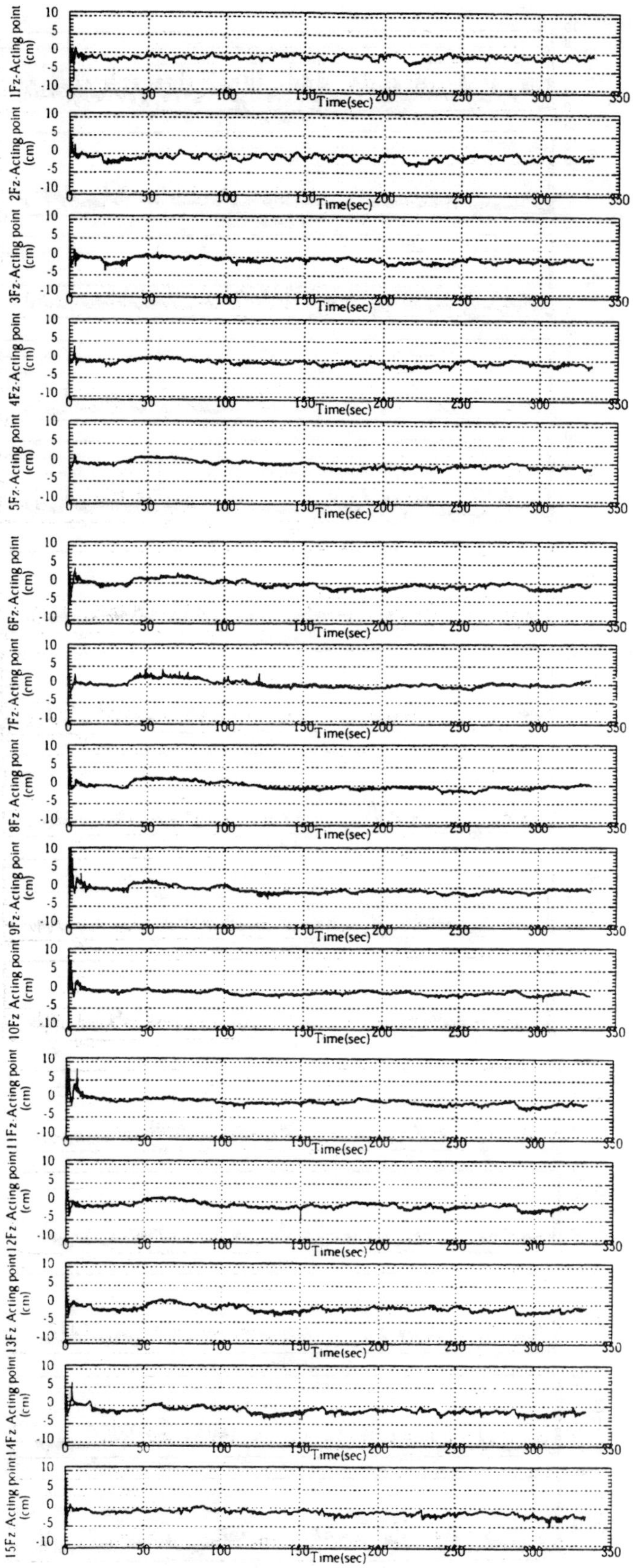

Fig.13 Plots of local ice loading points in the Z direction (for W=1.5m, h=12cm).

The acting position of local load for each panel was also calculated from the measurements of moment on Y-dir and ice force in Z-dir. The results are shown in Fig.13 for the test case of Feb. 20th (W=1.5m, h=12cm). The position of local load shows a variation of a few cm, and this is related to the nature of the contact between each panel and the ice sheet. In turn, the occurrences of both wedge-shape cracks may be non-simultaneous, during subsequent force ranges.

(3) Correlation coefficients between local ice loads on each panel

We first looked at the time course from the start of the indentation until the first peak loads occurred in the time series of the total ice loads, and then investigated the simultaneous and non-simultaneous failure of the ice sheet at the front surface of the panel by calculating the correlation coefficients between the local ice loads acting on each panel during this time duration. The absolute values of the correlation coefficients for simultaneous failure were around one. However, for non-simultaneous failure, the values were around zero (see Fig.14). Most of these values are larger correlation coefficients, indicating simultaneous failure of the ice sheet.

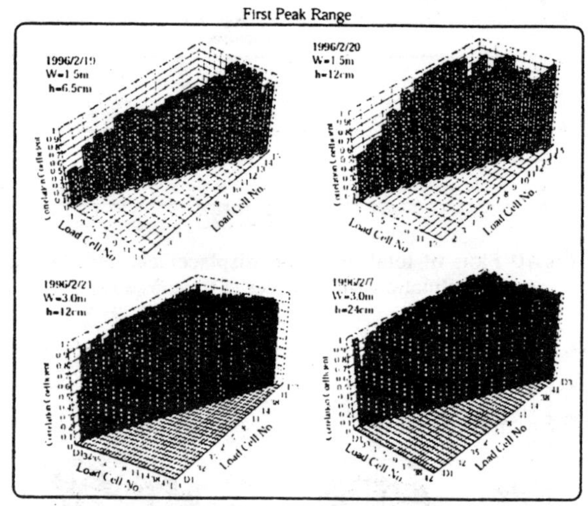

Fig.14 Correlation coefficients between local ice loads acting on each panel at first peak loading

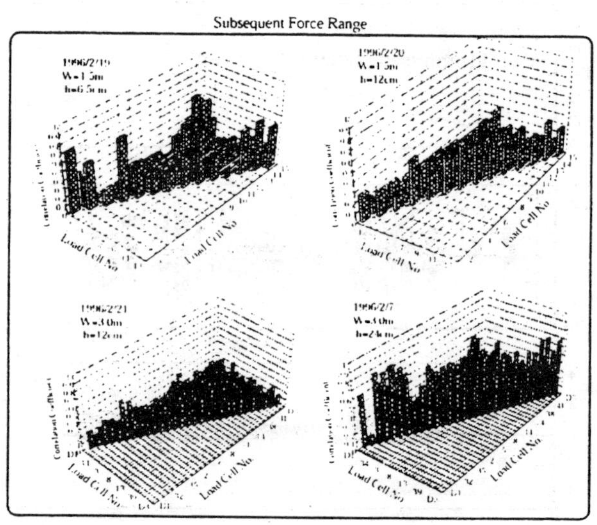

Fig.15 Correlation coefficients between local ice loads acting on each panel during subsequent loading

Similarly, the correlation coefficients were calculated between the local ice loads acting on each panel (see Fig.15). They were much closer to zero than those of Fig.14. The correlation coefficients of thicker ice (Fig. 15) were relatively large for the neighboring panels (diagonal parts of Fig.15, 1996/2/7), and the ice sheets at the front surface of the neighboring panels had almost simultaneous failure, indicating that the ice thickness is related to the width of the area where independent ice failure occurs.

3.2 Failure of Ice Sheet

In general, the failure modes of ice sheet can be classified according to the model structure velocity, the width of the model structure and the ice thickness (Timco 1986). Most of the ice sheet failure modes observed during continued ice forces in this test correspond to spalling (flaking, wedge) failure and crushing failure with horizontal cleavage cracks as observed in a previous study (Saeki and Ozaki, 1980). These failures occurred independently in each area in the directions of the width of the model structure, which is the same as the results of previous small-scale indentation tests (Takeuchi and Saeki, 1995, W is less than 1 m). Since the mode of ice sheet failure during the first peak force is influenced by the shape of the leading edge of the ice sheet at the start of the test, the failure starts to occur from the point where the model structure and ice sheet make contact. Fig.16 shows a vertical view of the ice sheet failure modes. The range of ice sheet destroyed by the indentation were the width of the model structure x ice thickness x the indentation depth. No other cracks were observed. Photograph 1 shows the ice sheet failure at the time of the test (W=3.0m, h=12cm). Under the test condition of T-1, because the ice sheet was thinner and weaker, flexural failure occurred. Figs 17, 18 and 19 show the shapes of the leading edges of the test ice sheets just after the ice indentation tests. The ice sheet in the direction of the model structure width had random roughness, and in the ice thickness direction had shapes corresponding to the spalling (flaking, wedge) failure of ice sheets. And, ratios of parts of the top and bottom wedges, and crushing zone were around one-third.

The size of an independent ice failure zone is evaluated under limited test conditions. Further, the relation between the size and velocity of the model structure and ice thickness should be investigated in later stages (Sodhi, 1997).

4. CONCLUDING REMARKS

The following is a summary of the experimental findings obtained from the indentation tests, using model structure widths of 1.5 and 3.0 m and ice with thicknesses of 6.5, 12 and 24 cm.

(1) Except for the ice of thickness 6.5 cm, the failure modes of the ice sheet could be classified as spalling (flaking, wedge) failure and crushing failure with horizontal cleavage cracks.

(2) The time series of total ice loads showed a high peak value initially (first peak load) followed by variable subsequent ice loads, the same as those of previous small-scale indentation tests (width of the model structure is less than 1 m).

(3) The time series of local ice loads showed a remarkable peak value at the outset followed by randomly spaced peaks. These random peaks for each panel were inconsistent, and corresponded to non-simultaneous failure in several areas of the ice sheet.

(4) From the calculations of the correlation coefficients between each local ice load, the size of the independent failure area is closely related to ice thickness. An apparent area of independent ice failure under subsequent force range is related to ice thickness under the velocities of the tests. However, an apparent area of independent ice failure has wedge-shaped cracks and crushing with horizontal cleavage cracks. Therefore, it is desirable to investigate whether the occurrences of these failure modes is simultaneous or non simultaneous.

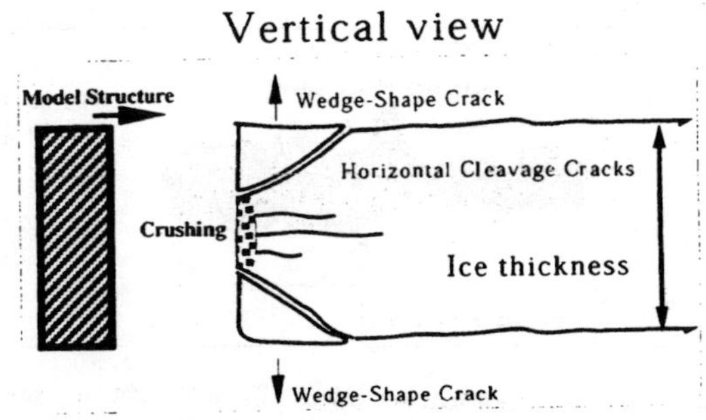

Fig.16 Vertical view of ice sheet failure modes(W=1.5m, h=12cm).

Photograph 1 Ice sheet failure at time of test (W=3m, h=12cm).

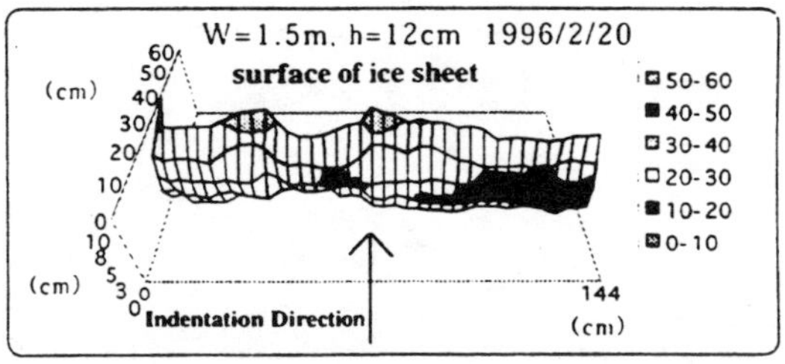

Fig.17 Shapes of leading edges of test ice sheets just after the ice indentation tests (W=1.5m, h=12cm).

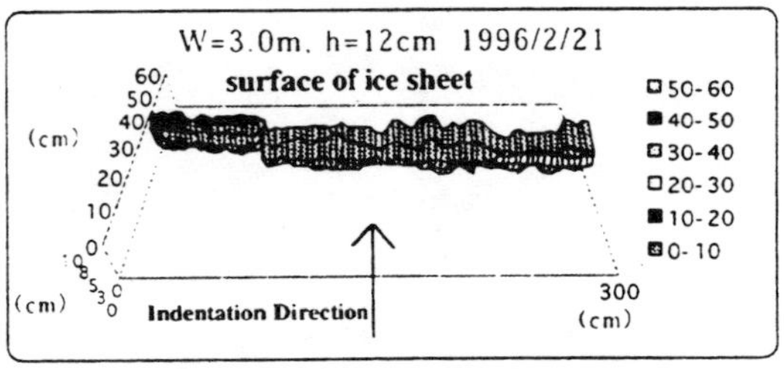

Fig.18 Shapes of leading edges of test ice sheets just after the ice indentation tests (W=3.0m, h=12cm).

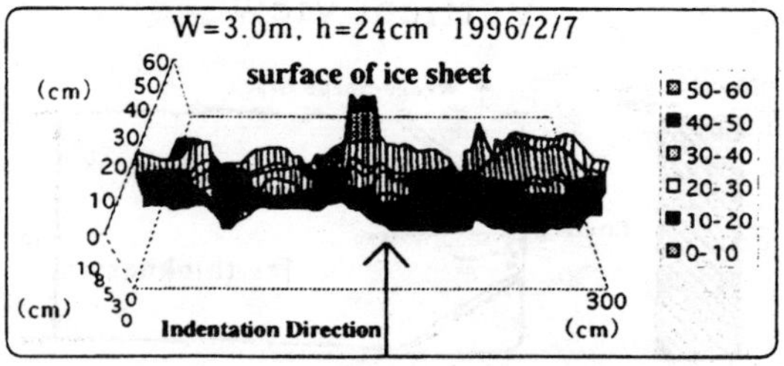

Fig.19 Shapes of leading edges of test ice sheets just after the ice indentation tests (W=3.0m, h=24cm).

ACKNOWLEDGMENTS

The authors sincerely thank all the people who participated in this project, especially those who worked in severe outdoor environmental conditions. Shimada Construction Co. Ltd collaborated in the the field test preparations. We couldn't have obtained the field data without their cooperation.

REFERENCES

(1) Saeki, H. and Ozaki, A. (1980): Ice Force on Piles, Physics and Mechanics of Ice, Ed. Per Tryde, Springer-Berlag, pp. 342 - 350.

(2) Takeuchi, T. and Saeki, H. (1995): Indentation Pressure in Ice/Vertical Structure Interaction. Int. Jour. Offshore and Polar Eng., Vol. 5, No. 4, Dec., pp. 279 - 285.

(3) Timco. G. W. (1986): Indentation and penetration of edge-loaded freshwater ice sheets in the brittle range. Proceedings of Offshore Mechanics and Arctic Engineering Conference (OMAE), Vol. 4, pp. 444-452.

(4) Sodhi, D. S. (1997): Correlation of ice crushing forces in segments of an indentor. Proceedings of Offshore Mechanics and Arctic Engineering Conference (OMAE) to appear.

Proceedings of the Seventh (1997) International Offshore and Polar Engineering Conference
Honolulu, USA, May 25-30, 1997

ISBN 1-880653-28-1 (Set); ISBN 1-880653-30-3 (Vol. II)

Medium-Scale Field Indentation Tests: Measurements of Ice Sheet Deformation Under Ice-Structure Interactions by Means of Photogrammetry

M. Sakai, M. Aoshima and H. Katsui
Taisei Corporation
Tokyo, Japan

M. Suzuki
Teikoku Construction Consultants
Gifu, Japan

ABSTRACT

The study on ice load under ice-structure interactions has been started since 1993 as Japan Ocean Industries Association (JOIA) project. Medium Scale Field Indentation Tests (MSFIT) are conducted in 1996 winter as part ofJOIA project and includes ice indentation tests, tests on physical properties ofice sheet and measurements ofice sheet deformation . This paper reports on the results of measurements of ice sheet deformation obtained from technique of stereo photogrammetry under ice - structure interactions in 1996 winter.

KEYWORDS: stereo photogrammetry, ice sheet deformation , strain range

INTRODUCTION

The ice load caused by interaction between the offshore structure and ice in icy water is known to be heavily dependent on the stain rate $\dot{\varepsilon}$ (denoting deformation of ice and defined by the ice moving velocity / the structure dimension).

The failure range of the ice sheet includes the following three ranges; (a) a range ofsmaller strain rate ($\dot{\varepsilon} \leqq 10^{-4}$, ductile failure range) , (b) a range ofgreater strain rate ($\dot{\varepsilon} \geqq 10^{-2}$, brittle failure range) and (c) a range of medium strain rate ($\dot{\varepsilon} \fallingdotseq 10^{-3}$, transition failure range). The result of research efforts made so far has revealed that the maximum ice load occurs in the transition failure range.

An effective evaluation technique to determine the strain range as one of the constituent elements of the strain rate, however, is not yet established. Evaluation of the strain rate is considered to have a serious impact on the evaluation of the ice sheet failure load, namely, evaluation of the structural design load. Accurate estimation and evaluation of the strain rate (strain range) is regarded as one of the major tasks.

Considering that the strain range of the ice sheet can be estimated from the displacement range of the ice sheet, the authors of this paper measured ice sheet displacement caused by the indentation of a structure into the ice in the Medium Scale Field Indentation Test (JOIA (1995), Kawamura et al (1996)) using the technique of photogrammetry, and estimated a way of determining the strain range.

EXPERIMENTAL SET UP AND PROCEDURES

The authors used stereo photogrammetry to measure the ice sheet displacement. Two cameras were used to stereo photograph the ice sheet displacement which occurred when the indentor entered the ice. Cameras were placed on the indentor support frame given in Fig.1. After deformation of the ice sheet was photographed, the displacements in three directions were obtained by the analytical plotter. For the three-axis coordinates, parallel direction (Y-axis), direction of indentation (Z-axis) and vertical direction (X-axis) were set with respect to the indentor, thereby calculating the displacements based on the differences from the coordinates prior to ice failure.

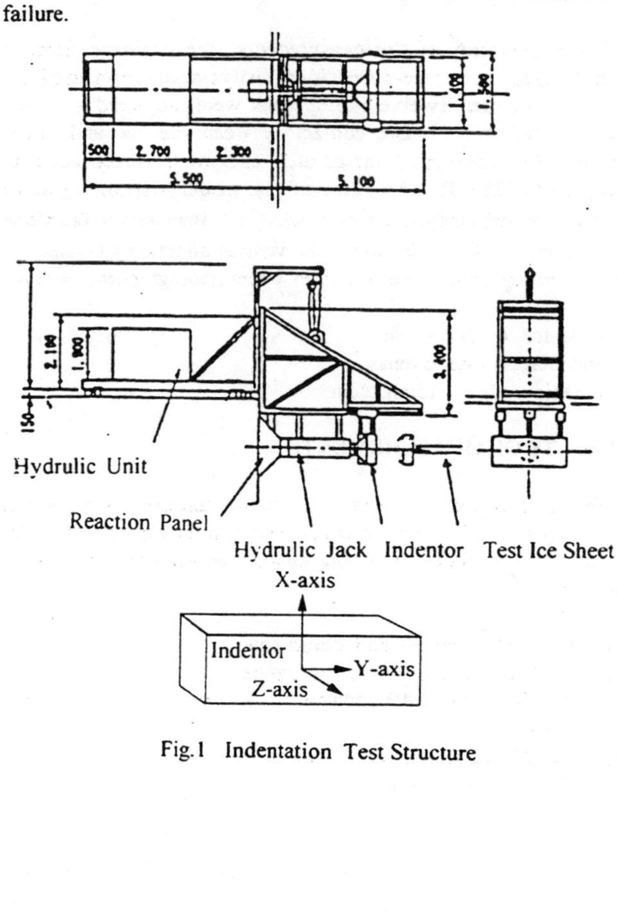

Fig.1 Indentation Test Structure

The experiments were taken at the Futamigaoka of the harbor of the Notoro fishing port in Notoro Lake in the suburbs of Abashiri City, in February, 1996.

The following shows the measurement procedures:

(1) Setting the measurement marks

To facilitate stereo photogrammetry, marks were placed at the measurement points of the ice sheet. These marks were put on 1 cm-square sheets of paper. To fix the marks on the ice sheet, about 10 cm long marking needles were used; they were pinned onto the ice. Marks were placed near the center of the indentor and about 50 mm away from the contact area between the indentor and ice sheet. Five rows were laid out at intervals of 50 mm parallel to the contact area, and three rows at intervals of 80 mm in the direction of indentation. (See Fig. 2.)

(2) Setting the control point

Four control points serving as references for relative displacement are placed on a rectangular aluminum frame of 2 m in the longitudinal direction and 1 m in the lateral direction. The aluminum frame is installed in overhung structure, supported by two round pipes provided in the forward portion of the test area. To eliminate impact on photogrammetry, the aluminum frame is designed to slide on the round pipes at the time of indentation test. (See Fig. 3.)

(3) Installing the cameras

The camera mount platform were fixed onto the support base of the indentor (about 70 cm high above the ice sheet surface), with optical axes of two cameras in the vertical direction. The effective photography area in this case was within the area of 25 cm × 35 cm, and layout was designed to ensure that marks and control points would be located within this area. (See Fig. 4.)

(4) Photographing and photogrammetry

Photographing was started immediately before the ice and indentor contacted each other. In one case, continuous shooting was performed at exposure intervals of 0.39 seconds per shoot, and measurement was conducted for 14 sec (for 36 frames) from the start of photographing.

For processing of the measurements, the negative films were directly used, and three-dimensional analysis was conducted by the analytical plotter. Twelve photographs were subjected to analysis; those immediately before contact between the ice and indentor (Model NO.1), and those immediately after the ice sheet was cracked (Model NO.12). The three dimensional model positioning accuracy obtained by the analytical plotter was ± 0.1mm within the plane (Y, Z-axis), and ± 0.2 to 0.3 mm in the vertical direction (X-axis).

The following shows two test cases where photographing was used:

Indentor width : 3.0 m
Indentation rate: 6 mm/s
Ice thickness : 15cm, 20cm

RESULTS OF MEASUREMENTS

The displacement obtained by photogrammetry was calculated according to the following formula, based on the mark position in the Model No.1 immediately before contact between the ice sheet and indentor:

Xn, Yn, Zn: No. n model mark coordinate
X1, Y1, Z1: No. 1 model mark coordinate
DX = Xn - X1 (upward +, downward -)
DY = Yn - Y1
DZ = Zn - Z1 (compression -, tension +)

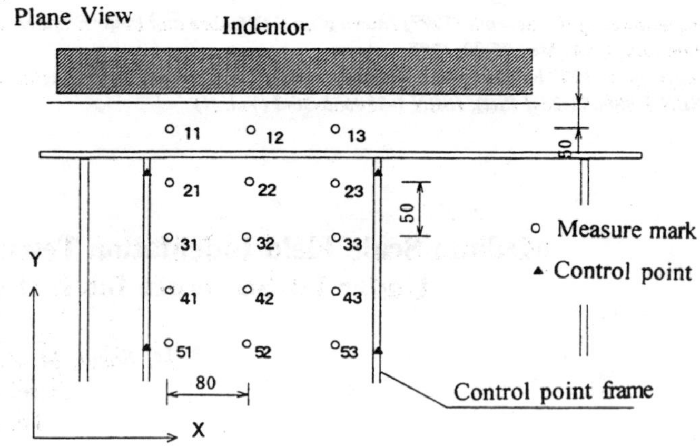

Fig.2 Installation of measure marks

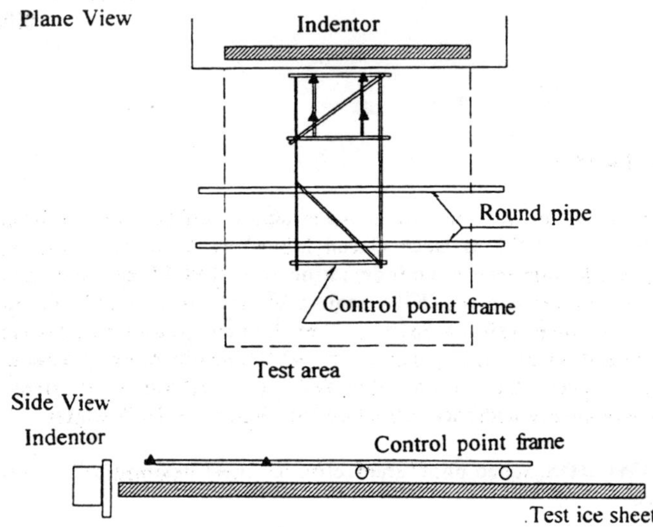

Fig.3 Installation of control point frame

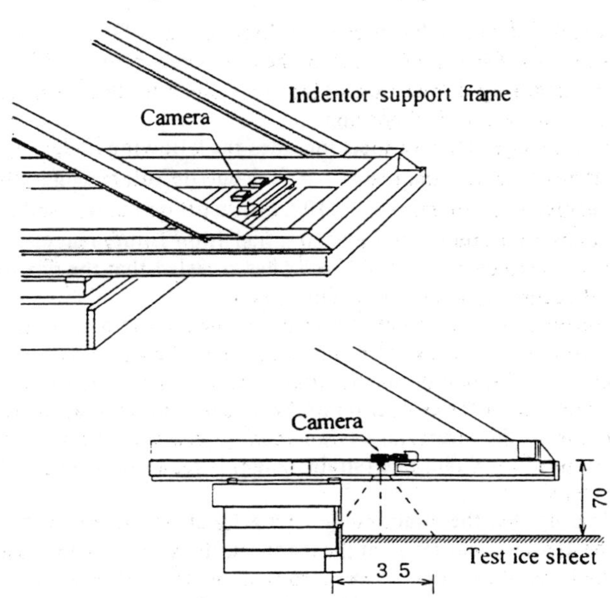

Fig.4 Camera Installation

Figs. 5 and 6 show the plots of ice sheet displacements in the indentation direction (Z-axis) and vertical direction (X-axis) in the case of ice thickness 15cm.

Figs. 7 and 8 show the plots of ice sheet displacements in the direction of indentation (Z-axis) and vertical indentation (X-axis) in the case of ice thickness 20cm.

In the indentation direction, a crack parallel to the indentor occurred on the mark in the foremost row of the ice (5 cm forward of the indentor) immediately after pressure was applied (2.73 sec. later / ice thickness 15cm , 2.32 sec..later / ice thickness 20cm). After that, multiple cracks parallel to the indentor occurred in the range from 10 to 15 cm in front of it, and upheaval and depression of the ice sheet were observed. Displacement in the direction of indentation occurred in the direction of compression. Changes of displacement were observed separately in two groups when the initial crack occurred. On the mark (first and second rows) close to the indentor, displacement was increased by the ice sheet failure, but on other marks, displacement was about 1 to 2 mm in both cases.

In the vertical direction, ice sheet was displaced in the upward direction immediately after the pressure was applied. Similar to the horizontal displacement, upward displacement was increased on the marks (first and second rows) close to the indentor when the initial crack occurred. After that, displacement on other marks (third and fourth rows) were changed within the range of about ± 2mm in the case of ice thickness 15cm, and about ± 2 to 3 mm in the case of ice thickness 20cm.

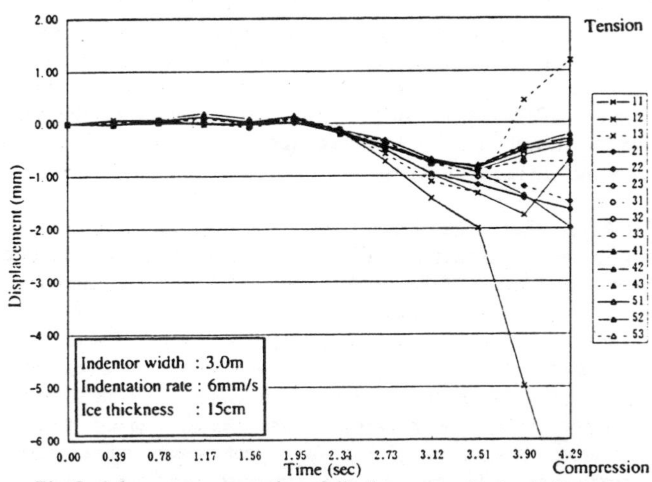

Fig.5 Measurement results of displacement in the Z direction

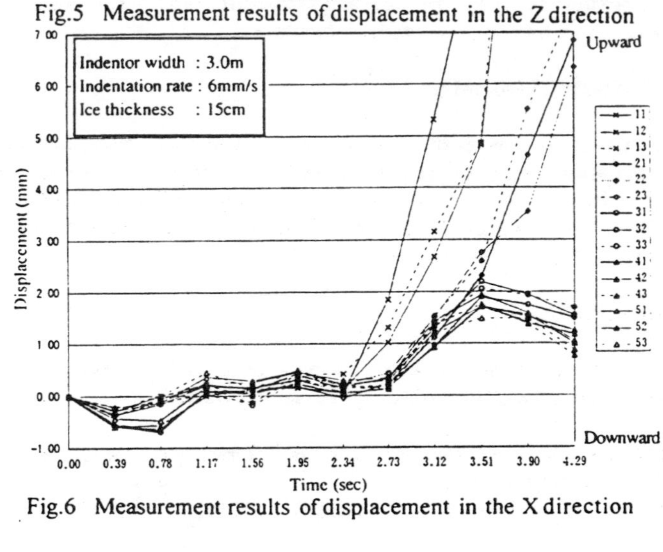

Fig.6 Measurement results of displacement in the X direction

CONSIDERATIONS

Figs. 9 and 10 show the temporal load data when the initial total ice load occurs.

Based on the load and the result of photogrammetry, the following conclusion has been obtained on the relationship between the ice load and ice sheet failure:

The ice sheet is deformed about one to two seconds after start of indentation. Of five measuring lines parallel to the indentor, the line of foremost row exhibits a gradual increase in deformation of the ice sheet as the indentor enters the ice. On other four lines parallel to the indentor, the horizontal displacement of the ice sheet tends to reach the maximum value in about 3.5 seconds. Almost simultaneously, the initial total ice load also tends to reach the maximum value at this time.

The occurrence of the horizontal displacement of ice sheet and the relationship between the horizontal displacement and initial total ice load have been found to show same tendencies in the measurements introduced in this paper.

Initial crack occurs on the mark immediately after indentation is started. After that, multiple cracks parallel to the indentor are observed to occur in the range of foremost line. As the indentor further enters the ice, multiple cracks parallel to the indentor are observed to occur in the forward range of other four lines. When these cracks have expanded to reach the range equivalent to the ice thickness, the initial total ice load tends to exhibit the maximum value, as has been made clear in this study.

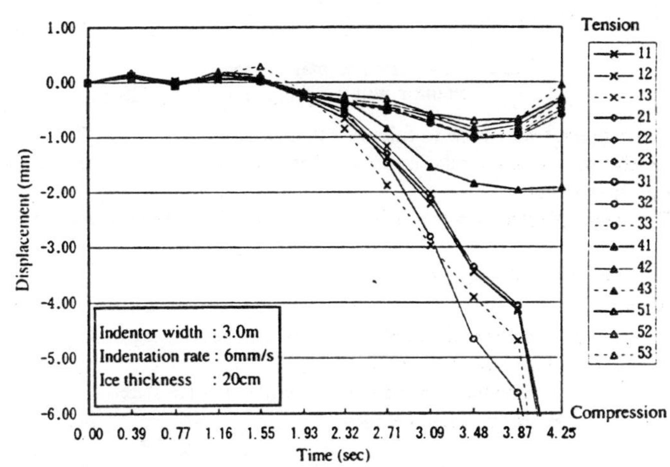

Fig.7 Measurement results of displacement in the Z direction

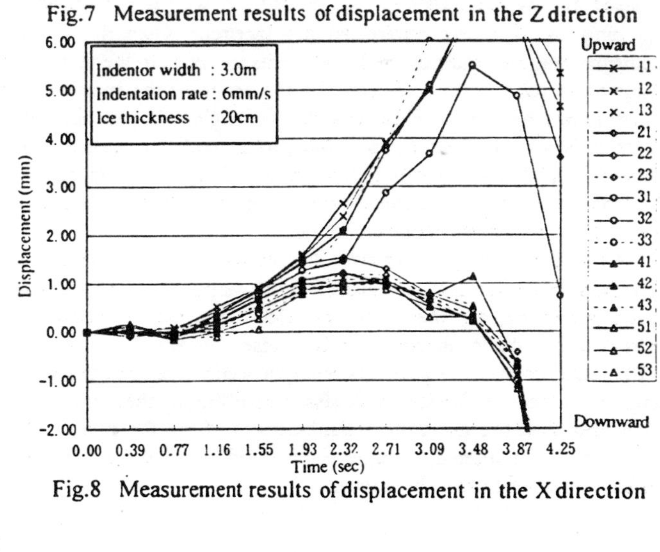

Fig.8 Measurement results of displacement in the X direction

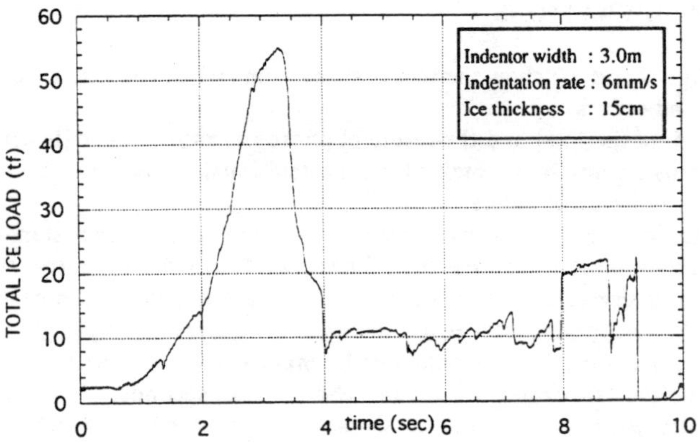

Fig.9 Time series of total ice load
(Indentor width 3.0m, ice thickness 15cm)

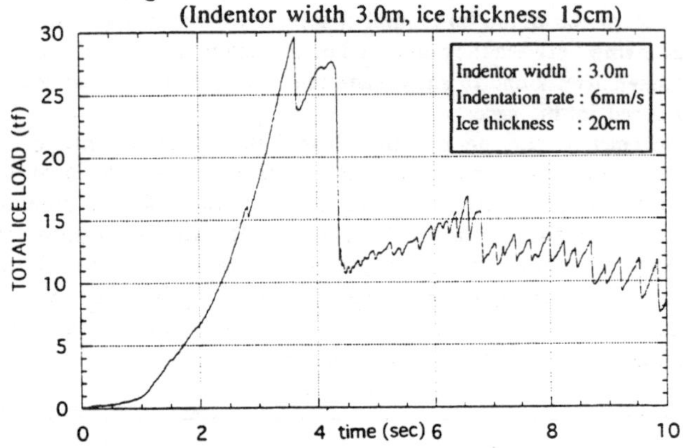

Fig.10 Time series of total ice load
(Indentor width 3.0m, ice thickness 20cm)

An attempt was made to calculate the range affected by strain, based on the stress-strain relationship when the ice is considered as an elastic material.
The range affected by strain L is calculated by the following formula:

$$L = \Delta L \cdot E \cdot A / P$$

where

Δ L: horizontal displacement of ice sheet
E:Young ratio of ice
A: Contact area between indentor and ice sheet
P: Initial total ice load

Based on the load data in Figs. 9 and 10, we made calculation assuming as Δ L the horizontal displacement when the initial maximumtotal ice load occurred. The measurements in the test were assigned to other values to find out the range of the influence of strain. It has been made clear that the value close to the indentor width are shown in both cases. This is the conclusion close to the conventional definition ofthe strain range of the ice sheet; namely, the conclusion is close to the view that the range ofthe influence is two to four times of the structure width.

Assuming the maximum horizontal displacement by photogrammetry as Δ L, we calculated the range ofthe influence by strain according to the same formula. The value shown is approximate to the ice thickness in both cases.

Thus, the result ofthe photogrammetry has confirmed that the strain range is related to the ice thickness, according to the conclusion calculated from the horizontal displacement of the ice sheet. The strain rate is considered to be more subjected to the influence of ice thickness than the structure width.

Actual ice has characteristics as visco-elastic material, so it will be not effective to handle the problems ofthe strain range for all states of the ice according to the theory of elasticity discussed above. Furthermore, it involves another problem that Young ratio and contact area in the above formula must also be handled as items subjected to the influence of ice crystal structure, ice temperature, indentation rate, etc.
The subsequent task will be to study the problems discussed above and to evaluate the characteristics ofthe strain occurring to the ice and the range ofthe strain with attention paid to the ice deformation mechanism.

CONCLUSIONS

The following summaries the result obtained from photogrammetry and load data with regard to the displacement resulting from the interaction between the ice and structure:

(1) The displacement occurring when the initial total ice load is applied to the ice sheet is about $\pm$2mm in the vertical direction (X-axis) with respect to the ice sheet, 1 to 2 mm in the direction of indentation (Z-axis) and almost no change in the parallel direction (Y-axis) in the test case (indentor width 3.0 m; ice thickness 15cm, 20 cm).
(2) The initial total ice load tends to exhibit the maximum value when the range ofthe crack occurring to the ice sheet has reached the range equivalent to the ice thickness in the direction of indentation.
(3) The measurement of the displacement when the initial total ice load exhibits the maximum value becomes the value closest to the displacement when the range of the strain is assumed as the ice thickness. This suggests that the range ofthe influence ofthe strain when the initial total ice load becomes the maximum is the range equivalent to the ice thickness.

However, these measurements have been gained from only the result of photogrammetry, and the number of tests was limited; therefore, evaluation of the displacements of the ice sheet and strain range remains only qualitative.

To ensure accurate measurement of the strain range of the ice sheet, it will be essential to study measuring methods over an extensive range including the technique of photogrammetry. At the same time, it is important to perform more systematic tests, thereby ensuring quantitative evaluation of the ice sheet strain range during interaction between the ice and structure, and verification in comparison with the past research results.

ACKNOWLEDGEMENT

This study has been made as part of the Medium Scale Field Indentation Test ofthe ice load consigned by Japan Ocean Industries Association (JOIA). Authors wish to take this opportunity to express their thanks for all the people who have kindly extended assistance and cooperation in this work.

REFERENCES

Japan Ocean Industries Association (JOIA)(1995). "Reports of the Research on Ice Loads Affecting Marine Structures."

Kawamura,M et al(1996)," Ice-Structure Interactions In Medium Scale Field Indentation Tests,"
proceedings of 13th IAHR'96 ICE SYMPOSIUM Vol.1

Proceedings of the Seventh (1997) International Offshore and Polar Engineering Conference
Honolulu, USA, May 25-30, 1997

ISBN 1-880653-28-1 (Set); ISBN 1-880653-30-3 (Vol. II)

Experimental Studies on Nonsimultaneous Failure Characteristics of Vertical Sided Indentors

Kazuhiko Kamesaki, Hiroyuki Tsukuda and Yutaka Yamauchi
NKK Corporation
Japan

ABSTRACT

A five year ice load research program is being executed to enhance design ice load prediction accuracy for vertical sided structures by the Japan Ocean Industries Association (Saeki et al., 1996). The program is comprised of extensive indentation tests and computer code development capable of simulating dynamic ice/structure response, local , global and foundation loads.

As a first year test program, four vertical rectangular indentors and one cylindrical indentor were tested in small scale with setting model stiffness relatively high to attain stable nonsimultaneous failure modes. Tests were conducted with varying ice thickness, indentation velocity and crushing strength. Segmented ice pressure panels were equipped to measure local and global ice pressure characteristics. This paper summarizes the experimental results with respect to the tendency of the global and local load. From the segmented panel data, local load characteristics regarding pressure area relation and a width of the independent zone are discussed.

KEY WORDS : Vertical sided structure, indentation, crushing, flaking, nonsimultaneous failure, independent zone, ice tank

INTRODUCTION

Ice load estimation on vertical sided structures is a classical problem, although the designer encounters many difficulties when selecting rational design loads. Ice interaction against a vertical sided indentor exhibits very complex local failure and that makes it difficult to develop mathematical models or to predict ice load using scaled ice tank tests. Croasdale(1988) reviewed the practical ice load estimation methods on vertical structures. He introduced three conventional methods, namely, Korzhavin's equation (1962), reference stress methods (Ponter et al., 1983) and cut-off stress methods (Walden et al., 1987). Sanderson (1988) claimed that a Korzhavin like equation predicts ice loads much higher applying a contact factor of 0.3 that is deduced from laboratory indentation tests. The contact factor required to match the full scale data would be less that 0.1 and typically the order of 0.07. This implies that the gap between the small scale experiments and the full scale data is considerably large. On the other hand, in addition to this scaling problem, after the lesson of Molikpaq in 1986 (Jefferied and Wright, 1988), it is well recognized that even massive structures may vibrate under certain conditions if a repeated load reaches a specific level during crushing. That would make the indentation problem more complex and difficult to predict ice loads on vertical sided structures.

To resolve these technical issues, the final goal would be to develop a computer code capable of predicting the ice/ structure interaction including dynamic response and local/ global loads. Some approaches have been proposed by Sodhi (1994) and Kärnä(1992) .

A series of indentation tests was performed in order to grasp the limitation of small scale model tests and to collect the validation data for the computer code as a first year test program. The tests were designed to attain nonsimultaneous failure modes. Test results are presented in terms of the global and local load. In particular, the analysis for the independent zone is reported for further application to computer code development.

TEST ARRANGEMENT

Model arrangement

The tests were performed in model ice at the Ice and Snow Engineering Laboratory of NKK Corporation. Fig.1 shows the side view of the test arrangement where models were towed against sheets of granular grained urea ice. Models were made of steel plates to attain sufficient stiffness, and supported by the four load cells to measure foundation loads through a model support structure sitting on an underwater carriage. The underwater carriage was connected to the main towing carriage through a hinged rod and pushed against the sheet ice. The models had sufficient stiffness, although the hinged rod had relatively weak stiffness. Thus the hinged rod stiffness governed the natural frequency for horizontal direction, and it is estimated to be approximately over 20 Hz. Four rectangular models were prepared named L, M, S, and N-model whose widths were 1500, 800, 400 and 100 mm respectively. Also a 800 mm diameter cylindrical model named CM-model was used to compare the pressure distribution with the rectangular model. Segmented pressure panels were arranged where the ice sheet interacted. Small load cells supported pressure panels and the local pressure was determined by dividing recorded load by nominal contact area multiplying panel width and ice thickness. Three types of pressure panels were used. Panels named A type measured 120 mm height and 100 mm wide, B type measured the same height and 50 mm wide, and C type was 1/6 area of A type measuring 40 mm height and 50 mm wide as to the rectangular models. Fig. 2 shows the panel arrangement and model sizes. N-model was not equipped with local panels. As for the cylindrical CM-model, the pressure panels aligned in a radial direction at an angle of 22.5° were arranged in the same manner as M-model as shown in Fig.2.

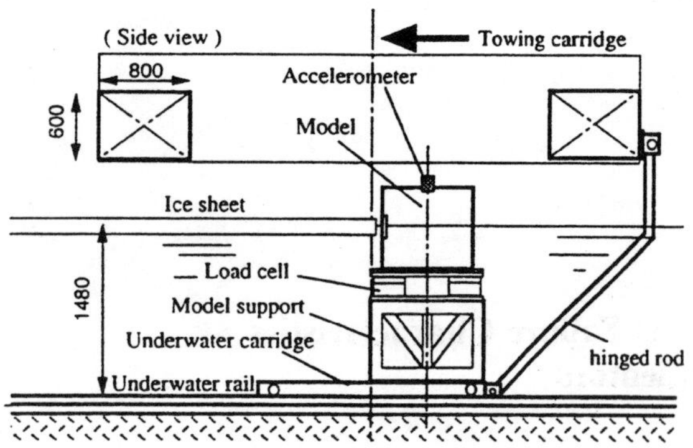

Fig.1 Test arrangement

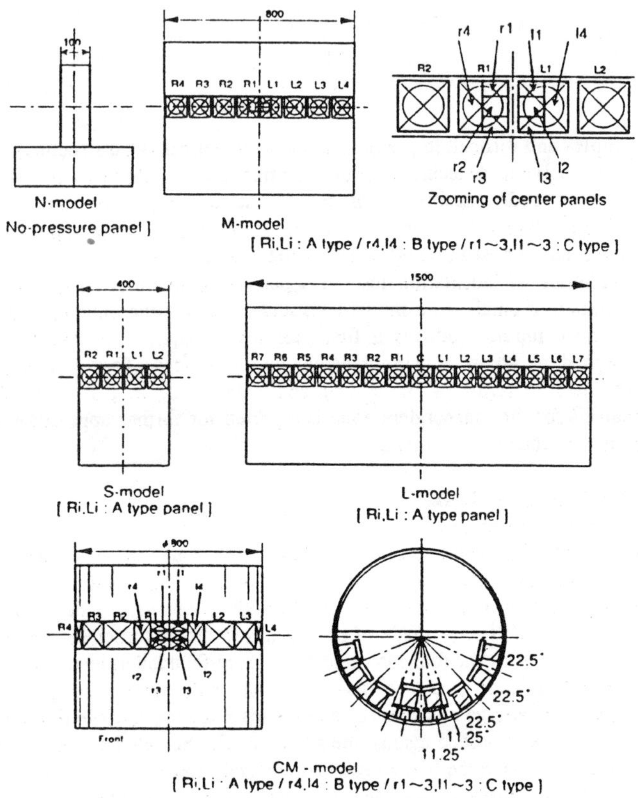

Fig 2 Size of Indentors and pressure panel arrangement

Table1 Test Condition

MODEL	Test ID	Ice thickness h(mm)	Compressive strength σ_{cc}(kPa)	Indentation velocity V(mm/sec)
Rectangular	1-4	63	103-105	6-212
N-model	5-8	99	123	6-208
Rectangular	1-4	62	105-113	6-210
S-model	5-8	98	94-98	6-207
Rectangular M-model	1-4	62	56-58	6-169
	5-8	64	129-135	6-168
	9-12	62	100-104	6-168
	13-16	42	96	6-169
	17-20	42	143-151	9-169
	21-24	64	121	6-168
	25-28	100	88-90	6-168
	29-32	102	99-103	6-167
	33-36	91	120-124	6-167
Rectangular L-model	1-3	93	107-110	6-105
	4-6	94	104-105	6-325
	7-9	65	121-123	6-316
	10-12	66	105-109	6-316
Cylindrical CM-model	1-4	42	87-88	6-169
	5-8	42	132-133	6-170
	9-12	66	109-110	5-170
	13-16	66	74-80	4-169
	17-20	64	58	5-168
	21-24	102	87-89	6-167
	25-28	103	104-105	5-167
	29-32	92	104-107	6-166

Test condition

In total, 96 tests were performed. Test conditions are summarized in Table 1. The test ranged in ice thickness from approximately 40 to 100mm, and in uniaxial crushing strength from 60 to 130 kPa. Three to four tests were performed per ice sheet with varying indentation velocity from 6 to 310 mm /sec. Before each test run, an edge of the ice sheet was cut by a saw parallel to indentors to attain perfect initial contact. Through the whole test series, the compressive strength was measured employing cantilever compressive tests proposed by Kato et al. (1982) in addition to the downward flexural strength to ensure index strength of ice sheets. The crushing strength determined by this test at the strain rate of 10^{-2} is adopted as the compressive strength value. For several ice sheets, the uniaxial compressive tests were additionally conducted.

TEST RESULTS AND ANALYSIS

Failure mode

The failure mode so called crushing with flaking was observed through the whole test series. It seems that the crushed ice was extruded upward and downward equally, and the considerable rubble was formed in front of the indentor.

Compressive strength test

Several ice sheets were dedicated to perform a series of the cantilever compressive tests. The test results are depicted in Fig. 3. The compressive strength by the cantilever test, σ_{cc}, slightly increased with increasing strain rate at the strain rate ranging from $3x10^{-3}$ to $3x10^{-1}$. As another test method, the test specimens were cut out from an ice sheet and the uniaxial compressive strength was measured using a universal testing machine with a room temperature of- 2 °C. The compressive strength measured by this test is denoted as σ_{cu}. The rate dependency was not found in these tests at a strain rate ranging 10^{-3} to $3x10^{-1}$ sec^{-1}.

Global load

Global load was determined as the sum of the segmented pressure panels, and then the global ice pressure Pg was calculated by dividing nominal contact area derived from indentor width and ice thickness. Relatively large peak pressure was observed at initial contact with the ice sheet and smaller fluctuated pressure was subsequent. The relation between the subsequent and initial peak pressures compared in Fig.4. $P_{g\text{-}INT}$ and $P_{g\ (peak)}$ means the initial contact peak pressure and the average of peak pressures picked in subsequent pressure time history exceed the simple mean value respectively. The ratio of $P_{g\ (peak)}$ to $P_{g\text{-}INT}$ shows the contact factor when the ice load is expressed in Korzhavin's equation form (1962). From this analysis, the contact factor is estimated to be 0.43. As mentioned earlier , this value is much larger than 0.07 obtained from the full scale data (Sanderson,1988). This discrepancy implies that ice load would be overestimated if the ice load is simply extrapolated applying the contact factors obtained from the small scale data. The trend of $P_{g\ (peak)}$ with respect to the ratio of the indentor width to ice thickness is plotted in Fig. 5. The indentation velocity ranged from 80 to 100 mm/sec. $P_{g\ (peak)}$ calculated by equation (1) is shown. The indentation factor for isotropic material given by Ralston (1978) is used in the equation. The test results agree well with the global pressure calculated the Korzhavin like equation.

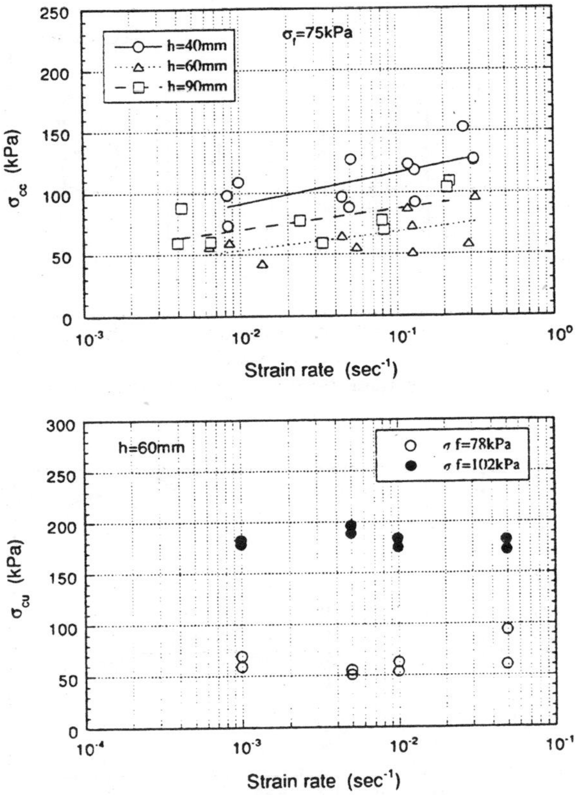

Fig.3 Uniaxial compressive strength as a function of strain rate

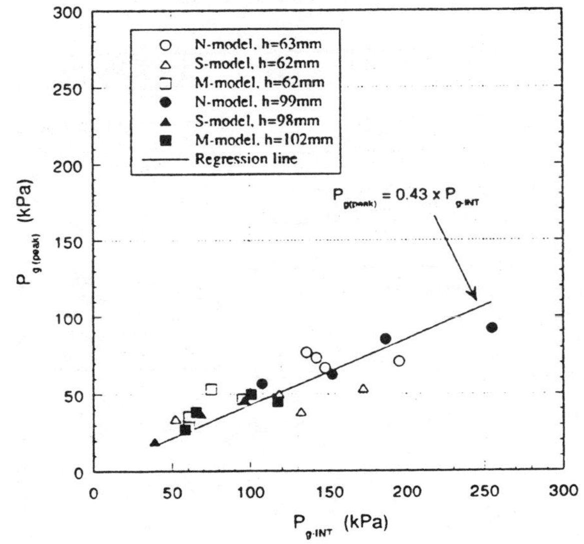

Fig.4 Relation between initial contact pressure and average of subsequent peak pressure (σ_{cc}=110kPa)

$$P_{g(peak)} = Fx / (B*h) = I\, fc\, \sigma\, cc \qquad (1)$$

where

Fx= global load, B= Indentor width, h= ice thickness

I= indentation factor, fc= contact factor

σ cc= crushing strength measured by the cantilever test

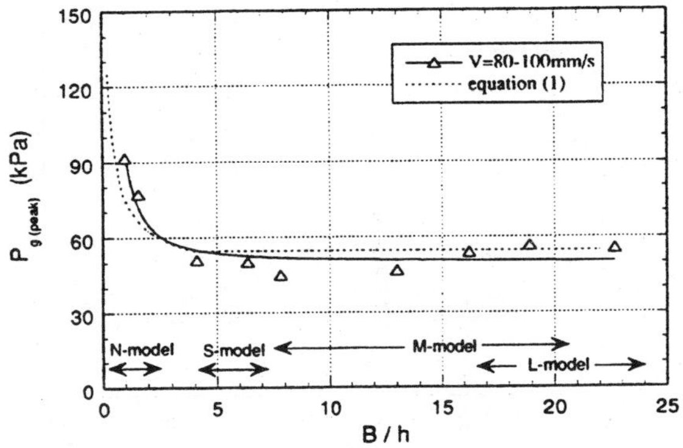

Fig.5 Relation of global ice pressure with the ratio of the indentor width to ice thickness

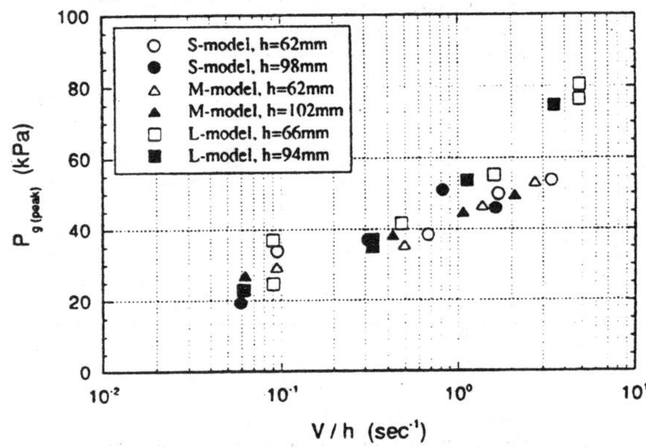

Fig.6 Global ice pressure as a function of strain rate (σ_{cc}=96~115kPa)

Fig. 6 shows the rate dependency on $P_{g\,(peak)}$ at the strain rate ranging from 6×10^{-2} to 5×10^{0}. Considerable rate dependency was observed. The indentation rate is defined as the ratio of indentor velocity to ice thickness (V/h). As indicated in Fig. 3, the compressive strength obtained by the cantilever test tends to increase slightly at the strain rate up to 3×10^{-1} sec^{-1}. The indentation tests were conducted with a much higher strain rate, therefore a direct comparison is not possible. This increase would be partly attributed to the characteristics of the model ice, however the increasing rate of $P_{g\,(peak)}$ is much higher than that of σ cc. The load obviously increased with increasing indentation velocity, although the reason for the increase is not clarified yet. Kärnä et al. (1993) found the same result for the global load in a series of field tests where a cylinder was pushed against 0.4 m thick sea ice at velocities between 0.2 m/sec. and 2.6 m/sec. Furthermore, a similar trend can be seen in the data of Sodhi (1992). $P_{g\,(peak)}$ and $P_{g(avr)}$ normalized by σ cc are plotted in the function of indentor width in Fig. 7 where V/h ranged 0.1 from to 1.0 sec^{-1}. $P_{g(avr)}$, this indicates that the simple mean value of the subsequent pressure curve. $P_{g(peak)}$ is decreasing with increasing indentor width although $P_{g(avr)}$ remains constant over 400 mm width. Fig. 8 shows the comparison between rectangular M-model and cylindrical CM-model. This difference is conventionally treated as the shape factor. The reduction factor depends on the indentation rate and ranges from 8 to 20 %. Significant reduction was not attained by changing the shape. It is considered that the global load mainly depends on the projection area.

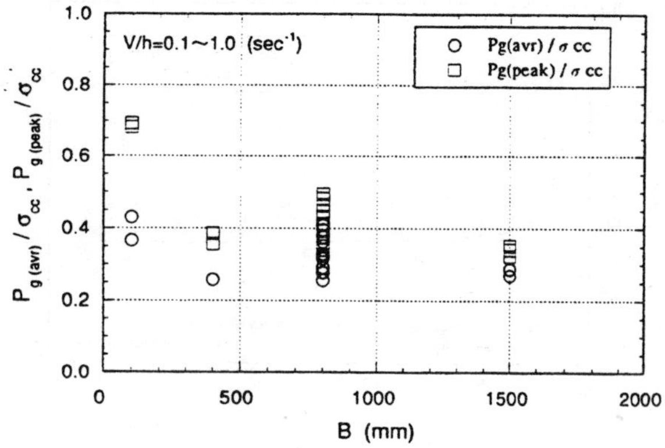

Fig.7 Average global ice pressure and peak global ice pressure as a function of width B

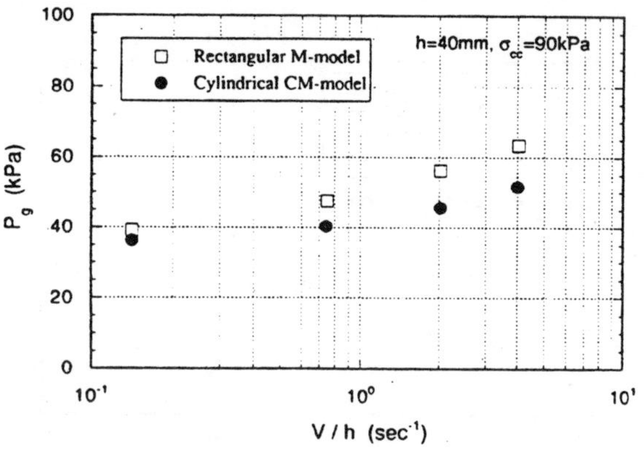

Fig.8 Global ice pressure comparison between rectangular M-model and cylindrical CM-model

Local pressure

Fig. 9 shows local pressure distribution for M-model and CM-model. In the figure, P_l means ice pressure range in each segmented panel. For the rectangular indentor , local pressure tends to slightly increase at both edges. On the other hand, local pressure is reduced for both edges for the cylindrical indentor. In both cases, the degree of the difference is sufficiently small and the results indicate that the approximation by the uniform pressure distribution is acceptable for structural calculation purposes.

The correlation factor in time basis is calculated to examine the degree of nonsimultaneous failure. The lower value indicates the sign of nonsimultaneous failure mode. The correlation factors are calculated against the combination of five different zones that are selected as symmetric with respect to the center. The zone definition and typical results are shown in Fig. 10. The correlation factors are plotted in the function of indentation velocity and referenced area. The correlating factors rapidly decrease as the reference area and indentation velocity increases. In this case, the test with 5.9 mm /sec. seems to exhibit simultaneous failure mode.

Local load time histories for l1, l2 and l3 panels are plotted in Fig.11 for 102 mm ice thickness. Panels l1 , l2 and l3 are located at upper, middle and lower levels in a vertical direction as shown in Fig.2. Each panel is 40 mm high, therefore l1 and l3 panels shared the load for 31 mm ice thickness and l2 panel shared the load for 40 mm ice thickness assuming that the center of the ice sheet adjusted exactly at the center of panels. The figure indicates that most of the pressure was absorbed by the center pressure panels except for initial contact. This implies that when the ice sheet fails associated with flaking , the load does not exert pressure on the top and bottom panels in the subsequent indentation.

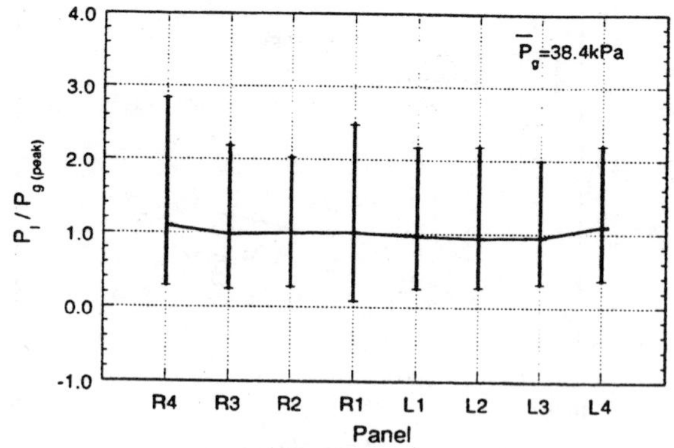

(a) Rectangular M-model (σ_{cc}=101kPa, V=85mm/sec)

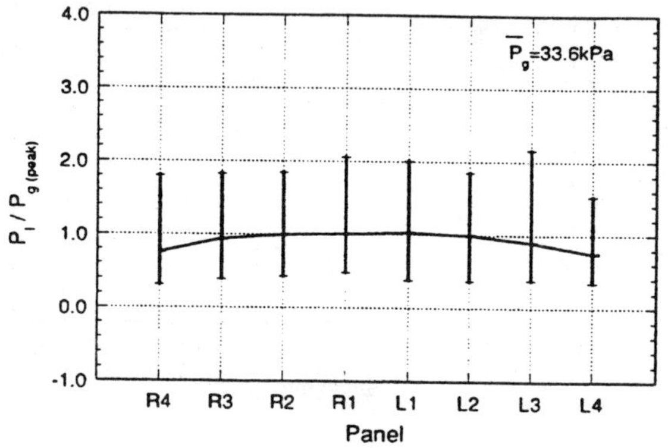

(b) Cylindrical CM-model (σ_{cc}=76kPa, V=85mm/sec)

Fig.9 Local ice pressure distribution

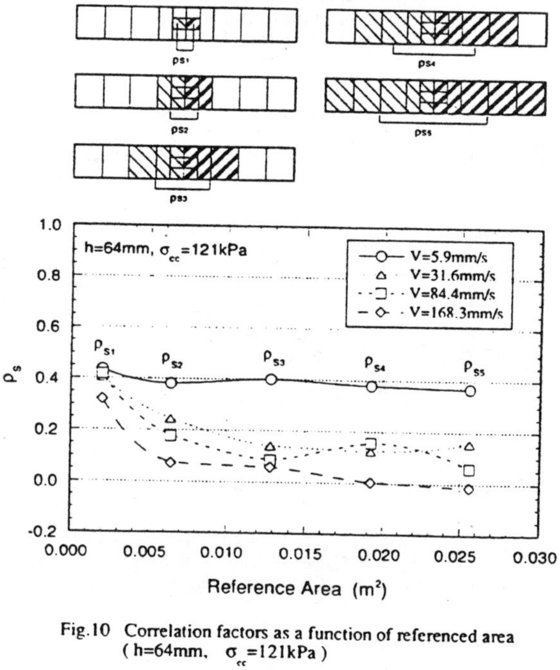

Fig.10 Correlation factors as a function of referenced area (h=64mm, σ_{cc}=121kPa)

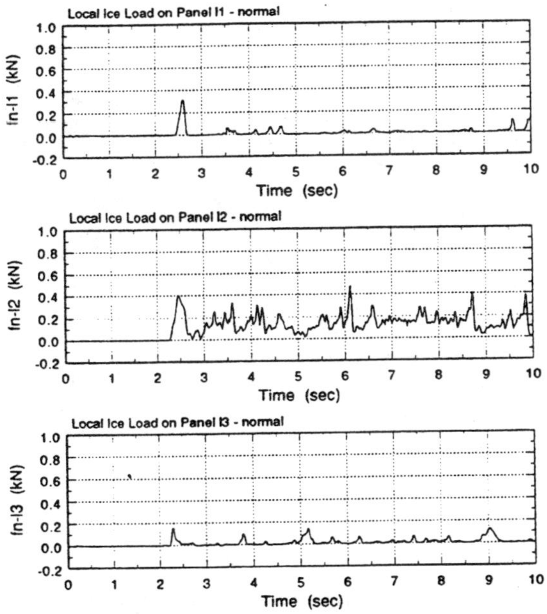

Fig.11 Local force time history — Panel I1, I2 and I3
(Test 31, h=102mm, σ_{cc}=100kPa, V=84mm/sec)

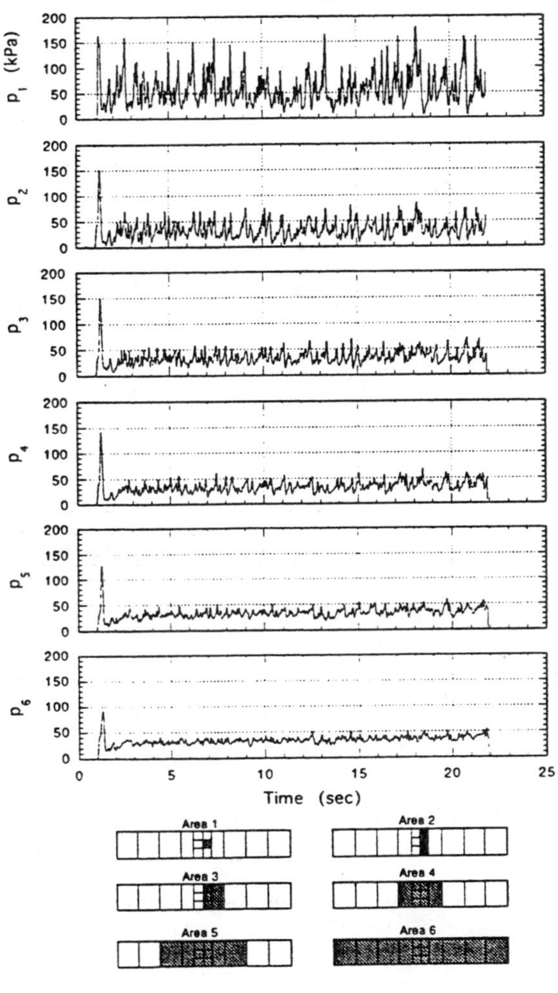

Fig.12 Typical pressure time history for each segmented area
(h=101mm, σ_{cu}=104kPa, V=84mm/sec)

Fig. 12 illustrates the typical pressure time histories for different zones. The fluctuation of the pressure, namely the variance of the pressure considerably decreased with increasing nominal area. Fig.13 shows the decrease of variance with increasing zone width for M-model. The pressure area curves synthesized in each indentation velocity is depicted in Fig. 14. The pressure decreasing rate increases with increasing the indentation velocity. This tendency can be well explained by the correlation factor depicted in Fig. 10. It means that the less correlated pressure time history, namely the more perfect nonsimultaneous indentation gives a larger pressure decreasing rate.

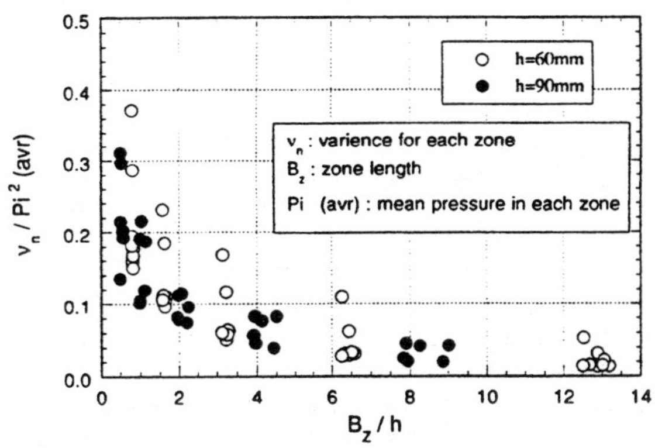

Fig.13 Relation between ice pressure varience and zone width

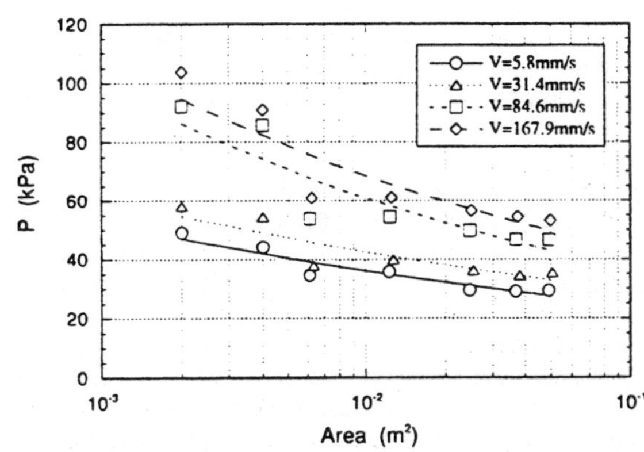

Fig.14 Pressure area relation synthesized for M-model
(h=62mm, σ_{cc}=102kPa)

Width of Independent Zone

Kry (1978) introduced the idea of the independent zone of crushing failure in front of a wide offshore structure and explained the pressure reduction with the increasing width of the structure. He deduced the independent zone as a width of four or five times of ice thickness. This zone is regarded as the minimum area where the failure occurs simultaneously. This concept is important to develop a mathematical ice/structure model employing a zonal local failure model (Kärnä,1992) or to synthesize full scale data employing the statistic assumption (Blanchet, 1996). An estimation of the width of the independent zone is attempted using the M-model ice pressure data recorded for over 90 mm ice thickness in each segmented zone. From Kry's assumption, the relation of ice pressure variance between different zones is expressed by;

$$b_I = b_n \times \frac{v_n}{v_I} \qquad (2)$$

where

b_I : independent zone width

b_n : zone width of independent zone

v_I : variance of ice pressure in b_I

v_n : variance of ice pressure in b_n

Assuming that the width of independent zone is given by bI = α h, the following simple relation is derived.

$$\frac{b_n}{h} = \alpha \frac{v_I}{v_n} \qquad (3)$$

Peak levels as high as initial contact pressure were frequently observed in the subsequent pressure curves in Area 1 as shown in Fig. 12. As for Area 2 , the pressure was mostly absorbed in the middle plate as explained in Figure 11, and the nominal contact area is simply different from Area 1. Thus, both Area 1 and Area 2 are regarded as the area within the independent zone. Assuming that the variance of Area 2 is representing v_I in this analysis, the variances for the tests over 90 mm ice thickness were plotted in accordance with eq.(3). The result is shown in Fig. 15. Regressed α ranges from 0.80 to 2.06 and gives the mean of 1.29. Thus, the independent zone is estimated to be approximately the order of ice thickness.

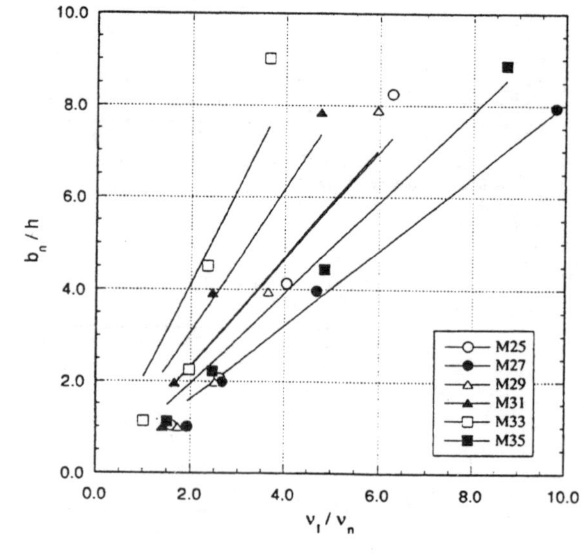

Fig.15 Relation between b_n / h and v_I / v_n

DISCUSSION

The ice tank test results demonstrated that Korzhavin's equation is still valid for small scale indentation tests for the wide range of B/h ratio. The authors expected that the contact factor would be slightly reduced at 1500 mm wide indentor tests, although it does not decrease at all in this relatively large size model indentor. It implies that the small scale test may over predict the ice load if the results are simply extrapolated with applying Korzhavin's equation. In another words, the current small scale test could not simulate the full scale indentation for wide offshore structures. The difficulty for correlation between full and model scale might arise from the discrepancy of flaw sizes and true contact area. Even for this test, as shown in Fig. 11 most of the ice load interacted with the middle part of the ice sheet. In full scale, the local load may concentrate a narrower area. Obviously interpretation or modeling other than Korzhavin's equation will be needed to extrapolate small scale data to full scale. Fortunately through this tests series, Kry's assumption was proved to be valid and the independent zone of failure is approximately the order of ice thickness. Fig.16 shows the plot of the variance and mean pressure for one panel located at near center corresponding to the data set shown in Fig.7. In this test series, the variance slightly decreased with increasing structure width. The mean pressures remain constant despite showing some scatter as assumed by Kry. The authors deduce that both the mean and peak pressures could be decreased if a structure width increases in full scale range. A numeric simulation can be possible if the proper zonal approaches and the statistics of local load is incorporated in the modeling.

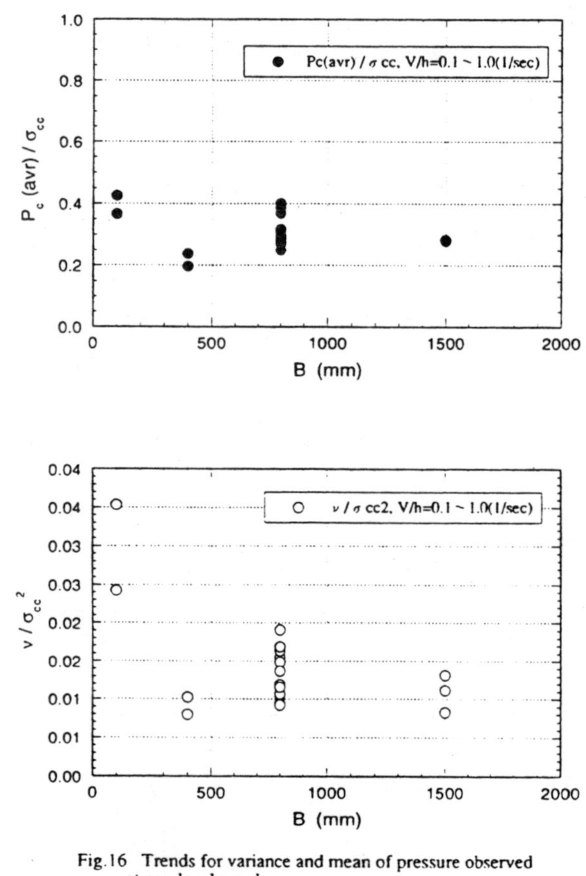

Fig.16 Trends for variance and mean of pressure observed at one local panel

ACKNOWLEDGMENT

This work was conducted by the Japan Ocean Industries Association, Ice Load Study Program. The author would like to acknowledge Japan Ocean Industries Association and the Minister of International Trade and Industry, for the permission to publish this paper.

REFERENCES

Croasdale, K. R. (1988) Ice Force: Current Practices, the International Conference on Offshore Mechanics and Arctic Engineering, Houston, USA, Vol.4, pp. 133-151

Jefferies, M.G., Wright, W.H. (1988) Dynamic response of "Molikpaq" to ice-structure interaction. Proceedings, 7th International Conference on Offshore Mechanics and Arctic Engineering (OMAE), Houston, TX, USA, Vol.4 pp.201-220

Kato, K., Sodhi, D.S., Haynes, F.D., Hirayama, K. (1982) Uniaxial In-situ compressive strength of model ice, CRREL internal report.

Kärnä, T. (1992) A Procedure for Dynamic Soil-Structure-Ice Interaction, Proceedings of 2nd ISOPE, San Francisco, Vol. 2, pp.764-770

Kärnä, T., Nyman, T., Vuorio, J. and Järvinen, E. (1993a). Results from indentation tests in sea ice. Proc. 12th OMAE Conf. Glasgow. Vol IV, pp. 177 - 185.

Korzhavin, K.N. (1962) Action of Ice on Engineering Structures, CRREL Draft Translation No.260

Kry, P.R. (1978) A Statistical prediction of effective ice crushing on wide structure, Proc. of the 5th International IAHR Conference, Lulea, Sweden, Part 1 pp.33-47

Ponter, A.R.S., Palmer, A.C., Goodman, D.J., Ashby, M.F., Evans, A.G. and Huchinson, J.W. (1983) The Force Exerted by a Moving Ice Sheet on an Offshore Structure., Part 1, The Creep Mode. Cold Regions Science and Technology 8, pp. 109-118

Ralston, T.D. (1978) An Analysis of Ice Sheet Indentation. , IAHR 78, Part 1, pp.13-31

Saeki, H. et al., (1996).JOIA PROJECT OF STUDY ON ICE LOAD, Proc. 13th IAHR Ice Symp. Beijing, China

Sanderson, T.J.O. (1988) Ice Mechanics, Risk to Offshore Structures, Graham & Trontman, ISBN 086010785X

Sodhi, D.S. (1992). Ice-structure interaction with segmented indentors. Proc. 11th IAHR Ice Symp. Banff, Alberta, Vol. 2, pp. 909 - 929

Sodhi, D.S. (1994). A theoretical model for ice-structure interaction. Proc. 13th international conference on OMAE, Houston , Vol 4, pp. 29 - 34

Walden. J.T., Hallam, S.D., Baldwin, J.T. (1987) An Explicit Technique for Calculating First-Year Ice Loads on Structures, Proceedings of ASME OMAE symposium. Houston

Proceedings of the Seventh (1997) International Offshore and Polar Engineering Conference
Honolulu, USA, May 25-30, 1997

ISBN 1-880653-28-1 (Set); ISBN 1-880653-30-3 (Vol. II)

Thermal Stressing of Pack Ice

James K. Lewis
Ocean Physics Research & Development
Long Beach, MS, USA

ABSTRACT

A thermodynamic formulation for pack ice is presented that includes a means of relating stresses in floes to thermally-induced strain rates. The formulation combines some basic thermodynamics as well as a mechanics component which includes viscous creep and mechanically-induced strain rates due to typical floes being too large to bend or twist in response to spatially varying strain rates. Moreover, the rheology quantifies the impact of existing cracks in the floe on the overall strain rate in the floe and the resulting stress state. A paradigm is put forward which specifies tensile fracturing and the resulting stress relief as being proportional to the amount of existing cracks and some measure of the tensile strength of the ice in the floe. This allows the thermo-mechanical equations to be used for both first-year and multi-year floes by differentiating the two by the extent of existing cracks.

KEY WORDS: Pack ice, thermodynamics, rheology, fracturing, modeling.

INTRODUCTION

In addition to the body of work on the thermodynamics of sea ice, a great deal of work has been performed dealing with the mechanics of pack ice. The two topics are closely related in that thermal forcing can cause significant stresses and fracturing in pack ice. In this work, an attempt is made to provide a thermo-mechanical formulation that combines the more basic thermodynamic processes in a floe with a rheology relating thermally-induced strain rates to the stress state within the floe. Crossing over from pure thermodynamics to thermo-mechanics is complicated by the fact that we do not have a definitive relationship between the stressing and fracturing of pack ice. In addition, observations of thermo-mechanical processes in first-year and multi-year floes are quite different even for identical thermal forcing. Here we will provide the basic formulation for the thermodynamics and then concentrate on the details of the mechanics.

Pertinent mechanical studies include the seismoacoustic investigations of Stein (1988) who studied fractures in pack ice, propagation paths, and the properties of the resulting ice- and water-borne elastic waves. His work showed that an event analysis of individual fractures could be used to determine the location of fractures in a floe. Xie and Farmer (1991) extended the work of Stein by determining the location of a host of individual cracks in adjacent first-year and multi-year floes in the Canadian archipelago. Their distribution of fracture events on a multi-year floe showed far more fractures over regions that were relatively smooth and covered with snow as compared to similar regions on an adjacent first-year floe. This indicated a fundamental difference between stressing and fracturing in first-year and multi-year floes for the same thermal forcing.

Related mechanical studies have dealt with understanding the stress state within a floe. Lewis (1993) put forward a formulation relating stresses to thermally-induced strains for multi-year floes. His formulation included the viscoelastic response of the ice (creep) as well as the effect of strains occurring elsewhere in the floe. Moreover, he included a very simple means of accounting for stress relief due to fracturing of ice under tension. His rheological model did a reasonable job in reproducing observed stresses in a multi-year floe in the eastern Arctic Ocean (Lewis, 1993). However, when the same model was applied to a first-year floe (Lewis et al., 1994), it was found that the rheological formulation could not reproduce the observed stresses except by excluding the component for fracturing and stress relief for ice under tension. Once again, this was an indication of a fundamental difference between the mechanics of first-year and multi-year floes.

In an attempt to address the rheological differences between first-year and multi-year floes, Lewis (1995) put forward a conceptual model of how existing flaws within a floe impact the stress state of the ice. A modified rheology was developed which quantifies the extent of existing cracks in the floe and how they impact the overall strain rate and resulting stresses within a floe. However, the question as to stress relief due to fracturing still needed to be addressed. Lewis (1995) concluded that, to make a rheology that is consistent between first-year and multi-year floes, stress relief during periods of tension should be specified as a function of the flaw characteristics of the floe.

In this paper, an enhanced thermo-mechanical formulation is presented that includes a rheology that relates stresses in pack ice to thermally-induced strain rates. As in Lewis (1995), the rheology quantifies the impact of existing cracks on the overall strain rate in the floe. But a paradigm has been added which specifies any stress relief as being proportional to the amount of existing cracks in the floe. Thus, we can simulate first-year and multi-year floes using the same thermo-mechanical model by differentiating the two based on the extent of existing cracks. The results of numerical simulations using the thermo-mechanical governing equations are presented.

THERMODYNAMICS

The thermodynamic component of the governing equations includes the more basic thermodynamics for a floe. The thermodynamic component is used to calculate vertical temperature variations as a function of the time-varying forcing at the top of a floe. Since lateral variations of temperature are relatively small, we use the governing equation for the vertical flux of heat within an ice layer of thickness H,

$$H \rho_I c_I (\partial T/\partial t) = (k_I \partial T/\partial z)_U - (k_I \partial T/\partial z)_L \quad (1)$$

where ρ_I is the ice density, c_I is the specific heat of ice, T is the ice temperature, t is time, z is positive upwards, k_I is the thermal conductivity, and $(k_I \partial T/\partial z)_U$ and $(k_I \partial T/\partial z)_L$ are the conductive fluxes of heat at the top and bottom of the layer, respectively. The model divides the floe into horizontal layers, the thicknesses of which do not vary with time but may vary with depth. The only exception is the bottom level which can become thicker or thinner with time depending on the net melting/growth of the floe at the ice-ocean interface.

In the model, the surface of the floe has an emissivity of 0.97 and an albedo of 0.65 (Maykut, 1986; Perovich, 1994). Moreover, a layer of snow on the surface of the floe can be specified, with its own characteristic thermal conductivity, specific heat, density, emissivity (0.99), and albedo (0.8). The conductive heat fluxes on each side of any snow/ice interface are made equal by calculating an appropriate interfacial temperature. This interfacial temperature is then used in the next time step for calculating heat fluxes in the snow above the interface and in the ice below the interface.

A finite difference scheme is used to approximate (1). The sequence of calculations is as follows. For a given snow cover, the albedo, emissivity, and surface heat fluxes are determined. Then the specific heats and thermal conductivities for each vertical level are calculated as functions of ice level temperatures, salinities, brine volumes, and densities. With the application of boundary conditions at the surface and bottom of the ice, we have a system of n equations for n unknown temperatures. These equations are solved using Gaussian elimination to give the vertical temperature profile at each time step.

To solve (1) we must determine the various parameters in the equation for each level (i.e., ρ_I and c_I) and at the interfaces of each level (k_I). The thermal conductivity of ice depends primarily on temperature and salinity. In this study, we use the relationship developed by Ono (1975): $k_I = k_o (1 - V_b) + k_b V_b$, where k_o is the thermal conductivity of pure ice at temperature T_i, k_b is the thermal conductivity of any brine in the ice, and V_b is the brine volume of the ice, a function of the bulk density of the ice ρ_b, the salinity of the ice S_i, the brine density ρ_{br}, and the salinity of the brine, S_b. Expressions for calculating these variables can be found in Wade (1993).

The specific heat of the ice is also expressed as a function of the salinity and temperature of the ice following Pounder (1965).

SURFACE BOUNDARY CONDITIONS

In the top layer of the model, there exist heat exchanges at the surface that include sensible heat flux (QSENS), absorbed short wave solar radiation (QSLR), longwave radiation from the floe to the atmosphere (QBI), and longwave back radiation from the atmosphere (QBA). We will be dealing primarily with winter-time conditions, so evaporative heat fluxes are neglected (Maykut, 1986). In addition, we neglect that component of solar radiation that is absorbed in the interior of the snow or ice.

The equation used for temperature changes in the top model layer is based on two assumptions. The first is that the top layer of snow or ice is specified to be thin enough that its temperature is indicative of the surface temperature of the ice/snow. This allows this temperature to be used in specifying QSENS and QBI.. The second is that the conductive flux of heat though the top of the surface layer can be approximated by QSLR + QSENS - QBI + QBA. We use QSLR = (1.0 - albedo) x QSOLAR, where QSOLAR is any measured incoming short wave radiation. We use QSENS = $\rho_a C_P C_S U (T_a - T_T)$, where the observed air temperature is T_a, ρ_a is the air density, C_P is the specific heat of air, U is the observed wind speed (m/s), and C_S is a bulk transfer coefficient (1.2×10^{-3}; Maykut, 1986). QBA is the observed longwave radiation (W/m^2). QBI is specified directly as a function of the temperature of the surface layer T_T: QBI = $e_L \Psi (273.16 + T_T)^4$ where e_L is the effective emissivity (for either ice or snow) and Ψ is the Stefan-Boltzmann constant.

BOTTOM BOUNDARY CONDITIONS

For this study, observations from studies on a first-year and a multi-year floe are used to specify a time-varying oceanic heat flux based on the calculated bottom ice temperature and a constant under-ice temperature (-1.9°C, first-year floe; -1.75°C, multi-year floe). Moreover, at the end of each time step the change in ice thickness of the bottom ice layer is calculated based on melting/freezing due to the net heat flux at the ice-water interface. We assume that the under-ice water temperature is always at the melting/freezing point, so

$$dH_{bottom} / dt = (\rho_I L)^{-1} k_{I,bottom} dT/dz \quad (2)$$

where H_{bottom} = z at the ice-ocean interface, z positive upwards, and L is the latent heat of fusion (a function of salinity and temperature; Pounder, 1965). The model keeps track of this moving boundary every time step for a given snow cover and redefines the thickness of the bottom layer.

If the ice is freezing ($dH_{bottom} / dt < 0$), the salinity of the bottom layer is adjusted based on the underlying water salinity (25 ppt for the simulations in this study). The salinity of the newly frozen water is assumed to be governed by $S = K_{eff}$ x 25 ppt where K_{eff} is an effective distribution coefficient, a function of the rate of growth, dH_{bottom}/dt as given by Cox and Weeks (1975).

ICE SALINITIES

Since salinity plays a significant role in determining the specific heat and thermal conductivity of pack ice, the thermodynamics includes a scheme to calculate a new vertical profile of salinity based on any brine expulsion or gravity drainage occurring within the ice. During episodes of cooling in a given ice layer, the expression developed by Cox and Weeks (1986) for gas-free ice is used to determine the change in salinity due to brine expulsion.

If the profile of brine volume is unstable, the expression of Cox and Weeks (1975) is used to determine the salinity change due to gravity drainage whenever the brine volume of a layer is >0.05. For either brine expulsion or gravity drainage, any salt removed from a given layer is added to the underlying layer.

To review, the thermodynamics includes the impact of snow cover (if any), temporally and spatially varying thermal conductivities and specific heats, varying atmospheric heat fluxes, growth/melting at the bottom of the floe, and temporally and spatially varying ice salinities.

MECHANICS

RHEOLOGICAL FORMULATION

The governing physics dealing with thermally-induced, uniaxial, normal stresses in pack ice consists of two components. The first component is based on thermally-induced strain rate information and represents the ice as a spring that is forced by strains which result from thermal and mechanical processes. The governing equation relating stress (positive being compression) and strain rates is

$$\sigma_t = E' (\varepsilon_t - \zeta_t - \gamma) \quad (3)$$

where σ_t is the time rate of change of stress, E' is the effective elastic modulus, ε_t is the thermally-induced strain rate, ζ_t is a mechanically-induced strain rate, and γ is the strain rate as a result of the viscous creep of a material under stress. As the thermal model marches forward in time, strain rates for various ice layers can be calculated using $\varepsilon_t = \beta\, dT/dt$ where $\beta = 51\times10^{-6}/°C$. The parameterization used for E' was derived by Lewis (1993) based on observed variations of E' with T, STR, and porosity P:

$$E'(T,P,STR) = \varpi \log(STR + 3) + 3.5\ \text{GPa} \times (1 - 7.5472 \times P) \times (1 - 0.0714 \times T) \qquad (4)$$

where $STR = |\varepsilon_t - \zeta_t - \gamma|$ is the overall strain rate for a given time and depth in the ice, and $\varpi = 0.1$ GPa per log(STR).

There are a number of parameterizations for the creep γ of freshwater ice (Nixon and McRoberts, 1976; Sinha, 1988; Sunder and Wu, 1990) as well as adaptations for saline ice (Lewis, 1993). This study uses a modification of the Nixon and McRoberts (1976) two-term viscous formulation for freshwater ice. The formulation was modified by Richter-Menge and Cox (1995) so that the creep of saline ice could also be specified by considering an effective stress which results from the porosity characteristics of saline ice:

$$\gamma = 1.86\times10^{-15}(1 - T)^{-5.25}(\sigma/(1 - P^{1/2})^2)^n + 3.63\times10^{-29}(1 - T)^{-2.09}(\sigma/(1 - P^{1/2})^2)^4 \qquad (5)$$

where the units are s^{-1}, σ is stress in Pa, and n has a value of $1.36 \times (1 - T)^{0.19}$. This creep parameterization gives results similar to that used by Lewis et al. (1994). The important exception is that the Richter-Menge and Cox formulation results in much greater creep at higher temperatures (>-5°C). This increased creep, previously seen in freshwater ice, has recently been verified for saline ice also (Richter-Menge and Cox, 1995).

The ζ_t term reflects strain transmitted throughout the ice as a result of the floe being too large to bend or twist in response to spatial variations of strain rates. As the upper layers of a floe attempt to shrink due to decreasing temperatures while the lower layers of ice are kept at a constant temperature, the floe, if small enough, will curl concave upwards in response to the vertical differences in strain. However, for typical floes (> 300 m in diameter), the restoring moments resulting from the gravitational pull on the outer edges of the floe and the buoyancy of the underlying water are too great to overcome. As a result, the ice will remain flat and not curl up (Evans and Untersteiner, 1971; Bogorodsky et al., 1972).

As the upper layers of a relatively large floe attempt to shrink in the horizontal, the lateral strains are transmitted downward, and the lower layers of the ice also sense shrinking to some degree. But since there is no associated thermal shrinking of the ice in the lower layers, the ice in the lower layers undergoes compression. Because the lower layers are not shrinking and the floe is too large to curl up, the physical connection between the layers does not allow the top layers to shrink completely either. Tensile stress thus develops in the upper layers of the floe in response to the thermal gradient. Overall, the vertically integrated stress is zero as a result of the ice not bending. (This is a well described process and was studied extensively during research into containment walls surrounding nuclear reactors; Hoyle, 1964).

A formulation for the above process was presented by Bogorodsky et al. (1972). Lewis (1994) extended the formulation to account for creep and the lateral variations in strain rates, and then again to account for the impact of existing flaws (Lewis, 1995). For ice of total thickness H, the mechanically-induced strain rate in the x direction (relative to a fixed coordinate system on the floe) is given by

$$\zeta_t = \int_{-L}^{L}\left[\int_{0}^{H} E'(\varepsilon_t - \gamma)(1 - 2N)\,dz\right]dy \Big/ \int_{-L}^{L}\int_{0}^{H} E'\,dz\,dy \qquad (6)$$

where 2L is the extent of the horizontal variations of the snow cover thickness and N is the fractional coverage of any cracks (whether continuous or not) in the x direction at a depth z in the ice. The stress in the y direction is calculated as above except the integrals are with respect to z and x. Explicit in the above expression is the assumption that the ice under all thicknesses of snow cover have the same relative fraction of existing cracks, namely N(z).

Sensitivity simulations were performed by Lewis (1993) for parameterizations of the effective elastic modulus, mechanically-induced strain rates, and the creep of the ice. Those and other results found in Lewis (1994) and Lewis (1995) indicate that 1) thermal stresses are not all that sensitive to the formulation for the effective elastic modulus, 2) various creep formulations give about the same results, but the Richter-Menge and Cox formulation performs better at the higher temperatures found in the lower layers of an ice floe, and 3) mechanically-induced strain rates due to lateral variations of snow cover and vertical variations of existing cracks can have a significant impact on the stress state of a floe.

The solution technique for the rheology given by equation (3) follows Cox (1984). For each time step, the first calculation of ζ_t is with a creep based on the stresses of the previous time step. After an updated vertical stress profile is calculated for different thicknesses of snow cover, a new ζ_t can be determined using the corresponding creep. Then new vertical stress profiles can be determined. In this study, this process is gone through three times per time step to allow the model to converge to an appropriate ζ_t, γ, and vertical stress profile for each thickness of snow cover.

In calculating an appropriate ζ_t term, the snow cover distribution information must be used. For example, the snow cover data in Table 2 must be interpolated to the time step of the thermal stress model and used as a multiplier in the routine to calculate the integrals in ζ_t. If a snow thickness of 15 cm covered 30% of the region, the E' and E' $(\varepsilon_t - \gamma)(1 - 2N)$ terms in ζ_t for the 15 cm snow cover regions are multiplied by 0.30. (Using these fractional coverages makes the specification of the distance 2L arbitrary.) Thus, even though stress calculations are always calculated for a variety of snow covers, only those with a non-zero fractional coverage are allowed to influence the value of ζ_t. In this way, the impact of the spatial and temporal variations of the observed snow cover is included.

Fracturing Paradigm

The second component governing thermally-induced stresses in pack ice deals with fracturing due to tensile stresses. This is required since any fracturing results in some reduction of the stresses that caused the fracturing. Thus, before we can use the above expressions to model the stress state of a floe, we must specify a fracturing paradigm tailored to the size and characteristics of typical floes.

There are a variety of factors that must be considered in the development of a fracturing paradigm. We know from under-ice acoustic data and in-situ stress observations that thermally-induced tensile fracturing can be initiated when the stress state is only of the order of 10's of kPa (Lewis, 1994; Lewis et al., 1994). This is considerably less than the intrinsic tensile yield strength of individual specimens from floes (-200 to -1000 kPa for typical floe temperatures and salinities). It is generally accepted that existing flaws within floes are responsible for the discrepancies between the measured tensile strength of ice specimens and the much lower stress levels at which tensile fracturing is initiated in nature. With pack ice, tensile fracturing most likely deals with stress states appropriate for propagating pre-existing flaws. Under uniaxial loading that is acting parallel to an existing crack, the sides of the crack respond in such a way that the crack takes on an elliptical shape (Dieter, 1986). The minor axis of the ellipse is parallel to the loading. The sides of the crack act as lever arms, and a torque is created which results in a stress concentration at the tip of the ice crack which is considerably larger than the background stress causing the crack to bend in the first place. As a result, the ice can fracture at the tips of the crack at relatively low background tensile loading. As the cracks are extended, the lever arms become longer, and the torque for the same environmental forcing becomes greater.

The term *fracture toughness* is used in fracture mechanics when referring to the yield strength of flawed materials as discussed above. Fracture toughness studies of ice deal with the prying open of a crack. The parameter worked with is the critical stress intensity factor, K_{Ic}, which is related to the overall tensile failure stress of the ice σ_{ten} by the relationship

$$K_{Ic} = \sigma_{ten} (\pi c)^{1/2} \quad (7)$$

(Mellor, 1986) where c is the half length of the crack. Studies of fracture toughness for sea ice have dealt with cracks though the entire ice thickness. The results of these studies show that, for $\varepsilon_t < 10^{-3}$/s, which is typical for thermally-induced strain rates, K_{Ic} is approximately constant. But observations also indicate that K_{Ic} decreases with increasing temperature (Mellor, 1986). This is likely a reflection of the fact that σ_{ten} increases as the ice becomes colder.

The development a fracturing paradigm for an ice floe is not straight forward. Firstly, cracks in floes commonly do not run entirely through the thickness of the floe. Secondly, for those cracks that only run partially through a floe, the controlling length scale c may be the depth of the crack, not the length of the crack. Thirdly, if we take 75 kPa $m^{1/2}$ as a reasonable value for K_{Ic} in a floe (Mellor, 1986), then for c = 1 m and c = 10 m, the corresponding σ_{ten}'s would be -42.3 kPa and -13.4 kPa, respectively. And although these are reasonable values for background tensile stresses based on under-ice noise and in-ice stress observations in the arctic (Lewis, 1994; Lewis et al., 1994), they are far too small based on laboratory tensile strength measurements (Dykins, 1970; Richter-Menge and Jones, 1993). Finally, recent measurements in first-year pack ice indicate that there is also a scale effect for through-ice cracks, with K_{Ic} decreasing with the size of the crack and the size of the floe (Dempsey, 1996). However, there is no clear means at this time of relating these results on size effect to actual thermally-induced fracturing in a first-year or a multi-year floe.

For this study we formulate a fracturing paradigm such that, if

$$\sigma(z) (\pi c)^{1/2} > \sigma_{ten} (\pi c)^{1/2} \quad (8)$$

at depth z for some c, fracturing is said to occur if N(z) is non-zero (meaning c is non-zero). If we make the assumption that c is relatively constant, the fracturing criteria reduces to $\sigma(z) > \sigma_{ten}$. This means that, for a given model-predicted $\sigma(z)$, we need only determine an appropriate σ_{ten} to ascertain if fracturing is going to occur. To do this, we consider the work of Dykins (1970) who provided data on the variation of the uniaxial tensile strength of saline ice as a function of temperature T and salinity S. Richter-Menge and Jones (1993) showed that similar strength levels are found for strain rates down to 10^{-5}/s, typical for the thermal stressing of floes in the arctic ice pack. These tensile strengths can be approximated over a range of T and S by

$$\sigma_{TEN} = (-348.9 + 19.8\,T + 17.9\,S - 0.9\,T\,S)\ \text{kPa}. \quad (9)$$

Previous studies indicate that $|\sigma_{ten}| << |\sigma_{TEN}|$; i.e., thermal fracturing is observed to occur at background stress levels within floes that are considerable less than the measured tensile strength of saline ice (Lewis, 1994; Lewis et al., 1994). Based on those previous studies, we chose $\sigma_{ten} = 0.05\ \sigma_{TEN}$, and thus allow for varying fracture toughness as the tensile strength of the ice changes. (Chosing a value larger than 0.05, say 0.1 or larger, results in the situation of $\sigma(z) < \sigma_{ten}$ during the simulations, so little or no fracturing occurs.)

Studies indicate that stress relief for fracturing that is approximately the scale of the ice being fractured results in a final stress that is ~45% less than the maximum stress (Mellor, 1986). But we can expect that, as the size of the ice specimen becomes larger with respect to the size of the flaw, the stress relief should become smaller. In terms of our parameter N, stress relief → 45% of σ_{ten} as N → 1.0, and stress relief → 0.0 as N → 0.0. We will assume a linear variation between these two extremes. Thus, when $\sigma(z) > \sigma_{ten}$, the model-calculated stress is reduced to

$$\sigma (1 - N) + \sigma_{ten} \times 0.55 \times N. \quad (10)$$

With our fracturing criteria and this expression for stress relief, we can simulate the thermally-induced stresses of an ice layer at a given time for a given snow cover given the vertically-varying N(z).

THERMO-MECHANICAL SIMULATIONS

The above thermo-mechanical governing equations were tested by simulating two scenarios using observations to drive the model. The following briefly describes the data collected during experiments on a first-year floe and on a multi-year floe. The model begins with an ice thickness of 1.55 m for all snow covers, with a vertical grid size of 8 cm for the ice and 3 cm for any snow cover as well as the top layer of ice. The time step of the model is 5 minutes.

FIRST-YEAR FLOE

Meteorological data over a first-year floe were collected during 22 March 1992 (Julian day 82) through 31 March 1992 in the Canadian archipelago (see Lewis et al., 1994, for complete details). These observations are shown in Fig. 1. A warming trend began on Julian day 84 and continued through Julian day 86. At the end of day 86, the air temperature dropped to approximately -30°C, the winds became calm, and the incoming atmospheric radiation decreased significantly. Over the next four days, the region underwent another period of gradual warming.

Snow depth measurements were obtained along a 200 m long line in the study area. Snow depths were measured at 5 m intervals along the line, and snow thicknesses ranged from 4 to 32 cm. Table 1 gives

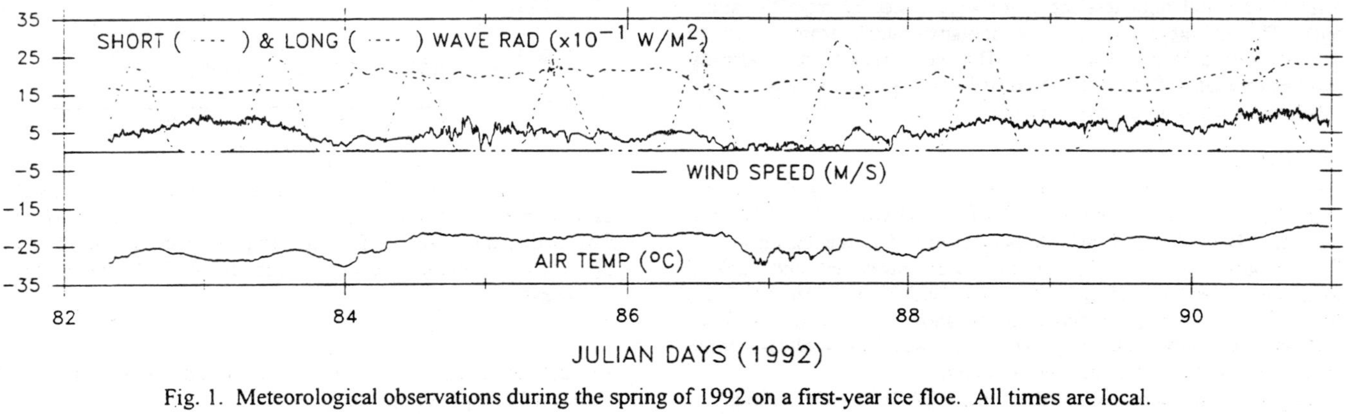

Fig. 1. Meteorological observations during the spring of 1992 on a first-year ice floe. All times are local.

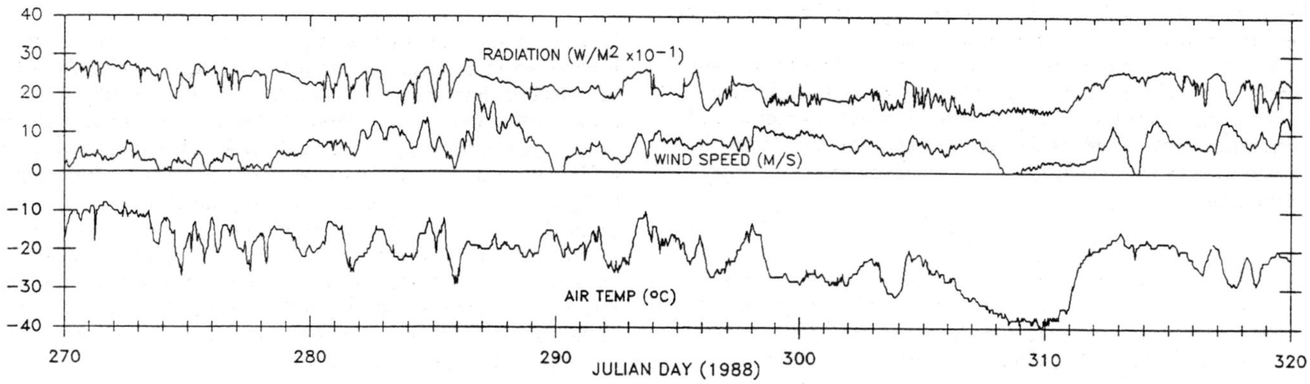

Fig. 2. Meteorological observations during the fall of 1988 on a multi-year ice floe. All times are local (shortwave radiation was negligible).

Table 1. Percent coverage of snow cover at 3 cm intervals based on the measurements on first-year ice, 1992.

Snow Cover Thickness (cm)											
0	3	6	9	12	15	18	21	24	27	30	33
0.0	0.0	17.1	21.2	9.8	9.8	17.1	12.2	7.3	0.0	2.4	2.4

Table 2. Percent coverage of snow cover at 3 cm intervals based on all measurements made during CEAREX.

	Snow Cover Thickness (cm)							
Jul Day	0	3	6	9	12	15	18	21
262.354	36	40	20	0	4	0	0	0
268.417	24	24	16	16	20	0	0	0
270.385	12	32	12	24	20	0	0	0
272.375	24	20	12	28	16	0	0	0
274.375	8	24	20	24	20	4	0	0
277.385	0	16	28	20	24	12	0	0
279.389	0	20	24	20	24	12	0	0
281.424	12	20	24	24	12	4	0	4
283.375	8	20	24	28	8	8	4	0
288.000	20	20	20	24	8	4	4	0
294.000	20	20	16	24	12	8	0	0
299.000	12	20	24	24	8	12	0	0
305.000	12	20	20	28	20	0	0	0
312.000	12	16	20	28	16	8	0	0
319.000	4	16	16	28	16	16	4	0
326.000	4	16	8	16	24	12	8	12

the percentages of the snow cover in bins of 3 cm intervals. Also, a total of nine full thickness ice cores were collected from the study area. The ice thicknesses at the coring sites ranged from 1.41 m to 1.71 m with a mean of 1.57 m. The mean salinity and porosity profiles through the ice were calculated using all cores (Lewis et al., 1994) and used as input to the model.

MULTI-YEAR FLOE

Conditions for a multi-year floe were collected during the Coordinated Eastern Arctic Experiment (CEAREX) (Lewis, 1993, 1994: the CEAREX Drift Group, 1990). Meteorological data are shown in Fig. 2. Temporal variations of snow cover were measured, and Table 2 gives the percentages of the snow cover with time in bins of 3 cm intervals. The average thickness of the snow cover is seen to have increased with time, but in general there was nearly always some fraction of the floe which had no snow cover.

Ice cores were also collected during CEAREX. As before, the

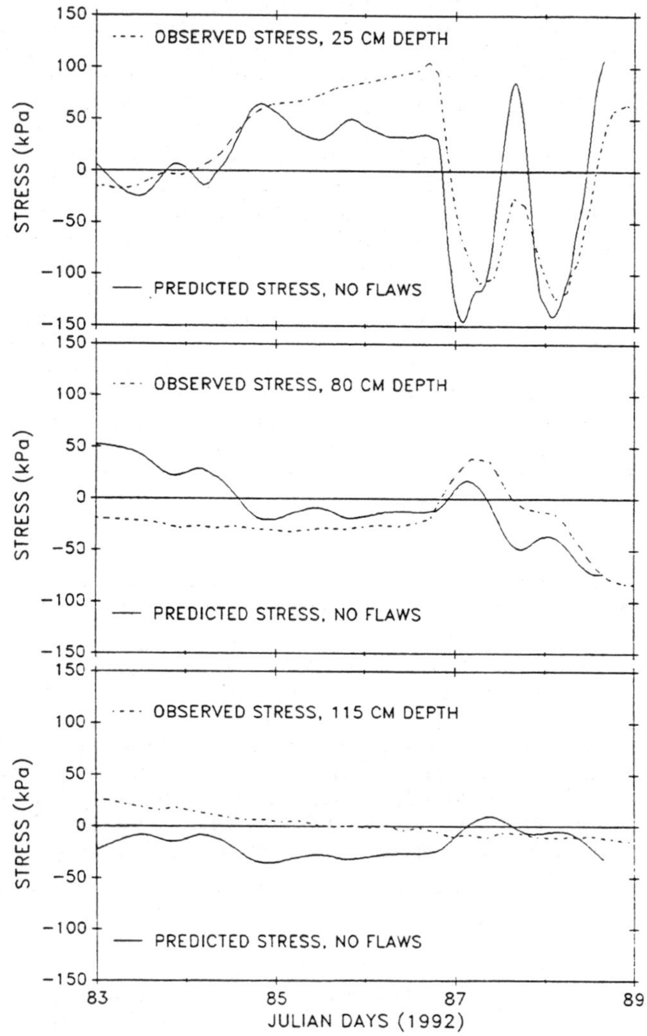

Fig. 3. Predicted and observed stresses for first-year ice based on an existing fracture profile of N = 0 for all levels in the ice (no flaws). The predicted stresses are for a region of ice with a 12 cm snow cover up to day 86.7 and no snow cover after day 86.7. Positive stresses are compressive.

vertical variations of the average salinity and porosity (Lewis, 1994) were used as input to the model.

Model Results, No Flaws

The stresses predicted for the first-year floe with N = 0 (no flaws) for all ice layers are shown in Fig. 3. Also shown are variations of the stress invariant $(\sigma_1+\sigma_2)/2$, where σ_1 and σ_2 are the principal stress components, determined by sensors in the floe (see Lewis et al., 1994, for details). The results of the simulation show that the model can reproduce many of the basic features of the observed stress variations. In general, there is a tensile (negative) bias in the bottom layers of the floe and a compressive (positive) bias in the upper layers of the floe during day 87. There is a dramatic change in the stress state of the upper layers of the ice during Julian days 86.7 to 87.4 after the passage of the cold front.

The stresses for the multi-year floe situation predicted with N = 0 (no flaws) are shown in Fig. 4 along with CEAREX stress data collected some 200 m from the edge of the floe. The dominant signal in the surface layer stresses occurred between Julian days 310-320 when compressive forces reached as high as 100 kPa. However, the strongest, most persistent signal near the bottom of the floe occurred between days 306-312. We see a negative bias for the upper layers of the ice and a positive bias for the bottom layers of the ice. For both the lower and upper layers of the ice, the model over-predicts the magnitude of the observed stress variations. A significant over-prediction occurs in the lower part of the ice. In response to the large compressive forces in the surface layers during days 312-317, the model-predicted stresses at 120 cm change by ~100 kPa (compression to tension). The observed stresses are seen to change by only 50 kPa.

Model Results, Flaws

To perform these simulations, we must specify the vertical variation of the relative amount of ice that is already cracked, an unknown that cannot be calculated or estimated using equations or existing observations. Thus, we can only guess at values of N and compare the resulting model-predicted stresses with observed stresses. Since each layer of ice can have its own value of N, we limited our test simulations in this study to those in which we approximated the vertical structure by assigning a particular value of N to the top third of the ice layers, another to the middle third, and still another to the bottom third. We then performed test simulations to determine which approximation of the vertical structure could best reproduce the observed stresses.

The predicted stresses for the first-year floe with N = 0.01 for the top third ice layers, N = 0.0 for the middle third, and N = 0.5 for the bottom third layers of ice are shown in Fig. 5. This would imply that the surface layers of the first-year ice had relatively few flaws, the central part of the ice was flaw-free, and the lower layers of the ice were considerably more flawed.

The multi-year floe conditions were modeled using N = 0.7 for the top third ice layers, N = 0.9 for the middle third, and N = 0.3 for the lower third. The results are shown in Fig. 6. This N(z) would imply that the surface layers of the multi-year floe were extensively flawed, considerable more so than the first-year floe. Such a flaw structure does a reasonable job in reproducing the observed stresses.

A final simulation was performed for the first-year floe scenario with the albedo set to 0.8 for ice surfaces as well as snow surfaces. This was done since most situations of a floe with "no snow cover" is actually one in which the snow cover is spotty or relatively thin. Thus, we might not have an albedo for a truly "ice surface" situation. The results are shown in Fig. 7. The higher albedo results in less of an excursion of the surface stress during days 87-89, and the predicted

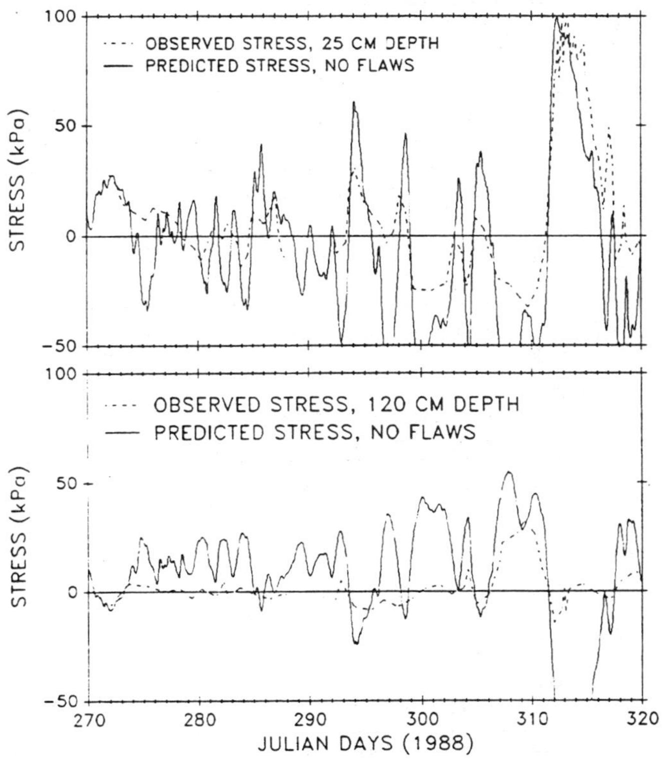

Fig. 4. Predicted and observed stresses for multi-year ice based on an existing fracture profile of N = 0 for all levels in the ice (no flaws). The predicted stresses are for a region of ice with a 12 cm snow cover. Positive stresses are compressive.

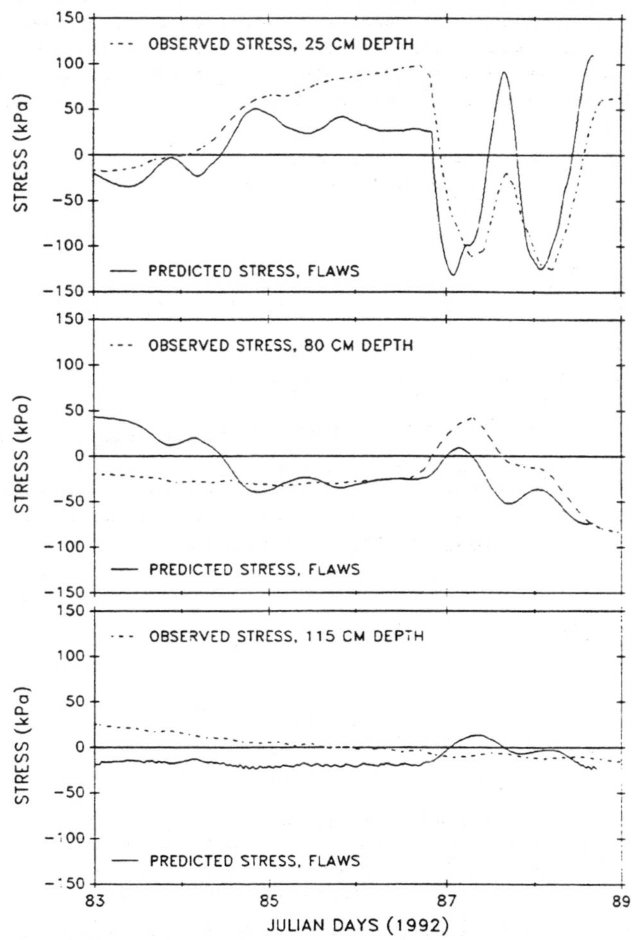

Fig. 5. As in Fig. 3 but with N = 0.01 for the top third ice layers, N = 0.0 for the middle third, and N = 0.5 for the bottom third.

stresses during these days are now in better agreement with the observations.

SUMMARY AND CONCLUSIONS

Expressions are presented that relate the thermal stressing of floes and the resulting stress relief for different characteristics such as salinity, porosity, and flaw structure. The basis of this thermo-mechanics is a heat flux model for pack ice that includes some very basic phenomena plus a rheology based on the fact that typical floes are too large to bend or twist in response to vertical or horizontal variations of strain rate. Along with the rheology is a paradigm used to determine the occurrence of fracturing when the ice is in tension and the amount of stress relief.

We see that our rheology and fracturing paradigm allow us to make reasonable predictions of the observed stresses in both first-year and multi-year floes by estimating the vertical variations of existing cracks in the ice. The simulations imply that the multi-year floe was significantly more flawed than the first-year floe, a logical result. In fact, the predictions would be considered surprisingly good considering all our assumptions, the fact that observed atmospheric forcing is in essence filtered through the thermodynamic and mechanical formulations to produce the stress predictions, and all variables in the formulations were based on information from the literature. However, the introduction of the unknown N(z) results in a parameter which might allow for tuning to fit observations without actually specifying the controlling physics exactly and completely.

It is obvious that the weak point in the formulation of the mechanics is the tremendously simplifying assumption that c is relatively constant. Certainly, there is no actual evidence that would support (or refute) that the controlling length scale parameter for thermally-induced fracturing in floes would be nearly constant. And, although our governing equations appear to be able to provide good predictions of observed stresses, the question of the traits of N(z) still remains. It is obvious that additional research is required to determine how flaws that extend only partially through a floe impact fracture toughness and to investigate the spatial and temporal variations of flaws in floes. Based on observations of thermally active flaws in first-year ice (Lewis et al., 1994), sampling flaw structure poses the practical problem in that the sizes of such flaws are apparently of the order of 5-10 m. Sampling at smaller scales would likely underpredict the true flaw structure of a floe.

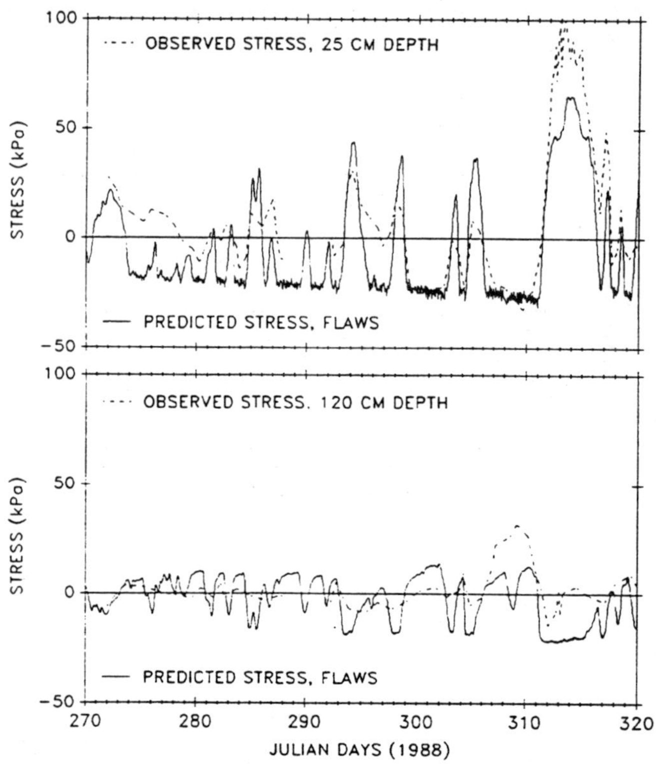

Fig. 6. As in Fig. 4 but with N = 0.7 for the top third ice layers of ice, N = 0.9 for the middle third (no observations available), and N = 0.3 for the bottom third layers of ice.

ACKNOWLEDGMENTS

This work was supported by the Office of Naval Research, Arctic Branch, through a contract to Scientific Solutions, Inc. The author wishes to thank Drs. P. J. Stein and E. Schulson for their helpful discussions and suggestions.

REFERENCES

Bogorodsky, V. V., V. P. Gavrilo, A. V. Gusev, Z. M. Gudkovich, and A. P. Polyakov (1972). "Stressed ice cover state due to thermal wave and related underwater noise in the ocean." Proc., Ice Symposium, 1972, Leningrad. AIRH Translation, pp 28-33.

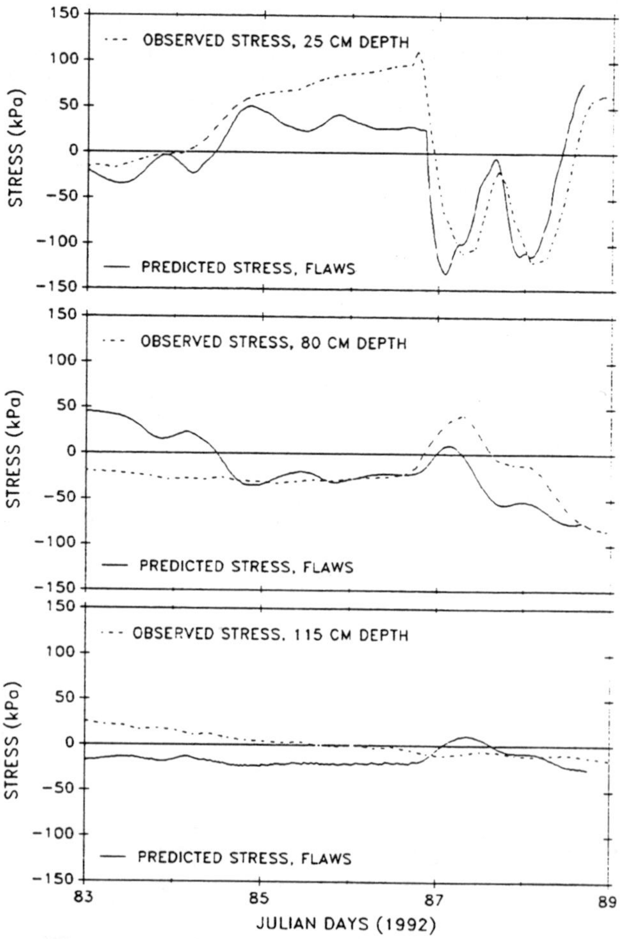

Fig. 7. As in Fig. 3 but with N = 0.01 for the top third ice layers, N = 0.0 for the middle third, and N = 0.5 for the bottom third and the albedo for all surfaces (ice or snow) was 0.8.

CEAREX Drift Group (1990). "CEAREX drift experiment." *Trans. Am. Geophys. Union*, 71, pp 1115-1118.

Cox, G. F. N. (1984). "A preliminary investigation of thermal ice pressures." *Cold Regions Sci. Tech.*, 9, pp 221-229.

Cox, G. F. N., and W. F. Weeks (1975). "Brine drainage and initial salt entrainment in sodium chloride ice." Tech. Rpt. 82-30, U.S. Army Corps. Eng. Cold Reg. Res. Eng. Lab., Hanover, NH. 85 pp.

Cox, G. F. N., and W. F. Weeks (1986). "Changes in the salinity and porosity of sea ice samples." *J Glaciol.*, 32, pp 371-375.

Dempsey, J. P. (1996). "Scale effects on the fracture of ice." In Johannes Weertman Symposium (R. J. Arsenault et al., eds.). The Minerals, Metals, and Materials Soc., TMS Annual Mtg., pp 351-361.

Dieter, G. L. (1986). *Mechanical Metallurgy*. McGraw-Hill, New York. 751 pp.

Dykins, J. E. (1970). "Ice engineering-tensile properties of sea ice grown in a confined system." U.S. Navy Civil Engineering Lab., Tech. Rpt. R689.

Evans, R. J., and N. Untersteiner (1971). "Thermal cracks in floating ice sheets." *J. Geophys. Res.*, 76 (3), pp 694-703.

Hoyle, Russell (1964). "Plane strain and plain stress." In *Thermal Stress* (P. Benham and R. Hoyle, eds.), Sir Isaac Pitman & Sons Ltd., London. pp 43-51.

Lewis, J. K. (1993). "A model for thermally-induced stresses in multi-year sea ice." *Cold Regions Sci. and Tech.*, 21, pp 337-348.

Lewis, J. K. (1994). "Relating Arctic ambient noise to thermally-induced fracturing of the ice pack." *J. Acoust. Soc. Am.*, 95 (3), pp 1378-1385.

Lewis, J. K. (1995). "A conceptual model of the impact of flaws on the stress state of sea ice." *J. Geophys. Res.*, 100(C5), pp 8829-8835.

Lewis, J. K., W. B. Tucker, III, and P. J. Stein (1994). "Observations and modeling of thermally-induced stresses in first year sea ice." *J. Geophys. Res.*, 99(C8), pp 16361-16371.

Maykut, G. A. (1986). "The surface heat and mass balance." In *The Geophysics of Sea Ice* (N. Untersteiner, ed.). Plenum Press, New York, pp 395-463.

Mellor, M. (1986). "Mechanical behavior of sea ice." In *The Geophysics of Sea Ice* (N. Untersteiner, ed.). Plenum Press, New York, pp 165-281.

Nixon, J. F., and E. C. McRoberts (1976). "A design approach for pile foundations in permafrost." *Can. Geotech. J.*, 13, pp 40-57.

Ono, N (1975). "Thermal properties of sea ice IV." Tech. Rpt, US Army Cold Regions Res. and Eng. Lab., Hanover, NH.

Perovich, D. K. (1994). "Light reflection from sea ice during the onset of melt." *J. Geophys. Res.*, 99(C2), pp 3351-3359.

Pounder, E. R. (1965). *The Physics of Sea Ice*. Pergamon, New York. 151 pp.

Richter-Menge, J. A., and G. F. N. Cox (1995). "Creep of saline ice at low stresses and high temperatures." Proc.'s, *5th Intl. Offshore and Polar Eng. Conf.*, The Hague, Netherlands, June, Vol. II, pp 312-316.

Richter-Menge, J. A., and K. F. Jones (1993). "The tensile strength of first-year sea ice." *J. Glaciol.*, 39, pp 609-618.

Sinha, N. K. (1988). "Crack-enhanced creep in polycrystalline material: strain-rate sensitive strength and deformation of ice." *J. Materials Sci.*, 23, pp 4415-4428.

Stein, P. J. (1988). "Interpretation of a few ice event transients. *J. Acoust. Soc. Am.*, 83(2), pp 617-622.

Sunder, S. S., and M. S. Wu (1990). "On the constitutive modeling of transient creep in polycrystalline ice. *Cold Regions Sci. and Tech.*, 18, pp 267-294.

Wade, R. H. (1993). "Studies of the geophysics of sea ice." Ph.D. thesis, Univ. Alaska, Fairbanks. 133 pp.

Xie, Y., and D. M. Farmer (1991). "Acoustical radiation from thermally stressed sea ice." *J. Acoust. Soc. Am.*, 89(5), pp 2215-2231.

Proceedings of the Seventh (1997) International Offshore and Polar Engineering Conference
Honolulu, USA, May 25-30, 1997

ISBN 1-880653-28-1 (Set); ISBN 1-880653-30-3 (Vol. II)

The Ductile Deformation of Columnar (S2) Saline Ice Under Triaxial Compression

Jeffrey S. Melton and Erland M. Schulson
Thayer School of Engineering, Dartmouth College
Hanover, NH, USA

ABSTRACT

The deformation of columnar (S2) saline ice under triaxial compression was investigated at a temperature of -10 °C and a constant strain rate of 3.9×10^{-5} s^{-1}, using a true multiaxial testing system. The ice was proportionally loaded under confinement ratios of $R_{21}=\sigma_{22}/\sigma_{11}=0.25$ and $R_{31}=\sigma_{33}/\sigma_{11}$ which varied between 0 and 1. The stresses σ_{11} and σ_{22} are normal stresses applied in two orthogonal directions across the columns and σ_{33} was applied along the columns. The behavior of the ice was macroscopically ductile and most of the deformation occurred perpendicular to the direction of the columns, which is consistent with deformation through basal slip. The flow curve is characterized by a linear region with an ill-defined yield point and by a slowly varying post-yield region. The ice exhibited a small amount of across-column cracking under all along-column confinements. Volume was generally not conserved. The triaxial yield stress $\sigma_{11,y}$ and the triaxial peak stress $\sigma_{11,p}$ can be described by Hill's (1950) criterion for the failure of plastically orthotropic materials.

1. INTRODUCTION

The prediction of maximum ice forces on offshore structures and on other features requires failure criteria for ice loaded under multiaxial compressive states. This work contributes to that need. It deals specifically with S2 ice because it is from such a material that first-year ice sheets are made.

Ice undergoes a ductile-to-brittle transition as the rate of loading increases. Smith and Schulson (1994) showed that a truncated Coulombic-type failure criterion describes the biaxial brittle failure stress of S2 saline ice when compressed across the columns, and Schulson and Nickolayev (1995) demonstrated that Hill's (1950) criterion describes its ductile failure stress under similar loading. Gratz and Schulson (1996) showed that the brittle compressive failure stress of S2 saline ice under moderate triaxial loading can be described by a similar Coulombic criterion, and both Häusler (1983) and Timco and Frederking (1986) attempted to describe its ductile failure stress under such loading in terms of *n*-type yield functions. The question we consider here, following our previous work (Melton and Schulson, 1995), is whether Hill's (1950) criterion can describe the ductile failure stress of S2 saline ice under triaxial loading.

Previous studies have reported that the failure strength of columnar saline ice is independent of the along-column confinement. Timco and Frederking (1983) performed biaxial confined compression tests on columnar sea ice at -11 °C and a nominal strain-rate of $\dot{\varepsilon}=2\times10^{-4}$ s^{-1}. The ice was ductile, and its failure strength measured across the columns increased to ≈4 times the uniaxial failure strength with increased across-column confinement. However, the failure strength did not change significantly when the ice was confined along the columns. Similar results were reported (Timco and Frederking, 1986) for columnar Beaufort Sea ice tested at -2 °C and $\dot{\varepsilon}=2\times10^{-4}$ s^{-1} under the same loading conditions. Häusler (1981) carried out true triaxial compression tests ($\sigma_{11}\neq\sigma_{22}\neq\sigma_{33}$) on laboratory-grown saline ice of high salinity (10.6 ‰), similar to tests performed at the Ice Research Laboratory. His samples were proportionately loaded at -10 °C and at a strain rate of 2.0×10^{-4} s^{-1} along the principal axis. Though the number of data points was limited, his results, like those of Timco and Frederking (1983, 1986), suggest that under moderate across-column confinement the ductile failure strength is independent of the along-column confinement.

This work presents new results on the ductile failure strength of columnar S2 saline ice proportionately loaded and moderately confined across the columns. The variable is the degree of along-column confinement.

2. EXPERIMENTAL PROCEDURE

The ice was grown in the laboratory using the procedure described in Kuehn and Schulson (1994). Briefly, a commercial brand of salts, "Instant Ocean", was mixed with filtered Hanover tap water in an insulated tank to produce a brine solution of ≈23 ‰. The solution was frozen vertically downward by placing an aluminum cooling plate on the surface of the brine. The plate was cooled to -12 °C by a continuously circulating refrigerated glycol solution. After approximately two weeks, a cylindrical puck of 915 mm diameter and of 300 mm thickness formed. It was removed from the

Figure 1a. Thin section of columnar saline ice showing irregular grain structure. The columns are perpendicular to the plane of the thin section. Scale: 1 cm = 5.8 mm.

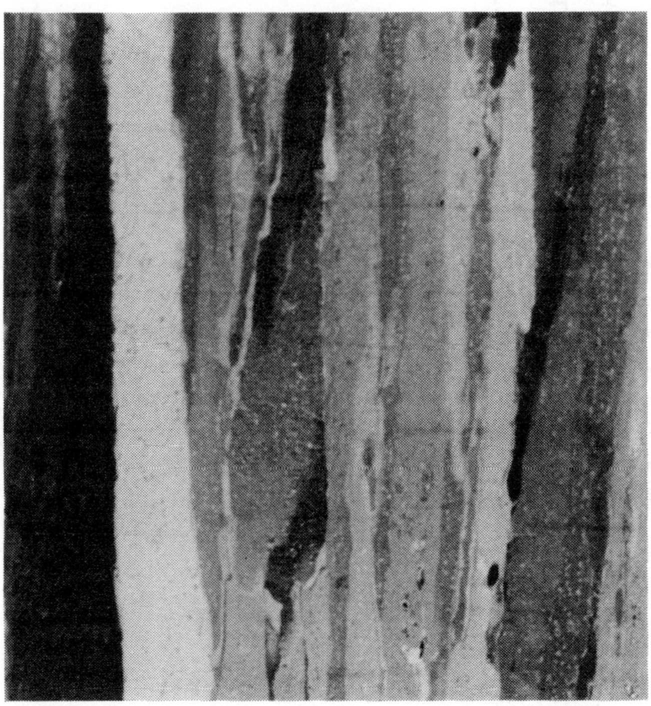

Figure 1b. Thin section of columnar saline ice showing the columnar grains. The columns are parallel to the plane of the thin section. Scale: 1 cm = 5.8 mm.

tank and cut into rough cubes 190 mm on edge using a chain saw. The structure of the top 20 mm or so of the ice was granular, giving way to a columnar structure; the bottom 50 mm or so had a mushy dendritic structure. When cutting the puck into cubes, care was taken to remove the top and bottom portions, leaving only the columnar ice. The samples were then cut into 165-170 mm cubes using a bandsaw and stored at -10 °C in insulated boxes until final milling and testing.

The cubes were composed of large, irregular shaped columnar grains, Figure 1, with an average diameter of ≈2 mm at the top of the sample and ≈6 mm at the bottom. The bulk density was 911± 9 Kg/m^3. The salinity of the ice was 4.3 ± 0.9 ‰, determined by conductivity measurements of the melt water. The *c*-axes were unaligned and confined to the horizontal plane within about ±15°. The grains contained plate-like regular arrays of trapped brine and gas spaced approximately 1 mm apart. The structure of the material was similar to the structure of first-year sea ice (see Weeks and Ackley, 1983).

Just prior to each test the samples were milled into 160 mm cubes. Care was taken to ensure that the vertical axes of the columnar grains remained normal to the top face of the cube. The maximum variation from parallel for the cube faces was 0.13 mm or 0.008 radians. Some cubes were tested within 24 hours after being milled, but most samples were tested within 6 hours to avoid sublimation.

Twelve tests were performed. The samples were loaded using the Ice Research Laboratory's multiaxial testing system (MATS). The MATS is a servo-hydraulic testing machine that is capable of independent triaxial loading along three orthogonal axes, and is housed in a dedicated cold room. The loads are applied through brass brush platens, similar to those used by Häusler (1981), which were designed to minimize transverse loading (see Gies, 1988). The tests were carried out at -10.0 °C with the temperature of the room within ±0.2 °C during the length of the test. The nominal strain rate was $\dot{\varepsilon}_{11}=3.9\times10^{-5}$ s^{-1} along the x_1 direction with σ_{11} and σ_{22} applied across the columns and σ_{33} along the columns. All twelve tests were terminated after 3% shortening in the x_1 direction. Figure 2 defines the co-ordinate axes with respect to the structure. During the experiments the across-column confinement, $R_{21}=\sigma_{22}/\sigma_{11}$, was kept at $R_{21}=0.25\pm0.02$ and the along-column confinement, $R_{31}=\sigma_{33}/\sigma_{11}$, was varied between 0 and 1 (±0.03). Two MTS extensometers per axis were mounted to the platens and used to measure the displacements along each axis. Load and sample displacement data were collected and stored on a personal computer using LabView software. When analyzing the displacement data, care was taken to remove the displacement of the platens, following the method given by Gratz (1996), leaving the true sample displacements.

3. RESULTS

3.1 General Features

In all tests the ice was macroscopically ductile, and in all but one test the deformation was characterized by a σ_{11}-ε_{11} curve that had an initial linear region followed by a peak and a slowly varying post-peak region. One curve did not have a peak. In nine tests the σ_{11}-ε_{11} curve exhibited a well defined peak and then strain-softening, with the post-peak branch of the curve having negative curvature, Figure 3a, which were designated as Type I curves. The σ_{11}-ε_{11} curves for two tests (SC 35, SC 36) had a broad, ill-defined peak and little or no strain-softening, Figure 3b, and were labeled Type II curves. Test SC 55 (R_{21}=0.27, R_{31}=0.53) had an initial steep linear region that progressed into a slowly rising linear region (Figure 3c) that did not peak before the test was terminated at ε_{11}=0.03. This σ_{11}-ε_{11} curve was designated a Type III curve. Table 1 lists the flow curve type for each test. The along-column confinement did not have any effect on the type of flow curve observed for a given test.

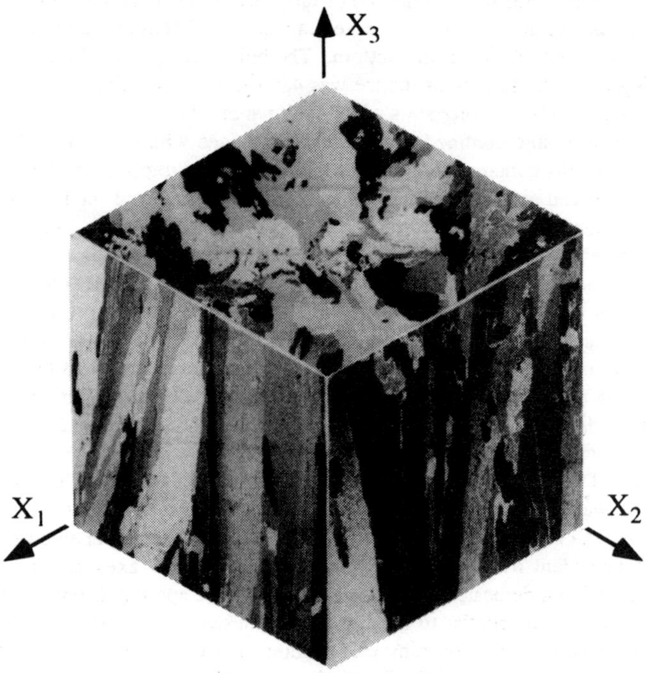

Figure 2. Definition of the co-ordinate axes with respect to the structure of the sample (Gratz, 1996). Cube edge length is 84 mm.

Audible reports were heard almost immediately upon loading the samples, owing to the formation of small cracks. The formation of the cracks was marked by flashes of reflected light, observed through the unloaded edges of the sample, though the cracks themselves were too small to be seen. A second type of crack was observed in the form of planar across-column cracks. The sound of the across-column cracks popping-in was louder and distinctly different from the sound of the smaller cracks forming, and they were not heard until 25-50 seconds (ε_{11}=0.001-0.002) into the test. The across-column cracks were easily seen with the naked eye after the sample was removed from the MATS (more below).

Post-test mass measurements showed that typically 3-5 grams (0.08%-0.13%) of the mass was lost during the tests, presumably owing to the expulsion of brine. The loss of the salts in the brine should cause a small decrease in the density of the sample, but post-test measurements of the samples showed the density to increase by 5±2 Kg/m^3, possibly due to the collapse of either gas pockets or brine filled pores during the test. It should be noted that these are preliminary observations due to the limited number of measurements made.

3.2 Strength

The strength of the ice is usually taken to be the peak stress, or highest stress that the ice can support. In this paper additional stress measurements are reported to better describe the behavior of the ice. The peak stress ($\sigma_{11,p}$) was defined to be the maximum stress sustained by the sample, and the corresponding strain was the peak strain. The yield stress ($\sigma_{11,y}$) was determined by finding the

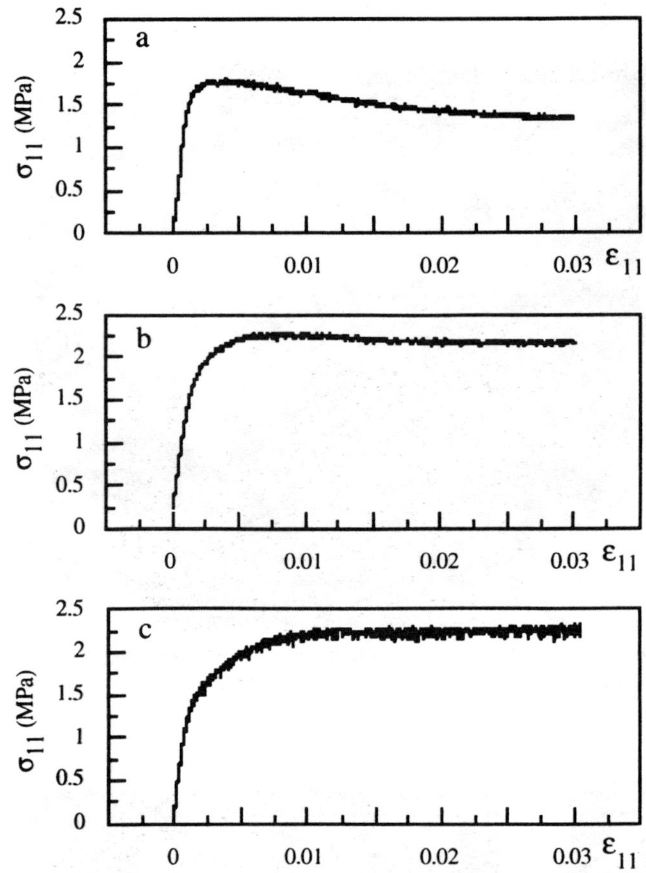

Figure 3. Ductile stress-strain curves for -10 °C and nominal strain-rate of 3.9×10^{-5} s^{-1}. (a) Type I stress-strain curve for SC 30 (R_{31}=0.10). (b) Type II curve for SC 36 (R_{31}=0.60). (c) Type III curve for SC 55 (R_{31}=0.53).

intersection of the σ_{11}-ε_{11} curve and a line with a slope equal to the initial tangent modulus ($\Delta\sigma_{11}/\Delta\varepsilon_{11}$), offset by 0.001 along the ε_{11} axis from the origin. The yield strain was then defined to be the strain at that stress. Two other stresses, $\sigma_{11,1}$ and $\sigma_{11,2}$, were the stresses at inelastic strains of 0.01 and 0.02, respectively. They were found by taking the value of σ_{11} from the σ_{11}-ε_{11} curve at 0.01 and 0.02 offset from the yield strain. Figure 4 illustrates the definitions of stress and strain used in this paper.

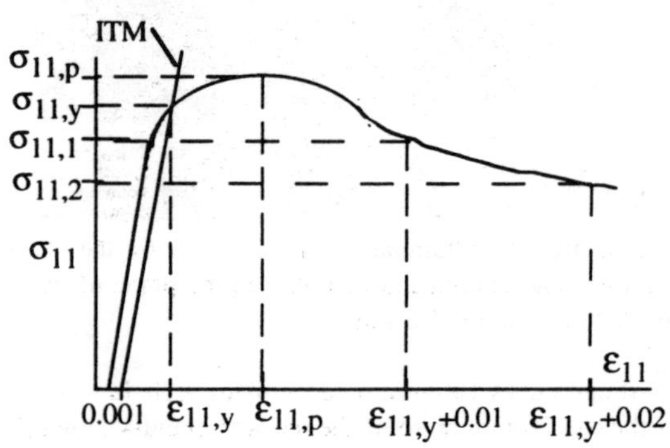

Figure 4. Sketch showing the definitions of the stresses and strains presented in this paper (not to scale).

Table 1 lists the values of the yield and peak stresses and the corresponding strains, as well as the stresses at 0.01 and 0.02 inelastic strain. Figure 5 shows the yield stress $\sigma_{11,y}$ versus the confining ratio R_{31}. To a good approximation the confinement along the columns has no effect on the yield strength. The slope of the fitted line, obtained through liner regression analysis, is 0.15 MPa, which is not statistically different from zero at confidence levels greater than 73%. The R^2 value is 0.11, which again suggests no significant effect. The best fit line predicts a rise in the yield stress of only 0.15 MPa over the range R_{31}=0 to R_{31}=1, while the standard deviation is 0.13 MPa. Thus, it appears that for the loading condition R_{21}=0.25 the yield stress can be represented by $\sigma_{11,y}=1.7\pm0.4$ MPa, independent of R_{31}. The regression co-efficients and critical values for the linear regression fit are given in Table 2.

Figure 5b shows the peak stress $\sigma_{11,p}$ plotted versus R_{31}. Test SC 55 (R_{31}=0.53) did not peak so the stress at the end of the test (ε_{11}=0.03) is plotted for reference. The stresses for both tests at R_{31}=0.6 were high compared to the rest of the data, but there were no unusual features found in the samples used for those tests, or in the test procedures, so the higher values are attributed to random scatter. Even with the SC 55 data included, the along column-confinement does not have any significant effect on the peak strength. The best fit line for the peak stress, with test SC 55 removed, has a slope of 0.26 MPa and an R^2 value of 0.18. The standard deviation of the peak stress, with SC 55 removed, is 0.17 MPa. Since the slope is not statistically different from zero for confidence levels greater than 82%, the linear trend is not considered significant compared to the scatter. Thus, the peak strength, like the yield strength, can the be expressed as a constant $\sigma_{11,p}=1.9\pm0.5$.

Figure 5 also shows the stress at 0.01 inelastic strain, $\sigma_{11,1}$, and the stress at 0.02 inelastic strain, $\sigma_{11,2}$, plotted versus R_{31}. The slopes of the best fit lines for stresses $\sigma_{11,1}$ and $\sigma_{11,2}$ are 0.42 MPa and 0.44 MPa, respectively, with critical confidence levels of 93% and 90%, respectively. At confidence levels below the critical levels the slopes are statistically different from zero. Even though the R^2 values for the fitted lines are low (0.24, 0.21 respectively), the high critical confidence levels suggest that there is an effect of increasing flow stress with increasing along-column confinement at these strains. This is reasonable because at higher strains the structure of the ice may begin to change, allowing the along-column confinement to have a greater effect on steady-state flow.

Direct comparisons of our strength values with the values from other studies is difficult because of the different test procedures used and the different types of ice tested. For instance, Häusler (1981) tested saline (10.6 ‰) ice at R_{21}=0.33 at a strain rate of $\dot{\varepsilon}_{11}=2\times10^{-4}$ s^{-1}. His strain rate is almost an order of magnitude larger than ours, which would give higher strengths. However, the higher salinity would weaken the ice somewhat. The across-column confinement at the peak for Timco and Frederking (1983, 1986) was different from the confinement in our tests, which makes it difficult to compare results. What is similar, though, is that we confirm the increase in the peak strength with increased across-column confinement. For the level R_{21}=0.25 used here, the yield stress increases by 21% and the peak stress by 36%, compared to uniaxial values of $\sigma_{11,y}=\sigma_{11,p}=1.4$ MPa. This is in agreement with the effect observed by Schulson and Nickolayev (1995), by Häusler (1981) and by Timco and Frederking (1983, 1986).

Test	R_{21}	R_{31}	$\sigma_{11,y}$ (MPa)	$\sigma_{11,p}$ (MPa)	σ_{11} @ ε_{11}=0.01 (MPa)	σ_{11} @ ε_{11}=0.02 (MPa)
SC28	0.23	0.00	1.6	1.7	1.5	1.4
SC30	0.24	0.10	1.7	1.8	1.6	1.4
SC31	0.25	0.21	1.7	1.8	1.6	1.5
SC58	0.23	0.29	1.6	1.8	1.6	1.4
SC32	0.25	0.41	1.7	1.9	1.9	1.7
SC55	0.27	0.53	1.6	2.3	2.2	2.3
SC34	0.25	0.60	1.9	2.1	1.9	1.8
SC36	0.25	0.60	2.0	2.3	2.2	2.2
SC57	0.24	0.70	1.7	2.0	1.9	1.7
SC37	0.25	0.79	1.8	2.0	1.9	1.8
SC56	0.23	0.90	1.9	2.0	1.9	1.8
SC35	0.24	1.01	1.6	1.8	1.8	1.6

Table 1a. The yield stress, peak stress, and stresses at 0.01 and 0.02 plastic strains are given. The confining ratios $R_{21}=\sigma_{22}/\sigma_{11}$ and $R_{31}=\sigma_{33}/\sigma_{11}$ are also provided. All tests were at -10.0±0.2 °C.

Test	R_{31}	$\varepsilon_{11,y}$ x10^{-3}	$\varepsilon_{11,p}$ x10^{-3}	Strain Rate $\dot{\varepsilon}_{11}$ x10^{-5} s^{-1}	ITM (MPa)	Curve Type
SC28	0.00	2.5	5.6	3.93	929	I
SC30	0.10	2.3	3.9	3.88	1227	I
SC31	0.21	2.7	5.4	3.45	996	I
SC58	0.29	2.4	5.9	3.92	1186	I
SC32	0.41	2.1	6.7	3.91	1425	I
SC55	0.53	2.1	29.9	3.97	1292	III
SC34	0.60	3.1	5.2	3.84	916	I
SC36	0.60	2.8	8.7	3.87	1148	II
SC57	0.70	2.1	6.8	3.94	1526	I
SC37	0.79	3.4	8.1	3.91	763	I
SC56	0.90	2.5	5.5	4.02	1269	I
SC35	1.01	2.9	7.0	3.91	901	II

Table 1b. The yield strain, peak strain and measured strain-rate are presented. The confining ratio $R_{31}=\sigma_{33}/\sigma_{11}$, the initial tangent modulus (ITM) and the curve type are also provided. All tests were at -10.0±0.2 °C.

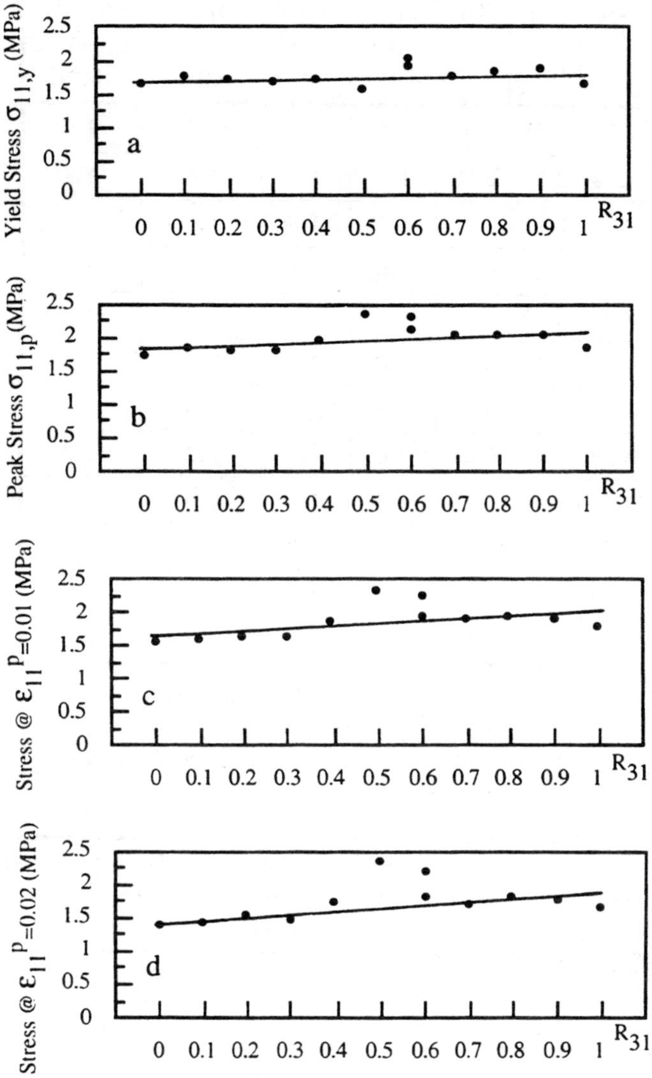

Figure 5. Plots of a) the yield stress $\sigma_{11,y}$, b) the peak stress $\sigma_{11,p}$, and c), d) the stresses at 0.01 and 0.02 inelastic strain, $\sigma_{11,1}$, $\sigma_{11,2}$, respectively, versus the along-column confinement ratio R_{31}. The slopes and intercepts of the fitted lines are given in Table 2.

3.3 Yield strain and peak strain

Table 1 lists the yield strains, which had an average value of 0.0026±0.0008, and lists the peak strains, which had an average value of 0.0083±0.0213. Figure 6 shows the yield and peak strains plotted versus R_{31}. It appears that the yield strains do not depend on the along-column confinement since the slope of the least squares fit line is 0.0005, which is not statistically different from zero at confidence levels greater than 73%. The peak strains show a slight dependence with a best fit slope of 0.0024 and a critical confidence level of 90%.

3.4 Inelastic deformation

All twelve tests showed the same strain-time dependence regardless of the shape of the σ_{11}-ε_{11} curve. Figure 7 shows typical plots of ε_{11}, ε_{22} and ε_{33} versus time (SC 36, R_{31}=0.6), and shows

Data	Slope	Intercept	R^2	Critical Confidence Level (%)
$\sigma_{11,y}$	0.15 MPa	1.65 MPa	0.11	73
$\sigma_{11,p}$	0.26 MPa	1.82 MPa	0.18	82
$\sigma_{11,1}$	0.42 MPa	1.63 MPa	0.24	93
$\sigma_{11,2}$	0.44 MPa	1.49 MPa	0.21	90
$\varepsilon_{11,y}$	0.0005	0.0023	0.13	73
$\varepsilon_{11,p}$	0.0022	0.0052	0.27	93

Table 2. The slope and intercept for the linear regression lines fitted to the data are presented. The abscissa is R_{31}. The R^2 value is given as an indication of how well the fitted line accounts for the scatter in the data. The critical confidence level is the level at which the slope becomes statistically different from zero. The peak strain $\varepsilon_{11,p}$ fit does not included the SC 55 data point.

Test	R_{31}	ψ_{21} @ $\varepsilon_{11}{}^{i}$=0.01	ψ_{31} @ $\varepsilon_{11}{}^{i}$=0.01	ψ_{21} @ $\varepsilon_{11}{}^{i}$=0.02	ψ_{31} @ $\varepsilon_{11}{}^{i}$=0.02
SC28	0.00	-1.02	-	-1.10	-
SC30	0.10	-1.08	-0.06	-1.13	-0.09
SC31	0.21	-1.05	0.00	-1.14	0.03
SC58	0.29	-1.03	0.00	-1.09	0.00
SC32	0.41	-1.00	0.01	-1.11	0.00
SC55	0.53	-0.98	0.02	-1.03	0.01
SC36	0.60	-1.12	0.01	-1.20	0.00
SC34	0.60	-1.08	0.00	-1.15	0.00
SC57	0.70	-0.99	0.03	-1.10	0.02
SC37	0.79	-1.02	0.01	-1.09	0.00
SC56	0.90	-0.99	0.03	-1.09	0.03
SC35	1.01	-1.06	0.03	-1.12	0.02

Table 3. The strain ratios $\psi_{21}=\varepsilon_{22}{}^{i}/\varepsilon_{11}{}^{i}$ and $\psi_{31}=\varepsilon_{33}{}^{i}/\varepsilon_{11}{}^{i}$ for 0.01 and 0.02 plastic strain. SC 28 was a biaxial test so there are no ψ_{31} strain ratios and the change in volume was not found.

σ_{11}, σ_{22} and σ_{33} versus time for the same test. The curve of ε_{11} versus time is a straight line with a negative slope equal to the applied strain rate; ε_{33} generally remained very close to zero. The ε_{22} plot initially remained close to zero and then began to increase rapidly in a nonlinear manner, Figure 7c.

In order to characterize the behavior of the inelastic strains the strain ratios $\psi_{21}=\varepsilon_{22}{}^{i}/\varepsilon_{11}{}^{i}$ and $\psi_{31}=\varepsilon_{33}{}^{i}/\varepsilon_{11}{}^{i}$ were found at $\varepsilon_{11}{}^{i}$=0.01 and 0.02, and are given in Table 3. The superscript i denotes the inelastic strain, found by subtracting the yield strain from the total strain. At $\varepsilon_{11}{}^{i}$=0.01 ψ_{31}= 0.007±0.072, and at $\varepsilon_{11}{}^{i}$=0.02 ψ_{31}=0.002±0.094. The ratio ψ_{21}=-1.035±0.085 at $\varepsilon_{11}{}^{i}$=0.01 and -1.113±0.088 at $\varepsilon_{11}{}^{i}$=0.02, which means that the inelastic deformation was almost completely plane strain.

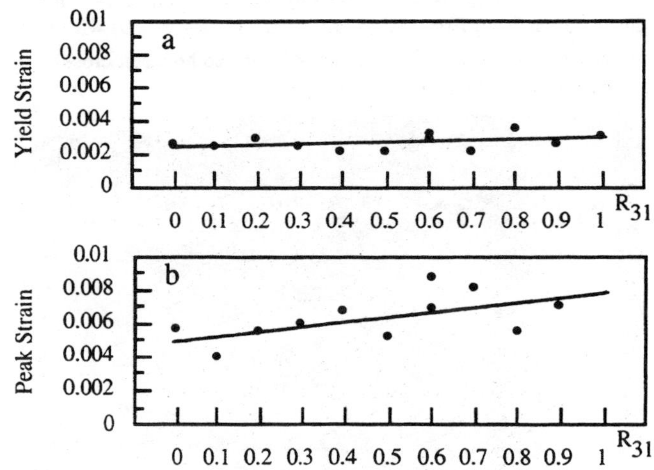

Figure 6. The yield strain $\varepsilon_{11,y}$ and the peak strain $\varepsilon_{11,p}$ versus the confining ratio R_{31}. The slopes and intercepts of the fitted lines are given in Table 2.

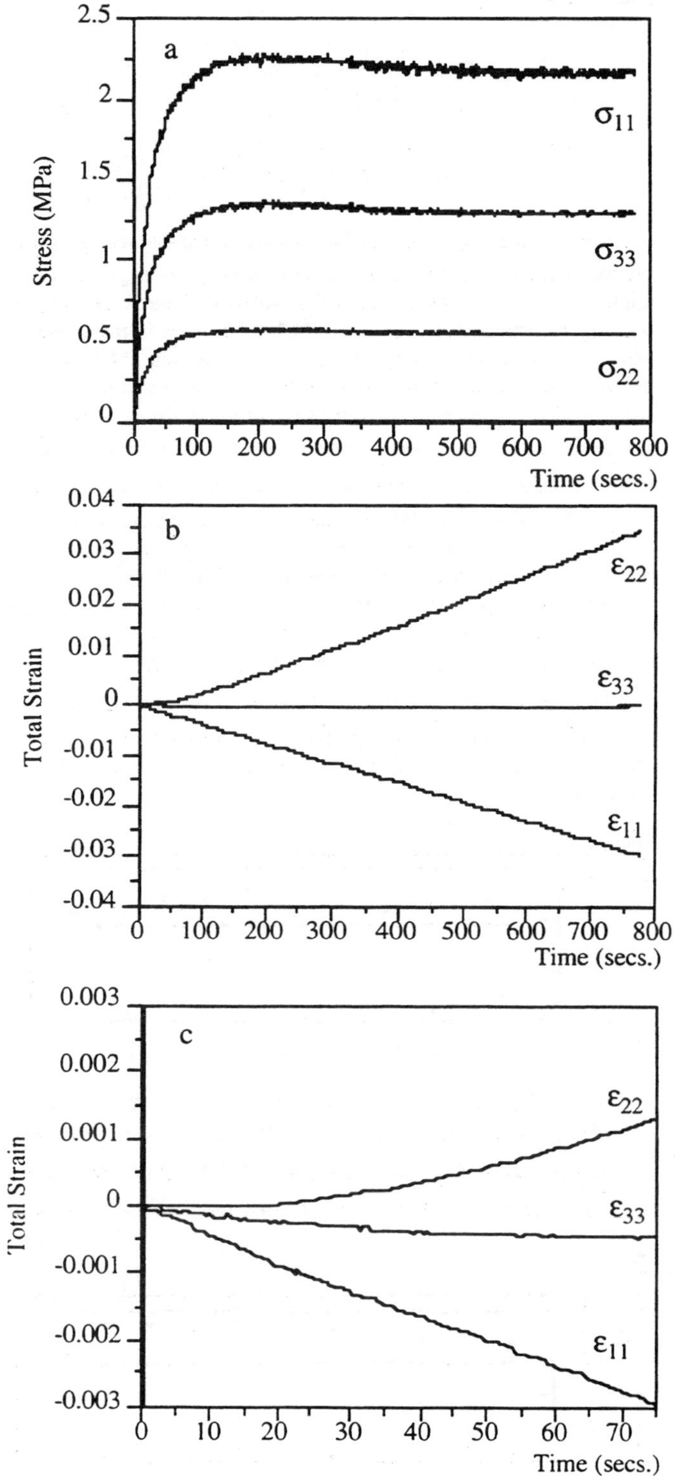

Figure 7. Plots of the stresses and strains versus time for SC 36 (R_{31}=0.60), tested at -10 °C and a strain rate of $\dot{\varepsilon}_{11}=3.87\times10^{-5}$ s^{-1}. (a) The three stresses σ_{11}, σ_{22} and σ_{33} plotted versus time. (b) The strains ε_{11}, ε_{22} and ε_{33} plotted versus time. (c) ε_{11}, ε_{22} and ε_{33} plotted for the first 75 seconds of the test.

3.5 Macroscopic features

In previous tests at R_{21}=1.0 marked patterns of bulges and depressions were found on the x_3 faces (Melton and Schulson 1995) the cubes. However at R_{21}=0.25 the x_3 faces remained essentially smooth and parallel during the tests, which suggests that the localized deformation observed earlier is a function of the across-column confinement. At the end of the tests, several samples had become much more opaque than they were initially, with parts of the sample becoming milky white, especially around brine drainage channels. These white sections are thought to be caused by clouds of microcracks popping in around the brine channels due to stress concentrations.

4. DAMAGE

Initial thin sections of ≈2 mm were made using a bandsaw so that the thin section could be viewed under cross-polarizing light within minutes of the sample being unloaded. Thus the observed microstructure was essentially free from post-test features. After the initial observations were complete, better quality thin sections were made for photographs and analysis.

All samples showed across-column cracks after testing to ε_{11}=0.03. As already noted, the cracks were visible at the surface of the sample and were easily seen in the interior by viewing under strong light. Viewed in thin sections (roughly 155 mm x 160 mm), the cracks were often uniformly distributed throughout the section with a density on the order of 400 cracks/m^2. In other sections they were less uniformly distributed and were confined to 1 or 2 grains, separated by ≈ 10 mm, still with ≈400 cracks/m^2. The crack lengths were typically 5 mm, usually with one tip originating at a grain boundary and the other terminating in the grain interior. The crack tips were generally sharp, though it was difficult to discern the shape of the crack tip at a grain boundary. The crack width was usually less than 1 mm, and many cracks did not appear to have opened. This differs from earlier triaxial tests on S2 columnar saline ice, carried out under the same conditions, except that the across-column confinement was $R_{21}=1.0$. In those tests the typical crack length was 20 mm, the typical crack width was approximately 1 mm and the crack density was an order of magnitude higher (Melton and Schulson, 1995). Figure 8 shows a typical thin section (≈ 2 mm thick), taken 5 cm from the x_2 face of SC 57 ($R_{31}=0.70$) within 10 minutes of unloading. The section was viewed between cross-polarizers to reveal the grain structure.

Figure 8. Thin section from sample SC 57 ($R_{31}=0.7$) taken 5 cm from the surface. The plane of the thin section is parallel to the x_2 face. The arrows indicate across-column cracks. Scale: 1 cm = 5.8 mm.

Another significant difference between the tests at $R_{21}=1.0$, mentioned above, and the present tests is the complete absence of recrystallization in the present tests. The tests at $R_{21}=1.0$ showed that extensive dynamic recrystallization, visible to the naked eye when seen in polarized light, accompanied the deformation (Melton and Schulson, 1995). Thin sections from the current tests were viewed with polarized light under an optical microscope at magnifications up to 50X and no recrystallization was seen, even five days after unloading. The cubes were shortened by 3% along the x_1 direction in both sets of tests.

5. ANALYSIS

5.1 Failure strength

Hill's (1950) criterion for the failure of a plastically orthotropic material can be expressed as

$$F(\sigma_{22}-\sigma_{33})^2+G(\sigma_{33}-\sigma_{11})^2+H(\sigma_{11}-\sigma_{22})^2=1 \qquad 1.$$

Where F, G and H are constants. For an orthotropic material with planar isotropy, such as S2 columnar saline ice, the constants F, G and H are given by:

$$F=G=\frac{1}{2\left(\sigma_{u,3}^{(d)}\right)^2},\qquad H=\frac{1}{\left(\sigma_{u,1}^{(d)}\right)^2}-\frac{1}{2\left(\sigma_{u,3}^{(d)}\right)^2} \qquad 2.$$

where $\sigma_{u,1}^{(d)}$ and $\sigma_{u,3}^{(d)}$ are the across- and along-column uniaxial ductile compressive strengths, respectively. This expression applies when the principal axes of anisotropy of the ice are aligned with the principal axes of stress, which is the case in these tests. This criterion is particularly appealing for columnar saline ice because there are only two parameters, which can be determined independently. Figure 9 shows upper and lower limits of Hill's (1950) criterion. The values used were $\sigma_{u,1}^{(d)}=1.4\pm0.3$ MPa and $\sigma_{u,3}^{(d)}=7.4\pm2.0$ MPa, determined at the same temperature and strain-rate, by Nickolayev (1995), by Kuehn (1993) and from uniaxial tests performed by the present authors. It is seen (from Figure 9) that Hill's (1950) criterion reasonably describes the yield strength measured in the present experiments. Similarly the peak stress can be described using Hill's (1950) criterion, using the separately measured uniaxial peak strengths $\sigma_{u,1}^{(d)}=1.4\pm0.3$ MPa and $\sigma_{u,3}^{(d)}=7.5\pm2.0$ MPa, Figure 10. Generally, the peak stresses lie well within the limits. Similar analysis shows that Hill's (1950) criterion also describes reasonably well the flow stresses at the higher strains.

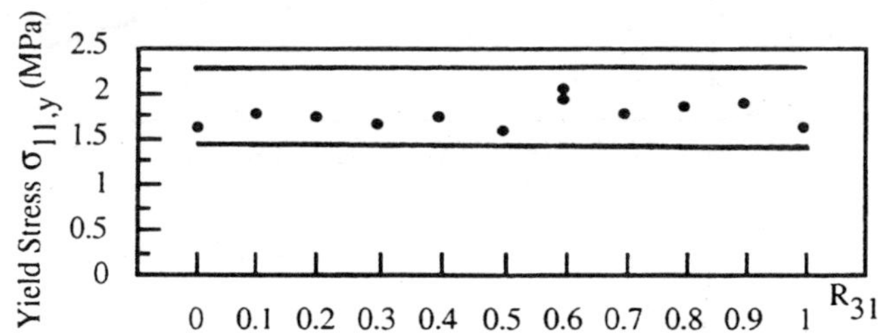

Figure 9. The yield stress $\sigma_{11,y}$ and the upper and lower limits of Hill's (1950) plotted versus R_{31}. The values used to plot the criterion were $\sigma_{1,u}^{(d)}=1.1$ MPa, $\sigma_{3,u}^{(d)}=5.4$ MPa for the lower limit and $\sigma_{1,u}^{(d)}=1.7$ MPa, $\sigma_{3,u}^{(d)}=9.4$ MPa for the upper limit.

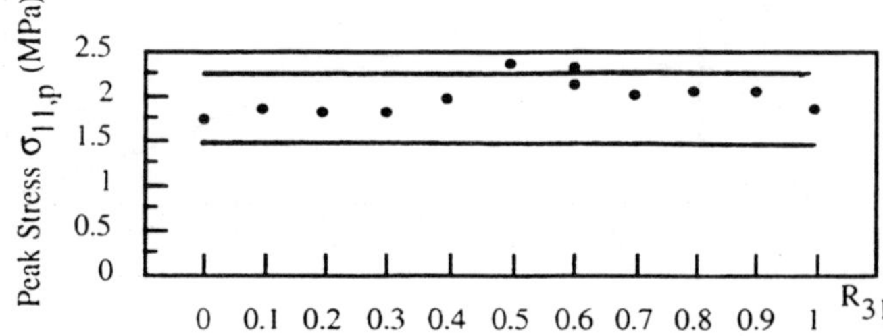

Figure 10. The peak stress $\sigma_{11,y}$ and the upper and lower limits of Hill's (1950) plotted versus R_{31}. The values used to plot the criterion were $\sigma_{1,u}^{(d)}=1.1$ MPa, $\sigma_{3,u}^{(d)}=5.5$ MPa for the lower limit and $\sigma_{1,u}^{(d)}=1.7$ MPa, $\sigma_{3,u}^{(d)}=9.5$ MPa for the upper limit.

5.2 Inelastic deformation

The flow rules for a non-dilatant plastically orthotropic material such as columnar saline ice (see Hill, 1950, p.320) are expressed as:

$$\begin{aligned} d\varepsilon^p_{11} &= dS\left[H\left(\sigma_{11}-\sigma_{22}\right)+G\left(\sigma_{11}-\sigma_{33}\right)\right], \\ d\varepsilon^p_{22} &= dS\left[H\left(\sigma_{22}-\sigma_{11}\right)+F\left(\sigma_{22}-\sigma_{33}\right)\right], \\ d\varepsilon^p_{33} &= dS\left[G\left(\sigma_{33}-\sigma_{11}\right)+F\left(\sigma_{33}-\sigma_{22}\right)\right], \end{aligned} \quad 3.$$

where dS is a constant and F, G and H are defined above. Using the values of $\sigma_{u,1}^{(d)}$ and $\sigma_{u,3}^{(d)}$ given above, the flow rules predict ψ_{21}=-0.97 and ψ_{31}=-0.03 at R_{31}=0, and that ψ_{21} will approach -1.02 and ψ_{31} will approach 0.02 as R_{31} goes to unity. As a first approximation, the flow rules agree with the measured inelastic deformation. However, the samples dilate somewhat, presumably due to the observed across-column cracks. It follows then that the flow rules might be used as a first approximation to predict the deformation behavior, but accurate predictions will not be possible until modifications are made to account for cracking.

6. CONCLUSIONS

From these studies of the ductile deformation of saline columnar ice at -10 °C and a strain rate of $\dot{\varepsilon}_{11}=3.9\times10^{-5}$ along the loading path $\sigma_{11}=0.25\sigma_{22}$, $0<R_{31}<1$, the following conclusions are made:

1. The yield strength $\sigma_{11,y}$ and the peak strength $\sigma_{11,p}$ are independent of the along-column confinement. The flow stresses at 0.01 and 0.02 inelastic strain show a dependence on along-column confinement, which may be attributed to a decrease in the anisotropy of the ice at higher strains.
2. The yield strength $\sigma_{11,y}$ and the peak strength $\sigma_{11,p}$ are reasonably described by Hill's (1950) criterion using appropriate values for the constants $\sigma_{u,1}^{(d)}$ and $\sigma_{u,3}^{(d)}$.

7. ACKNOWLEDGMENTS

This work was supported by the Office of Naval Research, grant no. N00014-92-J-1279. The work was performed in the Ice Research Laboratory at Thayer School of Engineering which is operated through additional funding provided by Chevron, Exxon, Mobil and Texaco.

8. REFERENCES

Gies, M (1988). "The Development Of Brush-type Platens For Loading Ice Specimens In Compression," M.E. Thesis, Dartmouth College, Hanover, New Hampshire.

Gratz, ET (1996). "Brittle Failure of Columnar S2 Ice Under Triaxial Compression," Ph.D. Thesis, Dartmouth College, Hanover, New Hampshire.

Gratz, ET and Schulson, EM (1996). "Brittle Failure Of Columnar Saline Ice Under Triaxial Compression," *J Geophysical Research, Solid Earth*, In Press.

Häusler, FU (1981). "Multiaxial Compressive Strength Tests On Saline Ice With Brush-type Loading Platens," *IAHR Symposium Proceedings*, Quebec, pp 526-536.

Häusler, FU (1983). "Comparison Between Different Yield Functions For Saline Ice," *Annals of Glaciology*, Vol 4, pp 105-109.

Hill, R (1950). "The Mathematical Theory of Plasticity," Oxford University Press, New York, pp 317-323.

Kuehn, GA (1993). "Laboratory-grown Saline Ice - A First Year Sea Ice Analog: The Uniaxial Compressive Properties And The Uniaxial Tensile Properties After Compressive Pre-strain," M.S. Thesis, Dartmouth College, Hanover, New Hampshire.

Kuehn, GA, and Schulson, EM (1994). "The mechanical properties of saline ice under uniaxial compression," *Annals of Glaciology*, Vol 19, pp 39-48.

Melton, JS and Schulson, EM (1995). "Preliminary Results On The Ductile Deformation Of Columnar Saline Ice Under Triaxial Compressive Loading," *Proc OMAE 95*, Vol 4, pp 139-143.

Michel, B (1978). "The Strength Of Polycrystalline Ice," *Canadian Journal of Civil Engineering*, Vol 5, No 3, pp 285-300.

Nickolayev, OY (1994). "Failure Envelopes And The Ductile To Brittle Transition In Columnar Saline Ice Under Biaxial Compression," M.S. Thesis, Dartmouth College, Hanover, New Hampshire.

Schulson, EM, and Nickolayev, OY (1995). "Failure of Columnar Saline Ice Under Biaxial Compression: Failure Envelopes and the Brittle-to-Ductile Transition," *J Geophysical Research*, Vol 100, No B11, pp 22,283-22,400.

Smith, TR, and Schulson, EM (1994). "Brittle Compressive Failure Of Salt-water Columnar Ice Under Biaxial Loading,"*J Glaciology*, Vol 40, No 135, pp 265-276.

Timco, GW and Frederking, RMW (1983). "Confined Compressive Strength Of Sea Ice," *Proc. POAC '83*, Helsinki, Finland, Vol 1, pp 243-253.

Timco, GW and Frederking, RMW (1986). "Confined Compression Tests: Outlining The Failure Envelope Of Columnar Sea Ice," *Cold Regions Science and Technology*, Vol 12, pp 13-28.

Weeks, WF and Ackley, SF (1982). "The Growth, Structure And Properties Of Sea Ice," USA Cold Regions Research and Engineering Laboratory, CRREL Report 82-1.

Proceedings of the Seventh (1997) International Offshore and Polar Engineering Conference
Honolulu, USA, May 25-30, 1997

ISBN 1-880653-28-1 (Set); ISBN 1-880653-30-3 (Vol. II)

Flexural Strength of Ice with Non-Uniform Thickness

G.W. Timco and A.M. Cornett
National Research Council
Ottawa, Canada

Abstract

A test program has been carried out to investigate the influence of non-uniform thickness on the flexural strength of ice. Over 250 cantilever beams were tested using model ice which had either uniform thickness, or non-uniform thickness. The results were analysed in two different, but complementary methods. In the first method, the flexural strength was calculated based on the maximum, minimum and average thickness of the beam. This was compared to the strength of the ice as determined in uniform ice. The second method computed the "effective" thickness of the non-uniform ice with a strength equal to the strength of the uniform ice. Both techniques clearly showed that the best correlation between the uniform and non-uniform ice was found by using the ***maximum*** thickness to characterise the non-uniform ice. An analysis of the data shows that a value of 95% of the maximum ice thickness provides the best representation of the thickness of non-uniform ice failing downwards in flexure.

Introduction

It is well known that ice thickness plays an extremely important role in the magnitude of the load that an ice sheet can exert on an offshore structure. Numerical and analytical models, which have been used to predict ice loads on sloping (or inclined) structures such as cones, usually have a functional dependence of ice load with a thickness-squared relationship for flexural ice failures (Edwards and Croasdale 1976; Croasdale 1980; Ralston 1977; Kato 1986; Hirayama and Obara 1986; Nevel 1992). For level ice conditions, these models can usually be used satisfactorily to determine the load that the ice sheet will exert on the structure. However, in most real situations, the ice thickness is not at all uniform - it can have a wide spatial variation. This is especially evident in the important situation of an ice ridge embedded in an ice floe. A recent analysis has shown that there can be variations by a factor of 4 in the thickness of the consolidated layer of a first-year ridge (Timco and Burden 1997). Given this variability, there is clearly a major uncertainty surrounding the choice of ice thickness to be used in load prediction models. One could argue that the *thinnest* ice should control the load since this is the "weak link" in the ice sheet. On the other hand, an argument could be made that the failure of the ice would be dominated by the *thickest* ice, since failure could not occur until the stress level was high enough to fail this ice. One could also argue that the *average* thickness could be the best representative value to use for non-uniform ice. To date, this issue has not been addressed.

This problem can be studied systematically in a laboratory environment. There, it is possible to produce well-defined conditions that can be used to understand the failure of ice in non-uniform thickness conditions. For this study, ice sheets were grown in a laboratory ice tank with two types of ice thickness - uniform ice and non-uniform, but well-defined, ice conditions. A larger number of flexural strength tests were performed in both types of ice. The tests in the level, uniform ice provided direct information of both the actual *strength* of the ice sheet, and the failure pattern with uniform ice. The tests with the non-uniform ice were compared to the uniform ice values, and an "effective" ice thickness was determined, for these controlled conditions. This paper provides information on the results and analysis of this systematic study of the flexural failure of non-uniform ice.

Experimental

The tests were carried out in the ice tank at the Canadian Hydraulics Centre of the National Research Council of

Canada in Ottawa. The ice tank is 21 m long, 7 m wide, and 1.2 m deep. It is housed in an insulated cold room with refrigeration that produces air temperatures of -25°C. For these tests, several sheets of PG/AD model ice were used. This type of model ice is a variation of the EG/AD/S ice developed by Timco (1986).

To produce the non-uniform ice, the following procedure was used:

1. Ice sheet growth was started using "wet-seeding" to nucleate the ice to produce uniform characteristics. This ice sheet was grown for approximately 3 hours.
2. Next, a number of 0.05 m by 0.03 m by 2.3 m strips of rigid Styrofoam insulation were placed in rows to produce a quasi-regular pattern of thickness variations.
3. The ice sheet growth continued until the desired level ice thickness was reached. At that time, the refrigeration was shut off, and the ice was allowed to temper at a temperature of +2°C.
4. After the tempering process, the flexural strength tests were performed.

This procedure produced ice which had a spatially-varying thickness. The Styrofoam strips insulated the growing ice so that the ice was always thinner under the Styrofoam strips. Figure 1 shows a photograph of a piece of the non-uniform ice. This ice had a flat top surface, and an irregular "undulating" bottom surface. The difference between the thickest and the thinnest ice ranged more than a factor of 2, depending upon the pattern of the Styrofoam strips.

Figure 1 Photograph showing a piece of the non-uniform ice.

Figures 2 and 3 show, respectively, typical thickness distributions for the uniform (i.e. level) and non-uniform ice. The uniform ice shows little variation in ice thickness, with an average thickness of 4.1 cm and a standard deviation of 0.1 cm. On the other hand, there is a wide variation in the thickness of the non-uniform ice. For this ice sheet, the average thickness was 3.3 cm, which is less than the thickness of the uniform ice. The standard deviation of the non-uniform ice thickness was 0.5 cm, which is significantly greater than that for the uniform ice.

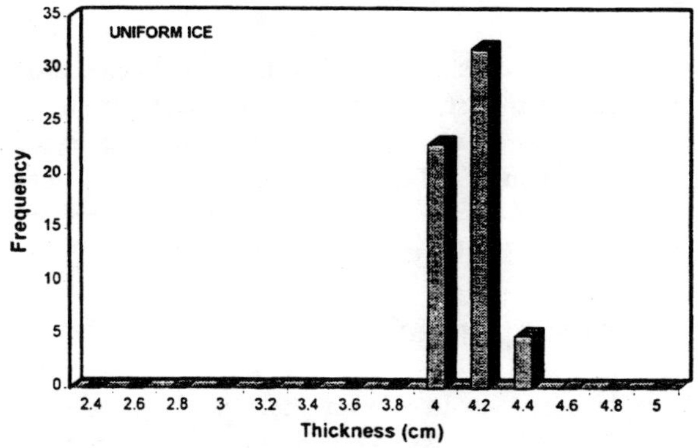

Figure 2 Histogram showing the variation in ice thickness for uniform ice.

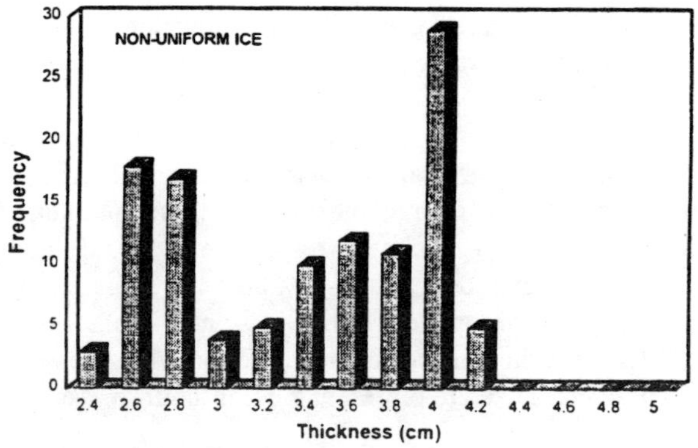

Figure 3 Histogram showing the variation in ice thickness for non-uniform ice.

Cantilever beam tests were performed *in situ* to measure the flexural strength of the ice cover. The beams were carefully cut in the intact ice sheet using a hand saw. The beam dimensions were chosen to provide the most accurate representation of the flexural strength of the model ice (Timco 1981). Care was taken to ensure that the sides of each beam were not in contact with the adjacent ice. Loads were applied using a calibrated push-pull gauge, with a loading time of 1 to 2 seconds. The flexural strength (σ_f) was determined from elastic beam theory according to

$$\sigma_f = 6\,P\,L\,/\,w\,h^2 \qquad (1)$$

where P is the applied load, L is the length of the beam, w is the width of the beam, and h is the thickness of the ice. For each beam, a number of thickness measurements were made along the root of the beam, and these measurements provided information on the thickness of the beam. Note that for all of the tests reported here, there was a significant thickness variation along the failure plane of the beam. That is, the beams were cut such that their long axis was parallel to the long direction of the linear styrofoam strips.

Results and Analysis

Four different ice sheets, each with different ice properties, were used for this study. From these sheets, more than 250 cantilever beams were tested. The full test results are presented in Timco and Cornett (1996). The results of the tests were analysed using 2 different, but complementary, approaches. Each will be discussed in turn.

"Strength" Approach

A simple approach to analyse the data involves computing several values of ice strength for each beam using different values of the ice thickness. In this "strength" analysis, the flexural strength of the beams was calculated using the *maximum*, *minimum* and *average* ice thickness along the root of the beam. Each of these calculated "strength" values is compared to the strength of the ice as measured in the uniform ice sheet. It is argued that the best correlation should reflect the thickness most representative of the non-uniform ice.

Figures 4 to 6 show the calculated strength of the non-uniform ice versus the flexural strength of the uniform ice for calculations based on the maximum (h_{max}), minimum (h_{min}) and average (h_{ave}) thickness respectively. In all figures, a dashed line is included which shows the line representing one-to-one correlation. Although there is scatter in all of the plots, it is clear that the best correlation is found when the *maximum ice thickness* is used to represent the non-uniform ice.

To quantify the agreement between the different approaches, the standard deviation ($s_{y\,x}$) about a one-to-one correlation was determined according to

$$s_{y\,x} = \sqrt{\frac{\sum (y - y_{est})^2}{N}} \qquad (2)$$

where y is the calculated value, y_{est} is the estimated value assuming a one-to-one correlation, and N is the number of data points. The standard deviation for each of the 3 cases shown in Figures 4 to 6 are as follows:

$s_{y\,x}$ (using the maximum ice thickness) = 12 kPa
$s_{y\,x}$ (using the minimum ice thickness) = 204 kPa
$s_{y\,x}$ (using the average ice thickness) = 47 kPa

Clearly, the difference between the measured and calculated values is significantly less if the *maximum* ice thickness is used to characterise the non-uniform ice.

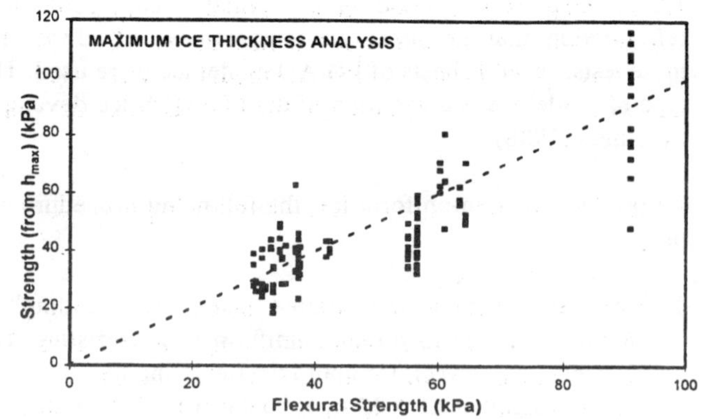

Figure 4 Strength calculated using the maximum ice thickness versus the flexural strength of the ice. The dashed line represents one-to-one correspondence.

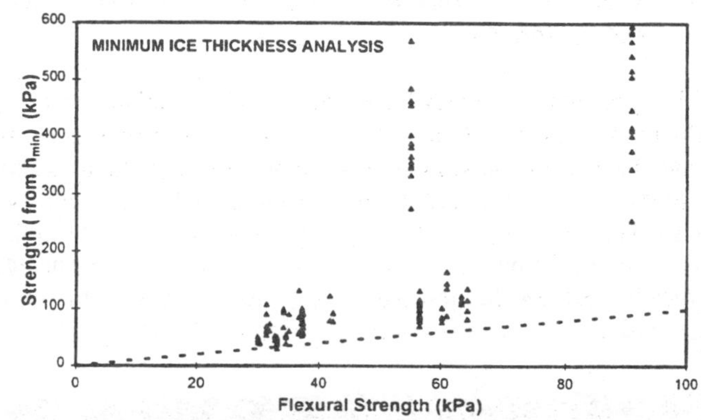

Figure 5 Strength calculated using the minimum ice thickness versus the flexural strength of the ice. The dashed line represents one-to-one correspondence.

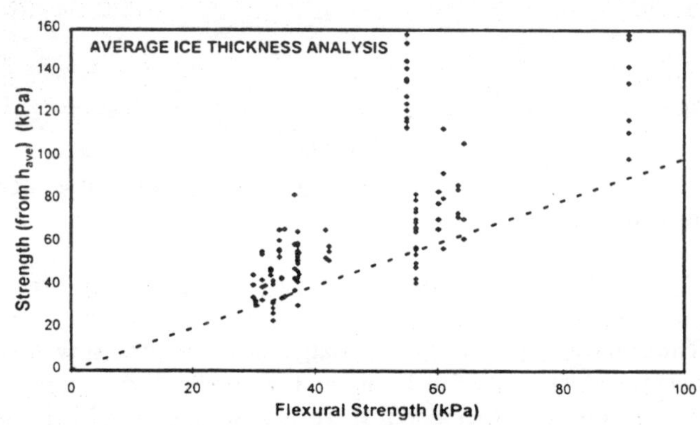

Figure 6 Strength calculated using the average ice thickness versus the flexural strength of the ice. The dashed line represents one-to-one correspondence.

"Effective Thickness" Approach

The second analysis approach uses the assumption that the true ice strength is constant and independent of ice thickness. Based on this assumption, an "effective" thickness for the non-uniform ice can be determined by computing the thickness of an equivalent rectangular beam that fails at the recorded load. For example, with *uniform* ice, the flexural strength (σ_u) is obtained from

$$\sigma_u = 6\,P\,L\,/\,w\,h_i^{\,2} \tag{3}$$

where σ_u is the flexural strength, P is the applied load, L is the length of the beam, w is the width of the beam, and h_i is the thickness of the ice. For tests with *non-uniform* ice, an "effective thickness" for the ice is computed according to:

$$h_{eff} = (6\,P\,L\,/\,w\,\sigma_u)^{0.5} \tag{4}$$

where the same strength (σ_u) is assumed. With this approach, quantitative information can be obtained on the effective thickness of non-uniform ice. By comparing the correlation between the *effective* thickness and the *maximum*, *minimum* and *average* thickness of the beam, information can be obtained on the thickness that best describes the strength of non-uniform ice.

Figures 7 to 9 show plots of the measured ice thickness versus the "effective" ice thickness for measurements representing the maximum, minimum and average thickness respectively. In all figures, a dashed line is included which shows the line representing one-to-one correlation. Although there is scatter in all of the plots, it is clear that the best correlation is found when using the *maximum ice thickness*.

The standard deviation ($s_{y.x}$) about a one-to-one correlation was calculated for each of the three cases shown in Figures 7 to 9. The results are as follows:

$s_{y.x}$ (using the maximum ice thickness)	= 0.39 cm
$s_{y.x}$ (using the minimum ice thickness)	= 1.21 cm
$s_{y.x}$ (using the average ice thickness)	= 0.66 cm

Clearly, the difference between the measured and calculated values is significantly less if the maximum ice thickness is used to characterise the non-uniform ice.

This analysis can be extended to quantify the agreement between the effective ice thickness and the maximum ice thickness. For this, a normalised thickness ratio (h_n) can be defined as

$$h_n = \frac{h_{eff} - h_{max}}{h_{eff}} \tag{5}$$

where h_{eff} is the effective ice thickness and h_{max} is the maximum ice thickness. The difference between the effective thickness and the maximum thickness has been normalised with respect to the effective thickness of the ice. Figure 10 shows a frequency histogram of this ratio for all of the tests with non-uniform ice. The average value of this distribution, which is -0.05, can be used in Equation 5 to determine a quantitative relationship between the effective thickness and the maximum thickness. Substituting a value of -0.05 for h_n, gives

$$h_{eff} = 0.95 \cdot h_{max}\,. \tag{6}$$

The foregoing analysis suggests that a thickness of 95% of the maximum ice thickness is the most appropriate thickness to chose for representing non-uniform ice.

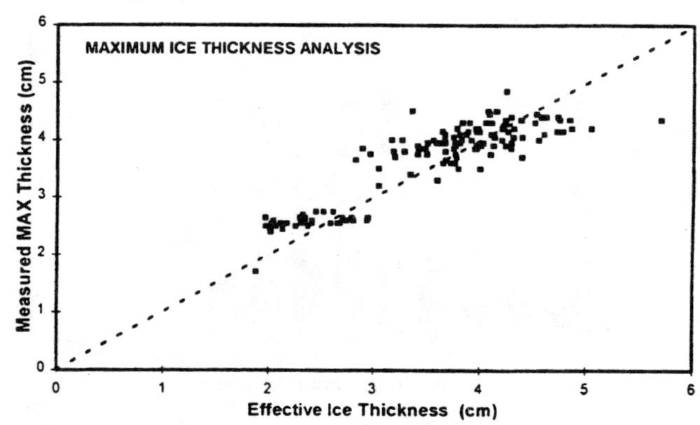

Figure 7 "Effective" ice thickness versus the maximum-measured ice thickness. The dashed line represents one-to-one correspondence.

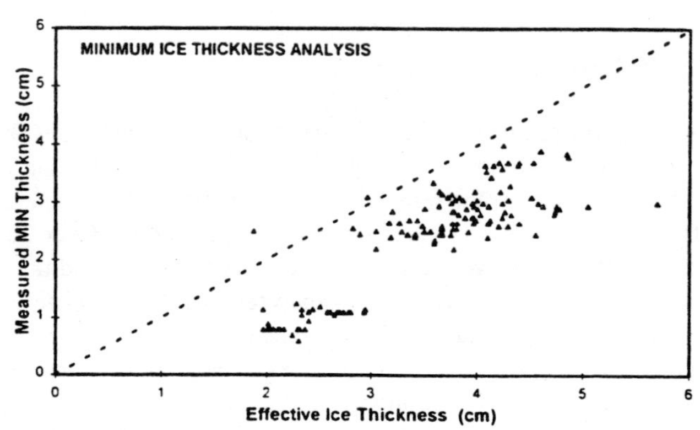

Figure 8 "Effective" ice thickness versus the minimum-measured ice thickness. The dashed line represents one-to-one correspondence.

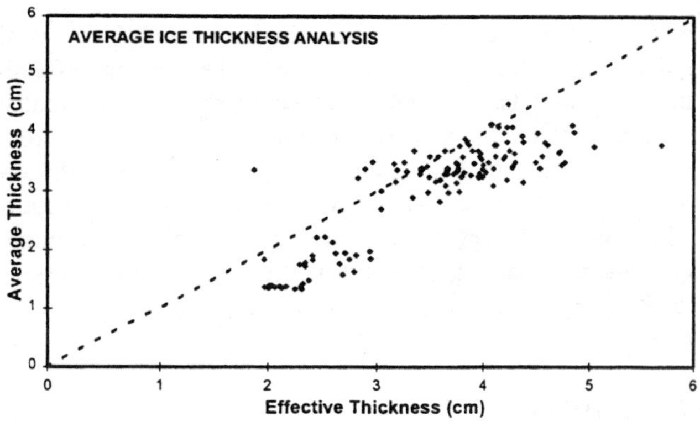

Figure 9 "Effective" ice thickness versus the average ice thickness. The dashed line represents one-to-one correspondence.

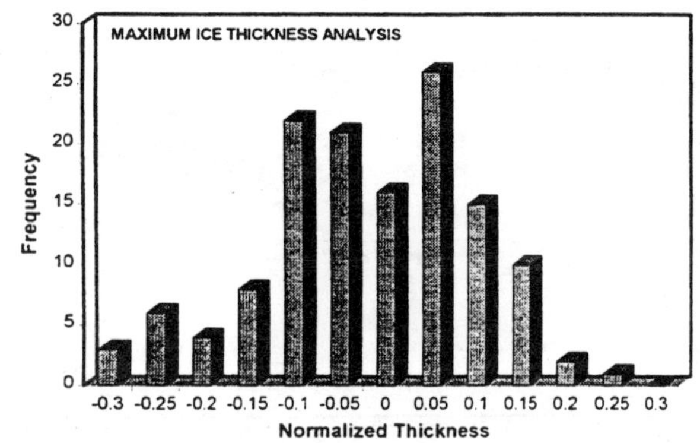

Figure 10 Histogram showing the distribution of the normalised thickness ratio for all tests with non-uniform ice. The analysis shows that a value of 95% of the maximum ice thickness provides the best characterisation of non-uniform ice failing in flexure.

Summary

A test program has been carried out to investigate the influence of non-uniform thickness on the flexural strength of ice. The results of the tests have direct implications to the load that an ice sheet can exert on a sloping structure. Over 250 cantilever beams were tested over a wide range of conditions. The results of the tests were analysed using two different approaches: a "strength" approach and an "effective thickness" approach. Both analyses clearly show that the *maximum* ice thickness along the failure plane controls the strength of the beam. An analysis of the data shows that a value of 95% of the maximum ice thickness provides the best representation of the thickness of non-uniform ice failing in flexure.

Acknowledgements

The authors would like to acknowledge the support of Dr. Ibrahim Konuk, and the financial assistance of Canadian Government Agencies, the Panel on Energy Research and Development, and the National Energy Board.

References

Croasdale, K.R. 1980. Ice Forces on Fixed Rigid Structures. in 1st IAHR State-of-the-Art Report on Ice Forces on Structures, pp 34-106, CRREL Special Report 80-26, Hanover, N.H., USA.

Edwards, R.Y. and Croasdale, K.R. 1976. Model Experiments to Determine Ice Forces on Conical Structures. Symposium of Applied Glaciology, International Glaciological Society.

Hirayama K. and Obara, I. 1986. Ice Forces on Inclined Structures. Proceedings OMAE Conference Vol. IV, pp 515-520, Tokyo, Japan.

Kato, K. 1986. Experimental Studies of Ice Forces on Conical Structures. Proceedings IAHR Ice Symposium, Vol. I, pp. 185-196, Iowa City, U.S.A.

Nevel, D.E. 1992. Ice Forces on Cones from Floes. Proceedings 11th IAHR Ice Symposium, Vol. 3, pp 1391-1404, Banff, Alta., Canada.

Ralston, T.D. 1977. Ice Force Design Considerations for Conical Offshore Structures. Proceedings POAC'77, Vol. II, pp. 741-752, St. John's, Nfld., Canada.

Timco, G.W. and Burden, R.P. 1997. An Analysis of the Shape of Sea Ice Ridges. Cold Regions Science and Technology 25, No. 1, pp 65-77.

Timco, G.W. and Cornett, A.M. 1996. Flexural Strength of Ice with Non-uniform Thickness. National Research Council Technical Report HYD-TR-008, Ottawa, Canada.

Timco, G.W. 1986. EG/AD/S: A New Type of Model Ice for Refrigerated Towing Tanks. Cold Regions Science and Technology 12, pp 175-195.

Timco, G.W. 1981. On the Test Methods for Model Ice. Cold Regions Science and Technology 4, pp 269-274.

Proceedings of the Seventh (1997) International Offshore and Polar Engineering Conference
Honolulu, USA, May 25-30, 1997

ISBN 1-880653-28-1 (Set); ISBN 1-880653-30-3 (Vol. II)

Temperature Effect on Strength of Ice Under Triaxial Compression

Anatoly M. Fish
U.S. Army Cold Regions Research and Engineering Laboratory
Hanover, NH, USA

Yuri K. Zaretsky
Institute of Geomechanics and Hydrostructures
Moscow, Russia

ABSTRACT

A temperature model has been developed that describes the ice strength in a multiaxial stress state over a wide spectrum of negative temperatures. The model takes into account the anomalous behavior of ice under high hydrostatic pressure, when its strength reaches a maximum, and then gradually decreases with the pressure increase. It has been shown that strength of ice under high hydrostatic pressure is described by an extended Drucker-Prager (parabolic) strength criterion with only three *fundamental* parameters, ice cohesion, internal friction angle, and ice melting pressure, which all have a *definite* physical meaning and are functions of temperature. The model has been verified using test data on the strength of iceberg ice and laboratory-made polycrystalline freshwater ice under triaxial compression at strain rates between 10^{-3} and 10^{-5} s^{-1} over the temperature range between –1°C and –40°C.

KEY WORDS: ice, triaxial strength, temperature, strain rate

INTRODUCTION

Studies of ice strength in the second part of this century attracted efforts of a number of researchers. Their attention was focused on investigation of ice strength as a function of the strain rate, temperature, grain size and other factors mainly in simple stress–strain, uniaxial compression. During the last two decades, however, the attention of researchers was shifted to investigation of the mechanical behavior of ice, particularly the strength of ice, in a complex stress-strain state such as triaxial compression.

In a general case, strength of ice in a multiaxial stress state, when other conditions (the type of ice, its structure, the grain size, etc.) are equal, can be presented as

$$\tau_i^* = \tau_i^*(\dot{\gamma}_i, p, T) \tag{1}$$

where $\tau_i^* = \tau_i = \sqrt{J_2}$ = octahedral (peak) shear stress
J_2 = second invariant of the stress deviator
$\dot{\gamma}_i$ = octahedral constant shear strain rate
$p = I_1/3$ = hydrostatic pressure
I_1 = first invariant of the stress tensor
T = ice temperature, K.

The first studies of the effect of low confining pressure on creep strength of ice under triaxial ($\sigma_2 = \sigma_3$) compression were carried out by Sayles (1974). The studies showed that the ice strength is a nonlinear function of the confining pressure. More detailed investigations of the effect of the strain rate and high confining pressures on the ice strength under triaxial compression were carried out by Jones (1978, 1982). The studies, carried out over a wide range of constant strain rates and confining pressures, revealed the anomalous behavior of ice when its shear strength, at a certain magnitude of confining pressure, reaches a maximum, attributable to melting of ice, and then gradually decreases with the pressure increase. Jones showed that the ice strength is a nonlinear (power) function of the axial (constant) strain rate (see Eq. 1c below). Further experimental studies with various types of ice revealed effects of temperature, salinity, structure and other factors on the ice strength (Hausler, 1983; Timco and Frederking 1986; Nadreau and Michel, 1986; Nadreau et al. 1991; Richter-Menge et al., 1986; Rist and Murrell, 1994; Gagnon and Gammon, 1995; Weiss and Schulson, 1995 and others).

At the same time mathematical models were also developed. Thus, Reinicke and Ralston (1977) and Hausler (1983) applied the equation of Smith (1974), developed originally for rocks, to describe the strength dependency of ice upon the hydrostatic pressure

$$\tau_i^* = \tau_i^*(p) \tag{1a}$$

The physical meaning of the parameters in the Smith equation have not been defined.

Nadreau proposed a model (Nadreau and Michel, 1986) that described the nonlinear dependency of ice strength on confining pressure ($\sigma_2 = \sigma_3$) by means of a third-order polynominal function with four phenomenological parameters, to be determined from a series of triaxial tests of ice at different strain rates, and the principal parameter—the ice melting pressure p^*. This pressure is suggested to be determined either from the ice state diagram or to be calculated by an empirical equation.

In the model of Fish (1991, 1992, 1993) the relationship

$$\tau_i^* = \tau_i^*(p, \dot{\gamma}_i); \qquad T = \text{Const.} \tag{1b}$$

was described by a product of two functions: an extended by the author Drucker-Prager (parabolic) yield criterion and a normalized (dimensionless) power function of the strain rate. The yield criterion describes the strength dependency on the hydrostatic pressure by means of three param-

* Denotes failure stress.

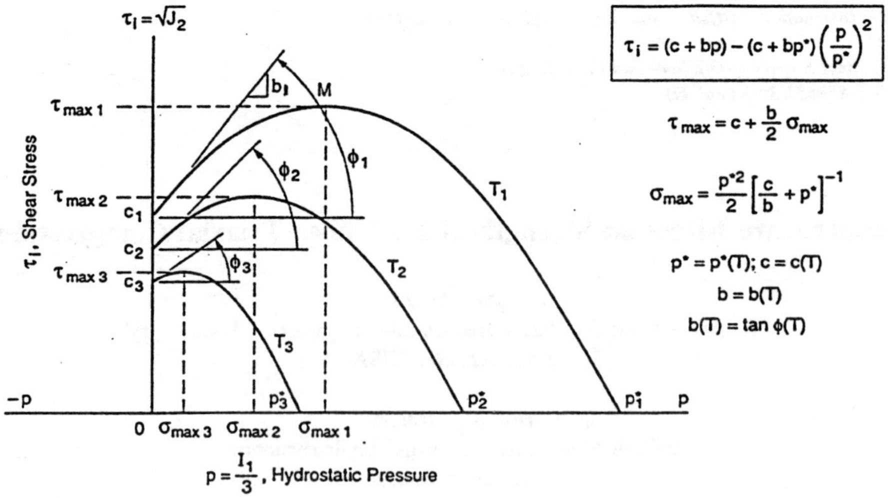

Fig. 1 Temperature model of ice strength in a multiaxial stress state: c, ϕ, p^* and T is the ice cohesion, internal friction angle, ice melting pressure and absolute temperature, respectively.

eters: the cohesion c, the friction angle ϕ and the magnitude of the hydrostatic pressure σ_{max} at which the shear strength of ice reaches a maximum τ_{max}, related to the ice melting pressure p^* (Fig. 1). The model described Jones' (1982) test data well. It was shown that the ice cohesion and the friction angle are functions of the strain rate.

The temperature effect on ice strength under triaxial compression was studied by Rist and Murrell (1994) and Gagnon and Gammon (1995) using the Norton-Glen equation

$$\dot{\varepsilon} = B\exp(-E/RT)\tau_i^n \quad (1c)$$

where $\dot{\varepsilon}$ = axial (constant) strain rate
B = empirical parameter
E = activation energy
R = universal gas constant
$n \approx 4$ = strain hardening parameter (Jones, 1982).

in which the temperature variations of the ice strength are characterized by only one parameter—the activation energy. The apparent activation energy magnitudes were found as E = 69 kJ mol^{-1} in the temperature range between −20°C and -40°C (Rist and Murrell (1994), and E = 101 kJ mol^{-1} in the temperature range between −1°C and −16°C (Gagnon and Gammon, 1995). The higher activation energy magnitudes were explained by grain-boundary softening associated with the presence of liquid on grain boundaries.

Equation 1c implies that the nonlinear (with regard to stress) viscosity coefficient of ice

$$\eta(T) = 1/B\exp(E/RT),$$

is independent of the hydrostatic pressure. In other words, ice was assumed to be incompressible.

It has been shown earlier (Fish, 1991) that ice strength in a multiaxial stress state at high strain rates is characterized by two or more temperature-dependent parameters which can affect the magnitude of the apparent activation energy: the ice cohesion, the friction angle, and the hydrostatic pressure. So the predicted values of the ice strength, calculated by Eq. 1c, may deviate considerably from those obtained in the tests.

A mathematical model (Zaretsky and Fish, 1996a, 1996b) takes into account the effect of all three variables in Eq. 1. The ice strength is described by three parameters: the ice cohesion, friction angle and ice melting pressure. The ice cohesion and the friction parameter are assumed to be proportional to temperature. It has been shown that the hydrostatic pressure being applied to ice decreases its melting temperature and thus its strength under triaxial compression.

In the present paper a different approach has been undertaken. The authors considered the ice strength as a function of two variables: the hydrostatic pressure and temperature, i.e.

$$\tau_i^* = \tau_i^*(p,T); \qquad \dot{\gamma}_i = \text{Const.} \quad (1d)$$

At a constant strain rate the strength of ice in a multiaxial stress state is described by the parabolic (extended Drucker-Prager) yield criterion with three parameters: the cohesion, and the friction angle which are different nonlinear functions of temperature, and the ice melting pressure. Then a strength criterion of ice, which takes into account the combined effect of all three variables in Eq. 1, the strain rate, hydrostatic pressure and temperature, is obtained by combining Eqs. 1b and 1d.

STRAIN RATE EFFECT

A constitutive equation for steady-state creep of homogeneous and isotropic of ice in a multiaxial stress state at constant strain rate and constant temperature takes the following form (Fish, 1991, 1992, 1993)

$$\dot{\gamma}_i = \frac{\bar{C}}{t_o}\left(\frac{\tau_i}{\tau_{io}}\right)^n \quad (2)$$

It should be emphasized that Eq. 2 is *fundamentally different* from the Norton-Glen power flow law (Eq. 1c), although it contains a power function of stress with exponent n.

In contrast to Eq. 1c parameters $\bar{C}$ and t_o in Eq. 2 have a *definite* physical meaning, and the denominator of the stress function is a temperature-dependent yield criterion (Eq. 5 below). This yield criterion, which is a function of the first invariant of the stress tensor and temperature, relates the minimum shear strain rate and the shear stress in the whole spectrum of hydrostatic pressures (mean normal stresses).

Equation 2 can be rewritten in terms of maximum (peak) shear strength $\tau_i^* = \tau_i$ and presented as a product of two independent functions: a yield function $\tau_{io}(p)$ and a nondimensional function $\Phi(\dot{\gamma}_i)$ of the constant strain rate

$$\tau_i^*(p,\dot{\gamma}_i) = \tau_{io}(p)\Phi(\dot{\gamma}_i) \quad (2a)$$

Function $\Phi(\dot{\gamma}_i)$ has been selected in its simplest form

$$\Phi(\dot{\gamma}_i) = \left(\frac{\dot{\gamma}_i t_o}{\gamma_{io}}\right)^{1/n} = \left(\frac{\dot{\gamma}_i}{\dot{\gamma}_{io}}\right)^{1/n} \tag{3}$$

where $\dot{\gamma}_i$ = applied octahedral constant shear strain rate
$\dot{\gamma}_{io} = \gamma_{io} / t_o$ = instantaneous (referenced) octahedral shear strain rate
$\gamma_{io} = \bar{C}$ = instantaneous octahedral shear strain
n = dimensionless parameter; for polycrystalline ice, $n \approx 4$ (Jones, 1982)
t_o = temperature-dependent time to failure, i.e., the time interval between the initiation of the test conducted at $\dot{\gamma}_i = \dot{\gamma}_{io}$ and the moment when the ice strength reaches a maximum (peak) value.

The temperature dependency of t_o is given by

$$t_o = \frac{h}{kT} \exp\left(\frac{E}{RT}\right) \tag{4}$$

where E = activation energy
R = gas constant
h = Planck's constant
k = Boltzmann's constant
T = absolute temperature, K.

In Eq. 3 the "instantaneous" (referenced) octahedral strain rate $\dot{\gamma}_{io}$ is defined as a strain rate at which the shear strength of ice reaches a maximum value, separating two different modes of failure: the brittle mode dominated by the cleavage mechanism of failure and the ductile mode dominated by the shear mechanism. At higher strain rates the ice strength either decreases (Gagnon and Gammon, 1995) or remains unchanged (Jones, 1982).

In experimental studies of ice under triaxial (biaxial) compression, the radial strains are assumed to be small and are usually ignored. In this case Eq. 3 can be replaced by

$$\Phi(\dot{\gamma}_i) = \Phi(\dot{\varepsilon}) = \left(\frac{\dot{\varepsilon} t_o}{\varepsilon_o}\right)^{1/n} = \left(\frac{\dot{\varepsilon}}{\dot{\varepsilon}_o}\right)^{1/n} \tag{3a}$$

where $\dot{\varepsilon}$ = axial strain rate
$\dot{\varepsilon}_o, \varepsilon_o$ = instantaneous (referenced) axial strain rate and strain respectively.
$t_o = \varepsilon_o / \dot{\varepsilon}_o$ = given by Eq. 4.

Thus, the strain rate function varies in the limits

$$1 \geq \Phi(\dot{\gamma}_i) = \Phi(\dot{\varepsilon}) \geq 0 \ . \tag{4}$$

The effect of the strain rate on the strength of ice can be excluded from consideration by selecting the strength test data of ice corresponding to $\Phi(\dot{\varepsilon}) = 1$.

EFFECT OF HYDROSTATIC PRESSURE

When applied strain rate is equal to the instantaneous strain rate $\dot{\gamma}_i = \dot{\gamma}_{io}$ or $\dot{\varepsilon} = \dot{\varepsilon}_o$, in Eqs. 2a and 3a function $\Phi(\dot{\gamma}_{io}) = \Phi(\dot{\varepsilon}_o) = 1$ and the strength dependency of ice upon the hydrostatic pressure is described by the parabolic yield criterion (Fish, 1991, 1992, 1993), which is presented in Figure 1:

$$\tau_{io}(p) = c + bp - \frac{b}{2\sigma_{max}} p^2 \tag{5}$$

or

$$\tau_{io}(p) = (c + bp) - (c + bp^*)\left(\frac{p}{p^*}\right)^2 \tag{5a}$$

where c = ice cohesion on the octahedral plane
$b = \tan \phi$, where ϕ is the angle of internal friction of ice on the octahedral plane
$p = (\sigma_1 + \sigma_2 + \sigma_3)/3$ = hydrostatic pressure (mean normal stress)
$\sigma_1, \sigma_2, \sigma_3$ = principal stresses
$\tau_i = \frac{1}{\sqrt{6}}\left[(\sigma_1 - \sigma_2)^2 + (\sigma_2 - \sigma_3)^2 + (\sigma_1 - \sigma_3)^2\right]^{1/2}$ = octahedral shear stress (resultant)
σ_{max} = magnitude of the hydrostatic pressure at which the shear strength reaches a maximum:

$$\tau_{max} = c + \frac{b}{2}\sigma_{max} \tag{6}$$

$$\sigma_{max} = \frac{p^{*2}}{2\left(\frac{c}{b} + p^*\right)} \tag{7}$$

p^* = ice melting pressure at which the shear stress of ice equals to zero. Subscript (o) at τ_{io} in Eq. 5 indicates that parameters c and b are referred to the instantaneous condition when the applied strain rate is equal to the instantaneous strain rate.

Note that the yield curve may also intersect the hydrostatic axis in the domain of the negative hydrostatic pressures ($-p$) at point h_2 (not shown in Fig. 1), the abscissa of which equals to

$$h_2 = \frac{-p^*}{1 + \frac{b}{c} p^*} \tag{8}$$

Equation 5 can be considered an extended Von Mises-Drucker-Prager yield criterion. At low stress level $p << \sigma_{max}$ the third term in the right side of Eq. 5 approaches zero and Eq. 5 transforms into the Drucker-Prager (1952) (extended Von Mises) yield criterion:

$$\tau_{io} = c + bp \tag{9}$$

For frictionless materials (b = 0) Eq. 5 reduces to the Von Mises yield criterion

$$\tau_{io} = c \tag{10}$$

Thus, in multiaxial stress state, the strength of ice as well as its strength characteristics are functions of only three physically well-founded parameters: p^*, c and b which can be determined from test data. Studies show that all these parameters are functions of temperature, i.e.

$$p^* = p^*(T), \quad c = c(T), \quad b = b(T)$$

Consequently, the shear strength of ice as well as the strength characteristics of ice are also functions of temperature:

$$\tau_{io} = \tau_{io}(T), \quad \tau_{max} = \tau_{max}(T), \quad \sigma_{max} = \sigma_{max}(T), \quad h_2 = h_2(T).$$

Thus, if a series of strength tests of ice is carried out at a constant strain rate instead of one yield curve, one obtains a family of curves for various temperatures (Fig. 1). It should be noted that parameters c, b and τ_{max} are also strain-rate dependent, while parameters σ_{max}, p^* and h_2 are independent of the strain rate (Fish, 1992, 1993).

TEMPERATURE EFFECT

Ice Melting Pressure

It is well known that the ice melting temperature, as well as the melting temperature of other crystalline materials, is a function of the hydrostatic pressure. This pressure can be determined (Zaretsky and Fish, 1996a) from the Clapeyron equation according to which a small change in the equilibrious melting temperature of a solid dT_m attributable to a small change in the hydrostatic pressure dp can be calculated from the relationship

$$dT_m = \frac{V_\ell - V_s}{S_\ell - S_s}dp = T_m \frac{V_\ell - V_s}{L_m}dp \quad (11)$$

where V_ℓ and V_s = unit mass volume of the material in the liquid and in the solid state, respectively

S_ℓ and S_s = unit mass entropy of the liquid and of the solid state, respectively

L_m = specific heat of melting of the unit mass.

Since at melting the volume $V_\ell < V_s, dT_m < 0$, i.e., the equilibrious melting temperature of ice T_m decreases as well,

$$dT_m = -Adp. \quad (12)$$

Thus, for ice when $p = 0$, $T_m = 273.1$ K, $V_\ell = 10^{-3}$ m³/kg, $V_s = 1.09 \times 10^{-3}$ m³/kg, $L_m = 3.336 \times 10^5$ J/kg, parameter A for ice at temperature $\theta = 0°C$ is equal to

$$A = 0.074 \text{ K/MPa}.$$

The magnitude of parameter A coincides with that calculated by Barnes et al. (1971). It is not difficult at this point to calculate the critical hydrostatic pressure p^* at which the shear resistance of ice equals zero, i.e.,

$$T_m + \theta = T_m - Ap^* \quad (13)$$

and

$$p^* = \frac{-\theta}{A} \quad (14)$$

where θ = ice temperature (°C) and $T_m = 273.1$ K, the ice melting temperature at atmospheric pressure. Note that the ice melting pressure is unrelated to the grain size or the structure of ice. The magnitudes of ice melting pressures p^* for various temperatures are presented in Table 1.

Ice Cohesion

The ice cohesion defines the ice strength when the hydrostatic pressure $p = 0$. Ice strength is a stochastic event, a culmination point of failure of intermolecular bonds and growth of cracks. A certain number of these bonds in the unit volume of ice are formed during freezing of water at temperature 0°C. Further temperature decrease brings about formation of new bonds, attributable to freezing of the liquid phase at grain boundaries, and a sharp increase of the ice strength (cohesion). Thus, one may conclude that the ice cohesion is a function of temperature and consists of a sum of two components:

$$c(T) = c_o + c_1(T) \quad (15)$$

where c_o = component of the ice cohesion brought about at the time of ice formation at 0°C

Table 1.[1] The strength parameters of ice at various temperatures.

θ°C	c (MPa)	b	~ φ°	σ_{max} (MPa)	τ_{max} (MPa)	p* (Mpa)
–40	12.93[2]	0.241	14	245.87	43.52	540.54
–16	9.43	0.125	7	80.15	14.44	216.22
–11.8	8.14	0.106	6	53.81	10.99	159.46
–11	7.91	0.103	6	49.0	10.43	148.65
–6	6.63	0.08	4°30′	20.05	7.43	81.08
–1	5.56	0.053	3	(4.79)[3]	(5.69)[3]	(36.85)[3]
0	5.37[4]	0.04[4]	2[4]			

1 – Data referred for strain rate, $\dot{\varepsilon} \approx 5 \times 10^{-3}$ s⁻¹
2 – Calculated by Eq. 17.
3 – From the best fit of Eq. 5.
4 – At freezing.

$c_1(T)$ = temperature-dependent component of the cohesion brought about by freezing of liquid phase at temperature below 0°C.

Since the physical nature of c_o and c_1 is identical, and taking into account that at temperature $T = T_m$, $c_1 = 0$, the temperature dependency of component c_1 can be presented in the form

$$c_1(T) = c_o\alpha(1 - T/T_m) \quad (16)$$

where α is a parameter. Combining Eqs. 15 and 16

$$c(T) = c_o[1 + \alpha(1 - T/T_m)]. \quad (17)$$

Equation 17 establishes a linear dependency of the ice cohesion upon temperature. Test data show that such a relationship is valid in the domain of relatively low temperatures below –20°C. In the range of moderate temperatures, this relationship becomes nonlinear and somewhat better described by

$$c(T) = c_o e^{\alpha(1-T/T_m)^\mu} \quad (18)$$

or

$$c(\theta) = c_o e^{\alpha|\theta|/T_m} \quad (18a)$$

in which $\mu = 1$ and $|\theta| = (T_m - T)$ is ice temperature (°C).

Equations 18 and 18a can be considered approximate forms of Eq. 17.

Angle of Internal Friction

The parameter of internal friction of ice $b(T)$ can also be presented as a sum of two components

$$b(T) = b_o + b_1(T) \quad (19)$$

where $b_o = \tan \phi_o$ and ϕ_o is the angle of internal friction of ice brought about at the time of ice formation at 0°C

$b_1(T) = \tan \phi_1(T)$ and $\phi_1(T)$ is the angle of internal friction attributable to the increased viscosity of the liquid phase (lubricant) on grain boundaries at temperatures below 0°C.

One may assume that the component $b_1(T)$ is proportional to b_o and to the temperature melting function, i.e.

$$b_1(T) = b_o\beta(1 - T/T_m) \quad (20)$$

where β is a parameter. Combining Eqs. 19 and 20

$$b(T) = b_o[1 + \beta(1 - T/T_m)]. \quad (21)$$

To simplify the parameter evaluation procedure, Eq. 21 can approximately be presented in the form

$$b(T) = b_o e^{\beta(1-T/T_m)^\nu} \quad (22)$$

or

$$b(\theta) = b_o e^{\beta_1|\theta|^\nu} \quad (22a)$$

where $\beta_1 = \beta / T_m^\nu$. It was found that $\nu = 1/2$ is in agreement with test data. In very low range of temperatures certain adjustments of Eqs. 15 through 22 will probably be required.

TEMPERATURE CRITERIA OF ICE STRENGTH

Since temperature dependencies of the ice strength parameters have been established, a temperature criterion of ice strength, taking into account Eq. 5, can be presented in a form

$$\tau_{io}(p,T) = c(T) + b(T)p - \frac{b(T)}{2\sigma_{max}(T)}p^2 \tag{23}$$

in which $c(T)$, $b(T)$ and $\sigma_{max}(T)$ are given by Eqs. 7 and 14 through Eq. 22a. Equation 23 is a temperature criterion of ice strength in a multiaxial stress state. To take into account the strain rate effect, Eq. 23 should be combined with Eqs. 2a through 4, i.e.

$$\tau_{io}(p,\dot{\varepsilon},T) = \tau_{io}(p,T)\Phi(\dot{\varepsilon})$$

$$= \left(c(T) + b(T)p - \frac{b(T)}{2\sigma_{max}(T)}p^2\right)\left(\frac{\dot{\varepsilon}}{\dot{\varepsilon}_o}\right)^{\frac{1}{n}} \tag{24}$$

In the range of the hydrostatic pressures $p << \sigma_{max}$ Eq. 24 takes the form

$$\tau_i^*(p,\dot{\varepsilon},T) = \left(c(T) + b(T)p\right)\left(\frac{\dot{\varepsilon}}{\dot{\varepsilon}_o}\right)^{\frac{1}{n}} \tag{25}$$

when $p = 0$ (pure shear)

$$\tau_i^*(\dot{\varepsilon},T) = c(\dot{\varepsilon},T) = c(T)\left(\frac{\dot{\varepsilon}}{\dot{\varepsilon}_o}\right)^{\frac{1}{n}} \tag{26}$$

Correspondingly

$$b(\dot{\varepsilon},T) = b(T)\left(\frac{\dot{\varepsilon}}{\dot{\varepsilon}_o}\right)^{\frac{1}{n}} \tag{26a}$$

where parameters $c(T)$ and $b(T)$ are corresponding to the strain rate $\dot{\varepsilon} = \dot{\varepsilon}_o$.

If the radial strain rates are taken into account function $\Phi(\dot{\varepsilon})$ in Eqs. 24, 25 and 26 should be replaced by function $\Phi(\dot{\gamma}_i)$.

TEST DATA

The validity of the temperature dependencies of the strength parameters of ice presented above was verified using test data of Gagnon and Gammon (1995). The triaxial ($\sigma_2 = \sigma_3$) compression tests were carried out using cylindrical specimens of Labrador iceberg ice, 9.58 cm in diameter and 26 cm long. Mean grain diameter $d \approx 8.1$ mm. The test temperature varied between −1°C and −16°C. The tests were conducted at constant axial strain rates between 10^{-2} and 10^{-5} s^{-1}. Test data corresponding to the strain rate $\dot{\varepsilon} \approx 5\times10^{-3} s^{-1}$ were selected in the following analysis to evaluate the strength parameters of ice. This strain rate was defined by the authors of this paper as the instantaneous strain rate $\dot{\varepsilon}_o$ because the test strength magnitudes at this rate reached their maximum values. The tests were carried out at four different confining pressures: 1.38, 3.45, 6.89 and 13.79 MPa. The test data replotted by the authors of this paper in terms of the shear strength $\tau_i = (\sigma_1 - \sigma_3)/\sqrt{3}$ versus the hydrostatic pressure $p = (\sigma_1 + \sigma_2 + \sigma_3)/3$ are presented in Figure 2. Note that each point in Figure 2 represents the mean of five tests.

In Figure 2 for comparison tests, data are presented on triaxial compression of freshwater randomly oriented polycrystalline ice obtained by Jones (1982). The laboratory-made ice samples were 20 mm in diameter and 60 mm long with mean grain diameter less than 1 mm. The test data selected for comparison refer to the axial constant strain rate $\dot{\varepsilon} = 5.4\times10^{-3} s^{-1}$. The test temperature was −11.8 ± 0.9°C. The confining pressure ($\sigma_2 = \sigma_3$) varied between ~5 and ~ 85 MPa.

In addition in Figure 5 (below) results of triaxial constant strain rate compression tests are presented of freshwater, randomly oriented, polycrystalline ice carried out by Rist and Murrell (1994). In these tests laboratory-made samples of ice were 40 mm in diameter and 100 mm long. Mean grain diameter was ~1.7 mm. These tests were conducted at constant strain rates from 10^{-2} to 10^{-5} s^{-1}; the test temperatures varied between −20°C and −45°C. The data selected for this analysis are referred to the axial strain rate $\dot{\varepsilon} \sim 10^{-3} s^{-1}$.

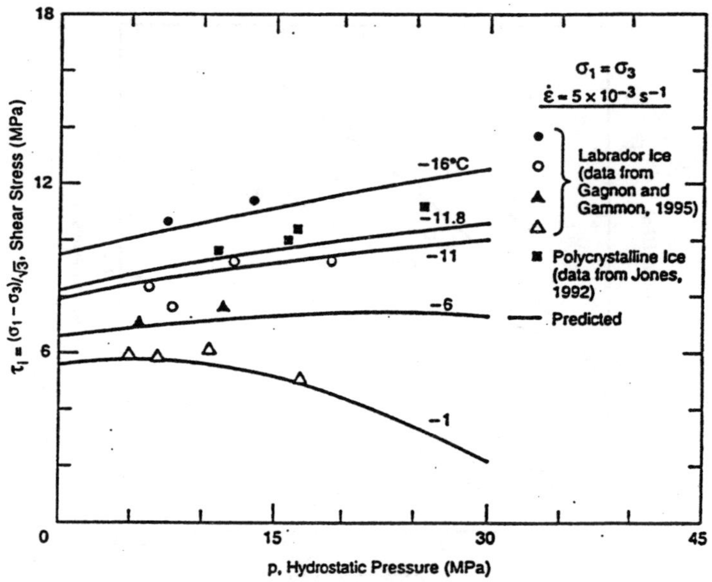

Fig. 2 Strength of ice under triaxial compression at various temperatures. Data from Gagnon and Gammon (1995) and Jones (1982).

PARAMETER EVALUATION

It was shown above that temperature variations of ice strength in a multiaxial stress state are defined by three temperature-dependent parameters: the ice melting pressure $p^*(T)$, ice cohesion $c(T)$ and friction parameter $b(T)$. The magnitudes of the ice melting pressure $p^*(T)$ are easily calculated by Eq. 14 for the test temperatures and are included in Table 1. The magnitude of p^* for $\theta = -11.8$ was found to be in agreement with $p^* = 123.7$ MPa obtained earlier (Fish, 1991) from the analysis of Jones' data and with that calculated by an empirical equation of Hallam and Nadreau (1988).

The temperature dependencies of parameters $c(\theta)$ and $b(\theta)$ can be determined based upon the considerations that in the range of low hydrostatic pressures $p << \sigma_{max}$ (test pressures in Fig. 2), as a first approximation, Eq. 5 can be replaced by Eq. 9. Then, test data in Figure 2 can be described by straight lines. The slopes of the straight lines and their intersects with the ordinate axis, obtained by a regression analysis of the test data, define friction parameter $b(\theta)$ and cohesion $c(\theta)$ for corresponding temperatures. The results of the analysis are presented in Figure 3.

One can see in Figure 3 that temperature dependencies of parameters $c(\theta)$ and $b(\theta)$ are indeed nonlinear and can be described by Eq. 18a and 22a. The evaluation procedures of parameters that enter these equations showed that for the Labrador iceberg ice

$$c_o = 5.37 \text{ MPa}; \quad b_o = 0.04$$

$$\alpha = 9.61; \quad \beta = 4.69$$

Using these parameters the magnitudes of the ice cohesion $c(\theta)$, the friction parameter $b(\theta)$ and the friction angle $\phi(\theta)$ calculated by Eq. 18a and 22a are included in Table 1. The magnitudes of $\sigma_{max}(\theta)$ and $\tau_{max}(\theta)$ in Table 1 were calculated by Eqs. 6 and 7 using corresponding values of $c(\theta)$, $b(\theta)$ and $p^*(\theta)$. Predicted temperature dependencies of parameters c and b are presented in Figure 3. From Table 1 and Figure 3 it follows that major changes of parameters b and c, and ice strength take place at temperatures below −10°C. Between −10°C and −12°C the ice strength changes insignificantly.

Note that the test point in Figure 3 corresponding to Jones' (1982) tests on polycrystalline ice with the grain size of less than 1 mm follows the same trend as the test data of Gagnon and Gammon on the Labrador ice with an average grain size of 8.1 mm, i.e., eight times larger than the ice in

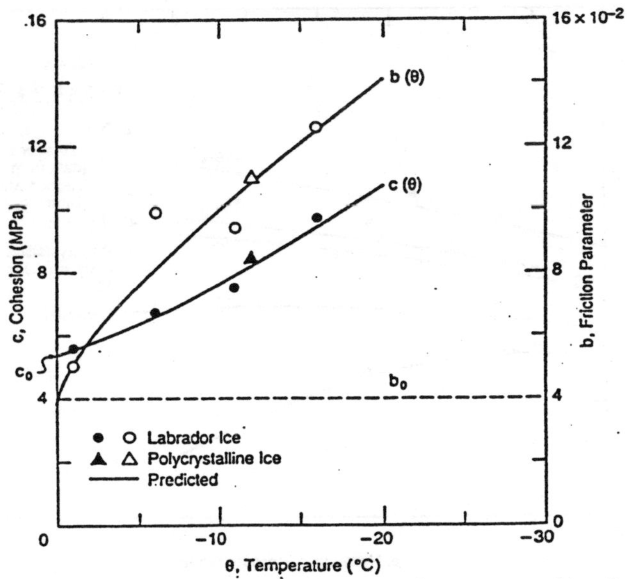

Fig. 3 Temperature dependencies of the strength parameters of ice. Data from Gagnon and Gammon (1995) and Jones (1982).

Jones' tests. Such correlation suggests that the ice cohesion may be unaffected by variations of the grain size.

To make sure that this conclusion extends beyond the temperature range between −1°C and −16°C, and that it is valid in the strain rate range different from ~5×10^{-3} s^{-1} used above in the evaluation of the cohesion, a comparative analysis of ice strength was carried out using test data of Rist and Murrell (1994). This polycrystalline ice had a grain size of 1.7 mm. The comparison was performed for the test temperature −20°C and the strain rates between ~10^{-3} s^{-1} and ~10^{-5} s^{-1}. The results of the analysis confirmed the above conclusion. At the same time our studies indicate that the ice cohesion magnitude is strongly dependent of the ice structure.

From the above studies an important conclusion can also be made regarding the physical nature of the angle of internal friction of ice. While the

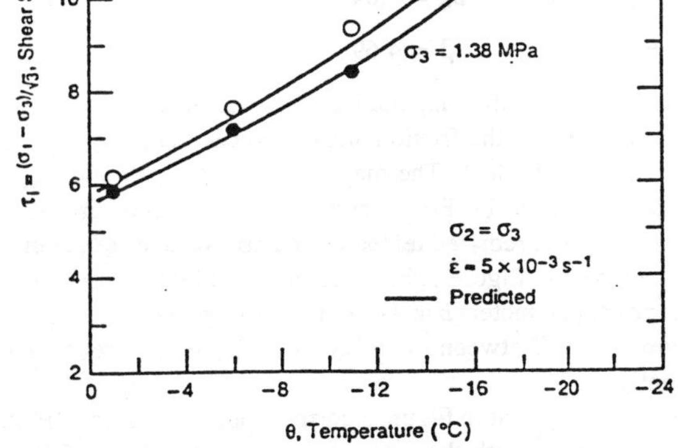

Fig. 4 Strength of ice as a function of temperature and confining pressure. Data from Gagnon and Gammon (1995).

temperature dependency of the friction parameter $b(\theta) = \tan \phi(\theta)$ is nonlinear (Fig. 3), the test data in Table 1 show that the friction angle is a simplest linear function of temperature:

$$\phi(\theta) = \phi_o + \omega|\theta| \qquad (27)$$

where $\phi_o = 2°50'$ and $\omega = 16.4'$ degree^{-1} and $|\theta| = (T_m - T)$ is ice temperature (°C).

Note that the magnitude of the internal friction angles ϕ in Table 1 were calculated based mainly upon test data on the Labrador iceberg ice ($d = 8.1$ mm) at the temperature range between −1°C and −16°C. These data were extrapolated to calculate the friction angle ϕ and the ice strength for temperature −40°C and compared (see Fig. 5 below) with test data of Rist and Murrell (1994) obtained for the grain size diameter of 1.7 mm.

One can see that despite the considerable difference (almost five times) in the grain sizes of these two types of ice and the difference in the test temperatures, the predicted angle of internal friction correlates well with the test data. We find that the friction angle magnitudes of ice are unaffected by variations of its grain size.

Since parameters c, b and p^* turned out to be independent of the grain size, Table 1 data can be used to predict strength of various types of iceberg and freshwater randomly oriented polycrystalline ice.

STRENGTH PREDICTIONS

Let us verify the accuracy of strength prediction by comparing the computed and the test results of ice strength as functions of temperature, strain rate and hydrostatic pressure.

For low hydrostatic pressures $p << \sigma_{max}$ and the strain rate $\dot{\varepsilon} = \dot{\varepsilon}_o$, ice strength can be calculated by Eq. 9 using data of Table 1 and the hydrostatic pressure magnitudes p from Figure 2, corresponding to confining pressures $\sigma_3 = 1.38$ MPa and 6.89 MPa. The results of such calculations for the temperature range between −1°C and −16°C are presented in Figure 4 together with the test data from Gagnon and Gammon (1995). One can conclude that the predicted ice strength values are in agreement with the test data.

The Table 1 data can also be used to evaluate the ice strength in those cases when the strain rate $\dot{\varepsilon} < \dot{\varepsilon}_o$. For $p << \sigma_{max}$ the ice strength is calculated by Eq. 25. To simplify the comparison with the published data the latter can be rewritten in terms of deviator stress

$$\sigma_1 - \sigma_3 = \sqrt{3}(c + bp)\left(\frac{\dot{\varepsilon}}{\dot{\varepsilon}_0}\right)^{1/n}. \qquad (25a)$$

In Table 2 the results of such calculations are presented for $n = 4$, $\dot{\varepsilon}_0 = 5 \times 10^{-3}$ s^{-1}, two different confining pressures assuming that $p = \sigma_3$ and two different strain rates together with the test results from Table 2 of Gagnon and Gammon (1995).

Table 2. Strength predictions of Labrador ice.

θ (°C)	Confining pressure (MPa)	Strain rate ($\dot{\varepsilon}$ s^{-1})	Strength (σ_1-σ_3) (MPa) Measured	Strength (MPa) Predicted	Ice type
−11	1.38	4.82 10^{-5}	3.50	4.36	Labrador ice
−11	6.89	5.96 10^{-5}	4.75	4.93	Labrador ice

From Table 2 it follows that predicted values of ice strength practically coincide with the test data. Unfortunately only two tests of this ice type were available for comparison.

In Figure 2 predicted dependencies of the ice strength are presented calculated by Eq. 5 using test magnitudes of hydrostatic pressures and the strength parameters from Table 1. One can see that in the moderate temperature range between −6° and −16°C the calculated values of the ice strength are in good agreement with test data. An exception is the test data for −1°C for which the magnitude of the ice melting pressure $p^* = 13.51$

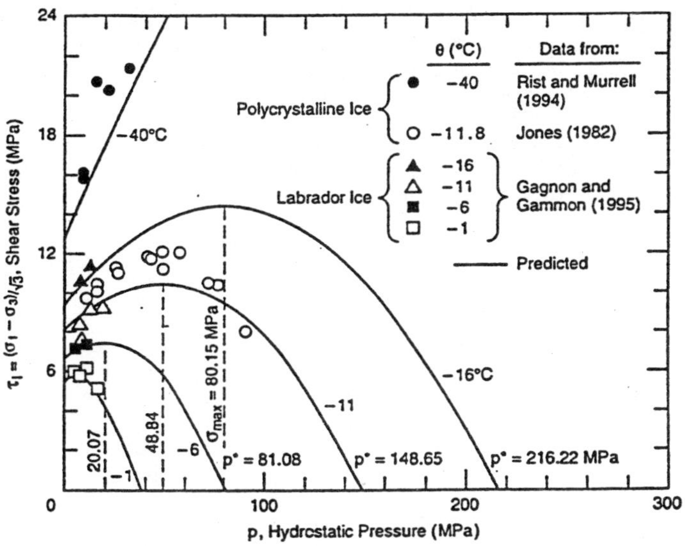

Fig. 5 Strength of ice in the range of high hydrostatic pressures at strain rate ~ $\dot{\varepsilon}$ 10^{-3} s^{-1}. Data from Jones (1982), Gagnon and Gammon (1995), Rist and Murrell (1994).

MPa calculated by Eq. 14 turned out to be considerably smaller than p^* = 36.85 MPa obtained from the best fit of the experimental points in Eq. 5. Apparently, as the ice temperature approaches the ice melting temperature, the relationship between the ice melting pressure and temperature becomes nonlinear. It is obvious that further studies of the temperature dependency of the ice melting pressure are necessary.

Equation 5 and the data in Table 1 were used to predict the ice strength over the range of high hydrostatic pressures. The results of such an extrapolation are presented in Figure 5 and compared with test data of Jones (1982) for temperature −11.8°C, and of Rist and Murrell (1994) for temperature −40°C. In the latter case parameter *c* was calculated by Eq. 17 to fit the test data, suggesting that further studies of the temperature dependency of ice cohesion in the temperature domain below −20°C are required. At the same time the general trend of predicted values of ice strength as functions of the hydrostatic pressure is well confirmed by Jones' test data.

CONCLUSIONS

1. A temperature model has been developed that describes the ice strength in a multiaxial stress state over a wide spectrum of negative (subfreezing) temperatures.

2. The strength dependency of ice on the hydrostatic pressure and temperature is well described by an extended Drucker-Prager (parabolic) yield criterion, three physically *fundamental* parameters of which are the ice cohesion *c*, angle of internal friction ϕ and ice melting pressure p^*. They all have a *definite* physical meaning and are functions of temperature.

3. The temperature model was developed based upon data on over 100 triaxial compression tests of iceberg ice and laboratory-made freshwater randomly oriented polycrystalline ice with a grain diameter between ~1 mm and ~8 mm in the temperature range between −1°C and −40°C at the range of strain rates between 10^{-3} s^{-1} and 10^{-5} s^{-1}.

4. Further studies are required of the temperature dependencies of the ice strength parameters for various types of ice, particularly the ice melting pressure over a wider spectrum of temperatures, strain rates and confining pressures.

Acknowledgments

Funding or this research was provided by the U.S. Army Corps of Engineers. The authors extend their appreciation to Drs. D.S. Sodhi, P. Black and S. Ketcham for their valuable comments.

REFERENCES

Barnes, P., D. Tabor, F.R.S. and J.C.F. Walker (1971) The friction and creep of polycrystalline ice. *Proc. Royal Soc. London, Ser. A*, 347: 493–512.

Drucker, D.C. and W. Prager (1952) Soil mechanics and plastic analysis of limit design. *Quarterly Journal of Applied Mechanics*, 10: 157–165.

Fish, A.M. (1991) Creep and yield model of ice under combined stress. U.S. Army Cold Regions Research and Engineering Laboratory. Special Report 91-31, Hanover, NH.

Fish, A.M. (1992) Three-dimensional visco-plastic flow model of polycrystalline ice. In *Proceedings 3rd International Conference on Ice Technology. M.I.T., Cambridge, Massachusetts, USA, 11–13 August* (T.K.S. Murphy, W.M. Sackinger and P. Wadham, Ed.). Computational Mechanics Publ., Southampton, U.K., pp. 193–207.

Fish, A.M. (1993) Combined creep and yield model of ice under multiaxial stress. *Journal of Offshore and Polar Engineering*, 12(3): 205–209.

Gagnon, R.E. and P.H. Gammon (1995) Triaxial experiments on iceberg and glacier ice. *Journal of Glaciology*, 41(139): 538–540.

Hallam, S. and J.P. Nadreau (1988) Failure maps for ice. In *Proceedings, 9th International Conference on Port and Ocean Engineering under Arctic Conditions (POAC '87) 17–22 August, Fairbanks, Alaska* (W.M. Sackinger and M.O. Jeffries, Ed.) Geophysical Institute of Alaska-Fairbanks, vol. 3: 45–55.

Hausler, F.U. (1983) Comparison between different yield functions for saline ice. *Ann. Glacial.*, 4: 105–109.

Haynes, F.D. (1973) Tensile strength of ice under triaxial stress. U.S. Army Cold Regions Research and Engineering Laboratory, Research Report 312.

Jones, S.J. (1978) Triaxial testing of polycrystalline ice. In *Proceedings, 3rd International Conference on Permafrost, Edmonton*, vol. 1, p. 670-674.

Jones, S.J. (1982) The confined compressive strength of polycrystalline ice. *Journal of Glaciology*, 28(98): 171–177.

Nadreau, J.P. and B. Michel (1986) Yield and failure envelope for ice under multiaxial compressive stresses. *Cold Regions Science and Technology*, 12(1): p. 75–82.

Nadreau, J.P., Nawwar, A.M. and Y.S. Wang (1991) Triaxial testing of freshwater ice at low confining pressures. *Transactions of ASME*, vol. 113, p. 260–265.

Reinike, K.M. and T.D. Ralston (1977) Plastic limit analysis with an anisotropic, parabolic yield function. *Int. Journal of Rock Mechanics and Mining Sciences*, 14(3): 147–154.

Richter-Menge, J.A., G.F.N. Cox, N. Perron, G. Durell and H.W. Bosworth (1986) Triaxial testing of first-year sea ice. U.S. Army Cold Regions Research and Engineering Laboratory, CRREL Report 86-16.

Rist, M.A. and S.A.F. Murrell (1994) Ice triaxial deformation and fracture. *Journal of Glaciology*, 40(135): 305–318.

Sayles, F.H. (1974) Triaxial constant strain rate tests and triaxial creep tests on frozen Ottawa sand. U.S. Army Cold Regions Research and Engineering Laboratory, Technical Report 23, Hanover, NH.

Smith, M.B. (1974) A parabolic yield condition for anisotropic rocks and soils. Ph.D. Thesis (unpublished) Rice University, Houston, Texas.

Timco, G.W. and R.M.W. Frederking (1986) Confined compression tests: Outlining the failure envelope of columnar sea ice. *Cold Regions science and Technology*, 12: 13–28.

Weiss, J. and E.M. Schulson (1995) The failure of freshwater granular ice under compressive loading. *Acta Metall. Mater.*, 43(6): 2303–2315.

Zaretsky, Yu.K. and A.M. Fish (1996a) Model of viscoplastic deformation of frozen and unfrozen soils and ice. In *Proceedings, 6th International Offshore and Polar Engineering Conference. Los Angeles, 26–31 May 1996*, vol. 2, p. 291–296.

Zaretsky, Yu.K. and A.M. Fish (1996b) Effect of temperature on the strength and viscosity of ice. *Journal of Soil Mechanics and Foundation Engineering* (Translated from Russian), vol. 33(2): 46-52.

Proceedings of the Seventh (1997) International Offshore and Polar Engineering Conference
Honolulu, USA, May 25-30, 1997

ISBN 1-880653-28-1 (Set); ISBN 1-880653-30-3 (Vol. II)

Ice Floe Collisions: A Neglected Attribute of Wave Dissipation in Ice Fields

Vernon A. Squire
University of Otago
Dunedin, New Zealand

Hayley H. Shen
Clarkson University
Potsdam, NY, USA

ABSTRACT

A model of a compact field of pancake ice and frazil acted upon by ocean waves is described, which includes collisions arising from the differential movement of separate ice cakes as a mechanism for energy dissipation. The model is based on the so called mass-loading theory, where the ice is assumed to behave as though composed of a dense cover of noninteracting points of mass with no inherent elasticity in flexure. Energy loss is found to depend upon the restitution coefficient of the floes, together with their impact velocities. For small dissipation, where the group velocity is unaffected by damping, the spatial attenuation of the wave energy is described roughly by a piecewise power law and not by the exponential function that is commonly hypothesized. With realistic values for parameters, the model predicts a highly dissipative band close to the ice margin, in agreement with some observational studies that suggest the presence of an edge zone where floes and cakes suffer significant buffeting by intense incoming seas.

Key Words. Pancake ice; ocean waves; collisions.

INTRODUCTION

Since the late 1950s there have been several experiments to measure the effect of pack ice on the passage of ocean waves, and a small number of theoretical studies. Rather than cite this work the reader is referred to the review by Squire et al (1995), which has a more or less comprehensive bibliography. The theoretical work has been based on empirical conclusions, on the thin elastic plate or the same with zero flexural rigidity, on scattering theory, and on parameterizations that can produce reasonable results but seldom have any physical justification.

There are various aspects of ocean wave behaviour in sea ice that theoreticians aim to reproduce in their models. They seek to correctly represent the way that the waves are attenuated by the ice field, how they are dispersed and how they are refracted at the edge—although data are equivocal on this feature, how their directional spread changes with penetration, and they seek this knowledge for each type of sea ice under a wide range of wave conditions. Only comparatively recently has scattering theory provided an answer to most of these questions (Masson and LeBlond, 1989; Meylan et al, 1997).

One facet of wave propagation in ice that has been omitted entirely from all the models, including those based on scattering theory, is the effect of floe collisions. And this is despite experimental evidence suggesting that it is important in many circumstances. Collisions are dealt with in Squire et al (1995) from an experimental perspective, where the work of Martin and Becker (1987, 1988), McKenna and Crocker (1992), Rottier (1992), and Rottier and Squire (1993) are discussed in some detail, but no theory of any sophistication exists to explain the data. Floes do collide regularly when moving under the action of ocean waves, and when they collide they lose energy. While this is only one of 4 possible processes by which dissipation occurs, the other 3 being due to: absorption from hysteresis as floes deform on the passing waves; absorption in the water column from wave breaking and viscous losses in the turbulent boundary layer; and redistribution of wave energy by scattering; collisions can dominate under certain conditions and, accordingly, should not be omitted. We find that they are most important for relatively short waves in highly compact sea ice.

We focus in this paper on pancake ice and frazil, as we wish to keep the sea ice model as simple as possible. First, the various ways waves can dissipate will be discussed briefly, drawing on the more detailed study of Shen and Squire (1997). Then, the energy density on either side of the ice edge will be found and, by means of the energy flux, a familiar relationship associating the incoming, reflected and transmitted wave amplitudes, i.e. $\mathcal{A}T^2 + R^2 = I^2$, will be re-established in a different way. The granular flow based collision model will be discussed next, and a simple differential equation written down and solved to allow the wave amplitude to be found at any distance from the ice edge. Finally, some conclusions will be drawn for short period waves, namely waves of about 5 or 6 s or less, where, as might be intuitively expected for small pancakes, the effect of collisions appears to be most significant.

DISSIPATION

Several mechanisms of energy loss act when ocean waves travel through pancake ice and frazil, with the relative importance of each process being dependent on the prevailing conditions. For compact fields we believe that energy dissipation associated with collisions between adjacent ice cakes is a significant contributor. As noted earlier, other ways of removing energy from the incoming sea are: hysteresis loss due to anelastic floe bending; water column losses; and redistribution of energy by scattering. For pancakes where the strains experienced by each cake are small, hysteresis loss is negligible. Likewise, scattering does not remove large amounts of energy. Water column losses, which arise in a compact ice cover from the rough undersides and edges of cakes and any rafting present, can be significant in some circumstances. We find that energy dissipation at short periods, particularly, is dominated by collisions. Referencing to 1 m amplitude waves, for example, at 5 s period the energy loss per unit area due to collisions is about 100 W m^{-2}, whereas that due to losses in the water column is 1 W m^{-2}. At longer periods energy loss by collisions becomes less important. Shen and Squire (1997) provide order of magnitude estimates for each mechanism.

ENERGY FLUX

Because we are considering ocean wave propagation in pancake ice and frazil, we suppose that the sea ice cover has no elasticity, i.e. we use the model of Wadhams and Holt (1991) based on the work of Weitz and Keller (1950), Peters (1950), and Keller and Weitz (1953). That model assumes that the ice behaves as a surface composed of disconnected particles of floating matter.

A simple relationship exists between the incident, reflected and transmitted wave amplitudes when waves penetrate a region covered with sea ice. This may be derived either by Green's theorem, as done quite generally by Fox and Squire (1994), or from first principles by considering the energy stored in the ice and the energy in the water column. Since we are most interested in physical quantities such as energy density and flux, the latter method will be used here. Also, to maintain generality we shall at first allow the sea ice to be elastic and then set the flexural rigidity to zero.

We suppose an incoming ocean wave of amplitude I, represented at the surface by $\eta_1 = I\cos(kx-\omega t)$, a reflected wave $\eta_2 = R\cos(kx+\omega t+\phi)$ and a transmitted wave $\eta_3 = T\cos(\kappa x - \omega t + \psi)$. Accordingly, the wave period is $2\pi/\omega$, the wavelength in open sea is $2\pi/k$, and the wavelength within the ice is $2\pi/\kappa$. The quantities ϕ and ψ are phase terms that have no relevance to the current work. The water is taken to be deep to keep the algebra uncluttered.

Assuming initially that the ice behaves as a thin elastic sheet, the time-averaged potential energy per unit area, hereinafter energy density, stored in the ice is

$$V_1 = \frac{1}{4}D\kappa^4T^2, \tag{1}$$

where D is the flexural rigidity of the sheet. The kinetic energy density is

$$K_1 = \frac{1}{4}m\omega^2T^2, \tag{2}$$

where m is the mass per unit area of the ice sheet. The kinetic and potential energy densities in the water column beneath the ice sheet are respectively

$$V_2 = \frac{1}{4}\rho g T^2 \quad \text{and} \quad K_2 = \frac{1}{4}\rho\frac{\omega^2}{\kappa}T^2, \tag{3}$$

where ρ is the density of the sea water and g is the acceleration due to gravity. The total mean energy density associated with the ice cover,

$$\mathcal{E}_i = V_1 + K_1 + V_2 + K_2,$$

is therefore given by

$$\mathcal{E}_i = \frac{1}{2}\rho g T^2\left(M\kappa^4+1\right), \qquad M = \frac{D}{\rho g}, \tag{4}$$

where the usual dispersion relation for ice-coupled waves, namely $\rho\omega^2/\kappa = D\kappa^4 + \rho g - m\omega^2$, has been invoked (Fox and Squire, 1994). Note also that

$$V_1 + V_2 = \frac{1}{4}D\kappa^4T^2 + \frac{1}{4}\rho g T^2 = \frac{1}{4}\rho g T^2\left(M\kappa^4+1\right) = K_1 + K_2,$$

(Wadhams, 1986), which is a statement of Rayleigh's principle that the total mean energy is twice the mean kinetic energy for a linear wave.

For a wave of amplitude A the energy density in open water, made up of its potential and kinetic parts, is

$$\frac{1}{4}\rho g A^2 + \frac{1}{4}\rho\frac{\omega^2}{k}A^2,$$

so using the dispersion relation for deep water waves, namely $k = \omega^2/g$, we find

$$\mathcal{E}_w = \frac{1}{2}\rho g A^2. \tag{5}$$

The energy flux $\mathcal{F}$, defined as the product of the energy density and the group velocity, must be conserved across the ice edge. Consequently,

$$\mathcal{E}_w U_w = \mathcal{E}_i U_i, \tag{6}$$

where U_w and U_i denote the group velocities for waves propagating in the open water and in the ice-covered region, respectively. Using

$$U_w = \frac{\omega}{2k} \quad \text{and} \quad U_i = \frac{\omega\left(5M\kappa^4+1-L\right)}{2\kappa\left(M\kappa^4+1\right)}, \tag{7}$$

where $L = m\omega^2/\rho g$, found from the deep water dispersion relations, (6) can be used to derive the same expression relating the amplitudes of the incident, reflected and transmitted waves derived by Fox and Squire (1994), i.e.

$$\mathcal{A}T^2 + R^2 = I^2, \qquad \mathcal{A} = \left(5M\kappa^4+1-L\right)\frac{k}{\kappa}. \tag{8}$$

The quantity $\mathcal{A}$ is called the intrinsic admittance.

Recalling (4), it is clear that if the ice has no elasticity, i.e. the flexural rigidity $D = 0$ and hence $M = 0$, then the energy density reverts to the usual expression for open water with the appropriate wave amplitude, namely $\frac{1}{2}\rho g T^2$. Further, because in this case $U_i = \omega(1-L)/2\kappa$, the energy flux in the ice is

$$\mathcal{F}_i = \left(\rho g - m\omega^2\right)\frac{\omega T^2}{4\kappa}, \tag{9}$$

and the intrinsic admittance in (8) becomes $\mathcal{A} = (1-L)k/\kappa$. For comparison, the energy flux in open water is

$$\mathcal{F}_w = \rho g\frac{\omega I^2}{4k}. \tag{10}$$

MODEL

The averaged energy equation within the sea ice when dissipation d is present is

$$\frac{\partial \mathcal{E}_i}{\partial t} + \frac{\partial \mathcal{F}_i}{\partial x} + d = 0. \tag{11}$$

We restrict our discussion to the steady state case, thereby omitting any time dependence, and assume that the dissipation is due principally to collisions between adjacent pancakes. Equation (11) then becomes

$$\frac{d\mathcal{F}_i}{dx} = \frac{d}{dx}\left(\mathcal{E}_i U_i\right) = -fec, \tag{12}$$

where f is the collision frequency per pancake, e is the energy lost per collision, and c is the concentration. The quantities f and e are themselves functions of T, ω, c, and the properties of the pancakes, which we denote by $\mathcal{P}$, i.e.,

$$f = f(T,\omega,c,\mathcal{P}), \qquad e = e(T,\omega,c,\mathcal{P}). \tag{13}$$

The determination of f and e is difficult, and this is where granular flow theory is employed to model the pancake ice field, as described in greater detail in Shen and Squire (1997).

An important facet of (9) and (10) is their respective dependence on the amplitude squared, an aspect natural to all linear wave theory. Introducing (9) into (12) we obtain

$$\frac{d}{dx}\left((\rho g - \rho' h c \omega^2)^2 T^2\right) = -4\rho\omega fec, \tag{14}$$

where m has been written as the product of the ice density ρ', the ice thickness h and the concentration c. This equation implicitly ignores any changes in the group velocity with penetration, i.e. it assumes that the dissipation is small enough that changes in the velocity at which the energy propagates are negligible. This is probably a reasonable assumption unless the waves are of comparable length to the ice cakes, and is analogous to wave propagation through a slightly lossy ionosphere (Budden, 1988). With constant thickness and concentration (14) becomes

$$\frac{dT^2}{dx} = -\frac{4\rho\omega}{(\rho g - \rho' h c\omega^2)^2} fec. \tag{15}$$

but (14) allows spatially variable ice covers to be considered also if the functions $c = c(x)$ and $h = h(x)$ are known.

When granular flow theory is used to find f and e at some c, a rough power law in T is found to hold piecewise over ranges of amplitudes, i.e. $fec = aT^{n+1}$, and

$$\frac{\partial T^2}{\partial x} \propto -T^{n+1} \quad \text{or} \quad \frac{\partial T}{\partial x} \propto -T^n = -ST^n. \tag{16}$$

A simple analytic general solution to this equation exists, namely

$$T^{n-1} = \frac{1}{(n-1)Sx + T_0^{(1-n)}}, \tag{17}$$

where T_0 is the amplitude at the ice edge. Because the magnitude of the reflection coefficient R will be very small for ocean wavelengths impinging on pancake ice, we may neglect it in the balance of energy flux at the edge. So with $M = 0$, (8) becomes

$$T_0 = \left(\frac{\rho g}{\rho g - \rho' h c \omega^2}\right) I. \tag{18}$$

This approximation, while convenient, is actually unnecessary as it is quite straightforward to find T_0 precisely. If this is done we find $T_0 = 2I/\left((2-L)\sqrt{1-L}\right)$. Comparison of (15) and (16) shows the quantity S to be given by

$$S = \frac{2\rho a\omega}{(\rho g - \rho' h c\omega^2)^2}. \tag{19}$$

It is also of some interest to point out that (16) and its solution (17) was derived by Wadhams (1973) using an entirely different dissipation mechanism, namely steady state creep. Equation (17), taken together with (18) and (19), provides the desired expression for energy decay when it is totally due to collisions between ice cakes.

RESULTS

The collision frequency per ice cake, the associated energy dissipation per collision, and the power density dissipation are presented against wave amplitude in Figure 1 for a wave period of 5 s. Note that because 5 s waves become unstable at 2.8 m amplitude, only points to the left of the vertical bracket at that amplitude are physically meaningful. Additional points are plotted to illustrate the stability of the procedure. Concentrating on Figure 1c, the power density dissipation is precisely the absolute value of the right hand side of Equation (12). As expected, when wave energy increases all three quantities increase, but the relationship between the dissipated power and the wave amplitude does not follow a simple power law.

The log-log plot of Figure 1d has been replotted as curve (a) in Figure 2 with corresponding curves at 10.3 and 14.6 s for comparison. All curves

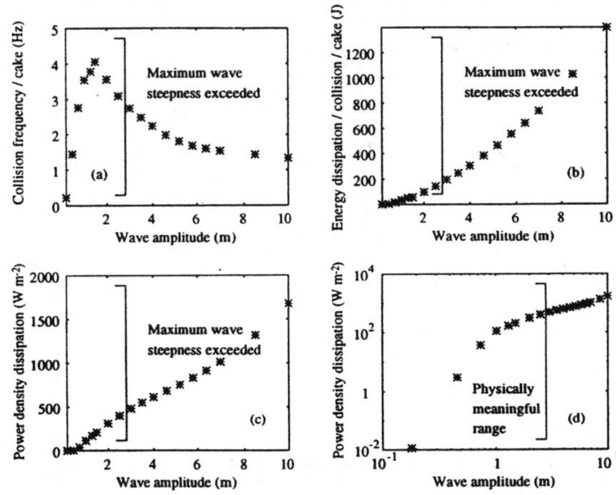

Figure 1: (a) Collision frequency, (b) energy dissipation per floe per collision, and (c,d) power density dissipation, each plotted against the normalized wave amplitude. Wave period is 5 s. Because 5 s waves above an amplitude of 2.8 m are too steep to exist, only points to the left of the bracket are physically meaningful in each plot. ($c = 0.9$, $d = 1$ m, $h = 0.3$ m, $E = 10$ MPa, $e = 0.2$, $C_m = 0.3$, and $C_w = 0.05$.)

increase monotonically. The points of both the 10.3 and 14.6 s curves, labelled (b) and (c), respectively, increase in a roughly linear manner at first from the lowest amplitude plotted, namely, 0.18 m. Power dissipation then begins to change rapidly at a specific wave amplitude that is different for each period. Finally, at a larger amplitude, the curves become less steep again, with a gradient similar to that at low amplitudes. The dissipation arising from collisions at 10.3 and 14.6 s is quite small except at the largest amplitudes, and consequently the energy decay experienced is likely to be dominated by other processes. This is not the case for 5 s waves, which suffer large dissipation. The results for 5 s waves, represented by +, appear at first sight to be different in form. However, the points trace out a curve that actually resembles the other curves shifted to the left. At very large amplitudes all curves tend asymptotically to one another, though this situation is physically meaningless as waves of such amplitude cannot exist. (A 5 s wave becomes unstable at 2.8 m, a 10 s wave at 11.1 m, and a 15 s wave at 24.9 m, as marked on the Figure.) A picture emerges then of an amplitude dependence that is qualitatively similar at each wave period, namely, a sheared, s-shaped curve that is positioned and sloped according to period. It is clear that Equation (16), which suggests a linear dependence between $\log T$ and the logarithm of the dissipation, cannot be used over the entire range of physically possible amplitudes.

The energy dissipated during each ice cake collision increases steadily with amplitude, apparently following a power law in a piecewise manner. As a result the velocity of each cake increases steadily, and the collisions, which consume energy proportionally, also increase steadily with amplitude. The unusual structure of the curves in Figure 2 is caused by differences in phase, which can prevent ice cakes from colliding and thereby cause changes in collision activity. For a real sea made up of waves of different periods with random phases, the s-shaped curves would be expected to be smoothed out to some degree.

Within prescribed ranges of amplitudes we may fit a power law piecewise to the points of Figure 2. This results in the linear fits shown, where (a) representing 5 s has piecewise slopes of 5.4 and 1.1 with intercepts of $10^{2.2}$ and $10^{2.1}$. Accordingly, the power law parameter n that appears in Equations (16)–(17) for 5 s is $n \approx 4.4$ when $T \leq 1$ m and $n \approx 0.1$ when $T > 1$ m, although these values will vary with ice properties, concentration, and floe size.

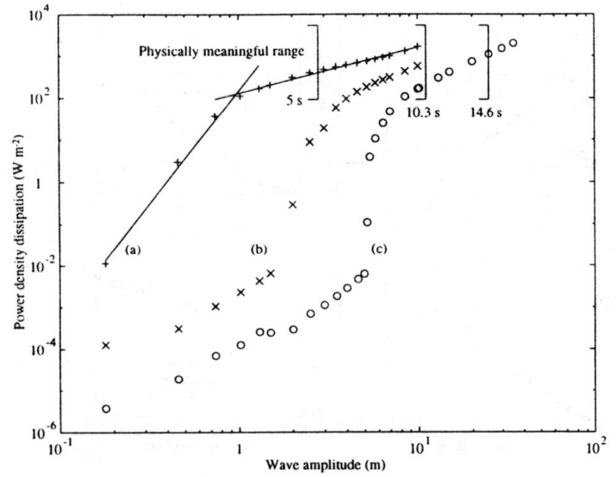

Figure 2: Curve (d) of Figure 1 with corresponding plots at longer periods. The 5 s curve is plotted with partial straight line fits of slope 5.4 and 1.1, and intercepts of $10^{2.2}$, and $10^{-2.1}$. The 10.3 and 14.6 s curves are drawn with × and o respectively. In each case the physically meaningful range lies to the left of the bracket.

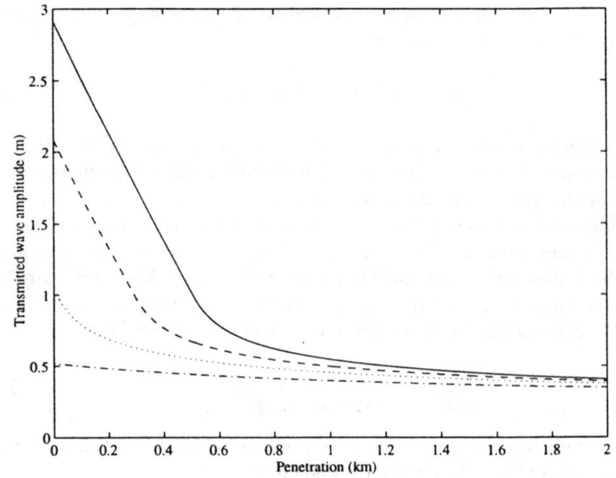

Figure 3: Four decay curves for the wave amplitude T, corresponding to incident wave amplitudes I of 2.8 m (solid), 2 m (dashed), 1 m (dotted) and 0.5 m (chained).

Limited by the largest open sea amplitude possible (2.8 m), Figure 3 presents decay curves for 5 s waves of amplitude T plotted against distance from the ice edge. Each curve is found using the piecewise linear fit described above if $T_0 > 1$, i.e. as the wave is attenuated there comes a point at which the power law (17) switches from one based on $n = 0.1$ to one based on $n = 4.4$. For $T_0 \leq 1$ $n = 4.4$ is used throughout. Consequently, when the wave amplitude is large, i.e. more than 1 m with the ice conditions considered, the effect of collisions is to produce a roughly linear decrease in wave amplitude in the vicinity of the ice edge, followed by higher order power law attenuation further in. Although a power law will always level off before the commonly assumed $T = T_0e^{-\alpha x}$, leading to greater eventual penetration of wave energy, near the edge the amplitude drop exceeds exponential decay for realistic attenuation coefficients α. This suggests a zone of substantial attenuation located near the ice edge where energy is quickly removed from the incoming sea before it enters a zone where the dissipation is less intense. While this may seem esoteric such a zone has indeed been observed.

The work of Dean (1966), for example, while appearing only as an abstract, reports that at short periods the energy decrease in the first 1 or 2 km is more severe than further into the ice field. At longer periods, i.e. above about 8 s, no difference between the region close to the ice margin and the interior is found. Wadhams (1986) reanalyzed Dean's data and notes the same effect, attributing it to the presence of brash between ice floes causing greater attenuation. The brash may have originated from collisions induced by intense wave activity, of course, and this would be an alternative explanation for dissipation in keeping with the present study. Squire and Moore (1980) in their experimental study of wave decay in the Bering Sea, divide up the marginal ice zone into three zones; an edge zone, a transition zone, and an interior zone. In the edge zone floes and cakes are buffeted together in the intense waves. Considerable rafting, ridging and fracturing takes place because of collisions between floes, and this leads to anomalously high attenuation across the zone. While the present model does not parameterize these processes in detail, it does include collisions as a source of energy dissipation. A comparative plot of power spectral densities at different wave periods with distance from the ice edge (Figure 4 of Squire and Moore, 1980) shows, similar to Dean's data, that whereas long periods decay systematically with penetration, the energy decay in the edge zone at the shortest period is more rapid than further into the ice interior. Although the current paper deals only with pancake ice where few data are available, similar results would be obtained for collisions between ice floes. This suggests that collisions may offer a plausible mechanism for augmented energy dissipation near the ice edge in penetrating, short period ocean waves.

SUMMARY AND CONCLUSIONS

We have presented herein a model for ocean wave dissipation in a field of sea ice that is based on collisions brought about by the differential movement of ice floes and ice cakes. Because various agencies act to remove energy from the waves, namely hysteresis losses, water column losses, and the redistribution of energy by scattering, we have chosen to examine a situation where the energy loss induced by collisions dominates other processes (Shen and Squire, 1997). Accordingly, short period waves entering a concentrated field of pancake ice have been considered.

For 5 s period waves, results suggest that collisions are responsible for power density dissipation values that range from about $10^{-2}\,\mathrm{W\,m^{-2}}$ for waves of amplitude 0.18 m to $500\,\mathrm{W\,m^{-2}}$ when their amplitude reaches a maximum pre-breaking value of 2.8 m. Such values are over an order of magnitude greater than those arising from other processes. But the amount of dissipation that occurs is also dependent on the wave energy in a nonlinear way. Indeed, a power law is insufficient to describe the functional relationship between wave amplitude T and the power density dissipation fec, and the piecewise linear fit shown in Figure 2 is more satisfying. This fit suggests that two dissipation rates exist, one in a band close to the ice edge and one further in. Waves entering an ice field are therefore rapidly attenuated until their amplitude is reduced to a particular value that depends on the nature of the sea ice, and thereafter the damping is more moderate. The band is of the order of a few hundred metres across for the ice conditions modelled. While this paper has focused on pancake ice, the phenomenon occurs with larger ice cakes and floes and has been observed on at least two occasions in a marginal ice zone made up of larger floes.

We believe we have demonstrated that ice floe collisions are an important, but neglected, facet of wave decay in ice fields. Furthermore, they may help explain why observations indicate very large attenuation in the near vicinity of the ice margin. Moreover, as Masson (1991) points out, the rapid decrease in short period wave energy causes a large compressive stress that also has the effect of compacting the pack ice at the extreme ice edge.

ACKNOWLEDGEMENT

The authors are grateful to the University of Otago, Clarkson University, the Royal Society of New Zealand, the NZ Foundation for Research, Science and Technology, and the US National Science Foundation for their continued generous support.

REFERENCES

Budden, KG (1988). *The Propagation of Radio Waves,* Cambridge University Press, New York, 669 pp.

Dean, CH (1966). "The attenuation of ocean waves near the open-ocean/pack-ice boundary," Symp. on Antarctic Oceanography, Santiago, Chile, pp 221–222.

Fox, C, and Squire, VA (1994). "On the oblique reflexion and transmission of ocean waves from shore fast sea ice," *Phil Trans R Soc A,* Vol 347, No 1682, pp 185–218.

Keller, JB, and Weitz, M (1953). "Reflection and transmission coefficients for waves entering or leaving an icefield," *Communs Pure Appl Math,* Vol 6, No 3, pp 415–417.

McKenna, RF, and Crocker, GB (1992). "Ice floe collisions interpreted from acceleration data during LIMEX'89," *Atmos-Ocean,* Vol 30, No 2, pp 246–269.

Martin, S, and Becker, P (1987). "High-frequency ice floe collisions in the Greenland Sea during the 1984 Marginal Ice Zone Experiment," *J Geophys Res,* Vol 92, No C7, pp 7071–7084.

Martin, S, and Becker, P (1988). "Ice floe collisions and relation to ice deformation in the Bering Sea," *J Geophys Res,* Vol 93, No C2, pp 1303–1315.

Masson, D (1991). "Wave-induced drift force in the marginal ice zone," *J Phys Oceanogr,* Vol 21, No 1, pp 3–10.

Masson, D, and LeBlond, PH (1989). "Spectral evolution of wind-generated surface gravity waves in a dispersed ice field," *J Fluid Mech,* Vol 202, pp 43–81.

Meylan, M, Squire, VA, and Fox, C (1997). "Towards realism in modeling ocean wave behavior in marginal ice zones," *J Geophys Res,* under revision.

Peters, AS (1950). "The effect of a floating mat on water waves," *Communs Pure Appl Math,* Vol 3, No 3, pp 319–354.

Rottier, PJ (1992). "Floe pair interaction event rates in the marginal ice zone," *J Geophys Res,* Vol 97, No C6, pp 9391–9400

Rottier, PJ, and Squire, VA (1993). "Collision events between adjacent ice floes driven by an ocean wave field," Proc. 3rd Int. Offshore and Polar Engin. Conf., Singapore, Vol. 2, pp 747–751.

Shen, HH, and Squire, VA (1997). " Wave damping in compact pancake ice fields due to interactions between ice cakes," Antarctic Science Research Series, AGU, in press.

Squire, VA, Dugan, JP, Wadhams, P, Rottier, PJ, and Liu, AK (1995). "Of ocean waves and sea ice," *Ann Rev Fluid Mech,* Vol 27, pp 115–168.

Squire, VA, and Moore, SC (1980). "Direct measurement of the attenuation of ocean waves by pack ice," *Nature, Lond,* Vol 283, No 5745, pp 365–368.

Wadhams, P (1973). "Attenuation of swell by sea ice," *J Geophys Res,* Vol 78, No 18, 3552–3563.

Wadhams, P (1986). "The Seasonal Ice Zone," *The Geophysics of Sea Ice* ed Untersteiner, N, NATO ASI Series, Plenum Press, NY, USA, pp 825–991.

Wadhams P, and Holt B (1991). "Waves in frazil and pancake ice and their detection on Seasat SAR imagery," *J Geophys Res,* Vol 96, No C5, pp 8835–8852.

Weitz, M, and Keller, JB (1950). "Reflection of water waves from floating ice in water of finite depth," *Communs Pure Appl Math,* Vol 3, No 3, pp 305–318.

Proceedings of the Seventh (1997) International Offshore and Polar Engineering Conference
Honolulu, USA, May 25-30, 1997

ISBN 1-880653-28-1 (Set); ISBN 1-880653-30-3 (Vol. II)

Discrete and Lattice Models of Floating Ice Covers

Mohamed Sayed
National Research Council
Ottawa, Canada

ABSTRACT

A numerical approach for modelling ice deformation, which has been used in solving several ice engineering problems, is presented. The formulation is based on tracking the movements and force equilibrium of discrete parts of the ice cover (e.g. floes). Contact models consist of arrangements of springs and dashpots which represent the required rheology (e.g. elastic or visco-elastic). Fracture is introduced by allowing tensile contact bonds to break; thus crack propagation and interactions are simulated. Examples of applications of the approach are presented, including a three-dimensional examination of ridge interaction with a bridge pier, and simulation of the evolution of open water wakes behind offshore structures.

KEYWORDS: - Ice-structure interaction, ice fracture, numerical simulations, discrete elements.

INTRODUCTION

Deformation of floating ice often involves complex processes whose description has been beyond the capabilities of numerical solutions. The present paper deals with the behaviour of floating ice at length scales from a few meters to hundreds of meters, which may be relevant to problems of ice interaction with offshore structures and ships, as well as in ice jam formation. In such situations, ice deformation may involve propagation of multiple large cracks, fragmentation, and rubble accumulation. Flow of broken ice floes can also take place.

Modelling crack propagation and interaction between cracks present the major difficulty in dealing with the above cases. Continuum approaches, such as the boundary element method (Tuhkuri, 1996), can only deal with idealized cases representing the initial stages of the propagation of a single crack. Obviously such solutions are not adequate for solving problems of practical interest. Continuum models also suffer other difficulties in dealing with large gradients (e.g. of pressure), definition of free boundaries, and uncertainty concerning the appropriate constitutive equations. Most such models employ finite element solutions. For example, Morsey and Brown (1993) modelled a few idealized situations of ice structure interaction. Continuum models also employ a damage variable to describe the intensity of small cracks (e.g. Wu and Shyam-Sunder, 1991), but they cannot model large cracks.

More recently, discrete element models have been introduced in ice forecasting, and in the study of ice jams. Shen et al. (1987) were the first to use a discrete formulation to model a marginal ice zone. A number of discrete models were later used to study some aspects of floating ice problems. For example, Hopkins (1993) developed a discrete element model to examine yield conditions at the meso-scale. Broken ice interaction with structures was examined by Løset (1994) using a discrete model. Also Yamaguchi et al. (1993) used a variation from traditional discrete element formulations to examine ice flow in channels and around structures. A new discrete formulation for meso-scale ice forecasting was also developed by Savage (1995). Discrete modelling was used as well in studies of ice jams by Shen et al (1993), and by Sayed et al. (1994).

Discrete models have several advantages, including the ability to handle discontinuities, accurate definition of ice edge, and elimination of numerical diffusion. They are also based on simple assumptions, which avoids the uncertainty regarding complex continuum constitutive equations. The major limitation, however, has been the need to use very large numbers of elements in order to deal with any problem of practical interest, which has been beyond most available models. As well, new models were needed for representing crack propagation.

Related developments of lattice models have made considerable progress in modelling the behaviour of large cracks, but these models are generally unknown in ice engineering literature. In such models, the material is discretized into a set of nodes. Each node can interact with its neighbours through contact elements that are allowed to break. The approach is suited for including homogeniety of the material and simulating the propagation of multiple cracks. Reviews of lattice models were given by Herrmann and Roux (1990), and Roux (1990). Applications of lattice models include studies of fracture of cement-based composites by Schlangen and van Meir (1992), and fracture of brittle materials by Dai and Frantziskonis (1994).

This paper presents a modelling technique which has been used to solve a number of practical problems. This approach is based on a discrete formulation, and introduces fracture according to lattice models techniques. The following sections describe the formulation of the model, and give examples of the applications. Those examples are intended to illustrate the manner in which simulations can be used. They do not, however, provide exhaustive conclusions for a particular case because of space limitations.

MODEL FORMULATION

Overview

Several ice cover types and regimes have been represented in the simulations. Broken ice floes, intact ice features, and floes with adhesive bonds between them have been modelled. Both two-dimensional and three-dimensional cases have also been addressed. For broken floes undergoing two-dimensional deformation, the ice cover is modelled as an assembly of random sized inelastic disks. At the contacts between disks, compressive normal forces and tangential forces may develop, but no tensile forces may exist. A "soft-particle" approach is used, where contacts between disks may persist for long durations. Groups of disks can also be linked to form a single rigid body as follows. The forces and moments acting on individual disks are used to calculate the sum values at the centre of the group. Next, the group translates and rotates as a single body. Finally new locations, and velocities of individual disks are calculated. Thus, irregularly shaped floes can be introduced. For three-dimensional deformation, spheres are used instead of disks. Compressive normal and tangential forces may develop at the contacts between spheres. Again, rigidly linked groups can form irregularly shaped ice blocks. For broken ice with cohesive bonds, tensile normal forces are allowed to develop at the contacts. The bonds break if the tensile force reaches a limit which corresponds to the continuum tensile strength.

Intact ice masses are modelled in a somewhat similar manner to broken ice. The ice mass is discretized into a set of nodes in a lattice. Each node is connected to its immediate neighbours by contact elements which consist of springs and dashpots. Normal, tangential, and torsional contact elements can be used. Only normal elements have been used in the present work, since they appeared to adequately produce the appropriate behaviour. Fracture is introduced by allowing the contact elements to break when the tensile force exceeds the strength. Broken elements do not transmit tensile forces, but they can transmit compressive normal and tangential forces.

Simulations keep track of the movements of individual floes or nodes. These movements are governed by the linear and angular momentum balance equations. Boundary conditions may consist of applied displacements, velocities, or forces. Body forces (e.g. gravity) can be applied to each discrete mass as well. For lattice models of intact ice, fragments consisting of one or more nodes may form and drift. Therefore, nodes may come in contact with new nodes other than their original neighbours. Simulations can be carried out to track new contacts, and interactions between fragments. As will be seen later, a node may be linked to its neighbours via elements of different lengths. The computations keep track of all such lengths. When a node starts to drift, however, a mean radius is assigned to it for use in new contacts calculations.

Contact Model

At the contact between two floes or blocks, normal and tangential forces increase their centres approach. The normal force, Fn is calculated using linear springs and dashpots as shown in Fig.1a , and may be expressed as follows

$$F_n = k\,\delta - b\,v_n \tag{1}$$

where k is a spring constant, b is a dashpot constant, d is the relative normal displacement (change in the length of the element), and vn is the relative normal velocity. This type of contact element represents elastic behaviour, with all deformation fully recoverable. The dashpot is included here to account for the viscous damping needed for energy dissipation. For broken ice, only compressive forces are allowed. In that case, a negative value for Fn according to Eq.1 (tension) is reset to zero. For intact ice and bonded floes, the normal force which is calculated by Eq.1 is allowed to have tensile values up to that of the tensile strength. The tangential force, Ft is calculated according to a Coulomb law

$$F_t = \mu\, F_n \tag{2}$$

where m is the coefficient of friction, and the tangential force acts in a direction opposite to that of the relative tangential velocity. This contact model is simple to implement and was found to give almost identical results to the more elaborate contact models of Walton and Braun (1986).

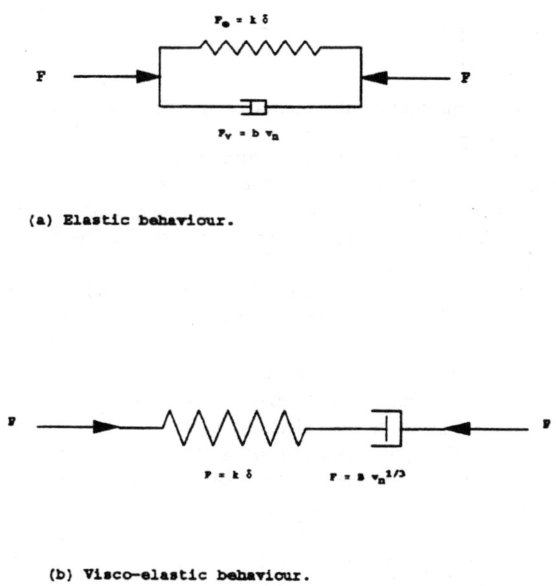

Figure 1 Contact Model.

Visco-elastic deformation is also introduced for intact ice in order to account for rate effects, and for irrecoverable deformation. A schematic of that model is shown in Fig. 1b. The spring is considered to be linear, but the dashpot was chosen to represent non-linear creep according to Glen's law. The force Fn between two nodes, would be equal to the force in the spring, and the force in the dashpot as follows :

$$F_n = k\,\delta = B\,v_n^{1/3} \tag{3}$$

where B is a constant for the non-linear dashpot. The present model corresponds to the simplest choice that gives the required visco-elastic behaviour. It accounts for secondary creep, but not other aspects of deformation such as primary and tertiary creep. Thus the model gives an upper limit for the stresses. The visco-elastic case requires additional steps for calculating contact forces. The relative displacement at each contact has to be divided into an elastic part and a viscous irrecoverable part. That involves solving the cubic equation which equates the elastic force of the spring to the viscous force of the dashpot.

Simulation Procedures

Linear and angular momentum equations govern the movements of discrete blocks either in two or three-dimensional coordinates, depending on the specific problem. The balance of linear momentum for each floe or node may be expressed as

$$m \frac{du}{dt} = \Sigma_{i=1}^{N} F_i \quad (4)$$

where m is the mass of the floe, u is the velocity, t is time, and $\Sigma_{i=1}^{N} F_i$ is the vector sum of contact forces due to N contacts with other floes. The balance of angular momentum is given by

$$I \frac{d\omega}{dt} = \Sigma_{i=1}^{N} T_i \quad (5)$$

where I is the moment of the inertia of the floe, w is the angular velocity, $\Sigma_{i=1}^{N} T_i$ is the sum of the torques due to N contacts. For two-dimensional deformation, the torque due to water drag on the floes may be significant. It is calculated using an integration of the local drag force, which varies from the centre to the edge of the block (see Savage 1995, and Renaut 1995).

Computations keep track of the positions and velocities of individual blocks. At each time step, the relative displacements and relative velocities between blocks are calculated. Following that, contact forces are calculated using the contact model described below. Contact forces are then summed, and boundary forces are added for each block. Accelerations are calculated from the resultant forces on each block. Velocities and positions are next determined by numerically integrating the accelerations. The time is then incremented.

Searching for neighbouring blocks is the most demanding part of the computations. If a straightforward search is done using N blocks at each time step, the computer running time would be proportional to N2. That drastically limits the number of blocks. Three approaches have been used to overcome that difficulty. In the first approach, neighbours' lists are maintained and frequently updated (Allen and Tildsley, 1987). The second approach is based on dividing the space into bins. Each block is assigned to a bin at each time step. Thus, only neighbouring bins need to be searched when looking for contacts (Løset, 1994). The third approach employs a tree-search which can quickly locate a block (e.g. Barnes and Hut, 1986).

PARAMETER VALUES

Several tests were done in order to correlate the local contact parameter values to those of the equivalent continuum parameters. For example, contact stiffness and block geometry should be related to the equivalent continuum elastic parameters. The tensile bond strength should also be related to the continuum tensile strength. For that purpose, uni-axial two-dimensional compression tests were simulated. A series of tests employed a 60 x 50 node lattice. The lattice was constructed by starting with equilateral triangles of length l. The nodes were next given random displacements of ±0.15 l along the horizontal and vertical directions (see Fig.2). The randomness was introduced in order to avoid anisotropy. Preliminary tests showed that a uniform lattice geometry can influence the directions of displacements and crack propagation.

Tests were done by holding the base of the rectangular sample fixed, while moving the top nodes at a constant downwards velocity until an axial strain of 0.025 was reached. The top nodes were then stopped. The sides of the sample were kept free. The normal contact model of Eq.1 was used, which corresponds to elastic behaviour. There was no fracture allowed in these tests.

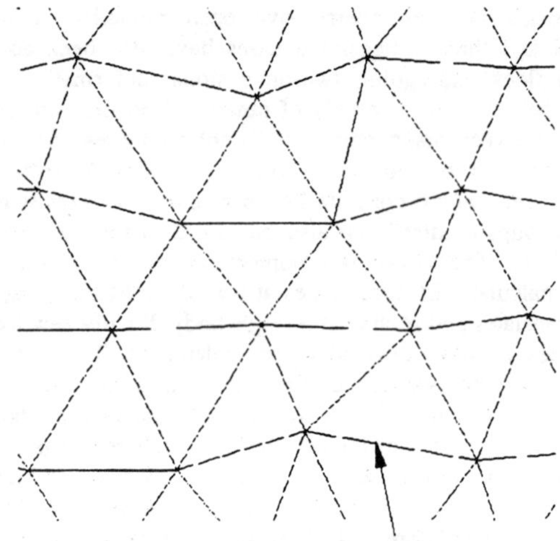

Figure 2 A portion of a lattice used in the simulations.

The resulting Poisson ratio was 0.33, which is a reasonable value for elastic behaviour. The calculated stress was used to calculate the elastic modulus E. The resulting relationship between E, and contact stiffness K can be expressed as E = 0.6 K. Note that both E and K have units of force/length2.

Several other simulations were done including uni-axial creep and plane indentation tests. The results were in good agreement with experimental data. Details of those results were reported by Sayed et al. (1995a). Another investigation by Sayed et al. (1995b) examined the continuum stress-strain relations of an assembly of floes with tensile bonds between them.

RIDGE INTERACTION WITH BRIDGE PIERS

The present model was used to examine failure modes and forces due to the impact of ice ridges on a bridge pier. Simulations determined the role of ridge shape, size and velocity, as well as rubble properties. Only an outline of that investigation and examples of the results are presented here. Details of the results were reported by Sayed (1995). The simulations started with a number of tests to verify model performance and parameter values. Those tests included simulation of plane strain bi-axial deformation of bulk rubble. The numerical results were in agreement with the experimental measurements of Løset and Sayed (1993). Those simulations also relate values of rubble bulk properties to contact parameter values. The other verification cases consisted of simulating indentation experiments. Those experiments measured indentation forces and displacements for cylinders into sand, and for small ice blocks, as well as model ice basin experiments. The numerical results were found to be in agreement with the experimental measurements of Timco and Cornett (1995).

Next, two-dimensional simulations were carried out in order to examine the modes of rubble failure and to give an upper bound on expected forces. Finally, three-dimensional simulations were made. The case presented here consists of a trapezoidal ridge of 20 m keel depth that impacts a cylindrical bridge pier of 10 m diameter. Geometry of the initial configuration is shown in Fig.3. Length of the ridge was 50 m. The ridge is represented by spheres which are initially placed in hexagonal packing. Then each sphere is given a random displacement of ±0.15 of its diameter. The consolidated layer at the water level is accounted for by using higher stiffness and tensile bond values than for the keel. The total number of spheres was approximately 7000.

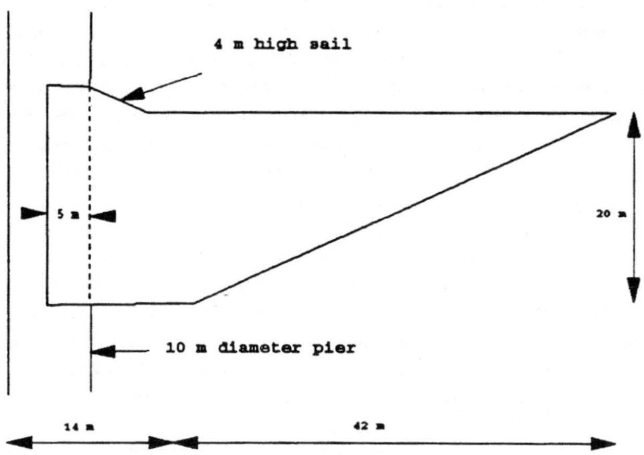

Figure 3 Initial ridge and pier geometry for a three-dimensional simulation.

The sides of the ridge are subjected to plane strain conditions; i.e. over the cross-section at each end of the ridge, velocity components perpendicular to the plane of the cross-section are set to zero. A 2 m wide part of the far end of the ridge moves at constant velocity, thus simulating a constant velocity of the pack ice. All spheres in the ridge start at the same initial velocity of the driving boundary (or pack ice). The consolidated layer thickness is approximately 2 m. The tensile strength of the consolidated layer is 400 kPa, and that of keel rubble is 5 kPa. Random components (20% of mean values) are added to spring constants and tensile strengths as in the above sections. Parameter values are : sphere diameter 2.0 m, velocity 0.5 m/s, spring constant (keel) 2.65 MN/m, spring constant (consolidated layer) 26.5 MN/m, spring strength (keel) 2.3 kN, spring strength (consolidated layer) 184.8 kN, dashpot parameter 12.9 kN s/m, friction coefficient (ice-ice) 0.2, friction coefficient (ice-pier) 0.1, and ice specific gravity 0.9. The corresponding bulk elastic modulus for the keel rubble is approximately 7 MPa, and that for the consolidated layer is 70 MPa.

The resulting forces are plotted versus time in Fig.4, for a number of velocities. Note that the velocities, ridge dimensions, and initial contact area were chosen to give conservative (high) forces. Fig.4 shows that the maximum force is reached after an indentation distance of between 0.5 m and 1m. The clearing and flow of rubble around the pier afterwards produces relatively small forces. A view of the bottom side of the ridge keel is shown in Fig.5. It shows ice blocks starting to break-out from the keel in front of the pier. Such a beak-out takes place after the force had considerably dropped. A parametric study examined the role of keel depth, ice-pier friction coefficient, and rubble stiffness. In other cases, a fixed boundary was used at the water level, instead of a breakable consolidated layer, to increase rubble confinement. Observed failure modes show that rubble flows around the pier without forming "fixed wedges". There was no evidence of the formation of shear planes, along which rubble deforms. This behaviour was also observed during the ice basin models of Timco and Cornett (1995) The conclusions indicate that the maximum forces are lower than those calculated using empirical soil mechanics formulas.

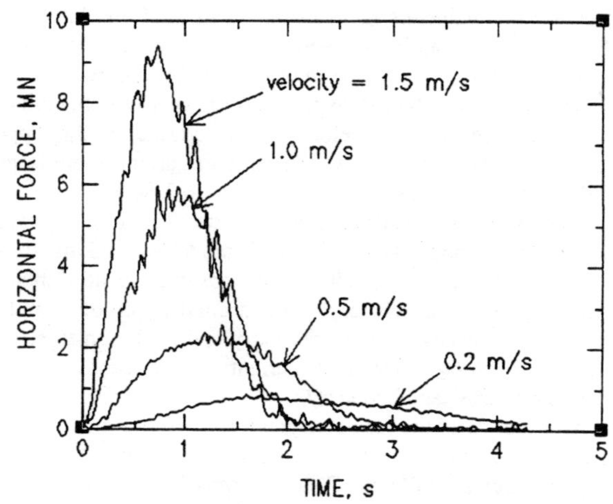

Figure 4 Total horizontal force on the pier.

OPEN WATER WAKE BEHIND THE OFFSHORE STRUCTURES

A project has been undertaken to examine the geometry of an open water wake behind structures, in the presence of ice covers. The objective was to determine the feasibility of mooring ships in such open water wakes. Obviously wake geometry depends on the environmental driving forces (e.g. wind and current), structure geometry, and ice cover type.

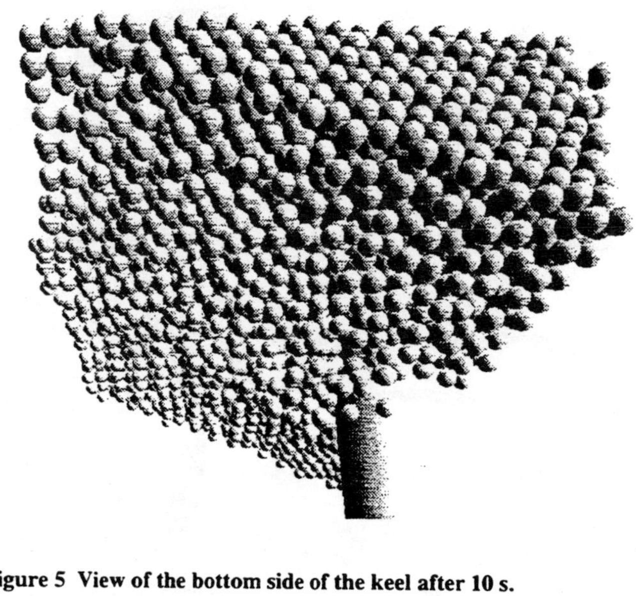

Figure 5 View of the bottom side of the keel after 10 s.

Simulations were done for an ice cover consisting of random sized floes, with no tensile bonds between them. The floes were driven by wind and current drag, and subjected to the boundary conditions imposed by the fixed structure. A relatively large study area was needed for the problem because of the length of the wake. Also ice floe radii had to be small compared to the dimensions of the structure in order to avoid spurious results. Preliminary runs were made in order to determine the size of the study area and the floe diameters. Square ice cover areas with lengths from 1.7 km to 2.3 km were used, depending on the case. Floe diameters ranged from 12 m to 20 m. Additionally, periodic boundary conditions were imposed at the sides of the rectangular ice cover. They are implemented by considering image floes at each edge. As a floe starts to exit (or is near) one boundary, an image of that floe with its velocity is introduced at the opposite boundary. Thus, the simulation would represent an infinitely large ice cover. The number of floes varied from 16000 to 20000.

Two types of simulations were done. The first reproduced specific field cases, and was used to verify model predictions. In those cases, comparison between predicted and observed wake dimensions and directions was satisfactory. The second type of simulation examined the sensitivity of wake geometry and evolution to structure dimensions, changing wind and current, ice thickness, and the existence of a nearby land boundary.

An example of the results is presented here for a case of a 100 m x 100 m structure, 1 m thick ice floes, and a wind of 10 m/s, changing direction at 300/hour. Contact parameters values are : spring constant 72 kN/m, dashpot constant 20 kNs/m, and friction coefficient 0.3. Air and water drag values are 0.0012 and 0.005, respectively. The resulting wake geometry is shown in Fig.6. The wake direction initially coincides with that of the wind, but lags with elapsed time because of ice cover inertia. In a similar manner, evolution of wake geometry was determined for several scenarios. Inspection of the predicted geometries enables the establishment of conditions under which ships can be moored, and the time they have available before the open water wake changes direction in response to environmental forces. More details about this project are given in Canmar (1996), and complete results will be published in the future.

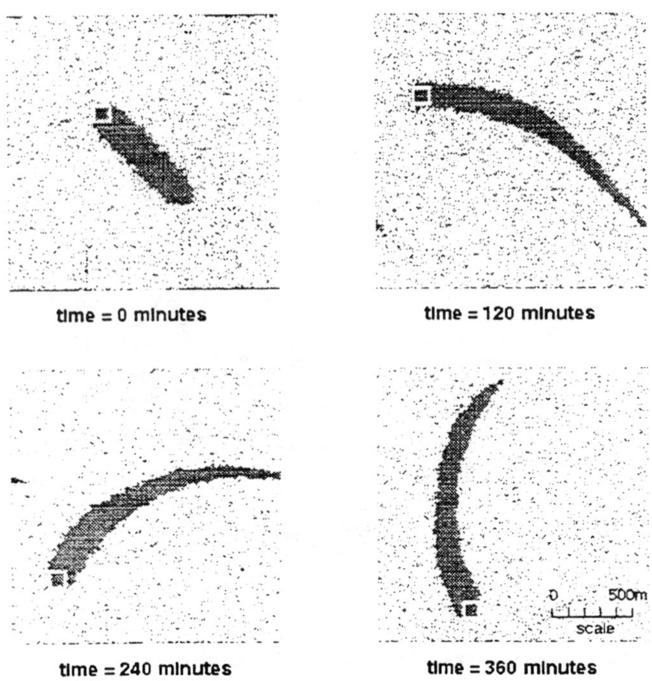

Figure 6 Evolution of open water wake geometry behind a structure in response to shifting wind.

CONCLUSIONS

A numerical approach, which has been used in solving several practical ice engineering problems, was presented. The approach is based on tracking the movements of discrete parts of the ice cover. Those parts can represent individual floes, or blocks of ice rubble. In addition to modelling the flow of broken ice, tensile bonds at the contacts were introduced. The same approach is further extended to deal with intact ice covers. In that case, the ice mass is discretized into nodes that are connected by breakable elements. Both elastic and visco-elastic contact models can be used. Fracture is introduced by assigning a tensile strength to the contact elements.

The preceding sections discussed formulation of the model and gave example applications. References to more complete records of specific results were given. The ability to solve practical design problems is a significant aspect of the present work. This was possible because the implementation allowed the use of large numbers of discrete elements (up to tens of thousands). Flexibility in choosing the rheology of the contact models and the introduction of fracture were necessary for treating many problems.

A major advantage of the present approach is the ability to model the initiation, propagation and interaction of multiple large cracks. Other advantages include the ability to accurately predict the location of the free ice edge, and to model large deformations of ice rubble. All of these problems had been beyond the capabilities of other numerical simulations. Furthermore, only a few well known parameters concerning ice rheology are needed. This reduces the uncertainty concerning the appropriate continuum constitutive equations.

Further work is needed in order to correlate the present discrete approach results to those of the well studied continuum formulations. Comparisons between model predictions and experimental measurements are also needed. Future work should address the computational aspects and examine ways to increase efficiency of the numerical schemes, as well.

ACKNOWLEDGEMENT

The present work has evolved from the results of a number of project. The author is grateful to Public Works and Government Services Canada and Dr. T. Brown for support of the study of ridge-pier interaction; and to CANMAR, Transport Development Centre of Transport Canada, and participants of a joint industry study, for support of the open water wake project. The author also thanks Drs. G.W. Timco and R.M.W. Frederking for useful discussions throughout the course of this work.

REFERENCES

Allen, MP, and Tildesley, DJ (1987). "Computer simulation of liquids," Clarendon Press, Oxford.

Barnes, J, and Hut, P (1986). "A hierarchical O(NlogN) force-calculation algorithm," Nature, Vol 324, p 446.

CANMAR (1996). "The simulation of ice drift behaviour behind offshore structures," Draft Rep, December 1996, 700-2nd St. W., Calgary, Alberta, Canada T2P 2H8

Dai, H, and Frantziskonis, G (1994). "Heterogeneity, spatial correlations, size effects and dissipated energy in brittle materials," Mechanics of Materials, Vol 18, pp 103-118.

Herrmann, HJ, and Roux, S (1990). "Modelization of fracture in disordered systems," Statistical Models for the Fracture of Disordered

Media, ed. H.J. Herrmann and S. Roux, Random Materials and Processes, Elsevier Science Pub., pp 159-188.

Hopkins, MA (1993). "A mesoscale simulation of the Arctic ice pack," ASME, AMD-Vol 163, Ice Mech, pp 85-96.

Løset, S and Sayed, M (1993). "Proportional strain tests of fresh-water ice rubble," J Cold Reg Eng, ASCE, Vol 7, No 2, pp 44-61.

Løset, S (1994). "Discrete element modelling of a broken ice-field-Part I: model development," Cold Regions Sci Tech, Vol 22, pp 339-347.

Morsy, UA and Brown, TG (1993). "Modelling for Ice-Structure Interaction," Proc OMAE'93, Vol 4, pp 119-126, Glasgow, Scotland.

Renaut, N (1995). "Granular dynamics simulations of wind-driven, broken ice fields," M Eng Thesis, Dept Civil Eng and Applied Mech, McGill Univ, April, Montreal, Quebec, Canada, 94 p.

Roux, S (1990). "Continuum and discrete description of elasticity and other rheological behaviour," in Statistical Models for the Fracture of Disordered Media, ed. H.J. Herrmann and S. Roux, Random Materials and Processes, Elsevier Science Pub., pp 87-114.

Savage, SB (1990). "Marginal ice zone dynamics modelled by computer simulations involving floe collisions," Mobile Particulate Systems, ed. E. Guazzelli and L. Oger, Chapter 18, pp 305-330, Kluwer Academic Publishers, The Netherlands.

Sayed, M, Serrer, M, and Arden, DA (1994). "Numerical simulation of ice jams in Lake St. Peter," Rep IECE-CRT-CTR-004, National Research Council, Ottawa, Ontario, Canada K1A 0R6.

Sayed, M (1995). "Numerical simulation of the interaction between ice ridges and bridge piers," IMD Rep TR-1995, February 1995, National Res Council, Ottawa, Ontario, Canada K1A 0R6.

Sayed, M, Arden, DA, and Timco, GW (1995a). "A lattice model of ice fracture," Rep Coastal Eng Prog, June, National Research Council, Ottawa, Ontario, Canada K1A 0R6.

Sayed, M, Neralla, VR, and Savage, SB (1995b). "Yield conditions of an assembly of discrete ice floes," Proc 5th Int Offshore and Polar Eng Conf (ISOPE), the Hague, 11-16 June, Vol 2, pp 330-335.

Shen, HH, Hibler, WDIII and Leppäranta, M (1987). "The role of floe collisions in sea ice rheology," J Geophys Res, Vol 92, pp 7085-7096.

Shen, HT, Chen, Y-C, Wake, A and Crissman, RD (1993). "Lagrangian discrete parcel simulation of river ice dynamics," Proc 3rd Int Offshore and Polar Eng Conf (ISOPE), Singapore 6-11 June, Vol. 2, pp 562-566.

Schlangen, E, and van Mier, JGM (1992). "Experimental and numerical analysis of micromechanisms of fracture of cement-based composites," Cement & Concrete Composites, Vol 14, pp 105-118.

Timco, GW and Cornett AM (1995). "Model tests of ridge interaction with a bridge pier for the Northumberland Strait Crossing," IMD Rep TR-1995-02, January 1995, National Research Council, Ottawa, Ontario, Canada K1A 0R6.

Tuhkuri, J (1996). "Experimental investigations and computational fracture mechanics modelling of brittle ice fragmentation," Acta Polytechnica Scandinavica, Me 120, Helsinki, Finland, 105 p.

Walton, O.R. and Braun, R.L. (1986). "Viscosity and temperature calculations for assemblies of inelastic frictional disks," J. Rheology, Vol. 30, pp. 949-980.

Wu, MS and Shyam-Sunder, S (1991)." Effects of Grain Size Variation on Damage in Polycrystalline Ice," Proc Cold Regions Eng, pp 542-553, West Lebanon, NH, USA.

Yamaguchi, H, Rheem, CK and Kato, H (1993). "Pack ice movement simulation using a distributed mass/discrete floe model," Proc 12th Int Conf POAC, Vol 2, pp 848-856.

Proceedings of the Seventh (1997) International Offshore and Polar Engineering Conference
Honolulu, USA, May 25-30, 1997

ISBN 1-880653-28-1 (Set); ISBN 1-880653-30-3 (Vol. II)

Assessment of the Wave-Iceberg Load Combination Factor

Ricardo Foschi and Michael Isaacson
University of British Columbia
Vancouver, Canada

Norman Allyn and Ibrahim Saudy
Westmar Consultant Inc.
Vancouver, Canada

ABSTRACT

The present paper describes an extension to a recent study (Foschi et. al., 1996), which was undertaken to determine the appropriateness of the recommended value of the load combination factor relating to the combined effects of wave and iceberg loads, as described in the Canadian Standards Association (1992) code for the design and construction of fixed offshore structures. The study examines the sensitivity of the load combination factor to various iceberg and wave parameters typical of three sites off the East Coast of Newfoundland. The methodology is based on a numerical analysis in which loads due to waves alone, an iceberg alone, and an iceberg and waves in combination, have been calculated for a range of iceberg and wave parameters, with the results applied to a first-order reliability analysis to study force levels corresponding to specified annual exceedence probabilities. The results indicate that the load combination factor is most sensitive to the wave angle relative to the current direction, and to the wave climate during the iceberg season - therefore, the load combination factor is site dependent. The load combination factor has been calculated as 0.25 on average, with a conservative value of 0.33. This compares with the Code values of 0.8 or 0.4, for icebergs and waves which are taken to be stochastically dependent or independent, respectively.

KEY WORDS: hydrodynamics, icebergs, ocean engineering, offshore structures, wave forces, waves.

INTRODUCTION

The selection of suitable environmental loads and load events is of critical importance in the design of offshore structures intended for operation in extreme environments. Such loads may include those due to wind, waves, earthquakes, ice floes and iceberg collisions. The CSA Offshore Structures Code CAN/CSA-S471-92 (S471) (Canadian Standards Association, 1992) describes the use of such loads in offshore design, and indicates the use of probabilistic methods on which the selection of load events and design loads should be based. The Code has been subjected to a comprehensive verification process, and this has identified several issues which warrant further study. One of these is an assessment of the combined effects of wave and iceberg collision loading. At present, this combination is treated by the use of a load combination factor which is used to determine a design value for the load due to a companion frequent environmental process (waves) acting in combination with a rare environmental event (iceberg collision). For combined iceberg-wave loading, the load combination factor γ is defined in the Code as follows:

$$E = E_r + \gamma E_f \qquad (1)$$

where E is the load due to an iceberg with waves with an annual exceedence probability of 10^{-4}, E_r is the iceberg alone load with an annual exceedence probability of 10^{-4}, and E_f is the wave load with an annual exceedence probability of 10^{-2}. Suitable values of γ are specified in the Code. The definition of load in Eq. 1 depends on the limit state under consideration, and includes the maximum iceberg load (in the presence of waves), applicable to a local damage limit state; and the maximum combined wave-iceberg load, applicable to a global sliding limit state.

Wave-structure interactions (in the absence of icebergs) and iceberg-structure interactions (in the absence of waves) have been studied extensively in the past (e.g. Sarpkaya and Isaacson, 1981, and Cammaert and Muggeridge, 1988, respectively), whereas the case of waves and an iceberg acting simultaneously on a fixed offshore structure has not been studied to the same extent. Isaacson (1987) considered the effect of waves on an iceberg up to the instant of impact, and described wave effects on the iceberg velocity and effective mass at the time of impact. Related studies include Lever, Attwood and Sen (1988), Lever, Colbourne and Mak (1990), and Isaacson and McTaggart (1990). In a recent study, Foschi et al. (1996) and Foschi and Isaacson (1996) described a numerical analysis in which loads due to waves alone, an iceberg alone, and an iceberg and waves in combination, have been calculated for a range of iceberg and wave parameters. The corresponding results were applied to a probabilistic study of the load event using the first-order reliability method (FORM), with the objective of determining suitable values of the load combination factor.

In the present study, the method of Foschi et al. (1996) is extended to examine the sensitivity of the load combination factor γ to wave and iceberg parameters typical of three sites off the East coast of Newfoundland. Loads due to waves alone, an iceberg alone, and an iceberg and waves in combination, have been calculated for a range of iceberg and wave parameters. These results have been used to develop expressions for wave and iceberg loads which are then used in a probabilistic study of the load event. The force exceedence probability is studied using FORM. For a given structure size, the influence of water depth, iceberg arrival rate, wave climate, wave direction, and mean current on the loads and on the load combination factor are examined. The study thereby makes recommendations for the load combination factor applicable to combined wave-iceberg loading under fairly general conditions.

In practice, the predominant wave direction may differ from the direction of iceberg motion, so that the effect of an oblique wave direction should be taken into account when calculating the combined wave and iceberg loads. In this work, the wave plus iceberg load results for various wave directions are considered and a parametric fit is made to the corresponding results. Different values of the relative angle between the incident wave direction and the iceberg trajectory are considered. In addition, a wave climate description based on the use of a specified recording interval is adopted. Finally, it is recognized that there is a seasonal difference between the annual wave climate and the wave climate occurring during the iceberg season.

The structure is taken as a fixed vertical circular cylinder of diameter 100 m and extending from the seabed to above the water surface. Water depths, iceberg conditions and wave conditions corresponding approximately to three sites off the East Coast of Newfoundland have been selected. The three sites are indicated in Fig. 1, and are denoted Sites H, A, and B. Site H corresponds to the location of the Hibernia platform; Site A is at the tail of the Grand Banks in 60 m water depth and located at 50° West, 43° 20' North; and Site B is near the north end of the Grand Banks in 100 m water depth and located at 50° West, 47° 47' North.

ENVIRONMENTAL CONDITIONS

In this study, data was required for both waves and icebergs for the three sites under consideration. Statistical distributions for the waves and the icebergs for the Grand Banks off the coast of Newfoundland are taken from Fuglem et. al. (1996), Jordaan (1996) and MacLaren Plansearch (1991), and are discussed below.

Icebergs

Following Det Norske Veritas (1988), each iceberg is assumed to be circular in plan and ellipsoidal in elevation as shown in Fig. 2. From statistical data for the Grand Banks off the coast of Newfoundland, all iceberg dimensions are expressed in terms of a single random variable L which is represented by a Gamma probability distribution. In particular, the iceberg diameter at the waterline D, the iceberg radius R, and the iceberg draft h may readily be related to L. However, icebergs capable of colliding must have a draft smaller than the water depth, and consequently h has been fitted by a suitable Beta distribution ranging from a minimum of 0 m to a maximum equal to the water depth. Statistics for each site are shown in Table 1. The table is based on an iceberg waterline length with a mean of 120 m and a standard deviation of 50 m, as provided by Jordaan (1996).

Waves

Distribution of Significant Wave Height. A simple representation of the long-term probability distribution of wave conditions is initially required. Various such representations are possible (e.g. Sarpkaya and Isaacson, 1981), and in the present study the significant wave height H_s based on a specified recording interval τ, is assumed to obey an Extreme Type I ("Gumbel") probability distribution as follows:

$$P(H_s) = \exp[-\exp(-a(H_s - b))] \tag{2}$$

where P() is the cumulative probability, and a and b are constants of the distribution. The constants a and b may be related to characteristic values of the distribution, such as the mean and standard deviation, or the slope and intercept of a straight line plot of the distribution.

Significant Wave Heights During the Iceberg Season. Since iceberg impacts are not expected to occur year-round, it is not appropriate to consider the probability distribution of wave heights on the basis of annual data, but rather to use conditions corresponding to a typical iceberg season. This distinction was considered by Fuglem et. al. (1996), who provided information on the distribution of significant wave heights near site H during the iceberg season. These results have been fitted by Eq. 2 and the resulting wave height distribution, representative of an iceberg season taken to last for four months, has been obtained. In order to extend this distribution to the other two sites, the relevant Wave Climate Atlas (MacLaren Plansearch, 1991) has been used to obtain the relative wave heights at the appropriate time of year for zones in which sites A, B and H are located.

Distribution of Wave Period. A suitable description of the peak wave period is also required, and is developed by adopting a simple assumption for the relationship between the peak wave period and significant wave height. In fact, it is assumed that T is fully correlated with the significant wave height and is given by:

$$T = 4.432\sqrt{H_s} \tag{3}$$

where T is the peak period in seconds and H_s is the significant wave height in metres. This relation has been found to be suitable for conditions in Canadian Atlantic waters (Neu, 1982).

Distribution of Annual Maximum Significant Wave Height. The corresponding probability distribution of the annual maximum significant wave height, denoted $P_a(H_s)$, is required and can be obtained from the distribution $P(H_s)$ given in Eq. 2 (based on a recording interval τ), by raising the latter to the power of N, where N is the number of such intervals per year. Two values for N are considered; one for the whole year [$N = 365 \times 24 / \tau$], and the other for a 4 month ice season [$N = 365 \times 24 / (3 \times \tau)$], where τ is in hours. It can thereby be shown that the probability distribution of annual maximum heights $P_a(H_s)$, is also given by Eq. 2, except that the constant b is replaced by $b + \ln(N)/a$.

Short-Term Distribution of Individual Wave Heights. Individual wave heights within a sea state (during which H_s is taken as constant) are assumed to possess a Rayleigh distribution.

RELIABILITY MODEL

Probabilistic analyses have previously been used in offshore engineering problems (e.g. Fuglem et. al., 1991, Det Norske Veritas, 1988, and Maes and Jordaan, 1984), including the calibration of the Canadian Code for offshore structures (e.g. Maes, 1986).

The estimation of conditional probabilities associated with a load event is conducted here using a computer program RELAN (Foschi et. al., 1990) which implements standard FORM and SORM algorithms (First and Second Order Reliability Methods) to calculate the probability that a "performance function" G of the vector of a set of random variables **x** is negative. In order to equate this result to a force exceedence level in the present context, the function G(**x**) is written as follows:

$$G(\mathbf{x}) = F - F_M(\mathbf{x})R_{nl} \quad (4)$$

where $F_M(\mathbf{x})$ is the maximum force developed on the structure due to waves, iceberg impact, or waves and iceberg impact in combination, as appropriate, **x** denotes a set of specified random variables characterizing the structure, the iceberg and the wave conditions, F is the sliding or local failure load level, and R_{nl} is a random variable associated with model inaccuracy in the calculation of F_M (Bea, 1992).

The probability of the event $G < 0$ corresponds to the probability that the load level F is exceeded. Three specific programs were extended for this study. The program ICELOAD is calculates forces due to iceberg collisions only; the program WLOAD calculates forces due to waves alone; and the program ICEWLOAD calculates forces due to an iceberg collision in the presence of waves.

In the case of an iceberg impact, with or without waves present, the force exceedence probability is first obtained conditional on the occurrence of an impact. In such a case, the programs allow for the estimation of the corresponding annual risk, denoted P_a, using the hypothesis that the events (i.e. iceberg impacts) follow a Poisson pulse process with a given mean rate of annual occurrence (events per year), denoted ρ. Thus, if the conditional exceedence probability of the event is P_e, the annual risk is given as:

$$P_a = 1 - \exp(-\rho P_e) \quad (5)$$

Collision arrival rates ρ have been calculated for combinations of three mean currents and the three water depths at the sites. The corresponding iceberg densities ν (icebergs / degree square) and arrival rates are shown in Table 2. In each of the three cases, the maximum force F_M developed during the event requires an appropriate mechanics model for its calculation. The approach that has been used to assess the loads due to waves alone, an iceberg alone, and waves and an iceberg acting in combination are described in detail by Foschi et al. (1996) and summarized by Foschi and Isaacson (1996) for the case of wave directions and iceberg trajectories which are colinear. For convenience, a brief summary is given here.

WAVE FORCE ON STRUCTURE

Regular Wave Forces. The loads due to regular waves interacting with a structure (in the absence of an iceberg) are obtained on the basis of linear diffraction theory. Results have been obtained using a computer program WELSAS, which is based on three-dimensional linear diffraction theory using a boundary element method (e.g. Sarpkaya and Isaacson, 1981). The program involves a discretization of the submerged structure surface into quadrilateral elements, and employs a suitable Green's function. Although a closed-form solution is available for a vertical circular cylinder (e.g. Sarpkaya and Isaacson, 1981), the program WELSAS is used for the case of waves alone, since it is needed for the subsequent case of combined wave-iceberg loading. For regular waves of height H and period T, the wave force on the structure varies sinusoidally in time with an amplitude F proportional to H. Since the wave forces are ultimately required for application in a probabilistic model, a simple expression has been fitted to numerical results obtained for a range of wave periods and for the three water depths.

Random Wave Forces. Loads due to random waves may be obtained as a direct extension to the case of regular waves. Since linear diffraction theory is used to calculate wave loads, the wave force amplitude is proportional to the wave height, so that the force amplitude F is taken as random with a Rayleigh probability distribution and with a significant value that can be obtained by applying diffraction theory to a regular wave train of height H_s.

ICEBERG FORCE ON STRUCTURE

Attention is now directed to the maximum force exerted on the structure due to an iceberg impact in the absence of waves. In this context, Sanderson (1988) provides a suitable background to ice mechanics and risks to offshore structures.

Ice Crushing Pressure. During the process of ice crushing against the structure, the area of contact A is continuously changing and the pressure p required to crush the ice depends on A. On the basis of previous studies and data, it is assumed that the crushing pressure p has a lognormal distribution, with mean μ_p and coefficient of variation $V_p = 0.50$. The size effect for the mean pressure μ_p is represented by:

$$\mu_p = \text{Max.}\begin{cases} \dfrac{9.0}{\sqrt{A}} \\ p_o \end{cases} \quad (6)$$

where A is in m^2 and μ_p in MPa. p_o represents a lower bound for μ_p which appears in observed data. In contrast to the earlier approach of Foschi et. al. (1996) in which specified values of p_o were used, p_o is instead represented here as a random variable to account for the uncertainty in ice crushing pressure for large areas A. Thus, p_o has been taken to possess a normal distribution with a mean $\mu_{po} = 4$ MPa and a standard deviation $\sigma_{po} = 1$ MPa.

Force-penetration Relationship. For a given iceberg penetration x due to ice crushing, the iceberg force F(x) acting on the structure can be calculated from an integration of the crushing pressure p over the area of contact A(x) on the basis of the specific geometry assumed here and assuming that the pressure is uniformly distributed over area. In general, it is found that, for the cases of interest, the iceberg will be stopped after a few metres of penetration, allowing for a linearization of the force-penetration relationship.

Maximum Iceberg Force. In calm water, the calculation of the maximum iceberg force F_M is implemented through consideration of an energy balance, which requires that the initial kinetic energy of the iceberg to be equated to the energy dissipated through ice crushing up to the time the iceberg is brought to rest. Thus:

$$\frac{1}{2} M (1 + C_a) V^2 = \int_0^{x_m} F(x)\, dx \quad (7)$$

where M is the iceberg mass, C_a is an added mass coefficient, and V is the impact velocity. The right-hand side corresponds to the energy dissipated through ice crushing up to a penetration x_m, obtained from the force-penetration relationship. An additional term on the right-hand side, corresponding to the energy dissipated through local structural damage was considered by Foschi et. al. (1996), but is omitted here. Given the iceberg geometry, its impact velocity, added mass, and the crushing pressure parameters, Eq. 7 can be solved to obtain x_m. Once this is found, the maximum force F_M is obtained from the force-penetration relationship.

Impact Velocity. As indicated in Eq. 7, the iceberg's impact velocity V influences the magnitude of the maximum iceberg force on the structure. V is generally influenced by both the prevailing current U and waves. However, in calm water (iceberg alone, no waves), the impact velocity V is taken to be equal to the current velocity U. This approximation is only needed with respect to the statistical descriptions of V and U, and is reasonably consistent with dynamic models of iceberg drift (e.g. El-Tahan et. al., 1986, and Isaacson, 1988) when applied in the absence of waves and wind. Following data from Det Norske Veritas (1988), the current U is assumed to possess a lognormal distribution. Three mean values for U are considered here: 0.14, 0.28 and 0.42 m/s, with corresponding standard deviations of 0.1, 0.2 and 0.3 m/s.

Added Mass. The added-mass coefficient at impact, C_a, needed in Eq. 7, is determined by solving the boundary value problem corresponding to an iceberg oscillating in otherwise still water. A description of the calculation procedure has been given by Isaacson and Cheung (1988). The added-mass coefficient depends on the submerged geometry of the iceberg and of any neighbouring structure (and thus it is a function of the iceberg distance from any such structure), the water depth and the oscillation frequency. The zero-frequency added-mass coefficient is suitable in the present context and has been estimated for a range of iceberg parameters, both in open water and when in contact with the structure. However, for the range of iceberg sizes of interest, the added mass is not strongly influenced by proximity to the structure. The numerical results have been used to develop a simple expression for the added-mass coefficient in terms of D/d and h/d for application to the probabilistic model.

Eccentric Collisions. The preceding description of the force-penetration relationship and the associated maximum force on the structure is based on the assumption of a head-on collision, whereas in practice an eccentric collision is likely to occur. In order to account for this possibility, the maximum force F_M calculated in the manner described above is multiplied by an eccentricity reduction factor $K_e \leq 1$. Data from Salvalaggio and Rojansky (1986) have been utilized to develop a suitable probability distribution of K_e, such that K_e varies from 1 for a head-on collision, to 0 in the limit of the iceberg just making glancing contact with the structure.

Modifications to the Probability Distributions for U and L. Finally, it is noted that available statistics for U and L correspond to all icebergs in open water. However, these differ from the corresponding statistics for impacting icebergs, since an iceberg's speed and size influence its probability of collision with the structure. Sanderson (1988) has investigated this difference and has shown that the speed and size probability distributions for colliding icebergs can be obtained as modifications of the corresponding distributions of all icebergs in open water. These modifications skew the original distributions so as to increase the chances of collision of bigger and faster icebergs.

COMBINED WAVE AND ICEBERG FORCES ON STRUCTURE

Attention is now given to an iceberg collision in the presence of waves. The preceding descriptions of iceberg shape and size, crushing pressure, and iceberg added mass continue to apply. However, the maximum iceberg force on the structure is altered, partly because the iceberg impact velocity is changed, and partly because the wave force on the iceberg influences the iceberg force on the structure. Also the maximum combined force on the structure, if required, should also account for the wave force which now acts on the structure. Furthermore, the description of wave parameters must now reflect the sea state at the moment of collision, so that more commonly occurring wave conditions should be accounted for, rather than wave conditions corresponding to annual maxima as is used in the case of waves alone. Finally, the influence of waves which are not colinear with the iceberg trajectory should be accounted for. These various aspects are now considered.

Wave Parameters. The description of waves under more commonly occurring conditions during which an iceberg impact occurs has already been summarized. The corresponding probability distribution of the significant wave height H_s, based on a recording interval of several hours, is assumed to be given by Eq. 2. As already indicated, the corresponding wave period is assumed to be proportional to $H_s^{1/2}$.

Impact velocity. The impact velocity is influenced by the presence of waves, and so is no longer taken to be equal to the current U. A simple formulation for the iceberg impact velocity V may be developed by taking V to be proportional to the iceberg drift velocity in open water, denoted V_o, and adopting a suitable expression for the latter. Following Isaacson (1988), an expression for V_o may be developed by equating the wave drift force to the current drag, taking the wave drift force coefficient to be proportional to D/L (see Isaacson, 1988), where L is the wave length, and taking T to be proportional to $H_s^{1/2}$. When the wave direction and the iceberg trajectory are colinear, the above approach leads to $V = U + V_r$, where V_r is the iceberg velocity relative to the current, given by $V_r = \alpha\ gT\ (D/h)^{1/2}$, where α is a constant and g is the gravitational constant. The value of the constant α has been estimated by examining previous results and data for the open water velocity V_o (e.g. Lever and Sen, 1987); and using a numerical model to relate the impact velocity V to the open water velocity V_o. Thus, a wave diffraction-radiation analysis has been carried out for a series of conditions corresponding to the iceberg approaching the structure, using an extension to the computer program WELSAS (see Isaacson, 1987). The foregoing procedure has indicated that $\alpha = 0.003$ should be suitable. When the wave direction makes an angle θ with the iceberg trajectory (see Fig. 2), an extension to the above approach yields instead:

$$V = V_r \cos\theta + \sqrt{U^2 - V_r^2 \sin 2\theta} \tag{8}$$

with V_r given as before.

Wave Force on Iceberg. The force due to waves on the iceberg is now required in order to treat the iceberg's equations of motion, and this must account for the presence of the structure. As a reasonable simplification to the problem, only the equation of motion of the iceberg in the x direction is considered, and the iceberg's transverse (y-ward) and rotational (yaw) motions are not treated. Thus, only the wave force in the in-line (x-ward) direction, denoted $F_{wx}^{(i)}$, is used and the force in the transverse (y-ward) direction is not needed. This wave force component, which is oscillatory, is calculated by treating the wave diffraction problem for the combined submerged surface of the iceberg in contact with the structure, with a pressure integration over the iceberg surface only used to provide the wave force on the iceberg. The force amplitude and phase angle have been calculated for the three water depths and a range of iceberg and wave parameters using an extension to the program WELSAS, and the numerical results are expressed in terms of relevant iceberg and wave parameters in order to be applied to the reliability model.

Equation of Motion of Iceberg. In the presence of waves, the iceberg force on the structure, F(x), is influenced in part by the in-line wave force on the iceberg, $F^{(i)}_{wx}$, and thus the calculation of F(x) is carried out by a direct integration of the equation of motion, rather than by a simple energy balance as in Eq. 7. Thus, dynamic equilibrium in the x direction of an iceberg of mass M gives rise to the following equation of motion:

$$M(1 + C_a)\ddot{x} = -F(x) + F^{(i)}_{wx} \tag{9}$$

with initial conditions x = 0 and $\dot{x}$ = V at t = 0, where x is the iceberg displacement in the x direction, corresponding to its penetration into the structure. In this equation, the steady forces due to current and wind drag are ignored. In order to obtain a solution to Eq. 9, estimates of the added mass coefficient, C_a, the iceberg impact velocity, V, and the in-line wave force on the iceberg, $F^{(i)}_{wx}$, are needed. These have been summarized above and are described in detail by Foschi et. al. (1996) for the case of colinear wave direction and iceberg trajectory. As already indicated, the iceberg force on the structure, F(x), is a nonlinear function of x, but can be linearized as F(x) = Kx. With this linearization, the equation of motion, Eq. 9, can readily be integrated in closed form to provide the iceberg force on the structure as a function of time.

Wave Force on Structure. In calculating the overall maximum force on the structure, needed for the limit state of global sliding, the force due to waves acting on the structure itself is also required, and this must now account for the presence of the iceberg. Because the wave direction and iceberg trajectory are not colinear, both the in-line and transverse components of the wave force on the structure, denoted $F^{(s)}_{wx}$ and $F^{(s)}_{wy}$ respectively, are required. As described earlier, these force components are obtained by treating the wave diffraction problem for the combined submerged surface of the iceberg in contact with the structure, with now a pressure integration over the structure surface only used to provide the wave force on the structure. Once more, the wave force amplitude has been calculated for the three water depths and a range of iceberg and wave parameters using an extension to the computer program WELSAS. The numerical results are used to express the x-ward and y-ward force amplitudes and phase angles in terms of relevant wave, iceberg and structure parameters. The phase angles, which control the effect of the force components at time t = 0 (i.e. at the beginning of the collision), are found to be important in the calculation of the exceedence probability.

Maximum Force on Structure. Once a numerical solution to Eq. 9 has been obtained, the required force on the structure may be obtained. Two limit states are considered here. The first is the local damage limit state, for which the maximum iceberg force on the structure in the presence of waves is needed. The second is the global sliding limit state, for which the iceberg force on the structure must be combined with the overall wave force on the structure, in order to develop the maximum combined force. Since the case of an oblique wave direction is being considered here, the maximum combined force is given as:

$$F_M = \mathrm{Max}\left\{\sqrt{(F(x) + F^{(s)}_{wx})^2 + F^{(s)2}_{wy}}\right\} \quad \text{for } 0 \le t \le t_o \tag{10}$$

The calculation of the maximum force F_M requires a search for the maximum combination of the iceberg force on the structure, F(x), the in-line wave force on the structure, $F^{(s)}_{wx}$, and the maximum transverse wave force on the structure, $F^{(s)}_{wy}$, from the instant of initial impact, t = 0, until the iceberg is stopped, t = t_o, bearing in mind that at maximum penetration the wave force components may not be at their peaks.

RESULTS

Table 3 shows a summary of the deterministic and random variables specified in the numerical model.

Waves Alone and Iceberg Alone

The program WLOAD was first run to obtain the forces for waves alone at an annual exceedence probability of 10^{-2}. The program ICELOAD was then run to obtain the iceberg alone forces at an exceedence probability of 10^{-4} for various values of the iceberg arrival rate μ, with the modifications to the U and L distributions described earlier. The results indicated that the iceberg collision forces are strongly dependent on both current speed and iceberg size (which is limited by water depth).

Combined Waves and Iceberg

The program ICEWLOAD was then run to obtain the total force on the structure as a result of an iceberg collision in the presence of waves, and the corresponding results based on the local and global damage limit states are shown in Tables 4 and 5 respectively. For the local damage limit state, F_M is the maximum iceberg force on the structure in the presence of the waves; and for the global damage limit state, F_M is the maximum combined iceberg-wave force on the structure. The annual exceedence probability was set at 10^{-4}, as required by the Code. The tables show results based on the three water depths, three values of mean current, and four values of mean wave direction, the iceberg parameters considered in Table 3 (collision rate, and with the modifications to the U and L distributions), and to the limit state under consideration. The wave climate considered is that which occurs during the ice season, and is relatively mild when compared to the year-round annual maximum.

Load Combination Factors

Finally, the preceding results may be used to determine the load combination factor γ used in the CSA Code. γ is used to determine a design value for the load due to a companion frequent environmental process (waves) acting in combination with a rare environmental event (iceberg collision), and is defined in the Code by Eq. 1. The Code recommends that γ = 0.8 for stochastically dependent events/processes, and γ = 0.4 for stochastically independent events/processes, and that iceberg impact and waves should be considered stochastically independent (i.e. γ = 0.4).

Values of γ can be calculated from the results obtained for waves alone, an iceberg alone, and those given in Tables 4 and 5, and are included in Tables 4 and 5. As expected, although the actual loads are substantially influenced by the various parameters that are considered, the results shown suggest that the load combination factor itself is much more stable. As the wave direction changes, the value of γ changes quite significantly, and eventually approaches zero for a wave direction α = 90° as expected. Finally, the results do not change substantially when the limit state is changed from the local damage state to the global damage state .

The values of γ were calculated with the 10^{-2} annual wave load, which may not occur during the ice season. As a result the values of γ are found to be smaller than the 0.4 level given in the code. If one were to use the 10^{-2} seasonal wave load, the value of γ would be larger (reaching the 0.8 level). Since the iceberg collision force is interrelated with the effect of the waves, it is recommended that the combined event should be classified as dependent rather than independent, and that a value of 0.25 could be used if a more detailed study cannot be conducted, and modified for different mean wave directions. From Tables 4 and 5, a conservative value of γ would be 0.33.

CONCLUSIONS

The analysis of loads due to an iceberg collision in the presence of waves, taking account of seasonal variations in wave climate and the influence of wave direction, is summarized. This analysis is based on mechanics models for the forces on an offshore structure due to waves alone, an iceberg alone, and the combination of waves with an iceberg, and these are combined with a reliability model in order to develop forces corresponding to specified risk levels.

A primary objective of this study has been to examine the load combination factor γ for combined iceberg-wave loads on large offshore structures. At present, the CSA Code recommends that iceberg impact and waves should be considered stochastically independent, with a load combination factor $\gamma = 0.4$. The results of the study suggest that iceberg impact and waves should instead be considered stochastically dependent, and that a value of γ of 0.25 could be used. This value may also be modified for wave direction as indicated here.

REFERENCES

Bea, R.G. (1992). "Evaluation of uncertainties in loading on offshore structures due to extreme environmental conditions." *Proc. Int. Symp. Offshore Mechanics and Arctic Engrg.*, ASME, Calgary, Vol. II, pp. 367-376.

Cammaert, A.B., and Muggeridge, D.B. (1988). *Ice interaction with offshore structures.* Van Nostrand Reinhold, New York.

Canadian Standards Association. (1992). *CAN/CSA-S471-92. General requirements, design criteria, environment and loads.* [Forming part of the code for the design, construction, and installation of fixed offshore structures.] Rexdale, Ontario.

Det Norske Veritas. (1988). "Probabilistic framework for ice loads on fixed marine structures." Calgary, Alberta. Rept. to Dept. of Public Works, Canada, Vol. 1.

El-Tahan, H., Venkatesh, S., and El-Tahan, M. (1986). "Evaluation of a model for predicting the drift of iceberg ensembles." *Proc. 5th Int. Symp. Offshore Mechanics and Arctic Engrg.*, Tokyo, Vol. IV, pp. 418-425.

Foschi, R.O., Folz, B., and Yao, F. (1990). *RELAN: User's manual.* Dept. of Civil Engrg., Univ. of British Columbia, Vancouver, B.C., Canada.

Foschi, R.O., and Isaacson, M. (1996). "A study of wave-iceberg load combination factors." *Proc. 6th Int. Offshore and Polar Engrg. Conf.*, Los Angeles, May, Vol. III, pp. 278-285.

Foschi, R.O., Isaacson, M., Allyn, N., and Yee, S. (1996). "Combined wave-iceberg loading on offshore structures." *Canadian J. Civil Engrg.*, 23(5): 1099-1110.

Fuglem, M.K., Duthinh, D., Lever, J.H., and Jordaan, I. (1991). "Probabilistic determination of iceberg collision design loads for floating production vessels." *Proc. IUTAM-IAHR Symp. on Ice-Structure Interaction*, St. John's, Newfoundland, Springer-Verlag, pp. 459-482.

Fuglem, M., Jordaan, I., Crocker, G., Cammaert, G., and Berry, B. (1997). "Environmental factors in iceberg collision risks for floating systems." *Cold Regions Science and Tech.*, in press.

Isaacson, M. (1987). "Ice mass motions near an offshore structure." *J. of Offshore Mechanics and Arctic Engrg.*, ASME, 109(2): 206-210.

Isaacson, M. (1988). "Influence of wave drift force on ice mass motions." *Proc. 7th Int. Conf. Offshore Mechanics and Arctic Engrg.*, Houston, Vol. II, pp. 125-130.

Isaacson, M., and Cheung, K.F. (1988). "Hydrodynamics of an ice mass near an offshore structure." *J. Waterway, Ports, Coastal and Ocean Engrg.*, ASCE, 114(4): 487-502.

Isaacson, M., and McTaggart, K. (1990). "Modelling of iceberg drift motions near a large offshore structure." *Cold Regions Science and Technology*, 19: 47-58.

Jordaan, I. (1996) - Personal Communication.

Lever, J.H., Attwood, D., and Sen. D. (1988). "Factors affecting the prediction of wave-induced iceberg motion." *Cold Regions Science and Technology*, 15: 177-190.

Lever, J.H., Colbourne, B., and Mak, L. (1990). "Model study of the wave-driven impact of bergy bits with a semi-submersible platform." *J. Offshore Mechanics and Arctic Engrg.*, 112: 313-322.

Lever, J.H. and D. Sen. (1987). "A method to upgrade iceberg velocity statistics to include wave-induced motion." *J. Offshore Mechanics and Arctic Engrg.*, 109(3): 278-286.

MacLaren Plansearch Ltd. (1991). "Wind and wave climate atlas. Volume I. The East coast of Canada." Rept. to Transportation Development Centre, Transport Canada, Ottawa.

Maes, M.A. (1986). "Calibration of partial factors in the new CSA Code for fixed offshore production structures." Technical Rept. No. 9, Environmental Protection Branch, Canada Oil and Gas Lands Administration, Energy Mines and Resources, Ottawa.

Maes, M.A., and Jordaan, I. (1984). "Probabilistic analysis of iceberg loads on offshore structures." *Proc. IAHR Ice Symp.*, Hamburg, pp. 175-188.

Neu, H.J.A. (1982). "11-year deep water wave climate of Canada Atlantic waters." *Canadian Technical Report of Hydrography and Ocean Sciences*, No. 13, p. 48.

Salvalaggio, M.A., and Rojansky, M. (1986). "The importance of wave-driven icebergs impacting an offshore structure." *Proc. 18th Offshore Technology Conf.*, Houston, Texas, Vol. 1, pp. 29-38.

Sanderson, T.J.O. (1988). *Ice mechanics and risks to offshore structures.* Graham and Trotman, London.

Sarpkaya, T., and Isaacson, M. (1981). *Mechanics of wave forces on offshore structures.* Van Nostrand Reinhold, New York.

TABLE 1. Summary of iceberg statistics at the three sites.

Site	Depth (m)	Max width (m)	Length parameter L		Draft	
		Mean	Mean (m)	S.D. (m)	Mean (m)	S.D. (m)
A	60	80.97	61.75	13.43	50.47	7.27
H	80	110.90	88.50	24.07	63.09	11.37
B	100	131.23	107.19	35.60	70.89	15.41

Note: maximum draft equals water depth

TABLE 2. Collision arrival rates at the three sites for various mean current values.

Mean current (m/s)	Site A (d = 60 m, ν = 0.31)	Site H (d = 80 m, ν = 0.74)	Site B (d = 100 m, ν = 0.74)
0.14	0.03	0.08	0.09
0.28	0.06	0.16	0.18
0.42	0.09	0.24	0.27

TABLE 3. Summary of specified variables and their statistics.

Variable	Description	Site A	Site H	Site B
Depth, d (m)	Deterministic	60	80	100
Structure radius, a (m)	Deterministic	50		
Minimum crushing pressure, p_o (MPa)	Normal	$\mu = 4, \sigma = 1$		
Iceberg density, ν (no / deg^2)	Deterministic	0.31	0.74	0.74
Current velocity, U (m/s)	Log-normal	μ = 0.14, 0.28, 0.42 σ = 0.1, 0.2, 0.3		
100 year sig. wave ht, H_s (m) - all year	Extreme Type I	19.70	19.50	19.50
100 year sig. wave ht, H_s (m) - ice season	Extreme Type I	12.60	9.80	9.80
Individual wave period, T (s)	Normal	$4.43 \sqrt{H_s}$		$\mu = 1.0$, σ= (input)
Wave direction, α (deg)	Normal	$\mu = 0$, $\sigma = 20$	μ = 0, 30, 60, 90, $\sigma = 20$	$\mu = 0$, $\sigma = 20$
Recording interval, r (hr)	Deterministic	6		
R_{n1}, associated with model uncertainty	Normal	$\mu = 1.0$, $\sigma = 0.1$		
R_{n2}, associated with ice crushing pressure	Normal	$\mu = 0.0$, $\sigma = 1.0$		
R_{n3}, associated with collision eccentricity	Uniform	Min. = 0.0, Max. = 1.0		
R_{n4}, associated with Rayleigh distribution for wave height	Uniform	Min. = 0.0, Max. = 1.0		

Note: Parameters shown refer to all three sites unless otherwise indicated.

TABLE 4. Maximum combined iceberg force F_M and load combination factor γ for an annual exceedence probability of 10^{-4} and for various currents and mean wave directions (local damage limit).

Mean current (m/s)	Wave direction (deg)		Site A		Site H		Site B	
	Mean	S.D.	F_M (MN)	γ	F_M (MN)	γ	F_M (MN)	γ
0.28	0	20	1,436	0.160	2,959	0.203	4,189	0.280
0.28	30	20	1,408	0.135	2,915	0.169	4,122	0.233
0.28	60	20	1,334	0.070	2,794	0.077	3,939	0.106
0.28	90	20	1,235	-0.017	2,630	-0.048	3,692	-0.067
0.14	0	20	734	0.176	1,469	0.235	2,097	0.326
0.42	0	20	2,226	0.152	4,623	0.185	6,529	0.254

TABLE 5. Maximum combined wave-iceberg force F_M and load combination factor γ for an annual exceedence probability of 10^{-4} and for various currents and mean wave directions (global damage limit).

Mean current (m/s)	Wave direction (deg)		Site A		Site H		Site B	
	Mean	S.D.	F_M (MN)	γ	F_M (MN)	γ	F_M (MN)	γ
0.28	0	20	1,439	0.163	2,960	0.204	4,190	0.281
0.28	30	20	1,419	0.145	2,916	0.170	4,123	0.234
0.28	60	20	1,345	0.080	2,794	0.077	3,948	0.112
0.28	90	20	1,248	-0.005	2,637	-0.043	3,695	-0.065
0.14	0	20	742	0.183	1,471	0.236	2,098	0.326
0.42	0	20	2,228	0.154	4,624	0.185	6,530	0.254

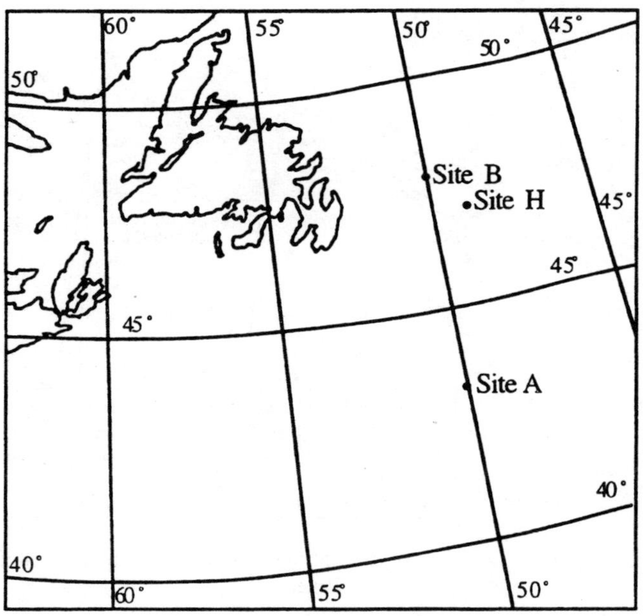

Fig. 1. Location sketch.

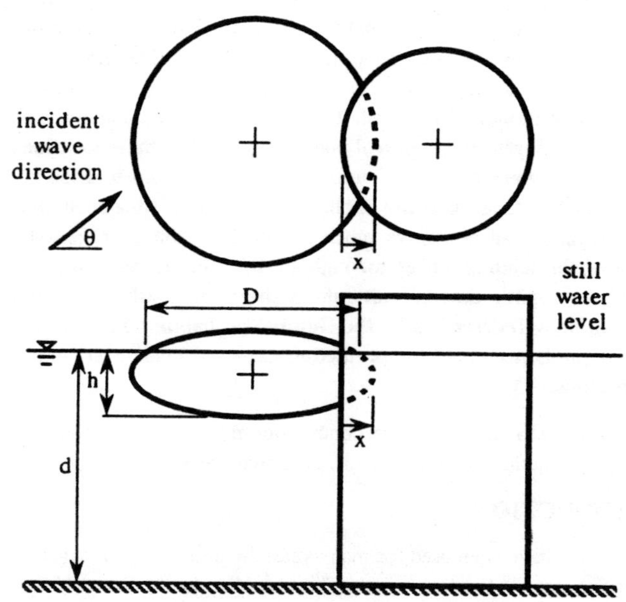

Fig. 2. Definition sketch.

Proceedings of the Seventh (1997) International Offshore and Polar Engineering Conference
Honolulu, USA, May 25-30, 1997

ISBN 1-880653-28-1 (Set); ISBN 1-880653-30-3 (Vol. II)

Ice Boom Loads in the St. Lawrence River, 1994-95 & 1995-96

A. Cornett and R. Frederking
Canadian Hydraulics Centre, National Research Council
Ottawa, Canada

B. Morse and S. Dumont
Canadian Coast Guard, Laurentian Region
Quebec, Canada

ABSTRACT

In recent years, a new type of ice control boom has been developed by the Canadian Coast Guard to improve the formation of a stable ice cover along the St. Lawrence River. The new boom uses cylindrical steel pontoons in place of the rectangular timbers that have traditionally been used. This paper presents an analysis of the loads measured in ice control booms deployed at three locations during the winters of 1994-95 and 1995-96. Loads up to 1250 kN were recorded. The loads are highly variable over time such that the maximum load is always much larger than the average. The nature of the loading varies throughout the winter, reflecting the type of ice regime that prevails along the river. Daily maximum loads can be well described by the Gumbel probability distribution. A technique is presented for predicting expected loads over longer time periods.

KEY WORDS: ice loads, ice control boom, probabilistic analysis, Gumbel distribution, long-term loads, St. Lawrence River.

INTRODUCTION

Ice booms have been used for many years by utilities and navigation interests to control the flow of ice. The Canadian Coast Guard (CCG) deploys booms at several sites along the St. Lawrence River to assist the formation of a stable ice cover during the winter months, thereby preventing ice floes from jamming the navigation channel (see Figures 1 and 2). Recently, efforts have been directed towards the development of more effective and durable booms for this purpose.

The CCG's ice control booms are typically deployed in the late fall and then removed at the end of the winter season so that they do not hinder the open-water uses of the waterway. Ice booms have many advantages over fixed ice control structures such as weirs, piers, piles, and man-made islands. Ice booms themselves have a negligible effect on the natural hydraulic conditions in the river. They have little influence on scour and siltation patterns, and do not impede fish migration (Foltyn and Tuthill, 1996).

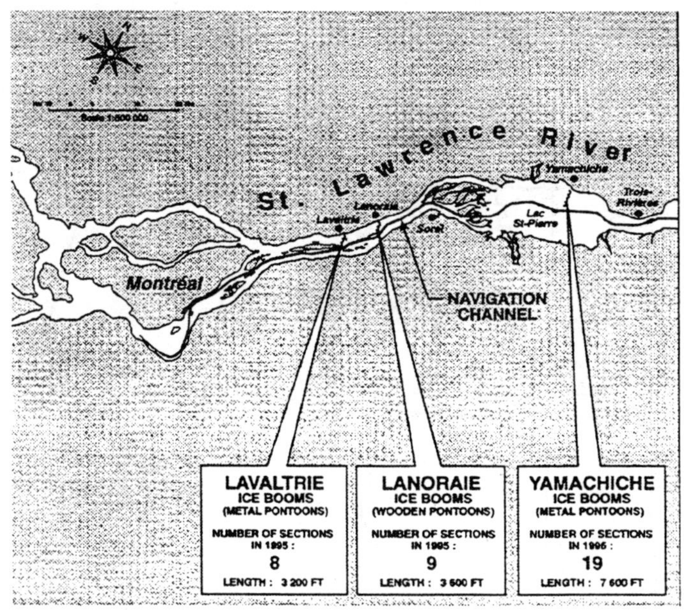

Figure 1. Locations of ice control booms along the St. Lawrence River.

During the winter of 1993-94, a new type of boom, with pontoons made from sections of steel pipe, was installed in the system as a field trial. During 1994, a series of physical model tests were conducted at the Canadian Hydraulics Centre (CHC) to gain information on the relative performance and loading of several different boom designs (see Timco and Cornett, 1994, 1996 and Cornett and Timco, 1995). During the winter of 1994-95, more boom sections with steel pipe pontoons were deployed, particularly at Yamachiche, where the entire boom was constructed using steel pontoons. Figure 2 shows an aerial view of the stable ice cover formed behind the Yamachiche boom. During the winter of 1995-96, steel pontoons were used in the booms at Yamachiche and Lavaltrie while the traditional timber pontoons were retained at Lanoraie only.

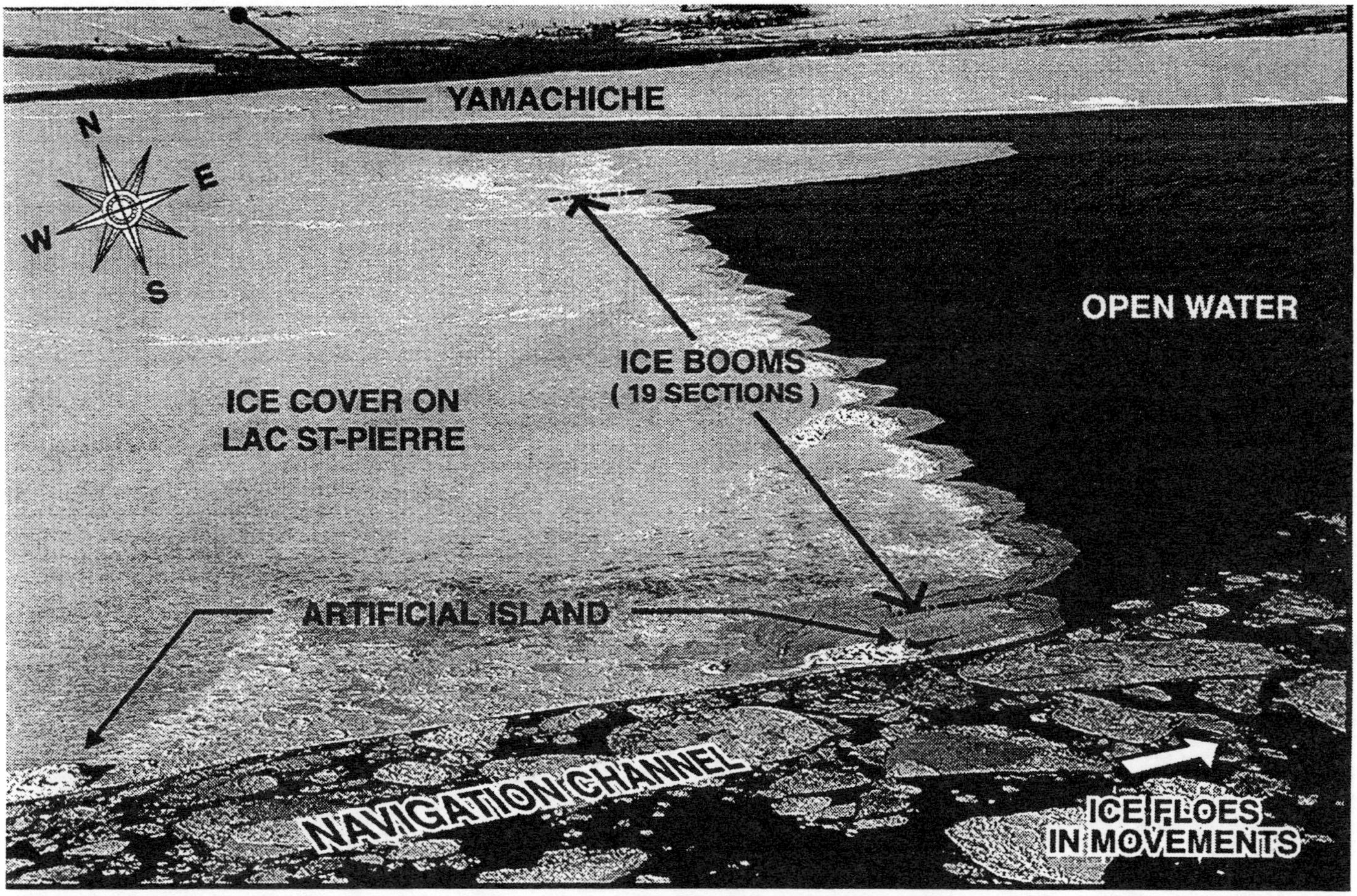

Figure 2. Aerial view of the Yamachiche boom in ice.

In order to assess the performance of the new booms, the CCG have made load measurements at these three sites during the winters 1993-94, 1994-95, and 1995-96. The CHC was asked to help analyse the loads measured during the winters of 1994-95 and 1995-96.

LOAD MEASUREMENTS

The ice control booms used in the St. Lawrence River consist of a number of semi-circular boom sections linked together in series through junction plates to form a linear structure. Each boom section consists of approximately thirteen pontoons (or floats) chained to a 150 m (500 ft) length of steel cable known as a section cable (see Figure 3). Longitudinal and lateral stability are provided through a system of anchors and anchor cables connected to the junction plates. The linear span of a boom section depends on the spacing of the anchors, but is close to 122 m (400 ft) on average. The new boom design uses pontoons made from 9 m (30 ft) lengths of circular steel pipe in place of the rectangular timbers which had traditionally been used.

The data acquisition systems used in the measurement program were set up with sampling intervals of 0.5 s and recording intervals ranging between 1 and 30 minutes. Two time series were recorded for each transducer: 1) the instantaneous value at the recording time; and 2) the maximum value sampled over the preceding recording interval. Subsequent analysis has been focused on the signal formed by the maximums, since this signal contains the most extreme recorded values.

The Lavaltrie boom incorporates 8 boom sections held in place by a network of 11 anchor cables. A single anchor cable, A3, was instrumented during both winters. The Lanoraie boom consists of 9 boom sections and 12 anchors. Load measurements were obtained in a single anchor cable, A4. The longer Yamachiche boom (shown in Figure 2) incorporates 19 spans fixed by a system of 28 anchors. During 1994-95, loads were measured in 3 anchor cables distributed along the boom and in 5 adjacent section cables, while 3 anchor cables and 6 section cables were monitored throughout 1995-96. In all cases, loads were measured where the cables (section or anchor) joined a junction plate.

ANALYSIS AND RESULTS

The anchor cable and section cable loads are highly variable over time, and can also differ widely along the length of a single boom. The temporal variations include: seasonal differences in the nature of the loading which reflect changes in the prevailing ice regime; diurnal oscillations caused by thermal stresses; and random high-frequency fluctuations associated with impacts from ice floes or sudden movements of the ice cover.

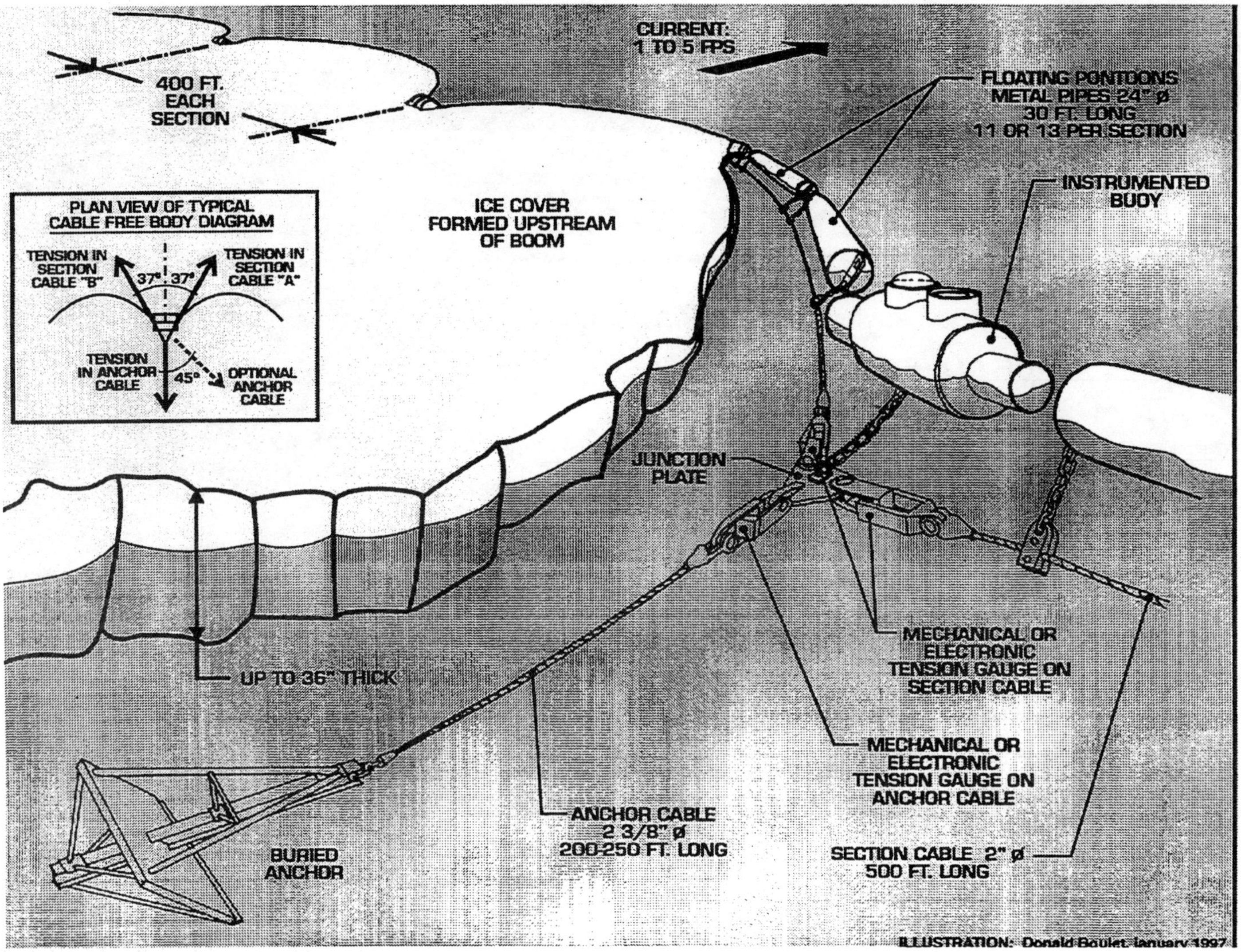

Figure 3. 3D sketch of a typical ice boom assembly.

Although the loads in each cable are unique, loads measured at junction plate 11 of the Yamachiche boom during 1994-95 are in many respects typical of those measured elsewhere. Junction plate 11 is located near the centre of the Yamachiche boom. Here, loads were measured in anchor cable A11S, section cable S10N (equipped with 0.76 m diameter pontoons) and section cable S11S (equipped with 0.61 m diameter pontoons). Loads in a second anchor cable were not recorded due to instrument failure. Force time series recorded at junction plate 11 of the Yamachiche boom during the winter of 1994-95 are plotted in Figure 4. All three transducers appear to give good data for the entire season from November through March. Peak loads recorded over selected time intervals throughout the winter are summarised in Table 1.

These results clearly illustrate that the larger 0.76 m diameter pontoons attracts higher loads than the smaller 0.61 m diameter pontoons. This behaviour is consistent with results from physical model tests reported by Timco and Cornett (1994, 1996).

Table 1. Peak loads over selected periods, junction plate No. 11, Yamachiche boom, 1994-95.

	S10N	S11S	A11S
Section/Anchor	section	section	anchor
Pontoon diameter	760 mm	610 mm	-
Dec. 19	1270	524	1010
Dec. 29 - Jan. 20	700	200	390
Mar. 1 - 12	420	380	490

The character of the loading tends to follow a general pattern in which the winter season can be sub-divided into the three periods described below.

1) **early winter freeze-up**: Ice floes accumulate behind the boom. Over time, they become partially consolidated into a semi-stable ice cover. Generally occurs for a few weeks in early- to mid-December.

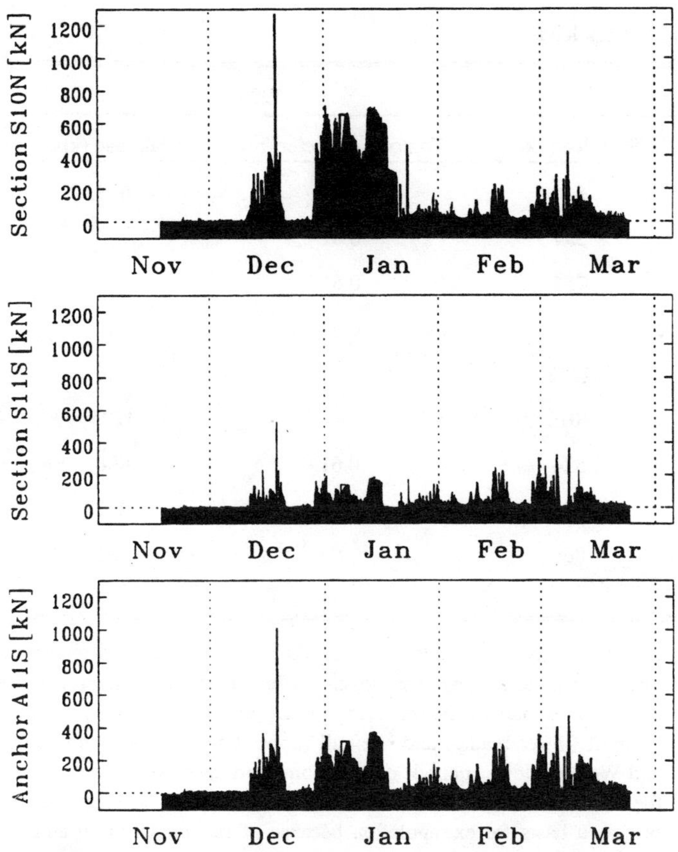

Figure 4. Loads at junction plate No. 11 (Yamachiche boom, 1994-95).

Produces wildly fluctuating boom loads that can attain large peak values. Loads on adjacent boom spans are poorly correlated.

2) **mid-winter**: Generally prevails from mid-December through early March. Characterised by lengthy periods of relatively stable, low level loads which may be interrupted by brief periods of more active loading. The ice cover is generally well consolidated and stable, but may become mobile at times. Loads on adjacent boom sections are generally well correlated throughout this period because the boom sections are frozen into the ice cover.

3) **spring break-up**: The stable ice cover breaks-up into floes which can impact against the boom, generating highly fluctuating loads similar to those associated with early winter freeze-up. Generally occurs during March. Loads on adjacent boom sections become un-correlated, as during the early winter freeze-up period.

Several periods were identified, each generally lasting less than a week, in which the loading was dominated by diurnal fluctuations. These diurnal fluctuations generate only moderate loads and invariably occur during the mid-winter period when the boom is well frozen into the ice cover. Figure 5 shows the daily fluctuations measured at junction plate No. 11 of the Yamachiche boom near the end of January, 1995. Due to their moderate load level and their relatively brief duration, these diurnal fluctuations represent a very small portion of the total energy in the load record.

All three signals recorded at junction plate No. 11 contain a dominant maximum load that occurs between 6 and 7 a.m. on the

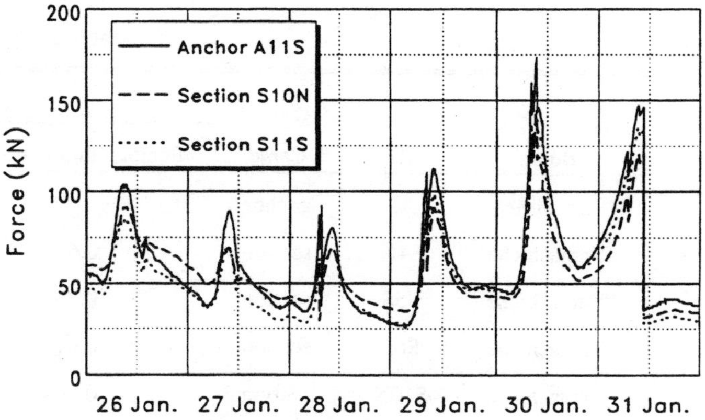

Figure 5. Diurnal loading at junction plate No. 11 (Yamachiche boom, 26-31 January 1995).

morning of December 19th. Figure 6 shows a detail of the raw maximum forces measured during this event. The loading event begins at 5:40 a.m. when the loads in all three components start to rise quickly. The maximum loads in the boom components occur sequentially rather than simultaneously over the next hour. The load in section cable S10N (with the larger pontoons) reaches 1270 kN at 6:26 a.m. while the load in section cable S11S (with the smaller pontoons) peaks at 524 kN a bit later at 6:42 a.m. During this event, the peak load in anchor cable A11S reaches 1010 kN at 6:22 a.m. It is quite likely that this event is caused by the arrival at the Yamachiche boom of a large and/or dense raft of ice floes.

At junction plate No. 11, the peak load for the entire winter occurs during early winter freeze-up, but this is not always the case. At other locations and in other years, the peak load occurs during the mid-winter period and even during spring breakup. The peak loads recorded at the various sites over the two winters are summarized in Table 2.

PREDICTING LONG TERM LOADS

Ice loading on a control boom is a random process that depends on many factors such as the ice thickness, the size and density of ice floes, the current speed, the boom configuration and geometry, the type of interaction that occurs (crushing, buckling, or flexural failures) and climatic factors including wind speed and direction as well as temperature. The peak loads summarized in Table 2 provide a concrete example of the wide range of loads generated through this random process. Because of this inherent randomness, ice loads are well suited

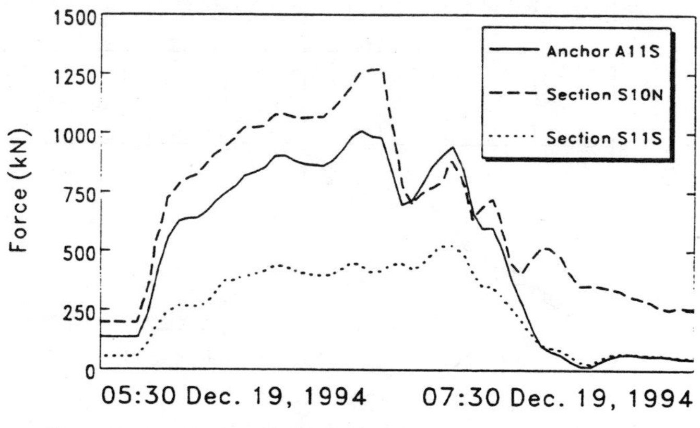

Figure 6. Loads at junction plate No. 11 (Yamachiche boom, 19 December 1994).

Table 2. Summary of peak loads.

Boom	ID	Cable	1994-95		1995-96	
			Pontoon diameter (m)	Peak load (kN)	Pontoon diameter (m)	Peak load (kN)
Lavaltrie	A3	anchor	-	646	-	673
Yamachiche	S4N	section	0.61	254	0.61	-
Yamachiche	S5S	section	0.76	755	0.61	-
Yamachiche	S6S	section	0.61	-	0.61	703
Yamachiche	S10N	section	0.76	1272	0.61	546
Yamachiche	A11S	anchor	-	1010	-	1225
Yamachiche	S11S	section	0.61	524	0.61	986
Yamachiche	S14N	section	0.61	524	0.61	212
Yamachiche	A15N	anchor	-	866	-	581
Yamachiche	S15S	section	0.76	-	0.61	752

to probabilistic analysis. In the following, a probabilistic analysis is developed to predict the peak loads that can be expected in the future based on the existing observations.

Anchor Cable Loads

Good quality measurements of anchor cable loads in the Yamachiche boom are available from anchors A5N, A11S and A15N over the winter of 1994-95, and from anchors A11S and A15N during the winter of 1995-96. In some cases, good data is available over the entire season from December through March, while in others, good data is only available for a fraction of the winter. First, these records were analysed to obtain the daily maximum force in each cable. Next, all the daily maximums from different cables and different years were combined into a single set of observations. After rejecting the very small forces recorded early and late in the season, when no ice interaction occurs, a total of 442 daily maximums were retained for further analysis. The largest recorded daily maximum was 1225 kN.

To obtain long-term design loads, one must fit empirically a convenient theoretical probability distribution which describes the

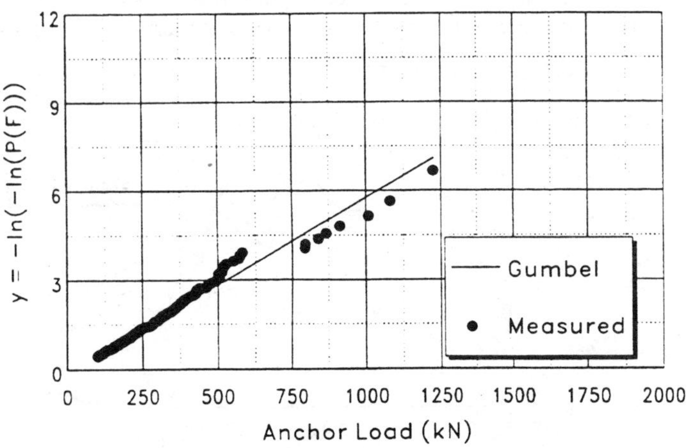

Figure 7. Daily anchor cable loads for the Yamachiche boom fitted to the Gumbel distribution.

observed daily maximums as closely as possible. The daily maximum loads were tested against seven different theoretical probability distributions, namely: Normal, Lognormal, Rayleigh, Gumbel, Fretchet, Weibull (lower bound) and Weibull (upper bound). Both the Gumbel and Weibull (lower bound) distributions were in close agreement with the observed data. The Gumbel distribution was selected as the preferred basis for extrapolation because of its simpler form and the fact that it has only two free parameters (compared to three free parameters for the Weibull distribution). The Gumbel distribution has been widely used in the context of flood prediction, and has also been applied to predict extreme winds, and extreme waves (Sarpkaya and Isaacson, 1981). The cumulative probability function for the Gumbel distribution is given by

$$P(F) = \exp\left\{-\exp\left[-\left(\frac{F-\varepsilon}{\theta}\right)\right]\right\} \quad (1)$$

where F denotes the random variate, in this case the daily maximum load in a single anchor, $P(F)$ denotes the probability that a general value $F(t)$ is less than or equal to F, ε denotes the location parameter and θ denotes the scale parameter of the distribution. Closely related is the exceedence probability $Q(F) = 1 - P(F)$, which denotes the probability that a general value $F(t)$ exceeds the value F.

The Gumbel distribution can be plotted as a straight line in the (x, y) plane under the transformation

$$x = F \ , \quad (2)$$

$$y = -\ln\{-\ln[P(F)]\} \ . \quad (3)$$

Figure 7 shows the good agreement between the Gumbel distribution and the observations of daily maximum anchor load for the Yamachiche boom. In this case, the best-fitting Gumbel distribution has a location parameter, $\varepsilon = 16.11$ and a scale parameter $\theta = 170.5$.

In order to extrapolate design loads using this analysis, it is necessary to express the exceedence probability $Q(F)$ in terms of a meaningful return period T_R. One possibility is to define the return period as a

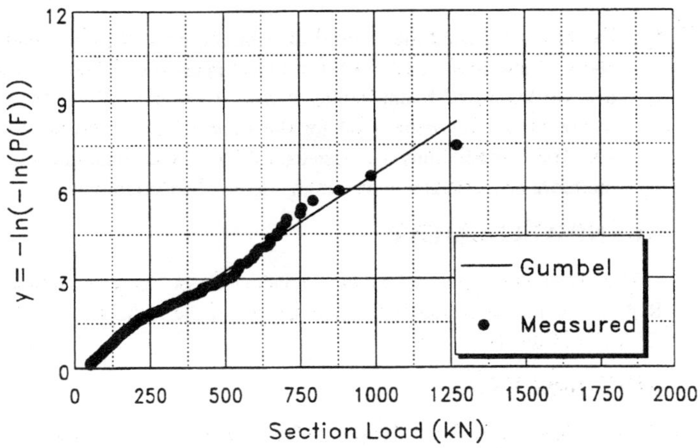

Figure 8. Daily section cable loads for the Yamachiche boom fitted to the Gumbel distribution.

combination of the total length of boom and the number of years it is deployed. This reflects the reality that the likelihood of experiencing larger ice forces in any single anchor cable increases with increasing boom length and with increasing deployment duration. Following this approach, a return period can be defined with units of kilometer-years (km-yr). Using this system, a 1 km long boom deployed for 10 years has a return period of 10 km-yr, while a 2 km long boom deployed for 5 years also has a return period of 10 km-yr.

Several assumptions have been made to obtain a relationship between return period and exceedence probability.

- The Yamachiche boom typically experiences ice loading for around 110 days each year, from early December through till late March
- There are approximately 10 anchors per linear kilometer of boom.

In the case of anchor loads, 442 good daily maximums are available from the 1994-95 and 1995-96 seasons. Assuming 110 days of ice loading per year on average, and 10 anchors per linear kilometer, this corresponds to 0.4018 km-yr of data. The recording interval represents the average number of km-yr between successive observations, and is given by

$$r = \frac{0.4018}{442} = 9.091 \cdot 10^{-4} \text{ km} - \text{yr} \quad (4)$$

The return period and the recording interval are related according to

$$T_R = \frac{r}{Q(F)} = \frac{r}{1 - P(F)} \quad (5)$$

This equation can be used together with the formula for the Gumbel distribution to estimate the expected anchor load for various return periods. Examples of this calculation are summarised in Table 3.

Table 3. Anchor cable loads predicted for various return periods.

T_R (km-yr)	1	2	5	10	20	50
Load (kN)	1210	1328	1485	1603	1721	1877

The table predicts a peak anchor load of 1721 kN for a 2 km long boom deployed over a period of ten years (20 km-yr). The table also shows how the expected load increases with increasing return period (boom length in kilometers and deployment duration in years).

It should be noted that these long-term loads have been estimated from quite a limited amount of data, and must therefore be associated with a fairly high degree of uncertainty. Much of the uncertainty stems from the fact that the extrapolation is based on data from only two winter seasons, neither of which featured particularly severe ice conditions. This sampling period is clearly insufficient to capture the natural annual variability of ice conditions in the St. Lawrence. A more severe winter with thicker and more abundant ice would likely generate higher loads. A second major source of uncertainty relates to the fact that (for the Yamachiche boom) all measurements of anchor loads were obtained at junction plates fitted with two anchor cables. It is quite possible that higher anchor loads prevailed at locations with only one anchor cable. A suitable factor of safety should be applied to account for these and other sources of uncertainty when computing factored design loads based on this analysis.

Section Cable Loads

A similar analysis has been applied to predict long-term loads for the section cables of the Yamachiche boom based on the loads measured during the winters of 1994-95 and 1995-96. During 1994-95, good measurements of section cable loads in the Yamachiche boom are available from cables S4N, S5S, S10N, S11S, and S14N. During 1995-96, good measurements were obtained in cables S6S, S10N, S11S, S14N, and S15S. Daily maximums from these cables were merged into a single data set with 989 observations. The largest recorded daily maximum was 1272 kN. As with the anchor cable loads, the Gumbel distribution provides a good basis from which to extrapolate extreme values for this data set. Figure 8 shows the good agreement between the daily maximum section cable loads and the Gumbel distribution with location parameter, $\varepsilon = -1.683$ and a scale parameter $\theta = 154.05$.

Since each boom section spans 122 m (500 ft), there are approximately 8.2 section cables per linear kilometer of boom. Considering that each section cable has two connection points, there are on average 16.4 section cable connections per linear kilometer of boom. Assuming 110 days of ice loading per year, the average recording interval for the section cable loads is

$$r = \frac{1}{110 \cdot 16.4} = 5.54 \cdot 10^{-4} \text{ km} - \text{yr} \quad (6)$$

Expected section cable loads for the Yamachiche boom are summarized in Table 4 for selected return periods.

Table 4. Section cable loads predicted for various return periods.

T_R (km-yr)	1	2	5	10	20	50
Load (kN)	1153	1260	1401	1508	1615	1756

For a return period of 20 km-yr (i.e. 2 linear kilometers of boom over 10 years), the expected section cable load is 1615 kN.

These estimates of long-term section cable loads are subject to a similar degree of uncertainty as discussed previously in the context of anchor loads. The major source of uncertainty here is the short 2-year sampling period. Again, an appropriate safety factor must be applied to account for this uncertainty when specifying design loads.

CONCLUSIONS

The main conclusions of this work are summarised below.

- A large quantity of good information on ice boom loads was collected during the winters of 1994-95 and 1995-96.
- Anchor cable loads up to 1225 kN and section cable loads up to 1272 kN were recorded. The loads are highly variable over time. The maximum load is always much larger than the average. The largest loads can occur at any time throughout the season.
- Section cables equipped with large 710 mm diameter pontoons attract larger loads than those with smaller pontoons.
- The ice loading season can be subdivided into three segments: early winter freeze-up; mid-winter; and spring break-up; based on the character of the boom loads which reflect the type of ice interactions which occur most often during each part of the season.
- The boom loads feature some diurnal periodic loading caused by thermal stresses within the ice cover.
- Daily maximum loads in the anchor and section cables tend to follow the Gumbel distribution, which can be used to extrapolate expected maximum loads over the longer term.
- Extrapolation using the Gumbel distribution indicates a maximum anchor cable load of 1721 kN for a return period of 20 km-yr (i.e. a 2 km length of boom deployed over 10 years). The expected maximum section cable load for the same return period is 1615 kN. These predictions are based on quite limited data and must therefore be associated with a high degree of uncertainty.

ACKNOWLEDGMENT

This work was supported by the Canadian Coast Guard, Laurentian Region and the Canadian Hydraulics Centre of the National Research Council of Canada.

REFERENCES

Cornett, A.M. and G.W. Timco, 1995. *Model Tests of Three Ice-Control Booms*, Proc. Canadian Society for Civil Engineering Annual Conference, Ottawa, Canada, Vol. 1, pp. 87-96.

Foltyn, E.P. and A.M. Tuthill, 1996. *Design of Ice Booms*. Cold Regions Research and Engineering Laboratory, Technical Digest No. 96-1, Hanover, USA.

Sarpkaya, T. and M. Isaacson, 1981. *Mechanics of Wave Forces on Offshore Structures*. Van Nostrand Reinhold Publishers, Toronto, Canada

Timco, G.W. and A.M. Cornett, 1994. *Laboratory Tests of Ice Interaction with the Booms in Lac St Pierre*, National Research Council Technical Report, IECE-CEP-TR-003 (NRC No. 38471), Ottawa, Canada.

Timco, G.W. and A.M. Cornett, 1996, *Laboratory Tests of Ice Interaction with Steel Booms*, CSCE, Canadian Journal of Civil Engineering. Vol. 23, No. 2, pp. 560-566.

Proceedings of the Seventh (1997) International Offshore and Polar Engineering Conference
Honolulu, USA, May 25-30, 1997

ISBN 1-880653-28-1 (Set); ISBN 1-880653-30-3 (Vol. II)

Ridge Ice Loads on Proposed Faceted Conical Structure

Zhiguo Wang
Memorial University of Newfoundland, St. John's, Canada
Derek B. Muggeridge
Okanagan University College, Kelowna, Canada
Ken. R. Croasdale
K.R. Croasdale and Associates Limited, Calgary, Canada

ABSTRACT

In order to simplify fabrication and reduce construction costs, the oil industry has proposed using a multifaceted shape to replace a smoothly curved surface for conical structures. An extensive test program was carried out to study the mechanism of ice interaction with this newly proposed structure. A total of five structural models at scales of 1:50 to 1:10 and 31 ice ridges were tested in three Canadian facilities. This paper presents the results of an extensive analysis of these tests. Previous tests with smooth conical structures showed that ridges failed by a central crack and a hinge crack, and that the maximum load occurred during the hinge crack of a ridge. It is shown that the ridges for the present tests also failed in bending mode, but with three distinct crack patterns varying from classical central and hinge cracks to local circumferential cracks. Moreover, our tests indicate that the maximum global loads on the tested faceted cone often occur during the ice clearing process. The effect of the ridge width and structure orientation on crack pattern and maximum ridge loads is analysed. The influence of neck size on global loads is also investigated.

KEY WORDS

Model testing, ice, faceted, conical, offshore structure.

1. INTRODUCTION

Conical shaped structures can induce ice bending failure and this mode will exert much lower loads on the structure compared to ice loads from a crushing failure mode. Therefore conical shaped structures are preferred by some oil companies for arctic oil and gas exploration/production operations. The difficulties encountered in manufacturing a smoothly-curved conical structure (SCS) lead to a higher cost of construction. For ease of fabrication and cost savings, Exxon Production Research Company proposed the development of a multifaceted surface to approximate the smooth surface (Weiss 1988, Croasdale and Muggeridge 1993). This proposed multi-faceted conical structure (MCS) is comprised of three sections: a main cone, a collar, and a vertical neck. Figure 1 shows the prototypes of the proposed large neck structure (LNS) and small neck structure (SNS) both of which had six facets in all their three sections.

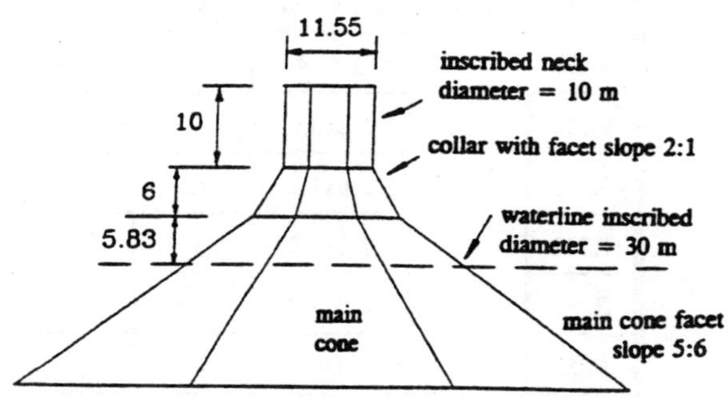

Prototype Small Neck Structure (SNS)

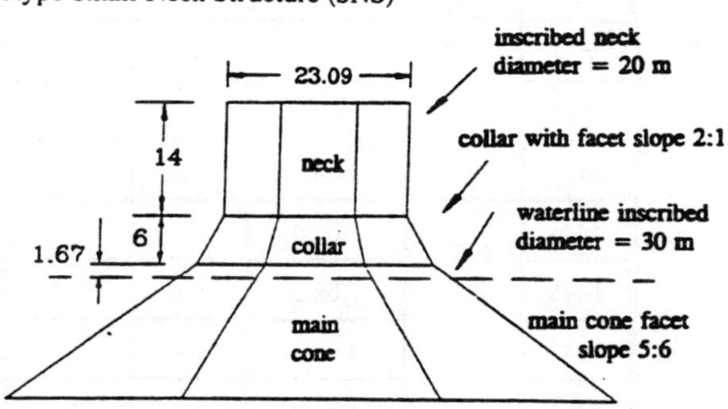

Prototype Large Neck Structure (LNS)

Fig. 1 Schematic of faceted cone (small and large neck structure). Typical full-scale dimensions are shown.

Five series of model tests were carried out in three Canadian test facilities: two series with structural models at scales of 1:10 and 1:20 in the outdoor ice basin of Esso Resources Canada (ERC), one series of tests with nominal scale of 1:50 at the ice tank of NRC-Ottawa, and another two series at scales of 1:25 and 1:50 in the ice tank of NRC's Institute for Marine Dynamics (IMD) in St. John's. Although the tests have been documented in several test reports (Metge and Weiss 1989, Metge and Tucker 1990, Irani *et al* 1992, Lau *et al* 1993a), only the data for IMD and NRC-Ottawa have previously

Tests		Ice Properties			Centre Crack Load (kN)		First Hinge Crack Load (kN)		Maximum Load and the Interaction Event		
No.	Stru. Scale	Dimen. (mxm)	σ_f (kPa)	E (MPa)	X_I	Z_I	X_H	Z_H	Load (kN) F_{mx}	Load (kN) F_{mz}	EML
R112	SNS 1:10	3.5x1.0	128	352	76	113	130	150	130	150	HC
R113		3.7x0.9	116	301	90	110	103	130	103	186	HC/R
R123	LNS 1:10	4.5x1.0	182	349	150	120	140	110	160	190	R/S
R124		3.5x1.0	162	295	-	-	135	145	149	170	R
R131		3.5x1.1	86	271	73	97	83	108	94.3	116	R/S
R133		3.5x1.3	150	281	56	70	80	91	125	138	R/S
R141		3.5x1.0	119	580	28	32	32	40	32	40	HC
R142		2.5x1.0	86	106	40	50	35	40	40	50	RC
R221	LNS 1:20	3.0x0.9	25	50	20	30	25	30	63	40	R
R222		2.5x1.90	40	80	32	25	28	35	72	60	R/FB
R231		2.8x.9	138	659	60	72	42	60	66.5	72	R/RC
R232		2.2x0.9	108	-	57	68	68	83	68	83	HC
R241		2.6x1.2	62	187	58	68	63	75	63	75	HC
R242		2.6x1.2	39	114	51	63	76	97	83	100	R
R251		2.0x0.3	20	-	7	8.5	-	-	7	8.5	RC
R252		2.0x0.3	41	263	12	15	-	-	12	15	RC
M33	SNS 1:25	1.0x0.4	99	77	21	21	10	13	21	21	RS
M34		1.0x0.3	49	75	15	15	-	-	15	15	RC
M36		1.0x.3 *	16	6	6	8	-	-	9	11	R/SC
M37		1x.4 *	33	5	10	11	12	13	12	13	RC/HC
M44	LNS 1:25	1.0x.4	112	105	26	24	-	-	26	24	RC/HC
M45		1.0x.3	69	207	14	12	14	13	14	13	RC/HC
M47		1x.3 *	27	7	10	9	15	14	15	14	HC
M53		1.0x.4	81	183	6	5	2	2	6	5	RS
M54		1x.3 *	11	2	8	6	-	-	8	8	RC/C
M61		1.0x.3	66	160	11	10	6	6	11	10	RC
M65		1x.2 *	12	11	7	8	7	6	8	9	3R/HC
M74	LNS 1:50	1.0x.3	136	282	15	15	6	5	15	15	RC
M75		3.0x.3	136	282	19	20	17	16	20	20	3R/HC
M77		1x.2 *	13	10	5	5	9	8	9	8	RC/C

Table 1 Ice Ridge Parameters and Maximum Loads for the ERC and IMD Tests

published (Timco *et al* 1993, Lau *et al* 1993b). The ERC's data are presented as new material and one of the core parts of this paper. In addition, the previous publications discussed only individual test series, this paper presents an analysis using all the ridge tests. Recently an extensive numerical simulation using the DECICE package has been undertaken (Wang *et al* 1997).

2. THE TEST PROGRAM

The ice and structures were supposed to be scaled down to the test scales mentioned above. For instance, ridge thickness and flexural strength should be one tenth of the full scale ridge thickness and flexural strength for the 1:10 tests. Croasdale and Muggeridge (1993) have summarized the full scale target for ice properties and achieved scaled values for all the test series. According to their analysis, the range of the scaled values roughly matched the range of the full scale target. The values presented here are those actually achieved in the tests. Details of the test techniques used were described in the individual test reports and related publications (Metge and Weiss 1989, Metge and Tucker 1990, Irani *et al* 1992, Lau *et al* 1993a, Timco *et al* 1993, Lau *et al* 1993b). An approach based on beam theory (Wood 1980) was applied in flexural strength and Young's modulus measurement and calculation. Global ice loads were measured using a set of load cells.

A total of 31 ice ridges were tested in the five test series. The main parameters of the ice ridges are listed in Table 1 for the ERC and IMD test series and in Table 2 for the IME series. The test numbers beginning with ``R'' (column 1 of Tables 1 and 2) denote the ERC tests, and those beginning with ``M'' and ``C'' are for the IMD and NRC-Ottawa tests, respectively. The structures used (LNS or SNS) and their model scales are identified in column 2 of Table 1 for the ERC and IMD tests. Only the SNS was used in the NRC-Ottawa test series, thus column 2 of Table 2 only gives the ratio of model diameter to prototype diameter at waterline level. The structural model used in the NRC-Ottawa tests was purposely raised out of the water to increase the loads on the structure, hence the actual scale at waterline varied from test to test, with average scale about 1:32 instead of 1:50 (Irani *et al* 1992).

Columns 3 of Tables 1 and 2 present ridge dimensions in form of width by thickness. The ridge length was 30 m, 12 m, and 7 m for the ERC, IMD, and NRC-Ottawa tests, respectively. Ridges M36, M37, M47, M54, M65, and M77 were constructed using a Dump Truck technique in which small ice pieces were dumped into ridge area to form a ridge (Lau *et al* 1993). These ridges which are to be referred to as ``Dump Truck'' ridges had a soft layer at their bottom. The total thickness including the soft layer was 50 cm for all these ridges, while the thickness of these ridges given in Table 1 is the thickness of the ridge core (specified with *). The remaining ridges used in the IMD tests were fully consolidated Split Layer (SL) ridges which were constructed by laying several layers of level ice strips one over the other (Lau *et al* 1993). A combination of these two techniques and others were used in construction of the ERC and NRC-Ottawa ridges. The ERC ridges can be considered to be fully consolidated, but most NRC-Ottawa's ridge also had a soft layer at bottom. The dimension given in column 2 of Table 2 is *width* x *total thickness (thickness of ridge core)* for the NRC-Ottawa ridges.

Columns 4 and 5 of Tables 1 and 2 present flexural strength σ_f and Young's modulus E, respectively. Saline ice was used in the ERC tests and EG/AD/S model ice (see Timco 1986 for details) was used for the IMD and NRC-Ottawa tests. The water density was about 1030, 1002, and 1002 kg/m^3 for the ERC, IMD, and NRC-Ottawa test series, respectively. The ice speed was approximately 6 cm per second for all the tests.

3. RIDGE FAILURE PATTERNS

Three ridge failure patterns have been identified from the video records and the information given in the test reports and were described in detail by Wang (1996). A brief summary is given in this section.

Ridge Failure Pattern I was characterized by distinct central and hinge cracks. A crack was initiated at or close to the centre line of the ridge. A pair of hinge cracks occurred a few seconds later at some distance away from the centre crack. It is believed that this would be the predominant failure pattern for multi-year ridges. The failure patterns of most of the ridges tested in the ERC and NRC-Ottawa series fall in this category. A typical final crack pattern is shown in Fig. 2A. Comparison of this failure pattern with that for a ridge against a SCS described by Lewis and Croasdale (1978) shows that they have many similarities. This means the faceted surface may not significantly affect the failure process of a ridge.

Ridge Failure Pattern II is illustrated by Fig. 2B. In this failure pattern, two radial cracks start at the edges of the structure's front facet. The hinge cracks that immediately follow the two radial cracks form in locations closer to the structure than those for failure pattern I. The pair of hinge cracks propagate and tend to converge toward one another till they meet in the sheet behind the ridge. The radial and hinge cracks break the central section of the ridge into three pieces, one with a trapezoidal shape on the front facet and the other two triangular pieces on either side of the front facet (Fig. 2B). This failure pattern was mainly observed from IMD's Split Layer ridges. ERC's ridges R251 and R252 as well as NRC-Ottawa's ridge C48 also failed in this failure pattern.

Ridge Failure Pattern III. The Dump Truck ridges tested in the IMD tank and NRC-Ottawa's ridge C64 failed in this pattern. As shown in Fig. 2C, a pair of radial cracks first form and then extend from the two corners of the front facet to the trailing edge of the ridge. At almost the same moment, a circumferential crack occurs in the ridge. These cracks break three relatively small pieces away from the ridge. A little later, the structure again contacts the new local front edge of the ridge and forms another circumferential crack which may still be within the ridge, or may extend into the interface of the ridge and the ice sheet behind. This crack breaks more pieces away from the ridge. This process may repeat once or twice, depending on the ridge's consolidation condition and its width.

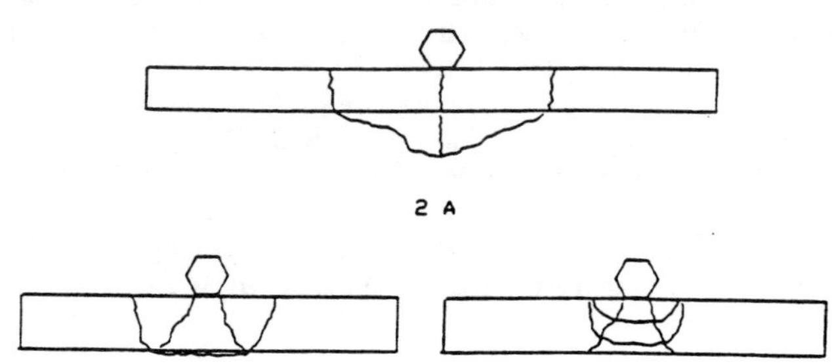

Fig. 2 Ridge Crack Patterns

4. WHAT CAUSES THE MAXIMUM RIDGE LOADS ?

Previous SCS tests (Lewis and Croasdale 1978) indicated that *the maximum force occurred when the ridge was cracked the second time (hinge cracks) at some distance away from the first crack.* The MCS tests showed that the maximum forces for a MCS model

Test No.	Model Scale	Ice Dimension (cm x cm)	σ_f (kPa)	E (MPa)	F_{mx} (kN)	F_{mz} (kN)
C45	1:32.3	58x16(9)	70	151	0.5	0.6
C46		72x15(10)	95	204	0.8	0.3
C47A		67x27(8)	265	570	0.8	0.7
C47B		109x25(8)	114	245	0.9	1.0
C48		67x13(7)	156	335	0.5	0.4
C49		105x16(8)	155	333	0.9	0.8
C51		63x17(8)	82	176	1.1	0.9
C52		60x15(6)	137	295	1.0	0.8
C58		64x25(6)	334	718	1.6	1.2
C59		100x15(10)	160	344	2.0	1.5
C63	1:31.2	64x18(5)	345	742	0.9	0.8
C64	1:32.3	100x20(8.5)	130	280	1.2	1.2
C65	1:31.2	62x35(9)	231	497	1.0	0.9
C66		100x42(12)	98	211	2.2	1.8

Table 2 Ice Parameters and Maximum Loads for the NRC-Ottawa Tests

occur during any of the three events: initial cracking, hinge cracking, and clearing process.

To clarify the relationship between the maximum loads and interaction events, the video records, the force records, and all the force plotting in the test reports have been reviewed and checked. The linking of the ice failure events and their corresponding loads through the recorded test time has been recognized for the ERC tests and the IMD tests. The initial crack and hinge crack loads together with the maximum ridge loads are listed in column 6 to 11 of Table 1 in which X_I and Z_I denote the horizontal and vertical initial crack loads, X_H and Z_H represent the horizontal and vertical hinge crack loads, while F_{mx} and F_{mx} denote the horizontal and vertical maximum loads, respectively.

The events causing the maximum loads (EML) were also identified and are given in the last column of Table 1. HC and RC mean hinge crack and radial crack, while C, R, S, FB, RS, 3R, and SC denote clearing of ridge segments (C), ridge fragment riding up (R), ridge pieces resting on the structure (S), ridge fragment breaking (FB), ridge separating from ice sheet (RS), the third radial crack (3R), and sheet cracking (SC), respectively. Generally, the events other than initial and hinge cracks (RC and HC) can be considered as particular events occurring during the ice clearing process. It can be seen in Table 1 that although radial and hinge cracks are still the main events causing the maximum ridge loads, nearly half of the ERC and IMD tests experienced their maximum ridge loads during the clearing process.

5. EFFECT OF VARIOUS PARAMETERS ON THE RIDGE FAILURE PROCESS AND MAXIMUM RIDGE LOADS

This section presents the analysis of the effects of three tested factors: structural or ridge orientation, size of the structure's neck, and ridge width. Two ridge orientations relative to the structure's facet face were tested, viz., (i) face-on orientation shown in Fig. 2A-2C, (ii) structural edge-on orientation and ridge oblique orientation showing in Fig. 3A and 3B, respectively. Since the structures used in the IMD tests were all in face-on orientation and the ridge widths were the same for all its tests except for M75, analysis of the effect of ridge orientation and width is based on the ERC and NRC-Ottawa tests; while the effect of neck size is analysed using the data from all the test series.

In the following analysis, the measured maximum forces are non-dimensionlized by $\sigma_f I/(z_t L_c)$ which is conventionally believed to be a combined parameter able to cover the effect of the members it contains (Abdelnour 1988). σ_f, I , z_t , and L_c are flexural strength, moment inertia of cross section, half thickness, and characteristic length of ridge, respectively.

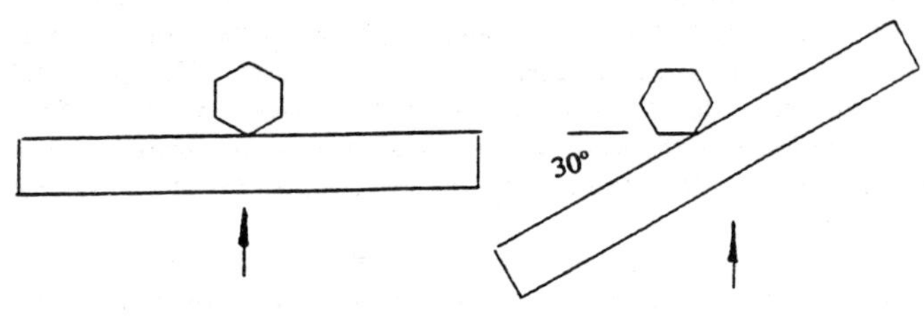

Fig. 3 Structural Edge-on Orientation (A) and Ridge Oblique Orientation (B)

5.1 Effect of Ridge Width

Of the five wide ridges (with width around 1 m) used in the NRC-Ottawa tests, three experienced a local circumferential crack formed within the ridges. In contrast to these wide ridges, the narrow ridges did not have local circumferential cracks. In the IMD tests, M75 was the only ridge that was wider than the others (three times wider than others, see Table 1). This ridge failed in failure pattern III and experienced a total of four circumferential cracks until it failed. In contrast, other ridges experienced no or just a few circumferential cracks. In the ERC tests, extremely wide ridges R123 and R221 were tested against 1:10 and 1:20 models, respectively. A common crack feature for these two wide ridges is that they had multiple radial cracks (near or within the edges of the front facet). This is very distinct from the other ridges's single radial crack pattern. Since the multiple radial cracks can also be considered as one type of local crack, we can say that the wider ridges had more local cracks than the narrower ridges.

To study the influence of ridge width on the maximum ridge

forces, the dimensionless forces are plotted against dimensionless ridge width (B_t/D_w) in Fig. 4 through 7 in which B_t and D_w denote ridge width and circumscribed waterline diameter of the structure, respectively. In these figures and those that follow, the points (x and o) are test data, the solid and dashed lines are from linear regression analysis. Since the effect of ridge width may depend on other parameters, viz. neck size, structural orientation, and force component, the data are sorted into eight groups, and two groups are plotted in each figure. For instance, the two groups of the data plotted in Fig. 4 are those vertical forces on LNS in face-on and edge-on orientations, respectively. It should also be noted that the edge-on orientation for the ERC tests is actually that in which the ridge was oblique at a 30° angle (Fig. 3B).

The data for the LNS are from ERC's tests, and those for the SNS are mainly from NRC-Ottawa's tests. The data from ERC's two SNS tests, R112 and R113, are also included in the face-on orientation for the SNS (Fig. 6 and 7). Because of the incompleteness in the data, tests C53, R251, and R252 are excluded from this analysis and from all the other analyses that follow.

Generally, the effect of ridge width is quite significant for the LNS but less significant for the SNS. For both edge-on and face-on orientations of the LNS, both the vertical and the horizontal components of the dimensionless force increase as the ridge width increases (Fig. 4 and 5). The vertical forces for the case of a face-on orientation of a SNS shows a similar trend (Fig. 6), whereas this effect is nearly negligible for the other cases, viz., the horizontal forces on a SNS in both orientations (Fig. 7) and the vertical forces for the edge-on orientation of the SNS (Fig. 6). A possible reason for the above trend may be that the wider ridges usually lead to large ice segments after their cracking, and these large segments are more likely to be stuck in front of the large neck, thus resulting in larger forces.

5.2 Effect of Structural Orientation

The location of the radial (basically, the initial crack) and the first pair of hinge cracks for the ERC and NRC-Ottawa tests are listed in Tables 3 and 4, respectively. In these tables, the dimensionless location of the cracks are expressed as the distance from the centre line of the ridge divided by the circumscribed waterline diameter of the structure, D_w. L and R in these tables denote ``on the left hand side'' and ``on the right hand side'' of the structure, respectively.

Tables 3 and 4 show that ridge orientation has an obvious effect on the location of the initial crack(s). In the ERC tests, every oblique ridge except R221 had a single initial crack starting from the contact point between the ridge and the structure's front corner. Even ridge R221 had one centrally located radial crack. On the other hand, only three (3) broadside ridges (which is equivalent to ``face-on'' orientation), out of eleven (11) in total, had initial cracks close to, but not at, the ridge's center line. Table 4 shows a similar trend in the NRC-Ottawa tests: every ridge in the edge-on tests except for C45 and C46 had a single central initial crack, whereas only four out of seven ridges, initially centrally cracked for the face-on orientation. Ridges C45 and C46 were not frozen completely. The location of the cracks were just the weak points where the ridge moulds were joined (Irani *et al* 1992). If these two tests are excluded from the present analysis, the initial cracks for NRC-Ottawa's edge-on tests would all occur at the centre line.

The structural orientation seems to affect the distance between the two first hinge cracks. The NRC-Ottawa tests indicate that this distance, on average, is smaller for the edge-on tests. As shown in Table 4, the average value of this distance for the face-on tests is about 33% larger than that for the edge-on orientation. However, the

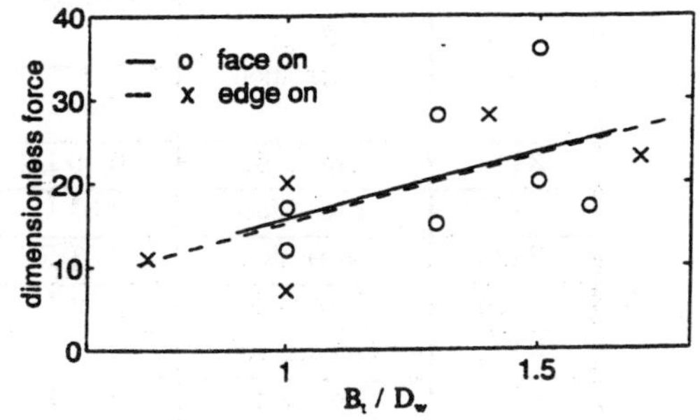

Fig. 4 Vertical Ridge Forces on the Large Neck Structure

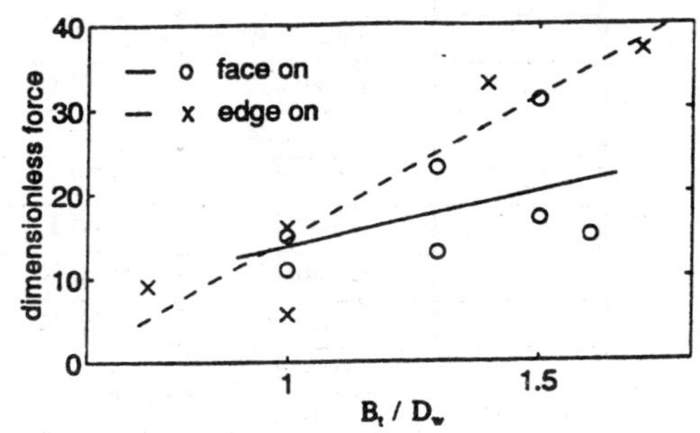

Fig. 5. Horizontal Ridge Forces on the Large Neck Structure

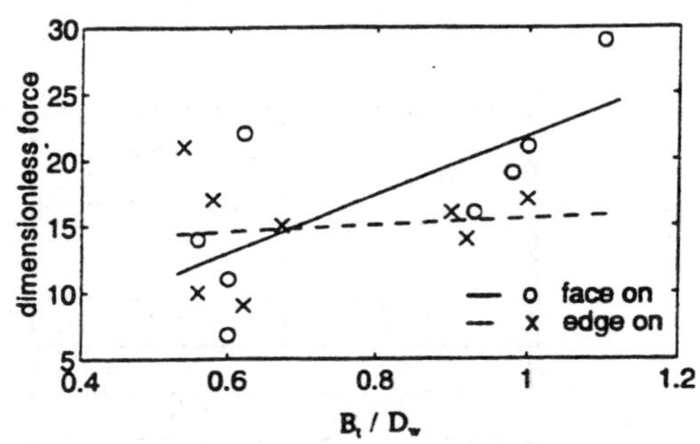

Fig. 6 Vertical Ridge Forces on the Small Neck Structure

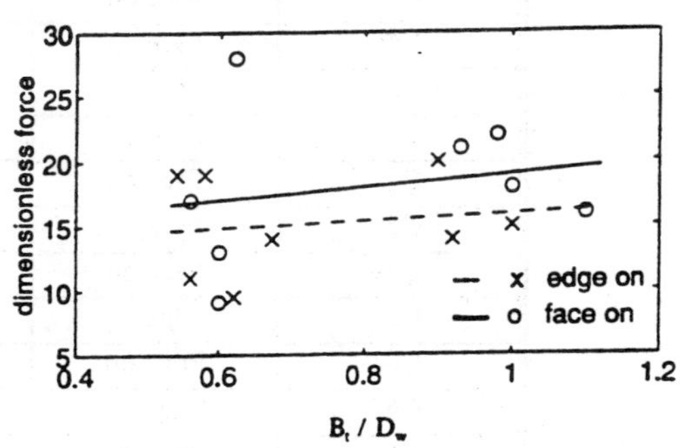

Fig. 7 Horizontal Ridge Forces on the Small Neck Structure

values for the ERC tests do not strongly support this trend. Table 3 shows this distance for ERC's broadside ridges on average is only 11% larger than that for the oblique ridges.

To identify its effect on ridge forces, the ridge (or structure) orientation must be separated from the other two factors: viz., neck

Test No.	Ridge Orientation	Dimensionless Radial Crack Location			Dimensionless First Hinge Crack Location	
		No. 1	No. 2	No. 3	Left	Right
R112	Broadside	0.20 L	-	-	1.76	2.17
R113	Broadside	0.03 L	0.12 R	-	2.05	1.59
R123	Broadside	0.03 R	0.25 L	0.25 R	2.48	2.48
R124	Broadside	0.23 L	-	-	2.45	2.51
R131	Oblique 30°	0	-	-	2.19	2.11
R133	Broadside	0.03 R	-	-	1.76	2.19
R141	Oblique 30°	0	-	-	0.95	1.96
R142	Oblique 30°	0	-	-	1.10	1.16
R221	Oblique 30°	0	0.87 L	0.35 R	4.22	4.39
R222	Oblique 30°	0	-	-	1.44	3.23
R231	Broadside	0.92 L	-	-	0.92	2.60
R232	Broadside	0.50 L	-	-	5.77	2.60
R241	Broadside	0.29 L	-	-	2.31	3.00
R242	Broadside	0.46 L	-	-	2.48	4.39
R251	Broadside	0.50 R	0.50 L	-	-	-
R252	Broadside	0.52 R	1.27 L	-	-	-
Average	Broadside	-	-	-	2.44	2.61
	Oblique	-	-	-	1.98	2.57

Table 3 Crack Location of the ERC Ridges

Test No.	Structure Orientation	Dimensionless Initial Crack Location	Dimensionless Hinge Crack Location	
			Left	Right
C45	Edge on	0.42 R, 0	1.12	1.45
C46	Edge on	0.42 R	1.12	1.45
C47A	Edge on	0	0.56	0.81
C47B	Edge on	0	0.80	1.03
C48	Face on	0.23 R, 0.23 L	0.93	1.17
C49	Face on	0.23 R, 0.32 L	1.73	1.40
C51	Face on	0	1.21	1.40
C52	Face on	0, 0.26 L	0.93	0.93
C53	Face on	0	1.63	1.82
C58	Face on	0	1.87	1.68
C59	Face on	0	1.59	1.87
C63	Edge on	0	0.90	0.63
C64	Edge on	0	1.31	1.12
C65	Edge on	0	0.72	0.72
C66	Edge on	0	1.76	1.80
Average	Face on	-	1.41	1.47
	Edge on	-	1.04	1.13

Table 4 Crack Location of the NRC-Ottawa Ridges

size and ridge width. The plots given in Fig. 4 through 7 were arranged to show the effect of orientation factor as well. Each of the figures is for a single neck size: viz., Fig. 4 and 5 are for the LNS, while Fig. 6 and 7 are for the SNS. In each of these figures, the forces for the two orientations can be directly compared for a given ridge width.

It can be seen from Fig. 4 that the orientation, on average, rarely has any effect on the vertical forces for the LNS because the regression lines for the two orientations nearly coincide. However, as indicated in Fig. 5, the edge-on orientation seems to induce larger horizontal forces on the LNS for $B_t/D_w > 1$. The SNSs show a reverse trend: the horizontal forces for the face-on orientation are larger than those for the edge-on orientation, see Fig 7. The case for the vertical forces on a SNS is a little complex. Fig. 6 shows that the forces for these two orientations are very scattered in the vicinity of dimensionless ridge width equal to 0.6 (narrow ridges). Though the regression line for edge-on orientation is above its counterpart for the face-on orientation, it is really hard to say which orientation experienced larger forces. In the region of $B_t/D_w = 1$ (wide ridges), the horizontal forces for the face-on orientation are obviously larger.

5.3 Effect of Neck Size

To analyze the effect of neck size, the mean values of dimensionless forces and their standard deviations are listed in Table 5. The data included in this table for the IMD series are those for the 1:25 scale models. The dimensionless ridge widths for these IMD tests were all the same: $B_t/D_w = 0.72$. In the ERC series, most tests were carried out with LNS in both face-on and edge-on orientations, and only two tests were run for the SNS. In the NRC-Ottawa series, only the SNS (in both orientations) was tested. Since the IMD tests cover both neck sizes and had a single value in ridge width, the results from this series are the most convenient and solid data for the study of neck size effect.

It is shown in Table 5 that the mean vertical force for the LNS in IMD's tests is smaller than its counterpart for the SNS, but the horizontal mean force for the LNS is slightly larger than (very close to) the forces on the SNS. Exactly the same trend is also found in the ERC tests for the face-on orientation in this table.

To utilize both the NRC-Ottawa and the ERC tests for this analysis, they are plotted in Fig. 8 and 9. The comparison of the dimensionless forces from these two series tests should be made at the same value of B_t/D_w. Since these tests share the values of B_t/D_w in the vicinity of $B_t/D_w = 1$, they are comparable in that region. These graphs show that the forces on the LNS in face-on orientation are quite close to (slightly smaller than) those on the SNS in the vicinity of $B_t/D_w = 1$. It should be noted that this trend is only in the average (or least square) sense and is supported by a small sample of the data. Looking at these figures more closely, one may note that the forces are quite close to each other for a few individual tests for both neck sizes.

One may argue that the scale factor played a role in the difference between the forces on the structures with different neck size. Although we cannot be completely sure that the scale factor does not play any role, the two points from tests R112 and R113 identified in Fig. 8 and 9 are somewhat reassuring that the role is small. These two ridges were tested with SNS at 1:10 scale. The two points for the LNS (represented with x) just below those for R112 and R113 in Fig. 7 and 8 are for tests R124 and R133 both of which were at 1:10 scale as well; thus they are directly comparable. It is shown in the figures that their forces are well above those for the LNS. The only exception is the horizontal force for R113 which might be smaller than its counterpart for the LNS if there was one.

The data for edge-on orientation are plotted in Fig. 10 and 11 for vertical and horizontal forces, respectively. Within the region of $B_t/D_w < 1.01$ in Fig. 10 and 11, the regression line for the SNS is higher than its' counterpart for the LNS. It is also shown in these figures that there is one case for the vertical and horizontal forces, respectively, wherein the force for the LNS is larger than the force for

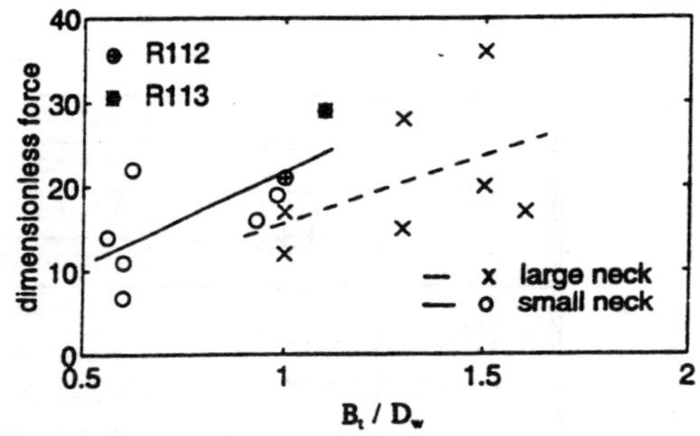

Fig. 8 Vertical Forces (Face-on Orientation)

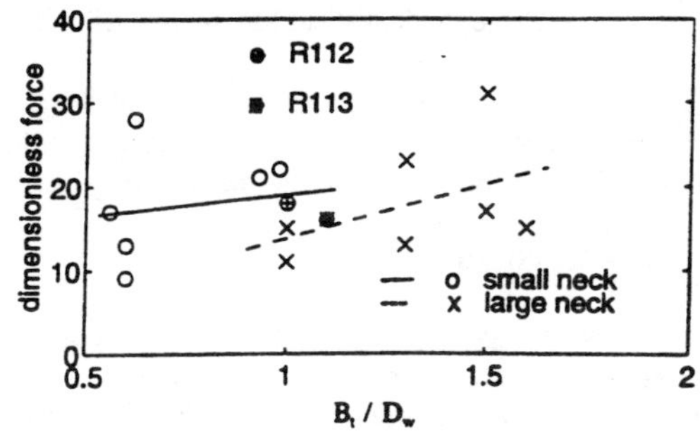

Fig. 9 Horizontal Forces (Face-on Orientation)

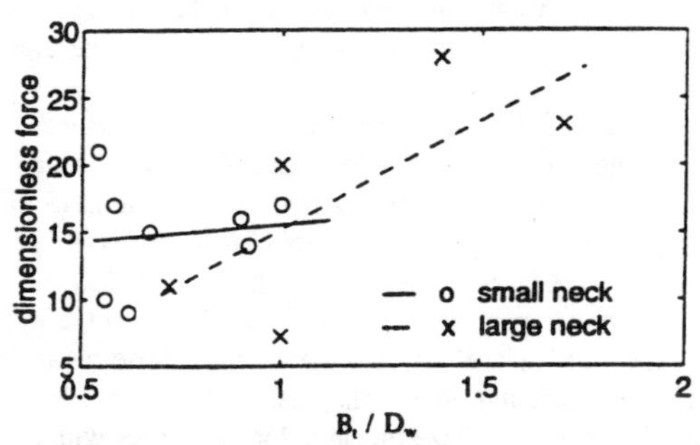

Fig. 10 Vertical Forces (Edge-on Orientation)

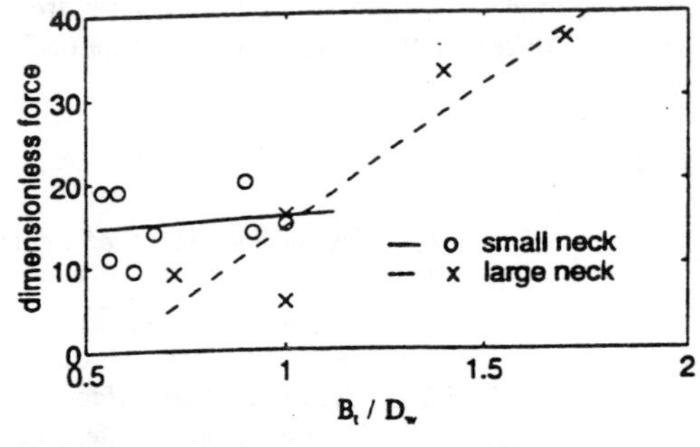

Fig. 11 Horizontal Forces (Edge-on Orientation)

Test Series	Structural or Ridge Orientation	Mean Value of Dimensionless Forces			Standard Deviation	
		Component	Large Neck	Small Neck	Large Neck	Small Neck
ERC	Face-on	Vertical	20.71	25.0	8.40	5.67
		Horizontal	17.71	17.0	6.60	1.41
	Edge-on	Vertical	17.84	-	8.58	-
		Horizontal	20.16	-	14.12	-
IMD	Face-on	Vertical	30.86	34.25	11.14	14.71
		Horizontal	32.29	32.25	11.00	14.59
NRC-Ottawa	Face-on	Vertical	-	14.80	-	5.48
		Horizontal	-	18.35	-	6.77
	Edge-on	Vertical	-	14.88	-	3.91
		Horizontal	-	15.19	-	3.87

Table 5 Statistical Measurements on the Neck Size Effect

the SNS whose value of B_t/D_w is the same as that for the LNS. Considering all the above, we would say that the forces on the edge-on LNS and SNS are relatively closer to one another.

6. CONCLUSION

The tested ridges failed in bending mode with three distinct patterns. Ridge width can also affect crack pattern. The wider a ridge, the more likely one or more local circumferential cracks will form. Ridge/ structure orientation affects the locations of cracks. The edge-on orientation more often induces a central crack along the centre line of the ridge and generates hinge cracks closer one another than face-on orientation. The maximum ridge load can be generated by any of three interaction events: initial crack, hinge crack, and clearing process.

Effect of ridge width, structural (or ridge) orientation, and neck size on the maximum ridge loads are analysed. As ridge width increases, both the vertical and horizontal loads on the LNS increase, the vertical loads on the face-on SNS also increase, while the horizontal loads on face-on SNS and the horizontal loads on SNS in both orientations are nearly unaffected. For a given ridge width, the loads on face-on SNS are basically larger than those on an edge-on SNS. Effect of orientation on the loads acting on LNS are not significant. Effect of neck size on ridge loads couples with structural orientation. For face-on orientation, the horizontal loads on LNS and SNS are close to one another, while the vertical loads on the LNS are slightly smaller than those on the SNS. If the structures are in edge-on orientation, the difference between the loads on the LNS and the SNS is small.

REFERENCES

Abdelnour, R., 1988, *Assessment of the Effect of Length and Orientation of Multiyear Ridge on the Force Acting on Conical Structures*, IAHR Ice Symposium, Sapporo, pp 59-75.

Croasdale, K.R. and Muggeridge, D.B., 1993, *A Collaborative Research Program to Investigate Ice Loads on Multifaceted Conical Structures*, Proceedings of the 12th International Conference on Port and Ocean Engineering under Arctic Conditions (POAC'93), Hamburg, August 17 - 20, pp 475 - 486.

Irani, M.B., Timco, G.W. and Muggeridge, D.B., 1992, *Ice Loading on a Multifaceted Conical Structure*, NRCC Report IME-CRE-TR-005, 615p.

Lau, M., Tucker, J.R., Jones, S.J. and Muggeridge, D.B., 1993a, *Model Ice Forces on an Upward Breaking Multifaceted Cone*, NRCC Report IMD-TR-1993-07.

Lau, M., Jones, S.J., Tucker, J.R. and Muggeridge, D.B., 1993b, *Model Ice Forces on an Upward Breaking Multifaceted Cone*, Proceedings of the 12th International Conference on Port and Ocean Engineering under Arctic Conditions (POAC'93), Hamburg, pp 537 - 546.

Lewis, J.W. and Croasdale, K.R., 1978, *Modelling the Interaction between Pressure Ridges and Conical Shaped Structures*, Proceedings of IAHR Ice Symposium, Part 1, pp. 165-196.

Metge, M. and Weiss, R.T., 1989, *Multifaceted Cone Tests, 1988-1989*, ERCL.RS.89.14.

Metge, M. and Weiss, R.T., 1990, *Multifaceted Cone Tests, 1989-1990*, ERCL.RS.90.11.

Timco, G.W., 1986, *EG/AD/S: a New Type of Model Ice for Refrigerated Towing Tanks*, Journal of Cold Region Science and Technology, Vol. 12, pp 175 - 196.

Timco, G.W., Irani, M.B., Funke, E.R., English, L.A., Carroll, L.B., and Arambarri, E., 1993, *Ice Load Distribution on a Faceted Conical Structure*, Proceedings of the 12th International Conference on Port and Ocean Engineering under Arctic Conditions (POAC'93), Hamburg, August 17 - 20, pp 607 - 616.

Wang, Z., Muggeridge, D.B., Prodanovic, A. and Chao, J.C., 1993, *Computation of Sheet Ice and Ridge Ice Forces on a Faceted Cone*, Proceedings of the 12th International Conference on Port and Ocean Engineering under Arctic Conditions (POAC'93), Hamburg, August 17 - 20, pp 627 - 638.

Wang, Z., Muggeridge, D.B., Jones, S.J. and Swamidas, A.S.J., 1997, *Numerical Simulation of Ridge and Sheet Ice Loads on a Proposed Faceted Conical Structure*, Proceedings of the 16th International Conference on Offshore Mechanics and Arctic Engineering. ASME Paper No. OMAE 97-207. Yokohama.

Wang, Z., 1996, *Ice Forces on a Multifaceted Conical Structure*, Ph.D. Thesis, Memorial University of Newfoundland, Canada, 248p

Weiss, R.T., 1988, *Proposed Ice Basin Model Test Program for a Multifaceted Conical Structure, to be Conducted by ERC, Internal Memo, Exxon Production Research to Esso Resources Canada Ltd.*, 24p.

Wood, K.N., 1980, *Conical Structure Test Program*, AOGA Project 113, IPRT. ME. 80.21. August, 1980.

Proceedings of the Seventh (1997) International Offshore and Polar Engineering Conference
Honolulu, USA, May 25-30, 1997

ISBN 1-880653-28-1 (Set); ISBN 1-880653-30-3 (Vol. II)

Vertical Ice Forces on Large-Diameter Marine Structures Under Water Level Changes

Takashi Terashima and Naoki Nakazawa
Pacific Consultants Co.
Sapporo, Japan

Akifumi Nishihata, Hideki Honda and Takaharu Kawai
Hokkaido University
Sapporo, Japan

Abstract

Vertical ice forces, generated by changes in water levels, often act on upright structures built in the ocean, in rivers, and in lakes, as well as on those constructed along beaches, in cold regions. They have pulled out piers of landing stages and bridges, and have collapsed navigation signs.

The theories and computation methods of how these vertical ice forces act on upright structures have been described by such references as Kerr, Christensen, Saeki, Nakazawa, and Terashima. However, few studies have been conducted on inclined large marine structures.

The goal of this study was to accurately compute ice forces to accommodate the design of large structures in frozen sea areas. We clarified the actions of ice forces depending on the form of a structure and on the presence or absence of ice adfreezing to it. We also suggested theoretical solutions and methods to compute vertical ice forces under various conditions.

1. Introduction

Offshore structures, constructed in cold regions, are affected by sea ice when the surrounding sea freezes or when ice floes reach them. Both vertical and horizontal forces can be exerted on these structures. Vertical (upward and downward) ice forces are caused by changes in tidal levels and sudden changes in water levels. Horizontal ice forces are created by such fluid forces as wind, current, and waves. Research on ice forces acting on structures in cold regions has been done by Kerr, by Christensen, and by Saeki, Nakazawa, and Terashima. However, little research clearly differentiated the influences of the following factors: the form of a structure and the presence or absence of ice adfrozen to a structure.

This paper examines methods to compute vertical ice forces on three special forms of large marine structures constructed in cold regions: cylindrical, conical, and inverted conical structures, incorporating the presence or absence of an ice sheet adfreezing to a structure and of flooding conditions.

2. Actions of Vertical Ice Forces

Vertical ice forces act differently depending on the form of a structure and by the presence or absence of ice adfrozen to a structure. Their actions can be summarized as follows:

① Cylindrical structures (Figure 1 (a) and (b)): when a structure is upright and is circular cross-sectional, vertical ice forces act only if an ice sheet is adfrozen to a structure.

② Conical structures (Figure 2 (a) and (b)): vertical ice forces act when an ice sheet adfreezes to a structure (a), and when an ice sheet is not adfrozen to a structure but the water level falls (b).

③ Inverted conical structures (Figure 3 (a) and (b)): vertical ice forces act when an ice sheet adfreezes to a structure (a), and when an ice sheet is not adfrozen to a structure but the water level rises (b).

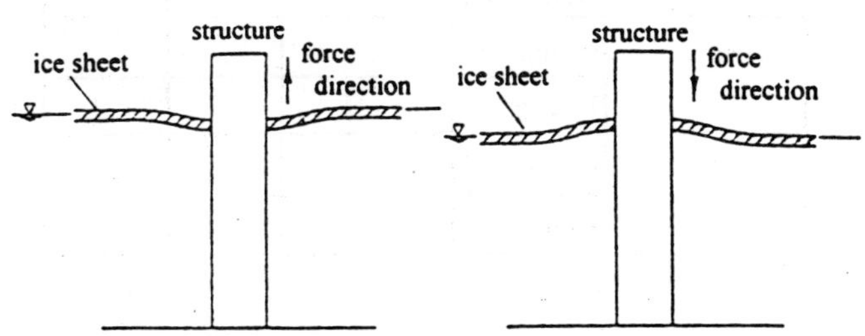

(a) Water level rising (b) Water level falling
Figure 1 Vertical ice forces acting on a cylindrical structure

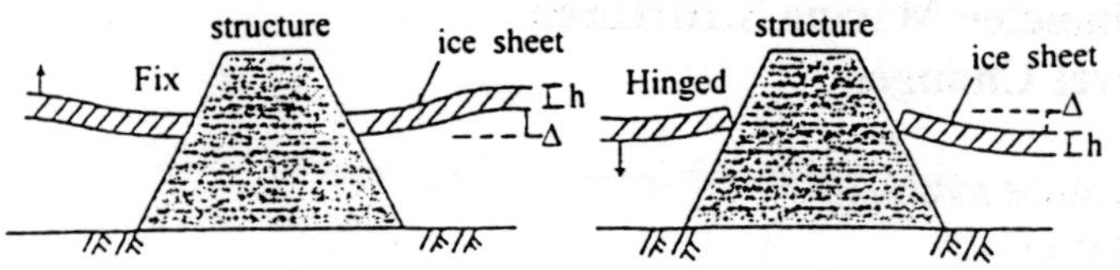

(a) Water level rising and ice adfrozen (b) Water level falling and ice not adfrozen

Figure 2 Vertical ice forces acting on a conical structure

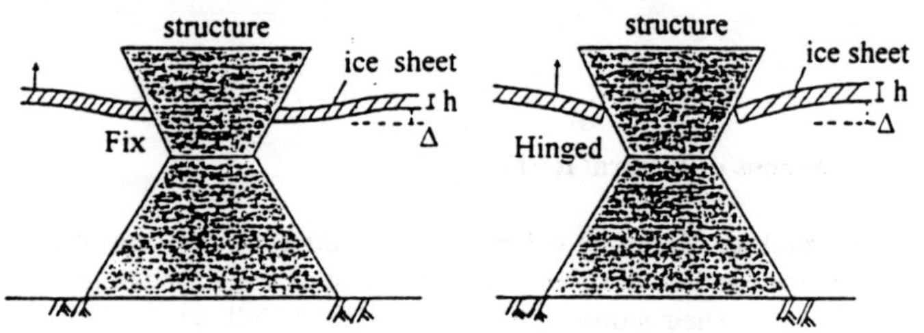

(a) Water level rising and ice adfrozen (b) Water level falling and ice not adfrozen

Figure 3 Vertical ice forces acting on an inverted conical structure

As shown above, the actions of vertical ice forces can be differentiated according to the form of a structure and the presence or absence of ice adfrozen to a structure.

Vertical ice forces act differently when the water level is rising and when it is falling. When it is rising, the occurrence of flooding also need to be considered.

The methods to compute ice forces under different conditions are presented in Table 1.

Table 1 Forms of structures and actions of vertical ice forces

Conditions \ Forms		Cylindrical	Conical	Inverted conical
Adfreezing ice	Water level			
Present	Rising	○*	○*	○*
	Falling	○	○	○
Absent	Rising	x	x	○*
	Falling	x	○	x

○: Ice forces act

x: Ice forces do not act

*: Flooding potential

3. Methods to Compute Vertical Ice Forces

3-1. Deflection of a floating ice sheet

The equation to compute the deflection of a floating ice sheet is:

$$D \bullet \nabla^4 W + Kw \bullet W = 0 \qquad (1)$$

where,
W is the deflection in distance r,
Kw is the unit weight of water, and
D is the flexual rigidity of an ice sheet, given by the following equation:

$$D = \frac{E \bullet h^3}{12(1 - v^2)} \qquad (2)$$

where,
E is Young's modulus of an ice sheet,
h is the thickness of an ice sheet, and
v is Poisson's ratio.

3-2. Vertical ice forces with ice adfrozen to a structure

Vertical ice forces are produced by changes in water level when an ice sheet is adfrozen to a circular cross-sectional structure (Figure 1). When the change is at a low speed, the structure is not subject to significant ice forces since an ice sheet exhibits viscoelastic properties. When the change is at a high speed, however, the structure is subject to considerable ice forces since an ice sheet displays elastic characteristics. Kerr [1)] formulated the following equation to compute the vertical ice force generated by a change in water level from the theory of elasticity of flat plates:

$$P = 2\pi a D \lambda^2 \Delta \frac{\{kei'(\lambda a)\}^2 + \{\ker'(\lambda a)\}^2}{k_1} \qquad (3)$$

where,
P is the vertical ice force exerted on a structure,
k_1=kei(λa)ker'(λa)-kei'(λa)ker(λa),
$\lambda=(Kw/D)^{1/4}$,
Kw is the unit weight of water,
a is the radius of the structure, and
△ is the change in the water level.

However, as the change in water level increases, either bending failure of an ice sheet is induced by the bending moment of the ice, or adfreeze failure is caused by shear forces acting on the area of contact between the ice sheet and the structure. Thus, both bending and adfreeze failures should be incorporated into the computation of vertical ice forces generated by changes in water level.

3-3. Vertical ice forces when ice is not adfrozen to a structure

Saeki et al. [4)] proposed the following equation to compute the vertical ice force when an ice sheet is not adfrozen to a structure:

$$P = 2\pi a D \lambda^2 \Delta \{F_1 kei(\lambda a) - F_2 \ker'(\lambda a)\} \qquad (4)$$

where,

$$F_1 = \frac{-\beta}{kei(\lambda a) \bullet \alpha - \ker(\lambda a) \bullet \beta}$$

$$F_2 = \frac{\alpha}{kei(\lambda a) \bullet \alpha - \ker(\lambda a) \bullet \beta}$$

$$\alpha = \frac{1}{a}\lambda \ker'(\lambda a) - V\lambda^2 kei(\lambda a) - \frac{v}{a}\lambda \ker'(\lambda a)$$

$$\beta = \frac{1}{a}\lambda kei'(\lambda a) + V\lambda^2 \ker(\lambda a) - \frac{v}{a}\lambda kei'(\lambda a)$$

3-4. Vertical ice forces under flooding conditions

As Figure 4 demonstrates, when the water level rises and the flooding conditions are present, the deflections of the ice sheet in flooded areas ($r \leqq r_0$, W_1) and in non-flooded areas ($r > r_0$, W_2) are described by:

$$\nabla^4 W_1 = -Kw \bullet \frac{nh}{D} \tag{5}$$

$$\nabla^4 W_2 = -Kw \bullet \frac{W_2}{D} \tag{6}$$

where,

$n = (1 - \frac{Ki}{Kw})$, and

Ki is the unit weight of an ice sheet.

The general solutions of equations (5) and (6) are obtained from equations (7) and (8), respectively:

$$W_1(r) = A_1 + A_2 \log(r) + A_3 r^2 \log(r) + A_4 r^2 - \frac{Kw \bullet nh}{64D} r^4 \tag{7}$$

$$W_2(r) = B_1 ber(\frac{r}{l}) + B_2 bei(\frac{r}{l}) + B_3 \ker(\frac{r}{l}) + B_4 kei(\frac{r}{l}) \tag{8}$$

where,

l (the characteristic length) = $(D/Kw)^{1/4}$

Since the boundary condition ($W(r) = 0$ as $r \to \infty$) gives $B_1 = B_2 = 0$, equation (8) can be simplified as:

$$W_2(r) = B_3 \ker(\frac{r}{l}) + B_4 kei(\frac{r}{l}) \tag{9}$$

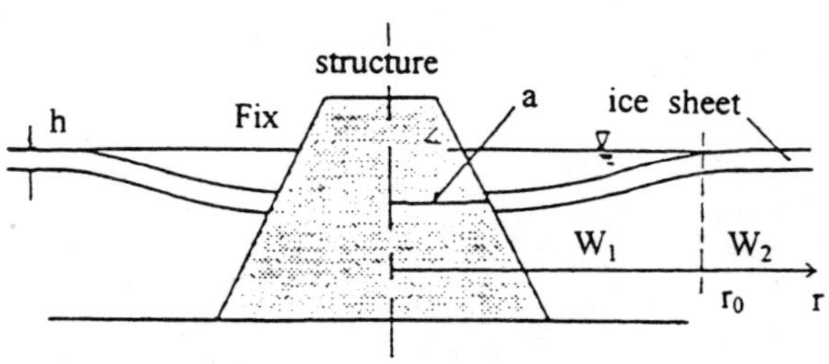

Figure 4 A conceptual figure to compute vertical ice forces when the water level is rising

When the water level is falling and flooding is present, general solutions for the deflection of the floating ice sheet in the flooded areas ($r \leqq r_0$, W_1) can be obtained from the following equation (for $r > r_0$, W_2, use equation (9)):

$$W_1(r) = A_1 + A_2 \log(r) + A_3 r^2 \log(r) + A_4 r^2 \tag{10}$$

(1) When ice is adfrozen to a structure

The following equations give the boundary conditions when an ice sheet is adfrozen to the structure, where a and r_0 are the radius of the structure and the distance to the boundary between the flooded and non-flooded areas, respectively:

$$W_1(a) = \Delta \tag{11}$$

(when r = a, the deflection is equal to the change in the water level)

$$W_1'(a) = 0 \tag{12}$$

(when r = a, the inclination is equal to zero)

$$W_1(r_0) = nh \tag{13}$$

(when $r = r_0$, the deflection is constant)

$$W_2(r_0) = nh \tag{14}$$

(when $r = r_0$, the deflection is constant)

$$W_1'(r_0) = W_2'(r_0) \tag{15}$$

(when $r = r_0$, the inclination is constant)

$$W_1''(r_0) + \frac{v}{r_0} W_1'(r_0) = W_2''(r_0) + \frac{v}{r_0} W_2'(r_0) \tag{16}$$

(when $r = r_0$, the r directional moment is constant)

$$\frac{d}{dr}\left[W_1''(r_0) + \frac{1}{r} W_1'(r_0)\right] = \frac{d}{dr}\left[W_2''(r_0) + \frac{1}{r} W_2'(r_0)\right] \tag{17}$$

(when $r = r_0$, the r directional shear force is constant)

(2) When ice is not adfrozen to a structure

When an ice sheet is not adfrozen to a structure, the following equation and equations (12)-(17) describe the boundary conditions:

$$W_1''(a) + \frac{v}{a} W_1'(a) = 0 \tag{18}$$

(when r = a, the bending moment is equal to zero)

The seven unknown variables, A_1 - A_4, B_3, B_4, and r_0, can be solved by the above equations for the boundary conditions. However, because r_0 is a variable of Kelvin's function, it seems rational to take a limit by assigning an approximate value of r_0. In this method, if seven unknown quantities are known, the shear force (Q) and the vertical ice force (P) under flooded conditions can be computed from the following equations:

$$Q = -D \bullet \frac{d}{dr}\left[\frac{d^2 w}{dr^2} + \frac{1}{r}\frac{dw}{dr}\right]_{r=a} = -D \bullet \left[\frac{4A}{r} - \frac{Kw \bullet nh}{2D} \bullet r\right]_{r=a} \tag{19}$$

$$P = -2\pi a \bullet Q \tag{20}$$

4. Conclusions

The computational results of equation (3) are demonstrated in Figure 5 (a). The vertical axis of Figure 5 (b) presents Pf/Pn, where Pn and Pf are the ice forces given by equations (3) and (20), respectively. The radius of the structure is 30-500 cm, and the ice thickness (h) is 50 cm. According to Figure 5 (a), the ice force from equation (3) is approximately 70 tf when the radius is 250 cm and the change in the water level is 10 cm. On the other hand, under flooded conditions the ice force from equation (20) is about 64% of the value from equation (3) according to Figure 5 (b). Thus the ice force is approximately 45 tf.

References

(1) Kerr A. D.: Ice Forces on Structure due to a Change in Water Level; Proceedings, IAHR Third International Symposium on Ice Problems, Hanover, 1975.

(2) F. T. Christensen: Interaction Between Floating Ice Sheets and Vertical Structures due to Water Level Fluctuations; Tech, Univ. of Denmark, S. Paper 5, No. 38, 1986.

(3) Nakazawa N., Yamada M., Hamanaka K., Saeki H.: Vertical ice Forces Exerted on Upright Circular Cross-Sectional Structures; Proceeding of Cold Region Technology Conference, 1995 (in Japanese).

(4) Enoki K., Terashima T., Yamazaki K., Nakazawa N., Saeki H.: Vertical Ice Forces Caused by Water Level Changes Exerted on Inclined Structures; Symposium on Okhotsk Sea and Sea Ice, 1991 (in Japanese).

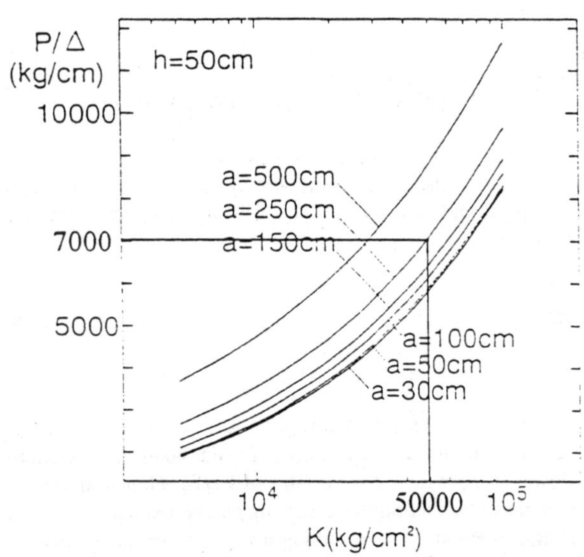

Figure 5 (a) The computational results of equation (3)
(h = 50 cm)

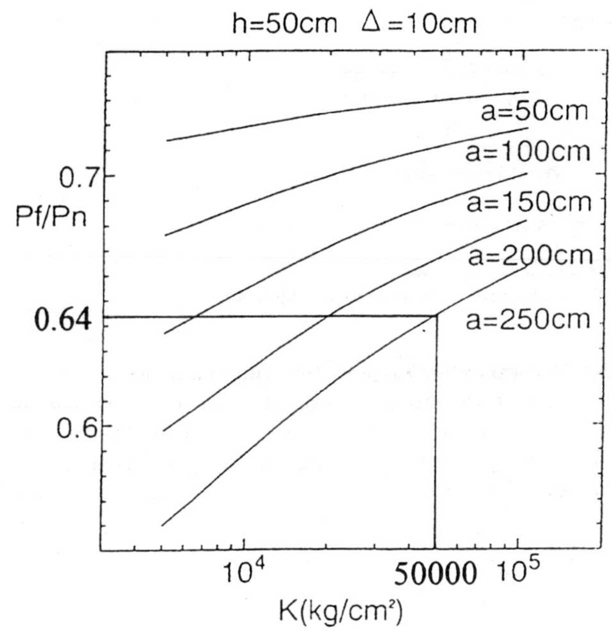

Figure 5 (b) The relationship between Pn/Pf and K
(the effects of a structure's radius)

Proceedings of the Seventh (1997) International Offshore and Polar Engineering Conference
Honolulu, USA, May 25-30, 1997

ISBN 1-880653-28-1 (Set); ISBN 1-880653-30-3 (Vol. II)

Method of Ridge Ice Force Analysis on Offshore Structures

Alexander T. Bekker, Olga A. Komarova and Sergey L. Vasiljev
Far-Eastern State Technical University
Vladivostok, Russia

ABSTRACT

The ice-structure interaction problem is investigated in terms of ridges action on oil and gas exploration structures with the aim of development of a reliability analysis method. Comparison analysis of forces from ridges according to the existing methods is made. Recommendations for the improvement of extreme ice force estimation procedure against offshore structures on northern seas are given.

KEY WORDS: Ice, ridge, force, analysis, Sakhalin.

INTRODUCTION

One of the important problems under exploration of freezing sea shelf is a supplement of exploitation reliability of offshore structures. The drifting ridge formations or ridges rows are most dangerous for structures in offshore zone of northern Sakhalin. Accounting for ridge influence is most complex problem in design of oil and gas fields. The existing recommendations for practice with semi-empirical nature have no strict theoretical base. It is explained, first, by lack of information about morphological and strength parameters of ice features, second, by complexity of determination and forecasting of integral physical and mechanical characteristics of ridges, third, complexity of ridge modeling and associated ice-structure interaction.

The following problems must be solved under definition of ridge ice forces. First, design scheme of ridge and its geometrical and strength parameters must be defined on the base of analysis of ice situation in the construction site. For definition of statistical relationship between ice topography of upper and bottom parts of ice fields with ridges the sonar survey and laser profiling are used. The thickness investigation of consolidated part of drifting ridges was made with a help of mechanical and thermal drilling. Data gathering about structure of bottom ice cover surface are expensive. This problem of internal ridge structure study is new and badly investigated accounting for high labor capacity and high cost of field works and test technology.

Large scale of ice ridges is a general complexity to obtain the integral strength parameters of these studied objects. The strength definition procedure of model ridges which are formed according to similarity criteria from ice pieces, has some disadvantages. They are consist in complexity of real modeling of cohesion between ice pieces, that is essentially influence on test results.

Second, the real available physical picture of ridge-structure interaction need to determine. Ridge failure scheme near structure depends on:

- dimensions and shape of structure and ridge;
- piece dimensions and their strength characteristics defining mainly the internal ridge structure,. i.e. density, nature and number of contacts between pieces;
- ridge consolidation level depending on time and temperature conditions ridge;
- drifting ridge kinematic parameters.

The investigation methods of ice ridges as objects acting on offshore structures at present time, as a rule, include the surface survey, underice works, drilling, ice sample study, level survey etc. But the expensive field works need to study the failure picture of ridges near structure.

The ice force field measurement is greatly complex accounting for small number of existing structures in freezing seas. Therefore the attention of researchers at last time is directed on creation of computer simulation ridge-structure interaction models on the base of field data about ice regime and mathematical ice-structure interaction models. The problem of ice-structure interaction development is sufficiently wide described in literature (Korzhavin, 1962; Croasdale, 1980; Nevel and Sisodiya, 1995).

At third, the probabilistic characteristics of extreme ice forces accounting for probability of ridge and structure collision need to receive for structure reliability estimation. The probabilistic distribution of ice forces is described by parameters of ice cover and it dynamics for all exploitation period of structure. Especially the 'tail" of drifting ridges distribution greatly influence on structure reliability evaluation. The ice force probabilistic characteristics of first-year ridges can be defined with probabilistic characteristics of parameters influencing on ice force value: dimensional, kinematic, physical and strength parameters. All these parameters can be considered as random values (Nevel and Sisodiya, 1995).

The obtaining of probabilistic characteristics of extreme ice forces at present time is complex problem and need additional special investigations. The accumulation of statistical information about geometrical and strength parameters of ridges, which influence on value of ice force, is difficult considering the absence of continuous field observations of these parameter changeableness. But at last years these investigations were managed in most perspective areas on shelf of northern seas.

The ridge-structure interaction simulation model need to develop for description of interaction. The simulation model allow to determine the probabilistic characteristics of extreme ice force. The field observations of geometrical, physical, mechanical and kinematic parameters are used for formation of ridge drift simulation model in-situ. It is based on probabilistic distributions of these parameters. But definition of impact probability of ridge with structure is important. Therefore the simulation model must account for probability of ridge impact with structure. The probability evaluation problem of impact of drifting ridge with structure is correctly given by (Nevel and Sisodiya, 1995).

So, the problems of failure mechanism definition of ridges near structure and probabilistic characteristics of geometrical, physical and strength parameters of ridge layers are most bad investigated for evaluation of offshore structures reliability.

The authors made an analysis of known ice force calculation procedures on the base of corresponding ridge-structure interaction models with different structural diameters. The suggestions were given for improvement of force calculation procedure for first-year ridges for northern Sakhalin conditions.

ICE FORCE EVALUATION FROM RIDGE FORMATIONS

Update there is no accurate mathematical solutions for definition of ice pressure on structure. The Building Codes (1989) recommend without sufficient basis to use the ridge factor which increase the force in comparison with the level ice. It is taken as 1.3÷1.5 of level ice thickness and does not depend from strength and dimensional ridge parameters. This is convenient but must be considered as tentative caused by lack of information about real nature of the ridge ice-structure interaction.

The analysis methods of ridge ice forces up to date allow only to approximately evaluate the force values. For example, in the works (Croasdale, 1980; Maattanen and Mauri, 1983; Mellor, 1980; Hoikkanen et al., 1984; Eranti and Lee, 1981; Prodanovic, 1981) a recommendations for practice are given mainly on the basis of static equilibrium equations of homogeneous non-cohesive - cohesive media of Mohr-Coulomb (scheme of Croasdale, 1980; Maattanen, 1983; Mellor, 1980; Hoikkanen, 1984; Eranti, 1981; Prodanovic, 1981). The obtained mathematical dependency on the basis of these schemes for ridge pressure definition on the offshore structure includes parameters which characterize the ridge strength properties: cohesion c and internal friction angle φ. This theoretical approach is well-founded because the material formed during sea ice failure relates to kind of discrete, non-cohesive media. The difference between calculation results by different researchers is explained by the choice of failure schemes. Disadvantages of this interaction scheme are, first, due to neglecting friction forces between structure and ice, second, the "two-dimensional problem" is considered although the stress-strain state is evidently three-dimensional.

SUGGESTED ICE RIDGE FORCE DEFINITION

The ridge observations in Sakhalin offshore allow to develop a structural ridge model with main geometrical characteristics. According to the study the thickness of ridge consolidated part can achieve of six thickness of pieces forming the ridge. Naturally to suppose that the ridge force main part will be depend on consolidated part with extreme strength characteristics in comparison with keel and sail. But the forces forming by keel and sail also need to consider. So, the total ridge force includes three components: sail force F_s, underwater (keel) force F_k and average (consolidated) part force F_c. The average force can be determined as force from level ice sheet during ice crushing near structure with vertical face according to well-known procedure (Korzhavin, 1962).

The force from ridge keel F_k can be defined on the base of admission that the ridge keel material is considered as cohesive homogeneous media. The force is defined as passive resistance according to Sokolovski (1960) method on the base of accurate theory of limit stress state. This method is based on admission that the shifting ground in any region before structure is fully in limit stress state (Sokolovski, 1960). The pressure calculation according to Sokolovski method essentially differ from Coulomb method which is adopted in all known procedures. As a result, the damage surface have curvilinear shape and force value is lower. The known solution from soil mechanics can be used for calculation of soil passive pressure on vertical structure face:

$$F_k = \sigma_{n.r.} = \sigma_y \lambda_p + \frac{c}{tg\varphi}(\lambda_p - 1) \qquad (1)$$

where $\sigma_y = \sum_1^n \gamma_B \Delta y_i + q$ is vertical stress on depth y; $\gamma_B = g(\rho_w - \rho_i)$; ρ_w and ρ_i are water and ice density relatively; λ_p is coefficient of horizontal component of passive pressure according to Sokolovski (1960) accounting for keel material friction on structure.

The three-dimensional work of ridge keel can be accounted for by a known procedure using design experience of offshore structures:

$$j = 1 + 2H_k/3B \qquad (2)$$

Then the ridge keel force is defined as:

$$F'_k = F_k \cdot j \qquad (3)$$

This procedure uses the strength parameters of ridge such as internal friction angle φ and cohesion c. But direct definition of ridge strength characteristics is greatly difficult because the characteristics φ and c are defined mainly from shear tests and depend on technical possibilities of experimental devices.

One of the available methods of strength parameters evaluation for solid ridge and its components (sail and keel), is calculation of so called integral strength of ridge in natural conditions (Truskov et al, 1992).

This work is the first attempt to determine the ridge integral strength. The compressive strength according to procedure of Truskov et al (1992) relating to drifting ridge near Sakhalin is 0.80 MPa, 0.34 MPa for sail and 0.30 MPa for keel.

CALCULATION RESULTS

The authors made an analysis of ice ridge force calculations with well-known interaction models of first-year ridges with structures of different diameters. The schemes of Croasdale (1980); Mellor (1980); Hoikkanen (1984); Eranti (1981) and also the scheme accounting and accepting the three-dimensional factor for ridge keel material j are taken. All calculations were made for one-year ridge 1% providence which drift on the Okhotsk sea with following parameters (see Fig.1): H_t = 24 m; H_s = 3 m; H_k =21 m; h_L = 3.45 m; c = 0.58 kPa; φ = 21°; B_s = 20 m; B_k = 57 m; n_k=0,26.

Fig. 2 shows the calculation results. The comparison analysis of calculation procedures show that the maximum force value for the same data give the scheme of Mellor (1980) and minimum - scheme of Croasdale (1980). According to results the total ridge force evidently mustn't to be equal the sum of maximum values of ridge components because it is unlikely that these values will reach the maximum simultaneously.

Secondly, the practice usage of suggested procedures difficulted by absence of necessary amount of initial data about geometrical, kinematic and strength ridge parameters. Besides it, the lack of this data make difficult the adequate ridge model (Croasdale, 1980; Hoikkanen et al., 1984; Eranti, 1981; Prodanovic, 1981).

CONCLUSIONS

1. The development of reliable recommendations in practice for definition of ice forces from the ice ridge is difficult by lack of field data about kinematic, geometrical and strength parameters of the ridges and also complexity of it modeling.
2. The existing analysis methods of forces on structures have some disadvantages: lack of theoretical basis of ridge models, does not reflect of physical interaction picture of ridges with structure, which doesn't account the properties relating to ridge ice.
3. The probabilistic characteristics of ice force from the drifting ridges must be obtained for reliability evaluation of offshore structures. It is necessary to define the probabilistic characteristics of kinematic, geometrical, physical and strength ridge parameters, develop the simulation model of ridge-structure interaction reflecting the real ridge failure picture near structure and accounting for probability of its impact with structure.

REFERENCES

Korzhavin, KN (1962). The Ice Action on Engineering Structures", Novosibirsk, *Siberian Div. of USSR Science Academy Publ.*, 203 p (in Russian).

Croasdale, KR (1980). "Ice Forces on Fixed Rigid Structures". New Hampshire, US Army Corp. of Engineers, *Cold Regions Research and Engineering Lab.*, Report 80-26, 145 p.

Nevel, D, and Sisodiya, R (1995). "Methods for Selecting Design Ice Forces". *The 2nd Int.Conf. on Development of the Russian Arctic*, Calgary, Canada.

Maattanen, M., and Mauri, P (1983). "Design Recommendations for Ice Effects on Aids-to-Navigation", *3rd draft,* Onlu, IALA Technical Committee to study the effect of ice on light-houses, 15 p.

Mellor, M (1980). "Ship Resistance in Thick Brash Ice", *Cold Region Sci. and Tech.*, pp 305-321.

Hoikkanen, J, Krankkala, T, Mattanen M, and Pulkkinen, E (1984). "Calculation Methods for Loads Against Offshore Structures". ATS, ESPOO, Report 4, 160 p.

Eranti E., and Lee, GC (1981). 'Introduction to Ice Problems in Civil Engineering". *Dept. of Civil Eng. and Techn., State Univ. of New York,* Buffalo, 194 p.

Prodanovic, A. (1981). "Upper Bounds of Ridge Pressure on Structures". *6th Int.Conf. on Port and Ocean Eng. under Arctic Conditions.* Quebec, Canada, Publ. Univ. Laval Quebec, pp 1288-1302.

Sokolovski, VV (1960). "The Static of Bulk Soils". Moscow, Physmatgis Publ. 243 p. (in Russian).

Truskov, PA, Beketski, SP, Surkov, GA, Polomoshnov, AM, and Bekker, AT (1992). "Strength Parameters of Hummocks". 2nd ISOPE Conf., San-Francisco, USA, vol.II, pp 783-789.

Building Codes (SNiP) 2.06.04-82* (1989). "The Forces and Actions (Waves, Ices, from Ships)". Moscow, Stroyizdat Publ. 39 p. (in Russian).

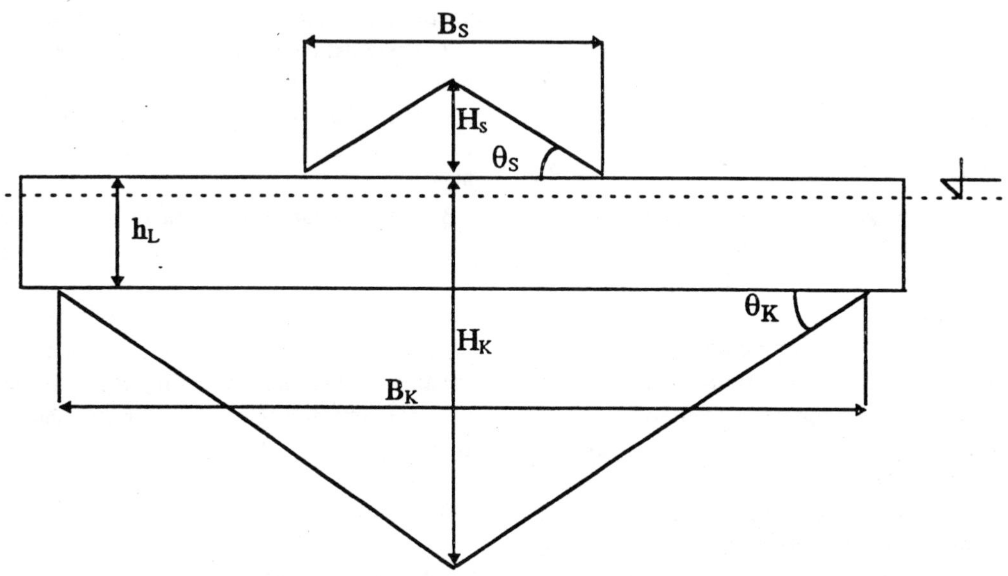

Fig. 1 The design scheme of drifted first-year ice ridge in Okhotsk sea.

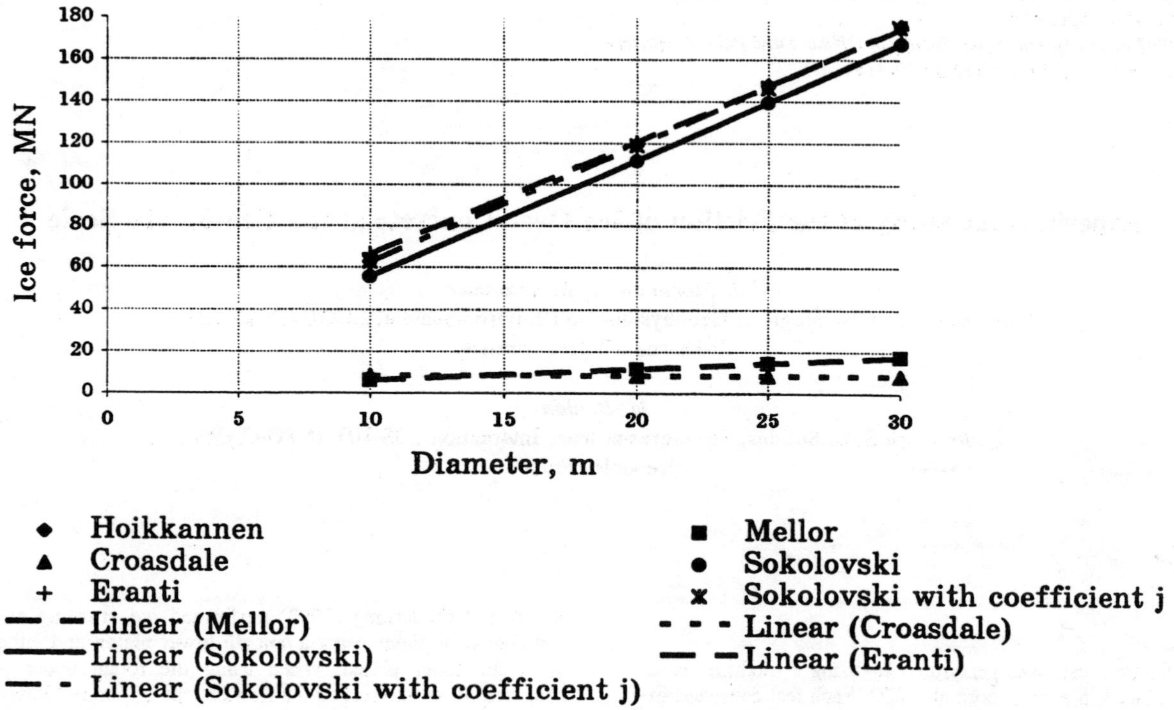

Fig.2. The chart of calculation results from different models.

Proceedings of the Seventh (1997) International Offshore and Polar Engineering Conference
Honolulu, USA, May 25-30, 1997

ISBN 1-880653-28-1 (Set); ISBN 1-880653-30-3 (Vol. II)

Experimental Study of the Friction of Ice Over Concrete at the Centimetre Scale

B. Fiorio and J. Meyssonnier
Laboratoire de Glaciologie et Géophysique de l'Environnement, LGGE-CNRS-UJF, St-Martin-d'Hères, France

M. Boulon
Laboratoire Sols, Solides, Structures et leurs Interactions, 3S-UJF-INPG-CNRS Grenoble, France

ABSTRACT

A series of friction tests was performed by using a modified shear-box apparatus placed in a cold room at -10°C. Each test consisted in a series of cycles during which a concrete plate was pulled then pushed at a constant velocity under a sample of pure ice, loaded normal to the concrete plate. The variables were the overall normal stress, in the range 25 - 800 kPa, the plate velocity varying from 0.1 to 10 mm.min^{-1}, and the conventional R_a roughness from nearly 0 to 0.6 mm. The analysis of the results shows that for given plate velocity, normal stress and plate roughness, the friction coefficient increases with time owing to the change in the real contact areas between ice and concrete. The stabilized friction coefficient increases with the plate velocity and roughness, and decreases with the normal stress as a result of the transition from adhesion to dynamic friction. Observed phases of stick-slip are discussed. Optical observation of the ice after the tests showed that the involved mechanisms are mainly visco-plastic deformation, recrystallization and micro-cracking of ice.

KEY WORDS: Ice, Concrete, Friction, Friction coefficient, Roughness.

INTRODUCTION

Offshore structures in polar regions are submitted to ice forces caused by the drifting of ice floes and ice fields under the action of winds and currents. The interaction between the ice and the structure is characterized by the fracture of ice and by the extrusion of small ice particles produced by crushing (Jefferies and Wright, 1988). The transmission of load from the ice to the structure can take place either by direct contact between undamaged ice and structure, or through a layer of crushed ice.

Experimental observations by Tuhkuri (1995) on the contact of a flat plate pushed against an ice sheet show the formation of a contact line which forms by spalling of ice at the edge of the ice sheet. Usually, on both sides of this line a layer of crushed ice is being pushed out of the contact zone. According to Zou and others (1996), spalling affects particularly the low confining pressure and the tensile zones, and friction between the end of the forming ice spall and the structure may be influential in the spalling process. Also friction could be partly responsible for the confining pressure observed near the ice-structure contact and could play a role in the extrusion of the crushed ice layer. Sayed and Frederking (1992) performed ice crushing and extrusion tests between plates using a smooth lower platen and either a smooth or rough upper platen. When using the rough upper platen, they observed different crushed ice flows at the upper and lower interfaces, caused by the difference in the boundary conditions. The pressure measured on the lower platen versus the thickness of the crushed layer was also greater under the same condition. A numerical simulation of these experiments confirms the role of friction on the confining pressure inside the crushed layer (Savage and others, 1992).

Studies of ice forces on structures at larger scales show that ice forces are generally cyclic (see for instance Kärnä and Turunen, 1989 ; Masterson and Frederking, 1993 ; Meaney and others, 1996). This is generally interpreted as a consequence of a cyclic damage regime of the ice cover upstream the contact zone (Jordaan and Timco, 1988 ; Kärnä and Turunen, 1989). However a stick-slip regime of friction at the ice-structure interface could explain changes in the contact pressure leading to cyclic ice forces. Ice is known to exhibit such a stick-slip behaviour under certain conditions (Bowden and Hughes, 1939 ; Barnes and others, 1971 ; Beeman and others, 1988).

As friction conditions at the interface appear to be an important factor in the ice-structure interactions, it was thought necessary to characterize more accurately the friction of ice against a structure by studying the different parameters which affect the ice-structure friction, and by achieving a better understanding of the physical mechanisms involved at the centimetre scale. To this aim a series of friction tests between ice and concrete was performed.

FRICTION TESTS

Experimental set-up

Ice-concrete friction tests were performed by using a modified direct shear box machine designed for soil mechanics studies. The apparatus, shown on Figure 1, was placed in a cold room at a constant temperature of -10°C ± 0.5°C .

The ice samples were given a cylindrical shape. Each ice cylinder was installed into a brass sample holder (60 mm in diameter, 60 mm height) then placed over a 175 x 150 mm plate of concrete. A small gap of 1 mm between the concrete plate and the sample holder was maintained by sustaining the sample holder with a vertical steel blade the ends of which were knife-edge shaped.

The concrete plate was fixed on a carriage which moved by

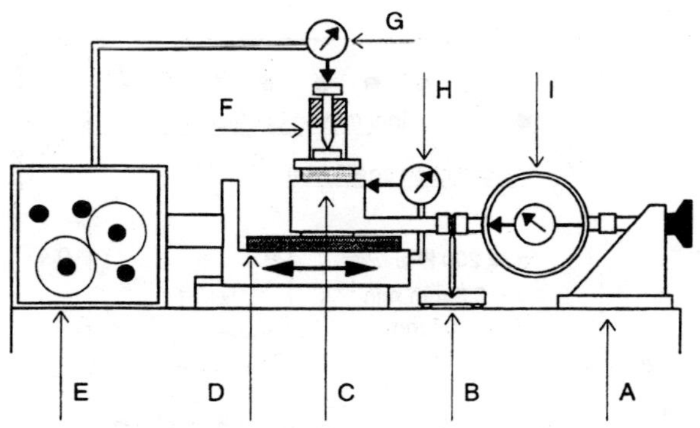

A : frame of the testing machine
B : steel blade
C : ice sample in sample holder
D : concrete plate on carriage
E : gearbox
F : application of normal load

Measurement of :

G : vertical displacement
H : horizontal displacement
I : tangential load

Fig. 1. Direct shear box machine.

horizontal translation along an approximately 30 mm stroke. This movement was reversed automatically when reaching the end of the travel in one direction. The carriage was driven at a constant velocity during the alternate translation. A gearbox allowed to select different plate velocities in the range 0.002 $mm.min^{-1}$ to 10 $mm.min^{-1}$, however the friction tests presented here were performed at five fixed velocity levels of 0.1, 0.5, 1, 5, and 10 $mm.min^{-1}$. A normal stress, measured with a load transducer, was imposed on the ice-concrete interface through a lever which was operated by an electromechanical device controlled by a computer.

Two spring type dynamometers were used to record the tangential and normal forces exerted on the ice specimen. The tangential stress was measured within ± 7.5 kPa, and could not exceed 880 kPa. The corresponding load transducer was inserted between the sample holder and the fixed frame of the machine. The maximum allowed normal stress on the ice-concrete interface was 1700 kPa. The accuracy of the corresponding transducer was within ± 4 kPa, but the computer control of the normal stress added a 4 kPa error owing to the oscillation of the actual stress around the prescribed value.

Normal and tangential displacements of the ice over the concrete plate were measured by means of two LVDT's, within ± 0.01 mm for the normal and ± 0.1 mm for the tangential displacements. The acquisition of the displacement measurements was made through a computer, at a rate up to approximately 20 Hz.

Concrete plate preparation and roughness measurement

Concrete plates were prepared by casting concrete, made of 15% water , 25% cement, 30% sand (grain size from 0.2 to 0.6 mm) and 30% small gravel (grain size from 3 to 5 mm) into a mould. A four weeks maturation period was imposed to ensure that concrete has reached its nominal resistance before performing the friction tests.

Denoting by z the altitude of the concrete surface along a longitudinal profile, and by $< z >$ the mean value of z over the length L of the plate, the average roughness of the concrete plate surface along the corresponding profile was defined as the mean value of $| z - < z > |$ over L.

The altitude z of the plate surface was measured by using a profilometer. This profilometer consisted in an hinged arm tapering down to a spherical tip 1 mm in diameter, which came into contact with the concrete surface. A displacement transducer recorded the vertical movement of the tip while the plate, attached to the machine carriage, was moved horizontally.

Mean values of the plate roughness R_a together with the average slope and the average plate thickness were obtained by an average over six longitudinal profiles chosen to represent the whole friction surface of the concrete plate.

Acceptable reproducibility of the initial shape and roughness of the plate surface was obtained by using three different specially designed moulds. One of them was used to obtain a smooth surface of reputed average roughness $R_a = 0$ mm, the two others were designed to obtain a random average roughness R_a of 0.11 and 0.28 mm. For a few preliminary tests, two other plates were used, one made of hand-shaped pure cement paste with a 0.05 mm roughness, the other made of concrete with a hand-made surface giving an average roughness of approximately 0.5 mm.

Preparation of the ice specimens

The ice samples were made of laboratory grown S2 columnar ice with grain diameter 5 mm. The samples were stuck on a metallic platen then machined on a lathe to obtain cylindrical specimens 60 mm in diameter and 95 mm long. The diameter was adjusted so that the specimen fit the sample holder while being allowed to slide along its axis direction. The tests were performed with columnar ice specimens machined in such a way that the long direction of the grains was perpendicular to the sample axis and to the direction of motion of the concrete plate.

The cylindrical face of the specimens was coated with silicon grease to prevent ice sublimation and water vapour recondensation between the ice specimen and the sample holder. The horizontal faces perpendicular to the cylinder axis were made perfectly smooth.

Testing procedure

Before installing the ice specimen, six profiles were measured to obtain the concrete plate average roughness. The mean slope was used to control that the mean plane of the plate surface was horizontal. The change in roughness of the concrete plate was measured at the end of each test.

The aim of this study was to observe the dependence of friction on the following parameters :

- Nominal applied normal stress, which can differ from the actual normal stress acting at the real ice-concrete contact areas. The tests were made to investigate the 25 to 800 kPa range.
- Nominal velocity of the ice relative to the concrete plate, which was varied from 0.1 to 10 $mm.min^{-1}$. This nominal velocity is only an average value because the stiffness of the dynamometric ring (1057 $N.mm^{-1}$), which measures the tangential load exerted on the ice specimen, causes the actual relative velocity to fluctuate around the value imposed to the carriage which supports the concrete plate.
- Roughness of the concrete plate. The whole series of friction tests was performed with a plate average roughness R_a in the range 0 to 0.6 mm. This parameter was partially controlled by using different moulds for preparing the concrete plates. Nevertheless, the wearing of the concrete surface led to small changes in the roughness of each plate from test to test. These changes were detected when measuring the plate roughness before each test.

At the end of each test photographs of the plate and of the ice surface were taken. Visual observations of the ice and the plate were done with an optical microscope, to study the different deformation patterns at the ice concrete interface.

Two types of tests, which differed according to the normal loading history, were performed.

Constant load tests

Constant load tests were run by prescribing the normal load on the ice specimen. When the value of the normal load was reached, the horizontal displacement of the concrete plate was started at the selected velocity. When reaching the end of its stroke, the movement of the concrete plate was reversed to come back to the initial position. One to twenty four cycles were done before stopping the experiment. Most of the tests were done within ten cycles. A typical friction curve is shown on Figure 2.

Level load tests

Level load tests were run with a 500 kPa normal stress prescribed for the first five cycles. This five cycles were necessary to reach a stable level of friction. Then the normal stress was decreased down to 25 kPa for one cycle, and the following cycles were performed under an increasing level of normal stress, respectively 50, 100, 200, 400 and 800 kPa. This latter cycle was repeated, and the next cycles were done following the reverse decreasing levels of normal stress, down to 25 kPa. The whole sequence, i.e. normal stress increase from 25 to 800 kPa then decrease from 800 to 25 kPa, was repeated twice for a few tests.

RESULTS

The results of 76 tests performed on S2 columnar ice are given in terms of the friction coefficient f, calculated as the ratio of the tangential stress σ_t to the normal stress σ_n.

Accuracy of the measurements and reproducibility of the tests

The accuracy of the measurements was good for the tests performed under a medium to high level of normal stress ($\sigma_n \geq 150$ kPa); the error on the friction coefficient was less than 10% for a normal stress higher than 300 kPa. For lower normal stresses, less than 100 kPa, this error increased rapidly with decreasing normal stress and a decreasing level of friction, the normal stress having the major influence. The error on the friction coefficient could reach 100% for the lowest normal stress and friction coefficient i.e. $\sigma_n = 25$ kPa and $f = 0.4$.

Tests reproducibility was generally good provided that the roughness of the concrete plate is sufficiently well controlled. However it was very difficult to perform two tests with the same plate roughness. Even if the initial roughness of a new plate was relatively well controlled by using the same mould, from the definition adopted for the roughness R_a and the way it was obtained, small difference in the real plate surface were unavoidable. Moreover, even if it were possible to achieve two identical plates, the evolution of the roughness during a test could not be exactly the same from test to test.

A compromise was achieved by performing both level load and constant load tests. Level load tests were necessary to ensure that the plate roughness changes were negligible for friction under different levels of normal load, while constant load tests were necessary to study the evolution of the friction coefficient with time and the plate roughness. It is worth to notice that for medium and high level of normal load, greater than 100 kPa, both types of test gave the same level of friction coefficient. For normal stresses less than 100 kPa, level load tests gave a higher friction coefficient than constant load tests. Such a behaviour can be explained by the differences in the histories of the actual contact areas between the ice and the asperities of the plate experienced during the two types of tests. In constant load tests the real contacts experienced only low normal stresses, whereas in level tests a higher normal stress was always imposed during the first five cycles.

Nevertheless the accuracy and reproducibility of the measurements were thought reasonably good enough to give reliable information on the evolution of the friction coefficient all along the life of the contact, and on the influences of the velocity, the normal stress, and the plate roughness on the friction coefficient.

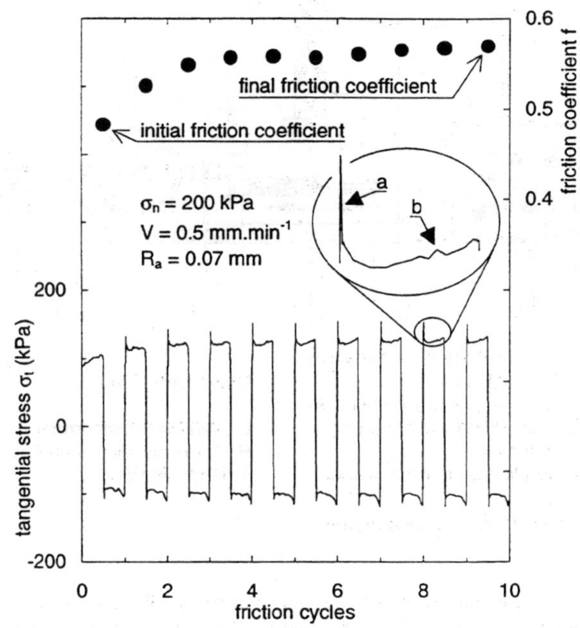

Fig. 2. Measured tangential stress and calculated friction ; a : stick peak ; b : local perturbation of friction by geometric defect of the plate.

Data analysis

Rough data gave information on the behaviour of the ice-concrete contact during the tests. A numerical treatment of the data was necessary to compare the different tests. The analysis of the friction coefficients computed for each cycle of a same test led to characterize each test by the average friction coefficient for the first and the tenth cycle. These values were obtained for each cycle by dividing the tangential stress averaged over the cycle by the nominal normal stress.

Evolution of friction over the cycles

Figure 3 shows the evolution of the average tangential stress versus the number of cycles for tests performed at different velocity levels under a normal stress of 200 kPa. The roughness of the plate was 0.07 mm. All the tests exhibit the same general behaviour, that is an increase of friction with the number of cycles (or with time). The friction coefficient increases from its initial value, computed with the value of the tangential stress averaged over the first cycle, up to a steady "final" value.

The increase of friction cycle after cycle can be explained by taking into account the visco-plastic deformation of ice under the imposed normal stress, possibly accompanied by micro-cracking. This deformation allows the asperities of the plate to penetrate into the ice and tends to increase the actual areas of contact between ice and concrete, leading to the increase in the tangential stress. At low velocity, more time is allowed to develop the deformation of ice in the direction perpendicular to the concrete plate, and the friction coefficient reaches its steady value during the first two cycles, whereas at the highest velocity this maximum is hardly obtained for the tenth cycle.

Influence of normal stress

Figures 4 and 5 show the friction coefficient f, obtained for level load tests, plotted versus the inverse of the normal stress, for two

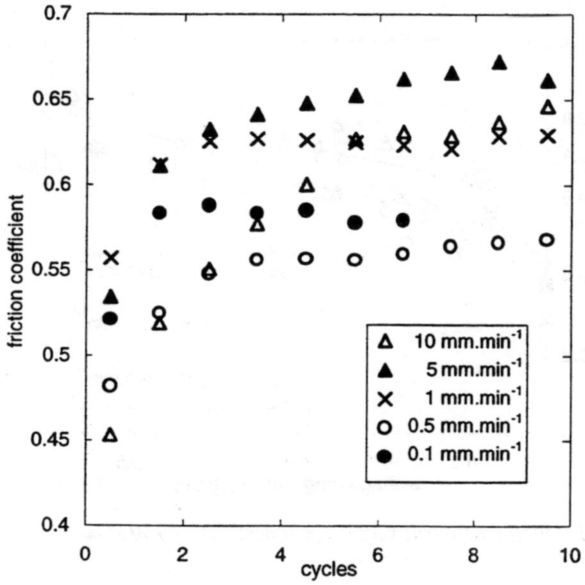

Fig. 3. Friction coefficient versus number of cycles. σ_n = 200 kPa, 0.06 < R_a < 0.08 mm.

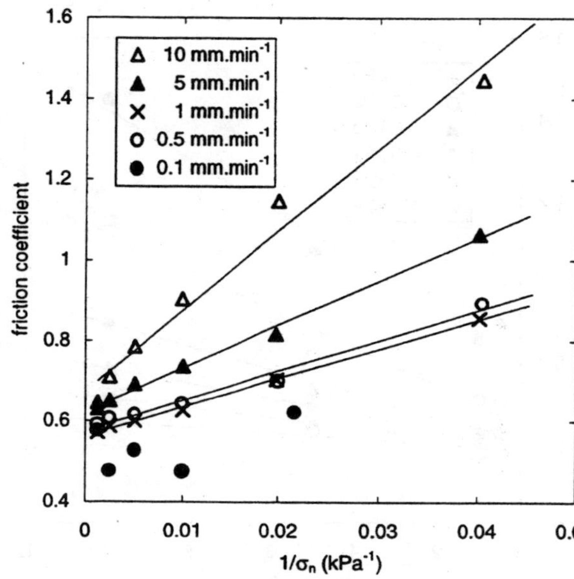

Fig. 4. Friction coefficient versus normal stress ; $f = C/\sigma_n + \mu$; 0.06 < R_a < 0.08 mm.

roughness ranges. Experimental data can be fitted by a law similar to Coulomb's law for internal friction of geomaterials, that is $\sigma_t = C + \mu.\sigma_n$, where σ_t and σ_n are the tangential and normal stress respectively, C is the adhesion or cohesive strength and μ is a friction coefficient. According to this relation the friction coefficient f defined as $f = \sigma_t / \sigma_n$ is given by $f = \mu + C/\sigma_n$, which means that f should be dependent on the normal stress. This dependency is really sensible at low normal stress, less than 150kPa. It was observed for all the tests except one (see Figure 5, dotted line) for which the change in roughness during the test was greater than that generally observed. Visual examination of the plate at the end of this test revealed an important wearing of the concrete plate, probably caused by a defect in the concrete composition. This resulted in a great number of torn up sand and cement particles which probably changed the nature of the contact by reducing the effect of the cohesive strength C. These particular data are given only for information.

Influence of the plate velocity

For constant load tests, the initial value of the friction coefficient, computed with the value of the tangential stress averaged over the first cycle, as well as the steady "final" value obtained after a number of cycles, are increasing functions of the velocity.

Figures 6 and 7 show the influence of the velocity on the friction coefficient calculated for level load tests. At low roughness (R_a < 0.08 mm), f is well represented by a linear function of the velocity in the range 0.5 to 10 mm.min^{-1}. For roughness greater than 0.1 mm, the linear approximation is acceptable for velocities less than 5 mm.min^{-1}. Data for 10 mm.min^{-1} plotted on Figure 7 correspond to the test shown by a dotted line on Figure 5 and cannot be considered as representative.

Influence of R_a roughness

Figure 8 shows the evolution of the friction coefficient with the roughness R_a. These results were obtained for the tenth cycle of the constant load tests performed under a normal stress greater than

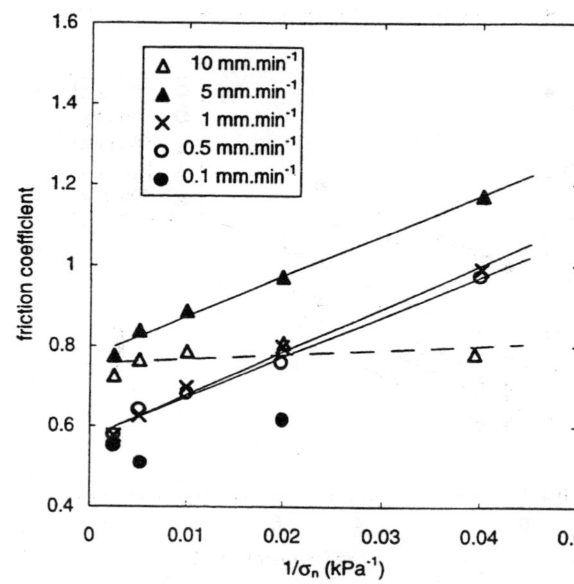

Fig. 5. Friction coefficient versus normal stress ; $f = C/\sigma_n + \mu$; 0.1 < R_a < 0.6 mm.

150 kPa. A strong increase of the friction coefficient was observed when the roughness passed from 0 (smooth plate) to R_a = 0.1 mm. For roughness higher than 0.1 mm, f kept increasing with R_a, but at a lower rate (0.3 mm^{-1} instead of 3 mm^{-1}). Increasing the velocity increases the friction coefficient, with no marked change in the rate of increase of f with R_a. As f becomes very sensitive to the normal stress when it is less than 150 kPa (see Figures 4 and 5), the corresponding values were not plotted on Figure 8, for clarity. However, the values of the friction coefficient derived from the constant load tests are consistent with that obtained in level tests, that is f increases with the velocity and decreases with σ_n.

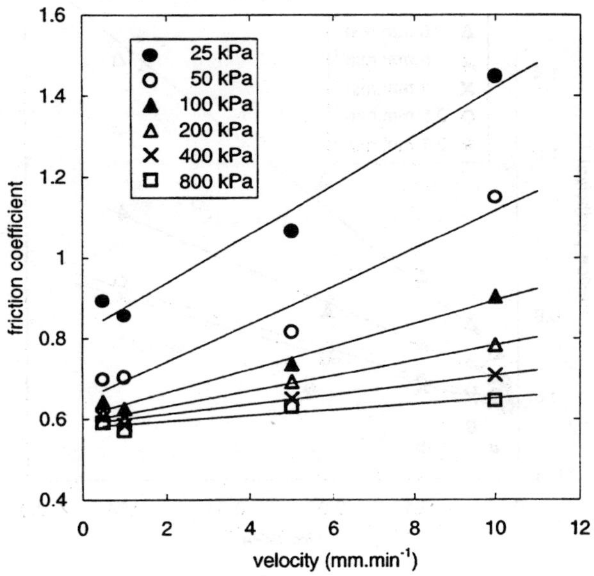

Fig. 6. Friction versus velocity ; $0.06 < R_a < 0.08$ mm.

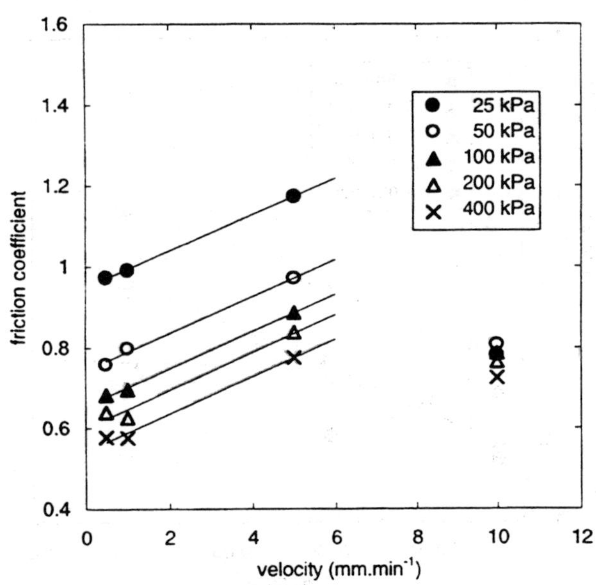

Fig. 7. Friction versus velocity ; $0.1 < R_a < 0.6$ mm.

Evolution of friction during a half-cycle

Figure 2 shows the typical evolution of the tangential stress versus time during a half-cycle. Stick peaks are visible at the beginning of each half-cycle (see Figure 9a). This is the consequence of the adherence of ice on concrete during the change in the direction of motion of the carriage. The compliance of the dynamometric ring and the small play between the ice and the sample holder allowed the ice to stop for a short lapse of time relative to the concrete plate, allowing adhesion to occur. The adhesion bond broke when the tangential stress reached a sufficiently high level. Bond breaking was brittle at high velocity and ductile at low velocity (see Figure 9a).

Stick-slip caused by momentary adhesion of ice over concrete was

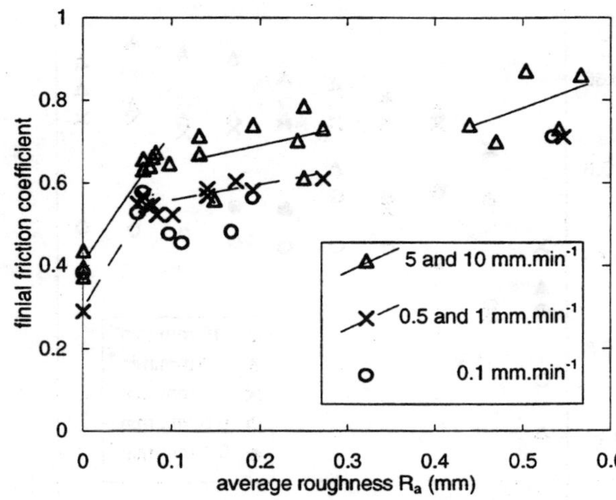

Fig. 8. Friction coefficient versus roughness ; $\sigma_n > 150$ kPa.

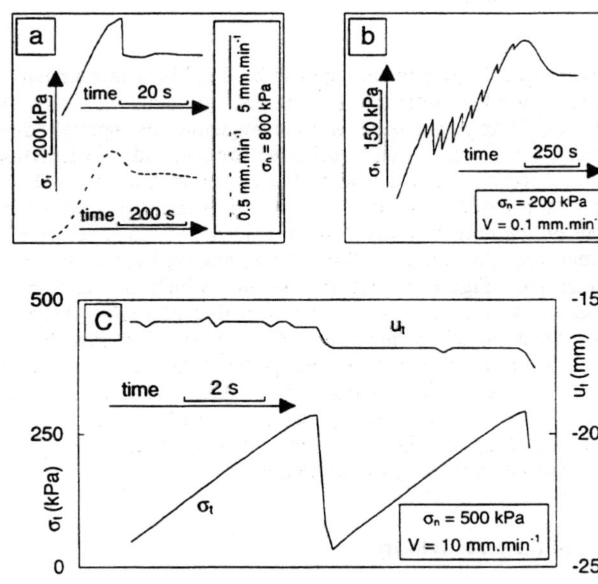

Fig. 9. Tangential stress σ_t and displacement u_t observed during stick-slip phases ; a : stick peak ; b : unstable ; c : stable.

not the only perturbation observed on the tangential stress. Local perturbations were observed which were the consequence of local perturbations in the shape of the plate (see Figures 2 and 9b). Geometric defects are the most influential at high velocities, and are smoothed at lower velocities owing to the visco-plastic deformation of ice.

Another type of perturbation of the friction curve is shown on Figure 2. It consists in an increase of the tangential stress at the end of each half-cycle. As this effect was observed during each test, and at the end of each half-cycle, this transient behaviour cannot be related to the geometry of the concrete plate. The most probable explanation is that the visco-plastic deformation of ice normal to the plate induces an increase in the actual contact areas between ice and concrete, which leads to the increase in friction. This is supported by the fact that the

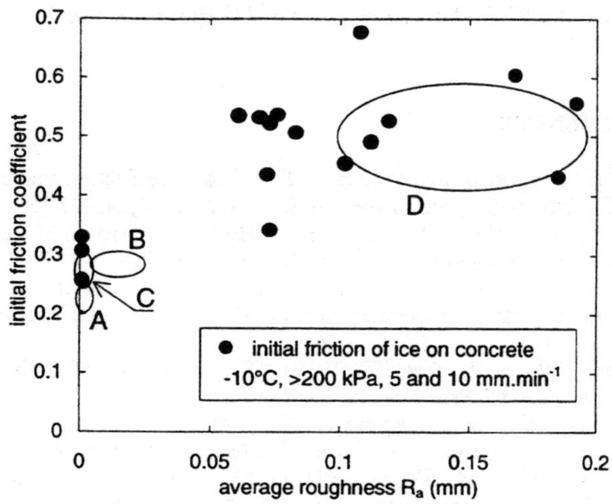

Fig. 10. Comparison between our tests and others ; A : Cammaert (1988), ice on concrete, -10°C, 200 kPa, 780 and 840 mm.min^{-1} ; B : Saeki and others (1986), ice on concrete, -8°C, 200 kPa, 10 and 5 mm.min^{-1} ; C : Tabata and Tusima (1981), ice on concrete, -2°C, <250 kPa, 5 and 24 mm.min^{-1} ; D : Shapiro and Metzner(1987), ice on beach gravel, near 0°C, a few hundred of kPa, 18000 mm.min^{-1}.

relative increase in the tangential stress at the end of the half-cycle is all the greater as the carriage velocity is lower. Moreover, evidence of ice visco-plastic deformation and micro-cracking was generally observed after the tests.

Stick-slip motion

Some tests exhibited a stick-slip phenomenon during their whole duration or only part of it. Stick-slip is characterized by the following phases : a) adhesion or anchoring of ice on concrete ; b) increase of the tangential stress while the plate carriage keep moving, until the ice-concrete bond reaches its failure point. During this phase, energy was stored in the dynamometer linking the sample holder to the frame of the machine ; c) breaking of the bond and rapid displacement of the ice relative to the plate together with a decrease of the tangential stress.

When there is full restitution of the elastic energy stored in the load transducer, the tangential stress reaches zero and so the velocity of ice relative to the plate. In this instance adhesion of ice is made possible and the full process can be repeated, leading to a stable regime of stick-slip during which the tangential stress is periodic on a time interval which can be part or the total duration of the test.

Unstable stick-slip (see Figure 9b) occurs when there is partial restitution of the stored elastic energy. It was generally observed during the transient phases, as the inversion of the direction of motion. In this case, there is a change in the configuration of the actual contact areas, and a mechanism of accommodation in the contact, as visco-plastic deformation or micro-cracking, is needed to limit the normal stress at points of real contact between ice and concrete.

Stable stick-slip (see Figure 9c) was influenced by the plate roughness, the velocity of ice relative to concrete and the normal stress. For high roughness, increasing the normal load and the velocity tended to favour stable stick-slip. For very low roughness, high normal stress favoured stable stick-slip, whereas the velocity was not influential on the occurrence of stick-slip, but was influential on the period of the stick-slip phases when they occurred.

Anchoring of ice on rough plates was caused by the indentation of ice by the asperities, under a high effective normal stress. This was confirmed by the observation of the friction surface of ice, which showed periodically spaced marks aligned in the direction of motion, the spacing of which was in agreement with the length covered by the ice sample during the energy release phase.

The arrest phases on smooth plates was explained by the adhesion of ice on concrete. The existence of fine crushed particles of ice, sintered in the hollow parts of the plate relief, could emphasize this mechanism, by favouring adhesion bond growth with time. The influence of the velocity and of the normal stress is consistent with this explanation.

DISCUSSION

Comparison with other studies

A comparison with a few other studies of friction of ice over concrete has been made in order to check the validity of our results. Figure 10 shows the initial friction coefficient corresponding to the first cycle of tests performed under normal stress greater than 200 kPa and for relative velocities greater than 5 mm.min^{-1}, compared to results found in the literature. For near zero roughness the friction coefficients of sea ice over concrete, under a 200 kPa normal stress obtained by Cammaert and Muggeridge (1988), Saeki and others (1986), and Tabata and Tusima (1981) are in good agreement with that of the present study. Friction coefficients for sea-ice seems to be 10 to 20% less than for pure fresh-water ice. However this deviation may be partly explained by the differences in temperature and velocity between the different tests. Shapiro and Metzener (1987) performed friction tests of sea ice over beach gravel and pebble under comparable normal loads. Their results are consistent with those found for high roughness.

No indication was found in the literature on the values of the stabilized friction coefficient measured after a number of cycles. However a reference to stabilized friction from Blanpied and others (1995) concerning the friction of granite on granite, in the temperature range 200 to 800°C , shows qualitatively the same increase of friction with time.

The present results were obtained for velocities and normal stresses which do not cover the full range of actual structure loading conditions. The relevant order of magnitude for the normal stress acting over an area comparable with that of our ice specimen should be 10 MPa (Cammaert and Muggeridge, 1988; Masterson and Frederking, 1993) rather than 1 MPa. However our results suggest that, with increasing normal stress, the friction coefficient tends towards a constant value μ which is very close to that obtained with the highest normal stress used in this study. Velocities less than 10 mm.min^{-1} correspond to quasi-static loading conditions, whereas typical ice-structure interaction velocity may be in the order of 1m.s^{-1}. The increase of the friction coefficient with velocity, found in the present study, results from the visco-plastic behaviour of ice. With increasing velocity, micro-cracking, then melting of ice, should produce a peak on the friction versus velocity curve, followed by a decrease of friction (Barnes and others, 1971; Saeki and others, 1986). Owing to the involved mechanisms, the velocity corresponding to the friction peak should be dependent on the normal stress and the concrete roughness.

Role of visco-plastic deformation

Evidence of visco-plastic deformation was given by grooves and swelled shoulders revealed by optical microscopy observation of the ice specimens after the tests. This was confirmed by the analysis of thin sections of ice, cut parallel and perpendicular to the interface. These thin sections observed between cross polarizers showed the occurrence of recrystallization characterized by the nucleation of new grains. These observations also showed that micro-cracking was involved in the process of friction. Visco-plastic deformation accompanied by recrystallization was present in each test, whereas micro-cracking concerned mainly high velocity tests.

As far as micro-cracking is not involved, the stress acting on the upstream face of an asperity in contact with the ice is an increasing function of the velocity, owing to the visco-plastic motion of ice past asperities. For a given plate roughness and given asperity slopes, the real area of contact must be such that the resulting vertical force equilibrates the prescribed force which is the product of the nominal normal stress by the nominal area of the ice specimen. Then the actual contact area must be a function of both the nominal normal stress and plate velocity. Assuming that the actual contact area is near zero at the beginning of a test, the time necessary to reach the contact area which corresponds to the prescribed normal stress and velocity is thought to correspond to the stabilization phase of the friction coefficient. This phase duration should be a decreasing function of the normal stress and of the plate velocity (the vertical strain rate which allows the asperities to penetrate into the ice increases with σ_n ; the stabilized contact area is a decreasing function of the plate velocity).

The increase of friction from the first cycle to the last cycle (see Figure 3) was found to be an increasing function of the velocity. This may be a consequence of the non-linearity of the ice flow law, as far as visco-plastic deformation is involved in the movement of ice past the asperities of the plate ; this may be also a consequence of the averaging procedure, as the tangential stress is averaged over one cycle, the duration of which is inversely proportional to the plate velocity : owing to the deformation normal to the plane of the plate, the actual roughness seen by the ice is evolving with time, then an average tangential stress for a low velocity test corresponds to a higher actual roughness than that computed for a cycle carried out at a higher velocity.

The fact that the plate roughness plays also a role in the evolution of friction, with a low roughness leading to a lower increase of the friction coefficient with the number of cycles, is consistent with the fact that the visco-plastic deformation of ice tends to increase the effective roughness. In the same way, the influence of the normal stress is consistent with this explanation.

CONCLUSION

The results of friction tests of ice over concrete at -10°C have been presented. In the range of variation of the normal stress σ_n (25 to 800 kPa), the plate velocity (0.1 to 10 $mm.min^{-1}$), and the average plate roughness R_a (0 to 0.6 mm), the friction coefficient f, defined as the ratio of the tangential stress σ_t to the normal stress σ_n, is an increasing function of the plate roughness and of the velocity of ice relative to the concrete plate, and a decreasing function of the normal stress. The transition from adhesion to dynamic friction is particularly sensible when the average plate roughness passes from 0 to a positive value. The role of the ice visco-plastic deformation, confirmed by thin section observation, appears to be very effective under the conditions which were imposed, even at the cold room temperature of -10°C. It can explain the variation of the tangential stress, then the friction coefficient, with velocity, as a direct consequence of the ice viscous behaviour. Moreover visco-plastic deformation in the direction perpendicular to the ice concrete interface, owing to the prescribed normal stress, tends to increase the real ice-concrete contact areas and to increase the effective roughness, seen by the ice, with time. This effect explains the increase of the friction coefficient with time under the conditions of constant normal load tests. Also, it makes a quantitative interpretation of the results very difficult.

ACKNOWLEGMENTS

The support from the technical staff of LGGE and 3S laboratories was greatly appreciated.

The first author had a financial support from the French MENESR (AMN grant).

LGGE is a CNRS laboratory, associated to Université Joseph Fourier - Grenoble.

REFERENCES

Barnes, P, Tabor, D, and Walker, JCF (1971). "The friction and creep of polycrystalline ice", *Proc Roy Soc Lond*, A 324, pp 127-155.

Beeman, M, Durham, WB, and Kirby, SH (1988). "Friction of ice", *J Geophysical Research*, V 93, No B7, pp 7625-7633.

Blanpied, ML, Lockner, DA, and Byerlee, JD (1995). "Frictional slip of granite at hydrothermal conditions", *J Geophysical Research*, V 100, No B7, pp 13,045-13,064.

Bowden, FP, and Hughes, TP (1939). "The mechanism of sliding on ice and snow", *Proc Roy Soc Lond*, A 172, pp 280-298.

Cammaert, AB, and Muggeridge, DB (1988). "Ice interaction with offshore structures", *Van Nostrand Reinhold*, New York, 432 p.

Jefferies, MG, and Wright, WH (1988). "Dynamic response of Molikpaq to ice-structure interaction", *Proc 7th Int Conf on Offshore Mech and Arctic Eng*, Houston, V 4, pp 201-220.

Jordaan, IJ, and Timco, G (1988). "Dynamics of the ice crushing process", *J Glaciology*, V 34, No 118, pp 318-326.

Kärnä, T, and Turunen, R (1989). "Dynamic response of narrow structures to ice crushing", *Cold Regions Sci and Tech*, V 17, pp 173-187.

Masterson, DM, and Frederking, RMW (1993). "Local contact pressures in ship/ice and structure/ice interactions", *Cold Regions Sci and Tech*, V 21, pp 169-185.

Meaney, R, Jordaan, IJ, and Xiao, J (1996). "Analysis of medium scale ice-indentation tests", *Cold Regions Sci and Tech*, V 24, pp 279-287.

Saeki, H, Ono, T, Nakazawa, N, Sakai, M, and Tanaka, S (1986). "The coefficient of friction between sea ice and various materials used in offshore structures", *J Energy Resources and Tech*, V 108, pp 65-71.

Sayed, M, and Frederking, RMW (1992). "Two dimensional extrusion of crushed ice. Part 1 : experimental", *Cold Regions Sci and Tech*, V 21, pp 25-36.

Savage, SB, Sayed, M, and Frederking, RMW (1992). "Two dimensional extrusion of crushed ice. Part 2 : analysis", *Cold Regions Sci and Tech*, V 21, pp 37-47.

Shapiro, LH, and Metzner, RC (1987). "Coefficients of friction of sea ice on beach gravel", *J Offshore and Artic Eng*, V 109, pp 388-390.

Tabata, T, and Tusima, K (1981). "Friction measurements of sea ice on some plastics and coatings", *Proc 6th Int Conf on Port and Ocean engineering under Arctic Conditions*, Quebec, V 1, pp 526-535.

Tuhkuri, J (1995). "Experimental observations of the brittle failure process of ice and ice-structure contact", *Cold Regions Sci and Tech*, V 23, pp 265-278.

Zou, B, Xiao, J, and Jordaan, IJ (1996). "Ice fracture and spalling in ice-structure interaction", *Cold Regions Sci and Tech*, V 24, pp 213-220.

Proceedings of the Seventh (1997) International Offshore and Polar Engineering Conference
Honolulu, USA, May 25-30, 1997

ISBN 1-880653-28-1 (Set); ISBN 1-880653-30-3 (Vol. II)

Abrasion of Steel Sheet Piles Due to Ice Flow Movements

Takashi Terashima
Pacific Consultants, Co., Ltd.
Sapporo, Japan

Masakuni Hanada, Takaharu Kawai, Kaori Oshima and Fumihiro Hara
Hokkaido University
Sapporo, Japan

ABSTRACT

Ice floes are widely known to abrade surfaces of structures constructed in sea water areas where ice floes move actively. According to Janson's research (1987)[1)] on concrete lighthouses in the Gulf of Bothnia, where sea ice movements are intense, the abrasion depth of several lighthouses was approximately 140 mm (approximately 7 mm per year). Also, since both corrosion and abrasion by sea ice cause a decrease in the thickness of the steel of marine structures in frozen sea water areas, differentiating these effects is difficult in practice.

The development of oil fields is just underway near northern Sakhalin in the Sea of Okhotsk. Sea ice is active in this area. Composite structures, in which the surfaces of structures are coated with steel, are believed to be effective for the future design of structures in frozen sea water areas. Hence, predicting the abrasion depth of surface steel due to sea ice movements is important for the design of these structures.

The area of our research was the water channel constructed in December, 1978, for the No. 2 estuary of Saroma Lagoon in Hokkaido, Japan. The channel connecting the lagoon with the Sea of Okhotsk is 50 m wide, 260 m long, and 5.0 m deep. We measured thickness of the steel sheet piles used for the side walls of the channel and for the piers of the bridge built over the channel. From the measurements we found the abrasion of the steel by sea ice movements and developed a technique to estimate the abrasion depth.

Keywords: frictional forces, abrasion.

1. LOCATION OF FIELD SURVEY AND METHOD OF STUDYING ABRASION

The No. 2 estuary of Saroma Lagoon was constructed in 1978 with two purposes; they are to improve the seawater quality within the lagoon and to make the estuary navigable. The water channel constructed for the estuary is about 260 m long, with two training jetties extending into the Sea of Okhotsk. The side walls of the channel are built of steel sheet piles, and the concrete used for the abutments is protected by steel pipe piles. The field survey was conducted at two locations on the steel sheet piles and at other two locations on the steel piles (Figure 1). The steel sheet pile is FSP IV A. The remaining thicknesses of the convex and concave parts of the sheet piles were measured at depth intervals of 50 cm from the upper end of the steel sheet pile. The survey showed the following.

(1) Steel sheet piles

① Observations of the area near the L.W.L. (low-water level) (+ 0.0 - − 1.0 m) showed that there were no organisms attached to the convex steel sheet piles, probably due to the movements of the drift ice.

② The area near the L.W.L. had no holes resulting from concentrated corrosion.

③ The convex steel sheet piles near the L.W.L. had a smooth outer surface because of the frictional action of drift ice.

④ Organisms attached to the piles were mainly seaweeds. They were observed from the depth of − 3.5 m to the sea bottom. This indicates that the drift ice also influenced the attached organisms.

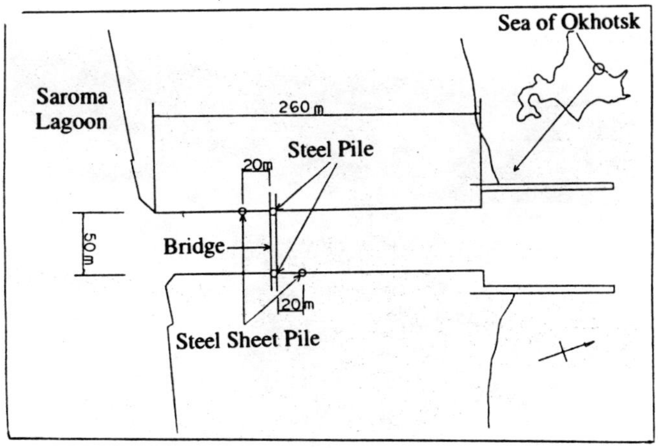

Figure 1. Location of Survey

(2) Steel piles

① In the area above the L.W.L., incrustations were formed in the splash zone.

② Organisms were not observed on the steel piles near the L.W.L., probably due to the abrasion by drift ice.

③ The area near the L.W.L. had no holes due to corrosion.

④ The area near the L.W.L. also had a smooth outer surface like that of the convex steel sheet piles, owing to the effect of drift ice. In this area, flattening of some piles was observed, which seems to have resulted from collision with the drift ice or the effect of ice forces. From visual observations, the range of the flattening of about 10 cm maximum was found to be from ± 0.0 - − 1.0 m, indicating that large ice forces acted on this section to a depth of up to − 1.0 m.

Figures 2 (a) and (b) are photographs of the steel sheet piles and the steel piles. The convex part of the steel sheet piles (Figure 2 (a)) had a smooth surface without any organisms attached, while many organisms were attached to the concave part and the sides. The convex part of the steel piles used for abutment protection (Figure 2 (b)) had little corrosion with no organisms attached. These observations showed that the steel sheet piles and the steel piles used for the water channel at the estuary were strongly affected by the ice floe movements, and that the ice floes affected only the convex parts.

Figure 2 (a). Steel Sheet Piles for the Side Walls of the Left Bank

Figure 2 (b). Steel Piles for the Abutment of the Right Bank

2. RESULTS AND DISCUSSION

Figures 3 (a) and (b) show the measurements of thickness and the corrosion rate (including abrasion) of the steel sheet piles used for the right bank of the water channel, Figures 4 (a) and (b) show these measurements for the left bank, and Figures 5 (a) and (b) and Figures 6 (a) and (b) show these measurements for the steel piles.

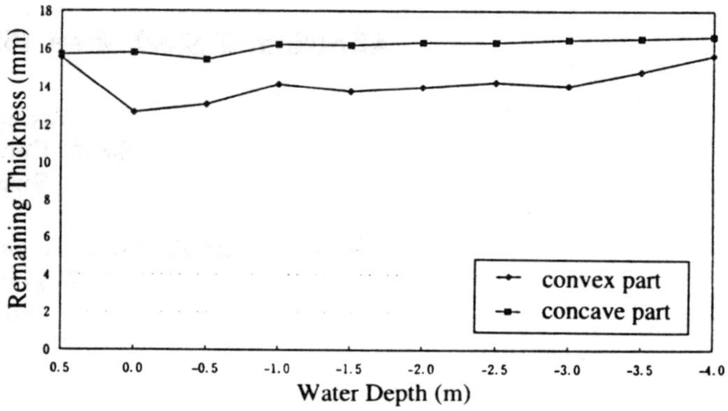

Figure 3 (a). Thickness of the Steel Sheet Piles for the Right Bank (during 16 years)

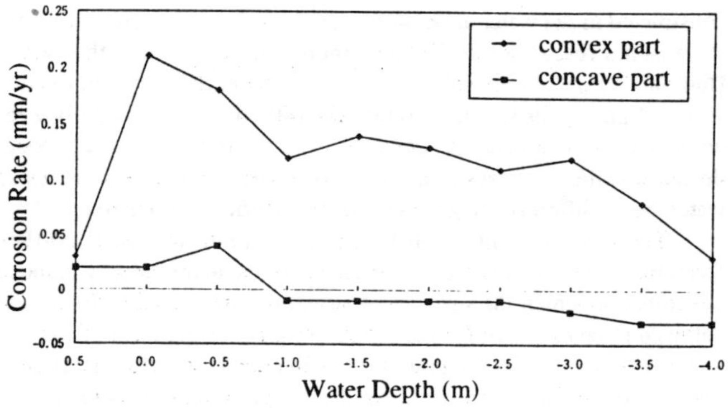

Figure 3 (b). Corrosion Rate of the Steel Sheet Piles for the Right Bank (Mean Rate during 16 years)

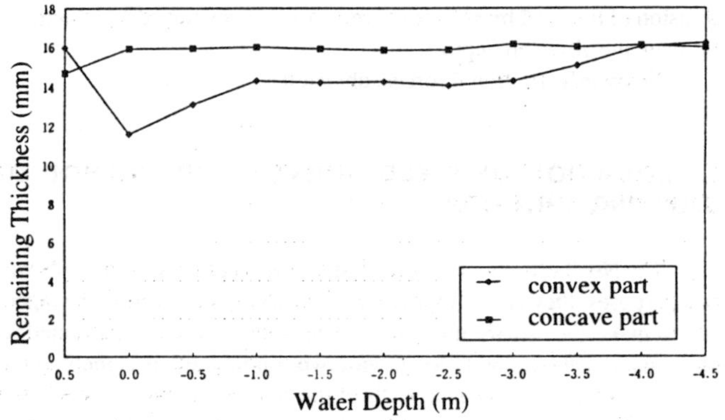

Figure 4 (a). Thickness of the Steel Sheet Piles for the Left Bank (during 16 years)

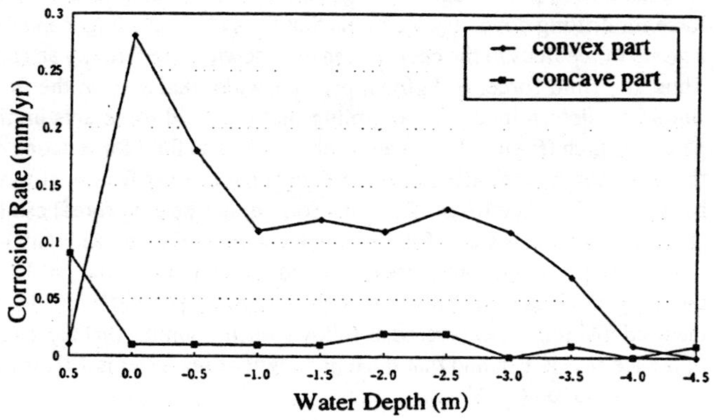

Figure 4 (b). Corrosion Rate of Steel Sheet Piles for the Left Bank (Mean Rate during 16 years)

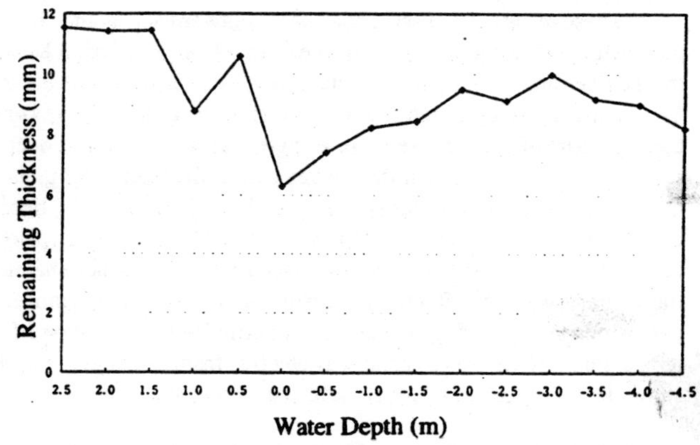

Figure 6 (a): Thickness of Steel Piles for the Left Bank (during 16 years)

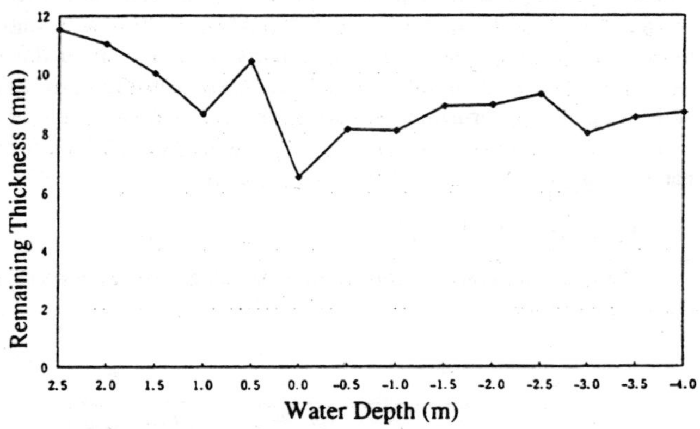

Figure 5 (a). Thickness of Steel Piles for the Right Bank (during 16 years)

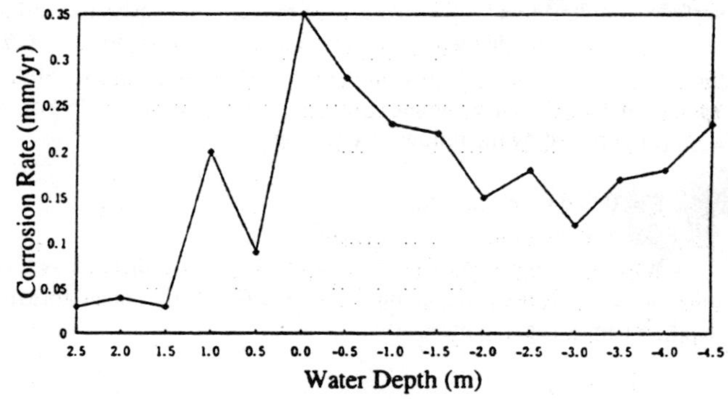

Figure 6 (b). Corrosion Rate of Steel Piles for the Left Bank (Mean Rate during 16 years)

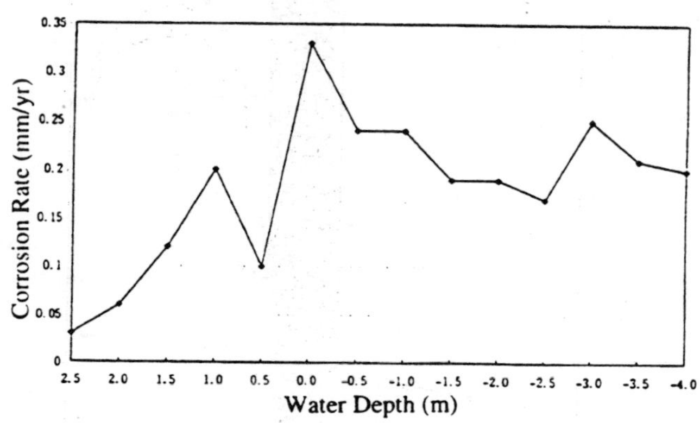

Figure 5 (b). Corrosion Rate of Steel Piles for the Right Bank (Mean Rate during 16 years)

(1) Steel sheet piles

① The right and left banks showed no significant difference in the tendency to corrode.

② The convex and concave parts showed large difference in the tendency to corrode; the amount of corrosion of the concave part was small, while that of the convex part was large.

③ Corrosion of the convex part varied with the depth. The largest amount of corrosion was recorded near + 0.0 m, and the amount of corrosion at the section deeper than + 0.0 m was about half of that at + 0.0 m.

④ The corrosion rate at this section near + 0.0 m was 0.2 - 0.28 mm/year, which was a much larger value than that at the concave part.

(2) Steel piles

① The right and left banks showed no significant differences in the tendency to corrode.

② Corrosion of the steel piles varied with the depth. The largest amount of corrosion was recorded near + 0.0 m, and the amount of corrosion at the section deeper than + 0.0 m was about half of that at + 0.0 m.

③ The corrosion of the body of the piles near + 0.0 m was as high as 0.35 mm/year.

These results and observations of the outward appearance indicate that drift ice flowing in and out of the channel causes friction between the ice floes and steel, such as the steel sheet piles, and abrades the steel. The measured amount and rate of corrosion described above for the concave part of the steel sheet piles agree with a general concept of corrosion. The corrosion of the convex part of the steel sheet piles and the steel piles seems to occur simultaneously with the abrasion because of the movements of ice floes. The vertical distribution of corrosion of the steel sheet piles suggests that the steel used for the water channel at the estuary was not affected by corrosion alone. Furthermore, the corrosion rate of 0.35 mm/year was abnormally high in view of low temperatures of the seawater and seawater free from contamination. This shows that the steel was strongly affected by abrasion.

3. METHODS OF ESTIMATING ABRASION OF STEEL DUE TO ICE FLOE MOVEMENTS – A CASE OF THE NO. 2 ESTUARY OF SAROMA LAGOON –

The mechanism of abrasion of armor materials, including concrete and steel, due to sea ice movements and their abrasion rate have been studied by Saeki et al.[2], Ito et al.[3] and Takahashi et al.[4][5]. Their studies showed that the abrasion rate (S') (the mean abrasion depth of material surface (mm/km) to a 1-km movement of the ice floe) varies depending on the material, and that when the sea ice temperature exceeds − 8 °C, the abrasion rate depends little on the ice temperature and the relative velocity but greatly on the contact pressure. The abrasion rate (S') of the steel is expressed by the following equation:

$$S' = 0.0003 \cdot \sigma_v \quad \text{.......} \quad (1)$$

S': mm/km $\quad \sigma_v$: kgf/cm²

When the contact pressure is σ_v and the abrasion distance is L (in other words, L is the distance moved by ice floes), the mean abrasion depth, $\bar{S}$ (mm), is given by:

$$\bar{S} = S' \cdot L$$
$$= 0.0003 \cdot \sigma_v \cdot L \quad \text{.......} \quad (2)$$

L: the abrasion distance (km)

The parameters σ_v and L must be determined to estimate the abrasion depth of the steel.

(1) Estimation of contact pressure (σ_v)

No measurements had been made on the contact pressure on the steel piles by ice floes passing through the channel at the No. 2 estuary, as well as on the amount of ice passing through the estuary. Therefore, estimation of the contact pressure was difficult. We made some assumptions to estimate the contact pressure (σ_v). We assumed that the channel is filled with ice floes drifting into the lagoon from the open sea and that the width of the channel ($2l$) is 50 m. From the typical sizes of ice floes in the Sea of Okhotsk and the flow velocity within the channel, we found the high possibility of ice floes forming an arch in the channel. Then we assumed that the floes drift into the lagoon in clusters forming an arch in the channel (Figure 7), and that the shape of an arch is a parabola. When the ratio ($f/2l$) between height of the arch (f) and the horizontal distance ($2l$) is set at 0.40 - 0.53, as obtained by Hara et al.[6],

$F_H = \frac{1}{2} F_v$ is given.

F_H multiplied by the friction coefficient between the steel sheet pile walls and the sea ice is the frictional force, F_F, which prevents ice floes from drifting down. The results of the study by Saeki et al.[7] have clarified that the coefficient of static friction is 0.23 or less and that of kinematic friction is about 0.1. Comparing this frictional force and the force for drifting down, F_V, we found that F_V is obviously larger and that a group of ice floes in the channel can drift down while forming an arch. Thus, the fluid force, q (kgf/m), per unit width that acts on the floes should be determined. By assuming that the fluid force acts on the parabolic arch (Figure 7), F_H can be obtained. The fluid force acting on the floes can be classified into the drag force and the frictional force between drifting ice floes. Since the form drag (the drag force) can be ignored when the inside of the channel is completely covered with ice floes, only frictional forces are taken into account. From the results of the study by Ueda et al.[8], the fluid force, q (kgf/m), per unit width due to the frictional force is expressed as follows when assuming that the length of the ice floe is 1 m and that the unit weight of sea water is 1.02 tf/m³.

$$q = 510 \cdot C_f \cdot V^2 \quad \text{.......} \quad (3)$$

where C_f is friction coefficient between water and ice and
V is the flow velocity (m/s).

Ueda et al.[8] suggested the friction coefficient (C_f) 0.007 for clusters of uniform ice floes = a considerably wide range of Reynolds numbers, and (C_f) 0.026 when a blunt comb-shaped roughness was provided for the bottom of the ice floe. Moreover, a friction coefficient (C_f) of 0.100 was obtained by McPhee from the measurements of ice floes in the Arctic Ocean. Enoki et al.[9] found C_f was 0.300 when ice floes with various thicknesses form ice jams. Since no reports exist on ice floes flowing through the No. 2 estuary of Saroma Lagoon, computed fluid forces (q) at the time of arch formation for various friction coefficients are shown in Table 1. The horizontal ice force, F_H (kgf), of floes which form a parabolic arch and when $f = 1.0\,l$ is expressed by:

$$F_H = \frac{ql^2}{2f} = \frac{1}{2} ql \quad \text{.......} \quad (4)$$

Table 2 is obtained by substituting the value from equation (3) for q and 25 (m) for l.

Figure 7. Assumption to Estimate Contact Pressure

According to the research by Kunimatsu et al.[10] on the size and thickness of drift ice in the Sea of Okhotsk, ice with a thickness of about 1 m is most frequently found. Therefore, we assumed in this study that the ice thickness was 1 m and that the ice forces (F_H) act on the area of 0.3 m² on the convex part of the sheet piles used for the side walls of the water channel. From this, the contact pressure (σ_v) in kgf/cm² was calculated as $F_H/3000$. Consequently, the abrasion rate (S') can be obtained by equation (1) when σ_v is determined. The abrasion rate depends on F_H, and F_H depends on the flow velocity and the friction coefficient. Table 3 shows the results of the computations of S'.

Table 1. Fluid Force (q) per Unit Length (kgf/m)

V(m/s) \ Cf	0.007	0.011	0.026	0.100	0.200	0.300
0.1	0.04	0.057	0.134	0.515	1.030	1.545
0.2	0.144	0.227	0.536	2.06	4.120	6.180
0.5	0.901	1.416	3.348	12.875	25.750	38.625
1.0	3.605	5.665	13.390	51.500	103.00	154.5
1.5	8.111	12.746	30.128	115.875	231.75	347.625
2.0	14.420	22.66	53.56	206.000	412.000	618.000
2.5	22.531	35.406	83.688	321.875	643.75	965.625

Table 2. Computation Table of F_H (kgf)

V(m/s) \ Cf	0.007	0.011	0.026	0.100	0.200	0.300
0.1	0.125	0.178	0.419	1.609	3.219	4.828
0.2	0.450	0.709	1.675	6.438	12.875	19.3125
0.5	2.816	4.425	10.463	40.234	80.469	120.703
1.0	11.265	17.703	41.844	160.938	321.875	482.813
1.5	25.347	39.831	167.375	362.109	724.219	1086.328
2.0	45.063	70.813	167.375	643.75	1287.5	1931.25
2.5	220.0	110.644	261.525	1005.859	2011.719	3017.578

Table 3. Computation Table of Abrasion Rate (S') (mm/km)

V(m/s) \ Cf	0.100	0.200	0.300
0.5	4.02×10^{-6}	8.04×10^{-6}	1.21×10^{-5}
1.0	1.61×10^{-5}	3.22×10^{-6}	4.82×10^{-6}
1.5	3.62×10^{-6}	7.24×10^{-5}	1.09×10^{-4}
2.0	6.44×10^{-6}	1.29×10^{-4}	1.93×10^{-4}
2.5	1.01×10^{-4}	2.01×10^{-4}	3.02×10^{-4}

(2) Estimation of the abrasion distance

To determine the abrasion depth by equation (2), an estimate of the abrasion distance (L) of ice movement is necessary. We provide two conditions for estimating L:

① The lagoon is frozen

When the lagoon is frozen before drift ice reaches the shore, a water area of some 160,000 m² remains open near the No. 2 estuary. Part of the drift ice which enters the lagoon is forced under the ice that was formed in the lagoon. When the tide is going out, the sea ice also drifts out of the lagoon. When the moving velocity of the floes (assuming it is the same as the flow velocity) is about 1 m/s, the open water areas inside and outside the lagoon are covered with drifting ice floes in about one hour. Therefore, when tides occur twice a day, ice floes move 12.8 km a day within the channel, when considering the flow of the ice in and out the channel.

② The lagoon is not frozen

When the lagoon is not frozen, we assumed that incoming and outgoing tidal currents move the drift ice outside the lagoon. According to past reports on drift ice, when the water area outside the lagoon was covered with drift ice and the lagoon is not frozen, the ice drifted into the lagoon and greatly damaged aquaculture facilities for scallops. Such damage was caused in 1988, 1989, 1990, 1991 and 1993 after the construction of the No. 2 estuary.

Assuming that the tide occurs twice a day and that changes in the flow velocity in the water channel are represented by sine waves, the flow velocity is expressed as:

$$V = V_{max} \sin \frac{2\pi}{T} t \qquad (5)$$

where V_{max} = the maximum flow velocity.

Also, the distance moved by the ice, L (km), is obtained by:

$$L = \int |V| dt \qquad (6)$$

Assuming that the period, T, is 12 hours, when V_{max} = 2.5 m/s, the distance moved by the ice was about 138 km a day; and when V_{max} = 2.0 m/s, the distance was about 110 km.

(3) Estimation of the abrasion depth

When the lagoon is not frozen, the contact pressure, abrasion rate and the distance moved by the ice vary due to changes in the flow velocity. Equation (2) shows that the abrasion depth, S (mm), is obtained by multiplying the abrasion rate, S' (mm/km), and the distance moved by the ice, L (km). When the flow velocity at the time of t (sec) is V (m/s), the abrasion depth, dS (mm), during the time of dt is expressed as:

$$dS = S'dL$$
$$= S'|V|dt \cdot 10^{-3} \qquad (7)$$

The abrasion rate, S' (mm/km), is expressed by using equations (1), (3) and (4).

$$S' = 6.375 \times 10^{-4} \times C_f \times V^2 \qquad (8)$$

Therefore, the total abrasion depth, S (mm), is expressed by integrating equation (7).

$$S = \int S' \cdot |V| dt \cdot 10^{-3}$$
$$= 6.375 \cdot Cf \int |V^3| dt \cdot 10^{-7} \qquad (9)$$

When the flow velocity is determined by the sine curve of semi-diurnal tide as shown in equation (5), the abrasion depth per day, S_{1day} (mm/day), is expressed as:

$$S_{1day} = 2.338 \cdot 10^{-2} \cdot Cf \cdot V_{max}^{3} \qquad (10)$$

Figure 8 is a graph drawn from equation (10). When the friction coefficient, C_f, of the ice floe in the state of an ice jam is 0.3, the abrasion depth is 0.11 mm per day when V_{max} = 2.5 m/s.

The abrasion depth was estimated for a case where the lagoon was frozen. When the flow velocity was 2 m/s and C_f was 0.3, the distance moved by the ice was 12.8 km per day. Hence, by using the abrasion rate (S') in Table 3, the abrasion depth per day was 12.8 km × 7.65 × 10^{-4} mm/km, or 9.80 × 10^{-3} mm. The water channel was constructed 16 years ago, but because the lagoon was not frozen during five winters, the abrasion can be considered to have been caused during 11 winters. Though correct data on how many days the drift ice flowed through the channel does not exist, the passage of drift ice through the channel averages 20 - 30 days a year. If we assume in this study that the period when the ice drifted into the lagoon was 20 days, the total abrasion depth in all the years when the lagoon was frozen (S_1) would be 9.80×10^{-3} mm $\times$ (11 $\times$ 20), or 2.156 mm, during the past 16 years. At the same time, to determine the total abrasion depth in all the years when the lagoon was not frozen (S_2), V_{max} and C_f were set at 2.5 m/s and 0.3, respectively, by using equation (10). When assuming that the days when the drift ice entered the lagoon was 20 days and, including the fact that the lagoon was not frozen in five winters during the past 16 years, the abrasion depth (S_2) would be 0.1096 (mm/day) $\times$ 5 (times) $\times$ 20 (days), or 10.96 mm. As a result of these computations, the total of these abrasion depths during 16 years was 10.96 mm + 2.16 mm, or 13.1 mm. At a level of + 0.0 m representing the location of the maximum abrasion, the maximum amount of corrosion (abrasion depth) during the past 16 years was 3.5 mm - 4.0 mm. Therefore, the calculated abrasion depth was much larger. The reasons for this seem to be: (1) the fluid force was the largest because the ice floe was in the state of an ice jam; (2) the largest flow velocity which can be observed in Saroma lagoon was used; and (3) the contact area between the ice and the structure was imagined to be rather small. In another words, the severest conditions thinkable were applied for the estimation. Therefore, the larger value obtained was natural. When the maximum flow velocity, V_{max}, is 2.0 m/s with C_f remaining 0.3, the total abrasion depth in 16 years would be 6.82 mm, which is closer to the calculated value.

The estimated value of the reduced thickness, assuming that steel is abraded by the ice floe movement, is almost equal to the measured value of the reduced thickness in the convex part of steel sheet piles and in the steel piles influenced by drift ice. From the results of the study on the thickness of the convex and concave parts of the steel sheet piles used for the channel, the corrosion of the convex part can be judged to be due to abrasion by drift ice, because the amount of corrosion of the concave part was extremely small.

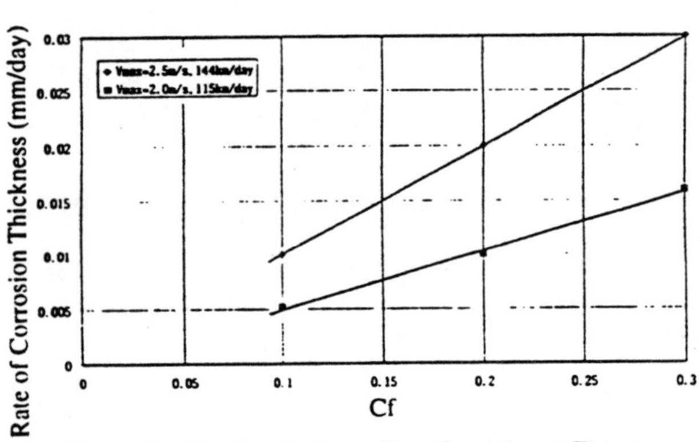

Figure 8. Abrasion depth per Day (Semidiurnal Tides)

(4) Corrosion rate

The amount and rate of corrosion of the steel used for the channel at the No. 2 estuary of Saroma Lagoon can be estimated from the measurements of the concave part of the steel sheet piles. In the case of the steel sheet piles, the measurements (Figures 3 and 4) show that the corrosion rate (mm/year) is 0.1 mm/year at the maximum. Compared with offshore structures in other water areas in Japan, this value is small. Water temperatures in Saroma Lagoon range from the highest 20 °C in August to below 0 °C from December through mid-March, which are lower than other water areas in Japan. In addition, because water free from contamination from the open sea flows in and out the estuary, the corrosion rate is extremely low. Therefore, for offshore structures constructed alone in water areas directly affected by drift ice, and for steel piles and the convex parts of steel sheet piles used for water channels where sea ice makes intense movements, careful thought is necessary about the effects of the abrasion caused by ice floe movements more than those of corrosion.

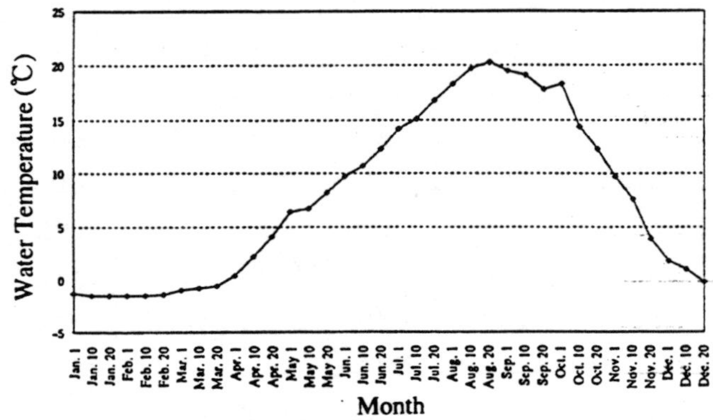

Figure 9. Mean Water Temperature in Saroma Lagoon (1986 - 1988)

4. CONCLUSIONS

The following conclusions were obtained from the results of the survey of the thickness of the steel sheet piles and the steel piles used to construct the No. 2 estuary of Saroma Lagoon and the analytic results:

1) The results of the study on the thickness of the concave part of the steel sheet piles not affected by drift ice show that the corrosion rate in the splash zone is 0.04 - 0.10 mm/year, and deeper sections have a rate less than 0.01 mm/year. Compared with similar structures in other water areas in Japan, these values are small. This can be due to such factors as low water temperatures, clean seawater and a salinity of 33 - 34 ‰.

2) In the splash zone, the corrosion and abrasion rate of the convex part of the steel sheet piles and the steel piles affected by the drift ice is 0.2 - 0.35 mm/year. In the deeper section of $-$ 1.0 - $-$ 4.0 m, the corrosion rate is a half of that of the splash zone, but the values were all much larger than the concave part. This indicates that while the concave part corrodes without being affected by the sea ice, the convex part shows the possibility of being abraded under the strong influence of sea ice.

3) By using the data from past research on the abrasion of steel, the abrasion depth of the steel sheet piles due to sea ice was estimated assuming that the ice floes form an arch in the channel. As a result, the total abrasion depth during the past 16 years was estimated, which agreed well with the measured value for the maximum amount of corrosion. From this, the amount of corrosion of the concave part can be considered to be caused by both corrosion and abrasion due to sea ice in the splash zone, while it is caused mainly by abrasion due to sea ice in the deeper section of $-$ 1.0 - $-$ 4.0 m.

REFERENCES

1) Janson, J. (1987), Long Term Resistance of Concrete Offshore Structures in Ice Environment, Proc. of 7th International Conference on Offshore Mechanics and Arctic Engineering (OMAE), Vol. III, P.225-231.

2) Saeki, H. et al. (1985), Study on the Abrasion of Concrete due to Sea Ice Movements, Proc. of Civil Engineering in the Ocean, Vol. 1, p.68-73.
3) Ito, Y. et al. (1988), Study on the Prediction Method of Abrasion Depth of Concrete Marine Structures due to Ice Floe Movements, Proc. of Civil Engineering in the Ocean, Vol. 7, p.221-225.
4) Takahashi, Y. et al. (1991), Study on the Abrasion of Surfaces of Structures due to Ice Floe Movements, Proc. of the 7th Cold Region Technology Conference, p.368-375.
5) Takahashi, Y. et al. (1990), Study on the Prediction Method of Abrasion Depth of Bridge Piers due to Ice Floe Movements, Proc. of the 6th Cold Region Technology Conference, p.367-373.
6) Hara, F. et al. (1993), Conditions of Arch Formation by Ice Floes at Bridge Piers, Proc. of Cold Region Technology Conference, Vol. 9, p. 654-659.
7) Saeki, H. et al. (1984), The Coefficient of Friction between Sea Ice and Various Materials Used in Offshore Structure, Jour. of Energy Resources Technology, Trans. of ASME, Vol. 108, p. 65-71.
8) Ueda, T. et al. (1992), Experimental Study on the Fluid Force Acting on Floating Ice Floes, Proc. of Civil Engineering in the Ocean, Vol. 8, p. 135-140.
9) Enoki, K. et al. (1992), Control of Ice Floe Movements by the Ice Boom, Proc. of Civil Engineering in the Ocean, Vol. 8, p. 135-140.
10) Kunimatsu, S. et al. (1993), Study on the Size of Ice Floes at the Okhotsk Sea Coast, Proc. of Civil Engineering in the Ocean, Vol. 9, p. 95-100.

Proceedings of the Seventh (1997) International Offshore and Polar Engineering Conference
Honolulu, USA, May 25-30, 1997

ISBN 1-880653-28-1 (Set); ISBN 1-880653-30-3 (Vol. II)

Discrete Element Modelling of the Local Interaction Between a Stationary Structure and a Moving Ice Pack

K. Sepehr and A.P.S. Selvadurai
McGill University
Montreal, Canada

G. Comfort
Fleet Technology Limited
Kanata, Canada

ABSTRACT

This paper presents the modification and application of a discrete element technique for the study of the two dimensional dynamic interaction between a moving pack ice and a stationary structure. The discrete element procedure accounts for the generation of fragmentation within the initially intact ice floes. The problems examined take into consideration brittle fragmentation in tension and viscoplastic flow in compression. The analysis also accounts for size dependency of the strength of the ice floe after fragmentation. The interaction between individual ice floes is modelled by nonlinear stiffness constraints with elastic-perfectly plastic behaviour. The methodology is used to examine both the fragmentation process in the intact ice floes which occur as a result of the interaction and to establish the time-dependent interactive stress development between the pack ice and a stationary structure.

Key words: Ice floes, Ice-structure interaction, Discrete element modelling, Numerical modelling

INTRODUCTION

Impact loads induced by storm-driven movement of extensive ice floes in open water is one of the ice load scenarios that needs to be considered in the design of arctic offshore structure. The dynamic interaction behaviour between the moving ice floes and the structure is a very complex process which is governed by the constitutive behaviour of ice, the configuration and flexibility of the structure, the interactions between the individual moving ice floes, local processes at the ice-structure interface and the speed of movement of the ice pack. The consideration of all the above factors in a dynamic ice-structure interaction scenario is a difficult exercise. As a result, the modelling of pack ice-structure interaction has been developed by considering a variety of simplifying assumptions and idealizations. In the most elementary of these treatments, the impact scenario was largely limited to the "limit stress" and "limit momentum" loads cases which neglect the possibility of initiation and propagation of cracks or fragment development in the impacting ice floe (Croasdale, 1985). An analysis of floe splitting based on the theory of plasticity, presented by Ralston (1981), is a first effort in this direction. The phenomenon of floe splitting was examined using continuum fracture mechanics considerations for the mechanical behaviour of the ice. Several studies on this subject have been reported, including the work of Palmer et al. (1983) who examined radial cracking in ice sheets of semi-infinite extent where the crack was modelled as an edge crack in a plate, wedged open by opposing crack opening forces and that of Hallam (1986) which presented a review of the role of fracture in limiting ice forces. Hamza and Muggeridge (1984) proposed an incremental numerical approach to propagation of cracks in a semi-infinite floes, and Cormeau et al. (1984) presented the results of a specific finite element analysis of the splitting of a finite ice floe with pre-existing cracks. The work of Bhat (1988) also considers the propagation of a crack in an impacting ice floe by applying conventional methods of computational linear elastic fracture mechanics, where the crack tip stress intensity factors were evaluated, numerically, using the finite element method. While the continuum based models described above can be used to examine the interaction between a single ice floe and a structure they are deficient in accommodating extensive brittle fragmentation, crushing and large scale movement of the ice fragments both at the ice-structure interaction region and at the boundaries of ice floes. The ability to model an ice floe which can accommodate a transition from an initially intact state to a fragmented transformed state is recognized as an essential step towards the realistic simulation of the ice-structure interaction processes. The application of discrete element modelling techniques to the study of ice floe-structure interaction is therefore considered a useful methodology which can address some of the deficiencies associated with the purely continuum - based procedures. The application of discrete element techniques to the study of ice-structure interaction is not entirely new; examples of the earlier studies are given by Hocking et al. (1985 a,b, 1987), Williams et al. (1986), and Mustoe et al. (1987). These discrete element procedures recently extended by Selvadurai and Sepehr (1996) to (i) identify stress states which promote either viscoplastic flow or a transition from a continuum to a fragmented discontinuum (ii) to incorporate influence of size dependency in the fragmentation strength.

In this paper, the basic concepts and methodologies are further modified and adopted to examine the problem of in-plane dynamic interaction between pack ice consisting of an assembly of close fitting individual floes and a stationary structure. The first modification relates to the concept of a size-dependent fragmentation stress. The second aspect examines the nonlinear interactive response between the ice floes. The results of the computational model presented in this paper illustrate the process of fragment development in an originally intact pack ice which consists of a number of single ice floes. The computational modelling also provides an estimate of the magnitude of the average maximum principal stress generated between the pack ice and a stationary structure, as a result of the movement of the pack ice.

CONSTITUTIVE MODELLING

Intact Ice

The constitutive modelling of the intact ice floe takes into consideration elastic and viscoplastic behaviour of the ice.. The elastic behaviour is characterized by isotropic linear elasticity and viscoplasticity is characterized via a Perzyna (1966), type model which has fluidity parameter (γ) and a yield function (F) and flow function Φ(F). In the present study the yield function is characterized by Mohr-Coulomb behaviour with cohesion (c) and internal friction (ϕ). The model also can accommodate limiting tensile strength defined by a tension cut off (σ_T).

Constitutive Behaviour of Intact Ice Floe:

The generalized constitutive model, for the intact geomaterial, adopted in the discrete element modelling is an isotropic elastic viscoplastic model with softening capabilities. In its rate form, the elastic constitutive model of the geomaterial is given by

$$\dot{\epsilon}_{ij}^{e} = (\dot{\sigma}_{ij}/2G) + (\dot{\sigma}_{kk}/9K)\delta_{ij} \qquad (1)$$

where (.) denotes the time derivative, $\dot{\sigma}_{ij}$ is the stress deviator tensor and G and K are, respectively, the linear elastic shear modulus and the bulk modulus.

The viscoplastic model proposed by Perzyna (1966) has been successfully applied by a number of investigators, including Zienkiewicz and Cormeau (1974), Cormeau (1975), Selvadurai and Au (1992) for the computational modelling of a variety of geomaterials. The viscoplastic strain rate is given by

$$\dot{\epsilon}_{ij}^{vp} = \gamma < \Phi(F) > \partial F/\partial\sigma_{ij} \qquad (2)$$

where γ is a fluidity parameter, F is the current yield function and Φ(F) is a flow function defined by

$$< \Phi(F) > = \Phi(F) \text{ if } F \geq 0\ ; < \Phi(F) > = 0 \text{ if } F < 0 \qquad (3)$$

As with other plasticity based models.the yield function F and the flow function Φ(F) are material properties which need to be determined from experiments. Two commonly used representations, however, are

$$\Phi(F) = \exp\{M(F-F_o)/F_o\}-1\ ;\ \Phi(F) = \{(F-F_o)/F_o\}^N \qquad (4)$$

where M and N are constants and F_o is a uniaxial failure stress. The Mohr-Coulomb yield function can be utilized to describe the behaviour of a wider class of geomaterials. i.e.

$$F = (I_1/3)\sin\phi + J_2^{1/2}\{\cos\Theta - [1/\sqrt{3}]\sin\Theta\sin\phi\} - c\cos\phi \qquad (5)$$

where c is the cohesion, ϕ is the angle of internal friction,

$$I_1 = \sigma_{kk};\ J_2 = (1/2)\sigma'_{ij}\sigma'_{ij}\ ;\ J_3 = (1/3)\sigma'_{ij}\sigma'_{jk}\sigma'_{ki} \qquad (6.a)$$

$$\Theta = (1/3)\sin^{-1}\{-(3\sqrt{3}J_3)/(2J_2^{2/3})\}\ ;\ (-\pi/6 \leq \Theta \leq \pi/6) \qquad (6.b)$$

ϕ = angle of internal friction

c = cohesion

The yield behaviour in the tension range can also be truncated by incorporating a tension cut-off σ_T. In this case the Mohr Coulomb yield criterion is a three parameter model which is defined by c,ϕ and σ_T.

Initiation of fragmentation

The process of fragment development during compressive shear failure is prescribed by
considering a Mohr-Coulomb criterion. This criterion can be written in terms of the principal stresses in the form

$$\sigma_1 \geq \sigma_c + \sigma_3\tan^2(45° + \phi/2) \qquad (7)$$

where σ_c is the unconfined strength in compression, ϕ is the angle of internal friction and σ_1 and σ_3 are, respectively, the maximum and minimum principal stresses. The unconfined compressive strength σ_c is related to the shear strength parameters c and ϕ associated with the Mohr-Coulomb failure criterion according to

$$\sigma_c = 2c\{(\tan^2\phi + 1)^{1/2} + \tan\phi\}. \qquad (8)$$

Considering the compression failure mode, there are two possible conjugate orientations of fragmentation inclined at equal angles to the direction of the stress on either side of it. In two dimensions, these are defined by

$$\alpha = \tan^{-1}\{(\sigma_1 - \sigma_{xx})/\sigma_{xy}\} \pm (90°-\phi)/2 \qquad (9)$$

where α is the angle between the fragmentation plane and the positive global axis, σ_{xx} , is the x-component of the stress tensor and σ_{xy} is the xy shear component of the stress tensor. In the computational scheme the sign of the second term on the right hand side of (9) is determined by using a random number generator. This allows for the development of a relatively random fragmentation pattern.

The tensile fragmentation criterion assumes that the material will fragment when the minimum principal stress σ_3 reaches the tensile strength of the material, σ_T. For two dimensions, the criterion can be written as

$$\sigma_3 \geq \sigma_T \qquad (10)$$

The orientation of the fragmentation plane is given by

$$\alpha = \tan^{-1}\{(\sigma_1 - \sigma_{xx})/\sigma_{xy}\} \qquad (11)$$

where, in two dimensions, α is the angle between the fragmentation plane and the positive global x-axis. In this case, the α-direction is aligned with the maximum principal stress direction.

Viscoplastic Flow vs. Fragmentation

The consideration of distinct stress states which will initiate either viscoplastic flow or fragment development can be ascertained only by appeal to experimentation. In the current study, fragment development and failure initiation for viscoplastic flow are described by the Mohr Coulomb criterion specified in terms of c and ϕ. The following procedure is adopted in the computational schemes:
The intact geomaterial will experience brittle fragmentation only for situations where any single principal stress component or both are in the tensile mode and viscoplastic flow will occur for only stress states involving purely compressive stress states for both principal stresses. In situations where viscoplastic flow occurs first, there is provision for subsequent fragmentation in tension; this subsequent fragmentation will be governed by the prescribed post peak strength characteristics of the geomaterial, which can include softening.

Size dependency of strength

Geomaterials have an inherent internal structure composed of crystals and grains and a fabric that can include random distributions of defects such as cracks and fissures. When the size of a specimen is small in comparison to a length scale of a grain or crystal, such that only relatively few defects are present, failure is forced to involve new crack growth; whereas a material loaded through a larger volume will encounter pre-existing flaws at critical locations. Thus, the geomaterial strength can be size-dependent. In particular, materials such as coal, granitic rocks and shale exhibit the greatest degree of size dependency; the ratio of laboratory to field strengths sometimes attaining values of

10 or more (Bieniawski, 1968; Johns, 1966; Pratt et al. 1972). Figure 1 illustrates the relationship between the tensile strength of ice and the largest specimen edge length, which is derived from experimental results available in the literature (Fleet Technology, 1996). As is evident, the tensile strength of ice (σ_T) increases with progressively smaller specimen size confirming the size dependency of ice strength. The relationship in Figure 1 is implemented in the discrete element model. The shear strength parameters (c,ϕ), are, however, assumed to remain constant at a prescribed reduced specimen size since there isinsufficient data available to postulate a conclusive relationship for size dependency of ice in compression.

Floe Interaction Responses

The interaction between individual floes is characterized by normal and shear interaction responses. The normal interaction response relates the normal forces to the differential normal displacements at the contacting surface. i.e.

$$dF_n = K_j\,(du_n^1 - du_n^2) \qquad (12)$$

where dF_n is the incremental change in the contact normal force per unit length between contacting floes, du_n^1 and du_n^2 are normal displacement increments at the contact plane between the regions (1) and (2) and K_j are the stiffness coefficients defined in the normal direction of the average plane at the point of contact. The shear interaction forces are related to the normal interaction forces through the coefficient of Coulomb friction μ. i.e.

$$dF_s = \mu\, dF_n \qquad (13)$$

where dF_s is the incremental change in the contact shear force unit length between contacting floes.

Constitutive Parameters

The constitutive response options include (i) elastic behaviour in the intact state, (ii) provision for fragment development in the tension and viscoplastic flow in the compression mode, and (iii) inter floe interaction responses which are characterized by an elastic-perfectly plastic response.

The constitutive parameters governing the elastic, viscoplastic and fragmented materials should be ideally selected by conducting a suitable material characterization of the intact and fragmented ice. For the purposes of numerical modelling, the material parameters are selected based on the best of experience and judgement. The values of the constitutive parameters are given below:

```
Ice Floe-Intact:
```

Young's Modulus	= 3500 MPa
Poisson's ratio	= 0.35
Cohesion	= 1.5 MPa
Angle of Friction	= 30 degrees
Tensile Strength	= 0.5 MPa

```
Ice Floe-Failed:
```

Young's Modulus	=3500 MPa
Poisson's ratio	= 0.35
Cohesion	= 1.5 MPa
Angle of friction	= 30 degrees
Tensile strength	= 0.3125/A*

```
Inter-Floe Interaction:
```

Normal stiffness coefficient	= 1000 MPa
Shear stiffness coefficient	= 1000 MPa
Cohesion	= 0.0
Angle of friction	= 30 degrees

```
Inter-Fragment Interaction:
```

Normal stiffness coefficient	= 1000 MPa
Shear stiffness coefficient	= 1000 MP
Cohesion	= 0.0
Angle of friction	= 30 degrees

```
Structural Properties:
```

Young's modulus	= 200000 MPa
Poisson's ratio	= 0.35
Proportionality constant	= 0.32
Damping ratio	= 0.2

* A = Area of the two dimensional fragment

Computational procedure for viscoplasticity:

The details of the computational procedures are documented by Selvadurai and Sepehr (1996). The accuracy of the computational scheme is governed by the time increments associated with the treatment of viscoplastic effects and dynamic phenomena. With viscoplasticity the time integration scheme is unconditionally stable if $\Omega \geq 1/2$; i.e. the procedure is numerically stable but does not ensure accuracy of the solution. Consequently, even for values of $\Omega \geq 1/2$, limits must be imposed on the selection of the time step to achieve a valid solution. For viscoplasticity problems which are based on an associative flow rule ($Q = F$); a linear flow for function $\Phi(F) = F$ and described by the Mohr Coulomb yield criterion, the recommended limit for the time increment is

$$\Delta t \leq [4(1+\nu)(1-2\nu)c\cos\phi]/[\gamma(1-2\nu+\sin^2\phi)E] \qquad (14)$$

The stability condition for the time increment Δt , which employs an explicit-explicit partition (applicable to linear systems), takes the form

$$\Delta t \leq 2\,\{(1 + D^2)^{1/2} - D\}/\omega_{max} \qquad (15)$$

where ω_{max} is the maximum frequency of the combined system involving both rigid body motion and deformability of the system; and D is the fraction of critical damping at ω_{max}. Other stability criteria have been proposed in the literature on computational methods for transient dynamic analysis; these include

$$\Delta t \leq 2/\omega_{max} \text{ and } \Delta t \leq \eta L[\rho(1+\nu)(1-2\nu)/E(1-\nu)]^{1/2} \qquad (16)$$

where η is a coefficient which depends on the element type used and L is the smallest length between any two nodes. This constraint is found to be suitable for application to non-linear systems.

PACK ICE - STRUCTURE INTERACTION

In this section, the discrete element technique is employed to examine the in plane problem of the interaction of a moving region of densely configured ice floes (pack ice) moving at a constant far field velocity of 0.1 m/sec. with a stationary, circular structure as shown in Figure 2. The basic objective of the analyses is to examine the influence of brittle fragmentation in the tension/compression mode, the size dependency of the tensile strength and the nonlinear inter-floe interaction responses on the average contact stresses between the pack ice and the stationary structure. Three geometric conditions for the moving pack ice are examined in order to investigate the influence of the location of the restrained upper boundary adjacent to the structure on the local fragmentation process and the interactive stresses. The influence of a converging pack ice channel on the local fragmentation process and the interactive stresses is also investigated.

Figure 3 shows the discrete element mesh for the case where the upper boundary is simulated at a some distance from the stationary structure. Figure 4 illustrates the extent of fragmentation that takes place after 5.0 seconds for this case. Fragmentation occurs in the region of the pack ice in the immediate vicinity of the stationary structure. The fragmentation of the ice floes in the vicinity of the stationary structure, however, proceeds indefinitely until the size of the fragments will result in the continual reduction in the time increments of the computational scheme necessary for its stability and convergence; this process will effectively terminate the computational procedure 5.04 seconds after the commencement of the computation. This is a limitation on the computing resources available rather than on the computational procedure. Figure 5 illustrates the time history of the average contact maximum principal stress between the pack ice and the stationary structure. A peak average contact maximum principal stress of 1420 kPa is predicted for the case for the duration of simulation. Figure 6 shows the discrete element mesh for the case where the stationary flexible structure is located in the immediate vicinity of the restrained upper boundary. Figure 7 illustrates the extent of fragmentation that takes place 7.4 seconds after the start of the computational process. Fragmentation occurs in the immediate vicinity of the stationary structure. Figure 8 illustrates the time history of the average value of contact maximum principal stress for this case. A peak average value of contact maximum principal stress of 1140 kPa is predicted for the duration of simulation. Figure 9 shows the discrete element mesh for the case where a pack ice channel converging on a stationary flexible structure is simulated. Figure 10 illustrates for this case the extent of fragmentation that occurs after 6.3 seconds. In this case fragmentation takes place in the region of the pack ice in the vicinity of the stationary structure. Figure 11 illustrates the time history of the average value of contact maximum principal stress for this case where a relatively higher peak average contact maximum principal stress of 2120 kPa is predicted.

CONCLUDING REMARKS

This paper summarizes the investigations that have been made to extend a two-dimensional discrete element model for the in-plane analysis of the interaction between a moving pack ice and a stationary flexible structure. Two major modifications have been implemented. The first modification accounts for the size dependency in the strength characteristics of the individual floes. The second modification accounts for the nonlinear interactive responses between the ice floes relating the normal forces to their normal differential displacements at the contacting surfaces as well as relating the contact shear force to the normal forces through a coefficient of Coulomb friction. The modified discrete element code is applied to examine the dynamic interaction between the storm-driven ice floes moving at a contant far-field horizontal velocity of 0.1 m/sec. and a stationary circular structure. The results of the numerical simulations are intended for demonstration purposes only. The accurate application of the modelling requires the identification of the elastic and strength parameters applicable to intact and fragmented ice. It is demonstrated however that the time history of the average contact stress between the moving ice pack and the stationary structure can be determined using the discrete element model. The modelling adopted in this paper considers for the first time the possible influences of fragmentation of the individual ice floes during the pack ice-structure interaction. Three different geometric conditions for the moving pack ice are simulated. The extent of fragmentation that takes place at the end of computational process as well as the time history of the average contact stress between the moving pack ice and the stationary flexible structure are determined for each case. In all cases fragmentation occurs mainly in the vicinity of the stationary structure where the pack ice-structure interaction takes plase. As expected, a relatively higher average value of contact maximum principal stress is predicted for the case of the converging pack ice channel than for the other two cases.

ACKNOWLEDGMENT

The acquisition of the DECICE Discrete Element Code was made possible by a Research Grant awarded to McGill University, by National Energy Board, Canada. The authors are grateful to Dr. Ibrahim Konuk of the National Energy Board for his continued interest in supporting research initiatives in Ice Mechanics.

REFERENCES

Bhat, S.U., (1988), Analysis For Splitting of Ice Floes During Summer Impact, Cold Regions Science and Technology, Vol. 15, pp. 53-63

Bieniawski, Z.T., (1968), The effect of specimen size on compressive strength of coal. Int. J. Rock Mech. Mining Science, Vol 5, pp. 325-335

Cormeau, A., Jordaan, J.J., Nassim, M. and Tomin, M., 1984, Ice-structure interaction: a fundamental energy based approach. Proc. IAHR Symp., Hamburg, Vol. 3, pp. 161-174

Croasdale, K., (1984), Sea Ice Mechanics : A General Overview, Marine Technology Society Journal, Vol. 18, No. 1

Fleet Technology Ltd., (1994), Private Communication

Hamza, H. and Muggeridge, D.B., (1984), A theoretical fracture analysis of the impact of a large floe with a large offshore structure. Proc. Third OMAE Symp., New Orleans, Vol. 3, pp. 291-297

Hallam, S.D. and Pickering, J.G., (1988), Modelling of Dynamic Ice Loading of Offshore Structures, Proc. POLARTECH, pp. 235-248

Hobbs, D.W., (1967), Rock Tensile Failure and its Relationship to a Number of Alternative Measures of Rock Strength, Int. J. Rock Mech. Min. Sci., Vol. 4, pp. 115-127

Hocking, G., Mustoe, G.G.W. and Williams, J.R., (1985a), Validation of the CICE Code for Ice Ride-Up and Ice Ridge Cone Interaction, Civil Engineering in the Arctic Offshore, ASCE Specialty Conference, San Francisco, pp. 962-970

Hocking, G., Mustoe, G.G.W. and Williams, J.R., (1987), Dynamic Analysis for Generalized Three-Dimensional Contact and Fracturing of Multiple Bodies, NUMETA 87, Swansea, U.K., A.A. Balkema, The Netherlands

Hocking, G., Williams, J.R. and Mustoe, G.G.W., (1985b), Dynamic

Global Forces on an Offshore Structure from Large Ice Floe Impacts, Civil Engineering in the Arctic Offshore, ASCE Specialty Conference, San Francisco, pp. 202-219

Jahns, H., (1966), Measuring the strength of rock in-situ at increasing scale. Proc. 1st Cong. ISRM(Lisbon), Vol. 1, pp. 477-482

Mustoe, G.G.W., Williams, J.R., Hocking, G. and Worgen, K., (1987), Penetration and Fracturing of Brittle Plates Under Dynamic Impact. NUMETA 87, Swansea, U.K., A.A. Balkema, The Netherlands

Palmer, A.C., Goodman, D.J., Ashby, M.F., Evans, A.G., Hutchinson, J.W., and Ponter, A.R.S., (1983), Fracture and its role in determining ice forces on offshore structures, Annals of Glaciology, Vol. 4, pp. 216-221

Perzyna, P., (1966), Fundamental Problems in Viscoplasticity. In: Recent Advances in Applied Mechanics, Academic Press, New York.

Pratt, H.R. Black, A.D. Brown, W.D. and Brace, W.F., (1972), The effect of specimen size on the mechanical properties of unjointed diorite. Int. J. Rock Mech. Mining Sci., Vol. 9, pp. 513-530

Ralson, T.D., (1981), Plastic Limit Analysis of Ice Splitting Failure, Proc. 6th POAC, Quebec, Vol. 1, pp. 205-215, Vol. 2, pp. 1434-1438

Selvadurai, A.P.S. and Sepehr, K., (1996), The role of discrete element modelling in the analysis of ice-structure interaction: Two-dimensional plane strain analysis, Department of Civil Engineering and Applied Mechanics, McGill University.

Williams, J.R., Mustoe, G.G.W. and Worgan, K., (1986), Force Transfer and Behaviour of Rubble Piles, Proc. IAHR Ice Symposium, Iowa, Iowa City.

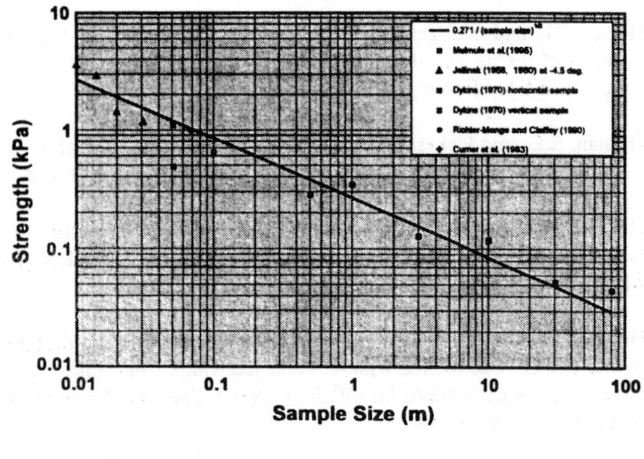

Figure 1. Size dependency on the tensile and fracture strength of ice: Experimental data.

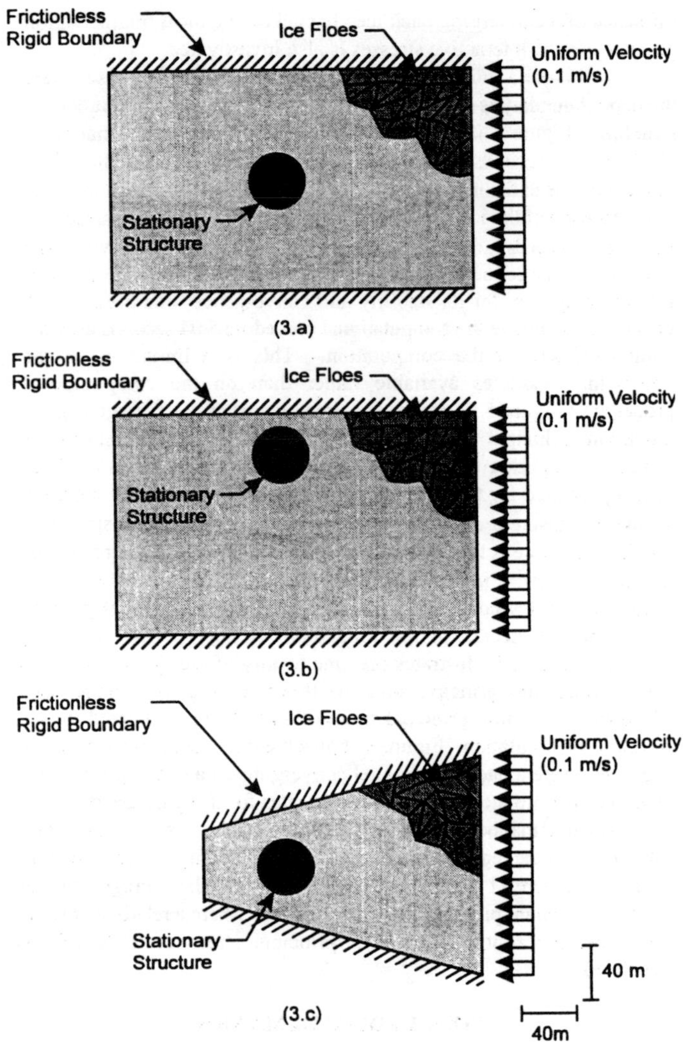

Figure 2: Schematic diagrams of pack ice stationary structure configuration.

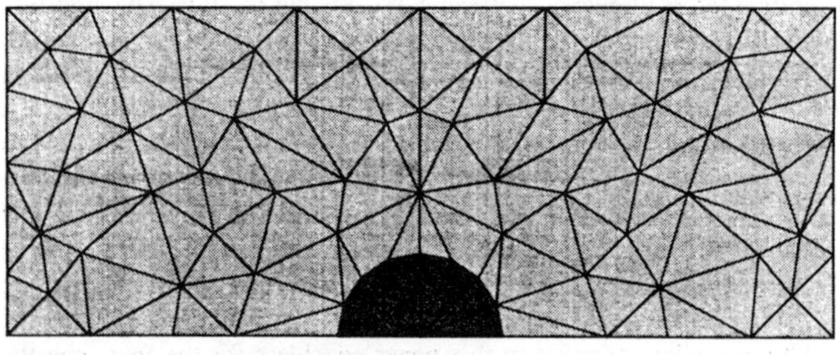

Figure 3: The discrete element mesh.

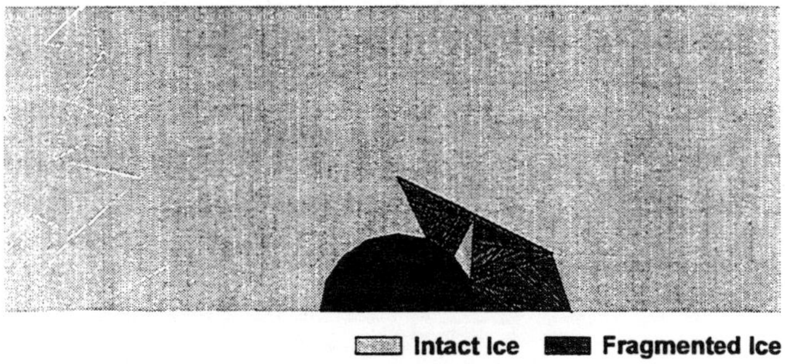

Figure 4: Fragmentation of the pack ice during the interaction process. Time = 5.0 seconds.

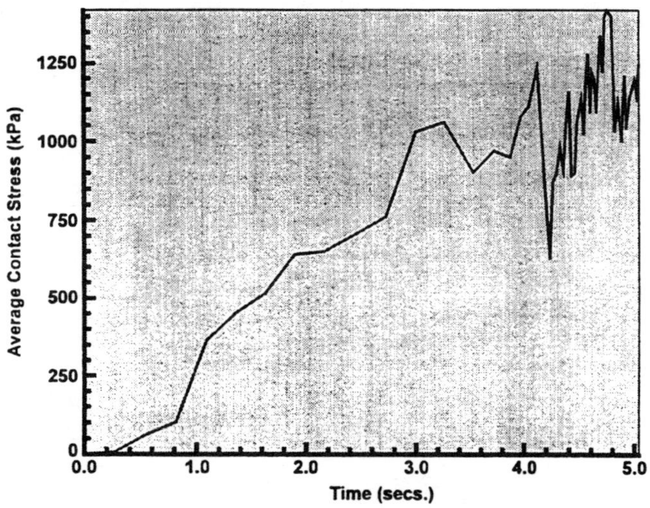

Figure 5: Time dependent variation of the average contact stress between the pack ice and stationary structure.

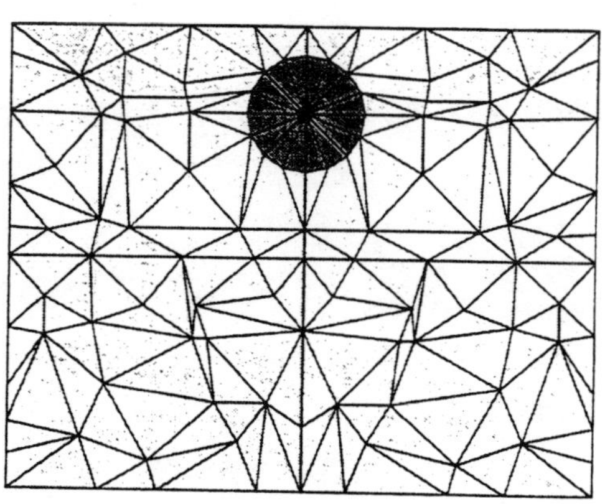

Figure 6: The discrete element mesh.

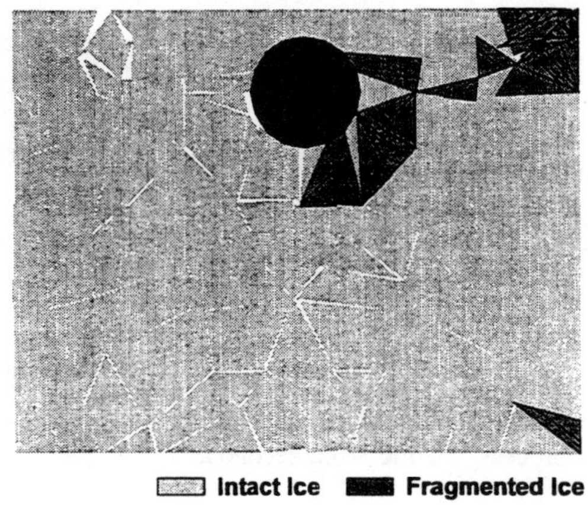

Figure 7: Fragmentation of the pack ice during the interaction process. Time = 7.4 seconds.

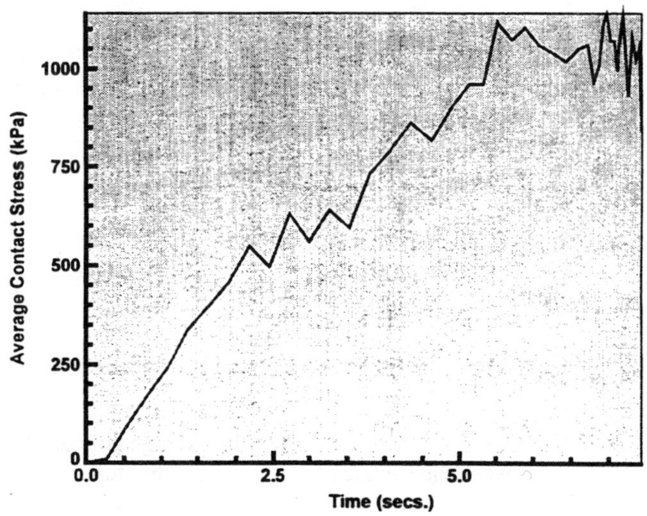

Figure 8: Time dependent variation of the average contact stress between the pack ice and stationary structure.

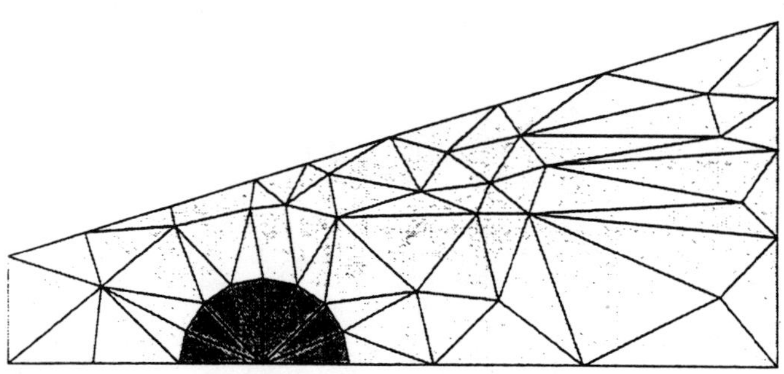

Figure 9: The discrete element mesh.

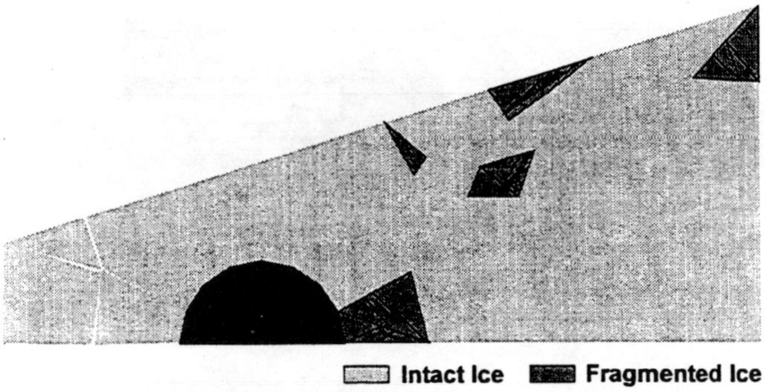

Figure 10: Fragmentation of the pack ice during the interaction process. Time = 6.2 seconds.

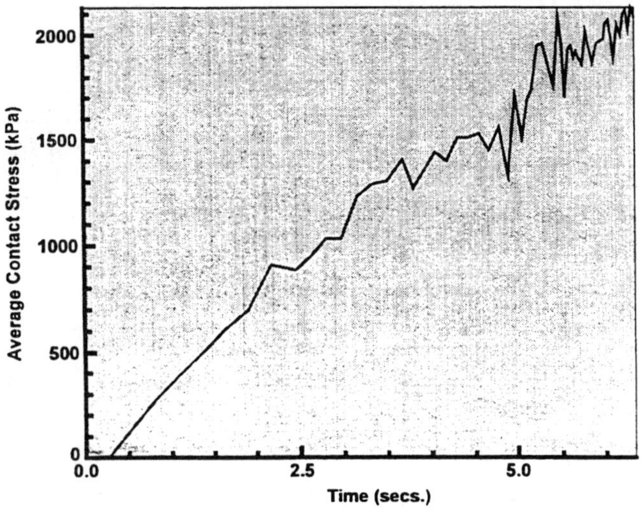

Figure 11: Time dependent variation of the average contact stress between the pack ice and stationary structure.

Proceedings of the Seventh (1997) International Offshore and Polar Engineering Conference
Honolulu, USA, May 25-30, 1997

ISBN 1-880653-28-1 (Set); ISBN 1-880653-30-3 (Vol. II)

Design Parameters for Hummocks and Grounded Hummocks in the Sea of Okhotsk

S.P. Beketsky, V.N. Astafiev and P.A. Truskov
Sakhalin Oil and Gas Institute
Okha, Russia

ABSTRACT

Consideration of hummock impact is a complex problem in designing oil and gas structures. Results of field studies of ice features serve as a basis for selecting design parameters. Data of hummock structure studies are further applied to develop "hummocks - oil and gas structure interaction" models.

This paper describes studies of both internal and external parameters of ice features on the northern Sakhalin offshore.

Based on the obtained data, design parameters for hummocks have been developed, which are necessary for adopting optimum technical solutions to provide for serviceability of future hydraulic structures.

KEY WORDS: Hummock, morphological parameters, strength parameters, ice load

INTRODUCTION

The analysis of the status of oil and gas field development in the Arctic seas indicates that offshore field development rates depend to a great extent on ice conditions and the level of their investigation. Data on ice conditions is required to ensure serviceability of hydraulic structures.

Hummocks and grounded hummocks will be the most dangerous for hydraulic structures on the northern Sakhalin offshore.

Consideration for ice hummock impact is a complex problem in designing oil and gas facilities. Design parameters are selected based on data obtained as a result of field studies of ice features. Data on hummock structure is further used to derive hummock - oil and gas structure interaction models. Basically, the available models describe hummock geometry, without representing the internal structure of a hummock. This is attributed to limited possibilities in obtaining data on internal structure of hummocks.

Ice hummocks are known to have heterogeneous vertical structure. Brash ice which forms due to hummocking by gravity and buoyancy moves towards the center of a hummock and fills voids between ice blocks. Therefore, porosity for middle part of a hummock (with a center being at the sea level) is lower than that for lower and upper parts of a hummock (Grishhenko, 1988).

Generally, a hummock model is currently used in engineering practice which includes upper part or "sail", middle or consolidated part and lower part or "keel". Occasionally, a hummock model is used which consists of two parts - "sail" and "keel". "Keel in this model is divided into consolidated layer and unconsolidated layer.

Field studies aimed at obtaining hummock structure and strength have been performed for fifteen years by Sakhalin Oil and Gas Institute on the northern Sakhalin offshore in the area of the discovered oil and gas fields. The studies cover visual observations of hummocks and grounded hummocks (stamukhas), geodetic serves and thermal drilling.

The northern Sakhalin offshore environment is noted for active ice cover dynamics, which means that ice cover suffers continuous deformation in winter and spring, such deformation being accompanied by hummocking and ice field failure.

MORPHOLOGICAL PARAMETERS OF HUMMOCKS

Three scenarios for hummocking on the northern Sakhalin offshore are distinguished based on hummocking conditions (Beketsky, 1996).

1. Formation of hummock ridges due to impact of drift ice floes on fast ice.

Under this scenario, the highest hummocking frequency is observed in case of combined impact of onshore winds, currents and tides resulting in total horizontal pressure of ice cover towards the coast. Sakhalin Bay is a region where a combination of the above-mentioned factors is in effect.

A zone of hummocking frequency equal to three points is present in Sakhalin Bay as early as in January. This zone develops in December when the highest daily tidal fluctuations of the sea level and maximum wind velocities are observed. Hummocks which are formed during this time period are basically 50-60 cm in thickness.

Formation of such hummock types on the north-eastern offshore shifts to the end of February and the beginning of April as compared to that in Sakhalin Bay. This is related to the average time periods for fast ice formation and failure in the region, which fall on middle ten days in June and last ten days in April respectively. Hummocking in fast ice edge occurs during this time period due to impact of drift ice floes under the influence of north-eastern, eastern and south-eastern winds. Ice blocks in such hummocks may be over 1m in thickness.

2. Formation of hummocks within boundary zones between drift ice and fast ice to the north of Schmidt peninsular.

In winter seasons, total water circulation in the Sea of Okhotsk, north-western monsoons and physic-geographic conditions in the region under study force movement of compact ice from north-western Sea of Okhotsk to the northern Sakhalin Bay and northern North Bay. This movement results in formation of permanent boundary zones between fast ice and drift ice which are noted for ice compression and hummocking.

Under this scenario, random piled up ice and hummock ridges are formed which are forced to the north-eastern Sakhalin offshore together with drift ice floes. Thickness of such hummocks varies from 20-30 cm to 1-1.2 m depending on the time of their formation.

3. The third scenario describes hummock formation due to interaction of two drift ice floes. As movement of two adjacent ice floes may differ both in velocity and direction, interaction of such ice floes results in ice hummocking. In this case, long straight hummocks are formed which are made up of ice cake, generally, no more than 70-80 cm in thickness.

Ice hummock age is very important in studying drift hummocks and analyzing their internal structure. The paper (Beketsky, 1996) gives the method for identifying hummock age based on difference in thickness of level unrafted ice at a hummock and thickness of ice blocks which make up a hummock. The method assumes that a hummock drifts together with ice the ice floe in which it was formed.

The sum of air temperatures, Σt_c for the design time period is the basic estimation factor. Special observations of ice accretion were performed to derive the design equation for ice thickness as a function of the sum of the degree-days of frost. Approximation of the obtained data resulted in plotting ice thickness against the sum of the degree-days of frost. The plot is a parabola (Beketsky, 1996):

$$h_L = 3.1\left(\sum t_c\right)^{0.47} \qquad (1)$$

where h_l is in (m) and t_c in (C°)

Sail height H_s, keel draft H_k, sail side slope angle θ_s keel side slope angle θ_k, sail width B_s, keel width B_k and thickness of level ice at the hummock are the basic parameters describing hummock geometry. Sizes of ice blocks that make up a hummock are also important.

Ice blocks in hummock sails were measured in the course of studies. The measurements covered ice block thickness(h_0), length (l), width (b) and dip angle (α). Based on these data, average l/ho = 3.1 and the average l/b =1.6. The average dip angle (α) for ice blocks was found to be equal to 35^0, which was close to the natural slope for piled up ice.

Hummock height depends on thickness of ice feature, in which it was formed. The analysis of relationship between sail height, H_s and ice block thickness, h_0 was performed for the northern Sakhalin offshore. This relationship is described by the following equation(Beketsky, 1996):

$$H_s = 4.7\sqrt{h_o} \qquad (2)$$

where H_s and h_l are in (m)

Two scenarios for hummock-hydraulic structure interaction are to be considered for the northern Sakhalin offshore environment.

1. Impact of a hummock which was formed from a grounded hummock. In this case a hummock may be considered as a homogenous body, its consolidated part being very large and its porosity being low. Data on size and integral strength of such a hummock is required to determine maximum load exerted by it.

2. Impact of a hummock having a three-layer structure. The thickest hummocks were found to develop within a permanent boundary zone between drift ice and fast ice. Such hummocks may be 35 m thick. In this case data on both hummock geometry and thickness distribution and strength for each layer are required.

It should be noted that fast ice fracturing and breaking off is very frequent on the Sakhalin offshore which is characterized by highly dynamic ice conditions. As a result, fast ice fields with hummocks and grounded hummocks are potential sources of impact on hydraulic structures.

Consolidated part of a hummock that has been present in fast ice for a long time period may reach 50 to 60% of total hummock size. This results in increasing design loads exerted by such hummocks on ice-resistant platform legs. Thus, hummocks and grounded hummocks in fast ice are very important subjects of studies.

Over thirty hummocks and grounded hummocks were drilled to study hummock morphology. The most representative results were obtained for three regions: Sakhalin Bay, North Bay and Odoptu field. The paper (Beketsky et al., 1996) gives data on hummock structure for the Sakhalin offshore.

In view of insignificant difference in total thickness of hummocks under study for various regions, the discrepancy in thickness for consolidated parts of hummocks is basically attributed to hummock age. Fig 1. gives a plot of h_c against H_t to determine vertical size of a consolidated part of a hummock. These relations were plotted for hummocks of various ages: 0,5 month, 1 month, 2 months, 3 months. The plots show considerable dependence of h_c/H_t relationship on hummock age.

The results of the studies performed on hummocks in fast ice were used to obtain the design model for a hummock in the Sea of Okhotsk offshore of Sakhalin (Fig.2) (Beketsky, 1996). This model describes both geometry and internal structure of grounded hummocks. As total thickness of a grounded hummock is not dependent on sea level variations, all other parameters are expressed in terms of this value.

Impact of drift hummocks and hummock ridges on offshore facilities will be one of the most probable cases of maximum design loads. More forty hummock ridges were studied on the northern Sakhalin offshore to investigate hummock structure. Analysis of the obtained profiles should be performed with consideration for highly complicated pattern of hummock ridge profiles. It is very difficult to perfectly identify a profile of a separate hummock. Generally, one hummock ridge merges with the other. As a result, one sail may correspond to two keels or two sails may correspond to one keel. Profiles were identified based on all of the obtained data to derive a design model for a hummock of a triangular cross-section, each of these profiles having one sail and one keel.

Fig.3 gives a design model of a hummock with triangular cross section for the Sea of Okhotsk (Beketsky, 1996). As sail height, H_s is the easiest to obtain, all other dimensions are expressed in terms of this value. This model has more gently sloping sail as compared to those used by other authors. This is primarily attributed to the age of the hummocks under study. Generally, the earlier models describe young newly formed hummocks. The age hummocks under study was over one month. High thickness of level ice near hummocks is also worth mentioning. This high thickness is attributed to rafted ice which is always present at hummock ridges. Thus, this paper gives rafted ice thickness which served as a reference for estimating side slope angles for a sail and a keel. Generally, hummocks in the Sea of Okhotsk attain their maximum height and draft when their sail width becomes as large as 20-25 m and keel width reaches 60 m. At this stage, hummock ridges begin to grow in width. Triangular profiles of such ridges transform to trapezoidal ones (Beketsky, 1996).

It should be noted that hummocks having large base width are of main interest for future studies as their load on offshore structures represent one of the design scenarios.

The analysis of all the profiles of hummock ridges under study resulted in obtaining a trapezoidal cross-section both for a sail and a keel. Sail axis and keel axis were found not to coincide; the displacement between them was equal to 5-10 m. Such displacement of a sail against a keel is equally probable in both directions (Beketsky, 1996).

A three-layer model including upper part, consolidated part and lower part was adopted for describing the internal structure of drift hummocks in the Sea of Okhotsk. The studies resulted in obtaining the design profile for a hummock ridge in the Sea of Okhotsk (Fig.4) (Beketsky, 1996). As sail height, H_s is the easiest to define, all other parameters are expressed in terms of this value.

STRENGTH PARAMETERS OF HUMMOCKS

Data on ice strength (integral hummock strength in particular) is required for determining loads exerted by such ice features.

The difficulty in obtaining such data is primarily attributed to large sizes of ice features under study. Conventional method for determining strength based on tests performed on drilled out ice cores do not give reliable strength data for a hummock as a whole. Method for determining strength of a simulated hummock made up of ice blocks on the basis of similarity criteria is associated with difficulty in reliable simulation of hummock genesis, which has considerable effect on test results. The paper (Truskov, 1992) gives the method for determining hummock strength based on study of grounded hummocks.

In the course of field study of grounded hummocks, broken grounded hummocks were encountered with tidal fractures going through them. Multi-year observations of grounded hummock formation as well as ice conditions around them indicate that breaking of such features is most probably caused by enlargement of a sail due to piling up of drift ice blocks. Based on this phenomenon and the strength theory, the integral strength of a natural hummock was estimated.

Tests on ice samples taken from a hummock simulated in fast ice in Chayvo Bay indicated that in case of complete filling of voids between ice blocks with congealed ice, confined compression strength for a hummock σ_t is essentially comparable to level ice strength σ_l. On this basis, strength of consolidated ice in a hummock may be assumed equal to strength of ice of thermal origin. Strength of upper and lower parts of a hummock depend to a great extent on porosity values.

Cubic blocks from simulated hummocks were tested for confined compression to obtain ice strength as a function of ice porosity. The procedure for simulating ice hummocks was developed by Surkov (Surkov, Truskov, 1993). Table 1 gives data on simulated hummock tests.

Table 1

Results of confined compression tests on cubic blocks of simulated hummocks

Parameter	№ sample				
	1	2	3	4	5
Initial porosity n_0, %	46	45	44	46	47
Porosity n_t, %	33	27	18	16	0
Ultimate strength σ_t, MPa	0.249	0.413	0.641	0.676	1.7
Relative ultimate strength σ_t/σ_l	0.146	0.243	0.243	0.4	1

CONCLUSIONS

The analysis of multi-year observations and performed studies resulted in obtaining morphology and strength for ice hummocks as well as their correlation.

One of the important results of the performed studies is the method which was developed for estimating maximum design parameters for hummocks required to design facilities for the Sakhalin offshore (Beketsky, 1996). The method is as follows:

1. Maximum thickness of level unrafted ice is calculated by eq.1 on the basis of the sum of the degree-days of frost for the region under study.
2. Maximum sail height is determined for maximum design ice thickness by the relation (2) defined for the region.
3. Hummock geometry including sail width, keel width and draft and total hummock thickness is obtained based on sail height (Fig.3).
4. Total thickness and age of a hummock serve as initial data for determining internal structure and strength of a hummock.
5. Consolidated layer thickness and total porosity of a hummock as a function of hummock age are determined based on total thickness of a hummock. (Fig.1).
6. Integral compressive strength of a hummock is calculated based on porosity value (Table 1).

Two nomograms were plotted (Beketsky, 1996) to apply the proposed method by designers. The first nomogram is aimed at obtaining morphometric parameters (Fig.5) and the second one to obtain internal structure and integral strength of a hummock (Fig.6).

This method may be applied for any particular case of offshore structure design for the Sea of Okhotsk which is insufficiently studied with respect to hummocks.

REFERENCES

Beketsky, SP (1996). "Morphology and strength of hummocks and grounded hummocks in the Sea of Okhotsk," *Synop. of thesis of Ph.D. of geogr.*, S-Petersburg: Arctic and Antarctic Institute, pp.1-25 (in Russian).

Grischenko, VD (1988). "Morphology of hummock ridges in ice of the Arctic basin," *Proc. of Arctic and Antarctic Institute*, L., Vol.401, pp.46-55 (in Russian).

Beketsky, SP, Astafiev, VN and Truskov, PA (1996). "Structure of grounded hummock in the sea of Okhotsk," *The 11-th International Symposium on "Okhotsk Sea and Sea Ice"*. Mombetsu, Hokkaido, Japan, pp. 339-342.

Surkov, GA and Truskov, PA (1993). "Method for evaluation of first-year ice hummock strength," *Proc. of the Third ISOPE Conf.*, Singapore, Vol.II, pp. 640-642.

Truskov, PA, Beketsky, SP, Surkov, GA, Polomoshnov, AM, Bekker, AT (1992). "Strength Parameters of Hummocks," *Proc. of the Second ISOPE Conf.*, San-Francisco, USA, Vol.II, pp. 783-789.

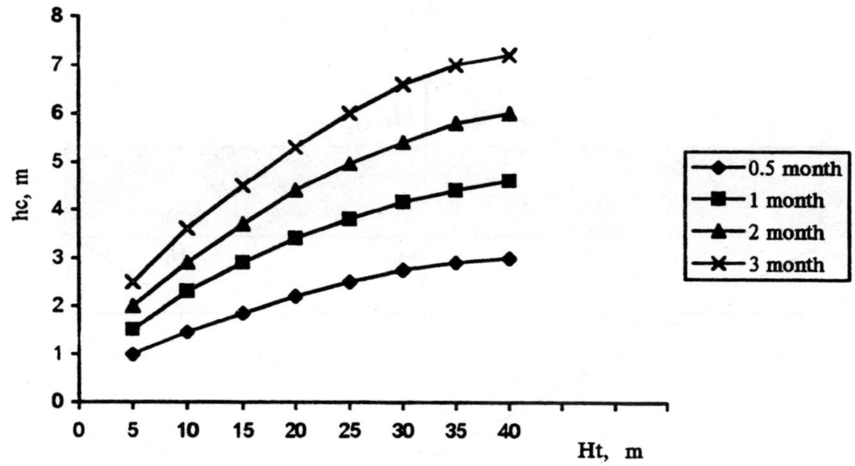

Fig. 1. Plot h_c as a function of H_t - for hummocks of different ages

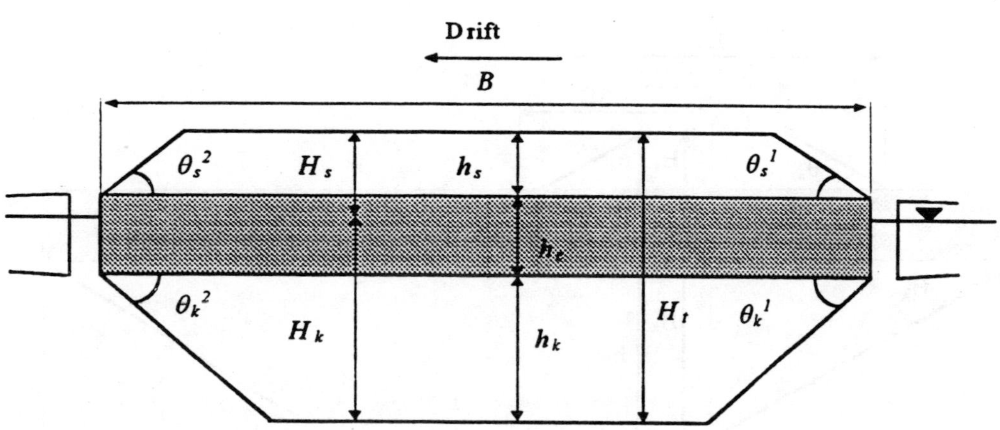

$h_s = 0.16\ H_t$ $\quad H_s = 0.23\ H_t$ $\quad \theta_s^1 = 23°$ $\quad B/L = 0.5$

$h_c = 0.32\ H_t$ $\quad H_k = 0.77\ H_t$ $\quad \theta_s^2 = 32°$

$h_k = 0.52\ H_t$ $\quad B = 2.7\ H_t$ $\quad \theta_k^1 = \theta_k^2 = 45°$

Fig. 2. Design model for grounded hummock in the Sea of Okhotsk

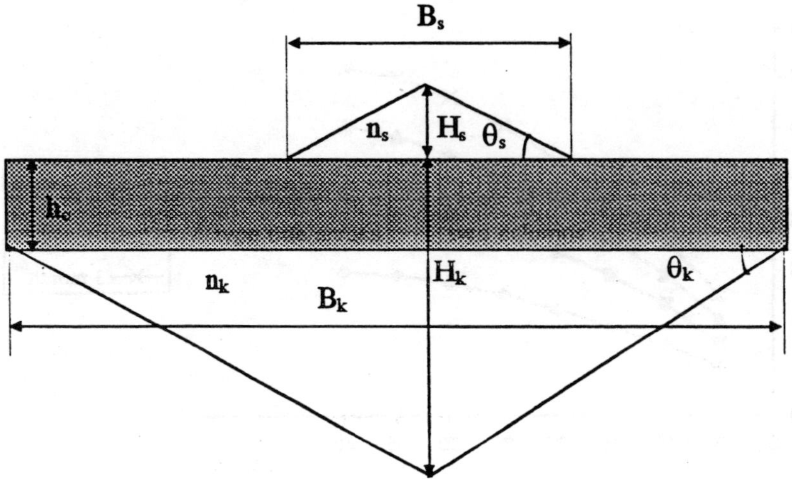

$H_k = 5\ H_s$ n_s=0.26 $h_c = 1.15\ H_s$
$B_s = 7\ H_s$ n_k=0.29 $\theta_s = 14°$
$B_k = 19\ H_s$ θ_k= 24°

Fig. 3. Design model for hummock in drift ice field in the Sea of Okhotsk

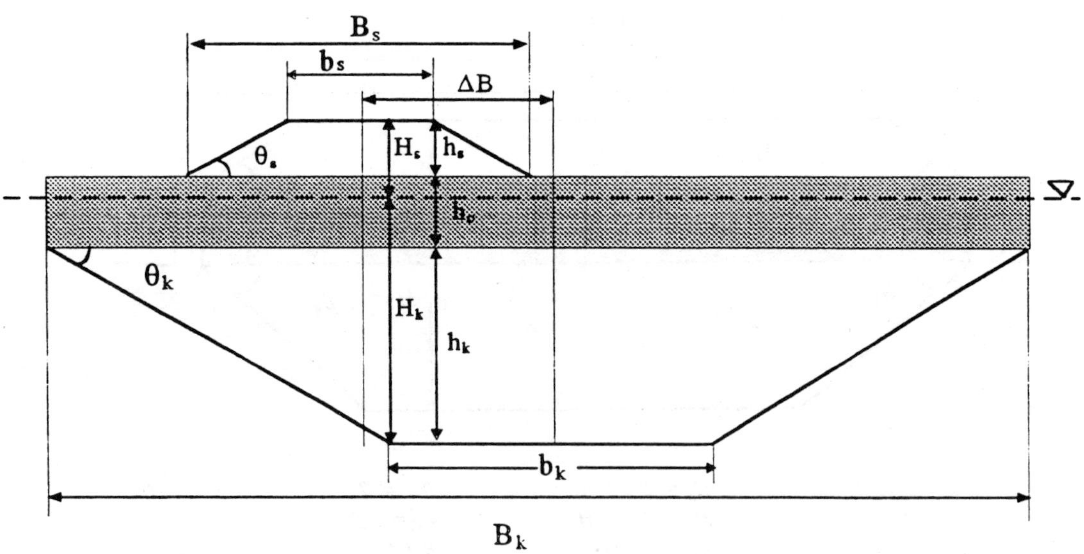

$H_k = 5.25\ H_s$ $b_k = 4.5\ H_s$ $h_c = 1.25\ h_s$ $\theta_s = 13°$ n_s=0.24
$b_s = 1.7\ H_s$ $B_k = 23\ H_s$ $h_k = 4.5\ h_s$ θ_k= 26° n_k = 0.26
$B_s = 9\ H_s$ $\Delta B = 3\ H_s$

Fig. 4. Design profile of a drift hummock ridge in the Sea Okhotsk

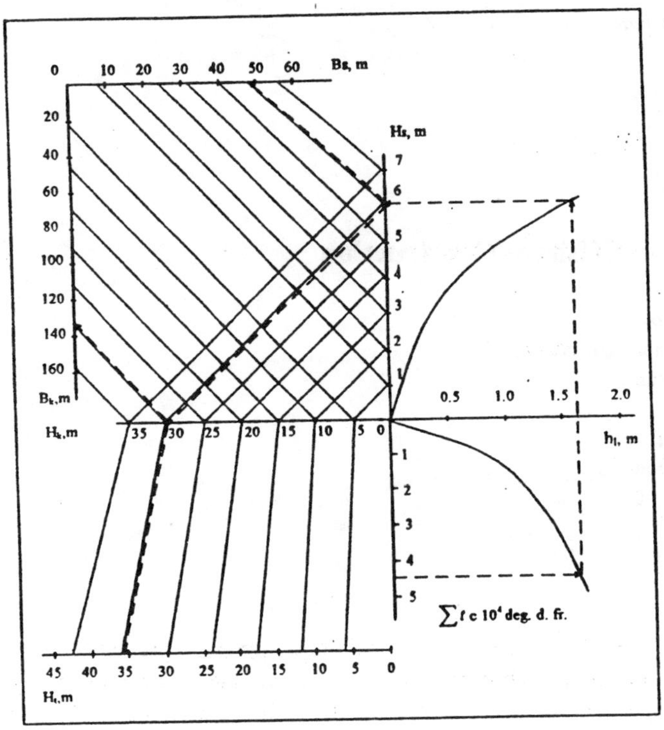

Fig. 6. Nomogram of determination of external parameters of hummocks

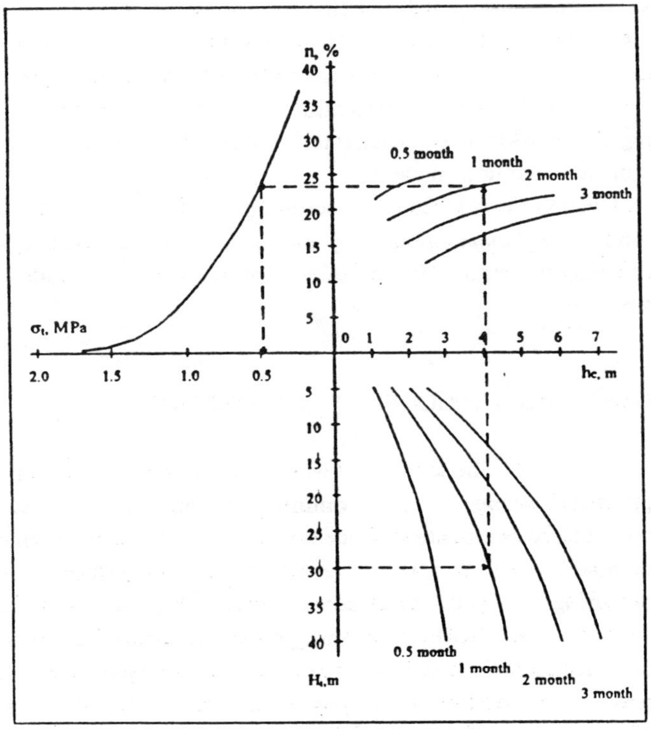

Fig. 7. Nomogram of determination of strength and internal structure of hummocks

Proceedings of the Seventh (1997) International Offshore and Polar Engineering Conference
Honolulu, USA, May 25-30, 1997

ISBN 1-880653-28-1 (Set); ISBN 1-880653-30-3 (Vol. II)

Description of Sea Ice Regime for Offshore Construction

A.T. Bekker
Far-Eastern State Technical University
Vladivostok, Russia

I.L. Appel
Fairweather Forecasting
Anchorage, AK, USA

INTRODUCTION

Increasing development of exploration of offshore fields of ice-infested seas is presents the modern tendency of sea oil and gas industry. The ice cover in winter season cause the several problems for offshore carbons fields exploration. The ice force analysis on offshore structures is greatly difficulty. The last ones are unique engineering structures. Therefore the increasement of ice force definition accuracy have the great importance accounting for influence on material expense and cost of these objects.

Numerical methods have found wide application analysis and forecasting of ice conditions. But the similar calculations were usually oriented on providing for navigation and description of processes of geophysical scale in the system "the atmosphere-ice cover-ocean". As a result of coarse resolution the corresponding mathematical models of ice cover evolution for large areas or whole seas cannot be directly used to calculate ice influence on offshore structures.

Existing approach to ice force analysis was based upon results of detailed observations of ice cover parameters and ice physical properties. But such an approach is very money consuming as need the continuous multi-year field observations series on ice regime variability at the site.

We propose the combining method of describing ice regime at the site that is based upon joining two approach together. In this case the parameters of ice regime of general character, for example, drift velocity, concentration, thickness distribution are calculated on the base of numerical model of ice cover of geophysical scale. More detailed features of ice cover state, for example, small-scale irregularities in spacechanges of ice characteristics, rigidity, ice cover shapes, are described by partial models received on the base of specialised observations on-situ.

The two models of different scales are successfully supplement each other and give the reliable information about all parameters of interest for ice force analysis. Proposed combined ice regime model at the site is considered to be effective means for analysis and forecasting of all range of available forces applied to different kinds of offshore structures.

So, the combined model describe the part of ice regime parameters in large scale with deterministic nature. The other parameters of common model are received on the base of treatment of field observations series with stochastic nature.

The combination of deterministic integral parameters of ice regime and stochastic characteristics of ice cover is done in range of imitation model.

This report includes the description of large-scale ice regime model and local model of ice cover features which are critically important for definition of ice forces on offshore structures.

GEOPHYSICAL MODEL OF ICE REGIME

In spite of abundance of works and advance in research, presently, there is no common notion even of some compulsory components of ice models. It is explained by more complexity of ice cover evolution in comparison with occurring in liquid or gaseous media. The ice cover has number of peculiarities governing its mathematical description. In mathematical model the more amount of parameters are need for description of ice cover state as for liquid or gas. Hence, a whole hierarchy of the models with different amount of parameters appears to be possible.

The characteristic features of different models greatly depend on the goals addressed by their authors and the temporal and spatial scales used. In this paper we lay emphasis on those model aspects that are important to interaction problems of ice cover with offshore structures.

Ice cover models as a geophysical phenomenon use typical spatial resolution of the order 10,000 m and bigger. The influence of minor ice cover features on its evolution is being parameterized. In particular, the similar approach is proposed to take into account narrow boundary layers of the ice shear deformations and subgrid inhomogeneity of the ice thickness. We need to remind that finer resolution is not reasonableness will notincrease accuracy of calculations at all.

Dynamic part of the model simulating forces and drift velocity affecting the ice cover, is of the great importance for the ice force analysis. The equation of momentum balance is the basis for drift velocity, internal stresses and interaction forces calculation. The tangential stresses of air and water on upper and bottom ice surfaces are the total result of tangential and form drag. The influence of the latter drag is clearly depends upon ice morphology that is calculated in the model using parameterization of existing observation on elements of ice topography.

The influence of internal stress in ice cover, accompanied by energy dissipation, produces a considerable effect on ice drift. T choose the way to describe the internal interaction appears to be quit complicated and important. Let us consider some general assumptions to model internal interaction in ice cover.

Many variants of internal interaction description were taken from mechanics of continuous media. But in difference from mechanics the coefficients of the constitutive law should not only reflect the ice physical properties. They should be considered as functions of thickness, concentration and type of deformations in ice cover.

One should emphasise the need to take into account some features of ice cover, differing its mechanical behaviour from the ice treated as a physical body.

First of all, the ice cover responses differently to the compression and tension. The tensile stresses can not exist in ice cover. Secondly, in the most cases, compressive and tensile strains are presented in ice cover simultaneously. The normal stresses are significantly anisotropy. Thirdly, ice cover response differently on shear and compression.

In the framework of an isotropic continuum description of ice cover it is expedient to use the principal axis of stress and strain tensor for division of shear and compressive stresses. But the ice cover is essentially anisotropy media. The approach to model ice cover anisotropy behaviour, first of all formation and evolution of leads, is proposed in another report on that Conference (Appel, 1997). The problem of leads modelling in ice cover is critically important for navigation but can be considered definitely as minor under study of ice cover influence on offshore structures. The marked difference is explained by different choice of ship motion but offshore structures have no this possibility.

We account for drift under pressure or opposite in description of boundary conditions for zone of internal stresses. The influence of narrow subgrid boundary layer is included to model for case of pressure drift near shore. The drift velocity on external bound of this layer is considered equal to zero as boundary condition for main calculation area if the drift is directed into zone but the internal stresses near bound is accounted for by usual manner.

If the model is created for practical usage than the heat factors need to account for even the short-time analysis. The modelling of ice growth and destruction under influence of heat processes is based on equation of heat balance accounting for heat streams of different sources: radiation, turbulence and connected with heat transfer.

The real processes of thermodynamic formation and destruction of ice in nature principally differ from evolution of hypothetically level uniform ice cover. The characteristic spatial scale of ice thickness inhomogeneity is much smaller than spatial resolution used in the model. Therefore we use a parametric description of subgrid changes in ice thickness and snow cover depth.

One should remember that at the present in choosing the number of calculated ice thickness categories for the practical purposes the observation data is not accurate in most cases. Therefore we guess that three or four categories of thickness seem to be maximum number for practical use. But the parameterization of thickness inhomogeneity including the pressure ridges for each separate ice category is included into mathematical description .

A more detailed description of the thermal processes inside ice massive is based on calculation of the vertical ice temperature profile and takes into account changes in ice salinity, density, thermal conductivity, heat of fusion, and penetration of short-wave radiation into ice.

We can greatly correct the calculations of ice thickness change in time accounting for the heat of fusion and thermal conductivity are not constant. Detailed calculations of heat flows through the ice massive gives us necessary information about vertical distribution of ice properties determining ice strength influencing on ice forces.

Mathematical description of ice regime implies simulation of ice cover redistribution for long time intervals. This circumstance need calculation of kinematics changes in ice cover state parameters under drift influence without accumulation of large errors. But problem solution out of kinematics changes definition of ice cover state parameters is not usual. In particular during calculation of advection influence the usage of an explicit-in-time finite difference schemes is connected with computational errors of order as from parameter change influence. The "viscosity" of numerical approximation come to pseudodiffusion which level the solutions. It is seemed that the any method accounting for only parameter values of ice cover in nodes of net area lead inevitably to great errors of calculations. We adopt to mind the local inhomogeneity of parameters in cells of net area with a help of calculation not only it average value but also the first moments of distribution space correspondingly to cell centre.

The suggested approach greatly increase the calculation accuracy of kinematics changes of ice cover parameters. The calculation errors of suggested method on two orders less than errors of evident finite difference approximation.

The numerical method of ice cover evolution modelling for geophysical scale was widely used for areas with different ice regimes. In particular method has been used to stimulate the interannual variations in ice cover state. The long-term calculations reliably average seasonal changes in ice cover state as interannual variability in ice distribution without accumulation of significant errors. An accuracy of calculation does not depend upon sign of value of ice anomaly or its magnitude. Atlases of ice regime and long-term ice simulations and forecasts can represent the basis for strategic planning and ice forces analysis.

The current supplement with ice information and short-term forecast includes satellite observations of ice state and numerical modelling of those parameters that can not be observed. To support offshore oil operations a three-to-four forecast is considered as a primary planning tool and one-to-two forecast allows us account extreme events such as hazardous pressure in ice cover or conditions leading to shutdown of rig operations.

Detailed knowledge about influence of ice conditions on the offshore structures is a mandatory requirement to create and utilise system of ice management for offshore activity on ice-infested waters.

The main parameters of offshore ice regime on concrete area (relating to characteristic spatial of grid discrete element of numerical model) and received from numerical model of geophysical scale are: concentration, drift velocity, ice cover thickness, stress distribution in it and etc. The concrete values, for example, concentration and drift velocity as time functions, can be used for description of ice regime accounting for that the numerical model reflect process in real time.

TRANSITION FROM GEOPHYSICAL CALCULATIONS TO LOCAL CONDITIONS

Large-scale ice cover models give valuable information on forces and deformation characteristics in ice cover. But those data correspond to large scale processes that differ significantly from real forces on structures. Thus, there is obvious necessity to realise transition from large-scale average parameters to local conditions. There are two different features of the transition. The first involves transition from a continuum medium to composite discrete elements and the second corresponds to transition from deterministic parameterization to a stochastic description.

The transition from continuum in global numerical model to discrete media in local model require a special solution relating to some parameters of ice regime. For example, the concentration of ice cover in-situ is estimated according to data of numerical calculation by special procedure. The joining of deterministic description in numerical model with stochastic model is done with a help of imitations modelling. It allow to receive the averaging of multi-year results.

The transition results must be verified and the real data play the important role in model calibration and examination of calculation reliability. The direct changes of local stresses and deformations together with field observations for ice cover characteristics are main information source about specific local conditions in considered area.

In addition to in-situ changes and observations, analyses of SAR imagery can be considered as a promising approach to study ice cover local properties in the vicinity of oil exploration and extraxtiob sites or along navigation routes. Using transition from large scale processes to local conditions allows us to predict ice forces and other parameters influencing on operations in ice-infested waters.

LOCAL MODEL OF ICE REGIME

At the present time our knowledge of many processes forming ice cover is insufficient to reliably it describe them. A number of the shortcomings of ice modelling is related to incorrect or insufficient representation of local specific features of physical processes. This explain an exceptional importance of experimental researches allowing us to more accurate define intercorrelations present in the phenomena under study.

At the same time, the observation results for specific local features of ice cover regime can be also used for parametrization of these observed features immediately including into the ice regime mathematical description.

Ice cover consists of basic discrete elements. The processes of ridging, rafting, sliding and others redistribute energy of large scale deformations to smaller scale. The latter determines design forces on offshore structures. The small-scale definition of ice media, in particular, the floe sizes, ice thickness, stress and pressure in ice cover is an important mandatory step in force analysis on offshore structures.

The following parameters of ice regime need to know for ice force analysis on stationary ice-resistant structure at concrete time moment: ice formation kind (field, ridge etc.) influencing on structure, it plan geometrical spatulas, ice thickness, drift velocity, concentration (interinfluence of neighbour ice formations), interference of drift velocity vector and axis of structure (for separate formations), strength characteristics of ice formation, it structure, added mass.

The ice regime description in concrete site of sea water area, an area of ice-resistant offshore structure construction, is based on multi-year field observations for ice cover parameters. In common case insufficiently continuos row of special observations (5-10 years) one can receive the basic information about ice regime for forecast of ice forces on structures.

The information about ice forces on structure need for based design and supplement of need level of reliability. As it is known, the reliability of engineering structures is characterised by probability of normal object exploitation for exploitation period without achievement of limit state (refuse). The refuses can be immediate or gradual (Bekker, 1995). Therefore the requirements to accuracy of ice regime information are sufficiently strict.

As a rule, the observation results for ice regime parameters are presented as histograms or probabilistic distributions of these parameters. The probabilistic characteristics of these distributions should be based on sufficiently continuous observation rows. The distribution data is basis for building the ice regime models in-situ of sea water area. The kind and character of such models are defined by taken purposes. For example, in case of evaluation of ice structure reliability on position of achievement of immediate refuse the special attention should be given to extremal values of ice forces which are defined by extremal values of ice regime parameters (ridges, giant fields etc.). In this the evaluation of extramal values distribution of studying parameters plays an essential role.

So, the mathematical model of ice regime will include the complex of ice regime parameters expressed as probabilistic characteristics of its distributions accounting for correlation connections.

Not only extremal force values but it average values must be known during consideration of reliability of structure on position of gradual refuse (destruction from fatigue damage accumulation in material). It is cause the necessity of such information obtaining and study of probabilistic distributions of ice regime parameters in all range of it changes. In this case the most good is to form the ice regime mathematical model in form of imitations model of processes of ice force formation on structure in all range of changes (Bekker, 1995). This model bear the more common character and can be used for the first case.

The creation of similar models need the great expenses of material means because for obtaining of initial information the full complex of expensive technical means is used: the Earth satellites, aircraft, helicopters, coastal radar's, floating and bottom devices, helicopter man groups, special equipment for research of ice properties and parameters of ice formations.

REFERENCES

Appel, IL (1997). "Numerical Modelling of Ice State for Ship Operation," *Proc. of the 7th Int. Offshore and Polar Eng. Conf* (ISOPE-97), Honolulu, Vol 2.

Bekker, AT, Perepelitsa, AN, Uvarova, TE (1989). "The Probabilistic Aspects of Hydrotechnical Structures Design on Ice Loads," *Proc. of all Union Conference "Ice-89,"* S-Peterburg, (in Russian).

Bekker, AT, Seliverstov, VI, Uvarova, TE (1995). "Definition of Loading Regime for Offshore Structure from Drifting Ice Cover," *Proc. of 7th Int. Symp. on Okhotsk Sea Ice*, Mombetsu, Japan.

Proceedings of the Seventh (1997) International Offshore and Polar Engineering Conference
Honolulu, USA, May 25-30, 1997

ISBN 1-880653-28-1 (Set); ISBN 1-880653-30-3 (Vol. II)

A Methodology for Developing a Scientific Basis for the Ice Regime System

G.W. Timco and R.M.W. Frederking
National Research Council
Ottawa, Canada

V.M. Santos-Pedro
Transport Canada, Prairie & Northern Region
Ottawa, Canada

Abstract

Transport Canada has developed an Ice Regime System which defines the conditions in which vessels can safely navigate in ice-covered waters. This System, which was developed in support of the Arctic Shipping Pollution Prevention Regulations, has important implications for ships operating in Canadian waters and potentially for polar navigation. To ensure that these regulations represent realistic guidelines for ship navigation in ice, it is necessary to have them based on a scientific approach. To this end, a logical approach has been developed to provide a platform for this scientific basis. The approach is based on an empirical evaluation of the behaviour of ships in ice-covered waters. It is developed in 7 different Tasks. This paper provides a short introduction to the existing Ice Regime System, and presents the methodology for developing a scientific basis for the Ice Regime System.

Introduction

Navigation in Canadian waters north of 60 degrees north latitude is regulated by Shipping Safety Control Zones. A Zone/Date matrix gives entry and exit dates for various ship types and classes. It is a rigid system with little room for exceptions, relying on nature to consistently follow a regulatory table.

The Ice Regime System is intended to introduce the necessary flexibility by making use of actual ice conditions, ultimately as seen by the Master. Responsibility to plan the route, identify the ice, and carry out a simple numeric calculation rests with the person in charge of the watch. Authority by the regulator to direct ships in danger or during an emergency remains unchanged. Due care and attention of the mariner, including avoidance of hazards (say, icebergs) is vital to the successful application of the Ice Regime System.

The need for a simple, effective, and fair system for shipowner and regulator alike, meant that it is empirical, based on the experience of navigators gained over the last 20 years using modern ships and advanced navigation equipment.

Accurate, reliable and timely ice information contribute greatly to introduction of a navigation control system that is responsive to rapid changes in climate conditions. As designed, the system will favor ships with the most knowledge and information.

Credibility of the new system has wide implications, not only for ship safety and pollution prevention but also in lowering ship insurance rates and predicting ship performance. Many verification voyages using the Ice Regime System attest to its validity and usefulness as well as some shortcomings - while certain sea passages may be safely done at low speed by a very knowledgeable and experienced navigator, the simplicity of the system may result in a negative entry.

Any changes to the Ice Regime System must be carefully done, in a controlled manner. Therefore, the need to identify the scientific basis of the system is of primary importance.

Transport Canada approached the National Research Council of Canada in Ottawa to assist them in developing a methodology for establishing a scientific basis for the Arctic Ice Regime System (AIRS). This lead to a "road map" approach which is based on 7 Tasks (Timco and Frederking 1996).

In this paper, the existing Ice Regime System is discussed, and shortcomings of the system are presented. The development and rational behind each of the 7 Tasks are discussed, along with the framework for the implementation of the proposed approach.

The Ice Regime System

An ice regime is any area composed of a relatively even distribution of any mix of ice types, including open water. The World Meteorological Organization (WMO) nomenclature is used to describe the ice types. These ice types are defined on the basis of both ***age*** and ***thickness***. Multi-year ice is usually thicker and stronger than first- or second-year ice. The ice thickness is inferred from the ice type.

In the proposed regulations, the ship's structure must be designed to withstand impacts with a maximum thickness of ice (ASPPR 1989). This is the thickest ice type in which a properly navigated ship may operate without risk of structural damage. The regulations define 9 categories of vessels such that each ship category is related to an ice type. Table 1 presents the nine ship categories along with the corresponding ice type for each category. Note that no equivalent Arctic classes of other Classification Societies are given in the table, since equivalency can only be done on a ship-by-ship basis.

An ice regime has two components: 1) ice that is above the "category ice" for the vessel; and 2) ice at and below the "category ice" for the vessel. These two components always total $10/10^{ths}$. In any given ice conditions, the ratio of the two components will differ between ship categories, since each ship category has a different ice type. The risk of damage, therefore will depend on the proportions of hazardous ice (i.e. above the category ice) and non-hazardous ice (i.e. at or below the category ice) in the regime, and on the ability of the ship to avoid the hazardous ice types.

Table 1. Ice Type for each Ship Category

WMO ICE TYPE	RANGE (m)	SHIP CATEGORY
Multiyear Ice (MY)	> 3	CAC 1
Second Year Ice (SY)	> 2	CAC 2
Thick First Year Ice (TFY)	> 1.2	CAC 3
Medium First Year Ice (MFY)	0.7 - 1.2	CAC 4
Thin First Year Ice - Second Stage (THFY2)	0.5 - 0.7	TYPE A
Thin First Year Ice - First Stage (THFY1)	0.3 - 0.5	TYPE B
Grey-White Ice (GW)	0.15 - 0.3	TYPE C
Grey Ice (G)	0 - 0.15	TYPE D
Open Water (OW)	0	TYPE E

To quantify the ice regime, a scheme has been developed which takes into account the relative amount of each type of ice, and relates it to the ship category. This is reflected through an **Ice Multiplier (IM)**. The value of the Ice Multiplier reflects the level of danger that the particular ice type poses to the particular category of ship with the larger negative numbers representing larger hazards. Table 2 lists the Ice Multiplier values for each ship category.

For any ice regime, an **Ice Numeral (IN)** is calculated by taking the sum of the products of the concentrations of the ice types present (in 10^{ths}), and their ice multiplier. The Ice Numeral is defined as:

$$IN = (C_a \text{ X } IM_a) + (C_b \text{ X } IM_b) + \ldots \qquad (1)$$

where IN is the Ice Numeral, C_n is the concentration (in tenths) of ice type "n", and IM_n is the Ice Multiplier from Table 2. The right side of the equation is expanded to include all the types of ice that are present, including open water. The Ice Numeral is therefore unique to the particular ice regime and category of ship operating within its boundaries. The system takes into account ice decay and ice ridging. For *decayed* ice, the multipliers are increased by 1 for second-year ice, and thick and medium first-year ice. The numbers for multi-year ice are not changed. If *ridges* are present in a concentration greater than $6/10^{ths}$, the multipliers are decreased by 1. This reflects the higher level of risk presented by the thicker ice.

Table 2 Ice Multipliers for each Ship Category

ICE MULTIPLIERS (IM)		SHIP CATEGORY						
Ice Type and Thickness		TYPE					CAC	
		E	D	C	B	A	4	3
MY	> 3	-4	-4	-4	-4	-4	-3	-1
SY	> 2	-4	-4	-4	-4	-3	-2	1
TFY	> 1.2	-3	-3	-3	-2	-1	1	2
MFY	0.7 - 1.2	-2	-2	-2	-1	1	2	2
THFY2	0.5 - 0.7	-1	-1	-1	1	2	2	2
THFY1	0.3 - 0.5	-1	-1	1	1	2	2	2
GW	0.15 - 0.3	-1	1	1	1	2	2	2
G	0 - 0.15	1	2	2	2	2	2	2
OW	-	2	2	2	2	2	2	2

For any given ship category, entry or non-entry into an ice regime is determined by the sign of the Ice Numeral. If the IN is *negative*, entry is not allowed. However, if the IN is *zero* or *positive*, then entry by the ship into that ice regime is allowed.

In order to use the Ice Regime System, a vessel must have access to relevant ice information that is both accurate and current, and as importantly, is reliably made available for shipboard decision making. Clearly, this type of ice information is required to define details of the ice conditions that will be encountered along a particular vessel's transit route and in turn, calculate Ice Numerals. The system is designed to enable vessels that have comprehensive information on the ice conditions to proceed, as long as it can be done safely. Vessels that have minimal information on the local ice regime may be hampered in certain circumstances.

In Canada, information regarding the ice cover is usually obtained from the range of products that are provided by the Atmospheric Environment Service's (AES) Branch of Environment Canada. The most important product that AES currently provides to vessels is the daily ice chart, which summarizes regional ice conditions in areas of interest to active marine operations. The information that is used to generate these ice charts is in transition. It includes visual and imaging radar observations (SAR and SLAR) acquired from periodic AES surveillance aircraft overflights, various forms of satellite imagery, and "as available" ship, helicopter reconnaissance and shore station reports. Although most mariners now feel that the level of ice information is adequate, there is no doubt that the use of new technology, such as RADARSAT, will significantly alter the amount, accuracy and timeliness of ice information. These changes, coupled with new commercial products and a willingness of the shipping community and AES to continue communicating regarding ice information needs and adequacy, should improve the level of ice information that is currently made available to support voyage planning and execution, and the application of the Ice Regime System.

For the past several years, there has been over 20 field trials to evaluate the suitability of the Ice Regime System (Norland 1994). In addition, there has been considerable discussion of the system, most notably at dedicated workshops on the topic (Dickins 1990; Norland 1991; Daley 1994). From these trials and workshops, a number of important points have been raised:

- On several voyages, negative ice numerals were calculated, at least for short segments, yet the ship traversed the ice safely;
- Different ice observers have estimated quite different concentrations of ice types for the same ice regime;
- The Ice Regime System does not take into account
    - vessel speed
    - visibility
- Ice information supplied by the AES often underestimated the initial growth of the ice, and was not informative on the details of second year ice distribution in a region;
- Ice information from the AES does not take into account the decay of first-year ice;
- Information from AES was not always presented in a clear fashion for the Ice Regime System. Ice charts are sometimes "cluttered" with information;
- The importance of ice ridging did not seem to be adequately addressed;
- Quantifying the effect of ship escort was difficult;
- The amount of time that a ship will be in a specific ice regime is important. This time directly relates to the risk of damage;
- The variance of experience of the various ship Masters is not considered in the System.

With the successes and concerns regarding the Ice Regime System, it is prudent to ensure that the system is based on a scientific foundation, both for reliability and confidence in its use. A proposed methodology is presented in the next section.

Overall Methodology for the Scientific Basis

It is important to have a logical strategy in order to develop the Ice Regime System in a scientific manner. This section will describe an overall methodology to implement this change. In developing the strategy, there were a number of key points that were used as guidelines:

- The System should be based on the ice conditions that cause problems for each class of ship - these have to be determined through specific events;
- The System must focus on addressing issues related to ship safety and pollution prevention, and not on the operational performance of the vessel;
- The System must be designed such that it is easy to apply by the ship operators;
- The System must satisfy the needs of Transport Canada in their application of pollution prevention measures;
- The System could also account for extrinsic conditions, other than ice characteristics - for example, ship speed, poor visibility, etc.;
- The System should try to make use of available information on the ice characteristics from both on-board and remotely-sensed systems, with its application in a very pragmatic manner.

The development of this system must include all of these factors. Figure 1 shows the context diagram for the proposed Arctic Ice Regime System. It shows the essential inputs into a scientifically-based system, and the important inter-relationship between the safety regulations of Transport Canada and the necessary operational efficiency of Ship Operators.

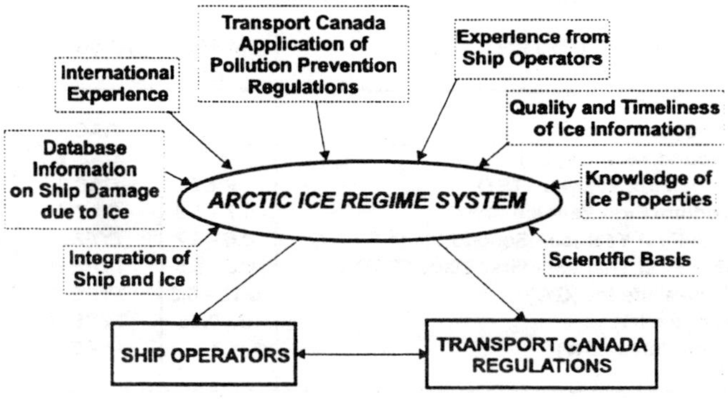

Figure 1 Context diagram for the Ice Regime System.

In developing the present methodology, a very simple strategy was employed. The system is based on asking and answering seven basic questions. Each question is a logical extension to the answer of the previous question. The basic questions are:

1. What problems can happen to a ship in ice?
2. What are specific examples of problems that have occurred which could be used for a deterministic development? i.e. specific case-histories that can be used to identify and understand the problems.
3. Would the current ice regime system predict these problems?
4. If not, how can the problem conditions be better defined?
5. Can the current ice detection methods identify the problem ice conditions?
6. If not, how can the ice detection systems be improved in a pragmatic manner to be able to detect the problem ice?
7. How can this information be communicated to the ship to implement the Ice Regime System?

The answer to each question provides the input into the next question in a logical fashion. Figure 2 shows a general flowchart for the proposed methodology. The work is divided into 7 Tasks that correspond to each of the questions, procedure and answers to the above 7 questions. Clearly, in answering the questions, important information will be required from a variety of sources. In addition, detailed knowledge of ice characteristics, ship behaviour and operational procedures will be key ingredients in the overall plan. There must be a number of verification procedures implemented to ensure that the methodology is based on actual field data. Each of the Tasks will be briefly discussed.

1. ***Define Safety-Related Issues*** - This task will systematically review historical data on the safety-related problems that a ship could encounter in ice. The output of the task is a classification of safety-related issues and identification of a comprehensive set of data sources.

2. ***Definition of Specific Problems with the Corresponding Ice Conditions*** - In this task, a large number of specific examples of ship-ice problems will be detailed. A major effort will be made to assess all available data sources including those developed by Transport Canada and international projects (*e.g.* Keinonen *et al.* 1992; Vefsnmo *et al.* 1995). In using these data sources, the events will be viewed to develop an understanding of the ice characteristics that caused the problem for the ship, as well as any other extrinsic circumstances such as vessel speed, visibility, etc. Cases where both damage and no-damage occurred will be identified. This data will provide a scientific guide to the type of ice characteristics that can cause problems for each class of vessel. A specific effort will be directed to *quantifying* the degree of damage based on the ship- and ice-related characteristics. Analysis will be carried out to investigate the damage as a function of ship speed, momentum and kinetic energy, with respect to the ice thickness, strength and concentration. The output from this task will be a series of "damage maps" that show the ice conditions that may cause damage for each vessel class.

   This information, organized in this manner is the key to developing a scientific basis for the ice regime system.

3. ***Assess Adequacy of AIRS*** - In this task, the current Ice Regime System will be used to assess how accurately it would have predicted damage (or no-damage) to the vessel for all cases developed in Task 2.

4. ***Definition of Problem Ice and Operating Conditions*** - In this task, the ice conditions that cause problems for each of the different class of vessels will be better defined. Detailed knowledge of ice properties and operators' experience are crucial to the successful completion of the task. New approaches to improve the definition of the ice regime will be explored. At the end of this Task, the Arctic Ice Regime System will be based on scientific evidence related to known ship behaviour in ice conditions.

5. ***Identification of Problem Ice*** - Once the "problem" ice conditions are known, it is necessary to see if they can be easily identified using the on-board and/or remote ice detection techniques. In this task, each of the techniques will be briefly reviewed with an eye towards the success of each technique for detecting the problem ice conditions. An important part of this process will be the essential ingredient of quick transmission of the ice conditions to the vessel. There will be a number of important contributors to this task, including ship operators, the Canadian Atmospheric Environment Service, and the Canada Centre for Remote Sensing.

6. ***Detection of Problem Ice*** - If the current detection techniques cannot provide the necessary information, a pragmatic study of the type of improvements that could be made will be undertaken. Developing expensive detection and/or analysis systems that may or may not work will not be part of this task. If current detection techniques are judged adequate, information from some of these systems will be analyzed in a different manner in order to provide the necessary information for the implementation of AIRS. To ensure the practicability of this task, the input of people with a knowledge of ship operations, remote sensing, and ice properties will be sought.

7. ***Implementation of AIRS*** - The final task will be to tie together the Ice Regime System with the available ice information in a user-friendly package for the vessel. A key ingredient will be the ease of use and general confidence in the methodology by the ship operators. There are several potential approaches that could be developed to perform this task. The most appropriate will have to be developed with input from the ship operators and the AES. At the conclusion of this task, there will be a viable ice regime system that is based on scientific data and is user-friendly to the end user. As an adjunct to this Task, the full implementation of the System will be achieved with adequate education and communication to those responsible for using and applying it.

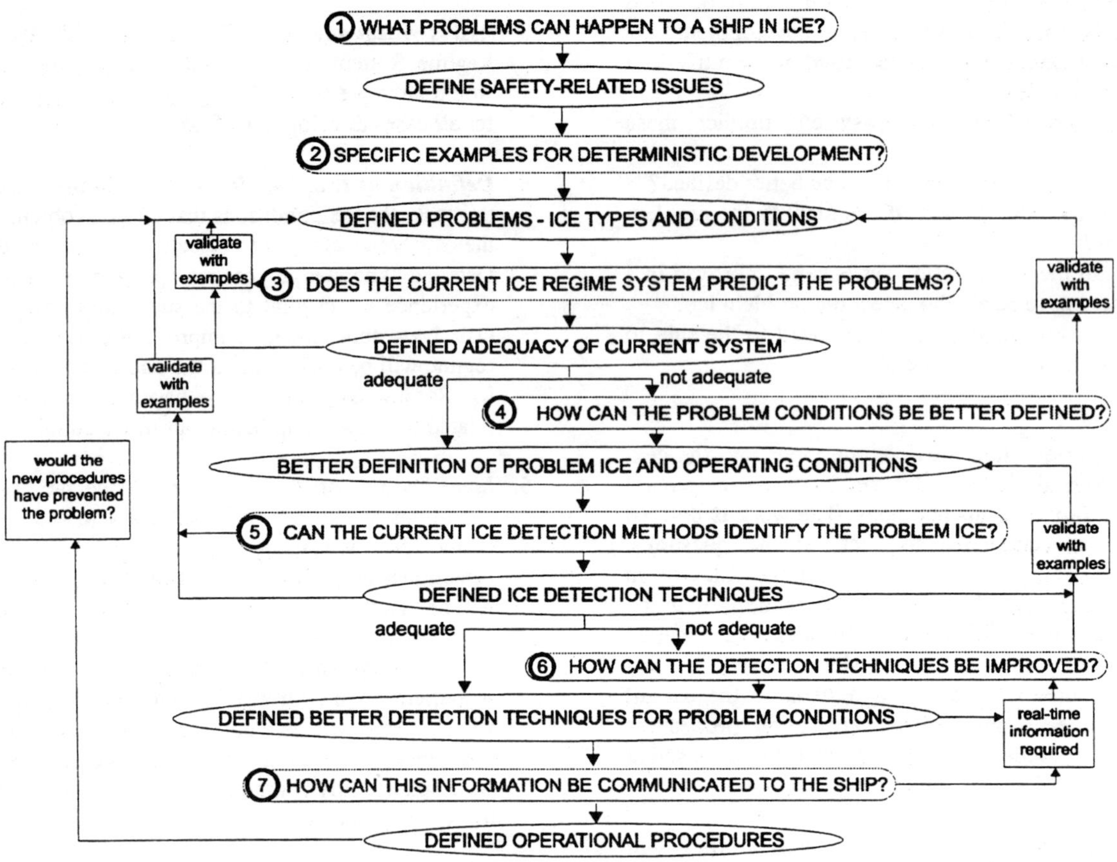

Figure 2 General flow chart showing the overall approach.

Relation to Other Ice Navigation Systems

In Russia the "Ice Passport" (Likhomanov et al. 1993) and the "QAD" system (Brovin et al. 1996) are used to manage ice navigation. The "Ice Passport" is unique to each vessel and sets out a safe operations envelope which relates speed limits to ice conditions. The "QAD" system is operationally oriented, taking information on ice thickness and length of transit in each ice thickness to optimize route selection. Developing a link between the Russian systems and the AIRS is a goal of the project.

The AIRS is based on ice information derives from the WMO ice classification. The classifications are based primarily on ice thickness as originally interpreted from aerial over-flights. Now, with remote sensing satellites, it is possible to obtain ice images in digital form in various spectrums. There is considerable effort underway to relate this digital information to the WMO classifications and also derive additional information about the ice conditions. The European Union is supporting one such project to improve the characterization of ice regimes; i.e. developing a more representative means of characterizing the "ice state" (Riska, personnel communication). The ice state could well be a more objective means of defining ice conditions and one which may be more relevant to defining safe and efficient operations of vessels in ice.

Summary and Conclusions

This paper has presented an overview of a proposed Canadian approach for developing a scientific basis for the Ice Regime System. The core of the approach is a verified relationship between the ice conditions and vessel category and low risk of damage. The approach is structured in 7 Tasks that each are formulated in terms of a question, a methodology for answering the question, and an answer to the question. When complete, this work should provide a valuable tool in support of the safe and efficient operation of ships in ice.

Acknowledgements

The authors would like to acknowledge the helpful comments on this work supplied by Brian Wright, Mohamed Sayed, Claude Daley, Andrew Kendrick and Bryan Kerman.

References

ASPPR, 1989. Proposal for the Revision of the Arctic Shipping Pollution Prevention Regulations. Transport Canada Report TP 9981, Ottawa. Ont., Canada.

Brovin, A., Tsoy, L. and others. 1996. Planning and Risk Assessment Volume 2 - 1994 Project Work. INSROP working Paper No.34,-1996, I5.5.

Daley, C. 1994. Workshop on the Scientific Basis of the Ice Regime System. Report to Canadian Coast Guard Northern, Transport Canada Report TP 12255, Ottawa, Ont., Canada.

Dickins, DF. 1990. Workshop Results: Experience with the Proposed Ice Regime Control System. Report to Canadian Coast Guard Northern, Transport Canada Report TP 10612, Ottawa, Ont., Canada.

Keinonen, A., Revill, C. and Brown, R. 1992. Review of the Proposed Revisions to CASPPR Study Area C: Damage. Report prepared for Memorial University, St. John's, Nfld.

Likhomanov, V.A. and others. 1993. Principles for the Development and Prospective Use of the Ice Passports. Proceedings POAC'93, Vol. 1, pp 219-226, Hamburg, Germany.

Norland Science & Eng. 1991. Workshop on the Ice Regime Shipping Control System. Calgary, Al. (unpublished).

Norland Science & Eng. 1994. Field Evaluation Trial of the Proposed ASPPR Ice Regime System CCGS Des Groseilliers. Report to Canadian Coast Guard Northern, Transport Canada Report TP 11921, Ottawa, Ont., Canada.

Timco, G.W. and Frederking, R.M.W. 1996. A Methodology for Developing a Scientific Basis for the Ice Regime System. National Research Council of Canada Report HYD-TR-009, Ottawa, Canada.

Vefsnmo, S., Lovas, S., Backlund, A. And Ranki, E. 1995. INSROP Natural Conditions and Ice Navigation: Content of Database, Planning and Risk Assessment. SINTEF Report STF60F95085, Trondheim, Norway.

Proceedings of the Seventh (1997) International Offshore and Polar Engineering Conference
Honolulu, USA, May 25-30, 1997

ISBN 1-880653-28-1 (Set); ISBN 1-880653-30-3 (Vol. II)

Real-Time Monitoring of River Ice Floes

Brian Morse, Marc Choquette and Marc Savard
Canadian Coast Guard
Quebec, Canada

ABSTRACT

Pertinent, timely information is a critical key to any modern, successful management system. Knowing when and where several thousand tons of ice wants to travel is a dramatic case in point. To provide real-time data on ice conditions and weather forcing factors along the St. Lawrence River, the Canadian Coast Guard (CCG) - Laurentian Region equipped critical key sites along the river with real-time cameras and load cells to measure ice forces. In the particularly critical portion of Lac St. Pierre, CCG installed a experimental, specially-instrumented, Ice Monitoring Station along with a permanent radar and a real time weather station. With the Ice Monitoring Station, the CCG can now measure ice thickness, concentration and speed in real-time. This paper describes what makes the station so successful and presents some preliminary measurements.

KEY WORDS: River ice, real time, mesurement, ice floe, ice thickness, ice concentration, ice flow rate, Saint Lawrence river, ice management

1. ICE MANAGEMENT ON THE ST-LAWRENCE RIVER

To see the true importance this experimental station, we can recap the history of ice management along the river. The Saint Lawrence River, downstream of Montreal, drains the Great Lakes, the Ottawa River basin and most of the southern portion of the province of Québec (Ellis, 1980). Historically, increases in water levels related to ice jams caused flooding more so than increases in spring runoff. Ice jamming happened twice each year: in late December, when ice cover on the river formed and later during spring break-up.

The first efforts at ice management on the river followed in the wake of particularly serious flooding at the turn of the century. The Canadian government decided to deploy ice-breakers. When the warm weather came, ice-breakers would steam up river from Québec City all the way to Montréal. This action avoided ice jams at spring break-up by helping regulate the orderly evacuation of ice.

The late 1950's and early 1960's saw the construction of hydro dams upstream of Montreal on both the Saint Lawrence River and the Ottawa River. There were plans to expand two islands in the river at Montréal to accommodate the "EXPO 67" world fair. The Canadian government initiated new studies in ice management to ensure that the expansion of the islands would not aggravate flooding.

Accordingly, the government developed and implemented a river ice management strategy. It built ice booms and artificial islands at strategic locations along the river to ensure formation of stable ice covers wherever possible. It deployed a new fleet of ice breakers to prevent flooding both during ice cover formation and spring break-up. These initiatives aimed to have a restricted channel within the river that would ensure that all ice formed upstream had an exit path. They also minimized the ice production potential by forming stable covers everywhere possible (except in this restricted channel). By the time Expo 67 opened, the ice management strategy was so effective that commercial shipping season to Montreal continued through the winter months.

What had started as a flood protection initiative led to the creation of a year-round, world-class port at Montreal. Nevertheless, for a few days every winter, bad weather conditions (temperature and wind) generated more ice upstream than the restricted channel was able evacuate downstream. From time to time, a persistent imbalance at lake Saint-Pierre (see Cornett et al... 1997 for location map) led to ice blockage and eventually to ice jams. Water levels rose and commercial navigation temporarily halted (Simard, 1993).

Traditionally, the bottle neck was "Lac Saint Pierre," a section of the St. Lawrence River, 70 km downstream of Montréal (Morse, 1994) Here, in this mild sloping portion of the river, there are relatively weak currents (Boudreau et al. 1993). In February 1993, there was a particularly severe jam. In the aftermath, the CCG decided to remedy and modernize ice management on the river (Clavelle et al. 1993):

at the structural level, we:
rebuilt artificial islands;

repaired weirs in the Sorel delta;

improved ice booms at Lavaltrie and Lanoraie;

constructed and deployed a new set of ice booms at Yamachiche (on Lac St. Pierre);

at the management level, we:

reviewed our ice-breaker deployment strategy

negotiated agreements among the CCG, the local ports, the river pilots, the industry and other CCG Regions to ensure coordination of all efforts;

at the technological development level, we:

we conceived and installed a real-time ice data information sites (cameras, radars, load cells, etc.) that will eventually be integrated into a global management system. Of particular interest for this paper is the experimental Ice Monitoring Station site.

2. REAL TIME ICE MONITORING STATION

The Ice Monitoring Station is located in the centre of Lac St. Pierre:

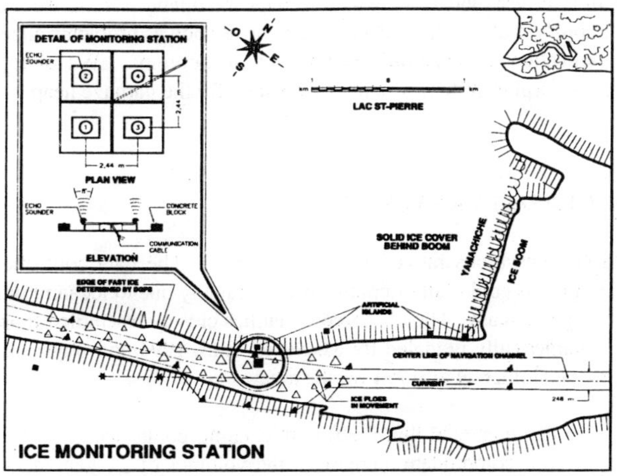

Figure 1. Sketch of Ice Monitoring Station

Four upward looking sounders are mounted on the corner a steel frame, spaced 2.44 m apart. A shielded cable joins the sounders to an instrument hut located on a nearby artificial island. The platform is weighted down with concrete blocks. The platform contained the sounders are placed in about 13 m of water on the bottom of the navigation channel.

The sounders and the associated electronic data acquisition system were surplus transducers retrieved from a retired sounding vessel. The sounder transducers ("Navitronic TC4") operate at a frequency of 200 Khz and have a footprint 1.8 m corresponding to a cone of 8 degrees. A Navitronic MCS-1 unit controls the transducers and sends data to the Navitronic Depth Pre-Processor from which it is captured by a PC 486 serial port. The signal resolution of the sounders is 1.5 cm.

Every twenty minutes, the data acquisition software on the PC calves a new raw data file and sends it to a second PC in a local network. The CCG has real-time access to the second PC via a cellular phone using REACHOUT software. In the winter of 1995-1996, we acquired data every second from each of the four transducers. In the winter of 1996-1997, we acquire data at 3 Hz. We developed special software in QBasic to analyze ice data and provide the water level, the ice concentration, thickness, speed and flow rate. This program now runs remotely at the site giving us easy real-time access to pertinent summarized information.

In the winter time, the site is used to record the passage of ice floes. Nevertheless, it does not rest in summer. Then, the site is used to record ship-induced and natural waves. For this purpose, CCG retained Carter and Anctil to develop software (in VBasic) to run on the second PC. It also analyses wave data in near real-time and creates an up to date output file every 20 minutes.

In winter, as the ice floes in the navigation channel pass over the instrumented platform, the sounders send out echoes. These echoes are reflected back from the bottom surface of the ice (or in the absence of ice, from the water/air interface). As long as the sounders are not covered by weeds or sand, most of the recorded data is accurate.

3.1 DATA ANALYSIS: DETERMINING THE WATER LEVEL

The first step of the data analysis determines water level. Normally, the greatest recorded value should correspond to the water/air interface. However, there are waves along the water and there are also data spikes. So, it is more pertinent to assume that the water surface is at a distance corresponding to a certain percentile (e.g. 99%) of the data.

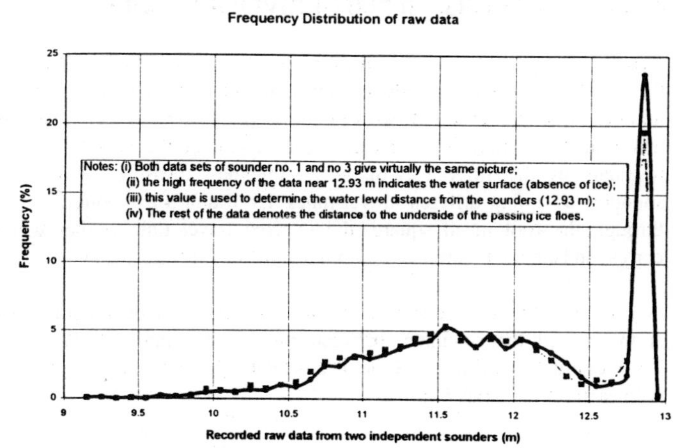

Figure 2. Frequency Distribution of Raw Data

Figure 2 shows the frequency distribution of the raw data for two sounders. Clearly, two distinct populations are present. The first has a very sharp peak frequency (about 22%) and corresponds to the echoes returned from the water/air interface. This peak frequency is easily detected unless there is 100% ice concentration. Under normal conditions, the water level estimate coming from each of the four sounders is within a range of 1 to 2 cm.

3.2 DATA ANALYSIS: ICE CONCENTRATION & THICKNESS

The second population is more largely distributed over a range of values. Once filtered and subtracted from the determined water level, it corresponds to a range in ice thickness as floes pass by (Figure 3). An important aspect of the data analysis is the consistency of the data from one sounder to the other. When the instruments are working properly, we find the same data distribution signature on each of the transducers. Normally, mean thickness vary by less than 5 cm from one sounder to the other. By keeping track of the respective times that the data represents ice rather than the water/air interface, the ice concentration is calculated. Usually it agrees to within 5% from one transducer to the other.

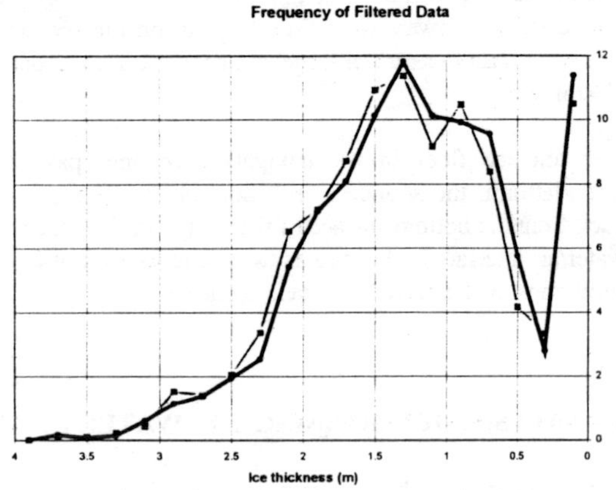

Figure 3. Frequency Distribution of Ice Thickness.

3.3 DATA ANALYSIS: SPEED OF RIVER ICE FLOES

Determining the speed of floating ice is a little more complex. Each of the ice floes that pass sounders number one or two pass over sounders number three or four about two to three seconds later. We calculate the travel time by analyzing the root mean square of the residuals. The residuals are the differences of the ice thickness measured at different sounders assuming different travel times. By fitting a parabolic function through the root mean square, the selected travel time is that which corresponds to the theoretical minimum value of the fitted parabola.

We compare the ice floe speed calculated between sounders number one and three with that calculated between sounders two and four. When the ice floes are well defined and the data is relatively clean, the speeds agree with each other within 5 cm/s.

3.4. DATA ANALYSIS: CHANNEL WIDTH

The restricted "open water" channel within which ice floes pass is confined by a solid fast-ice cover on each side (Figure 1). To calculate the volumetric ice floe throughput or "flow rate", we need to know the channel width:

Ice Flow rate (m3/s) = Ice Concentration x Average Ice Thickness x Ice Floe Speed x Channel Width

We cannot obtain channel width directly from the sounder instruments on the platform since they only see the ice passing a very localized area (2.44 x 2.44 m). Theoretically, we could use our real-time cameras to estimate channel width but their resolution for this purpose is not very good. The recent launch of RadarSat images could also now provide an interesting data source. In the meantime, we have devised an ingenious and precise way to plot and quantify restricted channel limits very quickly.

Daily, the CCG sends out helicopters on ice reconnaissance flights. After they complete ice mappings, using a Differential Global Positioning System, we request the helicopter to fly along the edge of the channel/fast ice interface at about 60 knots. During these flights, we capture the position of the helicopter (and hence the edge of the open water restricted channel). We overlay the DGPS data file on an existing AUTOCAD file of the lake. Within minutes, we can plot the location of the ice edge and calculate the channel width at strategic points within an accuracy of a few metres (note the ice edge denote in Figure 1 was created with this technique).

4.1. RESULTS: DATA ACQUISITION

During the 1995-1996 season, we had a fairly continuous data record except for a couple of weeks at the beginning of the season when the tape backups at the site were lost. This winter to date (March 4, 1997), the records also seem to be fairly. We attribute our excellent data capture success rate to the following:

- Fully armoured hydraulic hoses that contain and fully protect the cable between the acquisition units and the sounders.
- Redundancy in the instrumentation: sounders two and four provide independent confirmation of data originating from sounders one and three: We know to what extent our sounders are working and with what precision.
- Real time access and near real time data treatment. Thus, we can continuously monitor the site and intervene as soon as problems arise.
- An application of existing systems and technology. We simply adapted a surplus system which was destined for the garbage heap.

4.2 RESULTS: DATA ANALYSIS:

Measured concentrations ranged from 0 to 100%. The data confirmed that concentrations could fall dramatically and quickly due to ice jamming and loosening upstream. The observed dynamics closely resembled those previously numerically modeled by M. Sayed et al. 1994 and physically modeled by G. Holder et al. 1994 for the area.

The data also demonstrated that ice floe thickening could occur, up to a certain point, to accommodate increase in the volumes of ice generated. Data from 1995-1996 indicate that ice flow rate will increase up until an average thickness of about 0.4 m (Figure 4). When conditions force the floes to topple on top of each other and thicken beyond a mean va;ie pf 0.4 m, inter-ice floe pressures rise, ice floe speed reduces, and partial jamming occurs. This results in a drop in ice flow rate (channel evacuation capacity).

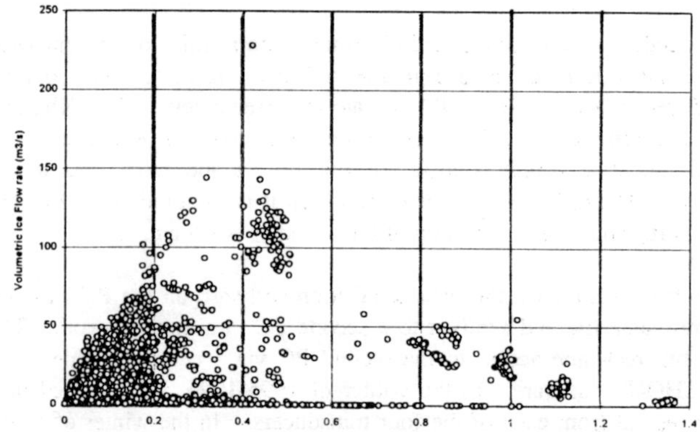

Figure 4. Mean Ice Thickness vs Ice Floe Flow Rate

This type of event was clearly demonstrated in the week of January 17-25 1996. Following balmy weather, on January 20th, temperatures dropped to -20 degrees and an adverse wind blew at speeds up to 20 knots. Some of the floes in the channel started to stick against the fast ice and channel width was reduced. The reduction in width caused ice floes to be crushed together and thicken, increasing friction between the ice moving in the

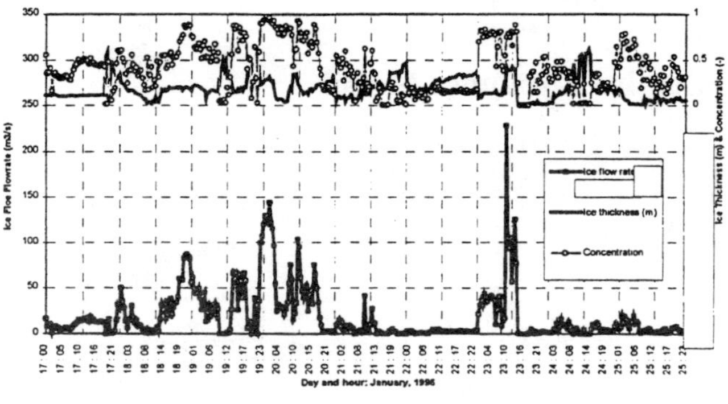

Figure 5. Record of Ice passage during January 17-25, 1996

channel and the edge of the fast ice. Speeds dropped and ice temporarily jammed.Downstream, the Ice Measuring Station recorded the subsequent drop in concentration from near 100% down to a low of 30% (Figure 5). Ice flow rate dropped from 150 m3/s down to 30 m3/s. This continued until commercial ships and an ice-breaker passing through the channel released the jam and widened the constricted channel. When the ice pack loosened, ice that was stored upstream was released and it passage was recorded as demonstrated for January 23 in Figure 5

4.3 RESULTS: TOWARDS AN INTEGRATED ICE MANAGEMENT SYSTEM:

Currently, we are comparing the measured ice flow rate at the ice station with a theoretically calculated ice volume production rate. We base the production rate on air temperature, wind conditions and hours of sunlight. Our application of the equation to the observed data at the Ice Station shows a good agreement between measured ice evacuation rates and theoretically calculated ice production rates. By further developing and refining our analysis, we hope to monitor and predict the potential for an imbalance between production and evacuation rates. With this kind of information, we hope to be able to act more pro-actively and hence avoid the possibility of ice jams developing even before they begin.

5. CONCLUSIONS:

Good ice management in the 1990's is based on having information at our finger tips. The ice station, in this paper, is an example of an interesting data source. The station works because there is redundancy in the instrumentation and because there is near-real time data analysis. Without analyzing data on a daily basis, it is impossible to know if the data acquisition systems are working or not and it is impossible to know if the data is representative or not. Furthermore, by treating data in real time, any unusual event can be readily linked to actual field conditions. By analyzing the data locally at the site prior to data retrieval, the amount of data management and data transfer time is significantly simplified.

So preliminary data interpretation shows that during severe temperatures ice can stack up and therefore more ice can be evacuated for the same surficial flow rate. However, the data show that there is a limit beyond which concentrations are so heavy that jamming occurs, reducing the speed of evacuation and therefore the ice flow rate. By comparing the ice evacuation capacity with available channel width, wind conditions, precipitation, temperature, we hope to use the data to develop a model to predict the river's vulnerability to ice jams in order to take pro-active remedial measures. The Canadian Coast Guard is presently developing just such a program and hopes to have it fully operational in the next three years.

REFERENCES:

Abdelnour,Razek & Cowper, Bruce. Investigation of Ice Thickness Measurement Techniques for Deployment at Laviolette Bridge. Report by Fleet Technology for the CCG. Nov. 1995

Boudreau, Paul, Dumont, Stéphane; Doyen, Bernard; Morse, Brian; Leclerc, Michel. Simulation numérique d'interventions visant l'amélioration du transit de la glace dans le lac Saint-Pierre. Report by INRS-Eau for the CCG. Dec. 1993.

Carter, Donald. Technical Assistance Ice Management of Lac St. Pierre. Report by Donald Carter Consultants for the CCG. Feb. 1995.

Clavelle, Jacques; Morse, Brian; D'Amours, Raymond; Provost, Roger. La Circulation Maritime en hiver sur le fleuve Saint Laurent - La situation de février 1993. A bilingual CCG report, July, 1993.

Cornett, Andrew; Frederking, Robert; Morse, Brian; Dumont, Stéphane. Ice Boom Loads in the St. Lawrence River, 1994-95 & 1995-96. ISOPE-97 conference paper May 1997.

Ellis, William. S. The St. Lawrence River. National Geographic. May, 1980.

Holder, Graham; Parkinson, Fred. Lac Saint-Pierre - Étude des glaces. Report no. 1194 Groupe Conseil LaSalle to CCG. Nov. 1994.

Morse, Brian. Embâcles sur le Lac Saint-Pierre - Rapport de synthèse. CCG report. Déc. 1994.

Sayed, Mohamed; Serrer, Martin, Arden, Deborah. Numerical Simulation of Ice Jams in Lake St-Peter. CNRC report for CCG. March, 1994.

Simard, Laurent. Rapport sur les embâcles dans le lac Saint-Pierre et en amont de Sorel. CCG report. Juin. 1993.

Proceedings of the Seventh (1997) International Offshore and Polar Engineering Conference
Honolulu, USA, May 25-30, 1997

ISBN 1-880653-28-1 (Set); ISBN 1-880653-30-3 (Vol. II)

Ice Conditions in an Anisotropic Sea Ice Dynamics Model

R.S. Pritchard
IceCasting, Inc.
Seattle, WA, USA

ABSTRACT

An oriented ice thickness distribution is presented. It describes sea ice as an oriented material and provides the description of ice conditions for an anisotropic plasticity model. This model describes the formation and evolution of individual lead and ridge systems. We can resolve ice behavior on scales as small as a few kilometers. Ship navigation and other offshore operations should benefit by the added information provided in simulations and forecasts of lead and ridge orientations.

KEY WORDS: sea ice dynamics, anisotropic plasticity, oriented thickness distribution

INTRODUCTION

Ice observations throughout the polar regions provide fractions of area covered by ice of different stages of development (thickness). Ice charts produced by the U.S. Joint Ice Center classify ice into the categories: New ice (0-10 cm), Young ice (10-30 cm), First year thin (30-70 cm), First year medium (70-120 cm), First year thick (120-200 cm), and Old ice (survived at least one summer's melt). The Russian ice observation system is similar. The Canadian Ice Centre uses similar classifications, but further divides New ice into New and Nilas ice, divides Young ice into Grey (10-15 cm) and Grey-white (15-30 cm). Additional information on the size of ice floes can also be added to an ice chart, but it will not be discussed here because sea ice dynamics models cannot now provide such information.

Thorndike, et al. [1975] introduced the ice thickness distribution to describe ice condition. They introduced a probability density function g(h, **x**, *t*) as the fraction of area with thickness greater than *h* but less than or equal to *h*+*dh* in the neighborhood of location **x** at time *t*

$$\int_{h_1}^{h_2} g(h)dh = \frac{1}{R}A(h_1,h_2)$$

where *R* is the area of the neighborhood and *A* is the area covered by this category of ice. The thickness distribution evolves thermally and mechanically according to

$$\frac{\partial g}{\partial t} = -\nabla\cdot(\mathbf{v}g) - \frac{\partial}{\partial h}(fg) + \psi \qquad (1)$$

where *f* is thermal growth rate and ψ is the redistribution function. The redistribution function is changed for oriented ice, but the essence of forming water during opening and ridging thinner ice into thicker ice during closing is preserved. This classical description includes the discrete classifications used by ice observers, and when numerical solutions are obtained the thickness categories can match those used by observers. The thickness distribution evolves as the ice cover deformations and by thermal growth or melt.

The thickness distribution was introduced as part of the Aidjex elastic plastic ice dynamics model [*Coon, et al.*, 1974]. *Hibler* [1979] used thickness distribution in a viscous plastic model, and *Pritchard* [1981] used it in a modified and extended Aidjex model. Most of the recent large scale sea ice behavior models use a reduced thickness distribution (two categories: open water and thick ice). A few exceptions appear: *e.g., Pritchard, et al.* [1990] described a coupled ice ocean model developed to forecast ice edge behavior; and *Flato and Hibler* [1995] extended the thickness distribution to describe ridged and undeformed ice separately. All of these models describe ice as an isotropic material.

More recently, *Coon, et al.* [1992, 1997a,b] introduced an anisotropic plasticity model that accounts for formation and evolution of individual lead systems directly. In their model, ice is composed of *oriented* ice and *isotropic* ice. The *oriented* ice is created when a lead system forms. The oriented ice condition is defined by its thickness distribution and its lead direction. This ice evolves mechanically and thermally. It can become a ridge if compressed. The *isotropic* ice is the surrounding older, thicker ice. It has no orientation. It is composed of floes and inactive ridges. It contains cracks from thermal stresses and isostatic imbalance. The efforts to develop the anisotropic plasticity model have focused on describing the yield surface and flow rule of a new lead, on the changes to the yield surface as ice grows in the lead and it is deformed, and on the rules for describing how a new lead forms. Less attention has been given to behavior of the isotropic ice. A yield surface and flow rule have been prescribed, but they have assumed that the yield surface for isotropic ice changes only a little and slowly.

In the present paper, we propose that all sea ice is oriented. It may begin as a lead system (either an individual lead or a set of parallel leads), which is obviously oriented, or it might begin as an open water area, which we interpret as equal amounts of ice having all orientations. This state is isotropic. We lay a mathematical framework needed to describe the formation of open water during isotropic opening. In addition, the new oriented thickness distribution provides a basis for estimating compressive strength in each direction consistently.

ANISOTROPIC ELASTIC PLASTIC MODEL

Coon, et al. [1992] introduced a method for incorporating the behavior of individual leading, rafting, and ridging events into the large-scale constitutive behavior of pack ice. The idea is that the large-scale stress state must lie within all of the yield surfaces that describe failure of individual ice features. Ice conditions are described by a thickness distribution for each lead and ridge system. Coordinate rotations provide the technique for bringing together the differently oriented surfaces and their flow rules [*Coon, et al.*, 1997a,b]. The new anisotropic model allows our research attention to focus on the individual small scale processes that contribute to leading, rafting and ridging. The model will resolve sea ice behavior on scales that may range from a few kilometers to hundreds of kilometers. Previous models [*e.g., Coon, et al.*, 1974; *Pritchard*, 1981] required that scales be large enough to ensure that enough randomly oriented features exist to make the ice behave as an isotropic material.

Much of the mathematical formalism of the plasticity model follows from the Aidjex model, with several crucial exceptions. First, the stress state in the isotropic Aidjex model had its principal directions aligned with principal directions of stretching, but in the new anisotropic model they are not necessarily aligned. Thus, solutions must be obtained for three stress components (σ_{xx}, σ_{yy}, σ_{xy}) in a general coordinate system, rather than the stress invariants. Second, the yield surface of the isotropic model was constructed from one tensile cutoff cone and a compression cap with strength dependent on the ice thickness distribution. The anisotropic yield surface is defined by the region in stress component space lying within the isotropic surface plus yield surfaces for each lead or ridge system. The numerical integration will be more complex because of the piecewise definition of this combined yield surface.

Assume the existence of a set of features: new deforming leads, ridges that have formed from older leads, or velocity discontinuities generated along a mathematical characteristic but have not yet opened. The stress state cannot violate the yield criterion of any of these features $\phi_j(\sigma, \kappa_j) \le 0$, where σ is the stress tensor, κ_j represents variables such as strength of the feature that depends on its thickness distribution. The stress state can be determined from the plasticity equations at any instant. If the stress state lies at the intersection of more than one of the individual yield criteria, then the plastic stretching will be composed of contributions from each of the deforming features

$$\mathbf{D}_p = \sum_{j=1}^{m} \lambda_j \mathbf{n}_j$$

where $\mathbf{n}_j$ is orthogonal to the yield surface ϕ_j if a normal flow rule is assumed, and m is the number of branches undergoing plastic deformation. The multipliers λ_j must be nonnegative. The normal tensor to each branch of the yield surface is

$$\mathbf{n}_j = \frac{\partial \phi_j}{\partial \sigma}.$$

and we also define $\mathbf{D}_j = \lambda_j \mathbf{n}_j$ as the plastic stretching of each lead system.

For an elastic plastic material, the stress satisfies a linear elastic law

$$\dot{\sigma} = \mathbf{M}\dot{\mathbf{e}}$$

where $\mathbf{M}$ is the elastic modulus tensor, $\mathbf{e}$ is the elastic strain, and the overhead dot indicates the material rate of change. The elastic strain rate and plastic stretching are related by the kinematic expression [*Pritchard*, 1975]

$$\dot{\mathbf{e}} - \mathbf{W}\mathbf{e} + \mathbf{e}\mathbf{W} = \mathbf{D} - \mathbf{D}_p$$

where $\mathbf{D}$ is stretching and $\mathbf{W}$ is spin, which are the symmetric and antisymmetric parts of the velocity gradient $\mathbf{L}$, respectively

$$\mathbf{L} = \frac{\partial \mathbf{v}}{\partial \mathbf{x}}.$$

Here $\mathbf{v}$ is ice velocity and $\mathbf{x}$ is position. We note that all rotational effects may be taken into account by introducing rotated stress and stretching tensors following *Colony and Pritchard* [1975]. Therefore, spin can be neglected in studying the material behavior of the anisotropic constitutive law.

This choice of kinematic variables provides a frame-invariant description that has important theoretical implications because the constitutive model describes specifically how rotations affect the solutions. This fact is crucial because other elastic strain measures are not necessarily independent of the way in which spin is decomposed into elastic and plastic parts [*Pritchard*, 1975].

ORIENTED THICKNESS DISTRIBUTION

The concept of an oriented thickness distribution that depends on thickness and orientation follows from the assumption that all ice is oriented. Thermal growth is isotropic and it reduces differences between ice having different orientations. Therefore, thermal changes tend to form isotropic ice conditions. Of course, some ice, *e.g.*, open water during the summer, can begin as isotropic ice, rather than result from deformation histories that become isotropic over time. The oriented thickness distribution can accept all these ideas. Furthermore, it includes the isotropic ice category introduced by *Coon, et al.* [1997a] to describe the older, thicker ice.

The oriented thickness distribution $g(h, \theta, \mathbf{x}, t)$ is a probability density function that describes all ice in the neighborhood R of location $\mathbf{x}$ at time t. The fraction of area covered by ice having thicknesses in the range $h_1 \le h < h_2$ and orientations in the range $\theta_1 \le \theta < \theta_2$ is

$$\int_{\theta_1}^{\theta_2}\int_{h_1}^{h_2} g(h,\theta)\,dh\,d\theta = \frac{1}{R}A(h_1,h_2,\theta_1,\theta_2)$$

where we have neglected position and time for simplicity. This definition follows from the isotropic definition presented by *Thorndike, et al.* [1975].

Since the oriented thickness distribution describes all ice, isotropic ice need not be defined independently. However, we must identify the properties of isotropic ice in this definition because much of the area is covered by isotropic ice, rather than by a lead or ridge system, which is strongly oriented. We find that isotropic ice appears naturally as that part of g that is independent of orientation.

The lead system shown in Fig. 1 has been spread over a range of orientations. If a finite area fraction were contained in one direction, its values of g would be infinite. The linear variation was used because of limitations in the plot package. Spreading the lead ice uniformly over a narrow band of orientations, say 1 degree, would be more meaningful. However, the essential features appear in the figure.

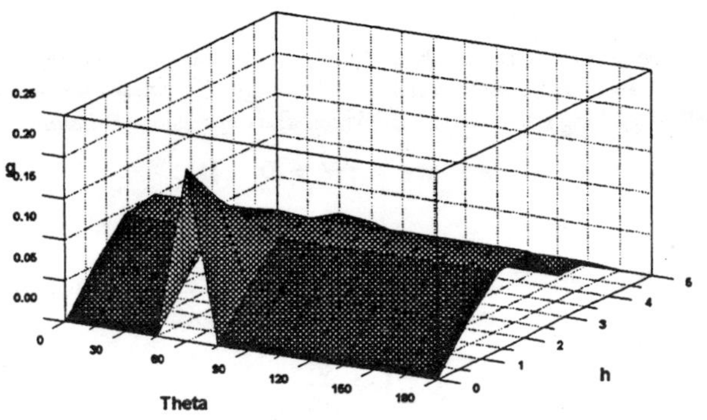

Fig. 1. A simplified oriented thickness distribution for an ice state that is isotropic except for a lead system oriented near 60 degrees.

The oriented thickness distribution evolves thermally and mechanically according to the same balance law derived by *Thorndike, et al.* [1975], and shown in equation (1). Introducing a Lagrangian description is useful in developing the constitutive law. Thus, changes in thickness distribution satisfy

$$\dot{g} + \frac{\partial}{\partial h}(fg) = \psi - g\nabla\cdot\mathbf{v}.$$

This description is not completely Lagrangian because the area fractions are referred to the final total area, not an initial reference area. This mixture is the reason that the $g\nabla\cdot\mathbf{v}$ term appears in the balance equation. The independent variable θ is implicit throughout the governing equation. There is no derivative with respect to θ. Thus, there is no communication between ice having different orientations, except that the total fraction of coverage by open water and ice of all thicknesses must be unity so that

$$\int_0^{\pi}\int_0^{\infty} g\,dh\,d\theta = 1.$$

The redistribution function is assumed to have the form

$$\psi(\theta_j) = \frac{1}{\pi}D^j_{\eta\eta}\begin{pmatrix}\delta(0) & \textit{if opening}\\ w_r & \textit{if closing}\end{pmatrix}$$

which is identical to the form introduced to describe uniaxial ridging behavior [*Thorndike, et al.*, 1975]. In both models, the Dirac delta function $\delta(0)$ describes formation of a finite amount of water. The isotropic model depends on the dilating $tr\mathbf{D}_p$, which equals $D^j_{\eta\eta}$ because stretching along the lead is zero ($D^j_{\xi\xi} = 0$). The coordinates (ξ, η) are measured along and across each lead, respectively.

The new formulation further assumes that shearing of the lead does not change its thickness distribution. The new redistribution function could be generalized to allow dependence on shearing, but such a formulation is beyond the scope of this paper.

Redistribution involves converting thin ice into thicker ice, with both the participating ice fractions and the resulting ice fractions having the same orientation. Thus, ice volume is conserved by requiring

$$\int_0^{\infty} h\psi\,dh = 0.$$

However, since the total area fraction must equal unity, the area fraction in all orientations must be normalized to satisfy

$$\int_0^{\pi}\int_0^{\infty} \psi\,dh\,d\theta = D^j_{\eta\eta}.$$

When isotropic opening or closing occurs, ice having each orientation changes its thickness distribution. This generalizes the model of *Coon, et al.* [1997a] who were concerned only with the behavior of discrete lead and ridge systems. An isotropic component of deformation occurs when the stress lies either at the origin or at the compressive tip of the cap. The plastic stretching lies in a fan with one component being parallel to the pressure axis σ_I. During opening, water forms in all orientations. During isotropic closing, ice in all orientations ridges by w_r.

The ridging function w_r remains unchanged from *Thorndike, et al.* [1975], except that it must be normalized over all orientations and over thickness. In summary, the new oriented thickness distribution is nearly identical to the isotropic thickness distribution of *Thorndike, et al.* [1975]. The sole difference is that it implicitly depends on orientation, and so a different thickness distribution is defined for ice of each orientation.

This oriented thickness distribution $g(h,\theta,\mathbf{x},t)$ is essential when the ice dynamics model is formulated in an Eulerian description. Advection of a field requires that the variable be defined everywhere. Therefore, having a thickness distribution for all orientations allows ice to be advected, though an adjacent area might have no ice of the same orientation. Of course, the oriented thickness distribution model of *Coon, et al.* [1997a] can be extended to allow such an interpretation, but in doing so we essentially introduce the present model.

STRENGTH

When a new lead forms, it contains open water and has no strength to resist closing or shearing deformations. The simplest way to describe this strengthless state is to introduce the yield surface that defines a line along the $\sigma_{\xi\xi}$-axis. This is a *pencil* [*Coon, et al.*, 1997b] or a *line* in stress space (Fig. 2). It has zero strength in tension, compression, and shear along the lead coordinate directions. Its strength along the lead direction is controlled by the

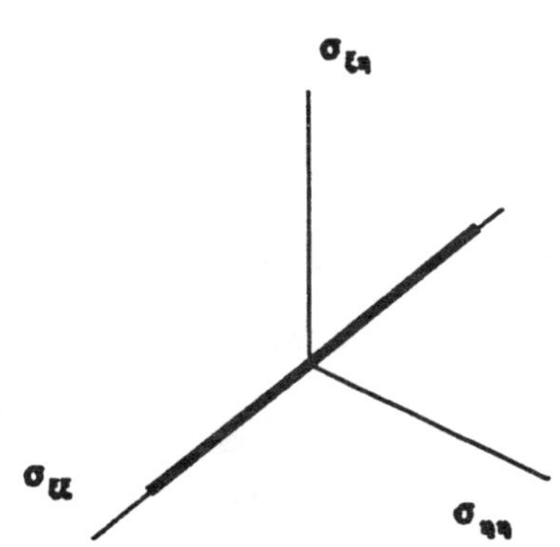

Fig. 2. Pencil-shaped yield surface of newly formed lead.

surrounding isotropic ice. Plastic stretching is allowed in all directions except along the lead system. This result is consistent with the mathematical characteristics of the system.

Ice may form in the lead, it may grow thicker, and the lead ice can raft and ridge when it closes. A thickness distribution defines the ice condition in the lead. This thickness distribution can be the complete formulation of *Thorndike, et al.* [1975] or a simpler four-component measure introduced by *Coon, et al.* [1997a]. Redistribution of ice can be assumed to depend only on the opening and closing of the lead system, not the shearing.

The yield surface for each lead system depends on how each lead fails in a combination of shearing and opening or closing. *Coon, et al.* [1992] chose a 2x4 to describe this behavior. This simple shape describes the three classical strengths independently: tensile t^*, shear s^*, and compression c^*. The fundamental yield surface is shown in Fig. 3. Initially, the 2x4 shape seems unreal. However, when combined with an isotropic yield surface, the extreme corners are clipped off, and the shape appears quite reasonable. An ellipse would also be a reasonable choice.

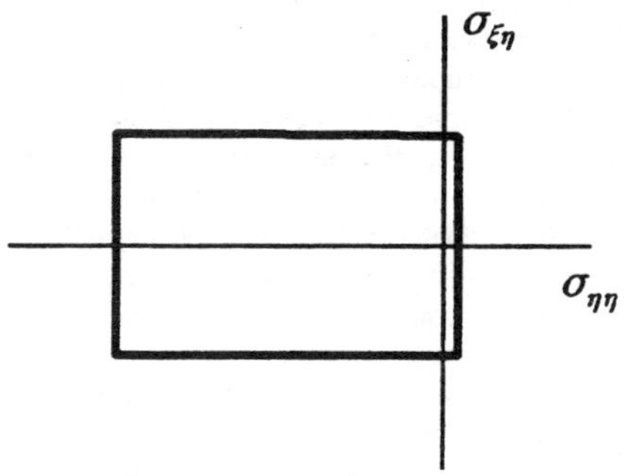

Fig. 3. Yield surface of evolving lead. A 2x4 shape is chosen for simplicity. Three strength parameters are used: tensile t^*, compressive c^*, and shear s^*.

Coon, et al. [1997a,b] described the older, thicker ice that surrounds younger, thinner leads as *isotropic ice*. This ice contains many leads that have been rafted and ridged and have grown to their (roughly) thermal equilibrium thickness. This ice is cracked from thermal stresses and isostatic imbalance. In the limit it has endured the formation of an infinite set of leads in all possible directions. It is not likely that this limit will be reached at any given location, but it is an essential limiting case. First, it is a conceptual limit. Second, it is needed to complete the definition of *isotropic ice*. Third, it provides a useful initial condition for conducting numerical simulations. It is the state that occurs before a new lead forms. Fourth, and finally, it approximates the ice state after several leads in different directions have evolved into thick ice.

The simplest way to understand the isotropic limit of the three-dimensional yield surface is to first describe the yield surface for a pair of similar orthogonal leads aligned with the x and y axes (Fig. 4). The yield surfaces in stress component space combine to form a box of height in the shear stress direction that equals the shear strength s^*. In the plane of normal stress components, a square box appears, with sides at the tensile and compressive strengths, t^* and c^*, respectively. The tensile and compressive faces of the lead yield surface aligned with the x-axis, cut off the infinitely long 2x4 yield surface aligned with the y-axis, and vice versa. This interaction is crucial to defining the isotropic yield surface.

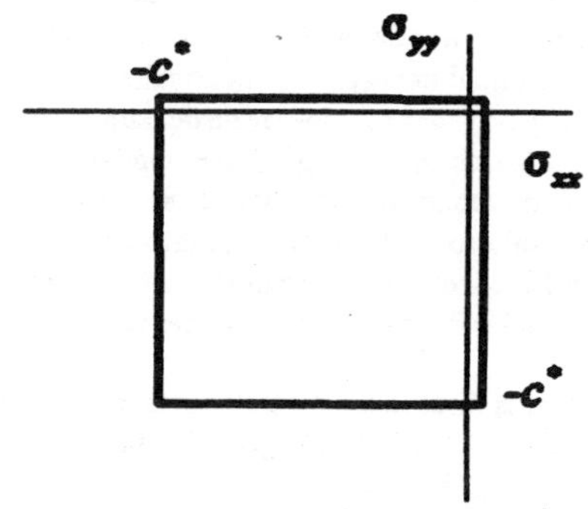

Fig. 4. Composite yield surface of two orthogonal 2x4 yield surfaces aligned along the x and y axes.

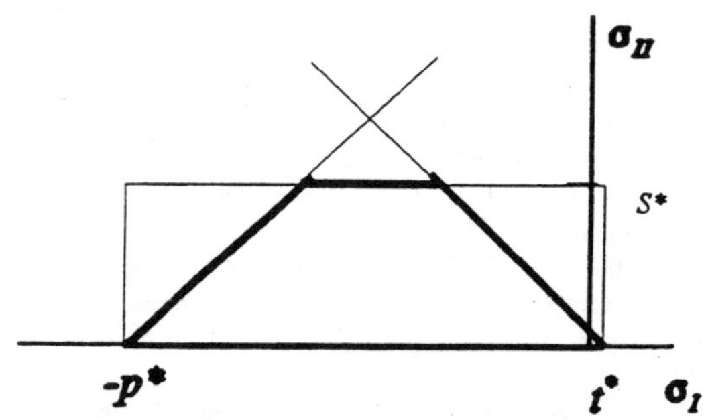

Fig. 5. Isotropic Yield Surface is a three-dimensional body of revolution, a belted diamond shape. The generic cross-section is shown in space of stress invariants.

To derive the isotropic yield surface, consider the yield surface generated by leads in all directions. This limit is most easily understood by considering the composite yield surface generated by pairs of orthogonal leads in all directions. Imagine spinning the yield surface for two orthogonal leads about the pressure axis. A body of revolution is formed, with the pressure axis as its centerline. This surface is presented in the space of stress invariants in Fig. 5. The isotropic compressive strength $p^* = c^*$.

Tensile strength is t^*. A nonzero tensile strength makes the behavior unstable if the tensile strength drops to zero whenever a failure occurs. This situation is undesirable because elastic wave energy is released and it propagates throughout the domain. We expect to assume that the small tensile strength is negligible, and may be assumed to be zero (or does not change from its small positive value after failure). This simple assumption removes the instability.

The isotropic yield surface is composed of three surfaces: (1) a tensile cutoff cone; (2) a compressive conical cap; and (3) a body of revolution that has the same cross-sectional shape as the yield surface of a single lead. If the shear strength equals or exceeds $p^*/2$, then the isotropic yield surface becomes a diamond, as the middle surface is eliminated.

Rothrock [1975] used an energetics argument to estimate compressive strength. The rate of energy dissipated by stress during plastic stretching, $tr\sigma\mathbf{D}_p$, was equated to power dissipated by known small scale sinks: change in gravitational potential energy, and frictional sliding. *Pritchard* [1981] extended this idea by considering more realistic parameter values and adding a sink to account for shear ridging. These approaches were complicated because they had to account for arbitrary deformations. For the anisotropic model, we need only to estimate the strength change for an opening or closing lead. In fact, Rothrock's original derivation described the energy dissipated in uniaxial compression, exactly the stretching state required for oriented ice.

Our attention here is focused on the yield surface. A normal flow rule and the elastic closure law remain unchanged from the isotropic ice.

APPLICATION TO OFFSHORE OPERATIONS

We must recognize that the redistribution function depends on the plastic stretching tensor for ice having each orientation, not just on the total plastic stretching tensor. The essential fact is that ice deforms in only a few orientations at any instant. For a given ice condition, stress tensor, and total (not plastic) stretching tensor, the plasticity constitutive law describes whether a new lead forms and it determines the rate of plastic stretching of each existing lead and of the isotropic ice [*Coon, et al.*, 1997b]. Therefore, the redistribution function is strongly constrained by the plasticity constitutive law. In fact we must look at the (instantaneous) plasticity solution to know which orientations of the thickness distribution are changing.

Although the oriented thickness distribution allows arbitrary variations with orientation, the fact that only a few leads are formed at each location, suggests that the oriented thickness distribution is uniform with respect to orientation, except for a few orientations. Therefore, we can describe these discrete directions separately, and approximate the remaining orientations as uniform. Thus, we can consider the oriented thickness distribution as a set of discrete orientations plus the fraction that is independent of orientation, namely the isotropic ice component. This was the approach used by *Coon, et al.* [1997a].

There are two important exceptions to the argument presented in the previous paragraph. When stress lies at either the tensile or the compressive tip of the yield surface, we cannot claim that evolution of the thickness distribution occurs only in a few discrete orientations. Instead it changes isotropically. During these deformations, the fraction of open water changes uniformly in all directions. This isotropic change can affect compressive strength in all directions, reducing it during opening and increasing it during closing. These special cases allow us to describe ice strength and ice condition throughout the year and under all conditions. This is an essential factor in developing a practicable constitutive law, one that can be used with confidence in a wide range of simulations.

With these special cases in mind, it is makes sense to describe the oriented thickness distribution as a finite set of discrete oriented thickness distributions plus an isotropic thickness distribution. The three-dimensional oriented thickness distribution can describe this model as a special case. The benefit of introducing the complete oriented thickness distribution model is that it can describe all possible ice conditions and deformations. This generality is necessary for a complete constitutive law.

In practice, however, the oriented thickness distribution is used in a computer model to simulate behavior over a selected time period and for specified initial conditions. These initial conditions are always idealized. We can imagine that the initial ice conditions over the problem domain would be specified as isotropic except for specifying lead systems in a few selected locations. Furthermore, *Coon, et al.* [1997b] have shown that the number of lead systems that may form at one location is limited to only a few. If two leads open at nearly the same time, the yield surface collapses to zero strength and free drift occurs. If, after these leads close and their strengths increase, another lead opens, the yield surface collapses to a line segment (we called it a *pencil*). Opening of a second lead shortens this line segment. Essentially, we do not expect very many leads to form at one location. As a result, it is reasonable to describe the three-dimensional oriented thickness distribution as a set (one for each lead system) of simple thickness distributions plus an isotropic thickness distribution for the remaining ice.

The first numerical approach is to divide the θ-domain into a grid of elements. A reasonable first approximation would be to divide into 12 elements having a width of 15 degrees. All lead and ridge systems would be assigned to one of these elements. This approach would likely be acceptable because adjacent locations would provide a better global estimate of lead orientation.

The second numerical approach requires that we look at the properties of the three-dimensional thickness distribution. Since only a few lead systems exist, tracking the formation and evolution of each lead separately is more efficient that the proposed first numerical approach. This result assumes that we need not track the small differences between old thick lead systems, or that we neglect them in the initial conditions. Here, we separate the three-dimensional thickness distribution into a simple thickness distribution for each lead system (as was done by *Coon, et al.* [1997a]), and the remaining part is isotropic. That is, the remaining part is independent of orientation. Using this second approach implies that we use both the isotropic thickness distribution of *Thorndike, et al.* [1975] to describe opening and closing that have large isotropic components, and the oriented thickness distribution of *Coon, et al.* [1997a] to describe the behavior of each lead. What is required of this second approach is that all components be properly normalized to conserve area and mass.

When a ridge forms, continued closing will eventually increase its strength to that of the surrounding ice. At this time, the ridge does not deform alone, instead the surrounding isotropic ice deforms along with it. If strength were to increase only in the ridge orientation, the yield surface would not remain convex. Therefore, we could either increase strength in a range of directions to ensure that the yield surface remains convex, or increase the isotropic strength using the rules developed for the isotropic models.

In practice, it is the amount of open water or width of a lead that is needed, not an area fraction. Lead width is hard to estimate because the continuum model uses stretching, which is an infinitesimal measure of deformation rate. The two can be converted if we introduce a length scale: multiply the area fraction by the length scale to estimate the amount of open water. The problem lies in deciding what is an appropriate length scale. We recommend using the numerical grid size Δx because open water areas in multiple adjacent cells can be added to estimate amounts of water in a fixed-size area.

CONCLUSIONS

The new anisotropic plasticity constitutive law of sea ice behavior provides the basis for the next generation of sea ice models. It describes the formation of individual lead systems and their evolution into ridges, which are the essential physical processes that affect large scale behavior. By identifying individual lead systems, smaller resolution can be achieved. This model can describe behavior averaged over only a few kilometers. In addition, the smaller scale processes that occur during a ridge-building event can be described directly, and their strengths can be incorporated directly into the anisotropic plasticity constitutive law.

Better estimates of large scale ice strength will give operators and designers better estimates of ice loads on their ships and structures. The ability to forecast the orientation of individual lead and ridge systems will give ships' masters an opportunity to plan faster and safer ship routing.

For the largest scale climate dynamics models, the more accurate description of lead systems provides a more accurate measure of the amount of open water in the ice cover. This variable improves the measure of heat transfer from ocean to atmosphere.

The oriented thickness distribution presented here provides a framework for using our knowledge derived from the isotropic models used for the past two decades, while adding the knowledge about individual oriented features such as leads and ridges.

ACKNOWLEDGMENTS

This work was funded under contracts N00014-94-C-0215 and N00014-96-C-0174 from the Office of Naval Research. I thank Drs. Thomas Curtin and Michael VanWoert and Cdr. Robert Edson for their support. Development of the anisotropic plasticity model is an ongoing joint effort with Drs. Max Coon and Stu Knoke and Mr. Skip Echert of NorthWest Research Associates, Inc.

REFERENCES

Coon, M. D., Maykut, G. A., Pritchard, R. S., Rothrock, D. A., and Thorndike, A. S. (1974) "Modeling the Pack Ice as an Elastic-Plastic Material," AIDJEX Bulletin No. 24, University of Washington, Seattle, WA., pp 1-106.

Coon, M. D., Echert, D. C., and Knoke, G. S. (1992) "Pack Ice Anisotropic Constitutive Law," in *IAHR 92, Proceedings of the 11th International Symposium on Ice*, Banff, Alberta, pp 1188-1205.

Coon, M. D., Knoke, G. S., and Echert, D. C., and Pritchard, R. S. (1997a) "An Oriented Thickness Distribution for Sea Ice," to be submitted to *J. Geophys. Res.*

Coon, M. D., Knoke, G. S., and Echert, D. C., and Pritchard, R. S. (1997b) "The Architecture of a New Sea Ice Mechanics Model," to be submitted to *J. Geophys. Res.*

Flato, G. M., and Hibler, W. D., III, "Ridging and Strength in Modeling the Thickness Distribution of Arctic Sea Ice," *J. Geophys. Res.*, vol. 100, no.C9, pp 18,611-18,626.

Pritchard, R. S. (1975) "An Elastic-Plastic Constitutive Law for Sea Ice," *J. Appl. Mech.*, Vol. 42, Series E, No. 2, pp 379-384.

Pritchard, R. S. (1981) "Mechanical Behavior of Pack Ice," in *Mechanics of Structured Media*, Part A, (ed) A. P. S. Selvadurai, Elsevier, Amsterdam, pp 371-405.

Pritchard, R. S., Mueller, A. C., Hanzlick, D. J., and Yang, Y.- S. (1990) "Forecasting Ice Edge Motion in the Bering Sea," *J. Geophys. Res.*, vol. 95, no.1, pp775-788.

Thorndike, A. S., Rothrock, D. A., Maykut, G. A., and Colony, R. (1975) "The Thickness Distribution of Sea Ice," *J. Geophys. Res.*, vol. 80, no. 33, pp 4501-4513.

Proceedings of the Seventh (1997) International Offshore and Polar Engineering Conference
Honolulu, USA, May 25-30, 1997

ISBN 1-880653-28-1 (Set); ISBN 1-880653-30-3 (Vol. II)

Numerical Modeling of Ice State for Ship Operation

I. Appel
Fairweather Forecasting Inc.
Anchorage, AK, USA

ABSTRACT

Sea ice is an important factor determining conditions of navigation in ice infested waters. The main parameters of ice cover influencing navigation are as follows: concentration (including partial), ice thickness (ice age), ice compacting, ridges, and leads. Proposed design of an improved scientific-operational system supporting navigation with information on Ice State includes numerical calculations of ice parameters and specialized recommendations for navigation. The collected onboard observations of ice cover and ship (icebreaker) velocity during polar navigation and scientific expeditions are used to study ice state inhomogeneity and create an empirical model of ship motion in ice cover.

KEY WORDS: Concentration, ice, lead, model, navigation, observation, thickness.

INTRODUCTION

Increasingly we see the industrialization of arctic and sub-arctic areas with mining, oil exploration, and fishing. Also, the possibility of using the Northern Sea Route to shorten international transportation routes is under active investigation. As a consequence of these developments, the improvement of scientific-operational systems to support ice navigation in arctic and sub-arctic seas has significant economic and social importance.

The modern system of navigation support includes the following constituent elements:

- observations on ice cover and other influencing factors;
- numerical methods of ice simulation and forecasting;
- specialized recommendations for navigation.

The structure of the paper is represented by three main sections which corresponds to aforementioned constituent elements of navigation support.

In the first section we consider the state of inhomogeneous ice cover based upon observation results. Consideration of the observation data is characterized by two important features. First of all, we limit our consideration to those parameters of ice state that exert the most significant influence upon ship motion in ice-infested waters.

Secondly, we place emphasize upon studying those ice conditions along the ships route that differ from conditions in the observable area around the ship.

We use a numerical model of ice evolution (Appel, Gudkovich, 1992) developed for time intervals from several days to months and for a spatial resolution of tens of kilometers. Section two includes a description of proposed constitutive law, input-output information, and analysis of results of the model validation.

Numerical methods have successfully found wide application in calculations of current and future ice conditions. Long-term ice forecast for up to the length of the navigation period is applied for strategic planning: scheduling of icebreakers and formation of freight traffic convoys. To support navigation, a three-to-four-day forecast is considered as a primary tactical planning tool and a one-to-two-day forecast allows us to take into account more detailed information about extreme events such as hazardous pressure in ice cover.

The spatially averaged geophysical ice state parameters (observed or calculated) cannot be used to directly determine conditions along the ships route. The use of calculated information to meet requirements of practical navigation is under consideration in the third section of the paper. There are two main features of using ice information that are critically important to practical navigation.

First of all, we would like to emphasize that even comprehensive information on distribution functions of ice parameters is not enough to describe the likelihood of encountering a specific ice conditions along the ship route. Results of existing observation show drastic difference between characteristics of ice condition along the ship route and in its observable area around the ship. This difference, representing an important aspect of ice inhomogeneity, is determined by the statistical relationship between conditions in the area around the ship and along the ship route.

The second feature of using ice information for navigation is connected with the need to estimate objectively the influence of ice cover state upon the velocity of ship motion. We believe that, despite availability of theoretical description of interaction between ship and ice, the data from onboard observations are the most reliable information about influence of ice parameters upon ship motion. The quantitative assessment of the influence is based upon an empirical model of ship motion in ice cover (Buzuev, 1981).

We propose an approach to support navigation in ice-infested waters with required information on ice cover state that is tailored to the specific requirements of the user. Therefore, this kind of service can be characterized as specialized information, as opposed to more generalized information on ice cover.

Both the numerical model of ice evolution and the empirical model of ship motion in ice cover undergo regular improvement. The latest results obtained during development of these models are the modeling of the appearance and transformation of leads and including ship "ramming" operation in the model of ship motion. These two modifications reflect extreme light and heavy conditions of navigation.

1. ICE COVER AS A SHIPPING MEDIA

1.1. ICE STATE

An increase of navigation in ice-infested waters and, in particular, use of the Northern Sea Route, will require more specialized, detailed, and higher accuracy ice information to support navigation in these waters.

On the whole, ice cover is a very complicated media consisting of numerous constituent elements. We agree with the predominant opinion that a realistic assessment of ice influence upon navigation can be obtained only on the basis of taking into account numerous features of ice cover. Therefore, the term "Ice State" implies a wide set of parameters characterizing properties of actual ice cover. A description of Ice State can be done for the local area or for the entire sea or any of its parts. In principal, ice conditions can be called Ice State or Ice Regime for any scale.

We believe that it is hardly possible to develop a comprehensive definition of Ice State, because ice cover causes different influence on different activity in ice-infested waters. Therefore, support of different activities with ice information must meet different requirements and deal with various properties and features of ice cover.

Thus, definition of Ice State should depend, at least to a certain degree, upon the task considered by researchers. For example, the process of interaction between ice cover and offshore structures, considered with participation of the author in another presentation (Bekker and Appel, 1997), is determined by another set of ice characteristics and therefore requires a different approach. In particular, an approach to determine design ice loads on structures and influence on ships should be different, if for no other reason than ships are mobile and are able to avoid most hazardous conditions.

Ice State depends upon numerous ice characteristics. Specifically, for ship operation we need to emphasize those parameters of ice that makeup the main contribution to resistance of ice cover to the motion of ships.

Considering ice cover as a shipping media, first of all, we will take the following parameters as defining Ice State:

- concentration (including partial);
- ice thickness (ice age);
- ice compacting;
- ridges;
- leads.

1.2. INHOMOGENEITY OF ICE STATE ALONG SHIP ROUTE

Spatial changes (as well as temporal) in all parameters of ice cover state are characterized by significant variability. And in turn irregularities in ice condition cause selectivity of ship motion in ice. The question of choosing a specific path among constituent small scale elements of ice cover is the responsibility of the ship's captain and is not considered in this paper. But the problem of greatest importance to our attention is the comparative analysis of ice parameter values along the actual meandering route of the ship and corresponding average values of the same parameters for the observable area around the ship.

Concentration of ice cover usually varies monotonously in space in the winter time. There is almost no alternating areas of different concentration along ship route (Table 1). There is also a very good correlation between information on concentration from different sources. In the winter time, parameters of leads exert greater influence upon ship motion in ice than variability of concentration. Therefore, questions about modeling of leads in ice cover are considered in a separate section of the paper.

Table 1 Probability (percent) of combinations of ice concentration (tenths) along the ship route and in the observable area around the ship for various concentrations around the ship.

Concentration in the area	Concentration along the ship way					
	winter			summer		
	0-3	4-6	7-10	0-3	4-6	7-10
0-3	100	0	0	100	0	0
4-6	5	95	0	70	28	2
7-10	0	0	100	27	45	28

In the summer time, inhomogeneity in ice concentration significantly increases. As a direct consequence of the increased inhomogeneity, difference between values of concentration on the ship route and average for the area around the ship also became significant (Table 1).

Thickness of ice is characterized by the most prominent inhomogeneity for all seasons. Considering the influence of different ice parameters on offshore activity, we obviously come to the conclusion that the assessment of idealized ice conditions will be most erroneous in the case of substituting real variability of ice thickness by a hypothetical mean value. Such a substitution can cause serious errors in calculation of ship velocity.

Those errors can be caused by the presence of ice of different age. It has been observed that velocity of movement in ice breccia of different components is larger than in relatively level ice of the same thickness. However, even ice of the same age can also be characterized by significant variability in thickness. In winter time, that variability usually increases with an increasing number of ridges (Fig.1, 2) and snow depth. In the summer period, inhomogeneity of ice thickness (Fig.3) is usually attributed to irregular melting, connected mostly with upper ice surface morphology and redistribution of melt water.

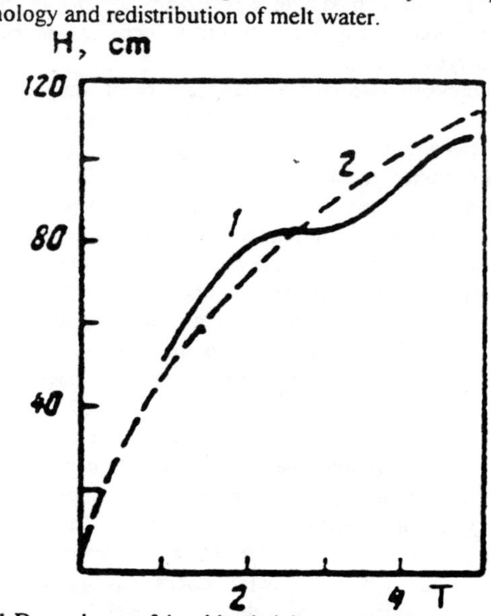

Fig.1 Dependence of the ridge height on hummocking for young ice: 1- observed data, 2 - their approximation

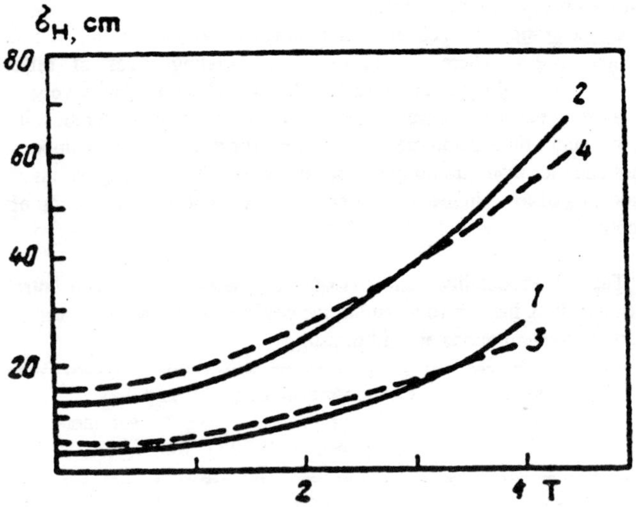

Fig.2 Standard deviation of ice thickness for level ice versus hummocking; 1,2 - observed for thin and thick first-year ice, 3,4 - modeled for thickness of 50 and 150 cm

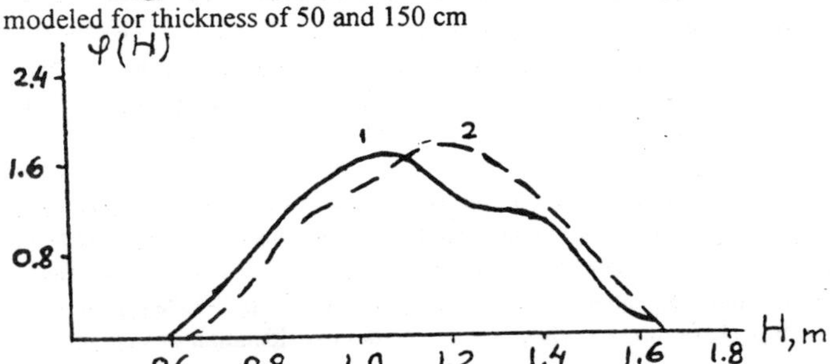

Fig.3 probability density of ice thickness distribution for fast ice of the Kara Sea in the middle of July for specific year: 1 - smoothed observation data, 2 - model simulation

Compacting in ice cover was also thoroughly studied on the basis of observation data collected onboard ships. That information has helped to study compacting in ice cover and include this phenomenon in the numerical model. The compacting corresponds to specific parts of atmospheric formations and its intensity - to velocity and trajectory of cyclones and anticyclones.

The results of onboard observations show that 90% of compacting cases correspond to the zones of significant negative air pressure tendency, as a rule in front of moving deep cyclones, or at a periphery of anticyclones. Theory explains this phenomenon by the convergence of isolabaric wind - the wind component, connected with movement of atmospheric synoptic formations. The greater the atmospheric pressure gradient and the velocity of pressure cell movement, the greater compacting in ice cover.

This description is typical for conditions outside of coast line influence. In the vicinity of islands and coast lines, compacting is determined by wind speed and its direction relatively to the coast. The strongest compacting is observed in a relatively narrow belt of ice cover along the coast line or fast ice.

Reliability of calculation of intensity and orientation of compacting has been confirmed by observation data from ships. We need to note that shipboard observations on ice compacting is the only source of reliable data. Data from visual air reconnaissance are essentially different (Fig. 4).

Ridge formation accompanies the processes of compacting in ice cover. The presence of ice ridges prominently influences ship speed. Influence of ridges has to be considered only in closed ice cover. The distribution of ridges is also characterized by noticeable irregularities (Table 2).

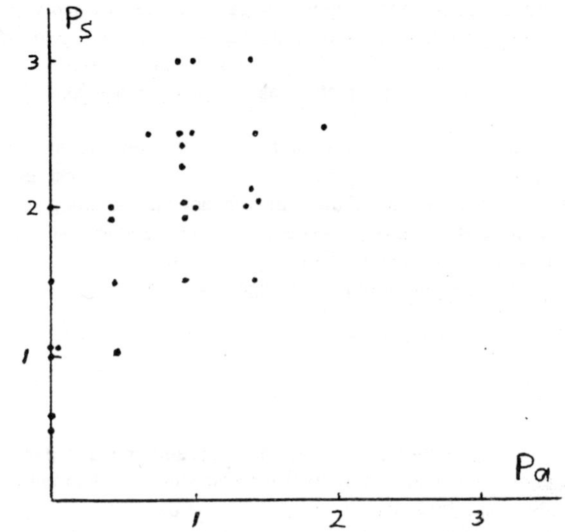

Fig.4 Ship observation on compacting in ice versus data of visual air reconnaissance

Table 2 Probability (percent) of combinations of hummocking along the ship route and in the observable area around the ship for various hummocking around the ship.

Hummocking in the area	Hummocking along the ship route 0-1	1-2	2-3	3-4	4-5
0-1	97	3	0	0	0
1-2	44	54	2	0	0
2-3	24	48	25	3	0
3-4	22	16	32	30	0
4-5	23	3	3	51	20

Selection of ship way in ice in the winter time is mostly determined by distribution of ridges. As a result, there is a significant disagreement between average amount of ridges in the area around the ship and their amount on the ship route. Due to selectivity of ship motion, the disagreement progressively increases with an increasing number of ridges.

2. NUMERICAL MODELING OF ICE STATE

Many features of a dynamic thermodynamic ice model used to simulate and forecast ice state are described in another paper presented at the same 7th International Offshore and Polar Engineering Conference (Bekker, Appel, 1997).

In the present paper, we limit our consideration of the model to discussion on mathematical simulation of internal stresses in ice cover. Our choice is explained by direct interrelation between process on internal interaction and compacting and ridges that significantly influence motion of ship in ice cover.

2.1. INTERNAL ICE STRESSES

General law for description of internal stresses in ice cover was proposed (Glen, 1970) and tested (Hibler, Tucker, 1977) on the basis of analogy with internal stresses in glaciers. In the law, difference between lateral and normal momentum exchange in shear and compression is determined by different dependence (or coefficients) for the spherical part of the stress tensor and its deviator. The approach can lead to the appearance of modeled tensile normal stresses that are impossible in pack ice consisting of discrete elements.

In the framework of a usual description of continuous ice cover it seems to be reasonable to divide the shearing and the compressive stress and strain on the basis of using the principle axes of respective tensor.

We include in the model only compressive components of internal stresses connected with compacting in close ice in directions of principle axes.

On the whole, effect of internal interaction in ice cover is accompanied by the dissipation of energy. Main losses of energy are connected with formation of ridges (or deformed ice in general case). For ice cover, expression determining the dissipation of energy due to processes of ridging and deformation is similar to description of "viscous" interaction. In the model, we state that principal values of internal stresses are proportional to corresponding strain rate.

Numerical realization of the model is based on the assumption that at each moment of time there is an exact correspondence between calculated field of ice drift and system of leads. Such quasi-steady solution is acceptable for periods of simulation exceeding time required to reach a steady solution.

Thus, internal stresses appear only in case of compression along one or two principle axes. Tension along principle axes is accompanying by formation of leads. In case of one system of leads, internal stress can exist only in a direction parallel to leads. In case of two lead systems there are no internal stresses. The latest conclusions of created model were confirmed by measurements during field experiment (SIMI, 1995).

Developed mathematical description of anisotropic ice was successfully applied to different ice regime, in coastal area and in the central part of the Arctic Ocean. Not only does model take into account possible presence of leads in ice cover, but also simulates typical orientation of leads existing in ice cover in the vicinity of coast line, islands, and arched leads at an approach to straits (Appel, Gudkovich, 1992).

2.2. RESULTS OF MODEL VALIDATION

The developed mathematical model serves as a basis for calculating ice evolution. Initial state of ice cover is mostly determined by its concentration and thickness as well as snow depth in grid cells.

Calculations are governed by current information on main influencing factors: surface pressure, air temperature, incoming short-wave solar radiation, and surface ocean currents. For different types of calculations the observed, predicted or mean multiyear values of influencing parameters can be used.

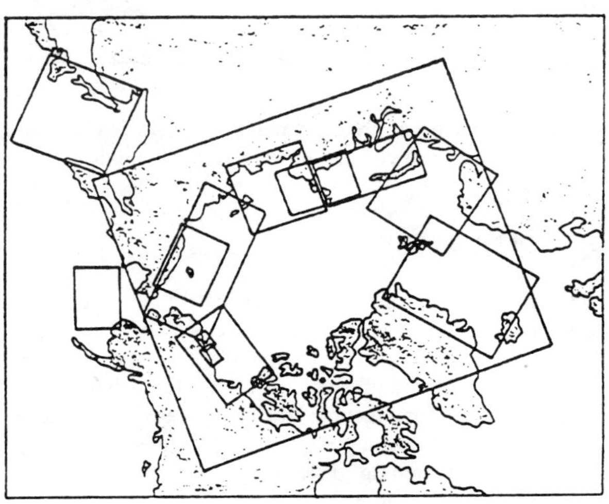

Fig.5 Location of various domains used for numerical modeling ice cover redistribution

Proposed numerical method has been widely used to decide different theoretical and practical problems. The method was applied to simulate and forecast ice conditions for the regions with essentially different ice regime (Fig.5) as along Russian arctic coast line and in other areas.

Numerical modeling of ice state has obvious advantages in comparison with other approaches, allowing calculation of spatial distribution of numerous ice state parameters and the use of calculated information to prepare specialized ice forecasts in support of navigation.

Results of using proposed method underwent detailed investigation. Distinctive feature of the present model verification was as follows. The quality of calculations was estimated not only for ice extent or relative coverage by ice of different concentration (or age) but also for values of concentration at each grid cell used for calculations.

The most thorough consideration has been applied to calculations for the summer period, although the model was used for all seasons as well as for climatic studies. The main conclusion is that long-term simulations reflect interannual changes in ice state for seas with different ice regime.

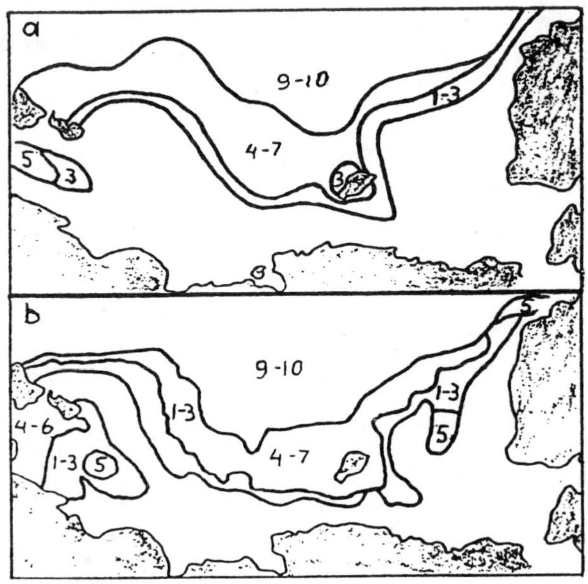

Fig.6 Distribution of ice concentration in the East Siberian and Chukchi seas in the end of August 1992: predicted two months in advance (a) and observed (b)

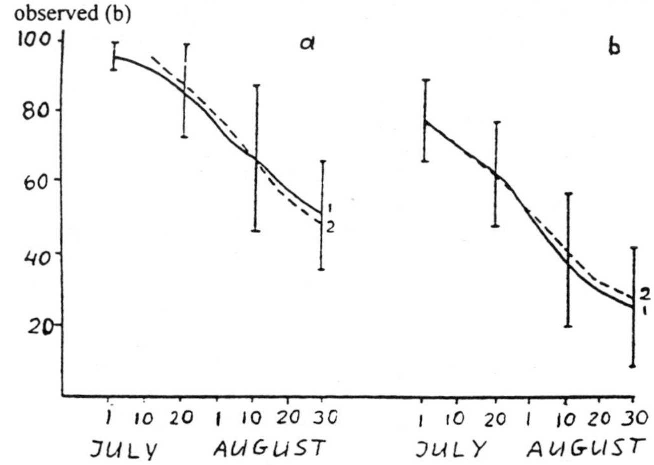

Fig.7 Mean actual (1) and calculated (2) changes in ice extent in the East Siberian (a) and Kara sea (b) for period of 1972-76 years. Error bars are twice standard deviation.

The model allows reliable simulation of redistribution of ice cover for the entire summer period (Fig.6) without accumulation of significant errors (Fig.7). The quality of ice simulation does not depend upon sign or magnitude of ice anomaly in every specific year.

Mean absolute error of total concentration for the second and third months of calculations (without any correction) is equal to 0.4 of standard deviation reflecting interannual variability. Only in less than 20% of all grid cells the errors exceed 0.8 of corresponding standard deviation.

Validation of calculating partial concentration of multiyear ice in the East Siberian and Chukchi seas for the end of the whole summer period shows that coefficients of correlation between calculated and observed distribution of the partial concentration exceed 0.8, the coefficient of proportionality in the regression equation is close to 1. Mean-square-root error of calculation is less than 2 tenths. We believe that similar results of simulation for the period of three months cannot be obtained if numerical realization does not use special algorithms effectively suppressing computational pseudodiffusion of finite-difference approximation (Bekker, Appel, 1997)..

This numerical method has being used for many years in operational activity of the Ice Information Center at Arctic and Antarctic Research Institute, St. Petersburg, Russia. The model is still used to support navigation in the Russian Arctic with information on current and future sea ice conditions (Fig.8). The developed ice forecasts include information about the ice edge configuration, concentration, thickness, ice compacting, choice of route and type of navigation (with or without icebreaker support).

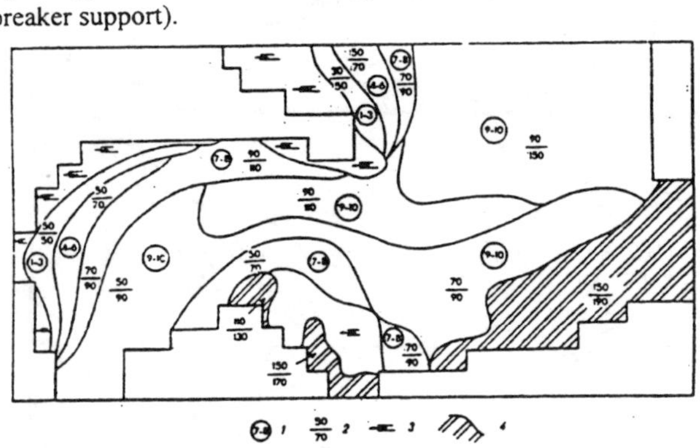

Fig.8 Sample of calculated ice distribution: 1 - concentration, 2 - thickness range for level ice, 3 - sea ice free, 4 - fast ice

In more than 80% of operational long-term ice forecasts, the error of predicting ice extent and area of heavy ice for 1 - 4 months in advance was less than 0.8 of standard deviation considered as a criterion for long-term forecasting.

Analysis of using numerical method for long-term ice forecasts indicates that the effective forecast of ice distribution for the entire summer season can be developed for the Barents Sea in the beginning of May, for the Kara and Chukchi seas in the beginning of June, while for the East Siberian Sea not until July.

The conclusion corresponds to existing comprehensive investigation of predictability for the summer ice processes (Kovalev, 1985). Obviously in this context we consider predictability of smoothed changes with typical temporal scale of 7-10 days. The magnitude of these changes is usually greater than day-to-day variations.

The method is also regularly used for short-term (up to eight days) ice forecasts of ice edge position, thickness, concentration, and the location of compaction areas (Fig.9). Ice information obtained on the basis of the numerical method has been used to prepare specialized ice forecasts. The forecasts allow one to choose the route of ship motion in ice infested water and estimate quantitatively the ship velocity and the time required for different marine operations.

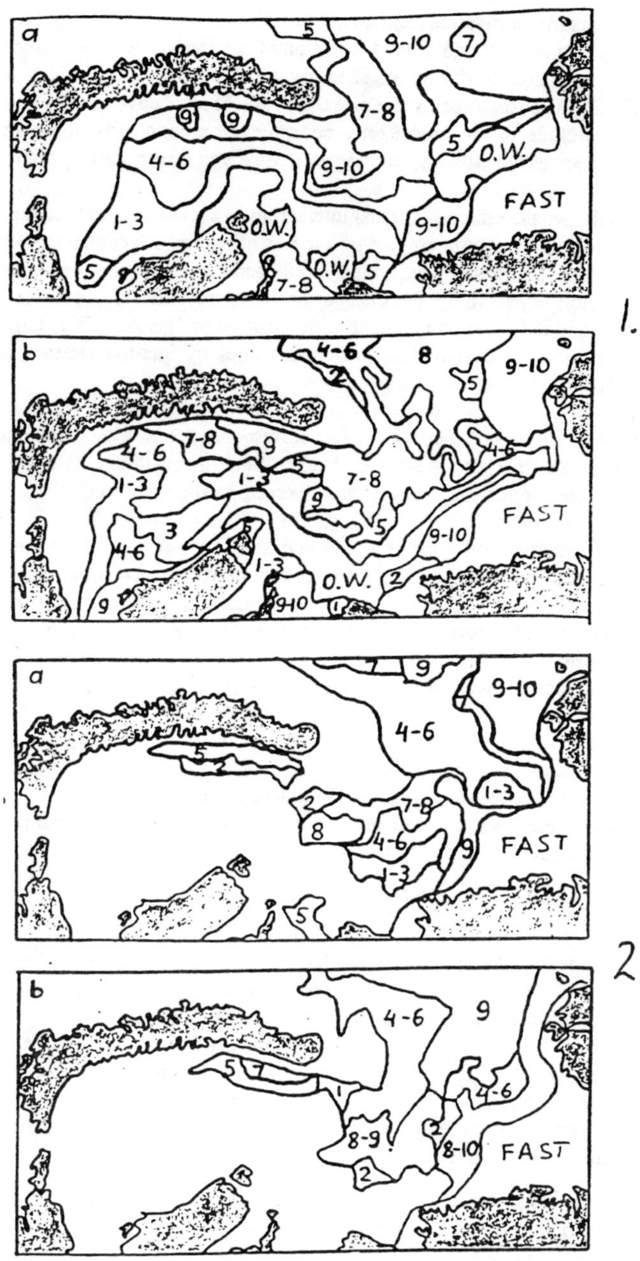

Fig.9 Distribution of ice concentration in the Kara Sea on July 23 (1) and August 6 of the same year (2): observed (a) and predicted eight days in advance (b).

2. 3. MODELING LEADS

If there is the system of leads approximately corresponding to the direction of ship movement, the resulting velocity of ship will be much higher than for the case of similar ice conditions but without leads. This clearly illustrates the significance of taking leads into account for operational support of navigation. Including leads in consideration to calculate ship motion does not present major difficulties. But the opposite statement is true regarding developing methods to calculate lead appearance and evolution in time.

Development of method allowing the simulation and forecasting of evolution of lead systems presents an attractive challenge. The primary problem is connected with the necessity to describe behavior of anisotropic ice cover.

The model of anisotropic ice cover, described in section 2.1., is based on an assumption about permanent exact correspondence between field of ice velocity and system of leads with their instantaneous adjustment to new influencing forces.

In the present section we illustrate natural extension of aforementioned model to the general case including simulation of lead appearance and its evolution under integrated influencing factors.

We consider shear as a major yield mechanism causing not only leads but also ridging. We use the following basic hypothesis about failure of ice cover: that irreversible plastic deformation (Coon et al., 1974) occurs when the maximum tangential stress reaches certain critical value.

Shearing causes appearance of linear discontinuities oriented in direction of the maximum tangential stress. Newly forming line of discontinuity in essence represent slip line (or crack) that can be transformed either in shearing ridge or lead depending on other influencing conditions.

The appearance of a lead is caused by tensile strain normal to the line of discontinuity. In this case only uniaxial compression (parallel to lead) can exist. Thus, creation of a lead is accompanied by abrupt changes in ice stresses: initial stress state causing shear motion immediately transforms into uniaxial compression corresponding to the direction of the new lead.

If the lead continues to open, normal to the lead stress remains equal to 0. If lead closes, resistance to normal compression in ice cover is determined by the lead's ice thickness distribution (or by thinnest ice in a lead). If there are no leads or slip line, we determine whether a new line of discontinuity is activated or not. If an active lead exists, shearing occurs along the lead or a new line of discontinuity appears. The choice of the weakest path from possible alternatives is determined by comparison of the maximum tangential stress in ice cover with shearing stress along lead.

In the event of favorable conditions for lead existence, we suggest a simple model to calculate typical distance between leads. The model is based upon mechanical consideration of separated blocks (floes) constituting ice cover. The essence of the developed approach is as follows.

The motion of each individual block is determined by integral balance of influencing forces. But forces are not balanced at every point of the block. The unbalance of the forces can be presented in the form of force couple of equal magnitude and opposite directions. If this couple causes shearing stress above shearing strength, the block will be broken in parts.

Characteristic changes in spacing between leads will depend upon changes of ice physical properties. As a result, this mechanical model explains structure of ice cover, size of constituting blocks, seasonal changes of size, and its dependence on ice thickness.

3. NAVIGATION RECOMMENDATIONS

3.1. EMPIRICAL MODEL OF SHIP MOTION IN ICE

Different ice parameters simultaneously influence ship motion in ice. This circumstance explains the complication of quantitative estimation of the dependence of ship velocity upon state of ice cover. There are two different approaches to study ship motion in ice cover.

The first uses theoretical consideration of ship motion, analytical description of ice resistance and elements of ice morphology. This direction of studies is promising, but many problems did not allow to get method applicable to practical calculations.

A different approach is based on accumulation of field observations, their statistical analysis and quantitative presentation of ship velocity as a function of ice state parameters.

More than 20 years ago, representatives of Arctic and Antarctic Research Institute, St. Petersburg, Russia, began specialized simultaneous observations on ice cover and velocity of ship motion in ice during regular polar navigation. Those observations, along with data from scientific expedition in Arctic and Antarctic waters, serve as a basis to create and validate an empirical model of ship motion in ice of different state.

First version of the empirical model was included in the book (Buzuev, 1981) that just recently became declassified and accessible for western scientists. Later, the model underwent comprehensive validation and has been used for routine scientific-operational support of navigation with specialized recommendations.

The empirical model gives an objective estimate of navigation difficulties represented by the velocity of ship motion in different kinds of ice conditions. The model can also be used to estimate errors in calculations of ship velocity, connected with error of determining ice state parameters by existing methods of calculation and observation.

Model includes concentration, thickness (Fig.10), and intensity of compacting as main factors determining ship velocity. The influence of other parameters is taken into account as a correction to the main functions.

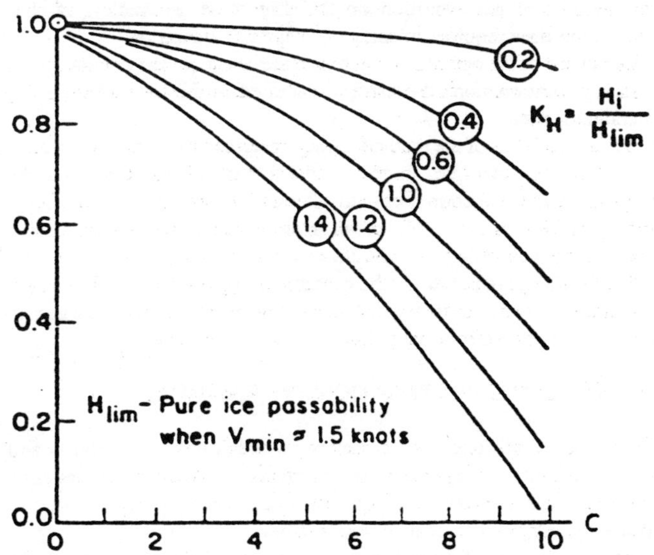

Fig.10 Characteristic dependence of ship velocity upon concentration, thickness.

Validation of the ship motion model shows its high reliability. The resulting time of icebreaker motion along ice route in 80% of cases can be calculated with the error of 10% or less. The quality of calculations for a convoy accompanying an icebreaker is worse: for 80% of cases the error of time estimation is 40% or less. This assessment has been obtained on the basis of actual operational calculations in support of navigation

3.2. TRANSITION FROM GEOPHYSICAL CALCULATIONS TO CONDITIONS ALONG ROUTE

Usually, available information on current or future ice state consists of average values or parameters for the area under consideration. We need to emphasize that the information on ice state is obtained not only on the basis of observations but as a result of numerical calculations. The calculations also give smoothed spatial changes of ice parameters, but this kind of information includes those parameters of ice state that at the present time cannot be obtained by observations.

Large scale ice models give valuable information on characteristics of ice cover. But those data differ significantly from actual conditions along meandering ship routes. Thus, there is an obvious necessity to realize a transition from average parameters to conditions along a selected ship route. Using the transition allows us to find ice parameters on the ship route, directly influencing ship motion in ice infested waters.

The same materials of field experiments that were used to develop model of ship motion in ice serve as a basis to study features of transition from geophysical calculations to conditions along the ship route. But in addition to the observation data described in the previous section, information on ice state in the observable area around the ship location was also taken into consideration.

Analysis of simultaneous field observations (Section 1.2.) gives quantitative statistical relationships between ice parameters along the ship route and generalized value of the parameters in the area around the ship.

3.3. OPERATIONAL RECOMMENDATION

At the present time scientific-operational support of navigation in ice infested waters includes recommendation (or specialized forecasts) about dates to begin specific operations, the most favorable route for ships, characteristics of ice condition on the ship route, estimation of ship velocity, and time required to carry out a specific operation.

The method of the optimal navigation route selection and calculation of the sailing duration along the route was created on the basis of the using ice modeling data.

Careful validation of the method during operational summer navigation shows that, for example, in the southern part of the Kara Sea the navigation route selection between alternative options was predicted correctly in 78% of cases. The mean error of predicting time for marine operations for typical light ice conditions in this area does not exceed 10% - 15% of real time required for the corresponding operations. For heavy ice conditions in different regions errors can be significantly larger, in some cases in the order of magnitude of calculated value.

3.4. MODELING OF OPERATIONS BY RAMMING

Heavy ice conditions force cessation of continuous ship motion and lead to necessity of operation by ramming. Analysis of amassed representative observations on operations by ramming allowed one to improve the empirical model of ship motion.

The calculations of ship speed consider all delays caused by ice conditions including operation by ramming (Table 3).

Table 3 Actual velocities (knots) of the "Arktika" icebreaker.

Velocity	Continuous motion in combination with ramming	Ramming in combination with continuous motion	Only by ramming
Average	5.5	3.6	1.6
Maximum	8.6	6.6	3.4
Minimum	2.6	1.4	0.2

Knowing distribution of ice conditions along ship route, we subdivide all route in two parts: one part can be passed without stops, another part requires operating by ramming. Algorithms of calculation takes into account velocity of motion for each of these two parts and their relative lengths.

Separation of the route into two parts is determined by limiting ice conditions that correspond to 50% probability of motion stop. Limiting values of ice thickness and other parameters for different types of ships and icebreakers were found on the basis of observations. Results and details of calculations for operation by ramming are included in Buzuev's presentation at POAC-97.

ACKNOWLEDGMENTS

The author would like to emphasize that the present paper in essence represents exposition of the approach to scientific-operational support of navigation with specialized ice information developed in close cooperation with Arkadiy Buzuev and Zalman Gudkovich, Arctic and Antarctic Research Institute, St. Petersburg, Russia, who can be considered as coauthors. The author and Dr. Gudkovich elaborated principal bases of dynamic thermodynamic model of ice evolution, Dr. Buzuew was responsible for creation of specialized navigation recommendations.

REFERENCES

Appel, IL, Gudkovich, ZM (1992) "*Numerical Modeling and Forecast of Ice Cover Evolution*," Leningrad, Gidrometeoizdat, 144 p. (in Russian)

Bekker, AT, Appel, IL (1997) "Description of Sea Ice Regime for Hydraulic Structures" *Proc 7th Int Offshore and Polar Eng, ISOPE, Honolulu.*

Buzuev AYa, (1981) "*Influence of Natural Conditions upon Navigation in Ice Infested Seas*," Leningrad, Gidrometeoidat, 200 p. (in Russian)

Coon, MD, Maykut, GA, Pritchard, RS, Rothrock, DA, and Thorndike, AS (1974) "Modeling the Pack Ice as an Elastic-Plastic Material," *AIDJEX Bulletin*, No 24, pp 1-106.

Glen, JW (1970) "Thoughts on a Viscous Model for Sea Ice," *AIDJEX Bulletin*, No 2, pp 18-27.

Hibler, WD, and Tucker, WB (1977) "An Examination of the Viscous Wind-Driven Circulation of the Arctic Ice Cover over a Two Year Period," *AIDJEX Bulletin* No 37, pp 95-134.

Kovalev EG, (1985) "Possible Limits of Predictibility for Ice Extent in Arctic Seas," *Trudy AARI*, Leningrad, vol 396, pp 152-160. (in Russian)

SIMI - Sea Ice Mechanics Initiative (1995) *Proceeding of the Sea Ice Mechanics and Arctic Modeling Workshop , Anchorage*, vol 1, 2.

Proceedings of the Seventh (1997) International Offshore and Polar Engineering Conference
Honolulu, USA, May 25-30, 1997

ISBN 1-880653-28-1 (Set); ISBN 1-880653-30-3 (Vol. II)

On the Damage of Skeg and Steering Gear of the M/V "Stepan Krasheninnikov" in the Antarctic Ice

N. Barabanov, A. Moskalenko and E. Lapin
Association of Independent Sea and River Surveyors
Vladivostok, Russia

ABSTRACT

The paper presents the data on the conditions of sailing of the m/v «Stepan Krasheninnikov» when being damaged in the Antarctic. Described are the damages of the skeg and the steering gear. Analysing this damage the authors made a conclusion on the possible causes of the accident and an estimation of the skeg construction.

KEY WORDS: Ice damage, skeg, steering gear

INTRODUCTION

By the order of the Owner the authors have studied the documents concerning the damage of the skeg and the steering gear of m/v «Stepan Krasheninnikov» occurring on February 28, 1994 in the Lazarev sea when the vessel was operating in the Antarctic in heavy ice. The diesel-electric Ro-Ro ship «Stepan Krasheninnikov» (of the «Vitus Bering» type) is of the special ice purpose class, which corresponds to the conditions of independent sailing during the summer-autumn period of navigation in all regions of the World Ocean, including the Arctic and Antarctic seas.

SAILING CONDITIONS

Before the accident resulting in damage of the steering gear of m/v "Stepan Krasheninnikov" the vessel was operating under the heavy ice conditions in the Lazarev sea near the Antarctic shore, where thickness of pack ice, individual ice pieces and fields reached 1,5-2,0 m, ice being packed at 10 balls. On February 27 and 28, 1994 heavy ice motion in the W direction opposite to the ship movement and strong compression of ice fields with the wind up to 17 m/sec were observed in the region of vessel operation.

DAMAGE OF THE RUDDER

Fig. 1 presents the character and size of damage. The ship being in the Lazarev sea on February 28, 1994, when the crew members made visual review of the steering gear. The results were as follows. The rudder blade was in an immovable state with its maximum movement off CL to the port side; the upper gudgeon of the rudder blade had two vertical fractures in the region of rotation axis of the rudder blade. On the starboard the fracture was through with an outer opening up to 80 mm and an inner opening up to 60 mm, the second fracture on the port side had an inner opening up to 30 mm (see Fig. 1), extended from the upper plate of the rudder blade vertically downward and did not reach 100 mm the bottom surface of the gudgeon. Partly there was no rudder stock key, the key way of the stock was cut off in the bottom part along the length of 90 mm with the cut depth about 15 mm. In addition, the sheets closing the nut holes of the stock and the middle pintle of the rudder blade were partly cut off along the weld, and there was a fracture of about 200 mm long in the upper plate of the rudder blade towards the stern because of the damage of the steering machine. However when working, the stock rotation could be observed.

After analysing the documents and examining the damages of the ship during its operation under heavy ice conditions we can state that the action of the crew members under the conditions of sailing near the Antarctic shore evidence that they are well-trained sailors and were guided by fine seamanship.

The main aim of examination of this casualty is to find out the true causes of damages as well as to present our considerations on the sequence of appearance of damages and their mutual influence upon each other.

On the basis of the materials available and evidences of the crew members we have made the following conclusions:

1. The steering gear and the sternframe have damages in the form of fracture in the upper gudgeon of the rudder blade and in the skeg.

2. The type and sites of damages (see Fig. 1) in the parts and structures evidence that all the damages appeared under the action of uncontrollable ice load which was frequently a cause of cracking in the hulls of ice ships.

3. The fracture in the skeg in the weld joint might arise under the action of ice loader on the rudder and impact loads which were due to the contact with ice under the ship hull.

4. As we suppose, the main ice load was applied at the level of waterline, which is confirmed by the absence of damages in the propeller blades.

5. The fracture in the skeg could not cause misalignment of the rudder blade and its jam, as no shifts in the welds were found out. And the weld itself throughout its full section was not broken. Consequently, the fracture could affect neither causes nor dimension of the damages in the upper gudgeon of the rudder blade.

6. The main damages in the rudder blade were due to heavy ice conditions of sailing and stress concentration available in the constructions and units of the key joint of the rudder stock with the rudder blade.

Propagation of the fracture in the vertical direction is explained by the presence of concentrations in the region of key way and ice action upon the rudder blade with its movement to the port side against the stop.

After weakening the considered construction (region of joint of the rudder stock with its upper gudgeon) the ice action caused cracking and damage of a part of the rudder blade gudgeon in the port side up to separation of the forward part of the gudgeon. This is confirmed by the illustrations presented to the surveyors in the place of repair of m/v "Stepan Krasheninnikov" in Durban

1. Simultaneously with the vertical fractures described above in (7) a longitudinal fracture appeared in the stern part of the upper horizontal plate of the rudder blade in the region of maximum concentration of stress (more than 5) in the round cross-section fractures (see Fig. 1).

CONCLUSIONS

1. All the above damages are due to heavy ice conditions of sailing and uncontrollable loading arising from these conditions of vessel operation.

2. The described damages of the steering gear and of the skeg did not cause misalignment of the gudgeon of the rudder blade and rudder post, which is confirmed by the long passage of the vessel to Durban for repair with the steering gear as it was.

After having analysed the presented materials we concluded that the most probable cause of damage of the stern frame bottom was ice action on the rudder blade on the port side resulting in the damage of the upper gudgeon of the rudder, residual sag of the rudder post and cracking at the beginning of the corner radius in the stern frame bottom half its breadth. Our conclusions are supported by the calculations.

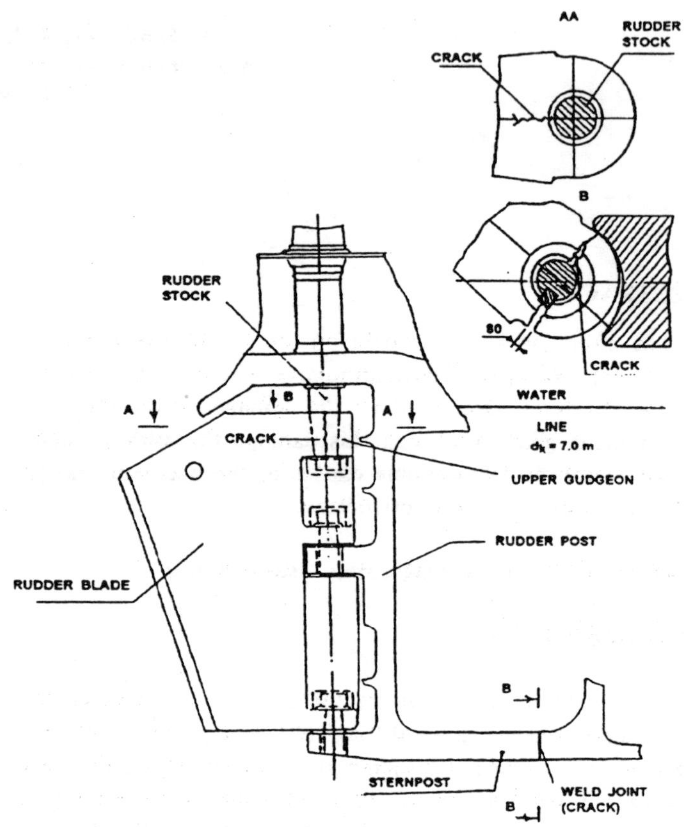

Fig. 1. The damage of skeg and steering gear

Proceedings of the Seventh (1997) International Offshore and Polar Engineering Conference
Honolulu, USA, May 25-30, 1997

ISBN 1-880653-28-1 (Set); ISBN 1-880653-30-3 (Vol. II)

Landing and Parking Curves for the C-17 Globemaster on Sea Ice: McMurdo Station, Antarctica

Joseph L. Barthelemy
Naval Facilities Engineering Service Center
Port Hueneme, CA, USA

ABSTRACT

The United States Antarctic Program operates a seasonal sea-ice runway at McMurdo Station that supports heavy-haul wheeled aircraft during the short but vital resupply window. The Naval Facilities Engineering Service Center provides criteria to the National Science Foundation for the safe operation of these aircraft. During 1996, landing and parking curves were developed for the newest Air Force transport, the C-17 Globemaster, which had been scheduled tentatively for service in Antarctica in the month of October 1996. Although this new transport weighs somewhat less than the C-5 Galaxy it is scheduled to replace, the C-17 also has just half as many wheels, and leaves correspondingly heavier footprints on the ice. In spite of somewhat greater localized stresses around the landing gear, the thickness of sea ice grown during the "typical" antarctic winter is sufficient to support the maximum aircraft weight of 2.6 MN on landing, takeoff and during turnaround parking.

INTRODUCTION

Scientific research in Antarctica is costly and conducted under great hardship to man and machine. As a result, it is a national priority to optimize research productivity there by ushering scientists and research equipment to the field at the earliest possible date. A key logistical strategy is to expedite delivery of high-priority cargo and personnel to the continent at the start of the brief summer season. The United States Antarctic Program (USAP) operates a seasonal sea-ice runway that is re-established yearly near McMurdo Station to support heavy-haul, wheeled aircraft during the vital re-supply window. From late September until early in the austral summer, around mid-December, the bulk of priority cargo and personnel is air lifted from New Zealand by heavy-haul transport, including C5 Galaxy, C-141 Starlifter and C-130 Hercules aircraft. A key logistical objective is to expedite the number of inbound flights and tonnage of goods delivered before the seasonal runway deteriorates and is closed. After the wheeled air strip is closed to intercontinental traffic, additional international flights as well as continued re-supply flights to inland stations are limited for the most part to smaller capacity LC-130 aircraft (C-130 Hercules equipped with retractable skis) operating with wheels down or skis, respectively, down from the hard and soft landing areas of the Ross Ice Shelf. By February, the continent is cut off to all air traffic as seasonal personnel and aircraft are evacuated for the duration of the long austral winter. Figure 1 shows an Air Force C5 Galaxy on the ice runway with Mount Erebus, an active volcano, standing in the background.

Figure 1. C5 Galaxy on the sea-ice runway at McMurdo Station.

A milestone in antarctic aviation was achieved in October 1989 when an Air Force C5 Galaxy fully loaded to 2.8 MN touched down on 2 m of annual sea ice. Since that date, C5 Galaxy flights have been used to launch most seasons, substantially accelerating the delivery of priority cargo and personnel. The C5 Galaxy is being phased out of production, however, and it is being replaced in the military by the C-17 Globemaster, a new rugged-terrain transport. Although the loaded Globemaster weighs about 10% less than the aircraft it is replacing, it has half as many wheels to spread the bearing load, making it potentially too heavy afoot for the ice foundation. The effect of footprint is particularly significant during the "elastic" landing event because localized stresses beneath the wheels determine the thickness of ice required. During the "visco-elastic" parking period, as an aircraft sits stationary on the ice, the particular wheel geometry is less of a factor in promoting progressive deflection than is the total dead load of the aircraft. Figure 2 shows the relative footprints of the C5 Galaxy and C-17 Globemaster.

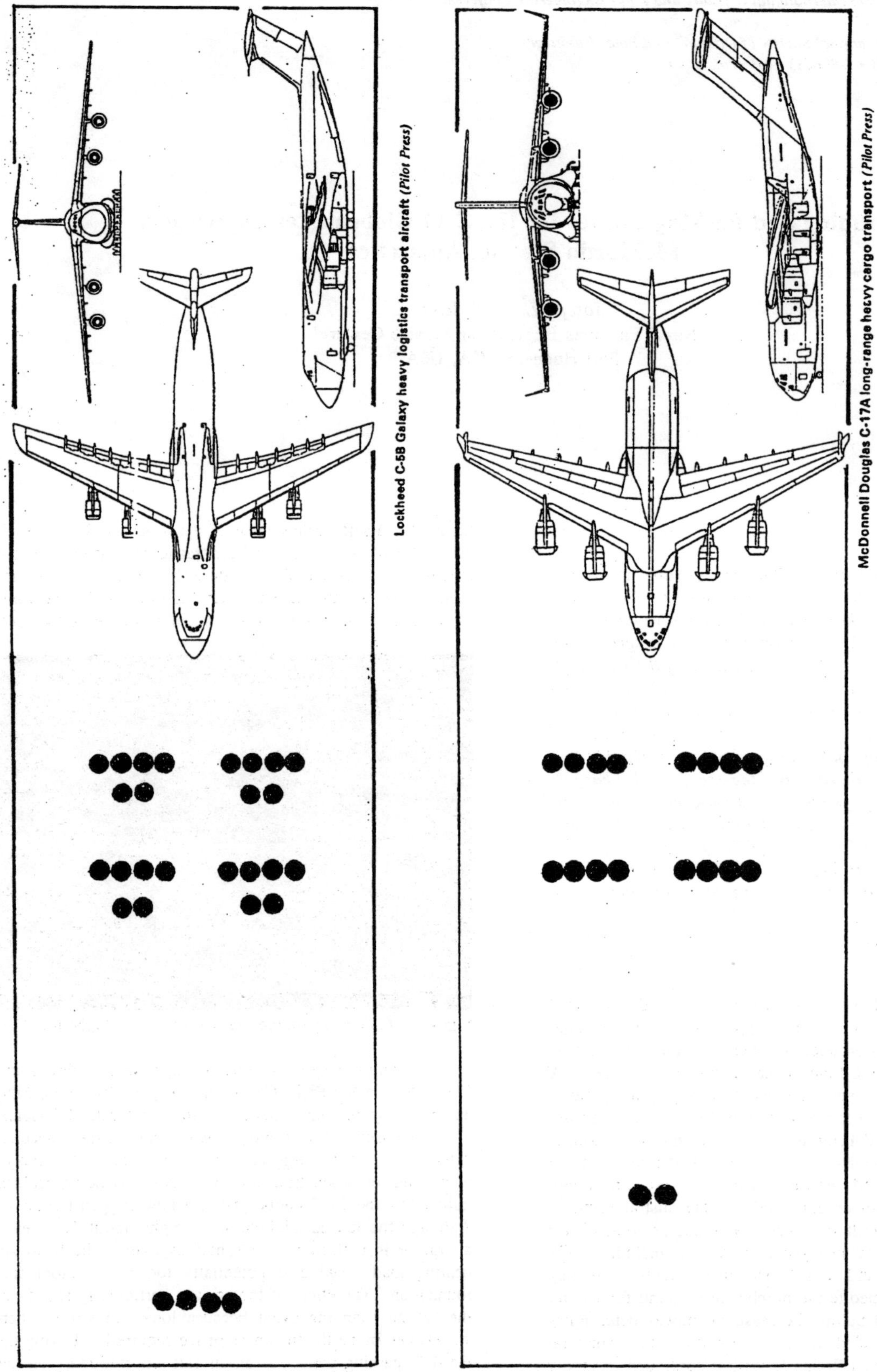

Figure 2. Relative footprints of C5 Galaxy and C-17 Globemaster (no scale).

BACKGROUND

The location that has evolved as McMurdo Station [77°51'S 166°40'E] is situated on bare volcanic rock at the barren end of the Hut Point Peninsula, an eighteen kilometer extension of Ross Island that marks the farthest south land mass accessible by ship. Although much of the shoreline area that defines Ross Island is held glacially captive by the impinging Ross Ice Shelf, the southern reaches of McMurdo Sound are free and expressed during the winter months as a continuous sheet of ice that breaks up during the spring. It is the close proximity of land, water, snow and ice that have enabled McMurdo Station to develop as the largest research center in Antarctica, and logistics hub for both air and sea replenishment of much of the continent. Figure 3 illustrates the confluence of *terra firma* (Ross Island), water/sea-ice (McMurdo Sound) and glacier (snow-blanketed Ross Ice Shelf).

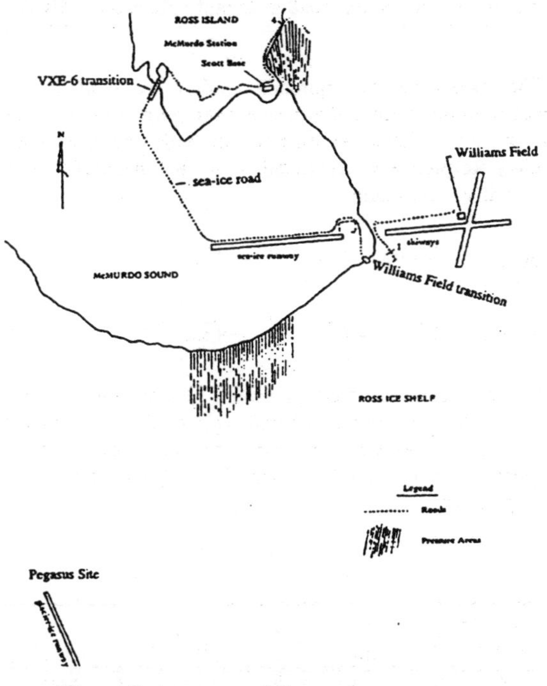

Figure 3. Area around the sea-ice runway.

From a humble beginning in 1955 as the Naval Air Facility, McMurdo Sound, to its opening as an American research site during the International Geophysical Year (1957), and subsequent renaming with full-station status in 1961, McMurdo Station has grown to house more that 80 buildings, including laboratories for biology, earth science, meteorology, and upper atmosphere physics. The National Science Foundation, Office of Polar Programs, is the United States agency that coordinates the science program, manages the budget and carries out congressional initiatives regarding construction.

The first ski-and-wheel equipped LC-130 was introduced to McMuro Sound in 1960. In 1967, the United States Air Force initiated deployment of the larger, jet-powered C-141 Starlifter, and has continued to operate those aircraft on a yearly basis. The latest aviation milestone was the October 1989 landing of the world's heaviest transport, the C5 Galaxy (Barthelemy, 1992)

The landing strip for wheeled aircraft is laid out each spring on the smooth annual sea ice that blankets McMurdo Sound. Beginning in early October, Air Force C5 Galaxy and C-141 Starlifter heavy-lift transports, operating from a staging area some 4,000 km away in New Zealand, deliver personnel, perishables, packaged goods and other priority cargo required to launch the science season. Just months later in January, a Coast Guard ice breaker removes fractured sea ice in McMurdo Sound that weeks earlier had accommodated the runway, and clears a turning basin and docking site for ship traffic along the Hut Point Peninsula. A tanker and cargo vessel deliver a year's supply of fuels, containerized goods, heavy equipment, and building materials for McMurdo Station and the inland locations. While shipping is required to import the bulk of goods necessary to sustain the continent through many months of isolation, it is air traffic that provides the essentials needed to expedite the start of the short science season.

The embayment south of McMurdo station is covered by sea ice for approximately 10 months during the year. The ice sheet appears generally as a flat, featureless plain with a snow cover of 15 to 20 cm, and wind-carved drifts rarely more than 45 cm in total height. In areas not totally cleared by ice breaker, old floes and small bergs pepper the low relief icescape. New ice attains a maximum thickness by mid-December. By late December, the ice sheet loses appreciable thickness and strength due to bottom melting and internal warming. If uncleared by Coast Guard vessel, cracks occur by early March and shortly thereafter the blanket breaks into drifting floes that are typically flushed from the Sound by wind and current. By April, as air temperatures continue to drop, new sea ice forms to begin the cycle anew.

Historically, after the sea-ice runway was closed in December, additional flights inbound and outbound were limited to LC-130 aircraft that could operate from a subordinate ski-way located on areas of deep snow blanketing the Ross Ice Shelf. The requirement for soft surface landing by ski reduced the allowable payload by an average of 3,600 kg (8,000 pounds) below that of the conventional C-130. Since 1993, however, a hard-surface, glacial-ice runway has been operated from a wind-swept region of the Ross Ice Shelf that is blown clear of deep snow. The Pegasus site, developed by the United States Army Cold Regions Research and Engineering Laboratory (USA CRREL), allows all season operation by wheeled aircraft, freeing valuable flight hours for the small fleet of LC-130 that must provide inland service to the Amundsen-Scott South Pole Station and other outlying locations. With the Pegasus site in operation, conventional C-130 Hercules and C-141 Starlifter airplanes provide increased service for passengers and goods headed north, while the LC-130s are able to "top off" with fuel and cargo before heading south. By the end of the current season, over 180 flights had used the new runway (Blaisdell *et al*, 1997). Figure 4 shows a C-141 Starlifter on the ice at the Pegasus site.

Figure 4. C141 Starlifter at the Pegasus runway site (*from George Blaisdell, USA CRREL*).

VISICE MODEL

The finite-element computer program VISICE was developed by the Naval Civil Engineering Laboratory [NOTE: on October 1, 1993, the Naval Civil Engineering Laboratory (NCEL) and five other Navy organizations were consolidated as the Naval

Facilities Engineering Service Center (NFESC) under the banner of the Naval Facilities Engineering Command (NAVFAC)] during the early 1970s to predict both the elastic ("landing") and linear visco-elastic ("parking") responses of a floating ice sheet to surface loads (Katona, 1979). The VISICE model assumes two criterion for failure: stress and deflection. In the short term, as an ice sheet is loaded suddenly by a landing aircraft, it behaves much like other elastic materials, where allowable stress dictates thickness required. On the otherhand, an airplane that lands successfully may not be able to remain stationary for prolonged periods. As an ice sheet relaxes beneath the overburden, allowable deflection between the main wheel assemblies limits the parking time permissible. The criteria applied by NFESC in plotting parking curves is that allowable deflection should not exceed the freeboard of the ice sheet, generally taken as 10% of thickness. The ten percent figure is considered conservative and safe since field tests have indicated that deflection beneath loads may exceed 25% of thickness without any major cracking or failure. The rationale behind the ten percent limitation is simply that once freeboard is exceeded, water can flood the surface through any cracks or holes, disrupting activities and eroding the confidence of attending personnel.

The carrying capacity of an ice sheet is a function of thickness, material properties, loading configuration, and duration of applied load. Material properties of sea ice, in turn, are highly dependent upon temperature and salinity, as characterized by time of year. The earliest load curves were developed on the basis of four general periods during the October-to-December operating season that reflected the progressive decay of material properties: mid-October to late November [period 1]; late November to mid-December [period 2]; mid-to-late December [period 3]; and late December through January [period 4]. That convention continues today, although safe operation of heavier aircraft such as the C5 Galaxy and C-17 Globemaster is restricted to period 1 only. Table 1 presents values of flexural strength and allowable stress assumed for each of the operating periods. The allowable stresses represent a conservative 25 - 30% reduction of flexural strength. Table 2 presents elastic material properties that are assigned as default values by VISICE.

Period of Operation	Surface Temp. °C	Flexural Strength kPa	Allowable Stress kPa
1	-20 to -10	482.6	338.0
2	-10 to -5	427.5	321.0
3	-5 to -2	400.0	300.0
4	isothermal	275.8	207.0

Table 1. Elastic flexural strength and allowable stress.

Temperature °C	Young's Modulus MPa	Poisson's Ratio
-27	4.1	0.30
-20	2.7	0.30
-10	1.7	0.30
-4	1.1	0.30

Table 2. Elastic properties assumed as default values by VISICE.

The finite-element approach applied by VISICE in the time-dependent mode requires that viscoelastic properties be given in the form of relaxation functions for both the bulk and shear moduli, K(t) and G(t), respectively, as modeled by Katona (1974) in the following simplified equations:

bulk modulus: $$K(t) = k_1 e^{-t/\alpha_1} + K_2 e^{-t/\alpha_2} \quad (1)$$

shear modulus: $$G(t) = G_1 e^{-t/\beta_1} + G_2 e^{-t/\beta_2} \quad (2)$$

Katona tacitly assumed Poisson's ratio as constant so that bulk and shear relaxation times would be identical and the modulus coefficients would be related in the same manner as an elastic solid. Table 3 presents values for the viscoelastic parameters at four different temperatures.

Temp °C	G_1 MPa	β_1 min	G_2 MPa	β_2 min	K_1 MPa	α_1 min	K_2 MPa	α_2 min
-27	1.50	4.50	0.060	2010.	3.25	4.50	0.130	2010.
-20	1.00	3.51	0.039	1350.	2.18	3.51	0.083	1350.
-10	0.61	3.58	0.028	2540.	1.33	3.58	0.061	2540.
-4	0.39	4.1	0.018	2400.	0.85	4.1	0.039	2400.

Table 3. Viscoelastic properties of sea ice assumed as default in the VISICE model.

The temperature-dependent material properties contained in Tables 1 through 3 are built into VISICE as default values. When operating in the default mode, the program assumes a linear profile between the assigned surface temperature and an isothermal (-3 °C) bottom condition, assigning material properties at intermediate nodes accordingly. The user need specify "elastic" or "viscoelastic" only, and the appropriate values are assigned automatically. In addition, a library subroutine for common aircraft allows the user to list a craft by name and percent of maximum weight; superposition of all wheel loads is provided automatically. Alternately, the user may enter his/her own values of material property and may specify an arbitrary pattern of loading. The modeling and mathematics used to develop VISICE are discussed in detail by Katona (1974), and by Katona and Vaudrey (1975).

RESULTS

Operation of the C-17 Globemaster will be limited to early in the season, during October before the sea-ice runway has warmed and weakened. The results of finite-element modeling by VISICE are contained in three charts plotted according to Period 1 conditions. The designation "early Period 1" refers to temperature conditions typical of the beginning of October, and "late Period 1" refers to conditions near the end. Figure 5 presents the results of elastic analysis in which the thickness of ice required in landing and taking off is shown as a function of gross aircraft weight. Figures 6 and 7 present the results of visco-elastic analysis with ice thickness shown as a function of required parking time and gross aircraft weight.

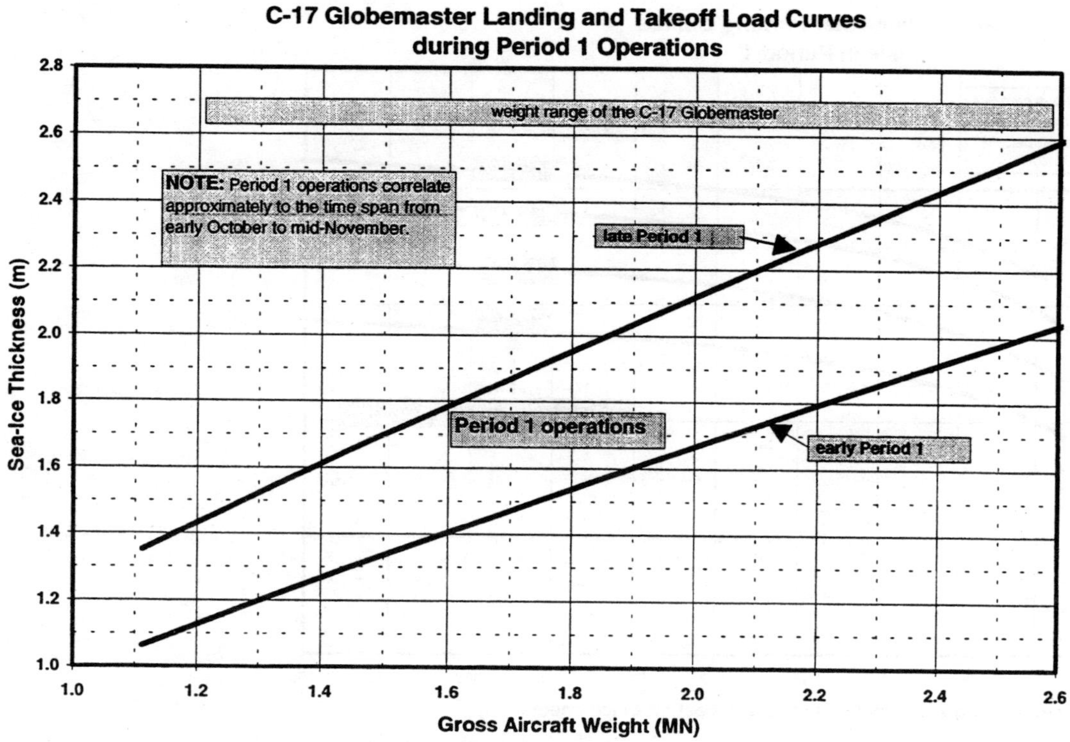

Figure 5. C-17 Globemaster landing and taking off curves.

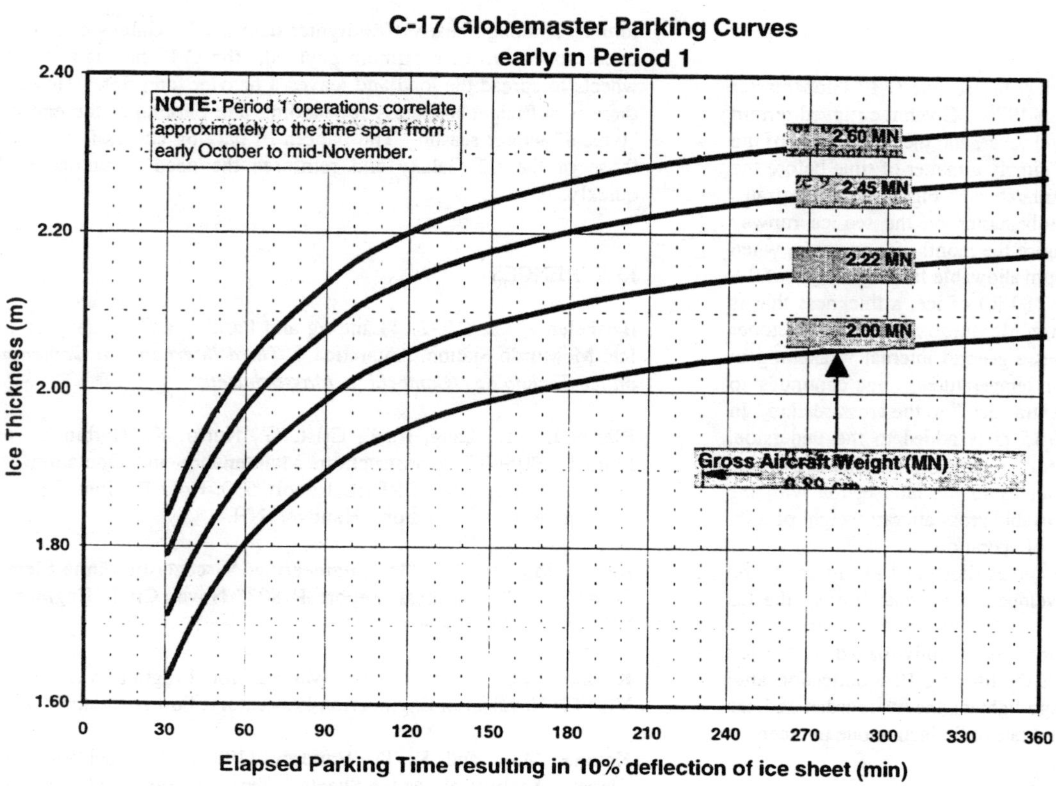

Figure 6. C-17 Globemaster parking curves (early Period 1)

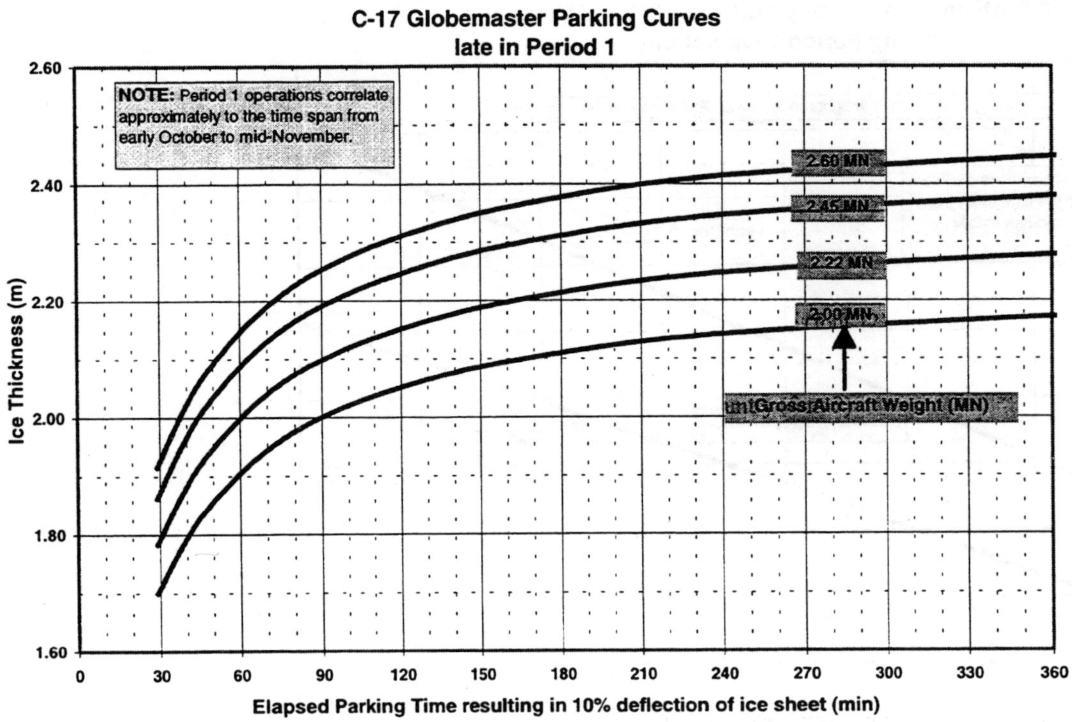

Figure 7. C-17 Globemaster parking curves (late Period 1).

DISCUSSION

The Air Force did not elect to use the C-17 Globemaster during the austral summer period of 1996. Given the rugged-terrain landing capability of this aircraft, however, and the phasing out of the C5 Galaxy from the military, it is simply a matter of time before the C-17 is pressed into service in Antarctica. Figure 5 demonstrates that it is feasible to land the Globemaster on the sea-ice runway during October. At the beginning of the month, for example, when the ice sheet is coldest, the maximum allowable landing weight of 2.6 MN (585,000 lbs) requires 2.0 m (80 in) of ice, a thickness that is typical of first-year annual growth in McMurdo Sound. As October progresses, the ice sheet undergoes a general internal weakening in response to warming ambient air temperatures, but continues to thicken even as it warms, adding extra "beef" to the cross-section. In a typical season, an additional 0.15 m is added to the underside, compensating somewhat for the reduced strength. Reentering Figure 5 at 2.15 m and traveling horizontally to the intersection with the "late Period 1" curve gives an allowable gross aircraft weight of 2.05 MN which is still within the range of operation.

The families of parking curves illustrated in Figures 6 and 7 are similar to those curves developed to validate leaving the C5 Galaxy on sea ice. As with the C5 Galaxy, the requirement is to unload the aircraft with dispatch, as a fully-loaded C-17 left stationary on the ice may produce the limiting 10% deflection after just one hour. Removing the cargo lightens the overburden and can significantly decrease the continued rate of deflection due to creep.

CONCLUSIONS

The C-17 Globemaster is the newest Air Force cargo transport and will likely see service in Antarctica in the not distant future. Although nearly 10% lighter than the C5 Galaxy counterpart under conditions of maximum payload, the C-17 has half as many wheels to spread the load and leaves a heavier footprint. However, there is sufficient thickness for landing and parking at the end of a "typical" winter season. The pattern of parking curves is similar to those of the C5 Galaxy and reinforce the need to unload cargo quickly.

REFERENCES

Barthelemy, J.L. (1992). "Landing and Parking a C5 Galaxy on Sea Ice: McMurdo Station, Antarctica," *Third International Conference on Ice Technology, Cambridge, Massachusetts.*

Blaisdell, G.L., Lang, R.M., Crist, G., Kurtti, K., Harbin, J., and Flora D. (1997). "Construction, Maintenance and Operation of a Glacial Ice Runway", CRREL Report 630, Cold Regions Research and Engineering Laboratory, Hanover, NH.

Katona, M.G. (1974). "Ice Engineering: Viscoelastic Finite Element Formulation," Technical Report R-803, Naval Civill Engineering Laboratory, Port Hueneme, CA.

Katona, M.G. (1979). User Manual for Program VISICE and VISFIT, Naval Civil Engineering Laboratory, Port Hueneme, CA.

Katona, M.G. and K. D. Vaudrey (1975). "Viscoelastic Finite Element Analysis of Sea Ice Sheets," *Third International Symposium on Ice Problems, International Association of Hydraulic Research, Hanover, NH.*

Proceedings of the Seventh (1997) International Offshore and Polar Engineering Conference
Honolulu, USA, May 25-30, 1997

ISBN 1-880653-28-1 (Set); ISBN 1-880653-30-3 (Vol. II)

Development of a Modern Heavy-Haul Traverse for Antarctica

George L. Blaisdell and Paul W. Richmond
Cold Regions Research and Engineering Laboratory, Hanover, NH, USA
Fred C. Kaiser
Caterpillar, Inc. Mossville, IL, USA
Russell G. Alger
Michigan Technological University, Houghton, MI, USA

ABSTRACT

An integrated, interdisciplinary study was completed to assess the feasibility of an oversnow route from McMurdo to the South Pole. Currently the only means of supplying the South Pole Station is by specialized aircraft. Air photo and satellite imagery were used extensively, along with ground-penetrating impulse radar (deployed from a helicopter) to make preliminary determinations of the suitability of glaciers in the Transantarctic Mountains for heavy tractor access from the Ross Ice Shelf to the polar plateau. These were followed by ground reconnaissance and data gathering. Modern traverse equipment was also developed and tested as part of this study. Tractor performance and terrain information were used to compare two potential traverse routes and to calculate delivered payload, fuel consumption, and travel time.

Key Words: **Oversnow vehicles, Antarctica, glaciers, remote sensing, ground-penetrating radar, route evaluation.**

INTRODUCTION

Aircraft are currently the only means of delivering large volumes of material to Amundsen-Scott South Pole Station. Cargo and fuel arrive in McMurdo, the main U. S. station in Antarctica, by conventional aircraft and ship. Supplies needed at the South Pole are moved with specialized ski-wheel Hercules (LC-130) aircraft, landing on a groomed natural snow surface. Alternatively, airdrop from C-141 Starlifters is sometimes used. Either method is generally expensive (more than US$6 per kilogram from McMurdo) and limiting (size and weight). Six-hundred to 1000 hours of flight time by the limited LC-130 fleet is currently consumed in delivering fuel alone to South Pole during its 100-day season of flight accessibility. Reliance on air delivery of all cargo has a strong impact on research and other logistics needs, and on aircraft availability for search and rescue missions.

Preliminary calculations suggest that an oversnow heavy traverse capability could provide considerable cost savings in the delivery of supplies to South Pole. Development of modern traverse technology could also make accessible new areas of the continent for scientific study. These factors prompted an integrated, interdisciplinary study to assess the feasibility of an oversnow route from McMurdo to the South Pole

In parallel with this feasibility study, modern traverse equipment was studied. Traditional heavy oversnow traverses have used steel-track tractors coupled with rugged sledges. These have many disadvantages, including slow speed, poor ride comfort (for both operator and cargo), poor maneuverability, significant terrain impact, and a high ratio of empty weight to delivered load. Currently, only the Russians routinely do long, heavy oversnow traverses (Mirny to Vostok, Antarctica). They use a variety of old tracked vehicles, most being modified versions of military vehicles. Other traverses in Antarctica have used custom Caterpillar D7 and Challenger C65 tractors coupled with heavy sledges (Australians), Hagglunds BV-206 articulated vehicles with light sledges (Scandinavian Antarctic programs), Caterpillar Challenger C65 tractors with rubber belted trailers and Kassbohrer PB tractors with Aalener sledges (joint Italian-French project), or specially built Ohara oversnow tractors towing heavy sledges (Japanese).

We tested a newly designed rubber-belted tractor and trailer system that was developed specifically for high-speed heavy-haul oversnow traversing. The results show significant improvements over traditional equipment. Measurements of speed, payload, fuel consumption, and oversnow mobility for the likely traverse routes to the South Pole show that each tractor-trailer system can deliver the equivalent of two LC-130 loads, for the same amount of consumed fuel.

CHALLENGES

The obvious hurdle to travel from McMurdo to the South Pole is climbing from the Ross Ice Shelf (essentially at sea level) up onto the polar plateau (elevation of 2000–3000 m) via a glacier in the Transantarctic Mountains. The last time the 1600-km overland trip from McMurdo to the South Pole was done (during the 1958-59 austral summer), the Skelton Glacier was used. Records of Sir Edmund Hillary's trip indicate that it was frequently slowed by deep snow, large sastrugi (hard, wind sculpted dunes), and tractors breaking through snow bridges over crevasses. History has also shown that travel a short way out of McMurdo can be challenging because of a deceptively hidden crevasse field between the McMurdo and Ross Ice Shelves. In 1990, the U. S. Antarctic Program lost a heavy tractor and two sledges in this area while attempting to follow Hillary's route. This area has recently been termed the McMurdo shear zone.

Our hope was to find a route that was reliable and robust enough to support frequent heavy haul tractor trains moving at good speed. We sought routes with grades of 4% or less, snow densities and hardness levels that would limit tractor train sinkage to less than 12 cm, and ones with minimal sastrugi or where roughness could be groomed for travel with minimal maintenance.

REMOTE STUDIES

Several preliminary and ad hoc studies were conducted by glaciologists familiar with portions of the Transantarctic Mountains that flank the Ross Ice Shelf and by air photo specialists. Their goal was to determine by remote means which glaciers in the range might be suitable for heavy tractor train travel. Information from notes made during assorted recent aerial reconnaissance flights within the Transantarctic Mountains were also used in analyzing the glaciers.

Initial work utilized existing air photos from the USGS map library (some dating back to the 1950's) and recent high-resolution satellite imagery to sort through the 40 largest glaciers that could provide access from the Ross Ice Shelf to the polar plateau. The principal concerns were to find routess with even, modest grades and firm, dry snow conditions and minimal crevassing. The air photo and satellite image study categorized the glaciers into general types based on their glaciological characteristics (Whillans et al., 1994). Ultimately, the glaciers were ranked on their apparent suitability for tractor travel. Only one glacier, the Leverett, appeared to satisfy the criteria desired for heavy tractor trains (Fig. 1).

Since the results of this initial study provided only one glacier that appeared suitable for travel by heavy tractor trains, we decided it would be prudent to include at least one other glacier in the follow-on aspects of this study. An additional candidate could provide a reference point for comparison to the Leverett Glacier and avoid leaving us with no options if the Leverett Glacier proved to have insurmountable problems. The Skelton Glacier (Fig. 1) was selected, partly in recognition of having supported the last successful tractor trip to South Pole.

In the McMurdo shear zone, a satellite image study was used to estimate the direction and rates of ice movement (Whillans and Merry, 1996). On the basis of the derived ice shelf motions, predictions were made of the areas where crevassing might occur and the direction and size of crevasses. This study also suggested a safe route from McMurdo through the shear zone and onto the Ross Ice Shelf.

LOW-LEVEL AERIAL RECONNAISSANCE

Low-level aerial reconnaissance flights were performed on the Leverett and Skelton Glaciers. These flights initially followed lines selected from the air photos and satellite image studies (Figs. 2 and 3). The lines were built from latitude and longitude coordinates estimated from the maps and photos and these were used to direct the pilots who were navigating with global positioning system (GPS) receivers. Observers on each flight noted problem areas along these flight lines and attempted to find more favorable alternative route segments. New GPS coordinates were established for these alternative segments.

Skelton Glacier

The Skelton Glacier (Fig. 2) was reconnoitered by Twin Otter aircraft in mid November 1995 at an elevation of 1500 m above the glacier. The route followed was roughly that of Hillary's expedition from the base of the glacier to just below The Portal. Landings were made just down-glacier of both Clinker Bluff and Stepaside Spur to place food and fuel caches for later ground and helicopter missions. During the flight it was difficult to assess grades or to guess at the existence of crevasse-free corridors. However, general observations of the possible tractor routes made it clear why the earlier studies did not consider the Skelton Glacier a good candidate. From Fishtail Point to a point north of Clinker Bluff and Mt. Tricouni, it appeared safe and easy to follow one of the central glacial "streets." As expected, the most challenging part of the glacier is between Clinker Bluff and a point north of Stepaside Spur. It was obvious that there was no suitable route to the west side of Clinker Bluff. Neither side of Twin Rocks appeared to be very attractive; however, passing to the east of the outcrop seems fraught with problems. Once a tractor train has made it safely north of Stepaside Spur, the terrain becomes considerably less challenging.

At the time of the reconnaissance flight a fresh covering of low density snow masked the glacier, but it could not hide the many open crevasses. Upon landing at the Stepaside Spur cache, the skis of the Twin Otter sank 15–20 cm in very dry, soft snow (less than 0.2 g/cm^3 density). Sinkage was even greater at the Clinker Bluff cache. A firm crust was present below the loose snow.

The Skelton Glacier was also reconnoitered at an elevation of 20–30 meters while performing helicopter-based ground-penetrating impulse radar (GPR) surveys at the end of November 1995. These flights were at very slow speed and stops were made at both caches for fueling, allowing much greater opportunity to make quantitative observations of the potential tractor routes. The general impression of the glacier in the region between just south of Twin Rocks to somewhat north of Stepaside Spur was that of a roller coaster. We estimated that the constantly rolling terrain had frequent grades from 4 to 6%, and up to 7 to 9% on occasion. Huge crevasses with partially open bridges were also common in the region between Twin Rocks and Stepaside Spur. Evidence of the maximum extent of a given crevasse in either direction from an opening was often non-existent, with no obvious sagging or subtle coloration change that can accompany large crevasses. Unlike the reconnaissance in mid November, there was a large area of exposed blue ice. This area spanned the entire width of the glacier and extended approximately between the entry points of the Baronick and Cocks Glaciers.

Leverett Glacier

The Leverett Glacier (Fig. 3) was viewed from an LC-130 Hercules aircraft during a dedicated low-level mission in mid December 1995. Observers were present on the flight deck and secured on the open cargo ramp at the rear of the aircraft. The flight took place with ideal visibility. The glacier appeared to have a very gentle slope from the ice shelf up to the area of Mt. Beasley. From Mt. Beasley, up-glacier to about 85° 56' S, the grade increases to perhaps 4%. Past this point, a tractor train would need to negotiate short, but very steep slopes in getting around the heart-shaped crevasse region up onto the plateau. Once over this hump, the terrain is virtually level all the way to the South Pole.

Crevassing could be seen in three areas on the Leverett Glacier that are important to the traverse. At the foot of the glacier, there are large open cracks approximately mid width. Crevassing could also be seen down-glacier and in line with Marsh Ridge. Finally, at the head of the glacier, there is an extensive network of crevasses, including the heart-shaped region, associated with ice spilling off the polar plateau and down the valley between the Stanford Plateau and the California Plateau.

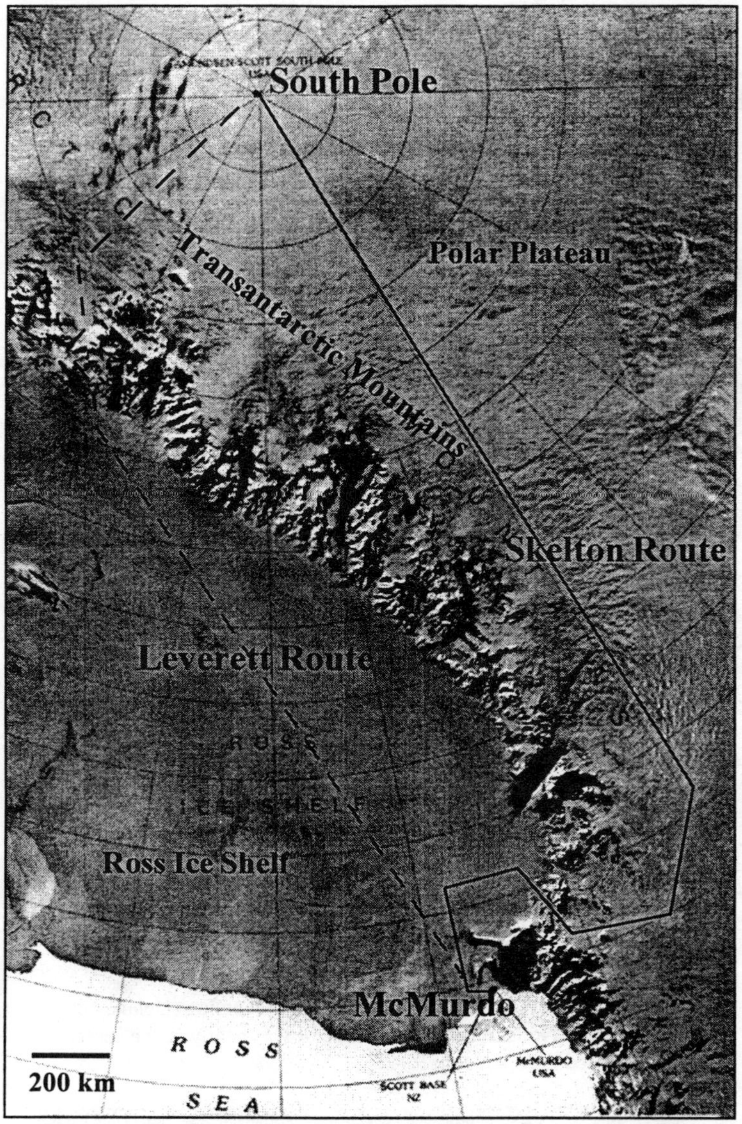

Fig. 1 Ross Ice Shelf and portion of Transantarctic Mountains showing potential traverse routes from McMurdo to the South Pole.

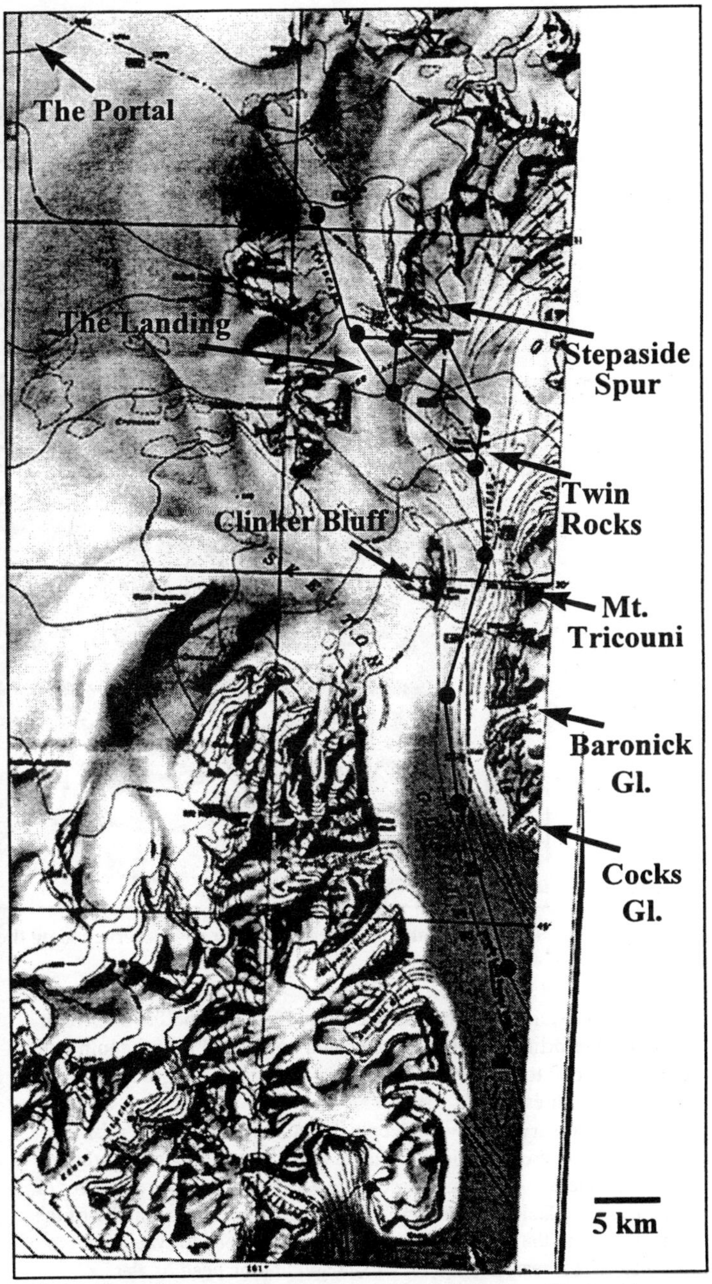

Fig. 2 Potential tractor routes surveyed on the Skelton Glacier (dots indicate major waypoints).

Viewed from the aircraft the snow surface appeared to be firm; it was very reflective and snow "snakes" were prevalent on the surface. Clearly, the surface of the Leverett Glacier was much stronger than what we saw on the Skelton Glacier.

A few weeks after the Hercules reconnaissance flight, low-level (25 m) GPR helicopter flights were made over the proposed traverse route on the Leverett. The impression of the observers on these flights was very similar to those who viewed the glacier from the Hercules.

GROUND-PENETRATING RADAR SURVEYS

Prior work has demonstrated that buried crevasses can be detected reliably by ground-penetrating radar (GPR) (Arcone and Delaney, 1995). Using the refined glacier route lines from the fixed-wing aerial reconnaissance flights, we completed helicopter-based GPR surveys of the selected glaciers to detect the presence of snow-bridged crevasses hidden below the surface. In the McMurdo shear zone the aerial radar surveys were followed by ground-based radar inspection. The radar equipment and techniques used for these surveys is described in Delaney et al. (1996).

Skelton Glacier

The GPR survey of the Skelton Glacier was performed in late November 1995. Detailed results from this survey are covered in Arcone et al. (1996). The results showed that most of the hidden crevassing is concentrated in the vicinity of Twin Rocks. Several lines were flown in the area in an unsuccessful attempt to find an uncrevassed route from just south of Twin Rocks to the center of The Landing (Fig. 2). Except for this 10-km segment of the glacier, and peculiar radar features that may indicate deeply buried crevasses or deep stratigraphic depressions in the snow seen at the very foot of the glacier, a crevasse-free route was mapped.

Leverett Glacier

The survey of the Leverett Glacier (Fig. 3) was performed in mid November 1995; detailed results are given in Arcone et al. (1996).

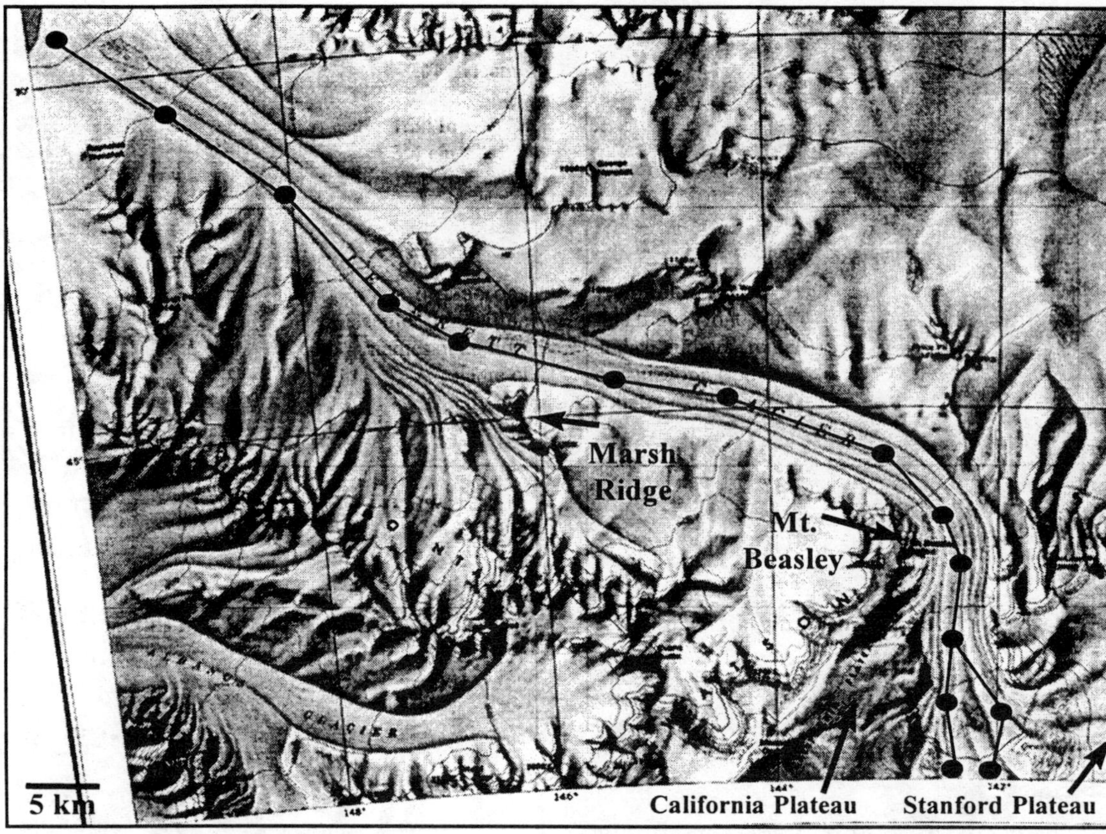

Figure 3. Potential tractor routes surveyed on the Leverett Glacier (dots indicate major waypoints).

The survey extended from well out on the Ross Ice Shelf to 86° south latitude. Unfortunately, the survey could not extend farther south onto the plateau due to poor visibility for low flying and lack of terrain reference features.

The radar survey showed deeply buried, narrow features near the foot of the Leverett Glacier, similar to those seen at the base of the Skelton Glacier. A crevasse-free route was found for the remainder of the Leverett Glacier, up to 86° S. The route keeps to the middle of the glacier around the Marsh Ridge area and a safe corridor was found to the east of the heart-shaped crevasse region at the head of glacier.

McMurdo Shear Zone

On the basis of the satellite imagery, we were concerned that the area where the McMurdo and Ross Ice Shelves intersect was heavily crevassed in a complex pattern. The GPR was flown along several lines from Williams Field out onto the Ross Ice Shelf. These lines included our best estimate of Hillary's route and the route suggested by Whillans and Merry (1996).

The radar results (Arcone et al., 1996) were encouraging; it appears that crevassing is limited to a distinct corridor. A strong correlation is suggested between the crevassed corridor and a narrow (4 km) band of tightly wrinkled flow traces seen in the satellite images (Fig. 4). This is a flow zone that Whillans and Merry (1996) identify as the Cape Crozier Shear Band. This band is thought to originate on the glacier-left edge of the Skelton Glacier where it enters the Ross Ice Shelf. The airborne radar records along all lines flown from Williams Field out onto the Ross Ice Shelf show no evidence of crevassing except in this 4-km band. We also plotted the position of a recent tractor fall into a bridged crevasse. Airborne radar investigation of this site in the past showed extensive chaotic crevassing (Delaney and Arcone, 1995). Not surprisingly, the tractor was nearly in the center of the proposed crevasse zone.

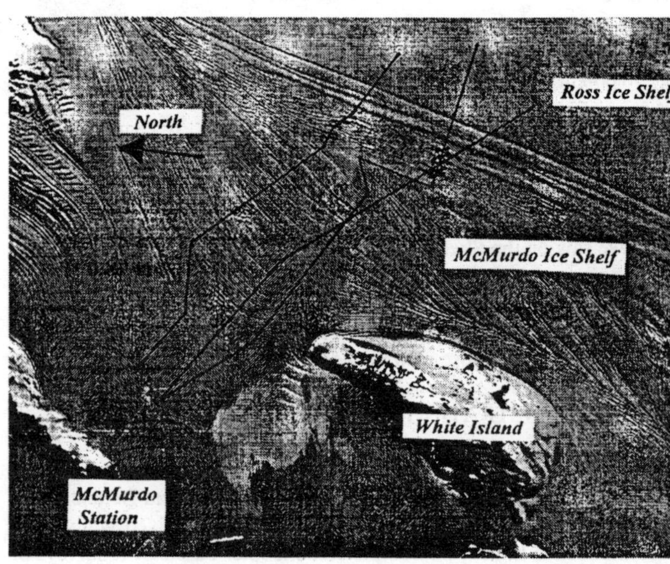

Fig. 4 Crevassed areas (hatched segments) detected in GPR survey lines overlaid on satellite image of McMurdo and Ross Ice Shelves (satellite image copyright 1994 SPOT Image Corporation. Scene ID: 30445569412011812251P).

To test the feasibility of using GPR as a warning system, the radar system was deployed as a ground-based unit by attaching it to a 3.5-m boom extending forward from a snowmobile (Delaney et al., 1996). The radar's data logging equipment and its operator rode in a sled being towed 12 m behind the snowmobile. The radar operator viewed the diffraction patterns in real-time and called out to a scribe when the antenna was passing over a crevasse. At the same time, snow bridge thickness was estimated. Data were stored and event markers were placed on the record to coincide with distance along the survey line.

More than 40 crevasses were identified along the 4 km radar survey line through the shear margin of the Ross and McMurdo Ice Shelves. Snow bridge thickness varied from 2 to 10 m and crevasse widths ranged from 1 to more than 10 meters.

GROUND PARTY SURVEY RESULTS

The results of remote data gathering for the proposed traverse routes were synthesized by a team of mountaineers, snow scientists, and engineers. This team deployed during the 1995-96 austral summer season by snowmobile to scout sections of the route. Being armed with such a large quantity of recent data, the ground team was able to operate efficiently and with confidence.

The ground team measured surface roughness, dug snow pits at regular intervals to determine snow properties, measured slopes, and probed crevasses identified by the GPR. In the McMurdo shear zone, four of the many crevasses identified by radar were opened and entered by the team to validate the radar results and to study the nature of the crevassing. These data, together with all of the remote data, were used to quantify the potential traverse routes (Table 1).

TRACTOR TRAIN EVALUATION

An extended-track redesign of the Challenger tractor has been developed by Caterpillar to meet the future logistic needs of the international Antarctic community. The tractor is designated the DV87 (Table 2, Fig. 5). The principal intent of this design was to reduce pitching at high speeds, thereby increasing driver comfort and facilitating efficient, long, oversnow traversing. The design also includes a two-person cab. Caterpillar has designed and built a rubber-belted tracked trailer (Fig. 5) to work with the DV87.

The purpose of our testing was to measure towing capacity and speed to allow estimates of travel time (loaded and unloaded), deliverable payload for candidate routes, and fuel consumption. We also performed some tests side by side with a standard Challenger tractor to show both objectively and subjectively the performance advantages of the DV87.

Table 1. Synopses of two potential heavy haul traverse routes from McMurdo to South Pole. (Location of route waypoints can be obtained from the authors.)

Skelton Route	Segment Length (km)	Maximum Grade (%)	Speed Limiter	Maximum Elevation (m)
Williams Field to Shear Zone	38	0	minor sastrugi	35
Shear Zone	22	0	minor sastrugi & crevasses	50
Z11 to S0 (Skelton base)	205	<2	big sastrugi	150
S0 to S3	40	2	big sastrugi & blue ice*	200
S3 to S5	12	-2 to 3	deep soft snow*	500
S5 to SB (Skelton top)	33	7-8, 10	deep soft snow, steep slopes*	1600
SB to Plateau	85	<4	unknown	2500
Plateau to South Pole	1473	<1	unknown	3000

Leverett Route	Segment Length (km)	Maximum Grade (%)	Speed Limiter	Maximum Elevation (m)
Williams Field to Shear Zone	38	0	minor sastrugi	35
Shear Zone	22	0	minor sastrugi & crevasses	50
Z11 to L00 (Leverett base)	990	<2	unknown	150
L00 to L4	57	4	large sastrugi*	1100
L4 to L7	21	5	hard snow & large sastrugi	1500
L7 to L8	6	-3 to 5	bumps	1600
L8 to L9	6	6	none	1850
L9 to L12	7	2-6	slopes	2100
L12 to L18 (Leverett top)	4	2, 10, 15	two 0.5 km steep slopes*	2400
L18 to South Pole	443	<2	moderate sastrugi	3000

* Crevasses appear in radar (GPR) records for some portion of route segment.

Table 2. DV87 Tractor Characteristics.

Weight (kg)	21 000
Track width (m)	0.90
Track length on ground (m)	3.58
Total track area (m²)	6.44
Contact pressure (kg/cm²)	0.326
Fuel capacity (L)	950
Maximum speed (km/hr)	35
Power (kW)	230

Fig. 5 Caterpillar DV87 traverse tractor in Antarctica towing two rubber-tracked trailers with 33 100 kg gross payload.

Drawbar and Towing Resistance

Using traditional drawbar-pull test protocol (SAE standard practice J872), the maximum available towing force for the DV87 was determined at steady speeds for gears 2 through 6 at full throttle. Tests were performed on firm snow (density 0.35–0.45 g/cm³), volcanic cinder soil, and glacial ice (Table 3).

Towing resistance was measured by pulling the vehicle or trailer forward at about 8 km/hr and using the load cell to obtain the force required. Towing tests of the tractor were done on ice and volcanic soil surfaces; towing resistance of the tracked trailers was done on a hard packed snow road and on the firm natural snow at several load configurations (Table 4).

We also performed several short traverses including up and down grades. With a single loaded tracked trailer (16 600 kg payload) the DV87 negotiated a 15% grade on a cinder road. On a longer traverse, over deep snowfields, the trailer was towed successfully up grades from 1 to 3%, but became traction limited when attempting 4% grades on untrafficked snow.

Table 3. Drawbar-pull test results (values are averages for all tests).

Surface	Drawbar-pull (kN) per Gear				
	2nd	3rd	4th	5th	6th
Firm Snow	80.95	80.06	75.62	71.17	71.17
Ice	70.28	64.50	63.61	66.72	68.94
Cinder road	*	*	*	*	*
Hard packed snow road	71.17	90.74	83.18	77.84	86.74

*All measurements exceeded the capacity of the load cell (111 kN).

Table 4. Towing resistance test results.

Equipment Towed	Surface	Towing Resistance
Tracked trailer; 16 600 kg	Hard packed snow road	13.3-17.8 (kN)
Tracked trailer; 16 600 kg	Firm snow	33.4-44.5
DV87	Glacial ice	8.45
DV87	Cinder road	12.45
Two trailers; 16 500 kg (front), 16 600 kg (rear)	Hard packed snow road	34.7-35.6
Two trailers; 12 410 kg (front), 12 500 kg (rear)	Hard packed snow road	26.7-31.1
Two trailers; 12 410 kg (front), 12 500 kg (rear)	Firm snow	48.9

Ride Comparison Test

Three operators were chosen to drive the DV87 over a round trip course with and without loaded trailers attached. The drivers were instructed to drive at comfortable speeds. Times were measured (Table 5) and driver comments recorded. This test was done on untracked surfaces along a 6.5 km loop. The surface was primarily ice with patches of crusted snow; little or no sinkage of the vehicles occurred. The loaded condition consisted of a tracked trailer carrying 16 600 kg of concrete. The operators indicated that speed was limited by terrain roughness (concern for equipment damage) rather than by driver comfort. Surface profiles were made to document roughness for comparison with Caterpillar's ride performance analysis.

Table 5. Travel times for ride test.

Operator	Time, min.	
	Loaded	Unloaded
1	38	37
2	34	39
3	46	30
Average	39.3	35.3
Speed (km/hr)	9.5	11.0

Vehicle Speed and Fuel Economy

Vehicle speeds obtained by the DV87 were measured by timing the vehicle over a straight course about 3.2 km in length. Two courses were used; a hard packed snow road and undisturbed firm natural snow. Successive tests in the natural snow were always in undisturbed areas. Flags were used to mark the start and end points where the tractor entered at full speed and maintained this speed throughout the course. Results of the speed tests while pulling various loaded trailer combinations are given in Table 6.

To determine fuel economy, the fuel tank was filled and the DV87, pulling a tracked trailer with a 16 600 kg payload was driven approximately 54 km. The course was around a measured loop and both clock time and engine hours were documented. The tank was then refilled and the fuel quantity measured. Fuel consumption was 200 L in 3.4 hours (60 L/hr). Engine hours were approximately equal to clock time. For comparison a Caterpillar D7 used 260 L in 6.4 hours (40 L/hr) while backfilling a snow trench under similar weather conditions (a slightly less strenuous task).

Table 6. DV87 tractor speeds while towing various loads.

Vehicle Configuration	Surface	Gear	Time (min:sec)	Speed (km/hr)
Two trailers; empty	Snow road	10th	6:34	29.4
	Firm snow	9th	10:48	17.9
Two trailers; 4160 kg each	Snow road	10th	8:00	24.1
	Firm snow	9th	12:12	15.8
Two trailers; 8320 kg each	Snow road	10th	8:05	23.8
	Firm snow[1]	8th	15:53	12.2
One trailer; 16 600 kg	Snow road	10th	6:26	29.9
	Firm snow	8th	14:25	13.4
Two trailers; 16 500 kg (front), 16 600 kg (rear)	Snow road	9th	10:50	17.9
	Firm snow[2]	5th	22:56	8.4
Two trailers; 12 410 kg (front), 12 477 kg (rear)	Snow road	10th	8:48	21.9
	Firm snow	6th	19:19	10.0

[1] In an additional test run, this vehicle configuration became stuck in snow that had been trafficked previously.
[2] The vehicle became stuck in a "pocket" of weak snow, was assisted back onto stronger snow, and timing resumed.

ANALYSIS

The two candidate traverse routes were divided into segments based on common features of the terrain (Table 1). This allowed us to use the performance data for the DV87 tractor train tests to estimate what speeds could be supported over each segment. By determining the most efficient load and gear combinations for the grades, surface roughness, and snow strength in each segment, it was possible to compile a likely scenario for a typical round trip traverse from McMurdo to the South Pole along the Leverett and Skelton routes. We also took into account the increased fuel consumption (8%) required when operating above 2300 m. Each tractor was responsible for carrying its own round trip fuel. Except when very concentrated loads are delivered, we assumed that each tractor would tow two trailers. To increase efficiency, we planned for return trip fuel for the segment from the base of the glacier back to McMurdo to be cached in stationary tanks at the foot of the glacier.

Both glaciers will require that each tractor tow one trailer at a time for the majority of its length. It also appears that both glaciers have short segments that would require winching loads up, or hitching two tractors to a single trailer, to get reasonable payloads over grades that exceed 5%. Other alternatives for navigating these segments exist, but were not included in our analysis (e.g., on-site heavy "assist" tractor, stationary tow winches, switchback trail preparation).

Calculations of driving time, delivered payload, and fuel consumed for the Skelton and Leverett Glacier routes, assuming minimal trail preparation, are given in Table 7. The Leverett route shows a 14% greater delivered payload while taking 16% less time. Average speed over the Leverett route is almost 1 km/hr faster. Equally important, the Leverett route consumes nearly 20% less fuel. These numbers, as well as equipment wear and tear, could probably be favorably affected if trail grooming was considered. For comparison, the data for an LC-130 Hercules delivery to South Pole is also shown in Table 7.

CONCLUSIONS

Our efforts using modern remote tools, such as air photos, satellite imagery, and air-deployed GPR provided a great advantage in quickly and efficiently determining the most promising corridors for tractor travel. This approach not only reduces cost, time, and numbers of personnel, it provides a ground reconnaissance party with valuable information that makes their task safer and more focused.

Clearly, the Leverett Glacier route is favorable to the Skelton route. This is not surprising, since the Skelton route is longer and enters the polar plateau early during the outbound journey (a greater quantity of fuel must be hauled up more than 2000 m in elevation, and considerably more time is spent operating at altitude). Table 7 shows that the only advantage to delivering materials by LC-130 is the vastly shorter time required for a round trip. A smaller payload (half) is delivered by aircraft for approximately the same amount of consumed fuel compared to either tractor train route. The LC-130 also suffers from payload weight limits (11 360 kg maximum delivered load from McMurdo) and volume limits (2.7 x 2.7 x 12 m). A further drawback is the greater cost of operating aircraft compared to ground vehicles (including more highly trained operators). These disadvantages must be weighed against the increased risk, exposure and other potential unknowns associated with a McMurdo to South Pole oversnow traverse.

Additional studies will be required to obtain greater refinement of the performance estimates given in Table 7. These studies will need to concentrate on quantifying surface conditions along the previously unexplored segments of the proposed routes, and performing longer trial traverses with the tractor train. This additional information will allow a reasonable estimate of the cost of delivery by tractor train.

Table 7. Comparison of potential traverse routes and aircraft.

	Leverett Glacier	Skelton Glacier	LC-130 Hercules
Round trip distance (km)	3450	3800	2060
Operating time (hrs)	336	402	6.5
Delivered load (kg)	25 910	22 780	11 360
Fuel used (kg)	12 727	15 855	14 960
Fuel used (L)	15 590	19 420	18 320
Overall speed (km/hr)	10.27	9.45	320
Efficiency 1 (kg/hr/L)	0.0049	0.0029	0.095
Efficiency 2 (tonne-km/hr)	266	215	3600

ACKNOWLEDGMENTS

This work was sponsored by the Operations Section, Office of Polar Programs, National Science Foundation. Steve Arcone and Allan Delaney (both of CRREL) performed all of the GPR work for this study. Field work was greatly assisted by Vernon Tejas, Oriol Sole-Costa, Quinton Rhoton, and Mike Darrah (all from Antarctic Support Associates). John Evans (Antarctic Support Associates) and Dave Bresnahan (National Science Foundation) performed much of the coordination for this work.

REFERENCES

Arcone, S. A., Delaney, A. J., and Blaisdell, G. L. 1996. "Airborne radar crevasse detection along the proposed South Pole inland traverses," In *Proc XXIV SCAR Conf*, 3-6 August, 1996, Cambridge, England.

Delaney, A. J., and Arcone, S. A. 1995. "Detection of crevasses near McMurdo Station, Antarctica with airborne short-pulse radar," In *Proc XXIII SCAR Conf*, 29 August - 2 September 1994, Rome, Italy, p 323-336.

Delaney, A. J., Arcone, S. A., and Blaisdell, G. L. 1996. "Ground-penetrating radar techniques for crevasse detection," In *Proc XXIV SCAR Conf*, 3-6 August, 1996, Cambridge, England.

Whillans, I. M., Hamilton, G., and Merry, C. J. 1994. "Search for a safe tractor route from McMurdo to South Pole," In *Proc Antarctic Traverse Workshop*, 2-4 May, 1994, Washington, DC, CRREL Tech. Note.

Whillans, I. M. and Merry, C. J. 1996. "Kinematics of the shear zone between Ross Ice Shelf and McMurdo Ice Shelf," Contract report OSU-10, Byrd Polar Research Center, Ohio State University, 17 p.

Proceedings of the Seventh (1997) International Offshore and Polar Engineering Conference
Honolulu, USA, May 25-30, 1997

ISBN 1-880653-28-1 (Set); ISBN 1-880653-30-3 (Vol. II)

The Polar Ice Prediction System (PIPS 2.0) - the Navy's Sea Ice Forecasting System

Pamela G. Posey and Ruth H. Preller
Naval Research Laboratory
Stennis Space Center, MS, USA

Abstract

The Polar Ice Prediction System (PIPS 2.0) is the operational model used by the Fleet Numerical Meteorology and Oceanography Center (FNMOC) for sea ice forecasting. The PIPS 2.0 coupled ice-ocean model forecasts ice conditions from the North Pole south to 30° N latitude. The PIPS 2.0 makes a 120-hour forecast each day of ice drift, ice thickness and ice concentration. Ice concentration data from the Special Sensor Microwave Imager (SSM/I) is used to initialize the system's forecast. Results from PIPS 2.0 forecasts with and without the use of SSM/I data for initialization will be discussed.

Key Words

Sea ice forecast, ice drift, ice edge and ice model

Introduction

Since the late 1980's, ice forecasting systems, developed by the Naval Research Laboratory, have been predicting conditions in the Arctic and several of its marginal seas at the Fleet Numerical Meteorology and Oceanography Center (FNMOC). Beginning in 1987, the Polar Ice Prediction System (PIPS) has produced forecasts over the central Arctic basin, the Barents Sea and the Greenland Sea using a coarse grid resolution of 127 km. In June 1989, the first regional ice forecast system, the Regional Polar Ice Prediction System-Barents Sea (RPIPS-B) was implemented. RPIPS-B, a higher resolution version of PIPS, forecasts over the Barents Sea and the western part of the Kara Sea using a grid resolution of 25 km. The second regional forecast system was implemented in October 1991, the Regional Polar Ice Prediction System - Greenland Sea (RPIPS-G). RPIPS-G forecasts over the region adjacent to the East Greenland coast using a grid resolution of 20 km. The forecast systems are driven by monthly mean ocean currents and deep ocean heat fluxes derived from the Hibler and Bryan (1987) coupled ice-ocean model. As of July 1996, a coupled ice-ocean model, the Polar Ice Prediction System (PIPS 2.0), was implemented operationally by FNMOC. PIPS 2.0 forecasts conditions in all sea ice covered areas in the northern hemisphere (down to 30°N in latitude). The horizonal grid resolution of the model is 0.28 degrees and uses 15 vertical levels. All four ice systems produce forecast fields of ice drift, ice thickness, ice concentration (ice edge) and the growth/decay of ice based on both dynamic and thermodynamic effects. The systems are driven by atmospheric forcing from the Navy Operational Global Atmospheric System (NOGAPS) (Hogan et al., 1991). At the present, all four systems (Figure 1) are currently running operationally, but within the next few months PIPS 2.0 will replace the other three systems.

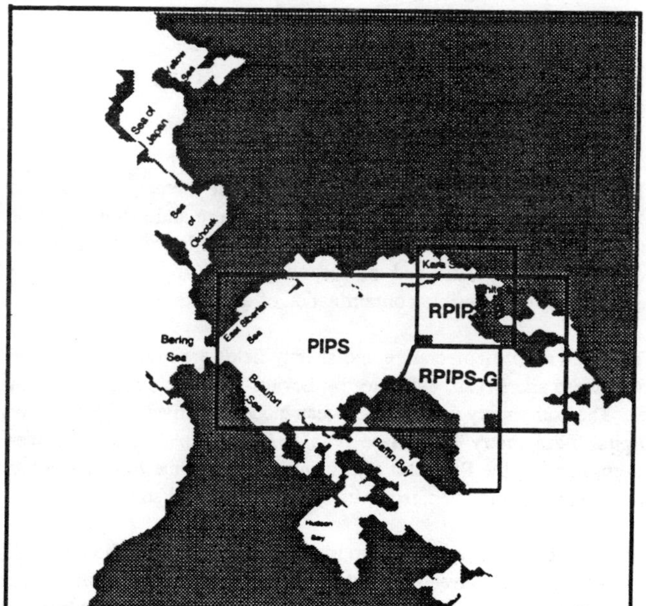

Figure 1. PIPS 2.0 domain with regional models included.

Model Description of PIPS 2.0

The coupled ice-ocean model used by PIPS 2.0 consists of the Hibler ice model (Hibler, 1979; 1980) and the Bryan and Cox ocean model (Cox, 1984). The Hibler ice model was rewritten in spherical coordinates to be consistent with the Cox ocean model and to define a more realistic grid at the lower latitudes (Cheng and Preller, 1996). The model domain extends from the North Pole south to approximately 30° N latitude and includes marginal seas such as the Sea of Okhotsk, the Sea of Japan and the Yellow Sea in the Pacific and the Gulf of St. Lawrence and the Labrador Sea in the Atlantic. The grid resolution is 0.28 degrees which varies from 17-33 km depending on the location of the grid square within the spherical coordinate system (Figure 2).

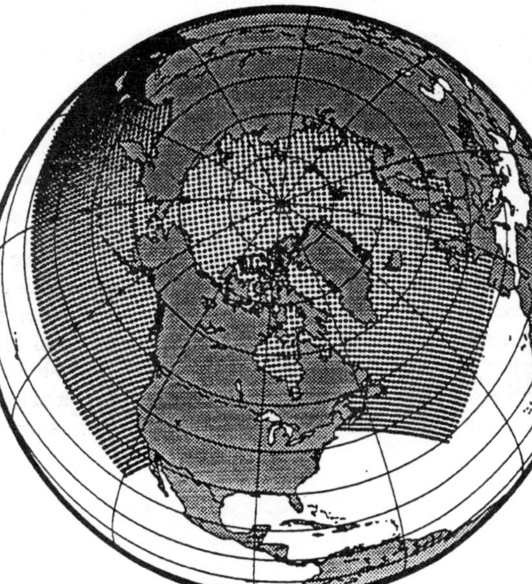

Figure 2. PIPS 2.0 model domain and grid. Every fourth point is plotted.

The ice model uses a time step of 2 hours while the ocean model uses a time step of 0.5/0.05 hours for the temperature and salinity/velocity equations. For both the ice and ocean models, the lateral boundaries are defined as solid walls and placed away from any sea-ice covered regions to avoid possible contamination of any forecasted fields.

The ocean model temperature and salinity fields are loosely constrained (Sarmiento and Bryan, 1982) to the Levitus (1982) climatological data set. The bathymetry used by the ocean model is derived from the Navy Digital Bathymetry Data Base 5' x 5' (DBDB5). These data were interpolated to the PIPS 2.0 grid using a cubic spline function, which was integrated and averaged over nine adjacent grid squares.

In this coupled system, the ocean model provides input to the ice model in terms of ocean currents, salinity and heat fluxes (temperatures). The ice model provides to the ocean model, surface stresses as well as salinity and temperature changes due to the growth and decay of sea ice. The ocean model contains 15 levels in the vertical, each level increasing in thickness with depth. Direct interaction between the ice and ocean model occurs in the first level which is 30 m deep and represents the ocean mixed layer.

Figure 3 is a schematic of the design of the forecast system. At the center of the system, is the coupled Hibler ice model/Cox ocean model. Atmospheric forcing is provided to the model by NOGAPS. The atmospheric forcing fields used by the forecast system are surface air temperature, surface pressure, surface vapor pressure, shortwave radiation, sensible plus latent heat flux and the total heat flux. The system produces a 120-hour forecast each day. Along with the output of the model (ice drift, thickness, concentration and growth/decay of ice), restart fields consisting of the model's 24-hour forecast are written to a file to be used in the next forecast. If the model restart fields are not available, a model derived climatology is used to restart the system.

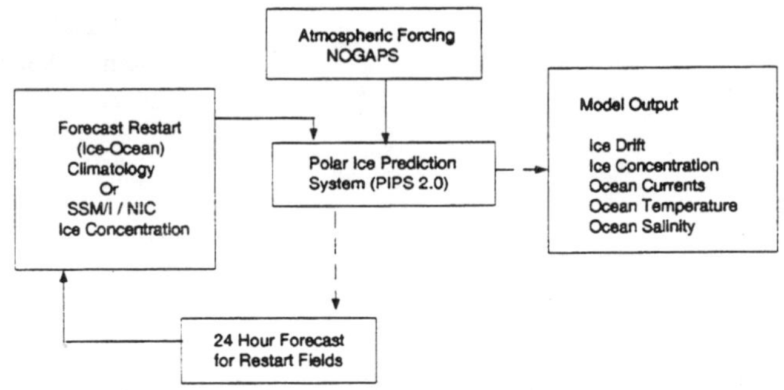

Figure 3. Schematic of the PIPS 2.0 sea ice forecast system.

PIPS 2.0 uses ice concentration data from the DMSP Special Sensor Microwave/Imager (SSM/I) for daily initialization of several model fields. There are 3 factors which make the SSM/I data appealing as a source of initialization data: the data are available in real time at the same site at which the forecast systems are run, the resolution of the data is similar to the resolution of the model and the data cover the entire model domain. During the past several years (1992-1996), FNMOC has been obtaining SSM/I data that have been converted into ice concentration data using the Navy algorithm (Hollinger, 1991) and are used by the NIC in their weekly analysis. The previous forecast systems were initialized once per week by the ice concentration data from the National Ice Center (NIC). The NIC's ice concentration analysis is a subjective analysis created from a combination of several sources of remotely sensed data (AVHRR, SAR and SSM/I). After the PIPS 2.0 restart fields are read in, the SSM/I (or NIC) ice concentration update data are read in. These real-time data are then used to replace the restart ice concentration field. The ice thickness field and the ocean temperature field are updated to be consistent with the data. The PIPS 2.0 products are reviewed regularly by NRL, FNMOC and NIC.

Forecast Products

Since the end of July 1996, PIPS 2.0 has been producing a 120-hour forecast daily. The forecast products from the model, in the form of either a graphic or printed numbers, are transferred to the NIC where they are used as part of the NIC forecasting analysis. Standard products from the system are ice drift, ice thickness and ice concentration. Along with these existing products, PIPS 2.0 also has the capability to forecast ocean currents, ocean temperature and salinity.

Figures 4a and 4b represent sample forecasts of ocean currents and temperature for winter and summer, respectively. The weak constraint on the ocean temperature and salinity (not shown) will not allow the model to forecast large temporal anomalies like the Great Salt Anomaly; however, the model does allow realistic daily, monthly and seasonal variability, as well as normal interannual variability.

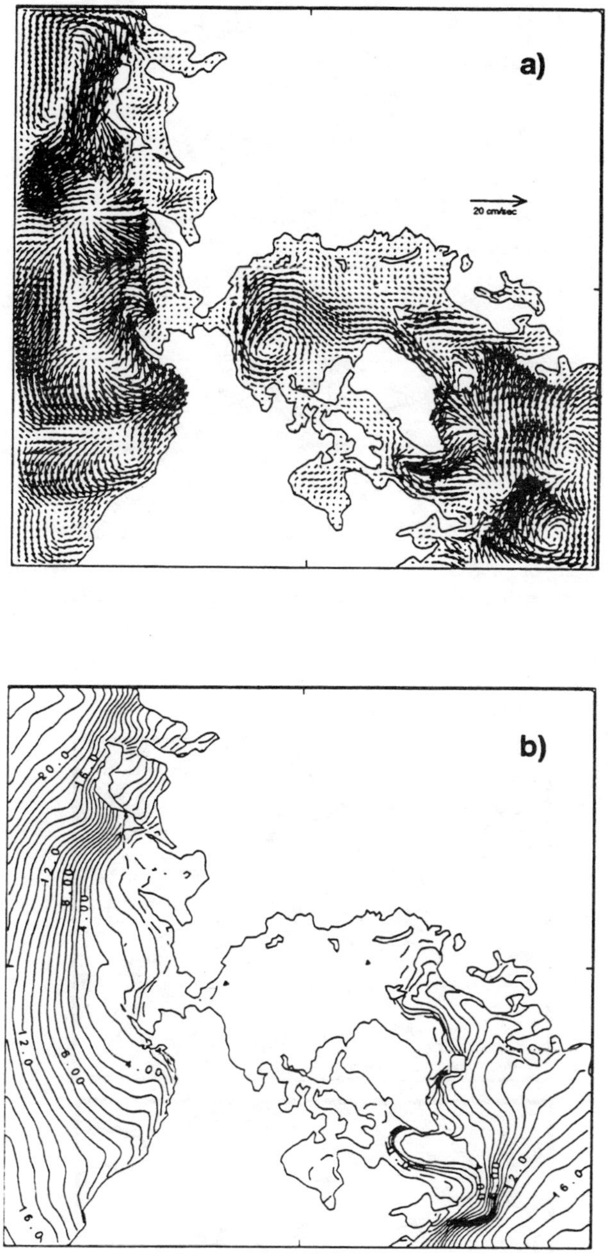

Figure 4a. Sample 24-hour forecast valid March 30, 1992 from PIPS 2.0 of a) Ocean currents (cm/sec) at 15 meters. Maximum scaled drift vector is 20 cm/sec, b) Ocean temperature (°C) at 15 meters. Contour interval ranges from -1°C to 24°C by 1°C at 15 meters.

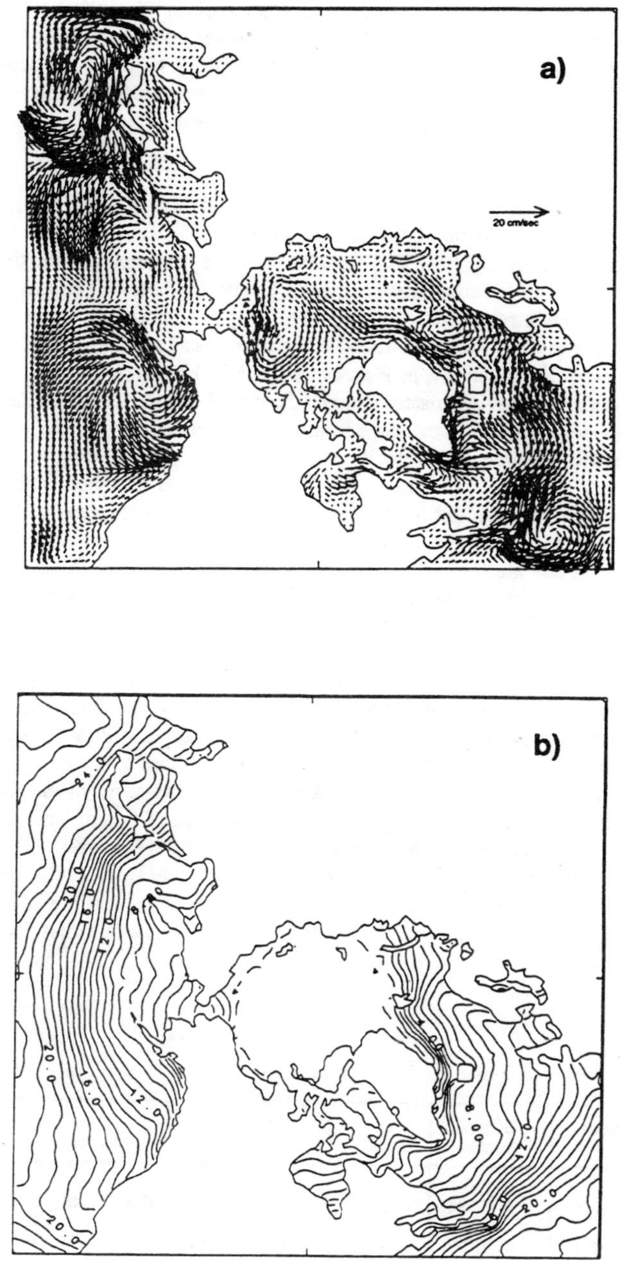

Figure 4b. Sample 24-hour forecast valid September 30, 1992 from PIPS 2.0 of a) Ocean currents (cm/sec) at 15 meters. Maximum scaled drift vector is 20 cm/sec, b) Ocean temperature (°C) at 15 meters. Contour interval ranges from -1°C to 27°C by 1°C.

Model Verification

Prior to operational implementation and verification, rigorous testing of a model is performed. During the last few years, the PIPS 2.0 coupled ice-ocean model, minus any data assimilation, was tested using several years of NOGAPS forcing. Ice motion is one of the few sea ice fields for which enough data exist to actually do a quantitative comparison between model and observations. Such a comparison was performed for the verification of the PIPS 2.0 ice drift (Preller and Posey, 1996). A statistical analysis performed on a year's worth of modeled ice drifts from 1992 versus ice drift derived from Arctic buoy positions, indicated that the PIPS 2.0 ice drift was an improvement over the existing operational products.

A qualitative comparison between model derived lagrangian trajectories and eight Argos drifting buoys was also performed. Table 1 shows the starting and ending time and location for each of the Argos drifting buoys used in this comparison. Figures 5a and 5b shows the tracks of these buoys through the Arctic in 1992. Figures 6a and 6b show the tracks of the eight drifters started at the same location and time as the observed drifters shown in Figures 5a and 5b. However, in this case, the drifter track is calculated based on the model-derived ice drift. Each day a new location is calculated based on the model-derived ice drift velocity. The ice-drift velocity is determined by interpolating to the drifter's location. Comparison of Figure 5a and 6a shows good agreement between the model data and observed data for buoy 12805 drifting through the Fram Strait. The two buoys that are placed in the Beaufort Sea are initially on a course in agreement with real data, but they have a strong westward component pushing them toward the Chukchi Sea during the last half of their track. Buoy 12807 also seems to move with good agreement during the first half of its track, but then it veers southward while the observed drifter moves poleward. Figures 5b and 6b indicate that buoy 12802, also exiting the Fram Strait, compares very well to the observed drifter's track. Buoy 11108, beginning north of the Barents Sea, tracks fairly close to the observed drifter along most of the pathway. Approximately two-thirds of the way through the track, however, it moves south instead of turning east and ends up farther east than the real buoy. Buoy 12800 tracks slightly farther north and west than the observed drifter, while Buoy 12801 gets caught in a relatively strong southerly flow not observed in the actual drifter's track. As a result, it is located closer to the Alaskan coast than the actual buoy.

Table 1. ARGOS buoy locations

Buoy	Starting Date	Ending Date	Starting Location	Ending Location
11108	01/01/92	12/31/92	84.996°N, 74.040°E	80.850°N, 16.381°E
12800	01/01/92	08/09/92	75.959°N,204.557°E	79.732°N,195.139°E
12801	01/01/92	12/31/92	79.855°N,222.181°E	72.243°N,217.321°E
12802	01/01/92	12/31/92	85.791°N,290.782°E	74.752°N,349.720°E
12805	01/01/92	12/31/92	86.608°N,327.827°E	71.241°N,341.248°E
12806	01/01/92	12/31/92	72.326°N,219.983°E	72.918°N,175.332°E
12807	01/01/92	10/22/92	79.647°N,192.552°E	80.595°N,227.456°E
12808	01/01/92	12/31/92	76.170°N,234.282°E	71.829°N,209.405°E

ARGOS Drifter Tracks

a)

b)

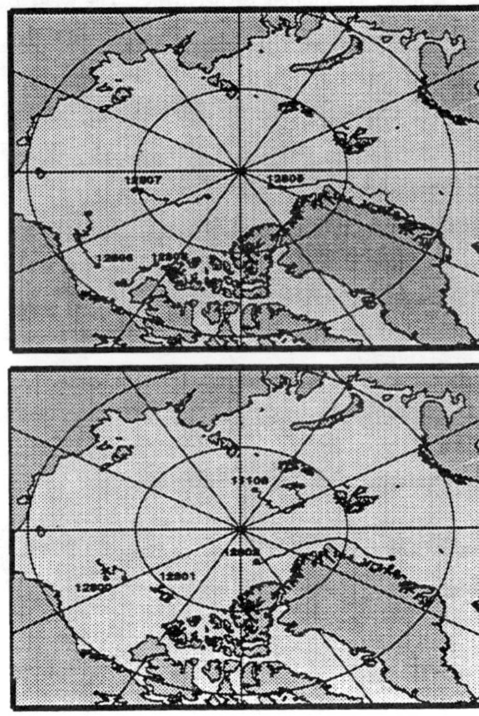

Figure 5. ARGOS Drifter Tracks from January 1, 1992 through December 31, 1992. a) Locates buoy 12805, 12806, 12807 and 12808 and b) locates buoy 11108, 12800, 12801 and 12802.

PIPS 2.0 Drifter Tracks

a)

b)

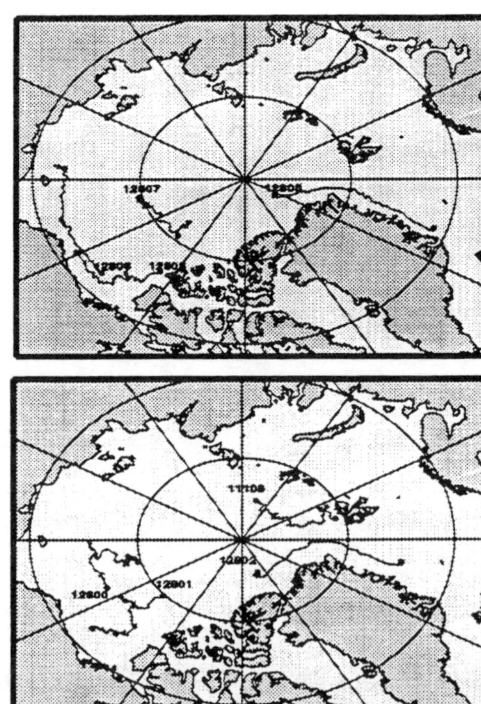

Figure 6. PIPS 2.0 Drifter Tracks from January 1, 1992 through December 31, 1992. a) Locates buoy 12805, 12806, 12807 and 12808 and b) locates buoy 11108, 12800, 12801 and 12802.

A final qualitative comparison was made between the modeled ice drift field and surface pressure fields derived by Colony and Rigor (1993) from the ARGOS buoy data. Figure 7 shows the ice drift field for each of four days representing the four seasons. Figure 8 shows the surface pressure fields for the same days derived from the buoy data.

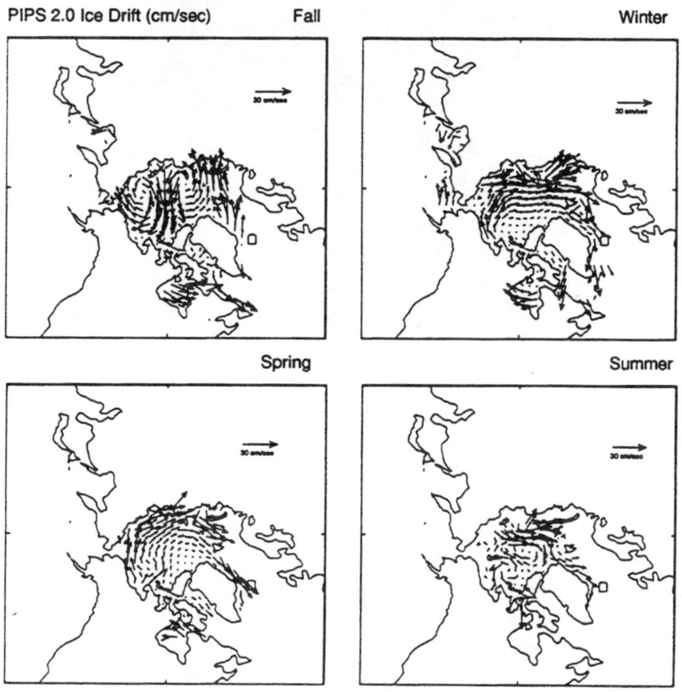

Figure 7. Daily ice drift velocities (cm/sec) from PIPS 2.0 representing the four seasons a) Fall, b) Winter, c) Spring and d) Summer. Maximum scaled drift vector is 30 cm/sec.

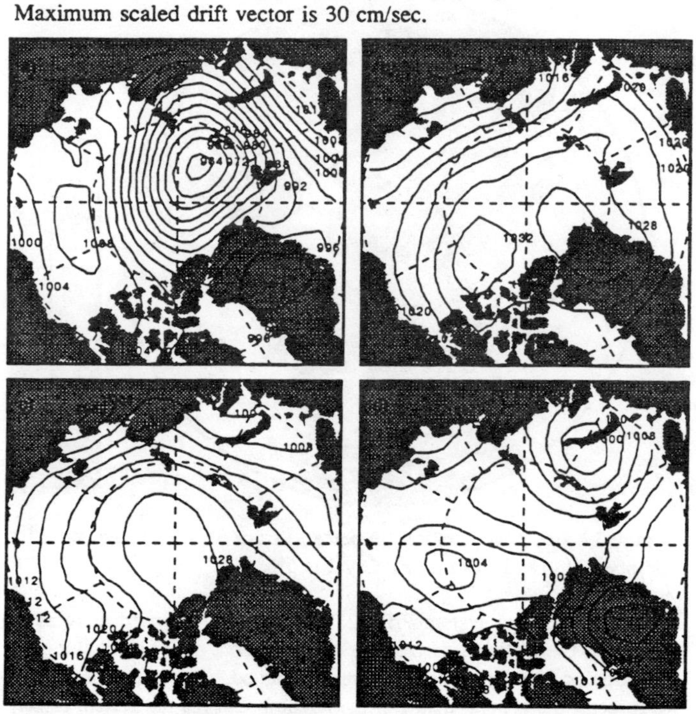

Figure 8. Daily surface pressure derived from Arctic buoys presenting the four seasons a) Fall, b) Winter, c) Spring and d) Summer.

The ice drift from the model clearly reflects the surface pressure fields for the same time. Surface pressure fields from the NOGAPS are generally in good agreement with the pressure fields produced by Colony and Rigor since NOGAPS uses the observed pressure data from these buoys in its daily analysis field. Therefore it is expected that the daily modeled ice drift would also agree well with these observations.

Figure 9 contains the daily mean ice concentration for these four days. The ice edge from the NIC analysis is drawn as a heavy dark line on this figure. PIPS 2.0 ice concentration and ice edge location forecasts were evaluated by comparing them to the digitized ice concentration analysis from the NIC on a weekly basis. Comparison of the modeled versus the observed ice edge indicates that the model does well in predicting the ice advance and retreat in the Pacific marginal seas, the Bering Sea and the Sea of Okhotsk. The model also does well in Hudson Bay, Baffin Bay and Davis Strait. However, the model has three inconsistencies: (1) in the Labrador Sea, the model underpredicts ice in the winter and fall; (2) in the marginal seas of the eastern Arctic Ocean, the Barents Sea and the Greenland Sea, the model overpredicts the amount of sea ice in winter, spring and summer, and (3) the model underpredicts ice in the Beaufort Sea, in part of the Chukchi Sea and in the East Siberian Sea in summer. Some of these inconsistencies between model and observations are due to the fact that the ocean model was not specifically designed to handle coastal and shelf processes; some are due to the model's treatment of all ice as a continuum and not as separate floes; and some are due to the inaccuracies in the atmospheric forcing.

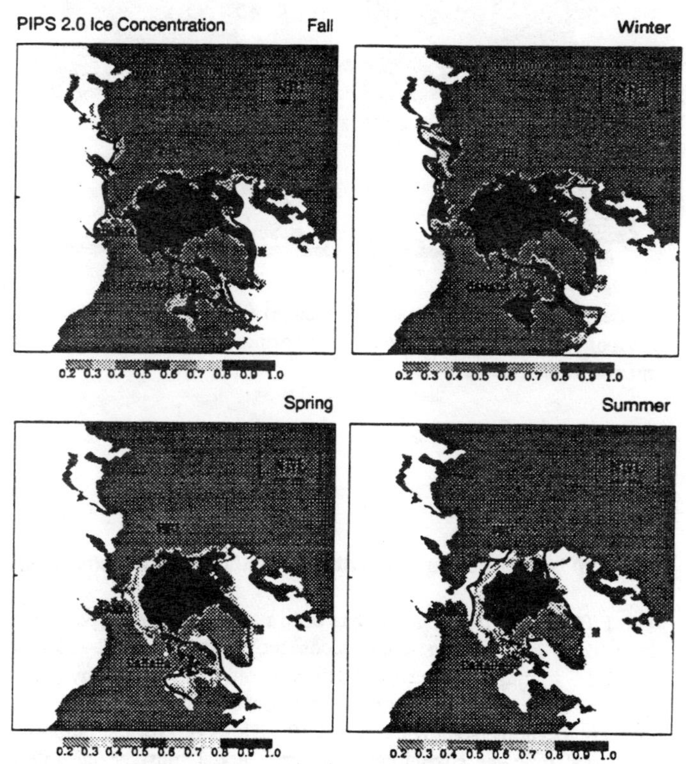

Figure 9. Daily ice concentration (in percent) from PIPS 2.0 representing the four seasons a) Fall, b) Winter, c) Spring and d) Summer. Solid black line represents the ice edge from the NIC weekly analysis. Contours are in percent ice concentrations with a contour interval of 10%.

Figure 10 shows the daily mean ice thickness from PIPS 2.0 representing the four seasons. With the harsh environmental conditions of the Arctic, insitu ice thickness observations are very difficult to obtain. However, PIPS 2.0 ice thickness has been compared qualitatively to seasonal and yearly averaged ice thickness values derived from submarine data (Le Shack et al., 1971: Bourke and Garrett, 1987). These comparisons showed that the PIPS 2.0 model developed a realistic ice cover with the thickest ice developing along the Canadian Archipelago, thinning towards the pole, and continuing to decrease from the North Pole towards the Soviet coast.

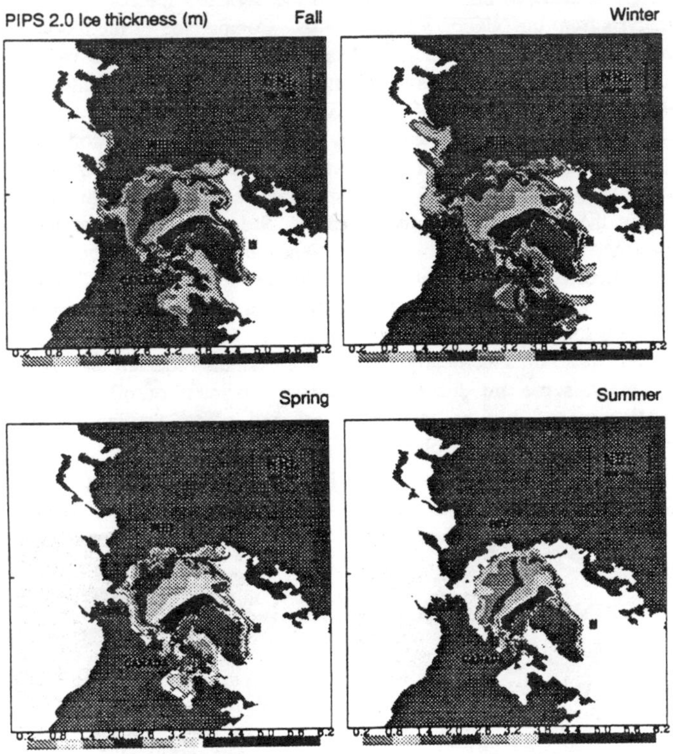

Figure 10. Daily ice thickness (in meters) from PIPS 2.0 representing the four seasons a) Fall, b) Winter, c) Spring and d) Summer. Contour interval is 0.6 meters.

PIPS 2.0 Ice Concentration With and Without Data Assimilation

The PIPS 2.0 ice-ocean model has been run without any type of real-time data initialization using consecutive years (1992-1996) of NOGAPS forcing. These results demonstrated that without any type of real time data initialization, the model produced realistic seasonal trends in many locations (see Figures 9 and 10). However, for the purpose of daily real time ice edge forecasting, the model must be capable of producing greater accuracy at the ice edge. Daily initialization from observations should enable the model to achieve far greater accuracy. Figure 11 shows a 24-hour forecast valid for October 31, 1996, of ice concentration from the PIPS 2.0 system without any type of real time data initialization. Ice concentration is less in coastal regional than in the central Arctic.

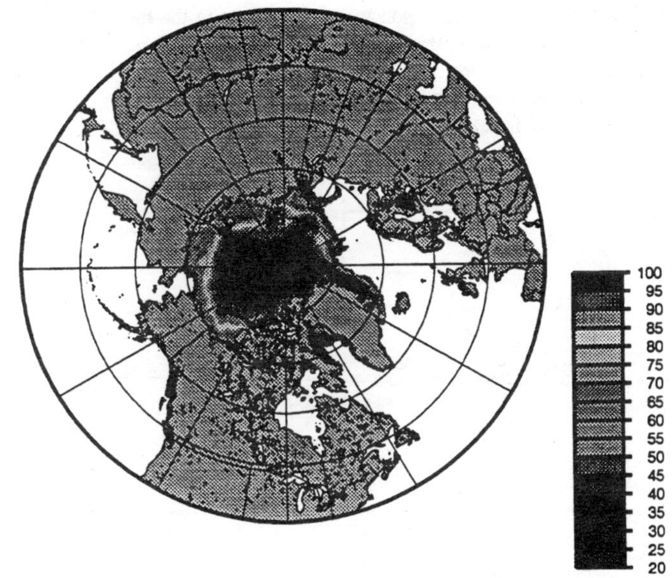

Figure 11. Ice concentration (in percent) from PIPS 2.0 without any initialization. Image valid for October 31, 1996. Contour interval is 5%.

Figure 12 shows an example of the SSM/I ice concentration image valid for October 31, 1996. Generally, most of the central Arctic is entirely (100%) covered with sea ice with a gradual reduction in the concentration towards the marginal ice zones. The SSM/I images along with weekly ice edge charts from NIC are extremely useful tools for the determination of ice edge locations. Comparison of the model result (Figure 11) to the SSM/I data shows that the modeled ice edge is generally in the correct location.

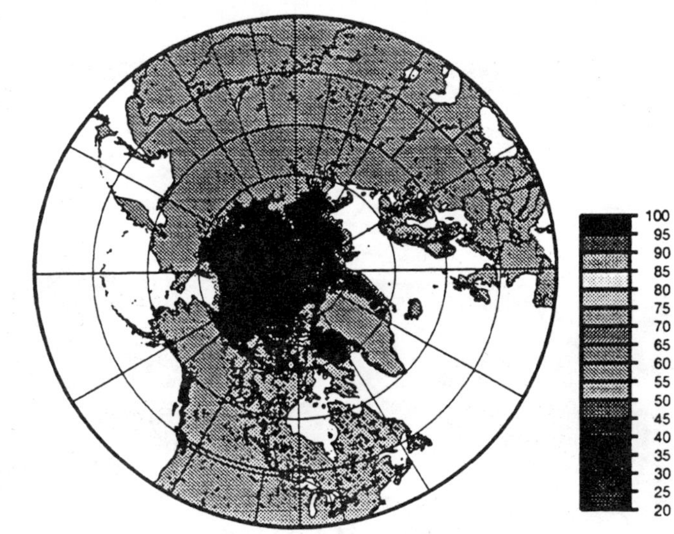

Figure 12. Ice concentration (in percent) from SSM/I. Image valid for October 31, 1996. Contour interval is 5%.

In the operational system PIPS 2.0, when SSM/I is used daily as an update to the modeled derived ice concentration, only values along the ice edge (< 50% concentration) are assimilated. This allows the model to forecast the general trends of the Arctic but keeps the system closely tied to the real time ice edge location. Figure 13 is the operational PIPS 2.0 24-hour forecast valid on October 31, 1996.

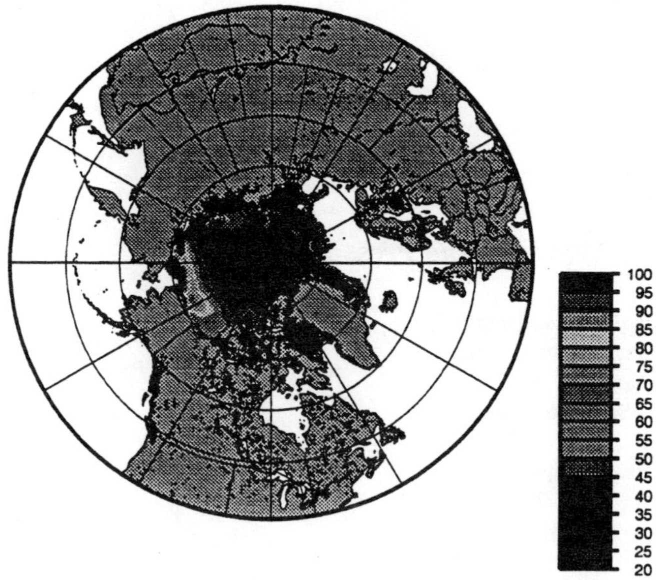

Figure 13. Ice concentration (in percent) from PIPS 2.0 initialized with SSM/I data. Image valid for October 31, 1996. Contour interval is 5%

There are slight differences along the ice edge between this model result (Figure 13) and the SSM/I data in the eastern Arctic (Figure 12). This difference is the development within 24 hours of thin ice along the ice edge. There are major differences between the two figures in the western Arctic near the ice edge and in the Beaufort Sea. Differences at the ice edge are again the growth of an area of very thin ice. In the Beaufort Sea, the modeled ice concentration is much less than that in the SSM/I data. This is because the operational forecast (Figure 13) allows the model to predict the concentration in areas 50% and greater, instead of using the real time data. During the summer of 1996, the NOGAPS heat fluxes were warm enough to cause a large region of low concentrations (between 50 and 90%) in the Beaufort Sea. The SSM/I data shows this region changing from low concentration values to 100% over a week while the model takes several weeks to develop such total ice coverage. The modeled ice concentration often seems more realistic in these situations than the ice concentration derived from the SSM/I data.

Conclusions

Since 1987, FNMOC has been running sea ice forecasting systems in various regions of Navy interest. PIPS 2.0, a coupled ice-ocean forecasting system, has been running operationally at FNMOC since July 1996 predicting ice thickness, ice concentration (ice edge location), ice convergence/divergence and the change in these fields over a forecast period. PIPS 2.0 provides each of these products in regions of Navy interest that were not covered by the three existing forecast systems (PIPS, RPIPS-B and RPIPS-G). PIPS 2.0 also provides fields of ocean temperature, salinity and ocean currents at each of its 15 vertical levels. Although these fields are weakly constrained to climatology, they should still provide the type of variability required of a forecast. Numerous studies have been performed to validate the system's forecast results. SSM/I ice concentration has been archived and evaluated by NRL for several years as a tool for initialization of the model ice concentration field. Several sensitivity studies have shown that daily initialization from SSM/I concentration can provide a valuable improvement to the existing operational sea ice forecasts.

Acknowledgements

This work was funded through the U.S. Space and Naval Warfare Systems Command, program element 0603207N. This paper, NRL contribution PP/7322-96-0031, is approved for public release; distribution unlimited.

References

Bourke, R.H. and R.P. Garrett (1987). "Sea Ice Thickness Distribution in the Arctic Ocean," *Cold Reg. Sci. and Technol.*, **13**, pp 259-280.

Cheng, A. and R.H. Preller (1996). "The Development of an Ice-Ocean Coupled Model in the Northern Hemisphere," NRL/FR/7322-95-9627, Naval Research Laboratory, Stennis Space Center, MS, p 61.

Colony, R.L. and I.G. Rigor (1993). "International Arctic Buoy Program Data Report for 1 January 1992 - 31 December 1992," APL-UW Technical Memorandum TM29-93, Applied Physics Laboratory, Seattle, WA.

Cox, M. (1984). "A Primitive Equation, 3-Dimensional Model of the Ocean," GFDL Ocean Group Technical Report, Geophysical Fluid Dynamics Laboratory, Princeton, NJ, p 1141.

Hibler, W.D. III (1979). "A Dynamic Thermodynamic Sea Ice Model," *J. Phys. Oceanogr.*, **9**, pp 815-864.

Hibler, W.D. III (1980). "Modeling a Variable Thickness Sea Ice Cover," *Monthly Weather Review*, **108**, pp 1943-1973.

Hibler, W.D. III and K. Bryan (1987). "A Diagnostic Ice-Ocean Model," *J. Phys. Oceanogr.*, **17**, pp 987-1015.

Hogan, T.F., T.E. Rosmond and R. Gelaro (1991). "The Description of the Navy Operational Global Atmospheric prediction System's Forecast Model," NOARL Report 13, Naval Research Laboratory, Stennis Space Center, MS.

Hollinger, J.P. (1991). "DMSP Special Sensor Microwave/Imager Calibration/Validation - Final Report, Volume II," Naval Research Laboratory, Washington, DC.

LeShack, L.A., W.D. Hibler, III and R.H. Morse (1971). "Automatic Processing of Arctic Pack Ice Obtained by Means of Submarine Sonar and Other Remote Sensing Techniques," *Propagation Limitations in Remote Sensing*, J.B. Lomax, Ed., AGARD Conf. Proc., No. 90, 5-1 to 5-19 {NTIS AGARD-CP-90-71}].

Levitus, S. (1982). "Climatological Atlas of the World Ocean," NOAA Prof. Pap., 13, p 173.

Preller, R.H. and P.G. Posey (1996). "Validation Test Report for a Navy Sea Ice Forecast System: The Polar Ice Prediction System 2.0," NRL/FR/7322-95-9634, Naval Research Laboratory, Stennis Space Center, MS, p 31.

Sarmiento, J.L. and K. Bryan (1982). "An Ocean Transport Model for the North Atlantic," *J. Geophys. Res.*, **87**, pp 395-408.

Proceedings of the Seventh (1997) International Offshore and Polar Engineering Conference
Honolulu, USA, May 25-30, 1997

ISBN 1-880653-28-1 (Set); ISBN 1-880653-30-3 (Vol. II)

Characteristics of Frequency Spectra of Wind Waves in the Presence of Sea Ice

Tetsuya Hayakawa and Takahiko Sasajima
Hokkaido Development Bureau, Sapporo, Japan
Masafumi Yoshino
Alpha Hydraulic Co., Ltd., Sapporo, Japan
Chiaki Goto
Tokai University, Hiratsuka, Japan

Abstract

In the presence of sea ice, wave forecasting and hindcasting is difficult because the relationship between the marine surface wind and wind waves has not been well investigated.

In this paper, a new spectral formula for wind waves in the presence of sea ice is proposed by the wave observation data at Mombetsu port as basic research for wave forecasting and hindcasting. The proposed spectrum form is able to calculate only of significant wave height, period and ice cover rate. The slope at high frequency band of this spectrum is sharper with the increase of ice cover rate, and the wave energy concentrates near the peak frequency band with the decrease of wave steepness.

Introduction

The Sea of Okhotsk near Hokkaido, Japan, is covered with sea ice during winter. It is known that, when sea ice is present, wave height is remarkably reduced. Also, in cases where the sea is almost entirely covered with sea ice, waves do not grow due to marine surface winds. In a case of wave observation at Mombetsu Port, the significant wave height only reached approximately 0.8 m, even when winds of 10 m/s were blowing coastward for over half of the day.

According to Aota et al. (1993), however, winter temperatures in this region have been increasing and the amount of sea ice has been decreasing in recent years. It is expected that such recent trends will lessen the effects of sea ice on wave, and wave heights in the Sea of Okhotsk are higher than ever. Therefore, in the design and management of port and harbor construction works, it is important to develop wave forecasting and hindcasting models in the presence of sea ice.

For sea areas where sea ice does not exist, various wave forecasting and hindcasting models have been developed. However, wave forecasting and hindcasting in areas with sea ice is difficult, because knowledge about correlation among wind waves, marine surface winds and sea ice is not sufficient.

In this study, as basic research for wave forecasting and hindcasting, the frequency spectra of wind waves in the presence of sea ice were investigated.

Analytical method

Observation data of sea waves obtained from 1985 to 1994 at Mombetsu Port, which sea ice approaches in winter, was used for the analysis. Observations were carried out every 2 hours for 20 minutes, at a depth of 52 m off Mombetsu Port, with a sampling frequency of 2 Hz using ultrasonic-type wave gages. However, this observation point has no wind profile. Thus, from the observation data, developing wind waves were extracted according to the following conditions:

(1) From the time series of significant parameters, significant wave height should be in their developing stage.

(2) The swell component is not included clearly. Thus, the spectral form should be a single peak spectrum.

(3) The JONSWAP spectrum parameter γ should be greater than 1.

The number of total observation data are 20 cases with the presence of sea ice and 4 with absence thereof, and frequency spectra were calculated using the FFT method.

Also, to consider the effects of sea ice on frequency spectra, relations between the spectral forms and ice cover rate Cv were examined. The range of Cv is from 1 to 10, and Cv is 10 when the entire area is covered with sea ice. Cv is the average of values measured every 20 km in a 200-km straight line, which is drawn from the wave observation point in the wind's direction, using sea ice distribution data from 1985 to 1993.

Observed frequency spectra

(1) Frequency spectra of wind waves with absence of sea ice

Goto and Aono (1993) proposed a standard form of wind wave frequency spectrum, which can be calculated only from significant wave data, by substituting values obtained from observation data for the generalized JONSWAP spectrum. Equation (1) shows this supectrum form.

$$
\begin{aligned}
&S(f) = \alpha (2\pi)^{-m+1} g u_*^{5-m} f^{-m} \exp\left\{-\frac{m}{4}\left(\frac{f}{f_m}\right)^{-4}\right\}\gamma^{\beta} \\
&\beta = \exp\left\{-\left(1 - f/f_m\right)^2 / 2\sigma^2\right\} \\
&m = 4,\quad \gamma = 6 f_{m^*}^{0.15},\quad \alpha = 0.17\gamma^{-1/3} \\
&f_m = 1/1.136 T_{1/3},\quad \sigma\begin{cases}\sigma_1 = 0.144 & (f < f_m) \\ \sigma_2 = 0.07 f_{m^*}^{-0.16} & (f \geqq f_m)\end{cases} \\
&f_{m^*} = f_m u_* / g,\quad u_* = H_{1/3} / g B^2 T_{1/3},\quad B = 0.067
\end{aligned}
\qquad (1)
$$

where, m is slope of wind wave spectrum at high frequency band, γ the peak enhancement factor, σ the band width near peak frequency, α the scale factor, f_{m^*} the non-dimensional peak frequency, f_m the peak frequency, u_* the friction velocity of wind, $H_{1/3}$ the significant wave height and $T_{1/3}$ the significant wave period.

When sea ice is absent, the observed spectra and spectrum calculated by Eq. (1) agreed well with each.

(2) Properties of frequency spectra in the presence of sea ice

Figure 1 shows an example of observed spectrum in the presence of sea ice and the spectrum, calculated by Eq. (1) using wave data at

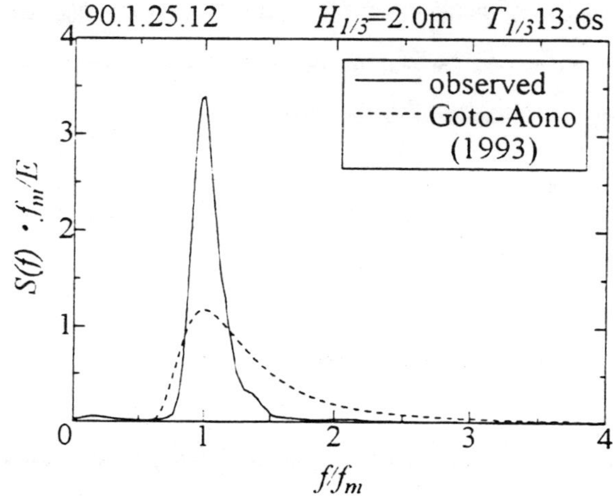

Fig.1 Example of observed spectrum in the presence of sea ice

that time. The x-axis is f/f_m, and the y-axis is $S(f)f_m/E$, which is obtained by making $S(f)$ non-dimensional, using f_m and the entire energy of the spectrum E. Miyabe (1987) pointed out that the form of frequency spectra in the presence of sea ice, compared with those in the absence of sea ice, is characterized by greater damping of energy at the high frequency band and more concentration of energy near the peak frequency. Similar results were obtained from the analysis of this study, and the form did not correspond with that of the standard spectrum calculated by Eq.(1).

Standard form of frequency spectrum in the presence of sea ice

In this section, the frequency spectrum in the presence of sea ice, as shown in the above Fig. 1, was examined to propose a standard form. A method similar to that of Goto and Aono (1993) was used, and each parameter was calculated by substituting values obtained from wave observation data for the generalized JONSWAP spectrum equation.

However, regarding the correlation between the marine surface wind, which was represented by the friction velocity of wind u_*, and wave data in the presence of sea ice, the 3/2 power law presented by Toba (1972) was not satisfied. Therefore, in this study, a new spectrum parameter A was applied to the part of the coefficient that did not include the frequency f of Eq. (1), as shown in Eq. (2). However, the band width near the peak frequency σ was not examined in this study, as the number of extracted disturbances was small.

$$S(f) = Af^{-m}\exp\left\{-\frac{m}{4}\left(\frac{f}{f_m}\right)^{-4}\right\}\gamma^{\beta} \qquad (2)$$

(1) Properties of the peak frequency

Figure 2 shows the relation between the observed significant wave period $T_{1/3}$ and the peak frequency f_m. The black circles in the figure indicate the average values of f_m and the vertical line shows the standard deviation of the data to indicate the amount of scatter. The solid line is the relational formula between f_m and $T_{1/3}$ using the method of least squares, as shown in Eq. (3).

$$f_m = \frac{1}{1.062T_{1/3}} \qquad (3)$$

The dotted line in the figure is the relational expression between $T_{1/3}$ and f_m in Eq. (1). It shows a tendency similar to that in Eq. (3). It was thought that the existence of sea ice would not affect the relation between significant wave period and peak frequency of the spectrum.

(2) Properties of slope of wind wave spectrum at high frequency band

A method similar to that of Goto and Aono (1993) was used to calculate the slope of wind wave spectrum m. It was calculated using the method of least squares for the range of 1.5 to 3 times of the peak frequency of the observed frequency spectra. In cases when sea ice did not exist, high-frequency properties of the average wind waves could be expressed by -4 power law (m = 4), as was in the results of field observations conducted by Goto and Aono (1993). However, in the presence of sea ice, the damping properties of the

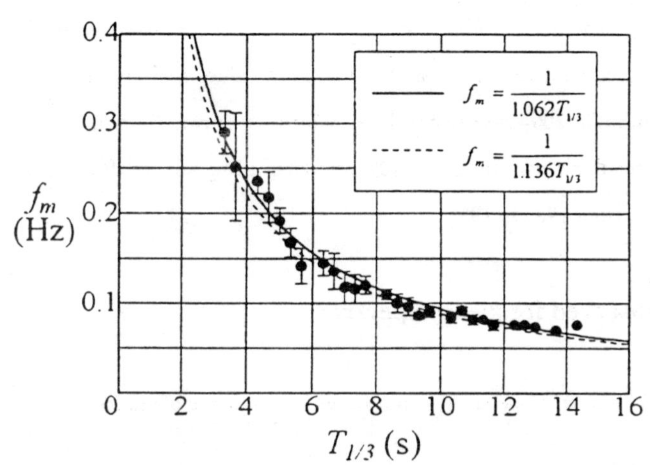

Fig.2 Relationship betweem $T_{1/3}$ and f_m

high-frequency energy density changed by the distribution condition of sea ice, and could not be expressed by a constant value.

Figure 3 shows the relation between the slope of wind wave spectrum at high frequencies m and ice cover rate Cv. m tended to increase in proportion to Cv. In other words, the damping of energy density at high frequency band increased with the increase of the area covered with sea ice. m is expressed by Cv using the method of least squares, and can be obtained, as shown in Eq.(4).

$$m = 4.52Cv^{0.0551} \tag{4}$$

However, while the wave observation was conducted every 2 hours, the data on sea ice distribution could be obtained only once a day. Also, as the effect of the thickness of sea ice was not taken into consideration, Eq. (4) could not sufficiently represent the effects of Cv on m. At this stage, m is expressed by Eq. (4), due to the limited number of cases and data on sea ice distribution.

(3)Properties of the peak enhancement factor

The parameter of the JONSWAP spectrum γ is called the peak enhancement factor and is a parameter representing the enhancement factor of wave energy near the peak frequency. When $\gamma = 1$, it describes wind waves that had fully developed. While γ is obtained from the friction velocity of wind u_* and other factors in the standard spectrum calculated from Eq. (1), this method cannot be used in the presence of sea ice, as the 3/2 power law is not satisfied.

Figure 4 shows the relation between the observed peak enhancement factor γ and the wave steepness $H_{1/3}/L_{1/3}$, where, the wave steepness was obtained by dividing the significant wave height $H_{1/3}$ by the wave length $L_{1/3}$, which was calculated from the significant wave period $T_{1/3}$. γ has almost a logliner relation to the wave steepness and tends to decrease as the wave steepness increases. In other words, the spectrum is the form with energy concentrated around f_m when the wave steepness is small. This is thought to be due to the damping of wave energy at high frequencies during the propagation of sea waves below sea ice. When the wave energy at high frequencies is damped, the wave steepness decreases because $H_{1/3}$ decreases and $T_{1/3}$ increases. Regarding the spectral form, the wave energy is concentrated near the peak frequency, due to the damping of wave energy. γ is expressed by the wave steepness, using the method of least squares, and γ can be obtained as shown in Eq.(5).

$$\gamma = 0.425\left(\frac{H_{1/3}}{L_{1/3}}\right)^{-0.581} \tag{5}$$

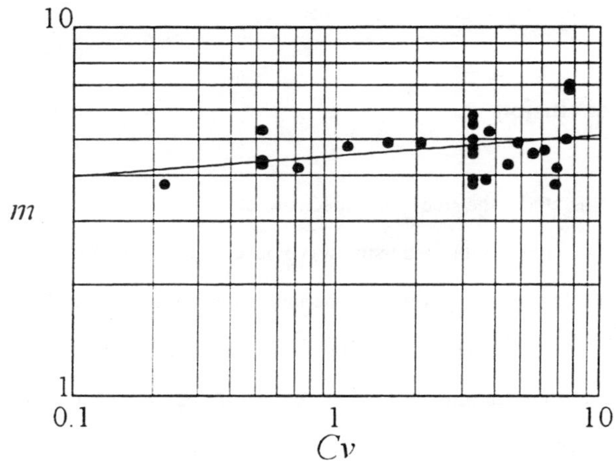

Fig.3 Relationship between Cv and m

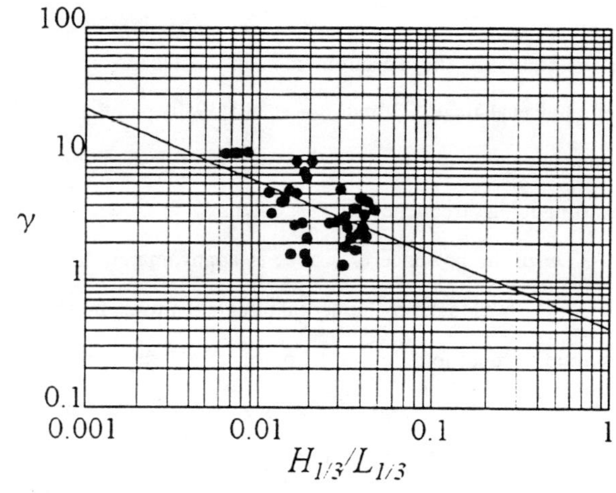

Fig.4 Relationship between $H_{1/3}/L_{1/3}$ and γ

(4) Properties of spectrum parameter

As shown in Eq. (2), spectrum parameter A was adopted as a new parameter. Although this spectrum parameter has no physical meaning, spectrum parameter A can be calculated from sea wave data as follows.

The wave energy E can be determined by integrating both sides of Eq. (3).

$$E = \int_0^\infty S(f)df = A\int_0^\infty f^{-m}\exp\left\{-\frac{m}{4}\left(\frac{f}{f_m}\right)^{-4}\right\}\gamma^\beta df \qquad (6)$$

A is expressed as shown in Eq.(7)

$$A = \frac{E}{f_m^{\ 1-m}M_0},\quad M_0 = \int_0^\infty f'^{\ -m}\exp\left\{-\frac{m}{4}f'^{\ -4}\right\}\gamma^\beta df' \qquad (7)$$

The significant wave height $H_{1/3}$ is known to be proportional to the square root of the wave energy E. This correlation was valid even in the presence of sea ice. $H_{1/3}$ is expressed by E, using the method of least squares, and E can be obtained as shown in Eq. (8).

$$H_{1/3} = 3.74\sqrt{E}\quad E = \left(\frac{H_{1/3}}{3.74}\right)^2 \qquad (8)$$

M_0 in Eq.(7) is the spectrum moment. Although it cannot be integrated analytically, Mitsuyasu et al. (1980) pointed out that it becomes proportional to the 1/3 power of γ, when the slope of spectrum at high frequency band, m equals 5. As shown in the above Fig. 3, m in the presence of sea ice is approximately from 4 to 7, where, if M_0 was found by numerical integration when m = 4 to 7, M_0 was almost proportional to the 1/3 power of γ, while errors became a little greater when m was larger. M_0 is expressed by m and γ, and M_0 can be obtained as shown in Eq.(9)

$$M_0 = 1.15\exp\{-0.336m\}\gamma^{1/3} \qquad (9)$$

(5) Standard spectrum in the presence of sea ice

From the examination of each spectrum parameter in previous sections, the standard spectrum in the presence of sea ice was proposed as Eq. (10). As for γ, however, the average of observation values by Goto and Aono (1993) was used.

$$\begin{aligned} S(f) &= Af^{-m}\exp\left\{-\frac{m}{4}\left(\frac{f}{f_m}\right)^{-4}\right\}\gamma^{\beta} \\ \beta &= \exp\left\{-(1-f/f_m)^2/2\sigma^2\right\} \\ A &= \frac{E}{f_m^{\ 1-m}M_0} \\ E &= \left(\frac{H_{1/3}}{3.74}\right)^2,\quad M_0 = 1.15\exp\{-0.336m\}\gamma^{1/3} \\ m &= 4.52Cv^{0.0551},\quad f_m = 1/1.062T_{1/3} \\ \gamma &= 0.119\left(\frac{H_{1/3}}{L_{1/3}}\right)^{-0.972},\quad \sigma\begin{cases}\sigma_1 = 0.144 & (f<f_m)\\ \sigma_2 = 0.200 & (f\geqq f_m)\end{cases} \end{aligned} \qquad (10)$$

Figure 5 shows the results of comparison of the observed and proposed spectra by Eq. (11). The spectrum of Goto and Aono (1993) by Eq. (1) is given in this figure for reference. The time series of $H_{1/3}$ and $T_{1/3}$ is also shown. It can be seen that the spectrum calculation formula proposed in this study better represents than conventional methods frequency spectral forms in the presence of sea ice, in which energy was concentrated near the peak frequency and the wave energy at high frequency band damped.

Conclusion

In this study, the frequency spectrum of wind waves in the presence of sea ice was proposed using wave observation data from Mombetsu Port. This spectral form can be calculated by using the significant wave data and the ice cover rate. The greater the ice cover rate, the greater the damping of the wave energy at high frequencies, and the smaller the wave steepness, the more energy was concentrated near the peak frequency. Also, it correlated better with observed one.

In the future, wave forecasting and hindcasting models in sea areas with sea ice will be examined by continuing sea wave observations and analysis and collecting more sea ice data on the basic of this research.

References

Aono, T, and Goto, C (1994)." On the Characteristics of One-Dimensional Spectra and Non-Dimensional Parameters of Wind Waves," *Proc. of 24th Coastal Engineering Conference*, pp.12-26

Aota, M, Ishikawa, M, Murata, K, and Hirata, T (1993)."Long Term Variation in the Sea Ice Concentration along Okhotsk Sea Coast of Hokkaido," *Proc. International Symposium on Ocean Energy Development*, pp. 3-8

Goto, C, and Aono, T (1993)."On the Characteristics of One-Dimensional Spectra and Non-Dimensional Parameters of Wind Waves -Wave Hindcast Model Using the Hybrid-parameter Methods (2nd report)- ," *Rep. of P.H.R.I*, Vol.32, No.1, pp. 53-99 (in Japanese)

Toba, Y (1972)."Local balance in the air-sea boundary processes, Ⅰ. On the growth processes of wind waves," *J. Oceanogr. Soc. Japam*, 28, pp.109-120

Mitsuyasu, H (1980)."Observation of the power spectrum of ocean waves using a cloverleaf buoy," *Jour. Physical Oceanography*, Vol.10, pp. 286-296.

Miyabe, S (1987)."Characteristics of Wave in Mombetsu," *Proc. 31th Conference of Hokkaido Development Bureau*, pp. 893-896

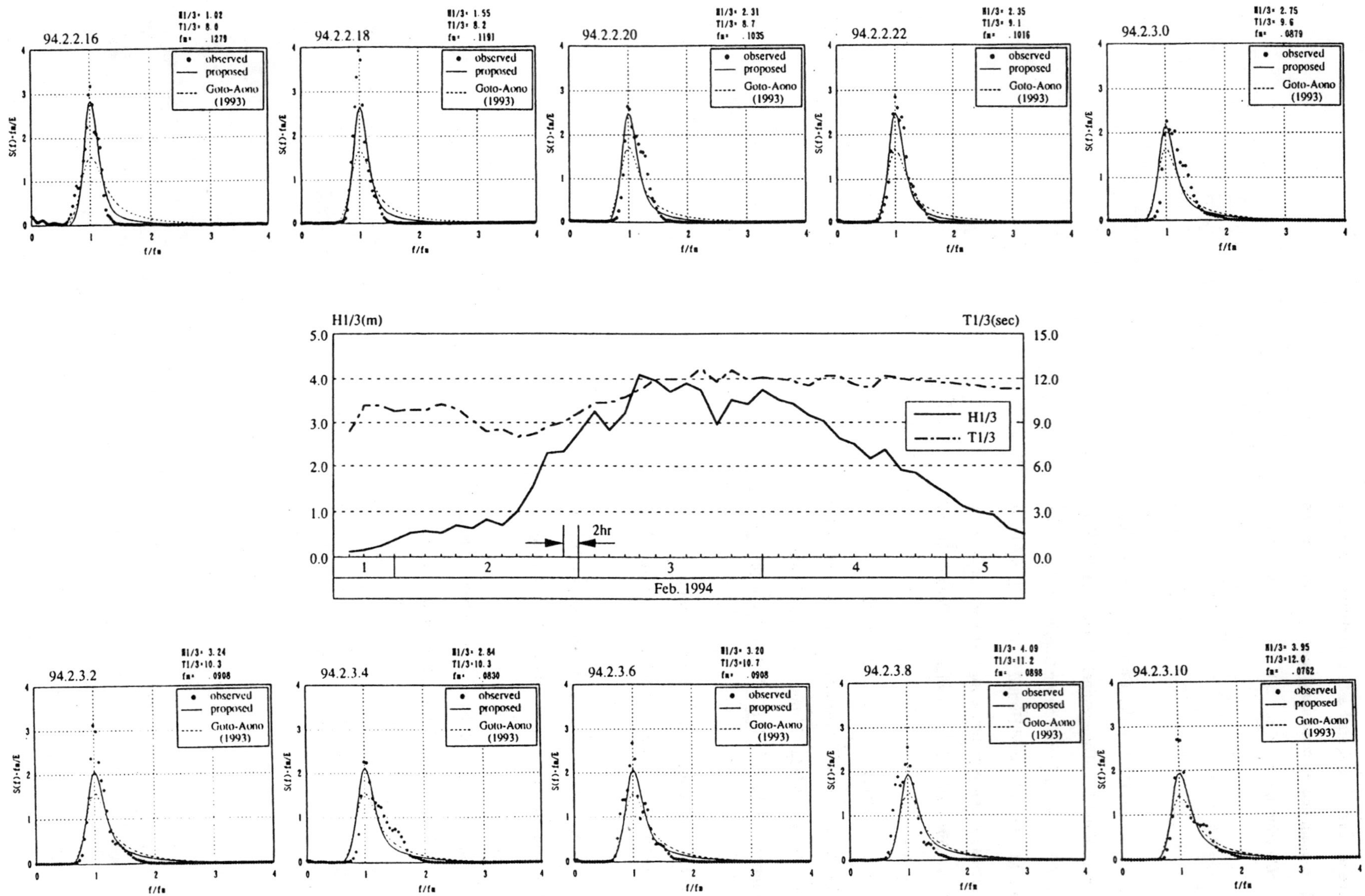

Fig.5 Results of the comparison of observed and proposed supectrum

Proceedings of the Seventh (1997) International Offshore and Polar Engineering Conference
Honolulu, USA, May 25-30, 1997

ISBN 1-880653-28-1 (Set); ISBN 1-880653-30-3 (Vol. II)

An Analysis of Wave Fields Around a Flexible Sheet Structure with Green Function Method

Ken-ichiro Hamanaka
Hokkaido Tokai University
Sapporo, Japan

Masaya Kato
Tokyu Construction Co., Ltd.
Tokyo, Japan

ABSTRACT

In the present paper, a new method is proposed to analyze the wave field around a submerged flexible sheet structure which was proposed to control the wave field. In this method, a doublet type of Green function is used. The necessary number of unknown variables can be reduced to one for each sheet-frame system.

KEY WORDS: Wave field, submerged structure, flexible sheet, Green function

1. INTRODUCTION

To satisfy the requirements of obtaining calm sea and exchange of sea water, several types of structure have been proposed; submerged plate and submerged breakwater, for example. The principle of these structures is basically based on the energy dissipation by wave breaking while the waves are propagating over the structures. Therefore, these structures need to have relatively wide width.

Kato and Hamanaka (1993) proposed an another type of submerged structure which is composed of textile sheets conjoined in frames with elastic springs. This structure is referred to as flexible sheet in this paper. They reported that the flexible sheet is more effective to control the wave fields than other types of submerged structures, because of resonant interaction between the waves and the motion of the sheets. This type of structure may be useful for temporary or seasonal use.

To analyze the wave fields around the flexible sheets, Kato and Hamanaka (1993) proposed a potential matching method in vertically 2-D cases. But a difficulty will arise in solving 3-D cases using the same potential matching method, because of the necessity of a large number of unknown variables.

In the present paper, a new method of analysis is proposed, in which a doublet type of Green function (John, 1950) is used. In this method, the necessary number of unknown variables can be reduced to one for each sheet-frame system, in both 2-D and 3-D case.

2. ASSUMPTIONS AND GOVERNING EQUATION

Irrotational fluid motion is assumed in a incompressible and non-viscous fluid. The following conditions are assumed.

(1) Wave motion is periodic in time.
(2) Water depth is constant.
(3) Amplitude of waves is sufficiently small.

From the assumption (1), the velocity potential $\hat{\Phi}$ may be expressed as;

$$\hat{\Phi} = \mathrm{Re}\left\{ \hat{\phi} e^{-i\hat{\omega}\hat{t}} \right\} \tag{1}$$

where the symbol Re represents the real part of a complex value, $\hat{\phi}$ is the reduced velocity potential, $\hat{\omega}$ is the angular frequency, $\hat{t}$ is time and the sign ^ denotes the dimensional variable.

Each variable is non-dimensionalized by $\hat{\omega}$ and the gravitational acceleration $\hat{g}$.

$$(x,z) = (\hat{x},\hat{z})\hat{\omega}^2 / \hat{g} \quad , \quad (h,d,B) = (\hat{h},\hat{d},\hat{B})\hat{\omega}^2 / \hat{g}$$

$$t = \hat{\omega}\hat{t} \quad , \quad \phi = \hat{\phi}\hat{\omega}^3 / \hat{g}^2$$

where h is the water depth, d is the submerged depth of the sheet and B is the width of the sheet (see Fig.1).

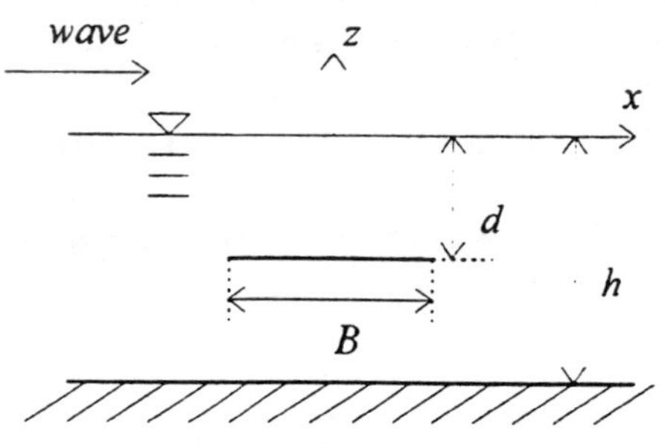

Fig. 1. Definition of coordinate system

The velocity potential is expressed in a non-dimensional form.

$$\Phi = \mathrm{Re}\left\{ \phi e^{-it} \right\} \tag{2}$$

The reduced velocity potential is governed by the Laplace equation.

$$\Delta\phi = 0 \tag{3}$$

where Δ denotes the Laplacian.
The linearized water surface condition and the bottom condition are,

$$\begin{aligned} \partial\phi / \partial z &= \phi \quad on \quad z = 0 \\ \partial\phi / \partial z &= 0 \quad on \quad z = -h \end{aligned} \tag{4}$$

3. GREEN'S INTEGRAL EQUATION

For simplicity, we start with one sheet-frame system in 2-D case and will discuss multiple sheet-frames system and 3-D cases in the later section.

To apply the Green's integral equation, we install two artificial vertical boundaries sufficiently away from the sheet structure; S_f in Fig.2.

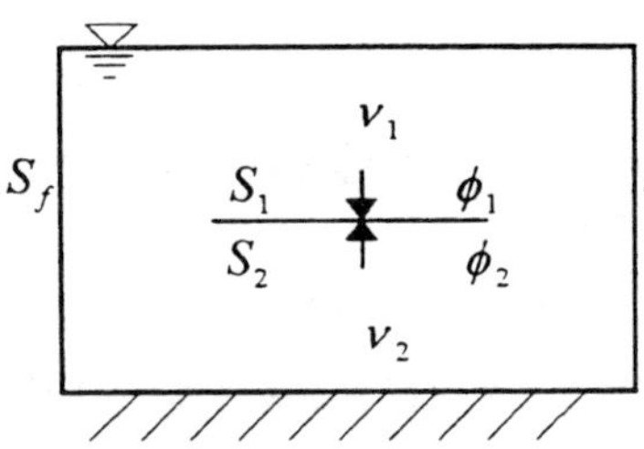

Fig. 2. Boundaries

The flexible sheet is horizontally installed in the water. Further conditions are assumed with the motion of the sheet.
(4) Motion of the sheet is periodic in time.
(5) Displacement of the sheet takes a half of sinusoidal form.
(6) Amplitude of sheet is sufficiently small.
(7) Pretension of the sheet is sufficiently large.
(8) Mass of the sheet is negligibly small.

Applying Green's integral equation to the region leads to;

$$-\phi(P) = \int_S (\phi\,\partial G/\partial\nu - G\,\partial\phi/\partial\nu)dS \tag{5}$$

where S denotes all boundaries of the region and G is John's Green function in 2-D (John, 1950) and ν is normal unit on the boundary outward from the region.

$$\begin{aligned} G \;=\; & C_0 e^{ik|x-x_0|} \cosh k(z_0+h)\cosh k(z+h) \\ + & \sum_{n=1}^{\infty} C_n\, e^{-k_n|x-x_0|} \cos k_n(z_0+h)\cos k_n(z+h) \end{aligned} \tag{6}$$

where

$$\begin{aligned} C_0 &= i/k(h+\sinh^2 kh) \\ C_n &= 1/k_n(h-\sin^2 k_n h) \end{aligned}$$

and the dispersion relations,

$$k \tanh kh = 1 \quad , \quad k_n \tan k_n h = -1$$

This Green function satisfies the conditions on the water surface and on the bottom, the integral on these boundaries is eliminated. Then,

$$S = S_1 + S_2 + S_f .$$

where S_1, S_2 denote the upper and lower surfaces of the sheet, respectively.

On the far boundaries, the scattered waves satisfy the radiation condition, therefore the integral on the far boundaries results in the potential of the incident waves $-\phi_I(P)$ on the singular point P.

$$-\phi(P) = \int_{S_1+S_2} (\phi\,\partial G/\partial\nu - G\,\partial\phi/\partial\nu)dS - \phi_I(p) \tag{7}$$

On the upper and lower surfaces of the sheet,

$$\partial G/\partial\nu_1 = -\partial G/\partial\nu_2,\quad \partial\phi_1/\partial\nu_1 = -\partial\phi_2/\partial\nu_2,$$
and $G|_{S_1} = G|_{S_2}$,

where ϕ_1 and ϕ_2 are the velocity potential on the upper and lower surface of the sheet, respectively.
Therefore, Eq. (7) becomes

$$-\phi(P) = \int_{S_2} (\phi_2 - \phi_1)\frac{\partial G}{\partial\nu_2}ds - \phi_I(P) . \tag{8}$$

Eq.(8) states that once the difference of potentials on the sheet is obtained, the potential at any point can be estimated by Eq.(8).

4. DIFFERENCE OF POTENTIAL ON THE SHEET AND AMPLITUDE OF THE SHEET

From the linearized equation of pressure,

$$p = -z - \partial\,\Phi/\partial\,t$$

where $p = \hat{p}\hat{\omega}^2/\hat{\rho}\hat{g}^2$.
The difference of pressure on the sheet is,

$$\begin{aligned} \Delta P = \; & p_1 - p_2 = \; -\partial\Phi_1/\partial t + \;\partial\Phi_2/\partial t \\ = \; & i(\phi_1 - \phi_2)e^{-it} . \end{aligned} \tag{9}$$

From the assumptions (4) and (5), the displacement of the sheet ζ is represented,

$$\zeta = \alpha\cos\beta x\; e^{-it} \qquad \beta = \pi/B \tag{10}$$

where α is the amplitude of the sheet.
From the assumptions (6) and (7), the vertical component of the tension of the sheet is,

$$F_z = T\partial^2\zeta/\partial x^2 = -T\alpha\beta^2(\cos\beta x)e^{-it}, \tag{11}$$

where T is the pretension of the sheet.
From the assumption (8), the difference of pressure Eq. (9) must be balanced to the vertical component of the tension, therefore,

$$\phi_1 - \phi_2 = iT\alpha\beta^2\cos\beta x \tag{12}$$

Eq. (12) states that once the amplitude of the sheet is obtained, the potential at any point can be estimated by Eq. (8) and Eq. (12).

Now the coordinate of the singular point is denoted by,

$$P = (x_0, z_0).$$

To obtain the amplitude α, Eq.(8) is differentiated with z_0,

$$\begin{aligned} \frac{\partial\phi(P)}{\partial z_0} \;=\; & \int_{S2}(\phi_1 - \phi_2)\frac{\partial^2 G}{\partial\nu_2\partial z_0}ds \\ & + \frac{\partial\phi_I(P)}{\partial z_0} \end{aligned} \tag{13}$$

Then the singular point approaches the center of the sheet, Q in Fig.3. In this limitation, the left hand side of Eq. (13) means the vertical velocity of the sheet at point Q. From Eq. (10),

$$\frac{\partial \Phi(P)}{\partial z_0} = \frac{\partial \phi(p)}{\partial z_0} e^{-it} = \frac{\partial \zeta}{\partial t} = -i\alpha e^{-it}, \tag{14}$$

then,

$$\partial \phi(P)/\partial z_0 = -i\alpha. \tag{14'}$$

Substituting Eq. (12) and Eq. (14)' into Eq. (13), and Considering

$$\partial / \partial v_2 = \partial / \partial z,$$

leads to;

$$-i\alpha = iT\alpha\beta^2 \int_{S_2} \cos\beta x \frac{\partial^2 G}{\partial z \partial z_0} ds + \frac{\partial \phi_I(P)}{\partial z_0}, \tag{15}$$

where $\partial G / \partial z_0$ is a doublet type of Green function.

In Eq. (15), the amplitude α is the only unknown variable, therefore when the integral in the right hand side is obtained, the amplitude is determined immediately.

For a multiple sheet-frame system, the amplitude of each sheet is an unknown variable. Locating the singular point at the center of each sheet, we obtain a set of equations like Eq. (15) and solve them simultaneously.

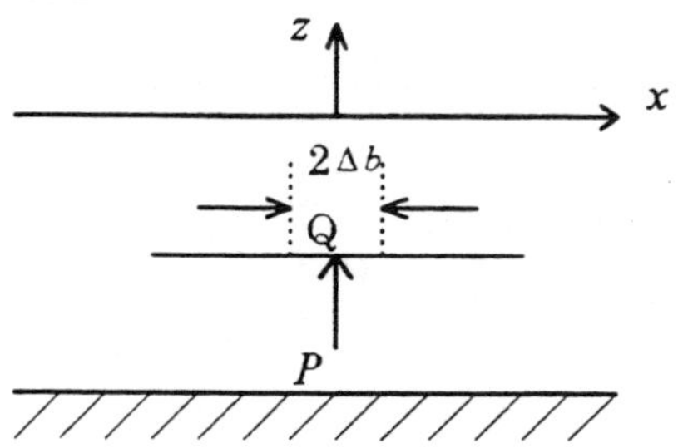

Fig. 3. Singular point and the center of sheet

5. INTEGRAL ON THE SHEET

The integral on the right hand side of Eq. (15) includes a singular function G, therefore special treatment is needed in the neighborhood of the singular point.

5.1 Principal-valued integral

Let us consider an integral on a small segment $-\Delta b \le x \le \Delta b$ at the limit $P \to Q$. Assuming $\cos\beta = 1$ (constant),

$$I_1 = \lim_{P \to Q} \int_{-\Delta b}^{\Delta b} \frac{\partial^2 G}{\partial z \partial z_0} dx.$$

The real part of Green function Eq. (6) can be approximated by a fundamental solution of 2-D,

$$\mathrm{Re}\{G\} \to -\frac{1}{2\pi} \log r \quad on \quad r \to 0,$$

$$r = \sqrt{(x - x_0)^2 + (z - z_0)^2}.$$

Therefore,

$$I_1 = -\frac{1}{\pi \Delta b} + 2(\Delta b) k^2 C_0 \sinh^2 (h - d) \tag{16}$$

5.2 Total integral

Except for the neighborhood of singular point, the integral can be obtained in the ordinal manner.

$$I_2 = \int_{\Delta b}^{B/2} \cos\beta x \; e^{ikx} dx$$

$$= \frac{ik}{\beta^2 - k^2} \left\{ \cos\frac{\beta B}{2} \exp(i\frac{kB}{2}) - \cos(\beta \Delta b) \exp(ik\Delta b) \right\}$$
$$+ \frac{\beta}{\beta^2 - k^2} \left\{ \sin\frac{\beta B}{2} \exp(i\frac{kB}{2}) - \sin(\beta \Delta b) \exp(ik\Delta b) \right\} \tag{17}$$

$$I_n = \int_{\Delta b}^{B/2} \cos\beta x \; e^{-k_n x} dx$$

$$= -\frac{k_n}{\beta^2 + k_n^2} \left\{ \cos\frac{\beta B}{2} \exp(-\frac{k_n B}{2}) - \cos(\beta \Delta b) \exp(-k_n \Delta b) \right\}$$
$$+ \frac{\beta}{\beta^2 + k_n^2} \left\{ \sin\frac{\beta B}{2} \exp(-\frac{k_n B}{2}) - \sin(\beta \Delta b) \exp(-k_n \Delta b) \right\} \tag{18}$$

With the notations Eq. (16), Eq. (17) and Eq. (18), Eq. (15) is represented as follows,

$$-i\alpha = iT\alpha\beta^2 \{ I_1 + 2k^2 C_0 I_2 \sinh^2(h-d)$$
$$+2\sum_n k_n C_n I_n \sin^2(h-d) \} + \partial\phi_I / \partial z_0$$
$$= iT\alpha\beta^2 I + \partial\phi_I / \partial z_0$$

Therefore, the amplitude α is,

$$\alpha = i \frac{\partial \phi_I(P)/\partial z_0}{1 + T\beta^2 I}. \tag{19}$$

6. POTENTIALS OF REFLECTED AND TRANSMITTED WAVES

Let us consider a single sheet-frame system submerged completely ($d > 0$). The potential on the water surface is obtained by Eq. (8) on $z_0 = 0$. Green function is regular on the sheet. Therefore the integral on the right hand side of Eq. (8) can be obtained analytically.

For $x_0 \le -B/2$, Eq. (8) is represented as,

$$\phi(P) = iT\alpha\beta^2 J k C_0 \cosh kh \sinh k(h-d)$$
$$-iT\alpha\beta^2 \sum_n J_n k_n C_n \cos k_n h \sin k_n (h-d) + \phi_I(P) \tag{20}$$

where

$$J = \int_{-B/2}^{B/2} \cos\beta x \; e^{ik(x-x_0)} dx$$

$$= \frac{2ik}{\beta^2 - k^2} \exp(-ikx_0) \cos\frac{\beta B}{2} \sin\frac{kB}{2}$$
$$+ \frac{2\beta}{\beta^2 - k^2} \exp(-ikx_0) \sin\frac{\beta B}{2} \cos\frac{kB}{2}$$

$$J_n = \int_{-B/2}^{B/2} \cos\beta x \; e^{-k_n(x-x_0)} dx.$$

$$= \frac{2k_n}{\beta^2 + k_n^2}\exp(k_n x_0)\cos\frac{\beta B}{2}\sinh\frac{kB}{2} + \frac{2\beta}{\beta^2 + k_n^2}\exp(k_n x_0)\sin\frac{\beta B}{2}\cosh\frac{kB}{2}.$$

For $x_0 \geq B/2$,

$$\phi(P) = iT\alpha\beta^2 J'kC_0 \cosh kh \sinh k(h-d) - iT\alpha\beta^2 \sum_n J_n' k_n C_n \cos k_n h \sin k_n (h-d) + \phi_I(P) \tag{21}$$

where

$$J' = \int_{-B/2}^{B/2} \cos\beta x e^{-ik(x-x_0)} dx = \frac{2ik}{\beta^2 - k^2}\exp(ikx_0)\cos\frac{\beta B}{2}\sin\frac{kB}{2} + \frac{2\beta}{\beta^2 - k^2}\exp(ikx_0)\sin\frac{\beta B}{2}\cos\frac{kB}{2}$$

$$J_n' = \int_{-B/2}^{B/2} \cos\beta x e^{k_n(x-x_0)} dx = \frac{2k_n}{\beta^2 + k_n^2}\exp(-k_n x_0)\cos\frac{\beta B}{2}\sinh\frac{kB}{2} + \frac{2\beta}{\beta^2 + k_n^2}\exp(-k_n x_0)\sin\frac{\beta B}{2}\cosh\frac{kB}{2}.$$

At a position sufficiently away from the sheet, the evanescent modes (the second terms of right hand side of Eq. (20) and Eq. (21)) are eliminated. Therefore, the potential of reflected waves $\phi_R(P)$ is,

$$\phi_R(P) = iT\alpha\beta^2 kC_0 J \cosh kh \sinh k(h-d), \tag{22}$$

and the potential of transmitted waves $\phi_T(P)$ is,

$$\phi_T(P) = iT\alpha\beta^2 kC_0 J' \cosh kh \sinh k(h-d) + \phi_I(P). \tag{23}$$

7. ENERGY LOSS MODEL

Actual energy loss may be caused in vortexes with separations from the two ends of sheet. But it is difficult to establish a suitable model of vortex in this complex flow. Therefore we propose a conventional model of energy loss in which all energy loss is assumed to occur inside the boundary layer on the sheet. When the difference of pressure on the upper and lower surface of the sheet is represented by Eq. (9) and the velocity of fluid on the sheet is represented by Eq. (14), total work is zero, which is done in one period by the fluid outside the boundary layer to the boundary layer. If there is $\pi/2$ phase sift in the vertical component of fluid velocity,

$$V' = \gamma \cos\beta x e^{-it}. \tag{24}$$

the work done by this component is,

$$W = V'P_2 - V'P_1 = -V'(P_1 - P_2)$$

$$= \gamma T\alpha\beta^2 \cos^2\beta x(e^{-2it} + 2 + e^{-2it}).$$

Therefore the fluid outside the boundary layer does positive work to the boundary layer in one period. This work is consumed as energy loss inside the boundary layer.

The coefficient γ in Eq. (24) is modeled in a similar manner to Kato and Hamanaka (1993),

$$\gamma = \alpha\beta^2 Tc.$$

Then Eq. (19) become,

$$\alpha = i\frac{\partial\phi_I(P)/\partial z_0}{1 + T\beta^2(I + ic)}. \tag{25}$$

8. EXTENSION TO 3-D

There is no change in the basic idea when the problem is extended to 3-D. For simplicity, the sheet is supposed to be square and the motion of sheet is assumed in two ways,

$$\zeta = \alpha\cos\beta x\cos\beta y\ e^{-it} \tag{26}$$

$$\zeta = \alpha\cos\beta x\ e^{-it}, \tag{27}$$

where Eq. (26) is the case of 4 conjoined sides, and Eq. (27) is the case of 2 conjoined sides.

For one sheet-frame system with 4 conjoined sides, the corresponding equation to Eq. (15) is,

$$-i\alpha = 2iT\alpha\beta^2 \int_S \cos\beta x\cos\beta y \frac{\partial^2 G}{\partial z\partial z_0} dxdy + \frac{\partial\phi_I(P)}{\partial z_0} \tag{28}$$

and with 2 conjoiened sides,

$$-i\alpha = -iT\alpha\beta^2 \int_S \cos\beta x \frac{\partial^2 G}{\partial z\partial z_0} dxdy + \frac{\partial\phi_I(P)}{\partial z_0} \tag{29}$$

Green function in 3-D is,

$$G = C_0 \cosh k(z_0 + h)\cosh k(z+h) H_0^{(1)}(kR) + \sum_{n=1}^{\infty} C_n \cos k_n(z_0 + h)\cos k_n(z+h) K(k_n R) \tag{30}$$

where

$$C_0 = i/2(h + \sinh^2 kh), \quad C_n = 1/\pi(h - \sin^2 k_n h),$$

$$R = \sqrt{(x-x_0)^2 + (y-y_0)^2}.$$

and $H_0^{(1)}$ is the Hankel function of the first kind and zero-th order and K is the modified Bessel function.

Once the integral in the right hand side of Eq. (28) or Eq. (29) is obtained, the amplitude of the sheet can be determined. The integral in the neighborhood of the singular point should be considered as the principal valued integral using the fact that the real part of Green function is approximated by,

$$\mathrm{Re}\{G\} \to 1/4\pi r \quad on \quad r \to 0$$

where

$$r=\sqrt{(x-x_0)^2+(y-y_0)^2+(z-z_0)^2}\,.$$

9. NUMERICAL RESULTS

Numerical experiments have been performed to examine the proposed method.

Fig.4 shows the reflection coefficients and the transmission coefficients for single flexible sheet in 2-D using the potential matching method by Kato and Hamanaka (1993). The circles show the experimental results. The water depth was taken as $\hat{h}=0.3$(m), the width of the sheet, $\hat{B}=0.26$(m) and the relative submerged depth, $\hat{d}/\hat{h}=0.3$. Fig.5 shows the results from the present method on the same conditions of Fig.4. In these figures, C denotes the coefficient of energy loss in Eq. (25). Fig.6 and Fig.7 show the case of two flexible sheets placed adjacently each other along the wave direction.

In each case, good agreements have been obtained, except for slight differences for large wave numbers. These differences are due to the assumption that the displacement of the sheet and the difference of pressure on the upper and lower surface of the sheet take the form $\cos\beta x$. On this assumption, the difference of pressure (difference of potential identically) is eliminated at the ends of the sheet. But the potential flow should be singular at these points, therefore the differences must not be zero.

Although there are slight differences between two method, the present method has a big advantage over the potential matching method. The necessary number of unknown variables is one for each sheet-frame system.

Fig.8 shows the reflection coefficients as a function of kh and the spacing of two flexible sheets. kh increases from the right to the lower left, and the spacing increases from the right to the upper left. The reflection coefficients on the upper right is in the case that the two flexible sheets are placed adjacent each other, and on the lower left the spacing of the center of sheets is about two times the wave length. From these results, the reflection coefficient takes maximum value when the spacing is zero.

3-D cases have been examined. 10 and 20 flexible sheets with 2 conjoined sides are placed side by side on a straight line. The wave fields near the center of the row of sheets almost coincide to that of the 2-D cases.

Fig.9 shows a example of 3-D case with 10 flexible sheets. The water depth, the size of the sheet and the relative submerged depth are the same of 2-D case. The incident waves propagate normal to the row of sheets, from the upper left to the lower right. (a) shows the water surface displacement. It can be seen there is a calm sea area behind the row of sheets and a reflection area in front of the row. (b) shows the diffracted and scattered waves after subtracting the incident waves from the total wave field. These results seem to be physically reasonable.

In these analysis, the integral on the sheet in Eq. (15) in 2-D cases was obtained analytically as explained in Section 5. In 3-D cases, the integral on the sheet was calculated numerically with Gaussian formula except in the neighborhood of the singular point where the principal-valued integral was used. Normally it is very difficult to obtain a precise value of numerical integral with a doublet type of singular function. In this analysis, we assume a suitable function for potential on the sheet, therefore it could be possible to obtain precise values.

10. Conclusion

A new method is proposed to analyze the wave field around the flexible sheet. In this method a doublet type of Green function is used. The necessary number of unknown variables can be reduced to one for each flexible sheet using a suitable function for the displacement of the sheet and the difference of pressure between the upper and lower surfaces of the sheet. But the function used in this paper does not express the singularity of flow at the ends of the sheet. It is needed to select another suitable function to improve this point.

REFERENCES

John, F. (1950) : On the motion of floating bodies II, simple harmonic motion, Comm. Pure Appl. Math., Vol.3, pp. 45-101

Kato, M. and K. Hamanaka (1993) : Analytical and experimental study of characteristics of textile sheet structure, 11th Australasian Conf. on Coastal and Ocean Engg., pp. 403-408

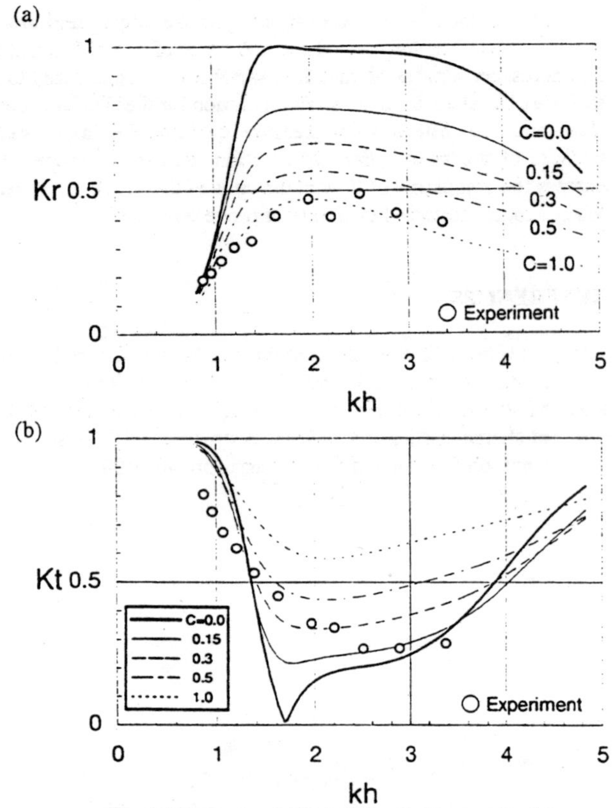

Fig.4 Single sheet (Potetial matching method)
(a) Reflection coefficient, (b) Transmission coefficient

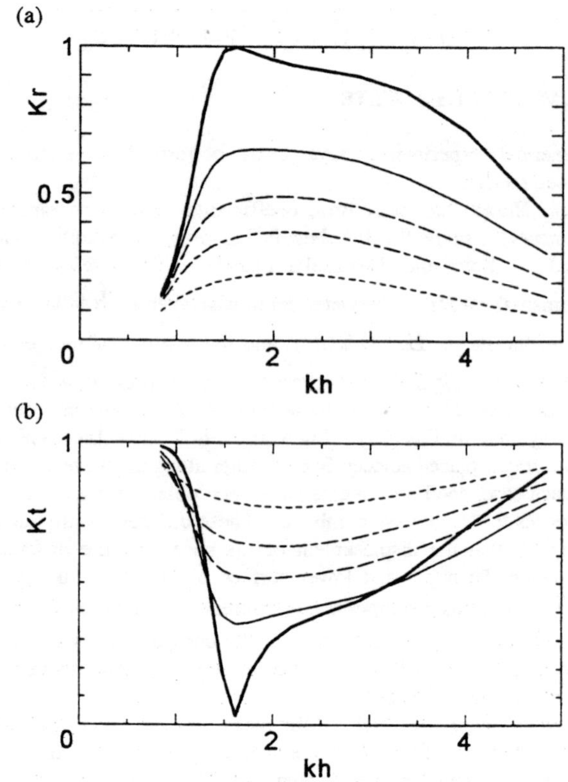

Fig.5 Single sheet (Present method)
(a) Reflection coefficient, (b) Transmission coefficient

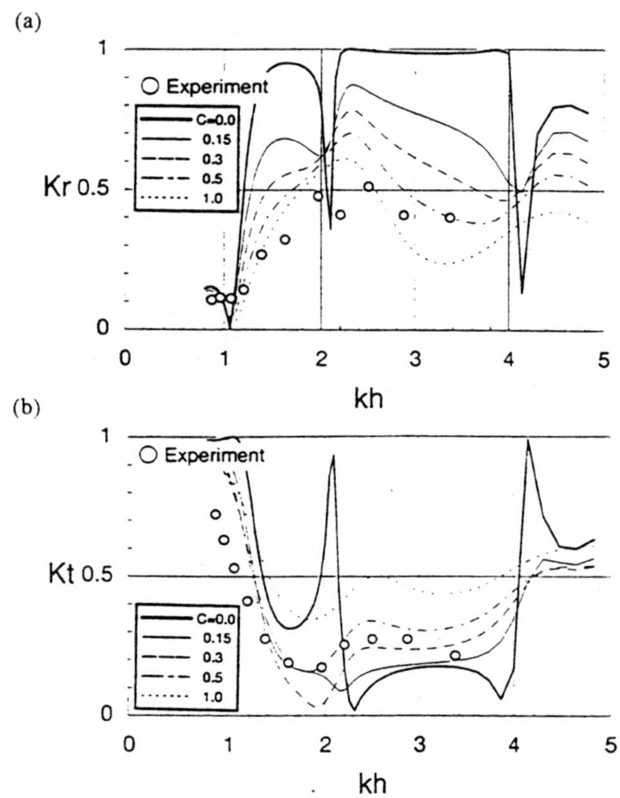

Fig.6 Double sheets (Potential metching method)
(a) Reflection coefficient, (b) Transmission coefficient

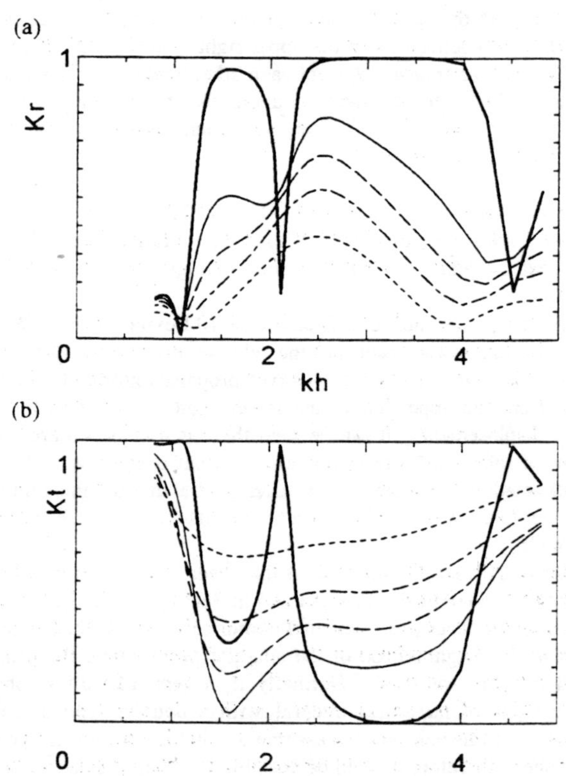

Fig.7 Double sheets (Present method)
(a) Reflection coefficient, (b) Transmission coefficient

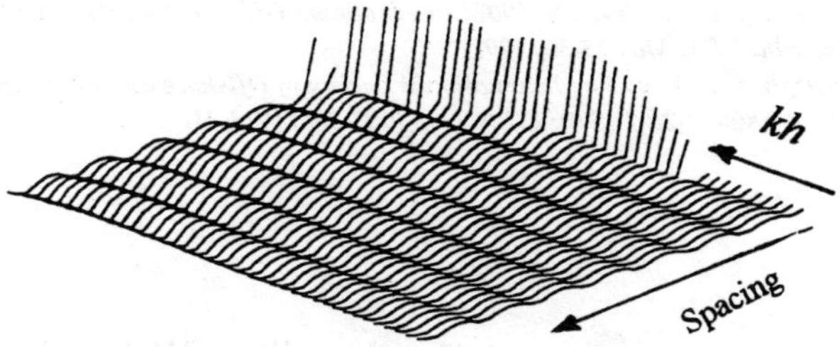

Fig.8 Reflection coefficient as a function of the wave number and the spacing of the double sheets

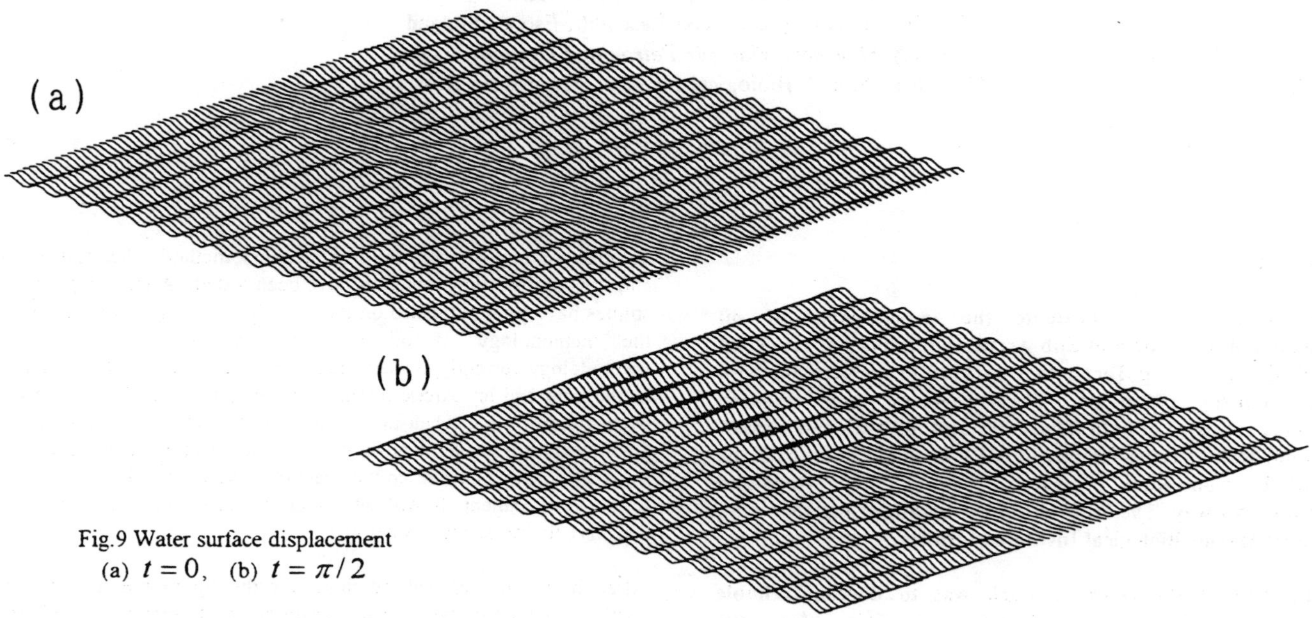

Fig.9 Water surface displacement
(a) $t=0$, (b) $t=\pi/2$

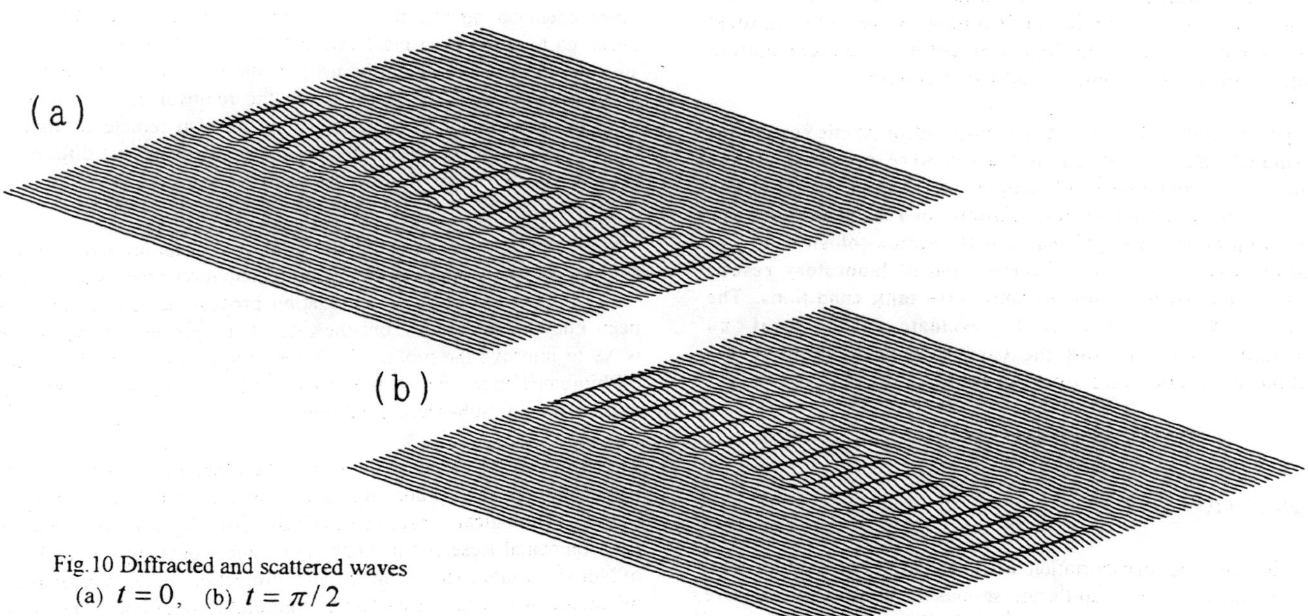

Fig.10 Diffracted and scattered waves
(a) $t=0$, (b) $t=\pi/2$

Proceedings of the Seventh (1997) International Offshore and Polar Engineering Conference
Honolulu, USA, May 25-30, 1997

ISBN 1-880653-28-1 (Set); ISBN 1-880653-30-3 (Vol. II)

Arctic Shoreline Oil Combatting Experiments in 1996 in Murmansk

Jorma Rytkönen
VTT Manufacturing Technology, Espoo, Finland
Reetta Piskonen and Merja Itävaara
VTT Biotechnology and Food Research, Espoo, Finland
Gennady Matishov, Vladimir Petrov and Gennady Iluyin
Murmansk Marine Biological Institute, Murmansk, Russia

ABSTRACT

An extensive study to evaluate the improvement of oil bioremediation in Arctic and Sub-Arctic conditions was launched by VTT Manufacturing Technology in 1995 in cooperation with several partner institutions. The two-years long research programme consisted of both laboratory and full-scale test phases. Laboratory tests were performed by VTT in cooperation with the Swedish Environmental Research Institute and Rogaland Research of Norway. The field tests were arranged together with Murmansk Marine Biological Institute of Russia.

The objective of the research work was to develop reliable analysing methods and evaluation principles for the bioremediation of oil spills in aquatic environment, especially in the shoreline conditions with tidal water variation. Two main categories of the research programme were the natural bioremediation of oil and the improvement of oil bioremediation by using biosorbent technology or added fertilizers.

This paper describes the field tests conducted in Arctic conditions in Murmansk, Russia. Several test areas were prepared on the shore-line. Oil, water and sand samples were taken and analyzed in order to study the oil bioremediation in real conditions, the impacts of oil on the environment and the ecotoxicological effects. The results were also used for verification of laboratory results achieved in micro-scale cylinders and wave tank conditions. The main task of this study is to evaluate the natural oil bioremediation potential and the various ways to improve the natural bioremediation , e.g. by using fertilizers or biosorbents.

INTRODUCTION

The production and transportation activities of oil in Arctic areas always contain a risk of oil spills and serious impacts on the sensitive nature. Methods for cleaning-up oil spills have been studied for decades. Among the conventional recovery methods chemical agents and various absorbing materials have been tested. A lot of reports and studies have been published on the oil behaviour, aging of the oil and the methodology of oil recovery techniques. However, the methodology currently available has been found inadequate, especially in the cold and icy Arctic marine environment. The clean-up task is even more difficult when the oil slick will reach the shore-line. The Russian statistics of large oil spills in the Barents Sea states clearly that oil will sooner or later reach the coast line and cause a lot of damages to the environment. It will also alter the ecological balance of the sensitive littoral shallow water area.

After the oil has reached the shore-line the countermeasures will be costly compared to the open water conditions. Removing the oil from the soil stratum is laborous work, and often the countermeasures can cause serious impacts on the environment. The use of dispersants or other chemical agents, the use of hot steam or water have been observed to cause even more serious problems than the oil itself. In-situ burning can be seldom be used in the shore-line conditions, and often the only possible measures are the removal of oil and/or oily sand by rather manual methods. There are also remote locations or shore-line conditions where effective countermeasures are difficult and even impossible to carry out without enormous costs.

The lessons learned from the shore-line oil clean-up activities have showed the importance of the oil bioremediation process as a natural 'soft' countermeasure. Oil degradation process, ie. oil alteration has been known for decades, but the rate of the process or the possible ways to improve the process have been rather unknown. The natural oil bioremediation rate has been found to be a slow process especially in the Arctic or Sub-Arctic conditions.

Therefore VTT's research divisions Manufacturing Technology and Biotechnology and Food Research together with the Murmansk Marine Biological Institute (MMBI) of Russia, the Swedish Environmental Research Institute (IVL) and Rogaland Research (RF) of Norway established a joint research project on the bioremediation of oil at the Arctic and Sub-Arctic shoreline conditions. The project entitled "The Improvement of Oil Bioremediation at the Shore Line" is scheduled for the years 1995-1997, and it consists of experimental studies supported by field experiments.

The preliminary tests were carried out in the Belokamenka Bay in Murmansk, May 1996. The goal was to study the behaviour of crude oil after oil spill had reached the shoreline in Arctic marine environment. This preliminary study was considered important for gaining practical information how sampling and analysis for microbial and oil analyses should be performed. Also it was considered important to find out the background oil level and the amount of oil degrading microorganisms before starting more wider field tests in Autumn 1996.

This paper describes the bioremediation field tests conducted in Murmansk. Oil bioremediation process on the test sections was first observed in May 1996 during a four weeks preliminary test period. The results of this test phase was used when designing a more wider field bioremediation experiment conducted Autumn 1996 - Spring 1997. In connection of the measurements of the oil quantity of the sand column measurements of the microbial development were carried out. The oil bioremediation was determined with an aid of adenosine triphosphate (ATP) content and substrate induced respiration (SIR). The influence of the oil on the microflora and bottom fauna of the littoral zone was studied, too.

TEST ARRANGEMENTS

Preliminary tests

Both the preliminary field experiment and wider field tests in Autumn were carried out on the sandy and silty littoral zone of the Belokamenka Bay. The preliminary test phase was carried out between 07- 28th of May 1996 in the conditions of the early Arctic spring, when the air temperature and wind regime were unstable. This test phase was then followed by a wider test phase which will be completed after snow melting, Spring 1997.

Altogether three test areas were established in the sandy beach, shown schematically in Figures 1 and 2.

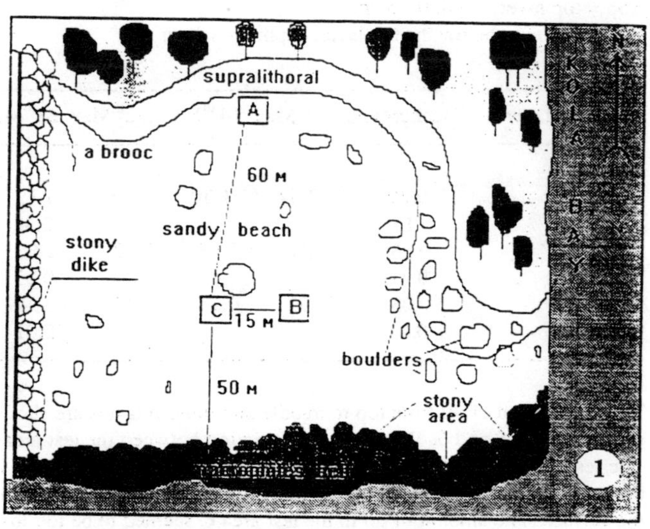

Fig. 1. The location of the preliminary test area in the littoral zone. Belokamenkaya Bay, Murmansk.

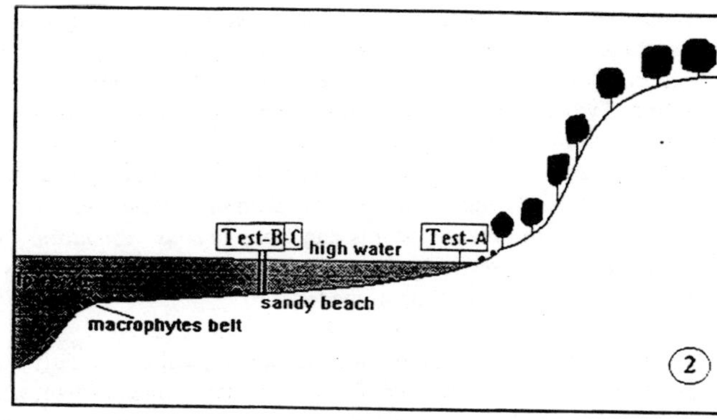

Fig. 2. The profile of the preliminary test area and the locations of the test sections A, B, C in the littoral zone of Belokamenkay Bay, May 1996.

The size of each test section was 2 x 2 m^2. Test section A was located in the upper part of the littoral zone near the highest mark of the tidal high water elevation. Two others, test sections B and C, were 60 meters offshore, in the downer part of the littoral zone. The distance between the areas B and C was 20 meters.

The test was started on the 7th of May 1996. Eight liters of the crude oil (Varandoy deposit) was equally spread onto each test section after the first sand samples were taken.

The sampling frequency was one week. Each time the test area was visited sampling, visual observations, meteorological and hydrographic measurements were carried out. Soil samples were taken with a core sampler and a metallic test spoon. Soil samples were taken both from the sand surface and deeper (down to the depth of 14 cm) of the sea bottom. The soil samples were frozen for oil analyses and determined later at VTT/Biotechnology and Food Research in Finland. The amount of oil degrading microorganisms was determined from the fresh samples stored at + 5^oC.

Water samples were also taken for the determination of background level of hydrocarbons and oil degrading microbes. Last samples were taken 28.5.1996.

The oil content of the shoreline sand was determined before and after addition of crude oil. The amount of hydrocarbons was determined after extracting and determination IR absorption maximum of C-H stretch of CH2-CH3 band at wavelength 2930 cm-1. This oil extraction method was performed according to the standard biodegradability test CEC L-33-T-82 test.

Test section C was also used for investigations of the impacts of oil on the littoral fauna. For this purpose 2 gages of blue mussels (*Mytilus edulis*), bivalves (*Macoma balthica*) and littoral macrophytes (*Fucus*) were added to the natural communities of the test section. Several middle sized stones with marine acorns (*Balanus balanoides*) living on them were put there, too. The most dominant bottom macrobenthos species was *Macoma balthica*, which had both the largest biomass and abundance. The population density was on the average 1733 pieces/m^2.

Altogether 800 additional mollusks (*Macoma balthica*) were collected and settled inside the test section C before pouring any oil into the section. Mollusks were put into a perforated cage settled on the sandy bottom. Two additional cages with blue mussels (*Mytilus edulis*) were mounted to the sides of the supporting wooden fences of the section. each cage contained 115 pieces of mussels. Furthermore small stones with blue mussels and marine acorns (*Balanus balanoides*), and four larger stones with marine acorns and algae (*Fucus vesiculosus*) were put inside the test section. All organism were kept in the test section three days before the start-up of the test.

During the preliminary test the weather was typical of the early Arctic spring in the Murmansk coast, i.e. very unstable. The air temperature fluctuated around the. Wind velocity changed from weak to strong and even storms were observed. Most part of the time the sky was cloudy, with fallouts-weak rain and snow showers. When starting the test the snow cover of the area started to melt. In the beginning of the test period the coast line around the bay and the littoral shallow water area was covered with a thick layer of snow. In the end of the test the coast line was completely free of the snow. Both the water and the ground temperature were always positive, i.e. the water temperature was stable in the range +3...+4 °C. The temperature of the surface layer of the ground up to 3 cm depth depended on the intensity of the solar radiation during the low water. Thus the ground temperature varied in the wider range- from+5.8 up to +9.0 °C. Figure 3 presents the temperature variation during the test.

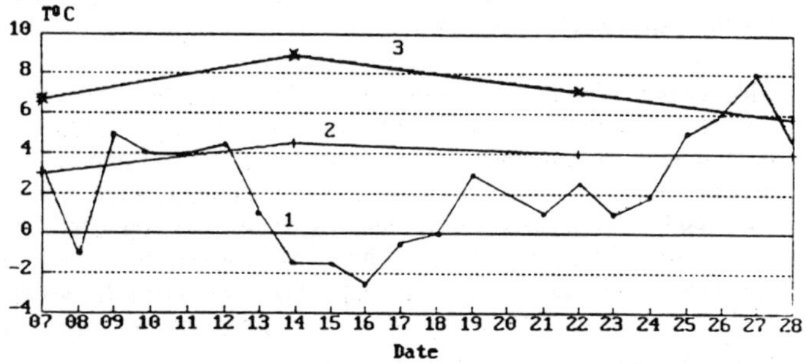

Fig. 3. The fluctuation of the air temperature (1), water temperature (2) and soil temperature (3) in May 1996.

Each of the test sections were in the beginning surrounded by a wooden low fence to protect the test area against floating ice floes during the high tide. Fig. 4. Fences were also used for marking the test area. and partly to protect the mussels, bivalves and macrophytes. The moving ice floes, however, partly removed the fences. Thus test sections A and B were left without the wooden shelters for the rest of the test duration. Due to the possible distortion effects of the fences on the local hydrodynamics of the test area, no samples were taken near the sheltered zones created by the fences.

The surface of the sea bottom of the littoral zone consisted mainly of find sand and silt. Stones having the diameter 0.1...1.0 m were rather uniformly spread on the area. The density of the stones was however scarce. No larger stones were located in the test sections. The sand surface of the test section was covered by small ripples.

During high tide test section A was completely covered with water for a short period of time. This upper part of the littoral zone can be defined as the transport zone and the area of the floating debris accumulation. This zone itself is the surfing zone. A great deal of small particles of debris was of the vegetable origin (relicts of sea weed and land-origined plants, parts and pieces of rind, chips etc.). These organic remains were distributed on the surface of the ground, partly mixed with sand, partly buried. Lower along the littoral shallow-water zone, where the test areas B and C were located, there are no debris on the surface.. Deeper under the surface of sand, however, buried debris was observed.

Fig. 4. Test section C was used for determining the impacts of oil on the littoral fauna.

PRELIMINARY TEST RESULTS

The amount of oil

The amount of oil decreased during the first week, after which only slight reduction was observed. The test areas A and B showed clear contamination of the shoreline sand with oil (Table 1). The amount of oil was still at the higher level than the background. The fast reduction of the amount of oil seemed due to the washing effect of seawater.

Table 1. The Oil content of shoreline soil (mg oil/ g sand) in test areas A. B and C.
Top = top layer, depth 0 - 5 cm ;
Bottom = layer below the top layer, depth 5 - 10 cm.

Test area-layer	Background	7th May	14 thMay	22thMay	28thMay
A TOP	0.30	9.84	1.30	1.38	1.19
A Bottom	0.15				2.38
B TOP	0.26	3.78	2.03	1.02	1.66
B Bottom	0. 03				0.03
C TOP	0.03	14.97	0.11	0.07	0.04
C Bottom	0.03				0.01

The penetration of oil from top to middle and bottom layers are shown in the Figures 5 and 6. There was big variation between the test areas. The test area A had the highest oil level 3-6 cm below the surface in the test area B the highest amount was on the top layer 0-3 cm from the surface. The amount of oil in the test area C seemed to be too low. No clear trend in the penetration of oil into soil was found. This may result from the big variation in the composition of the soil type and of the very few test areas and samples taken, Fig. 7.

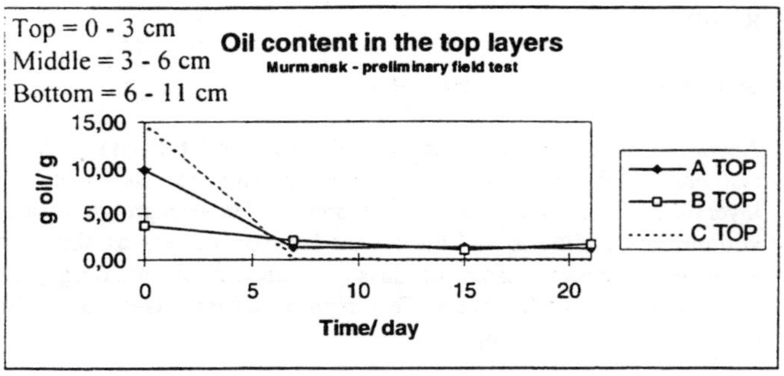

Fig. 5. The oil content extracted from the top of the shoreline soil in the test areas A. B and C. The results are shown as the function of time.

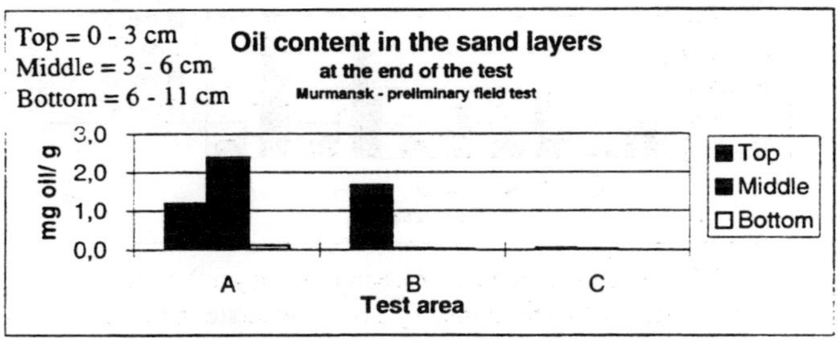

Fig. 6. The oil content in top. middle and bottom layers of shoreline soil in test areas A, B and C at the end of the test.

Fig. 7 Test section B after two weeks of test duration.

The amount of oil degrading microorganisms

The amount of the microbes were determined immediately after adding oil on the shoreline. In the test area A the amount of oil degrading microorganisms was originally higher than in the other test areas and no significant increase in oil degrading microbes was detected.

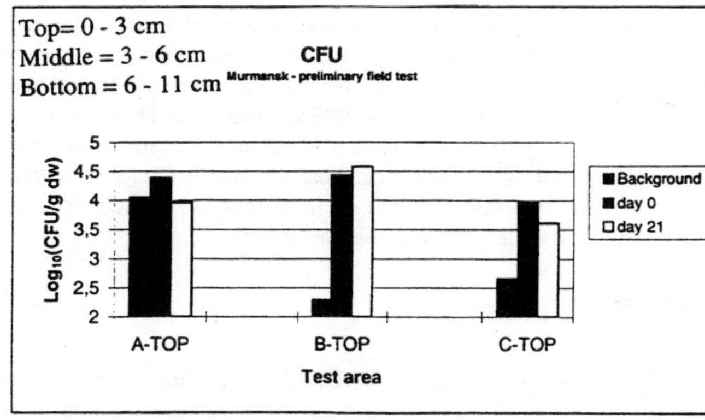

Fig. 8. The amount of microbes, expressed as colony forming units (CFU) in the test area A, B and C .

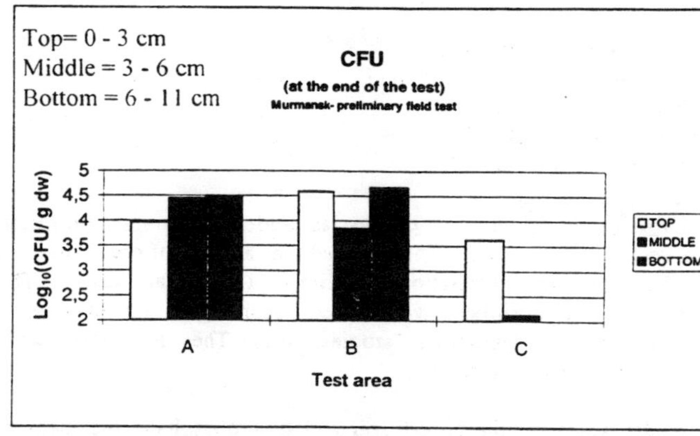

Fig. 9. The amount of microbes expressed as colony forming units (CFU) in different layers at the end of the test.

The impacts of oil

The oil had strong impact on the small crawfishes, *Gammarus genera*, in the beginning of the test phase. Crawfishes were located in a small puddle which was covered by fresh crude.

Crawfishes were observed to keep their mobility for half an hour, however partly stressed and losing their ability of orientation. The rise of the tidal water, however, postponed the subsequent observations. The animals were, however, completely covered by oil, which was impossible to remove with paper towel or rinsing with water.

One week later, on the 14th of May, all the animals and algae in the test section C and the stones and the netting of the cages were contaminated by oil. Only separate oil slicks were detected on the ground surface, under the stones, under the algae and under the cage.

The abundance of the mollusks, *Macoma balthica*, changed slightly in the test section C (Table 2) during the test execution. The reduction in the test section B was even more significant.

Table 2. Mollusks, *Macoma balthica*, abundance.

Date	Average abundance, pieces /m^2	
	Area C	Area B
07.05.96	1445	720
14.05.96	1474	---
22.05.96	2166	94
28.05.96	1550	42

Physiological examinations of the animals showed that after the high tide practically all blue mussels, marine acorns and macomas were actively feeding. In the end of the test the survival rate of blue mussels was 100%, marine acorns 98% and macomas 98% in the test section C. All the animals fed actively when put them into the fresh sea water: marine acorns were catching the surroundings with their legs, blue mussels opened their shells, macomas pushed forward their siphons and collected food from the ground.

FIELD TESTS IN AUTUMN

After the preliminary test results were analysed a more wider field test was established. Five new test areas were arranged in the Belokamenkaya Bay, near the previous preliminary test site. Four test sites (1,2,3 and 4) had a size of 2 x 4m^2, while as the test section 5 was smaller (size 1 x 3 m^2). The additives of the test sections were:

1. Background (without oil)
2. Russian oil (without additives)
3. Russian oil (fertilizer 1)
4. Russian oil (fertilizer 2)
5. Russian oil (biosorbent)

Approximately 3 liters/ m^2 of Russian crude oil was spread on each test section (2,3,4 and 5). This equaled the amount of crude oil used in the laboratory respiration experiment (microscale test). The fertilizers and biosorbent were added according to the experience gained in the laboratory studies, too. The fertilizers were manufactured by Kemira LtD.

The fast water soluble N-fertilizer and superphosphate, the medium soluble N-fertilizer and apatite and a Russian biosorbent were the additives added. The fertilizers were added in the following molar ratios C : N (10 :1) and C : P (30 : 1) as in the laboratory assays. In the laboratory test the carbon content of the oil was determined and the fertilizers were added in the above ratios. The biosorbent was added following the instructions of the manufacturer.
In the field test the carbon content of oil was unknown and the assumption of carbon content being approximately 85 % was made. The fertilizer were added (based on this assumption) in the molar ratios C : N = 10 :1 and C : P = 30 : 1. The carbon content of oil was later shown to be 87,7 % (Analysed by VTT Energy).

Every other week samples were taken for microbiological, nutrient and oil analysis as long as it was possible before the frost. The sampling dates were 17.9, 18.9, 30.9, 17.10, 4.11. In spring 1997 after snow melting the sampling will be continued in Murmansk.
The soil samples were taken from two layers (0-5 cm and 5-10 cm) by using a plastic tube (Ø 3 cm) especially prepared for sampling of sandy soil. The samples were collected from each area from several points and combined. The total amount of the samples taken from each test area was 3 x 300g. The samples were stored in the glass containers, cooled down and transported to Finland for microbiological analyses. The samples for oil modelling studies were freezed for transportation to Sweden.

The amount of microbial biomass in the soil samples were studied by three different microbiological method: SIR (substrate induced respiration) (Vanhala & Ahtiainen 1994), ATP (adenosine triphosphate) (Andersson & Domsch 1978), content and the total amount of bacteria (CFU).The amount of colony forming units (CFU) was counted after 3 weeks.

Results

SIR (substrate induced respiration)

According to the results from SIR, the microbial biomass increased fast after addition of oil to the shoreline. This was shown in both layers 0-5 cm and 5-10 cm (Figs. 10 and 11).The background area 1A without any additions had the lowest biomass content as should be expected. However, after 12 days the amount of microorganisms decreased in all the test areas. This might result from weather cooling down, winter was entering.

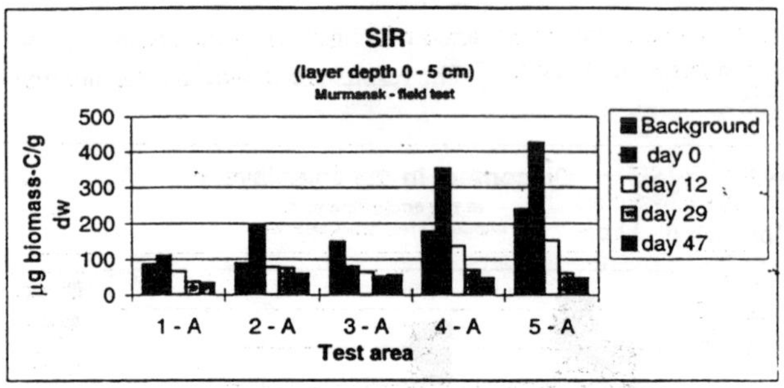

Fig. 10. The amount of biomass carbon in the layer 0-5 cm from the surface (A) determined by SIR (substrate induced respiration)

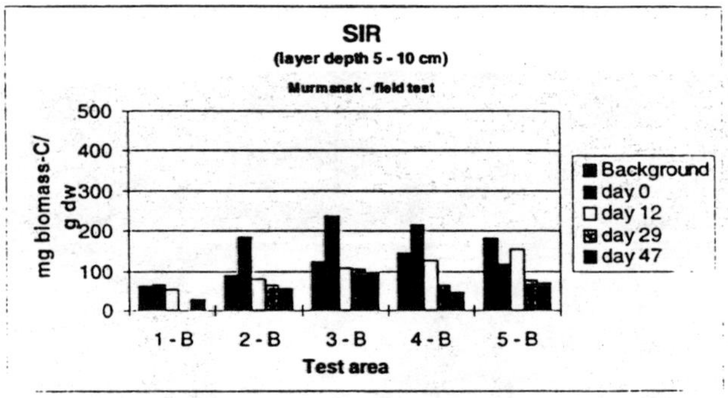

Figure 11. The amount of biomass carbon in the layer 5-10 cm from the surface (B) determined by SIR (substrate induced respiration)

Biodegradation of oil seem to occurr at least to 10 cm depth according to microbiological studies. In some test areas such as 4A and 5A a little higher microbial activity occurred near the surface (0-5 cm layer) than in 5-10 cm. Heterogeneity of the natural test area effects on the results.

ATP activity in the soil layers

When SIR and ATP are compared, the main difference between the results is that SIR indicates immediate increase in biomass after crude oil addition, but at the same time ATP content shows decrease or increase in biomass content depending on the test area.

No clear effect could be determined by ATP measurement (Figs. 12 and 13). ATP measurement is a very sensitive method and several agents may interfere the analysis eg. humus compounds. The peak in ATP activity on day 12 may result from errors in sampling and storage during the flight from Murmansk to Finland.

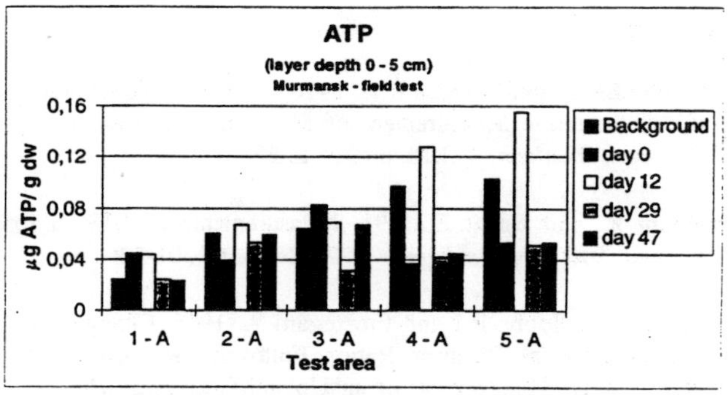

Fig. 12. Microbial biomass in the layer 0-5 cm (A) determined by ATP (adenosine triphosphate) activity.

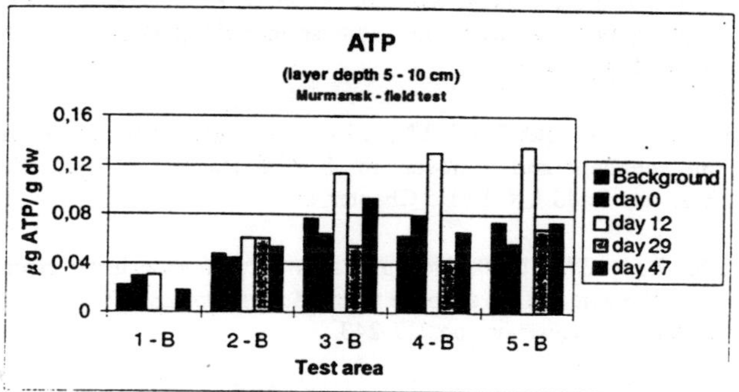

Fig. 13. Microbial biomass in the layer 5-10 cm from the surface (B) determined by ATP (adenosine triphosphate) activity.

CFU (colony forming units)

The amount of colony forming units (CFU) increased at first in the test areas 4 (fertilizer 2) and 5 (Russian biosorbent) and after 12 days the highest amount of microorganisms was counted in the test area 4 (Figs. 14 and 15). The number of CFU decreased towards the end of the test in all the areas. After 47 days the amount of cultivated microorganisms was lower than that of the background samples taken before addition of oil, except the area 4. The reason for this might be the cooling weather conditions as mentioned earlier.

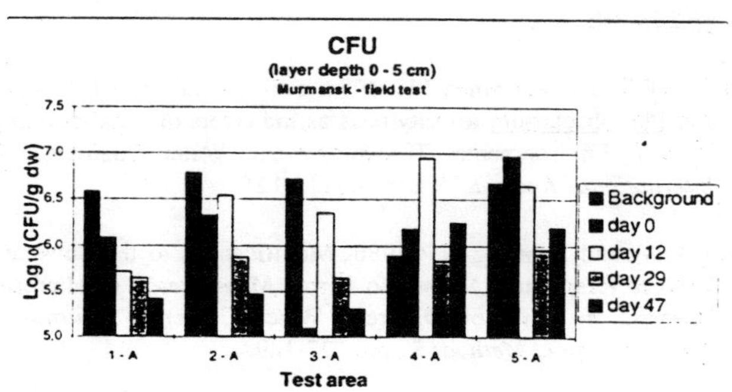

Fig. 14. The amount bacteria determined after cultivation on the seawater nutrient broth medium from the soil layer 0-5 cm (A).The amount of microorganisms is expressed as colony forming units (CFU).

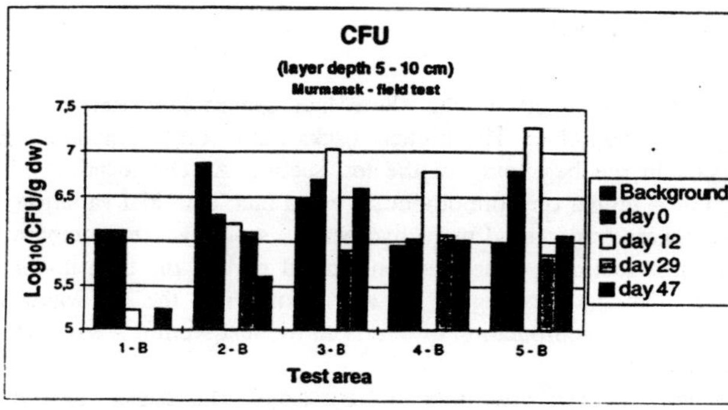

Fig. 15. The amount bacteria determined after cultivation on the seawater nutrient broth medium from the soil layer 0-5 cm (A).The amount of microorganisms is expressed as colony forming units (CFU).

pH in the sand layers

The addition of fertilizer caused decrease in the pH in the upper sand layer (0-5 cm) of the test area 3, but in two weeks increased to the background level. In general the pH of the test areas was very alkaline being about pH 9 at the highest, Figs. 16 and 17.

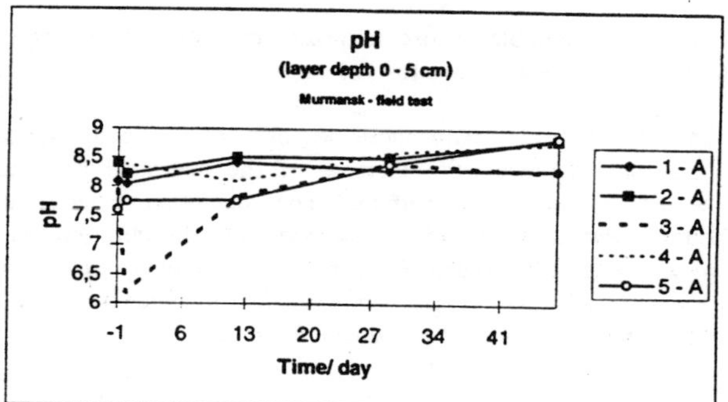

Fig. 16. The pH in the sand layers 0-5 cm (A).

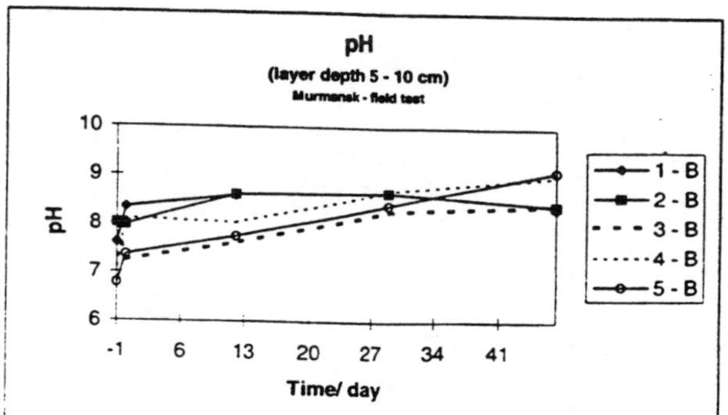

Fig. 17. The pH in the sand layers 5-10 cm (B).

CONCLUSIONS

The addition of oil on the shore line caused increase of the oil degrading microbes. The highest background level of hydrocarbons existed in the beginning in the test section A. The location of the section A higher onshore obviously was a fact, that tidal variation and waves were transported more hydrocarbons on this area compared to other test sections. Other sections stayed part of the time under the water, thus the flushing of oil and mixing into the sea water was evident. The accumulation of debris on the test section A support this explanation.
Oil will be sorb into pores of the sand. The degree of the oil penetration will depend significantly on the porosity, grain size and the water saturation degree of the soil. Oil will also lay under larger stones and stay a long period disturbing the normal existence of the biocenoses. The microphyte contamination of this type is especially dangerous for marine birds, catching contaminated amphipods, blue mussels and other source of the food. Here the direct contamination of oil on the birds or on other animals have not included to the subject of this study.

It may be supposed that bottom organisms can have an important role on oil degradation together with the natural physico-chemical weathering and bioremediation of oil. They will act like as a buffer for the coastal ecosystems, softening the sharp toxic impacts of oil.

It would be reasonable to use fertilizers preparations of long-term action on the adherent to the substrate basis.

The addition of crude oil to shoreline soils with lower background level resulted, however, the increase of the microbial population which can degrade hydrocarbons. The microbial determination method was very laborious and the colonies were difficult to detect on the oil-sea water agar plates. Furthermore the growing time is very long for three weeks at +5°C. Therefore other methods were used in the larger field study, started in Autumn 1996.

Literature

Andersson J.P.E. and Domsch K.H. 1978. A Physiological Method for the Quantitative Measurement of Microbial Biomass in Soils. - *Soil Biol. Biochem.* Vol.**10**. pp.215 to 221.

Arnebrant K. and Bååth E. 1991. Measurements of ATP in forest humus.- *Soil Biol.Biochem.* Vol. **23**, No.6, pp.501-506.

Bardgett R.D., Hobbs P.J and Frostegård Å. 1996. Changes in Soil Fungal:Bacterial Biomass Ratios Following Reductions in the Intensity of Management of an Upland Grassland. *-Biol. Fertil. Soils.* **22:**261-264.

Jenkinson D.S., Davidson S.A. and Powlson D.S. 1979. Adenosine triphosphate and microbial biomass in soil.- *Soil Biol.Biochem.* Vol. **11**, pp. 521-527.

Jenkinson D.S. and Ladd J.N. 1981. Microbial biomass in soil: measurement and turnover. In: Soil Biochemistry. Toim. Paul E.A. and Ladd J.N..Vol. 5 Chapter 10.

Lee K., Levy E.M. 1989. Biodegradation of petroleum in the marine environment and its enhancement.- *Aquatic toxicology and water Quality management.* pp.217-243.

Liukkonen, et al., 1994. Biosorbents and Absorbents in Oil Spill Combatting in Sea Environment. Test report of research phase I, Micro-Scale Tests. Finnish-Russian Offshore Technology Working Group. Report 17.89 p. + app.

Liukkonen, et al., 1994. Biosorbents and Absorbents in Oil Spill Combatting in Sea Environment. Test report of research phase II, Macro-Scale Tests. Finnish-Russian Offshore Technology Working Group. Report 18.54 p. + app.

Pritchard P.H. 1992. Oil spill bioremediation: experiences, lessons and results from Exxon Valdez oil spill in Alaska.- *Biodegradation* **3**. pp. 315-335.

Sparling G.P. 1985. The soil biomass. Teoksessa: Soil organic matter and biological activity. Toim. Vaughan D. and Majcom R.E. pp.223-262.

Vanhala P.T. and Ahtiainen J.H. 1994. Soil respiration, ATP content and Photobacterium toxicity tests as indicators of metal pollution in soil. Environmental Toxicology and Water Quality: - *An International Journal*. Vol. **9**. pp. 115-121.

West A.W. and Sparling G.P 1986. Modifications to the Substrate-Induced Respiration Method to Permit Measurement of Microbial Biomass in Soils of Different Water Contents.- *Journal of Microbiological Methods* **5** , pp. 177-189.

West A.W., Sparling G.P. and Grant W.D. 1986. Correlation between four methods to estimate total microbial biomass in stored, air dried and glucose-amended soils. Soil - *Biol. Biochem.* Vol. **18**. pp. 569-576.

Zelles L., Adrian P., Stepper K., Adrian M.V., Fischer K., Maier A. and Ziegler A. 1991. Microbial activity measured in soils stored under different temperature and humidity conditions. - *Soil Biol. Biochem.* Vol. **23**, No. 10pp. 955-962.

Proceedings of the Seventh (1997) International Offshore and Polar Engineering Conference
Honolulu, USA, May 25-30, 1997

ISBN 1-880653-28-1 (Set); ISBN 1-880653-30-3 (Vol. II)

The Improvement of Oil Bioremediation on the Shoreline

Jorma Rytkönen
VTT Manufacturing Technology, Espoo, Finland
Merja Itävaara
VTT Biotechnology and Food Research, Espoo, Finland
John E. Paulsen
Rogaland Research, Stavanger, Norway

ABSTRACT

The transportation of oil and fuel always poses a risk for the environment. Although means of combating marine oil spills have been studied for decades, once an oil spill reaches the shoreline there are very few effective methods available. As mechanical cleaning is laborious and hence expensive, other methods such as steam and emulsifying chemicals (e.g. surfactants) have been studied for shoreline treatment. Many of these methods have later been shown to have a detrimental and toxic effect on sensitive coastal wildlife. It is now-well known that sea water contains large numbers of oil-degrading microorganisms and, by improving the biodegradation conditions of these natural microbes, bioremediation can be speeded up.

During 1996 a research project on the improvement of oil bioremediation on the shoreline was started with funds provided by the European Commission and through local sources of from each country involved. The participating institutes in the EC project are VTT (Technical Research Centre of Finland), Swedish Environmental Research Institute, and Rogaland Research from Norway. Furthermore, a Finnish-Russian collaboration project is part of the overall consortium and is mainly concerned with field trials conducted in Murmansk by the Murmansk Marine Biological Institute in Russia.

This paper presents the basic test phases of the research programme. After the discussion on the laboratory tests, a description of the field tests has been carried out. The aim of the programme was to develop reliable testing methods for the evaluation of oil bioremediation. Thus different approach to study the oil bioremediation was followed.

INTRODUCTION

A lot of effort has been put into developing tools for cleaning the environment after oil spills of wrecked ships or failures in the offshore industry. Sometimes, however, the mechanical oil recovery methods are not sufficient, and more advanced technology is needed. For example, winter conditions with moving ice floes create a difficult environment for oil-combating authorities to implement countermeasures effectively. At the moment, small oil slicks can be dealt with during the winter time, but if a larger oil spill occurs in the winter time, these countermeasures will be difficult to implement and very expensive.

When the oil reaches the shoreline, usually the best recovery method has been to pick up oily spots with a spoon or a shovel. There are also remote areas and beaches where the effective recovery of oil is impossible. Oil can penetrate soil, and therefore the conventional methods have been the excavation of contaminated soil, burning of the soil or treatment with chemicals. This is the area where improvement through natural biodegradation can offer substantial advantages.

Biodegradation is the natural degradation process by microbes that consume hydrocarbons and produce carbon dioxide, water, biomass and weathered, biologically inert by-products. Bioremediation is the process that accelerates this natural biodegradation by microbes with the addition of nutrients. Nutrients, commonly nitrogen and phosphorous, are usually limited in availability in the areas of hydrocarbon supply. Nitrogen is an essential constituent of proteins and nucleic acids. Most bacteria require nitrogen in forms that are highly water-soluble. Ammonia, nitrate, urea and amino acids are among those which can be most readily assimilated by most micro-organisms.

Phosphorous is a vital component of nucleic acids and the lipids of cell membranes, and phosphorous plays central role in the biological energy transfer process (as phosphate) (Bragg, 1992). Normally in sea water, the biodegradation magnitude is in the order of 5 - -2 500 g/m^3/day. However, in- situ conditions of up to 300 times higher values have been observed in the nutrient-supplemented system. When considering the actual oil-combating situation and the sources of nutrients for speeding up the biodegradation process, a natural choice is to use fertilizers.

Several authors have found the use of nutrients, i.e. fertilizers, increases the efficacy of biodegradation. For example, in connection with the Exxon Valdez oil spill in 1989, the oil biodegradation rate increased to a level corresponding to 3 - 5 times the background level (Bragg, 1992). The biodegradation was also noted among all groups of hydrocarbons including the most polynuclear aromatics (PAH).

In 1992, VTT Biotechnology and Food Research and VTT Manufacturing Technology together with other research institutes of VTT started an international research programme on the use of biosorbents in oil-combating. Biosorbent can be explained as an additive (powder or liquid) consisting of three components: absorbent which captures the oil, natural fermented microbes and nutrients, usually a mixture of phosphorous and nitrogen to speed up the process.

Good results were achieved during the two-year-long study, which consisted of both laboratory simulations and full-scale experiments (Liukkonen, S. et al., 1995).

After the past biosorbent study a wider research programme was launched by VTT Manufacturing Technology in 1995. The goal of this study was to develop testing methods of oil bioremediation and to improve the natural oil degradation in the shore-line conditions. The work was started as a co-operation with the Murmansk Marine Biological Institute of Russia. Later, after the financial support from the EU, the programme was widened, and new research bodies participated in the programme: Swedish Environmental Research Institute (IVL) will use a special "Near-Infrared"-technique to detect the oil and develop a mathematical model for oil bioremediation. Norwegian Rogaland Research (RF) will conduct oil bioremediation simulation in special tailor-made biocylinders. VTT's task was to conduct parallel biocylinder studies, to arrange a macro-scale shore simulation test in VTT's 15 m-long simulation wave tank, and co-ordinate the programme. The Russian responsibility was to arrange the field tests.

This research programme is now underway. The micro-scale biocylinder simulations were conducted in Autumn 1996, and the shore-line simulations were completed in early Spring 1997. The field experiments in Murmansk were already conducted before the winter-time in 1996, but some sampling and analysis work is still to be carried out after the thaw in 1997. The project will be completed at the end of 1997. The supporting organisations involved in the study in Finland are the Ministry of the Environment, Finnish Environment Institute, Finnish-Russian Offshore Technology Working Group, Technological Development Centre of Finland (Tekes) and ST-Commission of the Ministry of Trade and Industry. The analysis work in the frames of this work has also been conducted by the Helsinki University of Technology and Kemira Ltd.

This paper describes the main test phases of the research programme.

DESCRIPTION OF THE CLOSED COLUMN TEST

The objective of the research programme was to develop reliable analysing methods and evaluation principles for the bioremediation of oil spills in the shore-line conditions. The goal of the work was to perform bioremediation of oil spills in three phases: the laboratory tests in both micro-and macro-scale and the field experiments for the verification of laboratory methods. The main task was to study the natural oil bioremediation in cold environment and to find out the suitability of different sampling and analysing methods. One important part is to find out possible means to improve natural bioremediation by using additives, i.e. fertilizers.

The goal of the micro-scale laboratory test was to ascertain the suitability of developed test cylinders for the reliable analyses of oil bioremediation in shore-line conditions. The main parameters studied in the tests were the effects of fertilizers and different amounts of oil on the bioremediation process. The basic idea was to determine the mineralisation of hydrocarbon compounds and to determine the carbon balance during the study. Test cylinders were closed during the test, i.e. closed column system (CCS) was used which ensures the reliable evaluation of the oil degradation due to the bioremediation process.

The test results conducted in the bioremediation tests in glass-walled test cylinders, each with a volume of 3 litres. Altogether 44 test cylinders were filled with real beach sand and brackish water. Oil and fertilizers or other additives were added to the top of the columns and incubated at 10 °C. Some of the cylinders were left without fertilizers, partly for comparison, and partly to study the natural oil biodegradation. Abiotic controls were included in the test matrix, too.

As biodegradation was determined as the evolution of carbon dioxide resulting from the degradation of the hydrocarbon chain, it was important to prevent the growth of algae and photo-oxidation by performing the studies in the dark. Biodegradation was determined as the total evolution of carbon dioxide measured from the gas and water phase. In addition, the fate of the oil was followed by extracting it from water and sand during the study. Microbiological activity was determined from sea water by measuring the ATP activity and SIR (substrate induced respiration). The test duration was 12 weeks.

The test vessel consisted of a glass cylinder partly filled with sand. The volume of the cylinder was 3 l, and the amount of sand 3 kg. Thus, the upper part of the cylinder had a certain space for air, Fig.1. The sand cylinder was connected by a tube to another flask, filled with brackish water initially taken from the Gulf of Finland. After oil and/or fertilizers were added on the sand of the sand cylinder, both vessels were sealed. The sand cylinder was assembled on the vertically moving table, while a sea-water vessel of 2 litres was placed on the stable support.

The vertically moving table was then run up and down simulating the tidal variation. Thus the water from the sea-water vessel periodically flushed the sand column. The maximum sea-water level inside the cylinder was 1 cm above the sand level. The whole tidal range was 20 cm.

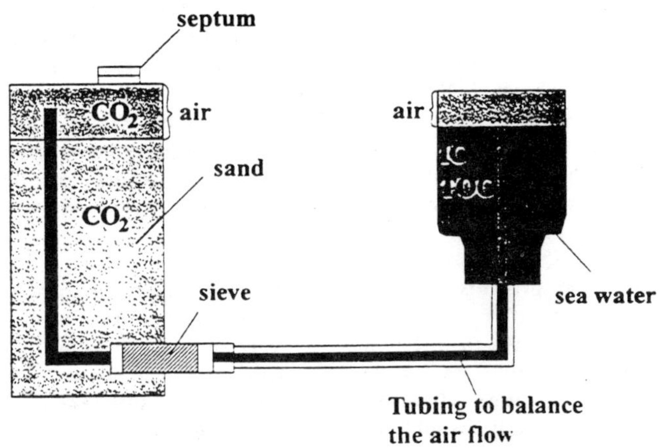

Fig. 1. Schematic presentation of the test cylinder and the source of water. Altogether 44 cylinders were prepared for the test.

Altogether 44 test cylinders were prepared for the test. The pre-altered crude oil, North Sea Brent, was poured on the sand cylinder during the simulated high tide. The water elevation was then lowered, and when oil reached the sand, fertilizers were added and the cylinders were closed. Some cylinders were also loaded with additional amounts of fertilizers or biosorbents. The basic aim in using fertilizers was to improve the natural biodegradation. Both fast and slow-releasing fertilizers were used. Fertilizers were added according to the following molar ratios: C : N = 10 : 1 and C : P = 30 : 1. The biosorbent used was originally designed by Russian Bios Group LTD for open-water oil-combating purposes. It contained a selection of microbes, nutrients and absorbing material, as described in more detail in (Liukkonen, et al., a and b, 1994).

The amount of oil was 25 g, which corresponds 2.8 kg/m^2 in full scale. One test series was performed with 5 g of oil (0.56 kg/m^2). Furthermore, positive control (biodegradable mineral oil) and abiotic control cylinders were performed. Table 1 shows the test matrix of the cylinder test.

After all the cylinders were loaded and closed, the vertically driven table was started. The movement of the table was adjusted to simulate the tidal cycle, i.e. two cycles per 24 hours. The cylinder arrangements, shown in Fig.2, were then kept in a room of constant temperature +10 ^oC for 10 weeks. During this period a lot of samples were taken both from the sand cylinder and the water vessel. Some of the cylinders were withdrawn and opened during the test in order to analyse the microbial development in the sand and the oil content.

Table 1. Test matrix for the biocylinder test.

Series No.	Parallel cylinders	Note
1.) Sea water	2	background
2.) Sea water + biodegradable oil + slow-releasing water-soluble fertilizer	2	positive control
3.) Sea water + Crude Oil, 25 g	6	
4.) Sea water + Crude Oil + Natrium Atsid	2	abiotic control
5.) Sea water + Crude Oil, 25 g + slow-releasing water-soluble fertilizer	6	
6.) Sea water + Crude Oil, 5 g + slow- releasing water-soluble fertilizer	6	
7.) Sea water − Crude Oil, 25 g − fast- releasing water-soluble fertilizer	6	
8.) Sea water − Crude Oil, 25 g + biosorbent	6	
9.) Sea water + biosorbent	2	background for the series 8
10.) Sea Water + fast-releasing water-soluble fertilizer	2	background for the series 7
11.) Sea Water + slow-releasing water-soluble fertilizer	2	background for the series 5
12.) Sea Water + Crude Oil, 25 g + slow-releasing water-soluble fertilizer	2	cylinders were open

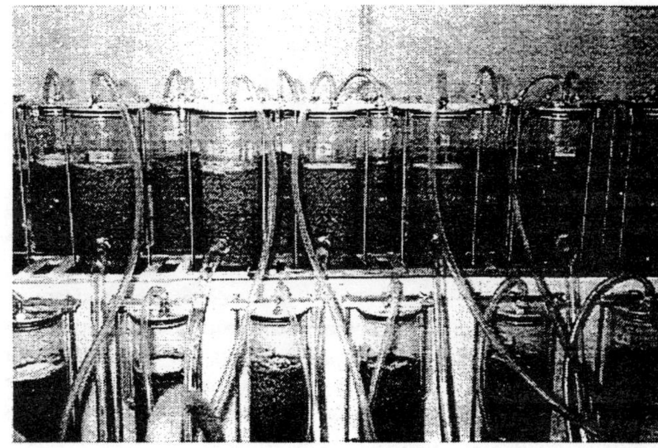

Fig. 2. The test arrangements.

Analysis

The sampling plan consisted of both sand and water samples. Furthermore, CO_2 was measured weekly from both the sand cylinder and sea-water vessel. After sample taking the test columns were aerated by fresh air.

Part of the sea water of the tidal vessels was changed weekly. The total organic carbon (TOC), inorganic carbon (IC), pH as well as nutrients and dissolved oxygen were measured from the sea water. The toxicity of the water or sand samples was determined on the basis of inhibition of luminescence of *Photobacter leiognathi*.

At the end of the test (or when opening some of the cylinders) the residual oil was extracted with carbon tetrachloride, and the absorbance of C-H bonds was measured using the FTIR method. The same method was used for the sea-water analyses.

After the test completion all the remaining cylinders were opened. The residual oil of sand was extracted and measured.

The amount of microbial biomass in the soil samples was studied by three different microbiological methods: SIR (substrate-induced respiration), ATP (adenosine triphosphate) content and the total amount of bacteria (CFU).

For pH measurements 20 g sand and 80 ml distilled H_2O were mixed and pH of the solution measured.

The changes in nutrient and oil content are not discussed in this paper. However, the determination of nutrients consisted of both nitrogen (N-tot, NH_4-N, NO_3-N) and phosphorous (P-tot, PO_4-P).

DESCRIPTION OF THE OPEN COLUMN TEST

The parallel sand cylinder test as described in the previous chapter was designed by Rogaland Research (RF) in Norway. The test arrangements of RF, however, differed slightly from the closed column system of VTT. The idea was to run both tests trying to establish system parameters that are identical in VTT's CCS test and RF's open colum system (OCS). Experimental parameters which had identical character in both tests were:

- sand sediment, grain size distribution profile,
- weight and volume of sand,
- same ratio of sediment-surface to added oil concentration,
- tidal frequency,
- oil source and the use of prealtered crude oil,
- additives and
- sampling frequency and monitoring methods.

The test apparatus designed by RF consisted of several identical plastic cylinders designed to contain sediments and to allow water circulation (tidal simulation). For the simulation of tidal effect fresh sea water was used taken up from the fjord near the laboratory. Fig. Dd Gives a schematic overview of the experimental apparatus. The cylinders were placed on the rigid table. while as fresh water cylinders were situated on the automatically operated telescope table. Sand cylinders and water distribution network were connected by tubes. Telescope table and the valve of the water distribution system were operating automatically. thus ensuring the correct tidal simulation.

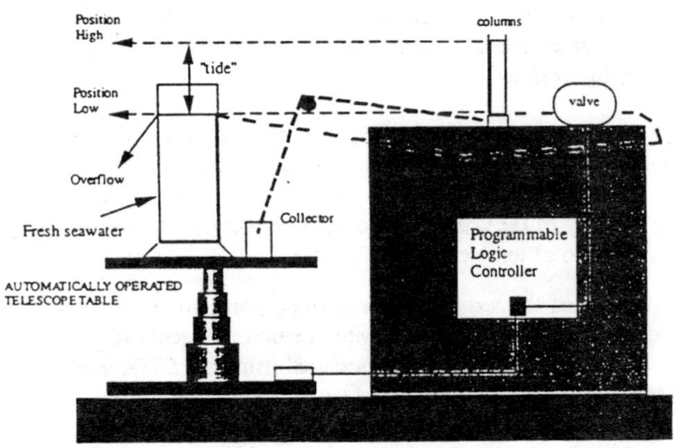

Fig. 3. Schematic representation of the OCS-system of Rogaland Research (Paulsen. J. E.. 1996).

Special attention was paid to the volumetric mass balance of the oil and the distinct detection of oil biodegradation. Conservative biomarkers were added to the prealtered (topped) crude oil before introduced to the columns. Biodegradation can then be evaluated by using GC-MS techniques and observing changes in the ratios of selected n-alkanes and added biomarkers.

The OCS-test arrangement enables the sampling through the whole test duration. Sediment samples were taken at different intervals both taken near the surface and deeper down to the depth of -10 cm. All the samples were mixed samples. thus representing mean values of the oil concentrations and microbial activity. By mixing the sediment the distortion effects of the uneven distribution of oil or possible effect of water channeling can be avoided.

In order to follow the oil bioremediation certain amount of biomarkers were added into the test oil.These can be defined as oil compounds which are reminiscenses of the organic material that once was turned into fossil fuel during diagenesis. Biomarkers are detected using GC-FID- GC-MS techniques. There are a lot of suitable biomarkers to be used such as androstane. pristane and phytane.

Oil extraction and clean up of sediment samples followed standard procedures described more detailed in (Paulsen, J.E., 1996). Samples were also taken for NIR-analyses taken care of Swedish IVL.

Microbial activity was determined by measurement of bacterial degradation of carbon sources, by enzymatic hydrolysis of 3',6'-diacetyl-fluorescein (FDA), modified by (Schnurer & Rosswall, 1982). In addition the numbers of oil-degrading micro-organisms were monitored by using a simplified most probable number technique, described as (Brown & Braddock, 1990).

Total soluble nitrogen and phosporous were also analysed during the tests.

MACRO-SCALE OIL BIOREMEDIATION TEST

After the cylinder test phase a new oil bioremediation test was carried out in the refrigerated test basin at VTT's Research Hall No. 1, Espoo, Finland. The test tank has the overall dimensions of 15 m x 3.2 m x 1.5 m. The air temperature in the tank hall can be adjusted in the range of +20...-25 °C.

For the oil bioremediation test, six individual and water tight compartments, called here the test basins. were built in the test tank. The wave maker of the previous biosorbent study was modified and assembled over the middle section of the test tank. The wave maker is a plunger-type wave maker having the vertically moved plunging unit both sides of the wall constructed cross the middle section of the basin.

An artificial shore-line was constructed in both ends of the test tank. shown in Fig. 4. The beach material was selected to correpond the sand grain size distribution curve of the Norwegian sand used in both cylinder tests. Thus the median grain size of the test sand was about 0.4 mm. The design height of the shore-line was 0.15 m above the water elevation. The design water depth in the tank was 0.75 m during the test execution.

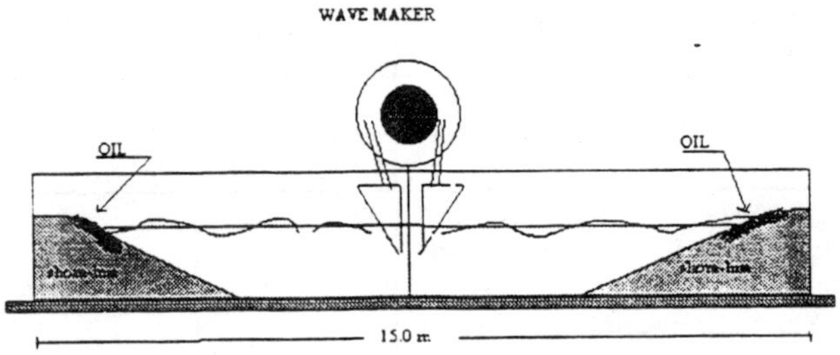

Fig. 4. The wave tank used in the oil bioremediation test.

Prealtered crude oil was spread on the artificial sand beach. thus five of the six compartments were treated with crude oil. The sixth compartment was left as a fresh water basin for verification. Some of the oily sand beaches were treated with additional fertilizers and/or biosorbents. The fertilizers selected were the same as used in the CCS-tests. see Table 2.

Table 2. Test matrix of the macro-scale test.

Test area	amount of oil (l/m^2)	Fertilizers /additives
1	-	fresh water basin
2	3.0	only oil
3	3.0	medium releasing fertilizer; N:0.74kg/m^2; apatite 1.30 kg/ m^2
4	3.0	fast releasing fertilizer; N:0.70 kg/ m^2 superphosphate: 2.56 kg/ m^2
5	3.0	Biosorbent I, 1.5 kg
6	3.0	Biosorbent II, 0.45 kg

The additives were added on the oily sand 30 minutes after oil was spread on the sand. The upper layer of sand of the test basin 5 was mechanically mixed with Biosorbent I according to the instructions of the manufacturer. The oil of the test basin 6 was also treated in the different way: oil was poured on the water surface, treated and let then drift to shore-line by the waves.

After all the test basins were loaded the wave maker was switched on, and the test was followed 10 weeks. The sampling and analysing plan was slightly modified from the procedure followed during the CCS-test.

FIELD TESTS

Both the preliminary field experiment and wider field tests in Autumn 1996 were carried out on the sandy and silty littoral zone of the Belokamenka Bay in Murmansk, Russia. The preliminary test phase was carried out between 07- 28th of May 1996 in the conditions of the early Arctic spring, when the air temperature and wind regime were unstable. This test phase was then followed by a wider test phase which will be completed after snow melting, Spring 1997.

The preliminary test phase consisted of three test areas, which were established in the sandy beach of the Belokamenkay Bay, see Fig. 5.

Fig. 5. Preparing the test section in the shore-line of the Belokamenkaya Bay.

The size of each test section was 2 x 2 m^2. Test section A was located in the upper part of the littoral zone near the highest mark of the tidal high water elevation. Two others, test sections B and C, were 60 meters offshore, in the downer part of the littoral zone. The distance between the areas B and C was 20 meters.

The test was started on the 7th of May 1996. Eight liters of the crude oil (Varandoy deposit) was equally spread onto each test section after the first sand samples were taken.

After the preliminary test results were analysed a more wider field test was established. Five new test areas were arranged in the Belokamenkaya Bay, near the previous preliminary test site. Four test sites (1,2,3 and 4) had a size of 2 x 4m^2, while as the test section 5 was smaller (size 1 x 3 m^2). The additives of the test sections were:

1. Background (without oil)
2. Russian oil (without additives)
3. Russian oil (fertilizer 1)
4. Russian oil (fertilizer 2)
5. Russian oil (biosorbent).

Approximately 3 liters/ m^2 of Russian crude oil was spread on each test section (2,3,4 and 5). This equaled the amount of crude oil used in the laboratory respiration experiment (microscale test). The fertilizers and biosorbent were added according to the experience gained in the laboratory studies, too. The fertilizers were manufactured by Kemira LtD.

More detailed description of this test phase is given in (Rytkönen, J.1997)

PRELIMINARY RESULTS

Both the end results and the measurements during the cylinder test period showed that the highest microbial activity exists on the cylinders loaded by fertilizers and biosorbent.

The highest activity of microbes was observed in the top layer of the sand column.

The SIR results correspond well the ATP results. The highest activity was observed in the biosorbent test cylinder (No. 8). In the biosorbent cylinder the differences between top and bottom layer were significant.

The enumeration of microbes, i.e. the density of microbes was determined from the samples by cultivating the bacteria via a suitable dilution series to a fixed agar plate. An example of the results of the analyses are shown in Fig.6.

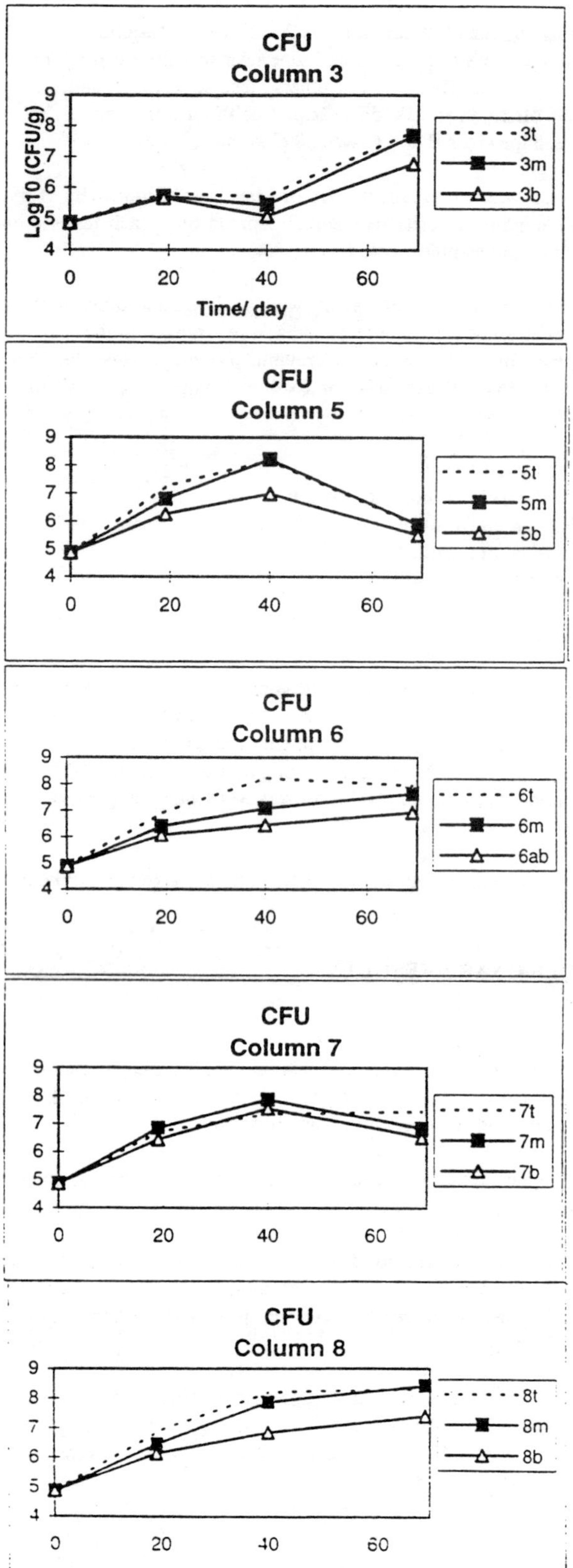

Fig. 6. The level of microbes at test completion.

The biodegradation percentage of the vertical column shows the estimated loss of oil by biodegradation based on CO_2 determination. The biodegradation percentage is calculated by comparing the evolved CO_2 to theoretically calculated maximal evolved CO_2 based on the carbon content of oil. Evolved CO2 in both test vessels were summed. The evolved IC in column water was added to this sum, and the following presentation of the biodegradation of oil was performed, shown in Fig.7.

The best oil degradation rate was observed in cylinder 6, where the amount of oil was 5 times smaller compared to other cylinders. The oil degradation was near 10 %, which was smaller than excepted. Other cylinders loaded with fertilizers and biosorbent showed rather constant growth, with the final value near 2.0.

The great difference between cylinder 6 and other cylinders can be explained only by the different amount of oil. Cylinder 6 contained only 5 g oil while other cylinders had an initial amount of 25 g. It is obvious that too large an amount of oil has altered the biodegradation process in the closed media. More tests should be made with even smaller oil content to find out the effect on the process. The knowledge of the optimum amount of oil in this kind of closed cylinder test will lead to better evaluation of the real oil degradation without any alteration of limiting factors.

In field conditions, however, it is more vital to know the correct amount of fertilizers needed and the releasing time in order to improve the bioremediation process effectively. For large amounts of oil, i.e. for a thick layer of oil, the oil bioremediation technique is ineffective without mechanical mixing: the oil degradation process takes place only on the surface of the oil, in thin biofilm, and therefore a large amount of oil will take a long time to degrade.

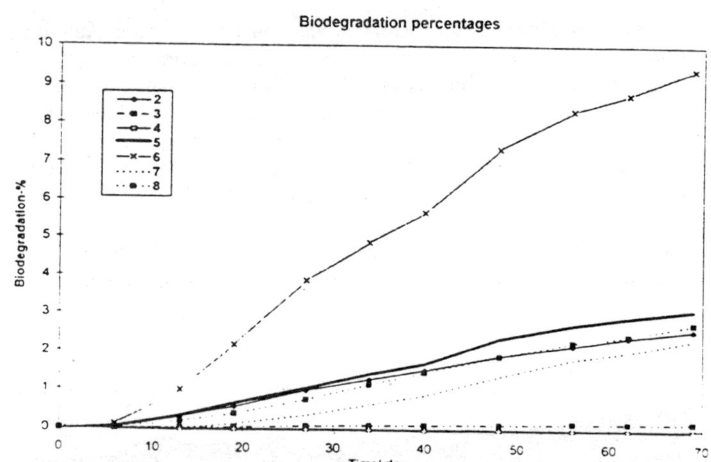

Fig. 7. Biodegradation percentage.

CONCLUSIONS

The observations gained from all the test phases showed, that fertilizers and biosorbents increased the biodegradation rate of crude oil.

The observed rate of oil bioremediation in CCS tests was in the order of 2...10 %. This is very small compared to several test results mentioned in the literature. During the biosorbent research programme, for example, the biodegradation percentages 20...40 % were observed (Liukkonen, et al, a, b, 1994) and an even higher figures, up to 50...75 % can be found in the literature. The results, however, are often carried out in full-scale conditions or in open-type test conditions where the loss of hydrocarbons due to other phenomena like bioremediation is evident. There are measurements conducted in open shore-line conditions, where the estimation of bioremediation is based on the hydrocarbon balance compared to the initial and end conditions.

The flushing effect of waves, oil dissolution and dispersion into the water, oil evaporation with photo-chemical oxidisation are processes that improve natural oil degradation, though cannot be taken into account as parameters of oil bioremediation. Moreover, microbes cut the more complex hydrocarbon chains that more easily evaporate or dissolute from the test area. As a conclusion it seems that primary oil biodegradation is not as high as is often mentioned in the literature. The 'secondary biodegradation rate', also including the accelerated physical or chemical degradation by the microbial activity can be very significant.

The amount of oil affected the observed biodegradation rate in closed test cylinders. Too high an amount of oil can decrease the process in a closed test system. Thus, for further testing and analyses, the amount of oil should be smaller than that used in the test system described in this paper.

An other interesting point observed was the decrease in pH value due to the use of fertilizers. In the closed test system this can alter the microbial activity unless some balancing activities are used. In natural conditions, however, the decrease in pH can decrease the oil degradation only in very limited conditions.

Laboratory tests can be used to determine the best additive and treatment. Field tests, however, are needed to verify the results and give a realistic impression of the improvement gained by the use of additives.

Next phase of this research programme will be the verification of each test phases and analyses of the test methods used. Decision will be made on the suitable and accurate analysing methods to be used when studying oil bioremediation in the shore-line conditions. Some discussions will also be carried out dealing with the use of biosorbent technology and the improvement of natural oil bioremediation as a part of total oil combating procedure.

LITERATURE

Andersson J.P.E. and Domsch K.H., 1978. A Physiological Method for the Quantitative Measurement of Microbial Biomass in Soils. *-Soil Biol. Biochem.* Vol.**10**. pp.215 to 221.

Arnebrant K. and Bååth E., 1991. Measurements of ATP in forest humus.- *Soil Biol.Biochem.* Vol. **23**. No.6. pp.501-506.

Bardgett R.D., Hobbs P.J and Frostegård Å., 1996. Changes in Soil Fungal:Bacterial Biomass Ratios Following Reductions in the Intensity of Management of an Upland Grassland. *-Biol. Fertil. Soils*. **22**:261-264.

Brown, E.J., & Braddock, J.F., Sheen screen, a miniaturized most-probable-number method for enumeration of oil degrading micro-organisms. Applied and environmental microbiology. Dec 1990, p. 3895 - 3896.

Bragg, et al., 1992. Bioremediation for Shoreline Cleanup Following the 1989 Alaskan Oil Spill. Exxon Research & Engineering Company. 94 p.

Jenkinson D.S., Davidson S.A. and Powlson D.S., 1979. Adenosine triphosphate and microbial biomass in soil.- *Soil Biol.Biochem.* Vol. **11**, pp. 521-527.

Jenkinson D.S. and Ladd J.N., 1981. Microbial biomass in soil: measurement and turnover. Soil Biochemistry. Edited by. Paul E.A. and Ladd J.N..Vol. 5 Chapter 10.

Lee K., Levy E.M., 1989. Biodegradation of petroleum in the marine environment and its enhancement.- *Aquatic toxicology and water Quality management.* pp.217-243.

Liukkonen, et al., a., 1994. Micro-scale tests. Test report of research phase I. Biosorbents and absorbents in oil spill combatting in sea environment. Finnish-Russian Offshore Technology Working Group, Report 17.

Liukkonen, et al., b., 1994. Micro-scale tests. Test report of research phase II. Biosorbents and absorbents in oil spill combating in sea environment. Finnish-Russian Offshore Technology Working Group, Report 18. 54 p. + app.

Liukkonen, et al., 1995., Bioremediation - An effective tool for marine oil spill clean-up. International Conference on Technologies for Marine Environment Preservation, MARIENV'95. Tokyo, 24.-29.9.1995. P. 360 - 367.

Pritchard P.H., 1992. Oil spill bioremediation: experiences, lessons and results from Exxon Valdez oil spill in Alaska.- *Biodegradation* **3**. pp. 315-335.

Scnurer, J. & Rosswall, T., Fluorescein diacetate hyrolysis as a measure of total microbial activity in soil and litter. Applied and environmental microbiology. Vol. 43. No. 6. June 1982. P. 1256 - 1261.

Sparling G.P., 1985. The soil biomass. Teoksessa: Soil organic matter and biological activity. Edited by Vaughan D. and Majcom R.E. pp.223-262.

Vanhala P.T. and Ahtiainen J.H., 1994. Soil respiration, ATP content and Photobacterium toxicity tests as indicators of metal pollution in soil. Environmental Toxicology and Water Quality: *- An International Journal*. Vol. 9. pp. 115-121.

West A.W. and Sparling G.P., 1986. Modifications to the Substrate-Induced Respiration Method to Permit Measurement of Microbial Biomass in Soils of Different Water Contents.- *Journal of Microbiological Methods* **5** , pp. 177-189.

West A.W., Sparling G.P. and Grant W.D., 1986. Correlation between four methods to estimate total microbial biomass in stored, air dried and glucose-amended soils. Soil - *Biol. Biochem.* Vol. **18**. pp. 569-576.

Proceedings of the Seventh (1997) International Offshore and Polar Engineering Conference
Honolulu, USA, May 25-30, 1997

ISBN 1-880653-28-1 (Set); ISBN 1-880653-30-3 (Vol. II)

Pollution Abatement at McMurdo Station, Antarctica

Erick Chiang, Shih-Cheng Chang and Arthur J. Brown
Office of Polar Programs, National Science Foundation
Arlington, VA, USA

ABSTRACT

The soil and marine sediment of McMurdo Station, the logistical hub of the U.S. Antarctic Program, are contaminated by both petroleum hydrocarbons and the residual products of waste disposal. These contaminants were released to the environment during earlier years of expeditionary operations. Since 1987, the National Science Foundation has initiated a number of measures in pollution abatement to correct past abuses and to protect the Antarctic environment. These measures include promulgation of waste regulations, development and implementation of a waste management program, cleanup of the station waste dump sites, and development of spill prevention and spill response plans.

KEY WORDS: McMurdo, Antarctica, contamination, pollution abatement, waste management

Introduction

McMurdo Station established in 1955 is the largest U.S. research station in Antarctica. This paper briefly reviews the U.S. Antarctic Program, its expeditionary operations and waste disposal practices at McMurdo and the resulting contamination at the station. The paper then describes the measures and investment in pollution abatement which the National Science Foundation has initiated since 1987. The national operators of other Antarctic Programs which have experienced similar conditions of contamination resulting from past practice have also implemented active pollution abatement programs. However, this paper is limited to the U.S. Antarctic Program pollution abatement at McMurdo Station.

Antarctica

Antarctica is the coldest and the driest continent on earth, and was largely unknown until four decades ago. Approximately 98 percent of the continent is buried under ice, averaging 2000 m thick, which represents 90 percent of the world's ice. The continent encompasses 14 million km^2, as large as the United States and Mexico combined. In the winter, the continent is surrounded by a belt of sea ice, 500 to 1500 km wide, making the continent inaccessible to ships, while erratic storm conditions combined with cold temperatures makes air access impractical.

The extreme cold and dry climate limit natural biodegradation. Pollutants and wastes released to the environment tend to persist. Therefore control of the release of pollutants and removal of waste are essential to the protection of the environment. This premise is embodied in the Antarctic Treaty. By international agreement, the Antarctic Treaty governs the conduct of human affairs south of 60 degree south latitude. The Antarctic Treaty assures free access to the region for scientific purposes, conserves the resources, and protects the environment. The Protocol on Environmental Protection to the Antarctic Treaty, Annex III, provides the basic framework for waste management and waste disposal in Antarctica.

U.S. Antarctic Program

The United States has strong interests in Antarctica and is a signatory to the Antarctic Treaty. The U. S Antarctic Program (USAP) is a national program administered by the National Science Foundation (NSF) to implement U.S. policy through the support of scientific research on the continent and in its surrounding oceans. The Program currently operates three year-around research stations in Antarctica: McMurdo, Amundsen-Scott South Pole, and Palmer. Additional seasonal field camps are established during the austral summer. The Program also charters two research vessels--Polar Duke and Nathaniel B. Palmer--to support research in Antarctic waters.

McMurdo Station

McMurdo Station (77 ° 51' S, 166 ° 40' E) (Figure 1) is located on the southern tip of Hut Point Peninsula on Ross Island, Antarctica, approximately 3860 km south of Christchurch, New Zealand. Ross Island is located in the western Ross Sea, separated from the Antarctic mainland by McMurdo Sound. The annual mean temperature at

McMurdo Station is -18 ° C: monthly mean temperatures range from -3 ° C in January to -28 ° C in August. The station is constructed on barren volcanic rocks consisting of pyroclastic deposits. The soil derived from weathering of volcanic rocks contains virtually no organic matter. The ground is frozen solid, permafrost underlying the station at a depth of 15 to 45 cm.

McMurdo has the world's most southern port accessible by ship. Winter Quarters Bay (WQB) is a small harbor of less than 4 hectares in area. The harbor has a floating ice pier, a solid 6 m thick ice platform reinforced with layers of steel cables. For most of the year, McMurdo Sound and the harbor are covered with sea ice about 2.5 m thick. Late in the austral summer, a U.S. Coast Guard (USCG) icebreaker opens a navigation channel to the harbor, which allows a fuel tanker and a supply vessel to replenish the station with a yearly supply of fuel and provisions. After unloading the cargo, the vessel is backloaded with science cargo and the annual accumulation of USAP wastes which are to be returned to the United States.

Starting as an expeditionary camp in 1955 in preparation for the International Geophysical Year, McMurdo expanded over the past forty years and has evolved into a modern research station currently having 54,000 m^2 of living and working space housed in more than 70 structures. McMurdo (Figure 2) has numerous storage yards, graded roads, a helicopter pad, a power plant, water plant, maintenance facilities, above ground water and sewer systems, above-ground bulk fuel storage tanks, and an extensive fleet of general purpose vehicles and heavy equipment.

McMurdo operates a fuel terminal which has a bulk fuel storage capacity of 34 million liters. Currently most of the fuel is AN-8 aviation turbine fuel (JP-8 with an icing additive inhibitor to depress the freezing point to -58 ° C) which is the primary fuel for power generation, space heating, and aircraft operation. Annual deliveries of AN-8 average 20 to 25 million liters. Fuel is transported to outlying aerodrome complexes by above-ground pipelines and hoses. Within the station proper, intermediate storage tanks are directly fueled through above-ground steel pipelines from the bulk storage tanks.

Most of the fuel tanks and delivery pipeline were built before 1970 and are antiquated compared to modern construction standards. Until recently, the steel pipes were jointed by flanges fitted with neoprene gaskets; the flanged joints were prone to leaks as the piping expanded and contracted with changing temperatures. Most of the fuel tanks are without secondary containment, and are not equipped with overfill protection devices such as high level alarms and indicators. This fuel system has traditionally been the major cause of fuel spills.

Early Practices in Waste Disposal

In the years prior to 1980, McMurdo disposed of its solid waste at a dump site located on the eastern shore of Winter Quarters Bay, approximately 2 hectares in area, which slopes to the bay from an elevation of 12 to 15 m. During winter months when the bay was frozen, the dump often extended out onto the sea ice. Combustible solid wastes were stock piled at the southern tip of the dump site. Periodically the piles of waste were doused with contaminated fuel and set on fire. Open burning lasted from hours to days. Afterward, the ash and residues were spread throughout the dump site.

Non-combustible wastes (scrap metals, construction debris, waste drums, unserviceable equipment) were frequently staged on the adjacent sea ice for eventual disposal at sea. During late summer months, the sea ice would break up and drift out to sea, dumping waste material into the deep waters of McMurdo Sound. However, the waste did not always make it all the way to the sea, and some would sink in the immediate vicinity of Winter Quarters Bay.

During a consecutive three-year period in the late 1970s, the sea ice failed to break away from the shoreline. As a result, wastes continued to build up and eventually encroached onto the shore. When the ice finally did break away in the fourth season, a significant amount of waste remained as an ice encased, consolidated mass on the shore. This mass was a source for continuous pollutant releases to the contiguous shoreline and adjacent marine environment. In 1980 the waste disposal activity was relocated from WQB to Fortress Rocks, a natural hollow on high ground on the northeast side of McMurdo Station. The dump site at Fortress Rocks was less than 2 hectares in size. Open burning of combustible waste continued to be practiced at Fortress Rocks until the end of 1990-91 field season when the dump was shut down. This marked the beginning of informed waste management practices. Today all USAP solid wastes are removed from McMurdo and returned to the United States for reuse, recycling or disposal.

Contamination

Petroleum hydrocarbon contamination is wide spread in the station soil and in the sediment of Winter Quarters Bay. The contamination mostly relates to accidental spills of fuel, hydraulic fluids and anti-freeze coolants (glycol). Some of the spills were caused by mechanical failures such as leaking pipe joints and fittings, rupture of fuel pipes or hoses, and the leaking from vehicles and heavy equipment of fluids or antifreeze coolants. Others were caused by human errors such as overfilling service tanks, opening wrong valves, overturning or puncturing fuel containers. Discharges in the early years of bilge water from ships into the bay are also suspect. The cold climate, high turn over of seasonal workers, and the expeditionary nature of early operations all played roles in generating human errors which caused spills.

Recently, preliminary investigations were undertaken at McMurdo Station to determine the extent and degree of fuel contamination. Surface soil samples were taken on 100-meter grids throughout most of the station, and on 25-meter grids for areas suspected of heavy contamination.

These investigations confirmed that fuel contamination was widespread. Approximately 85% of the samples tested showed a measurable level of total recoverable petroleum hydrocarbon (TPH). Over 45% of the samples measured TPH in excess of 100 parts per million (ppm); one sample which measured 24,000 ppm was collected near one fuel tank. Table 1 shows the distribution of TPH in the soil samples.

As suspected, areas in the vicinity of fuel systems and areas of fueling activity are the most contaminated. The station helicopter pad is one example. Most samples collected at the pad are contaminated with TPH at levels higher than 100 ppm. Nonetheless, level of TPH measured in the minuscule snowmelt runoff from the pad is very low, averaging 12 ppm, which suggests the contaminants are contained naturally by the underlying permafrost.

Table 1 Distribution of Hydrocarbon in Soil Samples

TPH Level (ppm)	Number of Samples
less than 10	35
10 to 100	86
100 to 200	27
200 to 300	15
300 to 400	11
400 to 500	6
500 to 1,000	6
1,000 to 5,000	20

A number of studies (Lenihan 1990, 1992; Risebrough 1990) have indicated that hydrocarbon contaminated marine sediments were largely confined to Winter Quarters Bay. Among the contaminants measured in the sediments in WQB were petroleum hydrocarbons, polychlorinated biphenyl's (PCBs) and terphenyl's (PCTs). Level of purgeable hydrocarbons (PHC) in sediments was measured as high as 4,500 ppm. Total polycyclic aromatic hydrocarbons (PAH) in the sediments varied from below the method detection limit to as high as 13 ppm. Level of PCBs and PCTs were measured in the range of 0.1 to 1.4 ppm and 0.03 to 1.2 ppm respectively. It was estimated that WQB contained 4 kg of PCBs and 1 to 2 kg of PCTs in its sediment. Certain trace metals were also elevated in the bay compared to sediments outside the bay.

WQB is enclosed at the bay entrance by a large submarine sill which greatly constrains the flow in and out of the bay. The inner part of the bay is 34-37 m deep and the submarine sill rises to a water depth of 18-20 m. Consequently, the contaminated sediments are naturally contained within the bay. Contaminants are found to drop precipitously away from the bay. The sharp pollutant gradient represents one of the steepest and claims to be the most well-defined sediment contamination in the world. Just a few hundred meters away, levels of PHCs in sediments were measured at only 3.5 ppm, about one thousandth of that in the bay sediment (Lenihan 1990, 1992).

Scattered around the floor of the bay and in the vicinity in McMurdo Sound are numerous debris; metal drums, vehicle frames, and discarded equipment. This debris was deposited as residuals from the former waste disposal activities.

Pollution Abatement

Since 1987, NSF has initiated a number of measures to abate pollution at the U.S. stations in Antarctica. These measures have included: termination of open burning and land disposal, surface cleanup of former dumps sites at WQB and Fortress Rock, promulgation of waste regulations for all U.S. citizens visiting or working in Antarctica, establishment of a waste permit system for the use and discharge of designated pollutants in Antarctica, development and implementation of a waste management program and a spill contingency and response plan, removal of all wastes except sewage from Antarctica, training of program participants, upgrade of fuel systems to prevailing standards, and investigation of soil remediation technologies applicable to the polar environment. A brief description of these measures follows.

Termination of Open Burning and Land Disposal and Cleanup of Dump Sites

Commencing in 1987 and extending over the next two field seasons, the residual surface waste and debris left at the former WQB dump site was removed and relocated to Fortress Rocks. The site was then covered with fill material and graded.

At the end of 1990-91 field season, the waste disposal operation-- open burning of combustible waste and land disposal of ash residue -- at Fortress Rocks was terminated. This decision in part resulted from concern for public health over the presence of friable asbestos found in the demolition waste.

During the 1991-92 field season, USAP, through a special Safety, Environment and Health initiative, launched a multi-million dollar operation to clean up the Fortress Rocks dump site and to remove all the waste backlog and accumulated surface debris (Reference 4). Specialized equipment for processing and packaging the waste and debris were purchased and flown in to McMurdo via U.S. Air Force C-5 Galaxies. Remediation specialists were subcontracted to undertake the actual hands-on cleanup. A temporary decontamination structure was erected on the site and air monitoring was set up in advance of remediation of the asbestos waste. Scrap metals were either baled or appropriately sized with cutting torches. Wood waste was chipped in a tub grinder to reduce volume. Other wastes were sorted and segregated into various waste streams -- cardboard, rubber, tires, glass --to facilitate packaging and shipment. The contents of 800 abandoned drums were sampled and tested to allow chemical characterization. Based on characterization, like substances were consolidated prior to packaging and manifesting for return to the U.S. All documentation was prepared in accordance with pertinent regulations of the Department of Transportation (DOT), the Environmental Protection Agency (EPA), the International Maritime Organization (IMO), and several state agencies.

The cleanup operation was a success and resulted in the recovery of 1800 bales of scrap metals, 1000 m^3 of large bulk metal, 57 m^3 of wood, 300 m^3 of cardboard, 40 m^3 of construction debris, 40 m^3 of asbestos waste, and 726 208-liter (55-gallon) drums of liquid waste and chemicals (Reference 4). The total amount of waste generated exceeded the capacity of the annual supply vessel and exhausted the available inventory of on-site sea worthy shipping containers. The backlog of waste took a number of seasons to be removed from the station but was all returned to the United States for disposition.

After all surface debris had been removed from Fortress Rocks, the entire area was covered with clean fill material. Backfill was layered over depressions to prevent thawing snow and ice from collecting in low lying areas. Surface samples were collected and analyzed in order to chemically characterize the entire dump site. The Fortress Rocks area was restored and made suitable for other uses provided that no excavation is involved.

Environmental Protocol and Waste Regulation

On June 29, 1993, NSF issued its final rule, 45 CFR Part 671, that amended its regulations to add provisions governing waste management and waste disposal practices in Antarctica. These changes implement the provisions of the Antarctic Conservation Act (ACA) of 1978 that require the Director of NSF to promulgate regulations designating, and governing the release of, pollutants in Antarctica. The waste regulation is consistent with the provisions for waste management and disposal, endorsed in October 1991 by the Antarctic Treaty nations, in the Protocol on Environmental Protection to the Antarctic Treaty. In 1996, the U.S. Congress adopted the legislation necessary to implement the provisions of the Protocol and on October 2, 1996, President Clinton signed into law the Antarctic Science, Tourism, and Conservation Act of 1996.

The waste regulation bans the use of certain substances in Antarctica. These include any PCBs, non-sterile soil, polystyrene beads, plastic chips or similar loose polystyrene packing material, pesticides (other than those required for scientific, medical or hygiene purposes). The regulation requires a permit for the use or release of designated pollutants, and for release of any waste in Antarctica. It specifies that certain wastes be removed from Antarctica, except when the removal of such waste would cause greater adverse environmental impacts.

The regulation sets forth requirements for waste storage and waste disposal. Maximum time limits are set for storage of hazardous wastes at U.S. stations: 15 months for McMurdo and South Pole Station, 28 months for Palmer Station, and 12 months for field camps. Hazardous waste storage facilities are to be inspected weekly.

The regulation requires phase out of open burning by March 1, 1994 and prohibits disposal of any waste onto ice-free areas, or into fresh water system. Sea disposal of sewage is allowed provided sewage is macerated before discharge at a location where dispersal is rapid.

The regulation establishes the requirements for waste permit application, and requires the tracking of the USAP wastes in five categories and the filing of an annual report on the USAP waste management.

Waste Management

Since the close down of Fortress Rocks dump site and phase out of open burning at McMurdo in 1991, USAP has launched an aggressive waste management program at all U.S. stations. All the solid wastes generated at South Pole and at the seasonal field camps are transported to McMurdo for further processing and preparation for removal from the continent. A brief attempt at incineration of food waste at McMurdo did not prove to be satisfactory, and incineration was terminated in 1993. The USAP no longer practices open burning, incineration or land disposal of wastes at any U.S. station in Antarctica. All wastes except domestic sewage are removed from Antarctica. Approximately 4000 metric tons of waste were removed from Antarctica in 1994, 3000 tons in 1995 and 2700 tons in 1996. The decline in the amount of the waste removed reflects a diminishing waste backlog at the station. Table 2 shows the annual production of USAP wastes in category and waste streams which are to be removed from the continent.

After discharging the annual resupply of equipment and material at McMurdo, the resupply vessel is backloaded with the USAP waste which has accumulated during the previous 12 months. On return to the United States, the vessel disembarks the waste at Indian Island, Washington; the State of Washington has one of the most aggressive waste recycling infrastructures in the nation. A waste disposal subcontractor takes custody of the USAP waste at the port and is responsible for its disposition in accordance with the USAP mandated priority of resale, reuse and recycling, with incineration or land disposal being the last consideration. For the past three years, USAP has achieved a recycling rate in excess of 60 percent, compared to the U.S. average recycling rate of 24 percent for municipal solid wastes.

Table 2

Annual Production of USAP Waste in 1995
(sewage excluded)

Category	Waste Stream	Quantity (1000 kg)	
Recyclables			
	aluminum	16	
	cardboard	149	
	cooking oil/lard	6	
	copper & wire	22	
	glass	12	
	scrap metals (heavy)	327	
	scrap metals (light)	108	
	paper, magazine	9	
	wood	298	
	subtotal	947	38%
Disposables	clothing	1	
	construction debris	587	
	greywater & urine	42	
	plastics	26	
	others	54	
	subtotal	710	29%
Combustible	food waste	163	
	food contaminated	8	
	human waste	8	
	trash	73	
	subtotal	252	10%
Used Equipment	tire & resalable	199	8%
Hazardous waste	hazardous waste	368	15%
Total waste produced		2,476	100%

In creating the necessary infrastructure to achieve its desired objectives, the USAP completely reordered the waste management organization and practices. Single point management was established, functions were consolidated, the organization streamlined, and an investment was made in staffing and in the procurement of capital equipment and supplies. Large capacity balers, compactors and shredding machines in addition to more than one hundred sea worthy containers--milvans--were purchased and shipped to McMurdo in 1994. This investment assured an adequate inventory of processing devices and containers for the storage and transportation of waste. Additionally, waste processing centers were established and waste management plans were developed and implemented for both hazardous and non-hazardous waste. These plans include the procedures for collecting, sorting, processing, and packaging waste, as well as the documentation required for transportation and importation of waste into the U.S. The procedures are revised in response to changes in state and federal regulations or in response to the introduction of new technologies. An educational program was launched to inform USAP participants about the waste management program and to instruct them in the need for compliance with established practices. This program was supplemented by special training provided to those contractors and researchers who generate or process hazardous waste or low level radioactive waste in the course of their normal duties.

Waste prevention and waste minimization are two keystones underlying the USAP waste management; the former is achieved through source reduction, and the latter through reuse and recycling.

Waste Prevention and Source Reduction

Education and employee training have proven to be essential in reducing the generation of waste. A waste management brochure (Reference 9) is distributed to each USAP participant before they start their journey to the continent. When packaging personal hygiene products, participants are asked to consolidate, use bulk containers, and to discard excess vendor packaging. The brochure encourages participants to observe the backpacker's rule (pack it in and pack it out) and further advises participants not to leave personal items behind when departing the station. Upon arrival at McMurdo all participants are required to attend a waste management orientation, which emphasizes each individual's responsibility for sorting and recycling of waste (Reference 10).

On-going measures are also being taken to improve material procurement practices and to streamline the consumable material inventory in order to reduce conditions of over stockage and the retention of products which have exceeded useful shelf-life. Vendors are notified, in the purchase order, of the ACA waste regulation which bans the use polystyrene foam "peanuts," plastic chips or similar loose polystyrene packing materials. Often, cargo is re-packaged in the U.S. before loading the supply ship, in order to reduce needless over-packaging which would end up in waste stream. Fresh fruits and vegetables which are purchased at Christchurch, New Zealand are pre-processed in New Zealand for the removal of excess leaf and stalks, then sealed in plastic bags which are filled with nitrogen gas. Such preprocessing significantly reduces the amount of food waste which would be generated from food preparation at McMurdo Station.

Product substitution is pursued as a means of reducing the on-hand inventory of highly toxic materials. For example, polypropylene glycol has been substituted for ethylene glycol. Polypropylene glycol unlike ethylene glycol is significantly less toxic. Since ethylene glycol is used as an antifreeze coolant in all vehicles and as a heat transfer fluid in the heating systems of most buildings at McMurdo, the on-hand inventory of the product is high. This substitution eliminates a major source of toxic waste.

Waste Minimization

Opportunities for reuse and recycling are continuously being assessed in order to minimize the quantity and volume of waste produced at McMurdo Station. Used pallets and crates which are in good conditions are reused. A reuse center has been opened for the distribution to the community of reusable discarded personnel items -- clothing, stereos, books. The procurement of a drum washing apparatus enables local cleaning and reuse of metal drums as they are returned to McMurdo from outlying camps. Off-specification aviation fuel is being burned in furnaces for space heating. Used lubrication oil is tested and, if found to be within the established standards, is then blended with furnace fuel and burned.

The USAP also maximizes reuse and recycling after the waste is retrograded to the United States. The terms of the waste disposal subcontract is designed to leverage the subcontractor to maximize reuse and recycling. The Request for Proposal requires each bidder to explicitly list how and where the bidder plans to dispose of individual waste streams, hazardous and nonhazardous, that are listed in the proposal. The criteria for evaluating the proposal explicitly assigns high scores to recycling and reuse, and low scores to incineration and landfill. This procurement mechanism has worked well.

Waste Segregation, Processing and Packaging

Reuse and recycling has been demonstrated to be most easily facilitated when the wastes are segregated at the source (Reference 10). This requires the active involvement of the entire McMurdo community. In the dormitories and in all maintenance shops and laboratory spaces, numerous receptacles are provided, each clearly marked and dedicated to a single type of waste. The receptacles are colored coded to indicate the type of waste acceptable: green for recyclable, gold for incineration, blue for disposable, and red for regulated waste. These receptacles are collected and the contents further sorted for ease of processing.

Aluminum, metal cans, and light-weight scrap metals are segregated and baled, as are cardboard and plastics. Glass is sorted by color, then crushed and stored in 208-liter (55-gallon) steel drums. Wood waste from carpentry projects and from unusable pallets and crates is shredded to reduce volume and then containerized in milvans.

Batteries are sorted so that wet-cell types can be packaged for recycling. Aerosol cans are rendered non-hazardous by use of a depressurizer that punctures the cans, drains the residual contents, and captures the propellants on a carbon filter. The containers are then recycled as aluminum or light metal.

The procedures for collection and documentation of the wastes are described in the waste management plans. A database utilizing commercially available software has been developed to keep track of wastes as they are generated and processed, and to prepare manifests for hazardous wastes. The waste management database provides useful information for identifying opportunities for improvement in waste management.

Fuel Spill Prevention and Contingency Planning

Fuel spills are the major source of hydrocarbon contamination at McMurdo. As such, a significant investment continues to be made in spill prevention, measures and projects, and spill response planning, training and procurements.

An Oil Spill Contingency Plan (Reference 6) has been developed following the requirements of DOT regulation as guidance. The plan describes incident notification procedures, mobilization of response teams, methods for containment, recovery and cleanup, as well as safety precautions, and formats of required incident reports.

The level of spill response is based on the scale of the spill. Level I are small spills of less than three 208-liter (55-gallon) drums of cleanup waste. Level II are spills that generate cleanup waste in excess of three 208-liter (55-gallon) drums. Level III are large spills that are beyond the capability of the local spill response teams and will require outside assistance.

It is the responsibility of the individual causing a Level I spill to clean it up. Level II and Level III spills are addressed through a team approach. There are four 5-person teams organized in the summer season to respond to Level II spills. The teams rotate the duty on a weekly basis. For Level III spills there are two 23-person response teams. During the winter, the response capability is reduced to two 5-person teams because the decreased level of winter activities limits the potential for spills to occur. Spill response teams are organized prior to deployment to the continent and team members receive hand-on spill response training at U.S. based training centers. An extensive inventory of equipment and supplies for containment and recovery of spills, both waterborne and terrestrial, have been procured. A representative amount of this inventory is staged in mobile containers which can be rapidly transported to the scene of a spill, if required.

Spill reporting procedures have been established for all levels of spill. The data are used to identify spill causes and trends so that preventive measures can be instituted.

A Spill Prevention, Control and Countermeasures Plan (SPCC) (Reference 7) has been developed following the requirements of the EPA regulations as guidance. The SPCC recommends measures deemed necessary or helpful to prevent and control spills. The recommendations include: replacement of the older bulk fuel storage tanks, replacement of all flanged pipe joints, and installation of secondary containment.

In response to these recommendations, in 1995 USAP completed the construction of two new 1,892,000 liters bulk storage tanks with secondary containment, initiated a multi-year program for the welding of all flanged type piping, and by 1996 completed the construction of secondary containment around 5 of the older bulk storage tanks (2-1.892 million liters, 2-946 thousand liters, 1-9.46 million liters).

Investigations Underway

The USAP continues to investigate the means to further reduce waste at McMurdo Station. One significant target for waste reduction is wood, a major waste stream of high volume. The feasibility of utilizing commercially available wood burners for space heating is being assessed. Such devices are clean burning and could potentially eliminate the need to annually retrograde approximately 110 milvans of waste wood. An added benefit would be realized in the reduction of heating fuel usage.

Soil cleaning technologies for remediating fuel contaminated soils at McMurdo are also being actively investigated. Technologies of interest include: soil washing with biodegradable detergents, in-situ bioremediation, thermal desorption, and steam cleaning.

At the international level, the USAP is currently in discussion with the national operators of other Antarctic Programs concerning regional based approaches to spill response. The capability to provide an increased level of response through a sharing of assets could be critical in limiting the extent and coincident damage to the environment of a large spill.

Conclusion

During the past several years, the NSF has taken aggressive measures to abate pollution at McMurdo. It has issued new waste regulations to govern waste management and disposal in Antarctica. It has banned open burning and closed down the waste dump. It has cleaned up former waste dump sites and removed the waste backlog from the station. It has developed and implemented spill prevention and spill response plans and continues to invest in the upgrade of its aging bulk fuel storage and distribution system. It has interacted with other national programs within the Antarctic Treaty System in sharing pollution abatement plans, procedures and technology. Opportunities for further reductions in waste are being pursued and technologies to remediate soil contamination are being explored. Perhaps the most significant improvement from past practices has been the shift in the culture of McMurdo community toward environmental responsibility. All these measures have enabled the USAP to assume a leadership role in the Antarctic Treaty System for the protection of the Antarctic environment.

REFERENCES

1. Lenihan, HS; Oliver, JS; Oakden, JM, and Stephenson, MD (1990). "Intense and Localized Benthic Marine Pollution Around McMurdo Station," *Marine Pollution Bulletin*, Vol 21, No. 9, pp 422-430.

2. Lenihan, HS (1992). "Benthic Marine Pollution Around McMurdo Station, Antarctica: A Summary of Findings," *Marine Pollution Bulletin*, Vol. 25, 9-12, pp 318-323.

3. Risebrough, RW; DeLappe, BW; and Younghans-Haug, C (1990). "PCB and PCT Contamination in Winter Quarters Bay, Antarctica," *Marine Pollution Bulletin*, Vol 21, No. 11, pp 523-529.

4. *Final Fortress Rocks Landfill Remediation Report for McMurdo Station, Antarctica*, 15 July 1992, AECOM, Fairfax, Virginia.

5. *Final Sampling and Analysis Report, Winter Quarters Bay Landfill, McMurdo Station, Antarctica*, Vol 1, April 1995, Consoer Townsend & Associates, AECOM Technology Corporation, Fairfax, Virginia.

6. *McMurdo Station Area Oil Spill Response Contingency Plan*, July 1995, National Science Foundation, Office of Polar Programs, Arlington, Virginia.

7. *U.S. Antarctic Program Spill Prevention, Control and Countermeasures Plan*, September 1994, National Science Foundation, Office of Polar Programs, Arlington, Virginia.

8. *U.S. Antarctic Program Participant Guide*, 1996 edition, National Science Foundation, Arlington, Virginia.

9. *Waste Management in the United States Antarctic Program*, NSF 93-128, National Science Foundation, Washington, D.C.

10. *Waste Management Plan and User Guidance*, October 1993, Antarctic Support Associates, Englewood, Colorado.

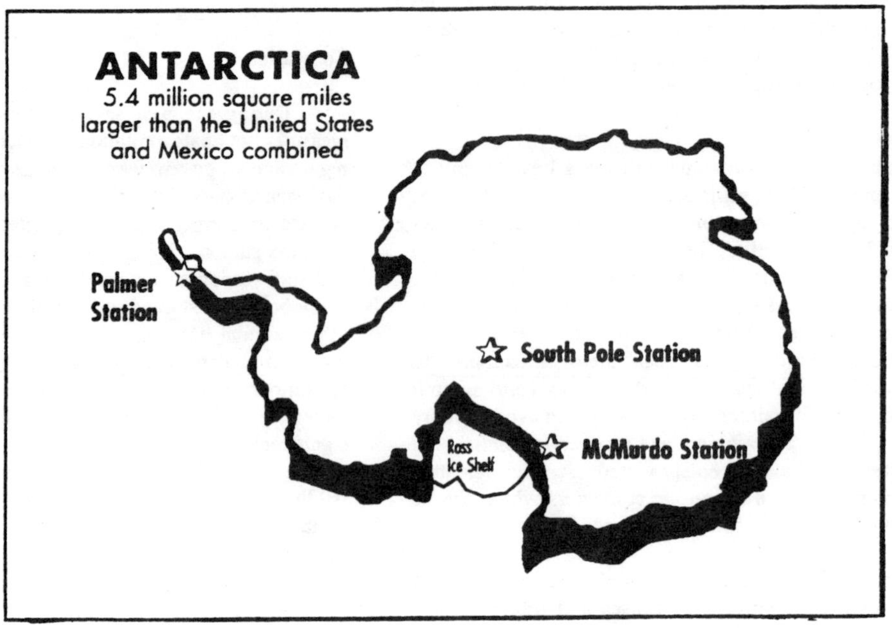

Figure 1 Location Map of McMurdo Station, McMurdo

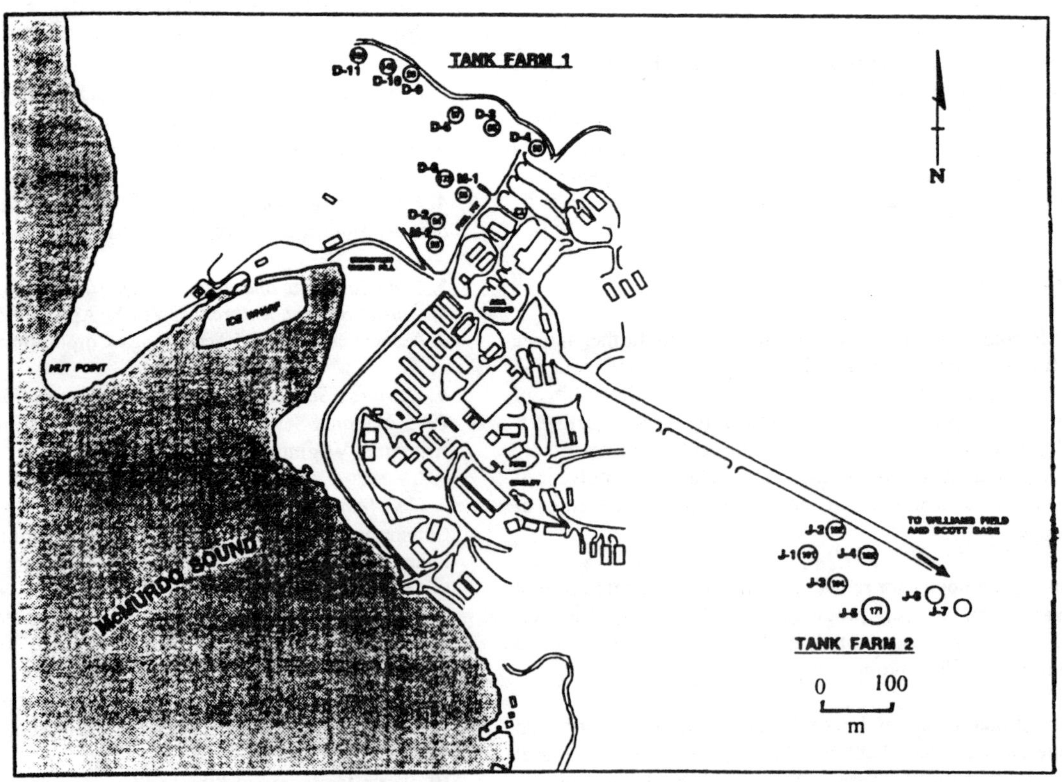

Figure 2 McMurdo Station Site Plan

Proceedings of the Seventh (1997) International Offshore and Polar Engineering Conference
Honolulu, USA, May 25-30, 1997

ISBN 1-880653-28-1 (Set); ISBN 1-880653-30-3 (Vol. II)

On the Adhesion of Oil to Ice

Seppo Liukkonen and Jorma Rytkönen
VTT Manufacturing Technology
Espoo, Finland

Alexander Alhimenko and Elena Kniazeva
St. Petersburg State Technical University
St. Petersburg, Russia

ABSTRACT

This paper describes a study on the adhesion between oil and ice and gives the main results. The experimental study consisted of contact angle and inclination angle measurements. Both measurements were taken both on ice in air and under ice in water. The test temperatures on ice in air were -1°C, -12°C and -18°C. Under ice the test temperature was the water temperature, i.e., 0°C. Three different oil types were included in the measuring program. North Sea Brent crude oil was tested as fresh and as artificially pre-altered. The third oil type was diesel fuel.

On the basis of the contact angle measurements, the surface energies at interfaces between oil, air, water and ice were investigated. In addition, theoretical assessments of the oil spreading on ice, the work of adhesion between oil and ice as well as the influence of the 'liquid-like layer' of ice on its adhesion to oil were performed. The inclination angle measurements yielded indications of the static and dynamic friction between oil and ice. Theoretical calculations were conducted on the motion resistance of oil drops sliding along the bottom of ice in water.

KEY WORDS: Adhesion, contact angles, oil spills

INTRODUCTION

General

As the unfortunate accidents of MT Antonio Gramsci in the Baltic Sea in 1987 and MT Exxon Valdez in the Prince William Sound in Alaska in 1989 have shown, the risk of a marine oil spill in an ice-covered sea is real. In addition, those accidents, among others, have shown that our knowledge of oil spill combating in cold and icy conditions is inadequate. Therefore, there is an obvious need for research on the subject. To meet this need, VTT Manufacturing Technology from Espoo, Finland, St. Petersburg State Technical University from St. Petersburg, Russia, and Murmansk Marine Biological Institute from Murmansk, Russia, have established a joint research project entitled 'Marine Oil Spill Combating in Arctic and Sub-Arctic Conditions'. In addition to the organizations above, the industry and oil spill combating authorities in both Finland and Russia take part in the project that has been scheduled for the years 1995-97.

As a part of the above research project VTT Manufacturing Technology and St. Petersburg State Technical University have carried out an experimental laboratory study on the adhesion between oil and ice in November - December 1995 in the VTT laboratories in Espoo, Finland. In the study, the contact angles of drops of different oils on the top surface of ice in air and under the bottom surface of ice in water were measured. In addition, special inclination angle tests were conducted. When the oil drop was lying on the ice piece, the ice was slowly inclined till the oil drop started to move on the ice surface. The angle initiating the oil movement was measured. In addition, the velocity of the oil drops was measured when the drops were sliding along the underside of the inclined ice piece in water.

Adhesion between oil and ice

Adhesion may be defined as the state in which two surfaces are held together by interfacial forces which may consist of valence forces or interlocking action, or both (Skeist, 1977, p. *xi*). Additionally, the adhesion between a solid and a liquid is sometimes also interpreted as the ability of the liquid to wet the solid (Zisman, 1977, p. 38).

When a sessile drop of a liquid lies on a solid surface, the Young-Dubrè equation defines the form of the drop using the three surface energies and the contact angle shown in Figure 1.

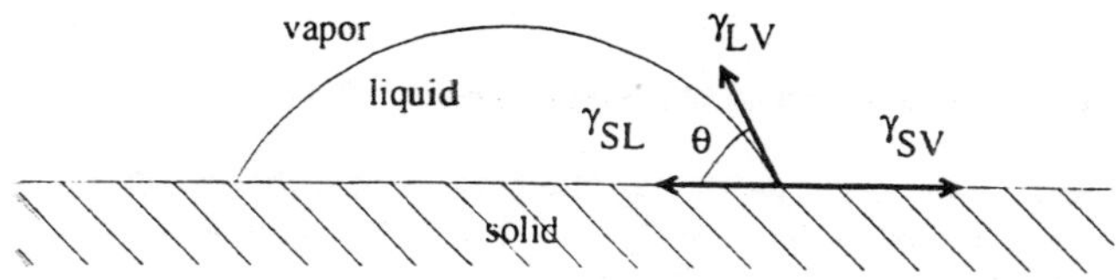

Fig. 1 Sessile drop of a liquid on a solid surface.

Using the notation of Figure 1, the Young-Dubrè equation is written (Zisman, 1977, p. 37)

$$\gamma_{SV} - \gamma_{SL} = \gamma_{LV} \cos\theta , \qquad (1)$$

where γ_{SV}, γ_{SL} and γ_{LV} are the surface energies at the solid/vapor, solid/liquid and liquid/vapor interfaces, respectively, and θ is the contact angle. Some of the surface energies above are difficult to measure in practice, but the contact angle is an easily measurable quantity. Therefore, the contact angle is a commonly used indicator of the adhesion between a liquid and a solid. The small contact angle indicates strong adhesion. In other words, the smaller the contact angle is the better the liquid wets the solid.

The Young-Dubrè equation holds strictly true only in a vacuum (Pocius, 1986, p. 48). In practice, however, the contact angle measurements are carried out in air, disregarding the possible small errors this may result in (Pocius, 1986, p. 48). Also, in this work the contact angle measurements on ice were conducted in air. For those tests the application of the Young-Dubrè equation is in accordance with common practice. Additionally, in this work contact angle measurements were taken also under ice in water. For those tests the Young-Dubrè equation was also applied, but the vapor phase in Figure 1 and equation (1) was replaced with water.

Some studies on the adhesion of oil to ice have been carried out. Mackay et al. (1976, pp. 72-74) have measured contact angles ranging from 140° to almost 180° for Norman Wells and Prudhoe Bay crude oils under the bottom surface of ice in water. The large contact angles indicate weak adhesion of oil to ice in water. However, contact angles smaller than 180° show that the oil comes into contact with ice and wets it to a certain extent. Malcolm & Dutton (1979, pp. 771-778) have measured the contact angle of 180° for Guanipa and Golden Eagle crude oils under the bottom surface of ice in water. Thus, their result is that the adhesion of oil to ice in water is extremely weak. In fact, Malcolm & Dutton (1979, p. 778) state that the oil does not come into contact with ice at all, but a thin water layer always exists between the oil and the underside of ice. Theoretically, the statement of Malcolm & Dutton means that either the work of adhesion between oil and ice equals to zero, or that the work of cohesion in oil is infinite (Adamson, 1960, pp. 266-267). Both alternatives above, however, are physically unreal (Adamson, 1960, pp. 266-267), and every liquid wets every solid to some extent, if brought into contact with each other (e.g., Zisman, 1977, p. 38). On the ice surface, however, there is the 'liquid-like' layer at the temperatures slightly below the melting point of ice (e.g., Hobbs, 1974, pp. 392-397 and Makkonen, 1994, pp. 40-42). This liquid-like layer may affect the adhesion of oil to ice as stated by Malcolm & Dutton (1979, p. 777), but further research on the subject is needed. Alhimenko (1989, p. 203) also assumes that the oil does not come into contact with the bottom surface of ice when drifting in a steady water flow under ice.

Liukkonen et al., (1995, pp. 69-90) have measured contact angles ranging from almost zero to 27° for North Sea Brent crude oil and light fuel oil on ice in air. Thus, these results indicate rather strong adhesion of oil to ice in air. The different adhesion behavior of oil on ice in air and under ice in water may be explained by the surface energies at the three interfaces included in the Young-Dubrè equation. Two of those interfaces with their surface energies change, when the vapor, i.e., air, in Figure 1 is replaced with water.

Influence of the adhesion on the oil spill behavior in ice

The adhesion of oil to ice may affect the oil spreading in an ice-covered sea as well as the recovery of oil out of the ice-covered sea in several ways:

i) The adhesion of oil to ice may be interpreted as a kind of static friction between oil and ice. This affects essentially the so-called threshold velocity, i.e., the velocity that the water flow under the ice cover has to exceed in order to set into motion an oil slick initially lying at rest under the ice cover (e.g., Uzuner et al., 1979, p. 5).

ii) The so-called net surface tension, calculated on the basis of the surface energies at the oil/ice, oil/water and ice/water interfaces and the contact angle of oil under ice, is commonly assumed to determine the final width of an oil slick under an ice cover (e.g., Uzuner et al., 1979, p. 18 and Yapa et al., 1993, p. 14). The adhesion between oil and ice affects the net surface tension at the edge of the oil slick.

iii) The adhesion of oil to ice may also be interpreted as the attraction force keeping oil stuck to ice. When oil is recovered out of the ice-covered sea, the separation of oil from ice is an important task, the success of which is highly dependent on the adhesion of oil to ice.

iv) The interaction between oil and ice is an important parameter for the mathematical modelling of oil spreading under ice. If the oil comes into contact with ice, a friction coefficient between the oil and ice has to be applied in the spreading calculations. If there is a water layer keeping the oil apart from the ice cover at all times, the direct oil/ice interaction may be neglected. Then, the oil spreading under ice may be calculated with the help of the water flow using the water as the carrying medium and the ice cover only as a boundary for the flow.

Target of the study

The target of the study was to find out numerical values for parameters affecting the adhesion of oil to ice, thus affecting also the behavior of oil spills in ice as explained above. Also, the influence of the liquid-like layer of ice on its adhesion to oil was studied.

DESCRIPTION OF THE TESTS

Test arrangements and procedure

The study consisted of two types of measurements, namely the contact angle and the inclination angle measurements. Both types of measurements were taken on the top of ice in air and under the bottom of ice in water. Figures 2-4 show the test arrangements.

Fig. 2 Test arrangements for the tests under ice in water.

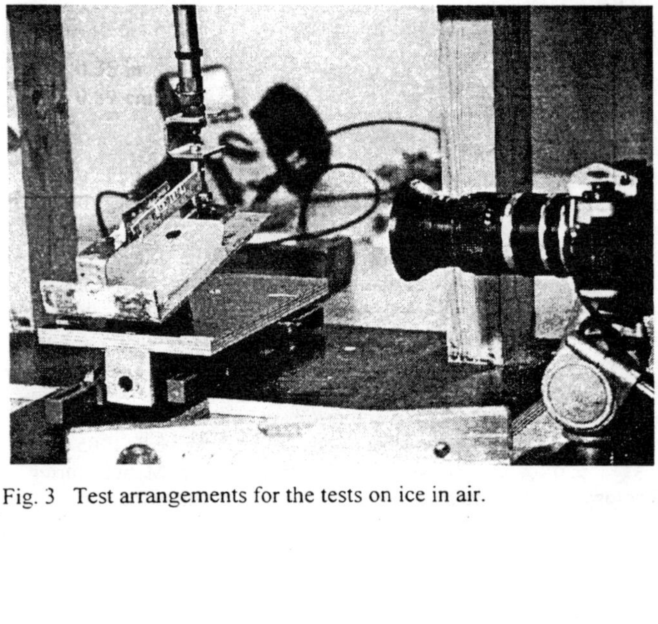

Fig. 3 Test arrangements for the tests on ice in air.

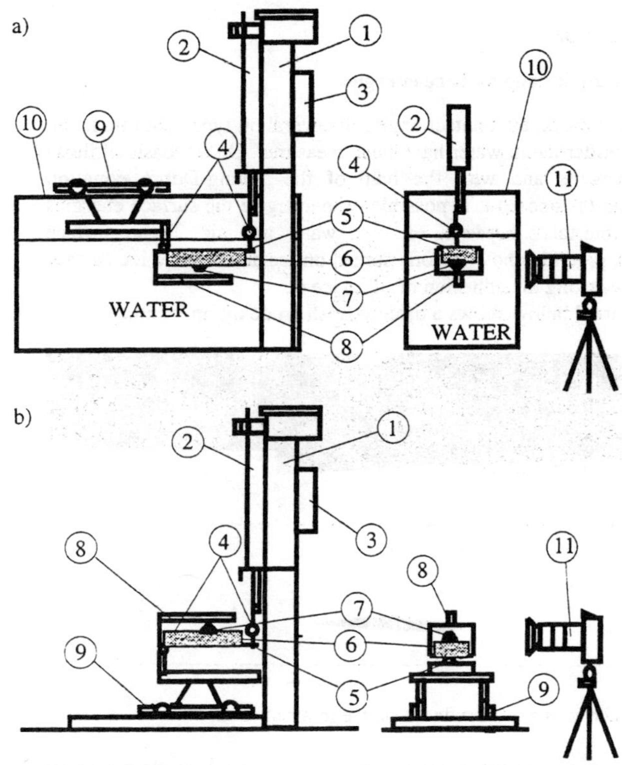

1: support frame
2: linear actuator; SKF CARR 22 x 200 x 1/D24B
3: actuator control box
4: swiwels
5: ice piece holder
6: ice piece
7: oil drop
8: ruler with centimeter scale
9: carriage with wheels to support the ice piece holder and to make the inclination smoother
10: glass-walled aquarium,35 liters in total volume
11: camera for contact angle photography, Olympus OM-4 with a zoom objective of f35-105 and extension rings of 14 mm and 21 mm. After the contact angle photography, the camera was replaced with a video camera, Panasonic S-VHS-C Movie camera NV-MS90, for video recording of the inclination tests

Fig. 4 Test arrangements for tests: a) under ice in water and b) on ice in air.

To start the tests, brackish Baltic Sea water was taken from the Gulf of Finland off the city of Espoo. This water was left to freeze naturally in a sub-zero temperature in a cold room. For the tests, the 2-cm-thick ice cover thus frozen was cut into pieces with the dimensions of 5 - 7 cm in width and 12 - 14 cm in length.

The surface roughness of the ice pieces was measured with a portable Taylor - Hobson Surtronic 3 surface roughness meter. For the top of ice, a total of 10 measurements yielded the mean center-line average, R_a, value of 22.03 µm with the standard deviation of 4.35 µm. The corresponding figures for the bottom of ice were 6.43 and 3.02 µm, respectively.

For the tests, an ice piece was installed in the test rig shown in Figures 2-4. The tests on ice in air were conducted with the test rig alone on a table (see Figures 3 and 4b). For the tests under ice in water, the ice piece holder with the ice piece was submerged into a glass-walled aquarium that was filled with Baltic Sea water (see Figures 2 and 4a). The total volume of the aquarium was 35 liters.

The oil drops were put on the ice piece with a pipette. To get the oil drops under the ice piece in water, a U-shaped pipe was connected to the pipette to lead the oil onto the underside of ice. The drop volume ranged from 0.01 to 0.5 ml. After an oil drop was put on or under an ice piece, it was photographed with a camera, and the contact angle was later determined from the photograph. After the contact angle photograph was taken, the initially horizontal ice piece was made to incline. The inclination was conducted with the help of a linear actuator connected to the ice piece holder (see Figures 2-4). The piston of the actuator moved with a rather slow speed of 0.07 centimeters per second yielding the angular inclination speed of 0.37 degrees per second for the ice piece. When the oil drop on or under ice was visually observed to start to move, the linear actuator was stopped, and the inclination angle of the ice piece was measured with an inclinometer connected to a water balance. In the tests under ice in water the moment when the whole drop started to move was easy to observe. In the tests on ice, the oil drop did not move as a whole, but it started to stretch downwards on the ice surface when the inclination angle increased sufficiently. The moment when the front edge of the oil drop started to move on ice was interpreted as the inclination angle and measured.

In addition, the velocity of the oil drops was measured in the inclination angle tests under ice in water. When the ice piece was inclined to the angle initiating the oil movement, further inclination of the ice piece was stopped. The oil drops, once they had started to move, however, continued sliding along the underside of ice at the constant inclination angle. The velocity of such sliding of the oil drops was measured from the video recordings taken from the tests.

The tests were conducted in altogether four temperatures. The under-ice test were conducted at the temperature of water, i.e., at 0°C. On ice in air, the test temperatures were -1°C, -12°C and -18°C.

Test oils

Three different oil types were used in the study. The North Sea Brent crude oil was used as fresh and as artificially pre-altered by 30 % of weight. The third test oil was diesel fuel. Table 1 presents the properties of the test oils.

Table 1 Properties of the test oils.

Property and measuring standard	North Sea Brent, fresh	North Sea Brent, pre-altered	Diesel fuel
Density[a] at +15°C [kg/m^3], ASTM D 4052	838	897	837
Density[a] at 0°C [kg/m^3], conversion from +15°C ASTM D 1250-80	848	907	847
Density[a] at -12°C [kg/m^3], conversion from +15°C, ASTM D 1250-80	856	915	855
Density[a] at -18°C [kg/m^3], conversion from +15°C, ASTM D 1250-80	860	919	859
Kinetic viscosity[a] at 0°C [m^2/s], ASTM D 445	$37.1 \cdot 10^{-6}$	*	$10.0 \cdot 10^{-6}$
Kinetic viscosity[a] at -12°C [m^2/s], ASTM D 445	*	*	$21.9 \cdot 10^{-6}$
Kinetic viscosity[a] at -18°C [m^2/s], ASTM D 445	*	*	$42.7 \cdot 10^{-6}$
Water content[a] [weight-%], ASTM D1744	0.07	0.01	0.04
Surface energy[b] at +25°C [mJ/m^2], ASTM D 971-91 (1992 mod.)	27.1	31.1	28.9
Surface energy[b] at ±0°C [mJ/m^2], ASTM D 971-91 (1992 mod.)	31.0	**	30.5

a measured by VTT Energy, b measured by Neste Oy
* viscosity of oil was too high for capillary viscosimeter
** viscosity of oil was too high for the surface energy measurements

RESULTS

Table 2 summarizes the numerical test results. Figures 5 and 6 show the results in graphical form.

Table 2 Test results.

Test conditions	Test type	Crude oil, fresh	Crude oil, pre-altered	Diesel fuel
Under ice, 0°C	Contact angle [deg.]	159.92 ± 7.83	165.00 ± 6.32	163.60 ± 5.77
	Inclination angle [deg.]	6.62 ± 3.25	5.40 ± 2.41	3.50 ± 2.22
	Sliding velocity [cm/s]	0.83 ± 0.75	0.51 ± 0.33	1.21 ± 0.43
On ice, -1°C	Contact angle [deg.]	15.75 ± 5.03	23.63 ± 5.38	*
	Inclination angle [deg.]	29.43 ± 7.63	9.22 ± 4.99	11.60 ± 3.20
On ice, -12°C	Contact angle [deg.]	30.69 ± 7.75	33.81 ± 10.80	*
	Inclination angle [deg.]	32.62 ± 7.09	27.50 ± 13.32	8.10 ± 2.28
On ice, -18°C	Contact angle [deg.]	**	**	*
	Inclination angle [deg.]	**	**	9.89 ± 2.47

* the contact angles of diesel fuel on ice in air were so small that they were impossible to measure (< 3 degrees).

** the fresh and pre-altered crude oil were too stiff to form a proper drop at -18°C. Thus, the measurements were not relevant.

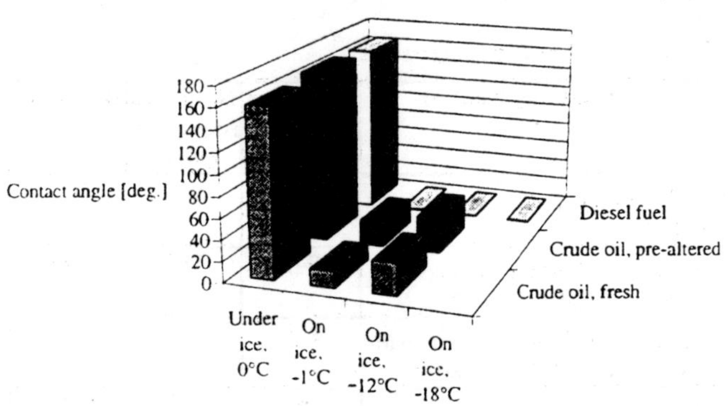

Fig. 5 Results of the contact angle measurements.

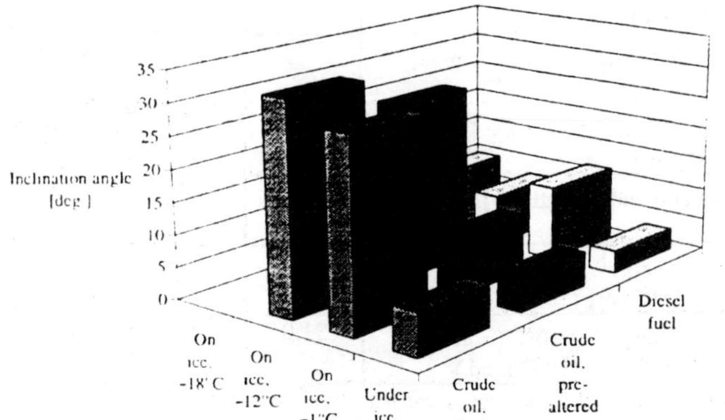

Fig. 6 Results of the inclination angle measurements.

DISCUSSION

Contact angles and surface energies

In this work, the contact angles of several oil types both on ice in air and under ice in water have been measured. On the basis of those measurements and with the help of the Young-Dubrè equation (equation (1) above) it is possible to investigate the surface energies at the interfaces between oil, ice, water and air. Those surface energies give idea how the oils spread on ice and wet the ice surface and how strong the adhesion of oil to ice is.

Figure 7 below shows a drop of fresh crude oil on ice in air.

Fig. 7 Drop of fresh crude oil on ice in air at -1°C.

The mean contact angle measured for the fresh crude oil on ice was 15.75° at -1°C (see Table 2). The surface energy of oil at ±0°C was measured to be 31.0 mJ/m^2 (see Table 1). Hobbs (1974, p. 440) gives the surface energy of 109 mJ/m^2 for the ice/air interface. By putting these values into the Young-Dubrè equation, the value of 79.2 mJ/m^2 for the surface energy at the oil/ice interface can be obtained. This value is not far from the results of Yapa et al. (1993, p. 51), who report the surface energy between oil and crushed ice to be 86.5 mJ/m^2. The difference may be due to the different oil type that Yapa et al. used in their tests.

The corresponding analysis can be made also for diesel fuel, even though its contact angle on ice was too small to be exactly measured. However, with good accuracy it can be estimated to be in the range from 0 to 3 degrees. The surface energy of the diesel fuel at ±0°C was measured to be 30.5 mJ/m^2 (see Table 1). With these values, together with the surface energy of the ice/air interface above, the Young-Dubrè equation gives the surface energies in the range from 78.00 to 78.04 mJ/m^2, thus practically 78.0 mJ/m^2, for the ice/diesel fuel interface.

There seems to be an apparent discrepancy above between the contact angles measured for the fresh crude oil and diesel fuel. In spite of the almost same surface energies at the oil/air interfaces, the contact angle of 15.75° for crude oil differs clearly from the almost zero contact angle of diesel fuel. The different contact angles, however, are possible. Even though the contact angles of liquids on a solid surface are considered to depend linearly on the surface energies of the liquid/air interfaces, the dependence is found to be valid with acceptable accuracy only for a homologous series of liquids (Pocius, 1986, p. 50). In spite of the fact that diesel fuel originates from crude oil, its molecular structure has changed so much during the refining process that it cannot be interpreted as homologous with the crude oil. Thus, the different contact angles of the crude oil and diesel fuel are theoretically quite possible. This fact indicates a difficulty in the practical oil spill combating work. The contact angle, i.e., the wettability of ice, with different types of oils may vary greatly and is poorly predictable without measurements with the particular oil type in question.

The picture of the oil/ice adhesion was clearly different under ice. Figure 8 shows the drop of fresh crude oil under ice in water.

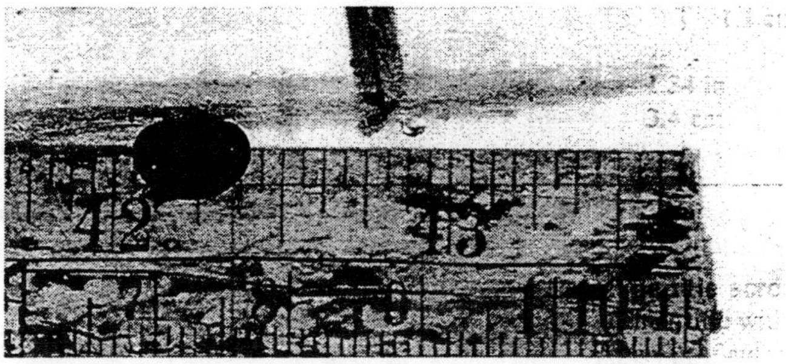

Fig. 8 Drop of fresh crude oil under ice in water.

The Young-Dubrè equation, written with the values of 33 mJ/m^2 for the water/ice interface (Hobbs, 1974, p. 440), 79.2 mJ/m^2 for the oil/ice interface (obtained above) and 159.92° for the contact angle (measured in these tests, see Table 2), gives the surface energy of 49.2 mJ/m^2 for the oil/water interface. In principle, this result seems to be correct, as the surface energy of a solid is usually expected to be higher than the surface energy of its melt (Adamson, 1960, p. 252). Here, the surface energy of oil/ice interface is 60 % higher than that of the oil/water interface. It is, however, true that the difference of 60 % is higher than expected by Adamson (1960, p. 252), who suggests the difference to be 20 - 30 %.

The comparison of the surface energy at the oil/water interface above with the values presented in the literature is difficult because of different oil types used in the measurements. The measuring temperatures also vary, but this is not so essential, because the temperature only slightly affects the surface energies. Chen et al. (1976, p. 81) report the surface energy of 21.0 mJ/m^2 for the interface between Norman Wells crude oil and water at the temperature of +24°C. Uzuner et al. (1979, pp. 37-38) report values in the range from 15.8 to 69.0 mJ/m^2 at the temperatures ranging from -2.3 to +5.5°C. The exact type of the crude oil they used has not been given. On the basis of the figures above, the value of 49.2 mJ/m^2 measured here for the North Sea Brent crude oil can only be stated to be of the expected order of magnitude.

The surface energy at the diesel fuel/water interface can also be calculated with the relevant values above. The calculations yield the value of 46.9 mJ/m^2. The result is well in harmony with those obtained above for the fresh crude oil.

Thus, the results from the surface energy investigations above can be summarized with the following table.

Table 3: Surface energies at different interfaces between oil, ice and water.

	North Sea Brent crude and ice	North Sea Brent crude and water	Diesel fuel and ice	Diesel fuel and water
Surface energy [mJ/m^2]	79.2	49.2	78.0	46.9

Surface energies and spreading of oil on ice

The surface energies at the three interfaces at the edge of the liquid drop lying on a solid surface (see Figure 1) define how the liquid is wetting the solid, i.e., how the liquid is spreading on the solid. A common factor describing the spreading is the spreading coefficient $S_{L/SV}$ defined as (e.g., Bateup, 1981, p. 233)

$$S_{L/SV} = \gamma_{SV} - (\gamma_{LV} + \gamma_{SL}). \qquad (2)$$

If the spreading coefficient is greater than, or equal to, zero, the liquid will spread spontaneously on the solid. In other words, the surface energies at the edge of the liquid drop generate a force trying to spread the liquid on the solid surface. In the case of a spreading coefficient smaller than zero, the surface energies at the edge of the liquid drop generate a force retarding the spreading of the liquid.

By substituting the surface energies in equation (2) with the values presented in the previous chapter for different interfaces between ice, oil, water and air, the following spreading coefficients can be calculated for different oils on ice in different environments:

Table 4 Spreading coefficients for different oils on and under ice.

	North Sea Brent crude on ice in air	North Sea Brent crude under ice in water	Diesel fuel on ice in air	Diesel fuel under ice in water
Spreading coefficient [mJ/m^2]	-1.2	-95.4	0	-95.9

The spreading coefficients above are smaller than zero in most cases. Thus, the oils are usually not spreading spontaneously on ice. In other words, the surface energies at the interfaces between oil, ice, water and air at the edge of the oil slick generate a force retarding the oil spreading. Only diesel fuel on ice in air is at the border line of spontaneous spreading.

When oil is released on ice in air or under ice in water, the retarding forces for the oil spreading both on and under ice are the inertial and viscous forces in the oil slick and interfacial forces at the edge of the oil slick. At the end phase of spreading, when the spreading velocity becomes very slow, the inertial and viscous forces vanish and the interfacial forces determine the final size of the oil slick. The spreading coefficients above are an indicator of the interfacial forces at the edge of the oil slick determining its spreading. Table 4 above shows that the spreading coefficients of the oils under ice are almost two decades smaller than those on ice. Thus, it can be expected that the oils would spread to a thinner film covering a wider area on ice in air than under ice in water. The expectation may be further emphasized by the fact that the driving force for spreading on ice, i.e., the gravity, is much stronger than the driving force under ice, i.e., the buoyancy of oil.

Work of adhesion

When a liquid drop is separated from a solid surface, as schematically shown in Figure 9, the liquid/solid interface (denoted with A-B in Figure 9) will be changed into one solid/vapor interface (C-D in Figure 9) and one liquid/vapor interface (E-F in Figure 9).

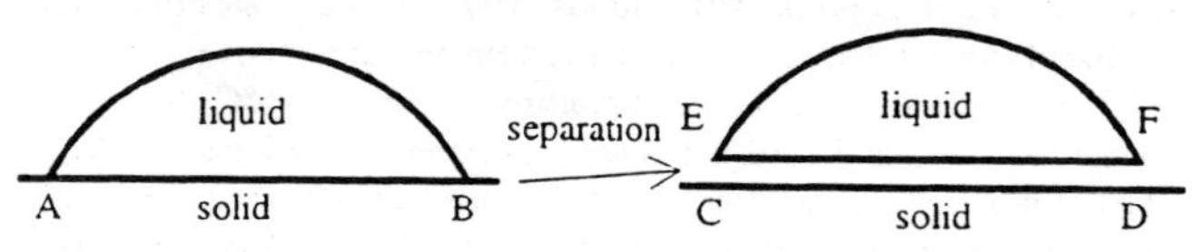

Fig. 9 Separation of a liquid drop from a solid surface.

The interface changes shown schematically in Figure 9 change the free energy, i.e., the total surface energy, in the liquid drop-solid system. To create this energy change, the work of adhesion must be done. The work of adhesion per a unit area of the solid/liquid interface, W_A is given by Bateup (1981, p. 233)

$$W_A = \gamma_{SV} + \gamma_{LV} - \gamma_{SL} \cdot \tag{3}$$

With the help of the Young-Dubrè equation, equation (3) can be derived to the form presented for instance by Zisman (1962, p. 180)

$$W_A = \gamma_{LV}(1 + \cos\theta) . \tag{4}$$

With the results from the contact angle measurements (Table 2 above) and the surface energy investigations (Chapter 4.1 above), equation (4) gives the following values for the work of adhesion between different oils and ice in different environments.

Table 5: Work of adhesion in different oil/ice contacts.

	North Sea Brent crude on ice in air	North Sea Brent crude under ice in water	Diesel fuel on ice in air	Diesel fuel under ice in water
Work of adhesion [mJ/m^2]	60.8	3.0	62.0	1.9

Recovery of oil out of the ice-covered sea

Regarding the recovery of oil out of the ice-covered sea, the oil spreading and adhesion work investigations above yield the clear conclusion that the oil and ice should be separated in the sea. If the oil and ice have been lifted out of the sea, the separation of oil from ice is much more difficult, because:

- On ice in air the oil tends to spread to a thinner layer than under ice in water. In other words, the same volume of oil covers a wider area of the ice surface, i.e., a wider area of the ice surface has to be cleaned.

- The work of adhesion to be done in order to separate the oil from the top surface of ice in air is roughly twenty to thirty times greater than that to be done to separate the oil from the underside of ice in water. The work of adhesion may be assumed to be proportional to the mechanical work, e.g., brushing, scraping, spooling with steam or water, etc., to be done to clean the ice.

Influence of the liquid-like layer on the adhesion of oil to ice

The liquid-like layer on ice is one of the arguments used to prove the assumptions that the oil does not come into contact with the underside of ice in water and that the contact angle between oil and ice in such conditions is 180° (Malcolm & Dutton, 1979, p. 777). Actually, the liquid-like layer exists also on ice in air, which means that the possible influence of the liquid-like layer on the oil/ice adhesion is not limited only to the case under ice, but should appear also on ice in air.

It is known that a liquid-like layer of a few tens of Ångstroms in thickness exists on ice at the temperatures close to its melting point. When the temperature goes down, the liquid-like layer disappears, but the exact temperature below which the liquid-like layer has totally disappeared is not known. Hobbs (1974, p. 743) states that the liquid-like layer on ice is stable at the temperatures above -2 ... -15°C, but a more accurate temperature of -3 ... -6°C for the disappearance of the liquid-like layer is given by Fletcher (1968, cited by Hobbs, 1974, pp. 393-395). Makkonen (1994, p. 42) defines -13°C as the lowest temperature at which the liquid-like layer exists on ice. Thus, -15°C seems to be a kind of a lowest temperature limit below which there is no liquid-like layer on ice anymore.

To study the influence of the liquid-like layer on the adhesion of oil to ice, contact angle and inclination angle measurements with diesel fuel on ice at the temperature of -18°C were conducted in this study. At such a low temperature there should not be any liquid-like layer on ice. Thus, by comparing the results of the measurements at -18°C with those of the measurements at -1°C and -12°C an idea can be obtained on how the liquid-like layer on ice affects the adhesion between oil and ice.

As Table 2 sets out, lowering the test temperature from -1°C to -18°C did not cause any discernible change in the contact angle of diesel fuel on ice. At all test temperatures the contact angle was very close to zero, actually smaller than what was possible to measure. Neither did the inclination angle nor the mode of oil drop movement, i.e., drop stretching, show any essential change when the test temperature was lowered to -18°C. These results seem to indicate that the liquid-like layer between oil and ice has no influence on the adhesion of oil to ice.

Inclination tests

In the inclination tests in air, the oil drop on ice did not start to move as a whole, but stretched downwards on the ice surface as shown in Figure 10.

Fig. 10 End of an inclination test with the oil drop on ice in air.

The behavior of the oil drops in the inclination tests on ice in air proved that in such conditions the viscous deformations inside the oil drops were more susceptible to taking place than the separation along the oil/ice interface. This is an evidence of strong adhesion between oil and ice in air and of the large work of adhesion needed to separate the oil from the ice surface. Thus, the behavior of oil drops on ice in air supported well the results obtained on the basis of the small contact angles measured for oils on ice.

When the ice piece in water, in turn, was inclined, the oil drop lying against the bottom of it was susceptible to moving. Figure 11 shows shematically the inclination test under ice.

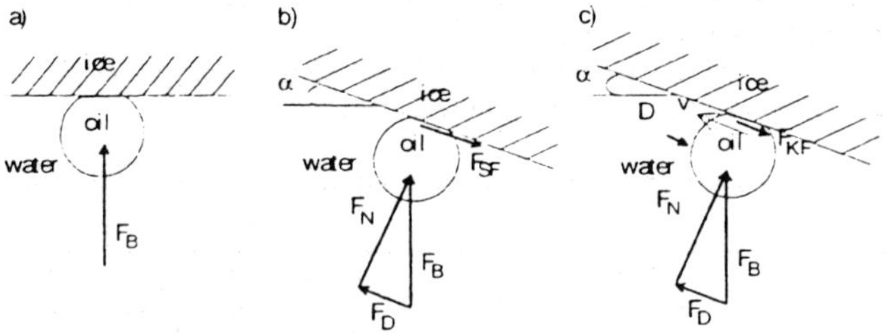

Fig. 11 Phases of an inclination test under ice: a) initial condition, b) start of oil movement and c) oil sliding along the ice surface.

In the initial condition (Figure 11 a), the oil drop was lying under the bottom of the ice piece. The buoyancy force, F_B, pressing the oil drop against ice is

$$F_B = gV(\rho_w - \rho_o) , \tag{5}$$

where g is the acceleration due to gravity, V is the volume of the oil drop, ρ_w is the density of water and ρ_o is the density of oil.

When the ice piece was inclined the oil drop lying against the bottom of it started to move very easily. The mode of motion was sliding along the underside of ice. The drop did not break or even deform discernibly during the sliding or its start. No tail or oily track on the ice surface behind the oil drop was observed during the sliding or its start. Thus, the behavior of the oil drop under ice in water showed weak adhesion of oil to ice in such an environment. This is also well in accordance with the large contact angles measured for the oil drops on the underside of ice in water.

Figure 11 b) shows the situation when the ice piece is inclined to the angle α that initiates the oil movement. Then, the force pressing the oil drop against the ice surface, F_N, is

$$F_N = F_B \cos\alpha . \tag{6}$$

The driving force, F_D, tending to move the oil drop upwards along the ice surface is

$$F_D = F_B \sin\alpha . \tag{7}$$

At the moment of the oil movement starting, the driving force F_D must be equal to the static friction force, F_{SF}, between the oil drop and ice. If an analogy is sought with the mechanical friction of solid materials, the static friction coefficient, f_{IS}, is

$$f_{IS} = \frac{F_B \sin\alpha}{F_B \cos\alpha} = \tan\alpha . \tag{8}$$

However, equation (8) must be considered as indicative only, because the analogy with mechanical friction is questionable. Most likely, the oil drops are not able to transmit the increasing normal force to the ice surface in a manner increasing the friction force according to the Coulomb law of friction, on which equation (8) is based.

In the tests reported here, the oil drops in water started to slide along the underside of the ice piece very easily. The inclination angles initiating the oil movement under the ice were measured as 6.62°, 5.40° and 3.50° for fresh crude oil, pre-altered crude oil and diesel fuel, respectively. With equation (8) the inclination angles above can be calculated to correspond to indicative static friction coefficients of 0.116, 0.095 and 0.061, respectively.

After the oil drop had started its movement, the inclination of the ice piece was stopped to the inclination angle α, and the oil drop slid upwards along the ice surface with a velocity v, shown in Figure 11 c). The force F_N pressing the oil drop against the ice surface and the driving force F_D still are the same as at the moment of the start of oil movement. The forces resisting the oil movement are the kinetic friction force, F_{KF} in the contact between oil and ice and additionally, the drag, D, i.e., the water resistance the oil drop experiences due to its motion. Then the force balance in the direction of the oil drop motion is

$$F_D = F_{KF} + D . \tag{9}$$

Thus, equation (9) shows the logical fact that the static friction is bigger than the kinetic one also in the contact between oil and ice.

The sliding velocities of oil drops along the underside of ice at the inclination angle that initiated the oil movement were measured as 0.83, 0.51 and 1.21 cm/s for fresh crude oil, pre-altered crude oil and diesel fuel, respectively.

To get an idea of the relations between the drag of the oil drop and the friction forces between the oil drop and ice, the following indicative study is made. The oil drop under ice is simplified as a sphere, of which the underside of ice cuts away a segment. The cut forms the contact angle θ at the point where the circumference of the sphere meets the bottom of ice. Then, the volume, V, of the oil drop can be calculated as

$$V = \frac{4\pi r^3}{3} - \pi[r - \cos(180-\theta)r]^2\left\{r - \frac{1}{3}[r - \cos(180-\theta)]r\right\} \tag{10}$$

where r is the radius of the sphere. Using the value of 0.2 ml for the volume of the oil drop (typical value in the tests) and 159.92° for the contact angle (measured for fresh crude oil under the ice, see Table 2), equation (10) gives the radius r of 3.6 mm for the oil drop.

Because the contact angle of the oil drop under the bottom of ice is large, the oil drop is close to a complete sphere. Therefore no great error is made if the drag of the oil drop is calculated as the drag of a solid sphere, however, taking into account that its frontal area is decreased with the area of the segment cut away by ice. Thus, the drag, D, of the oil drop may be calculated (Newman, 1982, p. 14)

$$D = 0.5\rho_w v^2 S C_D(R) , \tag{11}$$

where ρ_w is the density of water, v is the velocity of the oil drop, S is the frontal area of the oil drop and $C_D(R)$ is the drag coefficient that is a function of the Reynolds number. The frontal area S for the oil drop is calculated as

$$S = \pi r^2 - \left\{\frac{1}{2}\frac{2(180-\theta)}{180}2\pi r^2 - \frac{1}{2}r^2 \sin[2(180-\theta)]\right\} . \tag{12}$$

With r = 3.6 mm and θ = 159.92°, equation (12) yields the frontal area of 37.0 mm^2 for the oil drop.

The Reynolds number, R, for the oil drop is calculated as (Newman, 1982, p. 14)

$$R = \frac{vd}{\nu} , \tag{13}$$

where v is the velocity of the oil drop, d is the drop diameter and ν is the kinetic viscosity of water. With the values of v = 0.83 cm/s (= the mean velocity measured for the crude oil drops under ice, see Table 2), d = 7.2 mm and $\nu = 1.76 \cdot 10^{-6}$ m^2/s, equation (13) gives R = 33.95. With this Reynolds number, Newman (1982, p. 15) gives the drag coefficient for a sphere $C_D = 2$.

By substituting the values above to equation (11) the drag for the oil drop can be calculated to be $2.6 \cdot 10^{-6}$ N. This drag can be compared with the driving force of the fresh crude oil drop of 0.2 ml in volume under the ice piece at the inclination angle of 6.62°. Equations (5) and (7) give the driving force of $3.4 \cdot 10^{-5}$ N, when the densities of 1000 kg/m^3 and 848 kg/m^3 (see Table 1) are applied for water and oil, respectively. Thus, the drag of the oil drop is roughly 8% of the driving force. Due to the low sliding speed of the drop, the result sounds logical. The result also means that kinetic friction forms 92% of the motion resistance of the oil drop. Thus, the kinetic friction between the oil and ice is an important factor resisting the motion of oil under ice, at least at low speeds.

CONCLUSIONS

The adhesion of oil to ice has been studied with two types of measurements, namely the contact angle measurements and the inclination angle measurements. The measurements have been taken both on ice in air and under ice in water. The test temperatures in air were -1°C, -12°C and -18°C. The under-ice tests were conducted at the temperature of water, i.e., at 0°C. Three types of test oils were used. North Sea Brent crude oil was tested as fresh and as pre-altered. The third test oil was diesel fuel.

The contact angles measured for the oils on ice in air were small, ranging from almost zero to 34°. The almost zero value was measured for the diesel fuel at all test temperatures. The largest contact angle, 34° was measured at -12°C for the pre-altered crude. The small contact angles of oil on ice indicated strong adhesion between oil and ice in air.

Differing clearly from the situation on ice in air, the contact angles measured for the oils under ice in water were large, indicating weak adhesion between oil and ice in such an environment. The contact angles under ice ranged from 160 to 165 degrees. The oil type did not essentially affect these results.

The surface energies for different oil interfaces were calculated on the basis of the contact angle measurements. At the oil/ice interface, the surface energy was calculated to be 79.2 and 78.0 mJ/m^2 for fresh crude oil and diesel fuel, respectively. The corresponding numbers for the oil/water interface were calculated as 49.2 and 46.9 mJ/m^2.

On the basis of the surface energies, the spreading coefficients for the fresh crude oil and diesel fuel on ice were calculated to be -1.2 and 0 mJ/m^2, respectively. Under ice the corresponding numbers were -95.4 and -95.9 mJ/m^2. The negative spreading coefficients show that usually oils do not spread spontaneously on ice, but the interfacial energies at the edge of the oil drop generate a force retarding the oil spreading. Only diesel fuel on ice was at the border line of spontaneous spreading.

The work of adhesion, i.e., the work to be done in order to separate oil from the ice surface, was estimated on the basis of the surface energies calculated. In air, the work of adhesion per unit area of oil/ice interface was calculated to be 60.8 and 62.0 mJ/m^2 for the fresh crude oil and diesel fuel, respectively. Under ice in water, the corresponding figures were only 3.0 and 1.9 mJ/m^2, i.e., roughly one twentieth and thirtieth parts of those in air. These results gave a clear recommendation for the practical oil spill combating measures, if spilled oil must be recovered out of an ice-covered sea. From the adhesion point of view, the separation of oil from ice is much easier to do in water than in air, if practical methods to carry out the work in water only are available.

The inclination angle tests reflected the fact that the adhesion of oil iss strong on ice in air and weak under ice in water. In the inclination tests on ice in air, the oil drops did not start to move as a whole, but they just started to stretch downwards on the ice surface, when the inclination angle had grown sufficiently. Under ice in water, in turn, the oil drops were very susceptible to moving, and the dominating mode of oil movement under ice was sliding as a whole drop. The inclination angles measured to initiate the oil movement under ice were 7, 5 and 4 degrees for the fresh crude oil, pre-altered crude oil and diesel fuel, respectively.

When an oil drop was sliding at a speed of 0.83 cm/s along the bottom of ice in water, only 8 % of its motion resistance was calculated to be due to the drag in water. 92 % of the resistance was assumed to be formed by the friction between oil and ice.

A general conclusion of all test results was that when oils are lying at rest either on ice in air or under ice in water, they come to contact with ice or at least with the 'liquid-like' layer on ice. The 'liquid-like' layer on ice, however, was not observed to affect the adhesion behavior of oil on ice.

ACKNOWLEDGEMENTS

The following organizations are gratefully acknowledged for their financial support to the research and for the permission to publish the data:

Ministry of the Environment, Finland
Finnish Environment Institute, Finland
Kværner Masa-Yards, Finland
Neste Oy, Finland
TEKES, Finland
Ministry for Trade and Industry, Finland
Rosshelf, Russia
Marine Pollution Control & Salvage Administration of the Russian Federation, Russia

REFERENCES

Adamson, A.W., 1960. *Physical Chemistry of Surfaces.* Interscience Publishers, Inc., New York, USA, 629 p.

Alhimenko, A.I., 1989. *The Theory Development and Methods of Calculation of Oil Product Spreading at Aqua Medium Under Hydrometeorological Factors.* Doctoral Thesis, Leningrad Polytechnical Institute, Leningrad, Russia. 328 p. + appendices.

Bateup, B.O., 1981. Surface Chemistry and Adhesion. *International Journal of Adhesion and Adhesives*, Vol. 1, No 5, July 1981, pp. 233-239.

Chen, E.C., Keevil, B.E. & Ramseier, R.O., 1976. Behaviour of Crude Oil Under Fresh-Water Ice. *Journal of Canadian Petroleum Technology*, April-June, Vol. 15 - No. 2, pp. 79-83.

Fletcher, N.H., 1968. Surface Structure of Water and Ice. II. A Revised Model. *Phil. Mag. 18*, pp. 1287-1300.

Hobbs, P.V., 1974. *Ice Physics.* Clarendon Press, Oxford, Great Britain, 837 p.

Liukkonen, S., Koskivaara, R., Rytkönen, J. & Lampela, K., 1995. Adhesion of Oil to Plastics, Stainless Steel and Ice. *Proceedings of the Eighteenth Arctic and Marine Oil Spill Program (AMOP) Technical Seminar*, June 14-16, 1995, Edmonton, Canada, Environment Canada, pp. 69-90.

Mackay, D., Medir, M., & Thornton, D.E., 1976. Interfacial Behavior of Oil Under Ice. *The Canadian Journal of Chemical Engineering*, Vol. 54, No. 1/2, February/April, 1976, pp. 72-74.

Makkonen, L., 1994. The Mechanism of Surface Melting on Ice. *Proceedings of 12th Scandinavian Symposium on Surface Chemistry* (edited by Stenius, P. & Sarvaranta, L.). Helsinki University of Technology, Department of Forest Products Technology, Laboratory of Forest Products Chemistry, Report series C 6, Espoo, Finland, pp. 40-42.

Malcolm, J.D. & Dutton, C.R., 1979, The Interfacial Tension and Contact Angle of Crude Oil Under Ice. *The Fift International Conference on Port and Ocean Engineering Under Arctic Conditions*, August 13-18, 1979, Trondheim, Norway, Vol. 1, pp. 771-778.

Newman, J.N., 1982. *Marine Hydrodynamics*, fourth printing. The MIT Press, Cambridge, Massachusetts, USA, p. 402.

Pocius, A.V., 1986. Fundamentals of Structural Adhesive Bonding in *Structural Adhesives; Chemistry and Technology* (edited by Hartshorn, S.R.), Plenum Press, New York, USA, pp. 23-68.

Skeist, I., (editor), 1977. Standard Definitions of Terms Relating to Adhesives in *Handbook of Adhesives*, second edition. Van Norstrand Reinhold Company, New York, USA, pp. *xi - xviii*.

Uzuner, M.S., Weiskopf, F.B., Cox, J.C. & Schultz, L.A., 1979. Transport of Oil Under Smooth Ice. Report No. EPA-600/3-79-041, ARCTEC Incorporated, Columbia, Maryland, USA, 48 p.

Yapa, P.D., Weerasuriya, S.A., Belaskas, D.P. & Chowdhury, T., 1993. Oil Spreading in Surface Waters with an Ice Cover. Report No. 93-3, Clarkson University, Department of Civil and Environmental Engineering, Potsdam, NY, USA, 85 p.

Zisman, W.A., 1977. Influence of Constitution on Adhesion in *Handbook of Adhesives*, second edition (edited by Skeist, I.). Van Norstrand Reinhold Company, New York, USA, pp. 33-66.

Zisman, W.A., 1962. Constitutional Effects on Adhesion and Abhesion. In: *Adhesion and Cohesion* (Edited by Weiss, P.). Amsterdam, Elsevier Publishing Company, pp. 176-208.

Proceedings of the Seventh (1997) International Offshore and Polar Engineering Conference
Honolulu, USA, May 25-30, 1997

ISBN 1-880653-28-1 (Set); ISBN 1-880653-30-3 (Vol. II)

In-Situ Burning: An Alternative Approach to Oil Spill Clean-Up in Arctic Waters

Chantal C. Guénette
SINTEF, Applied Chemistry
Trondheim, Norway

ABSTRACT

In-situ burning is a spill countermeasure which involves burning the oil in place, usually at or near the spill site, under controlled conditions. During the 1980's in-situ burning emerged as a promising technology for dealing with oil spills in arctic waters, where ice cover can prevent deployment of conventional response equipment (skimmers and booms). In Norway, this method has been considered to be the most likely response option in high ice cover areas, for example in the Barents Sea. In light of this fact, an experimental Programme, spanning over four years, was initiated in 1990 to study in-situ burning. This research was carried out by SINTEF as part of a wider Programme "Oil spill response in Northern and Arctic waters" (ONA) sponsored by the Norwegian Clean Seas co-operative (NOFO). This paper describes the general principals of in-situ burning and presents some of the research carried out in Norway in this field.

KEY WORDS: oil spill, in-situ burning, countermeasure

INTRODUCTION

Background

The burning of spilled oil on the water surface, know as in-situ burning, has been recognised for many years as a potentially effective way of removing large quantities of oil in a relatively short time period. This technique has been successfully used in response to oil spills in broken ice in Alaska, Canada, Finland and Scandinavia on a number of occasions over the past few decades (Buist et al. 1994). In Europe, the first recorded use of in-situ burning was following the grounding of the *Torrey Canyon* of the coast of Great Britain in 1967. In 1969, two incidents occurred in Finland where in-situ burning was used as a response technique: the sinking of the cargo ship *Eiva*, which released 15,000 L of diesel, and the grounding of the tanker *Raphael*, releasing over 60 tons of crude. Some of the diesel was successfully burned using a paraffinic oil as an igniter while more than 90% of the crude oil was removed using peat moss, fuel oil and petrol as ignition sources (Coupal 1972, Haaktela 1970). More recently, tanker spills which have accidentally caught on fire (e.g. the *Agean Sea* and *Haven* incidents) have demonstrated the potential effectiveness of in-situ burning, particularly where nearby shorelines may have been threatened or further impacted by a oil coming to shore (Berguerio Lopez et al. 1993 and Turbini et al. 1993).

Advantages and disadvantages of burning

The two most commonly used oil spill response methods are mechanical recovery and use of chemical dispersants. Mechanical recovery involves collecting the oil in a containment boom and removing it from the water's surface using skimmers. Oil recovery can be quite effective under calm sea conditions, however, under more adverse conditions, recovery rates can be severely reduced. Skimming operations are labour intensive, require ocean going vessels and considerable heavy equipment and logistical support. Temporary storage is required for the recovered oil, which can contain over 75% water. Mechanical recovery in broken ice can be carried out when ice coverage is less than 20%, although recovery rates are reduced. Dispersants are used to enhance the natural dispersion of an oil spill at sea. They can be sprayed onto the oil slick from boats or aircraft. This technique is effective under certain conditions, however, some wave action is required to mix the chemical into the oil slick and dispersants have limited effectiveness once the oil become heavily weathered. Dispersants could be used in open waters or waters with low ice coverage (< 50%). Dispersants are less effective in waters with higher ice coverage in part due to the low wave energy in these conditions.

Burning of oil in broken ice can be accomplished when ice coverage is greater than 70%, in which case the oil is naturally contained and thickened between floes. Alternatively, fire proof booms can be deployed when ice coverage is less than 20%. As with any response method, there are a number of advantages and disadvantages for using in-situ burning in response to a marine oil spill, many of which are outlined in detail in Allen and Ferek 1993 and Buist et al. 1994. Some of the advantages of this method include, high efficiency and elimination rates of the spilled oil, minimal storage and disposal requirement for the recovered burn residue, and minimal environmental impact (Fingas et al. 1994). This method is also particularly suited for remote areas, due to the simple logistics and reduced storage requirements, and to clean-up of oil in ice where ice coverage may hinder or prohibit the use of booms and skimmers. The broken ice situation has the added advantage of providing natural containment. The most obvious drawback of this method is the production of black smoke when burning crude oil. There is also some concern regarding the fate and toxicity of burn residue.

Despite the promising results from spill cases where oil was burned either intentionally or accidentally, policy makers have been reluctant to approve in-situ burning as a tool for responding to oil spills. Only recently has burning been pre-approved for a number of specific regions in the United States, due primarily to the positive demonstration of this technique as an efficient cleanup method following the grounding of the *Exxon Valdez*, in Alaska (Allen 1991) and the ongoing research and development conducted to further develop this technique.

Research and development

Research and development efforts have helped define the capabilities and limitations of burning, have lead to improvements in the available technology and have addressed some of the concerns related to burning. In the 70's and 80's a number of experiments demonstrated the potential of this method for cleanup in remote areas and ice infested waters (Brown and Goodman 1986, Energetex Engineering 1977, Industry Task Force 1983).

Many different ignition devices have been tested over the last several decades, ranging in sophistication from something as simple as knotted diesel soaked rags and sorbents, to complex devices with time delay fuses or laser-based ignition devices. These are reviewed in detail in Buist et al. 1994. Effort were made to develop igniters which could be aerially deployed, and would be particularly suited for the ignition of oil in ice. The idea of using a Helitorch, a device which is slung below a helicopter, and remotely operated by the pilot to deploy globules of burning fuel was investigated for this purpose in the early 1980's. These devices are commonly used by forestry services in Canada and the United States during fire-control operations. Typically, the fuel used in this device consists of gelled gasoline and /or diesel. This concept has proven to be effective for the ignition of crude oils, but was limited by the degree of weathering of the oil. Suggested improvements to this concept were to replace the gasoline and diesel with fuels which burn at a higher temperatures and to add an emulsion breaker to the fuel for the ignition of emulsified oils (Bech et al. 1991 and 1993). These improvements have shown promise during recent field experiments (Guénette and Sveum 1995), and have lead to the development of an emulsion breaking igniter (EBI).

The development of fireproof oil containment booms in the early 1980's made it possible to confine slicks for burning, thus increasing the window of opportunity for in-situ burning to open sea scenarios. A number of fireproof booms are commercially available today (Buist et al. 1994).

The past decade has seen an increased interest to further develop in-situ burning technology. Recent experiments and spill incidents where burning was used have further demonstrated the effectiveness of in situ burning as a countermeasure to spills on water (Fingas et al. 1995, Guénette and Sveum 1995), in ice-infested waters (Bech et al. 1992, 1993, Brown and Goodman 1986, Guénette et al., 1995 Guénette and Wighus, 1996, Smith and Dias 1987), in snow (D'Atri et al. 1993, Sveum et al. 1991), in fresh and salt water marshes (Gonzales and Lugo 1995, Tunnell et al. 1995) and on land (Moir and Erskin 1994).

Considerable research efforts have been dedicated to quantitatively and qualitatively identifying the components of burning crude oil emissions and to developing models to predict plume structure, trajectory and composition. (Evans et al. 1994, Fingas et al. 1995, McGratton et al. 1995). Laboratory findings indicate that particulate emissions may account for 5 to 10 percent of the oil burned, with about half of this mass released as soot. Other emissions include CO, SO_2 and NO_x. Two full scale experiments were carried out off the coast of Newfoundland in 1993 involving a total release and burning of approximately 100 m^3 crude oil in a fireproof boom. An important finding from these experiments was one concerning the emissions from an in-situ burn: all compounds and parameters measured beyond about 150 m from the fire were below health concerns and very low emissions were detected beyond 500 m (Fingas et al. 1995).

In an effort to reduce emissions from an in-situ burn, an number of chemical additives have been investigated for their potential effectiveness (Mitchell 1990, Moir et al 1993, Guénette et al. 1994). Findings have shown that ferrocene and its derivatives can be very effective in reducing the amount of soot produced, by mixing it into the oil prior to ignition. Soot reductions of over 90% have been reported. However, due to the density and low solubility of this product, the problem remains to develop an effective means of applying ferrocene to a slick.

Burn residue tends to be a thick, tar-like substance with a very high viscosity and a density greater than that of the initial oil. It has been noted during a number of experimental burns that the burn residue generally remains neutrally buoyant (Energetex 1977, Evans et al. 1986 and Guénette et al. 1994) but has on some occasions and under certain conditions been known to sink. Chemical analysis and multi-species toxicity testing of water sampled below a burning slick, both in the lab and during full-scale burning indicated that in-situ burning did not adversely affect the underlying water column beyond those effects already associated with the unburned oil (Daykin et. al. 1995).

PRINCIPLES OF IN-SITU BURNING OF SPILLED OIL

Ignition and sustained burning

Ignition of an oil slick occurs when the surface temperature of the slick is raised to its flash point; the point where the hydrocarbons are vaporised in sufficient concentrations to support combustion. In order for flame spreading and sustained burning to occur, the slick surface temperature must reach its fire point, which is usually several degrees higher than the flash point. This is the temperature at which the rate of vaporisation is equal to or greater than the rate of combustion. Therefore for successful ignition of a slick to occur, the ignition source must be capable of raising the slick surface to its fire point. A minimum slick thickness (1 to 3 mmm for fresh crude oils) slick is required to achieve ignition. If the slick is too thin, heat escapes to underlying water, the slick surface never reaches it's fire point and ignition does not occur. Once a slick area of 1 m^2 is aflame, the burn is considered under way. Usually, a fresh or slightly evaporated oil can be easily ignited.

As crude oil weathers, it loses the lighter more flammable compounds by evaporation and begins to entrain water through emulsification. The burning process for water-in-oil emulsions is somewhat different than that of unemulsified oils. It was postulated by Bech et al. 1993 that it is not the emulsion that burns, but rather it is a layer of oil floating on top of the emulsion that supports combustion. The controlling factor in ignition and burning of emulsions appears to be the removal of water from the emulsion. Once the water is removed, oil is release, forming a layer of water-free oil on the emulsion surface. It is this water-free oil layer which can then be volatilised and ignited given that a sufficient heat source exists. The sustained burning of an emulsion slick will depend on a continuous and dynamic water removal process.

Extinction

As burning progresses the slick becomes thinner, therefore reducing the insulating capacity of oil layer. Extinction occurs when the slick thins to the point of no longer being able to support combustion. This occurs when the heat loss through the oil increases to the point where the achievable slick surface temperature is below that of the fire point. Radiant heat from the fire is no longer sufficient to volatilise the remaining oil, and the burning process stops. It has generally been found that a burning slick will self extinguish, once the slick thickness reaches about 1 mm.

Environmental conditions

The environmental conditions will impact both on the state of the oil and on the feasibility of carrying out the operational requirements for conducting a burn at sea, both of which will influence the success of an in-situ burn operation. Past experience has demonstrated that some wind can have a positive effect on flame spreading, particularly in broken ice. However, excessive winds can also make ignition and sustained burning difficult. This effect is dependant on the oil type and conditions. The presence of waves can prevent the ignition of marginally ignitable oils and stable emulsions, due to the temporary thinning of the slick at the crest of the swell.

Conditions required for burning

The following guidelines and limitations concerning the requirements for conducting a burn at sea have been established for Alaska (Anon 1995):

- minimum slick thickness:
  - 1 to 3 mm for fresh crude oil
  - 3 to 5 mm for diesel and weathered crude oil
  - 5 to 10 mm for Bunker C and emulsified oil
- state of the oil:
  - evaporative loss less than 30% for most crudes
  - emulsion water content of less than 25% to 50% depending on the emulsion stability and ignition source
- weather conditions:
  - winds speeds of less than 10 to 12 m/s
  - waves of less than 1 m for non emulsified oils, less for emulsions
  - current velocities of less than 0.5 m/s

Burn efficiency

One of the main advantages of in-situ burning is the potentially high oil removal efficiency which can be obtained. Efficiencies of up to 99% are reported in the literature. Burn efficiency, or oil removal efficiency, is defined as the percentage of oil ignited and burned minus the amount which remains upon extinction of the burn. This parameter is a function of initial slick thickness, the coverage of flame spreading and environmental conditions. Therefore, greater initial slick thickness will lead to higher oil removal efficiencies. Barriers and fire-proof booms are used to concentrate and contain an oil slick in order to increase the burn efficiency. The effect of wind, pushing the burning slick towards a barrier, known as "wind herding", results in increasing the slick thickness (Bech et al. 1993 and S.L. Ross and Energetex 1986). Currents pushing oil against a barrier can have a similar effects (Guénette et al. 1995). Broken ice or an ice edge can offer natural containment of an oil slick.

The high burn efficiencies observed during a number of spills which have accidentally caught fire have suggested the feasibility of burning slick uncontained (e.g.: Turbini *et al.* 1993). The concept of uncontained burning, investigated by Guénette and Sveum 1995, S.L. Ross and Energetex 1986, is based on the premise that the self-induced wind herding effect of the combustion air drawn in around the edges of a burning slick will prevent the slick from spreading as quickly as an unignited slick.

NORWEGIAN RESEARCH ACTIVITIES

Objectives

SINTEF Applied Chemistry has been working in the field of in-situ burning since 1988, beginning with the first open water testing of the 3M fire proof boom which took place on Spitsbergen. This was followed by a study of the feasibility of burning oil and diesel in snow in 1989, the results of which have been implemented into a Norsk Hydro drilling contingency plan. An experimental Programme was initiated by NOFO (Norwegian Clean Seas), and carried out by SINTEF starting in 1990 to study the in-situ burning of emulsions, as part of a wider NOFO Programme "Oil spill contingency in Northern and Arctic waters" (ONA). The Programme began with an extensive literature review of in-situ burning, and was followed by four years of laboratory and experimental field work. The main objectives of this Programme were to:

- define the limitations and effectiveness of in situ burning in terms of water content, evaporation and film thickness of crude oils
- study the effect of environmental conditions on burning i.e.: wind, waves, currents and ice
- study the processes governing burning emulsions
- improve the current technology available for igniting oil spills at sea and in ice
- study the feasibility of burning crude oil and emulsions uncontained
- test the effectiveness of ferrocene as a soot reducing agent

TEST FACILITIES

The field work for this research Programme took place at SINTEF Applied Chemistry's field laboratory on Spitsbergen, Norway. Two different test sites were used. In the winter/spring time, the test site was a frozen fjord, where experimental basins of any shape or size could be built by excavating the sea ice. The ice thickness in the fjord was on average one meter or more and 24-hour daylight prevails during the late spring. Temperatures typically ranged from -20°C to 0°C and wind speeds from calm to 15 m/s. The test basins ranged in size from 0.5 m^2 to 180 m^2, some of which were fitted with wave makers and current generators. Basin configurations included circular, square, rectangular and a circulating flume. In the summer and fall, experiments were carried out in a salt-water lagoon, measuring approximately 1300 m in length and 800 m at its widest point. Apart from a narrow channel, the lagoon was entirely enclosed, minimising the possibility of loss of oil or residue to the main fjord. The lagoon offered a calm, sheltered location for conducting burn experiments. Large scale uncontained burning of crude oil and emulsions was studied at this test site.

The majority of the experiments were carried out with Statfjord crude oil, a light paraffinic oil typical of North sea crudes.

MAIN FINDINGS

Effect of waves, currents, wind, and broken ice

Experiments were conducted to study the effect of environmental conditions on burning of crude oils and emulsions. These experiments also evaluated the feasibility of using burning in ice infested waters. Experiments in the order of 0.4 to 8 m^3 have been carried out in the field.

The general trends observed during these experiments were that the burn efficiency of oils and emulsions decreased with increasing degree of evaporation and increasing water content. The burn rate decreased both with increasing degree of evaporation and water content. For water free oil, the rate decreased from 3 mm/min for fresh crude to 1.4 mm/min for evaporated oil. Also, high winds (9 to 10 m/s) tended to decrease the burn rate. Fresh oil ignited instantaneously, whereas weathered and emulsified oil were more difficult and in some cases impossible to ignited. Flame spreading decreased with increasing evaporation and emulsification (Bech et al. 1992 and 1993, Guénette et al. 1995).

Small waves (0,3 m high, 3 m long) had little impact on the burning of fresh and evaporated oils, but made ignition and sustained burning of emulsions difficult to impossible. The effect of currents was studied by containing oil and emulsion slicks against a barrier in a 0.3 m/s current. Emulsions of weathered Statfjord crude oil burned with efficiencies of up 90% under the conditions tested.

For most oil conditions, burning was possible in winds of up to 12 m/s, although the ignition of some emulsions was difficult. Upwind flame spreading for emulsions was much slower than that observed with fresh and evaporated oils. Generally, wind enhanced flame spreading, especially when in broken ice. Flame spreading in broken ice to areas not directly connected to an ignited area was highly dependant on wind direction and flame deflection.

Experiments in broken ice suggested that burning can be an effective way of removing oil from water (Bech et al. 1993 and Guénette and Wighus, 1996). Three large scale experiments carried out in a 180 m^2 basin with up to 50% ice coverage, demonstrated that fresh crude and emulsions could be burned with efficiencies of up to 99%. Results and experimental set-up for these experiments are depicted in Figure 1. As indicated in this diagram, smaller basins were excavated in the ice downwind from the large basin. Each of these small basins contained a 10 mm slick of either fresh, 25% evaporated/50% w/o emulsion or 25% evaporated/65% w/o emulsion of Statfjord crude. The goal of these experiments was to study the factors affecting flame spreading from one burning area in broken ice to another separate pool of oil. The oil in the large basin was ignited and observations on flame spreading to the small basins were made. It was evident from these experiments that the wind velocity and direction was a very critical factor for flame spreading. Also indicated in this diagram are those basins in which the oil was ignited and burned during each experiment.

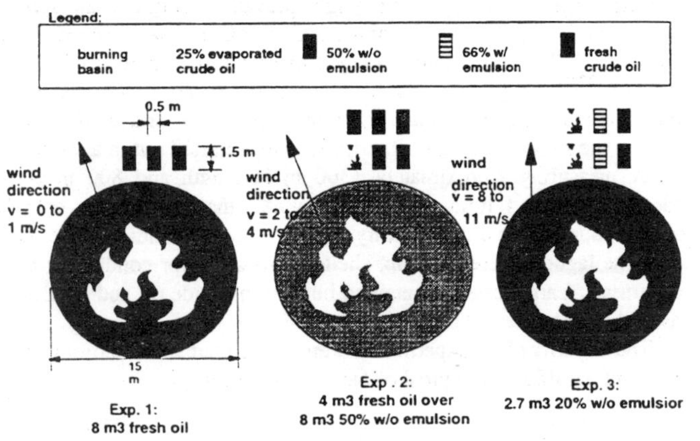

Figure 1. Layout of the test basins, excavated in the sea ice, for studying in-situ burning in ice. Several small basins were built in the ice 1.5 m downwind from the main burn.

In the first burn, carried out with 8 m^3 of fresh oil in the large basin, none of the oil in the small basins ignited. The wind speed during this experiment was low, ranging from 0 to 1 m/s, and caused very little deflection of the flames. Although a large amount of heat was generated from this fire, it was not radiated directly over the small basins.

In the following two experiments, the wind velocity was greater causing the flames to be deflected directly over some of the small basins. The flame angles were estimated to be between 30 and 35° from horizontal during the second experiment. This caused the oil in some of the small basins to ignite and burn. In this experiment, the contents of two basins, one with 25% evaporated oil and the other with a 50% water-in-oil emulsion ignited and burned with a very high efficiency (over 95%). These basins were located 1.5 m downwind from the main basin. In the third experiment, during which wind velocities were even higher, oil in four of the basins was ignited, including two which were further away from the burn (3.5 m). These results indicated that the wind direction, and more importantly the deflection of the flames due to wind have a significant impact on flame spreading in broken ice.

Emulsion breaking igniter research and development

Earlier work had demonstrated the effectiveness of gelled gasoline deployed from a Helitorch in igniting fresh oil slicks, however the effectiveness of this technique was limited with more weathered oils. The purpose of this part of the research Programme was to improve this igniter technology, and to develop an emulsion breaking igniter (EBI) deployable from a Helitorch for the ignition of heavily emulsified oils. This was achieved by assessing the effectiveness of using different fuels and the effect addition of demulsifiers and flame-spread/combustion promoting chemicals to the Helitorch fuel. Two concepts related to the application of the emulsion breaker were explored. One was to combine the chemical demulsifiers with the crude oil igniter, while the other was to apply the additives to the slick first, followed by the deployment of gelled fuel igniters. Both concepts showed promise and further work in this area was recommended. Figure 2 shows an experiment carried out to compare the effect of emulsion breaker concentration on the effectiveness of a gelled crude oil igniter. These igniters were tested on a 50% w/o emulsion of 25% evaporated Statfjord crude oil. A range of emulsion breaker concentrations, from 0.5% to 20% were evaluated. It was concluded that an emulsion breaker concentration of approximately 4 to 5% by volume in gelled Statfjord crude was most effective.

Figure 2. Testing of gelled fuel igniters with varying concentrations of emulsion breaker. The igniter to the right contained no emulsion breaker. The test oil was a 50% water-in-oil emulsions of 25% evaporated Statfjord crude oil.

Promising results obtained in 1992 (Bech et al. 1993) lead to a prototype emulsion breaking igniter which was successfully deployed from a Helitorch during the 1994 (Guénette and Sveum, 1994) field experiments. This igniter concept which consists of a gelled mixture of readily available fuels (Bunker, diesel and gasoline) and performance enhancing additives (an emulsion breaker and an anti-foaming agent) was deployed using the Helitorch from a height of approximately 4 m, and was capable of igniting a stable 50% water-in-oil emulsion of 25% evaporated Statfjord crude oil.

The main findings from this work was that the combination of an emulsion breaker with the gelled fuel igniter could effectively ignite otherwise unignitable emulsions. Regarding the deployment of igniters onto emulsified oil slicks, it was important that the gelled fuels be deployed upwind and directly onto the slick. Flame spreading over emulsified oils was almost entirely in the downwind direction.

Uncontained in-situ burning of crude oil and emulsions

A series of experiments was conducted in a lagoon on Spitsbergen during the fall of 1994 to study the feasibility, burn rates and efficiencies, and the behaviour of uncontained burning crude oil and emulsion slicks (Guénette and Sveum, 1995). The test oils were initially contained in a 10 m steel ring floating at the water's surface, ignited, and then released. Spill sizes ranged from 0,5 to 8 m^3 for fresh crude oil, and 1,2 to 4 m^3 for emulsions. The slick thickness varied from 6 to 100 mm depending on the initial volume of oil. The residue was collected and weighed so that burn efficiencies could be estimated.

The estimated burn efficiencies for fresh and emulsified Statfjord crude oil burning uncontained are given in Figure 3. The burn efficiency is given as percentage of oil or emulsion removed from the water surface. Fresh Stafjord crude oil burned with a maximum efficiency of 92%, while the emulsions of weathered Statfjord crude burned with a maximum of 75% oil removal efficiency. Burn efficiency increased with increasing slick thickness and the emulsions burned with lower efficiencies than the fresh crude oil. These results are consistent with previously reported results.

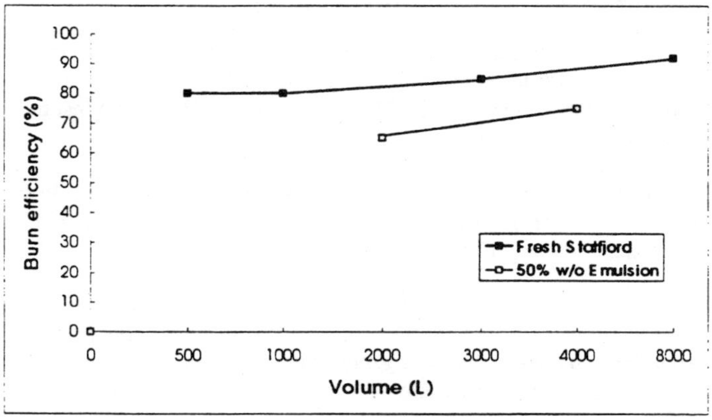

Figure 3. Burn efficiency of uncontained fresh crude and 50% w/o emulsion of 25% evaporated Statfjord crude oil. Initial slick diameter was 10 m.

The extent of flame spreading over the burning slicks is given in Figure 4, as the area of the slick covered in flames as a funtion of time. It is evident from this figure, that emulsions do not spread as quickly, nor as much as the fresh oil when burning uncontained. In the case of a fresh oil slick, as the oil layer is heated by the fire, the slick temperature increases and the viscosity of the slick decreases. The main resistance to spreading occurring during the fresh oil burns is the self-induced wind herding effect of the combustion air drawn into the slick. Because of the water in an emulsions slick, the temperature remains below that of the boiling point of water, until the water is evaporated. The viscosity of an emulsion is therefore likely to remain higher than that of an unemulsified oil and offer a greater resistance to spreading. This factor, combined with the self-induced wind herding effect may explain the limited spreading observed with the burning emulsions.

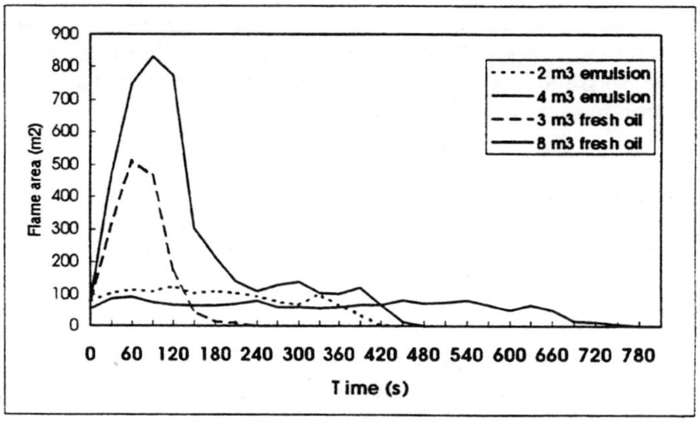

Figure 4. Flame area for fresh crude oil and 50% w/o emulsions burning uncontained

These experiments demonstrated that fresh crude oil and emulsions can be burned uncontained with reasonably high removal rates. The main conclusion from these experiments was that uncontained burning of crude oil and emulsions is feasible if the slick is sufficiently thick, and in the case of emulsions, if a large enough area can be ignited.

CONCLUSIONS

The in-situ burning of oil on water has been shown to be a response method worthy of consideration. Toxic effects and health concerns are not necessarily a major limitation. A deficiency still exists with respect to artificial containment mechanisms, however, if the oil is contained then high oil removal rates can be obtained. Natural confinement can be obtained under certain ice conditions and given the logistical difficulties of operating in arctic environments, in situ burning is a very promising response option for the cleanup of oil spills in broken ice. Important factors to consider are the wind velocity and direction. Ignition and removal efficiency of oil is a function of the weathered conditions of the spilled oil and the slick thickness. Recent advances in ignition technology, enabling the ignition of previously difficult or impossible to ignite emulsions, have widened the window of opportunity for using in-situ burning. The possibility of burning oils uncontained or in low ice coverage conditions further increases the opportunity for using this response technique.

The following areas of research, aimed at broadening the widow of opportunity for in-situ burning are suggested:

- Continue to improve the igniter technology for heavily emulsified oils.
- Further define the capabilities and limitations of burning in ice infested waters.
- Improve the possibility of burning under adverse environmental conditions (wind and waves).
- Uncontained burning of crude oil has shown promise and should be given further consideration.
- Develop effective methods of recovering the burn residue. Although the residue represents only a small percentage of the initial spill volume, for very large spills, or for emulsified oil which may not burn as efficiently, effective recovery techniques would be advantageous.

ACKNOWLEDGEMENTS

This Programme has been jointly funded by NOFO (The Norwegian Clean Seas Association for Operating Companies), Alaska Clean Seas (ACS), Marine Spill Response Corporation (MSRC), Chemische Betriebe PLUTO and AMOCO Eurasia. The author would also like to thank the Store Norske Spitsbergen Kulkompani and the Governor of Spitsbergen for their support during the experimental field work.

REFERENCES

Anon. (1995). In-Situ Burning: A Valuable Tool for Oil Spill Response. Alaska Clean Seas and Alaska Department of Environmental Conservation.

Allen, A.A. 1991. Controlled burning of crude oil on water following the grounding of the *Exxon Valdez. Proceedings of the 1991 Oil Spill Conference*, March 4-7, San Diego, California. American Petroleum Institute, Washington, D.C. pp213-216

Allen, A.A. and R. J. Ferek. 1993. Advantages and disadvantages of burning spilled oil. *Proceedings of the 1993 International Oil Spill Conference*, March 29-April 1, Tampa, Florida. American Petroleum Institute, Washington, D.C. pp765-772

Bech, C., P. Sveum, and I. Buist. 1992. In-situ burning of emulsions: The effects of varying water content and degree of evaporation. *Proceedings of the 15th AMOP Technical Seminar*, Environment Canada, Ottawa, Ontario. pp547-559

Bech, C., Sveum, P. and Buist, I.A. 1993. The effect of wind, ice and waves on the in situ burning of emulsions and aged oils. *Proceedings of the 16th Arctic Marine Oilspill Program Technical Seminar*, June 7-9, Calgary, AB. Environment Canada, Ottawa, Ontario. pp735-748

Bergueiro Lopez, J.R, N. Morales Corres and F. Dominguez Laseca. 1993. The *Aegean Sea* Incident: A quantitative evaluation of the fate of the oil. *Spill Technology Newsletter*. Vol.18(3). pp.1-6

Brown, H.M. and R.H. Goodman. 1986. In situ burning of oil in ice leads. *Proceedings of the 9th Arctic Marine Oilspill Program (AMOP) Technical Seminar,* June 10-12, Edmonton, AB. Environment Canada, Ottawa, Ontario. pp245-256

Buist, I.A., S.L. Ross, B.K. Trudel, E. Taylor, T.G. Campbell, P.A. Westphal, M.R. Myers, G.S. Ronzio, A.A. Allen, and A.B. Nordvik. 1994. The Science, Technology and Effects of Controlled Burning of Oil Spills at Sea. Marine Spill Response Corporation, Washington, D.C. MSRC Technical Report Series 94-013, 382p

Coupal, B. 1972. Use of peat moss in controlled combustion technique. Environment Canada, Ottawa, Ontario, Environment Canada Report No. EPS 4-EE-72-1, 32p

D'Atri, B. and T. King. 1993. Flicking your Bic at twenty-five below (or: The Atigun Pass Diesel Spill project). *Proceedings of the 16th Arctic and Marine Oilspill Program (AMOP) Technical Seminar*, June 7-9, Calgary, Alta. Environment Canada, Ottawa, Ontario. pp669-678

Daykin, M.M., P.A. Kennedy and A. Tang. 1995. Aquatic toxicity from in-situ burning. Emergencies Science Division, Environmental Technology Centre, Environment Canada, Ottawa, Ontario, 62p.

Energetex Engineering. 1977. Ignition and burning of crude oil on water pools under Arctic spring time conditions. Arctic Petroleum Operators Association (APOA) Project 141. Report to Canadian Marine Drilling Ltd. Calgary, AB. 70p

Evans, D.D. 1994. In situ burning of oil spills: Smoke production and plume behavior. *In Situ Burning of Oil Spill Workshop Proceedings*, January 26-28, Orlando, Florida. Minerals Management Service, U.S. Government Printing Office. pp29-38

Evans, D., H. Baum, B. McCaffrey, G. Mulholland, M. Harkleroad, and W. Manders. 1986. Combustion of oil on water. *Proceedings of the 9th Arctic and Marine Oilspill Program (AMOP) Technical Seminar*, June 10-12, Edmonton, AB. Environment Canada, Ottawa, On. pp301-336

Fingas, M.F., G. Halley, F. Ackerman, R. Nelson, M. Bissonnette, N. Laroche, Z. Wang, P. Lambert, K. Li, P. Jokuty, G. Sergy, E. J. Tennyson, J. Mullin, L. Hannon, R. Turpin, P. Campagna, W. Halley, J. Latour, R. Galarneau, B. Ryan, D.V. Aurand, R. R. Hiltabrand. 1995. The Newfoundland Offshore Burn Experiment -NOBE. *Proceedings of the 1995 Oil Spill Conference*, February 27-March 2. American Petroleum Institute, Washington, D.C., pp123-132

Fingas, M.F., Halley, G., Ackerman, F., Vanderkooy, N., Nelson, R., Bissonnette, M.C., Laroche, N., Lambert, P., Jokuty, P., Li, K., Halley, W., Warbanski, G., Campagna, P.R., Turpin, R.D., Trespalacios, M.J., Dickins, D., Tennyson, E.J., Aurand, D. and R. Hiltabrand. 1994. The Newfoundland offshore burn experiment - NOBE - Preliminary results of emissions measurement. *Proceedings of the 17th Arctic and Marine Oilspill Program (AMOP) Technical Seminar*, June 8-10, Vancouver, B.C. Environment Canada, Ottawa

Gonzalez, M.F. and G.A. Lugo. 1995. Texas marsh burn: removing oil from a salt marsh using *in situ* burning. *Proceedings of the 1995 Oil Spill Conference*, February 27-March 2. American Petroleum Institute, Washington, D.C., pp39-47

Guénette, C., P. Sveum, I. Buist, T. Aunaas and L. Godal. 1994. In-situ Burning of Water-in-oil Emulsions. SINTEF Report No. STF21 A94053 Trondheim, Norway. 139pp

Guénette, C. and P. Sveum. 1995. Emulsion breaking igniters: recent developments in oil spill igniter concepts. *Proceedings of the 18th Arctic and Marine Oilspill Program (AMOP) Technical Seminar*, June 14-16, Edmonton. Alberta. Environment Canada, Ottawa., pp1011-1026

Guénette, C. and P. Sveum. 1995. In-situ burning of uncontained crude oil and emulsions. *Proceedings of the 18th Arctic and Marine Oilspill Program (AMOP) Technical Seminar*, June 14-16, Edmonton, Alberta. Environment Canada. Ottawa., pp997-1010

Guénette, C.C., P. Sveum, C.M. Bech and I.A. Buist. 1995. In-situ burning of emulsions studies in Norway. *Proceedings of the 1995 Oil Spill Conference*, American Petroleum Institute, Washington, D.C., pp115-122

Guénette C.C. and R. Wighus. 1996. In-situ Burning of Crude Oil and Emulsions in Broken Ice. *Proceedings of the 19th Arctic and Marine Oilspill Program (AMOP) Technical Seminar*, June 12-14, Calgary, Alberta. Environment Canada, Ottawa., pp899-906

Haaktela, I. 1970. Oil spills off Finland. Marine Pollution Bulletin 1(1):19-20

Industry Task Group. 1983. Oil spill response in the Arctic. Part 2. Field demonstration in broken ice. Industry Task Force comprised of Shell Oil Company, Sohio Alaska Petroleum Company, Exxon Company, U.S.A., and Amoco Production Company, Anchorage, Alaska

McGratton, K.B., W.D. Watson, A.D. Putorti, W.H. Twilley, J. McElroy and D.D. Evans. 1995. Smoke plume trajectory from in-situ burning of crude oil in Alaska - field experiments. *Proceedings of the 18th Arctic and Marine Oilspill Program (AMOP) Technical Seminar*, June 14-16, Edmonton, Alberta. Environment Canada, Ottawa., pp901-914

Mitchell, J.B.A. 1990. Smoke reduction from burning crude oil using ferrocene and its derivatives. Combustion and Flame 86: 179-184

Moir, M. E., S. Charbonneau and J.B.A. Mitchell. 1993. Soot reduction chemicals for in-situ burning. *Proceedings of the 1993 International Oil Spill Conference*, March 29-April 1, Tampa, Florida. American Petroleum Institute, Washington, D.C. pp761-763

Smith, N.K. and A. Diaz. 1987. In-place burning of crude oils in broken ice. *Proceedings of the 1987 Oil Spill Conference,* April 6-9, Baltimore, Maryland. American Petroleum Institute, Washington, D.C. pp383-387

S.L. Ross Environmental Research Ltd. and Energetex Engineering. 1986. In-situ Burning of Uncontained Oil Slicks. Environment Canada Report EE-60. Environment Canada. Ottawa. Canada. 112pp

Sveum, P., C. Bech and M. Thommansen. 1991. Burning of oil in snow: Experiments and implementation in a Norsk Hydro drilling contingency plan. *Proceedings of the 14th Arctic and Marine Oilspill Program (AMOP) Technical Seminar*, June 12-14, Vancouver, B.C. Environment Canada, Ottawa. pp.399-410

Tunnell, J. W., B. Hardegree and D. W. Hicks. 1995. Environmental impact and recovery of a high marsh pipeline oil spill and burn site, Upper Copano Bay, Texas. *Proceedings of the 1995 Oil Spill Conference*, February 27-March 2. American Petroleum Institute, Washington, D.C. pp133-138

Turbini et al. 1993. The *Haven* incident: Lessons learned with particular reference to environmental damages. *Proceedings of the 1993 Oil Spill Conference*, American Petroleum Institute, Washington, D.C. pp179-183

Proceedings of the Seventh (1997) International Offshore and Polar Engineering Conference
Honolulu, USA, May 25-30, 1997

ISBN 1-880653-28-1 (Set); ISBN 1-880653-30-3 (Vol. II)

Modelling Oil Pollution Under Ice Cover

A. Alhimenko, A. Bolshev and A. Yakovlev
State Technical University, St. Petersburg, Russia
K. Klevanny
Flood Protection Agency "Morzashchita", St. Petersburg, Russia
S. Liukkonen
VTT Manufacturing Technology, Espoo, Finland

ABSTRACT

St. Petersburg Technical University and VTT Manufacturing Technology together with the industry and the oil spill combatting authorities in Finland and Russia are running a joint research project entitled "Marine Oil Spill Combatting in Arctic and Sub-Arctic Conditions".

This paper describes the objectives of the study and gives the main results obtained in the field and laboratory experiments and in the numerical modelling.

St. Petersburg Technical University and VTT Manufacturing Technology carried out a series of field experiments in the Gulf of Finland of the Baltic Sea during the winter season of 1995-96. The experimental study consisted of oil spilling under the ice cover and measuring diffusion coefficients under different water current velocities.

The first approximation of a numerical model of the process was developed and checked out. It is supposed that the numerical model will be used for oil combatting operations in ice conditions at the Prirazlomnoe oil field at the Barents sea.

1. INTRODUCTION

1.1. General

Field tests show rather high concentration of oil in water at the shelf zone of northern seas. The development of offshore oil and gas fields may increase the oil pollution in the seas. The problem is to provide companies responsible for oil and gas fields development with methods of oil spill combatting under ice conditions. The following typical situation may be considered. An under-water oil pipeline is damaged. Oil leaks out and floats under the ice cover. It is necessary to locate the oil slick in order to eliminate it. The search of oil spills under a thick ice cover by optical methods from aircraft and satellites is usually unsuccessful. So, the main instrument determining the locations and parameters of under-ice oil spills may become the numerical modelling of oil spreading under ice. Such a model may be useful for managing the spill combatting operations, selection of suitable clean-up equipment, etc.

The development of such a model for oil spreading under ice requires understanding the interaction processes between oil and ice. This interaction includes adhesion of oil to ice, diffusion processes under the influence of a water current, influence of the surface roughness of the ice bottom and other factors. The adhesion of oil to ice has been studied at VTT Manufacturing Technology. St. Petersburg State Technical University has conducted the laboratory and field tests and the theoretical investigations reported here.

1.2. Target of the study

At St. Petersburg State Technical University, research work on oil spreading in water has been conducted for more than 10 years (Alhimenko, 1989; Alhimenko, 1990). During the past five years processes of oil spreading under ice have been studied.

The most interesting results of studies on the process of oil spreading under ice have been described in works (Yapa, Weerasuriya, Belaskas, and Chowdhury, 1993; Usiner, Weiskopf, Cox, and Schultz, 1979; venkatesh, El-Tahan, Confort, and Abdelnour, 1990; Puskas, McBean, and Kouwen, 1987; Fingas, 1992). One detail is common for all these works: practically all experiments have been carried out in laboratories. However, it is known that the processes of oil spreading in water under ice in a real spill case in the field is difficult to predict only on the basis of the small-scale laboratory tests. Because real water and oil have to be used also in the laboratory tests, it is impossible to scale the spreading processes down to physical model tests. Therefore, field observations of oil spreading under ice after an accident or in an organized test are of essential importance. Monitoring the behaviour of oil in the sea after the accident of the MT Antonio Gramsci in the Baltic Sea in 1987 allowed to propose new type of idea for oil spreading under ice. Afterwards, the idea has been experimentally verified.

The main idea is that oil spreads in water under ice as a cloud drops of various sizes. The oil drops move due to the water current in the field of current boundary layer under the bottom surface of the ice cover. The oil drops are carried by the water flow separately from the ice cover. The velocity of the oil drop cloud depends on the velocity of the water current, sizes of the oil drops and roughness of the bottom surface of ice.

This idea had to be checked out in experimental way both in the laboratory and field conditions. According to (Liukkonen, 1992), no specially organized field tests on oil spreading under ice have been made. Some data (Liukkonen, 1996) concerning the monitoring of oil spills at the sea in broken ice are of an occasional character. Therefore, St. Petersburg State Technical University organized special field tests in order to check out the numerical results of the laboratory tests and the results of the numerical modelling.

The goals of the experiments were:

1) To obtain numerical values for the following parameters:

- dimensions of the oil slick ($V_s = L_x, L_y, L_z$) as a function of time under a constant water current velocity and ice roughness;
- velocity of oil slick movement;

2) To verify the results obtained in laboratories with the small volumes of spilled oil;

3) To develop numerical model using the results of experimental research;

4) To lay a ground for conceptual design of oil recovery methods with support systems for ice conditions.

At the same time, field tests on oil penetration through ice were conducted. The goals of that experiment were:

1) To evaluate the absorption of different oil types into ice;

2) To obtain data on the alteration of oil under ice.

2. EXPERIMENTAL METHODS

2.1. Field tests

2.1.1. Field tests on oil spilling under the ice

The experiments were conducted in January-February 1996 in the Gulf of Finland in the Baltic Sea, near St. Petersburg.

The water depth at the test site is about 4.0 m. The thickness of the ice cover was 0.4-0.6 m. The water current velocity varied from 0 up to 0.5 m/s. The background concentration of hydrocarbons in the water was 0.03-0.05 mg/l. Figures 1 and 2 show the procedure of the experiment.

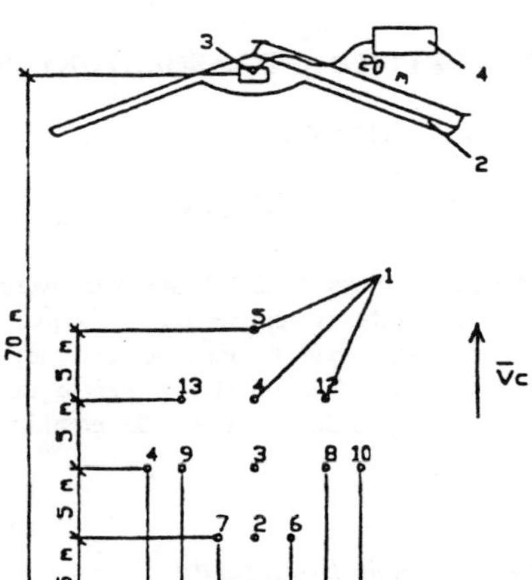

Fig.1. Procedure of field experiment

1 - boreholes; 2 - slots; 3 - booms; 4 - skimmer; 5 - barrel for oil emulsion.

The experiment consisted of spilling oil under the ice cover (Figure 2) from a barrel with a volume of 200 l (1). The oil was led under ice with a pump, through a special set of bore holes. Fourteen bore holes for oil monitoring and measuring and two additional bore holes for flow meters were drilled with a core drill through the ice cover. The water current velocities were measuring under ice cover and on the depth 1 m. The diameter of each hole was 20 cm. The monitoring of oil was conducted through the boreholes at the distance of 80 m in the direction of the current and about 20 m in a direction transverse to the current. The moment of pouring oil under the ice was the start of the experiment. After that, the personnel monitored visually the water surface in the bore holes. The time of the first appearance of oil on the water surface in the bore holes was recorded. Besides the oil samples, samples of water from a depth of 0.5 m and 1.0 m were taken. The samples were delivered to the laboratory and the concentration of hydrocarbons was measured.

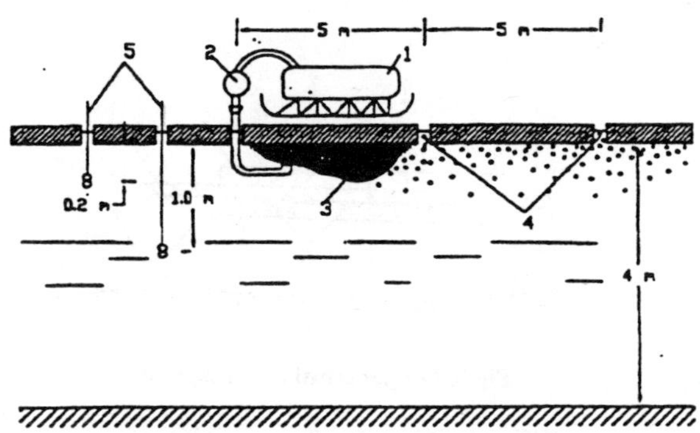

Fig.2. Procedure of the experiment

1 - barrel of oil; 2 - pump; 3 - initial lens of oil; 4 - boreholes; 5 - flow meters.

The residual oil in the sea ended up in a space restricted by booms (3) (Figure 1). The booms were installed in slots (2) made with ice saws. The total length of the boom was about 40 m. Oil was recovered by a skimmer (4). The oil-water emulsion was collected into a special barrel (5). After the completion of the experiment, the oil recovered was transported to a water treatment plant.

The experiments were carried out at three different locations for the following water current velocities: 0 cm/s; 10 cm/s; 25 cm/s and 50 cm/s. Two types of oil, North Sea Brent crude oil and motor oil were used.

2.1.2. Field tests on oil penetration through ice

The experiment started on January 20th, 1996, and was completed on April 8th, 1996. An open wooden grid (Figure 3) was installed not far from the coast line of the Gulf of Finland off the city of St. Petersburg. The grid consisted of 18 cells, each 1 m^2 in area. The grid floated on the water surface, and in order to keep it in its place it was anchored to the sea bottom. The water current velocity at the test site was close to zero. The water salinity was rather low, practically it was fresh water. The background concentration of hydrocarbons in the water was about 0.03 mg/l. An ice cover was let to freeze naturally in the grid. After the ice cover had reached a stable thickness, 0.2 liters of fresh North Sea Brent crude oil, diesel fuel and motor oil were injected into the cells under the ice.

2.2. Laboratory tests

Laboratory tests were organized in order to obtain data on the threshold velocity and relative velocity between ice and oil drops. Motor (lubrication) oil, crude oil and kerosene were used.

Figure 4 shows schematically the procedure of the laboratory experiment. The experiment was conducted in January 1995 at the open basin of St. Petersburg State Technical University.

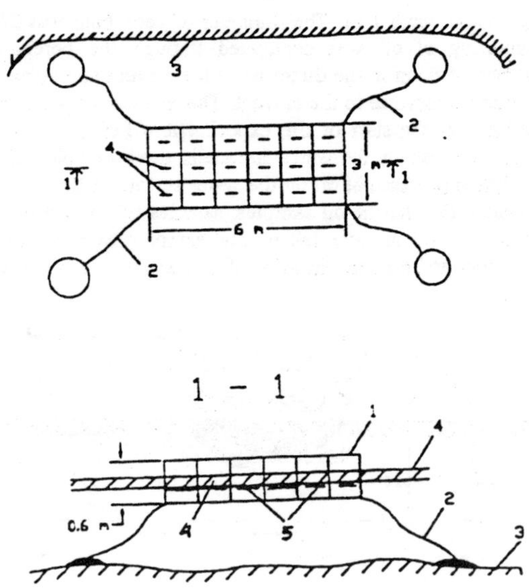

Fig.3. Oil penetration through ice

1 - grid; 2 - anchors; 3 - bottom; 4 - ice cover; 5 - oil layer.

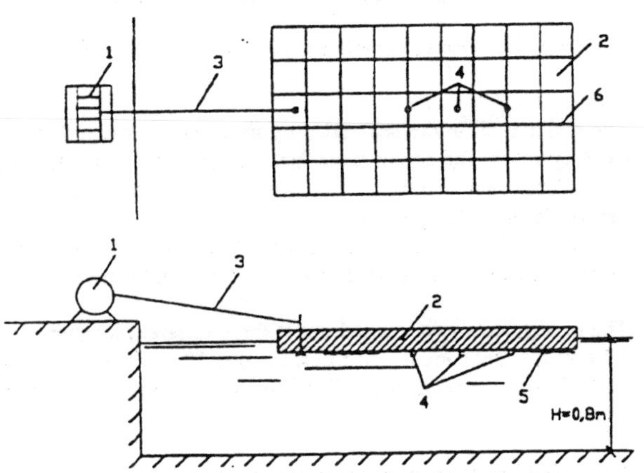

Fig.4. Scheme of laboratory experiment. Towing ice floes

1 - motor; 2 - ice floe; 3 - rope; 4 - oil drops; 5 - roughness bottom surface; 6 - coordinating grid.

To start the test, Baltic Sea water was taken from the Gulf of Finland. This water was left to freeze naturally in a sub-zero temperature at the open basin. After freezing, the 2-cm-thick ice cover was cut into pieces with the dimensions of 1.0 m x 0.8 m. The surface roughness of the ice pieces was measured. A total of 20 surface roughness measurements was taken with the following results. The mean center-line average, R_a, value measured for the bottom surface roughness was 6.43 μm with the standard deviation of 3.02 μm.

The ice pieces were put on the open water (2) and connected to the motor (1) with a rope (3). To get the oil drops under the ice in water, a U-shaped pipe was connected to the pipette in order to lead the oil onto the underside of ice. The drop diameters ranged from 1 mm to 5 mm. After putting the oil drops under the ice, the towing started. The towing velocity ranged from 3 cm/s up to 50 cm/s.

Sixteen samples were taken from the cells of the grid under the ice. The samples included: ice pieces from the cells of the grid, oil from the cells of the grid under the ice and control samples from clean surroundings. Three types of oil, fresh North Sea Brent crude oil, diesel fuel and motor oil were used in the tests. The samples were delivered to the laboratory of VTT Chemical Technology in Espoo, Finland, where they were analyzed. In the analysis, the oil content of the samples was determined with fluorescence spectrophotometry and the quality of oil with gas chromatographic analysis.

Samples were transferred from plastic to glass bottles and extracted with hexane, where after continuos spectra were measured from the extract using the two excitation wave length. Quantitation was made by using the oil that was most closely corresponding a sample.

The quality of oils which were extracted from the seven ice samples and nine oil samples were determined using gas chromatography.

The equipment used consisted of: GC- device, Hewlett-Packard 5890 with FI- detector, column HP-1, 25m, diameter 0.32 mm, liquid phase 0.17 µm. Temperature program: 40°C, 2 min, 5°C/min, 150°C, 0 min, 15°C/min, 300°C, 10 min.

3. RESULTS OF FIELD AND LABORATORY TESTS

3.1. Results of the field tests

3.1.1. Field tests on oil spreading under ice

Motor and crude oil slicks of 0.2 m^3 in volume were released under the ice cover. The water current velocity was equal to zero. The oil slicks were let to settle under the ice cover, and they were monitored after 2 and 27 hours (Figure 5) in order to define their final equilibrium thickness. Table 1 presents the equilibrium thicknesses observed.

Table 1: Equilibrium thicknesses of the oil slick

Oil Type	Dynamic Viscosity (mPa s)	Density (kg/m^3)	Observed equilibrium thickness (mm)	Test conditions source
Sea crude oil	500	915	3.03	Broken ice (Benoit, 1996) ice concentration (32%-62%) initial volumes of oil spills 1.136 m^3 and 1.09 m^3
	570	915	7.30	
ADGO	234	952	2.00	
Sea crude oil	450	915	1.30	
Motor oil	-	840	25	Continuous ice, our research initial volumes of oil spills 0.2 m^3
Crude oil	314	880	40	

Observed equilibrium thicknesses for motor oil and crude oil were rather high comparing to reported in literature. The explanation may be changing of ice thickness presented at Fig.6. Oil spreading was limited by depression of ice thickness. The difference of ice thickness is usual for Finnish Gulf of Baltic Sea. It is necessary to take into account this situation at the computer model.

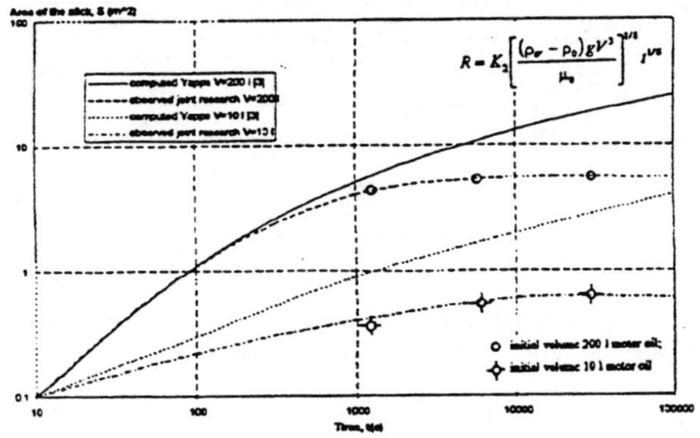

Fig.5. Comparison observed and computed data ($V_c = 0$)

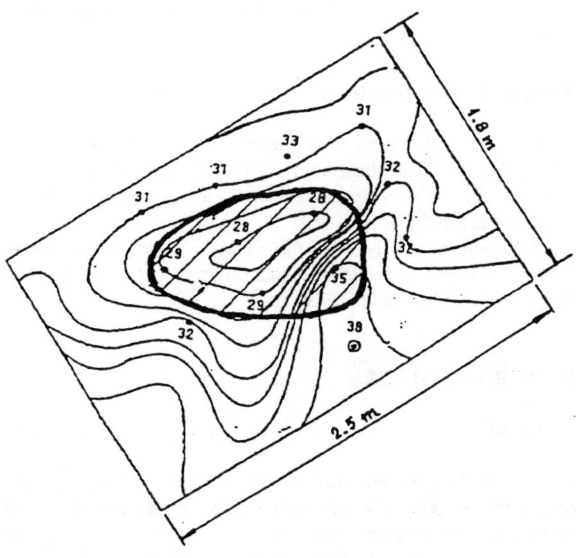

Fig.6. Spreading the crude oil under ice

—— - control of oil spill; 28...32 - ice thickness (cm)

In the tests, crude oil, motor oil and kerosene were released under the ice cover. The water current velocity varyed from 5 cm/s up to 50 cm/s. The bottom surface of the ice cover was rather smooth, but the waviness of the ice bottom resulted in some variations in the ice thickness. Figure 6 shows typical variations of the ice thickness at the test site. The ice thickness was measured by cutting sample cores out of the ice cover. The velocity of oil movement as a function of the water current velocity is presented in Figure 7. The velocity of the motor oil slick could be approximated with equation:

$$v_s = a v_c^{1.8}, \qquad (1)$$

were v_c is the water current velocity in cm/s and a is an empirical coefficient that was determined as 0.04.

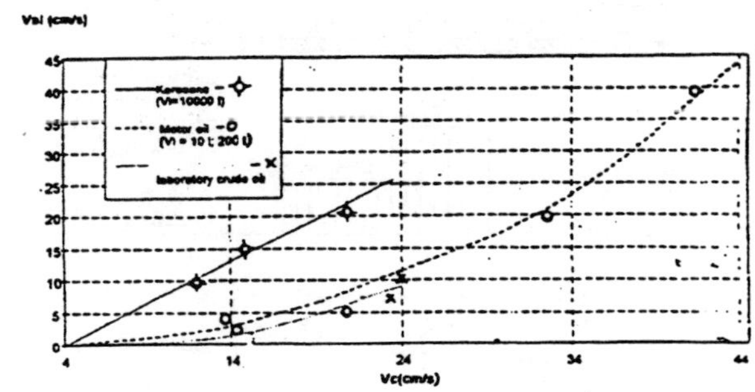

Fig.7. Experimental results: under-ice oil slick velocity as a function of water current velocity (smooth bottom surface)

The velocity of the kerosene slick may be assumed to be the same as the water current velocity. The drops of kerosene in the water could be detected only up to the distance of 50 m downstream. After that the kerosene dissolved in the water. The concentration of the kerosene distributed uniformly from the surface up to bottom after 300 m downstream from the source.

Using a lateral grid of bore holes, angles of lateral widening were measured. The areas of the oil slicks were computed. The influence of the current velocity of the area of the slick under the ice is presented in Figure 8. Also, the angles of lateral widening are presented.

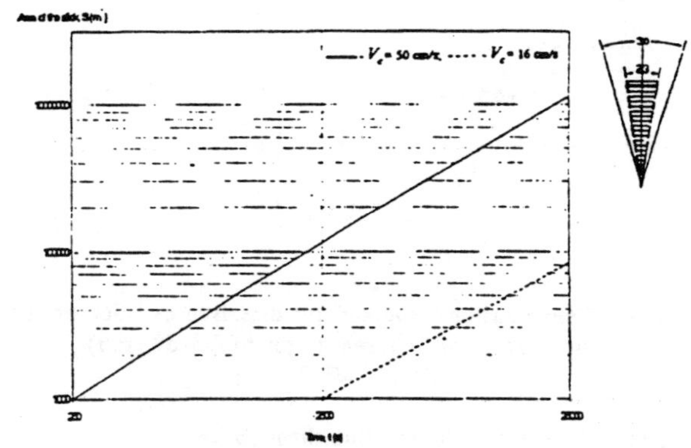

Fig. 8. Influence current velocity on the area of skick under ice. Initial volume of oil 10 litres.

Monitoring the water surface at the bore holes showed that the oil slicks spread under the ice as a cloud of oil drops of various sizes. The average oil drop size depends on the oil type, water current velocity (turbulence coefficient) and the distance from the initial source. In the experiments, the diameters of the crude oil drops were observed to range from 5 mm at bore hole No 2 (v_c = 16 cm/s) up to 1 mm at bore hole No 5, at the same velocity. The average size of the kerosene drops was about 2 mm in the beginning of the test, but at the distance of 20 m (v_c = 10 cm/s) it was impossible to define the droplet size anymore.

The concentrations of dispersed and dissolved hydrocarbons in water were defined. The numerical data are presented in Figures 9 and 10. The samples were taken with special samplers from a depth of 1 m and delivered to the laboratory for analysis.

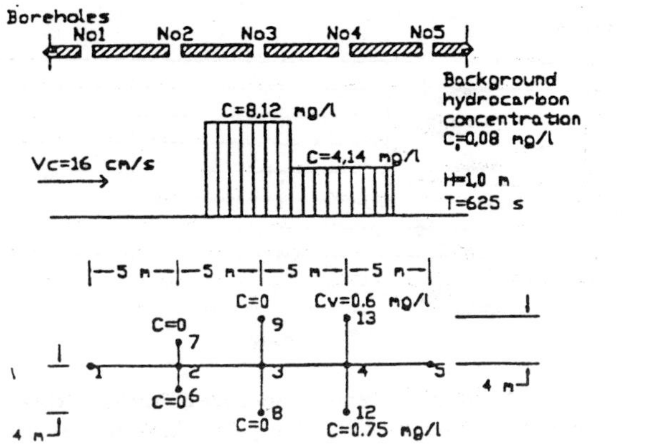

Fig.9. Concetration dispersed ans dissolved hydrocarbons in the water body current impact ($V_c = 16$ cm/s)

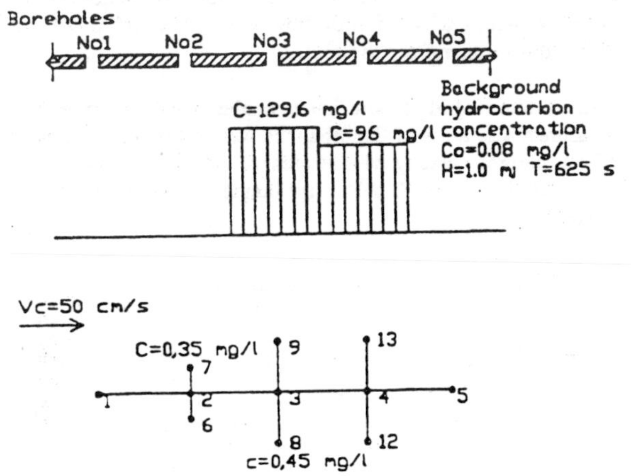

Fig.10. Concetration dispersed ans dissolved hydrocarbons in the water body current impact ($V_c = 50$ cm/s)

4.2. Field tests on oil penetration through ice

Sixteen samples were taken from the cells of the grid under the ice. The samples included: ice pieces from the cells of the grid, oil from the cells of the grid under the ice and control samples from clean surroundings. Three types of oil, fresh North Sea Brent crude oil, diesel fuel and motor oil were used in the tests. The samples were delivered to the laboratory of VTT Chemical Technology in Espoo, Finland, where they were analyzed. In the analysis, the oil content of the samples was determined with fluorescence spectrophotometry and the quality of oil with gas chromatographic analysis.

For the fluorescence spectrophotometry, proper dilutions in hexane were firstly prepared from calibration oils and their continuous fluorescence spectra were measured using two different excitation wave lengths (ex: 310 nm; em: 320-460 nm and ex: 230 nm; em: 240-460 nm).

Then, the oils from the samples from the grid were extracted with hexane, where after the continuous spectra of the extracts were measured using the two excitation wave lengths. The quantitation was made by using the oil that was most closely corresponding to the sample. Samples were transferred from plastic to glass bottles and extracted with hexane, where after continuous spectra were measured from the extract using the two excitation wave length. Quantitation was made by using the oil that was most clisely corresponding a sample. Table 2 shows the results for 7 samples of ice melted in laboratory.

Table 2: Results of the oil content analysis

Sampling date	Type of oil	Quantity (mg/l)
25.03.96	Crude	1300
25.03.96	Diesel	1600
25.03.96	Motor	17000
25.03.96	Crude	23
22.04.96	Motor	23000
22.04.96	Diesel	7000
22.04.96	Crude	15000

Two example gas chromatograms are shown in Figures 11 and 12. The main conclusions of the GC- analysis were:

1. On the basis of the gas chromatogrammes there are no dramatic changes in the quantity of low boiling hydrocarbons.
2. No changes in the ratio of C17 and C18 hydrocarbons have occurred, indicating that no biodegradation of oil has been taken place under ice.
3. Oil penetration into the ice takes the place. The distribution of penetrated into the ice is nonuniformly. More details will be presented in next paper.

4.2. Results of laboratory tests

The results of the laboratory tests demonstrated the influence of the oil drop diameter on the threshold velocity. In particular, our tests demonstrated that the oil drop diameter is a more important factor affecting the threshold velocity than the type of oil. The effect of the oil drop diameter is clearly visible in Figure 13 which also shows a comparison between our data and data from the literature (Usiner, Weiskopf, Cox, and Schultz, 1979).

4. NUMERICAL MODEL

4.1. Equations

For the calculations of oil propagation under ice during the most important phase, i.e., the gravity-viscous, spreading phase, 3D nonlinear hydrodynamic equations with a hydrostatic assumption for barotrophic conditions and 3D equations of advection and diffusion of dissolved substances may be applied

$$u_t + uu_x + uv_y + wu_z + gz_x = fv + (nu_z)_z + m_l Du - (P_a)_x r^{-1} \quad (2)$$

$$v_t + uv_x + vv_y + wv_z + gz_y = -fu + (nv_z)_z + m_l Dv - (P_a)_y r^{-1} \quad (3)$$

$$u_x + v_y + w_z = 0, \quad (4)$$

$$s_t + (us)_x + (vs)_y + [(w-w_0)s]_z = m_s Ds + (n_s s_z)_z - ls + s_l v_l \quad (5)$$

where **z** axis is perpendicular to the ice sheet, **x** and **y** are horizontal axises, **u,v,w** are the Cartesian components of water velocity vector **v**, **z** is the surface level, **f** is the Coriolis parameter, m_l is the horizontal eddy viscosity coefficient, **D** is the 2D Laplacian, P_a is the atmospheric pressure, **r** is the water density, **s** is the concentration of oil, w_0 is the vertical velocity of oil due to buoyancy, m_s is the oil diffusion coefficient , **l** is the coefficient of non-conservativity of oil, s_l is the oil concentration in internal sources, v_l is the discharges in

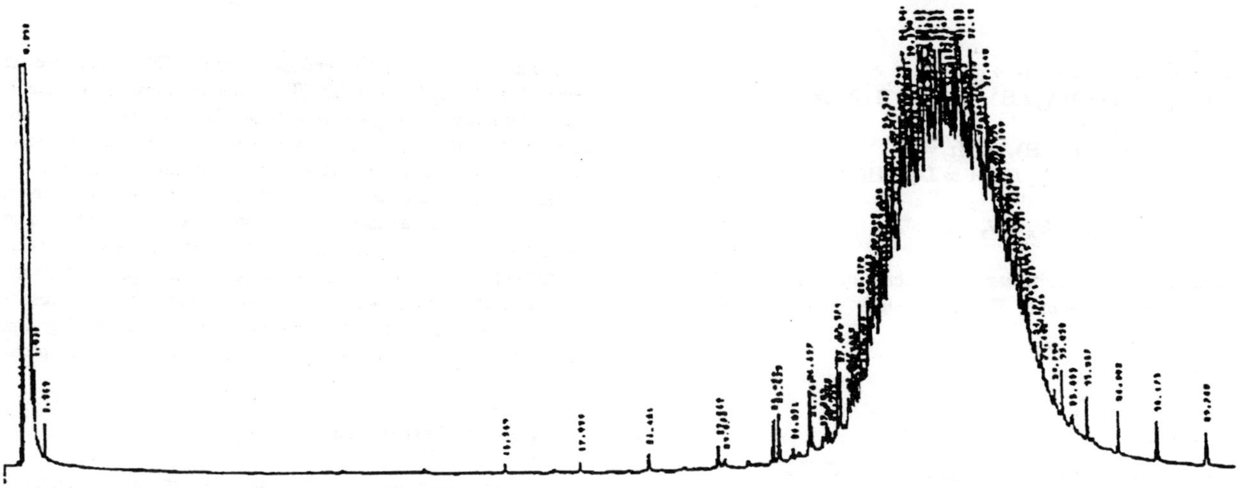

Fig.11. Chromatogramme of samples of motor oil extracted out of the water under the ice after 35 days

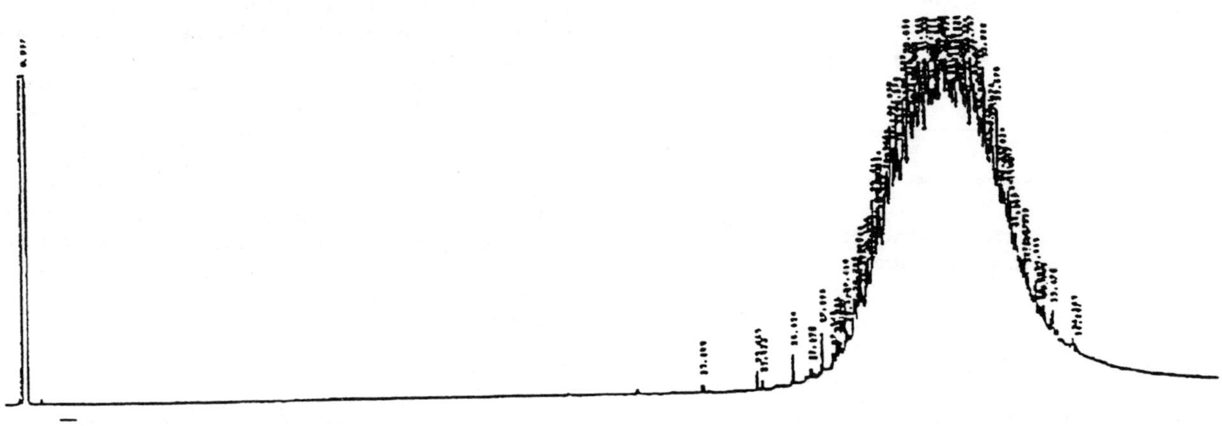

Fig.12. Chromatogramme of samples of ice pieces polluted by motor by motor oil during 35 days

internal water sources. Finally, n and n_s are the vertical eddy viscosity and diffusivity coefficients. They are determined from semi-empirical theories of turbulence. It may be Prandtl model for shallow seas (h < 50 m) and k-e model of turbulence for deeper regions, in which $n = c_m k^2/e$, where k is the kinetic energy of turbulence, e is the velocity of kinetic energy dissipation and c_m is the empirical coefficient. Two additional equations for k and e should be solved in the last case (Frost, and Moulden, 1977).

For the solid parts of the lateral boundaries, zero normal component of velocity vector was assigned. The tangential component is derived from the quadratic friction law with an arbitrary coefficient of wall friction. For the open parts of the lateral boundaries, u, v, z, or radiation ('free exit') relation may be specified. Concentration s is specified on the open parts of the lateral boundaries with ingoing fluxes. The surface velocities in ice-free parts of water surface (if any) are determined from the quadratic wind stress relation $nu_z = k_w|\underline{w}|w_{(x)}$, where $\underline{w}$ is the wind velocity vector with components $w_{(x)}$ and $w_{(y)}$, k_w is the wind stress coefficient. Zero turbulent fluxes of oil $n_s s_z$ are specified for the bottom and upper surfaces.

The bottom and under-ice velocities were determined in the same way using the quadratic friction law: $nu_z = -c_f u|\underline{v}|$, where c_f is the friction coefficient on the bottom or under ice, both determined from the logarithmic law:

$$c_f = [k/ln(z_b/z_0)]^2 \tag{6}$$

for hydrodynamically rough surfaces and

$$c_f = [k/ln[z_b u_b c_f^{1/2}/(0.11 n_{mol})]]^2 \tag{7}$$

for smooth surfaces (iterations should be used in this case), $k = 0.41$ is the von Karman constant, z_b is the distance between the first grid point and the bottom or ice surface, u_b is velocity at this grid point, z_0 is the roughness parameters of the bottom and ice surfaces that is functions of roughness height, length and shape and should be determined experimentally. It is known that for the bottom boundary layer z_0 is about 1/30 of the roughness height h_s, but for very rough surfaces z_0 may reach 1/5 of the roughness height (Monin, and Yaglom, 1965). It is supposed that during the future studies dependence of ice parameter z_0 from ice bottom surface characteristics would be revealed.

To find the water level on each time step with finite differences it is suitable to couple 3D equations with 2D depth averaged shallow water equations

$$U_t + (U^2/H)_x + (UV/H)_y + gHz_x =$$
$$= fV + k_w|\underline{w}|w_{(x)} - c_f U|\underline{V}| / H^2 + m_l DU - H(Pa)_x r^{-1} \quad (8)$$

$$V_t + (UV/H)_x + (V^2/H)_y + gHz_y =$$
$$= -fU + k_w|\underline{w}|w_{(y)} - c_f V|\underline{V}| / H^2 + m_l DV - H(Pa)_y r^{-1} \quad (9)$$

$$z + U_x + V_y = v_h \quad (10)$$

where U and V are the Cartesian components of water flux $\underline{V}$ through the unit width column, **H= h + z**, **h** is the undisturbed depth.

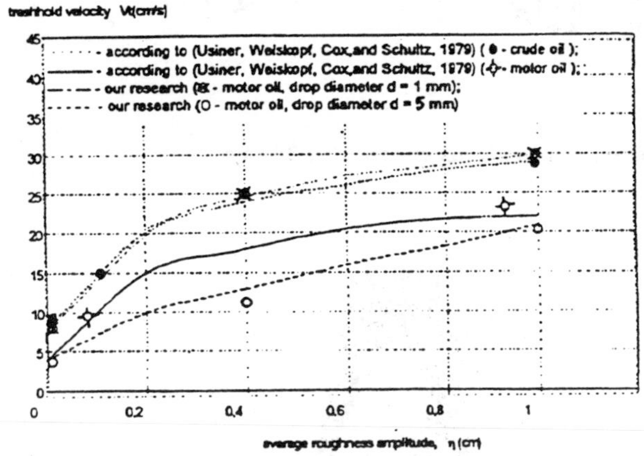

Fig.13. Influence of the ice surface roughness on the threshold velocity for an under-ice drops

4.2. Numerical method

For solving these equations in an arbitrary water domain it is reasonable to introduce new curvilinear co-ordinates **x(x,y)**, **h(x,y)** and **s(x,y,z,t)=(z+h)/H** concordant with the configuration of the given domain. On the lateral boundaries one of the co-ordinates **x** and **h** is fixed, and the other is distributed arbitrary but monotonically. The lateral boundaries are fixed, and the water thickness near them is non-zero. Introducing these co-ordinates leads to mapping of a given domain onto a cube. A drawback of using the Cartesian components of velocity vector **u**, **v** in curvilinear co-ordinates is the interpretation of the boundary condition of zero normal component of velocity on solid lateral boundaries: a complicated averaging procedure should be involved here. To simplify the lateral boundary conditions we introduce contravariant components of flux instead of the Cartesian ones.

A semi-implicit finite-difference scheme is used to solve the dynamics equations (Klevanny, Matveyev, and Voltzinger, 1994; Klevanny, and Matveyev, 1995). The advection, Coriolis and lateral eddy viscosity terms are approximated explicitly, while all other terms are approximated implicitly using the central differences in space and Crank-Nickolson approximation in time. The method of splitting gives a second-order accuracy both in time and in space. The space-staggered C-grid of Mesinger and Arakawa is used. After excluding the surface level from the first 2D motion equation using the mass conservation equation, a tridiagonal solver may be applied to find the first velocity component. After that the surface level and the second velocity component can be found explicitly. Then the vertical distribution of the first velocity component is obtained using a tridiagonal solver in the vertical direction. The procedure for the second half step is similar. The advection-diffusion equation is solved implicitly using a hybrid of the first- and third-order upwind finite differences. A pentadiagonal solver is used. This scheme does not diffuse the pollutant much, while oscillations are eliminated owing to the presence of the first order scheme. Velocity w_0 is assumed constant at internal grid points and equal to zero at the boundary top and bottom points. Thus, the mass of oil remains constant if l is zero. It should be mentioned that good results may be obtained only if the upper boundary layer has a fine grid resolution. Usage of the s transformation and uneven distribution of grid points in the vertical direction reveals the possibility to solve this problem.

4.3. Numerical experiment

A numerical experiment was carried out for the simulation of the propagation of oil under ice in a straight channel of 4.6 m in depth and 35 m in length, with a water discharge of 4.5 m^3/s. Roughness parameters z_0 equal to 0.5 mm were chosen both for the bottom and ice surfaces. In the vertical direction 15 layers were assigned, in the horizontal direction the grid step was equal to 1 m. The coefficient of the vertical eddy diffusivity was taken as constant and equal to $1.2x10^{-3}$ m^2/s. The vertical distribution of velocity is shown in Figure 14. The obtained velocity reduction due to the ice coincides with the observed during field experiments (19 cm/s in central parts and 16 cm/s under ice cover). At the initial moment a spot of oil was assigned in the surface layer of the channel axis. Propagation of the spill was simulated with zero lateral diffusion and with coefficient m_s equal to 0.1 m^2/s, w_0 was equal to 2 cm/s in both cases. In the first case, oil was detected only along the channel axis and in the second case, the influence of the lateral diffusion was rather big (Figure 15). A comparison with the experimental data (Fig.8) show that angles of oil slick borders rather close each other. Special analysis is needed to compare computed and observed details. It will be dine soon.

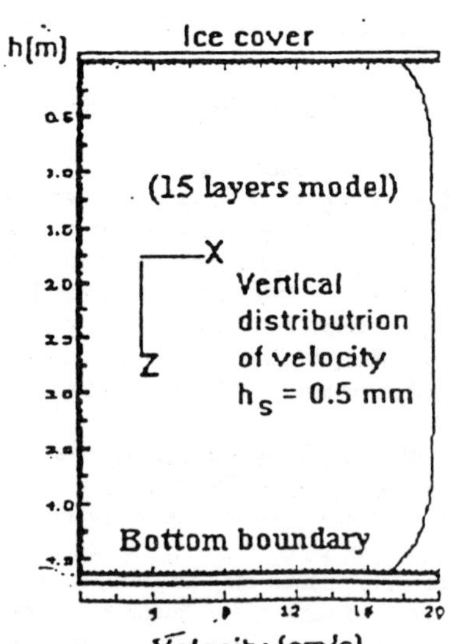

Fig.14. Hydraulic condition: a straight channel with uniform flow and two vertical boundary layers: top & bottom

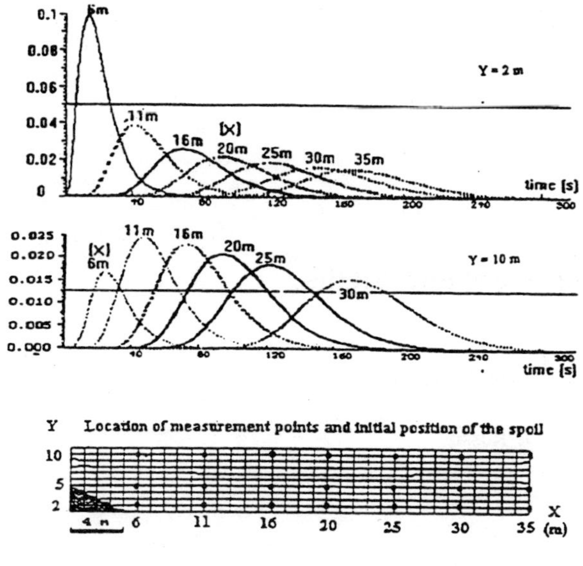

Fig.15. Propagation of oil spill with lateral diffusion

5. CONCLUSIONS

The paper presents field and laboratory experiments, theoretical research in order to create a general picture of fate and spreading of oil under ice.

The main distinction of this paper is results of field experiments. These field experiments were organized and conducted according special programme by contrast to occasional observations oil spills accidents. It showed that adequate describing of processes of oil spreading under the ice require more details. Such factors like roughness of bottom surface of ice cover, hummocks may substantially influence on these processes. The development of computer programme require also additional factors like oil drops sizes and changing it under influence water current and bottom surface ice rougness.

This problems needs additional experiments and farther development of computer model which will meet requirements of authorities and engineers responsible for practical oil combatting operations under ice conditions.

ACKNOWLEDGEMENTS

The following organizations are gratefully acknowledged for their financial support to the research and for permission to publish the data:
Rosshelf, Russia
Lencomecology, Russia
Ministry of the Environment, Finland
Finnish Environment Institute, Finland
Neste Oy, Finland
Kvaerner Masa-Yards, Finland
TEKES, Finland
Ministry for Trade and Industry, Finland

REFERENCES

Alhimenko, AI (1989). "The Theory Development and Methods of Calculation of Oil Product Spreading of Aqua Medium under Hydrometeorological Factors", Doctoral Thesis, Leningrad Polytechnical Institute, Leningrad, Russia, 328 p.

Alhimenko, AI (1990). "Matematyczny model rozprzestrzenia sie produktow naftowych w strefie przybrzeznej pod dzialaniem wiatrow, falowania oraz prdow". Inzynieria morska, Poland, ą 1, pp.5-7.

Benoit Petit (1996). Oil Spill Modelling in the Weddall Sea. Final report, Contract No A3/58/001, 1996, p.24.

El-Tahan, H, and Venkatesh, S (1994). Behaviour of Oil Spills on Cold and Ice-infested Waters. Analysis of Experimantal Data on Oil Spreading. Proc. 17-th Arctic and Marine Oilspill program Technical Seminar. June 1994. Vancouver, British Columbia, pp.337-354.

Fay, S (1969). The Spread of Oil Slicks on a Calm Sea. In: Oil in Sea, D.Hoult (editor), Plenum Press, pp.53-63.

Fingas, MF (1992) "The Behaviour of Oil in Ice", Proceedings of the Combatting Marine Oil Spills in Ice and Cold Conditions Seminar, Halsinki, 1992, Balttic Marine Enviroment protection Comission, pp.5-14.

Handbook of Turbulent (1977). Ads. Frost W, and Moulden T.N. Plenumpress, N.Y. and London.

Klevanny, KA, and Matveyev, GV (1995). 'Application of the integrated modelling system CARDINAL to the estimation of environmental impacts of coastal engineering in the Gulf of Finland', In: Proc. 4th Int. Conf. on Coastal and Port Engineering in Developing Countries, Rio de Janeiro.

Klevanny, KA, Matveyev, GV, and Voltzinger, NE (1994). 'An integrated modelling system for coastal area dynamics', International Journal For Numerical Methods in Fluids, v.18, p.181-206.

Liukkonen, S (1996). Behaviour and Spreading of Oil in Ice-Covered Waters. Literature study. Tech. report VALB 142, Espoo, July 1996, p.60.

Mackay, D, Paterson, S, and Trudel, K (1980). A mathematical Model of Oil Spill Behaviour. A report submitted to the Environmental Impact control Directorate, Environmental Protection Section, Ottawa, pp.39.

McNeill, MR, and Coodman, RH (1985). Motion of Oil on Leads. Proc. Eight Arctic Marine Oilspill Program Technical. Seminar. June 1985, Edmonton, Alberta. pp.42-52.

Monin, AS, and Yaglom, AS (1965). 'Statistical Hydromechanics', Nauka Publ.House, Moscow.

Ozer, J, and Geng, C (1991). PARCEL: a threedimensional oil spill model. Oil Pollution Enviromental Risk Assessment (OPERA), Proc. of the final workshop, Dalian, China, December 1991, pp.161-180.

Puskas, J, McBean, E, and Kouwen, N (1987). Behaviour and Transport of Oil under Smooth Ice. Canadian Journal of Civil Engineering, Vol.14, pp.510-518.

Reimer, E (1980). Oil in Pack Ice: The Kurdistan Spill. Proc. Third Arctic Marine Oilspill Program Technical Seminar, June 1980, Edmonton, Alberta, pp.529-544.

Usiner, MS, Weiskopf, FB, Cox, JC, and Schultz, LA (1979). "transport of Oil under Smooth Ice". Report ą EPA-600/3-79-041, ARCTECincorporated Columbia, Maryland, USA, 48 p.

Vefsnmo, S, and Johannessen, BO (1994). Drift and Spread of Oil in Broken Ice Field Experiment. IAHR Simposium 1994, Trondheim, Norway.

Venkatesh, S, El-Tahan, H, Confort, G, and Abdelnour, R (1990). "Modelling the Behaviour of Oil Spills in Ice-Infested Waters", Atmosphere Oceans, Vol.28, ą 3, pp.303-329.

Yapa, PD, Weerasuriya, SA, Belaskas, PP, and Chowdhury, T (1993) "Oil Spreading in Surface Waters with Ice Cover", Report ą 93-3, Clarcson University, Department of Civil and Environmental Engineering, Pottsdam NV, USA, 85 p.

Proceedings of the Seventh (1997) International Offshore and Polar Engineering Conference
Honolulu, USA, May 25-30, 1997

ISBN 1-880653-28-1 (Set); ISBN 1-880653-30-3 (Vol. II)

Study of Polluted Air Masses Coming Ways into Arctics

A.M. Prokhorenkov
Murmansk State Technical University, Murmansk, Russia
A.S. Sovlukov
Institute of Control Sciences, Moscow, Russia

ABSTRACT

Arctic region exerts essential influence on the climate and biosphere of the globe. This region is very subjectable to anthropogenic factors in different parts of the planet. Air pollution of Arctics is the most important problem and affects the interest of all the mankind. Polluted air masses coming ways into the arctic region are studied in the paper. Ways of polluting substances are analyzed, atmospheric transfer of these substances from world industrial regions into Arctics is considered. Influence of atmospheric precipitations on surface layer is investigated.

KEY WORDS: polluted air masses, coming ways, arctic region, precipitation, environment

1. INTRODUCTION

Influence of the arctic region on climate forming and of processes in high latitudes on the planet biosphere is very strong. Fragile biosphere at high latitudes is extremely subjected to the influence of anthropogenic factors. The problem of arctic environmental pollution is global and affects the interest of all the mankind. Environmental quality control is impossible without determination of the region and main factors of air pollution at high latitudes.

Atmospheric air pollution at high latitudes is the phenomenon of a global scale. Therefore it cannot be understood the aspiration of some nonrussian researchers to consider russian industrial regions as sources of arctic air pollution. As a rule such research works contain information about atmospheric air quality in a definite point that is placed in the arctic region: temporary concentration distribution of chemical substances Time interval includes only several days. On the base of trajectoric analysis is for instance discussed the influence of plants situated in Kolskiy peninsula or in the Norilsk town area, or in the Ural region, on air pollution in Arctics. It is considered only private case of the total process.

The purpose of this paper is determination of regions that influence most of all on air pollution in Arctics and of most important coming ways of polluted air masses into arctic latitudes. Also consiquence of precipitations influence on surface layer is analyzed here.

2. THE STATE OF KNOWLEDGE IN THE FIELD OF AIR POLLUTION MONITORING AT HIGH LATITUDES

There is a lot of investigations concerned the condition of atmospheric air pollution at high latitudes. First of all it is connected to fear that the extremely sensitive arctic biosphere can undergo irreversible destructions under the influence of unreasonable human economic activity. The natural structures and interaction between them were formed on the Earth during millions of years as results of long chemical, biological and geological evolution. In last century the influence of human economic activity has increased and accepted global scale. The phenomenon "Arctic smoke" described by many researchers and observed at high latitudes in "cold" period of a year, is defined by some scientists as an anthropogenic phenomenon. The severe climatic conditions of Far North do the environment of this region very subjected to the influence of anthropogenic factors. This region is industrially poorly advanced, therefore the influence of human economic activity directly in high latitudes is not great. The main amount of polluting substances comes to the region with air flows.

The study of arctic atmosphere includes the following directions:

- chemical structure definition of atmosphere aerosol;
- space-temporary distribution of aerosol;
- influence of aerosol on radiating balance;
- climatic research;
- clearing up of regions rendering the most considerable influence on the state of arctic air pollution.

In arctic latitudes a network of observation stations for monitoring the condition of atmospheric air and dislocation ways of substances polluting atmosphere to the regions of high latitudes is developed. It includes more than 10 points working within coordinated programs. For the last 20 years more than 10 scientific programs were devoted to the problem of arctic air pollution. Scientists in USA, Canada, Norway, Iceland, France were engaged in its realization. It is established that scale of the phenomenon connected with air pollution at high latitudes is significant. In cold period of a year it covers practically the whole arctic area. The essential concentration increase of some anthropogenic impurities is registered at the same period in various regions. This circumstance is caused by the throw out increase of polluting substances into atmosphere in winter from urban territories of the northern hemisphere, with peculiarities of atmosphere circulation over

Arctics and with properties of a bed surface in winter period. Practically the simultaneous concentration increase of sulphureous combinations in atmospheric air was registered over significant areas of territories.

The five-year climatic analysis of five-days trajectories of air masses dislocations curried out for station Borrow (USA) testifies that air masses came into this sector of Arctics by northern and southern ways. In cold period of a year the same regularity and speed of air dislocation increase were observed. Characteristic peculiarities of the weather in cold period of a year for arctic latitudes are high repeatability of cloudless atmosphere and decreased probability of precipitation fall. Substantial increase of carbonic acid, ozone, of optical density of atmosphere and of light scattering factors are observed.

3. ANALYSIS OF POLLUTING SUBSTANCES COMING WAYS INTO ARCTIC LATITUDES

There is no systematic approach now for information presentation about ways of polluted air masses displacement into Arctics that would cover all the urban areas of northern hemisphere. Also there is the lack of influence estimation for some urban areas of Barenz region on air pollution state at high latitudes.

Comparing the results of (Rahn et al., 1980) with our ones when back displacement routs of air masses are received there are found no contradictions. However, some cases of relatively high concentration registration of sulphureous combinations were not considered in detail in (Rahn et al., 1980). Here air masses into probes sampling (island Medvezhiy, Norwegian sector of Arctics) came by northern and north-west flows. It should be noted that back routs construction for areas of high latitudes is done much conditionally in view of the lack of exhaustive meteorological in formation for this region.

Consider now shortly the data that are known about the phenomenon "Arctic smoke". These available data show that there exists clearly expressed annual concentration behaviour for some anthropogenic impurities in atmospheric air at high latitudes. In Fig.1 annual concentration data for chemical combinations at Borrow Alaska (USA), in Fig.2 - the same data for Ny Alesund , Norge (Shaw, 1980; Joranger and Ottar, 1984).

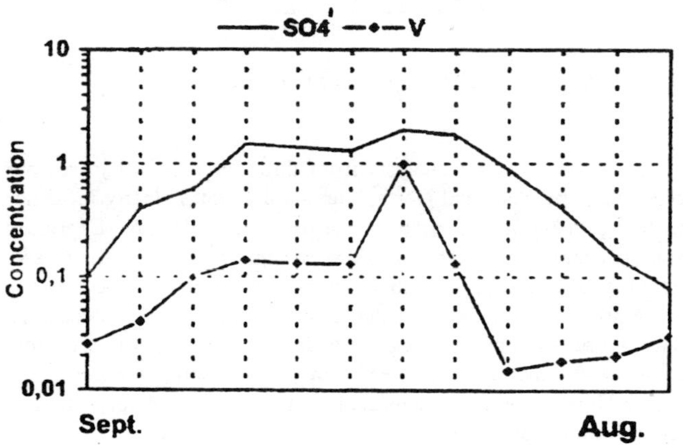

Fig.1. Seasonal SO_4' and V change in air probes from Borrow, Alaska (USA)

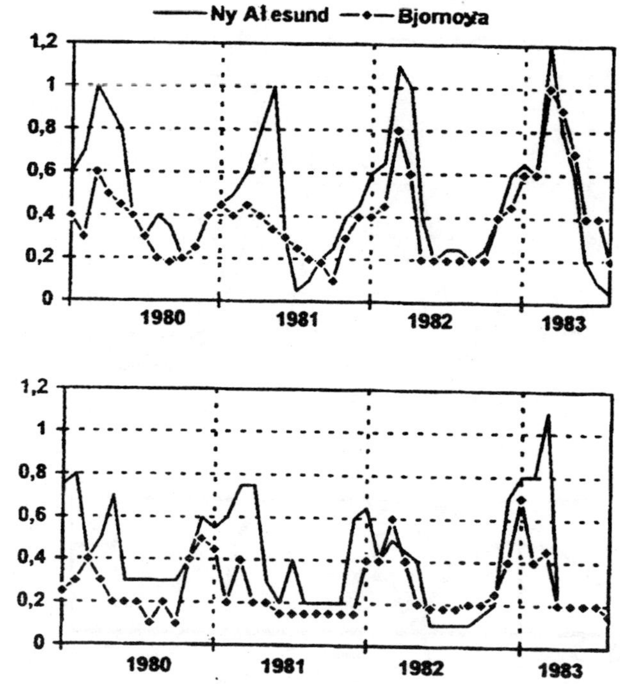

Fig.2. Monthly averaged concentration change of SO_2 and SO''_4 at arctic stations Bjornoya and Ny Alesund (Norge)

Expressed concentration increase of sulphureous combinations at winter period is caused by two reasons: firstly, by throw out increase of polluting substances into atmosphere at cold period of a year, secondly, by peculiarities of atmosphere circulation over this region in cold period of a year. Scale of "Arctic smoke" phenomenon is considerable. Increased anthropogenic impurities concentration is fixed at the same period of a year both in the western and in eastern hemispheres. Data in (Inversen, 1984) show that at three stations placed far each from the other, concentration increase of sulphureous combinations is registe red simultaneously (Fig.3).

Five year climatic analysis of back five-day routs for air masses displacement done for Borrow (Miller, 1981) shows that most often air masses come into this sector of Arctics by northern and southern ways. It corresponds to sectors 18 and 22 (Fig.4 and Fig.5).

For the period of a year when increased anthropogenic impurities concentration in atmosphere is observed, most of all air flows these sectors were also taken place.

Air masses transfer frequency from southern districts, sector 22, was considerably prevailed over air masses transfer frequency from northern districts, sector 18. For winter months some more high velocity of air masses transfer was fixed.

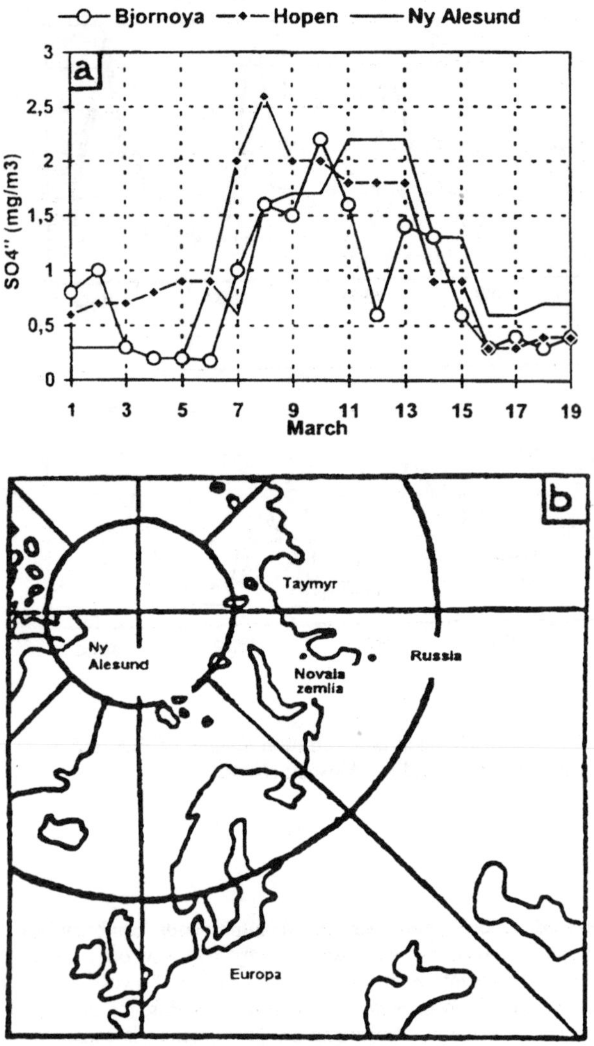

Fig.3. Concentration distribution of anthropogenic SO_4'' (a) at three stations in Norwegian sector of Arctics (b)

As a weather character peculiarities in cold period of a year for arctic latitudes is often repeatability of clear weather and smaller probability of precipitations. Besides of concentration increase of some chemical combinations in the most air layer in Arctics, clearly expressed increase of some physical parameters of atmosphere is also observed. Concentration of CO_2 in this layer, also of ozone, of optical atmosphere density at the wavelength 500 mkm and of light dispersion coefficient are considerably increased.

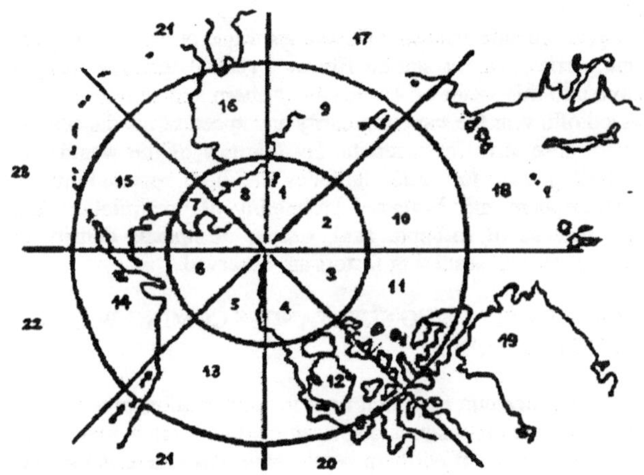

Fig.4. Choosen sectors for climatological analysis data averaging (station Borrow, Alaska)

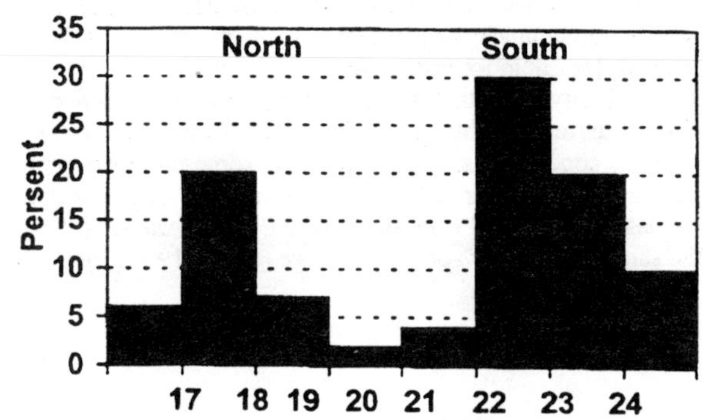

Fig.5. Percents of trajectoric types for five-years period

Air masses transfer analysis from three choosen urban regions placed in Europe, Asia and North America illustrates clearly influence of each region on air pollution in Borrow, Alaska (Fig.6). Expressed seasonal air masses transfer from choosen regions is fixed. Air masses transfer probability is considerably increased at the period since may up to september. Air masses transfer time is decreased in december, january, february from these regions to Borrow. In addition, trajectory analysis shows that for west sector of Arctics sources contribution placed in North America is considerable and is quite comparable with the contribution from industrial regions in european countries.

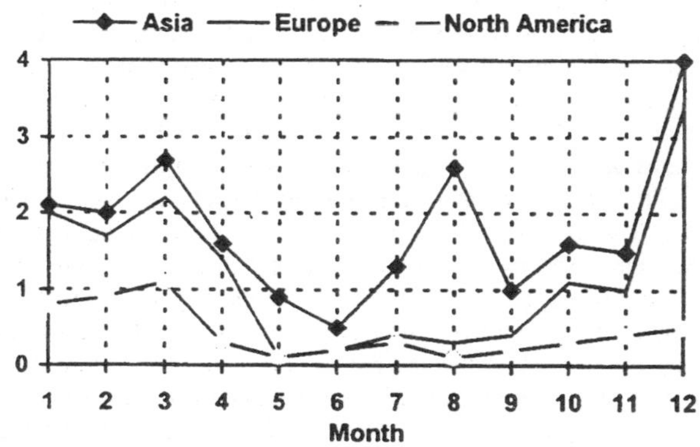

Fig.6. Relative monthly averaged contribution of each industrial region into atmospheric air pollution at Borrow, Alaska

4. ESTIMATION OF POLLUTING SUBSTANCES TRANSFER FROM INDUSTRIALLY DEVELOPED REGIONS INTO ARCTICS

In order to estimate contributions of various industrially developed regions into high latitudes atmospheric pollution based on the probe sampling data from stationary points located in nonindustrial regions of Kolskiy peninsula, and also on the probes sampled in expeditions (ships "O.Shmidt", "Arctics", airplanes-laboratories) was universal method designed. According to this method for every observation day back trajectory of air masses transfer was built on the level of 850 GPa for 36-120 hours. Concentration sets received for a definite period of time are presented as a frequency of data repeatability. Most often received concentrations are considered as quasi-backgrounded ones. Trajectories of air masses transfer are understood as "pure" ones for these concentration values. Back trajectories are called "dirty" ones for observation days when more than quasi-backgrounded concentrations are registered.

In order to represent back trajectories of air masses transfer in a compact way, all the analyzed region was divided into squares with sides of 450 km. Amount of trajectories coming through each square of a regular net was determined both for each choosen observation period and separately for cases of "dirty" trajectories registration.

Such an analysis gave ability to estimate contribution probability of any industrial region located in the european area and characterized by high throw out density of substances polluting the atmosphere.

Total probability P_t of pollution appearance is calculated so:

$$P_t = \frac{n_r}{n_t},$$

where n_r - amount of trajectories come through the region, n_t - total amount of observation days.

Transfer probability P_d of air masses with high SO_2 content and with anthropogenic sulphate is calculated according to this formula:

$$P_d = (SO_4^{''})\,SO_2 = \frac{n_d\,SO_4^{''}\,(SO_2)}{n_t},$$

where n_d $SO_4^{''}$ (SO_2) - amount of air masses transfer trajectories through the region with increased content of SO_4'' (SO_2).

As a criterium for considerability estimation of industrial impurities transfer from polluting regions the following value K is used:

$$K = \frac{P_d\,SO_4^{''}\,(SO_2)}{P_r},$$

where P_r - the value of P for a region.

For instance, results of K calculation for december of 1995 received for regions of St.Petersburg (Russia) and of London (England) can be presented:

	St.Petersburg	London
K SO_4''	0,49	0,33
K SO_2	0,33	0,69

Influence of throws out containing sulphreous combinations, from urban territories of England is more than that of St.Petersburg region industry under equal probability of air masses transfer (P_t = 0,196) through these regions. It takes place inspite of the fact that the distance from through out points into atmosphere till Kolskiy peninsula for the region in England 2,5 times exceeds that for the St.Petersburg region.

Observation analysis was done according to the data of the station located in nonindustrial region of Kolskiy peninsula. Back transfer trajectories of air masses with sulphur containing combinations and the data about their throughs out on european area were used. This analysis showed that the highest amount of anthropogeneous sulphate and of sulphureous gas comes to high latitudes with air flows from sectors "West" and "Europe", the least value of this amount - from sectors "East" and "North".

Here as flow F such a value was used:

$$F = \frac{8760\ cuLHP}{\alpha}$$

were c - averaged value for concentration of a chemical combination in air masses for the choosen sector, u - averaged value of air masses dislocation velocity for this sector, L - length of the path through that flow was calculated, it was taken equal to 1 km; H - height of dislocation layer, conditionally taken as equal to 1000 m; P - probability of air masses transfer from the sector; 8760 - amount of hours in a year; α - value of the sector angle.

For estimation of sulphureous combinations transfer into high latitudes through Kolskiy peninsula, characteristics of the most developed industrial regions located in northern areas of Europe and USA are needed. The chart was received for through out of sulphureous combinations from industrial regions located in european countries and in USA under generalization of published data. In Fig.7 is shown the chart for throws out of SO_2 on the territory of USA.

Here figures mean throw out of sulphureous combinations calculated to the values of SO_2 (thousands of tons/year).

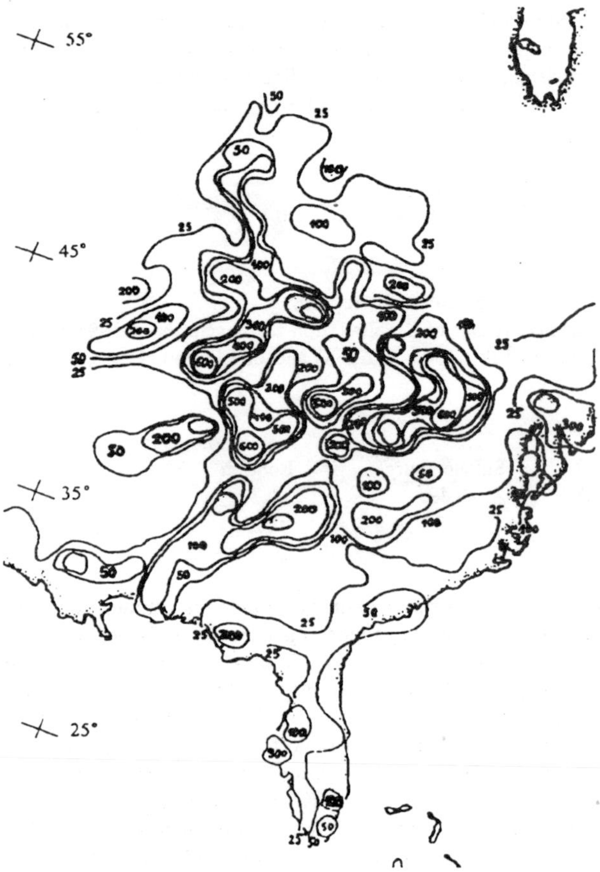

Fig.7. Throw out of sulphureous combination at the east territory of USA

Analysis of observation results for many years showed that the territory of Murmansk region is subjected to the pollution by sulphate sulphur considerably as a result of far transfer for neighbouring regions including transboundary transfer. There is no "reserve" for throw out from local sources of sulphureous gas. Precipitations of sulphate sulphur exceedes everywhere the value of 0,5 g/m^2 per year; this value was considered as threshold one for possible turning sour of surface waters at the Internat. conference for problems of environment souring. Reality of turning sour of basins in Kolskiy peninsula depends in such conditions only on buffer abilities of soils and basins themselves.

5. ATMOSPHERIC PRECIPITATIONS AS POLLUTION SOURCES

Pollution of atmospheric air is the most dangerous factor as compared with all the other components of environment. It is caused by penetrating abilities of air and of pollutants in it, of ability of air to dissolve in water, to be absorbed by alive organisms, plants and soils, to be brought through great distances, to form acid rains, smogs.

In the last case pollutants form as results of chemical reactions more toxic combinations than initial ones. It is also known that even during easy work a man consumes 45-50 kg of air, that is 15-20 times more than of food and water together. Even under comparable (quantitatively and qualitatively) pollutions, air pollution danger takes place rather significantly. Generalizing a lot of native and foreign papers concerning influence of industrial plants on northern regions, follow sequence of influences on their nature, for instance of two non-ferrous metallurgical plants and of one chemical plant, that are located in several tens of kilometers each from the other and of three heating plants, correspondingly. It should be reminded that flow of gases with the temperature 1000° and more carries away small coal particles, metal steams and their combinations and comes away in the form of aerosols. Simultaneously enterprices and heating plants give off a lot amount of SO_2 and dust. So, during one ton Ni, Cu reception, gases of 15-16 thousands of tons are received. Among them SO_2 is prevailed and are 300-400 kg of dust and aerosols containing up to 50% and more of said metals depending on used technology. In heating plants low quality coal is used. A heating plant with middle power throws out during a year SO_2 with amount of 25-30 thousands of tons, NOx (3-4 thousands of tons), ashes (15-20 thousands of tons). As a result during a year on 1 km^2 of earth surface falls 500-1000 kg of metals, 4000-6000 kg of sulphur, up to 10000 tons of dust, especially near heating plants and industrial enterprises. Events near industrial plants will develop approximately according the following scheme. Atmospheric precipitations will become sour, lowering up to pH = 4,0 - 3,5 from time to time. High-flood period for rivers and lakes will gain character of instantaneous throws out because of strong snow water decrease and of a lot of non-ferrous metals and other polluting substances in it that are accumulated for 6-7 winter period. Accumulation of metals in lakes because of their fall with precipitations takes place in such a way. At first in basins closely located with industrial plants, non-ferrous metals content begins to exceed background value. Concentration of each metal in first years is lower than permittable standards. But later they begin to exceed the value of one, then two and three. Then amount of each metal prevails already admittable standard value. Their dislocation from industrial plant to other areas is quickly increased not only because of precipitations including thawed waters but also of carrying out accumulated metals in plant surface layer under the action of acid rains. In basins their inhabitants perish because of considerable concentrations of pollutants. Simultaneously metals are accumulated in conifers, moss layers and foliage trees. When concentration of metals in conifers exceeds 1-5 mg/kg, it begins to perish. In pure air age of fir conifer its age shortens up to 10-8 years, further accumulation (40-100 mg/kg) leads to conifer age decrease up to 1-3 years and to a tree destruction. As a result, around industrial plants desolate areas at distances of 10-15 km from them curved birch and forest-tundra vegetations are formed.

If snow cover is concerned it should be noted that its some properties make it as suitable pollution indicator not only for precipitations and air themselves but also for the following pollution of surface water and soil. Under snow formation and precipitation, as a result of dry and wet processes of carrying out, concentration of polluting substances in snow becomes 2-3 times more than at atmospheric air.

Great industrial centres are basic sources of air and precipitations pollution by heavy metals in a studied region of the Far North. On snow cover around industrial centres halo of abnormal concentrations of heavy metals is formed. These metals are contained in air industrial throws out. Area of abnormal concentration fields and mass of polluting substances included there depend on throw out value and on accumulation period of time.

Under conditions of Far North, spring sewer of thawed waters takes place on still freezed through soil surface. Therefore impurities of snow cover don't take part in chemical reactions with soils and are carried out from water sampling areas proportionally to the water amount in sewers. In a studied region several tens of tons of heavy metals together with sewer of thawed waters are carried out into basins and water channels from the fields of abnormal concentrations.

Because microelements play an important part in biological processes, estimation of some metals storages in snow cover on ice of lakes placed along an axial track line of metals abnormal concentrations an axial track line of metals anormal concentrations was done. Tens of kilograms of various metals are contained in snow cover on the ice of lakes. Under snow and ice thawing a lot of metals comes into lakes, where they are accumulated by bottom deposits and living beings.

According to the data received, averaged concentration of some metals in water of these lakes after ice thawing was 60-70% from maximal value. Under investigation of microelements content in tissues of some representatives of hydrofauna it was discovered accumulation of biologically active metals in surface structures and inner organs of fishes. Accumulation degree of heavy metals by superiour water plants from bottom deposits of studied lakes appeared several times more than that from water.

Hence, industrial air pollution has direct influence on ecosystems of basins that are located at significant distance from large-scale industrial centres.

Chemical analysis of precipitation probes sampled at one of stations placed in nonurban region of Kolskiy peninsula showed that during summer months precipitation values of NO_3' ,NO_2' , PO_4''', Na^+ ,K^+ , SO_4'', NH_4 was increased as compared with winter period in 50 times. Monthly averaged value of sulphates precipitations (wet and dry precipitati on in total) being related to that value in atmospheric air for period july-august and november-march correspondes to values 6,3 and 2,4. This relation may be treated as intensity item of such impurities carrying out atmospheric air. It means that during warm period of a year process of sulphates carrying out from air proceedes 2,6 times quicker. Therefore using results of snow monitoring survey as the item that characterizes influence of industrial polluting throws out on bed surface, it is necessary to take into consideration precipitation velocity change of chemical combinations for various periods of a year. If there are no regime data about tendencies of polluting substances carrying out from atmospheric air, it is impossible to estimate trustworthy influence of anthropogenic factors on environment for long period of time (year and more than it).

Objective information about the influence and tendencies of environment pollution in the line "polluting substances atmosphere-bed surface" may be received only using the data of systematic determination of precipitations chemical composition.

6. CONCLUSION

The study carried out show that biosphere at arctic latitudes suffers from human industrial activity. Such a negative influence is caused not only by plants at northern regions but mainly remote industrial enterprises in Central Europe, America, Asia. As a result, because of precipitation falls of polluting substances coming into Arctics make significantly worse ecological items of northern regions of the globe. In order to improve ecological situation here it is necessary to control quality of environment in all regions of the planet.

REFERENCES

Joranger, E. and Ottar, B (1984). "Air Pollution Studies in Norwegian Arctic", Geophysical Research Letters, Vol 11, No 5, pp 365-368.

Inversen, T (1984). "On the Atmospheric Transport of Pollution to the Arctics". Geophysical Research Letters, Vol 11, No 5, pp 457-460.

Miller, IM (1981). "A Five-Year Climatology of Five-Day Back Trajectories from Borrow Alaska", Atmospheric Environment, Vol 15, No 8, pp 1401-1405.

Rahn, KA, Joranger, E, Semb, A, and Conway, TJ (1980). "High Winter Concentrations of SO_2 in the Norwegian Arctic and Transport from Eurasia", Nature, Vol 287, No 10, pp 824-826.

Show, G (1980). "Arctic Note", Weatherwise, Vol 33, No 10, pp 219-221.

Proceedings of the Seventh (1997) International Offshore and Polar Engineering Conference
Honolulu, USA, May 25-30, 1997

ISBN 1-880653-28-1 (Set); ISBN 1-880653-30-3 (Vol. II)

Source Locations of Contaminated Sea Ice Derived Using Backward Trajectories from an Ice-Ocean Model

Ruth H. Preller and Pamela G. Posey
Naval Research Laboratory
Stennis Space Center, MS, USA

Abstract

Results from the U.S. Navy's Polar Ice Prediction System 2.0 (PIPS 2.0), a coupled ice-ocean model, are used to determine source locations of sediment laden sea ice found in the Arctic. Model derived, daily ice velocities are used to calculate a "backward trajectory" for parcels of sea ice. The "end" of this trajectory represents the location at which the ice formed or detached from landfast ice. For ice parcels that contain contaminated sediment, this technique may provide a method for determining the source locations of these contaminants.

Key Words

Arctic contaminants, contaminated sediments, ice drift, contaminated sea ice, ice model

Introduction

Recent public attention has been drawn to the issue of the dumping of radioactive contaminants into the shallow waters of the Arctic Seas. In late 1991, the Greenpeace organization held a conference during which the location of several dumpsites for radioactive waste were identified in the shallow waters of the Kara and Barents Seas. Following this conference, the Russian Federation created a commission to report on these allegations. In 1993, the commission released a detailed report, now known as the Yablokov report or the White Book, documenting the dumping of both solid and liquid radioactive waste and the estimated radioactive inventories from the period 1959-1992. Figure 1 shows the dumpsite locations while Table 1 presents the contribution to the total radioactivity of each source in the Kara and Barents. Most of the dumpsites for solid radioactive waste were located in water with an average depth of 25 m. Liquid waste was dumped directly into the open ocean.

The largest contributor to the total inventory is the category of Solid Radioactive Waste (SRW) with Spent Nuclear Fuel (SNF). This source included containers, ships, barges and submarines containing nuclear reactors both with and without spent nuclear fuel. The majority of these objects were dumped into the shallow bays and inlets along the coast of Novaya Zemlya island in the Kara Sea. The second largest source, labeled Gulf Stream, comes from the discarded liquid waste released into the Irish Sea at the Sellafield nuclear reprocessing plant. The third largest source comes from atmospheric fallout directly onto the ice or ocean surface. The fourth source is the outflow from the Ob, Yenisei and Pechora rivers. Contaminated runoff from these rivers enters the Kara and Barents Seas in the shallow river mouth/estuary regions. The final source is low level liquid radioactive waste discarded mainly in the open ocean of the Barents Sea.

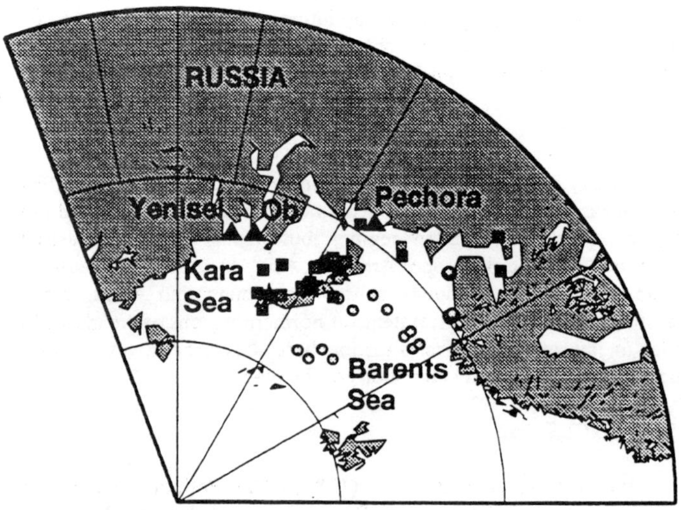

Figure 1. Dumpsites and sources of radioactive waste in the Kara and Barents Seas from Yablokov (1993). Circles represent liquid waste. Rectangles represent solid waste and stars represent nuclear reactors. River locations are indicated by triangles.

Table 1. Anthropogenic Radionuclide Budget of the Barents and Kara Seas Ecosystem, 1961-1990

Source of Long-Lived Radionuclides	Barents Sea Activity		Kara Sea Activity		Ecosystem as a Whole Activity		Ecosystem as a Whole Contribution
	kCi	TBq	kCi	TBq	kCi	TBq	%
Atmospheric fallout	100	3700	70	2600	170	6300	6.2
River runoff	6.0	200	33.0	1200	39	1400	1.4
Contributions from Gulf Stream System	200	7400	—	—	200	7400	7.3
Dumping of solid and liquid RW	13	480	16	600	30	1100	0.7
Sinking of SRW with SNF	—	—	2300*	85300*	2300*	85100	84.4
Underwater/surface nuclear explosions	... (No data) ...						
Total (upper limit)	319	11780	2419	89700	2739	101300	100

*--Expert estimate of the upper limit of activity at the time of disposal.

An issue of major concern is that any type of contaminated waste dumped into shallow water has a high probability of interaction with the ecosystem. Shallow seas are controlled by intense dynamic and thermodynamic processes. These are areas characterized by a strong and rapid response to synoptic weather systems producing variable winds and heat fluxes, by changes in density due to fresh water runoff and by the amplification of tides. An ice cover adds to this already complex system, a seasonal adjustment in the stratification and mixing due to the salination/freshening of the surface water resulting from the growth/decay of sea ice. Waste dumped into shallow ice covered seas therefore, has a high probability of interaction with the surrounding ice-ocean system.

Once contaminants are deposited into shallow seas, they may be transported out of the region in several different ways. First, they can act as passive tracers and be transported in the same way as the sea water through the mechanisms that drive the ocean circulation. In addition, the contaminated material may adhere to sediments and be transported along the sea floor by the mechanisms that govern sediment transport. A third method, and the one that will be the topic of this study, is the entrainment of contaminated sediments into sea ice and the resultant transport of these sediments by the ice. Sea ice may transport these contaminants a greater distance and to different final location than either the bottom sediment transport mechanisms or the ocean circulation. This paper will investigate the possible transport pathways of sea ice by using model derived ice velocities to calculate backward trajectories. This method can be used to estimate the source location of a parcel of sea ice. If the parcel contains contaminated sediment, the backward trajectory may be used to estimate the source location of the contaminants.

Mechanisms for Incorporating Sediments Into Sea Ice

A number of different processes have been suggested to play a role in the entrainment of sediments into sea ice. Several of these processes will be discussed briefly here. The interested reader is directed to a more detailed overview of these mechanisms found in Weeks (1993). Among the most important processes are 1) suspension freezing, 2) gouging of the sea floor by ice ridging, and 3) the flooding of sea ice by sediment laden river water. It should also be mentioned that sea ice and the sediments contained on the ice may become contaminated by airborne particles that settle on the ice surface.

Suspension freezing takes place in shallow water when there are strong winds, intense ocean turbulence and extreme subfreezing temperatures (Reimnitz et al., 1992). Under these conditions frazil ice forms and is distributed throughout the water column by turbulence. This saltwater frazil is sticky and can scavenge particles in the water column as well as adhere to bottom sediments when transported to the bottom by turbulent motion. Many frazil ice crystals can adhere to the same material forming a larger mass of ice called anchor ice. When the turbulence subsides, these buoyant ice masses float to the surface carrying material from the bottom as well as scavenging particles in the water column on the way up. These materials then become incorporated into a surface ice cover. Observations in the Arctic indicate that suspension freezing can carry material from the bottom to the surface in water depths of approximately 20 to 25 m (Reimnitz et al., 1992 and Reimnitz et al., 1987) and occasionally even in water as deep as 50 m (Reimnitz et al., 1993). These depths correspond to those of the dumpsites along Novaya Zemlya and to depths found in the vicinity of the river mouth/estuary systems implying that suspension freezing may provide a means for sediment and particle entrainment in these known regions of radioactive contamination.

The second method of sediment entrainment is ice gouging. As sea ice moves over shallow water, the sea floor is plowed by a possible combination of keels from pressure ridges, icebergs and ice islands. This interaction can cause sediments to be suspended in the water column or directly deposited into the ice cover. Finally the third method, the flooding of sea ice by sediment laden river water, occurs during the spring melt period, when the river runoff is increasing and the ice which formed near the river mouths begins to break up. Flooding can also occur during the fall when new ice begins to form in the shallow river estuaries.

The Ice-Ocean Model

The Polar Ice Prediction System 2.0 (PIPS 2.0) is used by the U.S. Navy to forecast daily ice conditions for the sea ice covered regions of the northern hemisphere. A coupled ice-ocean model consisting of the Hibler Ice model (Hibler 1979, 1980) coupled to the Bryan and Cox ocean model (Cox, 1984) forms the basis of this system. The models are coupled by exchanging both surface stresses and heat and salinity fluxes (Cheng and Preller, 1996).

The model domain extends from the North Pole to 30° N latitude in order to include the ice covered seas of the western Pacific. The model uses a horizontal resolution of 0.28 degrees on a spherical grid resulting in a grid spacing which varies from 17-35 km. Figure 2 shows the model domain with the horizontal grid overlaid. The model uses 15 levels, the thickness of each level increasing with depth, to define the vertical structure of the ocean. The model's bathymetry is defined from the Navy Digital Bathymetry Data Base 5° x 5° (DBDB5). The ocean model is weakly constrained to Levitus temperature and salinity climatologies (Levitus, 1982).

The coupled model is driven by the daily atmospheric stresses and heat fluxes from the Navy Operational Global Atmospheric Prediction System (NOGAPS) global model. The atmospheric forcing for each day is provided at six hourly intervals on a 1.25° grid. For this study the model was initialized with fall ice and ocean conditions from a previous model run and integrated using NOGAPS data from November 1991 through February 28, 1996. Daily averaged values of ice drift, ice thickness and ice concentration were produced for this four year period.

Figure 2. PIPS 2.0 model domain and grid. Every fourth grid point is plotted.

The Backward Trajectories

Backward trajectories are used to determine the source location of an observed parcel of sediment laden sea ice. The backward trajectory calculation is started from the known location of the parcel of sea ice. The "u" and "v" components of the ice drift are calculated for that location from model derived ice velocities using bilinear interpolation. The velocities are then "reversed" (u = -u and v = -v) and the backward motion of that day is calculated to find the initial location of the parcel on that day. The velocities from the previous day are then used to calculate the backward motion from the new location of the ice parcel. This procedure is repeated each day until the parcel either is in a liquid state indicating the location of ice growth or until the parcel is in contact with land. This process was verified by calculating forward trajectories from the source locations defined by the backward trajectories.

Several backward trajectories, calculated for this four year time period using the described method, are shown in Figure 3. Each trajectory is started on February 28, 1996, in the middle of the winter season. The starting locations for these backward trajectories (indicated by an "X" in the figure) were chosen to be in the Barents Sea, the Beaufort Sea and in the vicinity of the Fram Strait. These regions are all representative of marginal ice zones where sea ice is known to melt. It is in these regions that sea ice would be expected to release its load of contaminated sediment back into the surface of the water column.

The ice parcels found in the Barents Sea either originated at the northern boundaries of the Barents Sea or in the Kara Sea. Parcels B1 and B2 formed or grew the previous fall (October, 1995). Parcel B4 originated as fast ice or ice forming at the coast of Franz Josef Land. Trajectory B3 originated near the mouth of the Yenisei River, a possible source location identified by the Yablokov report for contaminated sediments. This ice parcel formed in the fall ice growth period (September 1994) and traveled for two years into the Barents Sea.

The ice parcels found in the Beaufort and Chukchi Seas, near the northern coast of Alaska, each traveled in a clockwise or east to west direction. Parcel BF1 formed in a region of ice growth in December of 1995. Parcel BF2 formed off of the northern coast of Alaska/Canada in the fall (October) of 1995 and traveled westward into the Chukchi Sea. Parcels BF3 and BF4 each traveled the entire four year period circling in a clockwise manner in the Beaufort Gyre. The trajectories indicate that these two parcels of ice were part of the Beaufort gyre circulation well before 1992.

The ice parcels found in the Fram Strait region came from two different locations. Parcel G1, located the farthest into the Greenland Sea, originated off the coast in the Canadian archipelago and traveled for nearly two and a half years before reaching the Greenland Sea. The ice parcels G2 and G3 both originated in the fall in the Laptev Sea. However, parcel G3 traveled for nearly two and a half years while parcel G2 traveled for a year and a half. Although these two parcels of ice have been displaced almost the same distance (from starting to finishing point) the route taken by parcel G3 was longer. This was due to several digressions from the direct path, such as brief periods when the parcel would travel in a circle.

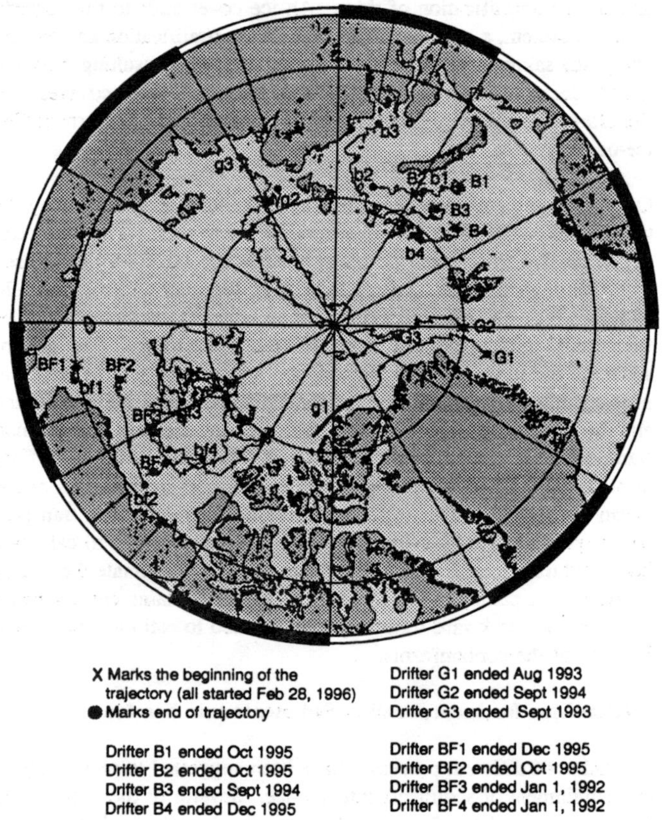

Figure 3. Backward trajectories calculated from daily model derived ice velocities. Upper case letters indicate the beginning and lower case letters indicate the end of the trajectory.

Discussion

For these model derived backward trajectories to be useful in determining the source location of sea ice parcels, they should compare reasonably well with the known circulation of sea ice in the Arctic. One such comparison is with the trajectories that lead back into the Laptev Sea. The Laptev Sea is a known region of sea ice generation for the Arctic. Strong average offshore winds in this region push existing ice away from the coast and allow new ice growth. The ice moves northward, often becoming part of the transpolar drift, and is carried out through the Fram Strait. Parcels G2 and G3 represent this type of motion.

Ice drift data from Arctic drifting buoys has been archived by the Applied Physics Laboratory at the University of Washington since 1979. Figure 4 from Pfirman et al. (1997) shows the annual mean ice motion in the Arctic from 1979-1990 as calculated from this Arctic buoy data. The mean ice circulation is represented by a large anticyclonic circulation in western Arctic connecting to a transpolar drift which exits the Arctic through the Fram Strait. The dark numbered lines indicate the length of time in years that it would take ice to travel from that general location out through the Fram Strait. Note that the buoy data indicates the average travel time from the northern Laptev Sea out through the Fram Strait is approximately 2-3 years, in agreement with the model travel times derived from the backward trajectories. This figure does not give a travel time for ice parcels originating in the central part of the anticyclonic circulation. It does however, indicate that the average travel time for a parcel of ice located off the northern coast of Alaska/Canada would be six years. Modeled ice parcels BF3 and BF4 are located in region 6. These parcels were originally located inside of the anticyclonic circulation and took a minimum of four years to reach region 6. Therefore a parcel of ice initially located inside the anticyclonic circulation could take a total travel time of 10 years or more to reach the Fram Strait.

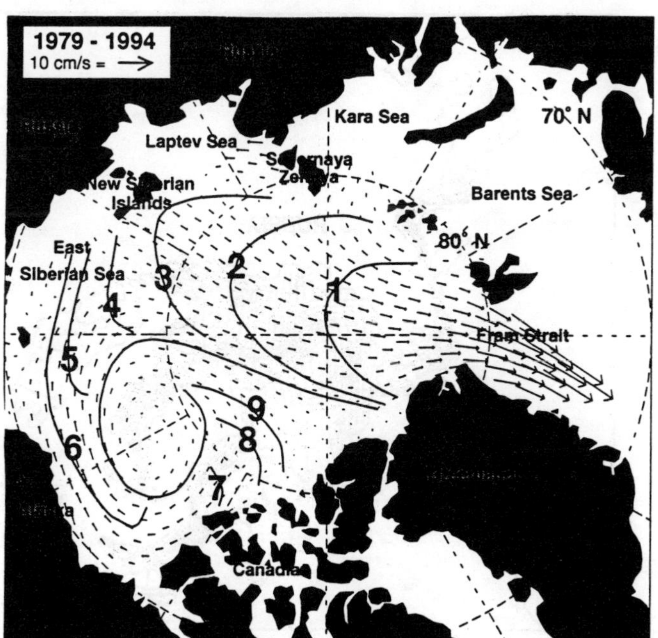

Figure 4. Mean Arctic ice drift derived from 12 years of buoy data (Pfirman et el., 1997).

To compare to the average buoy motion represented by Figure 4, model ice drift velocities were averaged over the four year study period and plotted in Figure 5. Note that this average field has the same general characteristics of the average buoy derived ice drift. This is not an unexpected result. Ice drift is dominated, even in the long term average, by wind forcing. NOGAPS uses the atmospheric pressure data from the buoy observations in its daily initialization. Therefore the wind fields in this area should be consistent with the observations as should the resulting modeled ice drift.

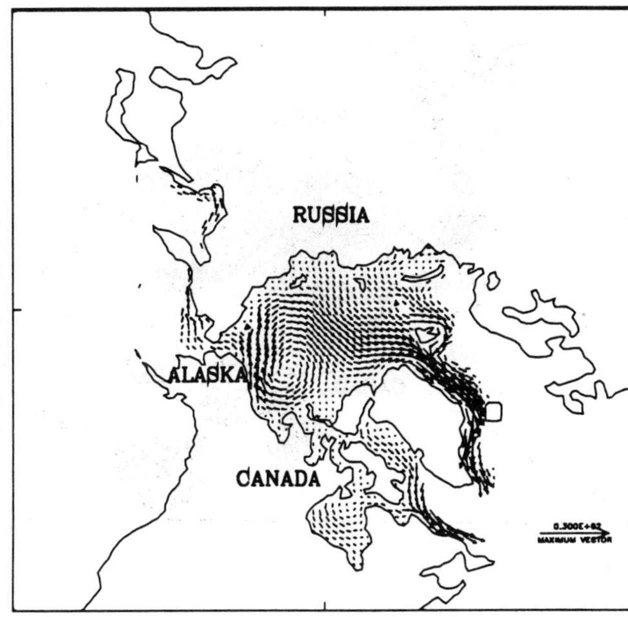

Figure 5. The four year mean (1992-1995) ice drift derived from PIPS 2.0. The maximum scaled vector is 30 cm/sec.

Pfirman et al. (1997) applied a technique for calculating backward trajectories using monthly mean ice drift derived from buoy data. However, when using the technique of backward trajectories to determine the source location of an ice parcel, it is important to note that the actual path taken by a parcel of sea ice may not be accurately represented by smoothed or averaged motion. The trajectories depicted by Figure 3 show that, depending on the variability in the daily wind forcing, a parcel of ice may oscillate, moving forward and backward or circulate several times or merely deviate to the right or left of its average path. Backward trajectories calculated using daily ice drift versus an average value such as a monthly mean may end up at different "source" locations. If this technique is to be used to determine the source location for contaminated material, this difference could be very important.

The difference in daily versus monthly ice drift is depicted in Figures 6 and 7. Figure 6 shows an example of two consecutive daily ice drift fields from the model while Figure 7 shows and example of two consecutive monthly mean ice drift fields from the model. Note the high degree of variability in ice drift fields from day to day. These values reflect the variability in the wind field due to the passage of synoptic pressure systems. The monthly mean ice drift fields are smoother and generally weaker due to the averaging. Note that the ice drift in Figure 7a looks very similar to the long term average circulation shown in Figures 4 and 5. The pattern shown in Figure 7b

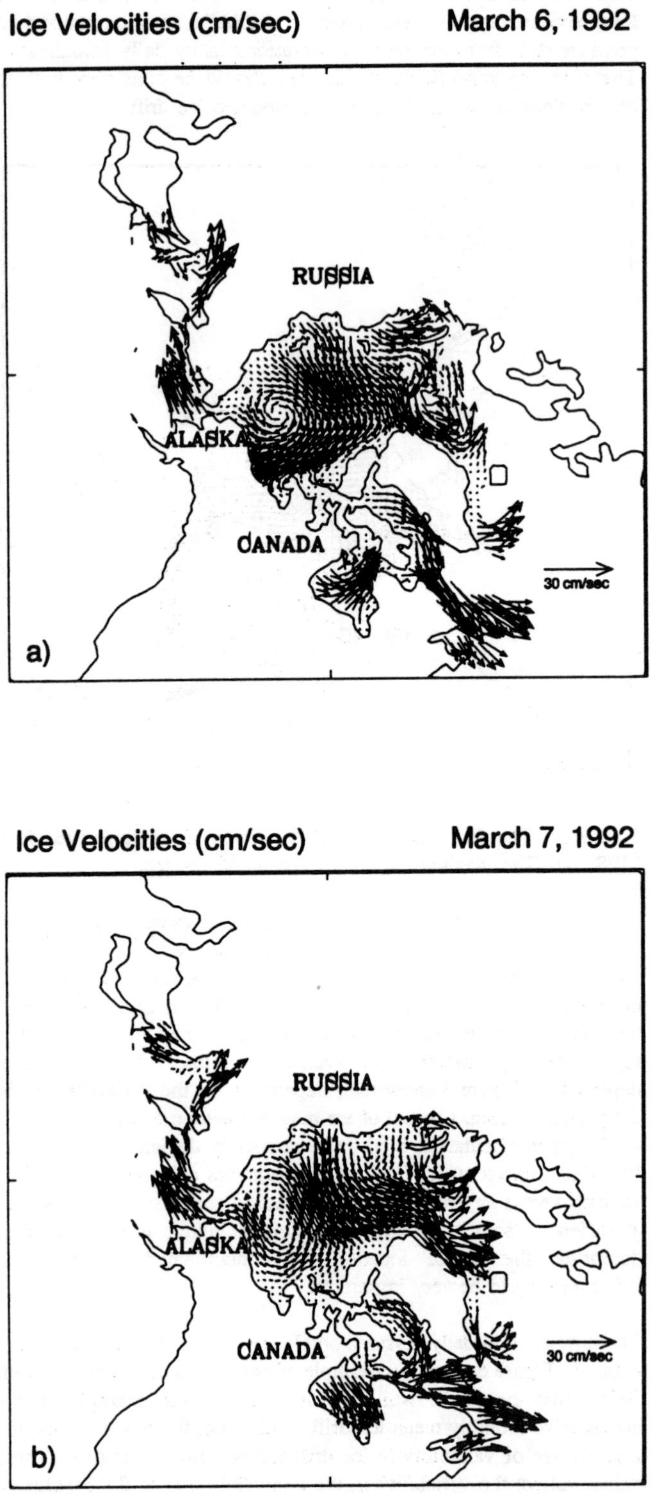

Figure 6. Daily averaged ice drift from PIPS 2.0 for a) March 6, 1992 and b) March 7, 1992. The maximum scaled vector is 30 cm/sec.

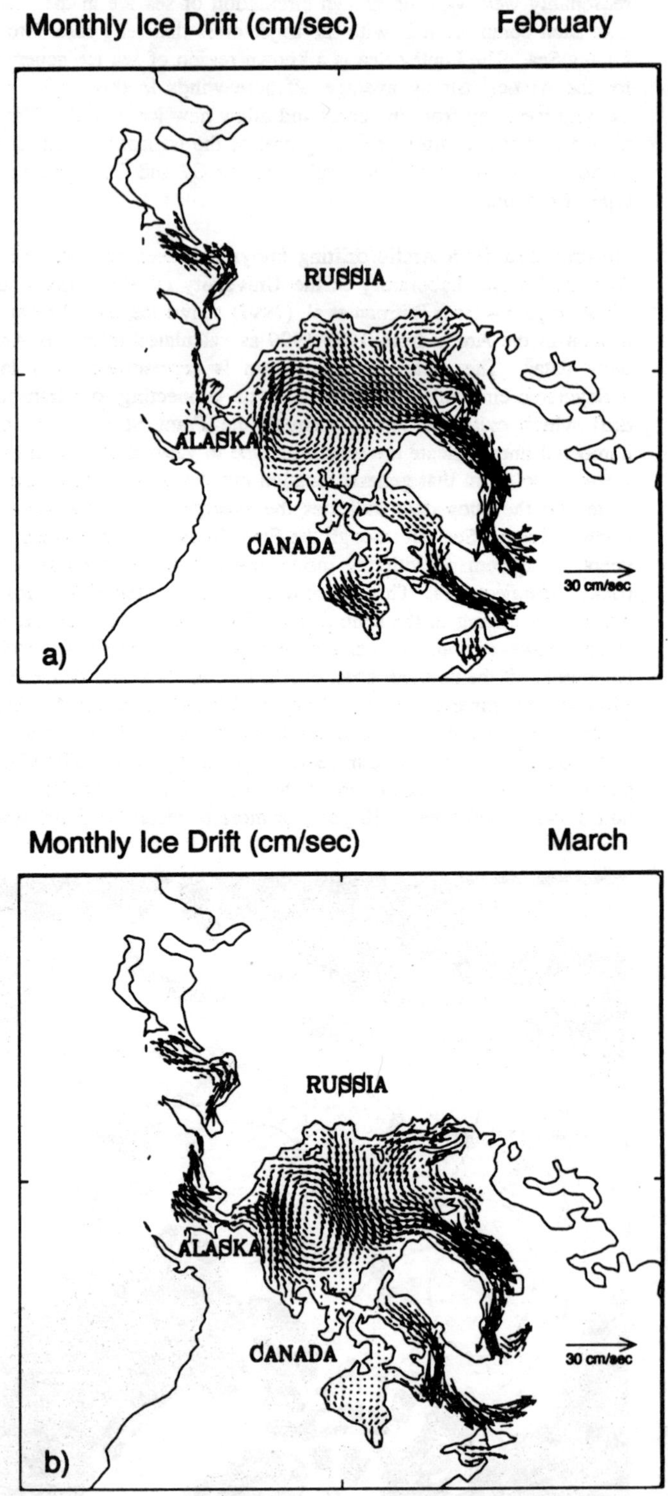

Figure 7. Monthly averaged ice drift from PIPS 2.0 for a) February 1992 and b) March 1992. The maximum scaled vector is 30 cm/sec.

is somewhat different showing a narrower anticyclonic gyre and a strong component of the ice drift from the pole to the eastern part of the Canadian archipelago and the northern coast of Greenland.

To show the difference between the average and daily ice motion on the calculation of backward trajectories, monthly mean ice drift values were derived from the daily ice drift values for each month during the model test period. The backward trajectories for each of the parcels shown in Figure 3 where again calculated, this time using monthly mean velocities (Figure 8). At first glance the trajectories look similar to those in Figure 3. In particular, ice parcels which did not travel great distances, such as B1, B2, B4, BF1 and even BF2 have similar trajectories in both cases. The "ending" times for these backward trajectories are also similar, differing by at most two months.

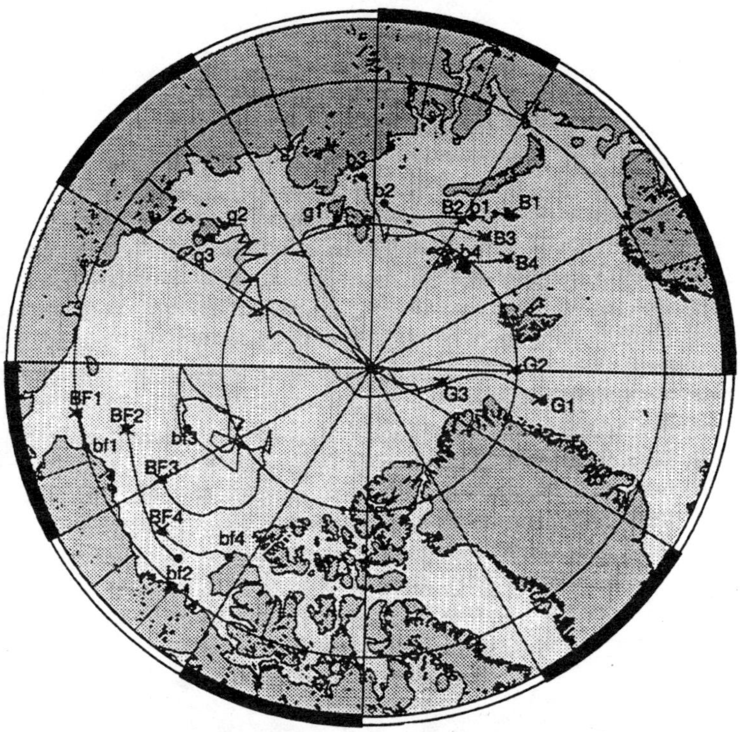

x Marks the beginning of the trajectory (all started Feb 28, 1996)
• Marks end of trajectory

Drifter G1 ended mid Feb 94
Drifter G2 ended end Dec 93
Drifter G3 ended Nov 93

Drifter B1 ended Aug 95
Drifter B2 ended Aug 95
Drifter B3 ended mid Jan 95
Drifter B4 ended mid Dec 95

Drifter BF1 ended mid Jan 96
Drifter BF2 ended Aug 95
Drifter BF3 ended Jan 92
Drifter BF4 ended Oct 95

Figure 8. Backward trajectories calculated from monthly averaged model derived ice velocities. Upper case letters indicate beginning and lower case letters indicate the end of the trajectory.

However, those ice parcels which traveled greater distances often "ended" at a much different location. The most drastic change in the "final" position for these trajectories is in the case of trajectory G1. Using daily ice drift, the source location for this trajectory was in the Canadian archipelago. Using monthly mean ice drift, the source location is along the boundary between the Kara and Laptev Seas, at Severnaya Zemlya! This indicates that the daily variability found in the ice drift, which is smoothed out in the monthly mean, is sufficient to cause this large change in trajectory. In addition, the length of this trajectory was six months shorter then the case using daily values. Ice parcels G2 and G3 have also shifted their "final" location substantially from the eastern Laptev Sea to the boundary between the Laptev and East Siberian Seas at the New Siberian Islands. Trajectory G3 is two months shorter in this case. However, trajectory G2 is nine months longer using the monthly means. This is not only due to the fact that the distance from starting to ending locations of this trajectory is longer, but also because this trajectory places the parcel in areas of slower moving sea ice.

An additional major change appears in the trajectory of BF4. Although the final locations don't appear that different, the main change is in the pathway taken by this parcel of ice to reach its final location. In the case of the daily ice drift, the parcel is caught in the anticyclonic circulation and makes one large clockwise rotation with some very small scale circulations along the way. The travel time for that trajectory was the full 50 months of model data. In the case of the monthly averaged trajectory, the parcel traveled a direct path from the Canadian archipelago to it's "final" location with a total travel time of 5 months.

Summary and Conclusions

This study presents a technique for using daily, model derived ice velocities to calculate the "backward" trajectory of a parcel of sea ice. Recent public and scientific interest in the ecological effects of contaminated waste dumped in the shallow waters of the Arctic or dumped in Arctic rivers make this technique of particular interest. Determining the pathway taken by a parcel of contaminated sea ice can provide information on the source location of the contaminants. Model derived ice velocities can also be used to determine the forward trajectory of an ice parcel which forms in a region of known contamination (Preller and Posey, 1997). However, these trajectories would be performed in a hindcast mode from known or archived data, not as a true forecast. The concept of the backward trajectory is appealing in that can provide real time information to the scientist, observing a parcel of sea ice, as to the source of that contaminated ice.

Backward trajectories created by this technique were compared to the observed average ice motion for the Arctic and produce realistic pathways for parcels of sea ice. However, it was noted that using daily model derived ice motion could produce a much different backward trajectory and therefore a different source location, than one produced using averaged ice motion which smooths out synoptic scale wind events. Numerical models, therefore, provide a unique tool for the determination of source locations. The models provide not only daily ice velocities, but they also provide them at a large number of point locations represented by the model grid. In addition, these models can be integrated for long periods of time using wind fields that have been shown to represent the Arctic very well. Daily ice velocities from these simulations can be saved and used to determine a multitude of backward trajectories.

Acknowledgements

This project was funded through the Office of Naval Research's Arctic Nuclear Waste Assessment Program and through the Naval Research Laboratory's Thermodynamic and Topographic Forcing in Global Ocean Models project. This paper, NRL contribution PP/7322-96-0032, is approved for public release; distribution unlimited.

References

Cheng A. and Preller, R. H. (1996). "The development of an Ice-Ocean Coupled Model in the Northern Hemisphere," NRL Formal Report 7322-95-9627, Naval Research Laboratory, Stennis Space Center, MS, p 65.

Cox, M. (1984). "A Primitive Equation, 3-Dimensional Model of the Ocean," GFDL Group Technical Report 1., Geophysical Fluid Dynamics Laboratory, Princeton, NJ, p 1141.

Hibler, W.D. III (1979). "A Dynamic Thermodynamic Sea Ice Model," *J. Phys. Oceanogr.*, **9**, pp 815-864.

Hibler, W.D. III (1980). "Modeling a Variable Thickness Sea Ice Cover," *Monthly Weather Review*, **108**, pp 1943-1973.

Levitus, S. (1982). "Climatological Atlas of the World Ocean," NOAA Professional Paper, 13, p 173.

Pfirman, S.L., R. Colony, D. Numberg, H. Eicken and I. Rigor (1997). "Reconstructing the Origin and Trajectory of Drifting Arctic Sea Ice", *J. Geophys. Res*, in press.

Preller, R.H. and P.G. Posey (1997). "Modeling the Dispersion of Radioactive Contaminants in the Arctic", 1997 NRL Review, Washington D.C., in press.

Reimnitz, E., E. W. Kempema and P.W. Barens (1987). "Anchor Ice, Seabed Freezing, and Sediment Dynamics in Shallow Arctic Seas," *J. Geophys. Res*, **92**, pp 14671-14672.

Reimnitz, E., L. Marincovich Jr., M. McCormick and W. M. Briggs (1992). Suspension Freezing of Bottom Sediment and Biota in the Northwest Passage and implications for Arctic Ocean Sedimentation," *Can. J. Earth Sci.*, **29**, pp 693-703.

Reimnitz, E., M. McCormick, K. McDougall and E. Brouwers (1993). "Sediment Export by Ice Rafting from a Coastal Polynya, Arctic Alaska, U.S.A.," Arct. Alpine Res.

Weeks, W.F. (1993). "Possible Roles of Sea Ice in the Transport of Hazardous Material," In: Proceedings of the Workshop on Arctic Contamination, Interagency Arctic Research Policy Committee, May 2-7, 1993, p 24.

Yablokov, A.V., Karasev, V.K., Rumyantsev, V.M., Kokeyev, M.Y., Petrov, O.I., Lystosov, V.N., Yemelyanenkov, A.F. and R.M. Rubtsov (1993). "Facts and Problems Related to Radioactive Waste Disposal in Seas Adjacent to the Territory of the Russian Federation," translated by P. Gallager and E. Bloomstein, Small World Publishers Inc., Albuquerque, NM, p 74.

Proceedings of the Seventh (1997) International Offshore and Polar Engineering Conference
Honolulu, USA, May 25-30, 1997

ISBN 1-880653-28-1 (Set); ISBN 1-880653-30-3 (Vol. II)

Intelligent Sensor Protection System for Polar and Marine Environments

Peter J. Stein, Armen Bahlavouni and Douglas W. Andersen
Scientific Solutions, Inc.
Hollis, NH, USA

ABSTRACT

An Intelligent Sensor Protection System is being developed under the U.S. Navy's Small Business Innovation Research Program. Its purpose is to provide a "smart" system to protect sensors during remote unmanned measurements. The first application is to perform solar and celestial radiation measurements under Arctic conditions. Here, the problem is frost, snow, and ice accretion on the sensor's optics. A prototype system was developed, tested, and patented. The system consists of an enclosure which houses and protects the sensor. Just before the measurement, the sensor is deployed while clearing snow and ice buildup. After the measurement, the data is stored, and the sensor is brought back within the enclosure for protection. A routine to determine sensor contamination is then executed and the sensor cleaned if necessary. An on-board computer controls all electro-mechanical and logic functions. Currently, this system is being improved under a Phase II SBIR supported by the Office of Naval Research, and will be expanded to underwater marine environment applications. This paper will discuss the system design and its applications.

KEY WORDS: Sensor Protection System, Harsh Environments, Computer Controlled, Remote Measurements, Pyranometer, Pyrgeometer, Solar Radiation

INTRODUCTION

Contamination of the sensing element of any sensor hinders its measurement ability, degrades its accuracy, and reduces the reliability of the measurement. Sensors are deployed in many locations; from climate-controlled clean rooms to the extreme harsh environments of the Arctic. In every situation, the sensor must provide accurate and reliable information. Here we report on the design, construction, and testing of a stand-alone, low-power, battery-operated system for recording radiometric data for periods of up to one year, without attention, in the Arctic environment.

Measurement of both long-wave and short-wave radiation is critical to understanding global environmental change. This is especially true in the Arctic, where the heat transfer among the ice, the atmosphere, and the ocean can alter the state of the ice pack. The ice acts as a thermal insulator keeping the warm ocean from transferring energy to the cold air mass. Changes in the ice cover could result in significant changes in the energy balance, affecting not just the Arctic, but potentially the entire biosphere. Therefore, it is necessary to understand environmental interactions in the Arctic. Long term autonomous systems are planned to be deployed in the Arctic to measure a variety of environmental parameters. It is currently not possible to include radiation measurements in these systems because there exists no portable, low-power-consumption system for keeping radiometer domes clean. Frost deposition and blowing snow and ice crystals form a hard crust-like surface on radiometer domes resulting in erroneous measurements. In recent field operations, radiometers needed to be cleaned as frequently as every hour.

Recent studies of the thermal stresses which often result in cracks in sea ice due to atmospheric heating and cooling exemplify recent work on this problem [1,2]. The cracks develop into leads (areas of open water) through which there is a tremendous heat flux. If climate is altered, the rate at which leads form and the overall rate of energy exchange between the ocean and atmosphere will change. This could act as an accelerator to climate change, or it may act as a negative feedback loop decelerating any change. However, it has been shown that calculation of the stresses developed in the ice, and modeling of the entire process, require short-wave and long-wave radiation data [1].

In a recent experiment in the Beaufort Sea, for the Sea Ice Mechanics Initiative (SIMI), a winter-over system remained on the ice from November 1993 to March 1994 to record stresses and ice cracking events. During the manned phase in November, radiation data was taken with the same difficulties described above. Radiation data could not be collected during the unmanned winter-over phase since there was no way to keep the radiometers clean. This was unfortunate in that it eliminated the possibility of performing any modeling to understand the thermally generated ice stresses and cracking. The experiment indicated that thermal-stress cracking started in the fall and continued into the spring. This continual cracking quite likely leads to the breakup of the ice floes.

In order to fully understand Arctic processes related to climate change and sea ice mechanics, long-term measurements are necessary, including autonomous radiation sensing. A measurement system must be portable and have low power consumption in order to operate on a self-contained power source for up to one year. The package must be small enough to be easily deployed. Furthermore, it would be best for the system to be stand alone, such that it could be deployed independently if required. Often, radiometers need to be deployed away from other equipment and structures to avoid shadowing, and cabling is not advisable since experience has shown it can be chewed by wildlife or easily broken by ice motion. Thus, the system should have its own data logging or satellite telemetry capability.

An Intelligent Sensor Protection System is being developed under the U.S. Navy's Small Business Innovation Research Program. Its purpose is to provide a "smart" system to protect sensors during remote unmanned measurements. Currently, this system is being improved under a Phase II SBIR supported by the Office of Naval Research, and will be expanded to underwater marine environment applications. This paper will discuss the system design and its applications.

IDENTIFICATION OF THE PROBLEM AND CURRENT TECHNIQUES

The most common difficulty with radiometers in the Arctic is the formation of frost and the accumulation of ice particles on sensor optics. In the absence of sufficient sunlight, the glass dome of the radiometer is cooled rapidly below ambient temperature by means of radiative heat transfer to space. The ambient air, due to it's lower emissivity, cools much slower. The cooler glass surface therefore becomes host to moisture condensation and frost formation [3]. The primary method currently used for keeping radiometer domes clear, next to manual cleaning, is to blow air around the dome (referred to as "ventilation"). The blowing air promotes mixing of the thermal boundary layer adjacent to the dome, which raises its temperature to near the ambient temperature. This in turn, delays the onset of frost formation. A number of ventilated units, some with heating capability, are available commercially from vendors. Both heated and unheated ambient air ventilators are acceptable techniques only in relatively modest frost conditions. In our experience, approximately the upper third of the surface of the dome is not directly affected by the flow of air since it falls within a separated airflow regime. Additionally, the relatively weak flow of air around the dome does not prevent freezing rain, heavy snow, or rime ice from accumulating on the dome and the surrounding shield. Accumulation on the shield can limit the field of view considerably, even if the dome remains clean. Furthermore, in the absence of logic control, the ventilator fan, and the heater (in the case of the heated unit) must remain turned on at all times, creating a sizable power liability for long term remote deployments. One must also be cautious of raising the dome temperature too high above the ambient, since the sensor is basically a thermopile and a warmer-than-ambient dome can result in erroneous data.

We found no performance data related to snow, rime ice or freezing rain conditions; none of the units of which we are aware have addressed these problems. The ventilated units deployed at Environment Canada weather station at Resolute Bay, are checked on a daily basis and require periodic cleaning in extreme conditions [4].

The results of this research strongly suggest that current techniques are inadequate for an unattended radiometer system and undesirable even for frequently maintained systems. Given the power constraints, and given the year-around conditions in which the radiometer must operate, the sensor requires some means other than, or in addition to, ventilation for keeping clean.

AN OVERVIEW OF THE PROTOTYPE SYSTEM DESIGN

SSI's technical approach to design of a prototype radiometer system was to protect the sensor from the harsh environment when not in use and mechanically deploy it for data acquisition; thus, conserving power by eliminating the need for continuous ventilation. Furthermore, the ability to customize the system operation via an onboard computer gives this design a great deal of flexibility and a number of distinct advantages over existing techniques. The prototype system depicted in Figure 1 was designed around the Eppley radiometer since both a short-wave and long-wave units were already available to SSI. Considering the environment and the necessary functions, the system was designed with the following features:

- A 10" diameter aluminum sphere, Teflon impregnated and hard coat anodized for ice release, houses the radiometer. The sphere is mounted on top of a cylindrical housing. When the radiometer is not in use, the sphere is turned so that the radiometer is protected inside the housing while the opposite side of the sphere closes off the opening. This arrangement eliminates the need for a separate cover and all the related linkages and hardware for it's operation. A circumferentially mounted knife-edge seal wipes the surface of the sphere as it rotates during deployment of the radiometer.
- A cylindrical housing open at the bottom to maintain the radiometer at ambient temperature, and prevent heat and humidity build up by providing natural ventilation to the enclosure.
- A light or heat source inside the housing which is energized to provide a calibrated signal for the sensor; thus determining the condition of the dome. A decision is then made whether to clean the dome. There is no added power consumption if the dome is clean.
- A hot air blower inside the housing which melts/evaporates frost or moisture from the dome if it should become contaminated.
- A fan which provides ventilation at ambient temperature when the radiometer is collecting data (similar to current commercial ventilated radiometers).
- A low-power computer which handles the data acquisition and logic control functions of the electro-mechanical system. Operational parameters such as frequency of data collection, conditions for cleaning, motor control and error checking, are all controlled through the software, and can be customized depending on the user and the environment requirements. Figure 2 shows a schematic of the electrical circuit. The housing is tied to the low side of the battery which is at ground potential. The relay board is powered and controlled by the computer, and provides isolated, automated switching of the loads. The gear-motor circuit is controlled by two DPDT electro-mechanical relays. Due to it's high current requirement at startup, the motor input power is buffered by a 27,000 mf capacitor.

The prototype Radiometer Control System (RCS) consisted of several single-board computer components purchased from Z-World Engineering of Davis, CA. The computer is a "Little PLC". Peripheral boards, connected to the computer by the Z-World PLC bus, are an A/D board, an RS-232 board, and a relay board. Z-World development software (Dynamic C for Windows) was used to create and download the control software. The RCS is rated for -40°C operation, and consumes an average of approximately 1 watt of power when awake and negligible power when asleep. The total power consumption thus depends upon the percentage of time that the system is awake, which for our testing was approximately 9%. (The total measurement cycle took approximately one minute, with 10 minutes sleep time). Figure 3 depicts the flow chart of the control logic for the prototype system.

To conserve power, the RCS is normally in a "sleep" mode. Controlled by the clock on the RCS board, the CPU is awakened once every minute. It quickly (50 milliseconds) checks whether to remain awake and proceed with a measurement or return to sleep mode. This cycle continues until a prescribed "sleep" time has elapsed, at which point the RCS begins execution of the main control and acquisition program. The RCS was programmed to control the system functions, including turning on the ventilation fan, rotating the sphere to expose the radiometer, acquiring and storing data, rotating the sphere back to hide the radiometer, turning on a calibration source to determine if the radiometer is frosted, and if frosted, turning on the heat source for a specified amount of time for clearing the radiometer. The frost-check/heating process is repeated until the check shows that the dome is clear, but for a maximum of 4 times to conserve power.

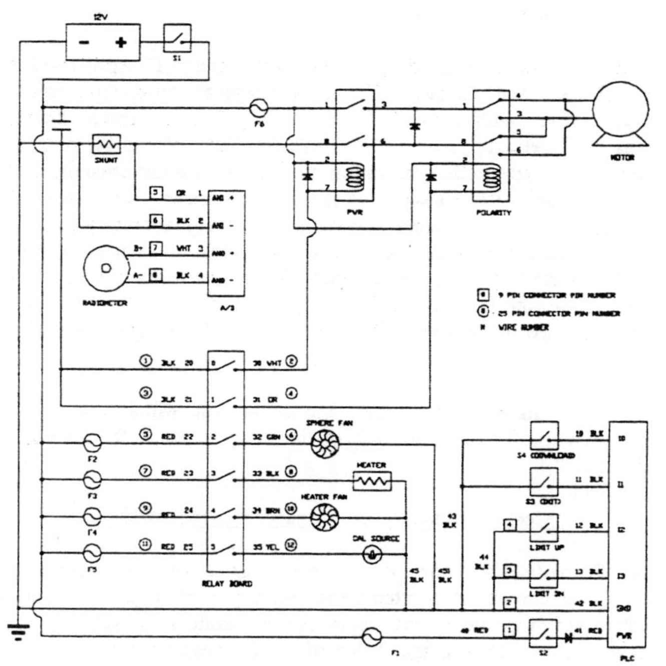

Figure 2. Schematic of the electrical system.

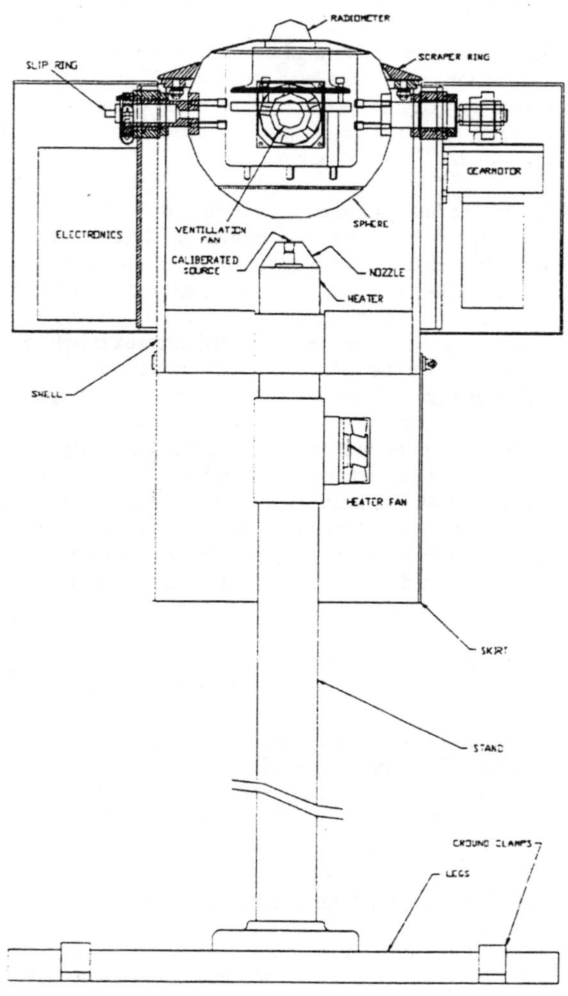

Figure 1. ARS Assembly, Front Cross-sectional view.

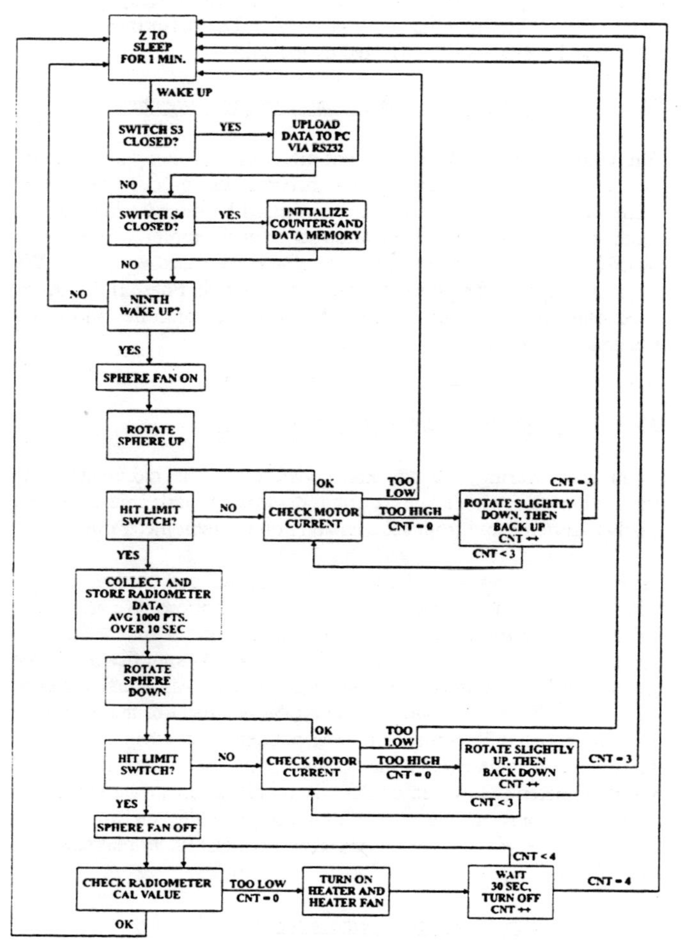

Figure 3. ARS Logic Flow Chart.

PROTOTYPE TESTING

A field test was conducted from 30 March through 12 April 1995, at the Environment Canada (EC) Meteorological station at Resolute Bay, NWT, Canada (74 43' N 94 59' W). The radiometer system was deployed approximately 24 m north-west of the Aerological building at the weather station, approximately 1.6 km on-shore, and approximately 100 m south of the EC radiometer field. The EC personnel gave us excellent support by providing laboratory space in the Aerological building, and providing access to their radiometer and weather data. Figure 4 depicts the radiometer deployed at the site. A second test was also carried out at US Army CRREL cold room located in Hanover, NH from January 15 through January 19, 1996.

The following is a description of the various tests which were carried out. The performance of each of the components of the system is discussed.

1) Mechanical system performance in Arctic weather.

At Resolute Bay, the overall performance of the mechanical components was tested in temperatures ranging from approximately -32 to -15 °C, with typical nighttime temperatures of -25 °C. On several days (and nights), sustained wind speeds were over 10 m/s, with blowing snow. All mechanical components worked very well in these conditions. There was no snow accumulation in either the motor or computer compartments. The system was subsequently tested in the CRREL cold room at -50 C. The performance of the mechanical system was unaffected by the cold.

2) Computer system performance in Arctic weather.

The cold weather had no apparent effect on the computer system. However, several computer failures occurred during our tests which were traced to power dips and spikes caused by the switching of the motor control relays and the startup surge of the motor. These were possibly more damaging as battery voltage decreased with use. We intend to improve the power distribution system in Phase II, by better grounding, filtering, and isolation between the controller and other systems.

3) Ability of the sphere to remove snow.

Light snow during one day had no effect on the radiometer; the ventilation fan which runs during deployment and measurement easily diverted snow from accumulating on the radiometer dome.

Heavier snow was shoveled from the ground and thrown on the sphere when the radiometer was in its down position. We attempted to simulate the amount of snow that would fall on the sphere during a typical 10 to 20 minute off time. This snow was significantly more granular and dense than newly fallen snow, and was actually a poor simulation. This snow built up around the perimeter of the sphere, at the surface of the scraper, and was not removed when the sphere turned up. Enough snow remained during the measurement to shield the radiometer at low sun angles. A modification to the scraper is planned to prevent the snow from accumulating on the scraper surface. However, we must also perform the test under real snowfall conditions.

4) Ability of sphere to remove ice.

Water at 0 C was sprayed on the sphere, with the radiometer in the protected position, in order to simulate freezing rain. A layer of ice approximately 0.8 to 3.2 mm thick was built up over the entire surface of the sphere. The motor was then activated to rotate the sphere. The motor current was measured as the motor broke the sphere free from the ice layer. The current peaked at 4.5 amperes, approximately 4 times the normal current required to rotate the sphere, and was significantly less than the full load capacity of the motor. Therefore, the motor could clearly handle this type of resistance, and could even be down sized to save power. Actual ice buildup in freezing rain conditions in the Arctic would probably be less than that in our test in a typical 10 to 20 minute measurement cycle. Even though the ice layer was cracked in various places due to induced stresses during the sphere rotation, it remained in place, of its own strength, when the radiometer reached the "up" position. The wind blew away the fragments before the computer actually started taking data, however, in the absence of wind, the ice dome would have remained, causing erroneous data. In fact, the ice dome would have continued to build up as long as freezing rain continued. Given the available excess torque of the motor, a simple modification was implemented and successfully tested later in the CRREL cold room. A sharp angle groove cut into the sphere surface causes stress concentration points, and forces the ice into the scraper, thus continually breaking it into small fragments during the rotation cycle.

5) Ability of the system to remove ice from the radiometer dome.

Two different tests were performed to investigate the ability of the system to clean the dome in the event that frost did form on it. Both tests involved spraying ice-water on the sensor dome to simulate frost or freezing rain, because no natural frost ever built up on the dome during the test.

a) With the heating circuit de-activated, the radiometer dome was sprayed with water to form a thin layer of "blotchy" ice droplets covering approximately 50% of the glass area (Figure 5). A similar layer was sprayed on a "control" dome (a small clear glass light bulb). The control bulb was left out in the sunshine and wind (approximately 5 m/s wind). The radiometer dome continued its normal cycle of 10 minutes down at rest and then, with the fan operating, 20 seconds turning up, 15 seconds measuring, and 20 seconds turning back down to rest position. After 1 hour the radiometer dome was significantly cleaner and the control bulb was unchanged. After 2 hours the radiometer dome was completely clean and the control bulb was only slightly improved. This is likely due to the effect of the fan in the immediate proximity of the dome which increases the rate of sublimation.

b) The radiometer dome and the control bulb were sprayed similarly to (a) above. The heater and the fan were then operated manually for the 10 minute period that the dome was down. When the radiometer turned up, it was nearly clean. The remaining small amount of ice on the dome would probably not affect the accuracy of the radiometer output. The condition of the control bulb however, was unchanged.

6) Ability of the system to prevent naturally occurring frost.

There was significant frost and ice buildup on the control bulb during the night of 7/8 April. This type of accumulation was also apparent on cables, and on the surface of the sphere as well. There was however, no trace of frost or ice on the radiometer dome with the system operating in its normal cycle as described in Sec. 5a above (the heater was never turned on). One of the EC radiometers (unventilated), which faces down to measure reflected energy, was frosted enough to prevent accurate data acquisition. Two EC ventilated radiometers had small traces of frost on their domes, perhaps enough to affect the accuracy of the data.

7) <u>Radiometer stabilization time.</u>

The objective of this test was to determine the minimum exposure time necessary for the radiometer output to stabilize after deployment. These tests were important in order to determine how quickly accurate data could be recorded after deployment, and to minimize the length of time the radiometer remained exposed to harsh elements. Furthermore, it would determine whether the shell enclosure adversely affected the sensor output by undesired heating of the sensor. It was found that the radiometer output stabilized within 5 seconds of being deployed, therefore, radiometer stabilization time was not an issue. However, this was a short-wave radiometer. Long-wave radiometers will be more sensitive to thermal changes and may require a longer stabilization time.

Figure 4. ARS deployed behind EC's Aerological Building.

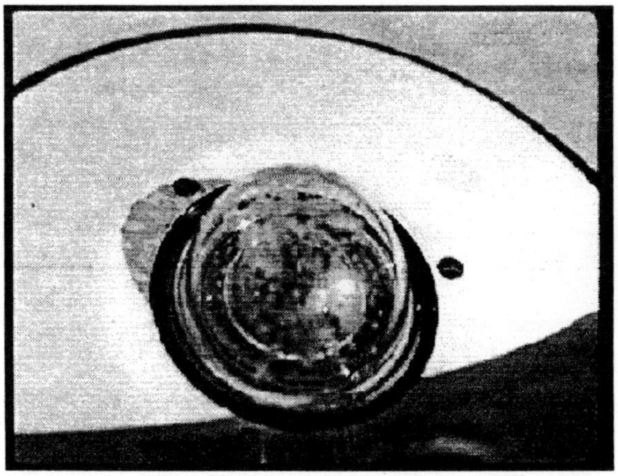

Figure 5. Frozen Dome after Exposure to "Freezing Rain".

8) <u>Multiple day short wave radiation data collection.</u>

This was perhaps the most important test to determine if the radiometer system could collect uncontaminated data in Arctic conditions. The radiometer computer was programmed to perform a nominal test cycle. The radiometer was kept inside for a period of 10 minutes. At the end of the 10 minute period, the motor and the ventilation fan were activated. After deployment, the sensor was allowed to stabilize for 5 seconds, followed by 10 seconds of data acquisition. During this time, 1000 readings were obtained, and averaged to produce a single value for that measurement. The motor was then reversed and the radiometer brought inside to close the cycle. This testing was performed for several multiple hour periods during the test and for one several day period.

The data for the several day period is shown in Figure 6. The drop outs occurring during the start of the rise each day were due to the shadow of a building passing over the radiometer (note the building shadow in Figure 4). The days associated with hours 72-84 and 96-108 had scattered clouds which leads to a noisy signal. The dropout from hour 114-119 was due to control hardware malfunction, and the dropout from 124-130 was associated with snow and ice tests being conducted on the radiometer as discussed above. The data compares well to that collected by the EC weather station.

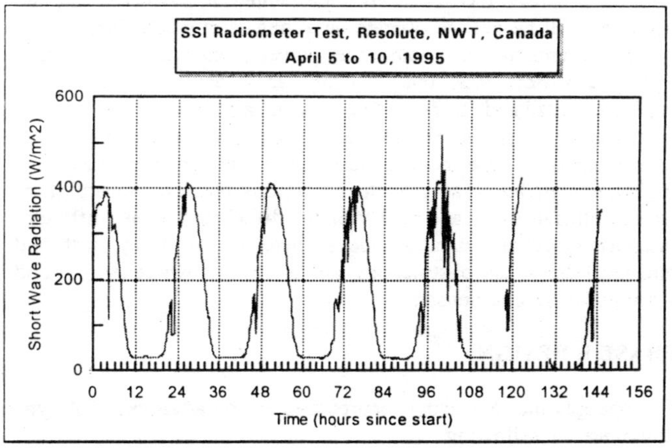

Figure 6. ARS data collected over 7 days, (mostly uniterrupted).

9) <u>Power drain.</u>

One of the goals of our system design is to develop a low power system capable of long term deployment. For the measurement condition described above, overall power drain was 2.5 amp-hours per day. This did not include any heating cycles. One heating cycle is expected to consume 0.3 amp-hours, however, since the need to heat the dome is most likely a seasonal requirement, it is not anticipated that it will significantly effect the power budget. (The fan alone of a ventilated radiometer system similar to the ones deployed by EC at Resolute consumes roughly 10 amp-hours per day.) This implies that our system would consume a large car battery once every 35 days (as compared to less than 9 days for a continuously ventilated radiometer). As most of the power is consumed by the motor, power consumption can be significantly lowered by a less frequent measurement cycle. The motor we employed was also over-powered given how easily it broke ice free. We may be able to lower power consumption with a smaller motor. Our Phase II effort also includes fabrication of a smaller system which will house a smaller radiometer, which in turn will require a smaller ball and smaller, lower power motor.

Features contributing to lower power consumption than continuously ventilated units are:

- The device is operated only when meaningful measurements are feasible. For instance, during dark hours, or when short-wave radiation measurements are meaningless, the system can remain idle conserving power.
- Operation of the ventilation fan only during the measurement cycle (approx. 1 min).
- Operation of the heater only as prompted by a calibration source indicating an iced or frosted dome.
- Intermittent operation of electrical and mechanical systems, since in the idle state, only the computer timer is on consuming minimal power.

CONCLUSIONS

The Phase I effort proved the feasibility of an unmanned, autonomous Arctic radiometer system. The testing showed that our design was successful in collecting uncontaminated radiometer data in an unattended mode for a multiple day period. With the exception of our deliberate icing of the radiometer, the sensor dome remained completely clean, even in the presence of frost and blowing snow. With the proposed modifications implemented, it is believed that the prototype will be able to operate on a long term basis. Performing six measurements per hour, the power consumption of SSI's prototype system was only roughly 25% of the commercially available ventilated units. The design also allows for simple modifications to protect various sensors in the Arctic and other harsh environments. These include measurements underwater where bio-fouling caused by algae growth is a serious problem, aboard ships where both salt spray and soot from the diesel stack cause difficulty in collecting reliable data, on buoys where freezing spray, dust, and bird fecal debris are obstructions, and at airports where jet fuel and urban pollution coat sensors.

PHASE II DESIGN

The changes that are being incorporated in the advanced prototype design are the following:

- Reduce size and weight of the ISPS system.
- A smaller and lower power consumption motor.
- Conditioned electrical power for the data acquisition and control circuitry.
- Software action based on battery voltage monitoring logic, which will safely shut down the system.
- Dual-axis inclinometer mounted within the sphere which indicates sensor orientation, and enables automatic leveling of the sensor.
- Improved heater design for better efficiency and lower power consumption.

ACKNOWLEDGEMENTS

Mr. Roy Bates, Cold Regions Research and Engineering Laboratory, Hanover NH.
Mr. Dave Beaubien, Yankee Environmental Systems Inc.
Mr. Joseph Carlin, Kipp and Zonen.
Dr. Thomas B. Curtin, Office of Naval Research, Arlington, VA.
Mr. John DeLuisi, NOAA/ERL/ARL
Mr. John Hickey, Eppley Laboratories, Newport, RI.
Mr. John MacIver, Manager, Environment Canada Meteorological Station, Resolute Bay, Canada.
Dr. Bruce McArthur, Head, National Atmospheric Radiation Center, Ontario, Canada.
Dr. Charles Ryerson, Cold Regions Research and Engineering Laboratory, Hanover NH.
Dr. Knut Stamnes, University of Alaska, Fairbanks.

REFERENCES

1. Lewis, JK (1993a). " A model for thermally-induced stresses in multi-year sea ice."*Cold Regions Sci. and Tech.*, Vol 21, pp 337-348.

2. Lewis, JK, Tucker, WB, Stein, PJ (1994) " Physical and mechanical properties of first year sea ice: Observations and modeling of landfast ice in Allen Bay, Canada." *J Geophysical Research*, Vol 99, No C8, pp 16361-16371.

3. Dr. Charles Ryerson, personal communication, CRREL, Hanover, NH.

4. Marie Gilkey, John MacIver, Managers, Resolute Bay Weather Station, Personal Communication.

Proceedings of the Seventh (1997) International Offshore and Polar Engineering Conference
Honolulu, USA, May 25-30, 1997

ISBN 1-880653-28-1 (Set); ISBN 1-880653-30-3 (Vol. II)

Application of the Pivot Point on FCP Diagram to Low-Temperature Fatigue of Materials

Meng-Lan Duan
China Classification Society
Beijing, China

James C.M. Li and Jing Li
University of Rochester
Rochester, NY, USA

ABSTRACT

By analyzing the test data of the fatigue crack propagation in offshore structural steel A537 at room and lower temperatures, a pivot point on the FCP diagram is presented, from which a relationship of $m-\ln C$ is derived, where m and C are the constants in Paris Equation. The derived $m-\ln C$ relationship for steel A537 is identified with a group of reports on the exist of experimental $m-\ln C$ relationship from literature. From the discussion and assessment on the $m-\ln C$ relationship, a critical stress intensity factor range named ΔK_{pc} is proposed. Hence, ΔK_{th}, ΔK_{pc} and ΔK_{fc} represent different fatigue crack propagation stages, and ΔK_{pc} can be used as a low-temperature fatigue safety index that is very important in monitoring the safe propagation of fatigue cracks and also important in evaluating the low temperature properties of materials.

KEY WORDS: fatigue crack propagation, low temperature fatigue, pivot point, Paris constants, stress intensity factor range.

NOMENCLATURE

a, b, B	material constants
da/dN	fatigue crack growth rate
ΔK	stress intensity factor range
ΔK_{th}	threshold stress intensity factor range
ΔK_{pc}	ΔK value at the pivot point
ΔK_{fc}	the critical ΔK value at which unstable propagation of a crack begins
m, C	Paris constants
m_i, C_i	Paris constants at temperature T_i
R_{pc}	fatigue crack growth rate at the pivot point
T, T_i	temperature
$U(\Delta K)$	a crack closure function

1. INTRODUCTION

The offshore petroleum platforms in Bohai Gulf, China, experience low temperatures that fall to as low as 238K for three months in winter. Cracks in the structures have to withstand low temperature fatigue and unstable propagation of fatigue cracks or brittle fracture would occur most possibly. Under ice loading, the shortest estimated fatigue life of an offshore platform by use of $S-N$ curves is only 2 years. The study of low temperature fatigue of offshore structures under ice loading is getting more and more urgent with the large scale exploitation of offshore oil in Bohai Gulf. Fatigue tests on the offshore structural steels are the basic part of this study. Steel A537 is the main structural steel of the offshore tubular joint platforms in Bohai Oil Field and has to be tested to determine if this steel can meet with the platform's withstanding both low temperatures and ice loadings. This paper is the extension of the work from the author (Duan, 1995; Fang and Duan, 1992).

2. RESULTS OF FCP TEST ON STEEL A537

Fatigue crack propagation tests on steel A537 at room and lower temperatures were partly reported (Duan, 1995; Fang and Duan, 1992). Here is a list of the test results, which are also schematically presented in Fig.1:

1) The threshold value of fatigue cracks in steel A537 and crack propagation resistance near the threshold region increase with the decrease of temperatures;

2) The stable fatigue crack propagation in steel A537 follows the Paris law, but the Paris exponent m varies with temperatures, presenting a much larger value at lower temperatures.

3) The curves on the FCP diagram as shown in Fig.1 tend to converge with the increase of stress intensity factor range and intersect at some high ΔK value. Before the cracks propagate to the intersecting point usually called the pivot point, the crack propagation rates at low temperatures are smaller than that at room temperature, but then get inversely higher until to the unstable propagation stage where the brittle fracture occurs.

This paper will investigate the pivot point as observed in Fig.1.

3. ANALYSIS OF THE PIVOT POINT

In 1963, Paris and Erdogan proposed a famous FCP formula to describe the fatigue crack growth rate which was later rewritten as:

$$da/dN = C(\Delta K)^m \quad (1)$$

It represents a linear relationship in FCP diagram as:

$$\ln(da/dN) = \ln(C) + m\ln(\Delta K) \quad (2)$$

Obviously, the fatigue crack propagation in steel A537 at the three tested temperatures, i.e., 238K, 273K and 303K, obeys the above law, as shown in Fig.1. For each temperature, Eq.(2) becomes:

$$\ln(da/dN) = \ln(C_i) + m_i \ln(\Delta K) \quad (3)$$

where C_i and m_i are the Paris constants at temperature T_i.

Let ΔK_{pc} and R_{pc} stand for the ΔK value and the crack growth rate at the pivot point respectively, and apply them to Equation (3), then we get:

$$\ln(R_{pc}) = \ln(C_i) + m_i \ln(\Delta K_{pc}) \quad (4)$$

or

$$C_i = \frac{R_{pc}}{(\Delta K_{pc})^m} \quad (5)$$

Therefore, Paris Equation takes the following form:

$$da/dN = R_{pc}\left(\frac{\Delta K}{\Delta K_{pc}}\right)^m \quad (6)$$

If suppose:

$$\begin{cases} a = \ln(R_{pc}) \\ b = -\ln(\Delta K_{pc}) \end{cases} \quad (7)$$

then from Eq.(1), (6) and (7), a linear relationship between the two Paris constants m and $\ln C$ can be easily gotten as the following form:

$$\ln C = a + bm \quad (8)$$

Evidently, a and b are material constants independent of temperatures. This is the recognized $m-\ln C$ relationship obtained from large amount of test data(Benson and Edmonds, 1978; Cortie and Garrett, 1988; Iost, 1991), it is derived from the pivot point through the above steps.

Then, we can conclude that if there exists the pivot point on the FCP diagram, there must be a $m-\ln C$ relationship as shown in Eq.(8), and conversely, if the $m-\ln C$ relationship is confirmed, there must be a pivot point on the diagram.

Least square regression of the test data for steel A537 is used to fit a line as:

$$\ln C = -11.4597 - 18.6597m \quad (9)$$

and

$$\begin{cases} a = \ln(R_{pc}) = -11.4597 \\ b = -\ln(\Delta K_{pc}) = -18.6597 \end{cases} \quad (10)$$

where the units of ΔK and da/dN are $MPa\sqrt{m}$ and $m/cycle$ respectively. Eq.(9) represents the correlation between the two Paris constants C and m for steel A537, with the test data of them being shown in Fig.2.

From Eq.(10), the constants ΔK_{pc} and R_{pc} at the pivot point for steel A537 are calculated as follows:

$$\begin{cases} \Delta K_{pc} = 127MPa\sqrt{m} \\ R_{pc} = 1.0546 \times 10^{-5}\, m/cycle \end{cases} \quad (11)$$

which are obviously independent of the environmental temperatures, taking notice that C and m are temperature dependent material parameters in Paris Equation, and form the pivot point which is fixed by the material itself and independent of environment conditions.

Furthermore, if considering the $m-\ln C$ relationship of Eq.(9), the fatigue crack growth rate of steel A537 can be expressed as:

$$da/dN = 1.0546 \times 10^{-5}\left(\frac{\Delta K}{127}\right)^m \quad (12)$$

Clearly, the fatigue crack growth rate is dependent on only one parameter m (besides ΔK). The two constants in Paris Equation are then reduced to one as shown in Eq.(6).

4. LITERATURE REMARKS ON THE PIVOT POINT

With the recognition of the Paris Equation after it was recommended as the basic law for fatigue crack propagation, the scientists have been trying to disclose its physical meaning by theoretical derivation, and a lot of work has been conducted on the two constants C and m since 1972, although theoretical results have not been reported. The following is a literature review on C and m the work on which are all test results.

Koshiga(1973) and Gurney(1979) collected a large amount of C and m data to investigate the $m-\ln C$ relationship in detail and assumed that the parameter C is fixed with all the experimental uncertainty being incorporated into the m parameter. In 1976, Niccolls reported that there is a very definite and precise relationship between the two parameters and pointed out that the fatigue crack propagation in Paris region is controlled by only one parameter m (besides ΔK). This conclusion is in agreement with the results shown above. Dimensional analysis by McCartney and Irving(1977) revealed that C is a function of m and a further material parameter B. But they disagreed with that the description of FCP in Paris region can be simplified into a problem of a single material parameter. Opposite to this contention, Cortie and Garrett(1988) made a conclusion that the correlation between $\ln C$ and m exists for a group of materials or test conditions only if B does not change.

Tanaka and Matsuoka(1977) examined the $m-\ln C$ relationship and concluded that all ferrous materials would have the same fatigue crack propagation rates at a certain ΔK value. They rewrote the Paris Equation in the form:

$$da/dN = C(\Delta K_{eff})^2 \quad (13)$$

by means of the empirical correlation where $\Delta K_{eff} = U(\Delta K) * \Delta K$ and $U(\Delta K)$ is a crack closure function yielding value between 0.1 and 0.2.

Radhakrishan(1980) also related C to m from dimensional analysis. He found that the Paris region fatigue data for all steel alloys lay on the same correlation line and confirmed that the pivot point for steels on the FCP diagram occurs at about $50MPa\sqrt{m}$ while Cortie and Garrett(1988) confirmed the point at about $60MPa\sqrt{m}$.

In 1978, Benson and Edmonds(1978) reviewed the correlation between C and m, showing that it was effected by the R-ratio but not by fatigue frequency, yield stress or specimen thickness. The variation of m with test temperatures was discussed by Radhakrishan(1980) where as the temperature rises, the fatigue crack growth rate increases and m correspondingly decreases and the following relationship was obtained as shown in Fig.3:

$$m = -0.004T + 4.8 \quad (14)$$

where T is in absolute degrees K.

In 1988, Cortie and Garrett(1988) studied in detail the correlation between C and m in the Paris Equation for fatigue crack propagation. They presented almost all the data of the literatures to confirm that the $m-\ln C$ relationship exists over a wide range of testing conditions and a computer simulation was described in that paper which generates random C and m values showing that these random values also exhibit the correlation. The simulation process was repeated 700 times and linear regression applied to the resulting $m-\ln C$ data, yielding an expression:

$$\ln C = -15.07 - 3.40m \quad (15)$$

with a correlation coefficient of 0.969. Fig.4 is the $m-\ln C$ data obtained from the simulation. As one of the conclusions, they pointed out that the pivot point is the result of the logarithmic technique commonly used when plotting FCP data and doesn't appear when data is plotted on linear axes, and that any models for fatigue crack growth which attempt to constrain m to values of, for example, 2 or 4, cannot be physically meaningful, for m can be any values from about 2 to more than 9.

In the recent years, A.Iost and J.Lesage(1990) discussed the existence of the pivot point for stage II fatigue crack growth and A.Iost(1991) the effect of load ratio on the $m-\ln C$ relationship. They linked the pivot point to the macroscopic crack growth rate, and the correlation to the transition between ductile striation and cyclic cleavage or other additional modes of propagation. M.L.Duan and H.C.Fang also found the pivot point when studying the low temperature fatigue of offshore structural steel A537(Duan,1995; Fang and Duan, 1992). The author has obtained some important results from the point, some of which will be further discussed in this paper.

5. SAFETY ASSESSMENT INDEX FOR LOW TEMPERATURE FATIGUE: AN APPLICATION OF THE PIVOT POINT

As has been pointed out above, there exists a pivot point on the fatigue crack propagation diagram in logarithmic axes for almost all materials including steels and their alloys, the position of the point being at around $60MPa\sqrt{m}$ (for ΔK value). Fatigue test results of offshore structural A537 steel show that this material has the same performance that the lines at different temperatures converge on one point on the diagram, but what is different from other literatures is the location of the pivot point which is at around $127MPa\sqrt{m}$, two times those of other materials. This is very important for A537 steel structures especially those served in low temperature environment, and the ΔK value at the pivot point is fatal to the fatigue performance of the material. As shown in Fig.1, fatigue cracks in steel A537 at lower temperatures are not so easy to initiate as those at room temperature because of the higher threshold stress intensity factor range ΔK_{th} and the crack growth rates are smaller than that at room temperature before the cracks propagate to the pivot point. But due to the larger exponent m which implies a higher acceleration of propagation at lower temperatures, the cracks propagate faster and faster and their propagation rates begin to exceed those at room temperature when the cracks pass the pivot point. Then, fatigue cracks at lower temperatures will enter stage III earlier which is called fast fracture region or unstable propagation stage. In other words, the critical ΔK value (ΔK_{fc}, marking the transition from stable propagation to unstable or uncontrolled one) at lower temperatures are smaller with brittle fracture faster. Anyway, it seems that, at lower temperatures, steel A537 has a higher threshold value which means that fatigue cracks initiate more difficultly, a lower growth rates during the most propagation period and a longer fatigue life as has been tested (most of the fatigue life is from initiation stage and earlier propagation stage) . This makes one think that the material has better fatigue performance at lower temperatures than that at low temperature, thus may behave more reliably and more safely when served in low temperature environment. But , considering the higher acceleration, lower ΔK_{fc} value and faster brittle fracture, and considering the rate transition point (i.e., the pivot point), there is a great potential unreliability and unsafety for the material to be served at low temperatures.

Now, we can make a conclusion from the above discussion. There are three transition points during the whole propagation stage of a fatigue crack. The ΔK_{th} indicates that the crack initiates and begins to propagate. The ΔK_{fc} means that it propagates fast from stable stage to unstable one or means that the brittle fracture begins. The ΔK value at the pivot point (marked by ΔK_{pc}) represents that the smaller crack growth rates at low temperatures begin to exceed that at room temperature with a higher propagation acceleration and afterwards the cracks enter the brittle fracture region earlier because of the higher acceleration. Therefore, ΔK_{pc} is a critical value for fatigue cracks showing a transition from a lower cracks growth rate to a higher one for A537 steel at low temperatures. Owing to the existence of ΔK_{pc}, an A537 steel an A537 steel structure which is considered reliable and safe at room temperature will become unreliable and unsafe once it is served in low temperature environment, and the value of ΔK_{pc} can thus be a reference for examining if the structure needs to be inspected or get repaired. It is advisable that if a fatigue crack propagates in a ΔK value less than ΔK_{pc}, the structure can be thought reliable and safe, or if the ΔK exceeds ΔK_{pc}, it is in a danger state and needs to be paid attention to or measures should be taken to resist the crack from propagating or to repair the crack. In one word, ΔK_{pc} can be a reference when determining the inspection period of A537 steel structures served in low temperature environment, and it can also be advised to be a low temperature fatigue safety index for A537 steel structures.

The only paper which dealed with temperature effect on the $m-\ln C$ correlation is from Radhakrishan (1980), but regretfully to say, the temperature range included in his paper is from 0°C to 550°C and temperatures less than 0°C have not been reported on the effect on the correlation. So, it is advised that low temperature effect on $m-\ln C$ relationship for materials used for structures that may be served in low temperature environment should be investigated to confirm if there are pivot points or ΔK_{pc} values as found above for A537 steel. If so, more attention should be paid to this specialty so as to avoid the structures from low temperature unstable fracture during the late crack propagation period, and the inspection period should be shortened to examine if the cracks have propagated near to the ΔK_{pc} values.

6. CONCLUSIONS

The above investigation on the pivot point for fatigue crack propagation in A537 steel and the literature review leads to the following conclusion.

1) Fatigue crack propagation in steel A537 at each temperature exhibits a linear relationship in logarithmic axes in stage II, and the lines at different temperatures tend to converge with the increase of stress intensity factor range and intersect at some high ΔK value. The intersection point is defined as a pivot point. Thus, there exist three transition points ΔK_{th}, ΔK_{pc} and ΔK_{fc} which have their own behaviours.

2) Theoretical analysis of the pivot point leads to the conclusion that the Paris parameters C and m yield a relationship as $\ln C = a + bm$ which was identified only by a large amount of test data rather than obtained from theoretical analysis. This conclusion theoretically supports the existence of the pivot point on the FCP diagram, hence, the ΔK_{pc} is further determined. For steel A537, the following expression of C and m at different temperatures is reached with the ΔK_{pc} value of $127MPa\sqrt{m}$:

$$\ln C = -11.4597 - 18.6597m$$

3) The proposed parameter ΔK_{pc} corresponding to the pivot point is meaningful and is very important for evaluating the mechanical properties of materials used in conditions of low temperatures. It is suggested to be a low temperature fatigue safety index of materials and it also should be refered when determining the inspection and repair period of the structures.

7. ACKNOWLEDGEMENTS

The author would like to thank Prof. Fang for his help and critical reading of this paper.

8. REFERENCES

Bailon, JP, Masounave, J and Bathias, C (1977), "On the Correlation between the Parameters of Paris Law for Fatigue Crack Growth in Aluminum Alloys", *Scripa Metall.*, Vol.11, pp1101-1106.

Benson, JP, and Edmonds, DV (1978), "The Relationship between the Parameters C and m of Paris Law for Fatigue Crack Growth in a Low-alloy Steel", *Scripa Metall.* Vol.12, pp645-647.

Cortie, MB and Garrett, GG (1988), "On the Correlation between the C and m in the Paris Equation for Fatigue Crack Propagation", *Engineering Fracture Mechanics,* Vol.30, No.1, pp49-58.

Duan, ML (1995), "On the Test of Low Temperature Fatigue of Offshore Structural Steels", *Oil Field Equipment,* Vol.24, No.1, pp14-17(in Chinese).

Fang, HC and Duan, ML (1992), "Temperature Dependent Crack Propagation in Steel A537 at Lower and Room Temperatures", *Proceedings of ISOPE-92*, Vol.4, June 14-19, San Francisco, pp130-136.

Gurney, TR (1979), "Fatigue of Welded structures", 2nd Ed., Cambridge University Press, Cambridge.

Iost, A (1991), "The Effect of Load Ratio on the $m - \ln C$ Relationship", *Int. J. Fatigue*, Vol.13, No.1, pp25-33.

Iost, A and Lesage, J (1990), "On the Existence of a Pivot Point for Stage II Fatigue Crack Growth", *Engineering Fracture Mechanics*, Vol.36, No.4, pp585-596.

Koshigs, D, and Kawahara, M (1973), "A Proposed Design Basis with Special Reference to Fatigue Crack Propagation", *J. Soc. Naval Architects Japan*, Vol.153, pp305-312(in Japanese).

Niccolls, EH (1976), "A Correlation for Fatigue Crack Growth Rate", *Seripa Metall,* Vol.10, pp295-298.

McCartney, LN and Irving, PE (1977), "Comments on: A correlation for fatigue crack growth rate", *Scripa Metall,* Vol.11, pp181-183.

Paris, P and Erdogan, F (1963), "A Critical Analysis of Crack Propagation Law", *Journal of basic Engineering, Transactions,* D85, pp528-534.

Radhakrishan, VM (1980), "Quantifying the Parameters in Fatigue Crack Propagation", *Engineering Fracture Mechanics*, Vol.13, No.1, pp129-141.

Tanaka, K and Matsuoka, S (1977), "A Tentative Explanation for Two Parameters, C and m, in Paris Equation of Fatigue Crack Growth", *Int. J. Fracture,* Vol.13, No.5, pp563-583.

Tanaka, K, Masuda, C and Nishijima, S (1981), "The Generalized Relationship between the Parameters C and m of Paris Law for Fatigue Crack Growth, *Scripa Metall*, Vol.15, pp259-264

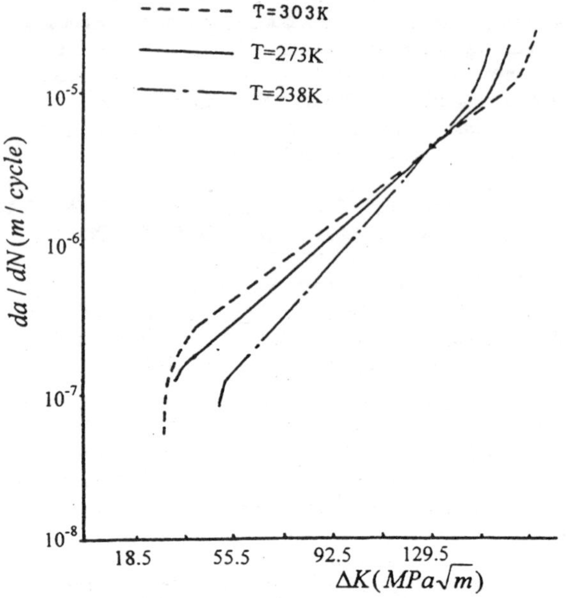

Fig.1 FCP Diagram of Steel A537 at Different Temperatures

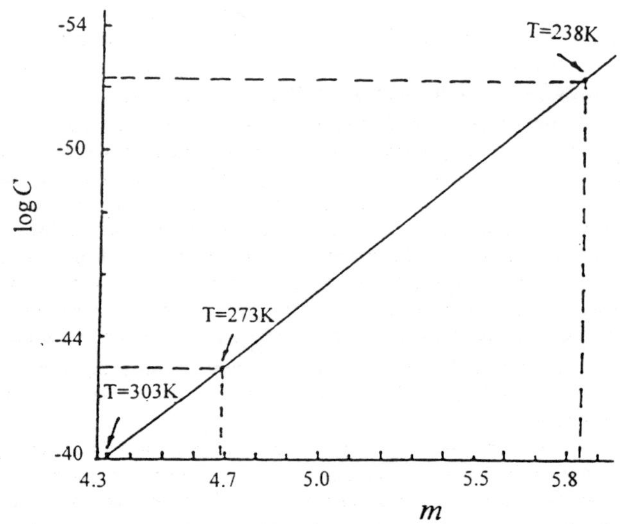

Fig.2 The $m - \log C$ Correlation of Steel A537

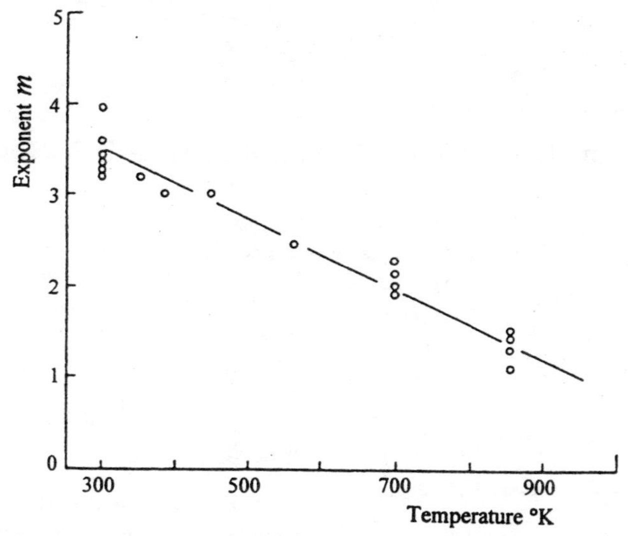

Fig.3 Relation between the Exponent m and Temperature(Radhakrishan, 1980)

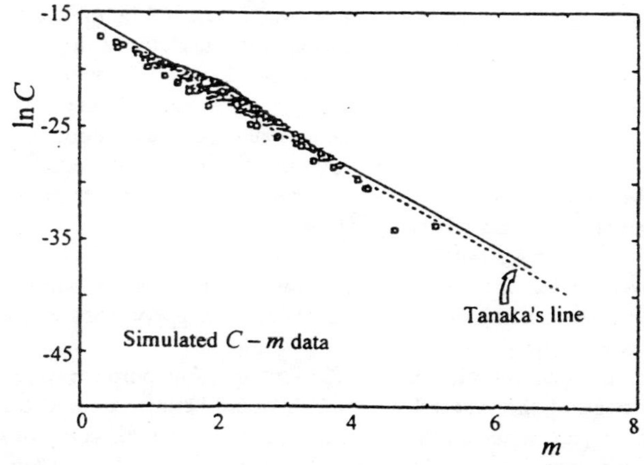

Fig.4 $C-m$ Data Generated by a simple Computer Simulation(Cortie and Garrett, 1988)

Proceedings of the Seventh (1997) International Offshore and Polar Engineering Conference
Honolulu, USA, May 25-30, 1997

ISBN 1-880653-28-1 (Set); ISBN 1-880653-30-3 (Vol. II)

A Laboratory Simulation of Wet Icing Build-up on H.V. Insulators

M. Farzaneh and J.-L. Laforte
Université du Québec à Chicoutimi
Québec, Canada

ABSTRACT

The objective of the present study is to propose a practical laboratory method to simulate wet atmospheric ice accretion on H.V. insulators. The relatively high liquid water fraction containing in such deposits is the principal cause of insulator flashover. The method developed is based on first accumulating a low density soft rime on energized insulators and then to heat the produced deposits. The liquid water fraction of the ice accretion is then determined using a centrifuge. This approach makes it possible to determine the performance of insulators under wet-snow-like conditions with relatively simple, low cost equipment.

*ISOPE Member
KEY WORDS: Atmospheric wet Icing, H.V. Insulators, Flashover.

INTRODUCTION

Over the last few years, several electrical power failures due to ice and wet snow accretion occurred on Hydro-Quebec overhead power lines (Drapeau, 1991; Hydro-Québec, 1995). These power outages were due, in part, to the occurrence of flashovers on ice and snow-covered insulators. In order to gain a consistent understanding of the roles of ice and snow in insulator flashover, a research program was established at the University of Quebec in Chicoutimi, in collaboration with Hydro-Quebec. The first stage of the research work involved ice accumulation on insulators. A part of the study, concerning the determination of atmospheric conditions favoring the accretion of the most severe type of ice , i.e. the type of ice associated with the highest probability of flashover, was reported in previous papers (Farzaneh et al, 1995; Farzaneh and Drapeau, 1995).

The second stage of the research, which is the subject of the present paper, deals with a practical laboratory simulation of wet icing accretion on H.V. insulators. The best example of such accretions is wet snow. This form of atmospheric wet icing is said to form when snow flakes, 1 to 5 mm in diameter, coalesce partially before touching the ground, if they come into contact with a positive temperature up to + 4°C. Depending on weather conditions, wet snow deposits may contain a fraction of liquid water up to 50% of their total weight (Mitsuru et al, 1990).

The problems caused for outdoor insulation by wet atmospheric accretions such as wet snow are, first, the bridging of insulator sheds, causing a notable reduction in leakage distance, and second, the presence of liquid fraction in the deposits. This causes a relatively high conductivity path between the high voltage and grounded sides of insulators. The combination of the two above effects causes a drastic drop in the leakage resistance of insulators, resulting in a significant increase in leakage current. In turn, the heating effect of the leakage current causes ice crystals to melt and provokes the deposits to fall off or melt completely in several places along the insulator or insulator string. Because of the low resistivity of the remaining deposits, the voltage drop along the airgaps caused by the melting or falling of deposits will be very high. Under such conditions, local arcs along the air gaps will occur and more ice crystals will melt. Such conditions could bring the insulators to the occurrence of a flashover arc (Farzaneh et al, 1995; Farzaneh et al, 1996).

In order to simulate wet snow accretion on H.V. insulators in the laboratory, two approaches may be considered:

The first approach purports to use a system for generating snowflakes which would be injected into the air flow that strikes the insulators. This approach requires complex and costly equipment (Mitsuru et al, 1990). Moreover, such a system produces small quantities of snow flakes and would be difficult to adapt to the cold room designed for insulator flashover tests.

The second approach consists of examining the possibility of obtaining wet ice accretion with the same device used to produce other types of atmospheric ice from supercooled droplets (Farzaneh and Drapeau, 1995).
The principal objective of the present study is to examine the practicality of the second approach and to identify a procedure by which the environmental conditions for obtaining ice accretion

producing bridging of insulators can be created. The approach retained consists of forming low density rime from supercooled droplets. The ice build-up is then heated to be practically melted. The study also calls for identifying the main characteristics of such wet ice deposits grown on insulators.

EXPERIMENTAL RESULTS

Facilities and Procedure

Ice was accreted on a short string of 5 Porcelain insulator units from supercooled droplets in a cold room 4.8 m x 2.8 m x 3.5 m (Fig. 1). The air temperature in the cold room was controlled by a proportional integral and differential system (PID), with an accuracy of ± 0.2°C.

Tap water, with a conductivity of about 80 $\mu S.cm^{-1}$ at 20°C, was used to spray the insulators, which were placed vertically in the center of the cold room. The spray system consisted of 4 air-atomizing nozzles mounted on a vertical support parallel to the axis of the insulator string, at a distance of 180 cm. In order to improve the uniformity of icing on the surface of the insulators, the nozzles oscillated along a vertical axis (see Fig. 1).

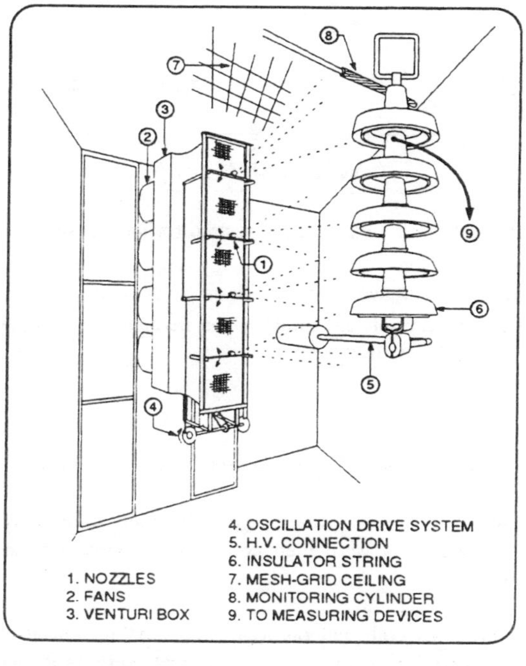

Fig. 1: Experimental Setup

The median volume diameter of the supercooled droplets was measured by exposing a glass slide coated with a solution of collargol to the flow of water droplets for a short time period (Godard, 1960). The samples were then examined under a microscope and the median volume diameter of these droplets was determined from the average diameter of approximately 1000 droplets.

A relatively uniform wind was obtained by using a system of eight fans placed in a tapering box with a honeycomb diffusing panel. The fans were mounted in a double stage arrangement enabling wind generation up to 50 $km.h^{-1}$ along the axis of the insulators.

The liquid water content of the air containing the supercooled droplets was measured by using the single cylinder method (Rush and Wardlaw, 1957; Stallabras, 1978), based on the weight of ice accreted on a cylinder with a radius of 3.15 mm, for a short time period (30 s to 1 min.).

Ice accretions were either accumulated without voltage or with the presence of an alternating voltage of 15 $kV_{r.m.s.}$ per insulator unit. This level of voltage applied per insulator unit is approximately equal to that used on Hydro-Quebec's 735 kV systems. The high voltage was supplied by a 240 kVA, 120 kV transformer with a short-circuit impedance of 5%. The primary coil of the transformer was connected to a 240 kVA regulator with a short circuit impedance of 2%.

The experimental conditions of forming low density ice deposits are characterized by the following parameters (Table 1).

Table 1: Atmospheric conditions for the formation of low density ice deposits

Air temperature	-12°C
Mean volume diameter of droplets	15 μm
Wind velocity	5.1 $m.s^{-1}$
Liquid water content	6.8 $g.m^{-3}$
Conductivity of freezing water	80 $\mu S.cm^{-1}$

The physical aspect of insulators covered with ice, accreted under the conditions reported in Table 1, is presented in Fig. 2. This type of accretion is soft rime, white and opaque with a density lower than 0.3 $g.cm^{-3}$. After 2 hours of accumulation, the ice deposits are let to warm up by rising the air temperature of the cold room from -12°C to a temperature above 0°C. During the warm-up period, the accreted ice partly melts. The fraction of liquid water inside the ice accretion increases with the rise in temperature.

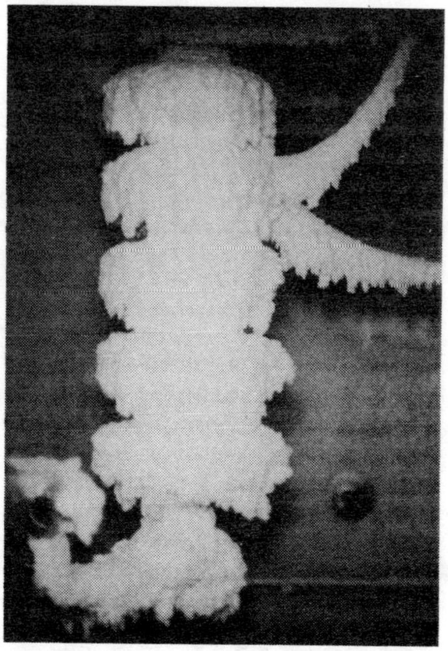

Fig. 2: Physical aspect of insulators covered with soft rime

Liquid Water Fraction Measurement

By definition, the liquid water fraction of ice crystals represents that portion of liquid water found in the deposits, and is determined by obtaining the ratio of the weight of the water contained in a given quantity of ice crystals to the total weight of the ice crystals. Establishing this parameter of laboratory produced deposits will make it possible to examine its effects on the flashover voltage of insulators, as compared to its effects on the liquid water fraction in natural wet snow.

A centrifuge was used in order to determine the liquid water content of the ice crystals. The principle of the measurement method is as follows: a sample of known weight and volume of ice crystal deposits is placed in a graduated test tube and submitted to centrifugal force. This pushes the water down to the bottom of the test tube where a mixture of water and ice crystals is formed. The weight of the water contained in the mixture can be determined from the results of a prior calibration.

The calibration procedure is as follows: First, a measured quantity of water is added to a test tube containing a known quantity of ice crystals previously cooled to -12°C (dry crystals). The mixture is then submitted to centrifugal force for a period of 5 minutes. The volume of the mixture of ice crystals and water (V_t) at the bottom of the test tube is then measured. Knowing the weight of the water added (P_e) and the total volume of the mixture (V_t), it is possible to calculate the index of water weight (η_e) from the following relation:

$$\eta_e = \frac{V_t}{P_e} \qquad (1)$$

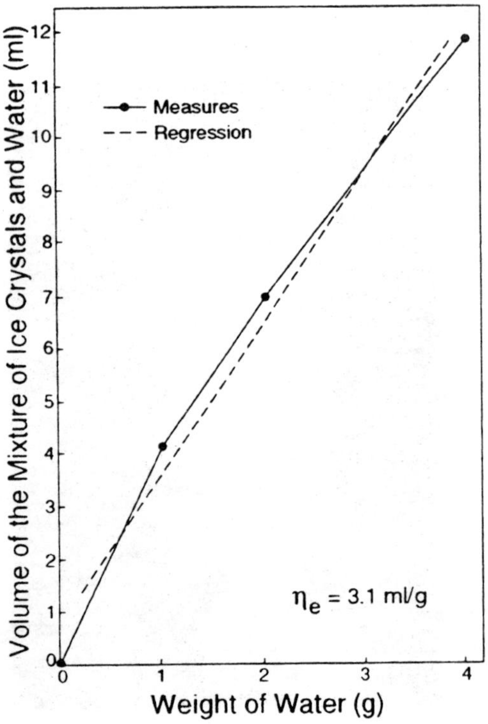

Fig. 3: Calibration of measurement system for soft rime

This ratio, supposed constant, should be used for the measurement of the liquid water fraction in any mixture. Calibration tests were made first using laboratory produced ice crystals and then natural snow for comparison purposes. The results obtained are presented in Fig. 3 and Fig. 4 respectively.

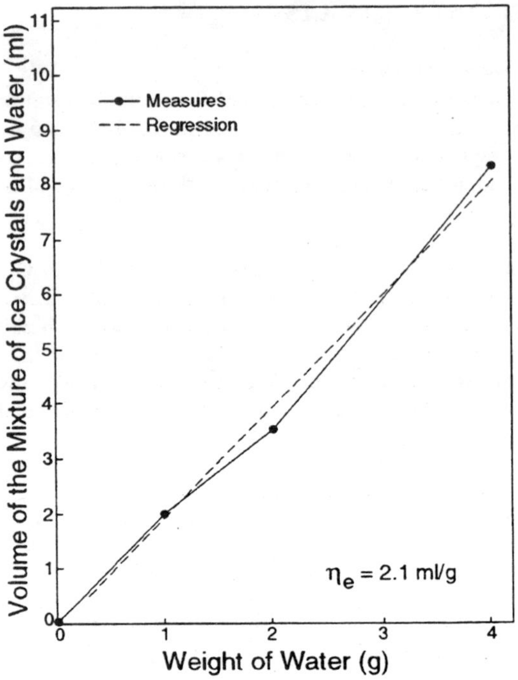

Fig. 4: Calibration of measurement system for natural snow

The η_e values calculated using the linear regression method were of 3.1 ml/g for ice crystals accumulated on insulators in the laboratory and of 2.1 ml/g for natural snow.

Experimental Results of Liquid Water Fraction Results.

The ice accretions were examined to ascertain their liquid water fraction. The results were then compared to those obtained from natural snow.

Natural snow was collected outdoors and kept in a heat-insulating container, while low density ice accretions were produced in the laboratory. Both samples were kept at a temperature of -12°C.

Samples were taken, placed into test tubes and were submitted to centrifugal force for a period of 5 minutes at varying temperatures. The weight of the water contained in the mixture was then calculated using relation (2).

$$P_e = \frac{V_t}{\eta_c} \qquad (2)$$

The liquid water content Ce, in Percentage, was calculated as follows:

$$C_e = \frac{P_e}{P_t} \times 100 \qquad (3)$$

where P_t is the total weight of the sample.

In order to minimize the margin for error, all the equipment used for the experiment was kept at the same temperature as that of the on-going test.

The results of liquid water fraction measurements corresponding to soft rime deposit grown, with no voltage applied to the insulators, and to natural snow flakes, are presented in Fig. 5. The outcome of the experiments presented in Fig. 5 leads to the following findings and results:

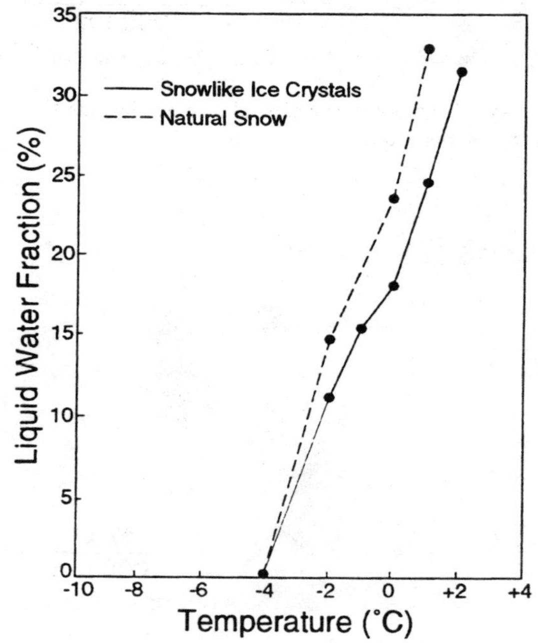

Fig. 5: Liquid water fraction of artificial soft rime deposits and natural snow as a function of temperature

- At temperatures below -4°C, no trace of water was discernible in either the laboratory produced ice accretions or natural snow.
- At a temperature of -2°C, the liquid water content of laboratory grown soft rime deposits and that of natural snow were about 11% and 14% respectively. These values increased with the increase in temperature. At + 2°C, the liquid water content for ice crystals was about 31.5%, with 32% for natural snow at +1°C.
- In accordance with the observations made during the tests, the liquid water content increases sharply at temperatures above + 2°C.

The influence of voltage on the liquid water content of ice accretion was also investigated by comparing the results obtained from two cases: in the first case, ice accretions were produced with an applied voltage of $15kV_{rms}$ per insulator unit and, in the second case with no voltage applied to the insulator string. The liquid water fraction for each type was then compared with the other. In both cases the atmospheric parameters were identical.

The results of the experiments are presented in Fig. 6. These results show that the liquid water fraction in both cases, with and without voltage, are practically of the same order of magnitude, and it may be concluded that the presence of an electrical field does not have any significant impact on the liquid water fraction of ice accreted on the insulators under laboratory conditions.

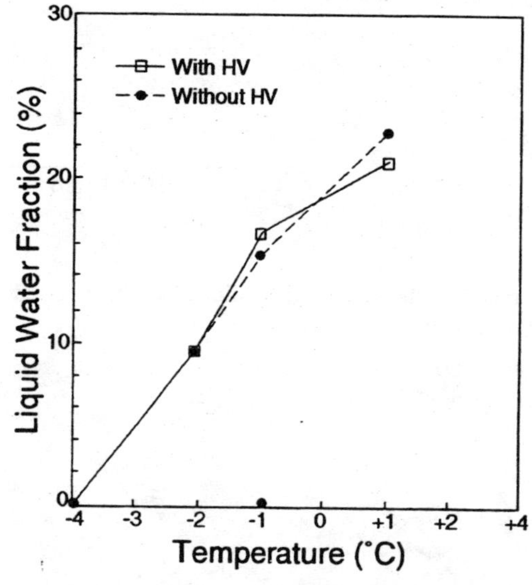

Fig. 6: Liquid water fraction of soft rime as a function of temperature

MICROSCOPE STUDY OF ICE ACCRETIONS

The objective of the microscope study was to compare the artificially produced soft rime to naturally occurring snowflakes. From among the different techniques in use, such as glass slides covered with an impressionable agent, e.g. formvar (Laforte and Du, 1988), collargol (Godard, 1960), or soot (Skidmore and Pavia, 1979), a clean glass slide for microscopic observation was chosen. This choice was based on the results of a series of preliminary experiments.

Characteristics of the soft rime accretion obtained in the Laboratory

Figure 7 shows ice crystals gathered on the glass slide, which had been exposed to the flow of droplets drawn to the ground below the insulator. No voltage was applied. The photograph shows a multitude of ice spheres, evenly distributed as a result of the spontaneous freezing of supercooled droplets upon contact with the slide. The diameter of the droplets, as shown, varies between 10 and 80 µm.

Figure 8 shows clusters of frozen droplets forming ice deposits on the insulator string during the tests at -12°C, with no voltage applied. In this case, the samples were removed from the insulator string using a spatula. The clusters were, in all likelihood, formed by spontaneous freezing of supercooled droplets upon impact with the insulator string. These clusters bear the characteristics of low density soft rime. The brittle, non-adhesive deposits come loose and fall off the insulators only to end up being dragged back into the air flow, then wind up in the air intake grid of the cooling system.

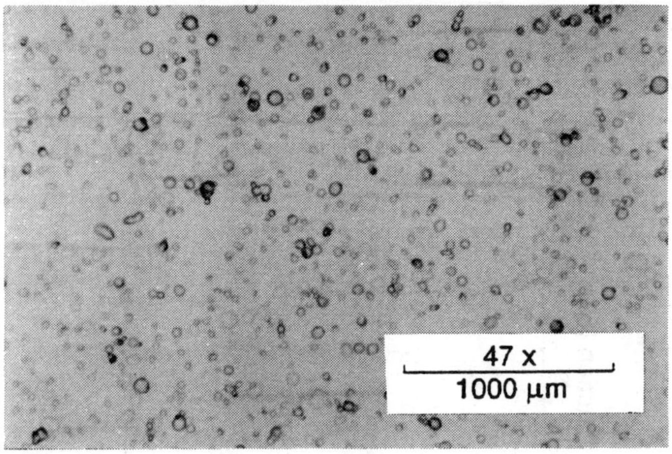

Fig. 7: Ice crystals gathered on a glass slide placed to received falling droplets

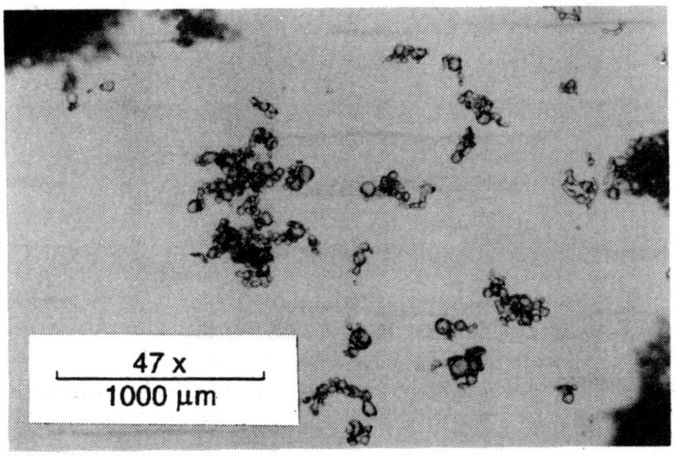

Fig. 8: Ice crystals gathered from the surface of non-energized insulators with a spatula

Figure 9 shows ice crystals formed on the glass slide which had been exposed to the flow of droplets moving towards the insulator string. During these tests, 15 $kV_{r.m.s.}$ was applied per insulator unit. In order to meet safety standards, and to achieve an accurate sampling of ice crystals formed in a high voltage environment, a suitable mechanism way designed, and then installed in the cold room (Farzaneh et al. 1992). The results presented in Fig. 9 show the presence of ice beads similar to those observed when the test was conducted in the absence of voltage (Fig. 7). Unlike the ice beads seen in Fig. 7 however, which are evenly distributed on the slide, these form several areas of frozen droplet clusters. The formation of these clusters indicates that the droplets tend to follow the same trajectory.

Characteristics of Naturally Occurring Snow

Figure 10 shows snowflakes gathered at -11°C on a clean glass slide while they are still falling, and before they hit the ground. These are relatively small flakes, all measuring less than 0.5 mm in diameter, which sets them apart from the more frequently observed large snow flakes of 1.5 mm. The structure of the flakes displays very fine hexagonal plates which form in the temperature range between -10 and -15°C.

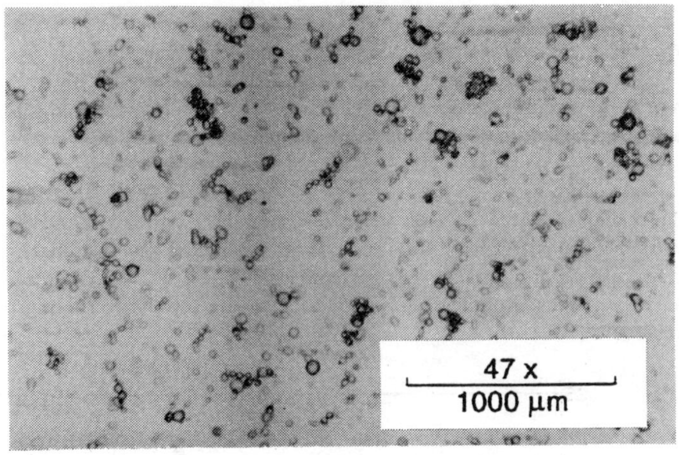

Fig. 9: Ice crystals gathered on a glass slide placed facing the energized insulators

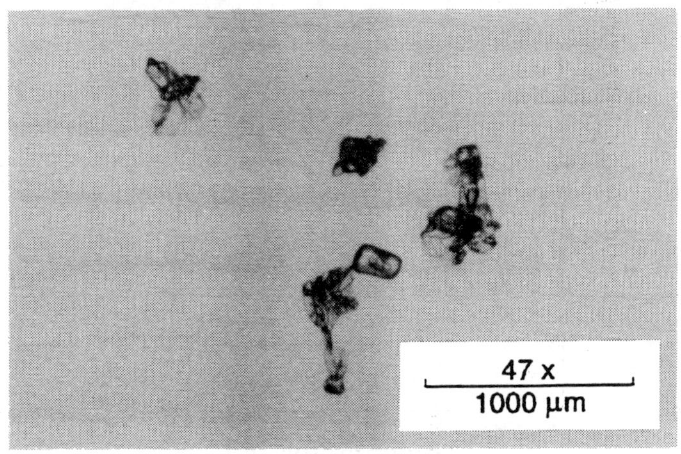

Fig. 10: Snowflakes gathered on a glass slide as they fall

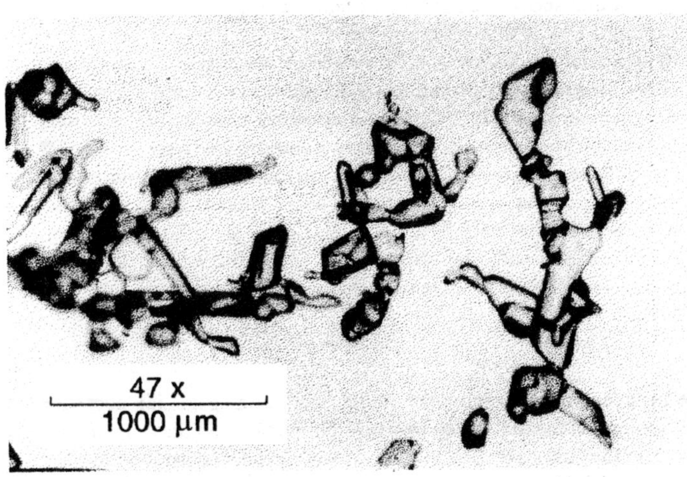

Fig. 11: Snowflakes gathered from the ground

The photographs in Fig. 11 show snowflakes gathered off the ground. Here again, the hexagonal symmetry of the flakes is in even sharper evidence. These, however, are partially fused together, a characteristic of snow that is said to have undergone a metamorphosis.

Examination under a microscope clearly reveals the difference between the characteristics of laboratory formed ice crystals (Fig. 8) and those of natural snow (Fig. 11). The first figure (Fig. 8) shows clusters of spheres formed by the spontaneous freezing of supercooled droplets, while the second (Fig. 11) displays the hexagonal symmetry, characteristic of having grown out of vapor.

Three hypotheses have been put forward to explain the features of ice accretion obtained in the laboratory.

First: Ice crystals are rerouted into the air flow system and injected into the flow of supercooled droplets. While this may be possible at the beginning of the experiment, the crystals rerouted get trapped in the air intake grid which has to be cleaned periodically. Those ice crystals that make it through the grid would be stopped in the heat exchangers. Thus, this hypothesis is unlikely.

Second: The ice crystals are formed through the spontaneous freezing of a portion of the supercooled droplets. This hypothesis is unlikely as well, since the microscope revealed traces of the impact of supercooled droplets on the glass slide.

Third: Deposits are formed from supercooled droplets, and thus constitute soft rime composed of clusters of ice beads. This hypothesis is supported by observation which reveals the low density of the clusters (0.1 to 0.2 g. cm^{-3}) and the size of the droplets produced by the spray nozzles (15 μm).

It seems feasible to consider using these soft rime deposits after heating them to a positive (°C) temperature in order to obtain a liquid water fraction that compares with those found in naturally occurring wet snow deposits.

In the event that the heated soft rime deposits show qualities acceptable for flashover testing, it would be possible to produce them within the same range of intensity of precipitation as that encountered in wet snowstorms. This approach presents the advantage of using the same setup that produces soft and hard rime, but in a specific microclimate. Another advantage is the possibility of producing high density precipitations using relatively simple and low cost equipment.

CONCLUSION

With the objective of proposing a practical method to simulate wet atmospheric icing with a liquid water fraction resembling that found in wet snow, an experimental study of the morphology of ice crystals formed in a cold room was made. The results obtained from these experiments lead to the following conclusions:

1- It is practically impossible, under the experimental conditions tested, to obtain ice particles from the spontaneous freezing of water droplets.

2- Ice accretions formed in the laboratory were soft rime deposits consisting of clusters of spheres resulting from the freezing of supercooled droplets, while those of natural snow present a hexagonal symmetry, characteristic of growth from vapor.

3- The liquid water fraction of laboratory produced soft rime accretions is virtually nil at temperatures below - 4°C, but increases rapidly over - 2°C, reaching 32% of total weight at + 2°C. The presence of high voltage on the insulators did not have a significant effect on the liquid water content of deposited ice crystals.

4- To determine the performance of insulator strings under wet snow conditions, it is possible to use soft rime deposits by heating them to a temperature of approximately + 2°C so that they present a liquid water fraction similar to that found in natural wet snow.

ACKNOWLEDGMENTS

This work was carried out at the University of Quebec in Chicoutimi. The authors are indebted to Mr. Du Nguyen-Dang for his assistance in performing a part of the experiments, and Mr. Claude d'Amours for his technical help. Our thanks are extended to Jean-François Drapeau and Jean Laflamme for discussions and helpful suggestions. The work reported here was sponsored jointly by Hydro-Quebec and the Natural Sciences and Engineering Research Council of Canada.

REFERENCES

DRAPEAU, J-F (1989). "État de l'art concernant les essais et le comportement des isolateurs sans glace," Hydro-Québec, Report No IREQ-4490.

FARZANEH, M, LAFORTE, J-L, KIERNICKI, J (1992). "Formation des cristaux de glace apparentés à la neige mouillée," UQAC, Research Report for Hydro-Quebec (IREQ), p54..

FARZANEH, M, AND KIERNICKI, J (1995). "Flashover Problems caused by Ice Build-up on Insulators," IEEE Electrical Insulation Magazine, Vol 11, no 2, pp 5 - 17.

FARZANEH, M, ZHANG, J, AND CHEN, X (1996). "Modeling of the AC Discharge on Ice Surfaces," IEEE Winter Meeting, Baltimore, Paper accepted for Publication in the Trans on Power Delivery (in Press).

GODARD, S (1960). "Mesure des Gouttelettes de Nuage avec un Film de Collargol," Bull Obsv Puy de Dôme, No 2, pp 41-46.

HYDRO-QUÉBEC (1995). "Impact du verglas et de la neige sur le réseau de transport d'Hydro-Québec - Analyse de la tempête du 13 - 15 janvier 1995 dans les régions Maisonneuve,Laurentides et Saguenay," Report no IREQ - 95 - 280 C (confidentiel).

LAFORTE, J-L, AND DU, N (1988). "Comparison of Three Slide Impact Method for Measuring the Size of Water Droplets," Proc of the 4th IWAIS, Paper No A10-3, pp 263-267.

MITSURU, Y, TOSIHIRO, O, AND TAROO, M (1990). "Experimental Study on Counter-Measure for Snow Accretion on Power Transmission Lines," Proc of the 5th IWAIS, Paper No B7-1, pp 1-6.

RUSH, CK, AND WARDLAW, RL (1957). "Icing Measurements with a Single Rotating Cylinder," Nat Aeronaut Estab Canada, LR-206, p 18.

SKIDMORE, FW, AND PAVIA, RE (1979). "Techniques for Examining Drop Size Spectra in Water Sprays and Clouds," Dept. of Defence, Australia, Aeronautical Research Laboratories, Report No ARL-Mech-Eng.-Note-375, 30 p.

STALABRASS, JR (1986). "Comparison of Droplet Size Measurements by Three Methods," Proc of 3rd IWAIS, pp 141-147.

STALABRASS, JR (1978). "An Appraisal of the Single Rotating Cylinder Method of Liquid Water Content Measurement," Nat Res Council Canada, Rept LTR-LT-92, p 36.

Proceedings of the Seventh (1997) International Offshore and Polar Engineering Conference
Honolulu, USA, May 25-30, 1997

ISBN 1-880653-28-1 (Set); ISBN 1-880653-30-3 (Vol. II)

A Theoretical Investigation of the Distribution of Freshwater Spongy Spray on a Vertical Cylinder

R.Z. Blackmore and E.P. Lozowski
University of Alberta
Edmonton, Canada

Abstract

A theoretical steady-state model of freshwater spray icing for a vertical cylinder gives insight into the icing of ship components such as masts or rigging. Even though salinity effects are neglected, a falling film of excess liquid and a surficial layer of growing dendritic ice called the freezing zone are included. The microphysics of the icing surface result in a nearly constant predicted sponginess. The model's prediction of icing rate agrees well with data. Depending on the spray temperature and the distribution of spray, the icing rate may increase or decrease with elevation.

Key Words

Spongy spray ice distribution, dendrite, falling film.

Nomenclature

a,b- empirical constants of the dendritic growth rate equation
c_r - freezing zone parameter
c_w - specific heat for water, $Jkg^{-1}K^{-1}$
g - acceleration due to gravity, ms^{-2}
L_f - specific latent heat of fusion for water, Jkg^{-1}
I_0 - mass flux of pure ice forming the ice matrix, $kgm^{-2}s^{-1}$
k - thermal conductivity of the freezing zone, $Wm^{-1}K^{-1}$
k_w - thermal conductivity of water, $Wm^{-1}K^{-1}$
q_{12}- heat flux from the freezing zone to the laminar layer, Wm^{-2}
q_{32}- heat flux from the laminar layer to the mixed layer, Wm^{-2}
q_a - heat flux from the outer surface of the falling film to the external environment, Wm^{-2}
q_c - conductive heat flux originating in the freezing zone and that is directed through the laminar layer into the mixed layer, Wm^{-2}
q_{i2}- sensible heat flux to warm the inflowing water of the laminar layer, Wm^{-2}
q_{i3}- sensible heat flux to warm the inflowing water of the mixed layer, Wm^{-2}
q_v- volumetric rate of heat evolution in the freezing zone, Wm^{-3}
r_c - radius of the tip of a hemispherical needle dendrite, m
R_0- mass flux of water entrapped in the growing ice matrix, $kgm^{-2}s^{-1}$
R_2- mass flux of water shed by the segment's laminar layer, $kgm^{-2}s^{-1}$
R_{21}- mass flux of water from the laminar layer to the freezing zone, $kgm^{-2}s^{-1}$
R_3 - mass flux of water shed by the segment's mixed layer, $kgm^{-2}s^{-1}$
R_{32}- mass flux of water from the mixed layer to the laminar layer, $kgm^{-2}s^{-1}$
R_4 - mass flux of impinging spray, $kgm^{-2}s^{-1}$
R_{i2}-mass flux of water entering the segment's laminar layer, $kgm^{-2}s^{-1}$
R_{i3}-mass flux of water entering the segment's mixed layer, $kgm^{-2}s^{-1}$
R_t -mass flux of water shed by the segment's falling film, $kgm^{-2}s^{-1}$
T_0 - temperature of the ice matrix, K
T_1 - temperature of the icing interface, K
T_2 - temperature of the laminar layer/mixed layer interface, K
T_3 - mixed layer temperature, K
T_4 - impinging spray temperature, K
T_a - air temperature, K
T(y)-temperature as a function of position normal to the icing surface, K
V_1 - rate of advance of the icing interface, ms^{-1}
V_c - linear rate of advance of a freely-growing ice crystal, ms^{-1}
y - coordinate normal to the icing surface, m
y_0 - location of the ice matrix/freezing zone interface, m
y_1 - location of the icing interface, m
y_2 - location of the laminar layer/mixed layer interface, m
y_3 - location of the outer surface of the falling film, m
y_t - falling film thickness, m
z_c - cylinder segment length, m
α_w- thermal diffusivity of water, m^2s^{-1}
ΔT- supercooling of bulk water, °C
η - non-dimensional thickness variable for the falling film model
η_1 - non-dimensional thickness at the outer surface of the laminar layer
μ_w - dynamic viscosity of water, $kgm^{-1}s^{-1}$
ν_w - kinematic viscosity of water, m^2s^{-1}
ρ_w - density of water, kgm^{-3}

1. Introduction

Accretion sponginess occurs in both freshwater and saltwater spray ice formation. Freshwater sponginess has been observed in hailstones (Knight, 1968; List, 1990), icicles (Knight, 1980; Makkonen, 1988), and experimentally grown ice accretions (List, 1959; Lesins et al., 1980). Saline spongy ice formation has also been observed in seaspray icing

(Makkonen, 1987), and in sea ice growth (Weeks and Ackley, 1982). The spongy formation mechanism is likely similar in each of these rather diverse instances of ice growth, namely, liquid is trapped in the vicinity of a dendritic icing interface as the interface advances into supercooled or constitutionally supercooled liquid.

The prediction of ice accretion loads has typically involved a thermodynamic calculation in which external heat transfers are translated into latent heat of freezing. Such prediction models can, by definition, forecast only the pure ice load and must inevitably ignore liquid inclusion. Over the years, various ad hoc and theoretical schemes have been produced to account for the liquid fraction of the ice accretion load. All such schemes proposed so far amount to making the liquid portion a fixed fraction of the total accretion load (typically about 25%). The phenomenon of spongy ice might be a scientific curiosity were it not that experiments show the entrapped liquid can more than double the ice load.

No model of spongy ice formation has been published that explicitly considers both the freezing zone and the surficial liquid layer. The model presented here is the first we know of that is intended for the prediction of ice accretion distribution with elevation while accounting for spongy growth with surficial microphysical modelling.

2. The model's concept and structure

The problem of icing on a vertical cylinder is a two-dimensional one, since the local icing rate is a function of height and azimuth around the perimeter (Fig. 1). We convert it into a one-dimensional problem by first calculating the total liquid flux impinging onto the cylinder, and then distributing this flux uniformly over the windward surface of the cylinder. Later, we use a similar procedure to sidestep the azimuthal dependence of the external heat transfer. Thus the icing problem is reduced to one which is a function of height only, and we discretize this problem by dividing the cylinder into a finite number of elements of height, z_c. The model begins by calculating the icing condition including the ice accretion flux and the flux of excess liquid shed from the uppermost cylinder segment (Fig. 1). This calculation is repeated for each segment down the length of the vertical cylinder to obtain the vertical distribution of the ice load.

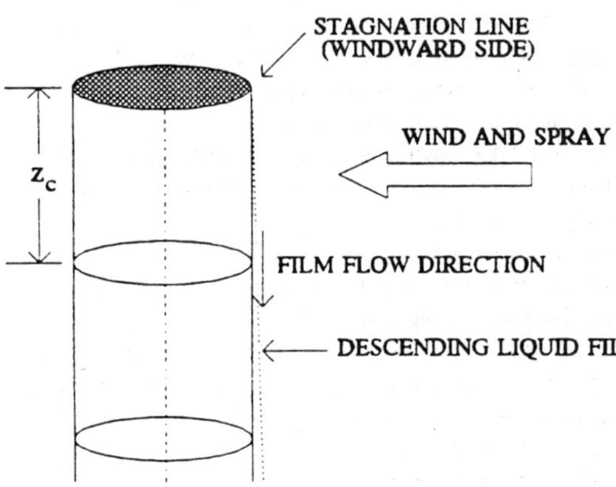

Fig. 1 Schematic of a vertical cylinder with the two uppermost cylinder segments shown. Cylinder segments of height, z_c, are used to discretize the distribution of impinging spray, and the accretion of ice with height. Only the windward half of the cylinder receives spray and exhibits a falling liquid film.

If there were no ice formation, the problem would reduce to that of a steady, vertically falling liquid film whose "classical" solution was formulated and experimentally verified by Dukler and Bergelin (1952). Ignoring wave and rivulet formation, we adopt their solution as the expression of the liquid film dynamics in our model. We also ignore any effect of dendritic ice growth on the film dynamics, except of course that the dendritic advance removes liquid from the film at the icing interface (Fig 2). The model's falling liquid film moves only vertically, and the accretion is assumed to be added uniformly to the windward side of the cylinder. Water either exits a segment as part of the falling film, or is accreted as a matrix of crystalline ice and entrapped liquid (Fig. 2). The ice crystal interface grows in response to the supercooling of the liquid in the falling film.

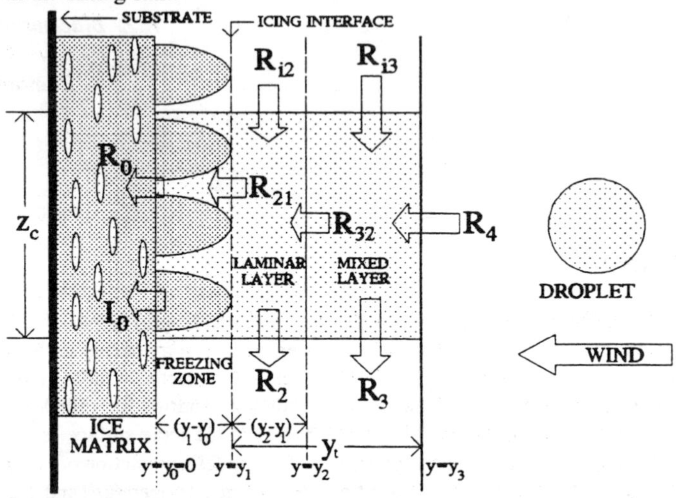

Fig 2. Schematic of the model's surficial structure with the various layers, interfaces and mass fluxes shown. The mixed layer may or may not be present, depending on the flux of excess liquid. The quantity, z_c, is the height of a cylinder segment.

The falling film model depends on the assumption that a laminar layer is formed next to the icing interface with buffer and turbulent layers (which together we call the mixed layer), forming only if there is a sufficiently large film flow. If present, the mixed layer is considered to be so strongly mixed that it offers no resistance to heat transfer and is isothermal. The vertical cross-section of Fig. 2 shows the ice accretion matrix adhering to the substrate. Transport of heat and mass occurs across the cylinder segment boundaries. The model consists of five regions. Proceeding outwards from the accretion towards the airstream, these are numbered from zero to four, with the numbers used as subscripts to designate each layer or region (Figs 2 and 3): 0. the ice matrix, 1. the freezing zone in which dendritic ice crystals grow outwards into the liquid film, 2. the laminar layer, 3. the mixed layer, and 4. the aerosol stream consisting of a monodisperse spray of liquid droplets (which may or may not be supercooled) in a subfreezing airstream that is directed normal to the film flow. Because the model assumes a steady-state ice growth condition with a constant impinging spray flux, the freezing zone, and the laminar and mixed layers maintain constant thicknesses with time.

The latent heat of fusion is assumed to evolve uniformly within the freezing zone and is conducted away from its outer surface through the laminar sublayer of the falling film. In turn, the outer surface of the falling film loses heat by convective, evaporative, and radiative heat transfer to the external airflow. The convective and evaporative heat loss are calculated with a heat transfer coefficient derived by Makkonen (1984). As in traditional icing models, this external heat transfer determines the rate of pure ice growth, but not the rate of liquid entrapment.

The flux of pure ice formed in the freezing zone along with the rate of advance of the icing interface determine two accretion regimes in the model: (1) the glaze ice accretion regime, and (2) the spongy ice accretion regime. Glaze ice accretion occurs when the growth speed of solid ice exceeds that predicted by a crystal growth model. In this situation, the ice accretion rate is based on the model's overall heat balance and the sponginess is zero. If, on the other hand, the crystal growth speed exceeds the thermodynamically predicted growth speed of solid ice, a portion of the liquid film is entrained by the freezing zone and then entrapped in the ice matrix. The model's spongy ice accretion regime occurs under these conditions. In both of these regimes there is a surficial liquid film.

3. The falling film

The mass fluxes and layer thicknesses of the falling film (shown in Fig. 2) are determined by the model. The falling film submodel provides the laminar layer thickness necessary to determine the conductive heat flux. It also provides the mass fluxes needed to ensure mass continuity in the film. Dukler and Bergelin (1952) describe fluid flow near the wall in terms of the non-dimensional variable, η, where

$$\eta = \frac{\sqrt{y_t^3 g}}{\nu_w} \tag{1}$$

The film's total mass flux (expressed in $kgm^{-2}s^{-1}$) is

$$R_t = \mu_w z_c^{-1} f(\eta) \tag{2}$$

where

$$\begin{aligned} f(\eta) &= 0.5\eta^2 && 0 \le \eta \le 5 \\ f(\eta) &= 12.5 - 8.05\eta + 5.0\eta \ln\eta && 5 \le \eta \le 30 \\ f(\eta) &= -63.8 + 3.0\eta + 2.5\eta \ln\eta && \eta \ge 30 \end{aligned}$$

The total mass flux shed from the cylinder segment, R_t, is an area-averaged flux density for the segment's forward surface ($kgm^{-2}s^{-1}$). For this reason, R_t is a function of both the film flow for the segment (kgs^{-1}), and the area of the forward surface of the segment. Since this area is a function of the segment's length it is not difficult to show that the film's total mass flux is also a function of the cylinder segment length, z_c. For the sake of brevity, the full derivation of Eq. 2 is not included here but it is given more completely by Blackmore (1996).

The ranges of the dimensionless coordinate, η, given in Eqn. 2, correspond to the laminar, buffer, and turbulent flow regions near the wall for the well-known von Kármán velocity profile (Holman, 1990; von Kármán, 1939). Dukler and Bergelin (1952) give experimentally verified film thickness, as a function of film flow rate and the physical properties of the fluid flow, over the viscous to turbulent range. These theoretical equations agree closely with measurements of film thickness for freshwater flow (without wind stress) on a vertical stainless steel plate into the turbulent regime, with a film thickness of up to 0.71 mm. The equations predict a laminar to turbulent film transition which agrees closely with observation.

The falling film model of Dukler and Bergelin (1952) predicts that a falling film of water at 0°C will form a steady-state laminar flow (i.e. in the range $0<\eta\le5$) up to a thickness of approximately 0.2mm at a film Reynolds number of up to 33. For greater film thickness, waves appear on the surface of the film with agitated waves beginning to form at the onset of turbulent flow with a Reynolds number of about 1000 for a film thickness of approximately 0.66mm.

Eq. 2 may be solved for η and then Eq. 1 may be solved for the film thickness, y_t, once R_t has been determined as the difference between the total incoming liquid flux and the ice accretion flux. Because of the influence of the laminar layer on the heat transfer to the external environment, the ice accretion flux depends non-linearly on the film flux and thickness. Consequently, the solution procedure is an iterative one. The non-dimensional variable used to describe the thickness of the laminar sublayer is defined in Dukler and Bergelin's model as

$$\eta_1 = \frac{\sqrt{y_t g}}{\nu_w}(y_2 - y_1) \tag{3}$$

where η_1 has a value 5 at the outer surface of the laminar layer. The laminar sublayer thickness, y_2-y_1, is calculated from Eq. 3 (setting η_1=5), and the flux of the laminar layer can then be calculated with

$$R_2(\eta_1) = 0.5\,\mu_w \eta_1^2 z_c^{-1} \tag{4}$$

with η_1=5. If a buffer layer exists over the laminar layer, then η_1=5 denotes the surface that separates the laminar layer from the mixed layer. If η_1=0, there is no film flow.

The mass flux in the mixed layer includes contributions from both the buffer and turbulent layers. Since the mass flux of the laminar layer, R_2, and that of the mixed layer, R_3, together constitute the total film flux

$$R_3 = R_t - R_2 \tag{5}$$

The conservation of water substance in the falling film (Fig. 2) may be expressed as

$$R_{i2} + R_{i3} + R_4 = I_0 + R_0 + R_2 + R_3 \tag{6}$$

All of the considerations to this point in the model's development give rise to an essentially conventional icing model with an explicit liquid film, of the type described by Lozowski et al. (1995). Such a model allows the prediction of the pure ice growth rate. If we wish to also predict liquid entrapment within the matrix, we will need to examine the freezing zone in more detail. Lacking experimental evidence as to the detailed morphology of the freezing zone, we rely on the ad hoc assumptions described in Section 4. The validity of these assumptions will require future experimental verification.

4. The freezing zone

Spongy growth results from dendritic ice growth at the ice surface (Knight, 1968; Makkonen, 1987; List, 1990; Lock and Foster, 1990). List (1990) refers to a layer of ice lying between the spongy ice matrix and the surficial liquid on a growing hailstone. We model this layer as a zone of dendritic ice and water of thickness, y_1-y_0 (Fig. 3). Referred to here as the freezing zone, it is assumed to contain an ensemble of dendritic ice crystals that are growing in close proximity to one another with a uniform tip growth rate.

4.1 The calculation of the freezing zone temperature

It is assumed that the latent heat of fusion produced by the growing ice is released homogeneously throughout the freezing zone. The conduction and evolution of heat in the freezing zone is described by the

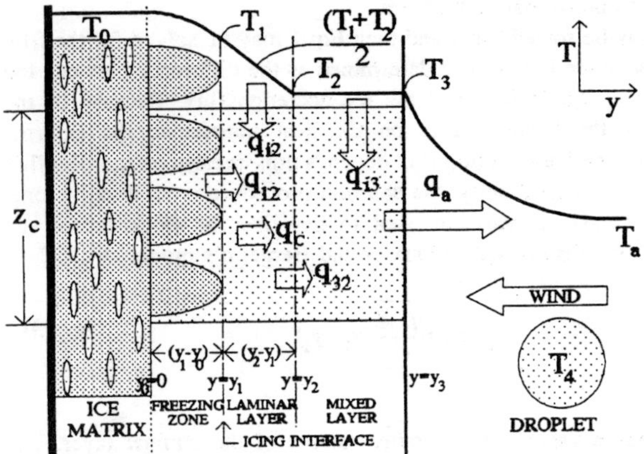

Fig 3. Schematic of the model's surficial structure showing the heat fluxes across the boundaries of the various layers, and the model's idealized temperature profile. The temperature axis is vertical with temperature increasing upwards.

one-dimensional steady-state heat diffusion equation

$$k\frac{d^2T(y)}{dy^2}+q_v=0 \tag{7}$$

If the thermal conductivity is independent of y, and q_v is constant, the solution to Eq. 7 is

$$T(y)=-\frac{q_v}{2k}y^2+c_1y+c_2 \tag{8}$$

Since the ice matrix is a freshwater accretion, T(0)=0. Because there is no heat conduction within the spongy ice matrix, dT/dy = 0 at y=0. Using both boundary conditions, the solution of Eq. 7 is

$$T_0-T_1=\frac{q_v(y_1-y_0)^2}{2k} \tag{9}$$

In the present work, the thermal conductivity of the freezing zone, k, is taken to be the average of the thermal conductivity of water and the thermal conductivity of the ice matrix (for details see Blackmore, 1996).

4.2 The rate of advance of the icing interface

Crystal growth theory has long recognized that the linear growth rate at the tips of dendrites is a function of the supercooling of the bulk liquid into which the dendrite is growing. This dendritic growth rate may be expressed as

$$V_c=a(\Delta T)^b \tag{10}$$

Tirmizi and Gill (1987) report values of a and b for ice crystal growth in pure water of 1.87×10^{-4} and 2.09, respectively. They measured the linear rate of advance of freely-growing ice crystals in bulk supercooled water in the range $0.05°C<\Delta T<1.1°C$. In addition, they show that natural convection in the liquid surrounding the ice crystal assists the growth of the crystal for $\Delta T<0.2°C$. For this reason, they recommend the values of a and b given above for $\Delta T>0.2°C$. We use these values of a and b and follow the example of List (1990). He extrapolates the formulation of Hobbs (1974) (valid for $2°C<\Delta T<6.5°C$) down to 0°C. He reasons that the equation gives an expected tendency in growth rate as the supercooling approaches 0°C. Therefore, we extrapolate Eq. 10 to lower values of ΔT on the very rare occasions when this is necessary. In addition, we follow Kachurin and Morachevskii (1965) and List (1990) with the assumption that the growth rate of the icing surface (that is the aggregate growth velocity of the dendrite tips) is a function of the temperature drop across the freezing zone, i.e. $\Delta T=T_0-T_1$.

4.3 The thickness of the freezing zone

It is proposed here that the freezing zone thickness is proportional to the radius of the tip of a hemispherical needle dendrite. This simple model of a growing dendrite consists of a hemispherical tip advancing on a cylindrical shaft (Kurz and Fisher, 1986). The hemispherical needle dendrite has the growth equation (Makkonen, 1990):

$$\frac{V_c r_c}{2\alpha_w}=\frac{c_w\Delta T}{L_f} \tag{11}$$

Eq. 10 may be substituted into Eq. 11 to give an expression for the tip radius of the needle dendrite:

$$r_c=\frac{2k_w}{a\Delta T^{b-1}\rho_w L_f} \tag{12}$$

In view of the above discussion, the freezing zone thickness in our model is given by

$$y_1-y_0=c_r r_c=\frac{2c_r k_w}{a\Delta T^{b-1}\rho_w L_f} \tag{13}$$

where c_r is a constant of proportionality referred to here as the freezing zone parameter. Eq. 13 is in qualitative agreement with the macroscopic experimental observations of Lock and Foster (1990) in which high air temperatures (low film supercooling) produced coarse columnar ice crystals.

The determination of c_r is possible if the ice growth rate and the ice fraction are known experimentally. Using a method of analysis developed by Blackmore (1996), nine spongy spray icing measurements made at the University of Alberta (J. Shi, pers. comm., 1995), were used to calculate values of the freezing zone parameter. These data yield a freezing zone parameter range of $2.73\leq c_r\leq3.00$, with a mean value of 2.9, which is used in the model.

5. The thermodynamics of the falling film

The assumption is made that conductive heat transfer occurs in the laminar layer of the falling film, and that the mixed layer offers no resistance to heat transfer. Under these assumptions, the heat balance for the mixed layer (or for the outer surface of the laminar layer if there is no mixed layer; cf. Fig. 3) is

$$q_{i3}+q_c+q_a=0 \tag{14}$$

The equation describing the heat balance for the laminar layer (cf. Fig. 3) is

$$q_{i2}+q_{12}+q_{32}=0 \qquad (15)$$

Once the thermodynamics of the falling film has been solved using Eqs. 14 and 15, Eqs. 9 and 13 along with the volumetric rate of heat evolution in the freezing zone, q_v, are used to solve for the temperature profile through the freezing zone and the falling film. For a full treatment of the model's algorithm the reader is referred to Blackmore (1996).

6. The ice growth regimes

The model's growth regime, glaze or spongy, is determined once the thermodynamics and the thermal profile of the model's surficial layers are known. The icing interface velocity is determined using Eq. 10 with $\Delta T=T_0-T_1$, viz

$$V_1=a(T_0-T_1)^b \qquad (16)$$

If the spongy ice accretion regime occurs in the model, the rate of advance of the icing interface is

$$V_1=\frac{I_0+R_0}{\rho_w} \qquad (17)$$

Eq. 17 is used in the model's algorithm to calculate the mass flux of liquid incorporated into the ice matrix, R_0. If the calculated value of $R_0 \leq 0$, the model is considered to predict glaze icing. In this regime, the accretion is solid and the ice accretion flux is I_0 with $R_0=0$.

7. Model performance

The model's performance has been examined by comparing the model's prediction of ice accretion flux with the results of an independent set of experiments performed by Stallabrass and Hearty (1967). They performed freshwater icing experiments in a closed circuit refrigerated wind tunnel with non-rotating cylinders ranging in diameter from 3.8 cm to 45.7 cm. The mass of ice and the accretion thickness were measured after one hour of ice growth for ten cases with the cylinders in a horizontal position. Eleven more cases were run with the same cylinders, of length 91.4 cm, in a vertical position. The liquid fraction of the accretions was not measured. The airspeed was 22.4 ms^{-1} with a range in air temperature between -4.5°C and -16.0°C. The median volume droplet diameter was reported to be 185μm. The cylinders were 5.2 m from the spray nozzles and the initial water temperature delivered to the nozzles was normally in the range 16°C to 20°C upon entering the spray rig. A hot wire liquid water content meter was used to confirm a spray flux of 0.112 $kgm^{-2}s^{-1}$ for the experimental runs.

A comparison of the modelled ice accretion flux (I_0+R_0) with the measured values for horizontal cylinders is shown in Fig. 4 as circles. Fig. 4 also shows eleven cases of vertical cylinder icing. The comparison of the model-predicted and measured values of height-averaged ice accretion flux for these cases is shown as squares. The model-predicted values were calculated using five cylinder segments. Since the spray impingement temperature was not measured, we have estimated a value of -4.1 °C, calculated using a model of the motion and cooling of a 185 μm spray droplet, initially at rest 5.2 metres from the icing cylinder. The initial droplet temperature was taken to be 18°C with a relative humidity in the airstream of 90%.

The model shows a reasonable ability to predict the ice accretion flux for the conditions measured by Stallabrass and Hearty (1967). For the 21 points shown in Fig. 4 the regression line has a y-intercept of -0.0061, a slope of 1.36, and a coefficient of determination, r^2=0.74. The Willmott index of agreement is d=0.88. This index is useful as an indicator of how closely the data lie to the diagonal, with a value of d=1.0 suggesting a perfect fit (Willmott, 1982). The model tends to underpredict ice accretion flux at lower values and overpredict at higher values. There seems to be no consistent difference between the horizontal and vertical cylinders.

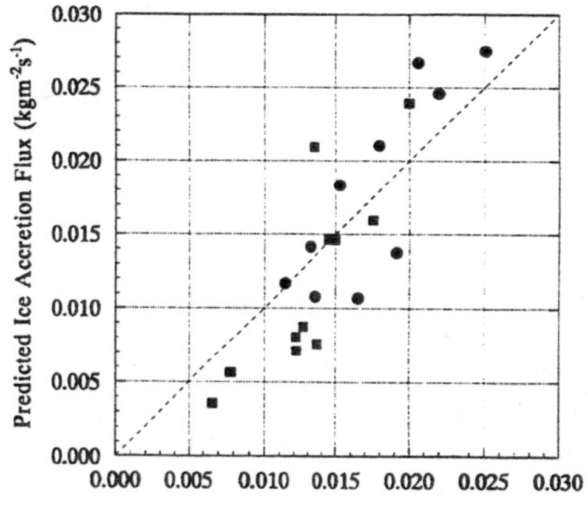

Fig 4. Model-predicted ice accretion flux compared to the experimental measurements of ice accretion flux by Stallabrass and Hearty (1967). The horizontal cylinder ice accretion data are represented by circles, and the vertical cylinder ice accretion data are represented by squares.

The model-predicted ice fraction for the 21 cases of Fig. 4 has a range of 0.59 to 0.64. This rather limited range in the predicted ice fraction is in qualitative agreement with Makkonen (1987) and Makkonen (1990), who hypothesizes that the ice fraction should be nearly constant over a significant range of spongy growth conditions.

8. Model sensitivity

In order to examine the icing distribution with elevation, we consider the conditions on the mast of a Soviet medium-sized fishing vessel. We consider a 5 m high mast, 40 cm in diameter. First, the distribution of impinging spray is calculated according to conditions experienced by the Soviet trawler "Academic Ber" in the Japan Sea on Jan. 12, 1969. Gurvitch (1972) reported an average windspeed of 12.5 ms^{-1} with a vessel speed of 4.6 ms^{-1}. Zakrzewski et al. (1993) give a formula describing the exponential decrease in spray cloud liquid water content with height above deck. This formula is used along with conditions reported by Gurvitch (1972) to calculate the distribution of time-averaged spray flux with height shown in Fig. 5 as curve "a".

Gurvitch (1972) reported an air temperature of -10.8°C and a sea-surface temperature of 0.7°C. We have estimated the impinging spray temperature to be -5.7 °C using a model of motion and cooling for a single droplet (1.25 mm diameter) over a distance of 11 metres (the approximate distance from the bow to the mast). The resulting ice accretion distribution curve is "b" in Fig. 5. Curve "b" shows an icing distribution that essentially "mimics" the impinging spray flux distribution. The model's ice fraction decreases from 0.71 to 0.58 from the bottom to the top of the mast.

For comparison, a height-averaged spray flux (0.1824 $kgm^{-2}s^{-1}$) was calculated from the exponential spray flux distribution. When this value is used in the model with the same impinging spray temperature as curve "b", the ice growth rate (not shown), like the spray flux is nearly constant with elevation at about 6.5 cmh^{-1}. Thus, it appears that the distribution of spray flux with height strongly controls the distribution of ice growth rate with height and with it the vertical distribution of ice load, which is important for vessel stability. On the other hand, the average ice accretion fluxes in the two cases differ by only a few percent, suggesting that the average impinging spray flux may yield a good estimate of the average accretion rate.

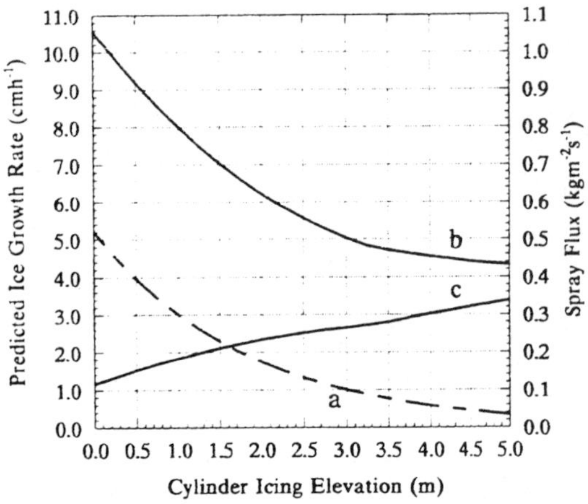

Fig 5. Spray flux (right ordinate) and model-predicted ice growth rate (left ordinate) as functions of height on a 5 m mast. The chain-dashed curve "a" shows an exponential vertical variation of spray flux. Curve "b" is for ice growth rate with a spray temperature of -5.7 °C. Curve "c" is for ice growth rate with a spray temperature of 0.7 °C.

To complete the picture, the model was run for the exponential variation of spray flux shown by curve "a" but with an impinging spray temperature equal to the sea-surface temperature of 0.7°C. In this case, curve "c" shows the ice growth rate increasing with elevation. This occurs because of the increasing contribution of sensible heat with higher fluxes of relatively warm impinging spray at lower elevation. The ice fraction increases with height from 0.58 to 0.70. This result suggests that the in-cloud cooling of the spray is very important.

9. Conclusions

1. A new theoretical model has been developed for simulating spray ice accretion on vertical stationary cylinders. It is the first spray icing model to predict both the growth rate of the accretion, the occurrence of spongy ice and its liquid fraction, and the variation of these quantities with height on the cylinder.

2. The model's spongy ice accretion regime shows that spongy growth can be successfully modelled with a traditional heat balance at the icing surface, and a surficial structure that includes a freezing zone and a surface liquid film. Both the surface liquid film and the freezing zone are important in predicting the sponginess of the model's ice accretion.

3. A comparison of the model's performance with the experimental data of Stallabrass and Hearty (1967) shows reasonable agreement for ice accretion flux. The success of the vertical cylinder model in predicting ice accretion flux for both the horizontal and vertical cylinder icing cases, concurs with the observation of Stallabrass and Hearty, that the orientation of the icing cylinders had little influence on the overall accreted ice loads. Much of the uncertainty in the comparison between the data and the model predictions arises from one principal factor, namely the impinging spray temperature, which is a critical factor in the model, but is a very difficult quantity to measure.

4. Experience with the present model suggests that the accretion sponginess remains relatively constant (around 0.6) over a wide variety of ambient conditions, in agreement with Makkonen (1987) and Makkonen (1990). The convective, evaporative, radiative and sensible heat fluxes together, drive the ice accretion process. However, if these heat fluxes produce a spongy accretion, the spongy accretion flux can be much larger than the solid ice accretion flux calculated in traditional models.

5. Liquid water content and spray temperature have a strong influence on the sensible heat transfer at the icing surface, the icing interface supercooling, and the thermal gradient in the surficial film, and consequently also on the accretion sponginess, the growth rate and the distribution of ice with height on a vertical cylinder.

6. Because of the expected similarity of the spongy ice formation mechanisms in freshwater and brine, it should be possible to extend the present model, "mutatis mutandis", to predict sponginess in saltwater spray ice accretions.

Acknowledgments

This research was supported through Research and Strategic Grants from the Natural Sciences and Engineering Research Council of Canada.

References

Blackmore, RZ (1996). "Vessel Icing and Spongy Accretion Modelling", Ph.D. Dissertation. University of Alberta, pp 231.

Dukler, AE, and Bergelin, OP (1952). "Characteristics of Flow in Falling Liquid Films," *Chem Eng Prog*, Vol 48, pp 557-563.

Gurvitch, LY (1972). "On Comparing the Observed Ice Loading of Fishing Vessels with USSR Norms," *Arkt. Antarkt. Nauc-Issled Inst. T298*, pp 137-148. [Brachnell, Met. Off. Trans. No. 876) and also, CRREL Draft Translation 411, ADA-003215, 140-152 pp.]

Hobbs, PV (1974). *Ice physics*, Clarendon Press, Oxford, 837 pp.

Holman, JP (1990). *Heat transfer*, McGraw Hill. New York, 714 pp.

Kachurin, LG, and Morachevskii. VG (1965). "Kinetics of Phase Transitions of Water in the Atmosphere," Leningrad State University, Leningrad, pp 59-73. (translated by Z. Lerman, Israel Program for Scientific Translations, Ltd., IPST Cat. No. 1807. 1966, 124 pp.)

Knight, CA (1980). "Icicles as Crystallization Phenomena," *J Cry Grow*, Vol 49, pp 193-198.

Knight, CA (1968). "On the Mechanism of Spongy Hailstone Growth," *J Atmos Sci*, Vol 25, pp 440-444.

Kurz, W., and Fisher DJ (1986). *Fundamentals of solidification*. Trans Tech Publications Ltd. Switzerland, p. 203, 244 pp.

Lesins, GB, List, R, and Joe, PI (1980). "Ice Accretions. Part I: Testing of New Atmospheric Icing Concepts," *J Rech Atmos*, Vol 14, pp 347-356.

List, R (1990). "Physics of Supercooling of Thin Water Skins Covering Gyrating Hailstones," *J Atmos Sci*. Vol 47, pp 1919-1925.

List, R (1959). "Wachstrum von Eis-wassergemischen im Hagelversuchskanal," *Helv Phys Acta*, Vol 32, pp 293-296 (in German).

Lock, GSH, and Foster, IB (1990). "Experiments on the Growth of Spongy Ice Near a Stagnation Point," *J Glaciol*, Vol 36, pp 143-150.

Lozowski, EP, Kobos, AM, and Kachurin, LG (1995). "Influence of the Surface Liquid Film on Cylinder Icing under Marine Conditions", *Proc 14th Inter Conf Offshore Mechanics Arctic Eng* (OMAE), Copenhagen, Vol 4, pp 115-122.

Makkonen, L (1990). "Origin of Spongy Ice," *Proc 10th Inter Assoc Hydraulic Res* (IAHR) Ice Symposium, Espoo, Finland, pp 1022-1030.

Makkonen, L (1988). "A Model of Icicle Growth," *J Glaciol*, Vol 14, pp 64-70.

Makkonen, L (1984). "Modeling of Ice Accretion on Wires," *J Climate Appl Meteor*, Vol 23, pp 929-939.

Makkonen, L (1987). "Salinity and Growth Rate of Ice Formed by Sea Spray," *Cold Reg Sci Tech*, Vol 14, pp 163-171.

Stallabrass, JR, and Hearty, PF (1967). "The Icing of Cylinders in Conditions of Simulated Freezing Sea Spray," *Mech Eng Rept MD-50, NRC No. 9782*, National Research Council of Canada, Ottawa, 10 pp.

Tirmizi, SH, and Gill, WN (1987). "Effect of Natural Convection on Growth Velocity and Morphology of Dendritic Ice Crystals," *J Cry Grow*, Vol 85, pp 488-502.

von Kármán, TH (1939). "The analogy between fluid friction and heat transfer," *Trans Am Soc Mech Engrs*, Vol 61, pp 705-710.

Weeks, WF, and Ackley, SF (1982). "The Growth, Structure and Properties of Sea Ice," *CRREL Monogragh 82-1*, U.S. Army Cold Regions Research and Engineering Lab., pp 136.

Willmott, CJ (1982). "Some Comments on the Evaluation of Model Performance," *Bull Amer Meteor Soc*, Vol 63, pp 1309-1313.

Zakrzewski, WP, Lozowski, EP, Thomas, WL, Bourassa, M, Blackmore, RZ, Szilder, K, and Kobos, A (1993). "A Three-dimensional Time-dependent Ship Icing Model," *Proc 12th Inter Conf Port and Ocean Eng under Arctic Conditions* (POAC), Hamburg, Vol 2, pp 857-873.

Proceedings of the Seventh (1997) International Offshore and Polar Engineering Conference
Honolulu, USA, May 25-30, 1997

ISBN 1-880653-28-1 (Set); ISBN 1-880653-30-3 (Vol. II)

Atmospheric Icing on Cables of Different Flexibilities

Pierre McComber
École de Technologie Supérieure, Montréal, Canada
Jacques Druez
Université du Québec à Chicoutimi, Chicoutimi, Canada
Jean Laflamme
Hydro-Québec, Montréal, Canada

ABSTRACT

In cold regions, power transmission lines are often exposed to atmospheric icing loads. It has been observed that smaller diameter ground wires tend to collect more atmospheric ice than conductors. These observations have been explained mostly by the twisting of the smaller cable under the accretion weight.

In this paper, a numerical model of the atmospheric icing of a cable, including cable twisting is presented for two cables of different flexibilities. The method simulates numerically in two dimensions the dry growth accretion on cables of different sizes and flexibilities. For each time step, the change in shape is calculated knowing the droplets size, its velocity and direction. The surface is then integrated to find the weight of accreted ice and the rotation of the shape is done numerically. The effect of cable height, wind velocity and drop size distribution are considered for freezing rain and in-cloud icing conditions. The wind speed and droplet diameter are treated as random variables and are used to determine the incidence angle of the impinging drops. In the case of freezing rain, cable twisting of the smaller cable is the result of its reduced flexibility. However, for in-cloud icing conditions, the larger collection efficiency of the smaller cable results in an even larger difference in the icing rate in favor of the small cable.

Keywords : Atmospheric icing, power transmission lines, cable torsional rigidity

INTRODUCTION

In recent years the understanding of atmospheric icing of structures has progressed [1]. Progress has been favoured where there exists an important economic incentive as in the case of overhead power transmission lines in northern regions. Presently, the usual design practice for most utilities is to adjust the strength of the structure of a transmission line to a maximum radial ice thickness applied to all cables and conductors irrespective of their size. However, it has been known for some time, largely due to the Mt. Valin on site testing [2,3], that a smaller diameter cable results in larger icing rate for the same icing conditions. This difference is even more important for power companies now that a smaller cable is used not only as a groundwire but also as an OPGW (**OP**tical Fibre **G**round **W**ire) for telecommunication purposes. In such a case, an icing overload of the OPGW could lead to a failure of the optical fiber [4] and hence of the telecommunication system itself.

In an attempt to quantify this phenomenon, measurements on an icing test site at Mt. Valin (altitude 902 m) Québec, Canada have been made on two cables of different sizes: a 12.5-mm (groundwire size) and a 35 mm ACSR cable (conductor size) on a non-energized test line since 1985. The larger icing rate on the smaller cable is observed most of the time [3] but it is not easy to quantify, since it varies during icing events as well as with different icing conditions. Icing events during the winter of 1991-92 are used to provide a basis for a comparison. It shows a larger maximum ice load of 10% to 40 % for the smaller cable depending on event duration and meteorological conditions [5].

Because of the numerous factors involved and the difficulty in observing their respective effect on icing at a test site, there is a definite advantage in using numerical simulation to determine the most important causes of the phenomenon and to know more about this hazardous situation. The behavior of cables as a type of flexible structure are even more complicated to model, since large displacements of the structure are possible during icing. In particular, cable twisting has been simulated, and the effect of the ice accretion is sufficient to make a significant difference in the icing rate on cables of different flexibilities [6].

The main purpose of the present paper is to verify both for freezing rain and in-cloud icing the effect of the variations of liquid water content, droplet size and wind velocity in a numerical simulation of the cable twisting on two cables of different size. This simulation also takes into consideration the random nature of the wind speed and droplet size.

DESCRIPTION OF THE NUMERICAL SIMULATION MODEL

The growth of the ice accretion is computed by making a time integration of the atmospheric icing on a smooth cylinder [6]. As shown in Fig. 1, rotation of an angle ϕ can modify the accretion rate by increasing the cross-section area of the iced cable.

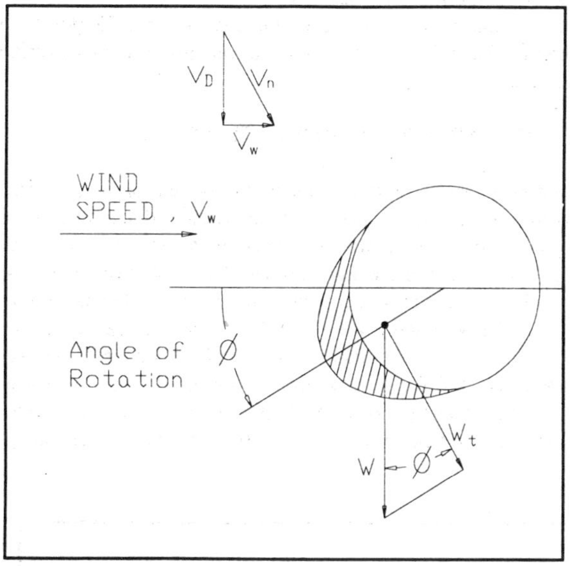

Fig. 1 Diagram of an ice accretion on a cable showing the angle of rotation ϕ.

At each time step, δt, the icing rate is determined by first calculating the portion of the water flux that hits the accretion surface. For dry growth the accretion rate is converted to a change in accretion volume using a typical atmospheric ice density. The model can be broken down in two parts.

First, the calculation of the water direction and intensity is computed from the wind speed and precipitation rate or liquid water content. The local collection is related to the local intensity of accretion by multiplying V_n, the drop velocity, by the liquid water content, and the collection efficiency E, w. I_n is the normal icing intensity, δt is the time step and ρ_i is the ice density:

$$I_n = EV_n \, w \, \delta t \, / \, \rho_i \tag{1}$$

The precipitation rate, drop size and wind velocity and direction are random variables and are treated as such in the present model.

The second part of the model determines the cable twisting under the calculated ice load. The ice accretion is computed locally on the accretion surface. Once the shape is obtained and the torque applied, the resulting twisting of the cable is computed. This portion of the model follows an approach similar to other existing models [6] and will not be reproduced here.

WIND VELOCITY AND DROP SIZE DISTRIBUTIONS

Wind and drop velocities

Wind speeds and directions are random variables in time and space. Also wind speeds vary with height in the atmospheric boundary layer close to the ground [8]. If a wind speed distribution is used, the average horizontal wind speed, U, is given by:

$$\frac{U(z)}{U_G} = \left(\frac{z}{z_G}\right)^{\alpha} \tag{2}$$

In Equation (2) z is the height from ground and G is a subscript indicating values taken outside the boundary layer. The distribution index is taken to be $\alpha = 0.28$ to account for the forest surroundings of Mt. Valin. There is also a component of turbulence or horizontal gust that must be considered. The wind velocity is considered to be a random variable described by a Gaussian distribution with a variance given by [8]:

$$S_U^2 = 5.5225 \; \kappa \; \mu_U^2 \tag{3}$$

with $\kappa = 0.28$. For the horizontal component of turbulence of the wind speed a normal or gaussian distribution is used with an average μ_U and the standard deviation σ_U. The wind velocity influences the direction of the cable, but mostly it interferes with the drop trajectory and the droplets angle of incidence.

The Mt. Valin test site [5] is located on the summit of a mountain. In such a case there are two effects on the velocity distribution [8]. The first one is an increase of wind speed and the second one is a decrease of the boundary layer thickness z_G. In order to estimate this effect, the upstream flow around half a cylinder is used with potential flow being applied to the distribution. This approximation gives a new $z_G = 221$ m instead of the $z_g = 410$ m for flat ground far from the mountain. The maximum speed is thus increased by a factor of 1.6.

Drop velocity field. For freezing rain and for a given drop size, D in mm, an expression given by Atlas et al. on the data of Gunn and Kinzer [9] is used to determine the terminal velocity, V_D, of a drop of diameter D:

$$V_D = 9.65 - 10.3 \, e^{-0.6D} \tag{4}$$

The droplet velocity is assumed to remain undisturbed by the presence of the cable, so that the collection efficiency is one. However, depending on the drop size, the terminal velocity changes and hence the incidence angle also changes with different drop diameter when there is a significant wind speed. If the wind speed is V_w in m/s, and the terminal velocity V_D is given by Eq. (4), the velocity V_n and the rain incident direction, α, are given by a vectorial sum. For in-cloud icing, droplet sizes are small and the impinging speed and direction is the same as the wind.

Drop size distribution

Liquid water content and droplet size are related in most meteorological models both for freezing rain and in-cloud icing. For rain, a Marshall-Palmer exponential distribution is used [10] for drop size. If the precipitation rate is R mm/h then the average drop diameter D mm is given by:

$$\mu_D = \frac{1}{4.1 \, R^{-.21}} \tag{5}$$

and the exponential number distribution is:

$$N(D) = N_0 \, e^{-D/\mu_D} \qquad \textbf{(6)}$$

where $N_o = 8000$. The liquid water content is found for each size D_j in g/m^3 by:

$$w = \frac{\pi \, N_j \, D_j^3}{6000} \qquad \textbf{(7)}$$

Best [11] suggests a two-parameter distribution to be used both for clouds and rain. However, he has shown that there is only a small difference when the Marshall and Palmer distribution is used for rain. Since the one parameter distribution is simpler to use, the Marshall and Palmer rain distribution as given by Eqs. 5 to 7 are used in the present model.

For in-cloud icing a two parameter distribution of the Weibull type suggested by Best [11] is used. For cloud or fog droplets, the collection efficiency is smaller than one and must be calculated for droplet sizes, wind velocities and collector size. The parametric equations used for the collection efficiency calculation can be found in the literature [12]

Point estimate method

To include in the model the effect of a wind speed and drop statistical distribution, the point estimate method [13] is used. Probably the most common approach to this type of calculation is the Monte-Carlo technique, but it is a rather innefficient method. For an approximate solution the point estimate method is better since its accuracy can be adjusted to the requirements of the problem. Details are given in Rosenblueth [13]. Basically, estimates of the function, in this case radii r_i (V_n, D) is made at pre-determined points, the number of which is related to the required accuracy. For a second order accuracy, two point estimates are used for each distribution: U_+, U_-, D_+, and D_- and the average and standard deviation of the resulting function r_i is estimated with a summation of these estimates with appropriate weights.

Point estimate method for wind. The two estimates of wind speed are taken in the following evaluation with a weight of $P_+ = P_- = 0.5$, corresponding to a wind speed of 5 m/s at 10 m, then for the 35-mm cable the values are $U_+ = 6.859$; $U_- = 2.987$ and for the 12.5-mm cable the corresponding values are $U_+ = 6.737$; $U_- = 2.934$.

Point estimate method for drop size. For the rain drop size distribution, the exponential distribution is a skewed distribution so that the weight and point location are no longer symetric. The two weights are $P_+ = 0.14644$ and $P_- = 0.85356$ respectively. For the rain drop (Marshall-Palmer distribution) $D_+ = 1.11414$ mm; $D_- = 0.19115$ mm.

For in-cloud icing, a Weibull distribution is used and the diameter is found from the distribution with parameters a and n determined. The value of n is taken as an average for all the distributions presented n = 3.26 and the value of a is evaluated from the liquid water content using a relationship suggested by Best [11]: $w = 1.1 \times 10^{-3} \, a^{1.79}$ (refer to the paper of Best [11] for the complete formulae). In the case of a LWC = 0.243 g/m^3 , a = 20.4 µm (230=µ) the median volumetric diameter is 18.23 µm and $D_+ = 24.77$ µm ; $D_- = 12.42$ µm.

The combined weight being given by $P_{++} = P_{-+} = 0.07322$ and $P_{--} = P_{+-} = 0.042678$ for rain and $P_{++} = P_{-+} = 0.4750312$ and $P_{--} = P_{+-} = 0.5249688$ for clouds.

Since the simulation results give the radius r_i for 32 points on a circumference, the radius at each point is a function of the two variables $r_i = r_i$ (V_w, D).

RESULTS OF CABLE ICING SIMULATIONS

A two-dimensional accretion shape was obtained for the two sizes of cables with a wind velocity and a median volume diameter associated with a given precipitation rate. The specific parameters used are the following. The two cables are a ground wire 12.5 mm in diameter and an ACSR conductor 35 mm in diameter. The 35 mm cable has a torsional rigidity of 0.378 N/m and a weight of 21.5 N/m whereas the 12.5 mm cable rigidity is .00383 N/m and a weight of 7.47 N/m [6]. The two ice shapes are calculated for a simulation period of 90 min, with a wind velocity of 5 m/s at a height of 10 m. This gives an average wind velocity of 4.923 m/s for the smaller wire and 4.836 m/s for the larger one.

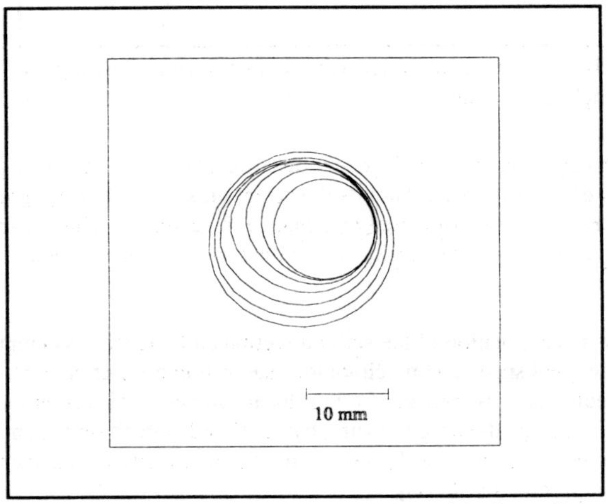

Fig. 2. Ice accretion on a 12.5-mm twisting cable, for uniform values of wind speed and drop diameter for glaze.

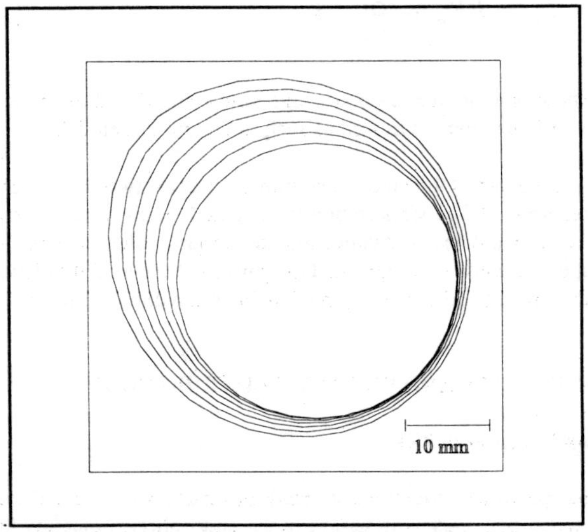

Fig. 3. Ice accretion on a 35-mm twisting cable, for uniform values of wind speed and drop diameter for glaze.

For freezing rain a precipitation rate of 4 mm/h was used and a corresponding liquid water content of w = 0.243 gm^{-3} is computed. The median volumetric diameter drop size is 1.52 mm. The average ice accretion density used is 850 kg/m^3 while the ambient temperature assumed is -3°C.

Figure 2 shows the shape of the simulated ice accretion on the smaller cable (12.5 mm), whereas Fig. 3 shows the ice accretion on the larger 35 mm cable. A comparison of the two figures (Figs. 2 and 3) shows that the smaller cable twists more and that this results in a more nearly circular cross-section. This increases the effective diameter of the cross-section of the smaller cable which in turn increases the icing rate.

Figures 4 and 5 show ice accretion on the same two cables (12.5 and 35 mm cable) but this time taking into consideration the difference in height and also the effect of horizontal turbulence and the droplet size distribution using the point estimate method. On each of these figures only the final accretion with the same duration simulation is shown. The average shape is given by the continuous line and the average plus and minus the standard deviation is shown in dotted lines. This shows the possible variation of shape resulting from the random nature of the phenomenon.

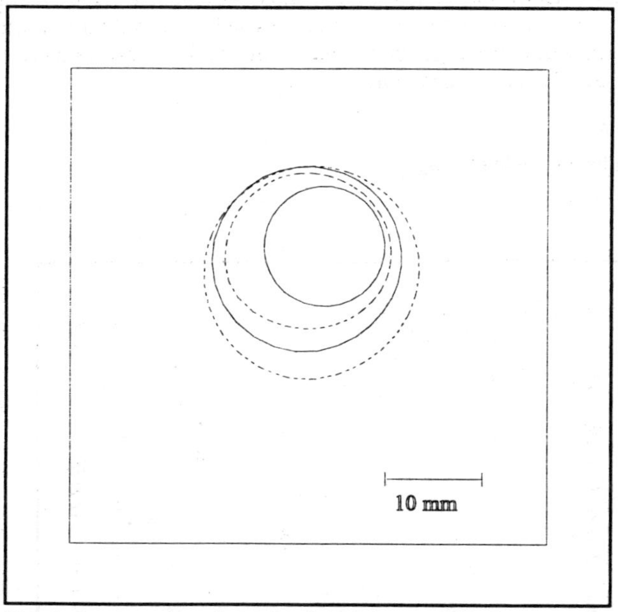

Fig. 4. Ice accretion shape for the 12.5 mm cable for a random distribution for droplet diameter and wind speed for freezing rain conditions.

Numerical simulation of cable icing was also performed for in-cloud icing conditions. To facilitate the comparison of the results with those of freezing rain, the following input conditions were used: liquid water content w = 0.243 g/m^3, median volumetric diameter = 18.3 µm, air temperature = -11° C and the same wind speed assumption. The next two figures, 6 and 7 show results of a simulation (for Figs. 4 and 5) with the same two cables diameters but this time for the case of in-cloud icing.

On each of these figures the average shape is also represented by a continuous line and the plus and minus standard deviation shown in dotted lines.

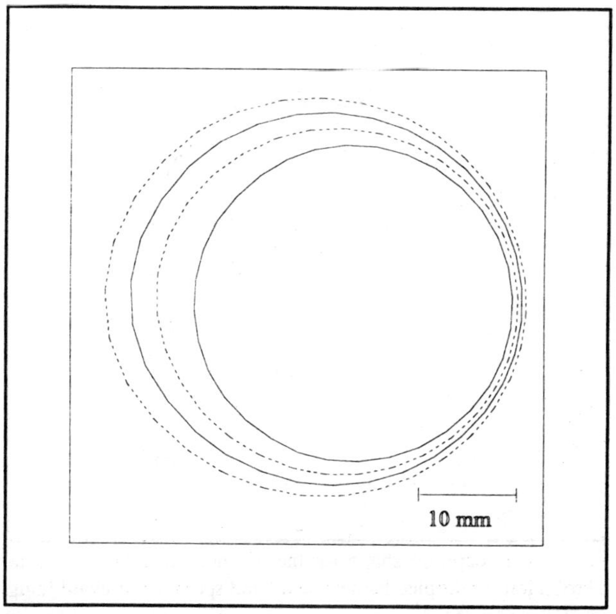

Fig.5. Ice accretion shape for the 35 mm cable for a random distribution for droplet diameter and wind speed for freezing rain conditions.

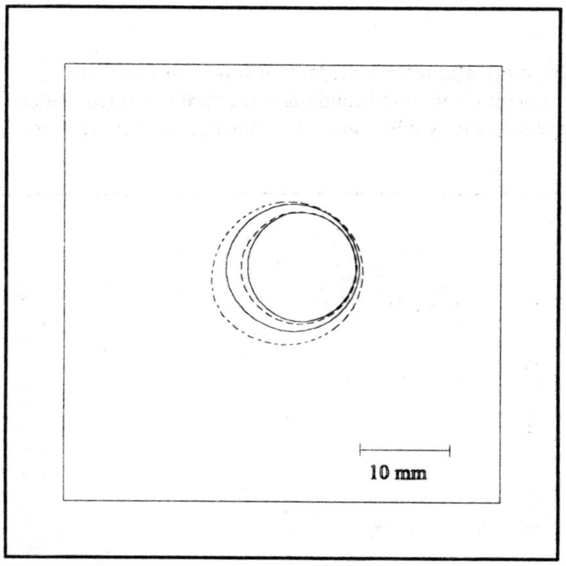

Fig. 6. Ice accretion shape for the 12.5 mm cable for a random distribution for droplet diameter and wind speed for in-cloud icing conditions.

Wind velocities and average precipitation rate have been varied, to see their respective influence on the icing rate. Figure 8 shows the results obtained for variation in wind velocity, and Fig. 9 for a variation in precipitation rate. The simulation results show that the wind velocity has

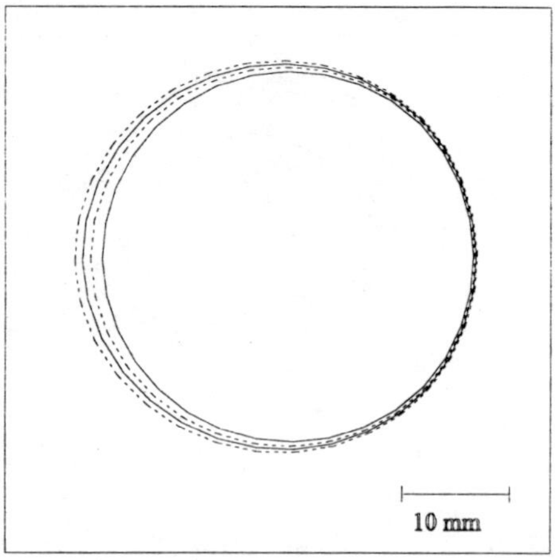

Fig. 7. Ice accretion shape for the 35 mm cable for a random distribution for droplet diameter and wind speed for in-cloud icing conditions.

a smaller influence on the glaze icing rate than the precipitation rate. When the precipitation rate increases, the rotation effect of the smaller cable becomes more important resulting in an increasing icing rate. If the trend persists the icing rate of the smaller cable should eventually become larger than the one for the larger cable as indicated by its more important slope.

Finally, in order to test the model's sensitivity to wind speed and liquid water content for in-cloud conditions, the total ice weight collected per unit length on each cable is given as a function of four velocities in Fig.

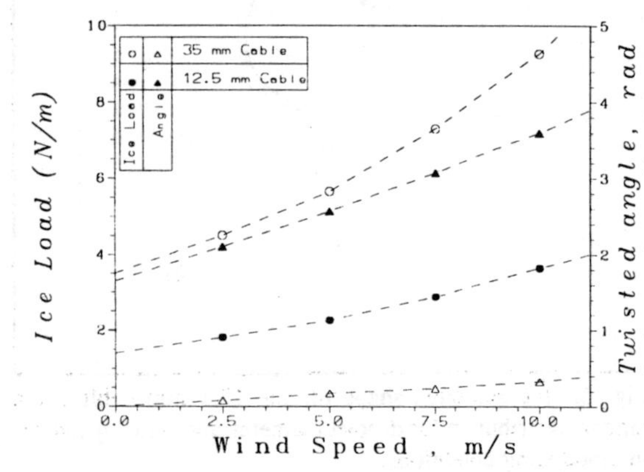

Fig. 8. Total ice mass collected on the cables as a function of average wind speed for glaze.

10 and as a function of the four liquid water contents plotted in Fig. 11.

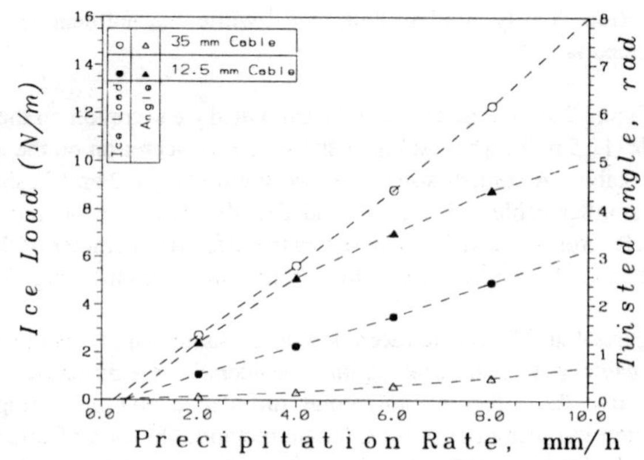

Fig. 9. Total ice mass collected on the cables as a function of precipitation rate for glaze.

The liquid water contents are chosen in Fig. 11 to correspond to the values obtained for the four precipitation rates plotted in Fig. 9. This gives a comparison of the behavior of the small cable as compared with the behavior of the large one and the sensitivity of the two variables in the case of in-cloud conditions.

Results for in-cloud icing

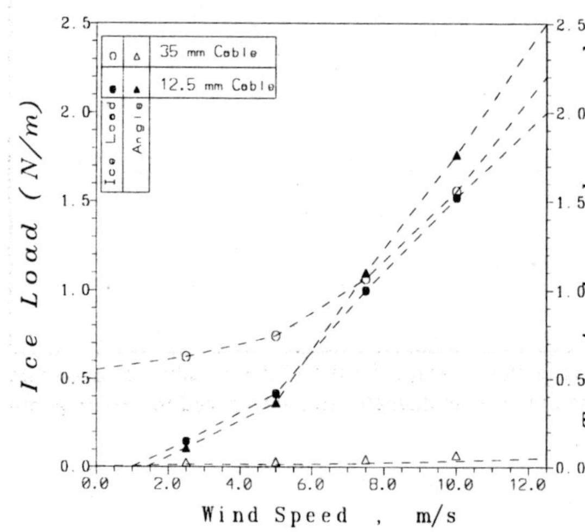

Fig. 10. Total ice mass collected on the cables as a function of average wind speed for rime

show only a small difference between icing rates for the two cables when compared with freezing rain results.

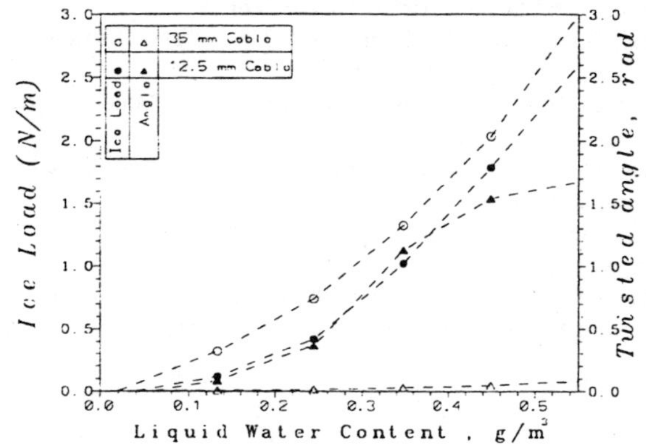

Fig. 11. Total ice mass collected on the cables as a function of liquid water content for rime

DISCUSSION

The change in precipitation rate has an important influence on the twisting of the 12.5 mm cable and much less influence on the 35-mm cable (Fig.9). If this is so, it would have to be explained by the influence of the accretion's shape. This would also indicate that the difference might be even larger in the case of a larger rain fall rate which in itself is a most dangerous condition because the maximum weight is achieved more quickly.

A comparison of the simulation of in-cloud icing with freezing rain shows that the collection efficiency is important for in-cloud icing and tends to increase the overall difference between the icing of the two cables. This is not the case for freezing rain where the size of the drops are sufficient to result in a collection efficiency of unity. Hence, the difference in icing of the two cables can be explained by the difference in collection efficiency as well as by the flexibility of the smaller cable. In fact, the two causes are always linked since for the same transmission line a smaller cable will collect more and at the same time will have a higher flexibility.

In the case of rime, two factors, i.e. cable flexibility and collection efficiency both increase the icing rate. There is a third factor, although smaller, smaller fog or cloud droplets follows the wind in an horizontal direction. This direction gives a maximum accretion thickness horizontally resulting in the maximum possible twisting torque. In the case of rain, usually falling close to the vertical line the resulting torque for the same ice weight is much smaller.

The effect of wind forces on flexible cables are not considered in the simulation. Wind forces increase with the square of the velocity so that these forces become important only for wind speed larger thant 10 m/s. The effect of these larger speeds would be to favor a more circular accretion shape. The smaller cable flexibility would again result in more twisting and swinging motions and hence in a more circular shape.

Recently, McComber et al. [7] have suggested that for stranded cable a change in mechanical tension could also result in cable twisting due to ice overloads. However, this occurs only on a transmission line with sufficient initial mechanical tension, and hence cannot explain the Mt. Valin data. For this reason, the mechanical tension was not included in this simulation since it is negligible at Mt. Valin. However, for a real transmission line it might be important. In such a case depending on the tension wind forces effect could be amplify and therefore more circular shape could be obtained for the ice accretion. However, the exact twisting charactreristics of various stranded cables are not well known so it is difficult to quantify this behavior [7]. Even the ice load changes the tension and therefore can cause a rotation of the cable.

Wet growth is not considered either in this paper. The influence of the wind on the heat transfer would mean that wind speed might become more important in the icing rate in a case of wet growth. [14]

The difference in cable height between the 12.5 and the 35 mm cables is not sufficient to result in a significant icing rate difference. This point had to be verified to make sure that the Mt. Valin results are not due to a difference in height of the two cables installation.

CONCLUSIONS

A simulation of cable icing was made on two cables of different sizes and flexibilities. Results show that for freezing rain the flexibility of the smaller cable makes possible twisting and a resulting icing rate increase. For in-cloud icing the same factors apply but in addition the smaller cable has a larger collection efficiency and also the maximum ice thickness occurs near the stagnation point on the windward side causing a larger twisting torque. Both added factors result in a larger icing rate for the smaller cable than would otherwise be expected. This combined effect should explain why the icing rate difference is more pronounced for in-cloud icing conditions than it is for freezing rain conditions. Both types of simulation were done for dry growth conditions. The results of the numerical simulations confirm the Mt. Valin icing load data at least qualitatively.

The effect of cable height, wind velocity and drop size distribution were also included in the numerical simulation. A larger precipitation rate has a more pronounced effect for the smaller cable than an increase in wind speed. The effect of a wind and droplet size increase is to cause the more flexible cable to grow a more circular accretion shape and hence tends to increase the overall atmospheric icing rate.

In summary if a smaller ground wire is used to carry an optical fibre for telecommunication purpose, the design criterion should be changed to account for an increase icing rate on the smaller cable. However, since the difference is not easy to establish quantitatively as it depends on many variables in a non-linear fashion, the solution might be to at least reduce the effect of cable twisting by installing pendulums at appropriate locations.

ACKNOWLEDGEMENT

This work is supported by the Conseil national de la Recherche en Science naturelle et en Genie du Canada. The Mt. Valin data were obtained in a contract from Hydro-Quebec.

REFERENCES

[1] Poots, G (1996). "Ice and Snow Accretion on Structures", *John Wiley and Sons*

Inc, New-York, 338 pp.
[2] McComber, P, Druez, J, and Laflamme, J (1995). "Icing Rate Estimation of Atmospheric Cable Icing", *IJOPE*, Vol 5 No 1.
[3] Druez, J, and McComber, P (1991). "A Comparison of the Atmospheric Icing of Two Different Size Cables" *Proc. of 1st Int. Conf. on Offshore and Polar Eng*, Edinburough (U.K.), August.
[4] Savadjiev, K and McComber, P (1995). "Mechanical Strength of OPGW under icing conditions", *Proc. Offs. Mech. and Art. Eng.*, Copenhagen, Vol. 4.
[5] Druez, J, and McComber, P (1993). "Étude de givrage au Mont Valin 1991-1992", Res. Rept. Hydro-Québec. (VPET-91-ÉTU-00-201), 358 p.
[6] McComber, P (1984). "Numerical Simulation of Cable Twisting due to Icing", Cold Reg. Sc. and Techn. Vol: 8
[7] McComber, P, Druez, J, and Savadjiev, K (1995). "Cable Twisting due to Atmospheric Icing", *Proc. 5th ISOPE-95, La Haye*, Vol. 2.
[8] Blevins, RD (1977). " Flow Induced Vibrations", *Van Nostrand Reinhold, Co.*
[9] Gunn , and G.D. Kinzer (1949). "The Terminal Velocity of Fall for Water Droplets in Stagnant Air" *J. Meteor.* Vol. 6
[10] Marshall, JS and W, M Palmer (1948). "The Distribution of Raindrops with Size" *J. Meteor.* Vol: 5.
[11] Best AC (1959). "The Size and Distribution of Raindrops" *Royal Met.* J.
[12] Horjen, I (1983). "Icing of Offshore Structures-Atmospheric Icing" *Norwegian Marine Research*, Vol: 3
[13] Rosenblueth, E (1975). "The Point Estimate Method" *Proc. Nat. Acad. Sci.*, Vol. 72 No 10,.
[14] Draganoiu, C, Lamarche, L, and McComber, P (1996). "A computer model of glaze accretion" *J. Offs. Mech. and Art. Eng*, Vol. 118, No 2.

Proceedings of the Seventh (1997) International Offshore and Polar Engineering Conference
Honolulu, USA, May 25-30, 1997

ISBN 1-880653-28-1 (Set); ISBN 1-880653-30-3 (Vol. II)

Ice Accretion and Shedding on Overhead Line Cables

Jacques Druez
Université du Québec à Chicoutimi
Chicoutimi, Canada

Pierre McComber
École de Technologie Supérieure
Montréal, Canada

ABSTRACT

In Canada as in many cold countries, atmospheric icing causes important damage to overhead lines and strongly reduces their reliability during the most critical periods of winter seasons. Ice accretion is a stochastic phenomenon and therefore probabilistic design is more and more used for overhead lines. Ice loads depend on ice accretion, ice shedding and persistence of ice on structures. A test site where icing is frequent, located on Mt. Valin (Québec) Canada, is used to obtain field data on icing events. This test site includes two non-energized test lines supporting three different single cables and one four conductor bundle, meteorological instruments, a data logger and a video recorder. Icing data recorded at the test site during the 1993-1994 winter season are presented and analysed. For the most interesting periods of this icing season, curves representing the ice load variation on the test cables and the corresponding signal for the meteorological instruments are displayed. It is shown that the persistence periods of ice on the cables can be very long, with a period lasting more than three weeks observed during November 1993. This period corresponds to large ice and wind loads reaching 3.51 kg/m, separated by partial shedding only of the ice. Results also indicate that the most dangerous types of atmospheric ice are wet growth accretions caused by freezing precipitations at temperature near the freezing point, dry growth accretions due to icing fog or in-cloud icing at lower temperature, and mixed accretions, at the same time or successively. Wet growth accretions are glaze ice or wet snow, and dry growth accretions soft to hard rime depending on temperature and droplets diameter. Different types of icing events can be distinguished in a data base by comparing the data obtained from a temperature probe, a heated rain and snow gauge, and an ice detector. For wet growth accretions, the ice mass reduction process on overhead lines is generally due to ice shedding caused by melting of the ice at the interface with structure and to ice melting at the external part of the ice accretion. For dry growth accretions, the ice mass reduction is generally due to cohesive failures inside the ice accretion, or to adhesive failures at the interface with structure, both under the effect of wind, gravity and cable movements. Over a longer period of time, the ice mass reduction can also be caused by ice erosion or sublimation of the ice. Temperature, wind velocity and shedding rate values can be used to distinguish different ice mass reduction modes in a data base.

KEY WORDS

Atmospheric Icing, Ice Load, In-cloud Icing, Precipitation Icing, Cables, Ice Accretion, Ice Shedding.

INTRODUCTION

In cold countries ice overloads cause important damage to overhead lines and strongly reduce the reliability of many vital networks during the most critical periods of winter seasons (Lott, 1996; Goia and Balan, 1996; Vuckovic et al., 1996). Icing events are stochastic phenomena and therefore probabilistic design is more and more used in the mechanical design of exposed networks such as power lines and communication networks (Laflamme and Périard, 1996; Raison et al., 1996). For strength and reliability analysis, atmospheric icing is an important design factor (Ervick and Fikke, 1996). Ice loads and wind forces depend on ice accretion and ice shedding, but also on the persistence period of ice on the structure (Druez et al., 1995; Hardy et al., 1996; Ryerson and Kenyon, 1996). Only few atmospheric icing measurements have been made on overhead lines or test cables (McComber et al., 1995) so many utility engineers need a more complete data base on ice storms (Shan et al., 1996). Therefore, a test site exposed to harsh icing events during about seven months each year is used since a few years to monitor different types of icing events, and to collect meteorological and icing data (Druez and McComber, 1988; Druez et al., 1991). This natural icing test site, located on Mt. Valin (Québec) Canada, includes two non-energized test lines supporting three different single cables and one four conductor bundle, meteorological instruments, a data logger and an automatic frame by frame video recorder (Druez et al., 1994; Druez, McComber and Laflamme, 1995).

Icing data recorded at the test site during the 1993-1994 winter season are presented and analysed in this paper . For the most interesting periods of the icing season, curves representing the ice load variation on the test cables and the corresponding signals for the meteorological instruments are displayed. It is shown that the most dangerous types of ice accretions are glaze and wet snow caused by precipitation, rime accretions due to in-cloud and fog icing, and mixed types of accretions. Different types of icing events can be distinguished by comparing the signals from a temperature sensor, a heated rain and snow gauge and an ice detector. Different ice mass reduction modes can also be distinguished using temperature, wind velocity and shedding rate values. Typical values of parameters related to the icing events are calculated from the field data and presented.

DESCRIPTION OF THE NATURAL ICING TEST SITE

The natural icing test site is located at an altitude of 902 m, on Mt. Valin, 45 km North-East of the city of Chicoutimi (Québec) Canada. Due to its northern climate and proximity of the Saguenay Fjord, the site is exposed to frequent icing events during about seven months every year, from October to April (Druez et al., 1996). The test site is mainly composed of two non-energized test lines supporting three different single cables and one four conductor bundle. The main test line, 96.5 m in span, consists of one stranded 12.5-mm steel ground wire, one stranded 35-mm ACSR conductor, and a bundle composed of four 35-mm ACSR conductors maintained at a distance of 406 mm using two cross-spacers. This test line orientation is North-North-East, almost perpendicular to the direction of dominant winds. The second test line is nearly perpendicular to the first one and used for comparison purpose. This second test line is 31.4 m in span and consists of one stranded 7.93-mm steel cable. At one end of each conductor, a load cell measures the mechanical tension which is converted into uniform ice mass per unit length, using the cable geometry and equilibrium equations. Due to this measurement method, the ice load includes the wind force on the ice accretion. This combined value is more significant in term of line reliability and moreover the separation of ice and wind loads would necessitate to make hazardous assumptions on ice density and ice accretion shape. Icing events are continuously monitored and recorded on the site using a video-camera and a frame-by-frame recorder. Figure 1 presents a diagram of the test site instrumentation.

The meteorological instruments are respectively a temperature sensor, an anemometer, a precipitation gauge and an ice detector. The thermistor which measures the air temperature is sheltered from wind and sun effects. The Hydrotech anemometer, model WS3/WD3, is electrically de-iced and measures the wind speed and the wind direction. This sturdy anemometer is reliable during intense icing events. The wind speed and the wind direction are averaged over a ten minute period called in meteorology "specified interval of time". From these averaged measurements, the wind velocity component normal to the main test line is computed, knowing the line orientation. The precipitation gauge is an electrically heated rain and snow gauge model Géneq P-1500, measuring the equivalent water precipitation in millimeters. The ice detector monitors ice accretion by measuring the shift in frequency, at resonance, of a vibrating cylindrical probe, under the effect of the mass of ice accreted. This instrument, developed by IREQ using a Rosemount probe, is electrically de-iced and mechanically shaked in order to facilitate water running-off after de-icing cycle.

DATA AQUISITION

A data logger, installed in a building close to the test lines, is linked by telephone to a computer located at Université du Québec à Chicoutimi. All the data are averaged for every 30-minute periods and are transmitted once a day, using a verification protocol to identify transmission errors. These data are averaged from samples taken every 10 seconds for each load cell, and from 200 samples in a ten-minute period, both for the wind speed and the wind direction. Only one sample is used for the temperature sensor, the precipitation gauge and the ice detector. Analytical and statistical analysis are performed on the data base obtained. Graphics showing the variation with time of the measured and calculated parameters are plotted as shown in Figures 2-4, and analysed. The video camera monitors ice accretion, ice shedding and cables vibration on the main test line. Its frame by frame automatic mode, generally adjusted at one picture every four seconds, gives an autonomy of three weeks before the videocassette must be replaced during a trip to the test site. The video camera can be remote-controlled only from the test site, and floodlightning allows video-recording 24 hours a day. Due to the severe atmospheric conditions prevailing at the test site, the experimental installation is protected against humidity, water infiltration, lightning and power interruptions, in order to improve its reliability.

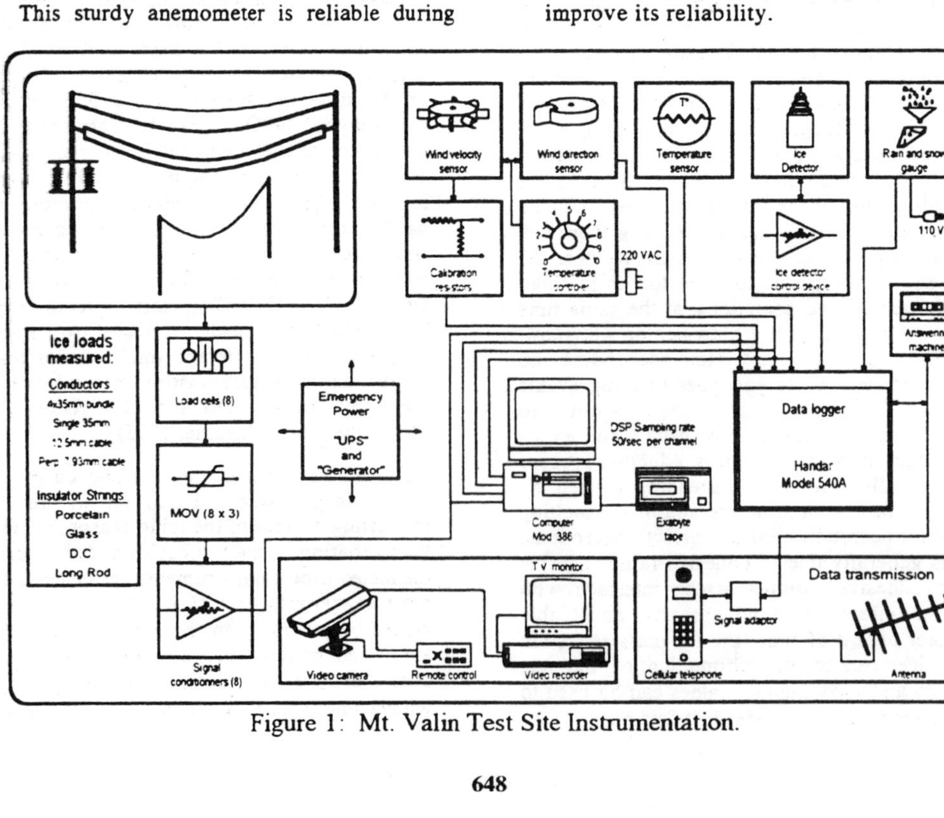

Figure 1: Mt. Valin Test Site Instrumentation.

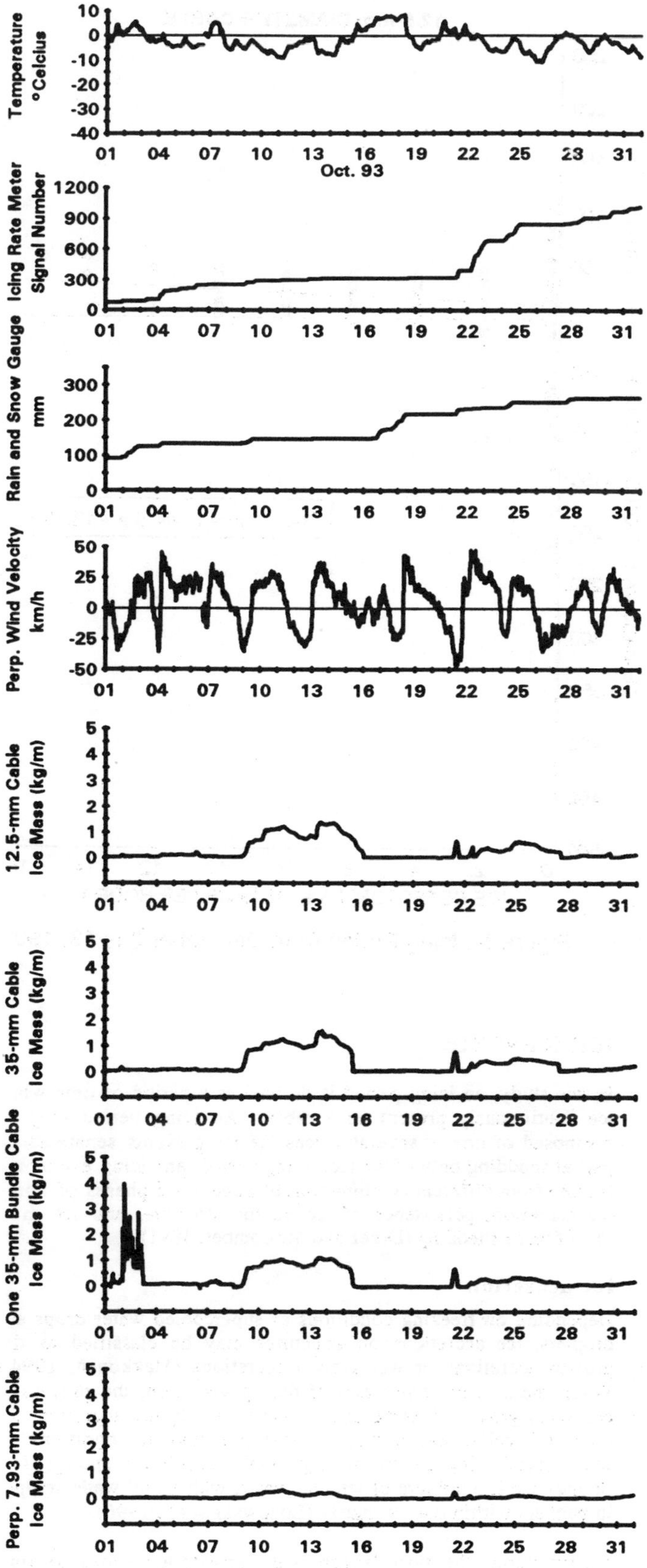

Figure 2: Icing during October 1993

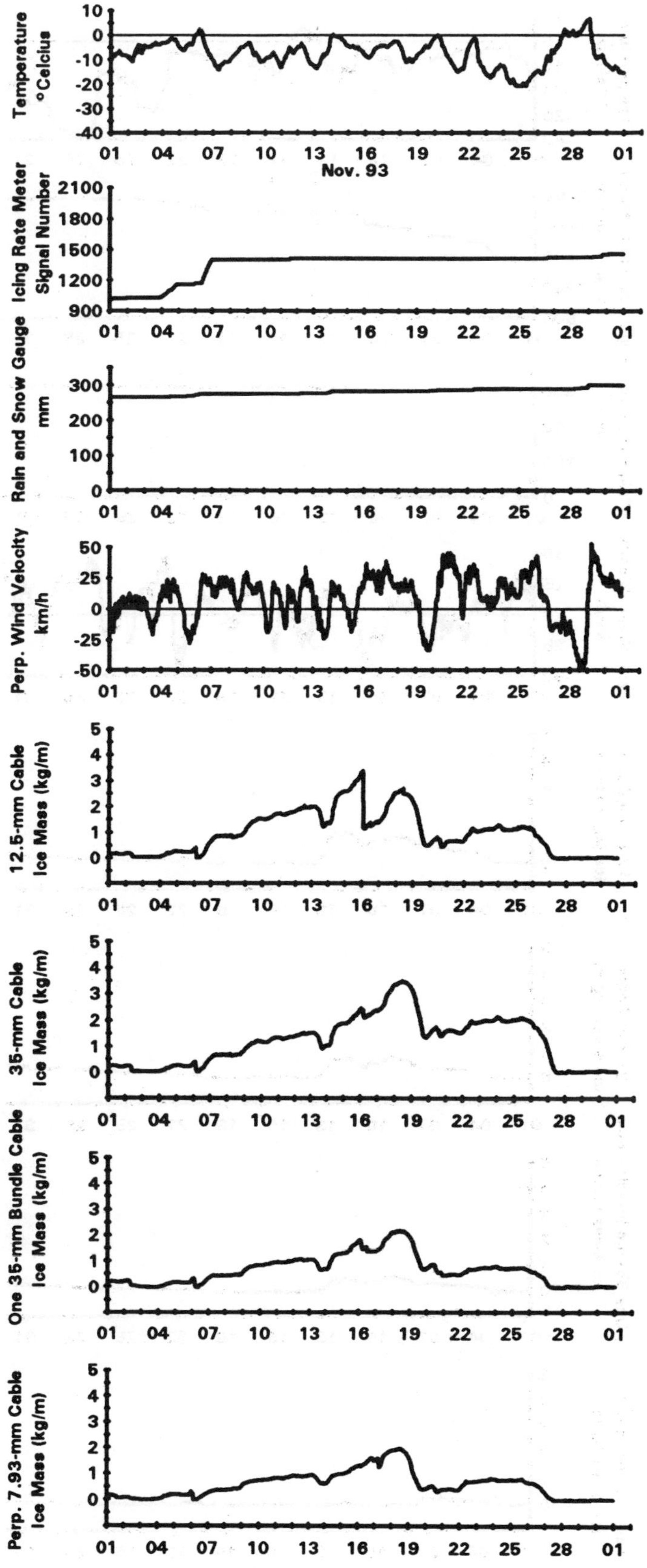

Figure 3: Icing during November 1993

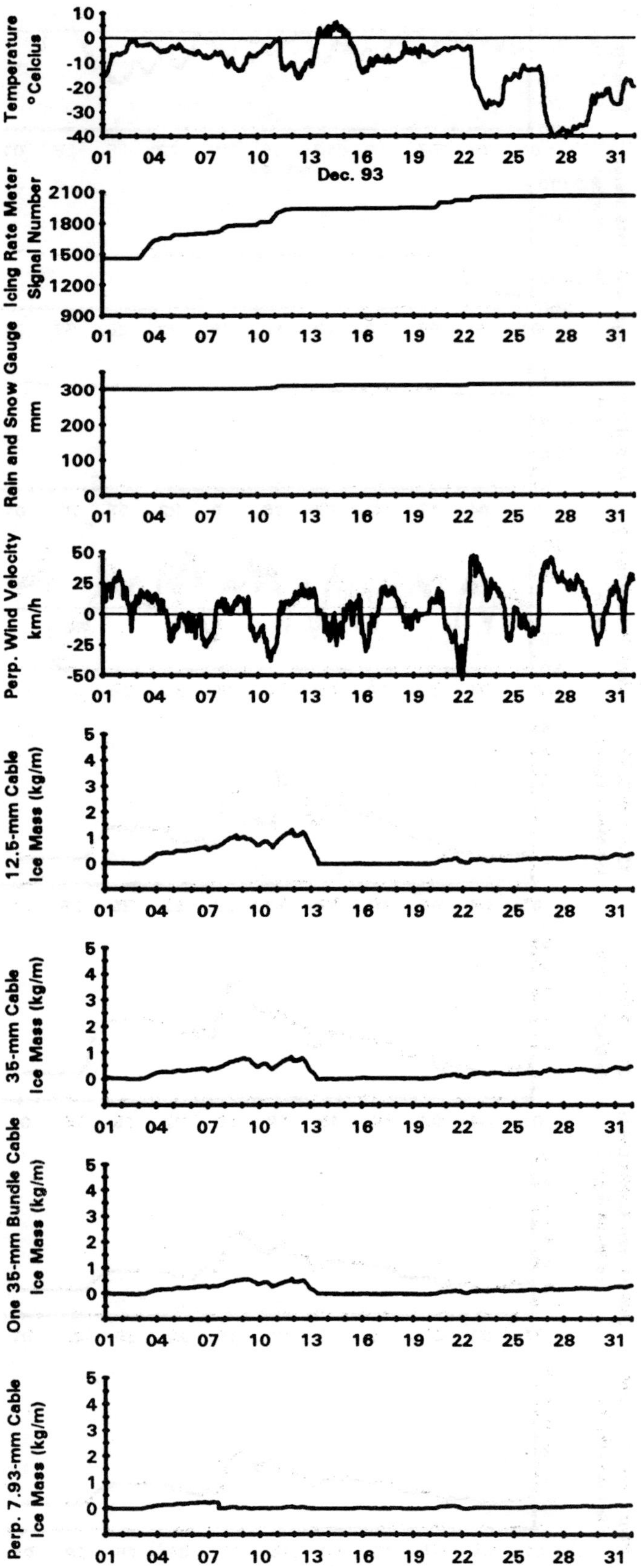

Figure 4: Icing during December 1993

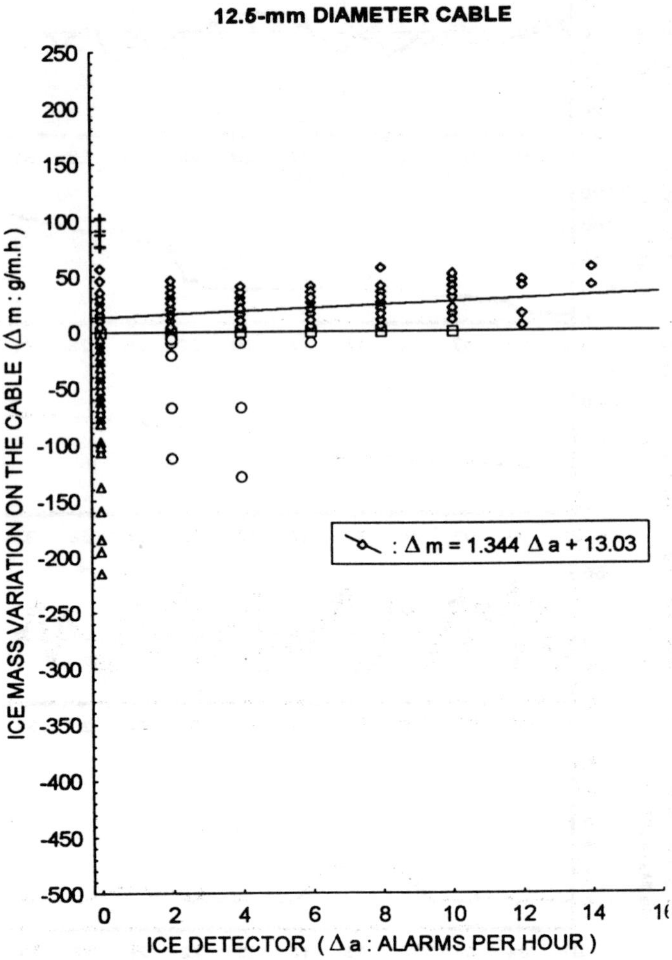

Figure 5 : Icing Period from December 3 to 13, 199

ICING EVENTS

In this study, an icing period is defined as a period of time whi ice continuously present on a cable. An icing period may l composed of one or several successive icing events separated t partial shedding only of the ice. Icing periods and icing events a formed from different combinations of three basic phases of icin ice accretion, persistence of ice on the structure, and ice ma reduction or shedding (Druez and McComber, 1991).

Ice accretion

Depending on freezing conditions of supercooled water drops ar droplets, ice accretions on structures may be classified as d growth accretions or wet growth accretions (Makkonen, 1994 When there is no water run-off during accretion, the process called dry growth. If some liquid water is lost during accretion, tl process is called wet growth. Under intermediate conditions, with mixed icing events, it is possible to obtain spongy ic Spongy ice is a mixture of ice and water, with liquid water trapp in pockets within the ice matrix (Lozowski et al., 1996).

For structures, the most dangerous dry growth accretions are rin accretions, formed at low temperature by impact of the sma supercooled droplets composing icing fog and icing clouds (Drue McComber and Laflamme, 1995). Depending on atmospher

Table 1 : Precipitation icing

DATE	TIME START	PHASE DURATION (h)			ICING RATE (g/m.h)		ICE LOAD (kg/m)	MEAN TEMP. (°C)		MEAN PERP. WIND (km/h)		PRECIPITATION (mm)		IRM SIGNALS	
		ACCR.	INTER.	SHED.	ACCR.	SHED.	MAXIMUM	ACCR.	SHED.	ACCR.	SHED.	ACCR.	TOTAL	ACCR.	TOTAL
							12.5-mm DIAMETER CABLE								
03 Oct.	23:00	7.5	45.5		11		0.12	-1.4		-23.0		7.00	7.25	49	115
08 Oct.	23:30	9.5	18.5		61		0.71	-4.1		-26.2		5.75	12.50	25	43
12 Oct.	17:00	6.0		7.5	25	11	0.87	-1.8	-3.6	-24.7	18.9	0.00	1.75	8	9
21 Oct.	07:00	5.5		4.5	132	160	0.65	-0.5	1.3	-42.8	-38.7	13.50	13.50	74	74
22 Oct.	04:30	6.5		1.0	65	384	0.42	-1.1	0.2	40.7	44.2	1.75	1.75	146	146
24 Oct.	01:30	25.5	58.0	1.5	17	152	0.64	-3.6	0.9	18.8	-18.0	17.00	17.00	159	160
03 Nov.	23:30	21.5	22.5		11		0.27	-2.5		15.2		2.75	4.00	120	122
05 Nov.	19:30	7.0		1.5	25	295	0.39	-0.6	1.1	-14.9	-6.0	3.25	4.50	15	15
19 Nov.	19:30	14.0		7.5	26	58	0.80	-1.1	-5.6	-17.4	30.7	1.50	1.50		
21 Nov.	22:00	14.0	80.5	35.5	38	35	1.28	-3.2	-10.1	29.1	17.7	3.75	3.75		
10 Dec.	18:00	26.5	16.5	22.0	23	56	1.29	-6.6	-9.4	-14.3	15.8	6.25	6.50	116	123
28 Jan.	14:30	7.0	239.5	3.0	19	53	0.41	-2.5	-32.5	-13.2	27.5	8.25	8.75	64	87
22 March	08:30	20.5		5.5	10	56	0.23	-3.3	-1.2	-10.2	-10.0	4.75	5.25	107	107
Average		13.2	68.7	9.0	35.6	126.0	0.62	-2.5	-5.9	-22.3	22.8	5.81	6.77	80.3	91.0
							35-mm DIAMETER CABLE								
03 Oct.	23:00	7.5	45.5		3		0.10	-1.4		-23.0		7.00	7.25	49	115
08 Oct.	23:30	9.5	18.5		85		0.88	-4.1		-26.2		5.75	12.50	25	43
12 Oct.	17:00	14.0			7		1.02	-3.0		-21.6		1.75	1.75	9	9
21 Oct.	07:00	5.5		4.5	148	170	0.76	-0.5	1.3	-42.8	-38.7	13.50	13.50	74	74
22 Oct.	04:30	6.5		2.5	48	62	0.32	-1.1	0.1	40.7	40.4	1.75	1.75	146	146
24 Oct.	01:30	25.5	58.0	1.0	9	250	0.51	-3.6	0.9	18.8	-17.4	17.00	17.00	159	160
03 Nov.	23:30	21.5	22.5		11		0.25	-2.5		15.2		2.75	4.00	120	122
05 Nov.	19:30	7.0		2.5	26	163	0.39	-0.6	1.4	-14.9	-6.3	3.25	5.25	15	15
19 Nov.	19:30	14.0		7.5	30	55	1.76	-1.1	-5.6	-17.4	30.7	1.50	1.50		
21 Nov.	22:00	14.0	80.5	41.0	34	54	2.14	-3.2	-8.8	29.1	17.2	3.75	3.75		
10 Dec.	18:00	26.5	16.5	22.0	16	37	0.82	-6.6	-9.4	-14.3	15.8	6.25	6.50	116	123
28 Jan.	17:00	7.5	237.5	4.0	21	83	0.69	-3.8	-33.4	-10.7	29.1	5.00	5.00	53	75
22 March	08:30	20.5		5.5	7	38	0.17	-3.3	-1.2	-10.2	-10.0	4.75	5.25	107	107
Average		13.8	68.4	10.1	34.2	101.3	0.75	-2.7	-6.1	-21.9	22.8	5.69	6.54	79.4	89.9
							ONE 35-mm BUNDLE CONDUCTOR								
03 Oct.	23:00	7.5	45.5		3		0.08	-1.4		-23.0		7.00	7.25	49	115
08 Oct.	23:30	9.5	18.5		81		0.75	-4.1		-26.2		5.75	12.50	25	43
12 Oct.	17:00	14.0			8		0.84	-3.0		-21.6		1.75	1.75	9	9
21 Oct.	07:00	5.5		4.5	143	160	0.65	-0.5	1.3	-42.8	-38.7	13.50	13.50	74	74
22 Oct.	05:00	6.5		2.0	31	27	0.13	-1.0	0.1	41.4	41.9	1.50	1.50	137	137
24 Oct.	01:30	25.5	57.5	1.5	7	97	0.33	-3.6	0.8	18.8	-16.4	17.00	17.00	159	160
03 Nov.	23:30	21.5	22.5		10		0.16	-2.5		15.2		2.75	4.00	120	122
05 Nov.	19:30	7.0		2.5	29	148	0.32	-0.6	1.4	-14.9	-6.3	3.25	5.25	15	15
19 Nov.	19:30	14.0		13.0	17	29	0.86	-1.1	-7.2	-17.4	34.4	1.50	1.50		
21 Nov.	22:00	12.5	82.0	35.0	17	20	0.79	-3.0	-10.3	29.0	17.7	3.75	3.75		
10 Dec.	18:00	26.5	16.5	22.0	10	25	0.57	-6.6	-9.4	-14.3	15.8	6.25	6.50	116	123
28 Jan.	14:30	9.5	237.0	6.0	13	14	0.37	-3.3	-33.6	-11.6	27.6	8.75	8.75	65	87
22 March	08:30	20.5		5.5	5	27	0.13	-3.3	-1.2	-10.2	-10.0	4.75	5.25	107	107
Average		13.8	68.5	10.2	28.8	60.8	0.46	-2.6	-6.5	-22.0	23.2	5.96	6.81	79.6	90.2
							PERPENDICULAR 7.93-mm DIAMETER CABLE								
04 Oct.	06:30	2.5	43.0		10		0.02	-1.6		38.2		0.25	0.25	30	66
08 Oct.	23:30	9.5	18.5		13		0.16	-4.1		-26.2		5.75	12.50	25	43
12 Oct.	17:00	14.0			3		0.14	-3.0		-21.6		1.75	1.75	9	9
21 Oct.	07:00	5.5		4.5	53	62	0.24	-0.5	1.3	-42.8	-38.7	13.50	13.50	74	74
22 Oct.	04:30	6.5		1.0	17	98	0.10	-1.1	0.2	40.7	44.2	1.75	1.75	146	146
24 Oct.	01:30	24.0	31.0	14.0	3	4	0.11	-3.4	-5.3	18.4	-27.5	17.00	17.00	159	159
03 Nov.	23:30	18.5	25.5		10		0.19	-2.5		15.0		2.75	4.00	104	122
05 Nov.	19:30	4.5		3.5	30	86	0.29	-1.0	0.6	-16.7	-10.1	3.25	3.25	15	15
19 Nov.	21:30	12.0		7.5	12	36	0.53	-0.9	-5.6	-15.2	30.7	1.50	1.50		
21 Nov.	22:00	11.5	83.0	41.0	28	19	0.82	-3.0	-8.8	28.7	17.2	3.50	3.75		
11 Dec.	05:30	14.5	17.5	21.0	3	3	0.07	-10.3	-9.5	6.5	16.0	0.00	0.25	25	34
28 Jan.	14:30	7.5	180.0	4.0	7	40	0.51	-2.7	-15.0	-12.6	-2.4	8.50	8.75	64	86
22 March	13:00	16.0		6.0	2	9	0.04	-3.0	-0.8	-7.3	-9.6	1.75	2.25	66	66
Average		11.3	56.9	11.4	14.7	39.7	0.25	-2.9	-4.8	22.3	21.8	4.71	5.42	65.2	74.5

conditions, these accretions range from soft rime to hard rime. In the case of wet growth, the most dangerous accretions are either glaze due to freezing rain, or wet snow. Glaze is formed from larger drops of water, at temperature just under the freezing point. Wet snow contains some liquid water and occurs when temperature is just over the freezing point. In the field, icing periods can last for several days or even several weeks, allowing successive observation of different types of icing events during a same icing period (Druez et al., 1996). Moreover, the variation of temperature during a given icing period can modify the physical, microstructural and mechanical properties of the atmospheric ice formed.

Persistence of ice

The persistence period of ice on a structure can vary from a few minutes to several weeks. This persistence period is generally much longer for dry growth accretions than for wet growth accretions, due to the corresponding air temperature values. During the coldest months of winter, rime accretions can remain on cables during many days before complete shedding of the ice can occur. This persistence period is very dangerous for an iced structure, because many cumulative icing events can be observed during such a long period. The probability for strong winds is also higher during a long period, resulting in a very severe combination of ice and wind loads. Figure 3 for example shows a persistence period of ice lasting about three weeks, from November 6 to 27. The maximum ice load for the season, 3.51 kg/m, was observed the 18 of November, during this period, on the 35-mm single cable. The ice load was also maximum for the three other conductors during the same icing period. The figure also shows a recurrence period of a few days for the wind peaks.

Ice mass reduction and shedding

An ice mass reduction on a cable can be caused by ice melting, ice sublimation, ice erosion, and mechanical failure of the ice under the combined effects of wind, gravity and cable displacements. Ice

Table 2 : In-cloud icing

DATE	TIME START	PHASE DURATION (h)			ICING RATE (g/m.h)		ICE LOAD (kg/m)	MEAN TEMP. (°C)		MEAN PERP. WIND (km/h)		PRECIPITATION (mm)		IRM SIGNALS	
		ACCR.	INTER.	SHED.	ACCR.	SHED.	MAXIMUM	ACCR.	SHED.	ACCR.	SHED.	ACCR.	TOTAL	ACCR.	TOTAL
12.5-mm DIAMETER CABLE															
06 Oct.	04:00	2.5		8.0	49	26	0.20	-4.4	-4.0	23.3	21.8	0.00	0.00	24	28
10 Oct.	03:30	5.0	27.5	28.0	77	16	1.20	-7.0	-3.4	20.2	-18.4	0.00	0.00	1	5
13 Oct.	06:30	7.0	17.5	43.5	92	27	1.37	-6.5	-0.6	31.9	7.2	0.00	0.00	4	4
22 Oct.	13:30	19.0	16.0		19		0.40	-3.3		27.0		2.00	2.00	144	144
28 Oct.	08:00	11.0	39.0	2.5	7	69	0.19	-3.2	-2.9	14.2	28.1	0.00	1.25	42	104
30 Oct.	12:30	66.0		3.5	2	41	0.17	-6.9	-8.5	7.8	3.1	0.00	0.00	60	60
06 Nov.	09:00	22.5	36.0		41		0.88	-8.4		20.0		0.25	0.25	231	231
08 Nov.	19:30	22.5	83.5	11.5	30	69	1.99	-6.5	-9.6	19.3	-12.6	0.00	1.50		
14 Nov.	03:00	9.0	37.0	1.0	124	2213	3.37	-0.7	-8.8	17.1	24.3	0.25	1.25		
16 Nov.	06:00	51.5	11.0	23.0	34	96	2.68	-5.4	-6.1	23.5	-18.0	0.00	0.75		
20 Nov.	17:30	7.0	21.5		35		0.68	-10.0		39.9		0.00	0.00		
03 Dec.	04:00	19.5	70.0	3.0	19	49	0.65	-2.9	-7.4	14.4	-25.1	0.50	1.75	164	238
07 Dec.	00:30	39.5	19.0	10.5	15	24	1.10	-8.1	-7.2	-9.6	-12.8	0.25	0.50	77	82
09 Dec.	22:30	8.0	4.5	7.0	21	28	0.85	-5.9	-3.4	-14.3	-31.6	0.75	0.75	25	29
20 Dec.	05:00	7.5	26.5	12.5	12	13	0.21	-6.4	-4.4	13.9	-39.3	0.00	1.00	48	71
22 Dec.	08:30	4.0	284.0	26.5	38	8	0.38	-7.1	-23.3	29.7	-10.3	0.25	4.00	26	90
13 Jan.	06:00	343.5		11.5	1	11	0.23	-26.0	-11.8	15.7	-14.8	0.00	0.00	18	18
21 Feb.	02:30	5.0	196.0	24.5	34	5	0.27	-5.3	-18.9	25.0	12.3	0.00	0.00	67	84
19 March	21:00	9.5	28.0	3.5	29	44	0.25	-9.0	-0.3	12.8	-13.8	0.00	0.00	83	86
25 March	08:00	12.0	11.5	6.5	13	35	0.20	-4.5	-3.9	11.5	12.8	0.00	0.00	62	66
Average		33.6	54.6	13.3	34.6	163.2	0.86	-6.9	-7.3	19.6	-18.0	0.21	0.75	67.3	83.8
35-mm DIAMETER CABLE															
06 Oct.	04:00	2.5		8.0	10	10	0.10	-4.4	-4.0	23.3	21.8	0.00	0.00	24	28
10 Oct.	03:30	6.5	26.0	28.0	34	9	1.25	-7.0	-3.4	20.8	-18.4	0.00	0.00	1	5
13 Oct.	07:00	5.5	45.5	3.5	102	235	1.53	-6.5	2.6	31.8	-2.1	0.00	0.00	3	4
22 Oct.	13:30	18.5	15.0		14		0.41	-3.2		26.9		2.00	2.00	144	144
28 Oct.	09:00	10.0	37.5	5.0	5	5	0.11	-3.3	-2.9	14.4	26.4	0.00	1.25	38	100
30 Oct.	13:30	65.0		3.5	2	70	0.25	-7.0	-8.5	7.5	3.1	0.00	0.00	60	60
06 Nov.	09:00	22.5	36.0		29		0.71	-8.4		20.0		0.25	0.25	231	231
08 Nov.	19:30	22.5	83.5	11.5	22	57	1.53	-6.5	-9.6	19.3	-12.6	0.00	1.50		
14 Nov.	03:00	9.0	36.0	4.5	89	66	2.46	-0.7	-8.9	17.1	27.5	0.25	1.25		
16 Nov.	06:00	51.5	11.0	23.0	31	99	3.51	-5.4	-6.1	23.5	-18.0	0.00	0.75		
20 Nov.	17:00	8.5	20.5		28		1.66	-10.1		40.2		0.00	0.00		
03 Dec.	05:30	18.0	70.0	3.0	13	31	0.40	-2.9	-7.4	14.4	-25.1	0.50	1.75	153	227
07 Dec.	00:30	51.0	7.5	11.5	10	25	0.78	-9.2	-7.2	9.9	-13.3	0.25	0.50	80	82
09 Dec.	22:30	8.0	4.5	7.0	16	28	0.57	-5.9	-3.4	-14.3	-31.6	0.75	0.75	25	29
20 Dec.	05:00	8.0	24.5	14.0	8	6	0.17	-6.4	-4.3	13.9	-37.8	0.00	1.00	48	71
22 Dec.	07:00	5.5	282.5	28.0	23	8	0.49	-6.3	-23.4	23.6	-10.4	1.00	4.75	28	92
13 Jan.	06:00	343.5		15.0	1	29	0.71	-26.0	-12.0	15.7	-17.0	0.00	0.00	18	19
21 Feb.	02:30	5.0	196.0	24.5	24	3	0.21	-5.3	-18.9	25.0	12.3	0.00	0.00	67	84
19 March	21:00	9.5	28.0	5.0	25	29	0.25	-9.0	-0.2	12.8	-16.2	0.00	0.00	83	86
Average		35.3	57.8	12.2	25.6	44.4	0.90	-7.0	-7.4	19.7	-18.4	0.26	0.83	66.9	84.1
ONE 35-mm BUNDLE CONDUCTOR															
06 Oct.	04:00	2.5		8.0	8	17	0.08	-4.4	-4.0	23.3	21.8	0.00	0.00	24	28
10 Oct.	03:30	5.0	27.5	28.0	36	8	1.03	-7.0	-3.4	20.2	-18.4	0.00	0.00	1	5
13 Oct.	07:00	5.5	45.5	5.0	44	145	1.09	-6.5	2.9	31.8	-3.6	0.00	0.00	3	4
22 Oct.	13:30	18.0	13.5		11		0.27	-3.1		26.9		2.00	2.00	144	144
28 Oct.	10:00	8.0	39.5	3.5	4	7	0.07	-3.2	-2.9	14.6	27.4	0.00	1.25	33	97
30 Oct.	13:00	65.5		3.5	3	52	0.20	-6.9	-8.5	7.7	3.1	0.00	0.00	60	60
06 Nov.	09:00	22.5	37.0		22		0.46	-8.4		20.0		0.25	0.25	231	231
08 Nov.	20:30	21.5	84.5	10.5	17	44	1.06	-6.6	-9.4	19.2	-13.4	0.00	1.50		
14 Nov.	02:30	9.5	37.5	3.5	59	105	1.82	-0.6	-9.2	17.1	30.3	0.25	1.25		
16 Nov.	06:00	51.5	11.0	23.0	18	70	2.19	-5.4	-6.1	23.5	-18.0	0.00	0.75		
20 Nov.	22:30	13.5	10.0		4		0.49	-13.4		35.4		0.00	0.00		
03 Dec.	06:30	17.0	70.0	3.0	10	24	0.31	-2.9	-7.4	14.1	-25.1	0.50	1.75	144	218
07 Dec.	00:30	51.0	7.5	11.0	7	16	0.56	-9.2	-7.2	9.9	-13.0	0.25	0.25	80	81
09 Dec.	23:00	7.5	4.5	7.0	11	24	0.45	-5.9	-3.4	-14.0	-31.6	0.50	0.50	22	26
20 Dec.	05:00	8.0	24.5	14.0	7	6	0.11	-6.4	-4.3	13.9	-37.8	0.00	1.00	48	71
22 Dec.	08:00	4.5	284.5	26.0	16	6	0.36	-6.9	-23.1	27.6	-10.2	0.25	4.00	27	91
13 Jan.	06:00	343.5		14.5	1	17	0.48	-26.0	-12.0	15.7	-16.7	0.00	0.00	18	19
21 Feb.	02:30	5.0	196.0	24.5	13	2	0.11	-5.3	-18.9	25.0	12.3	0.00	0.00	67	84
19 March	21:00	9.0	27.0	7.5	23	22	0.24	-9.2	-0.9	13.5	-16.8	0.00	0.00	83	86
25 March	08:30	11.5	11.5	3.5	2	17	0.05	-4.6	-4.8	11.4	9.0	0.00	0.00	66	66
Average		34.0	54.8	11.5	15.8	34.2	0.57	-7.1	-7.2	19.2	-18.1	0.20	0.73	65.7	81.9
PERPENDICULAR 7.93-mm DIAMETER CABLE															
06 Oct.	04:00	2.5		8.0	3	4	0.01	-4.4	-4.0	23.3	21.8	0.00	0.00	24	28
10 Oct.	03:30	6.0	26.5	28.0	12	6	0.31	-7.0	-3.4	20.7	-18.4	0.00	0.00	1	5
13 Oct.	07:00	7.5	43.0	1.0	10	38	0.21	-6.5	1.2	31.9	-1.5	0.00	0.00	4	4
22 Oct.	14:00	17.0	9.0		4		0.06	-3.1		26.7		2.00	2.00	140	141
28 Oct.	10:00	8.0	39.5	4.5	2	6	0.06	-3.2	-2.9	14.6	26.9	0.00	1.25	33	97
30 Oct.	14:00	42.5		5.5	3	11	0.16	-6.2	-7.2	8.0	-4.8	0.00	0.00	53	53
06 Nov.	09:00	22.5	37.0		15		0.41	-8.4		20.0		0.25	0.25	231	231
08 Nov.	20:30	21.5	83.5	14.0	13	26	0.94	-6.6	-9.2	19.2	-14.2	0.00	1.75		
14 Nov.	02:30	9.5	39.5	1.0	32	31	1.30	-0.6	-9.2	17.1	35.6	0.25	1.25		
16 Nov.	04:30	15.5	9.0	1.0	23	334	1.60	-6.7	-4.9	26.0	31.9	0.00	0.00		
17 Nov.	06:30	29.0	2.5	29.5	22	63	1.97	-5.5	-7.0	22.2	-18.2	0.00	0.00		
20 Nov.	17:00	3.5	25.5		21		0.41	-9.0		38.9		0.00	0.00		
03 Dec.	02:30	21.0	86.5	0.5	6	397	0.25	-2.9	-6.3	14.2	4.4	0.50	2.00	170	257
07 Dec.	20:30	10.5	8.0	2.0	5	19	0.06	-8.5	-11.6	6.9	8.8	0.00	0.00	48	52
20 Dec.	05:00	8.0	22.0	14.5	8	7	0.10	-6.4	-4.2	13.9	-34.9	0.00	1.00	48	71
22 Dec.	08:30	4.0	283.0	28.5	6	4	0.23	-7.1	-23.5	29.7	-11.1	0.25	4.00	26	90
13 Jan.	06:00	343.5		16.0	1	15	0.35	-26.0	-12.2	15.7	-17.3	0.00	0.00	18	19
21 Feb.	02:30	5.0	196.0	24.5	6	0	0.08	-5.3	-18.9	25.0	12.3	0.00	0.00	67	84
19 March	21:00	9.5	23.5	7.0	12	11	0.13	-9.0	-2.9	12.8	-10.7	0.00	0.00	83	86
25 March	07:30	12.5	11.5	3.5	5	27	0.07	-4.4	-4.8	11.2	9.0	0.00	0.00	63	67
Average		30.0	55.6	11.1	10.5	58.8	0.44	-6.8	-7.7	19.9	16.6	0.16	0.68	67.3	85.7

melting can occur at the external part of the ice accretion causing an ice mass reduction, and at the interface of ice with structure causing ice shedding. In the latter case, pieces of ice unstick from the structure and the shedding rate is generally larger than the rate observed for melting at the external layer of the accretion (Druez et al., 1990). Ice mass reduction and shedding by melting are usually observed with precipitation icing because the temperature is close to the melting point. Ice subblimation is a very low ice mass reduction process, typically observed during the coldest and dryest months of winter seasons (Druez et al., 1995; Ryerson and Kenyon, 1996). Ice erosion is a mechanical ice mass reduction process under the effect of wind. This is a type of mechanical failure at the external part of the accretion, mainly observed with the weakest types of atmospheric ice such as soft rime and snow. Two other modes of mechanical failure observed are the adhesive failure at the interface with structure, and the cohesive failure inside the ice accretion, both under combined effect of wind, gravity and cable movements (Druez et al., 1990). The corresponding shedding rates are generally smaller than for shedding due to ice melting.

ICING DATA

The most important icing events observed during the 1993-1994 season are presented in Figures 2-4. These figures correspond respectively to October, November and December 1993. This icing season was very distinctive because the larger icing events were regrouped during the first part of winter. During the second part of winter, beginning in January 1994, the weather became very cold and dry, and only very small and scarce ice accretions occurred. Figures 2-4 present the variation with time of the ice load on the test cables, expressed in kilogram per meter. Ice loads are shown for the four different types of conductors of the two test lines. For the bundle, the total ice mass has been equally divided between each of the four cables, and the ice mass is given for one cable only. The last graph in each figure corresponds to the 7.93-mm reference cable of the second test line. This cable however is nearly perpendicular to the other conductors, and hence parallel to the dominant wind direction.

The corresponding curves for the most interesting meteorological parameters are also given in the figures. These parameters are respectively the air temperature, the ice detector cumulative number of alarms, the equivalent cumulative precipitation in millimeters of water, and the averaged wind velocity component perpendicular to the main test line. The time scale on the figures refers to the begining of each day at midnight. All the scales in Figures 2 - 4 are uniform to allow direct comparison of the data presented. Wind velocities coming from the West-North-West sector are noted positive, and negative when coming from the East-South-East sector.

Table 1 presents values of typical icing and meteorological parameters calculated for the thirteen icing events of the season related to precipitation. The classification was made using mainly the level of precipitation and the air temperature. Table 2 gives the same data for the twenty in-cloud icing events. Due to space limitation, only the most significant icing data are presented in the tables. Each icing event is identified by the date and the beginning time. For each icing event and each cable, the data presented are the duration of each phase, the icing and shedding rates, the maximum ice load, the mean temperature during accretion and shedding, the mean wind velocity normal to the main test line during accretion and shedding, the precipitation and the ice detector number of alarms during the accretion phase only and during all the event.

The experimental installation is used most of the year in a very severe climatic environment, so it is possible to observe in Figures 2 and 3 some malfunctioning instruments. Problems generally take place at the beginning of the measurement season, when the temperature fluctuates around the freezing point. In order to avoid damage due to lightning and water infiltration, instruments are removed before summer. Other critical parts however, such as connecting devices and cables, must remain in place. Even if they are protected inside casing and pipes, they occasionally experienced some failures during summer, which are impossible to detect before the begining of the data acquisition process. Such case is apparent in Figure 2, for the bundle cable, during the period from October 1st to 3rd. Water infiltration caused instability of the signal during the period where the temperature was above the melting point. Also, on October 6, the data transmission was interrupted for the temperature and the wind velocity during one and half hour. More serious, the ice detector was malfunctioning during the period of October 9 to 15. In fact, the ice detector sended very few alarms during this period while the test cables accreted large amounts of ice. The vibrating probe of the ice detector was probably recovered with an ice cap obstructing new ice accretion on the probe. Since the ice melted during the period of October 15 to 18, the ice detector was operating again for the next icing event on October 21. Another breakdown arised on the ice detector on November 7 and was caused by a defective power supply. These two breakdowns of the ice detector were probably related and the power supply must be replaced on November 22.

ANALYSIS OF RESULTS

The four types of conductor are quite different in diameter, rigidity, height from the ground and orientation, and hence the ice loads vary on each cable as shown in Figures 2-4. However, the icing and shedding periods are very similar for the different test cables, except for the instrument breakdowns previously presented. The only significant difference is observed in Figure 3 for the 12.5-mm diameter cable. This cable is the only one to shed a large amount of ice, very rapidly, the 16 of November. With 2.21 kg/m of ice shedded, this shedding event was not due to ice melting because the temperature was about -10°C. This shedding can be explained by the impact of an ice chunk falling from a large antenna located close to the test line. A large amount of ice was accreted on the structures at this time and strong winds, after changing direction, were coming from the antenna location. The 12.5-mm cable is located above the others at the top of the main test line, and more exposed to the impact of falling ice chunks. This explanation is reinforced by the observations made during several icing seasons showing that this phenomena is extremely rare. The video camera however, settled at one frame every four seconds, did not recorded any impact on the cable, but many falling ice pieces are visible at this time.

The accretion phase duration is, on the average, 13.0 hours for precipitation icing and 33.2 hours for in-cloud icing. For the shedding phase, the values are very close, with respectively 10.2 hours and 12.0 hours. The longest accretion phase was 343.5 hours, beginning on January 13, and observed during in-cloud icing. The temperature was very cold, with -26°C on the average during this period. The longest persistence phase, 284.5 hours, was observed on the bundle conductor, during the December 22 in-cloud icing event. Another very long persistence phase, 239.5 hours on the 12.5-mm cable, was observed during the precipitation event of January 28. This phenomenon is very unusual because the temperature is generally close to the melting point during precipitation events. This time however, the temperature dropped very quickly from -2.8°C to -31°C in 24 hours, just after the accretion phase. The temperature remained very cold until February

18, preventing ice shedding. The shortest shedding periods for precipitation icing, 1 hour, correspond to ice shedding due to melting. The corresponding shedding rate was maximum on the 12.5-mm cable with 384 g/m.h. For in-cloud icing, shedding periods lasting 1 hour were also observed on November 14 and 16. On the 12.5-mm cable, the shedding rate was considerable with 2213 g/m averaged on one hour but mainly concentrated on a 30-minute data point. This very large value has been explained by the accidental impact of an ice chunk falling from the nearby communication tower.

On the average, the mean icing rate was larger, in decreasing order, on the 12.5-mm cable, the 35-mm cable, one 35-mm bundle cable and the 7.93-mm perpendicular cable. This was observed both for precipitation icing and in-cloud icing. This result confirms previous observations (Druez and McComber, 1991) and can be explained by the different torsional rigidities of cables. The torsional rigidity strongly reduces with cable diameter, increasing twisting for smaller cables during the accretion of an eccentric load. In fact, the rotation increases the effective cross-section of the iced cable and hence the icing rate. Icing rates however are larger for precipitation icing than for in-cloud icing, due to larger water droplets and larger liquid water content of air. For the shedding rate, values are generally smaller for in-cloud icing than for precipitation icing, except for accidental shedding episodes induced by impact of ice pieces.

The maximum ice load, observed the 18 of November on the 35-mm single cable, was 3.51 kg/m. However, the 12.5-mm cable was accreting more ice until the large accidental shedding episode of November 16 on this cable. The air temperature, averaged on the accretion phases, was -2.7°C during precipitation icing and -6.9°C during in-cloud icing. During the shedding phases, values were closer, with respectively -5.8°C and -7.4°C.

In Tables 1 and 2, the mean perpendicular wind velocity is calculated from the mean speed and the mean direction over ten minute periods. The sign indicates the direction with respect to the main test line: positive from WNW sector and negative from ESE sector. When averaged for each type of cable, the modulus is calculated using the absolute values and the sign obtained from the algebric values. That way the averaged wind velocity is representative of the mean perpendicular velocity during accretion or shedding, and the sign indicates the predominant wind direction during accretion or shedding. With the exception of the 7.93-mm cable perpendicular to the main test line, for precipitation icing the wind velocity is generally negative (from ESE) during accretion, and positive (from WNW) during shedding. The opposite is true for in-cloud icing. This shows that a change in the wind direction after an accretion facilitates ice shedding, probably because of the aerodynamical shape of the ice accretion in the direction of ice accretion.

From Table 1 and for precipitation icing, the equivalent level of precipitation, averaged over the complete events, is 4.68 millimeters of water. For in-cloud icing, the averaged level of precipitation is only 0.75 mm, as shown in Table 2. Concerning the number of alarms of the ice detector, the breakdowns observed on this instrument prevent a valid comparison.

Figure 5 presents for example the distribution of the ice mass variation on the 12.5-mm cable as a function of the ice detector number of alarms, for the icing period of December 3 to 13. The ice detector controller had been changed on November 22 and the instrument seemed to work properly since. Results are shown on the basis of one hour periods. Five types of data points are distinguished on the figure using different symbols. Triangles are used when the ice mass variation is negative and the number of alarms zero. This corresponds to shedding episodes. Circles are used when the ice mass variation is negative, but some ice detector alarms are observed. This indicates global shedding during relatively small accretions. In this case, ice shedding could be caused by mechanical failure of ice under the effects of wind velocity and cable vibration. Squares are used when there is no ice mass variation but ice detector alarms. This can be explained by some shedding combined with equivalent ice accretion. Depending on the icing rate, the equivalent shedding rate could be caused by mechanical failure of the ice, or ice erosion due to the wind velocity. Diamond-shaped points are used when the ice mass variation is positive and alarms are sent by the ice detector. This corresponds to an accretion episode. However, the ice detector sends an alarm only when a given amount of ice is accreted on the probe. So it is normal to observe small ice accretions on cable without any alarm of the detector. The corresponding points must be considered in a correlation between the ice mass accreted on the cable and the number of alarms of the detector. But if the ice accretion on the cable is very large and the number of alarms zero, breakdown of the ice detector must be suspected, such as jamming of the vibrating probe when too much ice is accreted, failure of the de-icing system, or insufficient de-icing power. Such points, represented in Figure 5 using a cross, should not be considered in the correlation previously described. This could be a convenient approach to detect ice detector failure, but the difficulty is to determine an applicable limiting value. What was done in Figure 5 was to calculate, for positive points corresponding to zero alarm, the mean corresponding ice mass on the cable, and the associated standard deviation. The above limiting value was chosen as the mean value augmented by three times the standard deviation in order to use most of the significant data points. It should be noted that the ice mass variation on the cable is a discrete mathematical distribution due to the digital data acquisition process, and that a given point on Figure 5 can represent more than a data point. The regression process has been made using all the data points corresponding to the diamond-shape symbols. The resulting line and the corresponding equation are shown in Figure 5.

CONCLUSIONS

1. Large ice loads over 1 kg/m on cables were observed both with precipitation icing and in-cloud icing.

2. The persistence of ice on a cable can last for several weeks and therefore could result in dangerous cumulative ice loads combined with large wind loads.

3. Precipitation icing and in-cloud icing can be distinguished in a data base using temperature, heated rain gauge and ice detector signals.

4. The accretion phase duration has been, on the average, 13.0 hours for precipitation icing and 32.2 hours for in-cloud icing. The corresponding mean temperatures were respectively -2.7°C and - 6.9°C.

5. For the three conductors of the main test line, and for precipitation icing, the normal wind velocity was generally from the East-South-East sector during accretion, and from the West-North-West sector during shedding. The opposite was true for in-cloud icing.

6. As an example, the correlation between the mass of ice accreted on the 12.5-mm cable and the number of alarms of the ice detector has been established using linear regression, for the icing period of December 3 to 13, 1993.

ACKNOWLEDGMENTS

The authors gratefully acknowledge the financial support from the Natural Sciences and Engineering Research Council of Canada, the Université du Québec à Chicoutimi Foundation and Bell Canada. They also would like to thank MM. Marco Boudreault and Serge Gauthier for their participation to this project, as well as Hydro-Québec and Radio-Québec for collaborating on the test site.

REFERENCES

Druez, J., McComber, P., Laflamme, J. (1996). "An Analysis of Selected Atmospheric Icing Events on Test Cables", *Proc 15th Int Conf on Offshore Mechanics and Arctic Engineering*, Florence, Italy, Vol IV, pp 1-9.

Druez, J., Louchez, S., McComber, P. (1995). "Ice Shedding from Cables", *Cold Regions Science and Technology Journal*, Vol 23, pp 377-388.

Druez, J., McComber, P., Laflamme, J. (1995). "A Comparison of Glaze and In-Cloud Icing Events on a Test Line", *Proc 14th Int Conf on Offshore Mechanics and Arctic Engineering*, Copenhagen, Denmark, Vol IV, pp 93-100.

Druez, J., McComber, P., Laflamme, J. (1994). "Field Data on Transmission Line Icing Collected at the Mt. Valin Test Site", *Proc 13th Int Conf on Offshore Mechanics and Arctic Engineering*, Houston (Texas), U.S.A., Vol IV, pp 51-56.

Druez, J., McComber, P. (1991). "A Comparison of the Atmospheric Icing of two Different Size Cables", *Proc of 1st Int Conf on Offshore and Polar Engineering*, Edinburgh, U.K., pp 528-534.

Druez, J., Laflamme, J., McComber, P. (1991). "Hydro-Québec Tests Cables and Insulators under Heavy Ice Loads at Mt. Valin", *Electricity Today*, vol 3, No 8, pp 22-23.

Druez, J., Louchez, S., Bouchard, G. (1990). "Study of Ice Shedding Phenomenon on Cables", *Proc 9th Int Conf on Offshore Mechanics and Arctic Engineering*, Houston (Texas), U.S.A., Vol IV, pp 143-148.

Druez, J. and McComber, P. (1988). "Atmospheric Icing Measurements on a Test Line in a Mountain Icing Site", *Proc IAHR Ice Symposium*, Sapporo, Japan, Vol 2, pp 477-485.

Ervik, M. and Fikke S. (1996). "IEC TC11 WG11 - Report on Meteorological Data for Assessing Climatic Loads", *Proc 7th Int Workshop on Atmospheric Icing of Structures*, Chicoutimi, Canada, pp 101-106.

Goia, L.M. and Balan G. (1996). "Romanian Experience Regarding Operational Behavior of HV OLs", *Proc 7th Int Workshop on Atmospheric Icing of Structures*, Chicoutimi, Canada, pp 216-221.

Hardy, C., Brunelle, J., Chevalier, J., Manoukian, B., Vilandré, R. (1996). "Telemonitoring of Climatic Loads on Hydro-Québec 735 KV Lines", *Proc 7th Int Workshop on Atmospheric Icing of Structures*, Chicoutimi, Canada, pp 169-174.

Laflamme, J. and Périard, G. (1996). "The Climate of Freezing Rain Over the Province of Québec in Canada, A Preliminary Analysis", *Proc 7th Int Workshop on Atmospheric Icing of Structures*, Chicoutimi, Canada, pp 19-24.

Lott, J.N. (1996). "The February 1994 Ice Storm in the Southeastern U.S.", *Proc 7th Int Workshop on Atmospheric Icing of Structures*, Chicoutimi, Canada, pp 259-264.

Lozowski, E.P., Blackmore, R.Z., Forest, T.W., Shi, J. (1996). "Spongy Ice in the Marine Environment", *Proc 15th Int Conf on Offshore Mechanics and Arctic Engineering*, Florence, Italy, Vol IV, pp 55-61.

Makkonen, L., Editor (1994). Ice and Construction. *RILEM. E and FN Spon*, London, UK.

McComber, P., Druez, J., Laflamme, J. (1995). "A Comparison of Selected Models for Estimating Cable Icing", *Atmospheric Research*, Vol 36, pp 207-220.

Raison, T., Albertin, J.C., Admirat, P. (1996). "A National Climatic Reference Base for Exceptional Wet Snow and Rime Ice Episodes", *Proc 7th Int Workshop on Atmospheric Icing of Structures*, Chicoutimi, Canada, pp 8-14.

Ryerson, C. and Kenyon, P. (1996). "Atmospheric Ice Ablation Process On Mt. Equinox, Vermont, U.S.A.", *Proc 7th Int Workshop on Atmospheric Icing of Structures*, Chicoutimi, Canada, pp 277-281.

Shan, L., Marr, L., McCafferty, R.M. (1996). "Ice Storm Data Base and Ice Severity Map", *Proc 7th Int Workshop on Atmospheric Icing of Structures*, Chicoutimi, Canada, pp 3-7.

Vuckovic, Z., Plazinic, S., Nikolic I. (1996). "Failures of Overhead Lines Due to Ice and Wet Snow in a Part of Balkan Peninsula (Serbia)", *Proc 7th Int Workshop on Atmospheric Icing of Structures*, Chicoutimi, Canada, pp 210-215.

Proceedings of the Seventh (1997) International Offshore and Polar Engineering Conference
Honolulu, USA, May 25-30, 1997

ISBN 1-880653-28-1 (Set); ISBN 1-880653-30-3 (Vol. II)

Dynamic Curvature in Catenary Risers at the Touch Down Point: An Experimental Study and the Analytical Boundary-Layer Solution

C.P. Pesce, J.A.P. Aranha and C.A. Martins
Escola Politécnica, University of São Paulo, Brazil

O.G.S. Ricardo and S. Silva
Instituto to de Pesquisas Tecnológicas do Estado de São Paulo, Brazil

ABSTRACT

Recently, the subcritical dynamics[1] of catenary risers in the vicinity of the Touch Down Point (TDP) has been studied in depth (Aranha, Martins & Pesce, 1995; Pesce & Pinto, 1996), by means of asymptotic methods and boundary layer theory. The major outcome of those previous studies is a simple mathematical expression that enables one to relate, in time, the dynamic curvature and TDP excursion when an oscillatory tension is applied to the riser. Such a dynamic asymptotic boundary-layer solution, though simple, contains all the non-linear features that result from the non-linear geometric boundary condition at the soil, of the 'unilateral type'. More recently, an experimental study has been conducted with a structural model, at IPT laboratory facilities. The riser model has been instrumented along the length by an array of strain-gages, of the resistive type, the static and the periodic dynamic problems being investigated in detail. Comparison with the analytical asymptotic (boundary-layer) solution has been performed showing an outstanding agreement. Not only time-average, rms and maximum amplitude variation of curvature over the riser length are recovered, but also are the corresponding curvature time-histories along the riser's span, within the boundary layer, validating the asymptotic solution in full.

KEYWORDS: catenary risers, dynamic curvature, experimental validation, asymptotic solution.

INTRODUCTION

Recent research work on the technical feasibility of the steel catenary riser alternative applied to the oil industry, see Phifer et al, 1994, gave rise to a number of questions concerning the local dynamics in the Touch Down Point (TDP) region. The precise evaluation of dynamic curvature in this region is mandatory, as fatigue life calculation is a crucial point in design. Driven by waves and by the floating unit, the riser undergoes oscillatory motions, causing TDP position to vary in time. In a purely two-dimensional problem, we can think of an "instantaneous TDP position" in close analogy to the instantaneous center of rotation in the kinematics of a rolling cylinder. Physically, the oscillatory variation of the TDP makes the riser to "lift off" and "lay down" cyclically on the sea floor, causing large variations of curvature in this region.

In fact, as flexural rigidity is rather small, considering the length of the riser, *bending stiffness plays a minor role* in the *global dynamics*, dominated by the geometric (catenary) rigidity. Close to the ends, particularly to the TDP, however, the flexural rigidity effect becomes important. This is an example of the so called "beam-string" problem; see, eg. Love, 1927, art. 273A, constituting itself into a singular perturbation problem, being generally and properly treated under the boundary-layer technique approach; see Kevorkian & Cole, 1981, art. 2.8, page 97. As the riser lays down on the floor repeatedly, the curvature, and so the bending stresses, vary from a very small value (exactly null if the soil is supposed to be rigid) to a somewhat relatively large one, of order of the catenary curvature. As well known, catenary curvature is discontinuous at TDP, since only the tangency boundary condition is satisfied, being lost the null curvature condition. This condition can only be recovered by introducing the bending stiffness effect.

The locally two-dimensional sub-critical[2] dynamic regime problem can then be addressed by a boundary-layer type technique, giving rise to an analytical asymptotic (non-linear) solution for the curvature problem in the TDP region (Aranha, Martins & Pesce, 1995). As a matter of fact, besides being remarkably simple in form, this analytical solution is only dependent on the dynamic tension and on the TDP instantaneous displacement, that can be easily evaluated from the catenary solution[3],

1 Shock condition not reached.

[2] (here defined as the absence of shock against the soil)

[3] We refer to the equations of a heavy inextensible cable under the action of steady current and dynamic loads due to waves and floating unit motions.

and can be used even in a random environmental condition, if small displacements around the static equilibrium configuration are considered, in a standard perturbation scheme. The local non-linearities are implied in the TDP oscillatory motion, and the curvature is written as a non-linear time-function of tension and TDP displacement. Making use of this boundary-layer solution, a number of qualitative and quantitative results could be obtained, as those presented in Pesce & Pinto, 1996. Its is possible, for example, to analytically determine the section corresponding to the maximum curvature variation, in a given environment condition. Fatigue evaluation becomes more tractable. And, as ever, an analytical solution provides insight to the designer in a rather rational manner.

Very recently, an experimental research program, aiming to deeply investigate the non-linear dynamics in the TDP region, has been conducted. A structural model was instrumented with an array of strain-gages, and static and dynamic experiments were performed. The static and dynamic experiments results confirmed, qualitative and quantitatively, the major features predicted by the boundary-layer model.

This work recovers the boundary-layer solution and presents some few comparisons with the dynamic experimental results. The non-linear boundary-layer solution is verified in space and time through a simple analysis procedure. Time-histories are shown to be in an outstanding agreement. RMS, peak-to-peak and average values calculated from the experiment and by the boundary-layer solution have been verified, along the structural model, within a very good agreement, as well.

THE TWO-DIMENSIONAL PROBLEM

The rationale for the theoretical results presented below have been essentially presented in Aranha, Pesce & Martins, 1995, but are here rederived and reinterpreted.

We shall restrict ourselves to the two-dimensional 'catenary' riser problem, dynamically formulated, under the usual small perturbation theory, around the static equilibrium configuration. The riser is suspended to a floating unit and hangs freely down to the sea-bed, under the action of current and waves. Boundary conditions at the upper end are supposed to be specified properly, considering the offset and motions of the floating unit and, possibly, end-fit effects. At the *touch down point (TDP)* the boundary condition is of the one-side type. For simplicity, soil is supposed to be rigid, but flexibility can be easily incorporated to the mathematical model. Figure 1 shows the considered problem.

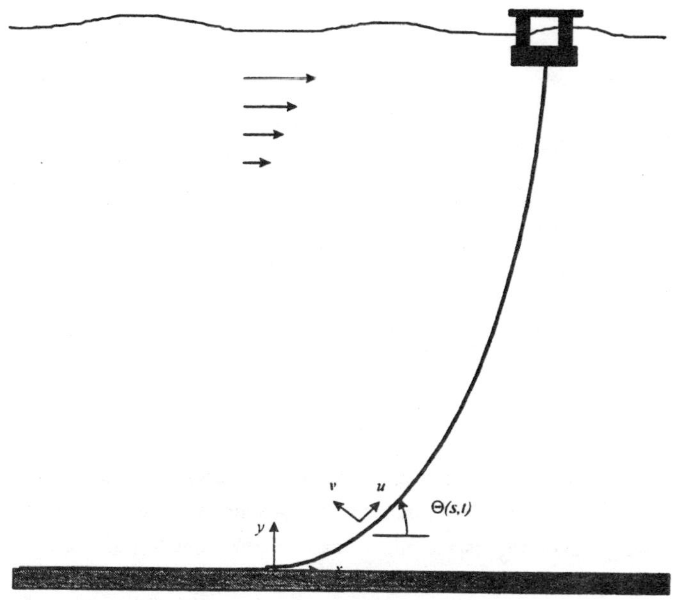

Figure 1 The two-dimensional 'Catenary" Riser problem

Governing Equations

Let $u(s,t)$, $v(s,t)$ be the axial and normal displacements along the arc-length, measured, for convenience, from the (catenary[4]) touch down point. Let

$$\Theta(s,t) = \theta(s) + \gamma(s,t)$$
$$\mathrm{T}(s,t) = T(s) + \tau(s,t) \qquad (1),$$

be, respectively, the total angle with respect to the horizontal and the effective tension along the line, being $\gamma(s,t)$ and $\tau(s,t)$ the corresponding dynamic perturbation around the static solution, the latter given by the functions $\theta(s)$, $T(s)$. The dynamic tension will be just enforced to satisfy $|\tau(s,t)|/T(s) < 1$, precluding dynamic compression. The dynamic perturbation $\gamma(s,t)$ is supposed to be much smaller than unit such that, up to third-order,

$$\frac{\partial v}{\partial s} = \gamma(s,t) \qquad (2).$$

Accordingly, let also

$$\mathrm{Q}(s,t) = Q(s) + \vartheta(s,t) \qquad (3)$$

be the shear force. Under standard Euler-beam theory, neglecting the rotatory inertia of the section, the following governing equations are easily derived.

[4] As if the corresponding catenary solution were determined

$$\frac{\partial T}{\partial s} - Q\frac{d\theta}{ds} - \frac{\partial}{\partial s}\left(Q\frac{\partial v}{\partial s}\right) + h_u + \varpi_u - q\sin\theta = m\frac{\partial^2 u}{\partial t^2}$$
$$\frac{\partial Q}{\partial s} + T\frac{d\theta}{ds} + \frac{\partial}{\partial s}\left(T\frac{\partial v}{\partial s}\right) + h_v + \varpi_v - q\cos\theta = (m + m_a)\frac{\partial^2 v}{\partial t^2} \quad (4).$$
$$\frac{\partial M}{\partial s} + Q \cong 0$$

Where (h_u, h_v) and (ϖ_u, ϖ_v) are, respectively, the static and dynamic parcels of the viscous hydrodynamic forces, q is the immersed weight and m_a is the added mass, all per unit length. Notice that equations (4) can be rewritten as

$$\left\{\frac{\partial T}{\partial s} - Q\frac{d\theta}{ds} - q\sin\theta + h_u\right\} + \left\{\frac{\partial \tau}{\partial s} - \vartheta\frac{d\theta}{ds} - \frac{\partial}{\partial s}\left(Q\frac{\partial v}{\partial s}\right) + \varpi_u\right\} =$$
$$= m\frac{\partial^2 u}{\partial t^2}$$
$$\left\{\frac{\partial Q}{\partial s} + T\frac{d\theta}{ds} - q\cos\theta + h_v\right\} + \left\{\frac{\partial \vartheta}{\partial s} + \tau\frac{d\theta}{ds} + \frac{\partial}{\partial s}\left(T\frac{\partial v}{\partial s}\right) + \varpi_v\right\} =$$
$$= (m + m_a)\frac{\partial^2 v}{\partial t^2}$$
$$\left(\frac{\partial M}{\partial s} + Q\right) + \left(\frac{\partial \mu}{\partial s} + \vartheta\right) \cong 0$$
(5)

the first brackets being identically null, since they correspond to the static equilibrium equations. This leads to,

$$\frac{\partial \tau}{\partial s} - \vartheta\frac{d\theta}{ds} - \frac{\partial}{\partial s}\left(Q\frac{\partial v}{\partial s}\right) + \varpi_u = m\frac{\partial^2 u}{\partial t^2}$$
$$\frac{\partial \vartheta}{\partial s} + \tau\frac{d\theta}{ds} + \frac{\partial}{\partial s}\left(T\frac{\partial v}{\partial s}\right) + \varpi_v = m\frac{\partial^2 v}{\partial t^2} \quad (6).$$
$$\frac{\partial \mu}{\partial s} + \vartheta \cong 0$$

If the classical constitutive relationship in the theory of the strength of materials, $M = B\chi$, is used, where $\chi(s,t) = \partial\Theta/\partial s$ is the (total) curvature, B the flexural rigidity, supposed constant, equations (4b) and (4c) can be combined to give

$$-B\frac{\partial^2 \chi}{\partial s^2} + T\chi + \frac{\partial T}{\partial s}\frac{\partial v}{\partial s} + h_v + \varpi_v - q\cos\theta = (m + m_a)\frac{\partial^2 v}{\partial t^2}$$
(7).

If flexural rigidity were totally neglected and only the static problem considered, from (5b) we would get the static curvature at the TDP ($\theta = 0$).

$$\chi_0 = \left.\frac{d\theta}{ds}\right|_{TDP} = \frac{q}{T_0} \quad \textbf{(8)}$$

being T_0 the static tension at TDP.

As mentioned before both the global static and dynamic problems can be treated under the classical 'beam-string' approach, see for instance Kevorkian & Cole, 1981, art. 2.8, or Love, 1934, art. 273A. Let T_0 be the typical effective tension scale. The small non-dimensional *flexural length* parameter $\varepsilon = \lambda/L = \sqrt{B/T_0}/L$ gauges the influence of the flexural rigidity in the overall statics and dynamics. In fact, equation (7) can be put in the non-dimensional form,

$$-\varepsilon^2\frac{\partial^2 \hat{\chi}}{\partial \hat{s}^2} + \hat{T}\hat{\chi} + \frac{\partial \hat{T}}{\partial \hat{s}}\frac{\partial \hat{v}}{\partial \hat{s}} + \hat{h}_v + \hat{\varpi}_v - \hat{\chi}_0\cos\theta = \frac{\partial^2 \hat{v}}{\partial \hat{t}^2} \quad (9),$$

where

$$\hat{t} = \frac{L}{c_0}t$$
$$c_0 = \sqrt{\frac{T_0}{m + m_a}} \quad (10)$$

are, respectively, the non-dimensional time and the cable wave velocity at TDP.

This is a singular perturbation problem and can be treated properly by means of boundary-layer technique (Kevorkian & Cole, 1981). For the static problem there exist two boundary-layers where flexural rigidity effect is important: the first one, at the TDP, with typical size λ; the second one close to the top, with a typical size given by $\lambda_T = \sqrt{B/T(L)}$. For the dynamic problem, flexural rigidity can be neglected along the riser span whenever the local flexure-length is much smaller than the typical semi-wave length-scale Λ_k ($\lambda/\Lambda_k \ll 1$), corresponding to the k-eigenmode, determined from the linearised equation.

So, the static *global* problem is dominated by the geometric rigidity, i.e., can be solved from the ideal cable equation. The same is true for the dynamic *global* problem, provided the condition above is satisfied. At both ends, boundary-layer technique is applied and solution can be matched in a standard way.

THE ASYMPTOTIC MODEL

We now focus our attention to TDP vicinity. Local analysis is performed and, assuming a sub-critical dynamic regime, boundary-layer technique is applied in order to construct an asymptotic expression for the curvature, valid inside the TDP boundary-layer.

The Sub-critical Dynamic Regime

Let $x_0(t)$ be the TDP instantaneous position. As presented in Aranha, Martins & Pesce, 1995, following Triantafyllou, 1985, it can be proved, for the ideal cable dynamic problem, that

$$\left(1 - M^2\right)\left.\frac{\partial y_c}{\partial x}\right|_{x_0(t)} = -\frac{F(t)}{T_0} \quad (11).$$

where $y_c(x,t)$ locally defines the '*catenary*'. $F(t)$ is a possible impact force provided by a shock against the soil and $M = \dot{x}_0/c_0$ is the local 'Mach' number. As, if only one TDP exists, $\partial y_c/\partial x|_{x_0(t)} \geq 0$, and as impact force, if it exists, is defined positive, then shock condition implies $M > 1$. Otherwise, $\partial y_c/\partial x|_{x_0(t)} \equiv 0$ and $M < 1$. We define here a *dynamic sub-critical regime* whenever $M < 1$. It can also be shown (Aranha, Martins & Pesce, 1995) that

$$\left.\frac{\partial y_c}{\partial x}\right|_{x_0(t)} \equiv 0; \quad M < 1 \qquad (12)$$

and

$$\left.\frac{\partial^2 y_c}{\partial t^2}\right|_{x_0(t)} = \frac{\dot{x}_0^2(t)}{\rho_c(x_0(t))} \cong \dot{x}_0^2(t)\left.\frac{\partial^2 y_c}{\partial x^2}\right|_{x_0(t)} \qquad (13)^5,$$

where $\rho_c(x_0(t)) = 1/\chi_c(x_0(t))$ is the curvature radius at the *instantaneous* TDP of the cable, leading to

$$\frac{y_c(0,t)}{\rho_{0c}} \cong \frac{x_0^2(t)}{\rho_{0c}}\frac{\partial^2 y_c}{\partial x^2}(0,t) \cong O(a_0\chi_0)^2 \qquad (14).$$

where a_0 is the typical amplitude of TDP motion and $x = 0$ is the static TDP position. In other words, for small TDP excursions, the static TDP can be treated as an articulation, in the ideal cable dynamics. However, the TDP excursion can be evaluated, from the ideal articulated cable dynamic solution as,

$$\frac{x_0(t)}{\rho_{0c}} = -\frac{\partial y_c}{\partial x}(0,t) = -\gamma(0,t) \qquad (15).$$

The Boundary-layer Solution in the *Sub-critical Regime*

In the vicinity of TDP, where $\theta << 1$, the hydrodynamic viscous forces can be neglected, correct to order θ^2. Accordingly, for the static tension we have $T(s) = T_0(1 + O(\theta^2)), s/\lambda << 1$. On the other hand, it was shown by Aranha et al., 1993, that the dynamic tension is weakly varying along the arc length, so that we can write,

$$T(s,t) \cong T_0 + \tau(t); \quad s/\lambda << 1 \qquad (16).$$

Equation (7) can then be written

$$-\frac{B}{T_0}\frac{\partial^2 \chi}{\partial s^2} + \left(1 + \frac{\tau(t)}{T_0}\right)\chi - \frac{q}{T_0} = \frac{(m+m_a)}{T_0}\frac{\partial^2 v}{\partial t^2} \qquad (17).$$

or,

$$-\lambda^2\frac{\partial^2 \chi}{\partial s^2} + \left(1 + \frac{\tau(t)}{T_0}\right)\chi - \chi_0 = \frac{1}{c_0^2}\frac{\partial^2 v}{\partial t^2} \qquad (18).$$

[5] Remarkable is the similarity with the acceleration of the instantaneous center of rotation, in the classical kinematics of a rigid disk rolling, without slipping, over a flat plane.

But, from (13), $\frac{\partial^2 v}{\partial t^2} \approx \dot{x}_0^2(t)\chi_0$, only valid in the TDP boundary-layer, and then,

$$-\lambda^2\frac{\partial^2 \chi}{\partial s^2} + \left(1 + \frac{\tau(t)}{T_0}\right)\chi \cong \chi_0(1 + M^2) \qquad (19).$$

In the *subcritical dynamic regime*, with an error of order $M^2 << 1$, *the inertia term can be locally disregarded*, and the curvature governing equation, in the TDP vicinity can be written as,

$$-\lambda^2\frac{\partial^2 \chi}{\partial s^2} + \left(1 + \frac{\tau(t)}{T_0}\right)\chi = \chi_0; \quad M^2 << 1 \qquad (20).$$

That is, inside the TDP boundary-layer, the curvature response, in the sub-critical regime, can be said to be governed by a quasi-static equation. Notice that all global static information, including current effect, is contained in the static parameter T_0 (so in $\chi_0 = q/T_0$). On the other hand, all global dynamic information is locally (inside the boundary-layer) represented by the functions $\tau(t)$ and $x_0(t)$. Moreover, under small perturbation (displacements) approach, the *global dynamics can be solved in the frequency domain, using the ideal cable equations* (neglecting shear efforts in equations(6a,b)), pinned at both ends.

The general solution for equation (20) is easily found as,

$$\chi(s,t) = \begin{cases} \dfrac{\chi_0}{1+\tau(t)/T_0} + C_1(t)e^{-\left(\sqrt{1+\tau(t)/T_0}\right)(s-s_f(t))/\lambda} + \\ + C_2(t)e^{+\left(\sqrt{1+\tau(t)/T_0}\right)(s-s_f(t))/\lambda}; \quad s > s_f \\ 0; \quad s \leq s_f \end{cases} \qquad (21).$$

where $s_f(t) = x_0(t) - \lambda$ gives the TDP instantaneous position if, with no loss of generality, the origin for the coordinate s is chosen as the static TDP for the corresponding 'ideal cable' case. As solution must be finite when $(s - s_f(t))/\lambda \to \infty$, then $C_2(t) \equiv 0$. Also, for rigid soil, $\chi(s_f(t)) = 0$, and then,

$$C_1(t) = -\frac{\chi_0}{1+\tau(t)/T_0} \qquad (22).$$

so that,

$$\frac{\chi(\xi,t)}{\chi_0} = \begin{cases} \dfrac{1}{1+\tau(t)/T_0}\left(1 - e^{-\left(\sqrt{1+\tau(t)/T_0}\right)\beta(\xi,t)}\right); & \beta(\xi,t) > 0 \\ 0; & \beta(\xi,t) \leq 0 \end{cases}$$

where

$$\beta(\xi,t) = \xi - \xi_0(t) + 1$$
$$\xi = s/\lambda$$
$$\xi_0(t) = x_0(t)/\lambda \qquad (23).$$

gives the local (boundary-layer) solution for the curvature in the sub-critical regime, considering the soil to be rigid. The static boundary-layer soution $\chi_e(s)$ is a particular case of (23), when $\tau(t) \equiv 0; \xi_0(t) \equiv 0$, and it is given by

$$\frac{\chi_e(\xi)}{\chi_0} = \begin{cases} \left(1 - e^{-\beta_e(\xi)}\right); & \beta_e(\xi) > 0 \\ 0; & \beta_e(\xi) \le 0 \end{cases}$$

with

$$\beta_e(\xi) = \xi + 1 \qquad (24)^6$$
$$\xi = s/\lambda$$

Equation (24) shows quite directly that the actual static TDP position, when flexural rigidity is taken into account, is given by $\beta_e(\xi) = 0$, or $s_{fe}/\lambda = -1$. In words, equation (24) shows that the actual static TDP position is displaced to the left (see figure 1), with respect to the TDP corresponding to the cable solution, by an amount equal to the local flexural length.

The asymptotic boundary-layer solution is rather powerful indeed, *since no assumptions have been made concerning the harmonic structure of the oscillatory (global) functions* $\tau(t)$ and $x_0(t)$, that are determined from the ideal cable dynamic equations. In fact, once the global dynamic solution is obtained, in the frequency domain, from the linearised dynamic equations, those functions can be easily constructed, for a given load condition (specified by a steady current, wave and floating unit spectra) by means of standard linear stochastic processes theory. Obviously, the global dynamics is supposed to satisfy small perturbation theory, in terms of displacements around the static equilibrium configuration. The linearised solution must be such that $|\tau(t)|/T_o < 1$ always, in order to preclude dynamic compression, what would cause the validity of the present asymptotic (dynamic) boundary-layer solution to break down.

As examples of the present theoretical results, figures 2 and 3 show time-histories for the curvature in the TDP region, normalized with respect to χ_0. We considered two cicles of monochromatic tension and TDP excursion functions, given by

$$\frac{\tau(t)}{T_0} = \frac{\tau_0}{T_0}\cos\omega t$$
$$\xi_0(t) = \frac{x_0(t)}{\lambda} = \frac{a_0}{\lambda}\cos(\omega t + \varphi) \qquad (25).$$

Similar examples have been published in Pesce & Pinto. 1996.

Figure 2 presents curvature, obtained from equation (23). corresponding to three different sections, identified by $\xi = s/\lambda = -0;5,0;+0,5$. and taking $a_0/\lambda = 1; \tau_0/T_0 = 0.2; \varphi = \pi$.

The intermediate section, defined by $\xi = \hat{\xi} = a_0/\lambda - 1$, is the 'critical-section' corresponding to the maximum curvature variation. This section just touches the soil once during a whole period. The first section rests on the soil periodically while the third one never touches it.

Figure 3 shows, for the critical-section, $\xi = \hat{\xi} = a_0/\lambda - 1$, three normalized curvature time-histories, obtained from equation (23) and corresponding to three values of the tension amplitude, $\tau_0/T_0 = 0.1;0.2;0.4$, keeping $a_0/\lambda = 1$ ($\hat{\xi} = 0$).

[6] A similar result can be found for a linerly elastic continuous support. Pesce. 1997.

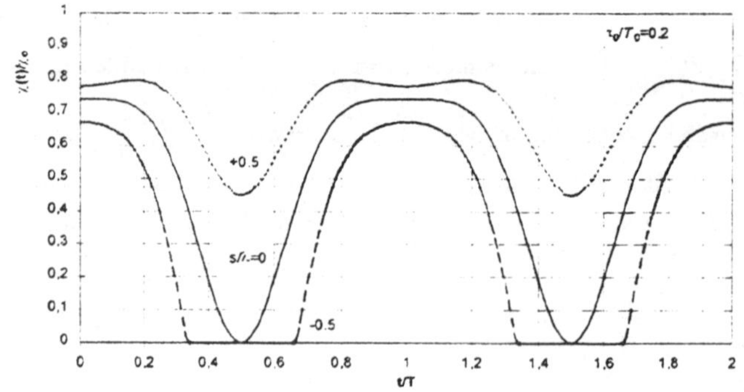

Figure 2. Nondimensional curvature $\chi(t)/\chi_0$, close to TDP. Boundary-layer asymptotic solution in sub-critical dynamic regime. $\tau_0/T_0 = 0{,}2; a_0/\lambda = 1; \varphi = \pi; s/\lambda = -0.5;0;+0.5$

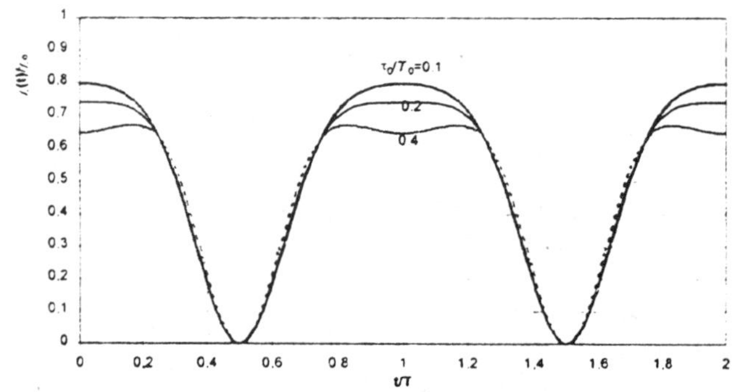

Figure 3. Nondimensional curvature, $\chi(t)/\chi_0$, close to TDP. Boundary-layer asymptotic solution in sub-critical dynamic regime. $\tau_0/T_0 = 0.1;0.2;0.4; a_0/\lambda = 1; \varphi = \pi; s/\lambda = 0$.

THE EXPERIMENTAL WORK

The experimental work was conducted at the São Paulo State Research Technological Institute (IPT), in the Ocean Engineering Laboratory Facilities, within a broader research program supported by PETROBRAS. Two types of experiments were performed: *static* and *dynamic* ones. A number of interesting results could be obtained, but only a few cases concerning the dynamic experiments are here presented. The agreement with the theoretical analytical results has been good enough to assert the model validity, as will be shown. It should be mentioned that the experiments were run in May-June, 1996, confirming a number of results previously predicted by using the boundary-layer solution by Pesce & Pinto. 1996[7].

The Experiment Set-up

A consistent structural modelling has been specially conceived to represent typical cases encountered in steel catenary risers designs. Making appeal to St. Venant's principle, a shortened length of tube

[7] See ISOPE'96. Final manuscript sent in February, 1996.

much larger than the boundary-layer can be chosen. As the theoretical model is generally applicable the structural model was chosen as a truncated round low-carbon laminated steel rod. This rod is about 19 m long (two bars of about 9.5m connected by an internal longitudinal pin and cemented), ½ inches (0.0127m) in diameter. The overall length was specified such that the suspended length was approximately10 times larger than the flexural (boundary-layer) length. This rod was laid on the towing-tank straight steel rail, experiments being conducted in the air[8]. The lower end was held by a longitudinal cable. The upper end was held by another cable, at a heigth of about 2.4m over the rail. The suspended span length was around 10m. Figure 4 shows a simplified view of the experiment set-up. Tangential forces stretched the rod, at will, at both ends, so various 'catenary' shapes could be simulated. These forces were measured by dynamometers. In the dynamic experiment a pin, eccentrically mounted on a pulley driven by an electric motor with constant speed, imposed a circular motion of radius R to the upper end, causing a (quasi-) sinusoidal dynamic tension on the rod. Scaling being under perspective, the following similarity non-dimensional parameters have to be considered,

$$\alpha_b = \frac{qL^3}{B} = \frac{q}{T_0}\frac{T_0}{B}L^3 = \chi_0 L\frac{L^2}{\lambda^2} = \frac{\alpha_c}{\varepsilon^2}$$

$$\alpha_c = \frac{qL}{T_0} = \chi_0 L$$

(26)

where b stands for *bending*, and c for *catenary. As an example*, in the experiment conditions, the rod scales a steel catenary riser whose particulars are shown in table 1.

Fifty sections were instrumented with 10mm common strain-gages (SG), two per section, diametrically opposed. So, the readings were duplicated. The initial (lower) six sections were 0.5m apart; then, four sections 0.3m apart from each other, followed by twenty-three sections 0.075m (3") apart, eleven sections 0.3m apart, and six sections 1.02m apart.

Table 1. Rod particular and scaled steel catenary riser

	Rod	Riser
External diameter (m)	0.0127	0.2731
Internal diameter (m)	-	0.2320
Linear weigth (q) (N/m)	9.807	1250.4
Young modulus (E) (GPa)	206	206
Flexural Rigidity (B) (kNm^2)	0.226	2.12E06

[8] Immersed experiments have been conducted as well, in order to confirm the hypothesis that neglects the local effect of the hydrodynamic forces. Results strongly indicate the consistency of such a hypothesis (IPT Report no. 34.463) and will be presented elsewhere.

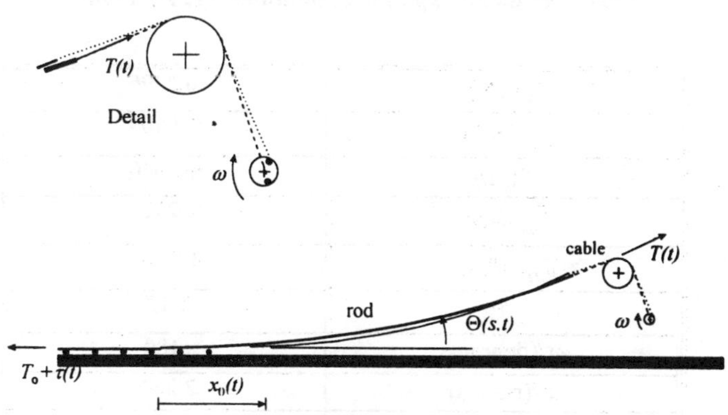

Figure 4. Sketch of the experiment set-up.

In order to avoid measure interference due to the dynamic contact of the SG's and wires with the rail, the rod was placed over small cylindrical wooden rods, disposed tranversally, about 0.5m apart from each other. Some doubts could be risen about the errors introduced by this supports. Neverthless, the readings were considered very much consistent as it will be seen. Consequently, in the final stages of the experiments, the downmost SG's (odd numbered) were taken off, the rod being laid directly over the rail, and some experiments repeated. The results were confirmed (IPT Report no. 34.643).

Dynamic Experiment Conditions

The results here presented refer to *experiment R21b-17/06/96*. The excitation period of motion imposed to the upper end was 2.66 seconds. Notice that three non-dimensional parameters control the local response: $a_0/\lambda; \tau_0/T_0; \mathrm{M}_0 = \omega a_0/c_0$. M_0 scaling the local 'Mach number', sub-critical dynamic regime being implied whenever $\mathrm{M}_0 < 1$. In the static equilibrium configuration, the tension at the upper end was $T_{upper} = 0{,}260\mathrm{kN}$, and the angle with respect to the horizontal $\theta_{upper} = 22.8''$. The corresponding flexural length is $\lambda = 0.964\mathrm{m}$. A (quasi-)sinusoidal dynamic tension resulted, as shown in figure 5. We may either input the actual measured force in the asymptotic model, given by equation (23), or represent the tension as a purely sinusoidal form, $T(t) = T_0 + \tau_0 \cos\omega t$, such that $\tau_0 = \sqrt{2}RMS(\tau(t))$ is determined from the measured signal. This latter choise is here taken. The TDP oscillatory motion is written, accordingly, as $x_0(t) = a_0\cos(\omega t + \varphi)$. The amplitude was observed within an error of $a_0/\lambda < 0.05$. Table 2 presents the main parameters calculated from the acquired data. The local 'Mach' number is about 0.0546, *so that the local sub-critical dynamic regime is assured.* As the suspended length was about 10.8m, the ratio between the imposed and the first natural frequency of the rod can be evaluated as $(\omega/\omega_1)^2 \cong (\omega l/\pi c_0)^2 = (\omega l/\pi)^2/(T_0/m) \cong 0.03$. In the experiment, the overall solution (and not the boundary-layer one) is quasi-static, and so $\varphi \cong \pi$. In actual applications, as those shown in table 3, the inertia term is important along the suspended length, and so, in general, $\varphi \neq \pi$; see, e.g., Pesce & Pinto, 1996. Notice also that $\tau_0/T_0 \cong 0{,}20$, a typical (though somewhat high) value encountered in steel catenary risers analysis in hard environmental conditions.

Table 2. Dynamic experiment conditions (r21b; 17/06/96)

T_0 (kN)	0.2396
χ_0 (m^{-1})	0.0409
λ (m)	0.9640
$\overline{T}_{top}$ (kN)	0.2727
$\overline{T}_{TDP}$ (kN)	0.2514
a_0 (m)	1.125
φ (degrees)	180
ω (rad / s)	2.3621
$v_0 = \omega a_0$ (m / s)	2.6456
c_0 (m / s)	48.4672
τ_0 / T_0	0.2004
$\tau_0 / \overline{T}_{TDP}$	0.1910
a_0 / λ	1.1619
M_0 ("Mach nr.')	0.0546
$M_0^{\ 2}$	0.0030

Table 3 Non-dimensional control parameters in experiment conditions, compared to typical situations encountered during steel catenary risers analysis in Campos Basin.

Riser/Sea	a_0/λ	τ_0/T_0	$M_0 = \omega a_0 / c_0$
16" - Sea 3	0,23	0,06	0,045
16"- Sea 4	0,55	0,08	0,099
8" - Sea 3	0,22	0,07	0,040
8" - Sea 4	0,39	0,10	0,064
Experiment	~1,20	~0,20	~0,055

Table 3 ilustrates, for two steel catenary risers (8" and 16" external diameters) the control parameters $a_0/\lambda ; \tau_0/T_0 ; M_0 = \omega a_0/c_0$ in two typical wave conditions. Sea-3 is the most likely to occur (probability of 39%) in Campos Basin, corresponding to a significant wave height of 1.75m and a zero up-crossing period of 5.8 seconds. Sea-4 condition (20%) correponds to 2.25m and 6.3 seconds, respectively. As observed from table 3, in the experiment conditions the most important control parameter, the 'Mach number', takes a value well inside the typical range. The values taken by the other two parameters are about twice the corresponding largest ones. From one side, measuring errors are reduced; from the other side, this hard experimental condition even provide us more confidence on the (asymptotic) boundary-layer solution

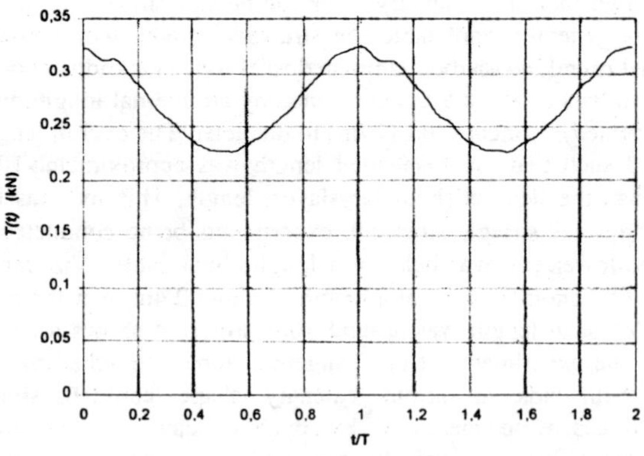

Figure 5. Tension measured at the upper end vs. non-dimensic time.

The signal from SG's numbered 05,07,09,17,41,61,81 were anal Sections corresponding to SG's 05,07,09 and 17 rested on the sup periodically. Sections corresponding to SG's 61 and 81 never tou the supports. It should be mentioned that SG'81 is circa 6 fle lengths apart from the static TDP. SG'41 was identified corresponding to the 'critical section', $\hat{s} = a_0 - \lambda$, i.e., the section just touches the soil once a cycle and is immediately risen up.

BOUNDARY-LAYER SOLUTION AND EXPERIM RESULTS

Analysis

The procedure below was followed:

- By inspection, the SG'41 was identified as the 'critical sect corresponding to the maximum curvature variation, such $s_{41}/\lambda = \hat{s}/\lambda = a_0/\lambda - 1$, where, for convenience, $s_f/\lambda =$ determines the TDP in the static equilibrium configuration. H all other SG's had their respective positions, s_k/λ, determine
- The asymptotic boundary-layer solution for the dynamic curva given by equation (23) was used, taking $T(t) = T_0 + \tau_0 \cos$ and $x_0(t) = a_0 \cos(\omega t + \varphi)$, with the set of parameters extr from table 2. The resulting analytical time-histories were gene and compared directly with the (measured) experimental ones.
- The rms, average, maximum, minimum, and peak-to-peak (do amplitude) values of curvature were calculated from experimental time-series, and compared to the correspon analytical results that were calculated continuously in the r $-3 < s/\lambda < 7$, thus covering the whole boundary-layer.

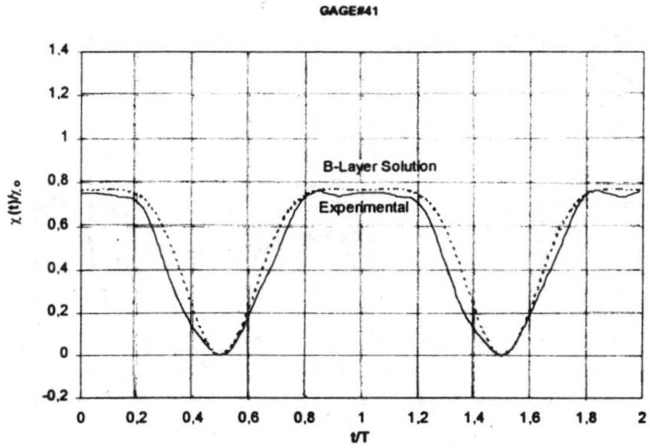

Figure 6. Nondimensional curvature $\chi(t)/\chi_0$. "Critical section": SG'41. $\chi_0 = q/T_0 = 0{,}0409m^{-1}; a_0/\lambda = 1{,}1619$ $s/\lambda = \hat{s}/\lambda = a_0/\lambda - 1 = 0{,}1619; \tau_0/T_0 = 0{,}2004; \varphi = \pi; \mathrm{M} = 0{,}055$

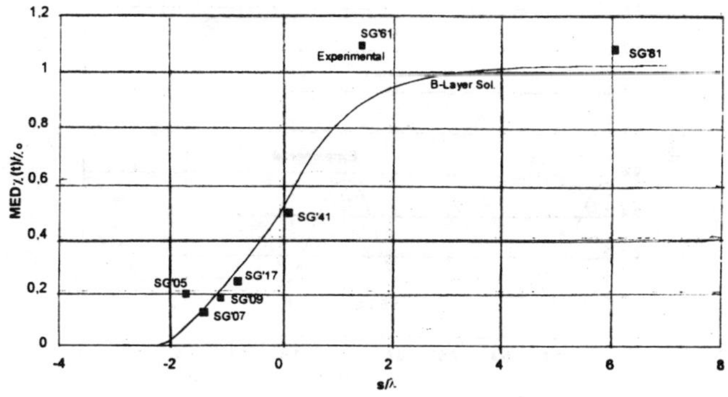

Figure 8. Average value for the nondimensional curvature $\chi(t)/\chi_0$ along the rod. $\chi_0 = q/T_0 = 0{,}0409m^{-1}; a_0/\lambda = 1{,}1619$; $\tau_0/T_0 = 0{,}2004; \varphi = \pi; \mathrm{M} = 0{,}055$

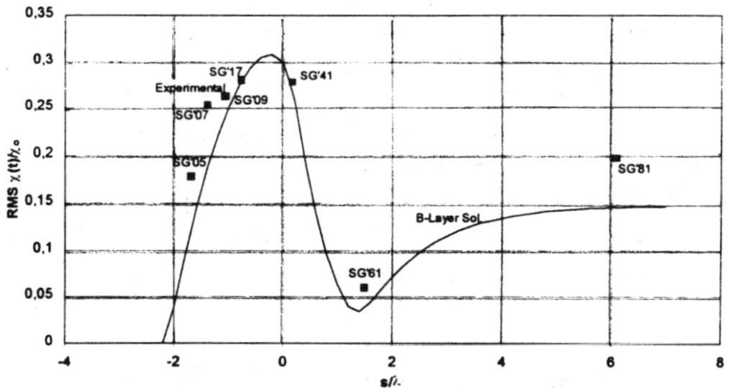

Figure 7. RMS value for the nondimensional curvature $\chi(t)/\chi_0$ along the rod. $\chi_0 = q/T_0 = 0{,}0409m^{-1}; a_0/\lambda = 1{,}1619$; $\tau_0/T_0 = 0{,}2004; \varphi = \pi; \mathrm{M} = 0{,}055$

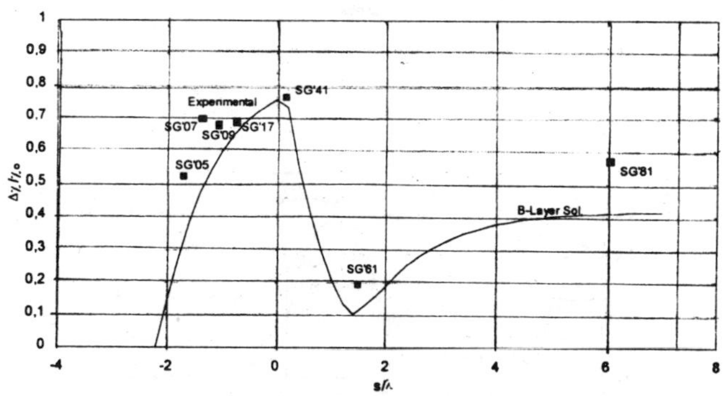

Figure 9. Peak-to-peak variation of the nondimensional curvature $\chi(t)/\chi_0$ along the rod. $\chi_0 = q/T_0 = 0{,}0409m^{-1}; a_0/\lambda = 1{,}1619$; $\tau_0/T_0 = 0{,}2004; \varphi = \pi; \mathrm{M} = 0{,}055$

Figure 6 shows the comparison for SG'41, at the 'critical section'. Curvature is normalized with respect to $\chi_0 = q/T_0$. Notice the remarkable agreement between the analytical solution and the measured signal. Figures 7 and 8 show, respectively, the rms and average curvature values, along the rod span. The solid line corresponds to the boundary-layer solution. The agreement is particularly good for the SG'41 ('critical section') and SG'17, in the region where the curvature variation attains a maximum. The local minimum in the curvature rms (SG'61) is remarkably predicted, as well. There is some discrepancy for the uppermost SG'81. However, it should be mentioned that SG'81 was placed outside the TDP boundary-layer, and close to the upper end, and should be affected by this end condition. The agreement for the average curvature is noticeable, as well, see figure 8, particularly inside the boundary-layer (SG'05,07,09,17,41,61). Figure 9 shows the peak-to-peak (double-amplitude) curvature variation. As mentioned before, SG'41 corresponds to the 'critical section', where variation attains a maximum. Figures 10, 11, 12, and 13 show the time-histories comparisons for SG'05,07,09,17.

The agreement can be said still remarkable, if considered that just the 'critical section' was identified, all the results in space being then directly determined. It should be noticed, for SG'07,09 and 17, the negative values attained by the curvature while touching the discrete supports, a local effect. For the SG'05, although curvature attains small values periodically, they are always positive, probably, perhaps due to some slight local imperfection of the rod. This kind of experimental results besides being explainable, are, as a matter of fact, quantitatively irrelevant and, by no means, affects the most relevant conclusions concerning the boundary-layer solution validity, or its predictability, either qualitative or quantitatively. Neverthless, the computed experimental rms values are affected, in a small amount, by such slight (higher-order) variations. The most important result concerns the 'critical section', where curvature variation attains its maximum. For this section the agreement is excellent, indeed. Figures 14 and 15 correspond to SG'61 and 81. The overall pictures are well predicted. For SG'61, the dynamic variation is quite similar and well predicted, despite some discrepancy on the average value. For SG'81, there are some quantitative discrepancies in average and rms values. Notice, once more, that SG'81 is relatively close to the upper end of the truncated rod.

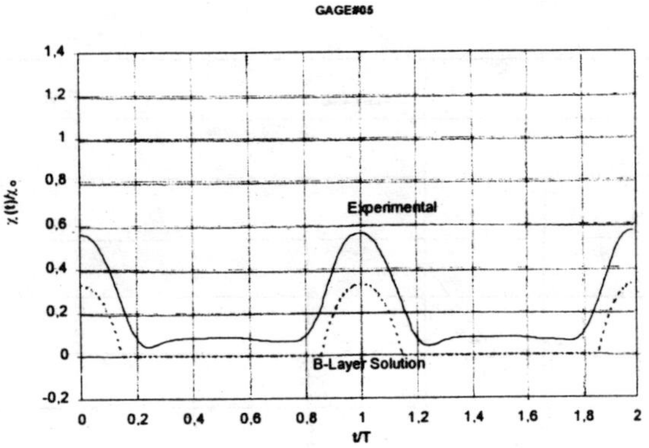

Figure 10. Nondimensional curvature $\chi(t)/\chi_0$. SG'05.

$\chi_0 = q/T_0 = 0{,}0409m^{-1}; a_0/\lambda = 1{,}1619$

$s/\lambda = -1{,}7054; \tau_0/T_0 = 0{,}2004; \varphi = \pi; \mathrm{M} = 0{,}055$

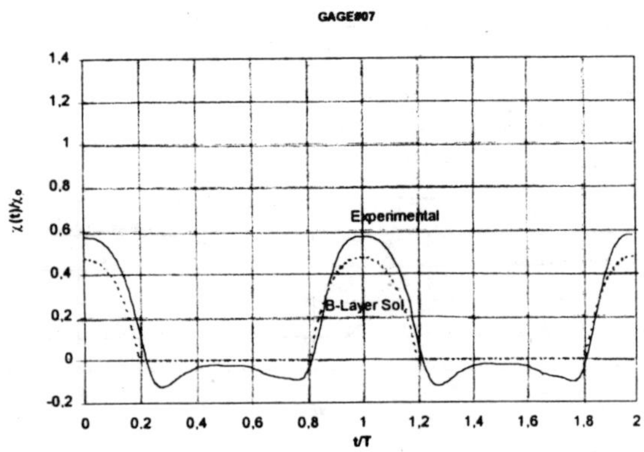

Figure 11. Nondimensional curvature $\chi(t)/\chi_0$. SG'07.

$\chi_0 = q/T_0 = 0{,}0409m^{-1}; a_0/\lambda = 1{,}1619$

$s/\lambda = -1{,}3942; \tau_0/T_0 = 0{,}2004; \varphi = \pi; \mathrm{M} = 0{,}055$

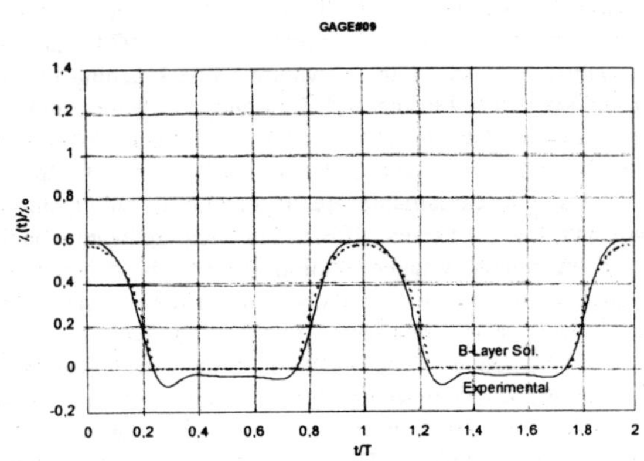

Figure 12. Nondimensional curvature $\chi(t)/\chi_0$. SG'09.

$\chi_0 = q/T_0 = 0{,}0409m^{-1}; a_0/\lambda = 1{,}1619$

$s/\lambda = -1{,}0830; \tau_0/T_0 = 0{,}2004; \varphi = \pi; \mathrm{M} = 0{,}055$

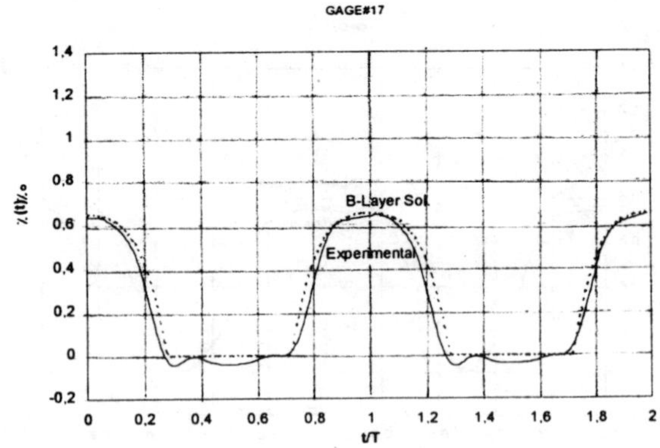

Figure 13. Nondimensional curvature $\chi(t)/\chi_0$. SG'17.

$\chi_0 = q/T_0 = 0{,}0409m^{-1}; a_0/\lambda = 1{,}1619$

$s/\lambda = -0{,}7718; \tau_0/T_0 = 0{,}2004; \varphi = \pi; \mathrm{M} = 0{,}055$

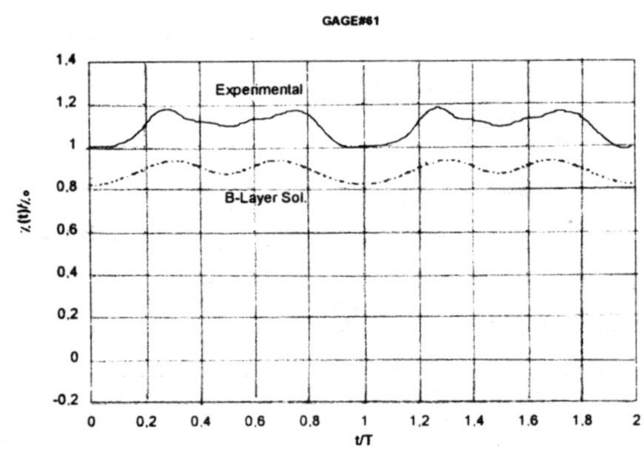

Figure 14. Nondimensional curvature $\chi(t)/\chi_0$. SG'61.

$\chi_0 = q/T_0 = 0{,}0409m^{-1}; a_0/\lambda = 1{,}1619$

$s/\lambda = 1{,}4846; \tau_0/T_0 = 0{,}2004; \varphi = \pi; \mathrm{M} = 0{,}055$

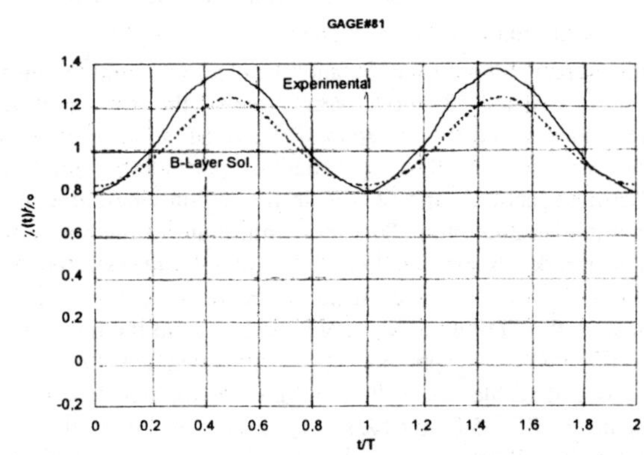

Figure 15. Nondimensional curvature $\chi(t)/\chi_0$. SG'81.

$\chi_0 = q/T_0 = 0{,}0409m^{-1}; a_0/\lambda = 1{,}1619$

$s/\lambda = 6{,}0491; \tau_0/T_0 = 0{,}2004; \varphi = \pi; \mathrm{M} = 0{,}055$

As a final example, ilustrating the power of the present non-linear boundary-layer dynamic solution, figure 16 shows the curvature time-histories for the SG'41, the 'critical section', but now taking as input the effectively measured tension $T(t)$. Notice that even the somewhat irregular small variations present in the flat part of the experimental curve are recovered. This "irregular" variations are due to the slight oscillations that can be observed in the measured tension signal, caused by a small "back-lash" in the pulley mechanism, see figure 5. Notice that as the tension decreases slightly, slackening the rod, curvature increases proportionally (look, e.g., at $t/T \approx 1.1$, in figures 5 and 16).

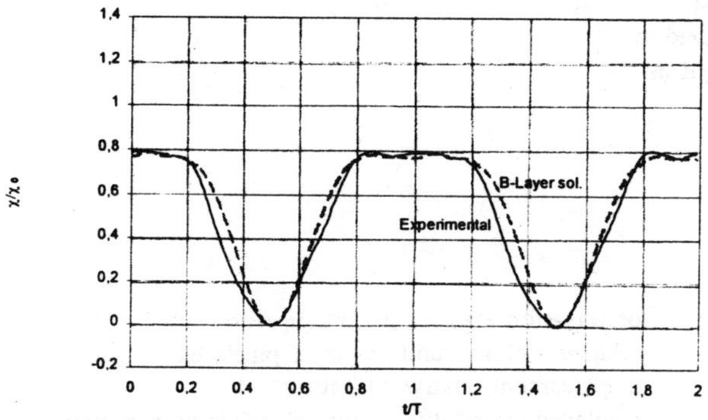

Figure 16. Nondimensional curvature $\chi(t)/\chi_0$. SG'41. Tension in eq. (23) as measured.

CONCLUSIONS

The present work gives an experimental support for an analytical solution for riser mechanics, previously published by Aranha, Martins & Pesce, 1995. The sub-critical dynamics of catenary risers in the vicinity of the Touch Down Point (TDP) has been addressed by means of asymptotic methods and boundary layer theory. A simple mathematical expression enables one to relate, in time, the *dynamic curvature* and *TDP excursion* when *an oscillatory tension* is applied to the riser. Such a dynamic asymptotic non-linear boundary-layer solution, though simple, contains all the non-linear features that result from the non-linear geometrical boundary condition at the soil, of the 'uni-lateral type'. The original analytical result, has been addressed and reinterpreted. The analytical boundary-layer non-linear solution is represented through a subcritical, *local* quasi-static expression, that depends on the *global* dynamics solution, represented by two time-functions describing the dynamic tension and the displacement of the 'instantaneous' TDP. There are no restrictions on these functions: they may be computed from the linearised global ideal cable equations, as a result from irregular waves action, for example, and then applied to the non-linear boundary-layer solution.

In order to enhance the understanding on TDP dynamics, and to validate the boundary-layer non-linear solution, *an experimental study has been specially designed and conducted with a structural model,* at IPT Ocean Engineering Laboratorial Facilities. The riser model has been instrumented along the length by an array of 100 strain-gages, of the resistive type, the static and the periodic dynamic problems being investigated in detail. The experiments were run in May-June, 1996, confirming results previously predicted by using the boundary-layer solution; see Pesce & Pinto, 1996. Comparison with the analytical asymptotic (boundary-layer) solution has been performed showing an outstanding agreement. Not only time-average, rms and maximum amplitude variation of curvature over the riser length are recovered, but also are the corresponding time-histories, along with their harmonic structures, validating the asymptotic solution in full.

REFERENCES

Aranha, J.A.P., Pesce. C.P., Martins. C.A. & Andrade, B.L.R. *Mechanics of Submerged Cables: Asymptotic Solution and Dynamic Tension.* 3rd International Offshore & Polar Engineering Conference, Singapore, Jun. 6-11, 1993. vol. 2, pp. 345-356.

Aranha, J.A.P., Martins. C.A, Pesce, C.P., *Dynamic Bending Moment near the Touch-Down Point of a Catenary Riser*, 1995, (to appear in IJOPE).

Instituto de Pesquisas Tecnológicas do Estado de São Paulo (in portuguese). *Ensaios com Modelo de Riser Rígido.* Rel. no. 34486, 1996.

Instituto de Pesquisas Tecnológicas do Estado de São Paulo Paulo (in portuguese). *Ensaios Complementares com Modelo de Truncado de Riser Rígido.* Rel. no. 34643, 1996.

Kevorkian, J. & Cole, J.D. *Perturbation Methods in Applied Mathematics.* (Applied Mathematical Sciences, v.34). Springer-Verlag, New York, 1981, 558 pp.

Love, AE.H., A *Treatise on the Mathematical Theory of Elasticity.* 4th (1927) edition., Dover Pub. Inc., N.Y., 1944, 643 pp.

Pesce, C.P., Aranha, J.A.P., Martins, C.A. & Pinto, M.M.O. *Steel Catenary Risers for Deep Water Applications.* 5th International Offshore & Polar Engineering Conference, The Hague, Jun. 11-16, 1995, vol. 2, pp. 190-202.

Pesce, C.P., & Pinto, M.M.O. *First-order Dynamic Variation of Curvature and Ttension in Catenary Risers.* 6th International Offshore & Polar Engineering Conference, Los Angeles, May 26-31, 1996, Proceedings, vol. 2 , pp 163-175.

Pesce, C.P. *Mechanics of Catenary Risers and Umbilical Cables.* (in portuguese) Monograph, University of São Paulo, 1997, 450 pp.

Phifer, E.H., Kopp, F., Swanson, R.C., Allen, D.W. & Langner (1994), C.G. *Design and Installation of Auger Steel Catenary Riser. Offshore Technology Conference*, 26th, Houston, Texas, May 2-5, 1994; pp. 399-408.

Triantafyllou, M.S & Bliek, A. & Shin, H: *Dynamic Analysis as a Tool for Open Sea Mooring System Design.* Anual Meeting of The Society of Naval Architects and Marine Eng., November 1985, N.Y.

Aknowlegdments

The authors are indebt to **PETROBRÁS**, Petróleo Brasileiro S.A for supporting the research program on catenary risers, for gently authorizing the use of experimental data and for original motivation on the analytical work. The authors also aknowledge the Technological Research Institute of S.Paulo State. The work of Pesce, Aranha and Martins has been partially supported by CNPq, National Research Council and FAPESP, S.Paulo State Research Foundation.

Proceedings of the Seventh (1997) International Offshore and Polar Engineering Conference
Honolulu, USA, May 25-30, 1997

ISBN 1-880653-28-1 (Set); ISBN 1-880653-30-3 (Vol. II)

A Simplified Analysis of Imperfect Thermally Buckled Subsea Pipelines

James G.A. Croll
University College London
London, United Kingdom

ABSTRACT

An alternative model for the one-way buckling of thermally compressed subsea pipelines is briefly described and shown to provide very close agreement with a new exact analysis. Compared with past analyses of the problem the present approach is shown to have considerable practical advantages. First, it is based upon staightforward physical concepts and involves little analytical complexity. Second, even for the situation in which the pipeline contains initial imperfections resulting from the lay process, the approach provides very simple and explicit analytical expressions for both upper and lower bounds of the buckling loads. As a consequence, this new and simplified method recommends itself as a practical basis for both future design and in-service strength assessment.

KEY WORDS

Analysis, buckling, design, imperfections, pipelines, thermal loading, upheaval

NOMENCLATURE

A_2, A_4	constants used in exact solution, eqn (24)
E	elastic modulus of pipeline
I	second moment of area of pipeline
k	parameter introduced for exact solution, eqn (24)
L	length of uplifted region of pipeline
L_f	length of continuous foundation imperfection eqn (11)
L_b	wavelength at maximum buckling load
P	axial load in pipeline
P_c	critical load of clamped column
P_b	maximum, buckling load for continuous imperfection
P_L	initial lift-off load
P_p	propagation load for perfect pipeline
q	relative weight / unit length of pipeline
w	displacement relative to flat seabed
w'	displacement relative to lift-off points on continuous foundation imperfection
$\bar{w}_l$	equivalent loading imperfection, eqn (4)
$\bar{w}_f$	foundation geometric imperfection, eqn (17)
$\bar{w}_g$	stress free geometric imperfection of pipeline, eqn (32)
x	position along pipeline axis
ξ	total equivalent imperfection, eqn (18) and eqn (34)
η	foundation and/or pipeline geometric imperfection parameter, eqn (20)
θ_0	end slope at lift-off point on foundation imperfection

INTRODUCTION

Problems of thermal buckling could be even more important in polar regions where the temperature differentials between pipe-lay and operational conditions are likely to be rather more extreme than in temperate regions. Furthermore, recently reported increases in the frequency of upheaval failures make the problem one of currently increased concern. While past work has been very successful in establishing the basic phenomena of thermal upheaval buckling, the sometimes complicated associated analysis has perhaps masked the inherent simplicity of the essential mechanics involved.

The following provides an alternative model for the one-way buckling associated with thermally loaded pipelines. The model is based upon the simple buckling of a clamped ended column. In its approximate form this alternative model has already been discussed in Croll, 1996 , and some of what follows is based upon the work of Beirne, 1995. In the following this approximate model will be briefly reviewed. It will first be used to reproduce the "classical" analysis (see Martinet, 1936 or more accessibly Hobbs, 1984), and in the process highlight the physics in a way that contrasts with the somewhat different approach employed in previous work. This approximate modelling will be shown to result in analytically explicit expressions for the classical, perfect, propagation uplift buckle behaviour, and, more importantly, the imperfection sensitive maximum buckling loads experienced when pipelines are lain with initial stress free out-of-straightness or over rough and undulating seabeds. The simplicity and explicit analytical expressions are particularly suited to the setting of imperfection tolerance limits, or to the provision of buckling estimates when distortions exceed design tolerance limits.

As confirmation of the simplified, clamped column modelling, an exact solution for both the perfect and the imperfect pipeline buckling is also presented. Although analytically less convenient, this exact solution provides reassurance of the accuracy of the simplified approach. While the following is presented in the context of thermally buckled pipelines, the new analysis could equally well be applied to the thermal upheaval of continuous concrete road pavements or steel railway track. Furthermore, although just a single mode of upheaval buckling is treated, a very similar approach could be adopted for other mode forms, and especially those associated with lateral buckling.

SIMPLIFIED UPHEAVAL BUCKLING MODEL

Perfect Buckling Behaviour :

The classical problem of thermal buckling (Martinet, 1936) is well established and has been used in many previous works (see for example: Kerr, 1974; Palmer and Baldry , 1974; Hobbs, 1984; Taylor and Gan, 1986; Maltby and Calladine, 1995). It basically involves the solution of the clamped column with the difference that a third boundary condition of zero curvature is added to the requirements of zero displacement and slope at the end boundaries. This has the effect of producing an eigenvalue problem which allows unique definition of the wave length characterising the buckling at a prescribed buckle amplitude. This same problem, with an arguably more direct physical rationale, can be alternatively formulated in the following way.

A stress and strain free straight pipeline of relative weight, **q** / unit length, rests upon a perfectly horizontal seabed.. Under the action of an increased temperature **T**, the restraint against the thermal expansion will induce a compressive force **P** = α**TAE**, where α is the coefficient of thermal expansion, **A** the cross-section area, and **E** the modulus of elasticity of the pipeline. Effects of differential internal and external pressure could be added to those of temperature induced axial compressions, to give a total compressive force **P**. Under the action of this induced axial load, pipeline at a state where the upheaval wavelength has reached a length, **L**, as shown in Figure 1(a), would possess equilibrium states governed by the simple imperfect column equation

$$\bar{w} = \frac{P_c}{(P_c - P)} \left(- \bar{w}_l \right) \tag{1}$$

where for the clamped column of instantaneous wavelength **L** the critical load P_c is

$$P_c = \frac{4\pi^2 EI}{L^2} \tag{2}$$

and the associated critical mode

$$w = \bar{w} \frac{1}{2} \left(1 + \cos \frac{2\pi x}{L} \right) \tag{3}$$

Equation (1) represents an approximation of the exact non-linear behaviour of the clamped column, and it is in this sense that the following analysis is approximate. It is perhaps worth noting that the approximation inherent in the use of eqn (1) is usually considered adequate for relatively small column buckle deformations; this will subsequently be shown to hold true in the present context. The negative loading imperfection w_l would represent the amplitude of the buckling mode eqn (3) contained within the downward deformation of a clamped beam also of length **L** and subject to a downward load of intensity **q** / unit length. This is the downward deformation that the pipeline would experience at this wavelength were it not for the restraint offered by the rigid seabed. Since the deformation form of the clamped beam is almost entirely in the form of the fundamental buckling mode of the clamped column, eqn (3), the amplitude of the equivalent load imperfection w_l may be taken as

$$\bar{w}_l = \frac{1}{384} \frac{qL^4}{EI} \tag{4}$$

Upheaval buckling states are of course in the opposite direction to the loading induced imperfection. In these circumstances it is the real , but unstable, equilibrium states associated with the so-called complementary paths, see Figure 2, that will determine the buckle propagation

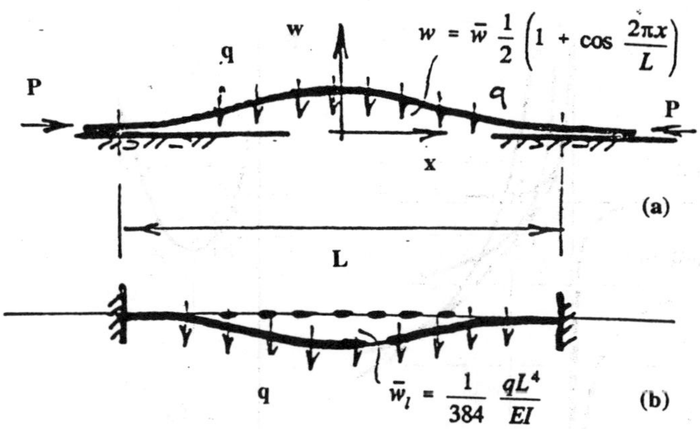

Figure 1 : Model for classical perfect uplift buckling
(a) propagation deformation mode,
(b) equivalent loading imperfection

characteristics. In most situations these complementary paths are of no practical significance. For the present one-way buckling phenomenon it is these complementary paths that control the post buckling loss of load carrying capacity of the pipelines.

To understand how this simple model of clamped column buckling reproduces the classical upheaval buckling (Martinet, 1936) consider the following. At a given level of maximum upheaval

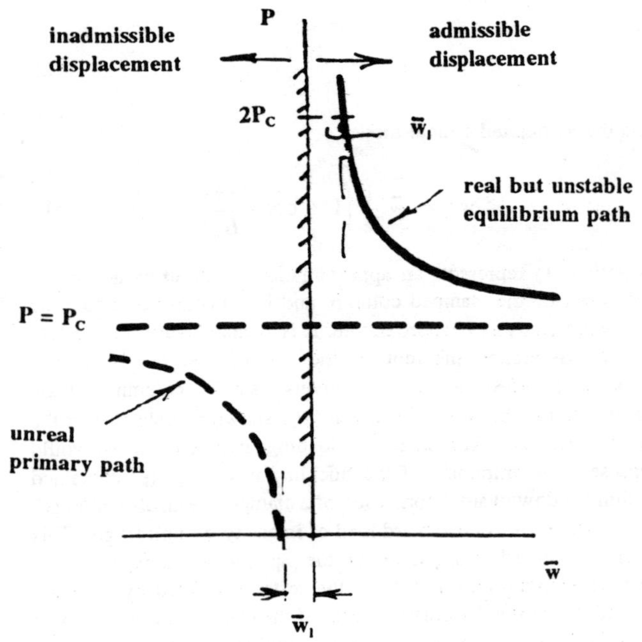

Figure 2: Relationships between one-way uplift buckle propagation response and the complementary path for the buckling of a clamped column.

displacement, $\bar{w}$, there exists an infinite number of possible equilibrium states associated with varying uplift wavelengths, **L**. Three of these infinite number of equilibrium states are depicted in Figure 3. For small values of **L** the forth power dependence of $\bar{w}_l$ on **L** will imply a very small

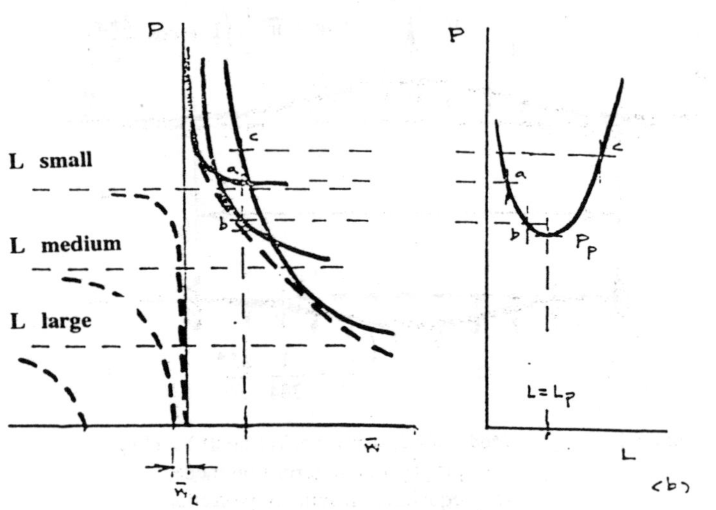

Figure 3 : Classical buckle propagation as the lower bound locus of the complementary equilibrium paths for the clamped column.

imperfection, but the inverse square dependence of P_c on **L** means a very high associated critical load. For small **L** the equilibrium state would be the point (a). As **L** increases the level of the imperfection rapidly increases, while the corresponding critical load decreases. There results a succession of equilibrium states denoted in Figure 3 by points (b), (c), ... At a particular wavelength, L_p, a minimum load will occur, as shown in Figure 3 (b), and it is this lowest possible propagation load, P_p, that will govern the propagation wavelength, L_p. Noting that

$$\frac{dP_c}{dL} = -\frac{2}{L} P_c \; ; \quad \frac{d\bar{w}_l}{dL} = +\frac{4}{L} \bar{w}_l$$

the condition $dP/_{dL} = 0$ may be easily shown from eqn (1) to occur when

$$\bar{w} = \bar{w}_l \tag{5}$$

at a load given by

$$P_p = 2P_c = \frac{8\pi^2 EI}{L^2} \tag{6}$$

This remarkably simple result is equivalent to the minimum post-buckled state at a maximum upheaval displacement, $\bar{w}$, occuring for that L_p which has a negative loading imperfection, $\bar{w}_l$, just equal to the upheaval displacement and at a load exactly twice the critical load for the clamped column having this same length L_p. Figure 4 summarises this situation which may be represented

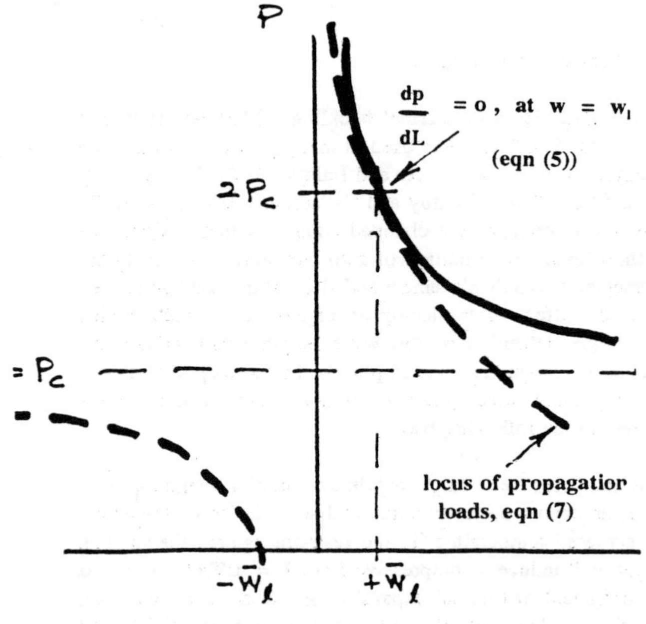

Figure 4 : Buckle propagation response for a perfect pipeline showing propagation load as twice that of the clamped column critical load at a wavelength given by eqn (8).

by the locus of such minimising states given by

$$P_p = 78.96 \left(\frac{EI}{L_p^2}\right) = 4.029 \left(\frac{EIq}{\bar{w}}\right)^{\frac{1}{2}} \tag{7}$$

where

$$L_p = \left(\frac{384\ EI\ \bar{w}}{q}\right)^{\frac{1}{4}} = 4.427 \left(\frac{EI.\bar{w}}{q}\right)^{\frac{1}{4}} \tag{8}$$

The very minor differences from the "classical " values (see Hobbs,1984)

$$P_p = 3.962 \left(\frac{EI.q}{\bar{w}}\right)^{\frac{1}{2}} \qquad (9)$$

$$L_p = 4.515 \left(\frac{EI.\bar{w}}{q}\right)^{\frac{1}{4}} \qquad (10)$$

can in part be explained by the minor errors introduced by the use of eqn (1) to represent the non-linear imperfect path. It is suggested that the very direct physical arguments used for the above derivation are somewhat more transparent than those normally used by the use of the third boundary condition to give the eigenvalue problem.

Imperfect Buckling Behaviour :

For the perfect uplift buckling analysis the interpretation based upon the above clamped column model provides a novel alternative, but adds little that is new or of practical significance to our understanding of uplift buckling. However, this new approach becomes much more useful when the buckling of imperfect pipelines is considered.

As a first example of the buckling of an imperfect pipeline consider the situation shown in Figure 5.

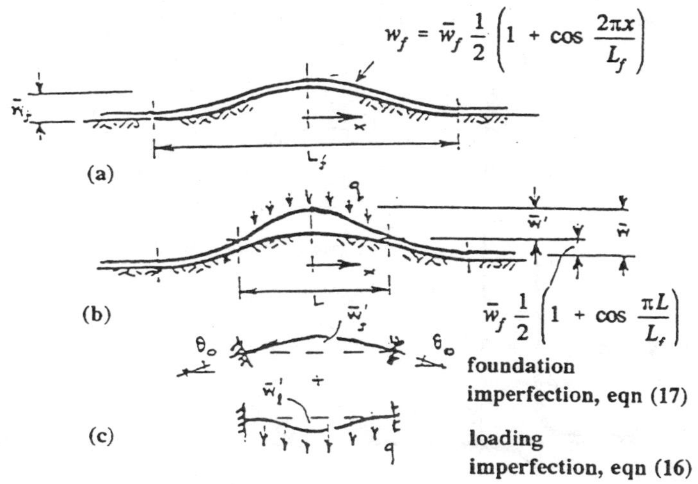

Figure 5 : Uplift buckle from a rigid seabed for a stress free pipeline when straight, showing (a) the initiation of uplift at the mid-point; (b) propagation over uplift length, L ;and (c) the associated geometric and loading imperfections.

It is assumed, for the present, that the pipeline is in a stress free state when straight, but laid onto a rigid seabed (or a trench) that has an out-of-straight profile taking the form

$$w_f = \bar{w}_f \frac{1}{2}\left(1 + \cos\frac{2\pi x}{L_f}\right) \qquad (11)$$

This represents the form of the prospective uplift buckling modes, and as such is likely to have the most damaging effects on the eventual buckling capacity. Once laid upon this imperfect seabed the pipeline will contain initial bending stresses consistent with its compatible deformed curvature. When subsequently subjected to an axial load, **P**, arising from the combined effects of differential temperatures and pressures, there will be an additional upward lineload exerted on the pipeline at the centre of the imperfection. Uplift at this mid-point will commence when the load reaches a level $\mathbf{P_L}$ such that at $\mathbf{x = 0}$

$$EI\frac{d^4 w_f}{dx^4} + P_L\frac{d^2 w_f}{dx^2} + q = 0 \qquad (12)$$

This occurs when

$$P_L = \frac{4\pi^2 EI}{L_f^2} + \frac{qL_f^2}{2\bar{w}_f\pi^2} \qquad (13)$$

For a fixed amplitude of initial out-of-straightness, $\bar{w}_f$, a minimum lift-off load will occur when

$$L_{fmin} = \left(\frac{8\pi^4 EI\ \bar{w}_f}{q}\right)^{\frac{1}{4}} \equiv 5.284\left(\frac{EI.\bar{w}_f}{q}\right)^{\frac{1}{4}} \qquad (14)$$

$$P_{Lmin} = 2\left(\frac{2EI.q}{\bar{w}_f}\right)^{\frac{1}{2}} = 2.828\left(\frac{EI.q}{\bar{w}_f}\right)^{\frac{1}{2}} \qquad (15)$$

Of particular importance is the nature of the subsequent load-deformation response accompanying the gradual growth of uplifted wavelength, **L** . The alternative approach based upon the simple column buckling model is as follows.

At the state where the uplift wavelength has reached the value **L** $< L_f$, the imperfect clamped column will be under the influence of two counterveiling imperfections. This is depicted in Figure 5(c), where it is indicated that there will still be the negative imperfection, $\bar{w}_l'$, resulting from the self-weight loading given by

$$\bar{w}'_l = \frac{1}{384}\cdot\frac{qL^4}{EI} \qquad (16)$$

However, to this negative loading imperfection must now be added a positive imperfection arising from the effects of the upward rotations ϑ_0 at the points of lift-off. It can be shown that the modal amplitude of the postive equivalent geometric imperfection, $\bar{w}_f'$, is given by

$$\bar{w}'_f = 0.829\frac{\theta_o L}{4} = 0.829\left(\frac{\pi}{4}\right)\left(\frac{L}{L_f}\right)\sin\frac{\pi L}{L_f} \qquad (17)$$

For both these imperfectiuons the (..)′ indicates that the imperfections are relative to the amplitude $\bar{w}_f$ ½(1 + cos ($\pi L/L_f$)) of the imperfection at the instantaneous point of lift-off. Relative to the points of lift-off, the deformation is then given by

$$\bar{w}' = \frac{P_c}{(P_c - P)}(\bar{w}'_f - \bar{w}'_l) = \frac{P_c}{(P_c - P)}\xi \qquad (18)$$

where ξ = (w_f' w_l') would represent the total equivalent imperfection for the clamped column of length, **L**.

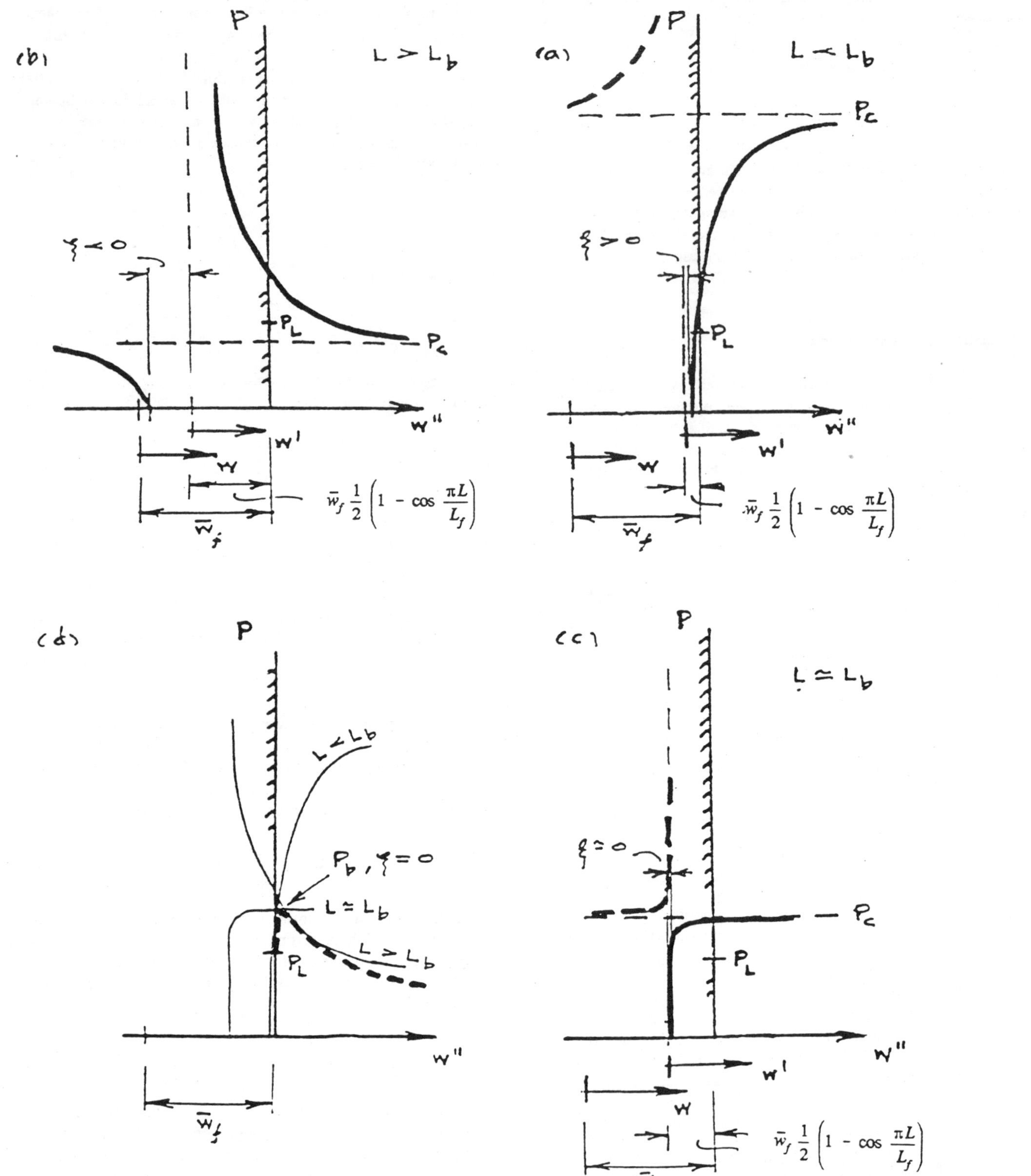

Figure 6 : Buckle propagation response from a rigid continuous foundation imperfection, showing : (a) stable, upward, rising path when ξ **>0;** (b) unstable, falling, path when ξ **< 0** ; (c) nuetral, horizontal, path when ξ **= 0** ; and, (d) their combination to define the propagation buckle locus.

For very small values of **L** the total equivalent imperfection will be dominated by $\bar{w}'_f$ and will consequently be positive. This is due to the geometric imperfection $\bar{w}'_f$ being dominated by L^2 terms while the loading imperfection $\bar{w}'_l$ of eqn (16) depends upon L^4. As illustrated in Figure 6(a) the imperfect paths for positive ξ will have positive slopes emerging from the initial lift-off state as $L \rightarrow 0$. For larger L the situation will be reversed. In this case shown in Figure 6(b) the total equivalent imperfection will be negative, and consequently the equilibrium paths will have negative slope associated with unstable equilibrium on the complementary paths. A transition state will occur when $\xi = 0$; this is shown in Figure 6(c). At this transition state the post-buckling will have an equilibrium path with an instantaneous zero slope. Superimposing these equilibrium paths, as shown in Figure 6(d), it becomes obvious that this transition state has special significance in that it represents the maximum load on the non-linear imperfection propagation path. It follows therefore, that this maximum or buckling load will be uniquely defined by the condition that $\xi = 0$, so that $\bar{w}'_f = \bar{w}'_l$. From eqns (16) and (17) this will occur at a lift-off wavelength, L_b, given by

$$\left(\frac{L_b}{L_f}\right)^3 = 0.6511\ \eta\ \sin\frac{\pi L_b}{L_f} \qquad (19)$$

where the non--dimensional composite imperfection parameter η is given by

$$\eta \equiv \left(\frac{384\ EI\ \bar{w}_f}{qL_f^4}\right) \equiv \frac{\bar{w}_f}{\bar{w}_{lf}} \qquad (20)$$

In this expression $\bar{w}_{lf}$ would represent the magnitude of the loading imperfection associated with the wavelenth L_f of the foundation imperfection.

Solution of eqn (19) is relatively straight-forward, and can be summarised as shown in Figure 7. Any values of the foundation imperfection ($\bar{w}_f$, L_f) for a given pipeline, defined by the quantities (E, I, q), allow definition of the composite parameter, η, which in turn allows from Figure 7, or the solution of eqn (19), the unique definition of the buckle wavelength L_b. Having defined L_b the buckling load P_b is simply the critical load of the clamped column having wavelength L_b, and is given by

$$P_b = \frac{4\pi^2\ EI}{L_b^2} \qquad (21)$$

For example, taking a pipeline having properties $E = 206\text{x}10^{+6}$ kN/m^2, $I = 48.402\text{x}10^{-6}\ \text{m}^4$ and $q = 1.144$ kN/m, as used in Taylor and Tran, 1996, and for an imperfection having $w_f = 0.1$ m, the buckle wavelengths, L_b, and associated buckling loads, P_b, are as shown in Figure 8. In this case the minimum buckling load $P_{bm} = 1.12$ kN would occur when $L_f = 30$ m which corresponds with the value of $\eta = 0.857$. The so-called 'empathetic' wavelength (see Taylor and Tran, 1996, or in greater detail, Tran 1994) would

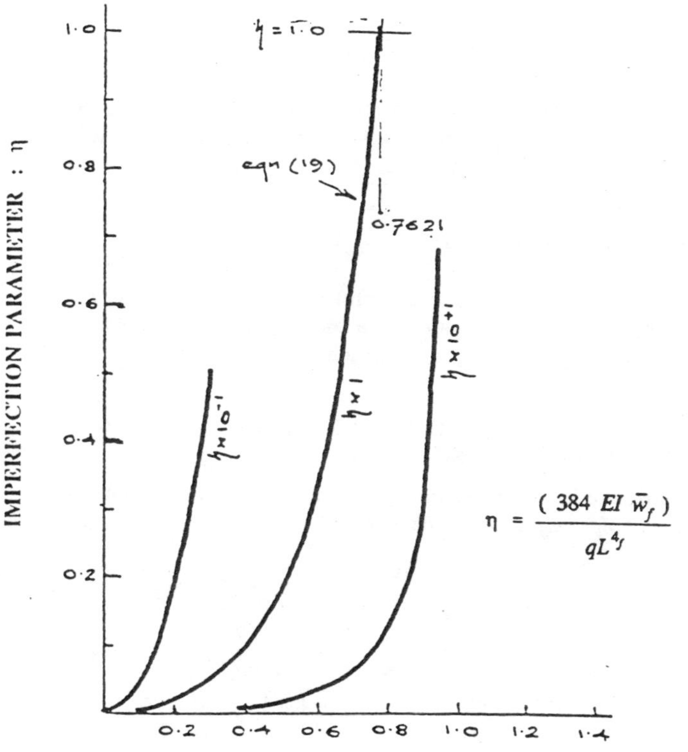

Figure 7 : Determination from eqn (19) of the unique wavelength, L_b, at the maximum buckling load for the case of a foundation imperfection and pipeline having imperfection parameter, η.

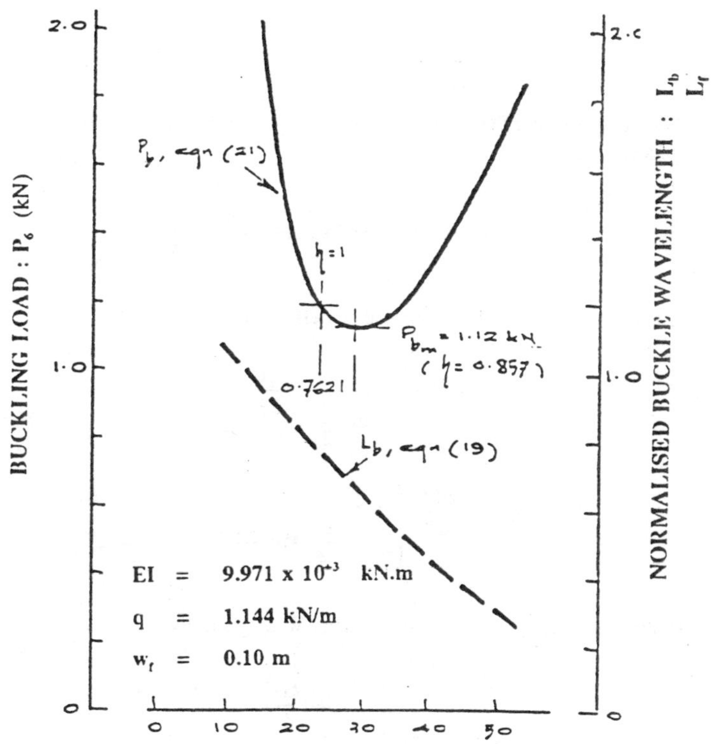

Figure 8 : Buckling loads P_b for the stressed foundation imperfection of varying wavelength L_f.

correspond with $\eta = 1$. From eqn (19) or Figure 7 this occurs when $L_b/L_f = 0.7621$, and has an associated maximum load

$$P_b = \frac{4\pi^2 \, EI}{(0.7621 \; L_f)^2} = 1.727 \; P_{cf} = 0.863 \; P_p \qquad (22)$$

where P_{cf} would represent the critical load for a clamped column having a buckle wavelength of L_f.

EXACT UPHEAVAL BUCKLING MODEL

With the above simplified model being based upon the approximate eqn (1), there must remain some uncertainty as to its relevance for the thermal buckling of subsea pipelines. Before extending the simplified model to other forms of imperfection it will be instructive to compare the predictions with an exact solution of this same problem.

Perfect Buckling Behaviour :

At an uplift wavelength, L, the pipeline of Figure 1 will be governed by the equation

$$EI \frac{d^4 w}{dx^4} + P \frac{d^2 w}{dx^2} + q = o \qquad (23)$$

With end boundary conditions $w = {}^{dw}/_{dx} = 0$ at $x = \pm 0.5 \, L$ the general solution of

$$w = A_2 \cos kx + A_4 - \frac{qx^2}{2P} \qquad (24)$$

where $k^2 = P / EI$, requires the coefficients A_2 and A_4 to be

$$A_2 = - \frac{1}{k \sin \frac{kL}{2}} \cdot \frac{qL}{2P}$$

$$A_4 = + \frac{qL^2}{8P} \cdot \left(1 + \frac{l}{\frac{kL}{4} \tan \frac{kL}{2}} \right)$$

In this case the maximum displacement, $\bar{w}$, is given by

$$\bar{w} = \frac{qL^2}{8P} \left[1 - \frac{\tan \frac{kL}{4}}{(\frac{kL}{4})} \right] \qquad (25)$$

For varying L the relationships between P and $\bar{w}$ are shown for the above pipeline example in Figure 9. Also shown is the locus of the lower bounds of these exact equilibrium states, which is identical to the previously reported solution of eqn (9) and (10). This exact lower bound locus can be seen to be very closely reproduced by the approximate model of eqn (7) and (8).

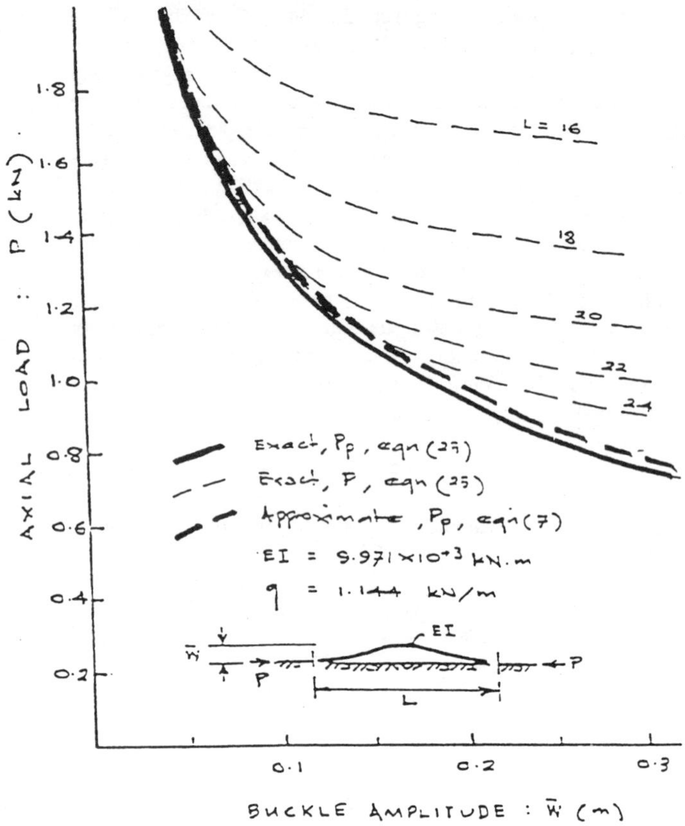

Figure 9 : Exact solution of perfect propagation behaviour using eqn (25), showing lower bound locus, compared with the approximate solution of eqn (7).

Imperfect Buckling Behaviour :

For the situation in which an uplift buckle has extended to a length, L, as shown in Figure 5, the total incremental displacement, w, will be governed by the equation

$$EI \, . \, \frac{d^4 w}{dx^4} + P \frac{d^2 w}{dx^2} + q = o \qquad (26)$$

In this case however, the kinematic constraints at the locations at which uplift begins would be given by

$$w = w_f \Big|_{x=\pm\frac{L}{2}} = \bar{w}_f \frac{1}{2} \left(1 + \cos \frac{\pi L}{L_f} \right) \qquad (27)$$

$$\frac{dw}{dx} = \frac{dw_f}{dx} \Big|_{x=\pm\frac{L}{2}} = - \bar{w}_f \left(\frac{\pi}{L_f} \right) \sin \frac{\pi L}{L_f}$$

The general solution of eqn (24) will, for the boundary conditions eqn (27), require

$$A_2 = \frac{1}{k \sin \frac{kL}{2}} \left[\bar{w}_f \left(\frac{\pi}{L_f} \right) \sin \frac{\pi L}{L_f} - \frac{qL}{2P} \right]$$

$$A_4 = \frac{1}{2}\,\bar{w}_f\left(1 + \cos\frac{\pi L}{L_f}\right) + \frac{qL^2}{8P}$$

$$- \frac{1}{k\tan\frac{kL}{2}}\left[\bar{w}_f\left(\frac{\pi}{L_f}\right)\sin\frac{\pi L}{L_f} - \frac{qL}{2P}\right] \qquad (28)$$

where again $k^2 = P / EI$. Maximum displacement at $x = 0$ will in this case be given by

$$\bar{w} = \frac{\tan\frac{kL}{4}}{\frac{kL}{4}}\left[\frac{1}{4}\,\bar{w}_f\left(\frac{\pi L}{L_f}\right)\sin\frac{\pi L}{L_f} - \frac{qL^2}{8P}\right]$$

$$+ \frac{1}{2}\,\bar{w}_f\left(1 + \cos\frac{\pi L}{L_f}\right) + \frac{qL^2}{8P} \qquad (29)$$

For the same pipeline example as that treated in Figure 9, and by Taylor and Tran, 1996, the various possible equilibrium states are shown in Figure 10 for a selection of buckle wavelengths, **L**. The locus representing the minimum loads for varying **L** are shown by the full bold lined path which will again represent the preferred equilibrium path followed by the buckling pipeline.

Comparing Figures 9 and 10 with the equivalent solutions obtained using the approximate modelling, summarised above but reported more fully in Croll, 1996, show that for practical purposes the approximate analytical model is an appropriate basis for design calculations. For this reason the approximate analysis will be used in subsequent discussions.

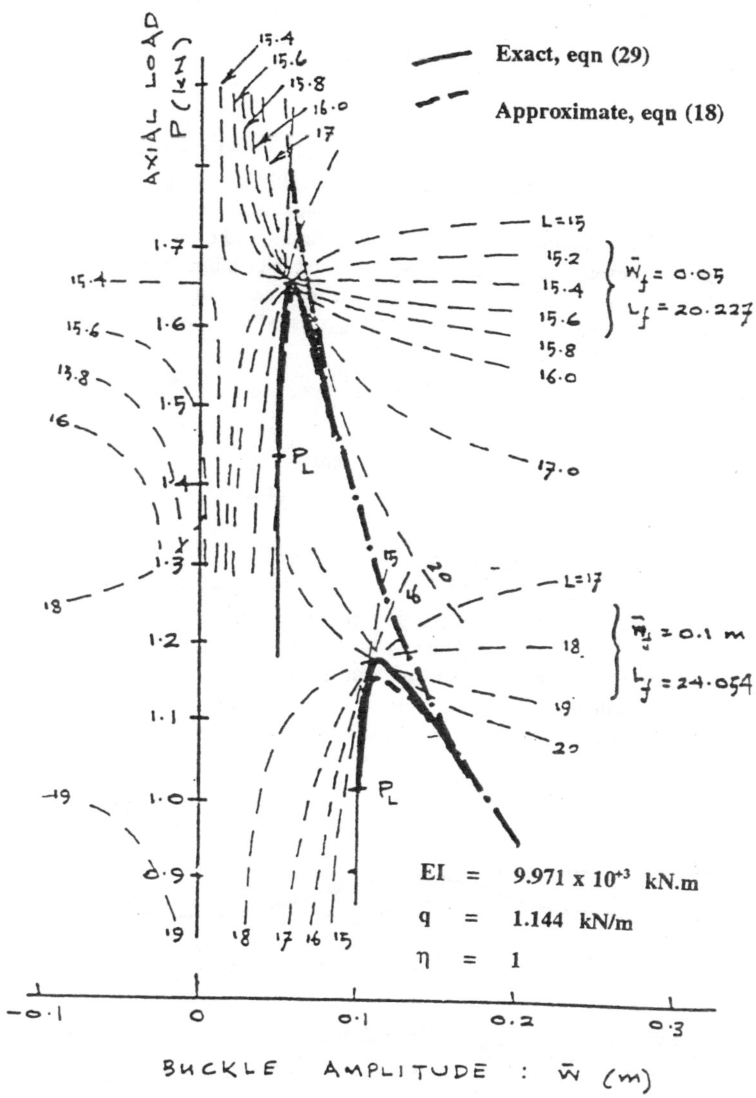

Figure 10 : Exact solution of the stressed foundation imperfect propagation behaviour using eqn (29), showing the lower bound locus, and comparison with the approximate solution of eqn (21).

OTHER FORMS OF GEOMETRIC IMPERFECTION

In the above analysis it was assumed that the pipeline when positioned on the the undulating seabed is internally stressed to a level required for the initially straight pipeline to take-up the imperfect seabed profile. There are circumstances in which a geometrically imperfect pipeline would contain lower stresses than those needed to take-up the out-of-straight profile; some of these have been discussed by Taylor and Tran, 1996, while the consequences have been highlighted by the major differences in buckling load as compared with the stress free imperfection considered by Croll, 1996.

If a pipeline is flexurally unstressed, but has an initial out-of-straightness taking the form

$$w_g = \bar{w}_g \cdot 0.5\,(1 + \cos 2\pi x/L_g) \qquad (29)$$

and if it is laid upon a seabed having an exactly matching form at the same location along its length, then under axial load, **P** , the initial lift-off at mid-length would be governed by

$$P_L \frac{d^2 w_g}{dx^2} + q = 0 \qquad (30)$$

In the mode of eqn (29) this indicates an initiation of lift-off when

$$P_L = \frac{qL_g^2}{2\bar{w}_g \pi^2} \qquad (31)$$

For a fixed amplitude of imperfection , $\bar{w}_g$, this suggests that the lift-off load, P_L , will monotonically increase as the imperfection wavelength increases. Once lift-off has started the non-linear equilibrium path associated with the propagating wavelength will again be governed by the lower bound to eqn (18), with the difference that now the geometric, $\bar{w}_g'$, and loading, $\bar{w}_l'$ imperfection at a lift-off length, **L** , will be given by

$$\bar{w}'_g = \bar{w}_g \frac{1}{2}\left(1 - \cos\frac{\pi L}{L_g}\right) \qquad (32)$$

$$\bar{w}'_l = \frac{1}{384} \cdot \frac{qL^4}{EI}$$

These are shown schematically in Figure 11(c), while Figure 12 shows the equilibrium states associated with varying uplift wavelengths, L , for two examples of 'empathetic' imperfection geometry. Pipeline properties are the same as those used in previous examples. For this imperfection the maximum, buckling load, P_b , will again occur when the total equivalent imperfection $\xi = (\bar{w}'_g - \bar{w}'_l) = 0$, which corresponds with a wavelength , L_b , that is governed by the condition

$$\left(\frac{L_b}{L_g}\right)^4 = 0.5\,\eta\left(1 - \cos\frac{\pi L_b}{L_g}\right) \qquad (33)$$

where $\eta = \left(\dfrac{384\,E\,I\,\bar{w}_g}{q\,L_g^4}\right) = \dfrac{\bar{w}_g}{\bar{w}_{lg}}$. Here $\bar{w}_{lg}$ would represent the downward deformation that would occur in a clamped beam having a length, L_g , under its self-weight, q . It might be observed that the solution of eqn (33) would become an approximation when $L_b/L_g > 1$, in the sense that the periodic imperfection form may not extend beyond one complete wavelength.

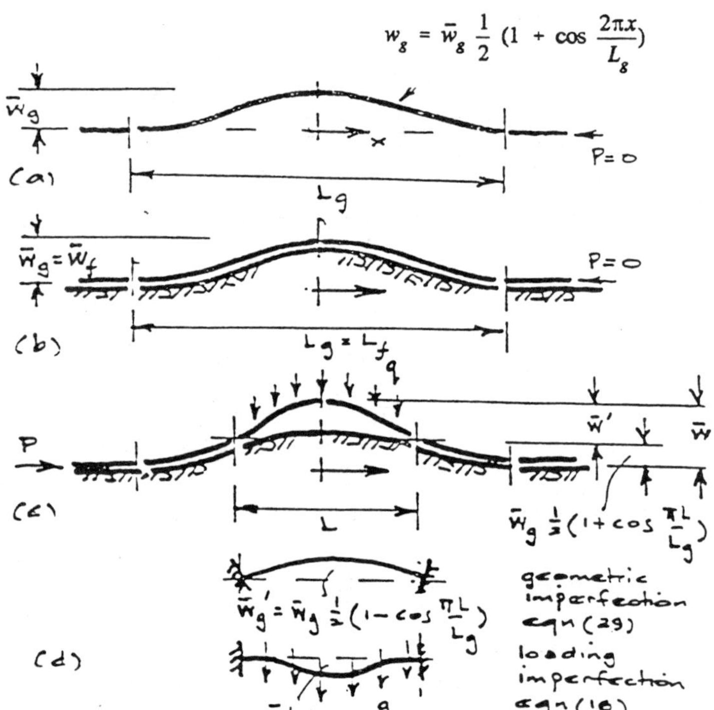

Figure 11 : Uplift buckle for a geometrically imperfect stress free pipeline, showing (a) stress free pipeline geometry ; (b) on rigid foundation geometry ; (c) propagation over uplift length, L; and (d) the associated imperfections.

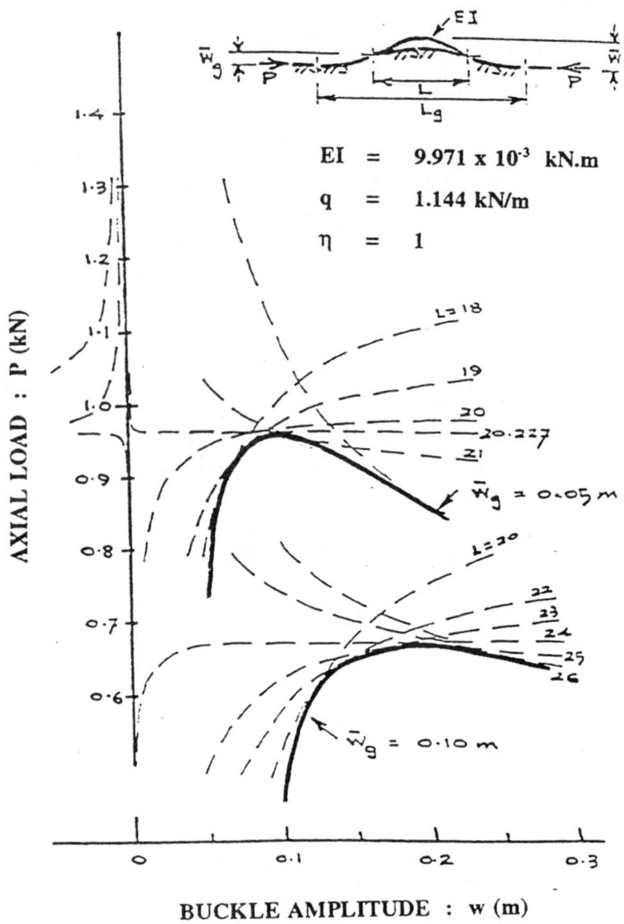

Figure 12 : Approximate solution of the stress free geometrically imperfect pipeline behaviour using eqn (18) and the imperfections of eqn (32).

Solution of the transcendental eqn (33) is again a very simple matter and may be summarised as shown in Figure 13. With η defined by the properties of the imperfection (w_g, L_g) and the pipeline (E.I.q) , the buckling wavelength, L_b , obtained from the solution of eqn (33) or more conveniently read off Figure 13 , may then be substituted into eqn (21) to give the buckle load, P_b .

For the same pipeline parameters as used in previous examples, an imperfection of amplitude $w_g = 0.1$ m and varying wavelength L_g will have buckle wavelength L_b from Figure 13 and buckling load P_b from eqn (21) as shown in Figure 14. What is at once striking is the severity of the reduction of the buckling load capacity for the stress free geometrically imperfect pipelines compared with those in a stressed imperfect state shown in Figure 8. What is also noteworthy is the existence of the minimum buckling capacity at the so-called 'empathetic' (see for example Taylor and Tran, 1996) mode for which η = 1 and the wavelength and amplitude of the imperfection are in the same relationship as the 'classical' propagation mode of eqn (8). To further highlight the increased imperfection sensitivity of the stress free imperfection of Figure 14 compared with the stressed foundation imperfection of Figure 8, the buckling loads, P , for 'empathetic' modes of varying amplitude of imperfection are shown in Figure 15. Also shown for

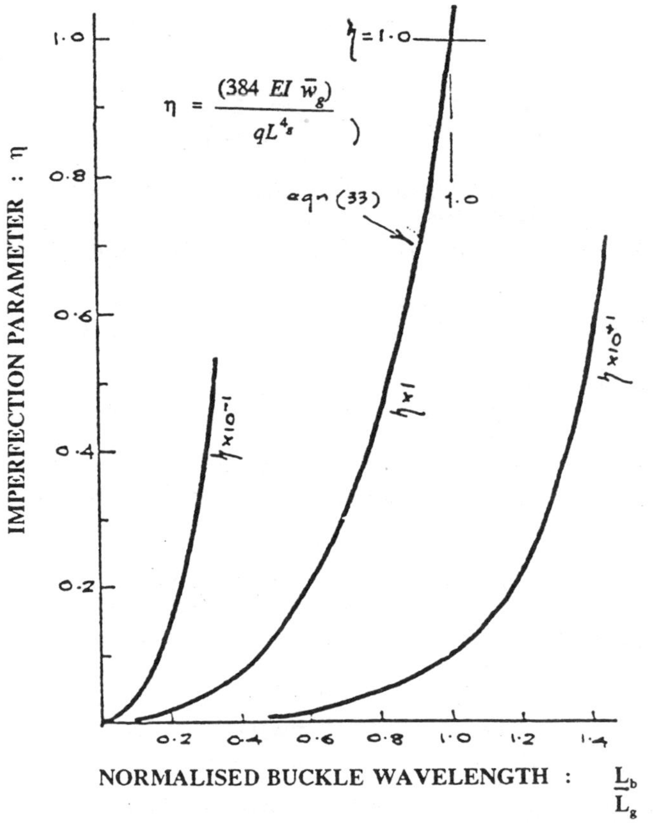

Figure 13 : Determination from eqn (33) of the unique wavelength, L_b, at the maximum buckling load for the case of a geometric imperfection and pipeline having imperfection parameter, η.

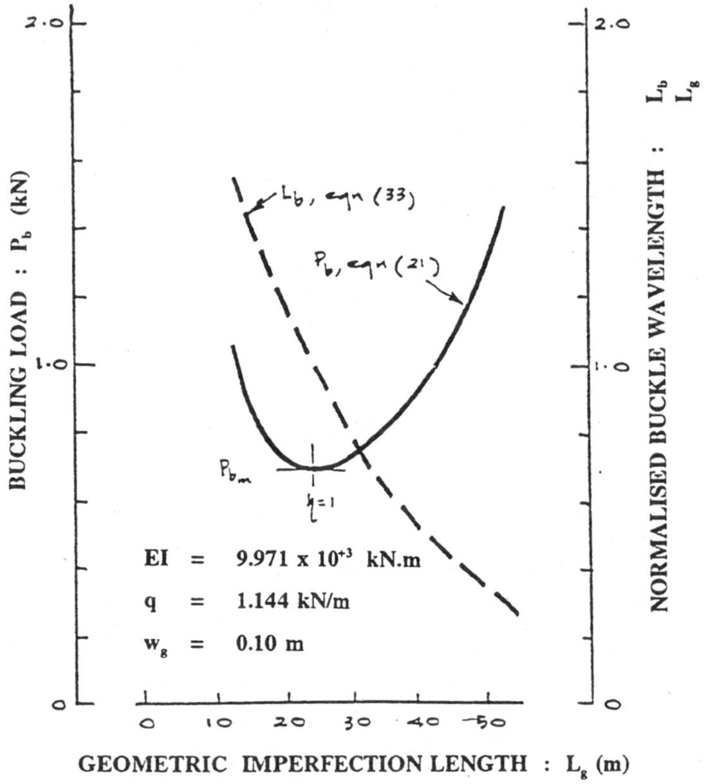

Figure 14 : Buckling loads P_b for the stress free geometric imperfection of varying wavelength L_f.

comparison are the related imperfection sensitivities reported by Tran, 1994, and referred to by Taylor and Tran, 1996. The potential consequences of stress free geometric imperfections are clear.

Although the above discussion of the stress free geometric imperfection has been undertaken using the approximate model, it is worth noting that the exact solution predicts effectively identical behaviour. By providing simple and explicit relationships for uplift buckling loads the simplified method has advantages in setting design tolerances or assessing the implications of in-service imperections. Some of these are touched upon in the following.

TOWARDS A PRACTICAL DESIGN ANALYSIS

It has been suggested that one of the possible sources of geometric error would be a less than perfect straightening process when pipelines are laid from spoolled pipe. There have even been suggestions that these lay induced distortions could have contributed to a number of recent uplift buckling failures. A pipeline having a stress free out-of-straightness when laid upon even a perfectly flat seabed, or trench, will exhibit all the characteristics discussed in the previous section. That is, there will be a stress free geometric imperfection superimposed upon the equivalent load imperfection, and the maximum buckling load will occur at that wavelength for which the effective geometric and loading imperfections are equal but opposite.

It would not be difficult to extend the above to the case where an out-of-straight stress free pipeline, having distortions defined by ($\bar{w}_g$, L_g), when laid upon a seabed having out-of-flat characteristics ($\bar{w}_f$, L_f) would have a total equivalent imperfection

$$\xi = \bar{w}'_g + \bar{w}'_f - \bar{w}'_1 \qquad (34)$$

where the effective imperfections at a buckle wavelength $L < L_f$ would depend upon the relative magnitudes of $\bar{w}_g$, $\bar{w}_f$ and L_g, L_f. This could of course include cases where $\bar{w}_g$ and/or $\bar{w}_f$ may themselves be negative. Although there would be greater numbers of independent parameters, the procedure could follow the lines discussed above for each case separately. As a worst possible case would be the assumption that both $\bar{w}_g$ and $\bar{w}_f$ have a wavelength corresponding with the 'empathetic' mode. As a conservative estimate of the combined effects would be the assumption that both $\bar{w}_g$, $\bar{w}_f$ are stress free. In this case the conservative estimate would be as outlined in the above section, but on the assumption that $\bar{w}_g$ and $\bar{w}_f$ are additive.

CONCLUSIONS

It has been shown that as an interpretation of the physical behaviour associated with one-way thermal buckling of subsea pipelines, the classical buckling of a clamped column provides a powerful alternative to more complex analytical procedures. This alternative involves the definition of a total equivalent imperfection resulting from possible geometric errors in either the pipelines geometry or the flatness of the seabed, combined with a loading imperfection resulting from the weight of the pipeline and/or the mobilised friction resisting buckling. It has been shown how the wavelength defining the maximum or buckling load is uniquely defined by the condition that the total equivalent imperfection is identically zero. At this wavelength the maximum or buckling load is controlled by the classical critical load of the clamped column having the same wavelength.

The directness of the approach, combined with the simplicity and explicitness of the analytical expressions, would seem to commend the method for routine design checking.

ACKNOWLEDGEMENTS

I am indebted to numerous former colleagues and collaborators, including Neil Taylor, Roger Hobbs, Chris Tam, Charles Ellinas, Alastair Walker, Andrea Kyriakou and many others for first drawing my attention to this fascinating little problem, and for many subsequent and stimulating discussions. Much of the work described resulted from final year projects; and I am indebted for the important contributions made by too many to list.

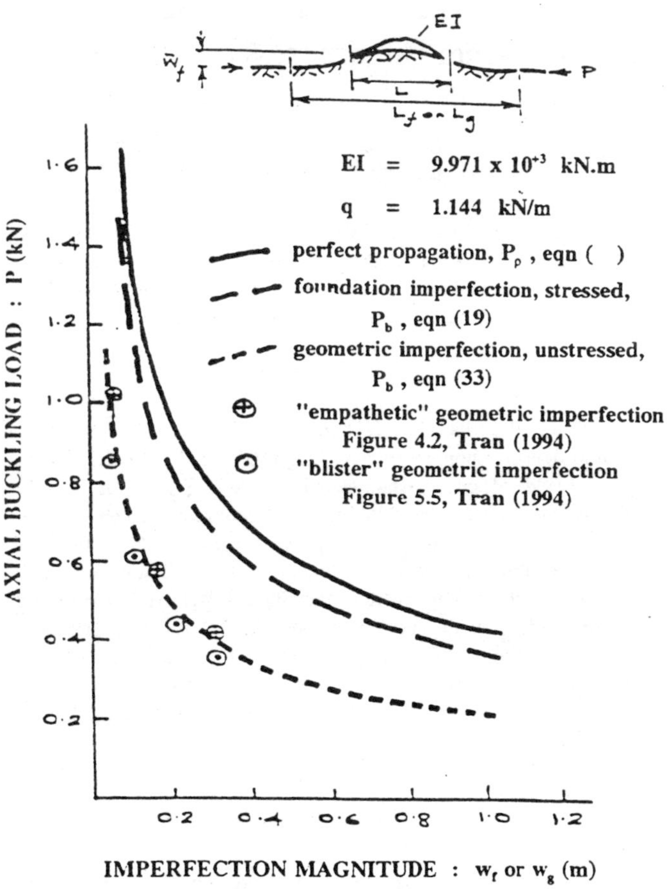

Figure 15 : Imperfection sensitivity for 'empathetic' continuous imperfections, comparing stressed foundation out-of - straightness and unstressed geometric errors.

REFERENCES

Bierne, C and Chen, P. L. (1995) "Thermal Buckling of Concrete Road Slabs" University College London. Final Year Thesis, 1-64.

Croll, J. G. A. (1996) "A Simplified Model of Upheaval Thermal Buckling of Subsea Pipelines", Bicentenary Conference on Thin Walled Structures, University of Strathclyde. 2-4 Dec. 1996.

Hobbs, R. E. (1984) "In-service Buckling of Heated Pipelines", ASCE J. of Transportation Engineering. vol 110. pp 175-189.

Kerr, A. D. (1974) "On the Stability of the Railroad Track in the Vertical Plane", Rail International. Feb. 1974, pp 131-142.

Maltby T. C. and Calladine. C.R. (1995) "An Investigation into the Upheaval Buckling of Buried Pipelines" Int J. Mech. Sci. , 37, pp 943-983.

Martinet, A. (1934) "A Flambement des Voies Sans Joints sur Ballast et Rails de Grande Longueur". Revue Generale des Chemins de Fer, vol 55/2. pp 212-230.

Palmer, A. C. and Baldry. J. A. S. (1974) "Lateral Buckling of Axially Compressed Pipelines" J. Petroleum Techn., 26, pp 1283-1284.

Taylor, N. and Tran, V. (1996) "Experimental and Theoretical Studies in Subsea Pipeline Buckling". Marine Structures, vol 9, pp 211-257.

Taylor, N. and Gan, A. B. (1986) "Submarine Pipeline Buckling-Imperfection Studies" Thin Walled Structures. 4, pp 295-323.

Tran, V. (1994) "Imperfection Upheaval Buckling of Subsea Pipelines". PhD Thesis. Sheffield Hallam University. May. 1994.

Tvergard, V. and Needleman, A. (1981) "On Localised Thermal Track Buckling" Int. J. Mech. Sci., 23, pp 577-593.

Proceedings of the Seventh (1997) International Offshore and Polar Engineering Conference
Honolulu, USA, May 25-30, 1997

ISBN 1-880653-28-1 (Set); ISBN 1-880653-30-3 (Vol. II)

Development of a Russian Standard for Submarine Pipeline Design, Installation and Operation

M.A. Kamyshev
AO VNIIST
Moscow, Russia

J.G. de Vries
TIDEWAY
Breda, The Netherlands

Abstract

At present, a large network of oil and gas pipelines is operated in the Former Soviet-Union (FSU). Large transmission lines with sizes up to 1,420 mm (56 inches) and infield lines have a total combined length of more than 200,000 km.
Except for a number of river crossings, these lines are all situated on land. Both are designed and constructed in accordance with existing FSU Standards and Regulations for land-lines.
The oil industry in the Russian Federation is now on the brink opening up their Continental Shelf, which covers an area of approximately 1/5th of the World's total Continental Shelf with large proven gas- and oil reserves in the Northern Arctic regions (85%) and in East Asia (15%). It is obvious that there is an urgent need for adequate standards and regulations for offshore installations and operations.

In a joint operation VNIIST (the All-Russian Pipeline Construction Institute) and Dredging International from Belgium, with its Dutch subsidiary Tideway, have developed a concept for a new Standard for the Design and Installation of Submarine Pipelines. It can be seen as a compromise between the existing FSU rules for land-lines and the Western experience with offshore pipelines.
The project was sponsored by the European Commission (EC) as part of the EC-Tacis program and by the Flemish Communities.
The proposed concept is now under review by other organisations in Russia for advice and comment, publication being expected later this year.

Key words: Russian Standard, SNiP, Submarine Pipeline System, Offshore, Shore Crossing, Design, Installation, Construction, Operation.

Introduction

Codes in Russia are generally divided into Codes for design and Codes for construction.
These exist at different levels of importance:

1. SNiP (= Construction Standards and Regulations) have the highest level of application. They define compulsory requirements with respect to design and construction of all kinds of Works, and are intended to ensure high quality design, to define design elements to be addressed, major design criteria, loads to be considered, quality and safety management, documentation. There are in total twenty such documents. The main codes for pipelines are:
   - SNiP 2.05.06-85: "Design of Long-distance Transmission Pipelines"
   - SNiP III-42-80 "Rules for the Construction and Acceptance of Works", Chapter 42: "Main Pipelines".

   Both documents refer to land-lines up to 1400 mm in diameter and with operating pressures between 1.2 MPa and 10 MPa.

2. GOST (= Federal specifications) and ONTP (= Federal inspection requirements). These documents are written to ensure standardisation of documentation, materials and material inspection. The SNiP documents, for instance, have to be written in accordance with requirements, specified by the GOST-documents

3. Svod Pravel (= Interdepartmental Regulations for Design and Construction). These documents give recommended methods and parameters but are not compulsory. The validity of a different approach must however be de-monstrated.
   Two such Regulations have been prepared in the past for offshore pipelines, including:
   - R 412-81: "Recommendations for the Design and Construction of Submarine Oil and Gas Pipelines" (Glavmorneftegasprom. 1981)
   - VSN 51-9-86: "Submarine Oil and Gas Pipeline Design" (Mingasprom. 1987).

   The two Regulations have only a limited application for small size pipelines.

4. Regulations, which may differ per territorial region in the Russian Federation, dictated by the local conditions, such as environmental restrictions, restoration of areas to be crossed, etc. Company Specifications are also supposed to be at this level.

Preliminary studies for offshore pipeline projects have been based on a combination of SNiP 2.05.06-85, particularly the articles concerning river-crossings, and current western methods and practice. However, the need for a new specific Russian Standard was felt to be urgently needed.

Direct application of Western Standards appeared not to be possible. Firstly, because it is generally not permitted to apply foreign standards, and secondly, because there is also in Russia extensive knowledge and experience with pipelines and pipeline installations which is worth to be implemented. This concerns, in particular, the harsh environments of arctic and earthquake areas.

During the study two new documents have been developed which cover the design and installation of submarine pipelines. In accordance with the outline given above, a SNiP-document was prepared which specifies the mandatory requirements of the Russian Federation for the design of the pipeline systems, the installation and the acceptance criteria for operation. The second document is a Guideline or Recommended Practice (Svod Privel), which presents the design elements to consider for safe installation and operation together with methods, parameters, quality aspects, documentation etc.

The two documents have been prepared in a joint project by the All-Russian Research Institute for Pipeline Construction AO VNIIST and Dredging International from Belgium with its Dutch subsidiary Tideway. The work was sponsored by the European Commission as one of their EC-Tacis projects and by the Flemish Communities.

AO VNIIST is a privatised, former governmental organisation, which is active in research, design and development of construction technology and certification of pipeline systems in the Russian Federation. In 1996, they have been assigned by the Ministry of Oil and Energy to operate a similar function in the future offshore industry. In this position, they have a co-ordination function in the development of Codes and Regulations for submarine pipelines.

Dredging International is a large international dredging organisation with its headquarters in Belgium. Tideway is an offshore contractor in the oil- and gas industry, specialised in rock-dumping, shore crossings of submarine pipelines and sewage outfalls and offshore dredging and pre-sweeping. The Company operates its own DP vessel for rock-dumping, the Rollingstone, with a loading capacity of 13.000 tonnes and an ROV controlled fall pipe system for very accurate dumping and back-filling.

This paper gives an overview of the proposed documents in their current status. Some major design elements are presented in more detail to give an impression of the forthcoming Svod Pravel Guidelines. The SNiP-document for the "Design and Installation of Offshore Pipelines" is already under review by several organisations in Russia for advice and comment. Official publication is expected later this year. The Guidelines will be probably agreed at a later date, as it is expected that acceptance by all parties involved will not be reached without extensive review, commentary and revision, especially where implementation of new techniques is proposed in lieu of familiar methods.

Area of Application

The new Standard covers all types of offshore pipelines including: carbon steel, stainless steel and flexible pipelines (irrespective of their size and internal pressure). It also covers all sorts of installation methods including: installation by specialised lay-vessel or lay-barge, utilising the S-curve (overbend-sagbend) or J-lay method (sagbend only), as well as reeling, towing and pulling methods. Shore crossing installations are also included.
The following Table of Contents may give an impression of the aspects of design, and installation that have been addressed. Broadly, a distinction can be made between General Provisions, Design and Installation aspects:

General Provisions

1. General Regulations. Some terms and definitions used in the Standards are shown in the Appendix to this paper.
2. Quality Assurance and Documentation.

Design Aspects

This part gives design guidelines and recommended practices to ensure safe installation and lifetime operation as well as consistency in design. It specifies design methods, criteria, quality control and documentation for pipeline and riser systems.

The following sections are included:
3. Design Data Requirements
4. Design Loads
5. Strength Analysis
6. Stability Analysis
7. Riser System
8. Corrosion Protection

Installation Aspects:

This part gives guidelines and recommended practices regarding work execution and acceptance for operation such as installation by various laying techniques, trenching, riser installation, cleaning, gauging and hydrostatic testing of the system, welding and weld inspection, shore approaches and landfalls, landfall restoration, survey and maintenance operations, prevention of environmental damage and safety during installation and operation of the pipeline and riser systems.

The following sections are included:
9. Pipeline and Riser Installation
10. Material Requirements and Welding
11. Hydrostatic Testing and (Pre)Commissioning
12. Environmental Protection and Safety.

It would be going too far to discuss all sections in detail, especially since most requirements will not be different from western methods and practice.
Some of the more interesting aspects will be discussed in this paper.

Pipeline Route Selection

As stated before, preliminary designs for offshore pipelines have been largely based on river crossings as they are specified in the SNiP 2.05.06-85: "Design of Long Distance Transmission Pipelines". This Regulation requires, for instance, the installation of spare pipelines in flooded areas of more than 500 m width. Minimum spacing between parallel lines ought to be at least 30 to 50 m for gas-lines. Parallel oil lines may be laid at closer distance and installation in a common trench is permitted.

Western Standards do not specify the necessity of spare pipelines. When a proper route selection is done with due attention to the system's safety, the probability of serious damage can be kept extremely low.
For the new Guidelines, it is recommended that the spacings between parallel pipelines are restricted by installation constraints only, irrespective the transported product. Such considerations concern laying and trenching tolerances, anchor patterns and anchor positions of the construction vessels, safety of a previously laid pipeline when a new pipeline is installed, interfaces with platform risers and land sections of the system, space for expansion loops, etc. This approach permits laying of lines in close proximity and the application of piggy-back lines.

In practice, the minimum distance between two (or more), not simultaneously installed pipelines, will be generally in the order of 5 to 10 m, when the lines remain resting on the seabed and 15 to 20 m, when the lines are buried afterwards with the use of post-trenching equipment. In shore approaches, where two or more lines are successively laid in a common trench (or as recently done in a common tunnel), the minimum distance can be kept as small as practically possible and only dictated by ensuring the structural integrity of the individual pipelines.

Allowable Stresses and Wall Thickness

A description of the method given in SNiP 2.05.06-85: "Design of Long-Distance Transmission Pipelines" to establish the wall thickness of a pipeline is presented in Appendix A.
Hereafter, a simple example is given to illustrate some of the differences in approach and results between this Code and some of the Western Standards. The simplification is that a straight pipeline section is considered with gas under high pressure. Bending is not addressed.

A pipeline with a pressurised product tends to expand circumferentially and, simultaneously, to reduce in length: the Poisson effect. However, when the pipeline is restrained in axial direction by the friction of the seabed, the line will be subjected to a tension force.

The example pipeline has the following main characteristics:
- outside diameter: D_H =1066.8 mm (=42 inch)
- gas pressure: p = 10 N/mm^2 (=100 bar)
- steel quality: API 5L Grade X60 or DIN StE 415.7 TM. Both steels have an SMYS (Specified Minimum Yield Strength) of 415 N/mm^2.

SNiP 2.05.06-85

With internal pressure in the pipeline and, due to pipe-to-soil friction, the pipeline being restrained and tensioned, the allowable stress in the steel is according to SNiP 2.05.06-85:

$$R_1 = \frac{R_1^H . m}{k_1 . k_H}$$

For the definition of the various design and reliability factors in this formula, reference is made to Appendix A
For an offshore pipeline with a size of 1066.8 mm, produced from high-strength steel, manufactured by a fully controlled process of rolling, double-sided longitudinal welding and 100 per cent non-destructive testing, the following values may be assumed:

m = 0.75
k_1 = 1.34
k_H = 1.10

The considered k_1 - value assumes a negative fabrication tolerance of not more than 5 per cent (API specifies 8 per cent).

The results are for API 5L Grade X60, with R_1^H=517 N/mm^2 (=SMTS):

$$R_1 = \frac{517 * 0.75}{1.34 * 1.10} = 263 N / mm^2$$

The nominal thickness according to SNiP 2.05.05-85 should be at least:

$$\delta = \frac{npD_H}{2(R_1 + np)} = \frac{1.1 * 10 * 1.0668}{2 * (263 + 1.1 * 10)}$$
$$= 0.02141 m. or 21.5 mm$$

with:

n = 1.1

The corresponding standard thickness is 22.22 mm (=0.875 inch).

A noteworthy aspect of a design method, which applies SMTS (R_1^H) instead of SMYS (R_2^H), is that materials, which are assumed to be equivalent (on the basis of an equal yield strength and elastic behaviour), like API 5L Grade X60 and DIN StE 415.7 result in a different wall thickness.

When DIN StE 415.7 is applied, the results are:

$$R_1 = \frac{550 * 0.75}{1.34 * 1.10} = 280 N / mm^2$$

with a minimum nominal wall thickness of:

$$\delta = \frac{npD_H}{2(R_1 + np)} = \frac{1.1 * 10 * 1.0668}{2 * (280 + 1.1 * 10)}$$
$$= 0.02016 m. or 20.2 mm$$

and a standard wall thickness of 20.62 mm (=0.812 inch).

ASME B31.8 "Gas Transmission and Distribution Piping Systems"

The nominal wall thickness is calculated from the following two formulae:

$S_h \leq F_1 ST$
$S_h = (P_i - P_o)D/2t$

The value of F_1 = 0.72 for pipelines.
Furthermore:
S = 415 N/mm^2 (=SMYS)
T = temperature derating factor, taken as T = 1
P_i = 10 N/mm^2
D = outside diameter
t = nominal wall thickness
P_o is conservatively taken as zero

From these figures, it follows that the nominal wall thickness should be at least:

t = 10*1.0668/2*0.72*415 =
0.01785 m =17.9 mm

This corresponds with a standard thickness of
t = 19.05 mm (= 0.750 inch)

BS 8010, Part 3: "Code of Practice for Pipelines, Pipelines Sub-sea"

This Code specifies the hoop stress as 72 per cent of the SMYS for pipelines.
The hoop stress is defined as:

$$\sigma_h = (P_i - P_o)D_o/2t$$

with:
D_o = outside diameter
t = minimum wall thickness

With P_o conservatively taken as zero, the minimum wall thickness is:

$$t = 17.9 \text{ mm}$$

Using a fabrication tolerance of 5 per cent, the nominal wall thickness should be at least

$$t_{nom} = 18.9 \text{ mm}$$

This corresponds with a standard wall thickness of 19.05 mm (=0.750 inch).

Summary

Except for the above mentioned codes, the wall thickness was determined for some more standards.

The results are given in Table 1.

Code	Pipe Specification	Nominal Wall Thickness (standard), mm
SNiP 2.05.06-85	API 5L Grade X60	22.22
ASME B 31.8		19.05
BS 8010, Part 3		19.05
DnV 1981 (Norway)		19.05
NEN 3650 (Netherlands)		20.62
SNiP 2.05.06-85*	DIN StE 415.7	20.62
NEN 3650* (Netherlands)		19.05

* These codes utilise the Specified Minimum Tensile Strength (SMTS or (R_1^H) as the basis for calculation of the pipeline wall thickness.

Table 1: Steel Wall Thickness according to Various Pipeline Codes

It is obvious that there is not really consistency between the various codes considered, also not between the western codes.

In our opinion, the thicker wall calculated with the SNiP-code is too conservative and we have proposed the following formulation for the calculation of the steel wall thickness in the new Guidelines for Offshore Pipelines in the Russian Federation:

$$t_{min} = \frac{(p_i - p_o)D_o}{2\sigma_A}$$

A combined stress level is calculated with the Von Mises criterion:

$$\sigma_{eq} = \sqrt{\sigma_h^2 + \sigma_L^2 - \sigma_h\sigma_L} \le \sigma_A$$

with:

t_{min} = minimum wall thickness in mm
p_i = internal design pressure in N/mm^2 or MPa
p_o = external hydrostatic pressure in N/mm^2 or MPa
D_o = outside diameter in mm

σ_A = allowable stress, which is equal to:

$$\sigma_A = \eta k_w k_t \sigma_y$$

with:

σ_y = Specified Minimum Yield Strength in N/mm^2
η = design factor, which is
- for the hoop stress:
  - 0.72 for pipelines
  - 0.60 for risers
- for the equivalent stress during construction and hydrostatic testing:
  - 1.0 for pipelines and risers
- for the equivalent stress during operation:
  - 0.96 for pipelines
  - 0.72 for risers.

k_w = weld joint factor, normally equal to 1
k_T = temperature derating factor. For temperatures between T= -25°C and +120°C, k_T is normally equal to 1.

The minimum wall thickness is chosen instead of the nominal thickness. Would the latter value be specified, yielding in a point with the minimum wall thickness would be possible during installation and/or hydrostatic testing.
For a straight pipeline, as considered in the above example, the result is a standard nominal wall thickness of 19.05 mm.

This approach is less conservative than as specified in the SNiP-code. However, based on the good record of offshore pipelines in Western offshore fields, there is no reason, why the suggested approach should not be introduced.

One last remark can be made regarding this subject:
SNiP 2.05.06-85 does not mention an extra allowance for internal steel corrosion or erosion, probably because the code is primarily concerned with larger trunk lines for the transport of treated, non-corrosive gases only. The proposed new Standard mentions the need to add an allowance for corrosion, because it is meant to be valid for the transport of all sorts of hydrocarbon products, gas and liquid, sweet, inert and sour, corrosive and non-corrosive.

Pipeline Stability

The equation to determine the minimum required submerged weight of the pipeline per meter length is:

$$f(W_{sub} - F_L) \ge n(F_D + F_I)$$

with:

f = pipe to soil friction coefficient
W_{sub} = submerged weight of the pipeline in N/m
n = safety factor between 1.0 and 1.1, depending on how accurate the hydrodynamic and seabed conditions around the pipeline are known.
F_D, F_L, F_I = drag, lift and inertia forces in N/m, which are calculated with the well known Morrison equations:

$$F_D = \frac{1}{2}\rho_w C_D D V_t |V_t|$$

$$F_L = \frac{1}{2}\rho_w C_L D V_t^2$$

$$F_I = \rho_w C_I \frac{\pi D^2}{4} \frac{dV_w}{dt}$$

with:

$\rho_w =$ density of seawater in kg/m^3
C_D, C_L, C_I = drag, lift and inertia coefficients
D = overall outside diameter of pipeline
V_t = water particle velocity, consisting of a quasi-static component of tidal and wind-induced current and an oscillatory wave-induced current
dV_w / dt = wave induced acceleration.

So far, the Russian and Western methods are similar.
However, there is a difference in the determination of the wave-induced water particle velocity and acceleration.
Where in Western design a most appropriate wave theory is applied, as a function of wave parameters and water depth (Linear, Stokes third or fifth order, Cnoidal or Solitary theory or Stream Functions), Russians utilise standard graphs showing the relations between the various factors from which the velocity and acceleration can be read off. This method specified in SNiP 2.06.04-82: "Loads and Forces acting on Marine Structures", is thus compulsory, but it gives conservative results. Although less conservative, Western approach is recommended as it has a proven record of pipeline safety.

Pipeline Burial

Besides the usual reasons of extra stability and protection against damage by anchors and fishing gear, pipelines in the Russian Arctic may require burial for safety against the risk of damage by floating ice, which 'ploughs' the seabed. Depth of burial can often be established from historical ice-gouges in the area. Burial is also required in shore approaches because of grounding of wind-driven ice sheets.

Permafrost Boundaries

Special care should be taken where a pipelines crosses a permafrost border, either close to or on shore, here suitable solutions and installation techniques should be employed to prevent serious pipeline damage and buckling due to successive thawing and freezing of the ground.
Excavation and filling the trench with granular back-fill material is one good solution.

Installation Operations

Offshore work periods in the Russian Arctic are rather limited because of the ice conditions not only at the work site itself, but also along the navigation routes to the site. Vessels which are mobilised for the work in the Arctic must be adequately ice-strengthened for the local circumstances and have the proper ice-classification.
Also onshore the working windows are limited, for instance in tundra areas with permafrost. During the summer period, a thin top layer will thaw and become very soft and thus seriously damaged by the pipeline installation equipment.

Anti-Corrosion Protection

This section of the new Standard has been written in line with Western design, manufacture, installation and operational practice. Sacrificial anodes have seldom been applied in the Russian Federation, whereas land-lines are usually protected by the different system of impressed current.

Quality Control

Proper welder and welding procedure qualifications have to be carried out as on Western projects. All welds have to be 100 per cent radiographically inspected with, at random, other non-destructive testing methods like by ultrasonics.
Weld acceptance criteria are in principle not much different from Western practice, but based on Russian research and experience. The new Standard considers Western criteria, except where Russian criteria are more stringent.

Safety of Pipeline Operation

Pipeline safety during operation should be ensured throughout the specified operational lifetime of the system and determined by properly addressing such items as:

- strength
- stability: weight coating and, if needed, burial
- burial for protection against seabed variations, fishing gear, anchors, floating ice etc.
- sufficient strength to prevent serious damage due to an earthquake
- prevention of collapse due to external pressure
- proper anti-corrosion protection by means of a suitable anti-corrosion coating, sacrificial anodes, internal corrosion allowance and/or protective internal lining
- over-pressure protection. As part of the review of Western Practice several project notes were exchanged between the parties involved. Appendix B gives, as an example, a project note on the matter of overpressure.

Conclusion

After a learning process, during which the parties had to acquaint themselves with each other, a good co-operation and working climate was established.
Despite the fact, that the communication was at times not easy and often required the assistance of an interpreter and translation of documents, the final product can be, in our opinion, of great value to further developments in the Russian Federation.
Our contributions to the new Standards are currently under review by various organisations with long experience in the Former Soviet Union. The exact form of the final versions is now in their hands.

References

1. SNiP 2.05.06-85: "Design of Long-Distance Transmission Pipelines"
2. SNiP III-42-80: "Rules for the Construction and Acceptance of Works"
3. R 412-81: "Recommendations for the Design and Construction of Main : Submarine Oil and Gas Pipelines"
4. VSN 51-09-85: "Submarine Oil and Gas Pipelines Design"
5. SNiP 2.06.04-82: "Loads and Forces acting on Marine Structures (Waves, Ice, Ship Loads)"
6. DnV 1981: "Rules for Submarine Pipeline Systems". Det norske Veritas
7. BS 8010: "Code of Practice for Pipelines, Part 3: Pipelines Subsea: Design, Construction and Installation"
8. ASME B 31.4: "Liquid Transportation Systems for Hydrocarbons, Liquid Petroleum Gas, Anhydrous Ammonia and Alcohols"
9. ASME B 31.8: "Gas Transmission and Distribution Piping Systems"
10. DIN 17172: "Steel Pipelines for Long-Distance Pipelines for Combustible Liquids and Gases"
11. API Spec 5L: "Specification for Line Pipe"
12. NEN 3650: "Requirements for Steel Pipeline Transportation Systems"

Appendix A: Allowable stresses and Wall Thickness Analysis
Appendix B: Project Note on Overpressure Protection
Appendix C: Short List of Definitions

Appendix A
Allowable Stresses and Wall Thickness Calculation

Allowable stresses

The SNiP 2.05.06-85 "Design of Long Distance Transmission Line is written for onshore oil and gas pipelines. Determination of the a lowable stresses is complicated because the formulae have to apply a large number of conditions. This is done by introduction of reliab ity or safety coefficients. Offshore pipelines are not addressed in th Code, but, for instance, river crossings are included.
The allowable stress follows from two expressions:

$$R_1 = \frac{R_1^H . m}{k_1 . k_H}$$

$$R_2 = \frac{R_2^H . m}{k_2 . k_H}$$

with:

R_1^H = Specified Minimum Ultimate (or Tensile) Strength

R_2^H = Specified Minimum Yield Strength

m = design factor for various functional properties, such as below or above ground, gas or oil transport, pipe size, area to be crossed etc. Values vary between m = 0.60-0.90

k_1, k_2 = reliability factors for the line pipe material. Values var between k_1=1.34-1.55 and k_2=1.10-1.20

k_H = reliability factor for the pipeline function. Values vary betwee k_H=1.00-1.40.

Wall-thickness

Two formulae are given for the calculation of the wall thickness:

$$\delta = \frac{npD_H}{2(R_1 + np)} \quad \text{for lines in tension}$$

and

$$\delta = \frac{npD_H}{2(\psi_1 R_1 + np)} \quad \text{for lines in compression.}$$

with:

δ = nominal wall thickness: the calculated value of the wall thickness is to be rounded off to the nearest larger value of the standard wall thickness. Negative tolerances on the wall thickness are not considered. So δ can be considered to be the minimum nominal wall thickness

p = working (standard) pressure in MPa

n = reliability factor for the internal working pressure of the pipeline

D_H= outside diameter of pipeline in mm

Finally:

$$\psi_1 = \sqrt{1 - 0.75\left[\frac{|\sigma_{loN}|}{R_1}\right]^2} - 0.5\frac{|\sigma_{loN}|}{R_1}$$

The longitudinal stress in this equation is given as:

$$\sigma_{loN} = -\alpha E \Delta T + \mu \frac{npD_{id}}{2\delta_H}$$

with:

α = thermal expansion coefficient ($11.7*10^{-6}$ m/m°C)
E = modulus of elasticity ($2.06*10^5$ N/mm^2)
ΔT= temperature difference between installation and operation
μ = Poisson's ratio (0.3 for steel)
D_{id}= internal pipe diameter
δ_H = minimum wall thickness.

The negative fabrication tolerance for welded line pipe should be not more than 5 per cent, so:

$$\delta_H = 0.95\,\delta$$

The longitudinal stress has to be verified against R_1 by:

$$|\sigma_{loN}| \leq \psi_2 R_1$$

in which:

$$\psi_2 = \sqrt{1 - 0.75\left[\frac{\sigma_h}{R_1}\right]^2} - 0.5\frac{\sigma_h}{R_1}$$

and the hoop stress:

$$\sigma_h = \frac{npD_{id}}{2\delta_H}$$

Bending stress

The Code gives also a formula for the longitudinal stress with a bending stress term included

$$\sigma_{lo}^H = \mu\sigma_h^H - \alpha E \Delta T \pm \frac{ED_H}{2R}$$

This longitudinal stress has to satisfy the condition that:

$$|\sigma_{lo}^H| \leq \psi_3 \frac{m}{0.9k_H} R_2$$

In axial tension: $\psi_3 = 1$

In axial compression:

$$\psi_3 = \sqrt{1 - 0.75\left[\frac{\sigma_h^H}{\frac{m}{0.9k_H}R_2^H}\right]^2} - 0.5\left[\frac{\sigma_h^H}{\frac{m}{0.9k_H}R_2^H}\right]$$

The term $\frac{ED_H}{2R}$ can be used to calculate the allowable bending radius (R) in horizontal and vertical curvatures.
This term can also be replaced by the more general term for bending stress, σ_b, and successively used to determine allowable free span lengths for a static loading condition (protection against oscillations caused by vortex shedding is calculated differently) and the maximum allowable size of seabed irregularities, such as solid outcroppings, sand waves, megaripples or scour holes. This information can be used to determine the best method and amount of intervention work to modify the seabed profile either by pre-sweeping or by installation of extra supports.

When the stresses in the pipeline approach the yield point, the stress-to-strain ratio (σ-ε) begins to deviate from the linear relationship:

$$\sigma = \varepsilon E.$$

In other words, the modulus of elasticity, E, is not constant anymore.
When the design is based on stress-limited values, Western practice is to use a constant value.
The SNiP-Code 2.05.06-85, however, gives formulae to calculate the real E-value as a function of the total equivalent stress and the consequential variation of Poisson's ratio, μ.

The influence on the stress-limited analysis, still the most common approach in offshore pipeline design, is relatively small and it would go too far to discuss this matter in more detail in this presentation.

Appendix B
Project Note: Pipeline Overpressure Protection

Introduction

A gas/oil production and transportation system consists of several components, each of them generally having its own design pressure.

Production wells will be exposed to pressures, which are directly related to the reservoir pressure. If a production wellhead assembly is located at some off-set from the receiving platform, the linking flowline has the same design pressure as the wellhead equipment.
Also, process installations on the platform may have their own design pressure, which is higher than the design pressure of the downstream export pipeline.

The sketched differences in design pressure require that at the boundary of two components a pressure control system is implemented which must protect the downstream component of the system against an overpressure. In addition, when the design pressure difference between the components is large, it may be needed to give the downstream section extra safety in order to prevent that its design pressure will be too far exceeded, should the pressure control system fail.
As stated elsewhere, the design pressure of a pipeline may be exceeded with 10 percent in case of an incidental pressure. Provided that the flow of gas or oil is rapidly interrupted in case of an upstream upset or failure, this kind of pressure exceedance can be seen as an incidental event.
So:

- if the design pressure in the process plant upstream of a pipeline is not more than 10 percent above the design pressure of the pipeline, it is not needed to install a system, which automatically stops or deviates (via an outlet) the flow of the commodity. This is under the proviso, that in case of malfunction of the pressure control valve at the boundary of plant and pipeline, an alarm signal is given and the reason for the upset relatively rapidly remedied. In this situation, the pressure increase can be considered to be an incident.
- if the design pressure in the process plant is more than 10 percent above the design pressure of the pipeline, an alarm system has to be installed, which activates interruption or deviation of the flow, when the pipeline design pressure will be exceeded by more than 10 percent. This can be an automatic system - direct closure of a valve - or a system with operates via human intervention.

Whatever approach is used, it has to be ensured that the response to the pressure upset is quick.

A code which discusses overpressure protection in more detail is the Dutch Code NEN 3650. This approach is further explained below.

It should be noted, that this code allows an incidental pressure of 15 percent above the design pressure, whereas most other regulations specify a maximum increase of 10 percent.

Overpressure Protection

The Dutch code NEN 3650 presents, briefly, the following formulation about the necessity and extent of an overpressure protection system:

- a pressure safety system is not required, if

  Pdv / Pdn $\leq$ 1.15

  in which: Pdv = upstream design pressure
  Pdn = downstream design pressure

- a single pressure safety system is required, if

  Pdv / Pdn $\leq$ 1.30

- two independent pressure safety systems are required, if

  Pdv / Pdn > 1.30.

The overpressure protection system consists of all installed equipment for pressure control, pressure alarm and pressure safety and can have the following components:

- the pressure control equipment which has to be set at such a point, that the design pressure (or the maximum operating pressure, if this is lower than the design pressure) in the export line is not exceeded
- the pressure alarm system, which needs to detect any malfunction of the pressure control system. Level and duration of pressure set-point exceedance have to be recorded via this system. An alarm system is not needed, if the pressure safety system is set at the design pressure, which is a lower setting and more conservative than follows from the above given ratios
- the pressure safety system, which has to be such, that exceedance of the set pressure level triggers automatically an operation, which prevents that the pressure in the downstream pipeline rises too high. It is obvious, that the set pressure may not be higher than the sum of the design pressure and the permissible incidental pressure, which is at maximum 1.15 times the design pressure.

The following pressure safety systems are possible:

- a pressure relief system. This combats the consequences of a pressure upset. The source upstream is not blocked and the pipeline system remains available. The excess product is released either to the atmosphere (gas) or into special buffers (oil). An example of this system is a spring loaded relief safety valve.
- a pressure non-relief system. This combats the cause of the pressure upset. The upstream source is blocked and the pipeline system is not longer available. An example is a block valve, which is activated by a 2 out of 3 choice system.

It could be added, that the pressure settings of the equipment have to take into account the accuracy of the instruments, normally in the order of $\pm$ 2 percent.

The export pipeline can be considered to include all elements of the system between the exit of the process plant on the platform and the block valve at the pipeline termination point on land. This includes such parts as the pig launcher with its appendages, piping components, valves, tees, reducers, blow-down system etc.

Emergency Shut-Down Valves

Another means of safety, particularly for the platform, is the Emergency Shut-Down Valve (ESDV). It is not really an overpressure protection system, but rather a tool to stop the gas/oil stream in an emergency situation. Its function is two-fold:

- to stop the flow, when there is a downstream damage of the system, either in the pipeline or the onshore plant

- to stop a return flow to the platform, when the platform system is out of order for any reason. This back flow is either caused by the fact, that the pressurised gas begins to expand when the pressure is dropped (this is less of a problem with the hardly compressible oil) or in a situation that the export line is also a trunkline for other fields with the risk of a reversed flow in the direction of the damaged platform.

Because of the latter possibility, ESDV's have been sometimes installed in the on-bottom section of pipelines at some distance from the platform within the 500 meter safety zone. This concept can be complicated when the valve is located in the expanding, moving part of the pipeline (near the expansion loop). Also, the installation with a protection frame against dropped objects can be rather difficult, while scour can undermine the installation. During operation, regular service of the valve, including functional testing and maintenance, will normally require diver's aid.

To avoid all these problems, many ESDV's have been integrated in the riser, at a safe location. BS 8010, Part 3 specifies that an ESD valve should be installed in the pipeline riser and it is added, that consideration should be given to the installation of a subsea isolation valve as this may contribute to the safety of offshore personnel and installations.

Conclusion

It is recommend to specify in a new standard the implementation of an overpressure protection system, with a pressure control system, and with one pressure safety system when the design pressure in the upstream platform process plant is more than 10 percent above the pipeline design system and a double safety system when the difference is more than 25 percent.

The latter value happens to be equal to the hydrostatic test pressure. It is important to specify an overpressure protection safety system, as it may prevent interface problems where the platform piping is often designed by others than the pipeline design team and in accordance with other -piping- standards.

Specifying a protection system is than a guarantee that the upstream facilities take into account the safety of the pipeline. Implementation of an ESD valve in the riser should be mandatory. Application of a bypass for the ESDV should be left to the operator and does not have to be mentioned in the Code.

Appendix C
Short List of Definitions

Submarine Pipeline (transportation) system
An interconnected system of submarine pipelines and pipeline riser system(s) for the transport of gaseous or liquid hydrocarbons or other commodities with associated safety systems, valves, tool launchers and/or receivers, corrosion protection systems and other appurtenances. It does not include the processing equipment piping systems on the offshore installations.

Submarine pipeline
The predominantly horizontal part of the pipeline system, normally resting in or on the sea bottom. It can be situated between risers on offshore platforms, between a riser and the first valve onshore or between a riser and the first valve on a subsea installation. In this Standard a submarine pipeline is referred to as a pipeline. The pipeline may include certain piping components.

Pipeline riser system
The predominantly vertical part of the pipeline system, which is supported by the platform.
The pipeline riser system consists of the pipeline riser, its supports and all integrated piping components.

(Pipeline) Riser
The part of the pipeline riser system which actually connects the pipeline on the sea floor with the processing equipment on the platform.. Depending upon whether a riser is exposed to or sheltered from the action of wind, waves and current, it is designated as an external or internal riser. The bottom bend is considered to be part of the riser.
Launching and receiving facilities installed on the platform for internal cleaning and inspection tools are generally considered to be a part of the riser.

Expansion device
A relatively flexible section of a pipeline system to reduce the direct transfer of loads and displacements from one section to another section. An expansion device is usually positioned at the transition of submarine pipeline and riser and investigated for stress and displacement in combination with the riser.

Piping components
Items integrated in the pipeline system such as flanges, tees, bends, reducers and valves.

J-tube system
J-shaped conduit with a riser inside. The steel or flexible riser is installed by pulling in the pipeline horizontally and converting into a vertical position. In case of a steel riser the bending usually causes the steel material to yield. The conduit extends to deck level as a permanent protection for the internal riser(s).

Corrosion protection system
The means to protect the pipeline system from corrosion. These can be corrosion protection coating and/or cathodic protection.

Cathodic protection
A technique to prevent corrosion of the pipeline system by making its surface the cathode of an electrochemical cell. This can be achieved by either impressed current or sacrificial anodes.

Weight coating
A coating applied around the pipeline to provide sufficient negative buoyancy. It may also be applied to serve as protection against mechanical damage to the pipeline and applied to the corrosion protection system.

Accidental loads
Loads to which the pipeline system may be subjected by accident or under abnormal circumstances.

Environmental loads
Loads caused by environmental conditions.

Functional loads
Loads caused by the physical existence of the pipeline system and by the operation and handling of the system.

Specified Minimum Yield Strength (SMYS)
The tensile stress at which the total elongation is not more than 0.2 per cent. For design purposes it is permitted to assume a linear relationship between stress and strain for stress values of less than or equal to SMYS.(Note: API specifies 0.5 per cent elongation)

Maximum operating pressure
Maximum pressure to which a pipeline system may be subjected during normal operation, which includes the static pressure and the pressure required to overcome the friction. It is essential that this pressure does not exceed the design pressure.

Design pressure
The pressure selected as the maximum sustained pressure exerted by the pipeline system contents to which the pipeline system is to be designed.

Transient pressure or surge pressure
An incidental pressure created by an upset in the steady state flow condition in a pipeline system, normally caused by valve operation, pump start or trip-out, fluctuation of a control valve or inaccuracies of instrument set-points.
Transient or surge pressure may not exceed the design pressure by more than 10 per cent.

Nominal wall thickness
The pipe wall thickness that is specified for the supply of pipes.

Minimum wall thickness
The nominal wall thickness less the maximum tolerance for under thickness of the wall.

Nominal pipe diameter
The pipe outside diameter that is specified for the supply of pipes.

Proceedings of the Seventh (1997) International Offshore and Polar Engineering Conference
Honolulu, USA, May 25-30, 1997

ISBN 1-880653-28-1 (Set); ISBN 1-880653-30-3 (Vol. II)

Limit Force Ice Loads and Their Significance to Offshore Structures in the Beaufort Sea

G. Comfort, S. Singh and A. Dinovitzer
Fleet Technology Ltd.
Kanata, Canada

ABSTRACT

The loads on an arctic offshore structure may be governed by the force necessary to fail the ice at the structure or the environmental forces available to drive the ice past the structure. The factors controlling limit-force loads were investigated, and the likely magnitudes of these loads were established. Limit-force and limit-stress loads were compared for a scenario in which an Extreme Ice Feature interacts with an offshore structure. Probabilistic descriptions were produced for all key parameters, and Monte-Carlo analyses were carried out. The analyses showed that, at exceedence probability levels of interest for design, extreme ice loads on the structure are likely to be reduced significantly by the available driving forces, in comparison to the case where only limit-stress loads are considered. However, the analyses can only be considered to be preliminary as more data are required to define the key variables, and other load-limits, such as limit-momentum, need to be considered.

INTRODUCTION

Ice loads on offshore structures in the southern Beaufort Sea may be limited by the force to fail the ice at the structure, (typically termed the limit-stress), the available environmental driving forces and the capability of the pack ice to transmit them (typically termed the limit-force), the momentum of the ice, and the force necessary to split floes contacting the structure. This paper summarizes analyses made by Comfort, Dinovitzer, and Gong, 1994; and Comfort and Singh, 1996 to quantify limit-force loadings, and to compare them to limit-stress loads.

APPROACH AND SCENARIO CONSIDERED

The distribution of limit-stress and limit-force loadings was predicted using Monte-Carlo analyses for the interaction of an Extreme Ice Feature (EIF) with a structure. This assumed scenario commences with EIFs being generated north of the southern Beaufort Sea, and then drifting towards the area of interest (which was considered to lie between latitudes 70° and 71° N, and longitudes 132.5° and 135°W). When an EIF was near the structure, it was assumed to move towards the structure, and to possibly contact it. If contact occurred, the EIF was assumed to lodge in front of the structure and stop there. Then, large pack ice driving forces were assumed to build up, and the resulting structure load was limited by either limit-stress or limit-force conditions. The annual load probability exceedence distribution was analyzed for structures at four locations covering the geographic area of interest for cases where only limit-force or limit-stress ice failures were allowed to occur (Cases A and B, respectively), and for the case where limit-stress and limit-force ice failures were both allowed to occur (Case C).

LIMIT-FORCE LOADS ON A STRUCTURE: CALCULATION APPROACH

Equation [1] was used to calculate the limit-force load, F_{lf}, during steady-state rubble building (depicted in Figure 1):

$$F_{lf} = DL_p + 2l\mu(L_q\cos\theta + L_p\sin\theta) \qquad [1]$$

where:
- l = the length of the rubble cone
- θ = the half angle of the rubble cone
- μ = the friction factor along the boundary between the stationary rubble cone and the moving pack ice. It is related to the internal friction angle of ice rubble, ϕ, as $\mu = \tan\phi$.
- D = the width of the floe lodged in-front of the structure
- L_p = the pack ice line load along the ice drift direction
- L_q = the pack ice line load normal to the ice drift direction

The shape and size of the rubble cone are related to the degree of confinement in the pack ice. Assuming that the rubble cone is formed with minimum force, L_p can be related to L_q by minimizing Equation [1] with respect to the half angle of the rubble cone. This gives:

$$L_p = L_q\tan\theta \qquad [2]$$

A value of 30° was used for the half-angle of the rubble cone based on available field data for large rubble fields (Figure 2). The friction angle was also taken to be 30°. This was selected by assuming that Coulomb's friction law is applicable, and thus, the rubble cone half-angle is related to the friction angle (equaling 45°-ϕ/2). The width of the rubble cone facing the pack ice (i.e., d in Figure 1) was taken as 0.5D. The EIF width is probabilistic. This was selected within each run of the Monte-Carlo simulation from the distribution shown in

Figure 3, which was truncated to produce a range of EIF widths from 0.5 to 20 km.

CALCULATION OF PACK ICE PRESSURE LINE LOADS

Data collected during previous field programs indicate that ridge-building pressures are related to the thickness of the ice from which the ridges were formed, termed the ridge ice thickness (Figure 4). Furthermore, the data suggest that the line load is relatively insensitive to the length of the ridging front, for ridging lengths greater than about 400 m (Comfort et al., 1994, and Comfort and Singh, 1996).

Croasdale et al., 1992 suggested that the interaction force would increase steadily with the ridge ice thickness raised to a power of 1 to 1.25. This is expected to be appropriate for ridge-building in thin ice where flexural failures are predominant. However, as the ridge ice thickness ice becomes greater, crushing becomes a more dominant part of the process. Hence, for thick ice, the contact pressure would be expected to decrease with the square root of the contact area (Sanderson, 1988). The equations below, which have this general form, were developed from the database (Figure 4), and used to predict pack ice line loads.

$$LL_{rb} = 100\ h_r^{1.25}, \quad \text{for } h_r < 2 \text{ m} \qquad [3]$$

$$LL_{rb} = 892 * h_r^2 - 3436\ h_r + 3540 \quad \text{for } 2 < h_r < 2.5 \text{ m} \qquad [4]$$

$$LL_{rb} = 90.6 * h_r^3 - 997 * h_r^2 + 3760 * h_r - 4030 \quad \text{for } h_r > 2.5 \text{ m} \qquad [5]$$

where: LL_{rb} = the ridge-building line load, in kN/m
h_r = the ice thickness, in m, from which the ridge was formed

Equations [3] to [5] allow the pack ice forces generated during an event to be calculated given that the ridge-building ice thickness is known. The ridge-building thickness, and hence, the ridge-building force is expected to increase during a storm event as ridge-building occurs progressively during an event, starting with the thinnest ice. For these analyses, it was assumed that all of the ice of one thickness in the pack is consumed by ridges before ridge-building shifts to ice of a greater thickness.

"Ice consumption," and hence the ridge-building ice thickness, during an event are controlled by: (a) the thickness distribution in the pack ice; and (b) the strain in the pack ice which was defined as:

$$\frac{\text{the ice movement that occurs during an event}}{\text{the overall distance scale over which pack ice processes act}} \qquad [6]$$

The ridge-building ice thickness and the pack ice strain are both probabilistic variables, and they were related to each other through "ice consumption" by ridge-building during an event. This relationship was established by analyzing ice thickness distribution data collected by Melling and Riedel, 1993; 1994, and by Wadhams and Horne, 1980. These analyses indicate that all of the relatively thin ice in the pack, i.e.. less than about 1 m thickness, would have been consumed by ridge-building after strains of about 5% had occurred (Figure 5).

A method for calculating pack ice strain was established by reviewing the pack ice dynamics that occurred during the 1986, 1989, and 1991 Pack Ice Projects (described in Croasdale et al., 1988; Comfort and Ritch, 1990, and Comfort et al., 1992, respectively). Comfort et al., 1994 showed that, to a first order, pack ice driving force events can be analyzed by assuming the polar pack to be a rigid boundary that is moved onshore and offshore by the prevailing wind and circulation patterns. Furthermore, Comfort et al., 1994 showed that the width of the shear zone was a reasonable choice for the denominator in Equation [6].

The pack ice strain is a probabilistic variable, and it was selected within each run of the Monte-Carlo simulation from the distribution shown in Figure 6, which was truncated for the analyses such that the range of pack ice strains considered was 0.75 to 60 %. This allowed the ridge-building ice thickness to be specified, and from that, the pack ice line load, using equations [3] to [5]. Figure 6 was established by analyzing ice movements recorded by drifter buoys placed in the southern polar pack through the International Arctic Buoy Program, ice charts prepared by the Atmospheric Environment Service, and wind data collected at Tuktoyaktuk, NWT. As there is insufficient space in this paper to describe the development of Figure 6, the reader is referred to Comfort, Dinovitzer, and Gong, 1994, and Comfort and Singh, 1996 for further information.

It should be noted that although the strain was calculated using Equation [6] by dividing the onshore ice movement, Δ, by the onshore width of the shear zone, W, the analysis is not restricted to onshore ice movements. If the ice movement is at an angle, α, to the landfast ice edge, the strain would still be Δ/W, as $(\Delta/\sin\alpha)/(W/\sin\alpha) = \Delta/W$. When the ice movement is along the landfast ice edge, where α=0, this formulation is not applicable.

LIMIT-STRESS LOADS: CALCULATION APPROACH

An empirical approach was adopted for predicting limit-stress loads, F_{ls}, as follows:

$$F_{ls} = \sigma_\varepsilon B H \qquad [7]$$

where σ_ε is the average ice pressure acting over the width of the structure, B, and the EIF thickness, H. Two methods were used to define the average ice pressure, as follows:

- Using the EIS Aspect Ratio Curve (EIS, 1982) which is an empirically-derived curve that relates the contact pressure to the aspect ratio (Figure 7).
- Using a constant contact pressure of 0.6 MPa for all cases.

The EIF thickness is the only probabilistic variable in this calculation. The EIF thickness distribution used was established by truncating an ice thickness distribution applicable to the area of interest to the range of 20 m to 30 m (Figure 8). Twenty meters were selected as the minimum EIF draft based on Croasdale, 1986 and Pilkington et al., 1992.

As the structure is wide, the average ice thickness "seen" across its width will be less than the maximum measured over smaller distance scales. This effect was quantified in Comfort et al., 1994 by analyzing raw ice draft data collected by Melling and Riedel, 1993 for multi-year ice conditions. For ice thicknesses more than 20 m, the average ice thickness seen by a 100 and 200 m wide structure was about 80% and 75% of the maximum ice thickness, respectively. This was accounted for in the analyses by adjusting the EIF draft shown in Figure 8.

ENCOUNTER FREQUENCY

As the EIF arrival rate varies with the geographical location (De Paoli et al., 1982), Cases A and C were evaluated for four locations (i.e., 70° N,132.5° W; 70° N,135° W; 71° N,132.5° W; and 71° N,135° W). Case

B was evaluated for two locations (i.e.,70° N,135° W and 71° N,132.5° W). De Paoli et al., 1982 developed equation [8] to predict the annual encounter frequency, F_{enc}, of an EIF with the structure during winter (which was taken to be 265 days long):

$$F_{enc} = N\, P_e\, P_{lc}\, (W + B) / B_s \qquad [8]$$

where: N = the number of ice features passing Cape Prince Alfred each year, which they found to be 0.1 for ice islands and 600 for multi-year hummock fields. "N" was taken as 600 for the analyses presented here.

P_e = an empirical factor developed by De Paoli et al., 1982 to account for the fact that most EIFs do not enter the southern Beaufort Sea. They determined "P_e" to be 7 %.

P_{lc} = the modified latitude crossing probability. Values were established for P_{lc} for each location by scaling Figure 4.10 in their report, as follows:

(a) 70°N;132.5°W 0.0010
(b) 70°N;135°W 0.0013
(c) 71°N;132.5°W 0.0033
(d) 71°N;135°W 0.0029

W = the width of the EIF

B = the width of the structure

B_s = the bin size used by De Poali et al., 1982 to develop the modified latitude crossing probabilities (which was 28 km for the analysis done with the 1/10 ice edge).

Equation [8] was modified as shown below in equation [9] to account for the following issues

- Peak pack ice pressures are only expected to occur during the March - May period as the ice is relatively thin in the early season, and ice movements are less frequent in the Jan.-Feb. period.
- The term $(W + B) / B_s$ in Equation [8], was accounted for within each Monte-Carlo run rather than explicitly as is done in Equation [8]. This allowed the EIF width, W, to be treated as a probabilistic variable.

$$F'_{enc} = N\, P_e\, P_{lc}\, F_t \qquad [9]$$

where F_t is a duration factor based on the proportion of the winter period over which high pack ice pressures are expected (taken as 92 days /265 days = 35 %).

RESULTS: EXCEEDENCE PROBABILITY LOAD DISTRIBUTION

The results for Case A (loads controlled entirely by limit-force), Case B (loads controlled entirely by limit-stress) and Case C (loads controlled by the combination of limit-force and limit-stress) are plotted in Figures 9-10, 11, and 12-13, respectively. As the limit-stress load is quite sensitive to the amount of open water in the pack ice initially, analyses were done for an extreme case representing a close-packed condition (with 0 % open water) and a looser initial pack ice condition (with 7 % initial open water). As expected, significantly higher loads were predicted for the "no open water case". Also, for these cases, the loads at an annual exceedence probability of 10^{-4} for the locations at 71° N latitude are much larger than those at 70° N latitude (Table 1). This reflects the fact that EIFs are much more likely to be present near the structure at the more northern locations. For the cases that include limit-stress loads, much higher loads are predicted when the EIS Curve is used to define the contact pressure, as expected.

For Case C, the load distributions at all four locations show a distinct change in slope at 1050 MN for the case where a constant limit-stress pressure of 0.6 MPa was assumed (Figures 12 and 13). This is a transition load between limit-force and limit stress, and it corresponds to the lower limit of the EIF draft used in this analysis. When the force on the structure is less than the transition load, the load is controlled entirely by limit-force, as the applied pack ice forces are insufficient to cause limit-stress failure of the minimum draft EIF. When the force is greater than the transition load, limit-stress and the limit-force failures can both occur. The transition load is not as clearly distinguishable for the analyses done using the EIS curve as it is a more complex relationship. However, slope changes are also evident (at 2300 MN) corresponding to this transition (Figures 12 and 13).

RESULTS: COMPARISON OF LIMIT-FORCE AND LIMIT-STRESS LOADS

Limit-force and limit-stress loads were compared at an annual exceedence probability of 10^{-4}, which is the rare end of the range specified in CSA code S471-92 (CSA, 1992) for Safety Class 1 for environmental loads.

Table 1: Predicted Loads For Cases A, B and C For a 100 m Wide Structure

Case	Location		Predicted Load (MN) At An Annual Exceedence Probability Of 10^{-4} For The Indicated Initial Open Water Amounts In The Pack And Limit-Stress Pressures			
	Lat. (°N)	Long. (°W)	7 % Initial Open Water		0 % Initial Open Water	
			0.6 MPa	EIS Curve	0.6 MPa	EIS Curve
A	70	132.5	575	Loads Are Not Affected By Limit-Stress	750	Loads Are Not Affected By Limit-Stress
	70	135	675		850	
	71	132.5	1200		1600	
	71	135	1050		1450	
B	70	135	1450	4025	Loads Are Not Affected By The Initial Open Water Amount	
	71	132.5	1525	4400		
C	70	132.5	575	575	750	750
	70	135	675	675	850	850
	71	132.5	1200	1100	1160	1500
	71	135	1050	1000	1140	1400

When limit-force and limit-stress ice failures were both allowed to occur, the load predicted at an annual exceedence probability level of 10^{-4} ranged from 575 MN to 850 MN for locations at 70°N (Table 1). Higher loads are predicted for a location at 71°N as EIF incursions are more likely to occur further north. For locations at 71°N, the load predicted at an annual exceedence probability level of 10^{-4} ranged from 1000 MN to 1500 MN.

The loads predicted when only limit-stress was allowed to occur were much greater than those where limit-force was taken into account. For locations at 71°N, and depending on the initial open water content in the pack and the method used to define limit-stress contact pressures, the limit-stress loads with an annual exceedence probability of 10^{-4} are 1 to 4 times higher than the corresponding limit-force loads. At the two 70° N locations, the predicted loads were reduced by factors of about 2 to 5 when limit-force was taken into account in comparison to limit-stress loads.

CONCLUSIONS AND RECOMMENDATIONS

The analyses indicate that extreme ice loads on an offshore structure are likely to be limited by the available pack ice driving forces at annual exceedence probability levels of interest. However, the analyses are subject to a number of limitations, and therefore can only be considered to be preliminary. A better definition of the key environmental variables is required. Also, the analyses should be expanded to include other load limits, such as limit-momentum.

ACKNOWLEDGEMENTS

The work was sponsored by the National Energy Board, with funding being provided through the PERD (Panel on Energy Research and Development) Program. Drs. I. Konuk and G. Timco were the scientific authorities.

REFERENCES

Abdelnour, R., Paterson, B., Comfort, G., and Singh, S., 1996, Pack Ice Pressure: A Review And Analysis Of Available Data, Fleet Technology Ltd. report 4368 prepared for the National Energy Board, the Canadian Coast Guard, and the Atmospheric Environment Service.

Canatec Consultants Ltd., 1994, Study of Ice Rubble Formation in the Beaufort Sea - Final Report, contractor report prepared for National Energy Board, Mobil R&D Corp., and Gulf Canada Resources Ltd.

Comfort, G. and Ritch, R., 1990, Field Measurements Of Pack Ice Stresses, proc. OMAE, Houston.

Comfort, G. Ritch, R. and Frederking, R., 1992, Pack Ice Stress Measurements, proc. OMAE, Calgary.

Comfort, G., Dinovitzer, A., and Gong, S., 1994, Limit-Force Ice Loads And Their Significance To Offshore Structures In The Beaufort Sea, Fleet Technology Ltd. report 4346, published by National Research Council.

Comfort, G., and Singh, S., 1996, Analyses Of Limit-Force Ice Loads And Their Significance To Offshore Structures In The Beaufort Sea, Fleet Technology Ltd. report 4491, submitted to the National Energy Board.

Coon, M.D., Lau, P.A., Bailey, S., and Taylor, B.,1989, Observations Of Floe Stresses In The Eastern Arctic, proc. POAC Conference, Lulea, Sweden.

Croasdale, K.R., 1986, Extreme Ice Feature - Structure Interaction, proc, Workshop On Extreme Ice Features, sponsored by the National Research Council, Banff, Alberta.

Croasdale, K., Comfort, G., Frederking, R., Graham, B. and Lewis, L., 1988, A Pilot Experiment To Measure Arctic Pack Ice Driving Forces, proc. POAC Conference, Fairbanks, Alaska.

Croasdale, K., Frederking, R., Wright, B., and Comfort, G., 1992, Size Effect On Pack Ice Driving Forces, proc. IAHR Symposium on Ice Problems, Banff.

CSA, 1992, General Requirements, Design Criteria, The Environment, And Loads, Part Of The Code For The Design, Construction, And Installation Of Fixed Offshore Structures, CAN/CSA-S471-92, prepared by the Canadian Standards Association.

De Paoli, S., Morrison, T., and Marcellus, R., 1982, Analysis Of Interaction Probabilities Between Large Ice Features And Offshore Structures In The Canadaian Beaufort Sea, Dome Pet. Ltd. report submitted with the Beaufort Sea Environmental Impact Statement.

EIS, 1982, An Initial Probabilistic Assessment Of Ice Loading On Beaufort Sea Structures. Prepared by CMEL for the Canadian Beaufort Sea EIS Partners, 1982.

Hallam, S., Duckworth, R, Pickering, J., and Westermann, P., 1985, Katie's Floeberg 1985: Pack Ice Driving Force Measurements, internal British Petroleum report prepared by the Technology Development Division.

Melling, H., and Riedel, D.A., 1993, Draft And Movement Of Pack Ice In The Beaufort Sea, April 1990 - March 1991, Canadian Technical Report Of Hydrography And Ocean Sciences No. 151.

Melling, H., and Riedel, D.A., 1994, Draft And Movement Of Pack Ice In The Beaufort Sea, April 1991 - April 1992, Canadian Technical Report Of Hydrography And Ocean Sciences No. 162.

Prinsenberg, S., Fowler, G., Van Der Baaren, and Beanlands, B., 1997, Ice Stress Measurements From Landfast Ice Along Canada's Labrador Coast, to appear in Cold Regions Science and Technology.

Pilkington, R., Hill, C., Metge, M., and McGonigal, D., 1992, Beaufort Sea Ice Design Criteria: Acquisition Of Data On EIFs, ESRF report 115.

Sanderson, T., 1988, Risks To Offshore Structures, published by Graham and Trotman, London.

Wadhams, P., and Horne, R.J., 1980, An Analysis Of Ice Profiles Obtained By Submarine Sonar In The Beaufort Sea, Journal of Glaciology, vol. 25, no. 93.

Wright, B., 1994, personal communication.

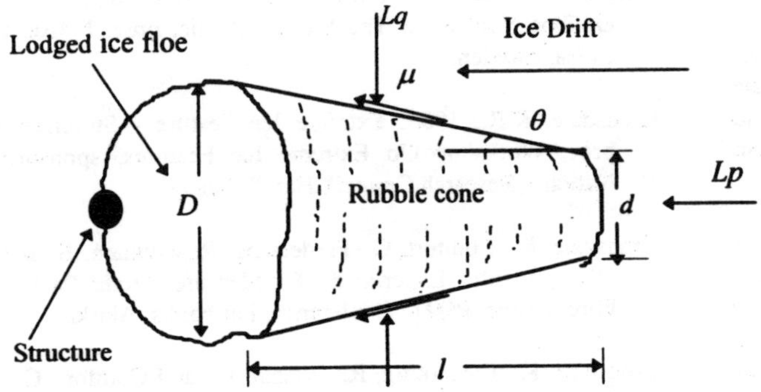

Figure 1: Development of rubble cone ahead of the lodged ice floe.

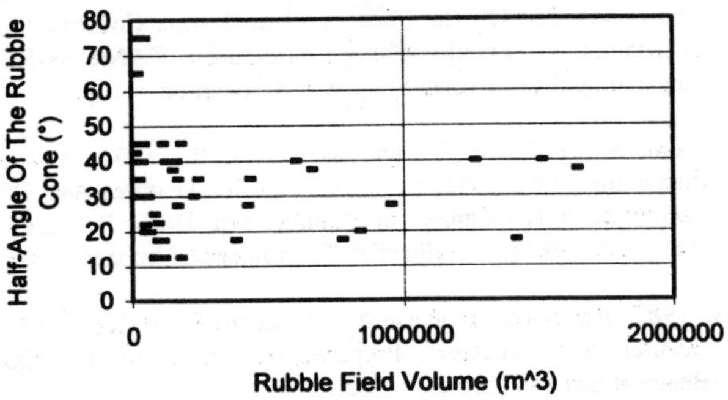

Figure 2: Geometry of rubble fields formed at structures in the southern Beaufort Sea (taken from data compiled by Canatec Ltd., 1994).

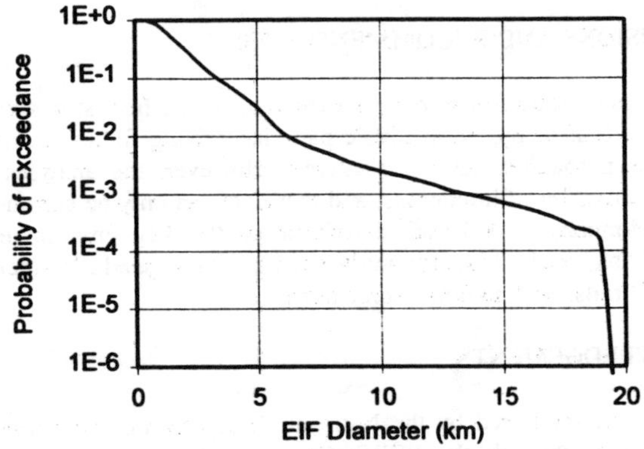

Figure 3: Distribution of EIF Diameter

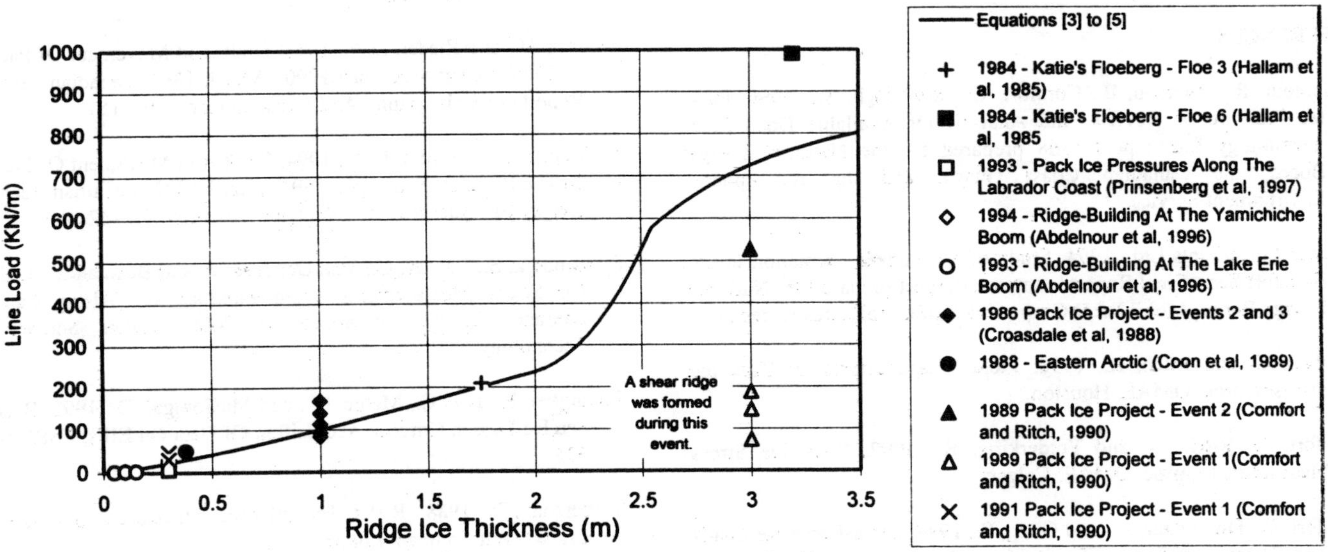

Figure 4: Ridge-building loads. The indicated data sources mainly present the raw stress data that were collected. These data were analyzed by Comfort et al, 1994 and Abdelnour et al, 1996 to produce the line loads plotted here.

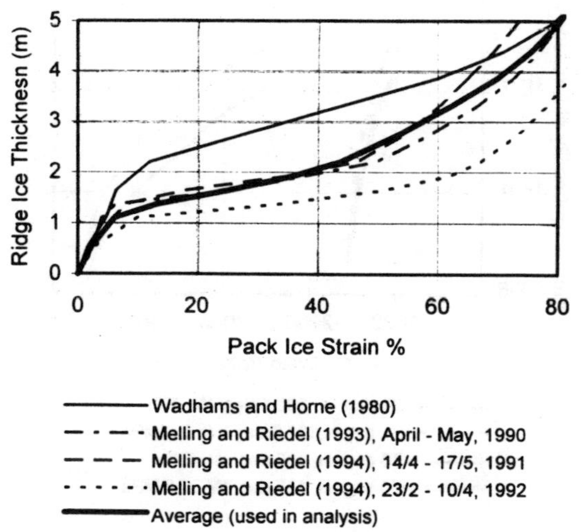

Figure 5: Progressive increase in the ice thickness available for ridge-building during an ice movement event.

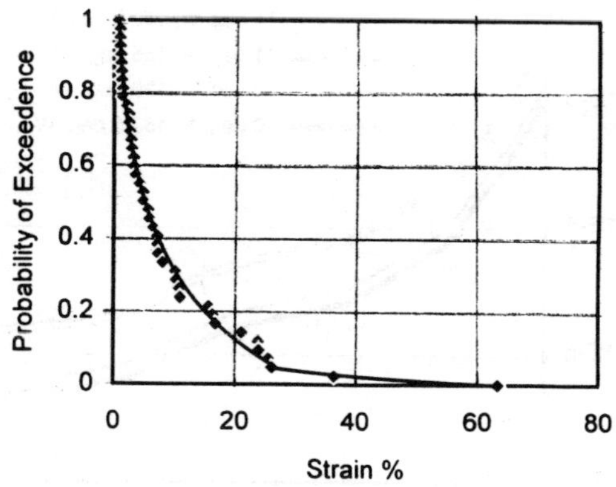

Figure 6: Onshore ice movement strains from buoy movement data and AES charts from period 1980 to 1991.

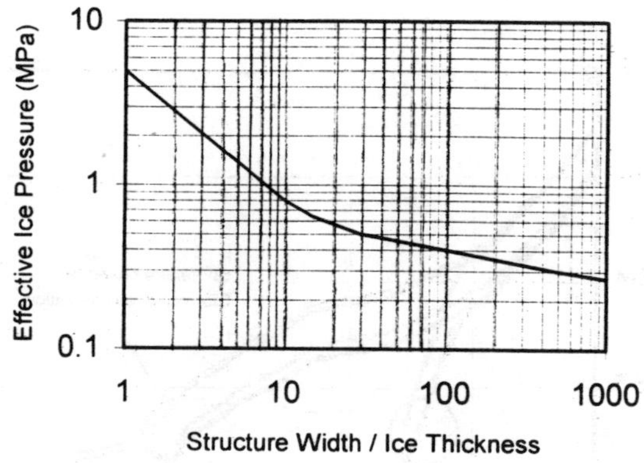

Figure 7: Effective ice pressure on large Arctic structures (EIS, 1982).

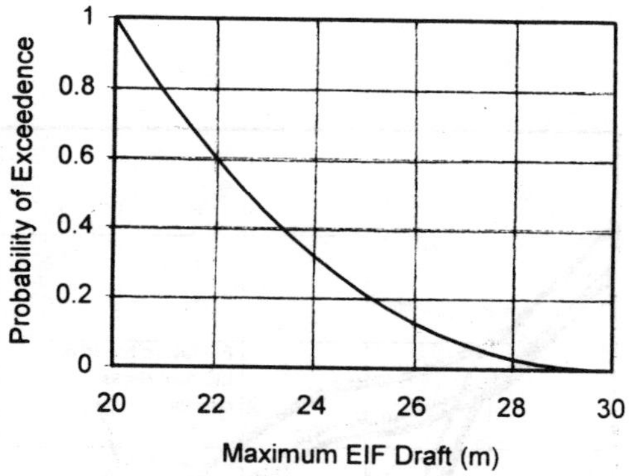

Figure 8: Maximum ice draft for ice floes > 470 m (Wright, 1994).

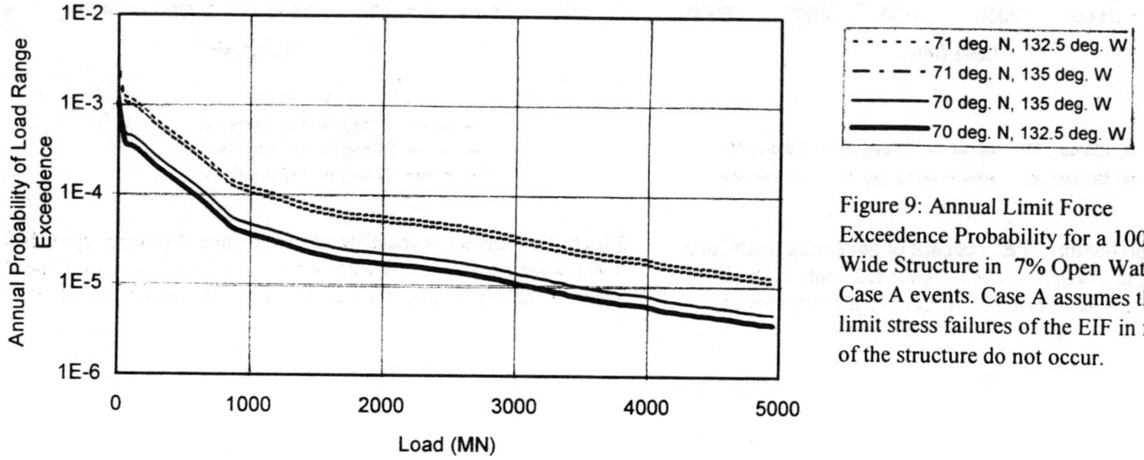

Figure 9: Annual Limit Force Exceedence Probability for a 100m Wide Structure in 7% Open Water for Case A events. Case A assumes that limit stress failures of the EIF in front of the structure do not occur.

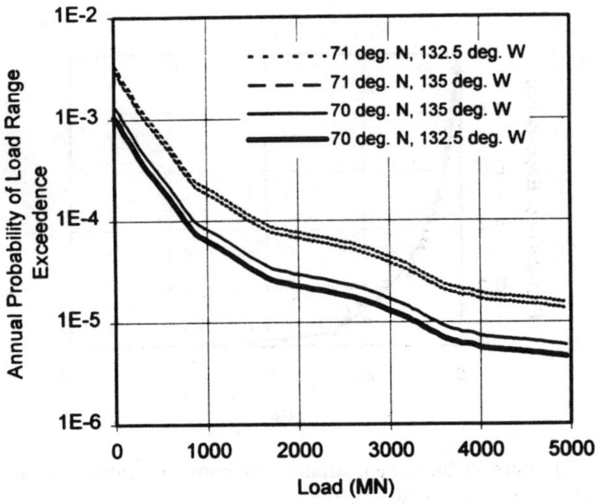

Figure 10: Annual limit force exceedence probability for a 100m wide structure in 0% open water for Case A Events. Case A assumes that limit stress failure of the EIF in front of the structure do not occur.

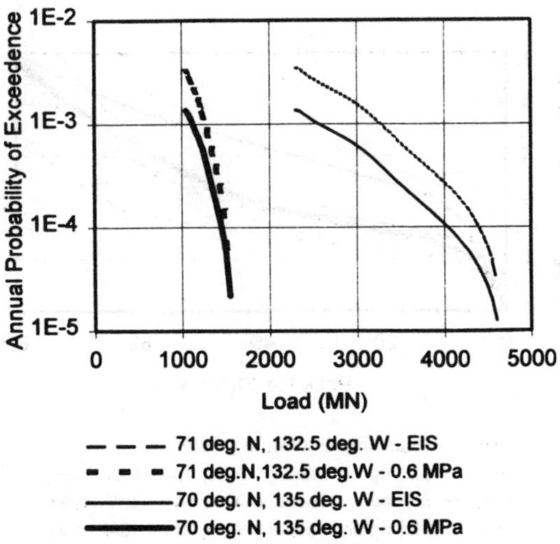

Figure 11: Annual limit stress force exceedence probability for a 100m wide structure for Case B events. Case B assumes that infinite pack ice driving forces are applied to the EIF in front of the structure.

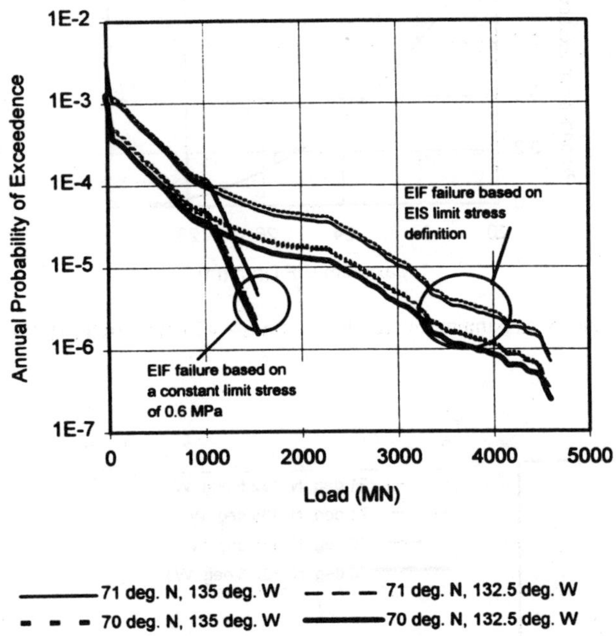

Figure 12: Annual probability of exceedence of governing loads for a 100m wide structure in 7% open water for Case C Events. Case C assumes that limit stress failures of EIF in front of the structure can occur.

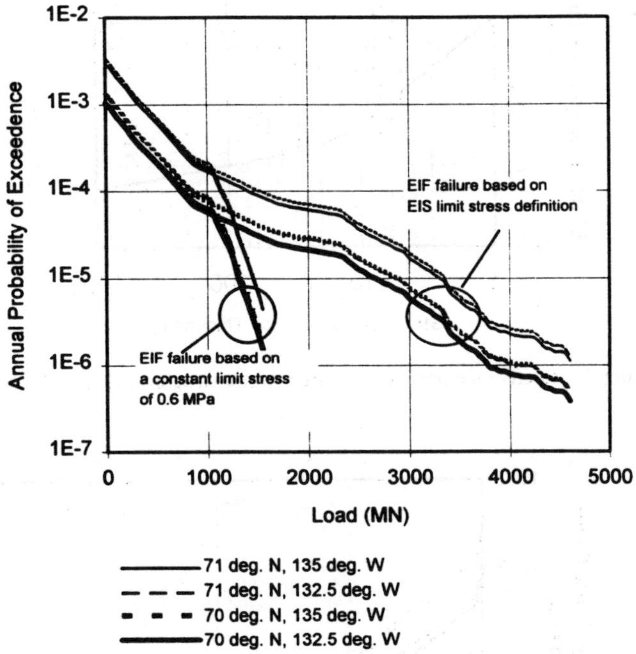

Figure 13: Annual probability of exceedence of governing loads for a 100m wide wtructure in 0% open water for Case C Events. Case C assumes that limit stress failures of EIF in front of the structure can occur.

Proceedings of the Seventh (1997) International Offshore and Polar Engineering Conference
Honolulu, USA, May 25-30, 1997

ISBN 1-880653-28-1 (Set); ISBN 1-880653-30-3 (Vol. II)

Wave-Induced Behaviour of a Submarine Pipeline Laid on an Erodable Bottom: Full-Scale Experiments

W. Dursthoff
University of Hannover
Hannover, Germany

W. Magda, T. Marcinkowski and B. Mazurkiewicz
Technical University of Gdansk
Gdansk, Poland

ABSTRACT

Tests were carried out in the *Großer Wellenkanal (GWK)* which is the Large Wave Flume in Hannover in order to investigate the degree of wave-induced lowering of a cylinder laid on a sandy seabed. These tests are part of further physical experiments with respect to significant influencing parameters. Therefore the results are representing the first step of subsequent tests under large-scale conditions.

NOMENCLATURE

D	=	diameter of the cylinder
e	=	gap between cylinder and bottom
f_{ha}	=	horizontal vibration frequency in air
f_{va}	=	vertical vibration frequency in air
f_{hw}	=	horizontal vibration frequency in water
f_{vw}	=	vertical vibration frequency in water
G_s	=	specific gravity of sand
H	=	wave height
H_s	=	significant wave height
I_D	=	relative density of sand
L	=	wave length
S_r	=	realative saturation of sand
T	=	wave period
T_p	=	peak period
t_{stb}	=	degree of lowering depth after stabilization
V	=	volume of samples

INTRODUCTION

From a review and resulting programme for further investigations on soil-water-pipeline interaction [*Dursthoff & Mazurkiewicz, 1996*] it can be concluded that the number of parameters which influence the soil-water-pipeline interaction is very high and that it is currently quite impossible to take into consideration not only all the parameters but also a significant part of them during model tests. Thus it was decided that the planned set of tests would concern the estimation of the influence of some parameters, i.e. of the permeable bottom, while the scale of the model tests was considered as a very important factor.

The investigations on submarine pipelines performed at the Large Large Wave Flume in Hannover can be classified into two typical sets. The first set concerned a horizontal pipeline fixed to the Large Wave Flume walls with a certain gap between the pipeline and the Flume bottom (e/D = 0.0 to 1.0), while the bottom was either smooth or rough. The pipe was also chosen as smooth or rough. The second typical set was performed for a pipeline connected with the Flume wall in such a way that the pipe could move vertically. The Flume bottom was covered by a layer of sand of the thickness of 1.0 to 1.3 m.

The results obtained from investigations performed within the framework of the first typical set were described in several publications, e.g. *Dursthoff & Mazurkiewicz, 1988; Dursthoff, Mazurkiewicz & Preser, 1989; Dursthoff, 1993; Preser & Dursthoff, 1993; Dursthoff & Sakout, 1995*. Thus it only has to be mentioned that as result a modified Morison equation was elaborated [*Preser, 1992*], followed by a theoretical solution based on the potential theory of incompressible perfect fluids and ideal boundary conditions [*Marcinkowski, 1993*].

This paper will focus on the results obtained during the investigations made within the framework of the second typical set, i.e. for a pipeline laid on a sandy bottom, while the defined subjects of the study were :

- development of erosion (scour) in the sandy bottom in the proximity of the pipeline under the influence of waves,
- self-lowering of the pipeline-permanent movements of the pipeline with respect to the initial position,
- development of soil resistance,
- hydrodynamic forces acting on the pipeline,
- loadings of pipeline span supports,
- velocity fields around the pipeline.

In this paper the two first subjects will be discussed. The other ones are curently under preparation.

EXPERIMENTAL ARRANGEMENTS

The experiments, as mentionned above, were made in the Large Large Wave Flume in Hannover. The main inside dimensions of the flume are 320 m length, 5 m width, 7.2 m total depth, maximum still-water depth 5 m.

Regular and random waves are generated mechanically by a vertical-plate wave generator. Maximum wave height 2.0 m, wave length H = 54 m, wave period 8 seconds (wave steepness 1 : 27). A special software programme has been designed and installed which controls the generation of waves with respect to the waves recorded at the profile where the tested structure is placed. With the aid of the process computer-based control of the wave machine, the re-reflections that can disturb the control of the generator are filtered out by means of a special software programme.

The test pipeline (Fig. 1) has a diameter of 800 mm, thus from the point of view of its diameter it lies within the range of diameter occurring in practice. The axis of the cylinder made from square steel elements with side dimension of 100 mm was equipped on both ends with special bearings housings which can be fitted to the clam-beam recesses in the flume walls. Due to the installation of hydraulically operated side profiles, the whole pipeline could be fixed at any desired point or height or left free, however without the possibility of horizontal movements.

The test pipeline was equipped with two types of measuring devices: strain gauges and pressure sensors. The strain gauges were attached to the axis of the cylinder for measuring the horizontal and vertical force components . The pressure sensors were placed in a measuring structure in the form of a ring located in the middle part of the pipeline. A total of 18 pressure sensors uniformly distributed around the circumference of the pipe were used for the tests described here. The sensors are of the piezoresistive type and are prepared for measurements of the absolute static and dynamic pressure occurring on the pipeline wall.

Several wave gauges were installed for the measurement of the changes in the water level in the Large Wave Flume. They consisted of two electrodes, one wave measuring string, one reference electrode and one pre-amplifier. The measurement principle is based on the assumption that the decrease in the current of the immersed measuring string is proportional to the immersion depth.

The calibration and estimation of the zero point of all the pressure sensors were made with the aid of an air pressure calibrator prior to the installation of the sensors in the pipeline using.

The strain gauges were calibrated dynamically on site at the installed pipeline in the Large Wave Flume. The installed wave gauges were calibrated for the water working levels during the filling and emptying of the Large Wave Flume. During the filling and emptying of the Flume the calibration factors of the pressure sensors could be checked for controlling purposes using the changeable height of the water column.

The mean values of several sensors were measured by measuring the initial state which lasted for 20 seconds at still water conditions and was taken into account during the final utilisation process by means of a computer programme.

After all the loading tests had been finished, the eigenfrequency of the vibration of the pipeline was estimated by plotting the vibration course due to the impulse and pointwise type of loadings. The following results were obtained:

- the horizontal pipeline in air (A), the gap between the pipeline and the bottom of the flume e = 0.5 m horizontal impulse:

  \+ 32.20 Hz = horizontal vibration frequency f_{ha}

+ 36.25 Hz = vertical vibration frequency f_{va}

- the horizontal pipeline under water (W) (water layer above the pipeline at a height of 0.2 m), gap between the pipeline and the flume bottom e = 0.5 m, horizontal impulse:

  + 22.40 Hz = horizontal vibration frequency f_{hw}

  + 22.50 Hz = vertical vibration frequency f_{vw}

The natural frequencies of the horizontal pipeline are significantly higher than those to be expected when induced by the vortex shedding transverse vibrations. Thus it was possible to cancel the resonance investigations.

The total arrangement of the pipeline and all the measuring sensors are shown in Fig. 2.

The sandy bottom consisted of a fine sand layer characterised by proper sieve curves. The thickness of the sandy layer was equal to 1.0 m during the first part of tests and 1.3 m during the second part (Fig. 3). Before the installation of the pipeline the sand was put in place by heavy machinery equipment and levelled roughly using the bulldozer. The final levelling was done by hand.

The first filling of the flume with water was done very slowly to a level roughly equal to the top level of the sand layer. The water level at this step was maintained through 48 hours. It has to be mentioned here that after each test, when the levelling and the compaction of the sandy layer had been done, the water level was always a little higher than the top of the sandy layer.

The measurements of the geotechnical parameters of the sand clearly indicate an increase in the degree of saturation of the sand with the increase of the number of tests performed. This conclusion is confirmed by tests made on samples of the volume V = 861 cm^3 of sand characterised fully by G_s = 2.65 g/cm^3 , n_{min} = 0.32, n_{max} = 0.43, e_{min} = 0.47, e_{max} = 0.75.

After each test and after levelling done by hand, the sandy layer was artificially compacted over an area within a length of 28.0 m and a width of 5.0 m (14 m in front of and behind the pipeline, Fig. 3). The compaction was done using deep vibrators which reached a depth equal to about 80% of the thickness of the sandy layer. The compaction time of the measuring area was equal to between 6 and 7 hours. The relative densities I_D and degrees of saturation S_r achieved, before and after experiments are shown in Fig. 4a and 4b for instance.

For the second typical set of investigations, the sand profile positioner and the velocity measuring technique were additionally installed as new measuring arrangements.

The sand profiler (Fig. 5) is a mechanically working system in which a profile roller connected to a contact sensor arm is pulled above the sand bed. It can be used either in water or in dry conditions over a length of 250 m (x-direction) and a height of 6 m (z-direction). The repeated accuracy is equal to ±10mm.

With a measuring velocity of 10 m/min it is also possible to detect very long sand profiles. The accuracy of the repeated measurements on a stiff bottom is ±2 mm. In sand, the profile roller compacts the surface layer in an insignificant way. With ±20 mm under water as well, the error of the final value is only ±0.5 %.

Generally speaking, through the use of a data acquisition system the bottom profiler provides the possibility of presenting the topographical changes caused by wave action on a horizontal pipeline.

Three types of velocity measuring gauges were tested: the micro-propeller gauge, the electromagnetic field gauge and the acoustic doppler velocimeter. It was decided after some trials to use the third type (Fig. 6), sensing three components of flow at velocities up to 5 m/s for current and wave measurements. It has to be mentionned that the velocity ranges from 1 mm/s to 5 m/s with a velocity resolution of 0.1 mm/s.

TEST RESULTS

Of 56 tests, 50 were performed with regular waves (R) and 6 with irregular waves (S). The irregular waves were generated on the basis of the JONSWAP spectrum with $\gamma = 3.3$, H_s.= 1.30 m and T_p = 8.0 s. The list of tests performed with the time added up and with the lowering values of the pipeline with respect to the initial position is given in Table 1.

The tests in group 1 to 5 were done for the sand layer thickness of 1.0 m, and in group 6 to 9 for the thickness 1.3 m.

From Table 1 and from Fig.7 it can be seen that a general conclusion can be drawn for the waves used, i.e. their parameters and time of action: after a certain time the pipe indicates stabilization which means that the self-lowering (settlement) is stabilized or stopped. As can be seen from the measurements of the geotechnical parameters of the sand on which the pipeline was laid, the increased number of waves caused an increase in the relative density I_D and the degree of saturation S_r.

This can mean that the resistance of the soil increased with time is independent of the erosion which took place at the pipeline.

Further phenomena were revealed:

- The stabilization of the self-lowering (settlement) of the cylinder took place roughly after the same time for the same wave height (e.g. for H = 1.0m, $t_{stb} \approx 40$ min for H = 1.5 m, $t_{stb} \approx 10$ min).

- The value of the self-lowering (settlement) of the pipeline at the moment of stabilization increased at the same wave height with the increase of the wave period.

T (sec)	SELF LOWERING AT STABILIZATION (cm)
4.0	~18
6.0	~26
8.0	~45

  This can mean that for longer waves the process of compaction is longer and that due to lower vibrations the pipeline itself has a smaller influence on the mobilization of the sand particles.

- The increase of the thickness of the sand layer from 1.0 to 1.3 m increased the final value of pipeline self-lowering (settlement), i.e. from 45.4 to 48.8 cm for T = 8.0 s and from 38.9 to 60.3 cm for T = 6.0 s. It is not a significant difference, particularly when the diameter of the pipeline is analysed. It indicates, however, that for a sandy layer of infinite thickness, the self-lowering (settlement) of the pipeline can be higher, even reaching a depth equal to the pipeline diameter. An additional parameter, which was not taken into consideration, is the weight of the pipeline and the resistance of the soil to vertical pressures. Assuming that the stability factor of a pipeline is near 1.0 for hydrostatic loads, one can consider that the vertical pressures of a pipeline acting on the sea bottom will be mainly caused by the water pressures acting on the pipeline and induced by waves.

- The cylinder could move only vertically. This means that horizontal motions of the pipeline, which could displace soil particles to the sides of the pipeline and thus increase the self-lowering (settlement) process, could not act. It also means that further tests should consider free movements of the pipeline.

The bottom profiles along the pipeline and perpendicular to the pipeline were measured during the investigations.

The profiles along the pipeline and on both sides of the pipeline are summarised for points indicated in Fig. 8a/8b and in Table 2.

The following interpretation can be made from the results presented, as well as from the results deduced from measurements of the longitudinal profile at a distance of 15 m from both sides of the pipeline (Fig. 9):

1. An increase of the embedding depth of the pipeline took place with the increased time and increased self-lowering of the pipeline (Figs. 10 and 11). This can be interpreted as an increase of the stability of the pipeline with the increased number of waves. Taking into consideration the increase of the relative density of the sandy bottom with time, the pipeline can be regarded as a time-dependent roughness element of the sea bottom, while the hydrodynamic loading decreases significantly.

2. The following statements can be made concerning the erosion:

   + no significant differences in the erosion depths could be observed in either side of the pipeline during the first phase of experiments;

   + the differences in the erosion depth increased after a longer period of wave impact, while it can be stated from results given in Table 2 that the lee erosion was greater than the luff erosion;

   + an increase in the erosion depth at both the ends of the pipeline, i.e. at the flume walls, was observed after almost all the tests; this aspect supports the previously mentioned recommendation concerning the necessity of the free placement of the cylinder on the sandy bottom during further experiments.

3. Measurements of the longitudinal bottom profile in the proximity of the pipe (Fig.9) indicate a significant influence of the pipeline on the shape of the bottom profile.

The bottom became very rough (Fig. 1), with several reefs which have a random distribution. Analysing the profile and the depths given in Table 2, it may be concluded, that, in the end, the pipeline will play the same role as the reefs which, according to the investigations, were formed from very compacted sand . This might mean that for sandy bottoms the main loadings on the pipeline are caused by sand waves.

4. The general assumption made during the preparation of the investigfation was to allow the pipeline free vertical movements. This was achieved, although it must be stressed that the cylinder indicated a slight inclination, which could be caused by uneven friction between both the side profiles of the pipeline and the dam-beam recesses. The limitation of the pipeline movements only to vertical ones significantly influences the development of an additionnal force, i.e. the active earth pressure from the luff side. Assuming that the pipeline could move freely in the horizontal direction, the active earth pressure could not develop. This also means that the erosion from both sides of the pipeline could reach the same depth. This statement can be supported by the fact that the transport of sand above the pipeline took place. This led to the assumption that the pipeline, instead of moving horizontally with the action of the waves, created a kind of obstacle behind which the sand was stored. The symmetrical development of erosion under the action of waves could be expected for a pipeline freely laid on the sandy bottom. Such symmetry was to be observed at the beginning of the experience.

ACKNOWLEDGEMENT

All tests carried out in the Large Wave Flume in Hannover were financially supported by the Commission of the European Union under the European Programme PECO. This enabled the authors to promote effective cooperation between the Universities of Hannover and Gdansk/Poland.

REFERENCES

Dursthoff, W./Mazurkiewicz, B. (1988): "Effects of seabed proximity on the hydrodynamic forces exerted on a submerged cylinder". *Proceedings 7th Conference on Offshore Mechanics and Artic Engineering, Houston, Vol. V*

Dursthoff, W./Mazurkiewicz, B./Preser, F. (1989): "Large-scale tests on smooth and rough cylinders placed horizontally on a plane bottom". *Archivum Hydrotechniki, T. XXXVI, Z. 3-4*

Preser, F. (1992): "Belastung horizontaler Rohrkörper in Meeresbodennähe durch Seegang im Übergangsbereich". *Mitteilungen des Franzius-Instituts für Wasserbau und Küsteningenieurwesen der Universität Hannover, Heft 73*

Wilde, P./Preser, F./Dursthoff, W. (1992): "A mathematical and observation model for waves and forces on a horizontal cylinder". *Proceedings 2nd International Offshore and Polar Engineering Conference, San Francisco, Vol III*

Marcinskowski, T. (1993): "Influence of the small gap between the pipeline and the sea bottom on the distribution and value of the wave-induced hydrodynamic forces on the pipeline". *Thesis Technical University of Gdansk*

Dursthoff, W. (1993): "The importance of large-scale tests in connection with the development of wave energy converters". *Proceedings of the 1993 European Wave Energy Symposium, Edinburgh, Scotland*

Preser, F./Dursthoff, W. (1993): "Loads on a prototype scaled horizontal pipeline in intermediate warter depth". *Proceedings 3rd International Offshore and Polar Engineering Conference, San Francisco, Vol III*

Dursthoff, W./Sakout, A. (1995)

Dursthoff, W./Mazurkiewicz, B. (1996): "Soil-wave-pipeline interaction - Review and resulting program for further investigation". *Mitteilungen des Franzius-Instituts für Wasserbau und Küsteningenieurwesen der Universität Hannover, Heft 79*

Fig. 1: Pipeline in the Large Wave Flume

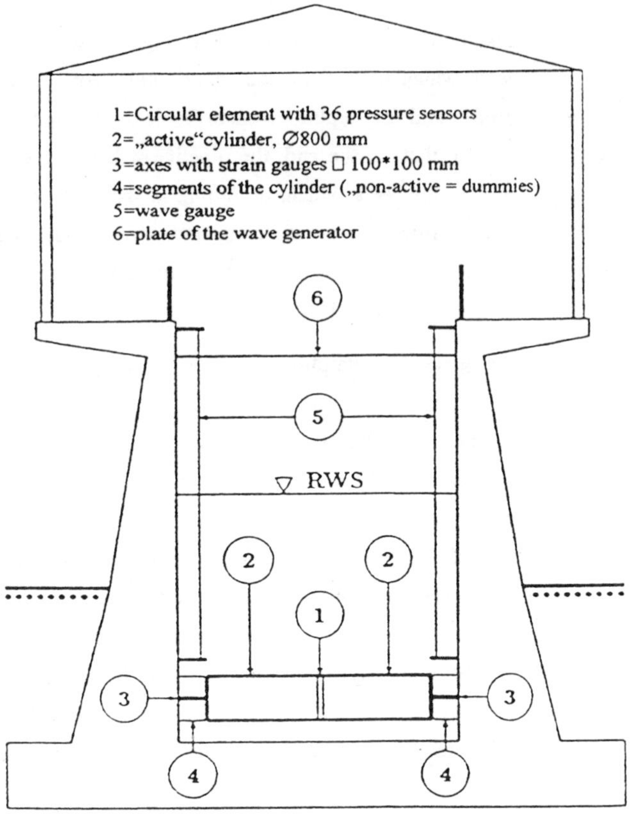

Fig. 2: Setup of the test cylinder

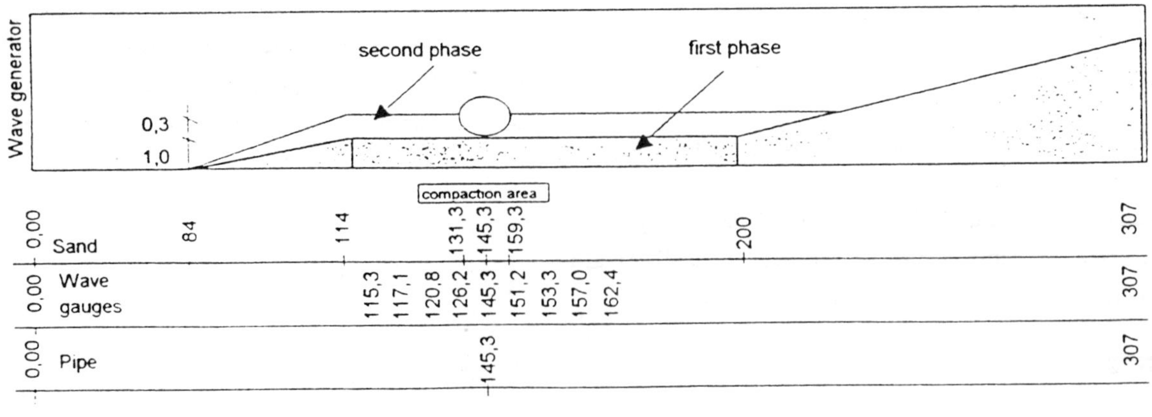

Fig. 3: Longitudinal section of the wave flume with the sand layer

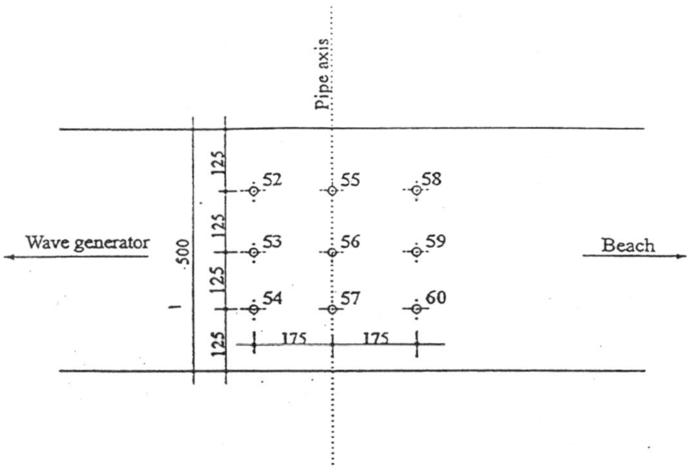

Nr	Box G_1 [g]	Wet probe + Box G_2 [g]	Dry probe G_3 [g]	I_D [-]	S_r [-]
52	123.1	1815.0	1441.2	0.60	0.79
53	122.7	1803.3	1425.6	0.54	0.79
54	123.2	1808.1	1436.6	0.58	0.78
55	122.9	1872.6	1491.8	0.79	0.87
56	123.2	1851.9	1477.0	0.72	0.83
57	122.8	1827.1	1452.8	0.64	0.80
58	123.5	1883.2	1490.7	0.78	0.90
59	123.3	1841.2	1465.2	0.69	0.82
60	123.5	1815.7	1439.5	0.59	0.80

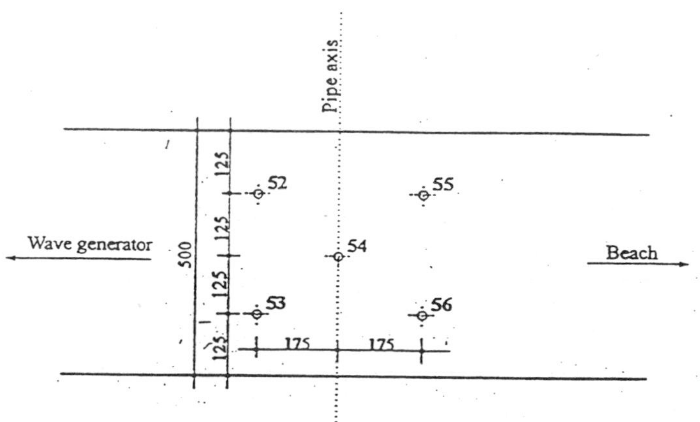

Nr	Box G_1 [g]	Wet probe + Box G_2 [g]	Dry probe G_3 [g]	I_D [-]	S_r [-]
52	123.1	1886.3	1502.0	0.93	0.89
53	122.7	1945.5	1545.2	0.98	0.99
54	123.2	1950.5	-	-	-
56	123.2	1931.0	1535.0	0.94	0.97
57	122.8	1924.3	1535.8	0.94	0.94

Fig. 4: Layout of the seabed sampling and relative density I_D and degree of saturation S_r before (Fig.4a) and after (Fig. 4b) the test runs

Fig. 5: Sand profiler

Fig. 6: Velocity measuring device

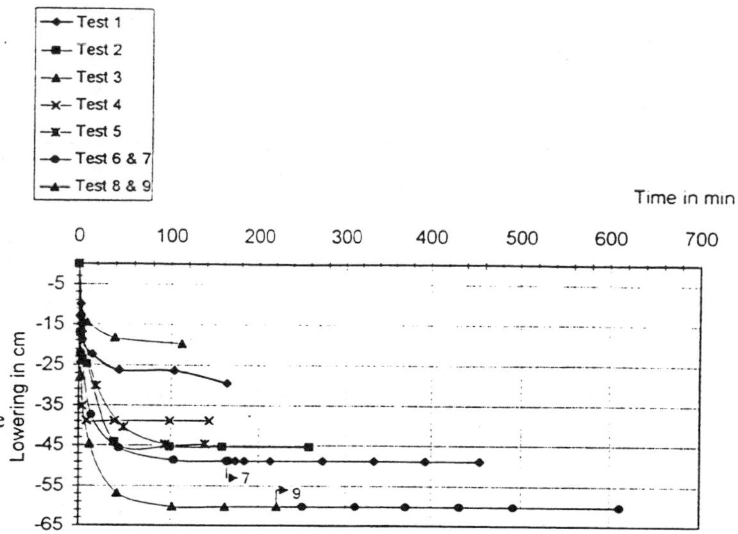

Fig. 7:Time dependent lowering of the cylinder under action of wave

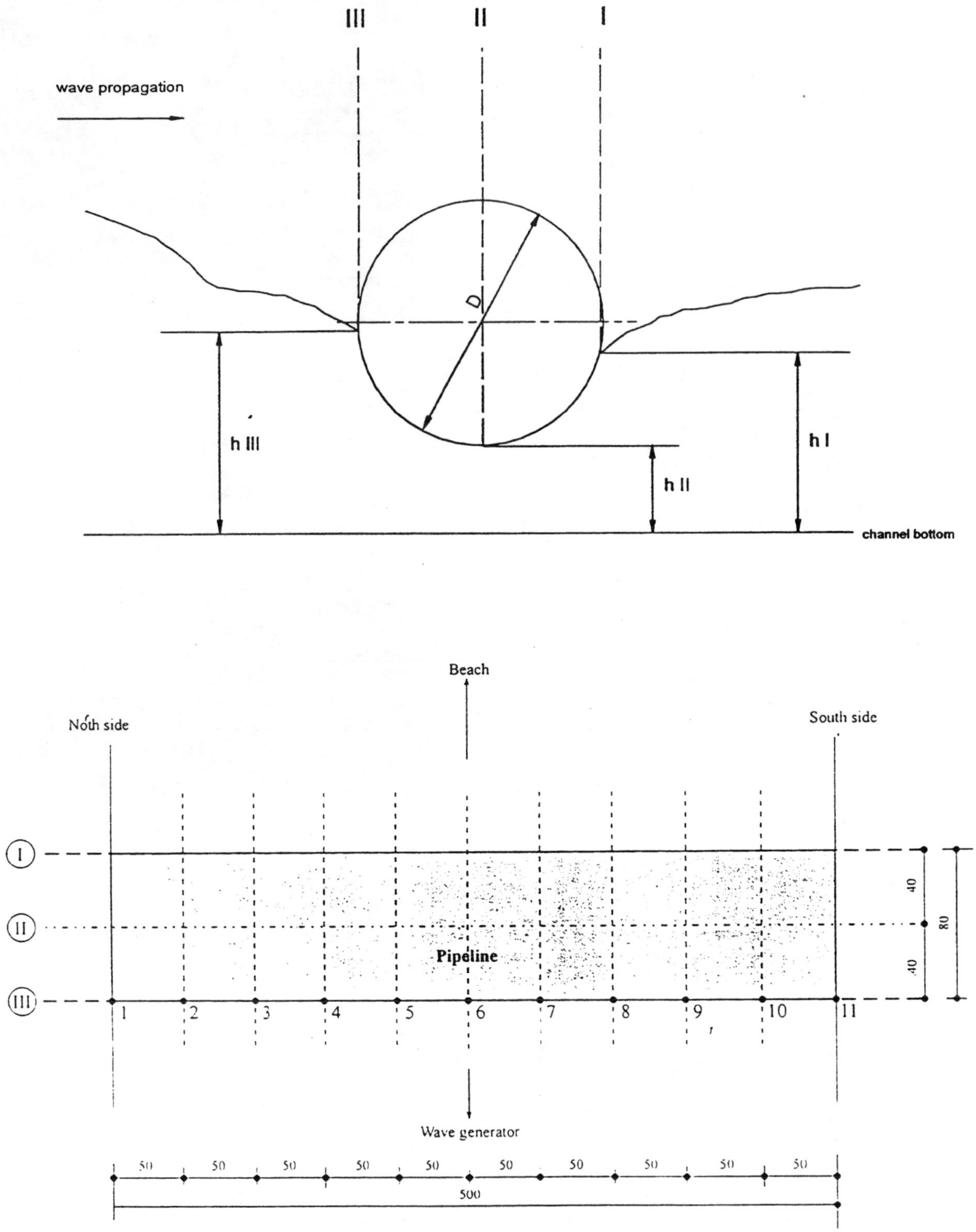

Fig. 8:Measurement layout of bottom profiles along the cylinder

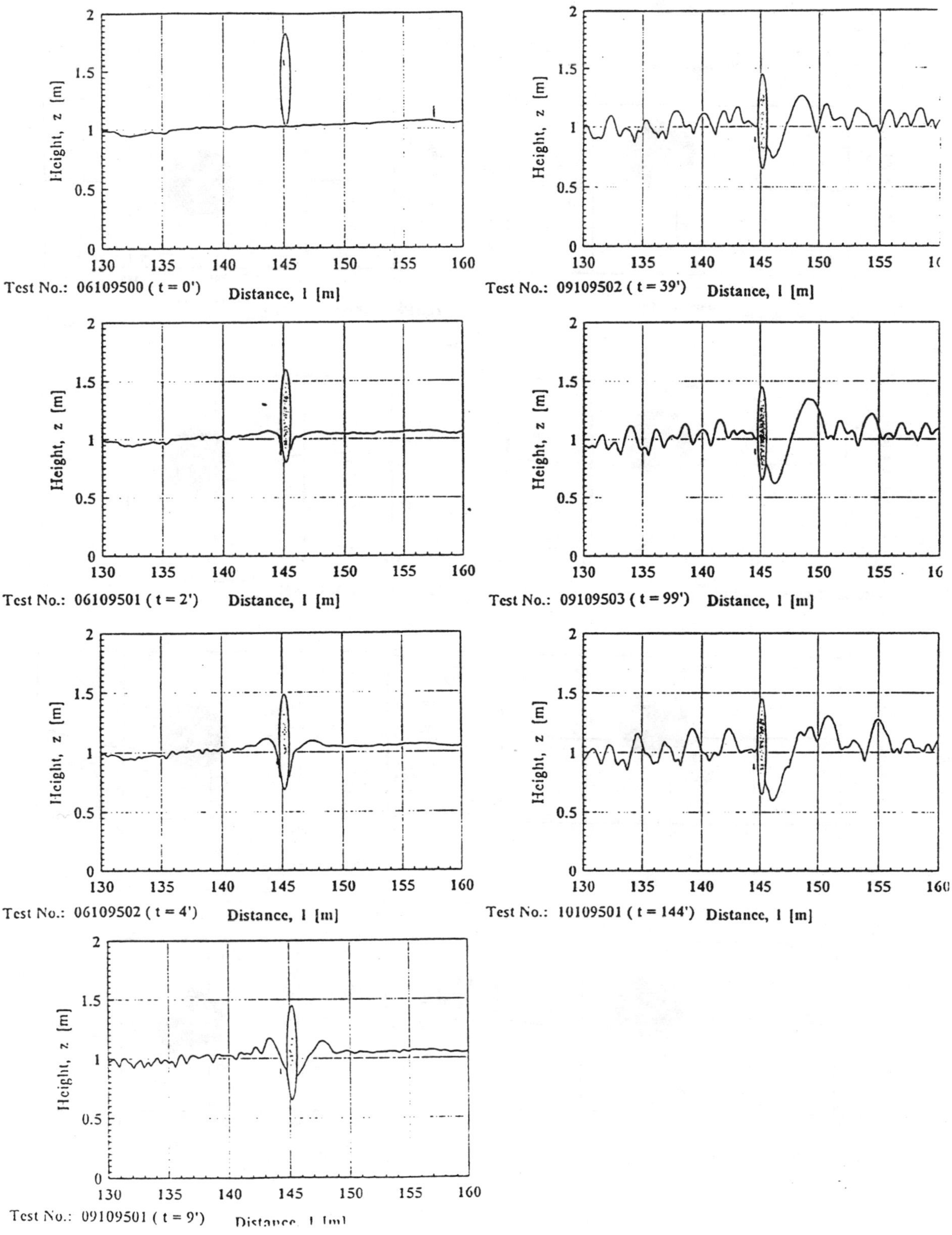

Fig. 9: Longitudinal bottom profiles 15 m in front of and 15 m behind the cylinder (with respect to the wave propagation)

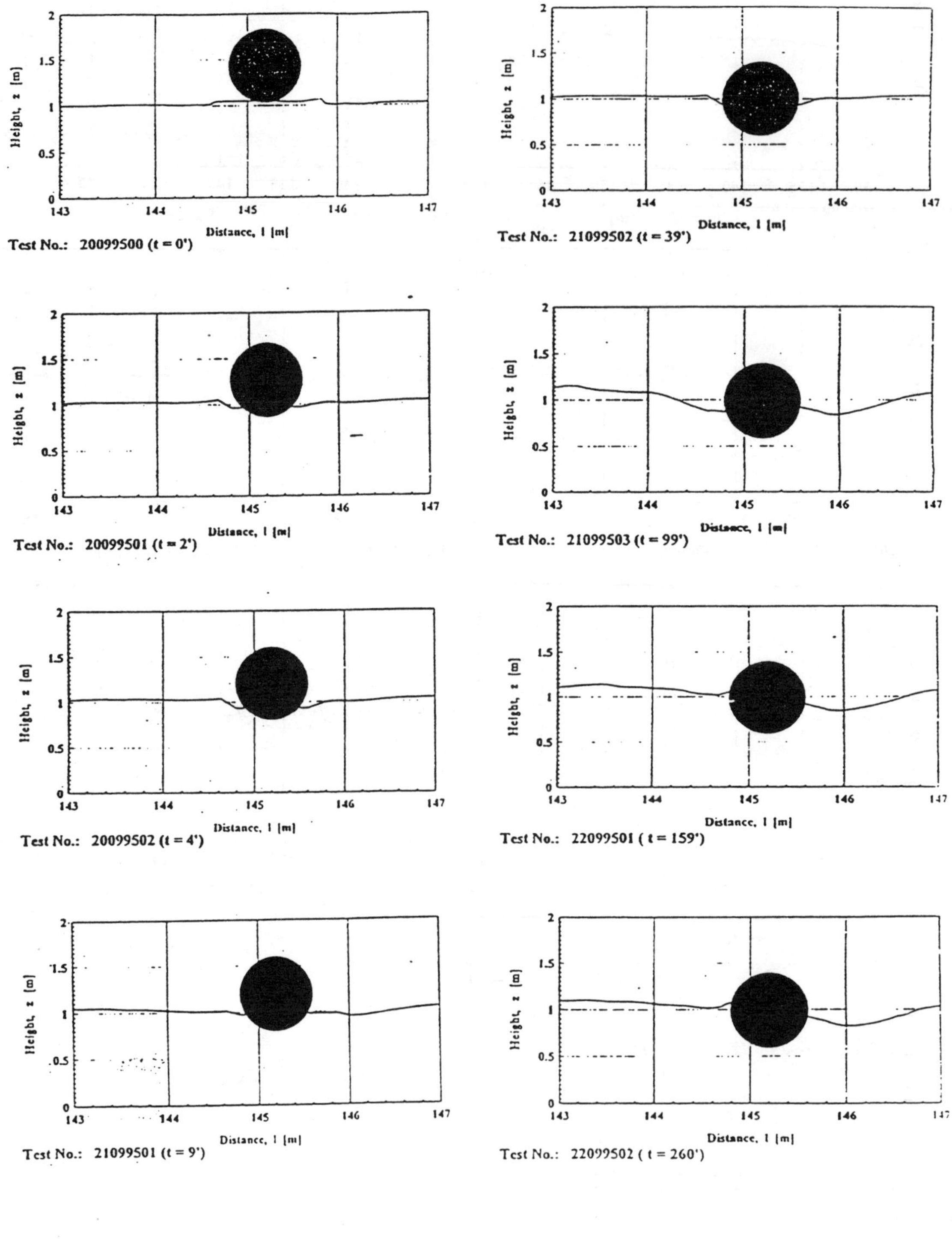

Fig. 10: Longitudinal bottom profiles in cylinder proximity

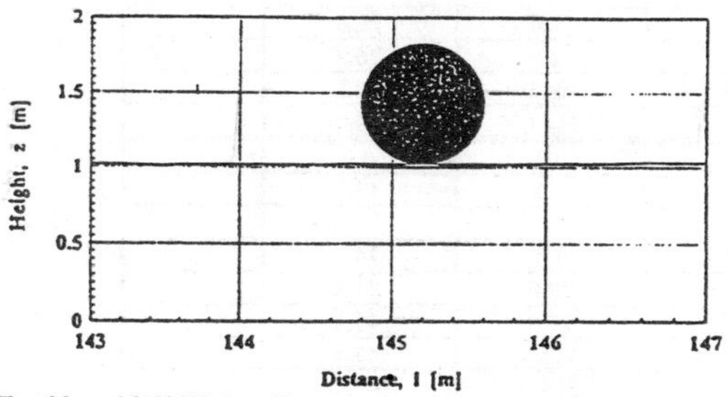

Test No.: 06109500 (t = 0')

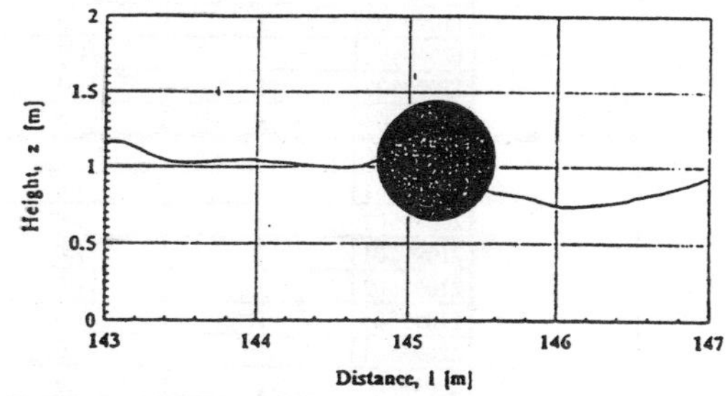

Test No.: 09109502 (t = 39')

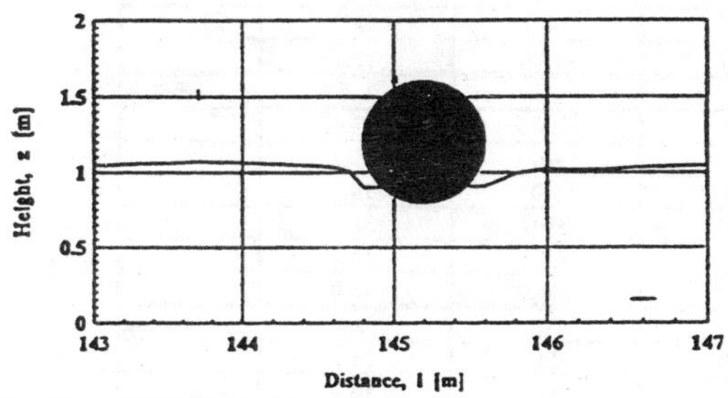

Test No.: 06109501 (t = 2')

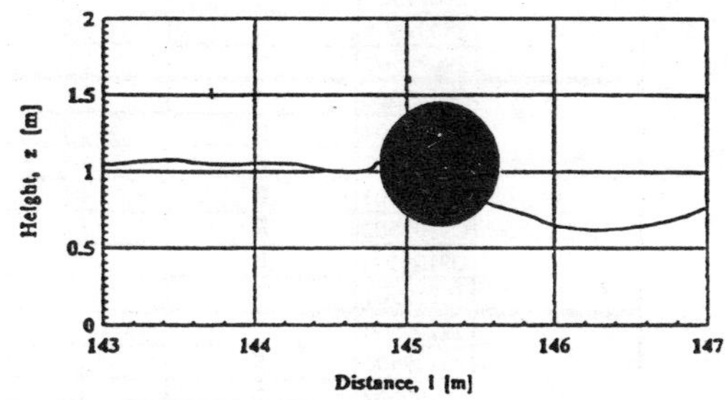

Test No.: 09109503 (t = 99')

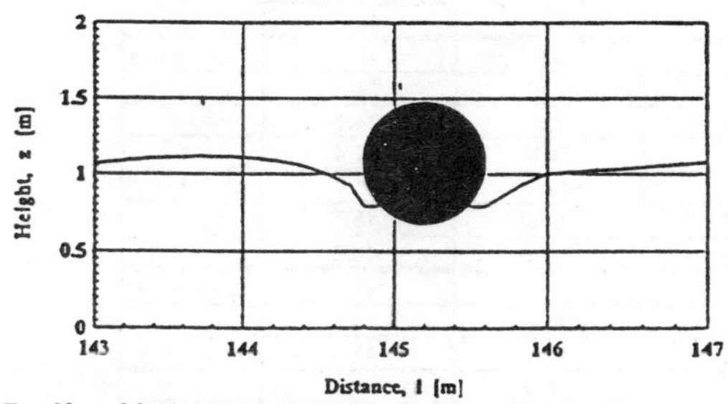

Test No.: 06109502 (t = 4')

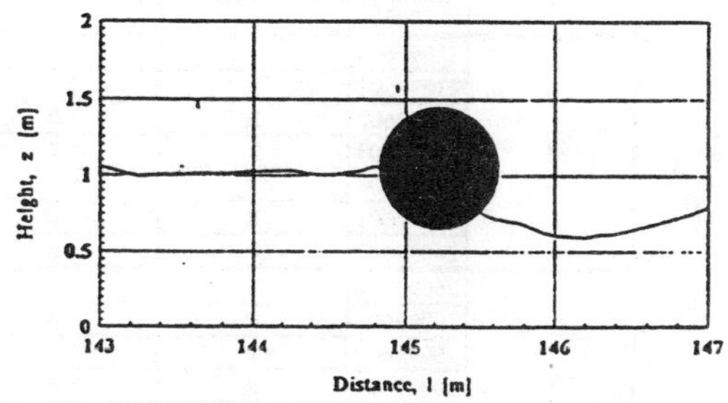

Test No.: 10109501 (t = 144')

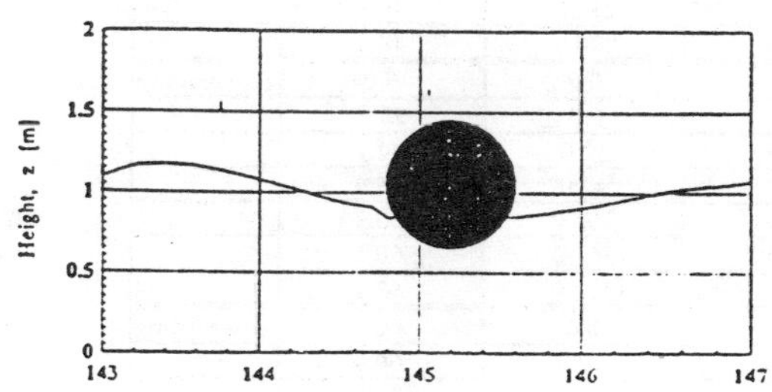

Test No.: 09109501 (t = 9')

Fig. 11: Longitudinal bottom profiles in cylinder proximity

Test Group No	Test No	Wave Regular /Spectrum	H, Hs [m]	T, Tp [s]	Summary Time [min]	Lowering of the pipeline [cm]	Remarks
	07099501	-	-	-	0.0	0.0	zero reading
	11099501	R	1.0	6.0	2.0	10.0	
	11099502	R	1.0	6.0	4.0	18.8	
1	11099503	R	1.0	6.0	14.0	22.3	
	12099501	R	1.0	6.0	44.0	26.2	
	12099502	R	1.0	6.0	104.0	26.3	
	12099504	R	1.0	6.0	164.0	29.4	
	20099500	-	-	-	0.0	0.0	zero reading
	20099501	R	1.0	8.0	2.0	17.1	
	20099502	R	1.0	8.0	4.0	23.6	
2	21099501	R	1.0	8.0	9.0	24.8	
	21099502	R	1.0	8.0	39.0	44.2	
	21099503	R	1.0	8.0	99.0	45.4	
	22099501	R	1.0	8.0	159.0	45.4	
	22099502	R	1.0	8.0	259.0	45.4	
	28099500	-	-	-	0.0	0.0	zero reading
	28099501	R	1.0	4.0	2.0	11.9	
3	28099502	R	1.0	4.0	4.0	14.2	
	28099503	R	1.0	4.0	9.0	14.5	
	28099504	R	1.0	4.0	39.0	18.3	
	29099502	R	1.0	4.0	112.0	19.7	
	06109500	-	-	-	0.0	0.0	zero reading
	06109501	R	1.5	6.0	2.0	23.5	
	06109502	R	1.5	6.0	4.0	35.2	
4	09109501	R	1.5	6.0	9.0	38.9	
	09109502	R	1.5	6.0	39.0	38.9	
	09109503	R	1.5	6.0	99.0	38.9	
	10109501	R	1.5	6.0	144.0	38.9	
	19109500	-	-	-	0.0	0.0	zero reading
	19109501	S	1.3	8.0	2.0	14.2	
	19109502	S	1.3	8.0	4.0	16.1	
5	19109503	S	1.3	8.0	19.0	30.1	
	20109501	S	1.3	8.0	49.0	40.4	
	20109502	S	1.3	8.0	94.0	44.7	
	23109501	S	1.3	8.0	139.0	44.7	
	18099600	-	-	-	0.0	0.0	zero reading
	23099501	R	1.0	8.0	2.0	22.0	
	23099602	R	1.0	8.0	4.0	24.1	
6	23099603	R	1.0	8.0	14.0	37.4	
	23099604	R	1.0	8.0	44.0	45.7	
	24099601	R	1.0	8.0	104.0	48.7	
	24099602	R	1.0	8.0	164.0	48.8	
	26099600	-	-	-	0.0	48.8	zero reading
	30099601	R	1.0	8.0	2.0	48.8	
	30099602	R	1.0	8.0	10.0	48.8	
	30099603	R	1.0	8.0	20.0	48.8	
7	30099604	R	1.0	8.0	50.0	48.8	
	01109601	R	1.0	8.0	110.0	48.8	
	01109602	R	1.0	8.0	170.0	48.8	
	01109603	R	1.0	8.0	230.0	48.8	
	01109604	R	1.0	8.0	290.0	48.8	
	09109600	-	-	-	0.0	0.0	zero reading
	09109601	R	1.5	6.0	2.0	27.7	
	09109602	R	1.5	6.0	12.0	44.7	
8	09109603	R	1.5	6.0	42.0	56.9	
	11109601	R	1.5	6.0	102.0	60.3	
	11109602	R	1.5	6.0	162.0	60.3	
	11109603	R	1.5	6.0	222.0	60.3	
	16109600	-	-	-	0.0	60.3	zero reading
	16109601	R	1.5	6.0	30.0	60.3	
	17109601	R	1.5	6.0	90.0	60.3	
9	17109602	R	1.5	6.0	150.0	60.3	
	17109603	R	1.5	6.0	210.0	60.3	
	17109604	R	1.5	6.0	270.0	60.3	
	18109601	R	1.5	6.0	390.0	60.3	

Table 1: Lowering of the cylinder after test ends (list of executed tests)

Group No	Test No	Point / Height	1	2	3	4	5	6	7	8	9	10	11
	11099503	h_I	0.82	0.90	0.90	0.90	0.90	0.91	0.91	0.94	0.92	0.92	0.88
	H=1.0m	h_{II}			0.77			0.78			0.79		
	T=6.0s t=14'	h_{III}	0.78	0.89	0.89	0.89	0.90	0.90	0.90	0.92	0.91	0.91	0.86
	12099501	h_I	0.79	0.85	0.86	0.85	0.86	0.87	0.89	0.88	0.89	0.89	0.85
	H=1.0m	h_{II}			0.73			0.74			0.75		
1	T=6.0s t=44'	h_{III}	0.79	0.89	0.85	0.87	0.86	0.87	0.89	0.88	0.88	0.91	0.82
	12099502	h_I	0.76	0.88	0.88	0.89	0.90	0.89	0.90	0.85	0.86	0.85	0.83
	H=1.0m	h_{II}			0.73			0.74			0.75		
	T=6.0s t=104'	h_{III}	0.80	0.90	0.91	0.90	0.92	0.91	0.92	0.85	0.84	0.86	0.85
	12099504	h_I	0.80	0.86	0.84	0.78	0.83	0.83	0.84	0.84	0.85	0.84	0.76
	H=1.0m	h_{II}			0.72			0.71			0.70		
	T=6.0s t=164'	h_{III}	0.81	0.87	0.85	0.84	0.83	0.88	0.84	0.84	0.85	0.87	0.77
	20099502	h_I	0.92	0.93	0.93	0.93	0.93	0.93	0.92	0.92	0.91	0.92	0.89
	H=1.0m	h_{II}			0.81			0.805			0.80		
	T=8.0s t=4'	h_{III}	0.90	0.93	0.93	0.94	0.94	0.93	0.93	0.93	0.91	0.91	0.91
	21099501	h_I	0.91	0.94	0.94	0.96	0.95	0.98	0.95	0.93	0.88	0.90	0.87
	H=1.0m	h_{II}			0.80			0.795			0.79		
	T=8.0s t=9'	h_{III}	0.86	0.98	0.96	0.99	0.95	0.97	0.94	0.93	0.80	0.90	0.90
	21099502	h_I	0.65	0.69	0.67	0.67	0.62	0.69	0.69	0.68	0.70	0.70	0.68
	H=1.0m	h_{II}			0.61			0.60			0.59		
2	T=8.0s t=39'	h_{III}	0.63	0.68	0.68	0.69	0.66	0.67	0.70	0.69	0.70	0.70	0.69
	21099503	h_I	0.75	0.88	0.88	0.90	0.86	0.92	0.86	0.90	0.90	0.89	0.73
	H=1.0m	h_{II}			0.60			0.59			0.58		
	T=8.0s t=99'	h_{III}	0.72	0.83	0.84	0.82	0.87	0.87	0.85	0.87	0.85	0.86	0.78
	22099501	h_I	0.85	0.85	0.94	0.92	0.86	0.92	0.87	0.91	0.92	0.92	0.82
	H=1.0m	h_{II}			0.59			0.585			0.58		
	T=8.0s t=159'	h_{III}	0.89	1.00	1.06	1.08	1.06	1.06	1.05	1.08	1.07	0.99	0.81
	22099502	h_I	0.89	0.88	0.93	0.92	0.91	0.92	0.89	0.92	0.92	0.91	0.83
	H=1.0m	h_{II}			0.60			0.59			0.58		
	T=8.0s t=259'	h_{III}	0.96	0.98	1.01	1.06	1.06	1.08	1.07	1.05	1.06	1.00	0.92
	28099501	h_I	0.96	1.03	1.03	1.05	1.05	1.05	1.05	1.04	1.05	1.03	1.00
	H=1.0m	h_{II}			0.93			0.93			0.93		
	T=4.0s t=2'	h_{III}	0.94	1.04	1.03	1.04	1.05	1.04	1.04	1.04	1.03	1.03	1.00
	28099502	h_I	0.95	1.04	1.04	1.04	1.04	1.04	1.04	1.04	1.05	1.04	1.00
	H=1.0m	h_{II}			0.91			0.91			0.91		
	T=4.0s t=4'	h_{III}	0.93	1.04	1.04	1.04	1.04	1.05	1.04	1.06	1.04	1.03	1.00
3	28099503	h_I	0.95	1.10	1.07	1.08	1.07	1.08	1.09	1.09	1.09	1.09	1.01
	H=1.0m	h_{II}						0.91					
	T=4.0s t=9'	h_{III}	0.92	0.99	1.05	1.07	1.07	1.08	1.07	1.07	1.07	1.06	1.00
	28099504	h_I	0.89	0.99	0.99	0.99	1.00	1.04	1.02	1.04	1.04	1.03	0.98
	H=1.0m	h_{II}			0.85			0.87			0.88		
	T=4.0s t=39'	h_{III}	0.88	1.01	0.98	1.01	1.00	1.00	1.04	1.02	1.03	1.00	0.97
	29099502	h_I	0.90	0.96	0.97	0.98	0.98	1.01	0.99	1.02	1.02	1.02	1.01
	H=1.0m	h_{II}			0.84			0.86			0.88		
	T=4.0s t=112'	h_{III}	0.84	0.96	0.95	0.97	0.97	0.99	0.99	0.99	0.99	1.00	0.95
	06109501	h_I	0.87	0.92	0.90	0.93	0.90	0.91	0.87	0.90	0.89	0.92	0.92
	H=1.5m	h_{II}			0.81			0.81			0.81		
	T=6.0s t=2'	h_{III}	0.76	0.92	0.90	0.90	0.91	0.90	0.90	0.91	0.89	0.89	0.93
	06109502	h_I	0.75	0.80	0.78	0.79	0.79	0.79	0.79	0.79	0.78	0.77	0.82
	H=1.5m	h_{II}			0.69			0.69			0.69		
	T=6.0s t=4'	h_{III}	0.74	0.80	0.80	0.79	0.81	0.79	0.79	0.80	0.79	0.78	0.79
	09109501	h_I	0.72	0.81	0.81	0.82	0.81	0.85	0.83	0.81	0.81	0.81	0.82
	H=1.5m	h_{II}			0.65			0.655			0.66		
4	T=6.0s t=9'	h_{III}	0.81	0.82	0.82	0.84	0.83	0.84	0.85	0.85	0.83	0.83	0.77
	09109502	h_I	0.74	0.85	0.87	0.87	0.83	0.83	0.83	0.87	0.87	0.87	0.83
	H=1.5m	h_{II}			0.65			0.65			0.65		
	T=6.0s t=39'	h_{III}	0.78	0.89	1.03	1.04	1.04	1.05	1.03	1.02	1.04	1.02	0.83
	09109503	h_I	0.71	0.82	0.82	0.79	0.75	0.78	0.76	0.80	0.80	0.80	0.79
	H=1.5m	h_{II}			0.65			0.655			0.66		
	T=6.0s t=99'	h_{III}	0.84	0.95	1.06	1.06	1.06	1.06	1.07	1.06	1.07	1.01	0.97
	10109501	h_I	0.67	0.78	0.77	0.76	0.71	0.72	0.73	0.77	0.75	0.77	0.75
	H=1.5m	h_{II}			0.65			0.65			0.65		
	T=6.0s t=144'	h_{III}	0.83	0.94	1.03	1.05	1.03	1.06	1.06	1.05	1.04	1.03	0.92
	19109501	h_I	0.95	1.04	1.05	1.03	1.03	1.02	1.03	1.04	1.02	1.05	0.99
	Hs=1.3m	h_{II}			0.90			0.905			0.915		
	Tp=8.0s t=2'	h_{III}	0.91	1.04	1.04	1.04	1.03	1.03	1.03	1.02	1.01	1.04	1.00
	19109502	h_I	0.94	1.03	1.02	1.02	1.02	1.00	0.98	0.98	0.95	0.99	0.99
	Hs=1.3m	h_{II}						0.89					
	Tp=8.0s t=4'	h_{III}	0.90	1.02	1.04	1.03	1.00	1.00	0.99	0.97	0.96	0.98	0.96
	19109503	h_I	0.76	0.83	0.81	0.85	0.88	0.88	0.89	0.89	0.90	0.89	0.82
5	Hs=1.3m	h_{II}			0.76			0.745			0.73		
	Tp=8.0s t=19'	h_{III}	0.89	0.85	0.82	0.80	0.88	0.88	0.88	0.90	0.88	0.87	0.82
	20109501	h_I	0.72	0.73	0.74	0.74	0.75	0.74	0.74	0.75	0.76	0.76	0.75
	Hs=1.3m	h_{II}			0.64			0.64			0.64		
	Tp=8.0s t=49'	h_{III}	0.68	0.74	0.74	0.73	0.74	0.74	0.73	0.75	0.74	0.74	0.73
	20109502	h_I	0.65	0.75	0.75	0.75	0.78	0.80	0.80	0.77	0.78	0.78	0.75
	Hs=1.3m	h_{II}			0.60			0.605			0.61		
	Tp=8.0s t=99'	h_{III}	0.76	0.80	0.81	0.83	0.81	0.82	0.84	0.82	0.82	0.81	0.81
	23109501	h_I	0.67	0.79	0.79	0.84	0.83	0.84	0.83	0.84	0.84	0.84	0.78
	Hs=1.3m	h_{II}			0.60			0.605			0.61		
	Tp=8.0s t=13[illegible]'	h_{III}	0.74	0.86	0.91	0.94	0.95	0.94	0.98	0.97	0.94	0.91	0.78

Table 2: Results of measurements after test runs

The Proceedings of the Seventh (1997) International OFFSHORE AND POLAR ENGINEERING CONFERENCE

VOLUME IV, 1997

Tubular Structures, Welding and Robotics, Fatigue and Fracture, Steel, Corrosion and NDT, Composite Materials, Mechanics and Structures, Reliability, Risk and Safety, Collision/Impact/Damage, Marine and Underwater Systems, Earthquake Engineering

edited by:

Ram S. Puthli, Universitat Karlsruhe, Karlsruhe, Germany
Yasuo Suga, Keio University, Yokohama, Japan
Nick Patrikalakis, Massachusetts Institute of Technology, Cambridge, USA

presented at:

The Seventh (1997) International Offshore and Polar Engineering Conference held in Honolulu, USA, May 25-30, 1997

organized by:

International Society of Offshore and Polar Engineers

sponsored by:

International Society of Offshore and Polar Engineers (ISOPE) with cooperating societies and associations

International Society of Offshore and Polar Engineers (ISOPE)
P.O. Box 1107 Golden, Colorado 80402-1107 U.S.A.

International Standard Book Number: ISBN 1-880653-28-1 (Set)
ISBN 1-880653-32-X (Vol. IV)

Library of Congress Catalog Card Number: 97-70292

Cooperating Organizations:

Canadian Association of Petroleum Producers (CAPP)
American Society of Civil Engineers (ASCE) - Engineering Mechanics Division
Korea Committee for Ocean Resources and Engineering (KCORE)
Canadian Society of Civil Engineers (CSCE) - Engineering Mechanics Division
Chinese Society of Ocean Engineers (CSOE)
Chinese Society of Naval Architects and Marine Engineers (CSNAME)
Chinese Society of Theoretical and Applied Mechanics (CSTAM)
Russian Academy of Sciences
Singapore Structural Steel Society (SSSS)
Norwegian Petroleum Society (NPF)
The Institution of Engineers Australia (IE Australia)
Kansai Society of Naval Architects, Japan (KSNAJ)
IRO (The Netherlands)
Technical Research Centre of Finland (VTT)
The Society of Materials Science, Japan (JSMS)
The Offshore Engineering Society (OES), United Kingdom
Ukraine Society of Mechanical Engineers (USME)
IFREMER, France
Scott Polar Research Institute (SPRI), UK
The Institute of Engineers Indonesia (PII)

Printed and bound in USA

FOREWORD

We greatly appreciate the excellent responses and help we have received from colleagues around the world in the successful organization of the 7th International Offshore and Polar Engineering Conference (ISOPE-97), Honolulu, May 25 -30, 1997. The Conference features **91** sessions **of 517** refereed papers and **10** plenary sessions from 40 countries.

The purposes of the ISOPE conference are to:

- promote technological progress and activities, international technological transfer and cooperation, and opportunities for engineers to maintain and improve technical competence; and
- provide a timely international forum for technical activities, cooperation, opportunity and fellowship among researchers and engineers,

by developing focused session topics with papers of high quality in both originality and significance of the results which were accepted after rigorous review. The conference promotes interdisciplinary interaction between academia and industry.

The International Society of Offshore and Polar Engineers (ISOPE) has already held 12 successful international meetings:

- the 1st (1990) European Offshore Mechanics Symposium (ISOPE EUROMS-90) in Trondheim, 1990;
- the 1st (1990) Pacific/Asia Offshore Mechanics Symposium (ISOPE-PACOMS-90) Seoul; PACOMS-94 (SOSC-94) Beijing; and PACOMS-96 Pusan.
- the Annual ISOPE conferences, starting in Edinburgh, 1991 were held in San Francisco, Singapore, Osaka, The Hague and Los Angeles. Since 1992, the annual ISOPE Conference has been held with the world's largest technical program of its kind with refereed papers.
- the 1st (1995) ISOPE Ocean Mining Symposium (ISOPE OMS-95), Tsukuba; the 2nd ISOPE OMS-97 will be held in Seoul, November 24-25, 1997.
- the 1st (1996) International Deep-Ocean Technology (IDOT-96) Symposium and Workshop, Los Angeles.

This year the ISOPE-97 TPC expanded greatly the number of papers on recent worldwide developments in environment, advanced marine technology, coastal engineering, Sakhalin oil and gas. We shall continue to add other emerging topics to the future conferences.

On behalf of all the individual TPC members (listed in this volume), we would like to thank the reviewers (listed in this volume) for their help in maintaining the technical accuracy and quality of papers. We greatly appreciate the help from colleagues around the world, particularly from the authors and reviewers in meeting the deadlines. The efficient handling of the conference organization by the ISOPE staff is greatly appreciated. TPC received 517 mats, examined individual papers, page by page, paginated, and retyped all individual titles according to the ISOPE format, all in 3 weeks, ready for the printer. Only about 5% of the papers were redone by the authors, following the ISOPE mat format and font size. It was possible with the help of fax, international express mail and wordprocessor.

It is our pleasure to welcome participants from all over the world to the ISOPE-97 Conference in Honolulu. We hope that the participants will also join us at the ISOPE-98 Conference to be held in Montréal, Canada, May 24-29, 1998.

Co-chairmen of the ISOPE-97 Conference:

Jin Chung, Colorado School of Mines, USA
Bob Frederking, National Research Council, Canada
Ron Knapp, University of Hawaii, USA
Michel Olagnon, Ifremer, France
Yukio Ueda, Kinki University, Japan

Acknowledgement

The success of the 7th International Offshore and Polar Engineering Conference (ISOPE-97), USA, May 25-30, 1997, is credited to the teamwork of the members of the Technical Program Committee (TPC), the conference cosponsors, the cooperating organizations and the reviewers (see the list).

The TPC would like to express its sincere appreciation to the TPC members' organizations for their support of the TPC members, and University of Hawaii TPC colleagues for their support toward the successful conduct of the conference events, and to the ISOPE staff for the efficient handling of the conference organization and proceedings.

ISOPE-97 Technical Program Committee (TPC)

Prof. A. Abel, Univ. of Sydney, Sydney, Australia
Prof. A. Akyarli, Dokuz Univ., Izmir, Turkey
Prof. A. Alhimenko, St. Petersburg Tech Univ., St. Petersburg, Russia
Dr. L. Andronicou, Lloyd's of London, UK
Dr. A.V.M. Arunachalan, WISER Associates, St, John's, Canada
Prof. K. Aso, Akita Univ., Akita, Japan
Dr. Y. Bai, JP Kenny, Stavanger, Norway
Dr. M. Ba, ENSMA, Futuroscope, France
Prof. R.G. Bea, Univ of California at Berkeley, CA, USA
Prof. A.T. Bekker, Far Eastern State Technical Univ., Vladivostok, Russia
Dr. S. Berg, Aker Verdal a.s., Verdal, Norway
Dr. T. Bjornsen, Det Norske Veritas, Norway
Dr. L.F. Boswell, City Univ., London, U.K.
Mr. C. Capanoglu, I.D.E.A.S. Inc., San Francisco, CA, USA
Dr. E.-S. Chan, National Univ. of Singapore, Singapore
Prof. J.R. Chaplin, City Univ., London, U.K.
Prof. Y.-P. Chin, National Cheng Kung Univ., Taiwan, China
Dr. Hin Chiu, Amoco Production Co., Houston, TX, USA
Dr. H.S. Choi, Hudson Engineering Corp., Houston, TX, USA
Prof. H.S. Choi, Seoul National Univ., Seoul, Korea
Dr. Y.S. Choo, National Univ. of Singapore, Singapore
Prof. Jin S. Chung **(Chairman)**, Colorado School of Mines, Golden, CO, USA
Prof. A.T. Chwang, Univ. of Hong Kong, Hong Kong
Dr. A. Clément, Ecole Centrale Nantes, Nantes, France
Dr. I.S. Cole, C.S.I.R.O., Highett, Australia
Mr. R.W. Copple, San Rafael, CA, USA
Prof. B.M. Das, California State Univ., Sacramento, CA, USA
Dr. Z. Demirbilek, Army Waterways Experiment Station, Vicksburg, MS, USA
Dr. J.F. Dos Santos, GKSS Research Center, Geesthacht, Germany
Dr. W. Dursthoff, Univ. of Hannover, Hannover, Germany
Dr. P. K. Dutta, USACRREL, Hannover, NH, USA
Dr. W. J. Easson, Univ. of Edinburgh, Edinburgh, UK
Dr. C.P. Ellinas, AME Ltd., Surrey, U.K.
Dr. J.M. Falzarano, Univ of New Orleans, New Orleans, LA, USA
Prof. M. Farzaneh, Universite du Quebec à Chicoutimi, Canada
Mr. J.A. Feltham, Northumberland Strait Crossing Project, Prince Edward Island, Canada
Prof. A.J. Ferrante, TSO s.r.l., Milan, Italy
Dr. A.M. Fish, USACRREL, Hanover, NH, USA
Dr. A. Francescutto, Univ. of Trieste, Trieste, Italy
Dr. R.M.W. Frederking, National Research Council, Ottawa, Canada
ir. A.M. Gresnigt, Delft Univ. of Technology, Delft, The Netherlands
Prof. P. Grundy, Monash Univ., Clayton, Australia
Dr. T. Hiraishi, Port and Harbour Inst., Yokosutka, Japan
Prof. Y. Hirose, Kanazawa Univ., Kanazawa, Japan
Prof. R.E. Hobbs, Imperial College, London, U.K.
Dr. S.W. Hong, Korea Research Inst. of Ship and Ocean Engineering, Daejeon, Korea
Dr. G. Horrigmoe, NORUT Teknologi, Narvik, Norway
Prof. T. Huang, Univ. of Texas, Arlington, TX, USA
Dr. Y. Ikeda, Univ. of Osaka Prefecture, Osaka, Japan
Prof. M. Isaacson, Univ. of British Columbia, Vancouver, Canada
Dr. C. H. Jo, Hyundai Heavy Industries Ltd., Ulsan, Korea
Dr. A. Judge, Geological Survey of Canada, Canada
Prof. K. Kaneko, Saga Univ., Saga, Japan

Dr. H. Karadeniz, Delft Univ. of Technology, Delft, The Netherlands
Dr. T. Karna, VTT Building Technology, Espoo, Finland
Prof. C.H. Kim, Texas A & M Univ., College Station, TX, USA
Dr. M.H. Kim, Texas A & M Univ., College Station, TX, USA
Prof. S.S. Kim, Chung-Ang Univ, Seoul, Korea
Prof. R.H. Knapp (**Co-chair**), Univ. of Hawaii, Honolulu, HI, USA
Dr. J. Koo, Exxon Research & Engineering Co., Annandale, NJ, USA
Prof. H.-J. Krock, Univ. of Hawaii, Honolulu, HI, USA
Prof. W. Koterayama, Kyushu Univ., Kasuga, Japan
Dr. B. Kvamme, Hogskole i Telemark, Porsgrunn, Norway
Prof. Y. Kyozuka, Kyushu Univ., Kasuga, Japan
Dr. I. Langen, Hogskolen i Stavanger, Stavanger, Norway
Dr. B.J. Leira, SINTEF, Trondheim, Norway
Dr. J. Leonard, M.I.T., Cambridge, MA, USA
Prof. Y.-C. Li, Dalian Univ. of Technology, Dalian, China
Dr. F.K. Lim, McDermott Marine Construction Ltd., Wembley, U.K.
Dr. T.S. Lin, US Naval Facilities Engineering Service Center, Port Hueneme, CA, USA
Prof. G.C. Ling, Chinese Academy of Science, Beijing, China
Dr. S. Løset, NTH, Trondheim, Norway
Prof. H. Maeda, Univ. of Tokyo, Tokyo, Japan
Dr. W. Magda, Technical Univ. of Gdansk, Poland
Dr. J.A. Maple, Maple & Associates, Houston, TX, USA
Prof. T. Matsui, Osaka Univ., Osaka, Japan
Prof. G.P. Miao, Shanghai Jiao Tong Univ, Shanghai, China
Prof. R.W. Miksad, Univ. of Virginia, Charlottsville, VA, USA
Dr. N. Mizutani, Nagoya Univ., Nagoya, Japan
Prof. V.J. Modi, Univ. of British Columbia, Vancouver, Canada
Prof. G. Moe, Norwegian Univ. of Science and Tech, Trondheim, Norway
Dr. H. Moshagen, Statoil, Stavanger, Norway
Dr. C.E. Myers, National Science Foundation, Arlington, VA, USA
Prof. S. Naito, Osaka Univ., Osaka, Japan
Dr. B.J. Natvig, Marine Technology Consultant, AS, Oslo, Norway
Prof. J.M. Niedzwecki, Texas A&M Univ., College Station, TX
Prof. S.-K. Oh, National Fisheries Univ. of Pusan, Pusan, Korea
Prof. T. Ohta, Niigata College of Technology, Niigata, japan
Mr. M. Olagnon, IFREMER, Plouzane, France
Dr. T. Onoufriou, Univ. of Surrey, Surrey, UK
Dr. I.R. Orisamolu, Martec Ltd, Halifax, Canada
Prof. J.K. Paik, Pusan National Univ., Pusan, Korea
Prof. S. Pamukcu, Lehigh Univ., Bethlehem, PA, USA
Prof. N. Patrikalakis, M.I.T., Cambridge, MA, USA
Dr. V. Pavlenko, Russian Academy of Sciences, Moscow, Russia
Prof. R.S. Puthli, Universitat Karlsruhe, Karlsruhe, GERMANY
Dr. Y.P. Pravdivets, Moscow State Univ. of Civil Engineering, Moscow, Russia
Prof. H.R. Riggs, Univ. of Hawaii at Manoa, Honolulu, HI, USA
Prof. K. Riska, Helsinki Univ. of Technology, Helsinki, Finland
Prof. J. Roesset, Univ. of Texas, Austin, TX, USA
Prof. H. Saeki, Hokkaido Univ., Sapporo, Japan
Dr. A.J.N.A. Sarmento, Lisbon Technical Univ., Lisbon, Portugal
Dr. M. Sayed, National Research Council, Ottawa, Canada
Prof. P. Sayer, Univ. of Strathclyde, Glasgow, UK
Prof. G. Scarsi, Univ. of Genoa, Genoa, Italy
Prof. E.M. Schulson, Dartmouth College, Hanover, NH, USA
Dr. J.G. Shin, Seoul National Univ., Seoul, Korea
Dr. I. Sibetheros, Univ. of Texas, Austin, TX, USA
Prof. R.A. Skop, Univ. of Miami, Miami, FL, USA
Dr. S. Soemantri, Bandung Inst. of Technology, Bandung, Indonesia
Prof. V.A. Squire, Univ. of Otago, Dunedin, New Zealand
Dr. M.-Y. Su, Naval Research Lab., Stennis Space Center, MS, USA
Prof. Y. Suga, Keio Univ., Yokohama, Japan
Dr. B.M. Sumer, Technical Univ. of Denmark, Lyngby, Denmark
Prof. H. Thiel, AWI fur Polar- und Meeresforschung, Bremerhaven, Germany
Prof. Y. Tomita, Osaka Univ., Osaka, JAPAN
Prof. M.S. Triantafyllou, M.I.T., Cambridge, MA, USA
Dr. A.L. Tunik, ABS, Paramus, NJ, USA
Prof. Y. Ueda, Welding Research Inst., Osaka Univ., Osaka, Japan
ir. R.J. van Foeken, TNO-BOUW, Delft, The Netherlands
Dr. K. Venkataranama, Kagoshima Univ., Kagoshima, Japan
Dr. P. Wadhams, Scott Polar Research Inst., Cambridge, UK
Dr. T. Yamazaki, National Inst. of Resources and Environment, Tsukuba, Japan

Prof. T. Yao, Hiroshima Univ., Higashi-Hiroshima City, Japan
Prof. R.W. Yeung, Univ. of California, Berkeley, CA, USA
Prof. S.C.S. Yim, Oregon State Univ., Corvallis, OR, USA

ISOPE-97 Executive Committee

Prof. Jin S. Chung (Chairman), Colorado School of Mines, USA
Dr. Robert M.W. Frederking, National Research Council, Canada
Prof. Ronald H. Knapp, University of Hawaii, USA
Prof.-Dr. Ram S. Puthli, Karlsruhe University, Germany

Conference Co-chairmen

Jin S. Chung, Colorado School of Mines, USA
Robert M.W. Frederking, National Research Council, Canada
Ron H. Knapp, University of Hawaii, Honolulu, USA
Michel Olagnon, Ifremer, France
Yukio Ueda, Kinki University, Wakayama, Japan

Hawaii Committee

Prof. R.H. Knapp (Chairman), University of Hawaii at Manoa, Honolulu, HI, USA
Prof. H.-J. Krock, University of Hawaii at Manoa, Honolulu, HI, USA
Prof. H.R. Riggs, University of Hawaii at Manoa, Honolulu, HI, USA

Partial List of Reviewers

Abe, N
Abel, A
Akyarli, A
Albert, D G
Alessandrini, B
Ali, P
Allyn, N
Andronicou, L
Aoki, S
Aoki, S I
Arai, M
Asai, K
Aso, K
Athanassios, T
Athanassoulis, G A
Bahlavouni, A
Bai, Y
Baier, V
Balasubramanian, S
Bangaru, N V
Barthelemy, J
Bea, R G
Bekker, A T
Berhault, C
Bertsimas, D J
Bitner-Gregersen, E
Bjørnsen, T
Blackmore, R
Boo, S Y
Borekci, O
Boswell, L F
Bouchard, G
Bowen, A
Brandes, H G
Bretz, G
Brown, A J
Bryden, I G
Burcharth, H F
Burgess, J J
Busch, W
Capanoglu, C
Caridis, P A
Carter, D J
Cavanié, A
Chan, D B
Chan, E S
Chandler, B
Chaplin, J R
Che, X L
Chen, H C
Cheung, K F
Chiang, E
Chien, L K
Chin, Y P
Cho, S R
Choi, H S
Choo, Y S
Chow, K W
Christensen, J
Chun, H H
Chung, J S
Chwang, A T
Clément, A
Cole, I S
Contento, G
Cornett, A M
Cox, D T
Croasdale, K R
Crookshank, N E
Dalane, J I
Damy, G
Das, B M
Deguchi, I
Demirbilek, Z
Den Haag, T
Dier, A
Dijkstra, G J
Dijkstra, O D
Dillon, W P
Doi, Y
Dos Santos, J F
Dursthoff, W
Dutta, P K
Duvat, G
Easson, W J
Engebretsen, T
Ertekin, R C
Fairchild, D P
Falzarano, J M
Farzaneh, M
Ferrant, P
Ferrante, A J
Fish, A M
Fithen, R
Fithen, R M
Ford, S F
Forristall, G Z
Francescutto, A
Frederking, R
Fujikubo, M
Fujino, M
Fujisaku, H
Fukumoto, M
Fuwa, T
Fylling, I
Garza-Rios, L
Goda, Y
Goel, A P
Gokce, T
Gokkus, U
Gouleilmos, A M
Grass, A J
Gresnigt, A M
Grilli, S
Grundy, P
Guerin, G
Gunbak, A R
Guo, T
Gutfraind, R
Handa, K
Hardy, C
Haritos, N
Harrison, G E
Haver, S
Herion, S
Hessler, R R
Higo, Y
Hiraishi, T
Hobbs, R E
Holder, G D
Hollebone, B
Hong, S W
Hong, S Y
Honji, H
Hopkins, M
Hosoyamada, T
Hover, F
Huang, T
Huang, T L
Huang, X L
Hyodo, M
Hyun, B S
Ifuku, M
Ikeda, Y
Isaacson, M
Iwata, K
Jacobsen, V
Jakobsen, J B
Jang, Y S
Jiang Q
Jin, D
Jo, C H
Johnson, B H
Jones, S J
Ju, F
Juhl, J
Kagawa, K
Kajiwara, H
Kaneko, K
Kao, C C
Kapdasli, S
Karahan, E
Karr, L A
Kashiwagi, M
Kawamura, Y

Kawano, K
Kershenbaum, N Y
Kielland, P
Kim, C H
Kim, D H
Kim, M H
Kim, S B
Kitamura, M
Kitamura, R
Kite-Powell, H L
Kiyokawa, T
Knapp, R H
Kioka, W
Kobayashi, A
Kobayashi, T
Kokawa, H
Kolle, J J
Koo, J Y
Koterayama, W
Krock, H J
Krogstad, H E
Kubo, M
Kujala, P
Kuriyama, Y
Kuroiwa, T
Kvamme, B
Kvenvolden, K A
Kyozuka, Y
Labeyrie, J
Laflamme, J
Laforte, J L
Lampela, K
Landrini, M
Langen, I
Le Boulluec, M
Le Goff, J P
Lee, S J
Lee, S K
Lee, Y G
Leira, B
Leonard, J
Li, J C
Li, Y C
Lian, D
Liu, Y H
Lin, G
Lin, M C
Lin, S S
Ling, S
Lipp, S
Liu, Y L
Liukkonen, S
Lozowski, E P
Luton, M J
Ma, Z L
Maeda, H
Magda, W
Makkonen, L
Manabe, Y
Mansard, E P D
Maple, J A
Marcus, H S
Martin, A J
Mase, H
Masson, D
Masuda, K
Masuda, Y
Matsui, T
McClimans, T
McClintock, F
McClure, G
McComber, P
Mebarkia, S A
Meon, G
Meylan, M H
Miao, G P
Mimura, N
Mitsuyasu, H
Miura, K
Mizutani, N
Modi, V J
Moe, G
Mogridge, G R
Mohr, W C
Molin, B
Moo-Young, H
Mori, K
Moshagen, H
Muggeridge, D B
Murakawa, H
Nabeshima, Y
Nadaoka, K
Naito, S
Nakai, T
Nakamura, M
Nam, D
Narayanan, R
Natvig, B J
Nerzic, R
Nestegard, A
Niedzwecki, J M
Niho, O
Nomura, S
Nwogu, N
O'Donoghue, T
Ogawa, K
Ogawa, Y
Oh, S K
Ohtaka, K
Olagnon, M
Onitsuka, K
Onoufriou, T
Oppenheim, B
Orisamolu, I R
Oshiwa, K
Otsuka, K
Oumeraci, H
Paik, J K
Pamukcu, S
Papanikolaou, A
Park, I K
Parsey, M R
Paté-Cornell, M E
Patrikalakis, N M
Pedersen, P T
Penna, R
Perkins, N C
Pervizpour, M
Pesce, C P
Powers, E J
Pradhan, B S
Prakash, S
Prasad, S
Prevosto, M
Puri, V K
Puthli, R S
Randall, R E
Raschke, S
Rigaud, V
Riggs, H R
Rijken, O
Riska, K
Roby, K
Rodd, J L
Roesset, J M
Ronold, K
Ryerson, C C
Saeki, H
Sakai, T
Sakakiyama, T
Sakamoto, Y
San, K
Sanni, S
Sarmento, A
Savenije, R Ph A C
Sayed, M
Sayer, P G
Scarsi, G
Schjølberg, I
Schriever, G
Senjanovic, I
Sergy, G
Setoguchi, T
Shen, F
Shen, H T
Shibayama, T
Shigematsu, T
Shin, J G
Shin, M S
Shinoda, T
Shogaki, T
Sibetheros, I
Sigbjörnsson, R
Silver, A L
Skjong, R
Sklenicka, V
Skop, R A
Skourup, J
Snijder, H H
Sodhi, D S
Sørensen, J D
Spencer, P
Squire, V A
Stansberg, C T
Stear, J
Strømsem, K
Su, M Y
Suga, Y
Sumer, B M
Sun, Z C
Suzuki, H
Suzuki, K
Szilder, K
Tabeta, S
Takagi, K
Takahashi, H
Takaki, M
Talley, W K
Tanaka, H
Tanaka, Y
Tanizawa, K
Tedeschi, R
Temperville, A
Teng, B
Thayamballi, A
Thiel, H
Timco, G W
Tomita, H
Tomita, Y
Toyosada, M
Triantafyllou, M S
Tsai, C P
Tsai, T K
Tsuboi, H
Tsunekawa, Y
Tsurusaki, K
Tuhkuri, J
Tulin, M
Uchida, K
Uchida, M
Ueda, Y
Utsunomiya, T
van der Vegte, G J
Vassalos, D
Venkataramana, K
Verley, R
Wadhams, P
Wang, B P
Wang, J
Wang, K H
Wang, Y S
Wang, Y X
Wang, Z Z
Watabe, T
Weave, T J
White, H B
Williams, A N
Williams, F M
Winterstein, S
Wright, B D
Wu, K Q
Xu, T
Yajima, H
Yamaguchi, S
Yamazaki, T
Yao, T
Yashima, A
Yasuzawa, Y
Yeung, R W
Yim, J Z
Yoneyama, H
Yoon, H K
Yoon, J H
Zdravkovich, M M
Zhang, D H
Zhang, H X
Zhang, J
Zhao, Y N
Zhou, L D

ISOPE Publication Order Information

International Journal of Offshore and Polar Engineers - Quarterly (March, June, September, December) (ISSN 1053-5381).

Editors: Jin S. Chung, J. Wardenier, R.M.W. Frederking, W. Koterayama. **Topics**: *offshore engineering, ocean engineering, environment, offshore mechanics, polar (Arctic and Antarctic) engineering, advanced marine technology, materials and technology and energy*: **Yearly subscription rates**: US$120. (Member, complimentary) and US$100 per a set of annual back issues.

Proceedings of the 1st (1990) European Offshore Mechanics Symposium (EUROMS-90), Trondheim, Norway, 20-22 August 1990. A single volume: (ISBN 0-9626104-4-5; LCCN 90-084691); approx. 590 pp., US$80. (Member, US$60.)

Proceedings of the 1st (1990) Pacific/Asia Offshore Mechanics Symposium (PACOMS '90), Seoul, Korea, 24-28 June 1990. For a set of 3 volumes, (ISBN 0-9626104-0- 2; LCCN 90-082039): approx. 1260 pp., US$150 per set and US$70 (Member, US$50) per volume.

Proceedings of Special Offshore Symposium China (SOSC-94) / 3rd (1994) Pacific/Asia Offshore Mechanics Symposium (PACOMS-94), Beijing, China, 16-18 April 1994. A single volume (ISBN 1-880653-15-X); 797 pp., US$100.

Proceedings of 4th (1996) Pacific/Asia Offshore Mechanics Symposium (PACOMS-96), Pusan, Korea, October 31-November 1, 1996. A single volume (ISBN 1-880653-27-3); 325 pp., US$60 (Member, US$50.)

Proceedings of 1st (1995) SOPE Ocean Mining Symposium, Tsukuba, Japan, November 21-22, 1995. A single volume (ISBN 1-880653-21-4, LCCN 95-77708): 238 pp., US$100 (Member, US$80.)

Proceedings of 2nd (1997) ISOPE Ocean Mining Symposium, Seoul, Korea, November 24-25, 1997. A single volume (ISBN 1-880653-33-8): 250 pp. est., US$100 (Member, US$80.)

Proceedings of the 1st (1991) International Offshore and Polar Engineering Conference (ISOPE-91), Edinburgh, U.K., 11-16 August 1991. For a set of 4 volumes, (ISBN 0-9626104-5-3; LCCN 91-071635): approx. 2190 pp., $320 (US$240; member) and $100 ($80; member) for a single volume.

Proceedings of the 2nd (1992) International Offshore and Polar Engineering Conference (ISOPE-92), San Francisco, USA, June 14-19, 1992. For a set of 4 volumes, (ISBN 1-880653-00-1; LCCN 91-78280): 2,899 pp. est., $340 (Member, $260) and $100 ($80; member) for a single volume. See below for postage and handling.

Proceedings of the 3rd (1993) International Offshore and Polar Engineering Conference (ISOPE-93), Singapore, 6-11 June 1993. For a set of 4 volumes, (ISBN 1-880653-05-2; LCCN 92-76219): 2,965 pp., $340 (ISOPE Member, $260) and $100 ($80; member) for a single volume. See below for postage and handling.

Proceedings of the 4th (1994) International Offshore and Polar Engineering Conference (ISOPE-94), Osaka, Japan, April 10-15, 1994. For a set of 4 volumes, (ISBN 1-880653-10-9; LCCN 93-80555): 2,700 pp. est., $340 (ISOPE Member, $260) and $100 ($80; member) for a single volume. See below for postage and handling.

Proceedings of the 5th (1995) International Offshore and Polar Engineering Conference (ISOPE-95), The Hague, June 11-16, 1995. For a set of 4 volumes, (ISBN 1-880653-16-8; LCCN 94-73796): 2,543 pp., $340 (ISOPE Member, $260). 2,589 pp., $340 (ISOPE Member, $260) and $100 ($80; member) for a single volume. See below for postage and handling.

Proceedings of the 6th (1996) International Offshore and Polar Engineering Conference (ISOPE-96), Los Angeles, May 26-31, 1996. For a set of 4 volumes, (ISBN 1-880653-22-2): 2,326 pp., $360 (ISOPE Member, $280). For a single volume, $110 (ISOPE Member, $80). See below for postage and handling.

Proceedings of the 7th (1997) International Offshore and Polar Engineering Conference (ISOPE-97), Honolulu, May 25-30, 1997. For a set of 4 volumes, (ISBN 1-880653-28-1): 3,548 pp., $380 (ISOPE Member, $300). For a single volume, $120 (ISOPE Member, $80). See below for postage and handling.

Postage (surface mail) and handling: add for each of ISOPE-92 through ISOPE-97 Conference Proceedings orders

A set	USA $15	Canada $20	Other countries $35
A volume	USA $8	Canada $10	Other countries $15

How To Order

To order, check or tick ☑ the appropriate boxes, fill in the information below, enclose the correct amount of payment or purchase order, and mail to ISOPE Order Department, P.O. Box 1107, Golden, Colorado 80402-1107 USA

❑ Member rate ISOPE Member No. _______ ❑ Non-member rate

❑ **International Journal of Offshore and Polar Engineering**
❑ **Proceedings of the 1st (1990) Pacific/Asia Offshore Mechanics Symposium**
❑ Vol. I ❑ Vol. II ❑ Vol. III ❑ A Set
❑ **Proceedings of the 3rd (1994) Pacific/Asia Offshore Mechanics Symposium (Special Offshore Symposium China)**
❑ **Proceedings of the 4th (1996) Pacific/Asia Offshore Mechanics Symposium**
❑ **Proceedings of the 1st (1990) European Offshore Mechanics Symposium**
❑ **Proceedings of the 1st (1995) ISOPE Ocean Mining Symposium**
❑ **Proceedings of the 2nd (1997) ISOPE Ocean Mining Symposium**
❑ **Proceedings of the 1st (1991) International Offshore and Polar Engineering Conference**
❑ **Proceedings of the 2nd (1992) International Offshore and Polar Engineering Conference**
❑ **Proceedings of the 3rd (1993) International Offshore and Polar Engineering Conference**
❑ **Proceedings of the 4th (1994) International Offshore and Polar Engineering Conference**
❑ **Proceedings of the 5th (1995) International Offshore and Polar Engineering Conference**
❑ **Proceedings of the 6th (1996) International Offshore and Polar Engineering Conference**
❑ **Proceedings of the 7th (1997) International Offshore and Polar Engineering Conference**

Proceedings of ISOPE-91~97 (*check*):
❑ Vol. I ❑ Vol. II ❑ Vol. III ❑ Vol. IV ❑ A Set

Method of Payment: Advance payment must accompany order (including the postage and handling charge above for 1992-97 Proceedings.) International money order in U.S. dollars, bank checks in U.S. dollars drawn on a U.S. bank, or credit card (Visa or MasterCard only) (*payable to ISOPE*) will be accepted. Remit with the mailing address including the postal (or zip) code for shipping.

❑ international money order in U.S. dollars is enclosed.
❑ bank check in U.S. dollars through a U.S. bank is enclosed.

❑ credit card no. ______________________ *Circle one*: Visa MasterCard American Express

Exp. Date ________ Cardholder's name ______________________

Signature ______________________ Date ______________

Ship To Name ______________________

Company ______________________ Dept. ______________________

Address ______________________

City ______________ State (Province) ______________

Country ______________ Postal (or Zip) Code ______________

Prospective authors are invited to submit abstract to

ISOPE-98 Montréal, Canada, May 24-29

The 8th (1998) International

Offshore and Polar Engineering Conference

Montréal, Canada, May 24-29, 1998

First

Call For Papers

Abstract Deadline	**August 1, 1997**
Manuscript (for review) Deadline	**November 30, 1997**

The **International Society of Offshore and Polar Engineers (ISOPE)** is organizing about 60 sessions of refereed papers in cooperation with international technical program committee (TPC) members and cooperating organizations. Delegates from more than 35 countries are expected. All manuscripts will be peer-reviewed prior to the final acceptance for the conference proceedings. Papers of archival value may be further reviewed for the *International Journal of Offshore and Polar Engineering*.

Main fields of interest are:

Offshore Technology	**Environment**	**Materials & Welding**
Offshore Mechanics	**Geotechnics**	**Tubular Structures**
Ocean Engineering	**Energy, Resources**	**Risers, Pipelines, Cables**
Polar Engineering	**TLP, SPAR, FPS**	**Comp Methods/CAD**
Deep-Ocean Tech	**Adv Marine Tech**	**Safety, Risk , Reliability**
Underwater Systems	**Hydrodynamics**	**Coastal Engineering**
Measurements	**Underwater Eng**	**Emerging Topics**

The abstract should stress the originality and the significance of the results, and include the complete names, addresses and fax numbers of all authors.

Send 2 copies of your abstract in 300-400 words to: (1) One of the TPC members (session organizers) or (2) **ISOPE-97 TPC,** ISOPE, P.O. Box 1107, Golden, Colorado 80402-1107 USA: FAX 1-303-420-3760.

Sponsors: *ISOPE.* **Cooperating Organizations and Societies**: *ASCE (USA), CAPP (Canada), CSCE (Canada), CSOE (China), CSNAME (China), CSTAM (China), IE Australia, IFREMER (France), IRO (The Netherlands), JSMS (Japan), KCORE (Korea), KSNAJ (Japan), NPF (Norway), OES (UK), PII (Indonesia), RAS (Russia), Scott Polar Res Inst (UK), SSSS (Singapore), and VTT (Finland).*

The Proceedings of
The Seventh (1997) International

OFFSHORE AND POLAR ENGINEERING CONFERENCE

Honolulu, Hawaii, USA, May 25-30, 1997

CONTENT

UNDERWATER WELDING

STEEL, CORROSION AND NDT

Proceedings of the Seventh (1997) International Offshore and Polar Engineering Conference
Honolulu, USA, May 25-30, 1997

ISBN 1-880653-28-1 (Set); ISBN 1-880653-32-X (Vol. IV)

Current Trends in the Safety of Offshore Structures

T. Moan
Norwegian University of Science and Technology
Trondheim, Norway

ABSTRACT

Recent developments in reliability analysis to aid structural design and reassessment during use as well as inspection planning are reviewed. Data and methods applied to calculate the probability of ultimate and fatigue failure of structural components and systems, considering the effect of inspection, are critically examined. Target reliability levels are briefly discussed.

KEY WORDS: Structural reliability, systems, design, code calibration, reassessment, inspection planning

INTRODUCTION

The harsh environmental conditions and other hazards associated with offshore structures as well as the significant consequences of failure make safety an important consideration for such structures.

Offshore platforms are designed for a service life of 20 years or more, with criteria for serviceability and safety against structural failure and overall instability or possible sinking (see e.g. ISO 1994-). Adequate structural safety is achieved by applying ultimate (ULS) and fatigue (FLS) limit state criteria for components. Fatigue is an important consideration for structures in areas with more or less continuous storm loading (such as the North Sea) and especially for dynamically sensitive structures. Progressive collapse limit state (PLS) criteria are applied to avoid catastrophic accidents, i.e., due to a small damage.

In addition to design measures inspection and monitoring, and repair, if necessary, are important measures for maintaining safety, especially with respect to fatigue, wear and other deterioration phenomena. But their effect on the reliability depends upon the quality of inspection, e.g., in terms of detectability vs. size of the damage. Hence, an inspection and repair measure can contribute to the safety only when there is a certain damage tolerance. This implies that there is an interrelation between design criteria (fatigue life, damage tolerance) and the inspection and repair criteria. Up to now, however, this interrelation has not been explicitly considered, due to lack of methods to deal with this problem in a rational way.

It is important to establish an optimal balance of the different safety measures at the design stage, when the costs of increasing the strength are small compared to what they are for an existing structure. However, the planning of structural layout and materials selection as well as inspection program made at the design stage can only be based on predicted resistances and loads, and not the additional information that become available during fabrication and use.

During operation a reassessment of the safety may be necessary due to

- required change of live loads
- occurrence of deterioration or overload damage
- extension of service life
- reduction of operational costs

Such assessments should take into account information about the geometry and material properties acquired during fabrication and observed response or damages during previous inspection, or condition assessments or survival of proof-loads or service loads. Observations both for the actual structure as well as for other relevant structures should be considered. In particular, it is essential to sequentially update the inspection plan made at the design stage as information is achieved during operation.

Predicted structural performance as well as inspection, maintenance and repair actions are subject to uncertainty. It was initially the significant fundamental variability of e.g. wave kinematics and wind speed and other physical phenomena that stimulated a rational treatment of uncertainty affecting the performance of marine structures. While this source of (aleatory) uncertainty cannot be reduced by collecting more data, other (epistemic) uncertainties due to lack of knowledge, can be reduced. Structural reliability analysis (SRA) in the classical sense deals with these uncertainties and especially provides the probability of failure - as a measure of reliability (see e.g. Melchers 1987). Human actions which influence design, fabrication and operation of structures normally follow defined procedures, which ensure that the corresponding uncertainties in structural resistances, loads and load effects follow certain patterns, which make it possible to measure and predict such uncertainties. However, sometimes the human actions depart from acceptable practice and result in so-called human errors, using design, fabrication or operation. Such errors and omissions are not handled by SRA, but by quantitative risk analysis (QRA). This also means that SRA can be applied to calibrate ULS and FLS requirements, while QRA is necessary to deal with PLS (see e.g. Moan 1994).

This paper addresses some of the recent theoretical developments and applications of SRA. The importance of safety for offshore structures and the significant engineering activity in the offshore petroleum industry in the last two decades have made this industry one of the important areas of recent SRA research and developments. This applies e.g. to Bayesian updating, techniques time-variant reliability as well as systems reliability, (see e.g. historical notes and survey by Moan 1994). This paper highlights some theoretical developments in these areas but especially applications to design code calibration, inspection planning and reassessment of existing structures. SRA has now become a mature engineering discipline by the availability of methods, computer tools and evidence of application by organizations other than research bodies and universities. Still, the pitfalls are many and standardization of methods and data is urgently needed. Besides observing the current capability of SRA, its limitations should be observed. Such issues will be addressed in this paper.

RELIABILITY ANALYSIS FOR CODE CALIBRATION OF ULTIMATE LIMIT STATES FOR COMPONENTS

Code calibration

The principal application of reliability methods so far has been to calibrate semi-probabilistic ultimate limit state criteria for components given in codes, to comply with a certain target reliability level. Previous efforts were recently critically examined (Moan 1995) to obtain a basis for comparing the safety level across different structural forms and material to provide a basis for harmonization design criteria in a new ISO Code for offshore structures (ISO 1994-). The overall framework of this work is ISO 2394 (1995), which is currently being revised to better reflect the new developments in reliability technology and the need for reassessment of existing structures.

Current ultimate limit state criteria in codes are formulated in terms of semi-probabilistic expressions, using defined characteristic values of load effects (S_i) and resistances (R_i) and partial safety factors for S_i and R_i. Such criteria (notably: the partial safety factors) are calibrated based on

- defined scope of the code (class of structures, load cases, failure modes etc.)
- relevant method of reliability analysis
- adequate uncertainty measures for load effects and resistances
- properly established target reliability level

The first reliability-based calibration of codes for offshore structures took place in the early 1970'ies (Fjeld 1977) only three years after the first calibration of a structural design code at all. An update of this early study relating to the Norwegian Petroleum Directorate's regulations was made in 1987-1988 (Moan 1988).

The significant effort by API (Moses 1986-1987; Lloyd and Karsan, 1988) to establish a load and resistance factor design code for steel jackets in U.S. waters is especially noteworthy. In parallel with the API work, the Canadian Standards Association conducted a study for fixed steel and concrete platforms (Jordaan and Maes 1991) which was implemented in a draft version in 1989. Theophanatos et al (1992), and Turner et al. (1992) have investigated the modifications of the API code which are necessary for implementing it for platforms in other geographical areas. Also, significant studies have been conducted and are going on to calibrate codes for floating platforms, especially for TLP's (Faulkner 1991; Mathisen et al. 1994; Banon et al. 1994), as well as for subsea pipelines (Jiao et al 1997). Much of the ongoing work is directed towards implementation in a new world-wide offshore code, being devloped under the auspices of ISO (1994-).

In certain situations when a new design falls outside the scope of existing codes, reliability analysis has been applied ad-hoc to establish design criteria. This was for instance the case when the planning of the first submerged tunnel bridges and offshore production ships was started some years ago. Clearly the novel design of a bridge subjected to stochastic traffic, sea and seismic as well as accidental loads required reassessment of the bridge design code. Hence, partial safety factors for ULS design checks were calibrated by Moan and Wu (1988) based on a more restrictive target safety level than for offshore structures. It was also apparent that the design of production vessels should differ from ocean-going merchant vessels because of,e .g., differences in mode of operation, implying different still-water and wave loads and consequences of failure. Moreover, it became clear that merchant ship rules and existing offshore codes differed significantly, implying a difference in steel weight of the order of 20-30 % for production vessels. Moan (1988) conducted a study to establish ultimate strength criteria for this type of vessel, which complied with the inherent safety level in the existing NPD code for offshore structures. The significant interest in production ships since then has stimulated further efforts to rationalize the design criteria for such vessels (Moan, 1995 and Wang, Jiao and Moan, 1996).

Since the first item of concern in the calibration is just a matter of defining a proper scope, the focus here will be on the latter three items. The following sections therefore provide a brief review of reliability methods, uncertainty measures and target levels primarily applied in the authoritative calibration studies mentioned above. The assessment is limited to jackets, the most widely used and studied type of structure. The purpose is especially to demonstrate differences in the premises used and, therefore, caution against the inconsistency in reliability estimates and target levels used in different studies.

Reliability analysis

The reliability problem for structural components is usually made time-invariant by replacing the stochastic load effect process by a random variable with a mean value equal to its expected value in the reference time period and appropriate measure for statistical and other sources of the uncertainty. If the time-invariant problem is described by a set, $\underline{Z}$ of random variables, the failure probability may be calculated as

$$p_F = \int_{g(\underline{z}) \le 0} f_{\underline{z}}(\underline{z}) d\underline{z} \qquad (1)$$

where $f_{\underline{z}}(\underline{z})$ is the probability density and $g(\underline{z})$ the failure function. The integral of Eq. (1) may be calculated by direct integration, simulation or FORM/SORM methods as described in, e.g. Melchers (1987).

While FORM and Monte Carlo type methods represent the state of the art approaches for reliability estimation, simple formulations have often been used in code calibrations. Lognormally distributed resistance R and load effect S are frequently assumed. Sometimes the corresponding safety index β_{LN} is approximated by

$$\beta'_{LN} = \ln(\mu_R/\mu_S)/\sqrt{V_R^2 + V_S^2} \qquad (2)$$

where μ and V denote mean value and COV, respectively. The difference between β_{LN} and e.g.β_N (based on lognormal R and normal S) amounts to about 0.5 to 0.9 units for a typical range of uncertainties.

The calibration of the API LRFD jacket code (Moses 1986-1987) in USA was based on a lognormal distribution of both the resistance and load effect. In the CSA work (Maes 1986) the resistance was taken to be lognormal while different probabilistic models were applied for the loads. The NPD code calibration Fjeld (1977) was based on log-normal and normal distributions for resistances and load effects, respectively. The study by Moan (1988) was based on similar distributions for consistency, but used updated uncertainty measures.

In calibration studies considered herein, the reliability analysis has been based on a load formulation similar to Eq. (3) given below. *If the airgap is sufficiently large, this assumption is justified.* However, if the *design value* of the wave height, E (as determined in the reliability analysis) exceeds E_c, Eq. (4) should be used to obtain a satisfactorily accurate estimate of the reliability. The practical implication of this feature for North Sea conditions is that an airgap on the 100 year crest of the order 1.0-2.0 m would be required to achieve annual failure probabilities of the order 10^{-3}, while the airgap needs to be 2.0-4.0 m to achieve a p_F of 10^{-4}; depending upon the magnitude of wave height, relative magnitude of resistance and load uncertainties. Using Eq. (3) when Eq. (4) is the correct formula, will obviously lead to significant errors in predicted p_F's.

Uncertainty measures

For jackets the uncertainties in load effect due to sealoads, wind, ice and earthquake predominate. They depend upon uncertainties in the environmental conditions as well as the force and response calculation. The uncertainty in environmental load effects for jackets with quasi-static behaviour, may be determined based on the following model (e.g. Moses 1986-1987 and Moan 1995):

$$S = K \cdot C_1 \cdot C_2 ... E^{\alpha} = C \cdot E^{\alpha} \qquad (3)$$

where

- K - a constant
- C_1 - transfer from environmental condition to load
- C_2 - transfer from load to load effect
- E - characteristic environmental parameter

Eq. (3) has typically been applied to model sea loads on jackets based on the design wave method (with α typically in the range 1.3 to 2.4) E represents a wave height. When Eq. (3) is applied to calculate the annual failure probability E should be taken to be the annual maximum value.

Systematic calculations of wave loads (and load effects) on jackets show that wave loads (and load effects) calculated by Eq. (3) provides a satisfactory representation of the load as long as the wave crest does not reach the platform deck. If the wave touches the deck at a wave height of E_c, the following formulation may be applied for $E \geq E_c$:

$$S = C \cdot E_c^{\alpha} + C'(E - E_c)^{\alpha'} \qquad for \;\; E \geq E_c \qquad (4)$$

where C' and α' represent a random model uncertainty and a constant, respectively.

The distribution of the relevant (annual maximum) E is expected to follow an (asymptotic) extreme value distribution, e.g. of Gumbel type. The bias of E corresponds to the ratio of expected annual maximum and the specified characteristic value. If C can be shown to be expressed as a product of several factors, it may be taken to be a lognormal distribution. Otherwise, a normal distribution may be most relevant.

The effect of different environmental climates, i.e. tropical vs. extratropical climate - on the inherent uncertainty of the environmental condition, specified by wave height and period, current velocity and the water depth (water level) is especially noted. Also the difference in approaches used to collect and implement data for these different wave climates contribute to the differences in uncertainty level (Moan 1995).

The uncertainty associated with the basic environmental data is commonly assumed to be negligible due to efforts made to avoid measurement errors etc. The statistical uncertainty of annual maximum significant wave height H_{mo} e.g. for extratropical areas with "continuous" storms, corresponds to a coefficient of variation, COV of 0.15. If an annual maximum individual wave height H is inferred from H_{mo} further uncertainty adds. The COV is smaller (approximately 0.07 - 0.08) for the maximum H when using the long-term distribution of H- as determined by using the scatter diagram and Rayleigh distribution of individual wave heights in storm sea. Both measures have been used in code calibrations (Moan 1995).

Since the wave height (E) is taken to a power α or α' (Eqs. (3-4)), the effect of using different measures of statistical uncertainty will be significant!

Extreme sealoads on jackets with a static behaviour, are universally estimated based on a regular design wave and Morison's equation, involving the following main features:

- waveheight and corresponding period, and a current velocity
- wave particle kinematics and current velocity profile
- values of hydrodynamic coefficients

Two main procedures which currently are used to estimate the sea loads on jackets include the API RP 2A (1993) approach and so-called standard North Sea practice. They differ with respect to wave kinematics, hydrodynamic force coefficients, as well as in treatment of current blockage and shielding effects. The model uncertainty (C in Eq.(3)) of the API load recipe has been determined by comparison with full-scale measurements, yielding a bias 1.06 and a COV of 0.25 (Heideman and Weaver 1992). In the initial API calibration (Moses 1986, 1987) a different bias of 0.93 was used. In an effort (Turner et al. 1992; Theophanatos, 1992) to develop a European API LRFD design procedure a more refined load model, explicitly based on the two terms in Morisons equation, was used. Uncertainties were assigned to the wave height, hydrodynamic coefficients and current velocity - resulting in total load bias and COV in the range of 0.71 to 0.88 and 0.19 to 0.29, respectively. The COV seems to be a bit small. This approach does not properly recognize the uncertainties in the design wave approach (vs. a stochastic wave model) itself and the kinematics. Other studies have been based on similar uncertainty models.

Methods which aim at predicting the most probable maximum load by considering a sea state, rather than a regular wave, exhibit a smaller model uncertainty - with a COV of the order 12-14 %. (Atkins JIP 1994 and Jonathan and Taylor 1996). This uncertainty measure obviously is applicable only if loads actually are calculated by a random wave approach and not a design wave approach.

Static load effects for ultimate strength design of jackets are generally determined by means of finite element frame models. The main uncertainties involved are associated with the flexibility of tubular joints and pile-soil interaction.

At a given site, current and wind conditions do not necessarily occur at the same time or in the same direction. The meteocean criteria in API RP2A permit the designer of a new platform to take full advantage of this fact. The guidelines of HSE in the UK, and the NPD regulations recognize this fact, but do not allow the designer to take advantage of it. However, this conservatism is partly compensated by the choice of hydrodynamic force coefficients.

Resistances for compression members, beam-columns and tubular joints are used in design and reliability assessment of jackets. For tubular members the bias and COV in ultimate strength is typically 1.0-1.1 and 5-10% respectively, implying relatively small variation between different approaches. For unstiffened simple tubular joints the uncertainties vary much more (see e.g. Moan et al 1994). The bias of different formulations for various joints may range from 1.0 to 1.4, while the COV may be in the range of 10 to 40%. For this reason care should be exercised in modelling such resistance uncertainties.

Target levels

The calibration of the API Code was undertaken using the same average target reliability as implied by the previous API Code and, hence, primarily aimed at a more uniform reliability level. A similar approach has been used in most calibrations so far.

If the same uncertainty measures and reliability methodology is used when determining the implied failure probability in an existing code (to establish the target level) as used when calculating the inherent failure probability in the new code, the sensitivity of the partial safety coefficients to the data and methods will be limited. However, demonstrated above different uncertainty measures are used by different researchers and organizations. The approach may hence not be consistent if target levels inferred by one organization is applied by another. Target levels should be tied together with the specific procedures for uncertainty and reliability analysis. Moreover, it is important that the reliability-based design is consistent with well established design practice for commonly occurring cases. It is, therefore crucial that uncertainty measures, reliability methodology and target values give designs that correspond to acceptable design practice. This also implies that a possible differentiation of target level according to e.g. the cause and mode of failure, should be consistent with design practice. For the above reasons the general guidance in NKB (1978, 1987) and DNV (1995), would, as demonstrated by Moan (1995), in general not be consistent with reliability measures quantified for different codes for various types of offshore structures by different organizations, and should not be used without validation.

FATIGUE RELIABILITY OF COMPONENTS

General

Fatigue and fracture are important failure modes for offshore structures as described,e .g., in Chapter 1 of Almar-Næss (ed.) (1985).

The first generation of jacket structures in the Norwegian part of the North Sea in the early 1970'ies was not designed against fatigue failure. The degree of detail and quality of the analysis has been significantly improved since 1975. The present tools (used after 1990) for predicting fatigue life or crack growth still imply highly scattered estimates. Vårdal and Moan (1997) present a method to utilise the inspection history/in-service observations to measure and improve the accuracy of analysis method and apply the method to approximately 4000 nondestructive testing inspection results obtained in the period 1972 to 1995, involving more than 600 platform years of operation. Table 1 gives an overview of the inspections. Fig. 1 shows the distribution of depth for the 690 cracks detected in a subset of 3411 inspections of the data base. Obviously a classification of cracks as propagating or non-propagating on a non-destructive basis will involve uncertainty. The cracks were repaired by grinding in steps of 0.5 mm until no indication of crack by Magnetic Particle Inspection, followed by a final 0.5 mm grinding to ensure complete removal of the crack. This repair procedure thus also provides a measure of the crack depth.

The importance of fatigue and the significant inherent uncertainties associated with fatigue imply that a probabilistic treatment is attractive. A reliability formulation for fatigue can be centred around design practice by using SN-curves and the Miner-Palmgren approach. However, a fracture mechanics model is required, especially to account for the effect of inspection; see, e.g. Kirkemo (1988). For these reasons, much attention has recently been paid to developing fatigue and fracture reliability methods, as e.g. reviewed by Moan (1994).

TABLE 1 Overview of inspection data (Vårdal and Moan 1997)

	Joint Category[1)]	First-time	Follow-nofind	Follow-repair	Follow-unrepair	Total
No. of insp.	primary	1983	525	210	167	2885
	secondary	626	94	107	53	880
	tertiary	140	2	58	12	212
	Sum	2749	621	375	232	3977
No. of cracks	primary	239	60	37	140	476
	secondary	118	30	54	36	238
	tertiary	65	0	25	11	101
	Sum	422	90	116	187	815
No. of cracks	primary	55	16	18	93	182
	secondary	72	27	47	28	174
	tertiary	54	0	25	11	90
	Sum	181	43	90	132	446

[1)] With respect to consequence of failure

By adopting a fracture mechanics model, the fatigue phenomenon may be described by the

a) crack occurrence in space
b) probabilistic distribution for the time to crack initiation T_0
c) a probability model for crack propagation, e.g., based on a deterministic Paris-Erdogan's law for the crack increment per cycle, da/dN:

$$da/dN = C(\Delta K)^m \tag{5}$$

where the material parameters C and m, and the stress intensity factor $\Delta K = S \cdot Y(a, P)\sqrt{\pi a}$, are taken as random variables. S is the nominal stress range and Y( ) is a function of a and $\underline{P}$ is a set of random variables.

Fatigue failure in a time period τ may be described by the following limit state function:

$$g(\underline{z}) = a_f - a(\tau) \leq 0 \tag{6}$$

where a_f and $a(\tau)$ are the crack sizes at failure and time τ, respectively.

Alternatively, the limit state function, g( ), may be written as:

$$g(\underline{z}) = T_0 + T - \tau \tag{7}$$

where the time, T from an initial crack size, a_0 to failure (crack size, a_f) can then be obtained by integrating Eq. (5) as

$$T = \frac{1}{C \cdot \nu_0 \cdot E(S^m)} \int_{a_0}^{a_f} \frac{da}{(\gamma_Y \cdot Y(a,\underline{P})\sqrt{\pi a}\,)^m} \tag{8}$$

where ν_0 is the mean zero upcrossing frequency of S, and $E(S^m)$ is the expectation of S^m. γ_Y is the model uncertainty of Y( ).

The failure probability can be calculated by Eq. (1).

Furthermore, because fatigue and fracture control relies on design as well as inspection and repair, much work has also been devoted to investigating the effect of inspection on the reliability by Bayesian statistics. The fact that inspection methods are not perfect is recognized by characterizing their quality by the probability of crack detection, POD, typically taken to be an exponential function; see, e.g., Silk et al. (1987) and Barnouin et al. (1993).

Updating of reliability can be made using the Bayesian approach in two ways, namely by updating of:

- the multivariant probability distributions and then recalculating the failure probability
- the probability of events, such as failure events

Probabilities may be updated by using Bayes theorem on discrete form as:

$$P(B_j|I) = \frac{P[I|B_j] \cdot P[B_j]}{\sum_j P[I|B_j] \cdot P[B_j]} \tag{9}$$

where B_j is the event updated upon the observed event I, while the failure probability is updated based on observed event (e.g. $H \leq 0$) by:

$$P[g(\underline{z}) \leq 0 \mid H \leq 0] = \frac{P[(g(\underline{z}) \leq 0) \cap (H \leq 0)]}{P[H \leq 0]} \tag{10}$$

Significant developments of the first type of approach were made by Shinozuka, Yang, Itagaki and others as reviewed by Yang (1993). Over the last decade FORM/SORM techniques have been used to develop general and efficient techniques for event updating, see Rackwitz and Schrupp (1985), Madsen (1985, 1987) and Skjong (1985).

Initial crack size

In probabilistic fracture mechanics analyses of tubular joints the initial crack depth is often modelled as an exponentially distributed variable with mean value equal to 0.11 mm. The source of this data may be Bokalrud and Karlsen (1981). However, in their paper the crack occurrence rate was 16 cracks per m. In a recent study (Moan et al. 1997) of 240 non-propagating cracks in 27 North Sea jackets, the mean a_0 was estimated to be 0.94 mm, however, at a rate of 0.36 cracks pr. joint, or one crack pr. three joints. It is noted that the expected maximum of the Bokalrud and Karlsen size cracks in three joints is 0.52 mm, i.e. smaller than that implied by Moan et al (1997).

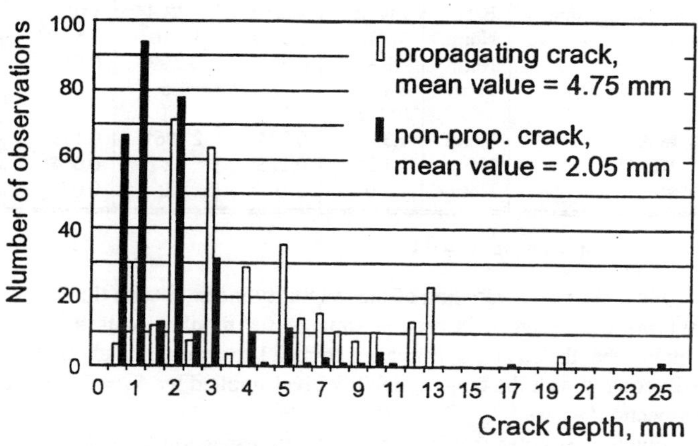

Fig. 1 Total number of 690 fabrication and fatigue cracks, as arranged by crack depth, detected in 3411 inspections

By representing the crack propagation (fatigue) in a tubular joint by a single crack, its equivalent size should account for the possible occurrence of crack in the hot spot area, which may be approximately 1/3 of the total weld length, the crack rate, and size distribution. The study by Moan et al (1997) indicates a mean equivalent crack depth of 0.38 mm.

Fracture mechanics model

Eq. (5) implies a linear relationship between da/dN and ΔK in a logarithmic plot. More general formulations, in which da/dN vs. ΔK is approximated by several line segments have been pursued by Kam and Dover (1989), and even by taking the effect of stress ratio into account (Hovde and Moan 1994). A particular challenge is associated with modelling crack growth in complex tubular joints. In principle, a multiparameter crack propagation law should be applied to describe the crack propagation of large cracks, including the effect of load shedding. Approximate formulations can be based on the assumption of a semi-elliptic crack with two parameters or a one parameter formulation in terms of crack depth, but with the aspect ratio as a function of crack depth; see e.g. Shetty (1992), Hovde and Moan (1994a) and Kam et al. (1995). The Paris equation for the two-parameter formulation is expressed by two simultaneous differential equations in terms of crack increments, see, e.g., Shetty (1992). The model uncertainty is represented by the random variable γ_Y. Vårdal and Moan (1997) show that the current approach used to predict through-thickness cracks is particularly conservative.

Fatigue loading

A crucial issue is the fatigue loading, represented by $E(S^m)$ in Eq.(8). Typically the distribution of stress ranges as determined by Rainflow or Local Range counting is taken to be a Weibull distribution

$$F_S(s) = 1 - \exp(-(s/A)^B) \tag{11}$$

implying $E(S^m) = A^m \cdot \Gamma\,(m/B + 1)$, where $\Gamma(\,)$ is the Gamma function.

The parameters A and B would then reflect the significant uncertainties in

- environmental conditions
- wave kinematics and hydrodynamic loading
- transfer from hydrodynamic loading to structural forces
- stress concentration factors, including "carry-over/influence function effects" between braces in multi-brace joints.

The implied model uncertainty is illustrated by Vårdal and Moan (1997), by comparing two alternative fatigue loading analyses procedures, indicating a ratio greater than 100 between the corresponding cumulative damages.

Fatigue-induced fracture

Steel structures subjected to dynamic loading may fail due to ductile collapse, fatigue or fracture. In particular fracture may occur after fatigue deterioration. Failure then occurs when the load X(t) first upcrosses the resistance $\xi(t)$ as indicated in Fig. 2. A time variant reliability procedure was established by Marley and Moan (1992) to calculate the failure probability with due account of the long-term variability of sea sates. This method was also used to demonstrate that a simplified time-invariant (TIV) and much less time-consuming method, yields accurate predictions (Marley and Moan, 1992, 1993). This TIV approach is based on a threshold of resistance, ξ at the end of the time period, τ and the loading, X is taken to be the maximum X(t) within the time τ.

Initial crack size and POD values for tubular joints of jackets

In a recent study by Moan et al (1997) the initial crack size distribution and probability of detection, POD curve (for MPI/Eddy Current inspections) were estimated based on 240 (non propagating cracks) detected in 3411 inspections of tubular joints in North Sea jackets. The initial crack size and POD curve are assumed to be given by:

$$f_{A_0}(a) = (1/\lambda_0)\exp(-a/\lambda_0) \tag{12}$$

$$P_D(a) = 1 - \exp(-\,a/\lambda) \tag{13}$$

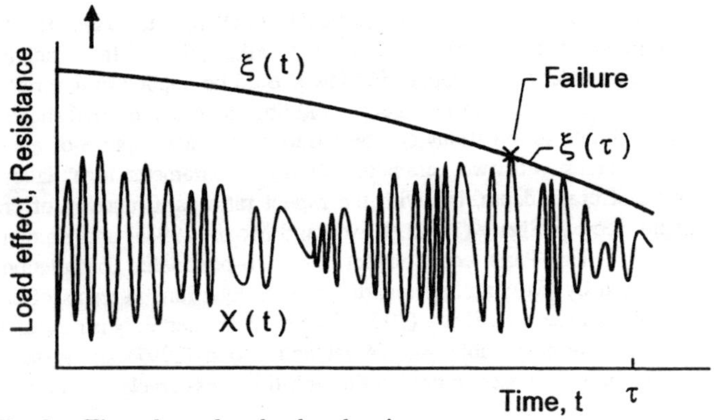

Fig. 2 Time dependent load and resistance

By introducing the event B_j in Eq. (9) as $f_{A_0}(a)da$ and the event I as $P_D(a)$ event, the distribution, $f_{A_0|D_1}(a)$ of the size of cracks detected in the first inspection, was obtained. Similarly, the distribution, $f_{A_0|ND_1}(a)$ of nondetected cracks was determined by taking I as $P_{ND}(a) = 1 - P_D(a)$. By updating $f_{A_0|ND_1}(a)$ conditional upon I taken to be a detection event, $f_{A_0|ND_1,D_2}(a)$ is computed. The unknown parameters λ_0 and λ are then determined by equating the mean crack sizes given by $f_{A_0|D_1}(a)$ and $f_{A_0|ND_1,D_2}(a)$ to the mean size of cracks observed in the first and second inspection, respectively. It was concluded that $\lambda_0 = 0.94$ mm and $\lambda = 1.95$. λ is sensitive especially to the mean crack size of the second inspection. However, if λ is fixed equal to the commonly used value of 1.2 mm, the λ_0 is found to be equal to be 1.02 mm, only slightly different from the value given above.

Calibration of fatigue design codes

Code calibrations described above refer to ultimate strength criteria. Because fatigue in offshore structures is mainly caused by one type of loading, there is limited merit of calibrating fatigue criteria, in the same sense as ULS criteria, relating to multiple loads. However, to achieve consistent design and inspection criteria, fatigue design criteria should be calibrated to reflect the consequences of failure and inspection plan. The purpose then may be to calibrate the design check: $D = \Sigma n_i/N_i \leq \Delta_d$ so that it implies a consistent life time failure probability When an inspection and repair strategy is implemented, the allowable cumulative damage Δ_d can be relaxed, as shown by Moan et al (1993), to obtain the target safety level equivalent to, say, a given Δ_d when no inspections are carried out (and the consequences of failure are the same).

To account for the effect of inspection and repair a fracture mechanics approach was used. Limit state functions g(z), here denoted by M(t), were based on Eqs. (7-8). The initial crack size, a_0, and fracture mechanics model were chosen so that the reliability level for the fracture mechanics model approach should be the same as that implied by $\Delta_d = 0.1$ or 0.3 for the SN approach when no inspection and repair are considered.

The major uncertainties were taken as shown in Table 2.

Inspections were assumed to be carried out every fourth year, and detected cracks were assumed to be repaired. The parameter λ in the POD curve was taken to be 1.5 mm.

The failure probability was calculated as follows (Madsen 1985, 1987)

$$p_{F1} = P[M(t) \leq 0] \qquad for\ 0 < t \leq \tau_1 (4\ years) \quad \textbf{(14a)}$$

$$\begin{aligned} p_{F2} = & P[M(\tau_1) \leq 0] \\ & + P[(M(\tau_1) > 0) \cap (H > 0) \cap (M^0(t) \leq 0)] \qquad \textbf{(14b)} \\ & + P[(M(\tau_1) > 0) > 0) \cap (H \leq 0) \cap (M^1(t) \leq 0)] \\ & for\ \tau_1 \leq t \leq \tau_2\ (8\ years) \end{aligned}$$

This means that the p_F for the time between the first and second inspection consists of two terms that depend upon survival in the time τ_1 and the outcome of the inspection - no crack detection (H > 0) or - crack detection (H ≤ 0) which is not known at time t = 0. $M^0(t)$ and $M^1(t)$ are expressed in the same way as M(t) by referring to an initial crack size of a_0 at t = 0 and a_R at time τ_1, respectively. The expression for H (at the time τ_1 of the first inspection) is given by the detectable crack size, a_D as: $H = a_D - a(\tau_1)$ and the distribution of a_D is given by the POD curve.

TABLE 2 Major probabilistic data for welded joints in tethers and jackets

Variable	Distr.	Jacket		Tether	
		Mean	St.dev.	Mean	St.dev.
$\log_{10}K$	Norm	12.52	0.248	14.03	0.204
Δ	Logn.	1.0	0.3	1.0	0.3
Initial crack depth, a_0 (mm)	Exp.	0.11	0.11	0.11	0.11
Crack depth after rep. a_R (mm)	Exp.	0.11	0.11	0.11	0.11
Detect.crack depth, a_D (mm)	Exp.	1.0/1.5	1.0/1.5	1.0/1.15	1.0/1.15
ν_0	Norm	$5.0 \cdot 10^6$	$3.5 \cdot 10^5$	$5.0 \cdot 10^6$	$3.5 \cdot 10^5$
$\ln C$	Norm	-29.97	0.514	-29.97	0.514
m	Fixed	SN:-3.0 FM: 3.1		3.5 3.1	
$\ln/A$	Norm	1.336[1)]	0.294	2.486[1)]	0.198
1/B	Norm	1.25	0.125	0.877	0.044
γ_Y	Norm	1.0	0.10	1.0	0.05

[1)] $\ln A$ corresponding to $\Delta_d = 0.1$

Fig.3 shows the variation of the reliability index, β with time. It also appears that assumed inspection strategy for tubular joints in jackets implies that the Δ_d could be relaxed from 0.1 to approximately 0.25 and still comply with the target reliability level implied by $\Delta_d = 0.1$ and no inspection/repair.

Finally it is noted that if we really do an inspection at the time $t = \tau_1$ and we know (for sure) that the outcome is no crack detection, the β varies as indicated in Fig. 3a just after the inspection. This implies that a much stronger statement of p_F is possible when there is no uncertainty about the inspection outcome.

SYSTEMS RELIABILITY APPROACH

General

Ultimate and fatigue design criteria in current codes are based on component failure modes (limit states) and a linear global model of the structure to determine the load effects in the components. However, an approach which is based on global (system) failure modes of the structure is desirable because significant consequences, e.g. fatalities will primarily be caused by global failure. A proper systems approach is also necessary to obtain the optimal balance between design and inspection plan since it is normally based on a certain damage tolerance, especially

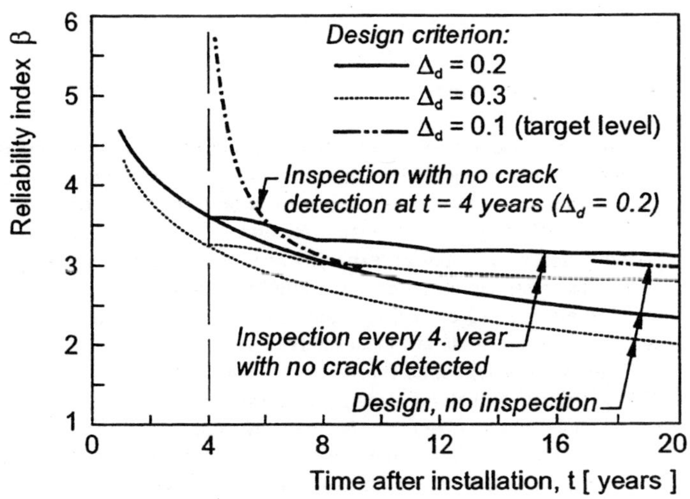

a) Tubular joint in jacket

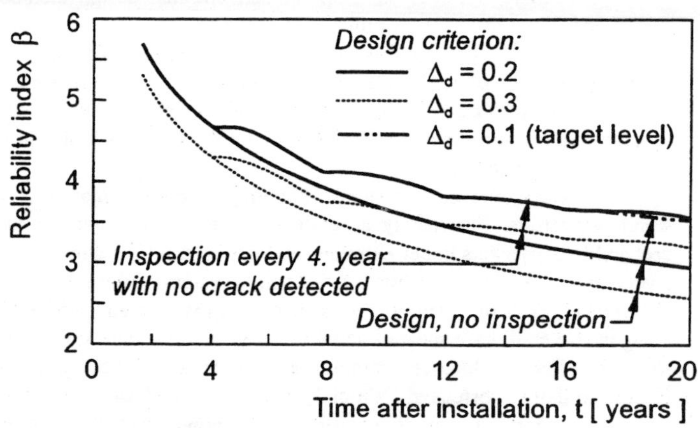

b) Welded joint in tether

Fig. 3 Reliability index for a welded joints as a function of time. Uncertainty measures given in Table 2. $E(a_D) = 1.5$ mm

when the inspections rely on detecting flooded or failed members.

While the structural mechanics analysis of e.g. a parallel series systems under tensile/bending loads is simple, prediction of the ultimate behaviour of framework structures like jacket is complex. Recent developments of efficient methods which account for large deflections and plasticity and even premature fracture (e.g. Moan and Amdahl, 1989, Ueda and Rashed 1991 and Stewart et al, 1993) facilitate systems reliability approaches. Significant progress has been made in reliability analyses of tether systems in TLPs and jackets over the last 15 years.

In the systems approach the relevant structure is assumed to be composed of different physical components (members, joints, piles, ..) which each may each have different failure modes, e.g. different collapse, fracture or fatigue modes.

System failure is then expressed mathematically by load- and resistance parameters relating to all failure modes for all components and calculate the systems failure probability by the probabilistic properties of these parameters. Broadly speaking, this may be achieved by a failure mode (or survival mode) analysis, or direct-simulation methods (Karamchandani 1990, and Moses and Liu 1992).

The failure mode analysis consists in

- identifying the sequences of component failures (system failure modes), E_i, considering members, joints and other components
- establishing a mathematical expression for the events of each sequence, E_i, based on structural mechanics. The event sequence (No. i) may involve failure of (n_i) components such that

$$E_i : E^2_{i_1} \cap E^1_{i_2} \cap \quad \cap E^{1,\dots,n_i-1}_{i_{n_i}} \tag{15}$$

where $E^{i,\dots,j-1}_{i_j}$ is the event that component no. j fails given that j-1 components already have failed.

- establishing probabilistic measures for the random variables involved
- calculating the failure probability of the system

The first two steps in this approach are particularly easily accomplished when ideal plastic behaviour can be assumed, see, e.g., Thoft-Christensen and Murotsu (1986). However, since buckling or collapse of members and fracture of joints are of concern for jackets (and other offshore structures), the more versatile member replacement or incremental approaches should be applied, to establish the failure function $g^{1,2,\dots,j-1}_{i_j}$ ( ) for the j'th failure given that (j-1) components already have failed in sequence i. Review of such approaches for sequences overload failures are given e.g. by Moses and Liu (1992) and Baker and Vrouwenvelder (1992) and Moan (1994).

Failure of redundant offshore structures may also be initiated by a fatigue failure or a fatigue induced fracture. If repair is not accomplished, a second fatigue or overload failure may occur. Even if the damage is detected, it may not be repaired until some time later, depending upon weather conditions. The increased stress in the remaining members will contribute to this second failure. While progressive overload failures are assumed to take place instantly (during the 18-20 sec period of a storm wave), the failures in the sequences of concern here, occur at different points in time.

For each failure sequence (i) of fatigue failures the failure function, $g^{1,2,\dots,i_n}_{i_j}$ ( ) may be established as follows: Consider a sequence with the first fatigue failure in section i_1, the second failure in joint i_2, etc. The first failure occurs when the time to failure, T_{seci_1}, is less than the service period, τ, i.e.

$$g^0_{i_1} = T_{seci_1} - \tau \le 0 \tag{16}$$

where T_{seci_i} is given by Eq. (8), and the initiation period is neglected. Similarly, the second failure occurs when

$$g^1_{i_2} = T_{seci_2/i_1} - \tau \le 0 \tag{17}$$

By assuming linear cumulative damage, it may be shown (Shetty 1992, Dalane 1993), that the conditional time T_{seci_2/i_1} to the second failure, in section i_2, given failure in section i_1, may be written as

$$T_{seci_2/i_1} = T_{seci_1} + (T_{seci_2} - T_{seci_1})\frac{d_2}{d_{2/1}} \tag{18}$$

where T_{seci_2} is the time to failure of section i_2 in the intact structure and

$$d_2 = \nu_{i02}E\left[(S_{i2}(\underline{P}))^m\right] \quad and \quad d_{2/1} = \nu_{i02/1}E\left[(S_{i2/1}(\underline{P}))^m\right] \tag{19a, b}$$

ν_{i02} and $\nu_{i02/1}$ being the zero-upcrossing frequency of the response at section i_2 for the intact structure and the structure with failure of section i_i; and S_{i2} and $S_{i2/1}$ are the corresponding stress ranges. $\underline{P}$ is a set of time-variant load and resistance parameters. The time to fatigue failure no. $k{:}T_{i_k|i_1 i_2 i_{k-1}}$ can be expressed in an analogous way as the time to the second failure (Shetty 1992, and Dalane 1993).

The probability of a given sequence of fatigue failures in the period τ may then be calculated by the intersection of events of the type $(g_{i_1 i_2 \dots i_k}^{1,2 \dots k-1}(\ ) \leq 0)$. The occurrence of an overload failure in the period τ, following the fatigue failures, may be included by referring the failure function for the overload to the period $\tau - T_{i_k/i_1 \dots k-1}$.

In computing the time to failure for the intact and damaged structure by using Eqs. (16-18), the main effort is associated with determining d_k and $d_{k/k-1}$. However, the calculation can be simplified by adopting a Weibull distribution, Eq. (11) and choosing A and B approximately (Shetty, 1992 and Dalane, 1993).

Modelling the uncertainties in loads and resistances in the components of the system is a crucial task. Measures for individual components used in code calibration are indicated above and in publications referred. The system approach in addition requires an estimate of the uncertainty of the system model as well as the correlation between variables in the different failure functions that represent the system. While the failure probability of a series system with n components may vary by a factor of n depending upon the correlation, the failure probability of a parallel system may vary even more - depending upon correlation and component characteristics.

Typically load uncertainties predominate in the calculation of the probability of wave overload, while load and resistance uncertainties are of the same order of magnitude in fatigue problems. Also, the correlation between component failure modes is less in fatigue. For fatigue failure modes the probability of first failure to the system failure probability will, hence, be large. Similarly, the small correlation between fatigue and overload failure events implies large systems effects for failure modes comprising fatigue events and a single overload failure.

Having established the limit states and uncertainty measures, the failure probability may be calculated by

$$p_{FSYS} = P[FSYS] = P\left[\bigcup_{i=1}^{N} \bigcap_{j=1}^{n_1} (g_{i_j}^{1 \dots j-1}(\ ) \leq 0)\right] \tag{20}$$

by FORM/SORM, bounding techniques or simulation methods. Due to the effort involved, it is important to apply some kind of technique to limit the number (N) of failure modes.

When the failure mode approach is applied, the updated failure probability for a system may be written as:

$$p_{FSYS} = P[FSYS|I] = \frac{P[FSYS \cap I]}{P[I]} \tag{21}$$

where FSYS is defined by Eq. (20) and $I{:}\bigcap_k (I_{\ell k} \leq 0)$ with ℓ and k referring to the inspection number and the type of inspection outcome, respectively. For a multi-component system and an inspection event tree with multiple branches, the computational efforts to calculate Eq. (21) are significant. Therefore, simplifications are necessary. For instance, simplified component limit state functions may be achieved by using time-invariant rather than time-variant approaches. The random variable T given by Eq. (8) in the failure function may be replaced by an equivalent one of the type $T = X_1 X_2 X_3$, where the X's are appropriately chosen (Hovde and Moan 1994a). Only the dominant system failure mode may be considered. Only significant branches of the inspection event tree may be considered.

In connection with FORM/SORM analysis, particular care needs to be exercised to eliminate non-active events (i.e.covered by other events).

The recent improvements of computer capability and simulation methods make it feasible to accomplish "direct systems reliability analyses" by simulation (Sigurdsson et al. 1993). Such methods are especially useful to achieve benchmark references for other approaches based on further approximations in the mechanics and probabilistic models.

Reassessment of jackets

There is an increasing need for reassessment of offshore structures. For instance almost one third of the world's 4500 fixed platforms are 20 years or more and an extended service life or reuse is called for in many cases. The ISO code for offshore structures (1994-) and other codes currently under development, explicitly address reassessment issues.

The reassessment is based on a condition assessment to identify possible presence of cracks, corrosion or other damages and involve analyses to demonstrate adequate safety.

While systems reliability approaches are not yet applied in the design of new structures, they are used in reassessment of existing structures when simple conservative methods are insufficient. (API, 1994; Digre et al., 1995). This is because more realistic and complex procedures would then pay in making decisions which affect safety and the significant costs of replacing or modifying the structure. Use of a system approach for some jackets is motivated by the significant reserve capacity beyond first component failure which is the basis for conventional design methods (Lloyd and Clawson 1983). A probabilistic system approach is attractive compared to a semi-probabilistic one, because it reflects more properly the uncertainties involved. Moreover, a semi-probabilistic approach based on safety factors does not represent the safety level for jackets with inadequate airgap (implying load calculation by Eq. (4) instead of Eq.(3)) in a consistent manner.

As indicated above, a systems reliability approach may require significant efforts. Fortunately, accurate estimates of the systems failure probability for jackets under extreme sea loading can be achieved with a very simple model, corresponding to a single systems failure mode, i.e. by referring both the load and resistance to a given load pattern and using the (overall) base shear as variable. This model has been validated for cases where the load uncertainties are dominant and the component forces are highly correlated (Wu and Moan 1989, and De et al 1989). Systematic Monte Carlo simulations have shown that the systems resistance can be accurately modelled by a mean value obtained by using mean values of all component resistance variables and a random uncertainty (COV) somewhere between zero- and the maximum component COV (Dalane, 1993 and Sigurdsson et al, 1993). Such a model is easily understood by designers and regulators and may hence contribute to a wider acceptance of systems reliability analysis. The simplified model needs further justification for systems with failure sequences involving collapse or fracture of tubular joints; or pile foundation failure, which are associated with a higher uncertainty and smaller correlation between components.

The calculated failure probability is compared with a target level. The target failure probability could be established as the component target failure probability multiplied by two correction factors. The first modification is to transform the target failure probability level for components to that for the system. The second factor is a cost-benefit factor to possibly permit the target p_F for an existing structural system to be higher than that for a new one. The balance between the two considerations would depend upon the regulatory regime These issues are discussed in detail by e.g. Moan (to appear).

Inspection planning relating to crack damage

Inspection, maintenance and repair (IMR) are important in ensuring continued integrity of offshore structures in the North Sea and other areas. Such actions should be directed at all types of damages (MTD 1989). While dents and gross corrosion are fairly easily controlled, particular attention should be paid to cracks, partly because cracks evidently occur, as demonstrated above, and the significant efforts devoted to IMR of cracks. Traditionally, inspection planning relating to cracks has been based on design fatigue life, although recently redundancy has become another parameter used in the planning process.

A number of limitations of inspection practice were identified in MTD (1989), especially the significant uncertainties in fatigue prediction, including design deficiencies, fabrication defects and crack detection probability, as well as the inadequacy of examining only a limited amount, typically 10-15%, of the joints in 4-5 year periods. The usefulness of probabilistic methods, especially Bayesian methods, to deal with uncertainties, and hence, increasingly recognized in inspection planning (Moan 1994).

A Design, Inspection, Maintenance and Repair (DIMR) strategy involves decision on

- design for a certain robustness against progressive failure
- identification of areas for surveillance, inspection and repair
- damages of concern (e.g. size of cracks to look for)
- frequency of inspection or monitoring
- inspection or monitoring method (and its quality)
- repair strategy (e.g. size of crack to be repaired, repair method, time required for repair)

A certain damage tolerance would be required because the damaged structure would need to carry the loads before the repair can be executed. Whether the inspection should be chosen to aim at detecting barely visible cracks,through-thickness cracks or member failures would depend on how much resources are spent to make the structure damage tolerant. This choice again would have implication on the inspection method; the main inspection methods being NDT methods, detection of through-thickness crack by e.g. member flooding, or failed members by visual inspection.

The most refined semi-probabilistic fatigue design practice is to let the allowable damage depend upon consequences of member failure and access for inspection (e.g. NPD 1992). Inspection requirements are not very specific. On the other hand, a systems reliability approach, considering the effect of inspection is extremely time-consuming even for a given structure and inspection program, and far more in case of optimization of design and inspection program.

A first step towards quantifying the effect of inspection is illustrated above (Fig. 3a) for a given tubular joint based on the component approach. If the same inspection program was implemented for all joints in the structure the target reliability will be achieved for all joints. However, even in the North Sea only a fraction of joints is inspected every year. For a redundant jacket such an inspection program will not be optimal. A more realistic approximate system approach would be to rank the inspection of a limited number of joints according to

$$p_j = p_{FFj} \cdot p_{FSYS|j} \quad (22)$$

where p_{FFj} is the probability of fatigue failure of joint j and $p_{FSYS|j}$ is the probability of system failure given failure of joint j. In principle $p_{FSYS|j}$ should be calculated by a formula like Eq. (20) for the structure less one component, and, hence, involve combined sequences of component failures due to fatigue and extreme loading. A further simplification would be to consider only a conditional ultimate failure, and calculate $p_{FSYS|j}$ by the simplified "single mode" model of the whole jacket mentioned above. The probabilistic inspection tool PIA (Kirkemo, 1989), which is widely used for North Sea platforms, is based on Eq. (22), using the mentioned "single mode" collapse model of the jacket to determine $p_{FSYS|j}$. It should be noted that for a given system and load conditions, the relative importance of progressive fatigue failure modes which are neglected in PIA when determining $p_{FSYS|j}$ increases with the service life of the structure (Moan 1994).

For systems with significant reserve strength after member failure, surveillance based on flooded member detection (FMD) may provide an opportunity for cost reductions, since flooding can be easily observed, when this feature exists. It is noted that 7 % of the cracks detected in North Sea jackets are through-thickness cracks (Vårdal and Moan 1997). Even if there is a residual fatigue life after the occurrence of a through-thickness crack, it may for simplicity be taken to be member failure. System reliability is governed by $p_{FF|j}$ and $p_{FSYS|j}$. FMD can be used to reduce the probability of progressive failure.

So far methods that may be used to plan inspections a priori. However, inspection plans needs to be continuously updated as information is gained during operation. As indicated in Fig. 3 no detection of cracks, reduces the failure probability, which in turn means that the time between inspections can be increased. This feature has been utilized in offshore activities in the North Sea as well as in other areas, as surveyed e.g. by Moan (1994) and Vårdal and Moan (1997). Since probabilistic fracture mechanics was introduced in the inspection planning for platforms in the Norwegian North Sea in 1990, the amount of inspections has been reduced by 70 to 80 percent. However, since the methodology on one hand is conservative by predicting 3 to 15 times more cracks than observed, and on the other hand fail to predict to 10 to 20 percent of the cracks observed, further developments are necessary (Vårdal and Moan 1997).

In the approach described above, inspection results are used to update the initial estimates of fatigue loading, crack growth parameters and initial crack size based on an assumed probabilistic nature of these variables. Since only fundamental randomness and "normal" uncertainties are considered, the updated hazard rate will be very conservative if, e.g., the initial crack size had been accidentally large due to a gross fabrication defect. Additional information about loads, may be a second measurement of the crack size, etc., would then be necessary to achieve a better estimate.

Inspection of tether systems

The tension-leg system is crucial for the safety of the TLP. Failure of all tethers in one of the four tether groups is considered to be catastrophic. Since a broken tether may hit other tethers and cause damage to or even failure of other tethers (Moan and Amdahl, 1989), it may be that even a single tether failure should be considered critical. A tether failure may occur due to tension or bending overload; or by going slack with excessive bending or subsequently experiencing a snatch load; or by fatigue. All these limit states are considered in the design. The tethers are normally designed with equal strength in all members, and in all joints.

The tether system is an ideal parallel system of series subsystems and is hence well suited for systems analysis. The tension in tethers in a group is highly correlated, while the possible bending in different tethers is less correlated.

The high correlation between component ultimate resistances makes the system analysis of overloads very simple and the systems effects relatively small (Guenard 1984). Fatigue is often the most critical failure mode for components and also for the system. It is noted that the fatigue probability of a single tether (a series system) may be several times higher than equally utilized and components with small correlation.

While the failure mode approach obviously can be applied to the tether system, simple and efficient alternative formulations have also been developed and applied.

Ximenes and Mansour (1991) applied a model based on extreme value statistics as well as a failure mode approach to determine the system fatigue reliability. Hovde and Moan (1994 a,b 1995)) extended the first approach by considering combined fatigue and overload failure modes, unequal loads and more general correlation properties between components as well as the effect of inspection. Their formulation is based on probability distributions and not events. Partial correlation between the loading and crack propagation properties at different crack sites are assumed, while the inspections are considered independent.

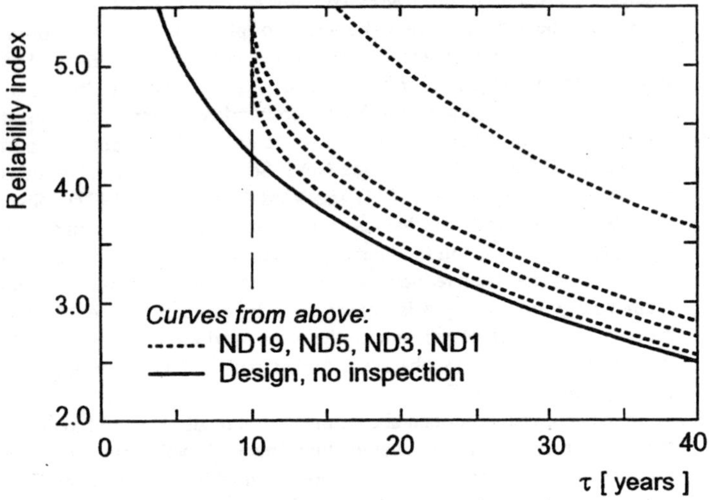

a) No cracks detected in the NDi joints inspected pr. tether

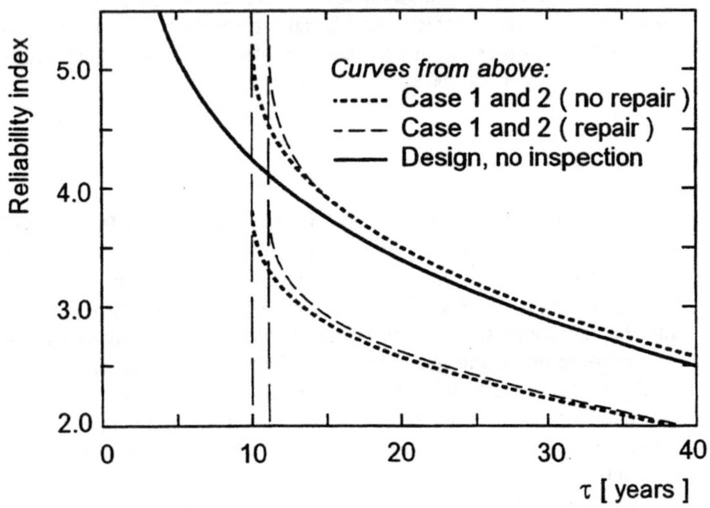

b) Three joints (out of 19) in each tether inspected. Crack detected and repaired for cases: 1) a single crack detected in one tether 2) one crack pr. tether in all legs. Repair period is 12 months.

Fig. 4 Updated system reliability for failure of one tether by inspection at τ_I = 10 years. $E(a_D)$ = 1.0 mm

Selected results of a case study of a tether system consisting of 4 tether groups, each with 4 tethers with 19 components in each tether. The results presented herein, refer to failure of one tether as the system failure mode. Fig. 4a shows the effect of the number of inspected joints in each tether. Considering the inspection costs, inspecting a few joints for each tether does not seem to be optimal. Fig.4b shows the reliability with respect to first tether failure after crack detection and repair. The repair period is conservatively assumed to be 12 months, and the number of inspected joints is 3 joints per tether. As expected and shown in the figure, the reliability decreases with the number of cracks detected. Further investigations (Hovde and Moan, 1994b) show that replacement of tethers instead of repair (by welding and grinding) is more efficient to increase the reliability.

CONCLUSIONS

In this paper an attempt has been made to demonstrate how some of the recent developments in structural reliability can aid decisions on structural design and modifications during use as well as inspection planning. However, the pitfalls are many when using advanced tools, especially because there is no standardization as yet of data and methodology. Examples are given, showing differences in approaches applied. Hence, such methods need to be used with caution, ensuring consistency with current design practice as well as consistency between reliability method, uncertainty measures and target levels applied. The early versions of guidelines for structural reliability analysis should therefore be further developed into standards.

REFERENCES

Almar-Næss. A. (ed.) 1985. *Fatigue Handbook for Offshore Steel Structures*, Trondheim: Tapir Publ.

API RP2A.WSD. 1994. Sect. 17.0, Draft, Dallas: American Petroleum Institute.

Atkins, W.S. 1994. Tern Structural Loading Study. JIP, Final Report, G3356/RPT/010, Atkins Oil & Gas Engineering, London.

Baker, M.J. & Vrouwenvelder, A.C.M. 1992. Reliability Methods for the Design and Operation of Offshore Structures. *Proc. 11th OMAE*, Vol. II: 123-132. New York: ASME.

Banon, H. et al 1994. Development of Reliability-based Global Design Equations for TLPs. *Proc. 13th OMAE*. Vol. II: 335-344. New York: ASME.

Barnouin, B. et al 1993. Underwater Inspection Reliability Trials for Offshore Structures. *Proc. 12th OMAE*, Vol. III-B: 883-890. New York: ASME.

Bea, R.G. 1993. Reliability-based Requalification Criteria for Offshore Platforms. *Proc. 12th OMAE*. Vol. II: 351-361. New York: ASME.

Bokalrud, T. and Karlsen, A. 1981. A Probabilistic Fracture Mechanics Evaluation of Fatigue Failure from Weld Defects, *Proc. Conference on Fitness for Purpose Validation of Welded Constructions*. Paper No. 8, London, UK.

Cornell, C.A. 1995. Structural Reliability - Some contributions to Offshore Technology, *Proc. 27th OTC*; 535-542, OTC 7753. Houston: Offshore Technology Conf.

Dalane, J.I. 1993. System Reliability in Design and Maintenance of Fixed Offshore Structures. Dr.ing.thesis. Trondheim. The Norwegian Institute of Technology.

De, R. et al. 1989. Study of Redundancy in Near-Ideal Parallel Structural Systems. *Proc. 5th ICOSSAR*. Vol.2: 975-982. New York: ASCE.

Digre, K.A. et al 1995. Modification to and Applications of the Guidelines for Assessment of Existing Platforms Contained in Section 17.0 of API RP2A. *Proc. 27th OTC*; 741-754, OTC 7779. Houston: Offshore Technology Conf.

Dharmavasan, W. & Dover, W.D. 1988. Nondestructive Evaluation of Offshore Structures using Fracture Mechanics. *Applied Mechanics Reviews* 41: 36-49.

DNV. 1995. Guidelines for Offshore Structural Reliability Analysis - General. Final Draft. Report No. 95-2018, Oslo: Det Norske Veritas.

Efthymiou, M. et al. Reliability-based Criteria for Fixed Steel Offshore Platforms. *Proc. 15th OMAE*: Vol. I:129-141, New York: ASME.

Faulkner, D. 1991. Unified Design Codes for Floating Systems. *Proc. Integrity of Offshore Structures*, Glasgow, 1990. London: Elsevier.

Fjeld, S. 1977. Reliability of Offshore Structures. *Proc. 9th OTC*: Vol. 2: 459-472, OTC 3027. Houston: Offshore Technology Conf.

Guenard, Y.F. 1984. Application of System Reliability Analysis to Offshore Structures. Report No. 71. Stanford, CA: John Blume Earth. Engng. Center, Stanford University.

Heideman, J.C. & Weaver. T.O. 1992. Static Wave Force Procedure for Platform Design. *Proc. Civil Engng. in the Oceans V*: 496-517. College station, Texas: ASCE.

Hovde, G.O. & Moan, T. 1994a. Fatigue Reliability of TLP Tether Systems. *Proc. 13th OMAE* Vol.II: 141-150, New York, ASME.

Hovde, G. O. and Moan, T. 1994b. Fatigue Reliability of TLP Tether Systems considering the Effect of Inspection and Repair. *Proc. Behaviour of Offshore Structures*, Vol. 3: 85-100. Pergamon, London.

Hovde, G.O. and Moan, T. 1995. Fatigue and fracture reliability of TLP Tether Systems before and after a tether failure. *Proc. 7th ICASP. Applications of Statistics and Probability*. Vol.1:319-326. Rotterdam: Balkema.

Jiao, G. et al. 1997. Limit State Philosophy and Reliability Based Design Recommendations for Pipelines. *Proc. 16th OMAE*, Paper No. 97-1812, Yokohama.

ISO 1994-. Standard for Offshore Structures for the Petroleum and Natural Gas Industries. ISO 13819-1, Part 1. General Principles (1994). Part 2 Fixed Steel structures (Draft B 1996).

ISO 1995. General Principles of Reliability for Structures, Revision of IS 2394. Draft No. 15, by ISO/TC98/SC2/WG1.

ISSC. 1988-94. Report of ISSC Committee IV.1. Design Philosophy. *Proc. 10th ISSC*, (1988), *11th ISSC*, Wuxi (1991) and *12th ISSC*, St. John's, Newfoundland (1994).

Jonathan, P. and Taylor, P.H. 1996. Wave and Force Variability on Offshore Structures. *Proc. 15th OMAE*:Vol.I:179-190. New York, ASME.

Jordaan, I.J. & Maes, M.A. 1991. Rationale for Load Specifications and Load Factors in the New CSA Code for Fixed Offshore Structures. *C.J. Civil Engng*. 18(3): 454:464.

Kam, J.C.P. & Dover, W.D. 1989. Advanced Tool for Fast Assessment of Fatigue under Random Wave Stress Histories. *Proc. Inst. Civ. Eng.*: Paper 9425. Part 2, Vol. 8. London: ICE.

Kam, J.C.P., Dover W.D. and Ma, C.N. 1995. The Prediction of Crack Shape Development for In-service Cracks in Offshore Welded Tubular Joints, *J. Marine Structures*, 8: 37-65.

Karamchandani, A. 1990. New Methods in Systems Reliability. Ph.D. Thesis. Stanford, CA: Department of Civil Engineering, Stanford University.

Kirkemoe, F. 1988. Applications of Probabilistic Fracture Mechanics to Offshore Structures. *Applied Mechanics Review* 41: 61-84.

Kirkemoe, F. 1989. PIA Theory Manual, Aker, Oslo.

Lloyd, J.R. & Clawson, W.C. 1983. Reserve and Residual Strength of Pile Founded Offshore Platforms. *Int. Symp. on the Role of Design, Inspection and Redundancy in Marine Structural Reliability*, Williamsburg, Va.

Lloyd, J.R. & Karsan, D.I. 1988. Development of a Reliability-based Alternative to API RP2A. *Proc. 20th OTC:* Vol. 4: 593-600, OTC 5882. Houston: Offshore Technology Conf.

Madsen, H.O. 1985. Random Fatigue Crack Growth and Inspection. *Proc. 4th ICOSSAR*, Int. Assoc. for Struct. Safety and Reliability: 1:475-484.

Madsen, H.O. 1987. Model Updating in Reliability Theory. *Proc. 5th ICASP*: 564-577. Vancouver.

Madsen, H.O., Krenk, S. & Lind, N.C. 1986. *Methods of Structural Safety*. Englewood Cliffs, MJ: Prentice-Hall Inc.

Maes, M.C. 1986. Calibration of Partial Factors in the New CSA Code for Fixed Offshore Production Structures. Technical Report No. 9. Calgary, Alberta: Environmental Protection Branch, Canada. Oil and Gas Lands Adm.

Marley, M.J. & Moan, T. 1992. Time Variant Formulation for Fatigue Reliability. *Proc. 11th OMAE*: Vol. II: 161-168. New York: ASME.

Marley, M.J. & Moan, T. 1994. Approximate Time Variant Analysis for Fatigue. *Proc. 6th ICOSSAR. Structural Safety and Reliability*. Vol. 2: 1099-1105. Rotterdam: Balkema.

Mathisen, J. et al. 1994. Reliability-based LRFD Code for TLP Hull Structure. Vol. II: 345-356.

Melchers, R.E. 1987. *Structural Reliability Analysis and Prediction*. Chichester, UK: Ellis Horwood Ltd.

Moan, T. 1983. Safety of Offshore Structures. *Proc. 4th ICASP*: 41-85. Bologna: Pitagora Editrice.

Moan, T. 1988. The Inherent Safety of Structures designed according to the NPD Regulations. Report No. STF71 F88043. Trondheim: Div. Structural Engineering, SINTEF.

Moan, T. and Wu,M.K. 1988. Calibration of Safety Coefficients for ULS Design Check of Submerged Buoyant Bridges. Report MK/R 100. Div. of Marine Structures, The Norwegian Institute of Technology, Trondheim.

Moan, T. & Amdahl, J. 1989. Catastrophic Failure Modes of marine Structures. In: T. Wierzbicki, & N. Jones (eds.), *Structural Failures*: 463-510. New York: John Wiley.

Moan, T. et al 1993. Reliability-based Fatigue Design Criteria for Offshore Structures Considering the Effect of Inspection of Repair. *Proc. 25th OTC*: Vol. 2: 591-599, OTC 7189. Houston: Offshore Technology Conf.

Moan, T. 1994. Reliability and Risk Analysis for Design and Operations Planning of Offshore Structures. *Proc. 6th ICOSSAR, Structural Safety and Reliability*. Vol. 1: 21-43. Rotterdam: Balkema.

Moan, T. et al. 1994. Applied Design-Strength Limit States Formulations. Report of ISSC Committee V.1, *Proc. 12th ISSC*, St. John's Newfoundland.

Moan, T. 1995. Safety Level Across Different Types of Structural Forms and Materials - Implicit in Codes for Offshore Structures. Report STF 70 A95210 Trondheim: SINTEF. Prepared for ISO/TC250/SC7.

Moan, T. Target Levels for Structural Reliability and Risk Analysis of Offshore Structures, in *Risk and Reliability in Marine Technology*, C.Guedes Soares (ed), Rotterdam: Balkema, to appear.

Moan, T., Vårdal, O.T., Hellevig, N.C. and Skjoldli, K. 1997. In-Service Observations of Cracks in North Sea Jackets. A Study on Initial Crack Depth and POD Values. *Proc.OMAE*, Paper No. 1335, Yokohama, Japan.

Moses, F. 1986-1987. Load and Resistance Factor Design-Recommended Practice for Approval. Final Report, API PRAC 86-22. Dallas: API. API PRAC 87-22. Dallas: API.

Moses, F. & Liu, Y. W. 1992. Methods of Redundancy Analysis for Offshore Platforms. *Proc.11th OMAE*:Vol.II:411-416. New York: ASME.

MTD 1989. Underwater Inspection of Steel Offshore Installations: Implementation of a new approach. Marine Technology Directorate Limited. Publication, London.

NKB 1978, 1987. Recommendations for Loading and Safety Regulations for Structural Design. Report No. 36, 55E. Copenhagen: The Nordic Committee on Building Regulations.

NPD 1992. Regulations Concerning Loadbearing Structures in the Petroleum Activity. Stavanger: Norwegian Petroleum Directorate.

Rackwitz, R. & Schrupp, K. 1985. Quality Control Proof Testing and Structural Reliability. *Structural Safety* 2: 239-244.

Shetty, N.K. 1992. System Reliability of Fixed Offshore Structures under Fatigue Deterioration. Ph.D. Thesis. London: Imperial College, Univ. of London.

Sigurdsson, G. et al. 1994. Probabilistic Collapse Analysis of Jackets. *Proc. 6th ICOSSAR*: Vol. 1: 535-544. G.I. Schuëller, M.Shinozuka and J.P.T. Yao (eds). Rotterdam: A.A. Balkema.

Silk, M.G., Stoneham, A.M. and Temple, J.A.G. 1987. *The Reliability of Nondestructive Inspection* - Adam Hilger, Bristol.

Skjong, R. 1985. Reliability-based Optimization of Inspection Strategies. *Proc. 4th ICOSSAR*. Vol. II: 614-618. Kobe: IASSAR.

Stahl, B. and Lloyd, J. 1995. Debate on the Application of Structural Reliability Techniques for Offshore Structural Engineering. London: The Institution of Structural Engineering.

Stewart, G., Moan, T., Amdahl, J. and Eide, O.I. 1993. Nonlinear Re-assessment of Jacket Structures under Extreme Storm Cyclic Loading. Part I - Philosophy and Acceptance Criteria. *Proc. 12th OMAE* Vol.II: 491-502. New York: ASME

Theophanatos, A. et al. 1992. Adaption of API RP2A-LRFD to Mediterranean Sea. *Proc. 24th OTC*. Vol. 2: 529-538, OTC 6932. Houston: Offshore Technology Conf.

Thoft-Christensen, P. and Murotsu, Y., 1986. *Application of Structural System Reliability Theory*. Berlin: Springer-Verlag.

Turner, R.C. et al. 1992. Towards the Worldwide Calibration of API RP2A-LRFD. *Proc. 24th OTC*: Vol. 2: 513-520, OTC 6930. Houston: Offshore Technology Conf.

Ueda, Y. & Rashed, S.M.H. 1991. Modern Method of Ultimate Strength Analysis of Offshore Structures. *Int. J. Offshore and Polar Engineering 1* (1): 27-41.

Vårdal, O.T. and Moan, T., 1997. Predicted versus Observed Fatigue Crack Growth. Validation of Probabilistic Fracture Mechanics Analysis of Fatigue in North Sea Jackets. *Proc. 16th OMAE*, Paper No. 1334, Yokohama, Japan.

Wang, X., Jiao, G. and Moan, T., 1996. Analysis of Oil Production Ships Considering Load Combination, Ultimate Strength and Structural Reliability. Proc. SNAME Annual Meeting, Paper No. 1, New York.

Wu, Y.L. & Moan, T., 1989. A Structural System Reliability Analysis of Jacket using an Improved Truss Model. *Proc. 5th ICOSSAR* Vol. 2: 887-894. New York: ASCE.

Ximenes, M.C.C. & Mansour, A.E. 1991. Fatigue System Reliability of TLP Tendons Including Inspection Updating. *Proc. 10th OMAE*: Vol. II: 203-211. New York: ASME.

Yang, Y.N. 1993. Application of Reliability Methods to Fatigue, Quality Assurance and Maintenance. The Freudenthal Lecture. *Proc. 6th ICOSSAR, Structural Safety and Reliability*. Vol. 1: 3 - 18. Rotterdam: Balkema.

Proceedings of the Seventh (1997) International Offshore and Polar Engineering Conference
Honolulu, USA, May 25-30, 1997

ISBN 1-880653-28-1 (Set); ISBN 1-880653-32-X (Vol. IV)

A New Ultimate Capacity Formula for Unstiffened CHS T-, TT- and K-Joints Under Axial Brace Loads

M. Morita, Y. Yamada, Y. Makino, S.R. Wilmshurst and Y. Kurobane
Kumamoto University
Kumamoto, Japan

M.M.K. Lee
Swansea University
Swansea, United Kingdom

ABSTRACT

In order to make simple design equations for CHS joints, a new ultimate capacity formula for K-joints under axial loads is derived by a multiple regression analysis method and using the database of Makino et al. (1996). Although there are several failure modes in K-joints, an ultimate capacity formula for joints failing only by local deformation of the chord wall is derived in this study. The derived ultimate capacity formula can be used for K-joints, T-joints, and TT-joints failing with no radial deformation of the chord wall between the braces. This formula represents the first stage of a continuous formula for hub joints where the braces are positioned around the chord circumference (e.g. T-, TT, X-, XT-, and XX-joints), and gap joints (i.e. joints with longitudinally spaced braces). The accuracy of the new ultimate capacity formula is slightly lower than the individual joint equations of Kurobane et al. (1984) and Paul et al. (1994), but it is simpler than the existing formulae.

KEY WORDS: Tubular joints, K-joints, Failure Type 1, Local deformation, Ultimate strength formula, Axial brace load

NOTATIONS

- D : chord diameter
- d : brace diameter
- e : eccentricity between point where two brace axes meet and chord axis
- g_c : circumferential gap between two braces
- g_l : longitudinal gap between two braces
- g_t : transverse gap between two braces
- L : chord length
- M_0 : bending moment in chord at center of joint
- M_p : full plastic moment capacity of chord member
- m_0 : bending moment ratio (M_0/M_p)
- N_u : ultimate axial force of brace
- $N_{u,c}$: estimated ultimate axial force of brace
- N_0 : axial force on chord (external force)
- n_0 : $\dfrac{N_0}{\pi(D-T)T\sigma_y}$
- $R_1, R_2, R_3 \cdots$: regression constants
- T : chord thickness
- t : brace thickness
- α : chord length ratio ($2L/D$)
- β : diameter ratio (d/D)
- γ : chord thinness ratio ($D/2T$)
- ε : error term in regression analysis
- ϕ : out-of-plane angle
- θ : in-plane angle between compression brace and chord
- σ_y : yield stress of chord material
- σ_u : ultimate tensile stress of chord material
- τ : thickness ratio (t/T)
- η : eccentricity ratio (e/D)
- ζ_l : longitudinal gap ratio (g_l/T)
- ζ_t : transverse gap ratio (g_t/T)

INTRODUCTION

Existing design formulae, such as AWS, API, CIDECT and AIJ Codes, are either too complicated or inaccurately estimate the ultimate strength for new other joints. To make simpler design equations than those existing, new ultimate capacity formulae should be proposed. New ultimate capacity formulae for X-, T- and TT-joints were already proposed as a first step by the authors (Morita et al. 1996). The new ultimate capacity formulae used no complicated geometric variables. Furthermore, the formulae for TT-joints were continuous as they related the ultimate capacity formulae for X- or T-joints in accordance with the two failure types of TT-joints.

It is desirable to make a new ultimate capacity equation which has following as requirements.

(1) The new ultimate formula should be simpler than existing equations. Even if it is as complicated as existing equations, the separate design formulae derived from the ultimate capacity formula should be simple. Furthermore, if the accuracy of the

ultimate capacity equation is lower than that of existing formulae, it should be simpler.

(2) It is better that new ultimate capacity formula is a continuous equation for various joints. For example, the ultimate capacity of a TT-joints with ϕ=180 degree is equal to that of a X-joint, that of a K-joint with large g_l should approximate to that of a T-joint (provided that shear failure of the chord cross-section is not the predominant failure mode) and so on, as shown in Figs. 1 and 2. The reason for continuous equations is the possibility to estimate easily the ultimate capacities of other types of joint.

DATA FOR REGRESSION ANALYSIS

Local deformation chord failures of axially loaded tubular joints can be distinguished into two types: Type 1 (Fig. 3 (a)) in which the two compression braces act together with no local deformation of the chord wall in the transverse gap area between them, and type 2 (Fig. 3 (b)) in which there is significant local deformation of the chord wall in the transverse gap area in the form of a radial fold. Failure type 1 occurs in joints with small transverse gaps while joints with larger transverse gaps yield failure type 2 .

If the ultimate capacities of all T-, K- and TT-joints for failure type 1 can be calculated by one equation as the former equation (Morita et al. 1996) for T- and TT-joints, it will be convenient for designers. In this paper, an ultimate capacity equation for T-joints, T-joints with failure type 1, and K-joints, is derived.

Data sets for regression analysis were selected from the database for the tubular joints (Makino et al. (1996)), and were screened according to the following conditions.

(1) Joints where the predominant failure mode was chord wall deformation were used.
(2) α of T-joint is larger than 10 and that of TT-joint is larger than 8.9.(According to van der Vegte (1995), α has no effect on the ultimate capacities when the α is greater than 12 and the effects due to overall chord bending moments have been excluded. However, the number of tests for $\alpha > 12$ is small.)
(3) Test results which do not reach the first peak load were omitted.
(4) Test results without measured yield and ultimate stresses of chord materials were omitted.
(5) Test results with unclear support conditions were omitted.
(6) Test results with partially thickened chord members were omitted.
(7) Test results with yield in both chord and brace members were omitted.
(8) Test results with local buckling in brace members were omitted.
(9) Test results with flexural buckling in brace members were omitted.
(10) Test results with concrete filled members were omitted.

As a result, the numbers of data sets and the ranges of data are as in Table 1, and the relationships between parameters are as shown in Figs. 4 (a) and (b).

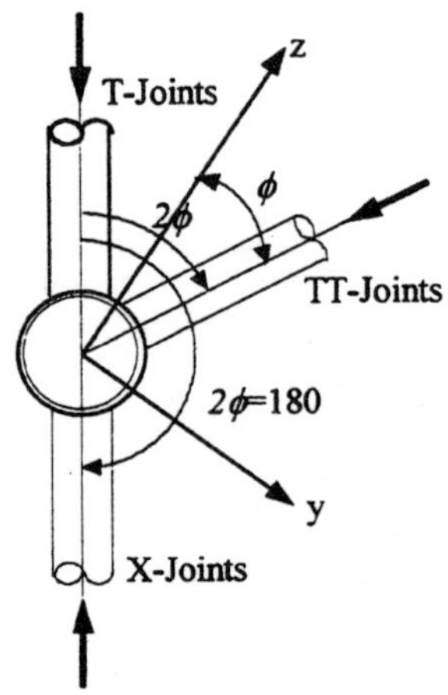

Fig. 1 Configuration Relationships between T, TT and X-Joints

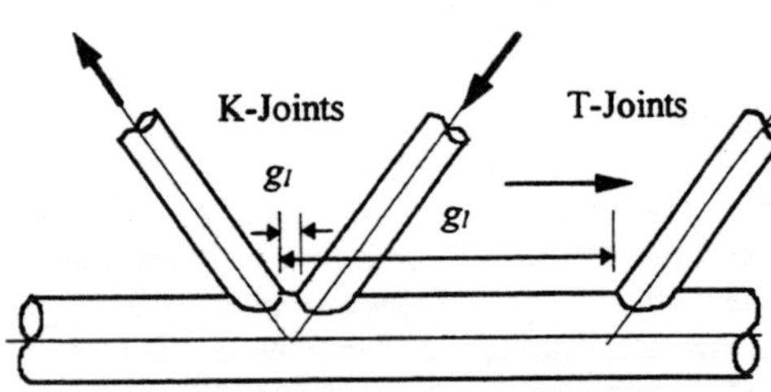

Fig. 2 Configuration Relationships between K and T-Joints

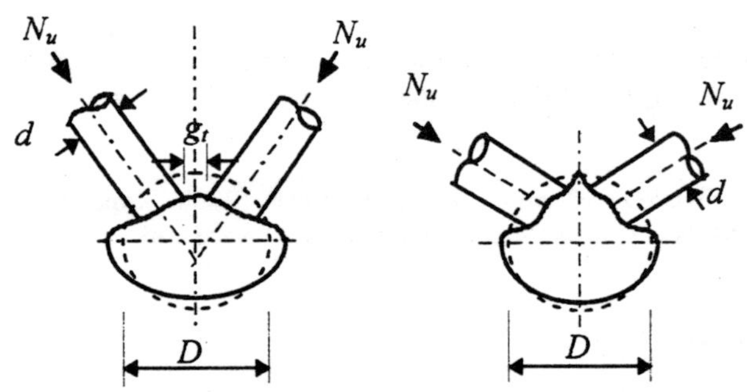

(a) Failure Type 1 (b) Failure Type 2

Fig. 3 Illustrations of joint failure for TT-joints

Table 1 Numbers and validity range of data on regression analysis

	number of data		validity range												
Joint	Test	FE results	D (mm)	T (mm)	d (mm)	t (mm)	θ (deg.)	φ (deg.)	β	γ	ζ_l	n_0	σ_y	σ_y/σ_u	m_0
K-joint	355	0	59.7～ 558.8	1.62～ 11.1	22.0～ 326.4	1.6～ 11.0	0～ 30	—	0.190～ 1.00	7.77～ 51.0	-14.4～ 66.2	-0.659～ 0.935	231～ 826	0.546～ 0.943	—
T-joint	28	40	113.9～ 1400	3.42～ 28.0	42.7～ 800	1.9～ 28.0	0～ 45	—	0.190～ 1.00	7.26～ 46.7	—	—	295～ 441	C.539～ 0.830	0.008～ 0.970
TT-joint	2	6	190.8～ 406.4	5.20～ 11.0	60.6～ 244.5	3.93～ 11.0	—	0～ 45	0.219～ 0.602	18.3～ 18.5	—	—	355～ 384	0.696～ 0.865	0.047～ 0.801

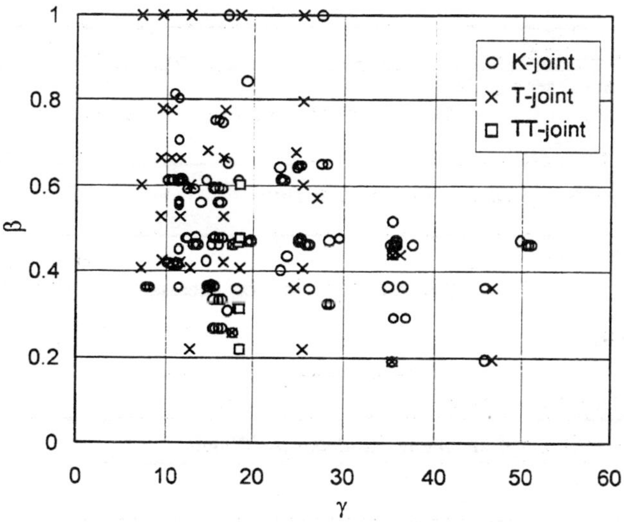

Fig. 4 (a) β versus γ. Geometrical parameter distribution of data for regression analysisζ_l.

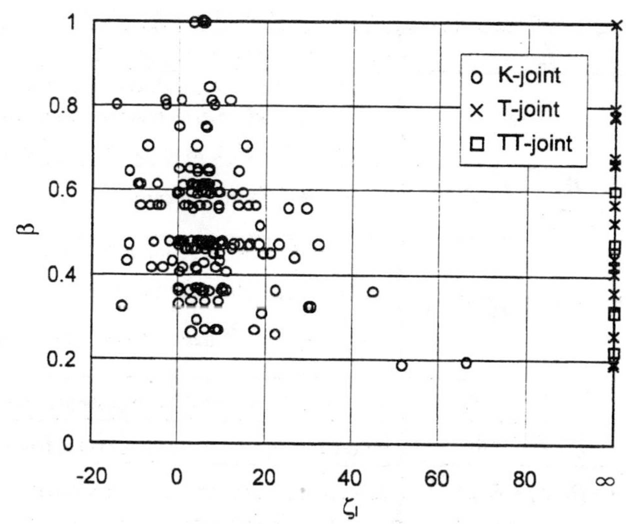

Fig. 4 (b) β versus ζ_l. Geometrical parameter distribution of data for regression analysis

MATHEMATICAL MODEL FOR REGRESSION ANALYSIS

In this paper, the ultimate capacity formula is derived by multiple regression analysis using test and numerical analysis results. Fig. 5 shows the relationships between the ultimate strength of K-joints from test and numerical analysis and the predictions according to the T-joint estimation equation of Morita et al. (1996). According to Fig. 5, a mathematical model for regression analysis was proposed as Eq. 1.

Final mathematical model is shown in Eq. 2, and the term meanings are shown in Table 2, since it was necessary to improve the accuracy of ultimate capacity equation.

$$_K N_{u.c} = \left(1 + R_1\, exp(R_2 \zeta_l) \cdot \beta^{R_3} \cdot \gamma^{R_4}\right)_T N_{u.c} \cdot \varepsilon \qquad (1)$$

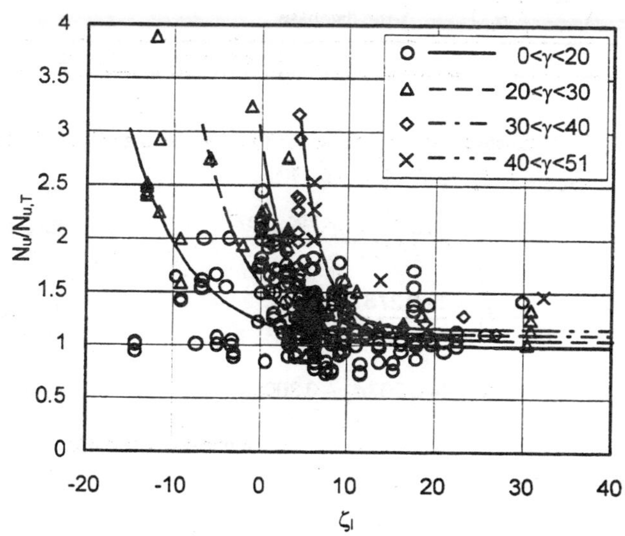

Fig. 5 Ultimate strength interaction between T-joints and K-joints

$$\frac{N_u \cdot sin\theta \cdot cos\phi}{T^2 \sigma_y} = \frac{R_1}{1+R_2\beta}\beta^{R_3}\gamma^{R_4}\left(1+R_5\, exp(R_6\zeta_l)\beta^{R_7}\right)(1+R_8\, cos\theta)(1+R_9\, sin\phi)\left(1+R_{10}n_0+R_{11}n_0^2\right)(1+R_{12}m_0)\left(1+R_{13}\frac{\sigma_y}{\sigma_u}\right)\varepsilon \qquad (2)$$

RESULT OF REGRESSION ANALYSIS

The equation is shown as Eq. (3) with mean and CoV values, and comparison between test and numerical results and predictions according to the equation are shown in Fig. 6, which are plotted on log-log axes. The dotted represents line the approximate 95% confidence limit, which is given by $N_{u.c} \cdot e^{\pm 1.96\sigma}$ (where σ is standard deviation of the ultimate capacity formula).

Test and numerical results to prediction ratios against the parameter β, γ and ζ_l, are shown in Figs. 7 (a), (b) and (c), respectively. The figures show there is no significant errors for parameters, and no residual trends between the capacity ratio and the parameters were found. Furthermore, mathematical model checks against the other parameters are shown in Figs. 8 (a), (b), (c), (d) and (e). It is confirmed that each mathematical model fits in with test and numerical results.

The mean and CoV values against each kind of joints are shown in Eq. 3. Since the total CoV value is equal to 0.132 and the approximate 95% confidence limit is between about 0.8 and 1.3, it is useful to make a design equation using this formula.

Table 2 Definitions of terms in the ultimate capacity formula

$N_u \sin\theta$	component force in the perpendicular axis to the chord for K- and Y-joints
$\cos\phi$	component force in the direction of z-axis for TT-joints (see Fig. 1) In a T-joint, this means that the braces of TT-joint act on one axis, the value of N_u represents a half of the ultimate capacity of T-joint, as in the former paper (Morita et al. 1996).
$\frac{R_1}{1+R_2\beta}\beta^{R_3}\gamma^{R_4}$	basic term for ultimate capacity equations for hub joints derived from ring-model analysis
$\left(1+R_5 \exp(R_6\zeta_i)\beta^{R_7}\right)$	correction factor for K-joints with a gap between two braces In Eq. (1), although there is a term γ^{R_4}, it was found that this term is not necessary as the term $\frac{R_1}{1+R_2\beta}\beta^{R_3}\gamma^{R_4}$ exists in the basic T-joint formula.
$(1+R_8\cos\theta)$	correction factor for the component force in the direction of chord axis
$(1+R_9\sin\phi)$	correction factor for the component force in the direction perpendicular to chord axis for TT-joints
$\left(1+R_{10}n_0+R_{11}n_0^2\right)$	function to account for chord pre-stress
$(1+R_{12}m_0)$	reduction factor for chord member global bending effects
$\left(1+R_{13}\frac{\sigma_y}{\sigma_u}\right)$	correction function for chord materials, since there is a mix of hot finished, cold finished, and high strength tubes in the database
ε	error term of the regression equation

Eq. 3 Ultimate capacity formulas for T and K-joints and TT-joints of Failure Type 1

$$N_{u.c} = f_1 \cdot f_2 \cdot f_3 \cdot f_4 \cdot f_5 \cdot f_6 \cdot f_7 \cdot T^2 \sigma_y \quad (3)$$

$$f_1 = \frac{11.6}{1-0.442\beta}\beta^{0.796}\gamma^{0.377} \qquad f_2 = 1+0.493\exp(-0.042\zeta_i)\beta^{-1.04}$$

$$f_3 = \frac{1-0.276\cos\theta}{\sin\theta} \qquad f_4 = \frac{1+2.03\sin\phi}{\cos\phi}$$

$$f_5 = 1+0.207n_0-0.300n_0^2 \qquad f_6 = 1-0.146m_0 \qquad f_7 = 1-0.556\frac{\sigma_y}{\sigma_u}$$

number of data=431 mean=1.009 CoV=0.132

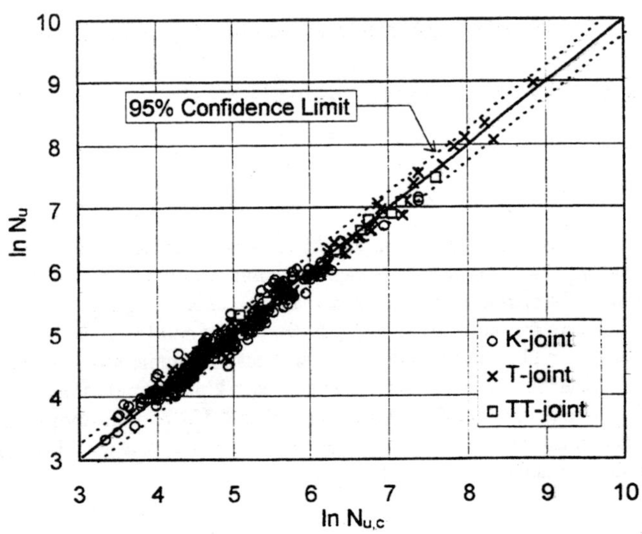

Fig. 6 Relationships between test and numerical results and prediction equation

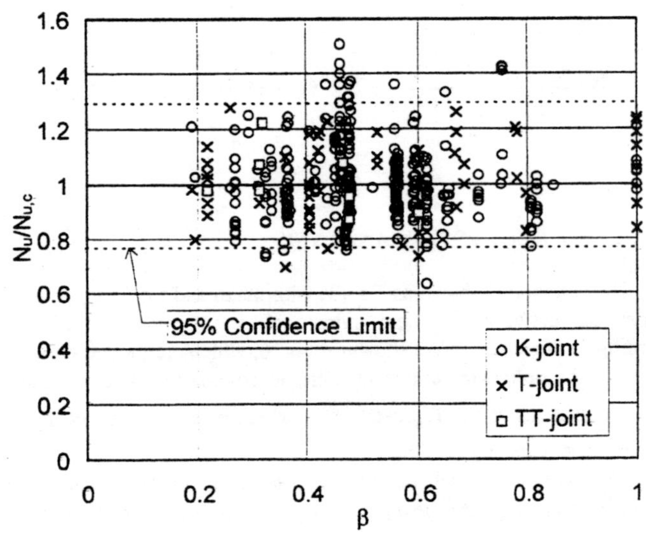

Fig. 7 (a) Test and numerical results to prediction ratios against β

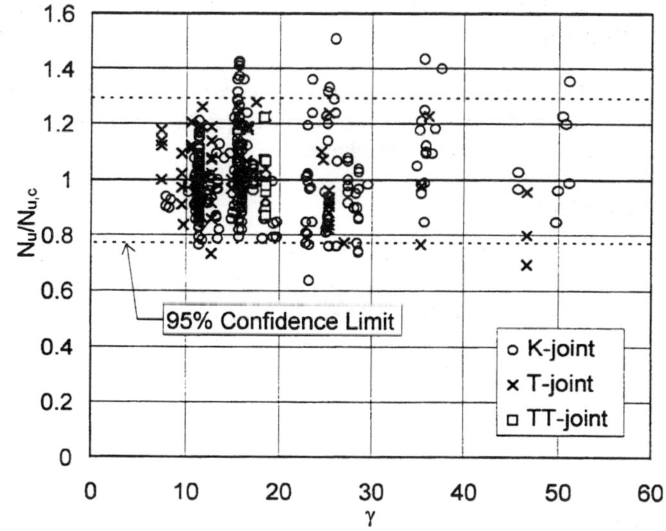

Fig. 7 (b) Test and numerical results to prediction ratios against γ

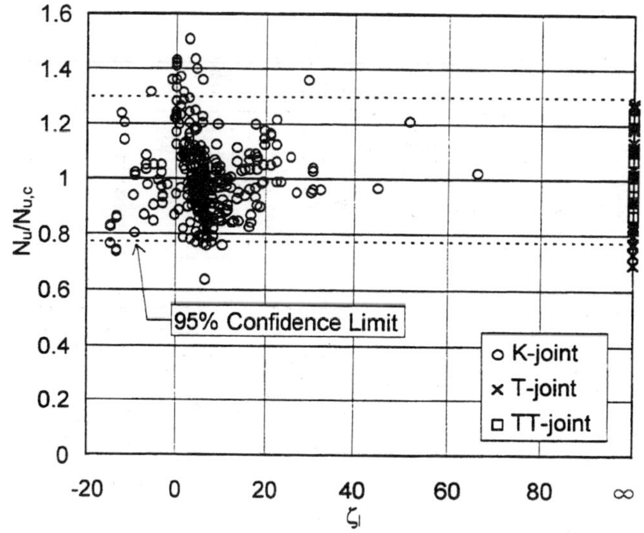

Fig. 7 (c) Test and numerical results to prediction ratios against ζ_l

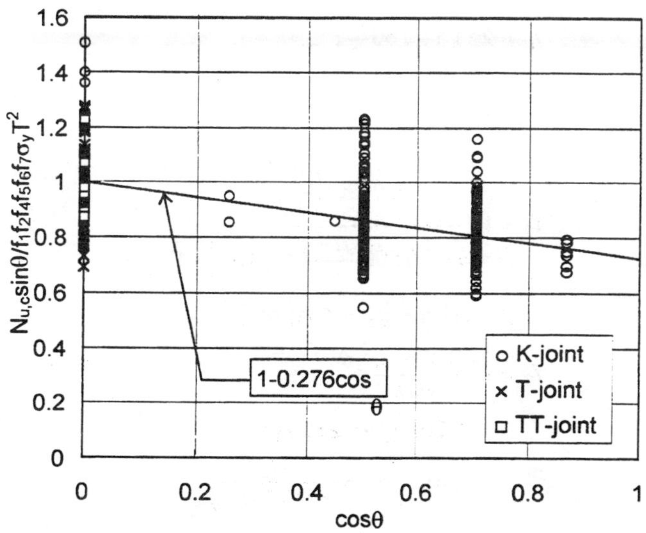

Fig. 8 (a) Test and numerical results to prediction ratios against cos θ

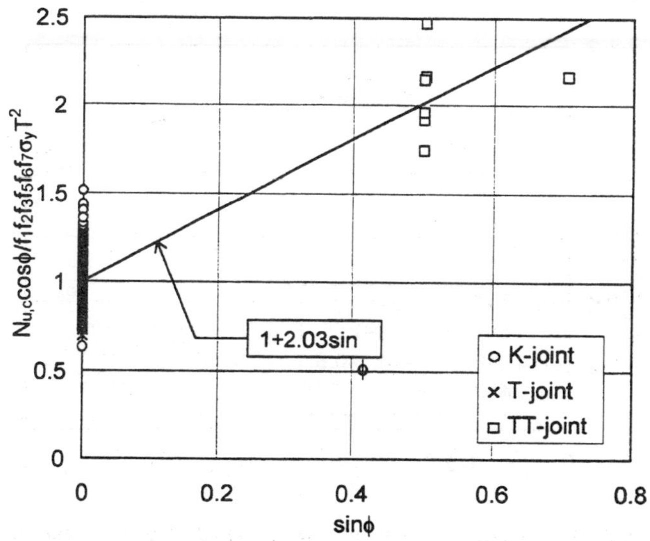

Fig. 8 (b) Test and numerical results to prediction ratios against sin ϕ

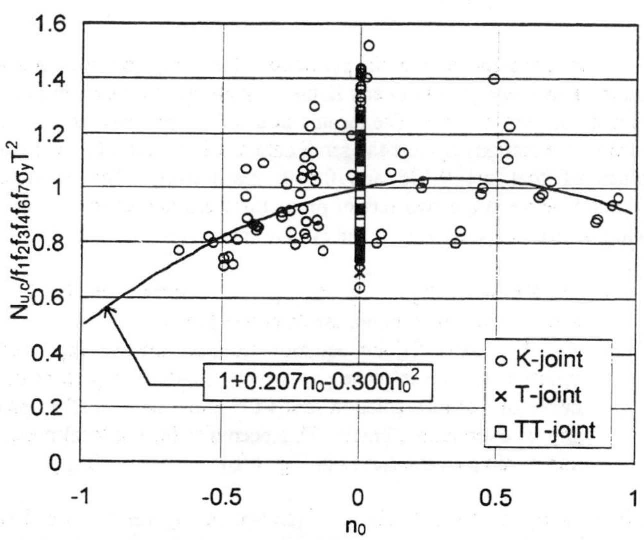

Fig. 8 (c) Test and numerical results to prediction ratios against n_0

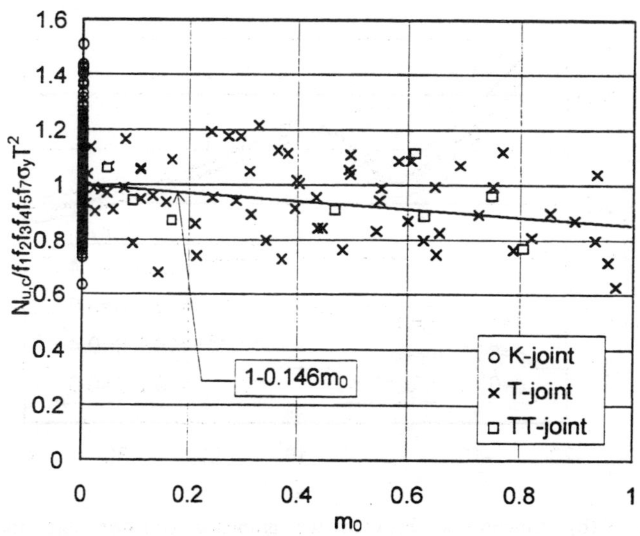

Fig. 8 (d) Test and numerical results to prediction ratios against m_0

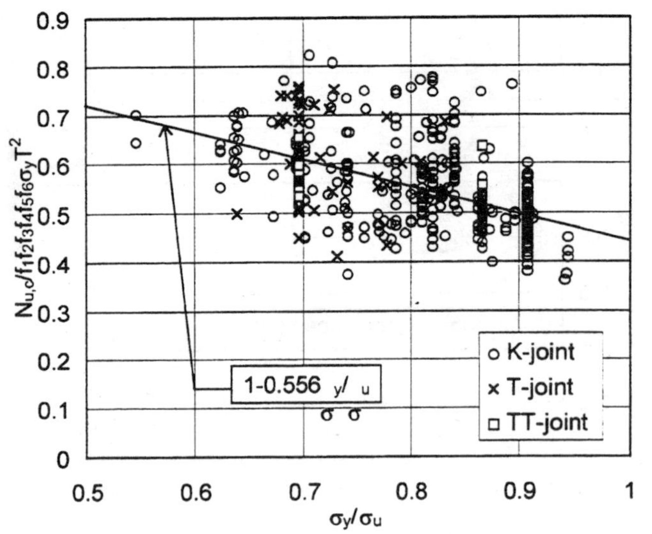

Fig. 8 (e) Test and numerical results to prediction ratios against σ_y/σ_u

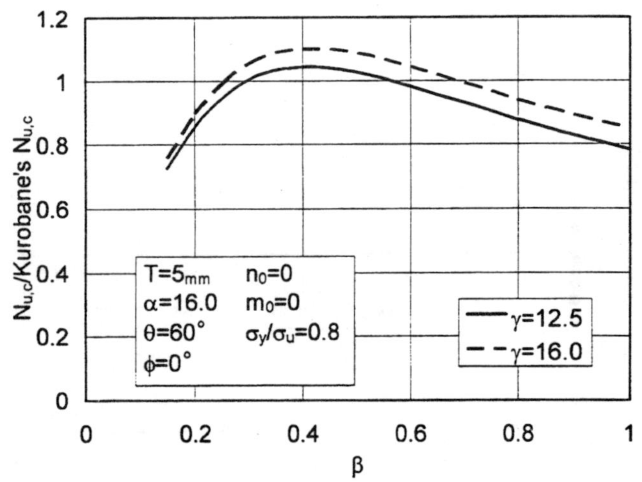

Fig. 9 (a) Comparison between the proposed equation and the Kurobane (1984) T-joint equation

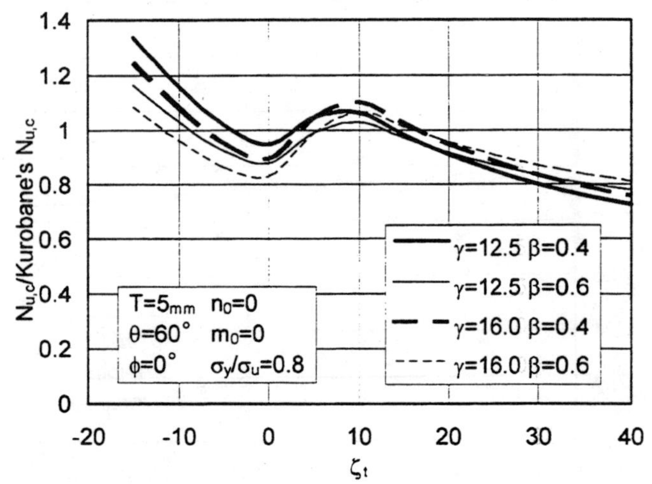

Fig. 9 (b) Comparison between the proposed equation and the Kurobane (1984) K-joint equation

COMPARISON WITH EXISTING ULTIMATE CAPACITY FORMULA

The formulae of Kurobane et al are recognized by other researchers as the most accurate prediction equations for tubular T-, and K-joints. The formulae are as follows:

For T-joints:

$$_TN_{u,c} = f_0 \cdot f_1 \cdot f_2 \cdot T^2\sigma_y$$

$$f_0 = 4.83\left[1+4.94\left(\frac{d}{D}\right)^2\right]$$

$$f_1 = \left(\frac{D}{T}\right)^{0.233}$$

$$f_2 = \left(\frac{L}{D}\right)^{-0.45}$$

For K-joints:

$$_KN_{u,c} = f_0 \cdot f_1 \cdot f_2 \cdot f_3 \cdot f_4 \cdot f_5 \cdot T^2\sigma_y$$

$$f_0 = 2.11\left(1+5.66\frac{d}{D}\right)$$

$$f_1 = \left(\frac{D}{T}\right)^{0.209}$$

$$f_2 = 1+\frac{0.00904\left(\frac{D}{T}\right)^{1.24}}{exp\left(0.508\frac{g_l-g_c}{T}-1.33\right)+1}$$

in here $g_c = 3.04(mm)$

$$f_3 = \frac{1-0.376\cos^2\theta}{\sin\theta}$$

$$f_4 = 1+0.305n_0-0.285n_0^2$$

in here $n_0 = \frac{N_0}{\pi(D-T)T\sigma_y}$

$$f_5 = \left(\frac{\sigma_y}{\sigma_u}\right)^{-0.723}$$

Comparisons between the predictions from the proposed equation and the Kurobane (1984) T- and K-joint equations are shown in Figs. 9 (a) and (b), and Table 3. The mean values of the predictions from the Kurobane equations against the same data are 1.08 and 1.00, and CoV values of that are 0.106 and 0.110, respectively. The non-linear variation between the two sets of predictions are due to the choice of functions for the parameters β and ζ_l. For example:

(1) The Kurobane T-joint equation uses a quadratic function for β, whereas the new formula uses a power function.
(2) The Kurobane K-joint equation uses an inverse exponential function for the in-plane gap which results in a plateau on a curve of test to prediction ratios for joints with small in-plane gap or overlapping braces. This accounts for the local maxima and minima on the curves in Fig. 9 (b).

A brief analysis of the T-joint data (shown in Figures 10(a) and (b), using the proposed formula and the Kurobane T-joint equation respectively, yields the following observations:

(1) When both prediction formulae are applied to the complete database the means are 1.015 and 1.248 and the CoV's are 0.137 and 0.183 respectively. The proposed formula appears to be more accurate. The Kurobane equation was originally derived from a database that used T-joints with simply supported chord ends (i.e. the data of Stamenkovic et al (1983) was excluded from the database). This would indicate that the proposed formula can be applied to T-joints where the actual chord end boundary conditions are not known; for instance in a frame or offshore jacket structure where the chord end conditions are often between fixed and simply supported.

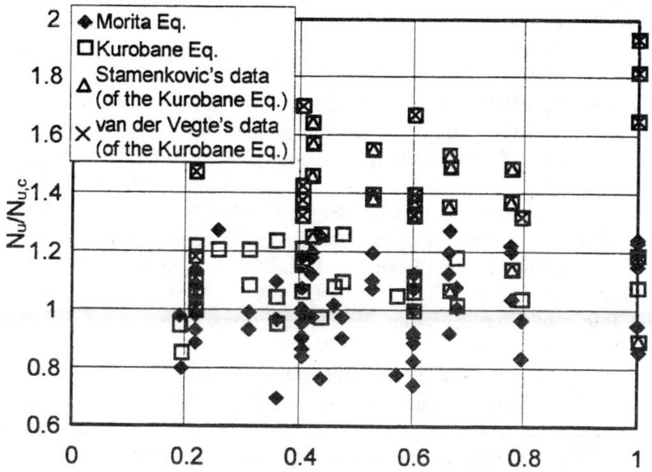

Fig. 10 (a) Comparison between the proposed equation and Kurobane (1984) T-joint equation

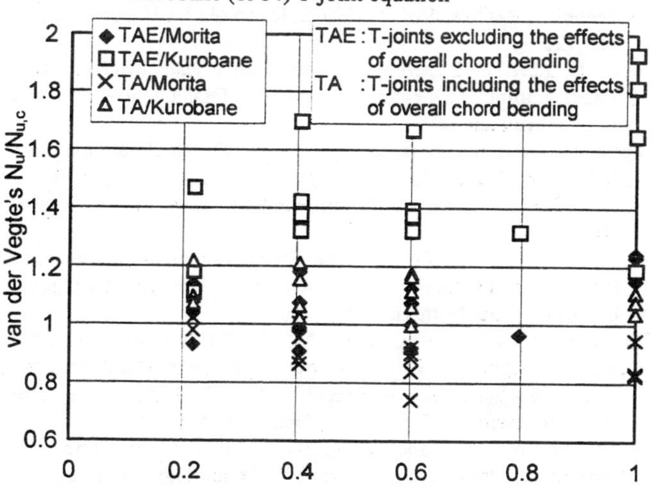

Fig. 10 (b) Comparison between the proposed equation and Kurobane (1984) T-joint equation

(2) When the data of Stamenkovic et al (1983) is excluded from the database the means are 0.987 and 1.213 and the CoV's are 0.135 and 0.185 respectively. This would indicate that the presence of the Stamenkovic data in the database does not significantly influence the accuracy of the predictions of either equation.

(3) When the formulae are applied to a database of joints with only simply supported chord ends and no applied counter moments at the chord ends the means are 1.017 and 1.256 and the CoV's are 0.135 and 0.199 respectively. Once more this indicates the accuracy of the proposed formula.

(4) From Figure 9(a) it appears that the proposed formula reduces the predicted strength of T-joints by 20% when compared to the Kurobane formula. An analysis of the data for joints with $\beta<0.25$ and $\beta>0.9$ showed that only 3 tests lie in this range of β. The predictions using both equations showed that both formula under-predicted the joint strength by between 3-21%, yet the maximum difference between the predictions from each formula was only 6%. The remaining data was numerically generated (van der Vegte (1995)). When both formulae are applied to the joints within the above range for β the means are 1.057 and 1.305 and the CoV's are 0.151 and 0.299 respectively. Therefore it can be concluded that the proposed formula accurately predicts the strength of T-joints at the extremities of the range of β used in this study.

A brief analysis of the TT-joint data, using the proposed formula and the Paul et al (1992) TT-joint equation respectively, yields the following observation:

(1) The formula of Paul et al (1992) was shown to accurate when applied to the database from which it was derived. However, it can be seen from Table 3 that when the equation is applied to the data other than that from which it was derived (i.e. data of van der Vegte (1995)) the equation can severely over-predict the strength of TT-joints. In contrast, the proposed formula can accurately predict the strength of TT-joints (mean=1.00 and CoV=0.117).

CONCLUSIONS

A new ultimate capacity formula for T-joints, TT-joints of failure type 1, and K-joints under axial loading was derived by a multiple regression analysis of a screened database. The characteristics of the ultimate capacity formula can be summarized as follows.

(1) The ultimate capacity formula is a continuous equation for T-, TT-, and K-joints, thus the ultimate capacities can be predicted by only one equation. The ultimate capacity formula consists of a basic equation and some multiplicative correction terms. The

Table 3 The mean and CoV values against each kind of joints

Joint Type	all data			the case of $\zeta_i>0$			the case of $\zeta_i \leqq 0$		
	numbers	mean	CoV	numbers	mean	CoV	numbers	mean	CoV
T-Joints	68	1.01 (1.25)	0.137 (0.183)	—	—	—	—	—	—
TT-Joints	8	1.00 (0.79)	0.117 (0.289)	—	—	—	—	—	—
K-Joints	355	1.00 (0.99)	0.140 (0.114)	307	0.99 (0.99)	0.126 (0.106)	48	1.05 (0.96)	0.197 (0.157)

Parenthesis values are for Kurobane's formulae (1984) and Paul's formula (1992) against same data.

each correction term has a clear meaning. As a result, ultimate capacities of new other joints could be estimated easily. There are no significant errors in the correction equations.

(2) The accuracy of the new ultimate capacity formula is higher than the Kurobane (1984) T-joint equation and TT-joint formula of Paul et al (1992), but is slightly lower than the Kurobane (1984) K-joint formula. The lower accuracy could be attributed to the simpler regression model used for the K- joint, than was used in the original equation of Kurobane.

In future studies, the ultimate capacity formula will be extended for various multiplanar joints. Finally, a continuous ultimate capacity formula for arbitrary joint types will be proposed

AKNOWLEDGEMENT

Financial supports for this project were jointly provided by the Grant-in-Aid for Scientific Research No.08044160 from the Japanese Ministry of Education, Science, Culture and Sports, and the Nohmura Foundation for Membrane Structure's Technology.

REFERENCES

Kurobane, Y. Makino, Y. and Ochi, K. (1984). "Ultimate Resistance of Unstiffened Tubular Joints," Journal of Structural Engineering, ASCE, Vol 110, No 2, pp 385-400

Makino, Y. and Kurobane, Y. (1986). "Recent Research in Kumamoto University in Tubular Joint Design", IIW Doc. XV-615-86, Tokyo, Japan, pp. 1-28

Makino, Y., Kurobane, Y., Ochi, K., van der Vegte, G. J. and Wilmshurst, S. R., (1996) "Database of Test and Numerical Analysis Results for Unstiffened Tubular Joints," IIW Doc. XV-E-96-220, Miskolc, Hungary

Morita, M., Makino, Y., Kurobane, Y. and van der Vegte, G. J., (1996) "A New Ultimate Capacity Formula for Unstiffened CHS Joints under Compression – Continuous Formula between T, TT and X-Joints –" Proceedings of 7th International Symposium on Tubular Structures, Miskolc, Hungary

REFERENCES OF DATABASE

Akiyama, N., Yajima, M., Akiyama, H. and Ohtake, A., (1974) "Experimental Study on Strength of Joints in Steel Tubular Structures," Journal of Society of Steel Construction, Vol. 10, No. 102, pp. 37-68, (in Japanese)

Hlavacek, V., (1970) "Strength of Welded Tubular Joints in Lattice Girders," Construzioni Metalliche, No. 6

Koning, C. H. M. de, and Wardenier, J., (1981) "The Static Strength of Welded CHS K-Joints," Stevin Report 6-81-13, Stevin Laboratory, Delft University of Technology, Delft, The Netherlands

Kurobane, Y., (1964) "Welded Truss Joints of Tubular Structural Members," Memoirs, Faculty of Engineering, Kumamoto University, Kumamoto, Japan, Vol. 12, No. 2, pp. 1-39

Kurobane, Y., Makino, Y., Honda, T. and Mitsui, Y., (1980) "Additional Tests on Tubular K-Joints -- with CHS Members under Static Loads -- " International Institute of Welding, Doc. XV-460-80, London, pp. 1-15

Makino, Y., Kurobane, Y., Takizawa, S., Yamamoto, N. and Oki, H., (1987) "Behavior of Tubular K-Joints - Part 3 : Test Results and Estimation of Ultimate Capacity," Research Report, Chugoku-Kyushu Branch of Architectural Institute of Japan, No. 7, pp. 321-324 (in Japanese)

Nakajima, T., Ishimaru, K., Shimizu, M., Koshihara, T. and Suzuki, K., (1972) "Study on Strength of Welded Joint in Thin-Walled Steel Pipe," Nihon Kokan Technical Report, No. 58, Nihon Kokan Co. Ltd., (in Japanese)

Nishida, Y., Sakamoto, S., Ohtake, A. and Ninoshima, N., (1978) "Method for Reinforcing Tubular Truss Joints (Investigation into Optimum Area of Chord Walls with Partially Increased Thickness)," Summary Papers, Annual Assembly of Architectural Institute of Japan, pp. 1303-1304, (in Japanese)

Ochi, K., Kurobane, Y. and Makino, Y., (1981) "Further Tests on CHS K-Joints with Various d/D ratios," Technical Reports, Vol.30, No.3, Kumamoto University, Kumamoto, Japan, pp. 119-124

Okujima S., (1984) "Actual Size Tests of Large Scale Tubular K-Joints - Main Roof Trusses of New Osaka Intenational Trade Center No. 3," Nihon Kenchiku Sougo Sikensyo, Report No. IIA-83-58 (in Japanese)

Paul, J. C., Makino, Y. and Kurobane, Y., (1993) "Ultimate Resistance of Tubular Double T-Joints under Axial Brace Loading," Journal of Constructional Steel Research, Vol. 24, No. 3, pp. 205-228

Stamenkovic, A. and Sparrow, K.D., (1983) "Load Interaction in T-Joints of Steel Circular Hollow Sections," Journal of Structural Engineering, ASCE, Vol.109, No.9

Stol, H. G. A., Puthli R.S. and Bijlaard, F. S. K., (1985) "Experimental Research on Tubular T-Joints under Proportionally Applied Combined Static Loading," Conference on the Behaviour of Offshore Structures (BOSS), Delft, The Netherlands

Takizawa, S., Yamamoto, N., Mihara, J. and Ohkata, S., (1979) " Full Scale Experiments of Tee and Cross Type Tubular Joints under Static and Cyclic Loading," Technical Reports, Kawsaki Steel Technical Review, Vol. 11, No. 2, Kawasaki Steel Corporation, pp. 115-125

Togo, T., (1967) "Experimental Study on Mechanical Behavior of Tubular Joints," Doctoral Dissertation, Osaka University, Osaka, Japan, (in Japanese)

Toprac, A. A. and Louis, B. G., (1970) "Research on the Fatigue Behaviour of Tubular Connections," Structures Fatigue Research Lab. (SFRL), Technical Report P550-13, Univ. of Texas, Austin, USA

Vegte, G. J. van der, (1995) "The Static Strength of Uniplanar and Multiplanar Tubular T- and X-Joints," Doctoral Thesis, Delft University, The Netherlands

Wardenier, J. and Koning, C. H. M., (1977) "Investigation into the Static Strength of Welded Warren Type Joints Made of Circular Hollow Sections," Stevin Report No. B1-77-19, Stevin Laboratory, Delft University of Technology, Delft, The Netherlands

Washio, K. and Kurobane, Y., (1963) "Truss Joint in Tubular Steel Structures (Preliminary Report)," Technology Reports, Osaka University, Osaka, Japan, Vol. 13, No. 553, pp. 147-163

Yura, J. A., Howell, L. E. and Frank, K. H., (1978) "Ultimate Load Tests on Tubular Connections," Civil Engineering Structures Research Laboratory, Univ. of Texas, Report No.78-1, Univ. of Texas

Proceedings of the Seventh (1997) International Offshore and Polar Engineering Conference
Honolulu, USA, May 25-30, 1997

ISBN 1-880653-28-1 (Set); ISBN 1-880653-32-X (Vol. IV)

Behaviour of Innovative Tubular KT-Joints Under Variable Repeated Loading

Nick K. Milani and Paul Grundy
Monash University
Melbourne, Australia

ABSTRACT

The behaviour of innovative flattened-end welded CHS KT-joints is the subject of an experimental program which consists of both cyclic and static loading programs. Tests show that under a limited number of cycles of high amplitude low frequency loading regimen, when the load levels exceed a certain limit, failure occurs by incremental collapse. This results in establishing an experimental shakedown limit.

Static testing of the same joints shows that the shakedown limit is significantly lower than the static collapse limit, proving a lower ultimate capacity for joints exposed to variable repeated loading, a loading case common to offshore and some land-based structures. The amount of reduction in ultimate capacity appears to be a function of the β-ratio, regardless of the loading history.

KEYWORDS : incremental collapse, innovative connections, shakedown limit, test, tubular KT-joint, ultimate load, variable repeated loading,

NOMENCLATURE

D	chord outside diameter
d	brace outside diameter
F_u	material ultimate tensile strength
F_y	material yield stress
P1	load from the 1000 kN actuator
P2	load from the 250 kN actuator
T	chord wall thickness
t	brace wall thickness
Δ_i	chord indentation at location i
δ_i	brace diameter at location i
β	= d/D
γ	= D/2T
τ	= t/T

1. INTRODUCTION

Steel tubular connections, in offshore structures, have traditionally used profile-cut joints where branch members are cut to a profile matching the contours of the through member, prior to welding.

Fabrication and welding of such connections, specially in overlapped and complex arrangements, is difficult and expensive. Avoiding complexity and adopting simple (gapped) connections, however, results in eccentricity in the line of action of forces and the introduction of bending stresses which in turn reduce the overall capacity of the joints. Added to this problem is the presence of high stress concentrations at the toe of welded members, an inescapable phenomenon of the geometry of profile-cut connections.

To improve the performance of tubular connections, an alternative connection detail (flattened-end) is considered for an experimental study. Cyclic loads of low frequency and high amplitude, as well as static loading, are used in the testing program to provide a better understanding of true capacity of the joints. This is due to the fact that with offshore structures, as well as some other land based structures, collapse under variable repeated loading can be more critical than collapse due to a single major loading.

Variable repeated loads have an unpredictable sequence of significant loads. The bounds of such loads for any given design life, however, can be predicted by using well-established statistical techniques. Wave loads, wind loads and earthquake loads are all examples of variable repeated loads.

Connection capacity has traditionally been based on the static and fatigue performance of the joints (Marshall, 1992), and current design guides recommend design equations based on experimental tests on joints under static and fatigue loading (API-RP 2A, 1991 and UEG, 1985).

Earlier studies (Milani, Grundy, 1996) have shown that conventional KT-joints fail at loads up to 17% below the static collapse load, when tested under variable repeated loading. The collapse mode in all cases has been established to be incremental collapse.

Incremental collapse occurs when a structure experiences progressive increments of deformation for every cycle of the same loads. This phenomenon occurs when the load exceeds a threshold known as the shakedown limit.

The shakedown limit lies between the elastic and static collapse limits. For loads below shakedown limit, the structure undergoes plastic deformation after each cycle of loads until it reaches a stage (shakedown) where the development of residual stresses allows the structure to respond elastically for further cycles of the same loads.

Alternating plasticity is another non-shakedown state and it occurs when the stress fluctuations at some point of the structure lead to generation of hysteretic inelastic strain reversals. This results in localised material exhaustion and consequently local failure which does not necessarily lead to structural collapse. This mode of failure has not been observed in this experimental program nor previous experimental programs conducted at Monash University (Goh, Grundy, 1994).

Incremental collapse seems to be the dominant non-shakedown state for tubular joints. It results in excessive deformation making the structure unserviceable for loads less than the static collapse load.

This paper provides the results of an experimental program on flattened-end CHS KT-joints under both variable repeated and static loading.

2. TESTING PROGRAM

An experimental program on flattened-end KT-joints with β-ratios of 0.64 and 0.73 was devised to investigate their behaviour under variable repeated loading, as well as monotonic static loading. Cyclic loads of low frequency were generated using a custom-built test rig which was fitted with two actuators with tension/compression capacities of 1000 kN and 250 kN, loading a Y-brace and the T-brace respectively.

In order to study the effects of loading history, two identical joints were exposed to different loading sequences. Load programs were applied for a limited number of cycles to establish whether shakedown had occurred. The amplitude of the load program was increased in stages until incremental collapse occurred, thereby establishing the experimental shakedown limit. As with previous tests, the static strength of the connections were not significantly affected by a modest amount of permanent plastic deformation due to the cyclic load, therefore, allowing usage of the specimens for static testing.

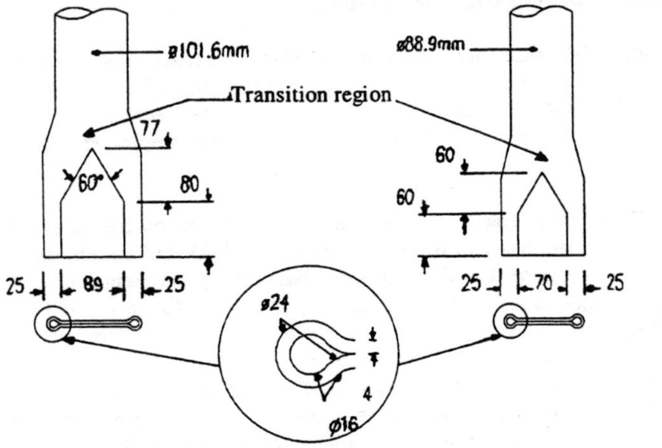

Figure 1. Flattened-end section details

2.1 FLATTENED-END KT-JOINTS

Flattened-end joints are an innovation in steel tubular connections. In these connections, the end of branch members (or braces) are flattened before cutting and welding to the through member (chord). The cross section of flattened-end brace, at the intersection with chord, has a dog-bone shape. This cross-section is gradually transformed, starting some distance from the joint, to fully circular section, in a transition region. Figures 1 illustrates the details of a flattened-end sections.

Fabrication of the required shape of the flattened sections was achieved by design and assembly of two sets of pressing dies in a 50-tonne Baldwin Pressing Machine. Pressing was completed in two steps and was based on a trial and error process to provide uniform and symmetric flattening and to avoid damaging or exhausting the material. Tubes were cold formed with electric resistance welding and their seams were placed at the middle of the fully flat section, away from the curves to prevent damaging the factory weld.

Although, the flattening process was done manually for the limited number of samples required, it was clear that an automated process could be used for a production-line case. The introduction of additional flattening process could be well justified by the time and cost savings in profile-cutting and welding of braces, in addition to increased capacity of the innovative joints.

Flattened end of braces were cut to a profile matching the contours of the chord, with the flattened surface normal to the plane of the joint, before welding. Welding was according to AWS D1.1-94 requirements, and sized to represent correct model scale. All flattened-end specimens had zero eccentricity in the line of action of their braces, eliminating the otherwise torsional effect of brace axial loads on the chord. This was achieved without brace overlapping, a complex and expensive solution in conventional profile-cut connections. Figure 2 shows the geometry of a completed KT-joint.

To investigate the effects of β-ratio on the behaviour of the joints, two groups of specimens were used. Table 1 shows the details of each group of specimens and material properties obtained from tensile coupon tests.

The free end of each member was fitted with an end-cap, consisting of a circular tube slightly larger than the member welded to a flat plate. The specimen was installed inside the testing rig by bolting the specimen's flat plates to the rig's supporting plates, providing a pin-joint type of connection at supports.

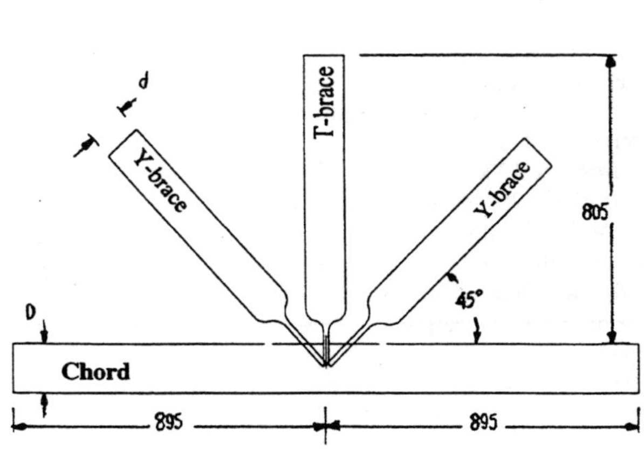

Figure 2. Joint geometry

Group	1	2
Quantity	2	2
D (mm)	140	140
T (mm)	5	5
d (mm)	89	102
t (mm)	4	4
β	0.64	0.73
τ	0.8	0.8
γ	14	14
$F_{y, chord}$ (MPa)	293	293
$F_{u, chord}$ (MPa)	369	369
$F_{y, brace}$ (MPa)	327	317
$F_{u, brace}$ (MPa)	399	381

Table 1: Specimen details

2.2 LOADING PROGRAM

Figure 3 shows a schematic diagram of the testing rig used in the experimental program. This custom-built rig was fitted with two hydraulic actuators. One of the actuators, with a tension/compressing capacity of 1000 kN, was positioned to apply force to a Y-brace, and the other actuator, with 250 kN capacity, was to apply force to the T-brace.

Both actuators applied compressive force and the ratio of the loads was maintained at 0.7 with the Y-brace experiencing higher loads. The load ratio was adopted arbitrarily, and is one of many possible combinations that could be the subject of a more comprehensive study.

Tests were conducted under axial load control, through a computer program specially prepared for the testing program, using DT-VEE version 2.

As the results of earlier experimental shakedown studies had shown, the static collapse limit of joints was not affected by the limited plastic deformation induced during controlled cyclic testing. Therefore, each flattened-end KT joint was tested under a low frequency, high amplitude cyclic test before being tested under static loading to determine its static collapse limit.

It was also intended to study the effects of loading history on the behaviour of the joints and therefore two identical joints were tested under two different cyclic loading programs.

2.2.1 CYCLIC TESTS

In Cyclic Program 1, each load cycle had four stages, as demonstrated in Figures 4 and 5a. Before the start of the first load cycle, both actuators were set to apply a compressive pre-load to the corresponding braces in order to avoid any vibration or fluctuation of the loading system once the cycle was completed. The value of pre-load was set to 30 kN for P1 (the load applied by the 1000 kN actuator) and 21 kN for the other actuator. Results of earlier tests on profile-cut joints under the same loading program had shown no sensitivity to the magnitude of this pre-load.

The first stage of the loading program 1 started by increasing the applied to a specified value. The second stage raised the load level on the T-brace

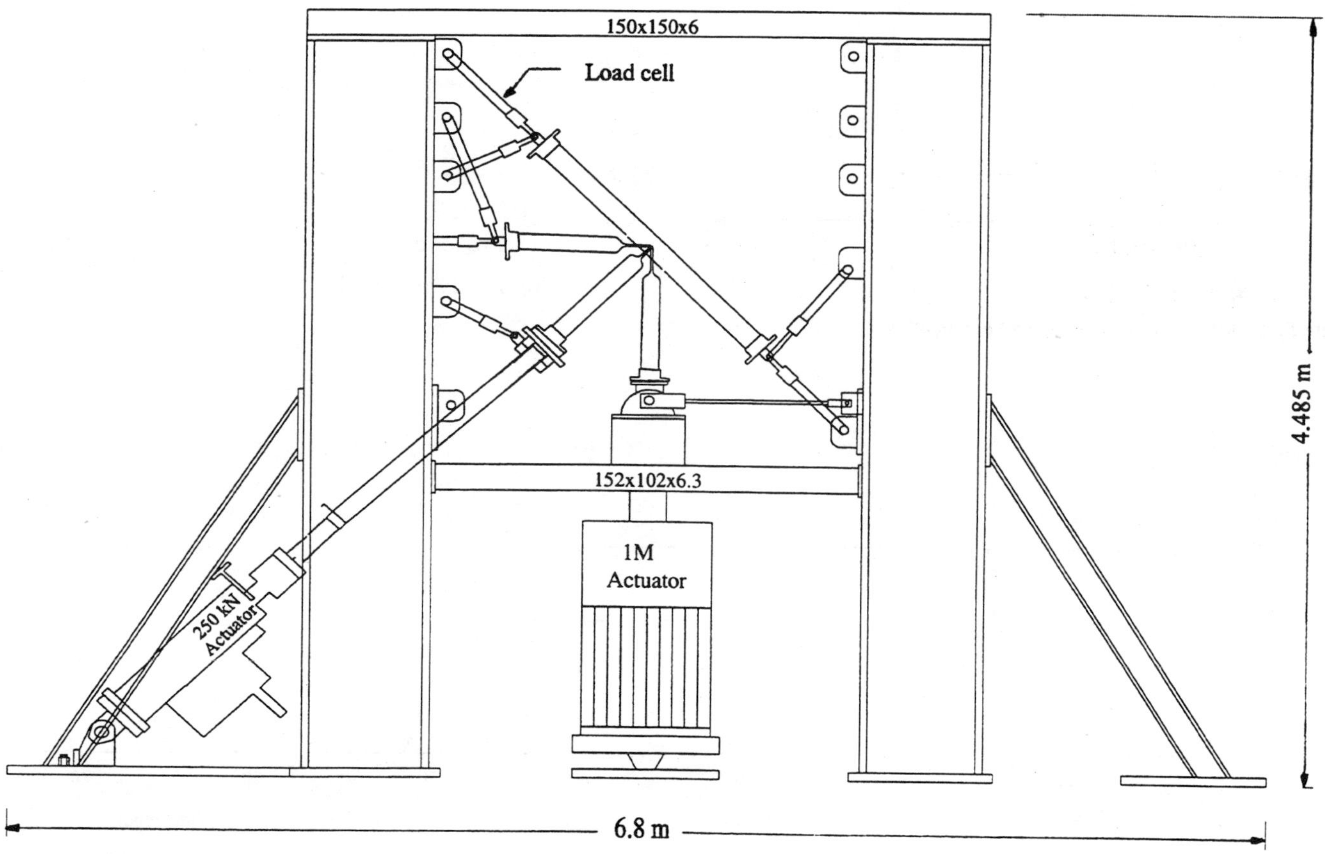

Figure 3. Testing rig

to 0.7 of that of the Y-brace, while maintaining the load in the Y-brace. Third stage followed by unloading the Y-brace to its pre-load level, and finally in fourth stage the load on the T-brace was reduced to its initial level.

The Cyclic Program 2 had two stages as illustrated in Figures 5b and 6. It started from a pre-load condition, similar to Cyclic Program 1. Stage one involved loading of both Y and T-braces before reducing them to previous level in stage 2.

In both loading programs, each load level was repeated for a finite number of cycles (between 5 to 15 cycles) and if shakedown was detected, load levels were increased. This process was repeated until the onset of incremental collapse. Once the non-shakedown response of the structure was detected, the test was terminated and the testing equipment and specimen was prepared for the following static test. In all the tests, the amount of permanent plastic deformation of the specimen was negligible and checking of the specimen and the supporting systems proved no significant alteration of the overall geometry of the specimens nor their supporting systems.

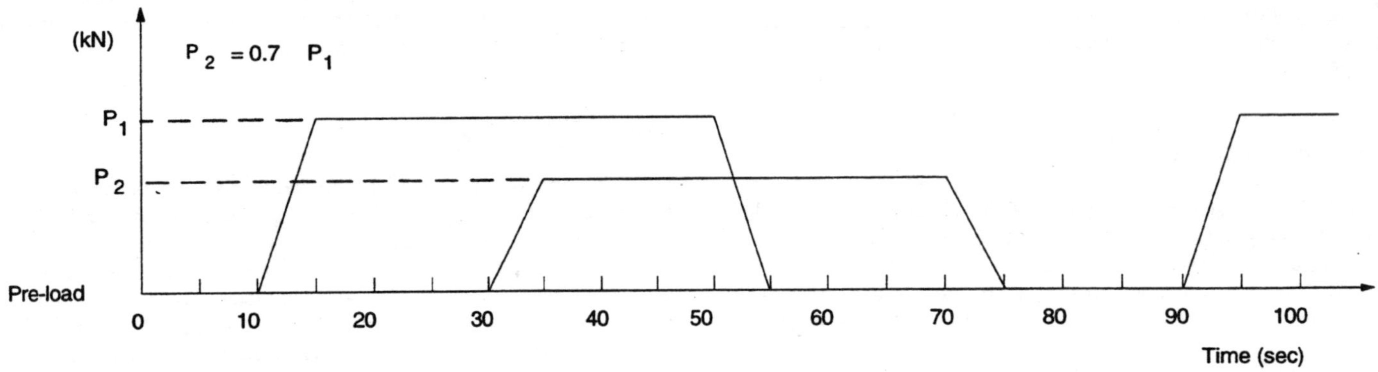

Figure 4. Cyclic Load Program 1

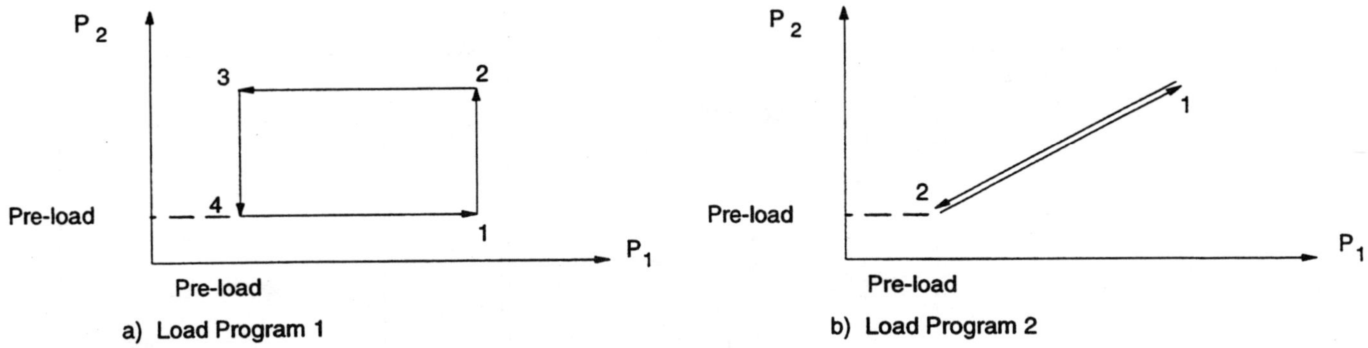

Figure 5. Sequence of loading for a single load level

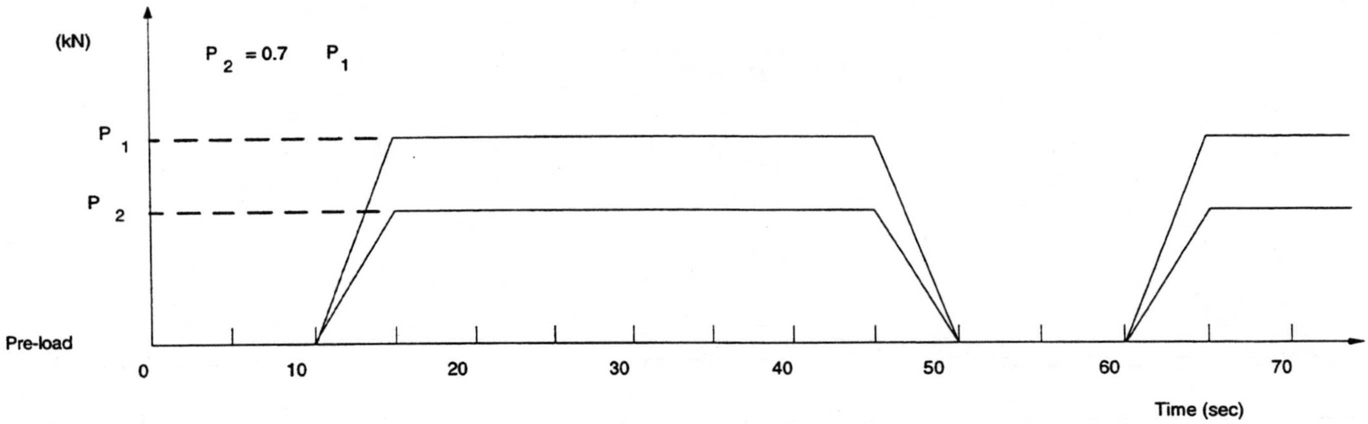

Figure 6. Cyclic Load Program 2

2.2.2 STATIC TESTS

Static tests on specimens were carried out by step-wise proportional compressive loading of both actuators, with the same 0.7 load ratio as for cyclic tests. Load increments were reduced when plastic deformation of the specimen was detected and more time was given for the system to stabilise before raising the load levels. This allowed correct determination of the static collapse of the specimen.

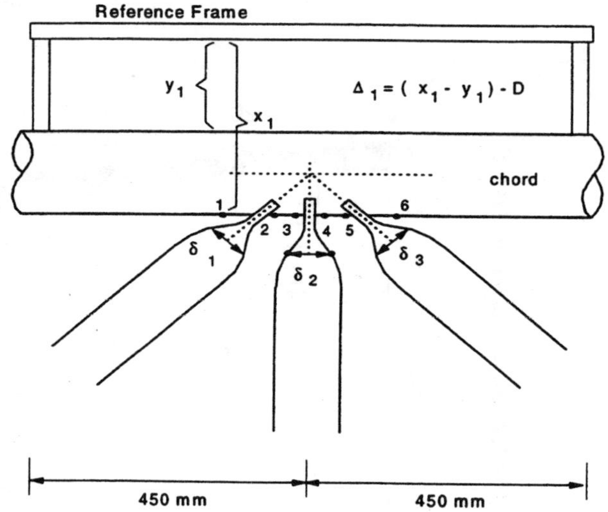

Figure 7. Location of chord deflection measurement

2.3 TEST MONITORING AND CONTROLLING

Real time measurement of a number of variables, in both cyclic and static tests, allowed careful monitoring and controlling of the tests.

The intended applied loads were measured through in-built actuator load cells and the results were progressively demonstrated on a PC screen, using the computer program prepared for the test program. It was observed that the applied loads were accurate to ±3 kN of the specified loads.

Moreover, having fitted the supporting systems with load cells, a global equilibrium check, with the possibility of halting the test and rectifying any undesired problems, was available in all tests.

In order to monitor the geometry changes of the joint, which was crucial in detecting both shakedown and static collapse limits, two deflection measurement systems were designed and implemented.

Chord deformation at the toe and the heel of all braces, as shown in Figure 7, was obtained using a number of linear measurement sensors in an adjustable set-up which measured the position of chord surface relative to a fixed reference frame, attached to the chord in a distance approximately 450 mm from the centre of the joint.

Deformation of the flattened section of brace members, in transition region, was also monitored using a scissors type set up around each brace with a linear sensor device attached to it.

All data was collected and stored automatically through the computer program, and selected information, in the form of graphs, was displayed on the screen enabling effective controlling of the test progress.

Delay times in data acquisition were appropriately specified, based on the elastic or plastic respond of the joint, to allow stabilisation of the system prior to data collection.

2.4 MODE OF FAILURE

In cyclic tests, once the shakedown limit was exceeded, the joint deformation grew after each cycle of loads, with a rate that was proportional to the amount the limit was exceeded. The tests were terminated before excessive permanent deformation of the joints, to preserve the joint for a further static test. However, if the cyclic tests were continued, joints would have failed from accumulated deflections, making them unserviceable. Figures 8,9 and 10 illustrate the onset of incremental collapse in the cyclic tests.

Results of an elastic analysis of a joint are tabulated in Table 2 and member numbering used in the analysis is shown in Figure 11. These results show that the non load bearing Y-brace carries the highest primary stress. It is at this point where the highest deflections in the chord and the brace were recorded.

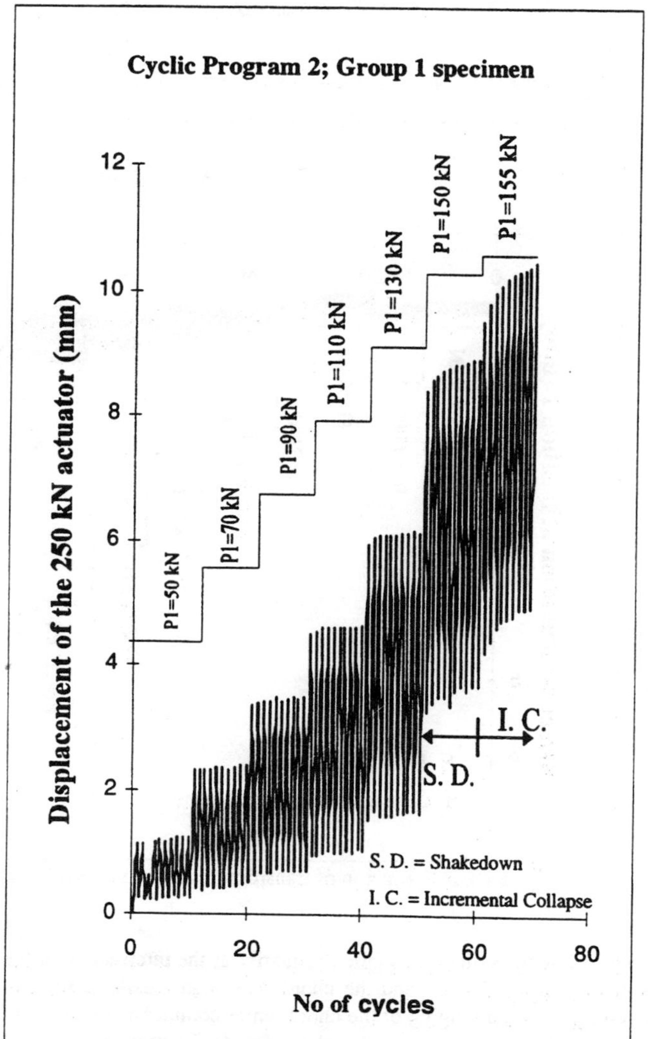

Figure 8. Actuator movement for applied loads(P2)

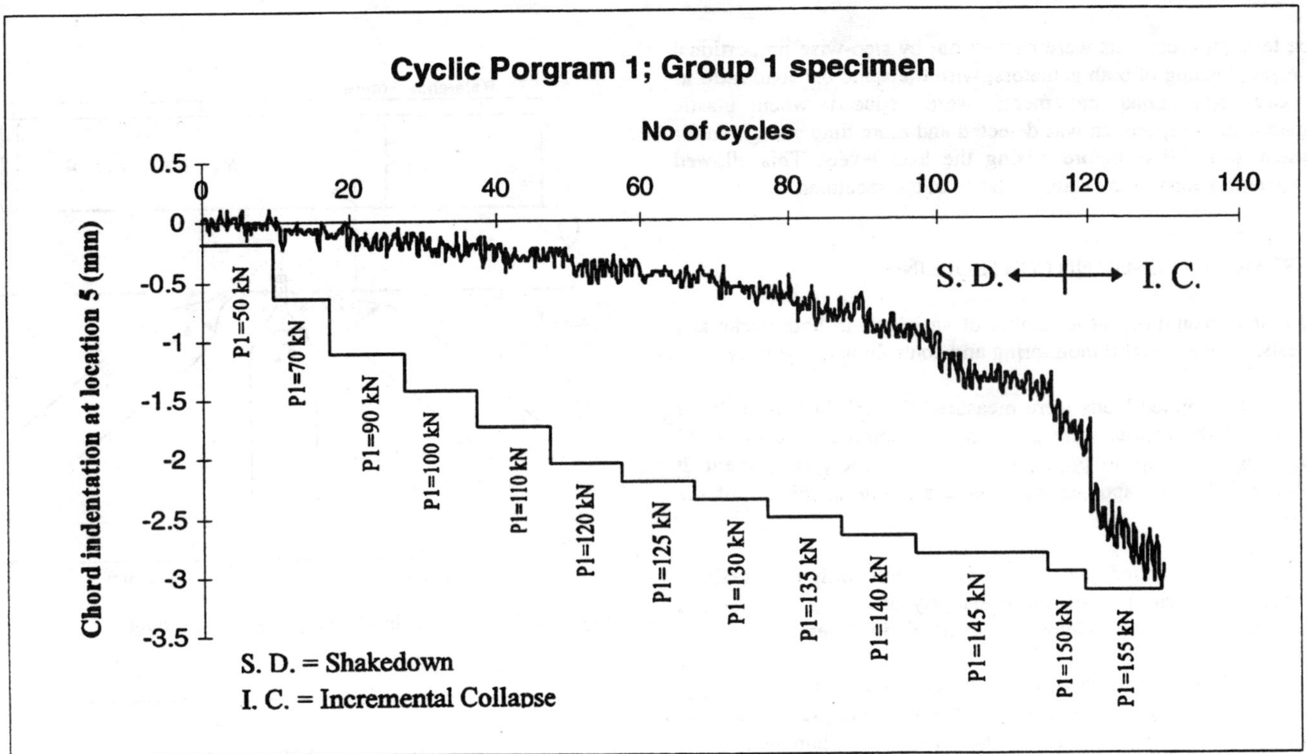

Figure 9. Chord indentation for cycle of loads applied by the 1000 kN actuator (P1)

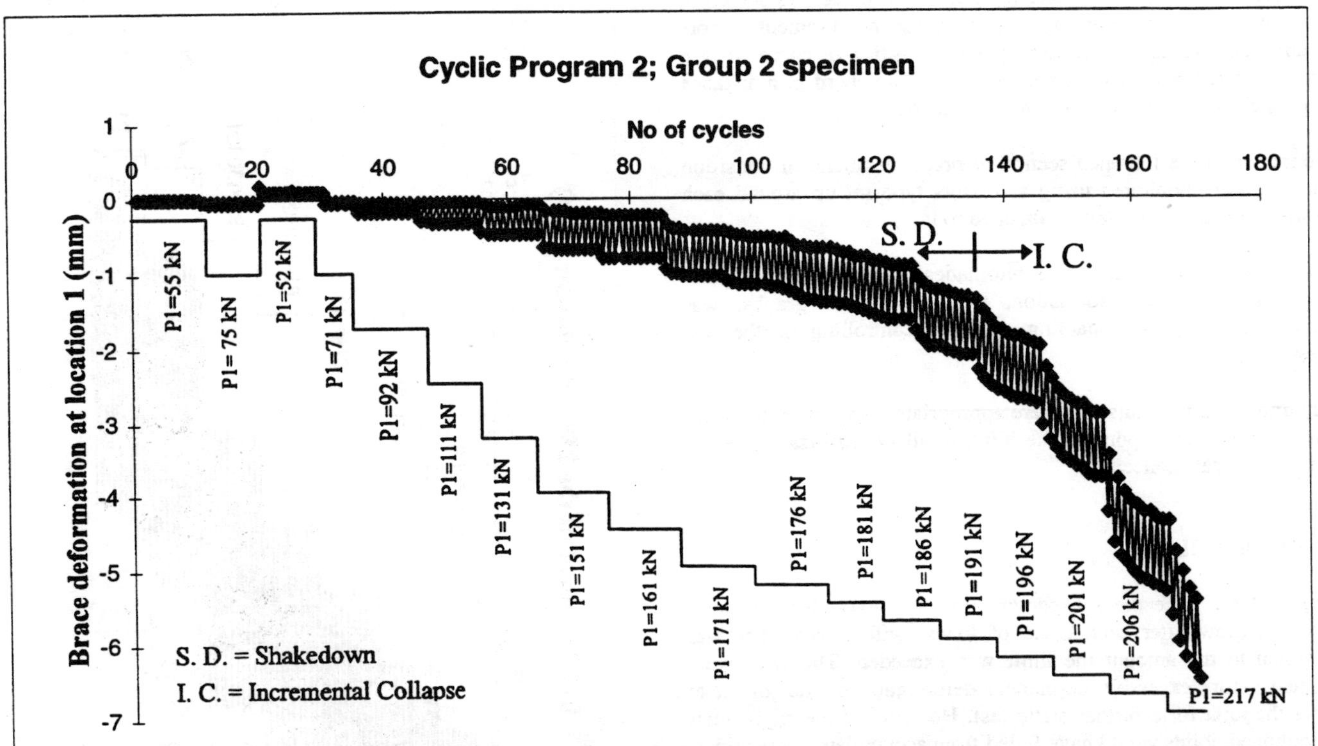

Figure 10. Deformation of transition region of non-load bearing Y-brace (δ_1) for applied loads (P1)

Under static tests, the joints failed primarily at the intersection of the non-load bearing Y-brace and the chord. The high tensile stresses in the Y-brace caused bulging of the chord while compressive stresses of the adjacent T-brace pushed the chord inside, causing a rotational effect in the chord wall in that region.

It was also observed that the transition region of the non-load bearing Y-brace, which had the highest axial force, was noticeably deformed for the larger β-ratio specimens. This deformation was in the form of reduction in diameter of the transition section in the plane perpendicular to the flattened surface. The amount of brace deformation in the transition region of other braces was negligible, due to lower levels of axial stress in them.

Failed samples showed no sign of weld cracks or material exhaustion, confirming the adequacy of material ductility and the distinct difference from that of fatigue failure. Figure 12 shows a schematic diagram of the failed joints under static loading tests.

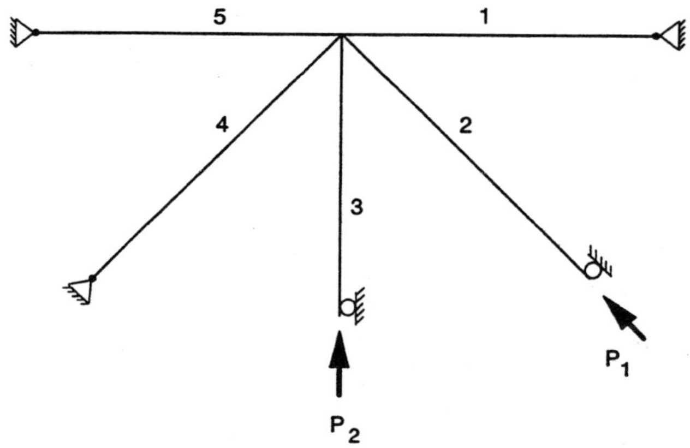

Figure 11. Member Numbering

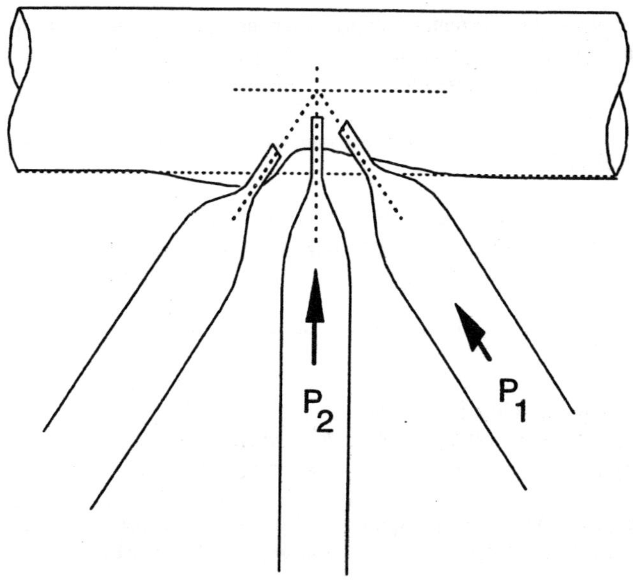

Figure 12. Connection collapse mode

Program 1 stage	Program 2 stage	P1 kN	P2 kN	A1 kN	A2 kN	A3 kN	A4 kN	A5 kN
1	---	*100*	*0*	*-68*	*100*	*0*	*-92*	*68*
2	1	*100*	*70*	*-100*	*100*	*70*	*-185*	*100*
3	---	*0*	*70*	*-32*	*0*	*70*	*-92*	*32*
4	2	*0*	*0*	*0*	*0*	*0*	*0*	*0*

Table 2. Nominal member axial forces obtained from an elastic analysis

Group	Cyclic Program	Shakedown Limit (P1) kN	Static Collapse limit (P1) kN	Reduction in capacity
1	1	151	179	16%
1	2	152	180	16%
2	1	>176	232	—
2	2	193	241	20%

Table 3. Test results for flattened-end KT-joints

3. TEST RESULTS

Table 3 summarises the results of the cyclic and static tests for shakedown and static collapse limits, showing the reduction of ultimate capacities of respective KT-joints under variable repeated loading regimen. The limit loads shown in this table are in terms of the loads applied by the 1000 kN actuator (P1).

In the Cyclic Program 1 on the Group 2 specimen, an accurate shakedown limit was not established since the cyclic test was terminated prematurely due to mal-functioning of a pair of linear positioning sensors which indicated the onset of incremental collapse in the chord. This was found to be erroneous after the static test was conducted on the specimen and the results were analysed. However, comparable joint deflection patterns under similar load conditions for the other specimen of this group and the fact that this specimen reached its shakedown limit at a higher load by only about nine percent, indicates that the true shakedown limit of Group 2 specimen under Cyclic Program 1 is not much different than that of the other member of this group

4. CONCLUSIONS

Results of this experimental study clearly show that flattened-end CHS welded KT-joints fail by incremental collapse under variable repeated loading at load levels below static collapse limit by as much as 20%.

These results also indicate that the level at which incremental collapse occurs, as well as the extent of chord indentation, is dependent on the β-ratio. With lower β-ratios, local buckling of the chord wall is more extensive. Joints with higher β-ratios, on the other hand, show better resistance against chord ovalisation. However, the presence of higher axial stresses in braces, results in a greater deformation of the transition region of flattened-end sections in a plane at right angles to the flattened surface. This brace deformation does not pose a structural problem to the integrity of the joint as much as chord deformation would, and therefore seems to increase the strength of the joint significantly.

Applying two different loading programs shows no variation in the behaviour of the joints, indicating the threshold of shakedown and ultimate static collapse limits are not affected by loading history.

5. REFERENCES

American Petroleum Institute, (1991). Recommended practice for planning, designing and constructing fixed offshore platforms(API-RP 2A). Washington D.C., API.

American Welding Society (1994). Structural welding code - steel (AWS D1.1-94). Miami, AWS.

Goh, TK, and Grundy, P (1994). Capacity of CHS YT- joints under high amplitude cyclic load. *Proc. 6th International Symposium on Tubular Structures,* Rotterdam: Balkema, pp 489-496.

Marshall, PW (1992). Design of welded tubular connections: Basis and use of AWS code provisions. Amsterdam: ELSEVIER.

Milani, NK, and Grundy, P (1996). Incremental collapse of KT-joints under variable repeated loading. *Proc. 7th International Symposium on Tubular Structures*. Rotterdam, Balkema, pp 293-300.

UEG Offshore Research (1985). Design of tubular joints for offshore structures. Great Britain, Page Bros (Norwich) Limited.

Proceedings of the Seventh (1997) International Offshore and Polar Engineering Conference
Honolulu, USA, May 25-30, 1997

ISBN 1-880653-28-1 (Set); ISBN 1-880653-32-X (Vol. IV)

Failure Assessment Diagram Method for Tubular Joints

Björn Klasén and Stig Wästberg
Det Norske Veritas
Høvik, Norway

ABSTRACT

The work performed was an investigation of stable crack growth failure assessment of cracked tubular joints similar to those used in offshore structures.

Three full scale experiments were performed where tubular X-joints with initial cracks developed by fatigue loading, were loaded till significant ductile crack growth occurred. The crack growth in the experiments at different load levels were beach marked and compared to crack growths predicted by Finite Element Analyses (FEA). The applied crack driving force in terms of the J-integral was used for the crack growth analyses.

Failure assessment procedures such as the PD6493 and the CEGB R6 were adopted to predict failure/crack growth and compared to the experimental results and FEA results illustrating the validity of the different failure assessment procedures for cracked tubular joints.

KEY WORDS

Fracture Mechanics, FAD, J-integral, Finite Element, Tubular Joint, Experiments

INTRODUCTION

The aim of the project was to provide recommendations and gain understanding of structure integrity assessment of tubular joints containing cracks by adopting Failure Assessment Diagram (FAD) based concepts. The following activities were performed:

Fracture testing of three large scale tubular X-joint specimens with fatigue precracks was carried out. Stable ductile crack growths during tests were measured. Limit loads were determined for all tests and strains and displacements were measured at a number of locations on the specimens.

Material characterisation for the chosen steel was conducted and both true stress-strain relation and fracture toughness data were determined.

Finite Element (FE) models were developed to determine quantities needed for evaluation of FAD based approaches. The FE results were checked against the test data.

Different FAD concepts were evaluated and recommendations and comment on FAD for tubular joints were worked out.

BACKGROUND TO FAILURE ASSESSMENT DIAGRAMS

Fracture mechanics based assessment procedures such as PD6493 /1/ and the CEGB procedure /2/, known as the R6 procedure, are based on the Failure Assessment Diagram (FAD) concept.

The approach suggested by Dowling and Townley in 1975 /3/ considered failure by linear elastic fracture as one limiting criterion and failure by plastic collapse as the second limiting criterion. A design curve is used to interpolate between the two failure criteria. When performing a structural integrity assessment of a flaw in a stressed structure an assessment point is derived from two separate calculations and plotted in the FAD. If the point falls inside the failure curve the structure is considered to be safe, while if the point lies on the curve or falls outside the curve failure is predicted.

The linear elastic fracture criterion is defined as failure will occur if the applied stress intensity factor, K_I, equals the critical stress intensity factor, K_{IC}, which is the characteristic material toughness. The fracture axis is defined as:

$$K_R = \frac{K_I}{K_{IC}} \tag{1}$$

The plastic collapse criterion is defined as the load, P, when plastic collapse of the structure occurs, either of the ligament or global plastic collapse, P_L. The load axis is defined as:

$$S_R = \frac{P}{P_L} \tag{2}$$

The first version of the failure assessment curve defined by Harrison et. al. in 1976 /4/ is shown in Figure 1, and the equation determining the shape of the curve is as follows:

$$K_R = S_R\left(\frac{8}{\pi^2}\ln\cdot\sec\left(\frac{\pi}{2}S_R\right)\right)^{-1/2} \tag{3}$$

The FAD curve above, equation (3), is commonly addressed as the R6 curve or the PD6493 Level 2 curve. The curve was derived for mainly non-workhardening steel materials.

The development of FAD based concepts continued by an examination of Chell in 1979 /5/. The fracture axis of the failure assessment curve was formulated for flawed structures where the J-solution was known as:

$$K_R = \sqrt{\frac{J_{el}}{J_{total}}} \tag{4}$$

J_{el} is the elastic part of the J-integral and J_{total} is the sum of the elastic and plastic parts of the J-integral. This formulation is adopted in this present work to determine the fracture axis, K_r, for the three specific tests performed. The J_{el} and the J_{total} values were determined by finite element analyses and, K_R, values were determined for the load range of interest for all three tests.

The J-integral based formulation above, equation (4), was used by Ainsworth 1984 /6/ to develop failure curves for different stress-strain curves. He derived the following assessment curve by introducing known J-integral solutions for different geometries:

$$K_R = \left(\frac{E\varepsilon_{ref}}{\sigma_{ref}} + \frac{1}{2}L_R^2/\left(1+L_R^2\right)\right)^{-1/2} \tag{5}$$

Where the load axis is defined by $L_R=\sigma_n/\sigma_{yield}$ and σ_n is the net section stress.

The FAD of Revision 3 option 2 was derived from Ainsworth's work with the second term modified to smooth out the transition from small scale yielding to extensive yielding:

$$K_R = \left(\frac{E\varepsilon_{ref}}{L_R\sigma_y} + \frac{L_R^3\sigma_y}{2E\varepsilon_{ref}}\right)^{-1/2} \tag{6}$$

The Revision 3 option 1 curve is an empirical lower bound fit to the option 2 failures lines for typical true stress-strain curves for a range of structural steels. The resulting equation for the option 1 failure curve is, Figure 1:

$$K_R = \left(1-0.14L_R^2\right)\left[0.3+0.7\exp\left(-0.65L_R^6\right)\right] \tag{7}$$

For all Revision 3 options the plastic collapse axis is normalised to yield load and not to flow stress. Thus the assessment curves extend beyond unity and represent material strain hardening as seen in Figure 1.

To produce FAD curves for the cracked tubular joint geometries of this study which could be compared to standard FAD curves, the load axis parameter, L_r, was determined as:

$$L_r = \frac{Applied\ Load}{Collapse\ Load} \tag{8}$$

Different collapse load definitions were used to normalise the applied load as will be shown in the proceeding text.

MATERIAL CHARACTERISATION

True Stress-Strain Curve and Fracture toughness

The material chosen was an ST52-3N steel with yield strength of 345 MPa and ultimate strength of 490 MPa. The true stress-strain curve for the material was identified.

Crack growth resistance tests, CTOD-R, J-R tests, were performed at room temperature. The J-R curves were determined in accordance with the standard method ASTM 1152-87 /8/, and the CTOD-R curves were determined in accordance with the BS 7448 /7/ standard method. The following crack growth initiation values and slopes of curves were found:

J_{INIT} = 153 [kN/m] dJ/da = 309000 [kN/m^2]

$CTOD_{INIT}$ = 0.49 [mm] d(CTOD)/da = 0.408

The specimens used for fracture toughness tests were taken from base material before post weld heat treatment.

EXPERIMENTAL VERIFICATION OF TUBULAR X-JOINTS

Experimental Verification of Tubular X-joints

Three large scale tests on X-joints with prefabricated cracks have been performed. The test specimen geometry is shown in Figure 2.

After welding the braces to the chord all specimens were postweld heat treated to minimise residual stresses. The postweld heat treatment was conducted at a temperature of 580 °C ± 20 °C for 2-4 hours.

A small notch, 5 mm deep and 15 mm long, was machined in all the specimens at one of the hot spot sites, on the side where the longer brace was connected to the chord. The weld toes at the other hot spot sites were ground to insure that the crack growth occurred only at the hot spot location where the notch was located.

The precracks were fatigued to three different crack lengths and crack depths for the three tests. The first and third tests contained cracks and the second test contained a through thickness crack. The crack depths were estimated by the potential drop method during the fatigue sequence.

For the fracture test the specimens were instrumented with strain gauges for measurements of strain/stress states and LVDT's for measurements of deformations. The locations of the gauges are shown schematically in Figure 3 for test 1 and test 3. Position 1 and 2 in the Figure shows the two clip gauges mounted at to positions at the crack mouth opening. Position 3 through 8 shows LVDT's used to measure ovalisaton of the chord and global displacements during tests. The rest of the positions are strain gauges.

The differences in instrumentation of test 2 compared to the instrumentation of test 1, were that there were additional strain gauges close to the crack tip and there were two LVDT's applied to measure crack mouth opening (Mode I and Mode III) during test.

During the tests, after some stable crack growth, the crack fronts were beach marked by fatigue loading of the specimen. The first and third test were beach marked once and the second test was beach marked by two fatigue load sequences.

In the first and the third test marking by use of ink was also used. It did not succeed in the first test, but it succeeded in the third test. The marking was conducted by spraying black ink into the crack at the first marking and then proceeding the test a bit further and then spraying red ink into the crack. After each marking compressed air was injected to dry the ink before continuing the test.

After the last beach marking, the specimens were loaded until the maximum point on the load deflection curve was passed in order to determine maximum load carrying capacity of the joints.

After the tests the cracks were cut out from the specimens and broken up to identify the initial crack geometry and the stable crack growths for each specimen.

Results from the First Test on Tubular X-joints

The initial surface precrack depth, was 16.5 mm at the deepest point, and the crack length at the chord surface was 210 mm measured straight between crack tips. The stable crack growth at the beach mark at the deepest point of the crack was measured to be 2.15 mm, and the total crack length increase along the weld toe was measured to be 2.0 mm. Maximum load was measured to be 3028 kN.

Results from the Second Test on Tubular X-joints

The initial through thickness initial precrack length on the inside of the chord was 110 mm and on the outside at the weld toe 260 mm. Both crack lengths were measured straight between crack tips. The stable crack growth at the first fatigue beach mark on the outside was 1.8 mm. On the inside the crack growth was extensive and measured to be 93 mm. Maximum load was measured to be 2292 kN.

Results from the Third Test on Tubular X-joints

The initial surface precrack depth was 10.6 mm, and the crack length at the chord surface was 152 mm measured straight between crack tips. All attempts to beach mark succeeded in this test, and the stable crack growth at the beach marks at the deepest point of the crack was measured to be 1.1 mm at the black ink marking, 2.0 mm at the red ink marking and 3.0 mm at the fatigue beach marking. Maximum load was measured to be 3423 kN.

HSE load carrying capacity to code

The Health and Safety Executive code, Guidance Notes 4th Edition /9/, have been used for calculation of load carrying capacity obtained from tests. For the HSE code the following equation is applicable for tensile loading of tubular joints:

HSE Guidance Notes (4th Edition)

$$Pu = P_{Basic} K_a = \frac{\sigma_y * T^2}{\sin\theta} Q_u Q_f K_a \qquad (9)$$

where: P_{Basic}= Basic ultimate load carrying capacity
σ_y= Chord yield stress
T = Chord thickness
$Q_u=(7+17\beta)Q_\beta$
$Q_\beta=1.0$ for $\beta<0.6$
$Q_\beta=0.3/(\beta(1-0.883\beta)$ for $\beta\geq0.6$
$K_a=0.5(1+1/\sin\theta)$

The following length and area corrections, due to the cracks, were adpoted when constructing the FAD curves:

$$\text{Length correction} = \left[1 - \frac{\text{Length area}}{\text{Intersection length} \times T}\right] \times \left[\frac{1}{Q_\beta}\right] \qquad (10)$$

$$\text{Area correction} = \left[1 - \frac{\text{Crack area}}{\text{Intersection length} \times T}\right] \times \left[\frac{1}{Q_\beta}\right] \qquad (11)$$

The predictions based on the HSE formula were safe for all cases except for the length corrected test 2. However, the prediction was only 1% off for that test. Several other predictions of the load carrying capacity was also tried but the HSE formula above gave the best agreement.

FINITE ELEMENT ANALYSIS OF TUBULAR X-JOINTS

Finite Element Models of the Tubular X-joint Test Specimens.

The three X-joint tests with cracks were analysed by linear and non-linear finite element analyses.

The FE-model of test 1 is shown in Figure 4. The meshes for test 1 and test 3 were identical except for the element geometry and node coordinates describing the crack tip and the vicinity of the crack tip. Figure 5 show the mesh of the crack in test 1. The chord was modelled with four elements through thickness and focused towards the crack tip. The crack tip elements had a significant height of 0.1 mm in all meshes adopted for the analyses.

The model of the through thickness cracked specimen, test 2, is shown in Figure 6.

The type of finite element used is a 20 node isoparametric brick element. For the linear analyses of cracked joints the crack tip elements had the midpoint nodes moved to the quarter points of the element length to obtain $1/\sqrt{r}$ stress singularities at the crack tip. The $1/\sqrt{r}$ stress singularities do not exist at the intersection of the crack front and the chord surface and results obtained are not valid close to or at the surface of the chord.

For the non-linear analyses of cracked joints ordinary wedge elements was used. The quality of the solutions was ensured by path independencies for calculated J-values.

The material models applied were a linear elastic and a multilinear plastic material model with isotropic hardening. Plasticity followed the von Mises criterion of yielding with the associated incremental flow rule.

The linear analyses were conducted with small deformations and small strain formulations

The non-linear analyses were performed with updated geometries, large strains and true stress strain curves. The elements used at the chord and the brace were the reduced integration order 20 node brick elements. The three layers of transition elements were modelled with full integration order 20 node brick elements and the crack tip was modelled by 15-node wedge elements in all analyses.

FINITE ELEMENT ANALYSES RESULT

Comparison of Analyses Results to Test Results

Comparisons of experimentally obtained load displacement curves for test 1 and 3 to those obtained from the analyses are shown in Figure 7. The displacements from the tests are the average of those measured at position 6 and position 8 as shown on the instrumentation plan, Figure 3. The analyses underestimate the load for a given displacement up to 9.0 mm for test 1 and up to 6.7 mm for test 3. This is due to the use of a symmetry model of the specimen, i.e. there are actually cracks modelled on both sides of the brace on the cracked side.

The model of test 2 was overstiff compared to the experiment, see Figure 8. The physical difference between the model and the experiment is that in the analyses the load is applied as a controlled displacement at the end of the brace on the cracked side of the specimen (i.e. the test rig is infinitely stiff) whereas the test rig is weaker and allows for some rotation of the brace.

J-integral calculations

The crack driving force, J-integral, was calculated at all node positions on the crack front by ABAQUS /10/. The method is a domain integral method described thoroughly in ABAQUS Theory Manual /11/, and the strain energy release per unit crack area increase is calculated by integration over a volume containing a cluster of finite elements surrounding the crack tip.

The J-values were calculated for all node positions along the crack front over five different volumes referred to as integration paths. The J-values obtained at the first and the fifth paths were excluded. The three values from path two, three and four were divided by the calculated average J-value from the same paths, and the deviations from the average J-value were calculated for the three paths. To show the confidence in the solution the maximum deviation from any of the three paths were plotted versus average calculated J-value for the positions of interest along the crack front. In Figure 9 the path independence for the deepest points of the crack for test 1 and 3 is shown. In Figure 10 the path independence for all points on the crack front is shown for test 2.

In Figure 11 the J-integral distribution along the crack front is presented for test 1 and 3 at the maximum loads. The variation of J between element midnodes and corner nodes is quite small at maximum load for both Test 1 and 3, whereas close to the intersection of the crack front to the chord surface where the midnode J-values deviates from the values obtained at the corner crack tip nodes. Position 12 and 14 in the Figures are such midnodes. It should be noted that the J-values close to the intersection and at the intersection of the crack front to the chord surface are not theoretically valid.

The variation of J-values along the test 2 through thickness crack front is extensive, not only close to the inner and outer chord surface, but for the entire crack front, see Figure 12. The position 1 in the Figure is the crack tip node at the intersection of the crack and the outer surface of the chord, and position 19 is the node at the intersection of the crack front and the inner surface of the chord.

FAILURE ASSESSMENT DIAGRAMS (FAD)

To evaluate different types of Failure Assessment Diagrams (FAD), and to determine how to normalise the load axis, the fracture axis for the FAD, K_r, was calculated as:

$$K_r = \sqrt{J_{el} / J_{total}} \tag{12}$$

J_{el} is the elastic part of J and J_{total} is the sum of the elastic and the plastic part of J. The values were plotted against the loads for all tests. The results from the analyses of the surface cracked specimens (Test 1 and 3) at the deepest point of the crack are shown in Figure 13. In Figure 14 the fracture axis values are plotted at three positions in the through thickness cracked test (Test 2).

In these Figures the test results are also plotted for the surface cracked specimens. The crack growth is known at the fatigue beach mark at the end of test 1 and at three beach marks for test 3. By introducing these crack growth values in the crack growth resistance curve, the value of J_{total} can be estimated from the experiment. The procedure is shown by the example in Figure 15. For the first measured crack growth of Δa_1=1.1 mm in test 3 the J_{total} value was estimated to be J_1=490 kN/m. The J_{elast} value was taken from the FE analysis, and K_r could be calculated according to equation (12) and plotted against the load in the FAD for test 3, see Figure 13.

The K_r values calculated from experimentally determined crack growths for test 1 and test 3 correlate excellent to those determined from numerical analyses.

The procedure of plotting results in the FAD constructed from FE analyses results was not carried out for the through thickness crack. This was due to very irregular crack growth along the crack front. However the crack growth qualitatively correlates to calculated J values.

A phenomenon observed when evaluating the analyses results of the through thickness crack test was the differences in K_r values for different positions at the crack front, see Figure 14. Position 2 is close to the outer surface of the chord, position 8 is in the middle of the chord thickness and position 18 is close to the inner surface of the chord. At the first analysis load step, F=350 kN, the K_r value is 0.91 for position 2, 0.99 for position 18 and 0.88 for position 6. When increasing the load, the differences in K_r for all the three positions at the crack front starts to increase up to the second load step. For further increase in load the difference between K_r at position 18 close to the inner surface of the chord and position 6 close to the middle of the chord thickness decreases. At the end of the analysis the values are almost the same for those two positions. The difference between position 2 close to the outer surface of the chord and the two other positions continues to increase during increased loading of the model. The phenomenon is believed to be an effect of mixed crack tip opening modes and different predominant modes at the three positions at the crack front. The Mode I opening mode seems to be dominating at the crack tip close to the outer surface, Mode II close to the middle of the chord and Mode III close to the inner surface of the chord. The crack tip opening Mode I is likely to generate a larger elastically strained volume yielding a relatively larger elastic part of J than the shear modes, Mode II and Mode III, and thus a higher K_r value.

In Figure 13 the K_r values for Test 1 and 3 are plotted versus load for comparison. The curves almost coincide at high load values, and this indicates that normalising the load axis by load estimates will separate the curves.

The FAD curves obtained for test 1 and 3, with the characteristic load (equation 9) used for calculation of the load axis, reduced by the crack area-correction (equation 11), are shown in Figure 16. The test results for both tests are included. The error bars applied on the test data points plotted demonstrates K_r values for approximate ±0.1 mm crack depths. The two FAD curves plotted in the figure are the R6/PD6493 Level 2 curve and the Revision 3 option 1 curve according the R6 document, see equation 3 and 7. All data points from the tests and the main part of both FAD curves determined from numerical analyses fall inside the failure curves.

The FAD curves obtained for test 1 and 3 with the characteristic load reduced by the crack length-correction (equation 10) are shown in Figure 17. Two of the test points, and a small part of the test 3 FAD curve obtained from analyses, still fall inside both the R6/PD6493 Level 2 curve and the Revision 3 option 1 curve.

The results indicate that the PD6493 Level 2 and the R6 document can give unsafe predictions of tearing of surface cracked tubular joints if the collapse load is calculated by the HSE characteristic strength formula. However, the HSE estimates predicted maximum load capacity rather accurately for the tested geometry and ductile material.

The FAD curves obtained from the analyses showed to be geometry dependent, and J-analyses of this type of structures, exhibiting high degree of load redistribution, is recommended.

When calculating the shape of the FAD curve it should be noted that the fineness of the finite element model is of critical importance. A too coarse model results in an underestimate of the total value of the J-integral, see equation (12). The position of the calculated FAD curve relative to the fracture axis will be overestimated, i.e. giving an overestimate of the acceptable K_r value used for fracture assessment.

CONCLUSIONS

The main findings and conclusions from the test results and numerical analyses are as follows:

1 The results show that the crack growths are J controlled for the material and geometry chosen for this investigation.

2 Results from numerical analyses and tests indicate that the PD6493 and the R6 document can give unsafe predictions of tearing of surface cracked tubular joints if the collapse load is calculated by the HSE characteristic strength formula. However, the HSE estimates predicted maximum load capacity rather accurately for the tested geometry and ductile material.

3 The FAD curves obtained from the analyses showed to be geometry dependent, and the J-analyses of this type of structures, exhibiting high degree of load redistribution, is recommended.

4 The fineness of the finite element model is of critical importance for the calculated shape of the FAD curve. A too coarse model results in an overestimate of the position of the curve relative to the fracture axis, i.e. giving an overestimate of the acceptable K_r value used for fracture assessment.

ACKNOWLEDGEMENTS

The sponsorship of CONOCO is gratefully recognised. We also wish to thank Mr Atle Johansen DNV for his skilful help during the experimental part of the work.

REFERENCES

/1/ PD6493:1991,"*Guidance on Methods for Assessing the Acceptability of Flaws in Fusion Welded Structures*", British Standard Institution, Second Edition, August 1991

/2/ Milne I., Ainsworth R.A., Dowling A.R. and Stewart A.T.,"*The Assessment of the Integrity of Structures Containing Defects*", CEGB Report R/H/R6, Revision 3, 1986

/3/ Dowling A.R. and Townley C.H.A.,"*The Effects of Defects on Structural Failure: A Two-Criteria Approach*", International Journal of Pressure Vessels and Piping, Volume 3, pp 77-137, 1975

/4/ Harrison R.P., Loosemore K. And Milne I.,"*Assessment of Integrity of Structures Containing Defects*", CEGB Report R/H/R6, Central Electricity Generation Board, United Kingdom, 1976

/5/ Chell G.G.,"*A Procedure for Incorporating Thermal and Residual Stresses Into the Concept of a Failure Assessment Diagram, in Elastic-Plastic Fracture*", ASTM STP 668, pp 581, 1979

/6/ Ainsworth R.A.,"*The Assessment of Defects in Structures of Strain Hardening Materials*", Engineering Fracture Mechanics, Volume 19, pp 633, 1984

/7/ BS 7448:Part1:1991,"*Fracture Mechanics Toughness Test: Part1. Method for determination of K_{IC}, critical CTOD and critical J values of metallic materials*", British Standard Institution, London, 1991

/8/ E 1152-87,"*Standard Test Method for Determining J-R Curves*", American Society for Testing and Materials, Philadelphia, 1987

/9/ Health and Safety Executive,"*Offshore Installations: Guidance on Design, Construction and certification*", HSE Guidance Notes, 4th Edition, Revised June 1992.

/10/ ABAQUS User's Manual (Version 5.4), Hibbitt, Karlsson and Sorensen, Provedence, USA 1994

/11/ ABAQUS Theory Manual (Version 5.4), Hibbitt, Karlsson and Sorensen, Provedence, USA 1994

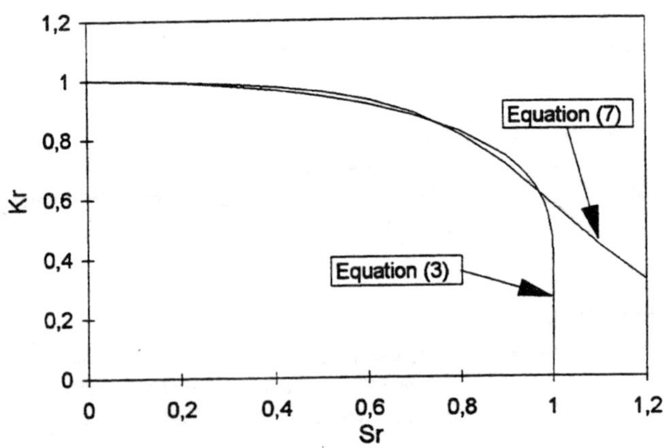

Figure 1 Failure Assessment Diagrams

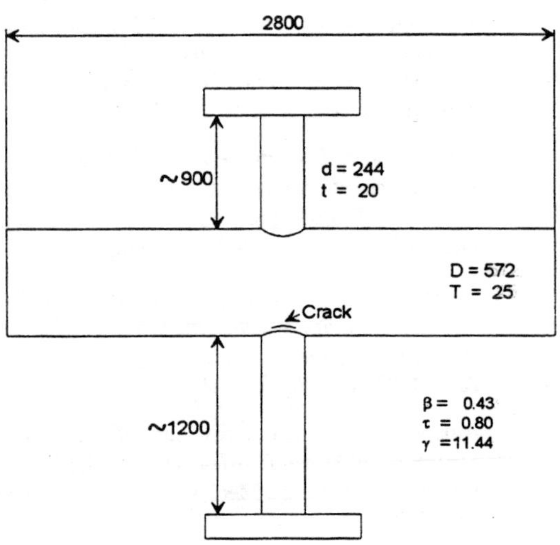

Figure 2 Test specimen geometry

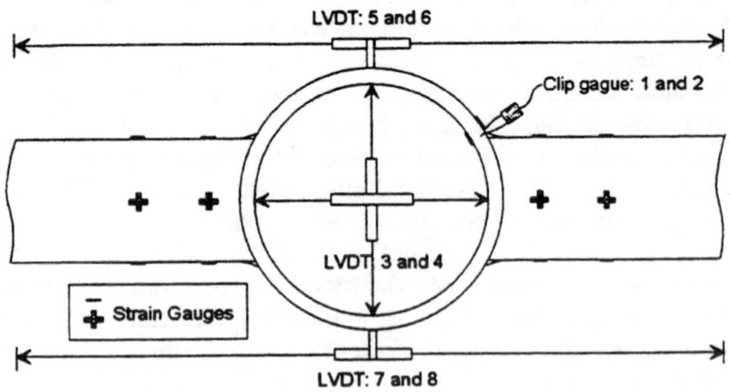

Figure 3 Overview of test instrumentation

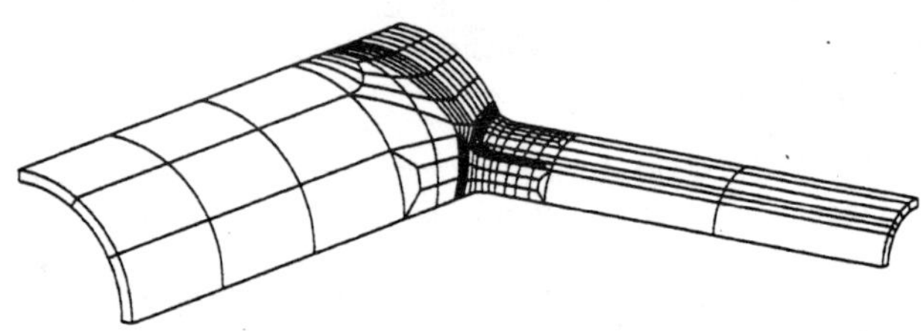

Figure 4 FE model of Test 1

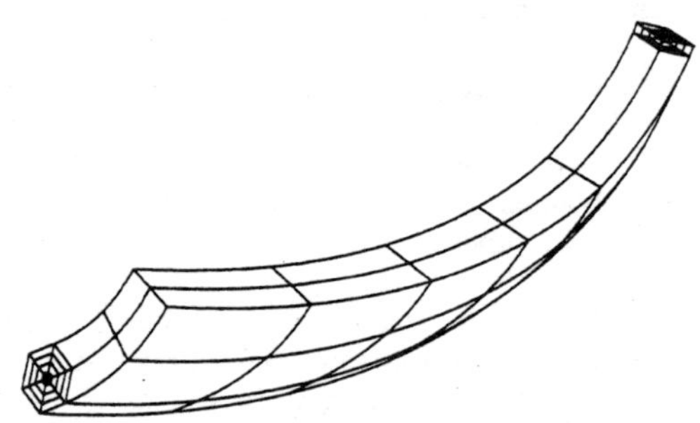

Figure 5 FE model of crack, Test 1

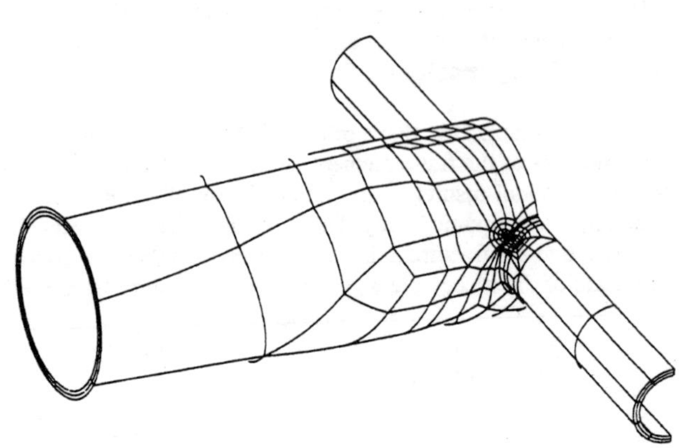

Figure 6 FE model of Test 2

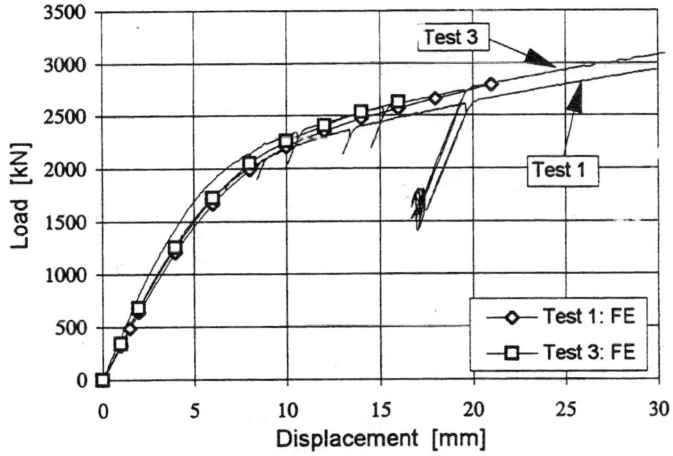

Figure 7 Load displacement curves for Test 1 and 3 compared to FE results

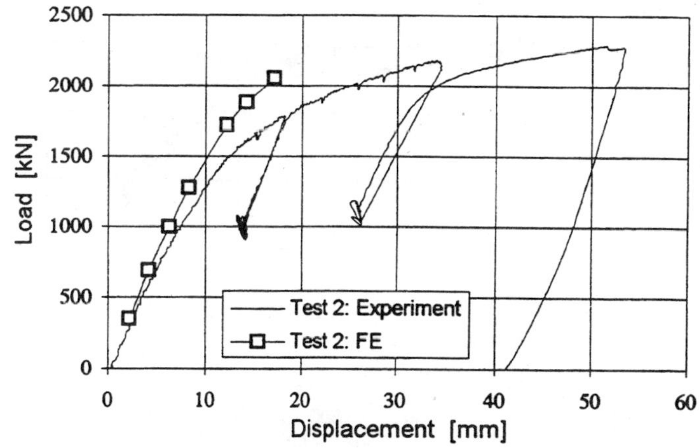

Figure 8 Load displacement curve for Test 2 compared to FE results

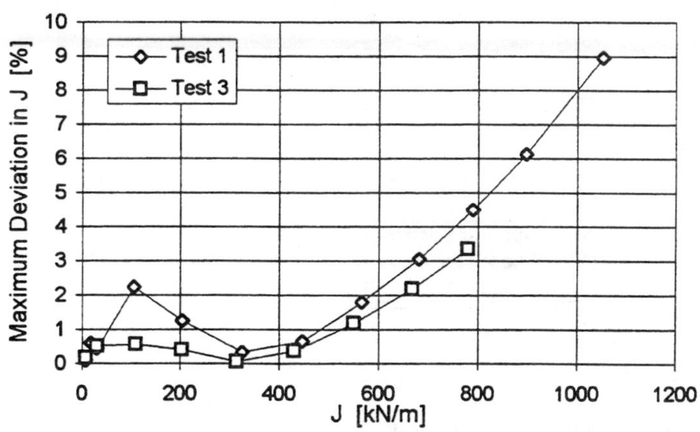

Figure 9 Path independence of J plotted versus J for Test 1 and Test 3

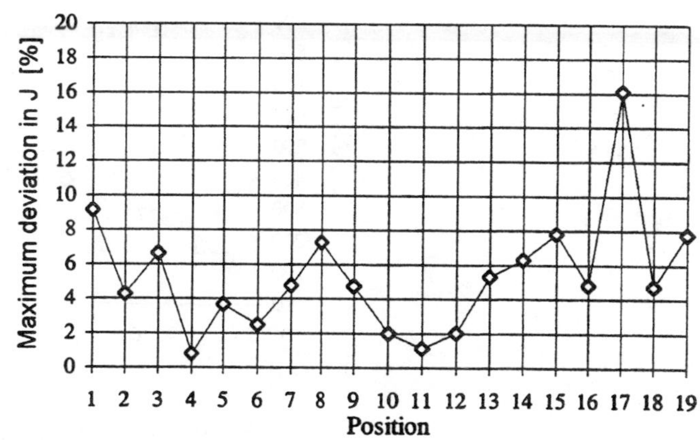

Figure 10 Path independence of J plotted along crack front for Test 2 at maximum load

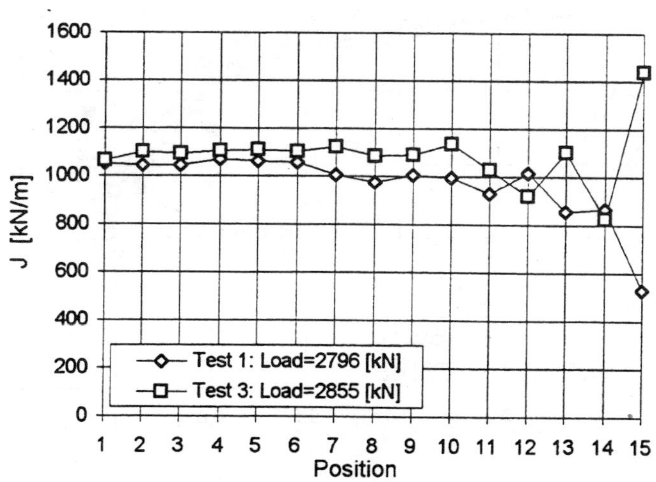

Figure 11 J distribution along crack front for Test 1 and Test 3

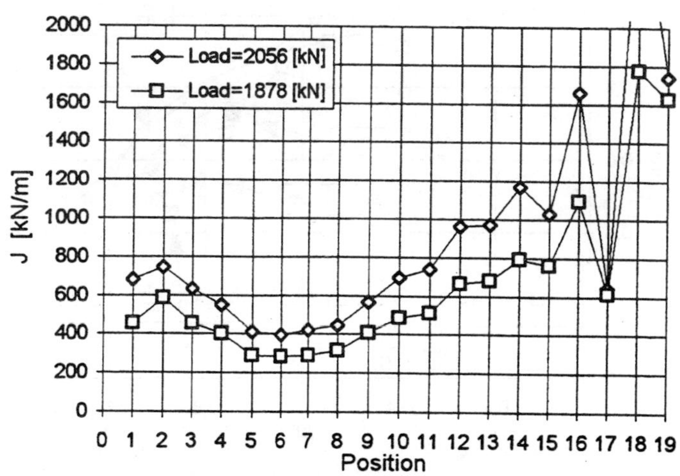

Figure 12 J distribution along crack front for Test 2 at two load levels

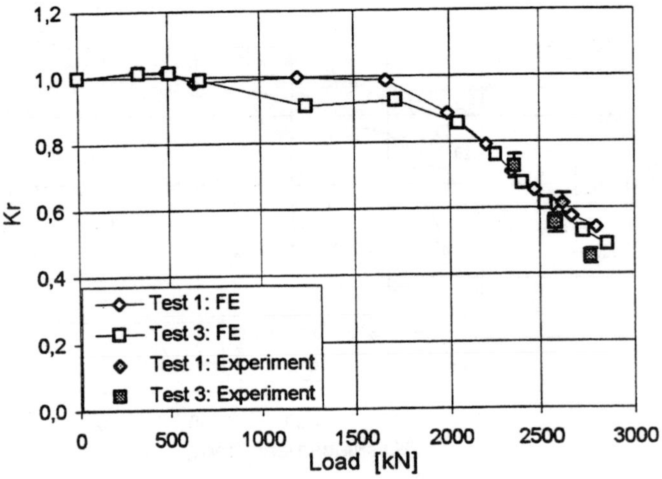

Figure 13 Kr versus load for Test 1 and 3

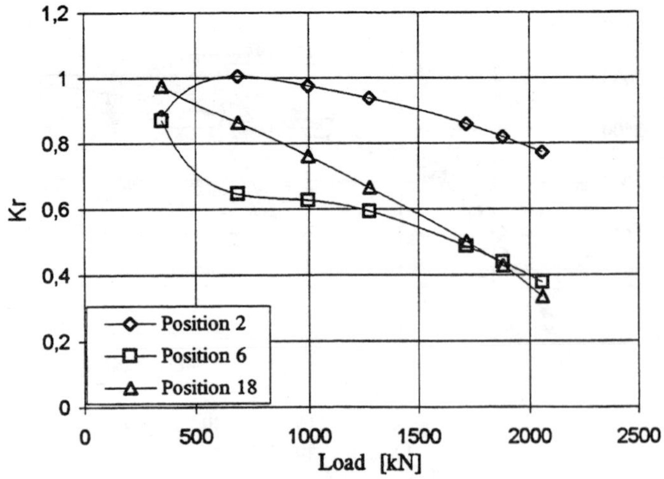

Figure 14 Kr versus load for Test 2

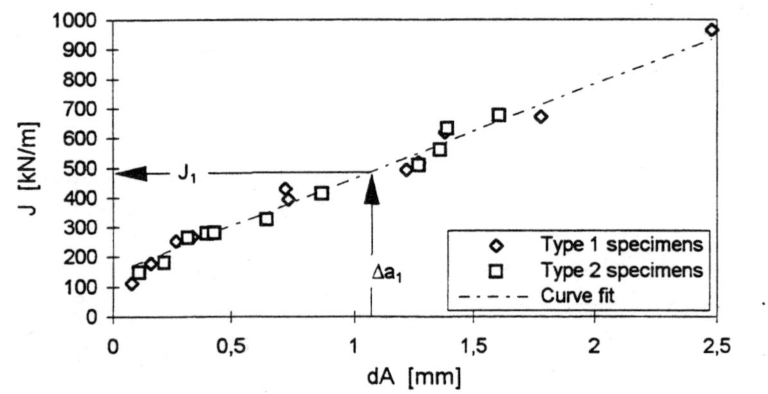

Figure 15 Procedure for J-estimate from experimental crack growth

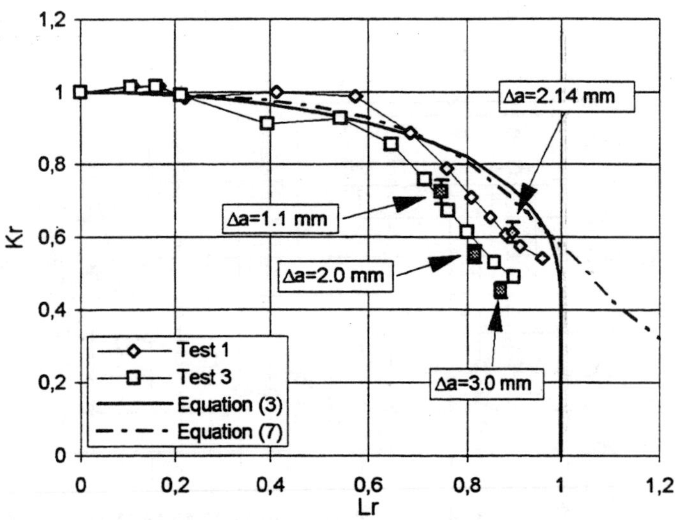

Figure 16 FAD curves for Test 1 and Test 3 compared to test results and two standard FAD curves. HSE area-corrected load estimate

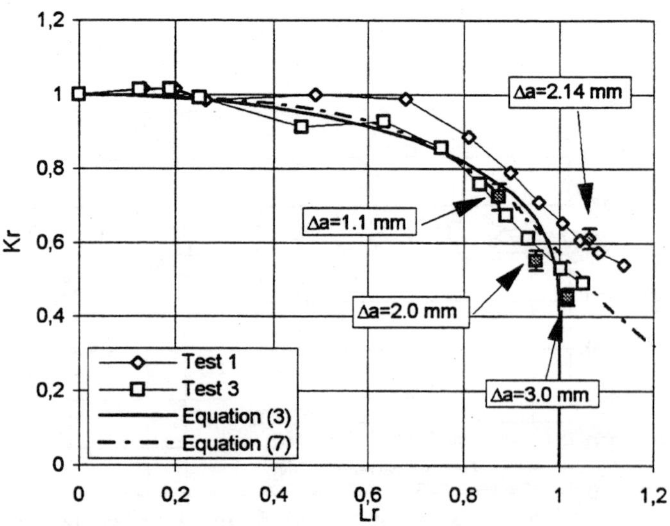

Figure 17 FAD curves for Test 1 and Test 3 compared to test results and two standard FAD curves. HSE length-corrected load estimate

Proceedings of the Seventh (1997) International Offshore and Polar Engineering Conference
Honolulu, USA, May 25-30, 1997

ISBN 1-880653-28-1 (Set); ISBN 1-880653-32-X (Vol. IV)

Mechanical Properties of Friction Welded Joints of AZ31 Magnesium Alloy

Kazuyoshi Katoh, Toshikatsu Asahina and Hiroshi Tokisue
Nihon University
Chiba, Japan

ABSTRACT

AZ31 magnesium alloy was friction welded using a brake type friction welding machine under argon shielding. Microstructures and mechanical properties of the friction welded joints were investigated. The fine grain structure resulting from disappearance of fibrous one is observed closely near the weld interface. The hardness of both the weld interface and heat affected zone are nearly equal to that of the base metal. The tensile strength of welded joints are improved with increasing of friction time, but it is lower than that of the base metal. In case of friction pressure of 50MPa and friction time of 6 seconds, the elongation of the joint shows the highest value, but it is inferior to that of the base metal. All the impact specimens with the notch into the weld interface showed impact strength inferior to that of the base metal, but impact strength increases with increasing notch distance from the weld interface. In case of smoothed specimen, fatigue strength of the welded joints are lower than that of the base metal irrespective of the number of cycles. Fatigue limit of the welded joint is 81.4% of the base metal. Fatigue strength of the welded joints with notch into the weld interface is slightly lower than that of the base metal and fatigue limit of the welded joint is 93% of the base metal. The notched specimens show fatigue limit lower values than smoothed specimen for welded joint.

KEY WORDS : Friction welding, AZ31 magnesium alloy, Microstructures, Temperature, Tensile strength, Impact value, Fatigue strength

INTRODUCTION

Magnesium alloy have superior to others materials in various properties, such as light weight, high strength in specific properties, excellent corrosion resistance, etc.. Those advantages allow the alloys to be increasingly used for automotive, spacecraft of aircraft components. The extended usage of magnesium alloys may present new problems in the selection of joining process. It is said that the magnesium alloy in general is readily welded because of the low heat capacity and low heat of fusion, means that power input can be low, welding speed can be high, and post heating for stress relief is not necessary. Despite some reports on the magnesium alloy, most of them are concerning to the fusion welding (Komatsu and Ito, 1976). The magnesium alloy joints, welded by fusion welding have lots of problems such as the welding metallurgy after welding. Meanwhile, a report says that the friction welding conducted in solid phase without melting a base metal makes it comparatively easy to produce joining free from the above-mentioned welding defects if proper welding conditions are given (Hazlett and Gupta, 1984). The others showed previously, upon examining the properties of friction welded joints of aluminum alloys, that they could be in practical use despite a certain decrease of ductility (Tokisue and Kato, 1978, Kato and Tokisue, 1990).

This paper deals with applicability of friction welding to the magnesium alloy by examining the structures, mechanical properties and fractography of tested specimens of friction welded joints.

MATERIAL AND EXPERIMENTAL PROCEDURE

For the work piece, the AZ31 magnesium alloy bar of 17 mm in diameter was used after machining it down to 150 mm in length and cleaning its welding surface with acetone. Chemical compositions and mechanical properties of base metal are given in **Table 1** and **Table 2**, respectively.

Table 1 Chemical compositions of base metal. (mass%)

Al	Zn	Mn	Fe	Si	Cu	Ni	Mg
2.8	0.81	0.36	0.003	0.01	0.002	Tr.	Bal.

Table 2 Mechanical properties of base metal.

Tensile strength (MPa)	Elongation (%)	Impact value (kJ/m^2)	Hardness (HV0.3)
295	22.7	223	59.6

Table 3 Friction welding conditions.

Rotational speed	n (s^{-1})	58
Friction pressure	P_1(MPa)	40, 50, 60
Friction time	t_1(s)	1, 2, 3, 4, 5, 6, 7
Upset pressure	P_2(MPa)	100
Upset time	t_2(s)	5
Shielding gas		Ar

Friction welding was performed by using a brake type friction welding machine under argon shielding. The welding conditions are given in **Table 3**.

JIS No.4 specimens were used for the tensile test. The impact strength was tested by Charpy impact tester, using the JIS No.3 specimens (U-notched, 2 mm) which the notch is put into the 0, 1, 2, 3, 4 and 5 mm from weld interface. The hardness was measured by the micro-Vickers hardness tester (load : 2.98N). The fatigue test was conducted with Ono's rotary-bending tester (capacity : 98N·m, rotational speed : $30s^{-1}$), using smoothed and notched specimens which is put into the weld interface. All the tests operated in the room temperature.

EXPROIMENTAL RESULTS AND DISCUSSIONS

Observation of macrostructures

The macrostructures of welded joints under condition of friction pressure of 50MPa which vary in friction time are shown in **Fig.1**. Regardless of the friction time, a symmetrical structure on the rotating axis of joint or on the weld interface appeared, and every joint clearly shows a weld interface taking the shape of a convex lens with its central part constricted, and the heat affected zone is seen taking the shape of concave lens enveloping the welding part. The original fibrous structures are pressed forward along the flux of burr in the heat affected zone, and they disappear near the weld interface. As the friction time increases, the welding layer in the shape of convex lens and heat affected zone become wider, and, naturally, the discharged burrs become large. The same friction time, the burr increased with increases of friction pressure. This tendency is observed on both carbon steel (Duffin and Bahrani, 1973) and aluminum alloys (Tokisue and Kato, 1978, Kato and Tokisue, 1990).

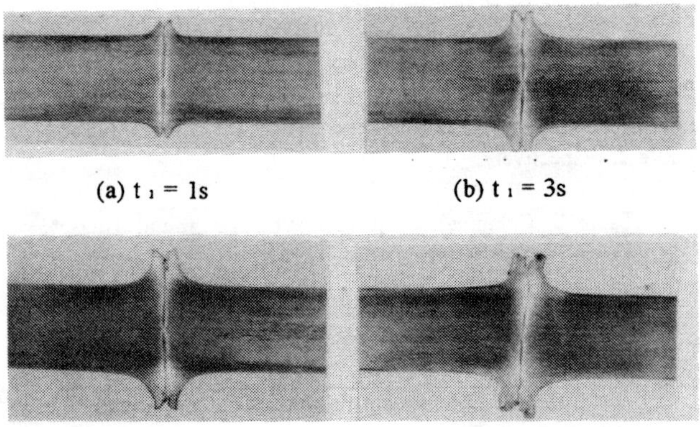

(a) t_1 = 1s (b) t_1 = 3s (c) t_1 = 5s (d) t_1 = 7s

Fig.1 Macrostructures of friction welded joints. (P_1=50MPa)

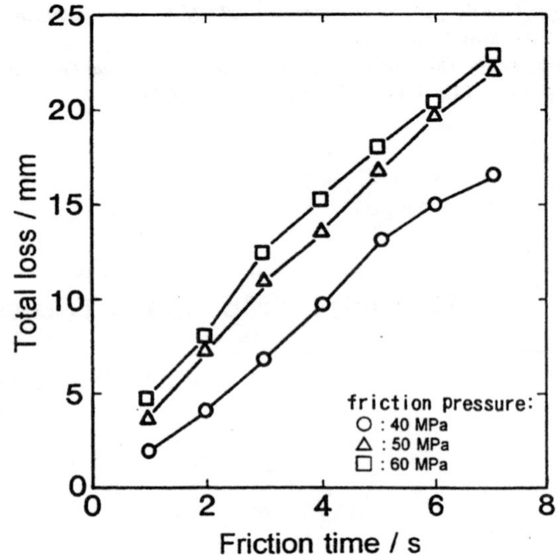

Fig.2 Relation between friction time and total loss.

Figure 2 shows the relation between friction time and total loss. The total loss (the difference of the total length of work piece before welding and length of the welded joint) increased almost lineally with the increase of friction time. The same friction time, total loss increased with increasing of friction pressure. The total loss with friction pressure of 50MPa are as same value as 5056 aluminum alloy joints (Kato and Tokisue, 1990).

Temperature-time history of the welding process

An example of the temperature-time history of a welding process, friction pressure of 50MPa and friction time of 6 seconds, is shown in **Fig.3**. The number in figure are shown the position of insert a thermocouple which is distance from weld interface. As shown in the figure, in the positions closer to the friction surface the temperatures rise rapidly. That is, in the position of 1 mm from

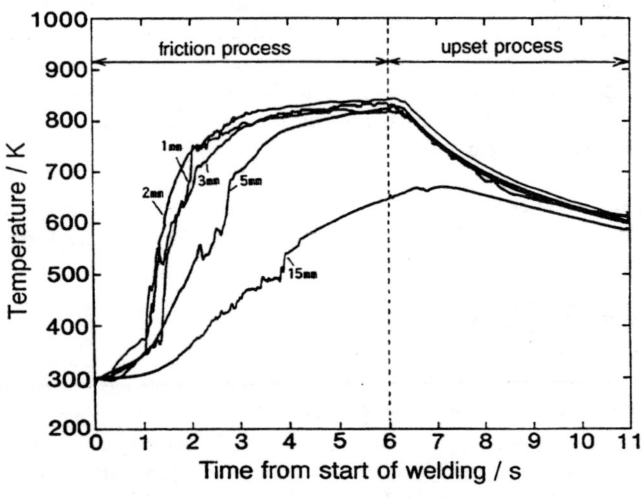

Fig.3 Temperature-time histories of friction welding process. (P_1=50MPa, t_1=6s)

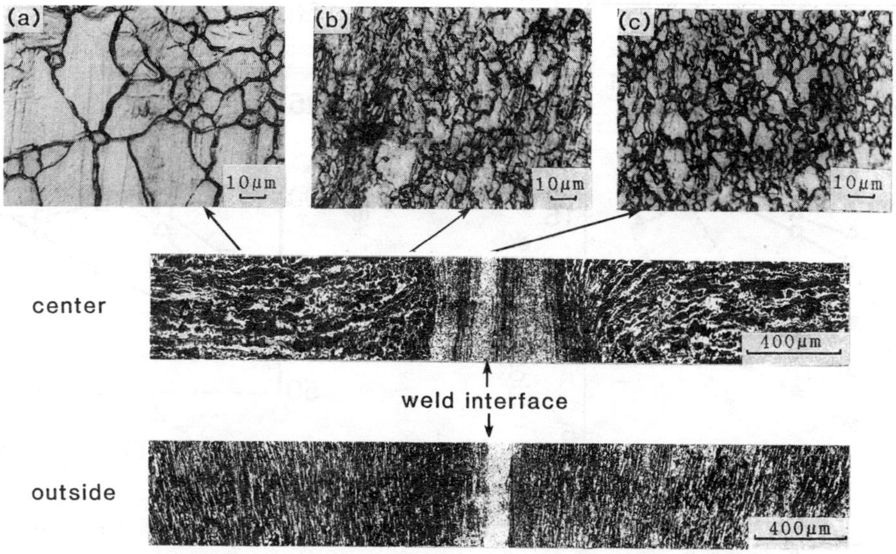

Fig.4 Microstructures of friction welded joint. (P_1 = 50MPa, t_1 = 6s)

weld interface, the temperature rise suddenly just after the start of welding, then, rises further at a slower constant speed, and reaches the highest value (823K) about 6.3 seconds after the start of welding slightly after it stops rotation. In the position of 2 and 3 mm from weld interface, the temperature rises sharply in less time than in position of 1 mm and reaches the highest value immediately after it stops rotation, through the nearly same process as that in position of 1 mm. In the position of 5 mm from weld interface, the temperature increases in less time than in position of 1 mm from weld interface and reaches the highest value immediately after it stops rotation through a little sharper temperature gradient than in the position of 1 mm from weld interface. In the position of 15 mm from weld interface, the initial temperature rises slowly where such a behavior as seen in the positions of 1 to 5 mm from weld interface is not found. The speed at which the temperatures rise in the positions of 1 to 5 mm from weld interface which are close to weld interface show a similar process. It is considered that the temperature grade moves at a constant speed acceding to the movement of the friction surface. The temperature-time history in the welding process displays the same tendency regardless of the friction time, however, in a short friction time the highest temperature is somewhat low. The same results are reported on the friction welding process of aluminum alloys (Tokisue and Kato, 1978, Kato and Tokisue, 1990).

Observation of microstructures

A microstructures of near the weld interface is shown in **Fig.4**. On the welding layer, a fibers structure which was observed on the base metal disappears and a non-directional fine grain structures where a deposit layer is observed along the boundary layer. The boundary layer is clearly observed regardless of the friction time, and its width has a tendency to expand with the increase of friction time. A fibers structure on the heat affected zone, a so-called plastic flow zone (Masumoto and Hira, 1979) which varies peripherally along the discharging direction of burr is clearly observed, especially clearly on the periphery of the joint. The variation of fibers structure such as this is also observed on the friction welded joint of mild steel (Duffin and Bhrani, 1973) and aluminum alloys (Hazlett and Gupta, 1984, Tokisue and Kato, 1978, Kato and Tokisue, 1990), which is considered to have resulted from the deformation of the heated area by the upset pressure. Obviously, the width of the plastic flow zone of the joint in short friction time has a tendency to become narrow.

Mechanical properties

The hardness distribution of the center of the welded joint under conditions of friction pressure of 50MPa and friction time of 5 seconds in the axial direction is shown in **Fig.5**. Since the hardness distribution of the joint is considered symmetrical on both the rotating and fixed sides with respect to the weld interface, the figure only shows the fixed side. The hardness is nearly equal on both the weld interface and heat affected zone to that of the base metal. It is observed similar tendency in TIG welded joint of AM60 magnesium alloy castings (Matsumoto, Kobayashi and Hotta, 1988).

Figure 6 shows results of tensile test. Regardless of the friction pressure, both tensile strength and elongation increase with increasing of the friction time. Particularly, all the welded joints in friction time of 7 seconds show the highest value in tensile strength.

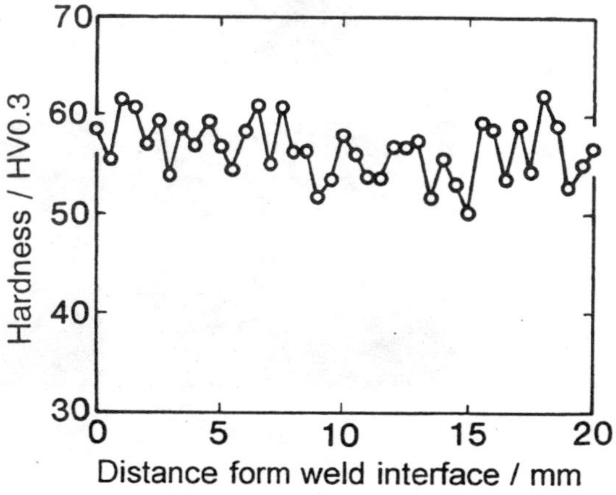

Fig.5 Hardness distribution of friction welded joints.
(P_1 =50MPa, t_1 =5s)

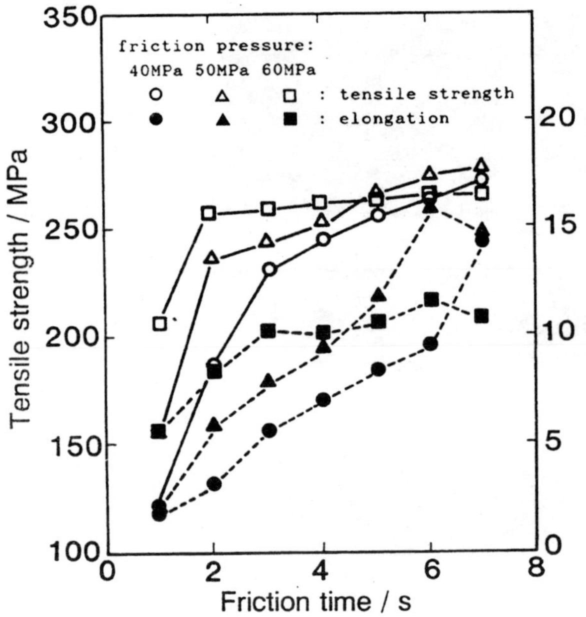

Fig.6 Relation between friction time and tensile strength and elongation.

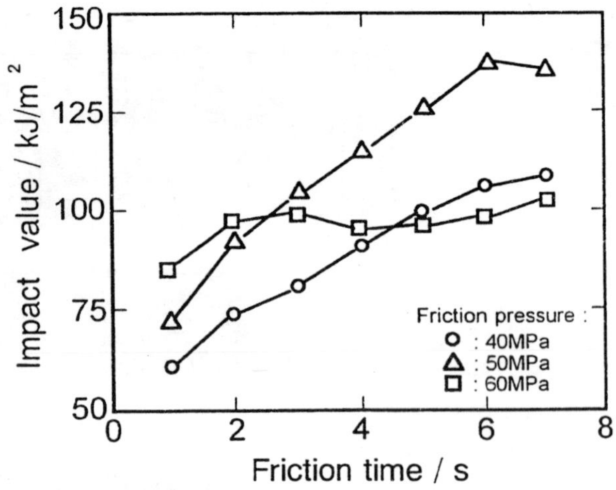

Fig. 8 Relation between friction time and impact value of welded joints with notch is weld interface.

The elongation tends to increase with the increasing of the friction time. However, elongation of the welded joints, with friction pressure of 50 and 60MPa, in friction time of 6 seconds show the highest value. The favorable results of tensile strength of joints such as about 90% of that of the base metal have been obtained. And, the maximum elongation is about 75% to that of the base metal. The welded joints, which were welded in the friction pressure of 50MPa and friction time of 5 seconds or over, fractured at the heat affected zone. But welded joints under the other welding conditions were fractured at the weld interface or the heat affected zone. In this case, tensile strength of welded joints show the same value, regardless of fractured position. but it's elongation of the welded joints fractured at heat affected zone are shown 2 times of the welded joint which are fractured at the weld interface.

Macro- and microfractographs of tensile fractured specimen of friction welded joints are shown in **Fig.7**. Every microfractographs are of the center of the fractured surface. Macrofractographs of welded joints which are fractured at heat affected zone show as same as base metal. Welded joints, which are fractured at weld interface, are fractured squarely an axis core and they showed spiral traces by the rotation in the welding process. The reason why, the weld interface received plastic deformation by twist moment in friction process. The fractured surface of welded joint which was fractured at heat affected zone were observed nearly the same equiaxed dimples which were observed on the base metal. But welded joint which was fractured at the weld interface observed brittle spiral traces.

Results of Charpy impact test of welded joints which the notch is put into the weld interface are shown in **Fig.8**. The impact value of welded joint are increased with increasing of friction time. The maximum impact value of the welded joints, under conditions of friction pressure of 50MPa and friction time of 6 seconds, is considerably lower than that of the base metal. On the weld interface, as clearly seen from the observations of the microstructures,

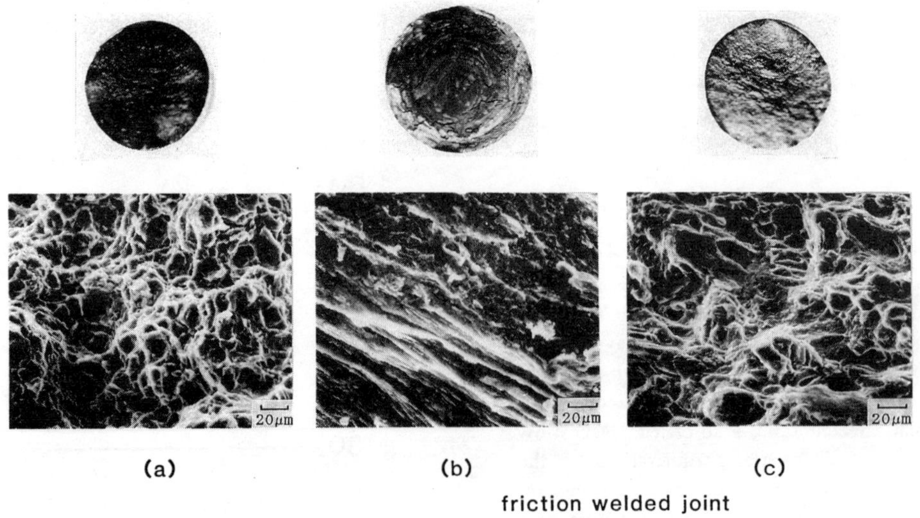

Fig.7 Macro- and microfractographs of tensile fractured surface of tested specimens.
(a) base metal, (b) P_1 = 40MPa, t_1 = 1s, (c) P_1 = 40MPa, t_1 = 6s.

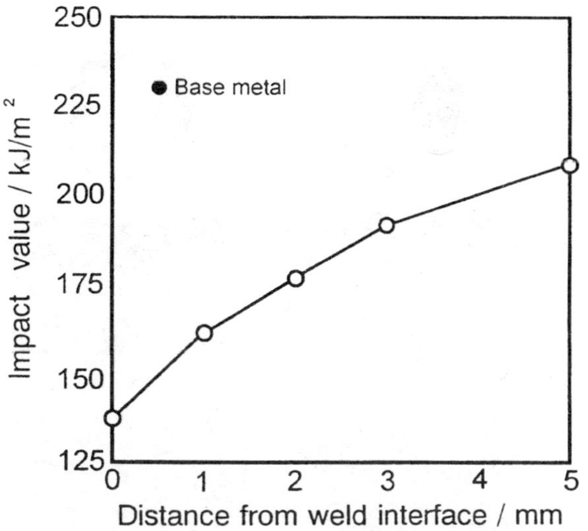

Fig.9 Relation between notch located position of distance from weld interface and impact value.
(P_1 = 50MPa, t_1 =6s)

the impact value decreases owing to disappearance of the fibrous structure. There is observed also both friction welded joints of carbon steel (Shioya, Yamada and Tarutani, 1966) and aluminum alloy (Tokisue and Kato, 1978).

Figure 9 shows results of the impact test with the welded joints, under conditions of friction pressure of 50MPa and friction time of 6 seconds, in which the notches are made in various position. The impact specimen of welded joint in which the notch put into the weld interface shows the lowest impact value of welded joints. The impact value of the welded joint increases of increasing of notch a distance from weld interface. The welded joint with the notch at 5 mm from weld interface shows about 90% of the base metal.

Macro- and microfractographs of the impact tested specimens on which the notches are made in various position are shown in **Fig.10**. Every microfractographs are of the center of the fractured surface and the upper side is the notch part. Macrofractographs of the

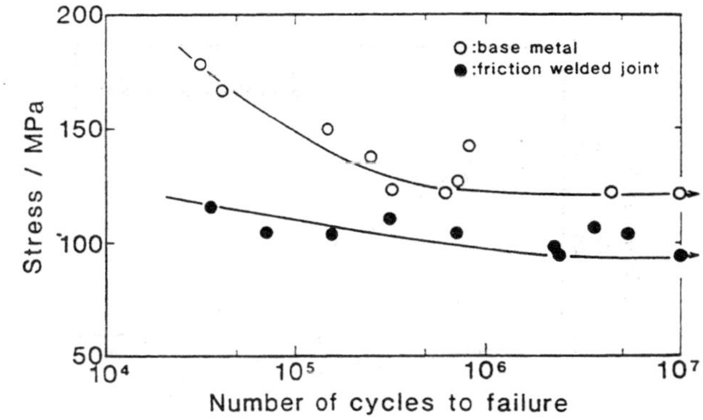

Fig.11 S-N curves of smoothed specimens.
(P_1 = 50MPa, t_1 =6s)

welded joint in which the notch is put into the weld interface dose not show a width spread under the notch and in relatively flat on the fractured surfaces, showing a brittle aspects. However, the more distance the notch is from the weld interface, the larger the width spread becomes under the notch; so-called widening phenomenon is observed, the unevenness of the fracture is large, and the impact value increases. The welded joints shows a dimple pattern like the base metal dose, regardless of distance from the weld interface of the notched position. The more distance is the notch from weld interface, the dimple becomes large and various size dimples are observed. When a notch is placed 5 mm from the weld interface, both size and depth of the dimples are nearly equal to those of the base metal.

The results of fatigue tests with smoothed specimen are shown in **Fig.11**. The fatigue limit of welded joint is 81.4% that of the base metal. In case of friction welded joint of aluminum alloys with same fatigue test, it is reported that fatigue limit of friction welded joints show about 90% of base metal and fractured at heat affected zone which is softened area (Kato and Tokisue, 1984, Kato and Tokisue, 1990). It is presumed that the smoothed specimen fractured at weld interface.

It is well known fatigue strength is relative to tensile strength. The ratio of fatigue strength to tensile strength is so-call fatigue ratio.

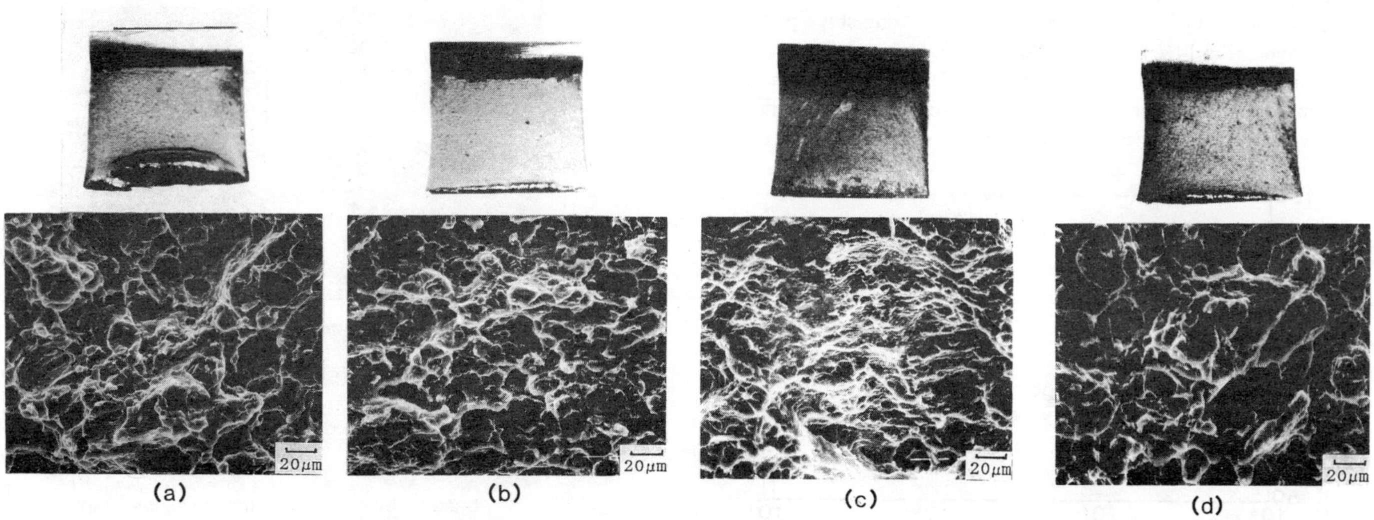

Fig.10 Macro- and microfractographs of Charpy impact fractured surfaces of tested specimens.
(a) base metal, (b), (c) and (d) are position of notch 0, 1 and 5mm from weld interface.
(P_1 = 50MPa, t_1 =6s)

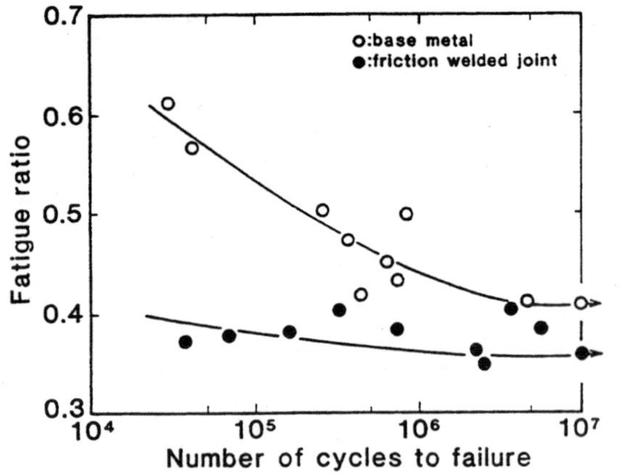

Fig.12 Relation between number of cycles to failure and fatigue ratio with smoothed specimens.

Figure 12 shows relation between number of cycles to failure and fatigue ratio with smoothed specimen. The fatigue ratio of welded joint is 0.36 which is lower than that of the base metal. However, it shows higher value than that of the friction welded joint of 2017 aluminum alloy which is 0.30 (Kato and Tokisue, 1984).

Figure 13 shows the results of fatigue test with notched specimen. The notched specimens of both base metal and welded joint show lower value than smoothed specimen. The fatigue limit of welded joint is slightly lower than that of the base metal. Inclinations of S-N curves of base metal and welded joint are nearly the same. The fatigue limit of welded joint shows 93% of base metal. The notch factor of base metal, which is calculated by results of this examinations, is 1.61 and welded joint is 1.42. The reason why, microstructures of the weld interface which is notched position shows a fine grain structure than that of the base metal.

Macro- and microfractographs of the fatigue tested smoothed specimens are shown in **Fig.14**. In all the microfractographs, left side is surface of specimen. Macrofractographs of smoothed specimens of both base metal and welded joint show flat fractured surfaces with small unevenness and beach marks are observed clearly. Microfractograph of base metal observed meandering stripe pattern at first stage (Fig.14(A)). But welded joint shows relatively flat on first stage showing a brittle aspect (Fig.14(D)). The second stage, a striation, typical pattern appearing on the fatigue fractured surface, is observed (Fig.14(B), (E)). A striation is clearly observed on the

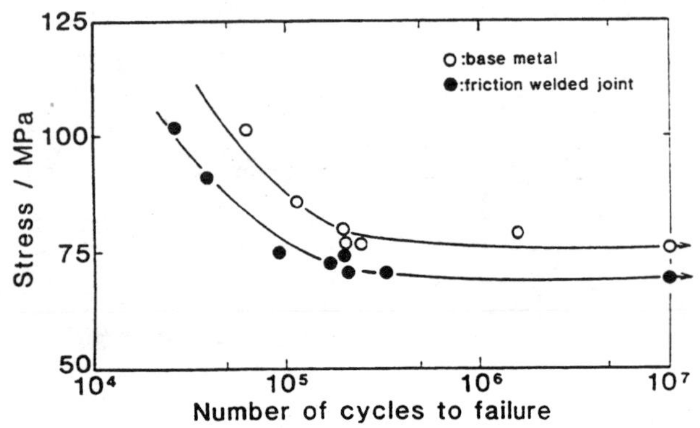

Fig.13 S-N curves of notched specimens. (P_1= 50MPa, t_1 =6s)

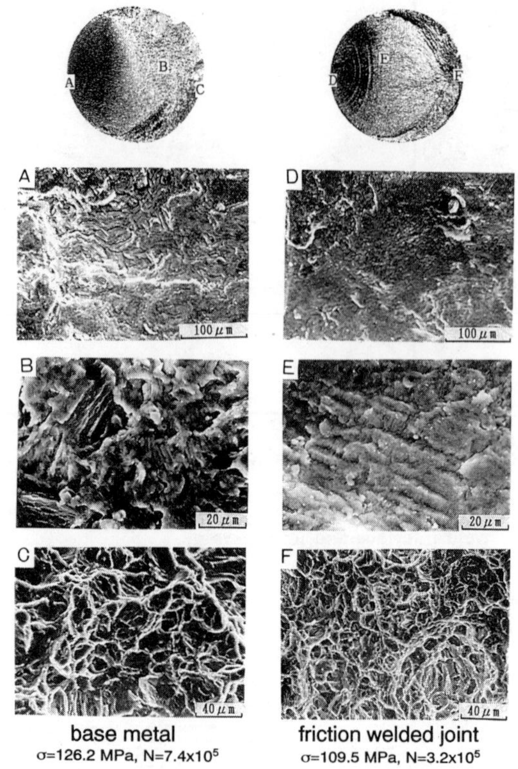

Fig.14 Macro- and microfractographs of fatigue fractured surfaces with smoothed specimens.

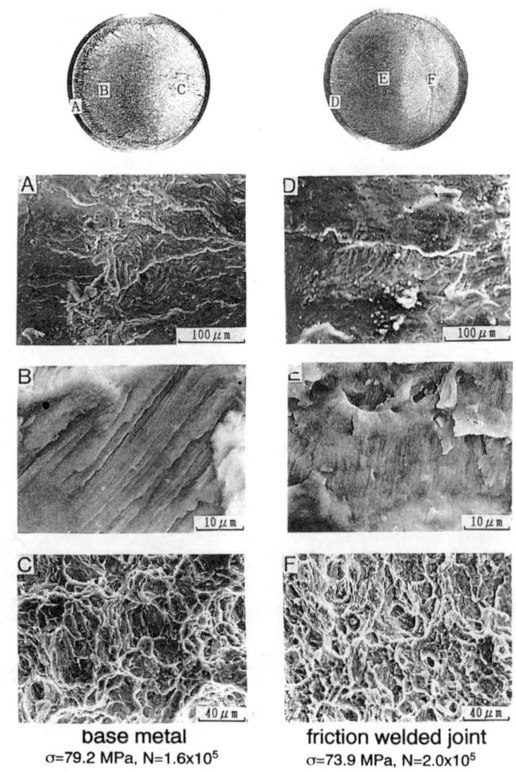

Fig.15 Macro- and microfractographs of fatigue fractured surfaces with notched specimens.

base metal, but the striation area of welded joint becomes narrower than that of the base metal. At the final stage, both base mental and welded joint show dimple, however, the dimple observed on welded joint is finer than that of the base metal (Fig.14(C), (F)). Because the weld layer show finer grain structure which is 3 to 10 μm than base metal which is 10 to 40 μm.

Figure 15 shows macro- and microfractographs of notched specimens of both base metal and welded joint. Macrofractograph of welded joint shows nearly the same and final fractured are observed clearly. Microfractographs of welded joint shows relatively flat than base metal. All the part of fractured surface of notched specimen of both base metal and welded joint show same appearance of smoothed specimens. This tendency are shown friction welded joint of aluminum alloys (Kato and Tokisue, 1984, Kato and Tokisue, 1990).

CONCLUSIONS

Upon examination of the structures, mechanical properties and fractography of the tested specimens of the friction welded joint of AZ31 magnesium alloy, the following results were obtained :

1) The fibers structure of the base metal are pressed forward to the outside along the flux of burr in the heat affected zone by the axial pressure, while they disappear near the weld interface.

2) The hardness of both weld interface and heat affected zone is nearly equal to that of the base metal.

3) The welded joints have lower tensile strength as the base metal, and have less elongation. All the friction welded joints are fractured at weld interface.

4) The impact specimen which is notched at weld interface shows the impact value lower than that of the base metal. Impact value of welded joint which is notch put into 5 mm from weld interface it shows about 90% of the base metal.

5) Fatigue limit of the welded joint is slightly lower than that of the base metal irrespective of the shape of specimen.

As above mentioned, it is considered that AZ31 magnesium alloy could satisfactorily be used when a certain decrease of both tensile strength and elongation are taken into consideration. In this examination, friction welding performed under argon shielding, however it is no flush, so friction welding is able to be welding in the air with risk of ignition.

REFERENCES

Duffin, FD and Bahrani, AS (1973), "Friction welding of mild steel . the effects of varying the value of deceleration", *Metal construction and British Welding Journal*, vol.5, pp 125-132.

Hazlett, TH and Gupta, KK (1984), "Friction welding of high strength structural aluminium alloy", *Welding Journal*, vol.42, pp 490s-494s.

Kato, K and Tokisue, H (1984), "Fatigue strength of friction welded joints of 2017 alloy", *Journal of Japan Institute of Light Metals*, vol.34, pp 325-330.

Kato, K and Tokisue, H (1990), "Friction Welding of 5056 Aluminium Alloy", *Journal of Light Metal Welding & Construction*, vol.28, pp 16-21.

Komatsu, R and Ito, S (1976), "Welding of magnesium ", *Journal of Light Metal Welding & Construction*, vol.14, pp 121-131.

Masumoto, I and Hira, H (1979), "Fundamental study of hot-heavy-worked and its properties of pure aluminum by friction welding", *Journal of Japan Welding Society*, vol.48, pp 626-631.

Matsumoto, J, Kobayashi, M and Hotta, M (1988), "Arc Welding of Magnesium Alloy AM60", *Journal of Light Metal Welding & Construction*, vol.26, pp 410-416.

Shioya, T, Yamada, S and Tarutani, Y (1966), "Cause of Impact Strength Decrease of Carbon Steel Welded by Friction Welding", *Journal of the Japan Welding Society*, vol.35, pp 65-71.

Tokisue, H and Kato, K (1978), "Mechanical properties of friction welded joints of aluminum alloy 2017", *Journal of the Japan Institute of Light Metals*, vol.28, pp 450-454.

Proceedings of the Seventh (1997) International Offshore and Polar Engineering Conference
Honolulu, USA, May 25-30, 1997

ISBN 1-880653-28-1 (Set); ISBN 1-880653-32-X (Vol. IV)

The Static Strength of Stiffened and Unstiffened L-Joints Made of Circular Hollow Sections

R. Puthli, F. Mang and D. Karcher
University of Karlsruhe
Karlsruhe, Germany

1 General

L-joints consisting of Hollow Sections are used in different structures. They are, for example, used in floating cranes and Offshore Structures. The load-bearing-behaviour of L-joints made of Rectangular Hollow Sections (RHS) has been already investigated (see Mang et al [1,2,3,4]). Examples of L-joints made of RHS are shown in figure 1. On the load-bearing-capacity of L-joints made of Circular Hollow Sections (CHS) however, only a few results are available. Design recommendations for L-joints on the basis of the research by Mang et al in RHS are given in DIN 18808 [8], EC 3 [9] and a CIDECT Publication [6]. In the scope of a CIDECT [7] and a DFG (German research foundation) research project the load bearing behaviour of CHS L-joints is being investigated. In this paper, the results of the existing investigations are presented.

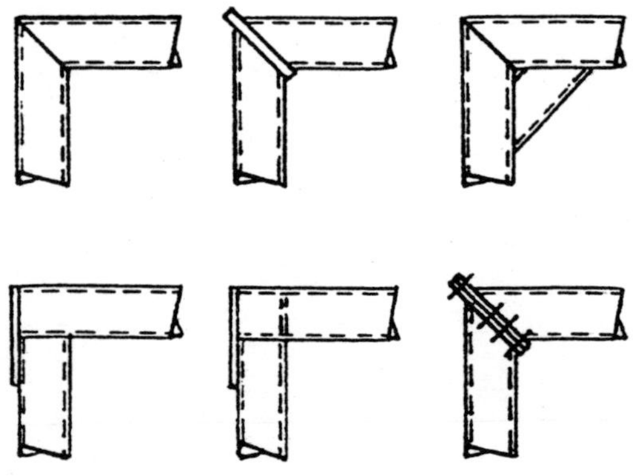

Figure 1: Different types of L-joints made of RHS

2 Symbols

a_w = weld throat thickness

α = Form factor: ratio of the load bearing capacity of unstiffened joint to the load bearing capacity of stiffened joint

D = Diameter of the circular hollow sections

b = width of Rectangular Hollow Sections

t = wall thickness of the tube

t_p = thickness of the stiffening plate

D/t = Diameter/wall thickness ratio

b/t = width/wall thickness ratio

P_{max} = Ultimate bearing load of the specimen

σ_{max} = maximum stress (determined from strain measurements)

σ_{FE} = maximum stress (determined from FE calculation)

fy = nominal yield stress determined from tensile tests

fu = nominal ultimate stress determined from tensile tests

E = Modulus of Elasticity (Young's modulus)

γ_F = partial safety factor for load

γ_M = partial safety factor for resistance

CHS = Circular Hollow Section

RHS = Rectangular Hollow Section

FE = Finite element

3 Test specimens

3.1 Dimensions

Altogether, 20 test specimens are fabricated. Ten different D/t ratios are used, so that ten specimens are fabricated with a stiffening plate in the joint area and the other 10 fabricated without a plate. Figure 2 shows these types of joints. A miter is cut and the tubes welded to each other. An overview of the dimensions is given in table 1. Test No. 10 is applied with strain gauges to determine local strains.

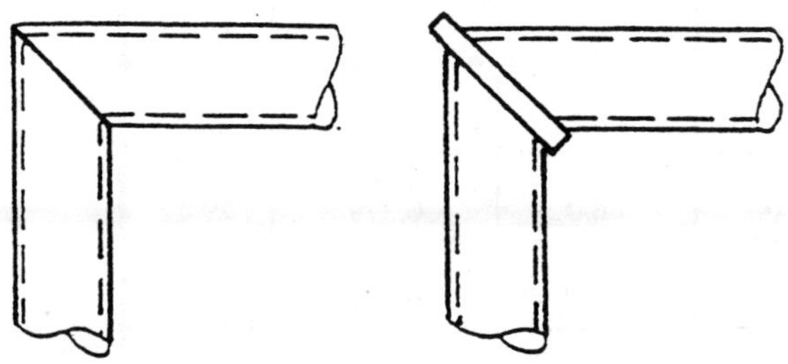

Figure 2: Types of tested L-joints made of CHS

The stiffening plate is fabricated according to the stipulations of DIN 18808. In clause 5.3 of DIN 18808 the plate size should comply with:

$t_p \geq 1.5 \times t_i \quad (i = 1,2..)$ and

$t_p \geq 10$ mm

A stiffener plate thickness of 15 mm is used for all test specimens so that the requirements in the DIN are fullfilled, since the maximum wall thickness of the tested tubes is 8.8 mm. The lateral dimensions of the stiffening plate are chosen on the basis of welding requirements. The distance from the outer edge of the tube to the edge of the plate should be at least two times the side length of the fillet weld to guarantee proper weld toes.

3.2 Weld details

The tubes of the stiffened specimens are welded to the plate by using fillet welds. The throat thickness of the welds complies with all important standards:

$a_w = t$ (wall thickness of the tube)

The miter is cut at an angle of 45°. Therefore, the angle between the stiffening plate and the tube varies from 45° to 135°. At sharp corners, the welding is to be carried out with care. In figure 3, the fillet welds at different positions can be seen.

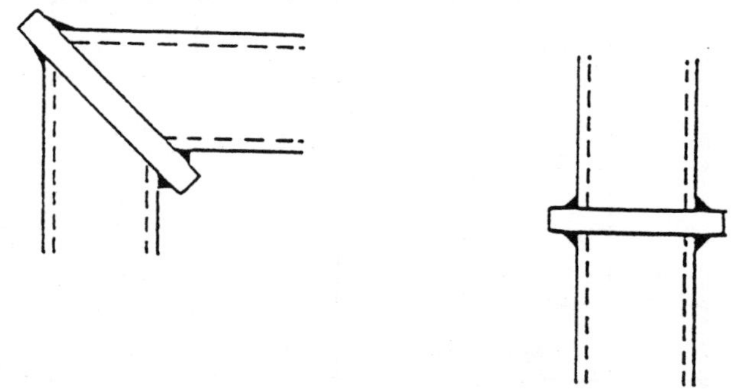

Figure 3: Fillet weld for stiffened L-joints

The unstiffened specimens are welded by a combination of butt and fillet welds with a smooth transition between the two. Edge preparation of the CHS wall as shown in figure 4 is carried out. The inner (sharp) corner is a fillet weld and the outermost corner a butt weld. There is a smooth and gradual transition between fillet and butt weld around the two half circumferences.

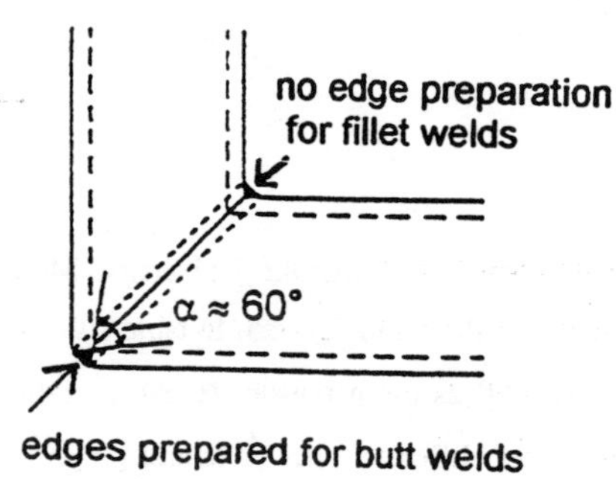

Figure 4: Edge preparation for welding specimens without stiffening plate

3.3 Material properties

Tensile test specimens are taken from all tubes and tensile tests carried out to determine the material behaviour. The yield strength and the ultimate strength of the material is given in table 1. The modulus of Elasticity E varies from 190.000 to 210.000 N/mm².

test no.	CHS mm x mm	length [mm]	$t_{measured}$ [mm]	f_y [N/mm²]	f_u [N/mm²]	$P_{ultimate}$ unstiffened [kN]	$P_{ultimate}$ stiffened [kN]	$\alpha = \frac{P_{unstiffened}}{P_{stiffened}}$	$\frac{D}{t}$
1	82.5x4.0	410	3.94	329.8	454.7	15.1	33.2	0.455	20.94
2	82.5x4.0	410	4.02	452.1	593.5	23.8	50.6	0.470	20.52
3	101.6x2.9	510	2.78	275.1	337.9	8.5	23.5	0.362	36.55
4	101.6x4.0	510	3.87	416.3	553.1	17.9	47.2	0.379	26.25
5	101.6x8.8	510	8.05	398.6	589.4	57.9	94.9	0.610	12.62
6	114.3x5.6	570	5.08	281.0	396.1	21.8	46.8	0.466	22.5
7	133.0x4.5	530	4.52	294.1	406.0	25.0	66.0	0.379	29.42
8	168.3x4.8	840	4.87	415.0	521.1	30.4	102.1	0.298	34.56
9	193.7x2.9	970	2.89	281.7	350.9	7.8	44.1	0.177	67.02
10	193.7x5.0	970	5.19	382.2	441.1	23.4	103.7	0.226	37.32

Table 1: Testing programme and test results

4 Tests

The tests were all performed at Karlsruhe University. Moment and axial loading was induced as shown in figure 5. In table 1 the maximum applied loads attained as well as the parameter α and the ratio D/t are given. In figures 6 and 7 the different types of failure for stiffened and unstiffened L-joints can be seen. The test specimens with the stiffener plate maintain the original section in the joint area due to the high plate stiffness. The failure is associated with plastic deformation in the compression zone. The unstiffened test specimen ovalizes and a buckling failure occurs. With increasing D/t ratios the buckling becomes more and more predominant.

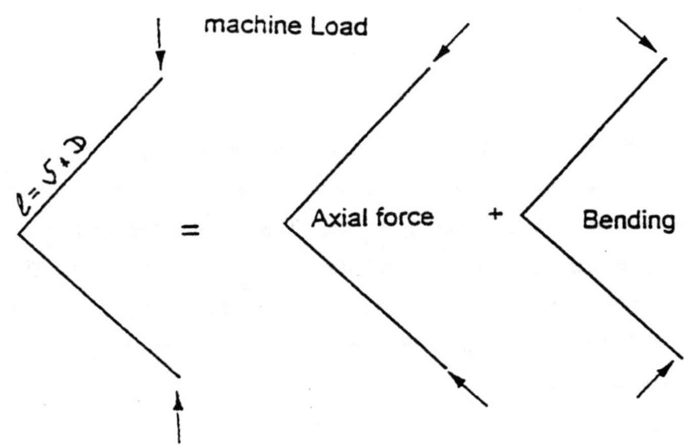

Figure 5: Components of the applied load

Figure 6: stiffened L-joint after failure

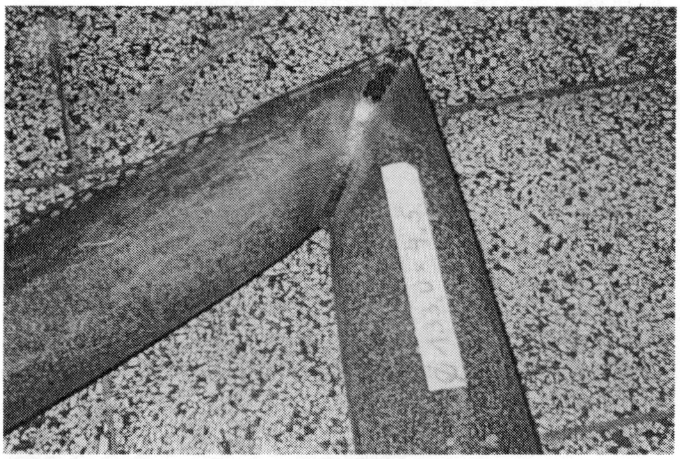

Figure 7: unstiffened L-joint after failure

5 FE - calculation

Most of the test specimens are also simulated by FE modelling. Figure 8 shows a typical FE mesh used for the investigations. Eight noded solid brick elements are used to model the complete geometry. However, only for the weld, 6 noded solid wedge elements are used. Figure 9 shows a cross section of a part of the joint. The measured dimensions as well as the material properties determined from the tensile tests are used to carry out the numerical investigations. A non-linear material behaviour is used. In table 2 a comparison of the maximum loads attained in the tests with the numerically determined maximum loads can be seen.

The FE ultimate loads of only 2 specimens, unstiffened test Nos. 9 and 10 do not give a good agreement with the test loads. There are two reasons for this fact:

- These tubes have a large D/t ratio and therefore the local buckling behaviour is very predominant, so that the mesh has to be refined further. In the further work [11], numerical investigations on the buckling behaviour are planned.

- On the other hand these test specimens have very small wall thicknesses (2.9 and 5.0 mm) in comparison to their diameters, so that fabrication (residual stresses) and local imperfections have a large influence in the ultimate loads.

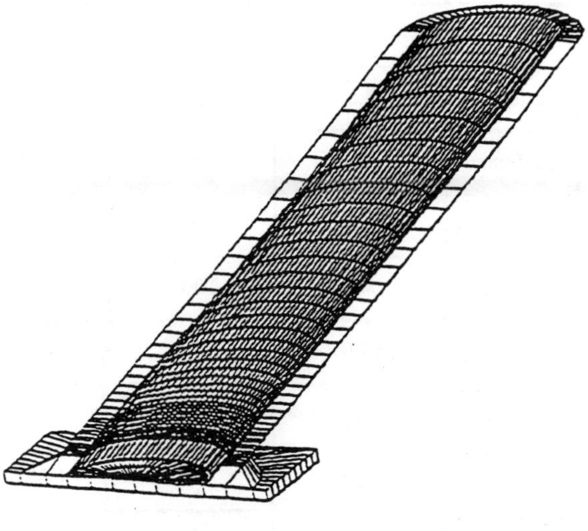

Figure 8: Example for an FE mesh

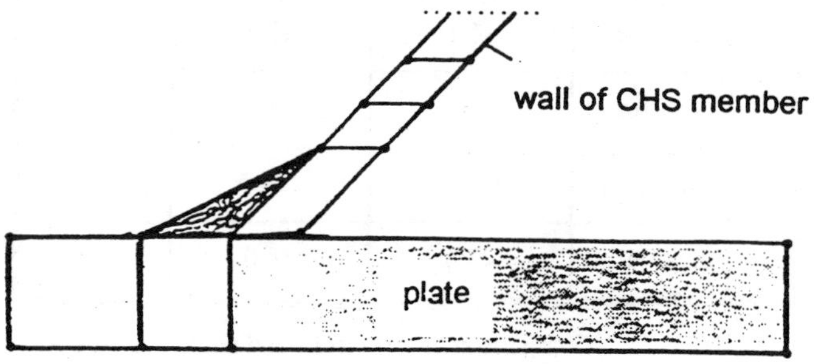

Figure 9: Cross section of a part of the joint

Typical examples of deformed mesh for the joint with and without a 8stiffened plate can be seen in figures 10 and 11. It coincides with the type of failure observed in the tests. The FE-analysis exhibited the same deformations as shown in all the tests.

After verification of the FE-models, parametric studies have been carried out. Altogether, 43 FE-models are generated and analysed. The

D/t ratio for these models ranges from 10 to 100. The maximum loads of the models under the described load conditions are graphically shown in figure 12 as a relationship between the nondimensional parameters α and D/t. As already mentioned, the factor α depends on the D/t ratio. This relationship is the main basis of the design proposal in chapter 6.

Test no.		max. load $_{test}$ [kN]	max. load $_{FE}$ [kN]	$\frac{load_{FE}}{load_{test}}$
1	stiffened	33.2	32.9	0.991
	unstiffened	15.1	14.4	0.963
3	stiffened	23.5	24.3	1.034
	unstiffened	8.5	9.1	1.071
4	stiffened	47.2	48.0	1.017
	unstiffened	17.9	18.0	1.000
5	stiffened	94.9	100.4	1.058
	unstiffened	57.9.	53.8	0.929
6	stiffened	46.8	53.2	1.137
	unstiffened	21.8	23.2	1.064
7	stiffened	66.0	68.4	1.036
	unstiffened	25.0	26.2	1.048
9	stiffened	44.1	47.3	1.073
	unstiffened	7.8	14.8	1.900
10	stiffened	103.7	109.6	1.057
	unstiffened	23.4	38.0	1.624

Table 2: Comparison of test results and numerical results

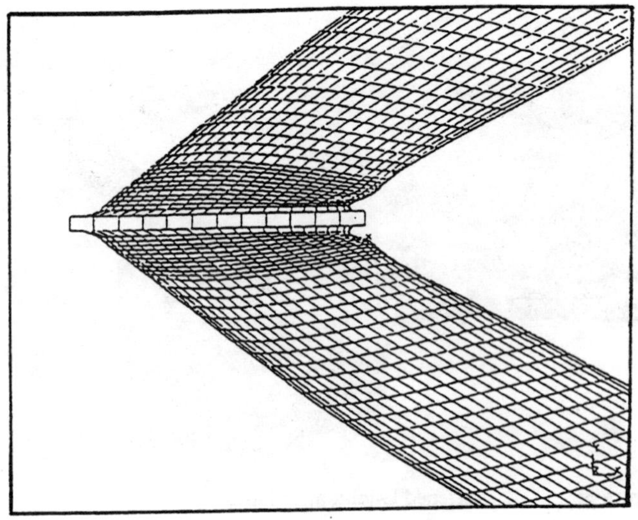

Figure 10: Example of a deformed FE-mesh with stiffening plate

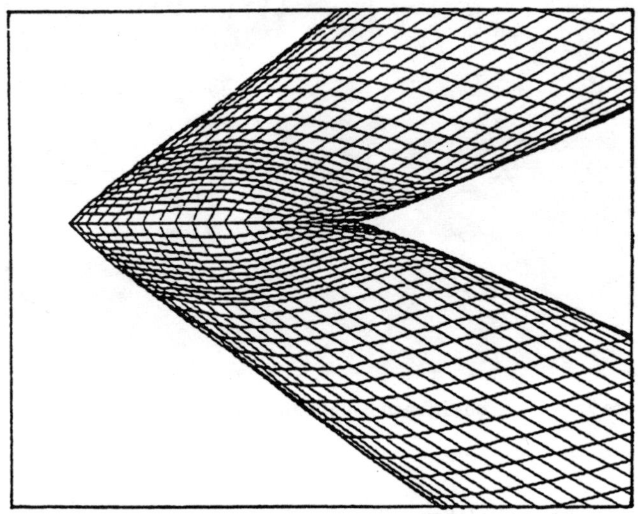

Figure 11: Example of a deformed FE-mesh without stiffening plate

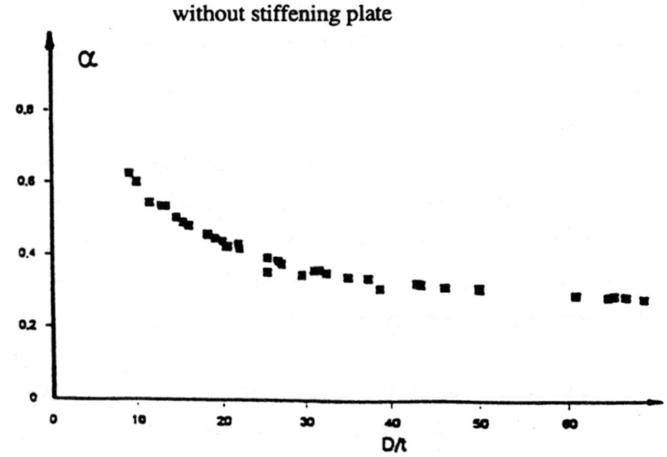

Figure 12: Dependence of the factor α on the ratio D/t (compressive load)

6 Design

Based on the results of the tests and the FE-analysis the following recommendations are made:

- members in stiffened joints can be designed as beam elements using elastic or plastic modulus depending on the ratio D/t
- a maximum limit of 40 for D/t is suggested for unstiffened joints, so that the design graph is cut off at D/t = 40 (figure 13)
- the bearing capacity of the unstiffened L-joints is the bearing capacity of the stiffened L-joint reduced by the factor α

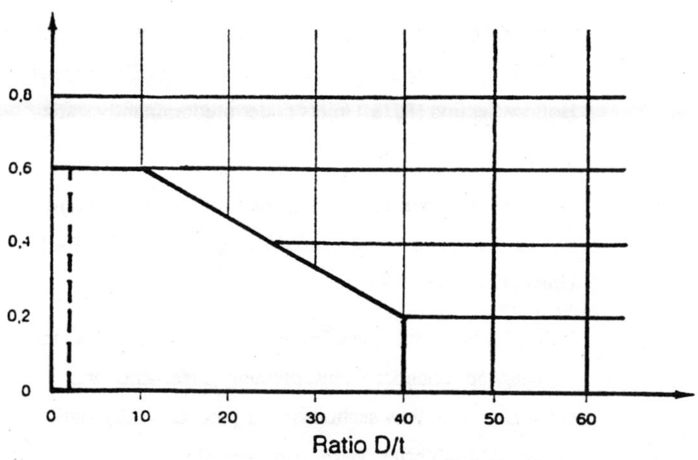

Figure 13: Design graph for unstiffened L-joints made of CHS

6.2 Design Example

$M_k = 2$ kNm (characteristic value)

CHS 193.7 x 5.0 (D x t) Fe 360 $W_{(Elastic)} = 136.0\ cm^3$

$$\Rightarrow \frac{D}{t} = 38.7$$

$\Rightarrow \alpha \cong 0.21$ (determined from the design curve in figure 13)

$$\sigma_{R,d} = \frac{f_{y,k}}{\gamma_M} = \frac{240}{1.1} = 218\ \frac{N}{mm^2}$$

$$\frac{M_k \cdot \gamma_F}{W} = \frac{M_d}{W} \qquad (\gamma_F = 1.35)$$

L - joint with stiffening plate:

$$\frac{M_d}{W} = \frac{2\ kNm \cdot 1.35}{136\ cm^3} = \frac{270\ kNcm}{136\ cm^3} = 19.9\ \frac{N}{mm^2} \le \sigma_{R,d} = 218\ \frac{N}{mm^2}$$

$\Rightarrow$ O.K.

L - joint without stiffening plate:

$$\frac{M_d}{W} = \frac{2\ kNm \cdot 1.35}{136\ cm^3} = \frac{270\ kNcm}{136\ cm^3} = 19.9\ \frac{N}{mm^2} \le \alpha \cdot \sigma_{R,d}$$

$$\alpha \cdot \sigma_{R,d} = 0.21 \cdot 218 \frac{N}{mm^2} = 45.8 \frac{N}{mm^2} \quad \Rightarrow \text{O.K.}$$

The index in the example are due to EC 3 [9].

7 Conclusions

The work presented describes FE studies, calibrated with experimental tests. A non-linear material law based on tensile test data is used. The ultimate loads of the FE and experimental models are determined and presented in Table 2 for comparison.

Parameter studies have also been carried out. A clear dependence between the parameters α and D/t can be observed in Figure 12. A simple (initial) design concept is developed based upon this relationship similiar to the design formulae for RHS L-joints. In a new DFG research programme, dynamic tests on L-joints have been proposed to determine the fatigue behaviour. Design recommendations using stress concentration factors (SCF) will be developed.

The results of the further work will clearify and perhaps modify the design proposal given in this paper for D/t ratios > 40. The FE-models are also analysed by using joint opening loads, as opposed to joint closing forces described upto now (see figure 5). The dependence between the parameters α and D/t is not so distinctive as for joint closing loads (see figures 12 and 14). Work is also progressing to obtain more information on this matter

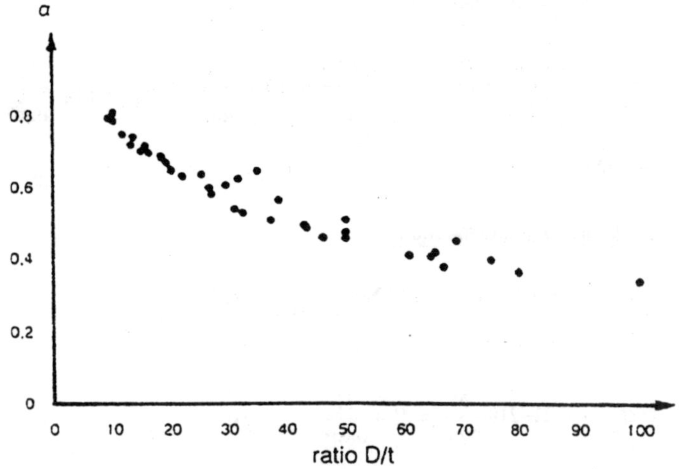

Figure 14: Dependence of the factor α on the ratio D/t (tension load)

8 References

[1] Mang, F., Bucak Ö., Knödel, P.: Ermittlung des Tragverhalten von biegesteifen Rahmenecken aus Rechteck-Hohlprofilen (St 37, St 52) unter statischer Belastung. AIF-research program No. 4117, Final report, April 1981, Studiengesellschaft für Anwendungstechnik Stahl und Eisen, Düsseldorf.

[2] Mang, F.: Tragverhalten von biegesteifen Rahmenecken aus Rechthohlprofilen der Werkstoffe AlZn4.5Mg1und AlMgSi0.5 unter statischer und schwingender Belastung. Final report of the AIF-research program, September 1986.

[3] Mang, F., Bucak, Ö., Herion, S.: Zum Tragverhalten von biegesteifen Rahmenecken aus Rechteckhohlprofilen der Werkstoffe AlZn4.5Mg1 und AlMgSi0.5. Publication No. 89 of the DVS (German welding association), pages 188-204, Düsseldorf.

[4] Mang, F., Steidl, G., Bucak, Ö.: Design of welded lattice joints and moment resisting knee joints made of hollow sections. International Institute of Welding Doc. XV-436-80, University of Karlsruhe, Germany, 1980.

[5] Dutta, Kurobane, Packer, Wardenier, Yeomans: Constructions with Hollow Steel Sections, Design Guide for Circular Hollow Section (CHS) joints under predominantly static loading. CIDECT, Köln 1991.

[6] Dutta, Kurobane, Packer, Wardenier, Yeomans: Constructions with Hollow Steel Sections, Design Guide for Rectangular Hollow Section (RHS) joints under predominantly static loading. CIDECT, Köln 1992.

[7] Mang, F., Karcher, D.: L-joints made of Circular Hollow Sections. Final report of CIDECT project No. 5 BE, Karlsruhe University, June 1996

[8] DIN 18 808: Stahlbauten, Tragwerke aus Hohlprofilen unter vorwiegend ruhender Beanspruchung (Steel structures consisting of hollow sections under predominantly static loads). Beuth-Verlag, Berlin, Oktober 1984.

[9] Eurocode 3: Design of steel structures, part 1.1: General rules and rules for buildings. 1992.

[10] Karcher, D.: Tragverhalten von biegesteifen Rahmenecken aus Rundhohlprofilen. PhD Thesis at the University of Karlsruhe, to be completed in 1998

Proceedings of the Seventh (1997) International Offshore and Polar Engineering Conference
Honolulu, USA, May 25-30, 1997

ISBN 1-880653-28-1 (Set); ISBN 1-880653-32-X (Vol. IV)

Additional Tests of CHS Planar KK-Joints Under Anti-Symmetrical Axial Loads

Y. Makino, S.R. Wilmshurst and Y. Kurobane
Kumamoto University
Kumamoto, Japan

A. Yoshida
Wakachiku Construction Co., Ltd.
Tokyo, Japan

ABSTRACT

In a former paper (Yonemura et al. (1996)), seven uniplanar KK-joints (the angle 2φ between two braces equal to 180 degrees) were tested under anti-symmetrical loads to expand the validity range for the ultimate capacity formula and to further investigate the ultimate behaviour of tubular joints. However, there were doubts about the validity of two of the tests due to welding defects. Two additional specimens were tested in this study, and the ultimate behaviour of incomplete welding joints is discussed.

The ultimate capacities of uni-planar KK-joints under anti-symmetrical axial loads were compared with those of planar K-joints. It was clear from the results that the ultimate capacity of uni-planar KK-joints is nearly equal to that of planar K-joints.

Furthermore, ultimate capacities of all KK-joint test results under anti-symmetrical loads, not only uniplanar but also multi-planar, were discussed and compared with the Kurobane (1984) K-joint formula and the KK-joint formula of Makino et al (1995).

KEY WORDS: Tests, Tubular joints, KK-joints, Anti-symmetrical loads, Ultimate strength

Introduction

In two of the tests of Yonemura et al (1996), it was reported that there was insufficient melting of the weld metal, and therefore it was doubted whether they developed their full joint and deformation capacities. The two new specimens, which are same size as the former tests, were manufactured by a different factory, and were tested and compared the ultimate behaviour between former and new tests. All tests, including the former seven tests, are reported to discuss the ultimate behaviour of the uniplanar KK-joints under anti-symmetrical axial loads in this paper.

Test Specimens

A uniplanar KK-joint specimen is shown in Fig. 1 with definitions

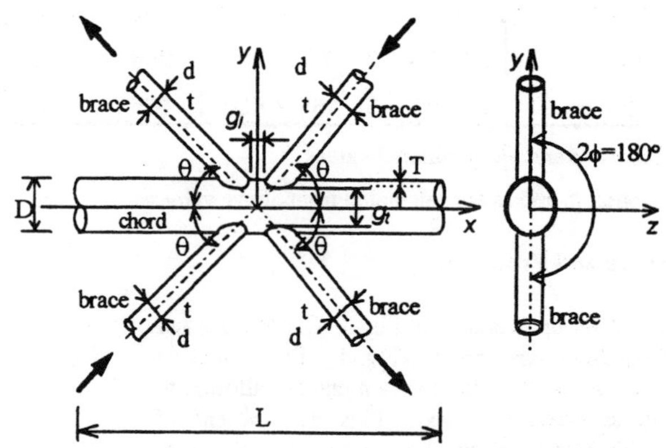

Fig. 1 CHS Planar KK-Joints under anti-symmetrical loads

DKA□ − □□ −H −1
- 1: Additional tests
- H: Hot finished tube
- d/D: none=0.36, 1=0.53, 2=0.65, 3=0.76
- D/T: L=50, M=36, H=27
- 5: 2φ=180°, θ=45°; 6: 2φ=180°, θ=60°
- DKA: Double K-joints under anti-symmetrical loads

Fig. 2 Definition of specimen name

of the symbols. The measured dimensions of the specimens are given in Table 1. In the specimen designations, the symbol DKA means double K-joints under anti-symmetrical loads. As shown in Fig. 2, the following number denotes the combination between angle θ. The following letters denote the chord thickness; L, M and H means light, medium and heavy chords, respectively. The next number denotes the β ratio. The next letter H shows if the chord is made of a hot rolled tube. When the letter H is not present, a cold-formed tube is used. The last number 1 shows the additional tests in this study.

To avoid effects of the chord length, L was designed to be 8 times the chord diameter. The brace length between the panel point and the brace end measured 500 mm. The panel point is the point where the brace centre-lines intersect the chord tube outer surface. In order that the braces did not buckle before joints reached the maximum capacities, thick-walled high strength steel tubes were used. The material properties of the chord and braces are given in Table 1.

The end profiles of the braces were prepared by automatic flame cutting, and the weld details were in accordance with Recommendations for the Design and Fabrication of Tubular Structures in Steel (AIJ, 1990).

Table 1 Test specimens (measured)

Specimen	$D \times T$	$d \times t$	Chord σ_y	Chord σ_u	Brace σ_y	Brace σ_u	2ϕ	θ	g_l	g_t	d/D	D/T
	(mm)		(N/mm^2)		(N/mm^2)		(degree)		(mm)			
DKA5-M2-H-1	216.1×6.4	140.3×4.6	342	485	366	647	180	45	16.5	165.1	0.65	33.8
DKA5-L3-1	216.8×4.3	165.2×5.6	379	474	427	625	180	45	-17.0	140.1	0.76	50.4
DKA5-M	216.8×5.5	76.9×4.0	330	436	494	631	180	45	108.1	202.7	0.35	39.3
DKA5-M1	216.3×5.8	114.4×4.5	367	474	438	596	180	45	54.5	183.5	0.53	37.0
DKA5-M2	216.8×5.5	139.5×4.1	330	436	366	647	180	45	19.5	165.9	0.64	39.3
DKA5-M2-H	216.5×6.3	140.3×4.6	369	504	366	647	180	45	18.2	165.0	0.65	34.3
DKA6-M	216.5×5.4	76.9×4.0	332	432	494	631	180	60	36.2	202.4	0.36	40.2
DKA6-M-H	216.5×6.3	76.3×4.0	369	504	518	658	180	60	36.9	202.6	0.35	34.3
DKA5-L3	216.0×4.3	165.2×5.6	405	532	427	625	180	45	-17.6	139.2	0.76	50.3

Note1: ϕ and θ are nominal values.

Note2: g_l and g_t are calculated from above values.

Test Set-Up and Procedure

The test set-up is shown in Fig. 3. The loading and support conditions are shown in Fig. 4. Pins at tension brace ends and a pin at tension ram end are allowed to rotate about x-axis and y-axis (Figs. 4). One end of compression braces are fixed to a ram end with a load cell. This compression ram at the brace end was supported by a loading rig, so that a compressive load was applied to the end of the other compression brace. One compression brace end was allowed to rotate in all directions and the other was fixed to the loading rig to supply load stability. The chord was connected by thin plate bracing at each end to the loading rig, to prevent rigid body rotations of the chord in the horizontal plane during testing and, also, to prevent out-of-plane buckling of the compression brace.

The loads were applied proportionally to the tension and compression braces until the maximum load by using 100 ton hydraulic rams. After the peak load, the displacement of the hydraulic ram of the tension brace was fixed, since proportional load control is difficult when negative stiffness occurs. As a result, the loads were not applied proportionally after the maximum load. The loads were measured by dynamometers mounted at the hydraulic ram ends. Relative displacements between the both tension and compression brace ends were measured, as a substitute for local deformations of joints (Fig. 3).

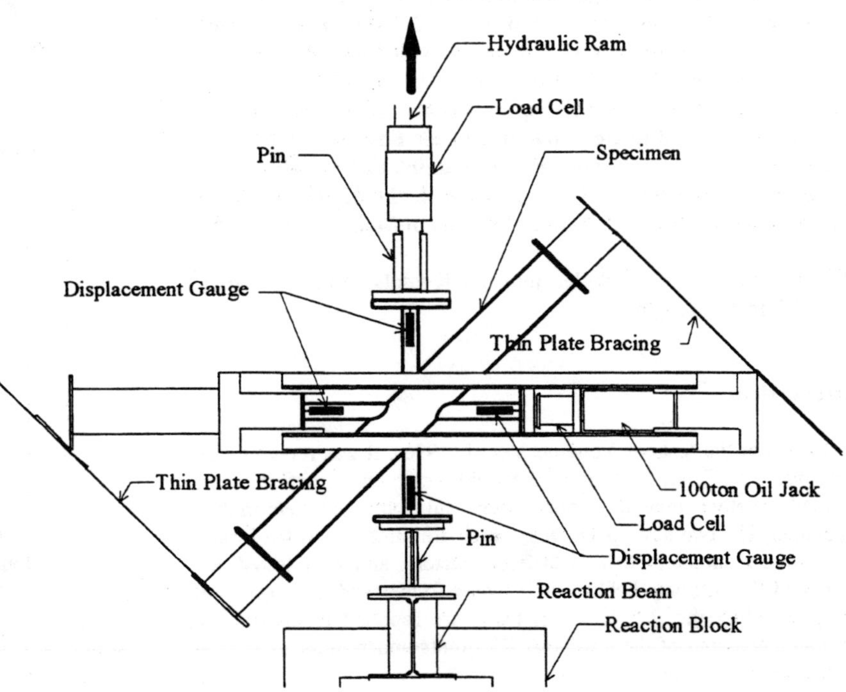

Fig. 3 Plan view of test set-up and measuring system

Summary of Test Results

The load-deformation relationships of all specimens except DKA5-M2-H and DKA5-L3 are shown in Fig. 5, and all test results are shown in Table 2.

The load-deformation curves for joints DKA5-M, DKA5-M1, DKA6-M and DKA6-M-H show that they reach the first peak loads, then become unstable, and deformations increase showing large deflections of the chord walls. Subsequently the joints show increased stiffness, as was observed in planar K-joint tests. Even if these joints sustained a further increase in load and reached a load greater than the first maximum load, the ultimate capacity is defined as the first maximum load (as in former research). The ultimate capacities are denoted by the circles in Fig. 5. One of the characteristics of these joints is seen in specimens that have the ratio (g_t/D) greater than 0.64 (see Table 2). In these specimens, cracks with certain depth occurring at weld toes on the chord surface were observed after first peak loads.

The failure of DKA5-M2 is caused by the crack growth with sufficient joint deformation as DKA5-M2-H-1, but the deformation at ultimate capacity of DKA5-M2 is smaller than that of DKA5-M2-H-1 in spite of same geometric and material properties. The capacity of DKA6-M is slightly lower than that of DKA6-M-H. In this comparison between hot-finished and cold-finished tubes (DKA5-M2-H-1, DKA5-M2, DKA6-M-H and DKA6-M), it is possible that the deformation at ultimate capacity of hot-finished tube is greater than that of cold-finished tube, due to higher ductility.

The pictures of joint deformation after testing are shown in Photos 1, 2 and 3. Although the centroidal axis of the chord was distorted and the brace axes were also distorted and curved in the joint area as shown in Photo 1, the local deformations in planar KK-joints after testing were similar to those in planar K-joints.

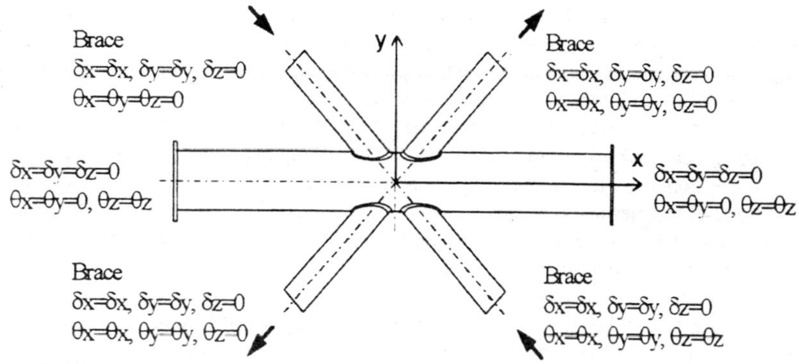

Fig. 4 Forces and support conditions in the anti-symmetrical loading test

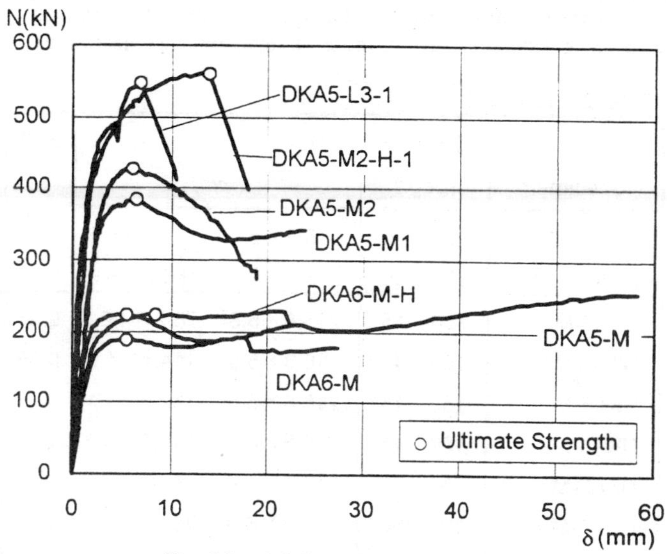

Fig. 5 Load-deformation relationships

Comparison with Former Test Results

The load-deformation relationships of the additional tests are shown in Figs. 6 and 7 and compared with the former tests. The load represents that recorded by the dynamometers at the compression brace ends, and the deformation represents the relative displacement

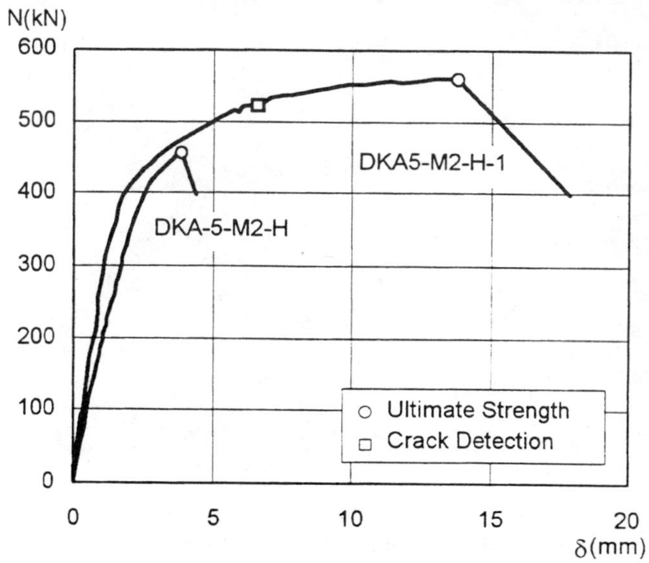

Fig. 6 Load-deformation relationships of DKA5-M2-H and DKA5-M2-H-1

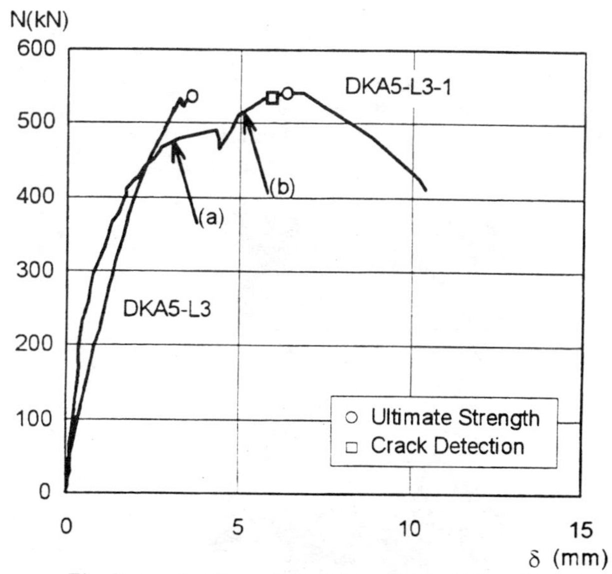

Fig. 7 Load-deformation relationships of DKA5-L3 and DKA5-L3-1

between the compression brace ends. In the load-deformation curve of DKA5-L3-1 (Fig.6), the sudden changes in the load-displacement behaviour were due to buckling of the thin-plate bracing (region (a) on the curve) followed by load redistribution with an increase in stiffness (region (b) on the curve). It should be noted that for this specimen γ=17 and g_l/D=-0.08 which indicate a very stiff overlapping joint.

As mentioned in the former paper, in specimens DKA5-L3 and DKA5-M2-H (former tests) cracks occurred at weld toes of the tension brace to chord welds before ultimate capacities were achieved, and failed suddenly without sufficient joint deformation. However, in this study, although cracks also occurred at weld toes in specimens DKA5-M2-H-1 and DKA5-L3-1, joint deformations at the peak load were longer than in the former tests. Crack initiation was carefully observed during these tests. The square symbol on DKA-M2-H-1 in Fig. 6 shows the observed crack detection point by eye. After the crack initiation, the load increased and the crack grew until the peak load same as some of planar K-joints. Although the final joint failure was due to crack extension, the ultimate strength was not necessarily controlled by cracks. In Wang et al. (1997), the relationship between the crack extension and the ultimate strength of specimen DKA5-M2-H-1 is discussed in detail using crack model FE analysis.

The failure cause of specimen DKA5-L3-1 (which is an overlapped joint) was chord deformation of saddle area as shown in Photo 2. The cracks initiated at the weld toes on the chord surface on the opposite side of lap area, and were detected just before the peak load. But the cracks did not extend as far as in specimen DKA5-M2-H-1. The ultimate strength was, once more, controlled by local deformation of joints.

The non-dimensionalised ultimate strengths ($N_u sin\theta / \sigma_y T^2$) of DKA5-M2-H-1 and DKA5-L3-1 are about 18% and 10% higher than those of DKA5-M2-H and DKA5-L3, respectively.

In the four test specimens, welding was difficult to perform owing to narrow space between the braces, and, as a result of insufficient melting of weld metal and defective welding, joint capacities can be lower than expected.

Table 2 Test results

Specimen	2ϕ (deg.)	θ (deg.)	$D \times T$ (mm)	$d \times t$ (mm)	g_l/D	${}_tN_u$ (kN)	${}_cN_u$ (kN)	δ_u (mm)	Failure Type	Comment
DKA5-M2-H-1	180	45	216.1×6.4	140.3×4.6	0.76	561	560	13.8	a , c	this test
DKA5-L3-1	180	45	216.8×4.3	165.2×5.6	0.64	543	543	6.4	a , b	this test
DKA5-M	180	45	216.8×5.5	76.9×4.0	0.93	253	223	4.8	a, b	
DKA5-M1	180	45	216.3×5.8	114.4×4.5	0.85	392	385	6.5	a, b	
DKA5-M2	180	45	216.8×5.5	139.5×4.1	0.77	437	431	6.0	a, b	
DKA5-M2-H	180	45	216.5×6.3	140.3×4.6	0.76	466	459	3.8	a, c	insufficient melting of weld metal
DKA6-M	180	60	216.5×5.4	76.9×4.0	0.93	210	190	5.5	a, b	
DKA6-M-H	180	60	216.5×6.3	76.3×4.0	0.94	247	224	8.4	a, c	
DKA5-L3	180	45	216.0×4.3	165.2×5.6	0.64	539	544	3.5	a, b/c	insufficient melting of weld metal

${}_cN_u$: Axial force in compression brace at ultimate load

${}_tN_u$: Axial force in tension brace at ultimate load

δ_u : Deformation at ultimate capacity

Failure type

a : Local deformation of chord wall
b : Crack at weld toe on the chord surface
c : Crack at weld toe on the tension brace surface in longitudinal gap region

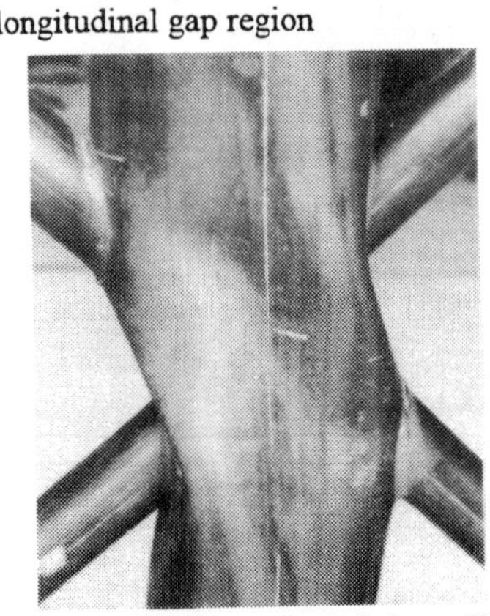

Photo 1 Overall deformation of planar KK-joints

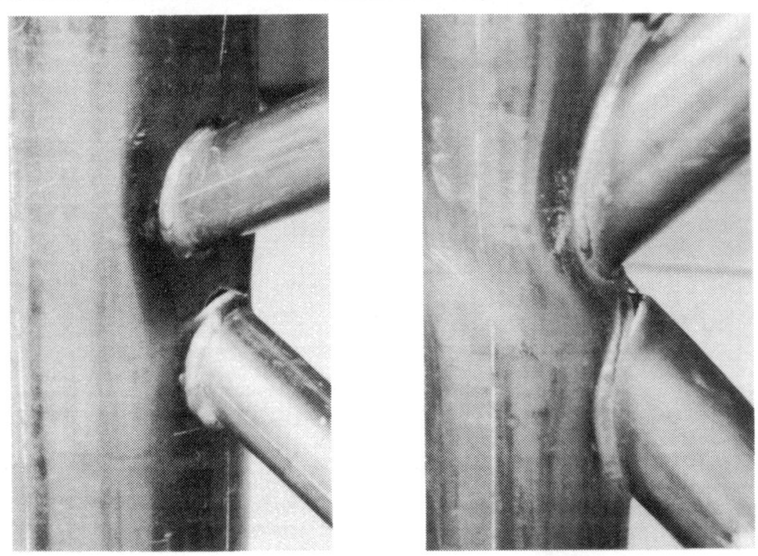

a and *b*
Local deformation of chord wall and crack at weld toe on the chord surface

a and *c*
Local deformation of chord wall and crack at weld toe on the tension brace surface in longitudinal gap region

Photo 2 Failure types

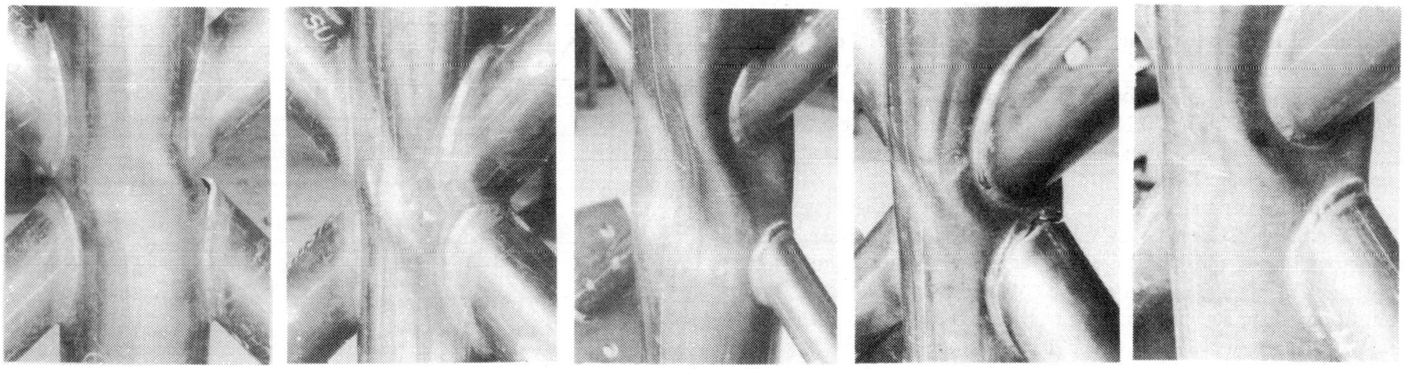

DKA5-M2-H-1 DKA5-L3-1 DKA5-M DKA5-M1 DKA5-M2

Photo 3 Joint deformation after testing

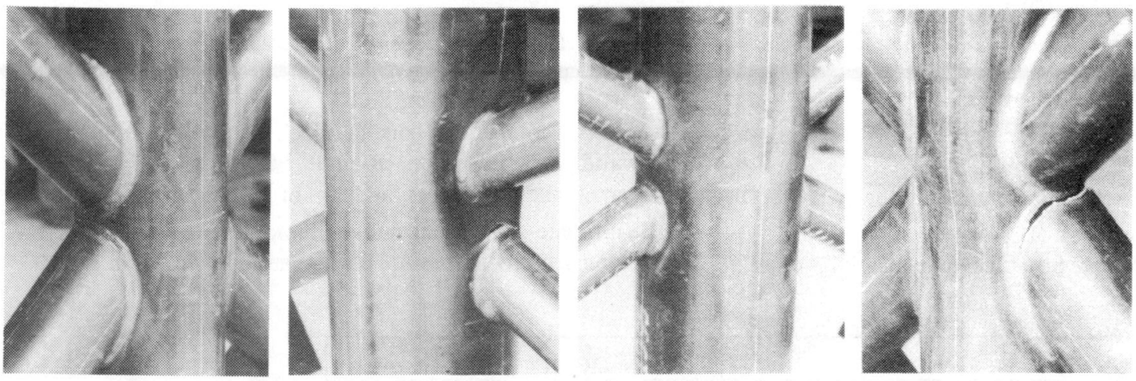

DKA5-M2-H DKA6-M DKA6-M-H DKA5-L3

Photo 3 (continued) Joint deformation after testing

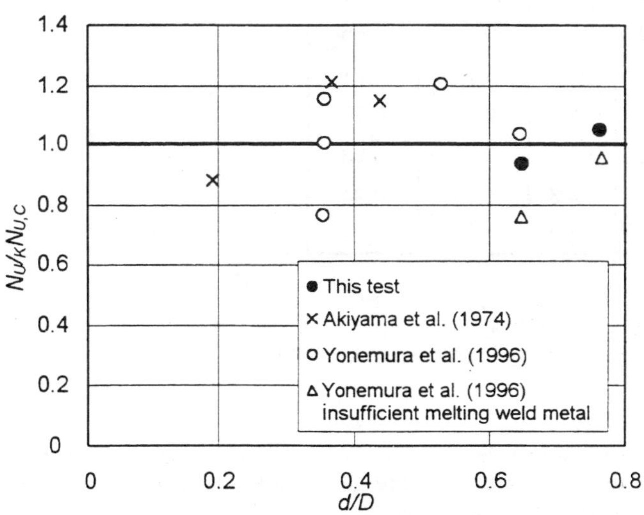

Fig. 8 Test to prediction ratio using planar K-joint formula plotted against d/D

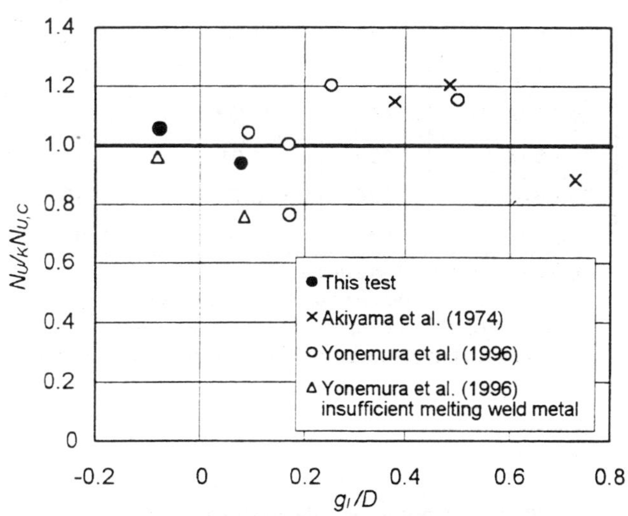

Fig. 9 Test to prediction ratio using planar K-joint formula plotted against g_l/D

Discussion of Ultimate Strength

In Table 3, test results are compared with the ultimate capacities predicted using the planar K-joint formula of Kurobane et al. (1984). The ratios of the test to prediction strength is shown as functions of d/D and g_l/D in Figs. 8 and 9, respectively. The test to prediction ratios according to planar K-joint formula vary from 0.76 to 1.21, with a mean of 1.01 and a COV of 0.154. It can be described that the ultimate strength of uniplanar KK-joints under anti-symmetrical loads is almost equal to that of planar K-joints.

Another comparison is shown as a function of g_t/D in Fig. 10, which includes test results for multi-planar KK-joints under anti-symmetrical loads (Makino et al. 1994), previous test results for uniplanar KK-joints (Akiyama et al. 1974) and numerical results for multiplanar KK-joints under anti-symmetrical loads (Wilmshurst 1995, Lee and Wilmshurst 1996). Crack locations are indicated by the symbols shown in the figure. Makino et al (1995) proposed an ultimate capacity formula for KK-joints under anti-symmetrical axial loads as follows, (also shown as the solid line in Fig. 10):

$$_{AKK}N_{u,c} = (1.36 - 3.17 \cdot g_t/D) \cdot {}_K N_u$$
$$g_t/D \le 0.157$$
$$_{AKK}N_{u,c} = 0.858 \cdot {}_K N_u$$
$$g_t/D \ge 0.157$$

In here, $_K N_u$ is the predicted ultimate capacity for planar K-joints according to Kurobane et al. (1984).

In the derivation of the above formulae, uniplanar KK-joints test results were not used, due to a lack of test data at that time. The validity range of out-of-angle between two braces was between 60 and 90 degrees, and that of g_t/D was less than 0.42. If the ultimate capacity formulae are applied to uniplanar KK-joint, strength predictions are conservative as shown in Fig.10.

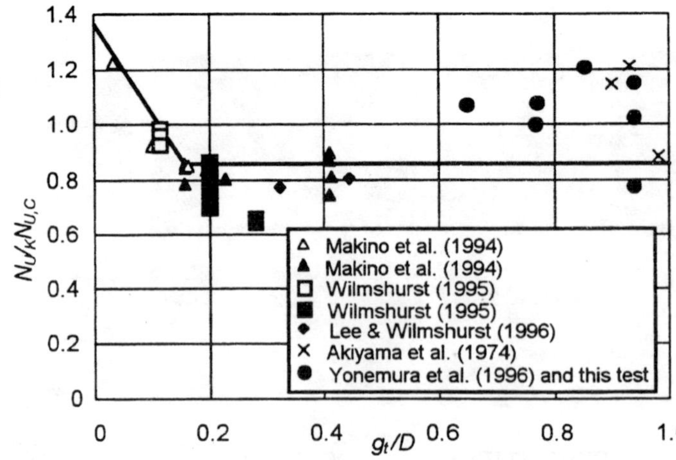

Hollow Symbol : crack occurred in transverse gap area
Solid Symbol : crack occurred in longitudinal gap area

Fig. 10 Test to prediction ratio using planar K-joint formula plotted against g_t/D

Table 3 Ultimate strength comparison between tests and predictions

Specimen	$N_u(kN)$	$N_u/_KN_{u,c}$	$N_u/_{AKK}N_{u,c}$	d/D	D/T	g_l/D	g_t/D
DKA5-M2-H-1	560	0.94	1.10	0.65	33.8	0.08	0.76
DKA5-L3-1	543	1.05	1.23	0.76	50.4	-0.08	0.64
DKA5-M	223	1.16	1.35	0.35	39.3	0.50	0.93
DKA5-M1	385	1.20	1.40	0.53	37.0	0.25	0.85
DKA5-M2	431	1.04	1.21	0.64	39.3	0.09	0.77
DKA5-M2-H	459	0.77*	0.90*	0.65	34.3	0.08	0.76
DKA6-M	190	1.01	1.17	0.36	40.2	0.17	0.93
DKA6-M-H	224	0.77	0.89	0.35	34.3	0.17	0.94
DKA5-L3	544	0.95*	1.10*	0.76	50.3	-0.08	0.64

* is insufficient melting of weld metal

N_u : Axial force in compression brace at ultimate load

$_KN_{u,c}$: Predicted ultimate capacity for planar K-joints according to Kurobane et al. (1984)

$_{AKK}N_{u,c}$: Predicted ultimate capacity for KK-joints under anti-symmetrical load according to Makino et al. (1995)

Conclusions

The two additional test results with the former tests are reported in this paper, and the joint deformation of the insufficient melting of weld metal was discussed. The ultimate capacities of uni-planar KK-joints under anti-symmetrical axial loads were compared with those of planar K-joints. The following conclusions may be summarized.

1. Narrow gaps that can not ensure an adequate space for welding are not recommended. It leads to a lack of penetration causing a crack failure before the ultimate strength is reached; also these joints have insufficient deformation capacities.

The ultimate capacities of these joints are much lower. The ultimate strengths decreased by 10% and 18% due to insufficient melting of the weld metal in these tests.

2. The local deformation patterns of uniplanar KK-joints under anti-symmetrical loads are similar to those of planar K-joints.

3. The ultimate strength for uniplanar KK-joints under anti-symmetrical loads can be predicted by the Kurobane (1984) K-joint formula.

4. In a comparison with two different material tube tests, it appears that the deformation at ultimate capacity of joints with cold finished chord tubes are lower than those of hot-finished sections.

5. When the Makino (1995) formulae for multiplanar KK-joints are applied to uniplanar KK-joints, the predicted ultimate capacities are conservative.

Acknowledgment

This study is supported by the Grant-in-Aid for Scientific Research No. 08044160 from the Japanese Ministry of Education, Science.

References

Akiyama, N., Yajima, M., Akiyama, H., and Ohtake, A. (1974). "Experimental Study on Strength of Joints In Steel Structures", Journal of Japanese Soc. of Steel Construction, Vol. 10, No. 102, pp. 37-68 (in Japanese)

Architecture Institute of Japan, (1990). "Recommendations for the Design and Fabrication of Tubular Structures In Steel", AIJ, Tokyo. Japan (in Japanese)

Kurobane, Y. Makino, Y. and Ochi, K. (1984). "Ultimate Resistance of Unstiffened Tubular Joints," *Journal of Structural Engineering, ASCE*, Vol. 110, No 2, pp. 385-400

Lee. M. M. K. and Wilmshurst, S. R., (1996). "Strength of Multiplanar Tubular KK-Joints under Anti-Symmetrical Axial Loading", Proc. of the 7th Int. Symposium on Tubular Structures, Miskolc, Hungary, pp.149-156.

Makino, Y. and Kurobane, Y., (1994). "Tests on CHS KK-joints under Anti- symmetrical Loads", Proc. of 6th Int. Symposium on Tubular Structures, Melbourne, Australia, pp. 449-456

Makino, Y., Kurobane, Y., Wilmshurst, S. R. and Lee, M.M.K., (1995). "Proposed Ultimate Capacity Equations for CHS KK-Joints under Anti- Symmetrical Loads", Proc. of the 5th International Offshore and Polar Engineering Conference, The Hague, The Netherlands, Vol. IV, pp. 6-11

Paul, J. C., Makino, Y. and Kurobane, Y., (1994). "Ultimate Resistance of Unstiffened Multiplanar Tubular TT- and KK-Joints", Journal of Structural Engineering, ASCE, Vol. 120, No. 10, pp. 2853-2870

Wang , B. , Kurobane , Y. and Makino ,Y. (1997). "Damage Criteria and Ultimate Strength for Tubular Joint", Proc. of 7th ISOPE, Honolulu, USA.

Wilmshurst, S. R. (1995) "The Ultimate Load Behaviour of Multi-planar KK-joints Subjected to Symmetrical and Anti-symmetrical Axial Loading", Doctoral thesis, Wales, Swansea University.

Yonemura, H., Makino, Y., Kurobane. Y. and van der Vegte, G. J., (1996) "Tests on CHS Planar KK-Joints under Anti-Symmetrical Loads", Proc. of the 7th Int. Symposium on Tubular Structures, Miskolc, Hungary, pp.189-195.

Appendix

Ultimate capacity formula for planar K-joint (Kurobane et al, 1984):

$$ {}_{K}N_{u,c} = f_0 \cdot f_1 \cdot f_2 \cdot f_3 \cdot f_4 \cdot f_5 \cdot T^2 \sigma_y $$

$$ f_0 = 2.11\left(1 + 5.66\frac{d}{D}\right) $$

$$ f_1 = \left(\frac{D}{T}\right)^{0.209} $$

$$ f_2 = 1 + \frac{0.00904\left(\frac{D}{T}\right)^{1.24}}{\exp\left(0.508\frac{g_l - g_c}{T} - 1.33\right) + 1} $$

in here $g_c = 3.04(mm)$

$$ f_3 = \frac{1 - 0.376\cos^2\theta}{\sin\theta} $$

$$ f_4 = 1 + 0.305 n_0 - 0.285 {n_0}^2 $$

in here $n_0 = \dfrac{N_0}{\pi(D-T)T\sigma_y}$

$$ f_5 = \left(\frac{\sigma_y}{\sigma_u}\right)^{-0.723} $$

Notation

D : Chord diameter
T : Chord thickness
d : Brace diameter
t : Brace thickness
ϕ : Out-of-plane angle between both braces
θ : Angle between chord and brace
g_t : Transverse gap between both braces
g_l : Longitudinal gap between both braces
${}_{K}N_{u,c}$: Predicted ultimate capacity for planar K-joint according to Kurobane et al.(1984)
${}_{AKK}N_{u,c}$: Predicted ultimate capacity for KK-joint under anti-symmetrical load according to Makino et al. (1995)

Proceedings of the Seventh (1997) International Offshore and Polar Engineering Conference
Honolulu, USA, May 25-30, 1997

ISBN 1-880653-28-1 (Set); ISBN 1-880653-32-X (Vol. IV)

Further Numerical Analyses of KK-Joints Under Anti-Symmetrical Axial Loading

S.R. Wilmshurst, Y. Makino and Y. Kurobane
Kumamoto University
Kumamoto, Japan

ABSTRACT

This paper presents the results of a parametric study of forty seven KK-joints subjected to anti-symmetrical axial loading using the finite element method. The data generation is a precursor to producing a continuous strength estimation formula that will be able to predict the strength of this and other more complex joint types. The modelling technique follows previously established guidelines and a check calibration of the model has been performed using test data for planar KK-joints. The failure modes and ultimate strengths were found to be accurately determined by the modelling method. The main parameters investigated for their effects on the ultimate capacity are the out-of-plane angle, ϕ (between 60° and 180°), the brace diameter ratio and the in-plane gap size. The results of this study show the importance of these parameters. Existing estimation formulae and the AWS strength formulation are assessed, and inaccuracies are exposed, in the light of these and previous results.

KEYWORDS: Multiplanar joints, axial loading, ultimate strength, finite element analysis.

NOMENCLATURE

Geometric parameters are shown in Figure 1. Other symbols used in this paper are given below:

P_u	Ultimate strength
P_{nd}	Non-dimensionalised strength ($P_u \sin\theta / \sigma_y T^2$)
$P_{u,K,Kur}$	Uniplanar K-joint strength according to the Kurobane (1984) K-joint formula
$P_{u,Mak}$	Ultimate strength of anti-symmetrically loaded KK-joints according to the formula of Makino *et al* (1995)
$P_{u,AWS}$	Ultimate capacity according to the AWS formulation
σ_y	Material yield stress
σ_u	Material ultimate tensile stress

INTRODUCTION

Multiplanar KK-joints are an unavoidable feature of many offshore and onshore structures. Their design traditionally follows guidelines (API (1993), HSE (1995)) adopted from the interpretation of the results of testing uniplanar joints. The standard procedure is to treat a multiplanar joint as a series of uniplanar joints and to consider each plane in isolation. This ignores the effects of the loading in, and the position of, out-of-plane members. The CIDECT Design Guide (Wardenier *et al* (1991)) and Eurocode 3 (1992), which are applicable to onshore structures, give some guidance for determining the strength of multiplanar joints. Such guidance takes the form of multiplying the uniplanar strength by appropriate correction factors, and is based upon a limited number of multiplanar joint test results. The only code that includes the effects of loading in the out-of-plane braces for an arbitrarily configured joint is the AWS code (1996). This is achieved by means of an 'ovalisation' parameter that attempts to relate the ovalisation of the chord cross-section with the loading pattern. However, the AWS formulation has been the subject of many investigations in recent years. Although the concept is a valiant attempt at reproducing the effects of loaded out-of-plane braces, it has been found that the strength predictions are inaccurate in the majority of cases. In the case of gap-joints with thick chord walls the strength predictions can be grossly unsafe.

For the KK-joint configuration (shown in Figure 1) anti-symmetrical axial loading can occur when wind or wave loading acts upon an offshore jacket or onshore truss structure. It has been found in previous work (Wilmshurst (1995), Lee and Wilmshurst (1996)) that the majority of KK-joints loaded in this manner are weaker than their uniplanar counterparts and, due to design practice, this can result in unsafe design. Makino *et al* (1995) proposed correction factors (Eqs. 1 and 2) that could be applied to the planar K-joint equation of Kurobane *et al* (1984) for estimating the ultimate capacity of KK-joints under anti-symmetrical loading:

$$P_{u,MAK} = \left(1.36 - 3.17\frac{g_t}{D}\right) P_{u,K,Kur} \qquad \text{for } \frac{g_t}{D} < 0.157 \qquad \text{(Eq. 1)}$$

$$P_{u,MAK} = 0.858\, P_{u,K,Kur} \qquad \text{for } \frac{g_t}{D} \geq 0.157 \qquad \text{(Eq. 2)}$$

However, the limited data upon which the correction functions were based dictated that the formula was only tentative, and that a more

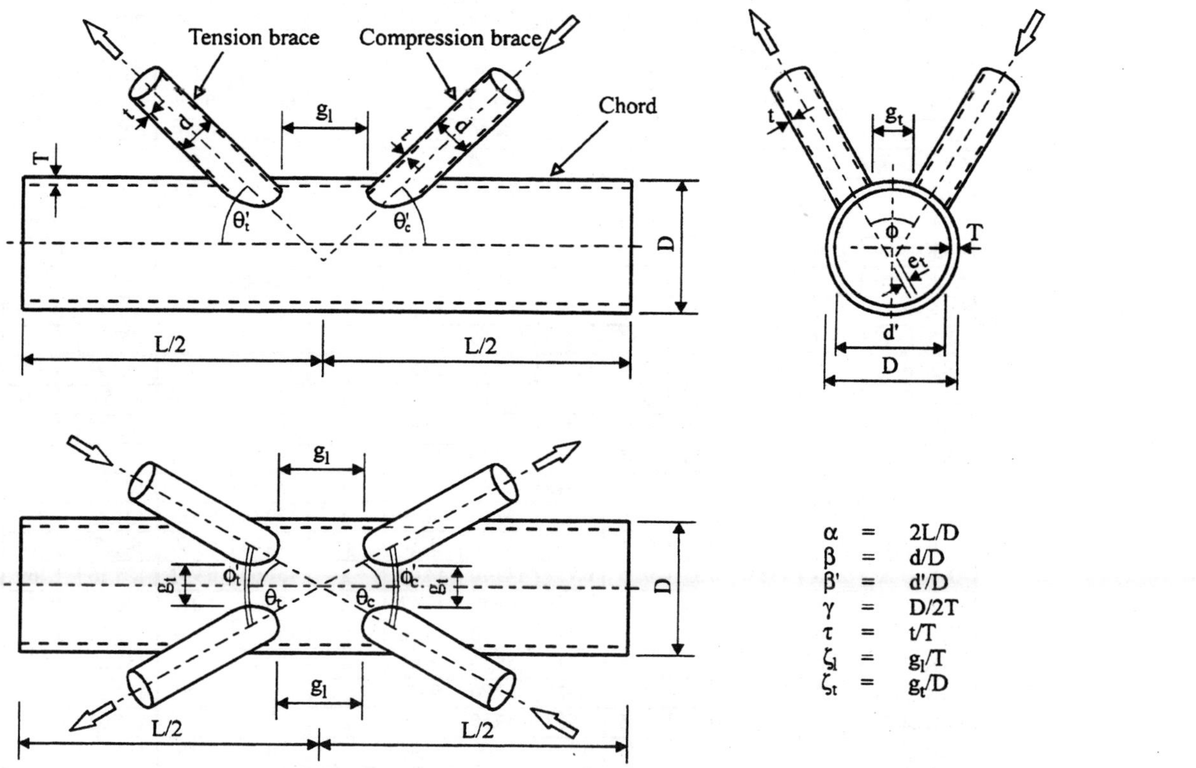

Figure 1 KK-joint configuration under anti-symmetrical axial loading

definitive formulation would be needed when more data became available. Lee and Wilmshurst (1996) found that, when $\zeta_t>0.2$, the capacity of anti-symmetrically axially loaded KK-joints increased with the out-of-plane angle, ϕ, in the range of $60°<\phi<90°$. Therefore, it would appear that any strength estimation formula for anti-symmetrically loaded KK-joints should include the parameter ϕ.

A study is being currently undertaken at Kumamoto University to derive a new continuous ultimate capacity formula for tubular joints. The formulation has, so far, related T-, TT-, and K-joints under compressive axial loading (Morita *et al* (1996,1997)). A future stage will be to include KK-joint configurations under symmetrical and anti-symmetrical loading. For KK-joints under anti-symmetrical axial loading, some 59 test and numerical data exist. However, this database does not include joints with $90°<\phi\leq180°$. This data is essential for a continuous formula to be derived, and for greater understanding of joints under anti-symmetrical loading. This paper presents numerically generated data to bridge the gap in the existing database. A total of 47 finite element analyses have been carried out. The new numerical data together with existing test and FE results are used to determine the effects of the geometric parameters on the static strength and failure modes. Existing design code formulae and estimation equations pertinent to multiplanar joints are also assessed in the light of the new data.

SCOPE OF STUDY

Table 1 shows the scope of this finite element study, in which effects of the parameters ϕ, γ, β, and g_t/D, were investigated . The ranges for each parameter are given below:

$60°\leq\phi\leq180°$ $12\leq\gamma\leq40$ $0.24\leq\beta\leq0.45$ $0.15\leq g_t/D\leq0.50$

The parameters β and g_t/D were investigated using three values of ϕ (90°, 135°, and 180°). The effect of γ was examined using these and three additional values of ϕ (60°, 112.5°, and 157.5°).

THE FE MODEL VERIFICATION

The FE model for KK-joints with $60°<\phi<90°$ has been extensively calibrated for both the symmetrically and anti-symmetrically axially loading cases (Lee and Wilmshurst (1995,1996), Wilmshurst (1995), Wilmshurst *et al* (1996)). However, with the availability of test data for anti-symmetrically loaded planar (*i.e.* ϕ=180°) KK-joints (Yonemura *et al* (1996)), it was possible to check the validity of the model across the entire range of ϕ considered in this study. Four specimens that failed by chord wall plastification were selected from this database. Table 2 shows the geometrical and material data for the selected test specimens. Finite element models of these specimens were generated according to the guidelines established in the literature referenced earlier in this section. The weld modelling was based upon the weld profiles given in AIJ (1990). As a plane of symmetry existed only half-models were used. A typical mesh used in the verification study is shown in Figure 2. The loading and boundary conditions were as per the test set-up of Yonemura *et al* (1996) (with local co-ordinate systems set up at the centre nodes of each end-plate to facilitate their application and the extraction of results). The force loading in the tests was applied proportionally to the tension and compression braces using 100 ton hydraulic jacks. This proportional force loading was achieved in the finite element analyses by means of the modified Riks method. The

Joint	γ	ϕ	g_t/D	β	ζ_t	P_u	P_{nd}	Failure Mode	$P_{u,K,Kur}$	[1]	$P_{u,AWS}$	[2]
AKK-B-01	12.0	90.0	0.15	0.35	0.415	541.80	12.07	a,c	673.81	0.804	457.72	1.184
AKK-B-02	12.0	90.0	0.35	0.24	0.517	363.80	8.11	a,b	458.44	0.794	357.78	1.017
AKK-B-03	12.0	90.0	0.35	0.35	0.415	483.10	10.77	a	579.46	0.834	435.10	1.110
AKK-B-04	12.0	90.0	0.35	0.45	0.313	669.10	14.91	a,b	689.48	0.970	500.46	1.337
AKK-B-05	12.0	90.0	0.50	0.35	0.415	454.90	10.14	a	566.56	0.803	422.43	1.077
AKK-B-06	12.0	135.0	0.35	0.35	0.732	520.40	11.60	a	579.46	0.898	514.40	1.012
AKK-B-07	12.0	180.0	0.15	0.35	0.937	583.60	13.01	a	673.81	0.866	655.48	0.890
AKK-B-08	12.0	180.0	0.35	0.24	0.971	382.60	8.53	a	458.44	0.835	497.32	0.769
AKK-B-09	12.0	180.0	0.35	0.35	0.937	515.10	11.48	a	579.46	0.889	655.48	0.786
AKK-B-10	12.0	180.0	0.35	0.45	0.893	641.20	14.29	a	689.48	0.930	799.26	0.802
AKK-B-11	12.0	180.0	0.50	0.35	0.937	504.70	11.25	a	566.56	0.891	655.48	0.770
AKK-B-12	12.0	180.0	0.50	0.45	0.893	626.60	13.96	a	674.13	0.929	799.26	0.784
AKK-B-13	25.0	90.0	0.15	0.35	0.415	149.90	14.50	a	170.18	0.881	112.93	1.327
AKK-B-14	25.0	90.0	0.35	0.24	0.517	98.32	9.51	a	119.94	0.820	87.53	1.123
AKK-B-15	25.0	90.0	0.35	0.35	0.415	130.40	12.61	a	151.60	0.860	107.41	1.214
AKK-B-16	25.0	90.0	0.35	0.45	0.313	172.00	16.64	a	180.39	0.954	124.09	1.386
AKK-B-17	25.0	90.0	0.50	0.35	0.415	126.60	12.25	a	151.47	0.836	104.14	1.216
AKK-B-18	25.0	135.0	0.15	0.35	0.732	157.00	15.19	a	170.18	0.923	128.23	1.224
AKK-B-19	25.0	135.0	0.35	0.24	0.805	103.20	9.98	a	119.94	0.860	98.28	1.050
AKK-B-20	25.0	135.0	0.35	0.35	0.732	145.10	14.04	a	151.60	0.957	124.16	1.169
AKK-B-21	25.0	135.0	0.35	0.45	0.653	182.20	17.62	a	180.39	1.010	146.55	1.243
AKK-B-22	25.0	135.0	0.50	0.35	0.732	141.20	13.66	a	151.47	0.932	121.64	1.161
AKK-B-23	25.0	180.0	0.15	0.35	0.937	155.20	15.01	a	170.18	0.912	151.02	1.028
AKK-B-24	25.0	180.0	0.35	0.24	0.971	102.70	9.93	a	119.94	0.856	114.58	0.896
AKK-B-25	25.0	180.0	0.35	0.35	0.937	143.50	13.88	a	151.60	0.947	151.02	0.950
AKK-B-26	25.0	180.0	0.35	0.45	0.893	184.00	17.80	a	180.39	1.020	184.15	0.999
AKK-B-27	25.0	180.0	0.50	0.35	0.937	139.90	13.53	a	151.47	0.924	151.02	0.926
AKK-B-28	40.0	90.0	0.15	0.35	0.415	66.04	16.35	a	67.28	0.982	45.97	1.436
AKK-B-29	40.0	90.0	0.35	0.24	0.517	44.33	10.98	a	51.64	0.858	35.47	1.250
AKK-B-30	40.0	90.0	0.35	0.35	0.415	61.14	15.14	a	65.28	0.937	43.83	1.395
AKK-B-31	40.0	90.0	0.35	0.45	0.313	72.44	17.94	a	77.67	0.933	50.85	1.425
AKK-B-32	40.0	90.0	0.50	0.35	0.415	59.29	14.68	a	65.28	0.908	42.53	1.394
AKK-B-33	40.0	135.0	0.35	0.35	0.732	64.23	15.91	a	65.28	0.984	49.89	1.287
AKK-B-34	40.0	135.0	0.15	0.35	0.732	68.58	16.98	a	67.28	1.019	51.39	1.334
AKK-B-35	40.0	180.0	0.35	0.24	0.971	44.05	10.91	a	51.64	0.853	44.76	0.984
AKK-B-36	40.0	180.0	0.35	0.35	0.937	64.54	15.98	a	65.28	0.989	58.99	1.094
AKK-B-37	40.0	180.0	0.35	0.45	0.893	84.05	20.81	a	77.67	1.082	71.93	1.168
AKK-B-38	40.0	135.0	0.50	0.35	0.732	62.22	15.41	a	65.28	0.953	48.93	1.272
AKK-B-39	12.0	60.0	0.35	0.35	0.165	453.00	10.10	a	579.46	0.782	469.87	0.964
AKK-B-40	12.0	112.5	0.35	0.35	0.584	512.70	11.43	a	579.46	0.885	454.48	1.128
AKK-B-41	12.0	157.5	0.35	0.35	0.850	514.70	11.47	a	579.46	0.888	603.94	0.852
AKK-B-42	25.0	60.0	0.35	0.35	0.165	113.00	10.93	a	151.60	0.745	114.92	0.983
AKK-B-43	25.0	112.5	0.35	0.35	0.584	141.40	13.68	a	151.60	0.933	111.63	1.267
AKK-B-44	25.0	157.5	0.35	0.35	0.850	144.10	13.94	a	151.60	0.951	141.61	1.018
AKK-B-45	40.0	60.0	0.35	0.35	0.165	50.21	12.43	a	65.28	0.769	46.58	1.078
AKK-B-46	40.0	112.5	0.35	0.35	0.584	64.93	16.08	a	65.28	0.995	45.39	1.431
AKK-B-47	40.0	157.5	0.35	0.35	0.850	64.47	15.96	a	65.28	0.988	55.88	1.154

$D = 216.5$ mm $\quad e_t = 0.0$ mm $\quad \theta_t = \theta_c = 45°$ $\quad \tau = 1.2$

$\sigma_{y,Chord} = 389$ Nmm^{-2} $\quad \sigma_{u,Chord} = 540$ Nmm^{-2} $\quad \sigma_{y,Brace} = 355$ Nmm^{-2} $\quad \sigma_{u,Brace} = 535$ Nmm^{-2}

[1] $\dfrac{P_u}{P_{u,K,Kur}}$ $\quad$ [2] $\dfrac{P_u}{P_{u,AWS}}$

Failure modes: a...Chord wall plastification b...Brace yielding c...failure by assumed crack

Table 1 Geometrical parameters, ultimate capacities, and failure modes of the specimens analysed in this study

Specimen	θ (°)	D (mm)	T (mm)	d (mm)	t (mm)	g_l (mm)	g_t (mm)	σ_y (Nmm^{-2})	σ_u (Nmm^{-2})
DKA5-M	45	216.8	5.52	76.9	3.98	108.1	202.7	330	436
DKA5-M2	45	216.8	5.52	139.5	4.13	19.5	165.7	330	436
DKA6-M	60	216.5	5.39	76.9	3.98	36.2	202.4	332	432
DKA6-M-H	60	216.5	6.32	76.3	4.03	36.9	202.6	369	504

Note: g_l and g_t are calculated values

Table 2 Geometric and material data for the specimens selected from Yonemura *et al* (1996)

Specimen	Ultimate capacity P_u (kN)		Ratio (FE / Test)	Deformation at ultimate capacity	
	Test	FE		Test	FE
DKA5-M	223.0	210.6	0.94	4.79	4.95
DKA5-M1	385.0	399.9	1.04	6.45	5.25
DKA6-M	190.0	184.6	0.97	5.53	6.33
DKA6-M-H	224.0	241.9	1.08	8.36	8.58

mean = 1.01 CoV = 0.061

Table 3 Results of the model verification for specimens with ϕ=180°

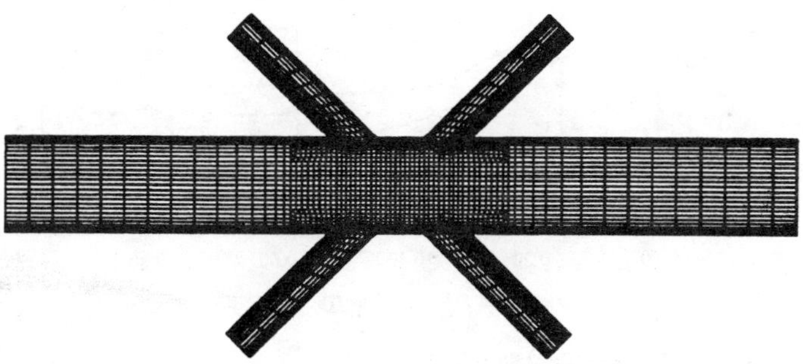

Figure 2 A typical mesh used in the planar KK-joint model verification

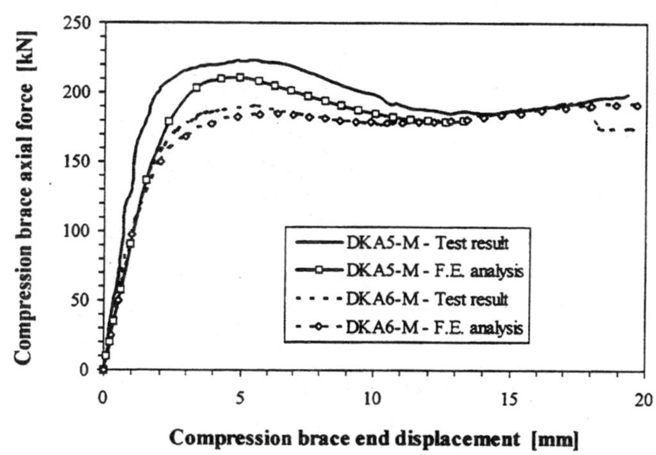

Figure 3 Typical load-displacement relationships from the model verification study

material data were those determined from tensile coupon tests during the testing of the original specimens. The results of the model verification are presented in Table 3, and the load-displacement relationships for specimens DKA5-M and DKA6-M are shown in Figure 3. It can be seen that there is very good agreement not only in the ultimate loads and the deformations at ultimate capacity, but also in the load-displacement behaviour. This confirms the validity of the modelling technique.

FINITE ELEMENT MODELLING OF THE JOINTS IN THE PARAMETRIC STUDY

The finite element modelling conformed to the guidelines applied in the verification study. However, the loading and boundary conditions were changed to match those applied to specimens analysed in Wilmshurst (1995). The tension brace ends were pinned, and guided displacement loading was applied in the direction of the brace axis at the compression brace ends. One chord end was pinned and the other chord end was supported by a pin-roller (*i.e.* displacements orthogonal to the chord centre-line were prevented and only displacements along the chord axis were allowed). The material properties of Grade BS50C cold formed sections were selected for the chord, and SM490B were chosen for the brace as data was available for a thick walled tube of this material. (The tubes from which the brace data was derived were machine formed from sheet coil and annealed). The material properties of the weld elements were chosen to match the brace material. The material stress strain curves are shown in Figure 4.

RESULTS AND DISCUSSION

Ultimate loads and failure modes

Table 1 shows the ultimate capacities and failure modes of the joints analysed in this study. The ultimate capacity was determined by the monitoring of the load-displacement history and the growth of the plastic zones in each analysis. As in previous research on the KK-joint configuration the ultimate capacity is defined as the first maximum axial load in the compression brace. However, where no maximum was

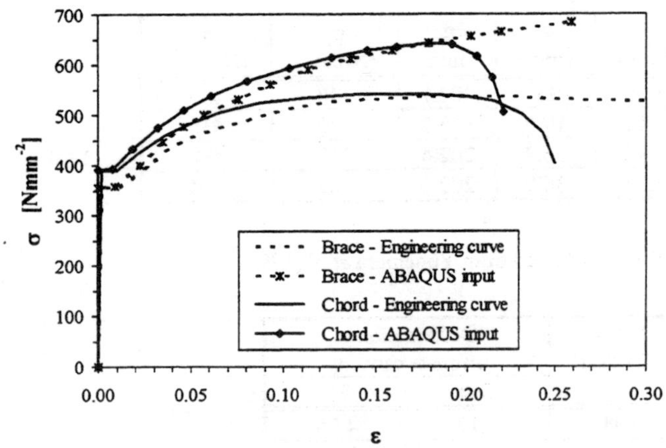

Figure 4 Stress-strain curves for the brace and chord materials

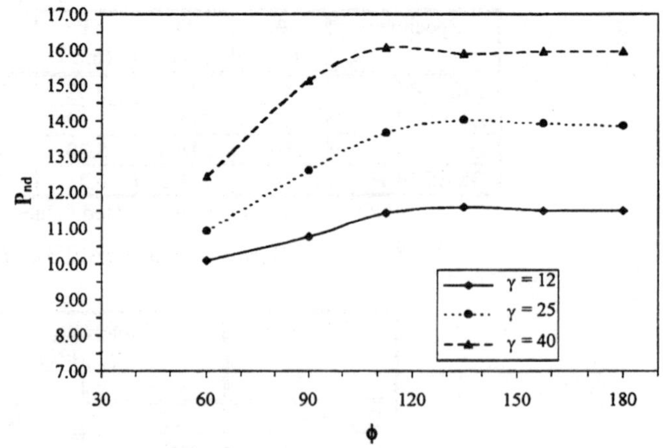

Figure 5 P_{nd} vs. ϕ as functions of γ

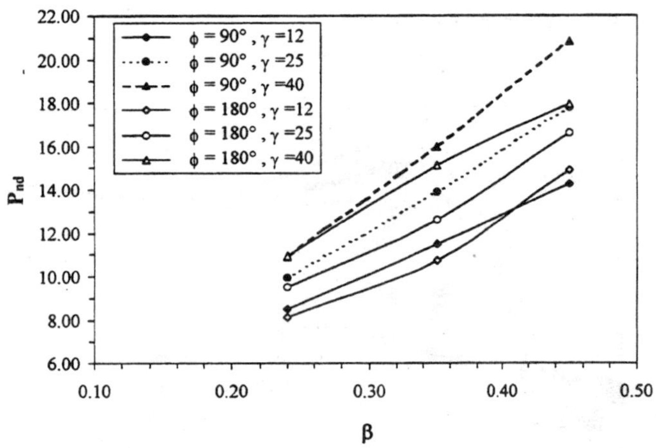

Figure 6 P_{nd} vs. β as functions of ϕ and γ

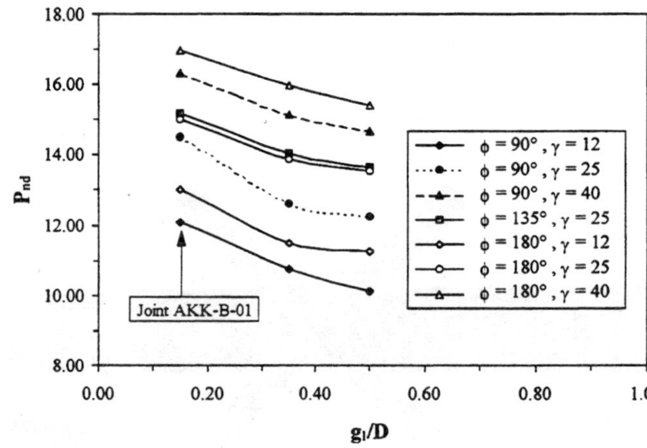

Figure 7 P_{nd} vs. g_l/D as functions of ϕ and γ

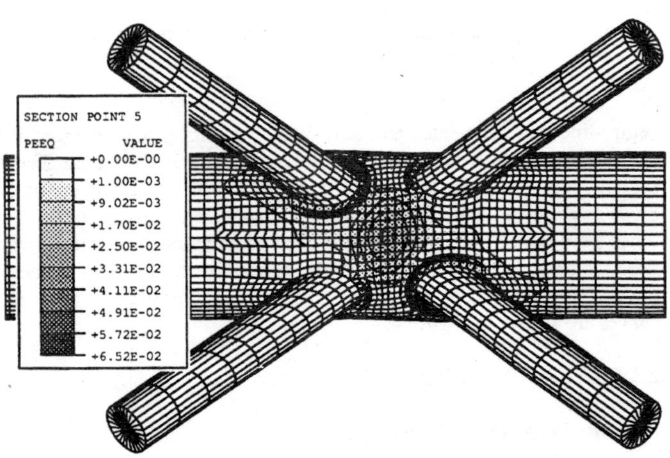

Figure 8 Equivalent plastic strain plot for a joint with ϕ=90° (AKK-B-15 - top surface at failure)

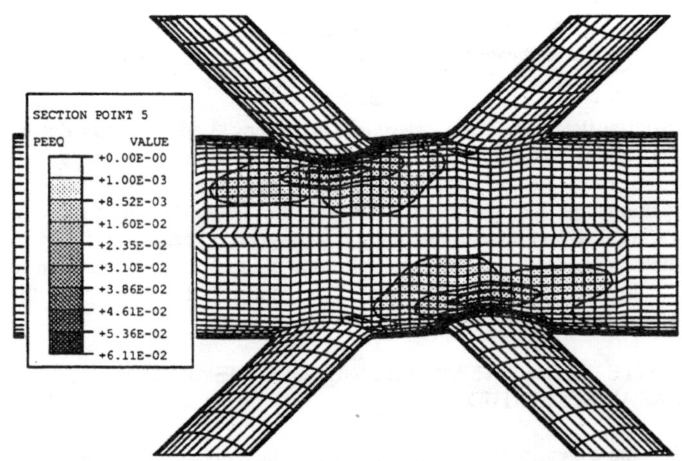

Figure 9 Equivalent plastic strain plot for a joint with ϕ=157.5° (AKK-B-44 - top surface at failure)

evident the following criteria were used to determine the ultimate capacity:

i) the first point of minimum gradient in the load-displacement history
ii) where there was a steady increase of the load after yielding, the load at which the plastic strain in the brace or chord material at the tension brace weld toes reached a value equal to the maximum strain in the material curve a crack was assumed to have formed. Although it is recognised as a only a crude method, it was found to give a reasonable indication of the ultimate capacity (see Figure 7). This method was only used for determining the strength of one specimen (AKK-B-01) which had low γ and small g_l/D.

All but three of the specimens failed by chord wall plastification alone. Two specimens, AKK-B-02 and AKK-B-04, failed by chord wall plastification with some yielding at the foot of the compression brace. Inspection of the equivalent plastic strains showed the maximum values of the strain in the compression brace to be 2.6% and 3.3% respectively. Although a low value it could still be possible that the failure mode was a combination of chord wall plastification and brace local buckling. The remaining specimen failed by combination of chord wall plastification and assumed formation of a crack.

The effect of the geometrical parameters on the ultimate capacity and failure mode

The influences of the geometrical parameters on the ultimate capacities and failure modes of the joints analysed in this study were investigated by plotting Figures 5 to 7, the non-dimensionalised strength P_{nd} against the individual geometrical parameters. These figures and Table 1 yield the following observations:

i) Figure 5 shows the effect of ϕ, the out-of-plane angle, upon P_{nd}, as functions of γ. It can be seen that initially P_{nd} increases almost linearly with ϕ, the increase being more prominent for higher γ ratios. When ϕ reaches a certain value (135° for γ=12 and 25, 112.5° for γ=40) the joint strength remains more or less constant. Inspection of the equivalent plastic strain distributions confirmed the suspicion that that there was a change of failure pattern. Figure 8 and Figure 9 show the plastic zones for two joints with γ=25 and where ϕ=90° and ϕ=157.5° respectively. The joint with ϕ=90° clearly shows the failure pattern observed in previous research where bending of the chord wall is evident in both the longitudinal and transverse gap areas. The joint with ϕ=157.5° appears to fail as two separate K-joints with the plastic zones localised at the base of the braces in each K-joint (*i.e.* chord wall bending in the longitudinal gap is predominant). This is similar to the observations of Yonemura *et al* (1996), who noted that the local deformations in the regions of the brace intersections for planar KK-joints were similar to that observed for K-joints. It would seem logical that there would be a transition stage between the two failure types.
ii) In accordance with the findings of a previous study (Lee and Wilmshurst (1996)) P_{nd} increases linearly with β, and the rate of increase becomes slightly greater as γ increases (see Figure 6).
iii) Figure 7 shows the non-dimensionalised strength plotted against g_l/D as functions of ϕ and γ. As observed in earlier studies, P_{nd} decreases linearly with increasing in-plane gap size for higher γ ratios. The apparent sharp increase of the strength when g_l/D decreases, for joints with $\phi\leq$90° and lower γ ratios, is thought to be the result of the increased bending stiffness of the chord wall in the gap regions.

Comparison with the Kurobane (1984) K-joint formula

Figure 10 shows the ratio of the joint ultimate capacity to the corresponding prediction using the Kurobane (1984) K-joint formula plotted against ζ_t, for the new data and the database of Makino *et al* (1996). Also plotted on the graph are the lines for Equations 1 and 2 (Makino *et al* (1995)). Yonemura *et al* (1996) found that the strength of KK-joints with ϕ=180° (*i.e.* high ζ_t)could be estimated by using the K-joint equation of Kurobane *et al* (1984). It is clear from the figure that Equations 1 and 2 do not accurately predict the strength of anti-symmetrically loaded KK-joints. Instead, the addition of the new data to the database shows that the trend for joints with $\zeta_t\geq$0.17 follows a power function (represented by the dashed line), and the equation for joints with $\zeta_t\leq$0.17 tends to overpredict the joint strength.

Comparison with the AWS design formulation

The AWS design code is the only code that attempts to incorporate the effects of out-of-plane loading in multiplanar joints as a function of the out-of-plane angle (which is incorporated into the α_o parameter). Figure 11 presents a comparison of the test and finite element strengths, for the new data and the database of Makino *et al* (1996), to the AWS code strength predictions plotted against ϕ. It is very clear from the figure that there is a residual trend of the capacity ratio with ϕ, indicating that ovalisation parameter does not adequately capture the effect of the out-of-plane angle on the joint strength. The strength predictions range from conservative for $\phi\leq$135° to unsafe when ϕ =180° (as the AWS equation is a lower-bound formulation).

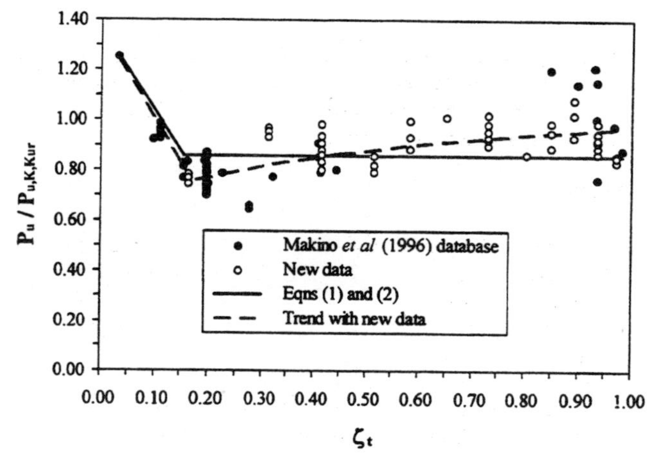

Figure 10 $P_u / P_{u,K,Kur}$ vs. ζ_t

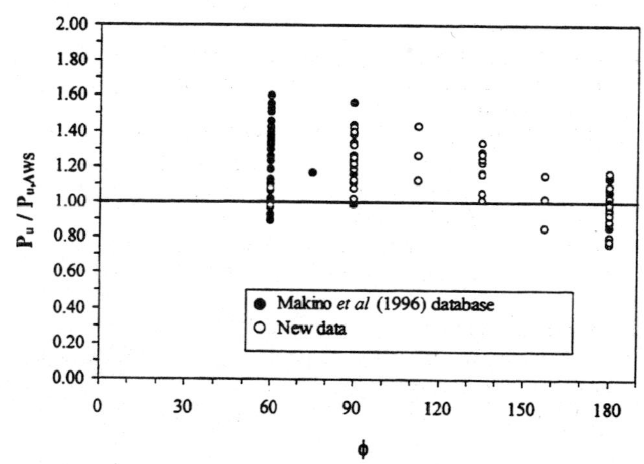

Figure 11 $P_u / P_{u,AWS}$ vs. ϕ

Proceedings of the Seventh (1997) International Offshore and Polar Engineering Conference
Honolulu, USA, May 25-30, 1997

ISBN 1-880653-28-1 (Set); ISBN 1-880653-32-X (Vol. IV)

The Axial Strength of Uniplanar X-Joints Reinforced by T-Shaped Ring-Stiffeners

G.J. van der Vegte
National University of Singapore/Delft University of Technology
Singapore/The Netherlands

D.H. Leray[1] *and Y.S. Choo*
National University of Singapore
Singapore

ABSTRACT

Since current design rules for ring-stiffened joints under static load are in-adequate, a numerical study has been carried out to quantify the effects of single (centred) T-shaped ring-stiffeners on the static strength of axially loaded uniplanar X-joints. The effects of the geometric parameters β and 2γ on the static strength have been determined. In addition, the influences of the thickness and the height of the T-shaped stiffeners have been investigated.

KEY WORDS : finite element analyses, static strength, stiffened tubular joints, T-shaped stiffener

INTRODUCTION

Un-stiffened tubular connections are widely applied in offshore structures, due to the easier welding process which can be done entirely from the outside of the tubular members. However, in order to avoid unacceptable wall thickness of the chord, as required from temporary loads during installation (for example, during launching of a jacket structure) or fatigue considerations, joints may be strengthened by adding internal or external stiffeners or diaphragms in the vicinity of the joints.

The design guidance for stiffened joints, however, is still in-adequate. For example, according to U.E.G (1985), the local ring stresses of ring-stiffened nodes can be determined using Roark's formulae (Young, 1989) which are based on an elastic "closed ring" approach, while the joint capacity can be derived using the ultimate (plastic) strength of the un-stiffened joints in combination with a contribution provided by the ring.

In this study, numerical analyses have been performed on ring-stiffened axially loaded uniplanar X-joints in order to determine the effects of the internal stiffeners on the strengths of the joints. Although various types of ring-stiffeners can be considered (e.g. flat-bar stiffeners or T-shaped stiffeners), this study is limited to single T-shaped stiffened joints. Various web height to thickness ratios of the T-shaped stiffeners are considered for different combinations of β and 2γ. A similar research programme has been carried out for axially loaded flat-bar stiffened uniplanar X-joints (van der Vegte *et al.* 1996).

[1] Visiting scholar

RESEARCH PROGRAMME

Fig. 1 shows the configuration of axially loaded uniplanar X-joints. The dimensions of the X-joints considered are summarized in Table 1. Three brace to chord diameter ratios β (0.25, 0.49 and 0.74) and two chord diameter to chord wall thickness ratios 2γ have been analyzed (36.4 and 48.8). The chord diameter d_0 is taken as 1854 mm. For all joints, the chord length parameter α has been set to 16.

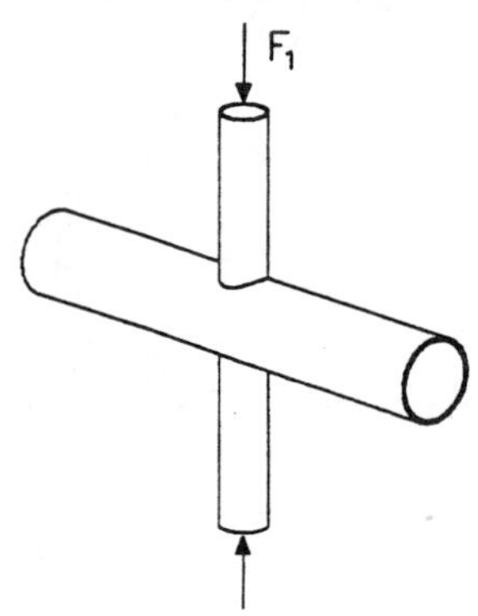

Fig. 1 Configuration of axially loaded uniplanar X-joints.

The configuration of the T-shaped stiffeners is illustrated in Fig. 2, while the dimensions are given in Table 2. The internal ring-stiffener is positioned in the centre of the X-joint. Each X-joint has been analyzed for each of the four different dimensions of the T-shaped stiffener. For reference, the un-stiffened uniplanar X-joints have been analyzed as well.

The steel grade of all chord and brace members is S355 with f_y = 355 N/mm^2 and f_u = 510 N/mm^2. The true stress - true strain curves have been modelled as step-wise linear relationships, including strain hardening.

Table 1 : Dimensions of the uniplanar X-joints

	$2\gamma = 36.4$ $d_0 \times t_0 =$ 1854 x 51 mm	$2\gamma = 48.8$ $d_0 \times t_0 =$ 1854 x 38 mm
$\beta = 0.25$ (d_1 = 457 mm)	-	X3 (t_1 = 32 mm)
$\beta = 0.49$ (d_1 = 914 mm)	X1 (t_1 = 44 mm)	X4 (t_1 = 38 mm)
$\beta = 0.74$ (d_1 = 1370 mm)	X2 (t_1 = 51 mm)	X5 (t_1 = 38 mm)

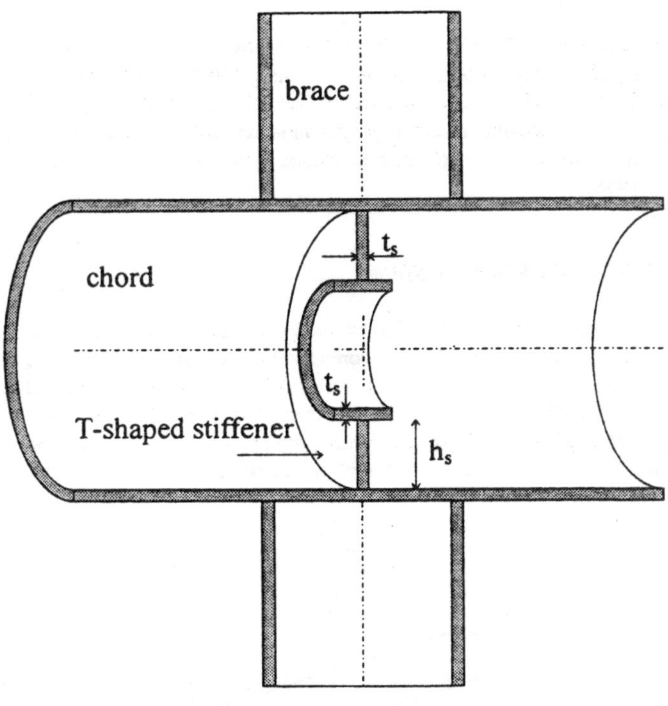

Fig. 2 The configuration of the T-shaped ring-stiffener.

Table 2 : Dimensions of the T-shaped ring-stiffeners

	$h_s = 5\ t_0$	$h_s = 10\ t_0$
$t_s = 0.5\ t_0$	$h_s/t_s = 10.0$	-
$t_s = 0.76\ t_0$	$h_s/t_s = 6.55$	-
$t_s = t_0$	$h_s/t_s = 5.0$	$h_s/t_s = 10.0$

FINITE ELEMENT ANALYSES

The most important features of the finite element analyses can be summarized as follows :

- the FE program ABAQUS has been used for all numerical analyses, whereas material and geometrical non-linearities have been taken into account.
- for axially loaded uniplanar X-joints, only one eighth of each joint needs to analyzed, due to symmetry in geometry and loading. However, to enable the ring-stiffeners to deform out-of-plane, one quarter of each joint has been modelled. The T-shaped stiffeners do not have initial geometrical imperfections.
- the joints have been analyzed with eight noded thick shell elements with reduced integration (ABAQUS element type S8R).
- loads have been applied by displacement control.
- the weld between the chord and the brace has been modelled in agreement with the weld sizes recommended by A.W.S. (1994), using shell elements (van der Vegte *et al.*, 1991).

NUMERICAL RESULTS

In Fig. 3, the numerically determined load-displacement curves are presented for each uniplanar X-joint. Since failure of the joints is due to plastification of the chord cross section, the load F_1 has been made non-dimensional according to $F_1/f_{y,0} \cdot t_0^2$ which is displayed against the (non-dimensional) displacement of the crown point δ_1/d_0. Ultimate load is considered to be reached if a maximum in applied force is observed. When no maximum in the applied load has been obtained (which holds for one of the curves of joint X3 only), failure load is defined as the brace force at a chord face indentation (at the crown point) of 0.03 d_0 (deformation limit of Lu *et al.*, 1994). The non-dimensionalized ultimate loads of the X-joints are summarized in Table 3.

From the load-deformations curves in Fig. 3 and the ultimate capacities presented in Table 3, it is observed that single T-shaped stiffeners significantly enhance the initial stiffness and ultimate strength as compared to the un-stiffened joints. This effect is most pronounced for the joints with β = 0.25, since for un-stiffened X-joints with small β-values, the load transfer through the joint is mainly governed by chord wall bending. Application of ring-stiffeners for X-joints with small β-values results in a large increase of the bending resistance of the "chord wall", thus resulting in significantly higher ultimate loads.

In line with the results for the flat-bar ring-stiffened joints (van der Vegte *et al.* 1996), it is observed that the influence of the thickness of the T-shaped stiffener on the ultimate capacity is relatively small (e.g. doubling the thickness of the web and the flange of the stiffener of joint X4 increases the ultimate strength from 1.42 to 1.86 times the ultimate strength of the un-stiffened joint).

As expected, enlarging the height of the T-shaped stiffener has a much larger effect on the ultimate strength as can be noticed from the results of the joints with $t_s = t_0$. From Table 3 it is observed that doubling the height of the web of the T-shaped stiffener of joint X4 (from 5 t_0 to 10 t_0) increases the ultimate strength from 1.86 to 2.79 times the strength of the corresponding un-stiffened joint.

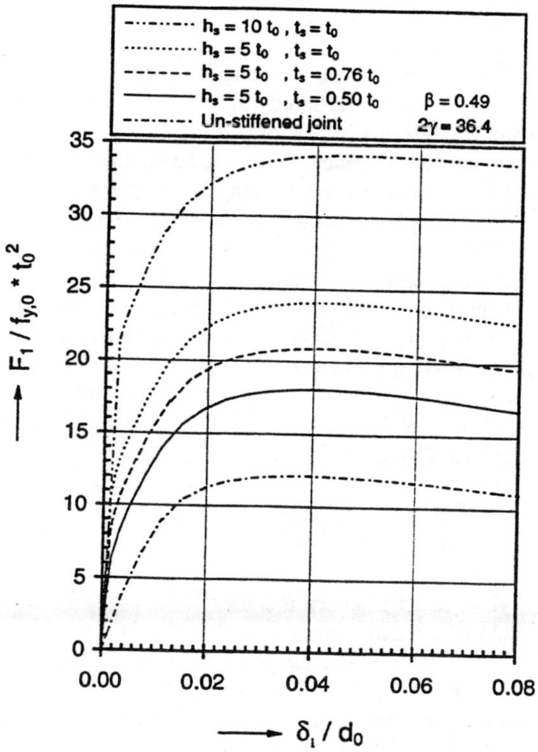

Fig. 3a : Joint X1 ($\beta = 0.49$, $2\gamma = 36.4$)

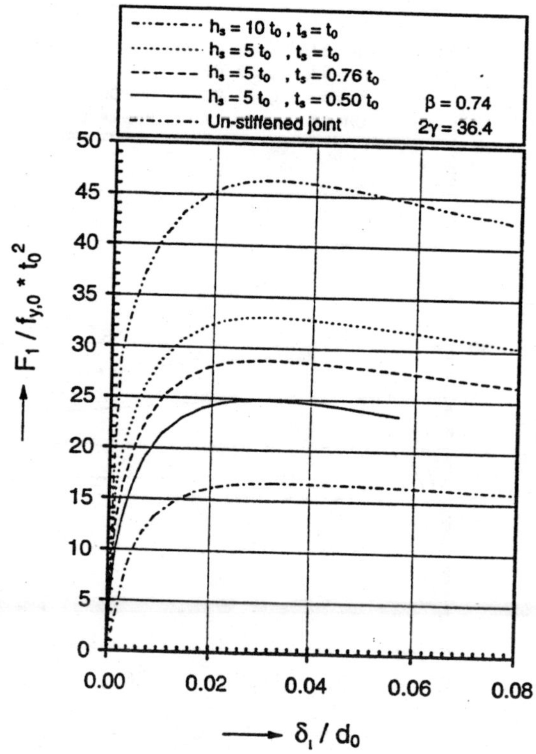

Fig. 3b : Joint X2 ($\beta = 0.74$, $2\gamma = 36.4$)

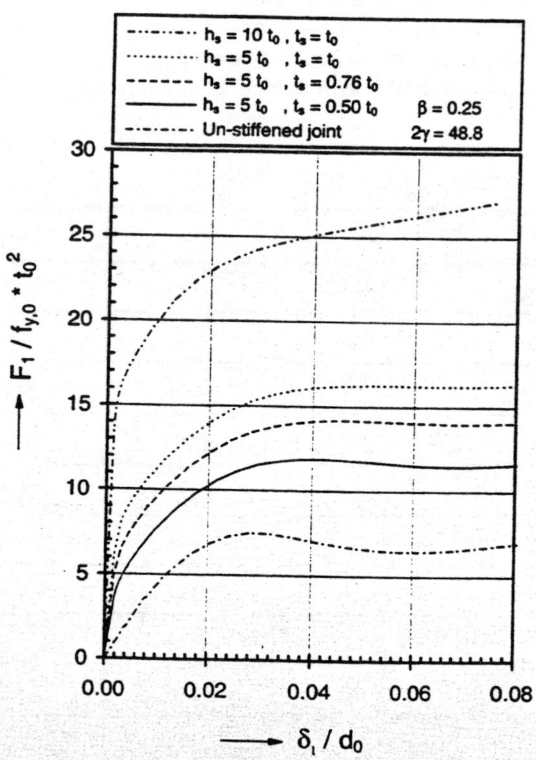

Fig. 3c : Joint X3 ($\beta = 0.25$, $2\gamma = 48.8$)

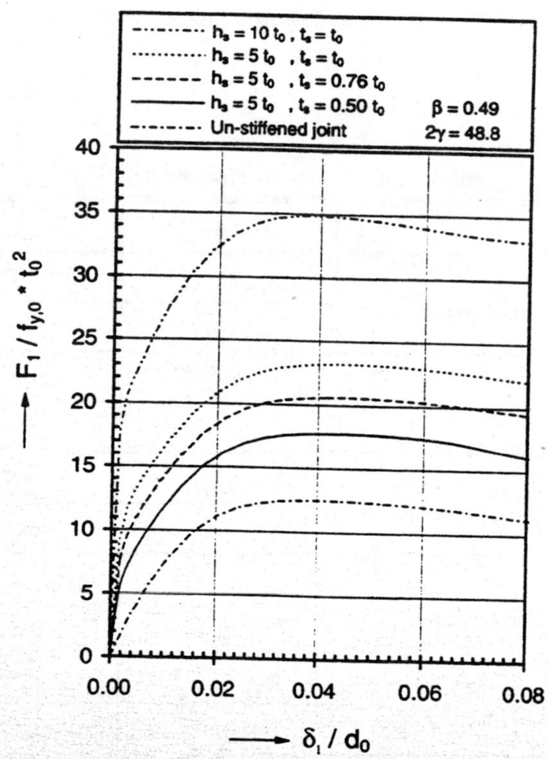

Fig. 3d : Joint X4 ($\beta = 0.49$, $2\gamma = 48.8$)

Fig. 3 Numerical load-displacement curves of axially loaded uniplanar X-joints reinforced by T-shaped ring-stiffeners.

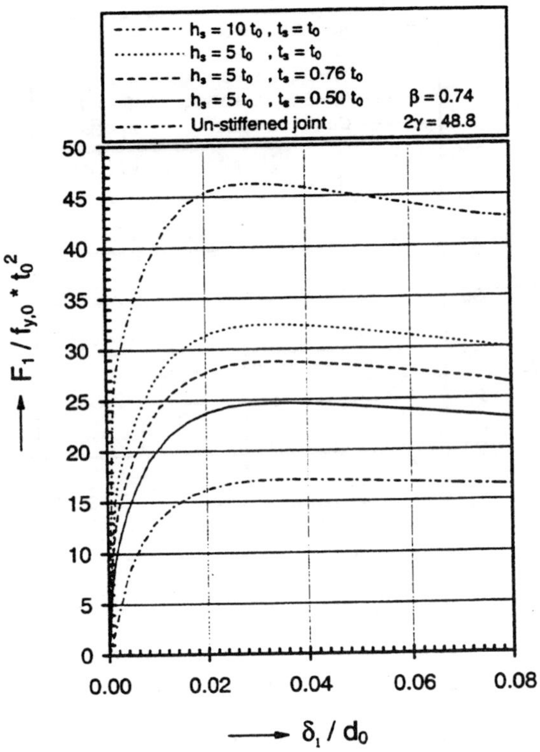

Fig. 3e : Joint X5 ($\beta = 0.74$, $2\gamma = 48.8$)

Fig. 3 Numerical load-displacement curves of axially loaded uniplanar X-joints reinforced by T-shaped ring-stiffeners (continued).

CONCLUSIONS

Based on the results of numerical analyses of axially loaded uniplanar X-joints reinforced by T-shaped ring-stiffeners, the following conclusions can be drawn :

- the strength enhancement due to a larger thickness of T-shaped stiffeners is relatively small compared to the increase in strength as a result of a larger height of the ring-stiffeners.
- the height of the T-shaped stiffeners has a significant influence on the ultimate capacity of the uniplanar X-joints.
- further numerical research has been planned on uniplanar X-joints reinforced by multiple ring-stiffeners subjected to axial loading and in-plane bending moments.

NOMENCLATURE

d_0 : outer diameter of chord
d_1 : outer diameter of brace
f_y : yield stress
$f_{y,0}$: yield stress of chord member
f_u : ultimate tensile stress
h_s : height of web of T-shaped ring-stiffener
l_0 : length of chord
t_0 : wall thickness of chord
t_1 : wall thickness of brace
t_s : wall thickness of flange and web of T-shaped ring-stiffener
FE : finite element
F_1 : axial force on in-plane brace
α : chord length parameter $2l_0/d_0$
β : diameter ratio d_1/d_0
2γ : chord diameter to thickness ratio d_0/t_0
δ_1 : indentation of crown point of in-plane brace

Table 3 : Numerical results for the X-joints with the T-shaped ring-stiffeners

	Joint :	X1	X2	X3	X4	X5
Un-stiffened joint	$F_{1,u}/f_{y,0} \cdot t_0^2$	12.13	16.74	7.46	12.49	17.14
	Ratio*	1.00	1.00	1.00	1.00	1.00
$h_s = 5\ t_0,\ t_s = 0.5\ t_0$	$F_{1,u}/f_{y,0} \cdot t_0^2$	18.06	24.87	11.88	17.76	24.64
	Ratio*	1.49	1.49	1.59	1.42	1.44
$h_s = 5\ t_0,\ t_s = 0.76\ t_0$	$F_{1,u}/f_{y,0} \cdot t_0^2$	20.92	28.73	14.18	20.60	28.71
	Ratio*	1.72	1.72	1.90	1.65	1.68
$h_s = 5\ t_0,\ t_s = t_0$	$F_{1,u}/f_{y,0} \cdot t_0^2$	24.09	33.00	16.24	23.20	32.39
	Ratio*	1.99	1.97	2.18	1.86	1.89
$h_s = 10\ t_0,\ t_s = t_0$	$F_{1,u}/f_{y,0} \cdot t_0^2$	34.25	46.43	24.30	34.90	46.24
	Ratio*	2.82	2.77	3.26	2.79	2.70

* Ratio = ultimate strength of each X-joint divided by the ultimate strength of the corresponding un-stiffened X-joint.

REFERENCES

American Welding Society, Structural Welding Code (1994). AWS D1.1-94.

Lu L.H., G.D. de Winkel, Y. Yu and J. Wardenier (1994). "Deformation Limit for the Ultimate Strength of Hollow Section Joints.", *Proc. Sixth International Symposium on Tubular Structures*, Melbourne, Australia, pp. 341-347.

U.E.G. (1985). *Design of Tubular Joints for Offshore Structures.*, UEG Publication UR33, London, U.K., ISBN 0-86017-231-7.

Vegte G.J. van der, C.H.M de Koning, R.S. Puthli and J. Wardenier (1991). "Numerical Simulation of Experiments on Multiplanar Tubular Steel X-Joints.", *International Journal of Offshore and Polar Engineering*, ISOPE, Vol. 1, No 3, pp. 200-207.

Vegte G.J. van der, D.H. Leray and Y.S. Choo (1996). "The Axial Strength of Ring-Stiffened Uniplanar X-Joints.", *Proceedings of the Fourth Pacific/Asia Offshore Mechanics Symposium*, Pusan, Korea, pp. 197-202.

Young W.C. (1989). *Roark's Formulas for Stress & Strain.*, Sixth Edition, McGraw-Hill International Editions, ISBN 0-07-100373-8.

Proceedings of the Seventh (1997) International Offshore and Polar Engineering Conference
Honolulu, USA, May 25-30, 1997

ISBN 1-880653-28-1 (Set); ISBN 1-880653-32-X (Vol. IV)

The Static Behaviour of Bolted I-Beam to RHS Column Connections with a Composite Floor

L.H. Lu
Public Works and Water Management
Utrecht, The Netherlands

J. Wardenier
Delft University of Technology
Delft, The Netherlands

ABSTRACT

Experimental and numerical investigations have been carried out on bolted I-beam to rectangular hollow section column connections with a composite floor, loaded by in-plane bending moments. From the experiments and the numerical modelling, a good insight into the static behaviour of these connections is obtained, such as: failure criteria of the connections, the effect of the concrete filling in the column and the influence of the multiplanar loading etc.

Based on the experimental and numerical results and the comparison with the analytical models, the proposed recommendations for the design of bolted I-beam to RHS column connections with a composite floor are given.

KEYWORDS

Bolted I-beam to RHS column connections, composite floor, static behaviour.

LIST OF SYMBOLS

b_0	:	width of RHS column
b_1	:	width of the flange of an I-beam
$f_{t,c}$	:	tensile strength of the concrete
$f_{y,0}$	:	yield stress of RHS column
$f_{y,cl}$	:	yield stress of the cleat flange
$f_{u,ø6}$	:	ultimate stress of ø6 reinforcement bars
$f_{y,ø8}$	:	yield stress of ø8 reinforcement bars
h_{cl}	:	height of the cleat flange welded to the column face
h_m	:	distance from reinforcements to bottom flange of an I-beam
t_0	:	wall thickness of RHS column
t_1	:	thickness of I-beam flange
t_{cl}	:	thickness of the cleat flange
$M_{u,exp}$	:	experimentally determined ultimate strength of the connections
$M_{u,num}$	:	numerically determined ultimate strength of the connections
RHS	:	rectangular hollow section
β	:	width ratio between I-beam's flange and RHS column b_1/b_0
2γ	:	width to thickness ratio of RHS column b_0/t_0
η	:	cleat flange height to RHS column width ratio h_{cl}/b_0

INTRODUCTION

Composite construction has been used for flooring system in buildings because of its overall economy of use of materials and ease of construction. However, there is still no standard design approach for the composite connections. Eurocode 4 (1992), for the design of composite construction, permits the general principle of using composite connections in frame design, but gives no application rules. The design is based on pin-ended connections and the reinforcements in the slab have been only used to control the crack width of the concrete. No structural action of the reinforcements in the slab has been taken into account in the strength design because of a lack of sufficient acknowledgements for these types of connections.

It is economically attractive to use the hogging bending moments of the connections. By taking the composite structural action through the reinforcements in the slab into account, a better stiffness and strength of a connection can be achieved even if relatively small steel members are used. With this aim, recently, extensive experimental and numerical research has been carried out on different types of composite connections (Ahmed, 1995).

In order to study the static behaviour of bolted I-beam to RHS column connections with a composite floor loaded by in-plane bending moments, experimental and numerical investigations have been carried out at the Delft University of Technology. The experiments have been carried out in the framework of ECSC project 7210-SA/611 and the numerical modelling have been carried out in the "Beek" programme of TU Delft.

RESEARCH PROGRAMME

The configurations of the bolted I-beam to RHS column connections with a composite floor is shown in figure 1. Four connections (4R1 to 4R4), as shown in table 1, have been considered. Connections 4R2 and 4R4 have a RHS column filled with reinforced concrete. Two load cases, i.e. J=0 and J=1 have been considered to investigate the influence of the multiplanar loading on the behaviour of the connections.

Each connection comprises of a 300x300x10 hotfinished RHS column, four IPE-240 beams and a composite floor. Angle cleats (160x80x10 mm, 120 mm long), which are welded to the column face, have been used to transfer axial compression loading to the column and web plates (180x81x6 mm) have been welded vertically to the columns for transferring the shear forces from the beams to the columns. 3 M16-8.8 bolts have been provided in the web plates to resist the web shear and 6 M12-10.9 bolts have been provided in the bottom flange of each I-beam. Each composite floor consists of a mesh of 16ø6 reinforcing bars, which have been distributed in the concrete slab at 150 mm centres. Additional reinforcements 8ø8 have been provided in the vicinity of the columns (over 300 mm from the column face), see figure 1b.

The nominal dimensions and the average values of the measured material properties of the connections have been summarized in table 1. The material properties of the reinforcements used in the composite floor are shown in figure 2.

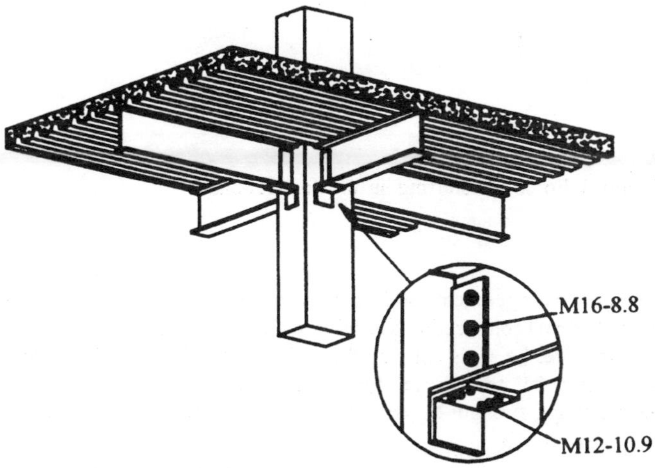

Fig. 1a Bolted connection

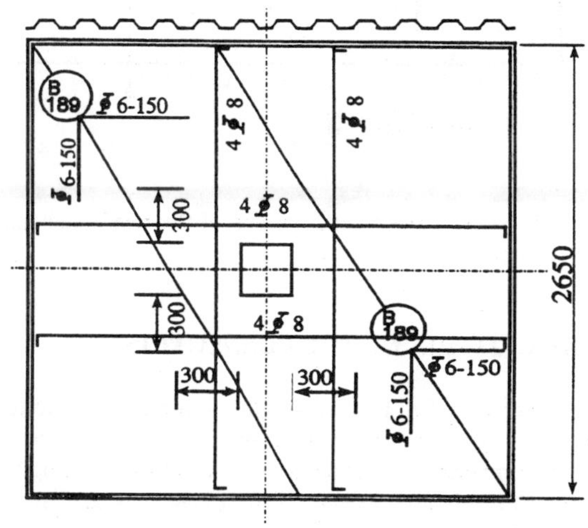

Fig. 1b Reinforcement arrangement

Figure 1 Configurations of the bolted I-beam to RHS column connections with a composite floor

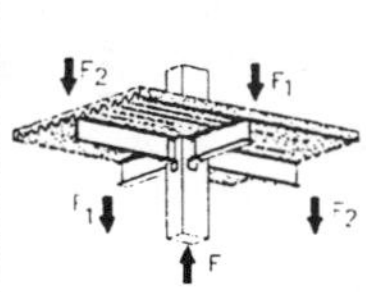	F_2/F_1	Concrete filling in column	Dimensions of the steel components [mm]					Material properties of steel and concrete [N/mm²]		
			b_0	t_0	b_1	t_1	t_{cf}	$f_{y,0}$	$f_{y,cl}$	$f_{t,c}$
4R1	0	no	300	9.76	119.6	9.72	10.4	439	377	3.58
4R2	0	yes	300	9.96	119.6	9.72	10.4	439	377	3.58
4R3	1	no	300	9.88	120.3	9.83	10.4	438	377	3.58
4R4	1	yes	300	9.88	120.3	9.83	10.4	438	377	3.58

Table 1 Average values of the geometries and material properties

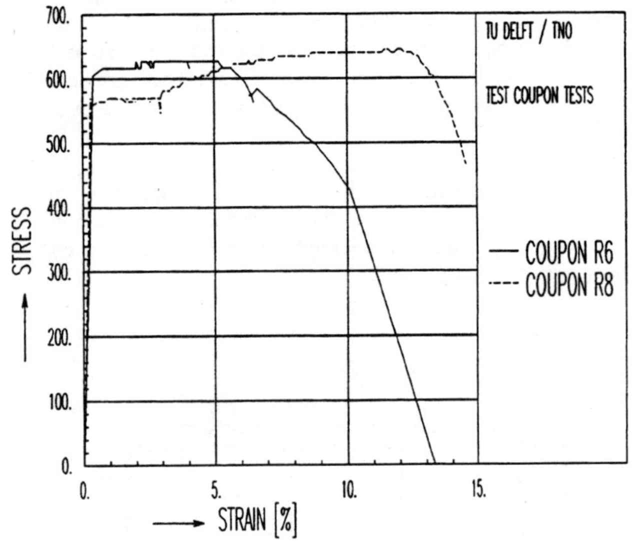

Figure 2 The stress - strain relationships of the reinforcements in the composite floor

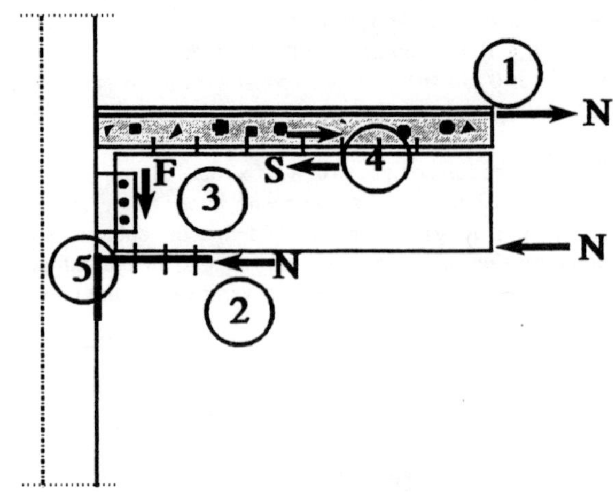

Figure 3 Force transferring in the connections

FAILURE CRITERIA OF THE CONNECTIONS

For the considered bolted I-beam to RHS column connections loaded by in-plane bending moments, an assumed model for force transferring through the components of the connections is illustrated in figure 3. The tensile forces resulting from the bending moments are taken by the reinforcements in the slab and the compression forces are transferred to the column through the bottom flanges and the cleat angles. The shear forces in the connections are transferred to the column by the web plates. Assuming that failure of the welds is not critical, the following failure modes in the connections can occur:

1. failure of the slab.
2. failure of the bolts in the bottom flanges, including bearing of the I-beam flange and the cleat flange.
3. failure of the bolts in the web plate under vertical shear force in combination with a moment due to the eccentricity of the loads.
4. failure of the studs between the slab and the I-beams.
5. failure of the column.

Failure modes 2 to 4 can generally be checked using the design formulae given in Eurocode 3 (1993). Failure of the slab and failure of the column will be considered in the present work.

Failure of the Slab

Failure of the slab depends on the arrangement and the properties of the reinforcements with the effective width b_c. According to Eurocode 4 (1992), the ultimate tensile capacity of the slab $N_{u,rf}$ can be given by the summation of the strength of the reinforcements within the effective width of the slab b_c at a particular strain. The effective width of the slab is related to the geometry of the construction and the loading cases. For the considered connections, assuming that all reinforcements of the slab are effective, i.e. b_c is equal to the width of the slab, and that failure occurs after failure of the reinforcements with the lowest failure strain, the following strength $N_{u,rf}$ of the slab can be expected (see figure 2):

$$N_{u,rf} = A_{\o 8} f_{y,\o 8} + A_{\o 6} f_{u,\o 6} \quad (1)$$

$A_{\o 8}$ total cross area of the reinforcing bars ø8.
$A_{\o 6}$ total cross area of the reinforcing bars ø6.
$f_{y,\o 8}$ the yield stress of the reinforcing bars ø8 when the reinforcing bars ø6 fail.
$f_{u,\o 6}$ the ultimate stress of the reinforcing bars ø6.

From figure 2 it can be seen that when the reinforcements reach a strain of about 5%, the bars ø6 break with an ultimate stress of $f_{u,\o 6}$=627 [N/mm^2] and the bars ø8 yield with $f_{y,\o 8}$=615 [N/mm^2]. With a h_m=332 [mm], the ultimate moment resistance $M_{u,rf}$ of the connection is obtained:
$N_{u,rf}$= 8ø8 *($f_{y,\o 8}$=615 N/mm^2) + 16ø6 *($f_{u,\o 6}$=627 N/mm^2) = 528.7 [kN].
$M_{u,rf}$ = 528.7 * 0.332 = 175.5 [kNm]

Column Face Yielding

Failure of the column under compression can be caused by: column face yielding, punching shear of the column face, chord side wall failure at β = 1.0 and concrete crash in the column. Punching shear and chord side wall failure of the column can be checked using the formulae given by Lu and Wardenier (1995a). Failure of the connection with concrete filling in the column subjected to compression loads has been investigated by Packer and Fear (1991).

Two simplified column face yield line models are considered for connection with on concrete filling of the column and shown in figure 4.

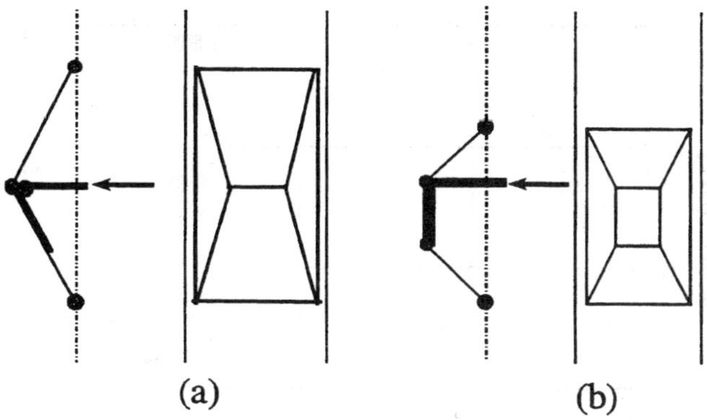

Figure 4 Column face yield line models

Model (b) may only occur if the cleat flange is so stiff that it does not deform. The following strength formulae have been derived using the same methode as used by Wardenier (1982) and Lu et al (1993, 1995a, 1995b):

For model (a):

$$\frac{N_{u,cy}}{f_{y0} t_0^2} = \frac{4}{\sqrt{1-\beta}} \sqrt{1+\frac{\beta f}{4}} \qquad (2)$$

with $f = \frac{f_{y,cf}\, t_{cf}^2}{f_{y,0}\, t_0^2}$.

For model (b):

$$\frac{N_{u,xb}}{f_{y0} t_0^2} = \frac{4}{\sqrt{1-\beta}} + \frac{2\eta}{1-\beta} \qquad (3)$$

with $\eta = \frac{h_{cl}}{b_0}$.

Using the values given in table 1, it can be shown that for the experiments model (a) governs over model (b). Note: for other combinations of β, 2γ and η values, this may not be true.

COMPARISON OF THE EXPERIMENTAL AND NUMERICAL RESULTS

The experiments (ECSC, 1993) have shown that for connections with concrete filling in the column (4R2 and 4R4) failure of the connections was caused by the reinforcements, while for connections without concrete filling in the column (4R1 and 4R3), failure of the connections was caused by column face yielding. It should be emphasized that for 4R1 loaded by J=0, although failure of the reinforcements occurred during the test, it happened after very large deformation. The indentation of the column face at the maximum moment had exceeded the deformation limit of $3\%b_0$ given by Lu et. al. (1994). For 4R3 loaded by J=1, no maximum moment has been obtained. Therefore, for determination of the connection strength $M_{u,exp}$, the deformation limit of $3\%b_0$ has been used. The corresponding rotation of the beam at this deformation limit was about ϕ=0.035 [rad.].

In order to obtain more information on the behaviour of these bolted connections, numerical modelling has been performed on the tested specimens. The details of the finite element modelling have been given by Lu and Wardenier (1997). In figure 5, the moment - rotation curves obtained from the experiments and the FE analyses are shown. The finite element analyses were terminated after the moment-rotation curves reached the points where failure of the reinforcements occurred during the testing. For connections with concrete filling in the column (4R2 and 4R4), failure of the slab is critical. The numerical results have been taken at the rotations where maximum loads have been obtained in the tests. The maximum connection strengths $M_{u,exp}$ and $M_{u,num}$ obtained from the experiments and numerical analyses have been summarized in table 2. For the comparison, the values determined by equations (1) and (2) have been also given.

From table 2, the effect of the concrete filling in the column and the influence of the multiplanar loading can be determined.

Connections	Concrete filling in column	Failure modes	$M_{u,exp}$ [kNm]	$M_{u,num}$ [kNm]	$M_{formula}$ [kNm]	$\frac{M_{u,exp}}{M_{formula}}$	$\frac{M_{u,num}}{M_{formula}}$
4R1	no	Column face yielding	78.0 *	76.2 *	75.1 (2)	1.039	1.015
4R2	yes	failure of the reinforcement	166.9	168.0	175.5 (1)	0.951	0.957
4R3	no	Column face yielding	72.4 *	76.2 *	76.7 (2)	0.944	0.993
4R4	yes	failure of the reinforcement	129.0	135.6	175.5 (1)	0.735	0.772

Table 2 Comparison of experimental and numerical results with the analytical models
Remarks:
* $M_{u,exp}$ and $M_{u,num}$ have been taken at ϕ=0.035 rad.
(1) values according to equation (1).
(2) values according to equation (2).

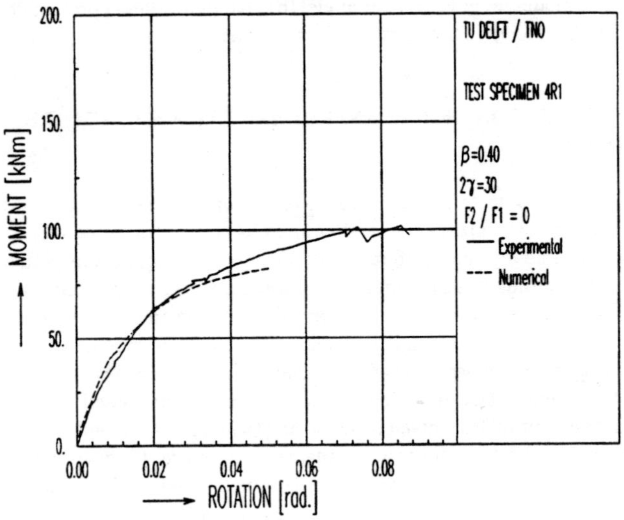

Fig. 5a Moment-Rotation Curves for 4R1

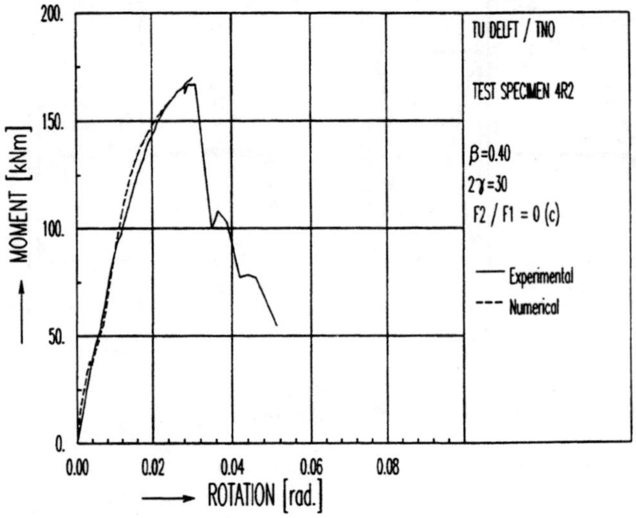

Fig. 5b Moment-Rotation Curves for 4R2

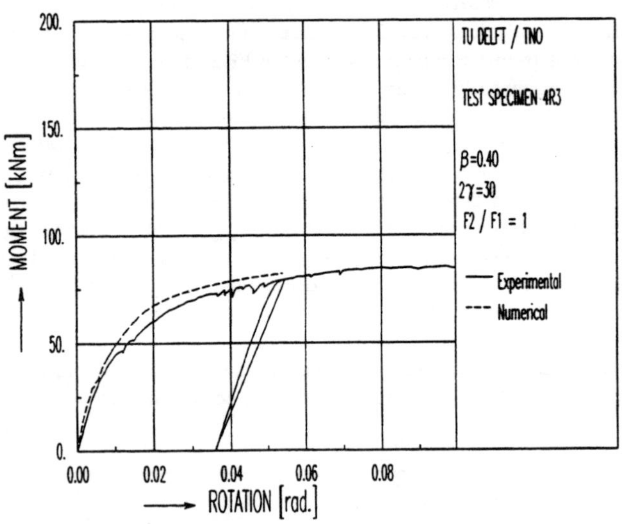

Fig. 5c Moment-Rotation Curves for 4R3

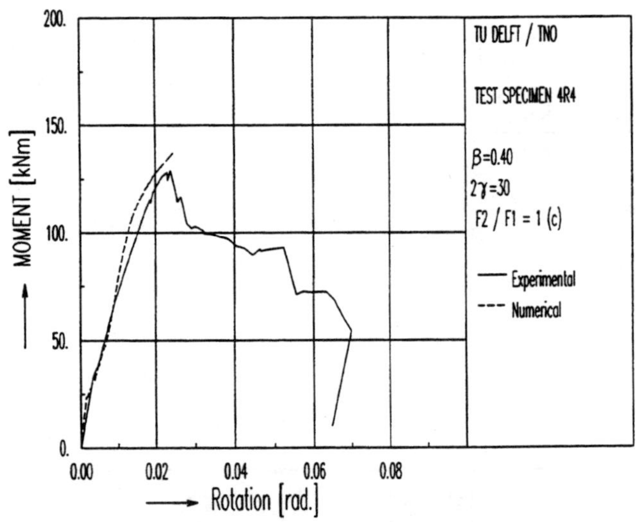

Fig. 5d Moment-Rotation Curves for 4R4

Figure 5 Moment-rotation curves obtained from the experiments and FE analyses

The Effect of the Concrete Filling in the Column

Figure 5 shows clearly that the stiffness and the strengths of the connections under the same load cases have been increased significantly by the concrete filling in the column, but the rotation capacities have been decreased considerably.

Comparison of 4R2 to 4R1 and 4R4 to 4R3 shows that for uniplanar loaded connections, the connection strength has been increased about 115% by the concrete filling in the column, while for multiplanar loaded connections, the increase is about 79%, see table 2.

The Influence of the Multiplanar Loading

For connections without concrete filling in the column (4R1 and 4R3), failure of the connections is governed by the column face yielding. The multiplanar loading has almost no influence on the load capacity at the deformation limit. The connection strengths at the deformation limit of ϕ=0.035 [rad.] are very close to the values determined according to the equation (2) for yield line model (a). However, the actual ultimate strength capacity for the multiplanar case is considerably lower than for the uniplanar case.

For connections with concrete filling in the column, failure of the slab is critical. The multiplanar loading J=1 (4R4) decreases the strength of the connection compared to the uniplanar loading J=0. During the testing, a progressive failure of the reinforcements has been observed. In test 4R4, the failure of the reinforcements started from the edge towards the column.

In order to understand this phenomena, the results of the numerical modelling have been further analyzed. In figure 6, the deformations of the slabs for 4R2 and 4R4 are shown. It can be seen that for 4R2 loaded by J=0, the slab rotates in one direction, about the non-loaded I-beams. The deformations of the slab above the loaded I-beams are larger than those at the edge of the slab. For 4R4 loaded by J=1, the slab rotates in two directions. The corners of the slab have larger deformations than the middle of the slab, see 6b. This is due to the deformation caused by the two sets of loads applied on the slab, see 6c.

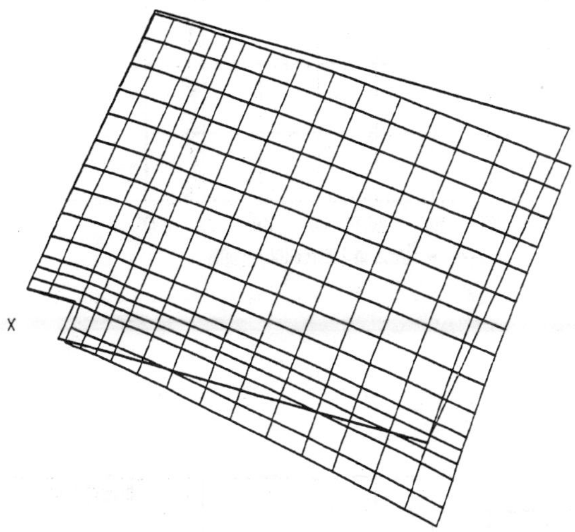

Fig. 6a J=0, 4R2

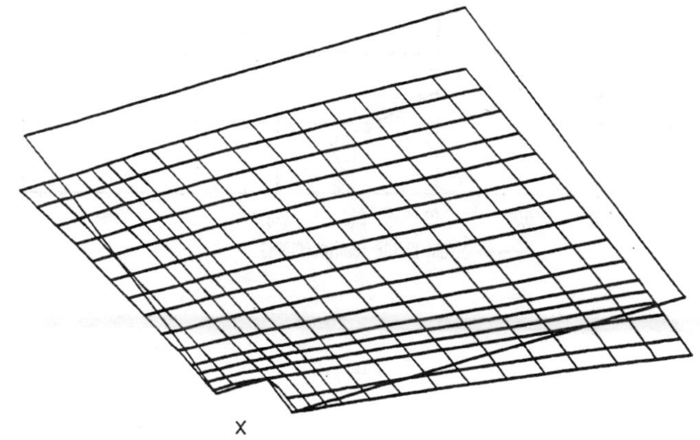

Fig. 6b J=1, 4R4

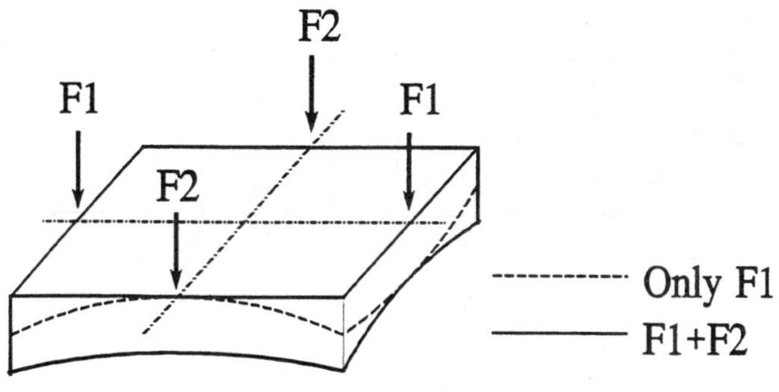

Fig. 6c The influence of F_2

Figure 6 Deformations of the composite floor under J=0 and J=1

The larger deformations result in larger strains in the reinforcements. In figure 7, the strains and the corresponding stresses in the reinforcements at the failure stage are shown at a distance of 105 mm from the middle line of the beams, where the main cracks of the slab were observed during the testing.

Figure 7a shows that for 4R2 loaded by J=0, the strains of the reinforcements nearby the column are larger than those of the reinforcements far away from the column. When the reinforcements ø6 nearby the column reach a certain strain, they fail. The stresses in the reinforcements nearby the edges of the slab are smaller than the yield stress of the reinforcement ø6, see figure 7c. It can therefore be said that those reinforcements nearby the edges of the slab are not effective.

For 4R4 loaded by J=1, the strains of the reinforcements at the edge are larger than those of the reinforcements close to the column because of the larger deformation of the corners, see figure 6c. Due to the small elongation at failure, the reinforcing bars ø6 failed starting from the edge of the slab towards the column.

CONCLUSIONS

Based on the experiments and the numerical analyses, the following conclusions have been made:

- For connections without concrete filling in the column, column face yielding is critical, while for connections with concrete filling in the column, failure of the slab governs.

- The concrete filling in the column increases significantly the stiffness and the strength of the connections, but decreases the rotation capacity of the connections due to the small elongation of the cold formed ø6 reinforcement bars.

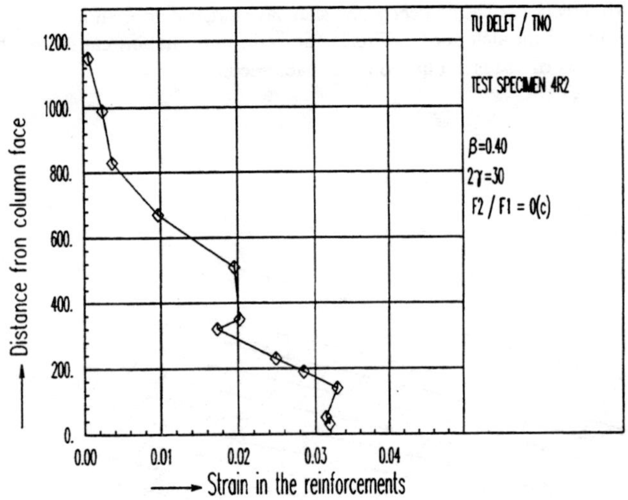

Fig. 7a

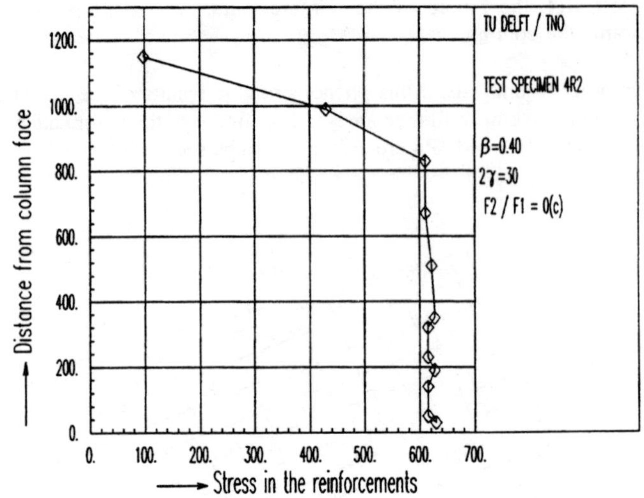

Fig. 7c

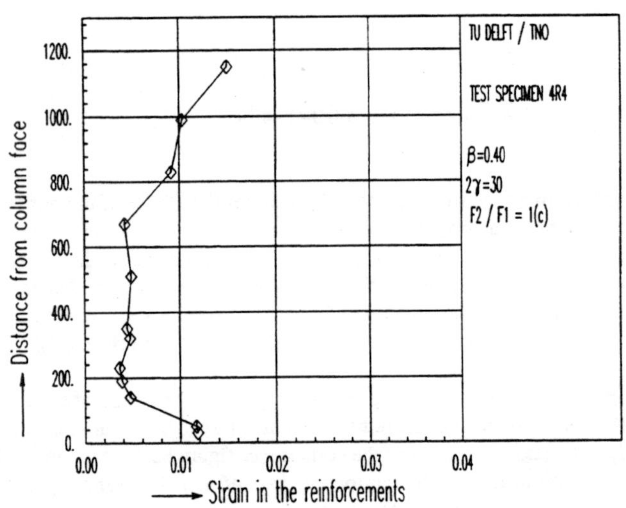

Fig. 7b

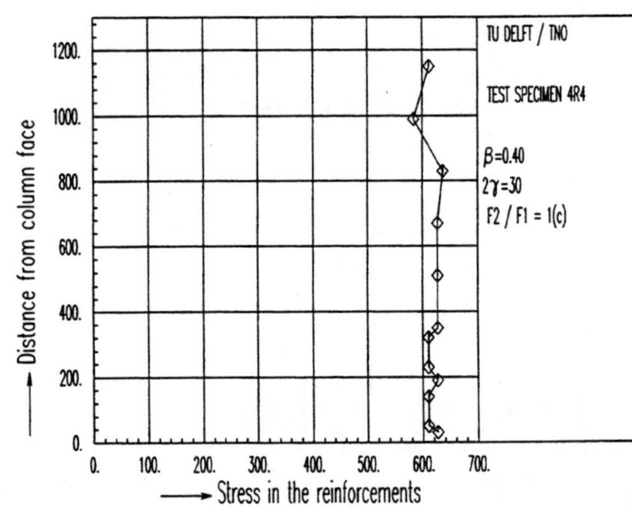

Fig. 7d

Figure 7 The strain and the stress of the reinforcements in x-direction

- For connections with β=0.4, 2γ=30 and η=0.27, without concrete filling in the column, the formula (2) derived from column face yield line model (a) can be used to predict the strength of the uniplanar loaded connections. The influence of the multiplanar loading is in this case negligible. For other combinations of β, 2γ and η, further parameter studies are needed.

- When failure of the slab is critical, the multiplanar loading has significant influence on the strength of the connections. Comparing the strength of 4R4 to that of 4R2, a decrease of about 22% in strength has been found.

RECOMMENDATIONS

The present experiments and numerical analyses show that the strength of semi-rigid connections can be considered in composite design. Use of the moment capacity of the connections results in more economical structures.

By filling concrete in the column, the stiffness and the strength of the connections can be significantly increased. However, the experiments have shown that the cold formed reinforcement bars do not have sufficient deformation capacity. In order to get a adequate warning of failure, hot rolled reinforcement bars are therefore recommended to use.

For the design of the composite connections, each structural part of the connections can be designed separately.

REFERENCES

Ahmed, B. and Nethercot, D.A. (1995). "Numerical modelling of composite flush end plate connections".

ECSC CONVENTION 7210-SA/611(F6.6/90) (1993). "Semi-Rigid Connections between I-Beams and Tubular Columns". Stevin report 6.93.27-A1/11.07, TNO-Bouw report 93-CON-RO832.

Eurocode 3 (1993). "Design of Steel Structures," Part 1: "General Rules and Rules for Buildings".

Eurocode 4 (1992), "Design of Composite Steel and Concrete Structures," Part 1: "General Rules and Rules for Buildings".

Lu, L.H., Puthli, R.S., Wardenier, J. (1993). "The Static Behaviour of Semi-Rigid Multiplanar Connections between I-beams and Rectangular Hollow Section Columns", Proc. Third International Offshore and Polar Engineering Conference (ISOPE-93) in Singapore.

Lu, L.H., Winkel, G.D. de, Yu, Y., Wardenier, J. (1994). "Deformation Limit for the Ultimate Strength of Hollow Section Joints," Proc. The Sixth International Symposium on Tubular Structures, 1994, Melbourne, Australia.

Lu L.H., Wardenier, J. (1995a). "Parametric Study on The Static Strength of Uniplanar and Multiplanar Plate to RHS Column Connections," Proc. Fifth International Offshore and Polar Engineering Conference (ISOPE-95), The Hague, The Netherlands.

Lu, L.H., Wardenier, J. (1995b). "The Static Strength of Uniplanar and Multiplanar Connections between I-beams and RHS Columns Loaded by Axial Compression," Proc. 14th International Conference on Offshore Mechanics and Arctic Engineering, Copenhagen, Denmark.

Lu, L.H., Wardenier, J. (1997). "The Numerical Modelling of Bolted Connections between I-beams with a Composite Floor and RHS Columns," to be presented at the Second International DIANA Conference on Finite Elements and Science Engineering, 11-13 June 1997, Amsterdam, The Netherlands.

Packer, J.A., Fear, C.E., (1991). "Concrete-Filled Rectangular Hollow Section X and T Connections," 4th International Symposium on Tubular Structures, Delft, The Netherlands. Proceedings pp. 382-391.

Wardenier J. (1982), "Hollow Section Joints", Delft University Press.

Proceedings of the Seventh (1997) International Offshore and Polar Engineering Conference
Honolulu, USA, May 25-30, 1997

ISBN 1-880653-28-1 (Set); ISBN 1-880653-32-X (Vol. IV)

Characteristics of Bending Strength and Residual Stress Distribution on High Thermal Cycle of Ceramic/Metal Joint

Young-Chul Park, Sun-Chul Huh and Hoo-Taek Koo
Dong-A University
Pusan, Korea

Kwang-Young Kim
Industrial Machinery Lab, KIMM
Korea

ABSTRACT

Since the ceramic has excellent qualities in light weight, abrasion resistance and heat reasistance etc., compared with the metal, it has been actively examined in order to apply for the structures such as gas turbine and turbo-charger etc., which require high strength and heat resistance. But it is not desirable to be used for the structural material since the ceramic is fragile, so the joint with the metal with abundant toughness has been studied. However, during the cooling process, the joint residual stress develops on the ceramic/metal joint by the difference in thermal expansion coefficient between two materials and this joint residual stress affects the bending strength significantly. Also in order to use the joint material as the structural material, the study about the fatigue of thermal cycle of actual use statement is necessary. Therefore, to ensure security and improvement of the bending strength of joint material, we studied the state of residual stress distribution to the high temperature-thermal cycle, and studied the effects of thermal cycle and state of residual stress distribution on the strength of joint material as well.

Key Words : Ceramic/Metal Joint, Residual Stress, Thermal Cycle, X-Ray Diffraction method, Bending Strength

1. Introduction

Recently, it is actively examined to use ceramic materal instead of metal material for the structures such as gas turbine, turbo-charger, which require light weight, high strength and heat resistance. And it is practically used in some cases. But, eventhough the ceramic is excellent in light weight, wear-resistance, and high heat resistance etc, it is very difficult to use the ceramic material for the general structural material because it has the fatal weakness in collision resistance, toughness strength and manufacturing process. Therefore using the ceramics/metal as the joint material with the metal with abundant toughness removing those weakness of ceramic, it can be used in practical structures as the material complementing good property and weakness.

In general, ceramics and metals are joined at a high temperature. The joint residual stress will develop during the cooling process by plastic constraint of both materials due to the difference of thermal expansion between ceramic and metal and at the same time the tensile residual stress occurring in ceramic lowers the fracture strength of joint material considerably.

For the ceramic/metal bonding material, as far as the bonding technology is concerned, the researches on the method lowering joint residual stress have been performed and especially understanding full information on the magnitude and distribution of joint residual stress. It can be regarded to obtain enough improvement of joint strength and reliance. The method measuring residual stress on the joint surface are supersonic, X-ray diffraction, and IF method etc.

However, X-ray diffraction which investigates the partial residual stress on material surface is the best way to measure non-uniform residual stress distribution, so in this study, measuring the joint residual stress grown on the joint surface by the X-ray method will be carried for the residual stress distribution of bonding material.

In addition, in order to use ceramic/metal material as a structural material the problems happening for the long time use such as the fatigue of heat cycle, stress corrosion crack, and creep should be solved prior to the use. In the general structures, the fatigue of heat cycle is the most wide affection on the strength of bonding material. The study to solve this point is necessary. Especially the repetitive heat cycle is considered of significant effect to lower the joint strength. The research for this problem has been rarely performed.

Si_3N_4/STS304 bonding material is employed for this research. After bonding, when the heat cycle of 1 cycle, 3 cycles and 10 cycles are loaded at 200℃, 300℃, 400℃ and 500℃ respectively, the residual stress distribution of joint surface is examined by PSPC type X-ray stress measuring device to study the variousness of residual stress distribution completely. By the 4-point bending, we carried the qualitative evaluation of heat state of the joint material due to repetitive heat cycle at each temperature as well as the complete understanding on the destructive device by analysis of fractural surface. Thus we have achieved full understanding about the temperature of heat cycle, the variousness of residual stress distribution by the number of repetition of heat cycle, and the effect of residual stress on joint strength.

2. Experimental Method

2.1 Specimen and Joining Method

A specimen is produced using the reactive metal method by joining nitriding silicon(Si_3N_4) and austenite stainless steel(STS304) with a copper thin layer between them. The shape and dimension of the specimen are shown in Fig. 1.

The material properties of specimen such as elasticity coefficient, Poisson's ratio, and thermal expansion coefficient are shown in Table 1. A reactive metal method is adopted as the joining method where relatively high strength and reliability can be obtained. The joining conditions are shown in Table 2.

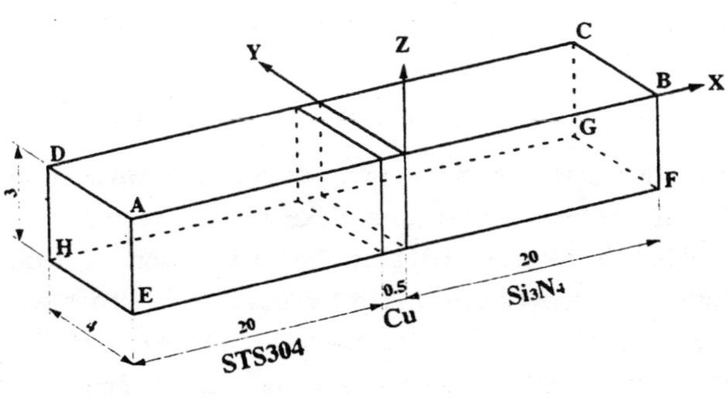

Fig. 1 Configuration of ceramic/metal joint specimen (unit:mm)

Table 1 Matreial properties

	Si_3N_4	Cu	STS304
E (GPa)	304	108	193
ν	0.27	0.33	0.3
α (/K)	3.06×10^{-6}	17.7×10^{-6}	16.5×10^{-6}

Table 2 Conditions of jointing

Brazing filler	Ti - Ag - Cu
Temperature	1073 K ~ 1123 K
Atmosphere	Vaccum, 1×10^{-5} torr
Inter layer	Cu (thickness 0.5mm)

2.2 Loading Condition of a Heat Cycle

The heat treatment of bonded material was performed at 200℃, 300℃, 400℃ and 500℃ using PIB silicone knit cylindrical electric furnace. Fig. 2 is the configuration of heat cycle. As seen in the picture the raising temperature-speed to reach each temperature is 1.74℃/mm.

After 8 hours of heat cycle temperature the specimen is cooled down to reach room temperature by furnace cooling and this process establishes one cycle. Repetition of this process provides the maximum 10 heat cycles.

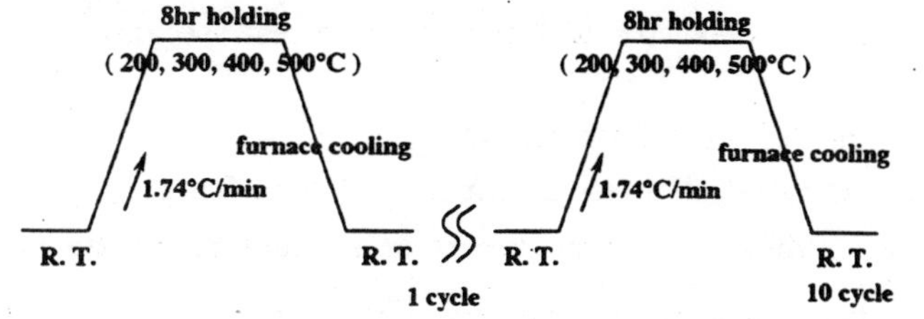

Fig. 2 Diagram of heat cycle

2.3 Measurement of Residual Stress by X-ray

Fig. 3 shows the measuring position to measure the residual stress of bonded material. As seen in the figure the residual stress is measured along three different directions. Along edge for the ceramic part of specimen surface, the vertical direction of joining surface(σ_x) is measured for x = 0.5, 1, 1.5, 2, 3, 5 and 7 mm along x line about y = 0.5 mm like ①. Along center line, the vertical direction of joining surface(σ_x) is measured for x = 0.5, 1, 1.5, 2, 3, 5 and 7 mm along x line about y = 2 mm like ②. Also, in order to have better investigation for the state of stress near the joining surface, the vertical direction of residual stress about joining surface(σ_x) is measured for x = 0.5, 1, 1.5, 2, 2.5, 3 and 3.5 mm along y line about x = 0.5 mm like ③.

2.4 4-Point Bending Test

At room temperature putting the joint at the center, a 4-point bending test was performed corresponding to JIS

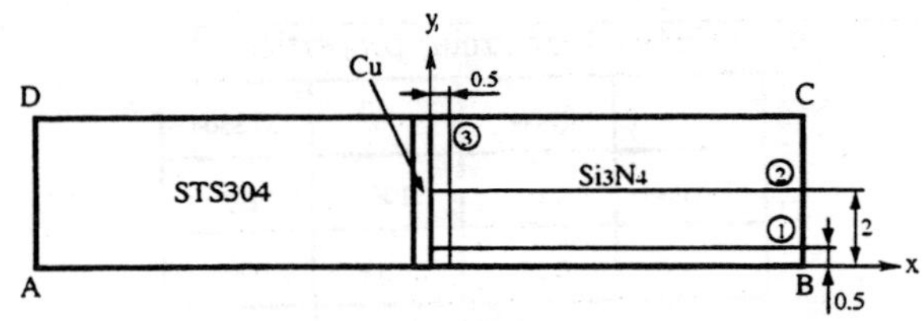

Fig. 3 Measurement positions of residual stress

R1601 on the specimens before a heat cycle and on the specimen after a heat cycle of 1, 3 and 10 at each temperature. Fig. 4 is the 4-point bending test and the speed of cross head is adjusted to 0.5 mm/min. From the maximum weight press P_{max} the bending strength σ_b is derived to the following.

$$\sigma_b = 3P_{max} (L-l) / 2WT^2$$

where W and T are width and thickness. L and I are lower and upper distance between two supports, respectively.

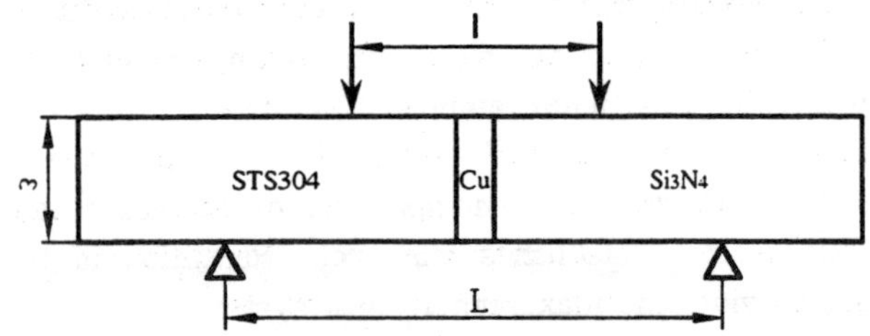

Fig. 4 Geometry of 4-point bending test specimen

3. Results and Discussion

3.1 The Residual Stress Distribution of Joint Surface

Fig. 5 shows the measurement of residual stress σ_x (x direction) distribution near the edge (by 0.5mm) in the vertical direction about the joining surface. As we can see in the figure, the maximum residual stress develops near the joining surface. However the stress reverses to compressive stress at about 2.5 mm from the joint and then the maximum residual stress is developed at about 4 mm from the joint. After that, the residual stress starts to increase again and compressive residual stress changes to tensile stress at about 5 mm from the joint, and then the residual stress becomes almost 0 constantly.

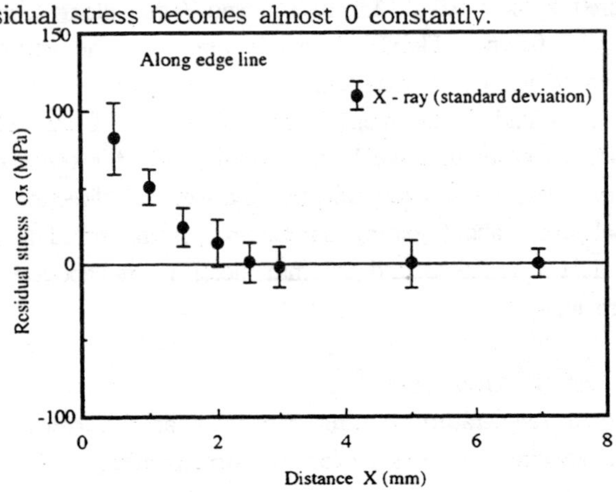

Fig. 5 Distribution of residual stress σ_x along the edge line of the specimen (X-axis)

Fig. 6 is the measurement of σ_x distribution along the center line (y = 2 mm). As seen in the figure σ_x distribution along the center line is almost same with that of the edge. Another word, σ_x becomes the maximum tensile stress around the joining surface and it decreases as the distance becomes far off from the joint. After σ_x becomes at the maximum compressive stress at about x = 2.5 mm, it approaches to 0 slowly. However σ_x at the

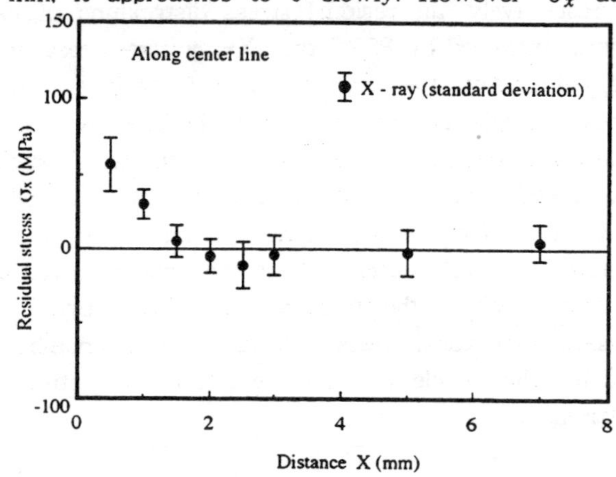

Fig. 6 Distribution of residual stress σ_x along the center line of the specimen (X-axis)

edge is much larger than that of the center line.

Fig. 7 is the measurement of residual stress σ_x distribution measured vertically (y direction) near the joint (x = 0.5 mm). According to the figure the distribution of σ_x has the maximum value of tensile stress at both edges of bounded material and the residual stress decreases as it approaches to the center. This distribution pattern is almost point symmetry to the center line. Since the tensile residual stress affecting fracture strength occurs maximum on the edges. It is presumed that the fracture point is near the joining surface. Therefore current research pays attentions to the edge parts for the variousness of residual stress by repetitive heat cycle.

Fig. 8 shows the variousness of residual stress (σ_x) corresponding to the heat cycle temperature when 1 cycle is loaded at each temperature. Like on the center line, for the edges, it is observed that the distribution of residual stresses obtained before and after heat cycle show almost the same trends. however, in the cases of 200℃, 300℃ and 400℃, the maximum tensile residual stress near of the joint increase due to the increasing heat cycle temperature.

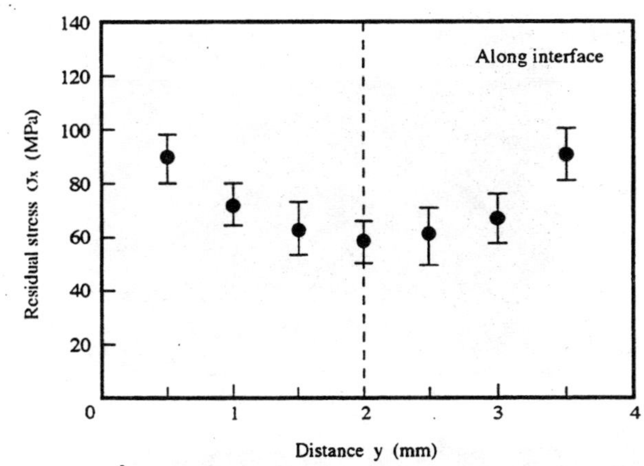

Fig. 7 The variation of residual stress σ_x at the interface along the Y-axis (x=0.5)

It is regared for the reason of residual stress that because the characteristic of carbonic isotropic, especially modulus of strain used as middle layer participates in the affection.

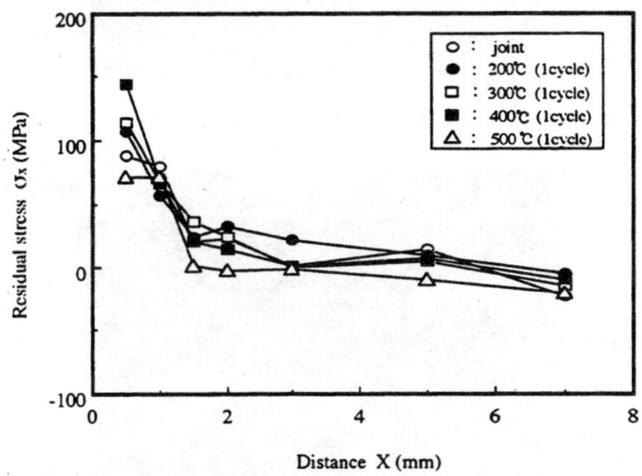

Fig. 8 Distribution of residual stress σ_x along the center line at various temperature

Fig. 9 is the result of the variousness of residual stress along the edge line corresponding to the number of heat cycle loaded when the heat cycle of 1, 3 and 10 are loaded at 400℃. As seen in the figure, it is known that the value of residual stress decreases by loading heat cycle repeatedly while the distribution pattern of residual stress remains constant by loading heat cycles. Especially the residual stress value near the joint showing the maximum tensile residual stress has a significant decrease.

Fig. 10 is the maximum tensile residual stress by loading heat cycle based on the measuring result so far, and near the joint the maximum tensile residual stress is the evaluation factor which shows the maximum tensile residual stress around the edge (y = 0.5 mm).

We see in the figure that the variousness of residual

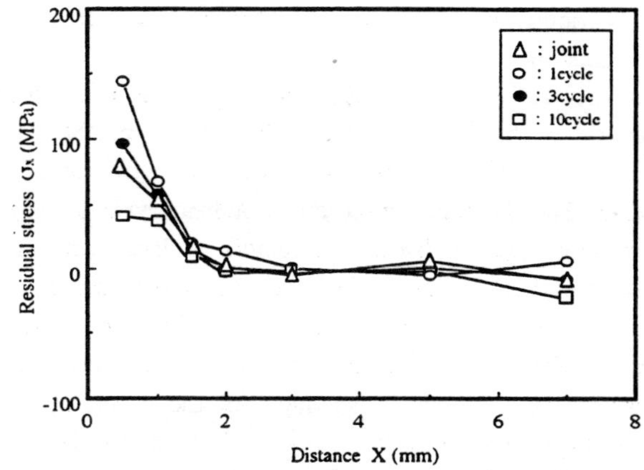

Fig. 9 Distribution of residual stress σ_x along the edge line at 400 ℃

stress by the number of loading heat cycle increases as well at every temperature of 200℃, 300℃, 400℃ and 500℃ for 1 cycle. however, it decreases for 3 cycles and 10 cycles, which means the number of heat cycle increses.

Also in the variousness of residual stress by the heat cycle temperature, for the case of low cycle loading such as 1 or 3, it shows the maximum residual stress at 400℃, while it has the maximum at 200℃ by increasing loading number. For the case of 500℃ of heat cycle loading, it has the minimum residual stress value regardless of the number of cycle. It is regarded that the reason for decreasing residual stress by increasing cycle is that the stress becomes free because there are low degree of plastic constraint due to heat of brazing filler along with the increase of plastic deformation of Cu. The study for this point will be carried with brazing filler as well as the arrangement of precipitation and compound by chemical reaction of ceramic.

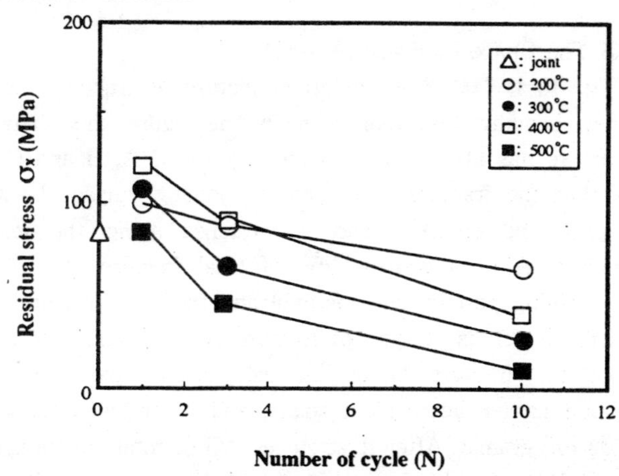

Fig. 10 Distribution of residual stress σ_x along the edge line at various cycle

3.2 Variousness of Fracture Strength by Heat Cycle Loading

Fig. 11 is the result for the variousness of fracture strength by heat cycle loading. In the figure the fracture strength of bonded material decreases by heat cycle loading. For the high temperature of heat cycle loading, the decreasing gap of fracture strength is large and as the number of heat cycle loading increases, decreasing gap of fracture strength is also large. Therefore, let's see the reason why the fracture strength of bonded material decreases by heat cycle loading. As the result in Fig. 9, for 1 cycle the residual stress increases by heat cycle loading which loads the decrease of fracture strength. For 3 cycles and more or on high temperature, brazing filler used in stress relaxation produces oxide and precipitation, so strength decrease largely occurs on the joint surface. Since the heat on the joint surface like that affects more than the relaxation of residual strength does, the fracture strength decreases.

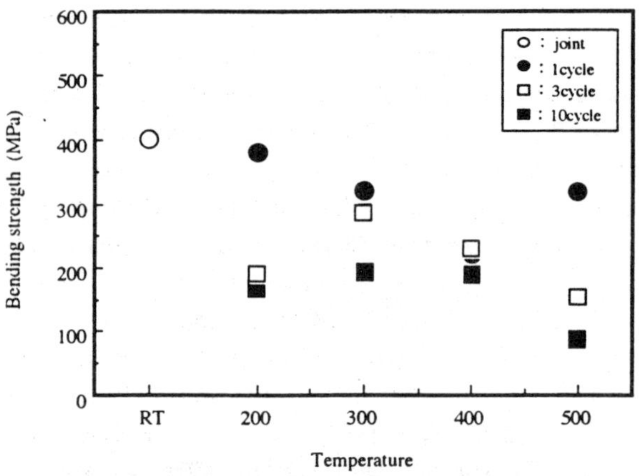

Fig. 11 Relation between bending strength and various thermal cycle temperature

3.3 The Shape of Fracture Surface

Fig. 12 is 200 times enlarged picture of fracture surface after bending test. As seen in the figure the shape of fracture has about three different types of A, B and C. (a) is that the fracture starts on the joining surface between copper and ceramic, and it stretches along the joining surface. After it moves 4/5 of total fracture, it turns to the joining part of two material and the final fracture ends there. Which is A type of fracture type. In contrary (b) is that the fracture states on the ceramic part with the maximum tensile residual stress, and it stretches along the side of ceramic. After it stretches 4/5 of total fracture, it turned to the border part and then establishes its last fracture surface on the border line between copper and ceramic. It is B type of fracture shape. And (c) is that the fracture states on the border between copper and ceramic,

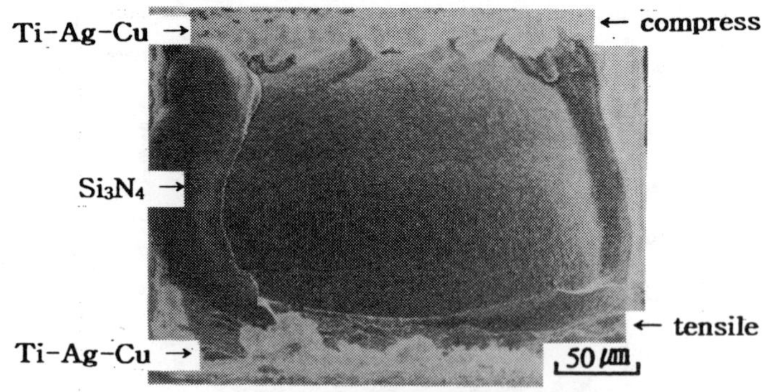

(a) a fracture mode

(b) b fracture mode

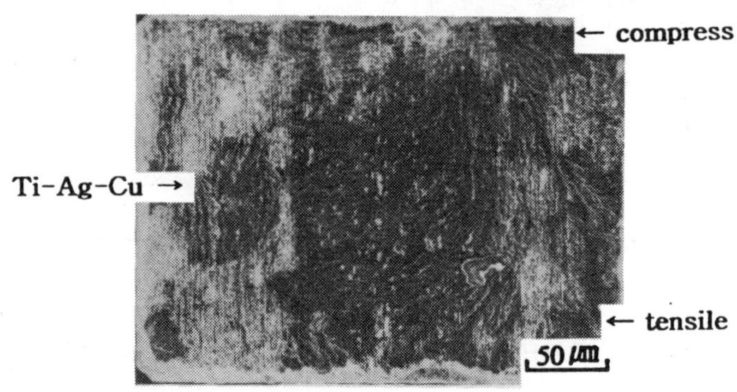

(c) c fracture mode

Fig. 12 Photographs of fractured plane by 4-point test of ceramic/metal joint

and it stretches linearly along the joint line. It is C type of fracture shape. For this type even the last fracture surface occures on the border part between copper and ceramic. The detailed explanation for this is in Fig. 13.

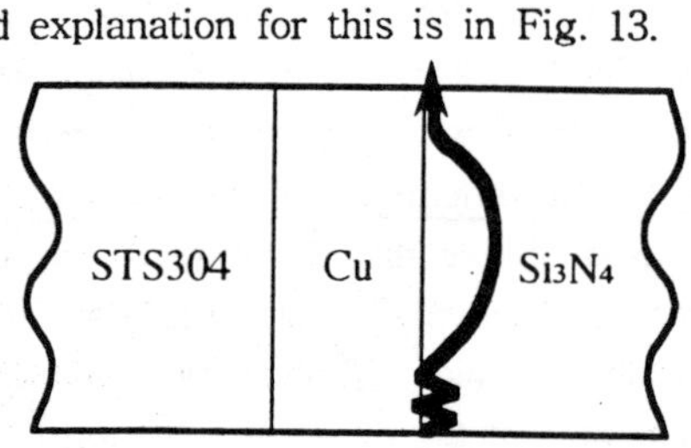

(a) a fracture mode

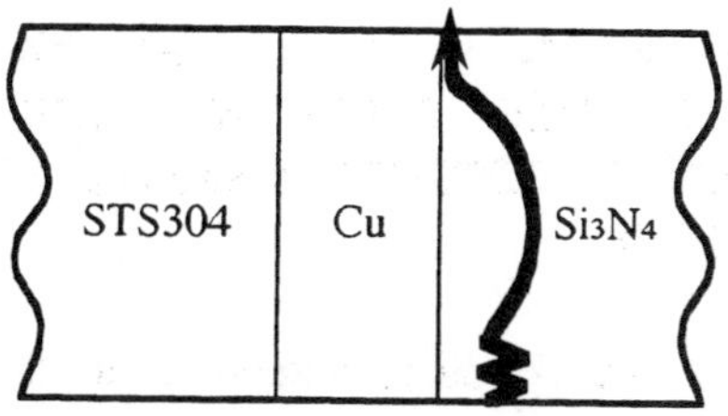

(b) b fracture mode

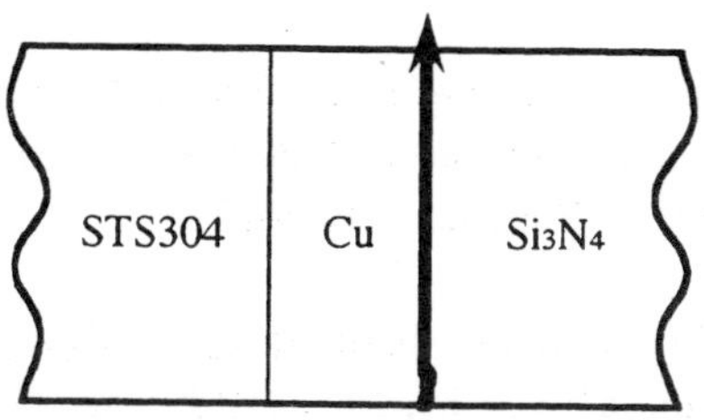

(c) c fracture mode

Fig. 13 Geometry of fractured plane of ceramic/metal joint

Fig. 14 is the result of different pattern of fracture by heat cycle loading at each temperature (c) for heat cycle loading at 200℃ all the shape of fracture is identical C type of fracture shape regardless of cycle number. (b) for heat cycle of loading at 300℃, it is B type for 3 cycles however for 10 cycles it is B type or A type. (c) for 400 ℃, as in 300℃, it is B type for 3 cycles and it is B type or A type for 10 cycles. (d) for 500℃ it is A type for both 3 and 10 cycles. By the above knowledge, it is understood that at the low temperature it is C type for somewhat low or high temperature it is B type, and for high temperature high cycle it is A type. So, the fracture shape varies to the heat cycle temperature and the number of repetition.

3 cycles 10 cycles

(a) 200℃

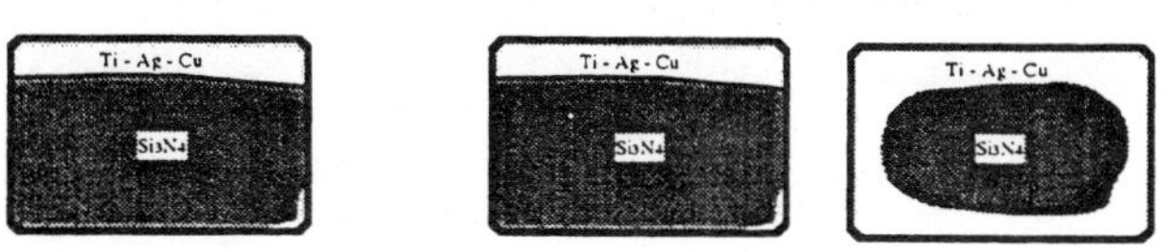

3 cycles 10 cycles

(b) 300℃

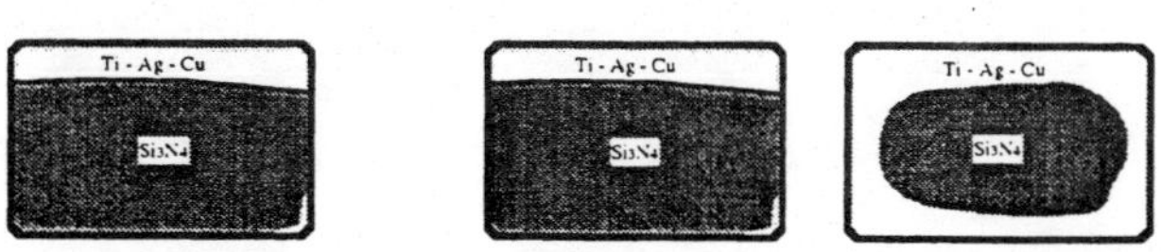

3 cycles 10 cycles

(c) 400℃

3 cycles 10 cycles

(d) 500℃

Fig. 14 Geometry of fractured plane in various temperature of ceramic/metal joint

Fig. 15 is the result that the maximum tensile stress applied on the fracture shape of C and A type when 4-point bending is carried out. It is 400 times enlarged picture for the fracture point (a) is the fracture shape of C type which the ceramic part becomes the fracture point. It is considered that the fracture being from the ceramic part by a residual stress and mechanical weight since the precipitation is not observed, but the oxide layer is found. (b) is the A type of fracture shape which the joint surface is the fracture point. For this type the oxide and precipitation layers formed around the joint surface of specimen can be observed and the parts like these are weak under the mechanical weight pressure. Therefore the fracture appears and grows at this part and it develops toward the joining part. And then it again turns to the ceramic part. For the case of high temperature and high cycle, the oxide and precipitation layers are produced by the fatigue of heat cycle. It is confirmed that these materials lower largely the fracture stress of joining surface.

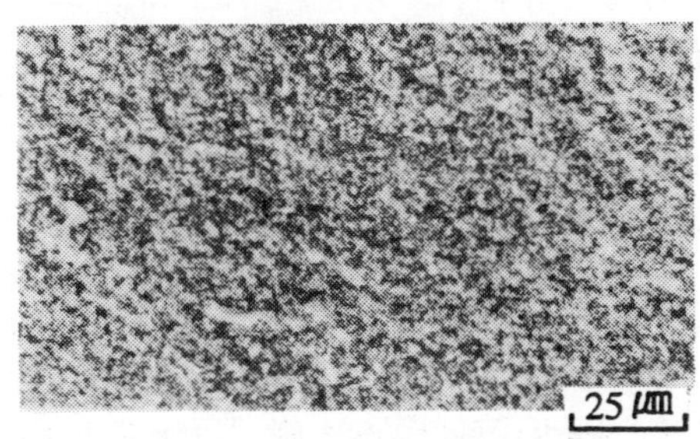

(a) c fracture mode

(b) a fracture mode

Fig. 15 Photo of fractured plane in b type and c type

4. Conclusion

1. High tensile residual stress is occurring on the edge near a joint of specimen.

2. The configuration of distribution of tensile residual stress does not change corresponding to a loading of single heat cycle, however the magnitude of maximum tensile residual stress changes. For 200℃, 300℃ and 400℃, it increases after 1 cycle, but the residual stress value decreases corresponding to the increase of the number of heat cycle such as 3 and 10 cycles.

3. By the heat cycle loading, the fracture strength decreases significantly. As the main factor to lower the fracture strength, the residual stress value affects directly to the fracture strength for the joint and 1 cycle. However, for the high temperature high cycle the oxide and precipitation rather than the residual stress directly affect to the decrease of fracture strength .

4. There are three types of fracture shape. For the case of low temperature and low cycle, it is C type for low temperature high cycles. It is B type, and for high temperature high cycle there are B and A types but it turns to A type consequently. Among these, A type is the lowest one for the fracture strength.

REFERENCES

Young Chul Park, Sae Wook Oh, Sun Chul Hue, Yomg Bae Cho, "Evaluation of Strength and Residual Stress in Si_3N_4 / SUS304 Joint", Trans. of the Korean Society of Mechanical Engineers, Vol. 18, No. 1, 1994, pp. 101~112

Young Chul Park, Sae Wook Oh, Sun Chul Hue, Kwang Young Kim, "Effect of Thermal Cycle on Strength of Ceramic and Metal Joint", Trans. of the Korean Society of Mechanical Engineers, Vol. 18, No. 7, 1994, pp. 1664~1673

Tanaka, s. Takahashi, Y. "Effects of X-ray beam collimation on the Measurement of Residual Stress Distribution in a Si_3N_4/Steel Joint", ISIJ International, Vol . 30, No. 12, 1990, pp. 1086~1091

D. B, Bogy and K. C. Wang, "Stress singularitys at interface corners in bonded dissimilar isotropic elastic materials", Int. J. Solids Structures, Vol. 7, 1971, pp. 993~1005

H. C. CAO and A. G. EVANS, "AN EXPERIMENTAL STUDY OF THE FRACTURE RESISTANCE OF BIMATERIAL INTERFACE", Mechanics of Materials 7, 1989, pp. 295~304

J. S. Wang and Z. Suo, "Experimental determination of interfacial toughness curves using brazil-nut-sandwiches", Acta metal, Vol. 38. No. 1 , 1990, pp. 1279~1290

F. Erdogan, A. C. Kaya, P. F. Joseph, "The crack problem in bonded nonhomogeneous Material", Transactions of the ASME, Vol. 58, 1991, pp. 410~418

Eager, T.W., "Ceramic-Metal Bonding Reseach of Japan", Welding Journal, Vol. 66, No. 11, 1987, pp. 35~37

B. J. Dalgleish, M. C. LU and A. G. Evans, "The Strength of Ceramics Bonded with Metals", Acta metal, Vol. 36, 1988, pp. 2029~2035

Proceedings of the Seventh (1997) International Offshore and Polar Engineering Conference
Honolulu, USA, May 25-30, 1997

ISBN 1-880653-28-1 (Set); ISBN 1-880653-32-X (Vol. IV)

Multiaxial Fatigue in Offshore Tubular Joints

Andras Abel and Xiao-bo Yu
University of Sydney
Sydney, Australia

ABSTRACT

A typical offshore steel platform structure may contain hundreds of welded tubular joints, the fatigue of which is a main concern in the design as well as during the operational life of the platform. Because of the geometry involved - a cylinder bisected by another cylinder for load transfer purposes - all three loading modes (axial, in-plane bending and out-of-plane bending separately and in any combination) will result in a biaxial stress state in the vicinity of the weld-line where fatigue is taking place.

Experimental and Finite Element Methods were used to evaluate the existing stress states by determining the so-called stress and strain concentration factors, SCF and SNCF respectively, for a number of large full-scale tubular joints. The results are presented with reference to the biaxial stress states existing in the vicinity of the weld lines

Experiments were also conducted on cylindrical specimen to investigate crack propagation tendencies when under controlled condition biaxiality can be established and varied during testing. Some of the findings are used to draw attention to the great differences arising in the fatigue performance when multiaxiality is involved.

KEY WORDS: Tubular Joints, Multiaxial Fatigue, Crack Propagation, Stress Concentration Factors (SCF)

INTRODUCTION

The advantages offered by tubular joints in onshore applications, such as warehouse, factory buildings, space frames, tower structures, bridges, etc., may be measured on terms of economy, depending on the skill of both designers and fabricators. In the case of offshore, two additional advantages are offered when circular tubular members are used. One relates to the drag characteristics, and the other to the buoyancy during the installation procedure. However, the welded tubular joints provide a few facets which are contributing heavily towards fatigue problems associated with offshore platforms.

The local maximum stress in a joint is the so called "hot-spot" stress, the magnitude of which is made up from the applied nominal stress multiplied by the SCF. The SCF, on the other hand, is influenced by the load transfer characteristics between the joined tubes and the notch stresses which are particularly influenced by the weld detail. As a result of these two factors plus the existence of a considerable residual stress state, due to weld effects, when a cylinder is bisected by another cylinder for load transfer purposes - all three loading modes, axial, in-plane bending and out-of-plane bending separately and on any combination, will lead to the biaxial stress state in the vicinity of weld-line where fatigue is taking place. A number of articles were dealing with this type of problems (Shinner 1983; Lee 1988; Abel 1989; Wu and Abel 1991a, 1991b, 1992; Abel and Wu 1993) and this communication is refocusing on the multiaxiality involved in offshore tubular joints.

EXPERIMENTAL RESULTS INDICATING BIAXIAL STRESSES

Nine large, full-scale specimens were fatigue tested in the as welded or residual stress relieved condition. Comprehensive strain gauging was used, including single element, rosette and crack propagation gauges and they were positioned on the outside and inside surface for both brace and chord. Two systems were used to monitor gauge response: a 200 channel Hewlett Packard automatic data acquisition, and a 200 channel Intercole Systems Compulog. A number of utility computer programs were written to process the data onto more meaningful form.

A specially designed loading rig allowed the simultaneous operation of three hydraulic loading jacks on the specimens which were bolted in place by flanges welded to the ends of the tubes.

Part of the obtained results due to static loading are shown in Figs. 1 to 4 and the fatigue results are indicated by Figs. 5 to 7.

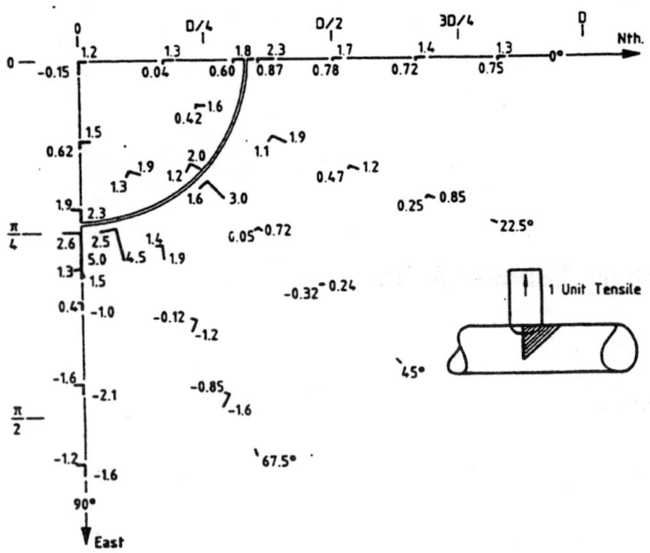

Fig. 1 Principal stresses on the outside surface of the chord. 1 unit tension is applied as indicated.

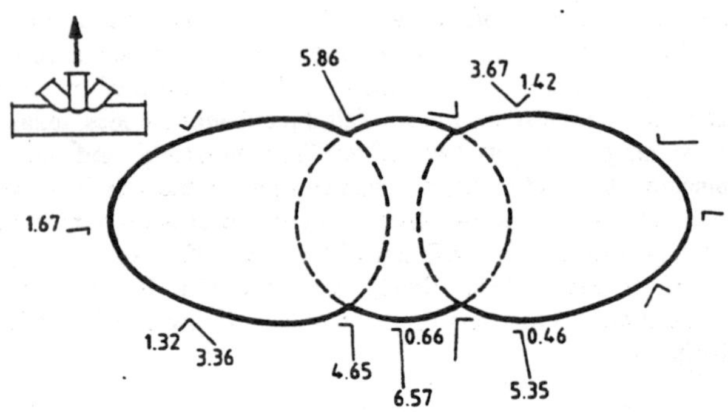

Fig. 2 Principal stresses on the inside surface of a TK joint. 1 unit tension is applied as indicated.

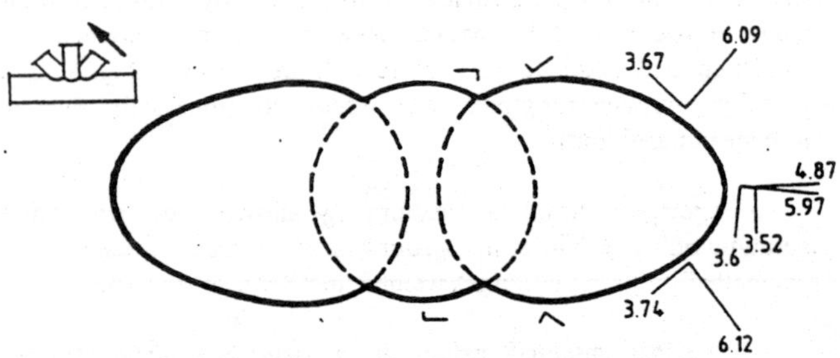

Fig. 3 Principal stresses on the outside surface of the chord. In-plane-bending is applied as indicated.

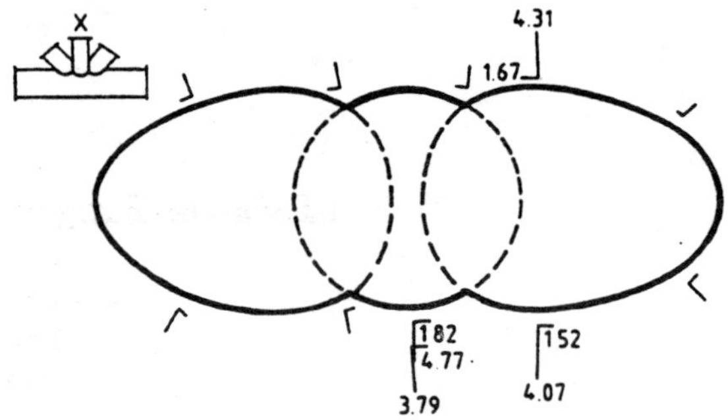

Fig. 4 Principal stresses on the outside surface of the chord. Out-of-plane bending is applied as indicated.

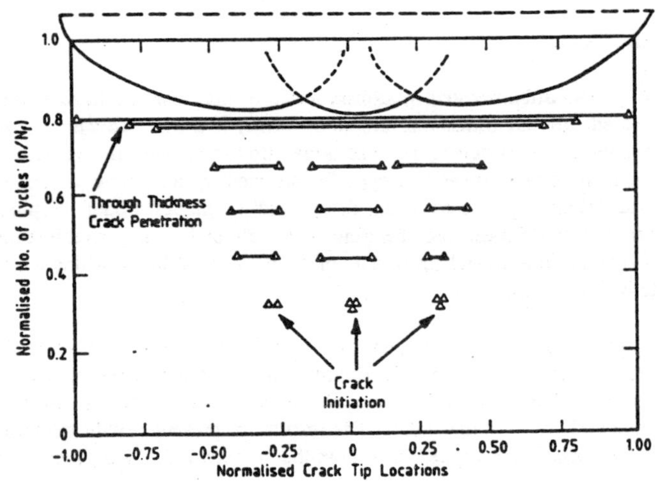

Fig. 5 Surface crack development

Number	1	2	3	4	5	6	7	8	9	10	11	12	13
n/Nf	0.21	0.30	0.35	0.38	0.41	0.44	0.48	0.54	0.73	0.83	0.89	0.95	1.00

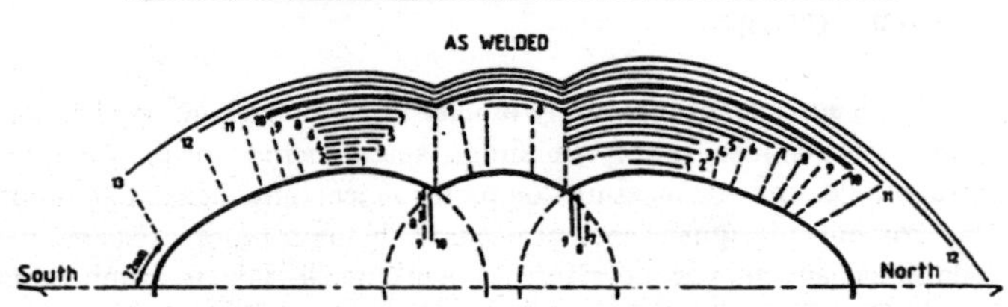

Fig. 6 Surface crack development. Through thickness crack achieved at 10.

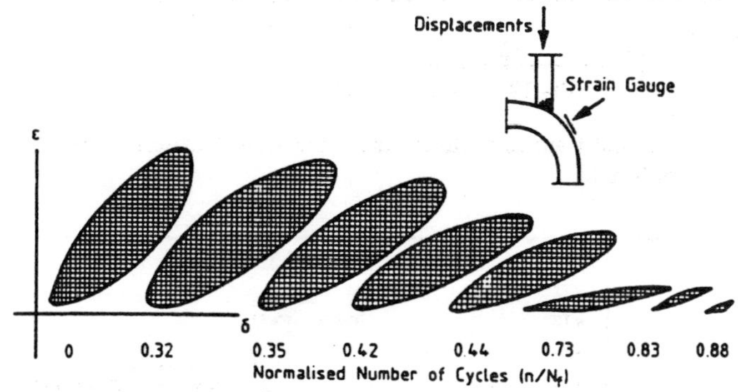

Fig. 7 Strain-displacement development.

EXPERIMENTAL RESULTS WITH BIAXIAL TESTING

Biaxial tests were carried out at room temperature on Instron 8503 electro-hydraulic machine.

Thin-walled tubular specimens with size of 57.5 mm outside dimension and 1.5 mm thickness were manufactured from 60.3 mm x 3.9 mm seamless carbon steel pipe (ASTM standard A106-93 Grade B), with the chemical composition of C 0.190, Si 0.240, Mn 0.56, P 0.018, S 0.008 and proof and ultimate strength of 399 MPa and 500 MPa respectively. A double-sided crack start notch, as shown in Fig.8(b), was introduced at the middle of the specimen's length by means of drill and electric discharge method.

The adaption of notched thin-walled tubular specimen is along the line of Yokobori *et al* (1985) and Gao and Chen (1989) but the present specimens have larger diameter-to-thickness ratio, which helps to eliminate the effects caused by unevenly distributed shear stresses under torsion.

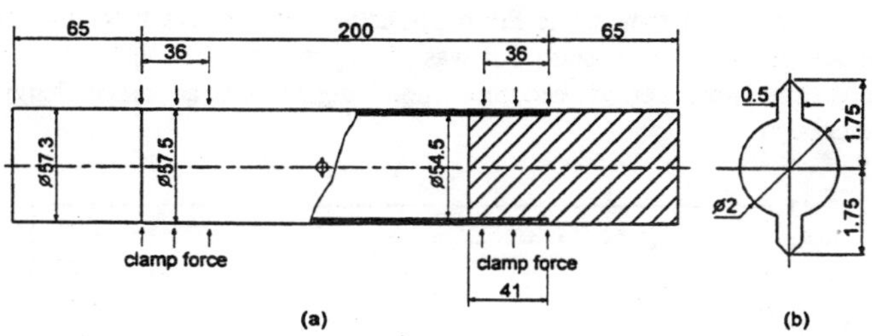

Fig. 8 Details of (a) thin-walled tubular specimen and; (b) double-sided crack start notch

Seven tests, numbered as T1 to T7, were performed under combination of axial cyclic loading and static torque. Axial loading was applied with 10Hz sinusoidal wave form for all tests with constant P_{min} and P_{max} values. Static torque, on the other hand, was only introduced after crack length reached a certain value (T1 to T6), or was applied from pre-cracking until crack length equaled to certain value (T7). Details of the test conditions are listed in Table 1, where 26.4 KN axial load and 360 Nm torque are equivalent to 100 MPa normal stress and 50 MPa shear stress respectively, as calculated for integral specimens. Symbols a_L and a_R stand for left-side crack length and right-hand crack length respectively, which are both measured from the centre of the notch. Crack lengths were measured by means of crack gauges and visual observation. Once the crack deviated from circumference (Fig. 9), crack lengths were mearsured by projection along the circumference.

Table 1. Details of Test Conditions

Test No.	Axial			Rotary (static torque)	
	P_{min}	P_{max}	R_{axial}	applied	when
T1	0	26.4 KN	0	360 Nm	$a_L \geq 5.3$ mm
T2	0	26.4 KN	0	-360 Nm	$a_L \geq 5.3$ mm
T3	0	26.4 KN	0	-360 Nm	$a_R \geq 8.3$ mm
T4	5.3 KN	31.7 KN	0.17	360 Nm	$a_R \geq 6.1$ mm
T5	-5.2 KN	21.2 KN	-0.25	360 Nm	$a_R \geq 5.3$ mm
T6	-5.2 KN	26.4 KN	-0.20	360 Nm	$a_R \geq 6.1$mm
T7	0	26.4 KN	0	360 Nm	$a_R \leq 6.4$ mm

Fig. 9(a) and (b) illustrate crack path deviations in tests T1 to T6 and test T7, respectively. Upon the introduction of torque, the crack propagation deviated in accordance to the tensile normal stresses arising from the applied torque. For tests T1 to T6, the deviated crack paths came out to be parabola-like curves and the largest tangential angles along the curves were found at the beginning of crack deviation.

For test T7, where static torque was superimposed from the very beginning, inclined crack propagation was observed from the crack start notch, however, unlike tests T1 to T6, no obvious curvature was observed. Once static torque was removed, cracks were observed to deviate to a direction opposite to their original inclination and parabolic-like curves were observed again and the largest tangential angles again occurred at the beginning of deviated curves. The deviation angels listed in Table 2 were obtained by use of a 50 times magnification.

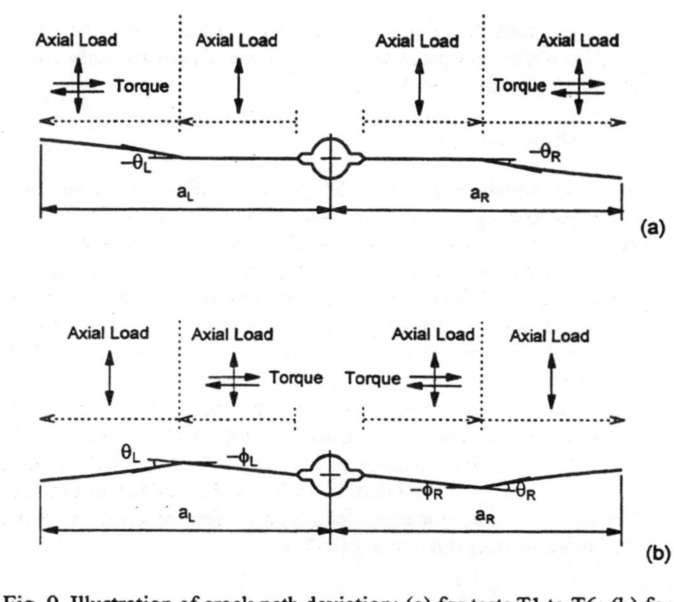

Fig. 9 Illustration of crack path deviation: (a) for tests T1 to T6; (b) for test T7.

Table 2. Crack Path Deviation Angles

Test	ϕ_L	ϕ_R	θ_L	θ_R
T1	—	—	-8.0°	-6.0°
T2	—	—	12.0°	13.5°
T3	—	—	7.0°	8.0°
T4	—	—	-10.5°	-9.5°
T5	—	—	-12.5°	-14.5°
T6	—	—	-12.0°	-12.5°
T7	-7.5°	-9.0°	23.0°	25.5°

Besides crack propagation paths, static torque also affected crack growth rates. Fig. 10 shows the relationships between crack length and relative cycle numbers for all seven tests. It can be seen that crack growth rate decreased with application of static torque. It should be also noted that crack propagation tendency is similar between left-side and right-side cracks.

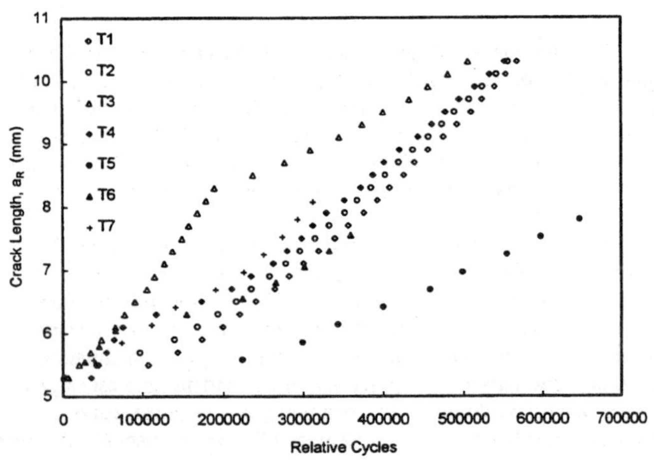

Fig. 10 Crack length versus relative cycle numbers

To evaluate the effects of static, the average crack growth rate da_m/dN is plotted against nominal mode I stress intensity range ΔK_I in Fig. 11. ΔK_I is calculated by Eq. 1,

$$\Delta K_I = Y \Delta \sigma_0 \sqrt{\pi a_m} \tag{1}$$

where $\Delta\sigma_0$ equals to $\Delta P / A_0$, ΔP is operational axial loading range and defined as $P_{max} - P_{min}$ for $P_{min} \geq 0$ or as P_{max} for $P_{min} < 0$ in accordance with ASTM standard E647-88a (1990) and, A_0 is transverse cross section area for a tube without crack and a_m is average length of left-side and right-side cracks. Y is shape function introduced to account for the geomatric difference of a crack with start notch and in cylinderical shell from one without notch and in infinite plane plate. While simplied, shape function Y is equal to multiplication of two parametres, Y_n for notch effects and Y_β for curvature effects. Results by Bowie (1956) and Newman (1971) showed that the effect of notch is less than 1% when crack length is over three times of the notch size so $Y_n \approx 1.0$ for the present case where $a_m > 5.3$ mm. For small length circumference crack, according to Lakshminarayana and Murthy (1976),

$$Y_\beta = 1 + \frac{\pi}{8}\beta^2 \tag{2}$$

where β is dimensionless curvature parameter and defined as,

$$\beta^2 = \frac{a_m^2}{8Rt\sqrt{12(1-\nu^2)}} \tag{3}$$

where R and t are respectively radia and wall thickness of tube, ν is Poisson's ratio.

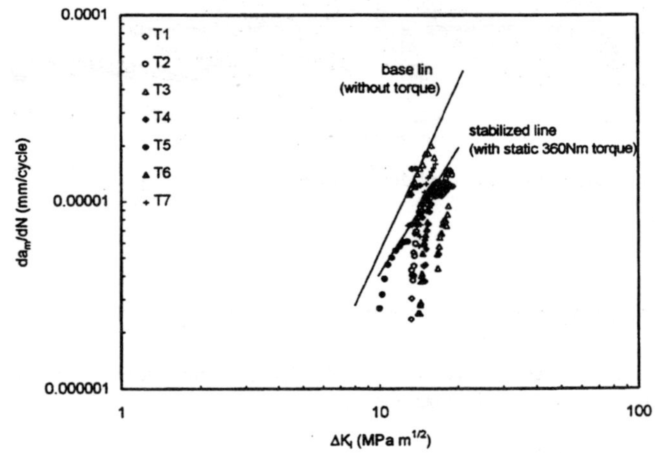

Fig. 11 Effects of static torque to crack propagation

It is shown from Fig. 11 that, upon the introducing of static torque, as the case of tests T1 to T6, significant reduction of crack growth rate follows and that such reduction gradually decreases with further crack growth. Crack propagation behavior within post-transition region can roughly be represented by a stabilized line as shown in the figure, which indicates lower crack propagation rates than base line without torque. Transient reduction of crack growth rate was also observed in tests T7 when the specimen was unloaded from static torque. Unlike test T1 to T6, crack growth rate of test T7 after the transition period is partially recovered.

Fig. 12 demonstrates the relationship between crack length and crack growth rate reduction. It was shown that torque's effect within post-transition period becomes more significant as crack length increases.

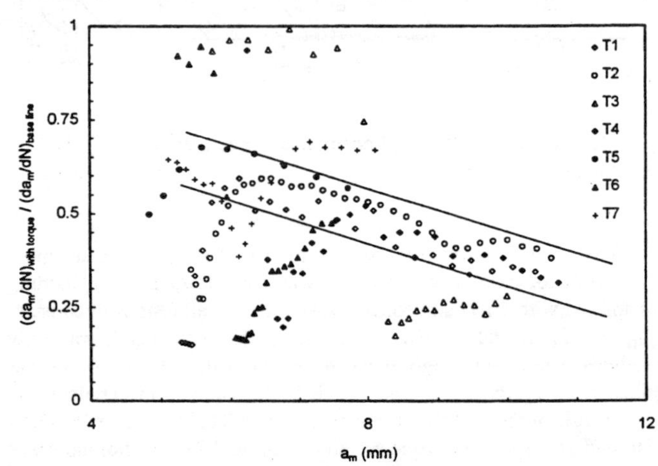

Fig. 12 relationship between crack growth rate reduction and crack length.

CONCLUSIONS

From the results of static and fatigue tests of tubular joints and biaxial tests of notched thin-walled specimens, it can be concluded that,

1) Axial loading or in-plane bending or out-of-plane bending or any combination of these leads to multiaxial stress state in tubular connections.

2) static torque affects axial load driven fatigue process in both crack path directions and growth rate: crack path deviation depends on the tensile normal stresses as arising from the applied torque; severe crack retardation occurs within transition period following the introduction or removal of static torque.

3) crack growth rate partially recovers with further crack propagation after introduction or removal of static torque but remains lower than those exhibited by axial loading only.

4) the effects of static torque after the transition period increase with the extension of crack length.

REFERENCES

Abel, A. (1989), "Fatigue of welded tubular joints," *3rd Int. Conf. on Biaxial/Multiaxial Fatigue*, April 3-6, Stuttgart, Germany, pp.66.1-66.20.

Abel, A and Wu, S. (1993), "Multiaxial fatigue study in tubular joints," *Proc. 3rd Int. Offshore and Polar Eng. Conf.*, Vol. **4**, pp. 56-61.

ASTM standard E647-88a (1990), "Standard test method for measurement of fatigue crack growth rate," *Annual Book of ASTM Standards*, Vol.**03.01**., pp.648-668.

Bowie, O. L. (1956), "Analysis of an infinite plate containing radial cracks originating at the boundary of an internal circlar hole," *J. Math. Phys.*, Vol. **XXXV**, No. **1**, pp.60-71.

Gao, Q. and Chen, X. (1989), "A fatigue life prediction methodology for cylindrical shell with an inclined crcak under torsion," *Theoretical and Applied Fracture Mechanics*, vol. **12**, pp. 155-160.

Lakshminarayana, H.V. and Murthy M.V.V. (1976), "On stresses around an arbitrarily oriented crack in a cylindrical sheel," *International Journal of Fracture*, Vol. **12**, No. **4**, pp.547-566.

Lee, P.K. (1988), *Fatigue Behaviour of Offshore Overlapped Tubular TK Joints, Ph.D Thesis*, School of Civil & Mining Engineering, The University of Sydney, Sydney 2006 Australia.

Newman, J. C. Jr. (1971), "An improved method of collocation for the stress analysis of cracked plates with various shaped boundaries," *NASA TN D-6376*, pp.1-45.

Shinner, C.D. (1983), *Fatigue Problems in Offshore Structures, Ph.D Thesis*, School of Civil & Mining Engineering, The University of Sydney, Sydney 2006 Australia.

Wu, S. and Abel, A. (1991a), "Analysis of fatigue surface crack growth in tubular joints used in offshore structures," *Proc. 1st Int. Offshore and Polar Eng. Conf.*, Vol. **4**, pp. 420-425.

Wu, S. and Abel, A. (1991b), "Application of line-spring element in analysis of stress intensity factor and fatigue crack propagation in tubular joints," *Proc. Int. Symp. Marine Structures*, Shanghai, China, pp. 345-350.

Wu, S. and Abel, A. (1992), "Fatigue crack stress intensity factor: the influence of residual stresses," *Proc. 2nd Int. Offshore and Polar Eng. Conf.*, Vol. **4**, pp. 312-317.

Yokobori, T. A. Jr., Yokobori T., Sato K. and Shyoji K. (1985), "Fatigue crack growth under mixed modes I and II," *Fatigue Fract. Engng. Mater. Struct.*, vol. **8**, n. **4**, pp.315-325.

Proceedings of the Seventh (1997) International Offshore and Polar Engineering Conference
Honolulu, USA, May 25-30, 1997

ISBN 1-880653-28-1 (Set); ISBN 1-880653-32-X (Vol. IV)

Effects of Joint Flexibility on the Fatigue Design of Welded Tubular Lattice Structures

A. Romeijn, S.A. Karamanos and J. Wardenier
Delft University of Technology
Delft, The Netherlands

ABSTRACT

The present paper investigates the effects of joint flexibility on the nominal stresses in tubular lattice structures. It is shown that these effects are quite significant for both axial and bending nominal stresses. Two simple formulae (for axial and bending stresses) based on finite element analyses, are proposed. Subsequently, the nominal stresses are "transferred" to the hot-spot stress level through appropriate stress concentration factors (SCFs). The results of the present study are summarized in a simple yet efficient method for calculating hot-spot stress in welded tubular structures. The method is of particular importance for the safe design of structures subjected to fatigue loading, such as offshore structures, bridges, cranes, masts and towers.

KEY WORDS: fatigue design, hot-spot stress, tubular joint, tubular structure, welded connection, joint flexibility.

INTRODUCTION - PREVIOUS WORK

The hot-spot stress method is used almost exclusively in the design of welded tubular connections. Hot-spot stress is calculated from nominal stresses which are combined with the corresponding stress concentration factors (SCFs). Significant effort has been devoted in the development of quite accurate SCFs for welded tubular joints under various loading conditions. On the other hand, a limited number of investigations have been conducted in order to determine the distribution of nominal stresses within a tubular lattice structure, with a high level of accuracy.

In tubular lattice structures, a simple framed-structure analysis (using beam elements) is conducted almost exclusively for practical design purposes, assuming rigid joints. Axial and bending stresses obtained from such an analysis are used as the nominal stresses for fatigue design. Nevertheless, there have been some doubts concerning the accuracy of such a model, especially regarding bending stresses at the brace ends. These doubts are motivated by the fact that the joint has a shell-like behavior with a certain local flexibility, resulting in nominal stresses quite different than those obtained through a rigid-joint assumption.

Det Norske Veritas rules (1977) was the first publication to report flexibility coefficients for tubular joints. These coefficients concerned in-plane and out-of-plane bending of T-joints and were based on FE analyses. The effects of tubular joint flexibility on structural response of offshore structures have been examined by Bouwkamp et al. (1980). The research was aimed at investigating deflections, force and moment distributions as well as dynamic characteristics of offshore structures taking into account that members are not rigidly connected. A computer program has been employed, using a nine-node doubly curved isoparametric shell element. The analyses were repeated using the assumption of rigidly connected beam elements. Results showed that deflections and distribution of bending moments within the structure obtained from the shell-element model are generally larger than those obtained from the beam-element model.

The UEG (1984) report presented a study on the effects of joint flexibility on the response of offshore jackets. Joint flexibility was examined with respect to deflections, axial force, buckling length and dynamic characteristics. The work by Fessler and Spooner (1981) on Araldite model joints, including seven multi-planer joints, constituted the basis for simulating joint flexibility. A methodology was proposed in the UEG report in order to form a stiffness matrix which takes into account the effects of joint flexibility.

Fessler et al. (1986a) presented improved equations for flexibility coefficients in terms of the joint parameters, regarding single-braced joints. In a subsequent publication, Fessler et al. (1986b) extended their research proposing coefficients for X and gap-K joints. The difference from the single-braced joints is the interaction between two braces, i.e. flexibility coefficients that relate displacement and rotations of a certain brace to the forces and moments at another brace.

Efthymiou (1985) published parametric equations for the local rotational stiffness of T, Y and K joints. Stiffness coefficients for rotational degrees of freedom were proposed in terms of joint parameters. Another interesting work on joint flexibility was reported by Buitrago et a. (1993). New equations for local joint flexibility coefficients were presented. The investigation was carried out using thick shell elements. The authors pointed out the difficulty of incorporating these coefficients in a structural analysis program. Instead, two models, referred to as the "spring model" and the "flex model", were proposed. It was the authors conclusion, that the consideration of joint flexibility in the analysis of offshore structures results in a redistribution of end-moments and axial loads, and in lesser structural demands on tubular joints.

The present study is aimed at providing evidence for the particularities of local joint behavior and investigating its effects on the overall behavior of a lattice structure. The effects are considered at both the nominal stress and the hot-spot stress level. A finite element formulation has been adopted in order to tackle the problem in a rigorous manner. Joints are modeled through shell finite elements, which can simulate local joint deformation quite accurately. Special emphasis is given on the reliable estimate of secondary bending stresses at brace and chord members.

The present investigation concerns girders with uni-planar concentric gap K-joints. A complete investigation for all possible structural configurations and loading conditions, and for all possible combinations of joint parameters would be an unrealistic goal for the purposes of the present study. However, the paper may be considered as a *pilot study on local joint effects*, which explores some aspects not investigated previously. The conclusions are of particular importance for assessing the integrity of lattice structures against fatigue loading.

FE MODELING - PARAMETERS

A typical gap K-joint girder has been considered for the purposes of our investigation. The girder, shown in Figure 1 has eight bays and its geometry is defined by specifying the following parameters (subscripts 0 and 1 refer to the chord and the brace respectively):

- brace over chord diameter ratio β (d_1/d_o)
- radius-to-thickness chord ratio γ ($d_o/2t_o$)
- brace over chord thickness ratio τ (t_1/t_o)
- brace angle with respect to the chord θ
- brace slenderness λ (L_1/r_1)

where L_1 is the brace length and r_1 is the radius of inertia of the brace (could be taken approximately equal to $0.354d_1$.). Furthermore, all joints of the girder are assumed to be identical.

It would be quite expensive from the computational point of view to examine all joints in the girder under consideration. Our attention is focused on Node 8, and all results which will be reported from the parametric study concern Node 8. The girder is loaded with a concentrated load at joint No. 9. Taking advantage of the symmetry with respect to the vertical axis through Node 9, only half of the girder needs to be analyzed.

The available numerical tools permit the systematic organization, solution and post-processing. A complete set of parameters has been selected and all possible combinations have been considered:

parameter	values		
β	0.25	0.40	0.60
γ	12	18	24
τ	0.25	0.50	1.00
θ	30°	45°	60°
λ	60	80	100

Table 1: Range of parametric study

No eccentricity has been considered at these joints. For a value of β equal to 0.60 and for brace angle θ equal to 60°, a concentric joint must be an overlap joint. In order to avoid overlap, a small eccentricity was assumed for this particular case, so that the gap between the two braces is equal to 8% of the chord diameter d_o. This small eccentricity, can be neglected for the purposes of the present study. All other joints considered in this parametric study are concentric. In total, a number of 243 ($=3^5$) analyses have been conducted in order to obtain a complete data base for all combinations of the joint parameters and slenderness stated above.

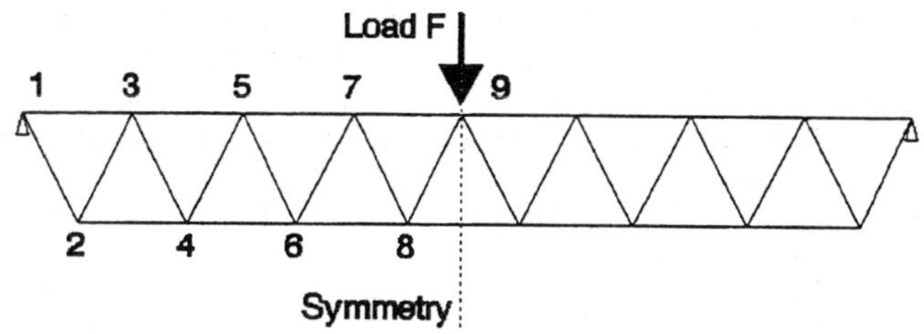

Fig. 1: Overall configuration of the K-joint lattice girder used for parametric study

Tubular joints were modeled using preprocessing package SESAM. This program generated nodes and elements to be used by FE package I-DEAS, through which the FE analysis of the girders was conducted. It was decided to use a combination of shell finite elements and regular beam-elements in order to model the girder efficiently. In addition, due to symmetry, only half of the girder needs to be analyzed.

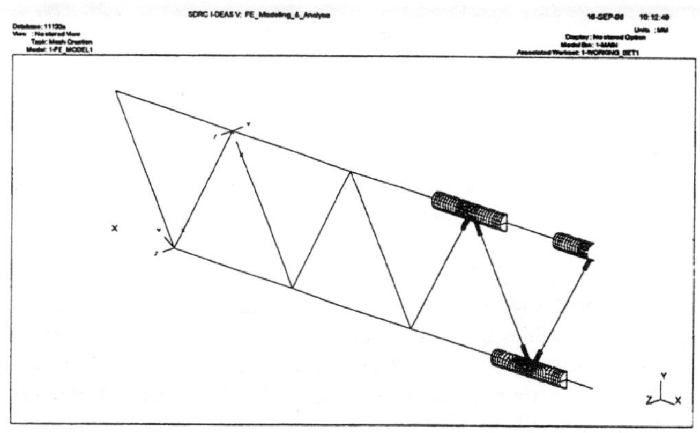

Fig. 2: Girder model used for parametric study (half-girder).

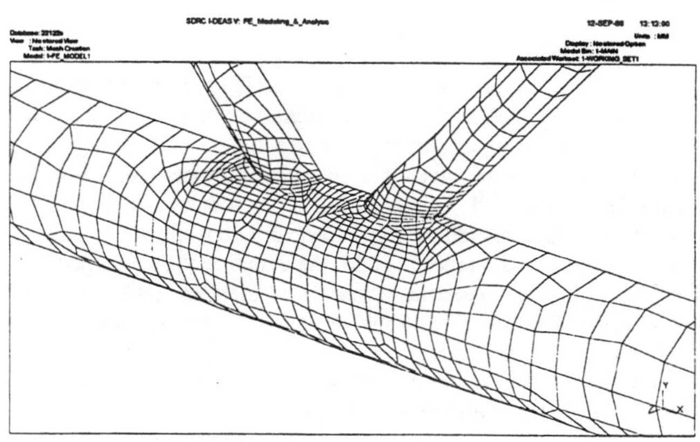

Fig. 3: Typical FE mesh for node modeling.

There are several issues regarding the numerical model such as the type of element to be used, mesh refinement, the number of nodes to be modeled with shell element as well as the length of braces and chords to be modeled with shell elements. These issues were resolved through a parametric study, described by Karamanos et al. (1997). Four-node thin shell elements were employed and three nodes of the half-girder were simulated with those elements (Nodes 7, 8 and 9 - Figures 1 and 2). A typical shell-element mesh used in this study is shown in Figure 3. Finally, the brace (or chord) part modeled with shell elements was taken equal to three brace (or chord) diameters. The model will be referred to as the "SHELL-model".

The following sign convention is adopted for the axial and the bending stresses:

- Axial stress is assumed to be **positive** if it is **tensile**, and negative if it is compressive.
- Bending stress is assumed to be positive if it corresponds to a bending moment having the direction shown in Figure 4. More specifically, it is assumed **positive** if it **"relieves" the stress** caused by axial forces at the weld **crown heel** locations.

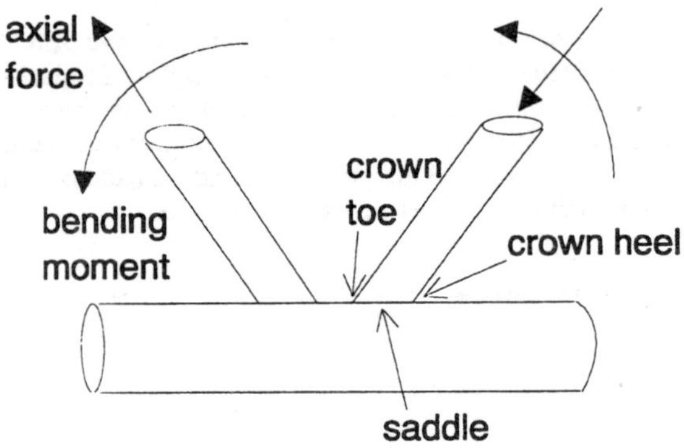

Fig. 4: K-joint configuration; definition of the points of interest of the connection; sign convention for bending moments.

For the purposes of our parametric study, the values of bending stresses or axial stresses are not reported in stress units, because they depend on the values of external load and the chord diameter. Instead, Bending-over-Axial-Stress (BoAS) ratios, indicating the relative magnitude of certain stresses, are reported, in a more "universal" format. More specifically,

$$BoAS = \frac{\sigma_{bend}}{|\sigma_{ax}|} \qquad ....(1)$$

The absolute value on the nominal axial stress is used so that the BoAS ratio has the same sign as the corresponding bending stress, regardless of the axial stress being tensile or compressive.

In order to compare with more conventional tools, another model has also been employed. This model uses beam elements exclusively. There is an issue concerning the implementation of this model: the "fictitious" parts of the brace "inside" the chord may be assumed rigid or to have the same properties as the brace element. Using infinitely rigid elements for these "fictitious" parts, the model is referred to as BEAM model. Several analyses were conducted using both the SHELL and the BEAM models regarding axial and bending stresses The following general conclusions were deducted:

- In general, axial forces (stresses) at chord and brace locations from the two models are quite close. However, there are some differences between the SHELL-model and the BEAM-model axial stresses for small brace size (i.e. β equal to 0.25) which are equal to about 8-10% in the average.
- There are some differences between the values of the bending moments (stresses) in the chord members. The difference between the two models is usually in the order of 10 to 50%, but in few cases it can be even larger (100-120% at most, i.e. an error factor of 2).
- Bending moments (stresses) obtained from the two models, at the brace members can be very different. Generally, the BEAM-model underestimates these stresses considerably. The ratio of the SHELL-model value over the BEAM-model value may be equal to 6 or more for several combinations of joint parameters. The direction of the predicted moment may also be opposite for the two models (Fig. 5).

It is the authors' opinion, that bending stresses at the brace provided by a simple framed-structure model that uses beam-elements exclusively may be *totally unreliable* for the general case. On the other hand, it would be impossible to use a model such as the SHELL-model for design purposes. The authors believe that a BEAM-model should be employed but its results should be corrected in order to provide reliable estimates of nominal stresses.

STUDY I: NOMINAL STRESS LEVEL

Axial stresses

When the axial forces (stresses) obtained from the SHELL and BEAM models were compared, it was found that there is a certain

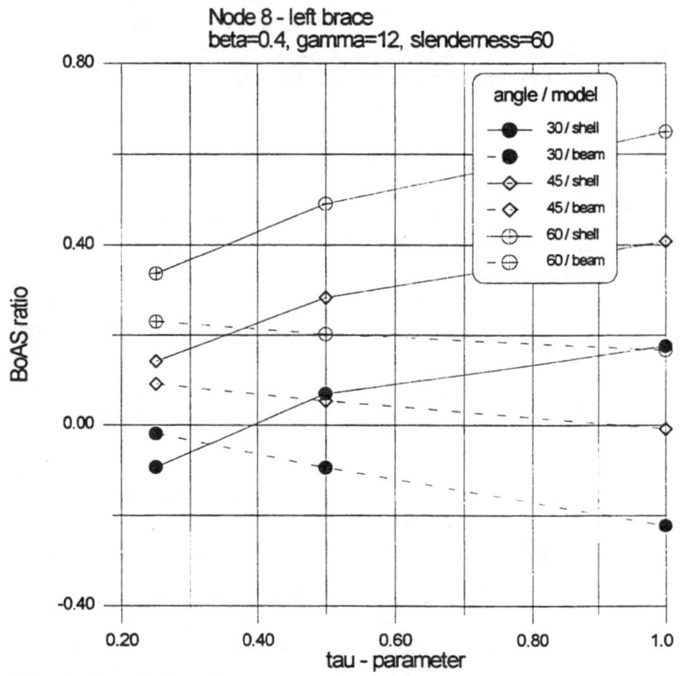

Fig. 5: BoAS (Bending-over-Axial Stress) ratio calculated form the BEAM model and the SHELL model.

difference especially for small β values. These differences were found less pronounced for β equal to 0.40 and 0.60. Nevertheless, the other joint parameters also appear to have an influence. The difference between SHELL-model and BEAM-model regarding axial stresses is explained by the fact that the chord ovalizes and cannot provide a rigid support to the brace. In fact, the tubular joint behaves similar to a spring (elastic support), which is responsible for reducing the value of the axial stress with respect to the corresponding value obtained from a rigid-joint BEAM model.

Based on the results of the rigorous analysis, the following parametric formula is proposed in order to take into consideration the effects of joint flexibility on the axial brace stress:

$$C_{ax} = C_o + C_1 \cdot \beta^{C_2} \cdot \gamma^{C_3} \cdot \tau^{C_4} \cdot (\sin\theta)^{C_5} \cdot \lambda^{C_6} \quad \text{....(2a)}$$

so that

$$\sigma_{ax} = (1 - C_{ax}) \cdot [\sigma_{ax}]_{beam} \quad \text{....(2b)}$$

The values of the seven constant coefficients C_i (i=0,1,...,6) are obtained from the results of the parametric study (243 cases) by means of a curve-fitting package, resulting in the following formula:

$$C_{ax} = 0.372 - 0.151 \cdot \beta^{0.322} \gamma^{-0.122} \tau^{-0.02} (\sin\theta)^{-0.137} \lambda^{0.306} \quad \text{......(3)}$$

The efficiency of the formula is shown in Figure 6. The inclusion of such a coefficient corrects the value obtained from a BEAM-model with respect to the more "rigorous value" obtained from the SHELL-model and its validity range is

$$0.25 \le \beta \le 0.60 \qquad 30^\circ \le \theta \le 60^\circ$$
$$12.0 \le \gamma \le 24.0 \qquad 60 \le \lambda \le 100 \quad \text{......(4)}$$
$$0.25 \le \tau \le 1.00$$

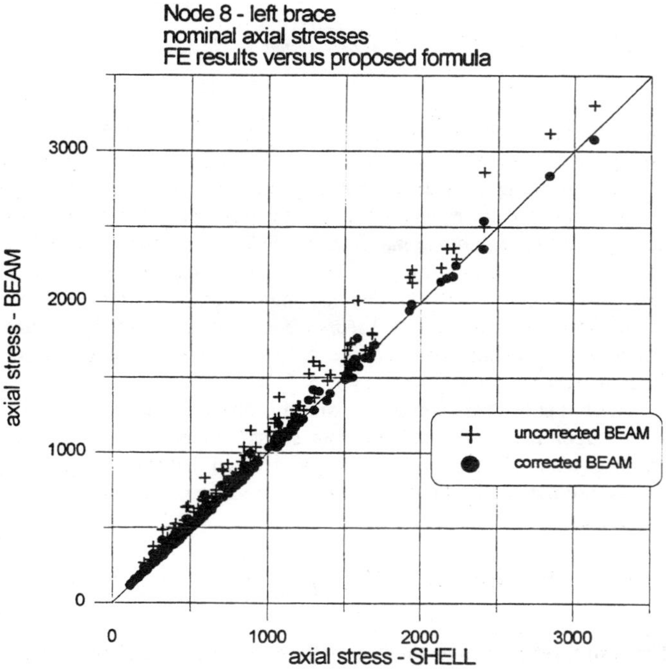

Fig. 6: Efficiency of the correction factor for axial stress, due to joint flexibility effects.

Bending stresses

Based on the FE results, it is possible to propose design guidance for design engineers to estimate bending stresses in lattice structures. In the course of the present study, one structural configuration was examined with one loading pattern. In order to make the results more "universal" and minimize the dependency on specific structural and loading configurations, the difference between the SHELL-model and the BoAS value from the BEAM-model is considered. More specifically,

$$BoAS = [BoAS]_{beam} + [DIFF] \quad \text{......(5)}$$

where *BoAS* is obtained from the SHELL-model, $[BoAS]_{beam}$ is obtained by the BEAM-model and *[DIFF]* denotes the difference between the results of the two models. The value of $[BoAS]_{beam}$ depends on the "overall" structural response of the structure (simulated with BEAM-model), whereas the value of *[DIFF]* would depend primarily on the local joint effects. Referring to Node 8, Figure 7 compares the *BoAS* ratio and the *[DIFF]* parameter at left and right braces. Clearly, the comparison between the *[DIFF]* parameter at the two locations is much more satisfactory than the comparison between the total *BoAS* ratio. This provides an additional indication that the value of the *[DIFF]* parameter depends on the local joint effects (i.e. on the joint parameters), and is rather independent on the "overall" structural behavior.

Using a curve-fitting package, it is possible to obtain a formula for *[DIFF]* in the form of equation (2a). After calibration the formula becomes:

$$[DIFF]_{predicted} = -0.618 + $$
$$+ 3.313 \cdot \beta^{0.0188} \cdot \gamma^{0.187} \cdot \tau^{0.30} \cdot (\sin\theta)^{0.219} \cdot \lambda^{-0.354} \quad \text{......(6)}$$

The value of *DIFF* should be superimposed to the value of $[BoAS]_{beam}$, obtained through a beam-type structural analysis of the lattice structure, considering joints to be rigid. The comparison between the *DIFF* value predicted by equation (7) and the rigorously computed *DIFF* value is shown in Figure 8. The comparison is very satisfactory. A nice feature of the above formula is the fact that it contains the five parameters (β, γ, τ, θ and λ) in an "uncoupled" form and, therefore, it is easily implemented. It is also assumed that every design engineer has the capability of conducting a beam-type analysis to obtain $[BoAS]_{beam}$. The total prediction of the *BoAS* ratio is computed as follows:

$$[BoAS]_{predicted} = [BoAS]_{beam} + [DIFF]_{predicted} \quad \text{...(7)}$$

where $[DIFF]_{predicted}$ is provided by equation (6). The range of validity is determined by equation (4).

Chord stresses

The comparison between the chord stress (axial and bending) obtained from SHELL-model and those obtained from the BEAM-model are in fairly good comparison. Axial stress from the two models are very close, their difference being within 2% for all cases examined. There were some differences regarding the value of the bending stresses, which range between 10-50%. Multiplication of the BEAM-model stresses with a "correction factor" of about 1.25 would improve the predictions of this model with respect to the more rigorous SHELL-model. Therefore, the following scheme is proposed, in order to obtain the total chord stress from a beam-type analysis to used in fatigue design:

$$\sigma_{chord} = [\sigma_{ch-ax}]_{beam} + 1.25 \cdot [\sigma_{ch-ben}]_{beam} \quad \text{......(8)}$$

where the axial and bending stresses are considered with their algebraic values in order to compute the total chord stress at the weld crown location.

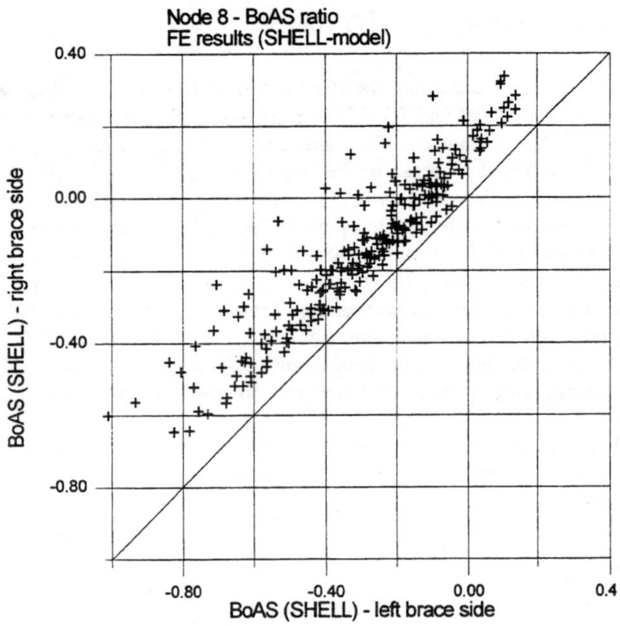

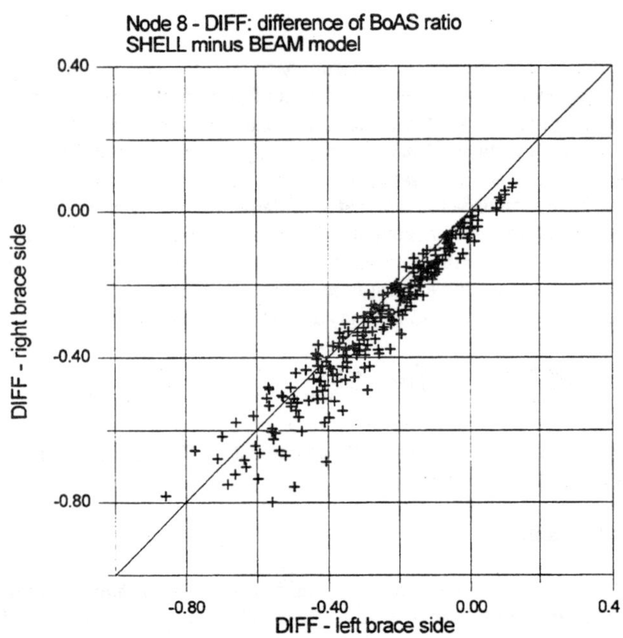

Figs. 7a & b: Comparison of *BoAS* ratio and *DIFF* parameter at the two braces (left and right) of Node 8.

STUDY II: HOT-SPOT STRESS LEVEL

The previous study was devoted on a development of design guidance for the reliable prediction of nominal stresses (axial and bending) for gap K-joint girders. This section is concerned with the "transfer" of these stresses to the "hot-spot" stress level through appropriate stress concentration factors (SCFs), investigating the effects of joint parameters on the hot spot stress of the chord and the brace. The following issues are of primary importance:

- Effects of joint parameters on the *location and magnitude* of the hot spot stress (saddle, crown-heel, crown-toe).
- Evaluation of the proposed design methodology with respect to the a rigorous hot-spot stress calculation.

The nominal stresses obtained in the previous section (STUDY I) are employed in order to compute the hot-spot stress of the joint. The hot-spot stress is to be used with an appropriate *S-N* diagram to obtain the joint fatigue life. Since *S-N* curves depend on the value of the tube thickness, it is necessary to compute hot-spot stresses at both the brace and the chord.

Hot-spot stress is computed rigorously, through the use of the TU Delft SCF data base for gap K-joints (see Romeijn 1994). Around each weld of the joint, ten locations are considered, five on the chord side and five on the brace side (crown heel, crown toe, saddle and two in-between points for the chord and the brace side). This implies a total of twenty locations at the joint, ten at the chord side and ten at the brace side. For the general case of combined loading conditions, the following superposition equations are used, for a specific location on the chord or on the brace (say location k):

$$\begin{aligned} S_{hs}(k) = \\ SCF^{(k)}{}_{ax-B1} \cdot \sigma_{ax-B1} + SCF^{(k)}{}_{ipb-B1} \cdot \sigma_{ipb-B1} + \\ SCF^{(k)}{}_{ax-B2} \cdot \sigma_{ax-B2} + SCF^{(k)}{}_{ipb-B2} \cdot \sigma_{ipb-B2} + \\ SCF^{(k)}{}_{ax-C} \cdot \sigma_{ax-C} + SCF^{(k)}{}_{ipb-C} \cdot \sigma_{ipb-C} \end{aligned} \quad ....(9)$$

where subscripts B1, B2 and C denote the two braces (left and right) and the chord respectively. Special care was paid on the sign of the axial and bending stresses so that they are compatible with the sign convention of the corresponding SCFs. It is possible to use equation (9) for any loading combination, since the SCFs are independent of the chord boundary conditions (see the "compensating moment" concept described by Romeijn 1994). The maximum geometric stress at the ten chord locations is the *chord hot-spot stress* (denoted by S_{chord}) and, similarly, the maximum stress concentration at the ten brace locations is the *brace hot-spot stress* (denoted by S_{brace}).

The hot spot stress computed from this analysis will be referred to as or "rigorously computed hot spot stress". In order to present the results in a more "universal" way, it was decided that the hot spot stress will be normalized by the average absolute value of axial stress at the two braces:

$$\sigma_{norm} = \frac{1}{2}\left(\left|\sigma_{ax-B1}\right| + \left|\sigma_{ax-B2}\right|\right) \quad ....(10)$$

The result, referred to as the "normalized hot spot stress" and denoted by NHSS, is:

$$NHSS = \frac{S_{hs}}{\sigma_{norm}}, \quad ....(11)$$

The following general observations concern the location and magnitude of hot spot stress, based on the results of the parametric study:

Chord side

- In the vast majority of the cases examined, hot-spot stress was located either in the crown toe or the saddle of the chord. Therefore, it is reasonable to examine stress concentrations in crown toe and saddle points for design purposes, neglecting all other points, thus simplifying joint design.

- Hot-spot stress value depends on the joint parameters. For the range considered, the *NHSS* value may vary from 1.4 to 14.8. The *NHSS* value has a strong dependence on the values of τ and θ. In the present study, the maximum values of *NHSS* were observed for τ and θ equal to 1.0 and 60° respectively.

Brace side

- In the vast majority of the cases, hot spot occurred at the crown heel, or at the saddle, or at the in-between point (i.e. the point between the saddle and the crown heel).
- There are very few cases where hot spot was located in crown toe, for high values of τ and θ.
- Based on the above observations, it is reasonable to examine stress concentrations in crown heel and saddle points of the brace for design purposes, neglecting all other points, thus simplifying joint design. The possibility of having S_{brace} on the toe is neglected for design purposes.
- Values of the hot spot stress may vary considerably with joint parameters (from 1.6 to 6.3 for the range considered). The value of τ appears to have a significant effect. The largest *NHSS* values were observed for τ equal to 1.0. The brace angle θ is not a crucial parameter for the value of *NHSS* at the brace side.
- In general, the *NHSS* values for the brace are smaller than the *NHSS* values for the chord. Only in very few cases, the opposite can occur (for θ equal to 30° and for small γ and τ values).

PROPOSED SIMPLIFIED METHODOLOGY

The following method is an alternative to the rigorous methodology described above but it is much more convenient to be used for design purposes. It is based on the simplified equations (3) and (7) for determining nominal stresses with a high degree of accuracy and the simplified SCF equations for gap K-joints presented in the Appendix.

- Solve the structure with beam elements and obtain axial and bending stresses in the chord $[\sigma_{ch\text{-}ax}]_{beam}$ and $[\sigma_{ch\text{-}bend}]_{beam}$ and the brace members $[\sigma_{ax}]_{beam}$ and $[\sigma_{bend}]_{beam}$. Obtain $[BoAS]_{beam}$ ratios at the brace members.
- Obtain total chord nominal stress σ_{chord}, from equation (8)
- Obtain correction factor C_{ax} from equation (3)
- Compute axial brace stresses σ_{ax} from equation (2)
- Assume balanced axial loads and obtain the average stress between the axial stress of the two braces.
- Obtain the predicted DIFF value from equation (6)
- Obtain the predicted BoAS ratio from equation (7)
- Compute bending stresses σ_{bend}

$$\sigma_{bend} = \sigma_{ax} \cdot [BoAS]_{predicted} \qquad \text{....(12)}$$

- It is noted that if the value of *DIFF* is positive, then the direction of the corresponding bending moment is in the direction shown in Figure 4, otherwise it has an opposite direction.
- Use σ_{ax} , σ_{bend} and σ_{chord} as nominal stresses to calculate geometric stresses at the locations of interest (at the two welds, at both the brace and the chord) through the corresponding SCFs for gap K-joints (see Appendix).
- Find maximum hot-spot stress for the chord and the brace (S_{chord} and S_{brace}) and calculate design life for the joint.

This procedure has been followed for the 243 girders of the present parametric study. In order to compare S_{chord} with S_{brace} in terms of their effects on the fatigue life and due to the fact that the S-N curve depends on the thickness, it is necessary to select a reference thickness. The chord thickness t_o is selected and S_{brace} is modified as follows:

$$S_{br-\text{mod}} = S_{brace}(\tau)^{0.30} \qquad \text{....(13)}$$

where a thickness correction with an exponent of 0.30 is assumed (see Reynolds and Sharp 1990). The maximum between the hot spot stress for the chord member (S_{chord}) and the modified hot spot stress for the brace ($S_{br\text{-}mod}$), denoted by S_{joint}, is the joint hot-spot stress with respect to the chord thickness t_o. The value of S_{joint} is computed from the proposed simplified methodology as well as through the rigorous FE analysis. The two values (both normalized by σ_{norm} defined by equation 10), are compared in Figure 9. The comparison is very satisfactory. Indeed, the simplified method is somewhat on the conservative side, which is always desired in design applications.

SUMMARY - CONCLUSIONS

An extensive parametric study was conducted in order to examine the effects of joint local flexibility on the fatigue design of typical lattice girders. A large number of gap K-joint girders, over a wide range of joint parameters and brace slenderness have been considered.

First, the nominal stress level was considered. In order to simulate accurately the effects of joint flexibility, three joints of the girder have been simulated with shell finite elements. The analyses have shown that the nominal stresses computed through a simple framed-analysis program, may be quite different than those obtained from a rigorous analysis which accounts for joint flexibility, especially for the brace members. Parametric formulae in terms of the joint parameters and the brace slenderness were proposed to take into account the effects of joint flexibility.

Subsequently, the nominal stresses were "transferred" to the hot-spot stress level. Conclusions regarding the location and the magnitude of the hot-spot stress were made. Finally, a design methodology based on the FE results of the present study was proposed for the determination of hot-spot stress in lattice girders. This methodology overcomes the problems involved using flexibility coefficients and semi-rigid connections for calculating nominal stresses and it requires only a simple framed-structure analysis.

It should be noted that the present study does not cover all possible configurations of lattice structures. In fact, it would be impossible to investigate all possible types of girders/joints with all possible types of loading in the course of the present study. Nevertheless, this study constitutes the first attempt to simulate rigorously the effects of joint flexibility on the fatigue analysis of K-joints and may be regarded as a "pilot" study for further investigation of this phenomenon. Repetition of the same methodology for other structural configurations would result in a more "universal" design guidance for tubular structures.

ACKNOWLEDGMENTS

The authors would like to thank Prof. R. S. Puthli and Dr. S. Herion (Univ. of Karlsruhe), and Ir. O. D. Dijkstra (TNO-Bouw) for their useful comments.

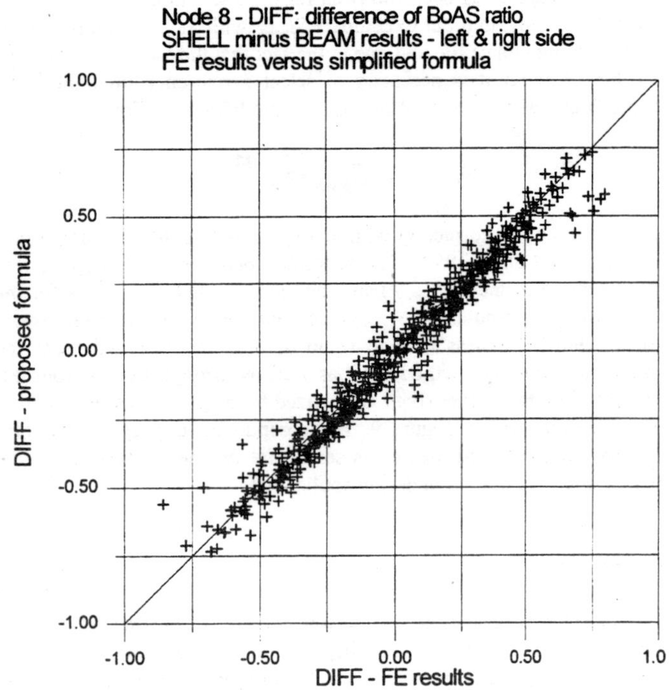

Fig. 8: DIFF values from rigorous analysis and DIFF values from simplified formula.

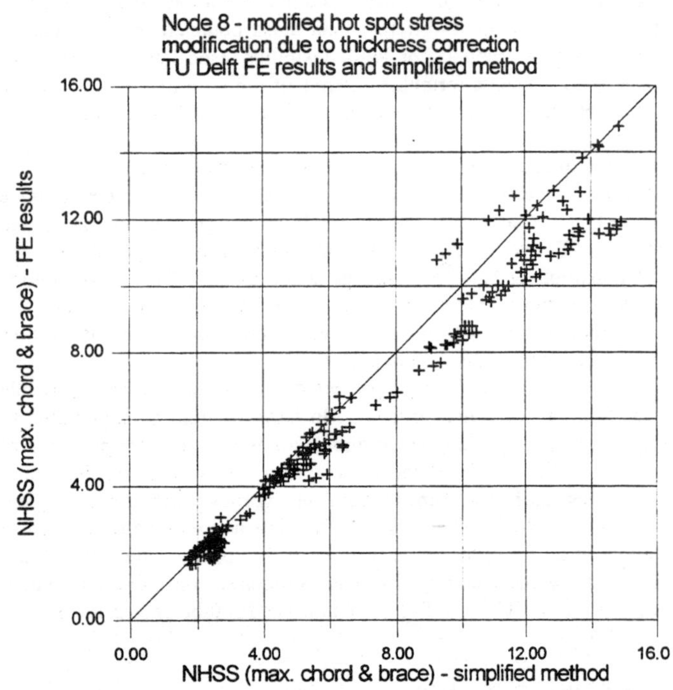

Fig. 9: Hot-spot stress for the joint from rigorous FE analysis and hot-spot stress from the proposed simplified formula.

REFERENCES

Bouwkamp, J. G., Hollings, J. P., Malson, J. P. and Row, D. G. (1980), "Effect of Joint Flexibility on the Response of Offshore Structures", *Offshore Techn. Conf.*, OTC 3901, Houston, Texas.

Det Norske Veritas (1977), "Rules for the Design, Construction and Inspection of Offshore Structures; Appendix C, Steel Structures", Oslo, Norway.

Efthymiou, M. (1985), "Local Rotational Stiffness of Unstiffened Tubular Joints", Report No. RKER.85.199, *Koninklijke Shell Exploratie en Produktie Laboratorium*, Rijswijk, The Netherlands.

Fessler, H., Mockford, P. B. and Webster, J. J. (1986a), "Parametric equations for flexibility matrices of single brace tubular joints in offshore structures", *Proc. Inst. Civil Engineers*, Part 2, No. 81, Paper 9089, pp. 659-673.

Fessler, H., Mockford, P. B. and Webster, J. J. (1986b), "Parametric equations for flexibility matrices of multi-brace tubular joints in offshore structures", *Proc. Inst. Civil Engineers*, Part 2, No. 81, Paper 9124, pp. 675-696.

Fessler, H., Mockford, P. B. and Webster, J. J. (1987), "An experimental technique for determining the flexibility of tubular joints in offshore structures", J. Strain Analysis, Vol.22, No.1, pp.7-15.

Karamanos, S. A., Romeijn, A. and Wardenier, J. (1997), "Stress concentrations and joint flexibility effects in the design of welded tubular connections", *Report, Faculty of Civil Engrg.*, TU Delft, The Netherlands.

Reynolds, A. G. and Sharp, J. V. (1990), "The fatigue performance of tubular joints - an overview of recent work to revise DEn Guidance", Integrity of Offshore Structures-4, Glascow, U.K., pp.261- 278.

Romeijn, A. (1994), *Stress and strain concentration factors of welded multiplanar tubular joints*, Doctoral Thesis, Delft Univ. of Technology, Published by Delft University Press.

Underwater Engineering Group (1984), "Node Flexibility and its Effects on Jacket Structures - A pilot study on two-dimensional frames", *UEG Report*, UR22, London, U.K.

APPENDIX

In this appendix, Stress Concentration Factors (SCFs) for CHS gap K-joints are presented. These SCFs were developed at TU Delft. They cover the needs for an approximate joint stress analysis and are based on extensive finite element results.

Extensive numerical work on tubular joints has been conducted at TU Delft, for the development of SCFs for axial load, in-plane bending, out-of-plane bending and chord loading. The numerical technique was supported by experiments and a summary of these results were reported by Romeijn (1994). Additional effort has been spent in order to obtain SCFs in simple graphs and formulae to be used in practical design purposes. This effort is described in detail in the report by Karamanos et al. (1997) and the simplified SCFs for gap K-joints under uni-planar action are summarized below:

Three loading conditions are considered:

- balanced axial load
- in-plane bending
- chord loading

Four locations around the weld are considered (see Figure 4):

- crown toe (chord side)
- saddle (chord side)
- saddle (brace side)
- crown heel (brace side)

The corresponding SCFs for all loading cases are obtained through the following power formula:

$$SCF(\beta,\gamma,\tau,\theta) = \left(\frac{\gamma}{12}\right)^{X1}\left(\frac{\tau}{0.5}\right)^{X2} SCF_0(\beta,\theta) \qquad ....(14)$$

where SCF_0 is the stress concentration factor for γ and τ equal to 12 and 0.50 respectively. The values of SCF_0 for balanced axial load, in-plane bending (ipb) and chord loading are obtained from Figures 10a, b and c, and from equations (15), (16), and (17). The following table summarizes the values of SCF_0 for critical locations:

SCF_0	Balanced axial (crown)	Balanced axial (saddle)	in-plane bending (crown)	chord load (crown)
chord side	Fig. 10a (toe)	Fig. 10a	Fig. 10b	eq. 17 (toe)
brace side	eq. 15 (heel)	Fig. 10c	eq. 16 (heel)	———

Table 2: Calculation of SCF_0 (for γ=12, τ=0.50).

were equations (15), (16), and (17) are given below

$$SCF_{0\,(bal.\,axial,\,brace\,crown)} = 2 \cdot \sin^{-0.4}\theta \qquad ....(15)$$

$$SCF_{0\,(ipb,\,brace)} = 1.85 \cdot \sin^{-1.2}\theta \qquad ....(16)$$

$$SCF_{0\,(chord\,load,\,crown)} = 1.2 \cdot \sin^{-0.9}\theta \qquad ....(17)$$

It is noted that the effects of chord loading on the brace side are neglected. The values of exponents *X1* and *X2* to be used in equation (14) are provided by the following Table:

Load Type	Location	*X1*	*X2*
bal.-axial	chord (ct & s)	0.40	1.10
	brace (ch)	0.00	0.00
	brace (s)	0.50	0.50
in-plane bending	chord (ct)	0.60	0.80
	brace (ch)	0.00	0.00
chord load	chord (ct)	0.00	0.30
	brace	n/a	n/a
Note: ct=crown toe, ch=crown heel, s=saddle			

Table 3: Values for exponents *X1* and *X2* to be used in equation (14) for gap K-joints.

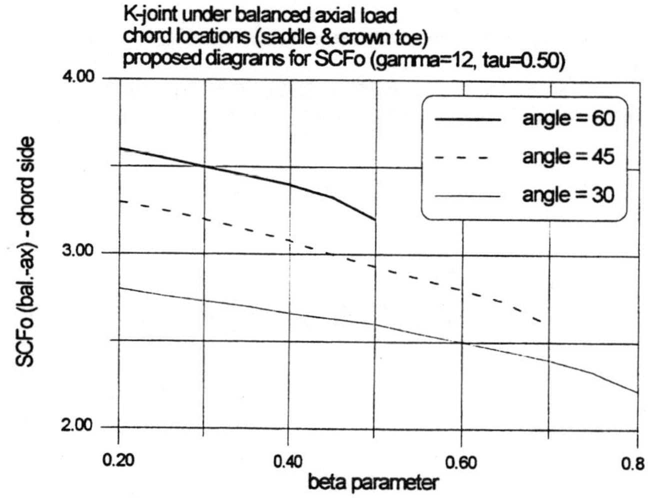

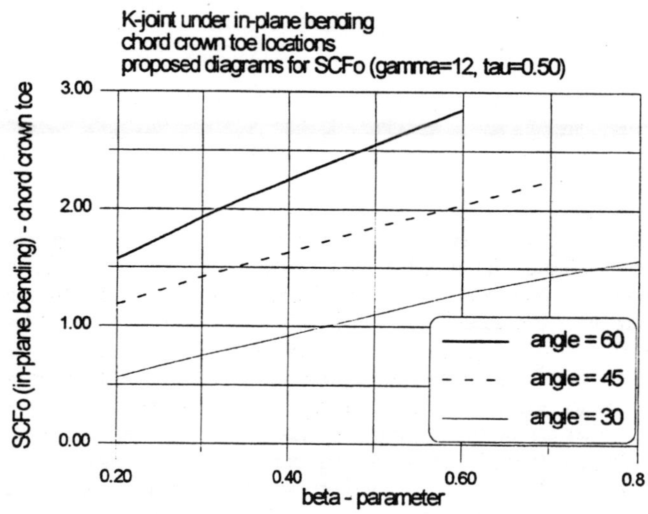

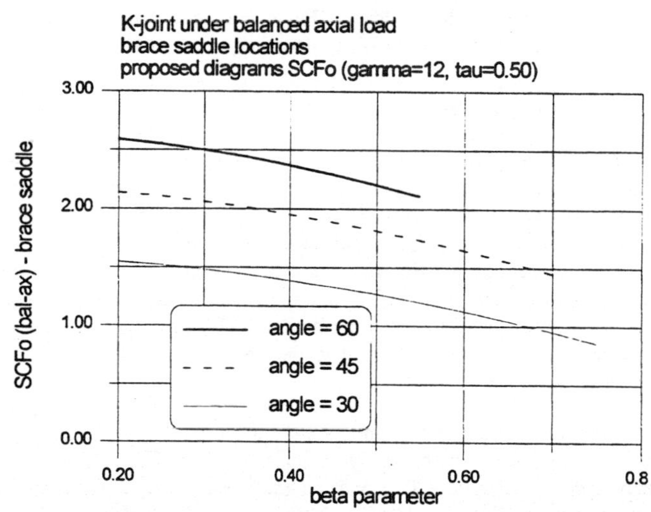

Figs. 10a, b & c: graphs for calculating SCF_0 for gap K-joints; simplified methodology developed at TU Delft.

Proceedings of the Seventh (1997) International Offshore and Polar Engineering Conference
Honolulu, USA, May 25-30, 1997

ISBN 1-880653-28-1 (Set); ISBN 1-880653-32-X (Vol. IV)

FE Mesh Generation for Circular Tubular Joints with or without Cracks

Junjie Cao, Guojing Yang and Jeffrey A. Packer
University of Toronto
Toronto, Canada

ABSTRACT

This paper concerns the mesh generation for different types of welded circular tubular joints with or without cracks for finite element analysis. Based on detailed geometry analysis of the intersection between a chord and a brace in a joint, formulae for mapping the chord/brace intersection on a plane have been derived. The mesh for the chord of a joint can be generated easily in a flat plane using circles and the circles can be mapped to form a mesh around the chord/brace intersection evenly. The flat plane can then be transferred to a tube and braces can be connected to the chord. The method can be applied to different types of joints, uniplanar or multiplanar and uncracked or cracked, and high quality meshes can be obtained by using the method.

KEY WORDS: Tubular Joints, Finite Element Analysis, Mesh Generation, Crack Modelling, Weld Modelling

INTRODUCTION

During recent years the application of the finite element (FE) method has become very popular in the analyses of different types of welded circular tubular joints. The rapid development of commercial FE programs and computation facilities has extended the analyses from elastic to elasto-plastic, from linear to non-linear, from uniplanar joints to multiplanar joints, from uncracked joints to cracked joints, from traditional study of stress distribution, SCF calculation and load-displacement behaviour of a joint to the assessment of the ultimate capacity of a joint.

A few years ago shell elements were used widely for modelling tubular joints, such as the analyses for a number of T/Y joints by Healy and Zettlemoyer (1993) and the analysis to a very complicated nine brace multiplanar joint by Ward and Billington (1992). Additional shell elements might be arranged around the chord/brace intersection of a joint to model the weld for the joint (Wilmshurst and Lee, 1993), but it has to be admitted that shell elements are not suitable to construct a model including welds (and also cracks). Although weld modelling might be thought to be not necessary in some cases by some researchers, such as the analyses of T joints by Lee *et al.* (1995) and the analyses of K joints by Healy (1994), the omission of weld shape in the models used in previous analyses was, mainly, due to the complexity in weld modelling, as pointed out by Romeyn *et al.* (1992). For an analysis for the calculation of SCFs around the weld toe of a joint or for an analysis for a joint with a crack around its weld toe, the weld of the joint has to be modelled carefully. A combination of solid elements and shell elements was used in some meshes but, based on a comparison of the results from different models for multiplanar joints, Herion *et al.* (1996) concluded that the best results were provided by models using 3-D quadratic solid elements for all tubes and welds.

More and more attention has been paid to the analysis of cracked tubular joints in recent years. Different models have been used to analyse the fracture behaviour of cracked joints by Kristiansen and Turner (1993), Burdekin and Yang (1997) and other researchers. Very fine meshes are needed for modelling cracks, so the transition between the fine mesh for a crack and the coarse mesh for the regions outside the crack is very complicated. For cracked joints 3-D quadratic solid elements should be used for all models.

Mesh generation is one of the key parts in a finite element analysis and a good quality mesh is the basis of the success of the analysis. Since full 3-D models are used for tubular joints, welds and cracks are included and more complicated multiplanar joints are concerned, the mesh generation for tubular joints become more difficult. Different programs have been used for creating meshes for different joints. Yang (1996) used PATRAN (PDA Engineering, 1987 and 1993) and PRETUBE (DNV, 1991) to produce a large number of models for uncracked and cracked tubular joints and other cracked components. It has been found that it is not easy to use the general mesh generation program PATRAN to generate meshes for tubular joints. As a special mesh generation program for tubular joints, PRETUBE is very efficient for some simple joints, but the program cannot generate good meshes for joints with high brace to chord width ratios and for cracked joints. Other programs have also been used for mesh generation for tubular joints. However, the mesh generation for a tubular joint is, perhaps, still the most time-consuming work for the FE analysis of a joint, due to the geometry complexity of tubular joints, especially the intersections between members and the difficulty in modelling welds and cracks.

It has been well known that the mesh for a circular surface can be generated in a plane and the plane can then be mapped to the circular surface very easily. The problem in applying this technique to a tubular joint is how to define the intersection between two members. In this paper detailed geometric analysis is carried out for the intersection curve between a brace and a chord and formulae for defining the curve on a plane, when the chord is opened out as a plane, are derived. Furthermore formulae for mapping a circle to the chord/brace intersection are developed. By using these formulae, the mesh for the chord of a joint can be generated in a plane using circles and these circles can be mapped to a series of curves around the chord/brace intersection evenly in the plane. The plane can then be mapped to the chord and the brace of the joint can be connected to the chord. Since the mesh for the chord can be generated in a plane using circles, the modelling of a weld or a crack becomes simpler. This approach can be implemented by using different programs for different types of joints, uniplanar or multiplanar and uncracked or cracked. Examples, which include a 12 brace member multiplanar joint and a DT joint with a surface crack, generated using ABAQUS (HKS Inc, 1995) are presented in the paper. All the examples in this paper were constructed using 20-noded quadratic solid elements. It can be seen that the approach is efficient and simple and the quality of the meshes generated by using the approach is high.

GEOMETRY ANALYSIS OF THE INTERSECTION IN A Y JOINT

For a circular tube it is very easy to generate a mesh in a flat plane and the plane can then be mapped to the tube very easily. Fig.1 shows the relationship between a circular surface ($X^2+Y^2=R^2$) and a plane ($X=R$). If the circular surface is cut by a longitudinal line and opened out along the line 1-1, the point a on the circle should be the point a′ on the plane. The coordinate system Y' - Z' is used to define the plane and it can be seen that the axis Z' is the same as the axis Z and the relationship between X, Y and Y' can be found to be:

$$X = R\cos\frac{Y'}{R} \qquad Y = R\sin\frac{Y'}{R} \tag{1}$$

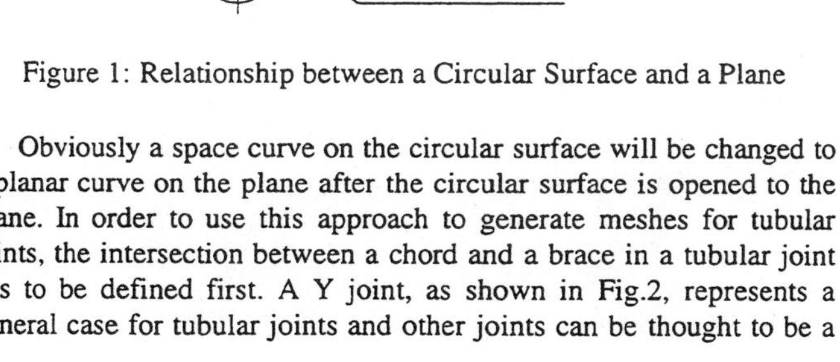

Figure 1: Relationship between a Circular Surface and a Plane

Obviously a space curve on the circular surface will be changed to a planar curve on the plane after the circular surface is opened to the plane. In order to use this approach to generate meshes for tubular joints, the intersection between a chord and a brace in a tubular joint has to be defined first. A Y joint, as shown in Fig.2, represents a general case for tubular joints and other joints can be thought to be a special case (e.g., a T joint) or a combination of several Y joints (e.g., K, KT, X and multiplanar joints). So the following geometry analysis for a Y joint can be also used for other types of joints.

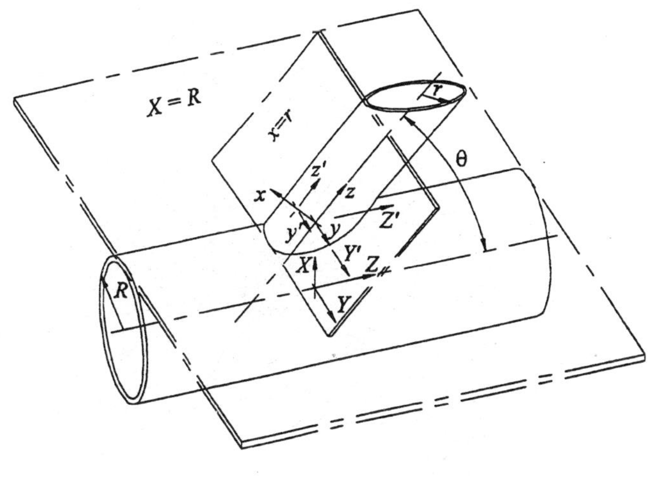

Figure 2: A Y Joint and Coordinate Systems

In Fig.2 the outside radii of the chord and the brace are R and r respectively, and the angle between the two axes of the tubes is θ. The axis Z is along the axis of the chord and the axis X passes through the origin of the coordinate system x-y-z, which is at the intersection point between the brace axis and the chord outside surface. The axis z is along the brace axis and the axes y and Y are in the same direction. Obviously, both the planes X-Z and x-z are co-planar with the two axes of the tubes.

The chord and the brace can be defined in the two coordinate systems as:

For the chord $$X^2 + Y^2 = R^2 \tag{2}$$

For the brace $$x^2 + y^2 = r^2 \tag{3}$$

The relationship between the two coordinate systems can be found to be:

$$\begin{aligned} X &= x\cos\theta + z\sin\theta + R \\ Y &= y \\ Z &= z\cos\theta - x\sin\theta \end{aligned} \tag{4}$$

or

$$\begin{aligned} x &= (X-R)\cos\theta - Z\sin\theta \\ y &= Y \\ z &= (X-R)\sin\theta + Z\cos\theta \end{aligned} \tag{5}$$

The intersection of the chord and the brace, which is a 3-D curve, is determined by Eqs.(2) and (3) and can be expressed in either of the coordinate systems. For the chord, the curve should be defined in the coordinate system X-Y-Z and Eq.(3) can be changed to Eq.(6) by using Eq.(5):

$$[(X-R)\cos\theta - Z\sin\theta]^2 + Y^2 = r^2 \tag{6}$$

As illustrated in Fig.1 (also in Fig.2) the chord outside surface can be changed to a plane ($X=R$) by inserting Eq.(1) into Eq.(6) so the intersection curve defined on the plane Y' - Z' can be obtained:

$$R^2\sin^2\frac{Y'}{R} + \left[Z'\sin\theta + R\left(1-\cos\frac{Y'}{R}\right)\cos\theta\right]^2 = r^2 \tag{7}$$

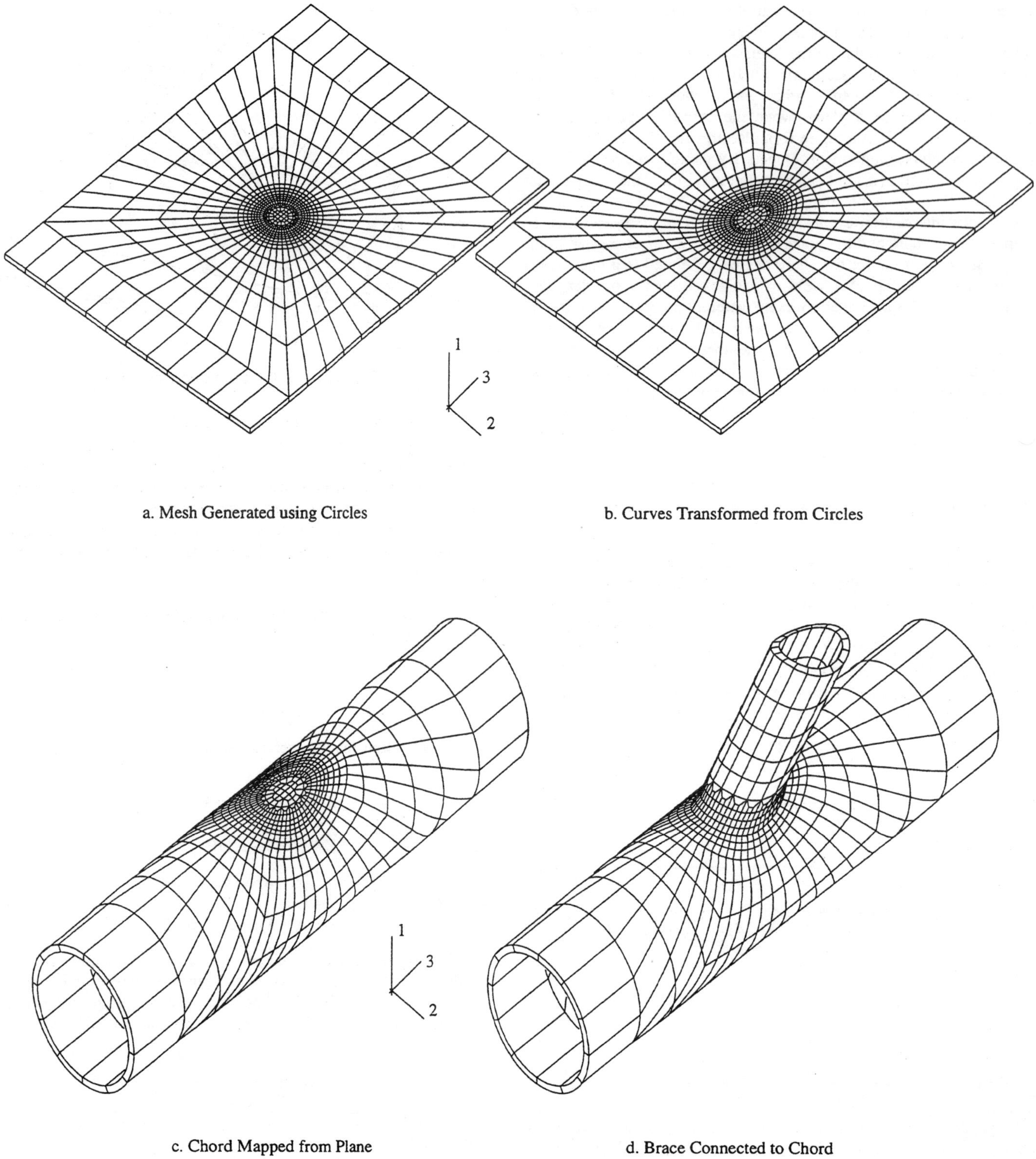

a. Mesh Generated using Circles

b. Curves Transformed from Circles

c. Chord Mapped from Plane

d. Brace Connected to Chord

Figure 3: Mesh Generating Procedure for a Y joint

For the brace the intersection should be defined in an x-y-z axis system. Inserting Eq.(4) into Eq.(2) gives:

$$(x\cos\theta + z\sin\theta + R)^2 + y^2 = R^2 \qquad (8)$$

A transformation similar to Eq.(1), where y' and z' are analogous to Y' and Z', $(x = r\cos(y'/r),\ \ y = r\sin(y'/r))$, can be made to change the brace into a plane ($x=r$ in Fig.2) so the intersection curve can be changed to a curve on the plane defined by y'-z':

$$r^2\sin^2\frac{y'}{r} + \left[z'\sin\theta + r\cos\frac{y'}{r}\cos\theta + R\right]^2 = R^2 \qquad (9)$$

The importance of determining the intersection curve has been emphasised by Suen and Wu (1993), who used discrete points to fit the intersection curve. The definition of the curve on a chord surface can be used for the control of an automatic welding process for the joint and the definition of the curve on a brace surface can be used for cutting the brace. The solutions can be also used to define intersection curves for generating FE meshes for different joints.

MESH GENERATION FOR JOINTS WITHOUT CRACKS

The mesh for the chord of a joint can be generated in a plane using any programs and the plane can be mapped to the chord according to Eq.(1). This can be done very easily in some commercial programs such as ABAQUS (HKS, 1995) and ANSYS (Swanson, 1992). It is very important to define the chord/brace intersection curve of the joint on the plane accurately and to keep the adjacent layers of elements around the intersection as even as possible. Although Eq.(7) can be used to define the intersection directly it is more efficient to design the mesh using circles and then transfer the circles to fit the intersection.

A circle of radius r can be expressed as:

$$u^2 + v^2 = r^2 \qquad (10)$$

Suppose:

$$Y' = R\arcsin\frac{u}{R}$$
$$Z' = \left[v - R(1-\cos\frac{Y'}{R})\cos\theta\right]\frac{1}{\sin\theta} \qquad (11)$$

If Eq.(11) is substituted into Eq.(10), Eq.(10) is actually changed to Eq.(7), which represents a chord/brace intersection on a chord surface for a joint. Hence the mesh around the intersection on the chord surface of a joint can be generated using circles on a plane first, which is the easiest way for mesh generation, and then Eq.(11) can be used to transfer the circles defined in a u-v coordinate system into a mesh in the Y'-Z' coordinate system. The plane in the Y'-Z' coordinate system can then be wrapped into a cylinder to form the mesh for the chord. The mesh for the brace connected to the chord can be generated very easily after the intersection curve on the chord is defined.

Fig.3 shows the mesh generation procedure for a Y joint. The first step (Fig.3a) is to generate a mesh for the chord in a plane and the elements around the chord/brace intersection are formed using circles. Then Eq.(11) is used to change the circles into the curves around the chord/brace intersection evenly, as shown in Fig.3b. The next step is to wrap the plane around the chord (Fig.3c). After the mesh for the chord is finished the brace can be connected easily (Fig.3d). The whole model includes 2212 20-noded quadratic solid elements (element type C3D20 in ABAQUS) and 12336 nodes.

Details of the mesh in Fig.3 around the chord/brace intersection are shown in Fig.4. Three layers of elements are used around the chord/brace intersection and one layer of elements is used for other regions. The weld shape is decided according to the AWS specification (1996) for Partial Joint Penetration (PJP) welds. Since the initial mesh for the chord of the joint was generated in a plane using circles, the definition of the weld toe of the joint becomes simpler and it is also easy to get a smooth transition between fine mesh regions and coarse mesh regions.

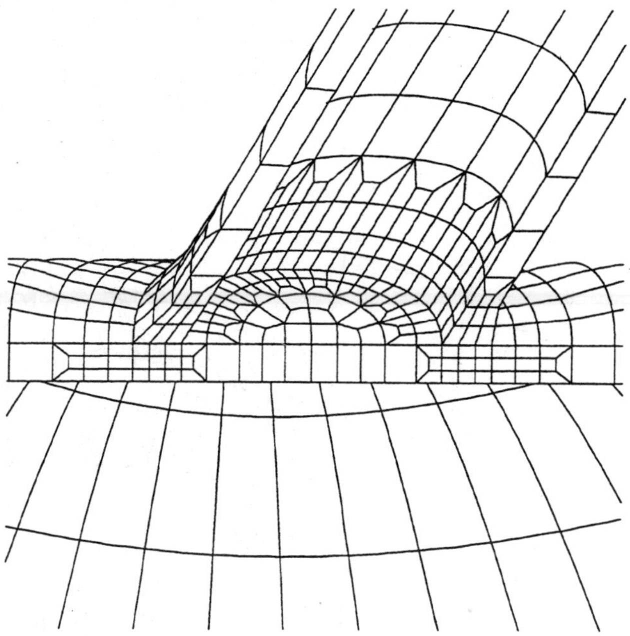

Figure 4: Mesh Details around Intersection for a Y joint

The procedure can be applied to K, KT, DT and multiplanar joints. For a general case of a joint with two or more braces, each brace can be treated separately to generate a part of the mesh for the chord of the joint and these parts can be combined to form the mesh for the chord in a plane. Because the mesh is generated in a plane the generation procedure is simple, even though several braces may join together with the chord. After the plane is wrapped around the chord the braces can be connected with their intersections on the chord. Fig.5a shows a mesh for a joint with 12 brace members joined with the chord, which includes 2736 20-noded quadratic solid elements (element type C3D20 in ABAQUS) and 20248 nodes. Different diameters and angles were used for these brace members and no symmetric conditions exist in the joint. The authors believe, based on the information in refereed publications, that up to now this mesh is the most complicated mesh developed for uncracked tubular joints. It is noted that this is not a real joint and the reason for using this joint is to show the generality of the method. Fig.5b gives more details around the intersections between members and it can be seen that all the welds for these braces are modelled. Most practical joints can be thought of as a part of this joint, so there is no concern in applying this method to different joints. If this method is used for an overlap joint the intersection curve for the overlapping member can be decided by its connections with the chord and the overlapped member.

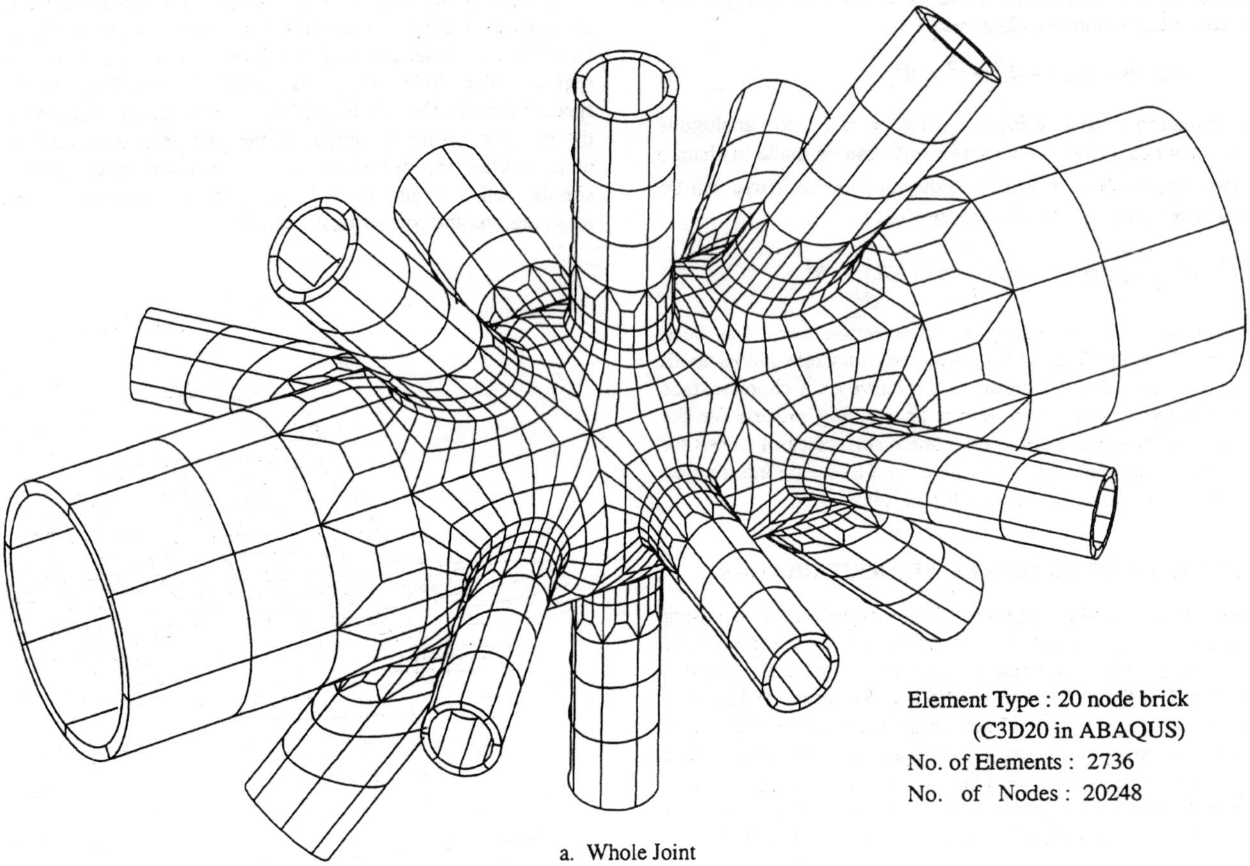

a. Whole Joint

b. Intersections between Members

Figure 5: Mesh for a 12 Brace Member Multiplanar Joint

MESH GENERATION FOR JOINTS WITH CRACKS

For tubular joints, two types of cracks, which are through thickness cracks and surface cracks, may be thought to exist in chords along chord/brace intersections. Different models for cracks have been investigated by Yang (1996) for the evaluation of J-integral and Crack Tip Opening Displacement (CTOD). In order to get reliable results for J and CTOD, several rings of focused elements around a crack tip (Yang, 1996) are needed. Because of the limit of space, detailed crack modelling will not be discussed here but the mesh models used in the examples in this paper have been used for different cracked objects successfully.

Because a very fine mesh is needed around a crack tip, the connection between crack tip elements and other elements is complicated. This results in the mesh generation for cracked joints being much more difficult than for uncracked joints. If the mesh generation method proposed in this paper is used, crack elements can be created during the first step of the mesh generation procedure shown in Fig.2. This means that the crack elements for a joint can be generated in a plane so the mesh generation for a cracked joint becomes relatively simple.

Fig.6 is a part of a mesh for a DT joint without a crack under the action of tension forces along the brace members. Because of symmetric conditions only half of the joint need be modelled. Fig.7 shows the case when a through thickness crack is included in the model. Twin nodes were created along the weld toe from the saddle to the crack tip, so the elements along the two sides of the weld toe were separated. A group of crack tip elements was used to replace the elements around the crack tip (4 elements for one layer) before the chord was wrapped. The crack tip is surrounded by three rings of elements in this mesh and other types of crack tip models can be also used for the mesh.

Compared with the case for a through thickness crack, the mesh for a surface crack is much more complicated. Fig.8 shows the mesh for the joint in Fig.6 with a surface crack. It can be seen that the crack tip in Fig.8 is different from that in Fig.7. This is a four crack tip model used for modelling crack growth and the crack tip consists of 348 3-D quadratic solid elements. Detailed discussion of this crack model is out of the scope of this paper and omitted here. The whole group of crack tip elements and the connecting elements with the crack tip elements were generated separately, and then these elements were used to replace some elements around the crack position for the mesh in Fig.6 before the chord was wrapped. The authors again believe that up to now this is the most complex model for cracked tubular joints, based on the information provided in refereed publications.

CONCLUSIONS

By using the formulae and procedure proposed in this paper, 3-D intersection curves between tubular members can be mapped easily using circles in a plane and thereby a complicated 3-D mesh generating procedure for tubular joints has been changed to a 2-D procedure. The method also makes the modelling of welds and cracks simpler so more detailed modelling of welds and cracks can be included. The examples show that the method is capable of generating meshes for complex joints and high quality meshes can be obtained by using the method.

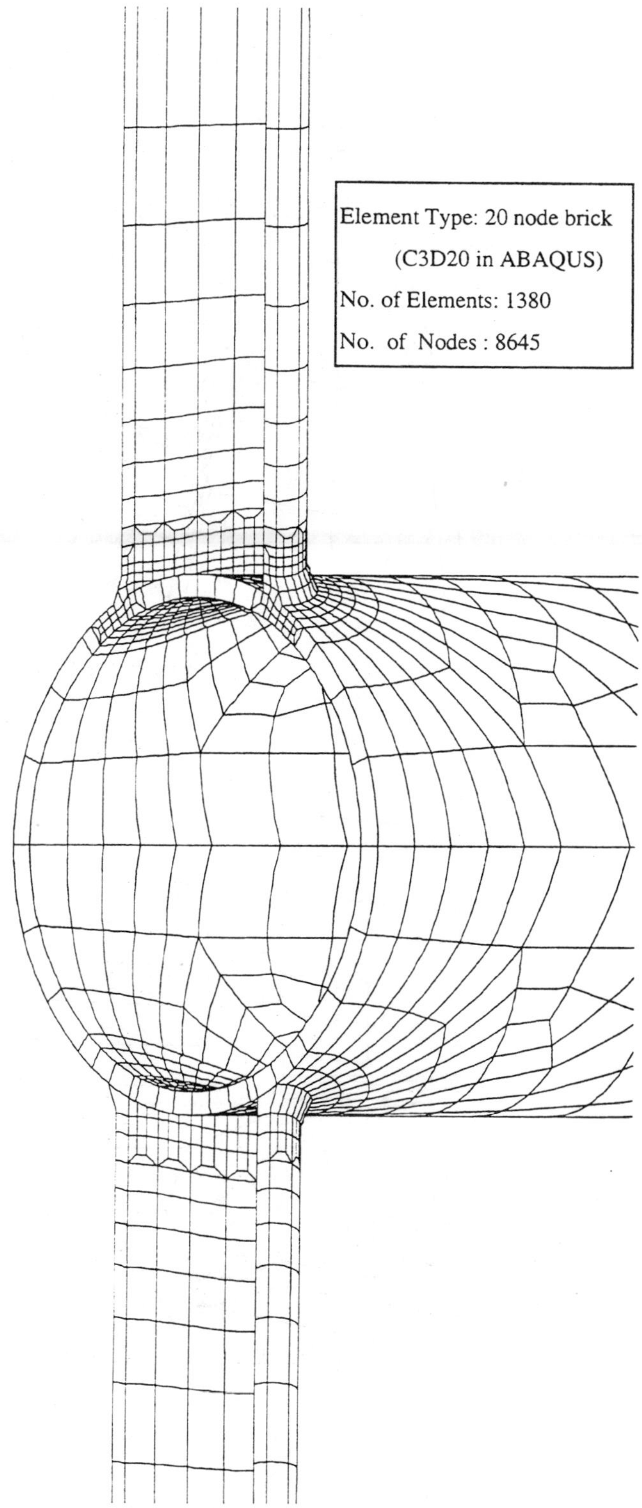

Figure 6: Mesh for a DT Joint without Cracks

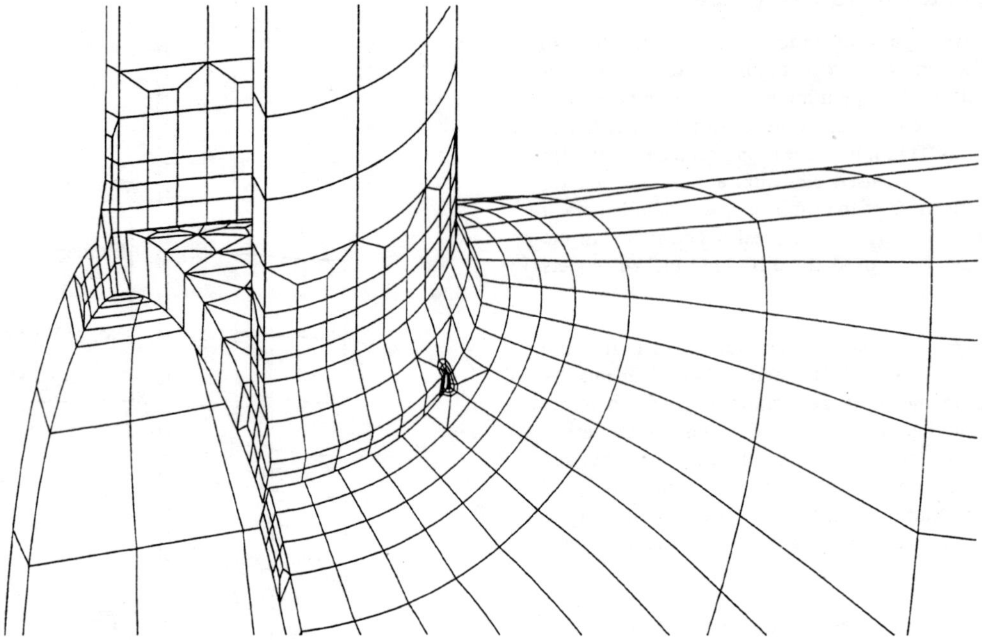

Figure 7: Mesh Including a Through Thickness Crack

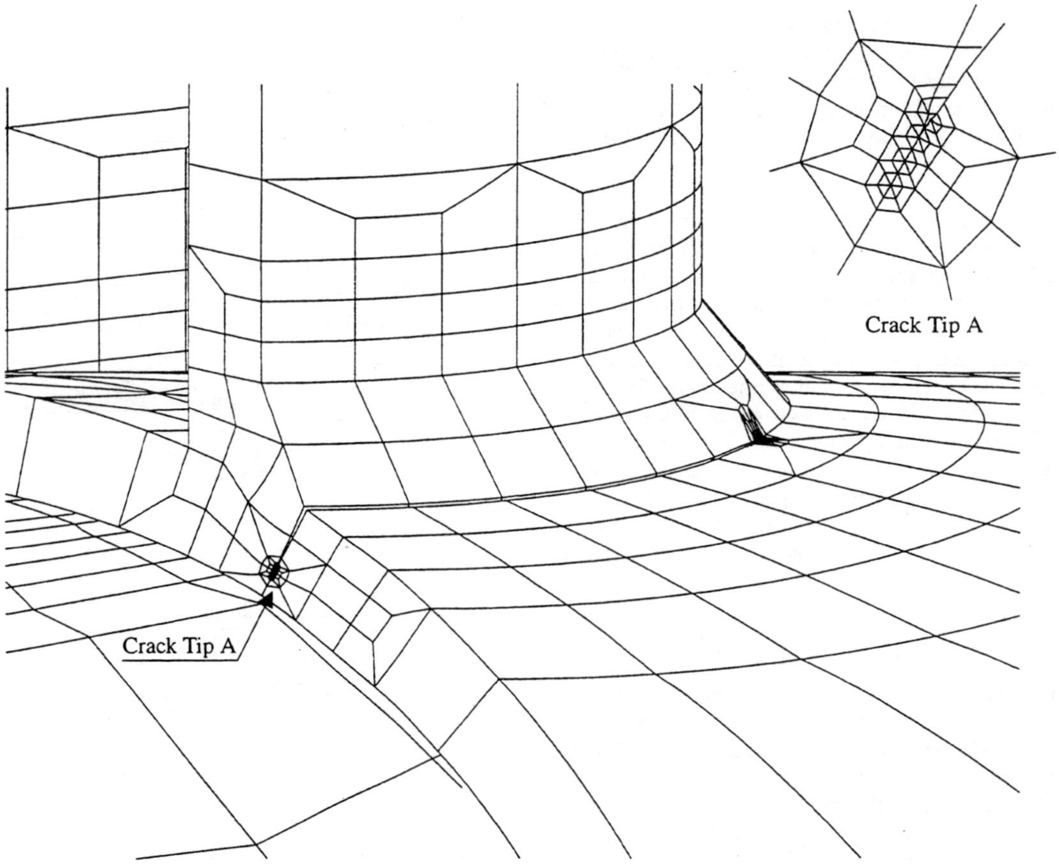

Figure 8: Mesh Including a Surface Crack

ACKNOWLEDGEMENTS

The authors are grateful to Prof. F.M. Burdekin, UMIST, UK for his advice for modelling cracks. Financial support from the Natural Science and Engineering Research Council of Canada (NSERC) is gratefully appreciated.

REFERENCES

American Welding Society (1996). *ANSI/AWS D1.1-96 Structural Welding Code-Steel*, Miami, USA.

Burdekin, F.M. and Yang, G.J. (1997). "Failure Assessment Diagrams for Mixed Mode Loading and Cracked Tubular Joints," *Proc. Ninth International Conference on Fracture*, Sydney, Australia.

Det Norske Veritas (DNV) Industry Ltd (1991). *PRETUBE User's Manual Veritas Sesam Systems*, Høvik, Norway.

Healy, B.E. and Zettlemoyer, N. (1993). "In-Plane Bending Strength of Circular Tubular Joints," *Proc. Fifth International Symposium on Tubular Structures*, Nottingham, UK, pp.325-344.

Healy, B.E. (1994). "A Numerical Investigation into the Capacity of Overlapped Circular K-Joints," *Proc. Sixth International Symposium on Tubular Structures* , Melbourne, Australia, pp.563-571.

Herion, S., Mang, F. and Puthli, R. (1996). "Parametric Study on Multiplanar K-Joints with Gap Made of Circular Hollow Sections by Means of the Finite Element Method," *Proc. Sixth (1996) International Offshore and Polar Engineering Conference*, Los Angeles, USA, Vol. IV, pp.68-73.

Hibbit, Karlsson & Sorenson Inc. (1995). *ABAQUS/Standard User's Manual Version 5.5*, USA.

Kristiansen, N. and Turner, C.E. (1993). "Defect Assessment of Tubular T-Joints Using R6 Failure Analysis Diagram," *ASME J. Pressure Vessel Technology*, Vol.115, pp.373-380.

Lee, M.M.K., Dexter, E.M. and Kirkwood, M.G. (1995). "Strength of Moment-Loaded Tubular T/Y-Joints in Offshore Platforms," *The Structural Engineer*, Vol.73, No.15, pp.239-246.

PDA Engineering (1987). *PATRAN Plus User Manual Version 2.5*, USA.

PDA Engineering (1993). *P3/PATRAN User Manual*, USA.

Romeyn, A., Puthli, R.S. and Wardenier, J. (1992). "Finite Element Modelling of Multiplanar Joint Flexibility in Tubular Structures," *Proc. Second (1992) International Offshore and Polar Engineering Conference*, San Francisco, USA, Vol. IV, pp.420-429.

Suen, D.S. and Wu, J.H. (1993). "Computer-Aided Design of Welded Tubular Connections," *Proc. Third (1993) International Offshore and Polar Engineering Conference*, Singapore, Vol. IV, pp.393-400.

Swanson Analysis Systems, Inc. (1992). *ANSYS User's Manual for Revision 5.0,* Houston, USA.

Ward, J.K. and Billington, C.J. (1992). "Strength Assessment of a Multiplanar Joint in a North Sea Jacket Structure," *Proc. Second (1992) International Offshore and Polar Engineering Conference*, San Francisco, USA, Vol. IV, pp.369-376.

Wilmshurst, S.R. and Lee, M.M.K.(1993). "Ultimate Capacity of Axially Loaded Multiplanar Double K-Joints in Circular Hollow Sections," *Proc. Fifth International Symposium on Tubular Structures*, Nottingham, UK, pp.712-719.

Yang, G. (1996). *Effect of Mixed Mode Fracture on Ultimate Strength of Cracked Tubular Joints*, Ph.D Thesis, Department of Civil and Structural Engineering, UMIST, UK.

Proceedings of the Seventh (1997) International Offshore and Polar Engineering Conference
Honolulu, USA, May 25-30, 1997

ISBN 1-880653-28-1 (Set); ISBN 1-880653-32-X (Vol. IV)

Numerical SCF-Analyses of Grouted Tubular T- and X-Joints

G. Partiman, J. Wardenier and A. Romeijn
Delft University of Technology
Delft, The Netherlands

R.J. van Foeken
TNO Building and Construction Research
Delft, The Netherlands

Abstract

Tubular joint geometry causes stress peaks in localised areas around the weld circumference. Chord grouting reduces stress peaks which results in longer fatigue life.
This study deals with numerical SCF-analysis of grouted T- and X-joints. TNO Building and Construction Research, Netherlands performed scale model experiments on (grouted) joints which allow for comparison with these numerical results.
The main effect of grouting is that chord ovalisation is reduced. Chord ovalisation is the main cause for stress peaks under (brace) axial- and (brace) bending loading.
SESAM Pretube and MARC were used for modelling and calculation. The grout was modelled by the MARC-related contact analyses and also by using an equivalent steel tube method.
Overall there was a good agreement between the numerical and the experimental results.

Keywords

Grouted tubular joints, numerical analysis, contact analysis

Nomenclature

d_0	: chord diameter
t_0	: chord wall thickness
E	: Young's module
I	: moment of inertia
ν	: Poisson's module
t (t/z)	: tension (tension zone)
c (c/z)	: compression (compression zone)
def	: deformable-deformable contact
rig	: deformable-rigid contact
bs	: brace saddle
bc	: brace crown
cs	: chord saddle
cc	: chord crown

Introduction

Offshore jacket structures are subjected to cyclic (e.g. wave) loading. These loadings change constantly in magnitude and direction, thus causing fatigue in the structure.
The fatigue behaviour of tubular joints is related to the hot spot stress ranges. Reduction of the hot spot stress will result in an extended life cycle. A method of stress reduction is grouting of the chord in tubular joints.
Up till now research on grouted joints has been limited to expensive scale model experiments that were carried out on specific configurations under specific circumstances. The results usually give a good indication on the behaviour of grouted joints, but are useless when an accurate stress analysis for another configuration is required.
TNO Building and Construction Research, Netherlands has conducted scale model experiments on a grouted T- and a grouted X-joint (see figure 1).

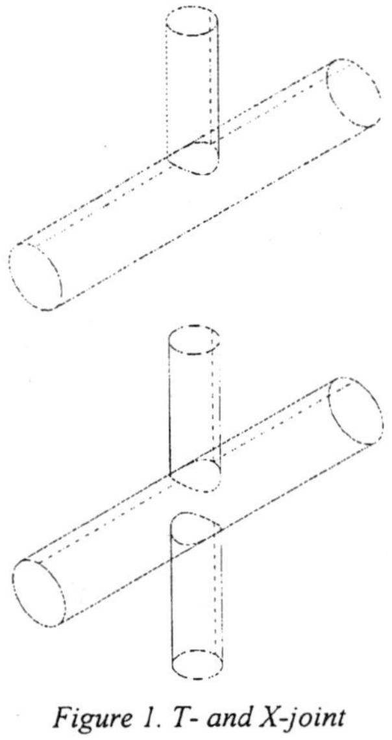

Figure 1. T- and X-joint

Parallel to that Finite Element analyses for these two particular joints were carried out. Finite Element analyses are cheaper than scale model experiments and may lower the financial threshold for proper application of grouted tubular joints.
The numerical results found in this work are valid for joints with fixed parameter values. The primary goal was to explore the possibilities and to point out the limitations of FE-calculations. The numerical results have been compared to the experimental data. However, because of the confidential nature of the latter these cannot be reproduced in this report. An attempt is made however, to pass a qualitative judgement on the results.

Survey of literature on grouted joints

Several papers on grouted joints were studied. Below an overview and a brief excerpt of the papers and other relevant literature is given. These articles deal only with experimental results.

- J.L. Trinh and Ph. Beguin [1] investigated the effect of the grout on static strength and fatigue resistance. Furthermore grout properties and bonding between grout and steel were investigated. Tests were carried out on X-joints of which the chord consisted of two coaxial tubes with grout in-between.

- I.E. Tebbett [2] reports of tests on static strength, SCF-measurements and local joint flexibility of grouted T- and DT-joints. The chords were assembled of two coaxial tubes. There are also graphs with ultimate strength for $F_{br;ax}$ and $M_{br;op}$ plotted against β and γ_p, where γ_p is radius/wallthickness ratio of the inner tube.

- M. Lalani and I.E. Tebbett [3] tested three K-joints with fixed joint parameters. Two as-welded- and one grouted specimen were tested. The results of the investigation consist of SCF- and ultimate load values. The load-displacement behaviour is also plotted.

- M. Lalani, D.J. Moran, R.J. van Foeken and J. Wardenier [4] performed tests on T- and DT-joints. SCF- and preload investigation were carried out. Actually measured SCF and SCF resulting from an equivalent steel tube method were compared.

- G.M. Brown, R. Holmes and J. Kerr [5] tested T-joints which had suffered damage while in service. By simulating circumstances of a joint in service they developed a method for grouting joints in service. After repair of fatigue damage SCF-values were determined.

Grouted joint behaviour

A quantitative description of composite action of grouted joints under various load types is given below. Grout length as well as joint flexibility will be briefly treated.

Joint behaviour under loading

The several types of brace loading will be treated separately in order to gain quantitative insight in the behaviour of grouted joints.

- Axial load (tension and compression)

In ungrouted joints ovalisation of the chord is the dominant factor for causing stress concentration. Under axial loading there is a relatively large ovalisation of the chord, which will be constrained by the grout mass.
Under axial tension local separation around the weld area may occur, while under compression the compressed area benefits from contact pressure between grout and steel.

- In plane bending

In case of in plane bending the chord wall in the tensile zone is locally separated from the grout. Due to the fact that contact pressure occurs in the compression zone, the expectation is that for brace in plane bending the neutral line of the reacting forces will shift to the compression zone. Grouting will result in minor reduction of these SCF.

- Out of plane bending

Out of plane bending can be roughly distinguished in axial compression and axial tension. It can be concluded that the SCF in this load case are generated by ovalisation of the chord. The grout may provide major improvement in stress peak behaviour. Here also the neutral line shifts.

- Behaviour under combined loading

Joints subjected to fatigue loading in general have stress peaks that, apart from some local plastification, remain in the elastic region. In overviewing the combined load cases, it will be assumed that there is neither plastic deformation in the steel nor significant cracking effects in the grout.
For instance, in case of bending in plane the neutral line will shift towards the compression zone when the load is increased. The stresses in the tensile zone will increase non-proportionally with the load. So the composite joint will act as a geometric non-linear substructure, which means that different loads cannot be superimposed in case of combined loading. Above all there is the complication of load level.
In fact the designer faces an almost infinite number of cases she/he has to investigate before maximum stresses can be determined. This is not a practical option, the number of calculations/experiments will be large which results in increasing computer costs. Also, a vastly growing amount of data has to be studied.
The conclusion is that calculation of separate load cases and/or several (discrete) load levels provides limited insight in the exact stress behaviour where combined loading is concerned.
More research will provide more data, but the designer's insight in composite joint behaviour is still the most important tool.

Grout length

Due to the fact that stress peak areas are (compared to the member length) of very local nature, increased grout length will yield little improvement in SCF-behaviour. For FE-calculations this conclusion can result directly in a reduced number of required elements for the grout mass. Reduction of the grout length results in the reduction of joint weight and grout costs (significant when often repeated).
In scale model experiments investigation of this parameter is of too little interest to study upon from an economical point of view.

Joint flexibility

Tebbett [2] and Lalani & Tebbett [3] deal with the joint flexibility aspect for respectively K-joints under balanced axial loading and T-joints under axial tension, compression, in plane bending and out of plane bending loads. In general real joint flexibility results in member and joint forces that are completely different from the forces that are

calculated when the joint is assumed to be rigid. So in fact the effect of the SCF reduction by grout injection could be partially lost by increased joint stiffness as a result of the grout.
Tebbett [2] also summarises several methods (such as substructuring) for inclusion of local joint flexibility in plane frame structural analysis programs.

Modelling

For the numerical analyses two FE-packages were used. SESAM Pretube for the global geometry and the element mesh. MARC was used for further refinement of the model, the actual calculation and for post-processing as well.

Modelling with Pretube

For this project the chord length was $6*d_0$. For the brace $3*d_i$ was used, where d_i is the diameter of the respective braces.
Superelement boundaries, so-called patches, are created. Inside these patches the element grid is defined. Figure 2 shows part of the joint with patches defined on the surface of the tubes.

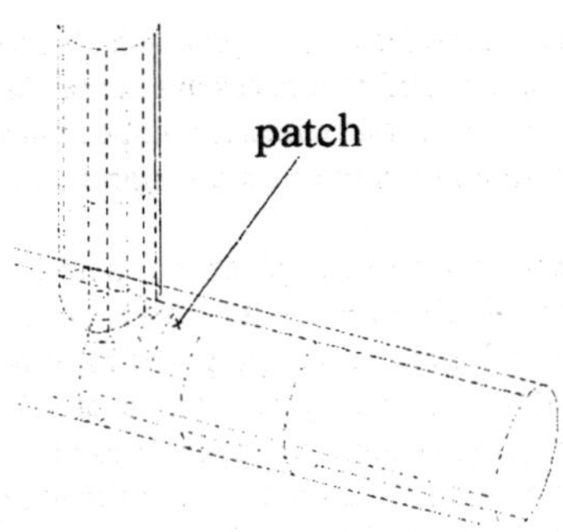

Figure 2. Joint with patches

For the SCF the stresses perpendicular to the weld toe are important. Therefore, patch boundary lines perpendicular to the weld toe are defined.

Element type

There are a number of possibilities regarding the choice of the element type. Fact is that the welds must be modelled with (8- or 20-node) solid elements if the AWS-code is applicable. For the tube there is a (realistic) choice between shell elements (6 DOF/node) and (8- or 20-node) solid elements (3 DOF/node). The attachment of the weld elements to the tube elements is shown in figure 3 for both alternatives.

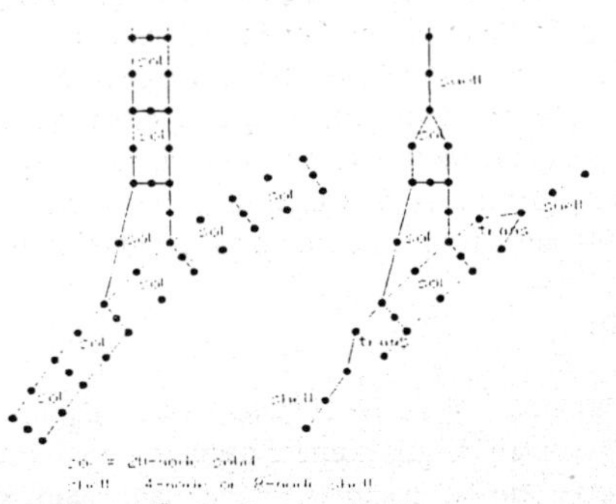

Figure 3. Solid- and shell modelling

The use of solid elements involves a larger amount of nodes, thus a larger amount of data. However, use of shell elements with transition elements increases the required computer resources. Table 1 gives an overview of all element types, integration schemes etc.

Table 1. Used element types.

Element type	20-node solid	8-node solid	4-node shell
Shape	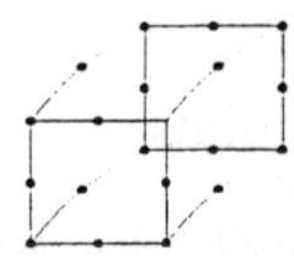	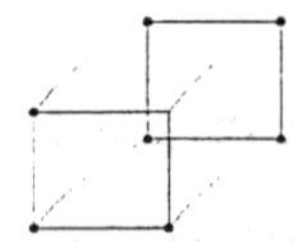	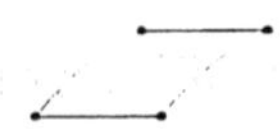
Integration scheme	3 x 3 x 3	2 x 2 x 2	2 x 2 x L (L = number of layers)
DOF/node	u_x, u_y, u_z	u_x, u_y, u_z	u_x, u_y, u_z ϕ_x, ϕ_y, ϕ_z
Applied for	Ungrouted steel tube	Contact analyses	Supports and grout

Modelling the composite joint

The modelling alternatives are:

1. Use of (Pretube-generated) 20-node elements for the modelling of the grout (grout modelling in Pretube) with interface elements between grout and steel. This method is too labour-intensive.
2. Use of MARC-related solutions such as contact analysis
3. The application of an equivalent steel tube method.

It was decided to concentrate on the second and the third alternative. These will be elaborated hereafter.

Contact analysis

The contact option is used to impose non-penetration constraints whenever two bodies come into contact with each other. Contact bodies can be either rigid or deformable. Rigid bodies are pure geometrical entities, deformable bodies are a collection of elements. The penetration constraint for both deformable-rigid and deformable-deformable contact can be formulated as in figure 4.

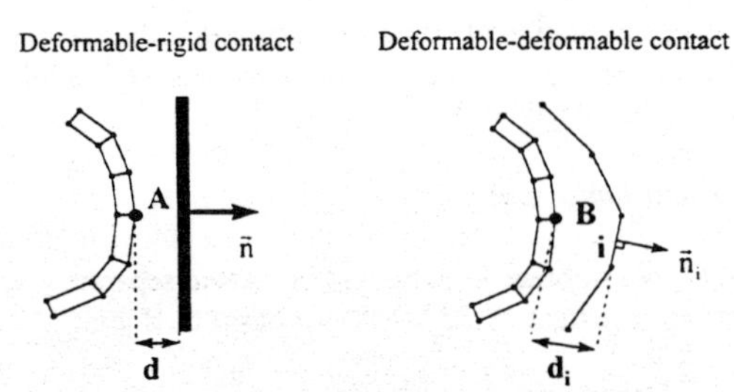

$$\bar{u}_A \cdot \bar{n} \le d \qquad |\bar{n}| = 1$$

$$(\bar{u}_B - \bar{u}_i) \cdot \bar{n}_i \le d_i \ \forall i \qquad |\bar{n}_i| = 1$$

Figure 4. Penetration constraint

These principles were implemented in the following way for the actual problem (figure 5).

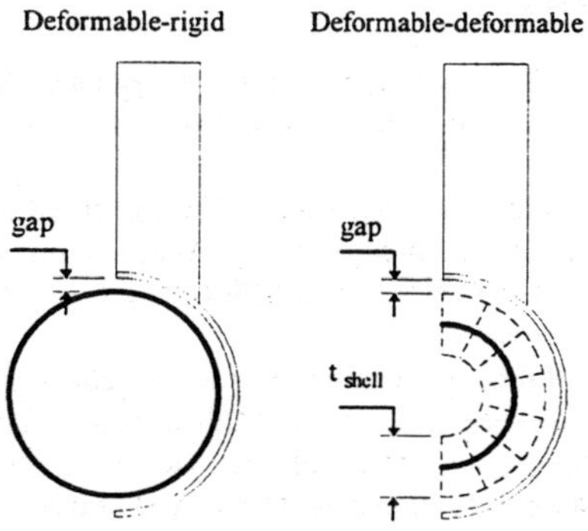

Figure 5. Implementation of contact analysis

It is possible to instruct MARC to check contact *only* for a collection of nodes that are likely to touch the other body. This reduces computational costs.
Rigid body surfaces may be described analytically or discretely (flat patches). The analytic (more accurate) form was used.
The contact algorithm degenerates 20-node solid elements to 8-node solid elements, which results in too stiff behaviour. This is also applicable for 8-noded shells. Logically, midside noded elements are not recommended in contact analyses. The model was rebuilt with addition of elements.

An equivalent steel tube method

The equivalent steel tube method assumes that the grout volume and the steel tube act as a perfect composite in case of *bending*. From this an equivalent wall thickness is calculated.
This method is best explained by the following elaboration:

$$E_{steel} I_{composite} = E_{steel} I_{steel} + E_{grout} I_{grout}$$

$$\Rightarrow t_e = \frac{d_0 - (d_0 - 2t_0)\sqrt[4]{\frac{n-1}{n}}}{2} \qquad \text{(Eq. 1)}$$

The stiffness ratio between steel and grout is n. Lloyds [7] recommends that the effective thickness t_e is limited to 1.75 t_0.
This method only changes the τ-parameter, which, in its turn, only depends on E_{grout}.

Analyses and results

In this chapter the results will be presented and discussed. First the calculation strategy will be summarised. After that all SCF at saddle- and crown locations are given.

FE-calculation set-up

All analyses treated here have been carried out with linear elastic material behaviour.
The material properties were

1. Steel	E = 210000 N/mm^2 ν = 0.3
2. Concrete (where applicable)	E = 30000 N/mm^2 ν = 0.16

Ungrouted joints

Both chord- and brace loads were applied to the ungrouted joints. This means that a total of six load cases per joint was calculated. Only one brace (the reference brace) of the X-joint was loaded. No distinction was made between axial tension and axial compression. For these analyses 20-node solid elements were used. SCF from chord loads were too small to be of further interest.

Grouted joint analysis set-up

The equivalent steel tube method and the contact analysis option were used.

1. Contact analyses have been conducted with the grout modelled both as a deformable and as a rigid body. Steel was modelled with 8-node solid elements. For the contact analyses the following subdivision was made:
   a. Contact analyses with a deformable grout volume modelled by a coarse mesh.
   b. Contact analyses with a deformable grout volume modelled by a refined mesh.
   c. Same as b, but with fixed chord ends, to test the influence of supports (axial load and in plane bending) in contact analysis.
   d. Contact analyses using a rigid cylinder touching the inner chord wall.

2. For the equivalent steel tube a value of 30000 N/mm^2 was appointed to E_{grout}. This resulted in an equivalent wall thickness of 17.5 mm.

In all cases the length of the grout entity was about 2 times the brace diameter. Gap between grout and steel remained 0 mm.

Discussion of SCF at reference braces

Table 2 and 3 contain the SCF from respectively the T-joint and the X-joint (reference brace).
Unless otherwise mentioned pin supports were used.

- Axial tension

For the brace saddle locations the several brace saddle SCF are in good agreement with each other. At the chord locations the first indications of support condition-dependency show at the chord saddle locations.
Notice that the equivalent tube method gives conservative results for brace locations.

- Axial compression

Contact analyses (for axial compression) with different meshes give results that clearly prove the
effect of mesh-refinement on stresses.
The scatter in SCF at the chord locations is worse than at the brace locations.
The trend however, is that compression SCF are lower than tension SCF, which was expected
Changing the chord supports from pinned ends to fixed ends also affects the SCF-values.

Table 2. T-joint SCF

SCF T-joint	Bs	Bc	Cs	Cc
Ax (ungrouted)	9.76	0.79	19.01	2.45
Ax (equivalent steel tube)	5.93	1.26	6.48	0.83
Axt (def - coarse mesh)	4.62	1.45	7.77	2.23
Axt (def - refined mesh)	4.56	1.41	7.67	2.18
Axt (def - refined mesh - fixed ends)	4.27	2.26	6.70	5.65
Axt (rig)	4.12	1.94	6.89	3.56
Axc (def - coarse mesh)	-3.00	-1.47	-1.86	1.12
Axc (def - refined mesh)	-2.00	-1.22	-2.78	0.32
Axc (def - refined mesh - fixed ends)	-2.64	-1.54	-2.32	-2.87
Axc (rig)	-3.23	-1.07	-5.36	-1.77
Ipb (ungrouted)	-0.01	1.99	-0.01	3.69
Ipb (equivalent steel tube)	0.00	1.95	0.00	2.09
Ipb (def -coarse mesh - t/z)	0.37	2.57	0.71	3.54
Ipb (def - refined mesh - t/z)	0.40	2.55	0.75	3.49
Ipb (def- refined mesh - fixed ends - t/z)	0.40	2.54	0.76	3.46
Ipb (rig - t/z)	0.30	1.81	-0.03	4.39
Ipb (def - coarse mesh - c/z)	0.37	-3.28	0.71	0.47 (!)
Ipb (def - refined mesh - c/z)	0.40	-2.99	0.75	-1.12
Ipb (def - refined mesh - fixed ends - c/z)	0.40	-3.00	0.50	-1.10
Ipb (rig - c/z)	0.03	-4.02	-0.03	-2.37
Opb (ungrouted)	7.55	-0.01	15.92	-0.01
Opb (equivalent steel tube)	2.29	0.00	1.79	0.00
Opb (def -coarse mesh - t/z)	5.23	-0.18	8.59	0.62
Opb (def - refined mesh - t/z)	3.98	-0.03	6.64	0.80
Opb (rig - t/z)	5.20	-0.04	8.53	1.01
Opb (def -coarse mesh - c/z)	-4.56	-0.18	0.18 (!)	0.62
Opb (def - refined mesh - c/z)	-2.16	-0.03	-0.42	0.80
Opb (rig - c/z)	-3.72	-0.04	-3.06	1.01

Table 3. X-joint SCF

SCF X-joint	Bs	Bc	Cs	Cc
Ax (ungrouted)	9.68	0.74	18.74	1.79
Ax (equivalent steel tube)	6.17	1.42	6.22	0.80
Axt (def - coarse mesh)	5.44	1.34	8.88	2.02
Axt (def - refined mesh)	4.23	1.53	7.26	2.27
Axt (def - refined mesh - fixed ends)	4.25	2.24	6.70	5.57
Axt (rig)	3.47	1.93	6.37	3.99
Axc (def - coarse mesh)	-2.98	-1.47	-1.81	1.23
Axc (def - refined mesh)	-1.99	-1.22	-2.77	0.38
Axc (def - refined mesh - fixed ends)	-2.63	-1.53	-2.41	-2.77
Axc (rig)	-2.46	-0.96	-4.03	-1.65
Ipb (ungrouted)	-0.01	1.98	-0.01	3.68
Ipb (equivalent steel tube)	0.00	2.63	0.00	2.06
Ipb (def -coarse mesh - t/z)	0.44	2.48	0.78	3.43
Ipb (def - refined mesh - t/z)	0.43	2.51	0.80	3.44
Ipb (def- refined mesh - fixed ends - t/z)	0.43	2.51	0.80	3.42
Ipb (rig - t/z)	0.45	2.55	0.87	4.03
Ipb (def - coarse mesh - c/z)	0.44	-3.50	0.78	0.43 (!)
Ipb (def - refined mesh - c/z)	0.43	-3.00	0.80	-1.07
Ipb (def - refined mesh - fixed ends - c/z)	0.43	-3.00	0.80	-1.06
Ipb (rig - c/z)	0.45	-4.42	0.87	-1.37
Opb (ungrouted)	7.57	0.00	6.15	-0.01
Opb (equivalent steel tube)	5.26	0.00	4.29	0.00
Opb (def -coarse mesh - t/z)	5.50	-0.25	9.08	0.52
Opb (def - refined mesh - t/z)	4.20	0.06	6.88	0.87
Opb (rig - t/z)	5.36	-0.14	9.15	0.97
Opb (def -coarse mesh - c/z)	-5.90	-0.25	1.8 (!)	0.52
Opb (def - refined mesh - c/z)	-2.49	0.06	-1.36	0.87
Opb (rig - c/z)	-4.15	-0.14	-2.53	0.97

- . In plane bending

In plane bending SCF are mainly the cause of very local chord bending. So in fact grouting has little effect, which is proven by the results from

the calculations. The compression zone SCF at the chord crown locations indeed show a mesh-dependency. Also, compression SCF at brace locations are larger than tension SCF which does not prove the benefit of contact pressure.
The equivalent steel tube method seems to work well for the brace locations, but at the chord locations the SCF-decrease is large compared to the SCF resulting from contact analyses. This is the cause of local bending of the thicker wall (thickness to the third power) of the equivalent tube.

- Out of plane bending

Application of the equivalent steel tube method gives lower SCF-values than the several contact analyses.
Mesh refinement causes tension zone SCF to decrease. Again the coarse mesh gives unrealistic values in the compressed crown area.
Overall tension SCF resulting from the several analyses are more consistent than compression SCF.
Notice the remarkably good agreement between coarse mesh SCF and SCF resulting from deformable-rigid contact in the tension zones.

Carry-over effects (X-joint only)

According to the results, carry-over effects (i.e. stresses around the weld area of the unloaded brace resulting from chord deformation) under axial loading and bending out of plane are virtually eliminated by the grout.

Experimental results

When a comparison was made to the (at the moment confidential) experimental results found by TNO Building and Construction Research (Netherlands), some remarkably good agreements were found in most cases.

Conclusions and recommendations

Grouting significantly reduces stress peaks of the joints and therefore extends the service life.
It is emphasised that the grout also increases the local joint stiffnesses, which may lead to increased bending moments. Finite Element calculation of grouted tubular joints proved to be a serious option when detailed stress information (SCF) is required.
MARC provides the contact analysis tool, which is powerful yet easy to use. However, in some cases the SCF results showed a strong mesh dependency.
The general trend is that the results are realistic and in agreement with experimental results (confidential). Further improvement of the contact analysis module should enable users to obtain results that are less mesh dependent.

The following recommendations apply:

- Research about the mesh dependency of SCF in areas that are in contact. Mesh densities of grout and steel should be in agreement with each other.

- The grout length is a parameter that has influence on the CPU-time and disk space. The results of several calculations with various grout lengths should recommend a value for this length that is both accurate and economical.

- The effect of more sophisticated material behaviour (local plastification, grout cracking) should be investigated.

- The influence of increased joint rigidity on load distribution in the structure should be investigated together with the actual SCF-reduction by the grout.

REFERENCES

[1] J.L. Trinh and Ph. Beguin (1994), 'Strength and Fatigue Behaviour of Grouted Tubular Joints Experimental Project'. Paper BOSS '94 p. 261 - 274. Tapir Publishers. Trondheim, Norway.

[2] I.E. Tebbett (1982), 'The Reappraisal of Steel Jacket Structures Allowing for the Composite Action of Grouted Piles'. Paper OTC 4194 of Offshore Technology Conference, Texas.

[3] M. Lalani and I.E. Tebbett (1985), 'Justification of enhanced Capacities for As-Welded and Grouted K-Joints'. Paper OTC 5025 of Offshore Technology Conference, Texas.

[4] M. Lalani, D.J. Moran, R.J. van Foeken and J. Wardenier (1996), 'Fatigue Behaviour and ultimate Capacity of Grouted Tubular Joints'. Tubular Structures VII, Farkas & Jármai (editors). Balkema, Rotterdam, ISBN 90-54-10-828-2

[5] G.M. Brown, R. Holmes and J. Kerr (1980), 'Fatigue Life Enhancement of Welded tubular joints by Injection of Grout'. Paper BOSS '88 p. 1081 - 1095. Pergamom, Massachusetts Institute of Technology. USA

[6] Lloyds Register of Shipping (1980), 'Fatigue Analysis of Fixed Steel Platform Tubular Joints'. Fatigue Appendix, Issue 5.

[7] Efthymiou, M (1988), 'Development of SCF Formulae and Generalised Influence Functions for Use in Fatigue Analysis'. Shell Internationale Petroleum Maatschappij BV The Netherlands.

[8] Romeijn, A. (1994), 'Stress and Strain Concentration Factors of Welded Multiplanar Tubular Joints'. Delft University Press -111, ISBN 90-407-1057-0. The Netherlands.

[9] Health and Safety Executive (1990), 'Offshore Installations: Guidance on Design and Construction'. Fourth Edition, HMSO, London. United Kingdom.

Proceedings of the Seventh (1997) International Offshore and Polar Engineering Conference
Honolulu, USA, May 25-30, 1997

ISBN 1-880653-28-1 (Set); ISBN 1-880653-32-X (Vol. IV)

Large Scale Testing of a Multiplanar Tubular DX-Joint

S.P. Chiew, C.K. Soh and T.C. Fung
Nanyang Technological University
Singapore

ABSTRACT

The ultimate capacities of multiplanar tubular DX-joints have be evaluated by the authors using the finite element method (Chong et al, 1995). To verify and support the accuracy of such numerical modeling and parametric studies, a test program was initiated where a multiplanar servo-hydraulic test facility was commissioned in the School of CSE, NTU Singapore in 1995. This paper first discusses briefly the general philosophy and special consideration behind the design of the NTU test facility. A large-scale tubular DX-joint specimen was subsequently tested to failure under static load, and this paper details its complete experimental strength test program involving the heat treatment and non-destructive testing of the weldments, test set-up, control and procedure, instrumentation and data acquisition, and discussion of results.

KEY WORDS: Ultimate strength, Multiplanar, Tubular DX-Joint, Failure mode, Design codes, Large Scale testing

INTRODUCTION

Steel structural hollow sections have been used extensively in both onshore and offshore structures. These sections have attractive engineering and architectural properties as structural members, however, they are more difficult to connect up to form complete multiplanar tubular joints. In the past, much research has been carried out on uniplanar joints leading to their present incorporation into design codes such as the API (1989) and AWS (1992) codes. Relatively less research and testing of multiplanar joints have been initiated mainly because of their higher costs and test complexities. To date, there are only 3 well-documented experimental test results on full-scale multiplanar DX-joints the authors could find in the literature for comparison and verification with numerical studies (Vegte et al, 1991). However, multiplanar joint is an intrinsic feature in steel tubular construction, and the understanding of its behavior is important for more efficient and economical design.

Supported by both NSTB Marine Technology Program and NTU Academic Research Fund, a new multiplanar test rig was erected to facilitate further test programs in this subject area. This paper first describes the special requirements for the design of the NTU multiplanar test facility, and then, the static strength test program and experimental results of a full-scale multiplanar DX-joint up to failure.

DESIGN REQUIREMENTS FOR TEST FACILITY

The test rig is housed in the Heavy Structures Laboratory of the School of Civil and Structural Engineering, NTU. The laboratory has a 1.3 m thick reinforced concrete strong floor with cellular box-type construction. The multiplanar test rig weighs about 50 tonne and occupies a total of 9 m^2 area of strong floor. It is a reactive load frame designed to the physical limitation of the laboratory as well as the strength of the strong floor. The design is governed by frame stiffness where there is a need to minimize overall frame deflections under critical load cases. Basically, the test rig consists of a main frame and two A-frames. The main frame is used to apply loading for the in-plane braces, and also to provide support for the chord of the tubular joint. Out-of-plane loading is applied from the two A-frames which sit opposite to each other at 90° from the chord axis. An overview of the test rig is shown in Figure 1.

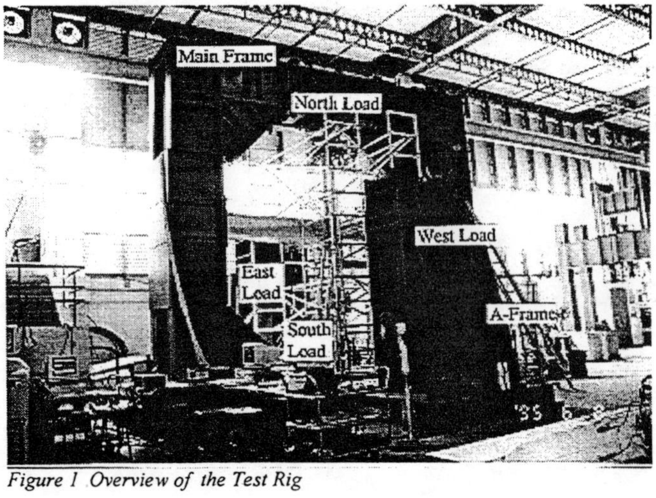

Figure 1 Overview of the Test Rig

The most important part of the design consideration is to figure out all the types of multiplanar joints that can be tested. It is not possible to design a test rig to test all possible joint configurations. However, it should at least cover a reasonable number of them. In this respect, 9 load cases for the main frame have been considered as shown in Figure 2. The loading points are fixed at 6 positions on the main frame and two positions on the A-frames. Two hundred metric tonne loading is considered as applied load. Due to the high load generated by the jacks and the self weight of the hydraulic jacks, it is not possible to move the jacks during testing. Also, changing the jack from time to time will add to the complexity of the test to be carried out. It was decided to fix the jacks at permanent locations of 90° and 45° with the chord.

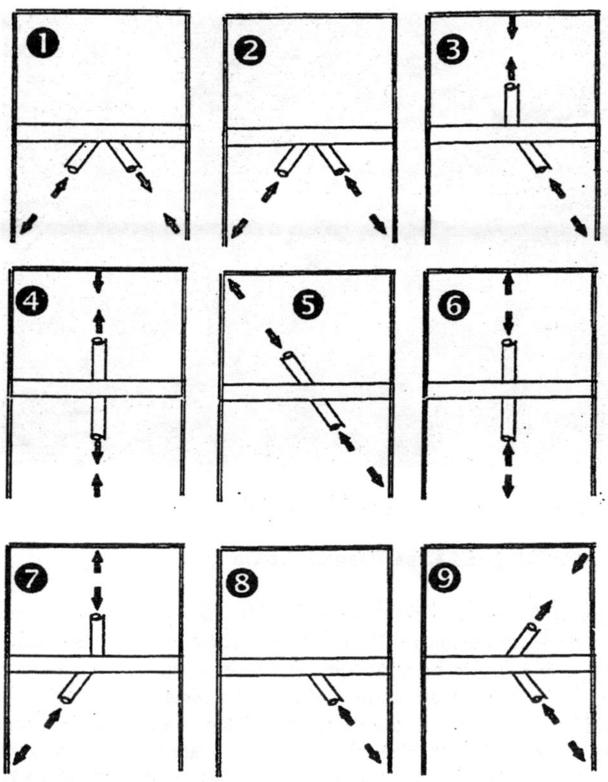

Figure 2 The Nine(9) Basic Load Cases for Main Frame Design

Another major consideration is the test rig configuration. The rig could be a self-contained equilibrium rig or reactive with the strong floor as the main support. In a self-contained rig, the loading will be balanced. By referring to Figure 2, it can be seen that Load Cases 3, 5, 7, 8 and 9 will cause much distortion to a self-contained frame if it is not properly braced. The increase of bracing would lead to an increase in the number of members and joints in the test frame. It will result in an increase of cost in terms of manpower and material for the manufacturing of the test rig. It will also be messy in appearance, and these will defeat the purpose of having the strong floor in the first place. An alternative solution is to bolt the main frame to the strong floor by means of holding-down bolts and anchor blocks. This system has the advantage of reducing the displacement of the overall frame subjected to loading comparable to those of the previously mentioned self-contained equilibrium frame.

The actuators are used to generate the required axial force during testing while the controller is used to control the movement of the actuators. A total of 8 actuators can be mounted on the multiplanar test rig, but only 4 actuators are installed at present. The 4 actuators installed are mounted on both the vertical and horizontal positions of the test rig. Where possible, front flange mounting are preferred over the usual pedestal-mounting to maximize headroom. The rate of loading is controllable in both position and load; with a minimum speed of 0.1 mm/min. All actuators have the capacity of generating a maximum static compression/tension force of +2000 kN/-1200 kN with a nominal rated stroke of ±100 mm. All the actuators are controlled by an Instron 8580+ digital controller. The controller essentially consists of one control tower and a computer. The stroke, load and other control parameters of the individual actuator are able to be viewed and programmed by the computer. More importantly, it can synchronize the action of all the 4 actuators or two pair of actuators. This synchronize action tends to instruct the actuators to response at the same time while each of them can have different waveform, either single ramp, double ramp, sinusoidal wave, etc. The load cell is capable of measuring up to 200 tonne and the swivel head is allowed to rotate 8° in every direction. An interesting design of the load cell is its integration with the swivel head which makes the total length much shorter than conventional design with separate units.

STATIC STRENGTH OF A MULTIPLANAR DX-JOINT

Details of the DX- Joint Specimen

Figure 3 shows the geometry as well as the dimensions of the multiplanar DX-joint. The geometric parameters for this joint are:

$$\beta\ (d/D)=0.6,\ \gamma\ (D/2T)=24.5,$$
$$\tau\ (t/T)=0.98,\ \alpha\ (2L/D)=21.86,\ \theta = 90°.$$

The β (d/D) value of 0.6 for this specimen is the upper range for non-overlapping multiplanar DX-joint based on the recommendation by API (1989) for a minimum of 50 mm clearance between the gap of the adjacent braces. The γ (D/2T) is also the upper range value used in the offshore structures. To prevent brace failure prior to the chord, a τ (t/T) of approximately 1.0 was chosen for the joint to be tested. The α (2L/D) used for the joint was sufficiently long to prevent any chord end effects. The difference in the length of the in-plane braces shown in Figure 3 was due to the necessary alignment of the chord ends which were mounted on the main frame of the test rig. The different length of the in-plane braces ensured the brace end touched the swivel head to the actuators and provides sufficient travelling distance throughout the testing process.

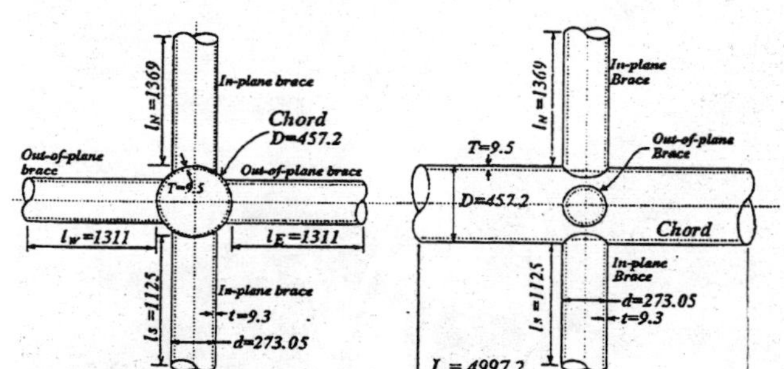

** All braces have the same diameter and thickness; all dimensions in mm*

Figure 3 Details of the DX-Joint Specimen

The standard API 5L Grade B pipe was used for the specimen. The brace and chord intersections were welded with full penetration weld conforming to the AWS (1992) code. At the free end of the braces, 25.4 mm thick flat plates were welded so that the load from the actuator could be transferred to the brace, and to serve as an attachment to the swivel head connected to the actuator. Table 1 shows the results of the coupon test for the brace material. The test coupon was machined according to the ASTM A370 (1992) specifications for the mechanical testing of steel products. The coupon was cut in longitudinal direction from a pipe section and the curve surface was flattened by machining. It has a gauge length of 50 mm, width of 12.48 mm and thickness of 7.70 mm.

Table 1 Material Properties

Results	Factory Cert.	Coupon Test	API Spec.
Yield stress (N/mm^2)	312.2	361.8	min. 250
Ultimate stress (N/mm^2)	482.8	478.5	min. 415
% elongation	36	33	min. 21

Stress-Relieving and Ultrasonic Testing

To eliminate or reduce the influence of residual stress on the stress concentration and the ultimate strength of multiplanar DX-joint, welding residual stress-relieving was necessary at the intersection region of the welded joint. Figure 4 shows the set-up for the heat treatment work. The braces and chord at the intersection region were covered by ceramic cells which functioned as heating elements. These elements were heated by nickel wire buried inside the cells that were connected to power supply. Sensor wires (thermocouples) stuck at several positions of the joint were used to feedback the temperature during the heating process. Finally, pieces of fiber glass sponge were used to cover the joint to prevent rapid heat emission. The feedback of the temperature and the power supply were controlled by a control unit. The whole processes of stress relief took about 8 hours. This included gradual increment of temperature up to 600° Celsius, maintaining at constant temperature for sometime before reducing it to the room temperature. The cooling process of the joint with the sponge wrapping around the intersection successfully reduced the difference of cooling time at different parts of the joint, thus reducing residual stress to a minimal.

Figure 4 Stress Relieving Of the DX-Joint Specimen

Before the ultrasonic testing was carried out on the multiplanar DX-joint, the surface of the steel was ground to smoothen the metal surface and to remove the paint as well as the protection layer for steel. Better transmission of the ultrasound was achieved by applying polycel couplant on the surface of steel. A picture of the test is shown in Figure 5. Defects such as cracks, incomplete fusion, inadequate penetration, slag and porosity are very common in the welds of tubular joints. From the results of the multiplanar DX-joint specimen, the discontinuity defects of inadequate penetration and porosity were found to be 14% of the butt welds around the intersection of the joint. These served as a useful piece of information for explaining the failure mode of multiplanar DX-joint if cracking is the controlling factor contributing to the ultimate failure of the specimen.

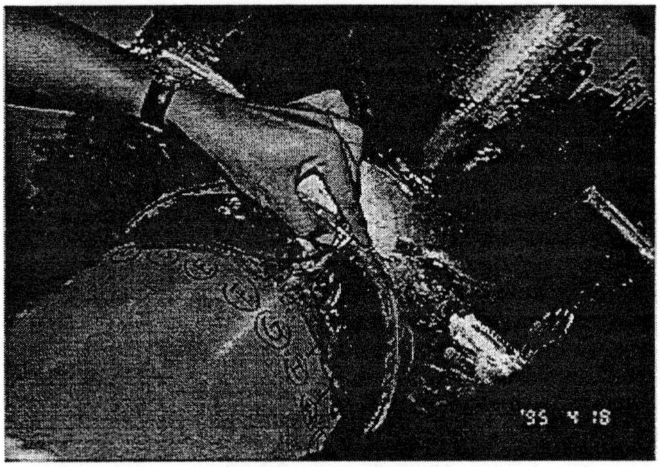

Figure 5 Ultrasonic Testing of the Weld Quality

Strain-Gauging and Experimental Set-Up

A total of 32 single element gauges with 2mm gauge length were placed on the multiplanar DX-joint for recording strain at a particular location. Strain can be converted to stress when the stress-strain curve of the material are known. Half of the gauges were placed at the mid-length of the 4 braces to record the strain resulting from externally applied load. The effect from the brace end and from the intersection of the joint to the strain was minimal at the mid-length of the braces. The other half of the gauges were placed at the intersection region of the North and West braces. The strain gauge positions are shown in Figure 6.

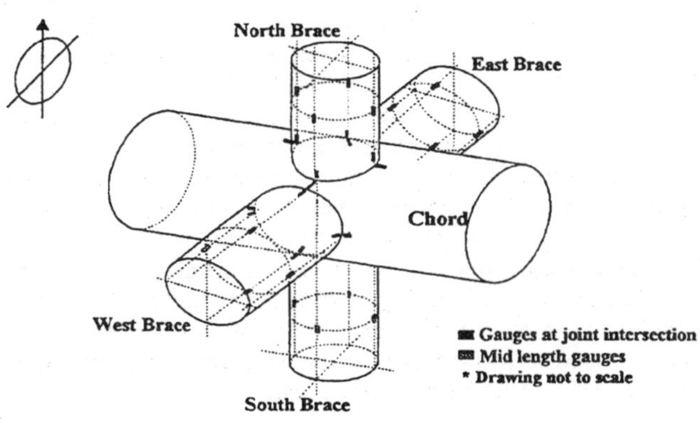

Figure 6 Strain Gauge Locations

Figure 7 shows the test set-up for the multiplanar DX-joint. The specimen chord ends were mounted by bolts on the main frame of the test rig. The self weight of the joint was thus taken by the main frame and transferred to the strong floor. With this arrangement, the chord centerline was maintained at the same position throughout the test. This facilitated the application of load for the four braces, and ensured minimum misalignment with the actuators. The free ends of the braces were connected to the actuators by swivel heads. Besides allowing for brace rotation during the application of loading, the swivel head also provided mounting surface for the end plates of the braces. Bolts were used to secure the brace end to the threaded holes at the swivel surface. This arrangement ensured no slippage at the connections during application of load while allowing for the rotation of brace at the swivel head.

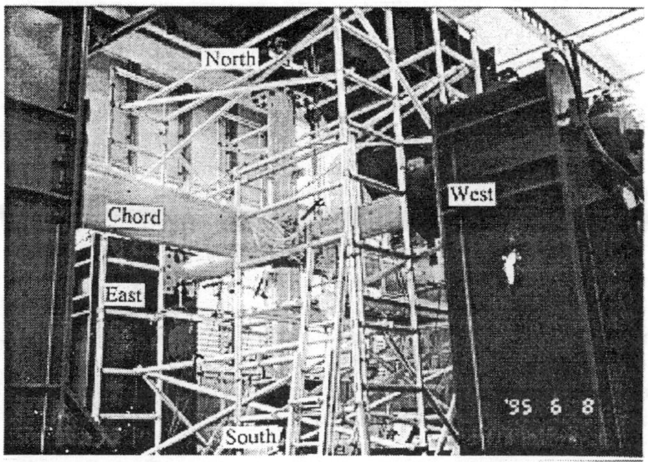

Figure 7 Test Set-Up for Multiplanar DX-Joint Specimen

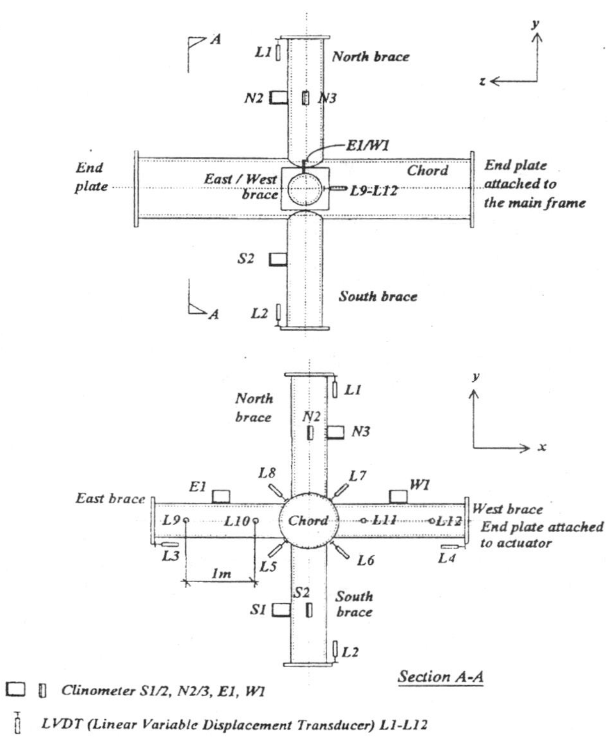

Figure 8 Locations of LVDTs and Clinometers

Instrumentation and Data Acquisition

Proper instrumentation is important because the interpretation of test results depends solely on the quality and accuracy of the data acquired. Thus, the equipment has been carefully checked before use and properly adjusted and placed at the locations shown in Figure 8.

The instrumentation comprises LVDT (Linear Variable Displacement Transducer), clinometer and strain gauges as illustrated earlier. LVDT is used for measuring displacement while clinometer is for measuring inclination. The primary objective of this experimental test is to establish a load-displacement curve for comparison with the numerical results, and to monitor the performance of the progress of the entire test system. The load reading can be directly obtained from the load cell integrated in the actuators. The mid-length gauges at the circumference of the braces can also be used to counter check the value of the applied load.

To measure the initial tilt of the braces and subsequent tilting during the progress of test, clinometers were placed at the locations shown in Figure 8. Each brace has two clinometers placed at 90° to each other, as indicated by N2 and N3 for the North brace shown in Figure 8. There are two locations at the East and West braces where horizontal rotation or rotation about the y axis cannot be measured by the clinometer. In this case, two LVDTs were used to replace the function of a clinometer. The location and orientation of these LVDTs are shown in Figure 8 as L9 to L12. The penetration of the braces due to axial compressive load was monitored by L1 to L4 as shown in Figure 8. The displacements were measured at the end of the braces because the measurement at the crown or saddle of the joint is prohibitively difficult to be carried out by using LVDT due to the limitation of space at the intersection. The small amount of elastic shortening of the braces due to the displacement of the joint can be ignored. Present loading condition with balanced axial load at opposite braces resulted in a more stable test with less rotation for the braces. These factors permitted the use of brace end as a replacement of crown or saddle point for monitoring of the displacement of the joint. To measure the bulging of the chord cross section, 4 LVDTs were placed at L5-L8 as shown in Figure 8. They were slanted at 45° to the horizontal line so that the displacement corresponds to the radial direction of the chord axis can be measured.

Figure 9 Set-Up for Data Acquisition

Figure 9 shows the set-up for data acquisition located beside the test rig. All readings were recorded automatically by computers except for the clinometer readings that have to be recorded manually at selected time interval. The LVDTs and strain gauges readings were channel through a switch box before connected to a data logger. A computer installed with LabWindows (1991) program was used to capture the data from the data logger. These data were stored in a file in the computer hard-disk and were printed out from the data logger as well. The load readings from the load cell were by default channeled to the console and computer. To facilitate that load and displacement readings can be monitored by only one computer, the load readings were also directed to the data logger. An amplifier was used to amplified the voltage analog load signal from the Instron console to the data logger.

Test Control and Procedure

Figure 10 shows the load pattern applied on the multiplanar DX-joint. The 350 kN pre-load at the out-of-plane braces (West and East braces) was maintained at constant value to simulate a multiplanar load condition. The in-plane braces (North and South braces) were loaded incrementally to failure.

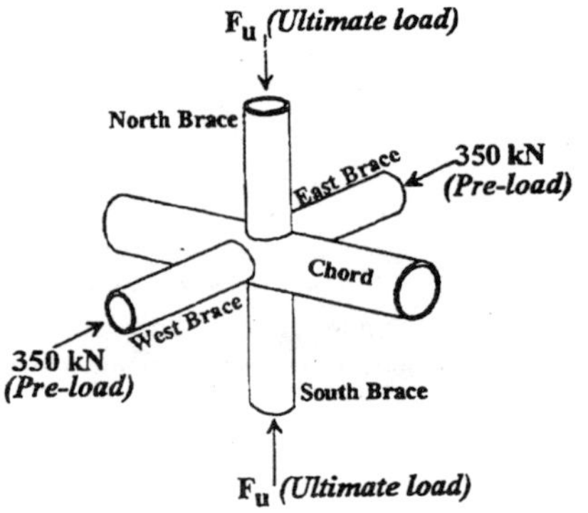

Figure 10 Load Pattern for Multiplanar DX-Joint Specimen

The test procedure can be divided into 3 stages. For the first stage, the 350 kN pre-load was applied at the out-of-plane braces by 10 increments/ramps. Each ramp of 35 kN was loaded at a rate of 10 kN/min. During the out-of-plane loading, the North and South braces were maintained at zero load by using the load control. These ensure no load interaction of the in-plane and out-of-plane braces during the pre-loading stage.

At the second stage, the in-plane braces were loaded to 400 kN at a rate of 20 kN/min. The out-of-plane braces were maintained at 350 kN. The second stage ensure the North and South braces were in contact with the swivel of the actuators and at equal load. At the final stage, the load control was switched to displacement control and the in-plane braces were loaded at a rate of 0.2 mm/min to joint failure. Displacement control can prevent sudden extraction of the piston of the actuator during instability or failure of the joint. The duration of the test was shortened by increasing the rate of displacement during the post-buckling failure of the joint.

Discussion of Results

The load-displacement curves for the multiplanar DX-joint tested experimentally are plotted in Figure 11. It can be noticed that there is a horizontal shift of the curve to the left at load equal to 0. This is caused by the pre-loaded force of 350 kN at the out-of-plane braces prior to the loading of the in-plane braces.

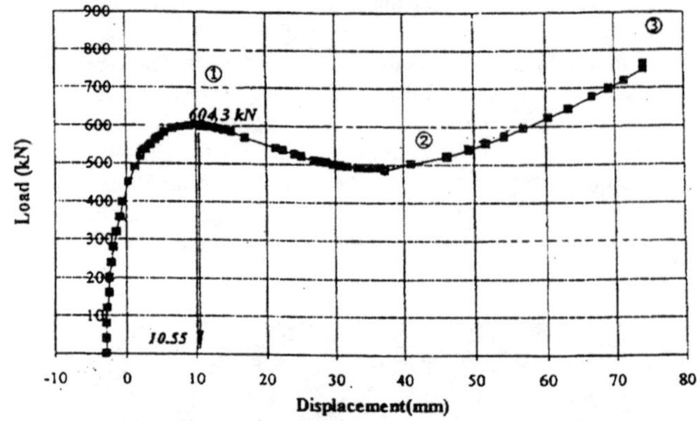

Figure 11 Load-Displacement Curves

The tilting and stresses of the braces were measured by the clinometers and mid-length strain gauges respectively. Results from the clinometers show that the East brace tilted upward at an angle of 1.01°. The other braces show tilting angle of less than 0.4°. Based on the applied load of 350 kN at the East brace, the angle of 1.01° has little impact on the loading. During the test, these tilting angles for the braces show a variation of ±0.02° only. Thus, the applied loads do not need to be corrected due to the small tilting angles. The amount of side load transferred to the actuators is also kept to the minimal. However, the strain gauge readings at the mid-length of the braces were very sensitive to the tilting angle of the braces. Only East brace shows a consistent amount of small bending moment at the location of the mid-length strain gauges. For the other braces, the 4 gauge readings around the circumference were almost the same.

The experimentally determined ultimate loads were summarized in Table 2 along with that predicted using the CIDECT design guide (Wardenier et al., 1991) and Equations 5.4 and 2.5 of the AWS (1992) design code.

Table 2 Tabulated Results for DX-Joint Specimen

Joint configuration and loading condition	d (mm)	D (mm)	t (mm)	T (mm)	L (mm)	l_{NORTH} (mm)	l_{SOUTH} (mm)
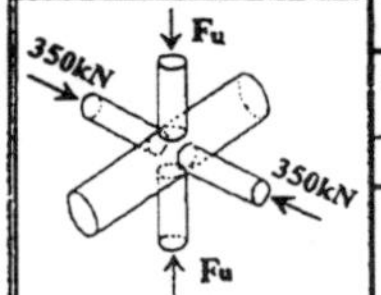	273.05	457.2	9.16	9.33	4,997	1,369	1,125
	$l_{WEST/EAST}$ (mm)	α	β	γ	τ	f_y (N/mm²)	f_u (N/mm²)
	1,311	21.86	0.60	24.50	0.98	361.8	478.5
	$F_{u,TEST}$ (kN)		$F_{u,AWS}$ (kN) Eq.2.5	$F_{u,CIDECT}$ (kN) Eq.5.4		$F_{u,AWS}/F_{u,TEST}$	$F_{u,CIDECT}/F_{u,TEST}$
	604.30		521.49	523.46		0.86	0.87

After achieving their respective peak load, both curves drop to their respective minimum value as indicated in the figure. After these minimum values, the load increases with increasing displacement. The final increment of load exceeds the peak load recorded earlier. For the experimental curve, the test stopped at a displacement of approximately 75 mm because of the limitation of the stroke of actuators.

The experimental ultimate load was compared to that predicted using the AWS code and CIDECT design guide. With the removal of code safety factors, both code predictions were conservative with the ratio of $F_{u,AWS}/F_{u,TEST}$ and $F_{u,CIDECT}/F_{u,TEST}$ equal to 0.86 and 0.87 respectively.

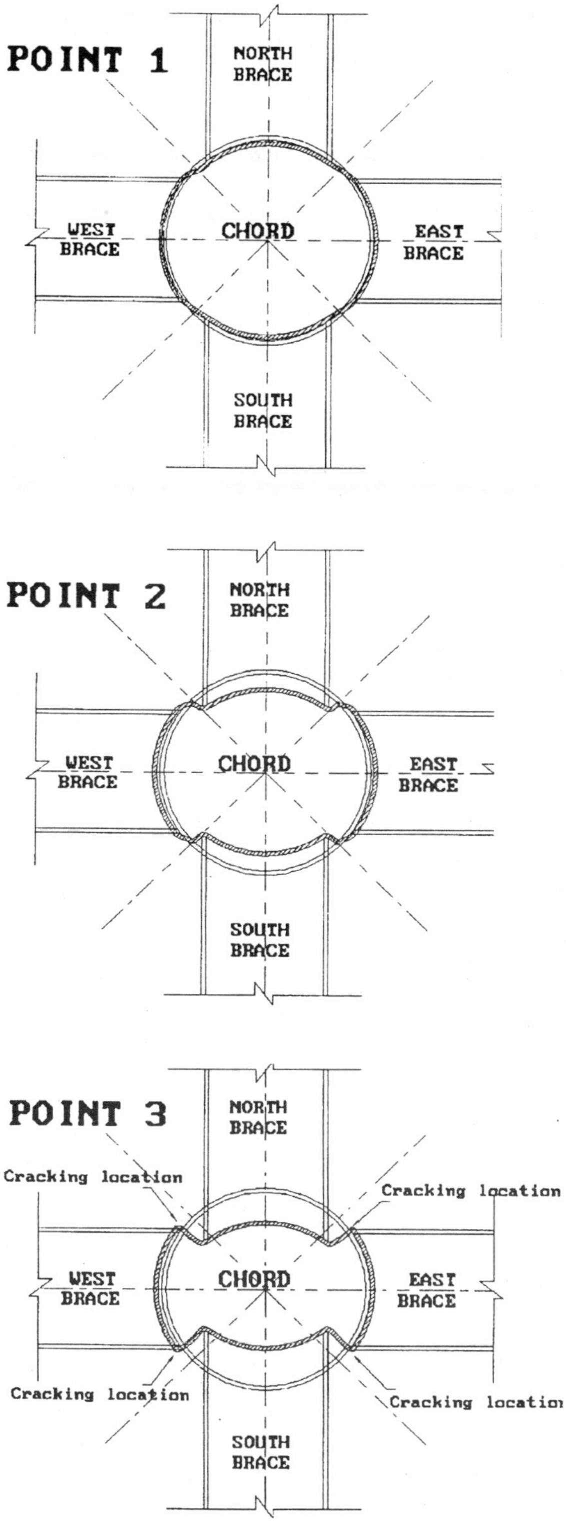

Figure12 Cross-Sectional View of Failure Mode

The failure mechanism throughout the test can be explained by Figure 12. The figure shows the cross-sectional view at the center of the multiplanar DX-joint according to the measured displacement during the experimental test at three stages i.e. Point 1 to Point 3 in the figure. At Point 1 (Figure 11), the in-plane (North and South) braces penetrated and buckled the chord wall at the gap areas. The out-of-plane (East and West) braces responded with slight bulge-out at the chord, although the braces were loaded with compressive load of 350 kN.

From Points 1 to 2, plastic hinges occurred around the circumference of the chord and were unable to sustain higher load. At Point 2, which is the lowest load after the peak load, the saddle of the in-plane braces is almost at the same level as the saddle points of the out-of-plane braces. At this stage, the gap area at the chord is almost parallel to the horizontal plane and looks like a flat plate. Further increase of load at the in-plane braces will hammock the chord and the increment of load is resisted by the tensile force of steel at the gap area.

Figure 13 Plastic Chord Failure and Cracking at Saddle Position of the Out-of-Plane Brace

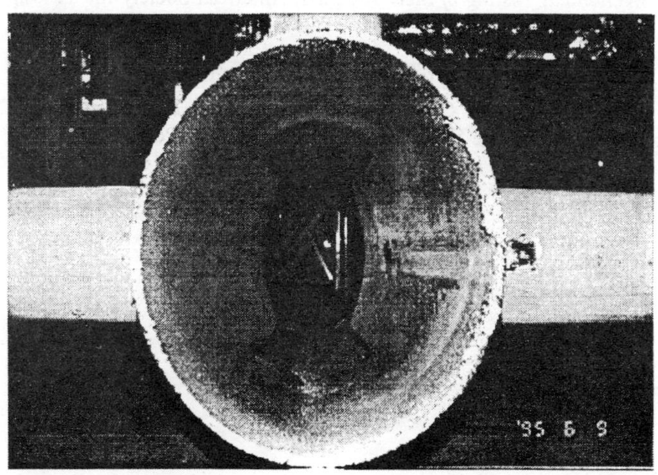

Figure 14 Cross-Sectional View of Plastic Chord Failure of the Multiplanar DX-Joint

During the experimental test, the formation of cracks can be observed at the saddles of the out-of-plane braces at the initial part of increasing load from Point 2 to Point 3. As the load increased, the cracks extended. At the final stage where the maximum load was reached, large cracks occurred at the weld toe of the four saddle points of the out-of-plane braces with an average of 11 cm long (see Figure 13). The failure mode at Point 3 is shown in Figure 12 and Figure 14. It can be concluded that plastic chord failure occurred for the multiplanar DX-joint and cracks initiated during the hardening part of the load-displacement curve.

CONCLUSION

This paper presented some important practical considerations associated with the design and fabrication of the NTU multiplanar test facility, and in some considerable details, the experimental test program of a full-scale multiplanar DX-Joint specimen. Further testing work is in progress, and other numerical modeling work and results will be reported later.

ACKNOWLEDGEMENT

The authors wish to thank the National Science & Technology Board and the Nanyang Technological University, Singapore for their financial support in this research project. The authors also wish to thank their former students, Mr Chong Chuan-Yi and Mr Lee Wooi-Min, and the technical staff of the CSE Structures Laboratory for their assistance in carrying out the experimental testing work described in this paper.

REFERENCES

American Petroleum Institute (1989). *Recommended Practice for Planning, Designing and Constructing Fixed Offshore Platforms*, API RP 2A, 18th Edition, Washington D.C., USA.

American Welding Society (1992). *Structural Welding Code - Steel*, 14th Edition, ANSI/AWI D1.1-92, Miami, USA.

ASTM A370 (1992). *Standard Method and Definitions for Mechanical Testing of Steel Products*, American Society for Testing Materials.

Chong, C.Y., Soh, C.K., Soh, A.K., Fung, T.C. and Chiew, S.P. (1995). *Ultimate Capacity of Multiplanar Double X-Joints.* Proceedings of the 5th International Offshore and Polar Engineering Conference, The Hague, The Netherlands, June 11-16, Vol. IV, pp. 109-115.

LabWindows (1991). *User Manual - Version 2.1*, National Instruments Corp., USA.

Wardenier, J., Kurobane, Y., Packer, J.A., Dutta, D. & Yeomans, N. (1991)., *Design Guide for Circular Hollow Sections (CHS) Joints under Predominantly Static Loading*, CIDECT - Comite International pour le Developpement et l'Etude de la Construction Tubulaire, Verlag TUV Rheinland, Koln, Germany.

Vegte, G.J Van Der, Koning, C.H.M.De., Puthli, R.S.and Wardenier, J. (1991). *The Static Strength and Stiffness of Multiplanar Tubular Steel X-Joints*, International Journal of Offshore and Polar Engineering, Vol.1, No.1, March, Golden, Colorado, USA.

Proceedings of the Seventh (1997) International Offshore and Polar Engineering Conference
Honolulu, USA, May 25-30, 1997

ISBN 1-880653-28-1 (Set); ISBN 1-880653-32-X (Vol. IV)

Tubular Joints Reliability and Fracture Analyses for Development of Underwater Inspection of Offshore Steel Structures

Wan Mahmood Wan Ab. Majid
Mara Institute of Technology
Kuala Lumpur, Malaysia

Mohamad bin Embong
Protek Engineers Sdn. Bhd.
Kuala Lumpur, Malaysia

ABSTRACT

Tubular joints of offshore structure are subject to cyclic loading of wave during the service life of the platform. This normally lead to fatigue failure, crack growth and fracture of the joints. All these failures will lead to reduction of platform reliability for safe operations. Operating oil company spent millions of dollar to carry out underwater inspection of tubular joints of offshore structures. Inspections are normally carried out in a fixed interval time period (i.e. 4, 5 years and etc.) determined by either company standard practice or international codes. The interval period may be increased or reduced if proper analyses of the joints reliability can be systematically established. This paper outlines the analytical basis and the calculation method used in the derivation of reliability indices, fracture assessment and the development of an outline inspection schedule for critical joints of offshore platform.

Keywords: cyclic, crack, fracture, reliability

INTRODUCTION

The emphasis of this paper is on the computation of the reliability index of steel tubular connections in plain cylindrical joint-cans (i.e. no ring stiffening present). There are two broad categories of joint that are considered:

- Where a joint is intact, the reliability index and an outline inspection schedule may be developed on the basis of the S-N spectral fatigue analysis.
- Where a joint is reported as cracked, the remaining life of the joint will be assessed by using fracture mechanics methods.

Overview of Analysis Procedure

The reliability index of a structural connection at a joint is a statistical measure of the acceptability of that connection for service in a marine environment. The calculation of the reliability index for joints in an existing structure involves a lengthy procedure. The procedure takes account of :

- The wave-loading environment for the structural location and orientation.
- The predicted fatigue life of the joint
- The expected remaining design life of the structure
- Previous and future inspection plans, including method and interval between inspections
- The occurrence of existing defects (cracks, repair grinding, etc.) in and adjacent to the weld forming the connection.

An ordered and systematic approach is required in order to take account of these factors.

In many cases, the reliability index will be calculated for major joints with known defects. There may often be a propensity for failure of the selected joint and, as a consequence, the calculations will be of a sensitive nature, often requiring scrutiny by third parties such as offshore certification bureaux.

The reliability index method described herein relates to joints which are uncracked: joints with known flaws are further evaluated using fracture mechanics (FM) techniques.

FM methods are used to calculate the remaining fatigue life of cracked joints, using a crack growth curve, a relationship for the stress intensity factor and a known defect size.

Preliminaries and Underlying Assumptions

A number of preliminary activities need to be completed before the calculations can be attempted:

1. Determination of the total and remaining service life of the platform.
2. Establish the underwater inspection history of the structure to date.
3. Existing anomalies database for the platform.

The preliminary work is conformed to the underlying assumptions which are used in the calculation of the reliability index. These are as follows:

1. An S-N spectral fatigue analysis is completed.

2. The API X' curve must have been used to compute the Damage Ratio of the joints.

3. The safety factor of 1.00 is used in the Spectral fatigue analysis.

4. Filters out the joints with a fatigue life < 1.2 times the specified service life. These are the reported low lives.

5. Produce a Reliability index for chord / brace joints.

Operation of Analysis

The bulk of the analysis may be undertaken using the simple program which can reads the long output file normally produced by spectral fatigue analysis.

The Fatigue detailed fatigue output file may be read in four stages:

1. The S-N data is assumed to be the API X' curve. The method used suits a one slope curve with an endurance limit, as used by the API X' curve.

2. Joint order summary report is scanned for low fatigue lives-results. Chord / brace failures are discriminated.

3. Wave spectra input is read.

4. For joints with fatigue lives less than 1.2 times the service life, the RMS stress and damage ratio per input wave spectra is read and stored for processing. Key data is output to a summary file.

The reliability index is computed for only the most damaging point of a brace/can intersection, which may be either on the brace or on the chord side of the joint.

The reliability index is calculated at discrete time intervals using the underlying principle that the Miner's Sum of damage (damage ratio) is directly related to the time elapsed.

The Reliability function is a statistical transformation, shown on figures 2.0 and 3.0. The probability of failure is computed from the damage ratio, and the reliability index is then derived from the probability of failure.

The probability of failure is affected by inspection [7,10,15]. If an inspection has been undertaken, and records no defects, then the maximum undetected defect can be estimated, and related to a revised probability of failure. If more than one inspection is undertaken, the probability of detecting a crack increases and the probability of failure decreases, as indicated by Lotsberg [15].

Probability of Failure

The relationship between P(f) and damage ratio is shown on Figure 4.0. It is based upon the curve developed by DnV, with modifications as required by the use of the API X' curve. The API X' S-N curve is a lower bound fatigue curve. API states (C5.2.5) that 97% of the experimental failures are on the safe side of the curve. This would indicate that:

P(f) = 0.03 when DR = 1.0

API adds that this value may need to be revised to 95% prediction interval, and so

P(f) = 0.05 when DR = 1.0.

API additionally advises that a safety factor of two be used in the fatigue calculations.

This is interpreted as: P(f) tends to the value 1.0 when DR=2.0.

The relationship between damage ratio and Probability of failure is charted by DnV [13] and by Lotsberg [10]. These two sources both show similar relationships to that used in this analysis. The major difference between this analysis and DnV's curve is that the DnV curve appears to relate to an S-N curve which is not a lower bound for fatigue life calculations in that the probability of failure is approaches a value of 1.00 when the damage ratio is 1.0.

Such a relationship indicates that the joint is "bound to fail" when the damage ratio is equal to 1.00, a proposition which is not upheld by the authors of the fatigue section of the API RP2A [14].

Reliability Index

The reliability index is based upon the computed probability of failure, P(f). The transformation used is of a standard statistical form, based upon the standard distribution curve. Because the standard distribution curve is not easily described in mathematical terms, the function is digitised and developed through the interpolation of a 34 point description of the curve.

A reliability index(β_R) that corresponds to an acceptable design limit, used to determine joint inspection programs from the result of joint reliability calculations. An index value of 1.47 (see figure 2.0) corresponds to a 7% probability of failure(P_F) and is used as a benchmark. The relationship of the joint reliability index and the probability of failure for joint is given in Figure 2.0.

Calculations undertaken shows that every joint reaches the base limit in a time interval which is equal to the calculated fatigue life. This result is identical to that returned by Madsen [16]. It is concluded that the description of the relationship between the reliability index and the damage ratio is correct for the framework within which the calculations are performed.

Treatment of Inspections

The reliability index is changed as a result of an inspection because the Probability of failure is altered. If there are no reported cracks on a joint with a low fatigue life, then there are three possibilities:

1. There is a significant crack, which has been missed

2. There are insignificant cracks which are too small to record or detect.

3. There is no crack at all.

The size of a remaining crack is a function both of the inspection technique, and of the history of inspections.

When a regular inspection has occurred, the Probability of failure may be updated, as follows:

After the 1st inspection, P(f) is set at 95.9% limit, which corresponds to a probability of failure of 0.041, and a damage ratio of 0.6.

A lower value would be justified at the 2nd, 3rd and subsequent inspections [7,15].

The damage ratio after an inspection is reset as the minimum of 50% of the existing damage ratio and 0.6^(number of inspections). See attached example figure 1.0.

Fracture Fatigue

The fatigue life of a joint can be estimated based on the results of fatigue analysis by the S-N method or by the FM (fracture mechanics) method.

Using either S-N or FM methods should be visualised as alternative post-processing methods applied to the results of a fatigue analysis.

The stress ranges per joint location relative to defined environmental descriptions are the starting point for both types of analysis (i.e. S-N and FM). This data is readily available from Fatigue analysis.

The fracture mechanics approach to fatigue analysis requires that the stress ranges be described in the "deterministic" format.

When a spectral fatigue analysis has been undertaken, the results can be transformed from the spectral format into a series of "Equivalent Stress Ranges" using a simple stress / damage relationship.

This operation results in a set of "deterministic" stress ranges that can then be used in the Paris Crack Growth equation.

Equivalent Deterministic Stress Range

The fracture mechanics approach demands the use of "deterministic" stress blocks which are normally obtained from a static deterministic fatigue analysis.

When a spectral fatigue analysis is run, the results can be transposed back into an equivalent set of deterministic stress blocks. This is because for defined S-N material curve, damage and number of cycles, there is one stress value which will return exactly the same damage ratio as the "probabilistic" stress range generated in the spectral fatigue analysis.

Stress ranges correspond to the description of the directional wave scatter diagram, and each stress / damage pair has an associated number of cycles, which is assumed to be directly related to the input Tz for the seastate.

This data can be transformed into Equivalent Deterministic stress ranges at the hot spot reporting the highest damage ratio only. Account is taken of:

- Zero damage Stress blocks
- Thickness Correction Factor
- S-N curve definition
- Re-computation of Number of Cycles where the Spectral stress range "straddles" the endurance limit.

Basis of Computation

The API X' curve is defined as follows:

$$Log_{10}(N) = 13.396 - 3.74 Log_{10}(S)$$

The equation is only valid for hot-spot stress range, S, in units N/mm^2

The equation is valid for hot spot stress ranges above the endurance limit, defined by the point N=200 Million Cycles, S= 22.02 N/mm^2 on the S-N curve.

When the stress is below the endurance limit, there is an infinite value of N with the result that no damage can be caused.

Crack Growth Curve

The crack growth curve used to predict propagation through the depth will be the linear Paris Relation

$$\frac{da}{dN} = C(\Delta K)^m$$

The crack growth curve to predict propagation through the thickness of the metal considered. The Paris Relation is used with a tubular forcing function.

For a < 2mm,

$$\frac{dA}{dN} = C(\Delta K)^3$$

For a > 2mm, the tubular forcing function is used:

$$\frac{dA}{dN} = 2.8x10^{-12}(\Delta K)^{2.5}$$

Crack growth calculation is terminated when the crack penetrates the wall thickness or when it becomes critical i.e. when the maximum stress intensity factor becomes equal to the fracture toughness.

Material Constants

The following material constants are used in the FM Module. They may be amended by any specific material information.

Paris Relation Constants

C	=	$5.793x10^{-12}$
m	=	3

Source of data: TNO [1].

Note the units for the constants given above is N and metre.

Threshold stress Intensity factor range is not required when used with the FM Module. The method basically returns a very high number of cycles when the stress is below 50 MPa

Fracture toughness

This constant is not used with the analysis method used because the tubular forcing function leads to a high rate of fatigue crack growth when the crack has grown to a depth equal to about 25% of the steel basic thickness.

Derivation of Stress Intensity Factors

The approach by Haswell [3], is used for the determination of SIFs for the tubular joints under consideration.

Haswell proposes a simple model relating the SIF, SCF, DOB, at each crack depth as follows:

$$\frac{SIF}{SCF.\sigma n} = (A + dA) + B.DOB$$

where A and B are constants derived from a linear regression analysis of the data performed by Haswell, dA is the maximum absolute scatter at each crack depth.

The deterministic stress range, including for SCF effects = SCF. σ_n. This value is computed by as defined previously.

Haswell lists the result of a parametric study of A+dA, B, DOB for a range of tubular joint geometries:

Haswell's results can be reduced to a set of simple formulae for use in plain tubular joints:

DOB =0.8

A+dA + 20.9 a/t

B= 4-22.35 a/t for a/t > 0.2 and B=-0.248 a/(0.20 t) for a/t < 0.2

The calibration of these parameters is shown on the attached Figure 5.0.

Note that the factor Mf is basically built into the procedure used, and is therefore not developed.

Haswell's method [3] is used because it summarises the general form of cracked joint behaviour over a range of tubular joint types.

FM Module

The FM module is based upon the results developed by TNO [3,4] in their fracture module. Because the rate of crack growth is highly non-linear, the selection of cycles between computation points is very significant. The cycle step size used in the spreadsheet is adjusted according to the rate of change of the a/T ratio.

As a result, the program uses relatively large solution step sizes when the crack growth rate is slow and much smaller steps when the crack growth rate is high.

Initial and Existing Defects - Intact Joints

An initial defect size of a=0.15 mm is used on for all joints which are assumed to be intact when performing FM check calculations. This size of initial defect has been used in the TNO comparison study. [See also TNO comparative study]

Flaws - Damaged Joints

The flaws is evaluated on a joint dependant basis, depending on location, and geometry of any existing crack locations.

The defects is extracted from the Anomalies database produced from underwater inspection report for the structure under consideration.

The defect size is used in the fracture mechanics evaluation of the joint.

Where only the defect length is known, a crack depth / length relationship shall be made for the crack under consideration on a joint by joint basis. Where possible, the recorded local grinding depths is used to define the current crack depth.

The rate of crack growth as determined by the analysis is only able to return an accurate value for the growth of a through thickness cracks.

For evaluating cracks other than those from 0.15 mm to through thickness, a correction factor is applied to the crack depth predicted by in the analysis. The correction factor is as figure 5.0.

TNO Comparative Study

Figure 6.0 shows some theoretical results obtained from the component fatigue analysis program developed by TNO under the aegis of THERMIE / RISC [1].

The TNO analysis shown in figure 6.0 is simply a Paris crack growth formulation for a tubular joint. It uses fixed stress ranges in discrete blocks to compute crack dimensions, making use of a multi-segment fracture growth curve. Both crack width and depth are computed, making use of the tubular forcing function.

The method of computing the crack growth has been used to process the data used by the TNO analysis.

Because the computation involves just one stress range, 180 N/mm^2 it was more practical to carry out the computation in a spreadsheet format.

The results of the comparison are shown on Figure 7.0. This indicates that analysis used and the TNO method are able to make good estimates of the total time to failure of the specimen joint.

CONCLUSION

An outline underwater inspection schedule may be determined for each platform upon completion of the reliability index and fracture calculations. The schedule shall be rationalised. The inspection schedule for a platform may be developed by referring to the Reliability graphs produced from analysis. The aim is to keep the RI value above the base index value of 1.47 by performing regular inspections [17]. This approach is only valid for uncracked joints.

The inspection schedule must be cost effective and there should be a reasonable interval between inspections [17]. Jacket structures are often given a thorough underwater inspection every 5 years. There will be cleaning and detailed surveys made of selected "major" joints.

It is common practice to specify that the divers grind out surface cracks as and when they locate them, provided that the depth of grinding does not exceed a locally agreed limit, often in the range 0.25 to 0.50 of the thickness of the virgin plate material.

For cracked joints, the expected time to failure may be calculated using fracture mechanics method. The detection of cracks in thin material is difficult and it is unlikely that cracks of a depth < 2mm can be detected. However a 2mm crack in thin steel plate, once detected, is likely to grow at a very high rate, with the possibility that the crack will go "through thickness" in less than one year. It needs to be repaired as soon as it is discovered.

The occurrence of an extremely short inspection return period (2 years or less) indicates an immediate need for repair. In some cases specific inspections may be required in order to monitor only certain joints - those with short predicted remaining fatigue lives, or known to contain cracks.

FIGURES

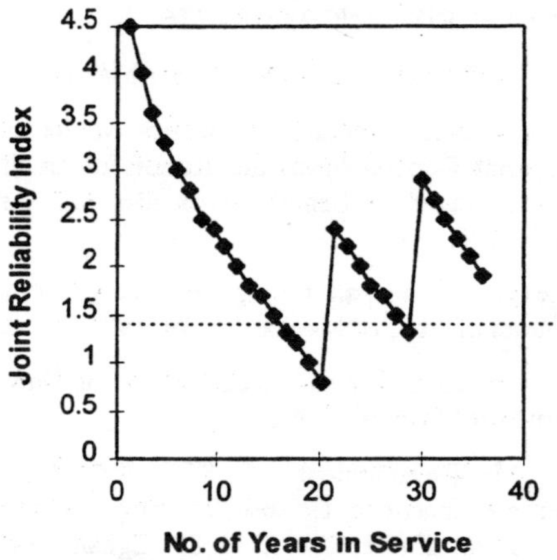

Figure 1.0 Example: Probability of Failure Reassessment

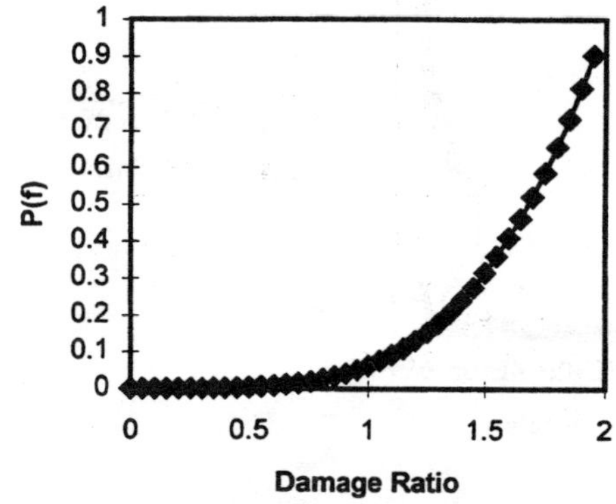

Figure 4.0 The Damage Ration/Probability of Failure Function

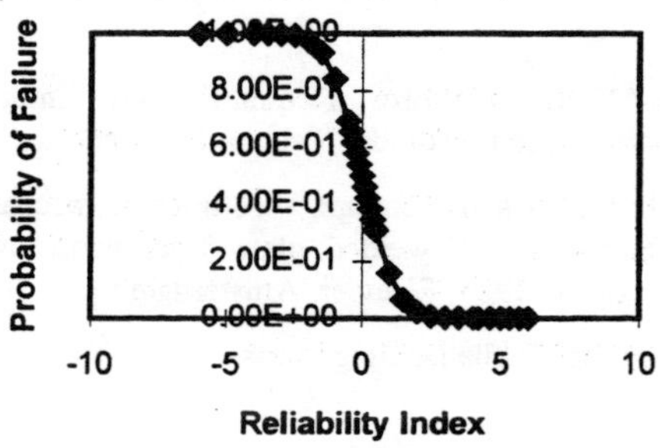

Figure 2.0 The Reliability Index Function / Probability of Failure Relationship

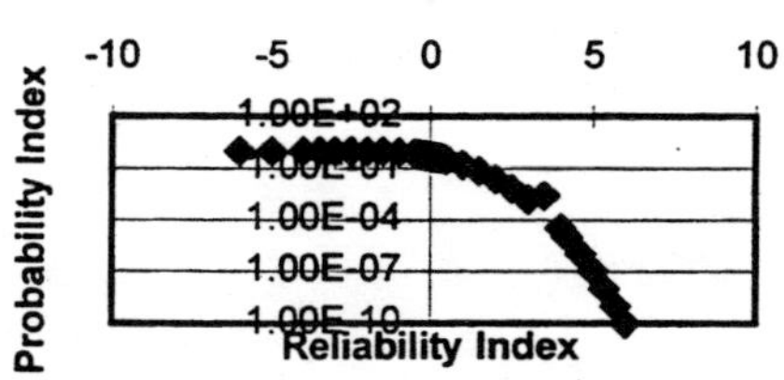

Figure 3.0 Joint Reliability Index Function - Log Scale

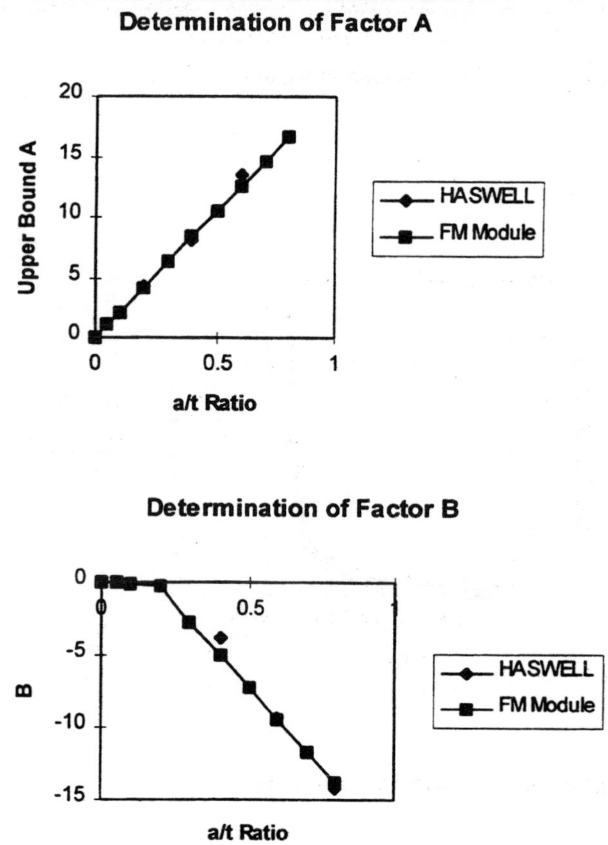

Figure 5.0 Derivation of Terms A and B in Haswell's Equation

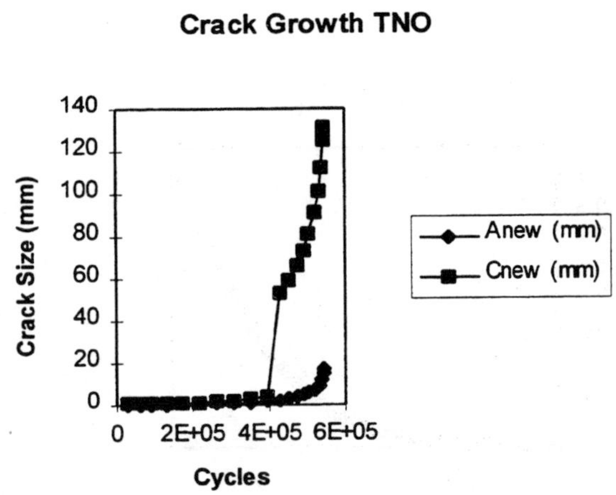

Crack depth is Anew and crack width is Cnew. A different forcing function is used when the crack depth exceeds 2mm. This is the tubular function. The above chart shows the results of a test case performed by TNO using RISC Component Fatigue Software. The data relates to a welded tubular T joint with chord 457 x 19 wt. And brace 324 x 19 wt. With a hot spot stress range of 180 Mpa. The experimental life was 540,000 cycles and good agreement is reached between the test and the software simulation.

Figure 6.0 TNO Theoretical Results

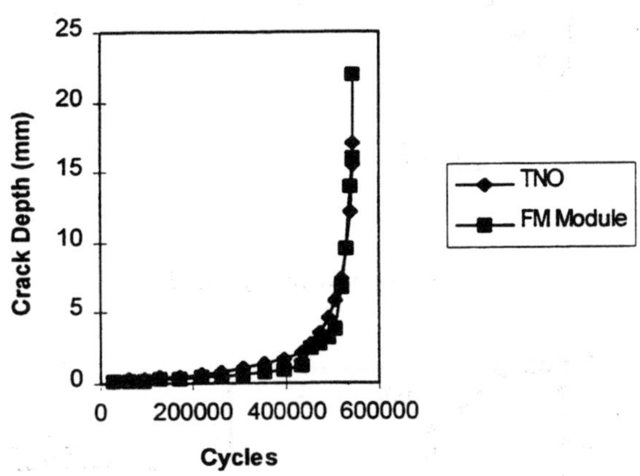

Figure 7.0 Comparison of Results

REFERENCES

1. TNO 'RISC Project, Component Fatigue analysis, Software User Manual' RISC Doc Ref. RISC/TNO/))/0794/116
2. 'RISC Project, Reliability analysis, Software User Manual'
3. Haswell, J.V ' A General Fracture Mechanics Model for a cracked Tubular Joint Derived From the Results of an Finite Element Parametric Study' Offshore Mechanics and Arctic Engineering Conference, 1992.
4. Rice J.R. and Levy. N. 'The part through surface crack in an elastic plate'. Journal of Applied Mechanics, 39, 1972.
5. BSI 'Guidance on methods for the acceptability of flaws in fusion welded structure ' PD6493, 1991
6. EFTHYMIOU M 'Development of SCF formulae and generalised influence functions for use in fatigue analysis', Offshore Tubular Joints Conference, OTJ 88, Egham, Surrey, UK, 1988
7. Marine Technology Directorate, UK. 'Probability based fatigue inspection planning'. MTD publication 92/100, London, 1992.
8. Peers, Tang and Dharmavasan. Knowledge based approach to inspection planning for offshore structures., OMAE , 1994.
9. Dijkstra, Foeken and Dharmavasan. Fracture Mechanics Limit States for Assessment and Maintenance Planning of Offshore Structures. OMAE, 1994.
10. INGE LOTSBERG AND FINN KIRKEMO (DnV) 'A systematic method for planning in-service inspection of Steel offshore structures.' OMAE Europe Conference, The Hague, March 1989.
11. ROBERTO BRANDI. 'Platform Requalification, integrity assessment and intervention methods. Tecnomare, Venezia.
12. Vosikovsky, Bell, Burns and Hohaupt. 'A fracture mechanics assessment of fatigue life of welded plate T-joints including thickness effect.' BOSS 1985, Elsevier, Amsterdam.
13. DnV : Regulations for Mobile Drilling Units
14. API RP2A.
15. Lotsberg, I : "Requalification of Offshore platforms" 1994 OMAE.
16. Madsen, HO Krenk, S and Lind, NC. Methods of Structural Safety. Prentice-Hall, 1986.
17. Dharmavasan et Al. "Reliability Based Inspection Scheduling for Fixed Offshore Structures", 1994 OMAE.

Proceedings of the Seventh (1997) International Offshore and Polar Engineering Conference
Honolulu, USA, May 25-30, 1997

ISBN 1-880653-28-1 (Set); ISBN 1-880653-32-X (Vol. IV)

A Modified CEGB Assessment Approach to Flawed Tubular Joints

*Bo Wang**
Tsinghua University
Beijing, China

Yoshiaki Kurobane and Yuji Makino
Kumamoto University
Kumamoto, Japan

ABSTRACT

The modified CEGB approach is applied for the evaluation of structural integrity for defected tubular joints. The guidance on prediction of ultimate load carrying capacity of flawed tubular joints is provided by means of the modified CEGB FAD. Cracked T, K, KK tubular joints were analyzed by nonlinear FE method. The crack driving force, J-integral, and ultimate collapse loads of cracked tubular joints subjected to axial brace loading were evaluated by FE analyses. From these results, two assessment parameters were available to construct the FADs for cracked tubular joints. The results show that the modified CEGB approach is suitable for the safety assessment to flawed tubular joints. Load-displacement curves and ultimate strengths for uncracked and cracked T, K and KK-joints were also obtained by FEM. Compared with test values, FE results show to be reasonable with regard to the effect of cracks.

KEYWORDS: Structural Integrity Assessment, Modified CEGB Approach, Tubular Joints, Crack, J-integral

1. Introduction

Recently, the crack failure has been gradually realized as one of the principal causes of damage to steel structures. Ductile cracks often initiated at weld toes in tubular structures (Wang et al, 1996, Makino et al, 1995) after structures sustain excessive plastic strain. These cracks frequently led to a premature failure of structures. Therefore, it is essential to understand crack growth behaviour and to be able to assess the significance of defects in steel structures. In elastic-plastic fracture mechanics, the J-integral has been proved to be a useful parameter to control the crack initiation and a small scale propagation for ductile materials. The fracture assessment procedures introduced by the CEGB (Harrison et al, 1976; Kumar et al, 1981) are based on the Failure Assessment Diagram (FAD) approach including the J-integral. Milne (1983) then developed this method into a new modified R6 approach based on the ratio of engineering ultimate strength divided by engineering yield strength. However, the modified CEGB R6 approach does not give optimal solutions for all types of welded joints. This paper outlines the CEGB R6 approach and the modified failure assessment curve. The modified CEGB R6 safety assessment approach will be applied to defected tubular joints.

The fracture behaviour of cracked tubular joints and the assessment of structural safety of these structures will be investigated according to the modified CEGB approach. The original EPRI handbook (Kumar et al, 1981) for the J-integral estimations did not include any through cracks in tubular joints. Hereafter, the two assessment parameters, the ultimate collapse load and the crack driving force of cracked T, K and KK tubular joints, which are loaded by axial forces on braces, will be obtained by nonlinear FE analyses. To calculate the plastic collapse parameter, plastic collapse loads of flawed tubular joints will be examined in terms of the ultimate capacity formulae for uncracked tubular joints combined with "area reduction factor". These numerical results will be compared with the modified failure assessment curve of the modified CEGB R6 approach. Load-displacement curves for uncracked and cracked tubular joints in the same load cases will be determined from numerical results.

2. The Modified CEGB R6 Approach

For ductile materials, the J-integral has been assumed as a fracture parameter. When the crack initiation is defined as the failure of structures, the equilibrium or energy balance equation at initiation will be

$$J(a,P) = J_{IC} \tag{2.1}$$

where a and P are respectively the crack length and the applied load, while the fracture toughness, J_{IC}, is the critical value at crack initiation.

In the linear elastic fracture mechanics (LEFM), the stress intensity factor is expressed as

$$K = \sigma\sqrt{\pi a} \tag{2.2}$$

*Formerly JSPS Research Fellow, Department of Architecture, Kumamoto University, Japan.

where σ is the uniform tensile stress at infinite boundary. Elastic J-integral can be written as

$$J^e = \frac{K^2}{E'} \tag{2.3}$$

where $E' = E$ for plane stress state and $E' = E/(1-\nu^2)$ for plane strain state.

In the situation of plane stress, when an infinite plate with a through crack is subjected to the uniform tensile stress σ, Dugdale solution of elastic-plastic J-integral is

$$J(a, \sigma) = \frac{8\sigma_y^2 a}{\pi E} \ln[\sec(\frac{\pi}{2}\frac{\sigma}{\sigma_y})] \tag{2.4}$$

where σ_y is the yield stress.

Two parameters are used in FAD. One is the fracture parameter J_r, which is the ratio of the applied crack tip driving force to the fracture toughness as the following:

$$J_r = \frac{J^e(a,P)}{J_{1C}} \tag{2.5}$$

Only the elastic component of the applied crack tip driving force is used in the calculation, since the effects of crack tip plasticity are allowed for by the shape of the failure assessment diagram. Another one, the plastic collapse parameter S_r, is the ratio of applied load to plastic collapse load of the flawed structure. S_r can be written as

$$S_r = \frac{P}{P_0} \tag{2.6}$$

where P_0 is the plastic collapse load of defected structures per unit thickness.

After applying Eq.(2.1) - (2.6), the equation of the R6 failure assessment criterion curve is described as (Kumar et al,1981):

$$\frac{S_r^2}{\frac{8}{\pi^2}\ln[\sec(\frac{\pi}{2}S_r)]} = J_r = K_r^2 \tag{2.7}$$

A defect is assessed by evaluating the fracture and plastic collapse parameters and plotting the corresponding points on the FAD. If the point lies within the assessment curve, the defect is deemed acceptable, if it falls outside the curve, the defect is unacceptable.

The original R6 diagram was shown (Wang et al, 1995; Cheaitini et al, 1994 Bloom et al, 1982) to be inadequate for representing geometry and strain-hardening effects for materials which have high work-hardening capacity. Therefore, Milne (1983) developed a modified R6 approach based on the ratio of engineering ultimate strength, σ_u, divided by engineering yield strength, σ_y. The strip yield model is retained with S_r introducing a new definition to account for high work-hardening, so that S_r and K_r are redefined by

$$S_r = \frac{\sigma}{\sigma_y} + (1-\sqrt{K_r})(\frac{\sigma_u}{\sigma_y} - 1) \tag{2.8}$$

and

$$K_r = \frac{\sigma/\sigma_y}{\sqrt{\frac{8}{\pi^2}\ln\ \sec(\frac{\pi\sigma}{2\sigma_y})}} \tag{2.9}$$

Note the $\sigma_u/\sigma_y = 1$ gives the original CEGB R6 procedure.

3. Plastic Collapse Load of Cracked Tubular T, K, KK-Joints

3.1 Ultimate Strength of Uncracked Tubular T, K, KK-Joints

Ultimate capacity formula for uncracked T-joint under the axial tensile load is given by (Department of Energy, 1990):

$$N_{u,T} = (7.57 + 17.95\beta + 1.18\beta\gamma)\sigma_y T^2 \tag{3.1}$$

where $\beta = d/D$, $\gamma = D/2T$.

The formula for ultimate strength of uncracked K-joint is recommended as (Kurobane et al, 1984):

$$N_{u,K} = f_0 \cdot f_1 \cdot f_2 \cdot f_3 \cdot f_4 \cdot f_5 \cdot T^2 \sigma_y \tag{3.2}$$

and ultimate capacity formula for KK-joints under anti-symmetrical axial loads is presented as (Makino et al, 1995):

$$\begin{aligned} N_{u,KK} &= 0.858 \cdot N_{u,K} & g_t/D \geq 0.157 \\ N_{u,KK} &= (1.36 - 3.17\frac{g_t}{D}) \cdot N_{u,K} & g_t/D < 0.157 \end{aligned} \tag{3.3}$$

where

$$f_0 = 2.11(1 + 5.66\frac{d}{D})$$

$$f_1 = (\frac{D}{T})^{0.209}$$

$$f_2 = 1 + \frac{0.00904(\frac{D}{T})^{1.24}}{\exp(0.508\frac{g_1+g_c}{T} - 1.33) + 1}, \qquad g_c = 3.04(mm)$$

$$f_3 = \frac{1 - 0.376\cos^2\theta}{\sin\theta}$$

$$f_4 = 1$$

$$f_5 = (\frac{\sigma_y}{\sigma_u})^{-0.723}$$

Numerical and test results have proved the accuracy of above formulae (Wang et al, 1997).

3.2 Plastic Collapse Load of Cracked T, K and KK Tubular Joints

We assume that a conservative estimation of the plastic collapse load of cracked T, K and KK tubular joints can be obtained by multiplying the ultimate strength of uncracked tubular joints from Eq.(3.1)-(3.3), with the same geometry, by an 'area reduction factor' to take account of net section area as follows:

$$\text{Area reduction factor } \alpha_0 = 1 - \frac{\text{crack area}}{\text{intersection length* T}} \quad (3.4)$$

The reduction in strength due to the through thickness cracks was lower than the associated reduction in the punching shear area. It should be stressed that the above assumption on the effects of cracks is independent of scale as far as plastic collapse behaviour is concerned (Cheaitini et al, 1994).

4. Structural Safety Assessment to Cracked Tubular Joints

4.1 FE Analyses on Cracked T, K and KK Joints

Non-linear FE analyses were performed by using ABAQUS FE package on cracked T, K and KK tubular models. Because of the symmetry, only one quarter part was analyzed for T-joint and one half structures were separately analyzed for K and KK-joints. The geometric parameters and mechanical properties of tubular joints are presented in Table 1. The original crack lengths for T, K and KK-joints are given in Table 2. Eight nodded thin isoparametric shell elements were used to model the chord, brace and weld. The Von Mises yield criterion and isotropic strain hardening have been used. Stress-strain curves were obtained from tests (Yonemura et al, 1996, Makino et al, 1996 and Ochi et al, 1981). The geometry of the weld elements is in agreement with the weld sizes recommended by the AWS (1994).

For cracked models, collapsed elements around the crack tip were used to reproduce the crack tip singularity, 1/r singularity, and the crack tip blunting at the crack tip, see Fig.1. We collapse one side (made up by nodes a, b, c) of an 8-node isoparametric element so that all three nodes have the same geometric location at the crack tip. When the 3 nodes (a,b,c) on the collapsed face can displace independently (by using different node numbers in the connectivity matrix) and mid-side nodes remain at mid-points, then $\varepsilon \propto r^{-1}$. For materials being in between linear elastic and perfectly plastic (most metal materials) it is better to have a stronger singularity than necessary if the correct singularity cannot be reproduced. The numerics in ABAQUS will force the coefficient of this singularity to be small.

Meshes have been refined sufficiently to model the very high strain gradients around the crack tip. The evaluation of fracture parameter, the J-integral, can be accurately obtained. A detailed mesh layout near the crack tip of cracked models is shown in Fig.2.

T-joint is subjected to the axial tensile load on the brace and is simply supported at the chord end.. An axial compressive load is applied to the compression brace and the top of tension brace and one end of the chord are fixed in K-joint. The uni-planar KK-joint is under the anti-symmetric axial load. In the process of calculation, the displacement control option was used to get the ultimate collapse load of structures for T and K-joint and the force control option was used for KK-joint. The contour around the crack tip was taken to calculate conservative estimates of the J-integral.

Additionally, nonlinear FE analyses of uncracked models, carried out to determine crack locations, indicated that the Mises stresses peak around the weld toes predicted the location of a likely crack initiation site. This conclusion is supported by test results (Wang et al, 1997).

In the finite element analyses, the following investigations were performed: (1) Elastic crack driving force J^e and elastic-plastic J-integral were respectively calculated aiming at defect assessment; (2) Load-displacement curves were evaluated to predict the ultimate collapse load of uncracked and cracked tubular joints; (3) Plastic collapse loads of uncracked and cracked models were analyzed. The ultimate capacity formula were used to predict the plastic collapse loads for uncracked T, K and KK tubular joints; (4) Mises stress field and equivalent plastic strain field were obtained to describe stress and strain field close the crack tip.

4.2 Failure Assessment Diagram

Plastic collapse loads of cracked tubular joints can be obtained by multiplying the ultimate strength of uncracked joints from Eq.(3.1)-(3.3), with the same geometry, by an 'area reduction factor'.

A FAD for cracked tubular joints is constructed by the fracture parameter K_r and the plastic collapse parameter S_r from the ABAQUS results. J_r is calculated as J^e/J, where J^e and J are the values of the J-integral obtained for an elastic analysis and an elastic-plastic analysis respectively, for the same load. S_r is evaluated as the ratio of the applied load to the plastic collapse load. Estimates for the latter were obtained by applying the area reduction factor α_0 to the uncracked plastic collapse load for T, K and KK-joints. The FADs for cracked T, K, KK joints lie within the criterion curve in the modified CEGB FAD (see Fig.6).

5. Discussions

In Fig.3 and Fig.4, it is clear that the crack tip is blunted after loading and there is a high plastic strain zone in small scale near the crack tip. It means that in the stress-strain field around the crack tip, the J-integral can be taken as a fracture parameter to govern the initiation and growth of the crack.

In Fig.5, the test curve is between the FE curve for the uncracked model and the FE curve for the model with an original crack. During the test process, the crack initiated and propagated in the chord near the weld toes. The stiffness will change because of the crack initiation and growth. Therefore, the crack initiation and growth should be considered in numerical models when test specimens are simulated by FEM. Ultimate strengths for uncracked and cracked tubular joints are reported in Table 2. The results are reasonable with regard to the effect of cracks.

In Fig.6, the results show that the FAD from FE analyses is close to the assessment criterion curve and FE values of two assessment parameters fall slightly inside the standard FAD curve. It means that the failure assessment curve described by Eq.(2.8) and Eq.(2.9) gives reliable prediction. The modified CEGB R6 safety assessment approach can be recommended to assess flawed tubular joints.

6. Conclusions

1) Nonlinear FE method by using the ABAQUS package has provided a useful and reliable tool for analyzing the mechanical and fracture behaviour of defected tubular joints. Compared with test results, numerical results for uncracked and cracked T, K and KK tubular joints show reasonable.

2) The J-integral can control the crack initiation and the estimates of the plastic collapse load of cracked tubular joints, which were obtained by multiplying the plastic collapse load of uncracked tubular joints with the 'area reduction factor', are conservative and suitable for engineering assessment.

3) The modified CEGB R6 type failure assessment procedure combined with the J-integral and plastic collapse parameter can be recommended to assess defected tubular joints. The modified safety assessment curve (Standard FAD) gives reliable prediction.

ACKNOWLEDGEMENTS

The first author wishes to thank the Japan Society for the Promotion of Science for a Post-Doctoral Fellowship. This project is partly supported by the Japanese Ministry of Education under Grant Nos. 07555178 and 94053. This work is also partly supported by the National Nature Science Foundation of China and Chinese Education Committee.

APPENDIX. REFERENCES

1. ABAQUS (1995) "ABAQUS v5.5 Manual (Users' Manual I and II, Theory Manual, Example Manuals I and II, Verification Manual), Hibbitt, Karlsson and Sorenson Inc.
2. American Welding Society, Structural Welding Code (1994), AWS D1.1-94.
3. Bloom, J. M. and Malik, S. N. (1982) "A procedure for the assessment of the integrity of nuclear pressure vessels and piping containing defects", EPRI Topical Report NP-2431, Research Project 1237-2, Palo Alto, CA, June.
4. Cheaitani, M. J. and Burdekin, F. M. (1994) "Ultimate strength of cracked tubular joints", Tubular Structures VI, Grundy, Holgate & Wong (eds), P607-616.
5. Department of Energy (1990) "Background to new static strength guidance for tubular joints in steel offshore structures", Offshore Technology Report OTH89 308, HMSO, Lodon, U.K..
6. Harrison, R. P., Loosemore, K. and Milne, I. (1976) "Assessment of the integrity of structures containing defects", CEGB R/H/6, Central Electricity Generating Board, UK.
7. Kumar, V., German, M. D. and Shih, C. F. (1981) " An engineering approach for elastic-plastic fracture analysis", NP-1931, EPRI.
8. Kurobane, Y., Makino, Y. and Ochi, K. (1984) "Ultimate resistance of unstiffened tubular joints", J. of Struc. Engrg., ASCE, Vol.110, No.2, P385-400.
9. Makino, Y., Kurobane, Y., Wilmshurst, S. R. and Lee, M. M. K. (1995) " Proposed ultimate capacity Equations for CHS KK-joints under anti-symmetric loads", Proceedings of 5th ISOPE conference, Vol.IV, P6-11.
10. Makino, Y., Kurobane, Y, Ochi, K., van der Vegte, G. J. and Wilmshurst, S. R. (1996) "Database of test and numerical analysis results for unstiffened tubular joints", IIW Doc. XV-E-96-220, Hungary.
11. Milne, I. (1983) "Failure assessment diagrams and J-estimates: A comparison for ferritic and austenitic steels", Int. J. of Pressure Vessels & Piping, Vol.13.
12. Ochi, K., Kurobane, Y. and Makino, Y. (1981) "Further tests on CHS K-joints with various d/D ratios", Research Report of Kumamoto University.
13. Wang B., Kurobane Y.. and Makino Y. (1996) "Stress analyses and safety assessment of welded T-joints", Memoirs of the Faculty of Engineering, Kumamoto Univ., Vol.40, No.3, P289-312.
14. Wang, B., Kurobane Y. and Makino Y.. (1997) "Damage criterion and ultimate strength for tubular joints", Proceedings of 7th ISOPE conference.
15. Yonemura, H., Makino, Y., Kurobane, Y. and van der Vegte, G. J. (1996) "Tests on CHS planar KK-joints under anti-symmetrical loads", Tubular Structures VII, JOZSEF FARKAS&KAROLY JARMAI (eds), P181-195,

Table 1. Test specimens (measured)

Specimen	$D \times T$ (mm)	$d \times t$ (mm)	L (mm)	Chord σ_y (N/mm^2)	Chord σ_u (N/mm^2)	Brace σ_y (N/mm^2)	Brace σ_u (N/mm^2)	θ	g_l (mm)	g_t (mm)	d/D	D/T
KK-joint	216.1×6.4	139.8×4.5	1978	342	485	366	647	45°	16.5	165.1	0.65	33.8
K-joint	217.4×4.4	140.3×3.3	1515	354	458	401	480	60°	30.0		0.65	49.6
T-joint	323.9×6.4	141.3×6.6	1219	274	451	274	451	90°			0.44	50.6

Note1: ϕ and θ are nominal values.

Note2: g_l : longitudinal gap between both braces; g_t : transverse gap between both braces.
g_l and g_t are calculated from above values.

Table 2 Numerical and Test Database

Specimen	crack (mm)	Ultimate Load, no crack, Test (kN)	Ultimate Load, no crack, FEM (kN)	Ultimate Load, with crack, FEM (kN)	Ultimate Load, with crack, Prediction* (kN)	Load Crack Initiation (kN)
T-joint	15	362	392	381	322	173
K-joint	25	225	232	216	227	172
KK-joint	25	560	595	525	554	505

*according to Eqs.3.1-3.4

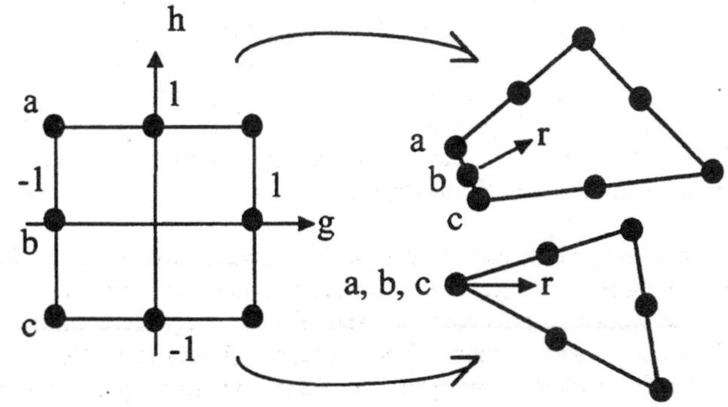

Fig.1 Collapse Elements around Crack Tip

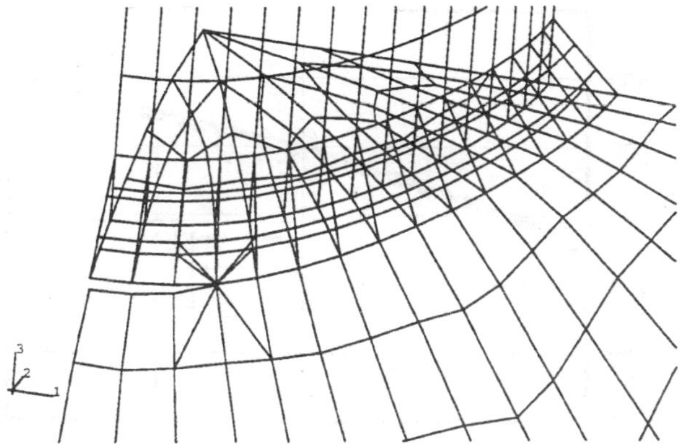

Fig.2a Deformed Cracked T-joint

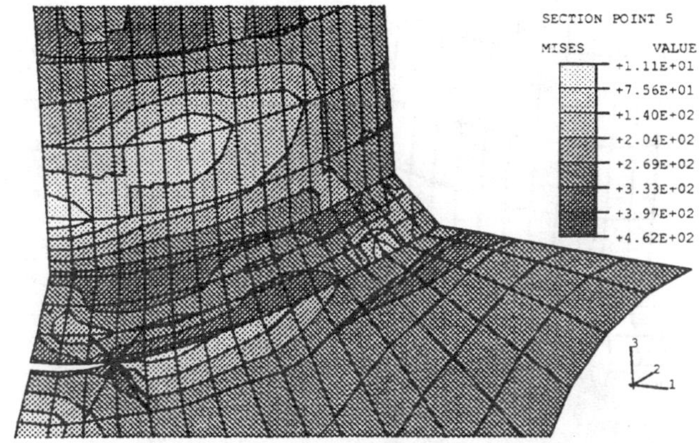

Fig.3a Mises Stress Distribution for Cracked T-joint

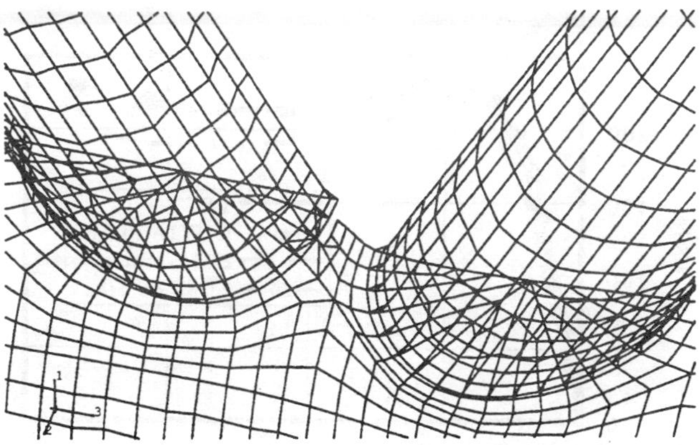

Fig.2b Deformed Cracked K-joint

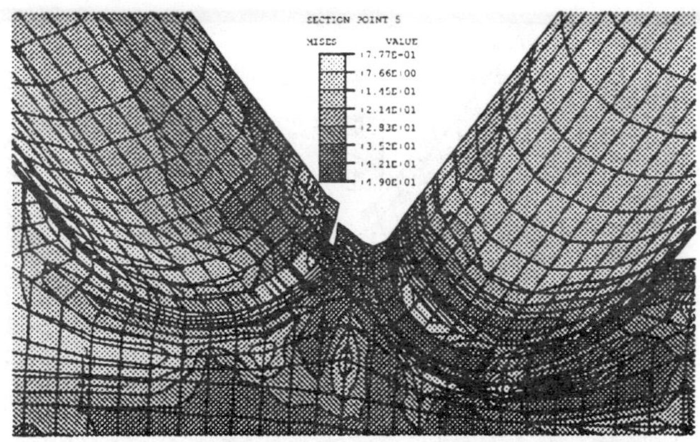

Fig.3b Mises Stress Distribution for Cracked K-joint

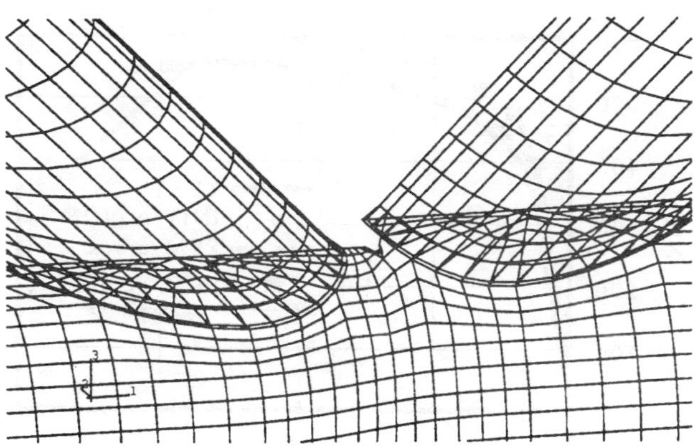

Fig.2c Deformed Cracked KK-joint

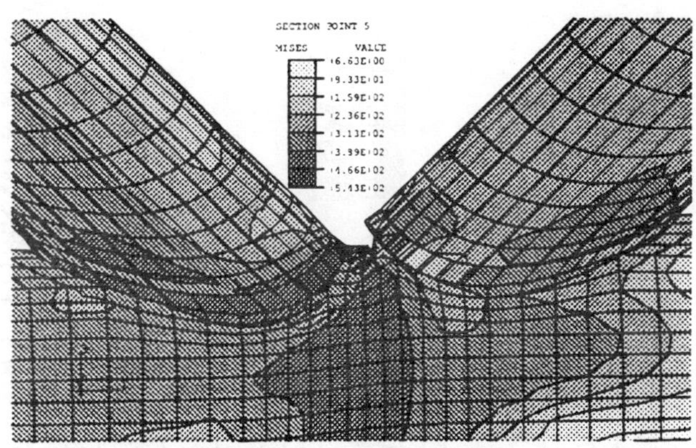

Fig.3c Mises Stress Distribution for Cracked KK-joint

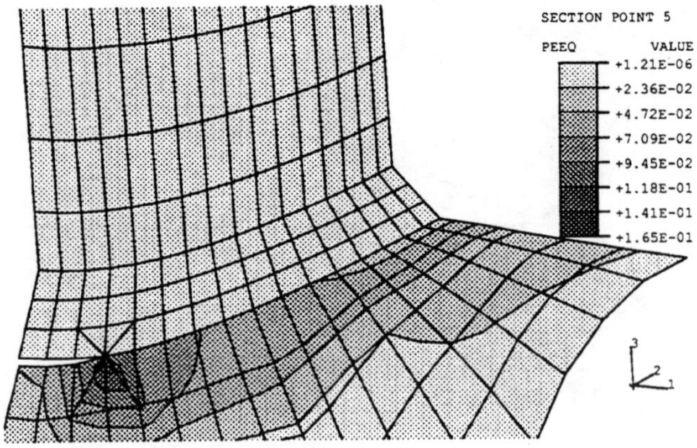

Fig.4a Equivalent Plastic Strain for Cracked T-joint

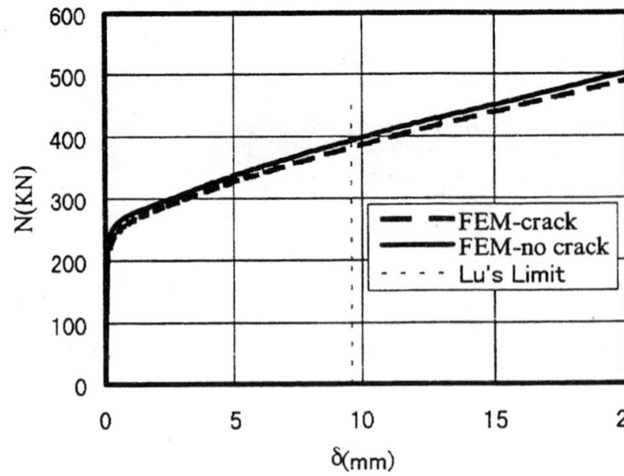

Fig.5a Load-displacement Curve for T-joint

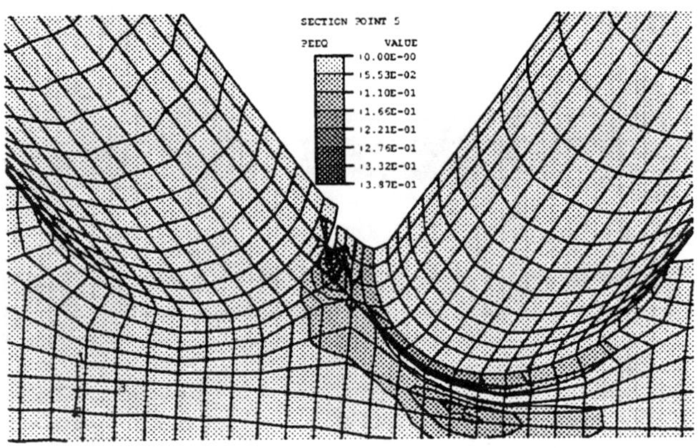

Fig.4b Equivalent Plastic Strain for Cracked K-joint

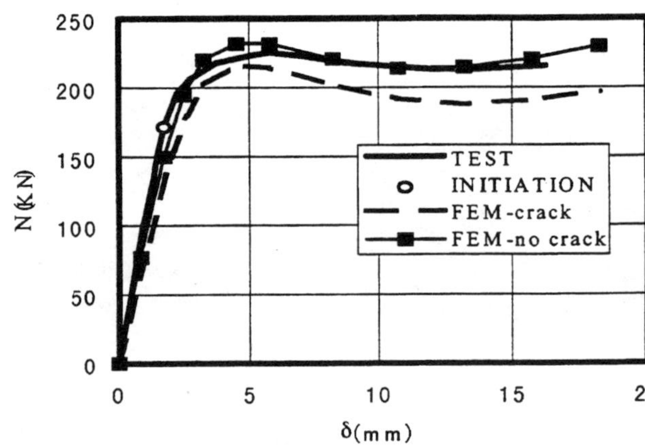

Fig5b Load-displacement Curve for K-joint

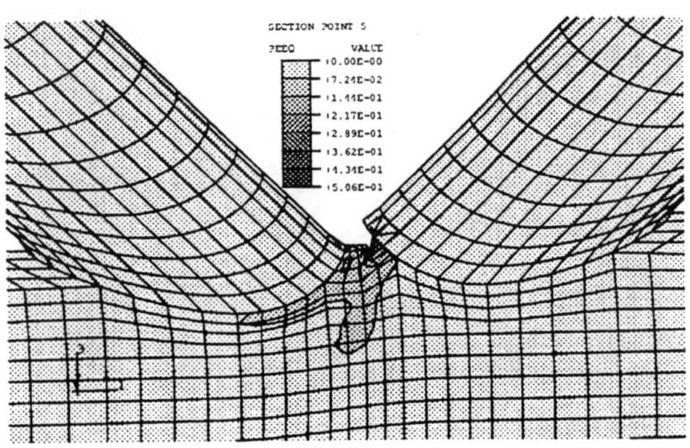

Fig.4c Equivalent Plastic Strain for Cracked KK-joint

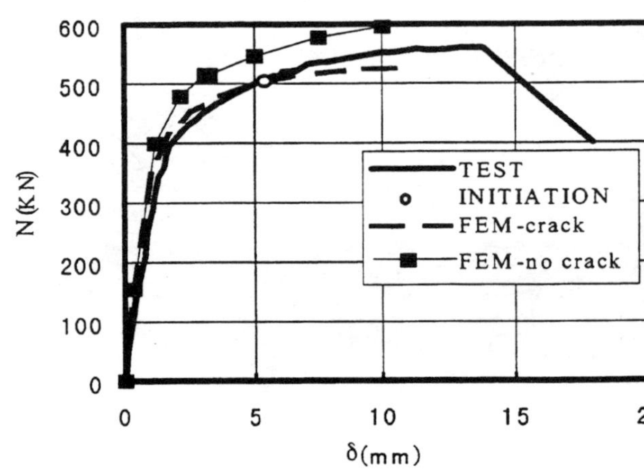

Fig5c Load-displacement Curve for KK-joint

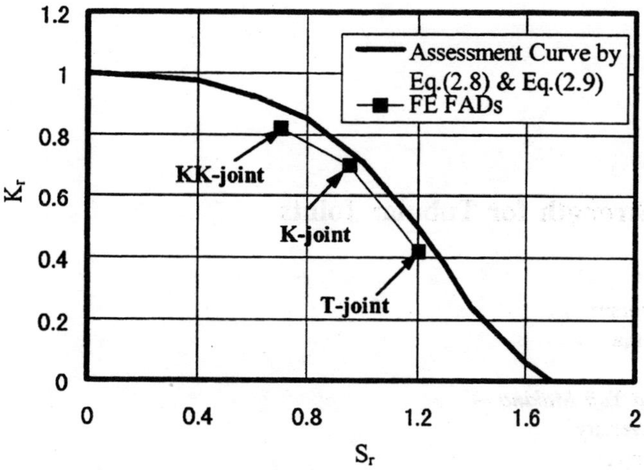

Fig.6 Failure Assessment Diagram in Modified CEGB

Proceedings of the Seventh (1997) International Offshore and Polar Engineering Conference
Honolulu, USA, May 25-30, 1997

ISBN 1-880653-28-1 (Set); ISBN 1-880653-32-X (Vol. IV)

Damage Criterion and Ultimate Strength for Tubular Joints

*Bo Wang**
Tsinghua University
Beijing, China

Yoshiaki Kurobane and Yuji Makino
Kumamoto University
Kumamoto, Japan

ABSTRACT

Based on the Continuum Damage Mechanics (CDM), a damage criterion is proposed to predict the macrocrack initiation for tubular joints. Nonlinear FE analyses on T, K and KK tubular joints were performed by using the ABAQUS package. Load-displacement curves and ultimate capacities for these joints were numerically determined. Triaxial stress distributions and equivalent plastic strains near weld toes in the chord around the brace were obtained from FE results. A uni-planar KK tubular joint specimen was tested under anti-symmetrical axial loads. During the test, the load-displacement curve and the ultimate strength were measured. Meanwhile, crack initiation and growth were observed until complete failure. FE results show that the damage criterion gives good prediction for macrocrack initiation of tubular joints; Analytical, numerical and experimental results for the ultimate strengths of tubular joints are consistent.

KEYWORDS: Damage Criterion, Macrocrack Initiation, Ultimate Strength, Tubular Joints.

1. Introduction

During the recent Kobe Earthquake, many steel structures were damaged owing to the initiation and growth of cracks. Ductile macrocracks often initiated at weld toes (Wang et al 1996, Makino et al 1995) after structures sustain excessive plastic strain. These cracks frequently led to a premature failure of structures. The previous T, K and KK-joint test results gave a conclusion that cracking in the chord wall is one of the important failure modes. The cracks often occurred along the weld toes between the tension brace and the chord and propagated through the chord wall to failure. In previous work (Makino et al, 1995), a crack was assumed to have formed when the tensile plastic strain reached a value corresponding to the ultimate tensile strength of material. However, This method to determine crack initiation is not very reliable without experimental verification, because crack initiation is not only dependent on the tensile plastic strain but also on other parameters.

*Formerly JSPS Research Fellow, Department of Architecture, Kumamoto University, Japan

From the microscopic point of view, damage is related to the process of nucleation and growth of micro-voids and cavities. During the damage process, the accumulated plastic strain and triaxial stress are two important factors affecting macrocrack initiation. In the present paper, based on the Lematre's damage model(Lematre 1985, 1990), the damage criterion for welded structures will be derived to predict the macrocrack initiation. A uni-planar KK tubular joint specimen will be tested to observe the crack initiation and to measure the load-displacement curve.

Nonlinear FE analyses on T, K and KK tubular joints subjected to axial brace loads will be conducted by using the ABAQUS general purpose FE package (ABAQUS 1995). The mechanical behaviour of tubular structures will be studied to simulate experimental observations. In order to verify the accuracy of the damage criterion proposed in this paper, the distributions of triaxial stress and equivalent plastic strain near weld toes will be calculated. Load-displacement curves and ultimate strengths numerically obtained will be compared with existing test results.

2. Damage Criterion for Macrocrack Initiation

Within the hypothesis of isotropy of damage and isotropy of plasticity, the constitutive equation for damage evolution is given by the normality property of the potential (Lematre 1990):

$$\dot{D} = -\frac{\partial \varphi^*}{\partial y} \tag{2.1}$$

where $\dot{D}$ is the damage growth rate, φ^* is the potential of dissipation and y is the damage strain energy release rate. The potential of dissipation can be written as follows (Lematre 1990):

$$\varphi^* = \frac{S_0}{(s_0+1)}\left(\frac{-y}{S_0}\right)^{s_0+1}\dot{p} \tag{2.2}$$

where s_0 and S_0 are the material constants, $\dot{p} = (2/3\ \dot{\varepsilon}_p : \dot{\varepsilon}_p)^{1/2}$ and ε_p is the Green Lagrange plastic strain tensor.

Based on the concept of the damage equivalent effective stress, the damage strain energy release rate y can be expressed as:

$$-y = \frac{\sigma_{eq}^2}{2E(1-D)}\left[\frac{2}{3}(1+\nu)+3(1-2\nu)\left(\frac{\sigma_H}{\sigma_{eq}}\right)^2\right] \tag{2.3}$$

where $\sigma_{eq} = (3/2\ s{:}s)^{1/2}$, s is the deviatoric stress tensor and σ_H is the hydrostatic stress.

For the three-dimensional case, the Ramberg-Osgood hardening law with damage leads to:

$$p=\left[\frac{\sigma_{eq}}{(1-D)K}\right]^M \tag{2.4}$$

where M is the hardening exponent of the material and K is a constant of the material.

From Eq.(2.1) to Eq.(2.4), the general constitutive equation for the plastic damage will be:

$$\dot{D} = \left(\frac{K^2}{2ES_0}\left[\frac{2}{3}(1+\nu)+3(1-2\nu)\left(\frac{\sigma_H}{\sigma_{eq}}\right)^2\right]p^{\frac{2}{M}}\right)^{s_0}\dot{p} \tag{2.5}$$

In the case of constant triaxiality ratio, a simple relation between the actual value of damage D and the equivalent plastic strain p can be obtained. p_D is defined as the damage strain threshold, that is, when $p < p_D$, $D = 0$. Therefore, Eq.(2.5) is integrated into

$$D = \left(\frac{K^2}{2ES_0}\left[\frac{2}{3}(1+\nu)+3(1-2\nu)\left(\frac{\sigma_H}{\sigma_{eq}}\right)^2\right]\right)^{s_0} \cdot \frac{M}{2s_0+M}\left(p^{\frac{2s_0+M}{M}}-p_D^{\frac{2s_0+M}{M}}\right) \tag{2.6}$$

Let's introduce the rupture strain p_R, which is a function of the triaxiality ratio σ_H/σ_{eq} corresponding to the intrinsic value of damage at failure D_c, where D_c is assumed to be a material property. Namely,

$$p = p_R \rightarrow D = D_c$$

we get

$$D_c = \left(\frac{K^2}{2ES_0}\left[\frac{2}{3}(1+\nu)+3(1-2\nu)\left(\frac{\sigma_H}{\sigma_{eq}}\right)^2\right]\right)^{s_0} \cdot \frac{M}{2s_0+M}\left(p_R^{\frac{2s_0+M}{M}}-p_D^{\frac{2s_0+M}{M}}\right) \tag{2.7}$$

For the large deformation of the material, the hardening exponent M is usually very high (a perfectly plastic material corresponds to $M = \infty$). s_0 is of the order of 1 and $(2s_0+M)/M \approx 1$. Physically, it is admissible to assume that the ratio p_D/p_R does not depend on triaxiality and is equal to its value in the one-dimensional case, that is

$$\frac{p_D}{p_R} = \frac{\varepsilon_D}{\varepsilon_R} \tag{2.8}$$

In the one-dimensional case, $\sigma_H/\sigma_{eq} = 1/3$ and the equivalent plastic strain at rupture p_R can be derived from Eq.(2.7) as follows:

$$p_R\big|_{\frac{\sigma_H}{\sigma_{eq}}=\frac{1}{3}} = \varepsilon_R = \frac{2ES_0D_c}{K^2\left(1-\frac{\varepsilon_D}{\varepsilon_R}\right)}\left[\frac{2}{3}(1+\nu)+3(1-2\nu)\left(\frac{1}{9}\right)\right]^{-s_0} = \frac{2ES_0D_c}{K^2\left(1-\frac{\varepsilon_D}{\varepsilon_R}\right)} \tag{2.9}$$

From Eq.(2.7) and Eq.(2.9), the rupture strain in three-dimensional case is

$$p_R = \varepsilon_R\left[\frac{2}{3}(1+\nu)+3(1-2\nu)\left(\frac{\sigma_H}{\sigma_{eq}}\right)^2\right]^{-s_0} \tag{2.10}$$

where s_0 is assumed as 1. This formula shows that stress triaxiality and equivalent plastic strain are two parameters governing the initiation of the macrocrack.

In welded structures, there exist stress triaxiality peak values at points close to weld toes for different plastic deformations. Voids or microcracks usually occur at the location where a stress triaxiality peak value takes place. Therefore, when the macrocrack initiates, σ_H/σ_{eq} and p_R should be replaced by the stress triaxialily peak value $(\sigma_H/\sigma_{eq})_{peak}$ and its corresponding equivalent plastic strain $(p_R)_p$ in Eq.(2.10), respectively. We assume that $\varepsilon_R = \varepsilon_u$, where ε_u is the maximum uniform strain of the material. Finally, the damage criterion at macrocrack initiation for welded structures can be expressed by Eq.(2.11).

$$\frac{(p_R)_p}{\varepsilon_u} = \left\{\frac{2}{3}(1+\nu)+3(1-2\nu)\left[\left(\frac{\sigma_H}{\sigma_{eq}}\right)_{peak}\right]^2\right\}^{-1} \tag{2.11}$$

Usually, for structural steel materials, Poisson's ratio ν is equal to 0.3. Therefore, the above expression will be simplified into:

$$\frac{(p_R)_p}{\varepsilon_u} = \left\{0.867 + 1.2\left[\left(\frac{\sigma_H}{\sigma_{eq}}\right)_{peak}\right]^2\right\}^{-1} \tag{2.12}$$

3. Test on a Uni-planar KK-joint

A uni-planar KK-joint specimen configuration under anti-symmetrical axial loads is shown in Fig.1. Geometric parameters are shown in Table 1. Mechanical properties of materials for the chord and braces are also shown in Table 1. Firstly, axial loads were proportionally applied to the tension and compression braces up to the maximum load by using the 100 ton hydraulic rams. After the peak load, the displacement of the hydraulic ram for the tension braces was fixed since the proportional load control was very difficult when negative stiffness occurred. As a result, the loads were not applied proportionally after the maximum load and only compression braces were loaded continuously until final failure.

The load-displacement curve for a uni-planar KK-joint subjected to anti-symmetric axial loads is reported in Fig.4c, where the load is recorded by the dynamometers at compression brace ends and the displacement represents the relative displacement between compression brace ends. Before the load reached the peak load, cracks were found at weld toes on the chord surface in longitudinal gap region. Cracks initiated at weld toes between the tension brace and the chord, and propagated in the chord wall. Then the ultimate capacity of the KK-joint, which is defined as the peak load, achieved and the structure failed suddenly.

4. FE Analyses on Tubular Joints

4.1 Model Description for T, K and KK-joints

The configurations for T, K and KK tubular joints are shown in Fig.1. Their geometric parameters and mechanical properties are shown in Table 1. Because of the symmetry of structures, one quarter was modeled for T-joint and one half was simulated as a numerical model for each of K and KK-joints. Typical meshes for T, K and KK-joints are shown in Fig.2. FE meshes near weld toes are refined to detect stress triaxiality around the brace. Four nodded thin isoparametric shell elements have been used to model the chords, braces and welds in tubular joints. The numbers of elements used to model T, K and KK-joints are respectively 544, 1812 and 2701. The Von Mises yield criterion and isotropic strain hardening were assumed. Stress-strain curves were obtained from tests (Yonemura et al, 1995, Makino et al, 1996, Ochi et al, 1981) and a piecewise multilinear relationship was approximated from those curves. The T-joint is simply supported at the chord end. For the K-joint, the top of tension brace and one end of the chord are free to rotate in the plane and fixed in other directions. For the uni-planar KK-joint, the tops of braces are free to displace in the axial direction and supported in other directions. Applied loads and boundary conditions are illustrated in Fig.1 and Fig.3. The geometry of the weld elements is in agreement with the weld sizes recommended by the AWS (1994).

4.2 Numerical Results

The numerically determined load-displacement curves are presented in Fig.4 separately for T, K and KK tubular joints, which were loaded by axial forces on braces. The ultimate strengths of tubular joints, which are defined as the peak load, are also presented in Table 2.

At the load when the first crack was found in the test, the distributions of triaxial stress near weld toes in the chord around the brace, which are from FE results, are shown in Fig.5 for T, K and KK joint models. Numerical results for K and KK-joints show that the high equivalent plastic strain zone is near weld toes in the chord between the tension brace and the chord, where the macrocrack initiates (see Fig.6b,c). For the T-joint, the high equivalent plastic strain zone is near the saddle point in the chord (see Fig.6a).

In order to apply the damage criterion derived in the above paragraph, the peak value of the stress triaxiality and its corresponding equivalent plastic strain near weld toes are used to predict macrocrack initiation. From the observation of tests (Ochi et al, 1981), macrocracks initiate at weld toes before structures reach the ultimate capacity. Therefore, during the process of numerical analyses, when the applied load is equal to the load at a macrocrack initiation in the test, the peak value of stress triaxiality and corresponding equivalent plastic strain near weld toes are adopted as critical parameters at crack initiation. These two critical parameters, $(\sigma_H/\sigma_{eq})_{peak}$ and $(p_R)_p$ from the external surface of the chord in tubular joints, are compared with the curve from the damage criterion according to Eq.(2.12) (see Fig.6).

5. Discussion

In Fig.4, there is a good agreement between FE and test results for load-displacement curves in the shape of the curves. Numerical results are a little bit higher than experimental results since the crack initiated and propagated in the chord during the test. The maximum error of both results is less than 10%. In Table 2, FE ultimate strengths for tubular T, K and KK-joints are also very close to test results. It means that adopted numerical models are accurate and valid to analyze the mechanical behaviour for tubular joints. Ultimate capacity formula for T-joint under the axial tensile load is given by (Department of Energy, 1990):

$$N_{u,T} = (7.57 + 17.95\beta + 1.18\beta\gamma)\sigma_y T^2$$

where $\beta = d/D$, $\gamma = D/2T$.

The formula for ultimate strength of K-joint is recommended as (Kurobane et al,1984)

$$N_{u,K} = f_0 \cdot f_1 \cdot f_2 \cdot f_3 \cdot f_4 \cdot f_5 \cdot T^2 \sigma_y$$

and ultimate capacity formula for KK-joints under anti-symmetrical axial loads is presented as (Makino et al,1995):

$$N_{u,KK} = 0.858 \cdot N_{u,K} \qquad g_t / D \geq 0.157$$

$$N_{u,KK} = (1.36 - 3.17\frac{g_t}{D}) \cdot N_{u,K} \qquad g_t / D < 0.157$$

where

$$f_0 = 2.11(1 + 5.66\frac{d}{D})$$

$$f_1 = (\frac{D}{T})^{0.209}$$

$$f_2 = 1 + \frac{0.00904(\frac{D}{T})^{1.24}}{\exp(0.508\frac{g_t + g_c}{T} - 1.33) + 1}, \qquad g_c = 3.04(mm)$$

$$f_3 = \frac{1 - 0.376\cos^2\theta}{\sin\theta}$$

$$f_4 = 1$$

$$f_5 = (\frac{\sigma_y}{\sigma_u})^{-0.723}$$

analytical values for ultimate strengths of tubular joints agree very well with numerical and test results (see Table 2).

In Fig.7, FE analyses indicate that the critical values of two governing parameters corresponding to the loads at crack initiation observed in T, K and KK tubular joint tests, exactly lie on the curve described by the damage criterion. After the comparison of these critical parameters and criterion curve from Eq.(2.12), it can be concluded that the damage criterion proposed in this paper gives good prediction for the macrocrack initiation of tubular T, K and KK-joints.

6. Conclusions

1) Load-displacement curves of T, K and KK-joints from FE results agree very well with those from tests. FE ultimate strengths for T, K and KK tubular joints are also close to analytical and experimental results.

2) Based on the Lematre's damage model, the damage criterion at macrocrack initiation is proposed in this paper. FE results by using ABAQUS have proved that this criterion gives good predictions for macrocrack initiation in tubular joints under triaxial stress state.

ACKNOWLEDGEMENTS

The first author wishes to thank the Japan Society for the Promotion of Science for a Post-Doctoral Fellowship and is also very grateful to Dr. S. R. Wilmshurst for his helpful discussions. He is also grateful for financial support from Nohmura Foundation For Membrane Structure's Technology. This project is partly supported by the Japanese Ministry of Education under Grant Nos. 07555178 and 94053. This work is also partly supported by the National Nature Science Foundation of China and Chinese Education Committee.

APPENDIX. REFERENCES

1. ABAQUS (1995) "ABAQUS v5.5 Manuals (Users' Manual I and II, Theory Manual, Example Manuals I and II, Verification Manual), Hibbitt, Karlsson and Sorenson Inc.
2. American Welding Society, Structural Welding Code (1994), AWS D1.1-94.
3. Department of Energy (1990) "Background to new static strength guidance for tubular joints in steel offshore structures", Offshore Technology Report OTH89 308, HMSO, Lodon, U.K..
4. Kurobane, Y., Makino, Y. and Ochi, K. (1984) "Ultimate resistance of unstiffened tubular joints", J. of Struct. Engng., ASCE, Vol.110, No.2, P385-400.
5. Lematre, J. (1985) "A continuous damage mechanics model for ductile fracture.", J. of Engng. Mater. and Tech. 107(1), p83-89.
6. Lematre, J. (1990) "Micro-mechanics of crack initiation", Int. J. of Fracture, 42, p87-99.
7. Lu, L. H., de Winkel, G. D., Yu, Y. and Wardenier, J. (1994) "Deformation limit for the ultimate strength of hollow section joints", Tubular Structures VI, Grundy, Holgate&Wong (eds), P341-347.
8. Makino, Y., Kurobane, Y., Wilmshurst, S. R. and Lee, M. M. K. (1995) " Proposed ultimate capacity equations for CHS KK-joints under anti-symmetric loads", Proceedings of the 5th ISOPE conference,The Netherlands, Vol.IV, P6-11.
9. Makino, Y., Kurobane, Y, Ochi, K., van der Vegte, G. J. and Wilmshurst, S. R. (1996) "Database of test and numerical analysis results for unstiffened tubular joints", IIW Doc. XV-E-96-220, Hungary.
10. Ochi, K., Kurobane, Y. and Makino, Y. (1981) "Further tests on CHS K-joints with various d/D ratios", Research Report of Kumamoto University.
11. Wang, B., Kurobane, Y. and Makino, Y. (1995) "Stress analyses and safety assessment of welded T-joints", Memoirs of the Faculty of Engineering, Kumamoto Univ., Vol.40, No.3, P289-312.
12. Yonemura, H., Makino, Y., Kurobane, Y. and van der Vegte, G. J. (1996) "Tests on CHS planar KK-joints under anti-symmetrical loads", Tubular Structures VII, JOZSEF FARKAS&KAROLY JARMAI (eds), P181-195.

Table 1. Test specimens (measured)

Specimen	$D \times T$ (mm)	$d \times t$ (mm)	L (mm)	Chord σ_y (N/mm^2)	Chord σ_u (N/mm^2)	Chord ε_u (%)	Brace σ_y (N/mm^2)	Brace σ_u (N/mm^2)	θ	g_l (mm)	g_t (mm)	d/D	D/T
KK-joint	216.1×6.4	139.8×4.5	1978	342	485	15.4	366	647	45°	16.5	165.1	0.65	33.8
K-joint	217.4×4.4	140.3×3.3	1515	354	458	14.5	401	480	60°	30.0		0.65	49.6
T-joint	323.9×6.4	141.3×6.6	1219	274	451	18.5	274	451	90°			0.44	50.6

Note1: ϕ and θ are nominal values.

Note2: g_l: longitudinal gap between both braces; g_t: transverse gap between both braces.

g_l and g_t are calculated from above values.

Table 2 Numerical and Test Database

Specimen	Ultimate Load Test (KN)	Ultimate Load FEM (KN)	Ultimate Load Prediction (KN)	Load Crack Initiation (KN)
T-joint	362	392	322	173
K-joint	225	232	227	172
KK-joint	560	595	554	505

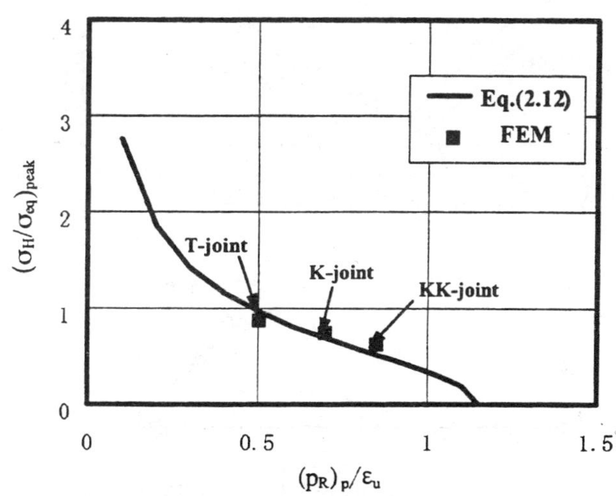

Fig.7 Comparison of Criterion Curve and FE Results

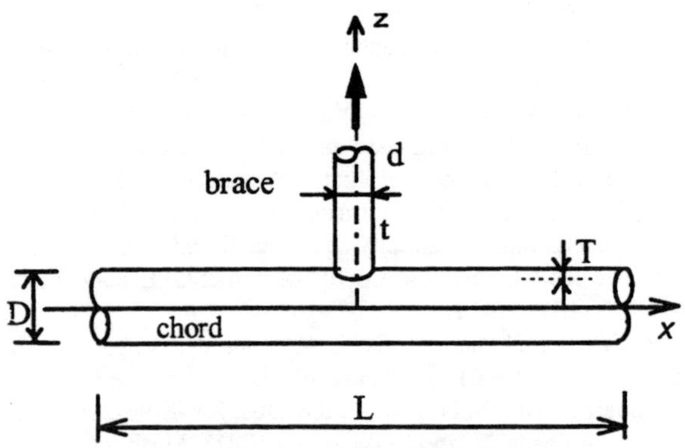

Fig.1a The Geometry of CHS T-joint

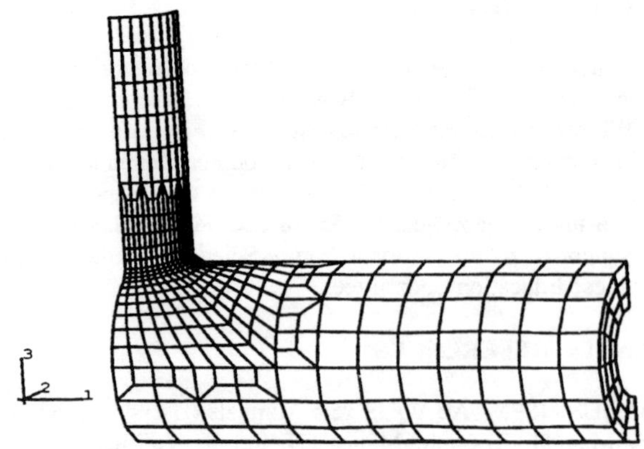
Fig. 2a Typical Meshes for T-joint

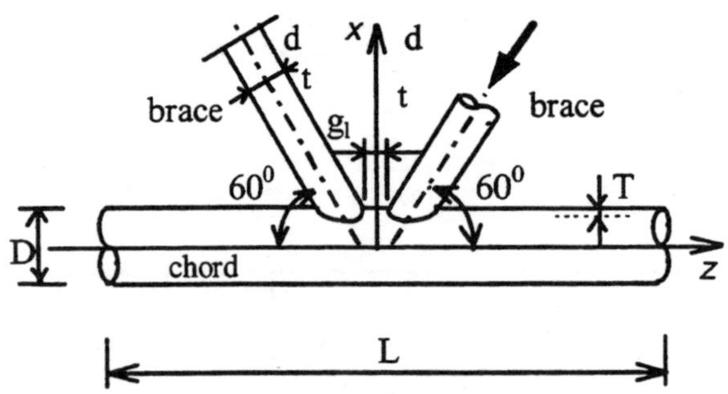

Fig.1b The Geometry of CHS K-joint

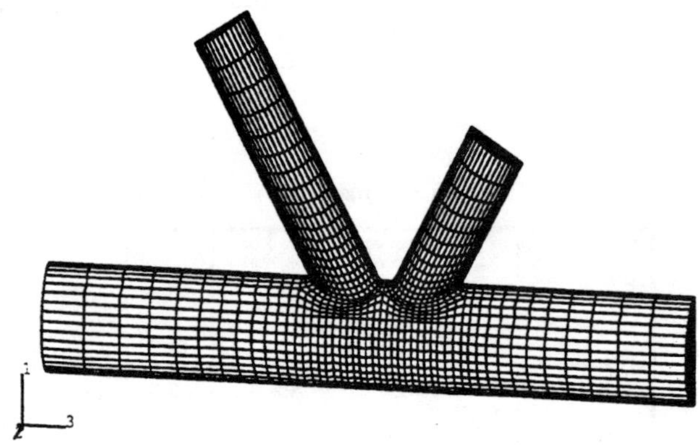
Fig.2b Typical Meshes for K-joint

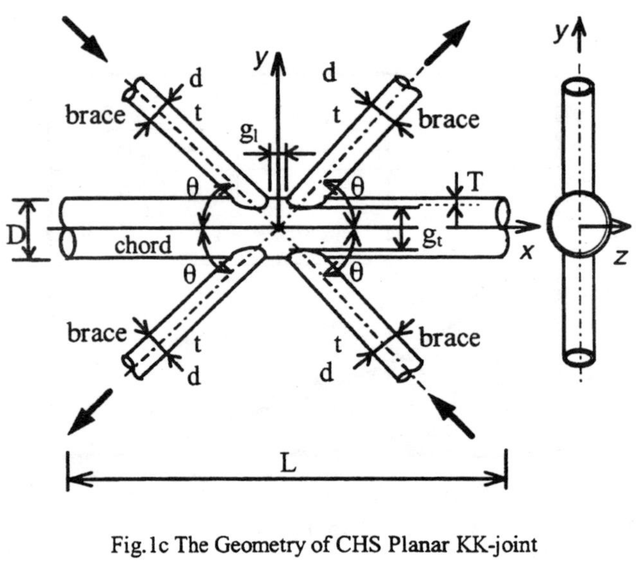

Fig.1c The Geometry of CHS Planar KK-joint

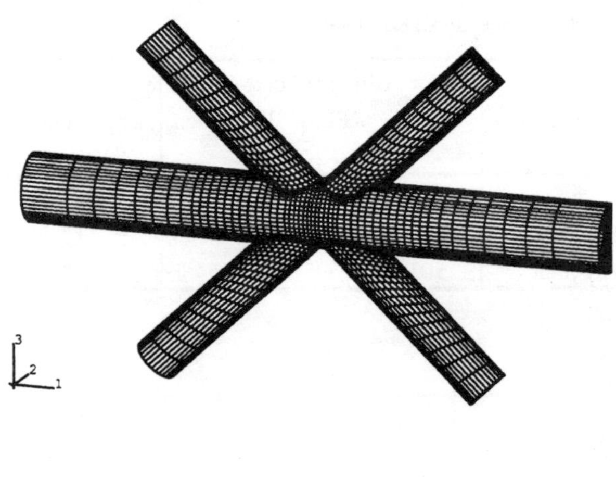
Fig.2c Typical Meshes for KK-joint

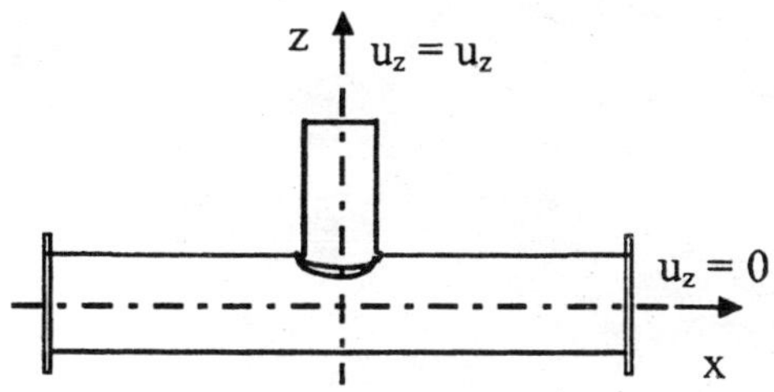

Fig.3a Boundary Conditions of T-joint

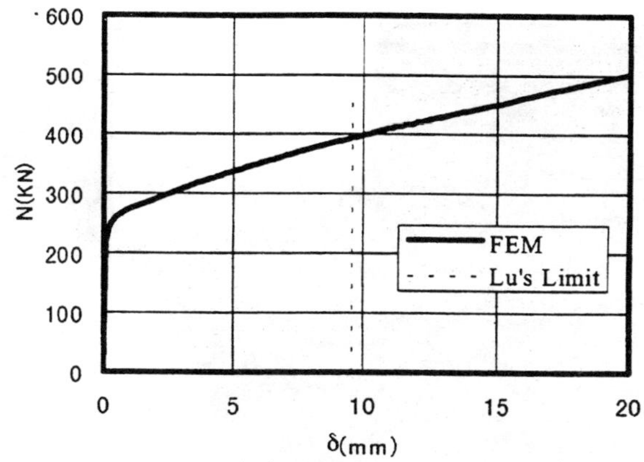

Fig.4a Load-displacement Curves for T-joint

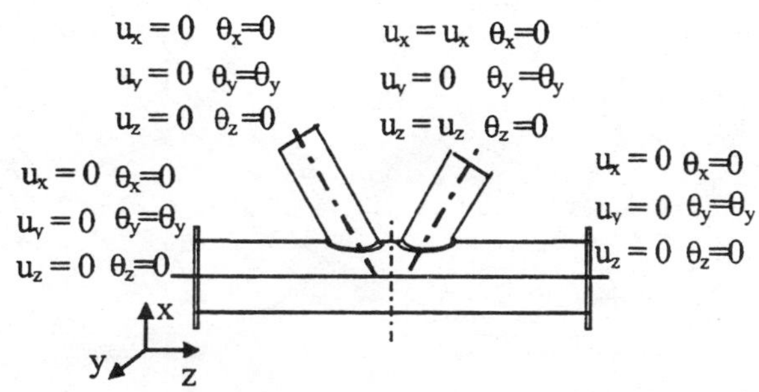

Fig.3b Boundary Conditions of K-joint

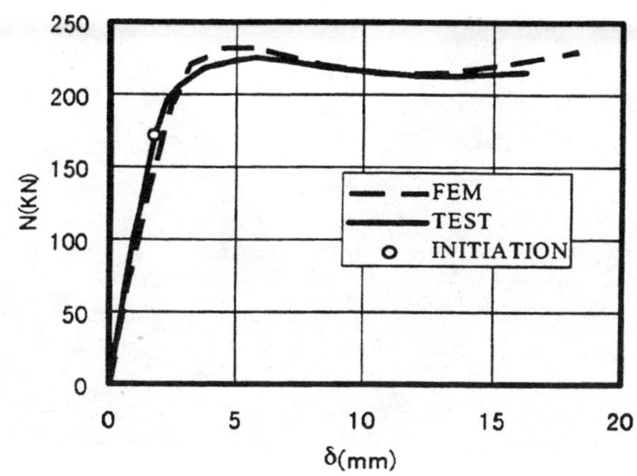

Fig.4b Load-displacement Curve for K-joint

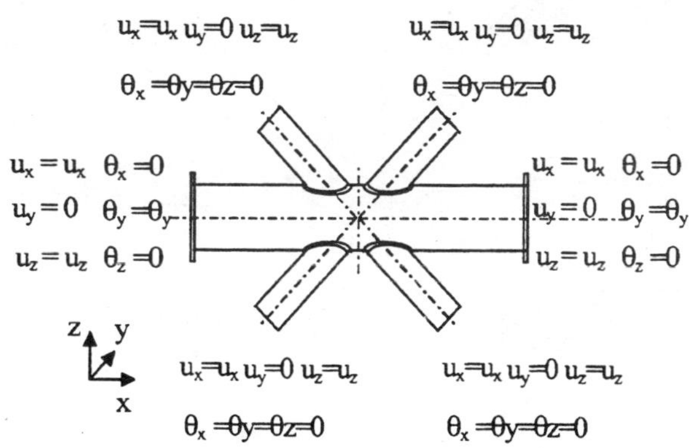

Fig.3c Boundary Conditions of KK-joint

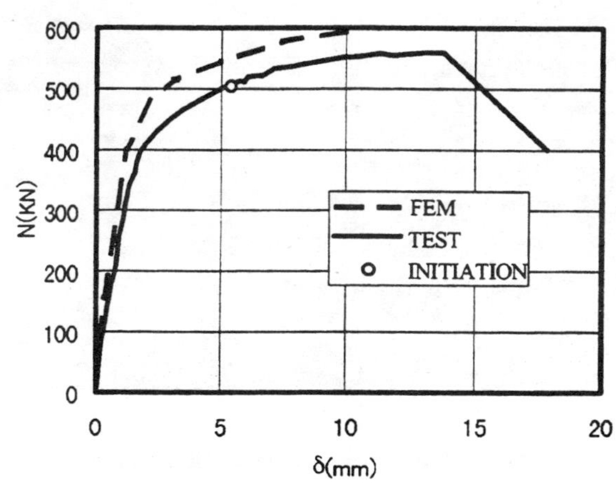

Fig.4c Load-displacement Curve for KK-joint

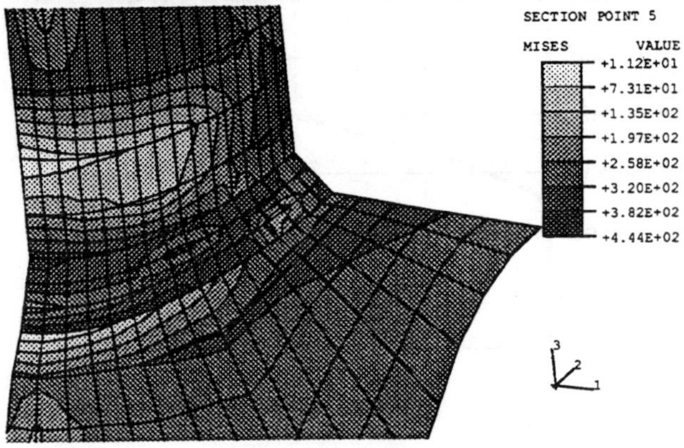

Fig.5a Mises Stress Distribution for T-joint.

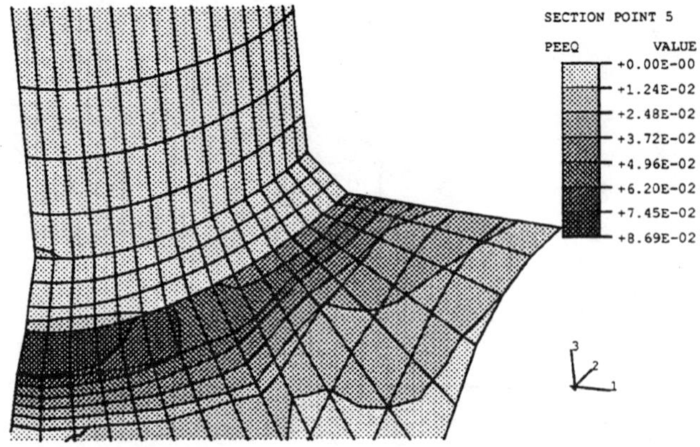

Fig.6a Equivalent Plastic Strain for T-joint

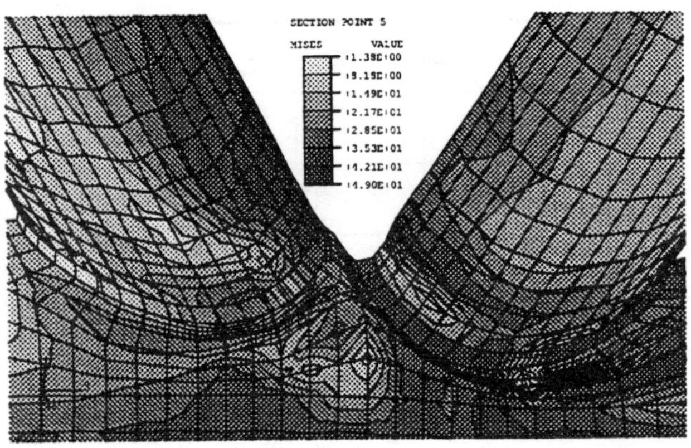

Fig.5b Mises Stress Distribution for K-joint

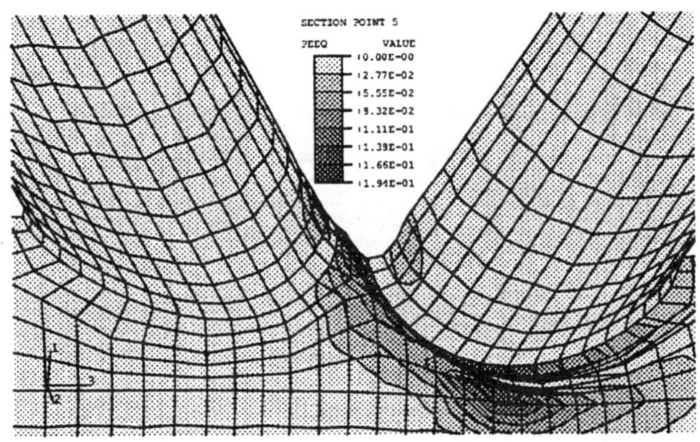

Fig.6b Equivalent Plastic Strain for K-joint

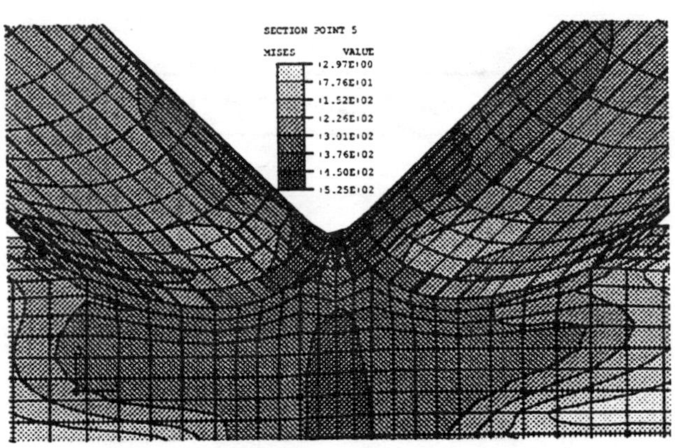

Fig.5c Mises Stress Distribution for KK-joint.

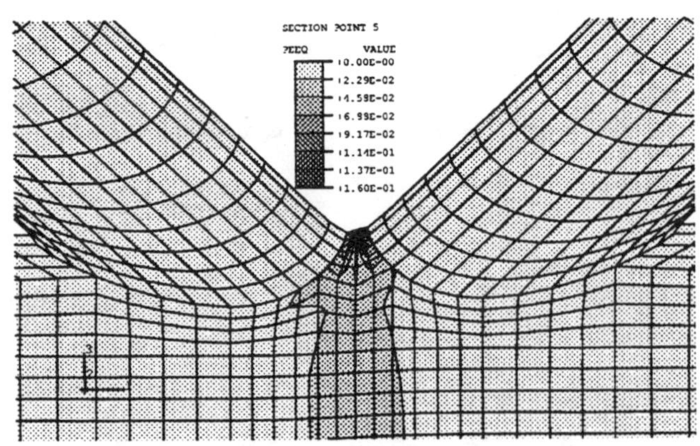

Fig. 6c Equivalent Plastic Strain for KK-joint

Proceedings of the Seventh (1997) International Offshore and Polar Engineering Conference
Honolulu, USA, May 25-30, 1997

ISBN 1-880653-28-1 (Set); ISBN 1-880653-32-X (Vol. IV)

Sensitivity of TLP Tendon Reliability Estimates to Excitation by Multi-Peaked Random Seas

J.W. van de Lindt and J.M. Niedzwecki
Texas A&M University
College Station, TX, USA

ABSTRACT

In this study a variation of the environmental contour technique is used to examine the sensitivity of basic reliability estimates for an idealized Tension Leg Platform (TLP) tendon subject to random sea excitation generated using the Ochi-Hubble and Torsethaugen spectral models. It is recognized that the response of offshore platforms is significantly influenced by the spectral shape as well as the storm severity. Often operational and storm seas are multi-peaked and this study examines the response of deepwater tendons whose natural period of vibration is in the range of peak spectrum energy. Initially, the primary peak is varied along the environmental contour, reflecting a specified return period, while the secondary peak is held constant at a natural period of vibration of the tendon. Reliability estimates are then computed for a deterministic displacement limit state using Monte Carlo simulation. The limit state is chosen with typical tendon spacing as a guideline. Then the most probable location of the secondary peak given the first is determined. Simulation is performed along the contour and a second reliability estimate made for this case. In the second part of the study the sensitivity of the reliability index as a function of the second peak is examined. Comparisons of these reliability estimates to estimates made utilizing typical single peaked JONSWAP spectrum generated seas are presented and discussed.

KEY WORDS: Environmental contours, Ochi-Hubble spectrum, TLP tendon, Reliability

INTRODUCTION

As the next generation of compliant platforms are being designed there is a need for more accurately characterized random seas. Compliant platforms are dynamically sensitive and their response can be significantly influenced by the shape of the wave elevation spectrum. The use of a spectrum with a single peak can be viewed as an over simplification of many offshore environments. Recent statistical work has shown that 'true' ocean spectra have as many as four peaks (Borgman 1993). In this study the affect of using a two-peaked Ochi-Hubble spectrum (Ochi and Hubble 1976) for generation of random sea waves in place of the usual single-peaked JONSWAP spectrally generated sea surface is investigated. The relatively new technique of environmental contours (Winterstein et al 1993) is used to estimate the reliability index of a tendon in 900 meters of water. The reliability index of the deepwater TLP tendon is estimated by combining the environmental contour technique with numerical simulation. It should be noted that the original formulation of an environmental contour (Winterstein et al 1993) was as a model describing the occurrence of peaks of storms exceeding a particular threshold. In this study the contour methodology has been adopted for illustrative purposes.

In the first scenario the location of the first peak is varied with respect to the significant wave height while the location of the second peak is set equal to the tendons second natural period of vibration. The critical seastate is then found by superimposing an environmental contour for the significant wave height, H_S, and the primary spectral period, T_P, onto a consistent response map for the tendon (Niedzwecki, van de Lindt, and Yao 1997). The probability of failure of the tendon was then determined using an allowable horizontal displacement failure threshold at a predefined depth as shown in Figure 1. The structural reliability index, β, which is a measure of the safety reserve in the system, is determined for the two-peaked seastate and compared to the reliability estimated from a 'single peaked' seastate of equivalent energy.

The second scenario involves a reliability based formulation that runs directly parallel to the original environmental contour formulation. The primary peak is set equal to it's most probable location conditional on the significant wave height (Ochi and Hubble 1976) and the second peak is varied. Similarly to the first case, an environmental contour for the significant wave height and the secondary spectral peak, T_{P_2}, is superimposed onto a consistent response map. The critical seastate can now be selected as a function of the second spectral peak by using the map. The reliability index was determined for this case and compared to the single peak JONSWAP results. The reliability indices are a direct function of the return period of the storm, or the probability of occurrence of the contour under investigation, warranting a brief theoretical description of the technique.

The final scenario compares estimates of the reliability indicies from waves generated using the two peaked Torsethaugen spectrum and the single peaked JONSWAP spectrum. Other two peaked spectral models for the North Sea have been developed (Soares 1984), but will not be a focus of this study. Because each of the parameters in the Torsethaugen formulation (Torsethaugen 1993) are a function of significant wave height and spectral peak period a typical North Sea environmental contour developed by Winterstein and Kumar (1995) is used.

ENVIRONMENTAL CONTOUR FORMULATION

The environmental contour approach initially requires one to separate the description of the environment from the platform response. Through the selection of (H_S, T_P) pairs and additional information about the storm, which includes T_R, υ and H_S one can now develop a solution plane of maximum tendon responses. Generating the contour mapping of the solution plane and overlaying the environmental contours one can determine the worst design sea

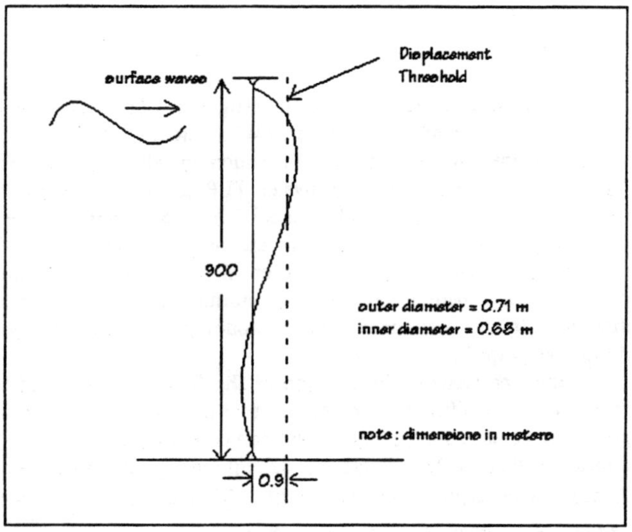

Fig 1 Illustration of limit state for tendon

condition specific to that structure. With this information one can then use Monte Carlo simulations to evaluate the probability of failure from which the reliability index can be estimated. It should be mentioned that although these responses are not deterministic they yield a reasonable approximation. In this procedure one must be careful to perform enough simulations to achieve the desired accuracy. To gain more confidence in the numerical results one may want to perform a sensitivity analysis.

The idea of approximately decoupling the uncertainty in the gross load environment and the extreme response given the loading was introduced through environmental contours (Winterstein et al. 1993). In that study it was shown that the storm with the largest significant wave height did not necessarily produce the greatest response for all structures. It follows that there exist a critical seastate that most probably produces the largest response with significant consistency (Niedzwecki, van de Lindt, and Rijken 1995). In this study the contours are used to determine the critical seastate for various types of spectra. The formulation for environmental contours was done using a single peaked JONSWAP spectrum that can be defined by two parameters, H_S, and T_P. Thus it is logical to formulate a joint statistical characterization of the environment based on these random variables.

Haver and Winterstein (1990) used a three parameter Weibull Cumulative Distribution Function to transform significant wave height to an equivalent standardized normal distribution to get the standardized normal variable U_1. The spectral peak period, T_P, was then assumed normally distributed with constant coefficient of variation. Because an environmental contour consists of all the possible points equidistant from the origin in the standard normal plane, forming a circle with center at the origin and radius β, specifically

$$\beta' = \sqrt{U_1^2 + U_2^2} \quad (1)$$

Thus, given β' and U_1 one can solve for the remaining standardized normal variable U_2. Transformation of the U_1 and U_2 vectors back to physical space, subject to the constraint given in Equation 1, results in the formation of an environmental contour. Because each point on the contour is equidistant from the origin in the standard normal plane, each point represents a storm with an approximately equal probability of occurring.

The probability of occurrence of a storm with a return period T_R, conditional on a storm with $H_S > H_{\min}$ is given as

$$p_{occ,event} = \frac{1}{\upsilon T_R} = 1 - \Phi(\beta') \text{ per storm} \quad (2)$$

where υ is the expected annual number of storms with $H_S > H_{\min}$ (Winterstein, Kumar, and Kleiven 1995; Winterstein and Kumar 1995). Since the occurrence of a storm does not necessarily constitute failure of the structure, Monte Carlo simulation is performed at the critical seastate to determine the probability of failure given occurrence. The critical seastate is found by superimposing the environmental contour on a structure specific iso-response map. These iso-response maps are formed using a load-response model for different combinations of H_S and T_P. The critical seastate can then be selected by inspection of the initial responses for various locations on and around the contour. The probability of failure at the critical seastate, p_f, of a structure can be expressed as the product of the probability of occurrence given an event and the probability of failure given occurrence (Niedzwecki, van de Lindt and Yao 1997), specifically

$$p_f = p_{occ,event} \times p_{f,occ} = [1 - \Phi(\beta')] \times \frac{f}{n} \quad (3)$$

where f is the number of failures and n is the number of independent storms simulated. $\Phi(\cdot)$ is the value of the standard normal distribution function. Once the probability of failure is determined the structural reliability index, β, is given as

$$\beta = \Phi^{-1}(1 - p_f) \quad (4)$$

(Madsen, Krenk, and Lind 1986). In the calculation of the $p_{f,occ}$, defined as the ratio of the number of failures to the number of simulations. The estimation of failure probabilities through Monte Carlo simulation is often limited to a single failure per storm as was done in this study.

OCHI-HUBBLE MULTIPLE-PEAK SIMULATION

The formulation by Ochi and Hubble (1976) defines the parameters for a most probable two-peaked spectrum given the significant wave height of a storm. Although there are currently other model spectra in use it is not the intention of this study to determine which of these multiple peak models is most accurate, but to determine if a more in depth investigation is warranted. In the first mutiple-peaked scenario of this study the primary and secondary significant wave heights were set equal to their most probable values for the '100 year' storm. It is assumed that both the high and low frequency components are narrow banded as is their sum. Thus, $H_S^2 = H_{S_1}^2 + H_{S_2}^2$, keeping the energy in the spectrum constant for a given significant wave height. The primary and secondary significant wave heights for the multiple-peaked spectrum are given here as

$$H_{S_1} = 0.84 \times H_S$$
$$H_{S_2} = 0.54 \times H_S \qquad (5)$$

The primary spectral period, T_{P_1}, was then varied while the secondary spectral peak, T_{P_2}, was set equal to the tendons second natural period of vibration of 7.36 seconds. It was hoped that the higher modes of vibration will be excited during simulation, resulting in exaggerated response and lower reliability levels. The critical seastate for this case was found to be H_{S_1} = 9.8m, T_{P_1} = 13.8sec, H_{S_2} = 6.3m, and T_{P_2} = 7.36sec. The results for the Monte Carlo simulation with a displacement limit state of 1.25 tendon diameters have been summarized in Table 1. It can be seen that the reliability is significantly lower for this multi-peaked case as compared to the JONSWAP case. Although the reliability index decreases by only 7%, it is important to recall that this index is a function of the tail of the normal distribution. It follows that the actual increase in the probability of failure is much greater, 130%.

The second scenario required a formulation parallel to the original environmental contour formulation. The total uncertainty in the system is given as β' in Equation 1. If H_{S_1} and H_{S_2} are set equal to their most probable values as given in Equation 5 they add no uncertainty to the system. The primary spectral peak period is set equal to its mean

$$T_{P_1} = \frac{2\pi}{0.7\exp(-0.046 H_S)} \qquad (6)$$

(Ochi and Hubble 1976). Thus, the system uncertainty not assigned to the significant wave height may now be given to T_{P_2}. This second spectral period is assumed normally distributed with a constant coefficient of variation of 0.06, giving

$$T_{P_2} = \frac{2\pi}{1.15\exp(-0.039 H_S)} \times (1 \pm 0.06 U_4) \qquad (7)$$

where U_4, the standardized normal variable for T_{P_2}, replaces U_2 in Equation 1. A flowchart showing the differences between a typical single-peak and a multiple-peak environmental contour formulation is shown in Figure 2. From this formulation an environmental contour for the significant wave height and the second spectral peak period is formed. The contour was superimposed over the iso-response map shown in Figure 3 to find the critical seastate of H_{S_1} = 9.24m, T_{P_1} = 14.82sec, H_{S_2} = 6.0m, and T_{P_2} = 8.0sec. A local peak can be seen at H_S = 11.0m and T_P = 13.0sec in Figure 3. The occurrence of this peak reinforces the existence of a basic assumption of the environmental contour formulation. Similar consistent response maps were developed for each seastate, but are not shown here. The purpose of these maps is not to directly read the maximum horizontal response of the tendon, but simply to determine the structure specific critical seastate. Since the structure is not changing, logic dictates that a significant change in the location of the critical seastate may be due to a change in the spectrum. Again the results of the Monte Carlo simulation at this critical seastate are summarized and compared to the single peak values under the heading 'Ochi-Hubble II' in Table 1. The reliability index for this scenario was only slightly lower than that for the JONSWAP case, indicating that the tendon models higher modes of vibration were not excited as they were in the previous scenario.

Seastate	Pf/occ	Pf	β (event)	β (year)
JONSWAP	0.016	0.0016	2.95	3.60
Ochi #1	0.037	0.0037	2.68	3.35
Ochi #2	0.020	0.0020	2.86	3.54

Table 1 GoM Simulation Results

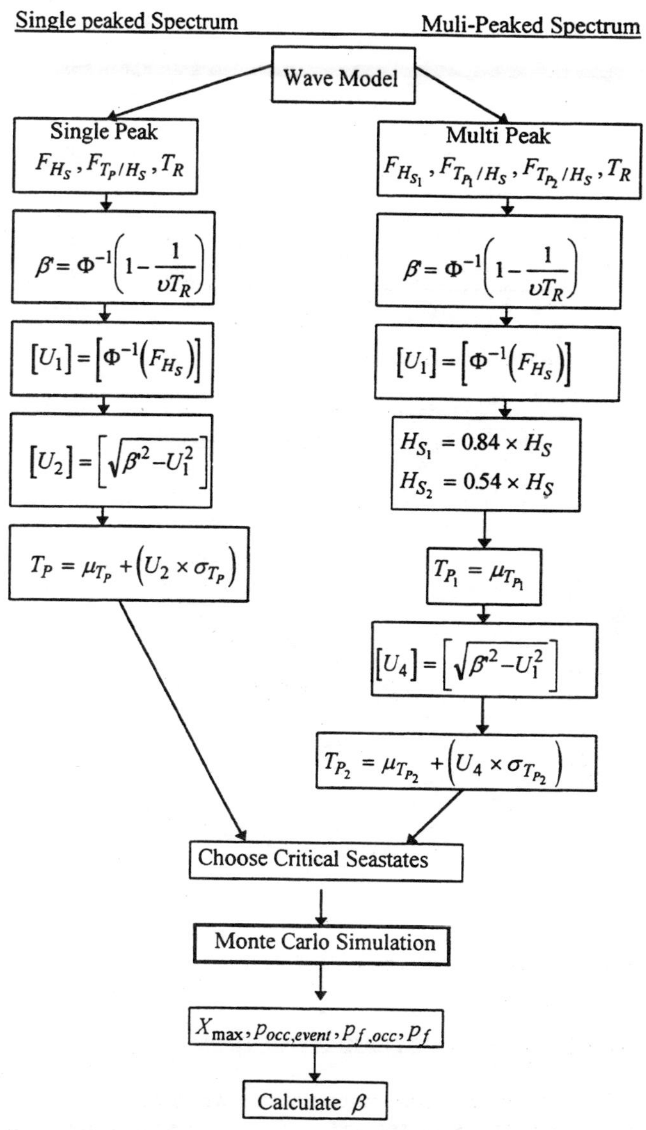

Fig 2 Flowchart

The three dimensional response surface in Figure 4 shows the locations of various response peaks for the second multiple-peak simulations. A similar plot for the responses generated from single peaked seastates tends to focus the maximum response near one particular seastate, resulting in a single peak on the surface. Thus a qualitative effect of using a multi-peaked spectrum for generation of random ocean waves can be observed. Table 1 shows the results for the reliability indices determined for the various spectral shapes. The reliability index for the 'Ochi-Hubble I' case, in which the secondary spectral peak was fixed at the second natural period of vibration of the tendon, was lowest. The probability of failure was more that twice that of the single peaked 'JONSWAP' case. The environmental contour formulated in the 'Ochi-Hubble II' example only gave a slight decrease in the reliability index.

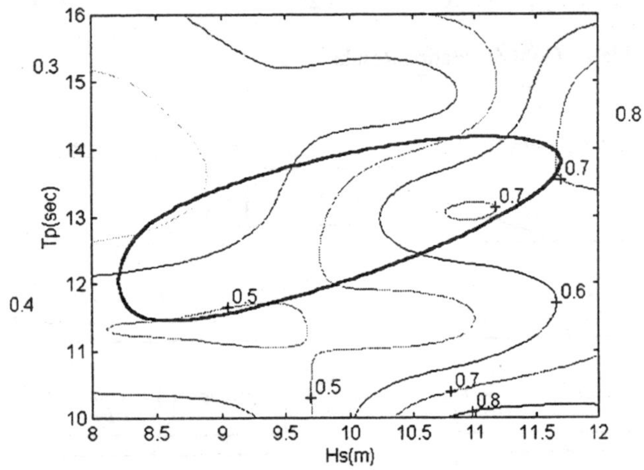

Fig 3 Ochi-Hubble Consistent Response Map

Figure 3: Hs-Tp Response Surface

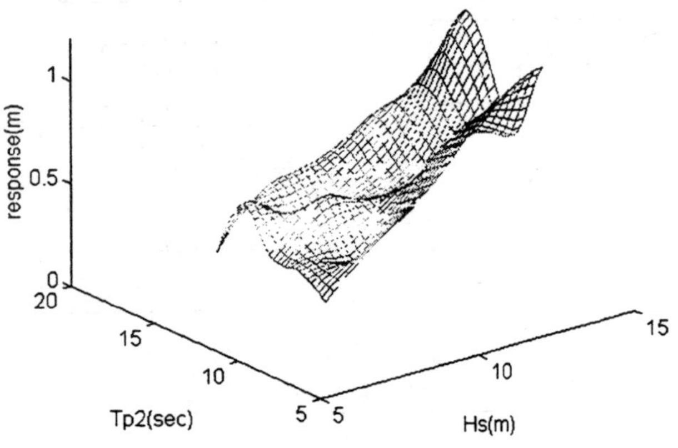

Fig 4 Ochi-Hubble Peaks for Gulf of Mexico

TORSETHAUGEN TWO PEAKED SPECTRUM

The two peaked Torsethaugen spectrum was developed to include locally generated wind sea and swell, with the individual peaks modeled as extensions of the JONSWAP spectrum. The separation between the two sea types is given as

$$T_f = a_f \sqrt[3]{H_S} \tag{8}$$

where a_f =6.6 for long fetches. The parameters for each of the peaks in this model are a function of the significant wave height and spectral peak period. Thus, the location of the design sea state is found exactly the same way as the original formulation. The reliability indicies for these cases were then compared for the idealized tendon. Figure 5 shows the response surface for the Torsethaugen case while Figure 6 shows the surface for the JONSWAP North Sea case both at the Statfjord site. Consistent response maps have been developed for each case but are not shown here. The critical or design sea state for the JONSWAP and the Torsethaugen sea states were both found to be H_S = 14.0m and T_P = 16.2 sec. indicating, as expected that the inline response of the tendon is wave height dominated. The Monte Carlo simulation was performed using the same displacement threshold as in the previous example was used and the reliability results are presented in Table 2. The reliability index for the idealized tendon model only slightly decreased for the Torsethaugen case. The probability of failure for the tendon was found to be about 20% greater for the Torsethaugen seas than for the JONSWAP seas. This is not significant when taking into account the order of magnitude of such calculations.

The spectral parameter, q, and the bandwidth parameter, ε, are both measures of bandedness of the time series and the spectrum from which they were generated. Both parameters are defined in terms of the spectral moments, λ_n.

The closer the parameters are to zero the more narrow banded the spectrum. These parameters are given as (Ochi 1990; Vanmarcke 1975).

$$q = \sqrt{1 - \frac{\lambda_1^2}{\lambda_0 \lambda_2}} \tag{9}$$

$$\varepsilon = \sqrt{1 - \frac{\lambda^2_2}{\lambda_0 \lambda_4}} \tag{10}$$

where

$$\lambda_n = \int_0^\infty \omega_n G(\omega) d\omega \tag{11}$$

The JONSWAP North Sea spectrum was slightly more narrow banded with q = 0.22 and ε = 0.47, while the Torsethaugen North Sea spectrum had q = 0.38 and ε = 0.76. It should be noted that first natural period of vibration of the pinned tendon is 3 to 4 seconds higher than the spectral peak in the JONSWAP spectrum. Because the Torsethaugen spectrum is less narrow there was more energy concentrated at this period. This may have been a contributing factor in the slight increase in the probability of failure.

Seastate	Pf/occ	Pf	β (event)	β (year)
JONSWAP	0.80	0.00114	3.04	2.41
Torsethaugen	0.94	0.00134	3.02	2.36

Table 2 NS Simulation Results

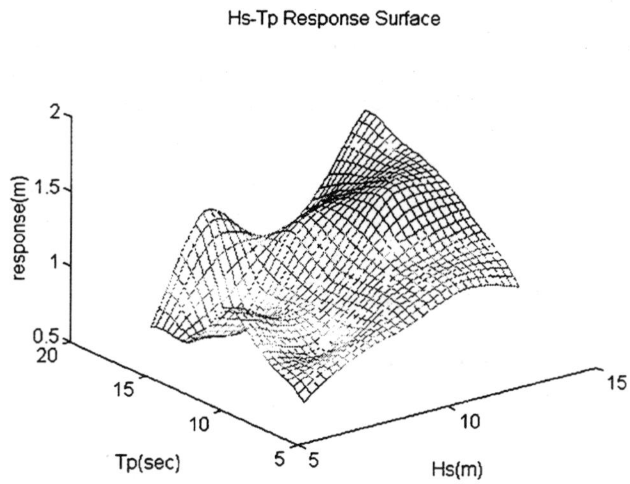

Fig 5 Torsethaugen Peaks for North Sea

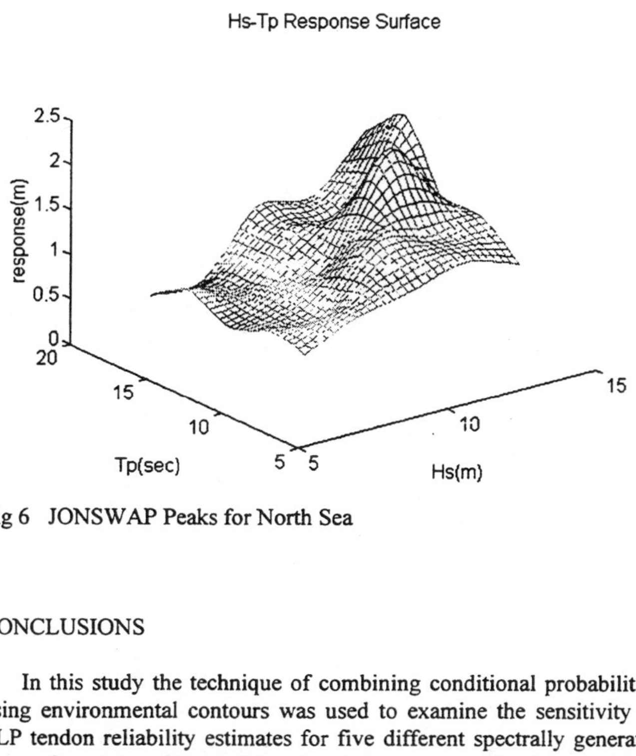

Fig 6 JONSWAP Peaks for North Sea

CONCLUSIONS

In this study the technique of combining conditional probabilities using environmental contours was used to examine the sensitivity of TLP tendon reliability estimates for five different spectrally generates types of seastates. Three of these seastates were examined for a generic Gulf of Mexico site using the JONSWAP and Ochi-Hubble spectral models while the remaining two seastates were examined for the Statfjord site in the North Sea using the JONSWAP and Torsethaugen models. The critical seastate in each case was determined by superimposing a contour onto an iso-response map. Monte Carlo simulation was performed for each of the five scenarios at the critical seastate and a conditional probability determined. The reliability index was determined with respect to a displacement limit state. It was determined that the critical seastate formed using logic parallel to the original environmental contour formulation did not cause a significant increase in the reliability of the tendon for the Gulf of Mexico example. However, the critical seastate found in the 'Ochi-Hubble I' scenario not only produced the largest extreme response, but more that doubled the tendons probability of failure. For the Statfjord cases the seas generated using the Torsethaugen spectrum resulted in a slightly lower reliability level. The spectral parameter and the bandwidth parameter were calculated and it was determined that the JONSWAP spectra was more narrowbanded in this North Sea case. This was sited as the possible reason for the slight change in reliability. Because reliability methods are based on extreme conditions, simulations of the critical seastate are imperative for accurate estimates. A design will not achieve its intended target reliability if there exists a seastate with an equal probability of occurring that is capable of consistently producing greater extreme responses. It follows that some of these critical seastates may be generated from spectra with two, three, or even four peaks. Thus, further investigation into reliability estimates using multi-peaked spectrally generated seastates is warranted.

ACKNOWLEDGEMENTS

Funding for this study was provided through the Offshore Technology Research Center Grant No. CDR8721512.

REFERENCES

Borgman, L.E., Bartel, W.A., and Shields, D.R. (1993). Simbat Theoretical Manual, *NCEL Technical Report R-941*, Port Hueneme, California.

Haver, S. and Winterstein, S.A. (1990). 'The effects of a joint description of environmental data on design loads and reliability.', *Proc. 9th Int. Offshore Mech. Arc. Eng. Sym.*, ASME, II, 7-14.

Madsen, H.O., Krenk, S., and Lind, N.C. (1986). *Methods of Structural Safety.*, Prentice-Hall, Inc., Englewood Cliffs, New Jersey, 07632.

Niedzwecki, J.M., van de Lindt, J.W., and Yao, J.T.P. (1997). "Estimating extreme tendon response using environmental contours.", *J. of Eng. Strut. In press.*

Niedzwecki, J.M., van de Lindt, J.W., and Rijken, O. (1995). "Behavior of Tendon Models in Random Seas.", *24th American Towing Tank Conference*, 157-161.

Ochi, M.K. (1990). Applied Probability and Stochastic Processes: In Engineering & Physical Sciences, Wiley Interscience, New York.

Ochi, M.K.,and Hubble, E. Nadine. (1976). "Six-Parameter Wave Spectra.", *Costal Engineering*, 301-328.

Soares, C.G. (1984). "Representation of a double-peaked sea wave spectra." , *Ocean Engineering*, Vol. II, No. 2, 185-207.

Torsethaugen, K. (1993). "A two peak wave spectrum model.", *Proc. 1993 OMAE*, Vol. II, 175-180.

Vanmarcke, E.M., (1975). "On the distribution of the First-Passage Time for Normal Stationary Random Processes." ASME *Journal of Applied Mechanics.*, March 215-220.

Winterstein, S.R. and Kumar, S. (1995). "Reliability of Floating Structures: Extreme Response and Load Factor Design." *Proc. 27th OTC*, Houston, Texas.

Winterstein, S.R., Kumar, S. and Kleiven, G. (1995). "Environmental contours and extreme response of deep-water floating strucutres." *Proc., 10th Eng. Mech. Specialty Conference,* Boulder, Colorado.

Winterstein, S.R., Ude, T.C., Cornell, C.A., Bjerager, P., and Haver, S. (1993). "Environmental parameters for extreme response : Inverse form with omission factors." *Proc., ICOSSAR-93*, Innsbruck, Austria.

Proceedings of the Seventh (1997) International Offshore and Polar Engineering Conference
Honolulu, USA, May 25-30, 1997

ISBN 1-880653-28-1 (Set); ISBN 1-880653-32-X (Vol. IV)

Fatigue and Overload Reliability of Mooring Systems

Geir Olav Hovde
Advanced Production and Loading AS
Arendal, Norway

Torgeir Moan
Norwegian University of Science and Technology
Trondheim, Norway

ABSTRACT

The methods of structural reliability, taking uncertainties related to both loading and strength into account, have been applied to mooring system design. Both the ultimate line tension and the fatigue failure mode are considered. One of the major concerns of the methodology is the combination of different response components, such as those due to low-frequency wind and wave motions, and wave-frequency wave motions. Different combination strategies have been compared. Efforts have also been devoted to obtaining a systems formulation for a single mooring line which accounts for correlation between individual components properly. The example application is concerned with a turret-moored vessel with 12 mooring lines.

KEY WORDS: Structural reliability, Turret-moored vessel, Catenary mooring, Extreme tension, Fatigue damage, Turkstra, Ferry-Borghes-Castanheta.

INTRODUCTION

Mooring systems for floating production systems can either be single point mooring or spread mooring. Single point mooring tends to be used more frequently for tankers and turret-positioned ships, while spread mooring is used mostly for semi-submersibles. A third type of station-keeping system is dynamic positioning, which often is used to assist catenary mooring. In this paper only a passive turret-positioned vessel is considered.

Mooring lines for permanently moored vessels may be made up of chain links, steel wire ropes, synthetic ropes, or a combination. In addition, clump weights or buoys may be added to achieve given mooring performance requirements. Connecting hardware such as shackles, swivels, triangular plates, sockets, and detachable links are used to connect together the main mooring line components. These components are however not considered to be critical for mooring systems, and is thus disregarded in the present study.

Mooring lines and mooring systems are commonly designed according to design codes, where uncertainties in applied loads and mooring line capacities are taken into account through safety factors or utilisation factors. A drawback with these codes is however that they may give different failure probability levels for different mooring systems.

Reliability methods provide a direct measure of the failure probability. This study therefore outlines a probabilistic approach for estimating the probability of failure directly, instead of assuming that an accepted probability level is achieved by the use of safety factors.

An important issue regarding the overload failure mode is how to obtain extreme values of combined low-frequency and wave-frequency responses. Commonly, either the Turkstra or the Ferry-Borghes-Castenheta rule are applied, i.e. the sine like response behaviour with positive and negative values are modelled very conservatively as rectangular processes with constant response values. In this study a simple modification is suggested to obtain less conservative estimates. Both extreme loads and fatigue are considered herein. Further details about the applied procedure are given by Hovde (1995). Some of the most recent publications by others on this subject are Larsen (1993), Mathisen and Mørk (1993), and Larsen and Mathisen (1996a,1996b).

SHIP MOTION

Environmentally induced responses of floating vessels can be divided into three frequency ranges, i.e. (1) steady state, which is response due to current forces, mean wind and mean wave drift forces, (2) low-frequency (LF) vessel motions due to wind and waves, and (3) wave-frequency (WF) vessel motions.

Conditioned upon an environmental state E specified by the significant wave height, H_S, peak wave period, T_P, wind velocity, U_W, current velocity, U_C, and their individual directions, the motion of the vessel can be formulated as

$$x(t;E) = \bar{x}(E) + x_{LF,WF}(t;E) \tag{1}$$

where t indicates time. Right hand terms are respectively steady state and time-varying displacement. The time-varying term is a coupling term of wave-frequency motion due to waves, and low-frequency motion due to waves and wind-gust.

Commonly, either frequency domain or time domain analyses are used to predict dynamic vessel motions and mooring line tension, but the frequency domain method is chosen for the present study. This method is always linear as the linear principle of superposition is used, implying that all non-linearities must be eliminated, either by direct linearization or by an iterative linearization procedure. The LF and WF motion components are thus computed independently. The wave part of the LF motion spectrum is herein calculated according to Pinkster's formula for spectral density (Pinkster, 1975), while the wind part is calculated based on Davenport's wind gust force spectrum (Davenport, 1978). Transfer functions for the vessel motions are found by linearizing both the non-linear damping mechanism and the restoring force characteristics of the mooring system. The consequence of the latter linearization may be an overestimation of the vessel offset.

Having determined the coupled standard deviations for surge, sway and yaw, extreme values of the motions can be determined in different ways, depending on the assumption of the motion processes. For motions with the characteristics of a narrow-banded Gaussian process the individual peaks will be Rayleigh distributed, but since slow-drift excitation of moored floaters due to waves usually lead to non-Gaussian processes another distribution proposed by Stansberg (1991) is applied herein. Stansberg (1991) investigated the extreme value behaviour of slow-drift motion due to waves, and found that the maxima of the LF motion due to waves may be assumed to asymptotically follow an exponential distribution, where the distribution parameters are determined by the bandwidth of the response spectrum. On the other hand, wind-generated LF motion is herein assumed to be Gaussian, implying Rayleigh distributed extremes.

The constant and slowly varying wave drift forces depend on current. This feature is accounted for by using frequency dependent wave-drift coefficients, using MIMOSA (Mo and Lie, 1990; Larsen and Sandvik, 1990).

EXTREME MOORING LINE TENSION

General

Tension responses of a mooring system due to steady state forces are predicted by a static approach. In addition, due to long response periods, static analyses can also be applied to predict responses due to low-frequency vessel motions. Responses due to the wave-frequency vessel motions should however be estimated by dynamic analyses; frequency-domain dynamic analysis is chosen herein. If non-linearities are properly linearized, then, in most cases, frequency-domain analyses will produce acceptable line tension and anchor load predictions. The non-linearities are among others: nonlinear stretching behaviour of mooring line, changes in geometry associated with changes in shape of mooring line, fluid loading such as the drag force in Morison's equation, and interaction between line and seabed. Model uncertainties related to the frequency domain approach is treated in Eq. (15).

The top tension of a specific mooring line no. l may then appropriately be divided into a quasi-static part (QS) due to mean displacement and low-frequency motion, and a dynamic part (DY) due to wave-frequency motion of fairlead, i.e.

$$T_l = T_{l,QS}(\overline{X} + X_{LF,wind} + X_{LF,wave}) + T_{l,DY}(X_{WF,wave}) \qquad (2)$$

Note that there is a strong coupling between the quasi-static and the dynamic part. Herein, this interaction is considered by computing the dynamic line tension at the same static vessel position as for the quasi-static tension component denoted as $T_{l,QS}$.

Dynamic analysis of mooring lines have shown that the amplitude of the dynamic tension tends to be relatively invariant with depth for a given excitation of fairlead (Mathisen and Mørk, 1993). On the other hand, due to the weight of the mooring line, the quasi-static tension falls off systematically. The tension at a specific point j of the mooring line is thus

$$T_l^{(j)} = T_{l,QS}^{(j)}(\overline{X} + X_{LF,wind} + X_{LF,wave}) + T_{l,DY}(X_{WF,wave}) \qquad (3)$$

where the dynamic tension is calculated at fairlead.

The dynamic line tension is in most cases a non-Gaussian process, implying non-Rayleigh local maxima distributions. Stochastic time domain simulations are therefore normally necessary for final design verification of catenary mooring systems, in order to obtain acceptable statistical confidence in estimated extreme values. However, for reliability analyses requiring results for a large number of short-term conditions, or in the early stage design, a far more efficient approach is needed. Lie and Sødahl (1993) therefore proposed a mixed distribution based on Borgman (1965) for predicting the dynamic tension from frequency domain analysis. The importance of this distribution is that the maxima distribution is described by a Rayleigh distribution for low probability levels and by an exponential distribution for high probability levels. This mixed model is herein considered to be a significant improvement from the simple assumption of Gaussian processes with Rayleigh distributed maxima. The difference between the models are demonstrated in the example application.

Combination of Response Components

Since the LF and the WF response components are stochastic processes with statistical coupling and different characteristic periods, problems related to their combination will arise. Each of the stochastic processes is associated by three distribution functions, i.e. the distribution of the continuous process, R, the distribution of the local maxima, $\tilde{R}$, and the distribution of the largest of N maxima,

$$\hat{R} = \max_{i=1}^{N}\{\tilde{R}\} \qquad (4)$$

where N relates to the short-term period.

Two strategies for combining the LF motion components and the WF tensile component are adopted in the present context. In the first strategy the response processes are initially represented by piecewise constant processes, commonly known as the Ferry-Borghes processes, where the magnitudes and durations correspond to the local maxima response values and the zero-crossing periods, respectively.

The second load combination strategy is not based upon the assumption of Ferry-Borghes processes, but on real process characteristics. This method thus requires the distribution function of the continuous process of the LF response, depending on whether it is the LF or the WF response that dominates. As discussed above, the continuous LF process may be non-Gaussian, and thus unknown in most cases. Herein, the distribution of local maxima for the wave-generated LF motion is chosen according to Stansberg (1991) for the wave-generated LF motion, according to Rayleigh for the wind-generated LF motion, and according to Lie and Sødahl (1993) for the dynamic line tension. Let

$\tilde{R}_1$ = Local maxima of either wind or wave induced LF vessel motion (motion with longest period)

$\tilde{R}_2$ = Local maxima of either wind or wave induced LF vessel motion (motion with shortest period)

$\tilde{R}_3$ = Local maxima for the dynamic tension process due to WF vessel motion

Let the number of cycles during a given period with stationary environmental conditions be $N_1 \le N_2 \le N_3$, where N_1, N_2 and N_3 relate to respectively R_1, R_2, and R_3, refer Figure 2. The tension maximum at fairlead of line no. k based on Ferry-Borghes processes is thus

$$\hat{T}_l = \max_{i=1}^{N_1}\left\{\max_{j=1}^{N_2/N_1}\left\{\max_{k=1}^{N_3/N_2}\left\{T_{l,QS}(\bar{x} + \tilde{R}_{1,i} + \tilde{R}_{2,j}) + \tilde{R}_{3,k}\right\}\right\}\right\} \quad (5)$$

In this equation both the negative and the positive part of the response cycle is represented by the response amplitude, which may be very conservative. A simple modification to reduce this overestimation, and thus the model uncertainty, is therefore to let the piecewise constant response processes be assigned the value of the response amplitude for one half of the response cycle, and zero for the other half. Extreme tension can thus be found by Eq. (5) with N_2 replaced with $N_2/2$, and N_3 replaced by $N_3/4$, see Figure 3.

It is difficult to predict extreme tension from Eq. (5) directly, but several simplifying methods, such as the Turkstra and the Ferry-Borghes-Castanheta combination rule, can be applied:

Turkstra:

$$\hat{T}_l = \hat{T}_{l,QS} + \hat{T}_{l,DY} = \max\begin{cases} T_{l,QS}(\bar{x} + \max_{i=1}^{N_1}\{\tilde{R}_{1,i}\} + E[\tilde{R}_2]) + E[\tilde{R}_3] & (6a) \\ T_{l,QS}(\bar{x} + E[\tilde{R}_1] + \max_{j=1}^{N_2}\{\tilde{R}_{2,j}\}) + E[\tilde{R}_3] & (6b) \\ T_{l,QS}(\bar{x} + E[\tilde{R}_1] + E[\tilde{R}_2]) + \max_{k=1}^{N_3}\{\tilde{R}_{3,k}\} & (6c) \end{cases}$$

$$= T_{l,QS}(\bar{x} + R_1^* + R_2^*) + R_3^* \quad (6d)$$

$$= T_{l,QS}(x_{QS}) + R_3^* \quad (6e)$$

In Eq. (6a) an extreme value of the first response is combined with the expected values of the second and third response. Eq. (6b) is similar to Eq. (6a), but with the extreme value of the second response combined with the expected values of the first and third response. In Eq. (6c) the expected values of the first and second response are combined with the extreme value of the third response.

Note that R_1^*, R_2^* and R_3^* in Eqs. (6d-6e) are only short notations for the actual figures given by the worst of Eqs. (6a-6c). For instance, if Eq. (6a) is the critical expression, then R_1^* refers to the extreme value of the first response, and R_2^* and R_3^* refer to the expected values of the second and third response, respectively.

Ferry-Borghes-Castanheta:

For the Ferry-Borghes-Castanheta rule the response combinations are based upon the worst combination of local maxima and extremes for the three response processes. For instance, in Eq. (7a) an extreme value of the first response, referring to the total short-term period, is combined with an extreme value of the second response, referring to one period of the first response, and an extreme value of the third response, referring to one period of the second response. Further, in Eq. (7b) an extreme value of the second response, referring to the short-term period, is combined with a random local maximum of the first response, and with an extreme value of the third response, referring to one period of the second response. Similar combinations yield for Eqs. (7c) and (7d).

$$\hat{T}_l = \hat{T}_{l,QS} + \hat{T}_{l,DY} = \max\begin{cases} T_{l,QS}(\bar{x} + \max_{i=1}^{N_1}\{\tilde{R}_{1,i}\} + \max_{j=1}^{N_2/N_1}\{\tilde{R}_{2,j}\}) + \max_{k=1}^{N_3/N_2}\{\tilde{R}_{3,k}\} & \\ T_{l,QS}(\bar{x} + \tilde{R}_1 + \max_{j=1}^{N_2}\{\tilde{R}_{2,j}\}) + \max_{k=1}^{N_3/N_2}\{\tilde{R}_{3,k}\} & (7b) \\ T_{l,QS}(\bar{x} + \max_{i=1}^{N_1}\{\tilde{R}_{1,i}\} + \tilde{R}_2) + \max_{k=1}^{N_3/N_1}\{\tilde{R}_{3,k}\} & (7c) \\ T_{l,QS}(\bar{x} + \tilde{R}_1 + \tilde{R}_2) + \max_{k=1}^{N_3}\{\tilde{R}_{3,k}\} & (7d) \end{cases}$$

$$= T_{l,QS}(\bar{x} + R_1^* + R_2^*) + R_3^* \quad (7e)$$

$$= T_{l,QS}(x_{QS}) + R_3^* \quad (7f)$$

The cumulative distribution functions for the largest of N independent local maxima are

$$\begin{aligned} F_{\hat{R}_1}(\hat{r}_1|E) &= F_{\tilde{R}_1}(\hat{r}_1|E)^N \\ F_{\hat{R}_2}(\hat{r}_2|E) &= F_{\tilde{R}_2}(\hat{r}_2|E)^N \\ F_{\hat{R}_3}(\hat{r}_3|E) &= F_{\tilde{R}_3}(\hat{r}_3|E, x_{QS})^N \end{aligned} \quad (8)$$

The distribution function of $\tilde{R}_3$ is conditioned upon the extreme displacement x_{QS} in order to account for variation in dynamic tension due to mean and LF vessel offset, confer discussion to Eq. (2).

The second alternative for combining the response components is to use an analogue to the Turkstra based combinations adopted by Larsen (1993), and Mathisen and Mørk (1993). Let us assume that the characteristic periods of the LF responses are almost identical, and significantly larger than the characteristic period for the WF response, i.e. $N_1 \approx N_2 \ll N_3$. Then, as long as one of the responses dominates, the extreme line tension can be approximated to

$$\hat{T}_l = \hat{T}_{l,QS} + \hat{T}_{l,DY} = \max\begin{cases} T_{l,QS}(\bar{x} + \max_{i=1}^{N_1}\{\tilde{R}_{1,i}\} + R_2) + \tilde{R}_3 & (9a) \\ T_{l,QS}(\bar{x} + R_1 + \max_{j=1}^{N_2}\{\tilde{R}_{2,j}\}) + \tilde{R}_3 & (9b) \\ T_{l,QS}(\bar{x} + R_1 + R_2) + \max_{k=1}^{N_3}\{\tilde{R}_{3,k}\} & (9c) \end{cases}$$

$$= T_{l,QS}(\bar{x} + R_1^* + R_2^*) + R_3^* \quad (9d)$$

$$= T_{l,QS}(x_{QS}) + R_3^* \quad (9e)$$

In Eq. (9a) an extreme value of the first response is combined with a random value of the continuous second response, and a random local maximum of the third response. Eq. (9a) is critical if the first response dominates the total tension, and if several local maxima of the third response are likely to occur in the immediate vicinity of the extreme maximum of the first response. Eq. (9b) is similar to Eq. (9a), but with the second response as the dominant component. If the third response dominates, then Eq. (9c), utilizing an extreme dynamic tension together with random values of the continuous LF responses, is more appropriate. Since the characteristic periods of the LF responses are much longer than for the WF response, local maxima of the LF responses are no more likely to occur together with the extreme dynamic tension than any other value. It is therefore reasonable to use the continuous distributions of the LF responses in Eq. (9c). An analogue explanation can also be given for the continuous process variables applied in Eqs. (9a) and (9b).

Distribution of Extreme Mooring Line Tension

By randomizing both the long-term distribution of the environmental state, E, as well as the conditional variables R_1^*, R_2^* and R_3^*, the extreme value distribution of the line tension during any stationary environmental state can be found by applying FORM or SORM, see e.g. Madsen et al (1986), on

$$F(\hat{t}_1) = P\left[\hat{T}_1\left(E, R_1^*, R_2^*, R_3^*\right) \leq \hat{t}_1\right] \tag{10}$$

where all the parameters within the parenthesis ( ) are random. This equation only gives the marginal tension distribution during a single random short-term condition, which means that for a long-term period of N_{ES} independent stationary environmental states, the following extreme tension distribution must be considered instead

$$F(\hat{t}_1) = P\left[\hat{T}_1\left(E, R_1^*, R_2^*, R_3^*\right) \leq \hat{t}_1\right]^{N_{ES}} \tag{11}$$

The quasi-static and the dynamic tension are evaluated by using MIMOSA (Mo and Lie, 1990) for a set of environmental states. Information such as standard deviations and quasi-static tension are then imported to the reliability analysis via a response surface interface based on natural cubic splines, providing a smooth approximation to all the necessary system response parameters.

Both the quasi-static and the dynamic model are subjected to uncertainties. Two bias factors are therefore introduced. The first bias, which is denoted as B_{QS}, relates to the quasi-static tension, while the second bias, B_{DY}, relates to the dynamic tension. These factors act as multiplicative factors to the quasi-static and the dynamic tension computed from Eq. (11), as further shown in Eq. (15).

FATIGUE DAMAGE

In contrast to extreme tension, which occurs during a short-term environmental state, fatigue damage accumulates over all environmental conditions. It is therefore important to have efficient computation tools that cover both wave-frequency and low-frequency ship motions in addition to line dynamics, such as the frequency domain analysis methods.

Even if the LF and the WF response spectra are both narrow-banded the combined loading will be wide-banded. Realistic fatigue damage predictions therefore require methods where the wide-banded nature of the load is accounted for. A review and test of such methods are found in Buoyssy et al. (1993), but among these methods the most convenient method for our purpose is the one proposed by Jiao and Moan (1990). This method is also recommended by Lie and Fylling (1994) in their paper on evaluation of methods for fatigue analysis of offshore mooring lines.

Stochastic analyses of fatigue under narrow-banded Gaussian processes are widely applied, and for such processes exact closed-form solutions to fatigue damage exist. Many structures are however subjected to two or more simultaneous dynamic loads acting in highly different frequency domains, such as the LF and WF tension components of mooring lines. When such processes with bimodal spectra are encountered, a narrow-band assumption of the combined spectrum may give too large fatigue damages. Jiao and Moan (1990) therefore presented a closed form approximation to fatigue damage of bimodal Gaussian response processes.

For a given environmental state, E, Jiao and Moan (1990) expressed the fatigue damage for a wide-band (WB) process, defined as the combination of two narrow-banded (NB) Gaussian processes, as

$$D_{WB|E} = \rho_{WB|E} \cdot D_{NB|E} = \rho_{WB|E} \cdot \frac{\upsilon_{0Z|E}\tau_E}{K} \cdot (8\lambda_{0Z|E})^{m/2} \cdot \Gamma(1+\frac{m}{2}) \tag{12}$$

where ρ is a damage correction factor, ν and λ are respectively the mean zero-upcrossing rate and the variance of the sum of the LF and WF tension processes, K and m are S-N curve parameters, and τ_E is the duration of the environmental state. Note that Eq. (12) is based on Rayleigh distributed local maxima. The damage correction factor is a function of several factors, of which Vanmarcke's bandwidth parameter is the only parameter not given from the MIMOSA analyses. Lie and Fylling (1994) recommended a fixed bandwidth parameter between zero and 0.1 for the present application. For further details of Eq. (12) confer Jiao and Moan (1990) and Hovde (1995).

Long-term fatigue damage is obtained by adding all short-term fatigue damages, i.e.

$$\begin{aligned} D_{WB} &= \sum_E \rho_{WB|E} \cdot \frac{\upsilon_{0Z|E}\tau_E}{K} \cdot (8\lambda_{0Z|E})^{m/2} \cdot \Gamma(1+\frac{m}{2}) \\ &= \frac{\tau}{K} \cdot \sum_E \rho_{WB|E} \cdot \upsilon_{0Z|E} \cdot (8\lambda_{0Z|E})^{m/2} \cdot \Gamma(1+\frac{m}{2}) \cdot p(E) \\ &= \frac{\tau}{K} \cdot D_{WB}^* \end{aligned} \tag{13}$$

where p(E) is the probability of the environmental state E, and τ is the service life of the mooring system.

PROBABILITY OF FAILURE DUE TO EXTREME TENSION

Mooring lines are made up of a large number of components, where the most obvious component is the chain link, but very often, mooring lines consist of both chain and wire rope components. Usually, a one to one relationship is used between the chain link and the chain component, while wire rope segments of length equal to the length of wire test pieces are used as wire rope components. For capacity testing, the length of wire rope test pieces is 32.5 times the nominal diameter (Mathisen and Mørk, 1993).

An individual mooring line has no redundancy, i.e. if one component fails, the line itself fails. The mooring line can thus be modelled as a series system. Let the line consist of m sections, where each section is made up of several components, i.e. for a line with three sections, the bottom section may consist of chain links, the centre section may be a wire rope, while the upper section may consist of additional chain links. The probability of line failure is thus

$$p_{f,l} = P\left[\bigcup_{i=1}^{m}\bigcup_{j=1}^{n_i}\left(C_1^{(i)} \cdot C_2^{(ij)} \leq \hat{T}_1^{(ij)}\right)\right] \tag{14}$$

where $C^{(ij)}=C_1^{(i)} \cdot C_2^{(ij)}$ is the capacity of component no. j of line section no. i. The capacity is split into two independent variables in order to consider correlation between components. The first variable is considered fully correlated between the section components, while the second variable is considered independent. Correlation between different sections can be considered by assuming full correlation between the variable C_1 for these parts.

As noted in Ch. 3.1, the dynamic tension is close to invariant along the line, while the quasi-static tension falls of systematically with increased distance to the fairlead. With full correlation between the tension at different positions, the component load may be written as

$$\hat{T}_l^{(ij)} = k_{ij} \cdot B_{QS} \cdot \hat{T}_{l,QS} + B_{DY} \cdot \hat{T}_{l,DY} \tag{15}$$

where k_{ij} is a deterministic scaling factor, while the quasi-static and the dynamic tension component are taken as the design point of Eq. (11). In order to consider correlation properties between different mooring line components, Eq. (14) is reformulated:

$$\begin{aligned} p_{f,l} &= P\left[g_S\left(B_{QS}, B_{DY}, C_l^{(1)}, \ldots, C_l^{(m)}, \hat{T}_l, U_S\right) \le 0\right] \\ &= P\left[U_S - \Phi^{-1}\left(1 - \prod_{i=1}^{m}\prod_{j=1}^{n_i}\left(1 - P\left[C_l^{(i)} \cdot C_2^{(ij)} \le \hat{T}_l^{(ij)} \mid B_{QS}, B_{DY}, C_l^{(i)}, \hat{T}_l\right]\right)\right)\right] \end{aligned} \tag{16}$$

where U_S is an auxiliary standard normal variate, and P[.|.] is the component failure probability conditioned upon the correlated random variables. For more details about this equation refer Hovde (1995).

PROBABILITY OF FAILURE DUE TO FATIGUE

The probability of a line failure due to fatigue can be found in a similar way as the probability of failure due to extreme tension, i.e.

$$p_{f,l} = P\left[\bigcup_{i=1}^{m}\bigcup_{j=1}^{n_i}\left(\Delta^{(ij)} \le \frac{\tau}{K^{(ij)}} \cdot B_\sigma^{m_i} \cdot B_\nu \cdot B_D^{(ij)} \cdot D_{WB}^{*(ij)}\right)\right] \tag{17}$$

where Δ is the Miner's sum at failure, B_σ is the bias of $\sqrt{\lambda}$, B_ν is the bias of ν, B_D is the bias of the fatigue damage model, and K and m are material parameters giving the S-N curve. B_σ and B_ν are assumed the same for all line sections since these parameters describe the uncertainties of the computed line tension response, which is more or less fully correlated along the line. Similar to the transformation of Eq. (14) to Eq. (16), Eq. (17) can be written as

$$\begin{aligned} p_{f,l} &= P\left[g_S\left(B_\sigma, B_\nu, U_S\right) \le 0\right] \\ &= P\left[U_S - \Phi^{-1}\left(1 - \prod_{i=1}^{m}\prod_{j=1}^{n_i}\left(1 - P\left[\Delta^{(ij)} \le \frac{\tau}{K^{(ij)}} \cdot B_\sigma^{m_i} \cdot B_\nu \cdot B_D^{(ij)} \cdot D_{WB}^{*(ij)} \mid B_\sigma, B_\nu\right]\right)\right)\right] \end{aligned} \tag{18}$$

where U_S is an auxiliary standard normal variate, and P[.|.] is the component failure probability conditioned upon the correlated random variables. For more details about this equation refer Hovde (1995).

EXAMPLE APPLICATION

General

The example application has been made to a turret-moored vessel at 150 m waterdepth exposed to North Sea environment. Principal parameters of the selected vessel, including a description of the mooring system are given in Tables 1 and 2, and Figure 1. The S-N curve parameters refer to API 2FP1 (1991) recommendations.

Due to the resonance characteristics of LF motions, the magnitude of the associated damping forces are key parameters to the LF motion. These forces are however difficult to predict accurately, and they may vary significantly, depending on the environmental conditions and on the characteristics of the mooring system. Herein, viscous damping due to the hull, mooring system damping, wave drift damping and aerodynamic damping are combined into single damping coefficients for surge, sway and yaw. These are respectively 1000 kNs/m, 500 kNs/m and 500 $kNms^2/m^2$; the first two being linear coefficients and the last being quadratic, refer definitions of Mo and Lie (1990).

Uncertainty Measures

The model uncertainty for both the quasi-static and the dynamic extreme tension are considered log-normally distributed with mean 1.0 and standard deviation 0.1 (Larsen and Mathisen, 1996a). Chain and wire rope strengths are assumed log-normal with mean 1.2 times the minimum breaking strength, and a coefficient of variation, COV, of 5% (Larsen and Mathisen, 1996a). Correlations between component capacities are initially neglected, but a parameter study of the importance of correlation is submitted.

The uncertainty in the Miner-Palmgren hypothesis for fatigue failure is treated as a log-normal variable with mean 1.0 and COV of 30% for the chain links, and 20% for the wire rope components. These figures are in the same range as in Engesvik (1987) and Larsen and Mathisen (1996b). Design S-N curves are based on statistical analyses of experimental data, and is usually defined as the mean minus two standard deviations of logN. In this study the mean of K is based upon an assumption of logN being normally distributed with standard deviation 0.4, and m being deterministic. The bias of the zero-upcrossing frequency is considered normally distributed with mean 1.0 and COV 7%. In order to be consistent with the extreme condition, the bias of the standard deviation, $\sqrt{\lambda_{0Z}}$, is considered to be log-normal with mean 1.0 and COV of 15%, i.e. approximately the uncertainty of the sum of two independent variables of COV 10%. The uncertainty of the fatigue damage model is assumed log-normal with mean 1.0 and COV 10%, which is the same as in Larsen and Mathisen (1996b).

Environmental Data

The joint environmental model selected for this study is based on experience gained from measurements and hindcast data from the Norwegian Continental Shelf. The model includes 1-hour mean wind velocity, current velocity, significant wave heights, spectral peak period. Wave direction is considered uni-directional with the wind and current direction co-linear to the wave-direction. Detailed description of this model is given in Bitner-Gregersen and Haver (1991). The two-parameter Pierson-Moskowitch wave spectrum and Davenport's wind-gust spectrum are used for response calculations.

Deterministic Design Check

Structural design is traditionally based on deterministic analysis. The safety factor according to the POSMOOR rules (DNV, 1989) is therefore determined before proceeding to the reliability analysis.

Deterministic analyses (Hovde, 1995) show that the 100-year seastate defined by H_S=17.9 m and Tp=18.4 s together with the 100-year wind velocity of 41.5 m/s, and the 10-year current of 0.59 m/s is the worst condition for the considered mooring system. For this loadcase the highest tension implied a safety factor of 1.59.

The estimated fatigue lives based on deterministic analyses are 320 years for the chain components and 8050 for the wire rope components. This implies a fatigue safety factor of 12.8 for a service life of 25 years.

Failure Probability due to Extreme Tension

Extreme line tension distributions are displayed in Figure 4, which shows that the response combination strategy based on real processes gives the smallest tension for a given probability level. It is also seen that the Turkstra combination rule for Ferry-Borghes processes gives smaller tension than the Ferry-Borghes-Castanheta rule, which is in accordance with the well known un-conservatism in the Turkstra rule. The simple modification of the Ferry-Borghes processes as proposed herein gives slightly lower extreme tension. Failure probabilities according to the unmodified and the modified Ferry-Borghes-Castanheta rule are respectively 4.40E-3 and 3.34E-3, i.e. almost identical. Note that the active response combination is given by Eq. (6a) for the Turkstra based rule, and Eq. (7c) for the Ferry-Borghes-Castanheta rule. The modified Ferry-Borghes-Castanheta rule is adopted for the further analyses. As a suggestion for future work it is important to compare these combination rules with measured or simulated time series.

Tables 3 and 4 display the results from the reliability analyses using FORM; note that for the present mooring system R_1^* is the LF wave motion. Table 3 presents the results for response maxima distributions according to Stansberg (1991) and Lie and Sødahl (1993), while Table 4 presents the results for Rayleigh distributed local maxima. As seen, Rayleigh distributed local maxima gives a failure probability more than 100 times smaller than for the distributions by Stansberg, and Lie and Sødahl. Hence, for catenary mooring systems it is very important to consider the non-Gaussian characteristics of the tension process.

The parameter study presented in Table 5 show that the uncertainty of the individual component strengths and their mutual correlations are of minor influence with respect to the failure probability. For the selected case studies the probability of failure only varied between 3.34E-3 and 6.40E-3. The reason for this insensitivity is the load distribution, which totally dominates the contribution to the failure probability, refer Table 3.

Failure Probability due to Fatigue

Analysis results for the fatigue failure mode is presented in Table 6. The results show that the probability of a fatigue line failure during a period of 25 years is 6.61E-3 when Vanmarke's bandwidth parameter is set to 0.1. The corresponding component failure probabilities are 3.19E-5 for the chain components and 7.27E-12 for the wire rope components. By reducing Vanmarcke's bandwidth parameter to 0.0 the mooring line failure probability decreases to 5.73E-3, which means that this parameter is insignificant for the present problem.

CONCLUSIONS

A framework for assessing the probability of failure due to extreme tension and fatigue is presented for catenary mooring systems, where the combination of response maxima for the extreme condition has been one of the major issues. It is however demonstrated that the assumption of local maxima distributions for the different responses is much more important than the response combination strategies. The analyses also show that for this particular mooring system, fatigue tends to be more critical than extreme tension.

REFERENCES

American Petroleum Institute (1991). Draft Recommended Practice for Design, Analysis and Maintenance of Mooring for Floating Production Systems - API RP 2FP1. Washington D.C., USA.

Bitner-Gregersen, E., and Haver, S. (1991). Joint Environmental Model for Reliability Calculations. Proceedings of the 1st ISOPE Conference, Edinburg, U.K., August 11-16.

Borgman, L.E. (1965). Wave Forces on Piling for Narrow-Band Spectra. *Journal of the Waterways and Harbors Division*, ASCE, August, pp. 65-90.

Bouyssy, V., Naboishikov, S.M and Rackwitz, R. (1993). Comparison of Analytical Counting Methods for Gaussian Processes. *Structural Safety*, Vol. 12, No. 1, pp. 35-57.

Davenport, A.G. (1978). The Prediction of the Response of Structure to Gusty Wind. *Safety of Structures under Dynamic Loading*, Vol. 1, pp. 257-84, Tapir Publishers, Trondheim, Norway.

Det Norske Veritas (1989). Rules for Classification of Mobile Offshore Units, Special Equipment and Systems Additional Class Pt. 6 Ch. 2 Position Mooring (POSMOOR). Det Norske Veritas Classification A/S, Høvik, Norway.

Engesvik, K. et al. (1987). Literature Survey on Parameter Uncertainties Related to Structural Analysis of Marine Structures. Report No. STF71-A87008, Sintef, Trondheim, Norway.

Hovde, G.O. (1995). Fatigue and Overload Reliability of Offshore Structural Systems, Considering the Effect of Inspection and Repair. Dr.Ing. Thesis, The Norwegian Institute of Technology, Trondheim, Norway.

Jiao, G. and Moan, T. (1990). Probabilistic Analysis of Fatigue due to Gaussian Load Processes. *Probabilistic Engineering Mechanics*, Vol. 5, No. 2, pp. 76-83.

Larsen, K. and Sandvik, P.C. (1990). Efficient Methods for the Calculation of Dynamic Mooring Line Tension. Proceedings of the 1st EUROMS, Trondheim, Norway.

Larsen, K. (1993). Reliability Analysis of Mooring System. Conference on "Posisjonering av Plattformer og Skip", Norsk Sivilingeniørers Forening, NIF Studiesenteret, Fagernes, Norway, November 1-3.

Larsen, K. and Mathisen, J. (1996a). Reliability-Based Mooring System Design for a Drilling Semisubmersible. Proceedings of the 6th ISOPE, Los Angeles, USA, May 26-31

Larsen, K. and Mathisen, J. (1996b). Reliability-Based Fatigue Analysis of Mooring Lines. Proceedings of the 15th International Conference on OMAE.

Lie, H. and Sødahl, N. (1993). Simplified Dynamic Model for Estimation of Exteme Anchorline Tension. Proceedings of the Offshore Australia 1993 Conference, Melbourne, Australia.

Lie, H. and Fylling, I.J. (1994). Evaluation of Methods for Fatigue Analysis of Offshore Mooring Lines. Proceedings of the 10th Offshore South East Asia Conference, Singapore, December 6-9.

Madsen, H.O., Krenk, S. and Lind, N.C. (1986). *Methods for Structural Safety*. Prentice-Hall Inc., Englewood Cliffs, New Jersey, USA.

Mathisen, J. and Mørk, J. (1993). Reliability-Based Design Methods for Mooring Lines. Proceedings of the 8th Int'l Conference on Floating Production Systems, London, U.K., December 9-10.

Mo, K. and Lie, H. (1990). USER'S Documentation MIMOSA-2, Version 2 (Additional theory notes of 1994 are submitted). Report no. 519616.00.01, MARINTEK, The Sintef Group, Trondheim, Norway.

Pinkster, J.A. (1975). Low-frequency Phenomena Associated with Vessels Moored at Sea. *Society of Petroleum Engineers Journal*, December, pp. 487-494.

Stansberg, C.T. (1991). A Simple Method for Estimation of Extreme Values of Non-Gaussian Slow-Drift Responses. Proceedings of the 1st ISOPE Conference, Edinburgh, U.K., August 11-16.

Table 1. Principal vessel and mooring system parameters. See Figure 1 for the mooring system layout.

Vessel:	
Displacement	112000 tonnes
Length between perpendiculars	254 m
Beam	46 m
Draught	12 m
Fairlead aft of fore perpendicular	5 m
Fairlead below vessel baseline	1 m
Mooring System:	
Number of evenly spaced mooring lines	12
Number of sections for each mooring line, m	3
Total length of mooring line	935 m
Horizontal distance to anchor	888 m
Line pretension	980 kN
LF damping:	
Linear damping coefficient in surge	1000 kNs/m
Linear damping coefficient in sway	500 kNs/m
Quadratic damping coefficient in yaw	500 $kNms^2/m^2$

Table 2. Mooring line characteristics. The S-N curve parameter, K, is $K=K_T \cdot MBL^m$, where MBL is the Minimum Breaking Load.

Section (i)	Lower (1)	Centre (2)	Upper (3)
Type of component	Chain	Wire Rope	Chain
Length	800 m	120 m	15 m
Nominal diameter	152 mm	140 mm	152 mm
Minimum Breaking Load	15590 kN	16040 kN	15590 kN
Submerged weight	4.34 kN/m	0.86 kN/m	4.34 kN/m
Drag coeff. on nominal dia.	2.5	1.3	2.5
Number of components, n_i	1316	27	25
T-N curve parameter, K_T	370	731	370
T-N curve parameter, m	3.36	4.09	3.36

Table 3. Results for the most loaded mooring line - extreme condition. Response maxima are distributed according to Stansberg (1991) and Lie and Sødahl (1993). Annual FORM failure probability is 3.34E-3.

Variabl.	Distrib.	Mean	St.D.	Des. Point	Imp. Fac.
B_{QS}	LogNor.	1.0	0.10	1.032	1.8%
B_{DY}	LogNor.	1.0	0.10	1.002	0.1%
$C_1^{(i)}$	Fixed	1.0	-	-	-
$C_2^{(ij)}$	LogNor.	1.2	0.06	-	Incl. in U_S
U_S	Normal	0.0	1.00	-0.093	0.1%
$\hat{T}_l$	Eq. (11)	-	-	16080 kN	98.0%
H_S				13.2 m	Incl. in $\hat{T}_l$
T_P				13.7 s	Incl. in $\hat{T}_l$
U_W				28.9 m/s	Incl. in $\hat{T}_l$
U_C				54.0 cm/s	Incl. in $\hat{T}_l$
$\bar{x}$				6.9 m	Incl. in $\hat{T}_l$
R_1^*				35.3 m	Incl. in $\hat{T}_l$
R_2^*				0.9 m	Incl. in $\hat{T}_l$
$\hat{T}_{l,QS}$				13350 kN	Incl. in $\hat{T}_l$
$\hat{T}_{l,DY}$				2730 kN	Incl. in $\hat{T}_l$

Table 4. Results for the most loaded mooring line - extreme condition. Response maxima are distributed according to the Rayleigh distribution. Annual FORM failure probability is 2.32E-5.

Variabl.	Distrib.	Mean	St.D.	Des. Point	Imp. fac.
B_{QS}	LogNor.	1.0	0.10	1.060	2.4%
B_{DY}	LogNor.	1.0	0.10	1.014	0.2%
$C_1^{(i)}$	Fixed	1.0	-	-	-
$C_2^{(ij)}$	LogNor.	1.2	0.06	-	Incl. in U_S
U_S	Normal	0.0	1.00	-0.176	0.2%
$\hat{T}_l$	Eq. (11)	-	-	15740 kN	97.2%
H_S				17.3 m	Incl. in $\hat{T}_l$
T_P				14.3 s	Incl. in $\hat{T}_l$
U_W				37.3 m/s	Incl. in $\hat{T}_l$
U_C				65.0 cm/s	Incl. in $\hat{T}_l$
$\bar{x}$				10.5 m	Incl. in $\hat{T}_l$
R_1^*				29.4 m	Incl. in $\hat{T}_l$
R_2^*				1.1 m	Incl. in $\hat{T}_l$
$\hat{T}_{l,QS}$				11970 kN	Incl. in $\hat{T}_l$
$\hat{T}_{l,DY}$				3770 kN	Incl. in $\hat{T}_l$

Table 5. Parameter study of the uncertainty of $C=C_1 \cdot C_2$.

Case	Distrib.	COV for C_1	COV for C_2	Pf,1year
1	LogNor.	0 %	5 %	3.34E-3
2	LogNor.	0 %	10 %	6.40E-3
3	LogNor.	5 %	5 %	3.69E-3

Table 6. Results for the most critical mooring line - fatigue condition. Vanmarcke's bandwith parameter is 0.1. Service life τ = 25 years. FORM failure probability is 6.61E-3.

Variable	Distrib.	Mean	St.D.	Des. Point	Imp. Fac.
B_σ	LogNor.	1.0	0.15	1.260	43.2 %
B_V	Normal	1.0	0.07	1.016	0.8%
$\Delta^{(1j)}$	LogNor.	1.0	0.30	-	Incl. in U_S
$\Delta^{(2j)}$	LogNor.	1.0	0.20	-	Incl. in U_S
$\Delta^{(3j)}$	LogNor.	1.0	0.30	-	Incl. in U_S
$\log K^{(1j)}$	Normal	17.5	0.40	-	Incl. in U_S
$\log K^{(2j)}$	Normal	20.9	0.40	-	Incl. in U_S
$\log K^{(3j)}$	Normal	17.5	0.40	-	Incl. in U_S
$B_D^{(1j)}$	LogNor.	1.0	0.10	-	Incl. in U_S
$B_D^{(2j)}$	LogNor.	1.0	0.10	-	Incl. in U_S
$B_D^{(3j)}$	LogNor.	1.0	0.10	-	Incl. in U_S
U_S	Normal	0.0	1.00	-1.858	56.0%

Figure 1. Horizontal mooring system layout.

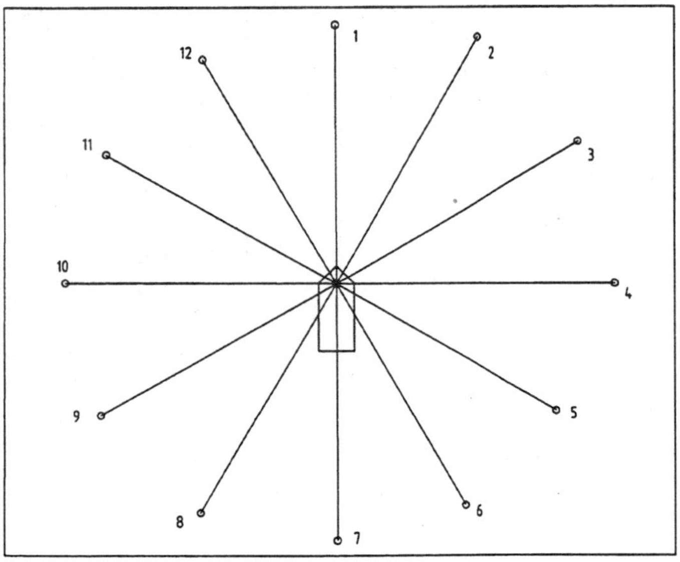

Figure 2. Piecewise constant response processes. Each response cycle is described by a constant value equal to the local maximum. Before modification.

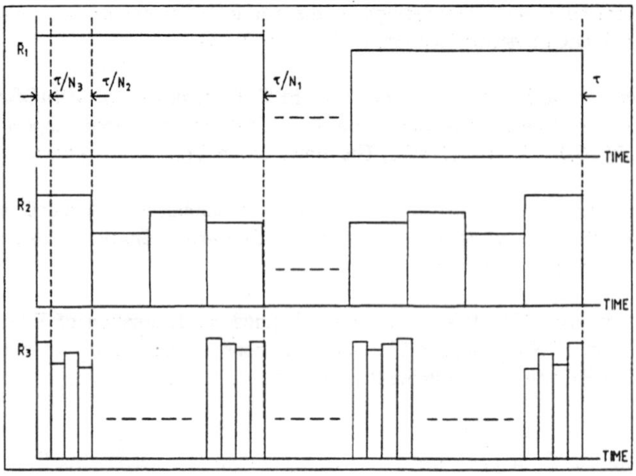

Figure 3. Piecewise constant response processes. Each response cycle is described by a constant value equal to the local maximum. After modification.

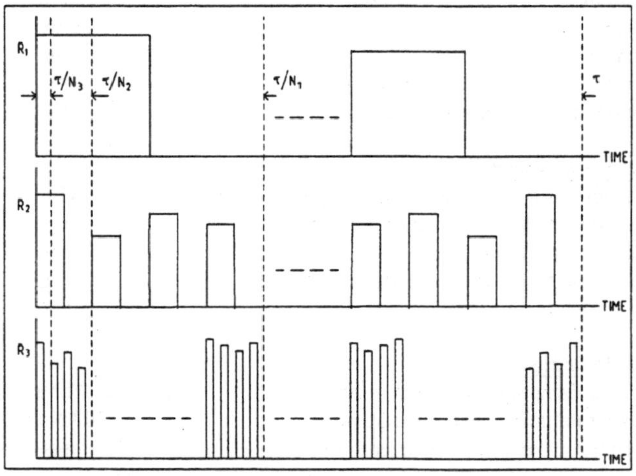

Figure 4. Cumulative distribution functions (FORM) for extreme tension at the upper end of the most loaded mooring line.

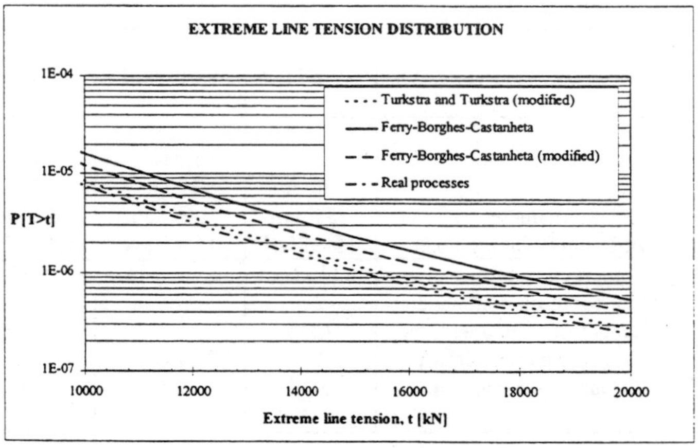

Proceedings of the Seventh (1997) International Offshore and Polar Engineering Conference
Honolulu, USA, May 25-30, 1997

ISBN 1-880653-28-1 (Set); ISBN 1-880653-32-X (Vol. IV)

Risk Analysis of the Vertical Offshore Breakwater at Gela, Italy

G. Passoni and A. Rigoni
Politecnico di Milano
Milan, Italy

L. Franco
3rd University of Rome
Rome, Italy

ABSTRACT

A reliability method (FOSM) has been used to assess the probability of sliding failure of the 33 years-old island caisson breakwater sheltering an oil terminal at Gela (Sicily).

The vertically composite structure is placed in just 13 m depth upon a weak silty seabed. It has experienced settlements of about 1 m, toe scour around 2 m and degradation of the slender reinforced superstructure, especially damaged after a severe storm on 24 November 1991.

On the basis of new detailed wave climate studies and field investigations a modern probabilistic analysis has been performed to verify the actual risk of failure, particularly against sliding over the rubble foundation due to wave loads.

A new formulation of the failure function as been derived introducing new expressions for the pressure coefficients in the Goda formula. A detailed sensitivity analysis of the failure probability with the caisson geometry has also been performed.

KEY WORDS: Vertical Breakwater, Stability, Reliability, FOSM.

INTRODUCTION

Conventional breakwater design in coastal engineering is mostly based on a deterministic approach, i.e. the design load is compared with the resistance of the structure. The assessment of the probability of failure is only performed by studying the probability distribution of the most important loading variable, the wave height (Franco et al., 1986; Lamberti, 1992; Burcharth, 1992; Van der Meer et al., 1995). On the contrary, the real behaviour of the structure is much more complex, mainly because other uncertainties are also due to those random variables which express the mechanical characteristics of the structure, hence its reliability. Therefore both a PIANC working group (PTC II-n. 28) and a EU-funded research project (MAST III Proverbs)are now being devoted to the probabilistic analysis of vertical breakwaters.

A first attempt was recently done by Christiani et al. (1996) assessing the reliability of a vertical breakwater as a series system of failures even if the geometrical parametrisation has been limited to the caisson width.

The possibility of an accurate assessment of the failure risk of a structure at sea is very difficult to be performed because all the joint probability density functions are generally unknown. Only for some of them the marginal probability density functions are known. When the random variables can be considered as independent (non correlated), the failure probability is obviously given by a n-fold integral, where n is the number of the independent variables. In practice, the evaluation of the failure probability via a direct multi-dimensional integration implies a huge amount of computations even for relatively small numbers n. In this respect an intermediate approach may be the so called FOSM (First Order Second Moment) methods, in which all the variables do exhibit a similar stochastic behaviour, i.e. reduced to the Gaussian distribution. This hypothesis is more reasonable for vertical type (caisson) breakwaters, at least when just considering the critical safety against sliding. The real occurrence of this characteristic failure case has been reported by Franco and Passoni (1994).

In the following section a brief description of the main probabilistic methods is given. Then a FOSM algorithm is applied to a specific real breakwater case which appears to survive with relatively small safety margin. Finally the numerical procedure has been iterated to explore the influence of the main geometric parameters on the risk of failure.

PROBABILISTIC APPROACH OF THE RELIABILITY ASSESSMENT

The evaluation of the safety of a structure, or its reliability, is accomplished by defining the failure modes and the corresponding "failure functions". Many failure events in marine engineering are characterised by different failure modes, but the attention can often been focused to the most important one, thus reducing to a single failure function. In the following the various approximations for the estimate of the failure probability are briefly summarised.

Level III Methods

Level III methods are based on the knowledge of the joint probability density function (jpdf) of the random variables which have to obey to the failure function $g(\underline{X})=\underline{R}-\underline{S}$. Denoting $f(\underline{X})$ the jpdf of a random variables vector $\underline{X}=(X_1, X_2, \ldots, X_n)$ the failure probability P_f is then given by the integral of the $f(\underline{X})$, over the subdomain in which the resistance R is lower than the load S, i.e.

$$P_f = \int_{R<S} f(x)dx \qquad (1)$$

For the simplest case of two independent random variables *R* and *S*, the previous integral (1) is reduced to

$$P_f = \iint_{R<S} f(r,s)drds \qquad (2)$$

In this case it is easier to indicate in a 3D plot the design point, as the most probable value of the variable $\underline{X}(\underline{R},\underline{S})$ (see Fig. 1).

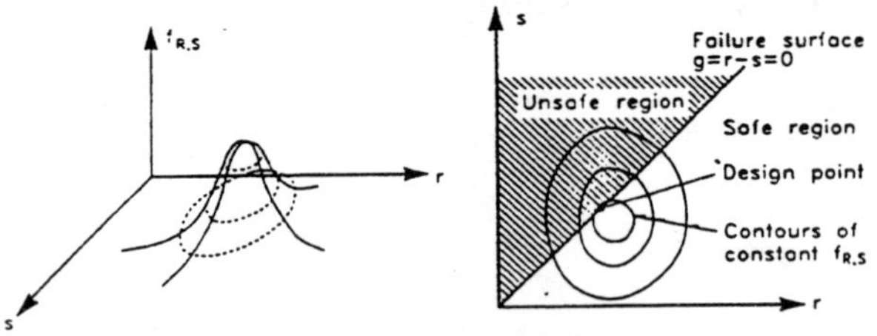

Fig. 1: Illustration of the two-dimensional *jpdf* for loading and strength (Burcharth, 1992)

Without sake of generality, let's consider a 2D case where the two random variables are independent, the previous integral may be casted in the form:

$$P_f = \iint_{R<S} f_R(r)\cdot f_S(s)drds \qquad (3)$$

which, after integration by parts, may be reduced to a 1D integral:

$$P_f = \int_0^\infty F_R(x)\cdot f_S(x)dx \qquad (4)$$

in which $F_R(x)$ is the cumulative distribution of *R*.

All the methods which fall into this class do express the reliability in terms of failure probability, which can be assessed by analytical solution for some simple cases, or numerical integration. Some examples are the First Order Reliability Methods (FORM) in which the failure hypersurface is approximated by the tangent hyperplane at a specific point, and Second Order Reliability Methods (SORM) in which a quadratic approximation of the failure hypersurface is adopted (Madsen et al., 1986).

Another simple way to compute the jpdf of the dependent variable g is the Monte Carlo method in which this goal is reached via the simulation of a large number of realisations of the stochastic process through a pseudo-random procedure applied to the random variables $\underline{X}=(X_1, X_2, \ldots\ldots, X_n)$, as reported by Ang and Tang (1984), Augusti et al. (1984), Casciati, Natale (1992), Lamberti (1992), Ditlevsen and Madsen (1996) and among others.

Level II Methods

In the engineering practice it is difficult to know the pdf of the vector $\underline{X}$, due to the lack of data, and furthermore the expression (2) is not trivial so that it can be integrated only by numerical quadrature. The most frequently known parameters of the distributions are the 1st (mean) and 2nd (standard deviation) order moments of the random variables, only. The methods which are based on this statistical information are the so called level II methods. Specifically they are referred to as reliability index methods. One such method is the First Order Second Moment method (FOSM) which has been used in this work.

Considering, for simplicity, a 2D case, and assuming as "failure" (or "performance") function the *safety margin* $g=R-S$, the limit state ($g=0$) between the safe ($g>0$) and the failure ($g<0$) regions is given by (see Fig. 2):

$$R'\sigma_r - S'\sigma_s + \mu_r - \mu_s = 0$$

where the primes indicate the normalised variables and σ, μ stand for standard deviation and mean values respectively.

In the normalised variables space the minimum distance β from the origin is a measure of the probability of failure (Cornell, 1969); in fact increasing β the area under the function f (r,s) will be larger, i.e. the probability P_f will be smaller.

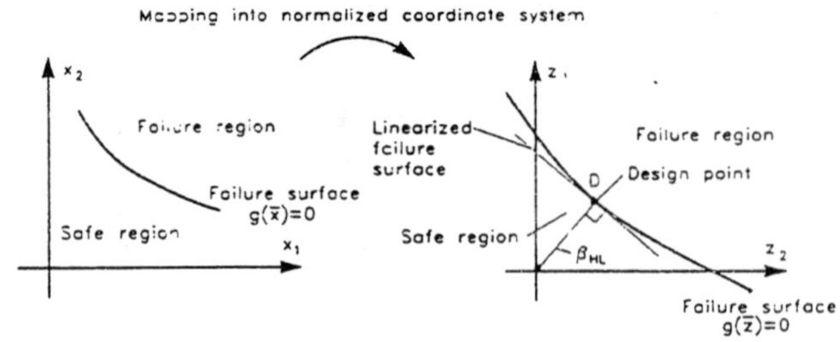

Fig. 2: Definition of the Hasofer and Lind reliability index β_{HL} (Burcharth, 1992)

From simple geometric considerations it's easy to verify that

$$\beta = (\mu_r - \mu_s)/(\sigma_r^2 + \sigma_s^2)^{1/2}$$

i.e. the ratio between the mean value of the function g and its standard deviation, assuming that g is a linear function of random independent variables. In this case the failure probability is given by:

$$P_f = \text{prob}(g \le 0) = \int_0^\infty f_g(x)dx = \qquad (5)$$

$$= \Phi\left(\frac{0-\mu_g}{\sigma_g}\right) = \Phi(-\beta)$$

Generally g is a function $g(\underline{X}) = g(X_1, X_2, \ldots, X_n)$ of several random variables, and $g(\underline{X})=0$ defines the limit of the safe state. If $\underline{X}=(X_1, X_2, \ldots, X_n)$ is a vector of n uncorrelated variables, after the normalisation in the n-space ($Z_i=(X_i-\mu_{Xi})/\sigma_{Xi}$) the reliability index becomes:

$$\beta_{HL} = min_{g(z)=0}\left(\sum_{i=1}^{n} z_i^2\right)^{1/2} \qquad (6)$$

i.e. the minimum distance of the limit state surface $g(\underline{X})=0$ from the origin of the normal space. The problem of finding the minimum distance of the function, subject to the constraint $g(\underline{X})=0$, has been approached by the method of Lagrange's multipliers (Harr, 1987; Ang and Tang, 1984) which, after some algebra, yields:

$$\beta = \frac{-\sum_i \left(z_i^*\right)\left(\frac{\partial g}{\partial Z_i}\right)_*}{\left(\sum_{i=1}^{n}\left(\frac{\partial g}{\partial Z_i}\right)_*^2\right)^{1/2}} \qquad (7)$$

where * stands for the point around which the function is approximated.

The same result may be obtained when looking at the first order term in the Taylor's series (8) of the function g where the mean value is given by the expression (9) and the approximation of the variance is given by expression (10) (Ang and Tang, 1984):

$$g(X_1, X_2, \ldots, X_n) \cong \sum_{i=1}^{n}(z_i^*)\left(\frac{\partial g}{\partial Z_i}\right)_* \qquad (8)$$

$$\mu_g \cong -\sum_{i=1}^{n}(z_i^*)\left(\frac{\partial g}{\partial Z_i}\right)_* \qquad (9)$$

$$\sigma_g^2 \cong \sum_{i=1}^{n} \left(\frac{\partial g}{\partial Z_i} \right)_*^2 \qquad (10)$$

and the ratio $\mu_g / \sigma_g = \beta_{HL}$ is the distance from the tangent plane of the failure surface at z_* to the origin of the normalised variates. The sensitivity of the reliability index with respect to the performance function (especially for non-linear functions g) has been overcome by first order approximations evaluated at a certain point (design point) on the failure surface (Hasofer and Lind, 1974). A similar procedure, may be applied also to correlated variates by mapping them into an uncorrelated set, after a coordinate transformation (Ang and Tang, 1984).

If some of the random variables are not normally distributed, the reliability can be assessed via the β_{HL} index, provided that a local equivalent Gaussian distribution can be found (Rackwitz, 1976). Such equivalent normal distribution can be also obtained using the Rosenblatt transform (Ang and Tang, 1984).

The numerical implementation of the procedure can be organised in different steps (Burcharth, 1992):

1) set some trial coordinates for the design point

$$Z^d = (z_1^d, z_2^d, \ldots, z_n^d)$$

2) computation of the derivatives at the design point $\alpha_i = \frac{\partial g}{\partial z_i}$

3) better estimate of the unknown vector

$$z_i^d = a_i \frac{\sum_{i=1}^{n} (a_i z_i^d) - g(Z^d)}{\sum_{i=1}^{n} (a_i)^2}$$

4) iteration of steps 2) and 3) until convergence in the coordinates is achieved.

5) estimation of the reliability index by the definition $\beta_{HL} = \left(\sum_{i=1}^{n} z_i^2 \right)^{1/2}$, as given in expression (6).

In order to avoid the convergence of the algorithm towards local minima, different sets of initial guesses for the vector Z^d have to be considered.

Level I Methods

The basic assumption in the Level I methods is the reduction of the system to a single random variable dependency. These are often quoted as deterministic/quasi probabilistic methods. With respect to maritime engineering application the wave height distribution is therefore taken as a direct measure of the probability of failure.

THE CASE OF GELA OFFSHORE BREAKWATER

The offshore breakwater of Gela industrial harbour is located inside the homonymous gulf on the Southwest coast of Sicily in the middle of the Mediterranean Sea (Fig. 3). It shelters a 2800 m long jetty and at the same time it is used as mooring berth for oil tankers. The design was carried out in 1962 and construction was completed in 1964.

This "island harbour" solution was selected due to the very shallow sandy seabed slope (0.4%). The water depth at and off the breakwater varies between 11 and 14 m MSL for a long distance offshore: a scour trench 150 m wide on average and up to 2 m deep now exists at the structure toe.

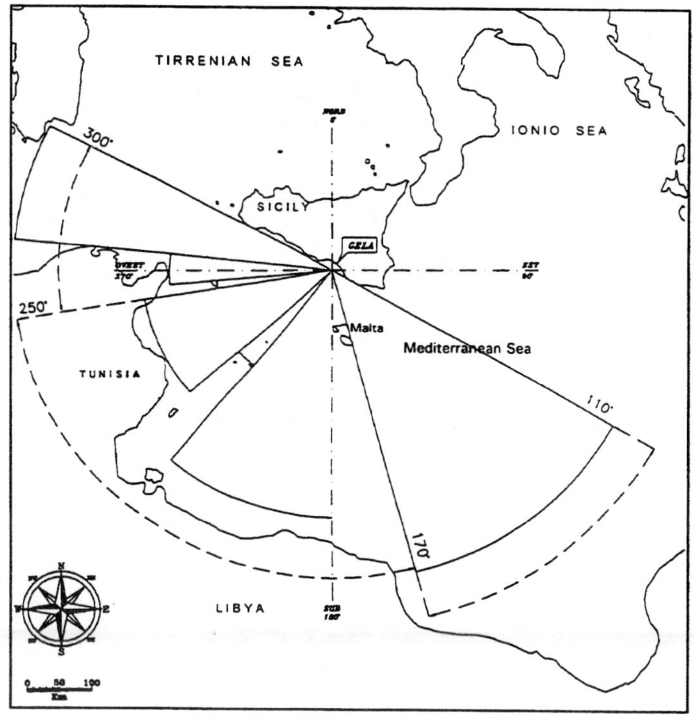

Fig. 3: Location map and fetches of Gela breakwater (AQUATER-Noli, 1996)

The breakwater planshape shows three differently aligned portions with a total length of 1160 m (Fig. 4). The breakwater is composed by 60 r.c. caissons sitting upon a thin (2 m) rubble foundation of 0.5-1.0 kN rock. The dimensions of each caisson are: length 20.4 m; width 14 m (b=17 m with the lateral bottom slab expansions); height hi=12 m (hi+Rc=17.5 m, including superstructure and parapet wall) (Fig. 5).

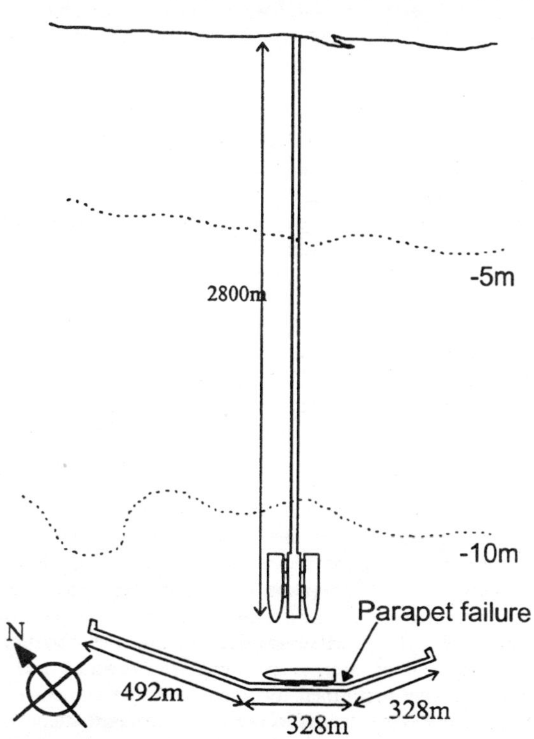

Fig. 4: Layout of Gela offshore breakwater

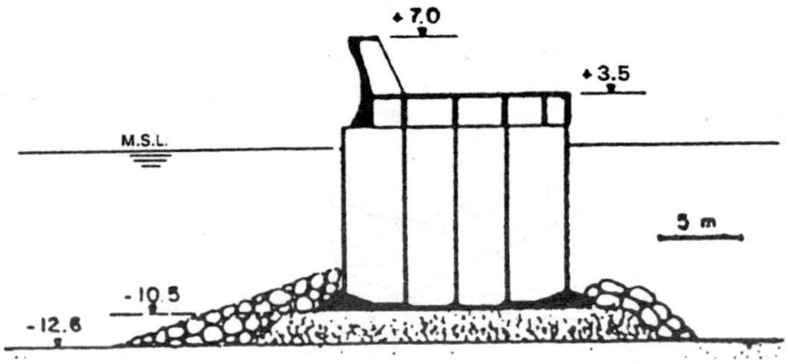

Fig. 5: Design cross section of Gela breakwater (Franco, 1994)

The caissons are divided in 24 cells, that are filled with a mixture of water and sand dredged from the seabed (unit weight of 19 kN/m^3). The thickness of the internal walls is 0.2 m, while the external ones are 0.3 m thick. Even the partly precast r.c. crown slab was made cellular and almost completely filled with sand, in order to reduce pressures on the seabed. The total volume per unit length is about 200 m^3/m. The total dry weighted averaged density results to be 20.2 kN/m^3 (2.06 t/m^3). Special attention was paid to the design of the parapet shape and elevation (originally +7.0 m MSL), in order to minimise wave overtopping and allow the loading and unloading operations on the rear quay. A thin curved r.c. crownwall is stiffened by light buttresses, again to reduce loads on the foundation.

The seabed stratigraphy shows up to 3-5 m sandy-silty soil; from 5 m to 40-45 m silty-clayey soil and below -45 m sandy-silty soil again (friction coefficient of 10° and shear resistance of 10 to 30 kN/m^2).

The geotechnical study made in 1961 gave a soil pressure from the structure of 170 kN/m^2. A load breaking-point of 145 kN/m^2 was obtained and a soil consolidation then suggested. A pretty uniform settlement of 0.7 - 1.0 m has recently been measured along the structure.

The original wave climate study and the model tests were performed by the French SOGREAH in 1963. Model tests with regular waves (T=7-11s) were conducted in 2-D (scale 1:50) and 3-D (scale 1:40) to check stability, overtopping and toe scour around the head.

The local spring tidal range is only 0.4 m. The wind setup is also small due to the narrow continental shelf. In storm conditions with strong persistent onshore winds and low atmospheric pressure a total water level setup of +0.5-0.6 m MSL can be assumed at the structure.

On 24 Nov. 1991 the breakwater superstructure (wave wall and pipelines behind) was damaged by a violent storm from SSW, when large overtopping occurred, although no caisson displacement has been observed. The estimated hindcasted deepwater significant wave height was around 7 m with a return period of about 50 years. The RON directional buoy off Mazara (some 150 km westward) recorded a maximum H_s = 6.15 m.

Some urgent repair work was carried out and a general rehabilitation project started up. A new detailed study of the local wave statistics has been recently performed (AQUATER-Noli, 1996) based on wave hindcasting models calibrated against buoy and satellite measurements. Wind input was gained from UKMO-Bracknell (10 years) and from Malta island meteorological station (28 years). Then the extreme wave height analysis has been carried out with the POT method (Goda, 1988) on the 28 years wave data set to produce directional and omnidirectional deepwater wave statistics: the Gumbel estimates obtained for a few selected return periods are listed in tab.1, together with the corresponding values of the standard deviation.

An advanced wave propagation model has been used to account for wave energy dissipation in the extended shallow water area (the isobath -50.0 m is about 10 km offshore) due to refraction, shoaling, friction and breaking and for energy spreading in frequency and direction. Computations were performed for three main offshore wave directions and three inshore water depths to derive wave height attenuation of 20-50%, mainly due to breaking dissipation.

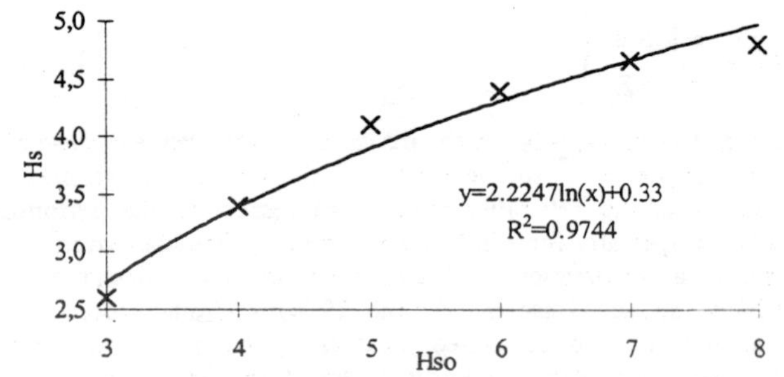

Fig. 6: Correlation between offshore and inshore significant wave height (in m) at 12m water depth

The resulting "transfer function" at a depth of -12 m is plotted in Fig. 6 and the actual values of Hs at the structure for the same return periods are also listed in tab.1, together with the corresponding standard deviations (reduced with the same ratio). Finally the corresponding mean values of the local Hmax (and their std. devs.) are obtained as 1.65 Hs, according to Goda, to account for a non Raleigh distribution of wave heights in shallow water (Table 1).

Tr	Hso	σHso	Hs	σHs	Hmax	σHmax
2	5.6	0.2	4.2	0.1	6.9	0.2
10	5.9	0.2	4.3	0.1	7.1	0.2
50	6.8	0.3	4.6	0.2	7.6	0.3
100	7.2	0.3	4.7	0.2	7.8	0.3

Table 1 Offshore (Hso) and inshore (Hs) extreme wave heights at Gela (all values in m except Tr in years)

FAILURE ANALYSIS AND RELIABILITY ASSESSMENT

A First Order Second Moment method has been used to assess the probability of failure of the vertical structure against horizontal sliding over the foundation. Other failure modes (e.g. overturning, slipping, settlement, scour etc.), which are usually less critical, are here neglected for sake of simplicity, even if they should be considered together their complex interactions in a full risk analysis of the breakwater.

The sliding performance function has been casted in the form:

$$g = f\left(\gamma_c V_{dry} - \gamma_w V_{wet} - UF_u\left(H_{max}\right)\right) - UF_h\left(H_{max}\right) \quad (11)$$

where:

f friction coefficient between caisson and rubble mound foundation (Gaussian variate)

γ_c specific weight of the caisson (Gaussian variate)

U uncertainty coefficient for the Goda wave pressure formula (Gaussian variate)

H_{max} maximum wave height (Gumbel variate)

are the four random variables, while F_u and F_h are expressions given by the complex formulae proposed by Goda (1985) for the vertical (i.e. uplift) and horizontal wave loads respectively and the remaining factors are known deterministically. The 1st and 2nd moments of the probability distributions are reported in the following Table 2:

random variables	μ	σ
f [#]	0.636	0.10
γ_c [kN/m³]	21.00	1.10
U [#]	1.00	0.20
H_{max} [m]	5.67	0.78

Table 2 Selected random variables for Gela caisson breakwater

The formal expressions of the functions F_u and F_h can include terms dependent on the wave length L, which is not known a priori. A linear approximation with respect to the wave height H_{max} is accurate enough (Rigoni, 1995), at least within a limited interval of practical interest (H_{max}=5-10 m, R^2>0.95) and the failure function becomes:

$$g = f\left(A\gamma_c - B\gamma_w - U\left(CH_{max} + DH_{max}^2 + EH_{max}^3\right)\right) + \qquad (12)$$
$$- U\left(FH_{max} + GH_{max}^2 + IH_{max}^3 + JH_{max}^4 + L\right)$$

where the coefficients A, B, C, D, E, F, G, J, I, L depend also on the geometrical parameters. The dependency on H_{max} is of the third order, with respect to the uplift force, and of the fourth order for the horizontal contribution. Different expressions are reported in the recent literature (Burcharth et al., 1994) mainly because in structural geometry and in the parametrisation of the Goda's pressure formulae were introduced (e.g. constant wave steepness).

The convergence of the numerical procedure, described in section 2, has been carefully tested with several trial values of the random variables, even if this convergence is not always achieved nor theoretically proved. Finally a threshold error for the g function was fixed of the order $|10^{-4}|$.

Some comparisons have been performed with other authors' results (Sorensen, private communication) and also with other numerical solvers for constrained non-linear optimisation problems (Microsoft-Excel 5.0, 1993; Wolfram, 1991).

The numerical procedure has been first used to determine the probability of failure for the present geometry of the Gela caisson breakwater (considering a storm water level setup of 0.6 m and a settlement of 0.7 m). The result of this Level II computations give Pf = 0.016 (1/year), which corresponds to a risk of failure R=55 % in a 50 years lifetime Tv (15 % in 10 years, 41 % in 33 years) since $R(Tv)=1-(1-Pf)^{Tv}$.

The same analysis has also been performed with a traditional Level I (deterministic) approach, but in this case no failure has been predicted because only a very high wave height (probably depth limited) could determine g=0 for the failure function.

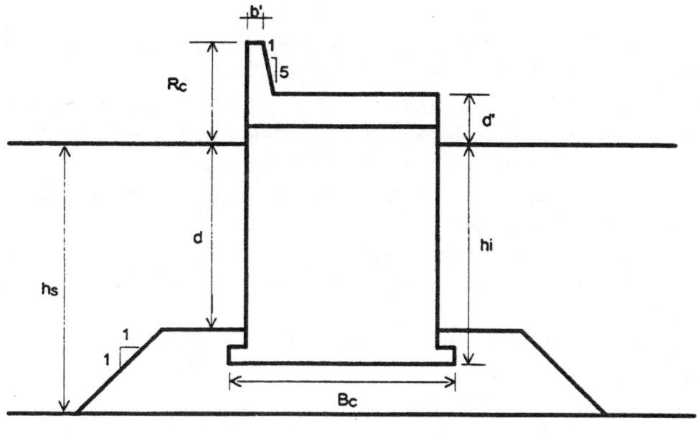

R_c=5.7m	B_c=17m	d=9.7m	h_s=13.1m
hi=11.8m	d'=2.2m	b'=2.0m	

Fig. 7: Cross section of the schematised caisson breakwater.

SENSITIVITY ANALYSIS OF Pf ON THE BREAKWATER GEOMETRY

A sensitivity analysis has then been performed in the space of the geometrical parameters in order to ascertain the influence of some possible design changes on the failure probability. For this purpose the structure has been schematised as indicated in Fig. 7 with the following set of non dimensional variables:

$$Rc^* = \frac{Rc - d'}{hs},\ Bc^* = \frac{Bc}{hs},\ d^* = \frac{d}{hs},\ d'^* = \frac{d'}{hs},\ hi^* = \frac{hi}{hs}$$

The intervals of the parameters have been chosen in order to include the actual geometry of Gel caisson breakwater, whose values are Rc^*=0.27, Bc^*=1.30, d^*=0.74, hs=13.10 m, d'^*=0.17, hi^*=0.90. The ranges are: 0.15<Rc^*<0.46, 1.15<Bc^*<1.45, while all the parameters within brackets have been set constant in order to detect the effect of the parapet height alone. Different sets of variables have been investigated.

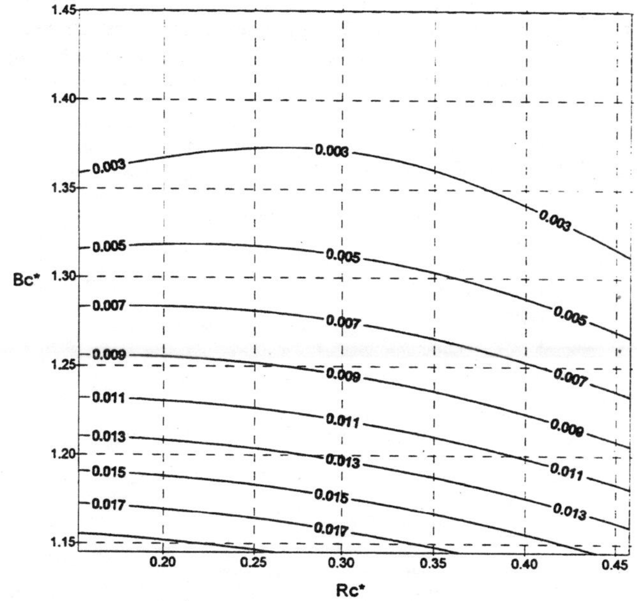

Fig. 8: Contour plot of P_f vs. Rc^* and Bc^* (d^*=0.74, d'^*=0.33, h_i^*=0.74)

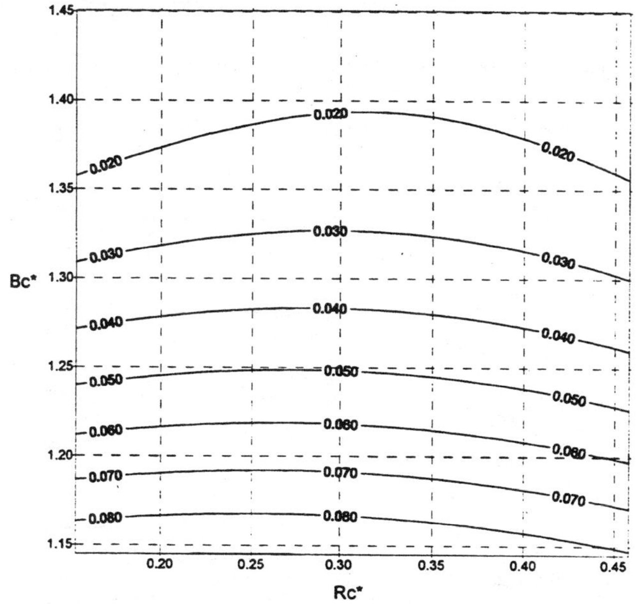

Fig. 9: Contour plot of P_f vs. Rc^* and Bc^* (d^*=0.74, d'^*=0.17, h_i^*=0.9)

First the failure probability has been computed for different values of Rc^* and Bc^* while the remaining ones at some intermediate values have been set constant (d^*, d'^* and h_i^*) for the case of the same caisson simply placed ($d^*=h_i^*$) over the rubble mound foundation (Fig. 8) or corresponding to the actual ones ($d^*=0.74$, $h_i^*=0.17$) of Gela breakwater (Fig. 9). The behaviour of the curves exhibits a trend of slightly increasing *Pf* for lower *Rc** and smoothly decreasing for larger *Rc**. Anyway the gradient of *Pf* exhibits a much larger component in the *Bc** direction, as obviously explained by the net weight increase with the caisson width. The negligible variability in the *Rc** direction is due to the slight increase of the horizontal load, which is counterbalanced by the increase of the total weight for larger *Rc**, thus determining a final reduction of *Pf*. The results presented in Fig. 9 refer to the actual overall geometry, i.e. a caisson whose bottom is lower than the berm elevation, while the other geometrical factors remain the same. The trend of the curves is similar, but the *Pf* values are much higher in this case which can be also explained by the increase of the buoyancy.

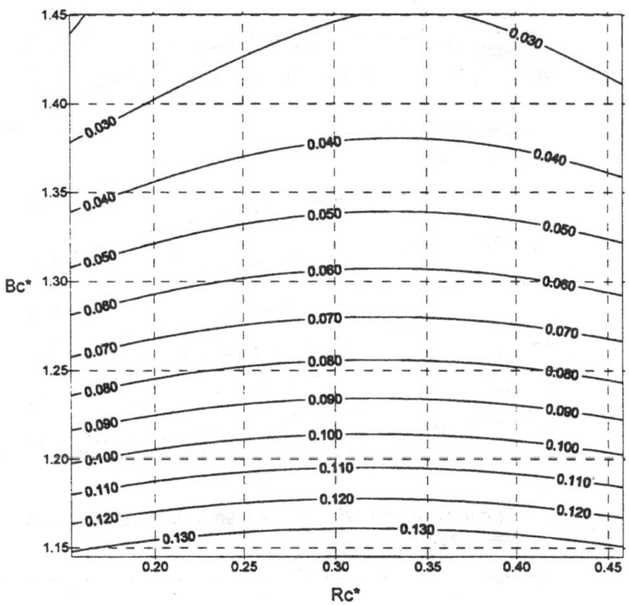

Fig. 10: Contour plot of P_f vs. hc^* and b^* ($d^*=1.0$, $d'^*=0.07$, $h_i^*=1.0$)

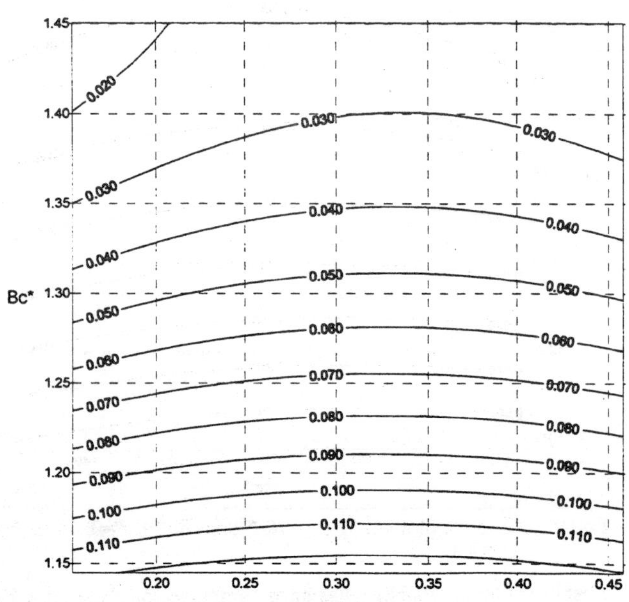

Fig. 11: Contour plot of P_f vs. hc^* and b^* ($d^*=1.0$, $d'^*=0.07$, $h_i^*=1.0$)

The horizontal loads are larger because of the additional contribution of the dynamic pressure acting in the zone between the berm crest and the caisson's bottom. Even larger *Pf* values result when the caisson is placed directly over the sea bottom without any rubble mound foundation, as indicated in Fig. 10 where $d^*=1.0$ and $h_i^*=1.0$.

The last geometry has also been tested using different expressions for the pressure coefficients (α_1 α_2 α_3) in the Goda's formula. Namely in Fig. 10, as also in the previous ones, the coefficients have been expressed as linear function of the design wave height as quoted in Rigoni (1995). The results in Fig. 11, on the contrary, have been obtained under the assumption of constant Goda's pressure coefficients for varying wave height and the difference of the results due to this approximation appears not negligible.

CONCLUSIONS

The application of a FOSM probabilistic computation for the stability against sliding of an existing 33-year old caisson breakwater off Gela (Sicily) has produced a quite large risk of failure (55% in 50 year lifetime).

The influence of some main geometrical parameters on the probability of failure has been studied with a parametric analysis. The elevation of the superstructure has a negligible influence on the probability of failure against sliding, under the particular environmental conditions considered in this case. This means that the safety against overtopping is not in contradiction with the global stability of the structure.

Other parameters, like the berm geometry, should be considered in a more general stability analysis of a caisson type breakwater together with the extension of the model to other failure modes. The dependency of the algorithm on the order of the failure function has also to be investigated.

ACKNOWLEDGEMENTS

This work has been carried out with the support of the EU funds within the MAST3-CT95-0041 PROVERBS research project. It was also stimulated by the PIANC-PTC II-Working Group 28 devoted to the analysis of vertical breakwaters (chaired by Prof. H.F.Burcharth).

The authors wish to thank Prof. J.D.Sorensen and Mr. E.Christiani (Aalborg University) for the fruitful discussions on the verification of the numerical procedure.

The permission to report the results of the new studies for the rehabilitation of Gela breakwater by AQUATER is gratefully acknowledged.

A special thank is also devoted to Ing. C.Restano for his kind assistance and advice.

REFERENCES

Ang A.H-S., Tang W.H.(1984) "Probability Concepts in Engineering Planning and Design", *Decision, Risk and Reliability,* Vol. II, John Wiley&Sons, New York.

AQUATER, Noli A., (1996) "Modelling studies for the rehabilitation design of Gela Breakwater", Int. Report (in Italian)

Augusti G., Baratta A., Casciati F. (1984) "Probabilistic Methods in Structural Engineering", Chapman and Hall, London-New York

Burcharth H.F. (1992) "Reliability Evaluation of a Structure at Sea", *Short Course on design and Reliability of Coastal Structures*, Ed. A. Lamberti, 23rd ICCE, October 1992, Venice.

Burcharth H.F., Sorensen J.D., Christiani E. (1994) "On the Evaluation of Failure Probability of Monolithic Vertical Wall Breakwater", *Intl. Workshop "Wave Barriers in Deepwaters"*, Port and Harbour Res. Institute, January 1994, Yokosuka (Japan).

Casciati F., Natale L. (1992) "Risk Analysis for Marine Systems: an Introduction", *Short Course on design and Reliability of Coastal Structures*, Ed. A. Lamberti, 23rd ICCE, October 1992, Venice.

Christiani E., Burcharth H.F., Sørensen J. D. (1996) "Reliability based optimal design of vertical breakwaters modelled as series system of failure", Proc. 25rd ICCE, Orlando (USA), ASCE NY.

Cornell C.A. (1969) "Structural Safety Specification Based on Second-Moment Reliability", *Sym. Int. Assoc. Bridge and Struct. Engr.*, London.

Ditlevsen O., Madsen H. O. (1996) "Structural Reliability Methods", John Wiley & Sons, Chichester (GB).

Franco L., Lamberti A., Noli A., Tomasicchio U. (1986) "Evaluation of Risk applied to the designed Breakwater of Punta Riso at Brindisi, Italy", *Coastal Engineering*, No. 10.

Franco L., Passoni G. (1994) "The Failure of a Caisson Breakwater in Neaples Harbour", *Bauingenieur*, Band 69, 73-80, Springer Verlag (in German), also in *Proc. 3rd Workshop MAST G6-S Project 2*, Franzius Institut, August 1992, Hannover.

Franco L. (1994) "Vertical Breakwaters: the Italian experience", *Coastal Engineering*, n.22

Goda Y. (1985) "Random Seas and Design of Maritime Structures", Tokyo University Press.

Goda Y. (1988) "On the methodology of selecting design wave height", *Proc. 21st ICCE*, Malaga, ASCE NY.

Harr M.E.(1987) "Reliability-Based Design in Civil Engineering", McGraw-Hill Book Company, New York.

Hasofer A.M., Lind N. (1974) "An Exact and Invariant First-Order Reliability Format", *J. of Engineering Mechanics*, ASCE, Vol. 100, EM1, February 1974, pp. 111-121.

Lamberti A. (1992) "Example Application of Reliability Assessment of Coastal Structures", *Short Course on design and Reliability of Coastal Structures*, Ed. A. Lamberti, 23rd ICCE, October 1992, Venice.

Madsen H. O., Krenk S., Lind N. C. (1986) "Methods of Structural Safety", Prentice-Hall, Inc., Englewood Cliffs, NJ 07632

Rackwitz R. (1976) "Practical Probabilistic Approach to Design", *Bulletin 112*, Comite European du Beton, Paris (F).

Rigoni A. (1995) "Probabilistic Analysis of Vertical Wall Structures", (in Italian) Master Thesis in Environmental. Eng., DIIAR-Politecnico di Milano, December 1995, Milan (Italy).

Takayama T., (1992) "Estimation of sliding Failure Probability of present Breakwaters for Probabilistic Design", *Report of Port Harbour Research Institute,* Vol. 31 N° 5.

Van der Meer J. W., d'Angremond K., Juhl J. (1995) "Probabilistic Calculation of Wave Forces on Vertical Structures", *Publication No. 487 Delft Hydraulics*, February 1995, Emmelord (NL).

Wolfram S. (1991) "Mathematica", Wolfram Res. Inc.©, Addison-Wesley Publishing Company Inc.

Proceedings of the Seventh (1997) International Offshore and Polar Engineering Conference
Honolulu, USA, May 25-30, 1997

ISBN 1-880653-28-1 (Set); ISBN 1-880653-32-X (Vol. IV)

Probabilistic Calibrated Design Equation for Burst Strength Assessment of Corroded Pipes

O.H. Bjørnøy, E.H. Cramer and G. Sigurdsson
Det Norske Veritas
Høvik, Norway

ABSTRACT

The commonly used criterion for corrosion damage assessment of pipes ANSI/ASME B31G is not in harmony with modern design philosophy for evaluation of the maximum allowable operating pressure, requiring uniform reliability against burst at a defined adequate level for different pipe materials, pipe geometries and corrosion configurations.

In this paper work conducted in the Det Norske Veritas (DNV) joint industry research project titled *Reliability of Corroded Pipes* is described. The goal of the project has been to develop a guideline with probabilistic calibrated design equations for determination of the allowable operating pressure of corroded pipes.

The proposed DNV code is based on an extensive number of FE analyses and full scale laboratory tests. The analyses and laboratory tests performed cover the internal pressure loading condition and combined loading from internal pressure and external loads. Both longitudinal and circumferential corrosion of the pipe have been considered.

The safety factor based design equations account for uncertainties associated with the determination of the largest annual operating pressure, the material properties and the assessment of the degree of corrosion in the pipe. As the developed design criteria account directly for the corrosion assessment accuracy, the benefit of improving the accuracy level of the inspections, for example intelligent pigging, can be utilised through increased allowable pressure.

The partial safety factors in the design criteria are defined for three reliability levels. The design criteria will therefore allow the pipeline operators to directly determine the maximum allowable pressure for a selected reliability level for the pipe in question, depending on the degree of corrosion.

KEY WORDS: pipeline, burst, reliability, code calibration, corrosion

BACKGROUND

A pipeline is a large financial assets for the pipeline operator and a safe operation of the pipeline is therefore of great concern. On the other hand, unnecessary repair and an over conservative operation of the pipeline may result in high costs and unexploded resource utilisation. As the pipelines are ageing and corrosion may develop, the economical consequences of reduced operation pressure, repairs, or replacements may become high. Available design equations for assessment of allowable operating pressures of degraded pipelines depending on the selected reliability level is therefore desirable.

When severe corrosion has been observed in a pipeline, the decision of necessary action to be carried out should be based on an overall assessment of the pipeline, where uncertainties associated with both the assessment of the degree of corrosion and the capacity evaluation should be considered. The cost of repair or replacement can be very high and should be avoided, or postponed in time, if this is possible within the safety requirements defined. Required actions should further also be initiated in order to maintain the integrity of the pipeline and to avoid an undesired risk exposure of the pipeline.

The commonly applied criterion for assessment of corroded pipes, ANSI/ASME B31G (ASME, 1991) was initially based on the NG-18 equation (Maxey et al, 1971) adjusted to account for available experimental data. The NG18 equation is defined as:

$$P = \frac{\sigma_{flow} \cdot 2 \cdot t}{D} \left[\frac{1 - \dfrac{A}{A_0}}{1 - \dfrac{A}{A_0} \cdot \dfrac{1}{M}} \right] \qquad (1)$$

where

$$A_0 = d \cdot t$$

and

M	:	Folias bulging factor, accounting for effect of stress-concentration at notch
P	:	failure pressure
σ_{flow}	:	flow stress
D	:	pipe outer diameter
A	:	projected corroded area
t	:	nominal pipe wall thickness
d	:	maximum corrosion depth

In the B31G criterion the projected corrosion area is assumed to be parabolic, and hence the projected corroded area is 2/3 $d{\cdot}t$. However, for long defects this assumption will obviously overpredict the capacity and a rectangular shape is assumed. The flow stress is limited to 10% higher than the specified minimum yield stress (SMYS). The B31G burst equation for Safe Maximum Pressure P' is defined as,

$$P' = 1.1P \cdot \left[\frac{1 - \frac{2}{3} \cdot \frac{d}{t}}{1 - \frac{2}{3} \cdot \frac{d}{t} \cdot \frac{1}{M}} \right] \quad \text{for } \sqrt{0.8} \cdot X \leq 4.0 \tag{2}$$

$$P' = 1.1P \cdot \left[1 - \frac{d}{t} \right] \quad \text{for } \sqrt{0.8} \cdot X > 4.0$$

where

$$P = \frac{SMYS \cdot 2t}{D} \cdot F \tag{3}$$

$$M = \sqrt{1 + 0.8 \cdot X^2} \quad \text{Folias bulging factor} \tag{4}$$

$$X = \frac{L}{\sqrt{D \cdot t}} \quad \text{Characteristic corrosion length} \tag{5}$$

and F is the Design factor normally equal to 0.72.

P' may, however, not exceed P (maximum allowable design pressure for uncorroded pipe).

No further evaluation is needed for corrosion less than 10 % of the wall thickness, and corrosion above 80 % is not accepted.

In Figure 1, the safe maximum pressure level based on B31G is shown for corrosion depth of 0.3, 0.5 and 0.7 of the wall thickness.

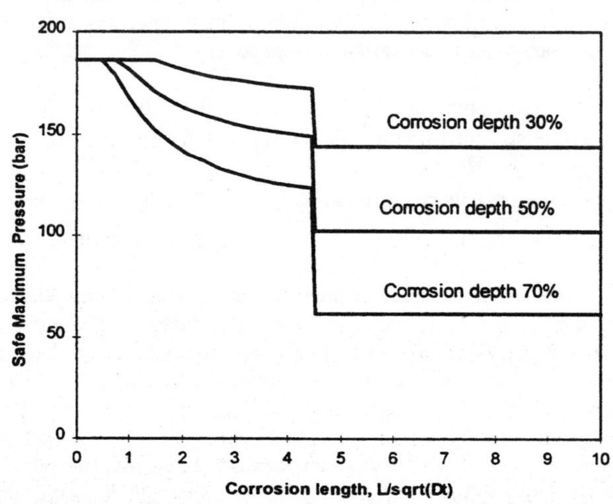

Figure 1 Safe maximum pressure according to B31G.

Several modifications have been proposed to improve the NG18/B31G criterion in order to better predict the actual failure pressure. These modifications have, however, mainly been based on modification changes of the equation parameters. An overview of some of these modifications to the flow stress, the bulging factor M and the procedure for estimating the projected corrosion area A is given below, (Denys, 1995).

Variations of proposed flow stress:

σ_{flow} = 1.1 SMYS
σ_{flow} = 1.15 SMYS
σ_{flow} = 0.5 (SMYS + SMTS)
σ_{flow} = SMYS + 69 MPa
σ_{flow} = a · SMTS, where a = 0.90, 1.0 or 1.1 and SMTS is the Specified Minimum Tensile Strength.

Variations of proposed projected corrosion area definition:

A = $d{\cdot}l$ (rectangular)
A = 2/3 $d{\cdot}l$ (parabolic)
A = 0.85 $d{\cdot}l$ (approx. average of rectangular and parabolic)
A = "exact" calculation (RSTRENG, Kiefner and Vieth, 1989)

Variations of proposed definition of bulging factor;

- Maxey et al (1971)

$$M = \sqrt{1 + 6.28 \cdot 10^{-1} \cdot X^2 - 3.38 \cdot 10^{-3} \cdot X^4} \tag{6}$$

- Kiefner (1974)

$$M = \sqrt{1 + 0.8 \cdot X^2} \tag{7}$$

- Kiefner and Vieth (1989)

$$M = \begin{cases} \sqrt{1 + 6.28 \cdot 10^{-1} \cdot X^2 - 3.38 \cdot 10^{-3} \cdot X^4} & X \leq \sqrt{50} \\ 3.3 + 3.2 \cdot 10^{-2} \cdot X^2 & X > \sqrt{50} \end{cases} \tag{8}$$

The drawback with modifying one or more parameters in the B31G equation in order to obtain a better adaptation to existing and newer results, is that it will most likely result in a negative effect for other design cases (geometries and corrosion configurations).

The inherent reliability level in B31G is very dependent on the length and shape of the corrosion. In Figure 2, the annual failure probability is estimated for rectangular corrosion shape for different corrosion depths as function of the corrosion length applying B31G. Uncertainties in the corrosion sizing accuracy are included using the specification from the British Gas 020 system. As can be seen in the figure, for a given measured corrosion depth the probability of failure changes significantly as function of the corrosion length. It should further be noticed the dramatic inconsistency in the reliability level for the different corrosion depths, e.g. for long corrosion the probability of failure is approximately 10-6 for d/t = 0.3, but more than 10-2 for d/t = 0.7.

The inconsistent behaviour of the inherent reliability level for various corrosion depths and corrosion lengths in existing codes has been the incitement for the DNV joint industry research project Reliability of Corroded Pipes, described in the following.

In order to obtain a better expression for the burst capacity for all actual design combinations, a new burst capacity equation has in this project been defined based on obtained data material on burst capacities and the use of advanced analysis procedures for transformation of these data to an applicable well behaving burst capacity formulations.

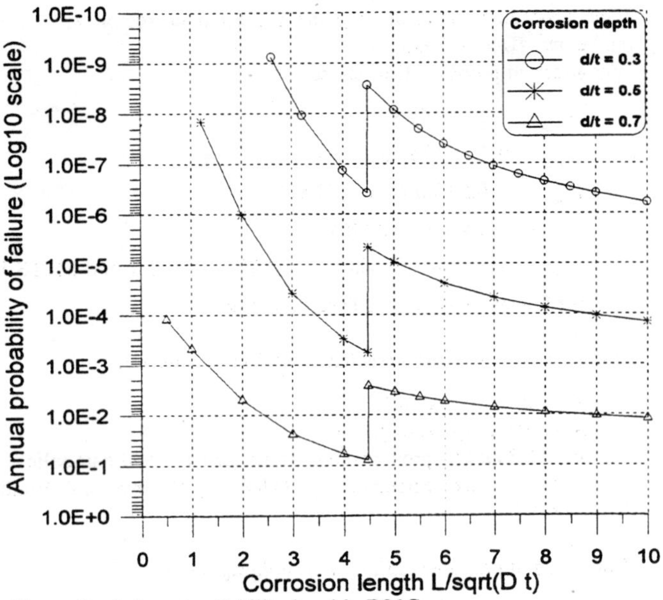

Figure 2 Inherent reliability level in B31G

With basis in the defined burst capacity equation and an understanding of the uncertainties involved in the determination of the pressure exposure, the burst capacity and the degree of corrosion, burst design equations with defined reliability levels for use in reassessment of corroded pipes have been developed.

BURST CAPACITY OF CORRODED PIPES

General

It was early recognised that the available published burst data were not alone sufficient as basis for an improvement of the existing burst capacity equations of corroded pipes. In order to establish an equation able to predict the burst capacity of corroded pipes for a wide range of pipe dimensions and corrosion conditions, new data material was therefore required.

In order to efficiently obtain reliable estimates for the burst capacities together with good predictions of the influence of parameter variations, the burst capacity was estimated based on a combination of a large number of finite element analyses and a series of full scale laboratory tests.

The finite element analyses were carried out using three-dimensional models with fully non-linear capabilities applying the program system ABAQUS, (Hibbitt, Karlsson and Sorensen, 1995). Approximately 200 analyses were carried in order to establish the burst capacity equation, where variations in pipe geometry *D/t*, corrosion length, corrosion depth and material characteristics were intensively studied. In addition to these parametric studies, the influence of various corrosion shapes, internal versus external corrosion, mesh density, solution strategy and 2D analyses for long corrosion were investigated. The failure criterion for the finite element analyses predicting the burst pressure was based on the work by British Gas (Fu and Kirkwood, 1995), and confirmed by laboratory tests.

The laboratory tests carried out within the DNV project cover variations in the corrosion geometries as well as different combinations of internal pressure loading and external axial and bending loads, where a total of 12 tests were initially carried out. The calibration of the expression for the burst capacity comprised both the outcome of the laboratory tests carried out within the project as well as available burst capacity results from the literature, (Chouchaoui and Pick, 1992), (Coulson and Wolthingham, 1991).

Burst Capacity Equation

The expression for the defined burst capacity was established with basis in the initial form of the NG-18 and the B31G equations in order to obtain continuity with previous work within the field, as the form of these equation was found suitable for describing the burst capacity. The expressions were, however, calibrated and adjusted in accordance with the results from the finite element analyses and the laboratory burst tests.

The established burst capacity expression is,

$$P_{fit} = P_0 \cdot \frac{1 - \frac{d'}{t}}{1 - \frac{d'}{t} \cdot \frac{1}{M}} \cdot H \tag{9}$$

where P_0 is a normalising tensile strength hoop pressure,

$$P_0 = \frac{2 \cdot t}{D - t} \cdot \sigma_u \tag{10}$$

and

- σ_u : ultimate tensile strength
- d' : equivalent corrosion depth, accounting for the corrosion shape
- t : pipe wall thickness
- D : pipe outer diameter
- M : bulging factor accounting for the length and the projected shape of the corrosion
- H : adjustment factor accounting for the influence of combined loading, the yield stress and tensile strength ratio. The adjustment for the ratio also dependent on the length and projected shape of the corrosion.

The burst capacity equation was defined based on a calibration to the results from approximately 200 finite element analyses, where the following parameter range variation were covered;

Characteristic corrosion depth, *d/t*	: 0.15, 0.3, 0.5, 0.7
Characteristic corrosion length, $L/\sqrt{D \cdot t}$	: 0.6 - 30
Tensile strength, σ_u	: 520 - 727 MPa
Yield stress tensile strength ratio σ_y / σ_u	: 0.7, 0.8, 0.9
Projected corrosion shape	: rectangular, parabolic

The burst capacity equation predicted the obtained finite element results extremely well, resulting in an unbiased factor with Coefficient of Variation (CoV) of only 1.5%. In Figure 3 the relationship between the capacity from the burst capacity equation and the finite element analyses is presented for internal pressure loading.

The burst capacity equation was also compared with the obtained laboratory test results and available published test results, as shown in Figure 4. The number of test results to be compared with the burst capacity equation is, however, limited as most of the published tests in the literature lack information on critical parameters, e.g. the actual yield stress or the ultimate tensile strength. The burst capacity equation is compared with tests from the AGA database (Kiefner and Vieth, 1994) and other published test results (Chouchaoui and Pick, 1992), (Coulson and Wolthingham, 1991). The comparison shows that the burst strength equation predicts the obtained laboratory capacity with good accuracy.

The burst capacity equation as function of the corrosion length and normalised with the tensile strength hoop pressure P_0 is shown in Figure 5 for different degrees of corrosion depth, having rectangular corrosion and a yield stress-tensile strength ratio of 0.8.

CALIBRATION OF CODE EQUATION

General

The burst capacity equation predicts the capacity with good accuracy for specified pipe material, pipe geometry, and corrosion characteristics as corrosion depth, length and shape. However, in real life some, or all, of these aspects are difficult to predict without some degree of uncertainty.

In order to establish a firm decision basis for the initiation of a possible action when corrosion has been observed in a pipeline, these uncertainties must rationally be accounted for.

The applicability of the design codes for corroded pipelines commonly used today are weakened due to an inconsistent reliability level against burst failure for various extent of corrosion and the accuracy of the corrosion assessment. It has, e.g. been observed that these codes are generally more conservative for ground corrosion than for deeper corrosion, see Figure 2.

In order to avoid an unpredictability in the reliability level applying the design codes, it is important that the codes reflect a known and consistent reliability against failure for all levels of corrosion.

Background for Code Development

The main requirements to a code equation is that is should reflect a defined and uniform reliability level for all design cases within the validity range of the code. As the different levels of uncertainty typically will have varying influence on the design from design case to design case, the development of code equations satisfying these reliability level requirements requires a detailed knowledge of the uncertainties involved and how these uncertainties affects the burst capacity of the pipe.

The choice of characteristic values representing the uncertain variables and the associated partial safety factors in the code equation must also be defined with carefulness in order to assure that the spreading in the uncertainty for the ranges of realistic design cases are contained within the requirements to uniformness in the resulting reliability level.

The uncertainties affecting the determination of the acceptable annual maximum operating pressure for a corroded pipeline is typically uncertainty associated with the determination of,

- the extent of corrosion (depth, length and shape),
- the pipe material characteristics (actual yield stress and tensile strength),
- the pipe geometry (e.g. pipe thickness)
- the operation pressure (daily variations and efficiency of pressure control systems),
- the goodness of the applied capacity prediction models.

The modelling of the uncertainties associated with the assessment of the degree of corrosion depends on the goodness of the applied inspection method and the type of corrosion. As these uncertainties to a large extend will affect the estimated burst capacity, the developed design code equations are defined for different levels of corrosion assessment accuracy, both with respect to the prediction of the corrosion depth and the corrosion length.

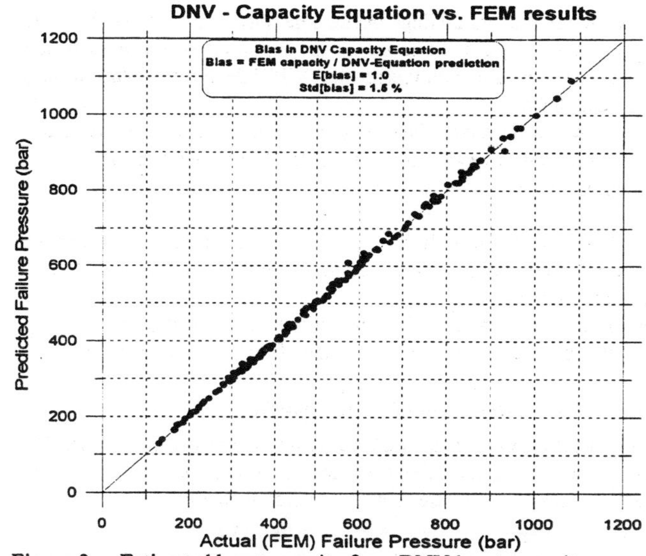

Figure 3 Estimated burst capacity from DNV burst capacity equation versus finite element results.

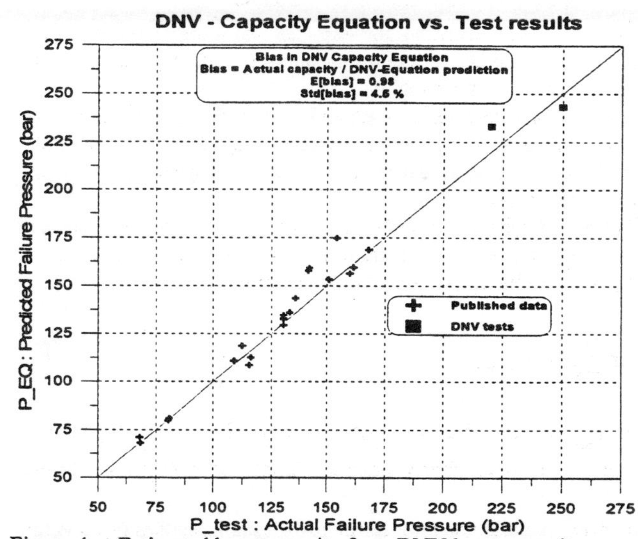

Figure 4 Estimated burst capacity from DNV burst capacity equation versus available laboratory test results.

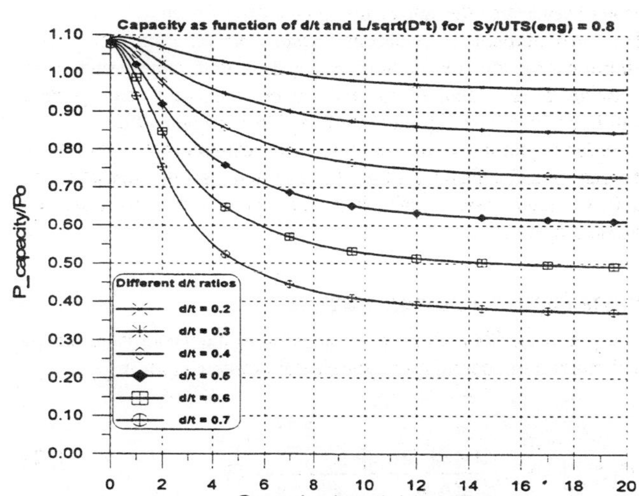

Figure 5 Burst capacity of depending on extent of corrosion.

The uncertainty modelling for the pipe material characteristics, the pipe geometry and the variations in the annual operation pressure are defined based on the findings in the SUPERB project (Jiao, Sotberg and Igland, 1995).

The uncertainties associated with the ability of the developed burst capacity prediction model to represent the "true" burst capacity were defined in two steps;

- Modelling of the ability of the defined burst capacity equation to represent the burst capacity obtained from the intensive 3D finite element analyses.
- Modelling of the ability of the finite element analyses to represent the burst capacity obtained from the laboratory tests.

Procedure for Calibration of Code Equation

In order to embrace all realistic design scenarios, more than 300 design cases were considered in the calibration of the burst code equation. These design cases covered variations in the:

- pipe material: X42 - X80
- measured corrosion depths: $d/t = 0.15 - 0.7$
- measured corrosion lengths: $L/\sqrt{D \cdot t} = 0.6 - 30$
- corrosion shape: rectangular and parabolic

The burst code equation for corroded pipes were developed for three different annual failure probabilities (reliability indices), 10^{-2} (2.32), 10^{-3} (3.09) and 10^{-4} (3.71).

In order to assure a uniform reliability level for the different design scenarios for each of the defined reliability levels, the partial safety factors in the code were determined based on advanced probabilistic code calibration procedures.

The set of partial safety factors to be applied with the selected characteristic values at each defined reliability level were obtained through an optimisation of a penalty function u,

$$\min \sum_{i=1}^{N_{des}} u\left(\beta_i - \beta_{\text{target}}\right) \tag{11}$$

where β_{target} is the target reliability index for the specified reliability level and β_i is the reliability index for design case i out of N_{des} design cases.

In order to avoid unconservatism in the established design code, the penalty function u were defined such that designs resulting in a reliability level lower than selected target reliability level where punished harder than designs resulting in a safer design.

The probabilistic evaluations were carried out using full distribution probabilistic methods and a first order approximation in the transformed standard normal space, applying the general probabilistic analysis program PROBAN (DNV Sesam, 1993).

In order to minimise the computational efforts in the optimisation of the partial safety factors, a five-dimensional response surface of the reliability index was established for three different corrosion assessment accuracy levels.

PROPOSED DNV CODE EQUATION

The proposed DNV code equation defining the mean value for the annual maximum operating pressure has the following form,

$$E\left[P_{\max}\right] = \gamma_M \cdot \frac{\text{SMTS} \cdot 2t}{D-t} \cdot \frac{1 - \dfrac{d'^{*}}{t}}{1 - \dfrac{d'^{*}}{t} \cdot \dfrac{1}{M^{*}}} \cdot H^{*} \tag{12}$$

where the subscript* defines the use of characteristic values and corresponding partial safety factors.

The proposed DNV code equation is based on the use of three partial safety factors,

γ_M: Model prediction
γ_d: Corrosion depth assessment
γ_L: Corrosion length assessment

Originally a separate partial safety factor for the material properties was incorporated. However, it was experienced that this partial safety factor could be set equal to one for all design conditions without affecting the reliability level uniformness of the code.

The partial safety factor values depend on the desired reliability level and the assessment accuracy for the observed corrosion.

The proposed DNV code is defined such that the decision maker may directly define the annual maximum operating pressure based on the accuracy of the corrosion assessment prediction and the selected reliability level for the pipeline.

The corrosion assessment accuracy will typically depend on the type and quality of the inspection carried out on the pipeline. The variation in the assessment accuracy is reflected in the code through the choice of partial safety factors, where an increased assessment accuracy results in the use of lower safety factors values for γ_d and γ_L. An increased effort to reduce the assessment uncertainties is then directly reflected through a reduced reduction in the annual maximum operating pressure for observed corrosion.

Dependent on the required reliability level of the pipeline, typically defined based on the estimated consequences of a possible burst failure, different selection of partial safety factors are made. An increased reliability level results in increased values for the partial safety factors and thereby an increased reduction in the annual maximum operating pressure for observed corrosion.

The safety factors and characteristic values for the uncertain variables in the design code are carefully defined such that the code reflects a uniform reliability level over the range of all possible corrosion extensions and corrosion shapes.

In Figure 6, the mean value of the annual maximum operating pressure based on the proposed DNV design code is compared with B31G and NG18 for a corrosion depth assessment of 30% of the thickness (measured 30% corrosion depth). The results are obtained assuming a typical estimate for the corrosion depth assessment accuracy of 80% probability within ±10% of the pipe thickness. For comparison purposes, only the internal pressure loads are here accounted for, assuming no possible influence from axial or bending loads on the pipe. The proposed DNV code is shown for three different annual failure probability levels, 10^{-2}, 10^{-3} and 10^{-4}.

It is observed that for the present example, the B31G gives an overall conservative estimate for the annual maximum operating pressure. This is also the case for no corrosion, where the allowable operating pressure is less in the ASME code compared to the DNV Pipeline Rules 96 (DNV'96, 1996). The horizontal curve defines the maximum annual operating pressure for uncorroded pipes from DNV'96, accounting for the influence of system effects. Allowable annual operating pressures above this limit are therefore not applicable.

For corroded pipes, the influence of possible system effects on the estimated burst probability is omitted in the evaluation of the allowable operating pressure, as the corrosion defect considered typically is the largest observed corrosion and thereby reflects the weakest burst capacity point along the pipeline.

In Figure 7, the mean values of the annual maximum operating pressures for 50% depth corrosion according to the proposed DNV burst code, B31G and NG18 are given. It is observed that for this more extensive corrosion, the degree of conservatism within the existing codes is not as significant as for 30% corrosion depth and is highly dependent on the corrosion length.

In Figure 8, the mean values of the annual maximum operating pressures for 70% depth are given. The figure shows that for deep corrosion B31G is unconservative for both short and long corrosion lengths with respect to a probability level of 10^{-3} and 10^{-4}.

The effect of the inspection accuracy on the allowable operating pressure using the proposed DNV code is shown in Figure 9 a), b) and c) for an annual failure probability of 10^{-4}, where different partial safety factors are used in the DNV code dependent on the inspection accuracy and the selected reliability levels.

The figure demonstrates the influence of the sizing accuracy of observed corrosion and highlights the necessity of including the quality of the inspections in the code formulation.

In the proposed DNV code, three levels of sizing accuracy of observed corrosion are considered;

- Exactly known corrosion
- The latest accuracy of intelligent pigging (80% within ±10%)
- The resent accuracy of intelligent pigging (80% within ±20%)

As can be seen in Figure 9, by increasing the inspection accuracy unnecessary repairs, operating pressure reductions or shutdowns can be avoided. The figure also indicates that B31G is rather conservative for shallow corrosion depths, but becomes unconservative for moderate and deep corrosion having lower inspection accuracy quality.

The estimated annual failure probability for rectangular corrosion shapes for different depths are shown in Figure 10 as a function of the corrosion length, where safety factors corresponding to an annual probability of failure of 10^{-4} is applied for the proposed DNV code. The figure presents thereby the inherent reliability level in the DNV code and B31G.

It is observed that the DNV code gives a very consistent probability of failure for various corrosion depths and corrosion length, while B31G varies dramatically. An inconsistency in the reliability level for the applied code dependent on the degree of corrosion depth and corrosion length makes it difficult to determine the inherent reliability level in a design based on this code.

CONCLUSIONS

In this paper the work conducted in the DNV joint industry research project *Reliability of Corroded Pipes* has been summarised. The goal of the project has been to develop a guideline containing probabilistic calibrated design equations against burst for determination of the allowable operating pressure of pipelines with smooth corrosion defects for selected reliability levels.

The main motivation for the initiation of the project was observed inconsistency in the reliability level in the commonly used B31G equation dependent on the degree of corrosion.

The proposed DNV code gives a defined and consistent reliability level for variations in the corrosion depths and corrosion length. The influence of corrosion assessment accuracy is further accounted for in the proposed DNV code, where an increase in the assessment accuracy of the observed corrosion will reduce the uncertainties and permit a higher allowable operating pressure.

The basis for the calibration of the design equations has been an extensive number of finite element analyses, a series of laboratory tests conducted within the research project and previously published test results on burst of corroded pipes.

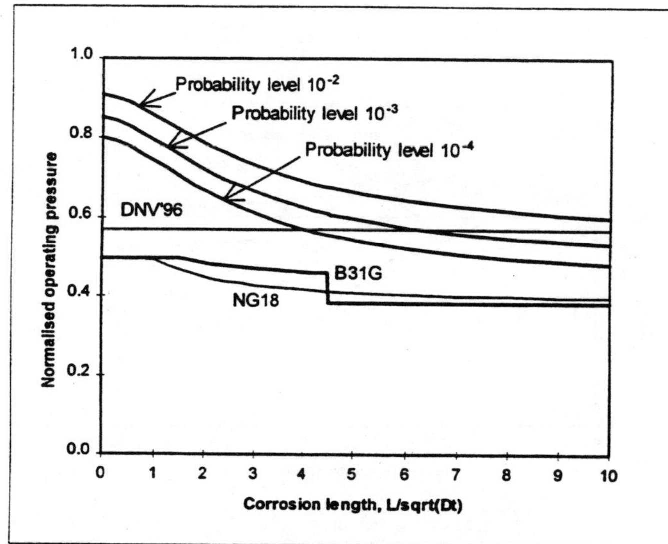

Figure 6 Normalised operating pressure for 30% depth corrosion.

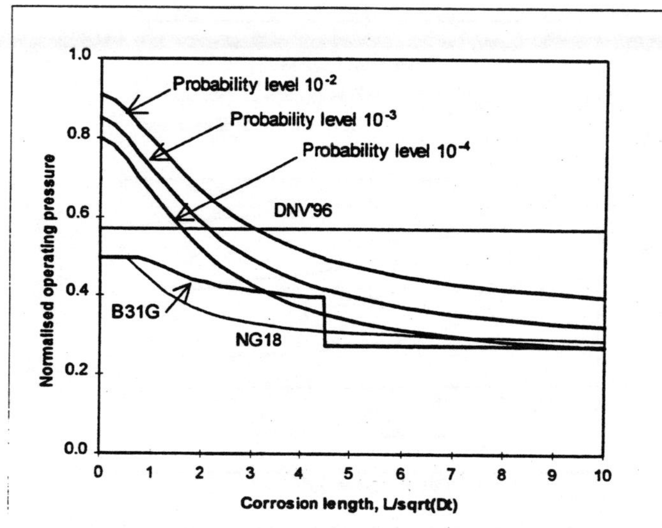

Figure 7 Normalised operating pressure for 50% depth corrosion.

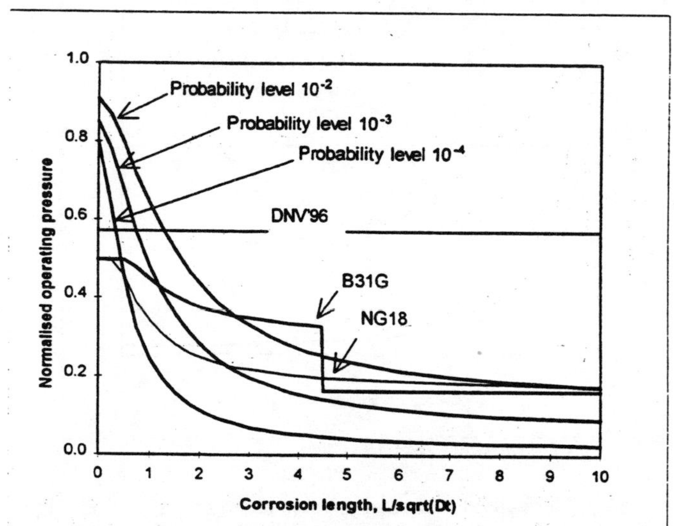

Figure 8 Normalised operating pressure for 70% depth corrosion.

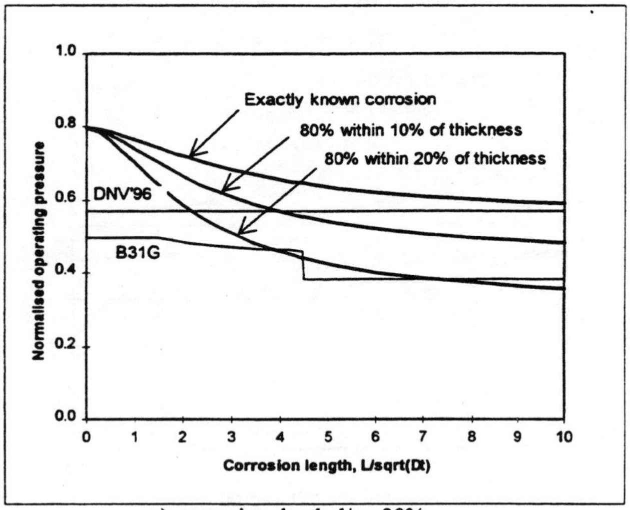

a) corrosion depth d/t = 30%

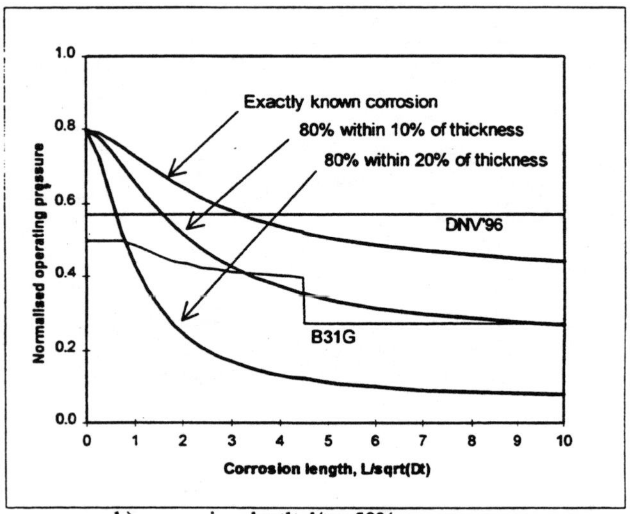

b) corrosion depth d/t = 50%

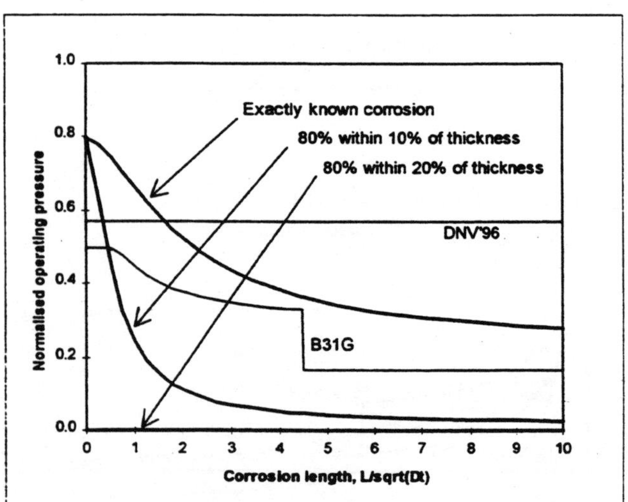

c) corrosion depth d/t = 70%

Figure 9 Normalised operating pressure based on DNV code (10^{-4}) for three levels of inspection accuracy having a) 30%, b) 50% and c) 70% measured depth corrosion.

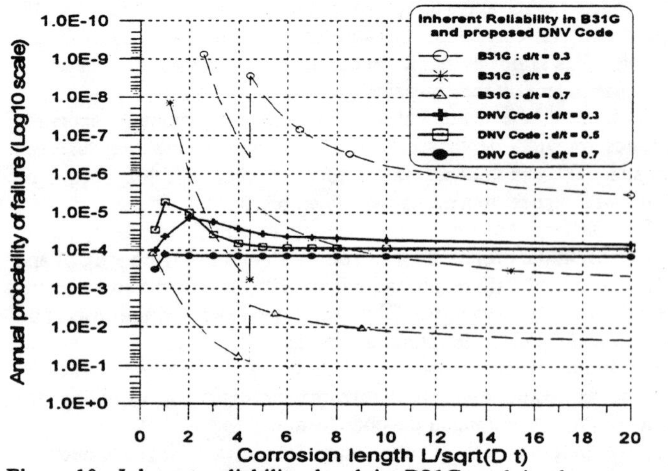

Figure 10 Inherent reliability level in B31G and in the proposed DNV code for target reliability level 10^{-4}.

ACKNOWLEDGEMENTS

The authors would like to thank Statoil, Amoco Norway Oil Company, Exxon Production Research Company, Norwegian Petroleum Directorate and Minerals Management Service for financial support of the *Reliability of Corroded Pipes* joint industry research project.

REFERENCES

ABAQUS v 5-4 (1995), Hibbitt, Karlsson and Sorensen, Providence, Rhode Island, USA

ASME, (1991), "Manual or Determining the remaining Strength of Corroded Pipelines, A supplement to ANSI/ASME B31 Code for Pressure Piping".

Chouchaoui, B. A. and Pick, R.J., (1992), "Residual Burst Strength of Pipe with internal Corrosion Pits", Proc. Int. Conf. On Pipeline Reliability, Calgary.

Coulson, K. E. W, and Wolthingham, R.G. (1990), "Pipe Corrosion-1, Standard Damage Assessment Approach is Overly Conservative", Oil and Gas Journal.

Denys, R. M. (1995), "Fitness for purpose assessment of corroded defects", The Pipeline Pigging Conference, Amsterdam.

DNV Sesam (1993), "PROBAN Version 4, Theory manual", DNV Research report no. 93-2056, Høvik, Norway.

DNV'96 (1996), "Rules for Submarine Pipeline Systems", DNV, Norway.

Fu ,B. and Kirkwood, M. G. (1995), "Predicting Failure Pressure of Internally Corroded Linepipe using the Finite Element Method", Offshore Mechanics and Arctic Engineering, OMAE'95, Copenhagen.

Jiao, G., Sotberg, T. and Igland, R. (1995), "SUPERB 2M - STATISTICAL DATA: Basic Uncertainty Measures for reliability Analysis for Offshore Pipelines", SUPERB project report.

Kiefner J.F. and Vieth P.H. (1989), "A Modified Criterion for Evaluating the Remaining Strength of Corroded Pipe, RSTRENG" Project PR 3-805 Pipeline Research Committee, AGA.

Kiefner J.F. and Vieth P.H. (1994), "Database of Corroded Pipe Tests" Pipeline Research Committee, AGA.

Maxey W.A., Kiefner J.F., Eiber R.J. and Duffy A.R. (1971), "Ductile Fracture Initiation, Propagation and Arrest in Cylindrical Vessels," Fracture Toughness, Proc. of the 1971 National Symp. on Fracture Mechanics, ASTM STP 514.

Proceedings of the Seventh (1997) International Offshore and Polar Engineering Conference
Honolulu, USA, May 25-30, 1997

ISBN 1-880653-28-1 (Set); ISBN 1-880653-32-X (Vol. IV)

An Application of Design of Experiments Concepts to the Inspection Planning of Structures

Alexandre Kawano
Escola Politecnica, University of São Paulo
São Paulo, Brazil

Abstract

For large structures under random loading such as ships and offshore installations, fatigue is often the governing limit state. Under repetitive loading, cracks initiate and grow in size reducing the strength of structures. If these cracks are detected during inspections, flawed parts are repaired or substituted, improving the overall system reliability. The contribution of this work in the field of inspection of fatigue sensitive structures is to consider in the inspection planning process, groups of structures of similar characteristics instead of isolated structures, as is common in the literature.

It is proposed a method to update probabilistic information about uncertain parameters by applying Bayesian methods using some concepts of classical design of experiments theory such as the factorial design.

Key words: Fatigue, Inspection Planning, Bayesian, Design of Experiments

1. Introduction

We live in a world that is increasingly sensitive to environmental damages of any kind and to losses of human lives due to disasters involving engineering systems. Among the causes of failures in dynamically excited structures, fatigue is among the most relevant ones. A very effective way to avoid failures of this kind is to inspect fatigue sensitive structures at critical spots. If a crack is detected in a structural element, it is repaired and the reliability level returns to acceptable levels. Besides, by using Bayesian methods, the results found in an inspection can be used to lessen the level of uncertainties about a given structure, thus increasing its reliability. However, since the number of critical spots is very large, and in general inspection procedures are very expensive, without planning the inspection process can be very cost ineffective.

There are several papers dealing with inspection of fatigue sensitive structures. For example, Yang & Trapp [Yang and Trapp, 1974] studied the effect of periodic inspections using classic probabilistic methods; Itagaki & Yamamoto [Itagaki and Yamamoto, 1985] realizing that the inspection results themselves furnish data that can be used to update the knowledge about uncertain parameters, applied Bayesian analysis into the problem in order to select inspection intervals; Sørensen at alli [Sørensen at alli, 1991] proposed a Method to minimize overall service costs; and Kawano [Kawano at alli, 1994] focused on the information gathered from the first inspection of a structure.

The contribution of this work is to consider in the inspection planning process, groups of structures with similar characteristics where it is expected some correlation between structural behavior and between inspection findings, instead of isolated structures, as is common in the literature.

It is proposed an application of the concepts of the design of experiments theory, such as the factorial design [Montgomery, 1976], to the inspection planning of a group of structures. The objective is to improve the absorption of information by Bayesian Methods. Specifically, it is considered two factors: different structures and different structural element types. The information about structural elements and structures can then be used subsequently in the planning of future inspections.

2. Mathematical Modeling

2.1 Crack initiation

Crack initiation time is predicted with the help of S-N curves, that relate a certain constant stress amplitude level to the corresponding number of cycles representing the number of cycles to crack initiation. The specimens used to obtain these S-N curves can be small standard specimens, scale models or even full scale structures. Moreover, the testing stress can simulated with several degrees of "realism". To relate constant amplitude data to the real stochastic process stress loading X(t), it is supposed that the Palmgrem-Miner rule applies. Conditions for the validity of this assumption is investigated in e.g. [Lutes et alli, 1984].

A crack is said to initiate when it reaches a macro size a_0 from which phenomenological fracture mechanics is applicable, i.e., stress intensity factor amplitude becomes meaningful for describing the crack propagation phase. In general, for common structures, a_0

is of the order of the size of detectable cracks by the naked eye, or more precisely, of the order of centimeters.

Mathematically, the crack starting time (number of cycles) is supposed to be governed by the Wöhler formula:

$$N(\Delta\sigma)^p = Q \quad (1)$$

where N is the number of cycles till crack initiation, $\Delta\sigma$ is the constant stress amplitude of the loading, and Q and p are material characteristics. As it is well known, fatigue test data show great scatter. It could be modeled by making p and Q random variables, but instead, what is usually done [Pittaluga, Cazzulo & Romeo, 1991] is to fix the value of p and model all randomness contained in these data by Q only.

According to the P-M rule, crack starts when the total damage Δ reaches the critical value Δ_{crit}. In the case of a stochastic stress process X(t), Δ is calculated as follows:

$$\Delta = \sum_j \Delta_j = \sum_j \frac{n_j}{Q(\Delta\sigma_j)^{-p}} = \frac{1}{Q}\sum_j n_j (\Delta\sigma_j)^p = \frac{1}{Q}\sum_j n_j [\Delta X_j]^p \quad (2)$$

where n_j is the number of cycles corresponding to the constant stress amplitude $\Delta\sigma_j = \Delta X_j$. Making the assumption that the stress process X(t) is a narrow band Gaussian process, and that the number of cycles is very high, the coefficient of variation (c.o.v.) of the summation factor is very small, and it is possible to approximate it by its mean value.

$$E\left[\sum_j n_j [\Delta X_j]^p\right] = T \times E[N_+(0)] (2\sqrt{2})^p [\sigma_x]^p \Gamma\left(1+\frac{p}{2}\right) \quad (3)$$

where in the formulas above, T is the time in years, σ_x is the standard deviation of the stress process X(t), and $E[N_+(0)]$ is the number of zero crossings from below per unit time. A crack is said to initiate when Δ reaches a critical value Δ_{crit}. Substituting (3) into (2) and solving $\Delta = \Delta_{crit}$ when $T = T_c$, the critical time can be written as:

$$T_c = \frac{Q\,\Delta_{crit}}{E[N_+(0)](2\sqrt{2})^p [\sigma_x]^p \Gamma\left(1+\frac{p}{2}\right)} \quad (4)$$

The fatigue life itself presents great variability. Besides, there are several uncertainties in the evaluation of Δ_{crit}, Q, and in other factors. To model them, a random factor Z that follows a 2 parameter Weibull distribution is introduced, so that

$$T_c = \frac{Q\,\Delta_{crit} \times Z}{E[N_+(0)](2\sqrt{2})^p [\sigma_x]^p \Gamma\left(1+\frac{p}{2}\right)} \quad (5)$$

The conditional probability density function of Z, given the knowledge of the values of the shape parameter α and the scale parameter β, is given by

$$f_Z(z|\alpha,\beta) = \left(\frac{\alpha}{\beta}\right)\left(\frac{z}{\beta}\right)^{\alpha-1} \mathrm{Exp}\left[-\left(\frac{z}{\beta}\right)^{\alpha}\right] \quad (6)$$

The conditional probability density function for the critical time T_c can be easily found by a simple transformation of variables:

$$f_{T_c}(t|\alpha,\beta) = J\left(\frac{\alpha}{\beta}\right)\left(\frac{t}{\beta/J}\right)^{\alpha-1} \mathrm{Exp}\left[-\left(\frac{t}{\beta/J}\right)^{\alpha}\right] \quad (7)$$

$$\text{where } J = \frac{E[N_+(0)](2\sqrt{2})^p [\sigma_x(h)]^p \Gamma\left(1+\frac{p}{2}\right)}{Q} \quad (8)$$

The probabilistic parameters α and β are considered uncertain in this work, and shall be updated by the Bayesian method.

Note that in the above model, no use has been made of a 'endurance limit'[1] that is observed in constant amplitude fatigue tests of steel specimens. In fact, the 'endurance limit' seems to vanish, or at least seems to be suppressed to a very low level under variable amplitude loading [Sarkani et al., 1994].

2.2 Crack propagation

After initiation, a crack is assumed to propagate according to the Paris law:

$$\frac{da}{dn} = C\left(\frac{\Delta K}{K_0}\right)^m \quad (9)$$

where a is the crack size, n is the number of cycles, ΔK is the stress intensity factor range, and C, m and K_0 are material properties.

In probabilistic reliability analysis the above formula is randomized in order to reflect the fairly high scatter present in crack propagation data see e.g. [Sobczyk K. & Spencer, 1992] [Itagaki, Ishizuka & Yan, 1993]. In this work, it is supposed that m is deterministic, and C is a lognormal random variable in line with the model proposed by Yang [Yang et alli, 1993]. The stress intensity factor ΔK is given by:

$$\Delta K = \Delta X\, Y(a) \sqrt{\pi a} \quad (10)$$

where Y(a) is a non-dimensional geometry factor, which depends both on the structural element and the shape and length of the crack. To simplify analysis, Y(a), in this work is supposed to be constant and equal to one, which corresponds to the infinite plate solution.

Supposing that the stress process $\Delta X = \Delta X(t)$ is narrow-band Gaussian and integrating (9), the crack length a as a function of time t is obtained:

$$a(t-t_c,c) = \left[a_0^{\frac{2-m}{2}} + (t-t_c)\, C\, \gamma(\bullet)\, \frac{2-m}{2}\left(\frac{\sqrt{\pi}}{K_0}\right)^m\right]^{\frac{2}{2-m}} \quad \text{if } m \neq 2 \quad (11)$$

where t_c stands for crack initiation time and:

$$\gamma(\bullet) = E[N_+(0)](2\sqrt{2})^m [\sigma_x]^m \Gamma\left(1+\frac{m}{2}\right) \quad (12)$$

In the above formula, a_0 is the initial crack size.

2.3. Crack detection

The probability of detecting an existing crack of size a is given by

$$D(a) = 1 - \mathrm{Exp}[-d(a-a_0)] \quad (13)$$

where d is a crack detection parameter.

[1] *For a few engineering materials such as steel and titanium, below a limiting stress called endurance limit, the material presumably can endure an infinite number of cycles without failure in the constant amplitude testing.*

2.4. Events at the inspection times and their respective probabilities

The Bayesian method is used to update the knowledge about uncertain probabilistic parameters present in the mathematical model. Specifically, in this work, by the knowledge of inspection results, the degrees of belief or probability distributions of the crack initiation parameter β (eq.(7)), the shape parameter α (also eq.(7)) and a measure of modeling error σ_e^2 (to be defined later) are updated via the Bayes method.

Here it is considered a group of similar, but different structures, that contain different structural element types. Since each member of the group of structures has been built with different construction teams, and is located at different places and therefore are subjected to different load histories, it is natural to imagine the influence of a particular structure on the value of β (eq.7). In the same manner, each structural element type considered has a particular geometry, and as a consequence, some influence on β must exist. Therefore, instead of a single β, a set of β_{ij} corresponding to the j-th element type of the i-th structure is considered.

The value of β_{ij} is partitioned into four components, or *effects*, as it is done in some design of experiments plans [Montgomery, 1976].

$$\beta_{ij} = \bar{\beta} + \beta s_i + \beta d_j + e_{ij} \tag{14}$$

where $\bar{\beta}$ is the characteristic value, βs_i is the effect of the i-th structure, βd_j is the effect of the j-th type of element, and e_{ij} is the modeling error that incorporates also a possible interaction effect between the structure i and the element type j.

To apply the Bayesian method, it is necessary first to describe what events are expected to be observed in the inspections. A structural element is defined so that it has only one fatigue critical location where a crack can initiate. Besides, in each structure i, only n_{ij} elements of type j that in principle are subject to a stress process that possesses the same RMS σ_x and the same mean number of zero crossings per unit time $E[N_+(0)]$ are inspected. The method presented here can be easily extended for the case when structural elements are subject to different loadings.

The following events form the sample space for inspection results concerning one element:

a) A crack of size a is found in the element[2].

b) No cracks are found in it.

To simplify the analysis, the event that an structural element fails is not considered. Further, if a crack is found in an element, it is repaired and the element is considered to return to a "as good as new" condition.

Consider the event

$$A_{ij} = \left\{ \begin{array}{c} (n_{ij})_1 \text{ elements out of } n_{ij} \text{ are found to have cracks of} \\ \text{sizes } a_1,\ldots,a_{(n_{ij})_1}, \text{ and } (n_{ij})_2 = n_{ij} - (n_{ij})_1 \text{ are found} \\ \text{to be intact (no cracks) in the j–th detail of the i–th strucure} \end{array} \right\}$$

Mathematically, the conditional probability of the occurrence of the event A_{ij} is given by:

$$P(A_{ij},T|\alpha,\beta_{ij})\, da_1 \ldots da_{(n_{ij})_1} = \binom{n_{ij}}{(n_{ij})_1} (P_2(T|\alpha,\beta_{ij}))^{(n_{ij})_2} \prod_{k=1}^{(n_{ij})_1} P_1(a_k,T|\alpha,\beta_{ij})\, da_k \tag{15}$$

In the above formula, T is the inspection time when the event A_{ij} takes place, $P_1(a_k,T|\alpha,\beta_{ij})\, da_k$ is the conditional probability of finding a crack of size a_k at time T in an element that were as good as new at $T = T_0 = 0$, irrespective of the number of times the element was repaired till T, and $P_2(T|\alpha,\beta_{ij})$ is the conditional probability of no crack detection at inspection at T.

Formulas for P_1 and P_2 used in (15) can be found by using the following procedure. Taking the example of what happens at the second inspection, $P_1(a,T_2|\alpha,\beta_{ij})$ can be calculated as the sum:

$$P_1(a,T_2|\alpha,\beta_{ij}) = p_1(a,T_2,T_0|\alpha,\beta_{ij}) + P_r(T_1|\alpha,\beta_{ij}) \times p_1(a,T_2,T_1|\alpha,\beta_{ij}) \tag{16}$$

where $p_1(a,T_j,T_k|\alpha,\beta_{ij})\, da$ is the conditional probability of finding a crack of size a at the inspection at T_j in any element that started service at T_k, and has never been repaired since then. Its formula is given in the appendix. $P_r(T_1|\alpha,\beta_{ij})$ is the conditional probability of finding any cracks in an element at $T = T_1$. It is given by

$$P_r(T_1|\alpha,\beta_{ij}) = \int_{a_0}^{\infty} P_1(a,T_1|\alpha,\beta_{ij})\, da \tag{17}$$

It can be readily seen that the fraction of elements found with a crack of size a at the inspection at T_2 is the sum of the fraction of the elements that started at $T = T_0$, but were not repaired at T_1, and the fraction of the elements that contains a crack of size a, but started service at $T = T_1$.

Analogously, at the third inspection:

$$P_1(a,T_3|\alpha,\beta_{ij}) = p_1(a,T_3,T_0|\alpha,\beta_{ij}) + P_r(T_1|\alpha,\beta_{ij}) \times p_1(a,T_3,T_1|\alpha,\beta_{ij}) + P_r(T_2|\alpha,\beta_{ij}) \times p_1(a,T_3,T_2|\alpha,\beta_{ij}) \tag{18}$$

The general formula can be written as:

$$P_1(a,T_\ell|\alpha,\beta_{ij}) = p_1(a,T_\ell,T_0|\alpha,\beta_{ij}) + \sum_{k=1}^{\ell-1} P_r(T_k|\alpha,\beta_{ij}) \times p_1(a,T_\ell,T_k|\alpha,\beta_{ij}) \tag{19}$$

The term $P_2(T|\alpha,\beta_{ij})$ can be calculated as

$$P_2(T_\ell|\alpha,\beta_{ij}) = 1 - \int_{a_0}^{\infty} P_1(a_i,T_\ell|\alpha,\beta_{ij})\, da \tag{20}$$

2.5. Information gathering from similar structures

The parameter β_{ij}, that governs the time for crack initiation, is divided into four components or effects: $\bar{\beta}$, βs_i, βd_j and e_{ij}, a modeling error term, according to eq.(14).

$$\beta_{ij} = \bar{\beta} + \beta s_i + \beta d_j + e_{ij} \tag{14 repeated}$$

The values of βs_i, βd_j and e_{ij} are assumed to be Gaussian random variables:

$$\beta s_i \sim N(0, \sigma_s^2) \qquad \beta d_i \sim N(0, \sigma_d^2) \qquad e_{ij} \sim N(0, \sigma_e^2) \tag{21}$$

That is,

$$\beta_{ij} = \bar{\beta} + \beta s_i + \beta d_j + e_{ij} \sim N(\bar{\beta}\, ,\, \sigma_s^2 + \sigma_d^2 + \sigma_e^2) \tag{22}$$

The values of σ_s^2 and σ_d^2 are selected based on previous experience. The greater their value, the more one expects the values of β_{ij} to deviate from the characteristic value $\bar{\beta}$, but not necessarily so.

---

[2] *In this work, "to find a crack of size a" means in fact "to find a crack of size between a and a+da"*

With the information about the distribution of β_{ij}, the conditional probability of the event A_{ij} at a inspection at time T_{ij} can be written as:

$$P(A_{ij}, T_{ij} | \alpha, \bar{\beta}, \sigma_e^2) = \int_0^{+\infty} P(A_{ij}, T_{ij} | \alpha, \beta_{ij})\, f_{\beta_{ij}}(\beta_{ij} | \bar{\beta}, \sigma_e^2)\, d\beta_{ij} \tag{23}$$

Suppose we wish to collect all information available to infer the probabilistic parameters α and β_{ab} about a specific type of element b in a specific structure a.

Consider the diagram shown in figure 1. It is divided into four areas I, II, III, and IV. Area I contains all combinations (i, j) of structures i and types of elements j such that $i \neq a$ and $j \neq b$. Area II corresponds to all cells of element type b, except the one on the structure a. Area III contains all elements of the structure a, except the element type b. Finally, the area IV corresponds to the element type b on the structure a. Note the resemblance to the structure of the factorial design [Montgomery, 1976]

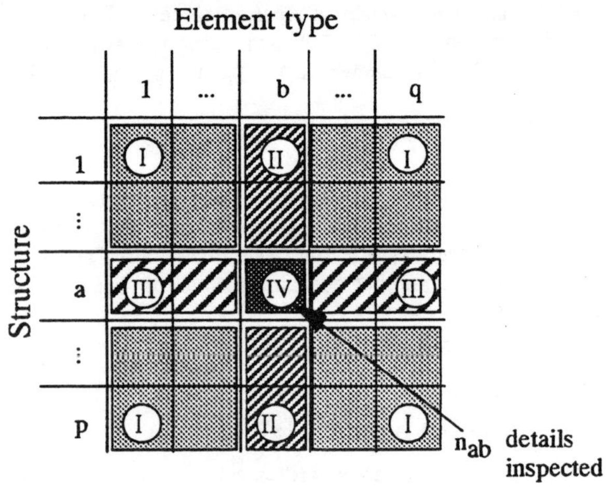

Fig. 1. Division into cells of structures and element types

By using the Bayesian method, it is possible to get information about the parameters α, $\bar{\beta}$, σ_e^2, βs_a, βd_b and e_{ab}. At the first level, to get information about α, $\bar{\beta}$ and σ_e^2, the following Bayesian updating relation is used [Sander & Bardoux, 1991].

$$f^{(1)}_{\alpha,\bar{\beta},\sigma_e^2}(\alpha,\bar{\beta},\sigma_e^2) = \frac{Like_{I,II,III,IV}(\alpha,\bar{\beta},\sigma_e^2 | A_{ij}) \times f^{(0)}_{\alpha,\bar{\beta},\sigma_e^2}(\alpha,\bar{\beta},\sigma_e^2)}{\int_{\sigma_e^2}\int_{\bar{\beta}}\int_{\alpha} Like_{I,II,III,IV}(\alpha,\bar{\beta},\sigma_e^2 | A_{ij}) \times f^{(0)}_{\alpha,\bar{\beta},\sigma_e^2}(\alpha,\bar{\beta},\sigma_e^2)\, d\alpha\, d\bar{\beta}\, d\sigma_e^2} \tag{24}$$

where $f^{(0)}_{\alpha,\bar{\beta},\sigma_e^2}(\alpha,\bar{\beta},\sigma_e^2)$ is the prior degree of belief distribution, $f^{(1)}_{\alpha,\bar{\beta},\sigma_e^2}(\alpha,\bar{\beta},\sigma_e^2)$, the posterior, and $Like_{I,II,III,IV}(\alpha,\bar{\beta},\sigma_e^2 | A_{ij})$ is the likelihood of obtaining the data of regions I,II, III and IV.

The likelihood function $Like_{I,II,III,IV}(\alpha,\bar{\beta},\sigma_e^2 | A_{ij})$ can be calculated by:

$$Like_{I,II,III,IV}(\alpha,\bar{\beta},\sigma_e^2 | A_{ij}) = \prod_{i=1}^{p}\prod_{j=1}^{q} P(A_{ij}, T_{ij} | \alpha, \bar{\beta}, \sigma_e^2) \tag{25}$$

where $P(A_{ij}, T_{ij} | \alpha, \bar{\beta}, \sigma_e^2)$ is given by eq.(23)

To get information about the parameter βd_b, another likelihood function $Like_{II,IV}(\beta d_b)$ based on eq.(15) is used. The conditional probability of getting a result in a cell (i,b) belonging to the regions II or IV is given by:

$$P(A_{ib}, T_{ib} | \alpha, \bar{\beta} + \beta d_b, \sigma_e^2) = \int_{-(\bar{\beta} + \beta d_b)}^{+\infty} P(A_{ib}, T_{ib} | \alpha, \bar{\beta} + \beta d_b + x)\, f_x(x | \sigma_s^2 + \sigma_e^2)\, dx \tag{26}$$

Above, $x = \beta s_i + e_{ij}$ follows the normal distribution $N(0, \sigma_s^2 + \sigma_e^2)$.

The likelihood function corresponding to the regions II and IV is given by:

$$Like_{II,IV}(\beta d_b) = \int_{\sigma_e^2}\int_{\bar{\beta}}\int_{\alpha} \left\{ \prod_{i=1}^{p} [P(A_{ib}, T_{ib} | \alpha, \bar{\beta} + \beta d_b, \sigma_e^2)] \right\} f^{(1)}(\alpha, \bar{\beta}, \sigma_e^2)\, d\alpha\, d\bar{\beta}\, d\sigma_e^2 \tag{27}$$

$f^{(1)}(\alpha, \bar{\beta}, \sigma_e^2)$ is the posterior distribution given by (24).

The posterior density function of βd_b is given by:

$$f^{(1)}_{\beta d}(\beta d_b) = \frac{Like_{II,IV}(\beta d_b)\, f^{(0)}_{\beta d}(\beta d_b)}{\int_{\beta d_b} Like_{II,IV}(\beta d_b)\, f^{(0)}_{\beta d}(\beta d_b)\, d\beta d_b} \tag{28}$$

where $f^{(0)}_{\beta d}(\beta d_b)$ is the prior distribution of the parameter βd_b. The consistent choice is the normal distribution $\beta d_b \sim N(0, \sigma_d^2)$ (see eq.(21)).

The posterior distribution $f^{(1)}_{\beta a}(\beta s_a)$ after observing the inspection results of the structure a (regions III and IV) can be obtained in a manner completely analogous as was done to obtain $f^{(1)}_{\beta d}$.

Finally, the posterior probability function for the non linear modeling term e_{ab} can be evaluated by observing the inspection results in the region IV.

$$f^{(1)}_{e_{ab}}(e_{ab}) = \frac{Like_{IV}(e_{ab}) \times f^{(0)}_{e_{ab}}(e_{ab})}{\int_{e_{ab}} Like_{IV}(e_{ab}) \times f^{(0)}_{e_{ab}}(e_{ab})\, de_{ab}} \tag{29}$$

To be consistent, the prior distribution $f^{(0)}_{e_{ab}}(e_{ab})$ is given by (21). The numerator of (29) can be given by

$$Like_{IV}(e_{ab}) \times f^{(0)}_{e_{ab}}(e_{ab}) = \int_{\alpha,\bar{\beta},\sigma_e^2,\beta s_a,\beta d_b} \{P(A_{ab}, T_{ab} | \alpha, \bar{\beta} + \beta s_a + \beta d_b + e_{ab})\, f_e(e_{ab} | \sigma_e^2) \times f^{(1)}_{\alpha,\bar{\beta},\sigma_e^2}(\alpha,\bar{\beta},\sigma_e^2)\, f^{(1)}_{\beta s_a}(\beta s_a)\, f^{(1)}_{\beta d_b}(\beta d_b)\}\, d\alpha\, d\bar{\beta}\, d\sigma_e^2\, d\beta s_a\, d\beta d_b \tag{30}$$

With more precise information about α, $\bar{\beta}$, βs_i, βd_j and e_{ij}, than in the beginning of the process, the planning of future inspections can be done more effectively [Deodatis, 1992].

3. Numerical example

In Table 1 it is shown the numerical values to be used in a numerical example.

There are inspections in the structures every two years, but not necessarily all element types of a structure are inspected, and the time when each structure is put into service is not fixed.

In order to test the Bayesian procedure described in this paper, hypothetical inspection results for all combinations of structures and element types were generated based on $\alpha = 5.0$ and $\beta_{ij} = \bar{\beta} + \beta s_i + \beta d_j + e_{ij}$ shown in tables 2 and 3.

Table 1. Physical parameters used

Physical Parameter	Symbol	Value	Unit
Mean value of the crack propagation rate C	μ_C	1.0×10^{-7}	m/cycle
K_0 parameter in the crack propagation formula	K_0	30.0×10^6	Pa $\sqrt{m}$
Standard deviation of lnC	σ_C	0.1	ln (m/cycle)
Minimum detectable crack size	a_0	0.01	m
Detection parameter d	d	10.0	m^{-1}
Paris law m exponent	m	3.0	
Mean Number of cycles per year	$E[N_+(0)]$	5.0×10^6	
Standard Deviation of the stress process	σ_X	21.0×10^6	Pa
Wöhler formula eq. (1)	$N(\Delta\sigma)^3 = 3\ 10^{30}$ $\Delta\sigma$ given in Pa		

In each cell of table 3, it is shown: when the inspection takes place (T=(insp. n.) x 2 years), the total number of elements inspected (tot. el.), the number of elements found without any cracks (intact) and the sizes of the cracks found (cr. sizes).

Note that the times when the inspections take place in each element type of a structure are not uniform (because in a inspection of a structure, which happens every two years, not necessarily all element types are investigated). Also, the number of inspected elements in each structure are not uniform. In the procedure used, the data can be analyzed in any chronological order. Further the Bayesian procedure is flexible enough to allow for "not balanced designs" typical of simple design of experiments plans [Montgomery, 1976].

Given the inspection results above, the joint posterior density distribution of the uncertain parameters $\bar{\beta}$, α, σ_ε^2 is evaluated by eq.(24). The prior distribution $f^{(0)}_{\alpha,\bar{\beta},\sigma_\varepsilon^2}(\alpha,\bar{\beta},\sigma_\varepsilon^2)$ used is uniformly distributed in the range $\alpha > 0$, $\bar{\beta} > 0$, $\sigma_\varepsilon^2 > 0$ and is therefore improper. The shape of the posterior distribution is exactly the same of the corresponding likelihood function (25). It is shown in figure 2 for $\sigma_\varepsilon^2 = 0.10$.

Now the element type number j=2 of the structure number i=4 is considered.

Observing the data corresponding to the element type j=2 across all structures considered, and using the posterior distribution $f^{(1)}_{\alpha,\bar{\beta},\sigma_\varepsilon^2}(\alpha,\bar{\beta},\sigma_\varepsilon^2)$ into (27), the numerator of (28) is obtained. Since the denominator is just a normalizing constant, the shape of the posterior distribution, shown in figure 3, is the same as of the numerator. Note that it gives us the shape of rational degree of belief about the effect of the element type j=2 on $\beta_{4,2}$ after observing the inspection results. As it could be expected from observing table 2, the effect of the element type j=2 should in fact tend to be negative.

Table 2. Set of β_{ij} used to generate the hypothetical inspection data set shown in table 3

	Structure						
element	1	2	3	4	5	6	mean
1	1.76	1.23	2.23	1.92	1.22	1.59	1.66
2	0.96	0.54	1.35	1.10	0.52	0.83	0.88
3	1.85	1.31	2.33	2.01	1.29	1.68	1.74
4	2.33	1.73	2.86	2.51	1.71	2.14	2.22
5	0.9	0.51	1.32	1.07	0.50	0.80	0.85
mean	1.56	1.06	2.02	1.72	1.05	1.41	**1.47**

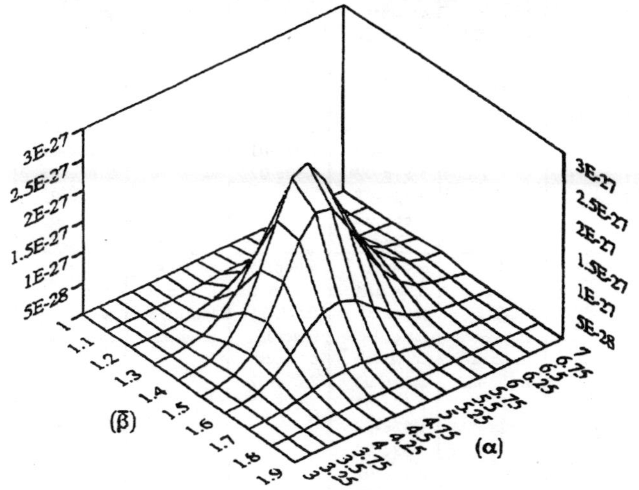

Fig. 2. Shape of the Posterior distribution of $\bar{\beta}$, α, σ_ε^2. (σ_ε^2 fixed at σ_ε^2=0.10)

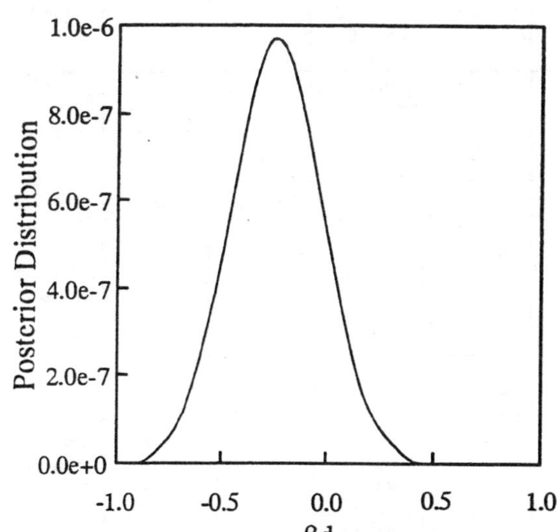

Fig. 3. Shape of the posterior degree of belief distribution of βd_2, the effect of the element type 2 on fatigue life parameter β_{ij}

Table 3. Hypothetical inspection results. Crack sizes given in meters

	Structure				
Elem	**1**	**2**	**3**	**4**	**5**
1	group 1 1 insp. n. 3 total 34 intact 34	group 2 1 insp. n. 3 tot.el. 29 intact 24 cr. sizes 1.73e-01 2.16e-02 1.47e-02 5.22e-02 1.57e-02	group 3 1 insp. n. 3 tot.el. 33 intact 33	group 4 1 insp. n. 1 tot.el. 31 intact 31	group 5 1 insp. n. 2 tot.el. 38 intact 38
2	group 1 2 insp. n. 4 tot.el. 31 intact 28 cr. sizes 2.21e-02 1.84e-02 1.71e-02	group 2 2 insp. n. 4 tot.el. 38 intact 28 cr. sizes 3.32e-02 2.78e-02 3.40e-02 6.02e-02 6.88e-02 3.05e-02 3.97e-02 4.71e-02 6.68e-02 2.97e-02	group 3 2 insp. n. 6 tot.el. 35 intact 27 cr. sizes 9.76e-02 1.08e-01 5.75e-02 3.87e-02 1.33e-01 1.83e-01 4.92e-02 3.79e-02	group 4 2 insp. n. 4 tot.el. 35 intact 31 cr. sizes 7.12e-02 1.26e-02 1.31e-02 2.08e-02	group 5 2 insp. n. 5 tot.el. 27 intact 17 cr. sizes 2.01e-01 7.73e-02 2.33e-01 7.39e-02 7.58e-02 1.18e-01 2.67e-01 9.37e-02 1.34e-01 2.27e-01
3	group 1 3 insp. n. 5 tot.el. 34 intact 30 cr. sizes 1.65e-02 5.25e-02 1.62e-02 2.52e-02	group 2 3 insp. n. 4 tot.el. 35 intact 29 cr. sizes 2.14e-02 2.54e-02 1.85e-02 1.39e-02 2.11e-02 2.15e-02	group 3 3 insp. n. 5 tot.el. 30 intact 28 cr. sizes 2.07e-02 1.67e-02	group 4 3 insp. n. 4 tot.el. 33 intact 33	group 5 3 insp. n. 5 tot.el. 34 intact 30 cr. sizes 1.07e-01 5.26e-02 3.89e-02 1.11e-01
4	group 1 4 insp. n. 2 tot.el. 29 intact 28 cr. sizes 5.19e-02	group 2 4 insp. n. 4 tot.el. 27 intact 24 cr. sizes 2.09e-02 3.52e-02 1.98e-02	group 3 4 insp. n. 3 tot.el. 30 intact 30	group 4 4 insp. n. 4 tot.el. 29 intact 29	group 5 4 insp. n. 3 tot.el. 40 intact 38 cr. sizes 1.36e-02 1.39e-02
5	group 1 5 insp. n. 3 tot.el. 29 intact 28 cr. sizes 1.58e-02	group 2 5 insp. n. 3 tot.el. 30 intact 24 cr. sizes 3.25e-02 2.70e-02 2.40e-02 2.92e-02 2.05e-02 2.16e-02	group 3 5 insp. n. 5 tot.el. 36 intact 29 cr. sizes 2.56e-02 8.95e-02 3.79e-02 5.36e-02 9.83e-02 2.85e-02 2.69e-02	group 4 5 insp. n. 4 tot.el. 33 intact 32 cr. sizes 1.50e-01	group 5 5 insp. n. 2 tot.el. 31 intact 26 cr. sizes 2.14e-02 3.36e-02 7.82e-02 1.16e-01 1.63e-02
6	group 1 6 insp. n. 3 tot.el. 33 intact 33	group 2 6 insp. n. 4 tot.el. 30 intact 22 cr. sizes 2.87e-02 9.50e-02 2.81e-02 3.94e-02 2.45e-02 2.20e-02 3.30e-02 3.40e-02	group 3 6 insp. n. 2 tot.el. 32 intact 32	group 4 6 insp. n. 4 tot.el. 33 intact 32 cr. sizes 2.67e-02	group 5 6 insp. n. 3 tot.el. 36 intact 33 cr. sizes 4.13e-02 1.49e-02 2.37e-02

Figure 4 shows the shape of the posterior degree of belief distribution of βs_4, the effect of the structure number 4 on fatigue life parameter $\beta_{4,2}$, after observing the data corresponding to all element types considered in the structure number 4.

Finally, the posterior degree of belief of the modeling error term, that includes the interaction effect between the element type and the structure it is embedded in, is evaluated by (29). Its shape is given by (30) and is shown in figure 5.

Note that there is a small (negative value) contribution of the error modeling term to $\beta_{4,2}$, that is compatible to the data used, shown in table 2, to generate the inspection results of table 3.

4. Conclusions

A method to infer the effects of the structure and the type of element on the fatigue life of structural members was proposed. It is based on Bayesian methods and on some concepts of classical designs of experiments such as factorial design and decomposition of a general factor (β) into a summation of effects, including an interaction, or modeling error term [Montgomery, 1976]. A numerical test was performed and showed satisfactory results. With the information collected by using the proposed method from the inspection results, it is believed that inspection planning of fatigue sensitive structures can be carried out more effectively.

5. Appendix

In order to evaluate $P_1(a_k, T|\alpha, \beta_{ij})$, which is the conditional probability of finding a crack of size between a_k and a_k+da at T in an element that were as good as new at $T = T_0 = 0$, irrespective of the number of times the element was repaired till T, it is necessary to know the formula of $p_1(a, T_j, T_k|\alpha, \beta_{ij})$ which is the conditional probability of finding a crack of size a at the inspection at T_j in a element that started service at T_k, and has never been repaired since then. It can be given by the following integral:

$$p_1(a, T_j, T_k|\alpha, \beta)\, da = \int_0^{+\infty} f(a|T_j, T_k, C, \alpha, \beta) \times \prod_{\ell=k+1}^{j-1} [1 - H(T_\ell - t_c(a, C))\, D(a(T_\ell - t_c(a, C), C))] \times D(a) \times f_C(C)\, dC\, da \tag{31}$$

In the formula above, H(x) is the Heaviside function defined so that H(x) = 1 if x>0 and H(x) = 0 otherwise.

$t_c(a, C)$ gives the deterministic time when a crack of size a observed at time T_j must have initiated, supposing that the crack propagation parameter C is known. It can be obtained from eq.(11), and is given by

$$t_c(a, C) = T_j - \frac{\left(a^{\frac{2-m}{2}} - a_0^{\frac{2-m}{2}}\right)}{\gamma \times C \times \left(\frac{\sqrt{\pi}}{K_0}\right)^m \times \frac{2-m}{2}} \tag{32}$$

$f_C(C)$ is the probability density function of the crack propagation parameter C, that in this work follows a lognormal distribution.

D( ) is the detection function given in eq.(13).

Finally, $f(a|T_j, T_k, C, \alpha, \beta)$ is the conditional probability density function of the crack size a at the time T_j in an element that started service at T_k, given the knowledge of the parameters C, α and β. It is given by

$$f(a|T_j, T_k, C, \alpha, \beta) = \begin{cases} f_{T_c}(t_c(a, C)|\alpha, \beta)\,\frac{dt_c}{da}, & \text{if } t_c(a, C) > T_k \\ 0 \quad , & \text{otherwise} \end{cases} \quad (33)$$

where $f_{T_c}(t_c(a, C)|\alpha, \beta)$ is given by eq.(7).

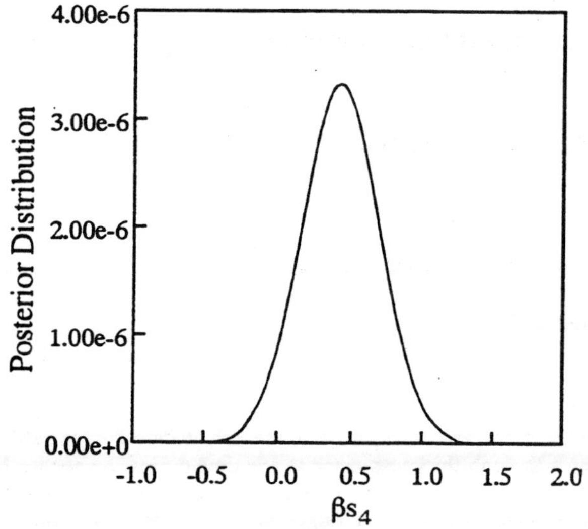

Fig. 4. Shape of the posterior degree of belief distribution of βs_4, the effect of the structure number 4 on fatigue life parameter βij

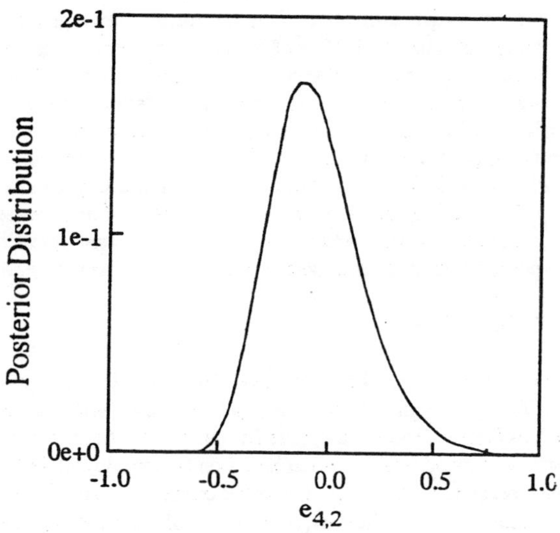

Fig. 5. Shape of the posterior degree of belief of the modeling error or interaction erm $e_{4,2}$

6. Acknowledgment

The author acknowledge the support given by FAPESP through the process number 95/8964-9.

7. References

[Deodatis et alli, 1992] Deodatis G., Fujimoto Y., Ito S., Spencer J., Itagaki H.: Non-Periodic Inspection by Bayesian Method I. *Probabilistic Engineering Mechanics* V.7, pp.191-204, 1992.

[Itagaki and Yamamoto, 1985] Itagaki H., Yamamoto N.: Bayesian Analysis of Inspection on Ship Structural Members. *Proceedings of ICOSSAR'85 4th International Conference on Structural Safety and Reliability*, pp.(III-533)-(III-542).

[Itagaki, Ishizuka & Yan, 1993] Itagaki H., Ishizuka T., Yan H. P.: Experimental Estimation of the Probability distribution of Fatigue Crack Growth Lives. *Probabilistic Engineering Mechanics*, V.8, pp.25-34, 1993.

[Kawano at alli, 1994] Selection of the First Inspection Time Based on the Maximization of Amount of Information. Kawano, A.;Itagaki H., Ishizuka T.: *Journal of The Society of Naval Architects of Japan* V176, pp. 597-602, 1994.

[Lutes et alli, 1984] Lutes L.D., Corazao M., Hu S-L.J., Zimmerman J.J.:Stochastic fatigue damage accumulation. *J. Struct. Engng. ASCE*, 110(ST11), pp. 2585-2601. 1984.

[Montgomery, 1976] Montgomery, D.C. - *"Design and Analysis of Experiments"*, Third ed., John Wiley & Sons, 1976.

[Pittaluga, Cazzulo & Romeo, 1991] Pittaluga A. Cazzulo R., Romeo P.: Uncertainties in the Fatigue Design of Offshore Steel Structures. *Marine Structures* V.4 pp. 317-332, 1991.

[Sander & Bardoux, 1991] Sander P., Bardoux R. (eds).: *Bayesian Methods in Reliability*. Kluwer Academic Publishers, Dordrecht, 1991.

[Sarkani et al., 1994] Sarkani S., Kihl D.P., Beach J.E.: Fatigue of Welded Joints under Narrowband non- Gaussian Loadings, *Probabilistic Engineering Mechanics*, V.9, pp.179-190, 1994.

[Sobczyk K. & Spencer, 1992] Sobczyk K., Spencer B.F. Jr.: *Random Fatigue*. Academic Press, Boston, 1992.

[Sørensen at alli, 1991] Sørensen J.D., Faber M.H. Rackwitz R., Thoft-Cristensen P.: Modelling in Optimal Inspection and Repair. *Proceedings of the OMAE1991*, V.2 Safety and Reliability.

[Yang and Trapp, 1974] Yang J.N. Trapp W.J.: Reliability Analysis of Aircraft Structures under Random Loading and Periodic Inspection. *AIAA Journal* V.12, N.12, 1974.

[Yang et alli, 1993] Yang J.N., Manning S.D., Norton J.M.: A Stochastic Crack Growth Analysis for Aircraft Structures. *Proceedings of the ICOSSAR'93*, Innsbruck, Austria.

Proceedings of the Seventh (1997) International Offshore and Polar Engineering Conference
Honolulu, USA, May 25-30, 1997

ISBN 1-880653-28-1 (Set); ISBN 1-880653-32-X (Vol. IV)

Using Structural Reliability Analysis in Inspection Planning of Offshore Structures

Luis Volnei Sudati Sagrilo and Edison Castro Prates de Lima
COPPE/UFRJ
Rio de Janeiro, Brazil

Carlos Cunha Dias Henriques and Sérgio Guillermo Hormazabal Rodriguez
PETROBRAS S.A.
Rio de Janeiro, Brazil

ABSTRACT

This paper proposes the practical application of a reliability-based procedure for fatigue inspection planning of fixed offshore structures. This procedure is based on a simplified probabilistic crack growth model for welded tubular joints available in the literature, which encompasses fracture mechanics and S-N curves and uses the results of the original fatigue design. Current techniques for reliability updating are implemented to re-evaluate the fatigue failure probability after inspection campaigns. Target reliability values are defined to be compatible with the simplicity of the procedure. Actual applications illustrate the inspection costs reduction that can be obtained.

KEY WORDS: Offshore Structures, Inspection Planning, Fatigue, Structural Reliability.

INTRODUCTION

Fatigue is an important limit state in the design and operation of steel offshore structures. The fatigue design is usually based on Miner's rule and S-N curves but alternatively a fracture mechanics approach can be employed. There are several uncertainties concerning its evaluation. In-service inspections, using nondestructive tests, are planned and performed in order to assure an adequate safety level and to gather more information about the fatigue process. Fatigue inspection results can be basically summarized in detection or no detection of cracks. Until recently, fatigue inspection planning was based mainly on engineering judgment and usually the results of previous inspections were not accounted for the next ones. However, since the structural reliability analysis has become a practical and widely spread tool this situation has changed.

Some important works on probabilistic inspection planning of offshore structures have been published recently (e.g., Madsen et al., 1987; Kirkemo, 1988, Madsen et al., 1989; Jiao, 1992). These works take into account inspections results and show that it is possible to establish a rational fatigue inspection planning. Relevant practical results of probabilistic inspection planning can be seen also in Pedersen et al. (1992). Most of these works employ a reliability method in connection with a fatigue crack growth model, based on linear fracture mechanics, to evaluate and update the probability of getting a through-thickness crack at any time during the service life of the structure. Inspections are indicated whenever this probability falls above a target value. This target must be established taking into account several issues such as consequence of failure, cost of repair and so on. Since this task is not straightforward, this topic has not been clearly defined. As a matter of fact this target should be stated by standard codes, but up to now there are few ones covering this topic.

This paper focus on practical inspection planning of fixed offshore structures using a probabilistic approach. It describes the theoretical topics employed in the development of a reliability-based procedure for fatigue inspection planning of fixed offshore structures and illustrates its practical utilization. This procedure uses a probabilistic mixed fracture mechanics/S-N model for crack growth in welded tubular joints proposed by Banon and Zimmermman (1994) and FORM (Madsen et al., 1996) to compute and update after inspections the probability of fatigue failure and its associated reliability index. Due to some approximations involved in the crack growth model and uncertainties in statistical data, these results and the target values are seen as qualitative indicators of reliability in the practical inspection scheduling process of individual tubular joints. Actual applications show the possible inspection costs reduction that can be obtained

FATIGUE ANALYSIS

Fatigue analysis can be performed basically by two methods (Naess, 1985): 1) Miner's rule and S-N curves (S-N model); 2) Fracture Mechanics (FM model). Traditionally, the fatigue design of welded offshore structures has been based on S-N model while the FM model has been most used in the maintenance of such structures. These two models and their equivalence will be briefly presented below.

S-N Model

The S-N model uses the well-known S-N curves which are obtained from fatigue experiments of tubular joints. An S-N curve is given by

$$N(S) = KS^{-m} \qquad (S > S_0) \qquad (1)$$

where N(S) is the number of cycles to failure under constant amplitude loading S, K and m are material parameters, and S_0 is a threshold level below which no fatigue damage is developed. For design purposes, generally, a characteristic S-N curve is defined by assuming m as a constant and setting K as the mean value of experiments minus two standard deviations (Naess, 1985; DnV, 1984). The Miner's law for fatigue damage is defined as

$$D=\sum_{i=1}^{N}\frac{1}{N_i(S_i)}=\frac{1}{K}\sum_{i=1}^{N}S_i^m \qquad (S_i>S_0) \qquad (2)$$

where N is the number of stress cycles in a reference period of time T_R and D is the fatigue damage accumulation. Assuming that fatigue failure occurs when D reaches an amount Δ, the fatigue design life T_L is given by

$$T_L=\frac{T_R\,\Delta}{D} \qquad (3)$$

In this procedure the stresses correspond to the local hot spot stresses, which are computed by multiplying the nominal stresses Sn, obtained from a global deterministic or stochastic structural dynamic analysis of the structure, by the stress concentration factor SCF. The SCF is obtained from standard parametric formulas or through a finite element analysis of the joint. The fatigue design life can thus be rewritten as

$$T_L=\frac{T_R\,K\Delta}{\sum_{i=1}^{N}\left(Sn_i\,SCF\right)^m} \qquad \left(Sn_iSCF>S_0\right) \qquad (4)$$

Fracture Mechanics Model

The most used FM model is the Paris-Erdogan crack growth equation defined by

$$\frac{da}{dn}=C\left(\Delta K(a)\right)^m=C\left(Y(a)S\sqrt{\pi a}\right)^m \qquad \left(\Delta K(a)>\Delta K_0\right) \quad (5)$$

where a is the crack size (in this paper it will refer to the crack depth only), n is the number of stress cycles, C and m are material parameters, K(a) is the stress intensity factor obtained from linear elastic fracture mechanics, Y(a) is the geometry function, S is the stress amplitude and ΔK_0 is a threshold below which no crack develops. Considering the mean frequency of long term stress cycles process equal to ν, the integration of equation (5) gives

$$\int_{a_0}^{a}\frac{da}{\left(Y(a)\sqrt{\pi a}\right)^m}=C\nu T\,E\left[S^m\right] \qquad (6)$$

where a_0 is the initial crack size, a is the crack size associated to time in service T and E[ ] is used to express the mean or expected value. The geometry function employed in this work corresponds to one that has been frequently used in offshore fatigue reliability analysis (Moan, et al., 1993; Zimmermann and Banon, 1994):

$$Y(a)=\left\{1.08-0.7\left(\frac{a}{t}\right)\right\}\left\{1+1.24\exp\left(-22.1\left(\frac{a}{t}\right)\right)+3.17\exp\left(-357\left(\frac{a}{t}\right)\right)\right\} \quad (7)$$

where a is the crack depth and t is the tube wall thickness.

The FM model can be employed whenever one needs to know the time interval for a crack to grow until a certain size. When using FM model for fatigue design purposes, similarly to S-N model, m is taken as a constant and C as a characteristic value (mean value of experiments minus two standard deviations) (DnV, 1984). The fatigue design life is calculated from equation (6) as the time in service until a critical crack size a_c is reached. Usually fatigue failures of tubular joints are defined by a through-thickness crack and in this case the critical crack size a_c is the tube wall thickness.

Equivalence between FM and S-N Models

A simplified calibration between the FM and S-N models can be obtained by assuming that the fatigue design life predicted by S-N model is the same as the time to obtain a through-thickness crack and the exponent m is the same in both models (Pittaluga et al., 1991; Jiao and Moan, 1992; Zimmermann and Banon, 1994). By this calibration it is possible to show that the time in service T(a) for a crack to grow from a_0 to a is given by (Zimmermann and Banon, 1994)

$$T(a)=\frac{\Psi(a)}{\Psi(a_c)}T_L \qquad (8)$$

where T_L is the fatigue life computed by equation (4) and the function $\Psi(x)$ for a particular value $x=b$ is given by

$$\Psi(b)=\int_{a_0}^{b}\frac{da}{\left(Y(a)\sqrt{\pi a}\right)^m} \qquad (9)$$

In this paper this calibrated model is identified as mixed FM/S-N model.

PROBABILISTIC FATIGUE ANALYSIS

As pointed out in detail in so many works, e.g. Wirshing (1983), Kirkemo (1988) and Zimmermann and Banon (1994), there are several uncertainties associated to the parameters of S-N and FM models. The rational way to address the effect of this uncertainties in the fatigue analysis is through structural reliability methods. For this purpose, a failure function or limit state function $G(\boldsymbol{X})$ must be established in such a way that it indicates a failure state whenever $G(\boldsymbol{X})\le 0.0$ and a safe state otherwise. $\boldsymbol{X}$ stands for a vector including all random variables considered in the analysis.

Probabilistic approaches have been developed for S-N, FM and mixed FM/S-N fatigue analysis models. As the S-N model does not take into account any parameter related to inspection results, i.e., detection or no detection of cracks, its probabilistic approach has been mainly used in the development of standard codes for fatigue design (Wirshing, 1983). On the other hand, FM probabilistic model has been widely used to develop rational fatigue inspection planning of offshore structures considering results of previous inspections (Madsen et al., 1987; Kirkemo, 1988; Madsen et al., 1989). The probabilistic mixed FM/ S-N model can be used for fatigue design criteria development (Pittaluga, et al., 1991; Jiao and Moan, 1992) or for probabilistic fatigue inspection planning (Zimmermann and Banon, 1994). Although some simplifications, the mixed FM/S-N model is attractive for practical purposes since it encompasses fatigue design and fatigue crack growth at same time. For this reason this model has been used in this work and it will be briefly explained in what follows.

Probabilistic FM/S-N Model

From equation (8) a failure function for the mixed FM/S-N model can be written as

$$G(X,T)=\frac{\Psi(a)}{\Psi(a_c)}T_L-T \quad (10)$$

where $P(G(X,T)\le 0.0)$ is the probability of having of crack of size a in T years in service. The random variables X considered in this work are the crack size a for the case when inspection results are considered (see next section), the Miner sum parameter Δ, the stress intensity factor SCF, the S-N curves variability and the uncertainty in the evaluation of the nominal stresses Sn (i.e., structural model idealizations, linearization of drag term in Morison's equation, etc.). Variability of S-N curves was accounted for by taking K as a random variable and m as a deterministic parameter. All random variables described above, but the crack size a, were represented by

$$X_i=\xi_{x_i}\overline{X}_i \quad (11)$$

where X_i stands for any of the parameters Δ, K, SCF or Sn, $\overline{X}_i$ its mean value and ξ_{x_i} is a random variable reflecting the bias inherent to it. In summary, for the present work $X^T=(a,\xi_\Delta,\xi_K,\xi_{SCF},\xi_{Sn})$. Substituting equation (11) into equation (10) one obtains

$$G(X,T)=\frac{\Psi(a)}{\Psi(a_c)}T_L\xi_\Delta\xi_K(\xi_{SCF}\xi_{Sn})^{-m}-T \quad (12)$$

Uncertainty in the geometry function, which is usually considered in many works (Kirkemo, 1988; Moan et al., 1993) by multiplying it by random variable having mean 1.0 and a certain COV, vanishes in the equation (12). Another variable usually considered is the initial crack size a_0. In this work it was taken as deterministic and equal to its mean value because it is assumed that implicitly its variability should be included in the S-N curve uncertainty. When using standard design S-N curves it is necessary to be careful in the evaluation of the bias ξ_K since generally for this curves, the parameter K is defined by a characteristic value instead of its mean value.

The probability of fatigue failure Pf(T) and its associated reliability index β(T) are defined by

$$P(G(X,T)\le 0.0)=Pf(T)=\Phi(-\beta(T)) \quad (13)$$

where Φ(.) is the standard normal probability function. Equation (13) can be evaluated by analytical reliability methods FORM and SORM (Madsen et al., 1986) or by Monte Carlo Simulation-based methods. In this work FORM has been used.

One should notice that in this probabilistic model, besides the statistics of the biases and crack depth, the only data needed is the fatigue design life. As this is usually available, no extra fatigue analyses are necessary.

RELIABILITY UPDATING THROUGH INSPECTION

Setting $a=a_c$ into equation (12), the probabilistic FM/S-N model reduces to the probabilistic S-N model and it is possible to establish the probability of fatigue failure at any time T. This probability can be updated by considering new information gathered from in-service inspections, i.e., no crack detection or detection of a crack with a measured size. The updated probability of fatigue failure $Pf(T)_{up}$ is obtained through the Bayesian approach

$$Pf(T)_{up}=P\left(G(X,T)\le 0.0/(I_1(X,T_1)\cap\cdots\cap I_k(X,T_k))\right)=\frac{P((G(X,T)\le 0.0)\cap I_1(X,T_1)\cap\cdots\cap I_k(X,T_k))}{P(I_1(X,T_1)\cap\cdots\cap I_k(X,T_k))} \quad (14)$$

where $I_i(X,T_i)=-G(X,T_i)\le 0.0/a=a_d$ for no crack detection after T_i years in service, and $I_i(X,T_i)=G(X,T_i)=0.0/a=a_i$ when a crack of size a_i is detected after T_i years in service and $i=1,\cdots,k$ refers to the inspections performed. a_d is the minimum crack size detected by the inspection method employed. It is a random variable modeled by the probability of detection (POD) curve (Madsen et al. 1987, Kirkemo, 1988, Moan et al. 1993). This formulation can be also extended for the case where a crack is detected and repaired (Madsen et al., 1989).

The expression (14) is usually evaluated using first order reliability method (FORM) for series systems. Further details can be found in Madsen et al. (1989) and Jiao (1992).

FATIGUE INSPECTION PLANNING

During their service life, offshore structures are inspected to gather more information and reduce the uncertainty in the fatigue design. In practice, members whose joints have the lowest calculated fatigue lives are those selected to be inspected. Until some years ago, the inspection intervals used to be based on engineering judgment and previous inspection results, mainly when no cracks had been found, used not to be accounted for the next ones. This has changed with the advent of probabilistic fatigue models and reliability updating methods.

Using probabilistic models (before and after updating) the strategy is to inspect a joint at the time in-service T that its probability of fatigue failure Pf(T) falls above a target value Pf_T. Using this strategy it is possible to take into account previous inspection results and calculate optimum inspection intervals, so that the number of inspections is minimum.

One of the most important points to elaborate a probabilistic-based fatigue inspection planning is to set up the target failure probability Pf_T (or its associated reliability index β_T). Setting up Pf_T is not an easy task since it involves considering consequences of failure, costs of maintenance and so on. Other important aspects are the several uncertainties associated to statistical parameters of the random variables involved in fatigue reliability analysis and simplifications in the methods to evaluate the crack growth in tubular joints. Due to these reasons, the use of an absolute value for Pf_T can be doubtful.

An approach for establishing Pf_T for a particular member *i*, which involves a nonlinear collapse analysis of the whole structure, is (Bjerager et al., 1991)

$$Pf_T\le\frac{Pf_{E_i}}{Pf_{E_T}} \quad (15)$$

where Pf_{E_i} is the probability of structural collapse considering the member failed and Pf_{E_T} is the target collapse probability for the structure under consideration. This approach automatically considers the member importance when setting up Pf_T but it requires besides a nonlinear collapse analysis for each member the setting up of Pf_{E_T}.

In this work a practical approach has been employed to set up the target probability Pf_T. When a joint is inspected at first time, according to an existing strategy, it implicitly establishes the target risk that the operator accepts. Therefore, the Pf_T for a given joint can be taken as the calculated probability of failure at the time of its first inspection. This approach has the advantage of not setting up an absolute target probability of failure. It establishes a qualitative target value that is in accordance with the method of analysis and the statistical description employed.

APPLICATIONS

A computer code called RLINSP, incorporating the topics described above, has been developed to be used in the re-analysis of the inspection plans of the fixed offshore platforms installed in Campos Basin, offshore Brazil. In what follows, re-evaluation of fatigue inspection planning of two joints belonging to Cherne-2 and Garoupa platforms will be presented. Table (1) shows the statistical data employed in the analyses. Most of them are taken from available literature on fatigue reliability analysis (Wirshing, 1983; Madsen et al. 1989; Zimmermann and Banon, 1994). Both structures were designed using API X' S-N curve and statistical bias for the parameter K of this curve was taken from Wirshing (1979). The mean initial crack depth a_0 was assumed to be 0.11mm for both joints. Table (2) shows the fatigue design lives and inspections results for both joints. The strategy of PETROBRAS for these joints is to inspect them each ¼ of their design life. The inspections performed do not correspond exactly to it because the fatigue lives shown in Table (2) are not those obtained in the original design. They were recently re-calculated when these two structures were analyzed considering new SCFs.

Figures (1) and (2) show the reliability index before and after the inspections for both joints. Through these results it is possible to observe that in order to maintain a reliability level above the target throughout the life time, the Cherne-2 joint should have been inspected after ≈13.00 years in-service and the Garoupa joint shall be inspected after ≈23.00.

The probabilistic approach can also be employed to define strategies for future inspections. For this purpose it was assumed that no cracks are found in the inspections and the life time of both structures is 30.00 years. Table (3) shows three strategies for each joint that maintain the reliability level above the target during the life time of the structures. For the Cherne-2 joint it was assumed that the next inspection will be in the beginning of 1997 (≈14.3 years in-service). Figure (3) illustrates graphically the three strategies for the Cherne-2 joint. It can be observed that, in case of no crack detection, one magnetic particle/Eddie current inspection or two visual inspections at time intervals shown in Table (3) will be enough to maintain the acceptable safety level during the service life of the platforms. It must be also observed that following the standard procedure and under the same conditions, i.e. no crack detection, at least three inspections would be performed in each joint without any sound basis to select the method. A potential inspection costs reduction is evident.

CONCLUSIONS

In this paper the utilization of a reliability-based procedure for fatigue inspection planning of fixed offshore platforms was presented. This procedure employs some techniques available in the literature in the fields of structural reliability analysis and reliability updating and uses a mixed Fracture Mechanics/SN model for crack growth. It is a very effective and practical procedure since it only needs the calculated fatigue design life of a joint and the inspection results to evaluate and update the probability of fatigue failure. The target probability of failure, which defines inspection intervals, is established by considering that the first inspection, according to an existing inspection plan, reflects the risk level that the operator implicitly assumes.

From the results presented it is possible to observe that inspections can be economically planned without any significant computational effort by simply retaining the same risk level of the first inspection.

Variable	Mean	Std. Dev.	Distribution
ξ_{SCF}	1.00	0.2000	Normal
ξ_{Δ}	1.00	0.3000	Lognormal
ξ_{Sn}	1.00	0.2500	Normal
a_d (mm)(1)	1.30(2) 2.00(3) 6.50(4)	1.30 2.00 6.50	Exponential Exponential Exponential

Notes:
(1) Minimum detectable crack depth;
(2) Underwater Magnetic Particle Inspection (MPI);
(3) Underwater Eddie Current Inspection (ECI);
(4) Underwater Visual Inspection (VI).

Table (1) - Random variables and their statistical description.

Platform	Brace/ Joint	Thickness (mm)	Fatigue Design Life (years)(1)	Inspection Results(2)
Cherne-2	5504/ 555	30.00	22	No cracks. Visual Inspection in February, 1989.
Garoupa	11023/ 1126	12.70	20	No cracks. MPI in March, 1992.

Notes:
(1) values obtained after reanalyzing both platforms using Efthymiou's SCFs.
(2) Cherne-2 installed in November, 1982 and Garoupa in May, 1980.

Table(2) - Fatigue design lives and inspection results.

Platform	Strategy(1)	Inspection Method	Years in-service
Garoupa	A	MPI	1st - 23.00
	B	ECI	1st - 23.00
	C	VI	1st - 23.00
		VI	2nd - 28.00
Cherne-2	A	MPI	1st - 14.30
	B	ECI	1st - 14.30
	C	VI	1st - 14.30
		VI	2nd - 22.00

Note:
(1) Assuming no crack detection

Table (3) - Inspection strategies for the joints.

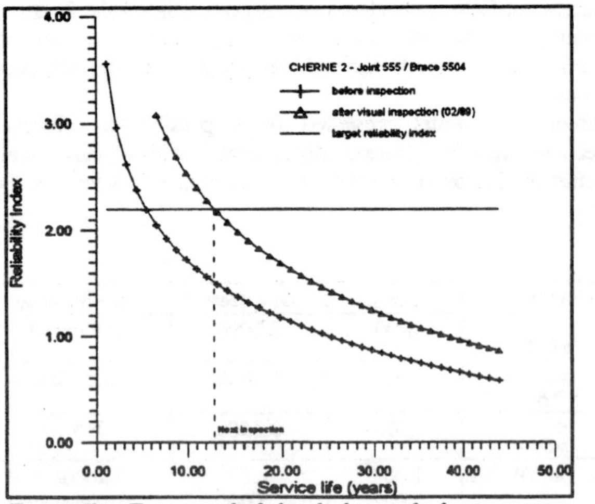

Figure (1) - Fatigue reliability before and after inspection of the Cherne-2 joint.

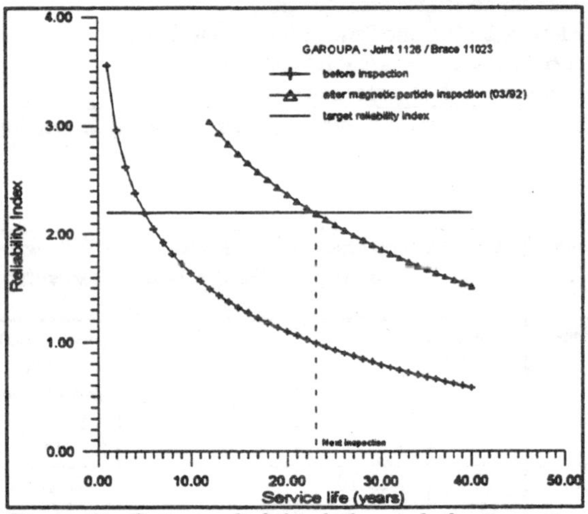

Figure (2) - Fatigue reliability before and after inspection of the Garoupa joint.

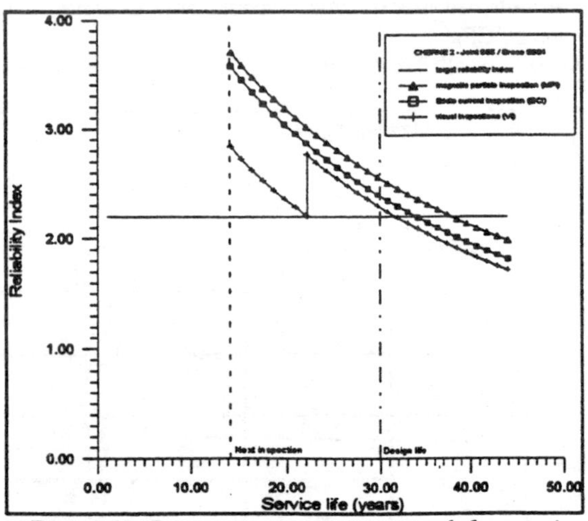

Figure (3) - Inspection strategies (no crack detection) for the Cherne-2 joint

REFERENCES

Bjerager, P., Langen, I., Winterstein, S.R., Marthinsen, T. and Karunakaran, D. (1991). "Application of Nonlinear Stochastic Mechanics in Offshore Engineering", *Proc. of IUTAM Symposium on Nonlinear Stochastic Mechanics*, Torino, Italy.

DnV (Det Norsk Veritas) (1984). "Fatigue Strength Analysis for Mobile Offshore Units", *Classification Notes*, No. 30.2.

Jiao, G. (1992). "Reliability Analysis of Crack Growth with Inspection Planning", *Proc. of the 11th Int. Conf. on Offshore Mechanics and Arctic Engineering (OMAE)*, Alberta, Canada.

Jiao, G. and Moan, T. (1992). "Reliability-Based Fatigue and Fracture Design Criteria for Welded Offshore Structures", *Engineering Fracture Mechanics*, Vol. 41, No. 2.

Kirkemo, F. (1988). "Applications of Probabilistic Fracture Mechanics to Offshore Structures" *Proc. of the 7th Int. Conf. on Offshore Mechanics and Arctic Engineering (OMAE)*, Houston, USA.

Madsen, H.O., Krenk, S. and Lind, N.C. (1986). *Methods of Structural Safety*, Prentice-Hall, Inc., Englewood Cliffs, New Jersey.

Madsen, H.O., Skjong, R. and Kirkemo, F. (1987). "Probabilistic Fatigue Analysis of Offshore Structures - Reliability Updating Through Inspection Results", *Proc. of Integrity of Offshore Structures Conference (IOS'87)*, Glasgow, Scotland.

Madsen, H.O., Sorensen, J.D. and Olesen, R. (1989). "Optimal Inspection Planning for Fatigue Damage of Offshore Structures", *Proc. of 5th Int. Conf. on Structural Safety and Reliability (ICOSSAR)*, San Francisco, USA.

Moan, T., Hovde, G.O. and Blanker, A.M. (1993). "Reliability-Based Fatigue Design Criteria for Offshore Structures Considering the Effect of Inspection and Repair", *Proc. of Offshore Technology Conference*, OTC Paper 7189, Houston, USA.

Naess, A. (1985) - *Fatigue Handbook*, Tapir, Trondheim.

Pedersen, C., Nielsen, J.A., Riber, J.P., Madsen, H.O. and Krenk, S. (1992). "Reliability Based Inspection Planning for The Tyra Field", *Proc. of the 11th Int. Conf. on Offshore Mechanics and Arctic Engineering (OMAE)*, Alberta, Canada.

Pittaluga, A., Cazzulo, R. and Romeo, P. (1991). "Uncertainties in the Fatigue Design of Offshore Steel Structure", *Marine Structures*, Vol. 4, No. 4.

Wirshing, P.H. (1979) . "Fatigue Reliability in Welded Joints of Offshore Structures", *Proc. of the Offshore Technology Conference*, Houston, USA.

Wirshing, P.H. (1983). "Fatigue Reliability for Offshore Structures", *Journal of Structural Engineering (ASCE)*, Vol. 110, No. 10.

Zimmermann, J.J. and Banon, H. (1994). "System Fatigue Reliability and Inspection Planning for Offshore Platforms", *Proc. of 13th Int. Conf. on Offshore Mechanics and Arctic Engineering (OMAE)*, Houston, USA.

Proceedings of the Seventh (1997) International Offshore and Polar Engineering Conference
Honolulu, USA, May 25-30, 1997

ISBN 1-880653-28-1 (Set); ISBN 1-880653-32-X (Vol. IV)

Sensitivity Analysis on Fatigue Reliability and Inspection of Ship Structural Members

Yukio Fujimoto, Sung Chan Kim and Eiji Shintaku
Hiroshima University
Higashi-Hiroshima, Japan

ABSTRACT

Sensitivity analysis on fatigue reliability and inspection is carried out for a typical structural member of bulk carrier. Fatigue properties of the member and probability of crack detection at the field inspection are obtained by the questionnaire asked to the naval engineers. Changing stress level(fatigue life) of the member, crack growth curve and inspection interval as parameters around a standard condition, failure probabilities of member during 20 years' service are calculated by the Markov chain model. The result of analysis can suggest the solutions for design requirements such as improvement of reliability level, reduction of structural weight and the simplification of inspection.

In the latter part of the paper, effectiveness of combined inspection method and sampling inspection method are discussed through the sensitivity analysis assuming member sets having different number of initiated cracks. It is found that combined inspection has more advantage than sampling inspection, when the number of members having initiated crack is a few in the inspecting member set.

1. INTRODUCTION

In-service inspection on ship structure plays an important role for maintaining safety against fatigue damage. However, when we want to perform reasonable inspection, it is necessary to clarify the present reliability level of structural members as well as the desirable level.

From this point of view, in the previous study, we conducted questionnaire about "Fatigue damage and inspection of bulk carrier" to naval engineers(Fujimoto et al, 1996, Kim et al, 1996 and SR226 Committee, 1996). In the questionnaire, six locations of bulk carrier reporting frequent damage occurrences were selected and the number of cracks initiated in the members, the fatigue characteristics of members, the probability of crack detection by field inspection, the target reliability level of members, etc., were asked.

Then, fatigue reliability analysis considering repeated inspections was carried out using the information obtained by the questionnaire. From the analysis, how much present inspections contribute to maintain the reliability level and what adequate inspection interval is to accomplish the target reliability level were made clear.

In this study, sensitivity analysis was carried out around the present inspection condition of a typical structural member. There exist certain uncertainties on stress level and crack growth curve for ship structural members. At the design stage, further, there are several requirements such as improvement of reliability, reduction of structural weight, and simplification of inspection. In order to respond to these problems, failure probabilities of the member were calculated by changing stress level(fatigue life) of the member, the crack growth curve and inspection interval as parameters around a standard condition. Markov chain model was employed in the reliability analysis.

From the limitation of cost and time at real inspection, an inspection is applied in the actual ships through long distance visual inspection, combined method with long and close visual inspection, or sampling inspection. The effectiveness of these inspection methods are discussed through the sensitivity analysis in the latter part of the paper.

2. OUTLINE OF QUESTIONNAIRE

The six locations of bulk carrier reporting frequent damage occurrences were the object of the questionnaire (M1 to M6 in Fig.1). In the questionnaire, answer was requested to reply subjectively the average condition of fatigue damage and inspection per a ship during 20 years' service on the basis of the damage examples he experienced in the past. Namely, frequent damaged ship, very sound ship, different sea route and different operating method were mixed in the average condition. To get more accurate answer, critical crack length was defined as 200mm or 500mm which is often experienced at ordinary inspection and damage examples. The crack exceeding the critical crack length was considered as failure event. The crack initiation was defined with 20mm length. The detailed content of questionnaire and its result are in References(Fujimoto et al, 1996, Kim et al, 1996 and SR226 Committee, 1996).

2.1 Fatigue Characteristics of Standard Member

Under the limitation of cracks which initiate within 20 years' service, their crack initiation time and crack propagation lives from 20 to 200mm and from 200 to 500mm were asked for the six members. In the results, the fatigue characteristics for the five members except member (M3) were similar. Fig.2 shows the distributions of crack initiation time, fatigue life to 200mm in length(crack initiation time is included) and fatigue life to 500mm obtained by averaging the properties of five members except member (M3). These distributions were used as the fatigue property of a standard member in the sensitivity analysis.

The crack growth pattern was asked, but the trend of crack growth pattern of each member was not made clear from the questionnaire. So, the sensitivity analysis was carried out for the four kinds of patterns shown in Fig.3. Each curve is expressed by the following equation.

$$a = (a_{cr} - a_0)(N/N_{cr})^k + a_0 \tag{1}$$

where the crack length at initiation time (a_0) is 20mm, critical crack length (a_{cr}) is 200mm or 500mm, N_{cr} is mean life to arrive to critical crack length and k is 4.0 for pattern①, 2.0 for pattern②, 1.0 for pattern③ and 0.7 for pattern④.

Failure was defined by the crack exceeding 200mm or 500mm in length as described above. So, the cumulative probability of failure is defined as follows.

$P_f(200)$: Cumulative probability of failure that a crack, which is initiated within 20 years and crack initiation time follows the distribution in Fig.2, propagates over 200mm in length (probability of failure per one member)

$P_f(500)$: Cumulative probability of failure that a crack, which is initiated within 20 years, propagates over 500mm in length (probability of failure per one member)

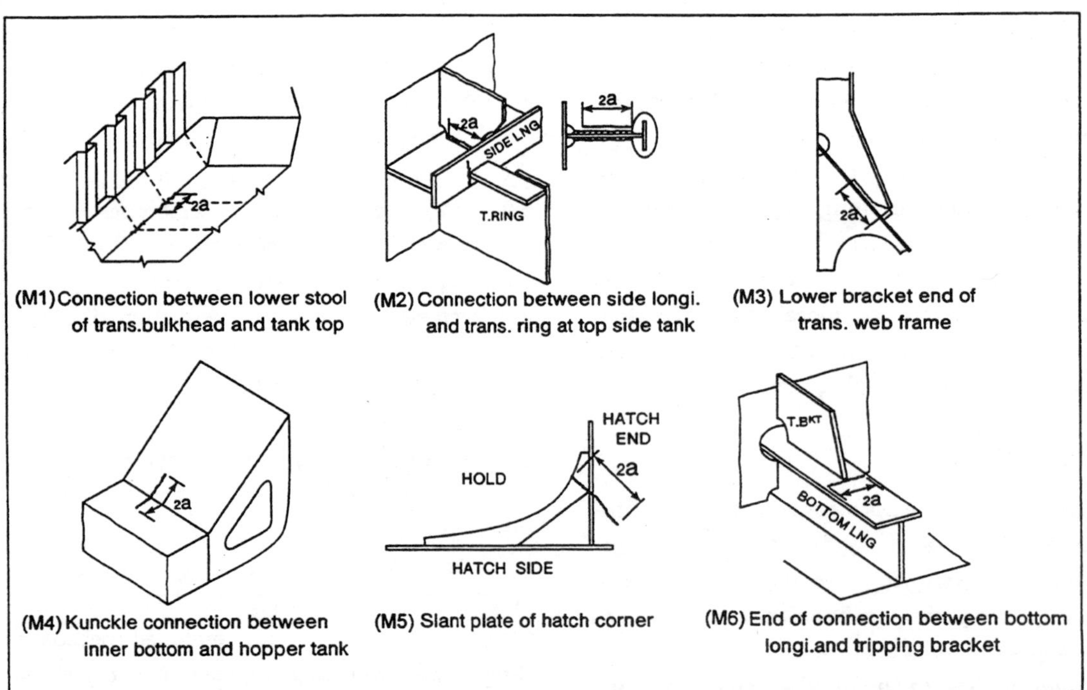

(M1) Connection between lower stool of trans.bulkhead and tank top

(M2) Connection between side longi. and trans. ring at top side tank

(M3) Lower bracket end of trans. web frame

(M4) Kunckle connection between inner bottom and hopper tank

(M5) Slant plate of hatch corner

(M6) End of connection between bottom longi.and tripping bracket

Fig.1 Structural members selected as the object of questionnaire.

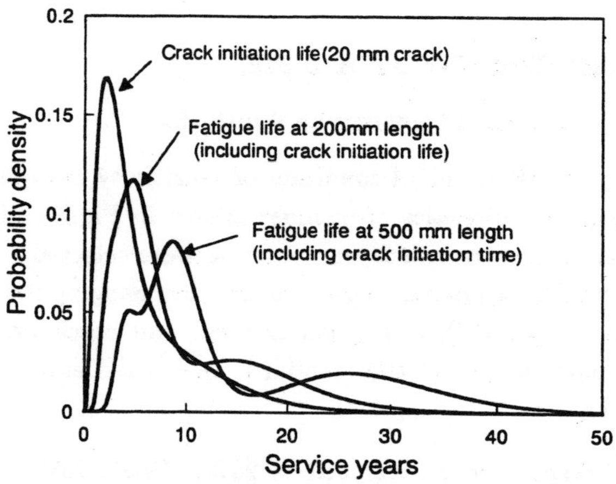

Fig.2 Fatigue life distributions of standard member.

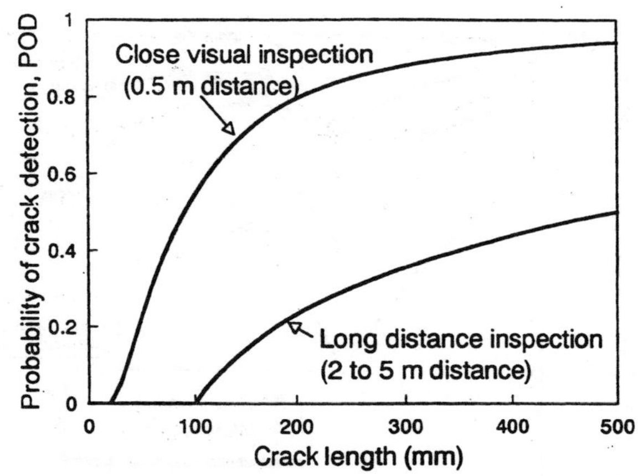

Fig.4 POD curves for close visual and long distance visual inspections.

2.2 Probability of Crack Detection

POD(Probability of crack detection) for visual inspection was asked under good condition without influence of corrosion and painting. In the results of reply, it was seen that detection probability was influenced by crack orientation, crack opening breadth, existence of hint for crack detection and existence of adjacent structural member hindering crack detection. However, average POD for close visual inspection and long distance inspection was obtained as shown in Fig.4. Close visual inspection is to inspect the member from about $0.5m$ distance and long distance visual inspection is from $2m$ to $5m$ distance. These POD's are expressed as

Close visual:

$$(POD)_1 = \frac{\exp\{-6.18 + 1.45\log(a-20)\}}{1+\exp\{-6.18+1.45\log(a-20)\}} \quad (2)$$

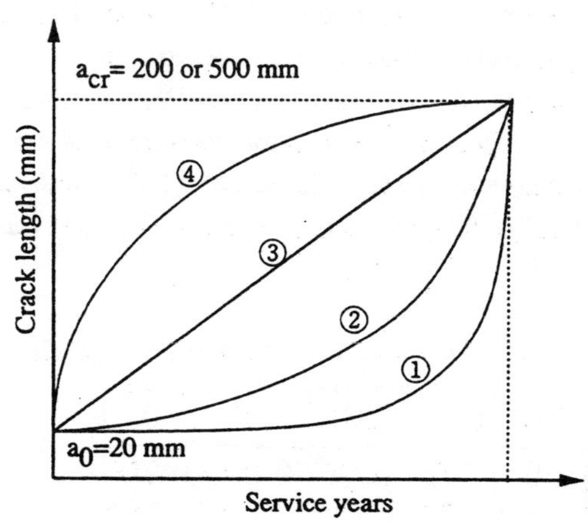

Fig.3 Crack growth pattern

Long distance:

$$(POD)_2 = \frac{\exp\{-5.18 + 0.86\log(a-100)\}}{1+\exp\{-5.18+0.86\log(a-100)\}} \quad (3)$$

where a is visible crack length.

2.3 Member Reliability under Present Inspection Condition

In the previous study, fatigue reliability analysis was carried out employing Markov Chain Model(Fujimoto et al,1989 and Fujimoto et al, 1996). Perfect repair model was assumed in the model. The fatigue life distribution and the POD curve of respective members were determined from the reply of questionnaire. The crack growth pattern③ was assumed for all the members. Five kinds of inspection interval as 3 months, 6 months, 1 year, 2 years and 4 years were examined.

Fig. 5 summarizes the calculated $P_f(200)$ and $P_f(500)$ at 20 year's service for the six members, in which the failure probabilities obtained from the questionnaire are also plotted for the comparison. The $P_f(200)$ of the questionnaire result is near to that of calculated result with 2 years inspection interval, and the $P_f(500)$ of the questionnaire result is near to the calculated result with 4 years interval. Four years and two years intervals correspond to regular or intermediate inspection interval of ship structure.

If inspections are not carried out during the entire life, about 90% of initiated cracks would propagate over $200mm$ and about 60% propagate over $500mm$. However, $P_f(200)$ and $P_f(500)$ is maintained in the level of 0.3 and 0.08, respectively, on the average of six members. These probabilities mean that if there are ten initiated cracks about three

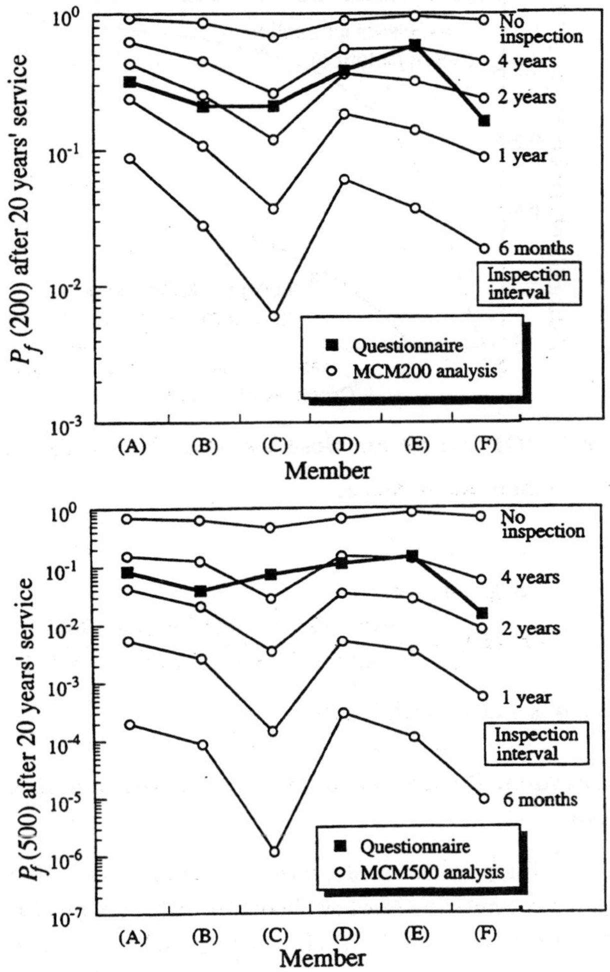

Fig.5 Failure probabilities of six members for respective inspection interval.

cracks among them would propagate over 200mm, and about one crack would propagate over 500mm under the present inspection. From the comparison with the case of no inspection, it is understood that the present inspection can reduce $P_f(200)$ to about 1/3 and $P_f(500)$ to about 1/7, respectively. In the sensitivity analysis 2 years interval of close visual inspection was chosen as the standard inspection condition.

3. SENSITIVITY ANALYSIS

3.1 Contents of Sensitivity Analysis

As the standard condition of sensitivity analysis, the fatigue characteristics of member shown in Fig.2 and close visual inspection with 2 years interval were selected. The sensitivity analysis was performed by changing the stress level (fatigue life), inspection interval and crack propagation pattern around the standard condition.Details are as follows.

1) Eight fatigue lives such as 25% , 50% , 75% , 100%, 150% , 200% , 300% and 400% of standard life.
2) Six inspection intervals such as 3 months, 6 months, 1 year, 2 years, 4 years and no inspection.
3) Four kinds of crack growth pattern in Fig.3.

Table 1 shows eight fatigue lives and corresponded equivalent stress levels which are calculated from $S = CN^{-m}$ $(m = 0.2)$.

3.2 Results of Sensitivity Analysis

Figs.6(a),(b) show one example of cumulative probability of failure to service years when inspection interval is 2 years, crack growth pattern is② and critical crack length is 200mm or 500mm. Figs.7(a),(b) show the cumulative probability of failure after 20 years' service to the ratio of fatigue life (or ratio of equivalent stress) for all the cases calculated.

It can be seen that cumulative probability of failure becomes to decrease with the decrease of stress level and inspection interval. For crack growth pattern①, failure probability is quite high, but big differences were not seen for the other patterns. The $P_f(200)$ and $P_f(500)$ of crack growth pattern① are about two times and five times of that of pattern④ respectively at the standard condition.

Table 2 is the classified level of $P_f(200)$ and $P_f(500)$ after 20 years' service. Level A means very safe state, level F means dangerous state. Tables 3(a),(b) summarize the $P_f(200)$ and $P_f(500)$ following the classification levels shown

Table 1 Eight fatigue lives and corresponded stress levels used in the sensitivity analysis.

Fatigue life ratio (N/N_0)	0.25	0.5	0.75	1.0	1.5	2.0	3.0	4.0
Stress ratio (S/S_0)	1.32	1.15	1.06	1.0	0.92	0.87	0.80	0.76

N_0, S_0 : Standard fatigue life and corresponded stress level

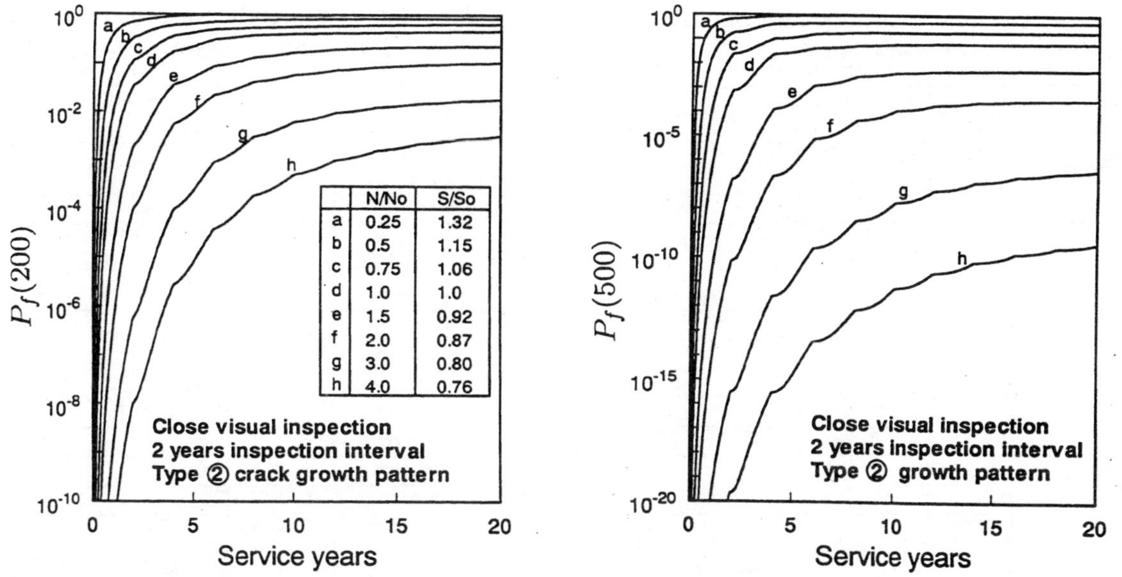

Fig.6 Relationship between failure probability and ser vice years with the change of stress level.

Table 2 Classified level of cumulative probability of failure after 20 years' service.

Level of reliability	Failure probability after 20 years $P_f(200), P_f(500)$	
A	Less than 0.0001	Very safe
B	0.0001 to 0.001	
C	0.001 to 0.01	
D	0.01 to 0.1	
E	0.1 to 0.5	
F	Larger than 0.5	Dangerous

in Table 2. . In the table it is understood that stress level influences most to structural reliability, but reliability level is also changed very much due to the change of inspection interval. In the case of standard stress level with 2 years inspection interval, $P_f(200)$'s almost belong to level E and $P_f(500)$'s belong to level D.

In this sensitivity analysis, the fatigue characteristics of the member and the inspection capability were determined based on the questionnaire asked to naval engineers. Therefore, if the analytical condition represents the fatigue

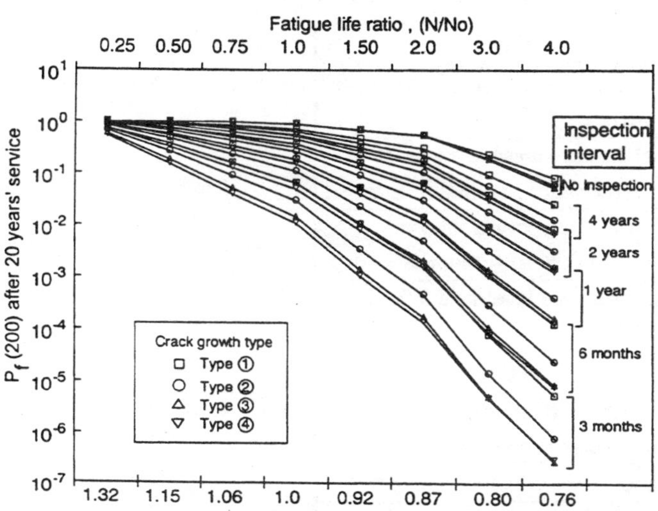

Fig.7(a) Cumulative failure probability, $P_f(200)$, after 20 years' service for several analytical conditions.

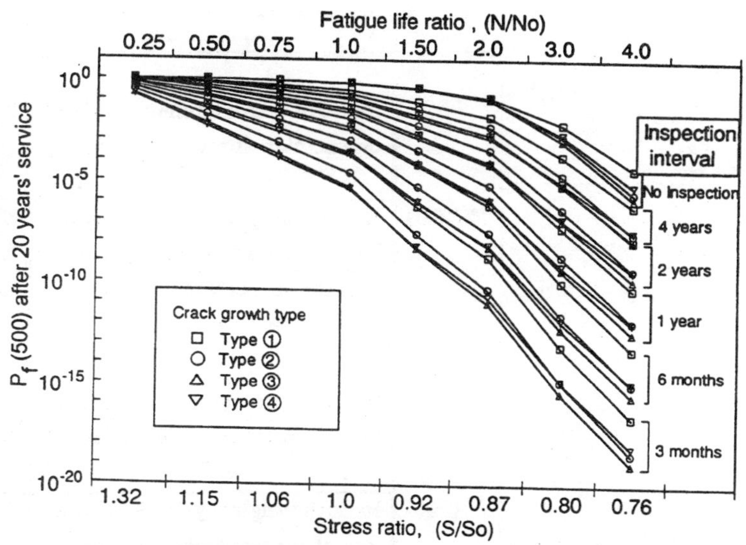

Fig.7(b) Cumulative failure probability, $P_f(500)$, after 20 years' service for several analytical conditions.

Table 3(a) Level of cumulative probability of failure, $P_f(200)$, after 20 years' service.

Crack Growth	Inspection Interval	Stress Ratio S/S_0							
		1.32	1.15	1.06	1.00	0.92	0.87	0.80	0.76
①	3 months	F	E	E	D	D	C	A	A
	6 months	F	F	E	E	D	D	C	B
	1 year	F	F	F	E	E	D	C	C
	2 years	F	F	F	(F)	E	E	D	C
	4 years	F	F	F	F	E	E	D	D
	NO insp.	F	F	F	F	F	F	E	D
②	3 months	F	E	D	D	C	B	A	A
	6 months	F	E	E	E	D	C	B	A
	1 year	F	F	E	E	D	D	C	B
	2 years	F	F	F	(E)	E	E	D	C
	4 years	F	F	F	F	E	E	D	D
	NO insp.	F	F	F	F	F	F	E	D
③	3 months	F	E	D	D	C	B	A	A
	6 months	F	E	E	D	D	C	B	A
	1 year	F	F	E	E	D·	D	C	B
	2 years	F	F	F	(E)	E	D	C	C
	4 years	F	F	F	F	E	E	D	C
	NO insp.	F	F	F	F	F	F	E	D
④	3 months	F	E	D	D	C	B	A	A
	6 months	F	E	E	D	C	C	A	A
	1 year	F	F	E	E	D	D	C	B
	2 years	F	F	F	(E)	E	D·	C	C
	4 years	F	F	F	F	E	E	D	C
	NO insp.	F	F	F	F	F	F	E	D

Table 3(b) Level of cumulative probability of failure, $P_f(500)$, after 20 years' service.

Crack Growth	Inspection Interval	Stress Ratio S/S_0							
		1.32	1.15	1.06	1.00	0.92	0.87	0.80	0.76
①	3 months	E	D	C	B	A	A	A	A
	6 months	F	E	D	C	A	A	A	A
	1 year	F	E	E	D	C	A	A	A
	2 years	F	F	E	(E)	D	C	A	A
	4 years	F	F	F	E	D	D	B	A
	NO insp.	F	F	F	F	E	E	C	A
②	3 months	E	D	B	A	A	A	A	A
	6 months	F	D	C	B	A	A	A	A
	1 year	F	E	D	D	B	A	A	A
	2 years	F	E	E	(D)	C	B	A	A
	4 years	F	F	E	E	D	C	A	A
	NO insp.	F	F	F	F	E	E	C	A
③	3 months	E	C	B	A	A	A	A	A
	6 months	E	D	C	B	A	A	A	A
	1 year	F	E	D	C	A	A	A	A
	2 years	F	E	E	(D)	C	A	A	A
	4 years	F	F	E	E	D	C	A	A
	NO insp.	F	F	F	F	E	E	C	A
④	3 months	E	C	B	A	A	A	A	A
	6 months	E	D	C	B	A	A	A	A
	1 year	F	E	D	C	A	A	A	A
	2 years	F	E	D	(D)	C	A	A	A
	4 years	F	F	E	E	D	C	A	A
	NO insp.	F	F	F	F	E	E	C	A

and inspection condition of real structure, it may be possible to suggest solutions for the several design requirements. The following solutions can be extracted from the Tables 3(a),(b).

(1) Present $P_f(200)$ belongs to level E. If we want to improve the reliability level to D, it is necessary to reduce inspection interval from 2 years to about 4 months, or to reduce stress level by 15% under 2 years inspection

Table 4 Solutions to achieve design requirements obtained by sensitivity analysis

	Requirements	Solutions
$P_f(200)$	Improve present reliability level by one step (from level E to D)	Shorten inspection interval from 2 years to 4 months
		(or) Lower stress level by 15 % under 2 years inspection interval
	Reduce weight by 10 % without loss of present reliability level	Shorten inspection interval from 2 years to 6 months
	Remove inspections without loss of present reliability level	Lower stress level by 20 %
$P_f(500)$	Improve present reliability level by one step (from level D to C)	Shorten inspection interval from 2 years to 10 months
		(or) Lower stress level by 10 % under 2 years inspection interval
	Reduce weight by 10 % without loss of present reliability level	Shorten inspection interval from 2 years to 8 months
	Remove inspections without loss of present reliability level	Lower stress level by 15 %

interval.

(2) If we want to reduce stress level by 10% with sustaining present $P_f(200)$, it is necessary to reduce inspection interval from 2 years to 6 months.

(3) If we want to sustain present $P_f(200)$ with no inspection, it is necessary to reduce stress level by 20%.

(4) Present $P_f(500)$ belongs to level D. If we want to improve the reliability level to C, it is necessary to reduce inspection interval from 2 years to about 10 months, or to reduce stress level by 10 % under 2 years inspection interval.

(5) If we want to reduce stress level by 10% with sustaining present $P_f(500)$, it is necessary to reduce inspection interval from 2 years to 8 months.

(6) If we want to sustain present $P_f(500)$ with no inspection, it is necessary to reduce stress level by 15%.

Table 4 shows the summary of these solutions.

4. COMBINED INSPECTION AND SAMPLING INSPECTION

4.1 Method of Analysis

As ship structure is composed of a lot of members, the application of close visual inspection for all the members is usually very difficult on the time and cost aspect. Therefore, long distance visual inspection, combined one or sampling one is often applied in the actual field. In this study, the effectiveness of combined and sampling inspection was studied assuming a member set.

The member set consists of two types of members; 'n' members will be cracked during 20 years' service and the rest members will not be cracked during this period. The cracked members('n' members) have the fatigue characteristics shown in Fig.2, and have the crack growth pattern 3. The critical crack length of the member was defined as $500mm$. The probabilities of crack detection for close visual and long distance visual inspections were evaluated by Eqs.(2) and (3), respectively. The analysis was carried out for the following four inspection methods changing the number of members having initiated crack.

Case A : Long distance visual inspection for all the members is carried out ($2m$ to $5m$ from inspector's naked eye).

Case B : Combined inspection is carried out. At first, long distance visual inspection for all the members is carried out. If at least one crack is detected, close visual inspection for all the members is carried out ($0.5m$ from inspector's naked eye). Otherwise, inspection is terminated.

Case C10 : Close visual inspection for 10% sampled members is carried out. If at least one crack is detected among the sampled members, close visual inspection for all the remaining members is carried out. Otherwise, inspection is terminated.

Case C20 : Sampling rate is selected as 20% and the inspection procedure is same as Case C10.

Case D : Close visual inspection for all the members is carried out.

The probability of crack detection for Case B, $(POD)_B$, is calculated by

$$(POD)_B = (1-Q_2)^n \times (POD)_2 + [1-(1-Q_2)^n] \times (POD)_1 \tag{4}$$

where n is the number of members having initiated crack during 20 years in the member set. Q_2 is the detection ratio of crack by the long distance visual inspection and is calculated by multiplying $(POD)_2$ by the distribution of crack length, $f_a(a)$ (crack state vector of Markov chain model), at each inspection time.

$$Q_2 = \int_{a_0}^{a_{cr}} (POD)_2 \times f_a(a)da \tag{5}$$

The probability of crack detection for Case C10 and Case C20, $(POD)_c$, is expressed by

$$(POD)_C = r \times (POD)_1 + (1-r) \times [1-(1-Q_1)^{rn}] \times (POD)_1 \tag{6}$$

where r is sampling rate and Q_1 is the detection ratio of crack by the close visual inspection and is calculated by the following equation.

$$Q_1 = \int_{a_0}^{a_{cr}} (POD)_1 \times f_a(a)da \tag{7}$$

4.2 Results of Analysis

Fig.8 shows cumulative probability of failure after 20 years for the case of n = 5, 10, 20, 50 and 100 members with respect to different inspection methods. Failure probability for long distance visual inspection is the highest among the five cases. The order of failure probability is 10% sampling inspection, 20% sampling inspection, combined inspection, and the lowest is the close visual inspection for all the members. Following results can be extracted from the Fig.8.

(1) The smaller the inspection interval is, the bigger the difference of cumulative probability of failure between Case A and Case D is.

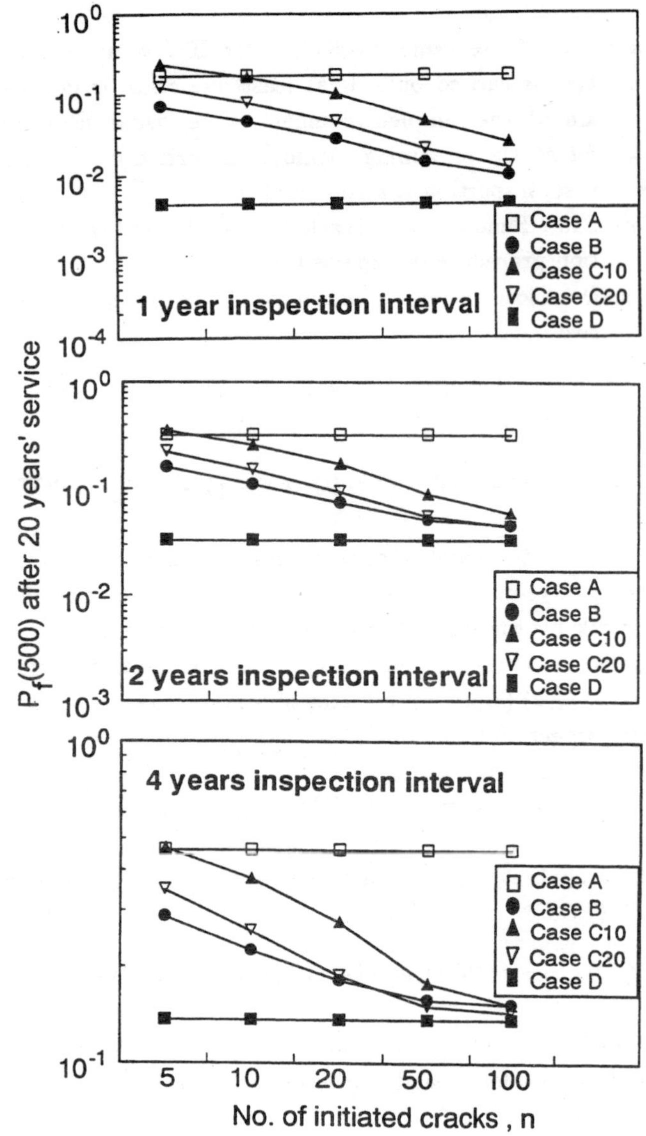

Fig.8 Effectiveness of combined and sampling inspection.

(2) Combined and sampling inspections become effective as the increase in the number of members having initiated crack in the member set.

(3) Combined inspection has more advantages than sampling inspection when number of members having initiated crack becomes small in the member set.

5. CONCLUSIONS

Hot spot stress amplitude and crack growth pattern are usually uncertain in ship structural members. Furthermore, there are several requirements such as the improvement of reliability, reduction of weight and simplification of inspection at the fatigue design. In order to respond to these problems, sensitivity analysis of fatigue reliability and inspection was carried out based on the information obtained from questionnaire. Also, the effectiveness of combined inspection method and sampling inspection method which are often used in the actual field was discussed through the sensitivity analysis. The major conclusions are drawn as follows:

(1) Under present inspection condition, it is thought that about 70% of initiated cracks are detected below 200*mm* and about 90% are detected below 500*mm*. This reliability level is maintained by the close visual inspection with 2 years interval.

(2) For the design requirements such as the improvement of reliability, reduction of weight and simplification of inspection, the solutions to respond to the requirements were extracted as shown in Table 4.

(3) Combined inspection, which is performed by both long distance and close visual inspection, has more advantage than sampling inspection, when the number of members having initiated crack is a few in the inspecting member set.

RERERENCES

Fujimoto, Y. et al, "Reliability Assessment of Deteriorating Structure by Markov Chain Model", J. of the Society of Naval Architects of Japan, 166, pp.303-314 (1989) (in Japanese)

Fujimoto, Y., Swilem, S.A.M. and Iwata, M., "Inspection Planning for Deteriorating Structures Based on Sequential Cost Minimization Method", J. of the Society of Naval Architects of Japan, 170, pp.755-768 (1991)

Fujimoto, Y., Kim, S.C., Shintaku, E. and Ohtaka, K., "Study on Fatigue Reliability and Inspection of Ship Structures Based on questionnaire Information", J. of the Society of Naval Architects of Japan, 180, pp.601-609 (1996) (in Japanese)

Kim, S.C., Fujimoto, Y., Shintaku, E., "Study on Fatigue Reliability and Inspection of Ship Structures Based on questionnaire Information",Proceedings of PACOMS'96 , pp.65-72

SR226 Committee, "An Investigation Study on Rational Structural Design for Bulk Carrier", Research Committee of the Shipbuilding Research Association of Japan (1996.3) (in Japanese)

Proceedings of the Seventh (1997) International Offshore and Polar Engineering Conference
Honolulu, USA, May 25-30, 1997

ISBN 1-880653-28-1 (Set); ISBN 1-880653-32-X (Vol. IV)

Use of Structural Simulation for Extension of Life of Offshore Structures

A.C. Walker and C.P. Ellinas
Advanced Mechanics & Engineering Ltd
Croydon, United Kingdom

W. Snedden
Shell Exploration and Production
Aberdeen, United Kingdom

ABSTRACT

The extension of life, or the change of use of offshore structures requires re-assessment of their capacity to carry the specified loads, which may have changed from the original design premise. The paper presents the methodology of using non-linear finite element analysis to simulate the limit states of response of a large offshore structural component to the imposed loads.

The work described here was initiated when linear analysis coupled with code check formulations showed that a significant degree of strengthening would be apparently required to enable the structure to function satisfactorily. However, structural non-linear simulation combined with a limit state assessment method proved conclusively the strengthening to be unnecessary, thus eliminating costly offshore fabrication and reducing shut-down time.

KEY WORDS: structures; non-linear; simulation; limit state; cost reduction; life extension.

INTRODUCTION

The first of the Brent concrete based Condeep platforms were designed in 1973 by Norwegian Contractors and when the Brent Field began production in 1976 it was estimated the field life would cease in 1988. Now, twenty years after initial production, the Brent topsides are being converted to low pressure gas production facilities for reservoir de-pressurisation - the first stage in the Brent Long Term Field Development. This development is the largest brownfield engineering project ever undertaken in the history of the North Sea at a cost of about $2.1bn.

The topside reconfiguration has required the re-assessment of the structural steel work to carry the process facility modifications, which include significant safety upgrades by relocation of the process equipment away from the living quarters. The re-assessment for structural strength made extensive use of the finite element (FE) method in a linear format coupled with strength checks using formulations from a relevant structural code (BS5400,1982). Essentially, this linear analysis forming the basis for the design encompasses the structural redundancies within the system but assesses the strength of the structure on the basis of the limiting strength of the individual structural elements and members. In this procedure, once the load in a single member reaches its ultimate capacity, as measure by the corresponding code formulation, the entire structure is deemed to have attained its maximum load capacity. This approach can be gross simplification of reality; most structures exhibit significant degrees of redundancy and have considerable capacity beyond first member failure by virtue of internal load re-distribution. In other words, further increases in external load beyond first member failure is carried by re-distribution of internal loads to adjacent members which still have reserves of strength.

The increased load capacity through re-distribution is well known but is seldom exploited in design analysis since, it is argued, the costs of analysis, about $60m is significantly increased. In the present case, the application of linear analysis and code check procedures resulted in requirements for very expensive offshore strengthening operations to be applied to the deck girders. Since these costs were estimated to be, far in excess of the cost of more extensive structural analysis, it was decided to base the assessment of the deck girders on results from non-linear finite element analyses.

The non-linear analyses encompass a number of features present in the practical steel plated structural elements, viz.,

- structural redundancy
- material yielding and plastic flow
- local and overall modes of buckling
- local strain concentrations at loading points
- initial deviation of the fabricated structure from its nominal design dimensions

In effect the non-linear analysis incorporates all the responses of the structure to the applied loads; that is, it is a numerical simulation of the behaviour of the structure. The simulation approach takes full benefit from the inherent multiple load path redundancy in the structure and enables the ultimate load and failure history of the complete deck structure to be evaluated accurately. The results from the simulation are assessed using requirement on the limit states of behaviour of the structure subjected to the specified loads.

The paper describes the non-linear simulation used for assessing the limit states of one of the deep web girders in the Brent D platform. Example results are presented and conclusions are drawn with regard to the use of structural simulation for a wide range structural re-assessment and design.

Table 1 Material Strengths

Component	Thickness (mm)	Yield Stress (N/mm^2)	Design Strength ULS (N/mm^2)
Flanges	20,25,3 8,50	265 355	252.4 338.1
Web	16,25,1 8,35	265 355	252.4 338.1
Flange Stiffeners	16,2525	265 355	252.4 338.1
Bulb flatsTee Stiffeners	11.5 - 30	235	195.8

Table 2 Loading Details

Vertical Loads in L1W Girder (MN)					
Load Point	Load Type	SLS		ULS	
		Factor	Load	Factor	Load
MSP 45	Gravity	1.0	-4.175	1.2	-5.01
	Environment	1.0	1.772	1.0	1.772
MSP 49	Gravity	1.0	-6.977	1.2	-8.372
	Environment	1.0	-1.773	1.4	-2.482
Floor beams *	Gravity	1.0	-2.403	1.2	-2.883
Shear at L1W/T1 Junction	Gravity	1.0	-1.682	1.2	-2.808

* Total load applied by five

STRUCTURAL MODEL

The Brent D deck structure is essentially a lattice of deep web steel girders sitting atop three concrete legs. The modules for accommodation, process equipment, drilling etc, are supported on the top flanges of the girders through individual load points. A representation of the deck is shown in Figure 1 which is a finite element modelling of the complete deck. The non-linear analysis reported here is concerned with the longitudinal girder L1W spanning between transverse girders T1 and T2. The junction is L1W and T2 forms part of the rigid box where the deck is supported by the leg.

Figure 1 shows the deformations obtained by the linear analysis, using the SESAM program, of the whole deck subjected to the localised vertical loads of the modules. It may be seen that L1W cantilevers out from T2 and indeed the detailed deflection results shown virtually no rotation in the vertical plane of junction of L1W and T1. The girder T1 is seen to displace vertically as a beam along its length but the linear analysis results show quite low values of bending and shear in the region of the L1W and T1 junction. Thus the structural model is taken to be a cantilever jointing on to a section of the T1 girder. The cantilever is subjected to local loading corresponding to the modules and specified shear on the T1 segments. This definition of the boundary conditions for the model are considered to provide a conservative basis for the analysis.

The structural outline for the girder is shown in Figures 2 and 3. Essentially, the girder consists of a plate web varying from 25mm thickness at the junction with T2 to 16mm thickness at the T1 junction. The web is stiffened horizontally by bulb-flats, 1m apart, and vertically by T beams 3.13m apart. The arrangement of stiffening at the module support points (MSP) is shown in Figure 3. Floor beams frame into the bottom flange of the girder and carry the vertical loads of equipment inside the depth of the deck.

The limit state design method requires factors to be applied to the material nominal yield stress to obtain the design stress. The factors used in the analysis were taken from the appropriate code (BS5400, 1982) and, for the analysis of the Ultimate Limit State (ULS) are γ_m = 1.2 for structural components which may fail by unstable buckling, such as the bulb-flat and Tee stiffeners, and γ_m = 1.05 for all others components, such as the main web and the flanges of the girder. The material yield strengths are shown in Table 1. The analysis for the Serviceability Linear State is performed with all factors equal to unity; that is the Design Strength is equal to the yield stress. It is evident a variety of steel grades had been incorporated in the original design. A uniform value of modulus, 205kN/mm2 was used in the analysis and the materials were modelled as being elastic - perfectly plastic.

The loading on the girder derived from its self-weight and the load transferred by the module sitting on the top flange. The latter load derives from the gravitational effect and the effects of the environment, i.e. wind loading. The factors which are specified to be applied to this loading depend on the limit state being analysed. Table 2 shows the load applied during the analysis; negative loads are vertically downwards. Note that the factor for ULS was not applied to environmental loads which reduced the total downwards load. The module support points (MSP) are shown in Figure 4 which presents the finite element model of the girder.

A major feature of structural simulation is that it must be capable of reproducing all the forms of structural response to the applied loading as would be observed in the actual structure. Extensive analysis, complementary to that described here, by Walker and Sneddon (1997) investigated the validity of finite element modelling to replicate plate and stiffened plate buckling failure modes. It is well recognised that such buckling is affected by the initial state of the deflections in the unloaded structure. Experience with testing and surveys of plated thin-walled structures have shown that several types of initial deformation may result from the fabrication process. These were incorporated in the finite element model of the girder, as shown in Figure 6. Comparison between this Figure and Figure 4 shows that the forms of initial deflections incorporated are

- local plate modes with a longitudinal wavelength equal to the width between bulb-flats
- column deflection modes in the horizontal and vertical stiffening element with a wave length equal to their length
- torsional deflection modes in all stiffening elements

The magnitudes of these deformations were at the maximum limit specified by BS5400 and the DnV Rules (1994) which is 10mm for the plates and stiffener elements. The initial deformations were incorporated in the plate in a stress free state. It is recognised that residual welding stresses can reduce the loads at which plate elements will fail by buckling. However, Walker and Sneddon (1997), and previous work, have shown that if the maximum initial plate deflections are used as a basis for the analysis, the additional effects of residual stresses in the plate failure load can be ignored.

ANALYSIS RESULTS

Summary results from the structural simulation process are presented in this Section and form the basis for the assessment of the limit states discussed in the next Section. Prior to initiating the simulation very careful checking is performed to ensure all geometries, loading and material properties were accurately incorporated in the model. A bench-mark test was performed in which the model, without initial local deflections, was run in a linear mode and the resulting stresses and girder deflections at a specified load, were compared to corresponding results from the SESAM model shown in Figure 1. A mesh convergence study was performed and it was shown that the mesh presented in Figure 5, compared to a model with half the element density, provide results less than 10% different for stresses and ultimate load. Only after the checking and valuation, also described in Walker and Sneddon (1997), had been completed were the simulation results accepted. Figure 6 shows a plot of the total vertical load applied to the girder, normalised by the specified ULS load, against the corresponding deflection MSP 49, i.e. at the extreme end of the cantilevered girder. The SLS load is 80% of the ULS, see Table 2. The plot shows that the load-deflection relationship is linear up to very near the ULS load and that the maximum load the girder can support is at least 25% greater than the ULS factored loads. Evidently at the ULS condition a temporary loss of stiffness occurs but is subsequently recovered.

Figure 7 shows the stress state in the girder at the specified SLS load. The stresses have been calculated using the von Mises formula and are at the mid-plane of the plates and stiffeners. The maximum stress occurs in the web plates around the openings and in the lower flange stiffeners at the junction with the T2 transverse girder. The maximum value of calculated stress is 265N/mm^2 compared to the corresponding yield stress of 355 N/mm^2 at these locations.

Figure 8 shows the deflected form of the girders at the maximum load calculated. The amplitudes of the deflections have been exaggerated for clarity. It may be seen that the web exhibits buckling in two main areas, both around opening. Detailed analysis of the results of the simulation show

- there is incipient buckling at the SLS load but the deflections, additional to the initial values, are less than the plate thickness
- the loss of stiffness of the girder shown in Figure 6 near the ULS load results from a rapid increase of the local buckling deflections.

However, the local buckles shown in Figure 8 are stable and the internal loading is re-distributed to enable the girder to continue to carry increasing external loading until a general failure mechanism is formed across the depth of the girder. The strain in the mid-planes of the plate and stiffener components were evaluated at a load 22% greater than the specified ULS loads. It was found that for almost all of the girder the strain was less than 0.12% and only in a few very localised areas, associated with the buckling deformation, was the strain in the range 0.35 -0.6%.

LIMIT STATE ASSESSMENT

The structural simulation procedures, described above, differs from the normal design approach in that it considers the complete response of the structural element to the applied loading and encompasses internal load re-distribution due to local plate or stiffener instability and material yielding. The normal linear approach relies on code formulations to indicate if the levels of stress calculated in the various structural components by the linear analysis are acceptable. Thus the usual approach is actually quite piecemeal in its treatment of structural behaviour in contrast to the holistic approach of structural simulation exemplified here.

The simulation of the L1W girder provided the deformation, stresses and strains at the specified SLS and ULS loads. However, it is necessary to specify acceptable conditions for the Serviceability and Ultimate Limit States to assess if the structure is fit-for-purpose at these loads. It is generally accepted by modern codes, e.g. Eurocode 3, (draft 1997) that computer analysis and limit state assessment may be used in design or re-design but no specific descriptions are provided to enable the assessment of particular structures. It was necessary therefore to prepare a set of acceptance criteria specific to the SLS and ULS conditions of the structural components in the Brent D decks. After a review of the background definitions of limit state in a variety of publications, a set of specific acceptance criteria were formulated for the assessment of the results from the structural simulation described here. The resulting criteria were:

SERVICEABILITY LIMIT STATE

General Requirements

- the analysis will be performed with unit load and material factors

and at specified serviceability load levels:

- the material should not have yielded in the mid-plane of the plates and the flanges and stiffeners of the girders
- the maximum out-of-plane deflection of the plates should not exceed a level at which confidence in the structure will be undermined or impede the use of the girder
- the non-linear deformation of the plates should not cause a significant departure from a linear relationship between the loads applied to the girder and the corresponding deflections at the loading points

These general requirements were formulated as specific criteria, viz:

- at the SLS load the maximum out-of-plane deflections of any plate between stiffeners should not exceed three times the plate thickness.
- the slope of the line relating the loads applied to the girder and the corresponding deflections at the loading points should not, at the SLS load level, be less than 97% of the original slope of the line at zero load.

ULTIMATE LIMIT STATE

General Requirement

- the analysis will be performed with the full appropriate material factors

and at the ultimate state load levels:

- the structure should be stable and robust
- the material should not have suffered local tearing

subsequently

- the structure should collapse in a ductile manner

The general requirements were formulation as specific acceptance criteria

- the slope of the graph of loads against corresponding deflections must be positive at the ULS load level
- the ratio of the maximum load calculated by the validated finite element model is to be not less than 1.1 times the specified ULS load level to accommodate modelling inaccuracies which may arise from the effects of mode density and other possible numerical uncertainties
- if the ratio of the maximum load to the ULS load is less than 1.15 it should be demonstrated that at the maximum load the structure is able to sustain deflections at the loading points up to 1.5 times the corresponding values at the ULS load without significant reduction in load carrying capacity
- if the ratio of the maximum to the ULS load is greater than 1.15 the ratio of the corresponding deflections should not be less than 1.2
- the strain levels in the plate and stiffeners should everywhere be less than 20% at the ultimate load levels

It is evident the results from the structural simulation, summarised in the previous Section, satisfy all the SLS and ULS criteria. It was concluded the girder was fit-for-purpose and any strengthening which appear to be required by the code checks was unnecessary.

CONCLUSIONS

The following conclusions have been drawn from the experience gained during the performance of the re-assessment project, part of which has been described above.

- It is concluded that structural simulation, coupled with limit state acceptance criteria, can be very effectively applied to the re-assessment of structures for extension of life and load upgrading.
- The process must be carefully validated and should involve engineers with experience of non-linear analysis and failure mechanisms, particularly those involving buckling.
- Programs and adequate computing power, are now available to enable the simulation of the response of large structures to complex systems of dynamic and quasi-static loadings.

The use of simulation will enable the implementation of a more rational limit state approach to design and reduce the reliance on a purely code based methodology which can result in unnecessarily conservative and costly structures.

REFERENCES

BS5400 (19982) Steel concrete and composite bridges Part 3

Walker, AC, and Sneddon,W (1997) Buckling of Plates Stiffened by Bulb-Flats, to be published

Det Norske Veritas (1994) Rules for Classification of Fixed Offshore Installations

Eurocode 3 (1993) Design of Steel Structures Part 2 Bridges (draft) CEN/TC280/SC3

Figure 1
SESAM finite element modelling of Brent D deck structure

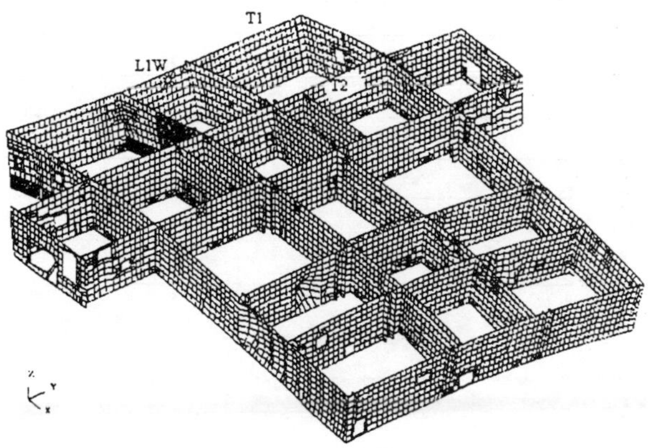

Figure 2
Structural details of L1W girder between T1 and T2 girders

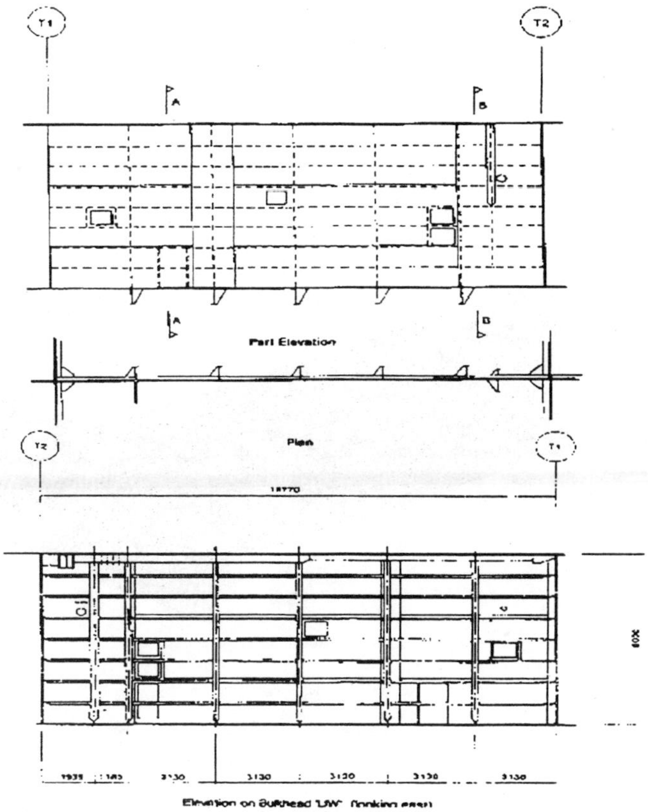

Figure 3
Sections through L1W girder

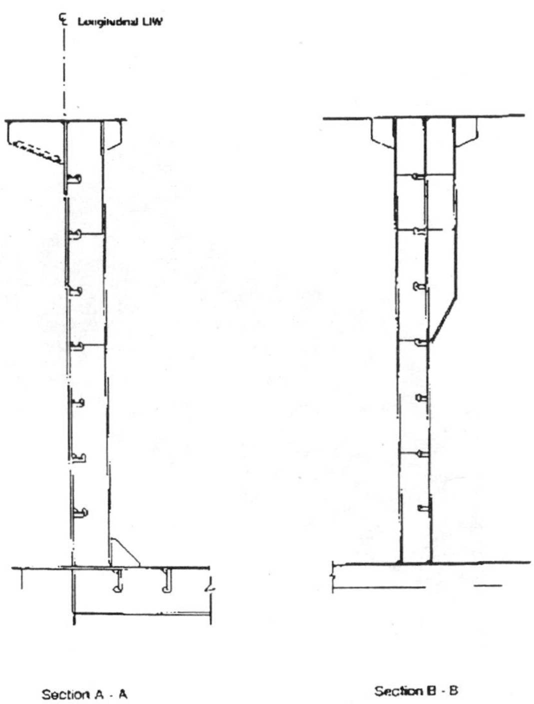

Figure 4
Across finite element model viewed from T2 end

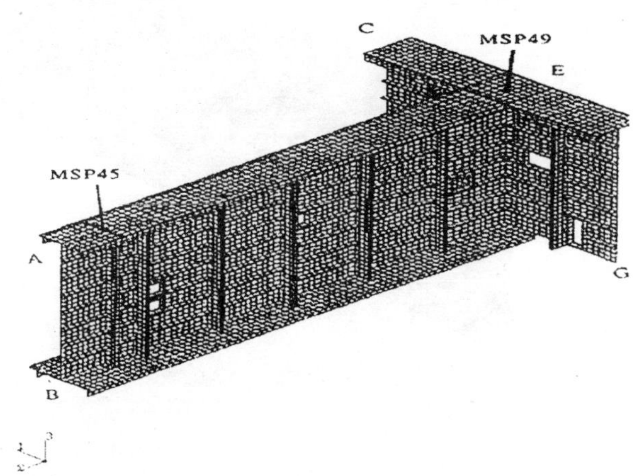

Figure 5
Finite element model showing imposed geometries deformations (at an exaggerated scale)

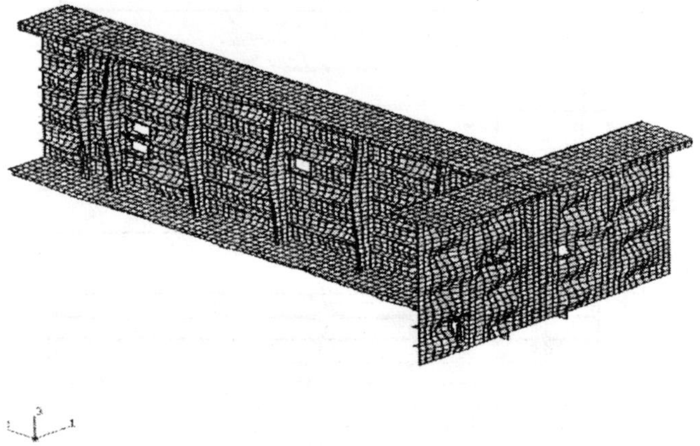

Figure 6
Plot of total applied load, normalised with respect to the ULS level, vs the vertical deflection at MSP49

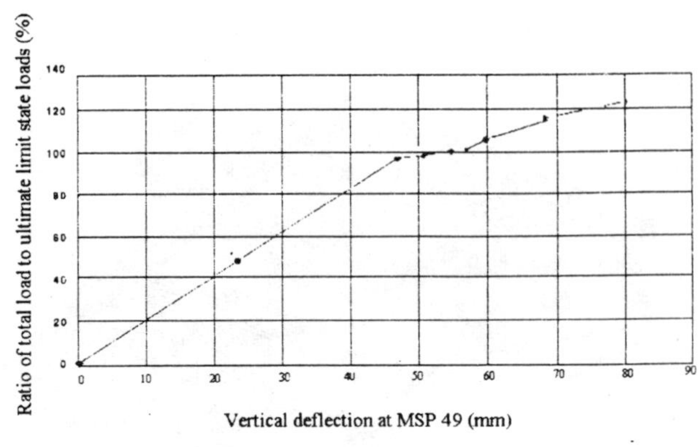

Figure 7
Mises mid-plane stresses (N/mm^2) for L1W and T1 girder at SLS load

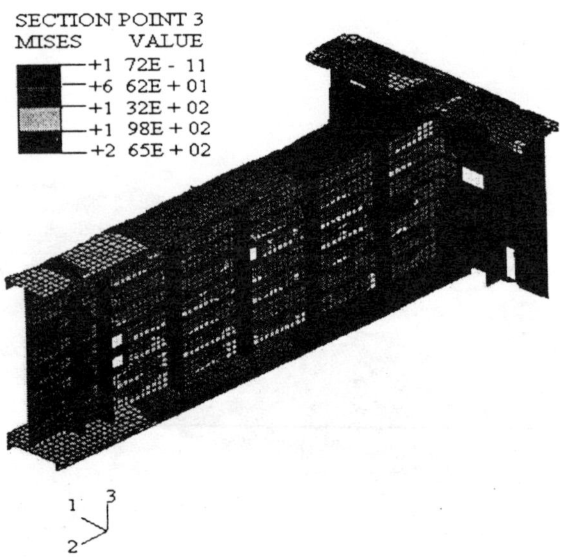

Figure 8
Deformed shape at an exaggerated scale of L1W and T1 girders at a total applied 1.22 times the ULS level

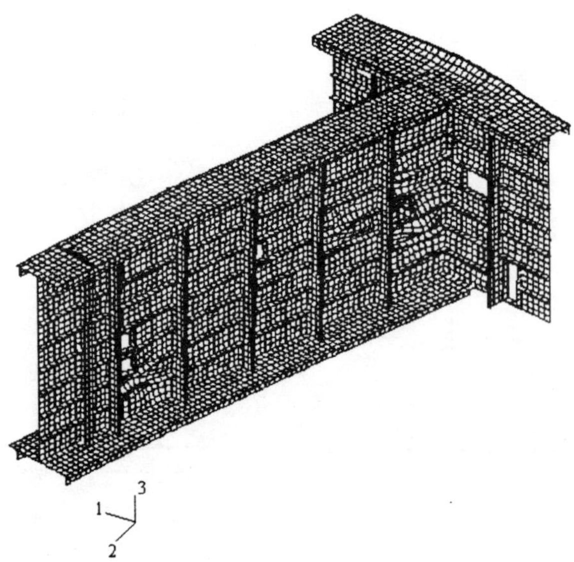

Proceedings of the Seventh (1997) International Offshore and Polar Engineering Conference
Honolulu, USA, May 25-30, 1997

ISBN 1-880653-28-1 (Set); ISBN 1-880653-32-X (Vol. IV)

Reliability-Based Design and Requalification Criteria for Longitudinally Corroded Pipelines

Yong Bai
J P Kenny A/S
Forus, Norway

Tao Xu and Robert Bea
University of California at Berkeley
Berkeley, CA, USA

ABSTRACT

Marine pipelines are a major component in the marine transportation infrastructure system. The consequences of pipeline failures may not only result in heavy economic losses due to the system repairs, but also constitute serious hazards due to the leakage of the transported gas and oil.

Corrosion is a major cause of leakage from marine pipelines. Repair of the corrosion damages, on the other hand, is expensive. Therefore, there are considerable economic incentives to continue the operation of the corroded pipelines, provided an acceptable safety level is maintained.

This paper develops reliability based criteria for design and requalification of the corroded pipelines. The existing criteria for corroded pipelines (ASME B31G) are reviewed. The new criteria are developed based on the analytical, experimental, and numerical studies. Safety factors for the new criteria are calibrated using reliability method to deliver the same safety level implied in the original criteria. The new criteria are applied in the requalification of an existing corroded pipeline.

1.0 INTRODUCTION

Because of economic considerations, marine pipelines are usually designed so that they can withstand some corrosion damages. These damages, however, must be controlled in order to prevent failures of the pipelines. Such failure can and has resulted in huge economic and environmental loss. Any economic savings made possible by allowing corrosion damages should be reconciled with potential failure of the pipeline. Therefore, proper selection of design criteria of corroded pipelines, is an important and complex task, and current design code recommendations only offer very general guidelines.

A number of studies about corroded pipelines have been conducted to develop better fitness for purpose criteria(Bai,et al 1994). Because of the variety of such studies and the many assumptions involved in their developments, it is not clear how much confidence an engineer can place them in a design or requalification process.

The purpose of this paper is to develop reliability based design and requalification criteria of corroded pipelines. Following a brief summary of available criteria, the technical developments and problems of corroded pipelines are illustrated by evaluating the reliability of ASME (B31G) design criteria. This evaluation focuses on three interrelated issues: 1) corrosion mechanism, 2) the accuracy of the maximum allowable corrosion length, the safe maximum pressure level, and 3) practical problems excluded in the B31G criteria. Based on this evaluation, new criteria are proposed to help overcome these problems. Reliability based safety factor calibration is conducted to assure that the safety level in the new criteria is the same as that in the existing criteria. The new criteria are then applied in the requalification of aging corroded pipelines.

2.0 REVIEW OF EXISTING CRITERIA

The design criteria for corroded pipelines is generally expressed as equations to determine the operating parameters: 1) the maximum allowable length of defects, 2) the maximum allowable design pressure for uncorroded pipelines, and 3) the safe maximum pressure. A number of criteria exist to determine these operating parameters. This section reviews the ASME B31 G Criteria.

The B31G criterion (ASME 1993) is widely used to assess corroded pipelines for fitness for purpose evaluation. The main equations in the ASME B31G criteria (1993) can be summarized as follows:

(1) Maximum Allowable Design Pressure P.

The maximum allowable design pressure in B31G criterion is expressed as:

$$P = \frac{2SMYS}{D} \bullet Ft \tag{1}$$

where, P is the maximum allowable design pressure, SMYS is the Specified Minimum Yield Strength, and F is the design factor which is normally 0.72.

(2) Maximum Allowable Defect Length and Depth.

In the B31G (ASME 1993), a criterion for the acceptable corroded length is given as below for a corroded area having a maximum depth 'd' in the range of $0.1 < d/t < 0.8$ where t is the nominal wall thickness

$$L_{allow} = 1.12B\sqrt{Dt} \tag{2}$$

where L_{allow} is the maximum allowable axial extent of the defect and B is:

$$B = \sqrt{\left(\frac{d/t}{1.1d/t - 0.15}\right)^2 - 1} \tag{3}$$

The Maximum Allowable Operating Pressure (MAOP) is defined to be less or equal to the maximum allowable design pressure P given by Equation (1).

$$MAOP \le P \tag{4}$$

Equating the Safe Maximum Pressure Level P' to the Maximum Allowable Operating Pressure (MAOP), the maximum allowable defect depth d_{allow} is:

(a) For $A \le 4$

$$d_{allow} = \frac{3t}{2}\left[\frac{1-\frac{MAOP}{1.1P}}{1-\frac{MAOP}{1.1P\sqrt{A^2+1}}}\right] \qquad (5)$$

(b) For $A > 4$

$$d_{allow} = \left[1-\frac{MAOP}{1.1P}\right]t \qquad (6)$$

(3) The Safe Maximum Pressure Level P'.

The safe maximum pressure level P' for the corroded area is:

$$P' = 1.1P\left(\frac{1-\frac{2}{3}\left(\frac{d}{t}\right)}{1-\frac{2}{3}\left(\frac{d}{t\sqrt{A^2+1}}\right)}\right); \; P' \le P, \; A \le 4 \qquad (7)$$

$$P' = 1.1P\left(1-\frac{d}{t}\right); \qquad P' \le P \text{ and } A > 4 \qquad (8)$$

where

$$A = 0.893\left(\frac{L}{\sqrt{Dt}}\right) \qquad (9)$$

3.0 EVALUATION OF EXISTING CRITERIA

The existing criterion ASME B31G (1993) for corroded pipelines was established based on the knowledge developed over 20 years ago. This criterion is re-examined to develop an improved criterion based on current knowledge. This evaluation is conducted based on the corrosion mechanisms, parameters in the existing criterion and the applications which are not included in the existing criterion.

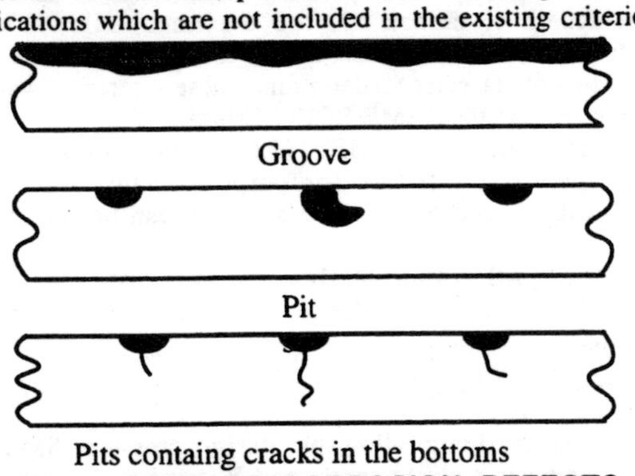

FIG. 1 TYPE OF CORROSION DEFECTS

3.1 Corrosion Mechanism

Figure 1 shows the types of corrosion defects. For marine pipelines, internal corrosion is a major problem (Mandke, 1990, Jones et al 1992). Many forms of internal corrosion occur, e.g., (a) girth weld corrosion, (b) massive general corrosion around the whole circumference, and (3) long plateau corrosion at about 6 o'clock position. External corrosion, on the other hand, is normally thought of as being local, covering an irregular area of the pipe. However, when the protective coating is failed, the external corrosion may tend to be pattern of long groove.

The B31G criterion has several problems for corrosion defects in real applications. It can not be applied to spiral corrosion, pits/grooves interaction and the corrosion in welds. For very long and irregularly shaped corrosions, the B31G criterion may lead to overly conservative results. It also ignores the beneficial effects of closely spaced corrosion pits.

Spiral Corrosion. For defects in other orientations, the B31G criterion recommends that the defect is projected on the longitudinal axis of the pipe to be treated as a longitudinal defect. This recommendation appears to be adequate for short defects. It is conservative for long spiral defects (Bai, et al 1994).

Mok (Mok, et al 1990, 1991) conducted extensive tests in the applicability of the B31G criteria to long spiral corrosion. For spiral defects with spiral angles other than 0 or 90 degrees, the study found that B31G underpredicted the burst pressure by as much as 50%. The effect of spiral angle is illustrated in Figure 2. (Mok, et al 1990).

Based on the experimental and numerical studies, Mok et al (1990) recommended the spiral correction factor in determining the burst pressure for $W/t \le 32$ as:

$$Q = \frac{1-Q_1}{32}\frac{W}{t} + Q_1 \qquad (10)$$

in which W is the defect width, and coefficient Q_1 is a function of the spiral angle φ ($\varphi = 90^0$ for longitudinal corrosion, $\varphi = 0^0$ for circumferential corrosion):

$$Q_1 = \begin{cases} 0.2 & \text{for } 0^0 < \varphi < 20^0 \\ 0.02\varphi - 0.2 & \text{for } 20^0 < \varphi < 60^0 \\ 1.0 & \text{for } \varphi > 60^0 \end{cases} \qquad (11)$$

for $W/t > 32$, the value of Q must be taken as 1.0

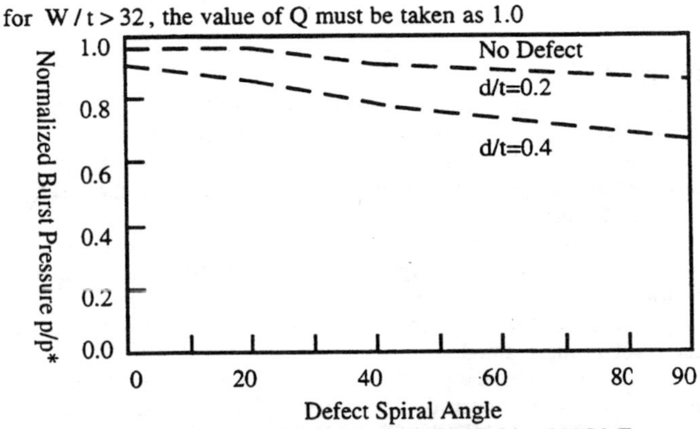

FIG. 2 THE EFFECT OF SPIRAL ANGLE

Pits Interaction. Corrosion in pipelines often results in colonies of pits over an area of the pipe. For closely spaced corrosion pits, a distance of t (wall-thickness) is used as a criterion of pit separation for a colony of longitudinal oriented pits separated by a longitudinal distance or parallel longitudinal pits separated by a circumferential distance.

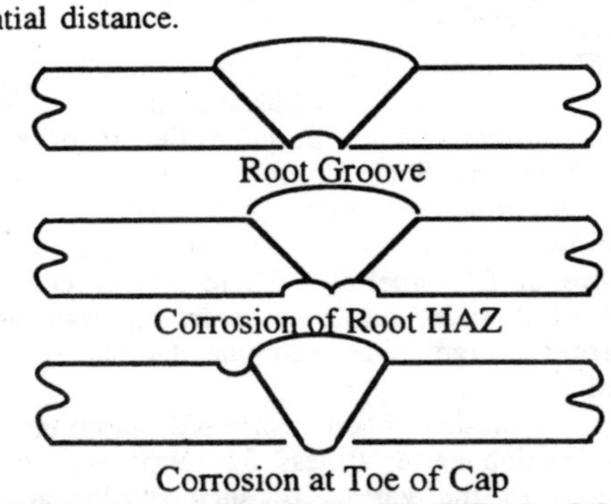

FIG. 3 TYPICAL PATTERNS OF WELD CORROSION

For circumferentially spaced pits separated by a distance longer than t, the burst pressure can be accurately predicted by the analysis of the deepest pits within the colonies of pits. For longitudinal oriented pits separated by a distance less than t, the failure stress of interacting defects can be predicted by neglecting the beneficial effects of the non-corroded area between the pits. For parallel longitudinal pits separated by a circumferential distance, experiments suggested that pits could be treated as interacting pits if the circumferential spacing is less than t (Bai, et al 1994).

Groove Interactions. For the interaction of longitudinal grooves, if the defects are inclined to pipe axis and the distance x between two longitudinal grooves of length L_1 and L_2 is larger than L_1 and L_2, the length of corrosion L is the maximum of L_1 and L_2. If the defects are inclined to pipe axis and the distance x between two longitudinal grooves of lengths L_1 and L_2 is less than L_1 and L_2, the length of corrosion L is the sum of x, L_1 and L_2, $L = L_1 + L_2 + x$.

Corrosion in Welds. One of the major corrosion damages for marine pipelines is the effects of the localized corrosion of weld on the fracture resistance. Figure 3 shows typical pattern of weld corrosion. The B31G criteria didn't cover the assessment of corroded welds. The existing fracture assessment procedures (BSI 6493) are recommended.

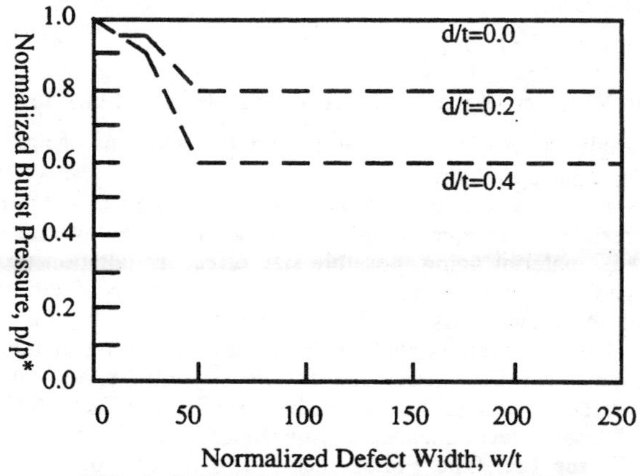

FIG.4 EFFECT OF DEFECT WIDTH

Effect of Corrosion Width. Figure 4 shows the effect of defect width on burst pressure with a longitudinal defect (Mok, et al 1991), for the case of X52, OD=508mm, t=6.35mm, d/t=0.4. It can be concluded in Mok's studies that the width effect is negligible on the burst pressure of pipe with long longitudinal defects.

Irregular Shaped Corrosion. The major weakness of the existing B31G criterion is its over-conservative estimation of the corroded area for long and irregular shaped corrosion (Kiefner and Vieth, 1990b, Bai, et al 1994, Hopkins and Jones, 1993).

Two shapes were considered in the development of the original B31G criterion. One was the rectangle area method. The other was the parabola area method. Tests of Hopkins and Jones (1992) indicated that irregularly shaped corrosion could be conservative assessed using the B31G criteria when the accurate cross-sectional area of the corrosion defect was used.

3.2 Material Parameters

The major material parameters in the B31G criterion are flow stress, Specified Minimum Yield Stress (SMYS), Folias Factor M.

Flow Stress and SMYS. In the B31G-1993 manual, the flow stress was defined as 1.1 SMYS which is an appropriate value for the new pipelines. However, the flow stress is influenced by a number of factors, fabrication process (e.g. hot rolled versus cold expanded) and material aging. Furthermore, the flow stress used in burst strength criteria is influenced by possible cracks in the pit bottom due to corrosion fatigue. Therefore, specific attention should be made for accurate estimate of flow stress for aging pipelines. Many researchers (Hopkins, Jones, 1992, Klever, 1992, Stewart, et al) indicted that the flow stress for base material could be estimated as ultimate tensile stress. An approximation of the ultimate tensile stress is the Specified Minimum Tensile Stress, a statistic minimum of the ultimate tensile stress:

$$\sigma_{flow} = SMTS \tag{12}$$

The value of SMTS are available in some design specification (API 5L, 1994).

Folias Factor M. The Folias factor M is a geometric factor developed by Folias (1964) to account for the stress concentration effect of a notch in the pipes. Recent studies (Kiefner and Vieth, 1989) recommended the following expression to improve the accuracy of the Folias factor:

$$M = \begin{cases} \sqrt{1 + \dfrac{2.51(L/2)^2}{Dt} - \dfrac{0.054(L/2)^4}{(Dt)^2}} & \text{for } \dfrac{L^2}{Dt} \le 50 \\ 0.032\dfrac{L^2}{Dt} + 3.3 & \text{for } \dfrac{L^2}{Dt} > 50 \end{cases} \tag{13}$$

3.3 Problems excluded in the B31G Criteria.

The ASME B31G criterion can not be applied in some practical corrosion problems including corroded welds, ductile and low toughness pipe, and corroded pipes under combined pressure, axial and bending loads. Recent studies concluded that the corrosion in submerged-arc seams (longitudinal welds) should be handled in the same manner as corrosion in the body of the pipe. Corrosion in Electric Resistance Welds (ERW) or flash-welded seams should not be evaluated on the basis of the existing B31G criteria. Corrosion in (circumferential) girth welds was recommended be evaluated using Kastner's local collapse criteria (Kastner, et al 1981).

A fracture mechanics approach (PD 6493) should be applied for assessing corroded welds, considering possible defects in the welds. The effect of material's fracture toughness (in ductile and low toughness pipe) is reflected by the critical fracture toughness of the material used in the fracture assessment criteria.

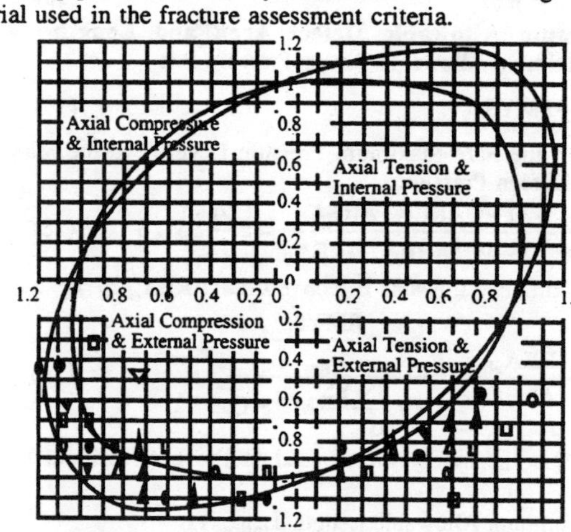

FUG. 5 EFFECT OF AXIAL LOAD ON THE COLLAPSE PRESSURE

In the B31G criteria, the effect of axial load is not discussed. In general, tensile longitudinal stress may delay yielding and pipe bursting. On the other hand, compressive longitudinal stress may accelerate yielding and result in reductions in bursting pressure. Figure 5 shows the effect of axial load on collapse pressure (Galambos, 1988). It is illustrated that:

- The internal burst pressure is largely reduced by axial compression
- The effect of axial tension is beneficial. The tension load is not significant when it is less than 60 percent of yield strength of the pipe section. This effect is significant when the axial tension is larger than 60% of the yield strength.

The dominant effect of bending stress is the reduction of the hoop stress in the corroded region. Therefore, the B31G criteria of burst pressure that considers internal pressure alone, may lead to unconservative results when large axial and bending stresses are coupled with corrosion.

4.0 DEVELOPMENT OF NEW CRITERIA

In this section, a new criterion is developed for longitudinally corroded pipelines. The criterion for circumferentially corroded pipelines will be developed on a companion paper.

Pit depth exceeding 80% of the wall-thickness is not permitted due to the possible development of leaks. General corrosion where all of the measured pit depths are less than 20% of the wall-thickness is permitted, without further burst strength assessment. If the ratio of maximum pit depth and wall-thickness is between 0.2 and 0.8, the following equations are recommended.

(1) The Maximum Design Pressure Level P

The maximum allowable design pressure in the new criterion is the same as that of the original B31G criteria:

$$P=\frac{2SMYS}{D}\bullet Ft \qquad (14)$$

where F is the usage factor for intact pipe which is 0.72 according to the B31G criterion.

(2) The Safe Maximum Pressure Level P'

$$P'=\frac{1}{\gamma}\frac{2\sigma_{flow}t}{D}\frac{1-QAREA/AREA_0}{1-M^{-1}AREA/AREA_0} \qquad (15)$$

where, P' = safe maximum pressure level, σ_{flow}=flow stress of the material, t = wall-thickness of the pipe, D = outside diameter of the pipe, AREA = Lt, $AREA_0$= original area prior to metal loss due to corrosion within the effective are which is Lt, L = defect length of corrosion profile, M = Folias factor, Q = Spiral correction factor, and γ=Factor of Safety.

The predicted bursting pressure level P_b is

$$P_b=\gamma P' \qquad (16)$$

(3) Maximum Allowable Defect Area and Length

The Maximum Allowable Design Pressure P is

$$P=F\frac{2t}{D}SMYS \qquad (17)$$

Equating the Safe Maximum Pressure Level P' to the Maximum Allowable Design Pressure:

$$\frac{1}{\gamma}\frac{2\sigma_{flow}t}{D}\frac{1-QAREA/AREA_0}{1-M^{-1}AREA/AREA_0}=F\frac{2t}{D}SMYS \qquad (18)$$

The Maximum Allowable Effective Area $AREA_{allow}$ is:

$$\frac{AREA_{allow}}{AREA_0}=\frac{F\gamma\frac{SMYS}{\sigma_{flow}}-1}{F\gamma\frac{SMYS}{\sigma_{flow}}M^{-1}-Q} \qquad (19)$$

in which γ is the safety factor used in the new residual strength criterion.

The Maximum Allowable Length L_{allow} is

For $M_{allow}\le 4.9$

$$L_{allow}=\sqrt{\frac{1.255}{0.0135}-\sqrt{\left(\frac{1.255}{0.0135}\right)^2+\frac{4\left(1-M_{allow}^2\right)}{0.0135}}}\sqrt{Dt}$$

and for $M_{allow}>4.9$

$$L_{allow}=\sqrt{(M_{allow}-3.3)/0.032}\sqrt{Dt} \qquad (20)$$

in which M_{allow} is solved by equating the Safe Maximum Pressure P' to the Maximum Allowable Design Pressure as:

$$M_{allow}=\frac{F\gamma\frac{SMYS}{\sigma_{allow}}}{Q+\frac{AREA_0}{AREA}\left(F\gamma\frac{SMYS}{\sigma_{flow}}-1\right)} \qquad (21)$$

Effective Area AREA. Two levels of AREA assessment are recommended. In the level 1, the AREA is estimated as:

$$L/(Dt)<30 \quad AREA=\frac{2}{3}L\bullet d$$
$$L/(Dt)>30 \quad AREA=0.85L\bullet d \qquad (22)$$

In the level 2, the exact area (AREA) of the corrosion profile is estimated by Simpson integration method.

Closely Spaced Corrosion Pits. A distance of t (wall-thickness) is used as a criterion of pit separation for a colony of longitudinal oriented pits separated by a longitudinal distance or parallel longitudinal pits separated by a circumferential distance.

Interaction of Longitudinal Grooves. For defects inclined to pipe axis, if the distance x, between two longitudinal grooves of lengths L_1 and L_2 , is greater than either L_1 or L_2, then the length of corrosion L is the maximum of L_1 and L_2; if the distance x, between two longitudinal grooves of lengths L_1 and L_2, is less than L_1 and L_2, the length of corrosion L is the sum of x, L_1 and L_2, $L=L_1+L_2+x$. For two longitudinal grooves separated by a circumferentail distance x, the wall thickness t is used as groove separation criterion.

Spiral Correction Factor. The spiral correction factor Q is determined as:

$$Q=\frac{1-Q_1}{32}\frac{W}{t}+Q_1 \qquad (23)$$

in which W is defect width, and coefficient Q_1 is a function of the spiral angle φ($\varphi=90^0$ for longitudinal corrosion, $\varphi=0^0$ for circumferential corrosion)

Flow stress. Consideration should be given to factors affecting flow stress, e.g., fabrication process (e.g. hot rolled versus cold expanded), material aging, possible size effect, installation process and possible crack in corrosion defect bottom. Use of the actual value of the flow stress is allowed provided the value has been obtained from a reliable approach (e.g., material testing of the pipe in situ. etc.). The flow stress can also be estimated as the ultimate tensile stress. For API 5L materials, SMTS (Specified Minimum Tensile Stress) is recommended as flow stress.

Folias Factor M. The Folias factor is estimated based on the following equations:

$$M=\begin{cases}\sqrt{1+\frac{2.51(L/2)^2}{Dt}-\frac{0.054(L/2)^4}{(Dt)^2}} & \text{for } \frac{L^2}{Dt}\le 50\\ 0.032\frac{L^2}{Dt}+3.3 & \text{for } \frac{L^2}{Dt}>50\end{cases} \qquad (24)$$

Corroded Welds. Corrosion in submerged-arc seams (longitudinal welds) should be handled in the same manner as corrosion in the body of the pipe. Corrosion in girth welds (circumferential) should be assessed using the Kastner's local collapse criterion. The level 2 (or level 3 analysis implemented in PD 6493 (1991) should be applied for assessing corroded welds. The corroded groove could be considered as a crack of the same depth and length. The effect of material's fracture toughness (in ductile and low toughness pipe) could be taken into account in the assessment procedure of the material fracture toughness.

Safety Factor. Traditional safety factors are given based on engineering experience and judgment. Within Kiefner and Vieth's studies (1989, 1993), several modified B31G criteria were developed. In all cases, the safety factors is assumed to be 1/0.72=1.39, as the original B31G criterion. However, the safety factor for the new criteria is calibrated based on reliability methods. It is around 1.8 and dependent on the accuracy of inspection tools and corrosion depth.

5.0 EVALUATION OF NEW CRITERIA

The evaluation of the new criteria is conducted in this section to compare with the test data from AGA database, NOVA tests, British Gas tests, and Waterloo tests. In the comparision, a model uncertainty parameter X_M is introduced as:

$$X_M=\frac{X_{true}}{X_{pred}} \qquad (25)$$

where X_{true} is the true strength in the tests, and X_{pred} is the capacity predicted by a given criteria (existing or new). Table 1 is the statistical parameters for X_M (mean μ and COV ν) which demonstrated that the uncertainty of the new criteria is smaller.

	B31G	NG18	New Criteria
μ	1.74	1.30	1.07
ν	0.51	0.19	0.18

TABLE 1 STATISTICS FOR DIFFERENT CRITERIA AND TEST DATA

6.0 RELIABILITY-BASED CRITERIA

The reliability-based design is to develop a Load Resistance Factor Design (LRFD) equation where bursting is taken as the failure criterion. This includes the following items:

- Specification of a target safety level
- Specification of characteristic value for the design variables
- Calibration of Partial Safety Factors
- Perform safety verification, formulated as a design equation utilizing the characteristics values and partial safety factors.

6.1 Target Failure Probability

The target failure probability is developed based on the historical failure data and the safety level implied in the existing B31G criteria. The target safety level should be determined considering the consequence of failure as well as the effects of inspection, maintenance, and repair. Based on the historical data, reliability analysis of the existing B31G criteria, and other factors, an annual target safety level of 10^{-4} is used in the development of the reliability-based criteria.

6.2 Design Equation and Limit State Function

For the sake of simplicity, only internal pressure is considered in the design equation. The LRFD approach leads to

$$P_R \geq \gamma P_L \tag{27}$$

where, P_R is the characteristic strength of the pipe, P_L is the characteristic load (internal pressure), γ is rthe partial safety factor.

A bias factor X is introduced to reflect the confidence in the criterion in prediction of burst strength.

$$X = \frac{\text{true burst strength}}{\text{predicted burst strength}} \tag{28}$$

Normalized random variables in the design equation are:

$$X_p = \frac{P_L}{\frac{2SMYSt}{D}} \tag{29}$$

$$X_f = \frac{\sigma_{flow}}{SMYS} \tag{30}$$

$$X_A = \frac{AREA}{AREA_0} \tag{31}$$

$$X_L = \frac{L^2}{Dt} \tag{32}$$

$$X_d = \frac{d}{t} \tag{33}$$

The design equation for corroded pipelines based on the B31G criterion is :

For $\frac{L^2}{Dt} \leq 20$

$$0.72\frac{1.1SMYS}{SMTS}X_t\frac{1-\frac{2}{3}X_d}{1-(1+0.79745X_L)^{-1/2}\frac{2}{3}X_d} - X_p \geq 0 \tag{34}$$

For $\frac{L^2}{Dt} > 20$

$$0.72\frac{1.1SMYS}{SMTS}X_t(1-X_d) - X_p \geq 0 \tag{35}$$

The limit state function is then expressed as:

For $\frac{L^2}{Dt} \leq 20$

$$g(\bar{x}) = 0.72\frac{1.1SMYS}{SMTS}X_M X_t \bullet \frac{1-\frac{2}{3}X_d}{1-(1+0.79745X_L)^{-1/2}\frac{2}{3}X_d} - X_p \tag{36}$$

For $\frac{L^2}{Dt} > 20$

$$g(\bar{x}) = 0.72\frac{1.1SMYS}{SMTS}X_M X_t(1-X_d) - X_p \tag{37}$$

The design equation for corroded pipelines, based on the new criteria is given by:

For $\frac{L^2}{Dt} < 50$

$$\frac{1-X_A}{1-\left(1+0.6275X_L - 0.003375X_L^2\right)^{-1/2}X_A} \bullet \frac{1}{\gamma}X_{flow}X_t - X_p \geq 0 \tag{38}$$

For $\frac{L^2}{Dt} > 50$

$$\frac{1}{\gamma}X_{flow}X_t\frac{1-X_A}{1-(0.032X_L+3.3)^{-1}X_A} - X_p \geq 0 \tag{39}$$

The limit state function is

For $\frac{L^2}{Dt} < 50$

$$g(\bar{x}) = \frac{1}{\gamma}X_M X_{flow} X_t \bullet \frac{1-X_A}{1-\left(1+0.6275X_L - 0.003375X_L^2\right)^{-1/2}X_A} - X_p \tag{40}$$

For $\frac{L^2}{Dt} > 50$

$$g(\bar{x}) = \frac{1}{\gamma}X_M X_{flow} X_t \bullet \frac{1-X_A}{1-(0.032X_L+3.3)^{-1}X_A} - X_p \tag{41}$$

6.3 Uncertainty

Bias for Criteria X_M. Model uncertainty X_M is introduced for the criteria to account for modeling and methodology uncertainties. It reflects a general confidence in the design criteria for a real life in-situ scenario.

The model uncertainty is calibrated from the 86 tests results in the AGA database (Kiefner and Vieth 1989). A Hermit model is applied to simulate the four lower moments. The mean bias and COV for the existing and new criteria is listed in Table 1.

Bias for Normalized Pressure X_p. The characteristic value of the normalized pressure X_p is obtained by substituting safety factors, characteristic values of the other parameters into the design equation. In general, the annual maximum operating pressure is higher than the nominal operating pressure. This is reflected by the mean bias in X_p. Sotberg and Leira (1994) assumed that the ratio of the annual maximum operating pressure to the design pressure followed a Gumble distribution. Its mean and COV is 1.07 and 1.5%. By further analyzing the data (Bai, 1994), a Gumbel distribution with a mean of 1.05 and a COV of 2% is used in this development.

Bias for Normalized flow stress X_f. The X_f mainly reflects the material property. Uncertainty of X_f is largely dependent of the material grade. A log-normal distribution is assumed to fit the data in the existing database. From the data analysis, the mean value and COV are selected as 1.14 and 6%, respectively.

Bias for Normalized Area X_a. The normalized defect area X_a is the ratio of metal loss area and its original area. Two kinds of inaccuracy are possible:

- inaccuracy in the calculation of the area of metal loss.
- inaccuracy due to use of measurement instruments.

A log-normal distribution with mean 0.8 and COV. 0.08 is recommended for X_a.

Bias for Normalized Depth X_d. The uncertainty in the corrosion depth is the combination of the uncertainties associated with pit separation, inspection, and future corrosion prediction. A lognormal distribution is thus assumed for X_d, and the mean value and COV are taken as 0.8 and 8% respectively based on the experimental data and expert judgment.

Normalized Length X_L. Similar to the discussion on corrosion depth, the uncertainty in normalized defect length X_L is the combination of the uncertainties associated with pit separation, inspections, and future corrosion. However, corrosion length is more easy to measure in inspections. Normal distribution is used to fit the X_L. Its mean value and COV are taken as 0.9 and 5% respectively.

6.4 Safety Level in the B31G Criteria.

Reliability methods are applied to estimate the implied safety level of the B31G criterion. The uncertainties described in the section 6.3 are used in the reliability analysis. The safety factor is taken as 1.4 in the B31G criteria.

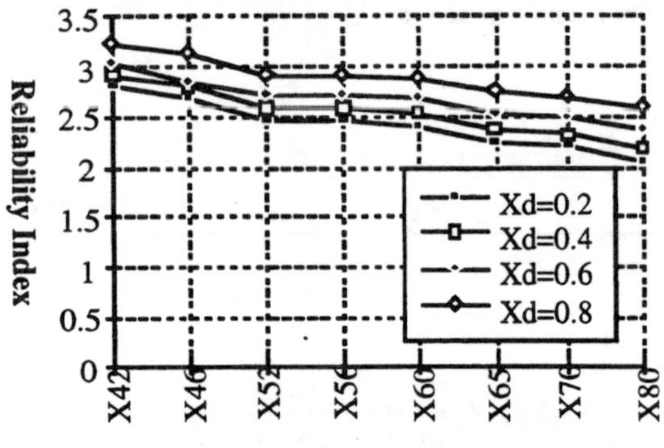

FIG. 6 IMPLIED SAFETY LEVEL IN THE B31G CRITERIA (SHORT CORROSION)

The obtained implied safety level (safety index) of the B31G criterion is shown in Fig. 6 for short corrosion $X_L = 10$ and Fig. 7 for long corrosion $X_L = 200$, as functions of defect depth X_d and material grade (SMYS). Due to large model uncertainty in the B31G criteria, the implied safety level in the B31G criteria is quite low. It is found for short ($X_L = 10$) and shallow corrosion defects ($X_d < 0.4$), the implied safety level is lower than 10^{-3}. For long ($X_L = 200$) and deep ($X_d > 0.4$) corrosion defects, the implied safety level is between 10^{-3} and 10^{-4}.

6.5 Reliability-based Calibration

The target safety level for the new criterion is between 10^{-3} and 10^{-4}, based on the implied safety level in the B31G criterion. This ensures that the safety level in the new criteria is higher than or equal to that of the original B31G criterion.

The relationship between the reliability index and the safety factor γ is shown in Fig. 8 for short corrosion defects ($X_L = 10$) and Fig. 9 for long corrosion defects ($X_L = 200$). The obtained reliability index for $X_A < 0.5$ was found to be close to the case $X_A = 0.5$. Comparing Fig. 8 with Fig. 9, it is obvious that the reliability index β for a given set of X_A and safety factor γ is only slightly different for short corrosion detects ($X_L = 10$) and long corrosion defects ($X_L = 200$). The sensitive analysis indicated that the model uncertainty of the criterion in questions was the dominantly important factor in the reliability analysis.

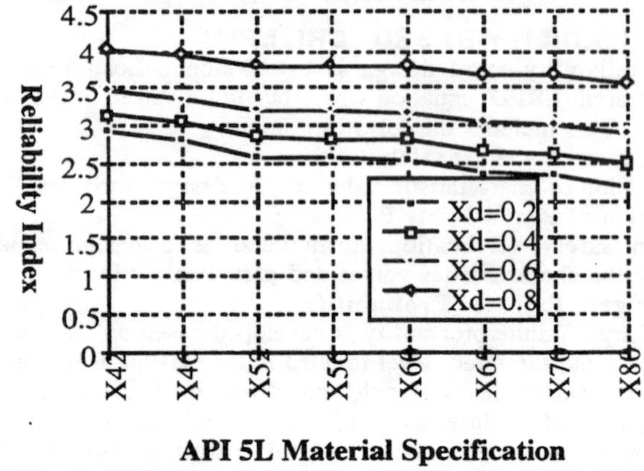

FIG. 7 IMPLIED SAFETY LEVEL IN THE B31G CRITERIA (LONG CORROSION)

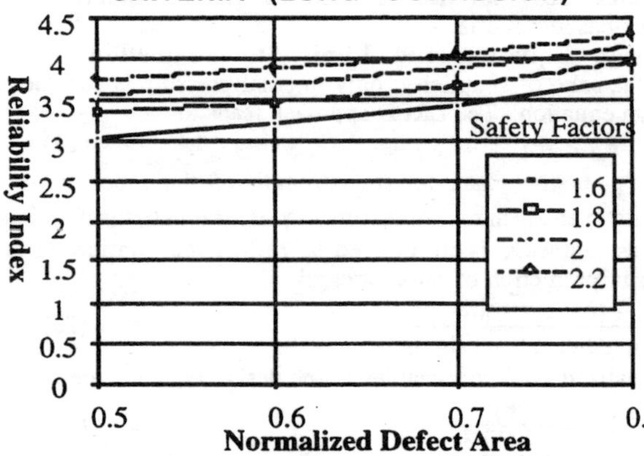

FIG. 8 RELIABILITY INDEX FOR DIFFERENT DEFECT AREA (SHORT CORROSION-NEW CRITERIA)

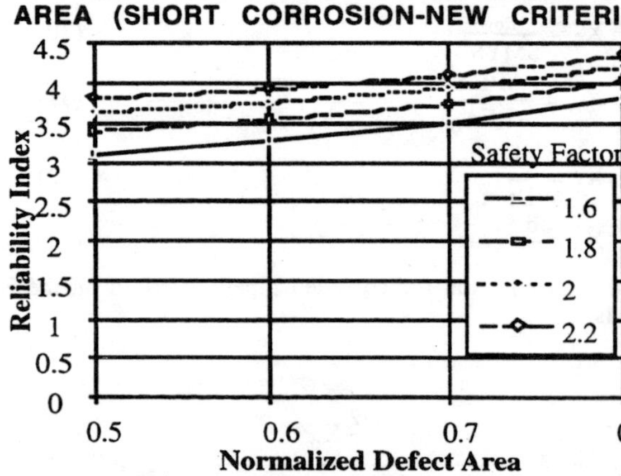

FIG. 9 RELIABILITY INDEX FOR DIFFERENT DEFECT AREA (LONG CORROSION-NEW CRITERIA)

7.0 EXAMPLE APPLICATIONS

An example is presented to illustrate the application of the new criteria in the pipeline requalification. As a result of a corrosion detection pigging inspection of a 10 year old offshore gas pipeline, grooving corrosion was found in the pipeline. The requalification of this pipeline is divided into following steps.

Requalification Premises. Extensive groove corrosion has been observed in a gas pipeline after 10 years of service. The observed grooving corrosion results in a reduced rupture (bursting) capacity of the pipeline, increasing the possibility for leakage with resulting possible environmental pollution, unscheduled down time for repair

Var.	Description	Distribution
$X_{\Delta p,max}$	Annual max pressure ratio	G (1.05, 2%)
n	Hardening index	N(0.2, 6%)
X_f	Burst capacity model	N(1.0, 10%)
X_{SMTS}	Ult. Tensile Strength Uncrt.	N(1.09, 6%)
SMTS	Ultimate Tensile Strength	517.0 N/mm2
X_t	Wall thickness uncrt	N(1.04, 10%)
X_{corr}	Corrosion model uncrt.	N(1.04, 10%)
ϕ	Degree of circum. corr.	LN(0.2, 20%)
n_{CO_2}	mol fraction of CO_2	0.17
α	Influence of inhibitor	0.02
d_{obs}	Observed relative corrosion	Beat(a, 50%)
X_a	Normalized Area Uncrt.	N(0.35, 14%)
X_d	Normalized Depth Uncrt.	LN(0.8, 8%)
X_L	Normalized Length Uncrt.	LN(0.8, 8%)
		N(0.9, 5%)

TABLE 2 UNCERTAINTY PARAMETERS IN THE ANALYSIS

The intended service life. The gas pipeline is scheduled for a life of 20 years, resulting in residual service life of 10 years after the observation of the corrosion. There is no intended change in the service of the pipeline within the residual life.

Available Information. The design and operation parameters and their uncertainties for the pipeline are given in Table 2. It is assumed that gas pressure and temperature linearly vary over the entire pipeline length based on the conditions specified at the inlet and outlet point. The gas pressure varies over the service life. The gas temperature, on the other hand, is assumed to be constant.

Service History. The pipeline is routinely inspected on 5 year interval with a conventional corrosion detection pig. The pipeline inspection after the first 5 years of service did not bring up any observed corrosion.

Present Conditions. The inspection after the first 10 years service resulted in the detection of grooving corrosion. The maximum measured corrosion was detected at 0.6 km from the inlet point with a corrosion depth of 35% of the wall thickness, $d_{corr} = 0.35t$, and a detection accuracy is represented through a COV of 5% of the wall thickness.

Bursting Model. The burst strength formulation is expressed as

$$M_{burst}(t) = \Delta p_{grv}(t) - \Delta p_{max}(t) \tag{42}$$

where, $\Delta p_{max}(t)$ is the annual maximum operating pressure, $\Delta p_{grv}(t)$ is the pressure resulting in bursting in year t.

The annual maximum occurring pressure in year t is expressed as a function of the operating pressure:

$$\Delta p_{max}(t) = X_{\Delta p,max}\Delta p_{oper}(t) \tag{43}$$

where, $X_{\Delta p,max}$ defines the relationship between annual maximum pressure and the average operating pressure.

The bursting capacity of the pipeline depends on the degree of grooving corrosion, and is modeled in accordance with the new criteria in section 4.

$$\Delta p_{grv} = \frac{1}{\gamma}\frac{1-QAREA/AREA_0}{1-M^{-1}AREA/AREA_0}\Delta p_f \tag{44}$$

where Δp_f is the burst pressure for uncorroded pipe, $AREA_0$ is the original area prior to metal loss due to corrosion Lt. AREA is the exact area of the metal loss due to corrosion in the axial direction of through-wall thickness. γ is the factor of safety, M is the Folias factor, Q is the spiral correction factor.

The burst pressure Δp_f for uncorroded pipe is

$$\Delta p_f = \frac{2\sigma_{flow}t}{D} \tag{45}$$

where D is the pipe diameter and t is the pipe thickness.

The flow stress is defined by Tresca or von Mises yield criterion as

$$\sigma_{flow} = X_f\left[\left(\frac{1}{\sqrt{3}}\right)^{n+1} + \left(\frac{1}{2}\right)^{n+1}\right]\sigma_u \tag{46}$$

where, X_f is the model uncertainty for predicting the burst capacity, n is the hardening index, σ_u is the ultimate stress.

Corrosion Rate. The corrosion rate, or the annual degree of grooving corrosion, is estimated based on the empirical "deWaard & Milliams" formula that the influence of the operating pressure and temperature on the corrosion rate is defined:

$$\log(v(t)) = 5.8 - \frac{1710}{T(t)} + 0.67 \bullet \log_{10}(n_{co_2}(t) \bullet \Delta p_{oper}(t)) \tag{47}$$

where T is the temperature in Kelvin, n_{co_2} is the mol fraction of CO_2 in the gas phase and $\Delta p_{oper}(t)$ is the operating pressure (bar).

The estimated degree of corrosion over a time period t can be derived by integrating the corrosion rate over the time period:

$$d_{corr}(t) = X_{corr} \bullet \int_0^t \alpha(t)v(t)dt \tag{48}$$

where parameter $\alpha(t)$ is the influence of inhibitors and X_{corr} is the model uncertainty associated with the empirical corrosion rate.

Basic Variables: The uncertainty defined in the Table 2 is introduced in the model, where the symbols N, LN, Beta and Gumbel indicated a Normal, Lognormal, Beta or Gumbel distribution. The first parameter is the mean value, the second is the COV, the third and fourth parameter are the lower and upper limits of the distribution.

7.1 Conditional Assessment

The first stage of the requalification process is an evaluation of the present state of the system. If the system satisfy the specified constraints, the system is continued to be operated as initially planned prior to the corrosion observation.

The specified constraints are summarized as

- Acceptable level of safety within the remaining service, or at least until next scheduled inspection.
- The annual bursting failure probability is less than 10^{-3} within the next 5 years.

Three level analyses are conducted: 1) simplified analysis, 2) deterministic analysis, and 3) probabilistic analysis in the conditional assessment.

For the simplified analysis, the observed corrosion is compared with the corrosion allowance.

Estimated corrosion: $0.35t = 7.8mm$
corrosion allowance: $1.6mm$

The observed corrosion is larger than the corrosion allowance.

For the deterministic analysis, the experienced degree of corrosion (stationary corrosion rate) is assumed to be valid over the remaining service life.

Corrosion after 15 years of service

corrosion rate: $\hat{v} = 0.35t / T = 0.78mm / year$

corrosion after 15 years: $\hat{d} = \hat{v}15 = 11.7mm$

Specified yield strength for X60 steel: $\sigma_y = 413MPa$

Acceptable hoop stress: $\sigma_{rules} = 297MPa$

Specified design pressure: $\Delta P = 13.4Mpa$

Hoop stress for uncorroded pipe:

$$\sigma_H = \frac{\Delta P \bullet D}{2 \bullet t} = \frac{13.4 \bullet 914.0}{2 \bullet 22.2} = 276\text{MPa}$$

Hoop stress for 11.7 mm corrosion:

$$\sigma_H = \frac{\Delta P \bullet D}{2 \bullet (t - \hat{d})} = \frac{13.4 \bullet 914.0}{2 \bullet (22.2 - 11.7)} = 583\text{MPa}$$

Based on the observed corrosion, the estimated hoop stress after 15 years service is larger than the acceptable hoop stress.

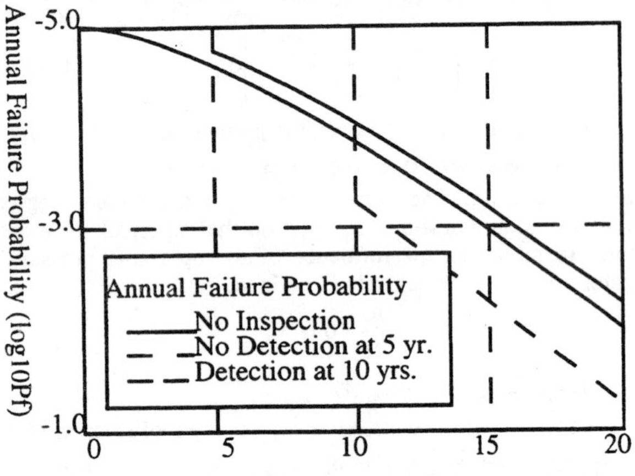

FIG. 10 ANNUAL BURSTING FAILURE PROBABILITY

For the probabilistic analysis, the following approaches are applied:

•The corrosion rate is based on the deWaard & Milliams formula,

•The reduced burst capacity is estimated by the new criteria,

•The design pressure for which the capacity model is to resist is developed over the service life as a function of he operating pressure.

Based on the capacity and loading model, the annual probability for bursting of the corroded pipelines is illustrated in Fig. 10. It can be seen that the estimated probability of failure increases slightly with time in spite of the reduced operating pressure due to the increase in the expected level of corrosion.

7.2 Evaluating of Repair Strategies

A minor repair/modification is recommended. Probabilistic analysis is conducted to evaluate different alternatives. The alternatives are summarized as:

- A reduction of the operating pressure, de-rating.
- Use of corrosion mitigation measures (inhibitors)
- Rescheduled inspection
- Combination of the above alternatives.

The life-cycle cost of mitigation measures and lost income is set as the evaluation criteria. The constraints are:

- Acceptable level of safety within the remaining service life, or at least until next inspection,
- The annual failure probability of the pipeline should be less than 10^{-3} with the remaining service life or until next inspection,
- Next inspection is scheduled for a service life of 15 years. Meanwhile, an early inspection can be recommended.

Two alternatives are studied: 1) derating, and 2) inhibitors.

De-Rating. The reduced operation pressure reduces the annual maximum pressure as well as, to some extent, reduce the additional corrosion growth.

In Fig. 11, the estimated annual bursting failure probability in the time period after the year 10 is shown as a function of the relative reduction in the operating pressure. It is illustrated in Fig. 11 that the time period until probability of failure 10^{-3} is approximately 14, 17 and 21 years when the operating pressure is reduced with 10%, 20%, and 30% respectively.

Inhibitors. The use of inhibitors reduces the additional corrosion growth over the remaining service life and thereby reduces the annual failure probability over time. Inhibitors resulting in 50%, 60%, 70% and 80% corrosion reduction are considered in the example applications. As the mitigation effects are uncertain, the influence of the inhibitors are modeled as Beta distribution with a median (50%) value as the specified corrosion reduction effect and a COV of 50%.

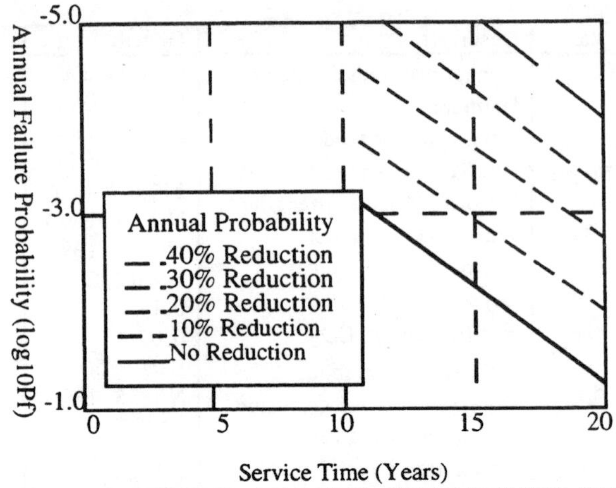

FIG. 11 ANNUAL FAILURE PROBABILITY FOR REDUCED OPERATING PRESSURE

The use of inhibitors greatly reduces the corrosion rate. Fig. 12 shows the estimated annual bursting failure probability in the time period after the 10 years service. The use of inhibitors reduces the failure probability.

Evaluation of Alternatives. The selection of the minor repair/modification alternatives (de-rating or inhibitors) satisfies the constraints. Table 3 summarizes the combination effects. It summarizes the operating time after 10 years service until the target probability 10^{-3} is reached.

Influence of Inhibitor

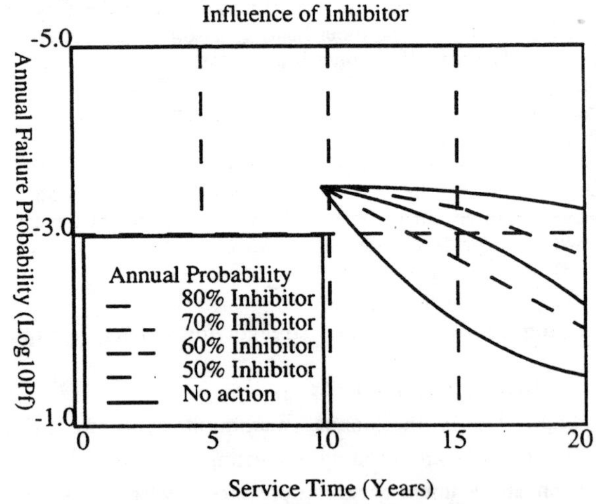

FIG. 12 ANNUAL FAILURE PROBABILITY FOR DIFFERENT INHIBITORS

	Effect of Inhibitors				
P Red.	0%	50%	60%	70%	80%
0%	1	3	4	6	12
5%	2	5	6	9	15
10%	4	7	9	12	18
15%	5	9	11	14	20
20%	6	11	13	17	23
25%	8	13	15	19	23
30%	9	15	17	21	29

TABLE 3 OPERATING YEARS INSPECTION UNTIL THE TARGET FAILURE PROBABILITY

If the next inspection is not scheduled prior to 15 years of service, the combinations of de-rating and inhibitors in the shaded area of Table 3 are the realistic decision alternatives. The darker shaded area indicates the most attractive combination of use of inhibitor with specified effect and degree of pressure reduction. If the pipeline inspection is rescheduled, the alternatives of upper left corner in Table 3 are recommended. However, in the evaluation of the alternatives incorporating a reduction of the time period until next inspection, the likelihood of possible major repair/mitigate measures at an earlier period should be addressed

7.3 Rehabilitation

A possible major repair alternative is replacement of a fraction or the whole pipeline. The major repair/modification can greatly reduce the estimated failure probability. However, as the observed damage can be effectively controlled by the minor repair/modifications, the major modification is not recommended in the requalification process of this pipeline.

8.0 CONCLUSIONS

This paper develops reliability based criteria for design and requalification of the corroded pipelines. The existing criteria for corroded pipelines (ASME B31G) are reviewed. The new criteria are developed based on the analytical, experimental, and numerical studies. Safety factors for the new criteria are calibrated using the reliability method. This calibration delivers the same safety level implied in the existing B31G criteria. The new criteria is applied in the requalification of the existing corroded pipelines.

9.0 REFERENCES

Bai, Y., et al (1994) "Probabilistic Assessment of Dented and Corroded Pipelines, International Conference on Offshore and Polar Engineering, Osaka, Japan

Bea, R.G., (1994) Evaluation of Alternative Marine Structural Integrity Programs, Journal of Marine Structures, Vol. 7, Elsevier Science Limited, England

British Gas Standard BGC/PS/P11, (1983) "Procedures for Inspection and Repair of Damaged Steel Pipelines".

British Standard Institute (BSI), (1991) PD 6493 - Guidance on Methods for Assessing the Acceptability of Flaws in Fusion Welded Structures".

Chouchaoui, B. A., Pick, R. J., and Yost, D. B., (1992), "Burst Pressure Prediction of Line Pipe Containing Single Corrosion Pits Using the Finite Element Methods", Proceedings of the 11th Offshore Mechanics and Arctic Engineering, Vol. 4

Chouchaoui, B. A., and Pick, R. J., (1992), "Residual Burst Strength of Pipe with Internal Corrosion Pits", Proceedings of International Conference on Pipeline Reliability, Calgary, Canada

Chouchaoui, B. A., and Pick, R. J., (1993), "Interaction of Closely Spaced Corrosion Pits in Line Pipes, Proceedings of 12th International Offshore Mechanics and Arctic Engineering Conference, Vol. 4

Darlaston, B. J. I., and Harrison, R. P., (1977) "Ductile Failure of Thin Walled Pipes with Defects under Combination of Internal Pressure and Bending", Proceedings of the 3rd International Conference on Pressure Vessel Technology, Tokyo, Japan.

Folias, E. S., (1965) "An Axial Crack in a Pressurized Cylindrical Shell", International Journal of Fracture, Vol. 1

Glover, A. G., (1962) "Assessment of Pipeline Girth Welds Subjected to High Longitudinal Strain", Proceedings of 11th International Conference on Offshore Mechanics and Arctic Engineering, Vol. 4

Hopkins, P., and Jones, D. G., (1992) "A Study of the Behavior of Long and Complex-shaped Corrosion in Transmission Pipelines", Proceedings of 11th International Conference on Offshore Mechanics and Arctic Engineering, Vol. 4

Kiefner, J. F., and Duffy, A. R., (1971) "Summary of Research to Determine the Strength of Corroded Area in Line Pipe", Presented to a Public Hearing at the U. S. Department of Transportation

Kiefner, J. F., Maxey, W. A., Eiber, R. J. H., and Duffy, A. R., (1973) "Failure Stress Levels of Flaws in Pressured Cylinders", Progress in Flaw Growth and Fracture Toughness Tests, ASTM STP 536s, American Society for Testing and Materials

Kiefner, J. F. and Vieth, P. H., (1989) "A Modified Criterion for Evaluating the Remaining Strength of Corroded Pipe, RSTRENG", Project PR 3-805 Pipeline Research Committee, American Gas Association

Kiefner, J. F. and Vieth, P. H. (1990) "New Method Corrects Criterion for Evaluating Corroded Pipe", Oil & Gas Journal

Klever, F. J., (1992) "Burst Strength of Corroded Pipe: 'Flow Stress' Revisited", Proceedings of Offshore Technology Conference, OTC 7029

Klever, F. J., Stewart, G., van der Valk, A. C., (1995) "New Development in Burst Strength Predictions for Locally Corroded Pipelines", Proceedings of the International Conference on Offshore Mechanics and Arctic Engineering, Vol. 4

Kvernvold, O., Johnson, R., and Helgerson, T., (1992) "Assessment of Internal Pipeline Corrosion", Proceedings of the 11th International Conference on Offshore Mechanics and Arctic Engineering, Vol. 4

Mok, D. H. B., Pick, R. J., and Glover, A. G., (1990) "Behavior of Line Pipe with Long External Corrosion", Material Performance, Vol. 29

Mok, D. H. B., Pick, R. J., Glover, A. G., Hoff, R., (1991) "Bursting of Line Pipe with Long External Corrosion", International Journal of Pressure Vessel and Piping, Vol. 46

Sotberg, T., and Leira, B. J., (1994) "Reliability-based Pipeline Design and Code Calibration", Proceedings of 13th International Conference on Offshore Mechanics and Arctic Engineering

Stewart, G., Klever, F., and Ritchie, D., (1994) "An Analytical Model to Predict the Burst Capacity of Pipelines", Proceedings of 13th International Conference on Offshore Mechanics and Arctic Engineering, Vol. 4

Sydberger, T., Edwards, J. D., and Mork, K. J., (1995) "A Probabilistic Approach to Prediction of CO_2-Corrosion, and Its Application to Life Cycle Cost Analyses of Oil and Gas Equipment", Corrosion 95, Paper No. 65

Proceedings of the Seventh (1997) International Offshore and Polar Engineering Conference
Honolulu, USA, May 25-30, 1997

ISBN 1-880653-28-1 (Set); ISBN 1-880653-32-X (Vol. IV)

Reassessment and Requalification of Two Gulf of Mexico Platforms

A. Sturm
Moffatt and Nichol Engineers, Long Beach, CA, USA
R. Bea
University of California at Berkeley, Berkeley, CA, USA
T. Miller
UNOCAL Corporation, Lafayette, LA, USA

ABSTRACT

The objective of the investigation summarized in this paper was to reassess the fitness for purpose of two template type steel platforms located in Block 209 of the Ship Shoal Area of the Gulf of Mexico. SS209B and SS209B-AUX were installed in 1961 and 1971, respectively. As a part of UNOCAL's loss control program, analyses were performed to evaluate the fitness for purpose of these field 'hub' platforms.

SS209B and SS209B-AUX were evaluated using the computer programs ULSLEA (Mortazavi, Bea, 1994, 1995) and StruCAD*3D (Zentech, 1995). The oceanographic loading was determined by use of the American Petroleum Institute Recommended Practice 2A WSD- Section 17 (Assessment of Existing Platforms) and confirmed with site specific oceanographic study results (Woodward-Clyde, 1982). The analysis was based on a comparison of the maximum base shear developed by intense hurricanes versus the capacity of the platform. The annual probabilities of failure for each of the platforms subjected to hurricane loadings were determined.

Using the criteria recommended by the API in Section 17 (API, 1996) and the annual probability of failure, the fitness for purpose of the system was determined. A potential structural upgrade (grouting leg - pile annuli) was analyzed to improve the reliability of the older of the two platforms (SS209B). This paper summarizes the results from the analyses of the platform loadings and capacities and the evaluations of fitness for purpose.

KEY WORDS: platforms, requalification, risk analysis, capacities, repairs, acceptable reliability

INTRODUCTION

The American Petroleum Institute (API) has developed guidelines to assist with the reassessment and requalification of steel template-type offshore platforms. These guidelines are designated as API RP 2A-WSD Section 17 (Assessment of Existing Platforms) (API, 1996). There are six components to the assessment process:

1. Platform Selection
2. Condition Assessment
3. Categorization
4. Design Basis Check
5. Analysis Checks
6. Consideration of Mitigation

The analysis checks consist of three levels of analysis:

1. Screening Analysis
2. Design Level Analysis (DLA)
3. Ultimate Strength Analysis (USA)

The three levels of analysis checks are typically performed sequentially, based on the outcome of the previous check. The screening analysis is typically performed first, with the design level analysis occurring only if the platform's capacitance fails to meet the requirements of the screening analysis. The same sequence follows for the ultimate strength analysis. The checks become more detailed and less conservative as one progresses.

In this study, two types of analyses were performed. The first type was DLA. The DLA were performed using the computer program StruCAD*3D (Zentech, 1995). The second type was USA. The USA were performed using the computer program ULSLEA (Ultimate Limit State Limit Equilibrium Analysis) (Mortazavi, Bea, 1995).

PLATFORMS

The two platforms (Figs. 1 and 2) are located in Block 209 of the Ship Shoal Area (Gulf of Mexico) in 94 ft. of water (MWL - Mean Water Level). Platform SS209B (Fig. 1), is a 4-pile jacket that was designed in 1963. It received minor damage in 1992 as a result of Hurricane Andrew. Platform SS209B-AUX (Fig. 2), was fabricated and structurally attached to SS209B in 1971. SS209B-AUX was not damaged by Hurricane Andrew. The two structures are connected with light bracing at the top of the jackets. Both structures were constructed using A 36 steel.

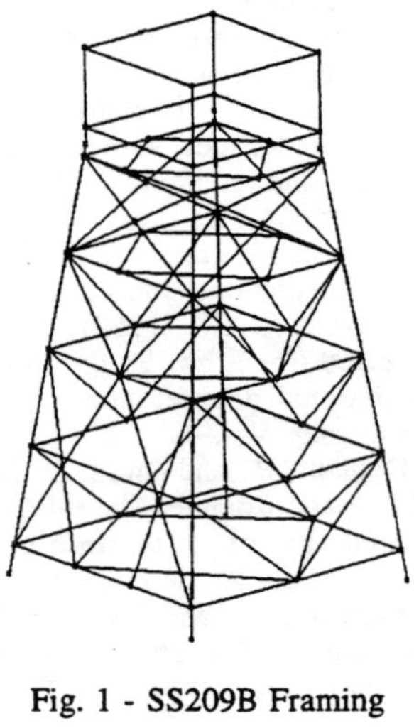

Fig. 1 - SS209B Framing

SS209B

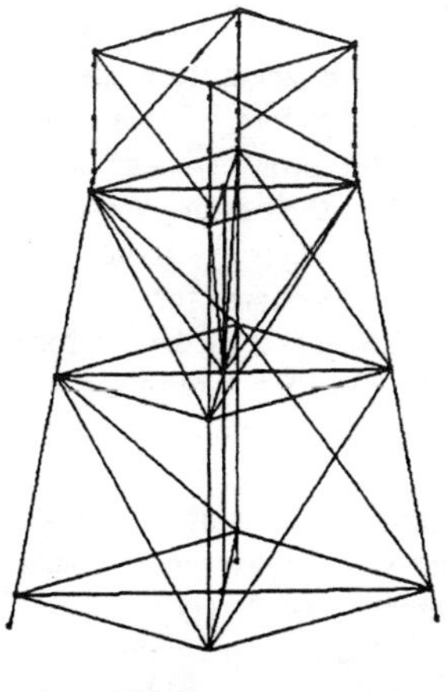

Fig. 2 - SS209B-AUX Framing

SS209B was originally fabricated with a deck, cellar deck and sub-cellar deck. The sub-cellar deck was removed to reduce the loads associated with hurricane storm wave crests that would reach this elevation. The cellar deck elevation is at +36 ft. MWL. This deck elevation exceeds the minimum deck height specified in API RP 2A-WSD Section 17 (API, 1996). A portion of the cellar deck was extended to allow for the placement of additional pumping equipment, while most equipment has been completely removed from the main deck.

Two of four walkways around the top of the jacket and the boat bumpers were removed to further reduce wave loads. The horizontal members in Bay 1[1] (elevation +7 ft. MWL) were modified at the same time that SS209B-AUX was installed; non-critical members were removed and the horizontal diagonals were replaced with stronger members.

SS209B is the only production platform of the two, with fourteen risers and five conductors. Following Hurricane Andrew in 1992, one of the conductors was found to have separated at approximately +3 ft. MWL and a dent was found in the northern horizontal at elevation -96 ft. MWL (bottom horizontal bay at the sea floor). The conductor has since been repaired.

Vertical cross-bracing in Bays 3 and 4 on the northern side differs from the vertical cross-bracing on the other three sides. This variation occurs because the platform was installed over an 8 ft. diameter caisson, that has since been removed. Hinged members were placed in the bottom two bays on the northern side, to allow for proper placement of the jacket. Based on results from a survey of the platform, deck and cellar deck loads were estimated at 50 psf (pounds per square foot) and 150 psf, respectively.

Piles and jacket legs are ungrouted. The 30 in. diameter piles range in thickness from 0.5 in. to 1.125 in. The pile penetration is 170 ft.

Above and below water inspections of this platform indicate that the structure has been well maintained. There is no significant corrosion, or member - joint damage.

SS209B-AUX

SS209B-AUX has not been known to have been damaged at any time. The deck height is at 45 ft. MWL, which exceeds the API RP 2A-WSD Section 17 minimum deck height requirement. SS209B-AUX has no conductors, but does support one 36 in. diameter sump that extends from the main deck to Bay 2 at -39 ft. MWL and two risers. Based on results of a field survey of the platform, deck loads were evaluated to be 150 psf on both the main deck and the cellar deck.

This platform consists of 2 bays with all joints grouted. The vertical cross bracing in the end-on direction is oriented in the same direction in both the north and south sides of the platform. This results in all of the cross-bracing acting in tension during a broadside loading condition that comes from the west and all of the cross-bracing acting in compression during a broadside loading condition from the east. Because the configuration of the structure is asymmetrical, analyses were conducted for all four load cases: north end-on and west broadside, and south end-on and east broadside.

The piles are 42-in. in diameter and range in wall thickness from 1.0 in. to 1.75 in. and penetrate to 180 ft. below the sea floor.

Above and below water inspections of this platform indicate that the structure has been well maintained. There is no significant corrosion, or member - joint damage.

Platforms API Classification

Both platforms are unmanned, have no significant oil storage, and are equipped with emergency shut-in equipment. These platforms would be classified by API as Unmanned, Insignificant Environmental Impact platforms. Therefore, because SS209B and SS209B-AUX are categorized as such and exceed the minimum deck height requirement in API RP 2A-WSD Section 17, both are considered to have minimum consequences.

ENVIRONMENTAL CONDITIONS & LOADINGS

Oceanographic criteria specified by API Section 17 for minimum consequence structures in the Gulf of Mexico for Design Level Analysis is based on winter storms and is outlined in Table 1. The associated wind force shape, and wave - current drag coefficients used to determine the hurricane forces acting on the structures are summarized in Table 2.

Table 1 - API Reassessment Environmental Conditions For Unmanned, Minimum Consequence Platforms (94 ft. MWL)

Surge Depth	3.5 ft.
Wind Velocity	45 mph
Wave Height	34 ft.
Wave Period	10.5 sec.
Current at SWL	0.9 fps
Current at Sea floor	0.9 fps

Table 2 - Force Coefficients

Drag Coef. -Jacket	1.05
Drag Coef. - deck	1.5
Wind Shape Coef.	1.0
Current Blockage	0.8
Directional Spreading	0.8
Global Load Factor	1.0

GEOTECHNICAL CONDITIONS & CHARACTERISTICS

Soil conditions at the platform locations were based on results from a 1995 geotechnical investigation (Fugro-McClelland, 1995). The site is composed primarily of soft to very stiff clay, with an undrained shear strength that varies from about 200 psf at the sea floor to approximately 2, 200 psf at 180 ft. below the seabed.

Skin friction versus deflection (T-Z) and end bearing versus deflection curves (Q-Z) were derived using data provided by Fugro-McClelland(1995). The static methods of analysis for pile foundations, as described in API 2A-WSD (1993), were used to calculate

[1] Jacket bays are numbered sequentially downward from the top of the jacket

the T-Z and Q-Z data. The lateral loading versus deflection curves (P-Y) were determined for cyclic loading using the procedures outlined in API 2A-WSD (1993). P-Y curves were determined for both a 42-in. diameter pile(SS209B-AUX) and a 30-in. diameter pile (SS209B) to a depth of 100 ft. in soft clay.

LOADING & CAPACITY ANALYSES

DLA with StruCAD*3D

StruCAD*3D is a three dimensional graphics oriented program used to design and analyze structures (Zentech, 1993). It has the capability to generate wind and wave loads as specified by the user. For this study, based on the water depth and storm wave characteristics, Stokes Fifth Order Wave Theory was used to determine the wave kinematics.

Structure loading and capacity were calculated using the API RP 2A-WSD (1993). Soil structure interaction was evaluated by providing the system with pile geometry and soil characteristics as represented by the T-Z , Q-Z, and P-Y curves.

An expected steel yield strength of 43 ksi (kips per square inch) was used (1.2 times nominal yield of 36 ksi). To develop an estimate of the ultimate bending capacity of the primary structure and foundation elements, the moments of inertia were increased by 1.3. The DLA capacity of a member in StruCAD*3D is determined by the creation of the first plastic hinge or member yielding.

USA with ULSLEA

ULSLEA derives its name from the ultimate limit state limit equilibrium analysis techniques employed to calculate a structure's capacity (Mortazavi, Bea, 1995). The program has been extensively verified with results from detailed nonlinear analyses of platforms and with performance information on platforms that have survived and failed during past hurricanes (Bea, et al. 1995).

Aerodynamic and hydrodynamic loadings are calculated according to API RP 2A-WSD (1993). The user defines the wind speed at +30 ft., the wave height and associated period, and current profile. Stokes Fifth Order Wave Theory is used to determine water particle velocities and accelerations (Preston (1994), Skjelbreia & Hendrickson (1961) and Fenton (1985)). Hydrodynamic loading is estimated using the first term of the Morison equation. The user is also provided with the opportunity to define directional spreading, current blockage factors as well as marine growth. All structural elements, including appurtenances are modeled as equivalent vertical cylinders that are located at the wave crest. For inclined members, the effective projected vertical area is used.

If the lower decks of the structure become inundated by the wave crest, only the horizontal hydrodynamic forces are considered. Drag coefficients for both decks and jacket are user specified. The methods used to compute hydrodynamic loadings in ULSLEA have been shown to over predict the maximum base shears by approximately 10 % (due to placement of all elements of the structure at the wave crest position).

The platforms lateral loading capacity is determined using plastic hinge theory. Limit equilibrium is developed by implementing the principle of virtual work, (Mortazavi and Bea, 1995). The structure is divided into its three primary components: the deck legs, the jacket, and the pile foundation. The ultimate shear capacity of a template-type platform is based on simplified analysis and experience, using assumed basic modes of failure in one or all of the three primary components. These modes of failure are as follows: plastic hinge formation in the deck legs and subsequent collapse of the deck portal, buckling of the primary vertical diagonal braces within the jacket, formation of plastic hinges in the piles resulting in lateral failure, or axial pile failure due to pullout or plunging.

The ultimate shear capacity of the deck portal is estimated based on the bending moment capacities of the deck legs. The contributing capacity of vertical bracing within the deck portal can also be accounted for.

The jacket of the platform is typically composed two or more bays. ULSLEA determines the capacity of each bay separately. The potential strength of the vertical bracing at each bay is estimated as the minimum capacity of the brace or the capacity of either of its joints. Two quantities represent the strength of each jacket bay. The first is the lower bound capacity as defined by the capacity of the member most likely to fail (MLTF) a.k.a. first member failure. The second is an upper bound capacity for the jacket bay. It is estimated by summing the residual strengths of all braces within the bay.

Brace capacity in ULSLEA is defined as yield when in tension and buckling when in compression. Joint capacity is defined as the first crack that develops in tension and in compression.

The pile foundation capacity is based on the assumption that the soil's undrained shear strength or angle of friction varies linearly along the length of the pile and that the diameter and wall thickness are constant. The pile lateral capacity is estimated based on the ultimate bending moment capacity of the pile (formation of two hinges), pile diameter and undrained shear strength of the soil. The axial capacity of the pile is determined from the end bearing capacity and /or the shaft frictional capacity. Scour depth can be defined by the user.

Once the ultimate lateral capacity is determined for all three primary components of the platform (deck legs, jacket bays, lateral and axial piles). A horizontal shear capacity profile for the platform is developed. This is compared with the storm shear profile. The static ultimate lateral capacity of the system is defined when the storm shear just exceeds the platform's horizontal capacity. The static capacity is then multiplied by a loading effects modifier, to account for transient wave loadings and nonlinear platform response (Bea, 1996a). The resulting capacity is the ultimate lateral capacity of the system. In these analyses, the loading effects modifiers were set at unity.

The structural reliability of the platform can be assessed by quantifying the uncertainties and biases that are associated with the loadings and capacities of the system (Mortazavi, Bea, 1995).

ULSLEA can also account for dented and corroded members and initial member out-of-straightness. Yield strength, modulus of elasticity, brace residual strength and brace buckling factors are user defined.

Differences

There are some important differences between the two types of analyses. ULSLEA and StruCAD*3D use different variations of Stokes Fifth Order Wave Theory to calculate the environmental loads due to waves. The ULSLEA program uses the methods discussed by Skjelbreia & Hendrickson (1961) and Fenton (1985). Fenton proposed a variation on the original formula by Skjelbreia & Hendrickson, that accounts for current-wave interactions. StruCAD*3D uses only the techniques of Skjelbreia & Hendrickson (1961). This variation results in ULSLEA calculating a slightly higher wave crest elevation that does StruCAD*3D.

There are differences in the definitions of brace failure. ULSLEA determines the expected ultimate strength of the member as a function of the joint capacity at each end and the brace tensile or compressive buckling capacity. StruCAD*3D bases the evaluation on the brace capacity only, and failure is defined as the formation of a plastic hinge or general yielding of the cross section.

There are important differences in how the foundation capacities are determined in StruCAD*3D and ULSLEA. In StruCAD*3D pile failure occurs when the first hinge forms in a pile. Further, the modeling of the soil resistance is based on traditional P-Y and T-Z characterizations. In ULSLEA pile failure occurs when two hinges form in all of the piles or pile plunging / pullout occurs. The soil is assumed to be at its maximum shear strength at all points along the pile. When traditional API static shear strength based pile-soil interaction characteristics are used (including cyclic P-Y) curves, the StruCAD*3D analyses will indicate foundation capacities that are substantially less than those determined by ULSLEA.

RESULTS - 209B

The API RP 2A-WSD Section 17 specifies a wave height of 34 ft. for requalification of minimum consequence structures in the location of SS209B. To gain insights into the performance of the two structures at wave heights above and below the API requalification wave height, ULSLEA and StruCAD*3D analyses were conducted using wave heights ranging from 20 ft. to 65 ft.

The DLA and USA maximum total lateral forces and the loadings at which first member 'failures' occur for SS209B are summarized Fig. 3. Because ULSLEA and StruCAD*3D differ in their methods for calculating loads and capacities, a slight discrepancy between the two programs is expected. Because ULSLEA calculates a slightly higher wave crest elevation than does StruCAD*3D, a slightly higher base shear is expected, especially at large wave heights when deck inundation occurs.

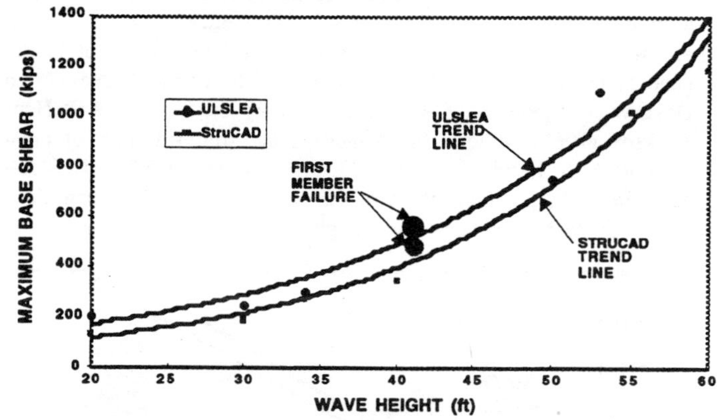

Fig. 3 - Wave Heights and Maximum Total Lateral Forces With Indicated First Member Failures

Another important difference is that StruCAD*3D assumes failure once the first plastic hinge is formed or yielding - buckling occurs in a member. The reserve strength of members is not accounted for. ULSLEA however determines the ultimate capacity of the member and the joints on both ends. The residual strength is also taken into account. ULSLEA assumes failure of a member only after specific failure modes occur.

In all cases both the maximum base shears were in very close agreement. At the 34 ft. wave height, both StruCAD*3D and ULSLEA results indicate a maximum base shear of approximately 300 kips. StruCAD*3D indicated that there was no inelastic action developed in any of the platform elements Therefore SS209B passed the requirements of API Section 17.

In ULSLEA, first member failure occurs in Bay 3 (Fig. 4) due to the cracking of a brace joint in tension at a maximum base shear of 550 kips (wave height of 41 ft). ULSLEA predicts that the storm shear will exceed the ultimate capacity of the platform at a wave height of 53 ft. with a maximum base shear of 1100 kips. At this load, there would be failure of all of the diagonal braces - joints in Bay 3.

StruCAD*3D analyses resulted in first member failure occurring due to plastic hinge formation in one pile at 30 ft. below the sea floor. This failure occurred at a wave height of 41 ft. with a corresponding maximum base shear of approximately 500 kips. StruCAD*3D also determined the diagonal braces in the third and fourth bays to be a weakness, with the failure of the first brace occurring within these bays at a wave height of 55 ft. Maximum base shear due to the 55 ft. wave height was 1010 kips.

Given the tendency for the StruCAD*3D analyses to underestimate the pile capacities relative to those from ULSLEA, if stronger foundation conditions were incorporated into the StruCAD*3D analyses there would be a consistent indication of first member failure in the jacket. In this instance there would be very good agreement between the StruCAD*3D and ULSLEA results. Both would indicate brace failure in the same location at a loading of approximately 1000 kips (wave height of 53 to 55 ft.).

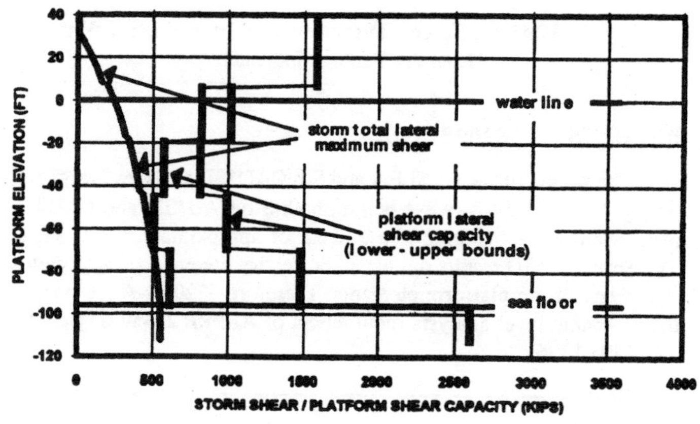

Fig. 4 - ULSLEA Lateral Forces (wave height = 41 ft) and Platform Lateral Capacities

Based on the StruCAD*3D results, a plot of total base shear versus deflection at the top of the structure was developed (Fig. 5). Because StruCAD*3D is not capable of determining an accurate load versus deformation curve after there is significant yielding of the structure elements, the number of failed members at each wave height is given as a qualifier. As already mentioned, first member failure occurs in a pile, with subsequent failure of diagonal braces, and ultimately hinging in all piles.

First member failure occurs in at a wave height of 41 ft. and with approximately equal maximum base shears in both analyses. However, the mechanism for first member failure is different. ULSLEA predicted the failure of a diagonal brace in Bay 3. This failure was not predicted by StruCAD*3D until a wave height of 55 ft. and a much higher base shear of 1010 kips. StruCAD*3D predicted first member failure due to the formation of a plastic hinge 30 ft. below the sea floor.

This failure was not predicted by ULSLEA because of the methods used to calculate pile capacity. ULSLEA does not assume lateral failure of the pile until two plastic hinges are formed. According to the StruCAD*3D results, the formation of two plastic hinges in the first pile does not occur until a wave height of 55 ft. ULSLEA predicts failure of the structure at a wave height of 53 ft.

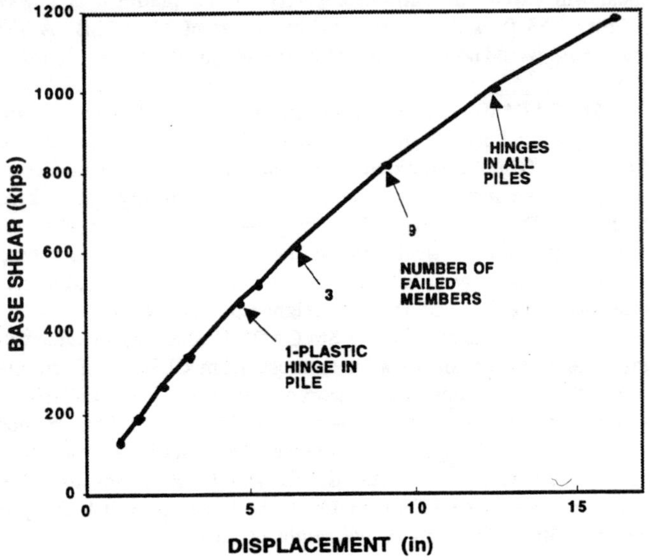

Fig. 5 - Platform Force - Displacement Characteristics With Element Failures Indicated

RESULTS - SS209B-AUX

The results of the ULSLEA and StruCAD*3D analyses are given in Fig. 6. At the 34 ft. wave height, both StruCAD*3D and ULSLEA results indicate a maximum base shear of approximately 300 kips. The StruCAD*3D results did not indicate the presence of any yielding in any of the platform elements. Therefore SS209B-AUX passed the screening level analysis requirement of API RP 2A-WSD Section 17 (API, 1996).

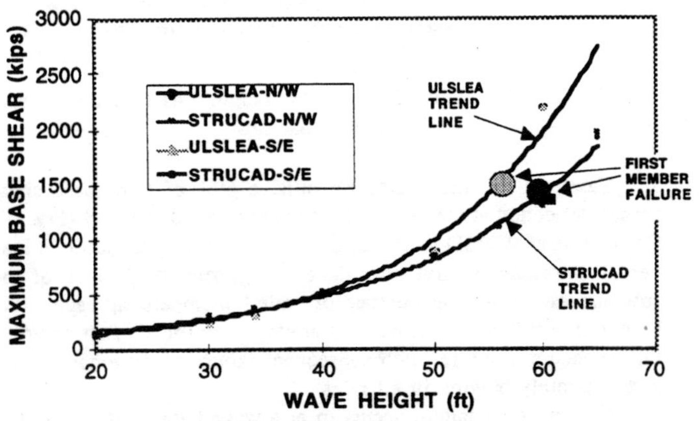

Fig. 6 - Wave Heights and Forces with First Member Failures Indicated

StruCAD*3D predicts first member failure at a wave height of 61 ft with a maximum base shear of approximately 1360 kips. StruCAD*3D was able to calculate base shears up to a wave height of 65 ft. At a wave height of 65 ft., plastic hinges formed in all four piles.

The maximum base shears calculated for SS209B-AUX are in close agreement up to the 56 ft. wave height. As shown in Fig. 7, it is at the 56 ft. wave height that first member failure occurs in ULSLEA at a maximum base shear of approximately 1500 kips. The structures ultimate capacity is reached at a wave height of 60 ft. with a maximum base shear of approximately 2200 kips.

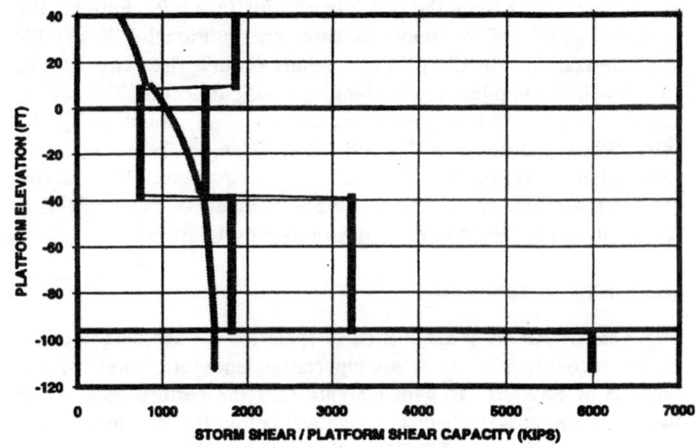

Fig. 7 - ULSLEA Lateral Forces (wave height = 56 ft.) and Platform Lateral Capacities

A maximum base shear versus deformation curve for SS209B-AUX is given in Fig. 8. Platform deformations were measured at the top of the deck portal (Fig. 3). The large differences in base shear at the 56 ft. wave height is due to a variation between ULSLEA and StruCAD*3D regarding in the amount of wave height above the still waterline. ULSLEA predicts the deck inundation by the wave crest at a wave height of 56 ft., while StruCAD*3D predicts deck inundation at a wave height of 61 ft. The reason for this disagreement is dependent on the method of calculation used to define Stokes Fifth Order Wave Theory. StruCAD*3D relies only on Skjelbreia & Hendrickson (1961) while ULSLEA incorporates the findings of both Skjelbreia & Hendrickson (1961) and Fenton (1985). ULSLEA calculates a higher percentage of the wave above MWL than does StruCAD*3D for the same wave height.

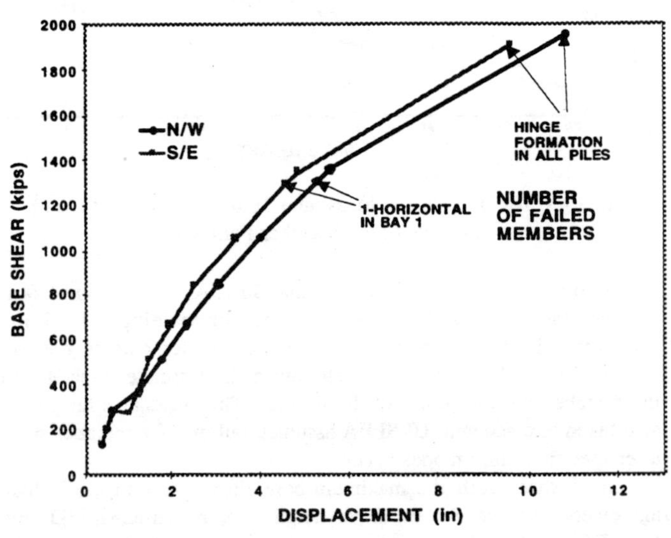

Fig. 8 - Platform Force - Displacement Characteristics With Element Failures Indicated

If maximum base shears are compared, regardless of wave height, then first member failure occurs as a result of similar stresses and both models predict first member failure will occur at the top of the jacket in Bay 1. StruCAD*3D specifically predicts failure in the northern horizontal member at the top of the jacket.

PROBABILITY ANALYSIS

A reliability analysis for hurricane storm loadings and platform capacities for SS209B and SS209B-AUX was performed as part of the ULSLEA analyses. A reliability study was performed to evaluate the implications of the uncertainties associated with the loadings and the various failure modes in the two platforms.

Modeling uncertainties (Type II) were based on characterizations of the 'biases' in the storm loadings and platform element capacities. Bias is defined as the ratio of the 'true' value to the predicted value. The mean biases and Coefficients Of Variation (COV) were used to characterize Type II uncertainties. The distributions of these uncertainties are assumed to be Lognormal (Mortazavi, Bea, 1996). The Type II uncertainties are summarized in Table 3 (Mortazavi, Bea, 1996)

Table 3 - Mean Biases and COV's for Model Elements

Hurricane Loadings	Mean Bias	COV
Wave in deck	1.0	0.4
Wave on jacket	1.0	0.3
Capacities		
Tubular Braces	1.4	0.3
Tubular Joints (209B)	1.6	0.2
Tubular Joints (209B-AUX	2.0	0.4
Piles Clay Lateral	2.0	0.3
Piles Clay Axial	3.0	0.3

ULSLEA was used to determine the probability of failure of the two platforms conditional on the maximum total lateral forces associated with the range of wave heights. Figures 9 and 10 summarize the results for the two platforms for wave heights of 41 ft. (SS209B) and 56 ft (SS209B-AUX), respectively. These wave heights produced lateral forces that brought these two platforms close to their ultimate limit state.

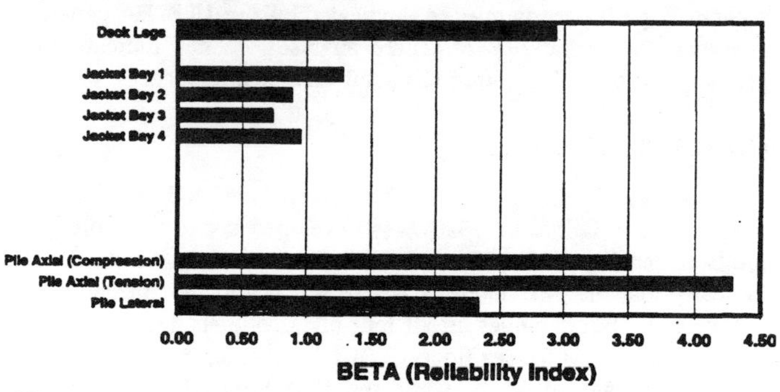

Fig. 9 - Safety Indices for SS209B Platform Components for Wave Height = 41 ft.

The ULSLEA results were used to define 'fragility' curves for the two platforms (Fig. 11). These fragility curves are based on the probability of failure of the first members in each of the structures and reflect the Type II uncertainties in both loadings and capacities.

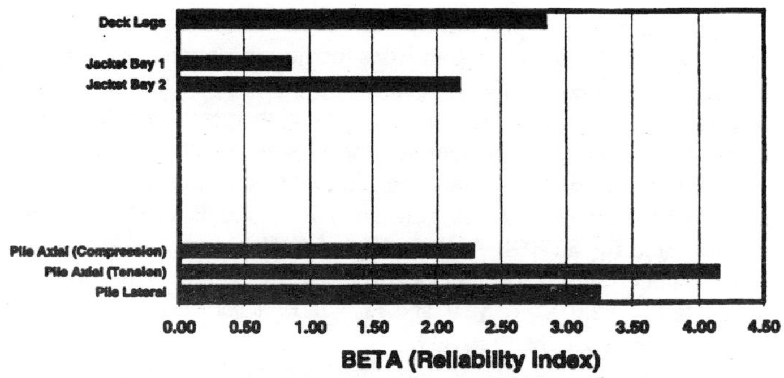

Fig. 10 - Safety Indices for SS209B-AUX Platform Components for Wave Height = 56 ft.

The long-term distribution of the expected maximum wave heights for this location is shown in Fig. 12. This characterization was based on site and regional specific oceanographic study results (Woodward-Clyde Consultants, 1982). Also shown are the wave heights that ULSLEA indicates would bring the platform to its ultimate limit state lateral loading capacity and the wave height specified by API for requalification of these platforms.

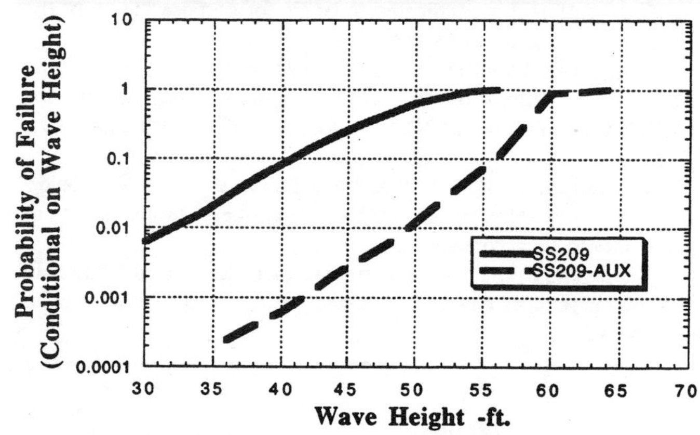

Fig. 11 - Probabilities of Platform Failure Conditional on Occurrence of Indicated Wave Heights (with Type II Uncertainties)

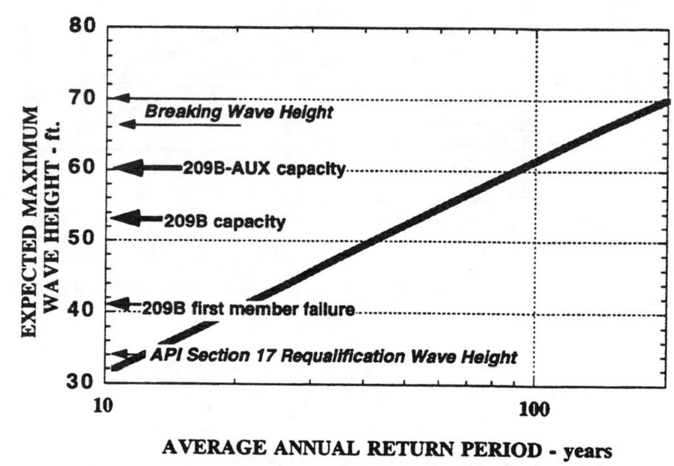

Fig. 12 - Average Annual Return Periods of Hurricanes Developing Sea States Having Given Expected Maximum Wave Heights

To compute the probability of failure of the two platforms, each of the conditional probabilities of failure summarized in Fig. 11 were multiplied by the probability of occurrence of sea states that would generate expected maximum wave heights equal to those shown in Fig. 12 and then summed for all wave heights.

The results of the calculations indicated that the annual probability of failure (Pf_E) due to hurricanes for SS209B is $Pf_E = 2.5$ % per year, and for SS209B-AUX is $Pf_E = 1.7$ % per year. When only Type I uncertainties are included, these probabilities of failure reduce to $Pf_E = 2.0$ % per year and $Pf_E = 1.1$ % per year for SS209B and SS209B-AUX, respectively.

Operations - processing hazards (fires, explosions) were estimated based on historic rates of failure for similar types of operations and platforms in the Gulf of Mexico (Visser, 1994). These hazards were estimated to develop probabilities of failure of the structures of $Pf_O = 0.5$ % per year. Due to the age of the reservoirs, minimum processing activities, and current well and pollution safety equipment onboard these platforms, the environmental probabilities of failure are much higher than the probabilities of failure related to operating hazards.

FITNESS FOR PURPOSE

Both platforms are indicated to be fit for purpose based on the API Section 17 guidelines and the results of the StruCAD*3D DLA.

Notional probabilities of failure that background development of the API guidelines for this class of structure indicate that the probabilities of failure due to storm waves should be acceptable (De, 1995).

In addition, economics based and Standard of Practice Guidelines (Fig. 13) (Bea, 1996) indicate that these probabilities of failure should be acceptable.

The results of the different approaches to the requalification of these two platforms develop consistent assessments of their fitness for purpose.

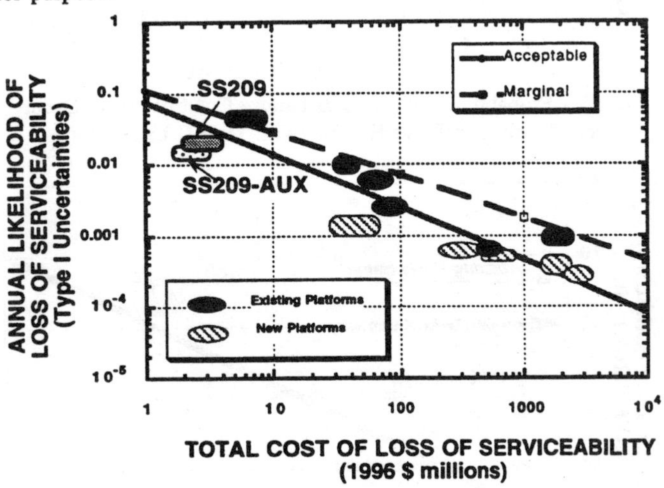

Fig. 13 - Standard of Practice Fitness for Purpose Guidelines

STRUCTURAL UPGRADE TO SS209B

A structural upgrade to SS209B was evaluated to improve the performance of the platform to be comparable with SS209B-AUX. The measures used to reduce the storm wave loadings on SS209B were discussed earlier.

Since initial failure of the structure is indicated by the analyses to be due to tensile joint capacities structural improvements investigated included grouting of the pile - leg annulus to improve the joint capacities.(UK Department of Energy 1988; Morahan, et al. 1996; Marshall 1992).

The tensile capacities of the joints were increased by a factor of 2.0 (Marshall 1992; Morahan, et al. 1996; U. K. Health and Safety Executive 1995). This improvement increased the capacity of Bay 3 (the bay where the first member failures occurs) by nearly 500 kips (Fig. 14).

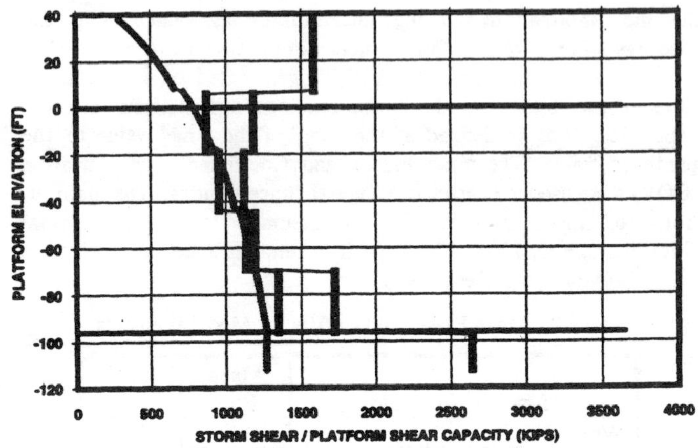

Fig. 14 - SS209B ULSLEA Lateral Loadings and Capacities for Wave Height = 54 ft. (Grouted Leg - Pile Annulus)

With the grouted pile - leg annulus, first member failure occurred at a wave height of 52 ft. The maximum lateral force capacity of the platform was mobilized at a wave height of approximately 54 ft. While the grouting option did have a substantial effect on the loading at which first member failure occurred, it did not have a large influence on the ultimate capacity of the platform. It is situations like this that results from linear analysis based evaluations can produce very misleading indicators.

Based on an ultimate capacity associated with a 54 ft. wave height, the new probability of failure of SS209B would be approximately $Pf_E = 2.1$ % with Type II uncertainties and $Pf_E = 1.6$ % without Type II uncertainties. As indicated by the ULSLEA capacity analyses, the probability of failure analyses do not indicate that grouting the pile - leg annuli has a substantial influence.

CONCLUSIONS

Both the StruCAD*3D and ULSLEA analyses indicate that the initial mode of failure in SS209B is developed due to the failure of joints in tension in Bay 3 at a wave height of 41 ft. StruCAD*3D indicated that the next mode of failure was due to the formation of two sets of plastic hinges in all four piles, one set below the sea floor and one set at the sea floor. ULSLEA indicated platform failure at a wave height of 53 ft. while the DLA StruCAD*3D indicated platform failure at a wave height of 55 ft.

SS209B-AUX StruCAD*3D and ULSLEA analyses indicate first member failure at a wave height of 61 ft. and 56 ft., respectively. The mechanism of first member failure is failure of the diagonal brace joints in tension. The StruCAD*3D and ULSLEA analyses indicate that the ultimate lateral load capacity of the platform occurs at a wave height of 65 ft. and 60 ft., respectively. It is at these wave heights that the decks become inundated resulting in a dramatic increase in the maximum base shear.

For both platforms, the DLA performed with StruCAD*3D and ULSLEA developed very comparable results. Similarly, the USA performed with StruCAD*3D and ULSLEA developed very comparable results.

Based on the criteria contained in API RP 2A-WSD Section 17, both SS209B and SS209B-AUX pass the screening level analysis. The probability of failure analyses and the economics and standard of performance based analyses indicated that both of the platforms are fit for purpose in their present conditions.

The analyses performed on SS209B based on grouting the pile - leg annuli indicated a substantial increase in the lateral loading required to cause first member failure. However, there was not a substantial increase in the total lateral load carrying capacity of the platform.

REFERENCES

American Petroleum Institute (API) (1993). *Recommended Practice for Planning, Designing, and Constructing Fixed Offshore Platforms - Load and Resistance Factor Design (RP 2A-LRFD)*, Washington, DC

American Petroleum Institute (1996). *API RP 2A-WSD 20th Edition, Section 17, Assessment of Existing Platforms*, Washington, D. C.

Bea, R. G. (1989). *Reliability Based Design Criteria for Coastal and Ocean Structures*, Institution of Engineers Australia, Barton, ACT.

Bea, R. G. (1996a). "Nonlinear Performance of Offshore Platforms in Extreme Storm Waves", Journal of Waterway, Port, Coastal, and Ocean Engineering, American Society of Civil Engineers, March.

Bea, R. G. (1996b). "Requalification of Infrastructure: Application to Offshore Platforms," Journal of Infrastructure, American Society of Civil Engineers, May.

Bea, R. G., Mortazavi, M. M., Loch, K. J., and Young, P. L. (1995). "Verification of a Simplified Method to Evaluate the Capacities of Template-Type Platforms," Proceedings of the Offshore Technology Conference, OTC 7780, Society of Petgroleum Engineers, Richardson, Texas.

De, R. (1995). "Risk Analysis Methodology for Developing Design and Assessment Criteria for Fixed Offshore Structures," *Proceedings of the Offshore Technology Conference*, OTC 7755, Society of Petroleum Engineers, Richardson, Texas.

Fenton, J. D. (1985). "A Fifth Order Stokes Theory for Steady Waves," ASCE Journal of Waterway, Port, Coastal and Ocean Engineering, Vol. 111, No. 2, pp. 216-234.

Fugro-McClelland Marine Geosciences, Inc., (1995). *Geotechnical Investigation - OCS-G-827, Platforms M & N - Block 209, Ship Shoal Area*, Report to Union Oil Company of California, Lafayette, Louisiana.

UK Department of Energy (1988). Grouted and Mechanical Strengthening and Repair of Tubular Steel Offshore Structures," OTH 88 283, HMSO, London.

UK Health and Safety Executive (1995), "Platform Strengthening and Repair using Grouted Joint Technology," *Offshore Research Focus*, No. 107, June, London.

Marshall, P. W. (1992). "Grouted Connections," *Design of Welded Tubular Connections*, Elsevier, London.

Morahan, D. J., van Foeken, R. J., Laiani, M., and Wardenier, J. (1996). "New Data on the Fatigue and Ultimate Limit State of Grouted Tubular Joints," *Proceedings of the Offshore Technology Conference*, OTC 8080, Society of Petroleum Engineers, Richardson, Texas.

Mortazavi, M., and Bea, R. G. (1995). *Screening Methodologies for Use in Platform Assessments and Requalifications*. Report to Joint Industry - Government Sponsored Project, Department of Civil Engineering, University of California , Berkeley, June.

Mortazavi, M. and Bea, R. G. (1996). "Reliability Based Screening Methodology for Template-Type Offshore Platforms," Proceedings of the Offshore Mechanics and Arctic Engineering Conference, American Society of Mechanical Engineers, Florence, Italy.

Preston, D. (1994). *An Assessment of the Environmental Loads on the Ocean Motion International Platform*, MS Thesis, Department of Naval Architecture and Offshore Engineering, University of California, Berkeley.

Skjelbreia, L., and Hendrickson, J.,1961, "Fifth Order Gravity Wave Theory." Proceedings, 7th Conference of Coastal Engineering, pp. 184-196.

Visser, R. C. (1993). "Offshore Platform Accidents, Regulations, and Industry Standards," *Proceedings of the Offshore Technology Conference*, OTC 7118, Society of Petroleum Engineers, Richardson, Texas."

Woodward-Clyde Consultants-Ocean Services Division (1982). *Shallow Water Wave Force Criteria Study - Gulf of Mexico*. Prepared for McMoRan Offshore Exploration Co., April.

Zentech (1993). *StruCAD*3D - User's Manual*. Distributed by Structural Software, Inc., Houston, Texas.

Proceedings of the Seventh (1997) International Offshore and Polar Engineering Conference
Honolulu, USA, May 25-30, 1997

ISBN 1-880653-28-1 (Set); ISBN 1-880653-32-X (Vol. IV)

Substructural Identification of Offshore Structures Using Neural Networks

Chung-Bang Yun and Eun Young Bahng
Korea Advanced Institute of Science and Technology
Taejon, Korea

ABSTRACT

A method is presented for estimating unknown parameters of offshore structures using a backpropagation neural network. Several techniques are employed to overcome the issues associated with a large number of degrees of freedom. They are the substructural identification and the submatrix scaling factor. The modal data such as natural frequencies and mode shapes are used as input to the neural network for effective element-level identification particularly for the case with incomplete measurement of modal data. A numerical example analysis on a jacket-type offshore structure is presented to illustrate the proposed procedure and to demonstrate the effectiveness of the method.

KEY WORDS: Substructural identification, neural networks, modal data.

INTRODUCTION

In relation to the problems of damage detection and maintenance of old structures, estimation of element-level stiffness parameters becomes an increasingly important issue (Natke and Yao 1986, Hong and Yun 1993, Ghanem and Sinozuka 1995). In recent years, the pattern matching techniques using neural networks have drawn considerable attention in the field of damage assessment. Several researchers have dealt with neural network approaches for damage estimation of structural models with a small degrees-of-freedom (Wu et al 1992, Tsou and Shen 1994, Pandey and Barai 1995). Although their results look promising, some issues related to a large number of degrees of freedom have to be resolved, before it can be a truly viable method for structural identification. In this study, the neural network-based approach is extended to the estimation of structural parameters of a complex structural system such as a jacket-type offshore structure. Several techniques, such as substructural identification (Koh et al 1991; Oreta and Tanabe 1994; Yun and Lee 1995), submatrix scaling factor (Lim 1990), and modal strain energy (Lim 1991) are employed to overcome the issues associated with a large number of unknowns. A numerical example analysis on a jacket-type offshore structure is presented to demonstrate the effectiveness of the proposed method.

NEURAL NETWORK-BASED STRUCTURAL IDENTIFICATION

Studies on neural networks have been motivated to imitate the way that the brain operates (Haykin 1994). It has recently drawn considerable attention in various fields of science and technology, such as character recognition, electro-communication, image processing, and industrial control problems. Many researchers have developed various neural network models for different purposes (Hush and Horne 1993). In this study, a model called backpropagation neural network (BPNN) is used for structural identification. The basic strategy for developing a neural network-based approach to identification of a structural system is to train the BPNN to recognize the element-level structural parameters from measurement data on the structural behavior, such as natural frequencies and vibration mode shapes.

Substructural Identification and Submatrix Scaling Factor

For the identification of a structure with many unknowns, it is not practical to identify all of the parameters in the structure at a time, because most of the identification techniques require expensive computation that would be even prohibitive, as the number of unknown parameters increases. Several researches were reported on identification of a part of a structure so as to reduce the size of the system under consideration. Those works were based on the reasoning that the expected damage of a structure occurs at several critical locations, hence it is more reasonable to concentrate the identification at critical locations of the structure.

The local-identification method is generally based on substructuring, in which the structure is subdivided into several substructures and the identification is carried out on a substructure at a time. In the present study, a substructure to be identified is called as the internal substructure, while the others as the external substructures as shown in Fig 1. Since the parameters to be estimated are limited to a substructure, it is expected that the numerical problems such as divergence or falling into local minima may be avoided. Another advantage is that this approach requires measurement data only on the substructure of interest, instead of on the whole structure.

The number of parameters to be estimated shall be kept reasonably small for successful identification of a structure. Reduction of parameters will improve the results of the identification and minimize the required measurement data. For the purpose of efficient parameterization of the structure, the submatrix scaling technique is adopted in this study (Lim 1990). Using the submatrix scaling technique, the stiffness matrix of the system can be described by introducing the submatrix scaling factors(SSF) corresponding to the element level stiffness matrices as

$$K = \sum_{j=1}^{p} s_j K_j^* \tag{1}$$

where K is the stiffness matrix of the structure in the present condition, K_j^* is the jth reference value of the stiffness submatrix for the jth element transformed into the global coordinates, which means the undamaged stiffness submatrix or the stiffness submatrix estimated by the previous identification, s_j is the jth SSF, and p is the total number of submatrices. Since the submatrix can represent a single element or a group of elements of the structure with the known geometry, material properties, and boundary conditions, a significant reduction of the unknown parameters can be achieved by grouping the structural elements with the same stiffness characteristics. Then, the identification of the stiffness matrix is performed by estimating the submatrix scaling factors instead of all of the stiffness matrix coefficients. Direct identification of the stiffness matrix coefficients may face with problems which violate the basic properties of the stiffness matrix; such as symmetricity and positive definiteness. However, the use of the SSF's described in Eq. (1) can avoid such problems.

Modal Data as Input to Neural Networks

A way of choosing the patterns representing the characteristics of the structure, which will be used as input to the neural network, is one of the most important subjects in this approach. Several researchers have used the various input patterns suitable for their purpose. For example, Wu et al. used the frequency spectrum for each DOF of the example structure for damage estimation (Wu et al 1992). This input has an advantage that does not need the modal parameter identification from measurements. Nonetheless, to characterize these spectral properties, tremendous amount of sampling data is required. Accordingly, a large number of input nodes is needed, which may reduce the efficiency and accuracy of the training process. Tsou and Shen used the dynamic residual vector that can be obtained from the modal data (Tsou and Shen 1994). It provides a simple and effective way to detect the damage and the length of the input pattern can be reduced significantly compared with the spectrum data. However, it still has the restriction that the modes should be measured at every finite element DOF's. In reality, the modes obtained from a typical modal survey are generally incomplete in the sense that the mode vector coefficients are available only at the test DOF's. Therefore, it is desirable to use the input patterns which are more suitable for the case with partial measurement data

Using the substructuring technique and referring to Fig. 1, the mode shape matrix Φ can be partitioned as

$$\Phi = \begin{bmatrix} \Phi_1^L & \Phi_1^H \\ \Phi_2^L & \Phi_2^H \\ \Phi_3^L & \Phi_3^H \end{bmatrix} \tag{2}$$

where the subscripts 1, 2, and 3 denote substructures, substructure 2 being the internal substructure; and the superscripts L and H represent lower and higher modes, where only the lower modes are generally used for the structural identification. Rewriting the ith mode in the partitioned modal matrix Φ_2^L as $\tilde{\varphi}_i$ for the simplicity, the input pattern can be defined as

$$\textit{Input Pattern Vector} = \{(f_i, \tilde{\varphi}_{1,i} \cdots \tilde{\varphi}_{n,i}),\ i = 1, \ldots, m\} \tag{3}$$

where f_i is the ith natural frequency, $\tilde{\varphi}_{ji}$ denotes the jth component of $\tilde{\varphi}_i$ which is normalized as $\tilde{\varphi}_i^T \tilde{\varphi}_i = 1$, n is the number of DOF's for Substructure 2, and m is the number of modes to be included in the identification.

A useful relationship between the submatrix for the element stiffness and the modes may be obtained by examining the strain energy of each submatrix with respect to the modes (Lim 1990). The modal strain energy distribution among the stiffness submatrices for the ith mode can be evaluated as

$$\beta_{ji} = \frac{\varphi_i^T \cdot K_j \cdot \varphi_i}{\varphi_i^T \cdot M \cdot \varphi_i \cdot \omega_i^2} \tag{4}$$

where β_{ji} is the modal strain energy(MSE) coefficient of submatrix K_j in the ith mode, K_j is the stiffness submatrix for the jth element transformed into global coordinates, φ_i is the ith mode shape, and ω_i is the ith circular natural frequency. The submatrices containing large values of the MSE's for a mode indicate that the corresponding structural elements are major load carrying elements for the particular mode. Stiffness changes in those elements will cause significant changes in the frequency and mode shape. Hence, detection of the stiffness changes may be more effective by examining the modal data for those elements. On the other hand, the structural elements having negligible MSE's reflect that it would be difficult to detect the stiffness changes in those elements using the corresponding modal data. They are used to determine what modes are better information for structural identification.

Generation of Training Patterns

Since the neural network-based structural identification is highly dependent on the training patterns, it is very important to prepare well-examined data sets. Consequently, it is required that the number of training patterns must be large enough to represent the structural system properly. However, for the computational efficiency, it must be limited to be reasonably small, because most of the computational time for this method is required for preparing the training patterns and training the network. When the number of the unknown parameters is N and each parameter has M sampling points, the size of whole population is M^N. Accordingly, a sparse sampling algorithm such as Latin hypercube is introduced in this study (Press et al 1992). The Latin hypercube sampling makes the required number of samples reduced from M^N to M.

The training patterns for the proposed neural network-based method consist of the modal data as input and the corresponding SSF's as output. To generate training patterns, a series of eigenanalyses are to be performed. However, iterative computations of eigenvalue problems for complex structures may be very expensive. Thus, the fixed-interface-component-mode-synthesis (CMS) method incorporating the substructuring technique is employed in this study (Craig 1968).

ILLUSTRATIVE EXAMPLE

A jacket-type offshore structure as shown in Fig.2 is chosen to demonstrate the applicability of the proposed technique for structural identification problems. The mass density and the elastic modulus of all members of baseline are 8000 kg/m^3 and $2.1 \times 10^{11} N/m$. The areas and the moments of inertia of the cross-sectional areas of vertical, diagonal and horizontal members are ($0.4m^2$, $0.12m^4$), ($0.1m^2$, $0.03m^4$), and ($0.02m^2$, $0.002m^4$), respectively. The substructure to be identified is called as the internal substructure, while the others as the external substructures. It is assumed that the unknown SSF's for the elements of the structure are between 0.5 and 1.5. The mode shapes are assumed to be measured only at 10 test DOF's, which include the displacements in x- and y-directions at 5 nodes of the internal-substructure, note that the rotational displacements are not included.

Input Patterns to BPNN

For the efficient training process, an appropriate number of modes as input to BPNN shall be selected. It is assumed that the first six modes of the internal-substructure at test DOF's are available in this study. The natural frequencies and mode shapes of the reference structure, in which all the SSF's are unity, are shown in Fig. 3. The fractional MSE distribution in the first six modes calculated for the reference state are shown in Fig. 4. Fig. 4 shows that the vertical elements such as 2 and 7 are carrying large fractions of the strain energy especially. It means that the modes contains more information concerning the vertical elements than the others. On the other hand, the MSE coefficients for the horizontal elements 1 and 8 are relatively small for all the modes, which indicates that the SSF estimation for the horizontal elements may not be effective.

Training Patterns and Testing Patterns

For generation of training patterns using the Latin hypercube sampling, 21 sampling points are selected between 0.5 and 1.5 with an equal interval of 0.05 for each SSF in the internal substructures, assuming each SSF is uniformly distributed, while 11 sampling points between 0.5 and 1.5 are taken with an interval of 0.1 for the SSF's in the external substructures. Finer sampling has been taken for the internal substructure than the external substructure, since the purpose of this study is to identify the SSF's in the internal substructure. Consequently, the number of the possible sample cases is 231 (= 21×11), which is not judged to be enough to represent the system adequately. Hence, eight iterations of the sampling process are taken, which results in 1848 sample cases. The elapse time for generating whole training data set is about 80 sec by the Alpastation 200[4/233] For the cases of testing patterns, the sampling points for the internal substructure are taken as the same to the training data, however the number of the sampling points for the external substructures is taken to be 10 in the range of 0.55 - 1.45 with an interval of 0.1. The difference in sampling is imposed, because the real testing patterns are not necessarily the same with the training patterns. Thus, the number of the possible sample cases for testing is 210 (21×10).

The patterns are generated by computing the natural frequencies and mode shapes using the CMS method. The training and testing data set for the noise-free and the 3% noise cases are prepared. The training data set is shown in Table 1.

Learning of Neural Network

It is important to choose the proper network size. In general, it is not straight forward to determine the best size of the networks for a given system. It may be found through trial and error process using knowledge about the system. A four-layer neural network as shown in Fig. 5 is selected for the present example. The numbers of neurons in the input, the first hidden, the second hidden and the output layer are 32, 20, 13, and 8, respectively. Numerical investigations show that small changes of the numbers of the neurons in the hidden layers have little effect on the estimation results.

The four-layer neural network is trained using 1,848 training patterns by the standard BP algorithm. The training process took about 2,000 iterations(epochs) to learn the pattern representation within tolerance accuracy, in which the elapse time is about 675 sec by an Alpastation computer system. One epoch means that all the 1,848 training patterns are used once for training. The tested outputs is listed in the Table 2. Fig. 6 shows the averages of the relative errors in the estimated SSF's for various cases with different number of modes included in the input patterns. It can be found that the accuracy of the estimation improves with increasing number of the modes included until the 6th modes.

Table 3 and Fig. 7 show the tested results for three cases with different noise levels in the training and testing data set. The average relative error of the tested outputs for the noise-free testing data, after the network is trained using the noise-free training data set, is 12.9%. But, the error for the 3% noise testing data using the same network is 23.3%. The result implies that the measurement error causes deterioration in the estimation. On the other hand, when the network is trained using the 3% noise training data set, the average relative error of the tested outputs for the 3% noise testing data is obtained as 15.8%, which is smaller that the average error of the test output (23.3%) using the noise-free training data. Fig. 8 is the learning curves to depict how the noise injection learning works. It is found that average relative error for the training case with 3% noise is smaller than that for the case without noise injection. This indicates that it is desirable to use the 3% noise data for training, if the measurement error is expected to be 3%. As anticipated in the previous section, the estimation errors of horizontal elements 1 and 8 are larger than those of the others.

CONCLUDING REMARKS

For the assessment of structural integrity, the neural network-based structural identification is applied to the estimation of the parameters of offshore structures. The substructuring technique and the concept of submatrix scaling factor are employed to reduce the number of unknown parameters for local identification. The proposed neural network-based method don't require any complicated formulation. Particularly the conventional approaches can not apply to substructural identification directly without any model reduction of a system. This approach depends upon how well and how compactly the training patterns represent the relationship between structural parameters and modal information. The Latin hypercube sampling and the fixed-interface-component-mode-synthesis(CMS) method provide the simple way to obtain such a training set.

A numerical example analysis on a jacket-type offshore structure is presented to demonstrate the effectiveness of the method. The average relative estimation error for the case with 3% noise testing data set is found to be about 16%. It has been found that the estimated results for the elements with larger MSE coefficients are better than those associated with smaller MSE coefficients.

REFERENCES

Craig, R. R., Jr. And Bampton. (1968). M. C. C., "Coupling of Substructures for Dynamic Analysis," *AIAA Journal,* Vol. 6, No. 7, pp. 1313-1319.

Ghanem, R. and Shinozuka, M. (1995). "Structural-System Identification I: Theory," *J. of Engineering Mechanics,* ASCE, Vol. 121, No. 2, pp. 255-264.

Haykin, S. (1994). *Neural Networks-A Comprehensive Foundation*, Macmillan College Publishing Company.

Hong, K-S. and Yun, C-B. (1993). "Improved Method for Frequency Domain Identification of Structures," *Eng. Struct.* Vol. 15, No. 3, pp. 179-188.

Hush, D. R. and Horne, B. G. (1993). "Progress in Supervised Neural Networks," *IEEE Signal Processing Magazine,* pp. 8-39.

Koh, Chan G., et al. (1991). "Estimation of Structural Parameters in the Time Domain: A substructure Approach," *Earthquake Engineering and Structural Dynamics*, Vol. 20, pp. 787~801.

Lim,T. W. (1990). "Submatrix Approach to Stiffness Matrix Correction Using Modal Test Data", *AIAA Journal,* Vol. 28, No. 6. pp. 1123-1130

Lim, T. W. (1991). "Structural damage detection using modal test data," *AIAA Journal*, Vol 29, No 12. pp 2271-2274.

Oreta, Andres W. C. and Tanabe, Tada-aki. (1976). "Element Identification of Member Properties of Framed Structures," *J. of Structural Engineering* ASCE, Vol. 120, No. 7, pp. 1961-1976.

Pandey, P. C. and Barai, S. V. (1995), Multilayer perceptron in damage detection of bridge structures, *Comput. Struct.* 54(4), pp. 597-608.

Press, W., Teukolsky, S.A., Vetterling, W. T. and Flannery, B. P. (1992). Numerical recipes in C-The art of scientific computing. Cambridge university press, New York, p.315.

Tsou, Poyu and Shen M.-H. Herman. (1994). "Structural Damage Detection and Identification Using Neural Networks," *AIAA Journal*, Vol.32, No.1, pp. 176-183

Wu, X., Ghaboussi, J., and Garrett, J. H. Jr. (1992). "Use of Neural Networks in Detection of Structural Damage," *Comput. Struct.*, 42(4), pp. 649-659.

Yun, C-B. and Lee, H-J. (1995). "Damage estimation of structures incorporating substructural identification," *Proceedings of APSSRA, Tokyo, Japan.*

Table 1. Training data set

Pattern 1	Input pattern without noise (f_i, $\tilde{\varphi}_{ji}$)							
	0.878	0.530	0.509	0.434	......	0.104	-0.068	
	2.608	0.460	0.458	0.457	......	-0.023	0.028	
	3.736	-0.011	0.085	0.036	......	-0.452	-0.403	
	4.250	0.068	0.070	-0.262	......	0.205	-0.237	
	Input pattern with3% noise in RMS							
	0.891	0.550	0.504	0.428	......	0.102	-0.066	
	2.772	0.469	0.450	0.458	......	-0.024	0.027	
	3.793	-0.011	0.083	0.035	......	-0.455	-0.407	
	4.166	0.063	0.072	-0.250	......	0.203	-0.241	
	Output pattern (SSF's)							
	0.60	1.50	0.50	1.25	1.75	0.60	1.35	1.35
⋮								
Pattern 1848	Input pattern without noise (f_i, $\tilde{\varphi}_{ji}$)							
	0.884	0.520	0.512	0.429	......	0.106	-0.069	
	2.618	0.449	0.459	0.450	......	-0.022	0.029	
	3.744	-0.012	0.076	0.041	......	-0.452	-0.402	
	4.273	0.003	0.088	-0.295	......	0.218	-0.277	
	Input pattern with3% noise in RMS							
	0.866	0.498	0.520	0.462	......	0.108	-0.072	
	2.576	0.437	0.462	0.455	......	-0.021	0.027	
	3.802	-0.012	0.077	0.040	......	-0.430	-0.403	
	4.102	0.003	0.087	-0.294	......	0.223	-0.283	
	Output pattern (SSF's)							
	1.10	1.40	0.80	0.75	1.00	1.20	1.50	1.45

Table 2. Test outputs for the testing data set with 3% noise when the network is trained with 3% noise injection.

Pattern 1	Input pattern (f_i, $\tilde{\varphi}_{ji}$) with 3 % noise							
	0.905	0.512	0.517	0.440			0.087	-0.105
	2.616	0.416	0.450	0.473			-0.029	0.034
	3.898	-0.018	0.034	-0.003			-0.360	-0.458
	4.393	0.119	0.056	-0.224			0.279	-0.275
	Actual output							
	1.10	0.90	1.38	1.31	0.89	1.40	1.21	1.07
	Target output (SSF's)							
	1.20	0.80	1.35	1.30	0.95	1.40	1.15	1.00
	Relative error* (%)							
	8.51	12.67	1.89	1.08	6.29	0.02	4.99	6.52
⋮								
Pattern 103	Input pattern (f_i, $\tilde{\varphi}_{ji}$) with 3 % noise							
	0.894	0.521	0.518	0.424			0.102	-0.060
	2.707	0.450	0.443	0.460			0.005	0.050
	3.797	0.040	0.159	0.102			-0.418	-0.330
	4.722	0.008	0.427	0.035			0.108	-0.070
	Actual output							
	0.84	0.57	0.98	0.94	1.20	1.24	0.64	0.80
	Target output (SSF's)							
	0.70	0.60	0.95	0.90	1.05	1.25	0.70	0.50
	Relative error (%)							
	19.5	4.70	3.30	4.77	14.6	0.48	8.91	60.4
⋮								
Pattern 201	Input pattern (f_i, $\tilde{\varphi}_{ji}$) with 3 % noise							
	0.816	0.513	0.520	0.440			0.111	-0.070
	2.625	0.439	0.470	0.461			-0.009	0.027
	3.393	-0.020	0.065	0.029			-0.434	-0.379
	4.082	0.055	0.184	-0.204			0.277	-0.234
	Actual output							
	0.98	1.13	1.31	0.67	0.99	0.93	0.84	1.05
	Target output (SSF's)							
	0.85	1.10	1.30	0.60	0.85	0.90	0.80	0.80
	Relative error (%)							
	15.2	2.94	0.83	12.3	16.5	3.51	4.53	31.1
	Average relative errors of each element for all the testing patterns							
	23.6	10.2	9.32	12.3	17.2	14.1	10.1	29.3
	Average error for 8 elements = 15.76 %							

* Note: $Relative\ error\ (\%) = 100 \times \frac{|Actual\ output - Target\ output|}{Target\ output}$

Table 3. Estimation errors (%) for different noise injection

Noise levels in RMS		in testing data 0 %	3 %	5 %
in	0 %	12.9	23.3	33.4
training	3 %	15.2	15.8	17.6
data	5 %	15.6	16.8	16.9

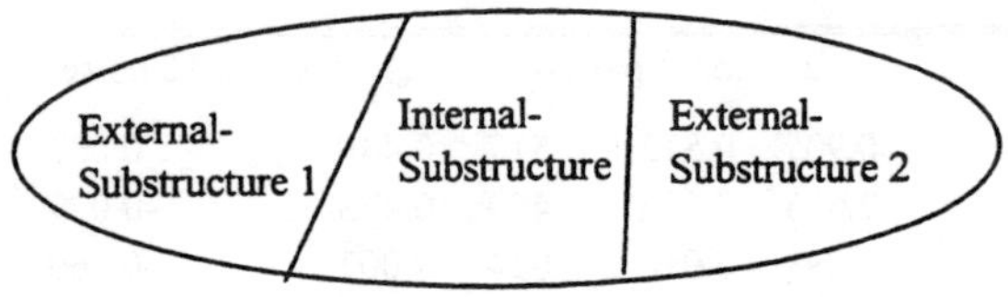

Fig.1 Substructuring for localized identification.

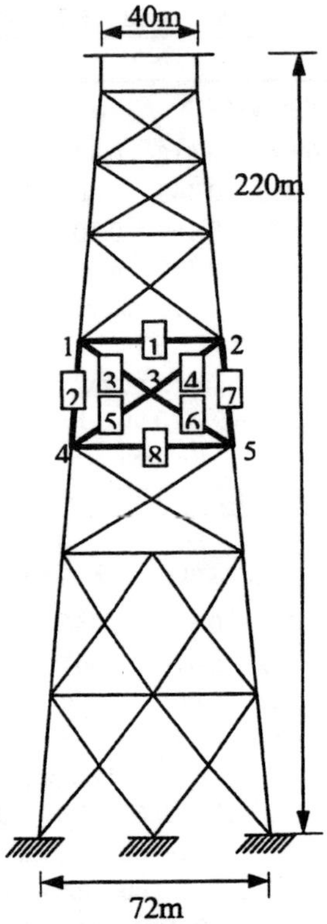

Fig.2 Example structure

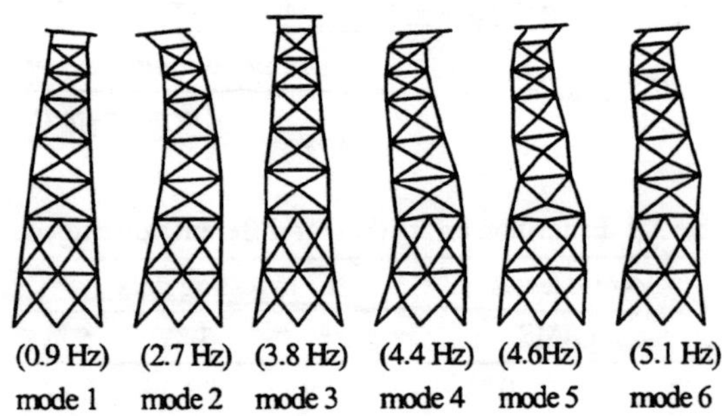

Fig.3. First six mode shapes of example structure

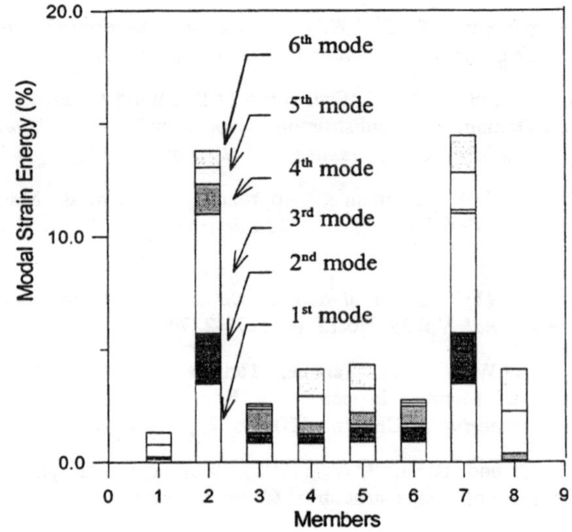

Fig. 4. Fractional modal strain energy distribution (%) for the reference structure

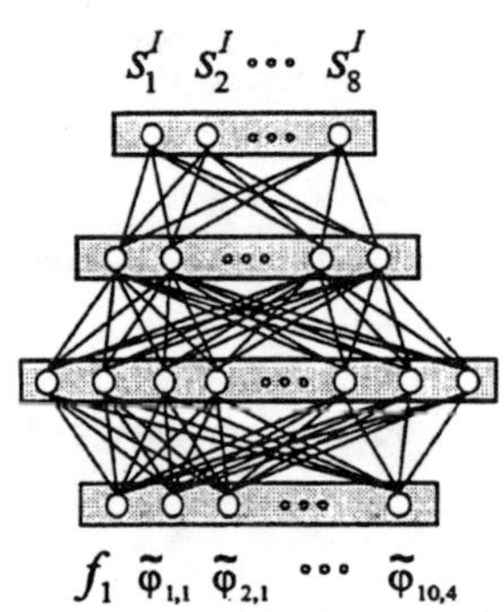

Fig.5. Backpropagation neural network

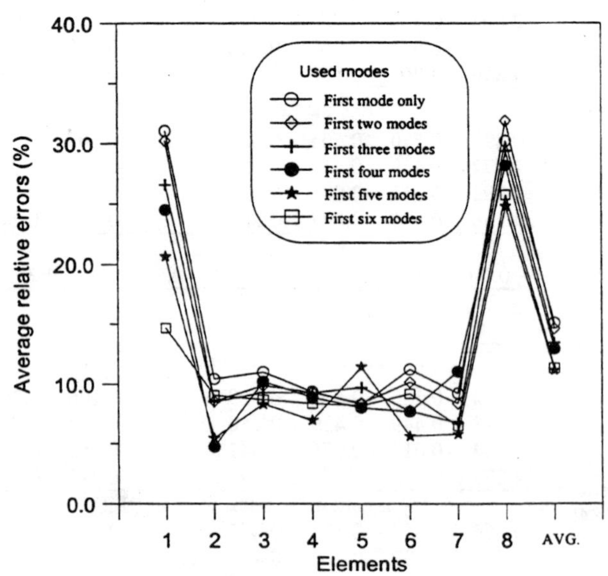

Fig. 6 Average estimation errors (%) for different modes used in input patterns (for testing data set without noise)

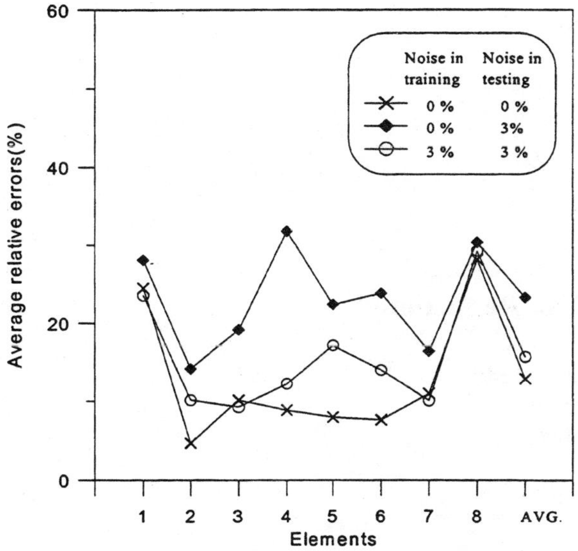

Fig. 7 Estimation error (%) for different noise injection learning

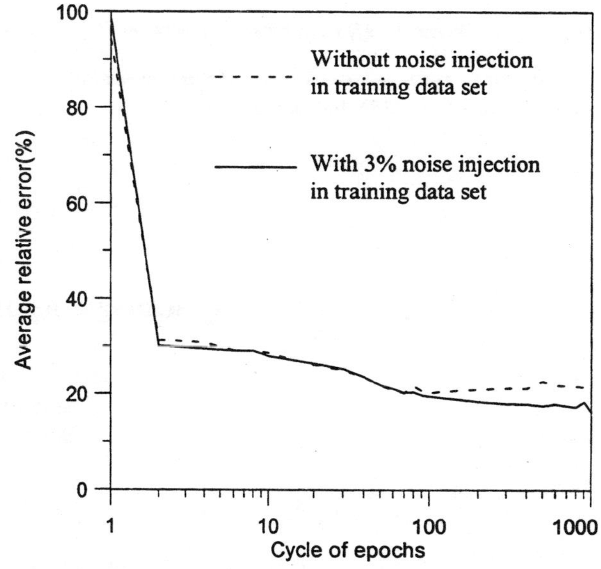

Fig. 8. Learning curves without and with noise injection levels.

Proceedings of the Seventh (1997) International Offshore and Polar Engineering Conference
Honolulu, USA, May 25-30, 1997

ISBN 1-880653-28-1 (Set); ISBN 1-880653-32-X (Vol. IV)

A Strategic Approach to Risk Reduction

Robert J. Simpson and Donald Smith
Health and Safety Executive
Bootle, United Kingdom

ABSTRACT

Strategic Planning has long been acknowledged by the Offshore Industry as an essential tool for the development of major projects. This paper outlines the approach adopted by the Offshore Safety Division of the Health and Safety Executive in developing its research strategy. Examples of the detailed strategies are presented as they relate to station keeping of floating systems and meteorological and oceanographic conditions in the North West Approaches region.

KEYWORDS

Offshore Research, Research Strategy, Risk Reduction, Station Keeping, Oceanography.

INTRODUCTION

It is widely accepted that science and technology has an important role to play in the reduction of risks in the work place. However, it is often difficult to identify the contribution to risk reduction which results from a specific application of science and technology. This is particularly the case where the risks being addressed have a low probability of occurrence, although their consequence, in terms of loss of life, may be severe (eg risks associated with environmental or seismic overload). Where the risks being addressed have a high probability of occurrence (eg risks associated with slips, trips and falls) it is usually possible to measure the benefits, in terms of reduced accident rates, which result from the application of science and technology.

Organisations involved in the exploitation of oil and gas reserves are, by necessity, 'high-tech' in nature. This is driven by a combination of the pressures which result from the competitive market place in which they operate, and the complexity and individuality of even the most basic of offshore operations. In addition, the ability to manage the myriad of risks associated with working offshore requires high level of in-house technical skill. For such organisations to be successful it is necessary for their technology function , including the element related to risk reduction, to be fully integrated into their business plan, and be accepted by the senior management as a real contributor to the organisations vision and mission.

Conventional approaches to strategy development have been based on two opposing schools of thought. The first is the traditional approach which was conceived as a rational, sequential process - define your objectives, analyse your position, evaluate alternative courses of action, develop the strategy and implement it. The second approach assumes that decisions in organisations are undertaken with only partial understanding of a situation, and that strategies emerge gradually over time. These approaches assume either that the future is understandable and predictable, or that strategy is reactive. Recognition that processes are not simple and linear, but complex and non-linear have already led to a rethink in business circles. Key issues for the future development of strategic planning are that there is a need to increase communications and remain flexible to accommodate organisational and market driven requirements.

This paper outlines how OSD has approached the problem of developing a research strategy which is consistent with its overall mission, and how it has reorganised the management of its research in order deliver a cost effective and user oriented research programme. The paper goes on to present details of how the strategy has been applied in the areas of station keeping, and meteorology and oceanography in the region of the North West Approaches to the UK.

OSD'S RESEARCH STRATEGY

The UK Health and Safety Executive (HSE) undertakes research, in part, to satisfy a duty placed on it through the Health and Safety at Work etc. Act 1974 (Health and Safety at Work etc Act, 1974) which requires HSC/E to *'make such arrangements as it considers appropriate for the carrying out of research, the publication of research, and the provision of training and information..'* The Offshore Safety Division (OSD) of HSE has, over the last few years, devoted significant effort in developing a strategic risk based approach to offshore oil and gas related research, in support of its mission; *'to ensure that risks to people from work activities in the upstream petroleum and diving industries are properly controlled.'*

The main objectives of OSD's research programme are to:

- Assess the extent and nature of the risks involved in work offshore;
- Undertake investigations to support specific enforcement activities, policy initiatives or accident investigations;
- Produce information to develop technical guidance to support legislation;
- Inform national and international standards making;
- Validate models (including computer models) enabling risk calculations to be made and inform regulatory decisions;
- Produce necessary health and safety information where the structure of the industry prevents it being produced by the industry (at least in an appropriate timescale).

The change in regulatory framework from a prescriptive approach (SI289, 1974) to a goal setting regime (Health and Safety Executive, 1992a) required that OSD refocused its research effort which had been largely orientated towards the development and updating of the Guidance Notes for the Design, Construction and Installation of Offshore Platforms (Health and Safety Executive, 1992b). As the main thrust of HSE's activity is the reduction of risk to the workforce, it is logical that research activity should be geared to:

- Identification of hazards, the risk posed by them and the consequence
- Awareness of existing knowledge on hazards and risks
- Provision of information both internally and externally on hazards and risks

The approach adopted by OSD to produce a research strategy consisted of identifying the major areas of risk offshore, prioritising these areas where possible, consulting with the offshore industry on key areas identified, and producing and openly publishing a strategy.

In the early development of OSD's research strategy a major risk study was undertaken (DNV Technica, 1994) which included a review of historical data for accident and incidents offshore. Where possible these data were taken from North Sea experience, but, in several cases, the data were so limited that other sources were used, particularly from the US where over 64,000 working platform years have elapsed.

The first version of OSD research strategy was published in 1995 and identified 17 key areas requiring research (HSE, 1995). The document consisted of a brief description of OSD's concerns in each of the key areas, and a list of research objectives specifically aimed at addressing the concerns.

It was recognised at the outset that the strategy would need to be dynamic and respond both to the information generated from new research projects and the changes in the offshore industry. A major source of information for identifying changes and shift in emphasis in the strategy has come from the data provided in the safety cases submitted as part of the UK Offshore legislative regime. A revised strategy was published in 1996 (HSE, 1996) extending the number of topic areas to 20. The areas included in the strategy are:

- Collisions
- Decommissioning
- Diving Operations
- Diver Physiology
- Evacuation, Escape and Rescue
- Fire and Blast
- Floating Stability
- Human Factors
- Material Performance
- Mechanical Equipment & Systems
- Moorings Failure
- Occupational Health
- Oceanography & Fluid Loading
- Organisation & Management
- Pipelines
- Process Systems & Process Control
- Safety Systems
- Structural Failure (Concrete)
- Structural Failure (Steel)
- Wells & Well Operations

In the first issue of its strategy document OSD attempted to prioritise the research effort in a risk based manner. However in the light of recent changes offshore, in particular the increase in activity in the North West Approaches region, it was felt that the original prioritisation was no longer valid. As part of OSD overall topic based strategy for risk reduction, key elements of research and new technology required to support OSD's core function will be developed as an integral part of the overall topic strategy.

OSD's intention is that the strategy will be revisited and revised in the light of the objectives which have been fulfilled, changing patterns of risk offshore, and the evolution of new operating practices.

Through its research strategy document OSD is attempting to encourage the offshore industry to form joint industry studies aimed at resolving the issues which will benefit the industry and reduce the risks to the offshore workforce.

Research Management within OSD

In order to improve the integration of the research function into its mainstream inspection activity, OSD has changed its approach to the development and management of research studies. Until recently OSD has had a dedicated research team whose responsibility was to develop medium and long term strategy across all technical areas, define and develop research projects, manage those projects, and disseminate the results. Although this was undertaken in close liaison with front-line inspectors, there existed a view that those who needed the information, and were best able to disseminate to the offshore industry, were always one removed from the source. As a result of this and a general restructuring of OSD's operations, it was decided that the responsibility for topic based research and technology should be devolved to teams across OSD who would have responsibility for delivering the overall strategy in a particular topic area. Individual teams would be encourage and expected to include the research function as a tool to be used alongside the more conventional assessment, inspection and enforcement options available to inspectors.

In order to manage and support the OSD's research activity, a small central team has been established whose responsibility will be to ensure that the overall aims and objectives of HSE are met, and that individual studies in which OSD intends to participate are consistent with the overall mission of the organisation. In addition, the team will take a very broad look at the key drivers behind new developments in the offshore industry, and inform inspectors of new technical challenges that they need to be equipped to assess. Individual topic teams will be able to use this information and their own market intelligence to develop appropriate topic based research programmes.

STATION KEEPING FOR FLOATING SYSTEMS

The ability to demonstrate that a mobile drilling, production or off-loading system is capable of maintaining station during predefined operating conditions, is an important factor in the overall demonstration of an installation's ability to operate in a safe manner.

There are two distinct requirements for moorings. The first is to maintain a vessel on station for a short period of time, such as a MODU; the second is to maintain a vessel on location for a long period of time in the case of a flotel or a floating production unit. The requirements of the mooring system are different in that for a floating production system there needs to be a higher reliability as the vessel is constantly in contact with hydrocarbons and does not have the options of moving off location on the same way as a drilling vessel. The performance of a mooring system is related to the assessment of the environmental conditions in which the vessel is to operate (HSE, 1997).

One of the difficulties associated with establishing the reliability of mooring systems is the lack of good data on both system and component performance. In developing its research strategy in the area of station-keeping, OSD undertook a detailed review of past incidents of loss of station keeping and the likely consequences. As a result of the greater availability of data, the work focused on drilling semi-submersibles rather than floating production (monohull) systems. However, some of the issues, particularly those relating to the failure rate of chain and wire rope, and the general design premise for mooring systems, are of relevance to other forms of moored systems such as FSUs and FPSOs. The initial study lead to the identification of the need for better quality data relating to both mooring failures and to the general operating conditions of floating systems which need to maintain station during a storm.

Analysis of Mooring Failures

A major review of mooring chain failures was undertaken on behalf of HSE by DNV. The Study was based on DNV's records and comprised some 183 incidents over the period 1976 to 1993. Although there has been a number of changes both in the use of materials for mooring lines and in operational procedures over the period, a number of important conclusion were derived. These included:

- the probability of failure per annum of an individual mooring chain manufactured prior to 1985 is about 0.04
- the probability of failure per annum of an individual chain manufacture in, or after 1985 is about 0.01
- over 50% of the failures occurred at between 10% and 30% of the maximum breaking load for the grade of material
- the reliability of joining shackles needed to be investigated.

Other investigations focusing on the UKCS with a smaller database have come up with similar conclusions. For example a study of the 8 failures that occurred in the October 1991 and January 1992 storms found that all the failures occurred at well below the rated line strength and half the failures were attributed to defective joining shackles. The differing performance of vessels under seemingly similar weather conditions is an issue that remains unresolved. Questions of different operating procedures, highly localized weather conditions, different equipment and mooring spreads, and human factors in the decision making process will all contribute to the reason for failure or non-failure.

Moorings Research Strategy

The information derived from the studies outlined above, and information available to OSD's inspectors as a result of their Safety Case assessment work, were used to help identify the areas in which research funds needed to be allocated for moorings related studies.

The overall strategy to be adopted was to attempt to reduce single-line failures in MODUs through an improved understanding of the underlying causes of the failures. This information could be used as an input to the development of reliability based approaches to design. A need for improved incident data, and a better analysis of the data, were identified as important inputs in prioritising whether work needed to be undertaken in better understanding the details associated with manufacturing faults, in-service inspection or fatigue damage. For multiple line failures the strategy to be adopted is one of improving knowledge in the area of the line tensions which can be experienced by a range of vessels, in a range of environments. Finally, the strategy recognised that line management has a major role to play in reducing the potential for a line to fail. An assessment needed to be made of the information available for decision making, and the associated management responsibilities. This information could be used to identify where improvement could be made.

A number of key, mooring related, research objectives were identified, which are listed below.

- To develop a database of moorings incidents
- To establish the criteria for risk assessment and analysis for vessels moored at close proximity.
- To investigate the risks to pipelines and SSIVs resulting from mooring failure and dragged anchors.
- To investigate the reliability of mooring chain and wire rope and to quantify the factors influencing reliability.
- To establish the accuracy with which mooring analyses predict line tensions for a range of vessels, line types, water depths, environmental conditions and mooring patterns.
- To establish with Industry inspection and discard criteria for mooring lines.
- To collate available experience on operational decisions during the onset of bad weather prior to a mooring failure.
- To consider the requirements on the reliability of moorings for a flotel in the event of an evacuation, escape and rescue scenario.

Work is already underway on many of the above issues as both joint industry initiatives and sole OSD funded projects. Recent major JIPs by both DNV (Larsen et al., 1996) and Noble Denton (NDA, 1996) have made a major contribution to the understanding of reliability based design.

METEOROLOGICAL AND OCEANOGRAPHIC CONDITIONS IN THE NORTH WEST APPROACHES REGION

Recent developments in the North West Approaches (NWA) region have resulted in the need for new information on both the environment and operating conditions of structures.

From meteorological, oceanographic (metocean) and fluid loading viewpoints, the NWA region presents considerable challenges for companies interested in operating there. In particular, the combination of water depth and severe environmental conditions makes the direct transfer of standard North Sea design technology into the area difficult. The severity and complexity of the metocean conditions in this region leads to difficulties in designing either fixed or floating structures capable of delivering adequate reliability against structural overload. New approaches, probably based on response rather than metocean criteria, require exploitation to assure designs have similar or improved levels of safety as achieved by structures located in the North Sea.

The wind, wave and current maps presented in the UK Guidance Notes includes the NWA region. Recent studies have shown that the wind and wave climate is not greatly dissimilar to that encountered in the Northern North Sea (Grant et al, 1994) and reasonably consistent with the information presented in the Guidance Notes. However, compared to the North Sea, the current climate is extremely variable, both temporally and spatially. This is a result of a variety of factors including; the water depth, the presence of the Continental Shelf, the complex bathymetry, and the variety of different water masses feeding and influencing the currents in the region. The current speeds measured tend to be similar in magnitude to those measured in the Southern North Sea, however, they are less predictable and act over considerable greater water depths. The current maps presented in the Guidance Notes cannot be used with confidence in the NWA region.

The problem associated with the severity of the metocean conditions in the NWA region are compounded by the fact that, compared to the North Sea, there is very little measured wind, wave and current data in the region. The only substantial measured datasets which are available were recorded by the UKOOA/HSE funded DB2 and DB3 buoys. These buoys were deployed in the NWA region for 27 months and 51 months respectively. Unfortunately, the analysis of date collected by the buoys show that the periods during which they were deployed were relatively benign. Hence, in addition to the datasets being relatively short, the data are not representative of the extreme conditions seen in the region.

The situation with respect to current data is equally as poor. Many of the datasets tend to be of a few months duration (usually just summer months), which, given the large temporal and spatial variability of the currents in the region, are of limited value. Many operators are now attaching Acoustic Doppler Current Profiles (ADCPs) to the vessels operating in the region. This should provide substantially more current data from which estimates of extreme current speed and profiles can be derived.

Metocean (NWA) Related Research Strategy

It is a combination of the severity and the poor understanding of the metocean conditions in the NWA which is presenting the industry with problems with ongoing and future operations in the region. The research strategy focuses on the need to increase the amount of data (in particular current data) available in the region, and to improve understanding of the types of phenomena which have the potential to result in occurrences of high current speeds (eg eddies, internal wave and slope current meandering).

A two key, NWA metocean related, research objectives were identified, which are listed below.

- To assess metocean parameters in the frontier region.
- To improve, with industry, the availability of metocean data in the West of Shetlands region.

In both of these areas OSD is working with industry through joint industry funded initiatives.

SUMMARY

The paper has reviewed OSD's approach to integrating the science and technology function, in particular the research function, in to its overall approach to achieving its mission of ensuring that the risk to the health and safety of people working in the offshore industry are properly controlled.

The major change in the way offshore research is identified, funded and managed from within OSD has been associated with the devolving of the management role from within a dedicated research oriented team., to mainstream inspection teams which have responsibility for delivering the overall strategy in specific topic areas.

At this stage it is too early to demonstrate the benefits which have resulted from these new working practices. However it is expected that the key benefits will be in the areas of better focused research and improved dissemination of finding of research studies both within OSD and to the industry as a whole.

ACKNOWLEDGEMENTS

The UK Health and Safety Executive Offshore Safety Division is thanked for granting permission to prepare and publish this paper. The views expressed are those of the authors and are not necessarily those of HSE.

REFERENCES

DNV Technica (1994). Update of the UKCS Risk Overview. HSE Books, ISBN 0-7176-1003-9.

Grant, CK, Dyer, RC and Leggett, IM (1995). "Development of a New Metocean Design Basis for the NW Shelf of Europe." OTC Paper No. 7685, Houston, Texas.

Health and Safety at Work etc Act (1974). HMSO 1989 ISBN 0 10 543774 3

Health and Safety Executive (1992a). The Offshore Installations (Safety Case) Regulations 1992. (SI 1992/2885) HMSO 1992 ISBN 0 11 025869 X.

Health and Safety Executive (1992b). Offshore Installations: Guidance on Design, Construction and Certification, 4th Edition HMSO, London, 1992.

Health and Safety Executive (1995). Strategy for Offshore Research. HSE, Report No. OTI 94 622, ISBN 0 7176 0916 2.

Health and Safety Executive (1996). Strategy for Offshore Research 1996/97. HSE, Report No. OTI 96 642, ISBN 0 7176 1299 6.

Health and Safety Executive (1997). Health and Safety Research needs arising from West of Shetlands Developments. HSE Report Number OTO96 900.

Larsen, K, Mathisen, J, (1996). "Reliability based mooring system design for a drilling semi-submersible". ISOPE, 1996, Los Angelos, USA.

NDA, (1996). "Review of Mooring Codes" Noble Denton and Associates, Joint Industry Project Final Report, January 1996.

SI289 (1974). The Offshore Installations (Construction and Survey) Regulations 1974.

Proceedings of the Seventh (1997) International Offshore and Polar Engineering Conference
Honolulu, USA, May 25-30, 1997

ISBN 1-880653-28-1 (Set); ISBN 1-880653-32-X (Vol. IV)

Jet Fire Resistance for Passive Fire Protection Materials on Tubular Sections

Tariq Al-Hassan
Health and Safety Executive, Bootle, United Kingdom
Alex B. Wenzel
Southwest Research Institute, San Antonio, TX, USA
Paul Mather
Lloyds Register, London, United Kingdom

ABSTRACT

A test development programme has been undertaken by Southwest Research Institute, BG Research and Technology, the Offshore Safety Division of the Health and Safety Executive in the UK (HSE), and Lloyd's Register, to validate the extension of the Interim Jet Fire Test Procedure (IJFTP) for testing passive fire protection (pfp) materials applied to tubular sections. The research programme was recommended by the International Jet Fire Working Group. In July '96, the Group published (through HSE books) the test procedure;' Jet-fire Resistance Test of Passive Fire Protection Materials' OTI95634. Initial work with tubulars sections had indicated that at the IJFTP-scale, heat fluxes could be achieved that were similar to those found in large-scale jet fires. To validate the IJFTP it was necessary to test pfp coated tubulars at both scales. Four tubular steel sections, of two sizes, were coated with the same thickness and type of proprietary epoxy intumescent pfp material system. One set of Tubulars were tested in accordance with the IJFTP at SwRI and the matching set were tested at Spadeadam, a BG test facility. This paper discusses the thermal response of the steel tubulars, reported by both test laboratories, and compares the results in terms of replicating the jet fire environment.

It is concluded that for tubulars at the IJFTP-scale, even though the heat flux are comparable, the erosion effects are not the same as produced in a large-scale jet fire. This is shown to be important when testing those pfp systems that may be vulnerable to erosion. Therefore, tubular tests at IJFTP-scale should be analysed with care, and further work will be required to replicate large-scale jet fires impinging on tubulars.

KEY WORDS

Jet fire resistance testing, passive fire protection materials, tubular protection.

INTRODUCTION

This paper discusses the results of a number of research and development projects conducted by Southwest Research Institute (SwRI), the Offshore Safety Division of the British Health and Safety Executive (HSE), BG, and Lloyds Register. These projects were recommended by the International Jet Fire Working Group consisting of the HSE, the Norwegian Petroleum Directorate (NPD), Lloyd's Register, UK Offshore Operator's Association (UKOOA), SINTEF-NBL, SwRI, Health and Safety Laboratory, Shell Research Limited (SRL) and BG. The scope of this programme was to conduct a series of medium-scale and large scale tests to validate and extend the Interim Jet Fire Procedure (IJFTP) described in reference [1], for evaluating passive fire protection (pfp) materials on tubular sections. The initial findings from this programme relevant to the test procedure were included in the recent report (9). A recent research programme (2,3,4) conducted by SwRI and HSE determined the medium-scale test parameters and procedures required to modify the IJFTP to reproduce the thermal profiles and erosive effects of a large-scale jet fire environment on tubular sections. This work was necessary because the existing IJFTP described in reference (1) is limited to planar geometries, i.e. bulkheads and I-beams. The earlier work on tubular specimens (2,3,4) required additional tests and had not been validated at large-scale. The large-scale jet fire environment was characterised by a programme conducted by Shell Research Limited and BG at the British Gas Spadeadam facility in Cumbria, England (5). This paper presents the modifications made to the IJFTP to test tubular sections, the large-scale test procedure used for the validation tests, a comparison of the medium and large-scale test results and conclusions.

BACKGROUND

The accidental release of contents from high-pressure hydrocarbon transfer, storage, and process plant may result in pool fires, explosions, and jet fires on offshore platforms, and these can pose a substantial hazard to personnel and property. Fires can escalate, thus increasing the potential for loss of life and damage to facilities. This kind of escalation was a major factor in the development of the Piper Alpha disaster in the North Sea, which resulted in the loss of 167 lives and the loss of the platform.

A jet fire is one of the most severe thermal threats to structures, and the overall integrity of offshore platforms. The avoidance of jet fire need to be a high priority, however, active and passive fire protection systems need to be developed and tested to mitigate the effect of jet fire, in the unlikely event that they do occur. The erosion and thermal loading from high-velocity jet impingement are different from those experienced in conventional fire resistance tests conducted in laboratory furnaces with low velocity flame impingement, and uniform heat flux. Jet fires are characterised by thermal effects consisting of combined radiative and connective heat transfer, in addition to erosion effects resulting from high velocity flow. The contamination of the gas with solid particles and liquid droplets, i.e. multi-phase release have not been evaluated, and therefore its effect on pfp is not known. The combined heat load and erosive effects of a multi-phase impinging jet fire would make it more severe than other fire threats. In addition, the weathering effects on the performance of pfp in jet fires is not known (15).

A jet fire transfers heat to an impinged object by both convection and radiation. The magnitude of these two components depends on many factors, including fuel type; release conditions; size and shape of the object and its relative location within the flame; and, at any instant, the surface temperature of the test specimen. Consequently, the heat flux to an object varies both over its surface and with time. An additional feature of a jet fire is the instantaneous attainment of the peak heat flux.

Large-scale jet fires, which are characterised by mass flow rates greater than 1 kg/s, have been studied since the mid 1980's by SRL and BG (5). Under this program, a large-scale jet fire facility was built at the BG Spadeadam test site in Cumbria, England. Tubular sections, and storage vessels were instrumented with total and radiative heat flux transducers and thermocouples, and exposed to jet fires with flow rates between 2 and 12 kg/s to measure the thermal environment and impact on the test section .

In the Public Inquiry into the Piper Alpha Disaster (6), the need to determine whether the currently available fire protective coatings could withstand a jet fire event was highlighted. To address this issue, in May 1989, the Shell Offshore Flame Impingement Programme, (SOFIPP) (7), was initiated. The response of full size unprotected and passively fire protected, structural steel members to impingement by a representative, large jet flame for one hour was determined. Tubular and I-beam sections were tested, resulting in incident heat fluxes of up to 300 kW/m^2, with substantial convective and radiative components, high gas velocities, and fire environmental temperatures.

LARGE-SCALE TESTING

The large-scale SOFIPP tests mentioned above were carried out at Spadeadam. These early tests formed the basis for many subsequent tests of pfp systems at large-scale at Spadeadam. In these tests full-sized tubular and I-beam sections are placed in a natural gas jet fire at a distance of 9m from the release point. This test configuration resulted in incident heat fluxes of up to 300 kW/m^2, with substantial convective and radiative components, gas velocities, and temperatures are in the order of 1200^{0}C. The jet fire is an ignited sonic release of with mass flow rate of 3 kg/s natural gas at 60 bar, from a 20mm diameter orifice. The unobstructed length of this high velocity jet fire is about 20m. It is about 2m wide at the point where the test object is placed and most objects are fully engulfed by fire.

INTERIM JET FIRE PROCEDURE (IJFTP)

Following the SOFIPP test program, SRL and the Norwegian Fire Research Laboratory (SINTEF NBL) embarked on a study to reproduce, on a smaller scale, a fire environment typical of the large-scale jet fires previously measured by SRL and BG (8, 10). Their objective was to develop a laboratory-scale jet fire test procedure for planar sections, which would produce the same thermal environment as the large-scale jet fire. This was achieved using 0.3 kg/s jet of propane and impinging on a 1500mm square steel box placed 1000mm away from the release point, produced an average heat flux within the box of 240 kW/m^2: Under these conditions, there is a maximum heat flux of 290 kW/m^2 (of which approximately 50% is radiative), and thermal environments of 1200^{0}C. The jet had to be directed into the 1500-mm square steel box, which was 500mm deep, in order to increase the heat flux environment up to the values of the large-scale jet fire at the 1000-mm stand-off distance, with the centre of the release nozzle one quarter of the height (h/4) from the base of the steel box.

In 1993, the International Jet Fire Working Group developed a medium-scale test procedure (1). This interim test procedure determines the jet fire resistance of pfp materials for planar sections and I-beam configurations following the pioneering work at SINTEF-NBL. The rationale behind the IJFTP is to subject pfp materials to fire conditions typical of those found in large-scale jet fire with the emphasis on reproducing those conditions to which pfp materials have been shown to be most vulnerable. This test is designed to indicate of how well passive fire protection materials will perform in a jet fire.

Currently, three test laboratories have the facilities that can undertake testing in accordance with this test procedure, they are Southwest Research Institute (SwRI) in San Antonio Texas, SINTEF-NBL in Trondheim, Norway and UK HSL laboratory in Buxton, Derbyshire. Reference (11) describes the results of uniformity tests among these test laboratories, and have showed that the test was reproducible. The IJFTP can be used to investigate the performance of pfp materials applied to structural sections. This involves testing the 1500-mm square box with an additional 20mm thick, 250mm deep, vertical web running its full height, simulating edge details such as flanges on I-beams. Details in using this procedure for testing panels, planar steelwork, and structural steelwork are given in Reference (9).

MODIFYING IJFTP PROCEDURES

Studies need to determine the modifications required to develop the IJFTP to evaluate pfp materials on tubular sections, and then to validated it in a large-scale jet fire environment. A research project conducted at SwRI consisted of testing two 3m long tubular steel sections (2,3), a 457mm and a 305mm diameter pipe each with a wall thickness of 9.5mm. These pipes were instrumented with 4 ellipsoidal radiometers, 8 total heat flux gauges and 12 Inconel sheathed thermocouples.

The tubular sections were oriented horizontally in front of the square steel box specified in the IJFTP, exposed to a direct flame from the 0.3 kg/s propane jet. The centre of the release nozzle was varied between one quarter of the height (h/4) and one half (h/2) from the base of the steel box. The position of the tubular section was varied in relation to the test box, always maintaining a distance of 1000mm from the front face of the pipe to the nozzle. Figure-1 shows the instrumented box and pipe test set up. Both the 457mm and the 305mm diameter pipes were tested flush with the front of the box, and recessed 100, 200, 300 and 400mm into the box. Details of the peak free jet heat flux levels and the peak free jet surface temperatures from the front, top, back and bottom positions of the three rings around the pipe are given in references (2,12). The principal results derived from these tests were:-

- The 457mm diameter pipe is too large for the current IJFTP procedure to yield equivalent results to the Spadeadam tests reported.
- The 305mm diameter pipe placed in front of the steel box, flush with the front face, produced a maximum total heat flux of 285kW/m^2, which is considered as being a representative reproduction of the Spadeadam jet fire environment (3,5). The highest convective heat flux for the 305mm diameter pipe was around (153 kW/m^2).

Based on these results, six additional tests were conducted using a 203mm diameter pipe with a wall thickness of 9.5mm placed in front of the steel box, flushed with the front face. The tubular sections were 3m long. Due to its internal space limitations, this pipe was instrumented with 3 ellipsoidal radiometers, 3 heat flux transducers and 6 Inconel sheathed thermocouples, as shown in Figure-2. The measurements of the heat fluxes and temperatures (12) showed an adequate correlation to that at large-scale. However, an important parameter; velocities of the jet at the two scales at the position of impingement have not been evaluated.

TESTS AT IJFTP-SCALE AND LARGE-SCALE

Three tests were carried out to provide information on the comparative performance of a pfp material applied to tubular sections, one test at large-scale and two at IJFT procedure. The objective was to demonstrate that the performance of pfp material when tested at large-scale was similar to that observed when tested using IJFTP.

Tubular Member	Length (m)	Diameter (m)	Thickness (mm)	Test Location
T-1	3.0	219.1	10.3	SwRI
T-2	3.0	219.1	10.3	SwRI
T-3	9.0	219.1	10.3	Spadeadam
T-4	9.0	518	9.5	Spadeadam

Table-1. Details of tubular sections

Four steel tubular sections, details shown in Table-1, were coated with a proprietary epoxy intumescent pfp material. The manufacturer specified the protection system so that it would protect the sections to limit the maximum temperature rise to 400^{0}C, for 40 minutes. The protection system included a galvanised steel mesh. The details of the system application are presented in reference (13). The tubular sections were instrumented with thermocouples, and the thickness of materials were measured as shown in Figure-3, and 4. In addition the Shore D hardness, and thickness were measured immediately before jet fire testing.

It was noted that the average thickness of the materials at the test sites were different to that measured at the applicator. The differences in the thicknesses were attributed to the difficulty in measuring partly cured pfp material. The average Shore D hardness of the pfp were higher at SwRI compared to that measured at Spadeadam. However, the manufacturer stated that they are within the acceptable limits, and that the material performance is not affected.

Two coated 9m long tubular sections were tested together at the Spadeadam facility described earlier. A 219mm diameter, 10.3m wall thickness and a 508mm diameter, 9.5mm wall thickness. Figure 5 gives a schematic of the test specimen layout. The test at Spadeadam (13) showed that sections T3, and T4 reached 400^{0}C at 26, and 25 minutes respectively. A post test inspection was carried out to give an overall picture of the protection system performance, including visual observation, char thickness measurements, and destructive examination. The test was witnessed by a representative from Lloyds Register (14).

Tests at SwRI showed that sections T1 and T2 reached 400^{0}C at 35, and 38 minutes respectively (12). A similar post test inspection was carried out to that performed at Spadeadam. The tests were also witnessed by the same representative from Lloyds Register (14).

The measurements of temperatures at the different positions shown in Figures 3, and 4 (details in references 12, 13, and 14), showed that the temperature rises were linear in all the three tests reaching around 200^{0}C in 15-20 minutes. Then, localised failures was noted in the two tubular sections T3 and T4 in the large-scale test. The post inspections showed that pfp material protection systems were severely damaged at large-scale compared to that at the IJFTP-scale. All the tubular sections did not perform to the required specifications, i.e. the protection system did not limit the maximum temperature rise to 400^{0}C, for 40 minutes.

The limited data indicate that the temperatures and velocities are lower at the IJFTP-scale compared those at large-scale, for the arrangements used in testing tubular sections. At large-scale, high velocities cause the relatively weak protective char, produced by the epoxy intumescent pfp materials, to erode rapidly and ultimately fail.

CONCLUSIONS

1. Measurements of the heat fluxes indicated that the Interim Jet Fire Test Procedure can be modified to test tubular specimens. However, validations tests with pfp coated tubulars has shown that despite the measured heat fluxes are comparable, the erosion effects are not the same as produced at large-scale.
2. None of the tubular specimens tested, at either scale, performed to the required specifications, i.e. to limit the maximum temperature of the steel to 400^{0}C for 40 minutes.
3. Preparation of the coated tubular sections highlighted the importance of the pfp material protection system application procedure, and quality control. Variation in thickness and hardness of the coatings were noted, although these were within the manufactures tolerances.
4. Further experiments will be required to replicate (at IJFTP), the full effects of large-scale jet fires impinging on tubulars.

ACKNOWLEDGEMENTS

The authors would like to acknowledge the contributions made by the members of the International Jet Fire Working Group, and in particular Mr Les Shirvill of Shell Research and Technology Centre, Thornton, UK.

REFERENCES

1. The Jet Fire Working Group, "Interim Jet Fire Test for Determining the Effectiveness of Passive Fire Protection Materials," OTO 93 028, Health and Safety Executive, UK, 1993.

2. Parker, A. J and Dodge, F.T, "Investigation of Critical Parameters Necessary to Simulate Large-scale Jet Fire Events Involving Tubular Sections," Southwest Research Institute Report No 01 9842, June 1995.

3. Parker, A.J, Wenzel, A.B, and Al-Hassan, T, "Characterisation of the Laboratory-Scale Jet Fire Environmental Involving Tubular Sections," 4th International Conference of Offshore Structures - Hazards, Safety & Engineering Working in the New Era, London, UK, December 1995.

4. Parker, A J, Wenzel, A B, and Al-Hassan, T,"Extension of the Interim Jet Fire Protection Test Procedure for the Evaluation of Passive Fire Protection Materials on Offshore Pipelines", Proceedings of the 29th Annual Loss Prevention Symposium- AIChE, Boston, Massachusetts, USA, 31 July- 2 August, 1995.

5. Crowley, L.T, and Pritchard, M.J, "Large-Scale Natural Gas and LPG Jet Fires and Thermal Impact on Structures", GASTECH 90 Conference and Exhibition, December 1990.

6. The Hon Lord Cullen, "The Public Inquiry into the Piper Alpha Disaster", presented to Parliament by the Secretary of State for Energy, November 1990.

7. Bennett, J.F, Cotgreave, T, Crowley, L.T, and Shirvill, L.C, "Shell Offshore Flame Impingement Protection Programme: Part 1. Programme and Test Description," Shell Research Limited, Chester, England, 1990.

8. Shirvill, L.C and Wighus, R, "A Test Method of Jet Fire Exposure," 7th International Symposium of Loss Prevention and Safety Promotion in the Process Industries, Taormina, Italy, May 4-8, 1992.

9. The Jet Fire Working Group, "Jet-Fire Resistance Test of Passive Fire Protection Materials", OTI 95634, Health and Safety Executive, UK, 1996.

10. Wighus, R, and Drangsholt, G, " Impinging Jet Fire Experiments- Propane 14 mW laboratory Tests", STF25 A92026, SINTEF NBL-Norwegian Fire Research Laboratory, Trondheim, Norway, 1993.

11. British Gas Research and Development, " Assessment of the Uniformity of the Interim Jet Fire Test Procedure", OTH 95477 , Health and Safety Executive, UK, 1995.

12. Southwest Research Institute, "Characterisation of the Laboratory-Scale Jet Fire Around 203mm inside Diameter Tubular Geometries", OTO 96048, Health and Safety Executive, UK, 1996.

13. British Gas Research and Development, "Validation of the Jet Fire Resistance Test Procedure Large-Scale Jet Fire Tests on Fire Protected Tubular Members", OTO report to be published, Health and Safety Executive, UK, 1997.

14. Lloyds Register, " Large-Scale and IJFTP-Scale Jet Fire Tests", OTO report to be published, Health and Safety Executive, UK, 1997.

15. CAPCIS, "Feasibility of Providing Accelerated Weathering test facility", OTO report to be published, Health and Safety Executive, UK, 1997.

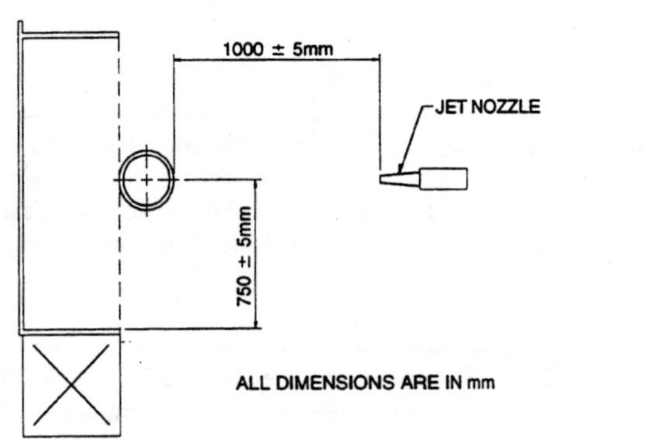

Figure 1 Layout of Test Facility with Tubular Section Test Specimen

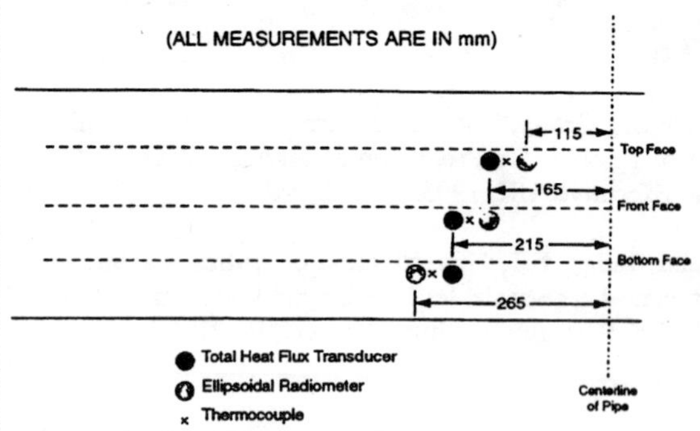

Figure 2 Instrumentation Location

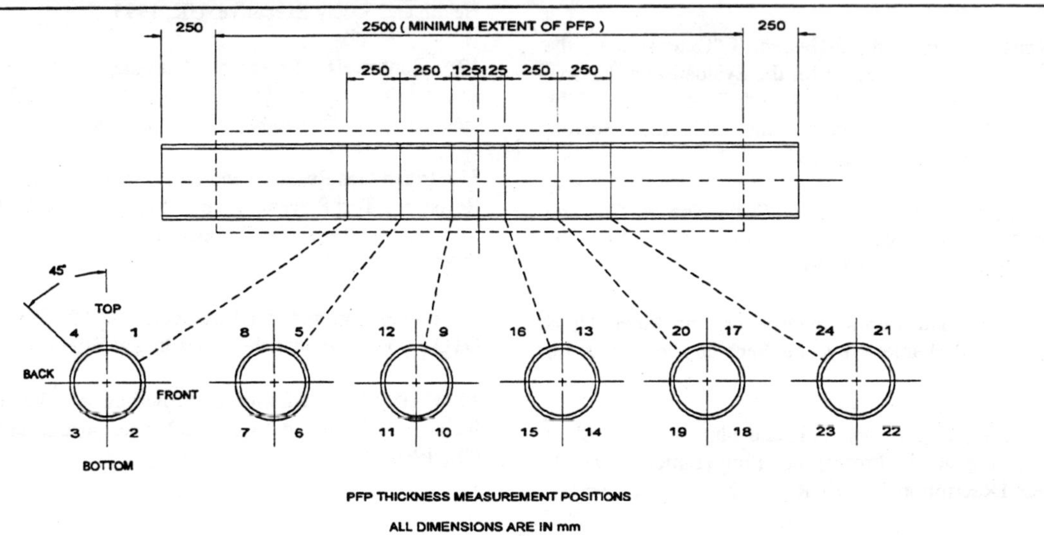

Figure 3a PFP Material Thickness Measurement Locations for Tubular Section Tests

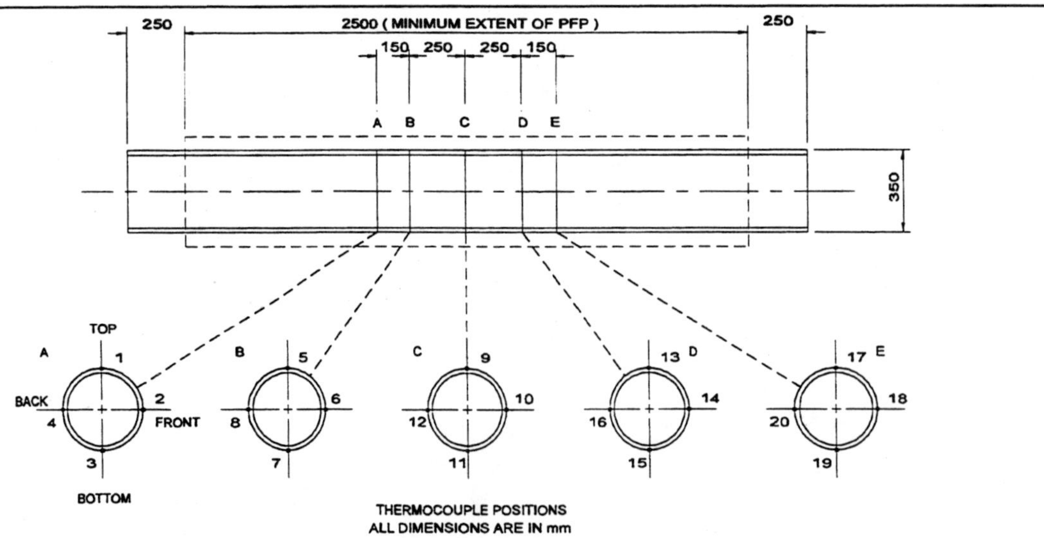

Figure 3b Thermocouple Positions for Tubular Section Test Specimen, SWRI Tests

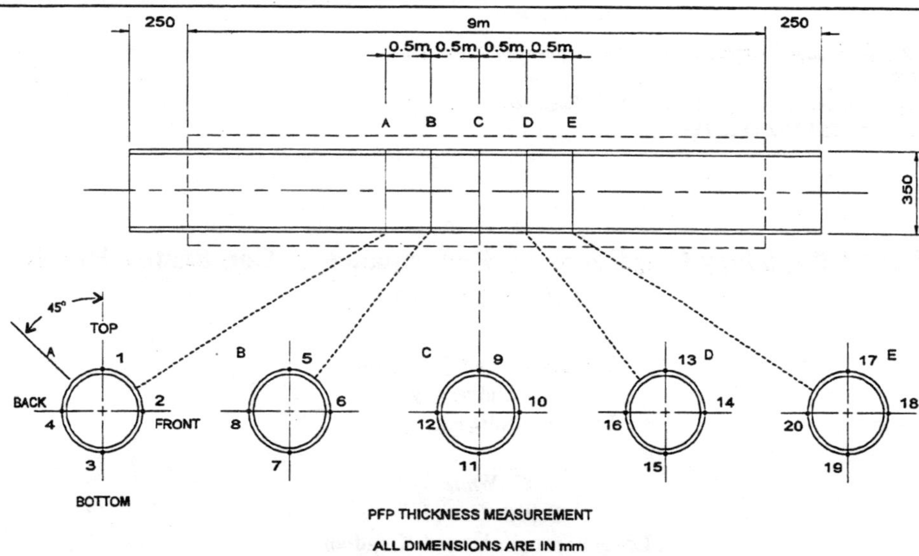

Figure 4a Location of PFP Thickness Measurements on Tubular Member T-3 and T-4, Spadeadam Test

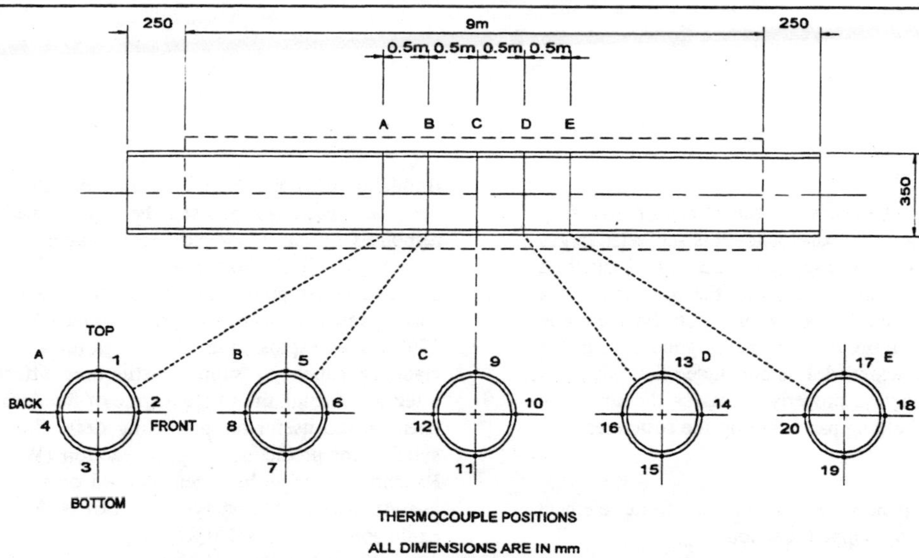

Figure 4b Location of Thermocouples on Tubular Member T-4 and T-3, Spadeadam Test

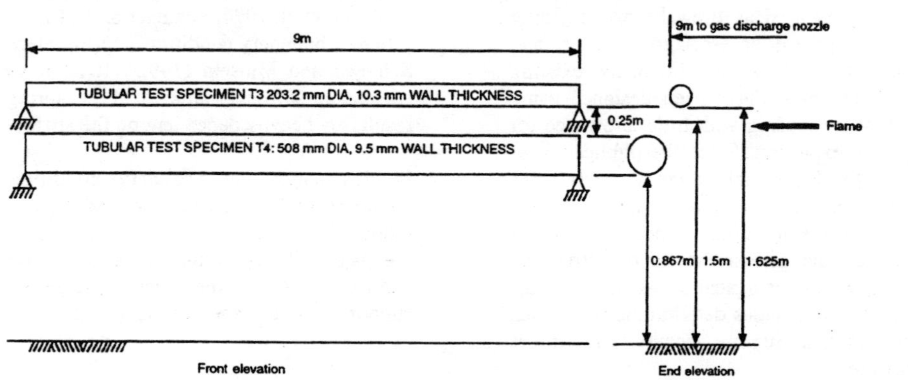

Figure 5 Orientation of Tubular Members in Jet Fire Test Rig

Proceedings of the Seventh (1997) International Offshore and Polar Engineering Conference
Honolulu, USA, May 25-30, 1997

ISBN 1-880653-28-1 (Set); ISBN 1-880653-32-X (Vol. IV)

Effect of Boundary Conditions on Performance of Corrugated Panels to Blast Loading

L.A. Louca and Y.G. Pan
City University
London, United Kingdom

G. White
British Gas plc
Loughborough, United Kingdom

ABSTRACT

Presented in this paper is the study on the effect of boundary conditions on the deformation of a panel when it is subjected to gas explosion. Also forces transferred to support structures are analysed in terms of the magnitude, distribution along the boundaries and their relationship with the overall panel deformation. By including the support structure in the finite element model good agreements with test results have been achieved. It is concluded that boundary conditions have to be modelled properly to assess the structural response of firewalls and cladding panels on topside structures.

Key words: corrugated panels, gas explosion, finite element analysis, boundary conditions, support reaction

1. Introduction

Corrugated panels are commonly used as firewalls and cladding panels on topside structures to prevent an escalation of a fire and provide protection to personnel. However, in many existing platforms, however, topside structures have been designed using codes of practice based on static loading with little allowance for any dynamic loading from an explosion. Since the publication of the Cullen report (1990) on the Piper Alpha tragedy, safety cases are required by the Health and Safety Executive in the United Kingdom to demonstrate that the installation meets acceptable safety criteria. This involves the assessment of the structural response of the topside, including corrugated panels in the event of an explosion. This includes determining how much the panel will deform, whether rupture occurs and forces transferred to support structures.

Both simplified and finite element methods have been explored in the past. Leach (1993) used the Biggs' method to predict panel responses by considering different shape functions in the elastic, elastic and plastic, and plastic stages. Boundary condition was assumed either pinned and fixed. The results are similar to the charts presented in the Interim Guidance Notes (Bowerman, et al, 1992). In an explosion incident, high loads can be transferred from cladding panels to support structures affecting their structural integrity. Though the single degree of freedom (SDOF) method has been proved useful for predicting deflection of a panel, it is not suitable for predicting support reaction (Wees and Peters, 1995). Recently, attention has been focused on using finite element (FE) analysis packages to analyse effect of both high explosion and gas explosion, such as ABAQUS (Louca, et al, 1996, Louca and Pan, 1996, Xu and Kirkvik, 1991, Walker, 1993, Fletcher and Milhsein, 1991, Wees and Peters, 1995), DYNA3D (Louca, et al, 1996, Pan and Watson, 1996). It has been observed that boundary conditions have significant effect on panel response (Louca and Pan, 1996, Schleyer, et al, 1994, Schleyer and Mihsein, 1992). To account for different boundary conditions, spring elements were introduced by Schleyer and Mihsein (1992). It is, however, difficult to assign reasonable stiffness values to these spring elements, and thus the results are heavily dependent on the stiffness selected.

In this paper, the boundary conditions were considered by including angle support attached to the corrugated panels in the FE model. Displacement at the centre of the panel is plotted to compare with test results obtained by British Gas. Displacement and rotation around the edges of the panel and forces transferred to support structures are also discussed.

2. Experimental set up

A test chamber approximately 3m by 3m by 10m long was constructed in which controlled gas explosion could be generated. Each panel was first welded to an angle frame which was then bolted to the test rig. Both short and long angle supports have been used (Figure 1a and 1b). Pressure time history, deflection history, etc. were recorded during each test. The length of the panel along the corrugation was 2.5 m and 1.87 m for the specimens with short and long angle support, respectively. The thickness of the corrugated panel was 2.5 mm. Details of the profile of corrugation and angle supports are shown in Figure 2. Three specimens have been included in this study due to that the displacement data was available.

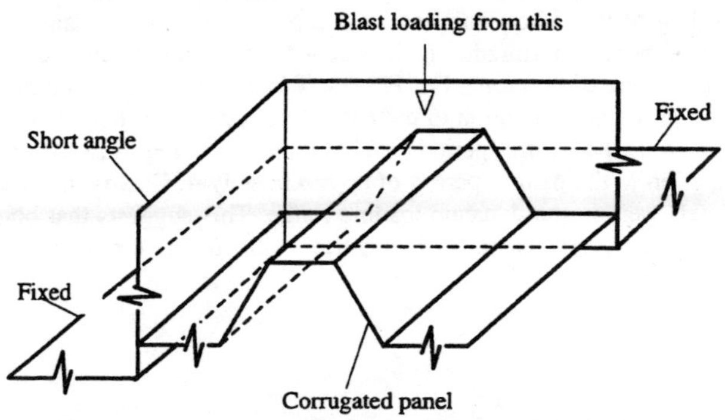

Figure 1a. A corrugated panel with short angle support

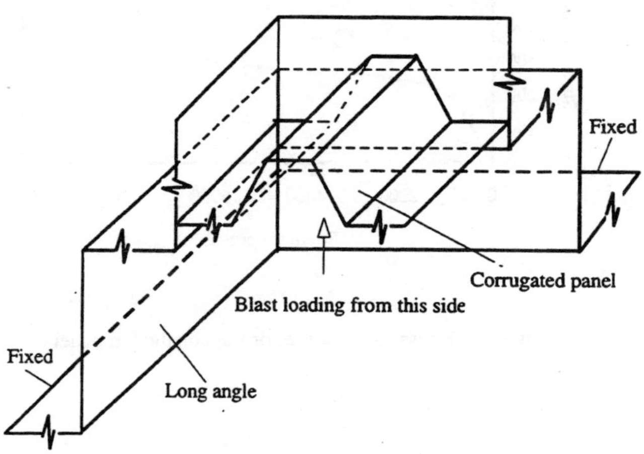

Figure 1b. A corrugated panel with long angle support

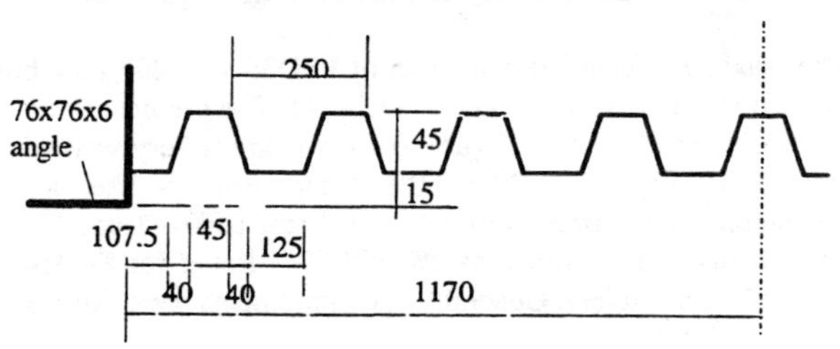

Figure 2a Details of specimen FFD39

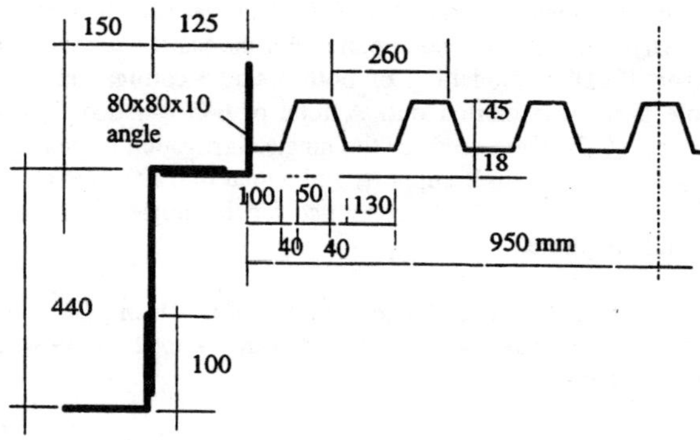

Figure 2b Details of specimens FFD141 and FFD142

3. Simplification of blast loading

As in any structural analysis, the determination of applied is critical. With blast walls, this is more apparent than in normal structures because the pressure from gas explosion is very difficult to predict. Even with identical experiments, a coefficient of variation of 10% has been observed (Wees, 1993). In order to make the FE results directly comparable with test results, pressure time history recorded with each panel was used and simplifications were made accordingly.

Seven pressure gauges were used for each test and positioned evenly over the panel. Generally, the peak overpressure recorded at the centre of the panel was the lowest compared with readings from other pressure gauges, and readings near the corners showed higher values. This is thought as the results of multi-reflection near the corner. In this study, however, the variation of pressure over the panel is ignored and the average value is adopted.

Another simplification made was to use a triangular form to represent the pressure time history. This is done in order to reduce the overall computing time. This implies that when the predicted displacement of the panel is to be compared with the test results, the time shift made during simplifying the pressure time history has to be included.

The simplified triangular pressure time histories for each specimen are given in Table 1.

Table 1 Simplified pressure time history

Specimen	Peak overpressure (bar)	Duration (ms)	Rise time (ms)
FFD39	1.0	50.0	25.0
FFD141	0.4	82.5	43.5
FFD142	0.7	90.0	45.0

4. Finite element modelling

The finite element code ABAQUS explicit was used to carried out the analysis of the corrugated panel. For the panel with short angle support (FFD39) modelling of both a single corrugation and the whole panel was carried out. A total of 645 4-noded thin shell elements (S4R) were used for the single corrugation to model both the panel and the angle support, and a total of 7200 S4R elements for the whole panel. Mesh generated for the single corrugation is shown in Figure 3.

For the panel with long angle support (FFD141 and FFD142), a quarter of the panel was modelled using a total of 1954 S4R elements (Figure 4).

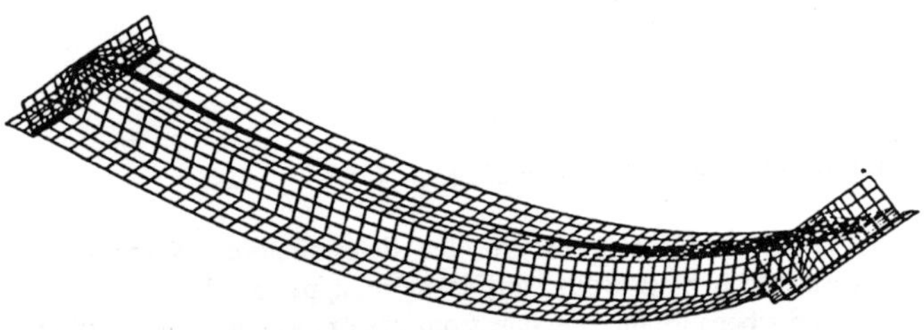

Figure 3 The FE model for FFD39

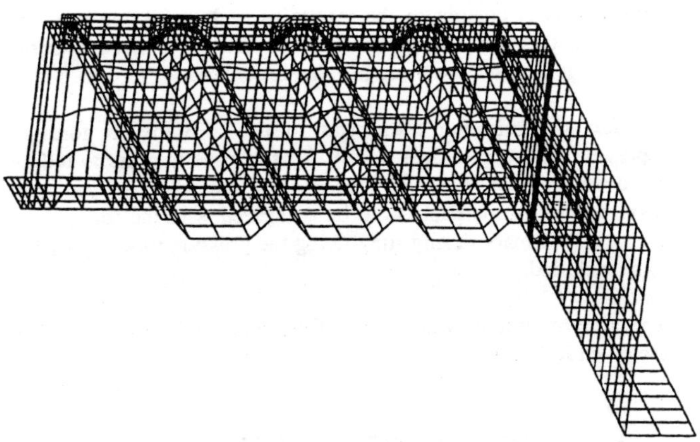

Figure 4 The FE model for FFD141 and 142

The corrugated panels were fabricated from stainless steel which, unlike most structural steels, does not exhibit a definite yield point. The stress strain relationship of the material was measured using coupons from the panel. Additional tests were carried out in City University which showed considerable elongation and an average maximum stress of 610 N/mm^2. In the FE analysis, the stress strain curve provided by the British Gas (Figure 5) is adopted with the assumption of an infinite plastic plateau after the maximum stress is reached.

A further assumption regarding the material properties is that the properties of both the short and long angle support are the same as the corrugated panel. The elastic modulus is assumed 200x10^3 N/mm^2 and the Poisson's ratio 0.3. In order to see whether the panel is influenced by the impulse or the peak overpressure, the natural period of the panel, T, is determined first. The vibration of the panel is initiated by applying a uniformly distributed blast loading of very short duration on the panel, and the fundamental natural period of vibration is determined from the deflection curve at the centre of the panel. For FFD39, T is about 30 ms, resulting a ratio of the blast duration to the natural period of 1.6. For FFD141 and 142, the natural period is 22 ms, resulting the ratio of blast duration to the natural period of approximately 4. Clearly all three panels were in the dynamic loading range. This suggests that both the peak overpressure and impulse will affect the structural response.

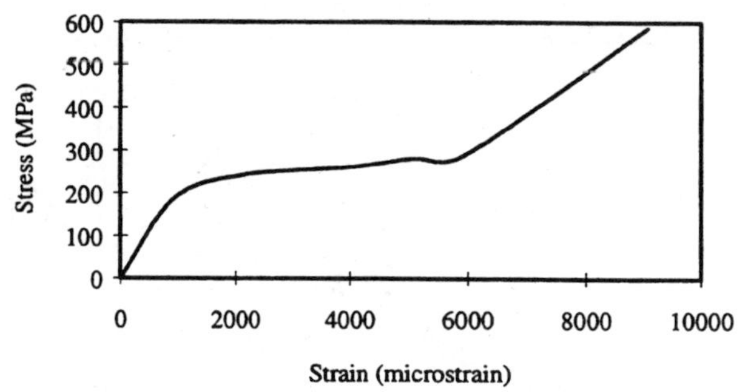

Figure 5 Stress strain curve for the corrugated panel

5. Comparison of FE analysis with experimental results

The maximum recorded deflection of FFD39 was 207 mm, but the deflection - time history was not available. The predicted maximum deflection was 203 mm (Figure 6) which agrees very well with the test result. Figure 7 shows the comparison of the predicted deflection - time history with test results for FFD141 and FFD142. The predicted deflection was 8% and 20% lower for FFD141 and FFD142, respectively. Generally, the agreements were fairly good.

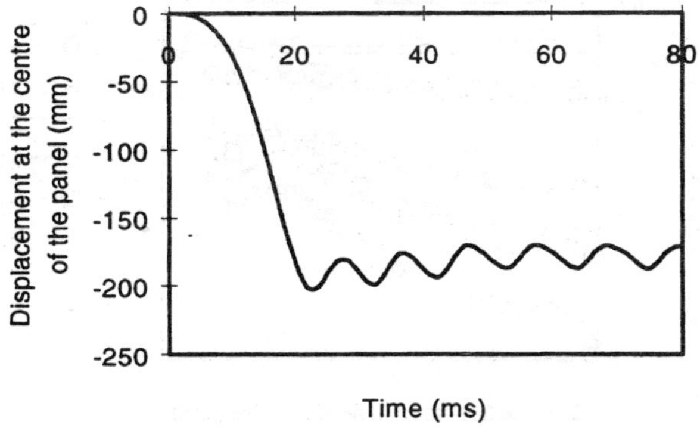

Figure 6 Displacement - time history of FFD39

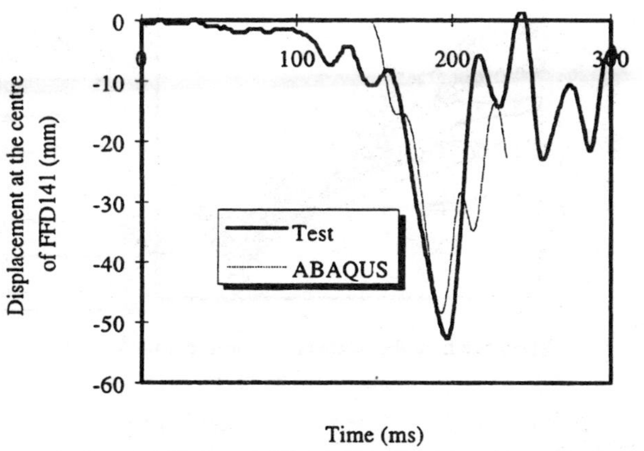

Figure 7 Displacement - time history of FFD141

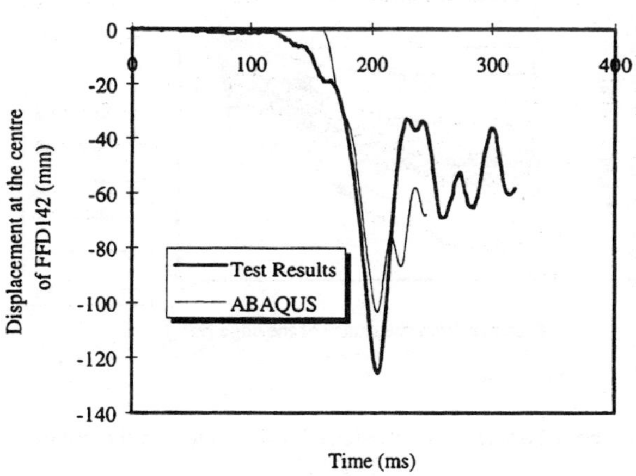

Figure 8 Displacement - time history of specimen FFD142

6. Discussion

The response of a panel when subjected to explosion can be divided into parts: the global response which controls the overall displacement of the panel, the local response which controls whether rupture of a panel occurs, and forces transferred to support structures which will influence the structural integrity of the support structures considering that they are already under service load. Because all three panels considered here remained attached o the test rig, the local response is not discussed in this paper.

6.1 The effect of angle support

Two types of angle support were used, the short one for the specimen FFD39, and the long one for FFD141 and FFD142. Under blast loading both vertical displacement and rotation at the interface between the panel and the angle support occurred.

For FFD39, the maximum displacement at the interface between the angle and the panel was about 20mm, approximately 10% of the total displacement. Assuming a half sine curve for the deformed shape, it can be worked out that the deflection at the interface position would be 9% of the total deflection, very close to the ABAQUS prediction. However, the rotation at the interface position based on the half sine curve assumption is only 50% of the ABAQUS result. This is thought as the influence of the fixed end of the angle because no rotation is allowed in the FE model and that the leg of the angle was only 76 mm considering that the deflection at the interface was about 20mm. This suggests that a half sine curve assumption might be used for determining the global response, but not for the local response at the connection area.

The maximum displacement at the edge parallel to the corrugation was 1.6mm for FFD141, whilst the maximum displacement at the edge normal to the corrugation was 4mm. The larger deflection at this edge suggests that it is mainly the angle support normal to the corrugation that undertakes the load. In addition to the vertical displacement at the edges, rotation also occurred at edges around the panel. Figure 9 shows the rotation at the edge normal to the corrugation. The maximum rotation was approximately 60 mrad.

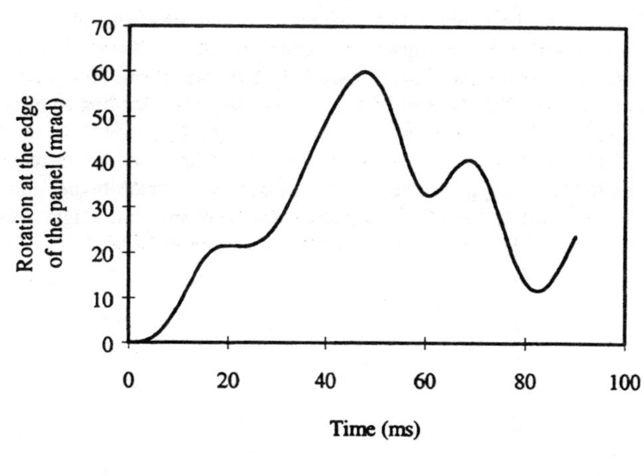

Figure 9 Rotation at the edge of the panel normal to the corrugation

When the peak overpressure was increased from 0.4 bar (FFD141) to 0.7 bar with roughly the same duration, the maximum deflection at the centre of FFD142 was increased by 110%. At the same time deflection and rotation around edges also increased. The deflection at the edge normal to the corrugation was increased by 85% whilst that at the edge parallel to the corrugation was increased by 145%. The increase in rotation was also higher at the edges parallel to the corrugation (130% compared with 115%). This suggests that with increasing levels of blast loading more load will be shared by the angle support parallel to the corrugation than that at lower levels of blast load. Because these long angles adopted were relatively large in size compared with the panel and were welded to form a much stiffer support structure the deflection at the interface between the angle and the panel was no longer fit with the half sine curve assumption.

6.2 Forces transferred to support structures

Forces transferred to support structures include shear force which is the result of direct application of blast loading and membrane force which is due to the in-plane restraint exerted on the panel by support structures. Quite often support structures are already under a certain level of service load, thus checking the safety of these support structures due to these extra forces will improve the reliability assessment of those structures involved.

It needs to be mentioned that in ABAQUS forces transferred through each node is the absolute value, thus depending heavily on the mesh size. So in order to make the prediction comparable with different mesh densities, all forces have been converted into stresses accordingly. Figures 10 - 13 show the distribution of shear and membrane stress transferred to support structures. Multiplying the stress with the thickness of the angle will give forces transferred per unit length. For FFD141 and FFD142 the thickness of the angle was 10mm, resulting the shear force transferred at the middle of the angle was about 30kN/m for FFD141 and 50kN/m for FFD142.

This is interesting to note that the both the peak shear and membrane forces transferred to support structure occurred at the middle of the angle. The decrease of membrane force from the centre of the angle was faster than that of the shear force.

Different angle support and different loading area also affect the forces transferred to support structure. It can be found that the loading area/boundary length ratio of FFD39 was about 60% higher than that of FFD141 and FFD142. Also the blast loading applied on FFD39 was much higher than that applied on FFD141 and FFD142. The result is that both shear force and membrane force transferred to support structure for FFD39 were much higher than the other two specimens. The peak shear force per unit length was 180kN/m, and the membrane for per unit length 965 kN/m.

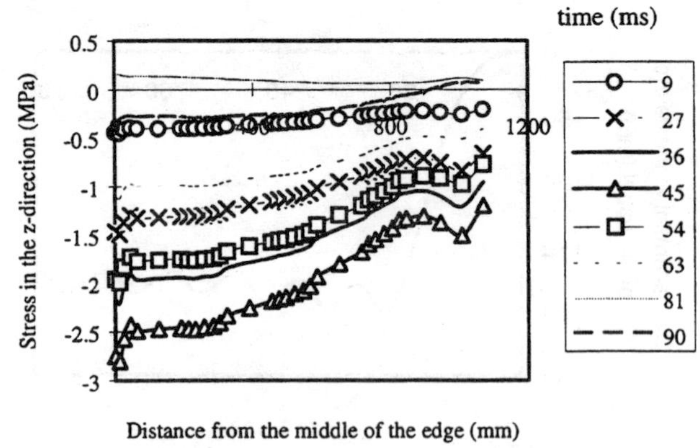

Figure 10 Shear stresses transferred to support structure of FFD141

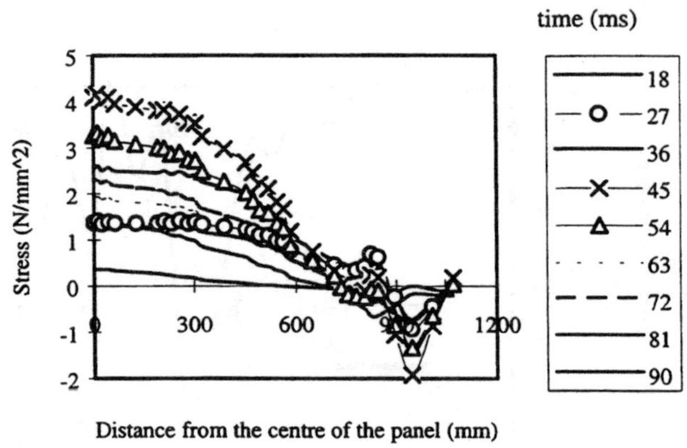

Figure 11 Membrane stress transferred to the support structure of FFD141

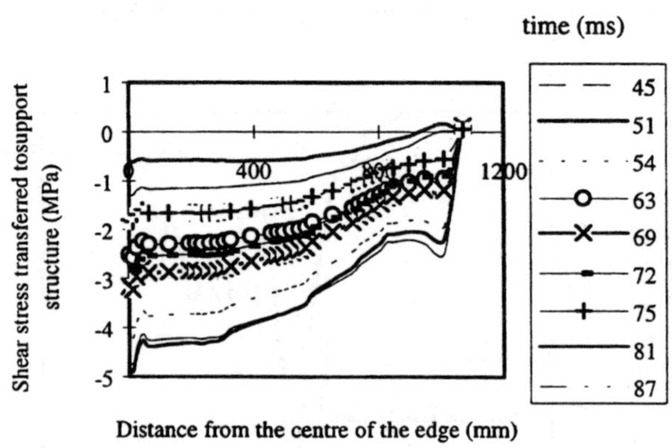

Figure 12 Shear force transferred to the support structure of FFD142

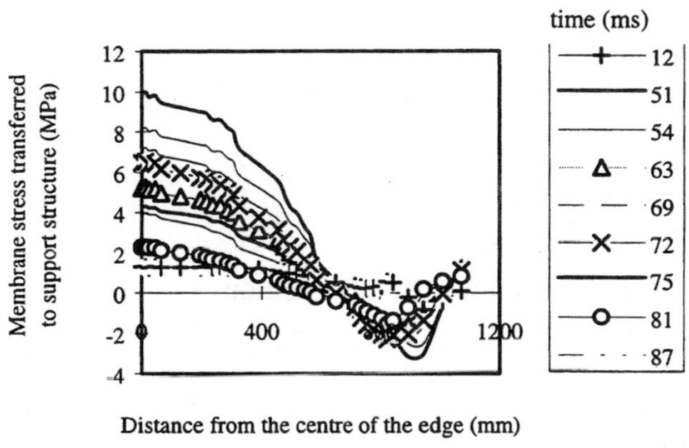

Figure 13 Membrane stress transferred to the support structure of FFD142

7. Conclusions

The finite element analysis package ABAQUS/Explicit was used to predict the response of corrugated panels subjected to gas explosion and good agreements in displacement has been achieved. The FE results showed considerable displacement and rotation at the interface between the angle support and the panel suggesting the effect of these angle supports have to be considered.

The response of the panel is mainly controlled by the stiffness along the corrugation and the load applied to the panel is transferred to the support structures through the two boundaries normal to the corrugation.

Forces transferred to the support structures include the shear force and the membrane force. At large deflection state the membrane force is much higher than the shear force. Its effect on the support structures has to be addressed properly.

Acknowledgements
The financial support from MTD and EPSRC of the United Kingdom is greatly appreciated (Grant No. MTD/EPSRC GR/L16804).

References

Bowerman, H. Owens, G.W., Rumley, J.H., and Tolloczko, J.J.A. (1992), "Interim Guidance Notes for the Design and Protection of Topside Structures against Explosion and Fire", *SCI-P-112/299, The Steel Construction Institute*, UK.

Fletcher, R. and Mihsein, M (1991), "The Dynamic Response of Firewalls under Blast Loading", *Proceedings of the ABAQUS users conference*, Oxford, England, 11-11 September, 1991

Leach, P. (1993), "Design Charts for Elasto-Plastic Design of Cladding Panels Subjected to Dynamic Loading", *Proc. Instn Civ. Engrs Structs & Bldgs,* Vol.99 Feb., pp.55-61.

Louca, L.A., Harding, J.E., and White, G.C. (1996), "Response of Corrugated Panels to Blast Loading", Proceedings of the International Conference on Offshore Mechanics and Arctic Engineering, Florence

Louca, L.A., Pan, Y.G. (1996), "Response of Offshore Topsides to Blast Loading", *Proceedings of the 1st Asia-Pacific Conference on Offshore Systems: Mobile and Floating Structures*, Kuala Lumpur, Malaysia, 10-11 December, 1996.

Louca, L.A., Punjani, M. and Harding, J.E. (1996), "Non-linear Analysis of Blast Walls and Stiffened Panels Subjected to Hydrocarbon Explosion", *J. Construct. Steel Res.* Vol.37 No.2 pp.93-113.

Pan, Y.G. and Watson, A.J. (1996), "Interaction between Concrete Cladding Panels and Fixings under Blast Loading", *Cement and Concrete Composite*, 18, 1996, pp.323-332

Schleyer, G. and Mihsein, M. (1992), "Development of Mathematical Models for Dynamic Analysis of Structures", Proceedings of the International Conference on Structural Design against Accidental Loads, London, ERA Technology Ltd

Schleyer, G., Galbraith, D., and Gillan, I., (1994), "An Integrated Approach to Optimising the Performance of a Blast Wall System", *Proceedings of the Third International Conference on Structures under Shock and Impact*, Madrid, Spain 1st - 3rd June, 1994, Ed. P.S. Bulson, pp.21-28

Walker, A.C. (1993), "Non-linear Analysis of Blast Walls to Evaluate Safe Ultimate Loads", Proceedings of the 2nd Int. Conf. on Offshore Structural Design against Extreme Loads, ERA, London

Wees, Rolf M. M. van and Peters, Dirk-Jan (1995), "Accuracy and Range of Applicability of the Single-degree-of-freedom Method", *Proc. of the 7th Int. Symposium on Interaction of the Effects of Munitions with Structures,* Mannheim, Germany, April 24-28 pp.371-380

Wees, R.M.M. van (1993), "Design of Stiffened Panels to Withstand Explosion Loads", *Proceedings of the 2nd International Conference on Offshore Structural Design against Extreme Loads*, ERA Report 93-0843

Xu, J. and Kirkvik, R.H. (1991), "Design against Explosion Loads in Offshore Structures", *Proceedings of the 1st International Offshore and Polar Engineering Conference*, Edinburgh, United Kingdom, 11-16 August, 1991 pp.183-189.

Proceedings of the Seventh (1997) International Offshore and Polar Engineering Conference
Honolulu, USA, May 25-30, 1997

ISBN 1-880653-28-1 (Set); ISBN 1-880653-32-X (Vol. IV)

Probabilistic Finite Element Methodology for the Reliability Assessment of Stiffened Plates

I.R. (Wally) Orisamolu and K-T. Ma
Martec Limited
Halifax, Canada

ABSTRACT

The present study reports a probabilistic finite element modeling and analysis procedure that is useful for the assessment of structural integrity and reliability. This procedure is efficient, accurate, and simultaneously accounts for the variabilities and uncertainties in the governing parameters as well as the nonlinear nature of the physical failure mechanisms. In the approach, nonlinear finite element methods are employed to model the physical response characteristics while an integrated stochastic finite element formulation that employs a FORM-based adaptive response strategy is used for probability calculations. Uncertainties in loads, material properties and structural properties are considered, and reliability levels for selected stiffened-panel configurations, with reference to some strain-based failure criteria, are determined. Probabilistic sensitivity studies are also conducted to study the importance of the influencing parameters as well as the uncertainty in these parameters. Results are compared to those obtained using approximate analytical models, and it is shown that the simplified models may predict reliability and sensitivity values that have significant errors. Finally, the paper suggests a strategy by which this methodology can be employed to improve the design of stiffened-plated structures.

KEY WORDS: stochastic, finite element, probabilistic, reliability, ductility, stiffened-plate, offshore

INTRODUCTION

Stiffened-plated structures have found wide applications in engineering construction. They have constituted the preferred mode of construction for ships, offshore structures as well as aircraft, for example. Thus, the assessment of the structural integrity and the residual strength of these structures require a thorough understanding of the behaviour of the basic structural elements which are essentially the main building blocks for the larger structural system. Traditionally, the assessment of structural integrity has been carried out via the application of simplified analytical models that are defined for various modes of failure for the stiffened panel configurations. Later on, sophisticated analytical techniques such as the finite element method were introduced to permit more realistic assessments of structural integrity in a manner that is more compatible with the true physical failure mechanisms. However, the application of such numerical techniques for gaining more accurate response and failure characteristics of stiffened plates have generally not been conducted in a framework that rationally and systematically accounts for the uncertainties in the governing parameters.

The engineering community is increasingly accepting the need to evaluate the integrity of engineering structures with due recognition of the uncertainties and variabilities that are associated with the various parameters involved in the assessments. Such parameters include loads, material properties, geometrical properties, strength or failure models, as well as structural analysis methods. The merit of the probabilistic approach for rationally and systematically accounting for the relevant epistemic and aleatory uncertainties have been well discussed and demonstrated in several engineering publications. For stiffened plates in particular, several efforts have been recorded in this connection. Notable examples in this regard include the works of Hart et al. (1985), Das (1986), and Ma and Orisamolu (1996). Most of these have relied upon the use of semi-empirical analytical models which characterize the response and capacity of stiffened-plated structures. In the previous work of the authors (1996), for example, probabilistic reliability assessments were reported for yielding in tension, flexural buckling, and ultimate failure to pressure loads.

The aforementioned simplified approaches have served as the basis for the design of several stiffened-plated structures over the years. However, it is well recognized that the failure modes of highly ductile stiffened panels occur via very complex nonlinear mechanisms which can only be adequately captured by more sophisticated analysis methods such as the finite element method. The application of sophisticated structural analysis methods within a framework that also duly accounts for the uncertainties and variabilities of the relevant parameters is, therefore, very desirable. This is the only strategy that can ensure that the introduction of probabilistic modelling/analysis concepts does not necessitate a compromise in the realistic characterization of the nonlinear physical deformation and failure mechanisms of the structure. The stochastic finite element method (SFEM) or probabilistic finite element method (PFEM) is emerging as a powerful framework for the practical implementation and application of this strategy. Several SFEM/PFEM methodologies and their limitations have been previously evaluated by the first author: Orisamolu and Liu (1994). In particular, a very versatile and robust methodology that is suitable for a wide variety of practical applications have been implemented in a computer program (STOVAST) which was employed in this study: Orisamolu and Liu (1994, 1996).

The present study has been motivated by the desire to develop capabilities and tools that can be applied for quantifying the reliability and ductility of monolithic steel platforms (such as Caisson-type structures) for oil production in arctic offshore environments. Research and feasibility studies have shown that such structural systems are the most cost effective conceptual structural design for arctic construction required in the offshore Canadian and US Beaufort Seas. Current designs are believed to be conservative primarily because of the uncertainties in the environmental load - particularly ice loads.

Accurate estimation of reliability and ductility levels has the benefit of providing quantitative information that can be employed for reducing design/construction costs without compromising the acceptable levels of structural reliability. In Section 2, a brief description of the stiffened-plated panel configuration and the finite element model is given. This is followed by the probabilistic finite element modelling/analysis methodology in Section 3, while Section 4 describes an illustrative application. Finally, some conclusions and recommendations are given.

PROBABILISTIC FINITE ELEMENT FORMULATION

The formulation and implementation of probabilistic finite element modelling and analysis (PFEM/PFEA) approaches must pay due recognition to several practical issues. First a PFEM/PFEA approach should be applicable to most classes of problems that are currently solved by conventional deterministic finite element methods. This implies the provision for modeling complex geometrical configuration, sophisticated constitutive equations describing mechanical material behaviour and large deformations. Such advanced modelling and analysis capabilities have become essential features of commercial and proprietary finite element modelling and analysis (FEM/FEA) computer programs. The compatibility of a PFEM/PFEA approach with the features of existing programs has obvious advantages. These include, among several other things, the benefit of utilizing well-tested algorithms for mechanical response computation. Equally important is the requirement for accurate characterization and representation of the true system uncertainties. In reality, these uncertainties could be random variables, random processes, or random fields which may have any arbitrary probability distribution and correlation structure. For arctic offshore structural systems fabricated from stiffened plates, any of the aforementioned types of uncertainties could be present. The structural material properties such as Young's modulus (E) could be random variables, the loading (such as wave loads) could be defined as random processes or local ice pressures could be properly characterized as random fields. Evidently, therefore, a true probabilistic treatment of the various possible types of uncertainties requires a robust and versatile probabilistic finite element modelling and analysis strategy.

Several PFEM/PFEA approaches have been proposed in the literature. These approaches deal with the representation and discretization of random fields and the estimation of the resulting probabilistic response or reliability. A comprehensive review of these approaches was carried out by Orisamolu and Liu (1996) in an earlier work. The approach utilized in the present work is a combination of a response surface methodology (RSM) and a fast probability integration scheme based on first-order reliability methods (FORM). The response surface methodology is used to approximate the nonlinear mechanical response of the stiffened-plate structure (obtained via nonlinear finite element analysis) while FORM algorithms are employed to compute the probability of limit state function violation. Brief descriptions for RSM and FORM computational procedures are provided in the following subsections.

Response Surface Methodology

The methodology applied herein involves the development of response surface based methods for probabilistic reliability analysis as well as the full characterization of probabilistic response. The response surface method represents an elegant, robust and efficient framework for developing closed-form mechanical models for predicting the behaviour of complex structural systems. For this reason several probabilistic analysis methodologies that utilized this methodology have been developed and implemented in STOVAST (Orisamolu and Liu, 1994, 1996).

The response surface method is a classical statistical technique in which a complex (computer) model is approximated by a simple functional relationship between the output quantities and the input (basic) variables. The approximation is usually based on polynomial functions and, often, linear or quadratic response functions are applied. Adopting the simpler response functions allows an efficient repeated computation, for example, as may be needed in simulations or parameter studies in structural reliability analysis. This is because the approximation to the response surface rather than the original limit state function is used in the calculation of failure probabilities.

The concept of response surface methods has been used when approximating costly to compute and/or non-differentiable limit state functions. Within the framework of the stochastic finite element method, the steps required for the implementation of the response surface technique were described by Faravelli (1989). This involves the application of regression analysis to obtain the polynomial coefficients involved in the representation of the limit state function using the results of several numerical experiments.

The explicit representation of the limit state function g(x), for the quadratic approximation for example, takes the form:

$$\bar{g}(x) = a + \sum_{i=1}^{n} b_i x_i + \sum_{i=1}^{n} \sum_{j=1}^{n} c_{ij} x_i x_j \qquad (1)$$

where n is the number of basic random variables (x_i) and the coefficients a, b, and c are to be determined from numerical experiments.

Response Surface Approximation Algorithm

In the finite element method, a response function (such as displacement, stress, strain, or buckling load) $\mathbf{Z} = Z(x)$ usually depends on a number of variables, X. The function Z(x) is implicitly built into the finite element modelling and analysis algorithms such that for any input, x, the output or response, **Z**, can be computed with a reasonable level of accuracy. The finite element computation process is, however, usually computation intensive. The response surface approximation seeks to develop an equivalent representation of the input-output relation, $Z = \tilde{Z}(x)$.

A computationally efficient response surface approximation procedure which has been previously implemented in STOVAST is employed in this work. First, only a subset of the input variables are used for the construction of the response surface. The subset utilized is the vector of random variables which are present in the response function. Secondly, the experimental design strategy utilized is to evaluate response values at a representative evaluation point such as the mean value, μ_i and at locations which are $\pm\gamma\sigma_i$ from the mean value, where γ is a positive real number ($0 < \gamma < R$) and σ_i is the standard deviation of the random variable. Reasonable values of the parameter γ are 1.0, 2.0, etc., but should be selected such that $|\gamma\sigma_i| < |\mu_i|$. This strategy provides experimental design points that are in the neighbourhood of the response surface evaluation point. The approximation scheme follows the recent work of Maymon (1993). Let **S** be the approximation of $Z = \tilde{Z}(x)$. Then a quadratic response surface of Z can be expressed as:

$$S = a_o + \sum_{i=1}^{L} a_i \cdot (X_i - X_{iep}) + \sum_{i=1}^{L} \alpha_{L+i} \cdot (X_i - X_{iep})^2 + \sum_{i=1}^{L-1} \sum_{j=i+1}^{L} \alpha_{(L-1/2)(i+1)+j} \cdot (X_i - X_{iep})(X_j - X_{jep}) \tag{2}$$

where **L** is the total number of random variables. The total number of terms in Equation (2) is, in the most general case of a complete quadratic function, given by k+1, where:

$$k+1 = \frac{(L+1)\,(L+2)}{2} \tag{3}$$

Based on earlier results which indicate that the contributions of mixed terms is usually negligible, the last set of terms in Equation (2), which represents the mixed terms, may be neglected. Thus Equation (2) reduces to:

$$S = \alpha_o + \sum_{i=1}^{L} \alpha_i (X_i - X_{iep}) + \sum_{i=1}^{L} \alpha_{L+1} (X_i - X_{iep})^2 \tag{4}$$

which may, in turn, be expressed as:

$$S = \alpha_o + \sum_{n=1}^{k} \alpha_n Y_n \tag{5}$$

where

$$Y_n = X_i - X_{iep} \quad , \quad n=1,2,...,L \quad i=1,2,...,L \tag{6}$$

while

$$Y_n = (X_i - X_{iep})^2 \quad , \quad n=L+1,L+2,...,2L \quad i=1,2,...,L \tag{7}$$

The parameter α_o is associated with the response surface evaluation point itself.

It can be shown that the optimal set of coefficients, $\{\alpha\}$, for Equation (5) that best approximates the actual (finite element) response function is given by the solution of the equation:

$$[Y]^T\,[Y]\,\{\alpha\} = [Y]^T\{\bar{S}\} \tag{8}$$

where the matrix [Y] is given by:

$$[Y] = \begin{bmatrix} 1 & 0 & 0\cdots0 & 0 & 0\cdots0 \\ 1 & +\Delta X_1 & 0\cdots0 & \Delta X_1^2 & 0\cdots0 \\ 1 & +\Delta X_1 & 0\cdots0 & \Delta X_1^2 & 0\cdots0 \\ 1 & 0 & +\Delta X_2\cdots0 & 0 & \Delta X_2^2\cdots0 \\ 1 & 0 & +\Delta X_2\cdots0 & 0 & \Delta X_2^2\cdots0 \\ 1 & 0 & 0\cdots0 & 0 & 0\cdots0 \\ \cdot & \cdot & \cdots & \cdot & \cdots \\ \cdot & \cdot & \cdots & \cdot & \cdots \\ 1 & 0 & 0\cdots+\Delta X_1 & 0 & 0\cdots\Delta X_1^2 \\ 1 & 0 & 0\cdots+\Delta X_1 & 0 & 0\cdots\Delta X_1^2 \end{bmatrix} \tag{9}$$

and $DX_j = \gamma_j \sigma_j$ in which γ_j is a suitably chosen factor.

The finite element experimental design response vector is:

$$\{\bar{S}\} = \begin{Bmatrix} S_o \\ S_1 \\ S_2 \\ \cdots \\ S_{M-1} \end{Bmatrix} \tag{10}$$

where **M** is the total number of finite element evaluations.

Fast Probability Integration via Advanced Reliability Analysis Methods

The general structural reliability problem is usually formulated in terms of a finite set of basic random variables $X = (X_1, X_2, \ldots X_n)$ and a limit state function g=g(x) in which g is defined by the event $\{g(x) \le 0\}$, while $\{g(x) > 0\}$ identifies a safe state. The probability of failure, P_f, is defined as:

$$P_f = Prob\ \{g(x) \le 0\} = \int_{\Omega}\cdots\int f_x(x)dx \tag{11}$$

$$where\ \Omega\ \{x{:}g(x) \le 0\} = failure\ region \tag{12}$$

and $f_x(x)$ is the joint probability density function of the random variables x.

The above integral is in general very difficult to evaluate and approximate procedures have evolved as practical tools for efficiently calculating the integral. Prominent examples in this connection include the first-order reliability methods (FORM) and the second-order reliability methods (SORM) which are well discussed in classical monographs on structural reliability as well as our earlier studies. These procedures are based on the evaluation of a reliability index (β) from which a failure probability (P_f) can then be computed. For the case of FORM, the failure probability is related to the reliability index (which is a widely accepted measure of safety in probabilistic analysis) through the expression:

$$P_f = \Phi\,(-\beta) \tag{13}$$

where ϕ is the standard normal cumulative distribution function.

The main steps involved in the application of FORM are well documented in several monographs; see, for example, Madsen et al.(1986). SORM and Monte Carlo simulation methods were also employed in this study.

PROBABILISTIC FINITE ELEMENT ANALYSIS APPLICATION

Stiffened-Plated Panel Configuration

A stiffened plated structure whose configuration is based on an exploration and delineation drilling Caisson Retained Island (CRI) located in the Canadian Beaufort Sea [Sandwell, 1991]. The CRI is primarily composed of stiffened plate elements which provide simplicity of construction using standard shipbuilding practices. The CRI (see Figure 1) is constructed with all-steel ice faces. This construction has been shown to be appropriate for exploration platforms which are designed for relatively short durations of exposure. For production platforms, the global ice loads may be significantly higher than those for exploration structures. Its ice face consists of a skin plate strengthened by a grid of longitudinal stiffeners and transverse frames, supported by deep horizontal girders and the watertight bulkheads. Two-by-two bay size of the stiffened panel is selected to be our model. Because of the symmetry condition, only a quarter of the two-by-two bay needs to be modeled. The finite element model was built by using HyperMesh, a commercial FE pre- and post-processor (see Figure 2). The length of the model (2500 mm) is taken as one span (the distance between two transverse frames). The selected model is typical of stiffened plated panel in offshore structures.

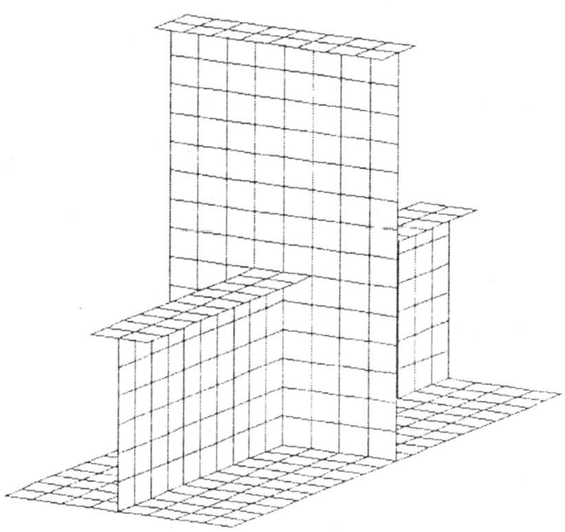

Figure 1. Model of a Stiffened-Plated Panel

Two main design issues must be considered to ensure the adequacy of an offshore structure against ice action, namely, local strength and global stability. Local design considerations such as steel plate thickness, stiffener dimensions, and concrete dimensions and reinforcement require the definition of ice pressure on the local area affecting the member being considered. On the other hand, the overall load on the structure is needed to ensure safety against global failure considerations such as sliding, overturning, and progressive collapse. Most of the emphasis in ice structure interaction research has been on global loads since these loads influence the overall dimensions and layouts of arctic offshore structures.

It is now being recognized that local ice pressures, which occur over smaller areas of the structure, are also very important. An ice feature moving against a structure will make initial contact over a small area. As the ice continues to move, the contact area progressively increases. The unit pressure over the initial small contact area may be quite high. Therefore the local design of the structure becomes very important. This paper concentrates only on local load effect to stiffened plated panels.

The uncertainty of ice formation, ice movement, and ice structure interaction demand that realistic models of ice loads be based on the probabilistic approach. The Canadian Standards Association Code: CAN/CSA S471-92 recommends this approach for the calculation of ice forces experienced by offshore structures. Other environmental loads that are of relevance to offshore structures were extensively reviewed by Orisamolu and Bea (1993). Pressure load due to ice impacts are the predominant loads considered in this study.

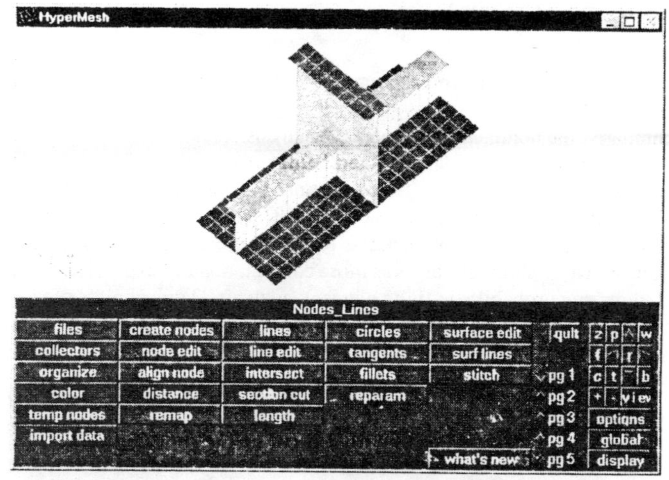

Figure 2. Finite Element Model of Stiffened-Plate Configuration

Probabilistic Reliability Analysis

A stochastic FEA requires a significant amount of computing time. One complete run of a nonlinear FEA usually takes 10 to 50 times of the computing time required by a linear FEA. This may be an obstacle to a stochastic FEA, and therefore the FEA needs to be carefully planned to minimize the computing time of a run. Before the actual stochastic FEA was performed, several tests were executed in order to better understand the influence on the results of several user-controllable factors such as material (plasticity) model, geometric displacement assumption and number of integration points used in the shell elements that were employed for discretization.

Deterministic nonlinear finite element analyses were first carried out to ensure the model integrity as well as understanding the physical failure/deformation mechanisms that were of interest. These analyses were performed using VAST (Martec Limited, 1996) and the results confirmed by ADINA (ADINA Inc., 1992). VAST and ADINA are both general purpose commercial finite element analysis computer programs. Elastic-perfectly plastic and bilinear plastic (strain hardening) models were used, and both gave essentially identical results as can be seen in the load-displacement curve shown in Figure 3.

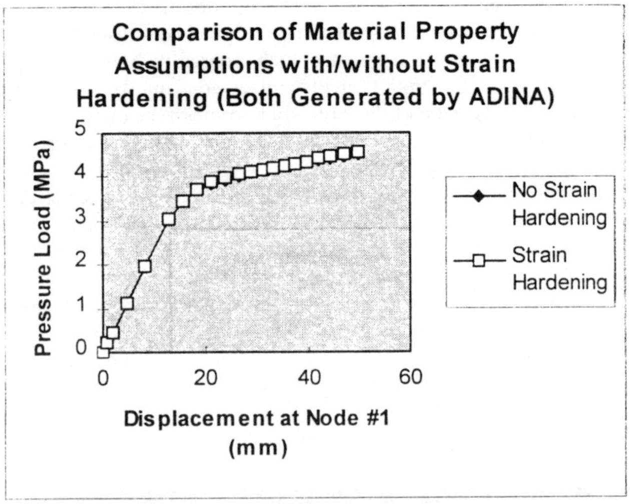

Figure 3. Load-Displacement Curves Using Different Plasticity Methods

Similarly, the nonlinear finite element analysis results were insensitive to small displacement or large displacement assumption. However, the results were very sensitive to the number of finite element integration points through the thickness of each element. The use of more integration points through the element thickness produced more accurate stress/strain results. Thus, although this was more computationally expensive, it was used for the probabilistic finite element analysis. Typical plots of some nominal deformed shapes and stress contours are shown in Figures 4 and 5, respectively.

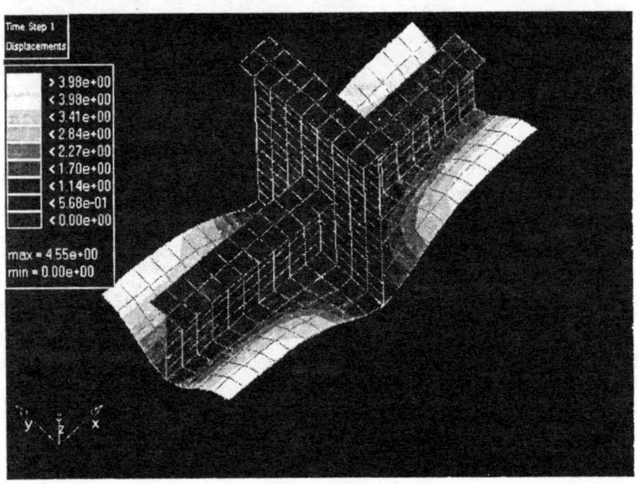

Figure 4. Exaggerated Deformed Shape Plot

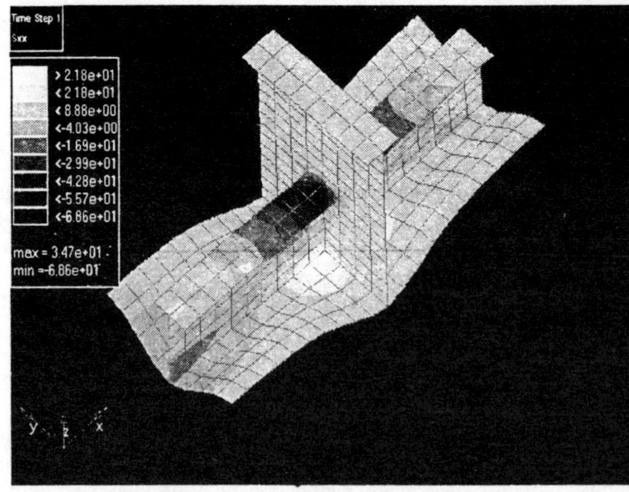

Figure 5. A Contour Plot of the X-Component of Strress σ_{xx}

The probabilistic finite element formulation that was briefly described in the previous section is applied here to assess the reliability of the stiffened-plate configuration shown in Figure 1. The structure is loaded by surface pressures acting on the elements. It is assumed that the load is deterministic for simplicity but the Young's modulus and the plate thickness are considered as random variables. The nonlinear limit state function is taken as the exceedance of the effective plastic strain, ϵ_p, beyond a particular threshold, ϵ_p^{THR}, that is:

$$g_{NL}(X) = \epsilon_p^{THR} - \epsilon_p \tag{14}$$

This limit state function is believed to be appropriate for demonstrating the concept of a probabilistic reliability analysis of a nonlinear structure in which the ductility of the structure is accounted for. The probability of failure, P_f, corresponding to the above limit state condition is computed for various probability models of the material property, E, and the thickness, t. These P_f values are computed for various uncertainty magnitudes.

In this demonstration application, a linear response surface is developed for approximating the effective plastic strain, ϵ_p, at the structural location with the most severe level of deformation. The response surface constructed for ϵ_p herein (using the procedure described above) is given by:

$$\epsilon_p = \alpha_0 + \alpha_E(E-E_m) + \alpha_t(t-t_m) \tag{15}$$

where α_0, α_E and α_t are the coefficients of the polynomial approximation, E_m is the mean value of the Young's modulus, and t_m is the mean value of the plate thickness at the deformation location of interest.

The coefficients α_0, α_E and α_t are determined via nonlinear finite element analyses of the structure described in the preceding chapter. Both E and t are taken to be random variables with lognormal probability distributions. Two representative sets of analyses are reported in Tables 4.1 and 4.2. The first set of results corresponds to the case in which the coefficient of variation (COV) of E and t are 5% and 2%, respectively. The second set of results corresponds to COVs of 5% for both E and t. Reliability results are presented for these cases for values of the threshold effective plastic strain as shown in Figure 6, while the corresponding failure probabilities are shown in Figure 7.

It should be noted that for current elastic designs, the probability of failure is assumed to be unity (i.e. 100%) as soon as yielding occurs. However, based on a limit state function formulation that exploits the ductility characteristics of the material and thereby allows for plastic deformation, acceptable reliability levels can be obtained for stiffened plated structures as illustrated in Figures 6 and 7.

Based on the results obtained in this work, further work will be conducted to select more representative values of the threshold value of the effective plastic strain to be used. We believe the plastic strain at rupture (which could itself be a random variable) is a very practical quantity that may be employed to represent the effective plastic strain. It should be noted that other nonlinear response measures such as ultimate load capacity could also be utilized as the basis for reliability analysis within the framework of the nonlinear probabilistic finite element analysis strategy applied in the present work.

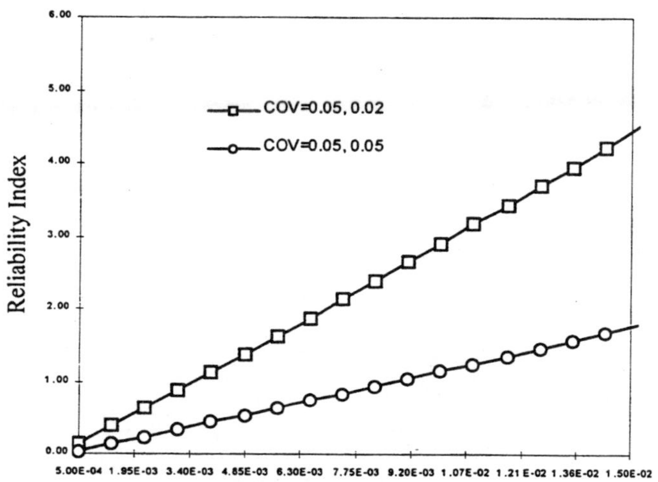

FIGURE 6: Reliability Levels for Different Values of the Effective Plastic Strain

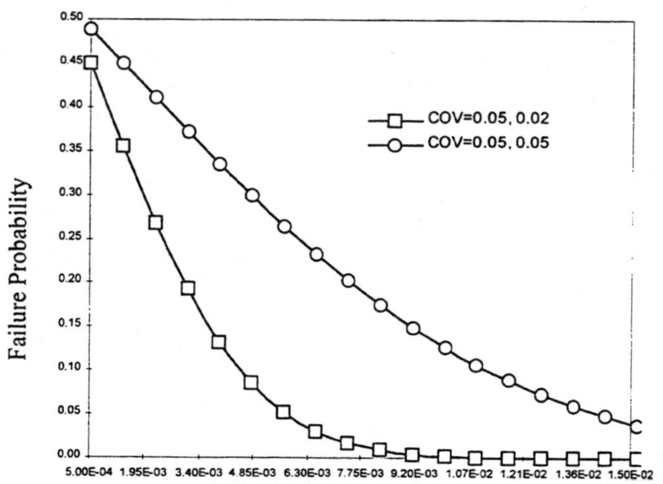

FIGURE 7: Failure Probabilities for Various Threshold Values of the Effective Plastic Strain

CONCLUSIONS AND RECOMMENDATIONS

The reliability of a stiffened-plated offshore structure has been studied on the basis of a nonlinear limit state function formulation that exploits the ductility characteristics of the material of construction. The methodology has been successfully demonstrated within the framework of nonlinear finite element analysis and advanced (Level III) probabilistic reliability analysis algorithms. It is shown that acceptable reliability levels can obtained for this class of structures well beyond the elastic design limits that are currently employed in practice.

Furthermore, the analysis demonstrated the feasibility of considering nonlinear deformation and failure mechanisms in the computation of probabilistic reliability levels without compromising the physical characteristics experienced in practice.

Based on the results of the present study, the application of a nonlinear probabilistic finite element modelling/analysis methodology for the reliability assessment of stiffened-plated structural systems appears to be promising. Work is, therefore, ongoing to further pursue the methodology development, including the implementation of additional nonlinear limit state functions that utilize ductility characteristics into the STOVAST (stochastic finite element analysis) system. An important limit state function in this connection is the ultimate strength. Realistic probabilistic models of ice loads are also being developed for use in such assessment. It is hoped that the capabilities being developed can be exploited for realizing cost savings in practical design of arctic (stiffened-plated) offshore structures.

ACKNOWLEDGEMENTS

Part of this work was funded by the National Energy Board of Canada through the Panel on Energy Research and Development (PERD) program. The authors express gratitude to Dr. Raymond Smith for his support and interest in this work.

REFERENCES

ADINA R&D, INC. "Automatic Dynamic Incremental Nonlinear Analysis (ADINA)". ADINA R&D, Inc., December, 1992.

ALTAIR COMPUTING, INC. "HyperMesh Documentation - Version 2.0". Altair/Finite Applications, 1995.

CSA (Canadian Standards Association). "CAN/CSA-S473-92: Steel Structures, Forming Part of the Code for the Design, Construction and Installation of Fixed Offshore Structures". CSA, Rexdale (Toronto), Ontario, 1992.

FARAVELLI, L. "A Response Surface Method for Reliability Analysis". ASCE, Journal of Engineering Mechanics, Vol. 115, No. 2, pp. 2763-2781 (1989).

MA, K-T and ORISAMOLU, I. R. "Reliability of Stiffened-Plated Panel in Offshore Structures". Proceedings of ISOPE-96, Log Angeles, May 1996, Vol. IV, pp. 423-430, The International Society of Offshore and Polar Engineers (1996).

MADSEN, H.O., KRENK, S. and LIND, N.C. "Methods of Structural Safety". Prentice-Hall Inc., Englewood Cliffs, New Jersey, USA (1986).

MAYMON, G. "Probability of Failure of Structures Without a Closed-Form Failure Function". Computers & Structures, Vol. 49, No. 2, pp. 301-313 (1993).

ORISAMOLU, I.R. and LIU, Q. "Development of Advanced Stochastic Finite Element Methods for the Analysis of Uncertain Structures". Martec Technical Report No. TR-95-26, February (1996).

ORISAMOLU, I.R. and LIU, Q. "Stochastic Finite Element Analysis Under Multiple Random Fields". Proceedings of the Second International Conference on Computational Stochastic Mechanics, pp. 451-460, Athens, Greece, June 1994.

ORISAMOLU, I.R. and BEA, R.G.. "Reliability of Offshore Structural Systems: Theory, Computation, and Guidelines for Application". Martec Technical Report No. Tr-93-33 to the National Energy Board, Martec Limited, Halifax, Nova Scotia, Canada, October (1993).

SANDWELL. "Verification Project G-1A Design Studies of Offshore Steel Structures". Prepared for Canadian Standards Association, June, 1991.

SNAME (Society of Naval Architects and Marine Engineers). "Principles of Naval Architecture - Volume One". Second Revision, 1989.

VAST: Vibration and Strength Analysis Program - User's Manual, Version 7.1, Martec Limited, Halifax, Nova Scotia, Canada, October 1995.

Proceedings of the Seventh (1997) International Offshore and Polar Engineering Conference
Honolulu, USA, May 25-30, 1997

ISBN 1-880653-28-1 (Set); ISBN 1-880653-32-X (Vol. IV)

Probabilistic Characterization of Pitting Corrosion Damage in Steel Panels

K-T Ma and I.R. (Wally) Orisamolu
Martec Limited
Halifax, Canada

ABSTRACT

This paper presents two probabilistic approaches that can be rapidly applied for characterizing the effective thickness of pitted plates. The approaches were formulated for compatibility with, and application to typical corrosion pit data that are collected via a standard procedure stipulated in a guide published by the Tanker Structure Cooperative Forum. The data items that are usually gathered in this connection include the depth of the deepest pit within a panel, a 300 mm x 300 mm representative sample square, the number of pits in the square, the average pit depth, the maximum depth, the average diameter, and the maximum diameter. Two probabilistic approaches are developed herein to model the effective plate thickness based on these data items. The first approach simulates the widths and depths of a sample of size N pits by using a known mean depth and an extreme depth, while the second utilizes the r deepest pit depths. Both approaches model the pit widths and depths as correlated random variables following certain probability distributions. The models are applied for the characterization of the effective thickness of typical pitted panels, and probabilistic reliability models that may be used for assessing the residual strength of such panels are also presented and demonstrated.

KEY WORDS: Pitting corrosion, pitted plate, random variable, probabilistic, simulation, residual strength.

INTRODUCTION

Pitting is one of the important damage categories for many metallic structures. The occurrence of pitting on any structural member, both in terms of spatial distribution and time, is a random phenomenon which is best modeled via a stochastic approach that duly accounts for the various uncertainties that are associated with the process. For ship structures in particular, pitting is one of the damage modes that constitute one of the main objectives of routine inspections and periodic surveys in order to assess the residual strength of structural members that have been affected by this damage mode. Pitting inspections are usually conducted using ultrasonic instruments or pit gauges. Although data gathered from pitting inspections serve as the basis for making replacement or repair decisions, there has been little research work directed toward the development of rational procedures by which the probabilistic character of typical field data can be systematically characterized and applied for structural integrity management.

The present paper is part of a research project directed at the realistic characterization of the residual strength of pitted ship panels. The ultimate objective of this Ship Structure Committee sponsored project is to develop guidelines that can be applied in practice for managing the structural integrity of pitted ship structural components.

Before presenting the approaches, the current inspection practice is reviewed here in order to identify the problem. Inspectors usually climb physically into tank bottoms to collect pitting data. According to a guide published by Tanker Structure Cooperative Forum (1986), seven items are recorded for each panel. A panel is usually defined as the steel plate bounded by longitudinal stiffeners and transverse frames. The intensity percentage is collected by visually comparing the pitting intensity with Pitting Intensity Diagrams. The depth of the deepest pit within the panel is measured and recorded. Then a 300 mm x 300 mm sample square which is regarded to be the most representative of the pitting on the panels being inspected is selected. The other items that are recorded for the sample square include the following:

- Frequency - The number of pits in the square.
- Average Depth - The depth in millimeters of the pit regarded as having the average depth for the square.
- Maximum Depth - The depth in millimeters of the deepest pit in the square.
- Average Diameter - The diameter in millimeters of the pit regarded as having the average diameter for the square.
- Maximum Diameter - The diameter in millimeters of the largest pit within the square.

Because of the large number of pits that could exist in a ship (see Table 1 for an example), it is almost impossible to measure and record every single pit. Following the above data collection procedures, there will be only seven pieces of data for each panel that is usually meters in length and nearly one meter in width in the case of Very Large Crude Carriers (VLCC).

Table 1: Summary of pit number in a ship.

Tank No.	0-7.9 mm	8-11.9 mm	12-14.9 mm	15-17.9 mm	18-Above	Total
1 STBD	47	0	0	0	0	47
1 PORT	63	3	0	0	0	66
2 STBD	87	0	0	0	0	113
2 PORT	141	2	1	0	0	144
3 STBD	1366	167	9	4	1	1547
3 PORT	1255	127	8	1	0	1691
4 STBD	166	12	2	1	1	182
4 PORT	140	2	0	0	0	142
5 STBD (Ballast Tank)						
5 PORT (Ballast Tank)						
6 STBD	152	20	0	0	0	172
6 PORT	124	1	0	0	0	125
7 STBD	1247	306	21	4	1	297
7 PORT	1075	251	15	14	0	1355
8 STBD	333	13	7	0	0	353
8 PORT	359	46	0	0	0	1708
	6555	950	63	24	3	7595

While one of the purposes of pitting surveys is to check if any pit has caused perforation (or potential perforation), another important purpose is to assess the residual strength of the pitted plate. If a pitted plate can be shown to have sufficient residual strength to continue service, the plate may not need immediate repair. On the other hand, a heavily pitted plate that is a potential hazard to the safety of the ship will need repair for the ship to continue service. To make a judgment on whether or not a pitted plate has sufficient residual strength, a criterion needs to be developed. An effective (or equivalent) thickness for the pitted plate needs to be defined. This paper presents two probabilistic approaches that have been developed to interpret the practical data format described above and determine the corresponding effective thickness for the panel.

EFFECTIVE THICKNESS

The scope of the first approach will be limited to a small 300x300 mm sample square identified by inspectors, assuming that the sample square is the most representative of the pitting in the plate panel that is surrounded by two transverse webs and two longitudinal stiffeners. A pitted plate inevitably has less strength than the original (undamaged) one because of reduced thickness. An effective thickness can be defined to be the thickness of a thinner non-corroded plate that has the same strength as the pitted plate. Note that the definition of effective thickness varies with failure modes. If the failure mode considered here is yielding in tension (see the following figure), then the effective thickness is the smallest cross sectional area divided by the plate width.

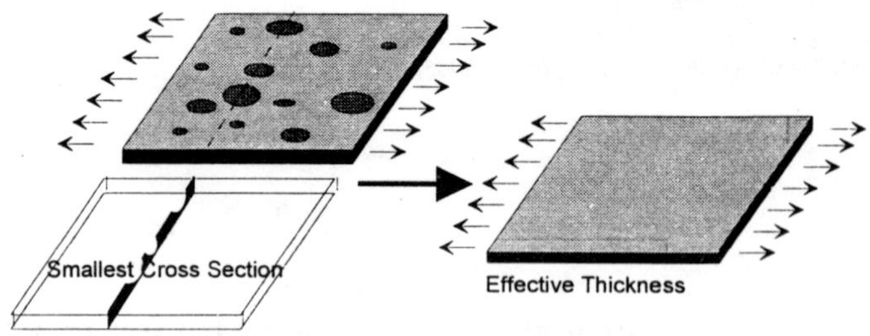

Figure 1: Definition of the effective thickness by equivalent cross section area.

While considering another type of failure mode such as buckling, a different definition of the effective thickness should be considered for the same pitted plate. The new definition has to be verified by experimental results. However, for simplicity, a universal definition of effective thickness will be defined so that it can be used for all kinds of failure modes. One can consider a thickness reduction Δt that is the average thickness loss in the sample square due to pitting. It can be expressed as:

$$\Delta t = \frac{V}{A} \tag{1}$$

where A = the area of the square, that is 300 mm x 300 mm according to the definition of the sample square. And V = the total steel volume loss due to pitting.

Then the effective thickness can be defined as:

$$t_e = t_o - \Delta t \tag{2}$$

where t_O = the original plate thickness. To determine t_e, the unknown V needs to be obtained. Two approaches have been developed to determine the steel volume loss V.

APPROACH USING AVERAGE AND MAXIMUM PIT DATA

To determine the effective thickness for a pitted plate, the volume loss of steel due to pitting, V, needs to be obtained. Ideally, the volume loss can be calculated by taking measurement on each pit and summing up all the pit volumes. The volume loss can be expressed as:

$$V = \sum_{i=1}^{N} c_i a_i d_i \tag{3}$$

where N is the number of pits in the 300x300 mm sample square. The quantities, a_i and d_i, are the area and depth of pit 'i', respectively. The quantity c_i is the cylinder coefficient defined as the actual pit volume divided by the corresponding cylinder of depth d_i and top area a_i. The parameter c_i has the range: $0 < c_i \le 1.0$, in which a value of c_i = 1.0 corresponds to a pit whose shape is actually (perfectly) cylindrical. Assuming that all pits have a semi-spherical shape, the value of the cylinder coefficient can be obtained as:

$$c = \frac{V_{semisphere}}{V_{cylinder}} = \frac{\frac{1}{2}\left(\frac{1}{3}\pi w^2 d\right)}{\frac{1}{4}\pi w^2 d} = 0.667 \tag{4}$$

Due to the large number of pits, it is practically prohibitive to measure all pits in a ship. There could be as many as seven thousand pits for an old tanker (see Table 1), or as high as one thousand in a single tank. In a panel, the number of pits may be several hundreds. Therefore, only a few data can be collected for one panel during a survey. As described previously, the data collected for a plate panel by the current inspection practice is limited to only seven data items. A probabilistic approach has been developed to make the maximum use of the limited data set.

Assuming the pit depth D and the pit width (diameter) W to be random variables following lognormal distributions (See Figure 2). The reason for choosing lognormal distributions to model pit depths and widths is because of its non-negative property and simplicity. Other probability distributions that deal with only positive values include Chi-square, Gamma, Gumbel, Rayleigh, and Extreme Type II. Any of these distributions can be a candidate to be applied to our mathematical model. It should be noted that no prior knowledge is available on which probability distribution produces the best fit for pitting data for ship panels.

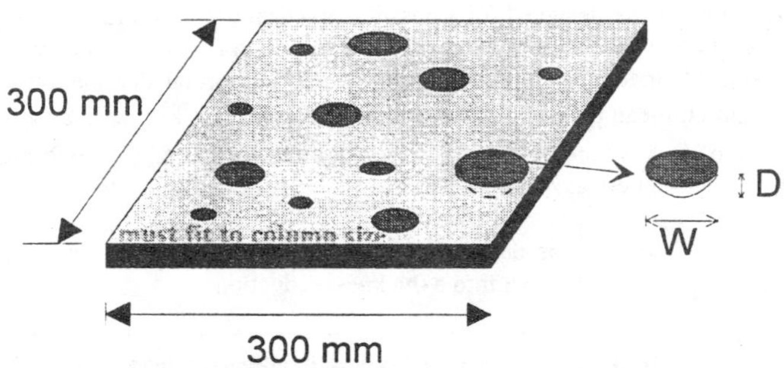

Figure 2: A sample square that is the most representative of the pitted panel will be selected to be inspected.

To determine which distribution better fits pit depth or width data, a sufficient number of pitting measurement on the depth and width of each pit in a field has to be supplied. Unfortunately, such data are expensive to gather and therefore are generally not available. Most of the existing survey reports record only the number of pits and the depth of the deepest pits. Some more detailed survey report listed the size of pits deeper than a threshold depth (for example, 12 mm) in order to monitor and track pit growth. For the shallower pits, no measurements were made mainly because of their insignificance and the large quantity. If a complete set of pitting measurements are available, the corresponding distribution can be determined using standard statistical inferencing, see Ang et al (1975). It is desirable that future work involving measurements from different types of ship tanks should be undertaken. However, in the current work only the lognormal distribution is used. The probability density function of the lognormal distribution is expressed as:

$$f_X(x) = \frac{1}{\sqrt{2\pi}\zeta x}\exp\left[-\frac{1}{2}\left(\frac{\ln x - \lambda}{\zeta}\right)^2\right], \quad 0 \le x < \infty \tag{5}$$

where $\lambda = E(\ln X)$ and $\zeta = \sqrt{Var(\ln X)}$ are, respectively, the mean and standard deviation of ln X, and are the parameters of the distribution. The goal is to determine the two parameters, that is, the mean and the standard deviation of the lognormal distribution. Once the two parameters are determined, a direct simulation can be performed on both pit depths and widths which will yield an estimate of the steel volume loss.

The average depth in a 300 mm x 300 mm square is one of the seven data items that were collected during inspection. Its value should be close to the mean of D. One can simply assume the averaged depth to be the mean, μ_D, of D. Then the only parameter left here to be determined is the standard deviation, σ_D. Note that this assumption may not be appropriate because the collected average depth comes from an inspector's human judgment. The error associated with this assumption may be very significant. This problem will be solved in the second approach that is discussed later in this paper.

Next, the second unknown parameter, the standard deviation needs to be determined. The ideal way of calculating the standard deviation is shown in the following equations:

$$Var(D) = \frac{1}{N}\sum_{i=1}^{N}(d_i - \mu_D)^2 \tag{6}$$

$$\sigma_D = \sqrt{Var(D)} \tag{7}$$

Since most of the pit depths in the sample square are not available from inspection data, σ_D cannot be calculated in this ideal manner. A alternative way of calculating σ_D has been developed by utilizing the extreme value theory.

The total number of pits in the square is available as Frequency N. The N pit depths in the sample square can be considered as a sample of N lognormal random variates. The maximum depth, χ_D, is therefore the extreme (largest) value of the N lognormal random variates. Note that the extreme pit depth in a sample of size N is a random variable with its own standard deviation. Assume that χ_D is the mean value of the extreme pit depth in a sample of size N. These two data together with the assumed mean depth can be used to predict the unknown standard deviation. In most engineering problems, the extreme value is the unknown to be solved (as shown in Equation 8). However, in this case, the extreme value is known, while the standard deviation σ_D is the unknown to be determined.

$$\chi_D = f_N(\mu_D, \sigma_D, N) \tag{8}$$

From Equation (8), the value of σ_D is obtained via an iterative numerical procedure. With a starting trial value of σ_D, a set of N lognormal random variates can be generated. The averaged extreme value of the set can be calculated. The trial-and-error iterations on σ_D are continued until its average extreme value matches the maximum depth. The estimate of the standard deviation σ_D is then determined. With both the mean and standard deviation obtained, the exact distribution function of D is determined. The distribution of the other random variable W can be determined in the same manner.

Based on these distributions, simulations on the pit depths and widths can be performed by generating N lognormal random vectors as $\{(d_1, w_1), (d_2, w_2), ..., (d_N, w_N)\}$ where (d_i, w_i) represents the depth and width of the i_th pit. To generate lognormal random vectors, one can first generate independent standard normal random variates, Press (1989), and then transform the independent standard normal variates to correlated standard normal variates as outlined by Chang et al (1994). There are several methods of transformations. Among those, Rosenblatt transformation and orthogonal transformation are very popular. However, another method which is developed by Rubenstein (1981), and further described by Melchers (1987), is used in this study. Correlated lognormal random variates can be obtained by transforming the correlated standard normal variates.

Before the simulation can be performed, the correlation coefficient, ρ_{DW}, needs to be obtained. It should be noted that the correlation coefficient, ρ_{DW}, between the depth and width is high (close to one), since a deeper pit tends to have a wider diameter. ρ_{DW} is a measure of the statistical dependence between two random variable D and W. It is defined as:

$$\rho_{D,W} = \frac{Cov(D,W)}{\sigma_D \sigma_W}, \quad -1 \leq \rho_{PD} \leq 1 \tag{9}$$

where $Cov(D,W) = E[(D-\mu_D)(W-\mu_W)]$,

E[ ]: expected value,

μ_D, σ_D : the mean and the standard deviation of D,

μ_W, σ_W : the mean and the standard deviation of W.

It would be valuable to gather actual measurements on pit depths and widths, and estimate their correlation coefficient. We searched existing survey reports in the industry for actual measurements, but only found an incomplete set of data from a special survey report. The survey report was aimed at monitoring and recording the size growth of deep pits. Only pits deeper than 12 mm were measured. The report summarizes pit numbers according to tank types and pit depth categories. The summary is reproduced in Table 1. The tanker has a total of 7595 pits. Ninety of them are deeper than 12 mm, and have measurements.

The correlation between the mean widths and depths of the 90 deep pits was analyzed, and found to be about 0.4. This value is lower than expected. The reason for this low correlation is due to the incompleteness of the data set. In order to find out the correlation coefficient for a complete set of data, we reproduced the data of pits shallower than 12 mm by re-scaling the sizes of the available 90 deep pits and distributing them uniformly to the range from 0 mm to 12 mm. The correlation coefficient was calculated again. The result turned out to be a correlation coefficient of 0.9. This number was used in this study. It is also recommended to be used in the pitting program developed in this study.

After ρ_{DW} is determined, a direct simulation can be performed. Equation 3 can be further simplified by assuming a constant cylinder coefficient and round shapes for all pits. Applying the random vectors from simulation, then the wasted steel volume, V, can be computed by the following equation:

$$V = 0.667 \cdot \sum_{i=1}^{N} \left(\frac{\pi}{4} w_i^2 \cdot d_i \right) \tag{10}$$

Note that since D and W are random variables, V is also a random variable. Each run of simulations may produce different values of V. Therefore, sufficient runs of simulation should be performed in order to determine the mean of V. Once V is obtained, the effective thickness can be computed as described in Equations 1 and 2.

Now, an approach has been successfully developed to compute the thickness reduction or the effective thickness. The advantage of this approach is that it utilizes all the pertinent information collected/recorded by surveyors under current practice. This approach has some disadvantages. The approach made a few assumptions that would cause uncertainties in the obtained equivalent thickness. The assumption which has the biggest impact is the one that assumes the average depth and width recorded by inspectors are indeed their true means. The measurement of the average depth and width are taken on a pit based solely on inspectors' human judgment. The variation in selecting the average depth and width by different inspectors can be large. The other sources of uncertainties include the nature of pitting randomness, data collection/recording, and the mathematical model employed.

In summary, care should be taken while using this approach. The accuracy of this approach depends on the assumptions. A computer program, PIT1, has been developed to demonstrate the approach. Using the program, a series of charts on thickness reduction were created for different values of coefficient of variations (COVs). The charts were designed for a possible practical use in the future. One sample chart with a COV value of 0.2 is shown in Figure 3. They were generated under the following assumptions: lognormal distribution of W and D; mean pit width is a function of mean depth as '$\mu_W \cong 2.5\mu_D$'; COV of D & W are both 20%; correlation between D & W is 0.9; pits have semi-spherical shape.

The chart is designed for an easy interpretation from a pit number and a mean depth into a thickness reduction.

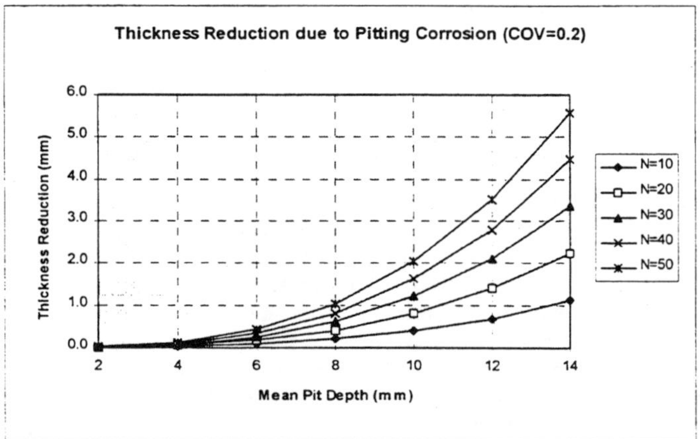

Figure 3: Thickness reduction as a function of mean pit depth and pit number (N = number of pits in a sample square of 300x300 mm^2)

APPROACH USING THE R DEEPEST PITS

In order to overcome the difficulty associated with the potential large variation of selected average depth and width, another approach has been explored. This second approach uses the number of pits and the depths of the r deepest pits as inputs to predict the average depth and width of the raw sample data. A similar concept of using the r deepest pit depths has been explored by Scarf (1992). The study had a different objective which was to extrapolate extreme pit depths in space and time.

In this second approach, inspectors do not have to define the most representative sample square or the average pit depth/width. Instead, the pit depths of the r deepest pits in the plate panel are measured and recorded. This approach minimizes the human judgment involved in pitting data collection.

Consider a plate with M pits. The plate for consideration could be either a 300x300 mm sample square or the whole plate panel bounded by longitudinal stiffeners and transverses webs. The first approach was demonstrated on a sample square, because the current practice collects data mostly on that basis. In the second approach, the entire plate panel will be used for demonstration. The approach can be applied to plate panels of any size. Since the considered area is larger than the one in the first approach, the value of M should be bigger than the value of N.

The depths and widths of the M pits can be denoted as a set of lognormal random vectors $\{d_1, d_2, ..., d_M\}$ and $\{w_1, w_2, ..., w_M\}$, assuming they are both lognormally distributed. The goal, again, is to find out the total wasted steel volume as:

$$V = c \cdot \sum_{i=1}^{M} \left(\frac{\pi}{4} w_i^{\,2} \cdot d_i \right) \tag{11}$$

The set of lognormal random vectors can be sorted in an ascending order so that (d_1, w_1) and (d_M, w_M) are the smallest and the largest respectively. The r largest pit depths and widths that will be recorded during inspections are known as $\{d_{M-r+1}, d_{M-r+2}, ..., d_M\}$ and $\{w_{M-r+1}, w_{M-r+2}, ..., w_M\}$. Assuming that the values of the COVs of D and W in any plate panels are the same, and are both fixed values, the means of the pit depth and width can be estimated.

By trial and error on the mean of D, a mean extreme (largest) value of the M pit depth can be found to be x_M. Continuing trial and error on μ_D until x_M matches d_M. This μ_D has been obtained as one estimate of the true mean pit depth. To better explain this approach, one can consider an extreme value as a function of the mean, the standard deviation (or COV) and the total number of samples:

$$x_M = f_M(\mu_D, \sigma_D, M) \tag{12}$$

which can be rearranged in the form:

$$\mu_D = g_M(d_M, \sigma_D, M) \tag{13}$$

in which x_M has been replaced by d_M. Equation (13) can be solved by an iterative procedure.

The obtained estimate of μ_D together with the assumed COV will determine the parameters of the lognormal distribution. A similar simulation can be performed on D as well as W. Then the wasted steel volume can be computed by using Equation 11.

The estimate of μ_D may be a bad estimate, if the input, d_M, is not close to the mean of the extreme value. This is often the case. Thus more extreme data should be used to make a better estimate of μ_D. The second extreme data can be used to predict the second estimate on μ_D. The function can be expressed as:

$$\mu_D = g_{M-1}(d_{M-1}, \sigma_D, M) \tag{14}$$

Using the similar procedures on all r extreme input data, r estimates of μ_D can be computed, and an overall better estimate can be obtained by averaging.

In order to demonstrate the approach and to illustrate the accuracy improvement by using more inputs, five tests were performed. Test #1 used only the deepest pit depth as input data. The test was conducted by generating 20 correlated lognormal random vectors to represent the depths and widths of 20 pits in a plate. The actual volume loss in the plate was calculated as V_{actual} using Equation 11. Then this approach was used to make predictions on V_{actual}. The largest depth and width were sorted out. They are used as inputs to determine the μ_D and μ_W.

Based on the μ_D, μ_W and assumed standard deviations, a simulation on lognormal random variates can be performed. Pits were generated to simulate the actual 20 pits in the plate. The predicted volume loss was calculated as $V_{predicted}$. To find the goodness of the prediction, a prediction error, R, is defined as:

$$R = \frac{V_{predicted} - V_{actual}}{V_{actual}} \tag{15}$$

R is also a random variable and should have a mean value of zero. Each run of the test produced a different error. Some of the errors can be higher than 100%. In each test, one-thousand runs were performed. The prediction error R in Test #1 is summarized in a histogram (Figure 4):

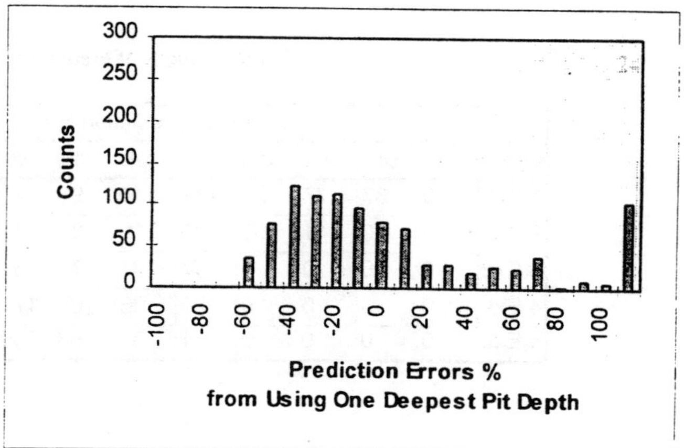

Figure 4: Histogram of prediction errors in percentage using one pit depth/width.

It can be seen that R falls mostly in the range from -60% to 100%. The errors are very large. More than a hundred of the errors in the 1000 runs are beyond 100%. This indicates that the prediction based on one extreme pit depth and width can not produce a reliable steel volume loss.

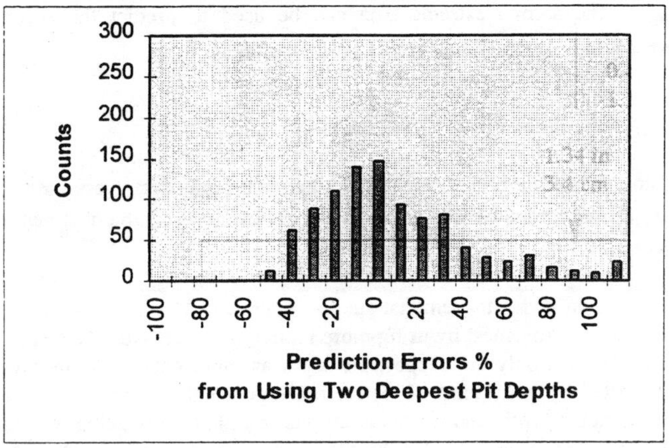

Figure 5: Histogram of prediction errors in percentage using two pit depth/width.

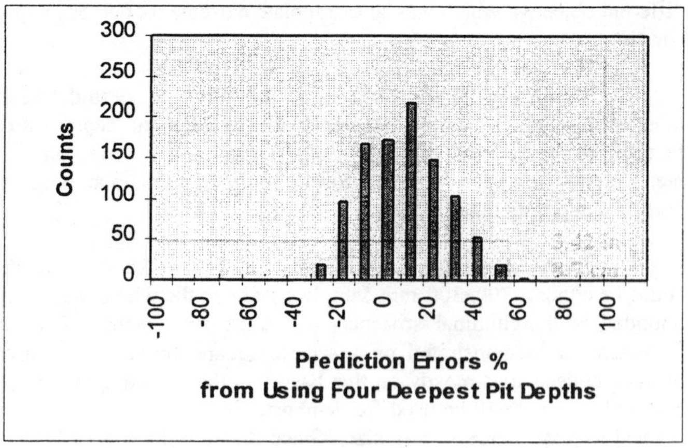

Figure 7: Histogram of prediction errors in percentage using four pit depth/width.

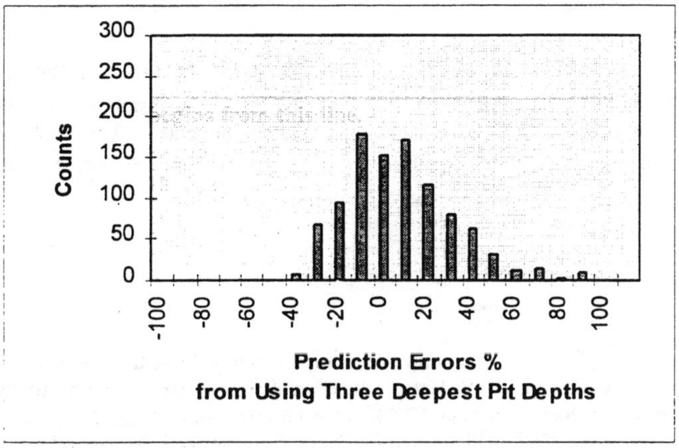

Figure 6: Histogram of prediction errors in percentage using three pit depth/width.

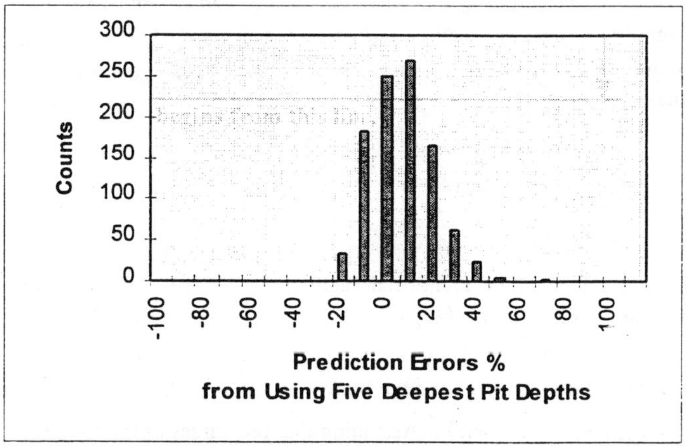

Figure 8: Histogram of prediction errors in percentage using five pit depth/width.

Table 2: Counts of Prediction Error Percentages for the five analyses.

			Counts of Prediction Error Percentages for Five Analyses															
Err %	-70%	-60%	-50%	-40%	-30%	-20%	-10%	0%	10%	20%	30%	40%	50%	60%	70%	80%	90%	100%	
1 Extr	0	37	77	122	111	114	96	79	73	29	28	20	27	24	39	3	10	8	103
2 Extr	0	0	11	62	90	110	139	146	94	78	81	41	30	24	31	16	13	10	24
3 Extr	0	0	0	8	67	94	179	153	171	116	79	64	31	12	14	2	9	1	0
4 Extr	0	0	0	0	20	96	166	171	218	148	103	53	20	3	1	1	0	0	0
5 Extr	0	0	0	0	1	34	183	250	270	165	63	25	6	0	3	0	0	0	0

Other four tests were performed by using the 2, 3, 4, 5 deepest pit depths and widths. The objective of the tests was to demonstrate that more input data would bring about a better prediction. Each test consists of 1000 trials. The prediction errors of the 1000 trials were again plotted on histograms (Figure 5 - 8). The raw data of the five tests are listed in Table 2.

From these figures, one can see that the prediction error R decreases when more extreme data is used as input. Test #5 which used five extreme pit depths/widths gave a very good prediction of the steel volume loss. The errors that lay in the range from -25% to 25% consists of 902 trials. In other words, 90.2% of the trials gave reasonable predictions. The tests showed a reduced trend on errors while using more extreme pit depths/widths as inputs. This indicates that this approach could be a useful method for a practical application on pitting data collection strategy in the future.

To further analyze the prediction errors in the five tests, a mean and a standard deviation of prediction error R can be calculated. The mean should have a value of zero, or it is biased. The means of the five tests are all slightly above zero. However, it approaches zero while more extreme pit depths/widths are used as inputs. The standard deviation also shows a decreasing trend thus indicating that better prediction of steel volume loss will be made when more inputs are provided.

In summary, an alternative approach has been developed. The approach eliminates the uncertainty caused by improper human judgment of selecting the most representative average pit depth in the first approach. The weakness of the second approach is that it still requires a computing facility to calculate the steel volume (and thus the thickness reduction). Another disadvantage of the second approach is that it may require five or more sets of input data to produce a reliable prediction. A computer program has also been developed to demonstrate and apply the approach.

CONCLUSION

Two probabilistic approaches have been developed to characterize the steel plate thickness reduction due to pitting corrosion. Both approaches assume pit widths and depths as correlated random variables following certain probability distributions. The first approach further assumes the sampled averaged pit depth and width to be the true mean depth and width. Values of thickness reduction versus mean pit depth and pit number were made into curves in a chart.

The second approach simulates the pitting by using the r deepest depths. This approach is more realistic, because it directly predicts the mean values of the depth and width instead of using the data collected by inspectors. A prediction error is defined and its mean and standard deviation are plotted against r. The plot result shows that when more data on the deepest depths are inputted (i.e., r is larger), the prediction on thickness reduction becomes more accurate.

Both approaches are more sophisticated than the current corrosion inspection practice which is based on inspectors' human judgment. However further verification based on real field inspection data may be needed before the approaches can be applied in residual strength assessment of pitted plates in marine structures. Further work is also underway to apply more rigorous random field models for the effective thickness characterization of pitted panels. Clearly, experimental strength data will be needed to validate the prediction of these models.

ACKNOWLEDGMENT

This work is part of a project supported by the Ship Structure Committee under Contract no. SR-1356. The authors wish to express their appreciation to its support. Also thanks to Maria Ximenes of Chevron Shipping for her technical advises and data input.

REFERENCES

Ang, Alfredo H-S.(1975), and Tang, Wilson H., "Probability Concepts in Engineering Planning and Design - Volume I, Basic Principles", John Wiley & Sons.

Chang, Che-Hao(1994), Tung Yeou-Koung and Yang, Jinn-Chuang, "Monte Carlo Simulation for Correlated Variables with Marginal Distributions", ASCE, Journal of Hydraulic Engineering, Vol. 120, No. 3.

Melchers, R.E.(1987), "Structural Reliability - Analysis and Prediction", Ellis Horwood Limited, John Wiley & Sons Press.

William H. and others(1989), "Numerical Recipes - The Art of Scientific Computing (FORTRAN Version)", Cambrige University Press.

Rubinstein, Reuven Y.(1981), "Simulation and the Monte Carlo Method", John Wiley & Sons.

Scarf, P. A., Cottis, R. A., and Laycock, P. J.(1992), "Extrapolation of Extreme Pit Depths in Space and Time Using the r Deepest Pit Depths", J. Electrochemical Society, Vol. 139, No. 9

TSCF(1986), "Guidance Manual for the inspection and condition Assessment of Tanker Structures"

Proceedings of the Seventh (1997) International Offshore and Polar Engineering Conference
Honolulu, USA, May 25-30, 1997

ISBN 1-880653-28-1 (Set); ISBN 1-880653-32-X (Vol. IV)

Integrity Up-Diving Down

M. Visser
Nederlandse Aardolie Maatschappij
Velsen, The Netherlands

ABSTRACT

The structural inspection cost of NAM's offshore installations has been systematically reduced by ca. 45 % since 1974, the year where NAM started their offshore operations.
Further potential cost reductions of ca. 19 % are identified by stretching the underwater inspection interval and changing of the contract strategy.
At the same time the level of confidence in the structural integrity of NAM's offshore installations improved as well.
Furthermore the spin-off of enhanced offshore safety to the workforce is discussed.

KEYWORDS

Structures, offshore, underwater, inspection, diving, contracts.

INTRODUCTION

The first NAM platform was installed in 1974 in the Dutch sector of the North Sea and since then the number of platforms has steadily increased to 30. Structural inspection costs of platforms could have increased proportionally with the number of platforms. However in the eighties NAM carried through a substantial rationalisation in their structural inspection program whilst maintaining the required level of confidence in the structural integrity of these platforms.
Since 1992 additional improvements have been achieved, mainly by research.
Further statistical and engineering work started in 1995, which has recently been completed successfully with the approval of the Certifying Authority Lloyd's and the Dutch Mining Authority S.o.d.M. (State Supervision of the Mines).
As a spin-off safety has been enhanced as the exposure of people to safety critical activities such as diving has been reduced.

PREVIOUS IMPROVEMENTS

Rationalisation of the structural inspection program of the NAM platforms has been achieved in the eighties in two steps.:
The first improvement step was carried out in 1982, when it was decided to review the inspection philosophy and program for the NAM offshore structures based on the inspection findings of the preceding years. The analyses of the date justified the increase of the interval for general underwater visual inspection of each platform from one to two years, .

The second improvement step took place in 1987. This time the cost reduction was basically achieved by structural redundancy considerations and improvement of the lifetime expectancy of fatigue sensitive welds as a result of updated fatigue analyses, carried out at that time.

These two improvement steps resulted in a cost reduction of ca. 30 % (figure 6).

RECENT IMPROVEMENTS

Objectives

In 1992 the structural inspection program for the NAM platforms for the years 1988 -1992 was nearing completion and a new five year cycle needed to be planned. The objectives of the update was to contain total inspection efforts with the aim :

- to reduce the annual inspection cost with approx. 15 %, whilst maintaining or improving the level of confidence in the structural integrity of the NAM platforms,
- to enhance the safety by reducing the exposure time to diving,

Approach

To come to a clear identification of cost reduction opportunities and improvement on safety a fishbone diagram was used (figure 1).

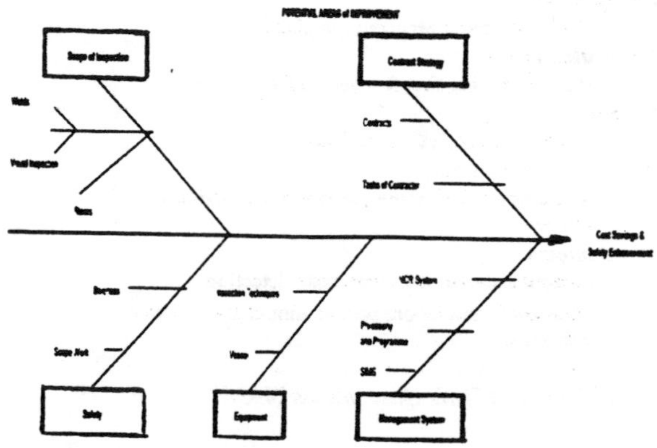

Figure 1 : Potential Areas of Improvement

The areas of improvement will be discussed hereafter.

Reduction of Scope of Inspection

The above water inspection of the structures is less than 10% of the total inspection cost and will not be further discussed in this paper. An analysis of the cost drivers of the under water inspection was made in 1992 (figure 2)

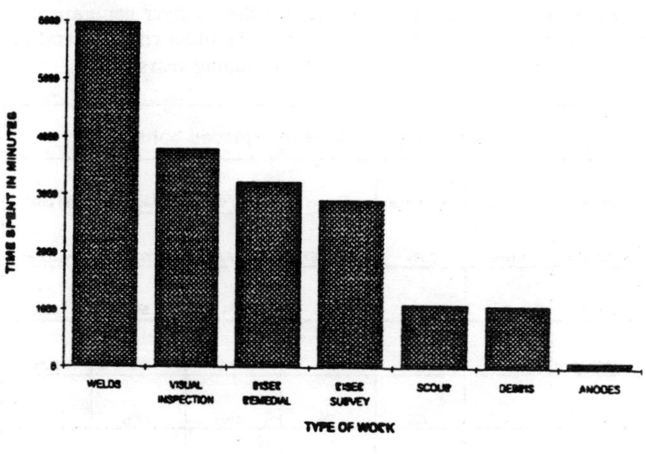

Figure 2: Breakdown of Underwater Inspection Time 1992

As can be seen from the figure 2 the two main cost drivers are

- the inspection of fatigue sensitive welds,
- the general visual inspection

The associated improvements will be discussed hereafter.

Fatigue sensitive welds

SHELL research developed an advanced inspection program for fatigue sensitive welds. This program is based on probabilistic techniques combinde with fracture mechanics. An important feature if this tool is that with feed-back from inspection results it is possible to update the probalbility levels of failure and the lifetime expectancy of these welds. Figure 3 demonstrates the further reduction of the number of underwater welds to be inspected as the result of this effort. An additional advantage of this program is that the level of confidence in the structural integrity of the NAM installations was enhanced.

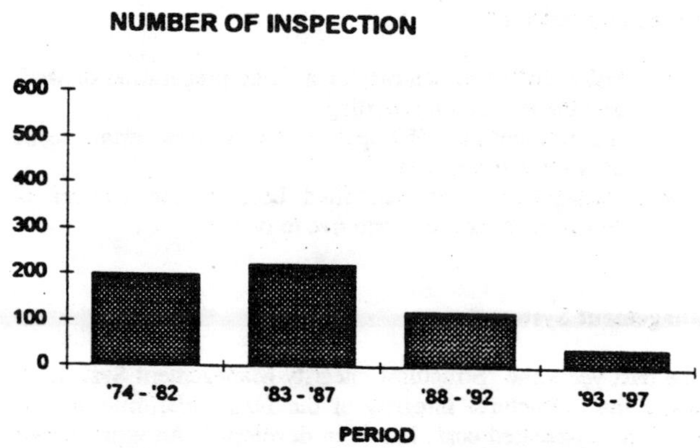

Figure 3: Number of Underwater Weld Inspections

General visual inspection.

The analysis of the inspection data revealed that cathodic protection and also scouring stayed within the original design envelope of the platforms. Occasionally the marine growth exceeded the allowable thickness. Re-analyses proved that this extra marine growth could be easily accommodated by the relevant platforms.
Similarities in the design and construction of the NAM platforms allows inspection to be carried out on a representative sample of platforms rather than of a full inspection of all platforms.

Safety

Opportunities were identified to shift inspection tasks from divers to R.O.V., particularly for general visual inspection and cathodic protection measurements.
Furthermore it became apparent that safety would be enhanced by the reduced number of underwater welds to be inspected. This was achieved by the introduction of the advanced inspection program for fatigue sensitive welds, mentioned above. The reduction in diving exposure time was substantial: 32 %.

Equipment

Improved inspection techniques e.g. cleaning methods and replacement of photo's by video were introduced. The alternative of using a more expensive and less weather sensitive semi-sub was considered but rejected on economic grounds.

Contract Strategy

In 1992 the underwater operations were covered by a total of five contracts, which would expire in 1993. Modest cost savings were achieved as a result of

- tasks shifted to contractors such as preparation of work and the inspection reporting,
- the discount of tariffs applicable in case of certain scope of work volumes and
- management was simplified because the number of contracts is reduced from five to two.

Management System

In the past years the "Structural Integrity Management System" to manage the structural integrity of the NAM platforms and to control the associated costs has been developed. An extra reason for doing so, is that such a system would also be beneficial towards Lloyd's and S.o.d.M. to demonstrate evidence of adequate control of the structural integrity of the NAM platforms. This system comprises:

1. the annual issue of an updated structural inspection philosophy and program for the next five year period ahead. The annual update has a built in feed-back loop of the inspection findings.

2. the monitoring and control of observed structural deficiencies by so called "Non Conformity Reporting" documentation. This constitutes the evidence of the control of the structural integrity of the NAM platforms and cost of deficiencies. Monthly reports and status lists are issued and discussed within the engineering department. These discussions result in corrective actions in consultation with Lloyd's.

Cost Reduction

The total cost reduction of 15% is largely due to reduction in the number of under water welds to be inspected and to a lesser extend to the improved contract strategy.

FURTHER IMPROVEMENTS

In 1995 it was realised that further improvements could be achieved by looking into technical and commercial aspects both.

5 Year under water inspection interval

General Visual Inspection

A special study was undertaken to investigate a data base on North Sea experience with platforms deficiencies. It covers 3850 platform years, whereas NAM exposure is 200 years. The main causes for defects are listed in figure 4, which also shows the position of NAM with respect to these causes.

Conclusions North Sea database analysis

- **design faults**
- NAM: simple, standard structures
- **dropped objects**
- NAM: recording procedure in place
- **fabrication faults**
- observed during first few years of operation
- **fatigue**
- subject to a separate investigation
- **pipeline pulling/pushing**
- probability small, consequences most onerous on pipeline
- **vessel impact**
- manned installations: immediate detection
- unmanned installation: part of annual above water inspection

Figure 4: North Sea Data Base versus NAM

The conclusion was that the 2 year interval for general visual underwater inspection could be extended to 5 years, without jeopardising the confidence level in the structural integrity of the offshore installations.

Risers

The second most time-consuming underwater activity is the inspection and repair of bolts of riser and J-tube clamps. So a statistical analysis was made of NAM's experience with the replacement of bolts. The outcome is interesting (figure 5). It shows that bolt replacement of the older riser generation, up to 1988, has not been effective, contrary to the performance of the younger generation of risers (as from 1988). Based on these assessments the inspection intervals for the younger generation of risers could be increased to 5 years. For the older riser generation a remedial program is scheduled for the coming years.

Annual percentage of replaced bolts

year of installation	bolted risers	number of bolts	annual percentage of bolts replaced			all bolts replaced
			1994-87	1988-91	1992-95	
1976-79	7	228	17%	6%	5%	28%
1980-83	3	132	5%	12%	5%	22%
1984-87	8	328	5%	1%	0.1%	5%
1988-91	17	594	-	0.4%	0.1%	0.5%
1992-95	5	508	-	-	0%	0%
total	40	1790	3%	2%	1%	6%

(riser = riser of J-tube)

Figure 5: Performance of Replaced Riser and J Tube Bolts

The expected cost reduction due to the introduction of the 5 year inspection interval is ca. 11%.

Bench marking

In 1996 a bench marking of underwater operations and the structural integrity process was completed of three Operators in the Southern part of the North Sea. The results of the past two years were compared and the outlook for the coming 5 years. The conclusion was clear: NAM could do better. This prompted NAM to look for improvements and to cut out chronic waste in the under water operations and structural integrity processes.

Partnering

To achieve these goals a project is now progressing on partnering. In essence this means that the current contractors and NAM both strive for a win-win situation with the aim to become "best in class". Our belief is that a cost reduction of 15% is possible.

Cost Reduction

The cost reductions of the underwater inspection over the past and coming years are shown in figure 6.

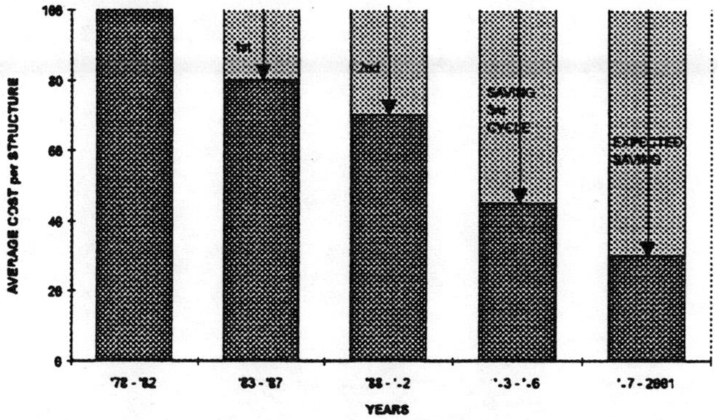

Figure 6: Cost Reduction Under Water Inspection

Structural Inspection Schedule

The structural inspection schedule for the under water part of the offshore installation is shown in figure 7.

NAM STRUCTURAL INSPECTION SCHEDULE		
Inspection Item	First inspection after installation	Intervals
general visual	year 1, 3 & 5	5 years (after 5 years)
NDT inspection	--	platform specific
debris	0	5 years and before any diver operation
scour	year 1, 3 & 5	5 years (after 5 years)
marine growth	--	5 years for appropriate structures
CP potential readings	1 year	5 years
anode dimensions	--	5 years
U/W inspection bolted clamps (risers and J-tubes)	--	old generation 2 years when modified 5 years new generation 5 years
visual inspection caissons	--	2 years
visual insp. exterior of risers	--	5 years

Figure 7: Structural Inspection Schedule

DISCUSSION WITH LLOYD'S AND DUTCH AUTHORITIES S.o.d.M.

This inspection schedule, based on a 5 year underwater inspection interval, was presented in 1996 to Lloyd's and the Dutch Authorities for their approval.

Lloyd's

Lloyd's were willing to accept NAM's line of thinking, provided that we could demonstrate that the offshore installations have adequate reserve strength at relevant damaged conditions. In co-operation with Lloyd's it was decided to select two representative platforms and to calculate the reserve strength. Push over analyses have been performed by Lloyd's on two older structures: a four pile platform (1977) and a six pile platform (1982). The main reasons to select these two platforms were:

- both platforms are manned.
- the original design shows high stress levels.
- those older platform have no properly documented ship impact analysis.

For each of those two platforms the intact and a damaged condition have been considered. For the damaged case a brace under water and below the Living Quarters was assumed to be missing. The outcome demonstrates that the reserve strength at damaged condition is more than adequate and is hardly effected by the missing brace.(figure 8).

RESERVE STRENGTH RATIO			
Condition	Norm	4 Pile Platform	6 Pile Platform
Int_ct	> 1.6	1.96	2.62
D_m_ged	> 1.6	1.94	2.5
Reserve Strength R_tion =	Ulitm_te Pl_tform Resist_nce / Design Lo_d		

Figure 8: Results Push Over Analyses

Dutch Authorities S.o.d.M.

Early 1996 discussions started with S.o.d.M. on NAM's philosophy of the structural integrity of the offshore installations. Special attention was paid to the technical justification, in which the encouraging results of the push over analyses of the four and six piles installations were crucial. It was also felt that the current Dutch prescriptive legislation will change over in due course to a goal setting legislation. NAM received concurrence from S.o.d.M. in February 1997.

Furthermore these discussions confirmed that the Safety Case Reports of the Installations are the key documents to manage the safety and structural integrity of the offshore Installations. So an update of these reports is essential to reflect the line of thinking with respect to the hazards in the structural sphere.
Enhanced attention was paid to the quality and the Structural Integrity Management System to safeguard proper execution of the works and to demonstrate an adequate confidence level in the structural integrity of the offshore installations. This resulted in following actions by NAM:

- development of a formal NAM policy for underwater inspections
- update NAM's Structural Integrity Management System.
- update the Safety Case Report of all Installations
- update the quality system for the execution of the underwater inspection work including internal audits.

CONCLUSION

The effort to reduce the underwater inspection cost, whilst improving the confidence in the structural integrity of offshore structures took many years of team work.

Dedication, creative thinking, sustained effort and constructive support from the Certifying Authority Lloyd's and the Dutch Authority S.o.d.M. proved to be essential.

The main results are

1. a 45 % cost reduction for the annual inspection is realised, compared to the 1992 cost level.
2. further cost savings of 19 %, compared to the 1992 cost level, are achievable.
3. an enhanced level of confidence in the structural integrity of all NAM platforms and
4. a reduction of safety critical diving activities by ca. 32 %.

ACKNOWLEDGEMENT

The author thanks the NAM for the opportunity to present this paper. Also he would like to thank his colleagues for their encouraging support.

Proceedings of the Seventh (1997) International Offshore and Polar Engineering Conference
Honolulu, USA, May 25-30, 1997

ISBN 1-880653-28-1 (Set); ISBN 1-880653-32-X (Vol. IV)

Risk-Based Decision Analysis for the Protection of Marine Pipelines from Dropped Objects

S. Yasseri
Haliburton Brown & Root Engineering Ltd
Surrey, United Kingdom

ABSTRACT:
The major threat to the spool piece of a pipeline, close to the platform, is from dropped objects. It is generally agreed that the spool piece should be designed so that it does not incur significant damage from the impact of small or moderate size dropped objects, and it will not rupture when impacted by major objects. This principle clearly implies that the risk of future loss must be balanced against the initial cost of providing stronger protection. As a result it is necessary to have a more explicit procedure for balancing cost and risk.

This paper describes a risk-based decision analysis procedure for the protection of pipeline spool pieces close to platforms from dropped objects. To illustrate the procedure, a pipeline near a typical North Sea platform is considered. The primary aim of the paper is to present and illustrate the procedure; however, some design conclusions, concerning design requirements applicable to the pipeline protection, are indicated.

1. INTRODUCTION

Producing a well-engineered solution to the problem of protecting pipelines against dropped objects involves the synthesis of many kinds of information. There is a requirement to analyse statistical information on the type and mass of objects dropped into particular areas of the sea around different offshore structures. There is also generally a choice of measures that can be used to protect the pipeline. These measures range from concrete coating (which is often required to prevent flotation) to very extensive pre-fabricated tunnels.

For each of the alternative protective measures, the associated total risk of pipeline failure needs to be assessed. This total risk is a composite of the various levels of threat from falling objects, and the uncertainty regarding the location of drop, current and finally the response of the pipeline. The choice of protective measure demands a weighing of costs, risks, and benefits and must also include any characteristic which leads to economic or other loss

This paper describes the decision process in choosing appropriate protective measures to protect pipelines against dropped objects in the vicinity of a platform. The implications and the underlying principles of each part of the decision process are also presented.

2. ANALYSIS OF TOTAL RISK

The calculation of the total risk for a particular protection scheme is based on the combination of two statistically varying characteristics. The first is the response of the particular protective scheme, which may be better or worse than expected. The second is the size and type of the dropped object, a factor that is obviously subject to statistical variation.

The determination of the total risk is in contrast to the usual design process undertaken in the design of the pipeline. In that case, the designer calculates the maximum (or equivalent) hoop stresses for a given pipe thickness because it is correlated with structural failure and economic loss. For the dropped object case, however, the type of response of interest is a dent or rupture in the pipe. A specific dent or a rupture may be caused by either an impact from a moderately sized falling object coupled with an unexpectedly bad response of protective system, or as a result of impact of an unusually large object combined with better than expected behaviour of the protective system. If both of these input-response combinations have the same net result in terms of the state of the pipeline (i.e. the size of dent or rupture), then they are considered equivalent within the selection process. That is in the analysis of the dropped object it is not necessary to distinguish further between any combinations that have the same result on the pipeline.

In estimating the risk that a particular level of response will be experienced by the pipeline's protective measure, it is necessary to determine the total probability associated with all possible combinations of impact due to falling objects and the deviations in the response of protective measure.

The total risk is determined by:

$$P[R_i] = \sum_{all\,j} P\left[R_i \middle| E_j\right] P\left[E_j\right]$$

in which $P[\bullet]$ represents the probability of the event indicated within the brackets, R_i denotes the event that the state of the pipeline is i, E_j indicates that the impact energy experienced is "level" j, and $P\left[R_i \middle| E_j\right]$ denotes the probability that the state (response) of the pipeline will be R_i given that the event E_j takes place.

In the context of protection of marine pipelines, R_i is the worst post-impact state (or state of damage, or response, or loss) of the pipeline during its design lifetime. E_j represents the entire range of possible objects which could fall and impact the pipeline during the lifetime of the platform. In this case, R_i is a unique description of the pipeline condition after each event.

The damage state which is assumed here to be the cost of repair or replacement may be represented in discrete form. Five general state of damage, R_i, were defined in terms of such loss-related factors as level of repair cost, degree of the pipe wall and concrete coating damage, the fraction of personnel injured and killed, etc. This discrete definition of damage states is described briefly in Table 1. Its use will be illustrated more fully later in this paper.

DAMAGE STATE AND SYMBOL	DESCRIPTION OF DAMAGE	INJURIES	FATALITIES
1. Negligible (N)	Minimal loss of asset	None	None
2. Marginal (M)	Single component damage	Potential	None
3. Substantial (S)	localised, but possibly severe damage	Few	None
4. Total (T)	Severe damage to parts of platform	Several	Up to 20
5. Catastrophic (C)	Widespread platform damage	Severe	Greater than 20

Table 1. Description of damage states

3. STATISTICS OF DROPPED OBJECTS

In the UK sector of the North Sea, detailed information on all crane accidents has been required to be reported for some years. This reporting requirement has become even more comprehensive in recent years. Table 2 shows statistics of crane accidents reported for UK fixed offshore platforms during 1980-1986 (Ref. 5). The data contained in this table is not intended to be exhaustive and is only reproduced here for illustration purposes. A distinction is made in the data between objects dropped onto the deck and those dropped into the sea.

DROPPED OBJECTS	ON DECK	OVER-BOARD	TOTAL	% OF TOTAL
Items up 1 tonne	11	9	20	22
Containers up to 5 tonnes	18	7	25	28
Crane Block	3	0	3	3
Crane test weight	1	2	3	3
Crane boom	3	3	6	7
Bundled pipes/casings 5-10 tonnes	9	1	10	11
Drill collar/casings	8	3	11	12
Hydril preventor	3	0	3	3
Riser/conductor parts	0	2	2	3
Mud pump	1	0	1	1
Well slot cover plate	2	0	2	3
Winch	0	1	1	1
Cable drum	0	1	1	1
Diving bell	0	1	1	1
Life boat	0	1	1	1
TOTAL	60	31	91	100
% OF TOTAL	66	34	100	

Table 2 Dropped objects in the British sector of North Sea 1980 to 1986

The above data has been obtained from a total of 825 crane-years. The adjusted frequencies of dropped objects per 100 crane years are given in Table 3. The data presented in this table was obtained by distributing the frequency of dropped objects between 34% over-board and 66% on deck for each type of object, irrespective of their actual recorded place of fall, see Table 1.

DROPPED OBJECTS	ON DECK	OVER-BOARD	TOTAL
Items up to 1 tonne	1.6	0.82	2.42
Containers up to 5 tonne	2.0	1.03	3.03
Crane Block	0.24	0.12	0.36
Crane test weight	0.24	0.12	0.36
Crane boom	0.48	0.25	0.73
Bundled pipes/casings 5 -10 tonnes	0.8	0.41	1.21
Drill collar/casings	0.88	0.46	1.34
Hydril preventor	0.24	0.12	0.36
Riser/conductor parts	0.24	0.12	0.36
Mud pump	0.08	0.04	0.12
Well slot cover plate	0.24	0.12	0.36
Winch	0.08	0.04	0.12
Cable drum	0.08	0.04	0.12
Diving bell	0.08	0.04	0.12
Life boat	0.08	0.04	0.12
TOTAL	7.36	3.77	11.13

Table 3 Adjusted frequency of dropped objects per 100 crane-years

4. IMPACT ENERGY

As a falling object moves through the water, the gravitational force is opposed by the drag force acting on the object. Terminal velocity is achieved when the gravitational force equals the total drag force opposing the object motion. For all objects listed in Table 1, the terminal velocity is reached within 30 metres of striking the water surface. Since this is much less than the water depth, all objects will achieve terminal velocity. Consequently, the velocity of the object at the water surface is irrelevant.

Gravitational force $F_g = mg$

Drag force $$F_d = \frac{1}{2}\rho C_d A V^2$$

where

m = mass of the falling object

g = gravitational acceleration

ρ = density of water

C_d = drag coefficient

A = cross sectional area of object

V = terminal velocity

Equating the gravitational and drag forces, $F_g = F_d$, gives:

$$mg = \frac{1}{2}\rho C_d A V^2$$

Hence

$$V = \sqrt{\frac{2mg}{\rho C_d A}}$$

Typical drag coefficients for the following objects are as follows:
bluff objects: 1.5 to 2.0
complex shapes and transverse tubulars: 0.7 to 1.5
spherical objects: 0.6 to 2.0
The impact energy, E, and terminal velocity of different dropped objects have been calculated by $E = mV^2/2$ with the results being noted in Table 4.

DROPPED OBJECTS	Terminal Velocity m/s	Impact Energy-$kN.m$
Items up to 1 tonne	2.8	3.9
Containers up to 5 tonnes	2.8	19.6
Crane Block	7.7	103.8
Crane test weight	12.2	372.1
Crane boom	2.5	31.3
Bundled pipes/casings 5-10 tonnes	5 Horizontal trajectory 33 Vertical trajectory	37.5 544.5
Drill collar/casings	5 Horizontal trajectory 33 Vertical trajectory	125 1210
Hydril preventor	10.3	265.2
Riser/conductor parts	3 Horizontal trajectory 14 Vertical trajectory	62.5 588
Mud pump	8.7	150
Well slot cover plate	5.5	38
Winch	10.0	200
Cable drum 18 tonne	9.5	812
Diving bell		
Life boat		

Table 4 Typical terminal velocities and impact energies

5. CLASSIFICATION (BANDING) OF IMPACT

The type of dropped objects, their shape and mass, and the terminal velocities are generally uncertain quantities. Many objects are grouped together for convenience which further adds to uncertainty. However, a continuous description of the full range of possible damage is neither necessary nor useful. There is a minimum level of energy below which the dropped object will not cause significant damage. On the other hand, when the impact energy exceeds a certain maximum level, then the damage is so extensive that any additional increase in impact energy can not aggravate the situation. It is appropriate to divide the range of energies into bands based on the level damage each would cause to the pipeline and its protection. Table 5 gives an example of such banding or classification. This classification is suitable for the example problem given later. The example pipeline is a 12 inch pipe with 45 mm concrete coating. This arrangement gives sufficient protection for impact energies up to 55 kJ.

6. EXAMPLE

The example is an typical North Sea platform with drilling and accommodation facilities in 110 metres of water with 12" (323.9 mm outside diameter) pipeline exporting oil from the platform. The pipeline wall thickness is 12.7 mm and the material is of API-X60. There are two purpose-built pedestal cranes on the platform. The maximum lift radius for the south crane (2) is 45 m and for the north crane(1) is 39 m. Their minimum lift radii are 7.5 m and 7.0 m respectively.

The type of objects dropped and frequency of such objects being transported are assumed to be equivalent to an average platform. The type of objects dropped into the sea and their adjusted frequencies for the example platform are given in Table 6. In this example, it is assumed that approximately 60% of all containers are 7.5 m^3 (2-3 tonnes), and the remaining 40% are 10 m^3 (4-5 tonnes).
By combining tables 5 and 6, the frequency of dropped objects in each impact energy band were obtained and are recorded in Table 7.

Energy Band $kN.m$	Criteria for banding
up to 55	This is a lower bound to the level of impact energy a pipeline with 40 mm concrete coating can take without sustaining damage
55 to 110	This is an upper bound to the level of impact energy a pipeline with 40 mm concrete coating can take without sustaining damage
110 to 200	A convenient band. With some additional protection measure above the basic concrete coating, a pipeline can be protected against this level.
200 to 400	A convenient band, but more substantial protection measure above the basic concrete coating would required for this level of impact energy.
400 to 800	This is an upper bound on the capacity of a typical protective scheme such as tunnel type protection
More than 800	This level of impact energy will require as extensive protection measure.

Table 5 Criteria for banding of impact energies.

DROPPED OBJECTS	FREQUENCY (per 100 crane-years)
Items less than 1 tonne	0.82
Containers 2-3 tonnes	0.62
Containers 4-5 tonnes	0.41
Crane Block	0.12
Crane test weight	0.12
Crane boom	0.25
Bundled pipes/casings 5 -10 tonnes	0.41
Drill collar/castings	0.46
Hydril preventor	0.12
Riser/conductor parts	0.12
Misc. Maintenance. Equipment	0.22
TOTAL	3.67

Table 6 Adjusted frequency of dropped objects for the Example platform

7. DEFLECTION OF DROPPED OBJECTS IN THE SEA

There are reports from experiments in the literature in which the time-displacement relations of rigid bodies passing through a wave/current environment were measured. The tests were done with lightweight tubulars falling into a test tank with currents ranging from 0.5 m/s to 0.25m/s. The results of these tests indicated that a 45 degree drop angle for all objects is conservative and that 30 degrees is a better value for lightweight tubulars.

Dropped Objects	Freq.(100 Crane year)						
	Total Imp act Energy Band $kN.m$						
	all Bands	0-55 $kN.m$	55-110 $kN.m$	110-200 $kN.m$	200 - 400 $kN.$	400-800 $kN.m$	>800 $kN.m$
Items less than 1 tonne	0.82	0.82	0	0	0	0	0
Containers 2-3 tonnes	0.62	0.62	0	0	0	0	0
Containers 4-5 tonnes	0.41	0.41	0	0	0	0	0
Crane Block	0.12	0	0.12	0	0	0	0
Crane test weight	0.12	0	0	0	0.12	0	0
Crane boom	0.25	0.25	0	0	0	0	0
Bundled pipes/casings 5 -10 tonnes	0.41	0	0	0.02	0.05	0.06	0.28
Drill collar/castings	0.46	0.04	0.05	0.12	0.17	0.08	0
Hydril prevented	0.12	0	0	0	0.12	0	0
Riser/conductor parts	0.12	0	0.02	0.03	0.05	0.02	0
Misc. Maintenancc. Equipment	0.22	0.11	0	0	0.01	0.01	0.09
TOTAL % of total	3.67 100	2.25 61	0.19 5	0.17 5	0.52 14	0.17 5	0.37 10

Table 7 Dropped object frequency in each energy band

Other experiments on the analysis of dropped pipes in water suggest that 60% of all dropped objects have a maximum deflection of 20° from the vertical, while the remaining 40% deflect between 20° and 45° from the vertical. This is valid for the present example where both general goods and piping are handled. Non pipe-shaped objects tends to fall more directly to the seabed, and do not deflect as much as the pipe-shaped objects.

With the water depth of 109 metres and the deflection angles outlined above, dropped objects land in three different zones. The distribution of objects in the zones is as follows:

- 59% inside the work area of the crane (Zone I)
- 29% outside the crane work area to 40 m beyond (Zone II)
- 12% from 40m to 109m outside the crane working area (Zone III)

The distribution is developed from geometric considerations of the deflection of objects dropped in every point within the work area of the cranes. The area covered by each zone is illustrated in Figure 2. It can be seen that the spool-piece is only subjected to impacts from the North crane (I). If the pipeline is routed differently it could be subjected from to impacts from more than one crane due to presence of an overlap area.

8. DAMAGE-PRONE AREA OF THE PIPELINE

Consideration of the average size of the dropped objects indicates that the pipeline may be damaged if the objects fall into an area whose width comprises the pipeline diameter plus 0.5 m. This is the damage prone area. The probability of impact of dropped object and pipeline is calculated by dividing the damage-prone area by the total area of the respective zone multiplied by the percentage probability of objects that falling in the zone. The results are given in Table 8.

9. TOTAL PROBABILITY OF IMPACT

The total probability of impact for each energy band on the pipeline around the platform from each crane is given in Table 9. The entries in this table are derived as follows:

The $\sum \text{Pr(impact:drop)}$ from Crane 1 which is equal to 3.94×10^{-3} is taken from Table 8. The probability of drop in the 0-55 $kN.m$ band is 2.25 (Table 7) for 100 crane years. For one crane year the probability is $(2.25/100) = 2.25 \times 10^{-2}$ and for entire design life of 20 years is $(2.25/100) \times 20 = 4.5 \times 10^{-1}$.

Crane	Zone	Damage Prone Area (m^2)	Total Zone Area (m^2)	Pr(zone)	Pr(impact:drop) $Pr=(3/4)\times 5$
1	2	3	4	5	6
A1	I	15.7	5202	0.59	1.78×10^{-3}
A1	II	62.6	9768	0.29	1.86×10^{-3}
A1	III	96.5	38250	0.12	0.30×10^{-3}
$\sum$ Pr(impact:drop)					3.94×10^{-3}
A2	I	0	5753	0.59	o
A2	II	0	11557	0.29	0
A2	II	0	48300	0.12	0
$\sum$ Pr(impact:drop)					0
$\sum$ Pr(impact:drop)					3.94×10^{-3}

Table 8 Annual probability of impact on the pipeline

Crane	1	2	Total from both cranes
Pr(impact: drop)	3.94×10^{-3}	0	
Pr (Drop: 0 - 55) Pr (Impact 0 - 55)	2.25×10^{-2} 0.89×10^{-4}	2.25×10^{-2} 0	0.89×10^{-4}
Pr (Drop 55 - 110) Pr (Impact 55 - 110)	1.9×10^{-3} 0.75×10^{-4}	1.9×10^{-3} 0	0.75×10^{-4}
Pr (Drop 110 - 200) Pr (Impact 110 - 200)	1.7×10^{-3} 0.67×10^{-4}	1.7×10^{-3} 0	0.67×10^{-4}
Pr (Drop 200 - 400) Pr (Impact 200 - 400)	5.2×10^{-3} 2.05×10^{-4}	5.2×10^{-3} 0	2.05×10^{-4}
Pr (Drop 400 - 800) Pr (Impact 400 - 800)	1.7×10^{-3} 0.67×10^{-4}	1.7×10^{-3} 0	0.67×10^{-4}
Pr (Drop: >800) Pr (Impact >800)	3.7×10^{-3} 1.46×10^{-4}	3.7×10^{-3} 0	1.46×10^{-4}
Total per year:			6.49×10^{-4}
Total for 20 year desig life:			1.3×10^{-2}

Table 9 Total probability of impact on the oil export pipeline

Thus the total probability of impact is:

Pr(impact: Drop) × Pr (Drop: 0 - 55 $kN.m$) $=3.94\times10^{-3}\times2.25\times10^{-2}=0.89\times10^{-4}$

For 20 years design life the total probability in this band is

$=3.94\times10^{-3}\times2.25\times10^{-2}\times20=1.78\times10^{-3}$

10. ENERGY ABSORPTION OF PROTECTIVE SYSTEM

Pipelines are typically installed with a concrete coating to provide the required stability against buoyancy. This concrete coating provides some energy absorbing capacity against impact.

The volume penetrated is related to the impact energy and the concrete cube strength by

$$\text{Volume Penerated} = \frac{\text{Impact energy } (J)}{5\times f_{cr}}$$

where f_{cr} is the cube strength in Pascal units of the concrete used.

For a typical value for the concrete coating, say $f_{cr}=38MN/m^2$, the energy required to give a given depth of penetrations is shown in Table 10.

Depth of penetration (mm)	Impact Energy ($kN.m$)
20	6
40	44
45	52
50	65
60	95
70	155
100	390

Table 10 Impact energy associated to varying depths of penetration.

11. ENERGY ABSORBED BY STEEL PIPE

Generally small dents in steel pipes do not necessarily lead to damage requiring repair or replacement. Typical intelligent PIGS can be reasonably expected to accommodate reductions in diameter, caused by denting, of up to 5%.

Various researchers (Refs. 3 & 4), have proposed relationships for the energy required to produce a permanent dent of a given size. In the same vein, Ellinas and Walker have proposed the following relationship for a knife edge indentor in a plane normal to the pipeline's axis, which gives a conservative estimate of the energy absorbed by the dent.

$$E_d = 25F_Y t^2\sqrt{\left(\frac{\Delta_d^3}{D}\right)}$$

where

$F_Y = 413N/mm^2$ $= X60\text{steel}$ = the specified minimum yield strength

$t = 12.7\ mm$ = the nominal pipeline wall thickness

$\Delta_d = 16\ mm$ = dent depth

$D = 12^{"}$ pipeline $= 323.9\ mm$ (the pipeline outside diameter)

Substituting these values gives

E_d = energy absorbed during denting = 6 kJ

The capacity of the coating and the pipe is summarised in the following table. These capacity estimate are not accurate and should only be used in the context of the present example.

Damage	Lower Capacity	Higher Capacity
Concrete coating only	52 $kN.m$	104 $kN.m$
Concrete coating only plus dent	58 $kN.m$	110 $kN.m$

Table 11 Summary of Energy Absorption capacity for the pipeline

12. RISK ANALYSIS

The concrete-coated pipeline (the base case) will be able to withstand up to 55 $kN.m$ impact energy with some local damage to the coat and a small dent in the pipe. Consequently, the conditional probability of the pipeline failing for impact energies within the 0-55 band is zero. The next band, 55-110, could cause damage ranging from the very small to the extensive requiring the shut-down of the pipeline. Considering the pipeline's capacity to withstand impact, it can be assumed that on average there is 50-50 chance that the damage would not lead to the pipeline's failure. In this case, the conditional probability is 0.5x0.5=0.25. It is therefore assumed 75% of objects with impact energy in 110-200 $kN.m$ band would lead to shut-down. For all energies above 200 $kN.m$ the conditional probability is 1.

As an alternative to the base case, it is assumed that the concrete coating has been increased to 65 mm which can withstand twice as much impact energy compared to the base case. The modified conditional probabilities are modified and given in Table 12. An impact energy in the range of 0-55 $kN.m$ would result a state of damage which can be classified as negligible as noted in Table 1. The second energy band, 55-110 $kN.m$ would cause significant damage that would probably be sufficient to require the pipeline to be shut-down. There would however be no injuries to platform personnel so this level of damage can be classified as marginal. The third band of energies, 110-200 $kN.m$, would require the pipeline's shut-down and may lead to injuries but is unlikely to cause fatalities. This damage scenario is therefore classified as substantial. Higher levels of impact energies should be classified as classification 4 (critical), or classification 5 (catastrophic).

The total probability of each band is given in Table 9. Thus each band has a classification and a probability associated with it.

Case	Protection Measure						
		0-55	5-110	10-200	200-400	400-800	>800
Base Case	45mm Concrete Coating	0	0.25	0.75	1	1	1
Alter native	65mm Concrete Coating	0	0	0.25	0.75	1	1

Table 12 Conditional probabilities factors.

Two requirements determine whether the level of protection provided by a scheme is adequate. The first and the most important requirement is the safety of the platform personnel. This requirement is evaluated by considering the level of risk to the personnel. The best way to achieve this is to plot the risk from each energy band on the hazard consequence chart as shown in Figure 2, using its classification and total probability as described earlier. If a point lies above the line separating lower and moderate risks, then increased pipeline protection is advisable in order to reduce the risk associated with that protection measure. When a point lies within the shaded area, then the adopted protection measure is adequate.

The second requirement is loss of revenue during the pipeline shut-down. Additional protection is considered justified when the additional cost of implementing extra protection is less than the total losses from pipeline failure resulting from dropped objects over the platform's life. These financial losses compromise the capital cost of repair and inspection of the damaged pipeline and the cost of lost or deferred production.

It should be noted that the hazard consequence chart depends on the risk aversion of the decision maker as well as legal requirements in some countries. Figure 2 is only an example of such a chart. A similar chart should be constructed, specifically adapted to reflect operator's approach to safety issues as well as accounting for any legal requirements.

13. CONCLUSIONS

A methodology was give here for deciding on the level of protection for a marine pipeline. This method accounts for the possible impact energies, the frequencies of dropped objects and the strength of the pipe and its concrete coating. The method allows various protective measures to be compared in terms of the residual risks.

14. REFERENCES

1. Hancock, J. R., The Impact Resistance of Concrete Weight Coating for Submarine Pipelines, 7th International Conference on the Internal and External Protection of Pipes, 1987.

2. Broyd, T. W., Macfarlane, C. R., and Moncaster, M. B., If it fall in the sea , The Oilman, July 1983, pp. 38-39.

3. Weirzbicki T, and Suh, M-S, Denting Analysis of Tubes Under Confined Loadings, Massachusetts Institute of Technology, MIT Sea Grant College Program, Report MITSG 86-5, 1986.

4. Ellinas C.P., and Walker A.C., Damage of Offshore Tubular Bracing Members, International Association for Bridge and Structural Engineering, IABSE Colloquium, Copenhagen, 1983.

5.World Wide Offshore accident Databank: Woad Statistical Report, 1988 : Statistics on Accidents to Offshore Structure Engaged in Oil and Gas Activities in the Period 1970-87.

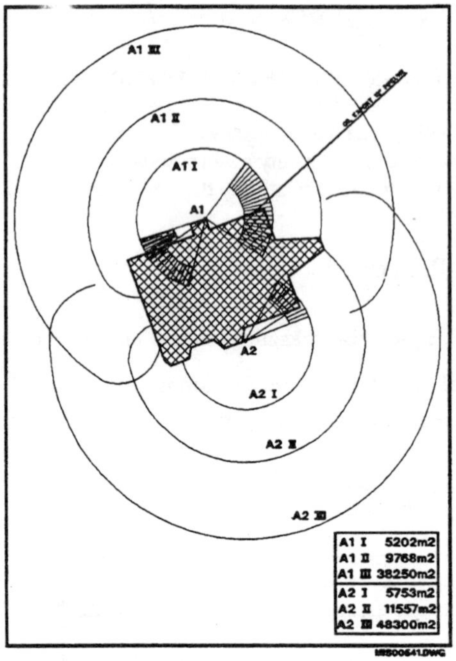

Figure 1

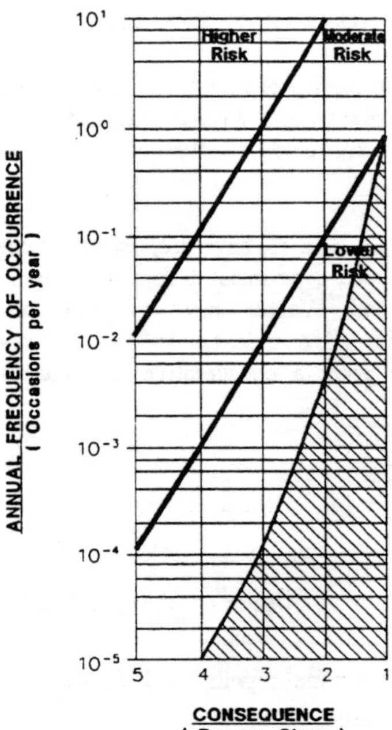

Figure 2

Proceedings of the Seventh (1997) International Offshore and Polar Engineering Conference
Honolulu, USA, May 25-30, 1997

ISBN 1-880653-28-1 (Set); ISBN 1-880653-32-X (Vol. IV)

The Relationship Between Public Policy and Risk Assessment in Tanker Design Regulation

Henry S. Marcus and Alan J. Brown
Massachusetts Institute of Technology
Cambridge, MA, USA

ABSTRACT

There is much room for improvement in the use of techniques involving risk assessment and technical analysis in the public policy decision making process as it applies to the marine field. The example of the regulation of tanker design, as mandated in the Oil Pollution Act of 1990, is utilized as a case study. Proper risk assessment for the complex tankers/waterway system requires a total-system approach, which is currently beyond the state-of-the-art, but the public demands an immediate response to what it perceives as a straight-forward problem and government must react. This paper discusses the further work needed in the areas of methodology, data bases, and communication/education to properly address these policy decisions.

INTRODUCTION

While techniques involving risk assessment and technical analysis are well respected and used in industry and academia, there often seems to be a dearth of such methodologies in the public policy decision making process as it applies to the marine field. This paper will use the example of the regulation of tanker design, as mandated in the Oil Pollution Act of 1990 (OPA 90), as a case study.

The tanker/waterway system is large and complex (and oil spills are a dynamic random-product). A valid assessment of risk for such a system cannot be made by analyzing a subset of the problem or by drawing conclusions from long-term oil spill statistics. Valid risk assessment and decision making requires a total-system model and analysis. When accidents occur, the public demands action, but the necessary tools to develop cost-effective solutions and to communicate the need and rationale for this process to the public do not exist. Often the solution in one scenario may not be the most cost-effective solution over all possible scenarios and vice versa.

Consider the Exxon Valdez oil spill in Alaska in 1989 which resulted in the Oil Pollution Act of 1990 that requires double hulls on all new tankers. Since Congress voted unanimously to pass this legislation, one might assume that it is a well thought-out law. One might consider the answers to the following questions:

- Will double hulls on tankers prevent an oil spill from an Exxon Valdez type accident?
- Are double hulls the best design to prevent or minimize oil spills in an Exxon Valdez type oil spill?
- Will new double hull tankers always spill less in accidents than the single hull tankers they replace?

The answers to all three questions are "No". There is no controversy among naval architects on this topic. First, a tanker weighing more than 200,000 tons at 12 knots hitting a reef will cause an oil spill no matter what commercial design is used. Unfortunately, many laypersons will be surprised after the first major oil spill in U.S. waters from a double-hull tanker that Congress didn't really "regulate" such spills out of existence. Second, to minimize an oil spill in a major "high energy" grounding, a design called mid-height deck would be more effective than a double hull. In the mid-height deck design, a horizontal, oil-tight deck through the cargo tank limits the outflow due to hydrostatic pressure. Most nations with the exception of the U.S., have declared that the mid-height deck design is equivalent or better than the double hull. Third, the maritime industry has determined that for tankers below 160,000 DWT, the least expensive double-hull design is one with only one tank across (rather than the normal two or three). When this tank is penetrated in a collision, more oil will spill than in a single-hull tanker with two or three tanks across. (U.S. laws and regulations requiring double hulls do not yet address the number of tanks across in the double-hull design.)

How can we be in this position where what might be thought of as the "obvious" objectives of OPA 90 are not being fulfilled. A simplified answer is "Too much emotion, not enough analysis." When seeing the pristine Alaskan environment covered with oil from the Exxon Valdez, the public demanded action. The double-hull design met the immediate emotion need. It is a quick "technology fix" that is easy for the layperson to understand. In addition, it should completely eliminate minor spills and reduce the impact of major spills. The cost to the U.S. of having double hulls on all tankers entering the U.S. is on the order of $700 million per year [1].

A MORE RATIONAL APPROACH

Since at least 80% of major tanker accidents (and probably the minor ones as well) are caused by human error [2], one might expect a formal risk assessment might have been made to determine the most cost effective way to reduce oil spills. In such a scenario, different "solutions" could be compared to see how many gallons of oil spills could be prevented by each solution (and at what cost).

However, it is clear that neither the industry nor academia was able to explain these issues to the general public or the Congress. Professors teaching the technical aspects of these issues had failed to communicate on a level that people with less specialized knowledge could comprehend.

To follow through with a rational approach to risk assessment, you need: a methodology, a credible data base, communication/education of the approach, and the political will to use it. The authors feel the political will can occur by itself if the other three needs are met.

METHODOLOGY

Once an accident has occurred it is too late to develop methodology, models, data bases, and tools to complete a total-system risk assessment, and select the most cost-effective response. The public demands that the government move more quickly than this. In such cases, the complexity of the tanker/waterway system forces industry, government and academia to focus on manageable pieces of the total problem, or to take a top-down statistical approach. It is difficult to draw cause and effect conclusions from statistics, particularly when data bases do not contain this type of information. Until there is a better understanding of accident mechanisms, any attempt to minimize accidents based on statistics alone is reactionary with questionable effectiveness. We must be working on a total-system methodology now if we expect to use it in crisis. This methodology should consider the basic physics and human factors of the problem and be sensitive to the impact of new regulations and technology.

The framework for a total-system methodology to assess tanker oil spill risk is presented in Reference [3]. The general structure of the oil spill model used in this methodology is illustrated in Figure 1. Simply stated, risk is the product of the probability of an accident and the consequence of an accident. Major disasters are rarely caused by any one factor. They arise from the enforceable concatenation of several diverse events, each one necessary but singly insufficient. Like the etiology of multiple-case illnesses due to resident pathogens, complex systems also breakdown due to resident latent errors. The challenge for this framework is to show how latent and active failures combine to produce accidents and to indicate where and how more effective remedial measures might be applied. The notional tanker risk calculation process is diagrammed in Figure 2. Once risk can be quantified, the cost/risk/benefit of various alternatives may be compared.

ACCIDENT PROBABILITY

Accident probability is calculated using Probabilistic Risk Assessment (PRA) techniques including fault trees, event trees and human reliability analysis (HRA)[4]. Fault tree analysis is very effective in analyzing complex systems that have multiple failure modes with physical and operational interactions, especially if the role of humans in the operation is a factor [5]. Fault tree analysis was developed for the aerospace industry for system safety analysis. Norman Rasmussen defined the essential tasks of a PRA in his safety study of the nuclear power industry [6]. This model follows his example. Human failure events are quantified using basic Human Error Probabilities (HEP's) that are modified by a series of Performance-Shaping Factors (PSF's)[4].

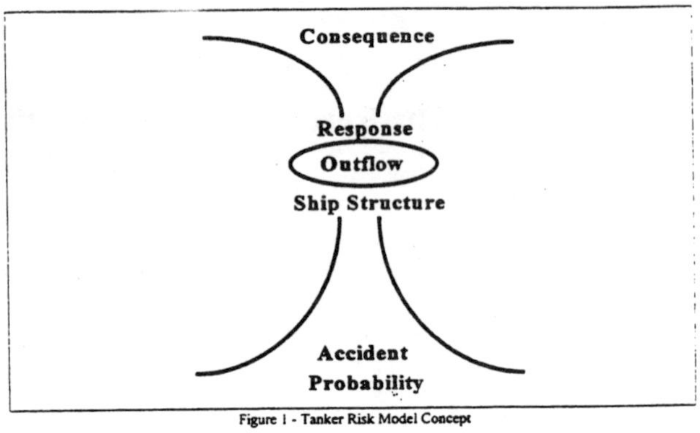

Figure 1 - Tanker Risk Model Concept

The fault trees become increasingly specific at lower levels where basic probabilities can be assigned reflecting individual error or other system failure. Standard HEP's are obtained from NUCREG/CR-1278, <u>Handbook for Human Reliability Analysis</u> [4]. Other probabilities are obtained from failure statistics, expert opinion and simulation studies. Event trees are used to further define tasks and develop fault probabilities. In this manner HEP's and other error probabilities are built up for each terminal event in the fault tree. Below this level performance shaping factors (PSF's) for the specific ship and port are used to adjust HEP's.

It is essential that important PSF's be identified to determine the effect external influences have on the basic HEP's. Table 1 shows PSF's from NUREG1278[4]. PSF's determine whether individual performance will be highly reliable, highly unreliable, or at some level in between. Using these generic PSF's as a starting point, specific ship and waterway shaping factors are identified. Their parametric relationship to the fault tree HEP's is determined by simulation and expert opinion.

The pending revision to the International Maritime Organization (IMO) Standards for Training, Certification and Watchstanding (STCW) and the new International Safety Management (ISM) Code are critical to the effective regulation, management, execution and monitoring of virtually all PSF's of concern to tanker safety. STCW and ISM certification and inspection records can provide an up-to-date status of human factors onboard ship and in the ship's management. This data can be used to update the total-system risk status for a specific ship for use by management or regulatory authorities.

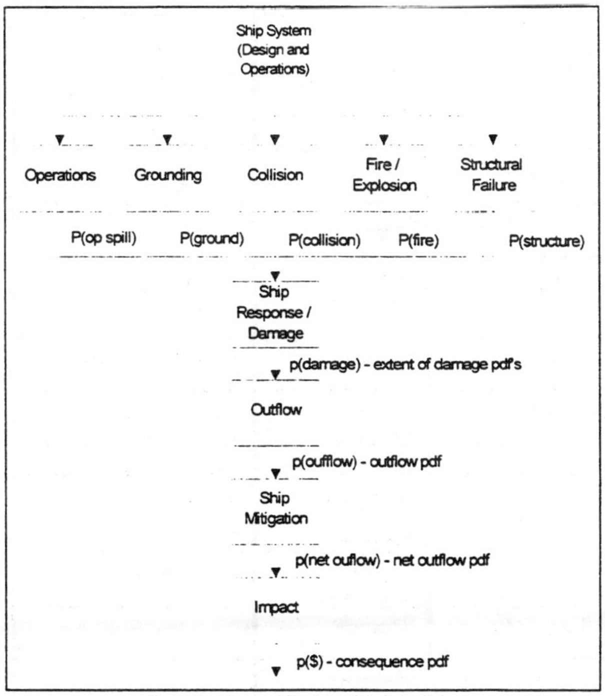

Figure 2 - Notional Tanker Risk Calculation

OIL OUTFLOW

Once probabilities for the major oil spill events (grounding, collision, fire/explosion and global structural failure) are determined, oil outflow is calculated using a probabilistic method based on the IMO "Interim Guidelines for the Approval of Alternative Tanker Designs Under Regulation 13F of Annex 1 of MARPOL 73/78" and a pdf is derived. This method calculates the probability of zero outflow, and values for mean outflow and extreme outflow for grounding and collision accidents using probability density functions (pdf's) for location and extent of damage.

One major shortcoming of the Regulation 13F methodology is that it does not consider the effect of structural design or crashworthiness on damage extent. The primary reason for this exclusion is that no definite theory or data exists to define this relationship. The solution of this problem must consider the interaction between local structural damage and global ship motion. The variety of structural details and potential accident scenarios makes this difficult. A methodology which considers crashworthiness should be sensitive to at least the basic parameters defining unique structural designs while maintaining sufficient generality and simplicity to be applied by working engineers in a regulatory content for a ship in a worldwide operation. It should not require detailed finite element analysis or be limited to a single accident scenario. The computational model developed at MIT by Professor Tomasz Wierzbicki under the Joint MIT-Industry Project on Tanker Safety is ideally suited to such an application [7].

SPILL RESPONSE

The options available for spill mitigation and cleanup include containment and elimination. Response to contain the spill is considered using a dynamic simulation model. Response is based on ship and waterway spill response assets. Probabilities are estimated using expert opinion. This calculation results in mean outflow after response.

Although the ship may have some onboard capability for containment, waterway assets, waterway management, and ship management are most important to the mitigation and cleanup function. Typically only 10-20% of the spilled oil is ever contained and recovered. The type and quantity of oil spilled, availability of personnel and equipment, environmental conditions and various human factors determine the effectiveness of the mitigation and cleanup effort. There is a room for great improvement in spill recovery and mitigation. Increasing recovery rates by 5-10% could significantly reduce spill consequences. Mitigation and cleanup should be actively investigated in the total system context.

SPILL CONSEQUENCE

The volume of oil spilled is not a valid metric for measuring the consequence of a spill. The impact or fate and effect of a spill is the correct metric and this is most conveniently quantified as a cost. There are five major categories of oil spill cost [8]: (1) commercial; (2) social and recreational; (3) ecological; (4) restoration; and (5) ship owners/cargo owners/insurance. Cost is extremely sensitive to the specific spill scenario. This model uses an average cost over many scenarios. Probabilities and pdf's are developed to identify potential spill locations in various waterways along primary tanker routes. A Monte Carlo simulation chooses the spill location and scenario based on these probabilities. The most recent Department of the Interior (DOI) Type A Damage Assessment Model is used to estimate primary restoration costs and compensation costs for various spill volumes (in U.S. waters). Regression analysis is applied to these results to calculate cost pdf's as a function of type and quantity of oil spilled.

DATA BASES

While methodologies that address such factors as human and organizational errors, such as the one described above, exist, they are rarely used. "The number one reason for this has been the almost total absence of any reliable data with which to do the modeling and calculations" [9]. Methodologies require data bases that explain "cause and effect" of previous accidents. Otherwise the methodology is nothing more than an academic exercise. Obtaining the right data bases is a challenging task. First, we have the "chicken and egg" problem. If no one is using comprehensive risk assessment methodologies, then why do we need the data (and how do we know what data we need). Of course, if there is not adequate data, it is impossible to apply the methodologies.

A researcher would like to have data bases with decades of information. The key assumption is that the "ground rules" for collecting the data have not changed over time. The U.S. Coast Guard's oil spill data base is the amalgamation of three different data systems: from 1975-1983 the Pollution Incident Reporting System (PIRS) was used, from 1986-1991 the Marine Safety Information System (MSIS) superseded the older system, and from 1992 until the present a modified MSIS is utilized. One must be careful in applying data across these time ranges.

One might think that everyone would be in favor of building comprehensive data bases; however, that is not always the case. In part due to accidents caused by pilots in U.S. waters, many persons have called for development of a rigorous relevant data base. The USCG's Merchant Personnel Advisory Committee and the pilots are against the idea of collecting such data on accidents "because the information collected might be used to pit one port against another" [10].

Table 1 - Performance Shaping Factors [4]

Procedures			Stressors		Internal
Situation Characteristics	**Job and Task Instructions**	**Task and Equipment Characteristics**	**Psychological Stressors**	**Physiological Stressors**	**Organismic**
Architectural Features	Required Procedures	Perceptual Requirements	Suddenness of onset	Duration of Stress	Training/Experience
Quality of Environment	Written or Oral Communications	Motor Requirements	Duration of Stress	Fatigue	Proficiency
Work Hours/Breaks	Cautions and Warnings	Control-Display Relationship	Task Speed	Pain or discomfort	Personality/Intelligence Variables
Availability/Adequacy of Equipment	Work Methods	Anticipatory Requirements	Task Load	Hunger or Thirst	Motivation/Attitude
Manning Parameters	Policies	Interpretation	High Jeopardy Risk	Temperature Extremes	Knowledge of Required Performance Standard
Organizational Structure		Decision-Making	Threats	G-force Extremes	Physical Condition
Actions by Supervisors		Complexity	Monotony	Atmospheric Pressure Extremes	Outside influences
Rewards, Recognition, Benefits		Narrowness of Task	Long, uneventful vigilance periods	Oxygen Insufficiency	Group Identification
		Frequency and Repetitiveness	Conflicts of Motives	Vibration	
		Task Criticality	Reinforcement absent or negative	Movement Constriction	
		Long and Short-Term Memory	Sensory Deprivations	Lack of Physical Exercise	
		Calculation Requirements	Distractions		
		Feedback	Inconsistent Cueing		

CHALLENGE FOR ACADEMIA[11]

Most college professors spend their time teaching to students that have chosen their area of specialty. Often the professor has been doing research on the leading area of specialization and is looking forward to sharing his or her new advanced knowledge with the students. In most colleges professors are also rewarded on their ability to perform research and publish the most sophisticated topics in their field.

Therefore, it is easy to forget that the students will often end up in environments where they must communicate their knowledge to people without their specialized backgrounds. In situations where the professor wants to effect a change in society, such as through the introduction of a new technology or with the passage of a new piece of legislation, the professor (or his or her students carrying out what they have learned) must be able to communicate the fundamental aspects of the topic to the general public, the media or the Congress. With these instances in mind the professor can plan in advance to be teaching for understanding on two levels: the basic level of understanding for the "layperson" and the more sophisticated student specializing in this topic.

Students today rarely comprehend that their specialized knowledge cannot be easily understood by the "layperson". Students, in effect, must be teachers to properly communicate their ideas. One by-product of the process of teaching for two levels of understanding is that the students will learn what it takes to communicate the specialized knowledge in their fields to "laypersons".

CONCLUDING COMMENTS

There is much room for improvement in the proper use of risk assessment in public policy decisions. More work must be done in the areas of methodologies, data bases, and communications/education. However, efforts have started in all these areas. While it will take considerable time and resources to effect proper risk assessments, one should remember that in the case of tanker design regulation, lack of such assessments have resulted in designs that are costing society hundreds of millions of dollars annually. It is the hope of the authors that these efforts to improve risk assessments will continue and will result in the political will to use such tools in public policy decision making in the future.

REFERENCES

[1] Marine Board, National Research Council, *Oil Spills: Prevention By Design*, Washington, D.C., 1991.

[2] U.S. Coast Guard, "Prevention Through People," Quality Action Team Report, Department of Transportation, Washington, D.C., 1995.

[3] Brown, A. and Amrozowicz, M., "Tanker Environmental Risk - Putting the Pieces Together", Joint SNAME/SNAJ Conference on Designs and Methodologies for Collision and Grounding Protection of Ships, San Francisco, 1996.

[4] Swain, A.D., Guttmann, H.E., *Handbook of Human Reliability Analysis with Emphasis on Nuclear Power Plant Applications*, NUREG/CR-1278, U.S. Nuclear Regulatory Commission, Washington, D.C., 1983.

[5] Modarres M., *What Every Engineer Should Know About Reliability and Risk Analysis*, Marcel Dekker, Inc., New York, 1995.

[6] WASH-1400 (NUREG 75/014), *Reactor Safety Study, An Assessment of Accidental Risks in US Commercial Nuclear Power Plants*, October 1975.

[7] Little, P., Pippenger, D. and Simonsen, B., "Development of a Computational Model for Predicting Damage to Tankers", Joint SNAME/SNAJ Conference on Designs and Methodologies for Collision and Grounding Protection of Ships, San Francisco, 1996.

[8] Astrup, O., Monnier, I., Sirkar, J. and Cojeen, P., "Framework for Evaluating Alternative Designs and Configurations for Tankers," 1995.

[9] Boniface, D. and Bea, R., "Assessing the Risks of and the Countermeasures for Human and Organizational Error," Annual SNAME Meeting, New York City, October 1996.

[10] "New Study Lights U.S. Pilots' Fire," *Seatrade Review*, Colchester, UK, November 1996.

[11] Marcus, H., "Why Professors Should Be Teaching for Understanding on Two Levels," Western Decisions Institute Twenty-Sixth Annual Meeting, March 1997.

Proceedings of the Seventh (1997) International Offshore and Polar Engineering Conference
Honolulu, USA, May 25-30, 1997

ISBN 1-880653-28-1 (Set); ISBN 1-880653-32-X (Vol. IV)

Total Loss Risk of Ship Accidents

Wayne K. Talley
Old Dominion University
Norfolk, VA, USA

Abstract

This paper investigates determinants of the total loss risk of ship (tanker, container and bulk) accidents. The results suggest that loss risk is greater for smaller than for larger sized ships and for human-caused and fire/explosion ship accidents than for other accidents. Also, it is greater when the weather is foggy, at higher wind speeds and in an ocean or gulf waterway, but less if the ship is manned by a licensed operator. The incremental loss risk (among binary variables) is greatest for a fire/explosion accident.

Key Words: accident, ship, risk, loss.

I. Introduction

In 1980, the number of ship accidents for which the ships were a total loss was 387 , representing 1.8 million ship gross tons; by 1990 these figures had declined to 188 losses and 1.1 million gross tons (Lloyd's Register, 1991). Although these data suggest that the safety of ocean transport has improved, safety concerns remain. Among marine insurers there is concern that the inspection of ships by classification societies has failed to spot defects and that maintenance of much of the world's aging ship fleet needs to be improved. Also, competition among open registries -- the registration of ships by owners in foreign countries -- has the potential to lower ship safety standards.

Do ships classified by a particular society or having a certain flag of registry have a greater risk of total loss in an accident? Does the age of a ship or the cause of its accident (e.g., human, environmental or ship) increase this risk? The purpose of this paper is to investigate the effects of these and other factors on the total loss risk of a ship accident, i.e., on the (conditional) probability that a ship accident (having occurred) will result in the total loss of the ship.

II. A Ship Total Loss Propensity Model

Suppose a ship total loss propensity index (STLPI) exists for commercial ship accidents that determines, in each accident, whether or not a ship is a total loss. I model this index as a function of ship characteristics, type of ship, type of accident, cause of accident, operating conditions, whether the ship is manned by a licensed operator and the price of ship building and repair.

Ship characteristics include a ship's age, size, flag of registry and classification society. A positive relationship is expected between ship age (SAGE) and STLPI, since ship structural failure is expected to increase with age.[1] The a priori sign of the relationship between ship size (SSIZE) and STLPI is indeterminate. Although larger ships are more seaworthy (i.e., less susceptible to windy and hazardous weather), it is unclear once an accident occurs whether they will be more or less susceptible to total loss than smaller ships.

Safety regulation of ships varies among countries and therefore by flags of registry. Since the United States (U.S.) has relatively high maritime safety standards (Gracey, 1985), I measure the variation in ship safety regulation among countries by whether a ship flies a U.S. flag (US-FLAG) or a non-U.S. flag. A negative relationship is expected between US-FLAG and ship total loss propensity.

While a country's ship safety standards may be high, the enforcement of these standards may be lax.[2] A function of ship classification societies is to inspect ships for adherence to safety regulations.[3] However, the thoroughness of these inspections for detecting safety violations may vary from one society to another. I measure the variation in inspection thoroughness by whether ships are classified (and therefore inspected for adherence to safety regulations) by the American Bureau of Shipping (ABS-SOCIETY) or by an alternative classification society. The relationship between ABS-SOCIETY and STLPI is indeterminate. Although the American Bureau of Shipping is one of the four largest classification societies in the world (the other three being the Lloyd's Register of Shipping, the Bureau Veritas of France and the Det Norske Veritas based in Norway), both the smallest and the largest classification societies have been criticized for lax safety inspections (see Loeb, Talley and Zlatoper, 1994).

The type of ship is either a tanker (TANKER), container (CONTAINER), or a bulk ship. It is unclear which ship (if any) will have the greater propensity to be a total loss in an accident, ceteris paribus.[4] The type of accident includes collision (COLLISION), a fire/explosion (FIRE-EXPLOSE), a material/equipment failure (MAEQ-FAILURE), or a grounding accident.[5] The a priori relationship between type of accident and total loss propensity is indeterminate.

The cause of an accident may be attributed to a human[6] (HUMAN) cause (e.g., stress, fatigue, lack of training or inattention to duty), an environmental (ENVIRON) cause (e.g., debris, adverse weather, unmarked channel hazard or an adverse current/sea condition) or a ship cause (e.g., corrosion, cargo shift, stress fracture or propulsion failure). The relationship between cause of accident and ship total loss propensity is indeterminate. Although maritime safety regulation has focused on the physical condition of ships (i.e., ship factors that cause accidents), "a series of new regulations representing a shift toward regulating human actions rather than the physical condition of vessels is creating hope that the number of serious accidents can be greatly decreased" (Abrams, 1996, p. 8B).[7] Under the International Safety Management code (which will become mandatory for many vessels in 1998) shipping lines will be required to document their management procedures for the purpose of detecting and eliminating unsafe human behavior. A positive relationship (if found) between HUMAN and STLPI would be supportive evidence for regulating human actions to improve ship safety (e.g., reducing total losses).

Operating conditions describe the circumstances of ship operation such as weather characteristics and type of waterway utilized at the time of the accident. Weather characteristics include fog (FOG) and wind speed (WIND). Although fog may increase the risk of a ship accident, its impact on the propensity of the ship to be a total loss is unclear. A positive relationship is expected between WIND and STLPI, since a higher wind speed may impair the ship operator's ability to lessen the damage severity of a ship accident.

I describe waterways as either river (RIVER), bay (BAY), or ocean or gulf waterways. The ship loss risk of collisions and groundings is likely to be greater in ocean and gulf than in river and bay waterways, since ships are often traveling at a higher speed in these waterways. Also, the ship loss risk for any type of accident is likely to be greater in these waterways given the battering from the sometimes larger waves. Hence, a negative relationship is expected between RIVER and STLPI as well as between BAY and STLPI.

I measure the experience of the manned operator of a ship by whether he is a licensed (LICENSE) or an unlicensed/undocumented operator.[8] A more experienced operator is not only expected to reduce the risk of an accident but also to lessen the damage severity of a disabled ship; therefore, a negative relationship is expected between LICENSE and STLPI.

The effect of the price of shipbuilding and repair (PRICE-BLDREP) on ship total loss propensity is not predictable: On the one hand, the more expensive repair services are, the greater the incentive to reduce this propensity (i.e., a positive safety effect); on the other hand, higher shipbuilding and repair prices may make some ship safety expenditures more expensive, thereby reducing the incentive to improve ship safety (i.e., a negative safety effect).

From the above discussion I obtain the following equation to be estimated:

$$\text{STLPI} = f(\text{SAGE, SSIZE, US-FLAG, ABS-SOCIETY, TANKER, CONTAINER, HUMAN, ENVIRON, COLLISION, FIRE-EXPLOSE, MAEQ-FAILURE, FOG, WIND, RIVER, BAY, LICENSE, PRICE-BLDREP}) \quad (1)$$

Since data on STLPI are not available, equation (1) is estimated utilizing the dependent variable LOSS, equaling one if the ship is a total loss and zero if the accident does not result in total loss.

III. <u>Data</u>

I estimate equation (1) utilizing detailed data of individual commercial ship (U.S. and foreign flag) accidents for which ship seaworthiness was affected. These accidents were investigated by the U.S. Coast Guard and occurred during the eleven-year time period 1981-91. Commercial ships include tanker, container and bulk ships. The foreign flag ship accidents occurred in U.S. waters; the U.S. flag accidents are not restricted to any body of water. Data (except for that of PRICE-BLDREP) were obtained from a computer tape of marine casualty information (i.e., the CASMAIN database) provided by the Coast Guard. Price index data for the variable PRICE-BLDREP were taken from various issues of <u>Producer Prices and Price Indexes</u> (Bureau of Labor Statistics, U.S. Department of Labor). The real price of shipbuilding and repair was determined by dividing PRICE-BLDREP by the U.S. Producers Price Index for all commodities (divided by 100) found in various issues of <u>Producer Prices and Price Indexes</u> (Bureau of Labor Statistics, U.S. Department of Labor).[9] Variables used in the estimation of equation (1) and their specific measurements appear in Table 1.

Variable descriptive statistics (mean and standard deviation) for both ship loss and no-loss accidents appear in Table 2. For the loss (no-loss) sample, the mean statistics reveal that the average ship age is 21.3 (15.7) years and the average ship size is 13,660 (25,069) gross tons; 70.2 (66.0) percent of the ships are U.S. flag; 36.8 (45.8) percent were classified by the American Bureau of Shipping; 38.6 (43.4), 14.0 (21.3) and 47.4 (35.3) percent are tanker, container and bulk ships, respectively; human, environmental and ship factors caused 48.2 (32.8), 16.7 (12.9) and 35.1 (54.3) percent of the accidents; and 19.1 (22.5), 28.6 (5.1), 21.4 (57.7) and 30.9 (14.7) percent of the accidents were collision, fire/explosion, material/equipment failure and grounding accidents, respectively. Also, for the loss (no-loss) sample, 9.8 (4.0) percent occurred in foggy weather; the average wind speed is 18.1 (12.2) knots; the accident location statistics reveal that 22.8 (36.4) percent occurred in rivers, 12.3 (12.3) percent in bays and 64.9 (51.3) percent in ocean or gulf waterways; and 55.3 (78.7) percent of the ships were manned by a licensed operator.[10]

IV. <u>Estimation Results</u>

I estimate equation (1) using the statistical technique, Probit analysis[11], where the dependent variable is LOSS. Since the Probit equation's coefficients are interpreted as the marginal effects of their explanatory variables on the propensity of a ship accident (having occurred) to be a total loss, the Probit equation may be interpreted as a ship accident total loss risk equation. I address possible estimation bias from omission of relevant explanatory variables by including a yearly time trend (YEAR) variable (having integer observations 1981 through 1991) to control for the exclusion of other causal factors (correlated with time).

Probit estimation results appear in Table 3. Among ship characteristic variables only SSIZE is statistically significant. Its negative coefficient suggests that ship loss risk is greater for a smaller than a larger sized ship. The coefficients of SAGE and US-FLAG have the expected positive and negative signs, respectively, but are not significant. The insignificance of TANKER and CONTAINER suggests that the loss risk is no greater for one type of ship than another (i.e., for tanker, container and bulk ships), <u>ceteris paribus</u>. This result may further suggest that the current concern for the loss of bulk ships relative to other types of ships may be exaggerated.[12]

The estimation results reveal that coefficients for both HUMAN and ENVIRON are positive, but only the former is significant, suggesting that ship loss risk is greater for an accident caused by a human as opposed to either an environmental or a ship factor. These results may also suggest that regulation of human activity aboard ship may be more effective than regulation of the physical condition of ships (which has been emphasized in the past) to reduce ship loss risk.

Among type of accidents COLLISION is significant with a negative coefficient; FIRE-EXPLOSE is highly significant, having a positive coefficient; and MAEQ-FAILURE is insignificant. These results suggest that a fire/explosion ship accident has a greater loss risk than a collision, a material/equipment failure or a grounding accident and the loss risk of a collision accident is less than that of a grounding accident. All of the operating condition variables are significant: The positive signs of the coefficients of FOG and WIND suggest that ship loss risk is greater when the weather is foggy and at higher wind speeds. The negative coefficients of RIVER and BAY suggest that loss risk is greater when a ship accident occurs in an ocean or a gulf waterway as opposed to a river or bay.

The variable LICENSE is significant with a negative coefficient, suggesting that the loss risk of a ship accident is less if a licensed as opposed to an unlicensed operator is manning the ship. The insignificance of the variable, the real price of ship building and repair, suggests that this price is not a determinant of ship loss risk. The coeffecient of the yearly time trend variable (YEAR) is positive but insignificant.

V. Incremental Loss Risk Estimates

The Probit estimation results may also be used to estimate incremental loss risks (or probabilities) associated with the explanatory variables, i.e., the incremental effects of explanatory variables on the probability of ship loss. They are calculated in either of two ways, depending on whether the explanatory variable in question is continuous or binary. The estimated incremental loss risk of a continuous variable is calculated by multiplying its Probit coefficient by the standard normal density function evaluated at the value of the Probit equation, determined by substituting mean values for the explanatory variables in the equation and then solving. The estimated incremental loss risk of a binary variable is calculated: by taking the difference between the value of the cumulative standard normal distribution function, where the binary variable is zero (and the values of the remaining explanatory variables are their means) and the value of the cumulative standard normal distribution function, where the binary variable is one (and the values of the remaining explanatory variables are their means).[13]

The incremental loss risk estimates appear in Table 3. The estimates indicate that the incremental loss risk (among binary variables) is greatest for a fire/explosion accident, followed by foggy weather. Specifically, a fire/explosion accident increases the risk (or the probability) that a ship will be lost in an accident by 0.0136, whereas foggy weather increases the risk by 0.0076. If the ship is manned by a licensed operator, the risk decreases by 0.0035.

VI. Conclusion and Policy Implications

This paper has investigated determinants of the total loss risk of ship accidents. Specifically, the effects of ship characteristics, type of ship, type of accident, cause of accident, operating conditions, whether the ship is manned by a licensed operator and the price of ship building and repair on ship loss risk were investigated. Probit estimation results of a loss risk equation suggest that ship loss risk: 1) is greater for a smaller than a larger sized ship; 2) is no greater for one type of ship than another (i.e., for tanker, container and bulk ships); 3) is greater for an accident caused by a human as opposed to either an environmental or a ship factor; 4) is greater for a fire/explosion ship accident than for a collision, a material/equipment failure or a grounding accident; 5) is less for a collision than a grounding accident; 6) is greater when the weather is foggy; 7) is greater at higher wind speeds; 8) is greater when a ship accident occurs in an ocean or a gulf waterway as opposed to a river or bay; and 9) is less if a licensed as opposed to an unlicensed/undocumented operator is manning the ship. The incremental loss risk (among binary variables) is greatest for a fire/explosion accident, followed by foggy weather.

Notes

1. Meek, Brown and Fulford (1985) conclude that for ships in general a positive relationship exist between ship casualty rates and ship age.

2. Although there is a worldwide parity trend in ship safety standards, there are significant differences in their enforcement across nations. "Disparities in enforcement do much more than economic damage -- they serve to undermine the safety system" (Gracey, 1985, p. 10C).

3. Ship classification societies (some dating back more than 200 years) inspect ships to ensure that: ships are seaworthy, they meet national-flag requirements and conform to international safety standards. No underwriter will insure a ship if it is not classed. For a discussion of current issues related to classification societies, see Boisson (1994) and Brooks (1996).

4. However, there is particular concern for the safety of bulk ships. In May 1995 bulk ship safety was a top agenda item at the London meeting of the Maritime Safety Committee of the International Maritime Organization. Between 1990 and 1994, 97 bulk ships were lost worldwide; ship structural damage and weather conditions contributed to approximately one-half of the losses; and the majority of the ships were over 15 years old (Porter, 1995).

5. The description of these types of accidents is as follows: 1) collision -- ship struck or was struck by another ship on the water surface, or struck a stationary object, not another ship (an allision); 2) grounding -- ship is in contact with the sea bottom or a bottom obstacle, struck object on the sea floor, or struck or touched the bottom; 3) fire and explosion -- the fire and/or explosion is the initiating event reported, except where the first event is a hull/machinery failure leading to the fire/explosion; and 4) material and equipment (or structured-machinery-other) -- hull/machinery damage, missing, and miscellaneous non-classified reasons (e.g., ships sunk due to either weather or break-up related to causes not covered by other casualty categories).

6. A common assertion is that nearly 80 percent of ship accidents are caused by human error (Goss, 1994). Human error is defined by the United Kingdom's (U.K.'s) P&I Club, the world's largest marine liability insurer, as any human action or omission identifiable as the immediate cause of the event from which the liability arises, including blame worthy behavior from simple mistakes in arithmetic, judgement and deliberate risk taking. For the period 1987-91 sixty percent of the claims of the U.K. P&I Club were attributed to human error (Tzannatos and Goulielmos, 1996).

7. The rationale for the shift toward regulating human actions as opposed to the physical condition of ships are: 1) it is less expensive to change human behavior than it is to redesign ships for safety and 2) "the fact that most accidents are caused by human error" (Abrams, 1996, p. 8B). For a discussion of maritime safety regulation, see National Research Council (1994) and Stewart (1996).

8. Unfortunately the data do not provide specific licensing information but only if the manned operator is licensed or unlicensed.

9. Since the accidents of non-U.S. flag ships (in the data) occur in U.S. waters, these vessels may have ship repairs done (especially for relatively major ship damage) in U.S. shipyards; if so, the U.S. price index of shipbuilding and repair is an appropriate ship-repair price index for these ships.

10. Some of the explanatory variables might be positively correlated, e.g., US-FLAG and ABS-SOCIETY and US-FLAG and LICENSE. A review of a matrix of simple correlation coefficients reveals coefficients of 0.401 and 0.765 for the latter pairs of variables, respectively. The former coefficient is not unreasonably high, but the latter is and may contribute to the insignificance of its variables in the estimation of equation (1).

11. The Probit probability model assumes there exists a theoretical (but not actually measured) index Z_i (for the ith observation) which is a function of an explanatory variable(s) X_i, i.e., $Z_i = g(X_i)$. This index is assumed to be a continuous variable which is random and normally distributed. Given that observations of Z are not available, Probit analysis utilizes data which distinguish only whether individual observations are in one category (for high values of the index Z) or a second category (for low values of the index Z). The problem which Probit analysis solves is how to obtain information about the underlying Z function utilizing categorical data: Let Y_i represent a binary variable which equals one when Z_i is greater than Z^*_I and equals zero when Z_i is less than or equal to Z^*_I. The Probit model assumes that Z^*_I is a normally distributed random variable, so that the probability that Z^*_I is less than or equal to Z_i can be computed from the cumulative normal distribution function. By construction, Probit analysis estimates a non-

linear probability function $P_i = h[g(X_i)]$, where P_i represents the probability of the selected event (or category) occurring. For further discussion of Probit analysis, see Berndt (1991).

12. See note #4 and Porter (1992).

13. For further discussion of the calculation of marginal probabilities in Probit models, see Greene (1990).

References

Abrams, A. (1996). "New Rules Put Focus on Human Factors," Journal of Commerce, May 2, p. 8B.

Berndt, E. R. (1991). The Practice of Econometrics: Classic and Contemporary, Reading, Massachusetts: Addision-Wesley.

Boisson, P. (1994). "Classification Societies and Safety at Sea," Marine Policy, Vol. 18, No. 5, pp. 363-377.

Brooks, M. R. (1996). "The Privatization of Ship Safety," Selected Papers from the 7th World Conference on Transport Research, Editor, H. Haralambides, a special issue of the journal Maritime Policy and Management, Vol. 23, No. 3, pp. 271-288.

Bureau of Labor Statistics, U.S. Department of Labor (various issues). Producer Prices and Price Indexes, Washington, D.C.: U.S. Government Printing Office.

Goss, R. (1994). "Safety in Sea Transport," Journal of Transport Economics and Policy, Vol. 28, No. 1, pp. 99-110.

Gracey, J. S. (1985). "Many Question Value of Drive for Vessel Safety Standards," Journal of Commerce, May 20, p. 10C.

Greene, W. H. (1990). Econometric Analysis, New York: MacMillan Publishing Co.

Lloyd's Register (1991). Casualty Return 1990, London: Lloyd's Register of Shipping.

Loeb, P. D., W. K. Talley and T. J. Zlatoper (1994). Causes and Deterrents of Transportation Accidents: An Analysis by Mode, Westport, Connecticut: Quorum Books.

Meek, M., W. R. Brown and K. G. Fulford (1985). "A Shipbuilder's View of Safety," Maritime Policy and Management, Vol. 28, pp. 251-262.

National Research Council (1994). Minding the Helm: Marine Navigation and Piloting, Washington, D.C.: National Academy Press.

Porter, J. (1992). "Bulk Carrier Losses Shrink, But Safety Concerns Remain," Journal of Commerce, August 5, p. 1A.

_______ (1995). "Bulk Carrier, Ferry Safety Discussed," Business Weekly: The Virginian-Pilot, May 15, p. 10.

Stewart, R. D. (1996). "Maritime Safety and Environmental Regulation," In Lovett, W. A. (Ed.), United States Shipping Policies and the World Market, Westport, Connecticut: Quorum Books, pp. 171-189.

Tzannatos, E. and A. M. Goulielmos (1996). "Management Information System for the Promotion of Safety in Shipping." A paper presented at the 1996 International Conference: International Association of Maritime Economists, Vancouver, Canada.

Table 1
Variables and Their Measurement

Variable	Measurement
Ship Age (SAGE)	Years
Ship Size (SSIZE)	Gross Tons
U.S. Flag Ship (US-FLAG)	1 if a U.S. flag ship; 0 if not a U.S. flag ship
American Ship Classification Society (ABS-SOCIETY)	1 if ship classified by the American Bureau of Shipping; 0 if classified by other than the American Bureau of Shipping
Tanker Ship (TANKER)	1 if a tanker ship; 0 if a container or bulk ship
Container Ship (CONTAINER)	1 if a container ship; 0 if a tanker or bulk ship
Human Cause of Accident (HUMAN)	1 if ship accident caused by a human factor, 0 if caused by an environmental or a ship factor
Environmental Cause of Accident (ENVIRON)	1 if ship accident caused by an environmental factor, 0 if caused by a human or a ship factor
Collision Accident (COLLISION)	1 if a collision; 0 if a fire/explosion or a material/equipment failure or a grounding
Fire/Explosion Accident (FIRE-EXPLOSE)	1 if a fire and/or explosion; 0 if collision or a material/equipment failure or a grounding
Material/Equipment Failure Accident (MAEQ-FAILURE)	1 if a material and/or equipment failure; 0 if a collision or a fire/explosion or a grounding
Fog (FOG)	1 if foggy weather; 0 if non-foggy weather

Table 1 (con't.)
Variables and Their Measurement

Variable	Measurement
Wind (WIND)	Knots
River (RIVER)	1 if a river, 0 if a bay or an ocean or a gulf
Bay (BAY)	1 if a bay; 0 if a river or an ocean or a gulf
Licensed Manned Operator (LICENSE)	1 if ship manned by a licensed operator; 0 if manned by an unlicensed/ undocumented operator
Price of Ship Building and Repair (PRICE-BLDREP)	Price Index

Table 2
Descriptive Statistics: Mean and Standard Deviation[a]

Variable	Loss Sample (n=25)	No-Loss Sample (n=984)
SAGE	21.321 (13.968)	15.681 (12.032)
SSIZE	13,660 (19,885)	25,069 (27,506)
US-FLAG	0.702 (0.462)	0.660 (0.474)
ABS-SOCIETY	0.368 (0.487)	0.458 (0.498)
TANKER	0.386 (0.491)	0.434 (0.496)
CONTAINER	0.140 (0.350)	0.213 (0.410)
HUMAN	0.482 (0.504)	0.328 (0.470)
ENVIRON	0.167 (0.376)	0.129 (0.335)
COLLISION	0.191 (0.397)	0.225 (0.418)
FIRE-EXPLOSE	0.286 (0.457)	0.051 (0.219)
MAEQ-FAILURE	0.214 (0.415)	0.577 (0.494)
FOG	0.098 (0.300)	0.040 (0.197)

Table 2 (cont'd.)
Descriptive Statistics: Mean and Standard Deviation[a]

Variable	Loss Sample (n=25)	No-Loss Sample (n=984)
WIND	18.100 (17.563)	12.191 (13.383)
RIVER	0.228 (0.423)	0.364 (0.481)
BAY	0.123 (0.331)	0.123 (0.329)
LICENSE	0.553 (0.504)	0.787 (0.410)
Real PRICE-BLDREP	95.408 (8.814)	94.802 (8.008)

[a]Standard deviations are in parentheses; n represents sample size.

Table 3
Ship Total Losses: Probit Estimation[a]

Variable	Estimated Parameter	Incremental Probability
Ship Characteristic		
SAGE	0.004 (0.008)	1.300×10^{-5}
SSIZE	-1.820×10^{-5}** (8.525×10^{-6})	-6.010×10^{-8}
US-FLAG	-0.253 (0.370)	-0.0009
ABS-SOCIETY	0.237 (0.270)	0.0008
Type of Ship		
TANKER	0.118 (0.250)	0.0003
CONTAINER	0.124 (0.296)	0.0005
Cause of Accident		
HUMAN	0.638** (0.272)	0.0004
ENVIRON	0.576 (0.390)	0.0026
Type of Accident		
COLLISION	-0.582* (0.329)	-0.0012
FIRE-EXPLOSE	0.988*** (0.347)	0.0136
MAEQ-FAILURE	-0.077 (0.316)	-0.0002
Operating Condition		
FOG	0.766** (0.357)	0.0076
WIND	0.015* (0.008)	5.000×10^{-5}
RIVER	-0.499* (0.273)	-0.0013

Table 3 (cont'd)
Ship Total Losses: Probit Estimation[a]

Variable	Estimated Parameter	Incremental Probability
BAY	-0.807* (0.431)	-0.0011
Licensed Manned Operator		
LICENSE	-0.619* (0.355)	-0.0035
Real Price of Ship Building and Repair and Time Trend		
Real PRICE-BLDREP	0.003 (0.013)	$1.000\text{x}10^{-5}$
YEAR	0.024 (0.039)	$7.900\text{x}10^{-5}$
Constant	-50.350 (76.692)	
-2 x Log Likelihood Ratio	351.830***	
n	1,009	

[a]Standard errors are in parentheses; n represents sample size.
***(**,*)Significant at the 1 (5,10) percent level.

Proceedings of the Seventh (1997) International Offshore and Polar Engineering Conference
Honolulu, USA, May 25-30, 1997

ISBN 1-880653-28-1 (Set); ISBN 1-880653-32-X (Vol. IV)

Probabilistic Analysis of the Effect of Resource Constraints on System Safety: Towing Failures During the Construction of Concrete Platforms

Elisabeth Paté-Cornell
Stanford University
Stanford, CA, USA

Wenche Rettedal
STATOIL
Stavanger, Norway

ABSTRACT

The failure of technical systems such as offshore platforms can generally be linked to some errors in design, construction or production. These errors, in turn, can often be related to features of the management or the organization. For example, time and budget constraints are unavoidable parts of industrial life, but the managers who set them are sometimes unaware of their actual implications. This is true because the technicians, operators and engineers that are subjected to these constraints generally try to satisfy them by taking shortcuts, sometimes without a clear idea of their long-term or short-term implications. Resource constraints thus have to be managed with care to avoid unanticipated catastrophic failures that can be the direct result of the shortcuts that people take to satisfy them. The objective of this study is to illustrate an analytical approach that allows improvement of the management of such constraints by anticipating their effects on the safety of marine systems. The framework is based on the SAM (system, actions, management) model that provides an analytical link between probabilistic risk analysis (PRA), human errors, and management and organizational factors. Its focus is on concrete platforms and on the problems that might occur in the maritime operations of the construction process. We present first a simple probabilistic risk analysis framework for a concrete structure. We then examine the towing process during construction to identify the major potential errors corresponding to each variable of the PRA model. Finally, we link those human decisions and actions to management features involving schedule and budget constraints.

KEY WORDS: Probabilistic risk analysis, management, human factors, human and organizational errors, resource constraints, concrete platforms

ORGANIZATIONS, RESOURCE CONSTRAINS, AND SYSTEM FAILURES

The failure of technical systems such as offshore platforms can generally be tracked back to errors in design, construction or production (Paté-Cornell and Bea, 1989; Paté-Cornell, 1990). These errors, in turn, can often be related to features of the management or the organization. For example, in the case of the Piper Alpha accident, problems of personnel management as well as couplings in design caused the catastrophic fire that engulfed the platform in 1988 (Paté-Cornell, 1993). In another instance, the case of the Sleipner structure, the failure of that concrete platform in 1991 was probably caused by an error of design. This design error was in part the result of a poor structural analysis probably induced by resource constraints (time and money): the mathematical model that was used was not sophisticated enough to represent the structural capacity with sufficient accuracy (Jacobsen, 1992).

Time and budget constraints are unavoidable parts of industrial life. The problem is that the managers who set them are not always aware of their actual implications: technicians, operators and engineers may try to satisfy them by taking shortcuts without a clear idea of their long-term or short-term implications (Paté-Cornell, 1996). Resource constraints thus have to be managed with care to avoid a significant increase of the risk of catastrophic failures.

In recent work (Paté-Cornell, 1993 and Paté-Cornell and Murphy, 1996), a method was developed to allow extension of the classic probabilistic risk analysis model (PRA) to include the effects of human and organizational errors on the probability of system failure. This paper represents one application of that work. It is based on a recent study by Paté-Cornell and Rettedal (1996). The focus is on concrete gravity base platform structures (GBS). We describe an analytical model linking failure risks and the decisions and actions that people may take under tight schedule and budget constraints. We present first a simple structure of probabilistic risk analysis for a GBS, then we illustrate this analysis by the case of towing failures offshore. We examine the towing process to identify, for each variable of the model, the major potential errors (human and organizational) and the role of resource constraints on the occurrence of these errors. We then link these errors to the PRA model in order to assess their influence on the overall probabilities of different outcomes.

THE CASE OF THE SLEIPNER PLATFORM

The Accident

This study was motivated in part by the case of the 1991 Sleipner accident. During a ballast test operation and prior to the installation of the deck, the Sleipner A1 (SLA1) Gravity Base Structure (GBS) sank in Gandsfjorden outside Stavanger. Sleipner was the twelfth in a series of Condeep GBS platforms built by Norwegian Contractors. As the structure was being submerged for testing and after a series of "bang-like" sounds, water started flowing into one of the shafts. About eight minutes after the first "bang", the order to abandon the platform was given. Five and half minutes later, the structure sank and disappeared from the sea surface.

Most of the basic events of the Sleipner accident sequence were caused or influenced by human decisions and actions that are either gross errors or errors of judgment (Paté-Cornell, 1990) and have been systematically identified by Jacobsen (1992). It was concluded that the incident probably had three main causes:

a. The global finite element analysis used to calculate sectional forces was faulty and yielded a 47% underestimation of shear forces in the tri-cell walls.

b. The reinforcement in tri-cell joints was improperly detailed.

c. The tri-cell joints were not designed or checked as separate components.

The role of resource constrains

These errors, in turn, seem to have been influenced by typical organizational factors that can be divided into four categories: personnel issues, time pressures, economic pressures, and uncertainties due to revisions in regulations, guidelines and standards.

Separately, each of these problems may not have caused significant damage. Yet, each of them influenced, to some extent, the probability of the events that eventually led to the Sleipner accident, for example, by decreasing the probability of detection of cracks in the concrete. Some of these problems may have been exacerbated by financial constraints induced by a drop in oil prices and by the bidding procedure that led to the contract award, which increased the pressures to reduce the cost of the Sleipner project.

SYSTEM FAILURE, HUMAN ACTIONS, AND MANAGEMENT FEATURES: THE STRUCTURE OF THE SAM MODEL

Quantification of the risks and, in particular, of the effects of the different factors (technical, human, and organizational) on the probability of failure is thus necessary to allow checking the "price" of each of the constraints. These constraint costs can occur both in the long term (increased overall failure probability) and in the short term (possibility of small problems that will delay the project). To link the technical, human and organizational aspects of a project in a quantitative (probabilistic) fashion, Murphy and Paté-Cornell have developed an analytical framework called S.A.M. (System-Actions-Management) that allows, for some of these factors, to assess the effect of variations of policies and practices on the safety of a system (Murphy and Paté-Cornell, 1996). The analysis is done in three steps:

Step 1: Technical PRA

The technical probabilistic risk analysis (PRA) framework has been well established, for instance, by the models developed in the nuclear power industry (NRC, 1975), and by those developed more specifically for civil structures. These models rely on the identification of the critical functions of a system and of its failure modes, and on the computation of the overall probability of failure accounting for the influence of external events. A convenient approach is to identify accident sequences beginning with the initiating events, and to use event trees to assess the probabilities of the different scenarios.

Step 2: Decisions and actions (including human errors)

The next phase is to assess the effect of human decisions and actions on the performance of the components and therefore, on the probability of failure of the overall system. Such actions (and errors) if they are directly part of a failure mode are already included in the PRA described above. Some of them, however, affect the probability of component failure but are not directly involved in an accident sequence. For example, maintenance problems may affect one or more components, but are not a critical element of a function of the system. To assess these problems, one needs to analyze the occurrence of human errors, both in the decisions that people make and in their actions. Two questions must be addressed: what is the probability of different types of errors? What is the effect of each type of error on the probabilities of the events of each accident sequence, and therefore, on the overall failure probability?.

Step 3: Management factors

Human decisions and actions (from engineers, technicians, operators, doctors, etc.) are in large part influenced by the organization and the management. The organization by its structure, its procedures and its culture determines, as mentioned above, the information available to the actors and the incentive or reward system (Paté-Cornell, 1990). The management factors on which we focus here are those related to both financial and schedule constraints. The problem is to make sure that shortcuts, when they occur, do not jeopardize the basic system safety, and that management understands the implications of the constraints that they set. To do so, they must understand how the base-line procedures were established, and what variations will take place if, suddenly, a project is to be executed in a fragment of the time that is normally required. It could be that there was a large safety factor to start from; it could also be that the base case was at the limit of acceptable safety and could not be tightened further without incurring serious risks.

The SAM model can be represented by three equations linking the basic PRA, people's decisions and actions, and management issues (Paté-Cornell, 1993; Paté-Cornell and Murphy, 1996). It can also be schematically represented by the influence diagram of Figure 1.

Notations:

$p(F)$: probability of system failure F

$p(IE_i)$: probability of the initiating event i

$p(L_h)$: probability of the different loss levels L_h

$p(DA_j)$: probabilities of the decisions and actions of the different actors

M_k: relevant management factors

The probability of failure is the sum of the probabilities of accidents triggered by each of the possible initiating events:

$$p(F) = \sum_i p(F / IE_i) p(IE_i) \quad \text{(Eq. 1)}$$

This equation can also be used to compute the probability of different levels of loss L_h (replacing F by each possible loss level).

The influence of the decisions and actions of the different actors on the probability of failure can be obtained by conditioning each term of the previous equation on individual actions, and multiplying by the probabilities of these actions:

$$p(F) = \sum_i \sum_j p(F / IE_i, DA_j) p(IE_i / DA_j) p(DA_j) \quad \text{(Eq. 2)}$$

The effects of the management factors on the probability of system failure are assessed through their influence on the decisions and actions of the people involved. The overall probability of failure conditional on a certain management system is thus:

$$p(F / M_k) = \sum_i \sum_j p(F / IE_i, DA_j) p(IE_i / DA_j) p(DA_j / M_k) \quad \text{(Eq. 3)}$$

In what follows, we chose to focus, among all management factors, on those that result in resource constraints: time, budget, and attention.

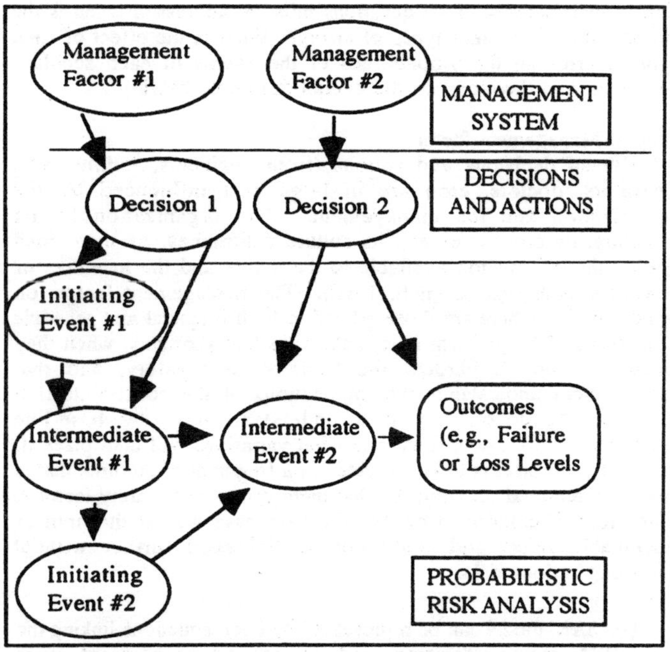

Figure 1: Influence diagram showing the structure of the SAM model

THE PROBABILISTIC RISK ANALYSIS MODEL

Equation 1 needs to be developed further to permit structuring the model in a manageable way. The structure of the PRA model used here is not the most classic one. It involves first, identification of different classes of *accident types* (e.g. structural failure of concrete structure). Second, for each class of possible accidents, a set of *initiating events* and the corresponding subsequent accident sequences are identified. The third step is to compute the probabilities of the different possible accident scenarios following each type of initiating event and to assess their consequences. This structure was dictated by the fact that for accident types, we may have global statistics in the first place. The PRA model is then used to allocate the overall probability of failure among different technical and organizational factors, rather than using a bottom-up structure to compute the probability of failure in the first place.

The accidents of interest in this analysis are those which could result (1) in damage to or loss of a gravity based concrete structure, (2) injury or death of personnel, or (3) delay in the project. To identify the initiating events of potential accident sequences, one needs to study the details of operations and systems (Paté-Cornell and Rettedal, 1996). In Table 1, *seven accident types* (AT_i) are identified. The probabilities shown here represent the opinion of one expert (Dr. Gudmestad from Statoil). They do not necessarily represent the situation of the whole industry but figures that he thought were relevant in the environment with which he is familiar. For each accident type, we then identify the set of potential initiating events IE_{ij} that represent the major accident initiators (Rettedal et al., 1994). To be usable in a probabilistic risk analysis, the failure scenarios have to be structured as a set of events that are collectively exhaustive and mutually exclusive. Although it is, by definition, impossible to guarantee that all possible events have been considered, it is important to ensure that the main accident sequences have been included.

For each accident type (e.g. failure of concrete structure during submergence test), all failure scenarios (accident sequences) are analyzed starting from each possible initiating event by developing the relevant influence diagrams. Influence diagrams are directed networks in which the nodes represent events or state variables. The links represent probabilistic dependencies (and not necessarily causality) among the variables and events. The analysis relies on the sequences of events that can follow the initiator, ranging from the detection and fixing of a minor problem to catastrophic event from which recovery is impossible. For each accident sequence, one then computes the probability that it results in each of the possible outcomes (consequence categories).

Notations:

AT_i:	Accident types
IE_{ij}:	Initiating events for each accident type
OC_k:	Outcomes (3 classes indexed in k)
p(.):	Probability of an event
p(.l.):	Conditional probability of second event given the first
p(.,.):	Joint probability of both events.
A,B,C,..:	Intermediate variables and events between initiating event and final state.

Assuming that the initiating events are structured as a set of mutually exclusive and collectively exhaustive elements, the overall probability of each outcome is the sum of the joint probabilities of accident scenarios and the outcome of interest for all such accident scenarios:

$$p(OC_k) = \sum_i \sum_j p(AT_i) \times p(IE_{ij} | AT_i) \times p(A | IE_{ij}) \times p(B | A, IE_{ij}) \times p(OC_k | A, B, IE_{ij}) \quad \text{(Eq. 6)}$$

(Paté-Cornell, 1993)

Figure 2 shows an influence diagram that represents this overall probabilistic model, based on accident types, their outcomes and their probabilities per operation. Three types of variables and events are represented here: accident types, final state of the key subsystems and final outcomes.

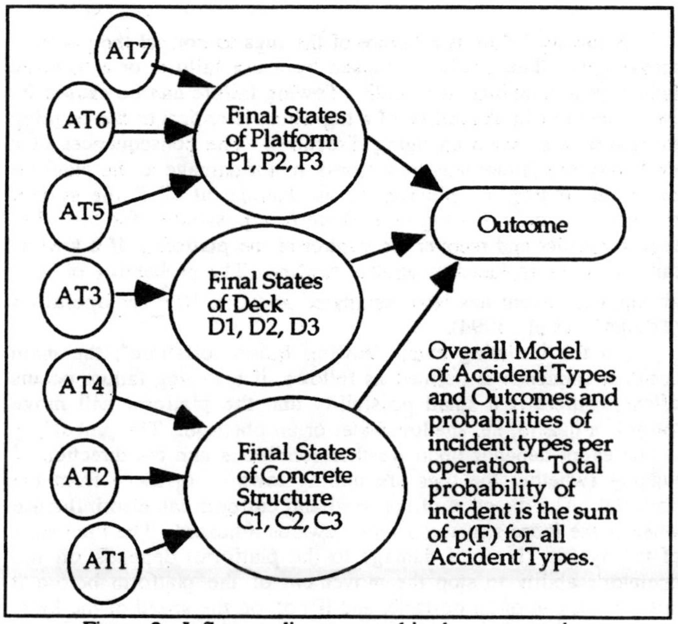

Figure 2. Influence diagram used in the computation of the different outcomes for all accident types

Class	Description of accident types	Marginal Probabilities per structure	Percentage of the Total Failure Probability	Initiating Events for each accident type
AT1	AT1 Failure of concrete structure during construction in dock, and out-of-dock operation	10^{-3}	10%	- Loss of air cushion (air below structure to create buoyancy) - Failure due to wrong estimate of the center of gravity - Miscellaneous
AT2	Failure of concrete structure during construction in the fjord	5×10^{-4}	5%	- Drop of conductor during installation - Collision with a passing ship - Mooring failure - Ballasting failure during construction - Miscellaneous
AT3	Failure of deck during transportation	10^{-3}	10%	- Towing line failure during deck transportation - Barge ballasting failure - Structural failure of a barge - Weather exceeds criteria - Miscellaneous

Table 1A. Accident Types 1-3

AT4	Failure of concrete structure during submergence and inclination test	5×10^{-3}	50%	- Ballasting failure during submergence test. - Failure in a temporary closure or around a penetration - Failure of concrete structure - Miscellaneous
AT5	Failure during deck installation and hook up inshore	5×10^{-4}	5%	- Load transfer monitoring failure. - Mooring failure - Ballasting failure during deck installation - Barge ballasting failure - Horizontal fine position failure - Miscellaneous
AT6	Failure of platform during tow to field	10^{-3}	10%	- Towing failure during tow to filed, inshore - Navigation error - Towing failure during tow to field, offshore - Passing vessel collision - Miscellaneous
AT7	Failure of foundation during installation	10^{-3}	10%	- Penetration failure - Tug failure due to extreme weather conditions during positioning - Positioning failure - Miscellaneous

Legend: AT: Accident Types

Table 1B: Accident Types 4-7.
Classification of accident types for the PRA Model and probability per operation at an event of each type and initiating events for each class of failure (the probabilities shown here are based on the opinions of our expert Dr. Gudmestad).

To obtain the overall failure probability, we sum the products of the probabilities of the different accident types and of the probabilities of the different classes of accident scenarios in each accident type. The probability of each scenario is obtained by multiplying the probability of the initiating event given the accident type, and the joint probability of different intermediate events of the scenario (A, B, C, etc.) conditional on the initiating event. The possible outcomes are subdivided into a set of mutually exclusive realizations that cover all the possibilities of loss of assets or delays in project. The probability of each outcome is computed by summing the joint probabilities of each accident sequence and that outcome. Table 2 shows the possible outcomes considered in this analysis. The final states are divided into different damage levels.

Possible outcomes (Consequence categories)	AT5, AT6, AT7	AT3	AT1, AT2, AT3
	Loss of Platform, OC_{p1}	Loss of Deck, OC_{D1}	Loss of Substructure, OC_{G1}
	Major damage to Platform, OC_{p2}	Major damage to Deck, OC_{D2}	Major damage to Substructure, OC_{G2}
	Minor damage to Platform, OC_{p4}	Minor damage to Deck, OC_{D3}	Minor damage to Substructure, OC_{G3}
		No damage to the Deck, OC_{D4}	No damage to Substructure, OC_{G4}
	No damage to Platform, OC_{p4}	No damage to the Deck, OC_{D4}	No damage to Substructure, OC_{G4}

Legend: AT: accident types
OC: Outcome
OC_P: Consequence categories for the platform
OC_D: Consequence categories for the deck
OC_G: Consequence categories for the substructure

Table 2: Outcomes of possible accident scenarios, (Rettedal et al., 1994).

ILLUSTRATION: TOWING FAILURE OFFSHORE (AT6)

Probabilistic risk analysis for this accident sequence

The object of this analysis is to allow linkage of the elements of the PRA with different relevant error types and management factors. Focusing on Accident Type 6 during the offshore and inshore tow of the platform, we consider here a specific initiating event: towing failure offshore. We construct a single influence diagram to analyze the probabilities of all possible outcomes of this initiating event. This influence diagram is shown in Figure 3.

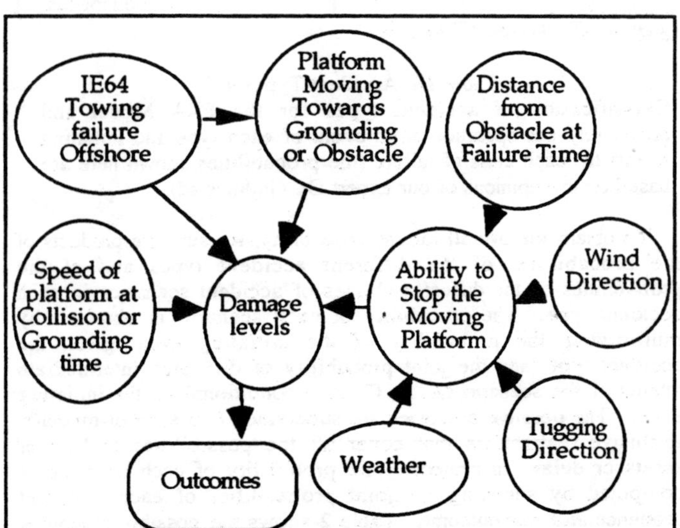

Figure 3. Accident Type: Failure of platform during tow to field. Initiating event: Towing failure offshore.

A towing failure is a failure of the tugs to control the platform movements. This could be caused by a tug failure, or a towline failure, or a combination of both. Towing failure can be caused by bad weather or by the failure of a tug which is critical to the ongoing maneuver (e.g. when changing direction). The consequences of a towline or tug failure may be major or minor damage to the platform due to grounding or collision, or no damage at all if the system recovers. If only one tug fails, there is a possibility for the other tugs to reorder and recover the control of the platform. If a towline fails, it can be replaced by another towline. The probability of such an initiating event has been estimated at 6.0×10^{-3} per operation (Trbojevic et al., 1994).

For the initiating event "towing failure offshore", the main accident sequence is defined as follows: if a towing failure occurs offshore, there is a small possibility that the platform will move towards a hazard, i.e. shallow water or an obstacle. The probability of that event depends upon weather conditions and the direction of tugging (whether the tugs are pulling towards or away from the obstacle). The direction of the wind and currents can also influence whether the loose structure moves towards a hazard. The final state of the system (level of damage to the platform) depends on the operator's ability to stop the movement of the platform before it hits the sea bed or an obstacle, and if not, on the speed of the loose structure at impact.

Effects of resource constraints on the risk of towing failure

The decisions and actions that influence these basic events are classified as maintenance errors, and errors of judgment in the decision to tow in uncertain weather. The maintenance level of a towline or a tug engine influences the probability of failure of the towing. The decision to tow in uncertain weather influences the probability of a towing failure and the probability that the platform is stopped in time given a towing failure. The organizational factors influencing these decisions and actions are limited in this analysis to schedule and budget constraints. A tight budget may force the management to choose the lowest bid which may include inexperienced crew and old tugs in bad condition. Further the schedule pressure which includes towing within a weather window may force management to take the tugs which are available at the moment. There is a probability that again these are old tugs with an inexperienced crew. To optimize cost, the management further may decide to place the backup tug far away from the tow if this turns out to be a cheaper solution. To increase the probability of controlling the platform movements if a tug or towline break, simulator training is important. Budget and time constraints may cut the scope of this training or prevent it completely (see Figure 4).

The influence diagram of Figure 4 was used to compute the effect of resource constraints in a specified environment for the chosen accident sequence, i.e., an increase of the probability of towing failure offshore due to time and budget constraints. This effect was computed by difference with a "base case" involving the probability of the various outcomes in the absence of exceptional constraints. The lowest level in the diagrams displays the basic events of the technical probabilistic risk analysis (accident scenarios). The second level shows the human actions and decisions and their influence on the basic events of the PRA. The top (third) level identifies the organizational factors that influence the human actions and decisions.

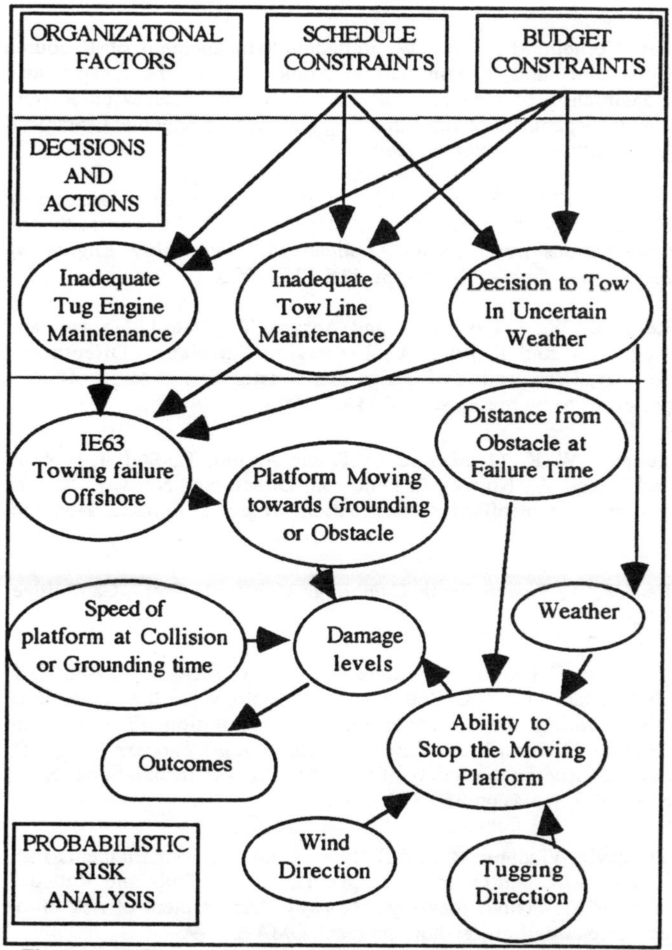

Figure 4. Influence diagram showing the effect of organizational factors on the risk of towing failure offshore.

The data were based in large part on the opinions of our expert (Dr. Gudmestad from Statoil). For example, given his experience, he estimated that under time and budget constraints, the probabilities of inadequate tug engine maintenance was 0.2 and the probability inadequate tow line maintenance was 0.1. As an example, his estimation for the probabilities of towing failure offshore per operation (IE63) given good weather and given the performance of the tug engine and the tow line are shown in Table 3. Given bad weather, he estimated these probabilities of failure to be twice as high. The results of the influence diagram analysis are shown in Table 4.

Tug Engine Failure	Tow Line Failure	Probability of Towing Failure
Yes	Yes	0.024
Yes	No	0.012
No	Yes	0.012
No	No	0.006

Table 3: Sample of data: expert opinion about the influence of tug engine failure and tow line failure on the probability of towing failure offshore, per operation and in good weather

Damage Levels	p(F), Base Case	Adding Potential Errors in Decisions and Actions	% Increase in P(F)
OC_{P2}	0.55×10^{-3}	0.9×10^{-3}	63%
OC_{P3}	2.59×10^{-3}	4.2×10^{-3}	62%
OC_{P4}	0.99685	0.99489	0.5%

Legend: OCpi: Outcome descriptor (different levels of overall platform failures)
i=2: Major damage to the platform
i=3: Minor damage to the platform
i=4: No damage to the platform

Table 4: Effects of resource constraints on the probability of a towing failure offshore

Again the figures shown in Table 4 are based on the opinions of our expert (Dr. Gudmestad from Statoil) and on the SAM model as displayed in the influence diagram of Figure 4. It is thus important to stress that they only give an indication of what the risk level *might* be in a specific situation. These figures probably vary considerably from case to case. These data should therefore be adapted to the situation at hand in order to make specific decisions in a given environment. The results indicate that, in the considered situation, the human errors (decisions and actions) potentially induced by time and budget constraints can increase the probability of total loss of the structure and the probability of significant damage to the platform by about 60%. The SAM model and the use of the influence diagrams corresponding to the different accident types thus permit a more detailed identification of the underlying root causes of accidents in the case of a specific organization, of their effects on the failure risk, and therefore, of the potential benefits of different policies affecting operators' decisions and actions.

CONCLUSIONS

The effects of human and organizational errors on the reliability of platforms is clearly significant. It should be noted, however, that almost all technical failures have a human component that is implicit in the record of system performance. Therefore, one should be careful when examining such effects to avoid double counting: human errors *do* enter past failure statistics simply because they influenced the frequencies of past accidents. In the absence of statistical data (which should be gathered when possible to obtain better results), we based our analysis on the work of Trbojevic (1994) and Rettedal (1994) and on the opinion of a Statoil expert (Gudmestad, 1996). Severe errors in design and construction are often due to shortcuts that people take under pressure, sometimes because they have no idea of the effects of what they are doing, sometimes because they underestimate the risk increase, and other times because they feel that even though the consequences could be severe, they have no choice. We computed that for the accident sequence that was examined here (towing failure offshore), these constraints, if not properly managed, could increase by 60% the probability of a severe outcome. This result was derived in part from expert opinions based on the environment with which the expert is familiar. Nonetheless, it suggests that for critical systems that could endanger human lives and environmental quality, resource constraints should be carefully set. A risk analysis model is the best way to assess the effects of potential shortcuts on the basic integrity of the system.

ACKNOWLEDGMENT

This study was supported in part by STATOIL and by the Stanford Department of Industrial Engineering and Engineering Management. Their support is gratefully acknowledged. This work was performed during Wenche Rettedal's stay at Stanford University during the Winter quarter of 1995-1996.

REFERENCES

Bellamy, L.J. and Geyer, T.A.W., "Organizational, Management and Human Factors in Quantified Risk Assessment," *HSE Contract Research Report*, Vol 33, 1991.

Bignell, V. and Fortune, J., *Understanding Systems Failures*, Manchester University Press, Manchester, UK, 1984.

Broughton P., Phillips Petroleum Company and K. Waagaard, Veritas, Norway; Civil Engineers, The Ekofisk Protective Barrier, Det Norske Veritas, Technical Report No. 541086 - E, 1977.

Gudmestad, O.T., Personal Communication, 1996.

Jacobsen, B., Norwegian Contractors, Stabekk, Norway, "The loss of Sleipner A Platform," *Proceedings of the 2nd (1992) International Offshore and Polar Engineering Conference*, San Francisco, CA, June 1992.

Murphy, D.M. and M.E. Paté-Cornell, "The SAM Framework: A Systems Analysis Approach to Modeling the Effects of Management on Human Behavior in Risk Analysis," *Risk Analysis*, Vol 16, No 4, pp 501-515, 1996.

Norwegian Petroleum Directorate, Data base for damage to load-bearing structures, CODAM, 1996.

Paté-Cornell, M.E., "Organizational Aspects of Engineering System Safety: the Case of Offshore Platforms," *Science*, Vol 250, pp 1210-1217, November 1990.

Paté-Cornell M.E. and R. G. Bea, "Organizational Aspects of Reliability Management: Design, Construction and Operation of Offshore Platforms," Research Report #89-1, Department of Industrial Eng and Eng Management, Stanfrod University, CA, April, 1989.

Paté-Cornell, M. E., "Risk Analysis and Risk Management of Offshore Platforms: Lessons From the Piper Alpha Accident," *J. of Offshore Mech and Arctic Eng*, Vol 115,179, August 1993.

Paté-Cornell, M.E., "Structure and Illustration of a Probabilistic Risk Analysis Model for Space Shuttle EVAs," *Proceedings of the ASME Int Mech EngiCongress and Exhibition (IMECE95)*, San Francisco, CA, November 1995.

Paté-Cornell, M.E. and P.S. Fischbeck, "Risk Management for the Tiles of the Space Shuttle," *Interfaces*, Vol 24, pp 64-86, January-February 1994.

Paté-Cornell, M.E., D.M. Murphy, L.M. Lakats and D. M. Gaba, "Patient Risk in Anesthesia: Probabilistic Risk Analysis, Management Effects and Improvements," in press, *Annals of Operations Research*, Vol 67, pp 211-233, 1996.

Paté-Cornell, M.E., and W. Rettedal, "Management of Resource Constraints and System Safety: Application to the Design and Construction of Concrete Offshore Platforms," Research Report #96-1, Department of Industrial Eng and Eng Management, Stanford University, CA, October, 1996.

Paté-Cornell, M.E. and D. M. Murphy, "Human and Management Factors in Probabilistic Risk Analysis: the SAM Approach and Observations from Recent Applications," *Reliability Engieering and System Safety*, Vol 53, pp 115-126, 1996.

Rettedal, W., Gillesvik, A and Aarum, T., Statoil, Norway and Vegge, A. and Kvitrud, A., Norwegian Petroleum Directorate, Norway, "Verification of design and construction of large offshore projects," to be presented at OMAE, 1996.

Rettedal, W. K., Gudmestad, O. T. and Aarum, T., Statoil, Norway and Vegge, A., Norwegian Petroleum Directorate, Norway, "Use of risk analysis in offshore construction projects," OMAE, Houston, TX, 1994.

Reason, J., *Human Error*, Cambridge University Press, Cambridge, UK, 1990.

Trbojevic,, Vladimir M., Linda J. Bellamy, P.G. Brabazon, O.T. Gudmestad, and WK. Rettedal: "Methodology for the Analysis of Risks during the Construction and Installation Phases of an Offshore Platform," *J. of Loss Prevention in the Process Industries*, Special Issue on the Safety of Offshore Process Installations: North Sea Vol 7, No 4, pp 350-359, 1994.

Trbojevic, Vladimir M. and Bellamy, Linda J., Four Elements Ltd., London, UK. and Gudmestad, Ove T., Aarum, Truls and Rettedal, Wenche K., Statoil, Stavanger, Norway, "Assessment of risk in the design phase of an offshore project," OMAE, 1996.

United States Nuclear Regulatory Commission (USNRC), *Reactor Safety Study, An Assessment of Accident Risks in U.S. Commercial Nuclear Power Plants*, WASH-1400 (NUREG-75/014), Washington D.C., 1975.

Vegge A. and A. Kvitrud, "Structural accidental events on offshore structures in Norway," Norwegian Petroleum Directorate, Norway, 1992.

Proceedings of the Seventh (1997) International Offshore and Polar Engineering Conference
Honolulu, USA, May 25-30, 1997

ISBN 1-880653-28-1 (Set); ISBN 1-880653-32-X (Vol. IV)

Risk of Collisions and Groundings in Bulk Carriers and Tankers

Alexander M. Goulielmos and Kostas Giziakis
University of Piraeus
Piraeus, Greece

Ernestini Giziaki
University of Athens
Athens, Greece

ABSTRACT

The authors have used 3575 ship accidents reported during 1994. The purpose was to calculate the risk of collisions and groundings among bulk carriers and tankers. This has been done by using the Odds Ratio and the Accident Rate as well as the Mantel- Haenszel techniques and chi-square trend statistic. The results are presented in tables (18 tables). Bulk carriers and container ships were found to be the most risky types of vessels (98 and 74 per 1000 ships). Bulk carrier accidents due to collisions and contacts are almost 28% of the total (59% of them occurred in N. America and Europe), while accidents due to groundings are 21% (54% have occurred in N. America and Europe). The USA flag holds the first position according to accident rate in collisions and groundings of bulk carriers. The influence of other factors such as age and size has been calculated and risk analysis applied. The above applications were also extended to tanker collisions and groundings.

KEY WORDS: Risk of collisions, Risk of groundings in Bulk carriers and Tankers- 1994.

INTRODUCTION

The purpose of this article is the calculation of the risk involved in specific types of accidents, those of collision and grounding, among specific types of ships, bulk carriers and tankers. Using a world-wide reliable marine accidents reporting (weekly) system, that of Lloyd's Register of Shipping, and the database created in the Marine Accident Laboratory in Piraeus (MAL), we processed 3575 ship accidents in 1994. Further, the fleet characteristics that were used were also provided by the Lloyd's Register of Shipping, World Fleet Statistics for 1994.

Our task was to quantify the risk of collisions and groundings and we thus decided to use the so called Odds Ratio (OR), i.e. the odds of failure of those with the risk factor compared with those without. This measure states by how much the probability that a specific event will happen changes, if a certain risk factor is present. Moreover, the accident rate is used, which indicates the probability of a specific accident, like collision or grounding, related to the rate of accidents per 1000 ships.

The statistical tools which were used for the analysis of data are the Mantel - Haenszel chi square, the chi-square trend statistic and the Mantel - Haenszel weighted OR to control for the effects of possible confounders.

In general, types of accidents will be first presented as well as the Accident Ratios by types of ships. We shall then analyse bulk carriers and tankers separately.

For bulk carriers the type of accidents will be given focusing on collisions and groundings. Moreover, the absolute number and percentage distribution of groundings and collisions by area will be given subject to certain limitations due to lack of rather expensive data.

As flag is the natural subject that concentrates attention, a flag analysis was attempted related to Accident Rate (i.e. per 1000 existing ships). The flag was ranked starting with that experiencing the highest accident rate.

Moreover, other factors will be examined, such as the influence of age on collisions and groundings. However, it will also be worthwhile to examine the possible association between age and risk of groundings and collisions as well as whether there is a linear trend between categories of age and risk of collision.

In addition, the size of ships was examined in relation to groundings and collisions. Here we tried to eliminate the possible confounding effect of age from size. A similar approach to that for bulk carriers was also attempted for tankers. 430 tanker accidents of all types have been reported and analysed.

DATA

The accident data file for the present study was extracted from the database created at the University of Piraeus. The main source of information for this database is the raw texts of Lloyd's Casualty Weekly. It contains all reported accidents all over the world for ships of over 100 grt, and includes data for total losses as well as for small incidents.

In the present analysis the most recent corrected and completed year was studied. There were 3575 reported ship accidents in 1994. The ship particulars were taken from Lloyd's Register of Shipping, that is name, type, age, flag, classification society, engine type etc.

A subfile was created for collisions and groundings which involved tankers and bulk carriers.
To calculate risks the world fleet characteristics are needed. These data were ordered from Lloyd's Register of Shipping and provided such information like type of ship, size, year of built and flag

METHODOLOGY

A proper risk analysis includes: (a) the risk assessment and (b) the risk management. Risk assessment quantifies the risk of failure and identifies the association of certain factors with this risk. Risk management takes specific measures to manage (prevent or reduce) the risk of accidents by using the results of risk assessment (E. Giziaki, 1997).

In order to quantify the risk of collisions / groundings we calculated a descriptive measure, the **Odds Ratio (OR)**. Odds Ratio is defined as the odds of failure among those with the risk factor in relation to the odds of failure among those without the risk factor. It measures the risk of an accident for those ships exposed to a specific factor compared to those ships not exposed to the factor under consideration. In other words, it expresses by how much the probability that a specified event will occur changes in the presence of a certain risk factor (Giziakis and Giziaki, 1996a) (Giziakis and Giziaki, 1996b). A value of OR equal to 1 indicates no change, a value greater than 1 denotes an increased risk, whereas a value less than 1 indicates a reduced risk.

Analysis of the data using the above descriptive measure provided an index that facilitates the interpretation of the data. In our presentation we also used another measure, that is the **accident rate**. The accident rate is an index of the probability of a specific accident, namely collision, and gives the rate of accidents per 1000 ships.

For the statistical analyses we also used : (a) the Mantel-Haenszel chi-square that tests the null hypothesis that the OR is equal to unity, (b) the chi-square trend statistic, (c) the Mantel-Haenszel weighted OR to control for the effect of possible confounders (Checkoway, Pearce and Crawford - Brown,1989) (Giziaki,1997).

RESULTS

Presentation of all accidents data.

The data file that includes all reported accidents for 1994 is presented in two tables. Table 1 shows an analysis of data by type of accident. It can be seen that the **collision** and contact types of accident amount to 27.8% and **groundings** to 15.4% of the total number of ships reported to have had an accident. These figures are for all types of ships reported to have had an accident.

TABLE 1 : ANALYSIS BY TYPE OF ACCIDENT FOR ALL REPORTED ACCIDENTS FOR 1994 .

TYPE OF ACCIDENT	Number of Accidents	% of Total
Foundered	438	12.3
Missing	29	1.0
Fires & Explosions	289	8.0
Collisions	566	15.8
Contact-struck	437	12.0
Grounded	551	15.4
Capsized-Listed-Shifted	87	2.3
Breakdownand Damaged	990	27.7
War Losses-Hostilities	24	1.0
Broke - Cracked	122	3.3
Other and Unknown	42	1.2
T O T A L	3575	100.0

Table 2 presents the Accident Ratio (AR) by type of ship. It expresses the number of accidents per 1000 existing ships . Comparisons can be made between all ship types since the population of ships in each type has been taken into consideration. The **bulk carrier and container types of ship seem to bear the two largest accident ratios for 1994, that is 98 and 74 per 1000 ships respectively.**

TABLE 2 : ACCIDENT RATES BY TYPE OF SHIP FOR 1994 .

TYPE OF SHIP	ACCIDENT RATE per 1000 ships
General Cargo	49
Bulk Carrier	98
Tanker	45
Ferry Ro/Ro	67
Fishing	15
Container	74
Passenger	22
ALL TYPES	44

Results for bulk carriers.

Descriptive Analysis.

An analysis by type of accident is shown in table 3. Collisions and contacts amount to 28.1% of the total number of accidents for bulk carriers. This type of accident is mainly due to human error. According to OECD, using only total losses from Lloyd's Register of Shipping, collision is essentially the result of short- comings in personnel, but also is due to congestion; moreover, the ship lost may not be the guilty ship (OECD 1974). Fatigue has also been incriminated for collisions (Goulielmos A- Tzannatos E, 1995). Using results of some 7000 major casualties over the period 1981- 1990 (Wijnolst N. and Wergeland

T., 1996), groundings (including stranding) and collisions (including contact with quay) were responsible for 34% of them. These two were the first and third most frequent causes (negligence of crew, repairs, owner etc. was the second cause with almost 19% of casualties). The human factor has been considered (US Dept. of Commerce 1976) responsible for 80% of all marine accidents over the last 31 years. Thus ISM Code and 1995 STCW amendments are hoped to become the means by which marine accidents will be effectively reduced as they concentrate on human factor. 9.7% of collisions and contacts were due to **mechanical failures** that resulted in collisions.

TABLE 3 : ANALYSIS OF BULK CARRIERS BY TYPE OF ACCIDENT .

TYPE OF ACCIDENT	Number of Accidents	% of total
Foundered - Missing	14	3.0
Fires & Explosions	31	6.5
Collisions	61	12.8
Contact	73	15.4
Grounded	98	20.7
Breakdown & Damaged	170	35.8
Broke	23	4.8
Other-Unknown	5	1.0
TOTAL	475	100.0

Groundings amount to 20.7% of the number of accidents for bulk carriers. 10% of groundings were due to **mechanical failures**. That is, 90% of groundings were probably due to human error. The areas in which a collision-contact has occurred have been reported and presented in table 4. The areas have been categorised subjectively into four groups. 58.6% of the total number of collisions took place in N. America and European waters.

In table 4, the absolute number and percentage distribution of groundings by area in which a grounding has occurred are presented. Again, about 54% of the total number of groundings took place in N. America and around Europe. A comparison among areas is not possible since the total number of bulk carrier ship movements in the specific area would be needed. To calculate a risk coefficient for the areas, the number of accidents and the ship movements for the area would be required. This in effect entails a cost that the authors cannot undertake without sponsorship.

TABLE 4 : DISTRIBUTION OF COLLISIONS-CONTACTS AND GROUNDINGS BY AREA OF OCCURRENCE FOR BULK CARRIERS .

AREAS	COLLISIONS		G ROUNDINGS	
	Number	%	Number	%
N. America	42	28.6	30	27.9
European Waters	44	30.0	28	26.1
Japanese Waters	16	10.9	8	7.5
Other Areas	45	30.5	41	38.5
TOTAL	147	100.0	107	100.0

An analysis by flag is presented in tables 5 and 6. The Accident Rate per 1000 existing ships is shown. Flags which have fewer than 10 ships in their bulk carrier fleet are excluded from the corresponding tables mainly because the AR can be extremely high only by chance. For example if 2 bulk carriers fly a certain flag and one of them had a small incident, then the AR will be 500, and this will be misleading. Tables 5 and 6 include ten flags ranked according to the highest Accident Rate. There are overall 39 flags, which have a bulk carrier fleet of over 10 ships, in which the loss ratio is lower than the last one shown in the corresponding tables. As shown, the ten most risky flags start with the USA, followed by Thailand and Canada in case of collisions. For groundings the USA is again first, followed by Canada and Chile.

TABLE 5 : RANKING OF FLAGS ACCORDING TO THE ACCIDENT RATE OF COLLISIONS FOR BULK CARRIERS .

FLAG (country)	Accident Rate (per 1000 ships)
United States	192
Thailand	111
Canada	95
Marshal Islands	87
Denmark	83
Vanuatu	75
Norway	59
Bulgaria	55
Greece	42
India	41

TABLE 6 : RANKING OF FLAGS ACCORDING TO THE ACCIDENT RATE OF GROUNDINGS FOR BULK CARRIERS .

FLAG (country)	Accident Rate (per 1000 ships)
United States	192
Canada	119
Chile	71
Bahamas	57
Thailand	55
Australia	48
Norway	44
Romania	44
Marshal Islands	43
Iran	41

The factor **age** and its influence on collisions and groundings for bulk carriers is shown in table 7. The Accident Rate is a measure which takes into account, as stated, the number of existing ships in each age category. It is perhaps strange to notice that the age factor can influence collisions and groundings. An explanation may be that these types of accidents are mainly due again to human factor. This finding agrees with the Analysis of Major Claims of UK P & I club (1992) , where human error was the main cause for 90% of the collisions reported . From the value of AR it can be concluded that there are two levels of AR for both collisions and groundings. Of the two levels for collisions, one corresponds to the age category up to 10 yrs old and the other to the category from 10 yrs and over. For groundings these two levels are ships up to 15 years old and ships above 15 yrs old.

TABLE 7: ACCIDENT RATE OF COLLISIONS AND GROUNDINGS FOR BULK CARRIERS BY AGE CATEGORY.

AGE CATEGORY	Accident Rate for Collisions (per 1000 ships)	Accident Rate for Groundings (per 1000 ships)
less than 5 yr.	15	13
5 - 10 yr.	18	16
10 - 15 yr.	33	13
15 - 20 yr.	32	24
20 - 25 yr.	32	29
25 or over	37	27

The influence of the **size** factor is presented in table 8. The AR in both types of accidents is reduced as the size increases. One explanation for this result is that large ships use pilot services in close waters or are monitored more closely from the onshore stations in these waters. Larger vessels, as being more costly, are entrusted to more experienced captains, and belong to better organised shipping companies which relay more accurate instructions to ship captains.

TABLE 8: ACCIDENT RATE OF COLLISIONS AND GROUNDINGS FOR BULK CARRIERS BY SIZE GROUP.

SIZE GROUP (in DWT)	Accident Rate for Collisions (per 1000 ships)	Accident Rate for Groundings (per 1000 ships)
less than 10000	57	48
10000 - 30000	34	22
30000 - 50000	30	20
50000 - 80000	27	18
80000 or over	10	10

Risk Analysis.

In table 9 we have examined the possible association between **age** and the **risk of an accident** (collision or grounding) in bulk carriers. We compare age categories from 5 years onwards to the reference smallest age category. The reference class is that of less than 5 years of age. A bulk carrier of age 10-15 yrs is subject to a 2.25 times higher risk for a collision as compared to a bulk carrier aged less than 5. The risk of a collision becomes 2.54 times greater in the age group 25 years and over as compared to the age group 0-5 yrs. We also examined whether there is a linear trend between age and the risk of collision. This test indicated that as the **age increases**, the **risk of collision gets higher**. When the age variable was grouped into two categories, that is one of ships less than 10 yrs old and another of ships over 10 yrs, the risk of collision became twice as great in the age category over 10 yrs (this result is not shown in any table). The risk of grounding for bulk carriers increases, from 1.18 to 2.09, as the ship gets older (table 9). Age as a secondary factor of collisions should be combined with a primary human error. Practical mechanical reasons for collisions have been advanced as secondary factors related to steering gear, radar reading, and engine response.

TABLE 9 : ASSOCIATION OF AGE WITH THE RISK OF COLLISIONS AND GROUNDINGS IN BULK CARRIERS

AGE GROUP (in years)	COLLISIONS Odds Ratio	(P- value)	GROUNDINGS Odds Ratio	(P- value)
- 5	1	reference	1	reference
5 - 10	1.22	(N.S.)	1.18	(N.S.)
10 - 15	2.25	(0.03)	1.01	(N.S.)
15 - 20	2.20	(0.038)	1.81	(N.S.)
20 - 25	2.17	(0.05)	2.22	(0.06)
25 or over	2.54	(0.04)	2.09	(N.S.)
Test for Trend	χ^2 = 5.96 (0.01)		χ^2 = 7.52 (0.0006)	

NS = Non - significant

In table 10 the effect of **size** of ship on the probability of **collision** and **grounding** is considered. The capesize bulkers have a reduced risk (OR=0.29) of collision as compared to the laker sizes. As the **size** of ship increases, the **risk of collision** is **reduced**, i.e from 0.86 to 0.29 (χ^2=8.19, P=0.004). Similarly for **groundings**, the **risk** is also **reduced** as the **size** of ship **increases**. The risk of a grounding in the capesize category is reduced to 0.45 as compared to the risk of grounding in a laker bulk carrier (table 10). We also attempted to eliminate the possible confounding effect of the age factor from the factor size. The ORs calculated from the stratified analysis are the adjusted OR. These OR follow the pattern of the crude OR presenting a reduced risk of collision from 0.94 to 0.27 as the size increases and a reduced risk of grounding for the capesize category (OR=0.62).

A possible explanation here is that smaller ships are obliged to make more journeys within a year due to distance, ceteris paribus, compared with large vessels. For groundings the nature of route should be taken into account (rivers, canals, channels) in connection with the breadth of the vessel.

Results for tankers.

Descriptive analysis

The subfile for tankers contains 430 accidents of all types. Table 11 presents the distribution of tankers by type of accident. The collisions and contact types of accident amount to 34.8% of all accidents to tankers. The grounding type includes 16.3% of the number of accidents.

TABLE 10 : ASSOCIATION OF SIZE OF SHIP WITH THE RISK OF COLLISIONS AND GROUNDINGS IN BULK CARRIERS

S I Z E group (in DWT)	COLLI SION S		GROUNDINGS`	
	O.R.	OR*	O.R.	OR*
-30000	1	1	1	1
30000 -50000	0.86	0.94	0.93	1.11
50000 -80000	0.78	0.72	0.83	1.00
80000 orover	0.29	0.27	0.45	0.62
Test for Trend	$\chi^2 = 8.19(0.004)$		$\chi^2 = 3.02\ (0.002)$	

OR*=Adjusted OR

TABLE 11 : ANALYSIS OF TANKERS BY TYPE OF ACCIDENT .

TYPE OF ACCIDENT	Number of Accidents	% of total
Foundered - Missing	8	2.0
Fires & Explosions	56	13.0
Collisions	86	20.1
Contact	63	14.7
Grounded	70	16.3
Breakdown & Damaged	127	29.5
Broke	10	2.2
Other-Unknown	10	2.2
T O T A L	430	100.0

The areas in which collisions and groundings have occurred are shown in table 12. Comparisons among areas cannot be made, as mentioned, since the traffic of vessels in the areas would be required. This table only give us an idea about the locations in which collisions and groundings have taken place. 22% of collisions took place in Japanese waters , while the corresponding figure for Europe is 31.4%. A percentage of 43.1 % of **groundings** occurred in Europe, while in Japanese waters this percentage was only 8.9%.

The ranking of **flags** according to the AR of collisions-contacts and groundings for tankers is shown in tables 13 and 14. The Accident Rate for flags having a tanker fleet of 10 ships or less has not been calculated, mainly because small numbers can distort the AR greatly. The Accident Rate has been calculated per 1000 existing ships and therefore comparisons can be made between flags. Tables 13 and 14 present the flags having the ten highest rates of accident out of the 75 flags with a tanker fleet of over 10 ships. As shown by these tables, Belgium holds the first place in collisions, followed by Greece and Vanuatu for 1994. In groundings Greece holds the first place, followed by Chile and Antigua for 1994.

TABLE 12:DISTRIBUTION OF COLLISIONS- CONTACTS AND GROUNDINGS BY AREA OF OCCURRENCE FOR TANKERS .

AREAS	COLLIS IONS		GROUNDINGS	
	Number	%	Number	%
N. America	18	11.8	15	19.0
European Waters	48	31.4	34	43.1
Japanese Waters	34	22.2	7	8.9
Other Areas	53	34.6	23	29.0
T O T A L	153	100.0	79	100.0

The ranking of **flags** according to the AR of collisions-contacts and groundings for tankers is shown in tables 13 and 14. The Accident Rate for flags having a tanker fleet of 10 ships or less has not been calculated, mainly because small numbers can distort the AR greatly. The Accident Rate has been calculated per 1000 existing ships and therefore comparisons can be made between flags. Tables 13 and 14 present the flags having the ten highest rates of accident out of the 75 flags with a tanker fleet of over 10 ships. As shown by these tables, Belgium holds the first place in collisions, followed by Greece and Vanuatu for 1994. In groundings Greece holds the first place, followed by Chile and Antigua for 1994.

TABLE 13 : RANKING OF FLAGS ACCORDING TO THE ACCIDENT RATE OF COLLISIONS FOR TANKERS .

FLAG (country)	Accident Rate (per 1000 ships)
Begium	90
Greece	75
Vanuatu	71
Antigua	71
Bermuda	64
Netherlands	63
United Arab Emirates	62
Argentina	62
Libya	62
Portugal	62

The accident rate per 1000 existing ships for several categories of **age** is shown in table 15. The rate for collisions and contacts does not present any constant pattern. Tankers of over 25 yrs have almost the same accident rate as tankers aged less than 5 yrs. There is only a peak in the age category 10-15 yrs. The AR for groundings is almost the same in ships up to 15 yrs old and is doubled for tankers of over 15 yrs old.

TABLE 14: RANKING OF FLAGS ACCORDING TO THE ACCIDENT RATE OF GROUNDINGS FOR TANKERS.

FLAG (country)	Accident Rate (per 1000 ships)
Greece	125
Chile	91
Antigua	71
Poland	67
Argentina	62
Ecuador	48
Finland	45
Cuba	45
Malta	42
Cyprus	31

TABLE 15: ACCIDENT RATE OF COLLISIONS AND GROUNDINGS FOR TANKERS BY AGE CATEGORY.

AGE CATEGORY	Accident Rate for Collisions (per 1000 ships)	Accident Rate for Groundings (per 1000 ships)
less than 5 yr.	9	5
5 - 10 yr.	13	3
10 - 15 yr.	23	5
15 - 20 yr.	12	10
20 - 25 yr.	19	10
25 or over	10	11

TABLE 16: ACCIDENT RATE OF COLLISIONS AND GROUNDINGS FOR TANKERS BY SIZE GROUP.

SIZE GROUP (in DWT)	Accident Rate for Collisions (per 1000 ships)	Accident Rate for Groundings (per 1000 ships)
less than 10000	12	6
10000 - 30000	15	12
30000 - 50000	20	12
50000 - 80000	25	21
80000 or over	20	5

In table 16 results are presented for collisions and groundings according to the size factor. The AR for collisions and groundings increase as the size of a tanker gets larger. Size can be related to the breadth of ships that sail on rivers and other narrow passages. This result is exactly the opposite of that for bulk carriers. The AR per 1000 existing ships in each size category for groundings indicates that small and very large tankers have almost the same rate. There is a peak of 21 accidents per 1000 existing ships in the category of 50000 to 80000 dwt.

Risk Analysis

In table 17 the effect of **age** on the risk of collision and groundings for tankers is considered. The risk of a collision is more than two times higher in the age categories 10-15 yrs and 20-25 yrs as compared to ships aged less than 5 yrs old. The risk of a grounding is more than two times higher in the age categories over 15 yrs as compared to the age group of less than 5 yrs old. There is a trend of an increased risk of grounding as a ship gets older ($\chi^2 = 10.1$, P=0.001).

TABLE 17: ASSOCIATION OF AGE WITH THE RISK OF COLLISIONS AND GROUNDINGS IN TANKERS

AGE GROUP (in years)	COLLISIONS O.R.	(P-value)	GROUNDINGS O.R.	(P-value)
- 5	1	reference	1	reference
5 - 10	1.49	(N.S.)	0.65	(N.S.)
10 - 15	2.69	(0.0008)	1.18	(N.S.)
15 - 20	1.42	(N.S.)	2.14	(0.06)
20 - 25	2.17	(0.01)	2.11	(N.S)
25 or over	1.18	(N.S.)	2.50	(0.03)
Test for Trend	There is no trend		$\chi^2 = 10.1$ (0.001)	

NS = Non - significant

In table 18 the effect of **size** has been further investigated. In the size group 50000-80000 dwt and the group over 80000 dwt the risk of collision is 2.47 and 1.7 times greater respectively, than in the group up to 10000 dwt. The risk of collision becomes higher as the size increases ($\chi^2 = 8.94$, P=0.003). The effect of size was greater when we examined the risk of a grounding. This risk becomes more than 3 times higher in the size group from 50000 to 80000 dwt, while it is reduced in the size category over 80000 dwt. When the odds ratios are adjusted to control the effect of age on size, the OR values were in the same direction. For groundings the values were higher than the crude OR, reaching the value of 5 for the size group 50000-80000 dwt. This size category, 50000-80000, has a risk almost twice as great as the size 30000-50000 dwt.

TABLE 18 : ASSOCIATION OF SIZE OF SHIP WITH THE RISK OF COLLISIONS AND GROUNDINGS IN TANKERS

SIZE group (in DWT)	COLLISIONS		GROUNDINGS	
	O.R	O.R.*	O.R.	O.R.*
- 10000	1	1	1	1
10000 - 30000	1.24	1.18	2.13	2.45
30000 - 50000	1.73	1.70	2.07	2.57
50000 - 80000	2.47	1.93	3.56	5.01
80000 or over	1.69	1.82	0.84	0.92
Test for Trend	$\chi^2 = 8.94$ (0.003)			

CONCLUSIONS

(A) Bulk Carriers

As has been presented in the general presentation (table 1), collisions (including contacts) cover almost 28%, and groundings almost 15%, of all reported ship accidents. To make figures comparable, Accident Ratio has been calculated by ship type expressing number of accidents per 1000 existing ships. It was found that bulk carriers and container ships share the largest accident ratio for 1994.

Bulk carriers were next examined. Collisions (including contacts) account for almost 28% of the total number of accidents in these ships. A search in the literature has led us to conclude that the human factor can be considered as the main cause for this type of marine accident. Mechanical failures explained only about 10% of collisions (and contacts). Groundings account for up to 21% of the accidents of bulk carriers. Mechanical failures again contributed to10% of groundings. By inference, 90% of groundings should be attributed to the human factor.

Our effort to gain insights into the geographical areas in which collisions took place has shown that N. America and European waters hosted about 59% of the total number of collisions. Groundings, too, followed the same pattern, with 54% of total marine groundings in these waters. This area needs more research if funds are available for it . For further investigation the total number of bulk carrier ship movements in these areas is required.

Flags of 39 nations attracted our attention (tables 5,6), but we concentrated on only the first ten flags in accordance with the level of Accident Rate. Risky flags were revealed to be the USA, Thailand and Canada for collisions. In groundings risky flags were the USA, Canada and Chile.

Next the influence of age on collisions and groundings was examined (table 7). **Indeed age influences collisions and groundings**, perhaps due to the predominance of the human factor (Goulielmos, 1996) combined with age. Age is examined further below.

Moreover, size was the next candidate for examination. Indeed, size is negatively related to both groundings and collisions (table 8). One tends to conclude that large bulk carriers may be preferable to small ones, if commercial criteria endorse this decision, as they seem to have fewer collisions and groundings. Size is further analysed below.

Is there any possible association between age and risk for collision or grounding in bulk carriers? (table 9). Indeed, a bulk carrier 10-15 yrs old **runs a 2.25 times higher risk for collision** than a bulk carrier of less than 5 yrs old .This risk remains at the same level (2.54 times) when the old bulk carriers (25 years plus) are considered. These differences are of interest to decision makers in order for the most appropriate policies to be applied. This result may be based on a specific generation of bulk carriers built to faulty specifications during a particular decade (e.g.1979- 1984), while another influence may be due to age as a secondary factor to the human one.

Insisting on the matter of age and risk of collisions in b.c., a test of linear trend was attempted (IMSOR, 1993). Indeed, as age rises, the risk of collision increases. It thus seems that bulk carriers older than 10 years are prone to more frequent collisions. It is not only the risk of a collision but also the risk of a grounding that rises (2.09 times) as ships become too old (table 9).

What is the probability of collision and grounding in B.C. due to size? Table 10 gives the following answers:

	Size of Bulk carriers	Risk (OR) :	Risk (Adj. OR)
1.	Capesize, collision	0.29	0.27
2.	Laker size, collision	0.86	0.94
3.	Capesize, groundings	0.45	0.62
4.	Laker size, groundings	0.93	1.11

B.C. between 30,000- 50,000 dwt have 3 times more risk for collision than the 80,000 plus ones. For groundings the risk is double between these size classes. Hence, the larger the B.C., the safer it is from groundings and collisions. The situation was no different when the possible age effect was removed.

(B) Tankers

Tankers had 430 accidents of all kinds. Collisions covered 35% of all accidents occurring to tankers, while groundings covered about 16%.

Areas where collisions took place were Europe with about 31% and Japanese waters with about 22%. Grounding percentages were: 43% in Europe and only 9% in Japan (table 12).

Out of 75 flags, ten were presented in tables 13 and 14 and among the 10 worst were those of Belgium, followed by Greece and Vanuatu, for collisions, while the ranking in groundings singles out at Greece, followed by Chile and Antigua.

As far as the age of tankers is concerned (table 15) we did not have a differentiating pattern. Old tankers (25 years and plus) behave accident-wise (collision, grounding) in the same way as newer tankers (5 yrs less). A peak, however, has been identified in tankers aged 10-15 yrs. In groundings there was an unvarying level of risk up to 15 yrs tankers, but AR doubles for those aged 15 yrs plus. Thus tankers in advanced age (15 yrs plus or 25 yrs plus) are more prone to grounding and collision . The human factor as a primary cause may be combined with problems due to the obsolescence of engine, steering gear and radar.

Size of tankers is the mysterious factor for both collisions and groundings (table 16). AR indeed increases as size gets larger. This tendency was not observed in bulk carriers. Both small and very large tankers have the same AR for groundings. A peak, however, has been identified (21 accidents per 1000 existing ships) in medium- sized tankers (50,000 to 80,000 dwt). The nature of the route should be examined here in relation to rivers and narrow passages. Smaller tankers are exposed to risks more often, since they carry out more voyages.

The risk analysis applied to tankers also stressed the relation between age and collision and grounding risk (table 17). Tankers aged 10-15 and 20-25 show a more than double risk for collisions (against the basic class of up to 5 yrs). Tankers aged 15 yrs and over are more than twice as prone to groundings. Aged ships present a higher probability of grounding. Age may be a secondary factor to human error.

Next, tanker size was investigated using risk analysis (table 18). Tankers 50,000 to 80,000 dwt and 80,000 plus have almost twice as much risk (2.47 to 1.70) as small tankers (up to 10,000 dwt). Size and collision show a positive relationship. For relatively large tankers (50,000 to 80,000 dwt) the risk of a grounding becomes three times higher, but it falls for large tankers above 80,000 dwt. When the probable age effect was eliminated, the influence of size did not alter, but it increased in the case of groundings for tankers sized 50,000 to 80,000 dwt, giving a two times higher risk compared to tankers of size 30,000 to 50,000 dwt. Possible explanations have been given above.

FURTHER RESEARCH

Information is needed about crew qualifications for ships that have successfully completed their voyage. For ships that fail information is usually available.

It is worth mentioning that the data used in this paper refer not only to total losses but also to minor marine incidents.

There are, of course, also other factors that should be investigated and their contribution to the risk of failure should be reckoned. Such factors are: classification society, type of engine, cargo, route, loading conditions. These can be tested either on their own or in combination. However, data about the population of ships having the characteristic under investigation would be required, and such can only be obtained at a high cost. Such data, if they were available, could allow the calculation of risk for marine accidents in each geographical area, testing also for the effectiveness of such policy measures as OPA etc

REFERENCES

Checkoway H., Pearce N.E. and Crawford-Brown D.J. (1989), "*Research Methods in Occupational Epidemiology*, Oxford Univ. Press.

Giziaki E., (1997), "Statistical Methodology for risk assessment in marine accidents", *Essays in memory of the late Prof. Emeritus of University of Piraeus B.N. Metaxas*.

Giziakis K. and Giziaki E., (1996a), "Analysing the risk of fires in shipping", *International Fire Safety Conference, Proceedings*, 24-25 May, 1996, pp.123-135.

Giziakis K. and Giziaki E., (1996b), "Risk Assessment Methods in Shipping Safety" , *The Journal of Science and Technology*, No.2,pp.17-28.

Goulielmos A.M., (1996), "The human factor and its role in Quality Shipping" , *Mare Forum Conference*, Erasmus University, Rotterdam, October 1996.

Goulielmos A.M. and Tzannatos E., (1996), "Management Information System for the promotion of safety in shipping", *IAME, 1996 International Conference -shipping, ports and logistics services: solutions for global issues* . Vancouver, B.C., Canada, June 26-28.

IMSOR, (1993), "Technical Report". Major marine casualties for vessels larger than 10000 dwt for 1984-1992 based on data from the Institute of London Underwriters were primarily attributed to age than flag or classification society.

Lloyd's Register of Shipping, "Weekly Casualty Returns".

OECD, (1974), "Marine Transport", ch.V., Loss records of ships flying flags of convenience 1964-1973.

U.S. Dept of Commerce, (1976), Maritime Transportation Research Board.

Wijnolst N. and Wergeland T., (1996), *"Shipping"* , p.411.

U.K. P & I club , (1992), " Analysis of Major Claims".

Proceedings of the Seventh (1997) International Offshore and Polar Engineering Conference
Honolulu, USA, May 25-30, 1997

ISBN 1-880653-28-1 (Set); ISBN 1-880653-32-X (Vol. IV)

Toward a Port-Level Model of the Physical Risk of Grounding

Hauke L. Kite-Powell and Di Jin
Woods Hole Oceanographic Institution
Woods Hole, MA, USA

Johan Jebsen, Vassilis Papakonstantinou and Nicholas Patrikalakis
Massachusetts Institute of Technology
Cambridge, MA, USA

ABSTRACT

We formulate a Bayesian model to estimate the physical risk of groundings during transits into and out of port as a function of explanatory factors. We assemble and analyze information on factors surrounding groundings in five U.S. ports between 1981 and 1995. Although the data are far from perfect, it is possible to establish associations between grounding risk and changes in factors such as vessel type and size, wind speed, and visibility. We apply the model to an investigation of the effect of visibility on the risk of grounding in the Port of New York/New Jersey.

KEY WORDS: risk assessment, vessel groundings, Bayesian statistics

1. INTRODUCTION

A collaborative research project on Formulation of a Model for Ship Transit Risk was started in late 1995 by researchers at the Massachusetts Institute of Technology's Ocean Engineering Department and the Marine Policy Center of the Woods Hole Oceanographic Institution. The project was developed as a three-year research activity. The focus of the first year was on collection and assimilation of data to support an improved understanding of factors that contribute to vessel groundings.

Groundings of commercial ships account for about one third of all commercial maritime accidents, including some of the most expensive in the United States' history, such as the *Exxon Valdez*. Many factors contribute to vessel groundings. Some of these factors are of particular concern to federal agencies responsible for aspects of the nation's marine transit routes. For example, the National Oceanic and Atmospheric Administration (NOAA) is responsible for the survey of U.S. waters and for the publication of nautical charts. The U.S. Coast Guard (USCG) and the U.S. Army Corps of Engineers (ACE) are responsible for navigation aids and for channel design and maintenance, respectively. All of these aspects may influence how likely groundings are to occur.

Like other kinds of casualties, groundings represent a risk because they expose vessel owners and operators, as well as the public, to the possibility of losses. These include vessel and cargo damage or loss, injuries and loss of life, environmental damage, and obstruction of the waterway, among others. Strictly speaking, the risk associated with groundings is a product of the probability of the vessel touching bottom and the probability, given that the vessel has grounded, of economic losses (where "economic" is broadly defined). This paper deals only with the physical risk component of grounding risk.

The focus of our work to date has been the association between historical groundings and factors such as environmental conditions and uncertainty in nautical charts. In our initial efforts to build a model of the physical risk of grounding, we paid particular attention to information that might inform the prioritization of port approaches for hydrographic re-survey. Our work over the past year suggests that, while the historical data on circumstances surrounding groundings in U.S. waters are neither perfect nor complete, they contain information useful to improving our understanding of why groundings occur. For a more detailed description of the project's first year research results, see Kite-Powell *et al.* (1996) and Jebsen and Papakonstantinou (1997).

Note that the work described in this paper investigates the *association* between the circumstances surrounding a transit and the occurrence of groundings, based on historical data. We emphasize here the word "association" as distinct from "cause." There is no intent in the present work to prove causal relationships between factors and groundings. Such proof is difficult to accomplish without conducting exhaustive controlled experiments; the data required exceed those available in the historical record. The present research aims only to establish associations, to identify and investigate factors that, on the basis of association and by rational consideration, appear to contribute meaningfully to risk. This approach has the advantage of permitting the inclusion of a wide range of potential contributing factors. Other methods, such as fault tree analysis (see Amrozowicz (1996)), provide a more detailed view of how groundings are caused but require a great deal of data beyond what is available in the historical record.

2. MODEL

The general hypothesis behind the physical risk model is that the probability of grounding on a particular transit depends on a set of explanatory variables. Formally, the model can be described as follows: let G denote the event that a transit results in a grounding, and let $X = (X_1, X_2, X_3, ..., X_p)$ be the vector of explanatory variables. These variables may be categorical (including binary) or continuous. The model attempts to estimate the conditional probability of G given a specified value x of X. By Bayes' Theorem, this probability is given by:

$$p(G|x) = l(x|G)\, p \,/\, (l(x|G)\, p + l(x|S)\, (1-p))$$

where p is the unconditional probability of G and where l(x|G) and l(x|S) are the likelihoods of x given G and S, respectively. S denotes the event that the transit is completed safely.

To implement this approach, it is necessary to select a set of explanatory variables that discriminates between G and S (Hand, 1981) and to estimate the unconditional grounding probability p and the likelihoods l(x|G) and l(x|S). The vector of explanatory variables X must capture the attributes of the transit that can be expected reasonably to contribute to the likelihood of a grounding. These attributes might include, among others:

- vessel characteristics (draft, beam, maneuverability)
- topography of the waterway (water depth, channel width, channel length, complexity of turns, traffic density)
- environmental conditions (wind, visibility, currents, waves)
- operators (experience with the vessel, training, local knowledge)
- information available to operators (quality of charts, quality of information about tide levels and currents, VTS guidance, navigation aids)

For each attribute, explanatory variables (x_i) must be extracted from historical data as numerical or categorical indicators.

3. DATA

Our analysis has focused on five study areas: San Francisco Bay, Houston/Galveston, Tampa Bay, Port of New York/New Jersey, and Boston Harbor. Data on historical groundings, transits, environmental conditions, and aspects of navigation infrastructure were collected for each study area for the period 1981 to 1995. Sources include national and local units of the Coast Guard and the Army Corps of Engineers; NOAA's National Ocean Service, National Climate Data Center, National Geophysical Data Center, and National Data Buoy Center; and local port authorities, marine exchanges, and pilots.

3.1. Underlying Probability of Grounding

Grounding data are drawn from the USCG's CASMAIN (1981-90) and MSIS (1992-95) databases. Data for 1991 are sparse and obviously incomplete in each dataset; we replaced the 1991 counts by averages of the four surrounding years for purposes of analysis. From the USCG data we selected for inclusion in this study only accidental, navigational groundings, and ignore those identified as intentional or due to mechanical failure or other, clearly non-navigational causes. We consider separately large (draft 30 feet or greater) and small tankers and dry cargo vessels, tank barge trains and dry cargo barge trains (a barge train is defined as a tug/towboat attached to one or more barges). Table 1 shows the number of groundings of ships in the data for each study port and year.

	Bos	**NY/ NJ**	**Tam**	**Hou/ Galv**	**San Fran**
1981	2	12	5	13	2
1982	1	2	1	5	5
1983	0	3	2	18	7
1984	0	6	9	13	3
1985	0	17	8	3	5
1986	0	8	3	2	5
1987	0	7	6	9	7
1988	0	6	15	13	4
1989	2	16	8	7	7
1990	1	8	3	9	6
1991*	0	2	0	1	1
1992	1	9	6	10	4
1993	0	7	3	5	5
1994	1	3	4	15	9
1995	1	1	1	11	6
1981-95	9	107	74	134	76

*1991 data are clearly incomplete

Table 1: Groundings of Ships, 1981-95, based on USCG data

Transit data are based on ACE Waterborne Commerce Statistics annual summaries, 1981-1994. For our purposes, "transit" is a vessel movement, so that a port call usually consists of two transits: one into and one out of the port. We assume that 1995 transits are the same as 1994, since 1995 data have not yet been made available. To avoid double counting, we base our transit counts on data for only one "waterway" in each port, as follows:

	waterway code
Port of Boston	0149
New York and New Jersey Channels	0388
Tampa Harbor	2021
Houston Ship Channel	2012
San Francisco Bay Entrance	4320

This procedure leads to underestimation of actual vessel movements, especially for Houston and New York. Unfortunately, there appears to be no simple way to build more accurate time series of transits for study port areas.

We examine transit data using the same vessel type and size breakdown as for grounding data. Table 2 shows average annual transits during the study period. The only explicit adjustment we made to ACE annual trip data is in the small dry cargo vessel category for the port of Boston, which includes a large number of very small vessels (local passenger ferries). These have been removed from the data. Barge train transits are estimated using an average number of barges per train for each study port.

	ships	barge trains
Boston	1,870	1,736
New York	10,193	30,050
Tampa	3,735	2,988
Houston	9,930	18,643
San Francisco	7,773	382

Table 2: Average Annual Transits, Ships and Barge Trains, 1981-95, based on ACE data

Grounding rates are calculated by dividing annual groundings by transits for each port. Average grounding rates are shown in Table 3. Caution is in order when comparing these grounding rates across ports. Local USCG offices may employ different reporting criteria from one port to another, and our procedure for building transit counts may underestimate actual traffic densities to varying degrees in different ports (which would lead to inflated grounding rates).

	groundings per 1000 transits	
	ships	barge trains
Boston	0.32	0.37
New York	0.72	0.18
Tampa	1.32	1.84
Houston	0.89	1.28
San Francisco	0.65	2.28

Table 3: Average Grounding Rates, Ships and Barge Trains, 1981-95

Given these caveats, it appears that the time-averaged grounding rate for ships is highest in Tampa, lowest in Boston, and clustered in between (around 0.75 groundings per 1000 transits, or one grounding in 1300 transits) for New York, Houston, and San Francisco. The most obvious temporal change occurred in Tampa, where ship grounding rates rose significantly from 1986 to 1990 and then declined again to pre-1985 levels.

Grounding rates for barge trains appear to be lower in Boston and New York than in other ports. The Houston barge train grounding rate declined during the early and mid 1980s but has risen again since then. In Tampa, the barge train grounding rate surged and then declined again, much like the ship grounding rate, but slightly earlier in time.

These underlying historical grounding rates form the basis for an estimate of p, the unconditional (underlying) probability of grounding, in the physical risk model.

3.2. Likelihood Function for Grounding Events

The distribution of explanatory factors associated with grounding events, l(x|G), is derived primarily from the USCG data on groundings. The USCG data include information about vessel type (ship/barge, oil/dry cargo), vessel size (tonnage, converted by an approximation to draft), and wind and visibility. The USCG reported wind speed and visibility data are not complete. Where these data are missing, we used NOAA's National Climatic Data Center (NCDC) hourly average wind speed and visibility data. If the time of the grounding is not reported in the USCG data, a daily average was used. The NOAA/NCDC data are taken from sensors located at airports near the study ports, and provide the only readily available proxy for on-site conditions.

Data on hydrographic surveys in grounding locations were obtained from NOAA's National Geophysical Data Center (NGDC). Grounding locations are provided in the USCG casualty data.

3.3. Likelihood Function for Safe Transits

The distribution of explanatory factors associated with safe transits, l(x|S), was derived from a variety of data sources. Vessel type and size counts were calculated from ACE transit data, as described above. The distribution of flags of registry for ships' safe transits is based on data provided by local marine exchanges and port authorities. The joint distribution of wind speed and visibility conditions was derived from NOAA/NCDC data for "safe" days (see sections 4.1.3 and 4.1.4 below). Lacking any specific information about the distribution of transits during the year, or about the joint distribution of ship size and flag, we assume no correlation between flag and ship size, and between safe transits and particular environmental conditions. These assumptions can be revised if additional data suggests the need to do so.

Data on hydrographic surveys for safe transit regions were obtained from NOAA/NGDC.

3.4. Data Issues

In the course of assembling and analyzing the historical data, we have identified certain deficiencies and limitations. We describe these briefly for each major data source.

3.4.1. <u>USCG Vessel Casualty Data</u> The USCG vessel casualty databases, CASMAIN and MSIS, are the most comprehensive source of commercial vessel casualty information for U.S. waters. However, these databases (particularly the older CASMAIN data) are occasionally difficult to work with because of missing data, duplicate entries, and inaccuracies. Also, some information useful for the analysis of casualties is not collected in these datasets at all.

The locations of groundings (and other accidents) are reported in theory to tenths of minutes latitude/longitude. In practice, this level of accuracy is not met for many entries; 18 percent of CASMAIN grounding records have no latitude/longitude information at all; and several groundings have erroneous location information (they plot on dry land). In several cases, a single casualty is described by two (slightly different) entries, one of which is probably erroneous and should have been removed from the database.

No data are presently collected on the actual draft or trim of vessels at the time of grounding; and it is difficult to reconstruct actual water depth at the time of the accident from the environmental data. No data are collected on the speed of the vessel at the time of the accident (making squat calculations impossible), or on wave states. Further, the presence and use of escort tugs is not quantified in the data, and the new MSIS data no longer include information about the presence of pilots on board vessels. These data could be usefully included in future USCG casualty data.

The USCG casualty dataset could be improved further by the adoption of a consistent set of criteria to govern what incidents are included in the dataset and how the information is to be obtained (i.e. wind speed, visibility, water level at time of the accident, etc.).

3.4.2. ACE Transit Data The ACE Waterborne Commerce transit data annual summaries are useful but suffer from several limitations for the purposes of our analysis. Dry cargo and passenger vessels are mixed in a single reporting category, requiring adjustments based on approximations. The breakdown of transits by specific waterways is useful, but makes the compilation of a composite "port region" transit history difficult because of potential double counting. Finally, barge movements are reported for individual barges, and there is no completely accurate way to determine the number of barge train movements.

3.4.3. NOAA Environmental Data Wind and visibility are available from NOAA as hourly averages. However, these are measured at airports located near each study port, and do not necessarily reflect conditions on the water. Historical water level (tidal and meteorological forcing) and current information is not always available with the same detail and consistency as wind and visibility, and water level and current conditions during historical groundings therefore cannot be reconstructed. This will change as real-time oceanographic data systems, such as PORTS in Tampa, become more common in U.S. harbors.

3.4.4. Port-Specific Traffic Data More detailed information about safe transits would be useful for several aspects of our investigation. Ideally, monthly (or even daily) counts by flag, vessel type, and vessel size, with tug escort and piloting information, could be used. Such data are only rarely available in usable, digital form from port offices or marine exchanges.

4. PRELIMINARY RESULTS

Promising explanatory risk factors identified by our research to date include vessel type, vessel size, wind speed, visibility, and (possibly) ship registry. We discuss briefly the intermediate results obtained in the individual analysis of these and other factors, and then describe an application of the model to a specific investigation of the effect of visibility on the risk of grounding in the Port of New York/New Jersey.

One factor we have not analyzed explicitly, but about which the data from Tampa may provide an interesting hint, is operators' quality of information about environmental conditions. Other things equal, better information about currents, tide levels, and winds may be expected to reduce the likelihood of grounding. This factor can be tested by distinguishing between study areas and time periods for which information from real-time monitoring systems was available to vessel operators (such as Tampa Bay in the 1990s) and those for which it was not. Our data show that grounding rates in Tampa have declined dramatically, for both ships and barge trains, in the 1990s. Further analysis is required to determine how much of this decline is attributable to the availability of real-time environmental information.

4.1. Intermediate Results

4.1.1. Vessel Type and Size Other things equal, a more maneuverable vessel may be expected to have a lower probability of grounding than a less maneuverable vessel. It is difficult to obtain meaningful summary measures of maneuverability. As a result, our analysis has to rely on proxies such as vessel type and size.

Our data suggest that, in most cases, barge trains are more likely to ground than ships (see Table 3). This is consistent with our expectations about maneuverability: barge trains are, in general, likely to be less maneuverable than ships. Our data also suggest that larger ships are consistently more likely to ground than small ships (draft less than 30 feet). This result is also consistent with our expectations about maneuverability.

These findings suggest that factors such as vessel type and size can be employed usefully in developing a physical risk model. We plan to investigate related factors, such as the presence and use of tugs during vessel transits.

4.1.2. Uncertainty in Chart Surveys Other things equal, a transit through an area for which perfect charts are available may be expected to involve less risk of grounding than a transit through an uncharted or poorly charted region. Historically, navigators' knowledge of their own position was uncertain enough to make them cautious of approaching charted hazards. The charts, and their underlying surveys, had generally greater accuracy, and were constructed with better instruments, than those available to the average mariner. Today, GPS users can position themselves with more accuracy than the surveyors who collected the data on which most existing charts are based. This may have eroded some of the safety margin previously incorporated into the charts. It is possible that uncertainty in paper charts based on older surveys has contributed to the incidence of groundings in U.S. ports. To investigate this, we analyzed hydrographic data for uncertainty using Hydrostat software (Kielland *et al.*, 1992) and combined the results with grounding location data to check for correlation between cartographic uncertainty and historical groundings.

Three main types of error arise when charts are designed from survey data: instrumental errors, interpolation errors, and design errors. Instrumental errors consist of positioning errors and depth measurement errors, and are assumed to be constant over the survey. The approximate size of these errors depends on the survey, but they are usually smaller than interpolation errors. Design errors arise from document handling and safety bias when data are transferred to navigational documents. Since both design errors and instrumental errors are well known and usually noted on the charts, they are of no particular interest to our investigation. Interpolation errors are bathymetric uncertainties that exist in the unsounded zones between measured soundings. They are the least controllable error for chart design and the focus of our investigation. If the surveyed depths are far apart, depending on the topography of the sea floor, these interpolation errors can be much greater than the instrumental errors in the measurements themselves. They vary continuously and are unique to every location on a chart.

The Hydrostat software computes the depth in the unsounded zones between measured soundings using a geostatistical depth interpolation algorithm, and also predicts the depth estimation errors inherent to each point on the interpolated bathymetric model. The results of the computation are two specific features: a bathymetric surface and a stochastic surface. The bathymetric surface is the digital terrain model interpolated from the observed sounding profiles, and is strictly a function of water depth. The stochastic surface is composed of the vertical error estimates for every point on the bathymetric surface. This surface is a function of both seabed texture and data sampling density. We used this stochastic surface to check for correlation between cartographic uncertainty and historical groundings.

Within the five study areas, for the period 1981-90 and 1992-95, we identified about 1000 navigational groundings. Of these, more than 100 records (primarily from the CASMAIN database, 1981-90) provide no latitude/longitude location information; others include an obviously faulty entry for location (they plot in marsh areas or on dry land). Of the remaining groundings, more than 80 percent occurred around dredged channels. Groundings around dredged channels generally are not due to the sort of bathymetric uncertainty that arises from sparse hydrographic surveys; and we eliminated them for the purpose of our survey investigation.

Unfortunately, not all of the remaining accidents are reported with sufficient accuracy to be useful for this analysis. Accident locations are reported in the USCG data (in theory) to an accuracy of +/-0.1 minutes latitude/longitude, or +/-150 to 200 meters in our study areas. However, many of the accident locations are reported without the last decimal, which results in an accuracy of +/- 1500 to 2000 meters and makes them worthless for an analysis of chart/survey uncertainty. This leaves us with 71 useful groundings for this part of the evaluation. Although we would prefer to have more data available, this is still a statistically useful sample.

NOAA's hydrographic survey data include depth measurements and bottom features. These data are part of the base from which NOAA charts are designed. We analyzed the most recent surveys for the areas of interest. These surveys differ greatly in age, quality, and density of data.

We examined charted depth uncertainty in detail for the ports of New York/New Jersey and Houston/Galveston. We compared the overall distribution of interpolated errors in estimated depth for the transit areas of each survey to the distribution of interpolated errors in the depth estimates at locations where accidents occurred. For all surveys examined in the New York and Houston regions, the results are consistent: there is no evidence that "open water" groundings tend to happen in high uncertainty areas.

Our analysis of uncertainty in charted depths and historical groundings suggests that while it is in theory a compelling factor, operators' uncertainty about topography appears to have made a negligible contribution to the incidence of groundings in U.S. ports in recent decades. Given this result, at this stage our model is not useful for the prioritization of navigable areas for re-survey.

<u>4.1.3. Wind Speed</u> Other things equal, a transit through an area characterized by unfavorable environmental conditions, such as high wind, may be expected to involve a greater risk of grounding than a transit through the same area under more favorable conditions.

To characterize wind conditions on "safe" days, we calculated the distribution of hourly mean wind speed (NOAA/NCDC data) on all days during the study period when no groundings took place in the port. (Another characterization considers a smaller sample of "safe" days that coincide in month and date with grounding days for the port. This approach reduces the influence of seasonal weather effects on the safe days wind speed distribution.) To characterize wind conditions when groundings took place, we used USCG reported wind speeds at grounding incidents. Where these data are not available, we used NOAA/NCDC hourly average data (if the time of the grounding is known) or daily average data (if it is not). We examined the cumulative distribution of mean wind speed w, $p(w>x)$, given either grounding or safe days. A higher "tail" of this distribution in the high wind speed range for grounding days indicates that higher wind speeds are more commonly associated with groundings. This higher "tail" was found consistently in the wind speed distributions for the study ports.

We have assumed here, with only annual transit data available, that safe traffic is uniformly distributed over the year. We will check this assumption against monthly transit statistics from ACE and will revise the associated distributions if seasonal transit data suggest otherwise. It may also be useful to refine the analysis by truncating the safe distributions on the right to reflect port closures (no safe transits) at times when winds are very high. These truncations will be based on information supplied by port operations personnel in each study port. It may be appropriate to use different high-wind cutoffs for ships and for barges. This refinement will amplify existing differences in the safe and grounding wind speed distributions.

<u>4.1.4. Visibility</u> The analysis of visibility data for "safe" and "grounding" days follows the approach described above for wind speed. We examined the cumulative distribution of mean visibility v, $p(v<x)$, given either grounding or safe days. A higher "tail" of this distribution in the low visibility range for grounding days indicates that lower visibility is more commonly associated with groundings. The expected higher "tail" was present in the visibility distributions for all study ports.

As with the wind data, we have assumed that safe traffic is uniformly distributed over the year. We will revise our estimates appropriately if seasonal transit data suggest otherwise. Here, too, we expect to truncate the safe distributions on the left to reflect port closure on days when visibility is very low, possibly using different cutoffs for ships and for barges. Again, this refinement will amplify differences in the visibility distributions.

<u>4.1.5. Operator Skill/Flag of Registry</u> Other things equal, more highly skilled or seasoned operators, and those with better local knowledge, may be expected to experience a lower risk of grounding. Unfortunately, no direct measures of operator skill are available in the historical data. One (admittedly rough) proxy for this factor that can be constructed readily from historical data is the flag of the vessel. This proxy is relevant only to ships; tugs/tows and barges in U.S. waters are almost without exception U.S.-registered.

	% transits	**% groundings**
New York	17	52
Tampa	16	36
Houston*	13	43
San Francisco	67	29

*based on incomplete data

Table 4: U.S. Flag Share of Ship Transits and Groundings

Our data (Table 4) suggest that during the study period, U.S.-registered ships were involved in groundings relatively more frequently than foreign-flag ships in New York, Tampa, and Houston; the reverse is true in San Francisco. Data for the safe transit flag distribution in Boston were not available, but U.S. flag ships accounted for 11 percent of groundings in Boston during the study period.

While there appears to be a signal in these data, the "flag" factor embodies many variables other than operator skill, and requires further exploration before it can be endorsed fully for the development of the physical risk model. For example, the size distribution of U.S. vs. non-U.S. ships in general is not known from available data.

4.2. Effect of Visibility on Grounding Risk in NY/NJ

The data on contributing factors, such as those discussed above, can be combined to build a Bayesian model of physical grounding risk, as described at the outset of this paper. We describe here the application of such a model to the analysis of the role of visibility in grounding risk for the Port of New York/New Jersey.

The underlying probability of grounding (p) for commercial vessels in the Port of New York/New Jersey during the study period was 0.000353, or 3.5 groundings per 10,000 transits. Explanatory factors included in this implementation of the model are: vessel type (ship or barge), vessel size, flag of registry, wind speed, and visibility. The likelihood functions l(x|S) and l(x|G) were derived from USCG and NOAA data, as described in section 3 above. Table 5 shows the model results for the effect of visibility on the probability of grounding for various vessels, in wind speeds under 10 m/s. (Poor visibility rarely occurs in New York during high wind.)

	probability of grounding	
	good visibility (>= 2km)	poor visibility (< 2km)
small ships (draft < 30 ft)		
U.S. flag	0.0018	0.0073
non-U.S. flag	0.0002	0.0015
large ships (draft >= 30 ft)		
U.S. flag	0.0047	0.0861
non-U.S. flag	0.0015	0.0147
barge trains	0.0002	0.0015

Table 5: Effect of Visibility on Grounding Risk in New York/New Jersey (wind speed < 10 m/s)

This simple example shows how the Bayesian model can be used to estimate the effect of a change in environmental conditions (or, equivalently, a change in operating procedure) on grounding risk and, by extension, on the number of groundings. In this case, the results indicate that poor visibility increases the risk of grounding by approximately one order of magnitude. The numbers also illustrate the higher risk associated with large ships and with ships registered under the U.S. flag, and the relatively low risk for barge trains, in the Port of New York/New Jersey.

5. CONCLUSIONS

We have illustrated the use of a simple Bayesian model of the risk of grounding, using historical data on casualties, safe transits, and environmental conditions in the Port of New York/New Jersey, and discussed data for similar investigations in other ports. Our analysis suggests that factors such as vessel type and size, and wind and visibility conditions, appear to contribute in measurable ways to the risk of grounding, and that historical data on these and other parameters, while far from perfect, can be used to develop models of grounding risk. Because uncertainty in charted depths appears not to have been a factor in port-approach groundings, the model is not directly useful to the prioritization of areas for re-survey.

The results described in this paper are preliminary, and we continue to develop and refine the models for the five study ports. We plan to supplement the "port level" analysis of groundings described here with a more detailed examination of groundings in particular channel segments, and to apply similar models to other accidents, such as collisions.

ACKNOWLEDGMENTS

This paper includes a condensed account of research findings described more fully in Kite-Powell *et al.* (1996) and Jebsen and Papakonstantinou (1997). The authors acknowledge the valuable assistance and helpful comments of Ric Walker, Larry Daggett, Stan Alper, Millington Lockwood, Peter Kielland, and Andy Solow. Many others at NOAA, the U.S. Coast Guard, and the U.S. Army Corps of Engineers, as well as port officials and pilots in the study ports, provided information to the project. In addition, the authors are grateful to Prof. M. Triantafyllou of MIT for arranging reviews of the paper, and to two anonymous referees for their comments and suggestions. Funding was provided by the MIT Sea Grant College Program under federal grant NA46RG0434 from the National Sea Grant College Program, NOAA, Department of Commerce; the U.S. Coast Guard; the U.S. Army Corps of Engineers; and the Marine Policy Center of the Woods Hole Oceanographic Institution.

REFERENCES

Amrozowicz, MD (1996). The quantitative risk of oil tanker groundings. Master's degree thesis, Ocean Engineering Department, Massachusetts Institute of Technology, Cambridge, Massachusetts.

Hand, DJ (1981). *Discrimination and classification*. John Wiley & Sons, New York.

Jebsen, J and Papakonstantinou, V (1997). Evaluation of the physical risk of grounding. Master's degree thesis, Ocean Engineering Department, Massachusetts Institute of Technology, Cambridge, Massachusetts.

Kielland, P, Burrows, K, Ward, B, Dagbert, M, and Velberg, R (1992). Toward "IHOstat": IHO approved software which evaluates the quality of bathymetric data. Hydro '92 Conference, Copenhagen, December 1992.

Kite-Powell, HL, Jin, D, Patrikalakis, N, Jebsen, J, and Papakonstantinou, V (1996). Formulation of a model for ship transit risk. MIT Sea Grant College Program Technical Report 96-19. Massachusetts Institute of Technology, Cambridge, Massachusetts.

Proceedings of the Seventh (1997) International Offshore and Polar Engineering Conference
Honolulu, USA, May 25-30, 1997

ISBN 1-880653-28-1 (Set); ISBN 1-880653-32-X (Vol. IV)

Admittance Policy Deep Draught Vessels and Safety

R.Ph.A.C. Savenije
Transposrt Research Center (AVV)
The Netherlands

Abstract

The design of the admittance policy of the Euro-Maas channel to Rotterdam and the Western Scheldt is based on a probabilistic method. With this method, instead if the previously implemented deterministic method, the accessibility of the harbours and safety during the channel transit have been increased without additional dredging costs.

Keywords

Harbour, Probabilistic Admittance Policy, Waves, Vessel Movements, Safety.

1 Introduction

The last few years, the nautical accessibility of the West-European harbours is again in the spotlights. During the sixties and seventies a clear separation developed between harbours who could and harbours who could not adapt to the progressive scale enlarging of the bulk carriers. For example, till the late fifties the maximum draught in Rotterdam, Antwerp as well as Hamburg was all the same: around 40 feet.

During the next decades the maximum draught with which the port of Rotterdam could be accessed, kept pace with the huge scaling up of the crude oil-tankers and bulk-carriers. At the moment, each year 350 channel-bound vessels (draught more than 17.40m) arrive at Rotterdam, with a maximum draught up to $22.55m$ (74 feet). Vessels heading for the Western Scheldt are limited to a draught of $15.00m$. Each year more than 400 vessels with a draught more than $11.00m$ bound for Flushing, Gent & Terneuzen or Antwerp.

The design of the admittance policy of the Euro-Maas channel at Rotterdam and the Western Scheldt is based on a probabilistic method. With this probabilistic method, a substantial improvement of the accessibility and safety of both channels has been achieved with only a minor change of infrastructure. A further scaling up of the bulk carriers and crude oil tankers is not expected. Nevertheless, from the market an urge exists for minimizing the accessibility restrictions.

The probabilistic admittance policy of Euro-Maas channel has proofed to enlarge the accessibility of the port of Rotterdam for already 10 years. For the Western Scheldt this method is recently introduced. It is also intended to recalculate the IJ-channel at Amsterdam in the near future using this method.

Contents

This paper gives an explanation of the implemented techniques, with which the accessibility and safety of the port of Rotterdam and the Western Scheldt has been improved. In the first section a brief introduction is given explaining some terms and expressions. After a short view on the previously used deterministic method, the design process using the probabilistic calculation method is explained in section 3. The base of the probabilistic design is the safety of the channel transit. Section 4 treats this subject. The schematization of all the information necessary for these calculations is treated in the next section (5). After a short explanation about some future developments in section 6 the paper is ended with a conclusion.

For a more thorough explanation of the probabilistic calculation method, or just out of interest or curiosity, the appended bibliography gives a list of papers and books which can be consulted.

2 Used expressions and terms

2.1 Tidal-Windows

Vessels with a draught of more than $20.00m$ with destination Rotterdam and vessels with a draught of more than $11.00m$ sailing at the Western Scheldt are *tidal-bound* and are provided with a *tidal-window* advice. For these vessels, the available water level is not sufficient. Only using the high tide they can reach the harbour.

A tidal-window consists of two times, an opening time and a closing time (figure 2). Between those two times a vessel is allowed to enter the channel. During the channel transit tidal-windows indicate the different opening and closing times at different locations. These set of tidal-windows shows during which time period at which location the tidal-bound vessels can safely sail the channel. The tidal-window depends on vessel type, dead-weight, draught, astronomical water level, meteorological water effects and wave conditions. Also the vessel speed and current are taken into account. The more unfavorable the situation gets (e.g. more wind

Figure 1: Major harbours in The Netherlands

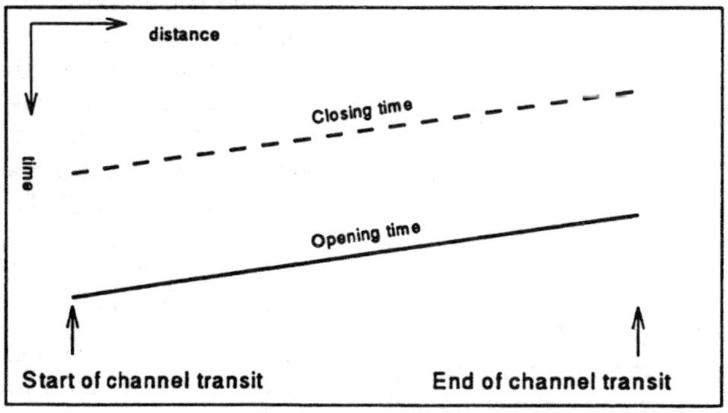

Figure 2: Tidal-Window

or waves) the narrower the tidal window becomes. In the most extreme situation no tidal window is available. In this case the vessel has to wait for the next tide and better circumstances.

2.2 Downtime and Inaccessibility-Percentage

The accessibility of a harbour can be determined using the *downtime* and the *inaccessibility-percentage*. The downtime equals the percentage of the *time* during which a vessel can not access the harbour because there is no tidal-window available. The inaccessibility percentage equals the percentage of the *tides* during which no tidal-window can be provided, due to unfavorable water level or extreme waves.

These two values are quite different. For a vessel it is most important that there is an opportunity each tidal cycle during which it can reach the harbour. It is less important how long this opportunity exist during the tidal cycle. For example, if during four tidal cycles a tidal window exist which is a continuous period of 1 hour each tide, the downtime equals 84%, but the inaccessibility percentage equals 0%!

3 Design of Tidal-Windows

3.1 Deterministic Method

Till 1985 a deterministic admittance policy has been used for the Euro-Maas channel. The Western Scheldt used this method till November 1995. The admittance of vessels was based on a fixed keel clearance percentage. The relation between the minimal keel clearance and the maximum draught was calculated by adding up the squat[1] vertical movements, sounding inaccuracies and sanding to the draught of the vessel, as explained in figure 3.

1. draught increase caused by the speed of the vessel (section 5.6.2).

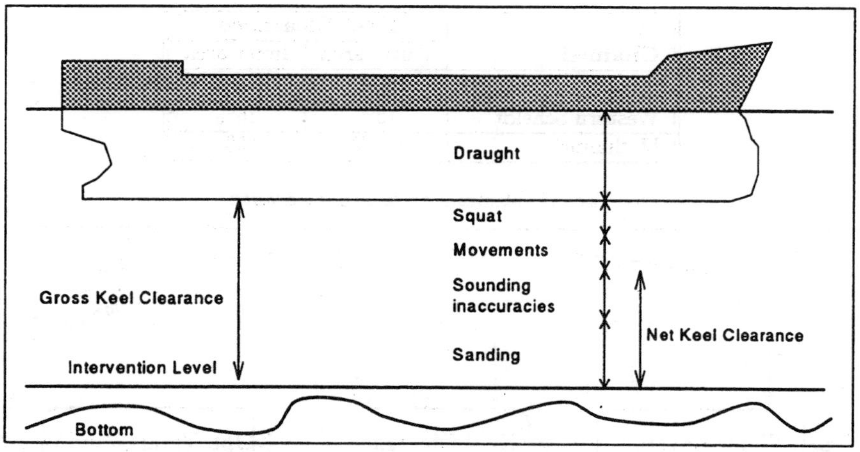

Figure 3: Keel-clearance and additions

These additions are equal to the maximum anticipated effects due to the factors mentioned before. The sum of all these additions determine the minimal gross keel clearance. The ratio between the gross keel clearance and the draught is called the keel clearance percentage. At table 1 the percentages of the major Dutch harbours are mentioned.

Using these keel clearance percentages the accessibility of the channel can be determined. At the Western Scheldt, the maximum draught was determined for the predicted water level at the most critical location along the channel. At the Euro-Maas channel and the IJ-channel tidal-windows were calculated at different locations along the channel, taking account of the changing water level during the channel transit.

3.2 Probabilistic Method

At this moment, the Euro-Maas channel [13] and the Western Scheldt have a probabilistic designed admittance policy, instead of a deterministic policy as described in section 3.1. All possible wave and water level conditions as well as ship characteristics are considered and used for determining an optimal accessibility. But instead of using discrete additions, each factor is translated into a probability distribution with a mean value and a variance. All the probabilistic distributions are combined and determine together the probability of touching the channel bottom.

The result of the probabilistic calculation are tidal-windows, inaccessibility percentages and downtimes. These tidal-windows and the according probability of touching the channel bottom conform to predefined safety criteria as described at section 3.5. Now it is possible, using these safety criteria, to weigh the dredging costs against accessibility of the port.

The reliability of the tidal windows depends on the input data. The predictions and measurements of the water and wave conditions which occur in the channel have a certain inaccuracy. The probabilistic design method takes full account of prediction reliability as well as spreading of measurements. The exact design process is described in section 3.3.

3.3 Probabilistic Design Process

The theory of the probabilistic design is explained in [12]. This theory has been used to develop a program called HARAP (HARbour APproach). HARAP calculates the probability of touching the channel bottom during a channel transit. It simulates a transit of a vessel. The process of determining the tidal-windows is explained in figure 4.

During such a transit first the water level is determined. Information about astronomical and meteorological water levels, current, vessel speed and wave climate is used. This results in the calculation of the keel clearance across the channel. Next the vertical movements of the different types of vessels are added which depend on the wave climate. The final result at this stage is one optimal tidal-gate.

This optimal tidal-gate consists out of *one* time: the optimal time to start the channel transit with minimal probability to touch the bottom of the channel given all the circumstances at that moment.

This process is repeated for all the different combinations of circumstances. As explained in [12], it is possible to divide all the circumstantial parameters into *regimes* or classes. Instead of calculating with all the different values only the mean or representative value of a regime is used. Regimes are used to characterize the different water level effects, vessel types and wave climates. The number of required calculations is significantly reduced by using this technique.

For each combination of regimes a tidal-window is calculated. All the probabilities for touching the bottom of the channel of all the transits are added up. This sum is compared to a safety criterion (section 3.5). Finally the different tidal-windows are enlarged, reduced or removed, changing the total probability of touching the bottom of the channel, till the required criterion has been reached.

Finally, a sensitivity analysis can be done to optimize the chosen regimes, as showed by the dashed arrow in figure 4. The result is a optimized set of tidal-windows for all the different combinations of regimes. For each vessel type a inaccessibility percentage and downtime is calculated to determine the resulting accessibility of the channel and port.

3.4 Advantages and Disadvantages Probabilistic Method

The deterministic method is based on additions which determine the keel clearance. Mostly these additions are at the outside cal-

Channel	Keel-Clearance	
	Outer area	Inner area
Euro-Maas channel	20%	10%
Western Scheldt	15%	15%
IJ channel	17.5%	15%

Table 1: Minimal gross keel-clearance

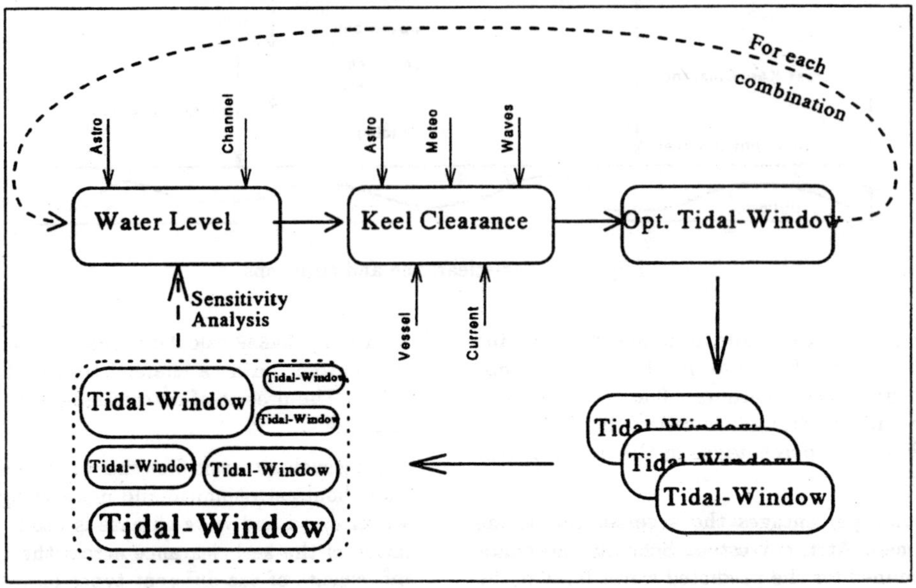

Figure 4: Process of calculating tidal-windows

culation, making sure that the safety is guaranteed. This results in a keel clearance which is at most times much to large. An advantage is the simplicity of this calculation method. It is possible to calculate the keel clearance by hand.

The probabilistic method, however, can not be calculated by hand. But the determined keel clearance is much more precise. With the same channel depth a much better accessibility can be achieved. Another advantage is the predetermined safety level. Now it is possible to weigh the accessibility, the required channel depth and the desired safety level.

The probabilistic calculations are based on the chance of touching the bottom of the channel. The result of the calculation is a predefined risk of the vessel entering the channel, and not a minimal keel clearance. This is sometimes quite difficult to accept and makes it impossible to compare the keel clearance of both calculation methods. However the accessibility percentages can be compared.

3.5 Safety Criteria

As explained in section 3.3, the calculation and optimization of tidal windows is done by determining the total probability of touching the channel bottom. This probability must be less or equal to a predetermined safety criterion. For the Dutch channels this safety-criterion is determined as:

- **During 25 years the chance of touching the channel bottom which maximum minor damage must not be more than 10%.**

Using a Poisson distribution, a chance of 10% equals a probability of 0.105 touches each 25 years [4]. This chance is equal to one touch of the bottom of the channel each 237 years (25/0.105 = 237). Taking into account the number of tidal bound vessels in the channel during 25 years, and the fact that only one out of ten occurrences results in more than minor damage, results in the criterion used by HARAP[2]

Besides the above mentioned safety criterion, two other criteria exist. The first criterion, the *manœuvering* criterion makes sure that:

- **The keel clearance never is less than 1 *m*.**

Finally the *single transit* criterion defines that:

- **The chance that a vessel during its transit touches the channel bottom must always be less than 1% at all (weather) conditions.**

Despite the fact that the last two criteria almost never limit the size of the tidal window, these criteria are necessary to guarantee a safe transit under all conditions, because the general criterion only limits the total probability of touching the bottom of the channel of all vessels together.

4 Safety

To ensure maximum and safe cargo deliverability in a given port, a proper balance must exist between excessive and insufficient

2. For the Euro-Maas channel 6250 vessels use the channel and are tidal bound each year. This results in $0.105 \times 10/6250 = 1.68 \times 10^{-4}$ as the safety criterion used by HARAP

required underkeel clearance. Historical data and new calculation methods have been used to determine the minimal underkeel clearance and the safety criteria used for the probabilistic calculation [10]. Also a study of the human factor has been done.

Not only the vertical movements of a vessel determine the safety of entering a channel. Also the horizontal movements are important. Cross current is a major problem for many harbours. For example, the tidal windows of Flushing, Rotterdam and IJmuiden are restricted for some type of vessels due to cross current.

For the Euro-Maas channel a full risk-analysis has been done for the entrance channel. General accepted safety criteria have been used for the first calculations. Taken into account was the fact that the chance of touching the channel bottom should be less than the chance of collision and running aground. Therefore the determination of the safety criterion can be split up in two parts:

- chance of collision,
- chance of running aground.

4.1 Collision

To determine the chance of collision the complete traffic information of one year is taken into account. All these vessels have a known origin and destination. With this information the chance of one vessel encountering an other vessel can be determined. Three different types of encounters are taken into account:

Head-on encounter A head-on encounter occurs when two vessels sail in opposite direction, while the course of both vessels have an angle between 150° and 180°.

Crossing encounter A crossing encounter occurs when the course of the two vessels have an angle between 60° and 150°.

Overtaking encounter When the angle between the course of the two vessels is between 0° and 60° an overtaking encounter takes place.

An encounter does not directly result in to an accident. The next step is to relate the database of real accidents with the information of encounters. This makes it possible to calculate how many encounters result into an accident or collision. An example of the calculations for the entry channel to Amsterdam, the IJmond-channel, is the following table (2). The risks for collision has been determined for different routes and has been split up to the different kinds of encountering. The figures are based on *Mean Collision Rates* which are calculated as explained before. More information about this subject can be read in [9].

4.2 Running Aground

Touching the bottom of the channel or running aground at the coast mostly does not result in any damage for the environment [15]. The Dutch coast is sandy and not very steep. In case a vessel runs aground it is not expected that any oil will leak out of the tanker. For this reason the risk of running aground is less important than the risk of collision.

The risk of running aground are expressed in *Area Stranding Opportunities, ASO* [9], which are calculated using the stranding opportunity of a single vessel and the stranding opportunity at a specific location. Table 3 gives an example of some values for the ASO at the IJmond-channel. Again these figures have been calculated for different routes. The fact that running aground occurs due to restrictions of the maximal draught is taken into account.

4.3 Cross Current

As mentioned in section 4, vessels with draughts of over $21.95m$ do have a cross-channel current restriction near the harbour entrance. This restriction has been implemented in the probabilistic admittance policy as a period during which no vessels with draughts of over $21.95m$ are allowed to enter the harbour. However, the risk which is being taken at the moment of entering the harbour is not taken into account. Instead of a probabilistic calculation a deterministic restriction is used. At the moment an investigation has been started to examine the relation of tide, wind and current to predict the cross-channel current. Also the risks are taken into account, which can be implemented for example as the other safety criteria (section 3.5).

4.4 Human Factor

Last but not least is the human factor. A lot of simulations have been done[11]. Also analysis have been done about the behavior of pilots on deep draught vessels. Finally, continuously a logbook is being kept up to date containing information about the piloting of the vessels to the Rotterdam harbour. This logbook can be used to examine specific situations during the navigation of the vessels.

5 Schematization

Using some kind of schematization the amount of data can be reduced, making it possible to calculate the keel clearance and the tidal-windows with a computer program.

As described in section 3.3 and [12] the meteorological and astronomical water levels can be divided in different regimes. Also the low frequency distribution of the wave energy is schematized into one or two values and divided into regimes. The different types of vessels can be characterized by a few parameters, and finally the channel has been divided into separate parts to simplify the input data.

The Transport Research Center, the North Sea Directorate and the Directorate General of Shipping and Maritime affairs of the Ministry of Transport, Public Works and Water Management have developed the computer program **HARAP**. This program uses the probabilistic theory and processes all the data. The following sections explain all the information necessary to calculate the tidal-windows. More detailed information can be read in [12].

5.1 The Channel

Because within the channel there are a lot of significant variance of the water level and channel depth, the channel has to be divided into smaller segments. The length of each segment must be chosen in such a way that the channel depth does not vary to much and that the water level variances during a segment transit are not more than the inaccuracy of other influences like depth inaccuracy.

Route	Meeting	Overtaking	Crossing	Total
1	1.55×10^{-6}	1.68×10^{-6}	22.93×10^{-6}	26.15×10^{-6}
2	4.14×10^{-6}	2.29×10^{-6}	29.71×10^{-6}	36.14×10^{-6}
3	4.63×10^{-6}	3.81×10^{-6}	55.25×10^{-6}	63.69×10^{-6}

Table 2: Risks of collision

Route	ASO
1	18.1
2	27.0
3	32.6

Table 3: Values of the Area Stranding Opportunities (ASO).

5.2 Astronomical Water Level

The astronomical water level is represented as astronomical curves, with each a determined frequency of occurrence. These astronomical curves are divided into regimes based on *tide-difference*. The tide-difference is the difference between the high water level and the previous low water level. The water levels are specified relative to the *stroke-middle*. The stroke-middle is the difference low water level and high water level (figure 5). The stroke middle is fixed with respect to a predetermined reference level. The advantage of this method is the significant smaller spreading within each regime, compared to the normally used method where the water level is determined directly relative to a predefined reference level.

The regimes of the astronomical water level are based on the frequency distribution of the tide-difference. The limits of each regime are chosen in such a way that the spreading is minimal within this regime. The values which the HARAP program use, are the mean values of the different regimes.

5.3 Meteorological Water Level

The meteorological water level is defined as the difference between the astronomical water level and the real water level measured during low and high water. These meteo effects are deviations of the astronomical water level as caused by wind and air pressure effects. Also the meteo effect is divided into regimes, based on the frequency of occurrence. Again the spreading is kept as low as possible by choosing the most optimal limits of the regimes.

Because the stroke-middle as described in section 5.2 is by definition constant during one tidal cycle, it is also incorporated in the meteorological regimes.

5.4 Current

For the HARAP calculations, the current can be defined using current speed and current direction. The current is important for determining a safe transit through the channel because:

- The cross-channel current limits the manœvrability near the harbour entrance, and
- The current is used for determining the squat of a vessel.

The cross-channel current must be determined with measurements and calculation models. The current in the channel mostly is extracted from tidal stream atlasses cause almost no current measurement time series are available. With this information the sailing speed of the vessel can be determined and the squat can be calculated.

5.5 Wave Climate

The wave climate is important for determining the vertical movements of the vessels, and therefore should be studied carefully. The wave climate can be divided into different wave directions and frequencies. For the Euro-Maas channel and the Western Scheldt the wave frequencies are projected to one wave direction which is the most critical one. The frequencies can be displayed using a wave energy density spectrum. For the Euro-Maas channel the wave frequency density spectrum has been schematized using the *lfe* parameter (figure 6).

The *lfe* parameter gives the amount of low frequency energy in the range between 0.03 and 0.1. This parameter is translated to a significant low frequency wave height, the He_{10}.

$$He_{10} = 4\sqrt{m_0}, \quad m_0 = \sum_{a}^{b}(S(f)\Delta f) \tag{1}$$

In this equation $a = 0.03$ and $b = 0.10$. The frequency is f and the spectral density equals $S(f)$.

For the Western Scheldt however a second parameter is used, the H_{res} value. The H_{res} parameter equals the amount of energy in the range between 0.1 and 0.5.

5.6 Vessels

Besides information about the waves, water level and current, also the specifications of vessels have to be determined. For the port of Rotterdam, two types of vessel are considered. The bulk carrier and the crude oil tanker. At the Western Scheldt container vessels and bulk carriers are considered. All the different types of vessels react differently on waves and swell. Also the squat depends on the type of vessel.

5.6.1 Vertical Movements

To determine the keel clearance of a vessel, one has to know the vertical movements of the vessels due to waves and swell. The relations which are necessary to know have been determined based on a lot of tests. An example is the test of a 200.000 DWT tanker [17]. Models have been made [16] of different types of vessel and put into a water basin, simulating all the different (wave) conditions according to reality.

The vertical movements of a ship as induced by the low frequency wave energy are a combination of heave, roll and pitch motions. Besides the wave energy spectrum, the vertical movements depends on:

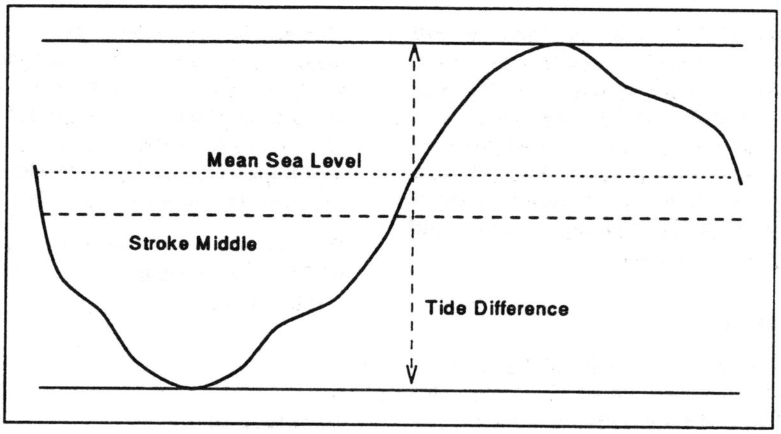

Figure 5: Astronomical water level

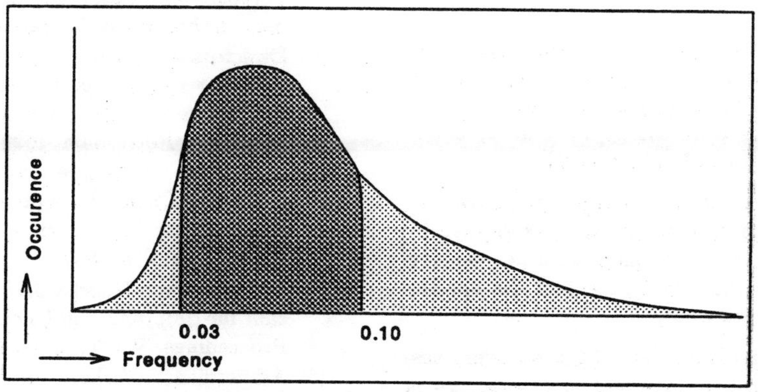

Figure 6: Wave frequency distribution,He_{10}.

- type, size, deadweight, cargo, center of gravity, etc.
- speed,
- wave angle,
- ratio between water depth and draught.

Low frequency waves (swell) are the most important factor for the vertical movements. For each wave spectrum a large data set of measured wave spectra is necessary. For each type of vessel a ship movement spectrum is determined, using the response characteristic of that vessel. This movement spectrum is then reduced to a single parameter, the H_mO_{sch}. With linear regression methods this value can be related to a characteristic value of the low frequency wave energy He_{10}:

$$Z_s = a \times He_{10} + b \tag{2}$$

Z_s denotes the vessel movement and a and b are regression parameters. The regression parameters depend on the speed, wave direction and the ratio of depth and draught.

5.6.2 Squat

The squat is the draught increase of a vessel due to the sailing speed. The squat depends on shape, water depth and speed. For the port of Rotterdam the simplified *Tuck-Taylor* [14] equation is used:

$$Squat = C \times \frac{F_{nh}^2}{\sqrt{1 - F_{nh}^2}} \times \frac{\Delta}{L_{pp}^2} \tag{3}$$

C denotes the squat constant, Δ the water displacement and L_{pp} the length of the ship. For the Euro-Maas channel a C value 1.75 has been used for the bulk carriers and 1.91 has been used for the oil tankers. The Froude number F_{nh} equals:

$$F_{nh} = \frac{v}{\sqrt{gh}} \tag{4}$$

In which v denotes the speed, h the water depth and g gravity acceleration. The speed is measured related tó the ground, so current is taken into account.

5.6.3 Speed and Vessel Regimes

At each segment of the channel a speed must be defined for each different type of vessel. All the information about the speed at the different segments together is called a speed regime. Totally three different speed regimes are defined for the Euro-Maas channel, a slow, medium and fast regime. With the information about the different speeds the position of the ship can be defined. For the Western Scheldt no different speed regimes are used.

6 Future Improvements

6.1 Neural Networks

In order to improve the rather coarse current restriction (5.4) in 1997 a permanent on-line current measurement location just next to the Maas channel will be installed. In order to obtain current predictions experiments are being carried out using *neural networks*. In the IJ-channel a neural network proved to be very well able to predict the current [5]. The Neural Networks are able to learn complex non-linear processes from examples. Their principle is based on the working of the biological brain. Numerical models are not yet able to provide operational predictions because of the large spatial variation of the current. The required fine grid models still ask too much computational time.

6.2 Individual Tidal Windows

At the moment, a tidal window is being calculated for a group of vessels, not for a individual vessel (section 5.6.3). Within this group, the most critical vessel is used for determining the vertical movements. In contradistinction to before, when only model tests were possible, now computer calculations can reliably determine the vertical movements.

In august this year a project has started to measure the exact movements of large vessels, in order to validate the calculations. Accelerations, speeds and displacements are measured, together with accurate information of the wave energy, in order to better tune the response functions to the individual vessel.

Large vessels like the *Berge Stahl* with a draught of 74 feet are calculated with their actual size, instead of the sizes defined by the group it belongs to. This way much more accurate calculations are possible. This means that the draught of the specific vessel must be given with great accuracy.

The goal is to determine an individual tidal window for every vessel, calculating the vertical ship movements of that vessel, given the predicted wave spectrum, besides using the predicted water level curve instead of the astronomical and meteorological water level regimes.

It is to be expected that these investigations will significantly increase the accessibility of the port of Rotterdam.

6.3 Safety Horizontal Manœuvering

As explained in section 4.3 the cross-channel current does limit the accessibility of the Rotterdam harbour. An investigation has been started to examine the risks related to the difficult manœuvres which have to be taken while entering the harbour. This includes a thorough study of the safety criteria used for the probabilistic calculation method.

7 Conclusion

The probabilistic calculation method is a great improvement compared to the deterministic method. Although, the process of schematizing the necessary information and optimizing the computer calculations is quite complex and time consuming.

However the resulting tidal-windows are very accurate and considers all the different weather and wave conditions the vessel can meat during the channel transit. The new probabilistic admittance policy for the Euro-Maas channel resulted in a very high accessibility of the port of Rotterdam.

The admittance policy design of the channel can be optimized using a pre-defined safety criteria, ensuring a safe channel transit. The vessels at the Western Scheldt profit by the use of the recently implemented probabilistic calculation method; the risks of touching the bottom of the channel have been minimized. With new improvements, being developed right now, even better results are expected for both channels.

Without any additional costs for dredging the safety and accessibility of a harbour can be increased using the existing channel configuration.

References

[1] Rijkswaterstaat Directie Noordzee, Adviesdienst Verkeer & Vervoer. *Informatie voor de vaart met diepstekende schepen naar de haven van Rotterdam*

[2] Directoraat Generaal Scheepvaart en Maritieme Zaken, Rijkswaterstaat Dienst Verkeerskunde, Directie Noordzee. *Informatie voor de vaart met diepstekende schepen naar de haven van Rotterdam*. 1990.

[3] J.Strating, T.Schilperoort, H.G.Blaauw.*Optimization of Depths of Channels*. September 1982.

[4] Adviesdienst Verkeer en Vervoer. *Ontwerphandleiding HA-RAP*. November 1994.

[5] J.C.Wüst, van Noort GJHL. *Neural Network Current reduction for Shipping Guidance*. IEEE OCEANS 94 Conference Proceedings, Vol. 1, pp 58–63.

[6] Adviesdienst Verkeer en Vervoer, *Onderzoek Tijpoorten Westerschelde*, 1 december 1993.

[7] *Invoer Randvoorwaarden "Evaluatie 74-voets geul"*. Rijswijk 15-8-1994.

[8] R. Wijnstra. *Onderzoek Tijpoorten Westerschelde Deelnotities*. Adviesdienst Verkeer en Vervoer.

[9] MARIN. *Aanvarings- en strandingskansen IJmond-Geul*, september 1987.

[10] P.M. Kimom, Exxon International Company. *Required Underkeel Clearance in Ports*, New Yersey 1982.

[11] Dienst Verkeerskunde. *Analyse gerealiseerde vaarten m.b.t. de bocht van het beerkanaal*, afdeling scheepvaart, april 1984.

[12] R.Ph.A.C.Savenije, Adviesdienst Verkeer en Vervoer. *Probabilistic Admittance Policy Deep Draught Vessels*, PIANC Bulletin n° 91, june 1996. 29 December 1995.

[13] R.Ph.A.C.Savenije, Adviesdienst Verkeer en Vervoer. *Probabilistic Admittance Policy Euro-Maas Channel*, proceedings 11th International Harbour Congress, Antwerp 1996.

[14] PIANC/IAPH WG30D. *Prediction of Squat*, 9 september 1996.

[15] R.Y. Schouten, J.v.d. Eikhoff, TNO. *Onderzoek naar schade welke schepen kunnen oplopen bij het raken van de bodem van de Eurogeul*, November 1983.

[16] MARIN. *Een model voor de berekening van verticale scheepsbewegingen*, februari 1994.

[17] NSP. *Proeven ter bepaling van het gedrag van een 200.000DWT tanker bij het aanlopen van Europoort*. report no. 68-179-GST.

Proceedings of the Seventh (1997) International Offshore and Polar Engineering Conference
Honolulu, USA, May 25-30, 1997

ISBN 1-880653-28-1 (Set); ISBN 1-880653-32-X (Vol. IV)

Houston/Galveston Safe Passage into the 21st Century, 1996

Stephen F. Ford
Texas A&M University
College Station, TX, USA

Ronald J. Bald
USCG Vessel Traffic Service
Galveston, TX, USA

ABSTRACT:

Mariners transiting the 70+ mile waterway complex bounded by the ports of Houston, Galveston, and Texas City are routinely affected by high winds, strong currents, low water levels, and sporadic shoaling. These conditions challenge the mariner's ability to control ships and barge-pushing towboats. Narrow, dredged channels provide the only means of transiting the area and do not offer abundant room to avoid other vessels. The combination of these and other factors led to a total of 1,525 groundings from 1986 through 1995 as documented by the U.S. Coast Guard Vessel Traffic Service (VTS) monitoring region. VTS loosely defines groundings for their statistical records as an interaction with the bottom which leads to the inability of the master to refloat the vessel at will. Wind, current, and low tide were given as the primary causes in 481 groundings equating to 31.6 percent of the 1986-1995 total of 1525 groundings. Very recent statistical data indicates a eleven year total (1986-1996) of 1,644 groundings in Galveston Bay. For the ten year period (1987-1996) of pro-active HOGANSAC-USCG waterways management of aids to navigation, there were 1,335 Galveston Bay groundings. The 1987-1991 total groundings were 932. The 1992-1996 total groundings were 403 wich reflects a 57% reduction between the five year periods of management.

Today, mariners are benefiting from the installation of a modern technological system. Since late-1995, National Oceanic and Atmospheric Administration (NOAA) personnel, members of academia, the maritime community, and the Coast Guard have acted in concert to acquire, deploy, and disseminate critical real-time weather, current, and tidal information from a Physical Oceanographic Real-Time System (PORTS). This intelligence can provide every mariner with crucial advance data regarding waterway and meteorological conditions that will affect their vessels. Application of this information can reduce the associated navigational risks which should lead to a reduction of maritime accidents. The area of greatest concern in this region is the entrance to Bolivar Peninsula along the Gulf Intracoastal Waterway (GIWW). This area is one mile east of the intersection of the Houston Ship Channel and the GIWW. About a mile on the opposite side of the Houston Ship Channel and across from Bolivar Peninsula is the Pelican Cut waterway which is of comparable concern to the maritime community. The next priority area, Morgan's Point, at the head of Galveston Bay is where the second current meter was installed. Strong broadside currents routinely affect towboats entering and departing the narrow cuts at Bolivar. VTS statistics rank Bolivar as having the highest occurrence of groundings over the last decade and the area has earned the highest percentage of overall groundings in during the 1987-1994 period. The 1996 year is dramactically improved. The number of groundings at the entrance to Bolivar Peninsula and Pelican Cut significantly declined and Morgan's Point Groundings remained level during a 1996 period of rising groundings per 1000 transits for the Galveston Bay region.

KEYWORDS:
Groundings, VTS, PORTS, Galveston Bay, NOAA, Safety

BACKGROUND:

Efficient and sustainable development of accurate charts and modern navigation systems are required for safe marine transportation systems. A strong National Oceanic and Atmospheric Administration (NOAA) and U.S. Coast Guard are important components of modern navigation safety and technology. As America and the Houston port complex advance into the 21st century, the international competitiveness of our local industries will depend upon the effectiveness of our marine transportation system and its associated port infrastructure. Safety and informed maritime decision-makers add a real cost to the final price of retail goods. However, the cost of a maritime catastrophe thrusts a much greater cost upon the maritime industry, the retail goods and ultimately, our society. The role of the Port Authorities is to optimize both the efficient transportation of water-borne commerce and the safety of all vessels operating on its waterway complex. As America experiences the globalization of trade, the port authorities of Houston, Galveston and Texas City are properly preparing for the 21st century. With the October, 1995 first current meter addition of the Houston/Galveston Physical Oceanographic Real-Time System (PORTS), NOAA, the Port Authorities and the local maritime community are demonstrating their concern for the environmental, ecological, and commercial sensitivities of their constituencies.

Every mariner fears being involved in a maritime accident. Groundings, collisions, and rammings can extensively damage the vessel, pollute the environment, or injure personnel. At a minimum, an incident will delay the vessel's transit which causes an unnecessary loss in efficiency which is certain to lead to a reduction in vessel revenue.

Figure 1 Two Ships Prepare to Meet in HSC {4}

Successful navigation requires the mastery of a vessel and her characteristics and must be accompanied with knowledge of waterway conditions. The fusion of these two components must be accompanied by foresight and pre-planning for the intended vessel transit in order to avoid an embarrassing accident. Professional experience plays the greatest role in recognizing potentially devastating factors for the voyage. But, experience must be accompanied with as much accurate and recent nautical information as possible if the vessel is to have an uneventful transit. In his opening paragraph of The American Practical Navigator, Nathaniel Bowditch states that a "good navigator gathers information from every available source...constantly evaluates the ship's position, (and) anticipates dangerous situations well before they arise..."

Figure 2 Two Ships Pass in HSC {4}

This paper investigates the relationships of the current meter installations on groundings at Bolivar Peninsula, Pelican Cut and Morgan's Point. The ten year period (1987-1996) of pro-active HOGANSAC-USCG waterways aids to navigation management is evaluated in terms of grounding frequency and compared to a unmanaged base year of 1986. The paper contrasts statistics between the two five year periods 1987-1991 vs. 1992-1996.

Narrow, dredged channels, high traffic density, and waterway characteristics which vary from 7 mile reaches to 20 miles of continuous turns welcome mariners transiting the 70+ mile waterway complex bounded by the ports of Houston, Galveston, and Texas City.
Even on a beautiful day with no wind, no current, and unlimited visibility, transits can be challenging. The only means for commercial ships and barge-pushing towboats to transit the area are the Army Corps of Engineers maintained channels. At a maximum width of 400 feet and a depth of 40 feet, these channels do not offer abundant room to avoid other vessels. Ships with beams exceeding 100 feet each routinely meet, and safely pass at a closest point of approach of fifty to seventy-five feet. This feat is accomplished through precision shiphandling. As shown in Figures 1 and 2, the vessels approach each other head-on, then simultaneously break to starboard when they reach a distance of approximately one-half mile apart. The ships use the hydrodynamic forces between the vessels and the banks to pass each other and regain their positions along the centerline. Shallows of less than ten feet lie within 600 feet of the channel centerline in most locations. Localized high winds, strong currents, low water levels, and sporadic shoaling routinely stress the Galveston Bay mariner's ability to control their ships and towboats.

The U.S. Coast Guard Vessel Traffic Services (VTS) Houston/Galveston monitors all waterborne traffic as required by federal law. In 1996, the VTS logged a daily average of 347 ships, towboats, and other vessels into the Vessel Movement Reporting System while monitoring 265 daily ferry boat transits. The level of VTS traffic has maintained a steady 2.5 percent annual growth (Figure 6) in vessel movements since 1986. The size of the vessels calling on the ports also continues to grow. Newer vessels carrying more cargo continue to expand the operating envelope of the waterways. However, throughout this period of increasing traffic, the physical waterway dimensions have not changed. Today, little room exists for vessels to maneuver out of harm's way in the event unexpected circumstances arise.

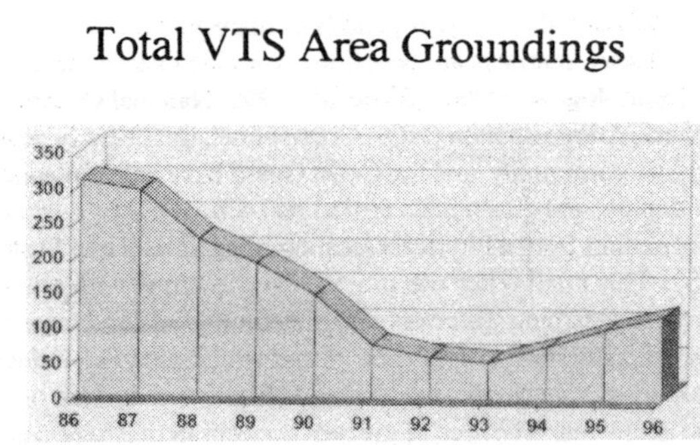

Figure 3 1986-1996 Total Galveston Bay Groundings {3}

Figure 3, Total Groundings, conveys the significance of the 1,644 Galveston Bay groundings between 1986 and 1996 as documented by the local VTS. VTS defines groundings for their statistical records as "an interaction with the bottom which leads to the inability of the master to refloat the vessel at will." Fortunately, the regional soft mud bottom contains an extremely low risk of adverse consequences resulting from the event. Wind, current and low tide were the primary grounding causes and typically represent about one third of total groundings from all causes. The 1987-1996 total (management period) groundings were 1,275. The 1987-1991 total groundings were 932 and 1992-1996 period was 403 for a 57% reduction.

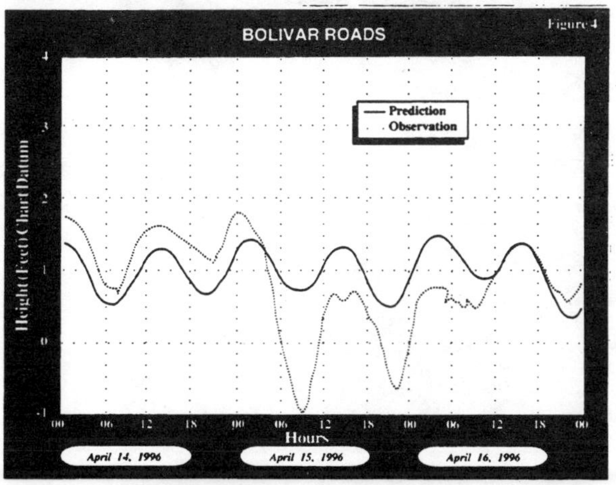

Figure 4 4/14-16/96 Bolivar Roads Water Levels {4}

The Galveston Bay basin is strongly influenced by the passage of local weather fronts. A 1987 NOAA evaluation, at the request of HOGANSAC, confirmed the local tidal predictions derived only from solar and lunar data had an accuracy level of only 50%. Figure 4, Galveston Bay Water Levels during northerly winds, illustrates their meteorological impact on the Bay on April 15, 1996 . Figure5, March18, 1996 Frontal Passage Impact, conveys the effect on the local water levels by the passage of a cold front that was involved in the break up of a barge and a significant fuel oil spill. The three foot drop in the water level in eight hours, observed through Figures 4 and 5, caused very high currents within the bay. The "zero mark" on the chart datum is Mean Lower Low Water (MLLW) which provides 40 feet depth in the HSC at this point. i.e. There was only 39 foot of water in the HSC

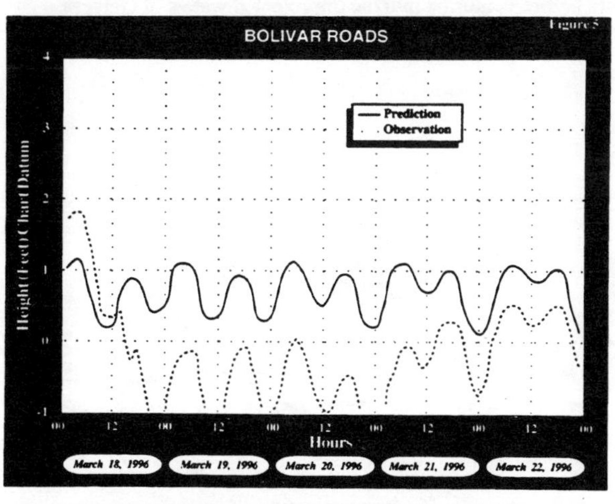

Figure 5 Northerly Frontal Passage Impact {4}

about 0900 4/15/96. From these examples, it is apparent that both the water level and current parameters have significant daily effect on the local marine transportation system.

Starting in 1987 and beginning in earnest in 1989, a cooperative effort between the local maritime industry's Houston Galveston Navigation Safety Advisory Committee (HOGANSAC) and the Coast Guard has yielded many positive changes in defining the waterway through aids to navigation placement. The late Coast Guard Captain John Witherspoon, while commanding officer of the Houston/ Galveston VTS, gathered regional statistics which provided a locational basis for cooperative industry-Coast Guard aids to navigation placements and changes. Over time HOGANSAC leaders, Mr. Milt Rose and Captain John Hoopaugh in concert with VTS representatives like Commander Michael Hunt and Commander Costello oversighted and fine-tuned aids to navigation placements based on statistics and vessel operator feedback. For the management period 1987-1996 the 1,275 groundings in 1,138,821 transits result in an average of a frequency of one grounding in 893 transits or 1.1 groundings per 1000 transits. This compares to the unmanaged 1986 base year of one grounding every 318 transits or 3.1 groundings per 1000 trransits. Over the ten year period

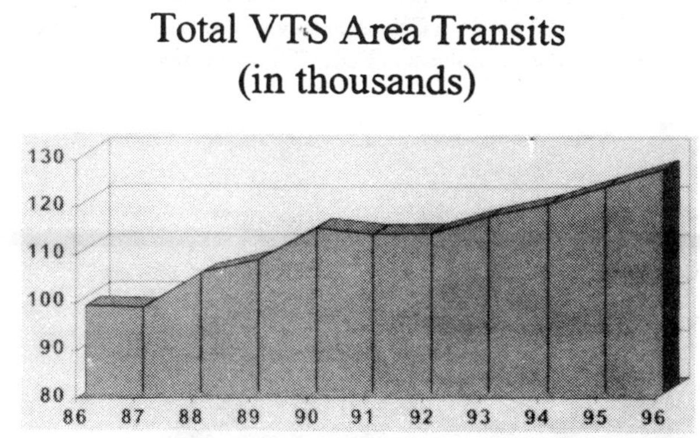

Figure 6 1986-1996 VTS Galveston Bay Transits {3}

strategicaly placed buoys and lights have virtually eliminated groundings in some areas of the bay.

The learning curve benefits are visible in a comparision of the two five year period of 1987-1991 vs. 1991-1996. In the first five year period, there was a grounding on the average of every 578 transits of the total 538,552 transits or 1.7 groundings per 1000 transits. In the more recent five year period (1992-1996), there was a grounding on the average of every 1,489 tranists of the total 600,269 transits or 0.67 groundings per 1000 transits. This was aproximate a 60% reduction between the two five year periods of waterways management. The 1992-1996 period statistic was an improvement of 469% over the 1986 unmanaged frequency of 3.1 groundings per 1000 transits or one grounding every 318 transits. While this is a tremendous improvement, every marine accident can be a catastrophe. Every member of the maritime community in the Houston, Galveston, Texas City complex continues to desire to give the mariner all the information and technology needed to conduct a safe transit. With the waterway boundaries better defined by 1993, it was apparent the mariner now needed real-time information on environmental factors in order to further reduce the incident statistics.

In 1989, HOGANSAC, under the leadership of Captain Jim Baker and Mr. Ted Thorjussen, convinced NOAA to evaluate the effectiveness of the local Tide Table predictions. This study confirmed local concerns about the accuracy of the table predictions. As a remedy, this HOGANSAC leadership sought the installation of current meters and water level devices in Galveston Bay.

During 1990-1991, a NOAA National Ocean Service (NOS) team had tested a prototype Physical Oceanographic Real-Time System (PORTS) of one current meter and of one water level sensor in Tampa Bay . Subsequently, NOAA prepared a report to the U.S. House of Representatives in June, 1994, discussing the benefits of a PORTS in several waterways including Houston/Galveston. By 1994, strong backing from local maritime industry officials, along with Congressional assistance, helped secure an initial NOAA investment of $750,000 for the procurement, installation, and one year of operation and maintenance of three Acoustic Doppler Current Profilers integrated with five meteorological and water level sensors in a real-time Galveston Bay system for the legislated purpose of enhancing marine transportation safety. These profilers combine to provide widespread coverage of 50 miles of area currents, tidal levels, and weather information. The water level sensors are part of the NOAA national water level observation network (NWLON) headed up by NOAA's Mr. Tom Mero and Mr. Jim Dixon of the National Ocean Service. Three of the water level sensors are Texas Coastal Ocean Observation Network (TCOON) sensors which are administered by Texas's General Land Office (GLO) and maintained by Dr. Manz of Lamar. In 1995, the Marine Transportation Department of Texas A&M University at Galveston (TAMUG) took the Galveston Bay project's lead and is now the base for the Houston/Galveston PORTS' initial period.

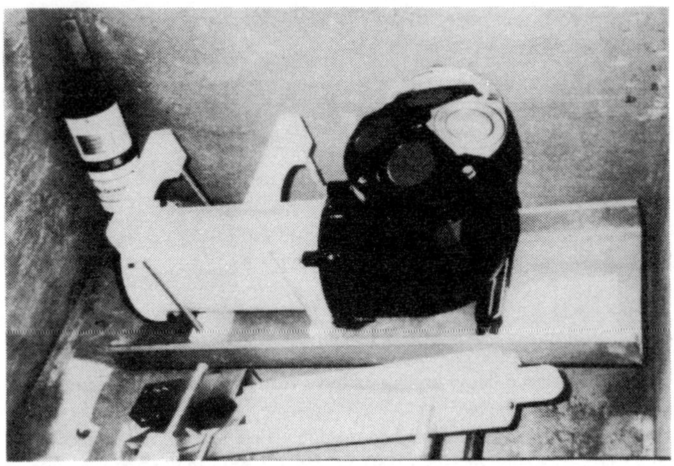

Figure 7 Acoustic Doppler Current Profiler (ADCP)

PRESENT PORTS ACTIVITIES:

An ADCP or Acoustic Doppler Current Profiler (Figure 7) is a sonic " look-up " device that is buried in the sea bottom to measure the doppler shift of sound waves and thus measure water current at different levels.

With the aid of the Galveston-based Coast Guard Cutter CLAMP, two profilers have been permanently installed near the intersection of the Houston Ship Channel entrance and the Gulf Intracoastal Waterway (GIWW) and near Morgan's Point at the upper end of Galveston Bay. As per Figure 8, Instrument Sites, a smaller, portable profiler will be rotated between several sites in the waterway to document conditions and changes over time. Its initial site is near Redfish Bar until sufficient data is acquired to permit "nowcasting".

Information from the three profilers is received every six minutes at TAMUG. A complex installation of four computers processes the data and makes it immediately available to the general public via telephone with the aid of a voice synthesizer or personal computer (PC) modem. The graphic data is also immediately available via Internet. Figure 9, PORTS Data Display, portrays the real-time information and graphics available from the system. For a real-time demonstration of PORTS technology, interested parties are invited to establish voice contact at (409) 740-4975 or internet contact at http://hgports2.tamug.tamu.edu or telnet ceob.nos.noaa.gov (login hgports). The Internet site is maintained by Mr. Todd Sanders of the TAMUG HGPORTS staff who welcomes your feedback.

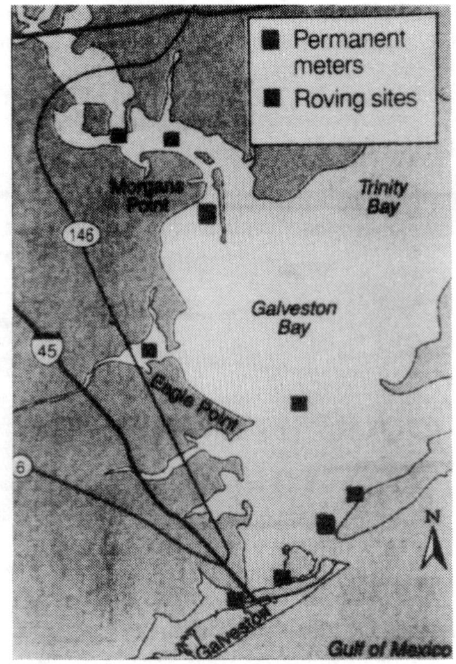

Figure 8 Galveston Bay PORTS Instrument Sites
Photo Courtesy of Galveston Daily News

Remote real-time PORTS displays are located at the Houston/Galveston VTS and Galveston Marine Safety Unit (MSU) offices and at both the Galveston-Texas City Pilot and Houston Pilot offices. The VTS disseminates the information to underway vessels while the pilot office continuously updates its members. MSU Galveston has already utilized the PORTS to predict and contain two significant spills during March, 1996. When necessary or desired, the data can also be accessed by underway pilots via cellular phone. Current speeds, water levels and wind data are critical parameters for mariner's avoidance of groundings and other incidents. Upon receipt of this PORTS information, many towboat flotillas are prudently awaiting more favorable operational parameters before venturing into the unprotected waters of Galveston Bay and the Houston Ship Channel.

Figure 9

```
                         Houston/Galveston PORTS
                       National Ocean Service/NOAA
                    at  9:57 am CST  March 18, 1996
.........................................................................
             TIDES                 :              CURRENTS
Pleasure Pier          0.6 ft.,Falling: Bolivar Roads    1.9 kts.(E), 140øT
Pier 21                0.5 ft.        : Redfish Bar      1.4 kts.(E), 146øT
Bolivar Roads          0.4 ft.,Falling: Morgans Point    1.1 kts.(E), 164øT
Eagle Point            0.7 ft.        : (F)lood,(S)lack,(E)bb,towards øTrue
Morgans Point          0.9 ft.,Falling:..................................
...................................:   Salinity & Specific Gravity
 Bottom Water Temp. (ADCP) :                       Sal.    S.G. Srfc Temp
Bolivar Roads      61øF    : Bolivar Roads      26.0 psu            63øF
Redfish Bar        66øF    : Eagle Point
Morgans Point      60øF    : Morgans Point      19.0 psu            65øF
...........................:.............................................
                            METEOROLOGICAL
                    Wind Speed/Dir                Air Pressure    Air Temp
Bolivar Roads   23 knots from WNW, gusts to  26   1006 mb,Rising    65øF
Eagle Point     13 knots from NNW, gusts to  27   1007 mb,Steady    62øF
Morgans Point   20 knots from WNW, gusts to  27   1006 mb,Rising    63øF
.........................................................................
For a description of PORTS, please contact Capt. Stephen Ford of TAMUG at
(409) 740-4471 or Dr. Wayne Wilmot of NOAA at (301) 713-2803
```

Figure 9 PORTS Data Screen on Telnet (PC Modem)

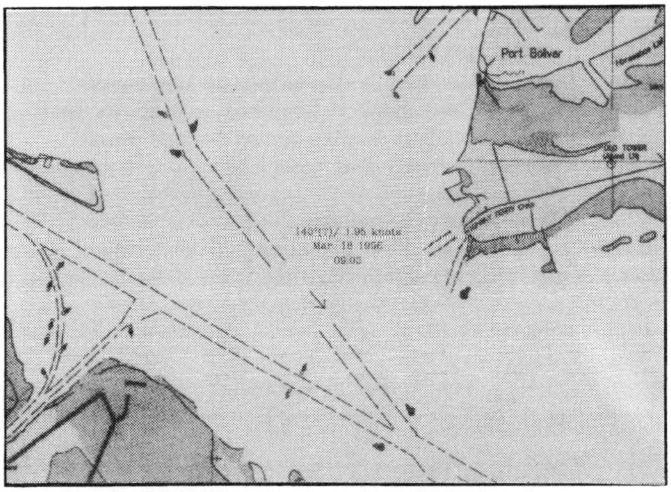

Figure 10 Bolivar Roads PORTS Display With
Current Data on Internet

As research at TAMUG has progressed, Internet, PC modem, and fax users are now able to receive pictorial displays of the numerical PORTS data in a fashion similar to Figure 10. Further research efforts envision that ships with the "right computer equipment" and vector electronic charts (Figure 11) or "Smart Charts" will receive real-time views of their nautical surroundings with complete details of the sea bottom including actual channel depth and real-time current velocity and direction from PORTS. Researchers at TAMUG are also investigating the utilization of PORTS data in ship simulators and in their development of virtual reality charts.

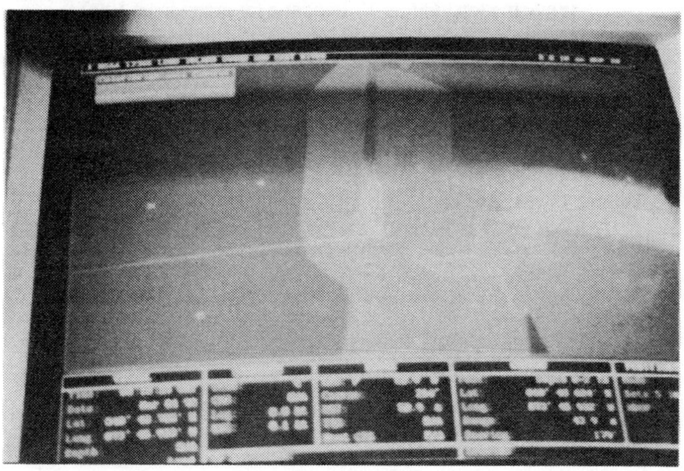

Figure 11 Vector Electronic Chart

The present PORTS installation is reducing navigational risks in the Galveston Bay complex by providing local mariners with accurate, real-time information. In essence, PORTS is adding more science to the seaman's eye in order to make waterborne commerce safer, easier and cheaper in a sustainable fashion.

The anticipated result of this new system was a reduction in the number of marine accidents in the waterway complex. Particularly, all parties expect to see a further decline in the number of groundings at the entrance to Bolivar Peninsula and Pelican Cut. This Bolivar area is one mile east of the intersection of the Houston Ship Channel and the Gulf Intracoastal Waterway (GIWW). Pelican Cut is about one mile west of the HSC. As depicted by Figure 10, PORTS' Bolivar Roads

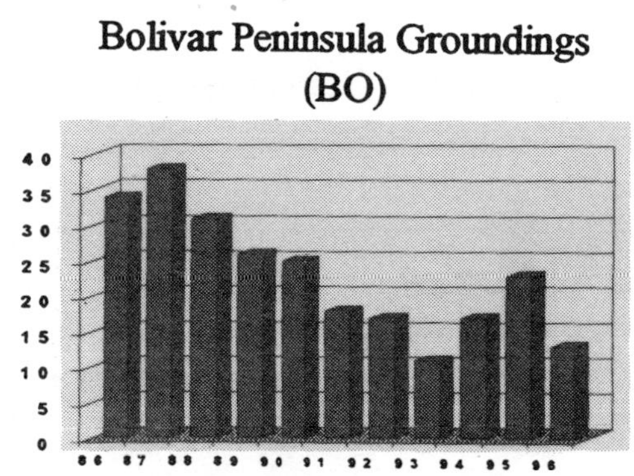

Figure 12 1986-1996 Bolivar Peninsula Groundings {3}

Display, strong broadside currents routinely affect towboats entering or exiting the narrow cuts on both sides of Bolivar Roads. As shown in Figure 12, Bolivar Roads Groundings, this area accounts for 219 groundings during the 1987-1996 period and Pelican Cut had 159 groundings for the same period for a total of 378 groundings in the locale.

Figure 13 Towboats "Crabs" Into Bolivar Peninsula Entrance
Photo by QM1 Scott Eriksen, USCG {4}

Figure 14 Gulf Intracoastal Waterway (GIWW)
Photo Courtesy of Kirby Corporation {4}

Pelican Cut Groundings (PE)

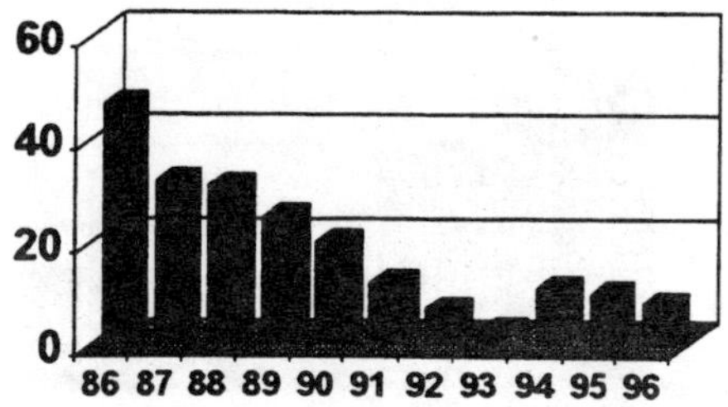

Figure15 1986-1996 Pelican Cut Groundings {3}

The 378 groundings represent 28% of the 1,335 total Galveston Bay groundings for 1987-1996. Figures 12 and 15 illustrate the annual grounding data for the Bolivar Road area.

The Bolivar area ranks as having the highest occurrence of groundings over the last decade. Bolivar Roads is the locale of the highest percentage of overall Galveston Bay groundings during the 1986-1995 period. Figures 13 and 14 display the local towboat operations on the GIWW. The GIWW is heavily traveled by towboats calling on the area facilities as well as those transiting to and from other ports of call. Groundings at the Bolivar Peninsula entrance periodically block the entire waterway which causes dozens of vessels to wait until the tow is able to work her way off the bottom. This channel blockage not only affects waterborne commerce and economics, but also its costs must ultimately be borne by the consumers of the impeded cargo. With PORTS as an added aid to marine transportation safety, fewer groundings should occur which should reduce the lost time of vessels and increase efficiency and productivity of local commerce.

Morgan's Point Groundings (MP)

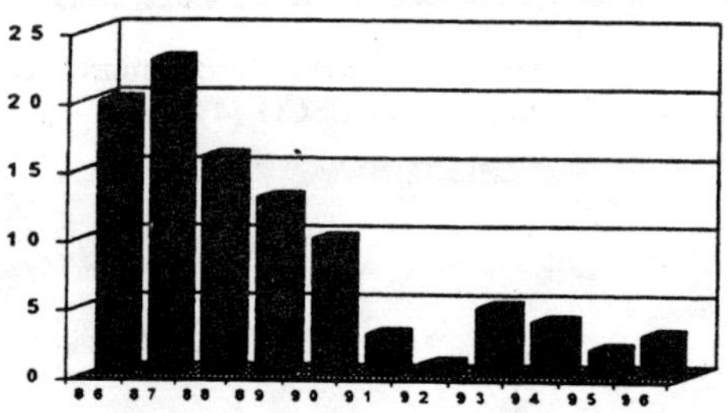

Figure 16 1986-1996 Morgan's Point Groundings {3}

After the October, 1995, Bolivar Roads current meter installation, VTS began communicating current velocities when they reached critical levels as determined by local towboat operating company representatives. Figure 16 illustrates the annual data for the less important locale of Morgan's Point at the head of the bay. This locale experienced 80 groundings during 1987-1996 which represents six percent of total groundings. However, a very large container terminal, Barbour's Cut, is located nearby and heavy course change traffic occurs in this locale. The curent meter supports continued safety in the area.

1996 GROUNDING IMPROVEMENTS AT CURRENT METER SITES:

Figure 17, Galveston Bay Groundings per 1000 transits, illustrates dramatic decline until 1993 when it begun an upward trend. It is postulated that groundings per 1000 transits over time would correlate with annual meterorological, water level and current parameters. It is hypothisized that unfavorable physical parameters will generate unfavorable operating performance (grounding) statistics. Hence,a rising trendline would indicate a period of comparatively adverse physical parameters. Nontheless, the declining trendline for the 1987-1993 period reflects the benefits of HOGANSAC - USCG waterway management efforts.

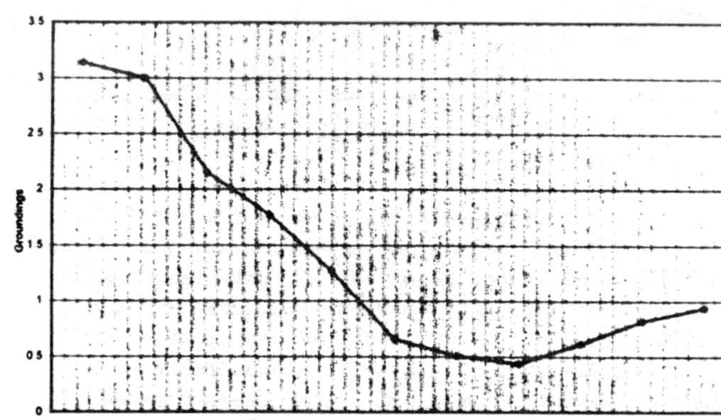

Figure 17 Galveston Bay Grounding Per 1000 Tranists

While the total Figure 17 trendline for groundings 1000 transits has been proceeding in an undesirable direction since 1994, the 1996 percentage of total groundings (Figure 18 and 19) at Bolivar Peninsula, Pelican Cut and Morgan's Point have been proceeding against the Galveston Bay trend of Groundings per 1000 transits.

Table 1, Grounding Statistics at Bolivar, Pelican, and Morgan's Point provides the underlying tabular information conveyed by Figures 17, 18, and 19.

TABLE 1. GROUNDING STATISTICS @ BOLIVAR . PELICAN, & MORGAN'S POINT

YEAR	GROUNDINGS PER 1000 TRANSITS	BOLIVAR % AGE OF TOTAL GROUNDINGS	PELICAN CUT % AGE OF TOTAL GROUNDINGS	COMBINED BOLIV/PELICAN % AGE OF TOT GRDGS	MORGAN'S PT % AGE OF TOTAL GROUNDINGS
1986	3.14	11%	15%	26%	6%
1987	3.00	13%	11%	24%	8%
1988	2.15	14%	14%	27%	7%
1989	1.77	14%	13%	27%	7%
1990	1.27	17%	14%	31%	7%
1991	0.66	24%	16%	40%	4%
1992	0.51	29%	12%	40%	2%
1993	0.44	12%	6%	27%	10%
1994	0.62	23%	15%	38%	8%
1995	0.82	23%	10%	33%	2%
1996	0.94	11%	7%	17%	2%

Table 1 Tabular Data For Fig. 17, 18. 19

For the 1987-1996 data period, Figure 18 illustrates the percentage of total Galveston Bay groundings reflected by Bolivar Peninsula (BP), Pelican Cut (PC), Morgan's Point (MP) and Bolivar Roads (BP+PC). In a less dramatic scale, these percentages are compared to the average annual groundings per 1000 transits for the entire Galveton Bay (VTS) region.

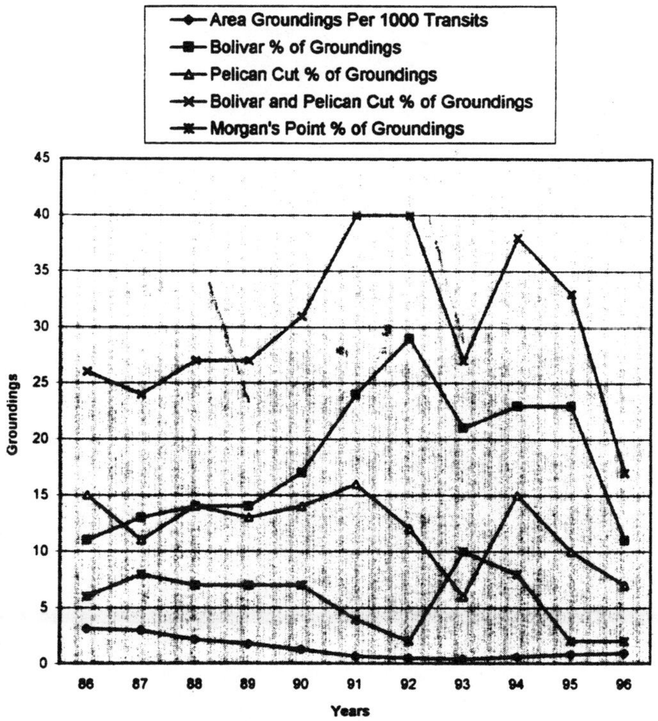

Figure 18 1986-1996 Bolivar, Pelican, Morgan Groundings Percentages

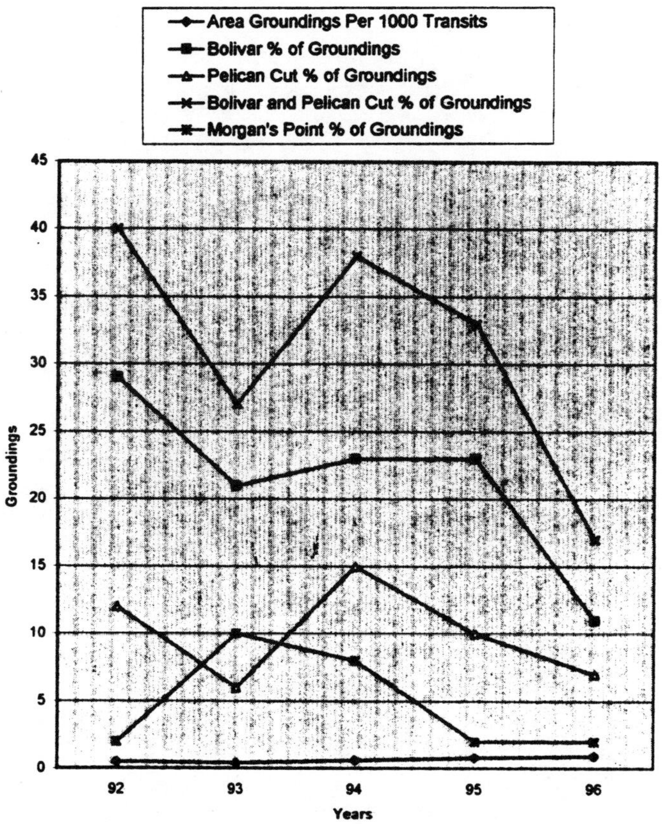

Figure 19 1992-1996 Groundings Statistics

Figure 19, 1992-1996 Grounding Statistics, more closely scrutinzed the most recent five year period. The improvements in 1996 statistics for Bolivar, Pelican, Bolivar Roads (BP+PC), is readily evident. This 1996 period was a significant improvement during a period of rising adverse meteorological conditions. Despite the fact the Morgan's Point data remains flat in 1996, this is regarded as an "improvement" for the locale in view of the rising statistic of groundings per 1000 transits for the region.

CONCLUSION:

It is the consensus of the author and local maritime personnel that the most significant impact on the Bolivar Roads and Morgan's Point locales during 1996 was the availability of the current meter, water level and meteorological information. It is hypothesized that the installation of the PORTS system is what explains the observed decrease in Bolivar Penisula and Pelican Cut groundings.

With continous recording of parametric data every six minutes by the PORTS system, it is anticipated that the individual contributions of the specific parameters (cuurent, wind, waterlevel) to the grounding statistics will be identified in the future.

Thus, it is the author's opinion based upon recent statistical data, the Houston/Galveston PORTS has made significant contributions to:

1. HOGANSAC - USCG Waterways Management Program
2. Improvements in Marine Transportation Saftey in Galveston Bay.

It will be of interest to the HOGANSAC-USCG and PORTS personnel to track how future data sustains this opinion over a more valid period of observation. In the interim, all participants in the HOGANSAC, USCG, PORTS groups deserve credit for their efforts and contributions towards the improvement in marine transportation safety in Galveston Bay.

Persuming favorable PORTS statistics in the future, other Port authorities should consider developing programs based upon the models of:

1. HOGANSAC - USCG
2. PORTS

to improve their local marine transportation systems. At the very least, all port complexs are encouraged to initate statistical tabulations of their groundings and accidents by geographic locations.

ABOUT THE AUTHORS:

Captain Stephen F. Ford is a 1970 graduate of the U.S. Merchant Marine Academy with a 1978 M.B.A. from the University of Houston. He is a Master Mariner of fifteen years seagoing experience in coastal and international tanker trades. Past vice-president of a U.S. flag shipping company and past Head of Marine Transportation of Texas A&M University at Galveston, he now serves as the PORTS Manager of the Houston-Galveston system. Eight years service on NAVSAC and almost twelve years on HOGANSAC plus three years on the Houston Pilot Review Board amply qualifies him for his PORTS participation and active Maritime Expert and Consultant assignments. Captain Ford was the key, lead team member in the construction of the first electronic chart in 1976 and continues to lead research and development in the dynamic field of Electronic Chart Display and Information Systems (ECDIS).

Lieutenant Ronald J. Bald of Closter, New Jersey, is a 1987 graduate of the U.S. Coast Guard Academy with a 1995 Master of Arts degree in Public Management from the University of Houston - Clear Lake. He has served on active duty for the last nine years in the Houston/Galveston region. His duty assignments include operations officer aboard the Coast Guard Sea-Going Buoy Tender BUTTONWOOD, Coast Guard Group Galveston aids to navigation officer, and operations officer for Vessel Traffic Service Houston/Galveston. He will depart the area in the summer of 1996 for Alameda, California, and a tour of duty aboard the Coast Guard High Endurance Cutter BOUTWELL.

REFRENCES:

1) USCG Vessel Traffic Service Houston/Galveston
1994 State of the Waterway Address, January 26, 1993
2) USCG Vessel Traffic Service Houston/Galveston
1996 State of the Waterway Address, January 18, 1995
3) USCG Vessel Traffic Service Houston/Galveston
1997 State of the Waterway Address, January 30, 1996
4) "Technology and Maritime Professionals Proactivaley Reduce Navigation Risks on Southeast Texas Waterways", by Ford, S.F. and Bald, R.J., Marine Safety Council Proceedigns, July-September 1996.
5) "Practial Navigator" by Nathaniel Bowditch, Defense Mapping Agency (DMA No. NVPUB9V1)

Proceedings of the Seventh (1997) International Offshore and Polar Engineering Conference
Honolulu, USA, May 25-30, 1997

ISBN 1-880653-28-1 (Set); ISBN 1-880653-32-X (Vol. IV)

Third Generation Electronic Charts: What They Will Provide Users and How to Produce Them

Peter Kielland
Canadian Hydrographic Service
Ottawa, Canada

Michel Dagbert
Geostat Systems International
Montreal, Canada

ABSTRACT:
The basic function of today's "Electronic Chart" (EC) is to continuously re-plot a vessel's changing position onto the digital facsimile of a traditional paper chart. While this plotting function is extremely useful, it is only a limited subset of the bathymetric display functions that could be supported if the survey data underlying the chart image were more realistically modeled. A continuous interpolated surface model describes the true seafloor more realistically than the discrete "points and lines" data model used to produce paper charts and their EC facsimiles. When supplied with a correctly computed surface model, present EC hardware can extract powerful, "third generation" views of the bathymetry, views such as: display of user defined contours and contour intervals, display of arbitrary "true depth" contours animated by real-time water level reductions, display of detailed sectional profile views along planned routes, display of 3D perspective views of the seafloor model and display of false colour seafloor imagery. Viewing the surveyed seafloor in this manner could easily give navigators a false and dangerous impression that; since the seafloor image appears "high tech" it must therefore be a complete and error-free spatial model. To avoid this danger, great attention must be paid to statistical modeling of the total probable depth error inherent to each estimated depth on the seafloor's surface model. The resulting "stochastic surface" is a critically important Quality Control tool since it enables fully automated safety biasing of whatever imagery an EC user might choose to display. This paper discusses a geostatistical surface modeling algorithm that supports the full spectrum of third generation EC viewing capabilities and demonstrates some of them using typical survey data.

KEYWORDS: geostatistics, kriging, risk analysis, bathymetry, hydrography, Electronic Chart

INTRODUCTION

Throughout history, nautical charts have been produced by compiling navigators' collective bathymetric knowledge then hand drawing it onto two dimensional media. The hallmark of this charting activity was, and still is, the interpretive mental judgments exercised by cartographers as they render the mass of surveyed data into an easily comprehended and "safe" image of the seafloor. Over the past 20 years, advancing computer technology has made it feasible to migrate the display of traditional chart imagery from two dimensional paper media into the cyberspace of the "Electronic Chart". The term "Electronic Chart" (EC) is a generic term used throughout this paper to refer to all navigation systems that display chart data. The terms "Electronic Navigational Chart" (ENC) and "Electronic Chart Display and Information System" (ECDIS) are also terms used in the paper. The ENC and ECDIS are evolved variants of the generic EC that are explained in context. The fundamental problem addressed in this paper is this: traditional paper chart data was never intended to be displayed on today's far more flexible electronic media and therefore cannot exploit its true potential. What follows is a discussion of that problem as well as the presentation of a data processing methodology that solves it. The proposed geostatistical data modeling algorithm employs objective statistical criteria to compute "third generation" bathymetric surface models whose structure is optimized to fully exploit the display capabilities inherent to the EC display media. Readers less interested in the mechanics of how this algorithm works may wish to jump to the section entitled "Safety Biasing The Seafloor Model" to see examples of the end-user products that the modeling process supports.

First and Second Generation ECs

The most striking benefit of presenting nautical chart imagery in an Electronic Chart environment is that the system's computer is used to automatically integrate the graphical chart information with real-time position information. This basic EC function automates the task of continually re-plotting the vessel's present position so that it's correctly displayed with respect to the backdrop of bathymetric information. By eliminating this manual drudgery, the EC enables navigators to concentrate more fully on analyzing all of the available data and making safer navigational decisions.

The first generation of Electronic Charts used raster scanned images of a paper chart to provide the visual backdrop of bathymetric information. Once the raster image is geo-referenced and overlaid with GPS position data (an icon of the vessel's current location), first generation ECs provide the basic automated plotting function which is the essence of any EC. Raster scanning paper charts is a fast and inexpensive way to create EC data, however, high resolution raster data files tend to be large and the raster images are difficult to maintain congruent with updates to the paper chart. However, their most severe limitation is that each pixel's value stores only the colour of the paper chart at that location. Raster chart data therefore really only contains 2D spatial information: the third "dimension" modeled by the raster is not the desired spatial parameter: "depth of water", it's the far less desirable graphical parameter: "color of paper".

This lack of true spatial information content limits the ability of first generation ECs to spatially decode the seafloor model (i.e. the EC's computer is unable to "know" the water depth at any location on the raster graphic).

To address this problem, second generation ECs were developed that display vectorized chart data. Vectorized chart graphics are described geometrically using a linked series of two-dimensional coordinates to define the various points and lines that make up a paper chart image. Vectorized descriptions of the chart's graphical elements are generally more compact than a rasterized facsimile of the same chart. Far more important however is the fact that, by vectorizing the chart's graphical "objects", spatial attributes can be attached to each one and thereby impart a limited degree of 3D spatial "intelligence" to the digitized chart data. For example, a vectorized 10 metre contour line is no longer simply a series of adjacent black pixels; it becomes a linked series of geographic coordinates that define the continuous geometry of the 10 metre contour object.

What is important about vectorized chart objects is that they can be intelligently managed within the EC's database, thereby providing the navigator with some degree of control over the data. For example, it enables the EC user to turn on or off the display of chart objects (e.g., "turn off" all contours deeper than 50 metres). Attributed chart objects can also be digitally manipulated to some extent. For example, by subtracting the current water level height from the depth attributes of vectorized contour and spot sounding objects, their displayed values are automatically reduced to chart datum. Object attribution also enables the EC's computer to "act behind the scenes" to automatically warn the user of impending danger. For example, if the vessel's position icon gets closer than a pre-determined distance from a specified depth contour then a grounding alarm can be activated to draw attention to the potential problem. Thus, second generation ECs are characterized by being able to actually apply some of the EC's computational power to augment the navigator's skills ... an obvious advance over simply using the EC as a display media for digitized paper charts. ECs that can display and exploit vector graphics are often referred to as ENCs (Electronic Nautical Charts).

Second generation ECs have other features that distinguish them from first generation raster plotters. One important added capability is that of overlaying a radar image onto the EC background to aid in collision avoidance. Other features exist, or are in the process of being developed, that can be added to the basic EC configuration. Some examples are:

- Hypertext links to List Of Lights information (e.g.: click on a buoy or light icon to display the light's characteristic),
- Hypertext links to Sailing Instructions (e.g.: click on marina to read how best to approach it or a list of its available services),
- Links to Notices to Mariners (to facilitate or even automate the task of maintaining the digitized chart data congruent with its paper chart equivalent),
- Links to tidal information (as simple as on-line tide tables or as sophisticated as real-time tidal modeling using broadcasts from digital tide gauges),
- Integration of images from weather maps and/or ice coverage maps to aid in route planning,
- Integration of hydro-dynamic modeling to predict the vessel's response to wind, currents, steering input and proximity to the seafloor.

Second generation ECs are not only characterized by additional convenience items and data integration features. Compliance with evolving international data format standards is also a fundamental characteristic. The International Hydrographic Organization is the body that works towards establishing a standard EC data format and the International Maritime Organization is the body the certifies standards and promulgates them to mariners. The IHO S-57 is the latest EC data standard which embodies object attribution conventions thereby insuring that all EC manufacturers can read and display the same digital chart files. A standards compliant, vector based EC that integrates external data and augments the user's decision making capability takes on the most sophisticated EC acronym: ECDIS (Electronic Chart Display and Information System).

Third Generation EC's

The line between "first" and "second" generation ECs is really quite fuzzy ... it's more a question of evolution rather than of revolution. Fundamentally, all current ECs are the same thing because they can only display the "points and lines" bathymetric data model that was developed long ago for the paper chart. When compared to a paper chart, today's EC systems have certainly improved the convenience of viewing traditional chart imagery. However, as long as the EC is limited to displaying paper chart graphics, the net bathymetric information content transmitted to navigators can be no greater than that conveyed by a paper chart. The real seafloor is not made up of discrete points and lines ... it's a continuous surface and it must be modeled as such if we are to achieve a quantum improvement in EC performance.

The required continuous (i.e. "very dense") 3D seafloor model can be produced by one of two methods:

1) By measuring all points on the seafloor using very dense depth soundings (such as provided by multibeam swath sounders). For display purposes, this ultra-dense volume of data can then be reduced by "binning" it onto a dense orthogonal grid.
2) By interpolating in between sparser (i.e. singlebeam) soundings to estimate a dense grid of depths for all points on the seafloor.

The seafloor model derived from either of these methods has a matrix data structure which is optimal for the EC display function. Most of the advantages of the gridded surface model are new EC display capabilities that are illustrated later in this paper. However, one computer related advantage that's often overlooked is the fact that the 3D model's "row and column" structure greatly compresses the size of the data file that must be manipulated and displayed by the EC. This data compaction benefit is due to the fact that the XY position of each gridded depth need not be explicitly written into the file. This is the same data storage efficiency that's exploited by raster graphic formats such as TIFF (Tagged Information File Format). A TIFF file uses a very compact header to define the row and column structure of the raster. This permits the XY position of every pixel value in the image to be implicitly decoded based solely on its sequential location within the file. This "value stream" data structure also lends itself to run-length encoding which further compresses the file. In the case of the gridded seafloor model being discussed here, each "pixel value" is actually an interpolated or measured depth rather than the "dumb" colour information normally associated with raster files. Since each grid node's value is a spatial parameter rather than an imaging parameter, the gridded 3D seafloor model is often referred to as an "intelligent raster" or "depth image". By exploiting the compressibility of its gridded data structure, a 3D depth image can approximate the high information density of the hydrographic survey record yet occupy only about the same file size as the dumb, raster-scanned image of a paper chart covering the same area.

Both multibeam surveys and interpolated singlebeam surveys are valid methods for producing a suitable gridded seafloor model, however, each method has distinct advantages and disadvantages. Since a multibeam sounder typically ensonifies the entire seafloor surface, the model derived from swath data has almost perfect spatial fidelity. Quality Control of such a model is simplified, since only positioning and depth measurement errors need to be considered. Swath data is generally dense enough that the roughness of the seafloor can be neglected as an error source (i.e. we are certain that all short wavelength shoal structures have been captured in the survey record). The disadvantage of swath sounded bathymetry is that it's very expensive, both to collect and to manage. Also, in terms of improving the fidelity of a seafloor model, sampling theory dictates that there is diminishing returns as more and more soundings are collected. Deriving seafloor models from multibeam data is therefore the ideal method, but one that, from an economic standpoint, should only be practiced in areas that are critical enough to shipping to warrant this ultimate expense.

The advantage of a gridded seafloor model derived by interpolating sparser singlebeam soundings is that it's much less expensive to survey, or can even be derived by re-processing existing survey data. The major disadvantage of interpolating between discrete sounding lines is that, due to incomplete sampling of the continuous and rough seafloor, a model interpolated from sparser data will invariably have inferior spatial fidelity with respect to one derived from full multibeam coverage. What is important from an economic standpoint is that in many geological and navigational situations the risk posed by interpolating a less than perfect seafloor model may be deemed both acceptable and cost effective. That's why objectively estimating the true spatial uncertainty of interpolated seafloor models is essential for adequate Quality Control of surveys. Without reliable error estimates for both the instrumental and interpolation errors that vary from place to place over the surveyed seafloor model, it's impossible to make informed logistical decisions about where the expense of continuous multibeam coverage is warranted and where discrete sampling presents an acceptable risk.

Regardless of the method used to derive a gridded seafloor model, its structure effectively permits users to access the very dense survey data that previously had to be discarded by cartographers. During cartographic data reduction, at least 90% of all surveyed soundings must be discarded and in many cases this ratio is over 99%. This high ratio of cartographic data reduction is necessitated by the fixed scale of a paper chart. If the complete survey data record were ever displayed at chart scale it would appear as an unintelligible mass of dense soundings. The survey data discarded by cartographers could only be displayed legibly at very large chart scales but for economic and logistical reasons these charts could never be compiled and published. The paper chart also needed to be uncluttered enough that a navigator could comfortably draw on it to record the vessel's planned and actual movements. Due to all these constraints, only soundings and depth contours deemed absolutely critical to navigators could be selected for portrayal at the judiciously chosen scale of the paper chart.

An EC does have the potential to eliminate the viewing constraints imposed by a paper chart's inflexible scale, however, first and second generation ECs have provided only very limited relief. If scale changes are applied to vectorized paper chart data, it merely enlarges the paper chart graphics. The same informational effect is experienced by a paper chart user simply moving closer to the chart. The visual effect of "zooming in" is even more noticeable when using first generation raster systems (since bit-mapped lines and numerals succumb to aliasing effects at larger scale). As more extreme magnification is selected, both rasterized and vectorized chart graphics become increasingly spread out making it increasingly obvious to the user that the data elements (and their spatial errors) are being viewed "overscale". Ironically, it's the lack of any increase in real spatial information content that serves to automatically alert users to overscale viewing conditions.

To eliminate the constraints of paper chart data, it's tempting to interpolate between the surveyed soundings thereby computing a gridded surface model of the seafloor. Such a model contains very dense spatial information so that zooming into it actually reveals more and more spatial information (until of course the spatial resolution of the hydrographic source data is reached). This data model also supports a variety of powerful EC viewing modes that are discussed below under "New EC display capabilities". Unfortunately this new bathymetry always appears "perfect" ... an unrealistic and dangerous situation for navigators. When displayed on an EC, it's not at all apparent to the user that the seafloor model's spatial fidelity in fact varies greatly from place to place. Some of these model errors will be on the side of safety (the model shows less water than reality) however some will be on the side of danger (the model shows more water than reality). Therefore, before mariners can be permitted to exploit the graphical flexibility of continuous seafloor models, this data quality problem must be addressed.

HOW TO COMPUTE A SAFE SEAFLOOR MODEL

There are pros and cons to all candidate algorithms for computing a continuous seafloor model from discrete surveyed soundings. One tempting methodology is to analytically define the surface by fitting 3D equations to the survey data. The most common implementation of curve fitting is the minimum curvature or "spline" fit. The minimum curvature approach is unsuitable for hydrographic applications because it defines a smoothed surface that doesn't pass exactly through the observed soundings. Smoothing is undesirable because the "noise" characteristics of natural seafloor terrain are often features that must be preserved for navigators; we cannot risk smoothing out shoal spikes that are obviously critical to safe navigation.

The most appropriate general approach to continuous surface modeling is to empirically define the seafloor by interpolating a dense grid of depths in between the observed soundings. All grid interpolation algorithms fall into one of two main camps: geometric algorithms and statistical algorithms. The geometric approach typically uses linear interpolation along a triangulated network whose nodes are the surveyed soundings. This approach mimics the mental process of a hydrographer drawing contours on a fairsheet and is thus easily understood. Another advantage of geometric interpolation is that the calculations execute very quickly on low powered computers. The big problem with geometric interpolation is that it is purely deterministic, i.e. it does not lend itself to estimating the probable error of each interpolated depth. Only a statistical approach to grid interpolation can provide the probable error estimates required for adequate QC of the seafloor model and its derived EC graphics.

Quality Control of the surface model

A Quality Control procedure for interpolated seafloor models must:

1) Provide a statistically valid estimate of the total depth uncertainty at each grid node.
2) Provide an intuitive means for transmitting that QC information to EC users which insures they cannot place undue confidence in the model's spatial fidelity.

Over the past 5 years, a software package has been developed by the Canadian Hydrographic Service that satisfies both of these QC requirements. The software, called "Hydrostat", is distributed without charge to anyone requesting a copy. Hydrostat 3.0 is currently being tested and runs on either Windows'95/NT PCs or on X-Windows UNIX machines. Hydrostat's spatial error estimates are computed at the same time as a seafloor model is computed. Each depth error estimate reflects the combined effect of both instrumental errors in the surveyed soundings and interpolation errors caused by under-sampling of seafloor roughness.

Hydrostat employs a statistical gridding algorithm known as "kriging". The kriging algorithm is essentially a least-squares depth interpolator that uses a "variogram" to determine each sounding's weight in the computation. The variogram assigns each sounding's weight based on the soundings

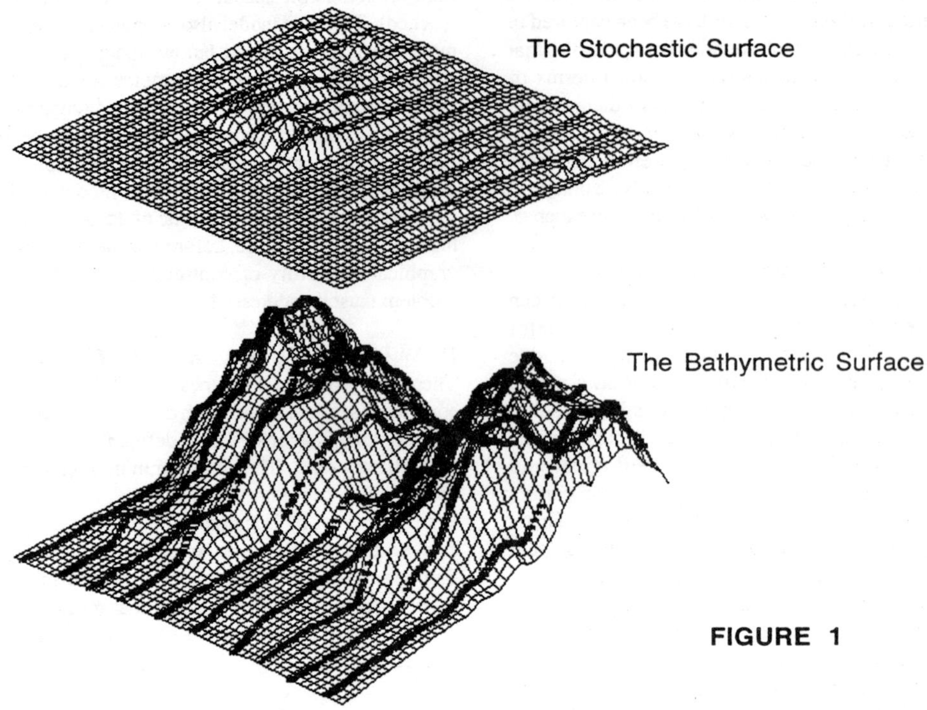

FIGURE 1

horizontal distance from the grid node being estimated. The variogram is ideal for this task because it's an objective statistical indicator of the spatial variability or "roughness" of the seafloor. For any population of depth soundings, the variogram relates the variability of the depth differences between all possible sounding pairs against all of the possible pair spacings. A variogram's slope and shape characterizes the roughness of the seafloor (a steeper variogram slope indicates a rougher seafloor). By employing the variogram as a least-squares weighting function, Hydrostat's kriging algorithm provides a sound statistical basis for estimating the probable error of depths interpolated at any distance from the surveyed soundings. Hydrostat models bottom roughness using many local variograms rather than using a single variogram to characterize the entire data set. It localizes the roughness analysis by using a small moving data window that traverses the whole data set. The local variograms define a "map" of changing bottom roughness which is subsequently used to interpolate the bathymetric model and its corresponding QC surface. At each grid location, the nearest local variogram is used to assign an appropriate weight to each sounding used to interpolate that grid point. Ten to fifteen of the nearest soundings are automatically selected for input to the grid node computation. The farther away each of these sounding is from the grid node being interpolated, the less its weight will be in the least-squares depth interpolation. If the seafloor is rough, the variogram insures that soundings further away are de-weighted more than if the seafloor is smooth. Regardless of the variogram model being used, the kriging algorithm insures that if an interpolated grid node is exactly coincident with a surveyed sounding then that sounding's weight will be 100% and the grid node will assume the sounding's exact value (i.e. kriging "honors" the data). The bottom line is that the stochastic nature of the kriging algorithm permits Hydrostat to estimate a standard deviation for each of the interpolated grid values.

Figure 1 illustrates Hydrostat's two data products. The lower surface is the "bathymetric surface" of gridded depths that were interpolated in between 10 singlebeam survey lines collected by CHS (the soundings are shown draped onto the model). The upper surface is the bathymetric model's "stochastic surface" that maps its estimated spatial fidelity. The height of each grid node on the stochastic surface is the estimated standard deviation of the corresponding depth on the bathymetric surface. For graphical clarity, the upper surface has been exaggerated 4 times (i.e. it's a 4 sigma stochastic surface). Each of the estimated depth interpolation errors on this surface is a function of its horizontal distance from the 10 to 15 nearby soundings that were used in the kriging computation. The closer an interpolated depth is to a measured sounding, the lower its estimated uncertainty. In Figure 1, this characteristic is visible in the "ridges" of depth uncertainty on the stochastic surface. These ridges grow (governed by the local variogram models) and peak midway between the surveyed depth profiles. The maximum predicted interpolation error within any zone of homogenous roughness is therefore a direct function of the sounding line spacing chosen for the hydrographic survey. Since the figure 1 data set uses a constant line spacing, the predicted interpolation errors correlate well to changes in the roughness of local seafloor topography. The smoother the seafloor, the lower the expected interpolation errors, and vice versa. Densifying the soundings over the rough shoal feature in the center of figure 1 would lower the corresponding peaks of uncertainty on the stochastic surface. Imagery of the stochastic surface is therefore a very useful Quality Control tool for hydrographers trying to decide where to densify sounding coverage and how dense those soundings must be in order to achieve a desired maximum uncertainty for the surveyed seafloor model.

"Data integration" is the task facing Hydrographic Offices when bringing together overlapping data sets to create a single bathymetric surface. The position and depth measurement errors of these different soundings can be highly variable (depending on the survey technology used to collect each data set). How can a single bathymetric model be optimally determined from all this non-homogenous survey data? An extension to Hydrostat's kriging algorithm has been implemented which effectively solves this problem. An estimate for each sounding's position and depth uncertainty must first be made, based either on a priori knowledge about the performance of the survey sensors or preferably on a more rigorous, sensor specific error analysis (e.g. error ellipses from over-determined GPS positions). These depth and position error estimates are then propagated through the kriging equations to appropriately de-weight less reliable measurements. The same procedure also correctly augments the interpolation error estimates to include the additional uncertainty caused by the imperfect surveyed measurements. The algorithm also has the desirable characteristic of propagating the vertical component of positioning errors into the stochastic surface. For example: over smooth seafloor, Hydrostat's positional error model correctly minimizes the effect that even large horizontal positioning errors will have on the interpolated depth values. Over more rugged terrain, the same level of positioning uncertainty will have a much larger effect on the interpolated depths and their estimated errors. By introducing instrumental error estimates into the kriging equations, Hydrostat is able to:

1) Produce a fully descriptive stochastic surface (i.e. one that includes the effect of both instrumental and interpolation errors).
2) Rigorously integrate diverse data sets into a single bathymetric surface model (by allowing each sounding's instrumental errors to also affect the sounding's weight in the depth interpolation).

Hydrostat insures the statistical validity of the stochastic surface by using an automated process to "calibrate" the variogram roughness models prior to using them to grid the seafloor model. The calibration procedure involves using about 10% of the surveyed soundings as control points. Using the remaining soundings, the algorithm then interpolates depths at the exact locations of the surveyed control soundings. The differences between the surveyed and interpolated depths (the real interpolation errors) are then compared to the estimated interpolation errors computed for those points. Statistical tests are then applied to determine if a bias exists between the real and predicted errors. If a systematic bias exists, then all of the interpolation error models (the local variograms) are adjusted proportionally to the observed bias. Only then are the variograms used to produce the final bathymetric and stochastic models. This validation and adjustment of the stochastic surface insures that the safety biased user products discussed below are truly safe and reliable.

SAFETY BIASING THE SEAFLOOR MODEL

This last section of the paper deals with how Hydrostat's seafloor model can be automatically safety biased and what the navigator might see when this new data model is exploited by an EC. Figure 1 shows Hydrostat's two outputs: a grid of estimated depths which model the seafloor (the "bathymetric surface") and a coincident grid of estimated standard deviations for each of the interpolated depths (the "stochastic surface"). The "one-on-one" relationship between the nodes on these two surfaces permits their values to be added and subtracted to form statistically biased seafloor models suitable for exploitation in an EC. For example, if a particular depth on the bathymetric surface has been estimated to be 52.7 metres and its 1 sigma uncertainty has been estimated to be 0.3 metres then it becomes easy to compute a "safe" depth at that location. By subtracting the depth's estimated standard deviation from the depth itself we derive a shallow biased depth estimate of 52.4 metres that has a 68% probability of actually being deeper than 52.4 metres. Similarly, at the 2 sigma (95%) confidence level, the true depth will not be shallower than 52.1 metres and not deeper than 53.3 metres, etc. When this arithmetic is applied to all of the nodes, a statistically biased surface model of the seafloor is rendered to whatever degree of confidence the user specifies.

New EC display capabilities

Together, the computed bathymetric and stochastic surfaces can support a variety of EC display modes that were impossible using traditional cartographic data. Nine such EC viewing modes are described below and many other variations are possible. The space and colour limitations for this paper permit only a few graphical examples to be presented. The sounding data used to generate these images was provided courtesy of the Voisey's Bay Nickel Co.

Option 1: Navigators could choose which depth contour values are extracted from the safety biased surface model and displayed on the EC. For example: if the vessel is drawing 12.2 metres of water, the user might choose to display only the 12.7 metre contour. Alternatively, if the user is navigating over a flat seafloor, he might choose to display contours every 0.5 metres. Customized contour views of the bathymetry can be rapidly reconfigured in response to changing conditions and user priorities. Once all the survey data has been integrated into a gridded seafloor model (a heavy computational task done at the Hydrographic Office), simply "joining up the dots" to draw arbitrary contour lines executes very quickly in real time, even if the EC has limited computing resources.

Option 2: Navigators could choose to apply changing water level data to animate the EC display. Since arbitrary contour intervals can be quickly extracted from the surface model, real-time, tidal reductions can be applied to the safety biased bathymetric model every few minutes and the screen refreshed with new depth contours. Automatic tidal reduction of the bathymetric surface model would for the first time provide navigators with a "true present depth" EC that models the total seafloor.

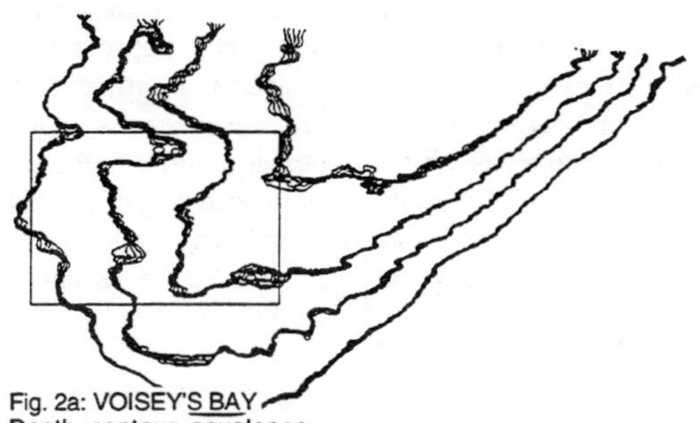

Fig. 2a: VOISEY'S BAY
Depth contour envelopes
(95% confidence level)

Option 3: Navigators could choose to apply different levels of safety biasing to any of the bathymetric views. The user need only specify the desired confidence level which in turn controls the grid differencing arithmetic between the underlying bathymetric and stochastic surfaces. In the case of displaying depth contours, this might take the form of a menu for selecting: "most probable contour locations", deep/shallow biased contour locations at 1 sigma confidence level, 2 sigma confidence level etc. or "all of the above". Figure 2a illustrates the contours for Voisey's Bay resulting from "all of the above". Each series of five contour lines was drawn using the most probable bathymetric surface and four statistically biased ones. The center line of this "contour envelope" represents the most probable location of the contour. The two contours on either side of it were drawn using gridded models that were biased by one and two standard deviations respectively towards both deep and shallow interpretations of the data. Together these five contours form a 2 sigma confidence envelope inside of which the true depth contour will lie with a 95% confidence.

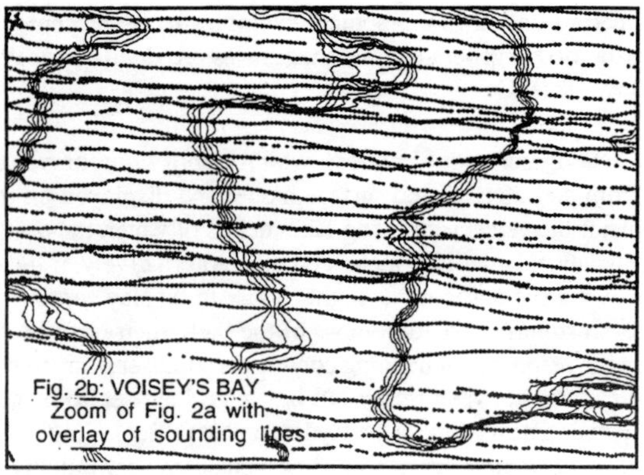

Figure 2b is a large scale inset of Figure 2a showing the western (rougher) portion of the bay. The surveyed soundings have been overlaid onto the contour envelopes so their approximately 100 metre line spacing is clearly evident. The effect of increased uncertainty in between the sounding lines is evident in the narrowing of the contour envelopes as they cross the survey lines.

Because both the spatial information and the QC information have been integrated by safety biasing the model, this image can be re-scaled to any degree and still maintain the user's desired level of confidence. Normally a prudent navigator would turn off all but the 95% shallow biased contour, however, any or all of the other interpretations could be instantly recalled. On the east side of Voisey's Bay, we see that the deep and shallow biased contours are virtually coincident. To some extent this appearance of higher spatial fidelity in the eastern portion of the model is an artifact of the slightly greater slope of the seafloor in that area. However, the main contributor to the tighter contour envelopes is the smoother seafloor in the eastern part of the bay. Figure 3a is Hydrostat's colour coded grid of estimated depth errors (1 sigma) and it illustrates the effect of changing bottom roughness. The maximum estimated interpolation errors in the western portion of the bay are about 1.5 metres whereas in the eastern portion of the bay (1 sigma), the maximum estimated errors are about half of that. Since interpolation uncertainty peaks mid-way between the lines, the east-west orientation of the survey lines is clearly visible in Figure 3a. Since the sounding vessel could not maintain perfect 100 metre spacing between the lines, we can also see the increased uncertainty that was introduced by inadvertent gaps in the sounding coverage.

Option 4: Navigators could choose to display an oblique perspective view of the bathymetric surface model (see Figure 1). A non-zenithal perspective view of the surface model permits the EC display to simulate a "fly through" of the seafloor topography. A logical configuration for this view would be to orient the viewer's line of sight pointing downward and forward of the vessel's present heading. As the vessel navigates across the bathymetry, the fly through view would refresh every few seconds to reflect the vessel's changing location and heading. The EC's screen image would simulate the view looking forward over the ship's bow however the ocean's surface would be perfectly transparent. An adjustable vertical exaggeration of the seafloor model would heighten the 3D effect. Optionally, a vertical offset equal to the vessel's draft would displace the navigator's viewpoint down to the keel of the vessel and an added artificial water surface would enhance the submarine viewing perspective.

Option 5: Navigators could choose to display a sectional profile view showing the depths that will be encountered along any planned route. Figures 3a, 3b and 3c illustrate how this might look. In figure 3a, four waypoints were chosen with a mouse to plan a voyage around Voisey's Bay, near its 40 metre depth contour. The planed "voyage" starts with a southerly leg down the west side of the bay then turns twice to port and finishes with north-east leg along the east side of the bay. Total distance traversed is 2.4 km (see X axis of Figure 3c). Normally a navigator would plan this route on a contoured depth image such as figure 2a. A colour coded "depth image" such that shown in Figure 3b would also be appropriate for route planning (figure 3b also shows an alternate planned route whose depth profile view is not shown here). Of course the route waypoints could also have been selected using a backdrop of conventional ENC paper chart data.

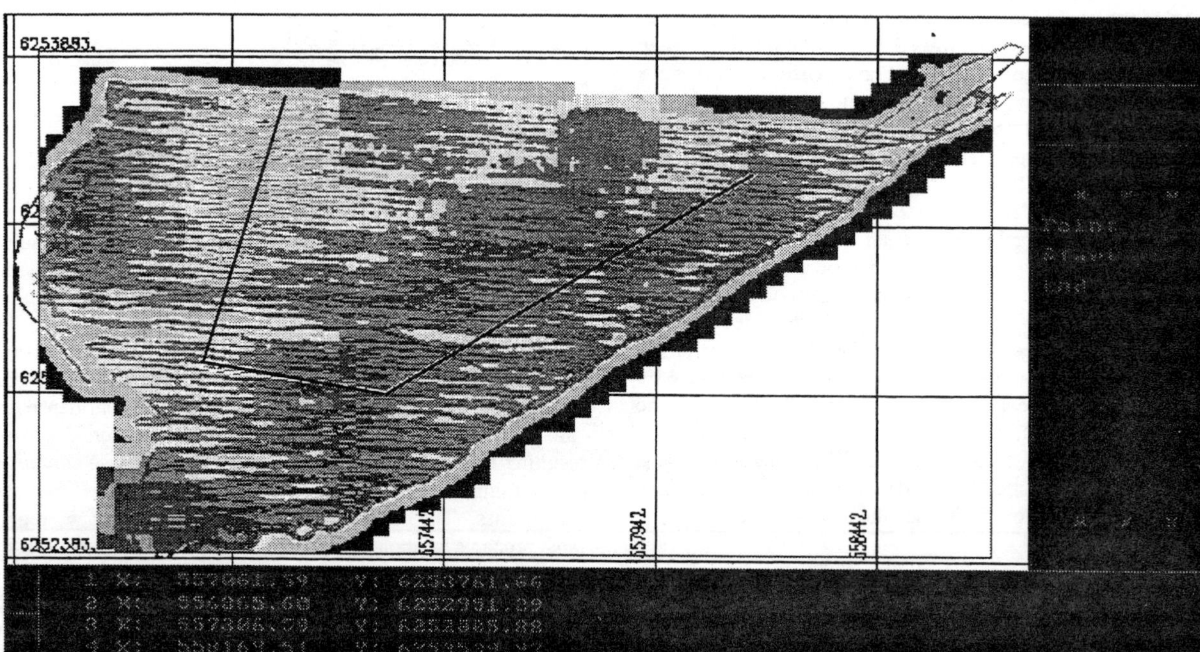

FIGURE 3a: Colour coded "error image" and proposed vessel track around Voisey's Bay

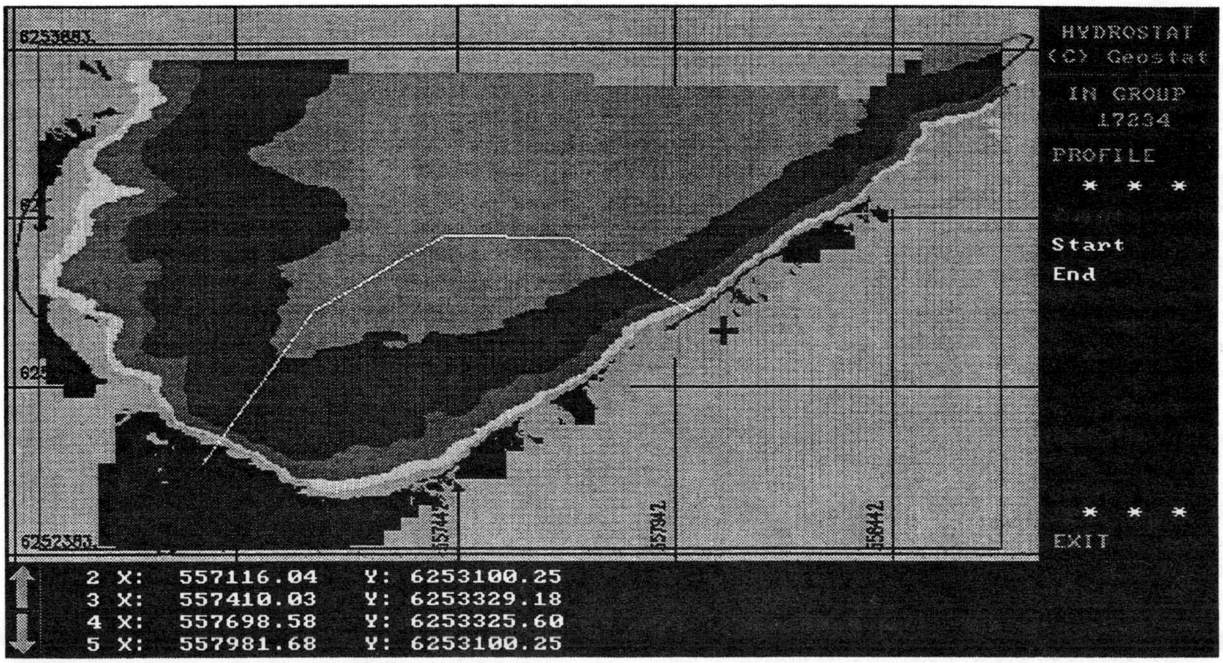

FIGURE 3b: Colour coded "depth image" of Voisey's Bay

Whatever their origin, the waypoints chosen from a 2D bathymetric display also define a sectional depth profile that can be instantly extracted from the 3D bathymetric surface model. The depth profile of the Voisey's Bay route planned on figure 3a is shown on Figure 3c. It took about 1 second for the prototype DOS version of Hydrostat to extract this profile and display it in its own window. Figures 3a, 3b and 3c are all screen snapshots taken of Hydrostat's processing screens. The two vertical lines shown on the depth profile correspond to the two southernmost turning points of the planned route. The horizontal resolution of this profile is 5 metres (the interpolated grid density that was chosen to approximate the surveyed sounding density along the lines). This high spatial resolution displays detail along the seafloor profile that could never be extracted from conventional chart data. Three depth profiles are shown on 3c; the most probable depths, the 95% shallow biased depths and the 95% deep biased depths. We see that the profile along the western leg of the route shows greater depth uncertainty than along the smoother east side of the bay. This correlates closely to the information shown on the error image of Figure 3a.

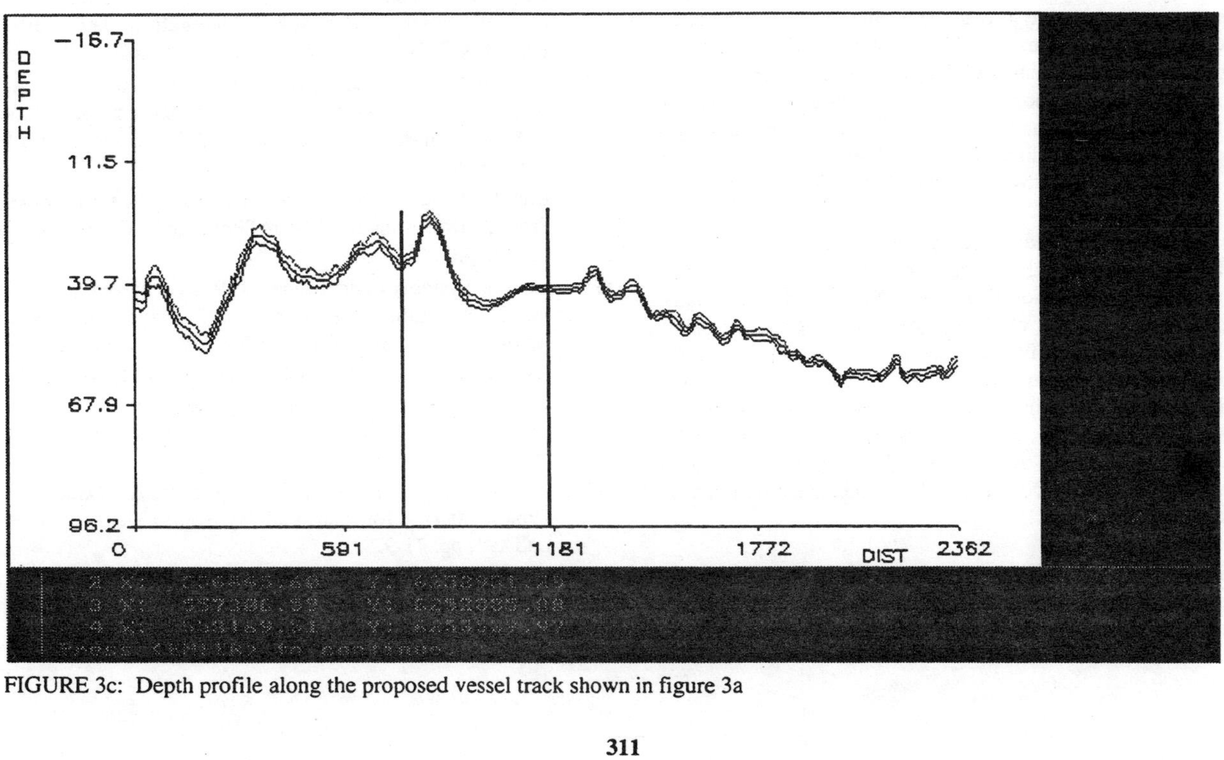

FIGURE 3c: Depth profile along the proposed vessel track shown in figure 3a

Option 6: Navigators could choose to display and manipulate a colour coded "depth image" of the seafloor model (example shown in figure 3b). Since such imagery is identical in structure to the raster imagery so prevalent in today's computer graphics industry, standard image processing algorithms can be used to render the bathymetric model easier to perceive. The simplest image processing approach is to re-assign the color of each depth in the image using only those few standard colors and contour intervals used on paper charts. This "stair step" color Look Up Table (LUT), would temporarily discard most of the vertical information in the depth image. When viewed zenithally, this processed depth image would take on an appearance fairly similar to that of a digitized paper chart. More sophisticated image processing algorithms such as edge finding and contrast stretching could also be applied by the navigator to temporarily enhance the appearance of the depth image.

Option 7: Navigators could choose to display a binary image showing only "safe" and "unsafe" areas. For example: if the vessel's current draft were 8.3 metres, that value could be entered into the EC which would then assign all depths less than 8.3 metres to red and all depths greater than 8.3 metres to green. This "go/no-go" depth image could be extracted from the "real present depth" surface model described in "Option 2", thereby presenting an extremely simplified steering display to the helmsman.

Option 8: Navigators could choose to display the error image shown in Figure 3a. The error image is simply the colour coded standard deviation values of the stochastic surface. This "Quality Map" shows at a glance the true spatial fidelity of the surveyed bathymetric model and is far more meaningful than the "source diagrams" shown on current ENCs. As seen in figure 3a, the surveyed sounding lines will be readily apparent in the error image (since the estimated errors are minimum where depths have been measured). In areas with sparse sounding coverage, the navigator could zoom into the error image to see where the hydrographic survey lines have been run. If keel clearance and data reliability were critical, the navigator could even use this image to steer along the exact same tracks taken by the survey vessels, thereby insuring that only the most reliable bathymetric data is traversed.

Option 9: Navigators could choose to set the colour of the entire depth image to transparent and overlay it onto the "points and lines" image of the digitized paper chart. The user could then continue to view familiar ENC chart graphics however the full resolution of the bathymetric surface could be accessed at any location by simply pointing there with an on-screen pointing device. For example a "magnifying glass" icon could be moved over the chart image to display patches of dense source soundings (shown either as numerical depths or as a false colour depth image).

The EC display scenarios described above are just some of the supplementary views that could be toggled on or off as required when navigating with a third generation EC. Their effect would be to greatly augment the safety and utility of the hydrographic information being transmitted by today's ENC and ECDIS systems. If Hydrographic Offices' digital survey data were appropriately modeled and furnished to EC manufacturers, these EC capabilities could be implemented immediately using simple display algorithms that run on modest EC hardware. HOs already have large holdings of suitable digital survey data and a geostatistical software tool that correctly analyzes and transforms it is now freely available.

CONCLUSIONS

The continuous surface of the seafloor can be modeled from discrete depth soundings using a geostatistical gridding algorithm. At each grid node the algorithm provides both a depth estimate and a statistically valid error estimate for that depth. At the location of each grid node, the error estimation algorithm models both:

1) The depth interpolation uncertainty caused by the spacing between the measured soundings with respect to the roughness of the local seafloor.
2) The effect that each sounding's measurement uncertainty has on the uncertainty of the interpolated depth.

The geostatistical gridding algorithm discussed in this paper has been implemented by the Canadian Hydrographic Service in a Public Domain software package called Hydrostat. The algorithm operates on any digital sounding data set but is particularly useful when operating on sparser data sets observed with single beam sounders. The statistical QC surface computed by Hydrostat can be used to automatically safety bias any of the bathymetric images that an EC user might choose to extract from the seafloor model. When exploited in an EC environment, these "third generation" seafloor models enable navigators to exercise much greater control over how the bathymetry is viewed than was possible when viewing digitized paper charts.

Bibliography

[1] Blais J.A.R. (1982): "SYNTHESIS OF KRIGING ESTIMATION METHODS", Manuscripta geodaetica Vol. 7, pp. 325-352.

[2] Herzfeld, Ute Christina, (1989): "VARIABILITY OF SUBMARINE MORPHOLOGY: PROBLEMS OF DEREGULARIZATION AND CARTOGRAPHICAL IMPLICATIONS", Math. Geol., Vol 21, No. 7, pp. 693-713.

[3] Royle A.G., Clausen F.L. and Fredriksen P. (1981): "PRACTICAL UNIVERSAL KRIGING AND AUTOMATIC CONTOURING", Geo-processing, (1), 1981, pp 377-394.

[4] Chiles J.P. and Chauvet P., (1974): "KRIGING: A METHOD FOR CARTOGRAPHY OF THE SEA FLOOR", International Hydrographic Review, pp. 25-41.

[5] Dagbert M., Kielland P., L'Esperance M. and Cowan A. (1992): "GEOSTATISTICS TO ASSIST HYDROGRAPHIC SURVEY DESIGN", Proceedings of Fourth International Geostatistical Conference, Troia Portugal, Sept 1992.

[6] Kielland P., Tubman T, (1995): "ON ESTIMATING MAP MODEL ERRORS AND GPS POSITIONING ERRORS", Navigation: The Journal of the Institute Of Navigation, Vol. 41, No. 4, pp. 479-499.

[7] Dagbert M., Kielland P., L'Esperance M. and Cowan A. (1992): "GEOSTATISTICS TO ASSIST HYDROGRAPHIC SURVEY DESIGN", Proceedings of Fourth International Geostatistical Conference, Troia Portugal, Sept 1992.

[8] David, Michel, (1986): "USING KRIGING TO MAP THE OCEAN FLOOR", Proc. of 18th Annual Offshore Technology Conference, Houston, pp. 433-439.

[9] Geostat Systems International (1992): "HYDROSTAT" users manual for Optimisation of Hydrographic Surveys using Geostatistics, written under contract for Canadian Hydrographic Service

[10] Kielland P., Burrows K., Ward B., Dagbert M., Velberg R., (1993) "TOWARDS IHOSTAT: IHO APPROVED SOFTWARE THAT EVALUATES THE QUALITY OF BATHYMETRIC DATA", Lighthouse #47-Spring 1993, pp. 19-32.

[11] Kielland P., Dagbert M., (1992), "THE USE OF SPATIAL STATISTICS IN HYDROGRAPHY", International Hydrographic Review, Monaco, LXIX91, pp. 71-92.

[12] Kite-Powell H., Jin D., (1996): "SAFETY BENEFITS OF DIGITAL NAVIGATION", International Hydrographic Review, Vol LXXIII, No. 1, pp. 65-74.

Proceedings of the Seventh (1997) International Offshore and Polar Engineering Conference
Honolulu, USA, May 25-30, 1997

ISBN 1-880653-28-1 (Set); ISBN 1-880653-32-X (Vol. IV)

A Probabilistic Analysis of Tanker Groundings

Michael D. Amrozowicz
Norfolk Naval Shipyard
Portsmouth, VA, USA

Alan Brown and Michael Golay
Massachusetts Institute of Technology
Cambridge, MA, USA

ABSTRACT

The culture, design, and operation of the maritime industry all contribute to create an error-inducing system. While risk acceptance and risky behavior are often attributed to the "traditions of the sea," the risks associated with sea transportation are no longer restricted to the domain of the seafarer. As oil tankers have gotten larger, the tolerance for error has decreased, and the consequences have increased. While the tanker industry has been identified by the USCG has a high-risk industry, the USCG has also stated that the industry has a high potential for improvement. A systematic approach must be undertaken to effectively identify tanker accident risks and consequences so that they can be minimized by appropriate safety measures. This paper concentrates on one aspect of a Level 1 Probabilistic Risk Analysis (PRA) for tanker groundings. The approach utilizes fault trees and event trees and incorporates The Human Error Rate Prediction (THERP) data to quantify individual errors.

KEY WORDS: Event tree, fault tree, grounding, oil tanker, probability risk analysis (PRA), risk assessment, tanker accidents.

The Motivation

Maritime oil spills are a significant international environmental problem. The culture, design, and operation of the maritime industry all contribute to create an error-inducing system (Perrow, 1984). Often, the consequence of these errors is the release of oil into the world's waterways. Since oil spills are low probability-high consequence events, and difficult to predict, prevention is the best response It is the risk of an oil spill that motivates further investigation.

The maritime industry has been identified as a high risk operation, requiring an active risk management program. (Based on roundtable discussion at the High Consequence Operations Safety Symposium, Sandia National Laboratories, Albuquerque, New Mexico, July 1984.) The U.S. Coast Guard (USCG) has expressed a commitment to reduce the risks of the maritime industry. Several major tanker owners have expressed the same commitment of cooperation with the USCG (USCG, 1995). Rear Admiral Card (Chief, Office of Marine Safety, Security, and Environmental Protection, USCG), has entrusted both industry and the USCG to make "prevention a strategic concept." Yet, in order to effect the appropriate risk management program, this risk must be understood.

While the possibility of an oil spill provides the impetus to investigate groundings, it must be remembered that the magnitude of oil outflow is a function of many unpredictable circumstances. There can be groundings that are preceded by marked and profound blunders, yet, the degree of oil spilled may be negligible. So while limiting oil outflow motivates the investigation of groundings, the scope is much broader and concerns itself with the nature of the events leading to the vessel's grounding. Hence, our goal is to understand the nature of the errors that lead to a grounding. This understanding can be an impetus for cultural change throughout the maritime industry–yielding a balanced approach to managing safety performance. The ultimate goal is to have safe and profitable operations balanced by the interaction of management, the work environment, human behavior, and technology, all supported on a firm foundation of sound rules, regulations, and standards (USCG, 1995).

The Means

Probabilistic Risk Assessment (PRA) is a technique for evaluating the hazards of a system, and for identifying areas offering the greatest risk-reducing potential. The approach has matured in the nuclear power industry's study of high consequence accidents. Many of the issues examined in the nuclear industry are germane to the oil tanker industry (Singh, 1993). Given the similarities, we apply the risk assessment methodologies of the nuclear power industry to the maritime industry.

Three levels of assessment have been defined as follows (Amrozowicz, 1996a):

1. *Level 1*: Determine the probability of ship damage and the damage extent.
2. *Level 2*: Given that an extent of damage has occurred, determine the probability of oil outflow.
3. *Level 3*: Given that oil has been released to the environment, determine the probability of consequences to the environment.
4. *Result*: The probability of oil pollution producing adverse economic and environmental consequences.

We focus on a Level 1 assessment for tanker groundings utilizing an event tree/fault tree approach. The event tree/fault tree approach employs discrete logic diagrams to show explicitly the causal relationships within the system model, and to determine the probability of accident scenarios. The methodology is widely used in technological systems applications (Siu et al, 1995), but it is also routinely performed to determine human reliability (Gertman and Blackman, 1994). Since over 80 percent of maritime casualties have been directly attributed to human error (NRC, 1976), it seems important to utilize a method that is consistent with both the technical and the human aspects of the system.

By conducting a PRA and integrating a Human Reliability Analysis (HRA), insight can be gained into the problems presented both to and by people aboard ships. The HRA allows the analyst to look at human failure and individual error as events whose causes can be investigated, rather than invoking stop rules at the events themselves and placing blame on the person or persons performing the events. Human failure factors are typically the largest source of uncertainty in a PRA, but they identify specific areas for potential risk reduction.

The most popular methods of analyzing individual reliability use decomposition. The basic technique is to break the system down into its constituent elements, or events, to assign reliability estimates to those elements, and compute the aggregated result (Hollnagel, 1993). The Technique for Human Error Rate Prediction (THERP) is the HRA device used in our analysis. THERP is a method to predict individual error rates and is the most widely used approach in HRA (Gertman and Blackman, 1994). The approach is similar to a traditional system reliability analysis modified to account for possible individual error. Rather than generate equipment system states, it produces possible human task activities and the corresponding error possibilities (Modarress, 1993). It allows the analyst to evaluate the degradation of the human-machine system likely to be caused by: either individual errors alone or with equipment functioning; operation procedures and practices; or other system and human characteristics that can influence system behavior (Swain and Guttmann, 1983). It combines a modeling method with a series of empirical data tables containing basic human error probability (HEP) rates that are modified by a series of performance shaping factors (PSFs).

The Grounding Fault Tree

Within the grounding fault tree (Amrozowicz, 1996a), accident faults are broken into two broad categories:

1. *Planning and piloting*: the vessel is able to follow a safe track, however, it proceeds down an unsafe track due to a planning or piloting failure.
2. *Equipment, assistance and environment*: the vessel is unable to follow a safe track because of mechanical failure, assistance failure and/or adverse environmental conditions.

Det Norske Veritas (DNV) defines the two categories as follows (DNV, 1995):

1. *Powered grounding*: An event type that occurs when a tanker collides with the shoreline while underway due to navigational error and lack of crew vigilance.
2. *Drift grounding*: An event type that occurs when a tanker loses its ability navigate, through loss of steering or propulsion, and is blown onto the shoreline before it is either taken in tow or is repaired.

The causality derived by the fault tree is consistent with the grounding definitions developed by DNV. For consistency and clarity the DNV terms are used to describe the two broad causal categories. For brevity, we will focus on the powered grounding accident to illustrate the concepts.

Powered Grounding

The fundamental failures resulting in a powered grounding (Fig. 1) are in the processes of planning and piloting. Voyage planning and piloting are essential skills required of any mariner. Those elements of the fault tree extending from "The Desired Track is Unsafe" constitute faults in the planning process. Likewise, those elements extending from "Course Deviates from a Safe Desired Track" are characteristic faults of the piloting process.

Event trees are used to quantify the probabilities of the faults identified in the fault tree. Event trees are developed by sequencing the fundamental events of each of the processes contributing to a failure event. From the event trees, the probabilities of either success, or failure of each of the processes, or elements of the processes are calculated. When events are human actions, HEP's are determined from Reference (Swain and Guttmann, 1983).

Passage Planning

The process of passage planning is analyzed using the event tree shown in Fig. 2 (Amrozowicz, 1996b). The values determined from Swain and Guttman (1983) are incorporated into the event trees.

The mariner has several sources of information available to ensure a safe and efficient passage. Prior to departure and arrival, publications must be corrected as necessary to reflect the most recent changes. The process can be tedious and time consuming. To determine the HEP value to apply to this task, the Table for "Estimated Probabilities of Error When Using Written Procedures Correctly" from Swain and Guttmann (1983) is used. It is assumed that the process for checking the navigation periodicals and messages is analogous to the HEP for following procedures with no check-off provision.

The HEP for correctly entering the changes in the appropriate charts and publications is taken from the same table. Since the mariner has developed a list of changes to make, the HEP is taken from the line item for procedures with check-off provisions.

The task of determining waypoints for the passage involves studying the charts to determine the track to take the vessel from origin to destination. It is assumed that the HEP is analogous to that of preparing written procedures.

The task of laying down the track involves the plotting of the waypoints and high-lighting any hazards to navigation. The process requires relatively precise use of dividers and simple mathematical calculations, analogous to a reactor technician's use of a micrometer. Swain and Guttman (1983) categorize these tasks under arithmetic computations.

It can be assumed that the first four events are independent of each other, since it is unlikely that the successive event will induce the operator to believe that the previous event was performed incorrectly. In other words, there is no mechanism for recovery. However, the performance of Event 5 does provide for recovery. It can be rationalized that in the process of plotting the track, the plotter has a general idea of the way the track will lay-out before actually plotting it, since the waypoints were determined from studying the charts. If this dependence is assumed, the event tree must model the recovery event.

The recovery event is the recognition of the faulty track after the track is laid-out. This presumes that the individual laying down the track is checking it for the specific purpose of meeting the constraints of a safe passage. This is analogous to checking displays for a specific purpose. This recovery event becomes Event 6 in the event tree.

The approval process requires the Captain to verify the validity of the track. A successful verification implies that the Captain has disapproved an improper plan. The analogous HEP from Swain and Guttmann (1983) corresponds to the table for "Estimated Probabilities of a Checker's Failure to Detect Errors." The final probability calculated for errors made in planning the track is **7.0049×10^{-5}**.

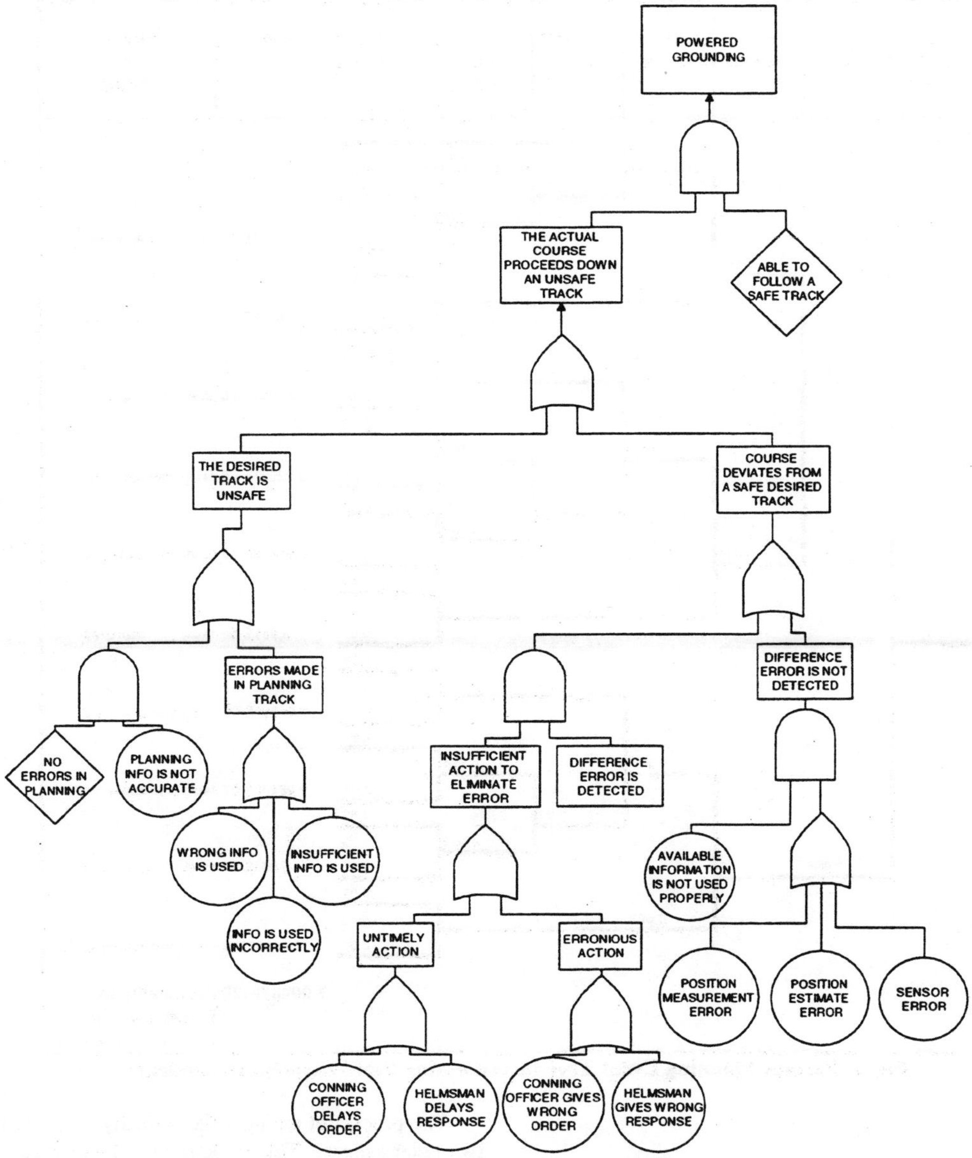

Fig. 1 Powered Grounding Fault Tree

Planning Information

Inherent in evaluating the probability of the desired track being unsafe, is the determination of the probability that the information used to plan the track is inaccurate. Only a small portion of U.S. waters have been surveyed using the most advanced techniques, and 60 percent of the soundings shown on nautical charts are based on lead-line surveys conducted over 45 years ago (NRC, 1994).

A query of the CASMAIN database was performed for the causes of vessel groundings. Specifically, the accidents where the vessels did not have the navigational information reflecting the actual environmental conditions. The database was queried for groundings that had the following causes attributed to the casualty:

1. Channel not maintained.
2. Unmarked channel hazard.
3. Inadequate weather information available.
4. Improper navigational aid location.

It was assumed that the results indicated cases where the mariner had inaccurate information:

The results of the query yielded 1,874 cases between the years 1980 and 1991 where vessels grounded due to false navigational information. Of the 1,874 vessel accidents identified, 298 involved tankers. The location for these accidents is dominated by those that occurred in inland waters. This illustrates the importance of understanding inland water and river dynamics and the increased caution that must be exercised when transiting inland waters and rivers.

Vessel transit data for four of the busiest ports in the U.S.--San Francisco Entrance, New Orleans, Baton Rouge, and Valdez was obtained from the Army Corps of Engineers (USACOE, 1986-1990). The accident quotient (the number of accidents due to faulty navigational information divided by the number of transits) is assumed to approximate the probability of grounding attributable to incorrect planning information. This data provides an estimate for the probability of inaccurate planning information causing an accident of $\mathbf{1 \times 10^{-4}}$ (Amrozowicz, 1996b).

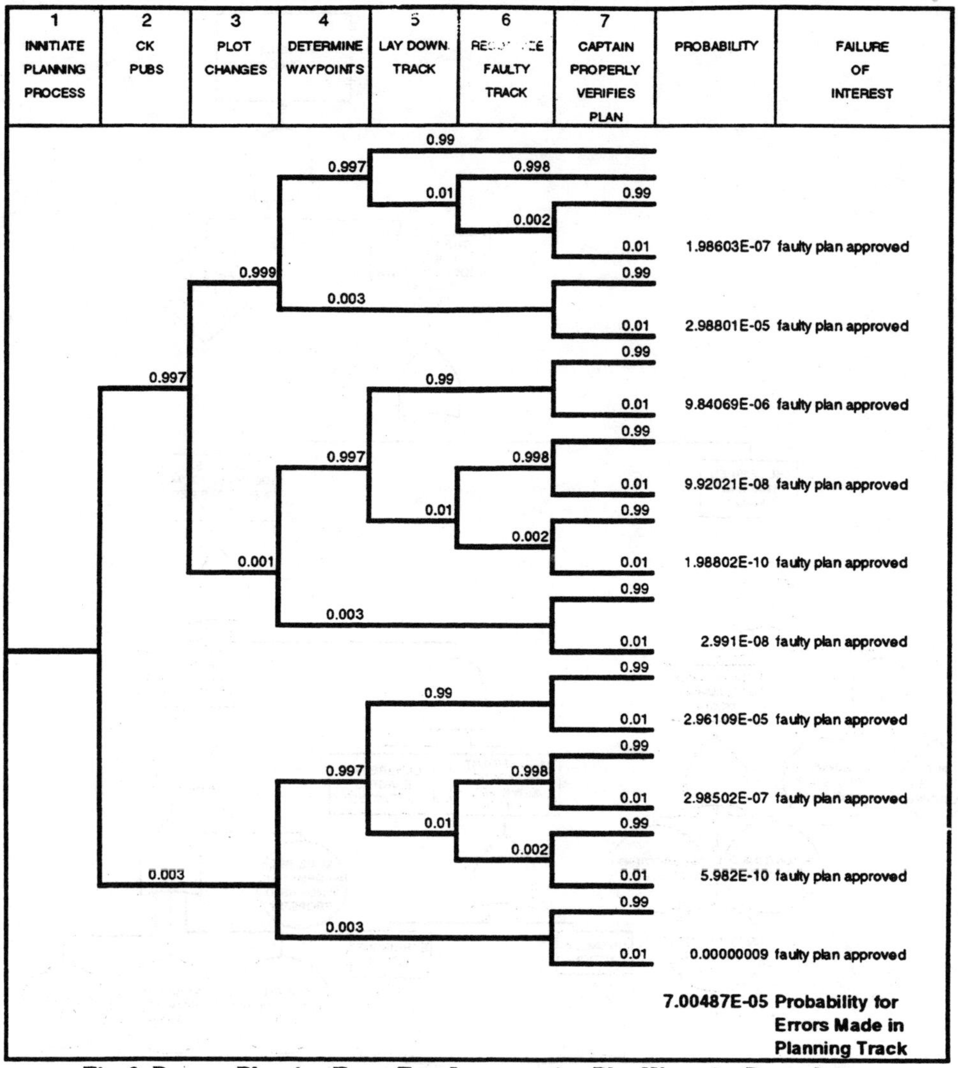

Fig. 2 Passage Planning Event Tree Incorporating Plot-Waypoint Dependency

Piloting

In the piloting event tree, the initiating event is the actual course deviating from the planned track. The subsequent sequence of events is as follows:

1. *The actual course deviates from the planned track.* This is the initiating event and the resulting probabilities are conditional upon this initial deviation.
2. *A difference error between the actual course and the planned track is generated.* To enable a detection of a deviation, the on board sensors detect and offer this information to the bridge team.
3. *A fix is taken and plotted.* Once the onboard sensors offer the information to the bridge team, the bridge team plots a fix on the chart.
4. *The difference error is detected.* When the fix is plotted, the bridge team evaluates the fix to detect that a difference exists between the actual position and the desired position.
5. *A correct course change is ordered.* Once the ship's deviation is recognized, a course change is given to negate further deviation.
6. *The helm responds correctly.* The helm responds with the proper rudder order to bring the ship's track back to the planned track.

Because there is no standard system installed on tankers, a value for the probability of generating a difference error is chosen based on Pederson (1992).

The process of taking a fix typically involves the taking of at least two radar ranges. This is done by selecting appropriate navigational aids, obtaining the ranges, and then plotting those ranges. The navigator must read the ranges off the radar and plot them correctly on the chart. The result is the estimated ship's position at the time the ranges were determined. The HEP is chosen from the table for "Probabilities of Errors of Commission in Reading Quantitative Information from Displays." The recording of the information obtained involves more than just writing down the information. Since some skill is required in using the dividers to plot the ranges at the correct scale, the HEP for recording is taken from the table for "Probabilities of Error of Commission in Recording Readings" and the higher HEP is used.

Once the fix is plotted, the navigator assesses whether the course is following the desired track. Singh (1995) refers to qualitative estimation and quantitative measurement as the methods mariners use to determine position. In many restricted waters, the pilotage of a vessel takes on many forms of comparing actual position to desired position. For this analysis, it is assumed that the process of computing actual position, regardless of whether there is a qualitative estimation or a quantitative measurement, takes on the HEP for plotting the actual position from a fix. This is analogous to a check-reading task where the navigator checks the plotted fix to ensure it is within tolerable limits of the desired track.

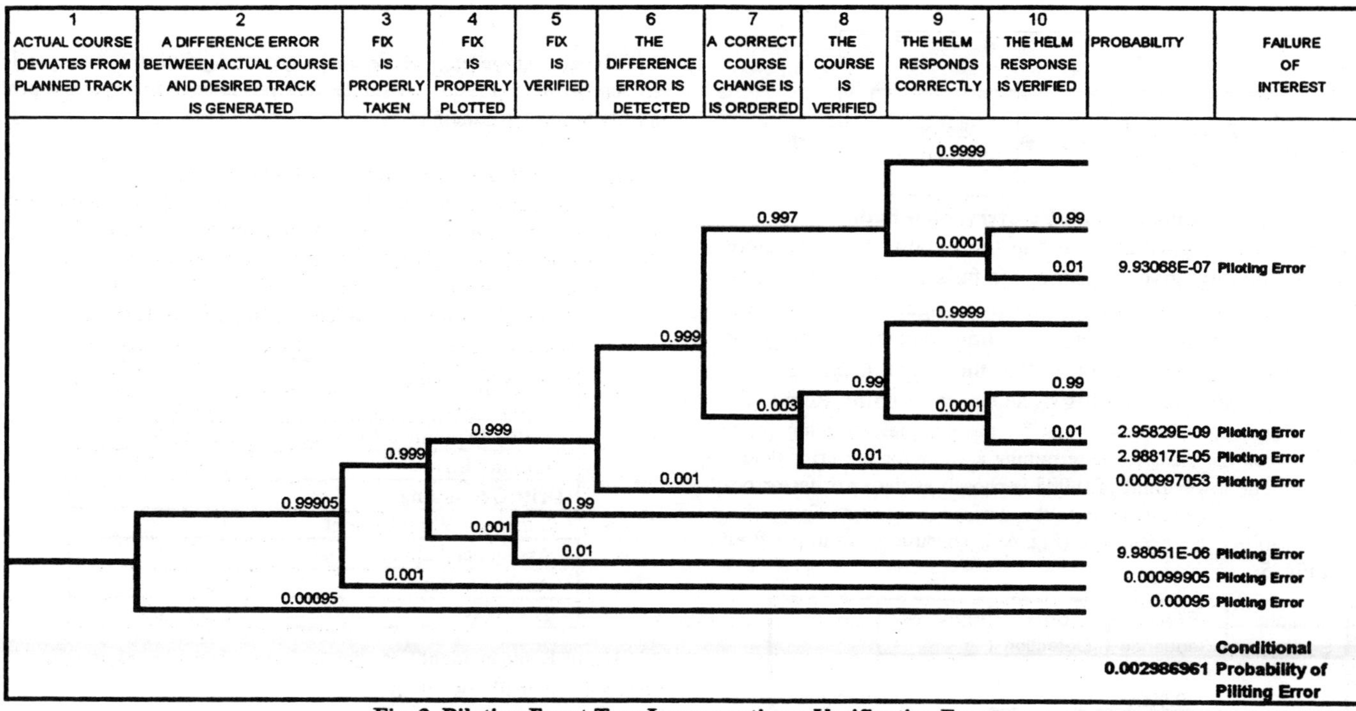

Fig. 3 Piloting Event Tree Incorporating a Verification Event

Given that the error in the course is detected, the conning officer determines the correct course change to order. This can be as simple as a rudder order. While there is no written procedure to follow, it is assumed that when a course deviation is detected the procedure is to order a course change. The corresponding HEP is taken from "Estimated Probabilities of Error When Using Written Procedures Correctly."

Once the order to change course is given, the helm must properly respond to the order. This involves turning the wheel while watching the rudder angle indicator and the gyro repeater until the ordered course is achieved. The helmsman must immediately respond and the procedure followed involves some skill. The standard order to the helm involves both a rudder angle order and a final course to steady on. The table "Estimated Probabilities of Errors in Recalling Special Instruction Items Given Orally" is used.

Once the helm responds to the order, the next event is to detect that the difference error is eliminated, which begins the sequence of events again. Therefore, the resulting probability is based on the number of fixes and assumes that the fix frequency is greater than the rate of departure from track. The resulting failure probability for this process is **9.91 x 10^{-3}** without recovery.

A more detailed analysis is done to determine the failure probability when recovery and redundancy are applied. Considering a verification role for the mate and pilot in taking fixes and ordering course changes, the failure probability is lowered. The analogous role in a nuclear power plant is that of either a second checker or an inspector. Fig. 3 is the event tree for this process. The resulting probability is **2.98 x 10^{-3}**.

Further verification must be included when the role of the Captain is considered. The Captain is responsible for the safe navigation of the vessel at all times. As such, the prudent Captain takes an active role in the piloting process. Incorporating the Captain's verification yields a more complex event tree with a final probability for piloting error of **1.95 x 10^{-3}** (Amrozowicz, 1996b).

The piloting event tree illustrates the importance of the verification role for each of the officers on the bridge. The results show that the additional verification role reduces the failure probability by an order of magnitude. Since the Captain is the individual who is responsible for the vessel, prudence dictates a verification role because it provides an additional recovery event for failure of either the mate or conning officer to perform their respective verification events. The Captain's verification role reduces the failure probability by an additional 35 percent. The Captain plays an integral role in the error detection cycle that is modeled to allow for a recovery event after each of the piloting processes.

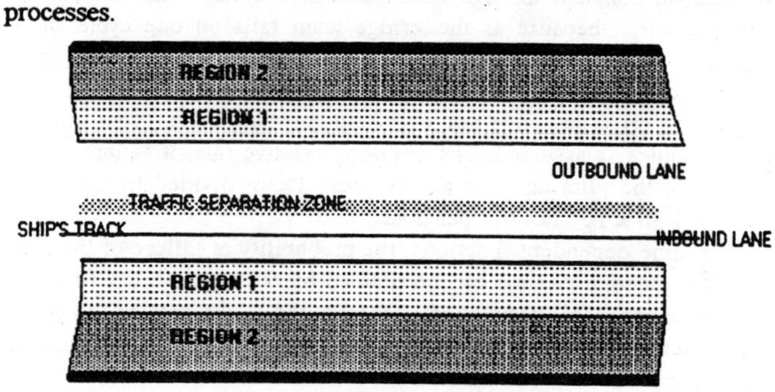

Fig. 4 Hypothetical Waterway

The piloting failure probability is time dependent; as the piloting process is periodic throughout the transit. Additional consideration must be made for recovery events after each of the piloting processes as the vessel transits the waterway. Consider the hypothetical waterway in Fig. 4. As the ship proceeds down the intended track, there can be errors in the piloting cycle that are not detected; however, the ship is not necessarily in a failure state. As the ship deviates from its intended track into Region 1, there exists the ability to recover. Once the ship enters Region 2, however, the ability to maneuver the ship to avoid grounding is lost, and avoiding grounding becomes impossible.

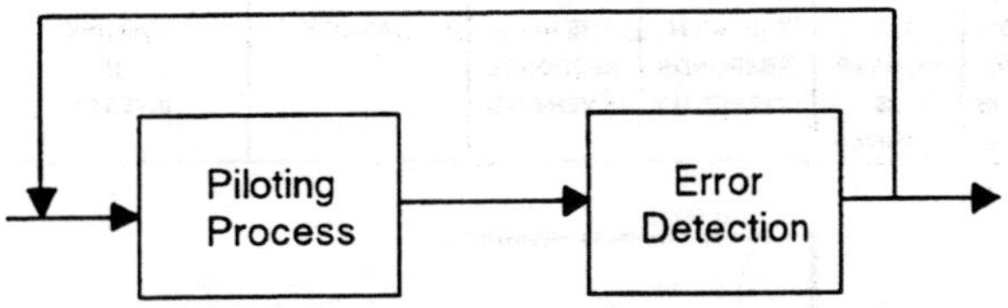

Fig. 5 Piloting, with Recovery, Flow Path

After each sequence of events in the piloting process, there is some probability, given that the ship fails to correct its course to the desired track, that the crew will recognize the error and implement correction in the next piloting sequence. The error detection factor is the probability that the bridge team will recognize its failure in the piloting cycle before reaching Region 2.

From Swain and Guttmann (1983), the nominal checking probability provides the basis for determining a value for the error detection factor. The lower limit of 0.005 is chosen as the error factor because of the many cues available to the mariner to recover. Given this failure probability, the event tree (Fig. 6) is constructed from the flow chart of Fig. 5.

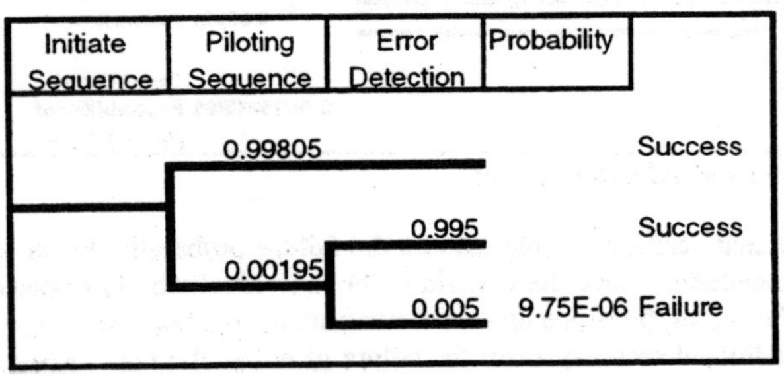

Fig. 6 Piloting and Recovery Event Tree

Implicit in the above model, is the assumption that the probabilities remain constant through each successive cycle. This is only an approximation, because as the bridge team fails on one cycle, it is plausible that the likelihood for failure on the next cycle is higher. The model uses a path dependent error detection factor. For this example, the error detection factor is held constant and a cycle time of three minutes is assumed. The resulting relative rate of failure is the product of the piloting error and recovery factor divided by the cycle time, or **1.95×10^{-4}** failures per hour.

For time dependent functions, the probability of failure of the system as a function of time can be defined by the unreliability function **F(t)**. The unreliability function is determined by integrating the probability density function (pdf) **f(t)**, which characterizes the behavior of the system.

The exponential distribution used to describe this pdf is:

$$f(t) = \lambda e^{(-\lambda t)} \qquad \text{Eq. 1}$$

where λ equals the relative rate of failure. The probability of piloting failure or unreliability function is determined as follows:

$$F(t) = \int_0^t f(e)\, d(e) \qquad \text{Eq. 2}$$

$$F(t) = 1 - e^{(-\lambda t)} \qquad \text{Eq. 3}$$

The probability of piloting failure along the track is determined by evaluating **F(t)** over the time of the transit. For small values of λt, **F(t)** is approximately equal to λt.

Drift Grounding

The drift grounding portion of the grounding fault tree is expanded by Amrozowicz (1996b). A similar analysis is done to develop a probability for drift grounding.

Summary of Probabilities and Sensitivity Analysis

The grounding fault tree has been inductively and deductively constructed for clarity in order to determine the basic faults of grounding accidents. The probability values determined from fault trees, event trees and historical data are summarized in Table 1.

Table 1 Summary of Grounding Probabilities (transit time = t hours, ship's speed = s knots)

Powered Grounding:	
Faulty Passage Plan	7.005×10^{-5}
Faulty Planning Information	1.0×10^{-4}
Piloting Error	1.95×10^{-4} t
Drift Grounding:	
Sufficient Wind/Current	1
Assistance Failure	0.25
Anchor Failure	0.25
Lost Way:	4.5×10^{-6} st

Using Boolean identities, the fault tree provides expressions for powered and drift grounding:

$$P(\text{powered grounding}) = 7.005 \times 10^{-5} + 1.0 \times 10^{-4} + 1.95 \times 10^{-4}\, t \qquad \text{Eq. 4}$$

$$P(\text{drift grounding}) = 1.0 \times 0.25 \times 0.25 \times 4.5 \times 10^{-6}\, st \qquad \text{Eq. 5}$$

The total probability of grounding is approximated as the sum the probabilities for drift grounding and powered grounding. For typical transit times, powered grounding dominates the contribution to the probability of grounding such that a final approximate expression for the probability of grounding is:

$$P(\text{grounding}) = 1.7 \times 10^{-4} + 1.95 \times 10^{-4}\, t \qquad \text{Eq. 6}$$

where t is the time of near-shore transit in hours.

A sensitivity analysis is done to determine those events that offer the largest potential for improving the failure probabilities. For the grounding event, powered grounding is shown to be the significant contributor. A sensitivity analysis of the event trees incorporated in the powered grounding analysis identifies the high-leverage factors. The high-leverage factors are determined by varying each of the probability events in the event tree over the range of uncertainty.

The effect of each high-leverage event on the probability for errors made in planning the track at the low-end and high-end of the uncertainty is summarized in Table 2. While these factors offer the greatest potential for improvement, over the range of uncertainty, they offer a greater potential for increasing the probability of failure (based upon the substantial increase in the probability of failure as the errors are increased to the upper end of their uncertainty values). For voyage planning, it is essential to begin with the correct information by checking publications, incorporating the changes on the charts, and determining the correct waypoints. The most important event is that of verification. The Captain's verification event has an impact on the complete planning process. This emphasizes the importance of navigation fundamentals and the Captain's role in verifying that the track meets imposed constraints.

When the probability of faulty planning information is included in the sensitivity analysis, Table 3 results. From Table 3, over the range of uncertainty, faulty navigational information offers the greatest potential for improving the failure probability (based upon the 53 percent reduction in the failure probability at the low end of uncertainty).

Table 4 shows the affect of the high-leverage piloting events on the overall probability for a piloting failure determined from the event tree that incorporates the captain's verification role. To gain further insight into the scope of the potential for the high-leverage factors, an additional sensitivity analysis for the event tree that does not incorporate the Captain's verification role is considered in Table 5. From Tables 4 and 5, the most sensitive events are also the most fundamental events to piloting and navigation:

1. Generating a difference error.
2. Properly taking a fix.
3. Detecting a difference error from the plotted fix.

Reductions in piloting error are dominated by the accuracy and reliability of the navigational equipment (a difference error is generated) and fundamental piloting techniques (fix is taken and a difference error is detected). Regardless of any verification processes, if these fundamental events fail, then there is a significantly higher probability of failure. Additionally, despite the methods used to determine the ship's position, if the conning officer is unable to detect that the ship has deviated from the desired track then the potential for grounding increases.

Conclusion

Based on the analysis of 100 accidents at sea, Groeneweg (1994) concluded that 96 of the accidents were preceded by human failures. There were 345 necessary human failures identified. Out of all the identifiable and necessary human errors, 76 percent of these errors occurred on the bridge. Since the bridge is the controlling station for the ship, it is not surprising that the majority of contributing events preceding an accident are attributable to the actions taken on the bridge. "Therefore, programs to improve safety should look carefully at what happens on the bridge" (Groeneweg, 1994). The significance of the bridge and the actions taken there, is reflected in the number of marine accident causal factors attributed to this controlling station of the vessel. This is substantiated by the grounding fault tree. The resulting probability for grounding is dominated by the piloting process in the powered grounding mode of failure vice the drift grounding mode of failure. This is confirmed by the CASMAIN database, which attributes only 15 cases of 716 tanker groundings to either steering failure, or propulsion failure (drift grounding).

Based on PRA and sensitivity analysis, critical tasks include:

1. *Planning:* Check publications for changes, determine waypoints properly, Master verify the plan.
2. *Planning Information*: must be accurate.
3. *Piloting*: Take fixes properly, recognize difference errors, provide accurate and reliable navigation equipment.

Effective implementation of the Standards for Training, Certification and Watchstanding (STCW) and the International Safety Management (ISM) Code can address the most important factors impacting these critical tasks including: bridge procedures, training, experience, morale, motivation, fatigue, substance abuse and maintenance. Besides sound seamanship, technology may also have a place in reducing tanker risk. The potential for Electronic Chart Display and Information Systems (ECDIS) to reduce planning errors is significant, if implemented properly and reliably.

Maritime culture, economics, and regulatory bodies all contribute to create a system that can be characterized as error-inducing. There has been little effort to characterize the system as a whole and to determine the areas that offer the greatest potential for risk reduction. The NRC has determined that maritime safety as a whole, could benefit from the increased use of quantitative and qualitative risk analysis to develop risk reduction strategies (NRC, 1994).

The outlined approach addresses the total tanker system (Brown and Amrozowicz, 1996) and has its foundations in the risk model and the event tree/fault tree methodology. Siu et al. (1995), have argued that the event tree/fault tree approach provides a natural framework for treating oil spill scenarios.

When developing a PRA for oil spills it is important to recognize the areas of uncertainty. PRA is a discrete analysis. Therefore, it is unable to account for the infinite number of possibilities. Ideally, a PRA considers all the important aspects that lead to the undesired event, but there is the possibility that important contributions have been overlooked.

Additionally, there are uncertainties due to the necessary approximations made in developing the model. Human failure factors, system complexity, and the subjective nature of the analysis, all present uncertainties that must be recognized. Despite the uncertainties, it is important to develop a PRA so that perceived risk does not produce either irrational behavior or reflex reactions. The performance of a PRA reduces the uncertainty concerning some of the elements of risk so that resources can be better allocated.

Table 2 Planning Failure Event Tree High-Leverage Factors

Event	Percent Deviation from the Mean Probability at the Low-End of Uncertainty	Percent Deviation from the Mean Probability at the High-End of Uncertainty
check publications for changes	-28%	100%
determine waypoints	-28%	100%
captain verify plan	-50%	4900%

Table 3 Planning Failure Event Tree (incorporating the probability for faulty information) High-Leverage Factors

Event	Percent Deviation from the Mean Probability at the Low-End of Uncertainty	Percent Deviation from the Mean Probability at the High-End of Uncertainty
check publications for changes	-12%	41%
determine waypoints	-12%	41%
captain verify plan	-21%	2018%
utilize faulty information	-53%	529%

Table 4 Piloting Failure Event Tree High-Leverage Factors (incorporating Captain's verification role)

Event	Percent Deviation from the Mean Probability at the Low-End of Uncertainty	Percent Deviation from the Mean Probability at the High-End of Uncertainty
a difference error is generated	-49%	3%
a proper fix is taken	-26%	205%

Table 5 Piloting Failure Event Tree High-Leverage Factors (without Captain's verification role)

Event	Percent Deviation from the Mean Probability at the Low-End of Uncertainty	Percent Deviation from the Mean Probability at the High-End of Uncertainty
a difference error is generated	-32%	2%
a proper fix is taken	-17%	135%
the difference error is detected	-17%	135%

The performance of a risk analysis of and by itself can reduce risk as knowledge and awareness are gained (Dougherty and Fragola, 1988). If the process of risk assessment is dynamic, the uncertainties can diminish with time as more knowledge is gained. Once the PRA is completed, allocating resources in the areas that are risk relevant, rather than trying to alleviate all conceivable hazards, allows for realizable risk reductions with limited resources.

There is momentum in the industry for change, and the outlined systematic approach offered by a PRA identifies the areas of change where industry can focus.

References

Amrozowicz, MD, "The Need for a Probabilistic Risk Assessment of the Oil Tanker Industry and a Qualitative Assessment of Oil Tanker Groundings," Master of Engineering Thesis, Department of Ocean Engineering, Massachusetts Institute of Technology, January, 1996.

Amrozowicz, MD, "The Quantitative Risk of Oil Tanker Groundings," Naval Engineer Thesis, Department of Ocean Engineering, Massachusetts Institute of Technology, June, 1996.

Brown, AJ and Amrozowicz, MD, "Tanker Environmental Risk - Putting the Pieces Together", Joint SNAME/SNAJ Conference on Designs and Methodologies for Collision and Grounding Protection of Ships, San Francisco, 1995.

DNV Technica, *Risk Analysis of Oil Spills from Tankers in UK Waters*, Prepared for Marine Pollution Control Unit, UK CoastGuard, Det Norske Veritas Industry LTD., May, 1995.

Dougherty EM, Fragola JR, *Human Reliability Analysis*, John Wiley & Sons, New York, 1988.

Gertman DI, Blackman HS, *Human Reliability & Safety Analysis Data Handbook*, John Wiley & Sons, Inc., New York, 1994.

Groeneweg J, *Controlling the Controllable*, DSWO Press, Leiden University, The Netherlands, 1994.

Hollnagel E, *Human Reliability Analysis: Context and Control*, Academic Press, San Diego, 1993.

Modarress, M, *What Every Engineer Should Know About Reliability and Risk Analysis*, Marcel Dekker, Inc., New York, 1993.

National Research Council, Panel on Human Error in Merchant Marine Safety, Maritime Transportation Research Board, *Human Error in Merchant Marine Safety*, National Academy of Sciences, National Academy Press, Washington D.C., 1976.

National Research Council, Committee on Advances in Navigation and Piloting, Marine Board, *Minding the Helm, Marine Navigation and Piloting*, National Academy of Sciences, National Academy Press, Washington D.C., 1994.

Pedersen PT, "Collision and Grounding Mechanics," Paper presented at WEMT 95 - Ship Safety and Protection of the Environment, 1992.

Perrow, C, *Normal Accidents*, Basic Books, New York 1984.

Singh SJ, "The Maritime Transportant and the Maritime Malaise," Unpublished, 1993.

Singh SJ, "The Aetiology of Grounding," Ph D. Thesis, University of Wales, College of Cardiff, Department of Maritime Studies & International Transport, March 1995.

Siu NO, Novack SD, Hill SG, Bhimavarapu KR, *Evaluating Risk Assessment Approaches in Oil Spill Prevention Applications*, INEL-95/0401, Idaho National Engineering Laboratory, Engineering Analysis Department, Idaho Falls, August, 1995.

Swain, AD, Guttmann, HE, *Handbook of Human Reliability Analysis with Emphasis on Nuclear Power Plant Applications*, NUREG/CR-1278, U.S. Nuclear Regulatory Commission, Washington D.C., 1983.

United States Army Corps of Engineers, *Waterborne Commerce of the United States, Waterways and Harbors*, (Excerpts from 1986 - 1990 received from U.S. ACOE Waterborne Commerce Statistics Center)

U.S. Coast Guard, Department of Transportation, *Prevention Through People: Quality Action Team Report*, U.S. Coast Guard Office of Marine Safety, Security and Environmental Protection, Office of Navigation, Safety and Waterways Services, July 15, 1995.

Proceedings of the Seventh (1997) International Offshore and Polar Engineering Conference
Honolulu, USA, May 25-30, 1997

ISBN 1-880653-28-1 (Set); ISBN 1-880653-32-X (Vol. IV)

Experimental Validation of the Ultimate Limit State Limit Equilibrium Analysis (ULSLEA) with Results from Frame Tests

M. Mortazavi
Exxon Production Research Company
Houston, TX, USA

R.G. Bea
University of California at Berkeley
Berkeley, CA, USA

ABSTRACT

The simplified structural analysis procedure to determine the ultimate limit state lateral load capacities of steel, template-type offshore platforms is identified as ULSLEA (Ultimate Limit State Limit Equilibrium Analysis) (Mortazavi, Bea, 1994). Simplified formulations have been developed to estimate the ultimate lateral shear capacity of the three primary structural components of a platform; deck legs, jacket, and foundation. The validation of ULSLEA using experimental results from laboratory tests on two-dimensional frames is described in this paper. Lateral loading capacity results developed by ULSLEA are compared with results from three sets of experimental frame tests (Grenda, et al. 1989; Bolt, et al, 1994; 1995). ULSLEA was able to predict the ultimate capacity of the six test frames very well. Defining a bias as the ratio of test over ULSLEA results, the mean and coefficient of variation of the bias were estimated to be 1.05 and 0.1, respectively, for the cases where brace failures governed the capacity. In cases where tubular joint failures governed the frame capacity, the mean and coefficient of variation of the bias were estimated to be 2.03 and 0.01 respectively. The latter set of results can be explained by the conservatism of the API joint capacity equations (API, 1993b), which were used in the ULSLEA analyses. If the bias in the conservative API joint capacity equations are recognized (approximately 2), the ULSLEA capacity bias is approximately 1.0.

KEY WORDS: platforms, capacities, ultimate limit state performance, failure, simplified methods

INTRODUCTION

The simplified nonlinear structural analysis procedure for template-type offshore platforms, developed previously by the authors, is identified as ULSLEA (Ultimate Limit State Limit Equilibrium Analysis) (Mortazavi, Bea, 1994). In this procedure, the structure is defined by specifying the following parameters: the effective deck areas, the proportion and topology of jacket legs, braces, joints, foundation piles (main piles through the legs and skirt piles), well conductors, and effective areas of appurtenances (e.g. boat landings, pipeline risers). Specialized elements are designated including grouted or ungrouted joints, braces, and legs. In addition, damaged or defective elements are included. Dent depth and initial out-of-straightness are specified for braces with dents and global bending defects. Element capacity reduction factors are introduced to account for other types of damage to joints, braces, and foundation (corrosion, fatigue cracks, etc.). Similar algorithms are included to characterize repaired elements (e.g. grout filled braces). Steel elastic modulus, yield strength, and effective buckling length factor for vertical diagonal braces are specified.

Simplified formulations are developed to estimate the ultimate lateral shear capacity of the three primary structural components of a platform; deck legs, jacket, and foundation. Elements within these components are identified first. These are deck legs, tubular braces and their connections (tubular joints) and main and skirt piles. For each component to fail, all elements within that component have to fail. Based on past numerical analyses and experience, failure modes are assumed for different platform components. Using the concept of plastic hinge theory, the principle of virtual work is utilized to formulate the component capacities; the virtual displacement is taken to be the actual collapse mechanism and an equilibrium equation is derived for each component at ultimate limit state. Where of significance, geometric nonlinearities are taken into account.

During the first generation verifications, ULSLEA was verified with resutls from detailed three-dimensional nonlinear pushover analyses performed on six Gulf of Mexico platforms (Bea, DesRoches, 1993; Bea, 1995). Defining the bias as the ratio of the nonlinear pushover analysis Reserve Strength Ratio (RSR = ratio of maximum lateral loading capacity to reference lateral loading) to the RSR determined using ULSLEA, the first generation verifications developed RSR biases in the range of $B_{RSR} = 0.80$ to 1.03 with a mean value of 0.95. Several of the verification platforms were subjected to intense loadings developed by hurricanes. The predicted and observed performance characteristics of the platforms were in good agreement.

During the second generation verifications (Bea, Mortazavi, 1995; Bea, Loch, Young, 1995; Bea, et al., 1995) ULSLEA was verified with results from detailed three-dimensional nonlinear pushover analyses performed on six Gulf of Mexico platforms. The detailed analyses were performed using the programs WAJAC (DNV, 1993) and USFOS (SINTEF, 1994).

Defining the bias as the ratio of USFOS to ULSLEA ultimate lateral loading capacities, the mean and the coefficient of variation of the bias are 1.03 and 0.09, respectively. These results indicate that compared with detailed nonlinear analyses, the simplified method predicts ultimate capacities that are practically unbiased. In addition, comparison of the results from the ULSLEA analyses with the performance characteristics of these platforms in past intense hurricanes indicates that the analyses produce results that are in agreement with the observed characteristics.

These biases and uncertainties are remarkably small. Experience during the past two years in analyzing more than 20 platforms indicates that the simplified ULSLEA analyses have taken from 1/30th to 1/60th of the time required to perform the complex USFOS analyses. In the majority of these analyses, ULSLEA results were able to identify errors in the initial USFOS results. The anticipated benefits of the simplified methods have been realized.

In the remainder of this paper we will describe the validation of ULSLEA that has been accomplished by comparing the capacities predicted by ULSLEA with those determined from experimental tests performed on two-dimensional test frames typical of those found in template-type platforms.

FRAME TEST PROGRAM I

Frame test program I was performed in 1986 in behalf of Esso Australia. The frame tests were performed to provide experimental background for upgrading some of Esso Australia's older platforms in the Bass Strait (Grenda et al., 1988). The program was conducted to support the results of nonlinear pushover analyses and the modeling assumptions upon which the analyses were based. Verification of these modeling assumptions were crucial to the validity of the nonlinear analyses results.

The test program included static testing of six two-dimensional single-bay K-braced frames (Fig. 1). Four of the frames were ungrouted with overlap-joints. In two other cases, the compression diagonals were grout-filled. The tubulars were fabricated from steel plate similar to steel conforming to the ASTM A36 specification (API Grade B/X42). The measured static yield strengths were 1.2 to 1.3 times the nominal yield strength. Two different K-joint can thicknesses were selected. The frames were heavily instrumented and sophisticated data acquisition systems were utilized. The 25 ft. by 28 ft. frames were loaded laterally using a 250 ton displacement-controlled hydraulic cylinder.

Fig. 2 shows the load-displacement response of test frames 1-6. The performance of the four ungrouted frames with different joint can thicknesses (tests 1-4) was reported to be similar, each resisting a peak lateral load of approximately 160 kips. In all four cases, the buckling of compression braces was reported to govern the behavior of the frames. The lateral load resisting capacity of the frames began to decrease rapidly after failure of the compression diagonals. This is a typical result for this brittle k-braced frame configuration. The measured effective length factors for the K-brace compression diagonals at peak load in the four ungrouted frame tests varied between 0.53 and 0.63 based on the node-to-node length of the members. The tests indicated an insensitivity of the measured frame behavior to the thickness of the K-joint cans. The apparent insensitivity to K-joint can thickness was attributed to the relatively high degree of brace overlap at the K-node (Fig. 1).

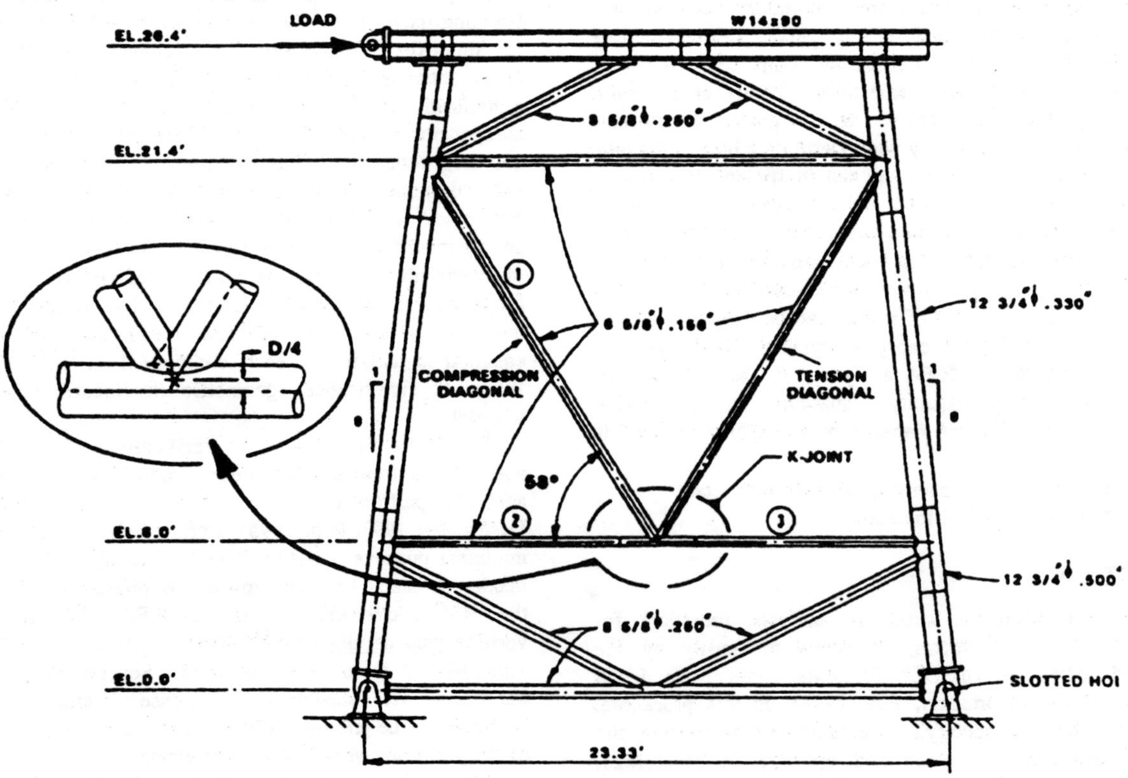

Fig. 1: Test Program I - Test Frame Geometry (After Grenda et al.,1988)

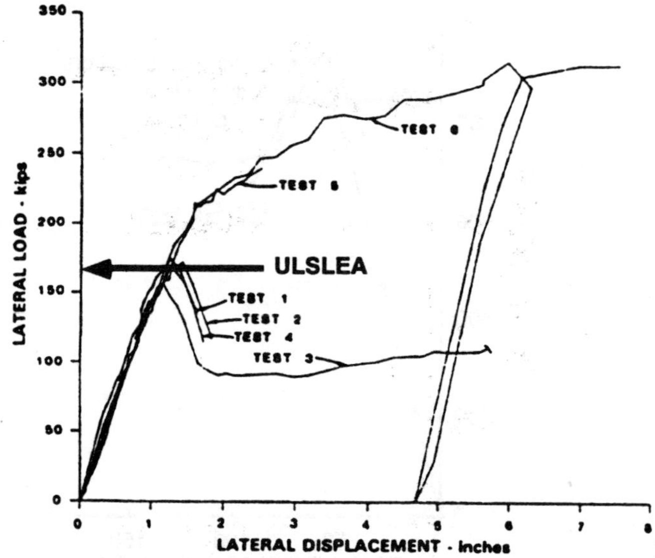

Fig. 2: Test Program I - Tests 1-4 Ungrouted Compression Diagonals (After Grenda et al., 1988)

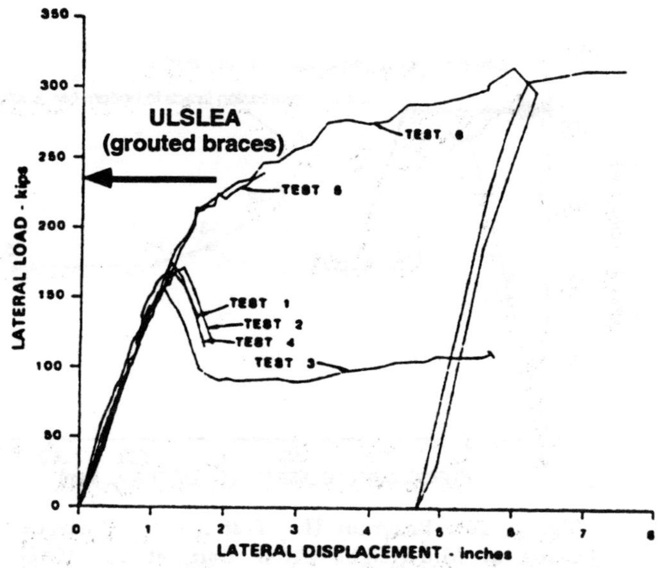

Fig. 3: Test Program I - Tests 5-6 (After Grenda et al.,1988)

In all four cases, ULSLEA was able to predict the frame failure at the same peak lateral loads (Fig. 2).

The lateral load capacity of the frames was significantly increased by grouting the diagonal compression braces. Grouting increased the buckling capacity of the diagonal braces in compression so that the diagonal braces in tension were able to fully develop their tensile capacity. This effect resulted in a redundant behavior of the frames. At the time the peak load was reached, both diagonal braces in tension and compression were contributing to the lateral shear resistance with their ultimate strengths. In addition, due to large displacement at collapse, the full portal strength of the frame legs were reported to be developed. However, the observed increase in lateral frame capacity due to frame action should not be overemphasized. In a real multi-bay offshore structure, frame action within one bay is likely to result in failure of diagonals in neighboring bays.

Frame test 6 resulted in an ultimate frame capacity of roughly 300 kips. The result gained using the simplified ULSLEA method indicated an almost simultaneous failure of two diagonal braces at a collapse load of roughly 240 kips (Fig. 3). This is an underestimation of 20% compared to the test results. A detailed comparison of predicted and actual member forces at collapse indicated that the effect of frame action, observed in the test, was the source of the difference in results. ULSLEA does not take the effect of frame action into account.

FRAME TEST PROGRAM II

Frame test program II was initiated in the U.K. in 1987. The program, Frames Project Phase I, was conducted by Billington Osborne-Moss Engineering Limited (BOMEL) as part of a joint industry project with the objectives of providing test data on the collapse behavior of jacket structures and developing a calibrated software for the nonlinear pushover analysis of framed structures (Bolt et al., 1994). Prior to release of the test data, the Health and Safety Executive in the U.K. invited interested companies in the U.K., Norway and U.S. to participate in a benchmarking effort. The results of this benchmarking exercise have been published by Nichols, et al. (1995).

In this phase of the project, four two-bay X-braced frames were pushed to collapse (Fig. 4). The effects of joint ductility and system redundancy on the ultimate and post-ultimate response of the frames were studied. The frames were heavily instrumented and tested in plane. The hydraulic actuators were located at the top of the frames and were operated in a displacement-controlled manner. The frames were pinned at the bottom. The four frames had virtually the same geometry but differed in the joint can thickness, horizontal bracing, initial imperfections and residual stresses. Frame 2 was similar to frame 1 with the exception of a reduced joint can thickness. Frame 3 was virtually the same as frame 1 with the horizontal brace at the mid height removed. Finally frame 4 had the same configuration as frame 3 with the difference of locked-in prestresses and initial imperfections. Frame test 4 was left out of the verification study.

The test results for frames 1 and 3 are shown in Fig. 5 in terms of lateral load displacement. In the case of frame 1, the buckling of the compression brace at the top half of the upper bay was reported to govern the ultimate lateral capacity of the frame which was reached at a lateral load of 920 kN. Due to redundancy of the X-bracing configuration and existing horizontal brace at the mid-height, a substantial residual strength in the system could be observed (Fig. 5). Portal action in the legs was also reported to have contributed to this residual capacity.

In the case of frame 3, a similar failure mode was observed. However, the peak lateral resistance was reached at a load level of 780 kN. The lower capacity as compared to the frame 1 was reported to be due to differences in the residual stresses and not a result of the missing horizontal brace at the mid-height. The lower residual capacity was, however, attributed to the reduced redundancy and lack of an effective load redistribution after first member failure. This resulted in a premature buckling of the compression brace in the lower bay and a consequent rapid fall of lateral resistance in post-ultimate regime (Fig. 6).

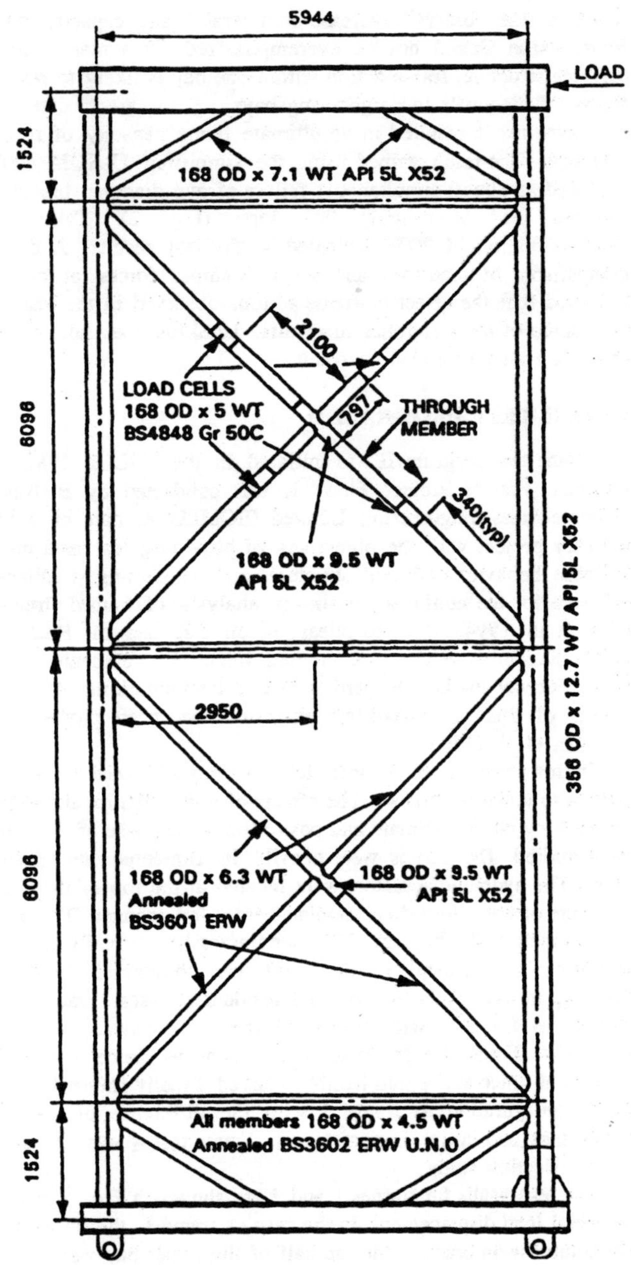

Fig. 4: Test Program II - Test Frame Geometry (After Bolt et al.,1994)

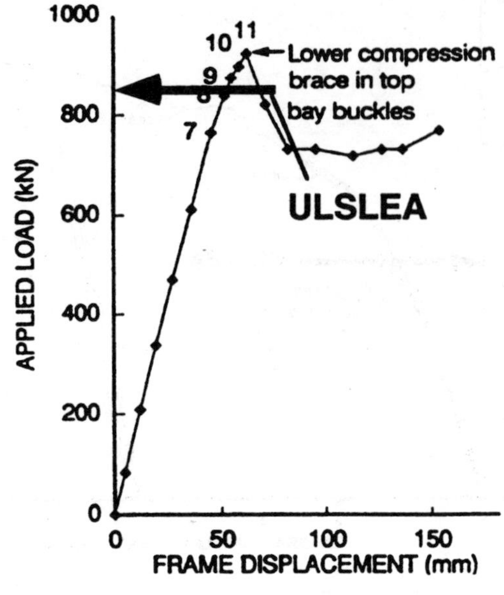

Fig. 5: Test Program II - Test 1 (After Bolt, et al., 1994)

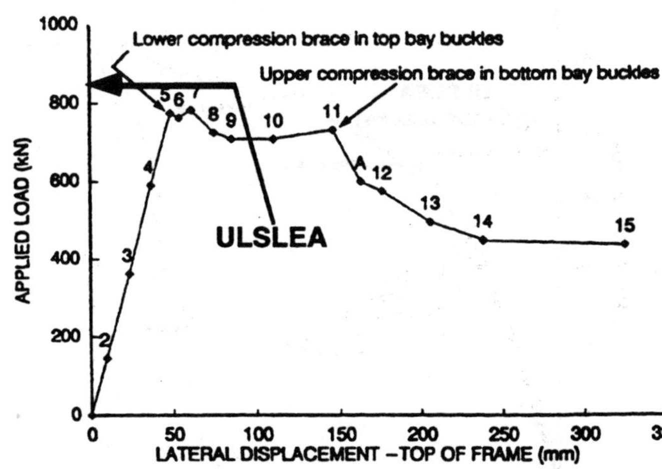

Fig. 6: Test Program II - Test 3 - No Horizontal Bracing at Mid-Height (After Bolt, et al., 1994)

ULSLEA was used to predict the ultimate lateral capacity of both frames 1 and 2. ULSLEA predicted almost simultaneous failure of the compression and tension diagonals at the upper bay at a lateral load of roughly 850 kN for both frames. This result is in excellent agreement with the test results. The minor difference is due to the fact that ULSLEA does not account for residual stresses. The fact that the absence of the horizontal bracing in frame 3 did not change the ultimate capacity of the system, confirms the assumption made in ULSLEA regarding horizontal framings. In ULSLEA, it is assumed that horizontal bracings exist and are rigid.

The results of the frame test 2 are plotted in Fig. 7. The X-joint in the upper bay with the reduced wall thickness was the weak link in the frame and started yielding at a lateral load of 689 kN. The flattening of the joint essentially postponed the buckling of the compression diagonal and the tension diagonal developed its full yield strength. After the X-joint was completely compressed, a new load path was created. With further increase in the lateral load, the compression brace in the upper bay buckled and a rapid load shedding followed. The large global deflection at collapse had also resulted in portal action. The peak capacity of 1,080 kN was reached.

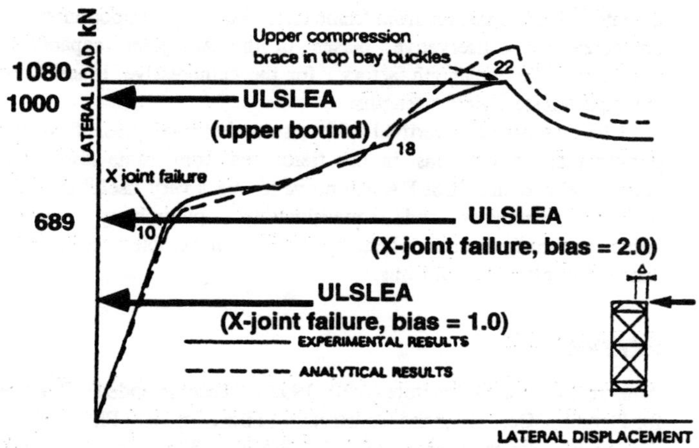

Fig. 7: Test Program II - Test2 (After Bolt, et al., 1994)

ULSLEA underestimated the lateral load at first member failure by 50%. The first member to fail was the X-joint in the upper bay. This underprediction is due to the conservatism of the joint capacity equations as given by API RP 2A-LRFD and used in ULSLEA. Based on use of a joint capacity bias factor of 2 resulted in an ULSLEA predicted lateral capacity of 689 kN; a very close prediction of the lower-bound capacity. An upper-bound capacity of 1,000 kN was predicted by ULSLEA. This peak lateral load was predicted to be associated with simultaneous failure of the tension and compression diagonals. The test indicated a peak lateral load capacity of 1,080 kN. In this case, the minor difference between ULSLEA and test results was traced back to the additional frame resistance due to portal action in the test.

FRAME TEST PROGRAM III

This frame test program was the second phase of the BOMEL's Frames Project (Bolt, 1995). Two objectives of this phase of the program were to investigate the effect of boundary conditions on joint ultimate capacity performance and to examine the collapse behavior of K-braced frames. In this phase four single bay K-braced frames were laterally pushed to collapse (Fig. 8). Gap and overlap K-joints were used. The frames were laterally pushed under displacement-control beyond the ultimate load and into post-ultimate regime in order to capture their residual strength.

Fig. 9 shows the load displacement behavior of one of the frames. The ultimate capacity was governed by the failure of the K-joint. With increasing lateral load, a crack was initiated at the chord side of the tension brace, which rapidly propagated and led to load shedding and a sudden reduction in frame lateral resistance.

Compared to test results on isolated joints, it was found that the joint capacity within frames was higher. ULSLEA also predicted K-joint failure, however at a lateral load 50% lower than reported in the test; 230 kN (Fig. 9). Again, the difference was found to be due to the conservatism in joint capacity formulations in API-RP 2A-LRFD (API, 1993b). Once the bias in the API joint formulation (= 2.0) was input to ULSLEA, the lateral load capacity was evaluated to be 460 kN. This was very close to the measured lateral capacity of 470 kN.

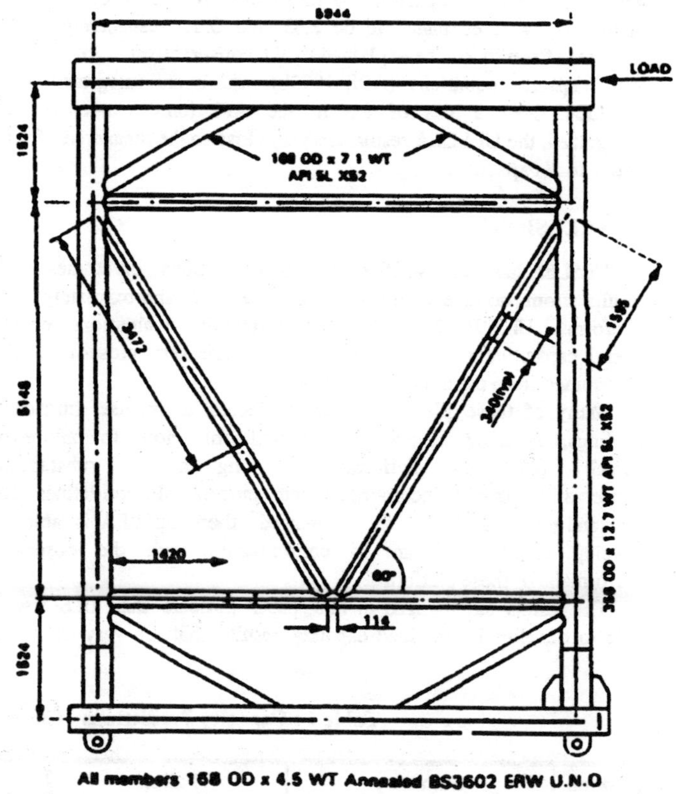

Fig. 8: Test Program III - Test Frame Geometry (After Bolt, 1995)

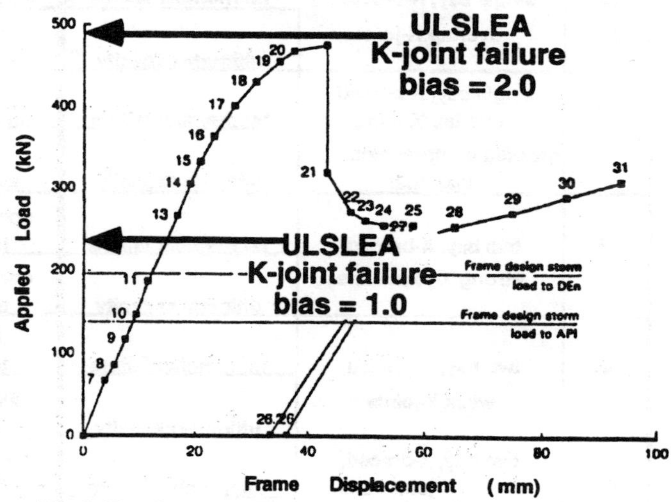

Fig. 9: Test Program III (After Bolt, 1995)

SUMMARY

Table 1 contains a summary of the verification studies based on frame test results. First member failures and ultimate lateral loading capacities of six test frames were predicted using the simplified method. These results were compared with the actual performance of the test frames. Defining a bias as the ratio of test over ULSLEA results, the mean and coefficient of variation of the bias were estimated to be 1.05 and 0.1 respectively for the cases where brace failures governed the capacity. In cases where tubular joint failures governed the frame capacity, the mean and coefficient of variation of the bias were estimated to be 2.03 and 0.01 respectively. The latter set of results can be explained by the conservatism of the API joint capacity equations (API, 1993b) which are integrated into ULSLEA. Once a bias of 2.0 in the API joint capacities are recognized, the ULSLEA results are very close to the measured frame lateral load capacities.

CONCLUSIONS

ULSLEA has been verified in two generations of studies with results from nonlinear inelastic analyses of approximately 30 platforms. ULSLEA has been able to develop evaluations of the lateral capacities of these platforms that are very close to those determined from the nonlinear analyses.

Many of these platforms were subjected to intense hurricane loadings, bringing several of the platforms close to failure or actually failed the platforms. As long as the substantial conservative biases incorporated into current API guidelines for pile and joint capacities are recognized, then ULSLEA is able to develop results that are in conformance with the observed performane of these platforms.

The results summarized in this paper indicate that ULSLEA is able to develop lateral load capacity results that are very close to the lateral load capacities from frame tests. Again, it is important to recognize the conservative biases in the API joint capacities, reasonable effective length factors for the compressive braces, and the expected steel yield strengths.

ULSLEA has been verified with results from analyses, from the performance of platforms in the field, and from frame tests. The authors believe that ULSLEA will prove to be a very useful tool for platform reanalyses and requalifications, preliminary design, parametric studies, and for checking results from complex nonlinear analyses of offshore platforms.

REFERENCES

American Petroleum Institute (API), 1993a. "Recommended Practice for Planning, Designing and Constructing Fixed Offshore Platforms - Working Stress Design (RP 2A-WSD)." 20th Edition, July, Washington, D.C.

American Petroleum Institute (API), 1993b. "Recommended Practice for Planning, Designing and Constructing Fixed Offshore Platforms - Load and Resistance Factor Design (RP 2A-LRFD)." First Edition, July, Washington, D.C.

Table 1 : Capacity Predictions - Comparison of ULSLEA and Test Results

Frame Test	Summary Description	Failure Category	Test Result		ULSLEA		Ratio of TEST/ULSLEA Base Shears
			Failure Mode	Base Shear	Failure Mode	Base Shear	
IA	single bay, K-braced, overlap K-joint	1st member failure	comp. brace	160 kips	comp. brace	160 kips	1.00
		ultimate capacity	comp. brace	160 kips	comp. brace	160 kips	1.00
IB	single bay, K-braced, overlap K-joint, grouted compression diagonal	1st member failure	tension brace	240 kN	all braces	240 kN	1.00
		ultimate capacity	all braces	300 kN	all braces	240 kN	1.25
IIA	two bay, X-braced, strong X-joint cans	1st member failure	comp. brace top bay	920 kN	comp. brace top bay	850 kN	1.08
		ultimate capacity	comp. brace top bay	920 kN	comp. brace top bay	850 kN	1.08
IIB	two bay, X-braced, weak X-joints	1st member failure	X-joint top bay	689 kN	X-joint top bay	345 kN	2.00
		ultimate capacity	all braces top bay	1080 kN	all braces top bay	1000 kN	1.08
IIC	two bay, X-braced, strong X-joint cans, no horizontal bracing	1st member failure	comp. brace top bay	780 kN	comp. brace top bay	850 kN	0.92
		ultimate capacity	comp. brace top bay	780 kN	comp. brace top bay	850 kN	0.92
III	single bay, K-braced, gap K-joint	1st member failure	K-joint	470 kN	K-joint	230 kN	2.04
		ultimate capacity	K-joint	470 kN	K-joint	230 kN	2.04

Bea, R.G., 1995. "Development and Verification of a Simplified Method to Evaluate Storm Loadings on and Capacities of Steel, Template-Type Platforms." Proceedings, Energy and Environmental Expo 95, American Society of Mechanical Engineers, Houston, Texas, Jan.

Bea, R.G., DesRoches, R., 1993. "Development and Verification of A Simplified Procedure to Estimate the Capacity of Template-Type Platforms." Proceedings, 5th International Symposium Integrity of Offshore Structures, D. Faulkner et al., Emas Scientific Publications, pp. 129-148.

Bea, R.G., Loch, K., Young, P., 1995b. "Evaluation of Capacities of Template-Type Gulf of Mexico Platforms." Proceedings, 5th International Offshore and Polar Engineering Conference, ISOPE-95, The Hague, The Netherlands, June 11-16.

Bea, R.G., Mortazavi, M., 1995a. "Simplified Evaluation of the Capacities of Template-Type Offshore Platforms." Proceeding of 5th International Offshore and Polar Engineering Conference, The Hague, The Netherlands, June 11-16.

Bea, R.G., Mortazavi, M., Loch, K., Young, P., 1995c. "Verification of A Simplified Method to Evaluate the Capacities of Template-Type Platforms." Proceedings, Offshore Technology Conference, OTC 7780, Houston TX, May.

Bea, R.G., Mortazavi, M., Loch, K., Young, P., 1995d. "Verification of A Second Generation Simplified Method to Evaluate the Storm Loadings on and Capacities of Steel Template-Type Platforms." Proceedings, Energy and Environmental Expo 95, American Society of Mechanical Engineers, Houston, Texas, Jan.

Billington, C.J., Bolt, H.M., Ward, J.K. 1993. "Reserve, Residual and Ultimate Strength Analysis of Offshore Structures: State of the Art Review." Proceedings, Third International Offshore and Polar Engineering Conference, Singapore, June, pp. 125-133.

Billington, C.J., Lloyed, J.R., Digre, K.A., 1993. "Structural Elements, Systems and Analyses." White paper prepared for the International Workshop on Assessment and Requalification of Offshore Production Structures, New Orleans, LA.

Bolt, H.M., 1995. "Results from Large-Scale Ultimate Strength Tests of K-Braced Jacket Frame Structures." Proceedings, Offshore Technology Conference, OTC 7783, Houston TX, May.

Bolt, H.M., Billington C.J., Ward J.K., 1994. "Results from Large-Scale Ultimate Load Tests on Tubular Jacket Frame Structures." Proceedings, Offshore Technology Conference, OTC 7451, Houston TX, May.

Cox, J.W., 1987. "Tubular Member Strength Equations for LRFD." Final Report API PRAC Project 86-55.

Grenda, K.G., Clawson, W.C., Shinners, C.D., 1988. "Large-Scale Ultimate Strength Testing of Tubular K-Braced Frames." Proceedings, Offshore Technology Conference, OTC 5832, Houston TX, May.

Mortazavi, M., Bea R.G., 1994. "ULSLEA, Simplified Nonlinear Analysis for Offshore Structures." Report to Joint Industry-Government Sponsored Project, Marine Technology and Management Group, Dept. of Civil Engineering, University of California at Berkeley, June.

Mortazavi, M., Bea R.G., 1995. "Screening Methodologies for Use in Platform Assessments and Requalifications." Final Report to Joint Industry-Government Sponsored Project, Marine Technology and Management Group, Dept. of Civil Engineering, University of California at Berkeley, June.

Nichols, N.W., Sharp, J.V., Kam, J.C.P., 1995. "Benchmarking of Collapse Analysis of Large-Scale Ultimate Load Tests on Tubular Jacket Frame Structures."

Titus, P.G., Banon, H., 1988. "Reserve Strength Analyses of Offshore Platforms." Proceedings, 7th Offshore South East Asia, Singapore, 2-5 February.

Yura, J.A., Zettlemoyer, N., Edwards, F.E., 1980. "Ultimate Capacity Equations for Tubular Joints." Proceedings, Offshore Technology Conference, OTC 3690, Houston TX, May.

Zayas, V.A., Popov, E.P., Mahin, S.A., 1980. "Cyclic Inelastic Buckling of Tubular Steel Braces.", University of California, Report No. UCB/EERC-80/16, June.

Proceedings of the Seventh (1997) International Offshore and Polar Engineering Conference
Honolulu, USA, May 25-30, 1997

ISBN 1-880653-28-1 (Set); ISBN 1-880653-32-X (Vol. IV)

An Empirical Formulation for Predicting the Ultimate Compressive Strength of Stiffened Panels

Jeom K. Paik
Pusan National University
Pusan, Korea

Anil K. Thayamballi
American Bureau of Shipping
New York, NY, USA

ABSTRACT

The aim of the present study is to empirically derive a simple and reasonably accurate expression for predicting the ultimate strength of longitudinally stiffened panels under uniaxial compression for use in the context of parametric structural reliability studies of ship hull girder global behavior. Existing test data on the ultimate compressive strength for stiffened panels were used as the starting point for the formulation. A study of the collected data indicated that the data needed supplementation in order for the derived formula to be applicable to a desired wide range of panel dimensions. Hence a series of 12 longitudinally stiffened plates having three different plate slenderness values were in the present study tested up to the collapse state under uniaxial compression. On the basis of the previously collected and newly developed test data, an useful empirical expression for predicting the ultimate compressive strength of stiffened panels, expressed in terms of the plate slenderness ratio and the column (stiffener) slenderness ratio, was derived. The formula implicitly includes the influence of initial imperfections at a moderately large level. All test data (previous and new) are documented.

KEY WORDS: Ultimate compressive strength, stiffened panel, empirical formula

NOMENCLATURE

A = cross-sectional area of one longitudinal stiffener including associated full plating
a = length of longitudinal stiffener between transverse support frames
A_{st} = cross-sectional area of one longitudinal stiffener exclusive of associated plating
B = breadth of the stiffened panel
b = breadth between longitudinal stiffeners (stiffener spacing)
b_f = breadth of stiffener flange
E = Young's Modulus
h_w = height of stiffener web
I = moment of inertia of one longitudinal stiffener including associated full plating
r = radius of gyration of one longitudinal stiffener including associated full plating $= \sqrt{I/A}$
t_p = thickness of plating
t_w = thickness of stiffener web
t_f = thickness of stiffener flange
w_{op} = initial deflection of plating
w_{os} = overall initial deflection of stiffener (column)
β = slenderness ratio of a plate between longitudinal stiffeners $= b/t_p \cdot \sqrt{\sigma_o / E}$
λ = slenderness ratio of a longitudinal stiffener including associated full plating considered as a column $= a/\pi r \cdot \sqrt{\sigma_o / E}$
σ_{rc} = welding-induced compressive residual stress
σ_u = ultimate compressive strength
σ_o = mean yield strength for the stiffened panel

$$= \frac{b\,t_p\,\sigma_{op} + A_{st}\,\sigma_{os}}{b\,t_p + A_{st}}$$

σ_{op} = yield strength of plate
σ_{os} = yield strength of stiffener

1. INTRODUCTION

Stiffened steel panels are the basic strength members in ship and offshore structures. Due to their simplicity in fabrication and excellent strength to weight ratio, stiffened steel panels are also widely used in land based structures such as box and plate girder bridges. During the 1970s, the unfortunate occurrence of a few failures in such bridge structures stimulated extensive theoretical and experimental research work on failure of stiffened panels. In ship design, the hull girder strength of ships is often governed by the collapse behavior of deck or bottom panels (Caldwell 1965, Paik et al. 1996). Hence the calculation of collapse strength of stiffened plate elements in deck and bottom panels is required for the reliability assessment of ship structures.

The stiffened plate elements are generally subjected to combined loads, and a primary load component is axial compression. Theoretically, the possible collapse modes of a stiffened panel subject to predominantly compressive loads can be categorized into the following five classes, namely

- overall collapse after overall buckling
- plate-induced (overall) collapse after local buckling between stiffeners
- stiffener-induced (overall) collapse after local buckling of stiffeners (without rotation of stiffener)
- torsional / flexural buckling (tripping) of stiffener (resulting in subsequent unloading of the stiffened panel)
- gross yielding

The collapse of stiffened panels can be postulated to occur at the lowest value of the various ultimate loads calculated for each of the above collapse patterns. A number of theoretical and numerical methods have been suggested for predicting the ultimate strength of stiffened panels. A benchmark study on the ultimate stiffened panel strength formulations was carried out by Das & Garside (1991), Hughes et al. (1994), Rigo et al. (1995) and Paik (1997). For convenience in design, most of those theoretical formulations have been derived for a single plate-stiffener combination that is representative of the stiffened panel. On the other hand, numerical procedures typically take an "implicit form", using a computer program. The latter are somewhat less useful than the former for efficiently calculating the ultimate strength of stiffened panels, either for use as a design equation, or as part of a failure function for reliability analysis.

The aim of the present study is to establish a reasonably accurate closed-form expression for predicting the ultimate strength of stiffened panels under uniaxial compression, using as few parameters as necessary (e.g., plate and column slenderness ratios, without explicit consideration of effective width) particularly for use in parametric reliability studies related to the ship hull girder. For this purpose, two approaches could be considered: One is to analytically derive a closed-form solution under a suitable collapse criterion for stiffened panels and the other is to empirically formulate a simple expression based on available experimental and / or numerical data. The present study uses the latter approach. All data used in the study are documented, which is a benefit in itself.

An useful formulation for predicting the ultimate compressive strength of stiffened panels using the latter approach was derived by Lin (1985), which was expressed in terms of slenderness ratios of plate and stiffener. Lin's formula is our starting point. Later in the paper, it is shown that Lin's formula can provide reasonably accurate solutions for stiffened panels with stiffeners of relatively large geometric dimensions, but can possibly seriously underestimate the ultimate strength for stiffened panels with stiffeners which have small geometric dimensions, i.e., with large equivalent column slenderness ratio.

With the possibility of higher speed vessels, and in the extreme, buckling-acceptable design, one cannot in principle rule out the larger column slenderness ratios in emerging merchant vessel concepts. In this situation, an empirical formula that can effectively cover a wide range of geometric dimensions of plates and stiffeners is needed. Thus the motivation in this study was to correct the potential weak point of Lin's formula such that the resulting new formula could be useful for stiffened panels with small and large geometric properties of stiffeners. Also, the use of an existing formula as the starting point enables us to reduce the number of new test data needed, as the new tests could be designed so that they intelligently supplement existing experimental data on ultimate strength of stiffened panels. Nevertheless, it is cautioned that the resulting formula is essentially empirical, and as such not suitable for use in the design of local structures.

2. A SUMMARY OF PREVIOUS COLLAPSE TEST DATA

Existing experimental data on collapse strengths of stiffened panels under uniaxial compression were first collected. Figure 1 shows a configuration of a stiffened panel between transverse support frames, subject to uniaxial compression. The panel has one-sided stiffeners and the stiffener spacings are the same. Collapse test data obtained only under nearly simply supported edge conditions were used, this being preferable for purposes of deriving a formulation useful for practical purposes.

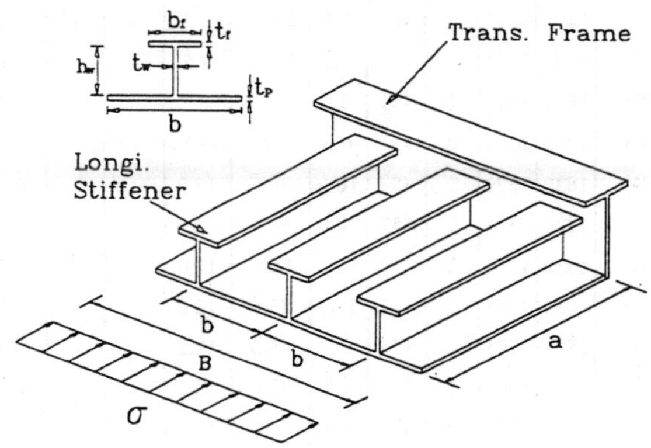

Fig. 1 Nomenclature: A stiffened plate structure under uniaxial compression

For many test specimen, the yield strength of the stiffener was different from that of the plating, and so a mean yield strength for the whole stiffened panel, see Nomenclature, was employed for deriving the present empirical formulation. Initial deflections and welding-induced residual stiffened panels, and thus the levels of initial imperfections for the test specimen were carefully surveyed. Also, the collapse modes of the various test specimen were reviewed.

An extensive set of collapse tests on longitudinally stiffened panels of varying dimensions of plate and stiffener were undertaken by Horne & Narayanan (1976) and Horne et al. (1977), see Table 1. Two longitudinal stiffeners were attached to the parent plate by welding, which was performed intermittently for some specimens or entirely for others. Only flat-bar type of stiffeners were attached to the specimen of Horne & Narayanan (1976), while bulb type as well as flat-bar type of stiffeners were used for the specimen of Horne et al. (1977). Table 1 also shows collapse modes of each test specimen as observed during and after the tests. The measured (maximum) initial deflections of plate part w_{op} / t_p were in the range of 5% to 97% and the normalized welding-induced compressive residual stresses $\sigma_{rc} / \sigma_{op}$ were in the range of 6% to 42%. Fixed as well as simply supported edge conditions were tested, but we used the results obtained only under simply supported boundary conditions for purposes of deriving the present empirical formulation.

Faulkner (1977) carried out a series of collapse tests on longitudinally stiffened plates, summarized in Table 2. Two separate series of tests are reported. Two T-type stiffeners were welded to the parent

Table 1 Experimental results of ultimate compressive strength for stiffened panels obtained by Horne et al. (1976, 1977)

Specimen No.	a (mm)	B (mm)	n_{st}	b (mm)	t_p (mm)	h_w (mm)	t_w (mm)	b_f (mm)	t_f (mm)	w_{op}/t_p	w_{os}/a ($\times 10^{-3}$)	λ	β	σ_{rc}/σ_{op}	σ_{op} (MPa)	σ_{os} (MPa)	σ_o (MPa)	σ_u/σ_o	Failure Mode	
3(FB)	915	1371	2	457	9.5	152.5	9.5	-	-	0.13	1.85	0.251	1.708	0.129	259.5	275.1	263.4	0.854	IW	Ⅲ
7(FB)	1830	1371	2	457	9.5	152.5	16.0	-	-	0.18	0.30	0.440	1.692	0.412	254.7	268.1	259.5	0.794	CW	Ⅱ
8(FB)	1830	1371	2	457	9.5	152.5	16.0	-	-	0.09	0.38	0.442	1.717	0.125	262.1	262.0	262.1	0.851	IW	Ⅱ
9(FB)	1830	1371	2	457	9.5	152.5	9.5	-	-	0.08	0.30	0.504	1.717	0.122	262.1	272.5	264.7	0.782	IW	Ⅲ
11(FB)	1830	1371	2	457	9.5	152.5	16.0	-	-	0.19	0.66	0.449	1.760	0.120*	275.5	262.0	270.6	0.791	IW	Ⅱ
12(FB)	1830	1371	2	457	9.5	152.5	16.0	-	-	0.27	0.52	0.440	1.692	0.418*	254.7	268.1	259.5	0.794	CW	Ⅱ
13(FB)	1830	1371	2	457	9.5	152.5	16.0	-	-	0.61	0.27	0.449	1.760	0.120*	275.5	262.0	270.6	0.750	IW	Ⅱ
14(FB)	1830	1371	2	457	9.5	152.5	16.0	-	-	0.63	0.46	0.440	1.6922	0.418*	254.7	268.1	259.5	0.832	CW	Ⅱ
A11(FB)	1830	1371	2	457	6.5	152.5	9.5	-	-	0.89	0.44	0.538	.971	0.120*	367.4	334.7	356.7	0.544	IW	Ⅱ
A12(FB)	1830	1371	2	457	6.5	152.5	9.5	-	-	0.23	0.66	0.534	2.897	0.066	349.3	354.5	351.0	0.569	IW	Ⅱ
A21(FB)	1830	1371	2	457	6.5	152.5	9.5	-	-	0.88	1.39	0.527	2.879	0.418*	345.2	334.7	341.8	0.642	CW	Ⅱ
A22(FB)	1830	1371	2	457	6.5	152.5	9.5	-	-	0.22	0.66	0.518	2.832	0.418*	334.0	323.9	330.7	0.564	CW	Ⅱ
A23(FB)	1830	1371	2	457	6.5	152.5	9.5	-	-	0.23	0.90	0.528	2.909	0.424	352.2	323.9	342.9	0.608	CW	Ⅱ
D11(FB)	1830	1371	2	457	10.0	80.0	12.0	-	-	0.54	0.68	1.096	1.694	0.120*	282.9	290.7	284.3	0.632	IW	Ⅱ
D12(FB)	1830	1371	2	457	10.0	80.0	12.0	-	-	0.31	0.60	1.001	1.540	0.132	233.6	252.3	236.8	0.648	IW	Ⅱ
D21(FB)	1830	1371	2	457	10.0	80.0	12.0	-	-	0.57	0.38	1.018	1.570	0.418*	243.0	256.0	245.3	0.574	CW	Ⅱ
D22(FB)	1830	1371	2	457	10.0	80.0	12.0	-	-	0.12	0.44	1.032	1.575	0.315	244.3	287.0	251.7	0.600	CW	Ⅱ
E11(FB)	1830	1371	2	457	6.5	76.0	12.5	-	-	0.97	0.44	1.148	2.840	0.120*	335.9	374.0	345.1	0.471	IW	Ⅱ
E12(FB)	1830	1371	2	457	6.5	76.0	12.5	-	-	0.26	0.49	1.148	2.835	0.088	334.7	377.9	345.2	0.476	IW	Ⅱ
E21(FB)	1830	1371	2	457	6.5	76.0	12.5	-	-	0.86	0.98	1.139	2.839	0.418*	335.6	353.3	339.9	0.443	CW	Ⅱ
E23(FB)	1830	1371	2	457	6.5	76.0	12.5	-	-	0.38	0.93	1.139	2.815	0.330	329.8	369.5	339.4	0.448	CW	Ⅱ
AF2(A)	3000	600	2	200	10.3	152.0	6.5	76.0	9.8	0.08	0.20	0.606	0.805	0.418*	353.5	410.2	379.4	0.892	CW	Ⅲ
AF3(A)	3000	600	2	200	10.1	152.0	6.5	76.0	10.1	0.11	0.40	0.593	0.779	0.594	318.4	426.2	368.5	0.739	CW	Ⅲ
AS2(A)	3000	600	2	200	10.35	152.0	6.6	38.0	12.0	0.10	0.43	0.659	0.816	0.418*	366.6	409.7	384.4	0.812	CW	Ⅲ
AS3(A)	3000	600	2	200	10.4	152.0	6.5	76.0	11.75	0.09	0.40	0.601	0.821	0.458	375.2	422.4	397.6	0.660	CW	Ⅲ
BF2(B)	3000	600	2	200	10.0	160.0	8.0	-	-	0.11	0.47	0.737	0.790	0.418*	320.8	302.8	313.8	0.613	CW	Ⅲ
BF3(B)	3000	600	2	200	10.1	160.0	8.0	-	-	0.08	0.30	0.759	0.796	0.545	332.4	329.3	331.2	0.546	CW	Ⅲ
FS4(FB)	3000	600	2	200	9.9	148.5	9.8	-	-	0.05	0.27	0.850	0.838	0.120*	353.7	410.3	377.7	0.681	IW	Ⅲ
FS5(FB)	3000	600	2	200	10.1	148.5	9.8	-	-	0.10	0.40	0.867	0.842	0.191	372.0	417.7	391.1	0.595	IW	Ⅲ
FS9(FB)	3000	600	2	200	9.9	148.5	9.8	-	-	0.09	0.57	0.840	0.828	0.418*	345.7	400.6	369.0	0.754	CW	Ⅲ
FS10(FB)	3000	600	2	200	9.9	148.5	9.8	-	-	0.09	0.80	0.844	0.832	0.568	348.8	403.7	372.1	0.709	CW	Ⅲ
PF2(FB)	2700	600	2	200	9.7	150.0	15.15	-	-	0.07	0.41	0.728	0.858	0.556	356.1	408.3	384.3	0.787	CW	Ⅱ
PF5(FB)	2700	900	2	300	10.0	150.0	15.15	-	-	0.14	0.26	0.790	1.344	0.329	413.3	415.6	414.3	0.791	CW	Ⅱ
PF11(FB)	2700	1050	2	350	9.8	150.0	15.15	-	-	0.16	0.39	0.784	1.532	0.219	378.8	410.1	391.3	0.750	CW	Ⅱ
SW1(FB)	2700	1440	2	480	9.7	150.0	15.15	-	-	0.19	0.41	0.839	2.132	0.102	382.1	428.3	397.3	0.717	IW	Ⅱ
SW3(FB)	2700	1440	2	480	9.7	150.0	15.15	-	-	0.28	0.44	0.838	2.138	0.177	384.2	421.8	396.5	0.714	IW	Ⅱ
SW5(FB)	2700	1440	2	480	9.7	150.0	15.15	-	-	0.20	0.30	0.851	2.204	0.269	408.4	408.8	408.5	0.636	CW	Ⅱ
SW7(FB)	2700	1440	2	480	9.7	150.0	15.15	-		0.14	0.30	0.866	2.229	0.067	417.7	434.1	423.1	0.693	IW	Ⅱ

Note: 1) Failure mode:

Ⅰ = overall collapse after overall buckling

Ⅱ = plate-induced (overall) collapse after local buckling between stiffeners

Ⅲ = stiffener-induced (overall) collapse after local buckling between stiffeners (without rotation of stiffener)

Ⅳ = torsional / flexural buckling (tripping) of stiffener

Ⅴ = full yielding

2) Welding type

CW = continuous weld

IW = intermittent weld

3) *Assumed value, others measured value

Table 2 Experimental results of ultimate compressive strength for stiffened panels obtained by Faulkner (1977)

Specimen No.	a (mm)	B (mm)	n_{st}	b (mm)	t_p (mm)	h_w (mm)	t_w (mm)	b_f (mm)	t_f (mm)	w_{op}/t_p	w_{os}/a ($\times 10^{-3}$)	λ	β	σ_{rc}/σ_{op}*	σ_{op} (MPa)	σ_{os} (MPa)	σ_o (MPa)	σ_u/σ_o	Failure Mode
P1(T)	244	530.4	5	88.4	3.07	17.4	4.88	12.7	6.17	0.13	0.20	0.314	1.004	0.178	250.0	283.0	262.4	0.976	II
P2(T)	384	882	5	147	2.62	30.4	4.83	12.7	6.22	0.47	0.10	0.330	1.956	0.178	250.0	262.0	254.4	0.733	II
P3(T)	638	1326	5	221	2.54	54.1	4.9	12.7	6.1	1.11	0.33	0.346	3.069	0.178	256.0	247.0	252.6	0.713	IV
P4(T)	523	1416	5	236	2.01	43.6	4.8	12.7	6.25	1.86	0.31	0.324	3.848	0.178	221.0	250.0	232.0	0.567	II
P5(T)	488	530.4	5	88.4	3.07	17.4	4.88	12.7	6.17	0.03	2.46	0.597	0.952	0.178	225.0	259.0	237.8	0.824	II
P6(T)	767	882	5	147	2.62	30.4	4.83	12.7	6.22	0.29	1.01	0.648	1.912	0.178	239.0	259.0	246.4	0.750	II
P7(T)	1275	1326	5	221	2.54	54.1	4.9	12.7	6.1	0.58	0.85	0.703	3.152	0.178	270.0	246.0	260.9	0.621	IV
P8(T)	1046	1416	5	236	2.01	43.6	4.8	12.7	6.25	1.36	0.47	0.675	4.068	0.178	247.0	259.0	251.5	0.515	II
P9(T)	732	530.4	5	88.4	3.07	17.4	4.88	12.7	6.17	0.12	2.43	0.918	0.963	0.178	230.0	283.0	249.9	0.716	II
P10(T)	1151	882	5	147	2.62	30.4	4.83	12.7	6.22	0.45	1.33	0.972	1.912	0.178	239.0	258.0	246.0	0.660	II
P11(T)	1913	1326	5	221	2.54	54.1	4.9	12.7	6.1	1.08	1.76	1.020	2.965	0.178	239.0	252.0	243.9	0.494	II
P12(T)	1570	1416	5	236	2.01	43.6	4.8	12.7	6.25	2.05	0.72	1.020	4.084	0.178	249.0	266.0	255.4	0.448	II
P13(FB)	262	530.4	5	88.4	3.1	26.4	3.1	-	-	0.12	0.15	0.405	1.000	0.178	253.0	261.0	254.8	0.988	II
P14(T)	244	1062	5	177	3.05	17.5	4.85	12.7	6.15	0.49	0.29	0.356	2.990	0.178	242.0	269.0	248.3	0.764	II
P15(T)	422	1590	5	265	3.07	33.9	4.95	12.7	6.2	1.03	0.36	0.376	2.867	0.178	227.0	267.0	236.3	0.569	II
P16(T)	384	1770	5	295	2.57	30.5	4.9	12.7	6.12	1.93	0.18	0.386	3.952	0.178	244.0	273.0	250.7	0.506	II
P17(FB)	523	530.4	5	88.4	3.1	26.4	3.1	-	-	0.03	0.71	0.777	0.951	0.178	229.0	256.0	235.2	0.822	II
P18(T)	488	1062	5	177	3.05	17.5	4.85	12.7	6.15	0.20	0.39	0.690	1.936	0.178	229.0	246.0	232.9	0.656	II
P19(T)	843	1590	5	265	3.07	33.9	4.95	12.7	6.2	0.65	0.42	0.781	3.027	0.178	253.0	266.0	256.0	0.563	II
P20(T)	767	1770	5	295	2.57	30.5	4.9	12.7	6.12	0.83	0.31	0.782	4.088	0.178	261.0	247.0	257.8	0.455	II
P21(FB)	785	530.4	5	88.4	3.1	26.4	3.1	-	-	0.12	1.12	1.223	1.010	0.178	258.0	262.0	258.9	0.696	II
P22(T)	732	1062	5	177	3.05	17.5	4.85	12.7	6.15	0.48	0.55	1.065	0.990	0.178	242.0	262.0	246.6	0.515	II
P23(T)	1265	1590	5	265	3.07	33.9	4.95	12.7	6.2	1.08	0.64	1.154	2.972	0.178	244.0	262.0	248.2	0.491	II
P24(T)	1151	1770	5	295	2.57	30.5	4.9	12.7	6.12	1.89	0.45	1.145	3.912	0.178	239.0	267.0	245.5	0.384	II
F1(FB)	348	1374	5	229	2.54	38.1	9.53	-	-	1.08	-	0.307	2.961	0.178	222.0	238.0	228.1	0.566	II
F2(FB)	653	1374	5	229	2.54	38.1	9.53	-	-	1.14	5.28	0.591	2.994	0.178	227.0	262.0	240.5	0.577	II
F3(FB)	958	1374	5	229	2.54	38.1	9.53	-	-	1.02	3.58	0.822	2.775	0.178	195.0	250.0	216.1	0.459	II
F4(FB)	1262	1374	5	229	2.54	38.1	9.53	-	-	0.93	-	1.030	2.725	0.178	188.0	208.0	195.7	0.339	II
FL(FB)	577	816	5	136	4.93	63.5	3.02	-	-	0.07	0.43	0.434	1.089	0.178	321.0	321.0	321.0	0.779	IV
FL1S(FB)	577	272	1	136	4.93	63.5	3.02	-	-	0.14	-	0.434	1.089	0.178	321.0	321.0	321.0	0.752	IV
FL2(FB)	577	816	5	136	4.93	63.5	3.02	-	-	0.11	0.35	0.376	0.956	0.178	247.0	219.0	240.8	0.787	IV
FL2S(FB)	577	272	1	136	4.93	63.5	3.02	-	-	0.11	-	0.376	0.956	0.178	247.0	219.0	240.8	0.723	IV
T1(T)	1224	1218	5	203	1.98	35.0	4.95	13.0	6.35	0.59	0.46	0.823	3.115	0.178	190.0	208.0	197.0	0.390	II
T2(T)	874	1014	5	169	1.98	25.4	4.95	13.3	6.35	0.27	0.09	0.797	2.580	0.178	188.0	278.0	222.7	0.352	II
T3(T)	986	1212	5	202	1.91	34.8	4.95	13.3	6.35	0.39	0.33	0.638	3.162	0.178	184.0	184.0	184.0	0.416	II
T4(T)	704	996	5	166	2.08	25.4	4.95	13.2	6.35	0.25	0.47	0.656	2.463	0.178	196.0	287.0	230.4	0.403	II
T5(T)	1019	954	5	159	2.41	35.6	5.08	13.3	6.35	0.01	0.95	0.714	2.062	0.178	201.0	267.0	228.0	0.619	II
T6(T)	1110	1278	5	213	2.54	50.8	4.45	12.7	6.23	1.01	-	0.626	2.905	0.178	247.0	247.0	247.0	0.610	II
T7(T)	775	942	5	157	2.41	35.6	4.95	13.3	6.25	0.02	0.79	0.573	2.257	0.178	247.0	262.0	253.1	0.558	II
T8(T)	546	696	5	116	3.09	25.4	4.95	13.23	6.25	0.21	-	0.532	1.308	0.178	250.0	267.0	256.3	0.744	II
T9(T)	673	1038	5	173	3.07	44.5	4.90	12.7	6.25	0.05	0.75	0.446	1.999	0.178	259.0	293.0	271.2	0.634	II
T10(T)	376	690	5	115	3.10	25.4	4.95	12.7	6.25	0.23	-	0.391	1.397	0.178	292.0	279.0	287.3	0.879	II
T11(T)	409	492	5	82	4.32	25.4	4.95	12.7	6.25	0.06	-	0.411	0.701	0.178	281.0	283.0	281.7	0.820	II

Note: * Assumed value, others measured value

Table 3 Experimental results of ultimate compressive strength for stiffened panels obtained by Niho (1978)

Specimen No.	a (mm)	B (mm)	n_{st}	b (mm)	t_p (mm)	h_w (mm)	t_w (mm)	b_f (mm)	t_f (mm)	w_{op}/t_p*	w_{os}/a ($\times 10^{-3}$)	λ	β	σ_{rc}/σ_{op}	σ_{op} (MPa)	σ_{os} (MPa)	σ_o (MPa)	σ_u/σ_o	Failure Mode
F20A(FB)	600	600	3	150	3.27	20	3.27	-	-	0.31	6.59	1.711	1.779	0.475*	309.7	309.7	309.7	0.550	III
F20R(FB)	600	600	3	150	3.27	20	3.27	-	-	0.31	9.86	1.737	1.807	0.460	319.5	319.5	319.5	0.386	III
F30A(1)(FB)	600	600	3	150	3.27	30	3.27	-	-	0.31	0.95	1.031	1.779	0.475*	309.7	309.7	309.7	0.644	II
F30R(1)(FB)	600	600	3	150	3.27	30	3.27	-	-	0.31	1.65	1.047	1.807	0.460	319.5	319.5	319.5	0.645	II
F30R(2)(FB)	600	600	3	150	3.27	30	3.27	-	-	0.31	2.74	1.047	1.807	0.460	319.5	319.5	319.5	0.595	II
F45R(FB)	600	600	3	150	3.27	45	3.27	-	-	0.31	1.09	0.629	1.807	0.460	319.5	319.5	319.5	0.633	II
T45R(T)	600	600	3	150	3.27	25	3.27	20	3.27	0.31	0.70	1.317	1.807	0.460	319.5	319.5	319.5	0.612	II

Note: * Assumed value, others measured value

Table 4 Experimental results of ultimate compressive strength for stiffened panels obtained by Yao (1980)

Specimen No.	a (mm)	B (mm)	n_{st}	b (mm)	t_p (mm)	h_w (mm)	t_w (mm)	b_f (mm)	t_f (mm)	w_{op}/t_p	w_{os}/a ($\times 10^{-3}$)	λ	β	σ_{rc}/σ_{op}*	σ_{op} (MPa)	σ_{os} (MPa)	σ_o (MPa)	σ_u/σ_o	Failure Mode
STY-1(FB)	500	500	1	250	3.13	18.50	3.13	-	-	0.07	0.44	2.021	3.244	0.330	339.4	339.4	339.4	0.305	III
STY-2(FB)	500	500	1	250	3.13	20.45	3.13	-	-	0.09	0.56	1.785	3.244	0.330	339.4	339.4	339.4	0.337	III
STY-3(FB)	500	500	1	250	3.13	22.40	3.13	-	-	0.06	0.38	1.592	3.244	0.330	339.4	339.4	339.4	0.336	III
STY-4(FB)	500	500	1	250	3.13	23.50	3.13	-	-	0.05	0.31	1.498	3.244	0.330	339.4	339.4	339.4	0.370	III
STY-5(FB)	500	500	1	250	3.13	25.0	3.13	-	-	0.10	0.63	1.383	3.244	0.330	339.4	339.4	339.4	0.382	III
STY-6(FB)	500	500	1	250	3.13	27.80	3.13	-	-	0.02	0.13	1.206	3.244	0.330	339.4	339.4	339.4	0.446	II
STY-7(FB)	500	500	1	250	3.13	38.34	3.13	-	-	0.15	0.94	0.792	3.244	0.330	339.4	339.4	339.4	0.449	II

Note: * Assumed value, others measured value

plate. The magnitude of welding-induced residual stress was not measured, but fairly large levels of initial imperfections were measured as indicated in Table 2.

Niho (1978) performed a series of collapse tests on longitudinally stiffened plates, summarized in Table 3. Three stiffeners of the flat-bar type or T-type were welded. Moderately large amounts of initial imperfections were observed.

Yao (1980) performed collapse tests for a total of 7 stiffened panels for which the geometric dimensions of the stiffener were relatively small, see Table 4. One flat-bar type of stiffener was welded in each case, and the magnitudes of initial deflection of the plate part w_{op}/t_p can be characterized as of an average level, i.e., 5% to 15%, while significantly large levels of welding induced residual stresses were predicted or assumed.

To investigate the tripping behavior of stiffeners, Tanaka & Endo (1988) carried out collapse tests for a total of 10 stiffened panels, see Table 5. The test models had two transverse frames (i.e. three spans) and three longitudinal flat-bar stiffeners each. The test specimen had small levels of initial imperfections. The buckling mode initial deflection normalized by the plate thickness for parent plate was in the range of 1.6% to 11.9% and the compressive residual stress normalized by the yield strength was in the range of 7.3% to 13.4%. As indicated in Table 5, all test models showed tripping failure of stiffener as was the intention.

3. PRESENT TESTS

Experimental data described in the previous section are quite useful as the starting basis for deriving the empirical formulation, but more data were needed in order to develop an expression applicable in a wider range, particularly for stiffened panels with small geometric dimensions of stiffeners. In this context, some additional collapse tests for longitudinally stiffened plates designed to have small geometric dimensions of stiffeners, i.e., with a large slenderness ratio for the plate-stiffener combination considered as a column, were undertaken in the present study.

3.1 Test Specimen and Procedures

Figure 2 shows schematic views of the test setup. Tests were conducted using the $1000\,kN$ loading machine at Department of Naval Architecture & Ocean Engineering, Pusan National University. Test specimen and the supporting jig are represented in Fig. 3. The rigid frame was attached to loaded edges to compress the test specimens uniformly. Because of the supporting jigs, the unloaded edges could

Table 5 Experimental results of ultimate compressive strength for stiffened panels obtained by Tanaka & Endo (1988)

Specimen No.	a (mm)	B (mm)	n_{st}	b (mm)	t_p (mm)	h_w (mm)	t_w (mm)	b_f (mm)	t_f (mm)	w_{op}/t_p	w_{os}/a ($\times 10^{-3}$)	λ	β	σ_{rc}/σ_{op}	σ_{op} (MPa)	σ_{os} (MPa)	σ_o (MPa)	σ_u/σ_o	Failure Mode
D0(F0B)	1080	1440	3	360	6.15	110	9.77	-	-	0.016	-	0.367	1.975	0.088	234.2	287.1	251.5	0.931	IV
D0A(FB)	1080	1440	3	360	5.65	110	10.15	-	-	0.044	-	0.343	2.220	0.088*	249.9	196.0	230.8	0.843	IV
D1(FB)	1080	1200	3	300	5.95	110	10.19	-	-	0.024	-	0.350	1.771	0.102	253.8	250.9	252.7	1.095	IV
D2(FB)	1080	1560	3	390	5.95	110	10.19	-	-	0.048	-	0.369	2.302	0.073	253.8	250.9	252.9	0.900	IV
D3(FB)	1080	1440	3	360	5.95	103.5	11.84	-	-	0.524	-	0.397	2.125	0.091	253.8	326.3	280.2	1.032	IV
D4(FB)	1080	1440	3	360	5.95	118.5	7.98	-	-	0.020	-	0.357	2.125	0.134	253.8	284.2	263.1	0.990	IV
D4A(FB)	1080	1440	3	360	5.65	118.5	8.08	-	-	0.067	-	0.349	2.220	0.134*	249.9	274.4	257.7	0.875	IV
D10(FB)	1080	1200	3	300	4.38	65	4.38	-	-	0.118	-	1.028	3.174	0.084	442.0	442.0	442.0	0.547	IV
D11(FB)	1080	1200	3	300	4.38	90	4.38	-	-	0.115	-	0.678	3.174	0.057	442.0	442.0	442.0	0.527	IV
D12(FB)	1080	1440	3	360	4.38	65	4.38	-	-	0.119	-	1.097	3.089	0.085	442.0	442.0	442.0	0.510	IV

Note: 1) * Assumed value, others measured value 2) w_{op}/t_p = buckling mode of initial deflection

remain straight, retaining a simply supported condition.

One flat-bar stiffener was welded to the parent plate in each case. The stiffener ends were connected to the rigid frame at the loaded edges such that the stiffener would also carry the applied axial compressive loading.

A total of 12 stiffened plate specimen of varying slenderness ratio of plate and stiffener, were tested up to collapse, see Table 6. The measured initial deflection of the plate part w_{op}/t_p was in the range of 10% to 53%. The magnitude of welding-induced residual stress was not measured, but fairly large levels of residual stresses are possible because of the thin plate thickness of the test models. The cross head of loading machine was quasi-statically pushed into the cross section of the test models. The loading speed was controlled to be less than $1mm/min$.

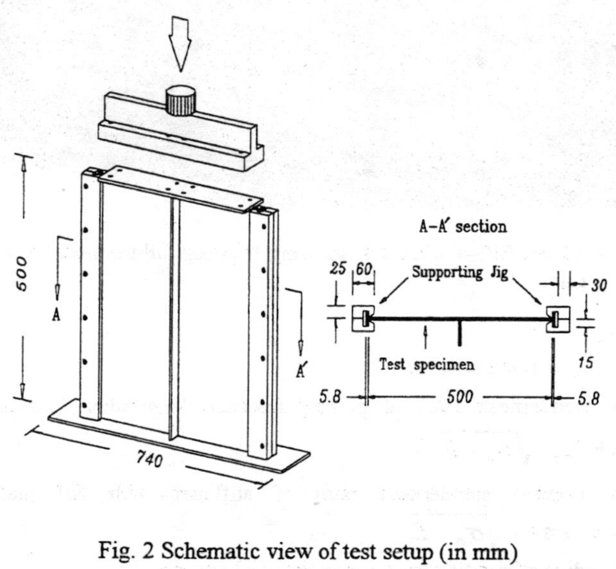

Fig. 2 Schematic view of test setup (in mm)

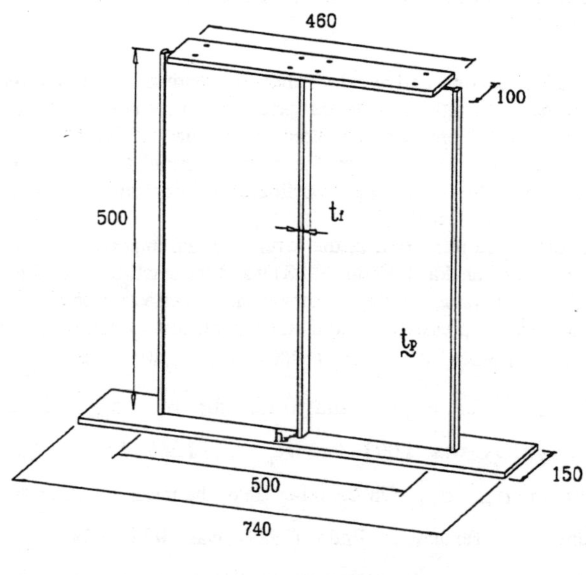

Fig. 3 Details of test specimen (in mm)

3.2 Analysis of New Test Results

During the test, two of the specimen, namely SP3-3 and SP4-2 fractured at one end of stiffener, and as a result, collapsed quite earlier than the other models. Therefore, the results of these two particular models were not used for deriving the present empirical formulation.

Table 6 indicates the collapse modes observed for each model. Most models showed a stiffener or plate induced collapse mode after local buckling between stiffeners, without rotation of stiffener, while model SP3-4 showed tripping failure of stiffener, see Fig. 4.

Table 6 Experimental results of ultimate compressive strength for stiffened panels obtained by the present experiment (1996)

Specimen No.	a (mm)	B (mm)	n_{st}	b (mm)	t_p (mm)	h_w (mm)	t_w (mm)	b_f (mm)	t_f (mm)	w_{op}/t_p	w_{os}/a ($\times 10^{-3}$)	λ	β	σ_{rc}/σ_{op}*	σ_{op} (MPa)	σ_{os} (MPa)	σ_o (MPa)	σ_u/σ_o	Failure Mode
SP3-1(FB)	500	500	1	250	3.5	17.5	3.5	-	-	0.47	3.12	1.987	2.729	0.330	300.5	300.5	300.5	0.399	III
SP3-2(FB)	500	500	1	250	3.5	19.5	3.5	-	-	0.44	2.94	1.747	2.729	0.330	300.5	300.5	300.5	0.447	III
SP3-4(FB)	500	500	1	250	3.5	37.0	3.5	-	-	0.53	1.49	0.775	2.729	0.330	300.5	300.5	300.5	0.522	IV
SP4-1(FB)	500	500	1	250	4.3	15.0	4.3	-	-	0.19	0.68	1.934	1.933	0.330	227.6	227.6	227.6	0.442	III
SP4-3(FB)	500	500	1	250	4.3	19.5	4.3	-	-	0.24	1.64	1.455	1.933	0.330	227.6	227.6	227.6	0.497	III
SP4-4(FB)	500	500	1	250	4.3	33.0	4.3	-	-	0.32	1.70	0.766	1.933	0.330	227.6	227.6	227.6	0.593	II
SP6-1(FB)	500	500	1	250	5.8	14.0	5.8	-	-	0.11	0.98	2.012	1.583	0.330	277.7	277.7	277.7	0.498	III
SP6-2(FB)	500	500	1	250	5.8	16.5	5.8	-	-	0.29	3.04	1.743	1.583	0.330	277.7	277.7	277.7	0.498	III
SP6-3(FB)	500	500	1	250	5.8	19.5	5.8	-	-	0.33	1.96	1.479	1.583	0.330	277.7	277.7	277.7	0.533	III
SP6-4(FB)	500	500	1	250	5.8	34.0	5.8	-	-	0.13	1.44	0.783	1.583	0.330	277.7	277.7	277.7	0.646	III

Note: * Assumed value, others measured value

4. DERIVATION OF THE EMPIRICAL STRENGTH EXPRESSION

Initial deflection and welding induced residual stress will reduce the ultimate strength of stiffened panels, and thus need to be considered in the development of any strength formulation. For practical purposes, however, we chose in the present formulation to implicitly rather than explicitly include the influence of initial imperfections (at a moderately large level).

As discussed in section 2, the level of initial imperfections for test specimens of Tanaka & Endo (1988) was very small compared to others. The mean value for the maximum initial deflection and the welding induced compressive residual stress of all test specimen except for those of Tanaka & Endo (1988) was 0.405 for w_{op}/t_p, 1.005×10^{-3} for w_{os}/a and 0.263 for σ_{rc}/σ_{op}. The corresponding COV was 1.091 for w_{op}/t_p, 1.369 for w_{os}/a and 0.490 for σ_{rc}/σ_{op}. On the other hand, the mean value for the test specimen of Tanaka & Endo (1988) was 0.110 for w_{op}/t_p (buckling mode initial deflection) and 0.094 for σ_{rc}/σ_{op} and the corresponding COV was 1.373 for w_{op}/t_p with 0.255 for σ_{rc}/σ_{op}. We thus excluded the test data of Tanaka & Endo (1988) in deriving the present formulation such that the resulting expression would implicitly include the influence of initial imperfections at a moderately large level.

As previously noted, Lin (1985) suggested a useful expression for predicting the ultimate compressive strength of stiffened panels, which is expressed in terms of the plate slenderness ratio and the column (stiffener) slenderness ratio as follows:

$$\frac{\sigma_u}{\sigma_o}=\frac{1}{\sqrt{c_1+c_2\lambda^2+c_3\beta^2+c_4\lambda^2\beta^2+c_5\lambda^4}} \quad (1)$$

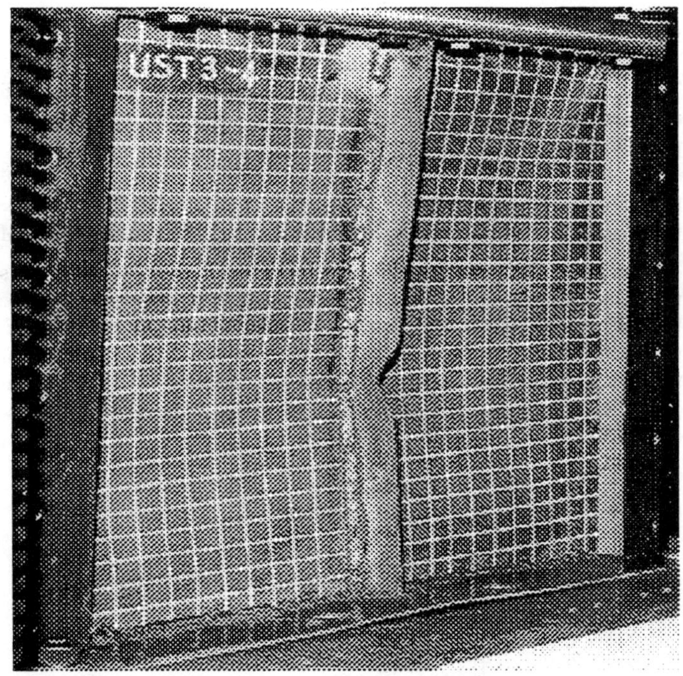

Fig. 4 Model SP3-4 after test showing tripping failure mode of stiffener

where

$c_1 \sim c_5 =$ test constants

$\beta =$ slenderness ratio of plating between longitudinal stiffeners $= b/t_p \cdot \sqrt{\sigma_o/E}$

$\lambda =$ column slenderness ratio of stiffener with full plating $= a/(\pi r)\cdot\sqrt{\sigma_o/E}$

$r =$ radius of gyration of stiffener with full plating $= \sqrt{I/A}$

σ_o = equivalent (mean) yield strength, see Nomenclature

I = moment of inertia of stiffener with full plating

$$= \frac{b\,t_p^{\,3}}{12} + b\,t_p\left(z_o - \frac{t_p}{2}\right)^2 + \frac{h_w^{\,3}\,t_w}{12}$$

$$+ h_w\,t_w\left(z_o - t_p - \frac{h_w}{2}\right)^2 + \frac{b_f\,t_f^{\,3}}{12}$$

$$+ b_f\,t_f\left(z_o - t_p - h_w - \frac{t_f}{2}\right)^2$$

z_o = neutral axis height

$$= \frac{\dfrac{b\,t_p^{\,2}}{2} + h_w\,t_w\left(t_p + \dfrac{h_w}{2}\right) + b_f t_f\left(t_p + h_w + \dfrac{t_f}{2}\right)}{A}$$

A = area of stiffener with full plating $= b\,t_p + A_{st}$

A_{st} = area of stiffener alone $= h_w\,t_w + b_f\,t_f$

Lin determined the value of test constants based on existing experimental (mainly Faulkner's data) and his own finite element analyses as:

$$c_1 = 0.960,\ c_2 = 0.765,\ c_3 = 0.176,\ c_4 = 0.131,\ c_5 = 0.146 \qquad (2.a)$$

or,

$$\frac{\sigma_u}{\sigma_o} = \frac{1}{\sqrt{0.960 + 0.765\lambda^2 + 0.176\beta^2 + 0.131\lambda^2\beta^2 + 1.046\lambda^4}} \qquad (2.b)$$

Equation (2.b) gives reasonably accurate predictions for stiffened panels with relatively large dimensions of stiffeners, but it can sometimes significantly underestimate the ultimate compressive strength of stiffened panels with small dimensions of stiffeners (as illustrated later in the paper).In our own present reanalysis of the existing and new supplementary collapse test data, we thus determined the following improved set of test constants by using the least squares method:

$$c_1 = 0.995,\ c_2 = 0.936,\ c_3 = 0.176,\ c_4 = 0.188,\ c_5 = -0.067 \qquad (3)$$

Substitution of equation (3) into equation (1) yields a new uniaxially stiffened plate ultimate strength formula given by

$$\frac{\sigma_u}{\sigma_o} = \frac{1}{\sqrt{0.995 + 0.936\lambda^2 + 0.170\beta^2 + 0.188\lambda^2\beta^2 - 0.067\lambda^4}} \qquad (4)$$

5. VERIFICATION AND DISCUSSION

The relative accuracies of Lin's formula (equation (2.b)) and the present formula (equation (4)) are now studied by comparing the approximate solutions with the experimental test data. Figures 5 to 9 show the relevant comparisons of the empirical strength expressions against experimental results.

It is seen that the approximate solutions obtained by Lin's formula and the present expression agree well with experimental results for stiffened panel models with large dimensions of stiffeners, i.e., small column slenderness ratio of stiffener.

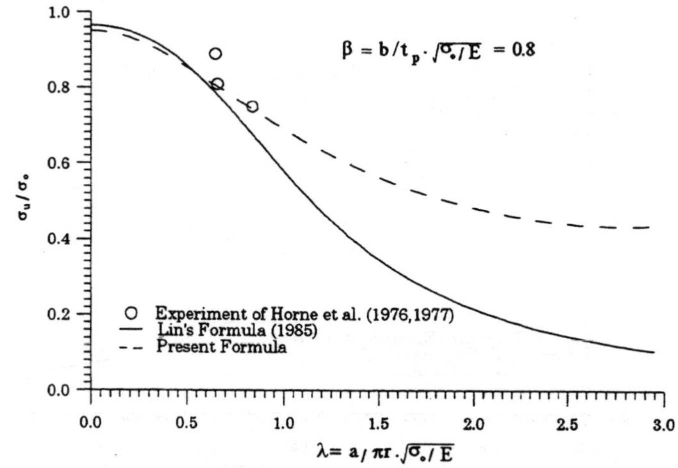

Fig. 5.a Comparison of ultimate strengths obtained by empirical formulas and experiment of Horne et al. For stiffened plates with β=0.8

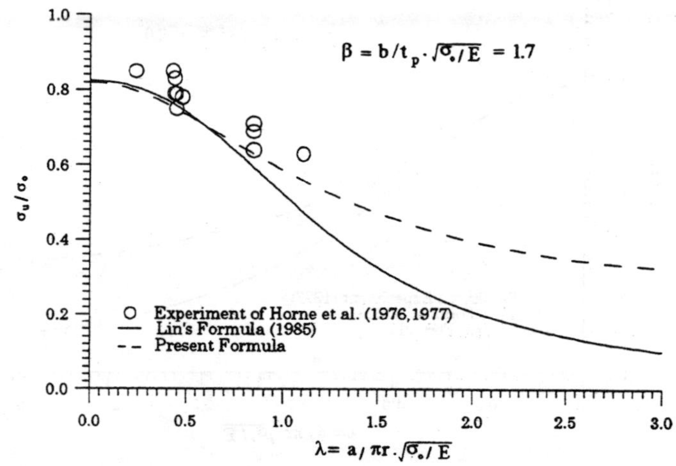

Fig. 5.b Comparison of ultimate strengths obtained by empirical formulas and experiment of Horne et al. For stiffened plates with β= 1.7

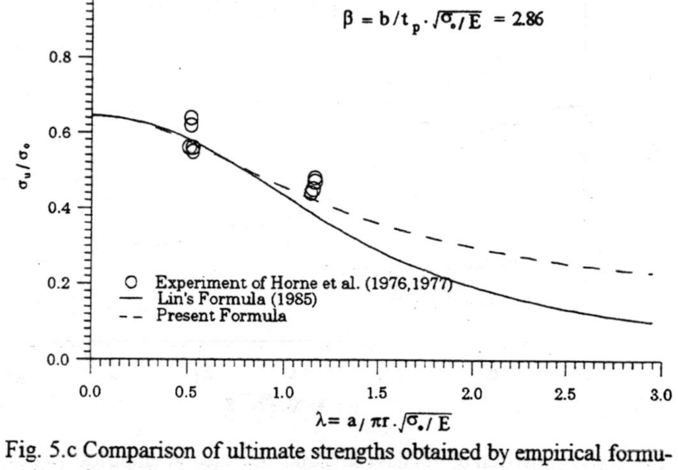

Fig. 5.c Comparison of ultimate strengths obtained by empirical formulas and experiment of Horne et al. For stiffened plates with β = 2.86

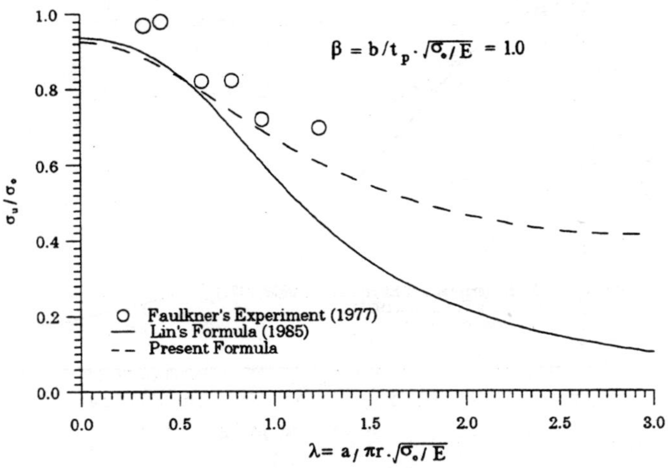

Fig. 6.a Comparison of ultimate strengths obtained by empirical formulas and experiment for stiffened plates with $\beta = 1.0$

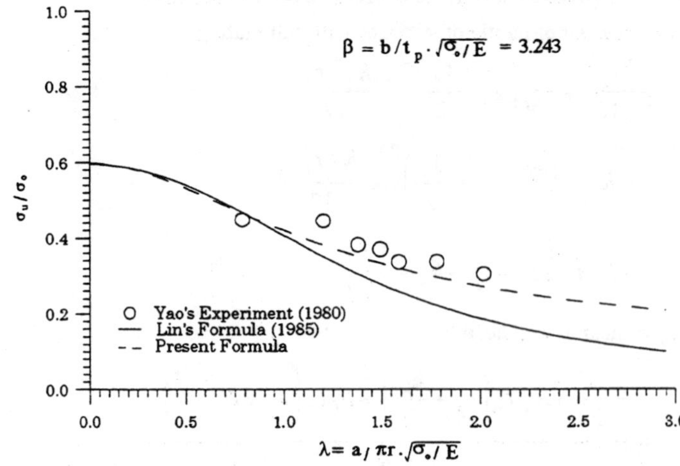

Fig. 7 Comparison of ultimate strengths obtained by empirical formulas Yao's experiment for stiffened plates with $\beta = 3.243$

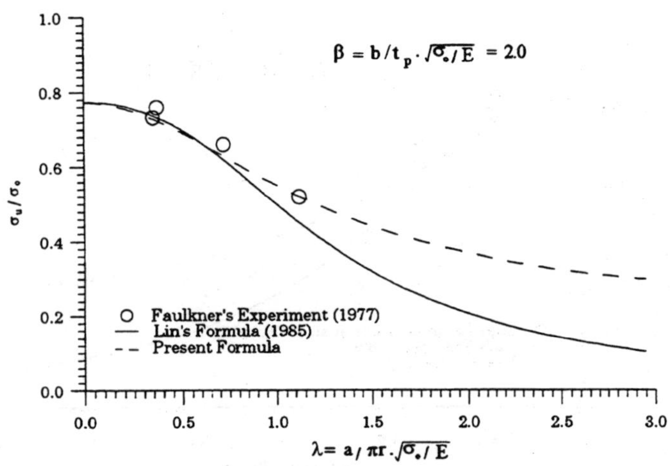

Fig. 6.b Comparison of ultimate strengths obtained by empirical formulas and experiment for stiffened plates with $\beta = 2.0$

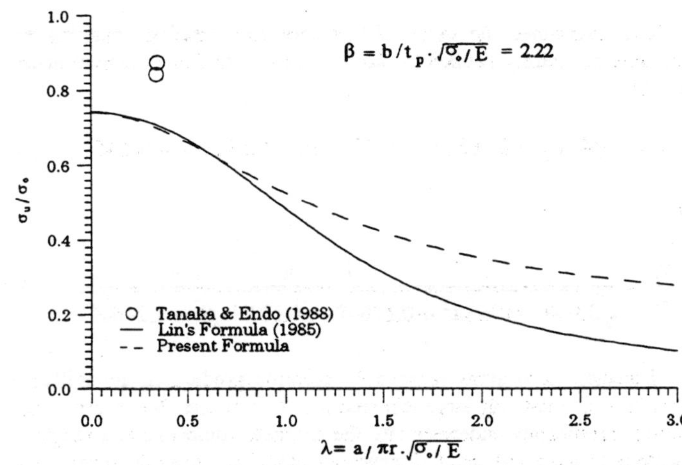

Fig. 8.a Comparison of ultimate strength obtained by empirical formulas and experiment of Tanaka & Endo for stiffened plates with $\beta = 2.22$

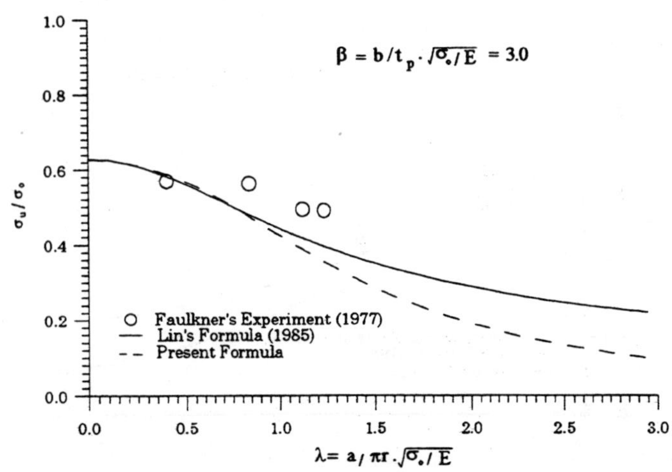

Fig. 6.c Comparison of ultimate strengths obtained by empirical formulas and experiment for stiffened plates with $\beta = 3.0$

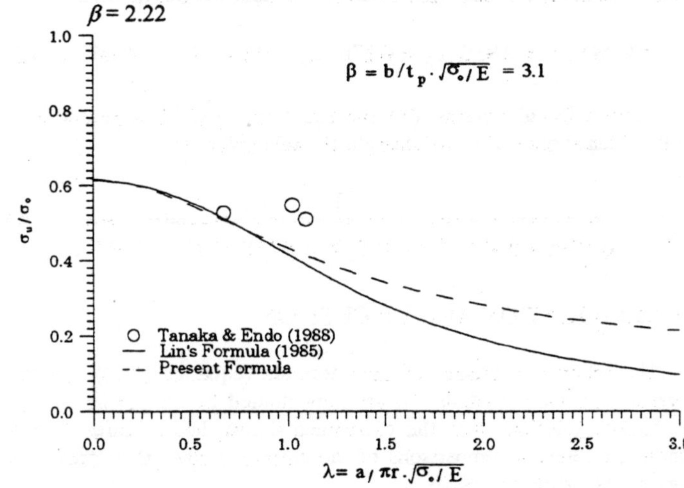

Fig. 8.b Comparison of ultimate strength obtained by empirical formulas and experiment of Tanaka & Endo for stiffened plates with $\beta = 3.1$

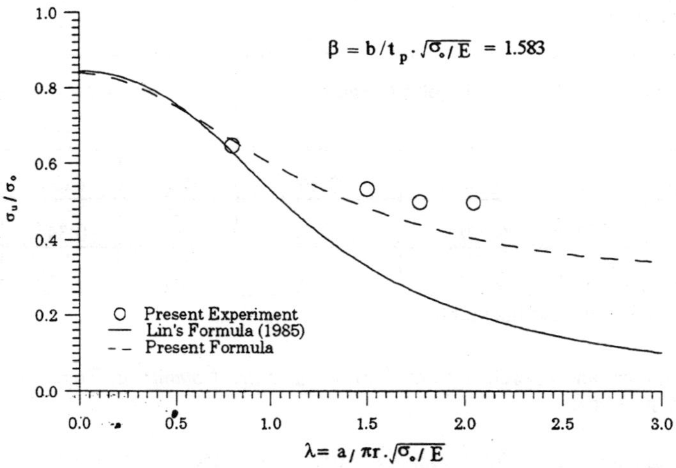

Fig. 9.a Comparison of ultimate strength obtained by empirical formulas and experiment of Tanaka & Endo for stiffened plates with $\beta = 1.583$

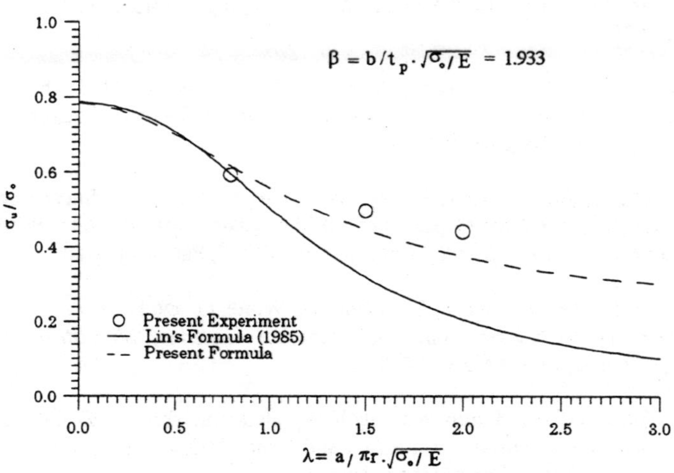

Fig. 9.b Comparison of ultimate strength obtained by empirical formulas and experiment of Tanaka & Endo for stiffened plates with $\beta = 1.933$

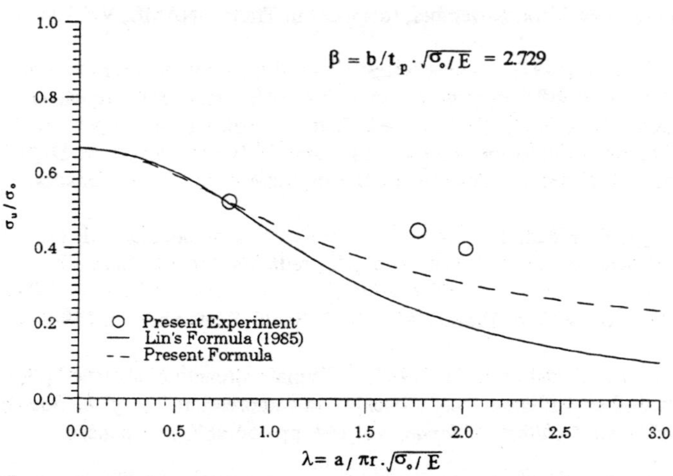

Fig. 9.c Comparison of ultimate strength obtained by empirical formulas and experiment of Tanaka & Endo for stiffened plates with $\beta = 2.729$

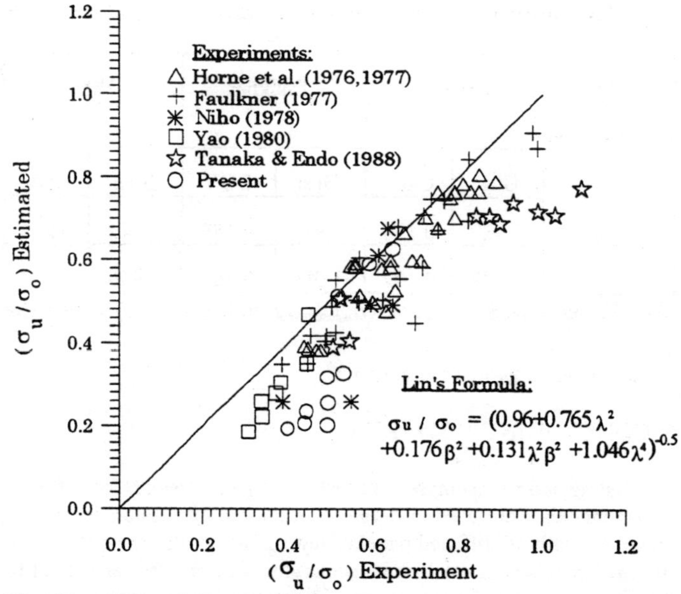

Fig. 10.a Comparison of ultimate strengths obtained by experiments and Lin's empirical formula

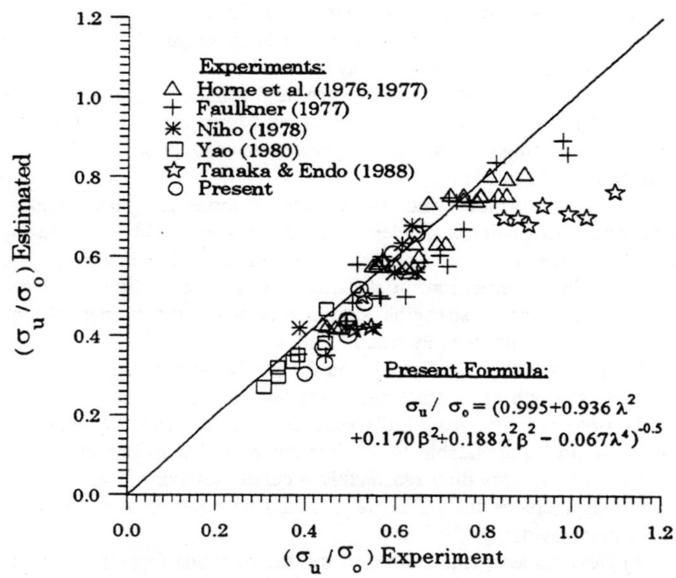

Fig. 10.b Comparison of ultimate strengths obtained by experiments and present empirical formula

However, for stiffened panel models with small dimensions of stiffeners, i.e., large column slenderness ratio of stiffener, which were obtained by Yao (1980) and the present study, Lin's formula tends to measurably underestimate the ultimate strength, while the present formulation agrees better with the experimental results as it should, see Figs. 7 and 9.

Table 7 indicates the results of the error analysis between experimental data and the approximate solutions. The test data by Tanaka & Endo (1988) which were obtained for test specimen with very small initial imperfections were not used in deriving the present formulation and thus a larger difference between approximate solutions and test results is given in that case. Comparisons of all test data with moderately large level of initial imperfection, against the Lin formula and the present expression are shown in Fig.10.a and 10.b, respectively. It is seen that the present formula gives reasonably accurate solutions for the

Table 7 Summary of comparison results between experiments and empirical formulas

Exp. / Formula	Horne et al.		Faulkner		Niho		Yao		Tanaka & Endo		Present	
	Bias	COV	Bias	COV	Bias	COV	Bias	COV	Bias	COV	Bias	COV
Lin	0.939	0.134	0.993	0.183	0.749	0.244	0.774	0.178	0.778	0.101	0.679	0.345
Present	0.979	0.132	1.011	0.162	0.956	0.123	0.921	0.066	0.766	0.110	0.903	0.114

Note: 1) Bias = Mean value of $(\sigma_u)_{estimated}/(\sigma_u)_{experiment}$ 2) COV = Coefficient of variation of the bias

cases of interest as was the intention.

6. CONCLUDING REMARKS

Existing and new compressive test data for the strength for stiffened panels models were used to derive an empirical expression for the ultimate strength of stiffened panels. New collapse tests for a total of 12 stiffened panels with small dimensions of stiffeners were carried out in the present study. Based on the old and new experimental data, an empirical expression for predicting the ultimate strength of longitudinally stiffened panels subject to uniaxial compression, which would be applicable in wide range of dimensions of stiffener and plating was derived. All test data used, old and new, are documented.

In conclusion, the following may be stated:

1) In the context of parametric structural reliability analyses of the hull girder global behavior of ships, closed-form expression for efficiently calculating he ultimate compressive strength of stiffened panels is desirable and needed.

2) The Lin formula which is expressed in terms of a plate slenderness ratio and a column slenderness ratio gives reasonably accurate solutions for the ultimate compressive strength of stiffened panels with relatively large dimensions of stiffeners, but it can potentially underestimate the ultimate strengths significantly when the proportions of stiffeners are comparatively small.

3) A new empirical formulation for predicting the ultimate compressive strength of longitudinally stiffened panels was hence derived in the present study, using Lin's work as a starting point. The new formula would be applicable to a wider range of dimensions of plating and stiffeners, providing reasonably accurate solutions. The formula implicitly includes the influence of initial imperfections at a moderately large level.

4) Nevertheless, it is cautioned that the resulting formula is essentially empirical, and while it may be suitable for application to the global ultimate strength behavior of ships, it is neither intended for, nor is it suitable for the compression design of local structures.

ACKNOWLEDGMENTS

The present research was undertaken with financial support from the American Bureau of Shipping which is thanked for their assistance. The views expressed in this paper are those of the authors, and not of the American Bureau of Shipping or Pusan National University.

REFERENCES

Caldwell, J.B. (1965). "Ultimate longitudinal strength," *Trans. RINA*, Vol 107, pp 411-430.

Das, P.K. and Garside, J.F. (1991). "Structural redundancy for continuous and discrete systems," *Ship Structure Committee*, Report No SSC-354.

Faulkner, D. (1977). "Compression tests on welded eccentrically stiffened plate panels," Steel Plated Structures, P.J. Dowling et al. (Eds.), Crosby Lockwood Staples, London, pp 581-617.

Horne, M.R. and Narayanan, R. (1976). "Ultimate capacity of stiffened plates used in girders," *Proc. Inst. of Civil Engrs.*Vol 61, Part 2, pp 253-280.

Lin, Y.T. (1985). "Ship longitudinal strength modeling," PhD. Thesis, Department of Naval Architecture and Ocean Engineering, University of Glasgow, Glasgow, Scotland.

Horne, M.R., Montague, P. and Narayanan, R. (1977). "Influence on strength of compression panels of stiffener section, spacing and welded connection," *Proc. Inst. Of Civil Engers.* Vol 63, Part 2, pp 1-20.

Hughes, O.F., Nikolaidis, E., Ayyub, B., White, G. and Hess, P. (1994). "Uncertainty in Strength models for marine structures," *Ship Structure Committee*, Report No SSC-375.

Niho, O. (1978). "Ultimate strength of plated structures," PhD. Thesis, Department of Naval Architecture and Ocean Engineering, University of Tokyo, Tokyo, December (in Japanese).

Paik, J.K., Thayamballi, A.K. and Che, J.S. (1996). "Ultimate strength of ship hulls under combined vertical bending, horizontal bending, and shearing forces," Presented at the *1996 SNAME Annual Meeting*, Paper No 2, New York, November, to appear in Trans. SNAME, Vol 104.

Paik, J.K. (1997). "A benchmark study of the ultimate strength formulations of stiffened panels," Joint PNU-ABS Research Project on A Benchmark Study on Stiffened Plate Strength, Final Report to the American Bureau of Shipping, by Pusan National University, Department of Naval Architecture and Ocean Engineering, Pusan, January.

Rigo, P., Moan, T., Frieze, P.A. and Chryssanthopoulos, M. (1995). "Benchmarking of ultimate strength prediction for longitudinally stiffened panels," *Proc. of the 6th Int. Symp. on Practical Design of Ships and Mobile Units (PRADS'95)*, Vol 2, Seoul, September, pp 869-882.

Tanaka, Y. and Endo, H. (1988). "Ultimate strength of stiffened plates with their stiffeners locally buckled in compression," *J. of the Society of Naval Architects of Japan*, Vol 164, pp 456-467 (in Japanese).

Yao, T. (1980). "Ultimate compressive strength of ship platings," PhD. Thesis, Department of Naval Architecture and Ocean Engineering, Osaka University, Osaka, Japan, May (in Japanese).

Proceedings of the Seventh (1997) International Offshore and Polar Engineering Conference
Honolulu, USA, May 25-30, 1997

ISBN 1-880653-28-1 (Set); ISBN 1-880653-32-X (Vol. IV)

A Posteriori Error Estimation and Improvement of Solution for Frame Structure Eigenvalue Problems

Mitsuru Kitamura, YuHua Chen and Hisashi Nobukawa
Hiroshima University
Higashi-Hiroshima, Japan

Abstract

A posteriori error estimation method for finite element solutions of natural frequencies and mode shapes of structures is proposed in this paper. The governing equations of a posteriori error for free vibration problems are deduced. The difficulty of error estimation for eigenvalue problems is that the error equations can not be solved directly. For this reason, a method used for solving the error equations is presented and a efficient solution strategy is developed. To reduce the size of matrix in error analyses, a technique of eliminating the degrees of freedom at the mid-node of element is developed. The finite element solutions of eigenvalues and eigenvectors are improved by adding the estimated error onto the original solutions.

Key Words: finite element method, *a posteriori* error analysis, eigenvalue problems, frame structure

Introduction

Finite element method is the most widely used and powerful tool for engineering structural analyses. However, it is well known that the accuracy of solutions obtained by the finite element method is heavily dependent upon the mesh subdivision used for simulating given problems. An improper finite element model for a given problem may produce a significant error in its solution. Hence, *a posteriori* error estimation has become a key feature of practical finite element analyses and has been a subject of considerable interesting for recent several decades. Considerable successes have been achieved on error estimations and adaptive finite element methods for elasto-plastic static problems and elastic dynamic problems (Zienkiewicz,1988; Ohtsubo ,1990; Ohtsubo, 1991; Ohtsubo, 1992; Wiberg, 1994; Babuska, 1994). However, the research on *a posteriori* error estimation for dynamic problems is far from completion and is very desirable today. Furthermore, to the authors' knowledge, the research on *a posteriori* error estimation for eigenvalue problems based on residual and recovery method have not been carried out.

In this paper, the method of *a posteriori* error estimation for free vibration problems is considered and the governing equation of the error is deduced. The error analysis for the eigenvalue problem of the framework in particular is under our investigation. Not only the energy norm of the error but also the error in the eigenvalue and eigenvector are estimated by the error analysis. If the error is estimated with enough accuracy, solutions for the eigenvalues and eigenvectors can be modified by adding the estimated *a posteriori* error to the original solutions.

The governing equations of error for eigenvalue problems

Without losing generality, we consider the longitudinal free vibration of a rod whose governing equation is given as

$$AE\frac{\partial^2 u(x,t)}{\partial x^2} = A\rho\frac{\partial^2 u(x,t)}{\partial t^2} \qquad \text{in } \Omega \tag{1}$$

with the following boundary conditions.

$$u = \bar{u} \qquad \text{on } \Gamma_u \tag{2}$$

$$\sigma = \bar{\sigma} \qquad \text{on } \Gamma_\sigma \tag{3}$$

Here $u(x,t)$ is the longitudinal displacement, E is a Young's elastic modulus, A is a cross section area and ρ is a mass density. The boundary of the rod is classified as Γ_u or Γ_σ where prescribed displacement $\bar{u}$ and surface traction force $\bar{\sigma}$ are given, respectively. Applying the principle of virtual work, the following weak formulation can be introduced.

$$\int_\Omega vA\rho\frac{\partial^2 u}{\partial t^2}d\Omega + \int_\Omega AE\frac{\partial v}{\partial x}\frac{\partial u}{\partial x}\,d\Omega - AE\left[v\frac{\partial u}{\partial x}\right]_{\Gamma_\sigma} = 0 \tag{4}$$

where v is virtual displacement. The finite element equations for free vibration problems can be derived from Eq.(4) and eigenvalues as well as eigenvectors are obtained by solving the finite element eigen-equations.

In order to obtain the governing equation for *a posteriori* error, it is assumed that the exact solution, $u(x,t)$, and the finite element solution, $\hat{u}(x,t)$, are expressed by the following equations, respectively.

$$u(x,t) = U(x)\sin\omega t \tag{5}$$
$$\hat{u}(x,t) = \hat{U}(x)\sin\hat{\omega} t \tag{6}$$

Here $U(x)$ and ω, and $\hat{U}(x)$ and $\hat{\omega}$ are the exact solutions and the finite element solutions of amplitudes of displacement and natural frequency, respectively. Introducing a vector which contains interpolation functions, $\{\hat{N}(x)\}$, the finite element solution $\hat{U}_i(x)$ for element i can be written as follows.

$$\hat{U}_i(x) = \{\hat{N}(x)\}^T\{\hat{U}\}_i \tag{7}$$

where $\{\hat{U}\}_i$ is the vector containing nodal displacements for element i.

Since the interpolation function for displacement is not exact, the finite element solutions can not satisfy the governing equation. Hence, the residual in the equation, r_i, is given as

$$r_i = A\rho\frac{\partial^2\hat{u}(x,t)}{\partial t^2} - AE\frac{\partial^2\hat{u}(x,t)}{\partial x^2} \quad \text{in } \Omega_i \tag{8}$$

for element i. Applying the principle of virtual work into Eq.(8) and assembling them over all the elements give

$$\sum_i\int_{\Omega_i} vr_i d\Omega = \sum_i\int_{\Omega_i}\left(vA\rho\frac{\partial^2\hat{u}}{\partial t^2} + AE\frac{\partial v}{\partial x}\frac{\partial\hat{u}}{\partial x}\right)_i d\Omega - AE\left[v\frac{\partial\hat{u}}{\partial x}\right]_{\Gamma_\sigma} - \sum_{i,j} v\lambda_{ij} \tag{9}$$

where

$$\lambda_{ij} = \left(AE\frac{\partial\hat{u}(x,t)}{\partial x}\right)_i - \left(AE\frac{\partial\hat{u}(x,t)}{\partial x}\right)_j \quad \text{on } \Gamma_{ij} \tag{10}$$

Here λ_{ij} is the traction jump along the boundary Γ_{ij} between elements i and j. Substituting Eq's.(5) and (6) into Eq's.(4) and (9) introduces

$$\left(-\int_\Omega vA\rho\omega^2 U d\Omega + \int_\Omega AE\frac{\partial v}{\partial x}\frac{\partial U}{\partial x}d\Omega - AE\left[v\frac{\partial U}{\partial x}\right]_{\Gamma_\sigma}\right)\sin\omega t = 0 \tag{11}$$

$$\left(-\int_\Omega vA\rho\hat{\omega}^2\hat{U}d\Omega + \int_\Omega AE\frac{\partial v}{\partial x}\frac{\partial\hat{U}}{\partial x}d\Omega - AE\left[v\frac{\partial\hat{U}}{\partial x}\right]_{\Gamma_\sigma} - \sum_{i,j} v\Lambda_{ij} - \int_\Omega vRd\Omega\right)\sin\hat{\omega} t = 0 \tag{12}$$

where

$$R \equiv -A\rho\hat{\omega}^2\hat{U} - AE\frac{\partial^2\hat{U}}{\partial x^2} \quad \text{in } \Omega \tag{13}$$
$$\Lambda_{ij} \equiv \left(AE\frac{\partial\hat{U}}{\partial x}\right)_i - \left(AE\frac{\partial\hat{U}}{\partial x}\right)_j \quad \text{on } \Gamma_{ij} \tag{14}$$

Since it is obvious that $\sin\omega t \neq 0$ and $\sin\hat{\omega} t \neq 0$, Eq's.(11) and (12) are equivalent to the following equations.

$$-\int_\Omega vA\rho\omega^2 U d\Omega + \int_\Omega AE\frac{\partial v}{\partial x}\frac{\partial U}{\partial x}d\Omega - AE\left[v\frac{\partial U}{\partial x}\right]_{\Gamma_\sigma} = 0 \tag{15}$$

$$-\int_\Omega vA\rho\hat{\omega}^2\hat{U}d\Omega + \int_\Omega AE\frac{\partial v}{\partial x}\frac{\partial\hat{U}}{\partial x}d\Omega - AE\left[v\frac{\partial\hat{U}}{\partial x}\right]_{\Gamma_\sigma} - \sum_{i,j} v\Lambda_{ij} - \int_\Omega vRd\Omega = 0 \tag{16}$$

Subtracting Eq.(16) from Eq.(15) results in the following weak formulation for *a posteriori* error analysis of eigenvalue problem in a global form.

$$\int_\Omega vA\rho e_a d\Omega + \int_\Omega AE\frac{\partial v}{\partial x}\frac{\partial e_u}{\partial x}d\Omega - AE\left[v\frac{\partial e_u}{\partial x}\right]_{\Gamma_\sigma} + \sum_{i,j} v\Lambda_{ij} = -\int_\Omega vRd\Omega \tag{17}$$

where e_u and e_a are the errors in the amplitudes of displacement and acceleration, respectively.

$$e_u(x) = U(x) - \hat{U}(x) \tag{18}$$
$$e_a(x) = (-\omega^2 U(x)) - (-\hat{\omega}^2\hat{U}(x)) = -\omega^2 U(x) + \hat{\omega}^2\hat{U}(x) \tag{19}$$

Errors in the amplitudes of displacement and acceleration for element i is interpolated as

$$e_u(x) = \{N(x)\}^T\{e_u\}_i \tag{20}$$
$$e_a(x) = \{N(x)\}^T\{e_a\}_i \tag{21}$$

where $\{e_u\}_i$ and $\{e_a\}_i$ are the vectors containing nodal errors for element i. Applying the finite element method with the above interpolations into Eq.(17) gives us

$$[M]\{e_a\} + [K]\{e_u\} = \{G\} - \{\Lambda\} \tag{22}$$

where $\{G\}$ is the nodal residual force vector corresponding to the residual force R and $\{\Lambda\}$ is nodal unbalanced traction vector assembled by λ_{ij} with considering the boundary conditions on Γ_σ.

The residual force $R_i(\hat{U})$ of element i is composed of the elastic force and the inertial force for the free vibration of structure and is expressed as

$$R_i(\hat{U}) = R_i^e(\hat{U}) + R_i^s(\hat{U}) \tag{23}$$

where $R_i^e(\hat{U})$ is the elastic residual force and $R_i^s(\hat{U})$ is the inertial residual force. With the expression of Eq.(23), the nodal residual force vector for element i becomes

$$\{G\}_i = -\int_{\Omega_i}\{N\}\left(R_i^e(\hat{U}) + R_i^s(\hat{U})\right)d\Omega_i = \{G^e\}_i + \{G^s\}_i \tag{24}$$

where $\{G^e\}_i$ and $\{G^s\}_i$ are element nodal residual force vectors corresponding to the elastic residual force and the inertial residual force, respectively. With this separation of $\{G\}$, Eq.(22) can be rewritten as

$$[M]\{e_a\} + [K]\{e_u\} = \{G^e\} - \{\Lambda\} + \{G^s\} \tag{25}$$

It should be pointed out that the error equations given by Eq.(25) deduced for the longitudinal free vibration of a rod is suitable for all structures. In other words, it is easily possible to apply the above theory of *a posteriori* error estimation into bending and torsional vibration problems. In this paper, we concentrated in *a posteriori* error estimation for the free vibration of framework. A 2-node (12 degrees of freedom) isoparametric element is used for the original finite element analysis and a 3-node (18 degrees of freedom) isoparametric element is used for the error analysis since the dominant term of *a posteriori* error is one order higher polynomial than that used in the original finite element analysis. In order to avoid locking, the reduced numerical integration technique is applied. Since the error equations of Eq.(25) can not be solved directly, the relation between the error acceleration and error displacement has to be found.

Energy norm of error

For free vibration problems, the total energy norm and the local energy norm of *a posteriori* error can be written as

$$||e|| = \sqrt{||e^s||^2 + ||e^k||^2} \tag{26}$$

$$||e||_i = \sqrt{||e^s||_i^2 + ||e^k||_i^2} \tag{27}$$

where

$$||e^s||^2 = \sum_i ||e^s||_i^2 \tag{28}$$

$$||e^k||^2 = \sum_i ||e^k||_i^2 \tag{29}$$

Here $||e^s||$ and $||e^k||$ are the total structural strain and kinetic energy norms of error, respectively. $||e^s||_i$ and $||e^k||_i$ are the local strain and kinetic energy norm of error for element i. $||e^s||_i$ and $||e^k||_i$ are defined as follows.

$$||e^s||_i^2 = \int_{\Omega_i} e_\sigma^T E^{-1} e_\sigma d\Omega_i \tag{30}$$

$$||e^k||_i^2 = \int_{\Omega_i} e_v^T \rho e_v d\Omega_i \tag{31}$$

where e_σ and e_v are the error in stress and the error in velocity written as

$$e_\sigma = E\frac{\partial e_u(x)}{\partial x} \tag{32}$$

$$e_v = \omega U(x) - \hat{\omega}\hat{U}(x) \tag{33}$$

Method of solving error equation

Since the consistent mass matrix is applied in the original finite element analysis, the finite element solution $\hat{\omega}$ is always larger than the exact solution ω. So it is assumed that

$$\hat{\omega} = \omega + \Delta\omega \tag{34}$$

where $\Delta\omega$ is the error of eigenvalue. Substituting Eq.'s (18) and (34) into Eq.(19) with the interpolations of Eq's (20) and (21) gives

$$\{e_a\} = -\omega^2\{e_u\} + (2\Delta\omega\cdot\omega + \Delta\omega^2)\{\hat{U}\} \tag{35}$$

Substituting Eq.(35) into Eq.(25) introduces the error governing equation with parameter β.

$$\left([K] - \omega^2[M]\right)\{e_u\} = \{G^e\} - \{\Lambda\} + \beta\{G^s\} \tag{36}$$

where

$$\beta = 1 - 2\frac{\omega}{\hat{\omega}}\cdot\frac{\Delta\omega}{\hat{\omega}} - \frac{\Delta\omega^2}{\hat{\omega}^2} \tag{37}$$

It should be emphasized that $\{e_u\}$ in the above equations is normalized to the mass matrix such that

$$\left(\{\hat{U}\} + \{e_u\}\right)^T [M] \left(\{\hat{U}\} + \{e_u\}\right) = 1.0 \tag{38}$$

However, it is difficult to determine the value of β since β is related to the exact eigenvalue ω. It is demonstrated that the same result is always obtained by solving Eq.(36) for error $\{e_u\}$ after normalization to the mass matrix without the exact β. According to the discussion in Appendix, the error equation is rewritten as

$$\left([K] - \omega^2[M]\right)\{e_u^0\} = \{G^e\} - \{\Lambda\} + \beta^0\{G^s\} \tag{39}$$

where $\{e_u^0\}$ is dependent on β^0 and is not normalized to the mass matrix. The difference of $\{e_u^0\}$ from $\{e_u\}$ caused by β^0 has effect only on the value of normalizing coefficient to mass matrix (seeing Appendix). Since the above are non-linear algebraic equations and the number of unknown is always one more than the number of equations, it is very difficult to solve Eq.(39) for $\{e_u^0\}$ and ω. For this reason, in this paper, a simple and efficient algorithm for solving this problem is presented.

A parameter α is introduced such that

$$\omega = \alpha\hat{\omega} \qquad (0 < \alpha < 1) \tag{40}$$

Substituting Eq.(40) into Eq.(39) rewrites *a posteriori* error equation with parameter α as

$$\left([K] - \alpha^2\hat{\omega}^2[M]\right)\{e_u^0\} = \{G^e\} - \{\Lambda\} + \beta^0\{G^s\} \tag{41}$$

Once α is given, the error equations of Eq.(41) become linear algebraic equations. Hence, $\{e_u^0\}$ can be computed easily and a modified eigenvector $\{\tilde{U}^0\}$ is given by the following formulation.

$$\{\tilde{U}^0\} = \{\hat{U}\} + \{e_u^0\} \tag{42}$$

This can be normalized as

$$\{\tilde{U}\} = \frac{1}{C_M}\{\tilde{U}^0\} = \frac{1}{C_M}\left(\{\hat{U}\} + \{e_u^0\}\right) \tag{43}$$

where

$$C_M = \sqrt{\{\tilde{U}^0\}^T[M]\{\tilde{U}^0\}} \tag{44}$$

A modified eigenvalue $\tilde{\omega}$ can also be given as follows by Rayleigh's quotient

$$\tilde{\omega}^2 = \frac{\{\tilde{U}\}^T[K]\{\tilde{U}\}}{\{\tilde{U}\}^T[M]\{\tilde{U}\}} = \{\tilde{U}\}^T[K]\{\tilde{U}\} \tag{45}$$

It is clear that the key of present method is how to determine the parameter α. The exact solutions of an eigenvalue problem satisfy

$$\left([K] - \omega^2[M]\right)\{U\} = \{0\} \tag{46}$$

However, for a given α which is different from the exact value, the modified results $\{\tilde{U}\}$ and $\tilde{\omega}$ do not satisfy Eq.(46) So,

$$\left([K] - \tilde{\omega}^2[M]\right)\{\tilde{U}\} \neq \{0\} \tag{47}$$

In order to reduce the unbalanced amounts in Eq.(47), an objective function F is introduce as

$$F = (A_1 - \tilde{\omega}^2 B_1)^2 + (A_2 - \tilde{\omega}^2 B_2)^2 + \ldots + (A_n - \tilde{\omega}^2 B_n)^2 \quad (48)$$

where

$$\{A\} = \left\{ \begin{array}{c} A_1 \\ A_2 \\ \cdot \\ A_n \end{array} \right\} = [K]\{\tilde{U}\} \quad (49)$$

$$\{B\} = \left\{ \begin{array}{c} B_1 \\ B_2 \\ \cdot \\ B_n \end{array} \right\} = [M]\{\tilde{U}\} \quad (50)$$

In the above, n is the number of DOF. It is clear that the proper α is determined such that the value of the objective function F becomes the minimum. In this paper, an incremental method is developed for finding a proper parameter α which is described as follows.

a) Set $\alpha^{(0)} = \tilde{\omega}_{k-1}/\hat{\omega}_k$ where $\tilde{\omega}_{k-1}$ is an error modified eigenvalue of the (k-1)-th order mode and $\hat{\omega}_k$ is the eigenvalue of the k-th order mode given by FEM, $\Delta\alpha$ be equal to a suitable given value and $m = 0$.

b) $m = m + 1$ and $\alpha^{(m)} = \alpha^{(m-1)} + \Delta\alpha$

c) Solve Eq.(41) for obtaining the error value of eigenvector $\{e_u^0\}^{(m)}$ and modify the eigenvector and eigenvalue from Eq.(42) to Eq.(45).

d) Calculate the objective function F by Eq.(48).

e) Repeat the process from step-b) to step-d) until the value of objective function F becomes minimum such that $F^{(m)} > F^{(m-1)}$.

f) $\{\tilde{U}\}^{(m-1)}$ and $\tilde{\omega}^{(m-1)}$ are a error modified eigenvector and eigenvalue of the k-th order mode.

If high accuracy is desired, additional repetitions of the above steps are performed with reduction of $\Delta\alpha$.

Reducing the size of matrix

Because, in error analysis, a one order higher polynomial function set than that used for the original finite element analysis is required, the size of the matrix in error analysis is much larger than that for the original finite element analysis. For reducing the size of matrix in Eq.(41), a technique of eliminating mid-side nodal degrees of freedom is developed. For element i, the error equation is written as

$$[\bar{K}]\{e_u^0\}_i = \{G^e\}_i - \{\Lambda\}_i + \beta^0\{G^s\}_i \quad (51)$$

where

$$[\bar{K}]_i = \left([K] - \alpha^2\hat{\omega}^2[M]\right)_i \quad (52)$$

It is supposed that $\{e_u^0\}_i = \{\{e_1^0\}^T\{e_2^0\}^T\}_i^T$ where $\{e_1^0\}_i$ and $\{e_2^0\}_i$ are the element nodal displacement vectors of the two side nodes and the mid-node, respectively. Eq.(51) is rearranged and repartioned as

$$\begin{bmatrix} [\bar{K}_{11}] & [\bar{K}_{12}] \\ [\bar{K}_{21}] & [\bar{K}_{22}] \end{bmatrix}_i \left\{ \begin{array}{c} \{e_1^0\} \\ \{e_2^0\} \end{array} \right\}_i = \left\{ \begin{array}{c} \{G_1^e\} \\ \{G_2^e\} \end{array} \right\}_i - \left\{ \begin{array}{c} \{\Lambda_1\} \\ \{\Lambda_2\} \end{array} \right\}_i + \beta^0 \left\{ \begin{array}{c} \{G_1^s\} \\ \{G_2^s\} \end{array} \right\}_i \quad (53)$$

Here $\{G_1^e\}$, $\{G_2^e\}$, $\{G_1^s\}$ and $\{G_2^s\}$ are known and $\{\Lambda_1\}$ is unknown as the exact allocation of jump force is not available. Because the mid-node of element is not connected with any other element nodes in frame problems, $\{\Lambda_2\}$ is equal to zero. Expanding the above equation and replacing $\{\Lambda_2\}$ with 0 give

$$[\bar{K}_{11}]_i\{e_1^0\}_i + [\bar{K}_{12}]_i\{e_2^0\}_i = \{G_1^e\}_i - \{\Lambda_1\}_i + \beta^0\{G_1^s\}_i \quad (54)$$

$$[\bar{K}_{21}]_i\{e_1^0\}_i + [\bar{K}_{22}]_i\{e_2^0\}_i = \{G_2^e\}_i + \beta^0\{G_2^s\}_i \quad (55)$$

Solving Eq.(55) for $\{e_2\}_i$ gives

$$\{e_2^0\}_i = [\bar{K}_{22}]_i^{-1}(\{G_2^e\}_i + \beta^0\{G_2^s\}_i) - [\bar{K}_{22}]_i^{-1}[\bar{K}_{21}]_i\{e_1^0\}_i \quad (56)$$

Substituting $\{e_2^0\}_i$ into Eq.(54) results in

$$[\hat{K}]_i\{e_1^0\}_i = \{\hat{G}_1^e\}_i - \{\Lambda_1\}_i + \beta^0\{\hat{G}_1^s\}_i \quad (57)$$

where

$$[\hat{K}]_i = [\bar{K}_{11}]_i - [\bar{K}_{12}]_i[\bar{K}_{22}]_i^{-1}[\bar{K}_{21}]_i \quad (58)$$

$$\{\hat{G}_1^e\}_i = \{G_1^e\}_i - [\bar{K}_{12}]_i[\bar{K}_{22}]_i^{-1}\{G_2^e\}_i \quad (59)$$

$$\{\hat{G}_1^s\}_i = \{G_1^s\}_i - [\bar{K}_{12}]_i[\bar{K}_{22}]_i^{-1}\{G_2^s\}_i \quad (60)$$

Because of eliminating the mid-node of an element, the total degrees of freedom in error analysis is the same as that in the original finite element analysis. Since solving only linear algebraic equations is required, the computational time of error analysis is very small compared to solving the eigenvalue problem.

Plane frame problem

A plane frame shown in Fig.1-a) is used as the calculation model of room of a deckhand and the mesh used for FEM calculation is shown in Fig.1-b). An isoparametric plane beam element with 2-nodes (6 degrees of freedom) is used for FEM calculation, while an isoparametric plane beam element with 3-nodes (9 degrees of freedom) is used for error estimation. The latter element also is used for FEM calculation to obtain referenced results signed as 'ref'. The referenced results are taken as the exact solutions in order to examine the accuracy of the error modified results which are signed as 'est'.

Table 1 shows the comparison of eigenfrequencies given by FEM, *a posteriori* error estimation and reference computation. It may be observed from Table 1 that the error modified results are almost the same as the referenced ones. Fig.2 shows the relationship between the objective function F and the parameter α for the 1-st mode, the 4-th mode and the 6-th mode. The relation of error modified eigenvalues to the parameter α is shown in Table 2. It may be observed from Table 2 that the error modified solutions are not sensitive to a parameter α. For example, the good error modified solutions can be obtained for the 6-th mode as parameter α is varied from 0.4 to 0.9 in spite of the proper value of 0.64. This phenomenon is very useful in the proposed error estimation.

In other words, an accurate error and a good error modified result are obtained even if the difference between the given parameter α and its proper value is large. Based on the above discussion, only solving the linear algebraic equations once is required with letting the parameter α be a given value. Hence, the computational time used in error estimation is very small compared with solving the eigenvalue problem.

The distributions of the strain and kinetic energy norms of error for every element are shown in Fig's. 3 and 4. It is found that the total strain energy norm of error is much larger than the total kinetic energy norm of error and the ratio of these norms is much smaller than 1.0, however; this ratio becomes larger with increasing of the mode order. In this numerical example, it is observed that the pattern of the distribution of error energy norm along elements is different for various modes. For example, the third element has the largest error energy norm at the 1-st mode, while the element has the smallest one at the 6-th mode.

The natural modes in this free vibration problem are shown in Fig.5 in which not only the original FEM solutions but also error modified solutions are plotted. The modes of referenced solution overlap with error modified solutions almost perfectly. The ratio between the estimated energy norm of error to the referenced energy norm of error is also shown in Fig.6. The ratios are almost 1.0 for all elements which means that very good agreement of $||e||_i^{est}$ with $||e||_i^{ref}$ is observed

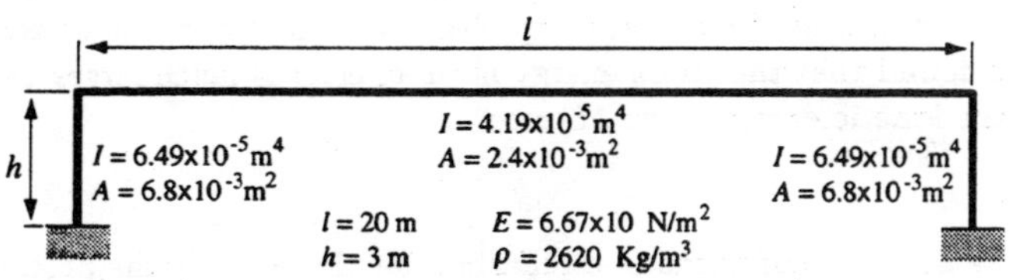

a) Strucutre model of plane frame

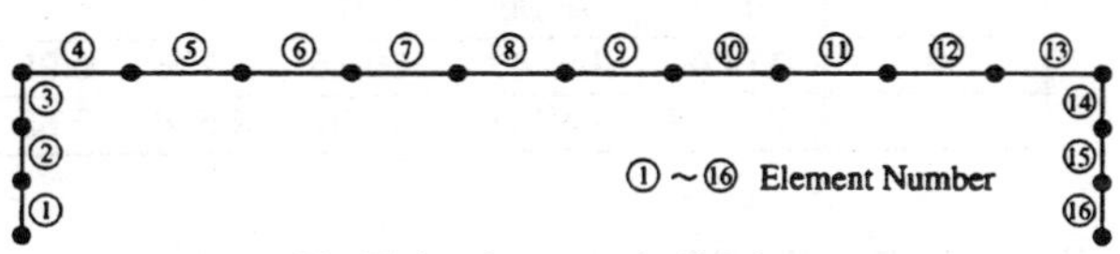

b) Finite element sub-division

Fig.1 Plane frame problem

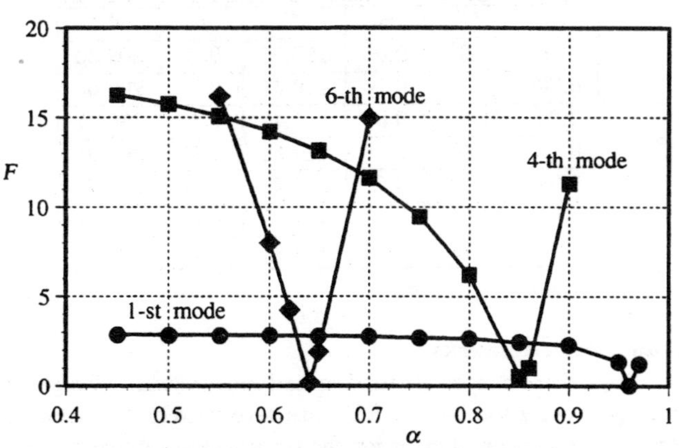

Fig.2 Relationship between F and α

Table 1 Comparison of eigenvalues for the plane frame problem

Mode #	$\hat{\omega}$	ω^{est}	ω^{ref}	F	α
1	15.87	15.35	15.35	0.003	0.967
2	24.45	23.28	23.28	4.689	0.952
3	56.97	52.77	52.77	5.444	0.927
4	131.87	112.58	112.58	0.587	0.850
5	260.72	196.80	196.80	2.132	0.755
6	475.82	304.88	304.88	0.159	0.640

Table 2 Relationship between the modified eigenvalue and parameter α for the plane frame problem

1-st mode		4-th mode		6-th mode	
α	ω^{est}	α	ω^{est}	α	ω^{est}
0.50	15.35	0.50	112.58	0.40	304.90
0.70	15.35	0.70	112.58	0.55	304.90
0.90	15.35	0.85	112.58	0.64	304.88
0.967	15.35	0.95	112.62	0.80	305.45
1.00	15.35	1.00	114.02	0.95	309.72

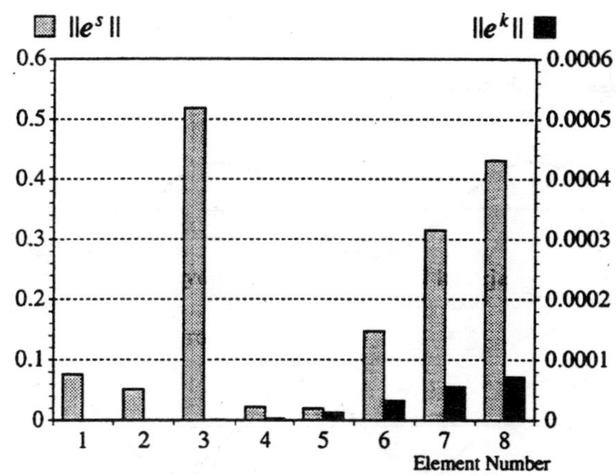

Fig.3 Strain and kinetic energy norms of estimated error for the 1-st mode

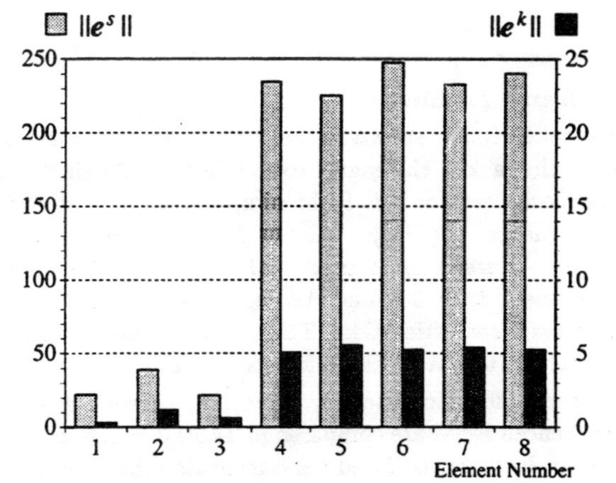

Fig.4 Strain and kinetic energy norms of estimated error for the 6-th mode

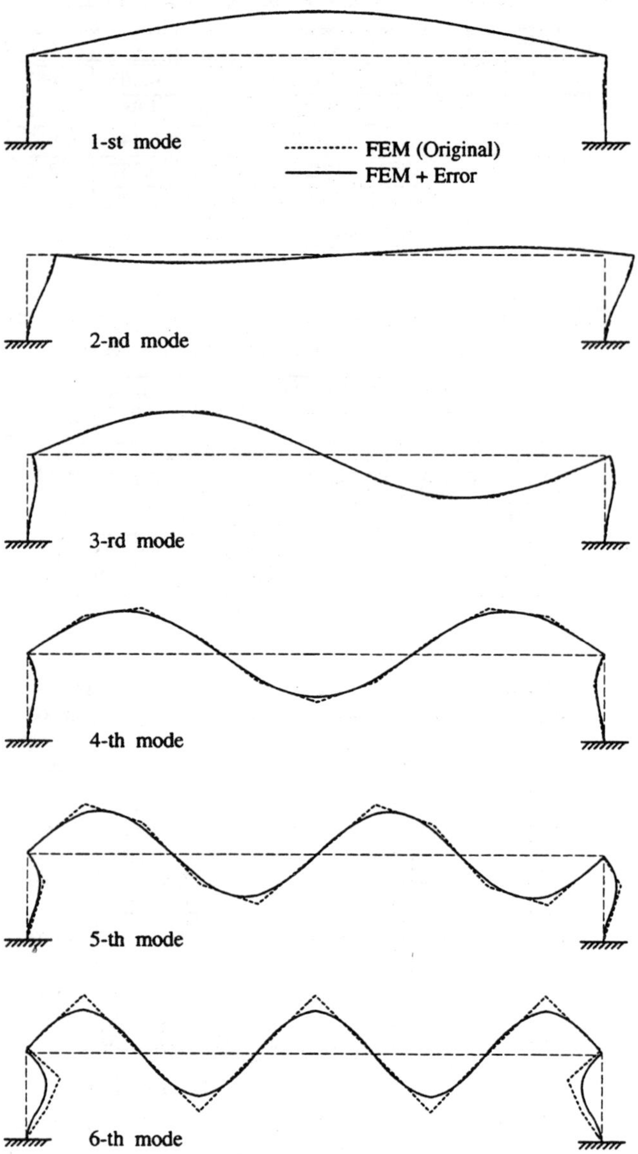

Fig.5 Natural modes of plane frame problem

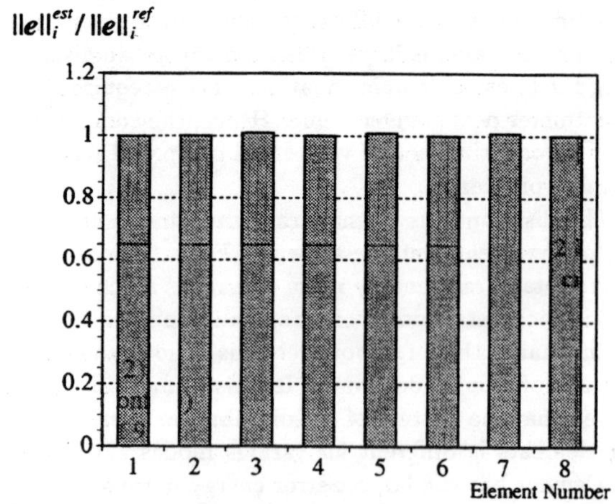

Fig.6 Ratio of estimated energy norm of error to referred energy norm of error

Space frame problem

A space frame shown in Fig.7-a) is used as the calculated model of the rack in the engine room of a ship. As shown in Fig.7-b), the mesh used for this FEM computation consist of 60 space frame elements. An isoparametric frame element with 2-nodes (12 degrees of freedom) is used for FEM, while an isoparametric frame element with 3-nodes (18 degrees of freedom) is used for *a posteriori* error estimation. The latter element is also used for FEM analysis to obtain the referenced results.

Eigenvalues computed by FEM, *a posteriori* error analysis and referenced FEM are compared in Table 3 which indicates that the eigenvalues are modified very accurately by this *a posteriori* error analysis. The relations of error modified eigenvalues to the parameter α is shown in Table 4 from which it is observed again that the error modified solutions are not sensitive to the parameter α. The good error modified solutions are obtained for the 1-st mode with α's varying from 0.60 to 0.98 (the most proper α is 0.86), for the 2-nd mode as α varying from 0.60 to 1.0 (the most proper α is 0.95), and for the 3-rd mode as α varying from 0.60 to 1.0 (the most proper α is 0.85).

Fig's. 8 and 9 show the distributions of the strain and kinetic energy norms of error for every element at the 1-st mode where it is found that the strain energy norm of error is much larger than the kinetic energy norm of error.

Table 3 Comparison of eigenvalues for the space frame problem

Mode #	$\hat{\omega}$	ω^{est}	ω^{ref}	F	α
1	88.16	75.90	75.90	0.313	0.86
2	119.07	110.45	110.45	2.796	0.95
3	240.53	209.01	209.01	5.616	0.85

Table 4 Relationship between the modified eigenvalue and parameter α for the space frame problem

1-st mode		2-nd mode		3-rd mode	
α	ω^{est}	α	ω^{est}	α	ω^{est}
0.60	75.98	0.60	110.56	0.60	209.30
0.70	75.91	0.70	110.47	0.70	209.05
0.80	75.90	0.80	110.45	0.80	209.01
0.86	75.90	0.85	110.45	0.85	209.01
0.90	75.90	0.90	110.45	0.90	209.01
0.95	75.90	0.95	110.45	0.95	209.01
1.00	75.90	1.00	110.45	1.00	209.02

Concluding remarks

A posteriori error estimation and improvement of eigenvalue and eigenvector solutions of FEM are developed in this paper. The governing equation of error for the eigenvalue-problem in a global form is deduced. A solution strategy for solving the governing equation of error is presented and Rayleigh's quotient is used for

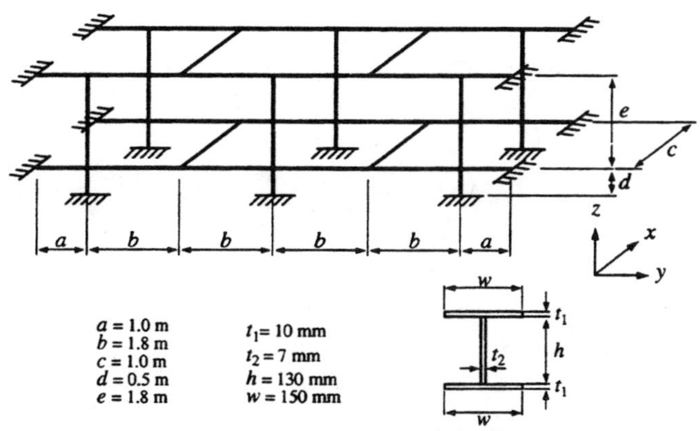

a) Structural model of space frame

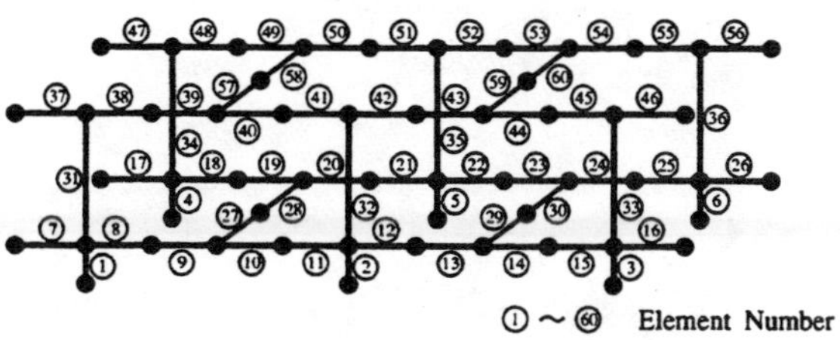

b) Finite element sub-division

Fig.7 Space frame problem

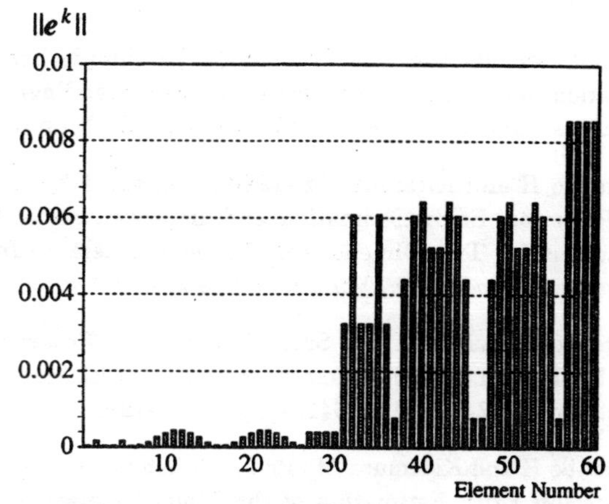

Fig.8 Kinetic energy norm of estimated error for the 1-st mode

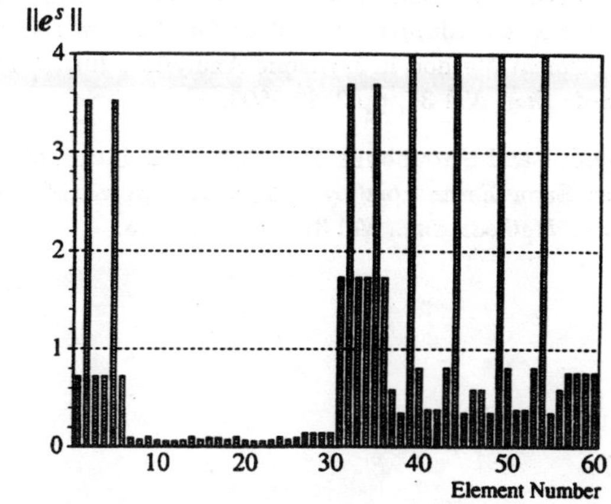

Fig.9 Strain energy norm of estimated error for the 1-st mode

obtaining the modified eigenvalues. To reduce the size of matrix in error analysis, a technique of eliminating the degrees of freedom at mid-node of element is utilized. It is observed from the selected numerical examples that the error modified results and the energy norm of error are not sensitive to the parameter α. Hence, in practical error estimation, giving an estimated parameter α can give us a good results. Since the computational time used in error estimation is very small compared with solving the eigenvalue problem, the proposed method is efficient to be used for any frame structures.

Numerical examples presented in this paper show that the proposed method can be successfully applied to *a posteriori* error estimation and improvement of eigenvalue and eigenvector solutions of FEM. The good error estimated results can be obtained by adding the error to the original FEM solutions and the good energy norm of error is given in the solved problems.

Acknowledgment

The authors appreciate many enjoyable discussions with Prof. H. Ohtsubo at the University of Tokyo during this research work. The support from Fundamental Research Developing Association for Shipbuilding and Offshore is gratefully acknowledged.

Appendix

For an arbitrary given β^0, the error displacement $\{e_u^0\}$ is obtained by solving the error equations of Eq.(39) and the modified displacement vector is given by Eq.(43). Hence, a normalized error displacement vector $\{e_u\}$ is as follows.

$$\{e_u\} = \{\tilde{U}\} - \{\hat{U}\} = \frac{1}{C_M}\left(\{\hat{U}\} + \{e_u^0\}\right) - \{\hat{U}\} \tag{61}$$

where $\{\tilde{U}\}$ is a modified displacement vector in normalization. Substituting Eq.(61) into Eq.(36) and doing some algebra give

$$\left([K] - \omega^2[M]\right)\{e_u^0\} = \{G^e\} - \{\Lambda\} + \left(\beta C_M - \frac{\omega^2}{\hat{\omega}^2}C_M + \frac{\omega^2}{\hat{\omega}^2}\right)\{G^s\} \tag{62}$$

Eq.(62) is equivalent to Eq.(39) with defining β^0 as

$$\beta^0 = \beta C_M - \frac{\omega^2}{\hat{\omega}^2}C_M + \frac{\omega^2}{\hat{\omega}^2} \tag{63}$$

If $\beta^0 = \omega^2/\hat{\omega}^2$, Eq.(63) is deduced

$$C_M\left(\beta - \frac{\omega^2}{\hat{\omega}^2}\right) = 0 \tag{64}$$

which results in $C_M = 0$, because $\beta - \omega^2/\hat{\omega}^2 \neq 0$. Since the normalizing coefficient C_M is never to be zero, $\beta^0 \neq \omega^2/\hat{\omega}^2$ is required.

References

1. Zienkiewicz OC and Zhu JZ (1988). "A Simple Error Estimation and Adaptive Procedure for Practical Engineering Analysis", *Int. J. Numer. Methods Eng.*, Vol 24, pp 337-357.

2. Ohtsubo H and Kitamura M (1990). "Element by Element *A Posteriori* Error Estimation and Improvement of Stress Solutions for Two-Dimensional Elastic Problems", *Int. J. Numer. Methods Eng.*, Vol 29, pp 233-244.

3. Ohtsubo H, Suzuki K and Sato K (1991). "*A Posteriori* Error Estimation in Finite Element Analysis for Elastodynamic Problem", *JSME*, Vol 57-541(A), pp 1979-1985.

4. Ohtsubo H and Kitamura M (1992). "Element by Element *A Posteriori* Error Estimation of the Finite Element Analysis for Three-Dimensional Elastic Problems", *Int. J. Numer. Methods Eng.*, Vol 33, pp 1755-1769.

5. Wiberg NE and Li XD (1994). "A Postprocessed Error Estimation and an Adaptive Procedure for the Semidiscrere Finite Element Method in Dynamic Analysis", *Int. J. Numer. Methods Eng.*, Vol 37, pp 3585-3603.

6. Babuska I and Strouboulis T (1994). "Validation of *A Posteriori* Error Estimators by Numerical Approach", *Int. J. Numer. Methods Eng.*, Vol 37, pp 1073-1123.

Proceedings of the Seventh (1997) International Offshore and Polar Engineering Conference
Honolulu, USA, May 25-30, 1997

ISBN 1-880653-28-1 (Set); ISBN 1-880653-32-X (Vol. IV)

Testing Ras Tanura Sea Island Breasting Dolphins

Abobakr M. Radwan and Ala A. Al-Sharif
Saudi Arabian Oil Company (Saudi Aramco)
Dhahran, Saudi Arabia

ABSTRACT

A unique field test was conducted in July 1995 to check the load-deflection characteristics of the breasting dolphins of Ras Tanura Sea Island No. 2. The purpose of the test was to confirm the assumptions used in the mathematical simulation and structural analysis of the dolphins. A secondary objective was to check whether there is a "play" in the structures, i.e. whether the dolphins deflected one way or the other under no significant load. Two opposing dolphins of different berths were tested simultaneously.

Deflections measured were in line with the expected deflections predicted by the computer structural simulation. The facility, test apparatus and procedure are described in this paper. Results of measured deflections and the corresponding computer simulation plots are also presented.

KEY WORDS: Dolphin, Sea Island, Loading Terminal, Field Test.

INTRODUCTION

Ras Tanura Sea Island 2 is a crude oil loading terminal which was constructed in 1969. The breasting dolphin structures consist of steel decks, supported by but not rigidly connected to, free standing vertical piles. The water depth at the Sea Island is about 90 feet.

It has been reported that large lateral deflections have been observed in relatively calm weather. The conditions have been reported to have deteriorated over the years and caused some concerns about the integrity of the structures. Therefore, it was the objective of the static field load test on the existing Outer Breasting Dolphin (OBD) structures at Sea Island 2 to measure:

1. Load/Deflection characteristics of OBD-7 of Berth-15 and OBD-8 of Berth-16 structures under lateral loads. This is to compare the results with the computer simulation and come to a conclusion regarding the relationship between the test results and computer modeling.
2. "Play" of the OBD structures, i.e. the deflections associated with small lateral load application. The "Play" measurements are to verify the structural behavior in opposite directions under small load.

The Load/Deflection characteristics was achieved by pulling the two dolphins towards each other. Unloading was done under controlled procedure to check the elasticity of the system. Coordinates of the top of all dolphin piles were precisely measured from the stationary loading platform. Applied load was monitored and recorded by a calibrated Portable/Battery operated, digital readout load cell, with a capacity of 100 Metric Tons. Waves, current, wind and direction measurements were recorded for the duration of the two-day test. Pile deflections near the mudline were recorded by divers. A repeat pull test was done the second day to check if any permanent deformation occurred in any of the dolphins. This was followed by the "Push" test to check the deflection characteristics of the two dolphins in the other direction, which determined the amount of "play" in the system.

The findings of the load/deflection test, simulating regular tanker berthing conditions in a static sense, were used to test the validity of the results of the computer analysis for the OBD structures. The correlation between the test measurements and the computer analysis will assist in the determination of the capacities of the dolphins.

The test was performed during the shut down period of the Sea Islands for Testing and Inspection (T&I) work.

TEST PROCEDURE

The detailed scope of work including material and equipment acquisitions and deliveries, fabrication, installation and miscellaneous construction requirements, personnel requirements, setup for test, testing procedure, reinstallation, cleaning and schedule are found in the complete report SAER-5534 (1996). A summary of the test procedure is described in Appendix A.

ENVIRONMENTAL PARAMETERS MEASUREMENT

The tests were to be conducted on relatively calm days. Wind speed and current velocities and profile were measured accurately during the course of the two day test to assess their contribution to the deflection of the structures. It was observed that wind and current changed their speed and direction during the test.

The Hydrographic Surveying Unit of Saudi Aramco continuously monitored the current profile, speed and direction. Average current speeds were in the range of 0.25 knot. Waves were about 2-3 ft. for the 2 days and wind speed was generally calm (about 5 knots) except for the second day when a wind gust of about 20 knots was observed.

Computer simulations to quantify the effects of currents and waves on the structural deflections were done. The results indicated that the pile deflections caused by current and wave loads were too small (1 mm from current alone and 5 mm from a 3 ft. wave) to affect the accuracy of this test, and were therefore ignored.

STRUCTURAL ANALYSIS

Structural Model:
Structural models using StruCAD*3D computer program were constructed for both OBD-7 and OBD-8. The pile-deck connections allowed the dolphins to deflect without imposing bending stresses to the piles at the deck elevation. Torsion bars were modeled carefully to distribute the loads to the various piles. Plate elements were used to model the decks. The loads were applied in the exact location as that of the field test. Model load cases depicted the same load intervals as those of the test.

Local Stresses resulting from the test frame and test loading:
Ring stiffeners were used around the pile top at OBD-8, where the load cell is attached to reduce local stresses and stiffen the pile against local bending. Stresses were calculated to be small (6 ksi) as a result of the maximum expected test load of 200 kips.

Test Frame Design:
The general arrangement of the test frame, its components and the attachment to the two dolphins are described in Appendix A. The test frame design and structural analysis are found in a separate report, (1995).

SOIL CONDITION AROUND THE PILES PRIOR TO THE TEST

A Seabed Penetration probe was used to check the condition of the soil around OBD-7 and OBD-8 dolphin piles prior to the test. The probe composed of a 60 inch long, 1 inch diam. pipe, with pointed end in a cone shape, equipped with a sliding weight of 44 lbs. that falls freely a distance of 12 inches driving the probe into the soil. The top layer soil is classified as granular, fine sand. The results of this test showed that penetrations near the piles (within 2 ft.) were in the order of 3 to 6 inches, while penetrations of more than 24 inches were measured further away from the piles (4 to 6 ft. away). This suggests that the cyclic pile movement, over the years, compacted the immediate granular soil around the piles. This was in line with the divers' observation of no voids adjacent to the piles.

RESULTS

Pile Top Displacements:
Test 1, Loading: (Pulling the dolphins towards each other) was performed on July 12, 1995. Unloading was done after a 3 hour pause, to check for any relaxation in the system; i.e. to check if the load will drop while holding the dolphins in the maximum deflected position for the test. The test was repeated on the next day to check if there are any changes in the deflection characteristics of the dolphins.

The coordinates of all eight pile tops were recorded by the surveying team before starting the test to mark the initial setting. Initial coordinates readings, and the readings at each load increment, were repeated 3 times, and the average of the 3 sets of readings was considered in calculating the displacements.

Displacements (deflections) were calculated from the coordinates readings for each dolphin separately. Graphs for each loading and unloading are shown in Figures 1 through 4.

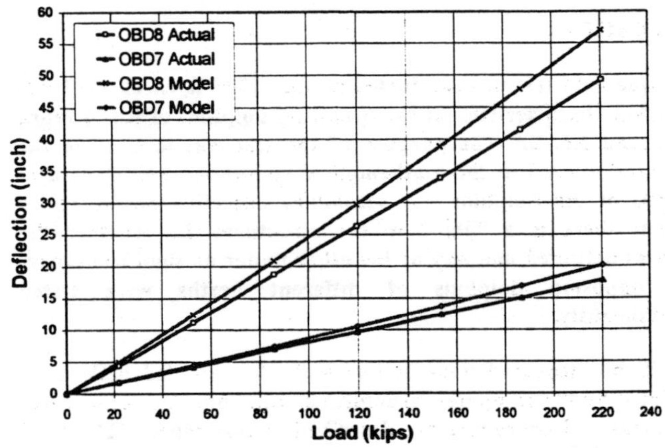

Fig. 1 "E/W" Deflection at Pile Tops
Loading Case on July 12, 1995

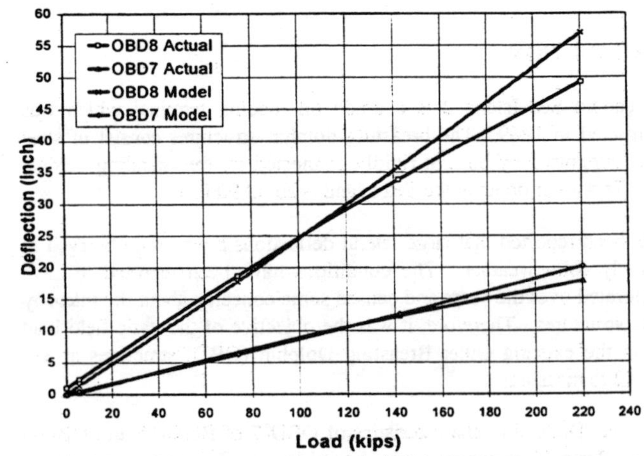

Fig. 2 "E/W" Deflection at Pile Tops
Unloading Case on July 12, 1995

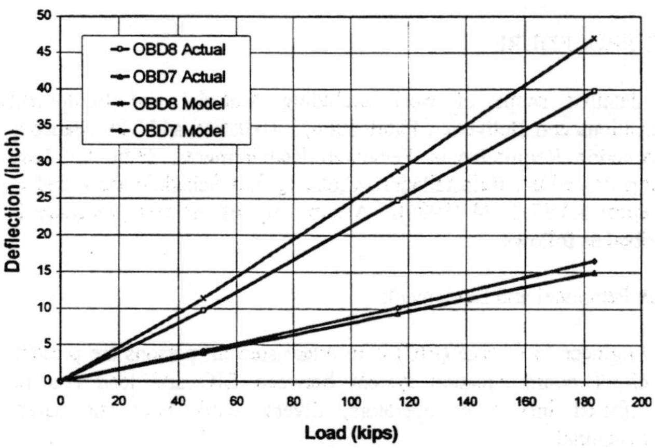

Fig. 3 "E/W" Deflection at Pile Tops
Loading Case on July 13, 1995

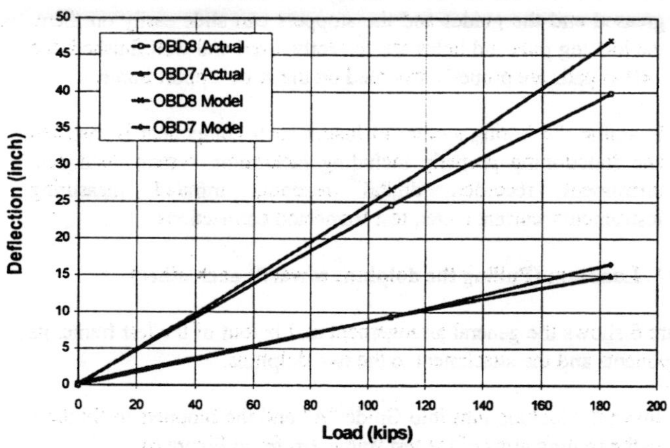

Fig. 4 "E/W" Deflection at Pile Tops
Unloading Case on July 13, 1995

Pile Displacements Near Mudline:

During the load test, a diving vessel was positioned at OBD 7 and a work boat at OBD 8 to provide diving support to divers monitoring the two dolphins pile movements near mudline. Displacements at mudline are hard to measure, since the bottom fine materials will be easily stirred by diver movement making it very difficult to read gauges at that location. After some trial and error it was determined that an optimum location to be away from the disturbed soil is 2 ft. above the mudline. Therefore, pile displacements under the various loading conditions at 2 ft above mudline were predicted using the results from the computer analyses for OBD 7 and 8 structures under the same test loads.

Actual pile displacements at the 2 ft above mudline elevation were measured using Starrett Dial Indicators Model 25-2041J. The dial gauges were mounted on stakes anchored to the ground about 3 ft. away from the 53" piles. To investigate the possible displacements of the dial gauge supports, a marker stake was driven at 10 feet west of Pile OBD8-2 (located at the north end of the dolphin). The distance between the marker and the gauge stakes was checked during the test. A movement of 3/4" at the maximum load (100 MT) was recorded, indicating that deflections obtained from the dial readings have to be modified to account for the movement of the gauge itself.

Play Test:

The test frame was constructed to have the capability to force deflections, and hence loading, in the other direction of the first load test, pushing the dolphins away from each other. Three sets of load versus deflection readings were recorded for the push test, up to a load of 33 Ton (73 kip). These values, along with the results from the original load test, were plotted on the same graph and are shown in Figure 5.

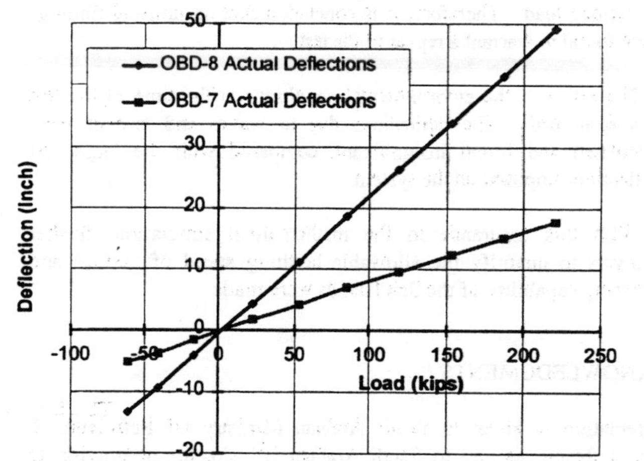

Fig. 5 Play Test Results

It is clearly demonstrated that all imposed deflections were associated with loads measured at each step. The load-deflection followed the linear relationship expected from the system as shown in Figure 5. No deflections were recorded under zero load. Therefore, it is concluded that there is no play in this system.

DISCUSSION AND CONCLUSIONS

The pile top displacements (deflections) were found to be 14% (on the average) smaller than the predicted values of the computer analysis for OBD-8, the three piled structure and 11% for OBD-7, the five piled structure. This difference is attributed mainly to uncertainties about the exact steel wall thickness and condition of the piles, especially after 26 years of service, and to uncertainties in the soil p-y characteristics that were predicted from the soil boring, refer to report on foundation investigation, 1970.

- Measured loading and unloading deflection results indicated almost linear deflection characteristics. Unloading to the same original position demonstrated the elastic behavior of the system.

- Reloading, the next day of the test, produced a slightly smaller deflection at the top of the pile at a load of 84 kips (40"- compared to 42" the previous day). This indicates no disturbance in the bond between the soil and piles even after a sustained static load at a high value (producing more than 4 ft. deflection).

- The play test, loading the dolphins in the reverse direction, demonstrated that there is no play in the system. Every movement of the piles is associated with an external load proportional to the elastic characteristics of each dolphin. Deflections observed at these dolphins, that were thought to be play in the system when no ship was moored to the facility, must have been the result of wind, wave and currents occurring at the time of the test.

- According to our analysis, diver readings for pile movement near the mudline were inconsistent, and no accurate conclusion can be drawn. Therefore, only the top of pile deflections were considered in the conclusion of this study. However, all diver readings were in the same order of magnitude of the expected deflections under the prescribed load. Therefore, it is concluded that no unusual findings were found to warrant a repeat of the test.

- The effect of the environmental conditions at the time of the test was examined. The deflections due to waves and current were calculated and found insignificant, compared with the large test deflections imposed on the system.

- With this assurance in the mathematical simulation, further analysis to quantify the allowable berthing speed of tankers and mooring capability of the Sea Islands were made.

ACKNOWLEDGMENTS

Appreciation is given to Saudi Arabian Ministry Of Petroleum & Mineral Resources and to Saudi Arabian Oil Co. for permission to publish this paper. The authors wish to express their appreciation to the Terminal Operations Department of Saudi Aramco for funding the project and providing various support to complete the test successfully. Special appreciation is extended to Khalid Addas,TOD, K. Yasin, I. Abusaai, and E. Brown, Offshore Projects Div.

REFERENCES

"Field Load Test, Outer Breasting Dolphins of Sea Island 2", Saudi Aramco Internal Company Publication SAER-5534, Jan. 1996.

"Foundation Investigation of Sea Island Loading Terminals Ras Tanura, Saudi Arabia", Saudi Aramco Internal Company Publication, May, 1970.

"Structural Calculations, OBD 7 & 8 Field Lateral Test", Saudi Aramco Internal Company Publication, May 22, 1995.

APPENDIX A

TEST PROCEDURE

The detailed scope of work including Material and Equipment Acquisitions and Deliveries, Fabrication, Installation and Miscellaneous Construction Requirements, Personnel Requirements, Setup for Test, Testing Procedure, Reinstallation, Cleaning and Schedule are found in the report SAER-5534 (1996). A summary of the test procedure is described as follows:

Check Personnel and Equipment:

- Engineer In Charge (EIC) is to make sure all persons are present, check communication system between EIC and load readers, infrared instrument operators, divers, work boat and safety personnel.

- Check the test frame is correctly aligned and connected, properly restrained laterally and vertically, the jacks are properly positioned and secured in the seats, the pulling arms and all the rollers are well greased and the guides and the stoppers can slide easily on them, the locking pins and holes are correctly sized and well finished, the Teflon pads are properly mounted on the strut support frame.

- Examine the testing system to insure each component is installed and functioning properly including jack/pump system, load cell, instrument recorder, digital readout, infrared measuring instruments, current meter, test frame and connections.

Test 1, Loading: (Pulling the dolphins towards each other)

Figure 6 shows the general arrangement and layout of the test frame, its components and the attachment to the two dolphins.

1. Insert the locking pins into Guide "A" and the Stoppers to fix these to the pulling arms of the test frame. (Refer to Figure 6)

2. Run the jacks slowly until the first pin is securely engaged and all the load is released from the jacks. This will cause westward movement for OBD 7 and eastward for OBD 8.

3. Load Reader to read and record the line tension load in both metric tons and kips (load cell readings are in metric tons) when the gauge digital readout is stabilized. Record the time of the reading in hours and minutes, and inform EIC of the line tension at same time.

4. Using the Test Record Form, the Record Keeper is to fill out the Test No./Date/Time/Pin Hole No. and record the load given by load reader, and read and record the pressure/force (in bars/psi and kips) from the gauge on the jack pump.

5. EIC is to instruct the Infrared Measuring Instrument Operators (IMIOs) and divers to take readings. The loading at the time shall be given to IMIOs also.

6. IMIOs to record the loading and time in hours and minutes, and take and record the angular and distance measurements for each pile.

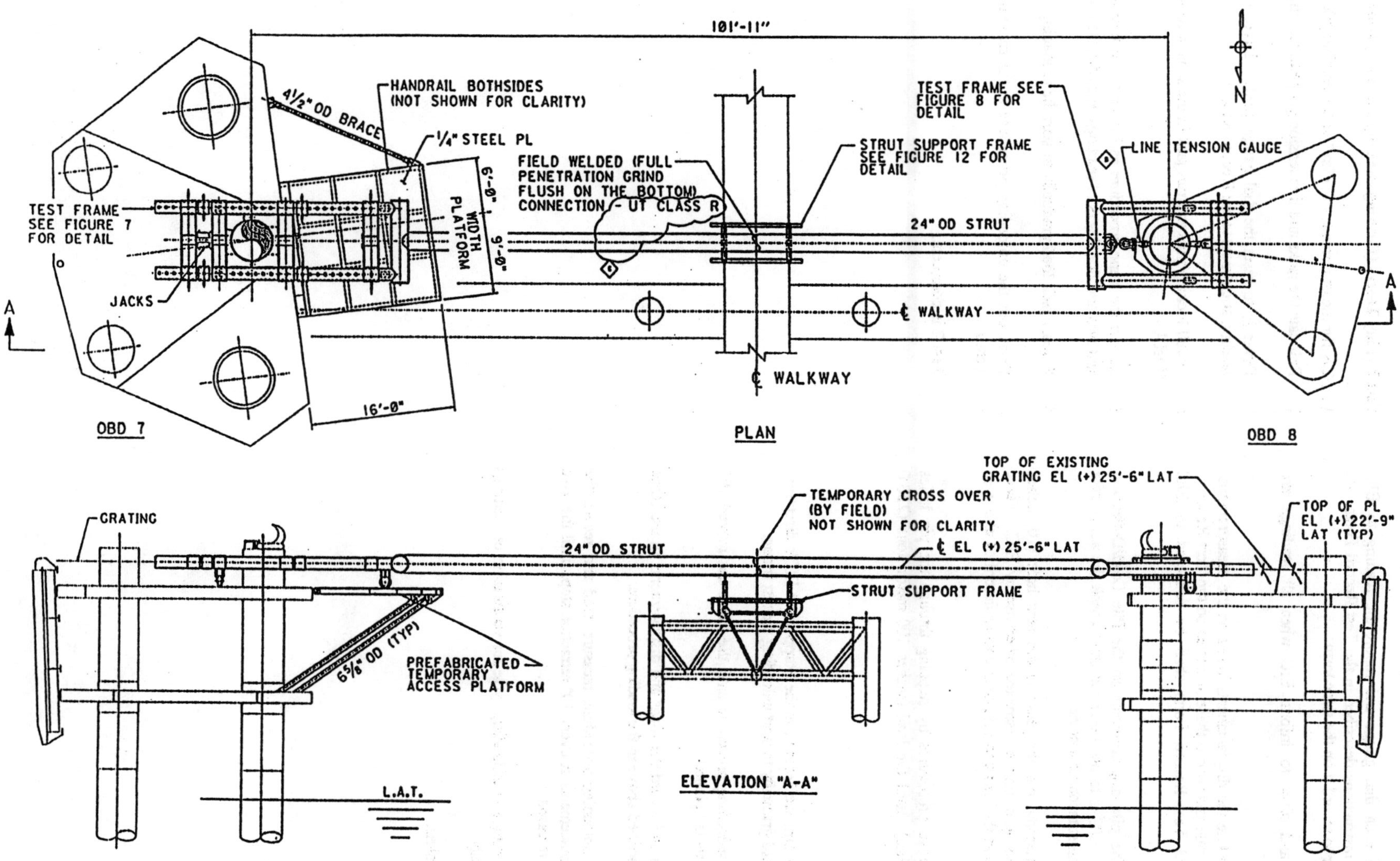

Fig. 6 Sea Island 2 Field Lateral Load Test
General Arrangement - Test Frame - Test 1

7. Diver to read each dial gauge and report the reading to EIC following the instructions described in the Test Record Form, for Record Keeper to record it on the same form.

8. Both IMIOs and diver to inform EIC when all readings are completed.

9. IMIOs to calculate the displacements based on the measurements taken for each pile and record these in inches on the Test Record. The completed Test Record Form shall be delivered to EIC immediately.

10. EIC to plot the pile displacements on the Pile Load/Deflection Curves forms. EIC is to use these curves to check the actual pile deflections against theoretical ones.

11. At EIC's instruction, run the jacks to the next hole (10 inches movement) and hold, unless instructed otherwise by EIC. Take, record and report the load and measurements as described above in this Section.

12. Repeat the above procedures by stepping through the pin holes (every 10 inches) until the load reaches 100 metric tons (220.0 kips).

Unloading:

1. When locking pins are secured in the Stoppers, at EIC's direction, release jack load gradually to start unloading the test frame.

2. Continue the unloading procedure until the frame has moved 10" and the Stoppers are pinned.

3. Measure, read and record the loads and displacements in the same manner as described above in the loading procedure.

4. Continue the unloading procedure, measure, read and record the loads and displacements at every 10" retraction stop, until the load is completely released.

5. Repeat the loading and unloading cycle for verification of readings taken previously.

6. Test 1 is complete.

Test 2 (Play Test) - (Pushing the dolphins away from each other)

Test 2 is for measurements of "Play" of the dolphin piles only.

1. Relocate the jacks and line tension gauge to starting locations.

2. Diver is to reset the dial gauge guide. Diver to take the initial reading and report to the Record Keeper.

3. Install locking pins and run jack in the same manner as described in Test 1.

4. Repeat the same procedure as for Test 1 to take an initial reading, and two additional sets of readings or as directed by EIC.

5. Release jack load gradually to start the unloading.

6. Carry out the unloading procedure in the same manner as that for Test 1.

7. Test 2 is complete.

Proceedings of the Seventh (1997) International Offshore and Polar Engineering Conference
Honolulu, USA, May 25-30, 1997

ISBN 1-880653-28-1 (Set); ISBN 1-880653-32-X (Vol. IV)

Buckling/Plastic Collapse Behaviour and Strength of Stiffened Plates Under Thrust

Tetsuya Yao, Masahiko Fujikubo and Daisuke Yanagihara
Hiroshima University
Higashi-Hiroshima, Japan

ABSTRACT

Series of elastoplastic large deflection analyses are performed on stiffened plates with flat-bar and angle-bar stiffeners subjected to thrust, and the influences of sectional geometries of stiffeners on the buckling/plastic collapse behaviours are investigated. It has been found that:

(1) Local panel buckling strength between stiffeners, post-buckling inplane rigidity and ultimate strength are significantly increased owing to the stiffener.

(2) Stiffened plates undergo elastoplastic secondary buckling after local panel buckling.

(3) Higher ultimate strength is attained by stiffeners with lower depth to thickness ratio of a web. However, the reduction in capacity after the ultimate strength is more rapid for stiffened plates with such stiffeners.

(4) When the stiffeners with relatively high depth to thickness ratio are provided, the ultimate strength is lower than the case with lower depth to thickness ratio, but the capacity reduction after the ultimate strength is rather moderate.

KEY WORDS : stiffened plates, local buckling strength, post-buckling inplane rigidity, secondary buckling, overall Eulerian buckling, ultimate strength, buckling/plastic collapse

1 INTRODUCTION

One of the critical failure events of a ship structure is the occurrence of overall buckling/plastic collapse at deck or bottom structure under longitudinal bending. So, the deck and the bottom plates are reinforced by a number of longitudinal stiffeners to increase their strength and rigidity. For a rational design avoiding such a failure, it is very important to know the buckling/plastic collapse behaviour of such stiffened plates under thrust.

Plenty of research works have been performed both theoretically and experimentally on buckling/ultimate strength of isolated plates as well as stiffened plates subjected to thrust. The results are put to practical uses establishing design rules or deriving design formulae.

However, in most of the previous research works, plates and stiffeners are separately considered. In general, stiffened plates are so designed that local panel buckling firstly takes place between stiffeners. So, buckling/ultimate strength of isolated plates are firstly investigated assuming that edges of the plate are simply supported. This boundary condition gives the lowest/conservative buckling strength. On the other hand, stiffened plates have been often modeled as a beam-column with plating between stiffeners as a flange.

In the actual ship structures, however, interactions between plates and stiffeners can not be ignored. The interactions depend on the sectional geometry of a stiffener. When stiffeners with high flexural-tortional rigidity are provided, the buckling strength of panels between stiffeners is expected to increase.

Here, the stiffeners in ship structures can be grouped into three types, which are flat-bars, angle-bars and tee-bars. As for stiffened plates with flat-bar stiffeners subjected to thrust, one of the authors performed series of FEM elastoplastic large deflection analyses and discussed the buckling/ultimate strength and the collapse modes of stiffened plates. He also found the minimum stiffness ratio of the stiffener to plate, γ^U_{min}, which guarantees the maximum ultimate strength [1-3]. In this investigation, however, the stiffener was assumed not to have resistance against torsion.

Caridis *et.al.*[4-6] considered both the panel and flat-bar stiffener as plates, and performed elastic large deflection analyses applying Finite Difference Method. In their analysis, flexural-torsional behaviour of a flat-bar stiffener was exactly simulated. Tanaka and Endo[7,8] proposed a simple approximate method to simulate collapse behaviour of stiffened plate with flat-bar stiffeners. The calculated results were compared with experimental results, and good correlation was observed between measured and calculated results although interaction between plate and stiffeners was ignored. Panagiotopoulos[9] discussed the interactions between plate and flat-bar stiffeners based on the results of FEM analyses.

Although useful information have been obtained on buck-

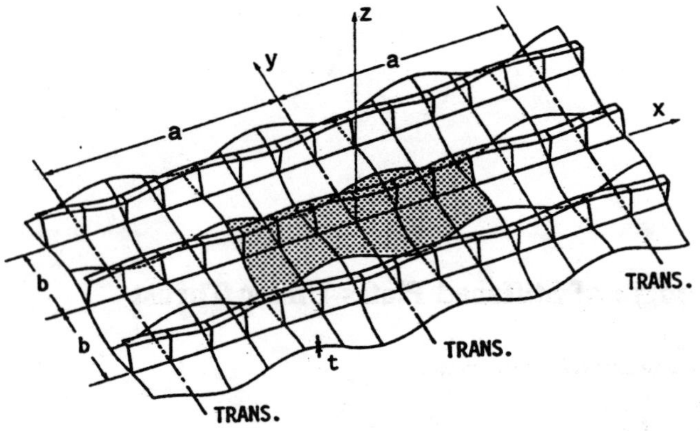

Fig. 1 Continuous stiffened plating

Table 1 Cross-sectional geometries of stiffeners

Model		type	h	t_w	b_f	t_f	h/t_w	I_s
Case 1	F1	flat	195	10	-	-	19.5	619
	A1	angle	150	12	90	12	12.5	619
	F4	flat	156	19.5	-	-	8.0	617
Case 2	F2	flat	322	10	-	-	32.2	2790
	A2	angle	250	12	90	16	20.8	2790
Case 3	F3	flat	395	20	-	-	19.8	10300
	A3	angle	400	11.5	100	16	34.8	10300

h : Height of stiffener web
t_w : Thickness of stiffener web
b_f : Breadth of stiffener flange
t_f : Thickness of stiffener flange
I_s : Moment of inertia of stiffener

h, t_w, b_f, t_f in mm I_s in cm^4

ling/ultimate strength of stiffened plates with flat-bar stiffeners, those for stiffened plates with angle-bar or tee-bar stiffeners are not so many[10].

In the present paper, series of FEM elastoplastic large deflection analyses are performed on stiffened plate with flat-bar and angle-bar stiffeners of the same flexural rigidity. Uni-axial thrust is applied, and the influences of cross-sectional geometries on collapse behaviour are discussed. Thin stiffened plates are considered to examine fundamentals in buckling-plastic collapse behaviour of stiffened plates under thrust.

2 MODELLING OF STIFFENED PLATES

2.1 Stiffened plates for analysis

In general, the deck and the bottom panels in ship structure are stiffened by equally spaced longitudinal stiffeners of the same size. Under longitudinal bending, compressive inplane force acts on the deck or the bottom plate in its longitudinal direction. In general, the plate and the stiffeners locally buckle under longitudinal thrust as illustrated in Fig. 1. In this case, considering the continuity and symmetry conditions, the buckling/plastic collapse bahaviour of a continuous stiffened plate can be simulated analyzing only the shaded part in Fig. 1.

Along the four edges, symmetry conditions are prescribed, and their inplane movements perpendicular to themselves are considered to be uniform. This symmetry condition is correct when flat-bars are provided as stiffeners, but is an approximation for stiffened plates with angle-bar stiffeners. The local plate dimensions are $a \times b \times t = 2,400 \times 800 \times 10$ mm. The thickness, $t = 10$ mm, is a little thin when the actual deck plate is considered, but it is chosen to get elastic local buckling of panels between stiffeners to investigate into the fundamental buckling behaviour of stiffened plates. The stiffeners shown in Table 1 are considered. In each group, flat-bar and angle-bar stiffeners have the same moment of inertia. Angle-bar stiffeners satisfy the requirement by DNV Rule [11] to prevent torsional buckling, but flat-bar stiffeners do not.

The material properties are as follows:

Yield stress: $\sigma_Y = 313.6$ MPa
Young's modulus: $E = 205.8$ GPa
Strain hardening rate: $H_k = E/65$

2.2 Initial imperfections

Initial deflection is assumed as the sum of local and overall buckling modes:

$$w_0 = A_0 \sin\frac{\pi x}{a} + B_0 \sin\frac{3\pi x}{a}\sin\frac{\pi y}{b} \qquad (1)$$

A_0/t and B_0/t are taken as 0.01. The stiffened plate is assumed to be free from welding residual stresses and their influences are not considered.

2.3 Method of analysis

Elastoplastic large deflection analyses are carried out using the computer code, ULSAS. In this code, incremental calculation is performed based on the updated Lagrangean approach. Four-noded isoparametric shell element with degenerated reduced integral[12] is used. This element has ten layers to examine if the yielding condition is satisfied, and to perform numerical integration toward thickness. The web and the flange of a stiffener are also discretised by shell elements as a plate part. Compressive load is applied by forced displacements.

3 RESULTS OF ANALYSES AND DISCUSSIONS

3.1 Buckling strength

For all the calculation models, eigenvalue analyses are performed to examine the influence of stiffeners on local buckling strength of stiffened plate under thrust. The results are summarized in Table 2. In this table, σ_{cr}^P (= 116.6 MPa) is the buckling strength when an isolated plate is assumed to be simply supported. $\sigma_{cr}/\sigma_{cr}^P$ represents the effect of a stiffener to increase local panel buckling strength. The inplane rigidity above buckling load, the initial yielding strength and the ultimate strength obtained by elastoplastic large deflection analyses are also given in Table 2.

For Case 2, the buckled cross-sections are illustrated in Fig. 2. It is seen that the flange resists against rotation of the web in case of angle-bar stiffeners. When a flat-bar stiffener is provided,

Table 2 Results of calculation for stiffened plate models

Model		$\sigma_{cr}/\sigma_{cr}^{P}$	E_T/E	σ_{iy}/σ_Y	σ_u/σ_Y
Case 1	F1	1.098	0.552	0.573	0.655
	A1	1.582	0.693	0.639	0.729
	F4	1.517	0.669	0.675	0.756
Case 2	F2	0.979	0.495	0.512	0.613
	A2	1.539	0.725	0.643	0.754
Case 3	F3	1.621	0.682	0.654	0.732
	A3	1.415	0.737	0.645	0.781

σ_{cr} : Elastic buckling strength of stiffened plate
σ_{cr}^{P} : Elastic buckling strength of unstiffened plate
σ_{iy} : Initial yielding strength
σ_u : Ultimate strength
E_T : Tangential modulus after elastic buckling

(a) Model F2 (b) Model A2
Fig. 2 Elastic buckling mode

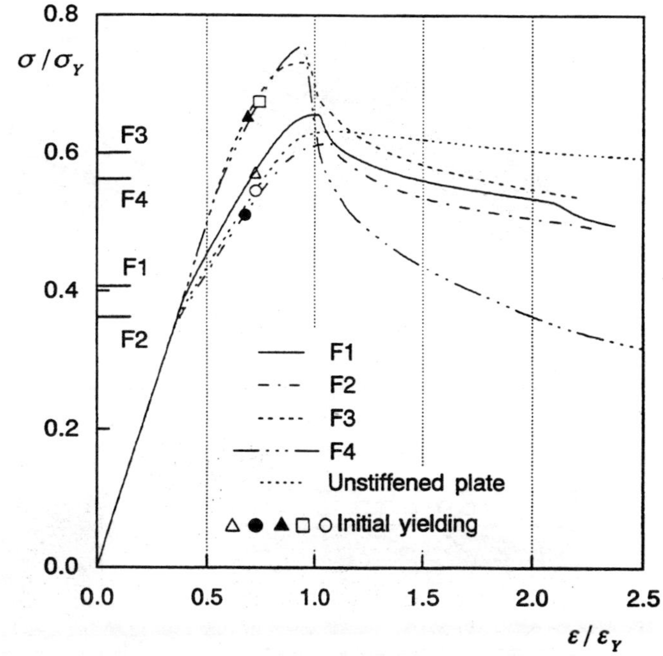

Fig. 3 Average stress-average strain relationships of stiffened plates with flat-bar stiffeners under thrust

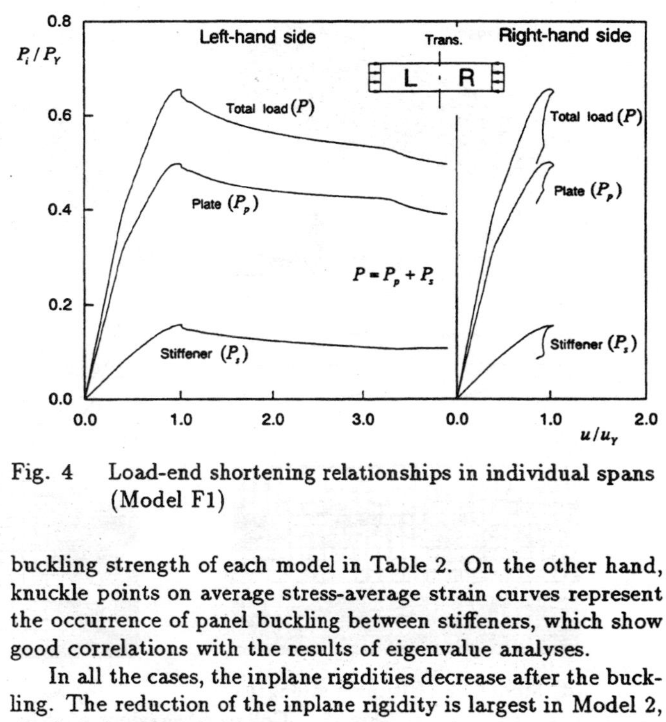

Fig. 4 Load-end shortening relationships in individual spans (Model F1)

it rotates according to the edge rotation of a plate keeping its cross-section straight. In all cases, the buckling mode is in three-half-waves in the loading direction.

Here, it should be noticed that the ratio, $\sigma_{cr}/\sigma_{cr}^{P}$, is less than 1.0 for Model F2. This is because the h/t_w ratio is very high and the local buckling strength of the web is lower than that of the plate. In case of Model F1, h/t_w ratio is relatively high, and the buckling strength is increased by 9.8%. The h/t_w ratio of Model F4 is low, and its buckling strength is increased by 51.7%. Model F3 also has high buckling strength.

In case of stiffened plates with angle-bar stiffeners, the horizontal movement of the top of the web is constrained by the flange. Consequently, the edge rotation of the plate is constrained, and a higher local buckling strength is attained. In this case, higher buckling strength is attained with lower h/t_w ratio.

The increase in buckling strength owing to stiffeners becomes smaller if the thickness of the plate becomes larger compared to the thickness of the web.

3.2 Compressive collapse behaviour

3.2.1 Stiffened plates with flat-bar stiffeners

(1) Inplane rigidity in post-buckling range and initial yielding strength

Figure 3 shows the average stress-average strain relationships for Models F1, F2, F3 and F4. The average stress is nondimensionalised by yield stress, and the average strain by yield strain. A square plate with the same breadth, thickness and material properties is also analyzed and the result is plotted by a dotted line. Short horizontal lines from the vertical axis indicate the buckling strength of each model in Table 2. On the other hand, knuckle points on average stress-average strain curves represent the occurrence of panel buckling between stiffeners, which show good correlations with the results of eigenvalue analyses.

In all the cases, the inplane rigidities decrease after the buckling. The reduction of the inplane rigidity is largest in Model 2, of which buckling has begun at a stiffener. In cases of Model F3 and Model F4 with thicker web plates, the reduction in inplane rigidity above buckling load is small.

In Models F1 and F2, initial yielding takes place at the bottom of the stiffener where local deflection becomes a node of deflection mode. It takes place also in the stiffener but at its top where the horizontal displacement of a web is maximum in Model F3. In Model F4, the initial yielding is at the bottom of the stiffener where horizontal displacement at the top is maximum. Model F4 has the highest initial yielding strength.

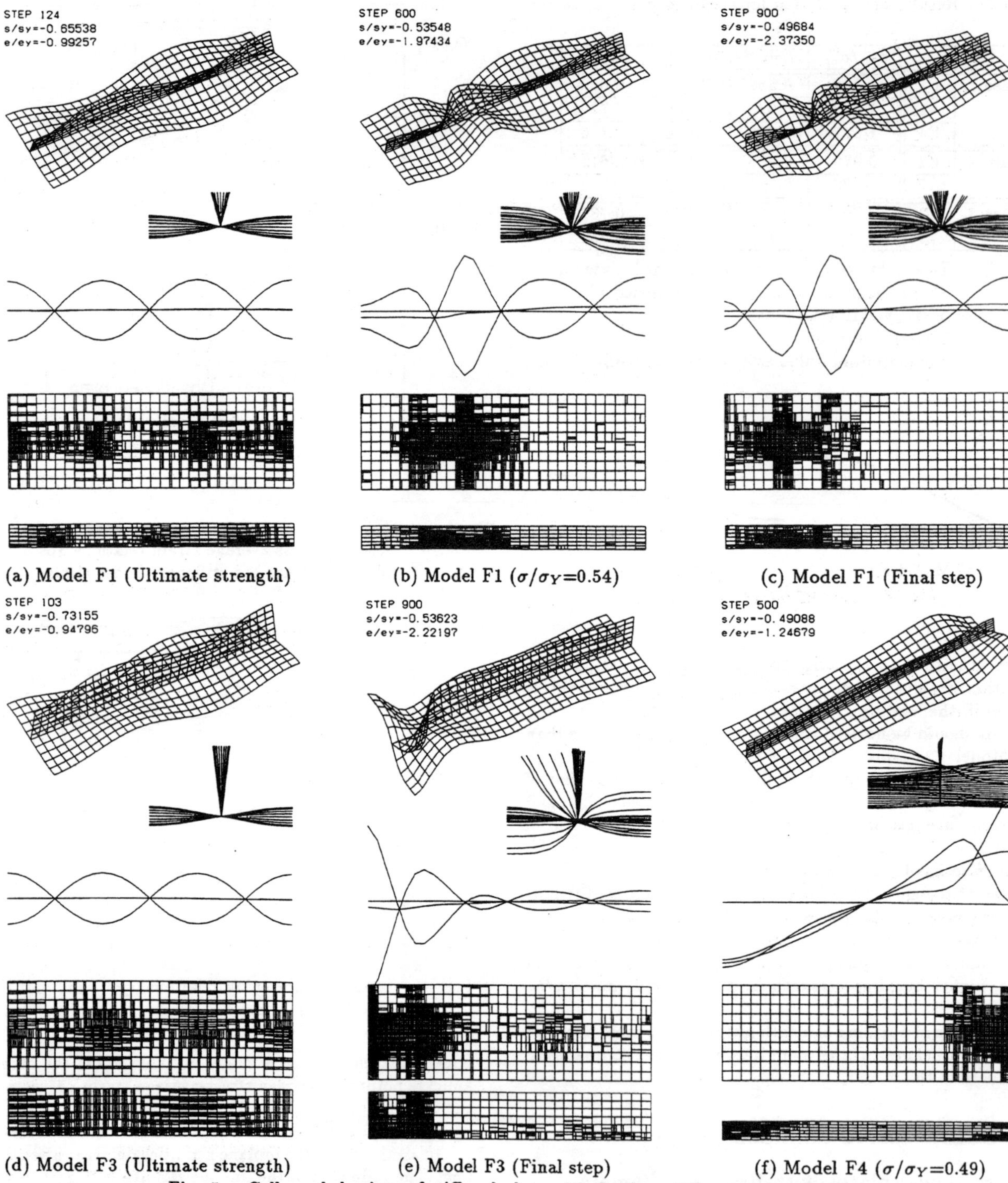

(a) Model F1 (Ultimate strength) (b) Model F1 (σ/σ_Y=0.54) (c) Model F1 (Final step)

(d) Model F3 (Ultimate strength) (e) Model F3 (Final step) (f) Model F4 (σ/σ_Y=0.49)

Fig. 5 Collapse behaviour of stiffened plate with flat-bar stiffener under thrust

(2) Buckling/plastic collapse behaviour

After initial yielding, the yielded zone spreads with the increase in applied displacement, and the ultimate strength is attained. Here, the collapse behaviour of Model F1 is firstly examined. The ultimate strength of Model F1 is about 4% higher than that of an isolated plate. After the ultimate strength, the capacity rapidly decreases because of the occurrence of the change in deflection modes due to the elastoplastic secondary buckling[13,14].

Figure 4 shows the load-end shortening relationships in the left-hand-side span and the right-hand-side span, respectively. P is the total load. P_p and P_s are the loads sustained by the plate and the stiffener. All the loads are nondimensionalized by the fully plastic axial load of the full cross-section, and the end shortening by the yield end shortening.

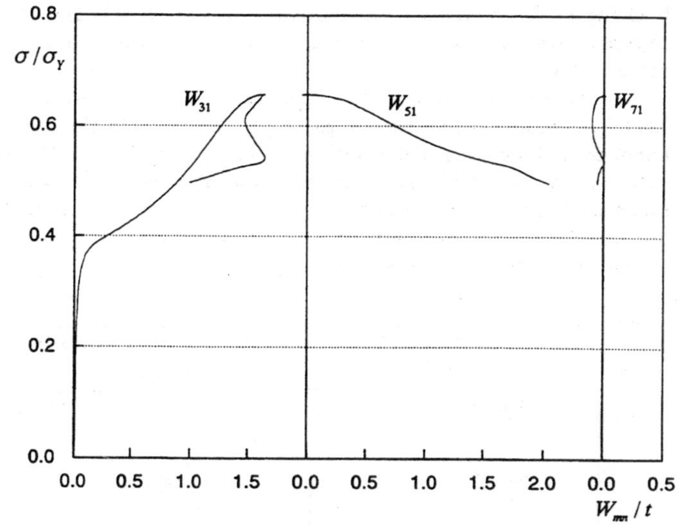

Fig. 6 Average stress-deflection component relationships in the left-hand-side panel (Model F1)

On the other hand, Figs. 5 (a) through (c) show the deflection modes and yielded zones of Model F1 at three loading steps. In the figures indicating the yielded zone, the vertical lines in each element represent the yielding from the back, and the horizontal lines from the face. The number of lines in each element indicates the number of yielded layers.

The capacity of the plate in the left-hand-side span rapidly decreases just after the ultimate strength, and then settles down to a certain level. During this strength reduction, the deflection mode changes from that in Fig. 5 (a) to that in Fig. 5 (b). This indicates that elastoplastic secondary buckling occurred in the left-hand-side panel, and the deflection mode changed from three-half-waves to five-half-waves. This is more clearly shown in Fig. 6, in which the changes in deflection components are indicated. It is known that the deflection component of five-half-waves begins to grow after the ultimate strength, although that of three-half-waves has been developed until the ultimate strength.

Due to the secondary buckling in the left-hand-side span, the capacity was decreased. Because of this, unloading took place in the right-hand-side span as is known from Fig. 4. This is also known from Figs. 5 (a) through (c) indicating that the yielded zone became elastic after the ultimate strength in the right-hand-side span.

This means that the plastic deformation is localized in the left-hand-side span, which results in the smaller increase in the average strain. Furthermore, compressive strain is decreasing in the right-hand-side span because of unloading. The steep decrease of the capacity after the ultimate strength in Fig. 3 is due to these two factors, that is the localization of yielding and the unloading in the right-hand-side span.

Model F2 shows almost the same collapse behaviour as Model F1. However, Model F2 has a lower buckling and ultimate strength than an isolated simply supported plate. This is because of local buckling at stiffener web as described in Section 3.1.

In Figs. 5 (d) and (e), the collapse behaviour of Model F3 is shown. The left-hand-side panel undergoes secondary buckling in Model F3, and the tripping of the stiffener takes place at the middle of the left-hand-side span. Similar collapse behaviour was observed in the experiment by Tanaka and Endo[7].

Model F4 has a stocky stiffener, and an overall buckling occurred as indicated in Fig. 5 (f). In this model, the plate in the left-hand-side span is completely unloaded. Comparing Models F3 and F4, the latter has a higher ultimate strength. This is because the stiffener has locally collapsed at the middle of the left-hand-side span in Model F3.

Local tripping takes place in Models F1, F2 and F3, and the deflection of the Eulerian buckling mode is not appeared. On the other hand, Model F4 undergoes Eulerian buckling. Model F4 has the highest ultimate strength, but the capacity after the ultimate strength rapidly decreases as indicated in Fig. 3.

It may be concluded that a stiffened plate can have higher ultimate strength but rapid decrease in capacity after the ultimate strength when stocky flat-bar stiffeners are provided

3.2.2 Stiffened plates with angle-bar stiffeners

(1) Inplane rigidity in post-buckling range and initial yielding strength

Figure 7 shows the average stress-average strain relationships of stiffened plates with angle-bar stiffeners. The angle-bar stiffener has high flexural-torsional rigidity since its flange contributes to the flexural rigidity for horizontal bending. Because of this, the buckling strength is high and so the inplane rigidity above the buckling load.

In all the models, initial yielding takes place at the bottom of the web where lateral deflection of the web is maximum. The initial yielding strength is almost the same regardless of the size of angle-bar stiffeners.

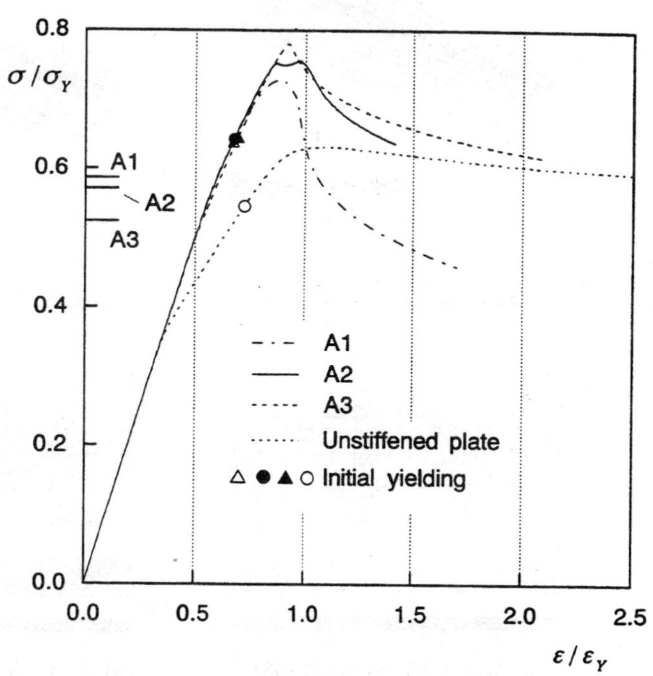

Fig. 7 Average stress-average strain relationships of stiffened plates with angle-bar stiffeners under thrust

(2) Buckling/plastic collapse behaviour

As indicated in Fig. 7 and Table 2, Model A3 shows the highest ultimate strength, and Model A1 the lowest. In case of Model A2, two load peaks exist.

Figure 8 shows load-end shortening relationships in individual spans of Model A2. For this model, deflection mode and spread of yielded zone at the first and the second load peaks are illustrated in Figs. 9 (a) and (b), respectively, and those at the last step in Fig. 9 (c).

At the first peak point of the capacity, the deflection mode is in three-half-waves in the loading direction as indicated in Fig. 9 (a). At this point, the plate in the right-hand-side undergoes secondary buckling, and the deflection mode changes to five-half-waves in the loading direction as illustrated in Fig. 9 (b). This secondary buckling phenomena can also be indicated in Fig. 10, where average stress-deflection component relationships are plotted. Above the initial buckling strength, the deflection component of three-half-waves in the loading direction increases. However, after the first load peak, this component begins to decrease and the component of five-half-waves starts to increase. After the second load peak, both components of three- and five-half-waves decrease. This is because higher order components appear to represent deflection mode with localized plastic deformation.

Figures 8 and 10 indicate that the secondary buckling strength is the ultimate strength for the plate between stiffeners, and the load increment is sustained by a stiffener after the secondary buckling. The ultimate strength as a whole is attained when the stiffener attains its own ultimate strength.

After the ultimate strength, the deflection of the Eulerian buckling mode is increased as indicated in Fig. 9 (c), and the capacity after the ultimate strength rapidly decreases as indicated in Fig. 7.

Model A1 shows almost the same collapse behaviour as in Model A2 as are known from Figs. 9 (d) and (e). On the other hand, overall torsional deformation is observed in Model A3 as indicated in Fig. 9 (f), and the deflection of Eulerian buckling mode is not appeared. Because of this, the strength reduction of Model A3 after the ultimate strength is rather rapid compared to that of Model A1 or Model A2.

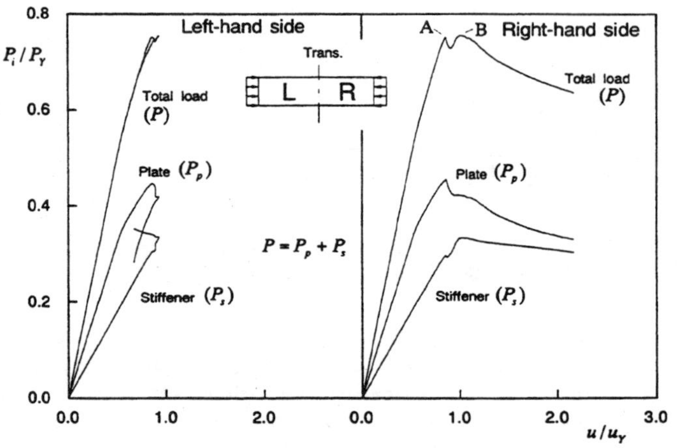

Fig. 8 Average stress-deflection component relationships in the left-hand-side panel (Model A2)

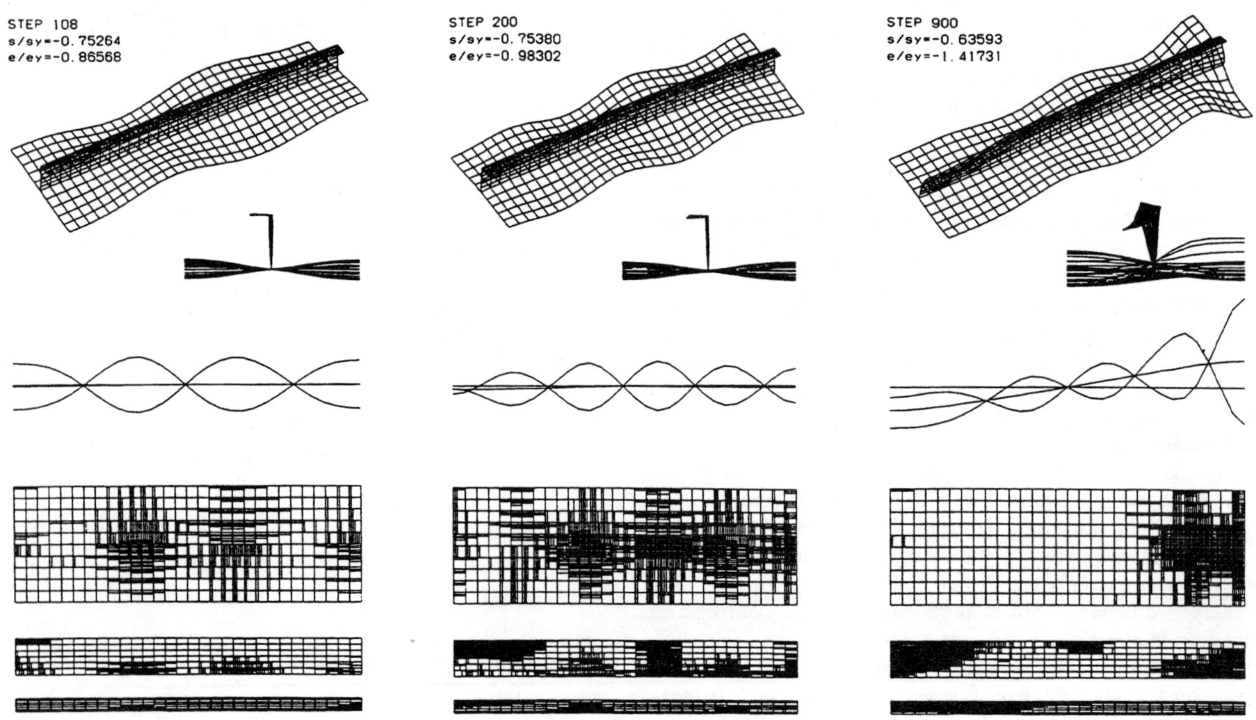

(a) Model A2 (σ/σ_Y=0.753) (b) Model A2 (Ultimate strength) (c) Model A2 (Final step)

Fig. 9 Collapse behaviour of stiffened plate with angle-bar stiffener under thrust

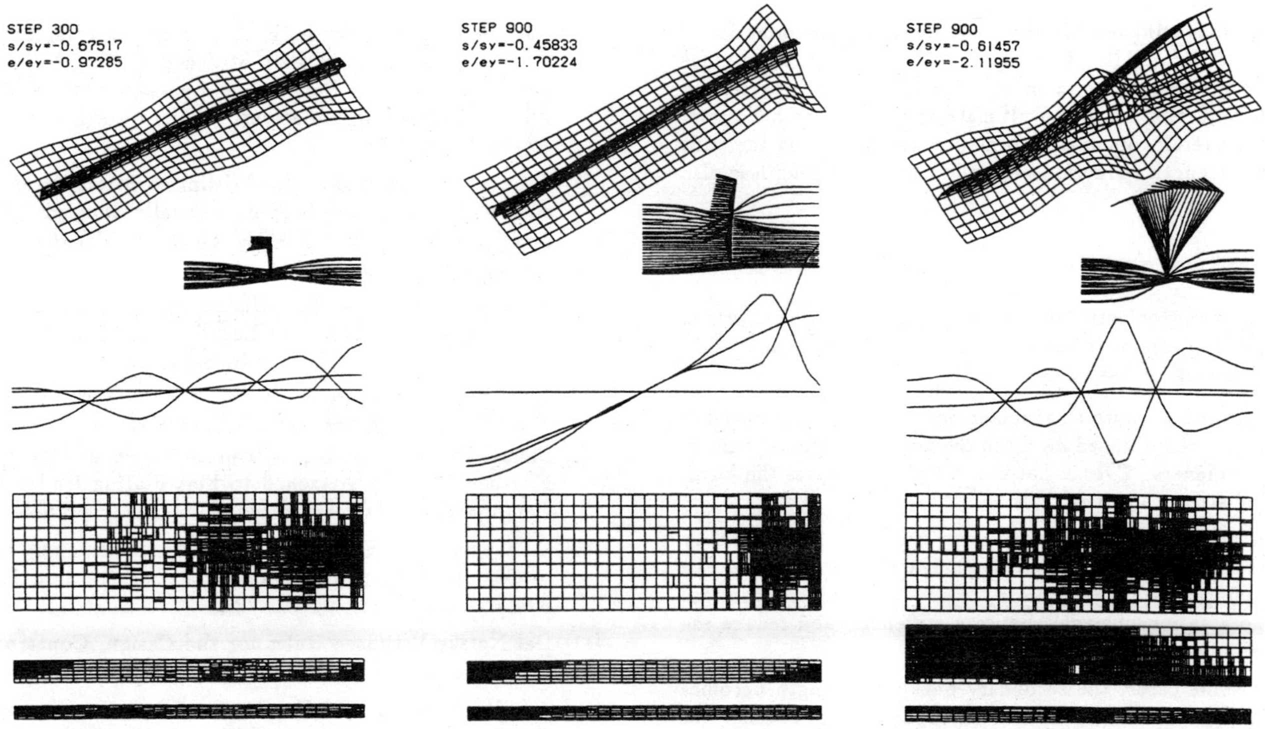

(d) Model A1 (σ/σ_Y=0.675) (e) Model A1 (Final step) (f) Model A3 (Final step)

Fig. 9 Collapse behaviour of stiffened plate with angle-bar stiffener under thrust (continued)

3.2.3 Influence of sectional geometry of stiffeners on collapse behaviour of stiffened plates

From the above observations, the influences of sectional geometries of stiffeners on buckling/plastic collapse behaviour of stiffened plates under thrust can be summarized as follows.

(1) Buckling strength, post-buckling inplane rigidity and ultimate strength

If stiffeners of the same flexural rigidity are provided, the panel buckling strength, inplane rigidity in post-buckling range and ultimate strength obtained by angle-bar stiffeners are higher than that obtained by flat-bar stiffeners. However, if thick flat-bar stiffeners having high torsional rigidity are used, the buckling strength is much increased.

(2) Secondary buckling

In general, the secondary buckling strength is very high compared to the primary buckling strength for simply supported plates, but is not so high for clamped plates[15]. In all cases analyzed here, the secondary buckling occurred. This is because the edge rotations of panels between stiffeners are constrained owing to the stiffeners. When stiffeners have high flexural-torsional rigidity, the secondary buckling takes place earlier, and the secondary buckling strength becomes the ultimate strength as for Models A1, A2 and A3.

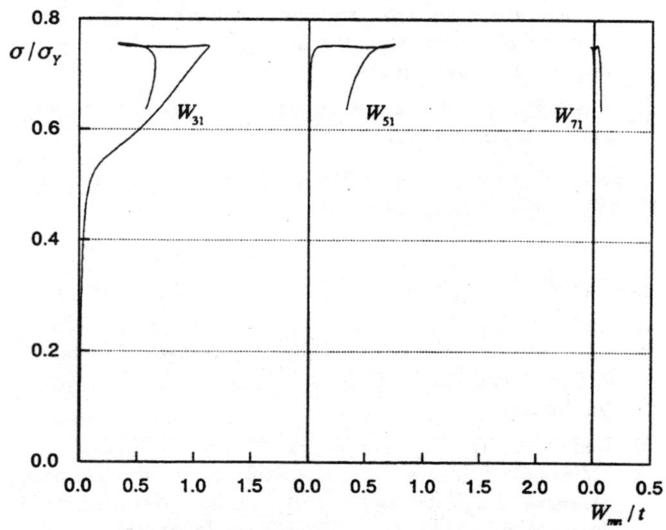

Fig. 10 Average stress-deflection component relationships in the left-hand-side panel (Model A2)

(3) Collapse mode

When a web has high depth to thickness ratio, the stiffener undergoes local tripping, and the plate also undergoes local collapse affected by local tripping. In this case, the overall Eulerian buckling does not occur, and the strength reduction after the ultimate strength is rather moderate.

Contrary to this, the stiffener undergoes overall Eulerian buckling in case of stiffened plates with angle-bar stiffeners or stocky flat-bar stiffeners. In this case, the plastic deformation is localized at the middle of the span, and unloading takes place in one of the two spans in many cases. Because of these, the strength reduction after the ultimate strength is rather rapid. When the overall torsional deformation takes place in the stiffener, the strength reduction after the ultimate strength is also rapid.

4 CONCLUSIONS

Series of elastoplastic large deflection analyses are performed on stiffened plates with flat-bar and angle-bar stiffeners subjected to thrust. It has been found that:

[1] Buckling strength and inplane rigidity in the post-buckling range are increased owing to the flexural-tortional rigidity of stiffeners. Even a flat-bar stiffener increases the buckling strength if it is thick.

[2] In all the stiffened plates analyzed here, the secondary buckling from three- to five-half-waves occurred in the elastoplastic range. This is because the stiffeners constrain the rotation of panels between stiffeners, and reduce the secondary buckling strength.

[3] In some cases, the secondary buckling strength becomes the ultimate strength, and the capacity does not increase after the secondary buckling because of the complex plastification during the mode change by secondary buckling.

[4] In the post-ultimate strength range, plastification in the panel is localized at one of the double spans. Consequently, the other span undergoes unloading. The rapid strength reduction after the ultimate strength is the results of these behaviours.

[5] The influences of sectional geometries on collapse behaviours are summarized in 3.2.3.

The authors are grateful to Messrs. Mizutani and Kanba for their help in performing calculations.

REFERENCES

[1] Ueda, Y., Yao, T. and Kikumoto, H.: "Minimum Stiffness Ratio of a Stiffener against Ultimate Strength of a Plate," J. Soc. Naval Arch. of Japan, Vol.140 (1976), pp.199-204 (in Japanese).

[2] Ueda, Y., Yao, T., Katayama, M. and Nakamine, M.: "Minimum Stiffness Ratio of a Stiffener against Ultimate Strength of a Plate (2nd Report)," J. Soc. Naval Arch. of Japan, Vol.143 (1978), pp.308-315 (in Japanese).

[3] Ueda, Y., Yao, T. Nakamine, M. and Nakamura, K.: Minimum Stiffness Ratio of a Stiffener against Ultimate Strength of a Plate (3rd Report)," J. Soc. Naval Arch. of Japan, Vol.145 (1979), pp.176-185 (in Japanese).

[4] Caridis, P. A.: "Interactive Flexural-Torsional Buckling Behaviour of Stiffened Plating," Marine Structures, 1 (198 6), pp.115-138.

[5] Caridis, P. A. and Frieze, P. A.: "Flexural-Torsional Elasto-plastic Buckling in Flat Stiffened Plating Using Dynamic Relaxation. Part 1: Theory," Thin-Walled Structures, 6 (1988), pp.453-481.

[6] Caridis, P. A. and Frieze, P. A.: "Flexural-Torsional Elasto-plastic Buckling in Flat Stiffened Plating Using Dynamic Relaxation. Part 2: Comparison with Test Results and Other Formulations," Thin-Walled Structures, 7 (1989), pp.37-72.

[7] Tanaka, Y. and Endo, H.: "Ultimate Strength of Stiffened Plates with Their Stiffeners Locally Buckled in Compression," J. Soc. Naval Arch. of Japan, Vol.164 (1988), pp.456-467 (in Japanese).

[8] Tanaka, Y. and Endo, H.: "Ultimate Strength of Stiffened Plates with Their Stiffeners Locally Buckled in Compression (2nd Report) - Effects of Initial Imperfections of Stiffeners due to Welding -," J. Soc. Naval Arch. of Japan, Vol.169 (1991), pp.391-402 (in Japanese).

[9] Panagiotopoulos, G. D.: "Ultimate Torsional Strength of Flat-bar Stiffeners Attached to Flat Plating Under Axial Compression," Marine Structures, 5 (1992), pp.535-557.

[10] GHavami, K.: "Experimental Study of Stiffened Plates in Compression up to Collapse," J. Constructional Steel Research, 28 (1994), pp.197-221.

[11] Det Norske Veritas: "Rules for the Design, Construction and Inspection of Offshore Structures, (1982).

[12] Toi, Y., Yuge, K. and Kawai, T.: "Basic Studies on the Crashworthiness of Structural Elements Part 1 Crush analysis by the finite element method," J. Soc. Naval Arch. of Japan, Vol.159 (1986), pp.248-257 (in Japanese).

[13] Yao, T., Fujikubo, M. and Ko, J.-Y.: "Secondary Buckling of Thin Plates with Initial Deflection," J. Soc. Naval Arch. of Japan, Vol.176 (1994), pp.309-318 (in Japanese).

[14] Yao, T., Fujikubo, M., Ko, J.-Y. and Yamamoto, M.: Influence of Secondary Buckling on the Buckling/Plastic Collapse Behaviour of Plates," J. Soc. Naval Arch. of Japan, Vol.177 (1995), pp.347-355 (in Japanese).

[15] Yao, T., Fujikubo, M. and Ko, J.-Y.: "Large Deflection Behaviour of Rectangular Plates under Inplane Compression," Trans. West-Japan Soc. Naval Arch., Vol.89 (1995), pp.179-190 (in Japanese).

Proceedings of the Seventh (1997) International Offshore and Polar Engineering Conference
Honolulu, USA, May 25-30, 1997

ISBN 1-880653-28-1 (Set); ISBN 1-880653-32-X (Vol. IV)

Buckling Strength Analysis of Ring-Stiffened Circular Cylindrical Shells Under Hydrostatic Pressure

Hai-Hong Sun
Wuhan Transportation University
Wuhan, China

Tie-Yun Chen
Shanghai Jiao Tong University
Shanghai, China

ABSTRACT

This paper concerns with the buckling strength problems of ring-stiffened circular cylindrical shells under hydrostatic pressure in both the deterministic and probabilistic analysis. A new singular perturbation method in deterministic analysis is put forward for solving the plastic non-dimensionalized nonlinear boundary-layer equations derived by the authors. The geometrical and physical reduction factors usually made by codes are regressed based on this method. A modified Monte Carlo simulation is used for the buckling reliability analysis of the shells with randomly geometrical and physical imperfections. The parametric sensitivities and failure mode correlation of inter-ring buckling and general buckling are examined and system reliability index is obtained. Conclusions are made from the asymptotic analysis and the modified Monte Carlo simulation.

KEY WORDS: plastic buckling; Monte Carlo simulation; circular cylindrical shells; system reliability analysis

INTRODUCTION

Buckling of stiffened and unstiffened circular cylindrical shells under hydrostatic pressure is the main failure mode in underwater vehicles. The initial investigations were based on linear theories which include many simplifications. The theoretical plastic buckling load is usually beyond the corresponding experimental one. This discrepancy can be attributed to 1) the effect of boundary layers (including pre-buckling non-uniform deformations and boundary conditions) and 2) the effect of initial imperfections (Huchison,1979;Simitses,1986). It was shown that the buckling load prediction in which the above effects are considered rigorously could be in well agreement with experimental one by using numerical methods in elastic range (Simitses,1979; Arbocz and Hol, 1990). The simplified nonlinear FEM was given by Yamamoto et al (1989) for fast analysis of the general plastic buckling of the shells. Little work has been done on comprehensively analyzing.the effects of boundary layers and initial imperfections on the load-carrying capacity of the cylindrical shells in plastic range. As the first part of this paper, a boundary-layer perturbation method with two small perturbation parameters is developed for the inter-ring and general buckling load predictions, in which the J_2 deformation theory of plasticity is adopted. An asymptotic evaluation is carried out in order to quantitatively effect the boundary layers and geometrical imperfections. It is found that these effects are relatively small and inversely proportional to Batdorf parameter Z. The calculating results from this method are in well agreement with the numerous experimental results. Based on this method, the geometrical and physical reduction factors usually made by codes are regressed.

The reliability theories of structures were vastly developed in the last several decades and widely used in the analysis of marine structures (Thoft-Christensen and Morutsu,1986; Mansour,1988). The first-order reliability method is quite suitable for the moderate curvatures of limit state surface but might not be appropriate for the limit state of the plastic buckling of ring-stiffened circular cylindrical shells with randomly geometrical and physical imperfections under hydrostatic pressure. A modified Monte Carlo simulation suggested by Harbitz (1986) with a little improvement in the formation of sampling points is suggested and the comparison is carried out between various methods for the buckling strength of ring-stiffened circular cylindrical shells under hydrostatic pressure. In this part, the parametric sensitivities and failure mode correlation of inter-ring buckling and general buckling are examined and system reliability index is obtained. It concluded that system reliability index is slightly reduced compared with the estimation only considering local buckling failure mode.

A BOUNDARY-LAYER PURTURBATION METHOD FOR THE PLASTIC BUCKLING ANALYSIS OF CIRCULAR CYLINDRICAL SHELLS UNDER HYDROSTATIC PRESSURE

Asymptotic Solutions

Let us consider a shell with radius R , wall thickness t, and length L. The location of a point on the shell surface is given by the axial coordinate X_1 and the circumferential coordinate X_2. The component of additional displacement W normal to the shell mid-surface is positive inward and the initial radial imperfection W^* is much less than R. We introduce the following non-dimensional quantities

$$\left.\begin{aligned} & x_1 = \frac{\pi}{L}X_1,\ x_2 = \frac{X_2}{R},\ \beta = \frac{L}{\pi R},\ Z = \frac{L^2}{Rt}\sqrt{1-\nu^2}, \\ & C = \sqrt{3(1-\nu^2)},\ \varepsilon = \frac{\pi^2}{\sqrt{12Z}},\ (w, w^*) = \frac{(W, W^*)}{t}2C\varepsilon, \\ & f = \frac{F}{D}\varepsilon^2,\ q_{cl} = \frac{\sqrt{2}\pi E}{3\sqrt{3}(1-\nu^2)^{3/4}}\frac{R}{L}\left(\frac{t}{R}\right)^{\frac{5}{2}},\ \lambda_q = \frac{q}{q_{cl}} \end{aligned}\right\} \quad (1)$$

where E =Young's modulus, ν =Poisson's ratio, D =bending stiffness, F = stress function, q =hydrostatic pressure. By introducing the Donnell-Mushtari-Vlasov strain-displacement relations and the J_2 deformation theory of plasticity for plane stress into the three equations of equilibrium, we can derive the following equations in terms of the additional outward normal displacement w and stress function f

$$\left.\begin{aligned} & L_h[f] + \varepsilon L_q[w] + \frac{\partial^2 w}{\partial x^2} = -\frac{1}{2}\beta^2(w; w + 2w^*) \\ & \varepsilon^2 L_d[w] - \varepsilon L_q[f] - \frac{\partial^2 f}{\partial x^2} = \beta^2(f; w + w^*) + \frac{4}{3^{3/4}}\lambda_q \varepsilon^{\frac{3}{2}} \end{aligned}\right\} \quad (2)$$

These are non-dimensionalized nonlinear incremental boundary-layer equations and are applicable to ring- and stringer-stiffened shells if the stiffener properties are "smeared out" to arrive at effective bending, stretching, and torsional stiffness for the skin-stiffener combination. The operators are defined as following

$$\left.\begin{aligned} & L_d[\ ] = h_1\frac{\partial^4}{\partial x_1^4} + 2h_2\beta^2\frac{\partial^4}{\partial x_1^2 \partial x_2^2} + h_4\beta^4\frac{\partial^4}{\partial x_2^4}, \\ & L_h[\ ] = s_1\frac{\partial^4}{\partial x_1^4} + 2s_2\beta^2\frac{\partial^4}{\partial x_1^2 \partial x_2^2} + s_4\beta^4\frac{\partial^4}{\partial x_2^4} \\ & L_q[\ ] = e_1\frac{\partial^4}{\partial x_1^4} + 2e_2\beta^2\frac{\partial^4}{\partial x_1^2 \partial x_2^2} + e_4\beta^4\frac{\partial^4}{\partial x_2^4}, \\ & (s;t) = \frac{\partial^2 s}{\partial x_1^2}\frac{\partial^2 t}{\partial x_2^2} - 2\frac{\partial^2 s}{\partial x_1 \partial x_2}\frac{\partial^2 t}{\partial x_1 \partial x_2} + \frac{\partial^2 s}{\partial x_2^2}\frac{\partial^2 t}{\partial x_1^2} \end{aligned}\right\} \quad (3)$$

The parameters $h_1, h_2, \cdots$ etc are given in the reference of Sun(1994) in detail. The boundary conditions and periodic condition are

$$\left.\begin{aligned} & x_1 = 0, \pi,\ w = \frac{\partial w}{\partial x_1} = 0;\quad \frac{1}{2\pi}\int_0^{2\pi}\beta^2\frac{\partial^2 f}{\partial x_2^2}dx_2 + \frac{2}{3^{3/4}}\lambda_q\varepsilon^{\frac{3}{2}} = 0; \\ & \int_0^{2\pi}[(s_1\frac{\partial^2 f}{\partial x_1^2} - \nu_{xy}\beta^2\frac{\partial^2 f}{\partial x_2^2}) + \varepsilon(e_1\frac{\partial^2 w}{\partial x_1^2} + e_y\beta^2\frac{\partial^2 w}{\partial x_2^2}) + w - \\ & \frac{1}{2}\beta^2(\frac{\partial w}{\partial x_2})^2 - \beta^2\frac{\partial w}{\partial x_2}\frac{\partial w^*}{\partial x_2}]dx_2 = 0 \end{aligned}\right\} \quad (4)$$

Because the numerical solutions have been found difficulty to treat the double singularities induced by bifurcation point and boundary layer near to the edges in these equations, a combined singularity perturbation method is proposed. For the perfect cylindrical shells, we can assume the solutions in the following form in the neighborhood of the critical state,

$$\left.\begin{aligned} & w = w_0(\lambda_p, x, y, \varepsilon) + \eta w_1(x, y, \varepsilon) + \eta^2 w_2(x, y, \varepsilon) + \cdots, \\ & f = f_0(\lambda_p, x, y, \varepsilon) + \eta f_1(x, y, \varepsilon) + \eta^2 f_2(x, y, \varepsilon) + \cdots, \\ & \lambda_q/\lambda_{qc} = 1 + \eta b_1 + \eta^2 b_2 + \cdots, \end{aligned}\right\} \quad (5)$$

where w_0, f_0 = axisymetrical prebucking quantities; w_1, f_1 = small incremental quantities during bifurcation buckling; w_k, f_k $(k \geq 2)$ = postbuckling quantities; λ_{qc} = load of bifurcation buckling; $b_k(k > 1)$ = load coefficients of postbuckling; η = a small number representing the contribution of w_1 to the buckled state. According to the boundary layer method in singularity perturbation theory (Nayfen,1981), w_k and f_k $(k \geq 0)$ can be further assumed as

$$w_k = w_k^{(o)} + w_k^{(i)} + w_k^{(I)}, \qquad f_k = f_k^{(o)} + f_k^{(i)} + f_k^{(I)} \quad (6)$$

where $w_k^{(o)}, f_k^{(o)}$ = outer or regular solutions; $w_k^{(i)}, f_k^{(i)}$ and $w_k^{(I)}, f_k^{(I)}$ = =inner or boundary-layer solutions at the edge x=0 and x=π respectively. The boundary layer variables ξ, ς are defined as

$$\xi = \frac{x}{\sqrt{\varepsilon}}, \qquad \varsigma = \frac{\pi - x}{\sqrt{\varepsilon}}$$

Substituting Eqs.(5) and (6) into Eq.(3), the regular equations and boundary layer equations and the corresponding plastic solutions of prebuckling, bifurcation buckling and post buckling can be obtained. The regular bifurcation buckling load is

$$\begin{aligned} \lambda_{qc}^{(o)} = & \frac{3^{3/4}m^2}{4\gamma^2}\left[\frac{m^2}{s_4\gamma^4} + \frac{\gamma^4}{m^2}(h_4 + \frac{e_4^2}{s_4}) - 2\frac{e_4}{s_4}\right] + \frac{3^{3/4}m^2}{2\gamma^2}\varepsilon^{\frac{1}{2}}\left[(h_2 + \frac{e_2 e_4}{s_4})\gamma^2\right. \\ & \left. - \frac{e_2 m^2}{s_4\gamma^2} - \frac{m^2 - e_4\gamma^4}{s_4\gamma^2}(\frac{s_2 m^2}{s_4\gamma^4} - \frac{e_4 s_2}{s_4} + e_2)\right] + \cdots \end{aligned} \quad (7)$$

where γ = parameter related to the stiffeners and m = half wave number in axial direction. The modified bifurcation buckling load accounting for the effect of the boundary layer is

$$\begin{aligned} \lambda_{qc} = & \left\{\left(1 + \frac{1}{2}\varepsilon^{\frac{1}{2}}\frac{m^2}{\gamma^2}\right) + \varepsilon\frac{m^2}{2\pi\gamma^2\alpha}\left(\frac{\gamma^2}{s_1}\frac{s_1 - 0.5\nu_{xy}}{\phi^2 + \alpha^2} + \frac{2\phi^2 - \alpha^2}{\phi^2}\right)\right\}^{-1} \times \\ & \left\{\lambda_{qc}^{(o)}\left(1 + \frac{1}{2}\varepsilon^{\frac{1}{2}}\frac{m^2}{\gamma^2}\right) + \frac{3^{3/4}}{4}\varepsilon\frac{m^2}{\pi\alpha}[(h_2 + 2\frac{e_1 e_2}{s_1} - \frac{e_1^2 s_2}{s_1^2})\right. \\ & \left. - 2\frac{e_2 - e_1 s_2/s_1}{s_1(\phi^2 + \alpha^2)} + \frac{s_2}{s_1^2}\frac{3\alpha^2 - \phi^2}{(\phi^2 + \alpha^2)^3}]\right\} \end{aligned} \quad (8)$$

The load-carrying capacity of the shells is

$$\begin{aligned} \frac{\lambda_q}{\lambda_{qc}} = & 1 - \eta^2\frac{3^{3/4}C^2 m^2}{2\lambda_{qc}}\{\varepsilon[2(\frac{m^2}{s_4\gamma^4} - \frac{e_4}{s_4}) + \\ & \frac{m^2}{4(h_4 + \frac{e_4^2}{s_4})\gamma^2 - \frac{4}{3^{3/4}}\lambda_{qc}}(\frac{2m^2}{s_4\gamma^4} - \frac{3e_4}{s_4})(\frac{m^2}{s_4\gamma^4} - \frac{3e_4}{2s_4})] + \cdots\} + \cdots \end{aligned} \quad (9)$$

where

$$\phi = \sqrt{(\sqrt{h_1 s_1 + e_1^2} + e_1)/2(h_1 s_1 + e_1^2)},\ \alpha = \sqrt{(\sqrt{h_1 s_1 + e_1^2} - e_1)/2(h_1 s_1 + e_1^2)}$$

According to Yamaki (1984), the effect of initial geometrical imperfection on the buckling load can be taken into account. The non-dimensionalized initial geometrical imperfection has the form

$$w^* = 2C\varepsilon\mu_{11}\sin mx \sin ny + 2C\varepsilon\mu_{20}\cos 2mx + 2C\varepsilon\mu_{02}\cos 2ny \quad (10)$$

Comparison with Experiments and Concluding Remarks

The plastic buckling behavior of ring-stiffened cylindrical shells under hydrostatic pressure is investigated analytically using a simple program developed for the purpose and numerous examples are solved numerically. The boundary condition for the experimental shells is realistic and is close to clamped edge condition. The results obtained are listed in Table 1 and Table 2. For the machine processing shells, the results obtained from the present method are in well agreement with the experimental results and are better than the theoretical results from Yamamoto et al (1989) and Zhou (1979) except for MYS-2 shell. However, the dimensions of MSY-2 shell are between those of MSY-1 and MSY-3 shells, the present result is reasonable. For the welded processing shells, the theoretical results obtained from the present method and Yamamoto et al (1989) are higher than the experimental results due to prominently geometrical imperfections. The imperfect sensitive analysis (Fig.1) show that the computational result becomes in consistent with the experimental one for WSY-1 shell when $\mu_{11} = 0.68$, which indicates that the ring-stiffened circular cylindrical shell is weakly sensitive to geometrical imperfections. Fig.2 shows the effect of ring-stiffening parameter $r = \sqrt{I_r/A_r}$ on the load-carrying capacity based on MSY-1 shell, which indicates that the increase of ring dimensions can significantly improve the load-carrying capacity in elastic range but is not effective in plastic range.

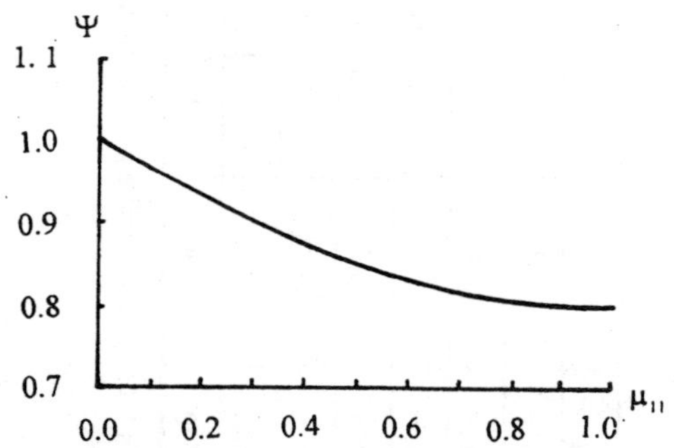

Fig.1 Imperfection sensitivity of WSY-1 shell

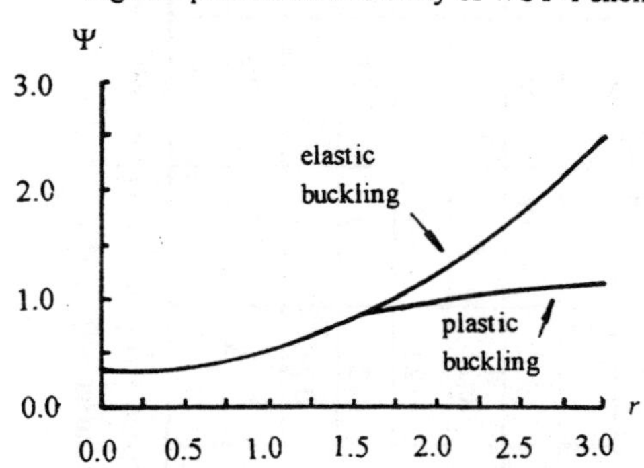

Fig.2 The effect of ring-stiffening parameter

In relation to ring-stiffened circular cylindrical shells subjected to hydrostatic pressure, the following concluding remarks can be directly made from the asymptotic analysis.

1) Eq.(7) shows that the bifurcation buckling loads of internally ring-stiffened shells ($e_1 = 0$, $e_2 > 0$ and $e_4 < 0$) are slightly greater than those of externally ring-stiffened shells if taking the eccentricity effect of rings into account;
2) Eq.(8) shows that the effect of boundary layer is the first order of ε, which implies that the effect of boundary layer is inversely proportional to Batdorf parameter Z;
3) Eq.(9) shows that the imperfection sensitive factor is the first order of ε, which also implies that the imperfection sensitive factor is inversely proportional to Batdorf parameter Z.

Regressive Formulas of the Geometrical and Physical Reduction Factors

The regressive analysis are carried out according to the least square method based on the above asymptotic analysis. The buckling strength of ring-stiffened circular cylindrical shells under hydrostatic pressure can be written as

$$\sigma_L = C_s C_g \sigma_E \tag{11}$$

The physical reduction factor is

$$C_s = \begin{cases} 1.0 & \sigma_E/\sigma_Y < 0.75 \\ 0.62806 + 1.25767(\sigma_E/\sigma_Y) - 1.24447(\sigma_E/\sigma_Y)^2 & \\ \quad + 0.30506(\sigma_E/\sigma_Y)^3 & 0.75 \le \sigma_E/\sigma_Y \le 1.8 \\ 1.15\sigma_Y/\sigma_E & \sigma_E/\sigma_Y > 1.8 \end{cases} \tag{12}$$

where

$$\sigma_E = \begin{cases} 0.60E\left(\frac{t}{R}\right) \Big/ \left(0.642\frac{d_r}{\sqrt{Rt}} - 0.37\right) \\ \quad \text{for the inter-ring buckling} \\ \frac{ER}{(t + A_r/d_r)[1+\beta^2(n^2-1)/2]}\left[\frac{t\beta^4}{R(\beta^2+n^2)^2(n^2-1)} + \frac{(n^2-1)I}{R^3 d_r}\right] \\ \quad \text{for general buckling} \end{cases} \tag{13}$$

n = wave number of general elastic buckling, d_r =spacing of rings, I = the inertia moment of ring with effective width of shell plating about the shell mid-surface and σ_Y = yielding stress.

The geometrical reduction factor for the inter-ring buckling is

$$C_g = \begin{cases} 1.0 & \frac{d_r}{\sqrt{Rt}} < 1.56 \\ 1.25 - 0.1605\frac{d_r}{\sqrt{Rt}} & 1.56 \le \frac{d_r}{\sqrt{Rt}} \le 3.12 \\ 0.75 & \frac{d_r}{\sqrt{Rt}} > 3.12 \end{cases} \tag{14}$$

and the geometrical reduction factor for the general buckling is

$$C_g = \begin{cases} 0.9 - 0.8333U & U < 1.2 \\ 0.8 & U \ge 1.2 \end{cases} \tag{15}$$

where

$$U = 0.535\frac{t^3 L d_r}{I\sqrt{R(t + A_r/d_r)}}$$

In the geometrical reduction factor for both inter-ring and general buckling, an admissible imperfection amplitude μ_{11} is 0.3.

RELIBILITY ANALYSIS OF RING-STIFFENED CIRCULAR CYLINDRICAL SHELLS UNDER HYDROSTATIC PRESSURE

Table 1 The theoretical and experimental results of the models presented by Chen & Shao(1979) and Zhou(1979)

Shells	Geometrical Properties							Chen&Shao(1979),Zhou(1979)			Present				
	L	R	t	d_r	e_r	A_r	I_r	q_{exp}	q	κ[b]	q	κ	μ_{11}	q^*	κ
Name[a]	(cm)	(cm)	(cm)	(cm)	(cm)	(cm^2)	(cm^4)	(kN/cm^2)	(kN/cm^2)		(kN/cm^2)			(kN/cm^2)	
MSC	22.0	16.5	0.14	4.4	-0.275	1.64×10^{-1}	2.3×10^{-3}	42.5(1,6)[c]	42.6(1,6)	—	46.0(1,6)	0.76	0.05	42.4(1,6)	0.70
MSZ-1	16.0	4.0	0.05	1.58	-0.08	6.76×10^{-3}	6.71×10^{-6}	28.0(1,3)	31.2(1,3)	1.00	26.8(1,3)	1.00	—	—	—
MSZ-2	16.0	4.0	0.06	1.60	-0.13	1.30×10^{-5}	4.31×10^{-5}	65.0(1,3)	61.6(1,3)	1.00	63.1(1,3)	1.00	—	—	—
MSZ-3	16.0	4.0	0.06	1.59	-0.18	1.89×10^{-5}	1.41×10^{-4}	82.0(1,3)	86.4(1,3)	0.66	84.3(1,3)	0.60	—	—	—
MAZ-1	16.0	4.0	0.06	1.59	-0.08	6.20×10^{3}	5.35×10^{-6}	5.4(1,3)	5.4(1,3)	0.80	5.4(1,4)	0.82	—	—	—
MAZ-2	16.0	4.0	0.06	1.58	-0.13	1.29×10^{-2}	1.81×10^{-5}	8.0(1,3)	8.7(1,3)	0.48	7.8(1,3)	0.44	—	—	—
MAZ-3	16.0	4.0	0.06	1.60	-0.18	1.95×10^{-2}	1.45×10^{-4}	9.4(1,3)	9.6(1,3)	0.25	9.5(1,3)	0.24	—	—	—

note: [a]MSC shell is quoted from Chen & Shao(1979), made of mild steel by machine processing.

MSZ and MAZ shells are quoted from Zhou(1979), made of mild steel and aluminum alloy respectively by machine processing;

[b] $\kappa = \frac{\text{plastic buckling load}}{\text{elastic buckling load}}$; [c]Buckling mode(m,n)

Table 2 The theoretical and experimental results of the models presented by Yamamoto et al(1989)

Shells	Geometrical Properties							Yamamoto et al(1989)			Present
	L	R	t	d_r	e_r	A_r	I_r	Ψ_{exp}	$\Psi_{end\ A}$	$\Psi_{end\ B}$	Ψ
Name[a]	(cm)	(cm)	(cm)	(cm)	(cm)	(cm^2)	(cm^4)				
MSY-1	30.0	14.9	0.23	5.0	-0.54	0.2125	0.01279	0.93(1,4)[b]	0.94(1,4)	0.97(1,4)	0.96(1,4)
MSY-2	45.0	14.9	0.23	5.0	-0.64	0.2625	0.02412	0.87(1,4)	0.90(1,4)	0.92(1,4)	1.00(1,4)
MSY-3	60.0	14.9	0.23	5.0	-0.69	0.2875	0.03619	1.01(1,4)	0.99(1,4)	1.05(1,4)	1.02(1,4)
MSY-4	60.0	15.2	0.23	5.0	-0.74	0.2275	0.05126	0.99(1,4)	0.95(1,4)	0.99(1,4)	0.98(1,4)
WSY-1	147.0	48.6	0.75	16.3	-2.28	3.1160	3.74959	0.83(1,3)	0.97(1,3)	1.09(1,3)	1.02(1,3)
WSY-2	196.0	48.6	0.75	16.3	-2.58	3.6080	5.82090	0.82(1,3)	0.91(1,3)	1.02(1,3)	1.05(1,3)

note: [a]MSY and WSY shells are quoted from Yamamoto et al(1989), made of mild steel by machine processing and welded processing respectively;

[b] $\Psi = qR/\sigma_y t$; [c]Buckling mode(m,n)

A Modified Monte Carlo Simulation

A modified Monte Carlo simulation is put forward in this section in order to examine the suitability of the first-order reliability method in the present analysis. According to Harbitz (1986), it can be done with the following two steps: calculating the Hasofer-Lind reliability index β by using first-order reliability method and restricting the sampling domain of simple Monte Carlo simulation to the points outside the n—dimensional β—sphere. A little improvement with Harbitz (1986) is in the formation of sampling points. In this method the random radius $R \geq \beta$ is directly sampled from a χ^2 distribution and the random direction from n independent (-1,1) uniform distributions and so the higher efficiency is achieved. Using this method to analyze the hyperspherical and hyperbolic limit surfaces as examples, we obtain the simulating results in well agreement with the exact results. In carrying out the reliability analysis of axially compressed circular cylindrical shells with random geometrical imperfections, we find that the limit surface of the shells is near to the hyperelliptical surface and the first-order reliability method is prone to considerable errors (Sun and Chen, 1996).

Reliability Results Based on the Asymptotic Analysis

A comparison between the first order reliability method (advanced first-order second-moment method, FOSM) and the modified Monte Carlo simulation (MMCS) is further dealt with for the reliability analysis of ring-stiffened circular cylindrical shells under hydrostatic pressure. The limit state function is selected as

$$G = \lambda_q q_{cl} - q \qquad (16)$$

where λ_q is directly calculated from the asymptotic analysis in the first part. The basic random variables , in which the mean values is in accordant with MSZ-3 shell and the coefficients of variation are assumed, and their probability density functions are listed in Table 3. Three cases of limit state functions are taken into account, they are 1) elastic buckling of perfect shells; 2) plastic buckling of perfect shells and 3) plastic buckling of imperfect shells. Reliability results are shown in Table 4. From Table 4, we can see that the difference of Case 1 between FORM and MMCS is the lowest, and of Case 3 is the greatest. Therefore, if simultaneously choosing the yielding stress and imperfections as random variables, FOSM is not appropriate for the theoretical reliability analysis of ring-stiffened circular cylindrical shells under hydrostatic pressure.

Table 3 Basic variables

Variable	Distribution	M. V.	C.O.V
q	Normal	65.00	10%
L	Constant	16.00	—
R	Normal	4.00	10%
t	Normal	0.06	10%
d_r	Constant	1.60	—
e_r	Constant	-0.18	—
A_r	Constant	0.0195	—
I_r	Constant	1.45×10^{-4}	—
E	Normal	2.06×10^{6}	5%
ν	Constant	0.3	—
σ_Y	Normal	3940	10%
μ_{11}	Normal	0.1	10%

Table 4 Results of Reliability Analysis

Description	FOSM		MMCS	
	β	P_f	P_f	β
Case 1	4.345	0.9^5297	0.9^5004	4.266
Case 2	0.916	0.82022	0.76613	0.726
Case 3	0.594	0.72357	0.66205	0.418

Reliability Results Based on the Regressive Formulas

In this section, the system reliability analysis are carried out which includes two failure modes of inter-ring buckling and general buckling based on the above regressive formulas. The limit state function is written as

$$G = C_s C_g \sigma_E \frac{t + A_r/d_r}{R} - q \qquad (17)$$

where the ring-stiffening parameters are selected zeros for the inter-ring buckling. The basic random variables and their probability density functions are listed in Table 5. Reliability results are shown in Table 6. From Table 6, we can find that the reliability result from FOSM is in reasonable agreement with the one from MMCS and so FOSM is applicable. Using FOSM, the correlation coefficient between the two failure modes is obtained

$$\rho = 0.354$$

According to the probability network evaluation technique, if we select the critical correlation coefficient 0.70, then the two failure modes can be considered incorrelate and the system reliability is

$$\beta = 2.59934 \qquad P_f = 0.46658\times10^{-2}$$

A slight reduction of the system reliability is obtained compared with the estimation only considering the inter-ring buckling failure mode.

Table 5 Basic variables

Variable	Distribution	M. V.	C.O.V
q	Normal	37.50	3%
L	Constant	900.00	—
R	Normal	300.00	—
t	Normal	3.00	—
d_r	Constant	60.00	—
e_r	Constant	11.45	—
A_r	Constant	17.96	—
I_r	Constant	468.00	—
E	Normal	2.02×10^{6}	1%
ν	Constant	0.3	—
σ_Y	Normal	6625	6.62%

Table 6 Results of Reliability Analysis

Description		β	P_f
Inter-ring buckling	FOSM	2.59961	0.46637×10^{-2}
	MMCS	2.63726	0.41788×10^{-2}
General buckling	FOSM	4.59736	0.21816×10^{-5}
	MMCS	4.34875	0.68458×10^{-5}

CONCLUSIONS

The following conclusions are based on our analysis for predicting the buckling strength of ring-stiffened circular cylindrical shells under hydrostatic pressure:

1) The increase of ring-stiffening dimensions can significantly improve the load-carrying capacity in elastic range but is not effective in plastic range;
2) The load-carrying capacity of internally ring-stiffened shells is slightly greater than that of externally ring-stiffened shells;
3) The effect of boundary layer including the boundary conditions and prebuckling deformations is the first order of ε, which implies that the effect of boundary layer is inversely proportional to Batdorf parameter Z;
4) The imperfection sensitive factor is the first order of ε, which also implies that the imperfection sensitive factor is inversely proportional to Batdorf parameter Z;
5) The reliability index calculating from the first order reliability method is theoretically higher than that from the present modified Monte Carlo simulation, which indicates that over estimation of the shell reliability will result if the first order reliability method is used, when taking the imperfection amplitude and yielding stress as random variables which is realistic in practice;
6) The system reliability including the two failure modes of inter-ring buckling and general buckling is slightly reduced compared with the estimation only considering the inter-ring buckling failure mode, in which the imperfection amplitude takes the admissible value from codes.

REFERENCES

Arbocz, J and Hol,J(1990). "Koiter's stability theory in a computer-aided engineering(CAE) environment", *Int. J. Solids Structures*, Vol.26,No.9/10, pp945-973.

Chen,TY and Shao,WJ(1979). "General Large-deflection Plastic Stability and Imperfection Sensitivity of Ring-stiffened Circular Cylindrical Shells under Hydrostatic Pressure", *Shipbuilding of China*, Vol 3, pp59-71.

Harbitz, A(1986). "An Efficient Sampling Method for Probability of Failure Calculation", *StructuralSafety*, Vol 3, pp109-115, 1986.

Hutchison, JW(1979) "Plastic buckling", *Advances in Applied Mechanics*, Vol 4, pp163-190.

Mansour,A(1988). "Reliability of Marine Structures", Draft Final Report, Contract No. DTCG23-86-C- 20054.

Nayfen, AH(1981). "Introduction to Perturbation Techniques", A Wiley Publication, John Wiley & Sons.

Simitses,GJ(1986)."Buckling and Postbuckling of Imperfect Cylindrical Shells: A Review", *Appl. Mech. Rev.*, Vol 39, No 10, pp1517-1524.

Simitses GJ et al(1979)."Nonlinear Stability Analysis of Pressure-loaded Imperfect Stiffened Cylinders", *J Ship Research*, Vol 23, No 2, pp123～126.

Sun,HH(1994). "Buckling and Reliability of Circular Cylindrical Shells in Ship and Ocean Engineering", Ph.D. Dissertation, Shanghai Jiao Tong University.

Sun,HH and Chen,TY(1996). "Reliability Analysis of Axially Compressed Circular Cylindrical Shells with Random Imperfections", *J Computational Structural Mechanics and Applications*, Vol 13, No 4, pp461-468.

Thoft-Christensen,P and Murotsu,Y(1986). "Application of Structural Systems Reliability Theory", Springer-Verlag.

Yamaki,N(1984). "Elastic Stability of Circular Cylindrical Shells", North-Holland Series in Appl Math & Mech, Eds. E.Becker, Vol 27, North-Holland, Amsterdam, New York, Oxford .

Yamamoto, Y et al (1989). "General Stability of Ring-stiffened Cylindrical Shells Under External Pressure", *Marine Structures*, Vol 2, pp133-149.

Zhou, CT(1979). "Elastic-plastic Stability Theory of Thin Shells", National Defense Industry Press.

Proceedings of the Seventh (1997) International Offshore and Polar Engineering Conference
Honolulu, USA, May 25-30, 1997

ISBN 1-880653-28-1 (Set); ISBN 1-880653-32-X (Vol. IV)

The Governfactor of Plastic Spin

Y.Y. Nam
Korea Institute of Machinery & Metals
Taejeon, Korea

S.G. Lee
Korea Maritime University
Pusan, Korea

ABSTRACT

Focusing studies on the plastic spin, three theories are extracted which have quite different concepts. The plastic spin is still an open research area. Though texturing is known as an origin of the plastic spin, the governing factor and its origin have not yet been studied thoroughly. It is, therefore, necessary to investigate and observe the fundamental behavior of plastic spin.

This paper provides some investigation for the behavior of plastic spin through the simulation of body centered polycrystal. The effects, such as texturing, hardening, deformation rate, and reverse loading, are investigated. In addition, texturing is emphasized and some significant governing factors are proposed.

Key Words: Plastic Spin, Finite Deformation, Texture, Rate-dependent Plasticity

INTRODUCTION

Since Kratochvil(1971) recognized a need for the constitutive equation of plastic spin, many researches have been studied on this subject. Summarizing them from the conceptual viewpoint, it can be found there are two divergencies.

First, as Kratochvil's statement, an additional constitutive equation has been widely introduced by Loret(1983), Dafalias (1983,1987), Pecherski(1988), Zbib and Aifantis(1988), Dafalias and Aifants(1990), Paulun and Pecherski(1992), Ning and Aifantis(1994), and so on. It has been accepted, in general, that the plastic spin can be described by a function of tensor, which may represent an aspect of internal state of a material. Dafalias (1983) introduced the isotropic function representation, and proposed the simplest one. The constitutive equation by the representation theory has an unknown coefficient. Paulun and Pecherski (1992) treated the coefficient more rationally. Besides the isotropic function approach, Gissen(1989) proposed a generalized flow rule where a microstress, in general unsymmetrical tensor, plays an important role.

Second, it was argued by some authors, eg. Schiek and Stumpf(1995), that there needs no additional constitutive equation for the plastic spin. According to their works, the additional constitutive equation was made to be a quite redundant one by deriving a kinematic function.

On the other hand, it is interesting that there are quite distinctive concepts each other in the above theories. It can be found that the plastic spin is still an opened subject to be dug out. Therefore, more investigation are needed on the plastic spin to understand its fundamental behavior and to develop its future researches.

The purpose of this paper is to provide a thorough investigation on the plastic spin. The simulation of plastic spin is conducted on a body centered polycrystal using a simple static accommodation model and the constitutive equation of Johnston-Gilman type(1965). Based on the present simulation, the effects of texturing, hardening, deformation rate, reverse loading and so on, are investigated. Though the texturing has been known as an origin of plastic spin, its significance is emphasized and the governing factors of the plastic spin are suggested.

SIMULATION MODEL

Kinematics

The multiplicative decomposition is applied to the deformation gradient tensor F.

$$F = F^e F^p \tag{1}$$

where F is evaluated from the reference configuration x^0

updated at previous step. The increment of Green strain tensor can be described as follows,

$$\Delta E=\frac{1}{2}(F^TF-I)$$

$$=\frac{1}{2}\{(F^p)^T(F^e)^TF^eF^p-(F^p)^TF^p+(F^p)^TF^p-I\} \quad (2)$$

where I is an unit diagonal tensor. With the definition of Green strain tensor, Eq. (2) is reduced to Eq.(3) as follows,

$$\Delta E=(F^p)^T\Delta E^eF^p+\Delta E^p \quad (3)$$

where ΔE^e is the elastic strain increment measured from x^p and ΔE^p, the plastic strain increment from x^o.

The velocity gradient of plastic deformation gradient is defined as follows.

$$L^p=\dot{F}^p(F^p)^{-1} \quad (4)$$

The plastic deformation gradient is approximately obtained by integrating its velocity gradient and under the assumption of constant velocity gradient during incremental deformation,

$$F^p=e^{(L^p\Delta t)} \quad (5)$$

where Δt is a small time increment. To avoid eigenvalue analysis, Eq.(5) is approximated as follow.

$$F^p=I+L^p\Delta t \quad (6)$$

From Eq.(3), (4) the elastic strain increment can be recast as follows.

$$\Delta E^e=(F^p)^{-T}(\Delta E-\Delta E^p)(F^p)^{-1} \quad (7)$$

Finally the plastic velocity gradient L^p is obtained from a constitutive equation, and the plastic spin is defined as follows.

$$W^p=\frac{1}{2}(L^p-(L^p)^T) \quad (8)$$

Formulation of Stress Increment

Notice the relation of the Cauchy stress to the second Piola-Kirchhoff stress,

$$\sigma=\frac{1}{J}F\sigma_p(F)^T \quad (9)$$

where σ is the Cauchy stress, σ_p the second Piola-Kirchhoff stress, J the determinant of F. Taking time derivative at both sides of Eq.(9) and rearranging it, the following objective rate is obtained.

$$\underset{\circ}{\sigma}=\frac{1}{J}F\dot{\sigma}_p(F)^T=\dot{\sigma}-L\sigma-\sigma(L)^T-tr(L)\sigma \quad (10)$$

The constitutive relation of the second Piola-Kirchhoff stress and Green strain tensor may be given, in incremental form, by

$$\Delta\sigma_p=C:\Delta E^e \quad (11)$$

where C is a forth order tensor. By substituting Eq.(7) into Eq.(11), the following constitutive equation is obtained in terms of total and plastic strain increments,

$$\Delta\sigma_p=C:\{(F^p)^{-T}(\Delta E-\Delta E^p)(F^p)^{-1}\}$$

$$=\overline{C}:(\Delta E-\Delta E^p) \quad (12)$$

where $\overline{C}=(F^p)^{-1}C(F^p)^{-T}$ ($\doteqdot C$ for small increment).

The updating of stress is approximately performed as follows,

$$\Delta\sigma=\Delta\sigma_p+L\sigma^o\Delta t+\sigma^o(L)^T\Delta t+tr(L)\sigma^o\Delta t \quad (13)$$

where σ^o is the Cauchy stress in the configuration x^o.

Simplified Averaging Model

Self-consistent models have been suggested to get macroscopic stress and strain from those of grains, but they are cumbersome because many iterations may be needed. Here an approximation will be dared to use. A grain stress σ_g is suggested by Bretheau and Zaoui(1979) as follows.

$$\sigma_g=\sigma+2\alpha\mu(1-\beta)(e^p-e^p_g) \quad (14)$$

By specifying a particular accommodation function α, many accommodation models can be obtained. Berveiller and Zaoui(1979) have shown that the accommodation function is small except at zero macro-plastic strain e^p, and rapidly decreases as the plastic deformation proceeds. Bretheau et al.(1984) have suggested further smaller accommodation function in the strain measurement of a grain. From the above context, it is plausible to assume $\sigma_g=\sigma$, namely, static model.

Under the assumption of static model, the plastic deformation gradient is approximated by the following procedure,

$$F=\frac{1}{n_g}\sum_i F_i=\frac{1}{n_g}\sum_i F^e_iF^p_i=\frac{1}{n_g}\sum_i R^rR^*_iU^e_iF^p_i \quad (15)$$

where F_i is the total deformation gradient of a grain, n_g the number of grains in the aggregate (assumed that all grains have the same volume), R^r the macroscopic rigid body rotation, R^*_i the relative rotation of grain. By assuming $R^*_iU^e_i \doteqdot U^e_i=U^e$ from the static model, the deformation gradient tensor can be decomposed as follows.

$$F=R^rU^e\frac{1}{n_g}\sum_i F^p_i=F^eF^p \quad (16)$$

Here it is assumingly confined that only dislocation movements of crystallographic slip systems are responsible for the plastic deformation. Then the plastic velocity gradient is represented by two vectors, slip direction(s) and its normal vector(n) to the slip plane, and the slip velocity $\dot{\gamma}$,

$$L^p=\frac{1}{n_g}(\sum_{i=1}^{n_g}\sum_{j=1}^{n_s}\dot{\gamma}_{ij}P_{ij}) \quad (17)$$

where P_{ij} is the dyad multiplication of two vectors($s_{ij}\otimes n_{ij}$), slip direction (s) and the normal vector (n), the indices (i, j) designate a grain and a slip, respectively (not components of tensor). Substituting Eq.(6) into Eq.(17), the plastic deformation gradient is expressed as follows.

$$F^p=I+\frac{\Delta t}{n_g}(\sum_{i=1}^{n_g}\sum_{j=1}^{n_s}\dot{\gamma}_{ij}P_{ij}) \quad (18)$$

And the plastic deformation gradient of a grain is, of course, given by the Eq.(19).

$$F_i^p = I + \Delta t \sum_{j=1}^{n_s} \dot{\gamma}_{ij} P_{ij} \tag{19}$$

The orientation of grain is updated by transforming two vectors s and n .

$$F_i^e = F(F_i^p)^{-1} \tag{20}$$

$$(s_{ij})^* = F_i^e s_{ij}, \quad (n_{ij})^* = n_{ij}(F_i^e)^{-1} \tag{21}$$

Slip Velocity Model

Gilman(1965) suggested the slip velocity model as shown in Eq.(22),

$$\dot{\gamma} = 2\,\mathrm{b}\, v_o(\rho_o + \rho\gamma) e^{-\frac{H_o + H_1\gamma}{\tau}} \tag{22}$$

where b is Burger's vector, v_o the limit velocity of dislocation, H_o material constant, ρ_o the initial dislocation density, ρ the dislocation increasing rate, and τ resolved shear stress on a slip. H_1 is concerned with a fixed portion of the generated dislocations (actually means hardening). This velocity model is adopted as the base of present model, but some modifications will be given to it.

$H_1\gamma$ is assumed to consist of two ingredients;

$$h_1|\gamma_{ij}^m| + H_i^s \tag{23}$$

where h_1 is the material constant, and γ_{ij}^m the maximum slip. H_i^s is given by the following evolution equation.

$$\dot{H}_i^s = c_s(h_s^u - H_i^s)\,\|\dot{\gamma}_i\| \tag{24}$$

where c_s is the material constant, h_s^u the saturation value of H_i^s, and $\|\dot{\gamma}_i\|$ is given by the following Eq.(2.26).

$$\|\dot{\gamma}_i\| = \sum_{j=1}^{n_s} |\dot{\gamma}_{ij}| \tag{25}$$

The evolution of internal stress due to dislocation substructure is described by

$$\dot{\alpha}_{ij} = c_b(h_b^u - \alpha_{ij})\,\dot{\gamma}_{ij} \tag{26}$$

where c_b is the material constant, and h_b^u the saturation value of internal stress. Finally introducing power n, the present slip velocity model can be described in detail as follows,

$$\dot{\gamma}_{ij} = sign(\tau_{ij} - \alpha_{ij}) 2\,\mathrm{b}\, v_o(\rho_o + \rho|\gamma_{ij}^m|)\cdot$$

$$\mathrm{EXP}\left[-\frac{(H_o)^n + (H_s + h_1|\gamma_{ij}^m|)|\tau_{ij} - \alpha_{ij}|^{n-1}}{|\tau_{ij} - \alpha_{ij}|^n}\right] \tag{27}$$

where the shear stress τ_{ij} resolved on the slip system (i, j) is expressed as follows.

$$\tau_{ij} = tr(P_{ij}\sigma) \tag{28}$$

A SIMULATION OF PLASTIC SPIN

Material Properties and Loading

The present simulation is performed in an aggregate of body centered polycrystal. The trial material constants are listed in Table 3.1. Initial orientations of grains are shown in Fig.3.1. The markings depict <100> directions of grains. Each point has three grains whose orientations are set by the method of Asaro and Needlmann(1985), thus 153 grains are considered in this study.

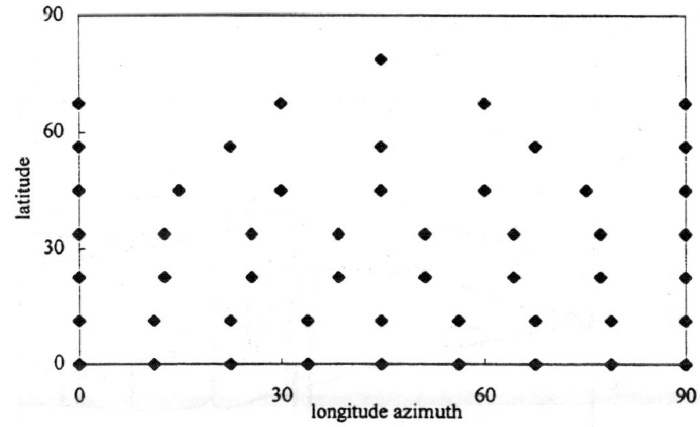

Fig.3.1 Initial orientation of grain(<001> direction)

The loading is applied by the prescribed increments of deformation gradient which is taken as simple shear,

$$F = \begin{vmatrix} 1 & 0 & 0 \\ \delta t & 1 & 0 \\ 0 & 0 & \delta_{33}^p \end{vmatrix} \tag{29}$$

where δ_{33}^p is the same to the plastic deformation, and t is a time incement. It can be nearly rate-independent plasticity with a small δ. At present, 1.0E-4(/s) and 2.0E2(/s) are used for δ, hereafter called as slow rate and fast rate, respectively.

Table 3.1 Traial material constants

b	2.5E-7 (mm)
ρ_o	375 (mm^{-2})
H_o	125 (kg/mm^2)
C_s	30
C_b	12.45
h_1	2000 (kg/mm^2)
v_o	3.2E+6 (mm/s)
ρ	1.0E+9 (mm^{-2})
n	1.35
h_s^u	100 (kg/mm^2)
h_b^u	26.5 (kg/mm^2)
Poisson Ratio	0.3
Young's Modulus	2.1E+11 (N/m^2)

Results and Discussions

Figure 3.2 shows a case of slow rate. The abscissa represents accumulated shear strain. It can be seen that non-dimensional plastic spin (W^p/W) is not asymptotically increasing. This tendency is not similar to the result of Taylor's model in which it is increasing toward 1.0, for instance Gissen(1992). It is manifest that the oscillatory nature of spin ratio depends on the competition between plastic spin and total spin.

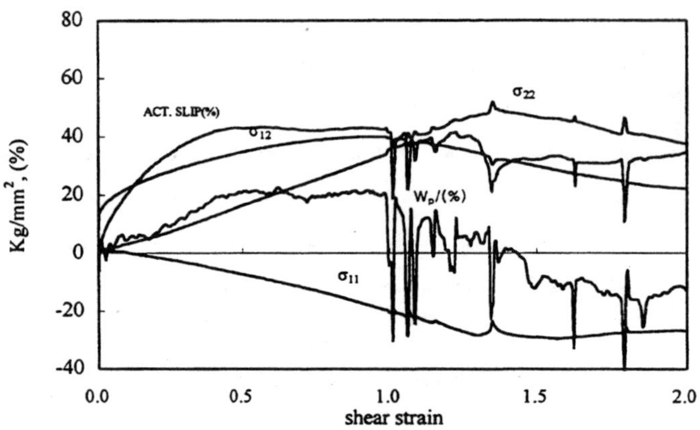

Fig.3.2 Simulation results for slow rate

Some sharp oscillations in curves of non-dimensional plastic spin and stress may be due to the abrupt changes of active slips. It is well explained by the active slip curve which represents the ratio of slip population currently active to the total slip, where the activation of slip is judged by the criterion $\dot{\gamma} > 10^{-6}(/s)$. Sharp oscillations appear to be well synchronized with those of the active slip curve. In the range of shear strain < 1.0 where there are no sharp fallings in the curve of active slip, such oscillations can not be seen. It is thought that texturing is responsible for those local sharp fallings of active slip population. No such fallings can be found in Fig.3.7 which is obtained under the condition that all grain rotations are forced to be fixed.

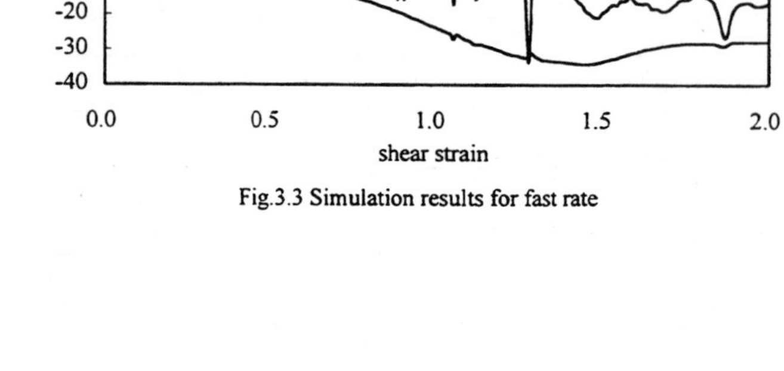

Fig.3.3 Simulation results for fast rate

On the other hand, Fig. 3.2 shows considerable axial stresses at large strain, especially, σ_{11} is a compression. In some torsion tests, for instances Montheillet et al.(1984), compressive axial stress can be observed while axial movement is prohibited. And it can be also seen the torque and axial stress are fluctuating. The present simulation shows similar behavior to those tests. Referring to Fig.3.7, it can be understood that texturing is an essential factor of those non-coaxial stresses. And the comparison of shear stress curves in Fig.3.2 and Fig.3.7 shows a geometrical softening effect.

Fig.3.3 of the case of fast rate shows similar trends to Fig.3.2, but has less sharp oscillations. Its non-dimensional plastic spin appears higher than that of slow rate. Fig.3.4 shows its comparison between two cases. The fast rate has more delayed point at which the plastic spin changes its direction. This rate dependency will be discussed later.

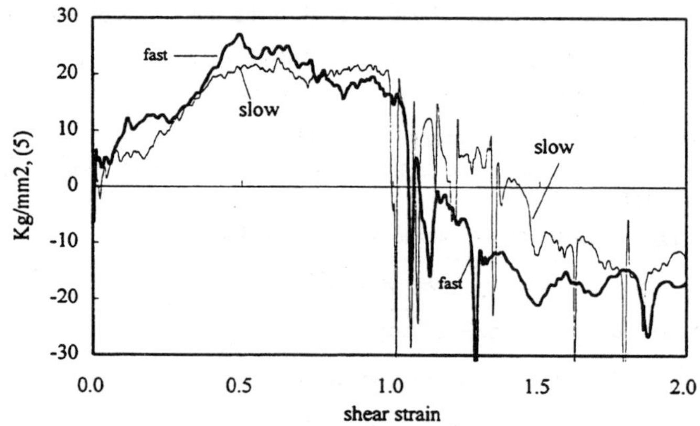

Fig.3.4 The rate effects

Hardening effect on the spin ratio is given in Fig.3.5. Dotted curves are reffered to the case of $h_1 = 1000(Kg/mm^2)$ and solid curves, to the case of $h_1 = 2000(Kg/mm^2)$. It can be observed the smaller hardening is, the larger non-dimensional plastic spin and rate effect are. Both effects, rate and hardening, can be interpreted in a quarry, overstress; the larger overstress, the larger spin ratio. How larger overstress causes larger non-dimensional plastic spin will be explained later.

The reverse loading effect is also investigated. Its result can be found in Fig.3.6. Just after loading reversed, the spin ratio jumps strikingly and goes down rapidly. While the tension σ_{22} is steeply decreasing to small level, the compression σ_{11} keeps its pace with further rate in a certain range of deformation. It is interesting in the behavior of σ_{11} just after loading is reversed. This behavior as well as the jumping of non-dimensional plastic spin can be explained such that, by back stress, the pattern of active slips is reconstructed to be more amiable to them.

If macroscopic plastic spin(MPS) is regarded as the average of grain spins and grain rotation is evaluated by subtracting grain plastic spin(GPS) from macroscopic total spin(MTS), it is natural that texturing can be the governing factor for MPS. Here an attention should be given to the fact that neither the conjugate

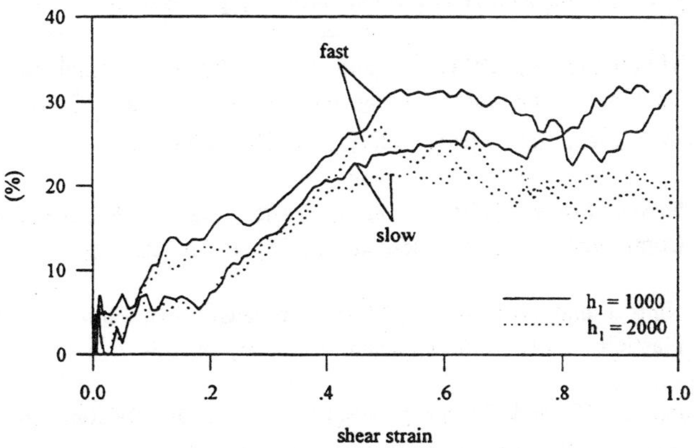

Fig 3.6 Hardening effects

force of texturing nor internal stress(e.g., back stress) is the driving force of plastic spin because it does not urged by such driving forces.

It is useful to classify two spins conceptually; superimposed spin and derived one. While the former is due to rigid body rotation, the latter, to symmetrization. In elasticity, this classification is useless because the distinction can not be measured by material itself and two spins have the same physical meaning. In plasticity, the classification does make sense. It can be deriven with a measurement, anyhow, because the microstructure of material has some information to tell about plastic spin. The plastic spin neither dissipates energy nor has physical meaning. Thus it is not subjected to thermodynamic constraint and there needs no thermodynamic conjugation force. However, texturing dissipates energy so that it has the conjugation force. In fact, the relative rotation of grain to the surrounding, actually giving texturing, is achieved by the deformations such as highly localized deformation in the vicinity of grain boundary and/or crystal twinning(see Jago and Hansen(1986)).

Returning to the origin of plastic spin, it can be considered in two steps; grain level and aggregate one. First, the plastic spin of a grain mainly comes from different levels of induced shear stresses on slips. Boukadia and Sidoroff(1988) showed this mechanism analytically. For comprehensive explanation, consider two groups of slip system giving plastic spins with opposite sign each other under the given loading. The resultant plastic spin, namely GPS, is the survivor in the competition between two groups. In this context, the induced shear stress on slip makes a crucial role. The shear stress level of a slip depends on the grain orientation and the combination of stress components. The deformation rate and hardening effect can be outlined as follow. These effects shift the overstresses of slips of both groups to higher levels, and the shifting effect enlarges the unbalance of slip velocity to yield the consequence because of highly accelerated growth in the stress-slip velocity curve.

Second, since MPS is got by averaging whole GPS controlled by grain orientation, the rearranged pattern of its orientationn might thought of it as a constitutive one. Recently, some authors(see Nemat-Nasser(1990), Schiek and Stumpf(1995)) argued that it is a kinematical function. Both seems to be still suspected in the following viewpoints. In spite of the nature of plastic spin discussed so far in this paper and the fact that IFR has a stretch rate, can the equation be regarded as a constitutive one? Rather a kinematical function? The present simulation and discussion states that the plastic spin can not be given by the complex of kinematic quantities only.

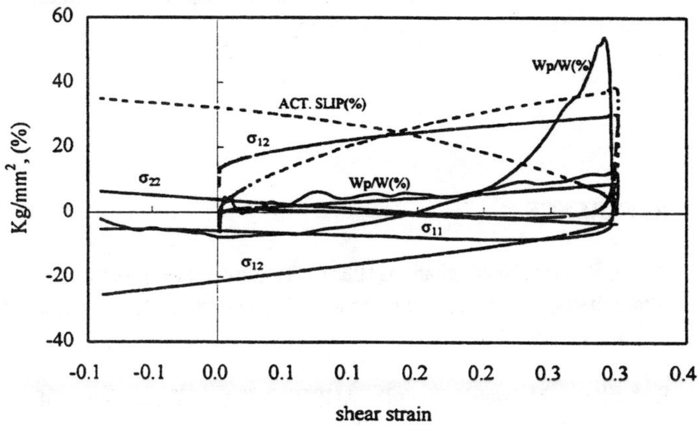

Fig.3.6 Reverse loading effects

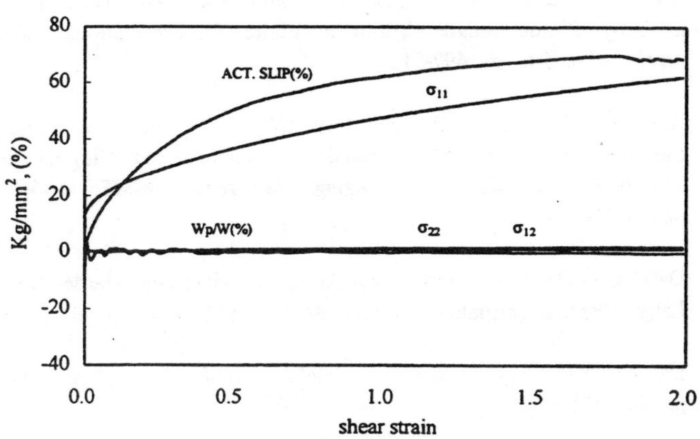

Fig.3.7 Simulation results for no grain rotation

SUMMARY AND CONCLUSIONS

Texturing itself is not effective on the plastic spin but the rearranged pattern of its orientation is. As a matter of fact, the orientation pattern of grains is a significant factor for the plastic spin. It is also the origin of the considerable non-coaxiality observed in the present simulation. The ingredient of stress tensor and the induced non-coaxiality are the important factors as well as the orientation pattern.

The non-dimensional plastic spin depends on deformation rate and hardening, but these effects are secondhand. The load reversing is followed by the jumping of non-dimensional plastic spin and steep rate of compression. These phenomena can not be described by the existing plastic spin models. Though conventional plastic spin models are thought to be effective to some extent, they are hardly regarded as general ones. Further more they can not be justified until a consideration is given to the induced non-coaxiality.

Finally, without a couple of stresse the plastic spin needs no conjugate force. It also needs not to be subjected to thermodynamic constraint as stated by Perzyna (1988). It should not be confused that the conjugate force of texturing has little effect to the plastic spin in spite of the significant relationship between the plastic spin and texturing.

REFERENCE

Asaro RJ and Needleman A(1985), "Texture Development and Strain hardening in Rate Dependent Polycrystals", *Acta Metall.*, 33, 6, pp 923-953.

Berveiller M and Zaoui A(1979), "An Extension of the Self-Consistent scheme to Plastically- Flowing Polycrystals", *J. Mech. Phys. Solids*, 26, pp 325-344.

Boukadia J and Sidoroff F(1988), "Simple Shear and Torsion of a perfectly Plastic Single Crystal in Finite Transformations", *Jrch. Mech.*, 406, 5-6, pp 497-513.

Bretheau T, Mussot P ,and Rey C(1984), "Microscale Plastic Inhomogeneities an Macroscopic Behavior of Single and Multiphase Materials", *J. Enging. Mat.Tech. ASME*, 106, 4, pp304-310.

Dafalias YF(1983), "Corotational Rates for Kinematic Hardening at Large PlasticDeformation", *J.Appl. Mech. ASME*, 50, pp 561-565.

Dafalias YF(1987), "The Plastic Spin", *J. Appl. Mech. ASME*, 52, pp 865-871.

Dafalias YF(1983) and Aifantis EC(1990), "On the Microscopic Origin of the Plastic Spin", *Acta Mechanica*, 82, pp 31-48.

Giessen EVD(1989), "Continuum Models of Large Deformation Plasticity Part I", *Eur. J. Mech., A/Solids*, 8, 1, pp 15-34.

Giessen EVD(1992), "A 2D Analytical Multiple Slip Model for Continuum Texture Development and Plastic spin", *Mech. Mat.*, 13, pp 93-115.

Gilman JJ(1965), "Dislocation Mobility in Crystals", J. *Appl. Phys.*, 36, 10, pp 3195-3206.

Jago RA and Hansen N(1986), "Grain Size Effects in the Deformation of Polycrystalline Iron", *Acta Metall.*, 34, 9, pp 1771-1720.

Kratochvil J(1971), "Finite-Strain Theory of Crystalline Elastic-Inelastic Materials", *J. Appl. Phys.*42, 3, pp 1104-1108.

Loret (1983), "On the Effects of Plastic Rotation in the Finite Deformation of Anisotropic
Elastoplastic Materials", *Mech. Mat.*, 2, pp 287-304.

Montheillet F, Cohen M and Jonas JJ(1984), "Axial Stress and Texture Development during the Torsion Testing of Al, Cu and $\alpha - Fe$", *Acta Metall*, 32, 11, pp 2077-2089.

Nemat-Nasser S(1990), "Certain Basic Issues in Finite-Deformation Continuum Plasticity", *Mechanica*, 25, 4, pp 223-229.

Ning J and Aifantis EC(1994), "On anisotropic Finite Deformation Plasticity Part I", *Acta Mechanica.*, 106, pp 55-72.

Paulun JE and Pecherski RB(1992), "On the Relation for Plastic Spin", *Arch. Appl. Mech.*, 62, pp 376-385.

Pecherski RB(1988), "The Plastic Spin Concept and the Theory of Finite Plastic Deformations with Induced Anisotropy", *Arch. Mech.*, 40, 5-6, pp 807-818.

Perzyna P, "Temperature and Rate Dependent Theory of Plasticity of Polycrystalline Solids", *Revue Phys. Appl.*, 23, pp 445-459.

Schiek B and Stumpf H(1995), "The Appropriate Corotational Rate, Exact Formula for the plastic Spin and Constitutive Model for Finite Elastoplasticity", *Int. J. Solids Structure*, 32, 24, pp 3643-3667.

Zbib HM and Aifantis EC(1988), "On the Concept of Relative and Plastic Spins and Its Implications to Large Deformation Theories", *Acta Mechanica.*, 75, pp 15-33.

Proceedings of the Seventh (1997) International Offshore and Polar Engineering Conference
Honolulu, USA, May 25-30, 1997

ISBN 1-880653-28-1 (Set); ISBN 1-880653-32-X (Vol. IV)

Development of the Buckling Strength Estimation System for Plate Members Considering Tensile Load Effects

Juh Hyeok Ham
Halla Institute of Technology
Wonju, Korea

Ul Nyeon Kim
Hyundai Heavy Industries Co.
Ulsan, Korea

ABSTRACT

Generally, a safety estimation based on the buckling strength is carried out to evaluate the strength of plate members in the design process of ship hull structures.

The design codes of classification societies do not consider the in-plane tension effects. However, in every loading conditions, many ship plate panels among enormous plate panels in a ship structure may be subjected to the in-plane tension with other loading components. Therefore, this in-plane tension effect gives the conservative value of buckling strength.

Firstly, in this study, to estimate the buckling strength based on the more real loading condition a simplified plate buckling formula considering in-plane tension is obtained on the theoretical basis and this formula needs not the number of buckling half waves for evaluation. The proposed formula is compared with the existing theoretical solutions, Ueda's formula, design code(LR) and the results which are obtained by numerical analysis. It has a good correlation with the present results and others.

Secondly, in order to increase the evaluation efficiency, simple estimation system of the buckling strength is developed using the window management system of engineering workstation based on the above plate buckling formula. This established system layout is simply explained.

From this evaluation system, more reliable buckling safety of plate panel will be guaranteed in the ship hull design stage of H.H.I. shipyard.

KEY WORDS: Elastic and critical buckling stress, Plastic correction, Elasto-plastic behavior, Large deflection analysis, Allowable design limit, Buckling design system

INTRODUCTION

Ship structure basically consists of plate members, and its overall strength depends on the stiffness and strength of each plate member. The buckling strength of plate is one of the most important design criteria when structural characteristic is investigated. Therefore, it is necessary to study a reasonable buckling formula in order to carry out a more efficient and reliable design.

Conventional buckling design codes can evaluate the buckling strength of the plate subjected to five load components, such as compression and bending in two directions and shear. Recently Hyundai Heavy Industry has developed the plate buckling estimation system in consideration of lateral pressure, residual stress and real edge condition (Paik & Ham 1992, Paik et al 1993).

In ship structures, structural components such as the transverse web, bulkhead & side stringer and enveloped plates of ship hull are under combined stresses. Therefore, buckling strength evaluation for these structural members is needed.

A buckling strength of the rectangular plate subjected to in-plane tension and compression generally increases according to the increase of tension loading. But there are few experiments or studies which examine into the relation between in-plane tension and buckling strength.

In this paper, a buckling formula for the simply supported plate subjected to the combined loads such as in-plane compression, tension and edge shear is proposed on the theoretical basis and this formula needs not the number of buckling half waves for evaluating buckling strength therefore, more simple application is expected together with the developed efficient evaluation system.

BUCKLING STRENGTH FORMULA

Fig.1 shows a simply supported plate($\beta=a/b\geq1$) under combined loads, such as in-plane tension, compression and shear. All plate edges are simply supported and remain straight after buckling. In this paper, a compression is defined as positive(+) and a tension is defined as negative(-).

Buckling interaction formula under in-plane compression

and tension

The plate buckling strength under the action of biaxial compression depends on the loading ratio, the aspect ratio and material properties.

The plate buckling strength equation can be derived accurately in this case. The differential equation governing the large deflection

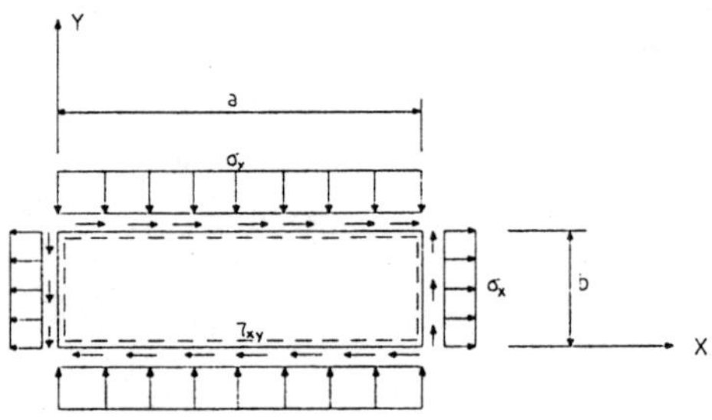

Fig. 1 Applied stress on the ship platings

behavior of plate is given by :

$$\nabla^4 w = \frac{t}{D}\left(\frac{\partial^2 F}{\partial y^2}\cdot\frac{\partial^2 w}{\partial x^2}+\frac{\partial^2 F}{\partial x^2}\cdot\frac{\partial^2 w}{\partial y^2}-2\frac{\partial^2 F}{\partial x\partial y}\cdot\frac{\partial^2 w}{\partial x\partial y}\right) \quad (1)$$

$$\nabla^4 F = E\left(\left(\frac{\partial^2 w}{\partial x\partial y}\right)^2-\frac{\partial^2 w}{\partial^2 x}\cdot\frac{\partial^2 w}{\partial^2 y}\right) \quad (2)$$

Since it is considered that boundary condition of the plate is simply supported and the aspect ratio is always more than 1, the buckled shape of biaxial compression can be idealized in the following form.

$$w = w_1 \sin\frac{m\pi x}{a}\sin\frac{\pi y}{b} \quad (3)$$

Substitution of Eq.(3) into (2) yields Airy's stress function(F). If appling the Galerkin's method for this stress function(F), w_1 will be found. Immediately before the buckling occurs, however, the plate deflection doesn't exist, that is, $w_1=0$. Finally the following equation can be obtained (Paik 1991).

$$\left[\frac{m_o b}{a}+\frac{a}{m_o b}\right]^2\frac{m^2a^2b^2}{(a^2+m^2b^2)^2}\left(\frac{\sigma_x}{\sigma_{xE}}\right)+\left[1+\frac{b^2}{a^2}\right]^2\cdot$$

$$\frac{a^4}{(a^2+m^2b^2)^2}\left(\frac{\sigma_y}{\sigma_{yE}}\right) = 1 \quad (4)$$

Eq.(4) developed by Paik et al. has experimentally good adaptivity and presents us with the exact solution for the relation of elastic buckling strength. But this formula needs buckling half-wave numbers(m & m_0) to obtain the buckling strength. Therefore, a simple continuous interaction equation is preferable for practical use. An elastic buckling equation of plate under in-plane tension in y-direction and compression in x-direction is approximately proposed as a function of the aspect ratio.

$$\left(\frac{\sigma_x}{\sigma_{xE}}\right)^2 + \alpha_5\left(\frac{\sigma_y}{\sigma_{yE}}\right) = 1 \quad (5)$$

where $\alpha_5 = 0.09^{\beta-1.47}-0.58$ for $1\le\beta\le\sqrt{2}$

$= 0.61^{\beta+0.54}+0.38$ for $\beta>\sqrt{2}$

Also, an elastic buckling equation of the plate subjected to in-plane tension in x-direction and compression in y-direction, is proposed as follows.

$$\alpha_6\left(\frac{\sigma_x}{\sigma_{xE}}\right) + \left(\frac{\sigma_y}{\sigma_{yE}}\right) = 1 \quad (6)$$

where $\alpha_6 = 0.6^{\beta-1}$

Buckling formula under in-plane tension and shear

The buckling interaction equation of plates subjected to uniaxial tension and edge shear is obtained from the curve shown in reference [5]. This curve is the plate buckling coefficient(k_s) versus plate aspect ratio(β) for various applied load ratio($\rho_{\sigma,\tau}$)(see Fig.2).

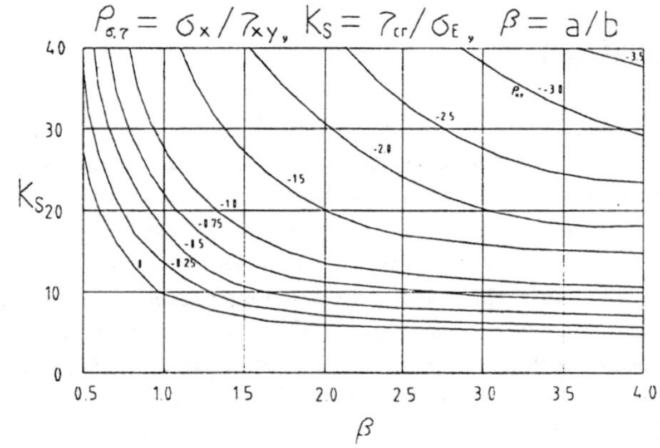

Fig. 2 Elastic buckling coefficient for aspect ratio under edge shear and in-plane tension

For the case of in-plane tension in x-direction and edge shear, the elastic buckling equation is approximated as follows.

$$\left(\frac{\sigma_x}{\sigma_{xE}}\right) + \alpha_7\left(\frac{\tau_{xy}}{\tau_E}\right)^{\alpha_8} = 1 \quad (7)$$

where $\alpha_7 = 0.004\beta^2+0.008\beta+1.009$

$\alpha_8 = 0.01\beta+1.81$

Also, for the case of in-plane tension in y-direction and edge shear, the buckling interaction equation is approximated as follows.

$$\left(\frac{\sigma_y}{\sigma_{yE}}\right) + \left(\frac{\tau_{xy}}{\tau_E}\right)^{\alpha_9} = 1 \quad (8)$$

where $\alpha_9 = -0.194\beta^2+1.314\beta+0.746$

Buckling formula under in-plane compression and shear

The following buckling equation for the plate under in-plane compression in x-direction and edge shear is proposed by Ueda et al

(1985).

$$(\frac{\sigma_x}{\sigma_{xE}}) + (\frac{\tau_{xy}}{\tau_E})^{\alpha_3} = 1 \qquad (9)$$

where $\alpha_3 = -0.160\beta^2+1.080\beta+1.082$ for $1\leq\beta\leq3.2$

$= 2.9$ for $3.2 < \beta$

Also, the buckling interaction formula for the plate subjected to in-plane compression in y-direction and edge shear is proposed by Ueda et al (1985).

$$(\frac{\sigma_y}{\sigma_{yE}}) + (\frac{\tau_{xy}}{\tau_E})^{\alpha_4} = 1 \qquad (10)$$

where, $\alpha_4 = 0.10\beta+1.90$ for $1\leq\beta\leq2$

$= 0.70\beta+0.70$ for $2\leq\beta\leq6$

$= 4.90$ for $6<\beta$

Plastic correction

For thin plate, buckling phenomenon usually occurs in the elastic range. But for thick plate, local plastic deformation may occur propotion as the load increases, and the bending stiffness of plate is depreciated abruptly. Buckling may take place in the elasto-plastic range before elastic buckling occurs. For this problem, Johnson-Ostenfeld parabola is used for the existing classification design codes(LR (1983) etc) to correct elastic buckling formula using plasticity correction factor. The well known Johnson-Ostenfeld parabola is as follows.

$$\sigma_{cr} = \sigma_E \qquad \text{for } \sigma_E/\sigma_o\leq0.5 \qquad (11)$$

$$= \sigma_o[1-\frac{\sigma_o}{4\sigma_E}] \qquad \text{for } \sigma_E/\sigma_o> 0.5$$

$$\tau_{cr} = \tau_E \qquad \text{for } \tau_E/\tau_o\leq0.5 \qquad (12)$$

$$= \tau_o[1-\frac{\tau_o}{4\tau_E}] \qquad \text{for } \tau_E/\tau_o> 0.5$$

In this paper, the following Paik's parabola (Paik & Ham 1992) is adopted for the plastic correction of proposed buckling formulas and Ueda's one (Ueda 1995).

$$\sigma_{cr} = \sigma_E \qquad \text{for } 0.5 > \sigma_E/\sigma_o \qquad (13)$$

$$= \sigma_o[-0.0847(\sigma_E/\sigma_o)^4+0.5857(\sigma_E/\sigma_o)^3-1.5631(\sigma_E/\sigma_o)^2+2.0784(\sigma_E/\sigma_o)-0.2164 \qquad \text{for } 0.5\leq\sigma_E/\sigma_o<1.9$$

$$= \sigma_o \qquad \text{for } 1.9\leq\sigma_E/\sigma_o$$

$$\tau_{cr} = \tau_E \qquad \text{for } 0.5 > \tau_E/\tau_o \qquad (14)$$

$$= \tau_o[-0.0847(\tau_E/\tau_o)^4+0.5857(\tau_E/\tau_o)^3-1.5631(\tau_E/\tau_o)^2+2.0784(\tau_E/\tau_o)-0.2164 \qquad \text{for } 0.5\leq\tau_E/\tau_o<1.9$$

$$= \tau_o \qquad \text{for } 1.9\leq\tau_E/\tau_o$$

Buckling formula under in-plane tension, compression and edge shear

The following buckling interaction equation for plate under tension in x-direction($\sigma_x<0$), compression in y-direction($\sigma_y>0$) and edge shear, is proposed by combination of Eq.(5), (7) and (9).

$$[\frac{\sigma_y}{R_{sx2}\,\sigma_{ycr}}] + [\frac{\tau_{xy}}{R_{sy2}\tau_{cr}}]^{\alpha_4} = 1 \qquad (15)$$

where $R_{sx2}=1-\alpha_6\frac{\sigma_x}{\sigma_{xcr}}$, $R_{sy2}=[(1-\frac{\sigma_x}{\sigma_{xcr}})/\alpha_7]^{1/\alpha_8}$

When the plate is subjected to compression in x-direction(($\sigma_x>0$), tension in y-direction($\sigma_y<0$) and edge shear, the following buckling interaction equation for plate is proposed by combination of Eq.(6), (8) and (10).

$$[\frac{\sigma_x}{R_{sx3}\,\sigma_{xcr}}] + [\frac{\tau_{xy}}{R_{sy3}\tau_{cr}}]^{\alpha_3} = 1 \qquad (16)$$

where $R_{sx3}=[1-\alpha_5(\frac{\sigma_y}{\sigma_{ycr}})]^{0.5}$, $R_{sy3}=[1-(\frac{\sigma_y}{\sigma_{ycr}})]^{1/\alpha_9}$

Recently, Ueda et al have presented explicit buckling design formula (Ueda 1995) and as follows.

- When σ_x is tension and σ_y is compression($\sigma_x<0$, $\sigma_y>0$)

$$\frac{(m_o^2+\beta^2)^2}{m_o^2(1+\beta^2)^2}(\frac{\sigma_x}{\sigma_{xcr}})+(\frac{\sigma_y}{\sigma_{ycr}})+(\frac{\tau_{xy}}{\tau_{cr}})^2=1 \qquad (17)$$

- When σ_x is compression and σ_y is tension($\sigma_x>0$, $\sigma_y<0$)

$$(\frac{\sigma_x}{\sigma_{xcr}})+\frac{(1+\beta^2)^2}{(m_o^2+\beta^2)^2}(\frac{\sigma_y}{\sigma_{ycr}})+(\frac{\tau_{xy}}{\tau_{cr}})^2=1 \qquad (18)$$

The above elastic buckling formula adopting the plastic correction parabola shown in Eq. (11)~(14) can be used not only in the elastic range but also in the elasto-plastic one.

APPLICATION

The large deflection analysis using finite element method(MSC/NASTRAN) is carried out in order to find out the accuracy and the applicability of proposed formulas.

It is assumed that plate has small initial deflection(w_o/t=0.001). In this case, the lateral deflection of the plate gradually increases with increase of the applied load and occasionally the clear buckling point is not appeared. Therefore, the so-called "P -W^2 method" is used to define the critical buckling point (Paik 1991).

It is assumed that all edges are simply supported and material follows perfect elasto-plastic behavior. In this study, non-dimensional term by the yield stress is introduced to compare the buckling interaction curves and the following characteristic properties are used.

b = 500 (mm)

E = 21000 (kgf/mm^2)

$\nu = 0.3$

$\sigma_o = 24 \quad (kgf/mm^2)$

Also, the allowable design limit is defined by the closed surface using the proposed buckling and von-Mises yield criterion. For the plate subjected to compressive stresses both σ_x and σ_y, the proposed formula employs Paik's equation (Paik & Ham 1991) in comparision with buckling interaction curves.

Interactive relation curve

The proposed interaction curve under in-plane compression, tension and edge shear is compared with theoretical solution, nonlinear analytical results, Ueda's formula and current design codes of class(LR).

These interaction curves shown in Fig.3 illustrate a failure criterion using the proposed buckling condition and von-Mises yield one. It is deemed to be failed when the value of von-Mises equivalent stress equals or exceeds the buckling and/or yielding strength of the plate.

Buckling interaction curve of the proposed formula is compared with other results for thin and thick plate and it is also compared with other results for combined loading condition.

Fig.3 a - d are buckling interaction curves under σ_x and σ_y. It is expected that elastic buckling will be occured for the thin plates shown in Fig.3 a & c. Also it is shown that proposed formula is a good correspondance with numerical or theoretical solution with a variance of the aspect ratio. Ueda's formula is in keeping with theoretical or analytical results for square plate but it gives underestimated buckling strength for rectangular plate. Design code of L.R. class is not considered the inplane tension effect, therefore it evaluates buckling strength conservatively.

Fig.3 b & d show buckling interaction curves for thick plate. From these curves it is shown that proposed formula produces somewhat overestimated results but it is expected that if more reasonable plastic correction effects are added, good results are expected.

Fig.3 e - h are buckling interaction curves under σ_x and τ_{xy}. In this figures the proposed formula is a good correlation with analytical results but Ueda's formula gives over estimated buckling strength as the aspect ratio increases. Fig.3 i - l are buckling interaction curves under σ_y and τ_{xy}.

From Fig.3 ,consequently, it is expected that the elastic buckling will be occurred in the thin plates (see Fig.3 a, c, e, g, i, k) and shown that the proposed formulas are a good correspondence with analytical results or theoretical solution, with a variance of the aspect ratio. Also Ueda's formula is in keeping with theoretical or analytical results for square plate, but it gives under or over estimated buckling strength for the rectangular plate. Also, the design code of L. R. class is not concerned about the increasing effect of buckling strength owing to in-plane tension. Hence, it evaluates buckling strength conservatively.

Also, for the thick plates, the proposed equation is in a good agreement with analytical results(see Fig.3 b, d, f, h, j, l).

Design results of the optimum plate thickness

Using each buckling formula, the optimum thickness of plate panels under the action of a certain combined stress is calculated.

If applied loads are slight, optimum plate thickness will be small, and elastic buckling will be occurred. But increasing of applied loads, optimum plate thickness will be large, therefore elasto-plastic or plastic buckling will be occurred. We compared with both slight and severe stress level (Ham et al 1995).

It is observed that the plate thickness against buckling is to be increased according to the increase of aspect ratio. For the applied loads of compression in x-direction and tension in y-direction, the thickness using proposed formula is larger than Ueda's one disregarding buckling strength change according to the effect of aspect ratio. Also, in case of acting edge shear and tension, the thickness from present formula is smaller than Ueda's one because Ueda's formula evaluates buckling strength of lower level. For the plate subjected to tension stress, it is found that LR's equation gives larger plate thickness compared with other results.

Additionally, to know the buckling strength according to the change of yield stress, derivation of optimum plate thickness is carried out for mild and higher tensile steel. For severe loads, high tensile steel(σ_o=36 kgf/mm^2) provides smaller plate thickness compared with mild steel(σ_o=24 kgf/mm^2) because of plastic correction effect.

CONSTRUCTION OF THE EVALUATION SYSTEM

Evaluation system at the engineering workstation is developed to easily and efficiently estimate the buckling strength of plate panel based on the above buckling strength formulas. First of all, construction layout is introduced and secondly evaluation steps are explained.

Outline of the System Construction

In this section we make an effort to prepare the advanced system in the respect to the accuracy and the efficiency of panel buckling evaluation in the initial ship design stage using above illustrated equations.

Actually evaluation for the enormous plate panel in ship structure is indispensable and many mistakes & fault factors by designers are potentially existed but in this advanced system time saving & automatical prediction of panel buckling are expected to keep with more accuracy.

In developing and processing this system in the SUN4 engineering workstation(EWS) interactive screen management technique is adopted and function of mouse is mainly used.

For the preparation of large windows in this system window setting function supplied at the internal supporting libraries in EWS is used. Icons are prepared using icon setting macro at the appropriate internal position of this large window and icon symbols are designed using icon image editing tool which is supporting tool in EWS.

At the sub-system or main system, after moving to next or previous windows each title window is needed to represent position of windows therefore large size characters matching for this 19" large size window are prepared using icon editing tool which

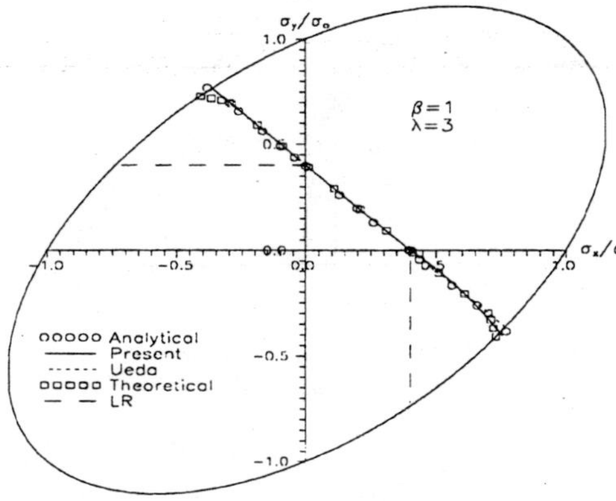

a) Thin plate under biaxial loads (σ_x, σ_y) β=1, λ=3

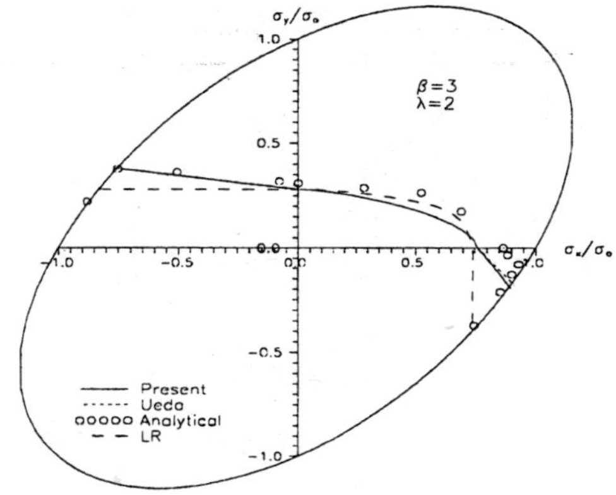

d) Thick plate under biaxial loads (σ_x, σ_y) β=3, λ=2

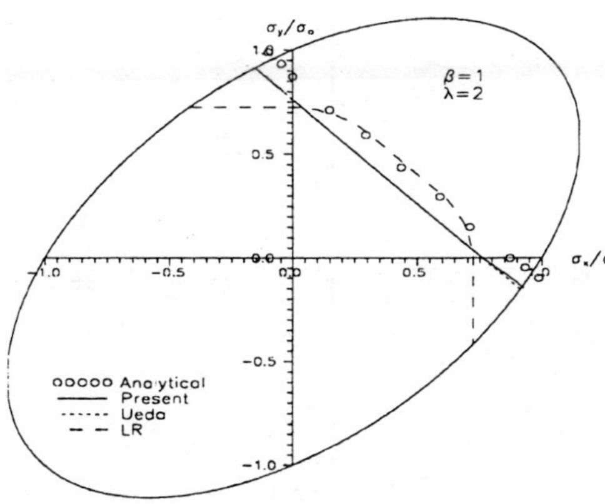

b) Thick plate under biaxial loads (σ_x, σ_y) β=1, λ=2

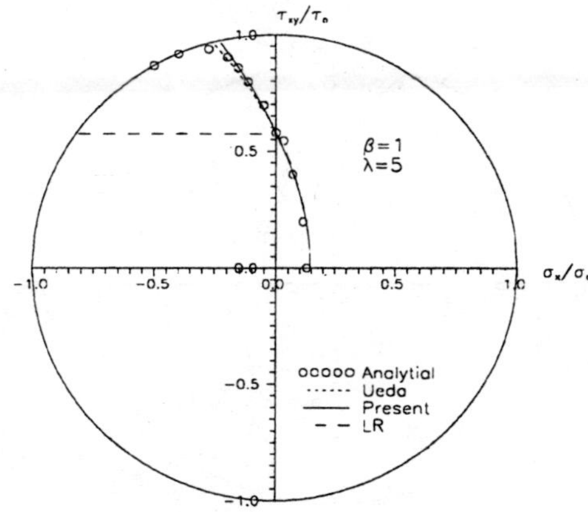

e) Thin plate under axial load and edge shear (σ_x, τ_{xy})
β=1, λ=5

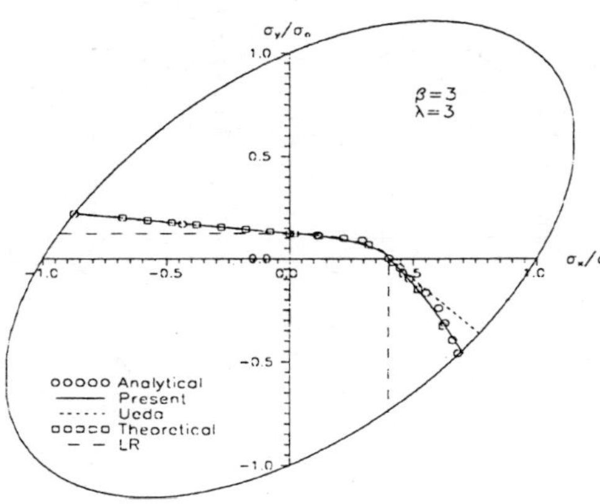

c) Thin plate under biaxial loads (σ_x, σ_y) β=3, λ=3

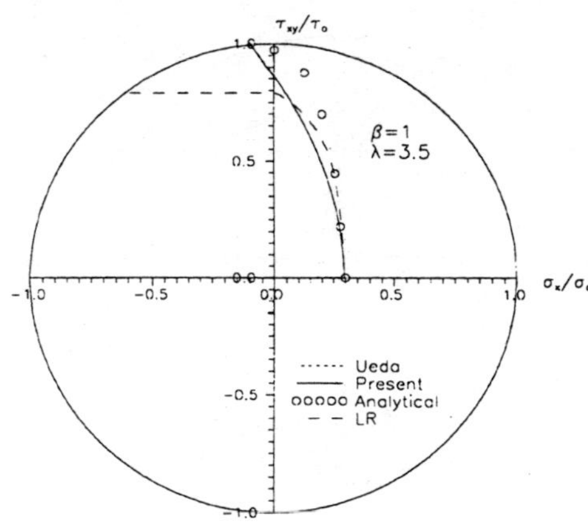

f) Thick plate under axial load and edge shear (σ_x, τ_{xy})
β=1, λ=3.5

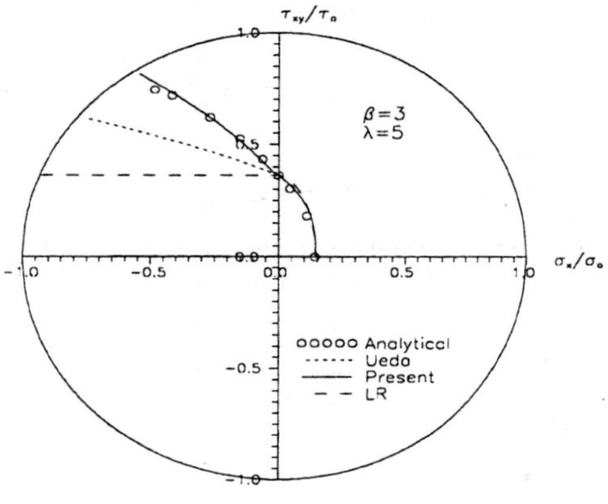

g) Thin plate under axial load and edge shear (σ_x, τ_{xy})
β=3, λ=5

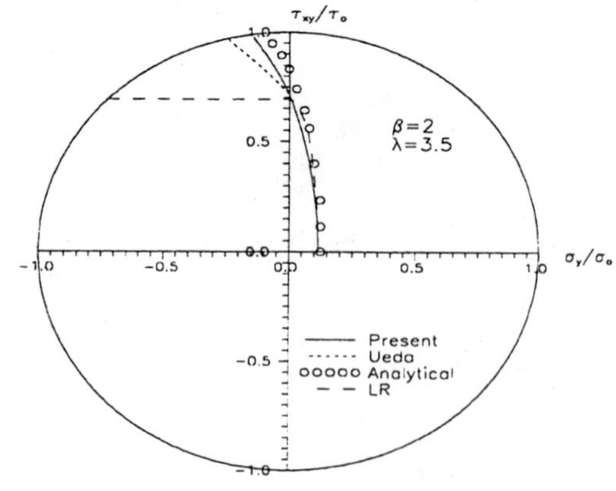

j) Thick plate under axial load and edge shear(σ_y, τ_{xy})
β=2, λ=3.5

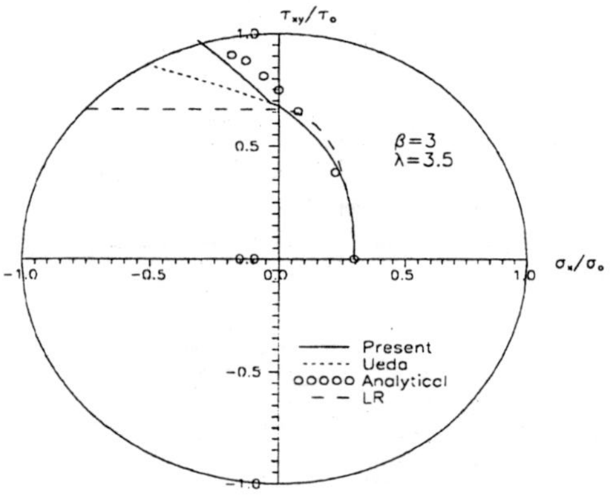

h) Thick plate under axial load and edge shear(σ_x, τ_{xy})
β=3, λ=3.5

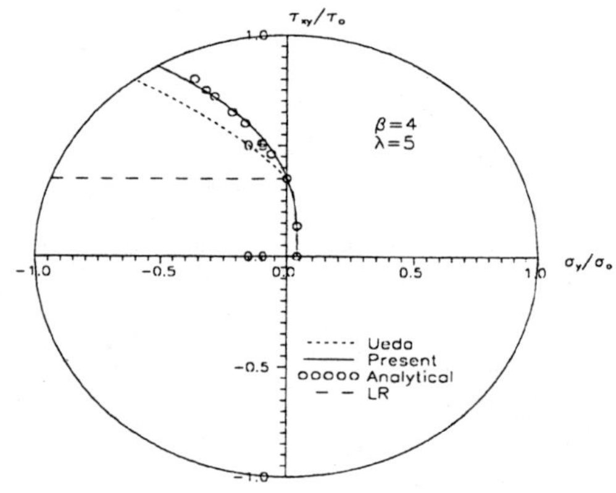

k) Thin plate under axial load and edge shear(σ_y, τ_{xy})
β=4, λ=5

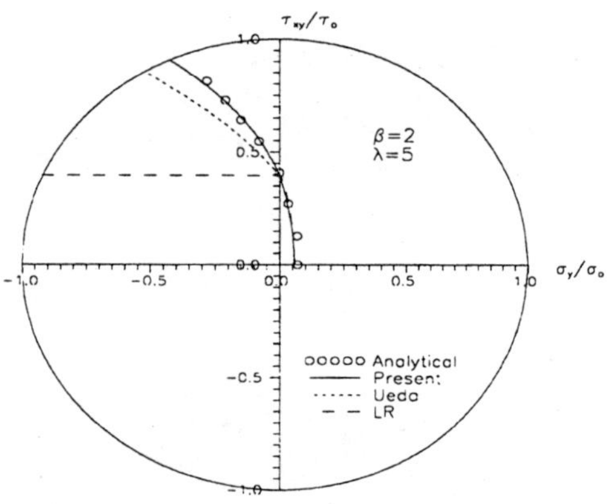

i) Thin plate under axial load and edge shear(σ_y, τ_{xy})
β=2, λ=5

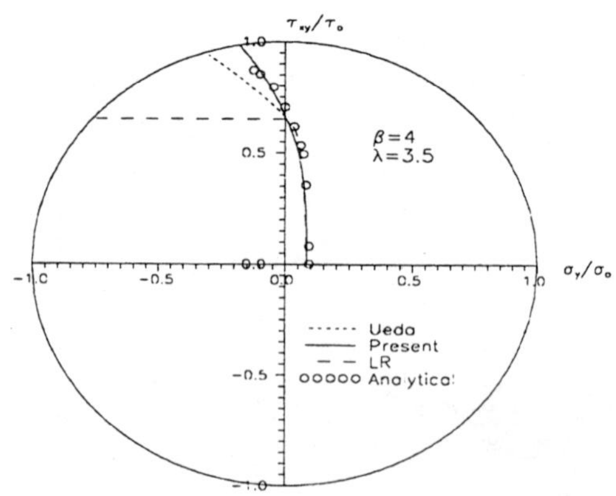

l) Thick plate under axial load and edge shear(σ_y, τ_{xy})
β=4, λ=3.5

Fig.3 Buckling interaction relationship curves.

construct alphabet and marks configuration using 64x64 array window of memory pixel in tool.

Each character panel automatically arranged on the title screen according to appropriate keyboard input string.

Operation for the excutable files and change of environments & working windows is controlled appropriately using the functions of system processing in C & C++ language.

After setting the output format, print system is adopted for the

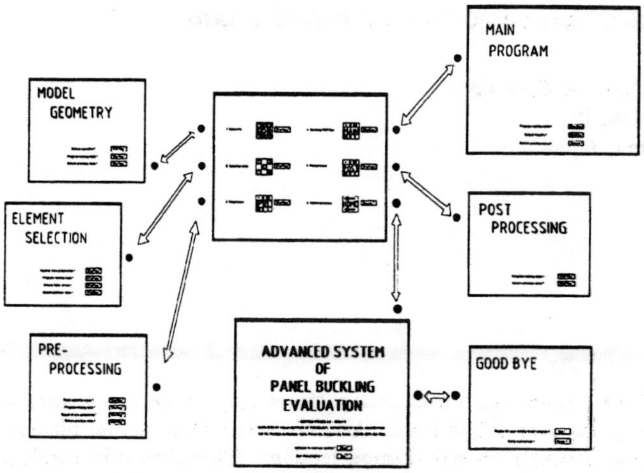

Fig. 4 Layout of buckling strength estimtion system

input & output list files and screen copy method for graphic results & design circumstance on the screen.

Print and screen copy buttons mentioned above are prepared in this developed system and automatic printing is available whenever being necessary.

Evaluation Steps of Buckling Strength

Over-all window management layout of this evaluation system is shown in the Fig. 4. At the lower middle part of this figure the main title window of this system is shown and main menu window is appeared by clicking the second button from bottom of main title window and the first one is used for system exit.

Main menu window located in the middle zone of above figure is prepared 6 main menus including return button to previous window. Left three buttons are pre-processing steps for buckling evaluation. First button is one of steps to confirm the mesh shape and element numbers of structural finite element model, second one is examination step for estimating the zone for buckling check including every loading cases of analysis results consequentially, therefore applied stress informations are obtained for considered elements or zones. Last button among three is used for automatic production of input data for the various buckling evaluations or formulas based on the above elements and the model information results.

Right upper part button is one of steps to easily evaluate the buckling strength based on the various clasificaton code, developed simple formulas and maximum deformed response as main procedure of panel buckling evaluation. Finally right middle button is prepared for the graphic presentation of evaluation results as the safety values on the considered each panel.

CONCLUDING REMARKS

In this study, a buckling design formula of plate panels considering in-plane tension effect was proposed on the theoretical basis.

The proposed formula shows a good correlation with the theoretical or the analytical results for various slenderness and aspect ratio. In consequence, for the strength evaluation of ship platings at initial design stage, the proposed formula can be utilized with reasonable accuracy.

Additionally, a simple estimation system of buckling strength is developed using the window management system of engineering workstation based on the above plate buckling formulas. This established advanced system is explained using system layout.

From this evaluation system, more reliable buckling safety of plate panel will be guaranteed in the ship design stage.

REFERENCES

J.K. Paik, J.H. Ham and U.N. Kim (1992). "A New Plate Buckling Design Formula(1st Report)", J. of the Naval Architects of Japan, Vol. 171, pp.559-566.

J.K. Paik, J.H. Ham and J.H. Ko (1992). "A New Plate Buckling Design Formula (2nd Report) - On the Plasticity Correction - ", J. of the Naval Architects of Japan, Vol. 172, pp.417-425.

J.K. Paik, J.Y. Kim, W. S. Kim and Y. S. Chung (1993). "A New Plate Buckling Design Formula(3rd Report) - On the Real Edge Condition - ", J. of the Naval Architects of Japan, Vol. 174.

J.K. Paik (1991). "An Analytical Solution of Nonlinear Behavior for Simply Supported Rectangular Plates Subjected to Biaxial Compressions Plates Subjected to Biaxial Compression", J. of the Naval Architects of Korea, Vol. 28(Korean).

C.R.C Japan (1971). "Handbook of Structural Stability", CORONA Publishing Co., Ltd., pp.3-37.

Y. Ueda, S.M.H. Rashed and J.K. Paik (1985). "Elastic buckling Interaction Equation of simply supported Rectangular plates Subjected to Five Load Components", J. of the Naval Architects of Japan, Vol. 157 (Japanese).

LR (1983). "Lloyd's Register's Plan Appraisal Systems for Ships", Ship Type Procedural Document for Bulk Carriers, January.

Y.Ueda, S.M.H. Rashed and J.K. Paik (1995). "Buckling and Ultimate Strength Interaction in Plates and Stiffened Panels under Combined in-plane Biaxial and Shearing Forces", Marine Structures, Vol. 8, No. 1, pp.1~36.

J.H. Ham, U.N. Kim and Y. S. Chung (1995). "The Development on the Buckling Strength Estimation Formula of Plate Members - in Consideration of In-Plane Tension (1) - ", Proc. of the Naval Architects of Korea, Fall Meeting (Korean).

Proceedings of the Seventh (1997) International Offshore and Polar Engineering Conference
Honolulu, USA, May 25-30, 1997

ISBN 1-880653-28-1 (Set); ISBN 1-880653-32-X (Vol. IV)

Buckling and Ultimate Strength of Plates Subjected to Combined Loads

Masahiko Fujikubo, Tetsuya Yao and Balu Varghese
Hiroshima University
Higashi-Hiroshima, Japan

ABSTRACT

Elastoplastic large deflection analyses of simply supported thick rectangular plates with initial deflection have been performed using nonlinear finite element method. The objective of the study is to check to what extent the plastic correction method used in the present classification rules is reliable for estimating buckling strength of a thick rectangular plate under combined in-plane loads. The buckling strength interaction curves of classification rules are compared with the ultimate strength interaction curves obtained from FEM analyses, and the degree of margin of safety is examined. A simple modification of plastic correction method is proposed for some combined loading cases for which it is revealed that the present plastic correction method estimates the buckling strength on the unsafe side.

KEY WORDS : plastic correction, ultimate strength, buckling strength, Johnson's formula, loading method, in-plane load, initial deflection

INTRODUCTION

Hull girder strength of a ship is derived from a lot of rectangular plates strengthened by stiffeners. These plates are subjected to many in-plane loads such as compression, shear, bending etc. and in the design of such plates it is inevitable to check their buckling strength. All the classification societies have prescribed standards for the evaluation of buckling strength of plates. At present, buckling strength evaluation of classification rule consists of estimating elastic buckling strength from elastic buckling interaction curves and then applying plastic correction using Johnson's formula etc. for the plates where plastic buckling may occur. A lot of researches have been done on elastic buckling interaction and reasonably good results have been obtained in this regard. On the other hand, plastic correction formulae are originally proposed as the safe side approximation of ultimate strength for column members. Although these formulae have been applied for plates straightforwardly, it does not have any physical meaning in itself. Also the strength of actual plates depend strongly on initial imperfections. Therefore it is absolutely necessary to check how far the estimated strength, using plastic correction, coincides with the buckling and ultimate strength of actual plates.

In this study, elastoplastic large deflection analyses using FEM have been performed on simply supported thick rectangular plates subjected to in-plane loads such as bi-axial compression and shear. The ultimate strength interaction curves obtained from these calculations are compared with the interaction curves of classification rules to check to what extent the plastic correction formulae are reliable when they are applied to the plates. In fact, the plastic correction is intended to estimate the plastic buckling strength of the plates and it cannot be called the ultimate strength. However for thick plates, the reserve strength after buckling is generally small and in the design of such plates buckling strength must be regarded as their ultimate strength. In this sense, here the ultimate strength obtained from FEM is compared with the buckling strength of classification rules.

FEM ANALYSIS AND MODEL

FEM analysis

Elastoplastic large deflection analyses of a rectangular plate having initial deflection, as shown in Fig. 1, are performed by using the non-linear FEM analysis code ULSAS developed by the authors. The accuracy of calculations by the ULSAS has been validated through a lot of analyses including the benchmark test performed by the ISSC (Dow, 1991). The following boundary conditions are assumed: regarding the out of plane deformations, simply supported conditions are considered and the in-plane loads are applied under load or displacement control keeping four edges of the plate straight.

Bi-axial compressive loads are applied in three different

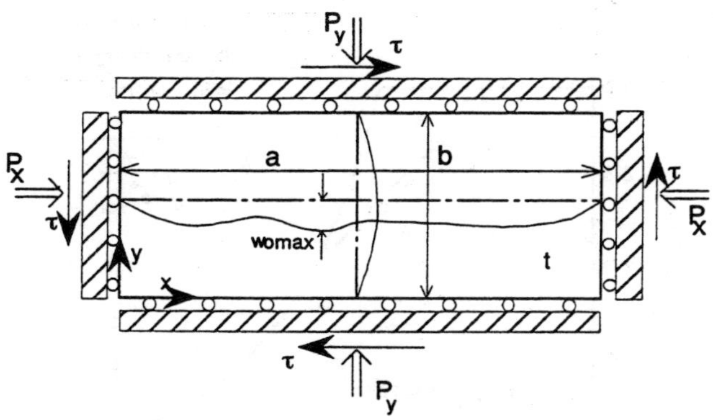

Fig.1 Rectangular plate under combine in-plane loads

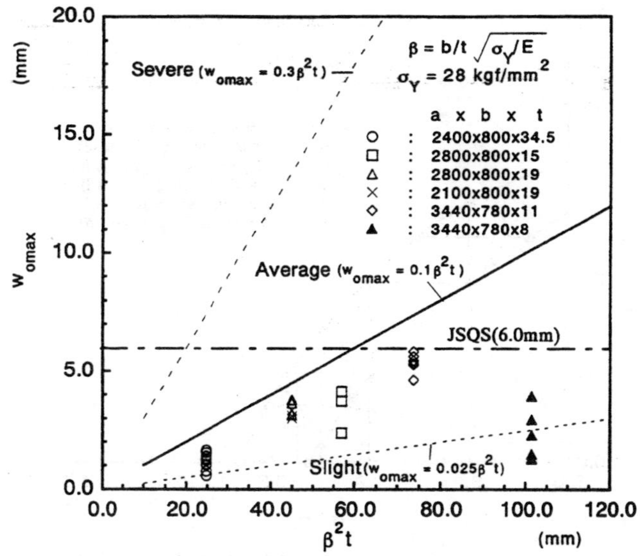

Fig.2 Maximum initial deflection of rectangular plates measured in ships

ways, namely d-d control, l-l control and l-d control. In d-d control, forced displacements are applied proportionately on both directions, whereas in l-l control loads are applied proportionately. In the case of l-d control, load is applied first in one direction and then keeping the load in that direction constant, forced displacement is applied in the other direction. It is studied how these loading methods influence the interaction relationship of initial yielding and ultimate strength. In the case of shear stress only load control is applied.

Model for analysis

The breadth, b, of the plates is fixed as 1000 mm and computations are done for different cases varying thickness, t, and aspect ratio, a/b. Yield strength of the material, σ_Y, is taken as 24 kgf/mm^2. Strain hardening effects of the material are not taken into account.

Initial deflection may have a large influence on the initial yielding/ultimate strength of thick plates. So it is necessary to consider the magnitude and shape of initial deflection as close as possible to the actual ones. In this study, the shape of initial deflection is expressed as a sum of sinusoidal components based on the measurement of initial deflection of deck platings on actual ships done by one of the authors (Yao, 1992). That is,

$$w_0 = \Sigma_m A_{0m} \sin\frac{m\pi x}{a} \sin\frac{\pi y}{b} \tag{1}$$

In Eq. (1), m is the number of half waves in the longitudinal direction and the coefficient A_{0m} depends on aspect ratio. Basically this equation gives the deflection shape of a hungry horse mode.

In Fig. 2, the maximum initial deflection measured by Yao et al. (1992) has been plotted against $\beta^2 t$, where β is the panel slenderness parameter defined as $\beta = b/t\sqrt{\sigma_Y/E}$. Smith et al. (1987) have categorized their measured data of maximum initial deflection, w_{0max}, by statistical analysis into three levels - 'slight', 'average' and 'severe' and formulated equations as shown in Fig. 2. 'Slight' and 'severe' correspond to 3% and 97% fractiles of probability distribution function of initial deflection, respectively. It can be seen from Fig. 2 that the measured data are distributed such that the 'average' line and the allowable limit prescribed by JSQS (Japan Ship Quality Standard) form an upper bound. In this study, the maximum initial deflection corresponding to the 'average' level, as per Eq. (2), is considered as the upper limit of actual initial deflection of thick plates.

$$w_{0max} = 0.1\beta^2 t \tag{2}$$

Effect of welding residual stress is not considered.

BUCKLING STRENGTH CALCULATION IN CLASSIFICATION RULES

As examples of ship classification rules, NK and DNV rules are considered. According to NK rules, σ^e_{xcr}, σ^e_{ycr} and τ^e_{cr} are calculated from elastic buckling interaction equation, and then equivalent elastic buckling stress, σ_{ecr}, is calculated using Eq. (3).

$$\sigma_{ecr} = \sqrt{\sigma^{e\,2}_{xcr} - \sigma^e_{xcr}\sigma^e_{ycr} + \sigma^{e\,2}_{ycr} + 3\tau^{e\,2}_{cr}} \tag{3}$$

If σ_{ecr} is greater than one half of the yield stress, equivalent plastic buckling stress, σ_{pcr}, is calculated using Johnson's plastic correction as

$$\sigma_{pcr} = \sigma_Y(1 - \frac{\lambda^2}{4}) \qquad \lambda \leq \sqrt{2} \tag{4}$$

where λ is a reduced slenderness ratio and is defined as follows:

$$\lambda = \sqrt{\frac{\sigma_Y}{\sigma_{ecr}}} \tag{5}$$

Check of buckling is done by comparing the buckling strength as calculated above with the equivalent applied stress, σ_{eq}, given by

$$\sigma_{eq} = \sqrt{\sigma_x^2 - \sigma_x\sigma_y + \sigma_y^2 + 3\tau^2} \tag{6}$$

In NK rule, plastic correction is performed on equivalent elastic buckling stress, σ_{ecr}. On the other hand, according to DNV, plastic correction is carried out for the elastic buckling stresses, σ^e_{xcr}, σ^e_{ycr} and τ^e_{cr}, separately. For combined loading cases, buckling stresses after the plastic correction, σ_{xcr}, σ_{ycr}

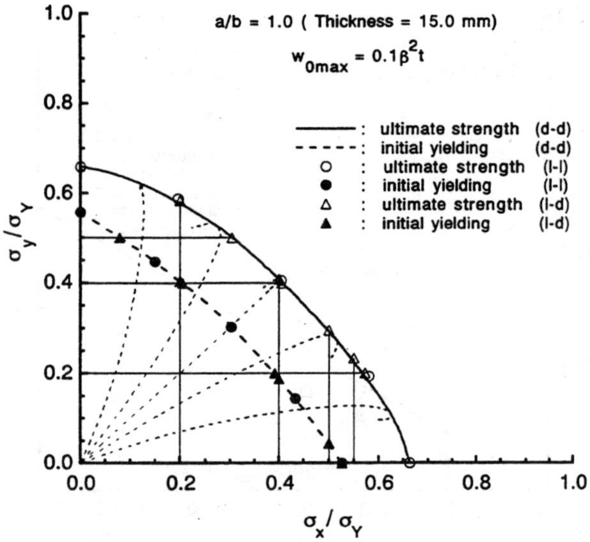

Fig. 3 Influence of loading method on initial yielding and ultimate strength interaction relationships (bi-axial compression)

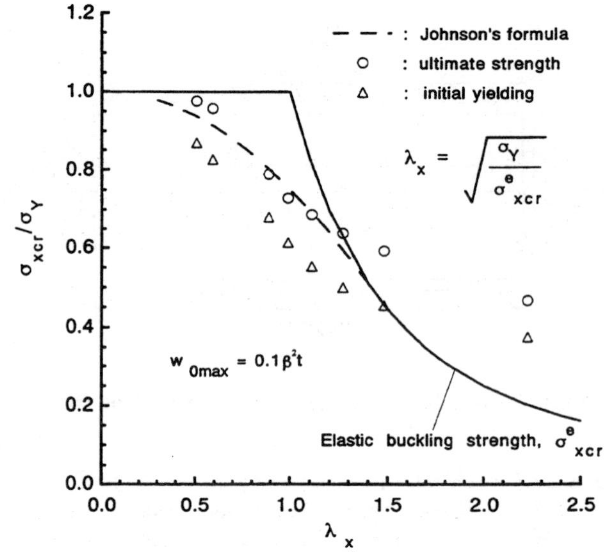

Fig. 4 Buckling/ultimate strength of a square plate under uni-axial compression

and τ_{cr}, are substituted into the following interaction relationship:

$$\frac{\sigma_x}{\eta_x \sigma_{xcr} q} - K\frac{\sigma_x \sigma_y}{\eta_x \eta_y \sigma_{xcr} \sigma_{ycr} q} + (\frac{\sigma_y}{\eta_y \sigma_{ycr} q})^n \leq 1 \qquad (7)$$

where, q represents the influence of applied shear stress and given by

$$q = 1 - \left(\frac{\tau}{\tau_{cr}}\right)^2 \qquad (8)$$

The parameters K and n in Eq. (7) are the functions of aspect ratio and slenderness ratio and η is the usage factor.

RESULTS AND DISCUSSION

Influence of loading methods

The influence of loading methods on initial yielding and ultimate strength interaction curves of a plate subjected to bi-axial compression is studied. The interaction curves obtained for a 15 mm thick square plate by d-d control, l-l control and l-d control are shown in Fig. 3. Thin dashed lines represent the path of average compressive stress obtained by d-d control and the ultimate strength interaction curve is obtained by drawing an envelope over many such paths. Circles show the results of l-l control in which σ_x and σ_y were increased proportionately. Thin solid lines represent the loading path in the case of l-d control. The initial yielding and ultimate strength interaction curves obtained by all the three methods are almost the same. Similar results were obtained for plates of other thicknesses and aspect ratios. It is understood from this that the effect of loading methods on initial yielding and ultimate strength interaction curves is almost negligible. So hereafter, only the results obtained by d-d control will be used for the case of bi-axial compression. On the other hand, for the case where shear stress is applied, the results obtained by l-d control in which shear stress is first applied will be used.

A square plate under uni-axial and bi-axial compression

The ultimate strength obtained by FEM analysis has been compared with the buckling strength obtained by using plastic correction formulae. First, as the most fundamental case, a square plate subjected to uni-axial compression is considered. The relationship between the buckling strength and the slenderness ratio λ is shown in Fig 4. In the FEM analysis, maximum initial deflection corresponding to the 'average' level, as per Eq. (2), is taken. It can be seen from Fig. 4 that the plastic correction curve represented by the dashed line almost coincides with the ultimate strength of the plate. It is to be noted that maximum initial deflection of 'average' level is nearly equal to the upper limit observed in actual ship platings as shown in Fig. 2. From this fact, it can be said that plastic correction gives an estimate of ultimate strength on the safe side for a square plate under uni-axial compression.

In Fig. 5(a), the ultimate strength interaction curve of a square plate of 15 mm thickness under bi-axial compression is shown. FEM analyses were performed with maximum initial deflections corresponding to 'slight', 'average' and 'severe' levels. Thin solid line represents the elastic buckling interaction curve. For intermediate load ratios, NK rule coincides with this linear curve, while DNV rule gives an outward convex curve. However, if we consider only 'average' level of maximum initial deflection, both NK and DNV rules give an estimate of strength almost on the safe side.

The results for 25mm thick plate is shown in Fig. 5(b). In this case, plastic buckling occurs for all the combinations of applied loads. The buckling interaction curve of DNV is in good correlation with FEM results for the average level of initial deflection, giving a lower bound buckling strength. In the case of NK rules, even though it estimates a buckling strength on the safe side for uni-axial compression, the estimate of strength for bi-axial compression is higher than the actual strength. The

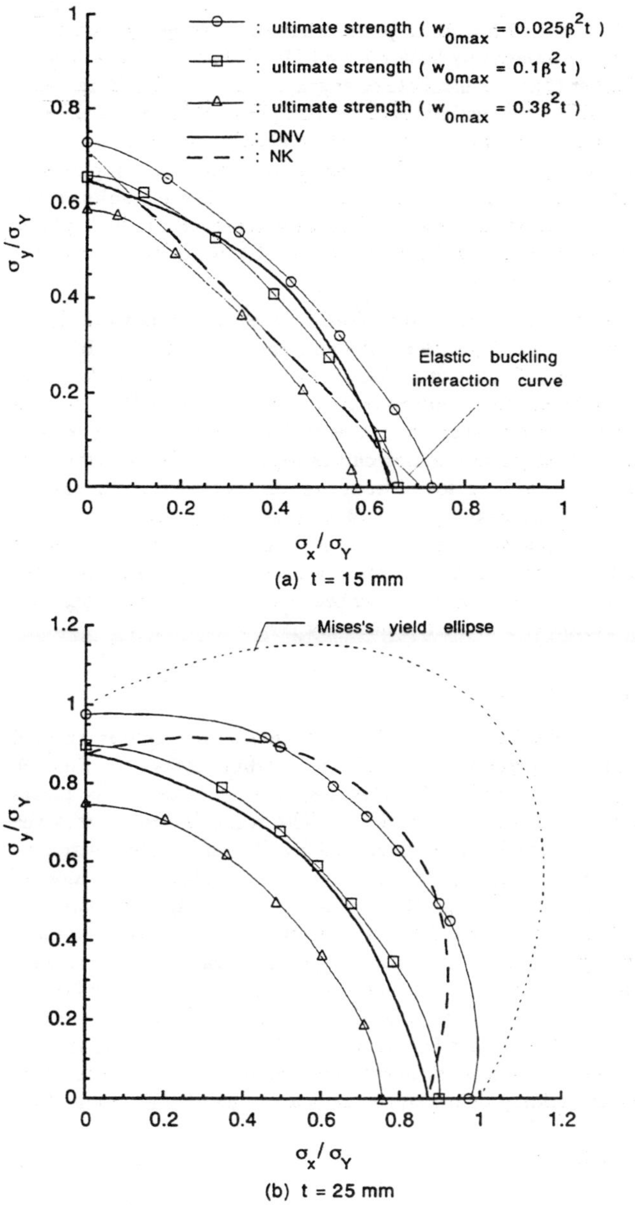

Fig. 5 Buckling/ultimate strength interaction relationships of a square plate under bi-axial compression

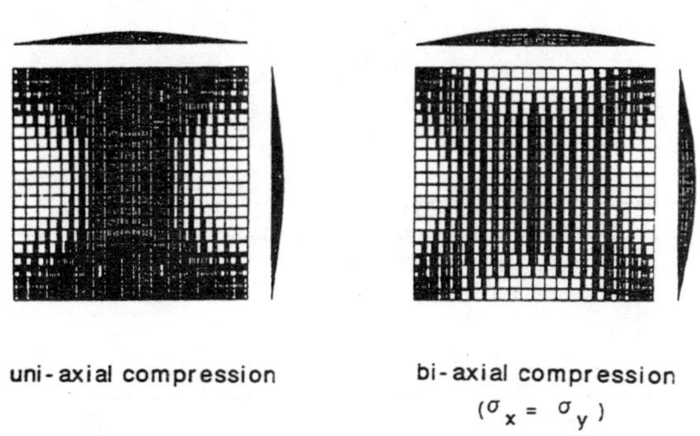

Fig.6 Spread of yielded zone of a square plate

reason for this can be attributed to the fact that in NK rule Johnson's correction is applied directly for bi-axial compression whereas DNV's interaction curve is derived by curve fitting based on FEM.

From the result by NK, it is known that Johnson's plastic correction method which depends only on slenderness ratio cannot reasonably evaluate a buckling collapse behaviour of thick plates under combined loads. This is basically because the collapse behaviour of plates depends not only on the slenderness ratio but also on applied load ratios. As a typical example, the collapse mode of a square plate under uni-axial compression and that under bi-axial compression where $\sigma_x = \sigma_y$ are compared. Figure 6 shows the spread of yielded zone at the ultimate strength for the respective cases. In the case of uni-axial compression, the in-plane load after buckling is carried by the effective width of the plate near the unloaded edges. In the case of bi-axial compression, bending is more dominant in comparison with uni-axial compression. As a result, reserved in-plane strength after buckling becomes small. Such factors are not considered in the plastic correction formulae.

Rectangular plate under uni-axial and bi-axial compression

Next, similar investigation about plastic correction has been carried out on a rectangular plate of aspect ratio, a/b = 3.0 and 5.0. In Fig. 7, the relationships between the ultimate strength and slenderness ratio obtained for the uni-axial compression in the longitudinal direction (x direction) and transverse direction (y direction) are shown. For a rectangular plate, elastic buckling

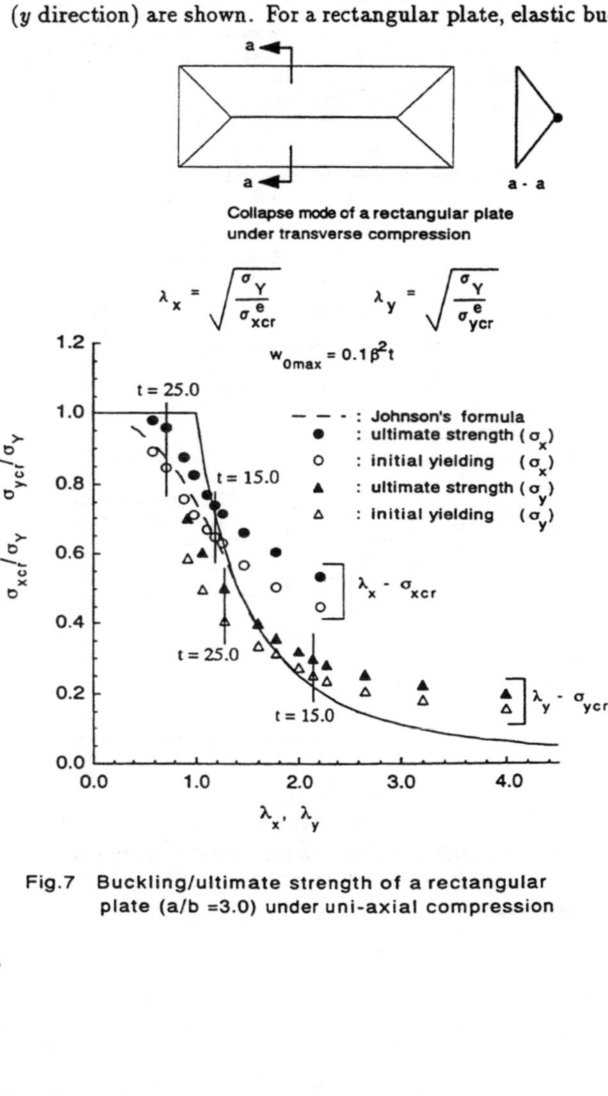

Fig.7 Buckling/ultimate strength of a rectangular plate (a/b =3.0) under uni-axial compression

strength is smaller for transverse compression than that for longitudinal compression. So, for the same thickness of plate, λ_y is greater than λ_x as indicated in Fig. 7. The ultimate strengths obtained for longitudinal compression lie above the plastic correction curve and they are even greater than the ultimate strength of a square plate. This is because the initial deflection in the shape of hungry horse mode resists the formation of three half-wave buckling mode, leading to a higher ultimate strength. In the case of transverse compression, if thick plates where $\lambda_y \leq \sqrt{2}$ are considered, the ultimate strengths obtained by FEM analyses lie below the plastic correction curve. The reason for this is that the middle part of a long plate under transverse compression fails like a strip with a mid-span plastic hinge and this collapse mode is further enhanced by the initial deflection of hungry horse mode. As a result, the reserve strength after buckling is less than that for longitudinal compression.

Figure 8 shows the ultimate strength interaction curves obtained for plates of 15 and 25 mm thicknesses and of aspect ratio 3.0. For 25 mm thick plate, when σ_y is dominant, buckling strength estimated by both NK and DNV are on the unsafe side, i.e. higher than the ultimate strength obtained by FEM analysis. However, the rule estimates of buckling strengths are on the safe side when σ_x is dominant or when elastic buckling occurs as shown in Fig. 8(a). Similar results are obtained for plates of aspect ratio 5.0 also. As can be seen from these results, there exists an uncertainty on the buckling strength of thick plates predicted by existing formulae, especially when the transverse stress is dominating.

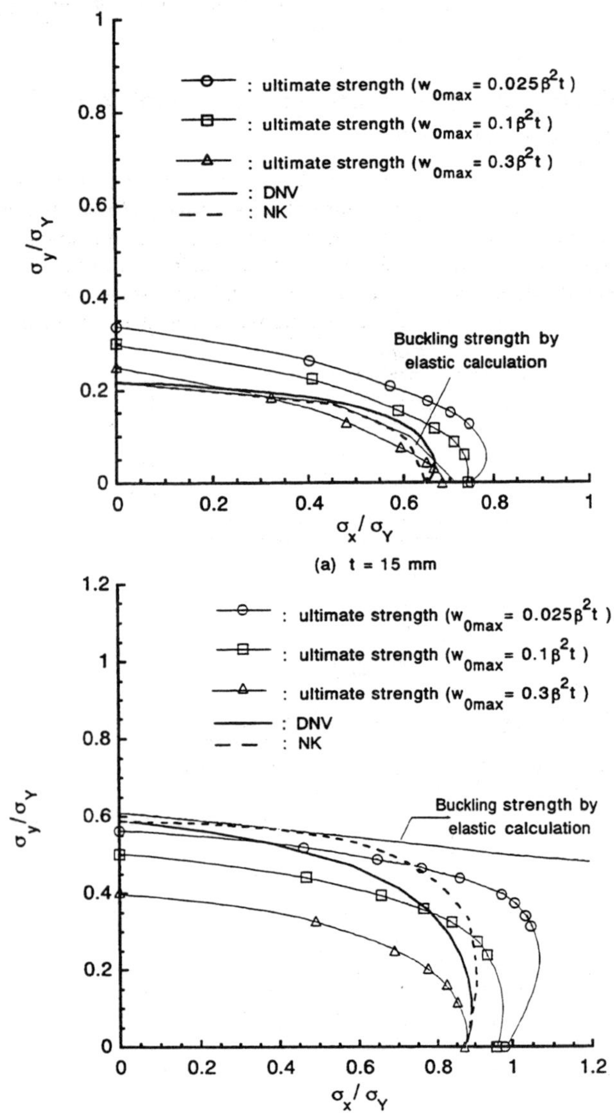

Fig.8 Buckling/ultimate strength interaction relationships of a rectangularplate (a/b = 3.0) under bi-axial compression

In this study, only unstiffened plates were analysed. However in a real case, plates are strengthened by stiffeners. In order to study the influence of stiffeners, in particular that of their torsional stiffness, on the ultimate strength of a plate, FEM analyses have been performed on the plates with and without stiffener and ultimate strength interaction curves were drawn for bi-axial compression. It has been found from the results that the influence of stiffener on ultimate strength of a plate is very negligible, especially when the transverse stress is dominant. This is primarily because the deformation of hungry horse mode is increased under the action of transverse compression and hence the torsional deformation of stiffeners is not very large.

Effect of shear

The effect of shear is discussed next. NK rule is applied for plastic correction for combined loading cases. In Fig. 9, τ_{cr}/τ_Y obtained by FEM analyses is plotted against λ_τ for plates of aspect ratios 1.0, 3.0 and 5.0. The solid line represents the buckling strength curve estimated by elastic buckling strength and Johnson's plastic correction, which is made for the range of $\lambda_\tau \leq \sqrt{2}$. It can be seen that for all cases the estimated strength by Johnson's formula is very much on the conservative side and the margin of safety increases as the aspect ratio increases. This is due to the reasons that for shear buckling load carrying capacity can increase after buckling because of the formation of a tension field, and that for the plate of larger aspect ratio the initial deflection of hungry horse mode more resists the formation of shear buckling deformation. It is also one of the reasons that

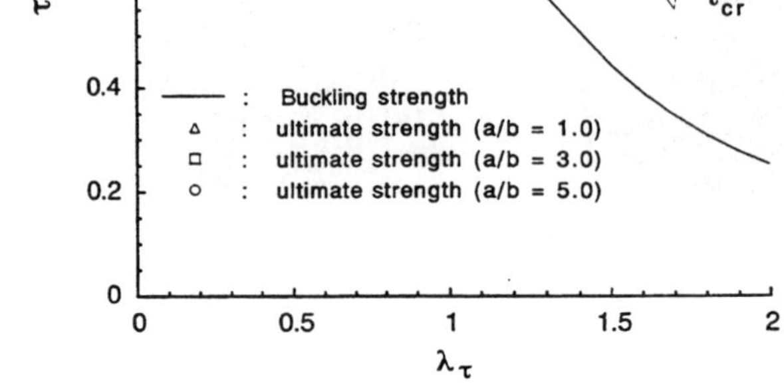

Fig. 9 Buckling/ultimate strength of a rectangular plate under shear

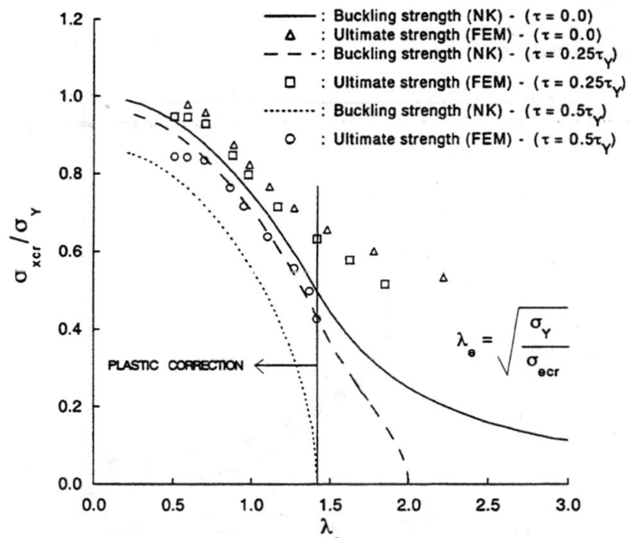

Fig. 10 Buckling/ultimate strength of a rectangular plate (a/b=3.0) under longitudinal compression and shear

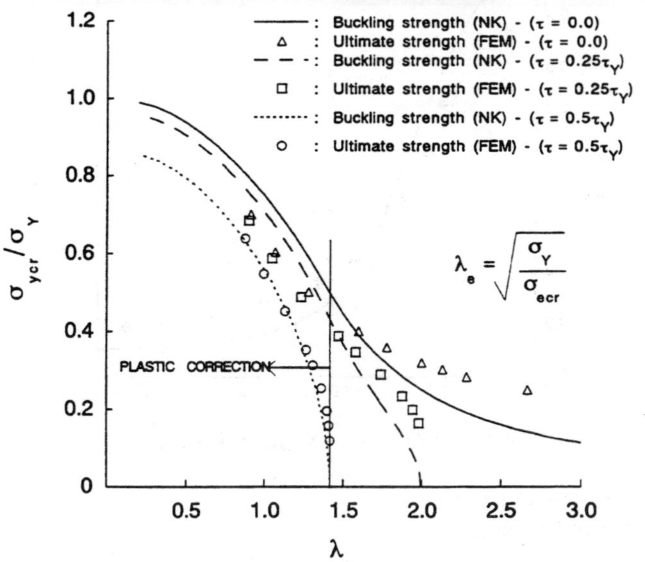

Fig. 11 Buckling/ultimate strength of a rectangular plate (a/b=3.0) under transverse compression and shear

for the plate of larger aspect ratio elastic buckling strength is small and λ_τ becomes large. In the case of a square plate, the formation of buckling mode was clearly observed for the applied shear stress above the buckling strength curve, whereas almost no buckling deformation was observed for the plates of aspect ratio 3.0 and 5.0 up to their collapse. It is known from these results that the existing plastic correction method gives over-conservative prediction of buckling strength for the plates under shear.

Figure 10 shows the case when longitudinal compression acts along with shear. In this figure, σ_{xcr}/σ_Y is plotted against λ_e for different intensity of applied shear stress. Plastic correction is performed for the range of $\lambda_e \leq \sqrt{2}$. Negative value of σ_{xcr}/σ_Y implies that the plate buckles under shear only. As can be deduced from the previous results, for this case also the estimate of strength by Johnson's formula is on the conservative side.

Figure 11 shows the result of a rectangular plate of aspect ratio 3.0 subjected to transverse compression and shear. When the transverse compression is dominant ($\tau = 0.0, 0.25\tau_Y$), the strength estimated by plastic correction is very much on the unsafe side. On the other hand, as mentioned above, when shear alone acts the estimate of strength by plastic correction is very much on the conservative side. So when the shear stress is increased both these effects get cancelled each other and for the case when $\tau = 0.5\tau_Y$, the plastic correction curve coincides well with the ultimate strength obtained by FEM. Similar results are obtained for plates of aspect ratio 5.0 also.

Proposed modification

As discussed above, it may be unsafe to use the present plastic correction method for thick plates subjected to transverse compression alone, transverse compression along with shear and bi-axial compression when transverse stress is dominating. Based on these results, simple modification has been proposed as an attempt to shift the predicted buckling strength towards the safe side. NK rule is considered as an example.

The elastic buckling interaction equation of NK rule is given by

$$\frac{\sigma_x}{\sigma_{xcr0}} + \frac{\sigma_y}{\sigma_{ycr0}} + \left(\frac{\tau}{\tau_{cr0}}\right)^2 = 1 \tag{9}$$

σ_{xcr0}, σ_{ycr0} and τ_{cr0} represent the elastic buckling strengths of the plate subjected to σ_x alone, σ_y alone and τ alone, respectively. Here, correction factors f_x and f_y have been introduced while estimating elastic buckling strength for longitudinal and transverse compression, as follows:

$$\frac{\sigma_x}{f_x^{n_1}\sigma_{xcr0}} + \frac{\sigma_y}{f_y^{n_2}\sigma_{ycr0}} + \left(\frac{\tau}{\tau_{cr0}}\right)^2 = 1 \tag{10}$$

$$\begin{aligned}
f_x &= \begin{cases} 1+\left(\frac{\beta}{2.5}\right)^2 - \frac{\beta}{2.5} \leq 1.0 & \text{for} \alpha \leq \sqrt{2} \\ 1 & \text{otherwise} \end{cases} \\
f_y &= 1+\left(\frac{\beta}{c}\right)^2 - \frac{\beta}{c} \leq 1.0 \\
c &= 1.5+\frac{1}{\sqrt{\alpha}} \\
n_1 &= 3.2(1-\frac{\beta}{2.5}) \\
n_2 &= 3.2(1-\frac{\beta}{c})
\end{aligned}$$

where $\alpha = a/b$ is the aspect ratio and $\beta = b/t\sqrt{\sigma_Y/E}$ the slenderness parameter. For the square plate, $f_x = f_y$ and $n_1 = n_2$.

The effect of the correction factor is to reduce the estimated elastic buckling strength under uni-axial compression. Even though the present estimate of strength for a square plate under uni-axial compression falls almost on the safe side, correction factors are applied for this case also, for the sake of making a safer estimate under bi-axial compression. It may be noted that for a rectangular plate under bi-axial compression, when the longitudinal stress is dominant, the present estimate falls on

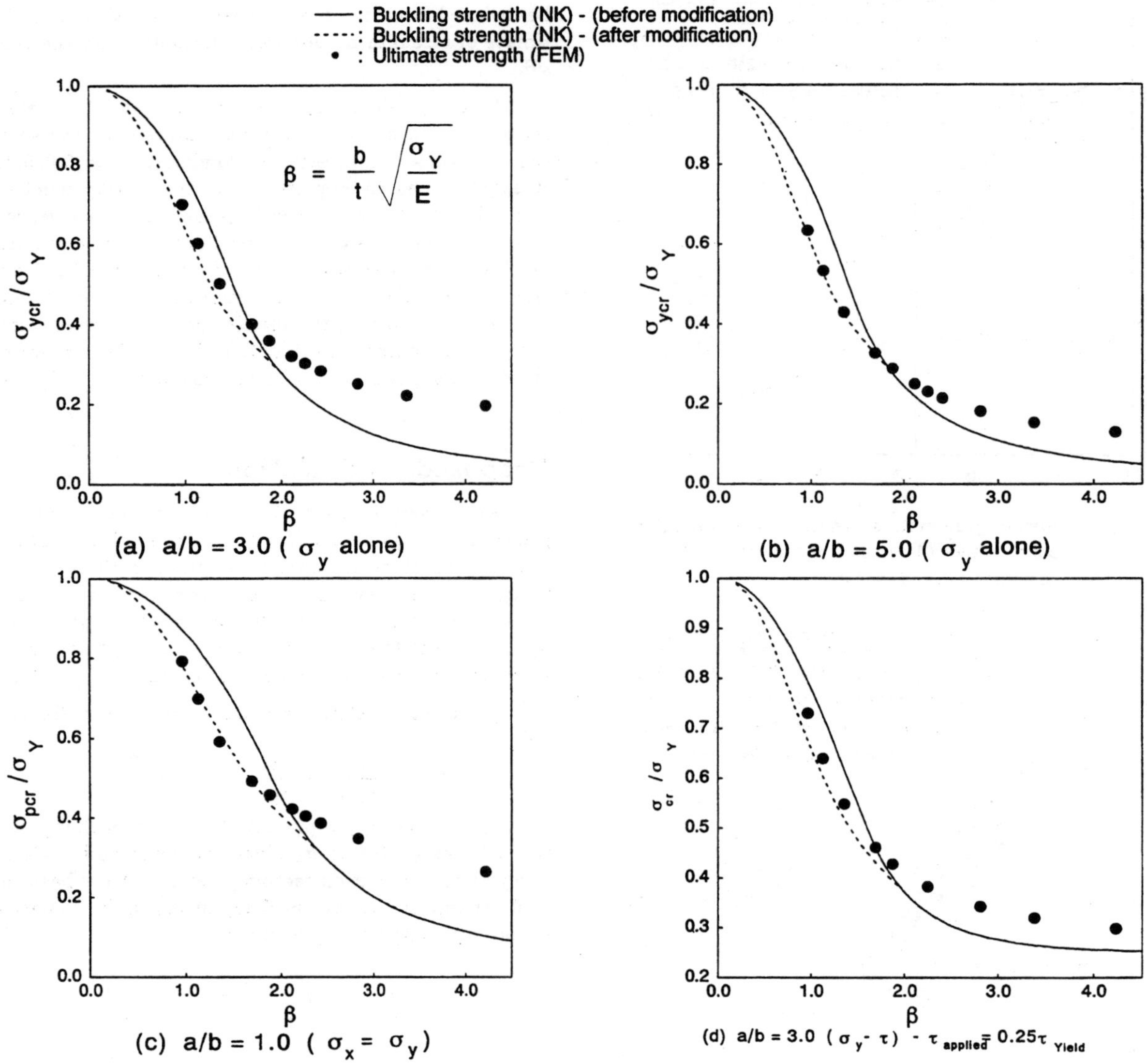

Fig. 12 Buckling/ultimate strength of a plate after the proposed modification

the safe side. So for plates which are likely to buckle with more than one half wave under longitudinal compression, it is sufficient to apply the correction only on uni-axial transverse buckling strength, for getting a safer estimate under bi-axial compression. Hence in the case of longitudinal compression, correction is applied only for plates of aspect ratio less than $\sqrt{2}$, whereas for transverse compression correction is applied for all cases depending on the slenderness ratio. Since the estimates of strength for shear are very much on the conservative side, no correction has been made for this case. The plastic correction is carried out in the normal way, using Johnson's formula.

The results after the proposed modification are shown in Fig. 12. It can be seen from these figures that the proposed modification gives an estimate of buckling strength on the safe side. Figure 13 shows the ultimate strength interaction curves for plates of aspect ratios 1.0, 3.0 and 5.0 and of thickness 25.0 mm under bi-axial compression. The estimated buckling strength interaction curves using the proposed modification coincide well with the ultimate strength of the plates with initial deformation of 'average' level.

CONCLUSIONS

In this study, ultimate strengths of plates with initial deflection subjected to bi-axial compression and shear were calculated using nonlinear finite element method and the results were compared with the buckling strength estimates of classification rules. The following conclusions are drawn from this study:

(1) Upper bound of the maximum initial deflection in the actual ship platings can be approximated by $w_{0max} = 0.1\beta^2 t$

(2) For the combined loading cases, loading method has only a negligible effect on the yielding and ultimate strength interaction curves.

(3) Plastic correction method used in the existing classification rules have applicability to thick plates as an estimation method of ultimate strength on the safe side within

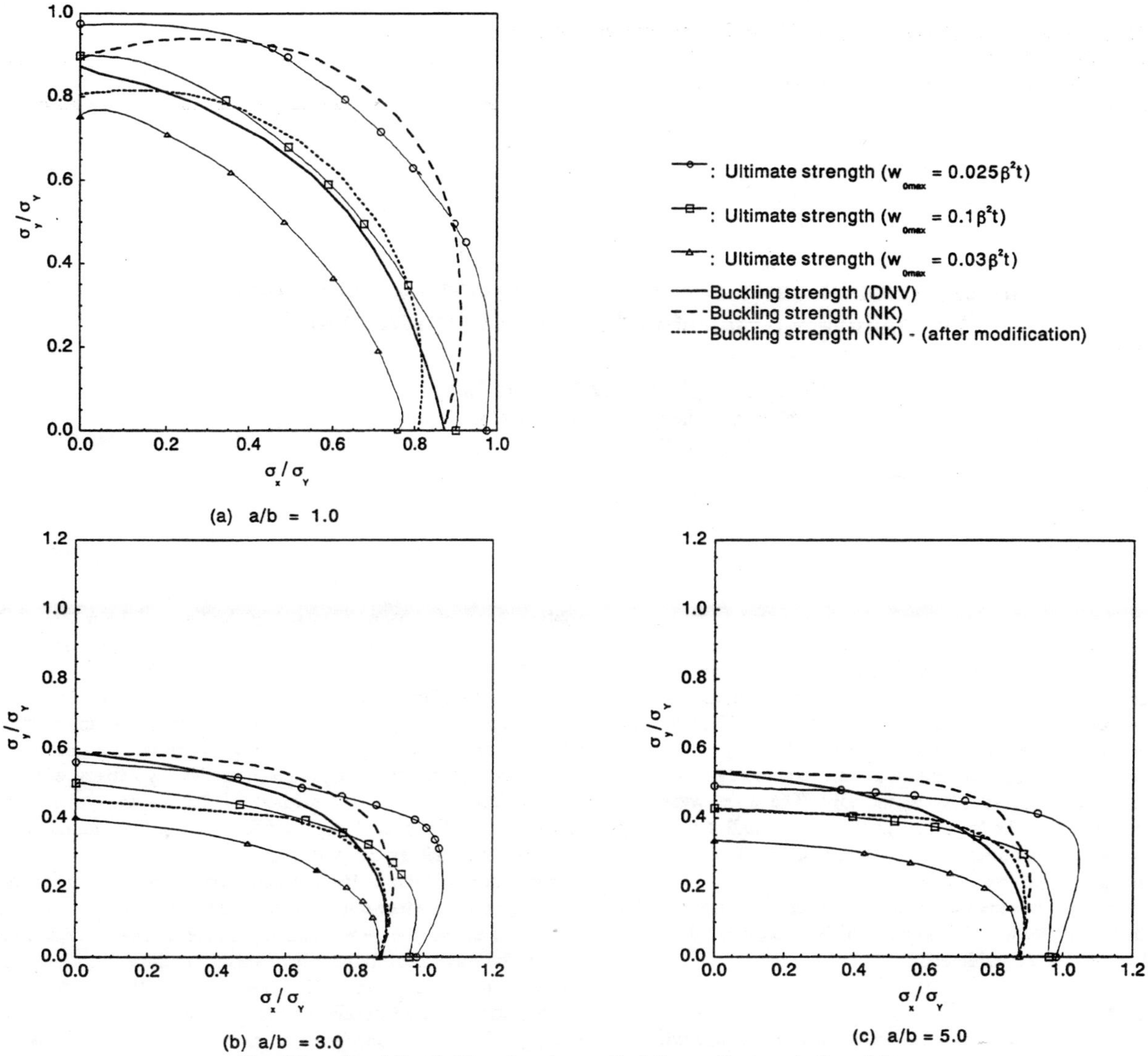

Fig. 13 Buckling/ultimate strength interaction relationships of a plate of thickness 25.0 mm after the proposed modification

the limit of upper bound of actual initial deflection. However in some cases, it may give buckling strength on the unsafe side depending on the ratios of applied loads.

(4) A simple modification has been proposed as an attempt to shift the estimated buckling strength toward safe side and its effectiveness has been demonstrated.

REFERENCES

DNV (1994). "Rules for Classification of Ships."

Dow, C.S. et al. (1991). "Report of Committee III.1 Ductile Collapse," *Proc. 11th ISSC*, Xusi, Vol 1, pp 319-403.

Smith, C.S., Davidson, P.C., Chapman, J.C. and Dowling, P.J. (1987). "Strength and Stiffness of Ships Plating under In-plane Compression and Tension," *Transactions of the Royal Institution of Naval Architects*, W6, pp 277-296.

NK (1993). "Rules and Guidance for the Survey and Construction of Steel Ships."

Yao, T., Nikolov, P.I. and Miyagawa, Y. (1992). "Influence of Welding Imperfections on Stiffness of Rectangular Plate under Thrust," *Mechanical Effects of Welding*, Eds. Karlsson, L. et.al, Springer-Verlag, pp 261-268.

Proceedings of the Seventh (1997) International Offshore and Polar Engineering Conference
Honolulu, USA, May 25-30, 1997

ISBN 1-880653-28-1 (Set); ISBN 1-880653-32-X (Vol. IV)

A Boundary Element Based Solution of an Inverse Elasticity Problem by Conjugate Gradient and Regularization Method

Cheng-Hung Huang and Wu-Yang Shih
National Cheng Kung University
Tainan, Taiwan, China

ABSTRACT

A boundary element solution to an inverse elasticity problem by utilizing conjugate gradient method is presented for estimating the residual or contact stresses on a body from the measurements of displacement at discrete locations either internal to the domain or on part of the boundary which already has traction specified.

The comparisons of the present method and commonly used regularization method are discussed systematically. The advantages and disadvantages in applying the large matrix (LM) and small matrix (SM) formulations are also examined. The present work is the first to the authors' knowledge to do above studies.

Finally the effects of the measurement errors to the inverse solutions are discussed. A statistical analysis which leads to a 99 % confidence bound will also be considered.

I. INTRODUCTION

In most boundary value problems of solid mechanics, one deals with solving for a governing system of equations (i.e. the balance of momentum, constitutive and kinematics equations), given sufficient initial condition and boundary traction or displacement conditions for the resulting displacement, stress and strain fields. This is the direct problem. The existence and uniqueness of such direct problem have been well established.

However many practical problems do not fall into this category. In particular, the boundary conditions are often incomplete and one wishes to determine them based on certain experimental measurements either in the form of overspecified boundary conditions on another part of the boundary or a part of the solution internal to the domain.

In general the inverse problems are ill-posed, this means their solutions do not necessarily satisfy conditions of existence, uniqueness and stability with respect to small change in input data. Instability implies that small measurement errors may result in tremendous errors in the calculated inverse solutions. With regard to uniqueness, there may be a variety of boundary conditions that result in nearly the same measured data at the sensor's locations. Therefore, in order to solve the inverse problem properly and exempt from the ill-posed phenomena, it is necessary to choose a suitable algorithm to obtain a stable solution.

Most of the literature on the solution of inverse problem has been devoted to inverse heat transfer, Beck et al (1985), research in the area of inverse elasticity problems has been limited. Inverse deformation problems have also been discussed in the field of thermoelasticity by Grysa et al. (1981), nonsdestructive measurement of plastic strains by Mura (1985), contact problem by Kuz'menko (1986).

Recently, Maniatty et al. (1989) used simple diagonal regularization in conjunction with the finite element method to determine unknown traction boundary condition; spatial regularization was introduced in conjunction with the boundary element method by Zabaras et al. (1989) and with the finite element method by Schnur and Zabaras(1990). Zhang et al. (1995) used constrained least-squares minimization to determine residual stress and contact stress. Martin et al. (1995) derived a non-iterative algorithm based on boundary element method to determine the surface tractions and deformations in elastostatics.

From the above review we learned that the method of regularization seems to be used most frequently to solve the ill-conditions sensitivity matrix in the recent years. However the smoothing effects of the regularization method depends on (1) the order of the regularization process, (2) the magnitude of *a priori* determined regularization coefficient and (3) the forms of the quadratic norm. The above three items are defined as the "regularization conditions", which are difficult to choose before the inverse calculations. To avoid such difficulties the method of conjugate gradient (or iterative regularization), Kammerer and Nashed (1972), should be used here.

In this paper, the inverse problem by using both conjugate gradient method and regularization method in estimating boundary traction from displacement measurements containing random errors will be addressed. The boundary element method (BEM) is proposed here for the direct problem solution. The boundary only discretization feature of the BEM is a distinct advantage over traditional finite difference or finite element methods which require whole domain discretization. As only boundary traction or displacement are needed to the inverse problem, the BEM is ideally suited for our purpose.

In the following sections a precise definition of the direct and inverse elasticity problem will be given. Then two different matrix systems (i.e. LM and SM formulations) containing sensitivity matrix and unknown boundary conditions that were derived by BEM discretization are developed respectively. The regularization and conjugate gradient methods are then applied to those two matrix systems to determine the traction boundary conditions by varying the measurement errors and the measured positions.

Finally, in order to include the effects, such as the measurement errors and the measured positions, to the inverse solutions, a statistical analysis is performed to derive the standard deviations for the estimated traction; thus 99% confidence bounds for the estimated functions can be determined.

II. THE DIRECT PROBLEM

Assuming that a two-dimensional body B with boundary $\Gamma = \{\Gamma_1, \Gamma_2, \Gamma_3\}$ is isotropic, homogeneous and linear elastic, the direct elasticity problem may be defined by the following equations in indicial form, Brebbia and Dominguez (1989)

$$s_{ij,j} = 0 \qquad (1)$$

$$s_{ij} = \lambda\varepsilon_{kk}\delta_{ij} + 2\mu\varepsilon_{ij} \qquad (2)$$

$$\varepsilon_{ij} = \frac{1}{2}(u_{i,j} + u_{j,i}) \qquad (3)$$

$$t_1(\overline{x}) = f(\overline{x}); \quad \overline{x} \in \Gamma_1 \qquad (4)$$

$$u_2(\overline{x}) = g(\overline{x}); \quad \overline{x} \in \Gamma_2 \qquad (5)$$

$$t_3(\overline{x}) = h(\overline{x}); \quad \overline{x} \in \Gamma_3 \qquad (6)$$

where s = stress tensor, ε = strain tensor, i, j = 1, 2 and traction $t = \{t_x, t_y\}$, displacement $u = \{u_x, u_y\}$. Equations (1) are the equilibrium equation without body forces, equation (2) is Hooke's law for an isotropic, elastic material where λ and μ are the Lame's constant. The relationship between infinitesimal strains and displacements is given by equation (3). The boundary traction on Γ_1 and displacement boundary condition on Γ_2 are specified in equations (4) and (5), another traction boundary condition on Γ_3 is given in equation (6). Please refer to Figure 1.

For a constant property, linear elasticity problem with a domain B and boundary Γ, the boundary integral equation for this problem without body force term can be derived as Brebbia and Dominguez (1989)

$$c^i_{lk}u^i_k + \int_\Gamma u_k t^*_{lk} d\Gamma = \int_\Gamma t_k u^*_{lk} d\Gamma \qquad (7)$$

where

i = point of Γ or B

u_k = displacement

t_k = traction

$c^i_{lk} = \delta_{lk}$, if i is in B; c^i_{lk} values will depend on type of corner if i is on Γ ($c^i_{lk} = 0.5\delta_{lk}$ if Γ is smooth at i)

u^*_{lk} = stationary fundamental solution for displacement

t^*_{lk} = stationary fundamental solution for traction

here for two-dimensional plane strain problem

$$u^*_{lk} = \frac{1}{8\pi\mu(1-\nu)}[(3-4\nu)\ln\frac{1}{r}\delta_{lk} + r_{,l}\, r_{,k}\,] \quad (8)$$

and

$$t^*_{lk} = \frac{1}{4\pi(1-\nu)r}\{\frac{\partial r}{\partial n}[(1-2\nu)\delta_{lk} + 2r_{,l}\, r_{,k}\,] + (1-2\nu)(n_l r_{,k} - n_k r_{,l}\,)\} \qquad (9)$$

where r = distance from i to a point of Γ.

When considering direct problems, if i is on Γ, one has with equation (7) the basic equation for linear elasticity problems. For i in B, the boundary integral equation which then connect internal traction and displacement fields are usually used for a posteriori computation.

Generally speaking, the discretization of Γ into n boundary elements allows substitution into the boundary integral equation (7) expressed for each boundary element of the algebraic linear system

$$\mathbf{C\,U + H\,U = G\,T} \qquad (10)$$

Where

U = vector of displacement over n boundary elements

T = vector of traction

H, **G** = geometry dependent matrices

C = diagonal matrix

In a similarly way, if there are n' internal points, one obtain

$$\mathbf{U}' = \mathbf{G}'\mathbf{T} - \mathbf{H}'\mathbf{U} \qquad (11)$$

where

U' = vector of displacement at the n' internal points

H', **G'** = internal points dependent matrices

Matrices **H**, **G**, **H'** and **G'** depend on the type of the boundary elements. Some elements of these matrices involve singular kernels as described in Brebbia and Dominguez (1989). The program for the above stated problem is modified based on the text book by Brebbia and Dominguez (1989) and the quadratic boundary elements are adopted for all the examples illustrated here.

The direct problem considered here concerned with the determination of the medium displacement or stress when the boundary conditions along Γ are known.

III. THE INVERSE PROBLEM

For the inverse problem, the boundary traction on Γ_3 is regarded as being unknown but displacement readings taken at some appropriate internal locations are considered available, i.e.

$$u(\overline{x}) = Y_m(\overline{x}); \quad \overline{x} \in (B-\Gamma), m = 1, M \quad (12)$$

Referring to Figure 1, we assumed that M sensors are used to record the interior displacement information in x and y directions $Y_m = \{Y_{mx}, Y_{my})$ to identify boundary traction on Γ_3 in the inverse calculations, where m = 1 to M represents the number of sensor. Then the inverse problem can be stated as follows: by utilizing the above-mentioned measured displacement data Y_m, estimate the unknown boundary traction $t_{\Gamma_3} = t_3$ on Γ_3.

The solution of the present inverse problem is to be obtained in such a way that the following functional is minimized

$$J(t_{\Gamma_3}) = \sum_{m=1}^{M}[(u_{mx} - Y_{mx})^2 + (u_{my} - Y_{my})^2]$$

$$= (\mathbf{u} - \mathbf{Y})^T(\mathbf{u} - \mathbf{Y}) \qquad (13)$$

where $u_m = \{u_{mx}, u_{my}\}$ are the estimated displacements in x and y directions at the sensor's locations. These quantities are determined from the solution of the direct problem given previously using the estimated $\hat{t}_{\Gamma_3}$ for the exact t_{Γ_3}. Here the hat '^' denotes the estimated quantities.

To solve the above inverse elasticity problem using BEM, there are two choices in obtaining the system of equations that contain the sensitivity matrix and unknown boundary condition matrix.

<u>(i) Large Matrix (LM) Formulation:</u>

Pasquetti and Niliot (1991) and Pasquetti and Petit (1995) suggested that the combination of equations (10) and (11) will yield the following basic discretized system

$$\begin{bmatrix}\mathbf{H+C}\\ \mathbf{H}'\end{bmatrix}\mathbf{U} = \begin{bmatrix}\mathbf{G}\\ \mathbf{G}'\end{bmatrix}\mathbf{P} + \begin{bmatrix}0\\ -\mathbf{U}'\end{bmatrix} \qquad (14)$$

Rearranging the unknowns in a vector $\mathbf{X}_1$ leads to the following system of linear equations

$$A_1 X_1 = B_1 \quad (15a)$$

or

$$X_1 = [(A_1^T A_1)^{-1}] A_1^T B_1 = [\Phi_1^{-1}] \Psi_1 \quad (15b)$$

here A_1 is the sensitivity matrix (or influence matrix) and X_1 (Large Matrix) is a column matrix that contains **all** the unknown surface conditions (i.e. not just the traction on Γ_3). For the case of Figure 1, $X_1 = \{u_1, t_2, u_3, t_3\}$.

(ii) Small Matrix (SM) Formulation:

Zhang et al. [8] suggested that in equation (10) by separating t_3 from other unknowns and moving t_3 and all known boundary conditions to the RHS results in the LHS an expression for the unknown surface conditions where t_3 is no included, let us denoted as **Q**.

Similarly, in the RHS of equation (11) the surface conditions can be regrouped as there parts, say t_3, **Q** and the known conditions. By inserting previous expression of **Q** (obtained from equation (10)) for the above **Q** (obtained from equation (11)), we finally obtain the following equation that contains sensitivity matrix and unknown traction t_3

$$A_2 X_2 = B_2 \quad (16a)$$

or

$$X_2 = [(A_2^T A_2)^{-1}] A_2^T B_2 = [\Phi_2^{-1}] \Psi_2 \quad (16b)$$

where A_2 is the sensitivity matrix (or influence matrix) and X_2 (Large Matrix) is a column matrix that contains **only** the unknown traction on Γ_3. For the case of Figure 1, $X_2 = \{t_3\}$.

The above mentioned matrix Φ ($\Phi_1$or Φ_2) is both symmetry and positive define matrix. We should note that equations (15) and (16) can be solved by the traditional way to obtain the solutions of **X** (X_1or X_2) provided that the matrix Φ is not singular. However for the inverse problem considered here the condition of matrix Φ will depend on the positions of sensors as well as the measurement errors. We observed from the numerical experiments that matrix Φ is more and more ill-conditioned (the determinant of Φ approaches to zero) as the distance between sensors and boundaries is increased or the displacement measurement errors are big, therefore the ill-posed phenomena occurs during the inverse calculations. To avoid this phenomena the following algorithms should be applied.

IV-1. THE REGULARIZATION METHOD

The regularization method is a modification of the sum of squares function with the addition of the regularization terms. These additional terms have a smoothing effect on the unknown function components. However the major difficulties in choosing a suitable quadratic norm, determining a proper regularization order and determining the optimal smoothing (or regularization) coefficient of the regularization method make it not so convenient to be used. From Pasquetti and Niliot (1991), the unknown vector **X** can be obtained as follows

$$X = [(A^T A + \Omega D^T D)^{-1}] A^T B = [(\Phi + \Omega D^T D)^{-1}] \Psi \quad (17)$$

where Ω is the unknown regularization coefficient, **D** is a square matrix associated with the regularization procedures. For the zero-order regularization process, the definition of **D** becomes an identity matrix and is given in equation (18)

$$D = \begin{bmatrix} 1 & 0 & 0 & 0 & \ldots\ldots & 0 & 0 \\ 0 & 1 & 0 & 0 & \ldots\ldots & 0 & 0 \\ . & . & . & . & \ldots\ldots & . & . \\ 0 & 0 & 0 & 0 & \ldots\ldots & 1 & 0 \\ 0 & 0 & 0 & 0 & \ldots\ldots & 0 & 1 \end{bmatrix} \quad (18)$$

One should note that the inverse solution obtained by the RM depends very strongly on how the quadratic norm is defined as well as how the regularization order and regularization coefficient are chosen. However the above stated inconvenient can be avoided if the iterative regularization method (i.e. conjugate gradient method) is used.

In spite of some existing techniques in finding the optimal value of regularization coefficient, Golub (1979), we adopt a general stopping criteria for the inverse problem, i.e. the discrepancy principle, Alifanov (1974), as will be discussed in text. The regularization coefficient Ω is initially chosen as a large number, then decreases gradually until the computed stopping criterion is less than the one that was calculated based on the discrepancy principle.

IV-2. THE CONJUGATE GRADIENT METHOD

Previous researchers have studied the iterative conjugate numerical methods for linear equations having symmetric and positive define matrix. They concluded that for the matrices which are very ill-conditioned, the conjugate gradient method is much the best. In the present study this is the case when the measured positions are too far away form the boundaries, therefore the conjugate gradient method is adopted here.

The detail descriptions and rigorous proof of conjugate gradient method can be found in references Kammerer and Nashed (1972), we are not going to repeat it here.

The algorithm of the iterative conjugate gradient method to solve equation

$$A X = B \quad (19)$$

is now stated briefly as:

(1). Chosen an arbitrary vector X_0 (the initial guesses to the solution).

(2). Set $E_0 = r_0 = B - AX_0$ (20)

(3). Determine successively

$$\alpha_i = \frac{(E_i, r_i)}{(E_i, AE_i)} \quad (21a)$$

$$X_{i+1} = X_i + \alpha_i E_i \quad (21b)$$

$$r_{i+1} = r_i - \alpha_i AE_i \quad (21c)$$

$$\beta_i = \frac{(r_{i+1}, AE_i)}{(E_i, \xi E_i)} \quad (21d)$$

$$E_{i+1} = r_i + \beta_i E_i \quad (21e)$$

Where

$r \equiv$ gradient

$E \equiv$ direction of descent

$\alpha \equiv$ search step size

$\beta \equiv$ conjugate coefficient

and subscript i represents the iterative index, (∗,※) denotes the inner product of the vectors ∗ and ※.

(4). Repeat step (3) with i+1 replacing i and continue until the functional J is minimized.

V. STOPPING CRITERION

If the problem contains no measurement errors, the traditional check condition is specified as

$J(t|_{\Gamma_3}) < \varepsilon$ (22)

where e is a small specified number. However, the observed displacement data may contain measurement errors. Therefore, we do not expect the functional equation (13) to be equal to zero at the final iteration step. Following the experience of Alifanov (1974), we use the discrepancy principle as the stopping criterion, i.e. we assume that the displacement residuals may be approximated by

$$u_{mx} - Y_{mx} \approx \sigma_1 \quad (23a)$$

$$u_{my} - Y_{my} \approx \sigma_2 \quad (23b)$$

where σ_1 and σ_2 are the standard deviation of the displacement measurements in x and y direction, respectively, which are assumed to be a constant. Substituting equation (23) into equation (13), the following expression is obtained for e:

$$\varepsilon = M(\sigma_1^2 + \sigma_2^2) \quad (24)$$

then, the stopping criterion is given by equation (22) with ε determined from equation (24).

VI. THE STATISTICAL ANALYSIS

Statistical analysis is important in determining the accuracy of the computed inverse solution. In this section we develop the confidence bounds analysis for the estimated unknown boundary condition **X** by assuming independent, constant-variance errors. We should note that **X** represents all the unknown boundary conditions in Large Matrix formula while denotes only the unknown traction along Γ_3 in Small Matrix formula. The variance-covariance matrix of the estimated parameter vector **X** is defined as (Farebrother (1988))

$$\text{Var-cov}(\mathbf{X}) \equiv E\left[\left[\mathbf{X} - E(\mathbf{X})\right]\left[\mathbf{X} - E(\mathbf{X})\right]^T\right] \quad (25)$$

where **X** = **X**(**Y**) is the unknown boundary condition vector, **Y** is the measured displacement vector, E(●) denotes the statistical expected value (averaging) operator, and superscript T refers to the transport. Equation (25) is a nonlinear system for the determination of the variance-covariance matrix.

The right hand side of equation (25) is then linearized by expanding **X** and E(**X**) in Taylor series and neglecting the higher-order terms. We obtain

$$\text{Var-cov}(\mathbf{X}) =$$

$$\left[\frac{\partial \mathbf{X}}{\partial \mathbf{Y}^T}\right] E\left[\left[\mathbf{Y} - E(\mathbf{Y})\right]\left[\mathbf{Y} - E(\mathbf{Y})\right]^T\right]\left[\frac{\partial \mathbf{X}^T}{\partial \mathbf{Y}}\right] \quad (26)$$

If $\sigma_1 = \sigma_2 = \sigma$ is assumed, for independent displacement measurements with constant variance σ^2, we have

$$E\left[\left[\mathbf{Y} - E(\mathbf{Y})\right]\left[\mathbf{Y} - E(\mathbf{Y})\right]^T\right] = \sigma^2 \mathbf{I} \quad (27)$$

Introducing equation (27) into equation (26), the variance-covariance matrix becomes

$$\text{Var-cov}(\mathbf{X}) = \sigma^2 \left[\frac{\partial \mathbf{X}}{\partial \mathbf{Y}^T}\right]\left[\frac{\partial \mathbf{X}^T}{\partial \mathbf{Y}}\right] \quad (28)$$

To express right-hand side of equation (28) in terms of Jacobian, the method of chain rule is applied, equation (13) is utilized and the high-order terms are neglected. After some length manipulations, Huang and Huang (1994), the following equation is obtained

$$\text{Var-cov}(\mathbf{X}) = \sigma^2 \left\{\left[\frac{\partial \mathbf{u}^T}{\partial \mathbf{X}}\right]\left[\frac{\partial \mathbf{u}}{\partial \mathbf{X}^T}\right]\right\}^{-1} \quad (29)$$

Here the Jacobian matrix $\partial \mathbf{u} / \partial \mathbf{X}^T$ and its transport $\partial \mathbf{u}^T / \partial \mathbf{X}$ are evaluated from the results of direct problem by perturbing the unknown boundary condition vector **X**.

If we assume independent errors, the nondiagonal elements in the variance-covariance matrix vanish and can be written in explicit form as

$$\text{Var-cov}(\mathbf{X}) = \begin{bmatrix} \sigma_{X_1}^2 & 0 & \cdots & 0 \\ 0 & \sigma_{X_2}^2 & & \vdots \\ \vdots & & \ddots & 0 \\ 0 & \cdots & 0 & \sigma_{X_{2k}}^2 \end{bmatrix} \quad (30)$$

where k represents the discrete number of node points along BC. Since there are x and y components of traction at each node, we have total of 2k unknowns. Then a comparison of equations (29) and (30) gives the standard deviation of the estimated parameters $\sigma_{\mathbf{X}}$ as

$$\sigma_{\mathbf{X}} = \sigma \sqrt{\text{diag}\left\{\left[\frac{\partial \mathbf{u}^T}{\partial \mathbf{X}}\right]\left[\frac{\partial \mathbf{u}}{\partial \mathbf{X}^T}\right]\right\}^{-1}} \quad (31)$$

where σ is standard deviation of the measurements and "diag" represents diagonal elements of the matrix. This equation is similar to the one that was shown by Maniatty and Zabaras (1994).

If we now assume a normal distribution for the measurement errors, the 99% confidence bounds are determined as

$$\left\{\mathbf{X} - 2.576\sigma_{\mathbf{X}} \le \mathbf{X}_{mean} \le \mathbf{X} + 2.576\sigma_{\mathbf{X}}\right\} = 99\% \quad (32)$$

This expression defines the approximate statistical confidence bounds for the estimated boundary condition vector **X**, i.e. the estimated values of the parameters are expected to lie between these two bounds with a 99% confidence.

VII. RESULTS AND DISCUSSIONS

To illustrate the validity and accuracy of the conjugate gradient and regularization methods using both LM and SM formulations in predicting boundary traction t_x and t_y from the knowledge of internal displacement recordings, we consider two specific examples. The technique illustrated here can be readily applied to the problems in estimating the residual stresses of a structure or the unknown contact stresses.

In order to compare the results for situations involving random measurement errors, we assume normally distributed uncorrelated errors with zero mean and constant standard deviation. The simulated inexact measurement data **Y** can be expressed as

$$\mathbf{Y} = \mathbf{Y}_{exact} + \omega\,\sigma \quad (33)$$

where $\mathbf{Y}_{exact}$ is the solution of the direct problem with an exact boundary condition **X**; σ is the standard deviation of the measurements; and ω is a random variable that generated by subroutine DRNNOR of the IMSL (1987) and will be within -2.576 to 2.576 for a 99% confidence bound.

One of the advantages of using the conjugate gradient method to solve the ill-conditioned matrix is that the initial guesses of the unknown quantities can be chosen arbitrarily. In all the test cases considered here using CGM, the initial guesses of $\hat{\mathbf{X}}$ used to begin the iteration are taken as $\hat{\mathbf{X}}_{initial} = 0.0$.

We now present below the numerical experiments in determining **X** by the inverse analysis using both the CGM and RM in both LM and SM formulations.

(A). Numerical test case 1

The first numerical test case involves a rectangular plate (1m×3m) assumed to be loaded in a state of plane stress as shown in Figure 2. The quadric elements along boundaries AB, BC, CD and DA are 5, 7, 5 and 3, respectively. 13 sensors located parallel to BC evenly and inside the plate are used to predict the traction at those 13 nodes on the boundary. The shear modules is taken as μ = 1000Mpa and Poisson's ratio as ν = 0.3. The boundary conditions are given as follows:

Along AB and CD:

$t_x = t_y = 0$; but $u_x = u_y = 0$ at the midpoint of each segments.

Along DA:

$u_x = 0,\ t_y = 0$

The prescribed traction condition on the right-hand side of the plate is

Along

$$t_x = 500 \times \mathrm{SIN}\left[\frac{2\pi}{13} \times \left(\frac{y - 0.214}{0.214}\right)\right] \mathrm{MPa}$$

$$t_y = 500 \times \mathrm{COS}\left[\frac{2\pi}{13} \times \left(\frac{y - 0.214}{0.214}\right)\right] \mathrm{MPa}$$

The inverse analysis is first performed by arranging 13 sensors in line along x = 0.9. This represents that the distance between boundary Γ_4 and the sensors is large. When assuming exact measurements, σ = 0.0, the exact and estimated boundary traction along BC using both the CGM and RM in LM formula with stopping criteria $\varepsilon = 10^{-3}$ is shown in Figure 3. It can be seen from these Figures that both schemes obtained good estimation of traction.

Now, let us discuss the influence of the measurement errors on the inverse solutions obtained by using CGM and LM in both LM and SM formulations.

Firstly, displacements are measured along x = 0.93 m with measurement error σ = 0.0001 (about 1 % of the average measured internal displacements), then the measured positions are increased to x = 0.92 m for the same measurement error σ = 0.0001, the exact and estimated traction with their 99 % confidence bounds are shown in Figures 4, 5 and 6. From those Figures we learned that the inverse solution become worse when the measured positions are increased. Moreover the 99 % confidence will also becomes wider. This implies that the computational procedure in Figures 4 and 5 provide more reliable solutions than in Figure 6 since the narrower bounds are desired. Finally the measured positions are further increased to x = 0.91 m for the same σ = 0.0001. The average errors for the inverse solutions obtained by the CGM and RM in using LM and SM formulations are listed in Table 1, where the definition of average error is given as

$$\sum_{i=1}^{2k}\left|\frac{X_i - \hat{X}_i}{X_i}\right| \div 2k \times 100\ \% \qquad (34)$$

here k represents the discrete number of node points along BC, i.e. k = 13 for the present case, while **X** and $\hat{\mathbf{X}}$ denote the exact and estimated values of boundary traction.

From the experiences of numerical experiments we found that the inverse solutions calculated by CGM seems better than that by RM and LM formulation is superior to SM formulation. However one should note that even though the LM formulation obtain more accurate solutions than SM formulation, LM requires more computational time than SM formulation since there are more unknowns in LM formulation than in SM formulation.

Secondly, the measurement error is increased to σ = 0.0004 (about 4 % of the average measured internal displacements), then the measured positions are varying from x = 0.93 m, 0.92 m to 0.91 m. The results (x = 0.91 m) by using CGM in both LM and SM formulations are shown in Figure 7. Same conclusions can be drawn, i.e. the inverse solutions become worse as either the measured positions or the measurement errors are increased. Moreover the CGM is superior to RM and LM formulation is better than SM formulation without the consideration of CPU time used in computer.

(B). Numerical test case 2

The second numerical test case is the same as the previous one except that the boundary traction along BC are now taken as

$$t_x = \begin{cases} 12 + 7(y - 0.214)\mathrm{Mpa} & \text{for} \quad 0.214\mathrm{m} \le y \le 1.5\mathrm{m} \\ 30 - 7(y - 0.214)\mathrm{Mpa} & \text{for} \quad 1.5\mathrm{m} \le y \le 2.782\mathrm{m} \end{cases}$$

$$t_y = \begin{cases} 20\mathrm{Mpa} & \text{for} \quad 0.214\mathrm{m} \le y \le 1.714\mathrm{m} \\ -20\mathrm{Mpa} & \text{for} \quad 1.714\mathrm{m} \le y \le 2.782\mathrm{m} \end{cases}$$

which represents the discontinuous functions.

The numerical experiments are first used σ = 0.00005 (about 0.5 % of the average measured internal displacements)and measured along x = 0.96 m. Then the measurement error is increased to σ = 0.0001(about 1 % of the average measured internal displacements) for same measurements. The exact and estimated traction with their 99 % confidence bounds are shown in Figures 8 and 9, respectively. We note that for the tractions with discontinuity considered here the positions of the sensors are required closer than that for continuous one as were estimated in case 1. The measurement errors are also required smaller than in case 1. Similar results as observed in case 1 are also observed in case 2.

VIII. CONCLUSIONS

The conjugate gradient method (CGM) and the regularization method (RM) using both Large Matrix (LM) and Small Matrix (SM) formulations were applied successfully for the solution of the inverse problem to determine the unknown boundary traction by utilizing internal displacement readings. Several test cases involving different functional forms of the boundary traction and measurement errors were considered. From the numerical test cases 1 and 2, we concluded that the CGM seems superior to RM and LM formulation is better than SM formulation without the consideration of computer time.

ACKNOWLEDGMENT

This work was supported in part through the National Science Council, R. O. C., Grant number, NSC-86-2611-E-006-002.

REFERENCES

(1). Alifanov, OM (1974). " Solution of an Inverse Problem of Heat Conduction by Iteration Methods," J. of Engineering Physics, Vol 26.

(2). Beck, JV, Blackwell, B and Clair, CR St. (1985). " Inverse Heat Conduction: Ill-posed Problems, Wiley-Interscience," New York.

(3). Brebbia, CA and Dominguez, J (1989). " Boundary Elements, An Introductory Course," McGraw-Hill, New York.

(4). Farebrother, RW (1988). "Linear Least Square Computations," Marcel Dekker, Inc., New York.

(5). Golub, GH (1979). " Generalized Cross-Validation as a method for choosing a good ridge parameter," Technometrics, Vol 21.

(6). Grysa, K, Cialkowksi, MJ and Kaminski, H (1981). " An Inverse Temperature Field Problem of the Theory of Thermal Stresses," Nucl. Eng. Des., Vol 64.

(7). Huang, CH and Huang, MC (1994). " Inverse Problem in Determining the Normal and Tangential Drag Coefficients of Marine Cables," J. Ship Research, Vol 38.

(8). IMSL Library Edition 10.0 (1987). User's Manual: Math Library Version 1.0, IMSL, Houston, TX.

(9).Kammerer, WJ and Nashed, MZ (1972). "On the Convergence of the Conjugate Gradient Method for Singular Linear Operator Equations," SIAM J. Numer. Analysis, Vol 9, No 1.

(10). Kuz'menko, VI (1986). " Inverse Contact Problems of the Theory of Plasticity," Prikl. Matem. Mekhan. U.S.S.R., Vol 50.

(11). Maniatty, A, Zabaras, N and Stelson, K (1989). " Finite Element Analysis of some Inverse Elasticity Problems, J. Eng. Mech.," Div. ASCE, Vol 115.

(12). Maniatty, A and Zabaras, N (1994). "Investigation of Regularization Parameters and Error Estimating in Inverse Elasticity Problems," Int. J. Numer. Methods Eng., Vol 37.

(13). Martin, TJ, Halderman, JD and Dulikravich, GS (1995). "An Inverse Method for Finding Unknown Surface Tractions and Deformations in Elastostatics," Computers and Structures, Vol 56, No 5.

(14). Mura, T (1985). "A New NTD: Evaluation of Plastic Strain in Bulk from Displacements on Surfaces," Mech. Res. Commun., Vol 12.

(15). Pasquetti, R and Le Niliot, C (1991). "Boundary Element Approach for Inverse Heat Conduction Problems: Application to A Bidimensional Transient Numerical Experiment," Numerical Heat Transfer, Part B., Vol 20.

(16). Pasquetti, R and Petit, D (1995). "Inverse Diffusion by Boundary Elements," Engineering Analysis with Boundary Elements, Vol 15.

(17). Schnur, DS and Zabaras, N (1990). "Finite Element Solution of Two-Dimensional Inverse Elastic Problems using Spatial Smoothing," Int. J. Numer. Methods Eng., Vol 30.

(18). Zabaras, N, Morellas, V and Schnur, D (1989). "Spatially Regularized Solution of Inverse Elasticity Problems using The BEM," Commun. Applied Numerical Method, vol 5.

(19). Zhang, F, Kassab, AJ and Nicholson, DW (1995). "A Boundary Element Inverse Approach for Determining Residual Stress and Contact Pressure," Seventeen International Conference on Boundary Elements, Madison, Wisconsin, USA.

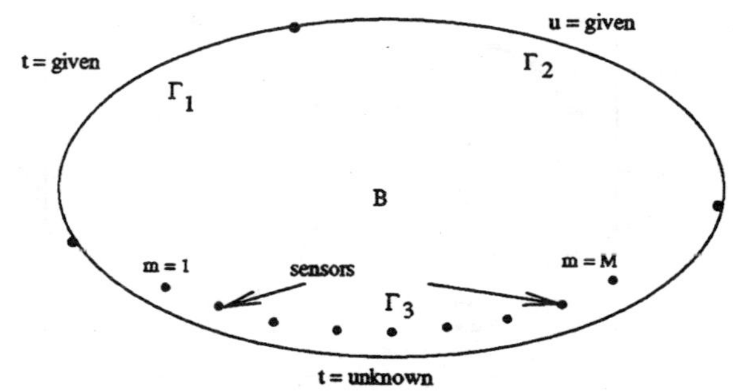

Figure 1. Schematic of an inverse elastic problem.

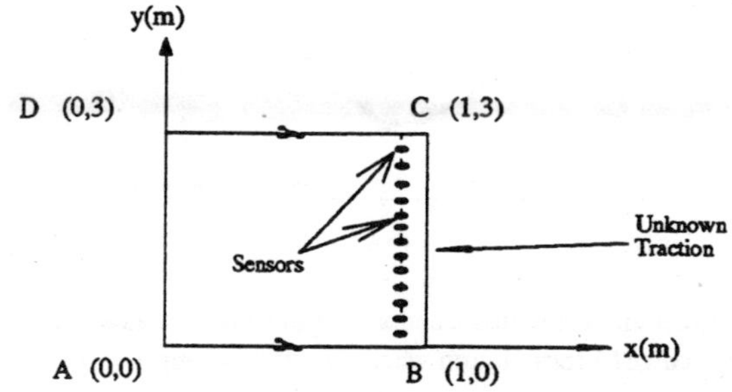

Figure 2. Geometry and coordinates for numerical test cases.

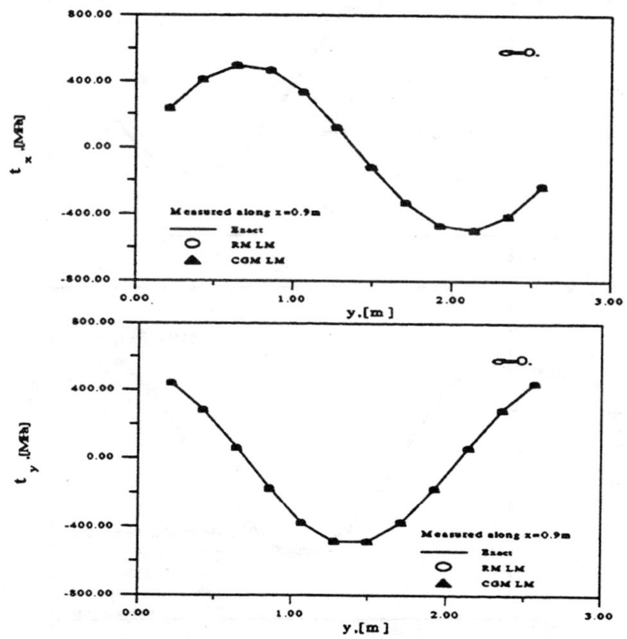

Figure 3. The exact and estimated traction with σ = 0.0 and LM formulation when measured along x = 0.9 m.

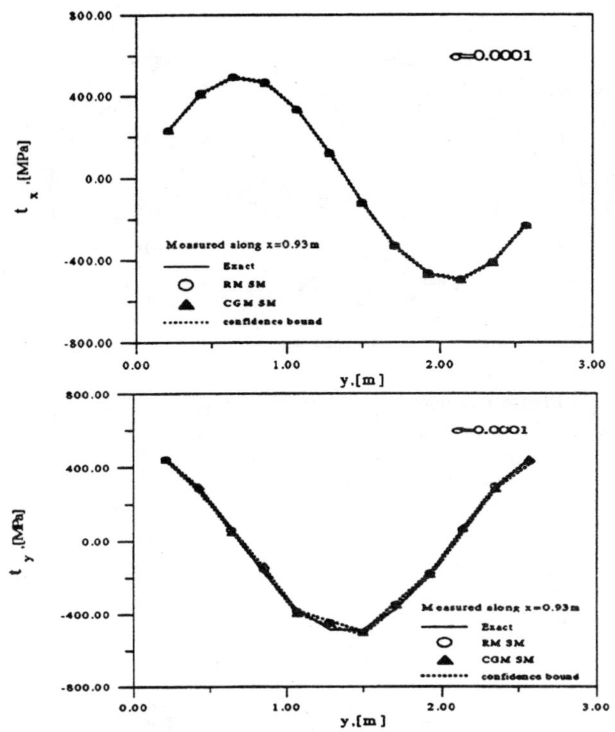

Figure 4. The exact and estimated traction and their confidence bound with σ = 0.0001 and SM formulation when measured along x = 0.93 m.

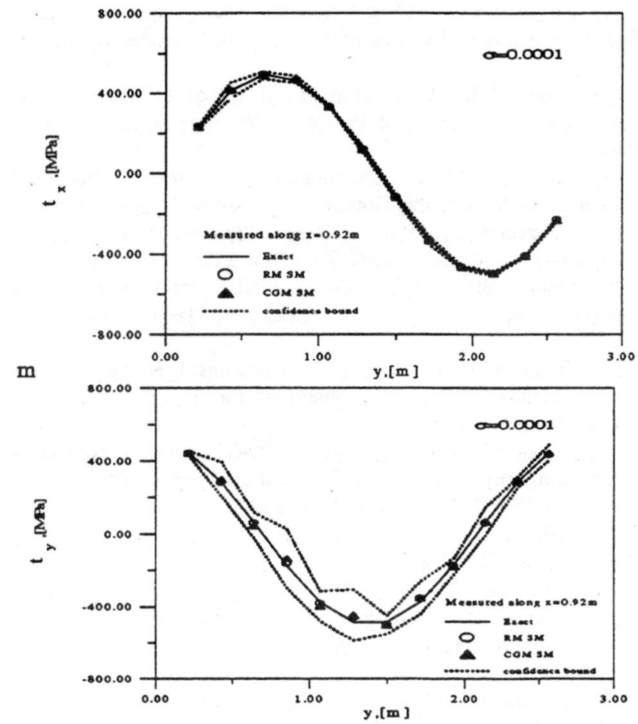

Figure 6. The exact and estimated traction and their confidence bound with σ = 0.0001 and SM formulation when measured along x = 0.92 m.

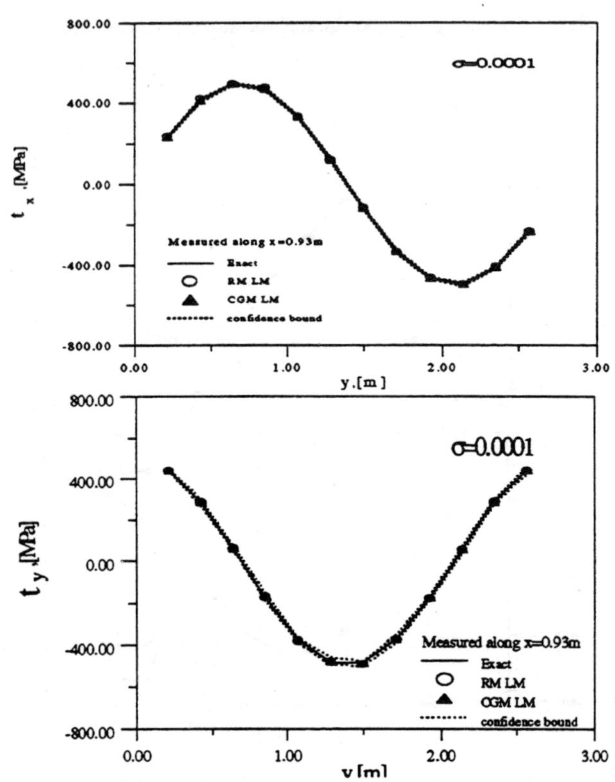

Figure 5. The exact and estimated traction and their confidence bound with σ = 0.0001 and LM formulation when measured along x = 0.93 m.

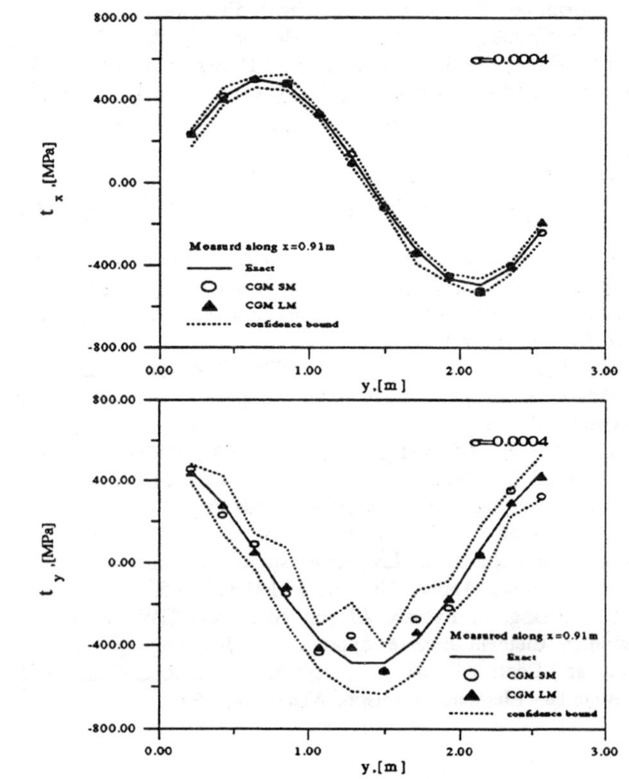

Figure 7. The exact and estimated traction and their confidence bound with σ = 0.0004 when measured along x = 0.91 m.

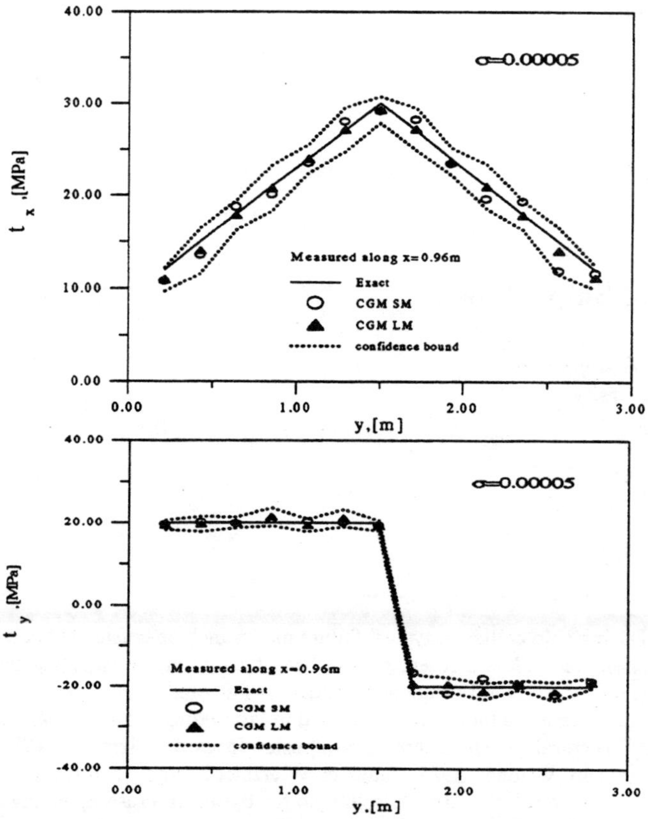

Figure 8. The exact and estimated traction and their confidence bound with σ = 0.00005 when measured along x = 0.96 m.

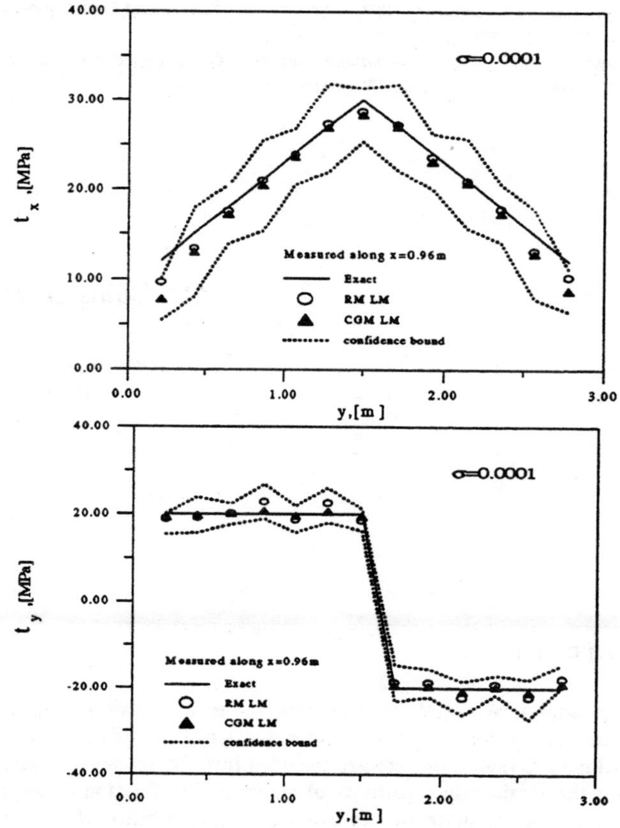

Figure 9. The exact and estimated traction and their confidence bound with σ = 0.0001 and LM formulation when measured along x = 0.96 m.

Proceedings of the Seventh (1997) International Offshore and Polar Engineering Conference
Honolulu, USA, May 25-30, 1997

ISBN 1-880653-28-1 (Set); ISBN 1-880653-32-X (Vol. IV)

On Collision Strength of Ship's Bow

Jeom Kee Paik and Jang Young Chung
Pusan National University
Pusan, Korea

ABSTRACT

An analytical model for predicting the mean crushing strength of stiffened plated structures under quasi-static crushing loads is developed. Dynamic effects are included into the model by considering the strain-rate sensitivity of material. Verification examples are shown to validate the accuracy and applicability of the proposed method. As an illustrative example, the model is applied to assessment for collision strength performance of bow structure for an actual chemical product carrier colliding with a fixed object.

KEY WORDS: Collision strength, crushing, individual plate unit method, intersecting method, effective crushing distance, mean crushing strength

INTRODUCTION

Ship collision and grounding can result in structural damage, marine pollution dependent on cargo type, and hull girder collapse after the accident.

The task of protecting against such environmental disaters or vessel loss due to potential accidents is complex, but can generally be divided into two major groups, namely active and passive safety methods (Pedersen 1995).

The active method is aimed at preventing an accident to occur. This approach includes auto-pilot system of the vessels, crew-training program for instance with realistic simulators for ship operations, and traffic control systems. The passive safety method works after an accident has occurred. It is designed to mitigate the risk of serious consequences, such as ingress of sea water, outflow of hazadous cargo, and hull girder collapse. The latter is related to the improvements of structural crashworthiness, and rapid salvage and rescue operations.

The present study is concerned with application of the passive method on the improvements of structural crashworthiness in ship's bow collision. For that purpose, it is necessary to better understand the mechanics of ship collision such that the structural safety of vessels against accidental loading can be improved.

During ship collision, typical failure modes include crushing of structural elements. Much research have been focused on investigating the behavior of structural components under crushing loads. For a review of relevant literature the reader is referred to Proceedings of three consecutive Structural Crashworthiness Symposia (Jones & Wierzbicki 1983, Wierzbicki & Jones 1988, Jones & Wierzbicki 1992). A benchmark study on several closed-form solutions for the mean crushing strength and the cutting resistance of plated structures during collision and grounding was carried out by comparing theoretical solutions with experimental data (Paik & Wierzbicki 1997).

This paper proposes a simplified analytical model for predicting the mean crushing strength of stiffened plated structures under quasi-static crushing loads. By applying the rigid-plastic collapse theory, the unstiffened plate unit is formulated by taking into account crushing behavior. Effects of the stiffeners on the crushing response are included in the formulation based on experimental results. Dynamic effects due to impact loading are incorporated by considering the strain-rate sensitivity in the material model.

The accuracy and applicability of the proposed model is verified by comparing the theoretical solutions with the existing experimental results for bow models under collision. As an illustrative example, the method is applied to evaluation of collision performance of bow structure for a chemical product carrier of 3,500 dwt.

ANALYTICAL MODEL FOR QUASI-STATIC MEAN CRUSHING STRENGTH

When thin-walled structrures are in such accidental loading situations as collisions, they can suffer deformations which are considerably larger than the wall thickness, and folding/contact of the walls occurs. In such cases, the safety evaluation of the structures is usually performed by utilizing the method based on energy absorption capacity. This method estimates damage by considering that the work absorbed must equal the initial kinetic energy.

Figure 1 shows the crushing response of thin-walled structures under predominantly compressive loads. In an usual loading situation where the magnitude of deformations is small, the internal force rapidly drops after the ultimate load is reached. As the penetration continues, local folding and internal contact of the walls occurs. If the walls contact

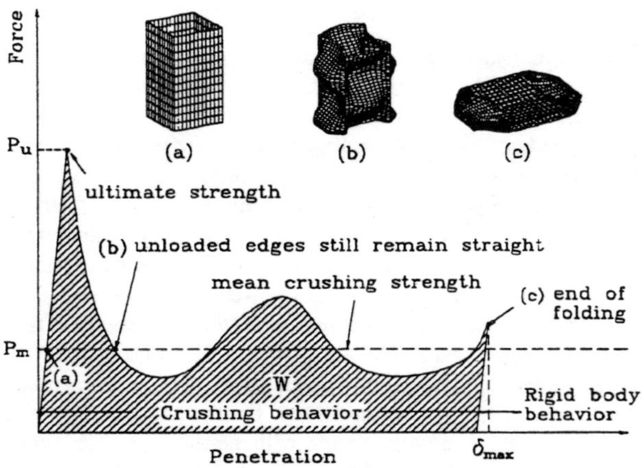

Fig. 1 Crushing response of a square tube under predominantly compressive loads

each other, the internal force rises until the adjacent walls fold. This response is repeated until the folding of walls ends. The completely folded structure then behaves as a rigid body.

To assess the crushing damage of structures, it is necessary to calculate the actual force-penetration relationship. Since the crushing behavior of structures is quite complex, it is normally very difficult to get the actual force-penetration relationship of structures by using an analytical or even numerical method. Only experiments will provide the actual force-penetration relationship of structures subject to crushing loads, but it is not possible for large sized structures.

However, in some cases, one does not need to calculate the actual force-penetration relationship because, if the mean crushing strength p_m is known, then the crushed length can be estimated by equating the absorbed energy to the initial kinetic energy, see Fig. 1. Previous studies for structural design against collision have focused mainly on the estimation of the mean crusing strength of structures instead of calculating the actual force-penetration relationships in detail.

Many simple formulations for predicting the mean crushing strength of thin-walled structures have been derived analytically or empirically in the literature. The analytical approach introduces rigid-plastic collapse mechanisms for the basic structural unit. Two kinds of thoughts currently exists depending on modeling technique of the structure. One called the intersecting unit method (Amdahl 1983, Wierzbicki 1983) is to model the structure by typical intersecting units, such as L, T, Y and X (or cruciform) sections, and the other called the individual plate unit method (Paik & Pedersen 1995) is to model the structure by individual plate units, see Fig. 2.

The former method admits a number of possible crushing mechanisms, and the crushing will occur at the failure mode which gives the lowest crushing strength. All analytical formulations except for one derived by Paik & Pedersen (1995) which used the individual plate unit method have employed the intersecting unit method.

The present study uses the individual plate unit method. Paik & Pedersen (1995) derived the mean crushing strength formula for unstiffened plate units by applying rigid-plastic collapse theory as:

$$\frac{\sigma_m}{\sigma_y} = \frac{1}{\eta}\left\{1.4245\left(\frac{t}{b}\right)^{0.5} + 0.2673\left(\frac{t}{b}\right)\right\} \tag{1}$$

where η = effective crushing distance normalized by the original unit

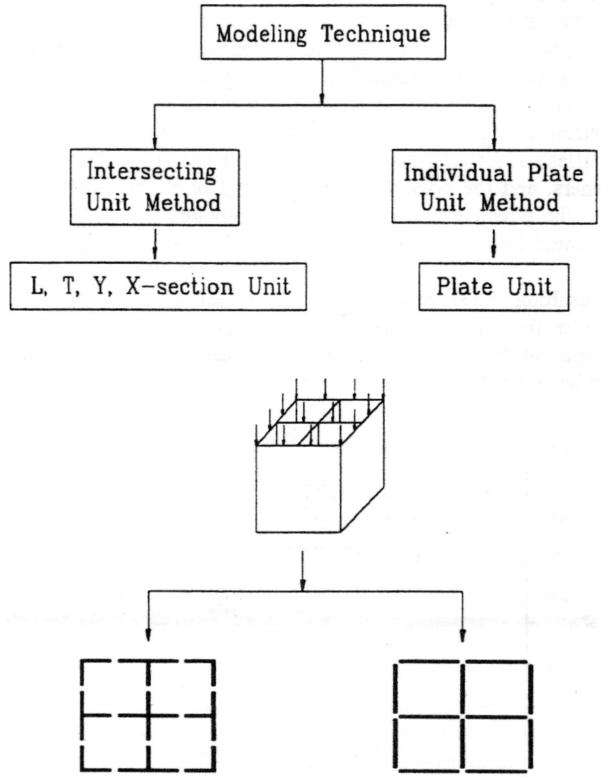

Fig. 2 Modeling of the plated structures under crushing loads

length, may be taken as 0.73 for unstiffened plate unit.

Recently, Paik et al. (1996a) performed a series of quasi-static crushing tests on longitudinally, transversely and orthogonally stiffened square tubes under axial crushing loads. Based on these test results, it was concluded that the longitudinally stiffened plate could be reasonably replaced by an unstiffened plate with the equivalent plate thickness. Also the effects of transverse stiffeners could be neglected if the spacing between transverse stiffeners is relatively large compared to the length of the folding wave.

The equivalent plate thickness t_{eq} can be expressed as

$$t_{eq} = t + k \cdot \frac{A_s}{b} \tag{2}$$

where k is a reduction coefficient accounting for the influence due to the type of longitudinal stiffeners and/or the welding method, which may normally be taken as less than 1.0. A_s is cross-sectional area of longitudinal stiffeners and is thickness of parent plate. b is plate width of loaded edges.

Applying the concept of the equivalent plate thickness, equation (1) would be applicable to stiffened plate units by replacing the plate thickness parameter t by the equivalent plate thickness parameter t_{eq} as:

$$\frac{\sigma_m}{\sigma_y} = \frac{1}{\eta}\left\{1.4245\left(\frac{t_{eq}}{b}\right)^{0.5} + 0.2673\left(\frac{t_{eq}}{b}\right)\right\} \tag{3}$$

Because of the finite folding radius, the maximum crushing distance is always shorter than the original unit length. This effect can be accounted for using the concept of the effective crushing length

(Wierzbicki 1983).

As the effective crushing distance becomes shorter the mean crushing strength becomes larger. It is known that for the unstiffened plates under crushing loading the penetration can continue only until 73% of the entire unit length (Abramowicz 1983, Paik et al. 1996a).

During the folding process of thin-walled structures with longitudinal stiffeners, the walls would come into contact much earlier due to the stiffeners, and the possible penetration will be further reduced. As a result, the effective crushing distance of longitudinally stiffened plate units could become much shorter than that for unstiffened structures, see Fig. 3. The transverse stiffeners could also disturb the folding process, resulting in additional shortening of the effective crushing distance. However, the test results obtained by Paik et al. (1996a) showed that the influence of transverse stiffeners on the effective crushing distance could be ignored.

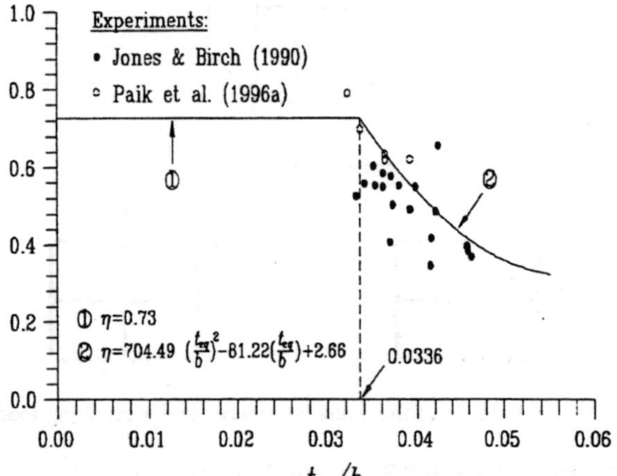

Fig. 3 The effective crushing distance normalized by the entire unit length for longitudinally stiffened units varying the equivalent plate thickness

Based on the test results, Paik et al. (1996a) empirically derived the formula for the effective crushing distance normalized by the entire unit length as

For unstiffened plate units:

$$\eta = 0.73 \tag{4.a}$$

For longitudinally stiffened plate units:

$$\eta = 0.73 \qquad \text{if } 0.0 < t_{eq}/b \le 0.0336$$

$$= 704.49\left(\frac{t_{eq}}{b}\right)^{0.2} - 81.22\left(\frac{t_{eq}}{b}\right) + 2.66 \qquad \text{if } 0.0336 < t_{eq}/b < 0.055 \tag{4.b}$$

$$= 0.36 \qquad \text{if } t_{eq}/b \ge 0.055$$

The mean crushing strength of unstiffened or stiffened structures under quasi-static crushing loads can then be predicted by using equations (2), (3) and (4).

DYNAMIC EFFECTS

So far the analysis was restricted to the quasi-static condition. However, actual ship collision is dynamic in nature which will affect the crushing response of structures.

Three aspects on dynamic loading situation are normally considered, namely material strain-rate sensitivity, inertia effects and dynamic frictional effects. With increase in the strain-rate, the yield strength of material increases. Due to inertia effects, deformation patterns may be varied. It is known that as the speed of dynamic loads increases the coefficient of friction becomes lower (Bowden & Tabor 1950). During collision, the first and second factors will raise the energy absorption capacity of structures, while the third factor will reduce it.

When the dynamic loads are applied, crushing response is mainly affected by material strain-rate sensitivity. During crushing response, the material strain-rate generally varies with penetration. For simplicity, however, the average value of strain-rate during the dynamic loading would be used. Using the formula for the folding wave distance of plated structures derived by Paik & Pedersen (1995), the average strain-rate for stiffened plate units can be estimated by

$$\dot{\varepsilon} = \frac{V_o}{2.2052\,\eta\sqrt{bt_{eq}}} \tag{5}$$

where V_o = initial loading speed

η = normalized effective crushing distance, equation (4)

b = plate width along loaded edges

t_{eq} = equivalent plate thickness, equation (2)

To estimate the dynamic yield strength of material, the Cowper-Symonds equation has been widely used (Jones 1989), namely

$$\frac{\sigma_{yd}}{\sigma_y} = 1.0 + \left(\frac{\dot{\varepsilon}}{d}\right)^{1/q} \tag{6}$$

where d and q are coefficients to be determined based on test data. For example, the parameters for the initial yield of mild steel are $d = 40.4\,sec^{-1}$ and $q = 5$.

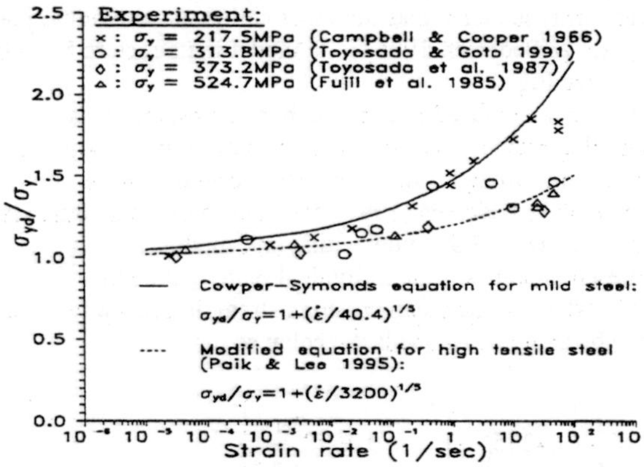

Fig. 4 Variation of yield strength with increase in the strain-rate for mild and high-tensile steel (Paik & Lee 1995)

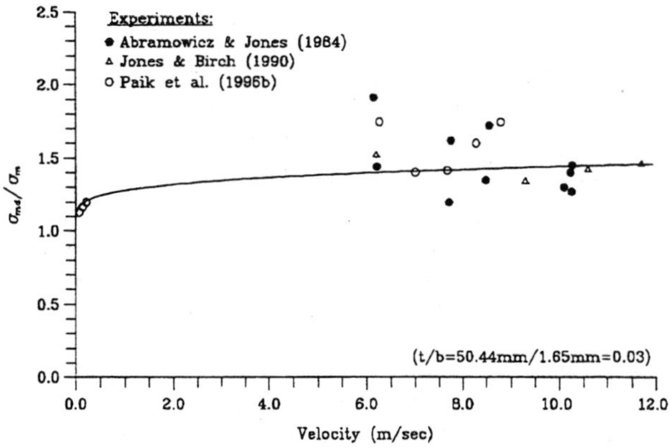

Fig. 5 Dynamic mean crushing strength for unstiffened square tubes varying the loading speed

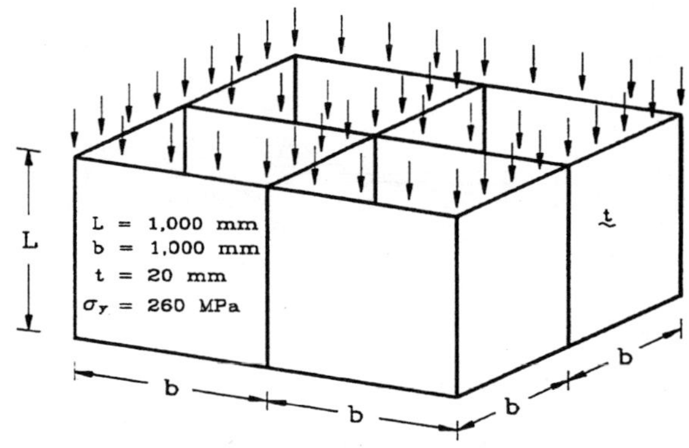

Fig. 6 Nomenclature: A plated structure under quasi-static crushing loads (Ohtsubo & Suzuki 1994)

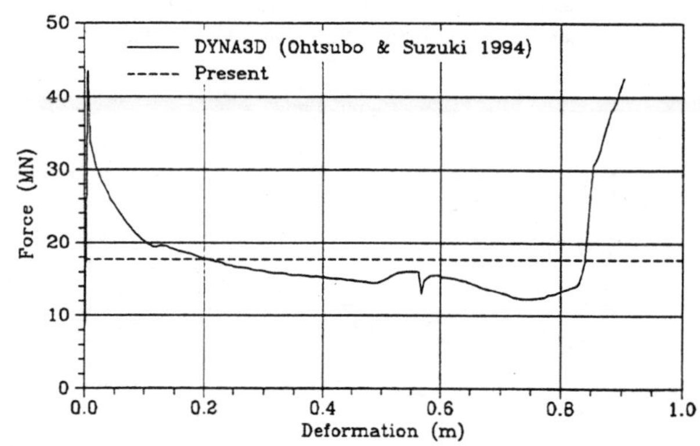

Fig. 7 Comparison of the present solution with the DYNA3D result

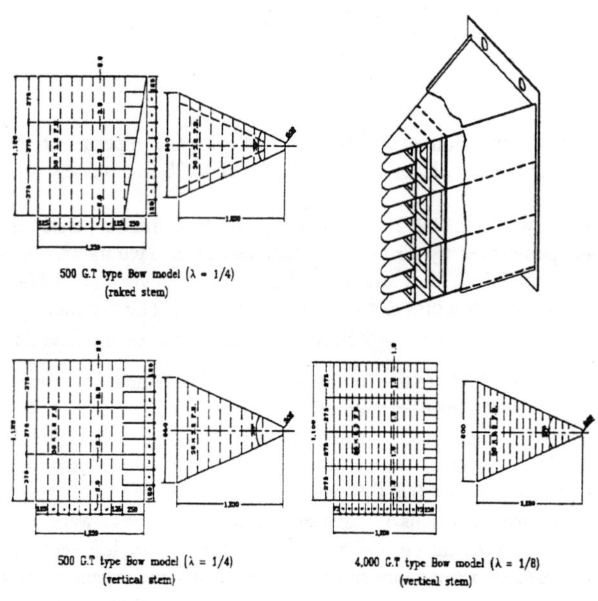

Fig. 8 Bow model of Nagasawa et al. (1977)

Marine structures are built by high-tensile steel as well as mild steel, and the strain-rate sensitivity of the high-tensile steel material should also be quantified. Based on existing test data for high-tensile steel materials, Paik & Lee (1995) determined the test coefficients as $d = 3200\, sec^{-1}$ and $q = 5$, see Fig. 4.

Due to inertia and the stress wave propagation phenomena within the structures during impact loading, the strain distribution (or deformation pattern) at any moment would be inhomogeneous. To investigate the characteristics of inertia effects on the crushing of structures, Paik et al. (1996b) carried out a series of experiments on square steel tubes under impact crushing loads. It was observed that as the maximum crushing indentation became larger and larger (which indicates that the mass of striking body and/or the impact speed would become larger resulting in increase of the initial kinitic energy) the inertia effects increased as well, but it was concluded that their effects could be ignored for the strain-rate less than about $50\, sec^{-1}$ (Note that the crushed strain-rate in ship collisions is normally less than $10\, sec^{-1}$).

During the crushing damage process, the influence of friction would be normally large when there is a relative velocity between the struck body and the indentor (or striking body). For crushing of structures, the influence of friction may be ignored because the relative velocity between striking and struck bodies is normally comparatively small.

The dynamic mean crushing strength of plated structures can then be estimated by replacing the static yield strength parameter σ_y by the dynamic yield strength parameter σ_{yd} in equation (6). Figure 5 shows change of the dynamic mean crushing strength normalized by the static mean crushing strength for unstiffened square steel tubes varying the initial loading speed.

VERIFICATION EXAMPLES

Recently, Ohtsubo & Suzuki (1994) calculated quasi-static crushing response of plated structures using the computer program DYNA3D, see Fig. 6. The present method models the object structure with a number of 12 individual plate units. Figure 7 represents the theoretical and finite-element solutions. The mean crushing strength for the structure obtained by the present method was $17.72 \times 10^7 \left(MN\right)$. The present method estimated the mean crushing strength by 11% higher than DYNA3D.

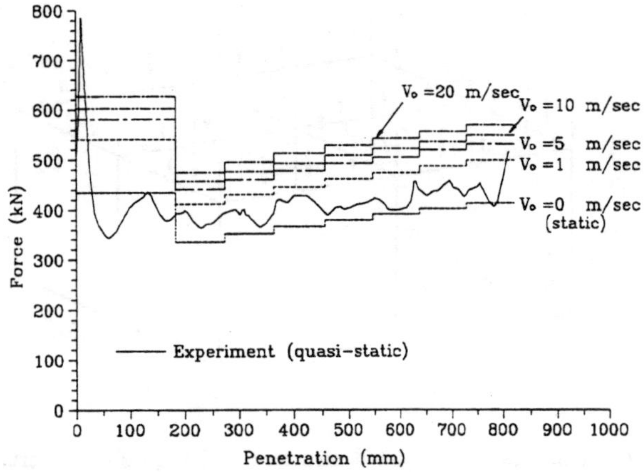

Fig. 9.a Comparison of the present method with experiment by Nagasawa et al. for 1/4 scale model of 500 G/T vessel

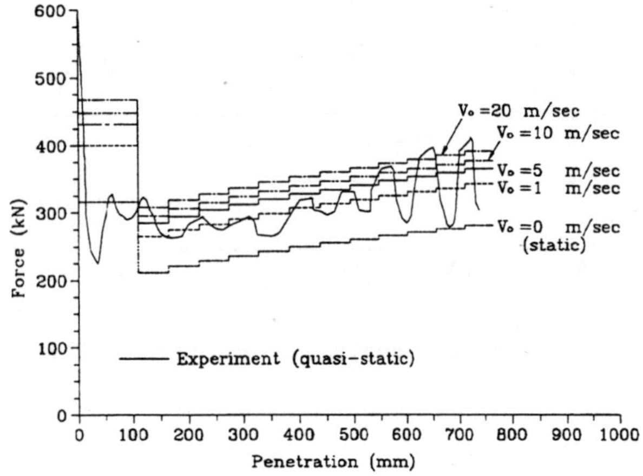

Fig. 9.b Comparison of the present method with experiment by Nagasawa et al. for 1/8 scale model of 4,000 G/T vessel

Nagasawa et al. (1977) carried out quasi-static crushing tests for bow models penetrated by a rigid indentor, see Fig. 8. Two models, namely 1/4 scale model for bow structure of 500 G/T class vessel and 1/8 scale model for bow structure of 4,000 G/T class vessel. The test models were made of mild steel ($\sigma_y = 260\ MPa$). Side shells of the models have inclination of 17.5° to the center line. Individual panels are modeled as plate units for the present theoretical analysis.

Figure 9.a and 9.b show the theoretical and experimental results for crushing responses of 1/4 scale bow model of 500 G/T vessel and 1/8 scale bow model of 4,000 G/T vessel, respectively.

Theoretical solutions for dynamic crushing response were also obtained varying the initial collision speed of the vessels. It is seen that the present method reasonably simulates the quasi-static crushing response compared to experimental results. With increse in the collision speed, the mean crushing strength increases. The trends of the mean crushing strength increase become moderate as the initial collision speed increases further. Examples for further verification of the present method are given by Chung (1996).

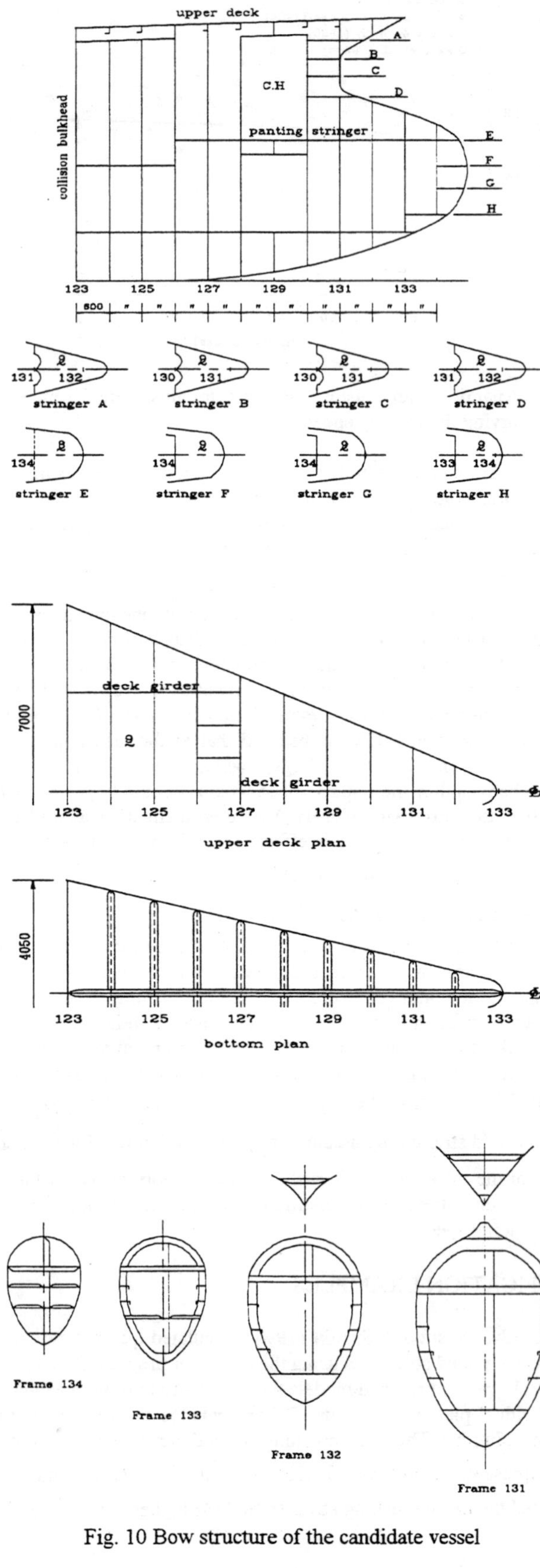

Fig. 10 Bow structure of the candidate vessel

APPLICATION EXAMPLE

The present method is now applied to assessment for collision performance of an actual chemical product carrier of 3,500 dwt. A collision scenario is assumed, where bow of the object vessel collides with a rigid fixed structure such as pier. A head-on collision with various speeds of the vessel in full loaded condition is analyzed. Table 1 indicates the principal dimensions of the vessel.

Table 1 Principal dimension of the candidate chemical product carrier

Dead weight		3,500 tonnes
Light weight		1,180 tonnes
Length	Loa	86.20 m
Length	Lbp	78.60 m
Breadth	B	14.00 m
Depth	D	6.60 m
Darft	d	5.80 m
Design speed	V	12.5 knots (6.43 m/sec)

Figure 10 shows the general arrangement for bow structure of the colliding vessel. The structure is made of mild steel ($\sigma_y = 235 MPa$). The spacing between transverse frames is $0.6 m$.

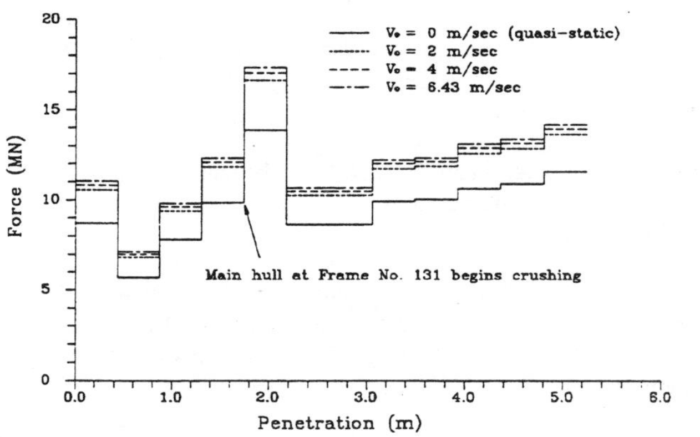

Fig. 11 Static and dynamic force-penetration curves for bow structure of the object vessel

Figure 11 represents the force-penetration relationships for the bow varying the initial collision speed. Since geometric properties of modules between transverse frames are different each other, different crushing responses are shown as the penetration continues, see Fig. 11.

The bulbous bow first comes into contact with the fixed object. As seen from Fig. 10, the tip of bulbous bow is quite stiff which is stiffened in the longitudinal direction and as a result, a large value of the mean crushing strength is provided. At the position of frame No. 131, main hull structure begins crushing, resulting in rapid increase of the mean crushing strength.

Based on the computed results, it could be evaluated whether cargo tanks of chemical products located after the collision bulkhead are damaged or not.

The initial kinetic energy may be estimated as

$$E_k = \frac{1}{2} m \left(1 + C_a\right) V_o^2 \tag{7}$$

where m = mass of the colliding vessel

C_a = added mass coefficient, may be taken as 0.1

V_o = initial collision speed

For an example, the initial kinetic energy for a head-on collision with full speed ($V_o = 6.43 m / sec$) in full loaded condition, can be calculated from equation (7) as

$$E_k = \frac{1}{2} \times (3{,}500{,}000 + 1{,}180{,}000) \times 1.1 \times (6.43)^2$$
$$= 106.42 \times 10^6 \ (N - m) = 106.42 \ (MJ) \tag{8}$$

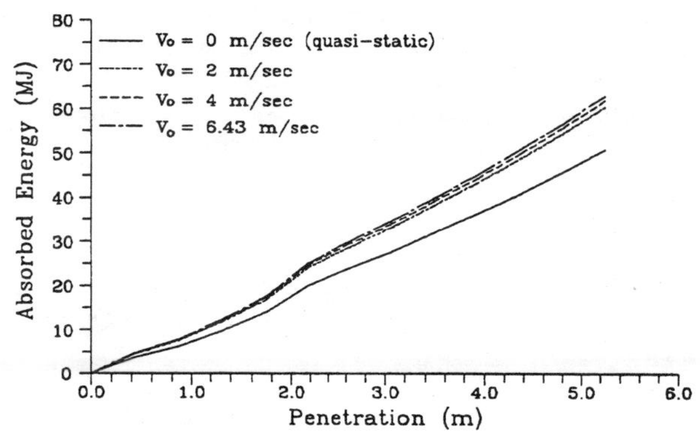

Fig. 12 Static and dynamic absorbed energy-penetration curves for bow structure of the object vessel

Table 2 Calculated results for penetration of the colliding vessel in full loaded condition

V_0 (m/sec)	E_k (MJ)	Penetration (m)	Remarks
2.0 (3.89 knots)	10.30	1.16	Non-damage of cargo tank
4.0 (7.78 knots)	41.18	3.67	Non-damage of cargo tank
6.43 (12.50 knots)	106.42	-	Damage of cargo tank

Figure 12 represents the absorbed energy of the bow structure which was estimated by integrating the area below the force-penetration curves shown in Fig. 11. By equating the estimated initial kinetic energy E_k to the absorbed energy, the possible penetration can be obtained, see Table 2. It is seen that cargo tanks would remain safe for the head-on collision with $4.0 m / sec$ of the initial collision speed (or 62% of the full speed), but they can be damaged for the full speed collision. This may imply that some improvements in structure crashworthiness of the bow structure are needed to prevent damage of tanks of hazadous cargo.

CONCLUDING REMARKS

Over the last decade there has been a growing interest in reducing marine pollution and vessel loss due to potential accidents such as ship collision and grounding. As a result, regulating for ship structural design have been strengthened. For instance, most oil tankers would have double-bottom and double-sides by the turn of the century in order to minimize the risk of oil outflow in collision and grounding.

Regardless of much efforts on preventing an accident to occur, potential accidents still have occurred. In this case, it is very important to mitigate the risk of serious consequences, such as outflow of hazadous cargo and vessel loss. For that purpose, the vessels must have sufficient performance against potential accidents such as collision and grounding, and structural crashworthiness of the vessels should be further improved.

The present study proposes a theoretical model for calculating the crushing response of stiffened plated structures in collision. Dynamic effects are accounted for by considering the strain-rate sensitivity in the material model. Verification examples for numerical and experimental models under quasi-static crushing loads are shown to validate the accuracy and applicability of the proposed method. As an application example, damage possibility of cargo tanks for a chemical product carrier of 3,500 dwt in head-on collision situation with various collision speed was assessed. It was seen that the cargo tanks of the object vessel would remain safe for a head-on collision with 62% of the design speed, but they could be damaged for full speed collision.

REFERENCES

Abramowicz, W. (1983). "The effective crushing distance in axially compressed thin-walled metal columns," *Int. J. of Impact Engng.*, Vol 1, No 3. pp 309-317.

Abramowicz, W. and Jones, N. (1984). "Dynamic axial crushing of square tubes," *Int. J. of Impact Engng.*, Vol 2, No 4, pp 179-208.

Amdahl, J. (1983). "Energy absorption in ship-platform impacts," Division of Marine Structures, University of Trondheim, Report No UR-83-34, Trondheim, Norway, September.

Bowden, F.P. and Tabor, D. (1950). "The friction and lubrication of solids," Oxford Press.

Campbell, J.D. and Cooper, R.H. (1966). "Yield and flow of low-carbon steel at medium strain rates," *Proc. of the Conf. on the Physical Basis of Yield and Fracture, Institute of Physics and Physical Society*, London, pp 77-87.

Chung, J.Y. (1996). "On bow collision strength of ships," Ph.D. Thesis, Department of Naval Architecture and Ocean Engineering, Pusan National University, Pusan, December.

Fujii, E., Ohkuma, I., Kawaguchi, Y. and Tsukamoto, M. (1985). "Effects of temperature and strain rate on dynamic fracture toughness of steel," *J. of the Society of Naval Architects of Japan*, Vol 158, pp 619-629.

Jones, N. (1989) "Structural impact," Cambridge University Press, pp 403-405.

Jones, N. and Birch. R.S. (1990) "Dynamic and static axial crushing of axially stiffened square tubes," *Proc. Inst. Mech. Engrs.*, Vol 204, pp 293-310.

Jones, N. and Wierzbicki, T. (1983). "Structural crashworthiness," *Proc. of the First Int. Symposium on Structural Crashworthiness*, University of Liverpool, Liverpool, September.

Jones, N. and Wierzbicki, T. (1992). "Structural crashworthiness and failure," *Proc. of the Third Int. Symposium on Structural Crashworthiness*, University of Liverpool, Liverpool, April.

Nagasawa, H., Arita, K., Tani, M. and Oka, S. (1977). "A study on the collapse of ship structure in collision with bridge piers," *J. of the Society of Naval Architects of Japan*, Vol 142, pp 323-332.

Ohtsubo, H. and Suzuki, K. (1994). "The crushing mechanics of bow structure in head on collision (1st Report)," *J. of the Society of Naval Architects of Japan*, Vol 176, pp 301-308.

Paik, J.K., Chung, J.Y. and Chun, M.S. (1996a). "On quasi-static crushing of a stiffened square tube," *J. of Ship Research*, Vol 40, No 3, September, pp 51-60.

Paik, J.K., Chung, J.Y. and Lee, T.K. (1996b). "Dynamic effects on crushing and cutting of thin-walled structures," (in preparation).

Paik, J.K. and Lee, T.K. (1995). "Cutting tests for a stiffened plate by a wedge," *Proc. of the Int. Conference on Technologies for Marine Environment Preservation*, Vol 1, Tokyo, September, pp 113-121.

Paik, J.K. and Pedersen, P.T. (1995). "Ultimate and crushing strength of plated structures," *J. of Ship Research*, Vol 39, No 3, September, pp 250-261.

Paik, J.K. and Wierzbicki, T. (1997). "A benchmark study on crushing and cutting of plated structures," to appear in *J. of Ship Research*, Vol 41, No 2, June.

Pedersen, P.T. (1995). "Annual Report for 1994 for the Research Programme on Marine Structural Design," Department of Naval Architecture and Offshore Engineering, Technical University of Denmark, Lyngby, January.

Toyosada, M., Fujii, E., Nohara, K., Kawaguchi, Y., Arimochi, K. and Isaka, K. (1987). "The effect of strain rate on critical CTOD and J-integral," *J. of the Society of Naval Architects of Japan*, Vol 161, pp 343-356.

Toyosada, M. and Goto, K. (1991). "Constitutive equation for mild steel involved the effect of strain rate and temperature," *Trans. of the West-Japan Society of Naval Architects*, No 81, pp 259-268.

Wierzbicki, T. (1983). "Crushing behaviour of plate intersections," *Proc. of the First Int. Symposium on Structural Crashworthiness*, University of Liverpool, Liverpool, September, pp 66-95.

Wierzbicki, T. and Jones, N. (1988). "Structural failure," *Proc. of the Second Int. Symposium on Structural Crashworthiness*, Massachusetts Institute of Technology, June.

Proceedings of the Seventh (1997) International Offshore and Polar Engineering Conference
Honolulu, USA, May 25-30, 1997

ISBN 1-880653-28-1 (Set); ISBN 1-880653-32-X (Vol. IV)

Experimental Study on Impact Force by Small Ship Collision to Fender System of Bridge

S. Shiraishi
Port and Harbour Research Institute
Yokosuka, Japan

R. Naitoh
Ministry of Transport
Yokohama, Japan

ABSTRACT

First of all, ship collisions with bridges in Japan are surveyed and their causes are analyzed. The results obtained by hydraulic model tests conducted on ship's impact force during collisions with the fender systems of bridge piers are discussed. The parameters in these hydraulic model tests are mainly the load-deflection characteristics and spring constants of the fenders. A numerical simulation to evaluate ship motions and fender deflections during ship collision are also presented and evaluated.

KEY WORDS : Fender System, Ship Impact Force, Simulation, Load-Deflection Characteristics, Bridge Pier

INTRODUCTION

Bridges spanning straights, canals, rivers have to be designed for safety against ship collision. To protect the piers of bridges against ship collision, fender systems are installed at the fronts of the piers of bridges. In the design of large ships, the impact energy by ship collision is considered to be absorbed both by deflection of the fender systems and deformation of the ship's bow or hull. For small ships, safety against deformation shall be preserved both by the piers of the bridge and the hull of the ships. Therefore, the impact energy by ship collision is considered to be absorbed only by deflection of the fender system.

In this paper, the results of both hydraulic model tests and computations for ship's impact force on fender systems which protect bridge piers against ship collisions are presented. The hydraulic model tests are carried out varying fender system characteristics such as, for example, linear-type, reaction increment-type, buckling-type and large hysteresis-type. Fender system spring constants are also changed in the hydraulic model tests. The reaction force and deflection of fenders, as well as shear forces are measured in the hydraulic model tests. Numerical simulations to evaluate ship motions and fender deflections against ship collisions are also presented and evaluated.

SHIP COLLISION WITH BRIDGE

Review of accident

Table 1 shows the number of ship collisions with marine structures in Japan between 1990 and 1995. These are from a judgment report by the Marine Accidents Inquiry Agency, Japan. **Table 2** shows the causes of ship collisions with bridges after 1974 in Japan. The most of ship collisions with bridge girders are caused by insufficient investigation of the clearance between the water level and the bottom level of bridge girders when ships sail under the bridge. Such errors are caused by insufficient investigation of tidal changes or insufficient observation of the height of the crane of working vessels before passing the bridge. On the other hand, the number of ship collisions with bridge piers are very small and they have various causes. Weather conditions such as dense fog and strong winds cause ship collisions. Ship collisions due to human factors are also common.

Table 1 Number of ship collisions with marine structures in Japan ('90-'94)

Type of Structures	'90	'91	'92	'93	'94	Total
Bridge pier	1	0	1	1	1	4
Bridge girder	0	1	2	3	3	9
Breakwater	18	24	17	27	22	108
Training jetty	0	0	0	2	0	2
Tide embankment	0	0	0	1	0	1
Detached breakwater	0	0	0	1	0	1
Light beacon	0	3	0	1	1	5
Light marking	0	2	0	0	0	2
Quaywall	5	14	11	10	7	47
Piled pier	1	2	1	4	2	10
Revetment	0	1	1	1	0	3
Dolphin	1	1	1	0	0	3
Mooring buoy	0	2	0	1	1	4
Others	0	0	3	2	2	7

Table 3 shows the distribution of the gross tonnage of ships which collided with bridge girders or bridge piers. In Japan, there have been no collisions with bridge girders or piers involving ships with a gross tonnage exceeding 1,000GRT. There are fewer ship collisions in Japan those in

Table 2 Cause of ship collisions with bridges in Japan ('74-'94)

Cause of Accidents	Number
Bridge Girder	16
Insufficient clearance	12
Engine trouble	1
Insufficient watching	1
Inadequate maneuvering	1
Strong winds	1
Bridge Pier	7
Strong winds	1
Dense fog	1
Inadequate maneuvering	1
Insufficient watching	1
Sleeping	1
Unlicensed (Pleasure boat)	1
Drunkenness (Pleasure boat)	1

Table 3 Distribution of gross tonnage in ship collisions in Japan ('74-'94)

Size of ship (GRT)	Bridge girder	Bridge pier
0 - 49	0	3
50 - 99	3	0
100 - 299	5	1
300 - 499	3	3
500 - 999	3	0
Others	2	0

Others: Tugboat and Floating crane

Europe or America because fewer large ships navigate on river in Japan. In the future, coastal development will progress and port traffic will increase, which means that more bridges spanning port areas will be constructed. If a bridge will be built near the entrance of a port, a ship will have to sail in more sivere condition than that of innner port. **Table 1** shows that ship collisions with breakwaters, where are more severe for ship sailing, are common even in Japan. Therefore, the possibility of ship collisions will increase making it more important to protect bridge piers against ship collisions.

Design problems

Fender systems are installed at the front of bridge piers or on the piers themselves to protect bridge piers against ship collisions. **Photos 1** and **2** show examples of such fender systems. The former type is directly installed to the pier, and the latter type is installed at the dolphins constructed at the front of bridges. Present design methods for fender systems are mainly aimed at protecting bridge piers.

One of design methods adopted for the fender systems of long span bridges constructed in straits is that the energy of ship collisions have to be absorbed by both the deflection of fenders and the deflection of the ship's bow or hull. This design method is broadly used to make fender systems that will protect bridges against relatively large ships navigating under bridges spanning straits in Japan.

Photo 1 Example of a fender system for a bridge pier

Photo 2 Example of a fender system for a bridge pier

To protect against small ship collisions, a system with dolphins and fenders is adopted in some cases. In this case, dolphins constructed at the front of bridge piers protect bridge piers gainst ship collisions. In this case, the energy of a ship collision is absorbed by fender deflection. The energy generated by a ship collision is computed by the Eq.(1) used for ship berthing force for mooring facilities at ports (Port and Harbour Bureau, 1989).

$$E_f = \frac{M_s V^2}{2} C_e C_m C_s C_c \qquad (1)$$

where,

E_f : Ship's berthing energy (Nm)

g : Gravitational acceleration(m/s^2)

M_s : Mass of ship(kg)

V : Approaching velocity of ship

C_e : Eccentricity factor

C_m : Virtual mass factor
C_s : Softness factor (1.0 as standard)
C_c : Shape factor (1.0 as standard)

Hereafter, we call this method the "Technical Standard". This method considers horizontal rotation (yawing) as the eccentricity factor C_e but not rolling. The eccentricity factor C_e is treated as 0.5 through 0.7 in the fender design at a harbour(Port and Harbour Bureau, 1989).

Therefore, in this paper, a numerical simulation method to compute the motions of a ship taking into account of rolling is discussed, and the results are compared with those obtained from model experiments and the "Technical Standard". In particular, to protect bridge piers against small ship collisions, it is expected that a fender system that protects not only the bridge but also the ship's hull shall be in demand in the future. To date, the load-deflection characteristics and energy absorption of fenders have not yet been clarified. Therefore, this study examines the relations between ship motions before and after a collision, and discusses the effect of the load-deflection characteristics of fenders and their spring constant s.

EXPERIMENTAL STUDY

Experimental method

Experimental studies of ship collisions against fender systems were carried out at the model basin of the Port and Harbour Research Institute (Naitho and Shiraishi, 1995; Shiraishi and Naitho, 1996). **Figure 1** shows the arrangement of the ship, fender and sensors in the model experiment. Five small electric lamps are set on the model fender and the model ship. The motions of these marker points are observed through a video camera, and their traces are analyzed by a tracking analyzer system to determine the ship's motion and its approaching velocity. The compression and shear forces acting on the model fender are measured by the use of load transducers. Fender deflections are also measured by the use of displacement sensors installed behind the model fender.

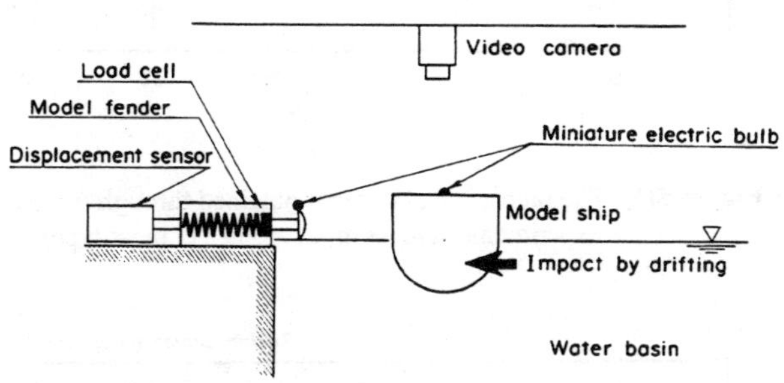

Figure 1 Arrangement of ship, fender, and sensors

Hydraulic model tests were carried out by the Froude similitude. The weight of the model ship was 26.3 kgf, and the model scale was 1/75 and 1/13 for a 5,000DWT general cargo ship and 10GRT fishing vessel, respectively. Although the model ship had a radio control system, it was difficult to control the ship's angle and the velocity when simulating a collision. Therefore, ship collisions in hydraulic model tests were controlled manually.

Model fender

Four types of model fenders were used in hydraulic model tests. There were a linear-type, reaction increment-type, buckling-type, and large hysteresis-type. It was very difficult to make a model fenders using rubber materials that satisfied the law of similarity. Therefore, model fenders with coil springs and weights were created to obtain similar load-deflection characteristics according to the law of similarity.

Coil springs of different spring constants were installed in the model fenders for each fender type. The spring constants at the initial linear part of the load-deflection characteristics were 0.2, 0.1, 0.05, and 0.025 kgf/mm for each type of fender, respectively. Coil springs for the buckling-type fender and the large hysteresis-type are the same. **Figures 2 (1)-(4)** show the

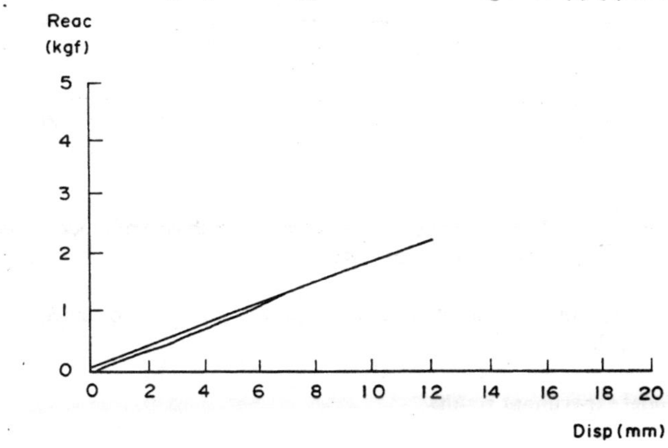

Figure 2 (1) Load deflection characteristics of each type of model fender (Linear-type)

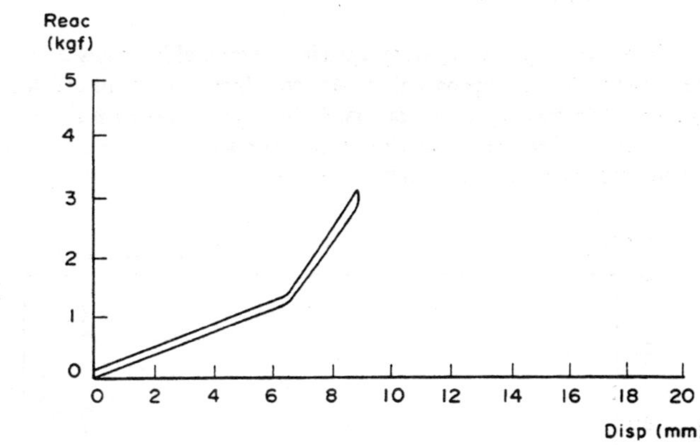

Figure 2 (2) Load deflection characteristics of each type of model fender (Reaction increment-type)

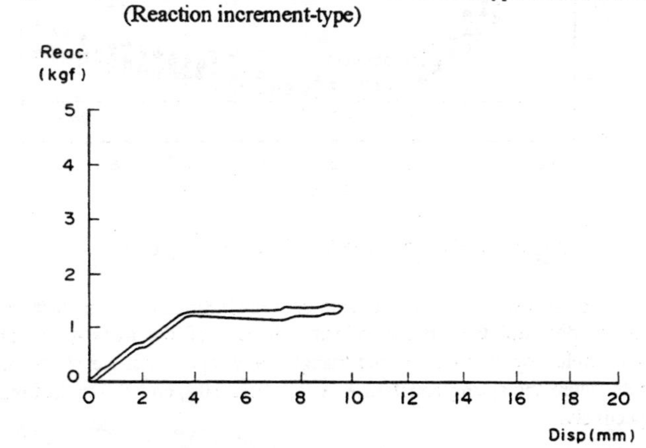

Figure 2 (3) Load deflection characteristics of each type of model fender (Buckling-type)

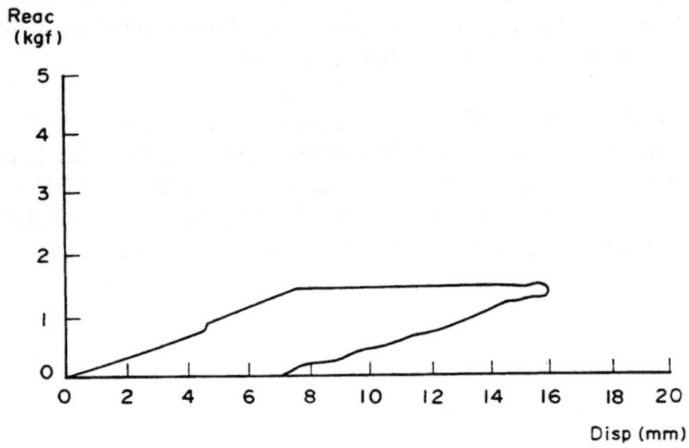

Figure 2 (4) Load deflection characteristics of each type of model fender (Large hysteresis-type)

load-deflection characteristics for each type of model fender obtained during static loading.

Model experiment results

Figure 3 shows the time histories of Y_G when a ship collides with the fender that has linear-type characteristics. In this figure, the time histories of the ship's motion are shown when the spring constant K is 0.2, 0.1, 0.05, and 0.025 kgf/mm, respectively.

Before the ship collides with the fender, the trace of Y_G shows a straight line. As a result, it is supposed that the approaching velocity of the ship is constant. The trace of Y_G is not straight because a swaying motion with rolling occurs after impact with the fender. The distance between the ship's hull and the fender after ship collision is greater when K is greater.

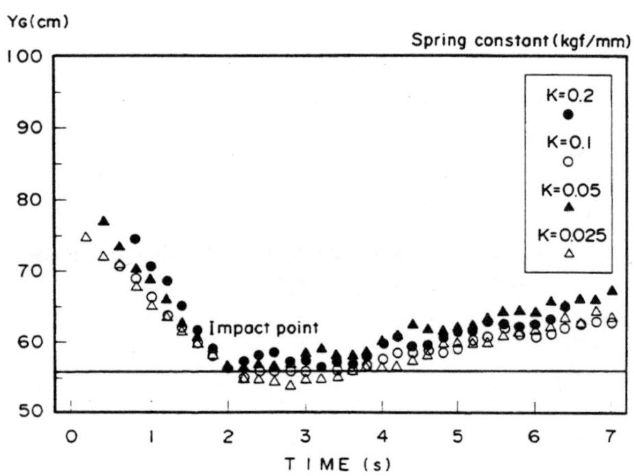

Figure 3 Time histories of Y_G when a ship collides

Figure 4 shows the relationship between the maximum displacement of the fender and the impact velocity in case of the linear-type. The closed circle, open circle, closed triangle, and open triangle show the results when the spring constant K is 0.2, 0.1, 0.05, and 0.025 kgf/mm, respectively.

The solid lines show the results computed by the "Technical Standard" method. In the computation, the eccentricity factor C_e is assumed to be 1.0 because no yawing motion was observed in hydraulic model tests. Maximum displacements become large when the impact velocities are high. However, computed displacements are greater than experimental displacements.

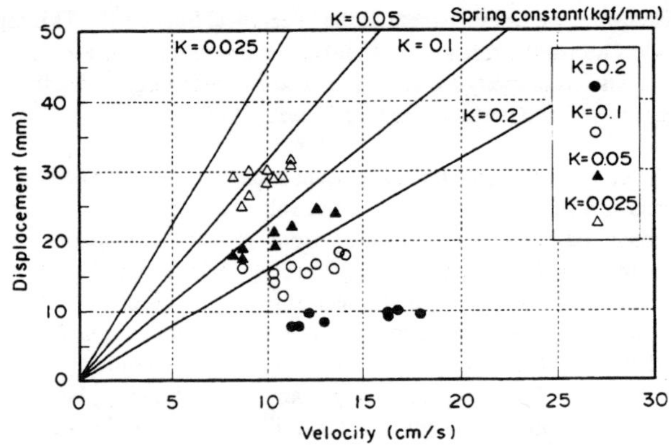

Figure 4 Relation between fender displacement and velocity

Figures 5 (1)-(4) show ratio δ_e / δ_c, where δ_e and δ_c are measured experimental displacement and computed displacement, when a ship

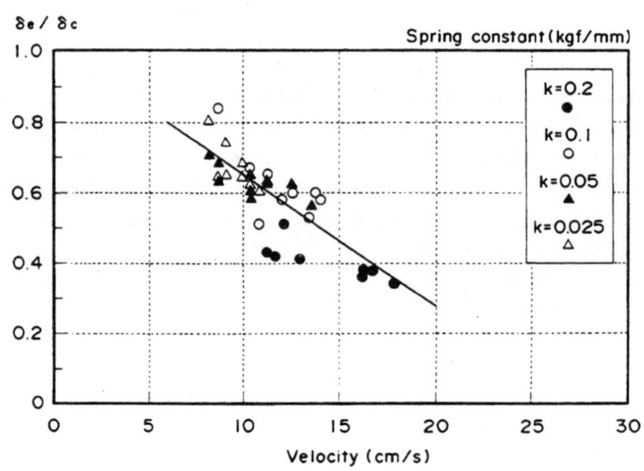

Figure 5(1) The ratio of displacement obtained through model experiments and computations (Linear-type)

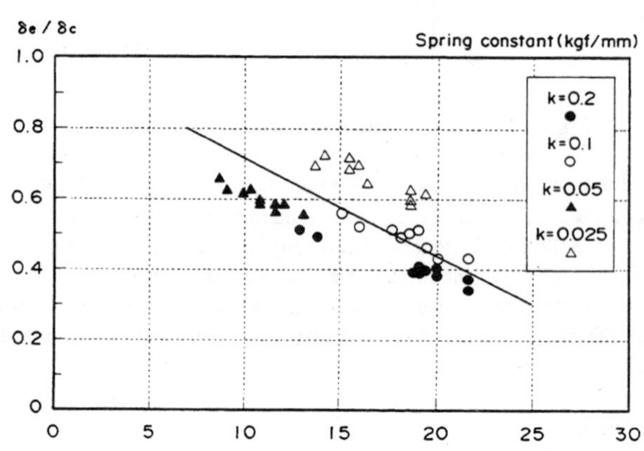

Figure 5 (2) The ratio of displacement obtained through model experiments and computations (Reaction increment-type)

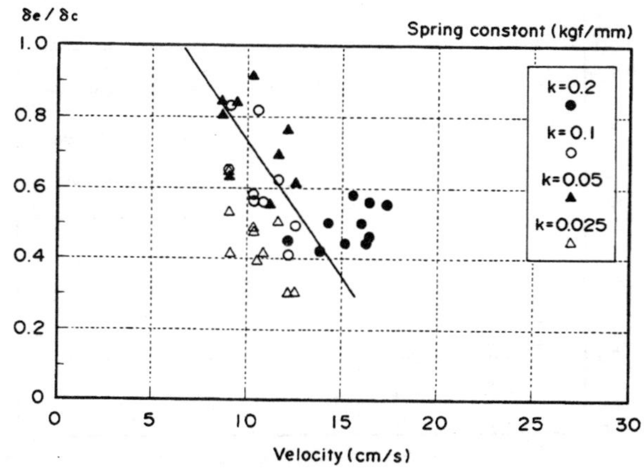

Figure 5 (3) The ratio of displacement obtained through model experiments and computations (Buckling-type)

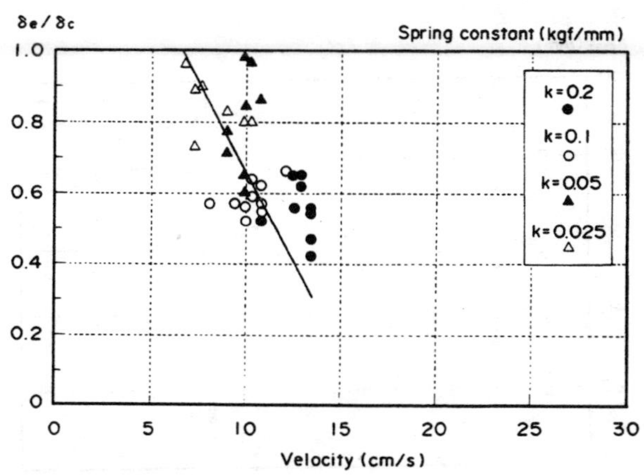

Figure 5(4) The ratio of displacement obtained through model experiments and computations (Large hysteresis-type)

collides with a fender at 90 degrees. These figures show the results when the fender characteristics are of linear-type, reaction increment-type, buckling-type, and large hysteresis-type, respectively. The ratio δ_e / δ_c decreases as the fender's spring constant increases, and the ratio is about 0.4 when the fender characteristics are linear-type and reaction-type and the spring constant K=0.2 kgf/mm. The results also show that the ratio δ_e / δ_c decreases when the velocity is high.

Fender displacements obtained by the "Technical Standard" method are larger than experimentally obtained displacements. The difference becomes especially large when the spring constant K is 0.2 kgf/mm. This difference is caused by the rolling motion of a ship when it collides with a fender. A large rolling motion was observed in hydraulic model tests, but this effect was not taken into account in the "Technical Standard" method. The results show that the ratio δ_e /δ_c decreases as the velocity increases when the reaction increment-type has the same spring constant.

The ratio δ_e/δ_c is almost 1.0 in case of the buckling-type fender when the spring constant K is 0.05 kgf/mm. On the other hand, the ratio δ_e /δ_c becomes about 0.4 through 0.6 when the fender characteristics are same and the spring constant K is 0.1 or 0.2 kgf/mm. Furthermore, the ratio δ_e/δ_c is about 0.4 when the spring constant K is 0.025 kgf/mm. In the other types of fender, the ratio δ_e /δ_c increases when the spring constant decreases, so the characteristics of the reaction increment-type differ from those of other types. The ratio δ_e /δ_c in the large hysteresis-type distributes around 0.6 through 1.0 when the spring constant K is 0.05 or 0.025 kgf/mm. These values are relatively large compared with those of the buckling-type fender.

The ship collision energy is well absorbed by the fender when the system of a large hysteresis-type with a small spring constant is installed. Ship rolling and yawing are also well controlled after impact with the fender, which means that the ship remains stable after collision.

NUMERICAL SIMULATION

Simulation method

In this study, the phenomena of ship collision is modeled by modifying the numerical simulation method of a moored ship in waves (Ueda and Shiraishi, 1983). In the analysis of a ship 's motion in waves, wave forces acting on the ship are considered. On the other hand, to analyze a ship collision, ship motions, fender displacements and reaction forces are obtained by solving the equation of motions considering the initial velocity of the ship and added masses as well as damping coefficients due to the radiation force of a ship in water.

In assessing ship collisions, the equation of motions can be simplified by ignoring the non-dominant motion components. In this study, six degrees of freedom are considered in motions and the equation is solved using the time domain method. Moreover, the non-linear characteristics and the hysteresis of the fender are also considered. The equation of motions is written as Eq.(2).

$$M\ddot{X} + N\dot{X} + C\dot{X}|\dot{X}| + BX + R(X) = F \qquad (2)$$

Where,
M : Virtual mass and inertia matrix
N : Linear damping matrix
C :Non-linear damping matrix
B : Hydrostatic restoring matrix
$R(X)$: Vector of reaction force with the mooring system
F : Fluctuating load vector
X : Displacement vector
$\dot{X}$: Velocity vector
$\ddot{X}$: Acceleration vector

In this paper, the fluctuating load is assumed to be zero because only ship collisions in calm sea are considered. The ship's initial coordinates are set as $X_{t=0} = (X_0, Y_0, Z_0, \Phi_0, \Theta_0, \Psi_0)$, and $X_{t=0} = (X_0, Y_0, Z_0, 0, 0, 0)$ when there is no angle eccentricity. The ship's initial velocity is $\dot{X}_{t=0} = (V_X, V_Y, V_Z, V_\Phi, V_\Theta, V_\Psi)$, and the terms of $V_Z, V_\Phi, V_\Theta, V_\Psi$ are usually set as zero.

Comparison with experimental data

Figure 6 shows the swaying, rolling, yawing, fender reaction force, and displacement of the ship's hull, all of which are obtained by numerical simulations when the initial velocity is 10cm/s and the spring constant is K = 0.05 kgf/mm. In the computation, the added mass coefficient for swaying is given by Ueda's formula (Ueda, 1981) , and the damping coefficient is 0.05. In the simulation, the coordinates for swaying and displacement at the ship's side are set as zero at the start of the simulation, so the "Impact point" is the time when the ship 's hull contacts the fender. Swaying (SWAY) moves to the fender side at first after ship collision and the ship moves again to the oppsite side by the reaction force from the fender. Rolling (ROLL) moves to the upper direction of the ship's hull at first and it shows the damped oscillation. The time history of displacement of ship's hull displacement (DISP-1) shows the composite motion of swaying and rolling, which is similar to the time history obtained by hydraulic model tests.

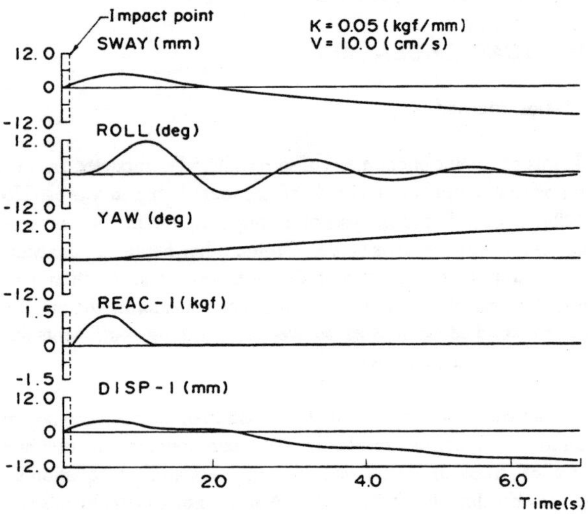

Figure 6 Time histories of computed results

Figures 7(1)-(4) show the maximum displacements of the fender (δ_{max}) obtained through model experiments, the "Technical Standard" method, and numerical simulations. These figures show the results in the linear-type, reaction increment-type, buckling-type, and large hysteresis-type, respectively. Because the approaching velocity in hydraulic model tests has dispersion, the average value of maximum displacements around 10cm/s are shown in these figures.

Results computed by the "Technical Standard" method are different from observed values. This difference is due to the rolling motion, which is

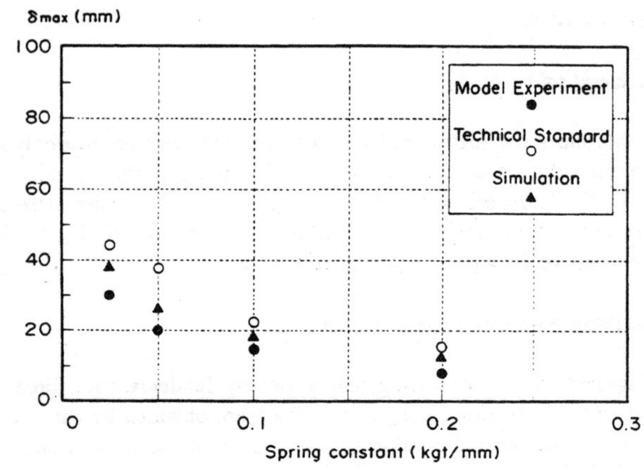

Figure 7 (1) Comparison of model experiments and computations (Linear-type)

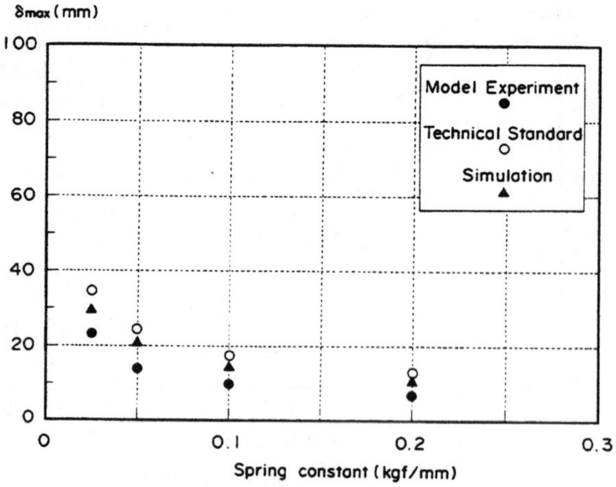

Figure 7(2) Comparison of model experiments and computations (Reaction increment-type)

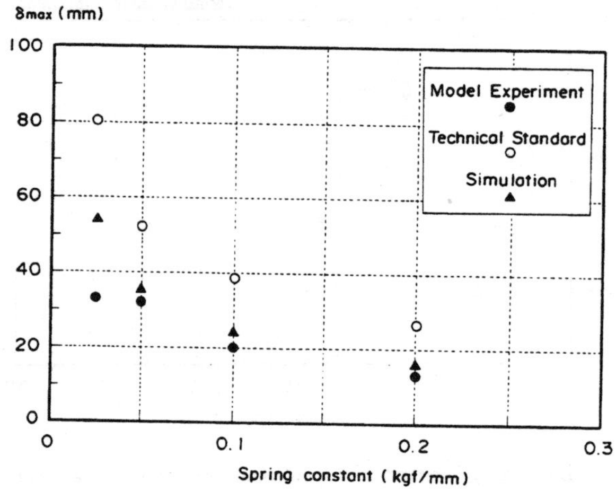

Figure 7(3) Comparison of model experiments and computations (Buckling-type)

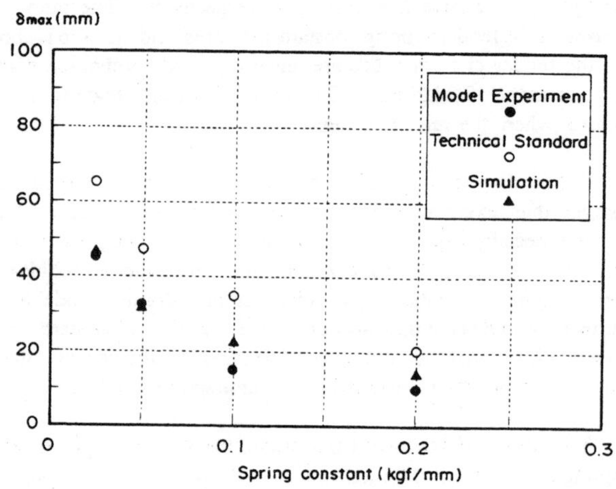

Figure 7(4) Comparison of model experiments and computations (Large hysteresis-type)

not considered in the computation using the "Technical Standard" method. On the other hand, simulation results agree well with observed values, because the analysis considers rolling motion. Maximum displacements obtained by numerical simulation for the reaction increment-type fender system are greater than those obtained by experiments. It is considered that the load-deflection characteristics obtained by a ship's collision are slightly different from those obtained by static loading. Because simulations were carried out by using the load-deflection characteristics obtained by static loading, there are some differences between computation values and the experimental values.

For other types of fenders, the difference between numerical simulation results and model experiment results is 10 through 20% at most. Therefore, it can be said that the numerical simulation method is adequate for evaluating fender reaction force and deflection when a ship collides with the fender of a bridge pier.

CONCLUSIONS

Major results obtained by this study are as follows.

1) Fender deflections computed by the ship impact force observed during hydraulic model tests are smaller than those computed by the "Technical Standard" method. These differences are caused by rotational motion such as rolling and yawing obserbed in hydraulic model tests.

2) Ship swaying and rolling in case of the large hysteresis-type fender are smaller than those of the buckling-type fender.

3) The validity of numerical simulation model used to estimate the ship motions and the deflection of the fender was verified by hydraulic model tests.

4) Fenders which have a large energy absorption capacity and a small reaction force such as the large hysteresis-type fender should be installed to protect bridge piers.

REFERENCES

Naitho,R. and Shiraishi,S. (1995): "Experimental Study on Impact Force by Small Ship Collision to Fender System of Bridge", *Tech. Note of Port and Harbour Research Institute*, No.808, 74p. (in Japanese)

Port and Harbour Bureau (1989): "Technical Standards for Port and Harbour Facilities with Commentary", Japan Port and Harbour association, (in Japanese)

Shiraishi,S. and Naitho,R. (1996): "Ship Collision Accidents and Simulation of Small Ship Collision System of Bridge", *Proc. of Civil Engineering in the Ocean*, Vol.12, pp.183-188. (in Japanese)

Ueda,S (1981): "Study on Berthing Impact Force of Very Large Crude Oil Carriers", *Report of Port and Harbour Research Institute*, Vol.20, No.2, pp.169-209. (in Japanese)

Ueda,S and Shiraishi,S. (1983): "Method and its Evaluation for Computation of Moored Ship's Motion", *Report of Port and Harbour Research Institute*, Vol.22, No.4, pp.181-218. (in Japanese)

Proceedings of the Seventh (1997) International Offshore and Polar Engineering Conference
Honolulu, USA, May 25-30, 1997

ISBN 1-880653-28-1 (Set); ISBN 1-880653-32-X (Vol. IV)

Assessment of Grounding Resistance of Oil Tankers

Sung Kyou Choi and Jae Hyung Park
Daewoo Heavy Industries
Changseungpo, Korea

Tomasz Wierzbicki
Massachusetts Institute of Technology
Cambridge, MA, USA

Abstract
A parametric study was performed for grounding damage calculation using the program DAMAGE developed by the MIT-Industry Tanker Safety Consortium, in support to structural development of pollution-preventive oil tankers. Four double hull and two single hull tankers of conventional type built by the Daewoo Heavy Industries were chosen as calculation models and damage predictions were made by varying parameters defining a grounding scenario. The grounding resistance was investigated by comparing the calculation results and the current design practices of the existing codes originated from international maritime agencies and classification societies. The result is that all four double hull tankers are "safe" against oil leakage caused by grounding accident; in other words their present structural arrangement and scantling are sufficient to carry the corresponding grounding loads. The results of single hull tankers were compared with those of double hull tankers and a concept of grounding resistance grade was discussed. The above mentioned computer program has proved to be a useful and convenient tool, especially in the initial stage of design when many alternative designs for grounding resistance should be considered with modifications of structural arrangement and scantlings. The program is, however, in its initial phase of development and has to be refined in many ways. Some comments on the program from a user's point of view are offered which will make it more robust.

Key words; tankers, grounding strength, oil-spill, single and double hull

1. Introduction
Much effort has recently been put worldwide to develop a robust tanker structure capable of minimizing oil spill with a considerable increase in the grounding resistance. Most of the activities has been focused on experimental studies and numerical simulations using commercial FEM codes. For a recent survey of methods and solutions the reader is referred to Ref. [1~12]. Despite numerous numerical analyses performed so far, these results can not be used for a systematic verification of the existing codes on structural arrangements. Furthermore, FE solutions are inadequate for any practical application to initial tanker design stage. The main difficulty is that a numerical simulation requires too much time and manpower in preparing a structural model, introducing relevant loading conditions and running the program. A simple and expedient usage of a program is a prerequisite for estimating grounding resistance when many runs are needed by varying grounding scenarios and structural arrangements and scantlings. The PC based programs have all the desired features, because they are conveniently based on relevant closed-form solutions and use the concept of the so called *Super-Elements*. However, such programs have also some problems to be refined, because they are in their initial stage of development.

Six tankers have been chosen for the grounding analysis. The results of calculations have been re-interpreted for the purpose of making a set of practical guidelines applicable to the actual tanker design. Then, a criterion of grading grounding resistance is discussed, which could be further improved by studying a large population of tankers.

Since structural arrangements and design characteristics vary from tanker to tanker, a high degree of stability, reliability and economy can not be achieved by adjusting the scantlings and arrangements of the double hull uniformly throughout the industry without taking characteristics of individual ships into consideration. Accordingly, an institutional measure should be prepared and put into practice to amend the current rules and regulations for the purpose of a proper and flexible application of rules to practical ship design. Such an action will ultimately result in reducing marine pollution efficiently and achieving cost-saving types of oil tankers. Thus, calculating the strength against grounding and collision with a suitable verification will help decide on the appropriate structural scantlings and arrangement inherent to each double hull tanker. The above design concept should also be used in conjunction with the so called performance standard of oil tankers reflecting the overall structural stability and reliability in order to evaluate tanker capability in preventing pollution.

2. Application to Tankers

2.1. Tanker Selection

The double hull tankers of three sizes are chosen here to characterize the grounding resistance depending upon their size. No. 1 tanker is of a small size, Nos. 2 and 3 are of a medium size and No. 4 is of a large size. Besides, grounding analysis of two single hull tankers Nos. 5 and 6 of almost the same size as the Tanker No. 4 has been performed in order to compare those two results derived for almost the same sized but differently structured vessels. This will enable to compare the grounding resistance of single hull tankers with those of double hull arrangement.

* # of compartments along ship's centerline

Applied Ships	No.1	No.2	No.3	No.4	No.5	No.6
LBP(m)	238	252	264	320	315	310
B(m)	43	46	48	58	57.2	56
D(m)	19.8	22.4	23	31	30.4	29.5
T(m)	13.5	14	15.22	22	20	19.82
Displacement (DWT)	115710 98K	141689 114K	162474 140K	320656 300K	310090 280K	292542 250K
Hull Type	D/H				S/H	
D/B height(m)	2.30	3.40	2.50	3.00	-	-
# of Holds*	6	6	9	5	6	6
Length of Holds(m)	181.44	184.0	207.36	254.3	239.0	251.3

Table 1. Principal Dimensions of Chosen Ships

2.2. Grounding Scenario

The grounding scenario is listed below. Several possible types of rock are presented in Fig. 1, and here a cone-shaped one is chosen. The ship's initial speed just before grounding is selected to be 12 knots. The assumption that the rock height from the baseline at the amidships is taken to be 2.0 meters.

Grounding Scenario

Ship's velocity	:	12.0 knots
Trim angle	:	0.1434 deg.
Frictional coefficient, f	:	0.3
Transverse rock eccentricity	:	10.0 m
Rock elevation, d	:	2.0 m
Spreading angle	:	45 deg.
Rock tip radius, r	:	1.0 m

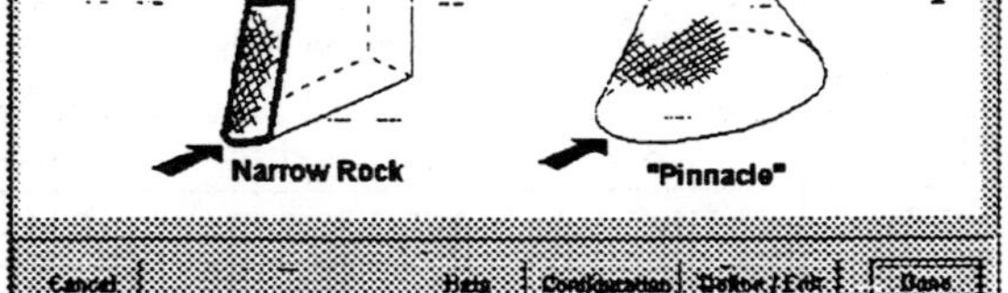

Fig. 1. Four Idealized Shape of Rock

2.3 Material Properties

There are two kinds of steels used in this ship, i.e., mild and HT32 steels with the following material constants :

$E = 200,000 MN/m^2$

$\nu = 0.3$ (Poison's ratio)

$R = 200 kJ/m^2$ (specific fracture work)

$\sigma_y = 262.83\ MN/m^2$ (for mild steel)

$= 340.00\ MN/m^2$ (for high tensile strength steel)

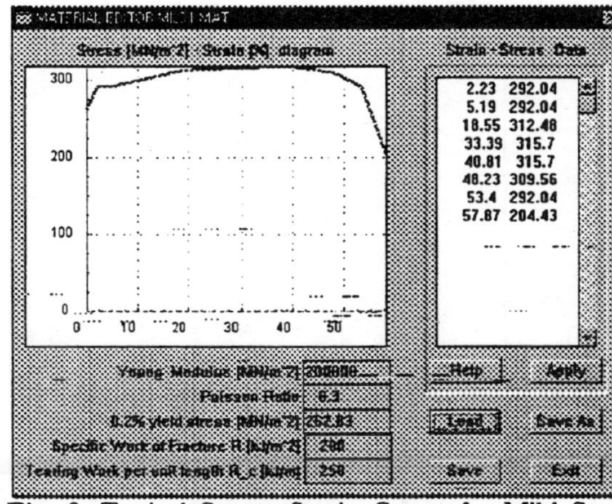

Fig. 2. Typical Stress-Strain Curve for Mild Steel

An example of stress-strain curve for mild steel is presented in Fig. 2, which was obtained in one of tests during the Tanker Safety Project in MIT.

2.4 Results of Analysis

The damage length of the hull is an increasing function of ship's initial speed, as shown in Fig. 3. It is interesting to observe that the damage length reaches a constant value, i.e., the whole hold length, from 8 knots on. This means that it is meaningless if the speed is reduced from 12 knots by less than 4 knots for the purpose of minimizing grounding damage. In other words, the ship should operate at a speed less than 8 knots in order to decrease a damage length. However, a damage length shorter than the length of all the cargo holds is not always desirable, because there is a possibility that the ship will enter the secondary grounding phase, which is called, stranding. In this phase, the damage of the inner bottom plate can be caused by a large concentrated force exerted on a stranded ship due to the effect of current, wave and tide with the resulting oil spill.

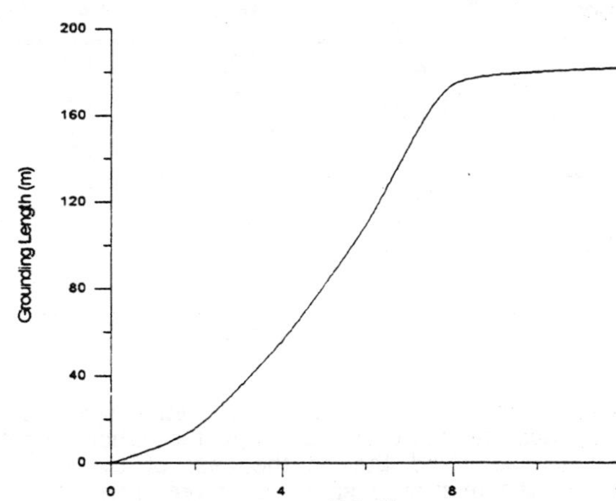

Fig. 3. Damage Length vs. Initial Speed

The maximum values of vertical penetration depending upon the rock eccentricity are presented for a typical tanker No.1 in Fig. 4, which shows that the vertical penetration is 1.45m at the zero eccentricity. It then increases stepwise in proportion to the eccentricity, reaches maximum 2.1m at $e=4$m, then it gradually decreases as the eccentricity increases. This interesting result can be explained by the interplay of global static equilibrium and local structural strength.

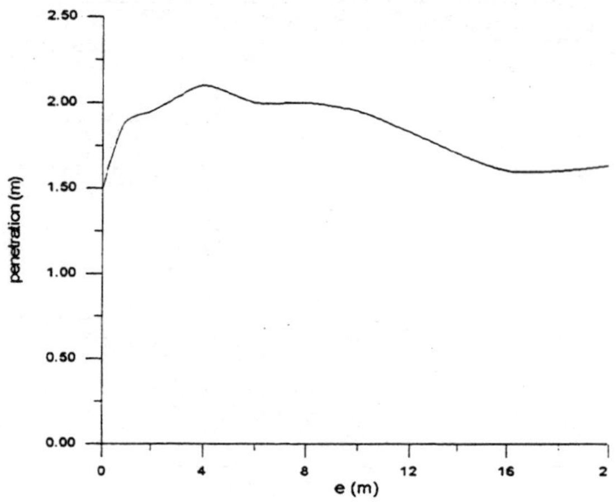

Fig. 4. Vertical Penetration vs. Rock Eccentricity

3. Some guidelines for Tanker Design

The results of calculations for six tankers will be re-interpreted emphasizing a number of different aspects. Let us consider the vertical penetration of rock into the ship hull which affects oil spill directly. Since there is no rupture in the inner bottom plate of all the four double hull tankers, no oil spill is likely to be initiated, see Table 1. In addition, it is also found that double bottom heights according to the regulations of IMO and OPA '90 are larger than the vertical penetration depth calculated for the assumed rock elevation. In other words, while it is true that the recommended double bottom heights higher than necessary will ensure a good ship structure from a potential spill and stability point of view, such a design will suffer from loss of hold volume.

Applied Ships	Requirement of D/B Height (m) OPA'90/ IMO	D/B Height (m)	Max. Vertical Penetration (m)
No.1	2.87/2.87	2.30	1.91
No.2	3.07/3.00	3.40	1.75
No.3	3.20/3.00	2.50	2.12
No.4	3.87/3.00	3.00	2.50
No.5	3.81/3.00	Single Hull	1.08
No.6	3.73/3.00		1.06

Table 1. Vertical Penetration vs. D/B Heights

Table 2 shows the needed ship's initial speed for the grounding damage to occur through the whole length of holds. It is also noted that in the case of the two single hull tankers the minimum speed is more than 12 knots in order to cover the grounding length over all the holds. It can be concluded that operating at more than 8 knots in the cases of Nos. 2 and 3, at more than 10 knots in the cases of Nos. 1 and 4 is preferable to operating at very low speeds, because stranding accidents can be possibly avoided. This is quite opposite to the case of single hull tankers where lower speeds are advantageous to minimize oil spill. Table 3 shows the results for the initial speed of 12 knots and the number of the corresponding ruptured holds.

Applied Ships	Length of Holds (m)	Grounding Length (m)	Ship's Speed (knots)
No.1	181.44	181.44	8.0
No.2	184.00	184.00	10.0
No.3	207.36	207.36	10.4
No.4	254.30	254.30	8.0
No.5	239.80	239.80	15.5
No.6	251.30	251.30	17.1

Table 2. Speed When Grounding Occurs over the Length of Whole Holds

The effect of the initial trim angle upon grounding damage is shown in Table 4, where the penetration ratio is defined as the ratio of the initial penetration to the maximum one. The trim angle has a positive correlation with the penetration ratio except the tanker No. 5. The penetration ratio is of course dependent upon the initial penetration depth and the trim angle. It may seem to be reasonable to mention with certainty that the trim angle suggested by the loading manuals and current practice plays a considerable role in determining the double bottom height, even if we do not rest upon theoretical estimation both qualitatively and qualitatively.

Applied Ship	H. No.	LBP (m)	B (m)	Grounding Length (m)	Hold Length (m)	Total Number of Holds	Number of ruptured Holds
1	5097	238	43.0	181.44	181.44	6	0
2	5016	252	46.0	184.00	184.00	6	0
3	5049	264	48.0	207.36	207.36	9	0
4	5091	320	58.0	254.30	254.30	5	0
5	5089	315	57.2	158.00	239.80	6	4
6	5029	310	56.0	161.00	251.30	6	4

Table 3. Grounding Length and the number of Corresponding Damaged Holds

H. No.	DWT	Penetration Ratio	Trim Angle (deg)
No.1	98K	1.91/1.10=1.74	0.1434
No.2	118K	1.75/1.37=1.28	0.1434
No.3	140K	2.12/0.80=2.65	0.4362
No.4	300K	2.50/1.24=2.02	0.1434
No.5	280K	1.08/0.25=4.32	0.2350
No.6	250K	1.06/0.25=4.24	0.4436

Table 4. Trim Angle vs. Penetration Ratio

Table 5 shows that the larger a ship is the greater is the grounding force. Furthermore resistance forces in the single hull ships are greater than those of the double hull ships. The fact that grounding occurs through the whole outer bottom plate without rupturing of the inner bottom plate can be thought as a limited capacity to resist grounding.

As pointed out earlier this may be advantageous in terms of reducing oil spill from the deferred damage. The fact that the grounding force in single hulls is higher does not mean that the grounding resistance is greater. One has to distinguish between the damage resisting capacity and of spill resistance. From this point of view, a single hull tanker poses a greater environmental threat, because oil spill, however small, is likely to occur, even though the damage length to the hull bottom is relatively short.

H. No.	DWT	Max. Grounding Force (kN)	Avg. Grounding Force (kN)	$10^9 F_{avg}/\sigma_u BT$
No.1	98K	0.63×10^4	0.59×10^4	0.317
No.2	118K	1.05×10^4	0.99×10^4	0.480
No.3	140K	1.68×10^4	1.17×10^4	0.500
No.4	300K	1.36×10^4	0.74×10^4	0.181
No.5	280K	5.84×10^4	3.95×10^4	1.079
No.6	250K	6.50×10^4	3.64×10^4	1.025

Table 5. Comparison of Grounding Forces

In this context, it is worth mentioning an assertion made by Böckenhauer [3] that the collision resistance of a double hull tanker is always greater than that of a single hull tanker and grading of the collision strength of double tankers is determined by considering how many times the energy dissipated by a reference single hull tanker a double hull tanker can absorb. For example, the capacity of a double hull tanker to resist collision impact is expressed by COLL1, COLL2, COLL3 when the energy of double hull tankers exceeds that of a reference ship single-hulled by more than 100%, 200%, 300%, etc. However, such a grading concept cannot be adopted in the current grounding analysis according to the foregoing discussion. It should be kept in mind that the grounding resistance and the corresponding energy dissipation themselves may increase when rupture happens in the inner bottom plate even in double hull tankers, but such a grounding scenario is undesirable from the point of view of preventing oil spill. Therefore, little meaning can be attributed to such an increase of grounding resistance from an oil-spill prevention viewpoint.

Another important result to be seen in Table 5 is that the grounding force should be normalized by a suitably chosen characteristic dimension of the vessel, not the absolute magnitude of the grounding force itself. The result of normalizing an average grounding force by $\sigma_u BT$, where σ_u is the ultimate strength of the material and B is the ship's beam and T is the draft, is shown in the column on the right hand side. According to this result, it is again pointed out that the grounding resistance of a tanker should be differentiated between a single and a double hull vessels. In the current grounding analysis, the tanker No. 3 possess the greatest resistance among the double hull tankers studied here and the next greatest resistance is offered respectively by the tankers No. 2, No. 1 and No. 4. In the case of single hulls, their grounding resistances are almost similar.

Table 6 introduces an index describing a capacity for the hull structure to absorb the grounding energy. Clearly single and double hull tankers should be distinguished in their energy dissipation grade. As discussed earlier, the energy dissipated by the single hull structure is greater than that by the double one in the same way as in the grounding force. The fact that the energy dissipation of the single hull tankers is greater than that of the double hull tankers bears serious practical consequences. Single hull tankers would stop while the vessel is still on the rock and accordingly the initial kinetic energy is fully dissipated during grounding. At the same time double hull vessel clears the rock with some remaining kinetic energy after the grounding process and accordingly the initial kinetic energy is only partly absorbed by the hull structure.

The absolute value of the average energy dissipation for the tanker No. 3 is the greatest of all. However, No. 2 has the largest dimensionless index and H5049 the next greatest when they are normalized by $10^8 E_{avg}/\sigma_u BTL$ in the case of the double hull tankers. The magnitudes of the dimensionless index for the single hull tankers are almost the same, as in the grounding force. A comparison of Tables 5 and 6 reveals that there is a direct correlation between them. It is natural, because the latter is obtained by integrating the former with respect to grounding length. It can be said that the average energy dissipation describes a global grounding resistance of a tanker through the whole grounding process, while the average grounding force represents an instantaneous grounding resistance. All the data obtained by those analyses so far will be used in developing practical guideline on how to use information from simplified calculations in real design and grading the grounding resistance.

H. No.	DWT	Max Energy Dissipation (kJ)	Avg. Energy Dissipation(kJ)	$10^8 E_{avg}/\sigma_u BTL$
No.1	98K	1.13×10^6	0.49×10^6	1.10
No.2	118K	1.80×10^6	0.87×10^6	1.68
No.3	140K	2.40×10^6	1.01×10^6	1.64
No.4	300K	2.74×10^6	0.80×10^6	0.61
No.5	280K	6.35×10^6	2.62×10^6	2.27
No.6	250K	5.80×10^6	2.00×10^6	1.82

Table 6. Comparison of Energy Dissipations

4. Criterion of Grading Grounding Strength

A need for introducing performance standards rather than design standards for oil tankers has been emphasized in Ref. [11] and [12]. In the present paper an attempt is made to re-interpret the grounding analysis made in Section 3 in a way of establishing some representative grading of the grounding performance. It should be remembered that our conclusions are tentative because of a small number of the studied ships. Therefore it is difficult to assign a definitive performance index to each of the studied ships. The methodology suggested here seems appropriate for the characteristic of the present task of gradation of grounding strength. That is, grounding analysis of as many tankers ever built worldwide as possible should be performed first for the purpose of using the results to get a better idea how grading of the grounding strength should be established. Perhaps, this may sound quite unrealistic but it is possible and practical when considering potential of simplified analysis tools.

An important step is to get blueprints of as many ships as we can. Then, a desirable gradation index can be derived from those results. If a consensus is drawn from all interested organizations including shipping companies, ship owners and operators, classification societies, concerned government agencies and shipbuilding companies, it will be possible to accomplish the task. The process of accumulating all the necessary materials may be long but a good data-base will successfully be established, so that performance standards for the grounding resistance of tankers could be established.

For example if a mean and standard deviation are obtained in a statistical way, we can grade the relevant ship by determining where the resistance belongs, for example, $\sigma/2$ or $\sigma/3$ in the whole statistical region.

5. Discussions

Suggestions for improvements of the present version of MIT computer program are given below.

(1) The effect of transverse floor or bulkhead should be estimated precisely. This task is included in the proposal of the Tanker Safety Project Phase II. The calculations can be verified by the test result performed at NSWC. See Ref. [1, 7].

(2) The initial amount of damage at the beginning of grounding should be estimated in a more rigorous way. That the height of the rock at midship is usually assumed to be 2.0m is just as good as the present statistical methodology based on the rule or regulations such as OPA '90. Therefore, the top priority in upgrading work of the Tanker Safety Project Phase II should be placed on the task of modeling the bow section of the hull including collision bulkhead. The Daewoo Heavy Industries shipyard undertake this research task, for which we will purchase a relevant universal tensile machine and a suitable numerical analysis tool. Since a full-scale test of grounding is almost impossible and the expense is enormous, the numerical simulation is very economic and efficient. We are planning to develope a theoretical model dealing with the initial vertical penetration based on the results of tests and numerical simulations.

(3) A grounding path should follow a curve not a straight line parallel to the ship's centerline due to the effect of moment arising from unbalanced force, when the rock eccentricity becomes larger. The corresponding theoretical model of external dynamics has not yet been verified by suitable experiments.

(4) Accuracy of calculating the horizontal damage width, especially in double hull tankers, should be increased. The results of grounding analysis of No.1 and No. 4 show too large magnitude of the horizontal damage width which is practically unrealistic, when we consider geometry of the cone-shaped rock. This deficiency could be corrected by a more detailed structural model.

(5) The effect of the longitudinal bulkhead in single hull tankers and center or side girders in double hull tankers should be estimated with more accuracy. Our experience is that, except a small number of cases, the effect of these structural members seem to be unrealistic from a view-point of structural strength.

(6) A competitive computer program should contain a module that facilitates function of the parametric studies such as performed in Section 3.

(7) As a vessel grows older, its structural strength becomes weak primarily due to the effect of corrosion of the plate, which, according to a recent statistics, is responsible for many accidents. Therefore, the corrosive effect should be taken into account by reducing the plate by an appropriate amount when need arises to analyze an old ship caught in a ground accident.

(8) There is a possibility that upon hitting a rock the vessel will be stranded. Such a scenario could be more devastating than one in which the vessel undergoes the accident of grounding through the whole length of cargo holds without fracturing in the inner bottom plate. The capabilities of estimating the so called deferred damage will be a nice feature of any PC based computer program. However, it is understood that wind, wave, current and tidal action will not be easy to deal with in a simple way. The loss of cargo can have an effect upon a grounding process, because the mass of the vessel is reduced with oil spill by a considerable amount, besides affecting the ship's speed etc.

(9) Plate thicknesses can vary with spatial structural positions even in the same structural member, for example, in a shell plating. Therefore, modification of input tool related with a variety of thickness should be reflected further.

6. Conclusions

The experimental grounding analysis program was implemented to several tankers built by the Daewoo Heavy Industries shipyard in order to develop a practical guideline which can be easily applied to the initial design of tankers and develop also a gradation index of grounding resistance of tankers.

We have selected a total of six tankers, and analyzed them to determine the grounding resistance. We have succeeded in developing a set of practical application guidelines and suggested a method of grading the grounding resistance. The original objective was to come up not only with the practical guidelines but also suggest the gradation method. Because the number of the studied ships was limited, the second objective could not be fully completed.

Some commercial numerical codes well known to naval architects are DYTRAN, LS-Dyna3D, Pam-Crash and the results of grounding and collisions analysis by using them are generated by many researchers around the world [4]. However, they are not suitable in a practical ship design and verification of the current rules and regulations. The computer program is based on the simpler concepts of Superfolding/Supertearing Element originally developed in automotive industry has great potential in the shipbuilding area in 1995. A limited experience with this program proves its usability and convenience in the design of oil tankers. This convenience and expedience can be very instrumental in examining the grounding resistance of a ship structure by checking various structural parameters in order to get the most optimum design. However, the program is in its initial stage of development, there are a lot of things to be refined and corrected. An extensive experience in using the program has identified a number of shortcomings in the current version and pointed out a direction of refinement and correction in order to attain our ultimate goal, i.e., a perfect program. We hope that when every suggestion of the ship designers and engineers is fed back to the present version for upgrade, it will become an efficient and state of the art tool for designing next generation of safe and economic ships.

Research activities in the field of grounding and collision should not be confined to crude oil tankers only but should be extended to LNG carriers and/or nuclear powered vessels. Especially, the Daewoo Heavy Industries shipyard in pursuit of LNG carriers as a main strategic product will focus activities on developing an essential technology needed of that field to build ships superior to other companies in terms of reliability, economy and stability.

References

[1] Choi, S.K. and Wierzbicki, T., "Correlation of NSWC Grounding Tests with Minorsky/Vaughan Method," Technical Report No. 39, Joint-Industry Program Tanker Safety, January 1995

[2] Daidola, J.C., "Tanker Structure Behavior During Collision and Grounding," *Marine Technology*, Vol. 32, No. 1, pp. 20-32, 1995
[3] Egge, E.G. and Böckenhauer, M., "Calculation of the Collision Resistance of Ships and its Assessment for Classification Purposes," Marine Structures Vol. 4, pp. 35-56, 1991
[4] Kuroiwa, T., "Dynamic Experiments and Numerical Simulations of Failure of Ship's Side Structure," Conference on Prediction Methodology of Tanker Structural Failure and Consequential Oil Spill, Association of Structural Improvements of Shipbuilding Industry, 1993
[5] McDonald, H.A., "Required Strength and Tear Resistance for Fillet Welds in Ships Exposed to Grounding or Collision Loads," Technical Report No. 14, Joint MIT-Industry Program in Tanker Safety, May 1993
[6] McNeill, D.C., Malmquist, D., Walter, D., Grothen, D. and Londot, K., "The Reconstruction of the Exxon Valdez," *SNAME Transactions*, Vol. 98, 1990, pp. 663-694
[7] Puente, I.J., "Correlation of NSWC Grounding Tests with MIT Theory," Master Thesis, Department of Ocean Engineering, MIT, May 1995; Technical Report No. 44, Joint MIT-Industry Program in Tanker Safety, May 1995
[8] Rodd, J.L. "Double Hull Tanker Grounding Experiment," *Advanced (Unidirectional) Double Hull Technical Symposium Proceedings*, Gaithersbrug, Maryland, Oct. 1994
[9] Rodd, J.L., Philips, M.P. and Anderson, E.D., "Stranding Experiments on Double Hull Tanker Structures," *Advanced (Unidirectional) Double Hull Technical Symposium Proceedings*, Gaithersbrug, Maryland, Oct. 1994
[10] Wierzbicki, T., Rady, R., Peer, D. and Shin, J.G., "Damage Estimates in High Energy Grounding of Ships," Technical Report No. 1, Joint MIT-Industry Program in Tanker Safety, June 1990
[11] Thomas, P. and Wierzbicki, T., "Performance Criteria vs. Design Standards for Commercial Tank Vessels," Proceedings of Prads '92, pp. 2.912-2.925, Newcastle upon Tyne, England, May 18-22, 1992
[12] Pippenger, D., Sirkar, J., Little, P. and Cojeen, P., "Design Performance Standards for Oil Tankers," Proceedings of International Conference on Design Methodologies for Collision and Grounding Protection of Ships, SNAME/SNAJ, August 1996, pp. 16-1 to 16-10.

Proceedings of the Seventh (1997) International Offshore and Polar Engineering Conference
Honolulu, USA, May 25-30, 1997

ISBN 1-880653-28-1 (Set); ISBN 1-880653-32-X (Vol. IV)

A Study on the Critical Collision Speed of Double Hull VLCC

Sang-Gab Lee and Young-Gu Chung
Korea Maritime University
Pusan, Korea

Abstract

This paper describes a series of numerical simulations of collision between a 310,000 DWT double hull VLCC (struck ship) and three 35,000, 70,000 and 105,000 DWT tankers (striking ships) using LS/DYNA3D. Collisions are assumed to occur at the middle of the VLCC with the striking ships moving at right angle to the VLCC centerline. Striking ship speeds are varied to find a critical collision speed without failure of inner side shell, and the informations of collision force and absorbed energy of each case are also reported.

Key Words: Collision, simulations, LS/DYNA3D, double hull VLCC, critical collision speed, collision force, absorbed energy

1. Introduction

From departure of the study on ship hull resistance to collision of nuclear powered merchant ships in the late 1950s, Minorsky[1] presented a formula to estimate absorbed energy in collision based on data of 25 collisions provided by the US Coast Guard. Studies have been extensively performed by many researchers. Since the grounding of Exxon Valdez in 1989 focused the international attention on the oil spills, the United States of America required a double hull for future tankers in 1990 and IMO adopted similar, though a little bit relaxed regulations in 1992. Further research is necessary to come up with design regulations and recommendations for tanker design to reduce the risk of oil spilling.

In structural design and optimization of a double hull tanker, it is important to have a good understanding of the behavior of its structure in collision and grounding, and to have an accurate prediction of its hull resistance against collision and grounding in terms of energy absorption capacity and resulting damage[2]. Since the ruptures of inner side shell and inner bottom shell due to collision and grounding and/or stranding are critical failure modes, the informations on the energy absorption capacity up to the above failures might be used as a basis to design and to optimize the structure of double hull tankers.

While simplified methods to estimate the energy absorption capacity in collision are proposed[3,4], since many years, hydrocodes such as MSC/DYTRAN, LS/DYNA3D and PAM-CRASH have been successfully used to simulate high-speed highly nonlinear phenomena in defense, safety and space development such as impact, crash, penetration and explosion. These hydrocodes were applied to ship collisions in inland water ways[5], and to simulate full scale collision experiments using a decoupled fluid-structure three step approach[6] and a coupled fluid-structure simulation[7]. Full scale numerical simulations of collision for double hull VLCCs were also carried out using MSC/DYTRAN or LS/DYNA3D to clarify the behavior of different structural members and to suggest a new design of VLCC side structures during failure as well as energy absorption mechanism[8-12]. Such a full scale numerical simulation may provide an accurate predictions on energy absorption capacity and failure modes of real-sized ship structures.

It is desirable to estimate the crashworthiness of ship structures against collision comparing their energy absorption capacities with each other. For this study, several sizes of striking ships with different collision speed are used to collide against each struck VLCC up to failure of inner side shell. This paper describes a series of numerical simulations of collisions between one type of 310,000 DWT VLCC(struck ship) and three types of 35,000, 70,000 and 105,000 DWT tankers(striking ships) using LS/DYNA3D. Collision are assumed to occur at the middle of the VLCC with the striking ships moving at right angle to the VLCC centerline. Collision speeds of each striking ship are varied to find a critical failure modes, and informations of collision force and absorbed energy of each case are also reported.

2. Validation of LS/DYNA3D

To validate LS/DYNA3D to use in this study, it is desirable to compare the result of numerical simulation of LS/DYNA3D with experimental one. Several large scale oil tanker grounding experiments have been performed by the Carderock Division of Naval Surface Warfare Center (CDNSWC) as part of the Advanced Double Hull Technology Project[13]. The first experiment represented a conventional T-5 Paul Buck double hull tanker bottom with transverse and longitudinal framing, which provided baseline information for comparison purposes in both numerical analyses and experiments.

Figure 1 shows geometry and construction details for the conventional double hull structure in the 30,000~40,000 ton range at approximately one fifth scale, which consists of the double bottom structure beneath one oil tank, the forward and aft transverse bulkheads, and heavy sideplates. The material used for the grounding model was ASTM A569 with the measured yield and ultimate stresses of 282 MPa and 344 MPa, respectively.

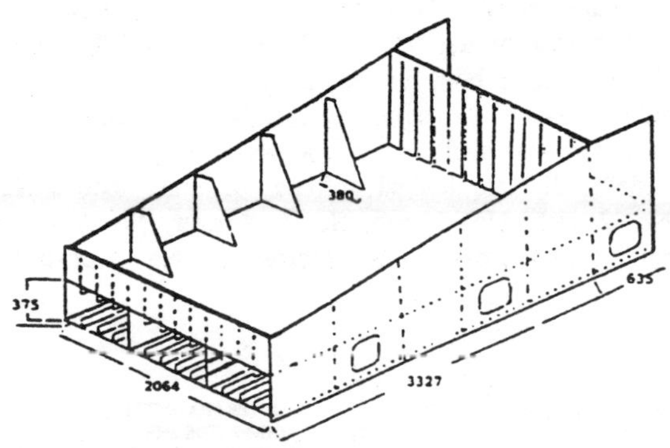

Fig. 1 Dimension of grounding model[13]

A series of large scale oil tanker grounding experiments have been conducted using the Grounding Test Machine, which consisted of a 227 tonne test vehicle (or sled) and impact area. Sled built on twin railcars carried the grounding model down an inclined set of railroad tracks to an impact area at a sled velocity of 6.1 m/sec (12 knots). A steel grounding rock included in the impact area was a 914 mm high, 90 degree cone with a 178 mm radius spherical tip. The installation and orientation of the inclined grounding model is shown in Fig. 2 with 7.4 degrees of attack angle and the entry/exit height of the rock tip.

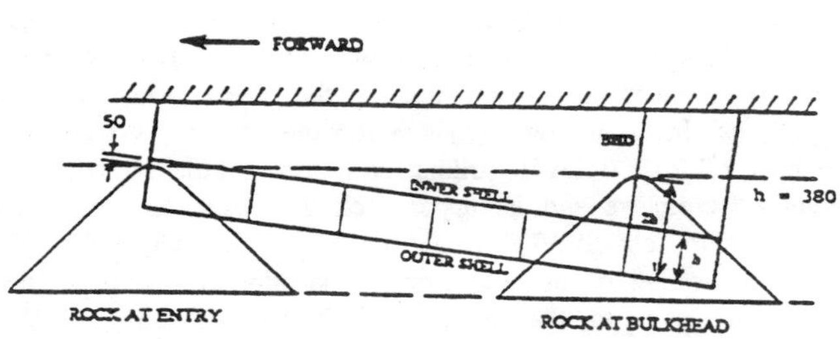

Fig. 2 Installation of grounding model[13]

Figure 3(a) shows the plots of the finite element mesh of grounding double hull structure and grounding rock. All members of inner and outer shells, transverse and longitudinal webs, forward/aft bulkheads, small stiffeners, and grounding rock are modeled using around 18,000 shell element. The mass of the sled is controlled by rigid plate on the top of the double hull model, which is invisible in Fig. 3. The grounding rock is modeled to be rigid and the double hull model is restrained to move straightly forward, and not upward/downward.

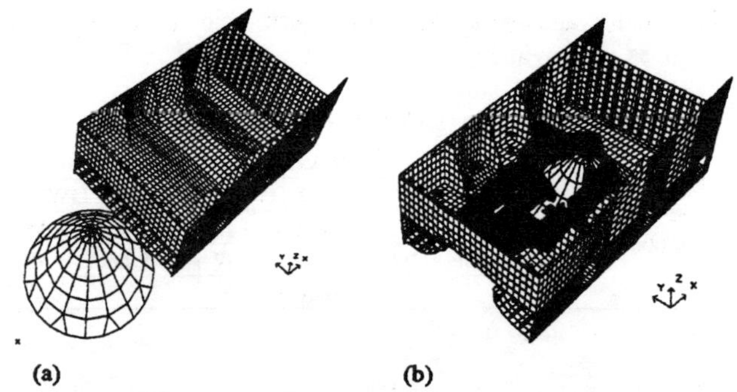

(a) (b)

Fig. 3 Finite element mesh and rupture configuration of grounding model

Two kinds of material types are employed for numerical simulation: strain rate dependent isotropic plasticity for the grounding double hull model and isotropic plasticity for the grounding rock. The former material type is suitable for the consideration of the material dynamic effect. Dynamic yield stress, $\sigma_Y = \sigma_0\{1+(\dot{e}/D)^{1/p}\}$, is used [14], where σ_0 is static yield stress, $\dot{e}$ is stain rate, and $D(=40.4\,s^{-1})$ and $p(=5)$ are constants. The density and Poisson's ratio to use here are 7850 kg/m^3 and 0.3, respectively. Every finite element is due to be eliminated when plastic strain reaches to the failure value, 0.25 in this study.

Figure 3(b) shows the configuration of rupture process of the grounding model by the grounding rock. Figure 4 shows the comparison of numerical simulation result with NSWC experimental one about the horizontal impact force (plate resisting force) vs. the sled position. Compared to the experimental result, it can be said that the numerical simulation result is satisfactory and that the already acknowledged tool LS/DYNA3D is sufficient to perform the following collision problems.

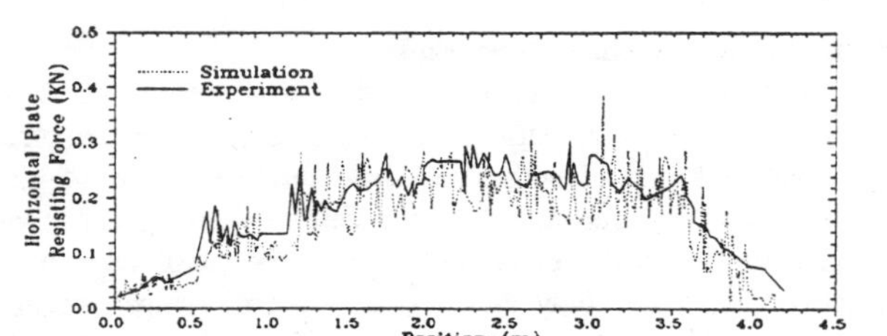

Fig. 4 Comparison of numerical simulation and NSWC experimental results

3. Collision Scenarios and Simulation Models

3.1 Collision scenarios

In the considered struck ship, 310,000 DWT D/H VLCC, three tanks are installed across its breadth inside the double hull, cross ties (struts) are fitted in the side tanks and three side stringers are also fitted in the double side, as shown Fig. 5. When the side of struck ship is hit by the striking ship, cross ties transfer the impact force to the side longitudinal bulkhead. It was reported that the side longitudinal bulkhead showed elastic deformation behavior even while cross ties failed[9]. Collision scenarios considered in this study are shown in Table 1, where striking ships collides against a stationary struck ship in each collision scenario as shown in Fig. 5.

Table 1 Collision Scenarios

Struck Ship	Striking Ship					
	35,000 DWT		70,000 DWT		105,000 DWT	
310,000 DWT	5.5 knots	I	4.5 knots	Ⅲ	4.00 knots	Ⅵ
	6.0 knots	Ⅱ	5.0 knots	Ⅳ	4.25 knots	Ⅶ
			5.0 knots(u)	V		

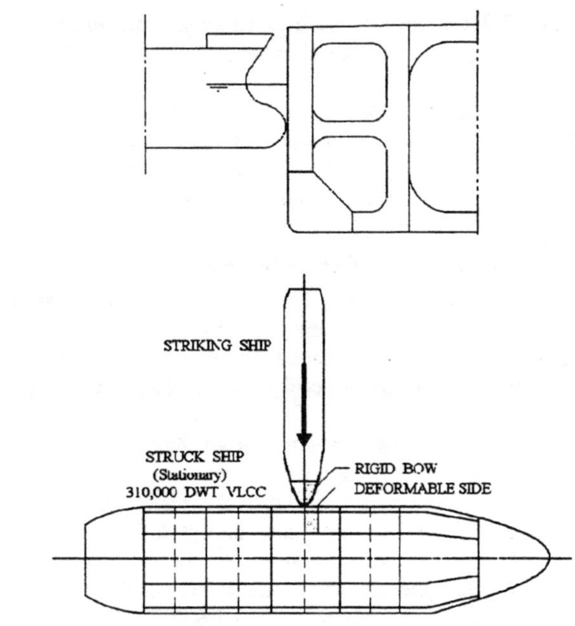

Fig. 5 Numerical model outline

3.2 Struck and striking ships models

In a high energy collision, large plastic deformations take place in the location of collision on the struck and/or the striking ships, while the rest of both ships show different levels of elastic deformations. Compared to the plastic deformations in the location of collision, the elastic deformations are very small and may be neglected. Therefore, only the regions in both ships containing members expected to show the plastic deformations are modeled as deformable structures, and the rest of both ships, as rigid bodies.

In full scale simulations during collision using MSC/DYTRAN [9], the side of the struck VLCC in the location of collision as well as the bows of the striking ships were modeled as deformable structures, and the reset of the ships, as rigid bodies. The water around the VLCC was modeled with an Eulerian finite volume mesh, and ALE(Arbitrary Lagrangian Euler) coupling was used to couple the water effect to the VLCC. Their results showed that the motions of both striking ship and struck ship were very small with heavy struck ship, and that the effect of the fluid around the struck ship was also small. Damage of the bow of the striking ship was also reported to be relatively smaller than that of the VLCC side[8,9].

By these results of full scale simulations, in this paper, the effect of the fluid around the struck ship is neglected. Only the deformable structure of the struck ship is employed and the rest of struck ship is neglected. On the assumption that the position of collision is the middle of cargo hold at the center of struck ship, in the longitudinal direction the deformable structure is modeled from the center of struck ship to the two web frames and a symmetric condition is established at the center of cargo hold, as shown in Fig. 5. In the transverse direction, the side longitudinal bulkhead is not modeled due to the level of elastic deformations until the failure of cross ties. Web frame and the shells of deck, bottom and inner bottom connected to the side longitudinal bulkhead are constrained to the transverse direction. Since damage of the bow of the striking ship was relatively smaller than that of the VLCC side, the bow of the striking ship is modeled as rigid body in this study.

Figure 6 shows the structural members of struck ship VLCC, such as outer side shell, inner side shell, longitudinal stiffener of outer side shell, longitudinal stiffener of inner side shell, No. 1 web, No. 2 web, No. 3 web, No. 1 stringer, No. 2 stringer, No. 3 stringer, No. 1 cross tie, No 2 cross tie, No. 3 cross tie, deck, etc..

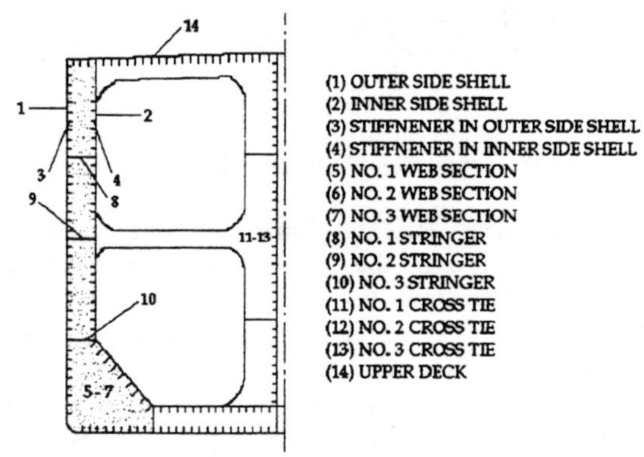

Fig. 6 Structural members of struck VLCC

Figure 7(a) shows the plots of the finite element mesh of the side of struck ship, where the breadth of side double hull is 3.38 m and the space of web section is 5.08 m. The bow of 35,000 DWT striking ship collides against the center of cross tie and its upper deck is located below that of the struck VLCC(scenarios Ⅰ, Ⅱ), as shown in Fig. 7(b). On the other hands, the center of the bow of 70,000 DWT striking ship hits a little bit below the center of cross tie and its upper deck is located the same level as that of the struck VLCC(scenarios Ⅲ, Ⅳ), shown in Fig. 7(c). For consideration of the effect on the location of the upper deck of striking ship to that of the struck VLCC, scenarios V is included in this study, such that the upper deck of 70,000 DWT striking ship is located above that of the struck VLCC, not shown in Figures. In the case of 105,000 DWT striking ship strikes much below the center of cross tie and its upper deck is also located above that of the struck VLCC(scenarios Ⅵ, Ⅶ), as shown in Fig. 7(d). All members of shells, web frames, longitudinal stiffeners, and stringers are modeled using shell element. Around 12,000 shell elements are used in this study.

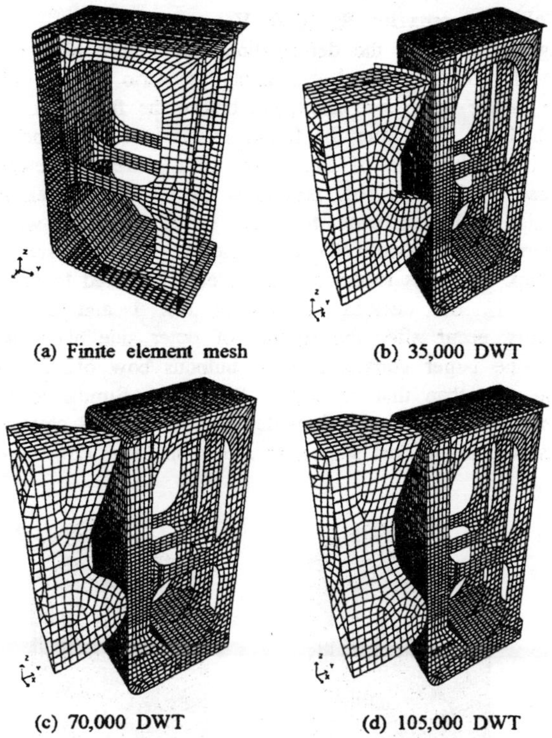

(a) Finite element mesh (b) 35,000 DWT

(c) 70,000 DWT (d) 105,000 DWT

Fig. 7 Finite element mesh of deformable side of struck VLCC and 3 striking ships

Elastoplastic material with nonlinear strain hardening and strain rate effects are used, and the properties are as follows:

Density	7850.0 Kg/m^2
Modulus of elasticity	210 GPa
Poisson ratio	0.3
Yield stress(mild steel)	250 MPa
Ultimate stress(mild steel)	400 MPa
Constants of Cowper Symonds Eq.	40.4, 5.0
Failure plastic strain	0.2

In collision, tearing of weld lines is usually observed in areas where large damage occurs. Modeling of weld lines and their failure may be necessary in order to accurately predict collision damage. When stresses acting on the weld satisfy a predefined failure condition, the joint breaks up and internal forces are unloaded to the surrounding structure. Since every shell element is complied with failure plastic strain, weld lines are to be neglected in this study.

In the plane of symmetry, grid points on continuous longitudinal members of the struck ship and the center plane of the striking ship are given symmetric boundary conditions ($U_x = \theta_y = \theta_z = 0.0$). The web frame in the symmetric plane of the struck ship model must contribute only half of its strength to the model. Instead of installation of the side longitudinal bulkhead of the struck ship as the boundary condition, all nodal points at that plane are constrained to the transverse direction. Because of symmetry, only one half of rigid body of bow is explicitly modeled, and the total mass of the striking ship together with 10% added virtual mass is distributed.

4. Results of Simulation

Simulations of the seven scenarios have been carried out, and their results are summarized as follows:

4.1 Collision scenarios Ⅰ & Ⅱ

Figure 8 shows the deformation and rupture plots of VLCC side with collision speed 6 knots (scenario Ⅱ) (a) after the failure in the outer side shell, (b) before the failure in the inner side shell, (c) after the failure in the inner side shell. Figure 8(d) is another one of Fig. 8(c) from different viewpoint. The bulbous bow penetrates the outer side shell, then the inner side shells at the cross tie of center web frame. It is interesting that the buckling mode of No. 1 cross tie in the tank changes during collision. At the initial stage No. 1 cross ties buckle mainly in bending mode as shown in Fig. 8(a), and at the last stages in Figs. 8(c) and (d) the torsional mode of No. 1 cross tie near the collision position can be observed. Since the deck of striking ship collides against the top outer side shell of struck VLCC, its deck penetrates the outer side and inner side shells without rupture of the upper deck of the struck VLCC during collision.

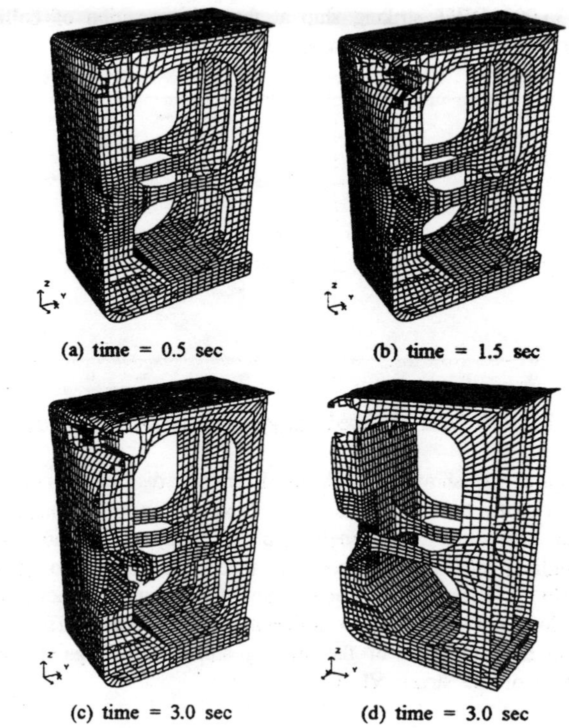

(a) time = 0.5 sec (b) time = 1.5 sec

(c) time = 3.0 sec (d) time = 3.0 sec

Fig. 8 Deformed side structure in scenario Ⅱ

The bulbous bow of striking ship with collision speed 5.5 knots(scenario Ⅰ) pierces the outer side shell and rebounds after the bow arrives at maximum 0.215 m ahead of the inner side shell without rupture of inner side shell. Scenario Ⅰ shows almost the same trend in the deformation and rupture configuration, and equivalent plastic contour plots as those of Figs. 8(a) and (b).

Figure 9 shows the variation of collision force with time and penetration in scenarios Ⅱ. It can be seen that failure of the outer side shell starts at time 0.15 sec and the inner side shell initiates to be torn away at time 1.79 sec with maximum deformation, 0.85 m, of the inner side shell.

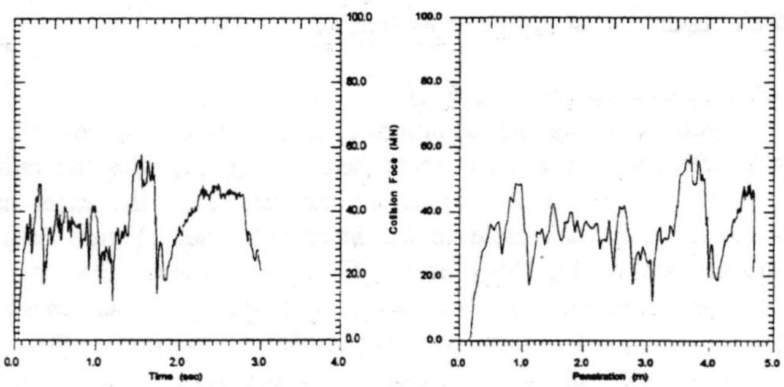

Fig. 9 Collision force variation in scenario Ⅱ

Figure 10 shows the variation of the total absorbed energy with time and with penetration in scenario Ⅰ and Ⅱ. It can be seen that the absorbed energy arrives at 159.6 MN-m when the inner side shell begins to be torn away in scenario Ⅱ. It might be thought, therefore, that the critical absorbed energy has to be larger than at least 159.6 MN-m to rupture the inner side shell with 35,000 DWT striking ship at the same position of collision to the struck VLCC as the scenarios Ⅰ and Ⅱ.

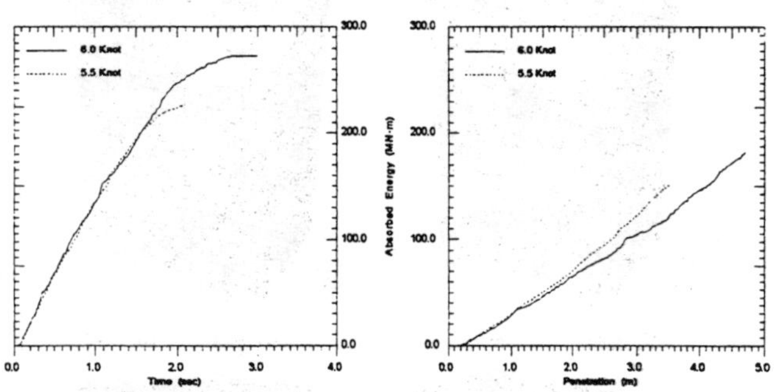

Fig. 10 Total absorbed energy in scenarios Ⅰ and Ⅱ

Figure 11 shows the absorbed energy distribution of each member with time in scenario Ⅱ. Most of absorbed energy are carried by outer side shell, No. 1 web, inner side shell, longitudinal stiffeners of outer and inner side shells, No. 2 web, and No. 2 stringer in capacity sequently during collision, while deck does not contribute to collision in this scenario. This might be due that the deck of the striking ship collides against below the deck of the struck VLCC.

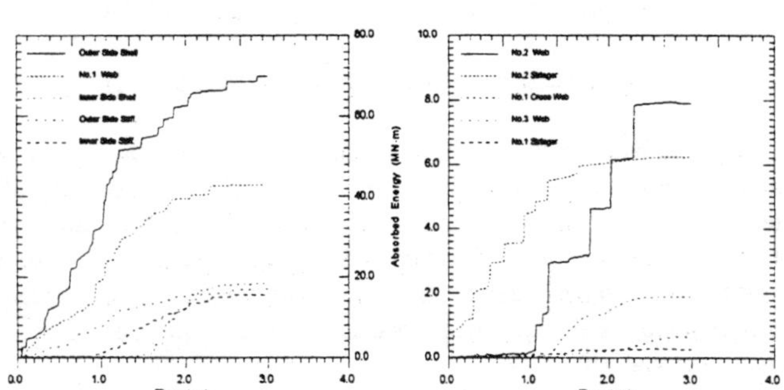

Fig. 11 Energy distribution in scenario Ⅱ

4.2 Collision scenarios Ⅲ, Ⅳ & Ⅴ

Figure 12 shows the deformation and rupture plots of VLCC side with collision speed 5.0 knots (scenario Ⅳ) (a) after the failure in the outer side shell, (b) before the failure in the inner side shell, (c) right after the failure in the inner side shell. Figure 12(d) is another one of Fig. 12(c) from different viewpoint. In this case the bulbous bow penetrates the outer side shell then the inner side shell just below the cross tie of the center web frame. The corner part between the deck and the top outer side shell of the struck VLCC is ruptured and folded. Compared to the case of scenario Ⅱ, the buckling modes of No. 1 and 2 cross ties apparently occur after the rupture of outer side shell, which is due to the larger contact area of bulbous bow of 70,000 DWT striking ship than that of 35,000 DWT one. Similar to the case of scenario Ⅱ, the torsional modes of No. 1 and 2 cross ties can be observed at the last stages, as shown in Figs. 12(c) and (d).

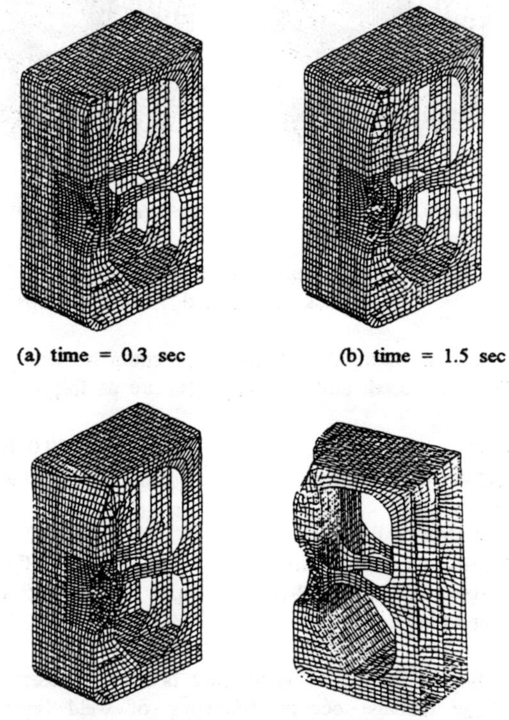

(a) time = 0.3 sec (b) time = 1.5 sec

(c) time = 2.5 sec (d) time = 2.5 sec

Fig. 12 Deformed side structure in scenario Ⅳ

The bulbous bow of striking ship with collision speed 4.5 knots (scenario Ⅲ) penetrates the outer side shell and rebounds after the bow arrives the inner side shell without rupture of inner side shell. Scenario Ⅲ shows almost the same trend in the deformation plots as those of Figs. 12(a) and (b). As can be expected from the scenario Ⅴ, the corner part between the upper deck and the top outer side shell of the struck VLCC is more ruptured and compressed than the case in scenario Ⅳ, while the bulbous bow of the striking ship penetrates the outer side shell and rebounds after the bow arrives the inner side shell without rupture of inner side shell like scenario Ⅲ.

Figure 13 shows the variation of collision force with time and penetration in scenario Ⅳ. It can be seen that the failure of the outer side shell starts at time 0.16 sec and the inner side shell initiates to be torn away at time 2.45 sec with maximum deformation, 1.5 m, of the inner side shell. Compared to scenario II in Fig. 9, it can be seen that the maximum collision force of scenario IV is larger than that of scenario II.

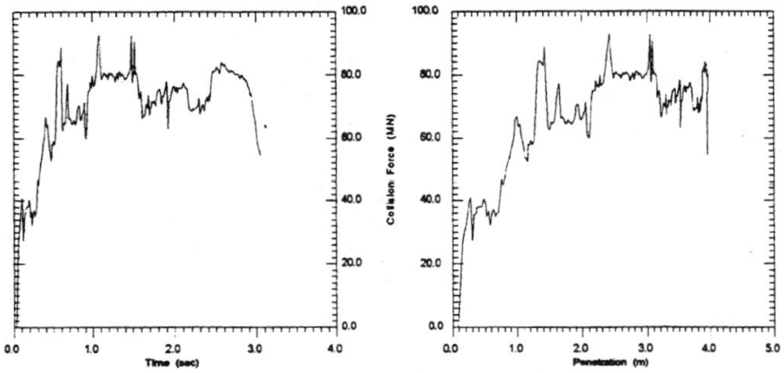

Fig. 13 Collision force variation in scenario IV

Figure 14 also shows the variation of the total absorbed energy with time and penetration in scenarios III to V. It can be seen that the absorbed energy arrives at 247.5 MN-m when the inner side shell begins to be torn away in scenario IV. The same conclusion as the scenario II can be drawn that the critical absorbed energy has to be greater than at least 247.5 MN-m with 70,000 DWT striking ship at the same position of collision to the struck VLCC as scenarios IV. Almost the same trends can be found between scenarios III and IV, as shown in Fig 14, while different tendencies, between scenarios IV(III) and V. The absorbed energy of scenario III is much smaller than those of scenarios IV and V. Even though the absorbed energy of scenario V is a little bit small than that of scenario IV, inner side shell of the former does not rupture. This might be the effect on the collision location of the upper deck of striking ship against struck ship. The collision of the upper deck of striking ship might reduce the portion of the absorbed energy of the bulbous bow of striking ship as its upper deck climbs the corner between the upper deck and the top outer side shell of struck ship.

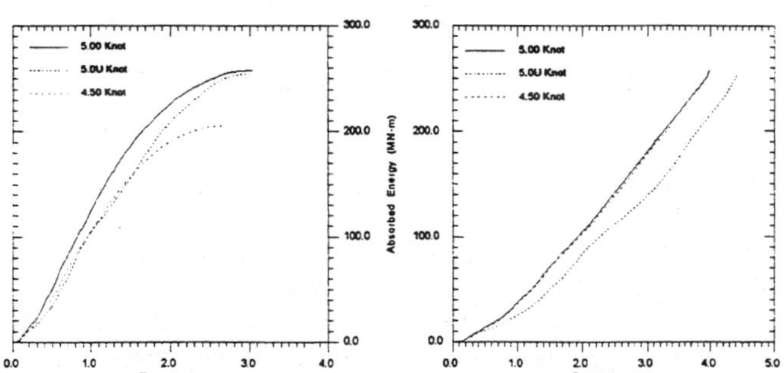

Fig. 14 Total absorbed energy in scenarios III to V

Figure 15 shows the absorbed energy distribution of each members with time in scenario IV. Similar to the trends of scenario II, most of absorbed energy is carried by outer side shell, No. 1 web, longitudinal stiffener of outer side shell, No. 2 and No 3 webs in capacity sequently. Compared to the case of scenario II, the inner side members, such as inner side shell and longitudinal stiffener of inner side shell, do not much contribute to the collision, while the deck member does, much. Outer side shell and No. 1 web in scenario IV is much larger absorbed energy (almost 1.5 times) than those of scenario II.

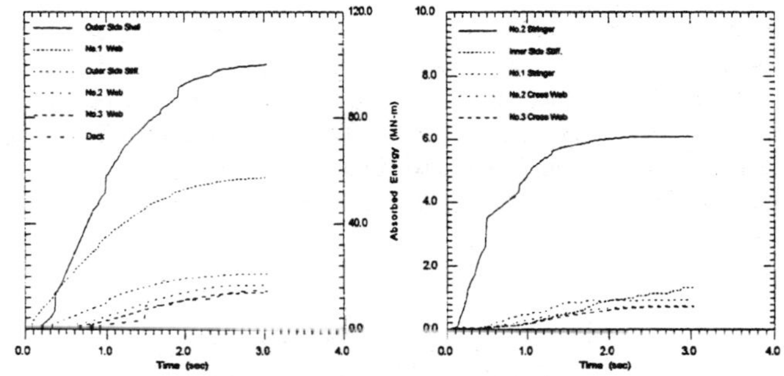

Fig. 15 Energy distribution in scenario IV

4.3 Collision scenarios VI & VII

Figure 16 shows the deformation and rupture plots of VLCC side with collision speed 4.25 knots (scenario VII) (a) at the onset of the failure in the outer side shell, (b) before the failure in the inner side shell, (c) right after the failure in the inner side shell. Fig. 16(d) is another deformation and equivalent plastic strain contour plots from different viewpoint of Fig. 16(c). The bulbous bow of striking ship crashes the outer side shell between cross tie and hopper tanker, and the upper deck of striking ship collides upward against the deck of struck VLCC like scenario V. The scope of rupture at the corner between upper deck and top outer side shell of the struck VLCC is much smaller than those of scenarios II and IV.

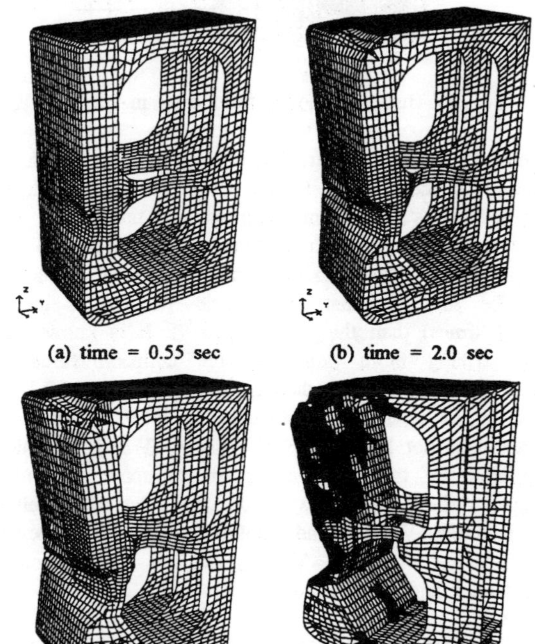

(a) time = 0.55 sec (b) time = 2.0 sec

(c) time = 2.3 sec (d) time = 2.3 sec

Fig. 16 Deformed side structure (scenario VII)

While the local rupture in the struck VLCC occurs in scenario II, the global damage and rupture happen to in scenario VII. Besides the failure in the inner side shell around the cross tie in scenario VII, the rupture can be found in the conjunction with hopper tank and inner bottom shell. All cross ties of the struck VLCC buckles more seriously than those of scenarios II and IV, as shown in Fig. 16. Compared to the cases of scenarios II and

IV, the buckling modes of all cross ties occur during rupture of the outer side shell, as shown in Fig. 17(a), which is due to the more larger contact area of bulbous bow of 105,000 DWT striking ship than those of 35,000 and 70,000 DWT ones. Extreme bending and torsional buckling modes in all cross ties can be found during the rupture of inner side shell, as shown in Fig. 17(b).

(a)

(b)

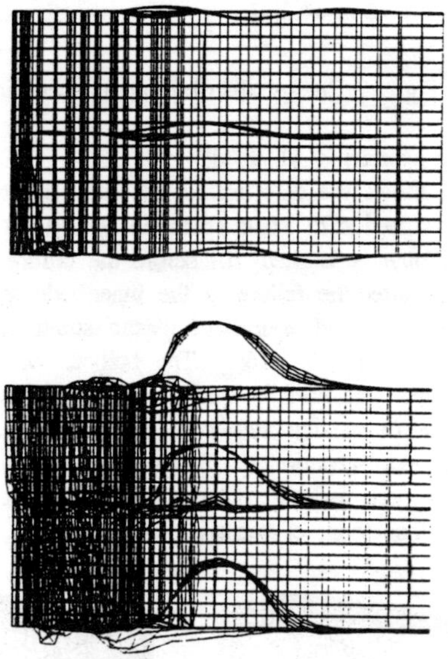

Fig. 17 Buckling modes from deck in scenario VII

Figure 18 shows the variation of collision force with time and penetration in scenario VII. It can be seen that the failure of the outer side shell starts at time 0.55 sec and that of the inner side shell, at time 2.28 sec with maximum deformation, 2.4 m, of the inner side shell. The maximum collision force in this case is almost the same as that of scenario IV, and is much larger (almost 2 times) than that of scenario II. It is known that mass of the striking more dominates the collision force than its collision speed does, even though scenarios IV and VII have about 1.4~1.5 times as large as scenario II in intial kinetic energy. This fact does not hold good in scenarios IV and V, which might be the effect of the collision location of upper deck of striking ship against the corner between the top outer side shell and the deck of struck VLCC. And another effect is the flexibility of impact position of struck VLCC, such as the location between hopper tanker and cross tie of the struck VLCC.

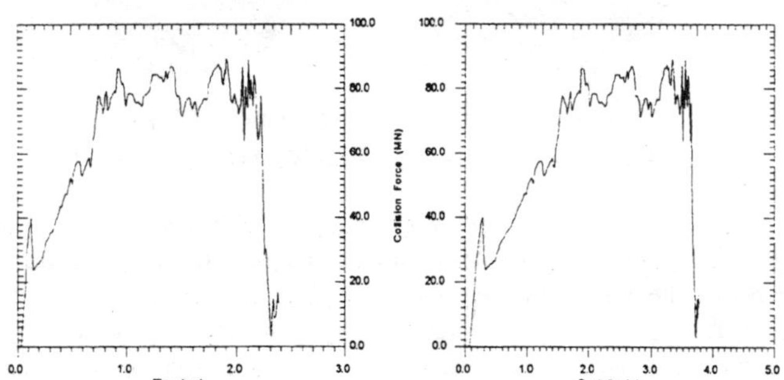

Fig. 18 Collision force variation in scenario VII

The variation of the total absorbed energy with time and penetration in scenarios VI and VII can be found in Fig. 19. In this case the bulbous bow of striking ship reaches the inner side shell with rupture beyond 230.0 MN-m.

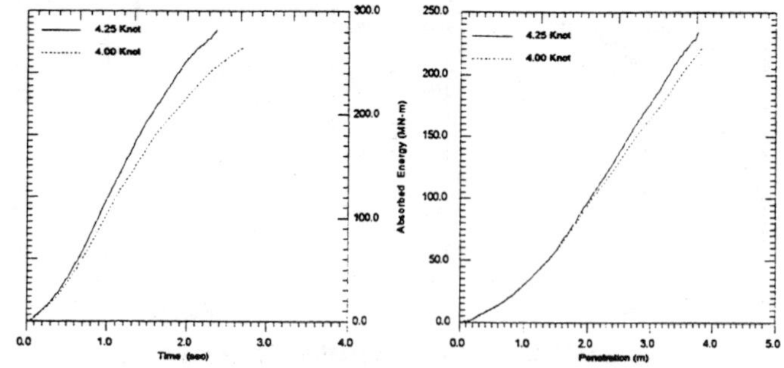

Fig. 19 Total absorbed energy in scenarios VI and VII

Figure 20 shows the absorbed energy distribution of each members with time in scenario VII. Most of absorbed energy are carried by outer side shell, No. 1 web, No. 2 web, No. 3 web, longitudinal stiffeners of outer side shell, deck, No. 3 stringer, No. 2 stringer, and longitudinal stiffeners of inner side shell in capacity sequentially. The contribution of outer side shell in scenario VII is less than that of scenario IV, and the others are almost the same level as those of scenario IV.

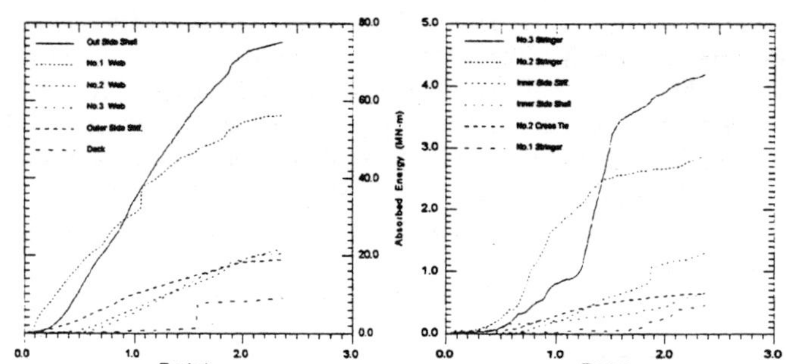

Fig. 20 Energy distribution in scenario VII

5. Conclusions

Full scale numerical simulations of collision for the struck 310,000 DWT double hull VLCC with three striking 35,000, 70,000 and 105,000 DWT tankers were presented using LS/DYNA3D, to figure out the critical failure modes and the informations of collision force and absorbed energy of each case. It might be thought that it was useful to use full numerical simulation to assess the structural failure and energy absorption capacity of VLCC in collision, for which large or full scale experiments are difficult and not practical.

It has been observed that the structural failure behavior and the absorbed energy greatly depend on the location of collision on the side structure(collision contact direction between striking and struck ships, flexibility of struck ship structure at collision region) and shape of sticking ship(collision contact area), besides the mass and speed of striking ship.

According to the numerical simulations of this study, the critical collision speeds of 310,000 DWT struck VLCC can be estimated just below 6.0 knots for 35,000 DWT, about 5.0 knots for 70,000 DWT, and about 4.25 knots for 105,000 DWT striking ships. Intial kinetic energy is increased as the mass of striking ship is increased, where the initial kinetic energy of 70,000 DWT of striking ship with speed 5.0 knots is about 1.4 times as large as that of 35,000 DWT one with speed 6.0 knots, and 105,000 DWT one with speed 4.25 knots, about 1.5 times.

As expected, the collision force of 35,000 DWT of striking ship is much less than those of 70,000 DWT and 105,000 DWT ones, and the absorbed energy of the former for fracture of inner side shell is much less than those of the latter. While the local rupture in struck VLCC occurs at the collision with relative small striking ship (35,000 DWT), the global damage and rupture happen to at the collision with large one (105,000 DWT). From these results, the following comment might be obtained that this type of 310,000 DWT struck VLCC has the failure modes in the inner side shell with lower absorbed energy from local rupture (30,000 DWT) than from global deformation and rupture (105,000 DWT) during collision.

Structural members, such as outer side shell and No.1 web of 310,000 DWT struck VLCC sustain almost the absorbed energy. As the local rupture occurs rather than the global deformation and rupture, the inner members, such as inner side shell and longitudinal stiffener of inner side shell, support the absorbed energy partly.

For the future study, realization of optimized double hull/ bottom as well as estimation of its critical (limit) collision/ grounding resistance in collision/grounding will be performed, through the simulations using more various cases and single parametric factor, while more complex factors are considered simultaneously in this study.

References

[1] V. U. Minorsky, "An analysis of ship collision with reference to protection of nuclear power plants", J. of Ship Research, 1959.

[2] J. Amdahl and D. Kavile, "Analysis and design of ship structures for grounding and collision", PRADS '92, 1992.

[3] H. Ito, K. Kondo, N. Yoshimura, M. Kawashima and S. Yamamoto, "A simplified method to analyze the strength of double hulled structures in collision(3rd report)", SNAJ, Vol. 160, 1986.

[4] S. Valsgard and E. Pattersen, "Simplified nonlinear analysis of ship/ship collisions", Norwegian Maritime Research, No. 3, 1982.

[5] H. Lenselink, K.G. Thung, H.L. Stipdonk and P.J. van der Weijde, "Numerical simulation of ship collisions", ISOPE-92, San Francisco, June 1992.

[6] H. Lenselink and K.G. Thung, "Numerical simulations of the Dutch-Japanese full scale ship collision tests", The conference on prediction methodology of tanker structural failure, Tokyo, July 1992.

[7] H. Lenselink et. al., "A 3-dimensional numerical simulation of the full scale Dutch-Japanese full scale ship collision tests with ALE fluid-structure interaction", FEM WORLD CONFERENCE, Monte Carlo, November 1993.

[8] J.Y. Kim, K.J. Lee, J.M. Kang, D.H. Kim, S. Rashed and D. Xiang, "Behavior of double hull VLCCs in collision", PRADS '95, September 1995.

[9] Jungsin Che and Gibok Jang, "Numerical simulation of structural response of D/H VLCC in collision", PRADS '95, September 1995.

[10] Atsushi Sano, et. al., "A study on the strength of double hull side structure of VLCC in collision", MARIENV'95, September 1995.

[11] Takao Kuroiwa, et. al., "Numerical simulation of collision and grounding of ships", MARIENV'95, September 1995.

[12] Ou Kitamura, "Comparative study on collision resistance of side structuer", Int. Conference on Designs and Methodologies for Collision and Grounding Protection of Ships, August 1996.

[13] James L. Rodd and Jerome P. Sikora, "Double hull grounding experiments", Proc. of the 5th ISOPE Conference, The Hague, The Netherlands, June 1995.

[14] G.R. Cowper and P.S. Symonds, "Strain-Hardening and Strain-Rate Effects in the Impact Loading of Cantilever Beams", Tech. Report No. 28, Brown University, Rhode Island, 1957.

Proceedings of the Seventh (1997) International Offshore and Polar Engineering Conference
Honolulu, USA, May 25-30, 1997

ISBN 1-880653-28-1 (Set); ISBN 1-880653-32-X (Vol. IV)

Impact Loading of Plates: Validation of Numerical Simulations by Testing

Magnus Langseth, Odd Sture Hopperstad and Torodd Berstad
Norwegian University of Science and Technology
Trondheim, Norway

ABSTRACT

Impact tests performed on steel and aluminium plates in the low velocity regime, i.e. v_i <50 m/, were used to validate numerical simulations performed with the computer code LS-DYNA3D. Good agreement was found for the steel plates grade St52-3N by using the quasi-static material properties and shell elements. For the aluminium plates, alloys AA5083-H112 and AA6082-T6, some difference between analysis and tests was found and more research has to be carried out before general conclusions and recommendations can be made.

LIST OF SYMBOLS

a - overhang
d - diameter of circular blunt ended projectile
F - interface force between projectile and plate
h - plate thickness
M_p - mass of projectile
R_i - isotropic hardening variable
t - time
v_i - impact velocity
w - displacement of plate
w_c - displacement of plate at the edge of projectile
X_i - kinematic hardening variable
ε - strain
$\dot{\varepsilon}$ - strain rate
σ - stress
σ_u - ultimate stress
τ_{cr} - mean shear stress at plugging
ε_p - plastic strain

INTRODUCTION

According to several design codes and regulations dropped objects shall be considered in the design of offshore structures. Based on a survey of possible dropped objects, long rods such as drill-collars are identified as the most dangerous item due to its high kinetic energy and small contact area with the structure during impact. A drill-collar has a mass of approximately 3200 kg, a length of 10 m, a diameter in the threaded pin end of 146 mm and may be dropped from heights up to 40-50 m, see Fig 1a.

The response of steel and aluminium plates subjected to a dropped drill collar is described by global deformations of the plates combined with the possibility of a local failure at the impact point. This combined mechanism makes the design very difficult and has therefore mainly been based on full scale testing. To improve the understanding of the response of steel and aluminium plates subjected to large mass projectiles in the low velocity regime (v_i < 50 m/s), a model test study has been carried out [Langseth and Larsen (1990, 1993, 1994)]. Rigid body blunt ended projectiles were fired against steel and aluminium plates in order to study the plugging capacity of such targets and to develop design recommendations. The primary variables were the target thickness, mass of the projectile relative to the mass of the target plate and the support conditions.

In order to extend the parametric range of the proposed design recommendations [Langseth and Larsen (1988, 1992), Veritec Report (1988)] and to get a better understanding of the physical mechanisms present during impact, numerical simulations can be used. However, before such analyses can be considered reliable the predictions have to be validated against test data. Therefore, the main objective with this paper is to investigate if the response of thin steel and aluminium plates subjected to large mass projectiles in the low velocity regime can be predicted by using the computer code LS-DYNA3D [Hallquist (1991)].

Even if the experimental results have been published elsewhere [Langseth and Larsen (1990, 1993, 1994)], a short summary will be given here as a basis for the comparison between the experimental results and the LS-DYNA3D calculations.

EXPERIMENTAL STUDIES

Test programme

All tests were carried out in scale 1:4 using replica modelling. As shown in Table 1, the structural model for the impacted steel plates was chosen as clamped plates with stringers (test series DM) and simply supported hatches (test series A1, B1-3, C1-2) of the type used in the wellhead area on a platform for simultaneous drilling and production. Test series A1 was designed to study the plugging capacity as afunction of the plate thickness, while series B1-3 were carried out

Table 1 - Test program and specimen geometry

Test series	Plate geometry		Projectile mass	Test specimen geometry and boundary conditions
	h (mm)	*a* (mm)	M_p *(kg)*	
A1	4,6,8,10	52	49,54	
B1	4,6,8	52	18,86	
B2	4,6,8	52	34,62	
B3	6,8	52	26,00	
C1	4,8	137	49,54	
C2	4,8	242	49,54	
DM	4,6		49,54	
E1	6,10,12,15,20	52	49.54	
E2	10,12,20,25	52	49.54	

Test series A1, B1-3, C1-2, E1-2

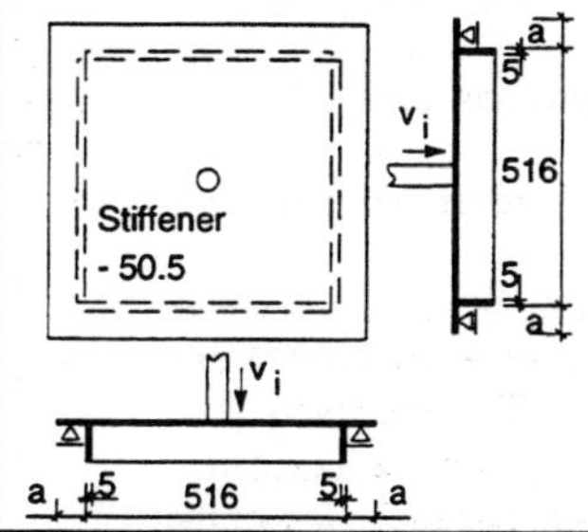

Test series DM

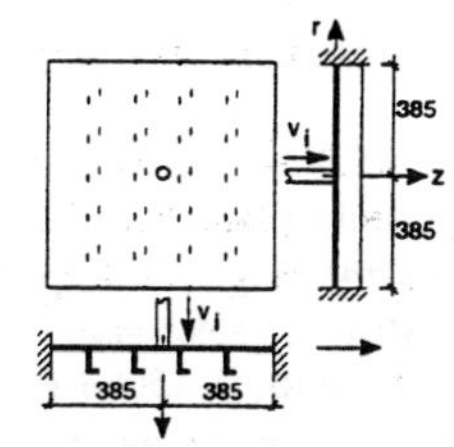

Boundary conditions in laboratory for test series A1, B1-3, C1-2, E1-2

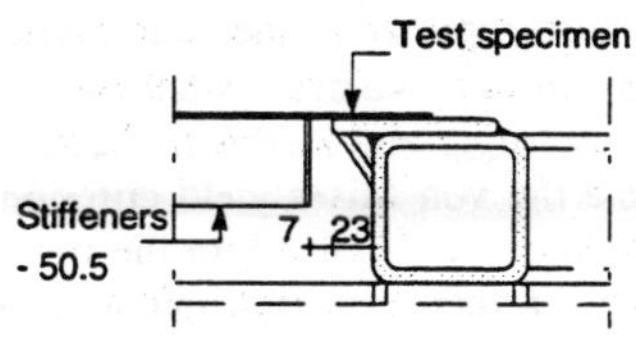

Boundary conditions in laboratory for test series DM

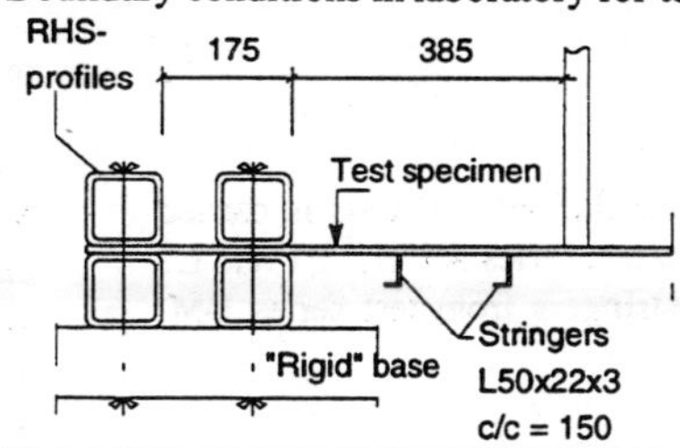

Test series A1, B1-3, C1-2 and DM: Steel grade St52-3N
Test series E1: Aluminium alloy AA5083-H112
Test series E2: Aluminium alloy AA6082-T6

to investigate the influence of the mass ratio of the projectile to target. Test series C1-2 were included to study the influence of an increased inplane stiffness, while test series DM was designed to study the effect of stringers in the free span. The increased inplane stiffness was obtained by increasing the plate overhang *a*. As structural model for the aluminium plates (test series E1-2) was chosen only simply supported hatches with the same geometry as that used for steel. Here, the primary variable was the plate thickness.

For the present tests the drill-collar was modelled as a rigid body, Fig 1, with a flat ended circular nose. Throughout the tests the diameter of the projectile was taken as $d = 36.5\ mm$, which is the scaled diameter of the threaded pin end.

The dynamic tests were performed by means of a pneumatic accelerator. In addition, a series of static tests were performed in order to study a possible relationship between the dynamic and static plugging capacities.

The material properties for the steel plates at elevated rates of strain were determined by tension tests [Langseth et al. (1991)], Fig 2. Only quasi-static tests were carried out for the aluminium plates as the material was assumed to be strain rate insensitive [Lindholm et al. (1971)], Fig 3.

The projectile was instrumented by strain gauges that were used to record the interface force between projectile and target. Based on this force-time history the displacement of the projectile and consequently the displacement at the center of the plates was generated by double integration. Furthermore, for test series, A1, B1-3 and C1-2 strain histories at different locations were measured using strain gauges, while a high speed camera shooting 10000 frames per second made it possible to determine the deformed shape of the target plate throughout the impact. The impact velocity of the projectile was measured by means of two photocells located immediately above the target.

Experimental results

The primary information gained from the experiments was the impact velocity at incipient plugging. This velocity was bracketed by two tests, one at an impact velocity where the projectile failed to remove a plug and a second at a slightly higher velocity where a plug was removed. Figs 4a and b depict typical time histories of the interface force between projectile and target for a hatch made of steel and aluminium respectively. As expected, no difference in behaviour is observed between a steel and an aluminium plate, and the impact can be divided into a transient and a global mode phase denoted *T* and *G* respectively. In the transient phase only inertia forces are present in the plate, as no forces are transmitted to the supporting structure. In the global mode phase the supports are fully activated, and the target behaves almost quasi-statically.

The most significant results from tests on the steel plates were as follows:

- The impact energy giving plugging was an increasing function of the plate thickness.
- Reducing the projectile mass from 49.54 kg to 18.86 kg, i.e. increasing the mass ratio between the target and the projectile, led to an increased impact energy at plugging. However, the critical kinetic energy seemed to approach the static plugging energy when the mass was increased.
- An increase in the inplane stiffness of the plate reduced the critical plugging energy.
- Plugging always took place at maximum interface force between plate and projectile, and the critical force for each plate thickness

was independent of the mass of the projectile and the support conditions.

- A comparison showed that the dynamic force at plugging deviated from the corresponding static one with ± 5 %.
- The energy absorption connected with the indentation of the projectile into the plate represented less than 10 % of the total absorbed energy at plugging.

From the aluminium plate tests the main conclusions were:

- Alloy AA5083-H112 was more efficient to prevent plugging in a homogeneous plate than alloy AA6082-T6.
- For the same height of drop, the weight saving of using aluminium alloy instead of steel grade St52-3N ranged between 31-38% and 12-15% for alloy AA5083-H112 and AA6082-T6 respectively.
- As for steel, the energy absorption connected with the indentation of the projectile into the plate at plugging was less than 10% of the kinetic energy at impact.

NUMERICAL SIMULATIONS

The comparison between numerical simulations and tests is carried out for the simply supported hatches from test series A1, B1, E1 and E2 and the clamped plates equipped with stringers from test series DM, Table 1.

Modelling

During an impact large elastic and plastic deformations take place which extend to the supports. Tests have shown that 90 - 95 % of the initial kinetic energy of the projectile is absorbed as strain energy in the plate and that only a small fraction is spent to push out a plug. This means that the test specimens can be modelled using shell elements and that no account needs to be taken to the local indentation of the projectile into the plate at the impact point when rebound takes place. The Belytschko-Lin-Tsay shell is the default element in LS-DYNA3D due to its computational efficiency and is chosen here. Five integration points through the thickness and one point in the plane of the element are used. The projectile is modelled as a rigid body using 8-node hexahedron elements.

Fig. 5 shows a view of the finite element model of the test specimens. As seen, the test specimens and projectile are both modelled using two symmetry planes. 144 solid hexahedron elements are used for the projectile while 1273 and 2067 shell elements are found to give sufficient accuracy for the loose hatches and clamped plates respectively. These element meshes require a time step of 1.3 μs for numerical stability. The support conditions of the test specimens from the series A1, B1, E1 and E2 (Table 1) are modelled as a stonewall that prevents inner penetration of boundary nodes while allowing for separation. Both the stringers and the plate are modelled as clamped for test series DM. Furthermore, the contact between the projectile and plate is modelled as a sliding interface without friction.

For both the steel and aluminium plates the material properties available are data obtained from tensile test specimens taken out normal and parallel to the rolling direction. For the steel plates insignificant difference is found between the two directions, while a variation of up to 10 % is found for the aluminium plates.

Due to the length of the yield plateau for steel grade St52-3N compared with the measured magnitude of the strains in the plates, the plate material is modelled as elastic-perfectly plastic using the von Mises yield criterion. Analyses with both the quasi-static material properties and the strain rate model of Johnson and Cook [Hallquist (1991)] will be shown.

The stress-strain curve of aluminium is non-linear, and a proper description of the strain hardening properties is necessary. In the present analyses of the aluminium plates a material model is used [Berstad et al. (1994), Berstad (1996)] that adopts a quadratic yield function and multi-component nonlinear isotropic and kinematic hardening rules. The yield function contains the von Mises yield criterion as well as Hill's criterion for initial anisotropic materials, whereas the combined nonlinear hardening rule makes the model suitable both for monotonic and cyclic loading conditions. In the uniaxial case, the stress-plastic strain curve is obtained as

$$\sigma = \sigma_0 + \sum_{i=1}^{n_R} R_i + \sum_{i=1}^{n_X} X_i$$

$$R_i = (1-\alpha)\, Q_i \left[1 - \exp\left(-C_i \varepsilon_p\right)\right] \qquad (1)$$

$$X_i = \alpha\, Q_i \left[1 - \exp\left(-C_i \varepsilon_p\right)\right]$$

The material constants are determined by curve fitting to uniaxial material data. In Eq. (1) σ_0 is the proportionality limit, C_i governs the rate of change in the isotropic and kinematic hardening variables, Q_i represents their asymptotic values, whereas α determines the relationship between isotropic and kinematic hardening. Based on the material data available the von Mises yield criterion with the average value of the stress-strain curves obtained for the two directions are used in the present analyses. Furthermore, isotropic hardening is chosen (i.e. $\alpha = 0$) as the numerical simulations show no difference in the plate response between analyses performed with kinematic and isotropic hardening.

Comparison and discussion

When comparing measured and computed time series, it should be noted that no attempt has been made to filter out the oscillations due to travelling stress waves in the projectile for the measured interface-force time series. As the computed values are only available at 10 μs time intervals, the computed histories are in effect filtered. Furthermore, the analyses shown in Figs 6-9 and 13 for the steel plates are carried out with the quasi-static material properties, while Fig 12 shows the effect of taking strain rate effects into account.

Test series A1 and B1: Figs. 6,7 and 8 show a comparison between the measured and calculated interface force-time curves for a representative case when a rebound occurs. Plotted is also the displacement of the plate at the edge of the projectile, and good agreement between the measured and calculated data is found when using the quasi-static material properties. However, the permanent displacement at the center of the plates is too large compared to the measured value. This means that the numerical model is too stiff during the unloading phase of the impact. High-speed camera measurements have shown that this effect may be related to a difference in deformation profiles for test and analysis during impact. Fig. 10 shows a comparison between the measured and calculated deformation profiles for test no A1-8-7. As seen, the deformation profiles are different during the loading period of the impact, Fig. 10a, as well as the permanent displacement profile of the plate, Fig. 10b.

Even if the magnitude of the deformations are different in analysis and test, the deformation patterns are similar during the deformation process. After initial impact, a deformation wave moves radially towards the support. Before this wave reaches the support both test and analysis show that the shape of the deformation pattern is approximately axisymmetric. After this time yield lines extend inwards from the corners. This is shown in Fig. 11, where contour line plots of the computed deformation pattern are given. Fig. 11a shows the deformation pattern before the waves reach the support, while Fig. 11b depicts the permanent deformation shape. Furthermore, it is worth noting that the impact event in the analyses, as for the tests, is divided into a transient phase and a global mode phase, and that the time when the supports are activated, is approximately equal for test and analysis, Figs. 6-9.

Fig. 9 shows a comparison for a case where a plugging failure occurs. As the local indentation of the projectile into the plate has insignificant influence on the global response, the measured and calculated curves are approximately equal until plugging takes place followed by a sudden drop in the measured force time curve. As plugging always takes place at maximum force, the shell analysis can be used to determine the onset of plugging provided the critical interface force between plate and projectile is known. In Ref. [Langseth and Larsen (1990)] the critical interface force for a circular blunt ended projectile is determined based on a mean critical shear stress, viz.

$$\tau_{cr} = \sigma_u(0.41\, h / d + 0.42) \qquad (2)$$

In the previous analyses, the quasi-static material properties were used. However, Fig. 12 shows a representative comparison between test and analysis when the latter is based on the Johnson and Cook plasticity model [Hallquist (1991)] and the strain rate data shown in Fig. 2. For comparison an analysis based on the quasi-static material properties ($\dot{\varepsilon} = 10^{-4}s^{-1}$) is shown as well. The inclusion of the strain rate effect increases the stiffness of the plate causing the maximum interface force to be overestimated, while the maximum displacement is underestimated compared with the test data. The increase in stiffness is evident as the response in the plate is mainly controlled by plastic membrane stresses. An increase in the yield stress will then imply an increase in the overall stiffness of the plate. The reason why the quasi-static material properties give the best fit to the test data is not understood and requires further study.

Test series DM: Fig. 13 shows a comparison between the measured and calculated interface force and displacement histories for a 4 mm thick plate equipped with stringers when a rebound occurs. Good agreement is also here found between test and analysis when using the quasi-static material properties.. However, as for the simply supported hatches, the numerical model is too stiff during the unloading period, resulting in a larger permanent displacement than for the test.

Test series E1 and E2: Figs 14 and 15 show a representative comparison between the measured and predicted force and displacement histories where rebound occurs for a 6 mm and 10 mm plate respectively. The shape of the measured and predicted curves is approximately the same, but the predicted maximum force is too large and the duration of the impact is too small compared with the measured curves. As for the steel plates, the response is dominated by membrane stresses, which means that the shape and magnitude of the input stress-strain curve to the analyses has a significant influence of the predicted interface force curves. It is a well known phenomena that aluminium can exhibit significant flow stress anisotropy even if the two directions tested do not show this behaviour (i.e. a significant reduction of the flow stress may be found in directions between those tested). This means that a better agreement between analyses and tests may be obtained by using models that are able to describe the anisotropy of the plate material with respect to the yield criterion and the flow conditions. However, even if the interface force as well as the maximum displacement are not well predicted, the numerical simulations seem to represent the permanent plastic deformations of the plates quite well.

CONCLUSIONS

Based on the numerical simulations carried out, good agreement is found between LS-DYNA3D and the test data when using the quasi-static material properties for the steel plates. The maximum interface force between projectile and plate as well as the maximum displacement at the impact point agree quite well with the test data. Hence, a shell analysis can be used for parametric studies to determine the plugging capacity of thin steel plated structures when a critical interface force between projectile and plate is used as a failure criterion. However, the permanent deformations are overestimated in the analyses. In a design situation these deformations will be of minor interest as a plated structure has to be replaced after an impact from a large mass projectile due to the large plastic deformations which take place.

The comparison carried out for the aluminium plates show that the interface force is overpredicted, which means that a mean shear failure criterion, as proposed for the steel plates, will give conservative estimates of the plate thickness. However, due to the observed deviations between test and analysis general recommendations regarding the use of shell elements and an isotropic material model cannot be given for the aluminium plates before further research has been carried out.

REFERENCES

Berstad, T. (1996). "Material Modelling of Aluminium for Crashworthiness Analysis". Dr ing Dissertation, Department of Structural Engineering, Norwegian University of Science and Technology.

Berstad, T., Hopperstad, O.S. and Langseth, M. (1994). "Elasto-Viscoplastic Constitutive Models in the Explicit Finite Element Code LS-DYNA3D". Second International LS-DYNA3D Conference, San Francisco, USA.

Hallquist, J.O. (1991).: "LS-DYNA3D, Theoretical Manual". Livermore Software Technology Corporation.

Langseth, M. and Larsen, P.K.(1988). "Dropped Objects. Plugging Capacity of Steel Plates". BOSS'88. Proceedings of the International Conference on Behaviour of Offshore Structures. Trondheim, Norway (Edited by T. Moan, N. Janbu and O. Faltinsen), Vol. 3, 1001-1014.

Langseth, M. and Larsen, P.K (1990). Dropped Objects -Plugging Capacity of Steel Plates - an Experimental Investigation". *Int J. of Impact Engng*. Vol. 3, 289-316.

Langseth, M. and Larsen, P.K. (1992).: "The Behaviour of Square Steel Plates Subjected to a Circular Blunt Ended Load". *Int. J. Impact Engng*. Vol 12, No. 4, pp 617-638.

Langseth, M. and Larsen, P.K.(1993). "Dropped Objects' Plugging Capacity of Stiffened Panels". Structural Dynamics - EURODYN'93. Moan et al. (eds). Balkema, Rotterdam, ISBN 90 5410 336 1.

Langseth, M. and Larsen, P.K.(1994). "Dropped Objects' Plugging Capacity of Aluminium Alloy Plates". *Int. J. Impact Engng* Vol. 15, No. 3 pp. 225-241.

Langseth, M. Lindholm, U.S., Larsen, P.K. and Lian , B. (1991). "Strain Rate Sensitivity of Mild Steel Grade, St52-3N". *American Society of Civil Engineers, Journal of Engineering Mechanics*. Vol. 117, 719-733.

Lindholm, U.S., Bessey, R.L. and Smith, G.V. (1971): "Effect of Strain Rate on Yield Strength, Tensile Strength and Elongation of three Aluminium Alloys" *J. Mater., JMLSA* 6(1), 119-133.

Veritec Report No. 88-3172, P12 (1988) "Design Guidance for Offshore Steel Structures Exposed to Accidental Loads".

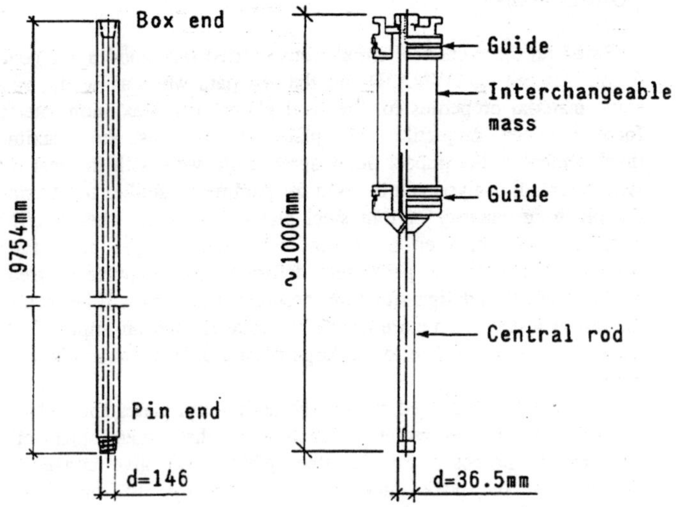

(a) Drill-collar in full scale M_p = 3171 kg

(b) Modelled drill-collar

Figure 1 Drill-collar modelling

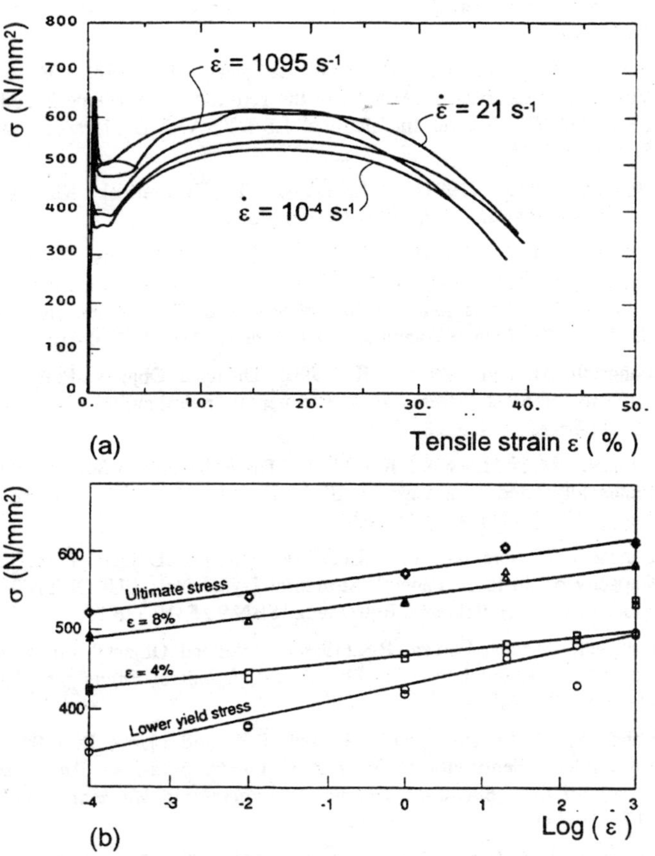

Figure 2 Material data for steel grade St52-3N
(a) Engineering stress-strain curve at elevated rates of strain
(b) Stress versus strain rate

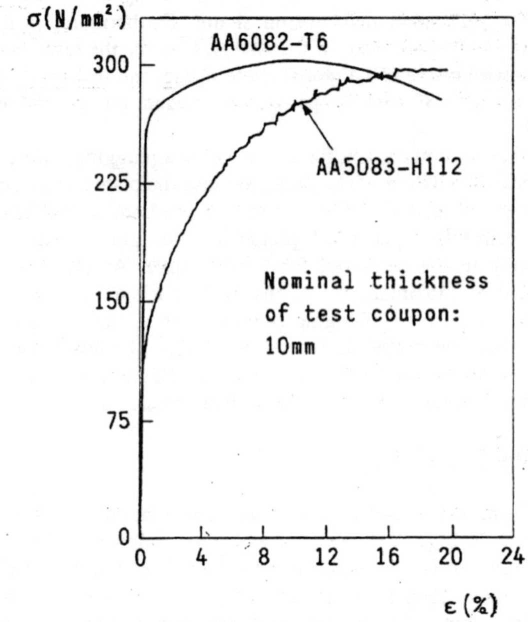

Figure 3 Typical engineering stress-strain curves

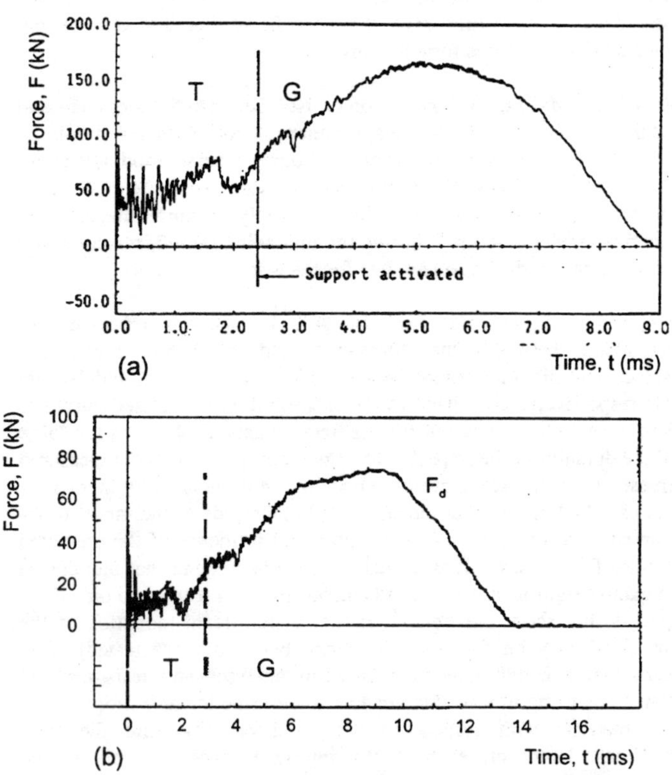

Figure 4 Force time histories
(a) Steel, h = 6.22 mm, v_i = 14.54 m/s
(b) Aluminium alloy AA5083-H112, h = 6.03 mm, v_i = 8.26 m/s

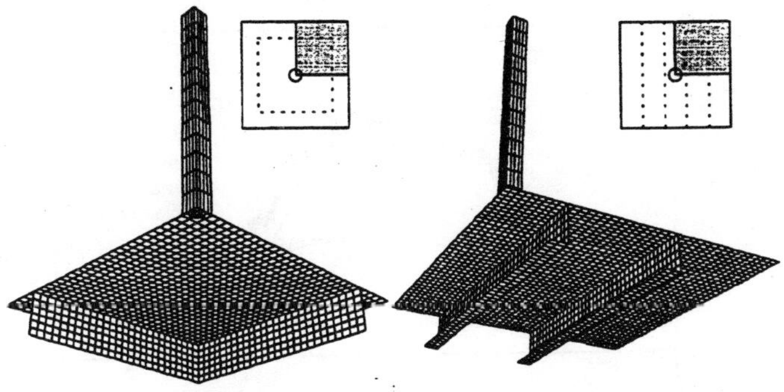

(a) (b)

Figure 5 Rear view of finite element models
(a) Hatch, test series A1, B1, E1 and E2
(b) Clamped plate equipped with stringers, test series DM

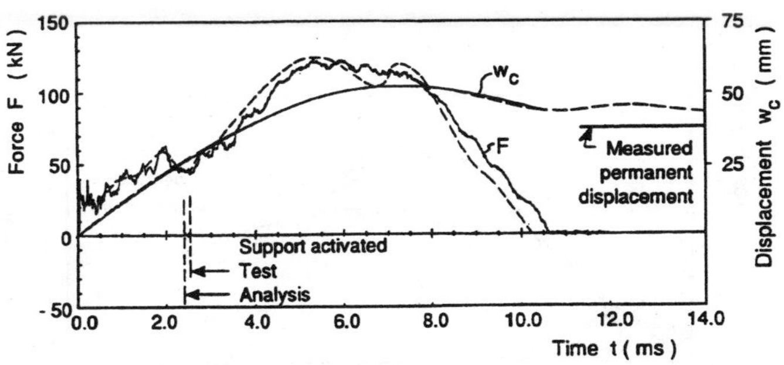

Figure 6 Force and displacement histories.
Test series A1, plate A1-4-5
v_i = 11.49 m/s, M_p = 49.54 kg, h = 4.54 mm
——— Test, -------- Analysis

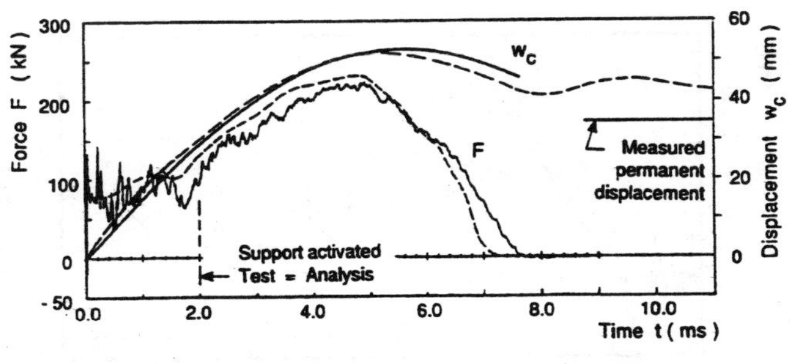

Figure 7 Force and displacement histories.
Test series A1, plate A1-8-7
v_i = 16.02 m/s, M_p = 49.54 kg, h = 7.88 mm
——— Test, -------- Analysis

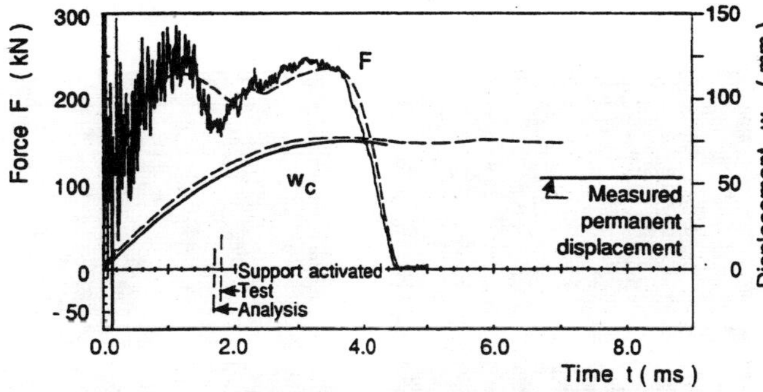

Figure 8 Force and displacement histories.
Test series B1, plate B1-8-7
v_i = 39.48 m/s, M_p = 18.86 kg, h = 7.94 mm
——— Test, -------- Analysis

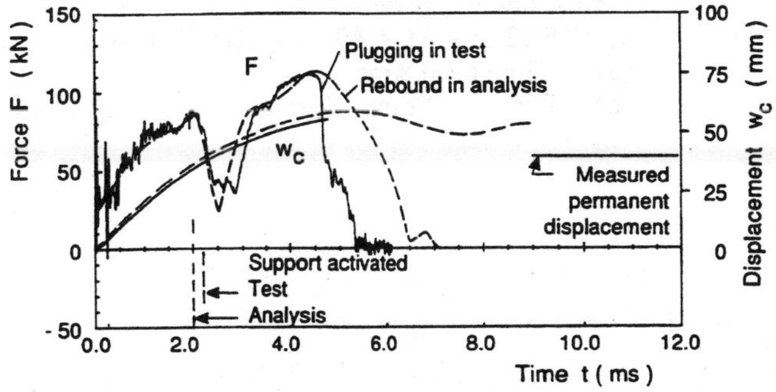

Figure 9 Force and displacement histories.
Test series B1, plate B1-4-6
v_i = 19.67 m/s, M_p = 18.86 kg, h = 4.14 mm
——— Test, -------- Analysis

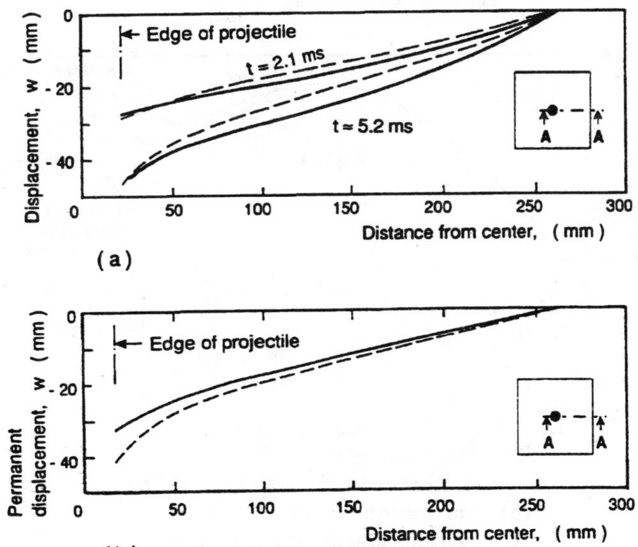

(b)

Figure 10 Deformation profiles. Test series A1, plate A1-8-7
v_i = 16.02 m/s, M_p = 49.54 kg, h = 7.88 mm
——— Test, -------- Analysis
(a) Deformation profiles at discrete times
(b) Permanent deformation profiles

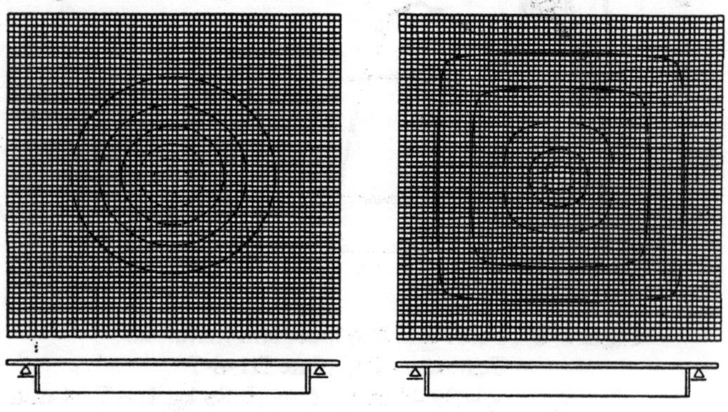

(a) (b)

Figure 11 Contour plots of deformation patterns.
Test series A1, plate A1-8-7
$v_i = 16.02$ m/s, $M_p = 49.54$ kg, h = 7.88 mm
(a) Time t = 0.8 ms
(b) Time t = 12 ms. Permanent deformation shape

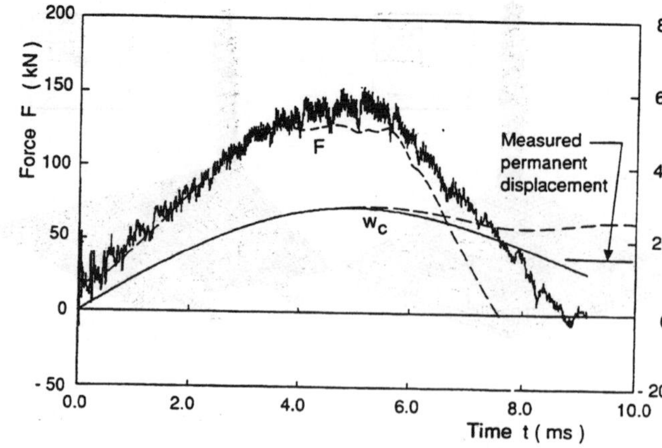

Figure 13 Force and displacement histories
Test series DM, plate DM-4-2
$v_i = 9.22$ m/s, $M_p = 49.54$ kg, h = 4.40 mm
——— Test, --------- Analysis

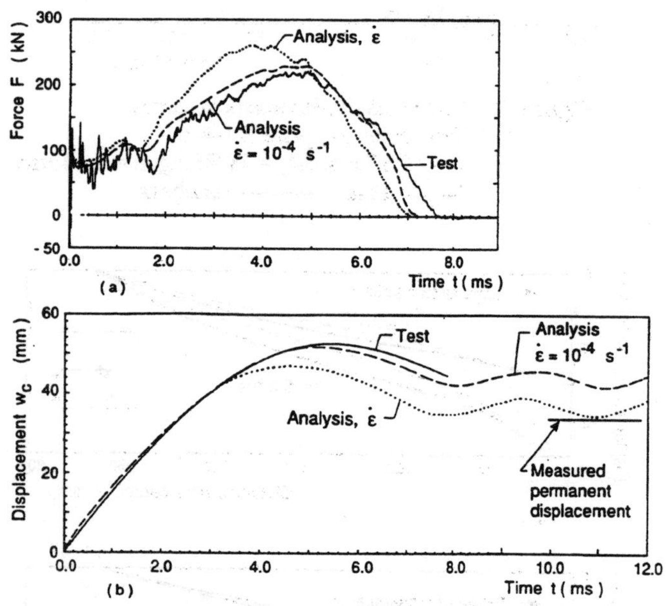

Figure 12 Force and displacement histories
Test series A1, plate A1-8-7
$v_i = 16.02$ m/s, $M_p = 49.54$ kg, h = 7.88 mm
(a) Force-time curves
(b) Displacement-time curves

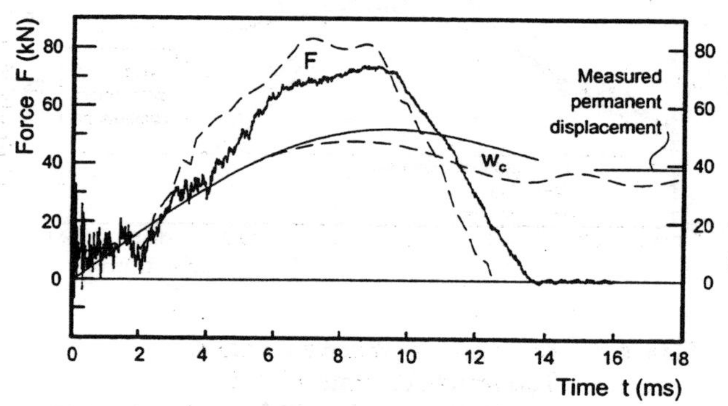

Figure 14 Force and displacement histories
Test series E1, plate I6-4
$v_i = 8.26$ m/s, $M_p = 49.54$ kg, h = 6.03 mm
——— Test, ----- Analysis

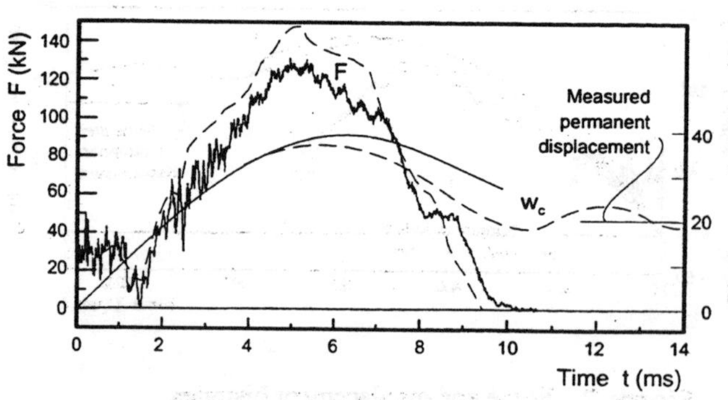

Figure 15 Force and displacement histories
Test series E2, plate B10-2
$v_i = 9.47$ m/s, $M_p = 49.54$ kg, h = 9.96 mm
——— Test, ----- Analysis

Proceedings of the Seventh (1997) International Offshore and Polar Engineering Conference
Honolulu, USA, May 25-30, 1997

ISBN 1-880653-28-1 (Set); ISBN 1-880653-32-X (Vol. IV)

Side Collision Resistance of Ship's Stiffened Panels

Sang-Rai Cho, Sang-Bock Lee and Il-Woong Kim
University of Ulsan
Ulsan, Korea

ABSTRACT

In order to investigate the collision strength of plates and stiffened panels three different approaches have been performed, namely experimental, analytical and numerical ones. Firstly, a collision testing machine of spring-roller conveyer type has been designed and fabricated. Using this machine, ninety-six unstiffened plates and twelve stiffened panels have been tested with different masses and collision velocities of the striker. Secondly, a simple analytical expression has been derived assuming that the kinetic energy of the striker can be dissipated by the formation of yield lines and membrane tensions of the impacted plate. Lastly, a simplified numerical method has been developed to predict the response of mass- impacted plates. The calculated predictions using these two theoretical methods have been substantiated with the test results. Reasonably acceptable correlations can be shown in this study.

KEYWORD: Plate, stiffened panel, collision test, simple analytical expression, simplified numerical analysis

INTRODUCTION

Recently, ship collision and grounding is more probable than before because ships are becoming larger and faster, especially those of tankers and passenger ships. These accidents cause economic loss and sea pollution and sometimes even human life loss. Generally, deformations of ship structures due to collision or grounding occur in a very short period remaining large permanent deformation and rupture. By the aid of computer, collision analysis is carried out by commercial computer codes, for instance, DYNA 3D and DYTRAN. But these codes are not yet fully substantiated with the results of experimental investigations and the economic aspect of using these commercial packages has not been justified yet.

In order to cope with the ship collision problems in more rational way, it is firstly necessary to quantify the possibility of collision occurrences. Secondly we need to predict the probable extent of damage due to the prescribed collision condition and then the residual strength of the damaged structure. Having tools to follow this collision analysis procedure the lifetime cost optimization can provide the guidance which way is more economical whether to pay more money in the initial construction stage or during the operation period for repair.

In experiment part of this study, the material properties of the models were quantified through tensile tests varying the strain rate. Collision tests were then performed on ninety-six unstiffened plates and twelve stiffened panels, changing the velocity and the mass of the striker. The collision testing machine used is of spring/roller conveyer type, where free rebound is allowed as in the real situation. The deformed shapes of the damaged models were measured afterwards.

In deriving the analytical expression to predict the damage of plates due to collision, kinetic energy of the striker is assumed to be dissipated by the formation of yield lines and membrane tensions of the impacted plate. A simplified numerical method was also developed to trace the dynamic behaviour of plate subjected to mass-impact loads. In the method in-plane deformation is treated as a plane stress problem using constant strain triangular finite elements and bending deformation is represented by torsional spring located at the edges of triangular elements. The calculate predictions using these two theoretical methods have been substantiated with the test collision results.

The test results and the analytical expression provided in this paper are expected to be useful to predict the situations when the tearings on plate occurs due to collision. The simplified numerical method may be helpful to efficiently trace the responses of plates subjected to collision loads.

COLLISION TESTS

Collision Testing Machine

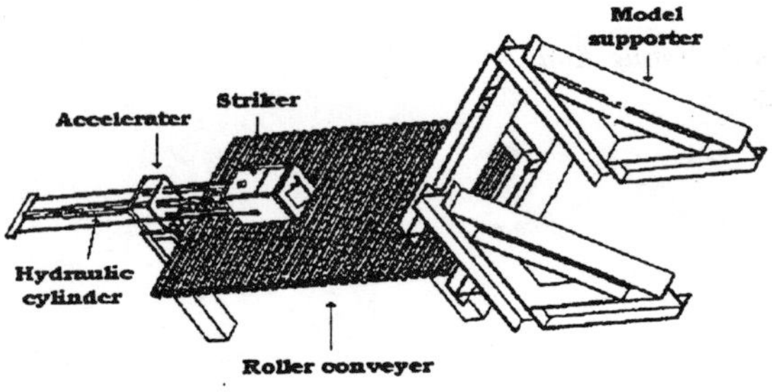

Fig. 1 Sketch of the collision testing machine

The collision testing machine is consist of a striker, an accelerator, a roller conveyer, a model supporter and two velocity measurement devices, which can be seen in Fig. 1. The shapes of striker header were of three type i. e. knife edge, rectangular and square. For the case of knife edge header the tip is blunt but having 4mm radius. The rear plate of the striker compresses the springs by the aid of a hydraulic cylinder operated by a hand pump. Then a stopper is placed to hold the striker in the position maintaining the springs compressed. The striker can be released free to run on the roller conveyer by sudden removing of the stopper. The velocity of the striker can be controlled by changing the compressed length of the springs. The mass of the striker can vary between 20 kg - 60 kg.

The accelerator consist of four steel bars each of which has a spring of 40 cm length. It is designed that the strain energy of the compressed spring is converted into the kinetic energy of the striker. The striker is positioned in front of the springs and these are compressed by a hydraulic cylinder up to the desired distance. Then a fixing pin is placed as a stopper to hold the striker in position and the striker can be released in sudden by removing the stopper. The maximum compression force of one spring is about 280 kgf.

The striker runs towards model on the roller conveyer, which minimizes the friction as low as possible. Two types of velocity measurement systems are equipped as shown in Fig. 2. One consists of a timer and two light senses. The first light sense starts the timer and the other stops it. The impact velocity of the striker is calculated dividing the distance between two senses by the measured time duration. The other velocity measurement system consists of an one-chip micro process and four light senses. Impact velocity is measured by nos. 1 and 2 senses and rebound velocity is measured by nos. 3 and 4 senses.

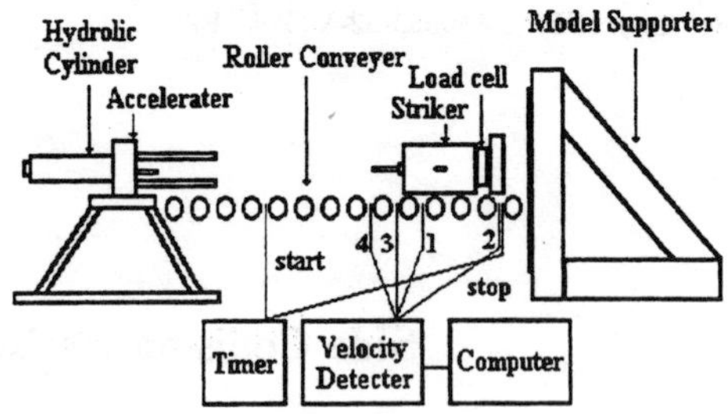

Fig. 2 Arrangement of velocity measurement system

Fabrication of the Specimens

For collision tests, a total one hundred and eight models were prepared of mild steel. Among them ninety-six were unstiffened but twelve were stiffened. The testing size of fifty-seven unstiffened models was of 500mm×300mm(the actual size was of 700mm× 500mm) and that of the other thirty-nine models was of 700×500mm(the actual size was of 900mm× 700mm). In the test the model was firmly fixed to the model supporter by bolting. The geometry and material properties of the unstiffend models are shown in Table 1. For twelve models one stiffener(flat bar or angle) was welded onto the plate in the middle lengthwise and the testing size was of 670mm×470mm(the actual size was of 900mm×700mm).

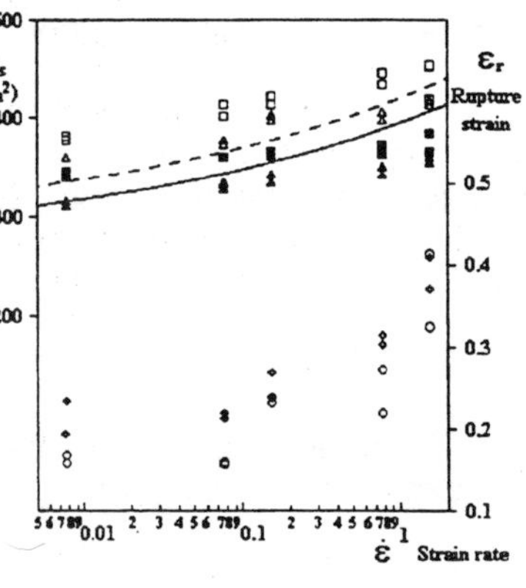

Note:

△ Tensile stress for D series($D_1 \sim D_{11}$)
□ Tensile stress for E series($E_1 \sim E_4$)
▲ Yield stress for D series($D_1 \sim D_{11}$)
■ Yield stress for E series($E_1 \sim E_4$)
○ Rupture strain for D series($D_1 \sim D_{11}$)
◇ Rupture strain for E series($E_1 \sim E_4$)
— Cowper & Symonds equation[2] for D series($D_1 \sim D_{11}$)
--- Cowper & Symonds equation[2] for E series($E_1 \sim E_4$)

Fig. 3 Effects of strain rate on the yield stress, tensile stress and rupture strain of mild steel plates

Tensile Test

For the tensile tests the specimens were machined

according to the Korean standard regulations. From the static tensile tests the yield stress and Young's modulus were obtained for A~E series. Average values of them are shown in Table 2, 3 for each series. For D and E series, dynamic tensile tests were also performed and the yield stress, the tensile stress and the rupture strain for different strain rates were obtained. The effects of strain rate on the yield stress, the tensile stress and the rupture strain of the material are illustrated in Fig. 3. As can be seen in the figure when increasing the strain rate all of the three quantities can be increased conspicuously.

Test Results

Table 1 Results of plate collision tests

Model	Width×length ×thick. (a×b×t) (mm)	Material properties		Striker			Depth of dent (mm)
		Yield Stress (N/mm^2)	Young's Modulus (KN/mm^2)	Impact velocity (m/s)	Mass (Kg)	Header type	
A_{1}	500×300×0.82	180.	210,	4.00	21.8	K(10)	16.50
$A_{2}^{\times}$	500×300×0.82	180.	210,	5.60	21.8	K(10)	24.50
$A_{3}^{\times}$	500×300×0.82	180.	210,	7.00	21.8	K(10)	29.50
A_{4}^{+}	500×300×0.82	180.	210,	8.62	21.8	K(10)	-
$A_{5}^{\times}$	500×300×0.82	180.	210,	5.60	21.8	K(10)	21.50
A_{6}^{+}	500×300×0.82	180.	210,	8.00	21.8	K(10)	-
$A_{7}^{\times}$	500×300×0.82	131.	208,	4.17	47.0	K(17)	24.70
A_{8}^{+}	500×300×0.82	131.	208,	5.66	47.0	K(17)	34.40
$A_{9}^{\times}$	500×300×0.82	131.	208,	3.88	47.0	K(17)	23.92
A_{10}	700×500×0.82	146.	205,	3.18	47.0	K(17)	22.13
$A_{11}^{\times}$	700×500×0.82	146.	205,	3.85	47.0	K(17)	29.94
$A_{12}^{\times}$	500×300×0.82	131.	208,	4.07	47.0	K(17)	25.17
A_{13}^{+}	500×300×0.82	131.	208,	5.15	47.0	K(17)	34.22
$A_{14}^{\times}$	500×300×0.82	131.	208,	3.22	47.0	K(17)	21.95
$A_{15}^{\times}$	500×300×0.82	131.	208,	3.80	49.5	R	18.68
A_{16}^{+}	500×300×0.82	131.	208,	4.78	49.5	R	24.10
A_{17}	500×300×0.82	131.	208,	3.24	54.0	R	18.44
$A_{18}^{\times}$	500×300×0.82	131.	208,	4.50	49.0	S	25.25
A_{19}^{+}	500×300×0.82	131.	208,	4.65	49.0	S	28.70
A_{20}	500×300×0.82	131.	208,	2.63	49.0	S	13.01
A_{21}	700×500×0.82	146.	205,	3.79	49.0	S	21.65
A_{22}	700×500×0.82	146.	205,	5.49	49.0	S	35.09
A_{23}	700×500×0.82	146.	205,	3.21	49.0	S	18.00
B_{1}	500×300×1.03	161.	200,	4.15	21.8	K(10)	15.50
B_{2}	500×300×1.03	161.	200,	5.33	21.8	K(10)	20.50
$B_{3}^{\times}$	500×300×1.03	161.	200,	6.59	21.8	K(10)	28.50
$B_{4}^{\times}$	500×300×1.03	161.	200,	8.62	21.8	K(10)	30.20
B_{5}	500×300×1.03	161.	200,	4.15	21.8	K(10)	16.50
B_{6}	500×300×1.03	161.	200,	5.60	21.8	K(10)	21.50
$B_{7}^{\times}$	500×300×1.03	140.	218,	5.43	47.0	K(17)	24.70
$B_{8}^{\times}$	500×300×1.03	140.	218,	3.75	47.0	K(17)	21.13
B_{9}	500×300×1.03	140.	218,	3.74	47.0	K(17)	18.20
B_{10}	700×500×1.03	136.	222,	3.84	51.5	K(17)	24.60
$B_{11}^{\times}$	700×500×1.03	136.	222,	5.25	40.0	K(17)	33.04
B_{12}	700×500×1.03	136.	222,	3.56	51.5	K(17)	20.90
$B_{13}^{\times}$	500×300×1.03	140.	218,	3.65	60.0	K(17)	23.70
$B_{14}^{\times}$	500×300×1.03	140.	218,	4.50	60.0	K(17)	27.67
B_{15}	500×300×1.03	140.	218,	2.63	60.0	K(17)	14.95
B_{16}	500×300×1.03	140.	218,	3.47	49.5	R	17.70
B_{17}	500×300×1.03	140.	218,	4.85	49.5	R	23.78
$B_{18}^{\times}$	500×300×1.03	140.	218,	3.28	49.5	R	16.65
B_{19}	500×300×1.03	140.	218,	3.57	62.0	S	16.75
B_{20}^{+}	500×300×1.03	140.	218,	4.34	62.0	S	25.85
B_{21}	500×300×1.03	140.	218,	2.61	62.0	S	14.43
B_{22}	700×500×1.03	140.	222,	3.75	49.0	S	14.43
B_{23}	700×500×1.03	136.	222,	3.91	49.0	S	24.68

(Cont'd)

Table 1 Results of plate collision tests(Cont'd)

Model	Width×length ×thick. (a×b×t) (mm)	Material properties		Striker			Depth of dent (mm)
		Yield Stress (N/mm^2)	Young's Modulus (KN/mm^2)	Impact velocity (m/s)	Mass (Kg)	Header type	
B_{24}	700×500×1.03	136.	222,	4.70	49.0	S	30.75
C_{1}	500×300×1.22	187.	202,	4.09	21.8	K(10)	11.50
C_{2}	500×300×1.22	187.	202,	5.29	21.8	K(10)	18.50
C_{3}	500×300×1.22	187.	202,	6.43	21.8	K(10)	22.50
$C_{4}^{\times}$	500×300×1.22	187.	202,	8.18	21.8	K(10)	29.50
C_{5}	500×300×1.22	187.	202,	6.43	21.8	K(10)	22.50
$C_{6}^{\times}$	500×300×1.22	187.	202,	8.18	21.8	K(10)	30.00
C_{7}	700×500×1.22	185.	209,	3.46	21.8	K(10)	12.60
C_{8}	700×500×1.22	185.	209,	5.29	21.8	K(10)	19.20
C_{9}	700×500×1.22	185.	209,	6.43	21.8	K(10)	26.20
C_{10}	700×500×1.22	185.	209,	4.20	40.5	K(10)	25.90
C_{11}	700×500×1.22	185.	209,	5.32	40.5	K(10)	34.00
C_{12}	700×500×1.22	185.	209,	5.61	40.5	K(10)	39.90
C_{13}^{+}	500×300×1.22	138.	203,	3.66	60.0	K(17)	20.57
C_{14}^{+}	500×300×1.22	138.	203,	4.42	60.0	K(17)	27.80
C_{15}	500×300×1.22	138.	203,	2.82	60.0	K(17)	16.97
C_{16}	700×500×1.22	134.	207,	3.98	51.5	K(17)	20.93
C_{17}	700×500×1.22	134.	207,	4.27	51.5	K(17)	26.60
C_{18}	700×500×1.22	134.	207,	3.52	47.0	K(17)	21.53
C_{19}	500×300×1.22	138.	203,	3.66	54.0	R	22.43
C_{20}	500×300×1.22	138.	203,	2.18	54.0	R	15.00
C_{21}	500×300×1.22	138.	203,	4.75	54.0	R	21.78
C_{22}	500×300×1.22	138.	203,	4.04	49.0	S	21.38
C_{23}	500×300×1.22	138.	203,	4.36	49.0	S	21.03
C_{24}	500×300×1.22	138.	203,	3.26	49.0	S	13.80
C_{25}	700×500×1.22	134.	207,	3.92	49.0	S	21.69
C_{26}	700×500×1.22	134.	207,	4.07	49.0	S	23.85
C_{27}	500×500×1.22	134.	207,	2.09	49.0	S	14.33
D_{1}	500×300×1.60	182.	206,	3.91	21.8	K(10)	8.30
D_{2}	500×300×1.60	182.	206,	5.63	21.8	K(10)	14.30
D_{3}	500×300×1.60	182.	206,	6.92	21.8	K(10)	18.50
D_{4}	500×300×1.60	182.	206,	8.18	21.8	K(10)	23.50
D_{5}	500×300×1.60	182.	206,	6.43	21.8	K(10)	18.80
D_{6}	500×300×1.60	182.	206,	8.11	21.8	K(10)	23.50
D_{7}	700×500×1.61	269.	214,	5.29	21.8	K(10)	10.70
D_{8}	700×500×1.61	269.	214,	6.92	21.8	K(10)	15.80
D_{9}	700×500×1.61	269.	214,	7.50	21.8	K(10)	21.50
D_{10}	700×500×1.61	269.	214,	4.39	40.5	K(10)	14.80
D_{11}	700×500×1.61	269.	214,	5.32	40.5	K(10)	22.90
E_{1}	700×500×2.02	286.	210,	4.74	21.8	K(10)	7.30
E_{2}	700×500×2.02	286.	210,	6.43	21.8	K(10)	10.50
E_{3}	700×500×2.02	286.	210,	8.18	21.8	K(10)	14.50
E_{4}	700×500×2.02	286.	210,	5.32	40.5	K(10)	15.60

Notes:
Header type:
K(17) ; knife (length 17cm)
R ; rectangular (10cm×17cm)
S ; square (10cm×10cm)
× denotes partially torn model
+ denotes completely torn model

In Table 1 the impact velocity and mass of the striker, non-dimensionalised depth of dent and energy ratio for each test are provided. The depth-of-dent was measured at the mid-point of the plate and is non-dimensionalised by the plate thickness. The kinetic energy of the striker(E_t), $(1/2)M\,V_i^2$, is non-dimensionalised by $E_p = \sigma_Y t^2\sqrt{ab}$ which may represent the capacity of energy of the plate dissipated by plastic deformation. Consequently the energy ratio, E_t/E_p, may indicate the severity of the damage due to collision. The test results of stiffened panels are summarised in Table 3, which will appear later.

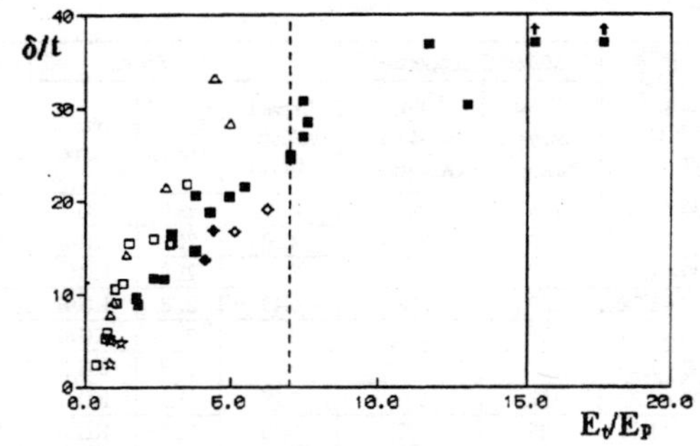

Note:

- ■ plate(500×300) , striker mass 21.8kg,
- □ plate(700×500) , striker mass 21.8kg,
- △ plate(700×500) , striker mass 40.5kg,
- ☆ stiffened plate , striker mass 40.5kg,
- ◆ box Girder[1] , striker mass 28.6kg,
- ◇ box Girder[1] , striker mass 39.9kg,

Fig.4 Non-dimensionalised test results of plate

It can be recognised in Fig. 4 that when the energy ratio is greater than 7.0 crack or partial tearing can be observed in the impacted plate while when the energy ratio is greater than 15.0 the plate can be completely torn and the striker pierces the impacted plate through.

In the figure, the symbol '◇' represents the test results of Samuelides[1] for box girder. Comparing with the test results of this study, relatively less severe damage can be seen in the results of Samuelides. It may be accused to the elastic behaviour of box girder which provides the boundary to the plate.

Another somewhat surprising finding from the results is that the more massive the striker, the larger the depth of dent of the impacted plate even though the energy ratio is the same. This may imply that the kinetic energy of the striker is not the most suitable physical quantity for collision problem.

In the static test of stiffened panel subjected to lateral concentrated loads at the mid-length, the stiffened panel shows bending only. However, in the stiffened panel collision test, where the striker impinges at the mid-length the stiffened panel undergoes not only the bending deformation but the tripping of the stiffener.

SIMPLE ANALYTICAL EXPRESSION

In order to derive a simple expression to predict thedepth of dent of the plate due to collision it is assumed that kinetic energy of the striker can be dissipated by the formation of yield lines and membrane tensions of the impacted plate. The deformed shape of the plate was then assumed as that shown in Fig. 5. The total plastic strain energy of the impacted plate may be written in the following form.

$$U_{total} = \int M_p \, \theta \, dL + \int \varepsilon \, \sigma_{Yd} \, dV \qquad (1)$$

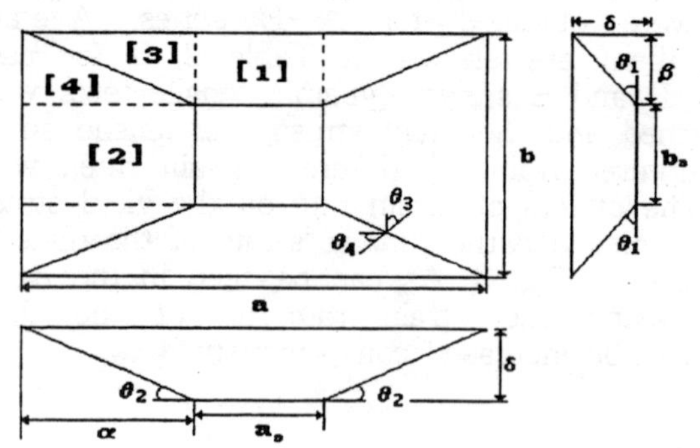

Fig. 5 Assumed deformed shape of a damaged plate after collision

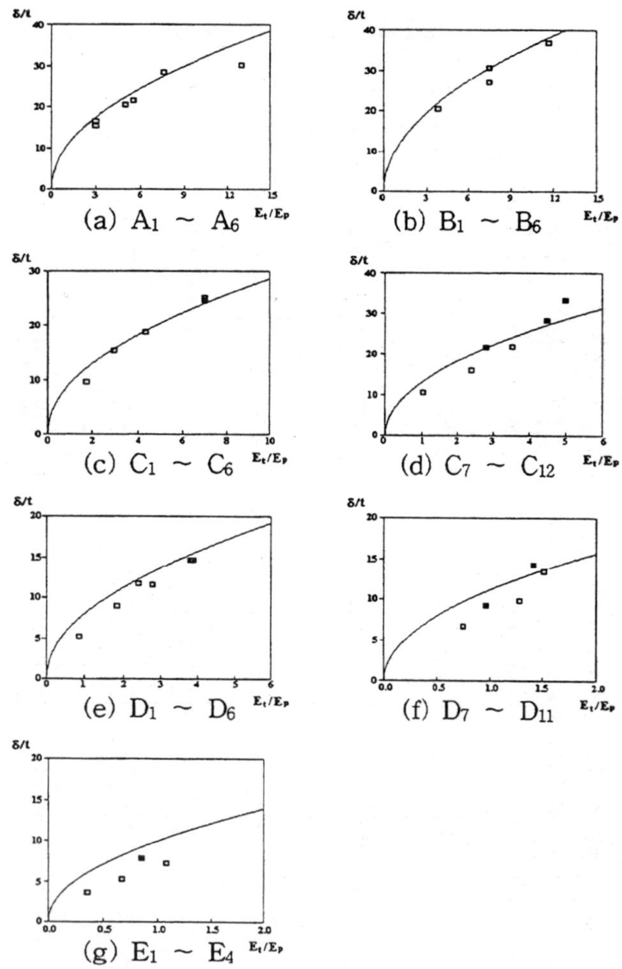

Notes:

analytical (mass 21.8kg) ———

experimental mass 21.8kg : □, mass 40.5kg : ■

Fig. 6 Comparison of analytical calculation results with test data where M_p : fully plastic moment

σ_{Ya} : dynamic yield stress

Energy Dissipated by Plastic Moment at Yield Lines

The energy dissipated by plastic moment at yield lines is expressed by the angle at yield lines, length of the lines and plastic bending moment.

$$U_p = M_p \{2\theta_1(a+a_s) + 2\theta_2(b+b_s) + 4l_\delta(\theta_3+\theta_4)\} \quad (2)$$

or

$$U_p = \frac{t^2\sigma_{Yd}}{2}\Big\{\tan^{-1}(\frac{\delta}{\beta})(a+a_s) + \tan^{-1}(\frac{\delta}{\alpha})(b+b_s)$$
$$2l_\delta(\tan^{-1}\frac{\beta\delta}{\sqrt{(\alpha\beta)^2+(\delta^2+\alpha^2)^2}} \quad (3)$$
$$+\tan^{-1}\frac{\alpha\delta}{\sqrt{(\alpha\beta)^2+(\delta^2+\beta^2)^2}})\Big\}$$

Energy Dissipated by Plastic Membrane Tensions

The energy dissipated by plastic membrane tension of the plate can be divided into those of region(1), (2), (3) and (4), as shown Fig. 5. The plastic strain energy of those three regions may be written in the following form.

$$U_1 = \sigma_{dY}\, t\, a_s(\sqrt{\beta^2+\delta^2}-\beta)$$
$$U_2 = \sigma_{dY}\, t\, b_s(\sqrt{\alpha^2+\delta^2}-\alpha)$$
$$U_3 = \frac{1}{2}\sigma_{dY}\, t\, \alpha\, (\sqrt{\beta^2+\delta^2}-\beta)$$
$$U_4 = \frac{1}{2}\sigma_{dY}\, t\, \beta\, (\sqrt{\alpha^2+\delta^2}-\alpha)$$

Accordingly the total dissipated energy by membrane tension can be written as the following equation,

$$U_s = 2U_1+2U_2+4U_3+4U_4 \quad (4)$$

where, dynamic yield stress, σ_{Ya}, can be obtained from the equation proposed by Cowper & Symonds[2]

$$\frac{\sigma_{Yd}}{\sigma_Y} = 1+(\frac{\dot{\varepsilon}}{Q})^{\frac{1}{p}} \quad (5)$$

where Q is 40 /sec and p is 5 for mild steel.

For the case of stiffened panels the energy dissipated at three hinges and the tensile elongation of the stiffener including the effective plating flange were considered using eqn(6).

$$4\theta' M_P + \int \varepsilon\sigma_y dv = \frac{1}{2}MV_i^2 \quad (6)$$

where M : mass of the striker

V_i : impact velocity of the striker

SIMPLIFIED NUMERICAL ANALYSIS

In order to rigorously trace the behaviour of collided plates a nonlinear elasto-plastic dynamic analysis should be employed. Futhermore if the yield stress is sensitive to strain rate, an elasto-visco plastic analysis needs to be performed. Generally finite shell elements, where each node has six degrees of freedom, are being used for static and dynamic analysis of plated structures.

The occurrence of yielding needs to be monitored throughout the thickness in elasto-plastic regime. The computing time, therefore, is enormous when performing the ordinary finite element analysis using shell elements in elasto-visco elastic regime.

In this paper a simplified numerical method has been developed adopting the concept proposed by Chan and Davis[3], where the plate bending deformation and the in- plane membrane action are decoupled. Hence only three degrees of freedom at each node are required to define the both deformation. Only a brief description of the analysis procedure is provided herein, but details can be found elsewhere[4,5].

Idealisation

The plate is divided into meshes of flat triangular elements. The in-plane deformation is represented by finite triangular elements of constant strain. While the out-of-plane deformation, i. e. bending, is assumed to be concentrated at the edges of the triangles. The rotations between the triangles about the edges are restrained by non-linear torsional springs representing the actual stiffness of the plate.

Only three translational degrees of freedom are contained in the analysis. The component of the resultant displacement at each node parallel to the element plane causes the in-plane internal forces and the bending moment can be obtained from the component normal to the element plane. The out-of-plane internal forces are, then, derived to balance the bending moment.

Formulation

Response of the impacted plate subject to dynamic load can be analysed by the step-by-step integration method. Using a central difference technique accelerations and velocities can be expressed in terms of the present displacements and the calculation time step. The displacements for the next time step can be obtained by the integration of the corresponding velocity.

(1) Equation of motion

For the case where the external actions are not existing the equation of motion may be written as follows :

$$[M]\{\ddot{D}\}_n + [C]\{\dot{D}\}_n + \{R^{int}\}_n = 0 \quad (7)$$

where

$[M]$: mass matrix

$[C]$: damping matrix

$\{R^{int}\}_n$: nodal internal force vector
$\{\ddot{D}\}_n$, $\{\dot{D}\}_n$: acceleration and velocity vectors respectively

As mentioned earlier a temporal finite difference technique and a spacial finite element method are adopted in this study. Using the finite difference method and providing that the displacements at (n-1)th and n th time steps are known, the displacements at (n+1) th time step can be easily obtained if the nodal internal force vector at 'n' th times, $\{R^{int}\}_n$ is known. Thus the only remaining task is to calculate the nodal internal forces.

(2) In-plane internal force
Constant strain triangular elements are adopted in the analysis to consider the membrane deformation. Therefore it does not take a long calculation procedure to obtain the membrane strains if the nodal displacements are provided. The next step is to calculate the stresses using the corresponding strains. In the elasto visco-plastic problems, the relation between stress and strain is nonlinear. Thus it is necessary to express these quantities in differential forms.

The differentials of strains can be obtained from the differences of displacements at each node. In this study, the relation between the differentials of the stress and those of the strain is obtained by the Prandtl-Reuss's flow rule. The yield condition is monitored by the von Mises criterion. The in-plane internal forces can, then, be obtained applying the virtual work principle[4].

(3) Out-of-plane internal force
In deriving the expression for the out-of-plane internal force the first thing to perform is to describe the folding angle between two adjacent elements in terms of the nodal translational displacements. Then the curvature can be approximated from the folding angle. Assuming that the plane remains plane after deformation the moment-curvature relation of the section can easily formulated[5]. From this relation the corresponding bending moment can be obtained. In order to satisfy the moment equilibrium conditions for two adjacent elements normal forces are required at the nodes off the common edge, which are the out-of plane internal forces.

VERIFICATION OF THEORETICAL METHODS

Simple analytical expression

In Table 2 and Fig. 6 the comparison with test results are provided of the prediction using the analytical expression derived in this study. When the energy ratio is large quite good agreement can be seen. However when the energy ratio is relatively small equation (1) can overestimate the depth of dent somehow. This is probably due to the assumption that all the kinetic energy of the striker can be dissipated by the plastic deformation of the impacted plate. But when the energy ratio is small the elastic deformation can not be neglected.

Table 2 Comparison of plate collision test results with theoretical predictions

Model	E_t/E_P	Non-dimemsionalised depth of dent (δ/t)		
		Exp.	Anal.	Num.
A_1	3.81	20.63	22.00	19.0
A_2	7.46	30.63	30.60	26.6
A_3	11.66	36.88	38.10	33.3
A_4	17.68	-	46.50	40.6
A_5	7.46	26.88	30.60	26.6
A_6	15.23	-	43.60	38.0
A_7	12.07	30.12	42.20	-
A_8	22.30	41.95	-	-
A_9	10.46	29.17	38.70	-
A_{10}	4.08	26.99	22.55	-
A_{11}	5.95	36.51	27.20	-
A_{12}	11.53	30.70	41.00	-
A_{13}	18.51	41.73	-	-
A_{14}	7.24	26.77	32.30	-
A_{15}	11.01	22.78	21.40	-
A_{16}	16.07	29.39	25.90	-
A_{17}	8.06	22.40	18.30	-
A_{18}	14.10	30.79	42.10	-
A_{19}	15.06	35.00	-	-
A_{20}	4.82	15.87	24.90	-
A_{21}	6.23	26.40	28.30	-
A_{22}	13.06	42.79	41.60	-
A_{23}	4.48	21.95	24.00	-
B_1	3.01	15.50	17.50	15.5
B_2	4.97	20.50	22.30	20.0
B_3	7.59	28.50	27.50	25.0
B_4	12.99	30.20	35.80	31.8
B_5	3.01	16.50	17.50	15.5
B_6	5.48	21.50	23.40	21.3
B_7	12.02	23.98	27.60	-
B_8	5.73	20.51	18.80	-
B_9	5.68	17.67	18.65	-
B_{10}	4.55	23.88	26.50	-
B_{11}	6.60	30.19	31.85	-
B_{12}	3.91	20.29	24.50	-
B_{13}	7.00	23.10	20.85	-
B_{14}	10.64	26.86	25.85	-
B_{15}	3.63	14.51	14.85	-
B_{16}	5.23	17.18	27.40	-
B_{17}	10.22	23.09	38.20	-
B_{18}	4.70	16.17	26.00	-
B_{19}	6.96	24.51	22.95	-
B_{20}	10.27	27.86	27.90	-
B_{21}	3.71	14.01	16.70	-
B_{22}	3.94	21.05	25.00	-
B_{23}	4.29	23.96	26.10	-
B_{24}	6.20	29.85	31.40	-
C_1	1.75	9.58	12.20	11.2
C_2	2.93	15.41	15.70	14.6
C_3	4.32	18.76	18.90	17.8
C_4	7.00	24.57	24.00	22.5
C_5	4.32	18.76	18.90	17.8
C_6	7.00	25.00	24.00	22.5
C_7	1.02	10.50	13.1	11.2
C_8	2.39	16.00	19.9	17.2
C_9	3.53	21.83	24.1	20.9
C_{10}	2.80	21.58	21.6	25.1
C_{11}	4.49	28.33	27.1	31.8
C_{12}	4.99	33.25	28.6	33.6
C_{13}	5.13	16.83	24.10	-
C_{14}	7.48	22.79	29.15	-
C_{15}	3.04	13.91	18.50	-

(Cont'd)

Table 2 Comparison of plate collision test results with theoretical predictions(Cont'd)

Model	E_t/E_P	Non-dimemsionalised depth of dent (δ/t)		
		Exp.	Anal.	Num.
C_{16}	3.16	17.16	25.60	-
C_{17}	3.63	21.80	27.50	-
C_{18}	2.25	17.65	20.00	-
C_{19}	4.62	18.39	18.20	-
C_{20}	1.64	12.30	10.60	-
C_{21}	7.73	17.85	23.70	-
C_{22}	4.98	17.52	20.90	-
C_{23}	5.82	17.24	22.60	-
C_{24}	3.24	11.31	16.80	-
C_{25}	3.17	17.78	24.20	-
C_{26}	3.41	19.55	25.15	-
C_{27}	0.90	11.75	12.70	-
D_1	0.88	5.19	7.5	6.8
D_2	1.83	8.94	10.7	9.9
D_3	2.77	11.56	13.1	12.2
D_4	3.87	14.69	15.4	14.2
D_5	2.39	11.75	12.2	11.3
D_6	3.81	14.69	15.3	14.3
D_7	0.75	6.69	9.7	7.8
D_8	1.28	9.88	12.6	10.2
D_9	1.51	13.44	13.7	11.0
D_{10}	0.96	9.25	11.0	11.9
D_{11}	1.41	14.31	13.2	14.4
E_1	0.36	3.65	6.0	4.1
E_2	0.67	5.25	8.2	5.6
E_3	1.08	7.25	10.3	7.1
E_4	0.85	7.80	9.2	8.5

Table 3 Results of stiffened panel collision tests results with theoretical predictions

Model	Geometry		Striker		Depth of dent (mm)	
	Plate(mm)	Stiffener(mm)	Impact velocity (m/s)	Mass (Kg)	Exp.	Anal.
AS_1	670×470×1.2	40×1.5	3.91	42	14.54	36.35
AS_2	670×470×1.2	40×1.5	2.57	51	14.14	30.60
AS_3	670×470×1.2	50×1.5	4.09	42	11.97	33.70
AS_4	670×470×1.5	40×1.5	5.63	42	14.23	53.35
AS_5	670×470×1.5	40×1.5	4.50	51	12.32	48.15
AS_6	670×470×1.5	50×1.5	6.00	42	10.11	57.75
BS_1	670×470×1.2	30×1.5+20×1.5	3.91	42	15.04	31.70
BS_2	670×470×1.2	30×1.5+20×1.5	2.65	51	13.12	22.45
BS_3	670×470×1.2	40×1.5+20×1.5	3.91	42	13.31	27.50
BS_4	670×470×1.5	30×1.5+20×1.5	5.00	42	10.15	26.80
BS_5	670×470×1.5	30×1.5+20×1.5	4.74	51	12.21	36.00
BS_6	670×470×1.5	40×1.5+20×1.5	5.29	42	13.92	32.80

Notes:

type of striker header ; knife edge (17cm)

type of stiffener for AS series ; flat bar

type of stiffener for BS series ; angle

In Table 3, the results of the collision tests on stiffened panels are provided together with the analytical predictions obtained assuming that the collision energy can be dissipated at three hinges and tensile elongation of the stiffener. However, the analytical predictions overestimate the depth of dent of the impacted stiffened panel. It seems that if the energy dissipation in the plate is considered the predictions can be better.

Simplified numerical analysis

A computer program has been written following the procedure described herein to trace the behaviour of plates subjected to mass impacts. Using this program a rigorous computation has been performed and followings are the findings during the computation The calculation time interval hardly effects accuracy of the results, but if the interval is longer than the certain value the calculation may diverge.

Five mesh types, namely a, b, c, d and e type, has been investigated, whose element sizes are $625mm^2$, $187.5mm^2$, $156.25\ mm^2$, $100mm^2$ and $39.0625mm^2$ respectively for the case of the plate of 700mm x 500mm . The meshed model of 'd' type can be seen in Fig. 7. The effect of mesh type on the response of the plate is not significant.

One of the most influential parameter on the response is the damping coefficient, which sensitively affect the maximum displacement, the occurrence of separation and the shape of displacement history. Therefore, the adequate of prediction is strongly depend upon the choice of the damping coefficient. In this study, the most suitable value of the damping coefficient is determined for each test. Using this results a regression has been performed and an equation has been derived to approximate the damping coefficient as follows :

$$c = 3.2\left(\frac{\sigma_Y}{\rho V_i^2}\right)^{0.2}\left(\frac{a}{d}\right)^{1.3}\left(\frac{t}{\sqrt{ab}}\right)^{1.0}\left(\frac{\sigma_Y}{E}\right)^{0.02} \quad (8)$$

where d: width of the smallest mesh

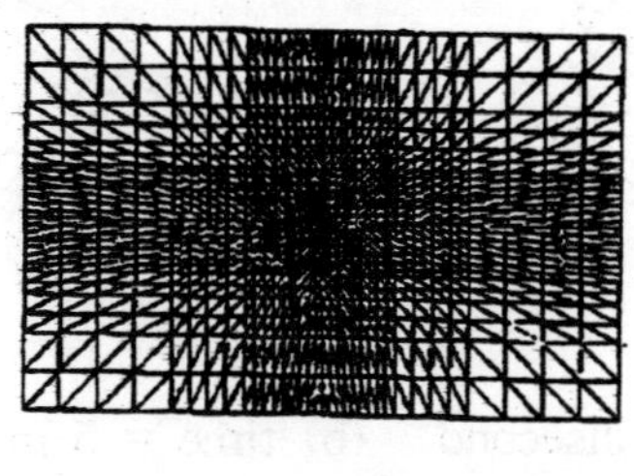

700mm×500mm Model

Fig. 7 Example of meshed model of 'd' type

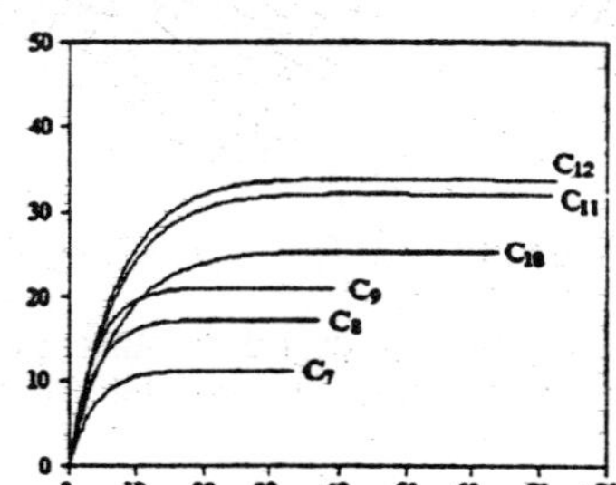

Fig. 8 Deflection history at the mid-point of the plates

The deflection history at the middle point of Model C_2 is demonstrated in Fig. 9. As can be seen in the figure the shape of denting is very sharp. Comparing with the deformed shape of the tested plates, the shape predicted by the numerical method is much sharper and very localized. It seems that some modification is necessary in the estimation of the bending rigidity of the plate.

CONCLUSIONS

In this study, a handy collision testing machine has been designed and fabricated. Using this machine one hundred and eighty collision tests have been successfully performed on unstiffened steel plate and stiffened panel. A simple analytical expression and a simplified numerical method have been developed to predict the depth of dent of plates subjected to mass-impacts. The correlation study of the theoretical methods with test data showed reasonably good agreements. However, in order to improve the accuracy of the prediction some modification is necessary especial on the bending rigidity estimation.

Followings are the findings from the experimental works of this study:

- For the case of mild steel not only the yield stress but the tensile stress and rupture strain can be conspicously increased for higher strain rate.
- Even for the same amount of collision energy the more massive the striker, the larger the dent of depth of the impacted plate. This may imply that the kinetic energy of the striker may not be the most suitable design parameter for collision problem.

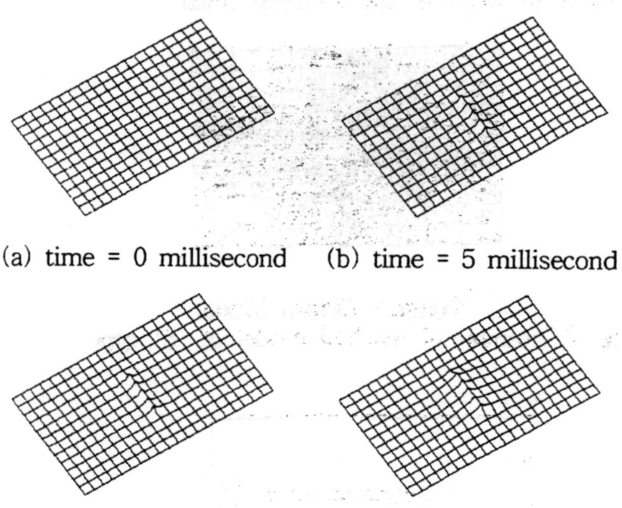

Fig. 9 Deformation history of collided plate predicted by the numerical method : model C_2

REFERENCES

[1] Samuelides, E. 'Structural Dynamic and Rigid Body Response Coupling in Ship Collision', PhD Thesis, Glasgow Univ., 1984.

[2] Cowper, G.R. and Symonds, P.S. "Strain-Hardening and Strain-Rate Effects in the Impact Loading of Cantilever Beams", Rep. No. 28 Brown Univ. 1957.

[3] Chan, A.S.L. and Davis, G.A.O., "A Simplified Finite Element Model for the Impact of Thin Shells", in 「Structures under Shock and Impacts」, Proceeding of the 1st International Conference on Structures Under Shotck and Impact, Ed. Bulson, P.S., Elseiver, Amserdam, pp361-380, 1989.

[4] Park, J.W. "Development of a simplified technique to analyse plated structures under impact loadings" Msc Thesis university of Ulsan 1993.

[5] Cho, S.-R., Lee, S.-B., and Kim, I.-W., "Experimental and theoretical investigates on the collision strength of plates", proc. of the Tenth Asian Technical Exchange and Advisory Meeting on Marine Structures(TEAM 96)', Pusan, pp 287~305 1996.

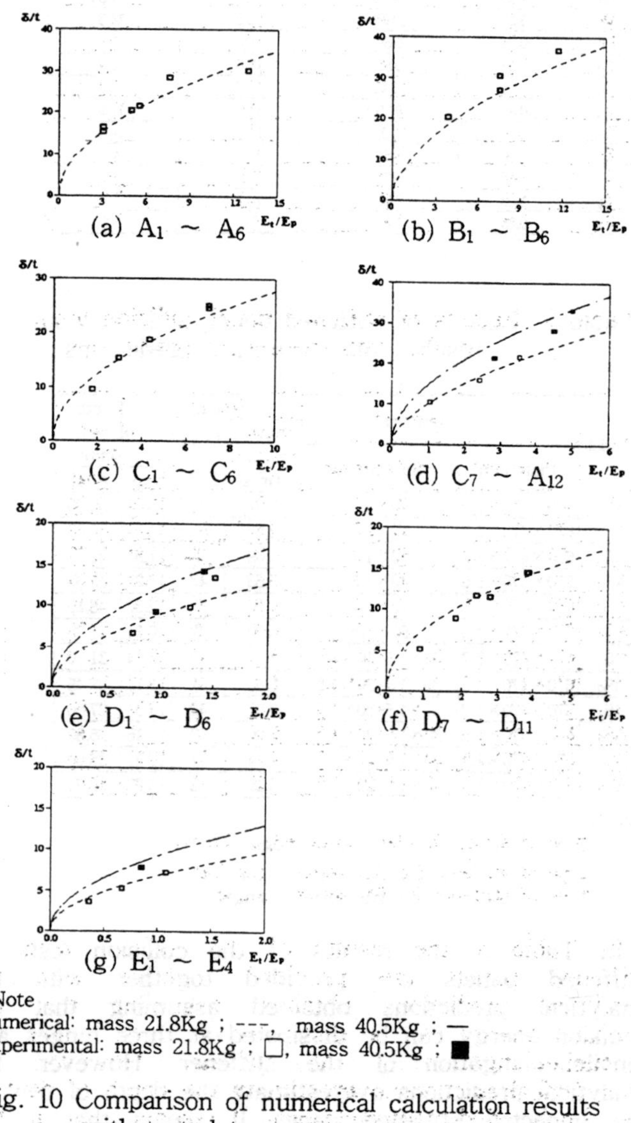

Fig. 10 Comparison of numerical calculation results with test data

Proceedings of the Seventh (1997) International Offshore and Polar Engineering Conference
Honolulu, USA, May 25-30, 1997

ISBN 1-880653-28-1 (Set); ISBN 1-880653-32-X (Vol. IV)

Integrity Monitoring of Offshore Platforms Using Artificial Neural Nets

Pradipta Banerji and Tushar K. Datta
Indian Institute of Technology
Bombay, India

ABSTRACT

This paper explores the possibility of using Artificial Neural Nets (ANNs) for monitoring the integrity of offshore structures. The methodology consists of comparing measured rms displacement responses at different elevations of an offshore structure, for a given sea state, with a datum set, obtained using a number of trained feedforward ANNs. Each ANN, given the significant wave height and average wave time period that characterise a sea state, provides structural response estimates for a pre-defined damage state. These estimated responses are suitably scaled to generate the datum set. If the measured response pattern does not match the undamaged state ANN response pattern, the structure is damaged. The region of damage is given by the damage state whose ANN response pattern matches the measured response pattern. In this paper, analytically determined rms displacements from a simulation study are used, instead of measured displacements from the platform structure, to train the ANNs.

KEY WORDS: ANN, monitoring, maintenance, integrity, damage, jacket platform, offshore

INTRODUCTION

Collisions with vessels and material fatigue among various reasons, may cause cracking or severance of structural members in offshore jacket platforms, which compromise the integrity of these structures. The standard approach for integrity monitoring is a regular inspection of all members in an offshore structure. Operational and environmental factors, however, make on-site inspection of underwater members extremely difficult for current platforms in deep waters. The main thrust, therefore, of research in this area has been in the development of other approaches for integrity monitoring. Most of these approaches evaluate structural integrity of offshore platforms by monitoring their vibrations.

The basis for one approach is the assumption that natural vibration frequencies of an offshore platform shift measurably due to member damage. Vandiver (1975) used this approach to confirm the lack of member failure in a Coast Guard tower after a boat collision. This concept is also used in a monitoring system for an existing platform, where changes in the first two platform frequencies are monitored (Chiostini and Vignoli, 1994) to check structural integrity. A comprehensive study reported by Duggan et al (1980), however, has shown that changes in natural frequencies of typical jacket platforms due to removal of bracing members, representing structural damage, could not be distinguished from shifts caused by normal operational changes. Structural frequency monitoring, therefore, has limited application in integrity monitoring of offshore platforms.

Another approach is broadly based on monitoring the vibration response of structures to ambient excitations and using system identification procedures (Agbabian et al, 1991) to determine damage, if any. Kenly and Dodds (1980) conducted detailed experiments with an actual decommissioned platform to establish that the vibration signature of an offshore platform is measurably affected by damage in the structure and can be used for integrity monitoring. Shahrivar and Bouwkamp (1984) showed that vibration responses from nodes above water only could be used to monitor changes in the lower mode characteristics of offshore platforms to detect damage. Actual monitoring methods using the system identification approach are currently of two types: modal analysis methods, and flexibility monitoring methods. The changes in modal characteristics of a structure — natural frequencies, mode shapes, and modal damping (the last two are sensitive to even low levels of damage) — are monitored (Hearn and Testa, 1991; Chen et al, 1995) to indicate damage in the modal analysis methods. The changes in flexibility ratios at different levels of an offshore platform are monitored (Rubin and Coppolino, 1983; Shyamsunder and Ting, 1985) for damage indication in the flexibility monitoring method. Both types of methods can also be used to determine the approximate location of the damage.

Since integrity monitoring approaches depend on system identification, artificial neural networks (ANNs) can be used for damage detection. This is because neural networks are signal processing devices that can be used to represent transfer functions between input and output processes in a non-functional form, approximating the neural processes in a human brain. They can, therefore, be trained to solve problems such as the system identification process without resorting to approximate functional representations of the structure.

Wu et al (1992) applied this approach to damage detection in a model structure subjected to earthquake ground motion. They were partially successful in using a multi-layered perceptron feedforward ANN to predict damage. The input was taken as the Fourier spectra of acceleration time histories measured at the top of the structure only, and the output was the level of damage in each member of the structure. This study established the efficacy of using ANNs in integrity monitoring of structures. Later work by Elkarody et al (1993) and Barai and Pandey (1995) reinforced the applicability of ANN in damage detection and integrity monitoring of skeletal structures, such as offshore jacket platforms. One of the main problems in using ANNs for such a purpose is the requirement of a large number of sample cases simulating all possible damage states to train the ANNs to monitor damage. Elkarody et al (1993) have, however, shown that this training can be done very well using numerically simulated sample cases for a particular structure.

In this paper an attempt is made to illustrate the usefulness of ANNs for integrity monitoring of offshore jacket platforms. The root mean square (rms) displacement responses at different levels of a particular jacket platform to different sea states are compared with those predicted by a set of trained ANNs. In order to estimate the damage state of the platform, the ANNs are trained for a number of input-output pairs; inputs are the significant wave height and average wave time period, while the outputs are the rms displacements at different elevation levels of the platform. The platform structural response to each random sea state is determined, for training the neural nets, using a numerical simulation procedure. The application of this concept to an actual offshore jacket platform requires the concurrent measurement of the time histories of the sea-surface elevation and displacements at different platform levels over a period of a few minutes at certain intervals during a day.

BRIEF BACKGROUND OF FEEDFORWARD ANN

Multi-layered neural networks are an efficient way to map transfer functions between an input to a system, for example, the wave forces on an offshore structure, and the output of the system, which in the above case would be the response of the structure (Haykin, 1994). Multi-layered neural networks usually consist of many neuron units, where the units are arranged into several layers, as shown in Fig. 1. The input layer receives the input parameters from the outside, and sends its output to the first hidden layer. The output layer receives the signal from the last hidden layer, and produces the final output. The hidden layers are important in a neural network, because they allow the transfer function mapping to be quite complex.

Each neuron unit has an activation function, which represents the output value o_j of that neuron j, that is given by the sigmoid function:

$$o_j = \frac{1}{1+e^{-tl_j}} \qquad \text{Eq. 1}$$

where tl_j denotes the site function for that neuron, which is given by

$$tl_j = \sum_i w_{ij} o_i + bias_j \qquad \text{Eq. 2}$$

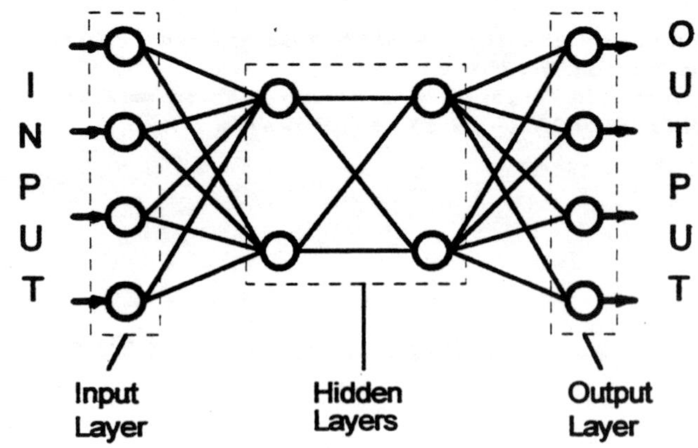

Fig. 1. Multi-layered perceptron model of a feedforward ANN

in which o_i denotes the input to the neuron from the neuron i of the previous layer, w_{ij} represents the weight of the connection between the neurons of the two layers, and $bias_j$ denotes a bias for the unit, which is a threshold in the site function. The output value for an input layer neuron is directly an appropriately normalized input value, and the output value for an output layer neuron is directly the activation function for that neuron.

For a neural network to be able to represent the transfer function for a particular system, it has to be trained first with many sample input-output pairs, so that it could represent the relationship between the input and output with a reasonable degree of fidelity. The training algorithm used for feedforward neural nets is the back-propagation rule. It is basically an error minimization procedure that consists of the following steps:

- Construct the network, initialize weights and threshold values with small random values, and choose an error tolerance value.
- Randomly present training sets of input and output values to the network, and calculate the error norm in the output values.
- Adapt weights of the network connections, starting from the output layer and then proceeding backwards through the hidden layers, using any error minimization scheme. Here the generalized delta rule with momentum term scheme (Haykin, 1994) is used.
- Repeat the training process until the error norm from the output units is less than the error tolerance value.

Once this process is completed, the feedforward network is said to be a trained network. To study whether this trained net can indeed represent the transfer function map for a particular structure, some input sets not used in the training process, popularly defined as "unseen" examples for the net, are fed to the network. If the trained network provides output values that are comparable to the true output set corresponding to a given input set, it can be said to have become a system identification tool for the structural system.

STRATEGY FOR MONITORING USING ANN

In an offshore platform, the structural response over any pre-determined period of time can be recorded using vibration measurement devices. These devices may be placed at different locations. In this study it is assumed that they are placed at different elevation levels of the structure. Similarly the sea state, that is, the root mean square (rms) value of the sea-surface elevation and significant wave height and average wave time period, can also be obtained from a record of the sea-surface elevation during the above period of time.

A neural network is trained to map the transfer function of an undamaged offshore platform using the inputs as the significant wave heights and average wave time periods for particular sea states and the outputs as the rms displacements at different elevations of the platform. The outputs of the training sets required for this process are obtained either from actual measured data or by numerical simulation of the jacket platform. It should be noted here that the monitoring scheme proposed here depends only on matching the pattern of the displacements, and thus the absolute response values have little relevance. Furthermore, the scheme involves processed information from raw recorded data, and is, therefore, an off-line scheme, although the monitoring can be done in real-time. Once the ANN is trained it acts as a monitoring system in the following manner:

- The structural responses and sea surface elevation are recorded over a pre-determined period of time, say 15 minutes.
- These are then processed to determine the significant wave height, H_s, and average wave time period, T_0, for the sea state and the rms structural responses, given by the vector **v**.
- The significant wave height, H_s, and average wave time period, T_0, are used as inputs for the trained ANN to determine the output rms structural responses, given by the vector **x**.
- The normalized correlation given by the similarity function $S(\mathbf{x},\mathbf{v})$

$$S(\mathbf{x},\mathbf{v}) = \frac{\mathbf{x}.\mathbf{v}}{|\mathbf{x}||\mathbf{v}|} \quad \text{Eq. 3}$$

is determined, where the term in the numerator denotes the dot product of the two vectors, and the terms in the denominator denote the magnitudes of each vector. If this function is very close to 1, the platform is considered to be undamaged. If the value is measurably less than 1, then there is a strong possibility that the structure is damaged.

In this study a number of pre-determined damage states are also considered. The ANNs for different damage states are trained using numerically simulated data only, since it is not possible to obtain measured data for a damaged platform before hand. The response vector from each ANN for a given sea state is obtained. The response vector that gives the largest similarity function value from Eq. 3 with the measured structural response identifies the likely damage state, and, therefore, the damage location. For an actual platform it may not be possible to generate an ANN for each damage state of the structure. However, since the similarity function given in Eq. 3 is basically a comparison of response patterns, a few damage states for different damage locations would suffice, even for the actual structure.

Because of the non-availability of data from an actual platform, this strategy is illustrated using an example jacket platform with numerically simulated data sets to train the ANNs for different damage states, and evaluate their performance in integrity monitoring.

RESPONSE SIMULATION FOR EXAMPLE JACKET PLATFORM

A 60-meters high 2-dimensional jacket platform, which is assumed to be standing in a mean water depth of 50-meters, shown in Fig. 2, is considered to illustrate the integrity monitoring approach proposed in this paper. All the members of the platform, including the bracing members, are annular pipes with an outside diameter of 1-meter and a wall thickness of 5-centimeters. The elastic modulus of the steel pipe section is considered to be 245 GPa, and the mass density is taken as 7.86 tonnes/cubic-meter. To simulate realistic conditions, a lumped mass of 400 tonnes is considered at the top of the platform, where large masses are normally present due to operational reasons.

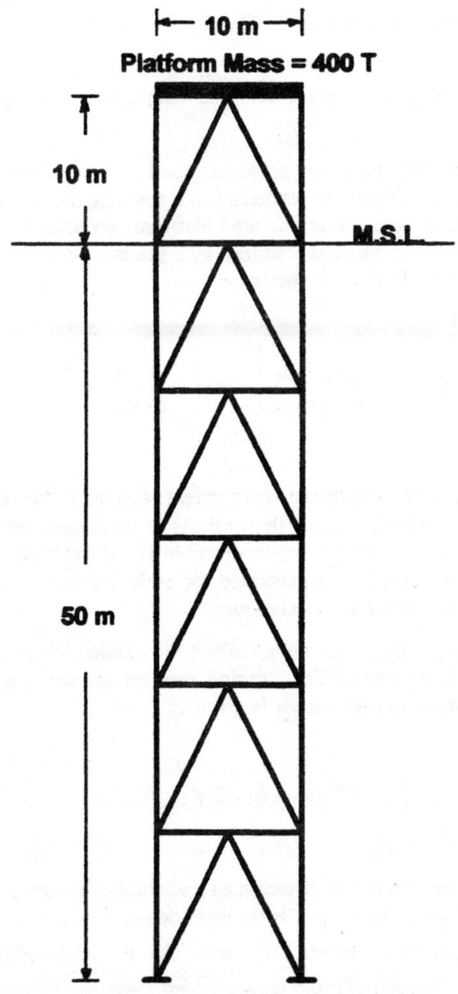

Fig. 2. Example jacket platform considered in this study

To simulate typical damage that can occur in this platform, four different damage states are considered, apart from the undamaged state. These five damage states, including the undamaged state, are schematically represented in Fig. 3. In this preliminary study, each damage state is simulated by assuming one or more bracing members to be severed or unable to contribute as load carrying members due to extensive cracking across the entire cross-section. Later studies will consider members with partial load carrying capacity due to low levels of cracking, which will include more realistic damage scenarios.

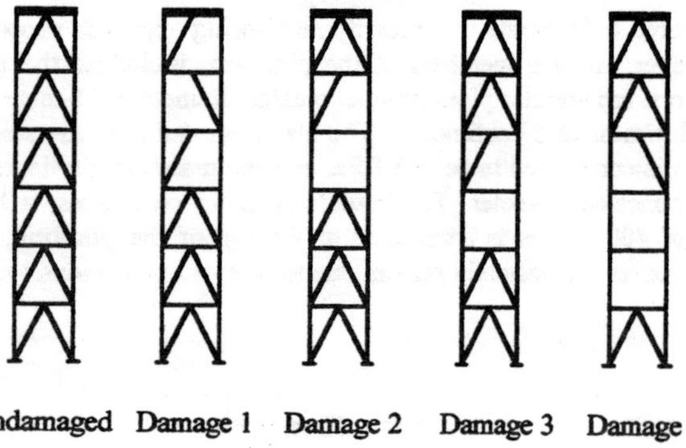

Fig. 3. Representation of damaged members in five damage states

The response analysis procedure used to generate the training set for each condition of the platform is the standard method for analyzing stochastic response of an offshore platform subjected to wave forces. The wave forces are characterized by a sea state defined by the DNV version of the P-M wave spectrum:

$$S_\eta(\omega) = \frac{H_s^2}{\pi\omega_0}\left(\frac{\omega}{\omega_0}\right)^{-5}\exp\left[-\frac{4}{\pi}\left(\frac{\omega}{\omega_0}\right)^{-4}\right] \qquad \text{Eq. 4}$$

where H_s is the significant wave height, and ω_0 is the average wave frequency ($=2\pi/T_0$). Since the analysis is non-linear, because of the wave forces, a P-M spectrum-compatible 15-minute time history of the sea-surface elevation is generated for each sea state, using standard Monte-Carlo simulation techniques.

The wave forces on the members are obtained from the modified Morrison's equations. For a typical member the wave force per unit length normal to the member is given by

$$\bar{p}(t) = \frac{1}{2}\rho_w A C_d(\dot{u}-\dot{x})|\dot{u}-\dot{x}| + \rho_w V C_m(\ddot{u}-\ddot{x}) + \rho_w V\ddot{x} \qquad \text{Eq. 5}$$

where A and V are the surface area and volume, respectively, per unit length of the member, ρ_w is the mass density of water that is taken to be 1 tonne/cubic meter, C_d and C_m are, respectively, a drag coefficient taken to be 0.6 and an added mass coefficient taken to be 2.0, and u and x are, respectively, the water particle and structural displacements. These forces are lumped at member nodes by assuming a trapezoidal distribution of the wave forces over the member length.

A lumped mass model is used for the dynamic analysis of the structure, where the mass terms include the added hydrodynamic mass coming from the last two terms in Eq. 5. The members are assumed to be axially inextensible to reduce the computational effort required for analysis. There are, therefore, six dynamic degrees-of-freedom, given the vector **v**, for the platform shown in Fig. 2, the lateral displacements at the six horizontal member levels. The governing equations of motion can be represented by

$$\mathbf{M}\ddot{\mathbf{v}} + \mathbf{C}\dot{\mathbf{v}} + \mathbf{K}\mathbf{v} = \mathbf{F}(t) \qquad \text{Eq. 6}$$

where **M**, **C**, and **K** are, respectively, the structure mass, damping, and stiffness matrices, and **F**(t) is the hydrodynamic force vector corresponding to the dynamic degrees-of-freedom. It should be noted that this force vector is obtained from Morrison's equations for each structural member, given in Eq. 5, and is, therefore, non-linearly dependent on the water particle and structural velocities. The equations of motion in Eq. 6 are solved using the normal mode superposition principle, so that the left-hand-side of the equations are decoupled. The right-hand-side remains non-linearly coupled.

An iterative procedure is used at each time step to solve the modal equations of motion. The Newmark-beta time-marching numerical integration method is used to solve these equations. At each time step the iteration starts with the structural velocity considered to be identical to that for the previous time step. The right-hand-side of the modal equations are then completely defined, and the modal responses are consequently determined. These computed responses are then used to get a revised value of the loads. The iteration continues till the responses in two successive iterations are within a tolerance value. The rms displacement at any level is obtained by integrating the square of the displacement time-history at that level over the entire 15-minute duration and taking its square root.

NUMERICAL STUDY AND RESULTS

For training the ANNs 22 input-output pairs were required, and 5 sets were analyzed to be kept as "unseen" examples. Therefore, 27 sea states were defined, and the responses of the structure for each of the five damage states were numerically simulated. The significant wave height and the average wave time period for the sea states considered are given in Table 1. A typical set of input-output pairs used in the training of the 5 neural nets are presented in Table 2.

Table 1. The significant wave height, H_s , and average wave time period, T_0 , for the sea states considered in this study

H_s (m)	T_0 (sec)	H_s (m)	T_0 (sec)	H_s (m)	T_0 (sec)
2	4	2.5	6	3	7
3	8	3*	9*	4	9
4	10	5	11	5	12
6*	12*	6	12.5	8	13
8	14	8	15	9	15
10	15.5	10*	16*	11	16
12	16.5	12	17	12*	17.5*
13	18	14	18	14	18.5
15*	19*	16	19	16	20

*"Unseen" problem sea states for verifying the trained ANNs

Table 2. Rms displacements at different elevation levels of the example platform for a sea state (H_s = 3 m, T_0 = 8 secs) used in training the ANNs for the five damage states

RMS displ.	Undam. state	Damage state 1	Damage state 2	Damage state 3	Damage state 4
1st Level	0.193	0.193	0.194	0.180	0.238
2nd Level	0.382	0.382	0.383	0.393	5.879
3rd Level	0.551	0.552	0.577	4.230	11.33
4th Level	0.688	0.815	3.667	4.598	11.74
5th Level	0.763	0.962	3.898	4.669	11.81
Top Level	0.764	0.965	3.898	4.669	11.82

In this study each neural net consists of two nodes in the input layer and six nodes in the output layer with one hidden layer of six nodes. The networks are fully connected (the input layer nodes are connected to both the hidden and output layer nodes) and are trained using the SNNS (Studgurd Neural Network Simulator) package. Approximately two hundred thousand training cycles were required to train each ANN. A schematic layout of a typical trained net with the node numbers and threshold values is shown in Fig. 4, and the final connection weights between the nodes are given in Table 3.

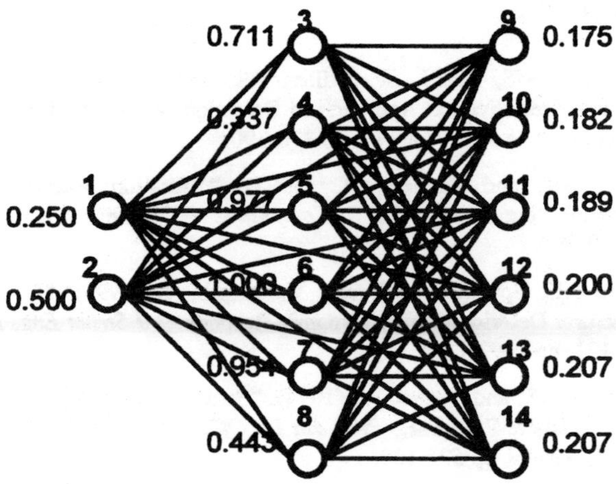

Fig. 4. Schematic layout of a typical integrity monitoring ANN with nodal threshold values for undamaged example jacket platform

Table 3. Connection weights for trained undamaged structure ANN

Target	Source : Weight			
3	1 : 3.621	2 : 0.136	-	-
4	1 : 0.868	2 : -1.698	-	-
5	1 : 12.221	2 : -7.024	-	-
6	1 : -9.212	2 : -6.533	-	-
7	1 : 4.039	2 : -4.060	-	-
8	1 : 0.275	2 : -3.067	-	-
9	1 : 1.243	2 : 3.302	3 : 3.228	4 : 0.756
	5 : -3.053	6 : -3.957	7 : -0.719	8 : 0.137
10	1 : 2.404	2 : 3.688	3 : 1.325	4 : 1.296
	5 : -3.229	6 : -5.007	7 : -0.323	8 : 0.767
11	1 : 1.951	2 : 3.424	3 : 2.419	4 : 1.198
	5 : -3.148	6 : -3.764	7 : -1.128	8 : 0.824
12	1 : 2.122	2 : 3.148	3 : 2.241	4 : -0.022
	5 : -3.536	6 : -3.766	7 : 0.211	8 : 1.098
13	1 : 2.421	2 : 2.704	3 : 2.043	4 : 0.312
	5 : -3.546	6 : -3.786	7 : -0.287	8 : 0.751
14	1 : 2.540	2 : 3.620	3 : 1.405	4 : 0.803
	5 : -3.196	6 : -3.689	7 : -0.859	8 : 1.640

A comparison of the response values "predicted" by neural nets with analytical values for one "unseen" sea state is given in Table 4. This shows that responses "predicted" by neural nets are very close to actual response values. A comparison for all "unseen" examples shows that the errors in ANN response estimates are less than 5%, which is acceptable. This illustrates the applicability of neural nets as fast response analysis tools for offshore jacket platforms.

Table 4. Comparison of analytical (AN) and neural net "predicted" (PR) rms displacement values for all five damage states considered of the example jacket platform structure for an "unseen" sea state ($H_s = 6$ m, $T_0 = 12$ secs)

Rms disp. vector (in mm)	Undamaged state		Damage state 1	
	AN (v)	PR (x)	AN (v)	PR (x)
1st level	0.527	0.520	0.525	0.520
2nd level	1.006	0.978	1.005	0.997
3rd level	1.396	1.367	1.401	1.392
4th level	1.676	1.649	1.933	1.916
5th level	1.801	1.776	2.183	2.161
Top level	1.802	1.764	2.184	2.168
Rms disp. vector (in mm)	**Damage state 2**		**Damage state 3**	
	AN (v)	PR (x)	AN (v)	PR (x)
1st level	0.527	0.528	0.500	0.491
2nd level	1.004	1.000	1.031	1.053
3rd level	1.446	1.440	9.827	9.772
4th level	7.721	7.705	10.60	10.55
5th level	8.135	8.116	10.72	10.69
Top level	8.131	8.112	10.71	10.73
Rms disp. vector (in mm)	**Damage state 4**			
	AN (v)	PR (x)		
1st level	0.639	0.641		
2nd level	14.60	14.71		
3rd level	27.16	27.31		
4th level	28.00	28.22		
5th level	28.12	28.34		
Top level	28.13	28.26		

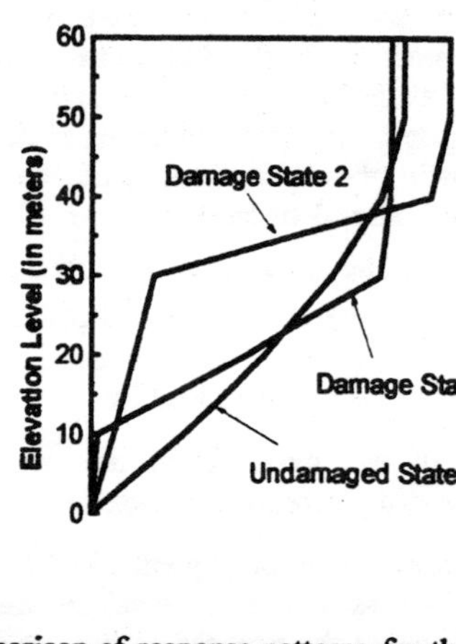

Fig. 5. Comparison of response patterns for three damage states of the example jacket platform subjected to a particular sea state ($H_s = 6$ meters, $T_0 = 12$ seconds)

For integrity monitoring, however, the actual response values have no significance, as only response patterns should be determined. Typical analytical and "predicted" response patterns for the undamaged and two other damage states are shown in Fig. 5, which shows that the two patterns coincide for a given damage state. A numerical comparison of these response patterns is done by evaluating the similarity function given by Eq. 3 for the two different patterns, which is presented in Table 5.

Table 5. Similarity function values, given by Eq. 3, for comparing the datum set of "predicted" and analytical/measured response patterns for different damage states of the example jacket platform subjected to a sea state ($H_s = 6$ m, $T_0 = 12$ secs)

Anal. Sim. Damage state	Datum Damage State from ANN				
	0	1	2	3	4
0	1.0000	0.9971	0.9204	0.9616	0.9866
1	0.9971	1.0000	0.9463	0.9664	0.9826
2	0.9206	0.9462	1.0000	0.9274	0.9046
3	0.9618	0.9661	0.9268	1.0000	0.9784
4	0.9864	0.9826	0.9043	0.9785	1.0000

The values in the first column of Table 5 clearly show that similarity function values are less than 1 when the analytical response patterns of a damaged structure is compared with the ANN response pattern estimates for an undamaged structure. This illustrates the point that a measurable reduction in the similarity function value can give an alert that the structure may be damaged. For all the different damage states the function has the largest value (equal to 1 here) when the analytical response pattern and the ANN response pattern estimate are for the same damage state. This shows that the integrity monitoring ANN does indeed predict the damage state of the platform accurately.

SUMMARY AND CONCLUSIONS

The efficiency of artificial neural networks in mapping the transfer function for the rms displacement of an offshore structure, for specific sea states, is investigated in this paper. It should be noted that, although, the rms displacement is the response quantity considered in this study, the conclusions given should be equally valid for any other response quantity, for example, stress level. In addition, the possibility of using such neural nets for integrity monitoring of jacket platforms is examined. The following broad conclusions can be made from the results of the numerical study done on an example jacket platform:

- A fully connected feedforward neural network can efficiently map the non-linear transfer functions for rms displacements at various elevation levels, or any other response quantities, of a jacket platform subjected to random wave loading.
- With the help of the above neural nets and the similarity functions, damage to a platform can be detected.
- This procedure can also be usefully employed for the location of the damage, as long as a suitable neural net can be trained to approximately represent the response pattern for a corresponding damage state.

Thus artificial neural networks show great promise in integrity monitoring of offshore structures. Further investigations, however, are required to produce a robust scheme that can be easily implemented for practical platforms.

ACKNOWLEDGMENT

The authors gratefully acknowledge the assistance provided by Mr. Mallikarjuna, Research Scholar, IIT, Delhi, in training the ANNs used in this study.

REFERENCES

Agbabian, MS, Masri, SF, Miller, RK, and Caughey, TK (1991). "System Identification Approach to Detection of Structural Changes," *J Eng Mech*, ASCE, Vol 117, pp 370-390.

Barai, SV, and Pandey, PC (1995). "Vibration Analysis using Artificial Neural Networks," *J Computing in Civ Eng*, ASCE, Vol 9, pp 259-265.

Chen, HL, Spyrakos, CC, and Venkatesh, G (1995). "Evaluating Structural Deterioration by Dynamic Response," *J Struct Eng*, ASCE, Vol 121, pp 1197-1204.

Chiostrini, S, and Vignoli, A (1994). "Structural Integrity Monitoring of an Offshore Platform," *Proc 4th Int Offshore and Polar Eng Conf*, ISOPE, Osaka, pp 528-533.

Duggan, DM, Wallace, ER, and Caldwell, SR (1980). "Measured and Predicted Behavior of Gulf of Mexico Platforms," *Proc Offshore Tech Conf*, OTC 3864, Houston, pp 91-100.

Elkarody, MF, Chang, KC, and Lee, GC (1993). "Neural Network Trained by Analytically Simulated Damage States," *J Computing in Civ Eng*, ASCE, Vol 7, pp 130-145.

Haykin, S (1994). *Neural Networks - A Comprehensive Foundation*, Macmillan College Publishing Company, Inc.

Hearn, G, and Testa, RB (1991). "Modal Analysis for Damage Detection in Structures," *J Struct Eng*, ASCE, Vol 117, pp 3042-3063.

Kenley, RM, and Dodds, CJ (1980). "West Sole WE Platform: Detection of Damage by Structural Response Measurements," *Proc Offshore Tech Conf*, OTC 3866, Houston, pp 111-118.

Rubin, S, and Coppolino, RN (1983). "Flexibility Monitoring of Offshore Jacket Platforms," *Proc Offshore Tech Conf*, OTC 4535, Houston, pp 201-208.

Shahrivar, F, and Bouwkamp, JG (1984). "Damage Detection in Offshore Structures using Vibration Information," *Proc ASME Symp on Offshore Mech and Arctic Eng*, pp 174-185.

Shyamsunder, S, and Ting, SK (1985). "Flexibility monitoring of Offshore Platforms," *J Applied Ocean Res*, Vol 7, pp 14-23.

Vandiver, JK (1975). "Detection of Structural Failure on Fixed Platforms by Measurement of Dynamic Response," *Proc Offshore Tech Conf*, OTC 2267, Houston.

Wu, X, Ghaboussi, J, and Garret Jr., JH (1992). "Use of Neural Networks in Detection of Structural Damage," *Computers and Structures*, Vol 42, 649-659.

Proceedings of the Seventh (1997) International Offshore and Polar Engineering Conference
Honolulu, USA, May 25-30, 1997

ISBN 1-880653-28-1 (Set); ISBN 1-880653-32-X (Vol. IV)

Three-Dimensional Characteristics of Seaquakes

T. Kiyokawa
Shimizu Corporation
Tokyo, Japan

ABSTRACT

Three-dimensional potential function is developed to investigate seaquakes. A Green function describing three-dimensional waves in the compressible fluid is derived and employed to establish the seaquake potential. An approximate solution of the seaquake potential is also derived for the case of a flat rectangular wave source on the sea bottom. Numerical results are presented to show the three-dimensional characteristics of seaquakes. The results agree well with the reported fact that seaquakes can propagate in a long distance.

KEY WORDS: Seaquakes, hydrodynamic pressure, Green function, three-dimensional theory, seaquake potential, numerical simulation, attenuation relationship

INTRODUCTION

Seaquakes are known as a phenomena such that vessels traveling in oceans suddenly feel a very strong shock as if they are running ashore or be attacked by a mine. There are several reports that reported ships subjected to seaquakes had severe damages. Hove et al. (1983) reported such a shock motion of ships during seaquakes. Seaquakes may occur by oceanic earthquakes. Therefore, floating structures mooring in a sea nearby earthquake zone may have a high potential to take damages. The potential damages caused by seaquakes for floating structures should then be discussed in the design stage.

The characteristic facts of seaquakes are: seaquakes propagate in a long distance (order of several hundred kilometers) and cause high frequency response motions of the vessels (Ex. Hamamura, 1987). Baba (1987) investigated seaquakes analytically using velocity potential theories. However, the theory could not explain that seaquakes can propagate in a long distance. According to the theory, hydrodynamic pressure caused by seaquakes should decrease exponentially as the observation point goes away from the source area. Iwasaki (1988) assumed seaquakes to be sound waves and carried out one-dimensional unsteady analysis, in which multiple reflection of waves between free surface and sea bottom was neglected. Therefore, the theory used by Iwasaki seems to be unrealistic and not sufficient for the design stage. Kiyokawa and Inada (1989) proposed a vertically two-dimensional theory assuming fluid compressibility. They have then shown that the compression waves generated under certain circumstances can propagate in a long distance without decay as well as the generated compression waves generally have high frequencies. Shingu et al. (1990) carried out an experimental study on response of a floating structure subjected to seaquakes. However, it is almost impossible to generate seaquakes in a model tank because generating a phenomenon similar to seaquakes requires deep water. Therefore, Shingu's work may be related to the case of direct input of earthquakes. Kawanishi et al. (1991) also studied the response of a tension leg platform during earthquake. The author has shown that high frequency response motions of floating structures, as a typical phenomenon of seaquakes, can be obtained using the seaquake potential derived by Kiyokawa et al. (Kiyokawa, 1991, 1995). However, these studies can only be applied to a vertically two-dimensional problem. Hamamoto et al. (1993) investigated analytically the response motion of a floating elastic circular plate subjected to seaquakes. Although their solution is derived using 3D theory, the structure profile is a very restricted circular plate. It can then be concluded that there are no 3D theories available to predict the hydrodynamic forces acting on arbitrary floating structures during seaquakes so far.

In the present paper, a 3D seaquake potential which can be used to evaluate hydrodynamic forces acting on arbitrary floating structures during seaquakes is presented. The seaquake potential is developed using a Green function which describes the wave motions in compressible fluids. At first the Green function is derived using an eigenfunction expansion method. The derived Green function satisfies the corresponding boundary value problem so that we can obtain the seaquake potential in a closed form. The seaquake potential is derived by applying the Green's theorem to the fluid domain with the Green function. The exact expression for the seaquake potential is obtained in integral form. However the analytical integration over the wave source area may be difficult in general. Discretization of the wave source area is then used to study seaquakes numerically. Consequently, the solution is expressed as a sum of contributions from each sub wave source area. The solution finally obtained is physically composed of three components, that is, surface wave component, evanescent mode component, and compression wave component. Most essential part for seaquakes is the compression wave component for the case of deep water because it makes seaquakes propagate in a long distance without losing much of energy. The numerical results show that seaquakes may propagate over several hundred kilometers and the result agrees well with the reported fact.

THEORY

Governing Equation and Associate Boundary Conditions

We consider a wave source that is a part of a flat seabed of depth h and oscillates in the vertical direction with angular frequency ω as shown in Fig.1. The coordinate system is also defined as shown in Fig.1. We assume that the fluid under consideration is inviscid and compressible as well as its motion is irrotational. Both amplitudes of the wave source and the surface wave are also assumed to be small enough compared with their horizontal scale. The fluid motion can then be described using linear potential theory. The velocity potential $\phi(x,y,z;t)$ is assumed to be in the form of $\phi(x,y,z)e^{-i\omega t}$ for the frequency domain analysis, in which $i=\sqrt{-1}$, ω is the angular frequency, and t is the time. The velocity potential may be obtained as a solution of the following boundary value problem.

$$L[\phi]=0 \qquad (in\ \Omega) \qquad (1)$$
$$\partial\phi/\partial z - \nu\phi = 0 \qquad (\nu \equiv \omega^2/g) \qquad (on\ S_F) \qquad (2)$$
$$\partial\phi/\partial z = 0 \qquad (on\ S_B) \qquad (3)$$
$$\partial\phi/\partial z = V_0 \qquad (on\ S_W) \qquad (4)$$

where, $L[\,]$ is the three-dimensional Helmholtz operator $(\equiv \partial^2/\partial x^2 + \partial^2/\partial y^2 + \partial^2/\partial z^2 + \omega^2/c^2)$, c is the sound speed in the water, g is the acceleration of the gravity, V_0 is the complex velocity amplitude of the wave source.

Equation (1) is the three-dimensional Helmholtz equation, which is the governing equation for the problem and should be satisfied in the whole fluid domain Ω. Equation (2) describes the kinematic condition at the free surface S_F, Eq.(3) describes the impermeability condition at

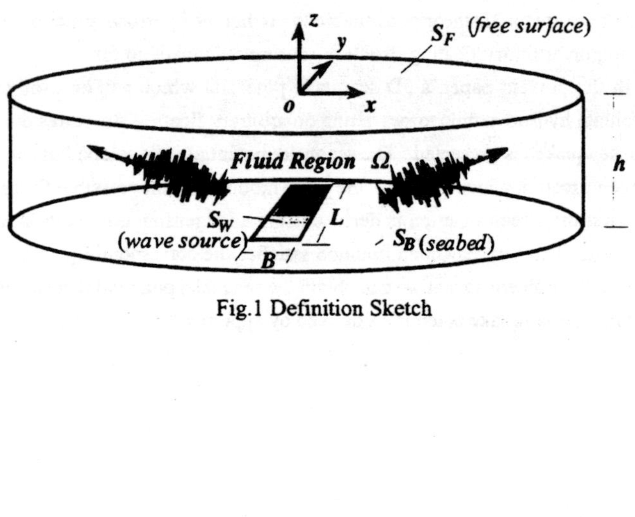

Fig.1 Definition Sketch

the seabed S_B, Eq.(4) describes the continuity between the fluid motion and the wave source motion at the surface of the wave source S_W. In addition, a radiation condition must be specified at large distance from the wave source. The radiation condition allows only outgoing waves of infinitesimal amplitude from the fluid domain including the wave source.

Green Function

We derive a Green function to solve the boundary value problem defined by Eqs.(1)-(4). The Green function G corresponding to the boundary value problem is defined as a function satisfying the following boundary value problem.

$$L[G] = -\delta(x-\xi)\delta(y-\eta)\delta(z-\zeta) \qquad (in\ \Omega) \qquad (5)$$
$$\partial G/\partial z - \nu G = 0 \qquad (on\ S_F) \qquad (6)$$
$$\partial G/\partial z = 0 \qquad (on\ S_B \cup S_W) \qquad (7)$$

where, $\delta(\,)$ is the Dirac's delta function, (ξ,η,ζ) is a source point in the fluid domain. The Green function should also satisfy the radiation condition similar to that for the potential function.

We derive the solution of the above boundary value problem using an eigenfunction expansion method. The eigenfunction satisfying both Eqs.(6) and (7) can be expressed as follows;

$$z_n(z) = \begin{cases} \sqrt{\dfrac{2k_0^2}{h(k_0^2-\nu^2)+\nu}}\,\dfrac{\cosh k_0(h+z)}{\cosh k_0 h} & (n=0) \\ \sqrt{\dfrac{2k_n^2}{h(k_n^2+\nu^2)-\nu}}\,\dfrac{\cos k_n(h+z)}{\cos k_n h} & (n=1,2,\cdots) \end{cases} \qquad (8)$$

where, $z_n(z)\ (n=0,1,\cdots)$ are the eigenfunctions with respect to the z-coordinate, k_0 is the wave number, $k_n\ (n=0,1,\cdots,\ k_1<k_2<\cdots)$ are the eigenvalues. The wave number k_0 and the eigenvalues $k_n\ (n=1,2,\cdots)$ are obtained as real and positive solutions of the following equation.

$$k_0 \tanh k_0 h = -k_n \tan k_n h = \nu \qquad (9)$$

Equation (8) is a set of orthonormal functions and therefore the Dirac's delta function can be expanded using Eq.(8) as follows:

$$\delta(z-\zeta) = \sum_{n=0}^{\infty} z_n(\zeta) z_n(z) \qquad (10)$$

Corresponding to Eq.(10), we then assume the Green function to be in the following form;

$$G(x,y,z;\xi,\eta,\zeta) = \sum_{n=0}^{\infty} g_n(x,y;\xi,\eta) z_n(\zeta) z_n(z) \qquad (11)$$

where, $g_n(x,y;\xi,\eta)\ (n=0,1,\cdots)$ are unknown functions of x and y with parameter (ξ,η).

Substituting Eq.(11) into Eq.(5), we obtain the following partial differential equation to be solved for $g_n(x,y;\xi,\eta)\ (n=0,1,\cdots)$;

$$\frac{\partial^2 g_n}{\partial x^2} + \frac{\partial^2 g_n}{\partial y^2} + \lambda_n^2 g_n = -\delta(x-\xi)\delta(y-\eta) \qquad (12)$$

in which, λ_n is a new parameter defined by

$$\lambda_n^2 \equiv \omega^2/c^2 \pm k_n^2 \quad (n=0,1,\cdots) \qquad (13)$$

where, the plus sign is used for n=0 while the minus sign is for n=1,2,..., respectively.

Equation (12) is the well known equation to obtain the fundamental solution for the two-dimensional Helmholtz equation. The solution of Eq.(12) satisfying the radiation condition mentioned before can be written as

$$g_n(x,y;\xi,\eta) = (i/4)H_0^{(1)}(\lambda_n R) \tag{14}$$

where $H_0^{(1)}(\,)$ is the first kind and 0-th order Hankel function. R is the horizontal distance between the source point and the observation point ($\equiv\sqrt{(x-\xi)^2+(y-\eta)^2}$).

When the right hand side of Eq.(13) becomes negative, λ_n becomes to be a pure imaginary number. Then introducing a new parameter λ_n' defined by $\lambda_n = i\lambda_n'$, We rewrite Eq.(14) in the following way.

$$g_n(x,y;\xi,\eta) = (i/4)H_0^{(1)}(i\lambda_n' R) = (1/2\pi)K_0(\lambda_n' R) \tag{15}$$

where $K_0(\,)$ is the second kind and 0-th order modified Bessel function.

Substituting all the results into Eq.(11) and performing an algebraic calculation for a simple expression, we finally obtain the Green function as follows;

$$\begin{aligned} G(x,y,z;\xi,\eta,\zeta) = {} & \\ & \frac{i(k_0^2-\nu^2)}{2\{h(k_0^2-\nu^2)+\nu\}}\cosh k_0(h+\zeta)\cosh k_0(h+z)H_0^{(1)}(\lambda_0 R) \\ & +\sum_{n=1}^{s}\frac{i(k_n^2+\nu^2)}{2\{h(k_n^2+\nu^2)-\nu\}}\cos k_n(h+\zeta)\cos k_n(h+z)H_0^{(1)}(\lambda_n R) \\ & +\sum_{n=s+1}^{\infty}\frac{(k_n^2+\nu^2)}{\pi\{h(k_n^2+\nu^2)-\nu\}}\cos k_n(h+\zeta)\cos k_n(h+z)K_0(\lambda_n' R) \end{aligned} \tag{16}$$

in which, s is the maximum value of integer n which makes the right hand side of Eq.(13) to be positive.

The Seaquake Potential

To obtain the seaquake potential, we apply the extended Green's theorem to the fluid domain Ω with the Green function (16). The extended Green's theorem is written as

$$\iiint_\Omega \{G\cdot L[\phi]-\phi\cdot L[G]\}d\xi d\eta d\zeta = \oiint_S \left\{G\frac{\partial\phi}{\partial n}-\phi\frac{\partial G}{\partial n}\right\}dS \tag{17}$$

in which, the left hand side of the above equation denotes a volume integral over the fluid domain Ω, S is the total surface enclosing the fluid domain, and differentiating with respect to n means a normal differential at the surface.

Substituting Eqs.(1)-(7) into the above equation and considering the radiation conditions for each the potential function and the Green function, we may derive the following expression of the seaquake potential after several algebraic calculations.

$$\phi(x,y,z) = -\iint_{S_w} V_0(\xi,\eta)\cdot G(x,y,z;\xi,\eta,-h)d\xi d\eta \tag{18}$$

It should be noted that this expression is the exact solution to the boundary value problem (1)-(4). When the complex amplitude $V_0(\xi,\eta)$ is given from earthquake science like a fault theory, the integrand of Eq.(18) becomes all known functions and we can then obtain the potential value at an arbitrary point in the fluid through the integration. However to perform this integration analytically seems to be almost impossible in general. Thus, we use the following approximation instead of Eq.(18).

$$\phi(x,y,z) = -\sum_{m=1}^{M} V_0(\xi_m,\eta_m)\cdot G(x,y,z;\xi_m,\eta_m,-h)\cdot\Delta_m \tag{19}$$

where M is the number of division to the wave source element, (ξ_m,η_m) is the center coordinate of m-th source element, Δ_m is the area of m-th source element.

The approximation (19) may be sufficiently accurate for practical use. We assume much more simple case for a fundamental study, that is, the velocity amplitude is assumed to be a constant over the source area ($V_0(\xi,\eta) = V_0 = const.$). For such a case, Eq.(19) yields

$$\begin{aligned} \frac{\phi(x,y,z)}{-V_0} = {} & i\frac{(k_0^2-\nu^2)}{2\{h(k_0^2-\nu^2)+\nu\}}F_0(x,y)\cosh k_0(h+z) \\ & +\sum_{n=1}^{s}\frac{i(k_n^2+\nu^2)}{2\{h(k_n^2+\nu^2)-\nu\}}F_n(x,y)\cos k_n(h+z) \\ & +\sum_{n=s+1}^{\infty}\frac{(k_n^2+\nu^2)}{\pi\{h(k_n^2+\nu^2)-\nu\}}F_n(x,y)\cos k_n(h+z) \end{aligned} \tag{20}$$

where the functions $F_n(x,y)$ $(n=0,1,\cdots)$ are given by

$$F_n(x,y) = \begin{cases} \sum_{m=1}^{M} H_0^{(1)}(\lambda_n R_m)\cdot\Delta_m & (n=0,1,\dots s) \\ \sum_{m=1}^{M} K_0(\lambda_n' R_m)\cdot\Delta_m & (n=s+1,s+2,\dots) \end{cases} \tag{21}$$

in which, R_m is the horizontal distance between the observation point and the center coordinate of the m-th source element ($\equiv\sqrt{(x-\xi_m)^2+(y-\eta_m)^2}$).

It should be noted that the first term of Eq.(20) describes the surface wave component, the second term is the compression wave component, and the last term, the evanescent mode component. In the numerical calculation, the surface wave component should be omitted for high frequency and deep water cases because it becomes $0\times\infty$ type calculation. It can be easily shown that the surface wave component should vanish when kh becomes large.

The hydrodynamic pressure p due to seaquakes is obtained from the linearized Bernoulli's equation: $p(x,y,z) = i\rho\omega\phi(x,y,z)$, in which ρ is the fluid density.

NUMERICAL RESULTS AND DISCUSSION

Distribution Characteristics of Hydrodynamic Pressures

We assume the water depth h to be 1000m, the wave source width B to be 1km, and the wave source length L to be 20km, respectively for the following numerical studies. The oscillating acceleration of the wave source is assumed to be a unit acceleration (=1m/s²=100gal). Figure 2 presents vertical distributions of the hydrodynamic pressure for y=0m, 5000m, 10000m in the case of oscillation frequency f=1Hz. In this case

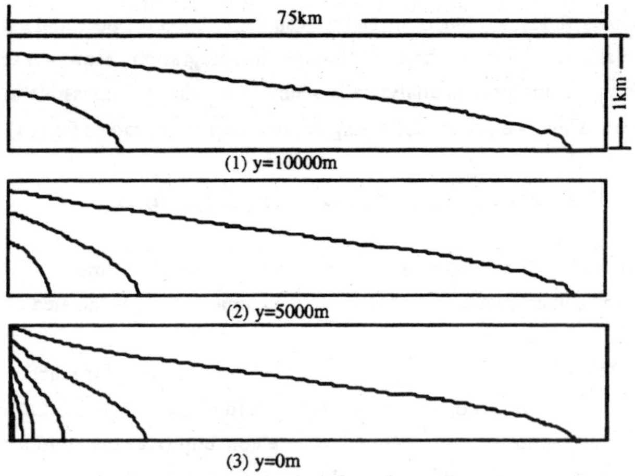

Fig.2 Vertical Distribution of Hydrodynamic Pressure (1)
(Frequency f=1.0Hz, Number of generated compression waves is one)

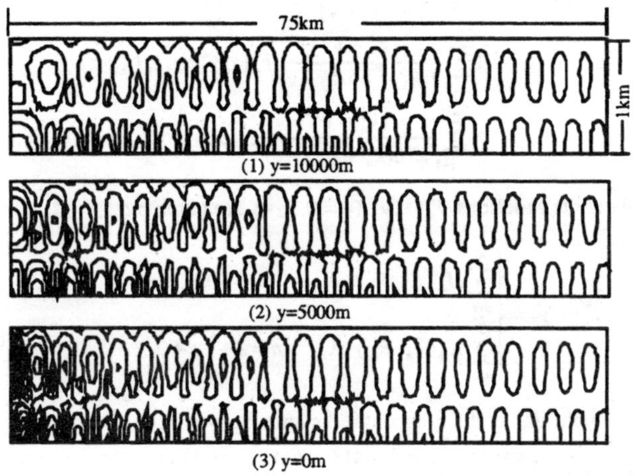

Fig.3 Vertical Distribution of Hydrodynamic Pressure (2)
(Frequency f=1.5Hz, Number of generated compression waves is two)

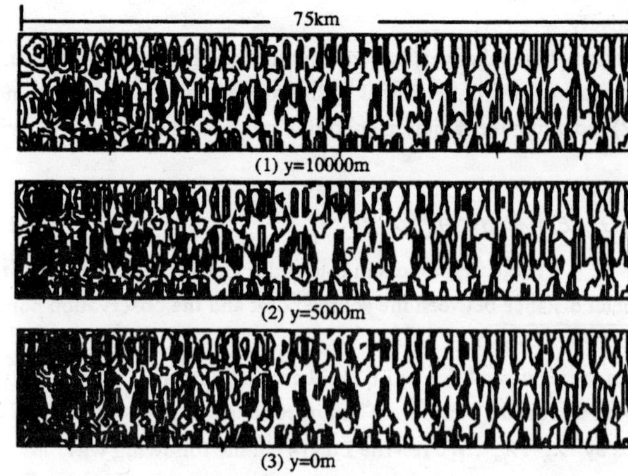

Fig.4 Vertical Distribution of Hydrodynamic Pressure (3)
(Frequency f=2.0Hz, Number of generated compression waves is three)

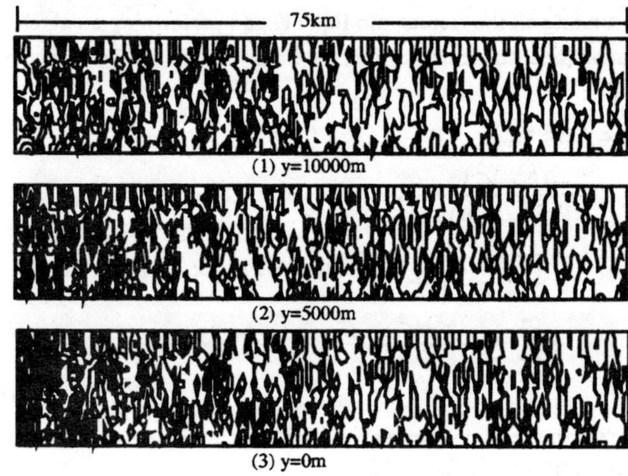

Fig.5 Vertical Distribution of Hydrodynamic Pressure (4)
(Frequency f=3.0Hz, Number of generated compression waves is four)

the number of generated compression wave component is one. The left down corner of each figure is a edge of the wave source. The distributions are shown within the horizontal distance of 75km from the edge. The hydrodynamic pressure becomes large as the observation point approaches to the center of the wave source. In this case, we can easily imagine the three dimensional structure of the pressure distribution. Figures 3-5 also show the vertical distributions of the hydrodynamic pressure for the cases where the oscillation frequencies are 1.5Hz, 2.0Hz, 3Hz, respectively. These cases correspond to the cases where the number of the generated compression wave component are two, three, four, respectively. It is found that the distribution pattern becomes complicated more and more as the number of compression wave component increases. This is due to the interaction of the compression waves and of the multiple reflected waves between the free surface and the bottom. We can hardly imagine three dimensional structure of the pressure distribution when the number of generated compression waves becomes more than two components.

Figures 6-9 present the horizontal distribution of the hydrodynamic pressures related to Figs.2-5, respectively. The figures show a quarter plane because of symmetry, in which the upper end of each figure is the x-axis and the upper left is the center of the wave source. The distributions are shown within an area of 75km in the x-direction and 10km in the y-direction. The water depth are chosen to be 500m and 1000m, respectively. In all the cases, it is found that the pressure takes the largest value around the center of the wave source and decreases radiately as the observation point moves far from the center. In addition, the result shows that there are many local peaks when the number of the compression waves is more than two, and the peak values decrease as the observation point moves far from the center. The distribution characteristics quite differ from those of the vertical distribution, that is, the vertical distribution varies complicatedly in the whole fluid domain while the horizontal distribution simply decreases as the observation point moves far from the center of the wave source.

Frequency Response of Hydrodynamic Pressures

Figure 10 shows the frequency response characteristics of hydrodynamic pressures. The calculation points are chosen as 1km, 10km, 100km, 1000km

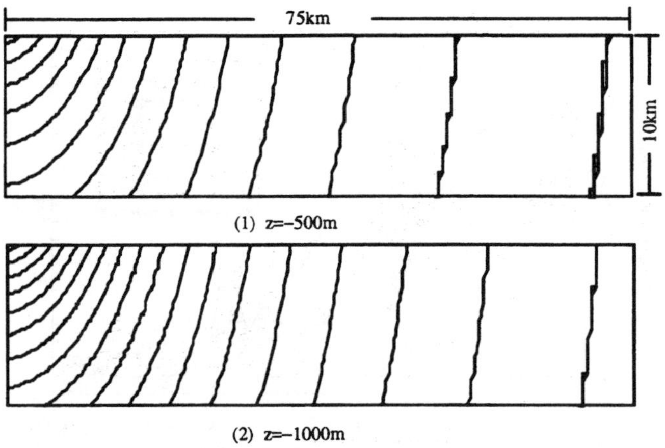

Fig.6 Horizontal Distribution of Hydrodynamic Pressure (1)
(Frequency f=1.0Hz, Number of generated compression waves is one)

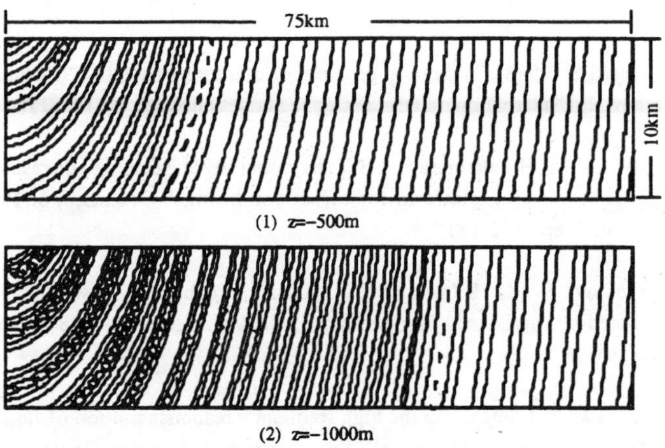

Fig.7 Horizontal Distribution of Hydrodynamic Pressure (2)
(Frequency f=1.5Hz, Number of generated compression waves is two)

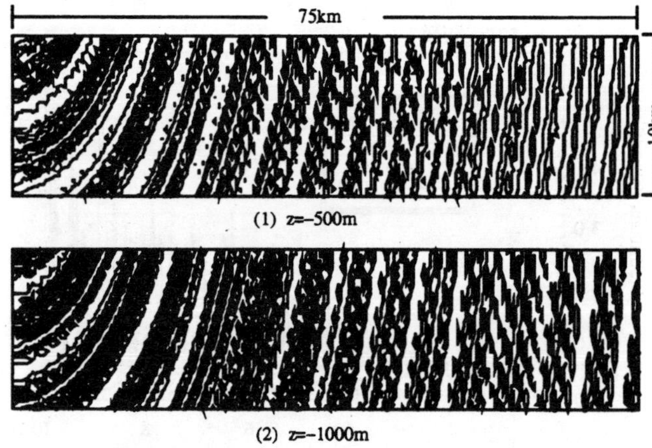

Fig.8 Horizontal Distribution of Hydrodynamic Pressure (3)
(Frequency f=2.0Hz, Number of generated compression waves is three)

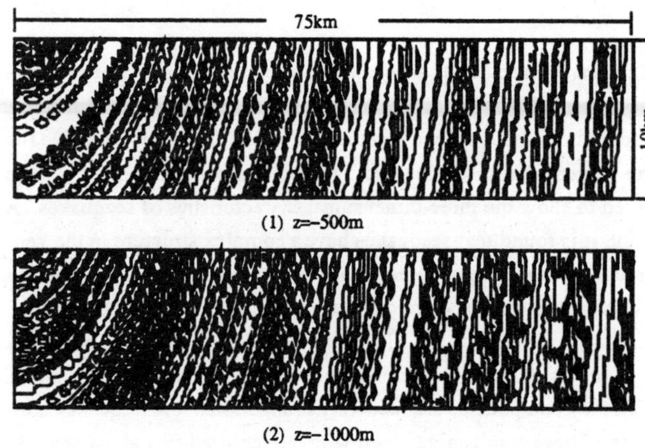

Fig.9 Horizontal Distribution of Hydrodynamic Pressure (4)
(Frequency f=3.0Hz, Number of generated compression waves is four)

from the center of the wave source and 10m beneath the x-axis. Behavior of the hydrodynamic pressure as a function of frequency is found to be very complex. There are several points where the pressure values for the long distance point are larger than those for near point. This may be due to the effects of interaction among the compression waves. This result may also suggest that response motions of floating structures subjected to seaquakes depend significantly on its location. It is also found that high frequency components tend to be strong. Consequently, people on ships subjected to seaquakes may feel strong shock as reported.

Attenuation Relationship of Hydrodynamic Pressures

As mentioned before, compression waves which are main part of the seaquakes propagate from the wave source to far field with decreasing strength. The attenuation of energy in propagation may be mainly due to three-dimensional effects even in an actual case. To know the attenuation relationship is of importance to design floating structures in a potential risk zone of seaquakes. We discuss the attenuation relationship in propagation of seaquakes. Figure 11 presents an attenuation relationship as a function of distance from the center of wave source for different frequencies as shown in the figure. In the calculated cases, the hydrodynamic pressures tend to attenuate from roughly over 10km though the pressure values still move up and down. This fact may suggest that the hydrodynamic pressure dose not decrease within the area of half length of the wave source ($=L/2$). The hydrodynamic pressures begin to decrease as the observation point moves out of the area. It should be noted that the hydrodynamic pressures are still substantial even at a location several hundred kilometers far from the center of wave source, and this result agrees well with the reported fact (Ex. Hamamura, 1987). Such a result that the hydrodynamic pressure is substantial even in the far field from the source area can be obtained only when the fluid compressibility is taken into account. Thus, the results may have given a qualitative proof to the author's hypothesis that principal part of seaquakes are the compression waves.

CONCLUSIONS

The three-dimensional seaquake potential is presented in the paper and which enable us to predict the external loads acting on floating structures

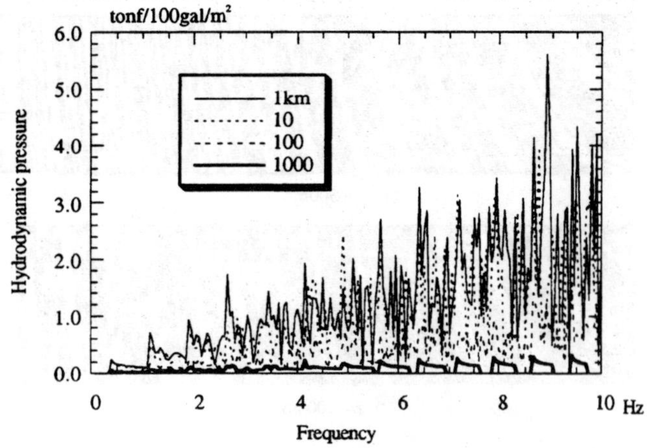

Fig.10 Frequency Response Characteristics of Hydrodynamic Pressure

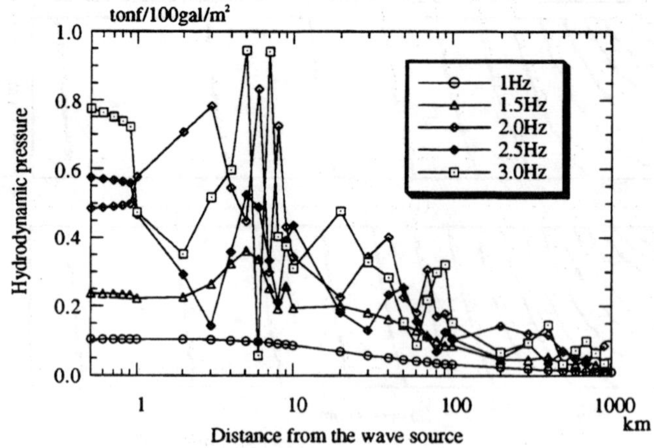

Fig.11 Attenuation Relationship of Hydrodynamic Pressure

of arbitrary shape during seaquakes. The corresponding Green function is also presented to derive the seaquake potential analytically, which describes fluid motions induced by a point source oscillating in compressible fluids. The numerical results obtained by using the seaquake potential are presented to show the three-dimensuinal characteristics of seaquakes. As a result, it is found that seaquakes have a complex structure in the vertical direction and propagate radiately in the horizontal direction, in the propagation process hydrodynamic pressures reduce the strength because of the three-dimensional effects. It is also found that seaquakes propagate at least several hundred kilometers without losing much energy, and high frequency components are relatively stronger than low frequency components. These results agree well with the reported fact about seaquakes. Therefore, it can be concluded that the fundamental mechanism of seaquakes is made clear in the paper. Namely, the generation of compression wave components, which are a kind of progressive waves and be generated when the water depth is deep and exciting frequency is high, is essential for seaquakes.

REFERENCES

Baba E. (1978). "A study on the effect of seaquakes on a floating body," Jour. of Japan society of naval architecture, No.162, pp.90-98.

Hamamoto,M.and Y.Tanaka (1993). "Fluid-structure interaction analysis of floating elastic circular plate Part.3 Response behavior of artificial floating islands subjected to seaquakes," Jour. of structural and construction Engineering, Trans. of AIJ, No.448, pp.173-185.

Hamamura, K (1987). "Is the ocean isolated seismically? Investigation on sea shock," Jour. of ship science, Vol.40, No.10, pp.58-61 (in Japanese).

Hove K., P. B, Selnes and H. Bungum (1983). "Seaquakes: A Potential threat to offshore structures, " Behav. Offshore Structure, vol.2, pp.561-571.

Iwasaki S. (1988). "Vibration problem of offshore structures," Technical report, Ed. International marine science and technology, pp.111-112.

Kawanishi T, H.Takamura and H.Kobayashi (1991). "Earthquake response of the tension leg platform for offset condition, " Oceans, Vol.1991, No.Vol.2, pp.1112-1117.

Kiyokawa T. and H. Inada (1989). "On the mechanism of seaquakes, " Proceedings of Coastal Engineering, JSCE,Vol.36, pp.734-738 (in Japanese).

Kiyokawa T. (1992). "On the high frequency response motion of floating structures during seaquakes," Proceedings of Coastal Engineering, JSCE,Vol.39 (2), pp.816-820 (in Japanese).

Kiyokawa T. (1995). "Numerical prediction for response motion of floating structures subjected to seaquakes," Proceedings of the Fifth (1995) International Offshore and Polar Engineering Conference, pp.429-434.

Shingu k.and D. Funamoto (1990). "An experimental study on responses of a floating structure subjected to seashock," Proceedings of 4th Pacific

Proceedings of the Seventh (1997) International Offshore and Polar Engineering Conference
Honolulu, USA, May 25-30, 1997

ISBN 1-880653-28-1 (Set); ISBN 1-880653-32-X (Vol. IV)

Analytical Solution for the Earthquake Response of a Cylindrical Structure Considering Hydrodynamic and Soil Dynamic Interaction

Yoshihiro Tanaka
Taisei Corporation, Tokyo, Japan
Takaaki Nakamura
Shinozuka Research Institute, Tokyo, Japan
Kyohei Ito
Civil Engineering Research Laboratory, Tokyo, Japan

ABSTRACT

An analytical solution for the earthquake response of a rigid cylindrical offshore structure has been developed considering both hydrodynamic forces-structure and soil dynamic forces-structure interactions. The solution is useful for the practical earthquake design of the cylindrical type pier foundation in deep water such as Honshyu-Shikoku suspension bridge. The radiation damping effect due to water compressibility on the earthquake response is more important than that due to ground motion.

KEY WORDS : Eigenfunction expansion method, Hydrodynamic and soil dynamic structure interaction, Sway and rocking motions

INTRODUCTION

A rigid gravity type concrete cylindrical structure is a typical main pier foundation constructed in deep water for suspension bridges such as Honshyu-Shikoku connecting bridges and newly planned Kitan Strait bridge. For the purpose of earthquake design for such kind of huge foundations, it is necessary that the swaying and rocking motions of the foundation settled on the soil ground should be taken account into the coupled both hydrodynamic and soil dynamic interactions. Quick and accurate prediction method of earthquake response for such a simple rigid cylindrical structure is required for parametric studies of earthquake performance to determine the geometrical configuration of the foundation and the stiffness of the seabed soil ground.

Tanaka and Hudspeth (1988) have developed the analytical solution using eigenfunction expansions to investigate the hydrodynamic interaction between a flexible cylindrical structure and the surrounding compressible fluid and demonstrated the importance of the relative contribution on the fluid compressibility to the dynamic response for the flexible squatty cylinders. Kiokawa (1984,1985) have been investigated the hydrodynamic restoring forces on the axis-symmetric and plane-symmetric rigid bodies using the boundary element method. A new Green's function to satisfy the three-dimensional Helmholtz equation and boundary conditions has been derived by Nakamura and Tanaka (1994) to simulate the dynamic response of the arbitrary shaped three-dimensional flexible structures to earthquake excitation. Hitherto, the generation of the free surface waves and interaction between the structure motions and the ground deformations have not been considered yet, although the fluid compressibility effect on the hydrodynamic restoring forces has been considered.

In this paper, an analytical solution using eigenfunction expansions has been derived which considers both the hydrodynamic interaction between the cylindrical structure and the surrounding fluid, and the soil dynamic interaction between the structure and the soil ground. Through the numerical implementation for the similar pier foundation to the Akashi suspension bridge, the radiation damping due to the hydrodynamic effects by free surface wave and the fluid compressive wave is compared with the radiation damping due to the soil dynamic effect by the progressive shear wave into the foundation ground.

MATHEMATICAL FORMULATION

Governing Equation and Boundary Conditions

Let the vertical circular cylinder of height, h_0, in water of constant depth, h, settled on the semi-infinite elastic soil ground shown in Fig. 1 have a harmonic horizontal motion with frequency ω, at the infinitely deep ground. Assuming that the fluid is compressible and inviscid and that the fluid motion is irrotational, the scalar velocity potential $\Phi(r,\theta,z,t) = \phi(r,\theta,z)\exp(i\omega t)$ can be defined in the fluid domain Ω. The small amplitude fluid motion is governed by the wave equation in cylindrical coordinates. The spatial part of the velocity potential $\phi(r,\theta,z)$ should satisfy the Helmholtz equation as well as free surface, bottom and kinematic body surface boundary conditions;

$$\frac{\partial^2\phi}{\partial r^2}+\frac{1}{r}\frac{\partial\phi}{\partial r}+\frac{1}{r^2}\frac{\partial^2\phi}{\partial\theta^2}+\frac{\partial^2\phi}{\partial z^2}+\frac{\omega^2}{c^2}\phi=0 \quad ;\text{ in } \Omega \tag{1}$$

$$\partial\phi/\partial z-\mu\phi=0 \quad (\mu\equiv\omega^2/g) \quad ;\text{ on } S_f \tag{2}$$

$$\partial\phi/\partial z=0 \quad ;\text{ on } S_b \tag{3}$$

$$\partial\phi/\partial r=i\,\omega\,S(z)\cos\theta \quad ;\text{ on } S_c \tag{4}$$

where c is the acoustic speed in water, g is gravitational constant, a is radius of the cylinder and S(z) is the shape function corresponding to the swaying and rocking motions. In addition, a radiation condition is required for the velocity potential $\Phi(r,\theta,z,t)$ to insure that propagating eigenmodes are only outgoing and that evanescent eigenmodes are bounded. The shape function S(z) can be defined as S(z) = s and S(z) = b(h+z) for swaying and rocking motions of the cylindrical structure, respectively; where s is swaying amplitude and b is rocking amplitude which can not be known in a prior.

When the rigid cylinder settled on the semi-infinite soil ground have the swaying and rocking motions by harmonic earthquake excitation, a mathematical model of two degree of freedom can be defined as a swaying-rocking model (S-R model) as shown in Fig. 2. Taking the equilibrium conditions of the horizontal forces and the moment around the horizontal axis which is point O in Fig. 2, which are made up of components due to the hydrodynamic pressure on the body and due to the soil dynamic pressure on the bottom, the equations of motion can be obtained by

$$m\ddot{u}+C_s(\dot{u}-\dot{x})+K_s(u-x)+mL\ddot{\varphi}=-F_s u-F_r\varphi \tag{5}$$

$$I_0\ddot{\varphi}+C_r\dot{\varphi}+K_r\varphi+mL\ddot{u}=-M_s u-M_r\varphi \tag{6}$$

where x is the absolute displacement of ground at infinite depth, u is the swaying response amplitude, φ is the rocking response amplitude, m is the total mass of the cylinder, I_0 is the mass moment of inertia, K_s and K_r are the dynamic stiffness, C_s and C_r are the damping coefficients in swaying and rocking modes, respectively, F_s and F_r are the horizontal restoring forces due to the unit swaying and rocking motions and M_s and M_r are the rotational restoring moments due to the unit swaying and rocking motions. Because the ground motion at infinite depth x and the cylinder responses u and φ are harmonic, x, u, φ are described by $x=x_0\exp(i\omega t)$, $u=u_0\exp(i\omega t)$ and $\varphi=\varphi_0\exp(i\omega t)$, where x_0 is real amplitude of ground displacement, u_0 and φ_0 are complex amplitude of swaying and rocking modes, respectively. The equations (5) and (6) can be reduced into the following matrix form.

$$\begin{bmatrix} -m\omega^2+i\omega C_s+K_s+F_s & -mL\omega^2+F_r \\ -mL\omega^2+M_s & -I_0\omega^2+i\omega C_r+K_r+M_r \end{bmatrix}\begin{Bmatrix} u_0 \\ \varphi_0 \end{Bmatrix} = \begin{Bmatrix} (K_s+i\omega C_s)x_0 \\ 0 \end{Bmatrix} \tag{7}$$

Solution Using Eigenfunction Expansions

Applying the separation of variable method, the velocity potential solution which satisfies the governing equation and the boundary conditions except the kinematic boundary condition on the body surface given by Eq. (4) may be given by

$$\phi(r,\theta,z)=[A_0H_1^{(2)}(\lambda_0 r)\cosh k_0(h+z)+\sum_{n=1}^{N}B_nH_1^{(2)}(\lambda_n r)\cos k_n(h+z)+\sum_{n=N+1}^{\infty}C_nK_1(\lambda_n' r)\cos k_n(h+z)]\cos\theta \tag{8}$$

where $H_1^{(2)}$ is the Hankel function of the second kind of order one and K_1 is the modified Bessel function of the second kind of order one. Also k_0 indicates the wave number and k_n is eigenvalues which is given as positive real number solutions by the following dispersion equation.

$$k_0\tanh k_0 h=-k_n\tan k_n h=\mu \tag{9}$$

where N in Eq. (8) is the maximum integer value of n which satisfies $\omega^2/c^2 - k_n^2>0$. Also λ_0, and λ_0 are the eigenvalues which should satisfy the following equations.

$$\lambda_0^2\equiv\omega^2/c^2+k_0^2,\quad \lambda_n^2\equiv\omega^2/c^2-k_n^2,\quad \lambda_n'=i\lambda_n \tag{10}$$

The unknown coefficients in the series for velocity potential A_0, B_n, and C_n in Eq. (8) may be derived by satisfying the kinematic boundary condition on the body surface using the orthogonality of the depth-dependent functions, with the result that

$$A_0=\frac{i\omega\int_{-h}^{0}S(z)\cosh k_0(h+z)\,dz}{\lambda_0H_1^{(2)\prime}(\lambda_0 a)\int_{-h}^{0}\cosh^2 k_0(h+z)\,dz} \tag{11}$$

$$B_n=\frac{i\omega\int_{-h}^{0}S(z)\cos k_n(h+z)\,dz}{\lambda_nH_1^{(2)\prime}(\lambda_n a)\int_{-h}^{0}\cos^2 k_n(h+z)\,dz} \tag{12}$$

$$C_n=\frac{i\omega\int_{-h}^{0}S(z)\cos k_n(h+z)\,dz}{\lambda_n'K_1'(\lambda_n' a)\int_{-h}^{0}\cos^2 k_n(h+z)\,dz} \tag{13}$$

where [Bessel function]' denotes the derivative with respect to the argument of the function. The solution given by Eq. (8) give the velocity and pressure field for the swaying and rocking motions of the rigid cylinder. The first term in Eq. (8) denotes the component of the progressive free surface wave, the second term denotes the component of the compressive propagating wave and the third term denotes the component of the evanescent wave existing only near the body surface.

Restoring Force and Moment Coefficient

In order to solve the equation of motion given by Eq. (7), the hydrodynamic restoring force and moment coefficient such as F_s, M_s, F_r, M_r need to be derived. The horizontal hydrodynamic restoring force $f_s(z)$ acting on the cylinder per unit length due to the unit swaying motion can be obtained by integrating around the circumference of the dynamic pressure $p(r,\theta,z)$.

$$f_s(z) = a\int_0^{2\pi} p(r,\theta,z)\cos\theta\, d\theta \qquad ; r = a \qquad (14)$$

where the dynamic pressure $p(r,\theta,z)$ can be obtained by the linearized Bernoulli equation. The hydrodynamic restoring force and moment coefficient F_s and M_s are computed by integrating $f_s(z)$ over the water depth.

$$F_s = \int_{-h}^{0} f_s(z)\,dz \quad , \quad M_s = \int_{-h}^{0} (h+z)\, f_s(z)\,dz \qquad (15)$$

The hydrodynamic coefficient F_r and M_r can be obtained in the same way by using the shape function of rocking mode.

$$F_s = 4\pi a\omega^2\rho \Big[\frac{D_0 k_0 \sinh k_0 h}{k_0 h \sinh k_0 h - \cosh k_0 h + 1} + \sum_{n=1}^{N} \frac{E_n k_n \sin k_n h}{k_n h \sin k_n h + \cos k_n h - 1} + \sum_{n=N+1}^{\infty} \frac{F_n k_n \sin k_n h}{k_n h \sin k_n h + \cos k_n h - 1}\Big] \qquad (16)$$

$$M_s = F_r = 4\pi a\omega^2\rho\,(D_0 + \sum_{n=1}^{N} E_n + \sum_{n=N+1}^{\infty} F_n) \qquad (17)$$

$$M_r = 4\pi a\omega^2\rho \Big[\frac{D_0(k_0 h \sinh k_0 h - \cosh k_0 h + 1)}{k_0 \sinh k_0 h} + \sum_{n=1}^{N} \frac{E_n(k_n h \sin k_n h + \cos k_n h - 1)}{k_n \sin k_n h} + \sum_{n=N+1}^{\infty} \frac{F_n(k_n h \sin k_n h + \cos k_n h - 1)}{k_n \sin k_n h}\Big] \qquad (18)$$

$$D_0 = \frac{\sinh k_0 h(k_0 h \sinh k_0 h - \cosh k_0 h + 1)}{k_0^2 \lambda_0 (\sinh 2k_0 h + 2k_0 h)} \frac{H_1^{(2)}(\lambda_0 a)}{H_1^{(2)\prime}(\lambda_0 a)} \qquad (19)$$

$$E_n = \frac{\sin k_n h(k_n h \sin k_n h + \cos k_n h - 1)}{k_n^2 \lambda_n (\sin 2k_n h + 2k_n h)} \frac{H_1^{(2)}(\lambda_n a)}{H_1^{(2)\prime}(\lambda_n a)} \qquad (20)$$

$$F_n = \frac{\sin k_n h(k_n h \sin k_n h + \cos k_n h - 1)}{k_n^2 \lambda_n' (\sin 2k_n h + 2k_n h)} \frac{K_1(\lambda_n' a)}{K_1'(\lambda_n' a)} \qquad (21)$$

On the other hand, the final velocity potential when the circular cylindrical structure is responding for the earthquake excitation, may be obtained by plugging the swaying and rocking complex amplitude including the phase information into the shape function and superimposing the each velocity potential.

NUMERICAL IMPLEMENTATION

Verification of Analytical Solution

Let us make the verification of the analytical solution taking the example circular cylindrical foundation with the diameter a = 35 m, the water depth h = 70 m (a/h = 0.5) and the height h_0 = 80 m. Three different analytical solutions of the dimensionless restoring force coefficient F_s are compared with each other in Fig. 3; i.e., one solution by Tanaka and Hudspeth (1988) considering the water compressibility but neglecting the gravity wave, another solution by Dalymple and Dean (1972) considering the gravity wave but neglecting the water compressibility and the last present solution considering both effects. For the range of forcing frequency f > 1 Hz, the restoring force increases because of the water compressibility effect, therefore the present solution agrees well with Tanaka and Hudspeth's solution. When the dimensionless frequency ($\Omega = 2h\omega/\pi c$) becomes unity, which is called "First cut-off frequency", the compressive propagating waves shown in the second term in Eq. (8) become significant. At this frequency (f = 5.2 Hz), the dimensionless restoring force coefficient F_s indicates peak values.

On the other hand, for the rage of f < 1 Hz, the present solution agrees well with Dalrymple and Dean's solution. The restoring force decreases as the frequency decreasing and has a minimum value at f = 0.2 Hz because the added mass component decreases for this range of frequency, however for f < 0.2 Hz the radiation damping component due to the gravity wave increases hence the restoring force increases.

Frequency Characteristics of Restoring Forces

Let us investigate the restoring forces coefficient for the same configuration cylindrical foundation of the previous section. The real and imaginary parts of the swaying, rocking and swaying-rocking dimensionless restoring force coefficients are illustrated in Figs. 4, 5, 6, respectively. The real part of the dimensionless restoring force coefficient is defined as added mass coefficient because it is the same phase of the body acceleration, while the imaginary part of that is defined as radiation damping coefficient because it is the same phase of the body velocity.

The real part of the swaying dimensionless restoring force coefficient shown in Fig. 4 demonstrates that the peak value exists near the frequency f=5.8 Hz where the imaginary part of that is gradually growing up because the compressive propagating waves appear. At the range of the frequency $f < 0.4$ Hz, the imaginary part of F_s is indicated small because the radiation surface waves appear.

The rocking dimensionless restoring force coefficient illustrated in Fig. 5 demonstrates almost same tendency as the swaying dimensionless coefficients except that the damping due to the radiation surface waves in rocking mode is larger than that in swaying mode. Since the hydrodynamic pressure due to the radiation surface waves in rocking mode is distributed around the cylinder near the free surface, the dimensionless restoring moment coefficient results in significant for the unit rotation.

The swaying-rocking coupled dimensionless restoring force coefficient is illustrated in Fig. 6 which also demonstrates same tendency as the swaying and rocking dimensionless coefficients. It should be noted that the magnitudes of those coupled coefficients are extremely smaller than that in swaying or rocking mode.

Earthquake Response Characteristics

The dynamic stiffness which Kausel et al. (1975) derived for the rigid circle plate settled on the semi-infinite soil ground considering the interaction between the soil ground and the cylindrical foundation are given by

$$K_s = \frac{8}{2-\nu} Ga \;, \quad C_s = 0.4 \frac{Ga^2}{V_s} \tag{22}$$

$$K_r = \frac{8}{3(1-\nu)} Ga^3 \;, \quad C_r = 0.3 \frac{Ga^6}{V_s} \frac{\omega^2}{V_s^2 + \omega^2 a^2} \tag{23}$$

where ν (=0.35) is the Poison's ratio, V_s (=1000m/sec.) is the shear wave velocity in soil and G (=1800MPa) is the shear modulus in soil. It is noted that the damping coefficient C_r in rocking motion is a function of the frequency ω. The natural frequencies in swaying mode and rocking mode in air calculated using given soil and structure conditions become $f_s = 3.2$ Hz and $f_r = 2.1$ Hz, respectively without considering the coupling effects.

In order to investigate the fluid-soil-structure interaction in swaying and rocking modes, first of all the response in air is illustrated in Fig. 7. The response functions in air both for swaying and rocking modes demonstrate the first peak at the frequency f = 1.841 Hz and the second peak at f = 6.30 Hz. The second peak may be caused by the eccentric moment of the cylinder including in the coupled terms in Eq. (7). The response functions for both modes in incompressible water are shown in Fig. 8. The peak frequencies in incompressible water shifts to smaller frequencies (f = 1.735 Hz and f = 5.90 Hz) compared with that in air, because of the added mass due to evanescent mode. The response functions for both modes in compressible water are illustrated in Fig. 9. Since the compressive propagating waves are generated for f>4 Hz, the added mass component (the real part of F_s and M_r) makes the second peak frequency smaller, i.e. f = 5.9 Hz reduces to f = 5.50 Hz. In addition the radiation damping component (the imaginary part of F_s and M_r) effects on 25% reduction of the peak values of swaying and rocking responses at f = 5.50 Hz.

The rocking motion has significant contributions on the response because this model does not include the internal energy dissipation damping in the soil. Experiments seem to indicate the internal damping is independent of frequency but is dependent of shear strain of the soil. Kausel et al. (1975) indicated that the effect of the internal damping on the damping coefficients is very small for rocking motion. It is noted, however, the internal damping which increases as the shear strain increases, must smooth out the peak at the first natural frequency.

Damping Factor of Soil Ground and Fluid Domain

The radiation damping for both soil ground and fluid domain exists analytically in our proposed S-R model. The damping factors for the swaying and rocking motions are compared with by soil ground and by fluid domain in Figs. 10 and 11, respectively. The normalization of the swaying and rocking modes is carried out by the critical damping coefficient without considering the coupled effect.

For both cases of the swaying and the rocking motion, the radiation damping due to the water compressibility (the imaginary part of F_s and M_r) grows up for over the first cut-off frequency (f > 5 Hz). The ground radiation damping of rocking motion is an increasing function of frequency as shown in Fig. 11. Comparing the radiation damping due to soil motion and due to fluid motion, it is noted that the radiation damping caused by the compressive propagating waves effects mainly on the earthquake response of the structure for over the first cut-off frequency.

CONCLUSIONS

A mathematical S-R model to simulate the earthquake response of a rigid cylindrical offshore structure is proposed and an analytical solution considering fluid-soil-structure interaction has been developed. This model is practical for the earthquake design of the cylindrical foundation because of easy calculation. Through the numerical implementation of a typical cylindrical foundation with a/h = 0.5, the following results are obtained.

(1) The radiation damping force due to gravity waves are generated near the free surface for earthquake excitation with long period (3 - 5 seconds), however it effects little on the response.

(2) The added mass and radiation damping restoring force components over the first cut-off frequency effect a lot on the response as well as on the resonance frequencies.

(3) The radiation damping due to the compressive propagating waves of fluid is much larger than that due to the shear propagating waves of soil over the first cut-off frequency. The radiation damping component effects on 25% reduction of the peak values of swaying and rocking responses at f = 5.50 Hz for a typical cylindrical foundation with a/h = 0.5.

REFERENCES

Dalymple, RA, and Dean, RG (1972). "The Spiral Wavemaker for Littoral Drift Studies," *Proc 13th ICEE*, pp.689-705.

Kausel, E, and Rosset, JM (1975) "Dynamic Stiffness of Circular Foundations," *Proc ASCE*, Vol. 101, No. EM6, pp.771-785.

Kiokawa, T (1984) "Added Mass of Axis-symmetric Structure Considering Water Compressibility," *Proc 31th coastal engr in*

Japan, pp.471-475.

Kiokawa, T (1985) "Added Mass of Plane-symmetric Structure Considering Water Compressibility," *Proc 32th coastal engr in Japan*, pp.560-564.

Nakamura, T, and Tanaka, Y (1994) "Twist Earthquake Response of 3-D Offshore Structures Considering Compressive Fluid and Flexible Structures Interaction," *Proc JSCE Journall*, No.485/II-26, pp.366-370.

Tanaka, Y, and Hudspeth, RT (1988) "Restoring Forces on Vertical Circular Cylinders Forced by Earthquake," *Earthquake Engr and Structural Dynamics*, Vol.16, pp.99-119.

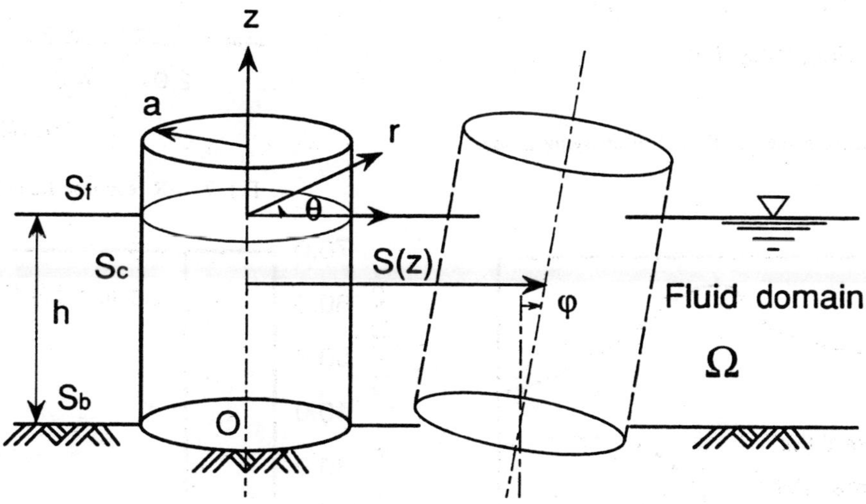

Fig. 1 Definition of fluid domain and coordinate system

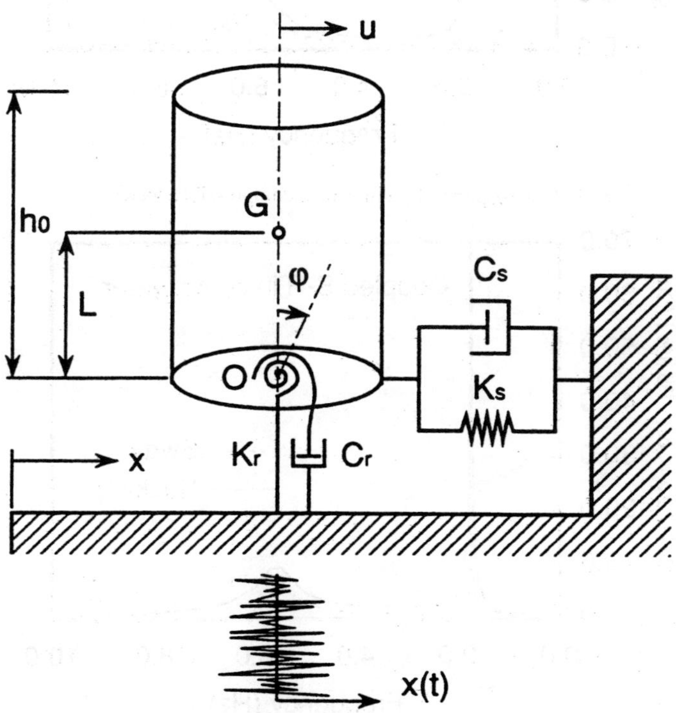

Fig. 2 Swaying - Rocking model

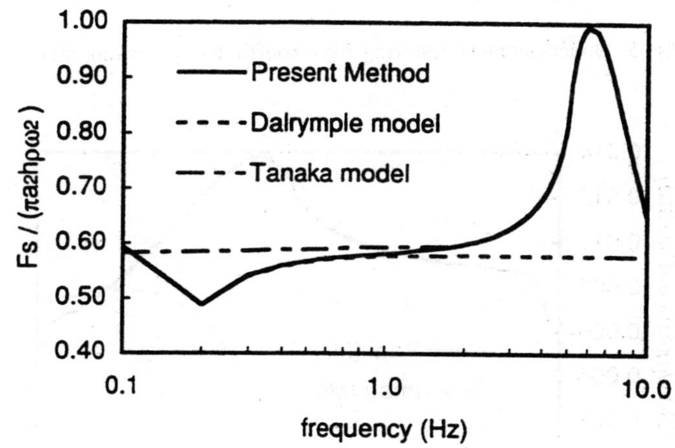

Fig. 3 Comparison of three different analytical solutions

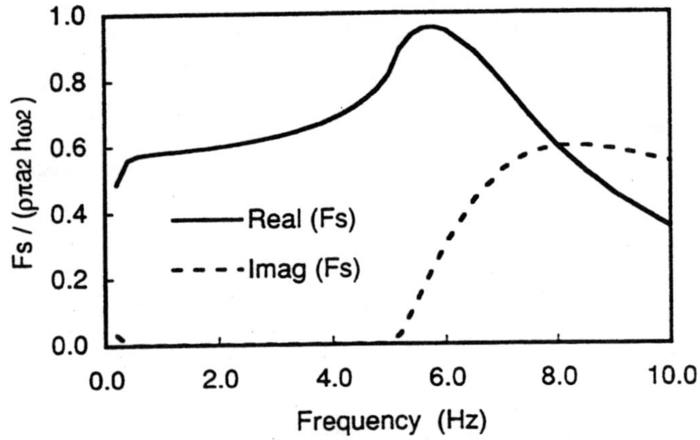

Fig. 4 Dimensionless restoring force coefficient of swaying mode

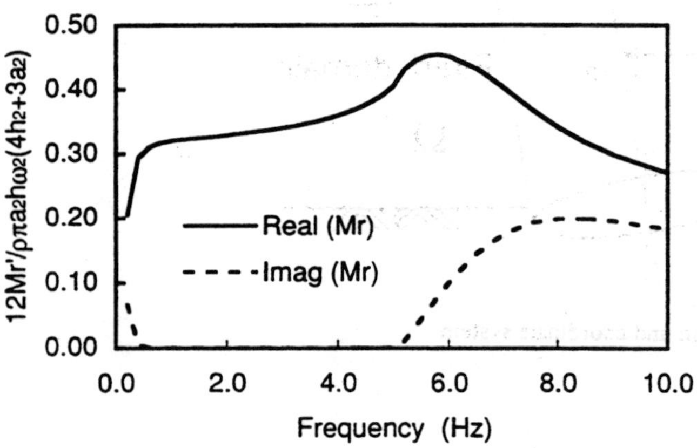

Fig. 5 Dimennsionless restoring force coefficient of rocking mode

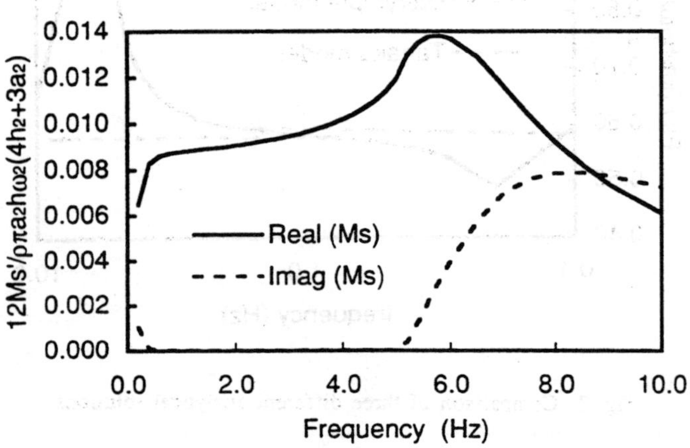

Fig. 6 Coupled restoring force coefficient of swaying-rocking modes

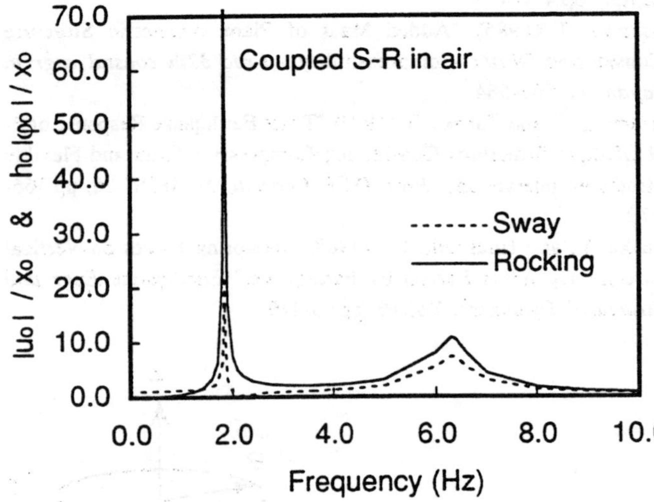

Fig. 7 S-R response function in air

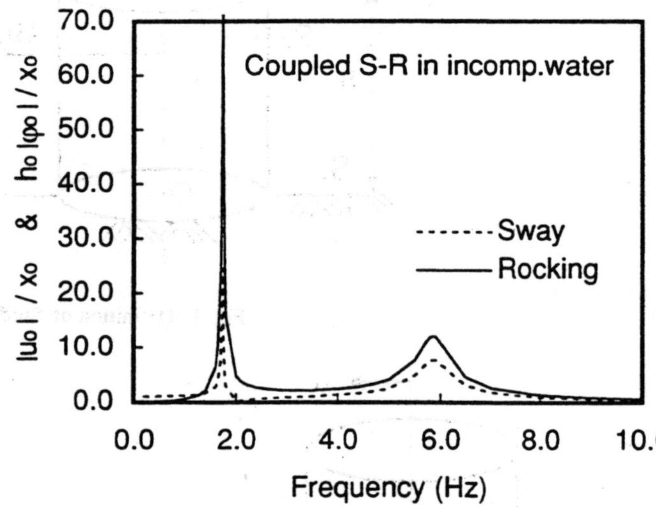

Fig. 8 S-R response function in incompressible water

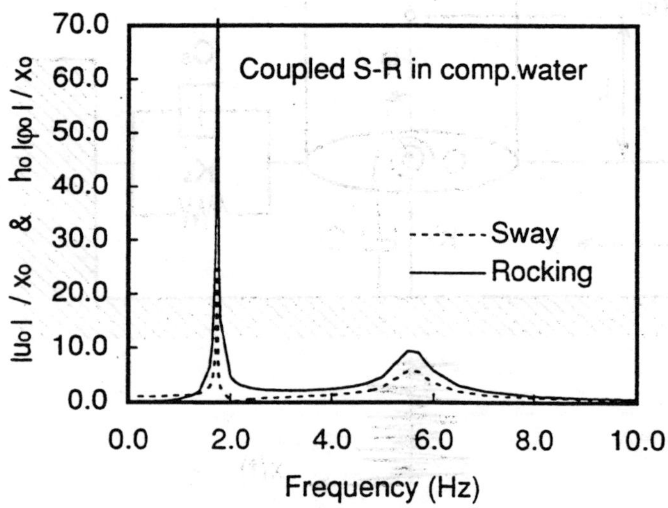

Fig. 9 S-R response function in compressible water

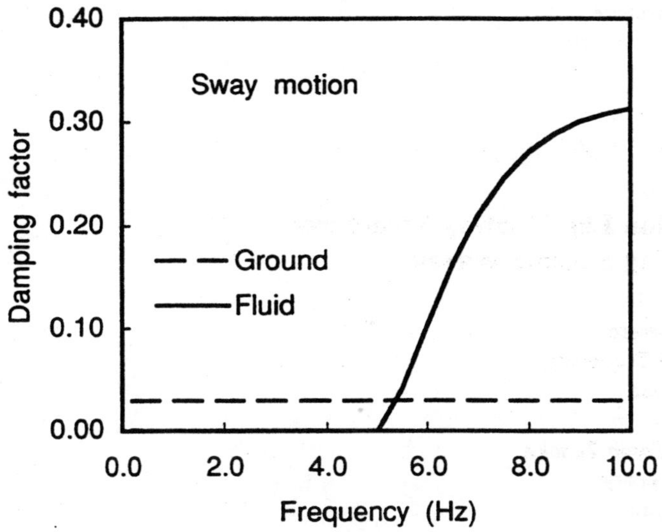

Fig. 10 Radiation damping due to ground and fluid in swaying motion

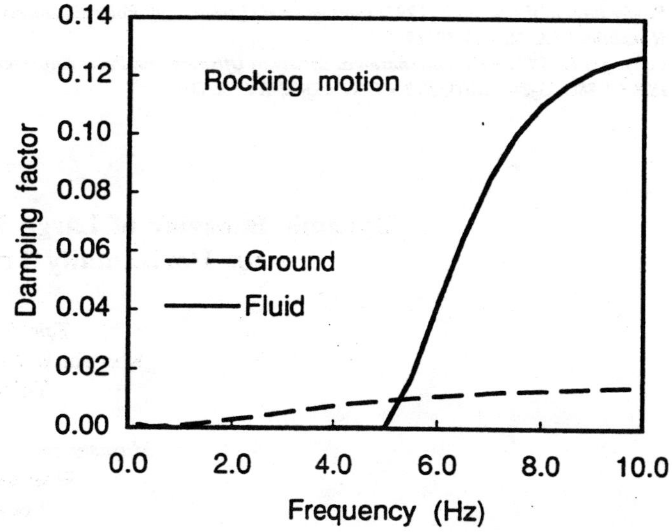

Fig. 11 Radiation damping due to ground and fluid in rocking motion

Proceedings of the Seventh (1997) International Offshore and Polar Engineering Conference
Honolulu, USA, May 25-30, 1997

ISBN 1-880653-28-1 (Set); ISBN 1-880653-32-X (Vol. IV)

Dynamic Behavior of Large Tension Leg Floating Structures for Horizontally Traveling Seismic Waves

Takuji Hamamoto
Musashi Institute of Technology
Tokyo, Japan

Masashi Inoue and Yasuo Tanaka
Waseda University
Tokyo, Japan

ABSTRACT

Stochastic response of large tension leg floating structures for horizontally traveling seismic waves on sea bed is formulated, taking account of structural flexibility and fluid structure interaction. The floating structure is idealized as an elastic circular plate with a series of linear springs. Hydrodynamic pressure acting on the floating structure is obtained in closed form on the basis of a linear potential flow theory. Response quantities are evaluated by a wet mode superposition approach and stationary random vibration theory. Based on the numerical results, the dynamic behavior of large tension leg floating structures is discussed for horizontally traveling seismic waves.

KEY WORDS: Large floating structure, tension leg mooring, fluid structure interaction, seaquake, horizontally traveling seismic wave, potential flow theory, elastic circular plate

INTRODUCTION

Tension leg mooring is one of the promising methods to reduce the elastic deformation of large floating structures for wind induced waves (Hamamoto *et al.*, 1991). However, in the seismically active regions, earthquake-related forces are another threat to such structures. In the case of tension leg mooring, seismic waves directly propagate through tension legs (Kawanishi *et al.*, 1987) in addition to seaquake phenomenon (Baba, 1988 ; Liou *et al.*, 1988). Taking account of structural flexibility, Babu and Reddy (1986) have formulated a two dimensional finite element analysis of a floating plant and Hamamoto (1994a) have formulated a three dimensional response analysis of a large circular floating island. However, these studies have only concerned with vertically incident seismic waves.

When subjected to horizontally traveling seismic waves with apparent velocities, the sea bed moves with phase difference in the vertical direction. Consequently, the dynamic tendon force as well as hydrodynamic pressure due to seaquakes act on the lower surface of moored floating structures with phase difference. The frequency content of vertical ground motion has wide band nature. As a result, the spatial response distribution of large tension leg floating structures may be considerably complicated according to the excitation of a number of higher modes of vibration for horizontally traveling seismic waves.

In this study, the dynamic behavior of large tension leg floating structures subjected to horizontally traveling seismic waves is formulated within the framework of hydroelasticity. The floating structure is idealized as an elastic circular plate. The mooring system is assumed to be composed of a series of tension legs which connect from the lower surface of the structures to the sea bed. The hydrodynamic pressure acting on the floating structures is evaluated by a linear potential flow theory. The fluid coupled modal equations of motion of the floating structures are derived by energy method. Structural responses such as displacement, acceleration and internal forces are evaluated by a wet mode superposition approach and stationary random vibration theory. Based on the numerical results, the emphasis is placed on the difference in the dynamic behavior of the floating structures with and without tension legs for horizontally traveling seismic waves.

ANALYTICAL MODEL AND ASSUMPTIONS

The analytical model of a large tension leg floating structure subjected to horizontally traveling seismic waves on sea bed is shown in Fig.1. In the figure, a is the radius of the floating structure, h is the thickness, d is the water depth, $\bar{d}$ is the distance from the sea bed to the lower surface of the floating structure, U_g is the vertical displacement of the sea bed, and C_a is the apparent velocity of horizontally traveling seismic waves. Cylindrical coordinate (r, θ, z) is used. The origin is located at the projected point of the center of the floating structure on the sea bed. As shown in Fig.2, three mooring states of the floating structure are considered : (a) freely floating case without tension legs, (b) circumferential mooring case in which tension legs are attached around the circumference, and (c) distributed mooring case in which tension legs are attached over the lower surface. The following assumptions are introduced in this study:

1. The floating structure is isotropic and elastic.
2. The sea water is irrotational, inviscid and incompressible.
3. The tendon force of mooring system is linearly elastic.
4. Mass effect of the mooring system is ignored.
5. Tension legs transfer only vertical tension force.
6. The vertical ground motion is caused by horizontally traveling seismic waves with apparent velocities.
7. The stationary part of the vertical ground acceleration at sea bed is ergodic and zero-mean Gaussian process.
8. The water depth is constant and sea water extends to infinity.

HYDRODYNAMIC PRESSURE

The hydrodynamic pressure acting on the floating structure is evaluated by a linear potential flow theory. The hydrodynamic pressure may be expressed as a linear combination of the pressure component p_f induced by seaquakes on the motionless floating structure and the pressure component p_m induced by the motion of the floating structure, respectively, as shown in

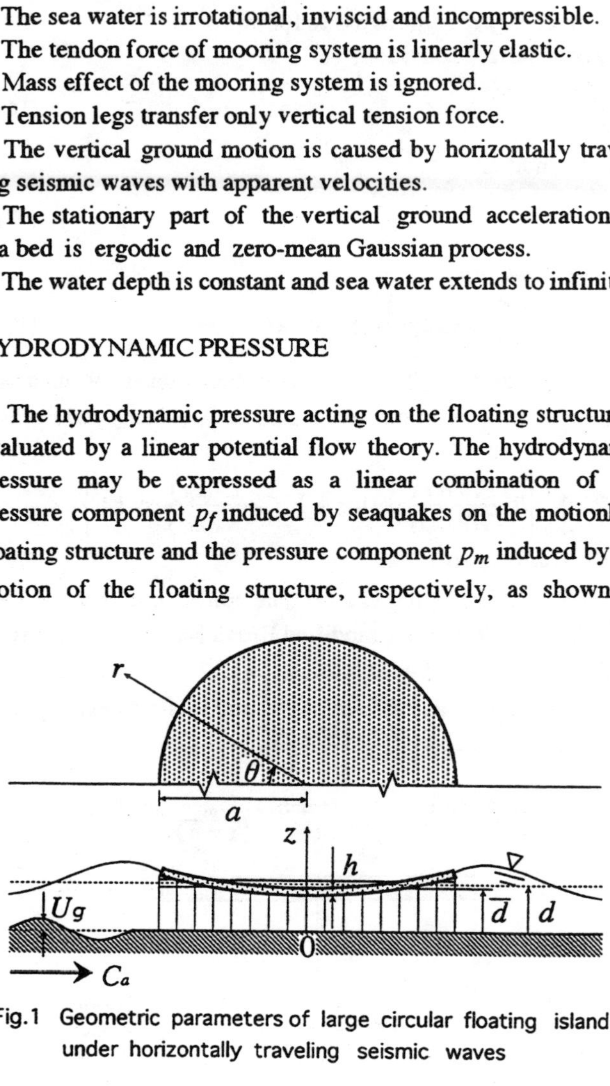

Fig.1 Geometric parameters of large circular floating island under horizontally traveling seismic waves

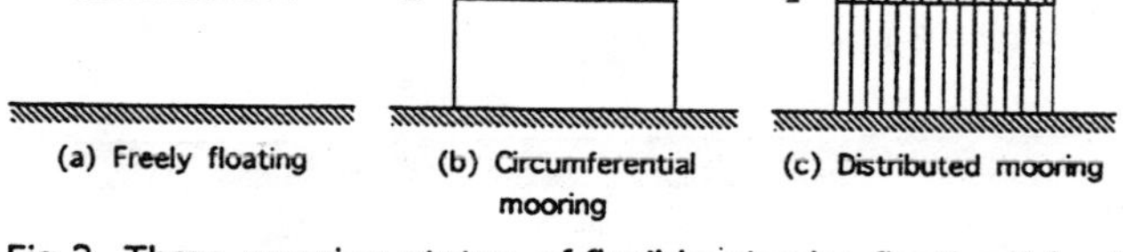

Fig.2 Three mooring states of flexible circular floating island

Fig.3. Moreover, the pressure component p_f may be divided into two components, p_i and p_s which are generated by incident and scattered waves of seaquakes, respectively, as shown in Fig.4. Because of the geometric simplicity, each pressure component can be obtained in closed form in terms of Bessel function families.

Incident wave

The hydrodynamic pressure acting on the lower surface of the motionless structure due to incident wave of seaquakes may be evaluated in the free field without the floating structure. The linearized governing equation for an incompressible and inviscid fluid is given by

$$\nabla^2 \phi_i = 0, \qquad (1)$$

in which ϕ_i is the velocity potential function of incident wave and ∇^2 is Laplace operator.

The boundary conditions are as follows :

Free water surface condition,

$$\frac{\partial \phi_i}{\partial z} + \frac{1}{g}\frac{\partial^2 \phi_i}{\partial t^2} = 0 \qquad \text{at} \quad z = d. \qquad (2)$$

Sea bed condition,

$$\frac{\partial \phi_i}{\partial z} = \dot{U}_g \qquad \text{at} \quad z = 0, \qquad (3)$$

in which g is the gravitational acceleration, t is time, and $\cdot = \partial / \partial t$. If seismic waves propagate with apparent velocity C_a on the sea bed, the vertical ground acceleration with phase difference may be simulated by the superposition of harmonic wave components as follows:

$$\ddot{U}_g(r, \theta, t) = \sum_{l=1}^{\infty} \sqrt{2 S_{\ddot{U}_g \ddot{U}_g}(\sigma_l) \Delta\sigma}\, \exp\left[i(\sigma_l t - k_l r \cos\theta + \varphi_l)\right], \qquad (4)$$

in which $S_{\ddot{U}_g \ddot{U}_g}(\sigma_l)$ is the power spectral density function of the vertical ground acceleration, σ_l, k_l, and φ_l are the circular

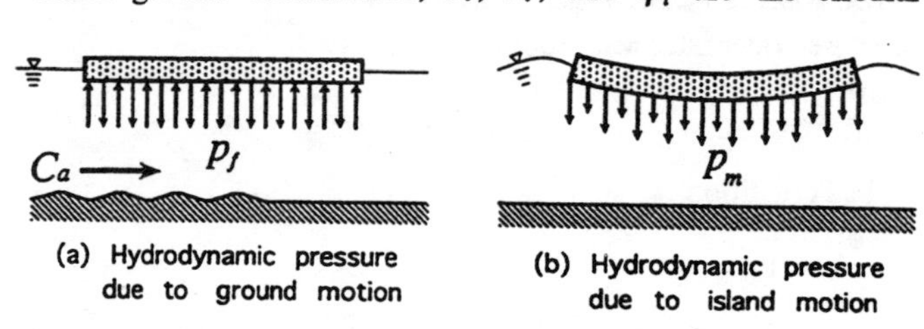

Fig.3 Two components of hydrodynamic pressure when subjected to seaquakes

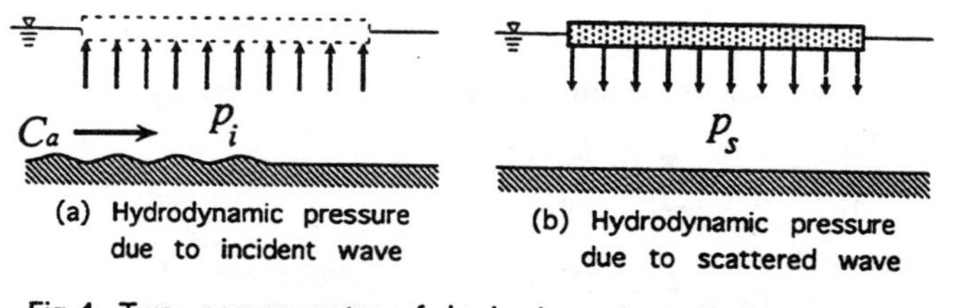

Fig.4 Two components of hydrodynamic pressure due to ground motion

frequency, wave number and phase of the l -th component wave, respectively, and $\Delta=(\sigma_{l+1}-\sigma_{l-1})/2$. The relationship between circular frequency and wave number is given by

$$k_l=\frac{\sigma_l}{C_a}. \tag{5}$$

The following Fourier-Bessel series expansion is introduced,

$$\exp\left[-ik_l r\cos\theta\right]=\sum_{n=0}^{\infty}\gamma_n(-i)^n J_n(k_l r)\cos n\theta, \tag{6}$$

in which $J_n(k_l r)$ is the Bessel function of n-th order and $\gamma_0=1$ and $\gamma_n=2$ $(n\geq 1)$. Incorporating Eq.(6) into Eq.(4) yields

$$\ddot{U}_g(r,\theta,t)=\sum_{l=1}^{\infty}\sum_{n=0}^{\infty}\sqrt{2S_{\ddot{U}_g\ddot{U}_g}(\sigma_l)\Delta\sigma}\,\gamma_n(-i)^n J_n(k_l r)\cdot\cos n\theta\left[i(\sigma_l t+\varphi_l)\right]. \tag{7}$$

Solving Eqs.(1)-(3) yields the velocity potential function of incident wave as

$$\phi_i=\sum_{n=0}^{\infty}\frac{V_0\gamma_n(-i)^n}{k}\frac{-kg\cosh k(d-z)+\sigma^2\sinh k(d-z)}{kg\sinh kd-\sigma^2\cosh kd}\cdot J_n(kr)\cos n\theta\exp(i\sigma t), \tag{8}$$

in which $V_0=A_0/i\sigma$ and $A_0=\sqrt{2S_{\ddot{U}_g\ddot{U}_g}(\sigma_l)\Delta\sigma}$. The subscript l is omitted in Eq.(8) for simplicity.

Scattered wave

Due to the existence of the floating structure, the incident wave is reflected and scattered at the lower surface of the floating structure. To solve a scattering problem, the fluid domain is divided into interior and exterior regions as shown in Fig.5.

The governing equation in the interior region is given by

$$\nabla^2\phi_s^{(i)}=0, \tag{9}$$

in which $\phi_s^{(i)}$ is the velocity potential function of scattered wave in the interior region.

The boundary conditions are as follows :

Structure-water interface condition,

$$\frac{\partial\phi_s^{(i)}}{\partial z}=-\frac{\partial\phi_i}{\partial z} \quad \text{at} \quad z=\bar{d}, \tag{10}$$

Sea bed condition,

$$\frac{\partial\phi_s^{(i)}}{\partial z}=0 \quad \text{at} \quad z=0. \tag{11}$$

Solving Eqs.(9)-(11) yields the velocity potential function of scattered wave in the interior region as

$$\phi_s^{(i)}=\phi_{sh}^{(i)}+\phi_{sp}^{(i)}, \tag{12}$$

in which $\phi_{sh}^{(i)}$ and $\phi_{sp}^{(i)}$ are the homogeneous and particular solutions of Laplace equation, respectively, and given by

$$\phi_{sh}^{(i)}=\sum_{n=0}^{\infty}\left\{D_{n0}\left(\frac{r}{a}\right)^n+\sum_{s=1}^{\infty}D_{ns}\frac{I_n(l_s r)}{I_n(l_s a)}\cos l_s z\right\}\cos n\theta\exp(i\sigma t), \tag{13}$$

$$\phi_{sp}^{(i)}=\sum_{n=0}^{\infty}\sum_{s=1}^{\infty}\gamma_m(-i)^n\frac{2V_0}{a\lambda_{ns}}\frac{kg\sinh k(d-\bar{d})-\sigma^2\cosh k(d-\bar{d})}{kg\sinh kd-\sigma^2\cosh kd}\cdot\frac{J_n(\lambda_{ns}^* r)}{J_{n+1}^2(\lambda_{ns})}\frac{\cosh\lambda_{ns}^* z}{\sinh\lambda_{ns}^*\bar{d}}\cdot\int_0^a J_n(kr)J_n(\lambda_{ns}^* r)rdr\cdot\cos n\theta\exp(i\sigma t), \tag{14}$$

in which $I_n(l_s r)$ is the modified Bessel function of n-th order of the first kind, $l_s=s\pi/\bar{d}$, D_{n0} and D_{ns} ($s=1, 2, \cdots$) are unknown coefficients, λ_{ns} is the s-th positive root of $J_n(\lambda_{ns})=0$, and $\lambda_{ns}^*=\lambda_{ns}/a$.

On the other hand, the governing equation in the exterior region is given by

$$\nabla^2\phi_s^{(e)}=0, \tag{15}$$

in which $\phi_s^{(e)}$ is the velocity potential function of scattered wave in the exterior region.

The boundary conditions are as follows :

Free water surface condition,

$$\frac{\partial\phi_s^{(e)}}{\partial z}+\frac{1}{g}\frac{\partial^2\phi_s^{(e)}}{\partial t^2}=0 \quad \text{at} \quad z=d, \tag{16}$$

Sea bed condition,

$$\frac{\partial\phi_s^{(e)}}{\partial z}=0 \quad \text{at} \quad z=0, \tag{17}$$

Radiation condition,

$$\lim_{r\to\infty}\sqrt{r}\left(\frac{\partial\phi_s^{(e)}}{\partial z}+ik\phi_s^{(e)}\right)=0 \quad \text{at} \quad r\to\infty. \tag{18}$$

Solving Eqs. (15)-(20) yields the velocity potential function of scattered wave in the exterior region as (MacCamy and Fuchs, 1954)

$$\phi_s^{(e)}=\sum_{n=0}^{\infty}\left\{B_{n0}\frac{H_n^{(2)}(kr)}{H_n^{(2)}(ka)}\cosh kz+\sum_{j=0}^{\infty}C_{nj}\frac{K_n(k_j r)}{K_n(k_j a)}\cos k_j z\right\}\cdot\cos n\theta\exp(i\sigma t), \tag{19}$$

in which $H_n^{(2)}(kr)$ is the Hankel function of n-th order of the second kind, $K_n(k_j r)$ is the modified Bessel function of n-th order of the second kind, B_{n0} and C_{nj} ($j=1, 2, \cdots$) are unknown coefficients, and k and k_j are wave numbers satisfied by the following equation:

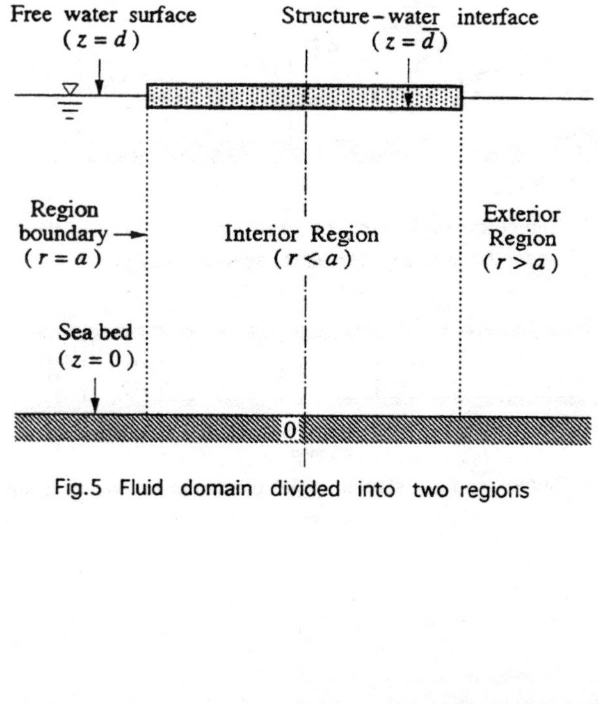

Fig.5 Fluid domain divided into two regions

$$\sigma^2 = kg\tanh kd = -k_j g \tan k_j d. \tag{20}$$

The unknown coefficients can be determined by imposing the following continuity conditions between interior and exterior regions:

Kinematic continuity condition,

$$\frac{\partial \phi_s^{(e)}}{\partial r} = \frac{\partial \phi_s^{(i)}}{\partial r} \quad \text{at} \quad r=a,\ 0 \le z \le \bar{d}, \tag{21}$$

$$\frac{\partial \phi_s^{(e)}}{\partial r} = 0 \quad \text{at} \quad r=a,\ \bar{d} \le z \le d, \tag{22}$$

Pressure continuity condition,

$$\phi_s^{(e)} = \phi_s^{(i)} \quad \text{at} \quad r=a,\ 0 \le z \le \bar{d}. \tag{23}$$

The hydrodynamic pressure component p_f may be obtained by the linearized Bernoulli equation:

$$p_f = -\rho_w \left(\frac{\partial \phi_i}{\partial t} + \frac{\partial \phi_s^{(i)}}{\partial t} \right) \Bigg|_{z=\bar{d}}, \tag{24}$$

in which ρ_w is the mass density of sea water.

Radiated wave

When subjected to horizontally traveling seismic waves with apparent velocity, the sea bed moves with phase difference in the vertical direction. Different from the axisymmetric response for vertically incident seismic waves, however, the hydrodynamic pressure associated with horizontally traveling seismic waves acts on the lower surface of the floating structure with time lag in the propagating direction. As a result, the floating structure responds with rigid body motion of heave and pitch as well as elastic deformation of higher order circumferential Fourier wave number. Making using of a wet mode superposition approach, the total displacement of the floating structure can be expressed as

$$\zeta = \sum_{n=0}^{\infty} \sum_{m=1}^{\infty} \zeta_{nm}(r) \cos n\theta\, q_{nm}(t), \tag{25}$$

in which n is the circumferential Fourier wave number, m is the radial mode number, $\zeta_{nm}(r)$ is the nm-th wet mode shape function in the radial direction along $\theta=0$, and $q_{nm}(t)$ is the nm-th generalized coordinate. The wet mode shape function of the floating structure can be obtained by fluid coupled free vibration analysis (Hamamoto *et al.*, 1994b).

The hydrodynamic pressure component p_m induced by rigid body motion as well as elastic deformation of the floating structure can be obtained for the motion of the floating structure in still water. The fluid domain is divided into interior and exterior regions in the same way as a scattering problem.

The governing equation in the interior region is given by

$$\nabla^2 \phi_m^{(i)} = 0, \tag{26}$$

in which $\phi_m^{(i)}$ is the velocity potential function of radiated wave in the interior region.

The boundary conditions are as follows:

Structure-water interface condition,

$$\frac{\partial \phi_m^{(i)}}{\partial z} = \frac{\partial \zeta}{\partial t} \quad \text{at} \quad z=\bar{d}, \tag{27}$$

Sea bed condition,

$$\frac{\partial \phi_m^{(i)}}{\partial z} = 0 \quad \text{at} \quad z=0. \tag{28}$$

Solving Eqs.(26)-(28) yields the velocity potential function of radiated wave in the interior region as

$$\phi_m^{(i)} = \phi_{mh}^{(i)} + \phi_{mp}^{(i)}, \tag{29}$$

in which $\phi_{mh}^{(i)}$ and $\phi_{mp}^{(i)}$ are the homogeneous and particular solutions of Laplace equation, respectively. $\phi_{mh}^{(i)}$ may be expressed by replacing the subscript s by m in Eq.(13). $\phi_{mp}^{(i)}$ is expressed as follows:

$$\phi_{mp}^{(i)} = \sum_{n=0}^{\infty} \sum_{m=1}^{\infty} \sum_{s=1}^{\infty} \frac{2i\sigma}{a\lambda_{ns}} \frac{J_n(\lambda_{ns}^* r)}{J_{n+1}^2(\lambda_{ns})} \frac{\cosh \lambda_{ns}^* z}{\sinh \lambda_{ns}^* \bar{d}} \int_0^a \zeta_{nm}(r) J_n(\lambda_{ns}^* r) r dr \cdot q_{nm}(i\sigma) \cos n\theta \exp(i\sigma t), \tag{30}$$

in which $q_{nm}(i\sigma)$ is the complex amplitude of the generalized coordinate and is defined as $q_{nm}(t) = q_{nm}(i\sigma)\exp(i\sigma t)$.

On the other hand, the linearized governing equation in the exterior region is given by

$$\nabla^2 \phi_m^{(e)} = 0, \tag{31}$$

in which $\phi_m^{(e)}$ is the velocity potential function of radiated wave in the exterior region.

The boundary conditions are as follows:

Free water surface condition,

$$\frac{\partial \phi_m^{(e)}}{\partial z} + \frac{1}{g} \frac{\partial^2 \phi_m^{(e)}}{\partial t^2} = 0 \quad \text{at} \quad z=d, \tag{32}$$

Sea bed condition,

$$\frac{\partial \phi_m^{(e)}}{\partial z} = 0 \quad \text{at} \quad z=0, \tag{33}$$

Radiation condition,

$$\lim_{r \to \infty} \sqrt{r} \left(\frac{\partial \phi_m^{(e)}}{\partial z} + ik\phi_s^{(e)} \right) = 0 \quad \text{at} \quad r \to \infty. \tag{34}$$

Solving Eqs. (31)-(34) yields the velocity potential function in the exterior region $\phi_{mh}^{(e)}$. $\phi_{mh}^{(e)}$ may be expressed by replacing the subscript s by m in Eq.(19).

The unknown coefficients can be determined by imposing continuity conditions at interior and exterior region interface.

The hydrodynamic pressure component p_m may be obtained by the linearized Bernoulli equation:

$$p_m = -\rho_w \frac{\partial \phi_m^{(i)}}{\partial t} \Bigg|_{z=\bar{d}} - \rho_w g \zeta. \tag{35}$$

DYNAMIC TENDON FORCE

When tension leg mooring is attached to large floating structures, seismic waves directly propagate through the mooring system in addition to hydrodynamic pressure due to seaquakes.

In the case of circumferential mooring, the dynamic tendon force acting on unit length of the circumference of the floating structure is given by

$$p_c(r=a,\theta,t)=-k_c(r=a,\theta)\{\zeta(r=a,\theta,t)-U_g(r=a,\theta,t)\}, \quad (36)$$

in which $k_c(r=a,\theta)$ is the stiffness per unit length of the circumference and $U_g(r=a,\theta,t)$ is the displacement of vertical ground motion at sea bed.

In the case of distributed mooring, the dynamic tendon force acting on unit area of the lower surface of the floating structure is given by

$$p_d(r,\theta,t)=-k_d(r,\theta)\{\zeta(r,\theta,t)-U_g(r,\theta,t)\}, \quad (37)$$

in which $k_d(r,\theta)$ is the stiffness per unit area of the lower surface.

In this study, the mooring stiffness is assumed to be independent and constant in space. Consequently, the stiffnesses of circumferential and distributed moorings are given by $k_c(r=a,\theta)=k_c$ and $k_d(r,\theta)=k_d$, respectively.

EQUATION OF MOTION

Having obtained the hydrodynamic pressure components, p_f and p_m, and dynamic tendon force, the equation of motion of the floating structure may be derived. The dynamic behavior of the floating structure is governed by the Lagrange equation:

$$\frac{d}{dt}\left(\frac{\partial T}{\partial \dot{q}_{nm}(t)}\right)-\frac{\partial T}{\partial q_{nm}(t)}+\frac{\partial S}{\partial q_{nm}(t)}=Q^L_{nm}(t)+Q^D_{nm}(t), \quad (38)$$

in which $Q^L_{nm}(t)$ and $Q^D_{nm}(t)$ are the nm-th generalized force and damping force, respectively, and T and S are the kinetic energy and strain energy of the floating structure, respectively, given by

$$T=\frac{\rho_P h}{2}\int_0^a\int_0^{2\pi}\left(\frac{\partial\zeta}{\partial t}\right)^2 r\,d\theta\,dr, \quad (39)$$

$$S=\frac{D}{2}\int_0^a\int_0^{2\pi}\left[\left(\frac{\partial^2\zeta}{\partial r^2}+\frac{1}{r}\frac{\partial\zeta}{\partial r}+\frac{1}{r^2}\frac{\partial^2\zeta}{\partial\theta^2}\right)^2 -2(1-\nu)\left\{\frac{\partial^2\zeta}{\partial r^2}\left(\frac{1}{r}\frac{\partial\zeta}{\partial r}+\frac{1}{r^2}\frac{\partial^2\zeta}{\partial\theta^2}\right) -\left(\frac{1}{r}\frac{\partial^2\zeta}{\partial r\partial\theta}-\frac{1}{r^2}\frac{\partial\zeta}{\partial\theta}\right)^2\right\}\right] r\,d\theta\,dr, \quad (40)$$

in which ρ_P is the mass density of the structure, $D=Eh^3/12(1-\nu^2)$ is the flexural rigidity of the structure (E is Young's modulus and ν is Poison's ratio).

Making use of orthogonal properties of wet mode shapes of the floating structure, the nm-th uncoupled modal equations of motion can be obtained as follows:

$$(M^P_{nm}+M^W_{nm})\ddot{q}_{nm}(t)+(C^P_{nm}+C^W_{nm})\dot{q}_{nm}(t) + (K^P_{nm}+K^W_{nm}+K^T_{nm})q_{nm}(t)=Q_{nm}(t), \quad (41)$$

in which M^P_{nm}, C^P_{nm} and K^P_{nm} represent the nm-th generalized mass, generalized damping and generalized stiffness of the floating structure, respectively, given by

$$M^P_{nm}=\varepsilon_n\pi\rho_P h\int_0^a\zeta^2_{nm}(r)r\,dr, \quad (42)$$

$$C^P_{nm}=2\xi_{nm}\omega_{nm}M^P_{nm}, \quad (43)$$

$$K^P_{nm}=\omega^2_{nm}M^P_{nm}, \quad (44)$$

in which ξ_{nm} and ω_{nm} are the nm-th dry mode material damping ratio and natural circular frequency of the structure, respectively, and $\varepsilon_0=2$ and $\varepsilon_n=1$ $(n\geq 1)$. M^W_{nm}, C^W_{nm} and K^W_{nm} represent the nm-th generalized added mass, generalized added damping and generalized added stiffness, respectively, given by

$$M^W_{nm}=\frac{\varepsilon_n\pi}{\sigma^2|q_{nm}(i\sigma)|^2}\left\{Re\left[\int_0^a p_{mh}(r,\sigma)\zeta_{nm}(r)r\,dr\right]\cdot Re\left[q_{nm}(i\sigma)\right] + Im\left[\int_0^a p_{mh}(r,\sigma)\zeta_{nm}(r)r\,dr\right]\cdot Im\left[q_{nm}(i\sigma)\right]\right\} + \frac{\varepsilon_n\pi}{\sigma^2 q_{nm}(i\sigma)}\int_0^a p_{mp}(r,\sigma)\zeta_{nm}(r)r\,dr, \quad (45)$$

$$C^W_{nm}=\frac{\varepsilon_n\pi}{\sigma|q_{nm}(i\sigma)|^2}\left\{Re\left[\int_0^a p_{mh}(r,\sigma)\zeta_{nm}(r)r\,dr\right]\cdot Im\left[q_{nm}(i\sigma)\right] + Im\left[\int_0^a p_{mh}(r,\sigma)\zeta_{nm}(r)r\,dr\right]\cdot Re\left[q_{nm}(i\sigma)\right]\right\}, \quad (46)$$

$$K^W_{nm}=\varepsilon_n\pi\rho_w\int_0^a\zeta^2_{nm}(r)r\,dr. \quad (47)$$

$Q_{nm}(t)$ is the nm-th generalized force given by

$$Q_{nm}(t)=Q^F_{nm}(t)+Q^T_{nm}(t), \quad (48)$$

in which $Q^F_{nm}(t)$ is the nm-th generalized force associated with the hydrodynamic pressure p_f given by

$$Q^F_{nm}=\varepsilon_n\pi\int_0^a p_f\zeta_{nm}(r)r\,dr. \quad (49)$$

$Q^T_{nm}(t)$ is the nm-th generalized added force which is caused by the dynamic tendon force due to direct propagation of seismic waves. Corresponding to mooring states, the generalized added force is given as follows:

Freely floating case,

$$Q^T_{nm}(t)=0, \quad (50a)$$

Circumferential mooring case,

$$Q^T_{nm}(t)=\varepsilon_n\pi a k_c U_g(r=a,t)\zeta_{nm}(a), \quad (50b)$$

Distributed mooring case,

$$Q^T_{nm}(t)=\varepsilon_n\pi k_d\int_0^a U_g(r,t)\zeta_{nm}(r)r\,dr. \quad (50c)$$

K^T_{nm} is the generalized added stiffness which is caused by the restoring force of mooring system. Corresponding to mooring states, the generalized added stiffness is given as follows:

Freely floating case,

$$K^T_{nm}=0, \quad (51a)$$

Circumferential mooring case,

$$K^T_{nm}=\varepsilon_n\pi a k_c\zeta^2_{nm}(a), \quad (51b)$$

Distributed mooring case,

$$K^T_{nm}(t)=\varepsilon_n\pi k_d\int_0^a\zeta^2_{nm}(r)r\,dr. \quad (51c)$$

Making use of the relation:

$$K^P_{nm}+K^W_{nm}+K^T_{nm}=\bar{\omega}^2_{nm}(M^P_{nm}+M^W_{nm}), \quad (52)$$

in which $\bar{\omega}_{nm}$ is the nm-th wet mode natural circular frequency, and dividing both sides of Eq.(41) by $(M^P_{nm}+M^W_{nm})$ yields the following equation,

$$\ddot{q}_{nm}(t)+2(\bar{\xi}_{nm}+\bar{\xi}^{*}_{nm})\bar{\omega}_{nm}\dot{q}_{nm}(t)+\bar{\omega}^2_{nm}q_{nm}(t)=\frac{Q_{nm}(t)}{M^P_{nm}+M^W_{nm}}, \quad (53)$$

in which $\bar{\xi}_{nm}$ and $\bar{\xi}^{*}_{nm}$ are the nm-th wet mode damping ratio and hydrodynamic radiation damping ratio, respectively, given by

$$\bar{\xi}_{nm}=\frac{1}{2\bar{\omega}_{nm}}\frac{C^P_{nm}}{M^P_{nm}+M^W_{nm}}=\frac{\omega_{nm}}{\bar{\omega}_{nm}}\frac{M^P_{nm}}{M^P_{nm}+M^W_{nm}}\xi_{nm}, \quad (54)$$

$$\bar{\xi}^{*}_{nm}=\frac{1}{2\bar{\omega}_{nm}}\frac{C^W_{nm}}{M^P_{nm}+M^W_{nm}}. \quad (55)$$

STOCHASTIC RESPONSE

The vertical ground acceleration at sea bed is assumed to be characterized by the modified Kanai-Tajimi power spectral density function as

$$S_{\ddot{U}_g\ddot{U}_g}(\sigma)=S_0\left[\frac{1+4\xi_g^2(\sigma/\omega_g)^2}{\{1-(\sigma/\omega_g)^2\}^2+4\xi_g^2(\sigma/\omega_g)^2}\right]\cdot\left[\frac{(\sigma/\omega_k)^4}{\{1-(\sigma/\omega_k)^2\}^2+4\xi_k^2(\sigma/\omega_k)^2}\right]$$
$$(-\infty<\sigma<+\infty), \quad (56)$$

in which the first and second brackets on the right hand side represent the Kanai-Tajimi low pass filter (Tajimi, 1960) and the Clough-Penzien high pass filter (Clough and Penzien, 1975), respectively, S_0 is the spectral intensity, ξ_g and ω_g are the parameters of the low pass filter (ξ_g is the equivalent damping ratio and ω_g is the predominant circular frequency of local soil deposit, respectively), and ξ_k and ω_k are the parameters of the high pass filter. Generally speaking, S_0 and ω_g are larger than those of horizontal ground acceleration.

On the basis of a linear random vibration theory, the power spectral density function of displacement response ζ may be obtained as

$$S_{\zeta\zeta}(r,\theta;\sigma)=\sum_{n=0}^{\infty}\sum_{m=1}^{\infty}\zeta^2_{nm}(r)\cos^2 n\theta\,|H_{nm}(\sigma)|^2 S_{Q_{nm}Q_{nm}}(\sigma), \quad (57)$$

in which $S_{Q_{nm}Q_{nm}}(\sigma)$ is the power spectral density function of the generalized force $Q^F_{nm}(t)$. $|H_{nm}(\omega)|^2$ is the nm-th transfer function given by

$$|H_{nm}(\sigma)|^2=1/(M^P_{nm}+M^W_{nm})^2\left[(\bar{\omega}^2_{nm}-\sigma^2)^2+4(\bar{\xi}_{nm}+\bar{\xi}^{*}_{nm})^2\bar{\omega}^2_{nm}\sigma^2\right]. \quad (58)$$

In Eq.(58), cross spectra related to coupling between different modes of vibration are disregarded.

The variance of displacement response can be obtained by (Hurty and Rubinstein, 1964)

$$\bar{\zeta}^2(r,\theta)=\int_{-\infty}^{+\infty}S_{\zeta\zeta}(r,\theta;\sigma)d\sigma$$
$$=\sum_{n=0}^{\infty}\sum_{m=1}^{\infty}\zeta^2_{nm}(r)\cos^2 n\theta\cdot\bar{q}^2_{nm}, \quad (59)$$

in which $\bar{q}^2_{nm}$ is the mean squared value of the nm-th generalized coordinate given by

$$\bar{q}^2_{nm}=\int_{-\infty}^{+\infty}|H_{nm}(\sigma)|^2 S_{Q_{nm}Q_{nm}}(\sigma)d\sigma. \quad (60)$$

Other responses, such as acceleration and internal forces, can be obtained in the same manner.

NUMERICAL RESULTS AND DISCUSSION

Numerical examples are presented to discuss the difference in the response behavior of circular floating structures with and without tension legs for horizontally traveling seismic waves. The floating structure is considered as a large concrete barge with a number of bulkheads. Geometrical parameters and material properties are characterized as follows: radius of structure = 1000m, thickness of structure = 50m, Young's modulus of structure = 2.0×10^{10} N/m^2, Poison's ratio of structure =0.15,

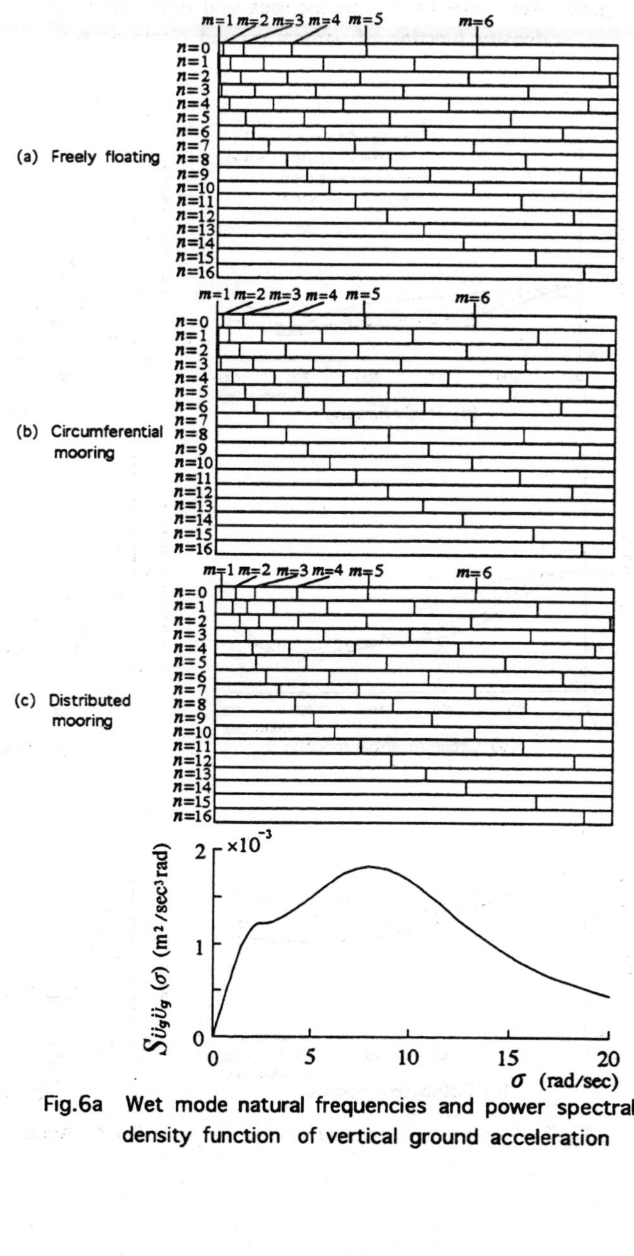

Fig.6a Wet mode natural frequencies and power spectral density function of vertical ground acceleration

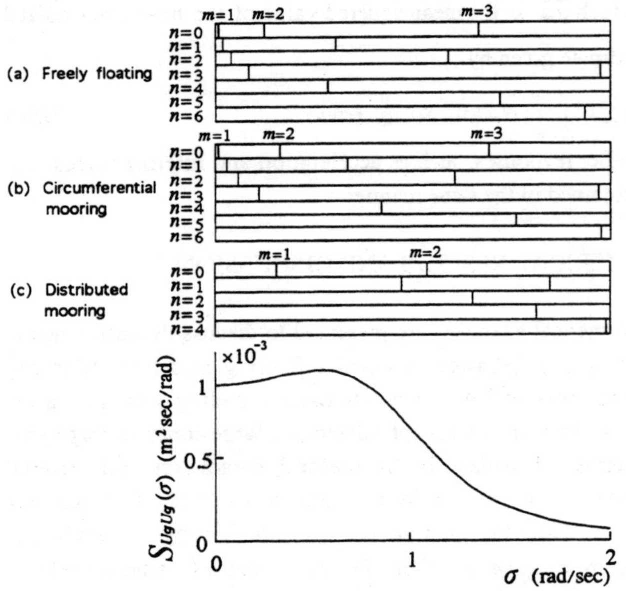

Fig.6b Wet mode natural frequencies and power spectral density function of vertical ground displacement

mass density of structure = 0.4×10^3 kg/m^3, material damping ratio in air = 0.05, water depth = $200m$, stiffness of circumferential mooring $k_c=5.0\times10^6$ N/m² and stiffness of distributed mooring $k_d=5.0\times10^5$ N/m³. The floating structures with mooring system are pulled from the location of the freely floating case by $5m$. Three different apparent velocities are considered for horizontally traveling seismic waves: $C_a = 1\times10^3$, 3×10^3 and 5×10^3 m/sec. Vertically incident seismic waves correspond to $C_a=\infty$. In Eq.(56), the spectral intensity $S_0 = 1.0\times10^{-3}$ m^2/sec^3, parameters of the low pass and high pass filters are $\omega_g = 10$ rad/sec, $\xi_g = 0.6$, $\omega_k = 1$ rad/sec and $\xi_k = 0.6$.

Figures 6a and 6b show the relationships between the location of wet mode frequencies for each mooring case and the power spectral density functions of vertical ground acceleration and displacement, respectively. A number of higher order wet modes are located in the central region of the power spectrum of vertical ground acceleration. On the other hand, only several

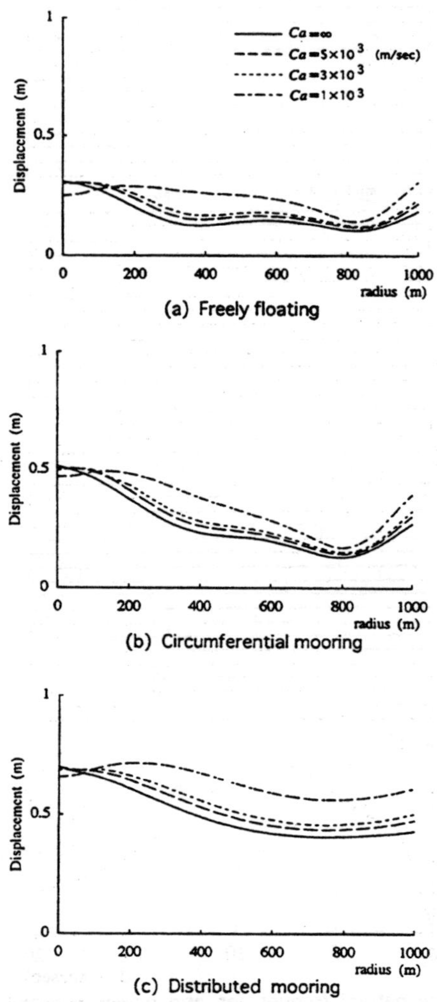

Fig.7 Displacement responses

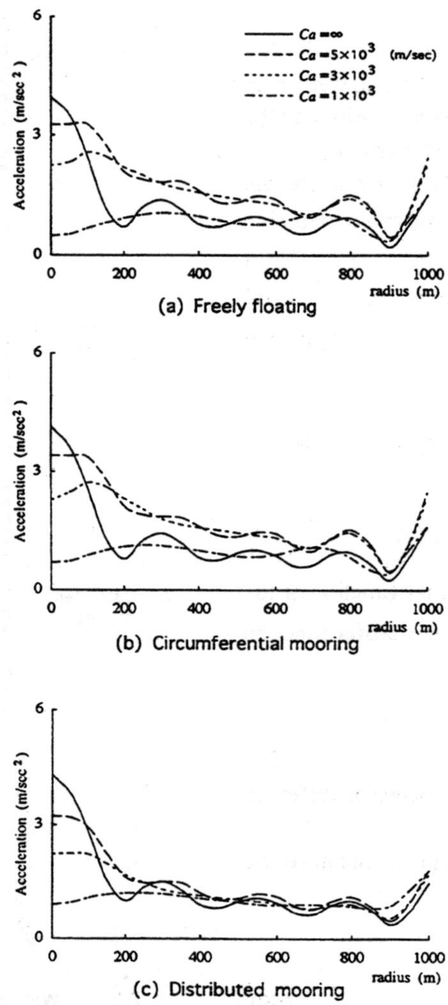

Fig.8 Acceleration responses

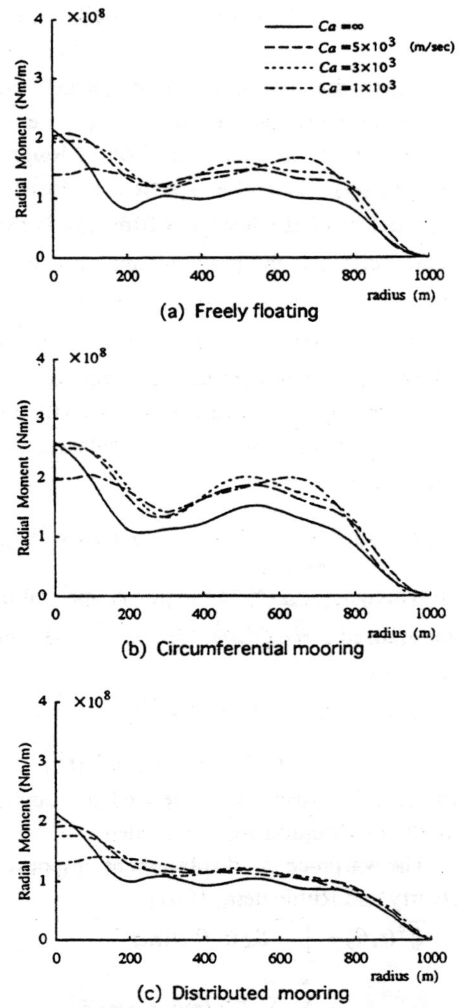

Fig.9 Radial bending moment responses

lower order wet modes enter the central region of the power spectrum of vertical ground displacement. The location of wet mode frequencies of circumferential mooring case is similar to that of freely floating case. However, the location of wet mode frequencies of distributed mooring considerably shifts to the region of higher frequency. 55 wet modes are considered in the analysis, which range from 0 to 16-th mode for circumferential Fourier wave number n and from 1 to 6-th mode for the radial mode number m.

Figure 7 shows the standard deviations of displacement response along $\theta=0$. The origin, $r=0$, corresponds to the center of the floating structure. The responses of both circumferential and distributed moorings become larger than that of freely floating case. In the response of distributed mooring, the structural deformation is significantly suppressed and rigid body motion becomes predominant. With the decrease of apparent velocity of horizontally traveling seismic waves, the displacement responses increase except for the central region of the floating structure for each mooring case.

Figure 8 shows the standard deviations of acceleration response along $\theta=0$. Higher order wet modes are remarkably excited for each mooring case. Consequently, the overall shape of response distribution becomes considerably complicated. The difference in the response distribution is relatively small for each mooring case. The response in the central region tends to decrease with the decrease of apparent velocity. However, the response except for the central region tends to increase with the decrease of apparent velocity.

Figure 9 shows the standard deviations of bending moment response along $\theta=0$. Compared to displacement response, the contribution of higher modes of vibration is observed. The response of circumferential mooring case is the largest among different mooring states. The large response may be due to local constraint around the circumference. The response of distributed mooring case is similar to that of freely floating case. The response except for the central region of the floating structure becomes large with the decrease of apparent velocity for each mooring case.

CONCLUSIONS

Based on a linear potential flow theory, the hydrodynamic pressure acting on the lower surface of a large tension leg circular floating structure subjected to horizontally traveling seismic waves on sea bed is obtained in closed form. Then, the seaquake response is calculated by using a wet mode superposition approach and stationary random vibration theory. Within this study, the following conclusions can be obtained.

1. The displacement response is amplified by attachment of tension leg mooring because of direct propagation of seismic waves through mooring system. The response of the floating structure with distributed mooring is the largest among different mooring states. Only lower modes of vibration are excited for each mooring case. With the decrease of apparent velocity of horizontally traveling seismic waves, the displacement response tends to increase for all mooring cases.
2. The difference in the spatial distribution of acceleration response is not remarkable for each mooring case. The acceleration response in the central region of the floating structure tends to decrease with the decrease of apparent velocity. However, the response except for the central region tends to increase with the decrease of apparent velocity.
3. The bending moment response of circumferential mooring case is the largest due to local constraint at the edge. The bending moment response except for the central region of the floating structure becomes larger with the decrease of apparent velocity for each mooring case.

REFERENCES

Baba, E., (1988), "A Study on the Effect of Seaquakes on a Floating Body", J. of the Society of Naval Architects of Japan, Vol.162, pp.90-98.

Babu, P.V.T., Reddy, D.V., (1986), "Dynamic Coupled Fluid-Structure Interaction Analysis of Flexible Floating Platforms", J. of Energy Resources Technology, ASME, Vol.108, pp.297-304.

Clough, R. W., Penzien, J., (1975), "Dynamics of Structures", McGraw-Hill.

Hamamoto, T., Takahashi, H., Tanaka, Y., (1991), "Stochastic Response of Anchored Flexible Floating Island Subjected to Wind-Waves and Seaquake", Proc. of the 1st Int. Offshore and Polar Eng. Conf., Edinburgh, Vol.3, pp.391-398.

Hamamoto, T., (1994a), "Dynamic Response of Flexible Circular Floating Islands Subjected to Stochastic Waves and Seaquakes", Proc. of Int. Conf. on Hydroelasticity in Marine Technology, Trondheim, pp. 433-445.

Hamamoto, T., Asami, T., (1994b), "Fluid-Coupled Free Vibration of Floating Elastic Circular Plates with Elastically Supported Edge", Theoretical and Applied Mechanics, Vol. 43, pp.171-179.

Hurty, W. C., Rubinstein, M. F., (1964), "Dynamics of Structures", Prentice-Hall.

Kawanishi, T., Ikeda, N., Kato, A., (1987), "Response of Tension Leg Buoy to Vertical Seismic Excitation", J. of the Society of Naval Architects of Japan, Vol.161, pp.307-315.

Liou, G.S., Penzien, J., Yeung, R.W., (1988), "Response of Tension leg Platforms to Vertical Seismic Excitations", Earthquake Engineering and Structural Dynamics, Vol. 16, pp.147-182.

MacCamy, R. C., Fuchs, R. A., (1954), "Wave Forces on a Pile: A Diffraction Theory", Tech. Memo 69, U. S. Army, Corps of Engineers Beach Erosion Board.

Tajimi, H., (1960), "Statistical Method of Determining the Maximum Response of Building Structure During an Earthquake", Proc. of the 2nd World Conf. on Earthquake Eng. Tokyo, Vol. 2, pp.781-798.

Proceedings of the Seventh (1997) International Offshore and Polar Engineering Conference
Honolulu, USA, May 25-30, 1997

ISBN 1-880653-28-1 (Set); ISBN 1-880653-32-X (Vol. IV)

Seismic Response of TLP Under Offset Condition

Katta Venkataramana, Kenji Kawano and Susumu Yoshihara
Kagoshima University
Kagoshima, Japan

ABSTRACT

The dynamic response of tethers of tension leg platforms is investigated for the inputs of current and earthquake excitations. The offset condition of the tether is caused by current flow. The static deflected shape of the tether under this current load is firstly identified. The dynamic analysis for earthquake input is then carried out for this deflected configuration of the tether. The fluid loading due to the surrounding water is taken into account in the form of an added mass term and a hydrodynamic damping term. Input ground acceleration is represented using the power spectral density function of the modified Kanai-Tajimi type. The coupled axial and lateral responses are evaluated for both horizontal as well as vertical ground excitations.
KEY WORDS: TLP, current force, earthquake force, dynamic response

INTRODUCTION

A Tension Leg Platform (TLP) consists of buoyant platform moored to the sea bed through vertical tethers. The tethers are kept in tension by extra buoyancy of the platform. The main advantage of the TLP is that the heave motion of the platform is suppressed due to the axial stiffness of tethers. Thus the platform acts like a moored semi-submersible vessel with great flexibility in the horizontal direction but is quite rigid in the vertical direction. TLPs are being recommended for oil and gas production works in deep seas based on economical considerations. Their behavior under dynamic loads is of main concern to offshore engineers and has attracted wide attention.

In a seismically active region, analysis for earthquake excitations is a major design requirement. Liou et al (1988) have presented a method for analyzing the vertical motion, tether force and foundation uplift of a TLP when the subsea soil is excited by vertical seismic waves. The multiple interactions among the soil, foundation, fluid and platform were mathematically modeled using the substructuring technique. Their study showed that the maximum values of the response are controlled primarily by the earthquake ground motions in a narrow band centered on the vertical mode frequencies of the TLP. In their analysis, the tethers were considered as massless springs and the offset condition was not included.

Since the tethers of TLPs are highly flexible in the horizontal direction but are very rigid in the vertical direction due to pretension, their response behavior in the vertical direction is of main interest to structural engineers. A horizontal load such as wind, wave or current will create an offset of the platform. When this platform under offset condition is subjected to a seismic excitation, the response characteristics of the platform and the tethers may deviate from their originally expected values based on a no-offset assumption. When the tethers are perfectly vertical, they can be treated as aseismic structures for vertical excitations whereas when offset condition is created, tethers undergo elongation under earthquake loading. Kawanishi et al (1991, 1993) investigated theoretically as well as experimentally the earthquake response of a fixed-base TLP under an offset condition. They examined the tension variation in the tethers and found that the minimum tension of a downstream tether falls remarkably under the offset condition. In their analysis weight, inertia force and hydrodynamic force of tendons were neglected.

In the present study, the offset of the tether of TLP is caused due to the action of current force. The dynamic behavior of this tether, which is in the offset condition, for the inputs of horizontal and vertical seismic excitations is investigated. The bending stiffness as well as the hydrodynamic loading due to surrounding water is included in the analysis. Table 1 lists the details of a typical tether used for study (Aker Engineering, 1989). The depth of water is 300m, the length and diameter of each tether are 262.5m and 0.81m respectively, and the pretension in tether is 15205.5kN. The static deflected shape of the tether under steady current load is firstly identified. The dynamic analysis for earthquake input is then carried out for this deflected tether. The fluid loading due to the surrounding water is taken into account in the form of an added mass term and a hydrodynamic damping term. The tether is discretized by lumping masses at selected nodal points. At each node three degrees of freedom corresponding to surge, heave and pitch motion are considered.

STATIC DEFLECTED SHAPE (INITIAL OFFSET) OF THE TETHER UNDER STEADY CURRENT

Figure 1 shows the schematic diagram of a tether-TLP system under steady current . Let N be pretension in the tether, l be the length of the tether and p be the horizontal load on the tether due to current.

The tether is modeled as a cantilever beam under axial and lateral loads. The loading p due to the current per unit length of the tether is

$$p = 0.5 C_{dc} \rho A V_c^2 \quad (1)$$

where C_{dc} is a drag coefficient, ρ is the mass density of water, A is the area projected in the direction of flow and V_c is the velocity of steady current. For unidirectional steady flows, V_c is a function of the Reynolds number. When the current velocities vary from 0.5m/s to 2m/s, for a tether of 0.81m in diameter, C_{dc} lies between (Schlichting, 1968) 1.0 and 0.7. In the present study, $C_{dc}=1$ is assumed.

The governing equation of deflection of tether may be obtained as

$$EI\frac{d^4x}{dz^4} - N\frac{d^2x}{dz^2} = p \quad (2)$$

where x is the deflection at distance z measured up from the base, E is the modulus of elasticity of the tether and I is the moment of inertia of the tether. The general solution of the above equation is

$$x = C_1 \exp(kz) + C_2 \exp(-kz) + C_3 z + C_4 - \frac{p}{2N} z^2 \quad (3)$$

where $k = \sqrt{N/EI}$. The constants C_1, C_2, C_3, C_4 are determined using the following boundary conditions at the ends of the tether:

$$\left.\begin{aligned} &[x]_{z=0} = 0 \\ &\left[\frac{dx}{dz}\right]_{z=0} = 0 \\ &\left[\frac{d^2x}{dz^2}\right]_{z=l} = 0 \\ -EI&\left[\frac{d^3x}{dz^3}\right]_{z=l} + N\left[\frac{dx}{dz}\right]_{z=l} = 0 \end{aligned}\right\} \quad (4)$$

Finally the expression for the deflection is

$$x = \frac{p(1 - kl\exp(-kl))}{Nk^2 \exp(-kl)(1+\exp(2kl))}\exp(kz) + \frac{p(1 + kl\exp(kl))}{Nk^2 \exp(-kl)(1+\exp(2kl))}\exp(-kz)$$
$$+ \frac{-2p + plk(\exp(-kl) - \exp(kl))}{Nk^2 \exp(-kl)(1+\exp(2kl))} + \frac{pl}{N}z - \frac{p}{2N}z^2 \quad (5)$$

The last two terms of the above equation correspond to the solution for a taut tether without bending.

The examples of the deflected shapes are shown in Fig.2. They are proportional to the square of the current velocity, as is clear from Eqs.(1) and (5). The maximum deflection of the tether, against a current velocity of 2m/s, is 3.4m which is about 1.3% of its total length. The pretension in the tether has prevented it from undergoing large deflection.

DYNAMIC RESPONSE ANALYSIS METHOD

The static deflected shape of the tether under steady current is obtained as mentioned above. The dynamic analysis for earthquake input is then carried out for this deflected configuration. For this purpose the deflected tether is represented by finite element method by lumping masses at selected nodal points and using the resultant forces over the corresponding projected areas and volumes. The hydrodynamic loading due to surrounding water is taken into account in the form of an added mass term and a hydrodynamic damping term. This damping term is nonlinear as it contains the squared term of structural velocity. Assuming that the distribution of the probability density function of the structural velocity is Gaussian process, using the equivalent linearization technique, the equation of motion can be linearized.

The linearized equation of motion, for the tether under earthquake excitation, can be obtained as:

$$[\tilde{M}]\{\ddot{u}^t\} + [\tilde{C}]\{\dot{u}\} + [K]\{u\} = \{0\} \quad (6)$$

where

$$[\tilde{M}] = [M] + \left[\ddots\, \rho C_m V \,\ddots\right], \qquad [\tilde{C}] = [C] + \left[\ddots\, \rho C_d A \sqrt{2/\pi}\,\sigma_{\dot{u}} \,\ddots\right],$$
$$[K] = [K_s] + [K_g], \qquad \{\ddot{u}^t\} = \{\ddot{u}_g\} + \{\ddot{u}\}$$

in which $[M]$ is the lumped mass matrix, $[C]$ is the structural damping matrix, $[K]$ is the stiffness matrix, $[K_s]$ is the elastic stiffness matrix, $[K_g]$ is the geometric stiffness matrix which takes into account the pretension in the tethers, $\{u\}$ is the displacement vector of the tether, V is the volume enclosed with respect to flow, A is the projected area in the direction of flow, C_m is the added mass coefficient, C_d is the drag coefficient, $\sigma_{\dot{u}}$ is the root-mean-square (rms) value of the structural velocity which is obtained by an iterative procedure, the superscript t denotes the total acceleration, $\{\ddot{u}_g\}$ is the ground acceleration due to earthquake. The displacement vector $\{u\}$ consists of horizontal displacements (surge) u_x, vertical displacements (heave) u_z and rotational displacements (pitch) u_θ.

The effective earthquake force which produces the dynamic response of the system results from the fact that inertia force term in Eq.(6) depends on the total motion, while the damping and elastic force depend only on relative motion. Therefore Eq.(6) can be expressed in terms of relative displacements as

$$[\tilde{M}]\{\ddot{u}\} + [\tilde{C}]\{\dot{u}\} + [K]\{u\} = -[P]\{\ddot{u}_g\} \quad (7)$$

where

$$[P] = [\tilde{M}]\{C_g\}$$

in which $\{C_g\}$ is a vector representing a column of ones and zeros depending on the type of excitation.

Now using eigenvalue analysis of an undamped vibration system and assuming Rayleigh-type damping, Eq.(7) can be transformed as

$$\left[\ddots\, 1 \,\ddots\right]\{\ddot{q}\} + \left[\ddots\, 2\beta_j\omega_j \,\ddots\right]\{\dot{q}\} + \left[\ddots\, \omega_j^2 \,\ddots\right]\{q\} = -[\Phi]^T[P]\{\ddot{u}_g\} \quad (8)$$

where

$$\{u\} = [\Phi]\{q\}, \qquad [\Phi]^T[\tilde{M}][\Phi] = \left[\ddots\, 1 \,\ddots\right],$$
$$[\Phi]^T[\tilde{C}][\Phi] = \left[\ddots\, 2\beta_j\omega_j \,\ddots\right], \qquad [\Phi]^T[K][\Phi] = \left[\ddots\, \omega_j^2 \,\ddots\right]$$

in which ω_j is the natural frequency for the jth vibration mode of the structure, β_j is the corresponding damping ratio which includes both the structural damping and the hydrodynamic damping, $[\Phi]$ is the undamped eigenvector, $\{q\}$ is the corresponding modal displacement vector and the superscript T denotes the transpose of a matrix.

Applying the Fourier transformation to Eq.(8), the response is obtained as

$$\{\bar{q}(\omega)\} = -[H(\omega)][\Phi]^T[P]\{\bar{\ddot{u}}_g(\omega)\} \quad (9)$$

where

$$\{\bar{q}(\omega)\} = \int_{-\infty}^{\infty}\{q\}\exp(-i\omega t)dt, \qquad \{\bar{\ddot{u}}_g(\omega)\} = \int_{-\infty}^{\infty}\{\ddot{u}_g\}\exp(-i\omega t)dt$$

$$[H(\omega)] = \left[\ddots\, \omega_j^2 - \omega^2 + 2i\omega\omega_j\beta_j \,\ddots\right]^{-1}$$

in which $[H(\omega)]$ is the complex frequency response function and $i=\sqrt{-1}$. Due to the effects of hydrodynamic damping, $[H(\omega)]$ depends on the magnitude and the nature (horizontal or vertical) of the ground excitation.

The earthquake ground motions are highly random in nature and hence the response analysis is carried out using the random vibration approach. The random earthquake ground motions may be represented with a stationary filtered white noise, given by Kanai and Tajimi (This is referred to as KT spectrum in this paper) as

$$\left[S_{\ddot{u}_g\ddot{u}_g}(\omega)\right]=\left[S_1(\omega)\right]S_0 \qquad (10)$$

where

$$\left[S_1(\omega)\right]=\frac{1+4h_g^2(\omega/\omega_g)^2}{\left(1-(\omega/\omega_g)^2\right)^2+4h_g^2(\omega/\omega_g)^2},$$

$$S_0=\frac{\sigma_{\ddot{u}_g}^2}{\pi\left(1+4h_g^2\right)\omega_g/(2h_g)}$$

in which $\left[S_{\ddot{u}_g\ddot{u}_g}(\omega)\right]$ is the power spectrum of ground acceleration, ω_g and h_g are the filter parameters (characteristic ground frequency and characteristic ground damping ratio respectively), $\sigma_{\ddot{u}_g}$ is the rms value of ground acceleration, S_0 is the intensity of white noise at the base.

But the above formulation does not change the amplitudes as the frequency approaches zero and some difficulty may arise with very-low frequency components. In particular, the natural frequencies of TLPs have generally lower values and hence proper selection of ground acceleration spectrum is necessary for reliable designs. Therefore Eq.(10) is modified by including another filter which attenuates the very-low frequency components as

$$\left[S_{\ddot{u}_g\ddot{u}_g}(\omega)\right]=\left[S_1(\omega)\right]\left[S_2(\omega)\right]S_0 \qquad (11)$$

where

$$\left[S_2(\omega)\right]=\frac{(\omega/\omega_f)^4}{\left(1-(\omega/\omega_f)^2\right)^2+4h_f^2(\omega/\omega_f)^2}$$

in which ω_f and h_f are the filter parameters (frequency parameter and damping parameter respectively) of a second filter, introduced to overcome the limitations of Eq.(10). Eq.(11) is referred to as the modified Kanai-Tajimi spectrum (MKT spectrum) in this paper.

The power spectrum of the modal responses is

$$\left[S_{\bar{q}\bar{q}}(\omega)\right]=\left[H(\omega)\right][\Phi]^T[P]\left[S_{\ddot{u}_g\ddot{u}_g}(\omega)\right][P]^T[\Phi]\left[H(\omega)^*\right] \qquad (12)$$

where $\left[H(\omega)^*\right]$ is the complex conjugate of the frequency response function.

Now the variance of the relative displacement responses is

$$E\left[\{u\}\{u\}^T\right]=[\Phi]^T\left[\int_{-\infty}^{\infty}\left[S_{\bar{q}\bar{q}}(\omega)\right]d\omega\right][\Phi] \qquad (13)$$

The variance of the axial stress, shear stress and the bending moments in the tether elements can be obtained by multiplying the variance of the displacement responses by their stiffness matrices as

$$E\left[\{f\}\{f\}^T\right]=[K]^T[\Phi]^T\left[\int_{-\infty}^{\infty}\left[S_{\bar{q}\bar{q}}(\omega)\right]d\omega\right][\Phi][K] \qquad (14)$$

NUMERICAL RESULTS AND DISCUSSIONS

The dynamic response analysis is carried out for the tether model shown in Fig.1. The depth of water is 300m from mean sea level. The structural details of the tether are given in Table 1. Its static deflected shape for different current loads is given in Fig.2. The dynamic analysis is carried out for this deflected tether. The bottom end of the tether is assumed to be fixed to the sea bed whereas the top end is free to move laterally. For analysis purposes, this tether is discretized by lumping masses at selected nodal points into 51 nodes and 50 elements. The effect of platform is included in the top node as a concentrated mass. The displacement of each node has horizontal (sway), vertical (surge) and rotational (pitch) components in plane. The pretension in the tether is taken into account in the form of a geometric stiffness term.

The natural frequencies and vibration mode shapes are computed by eigenvalue analysis. The structural damping is assumed to be 2% for the first mode of vibration. Table 2 and 3 show the natural frequencies and the structural damping ratios respectively for the first five modes for the five cases of current loading i.e., current velocity $V_c=0$ (undeflected position), 0.5m/s, 1m/s, 1.5m/s and 2m/s. The first natural frequency for all the cases is lower than the dominant wave frequency range (0.314 to 1.256rad/s or 5 to 20sec) whereas the frequencies corresponding to second and higher modes are higher than the dominant wave frequency range for a water depth of 300m, assumed for the TLP.

The hydrodynamic force on the tether due to the surrounding water is expressed as the sum of an added mass term and a linearized hydrodynamic damping term based on Morison equation. The values of the added mass coefficient C_m and the hydrodynamic damping coefficient C_d are taken as 1.0 and 1.0 respectively. Since the hydrodynamic damping force is coupled to the dynamic response, the linearized damping term is calculated by a cyclic procedure and, in this study, reasonable convergence was attained in about four cycles of iteration. Table 4 shows the examples of hydrodynamic damping ratios up to the 4th mode of vibration for various current velocities for horizontal earthquake excitations with *rms* acceleration of 100gal.

Earthquake ground motions are modeled stochastically using the MKT spectrum (Eq.11) for stationary conditions. Numerical results are presented for *rms* ground accelerations of 100gal (severe earthquake situation) and 30gal (moderate earthquake situation). Fig.3 shows the ground acceleration spectrum for *rms* ground accelerations of 100gal. The values obtained using the KT spectrum are compared with those obtained using the MKT spectrum. The difference is prominent in the low-frequency region which includes the natural frequency of the present tether model.

Fig. 4 shows the variation of *rms* displacement of node 1, in the horizontal dicrection) as a function of frequency parameter of the earthquake spectrum for horizontal excitations with no-offset condition. As mentioned earlier, the response value is strongly dependent on the frequency parameter ω_f chosen for the calculation as the present offshore structural model has lower natural frequencies compared to onshore structures. Therefore, for reliable response evaluation for earthquake excitation, it is suggested that the value of the parameter must be chosen by studying many typical earthquake records and their characteristics. The value of the frequency parameter usually chosen for earthquake response of onshore structures is around 1.5 rad/s (with characteristic ground frequency ω_g of around 15 rad/s representing firm soil conditions). Since the marine soil is generally soft, in the present study, the value of ω_f of 1.0 is assumed (with characteristic ground frequency ω_g of 10 rad/s.

Numerical computations were carried out for horizontal as well as vertical ground excitations and *rms* responses of structural nodes were obtained for *rms* ground accelerations representing severe ground motions. Fig.5 and 6 show the horizontal and vertical displacements respectively of the tether for horizontal ground excitations. The distribution of displacements for different current velocities and *rms* ground accelerations are compared. It is observed that the horizontal displacements increase with the increase in ground acceleration, but are nearly the same for all the cases of current velocities for a given value of ground acceleration, indicating very little effect of initial offset due to current on the earthquake response characteristics in the horizontal direction. On the other hand, vertical displacements, which are zero in the absence of initial offset with no current, increase rapidly with the increase in initial offset and earthquake forces. Excessive vertical displacements may cause slackening of tethers and may cause serious operational problems.

Figs. 7 and 8 show the examples of horizontal and vertical displacements respectively of the tether for vertical ground accelerations. Both the horizontal and vertical displacements, which are zero in the absence of currents, increase rapidly with the increase in current velocity which causes the initial offset and the intensity of earthquake loading which causes the dynamic response of the tether.

CONCLUSIONS

Since the tethers of TLPs are highly flexible in the horizontal direction but are very rigid in the vertical direction due to pretension, their response behavior in the vertical direction is of main interest to designers. When the tethers are perfectly vertical, they can be treated as aseismic structures whereas when offset condition is created, tethers undergo elongation under earthquake loading. The horizontal components of response displacements for horizontal excitations are mainly influenced by the intensity of earthquake loading only and not much by the amount of initial offset. On the other hand, vertical components of response displacements for horizontal excitations, and both horizontal and vertical components of displacements for vertical excitations are strongly influenced both by the amount of current loading which causes the initial offset, and by the intensity of earthquake loading which causes the dynamic response.

REFERENCES

Aker Engineering (1989). "Feature of the month - TLP technology", *Noroil*, p. 51.

Kawanishi, T et al (1991). "Earthquake response of Tension Leg Platform under offset condition", *Proc. of the 1st Int. Offshore and Polar Eng. Conf.*, ISOPE, Vol.1, pp. 80-86.

Kawanishi et al (1993). "Earthquake response of the Tension Leg Platform under unbalanced initial condition", *Proc. Of the 3rd Int. Offshore and Polar Eng. Conf.*, ISOPE, Vol.1, pp. 319-325.

Liou, GS et al (1988). "Response of tension leg platforms to vertical seismic excitations", *Earthquake Engineering and Structural Dynamics*, Vol.16, pp.157-182.

Schlichting, H (1968). *Boundary Layer Theory*, McGraw-Hill, New York.

Table 1 Structural details of the tether

Outside diameter of the tether	0.81m
Wall thickness	38mm
Length of tether	262.5m
Pretension	15205.5kN
Weight of platform	14715kN
Self-weight	7.081N/m
Modulus of elasticity	$2.06 \cdot 10^{8}$ kN / m^2
Modulus of rigidity	$7.94 \cdot 10^{7}$ kN / m^2
Water depth	300m

Table 2 Natural frequencies of vibration (rad/s)

Vibration mode	Current velocity (m/s)				
	0	0.5	1.0	1.5	2.0
first	0.190	0.190	0.192	0.198	0.210
second	1.568	1.568	1.575	1.585	1.614
third	3.160	3.160	3.162	3.162	3.162
fourth	4.885	4.885	4.886	4.886	4.885
fifth	6.788	6.788	6.789	6.789	6.790

Table 3 Structural damping ratios (%)

Vibration mode	Current velocity (m/s)				
	0	0.5	1.0	1.5	2.0
first	2.00	2.00	2.00	2.00	2.00
second	16.51	16.51	16.37	16.05	15.40
third	33.28	33.27	32.87	32.01	30.27
fourth	51.45	51.43	50.79	49.47	46.80
fifth	71.50	71.47	70.58	68.75	64.62

Table 4 Hydrodynamic damping ratios (%)
(Horizontal excitations, rms ground accn=100gal)

Vibration mode	Current velocity (m/s)				
	0	0.5	1.0	1.5	2.0
first	0.0039	0.0039	0.0040	0.0039	0.0035
second	0.0091	0.0091	0.0091	0.0092	0.0094
third	0.0046	0.0046	0.0046	0.0047	0.0049
fourth	0.0021	0.0021	0.0021	0.0021	0.0022

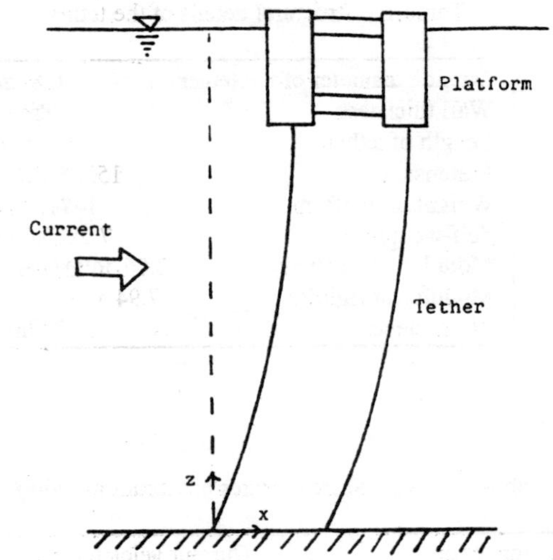

Fig.1 Schematic diagram of a TLP-tether system

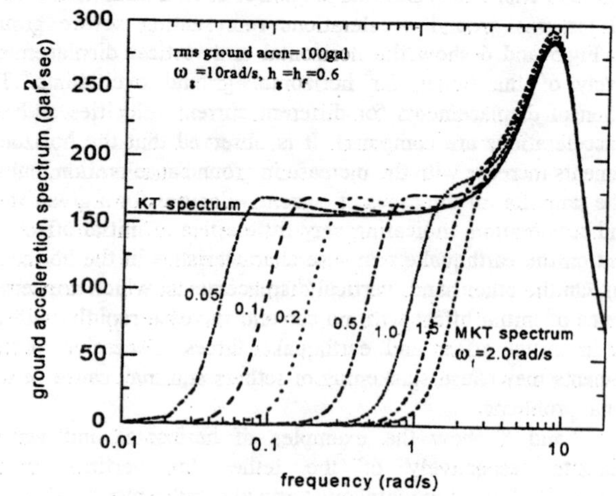

Fig.3 Ground acceleration spectrum

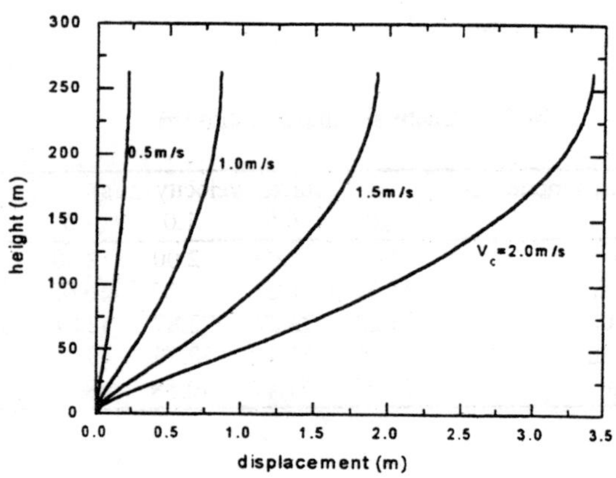

Fig.2 Static deflection of the tether

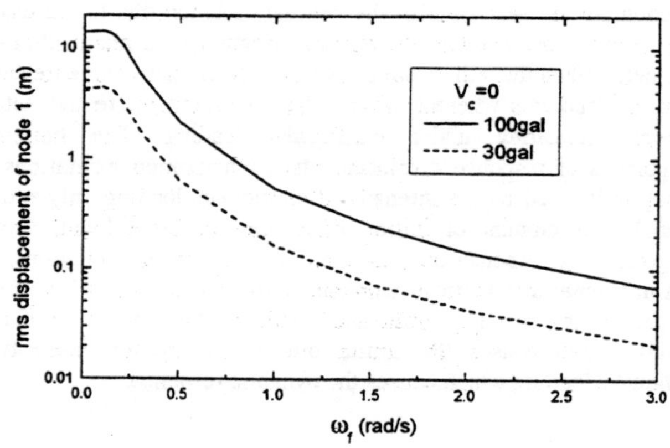

Fig.4 rms displacement at node 1 against frequency parameter ω_f

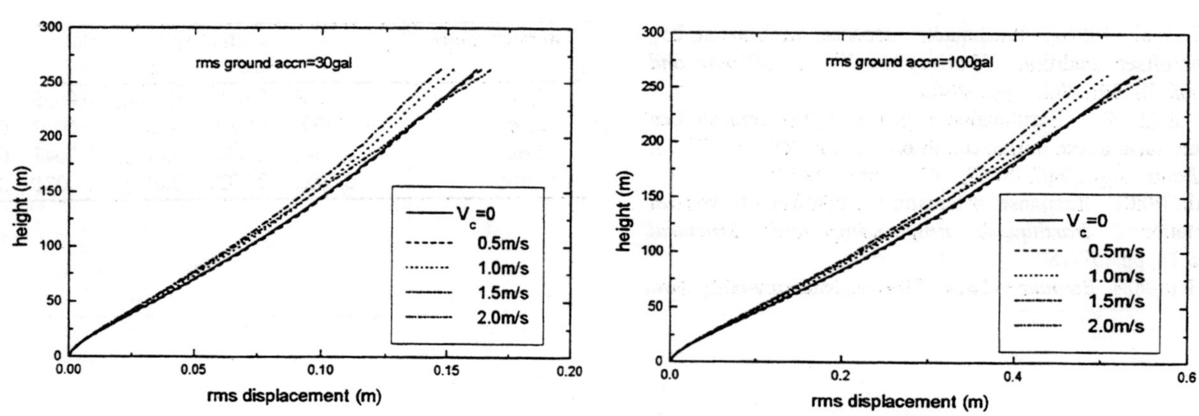

Fig.5 Horizontal displacements (for horizontal excitations)

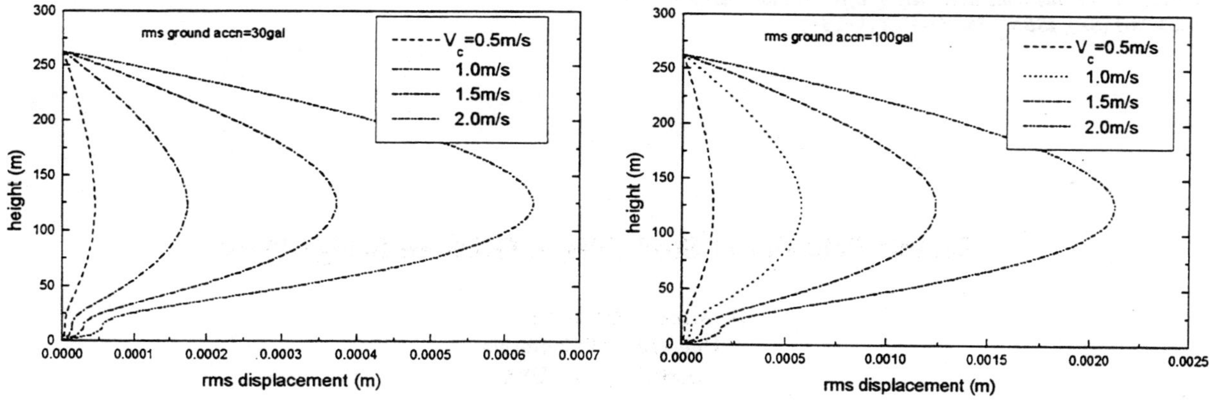

Fig.6 Vertical displacements (for horizontal excitations)

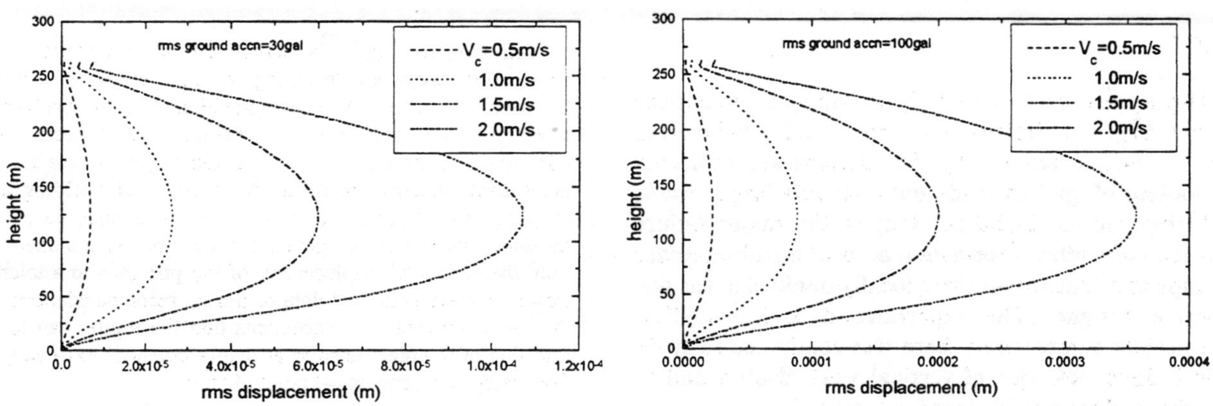

Fig.7 Horizontal displacements (for vertical excitations)

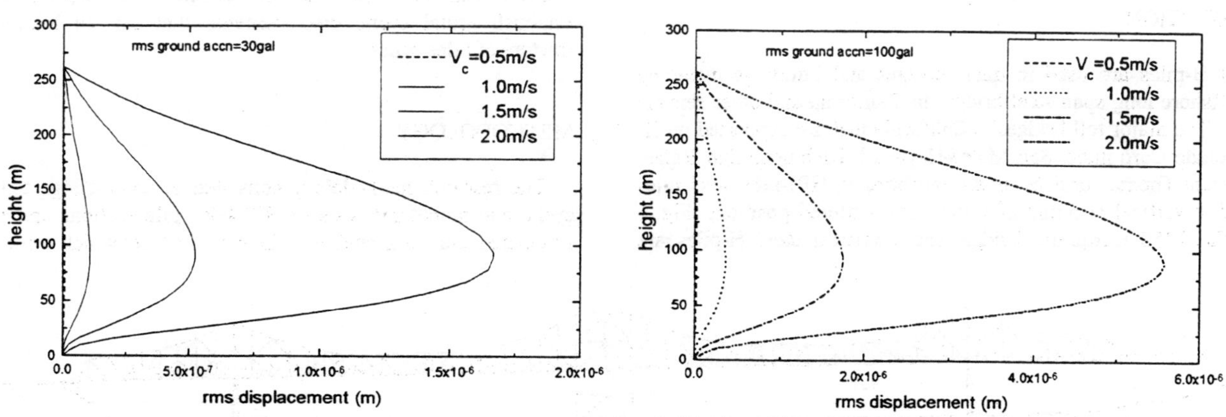

Fig.8 Vertical displacements (for vertical excitations)

Proceedings of the Seventh (1997) International Offshore and Polar Engineering Conference
Honolulu, USA, May 25-30, 1997

ISBN 1-880653-28-1 (Set); ISBN 1-880653-32-X (Vol. IV)

Seismic Behavior of Steel Piles in Offshore Bridge Piers

A. Astaneh-Asl and S. Ravat
University of California
Berkeley, CA, USA

ABSTRACT

Steel H-piles are used extensively to support foundations of major bridges, buildings and other civil engineering facilities. In seismic regions, the foundations are subjected to combinations of gravity loads and relatively large seismic effects during the earthquakes. One of the major failure modes of the steel piles is formation of local buckling under the pile caps and fracture of these locally buckled areas due to low-cycle fatigue. The experimental and analytical research project summarized here was conducted to study low-cycle fatigue behavior of vertical steel H-piles and to develop seismic design guidelines for these piles.

KEYWORDS: Steel piles, seismic design, cyclic tests, bridges, foundations, offshore piles

INTRODUCTION

Steel H-piles are used in many bridges and buildings including major offshore long span steel bridges in California and other seismic regions. The major toll bridges in California that are supported on H-piles include Carquinez, San Mateo-Hayward, Richmond-San Rafael and Vincent Thomas bridges. In all four bridges, HP14x89 steel piles are used in vertical or a mix of vertical and battered positions. Fig. 1 shows the 1958 Carquinez bridge where vertical steel H-piles are used.

Currently, the California Department of Transportation and bridge engineering firms are designing seismic retrofit of toll bridges in California including the above mentioned major bridges. In many cases, the piles supporting the bridges, are expected to undergo relatively large horizontal as well as axial cyclic displacements. Such large displacements can result in severe local buckling and overall buckling in steel piles. To prevent fracture of piles, extensive retrofit measures are being developed for the above-mentioned bridges to limit the horizontal displacement of the pile caps to a tolerable level. However, there is no test data or design recommendations on what is the horizontal cyclic displacements that these piles can tolerate? The selection of tolerable horizontal displacement is extremely important from safety and economical point of view.

OBJECTIVES

The main objective of the research project summarized here was to conduct realistic cyclic tests of vertical steel H-piles in order to establish actual cyclic displacements that can be tolerated before fracture of piles occurs.

METHODOLOGY

The research methodology consisted of conducting ten full-scale cyclic and monotonic tests of HP14x89 pile sections and conducting associated analytical studies. Three cyclic tests which were conducted

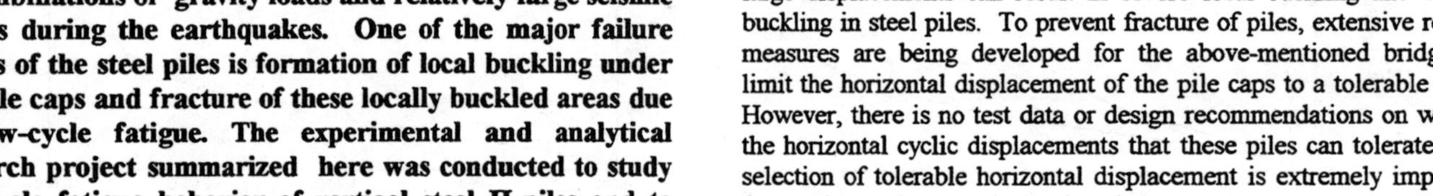

Fig. 1 A typical offshore bridge with steel H-piles

on vertical piles are summarized in this paper. Three more cyclic tests were conducted on battered piles. In addition to cyclic tests, four monotonic tests of pile sections were conducted. The main purpose of monotonic tests was to establish M-P curves for the pile section. For the information on battered pile tests and monotonic M-P tests, the reader is referred to Astaneh-Asl and Ravat (1997).

TEST SPECIMENS

The test specimens for cyclic tests of vertical piles consisted of 20 feet long steel HP14x89 piles embedded in a reinforced concrete pile caps. Fig. 2 shows the specimens. The 20-ft length of pile was selected to represent the typical location of point of inflection of the piles in major bridges. The orientation of H-piles inside the pile caps was such that the piles could be subjected to combinations of cyclic axial force and cyclic bending moment about strong axis, 45-°-angle axis and weak axis as shown in Fig. 2. Table 1 shows properties of test specimens and final failure mode

Table 1. Test specimens

Speci -men	Pile section	Pile length	Axis of bending	Failure mode
A	HP14x89	20 feet	Strong Axis	Local buckling
B	HP14x89	20 feet	45°-Axis	Local buckling and overall buckling
C	HP14x89	20 feet	Weak Axis	Local & overall buckling

Note: Material of piles was A36 steel with yield point of 38 ksi.

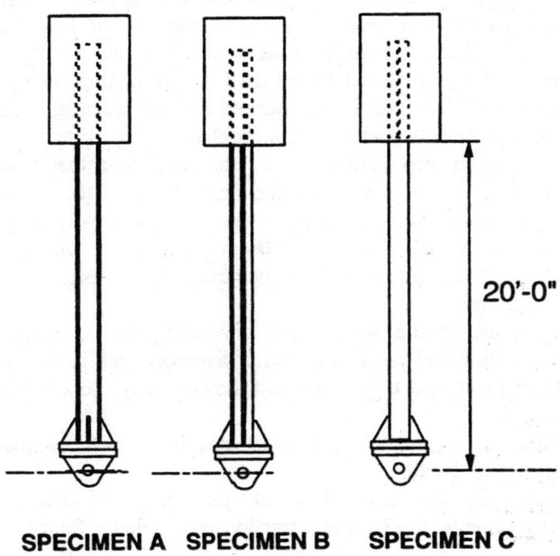

Fig. 2 Specimens for cyclic testing of vertical steel H-Piles

TEST SET-UP

Fig. 3 shows test set-up used to apply cyclic axial and horizontal displacements to the pile specimens. A relatively large actuator, with a capacity of ±1500 kips force and ±12 inches of displacement was used to apply the bulk of axial cyclic displacements. The horizontal cyclic displacements were applied by using two parallel actuators, each with a capacity of ±120 kips force and ±18 inches of displacements. The specimens were instrumented using strain gages and rosettes as well as displacement transducers. Each steel pile was embedded in a reinforced concrete block.

The pile-cap was post-tensioned to the floor of the laboratory as well as to six other pile caps and reinforced concrete blocks to be able to transfer the relatively large axial load of the steel pile to the laboratory floor with sufficient factor of safety. The horizontal and vertical displacements were applied to the free end of the pile. The free end of the pile represents most probable location of inflection and was chosen to be about 20 feet below the pile cap as the most common depth in above-mentioned bridges.

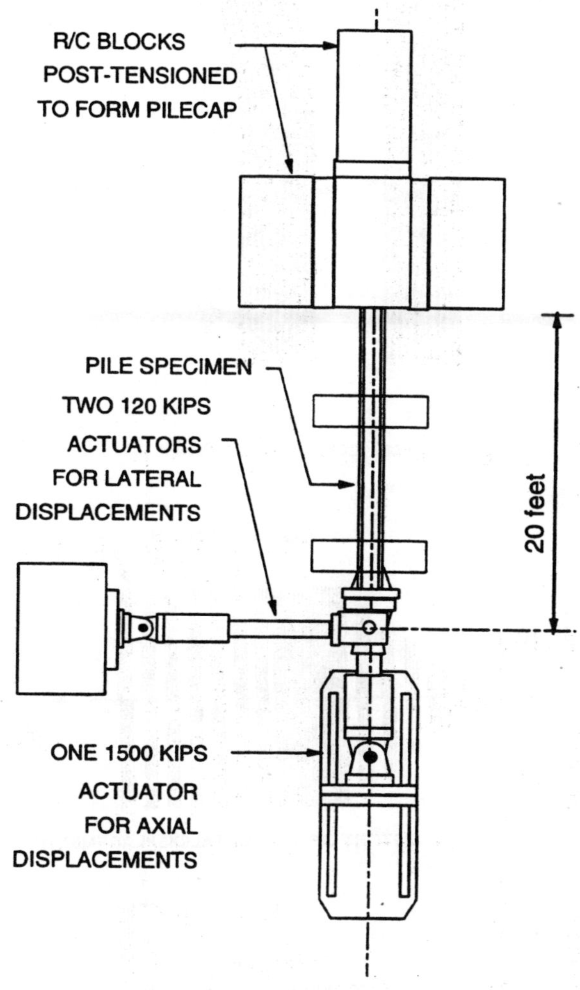

Fig. 3 Test set-up used in the tests

CYCLIC DISPLACEMENT HISTORY

Fig. 4 shows movements of a typical bridge pier and its pile foundation during a major earthquake. The pile cap foundations normally experience horizontal and vertical movements as well as rotations. As a result of these movements, each individual pile in the group experiences primarily a combination of axial and horizontal displacements. To establish realistic axial and horizontal displacement time histories to be applied to pile specimens, the results of dynamic analyses of four major bridges in California were collected and studied. The displacement time history shown in Fig. 5 was established and applied to test specimens.

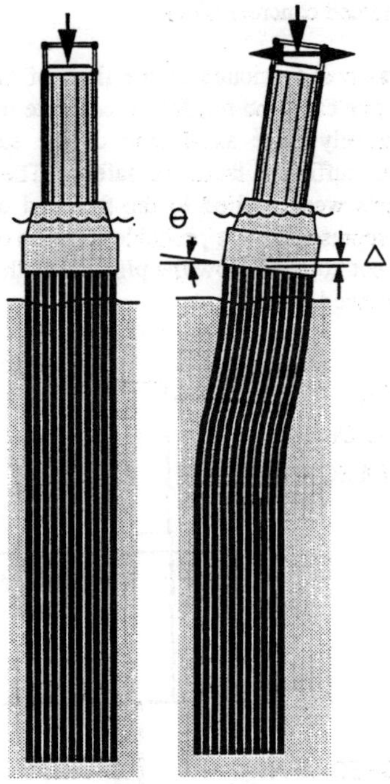

Fig. 4 Typical movements of a bridge pier

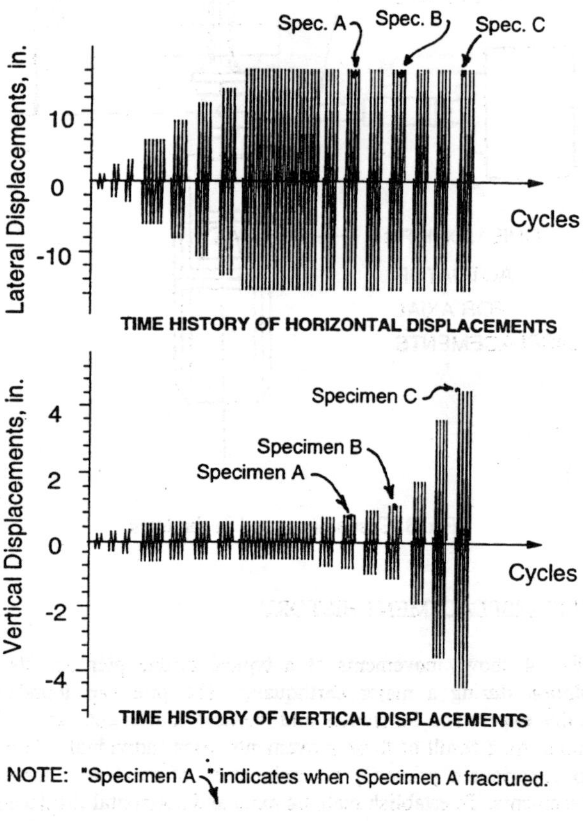

NOTE: "Specimen A" indicates when Specimen A fracrured.

Fig. 5 Time histories of axial and horizontal displacements

CYCLIC TEST PROCEDURES

The major steps taken before, during and after cyclic testing of the specimens are summarized herein:.

1. The specimens were fabricated. During fabrication, cylinder samples of concrete were taken for testing to establish strength of concrete in pile caps. The concrete that was used in pile caps was 5000 psi. After completion of fabrication of piles and casting of pile caps, the specimens were left to cure.
2. The material for tension coupon tests were machined and standard coupons were prepared according to ASTM E6 standards. At least a total of three coupons of steel sections from the flanges and three from the webs were taken. The yield stress of steel was established as 38 ksi.
3. The steel piles were cleaned and relevant dimensions of the specimens were measured.
4. The specimens were installed inside the test-set-up. The support brackets for installing displacement transducers were added.
5. The specimens were prepared and strain gages were installed. A total of 52 strain gages (single or in a rosette form) were installed on each specimen. The location and type of strain gages were chosen such that the most useful strain measurements could be made especially in the areas of local buckling and plastic hinge formations. The strain gages were covered with standard protection to prevent moisture and physical damage.
6. Displacement transducers were added.
7. White wash paint was applied to the piles.
8. A small load (less than 10% of the buckling capacity of the specimen) was applied to check the test apparatus and data acquisition system. During the application of small load, the data was recorded. The small load was removed and data was processed and analyzed to ensure that all instrumentation and load application systems are checked and are working perfectly. If any problem was detected, the problem was corrected.
9. The Sponsor was notified as to the date of actual tests and Sponsor's engineers attended the final stages of testing. Due to extra-ordinary large and unprecedented loads involved in these tests, and to secure safety of observers, only the final stage of testing when actual failure occurred was open to public observation.
10. The actual cyclic tests were conducted by applying axial compression and horizontal cyclic displacements in increments. The displacement time histories followed in the tests are shown in Fig. 5.
11. Cyclic testing continued until specimen was significantly damaged or fractured.
12. Throughout the test, data in the form of strain gage measurements, displacement transducers, actuator forces, visual observations, video and photo recordings were collected.

The following sections explain behavior of each specimen and provide a summary of test data. A complete set of tests data can be found in Astaneh-Asl and Ravat (1997).

BEHAVIOR OF SPECIMENS DURING CYCLIC TESTS

The time histories of horizontal and vertical displacements that were applied to the specimens are shown in Fig. 5. Also shown in the figure are points where each specimen, A, B and C fractured. Fig. 6 shows appearance of specimens at the end of cyclic tests. In the following, behavior of each test specimen is summarized.

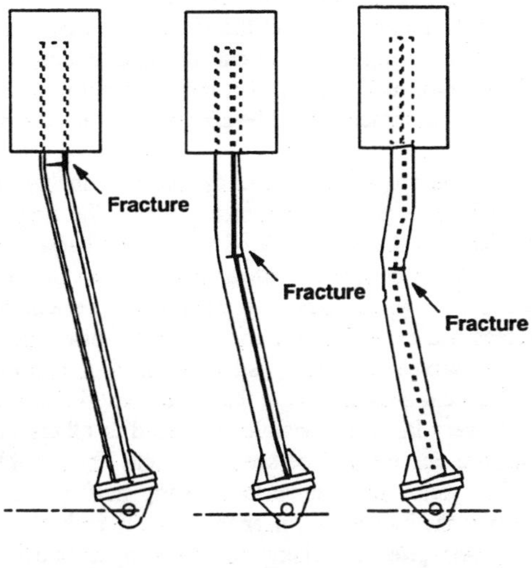

Fig. 6 Test Specimens at the end of tests

Cyclic Behavior of Specimen A

The total number of inelastic cycles that Specimen A could tolerate prior to developing cracks was 49. The maximum cyclic horizontal displacement was ±16 inches. In the following a summary of behavior of Specimen A during the cyclic loading is provided.

The specimen behaved almost elastically during the cycles up to the horizontal displacement of ±2.5 inches and vertical displacement of ±0.2 inch. During these cycles, some yield lines could be seen on the south flange over a length of about 40 inches just below the pile cap. During the following cycles of the horizontal displacement of ±3.5 inches and vertical displacement of ±0.25, yielding increased and the south flange of the pile showed vertical and diagonal yield lines over a length of about 55 inches of the pile just below the pile cap. During the following cycles yielding spread over longer area of the south flange.

During the horizontal displacement cycles of ±2.5 and vertical cycles of ±0.3, the severely yielded area of the south flange was about 55 inches long. However, sporadic yield lines could be observed on this flange over a length of about 10 feet from the pile cap. During the cycles of ±6 inches of horizontal displacements, local buckling of plastic hinge area just below the pile cap was quite visible. The strain gage readings indicated that local buckling of this flange was initiated during the last cycle of ±6 inch horizontal displacement. The deflected shape of locally buckled flange was in the form of a half-sine curve with the length of locally buckled area about 18 inches and the out of straightness of about 1 inch. At the end of these cycles, there was no other signs of yielding or local buckling on the web or the north flange.

During the following horizontal displacement cycles of ±8.5 and vertical cycles of ±0.45, local buckling of south flange was very visible. The length of locally deformed area of flange was still about 18 inches. However, the out-of-straightness was about 1.2 inches. During these cycles, the north flange also showed a few diagonally drawn yield lines over a 30 inch length of pile from the pile cap. Maximum horizontal force during these cycles was about 125 kips and maximum vertical force was about +300 kips in tension and -550 kips in compression.

The ±8.5 horizontal displacement cycles were followed by horizontal displacement cycles of ±11 and vertical cycles of ±0.45. At the end of these cycles, the out-of-straightness of locally buckled area was about 3.5 inches. The local buckling had a deformed shape of V instead of sine-wave. Yielding of north flange increased during these cycles. However, the yielding occurred during only axial tension cycles and was in the area of plastic hinge just below the pile cap.

During the horizontal displacement cycles of ±13.5 and vertical cycles of ±0.45 that followed, the north flange also buckled locally, although with less severity than the south flange. At the end of these cycles, the shape of locally buckled areas of both flanges was close. The out-of-straightness of locally buckles areas of the south and north flanges were about 4.5 and 2.5 inches respectively. Strain gage readings indicated that during the first cycle of these series, the web within the plastic hinge area also buckled locally. The length of locally deformed area of flange was still about 18 inches. Maximum horizontal force during these cycles was about 125 kips and maximum vertical force was about +620 kips in tension and -800 kips in compression.

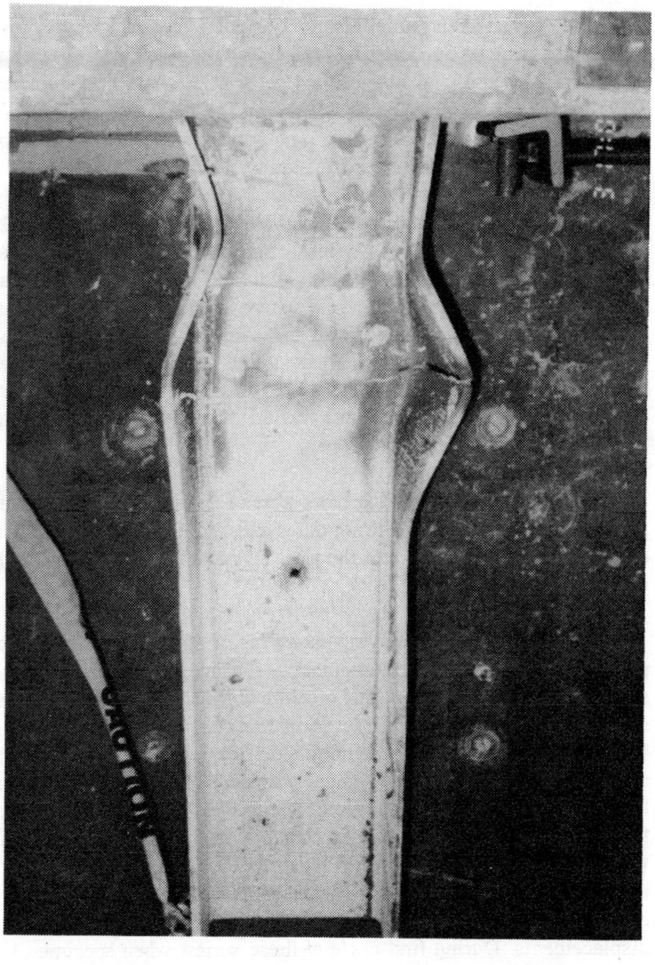

Fig. 7 shows Specimen A after fracture through locally buckled

The horizontal displacement cycles of ±13.5 and vertical cycles of ±0.45, were followed by horizontal displacement cycles of ±16 inches. The vertical displacements were ±0.45. During these cycles, the inelasticity was primarily in the area of plastic hinge below the pile cap with local buckling deformations becoming larger. After applying 27 cycles of ±16 inches of horizontal displacement along with ±0.45 inches of vertical displacement, two hair cracks were observed at the tip of the locally buckled area of the south flange. As cyclic displacements at ±16 inches continued, after 3 more cycles the crack propagated throughout the south flange and half of the web.

Maximum horizontal force during these cycles was about 140 kips and maximum vertical force was about +620 kips in tension and -800 kips in compression. After fracture, the vertical compression dropped to about -400. Local buckling and fracture occurred at a cross section approximately 16 inches below the pile cap.

Cyclic Behavior of Specimen B

The total number of inelastic cycles that Specimen B could tolerate prior to developing cracks was 59. The maximum cyclic horizontal displacement was ±16 inches. In the following a summary of behavior of Specimen B during the cyclic loading is provided.

The displacement time histories applied to this specimen was exactly the same as Specimens A and C as shown in Fig. 5. The cyclic test started by applying two initial cycles of ±1 inch horizontal and ±0.15 inch axial displacements. At the end of these two cycles, there was no signs of yielding visible. The specimen during these initial cycles behaved in an almost elastic manner.

The next step in loading history was applying two cycles of ±2.5 inches of horizontal and ±0.25 inch of axial. During this cycle when displacement was about 2 inches, yield lines occurred on the south half of the top flange of the pile where stresses due to axial and bending forces are combined with the same sign. These yield lines were spread over a length of about 7 feet just below the pile cap. There was also some localized yield lines at the end of pile just outside the clevis end plate. This was an indication of yielding being the result of axial effects.

The next step was applying two cycles of ±3.5 inches of horizontal and ±0.25 inch of axial. During this cycle, more yield lines occurred on the south top half flange of the pile. There was no yield lines on the other parts of the pile.

The next step was applying six cycles of ±6 inches of horizontal and ±0.3 inch of axial displacements. During the first cycle of this step, yield lines on the top half of south flanges became more visible and thicker. At the end of these ±6 inch cycles, there was some yield lines on the bottom half of the north side flange of the pile. There was no yield lines on the web. There was no signs of local or overall bucking yet and other than flange yield lines as explained earlier, there was no other damage. The concrete in the pile cap was intact with no hair crack. But, no yield lines could be seen on the web or on the north half of the flanges. The ±6 inches horizontal displacement cycles were followed by four cycles of ±8.5 inches of horizontal displacements. During first cycle of these series, when specimen had about 7.5 inches of horizontal displacement and the pile was in compression, very small local buckling waves were observed over a length of 6 feet of pile from the pile cap end.

The local buckling resulted in the axial load axial displacement curve to show a flat plateau. Also some additional yield lines occurred. When specimen was subjected to tension, more 45 degree yield lines occurred. The yield lines were on about 6-7 feet of pile length almost uniformly distributed and mostly on to north and some on bottom south flanges. Some of the yield lines were extended to the bottom half south flange. At the end of these cycles, local buckling waves had an amplitude of about 1/8 inch. No yield lines were visible on the web. No hair cracks could be observed in the pile cap.

The next step was applying four cycles of 11 inches of horizontal and 0.45 inch of axial displacements. During the first cycle of this step, yield lines on the south half of top flanges became more visible and thicker and extended through a length of about 10 feet of pile from pile cap. Also yield lines on the north side became more visible, all 45 degrees and extended over about 4 feet of pile cap. At this time, local buckling waves were very visible and had an amplitude of about 3/4 inch. The local buckling wave that was about 4 feet away from the pile cap was very large and visible compared to others. Some minor overall bending curve could be seen. No yield lines could be seen on the web. The yield lines on north side top half of the section had penetrated into the bottom half near the pile cap and at a section about 5 feet away from pile cap. There was some signs of overall buckling although the curvature was small. There was no other damage. The concrete in the pile cap did not have any damage.

It was clear now that the pile was yielding inside the pile cap since anytime tension was applied, the pile would stretch inside the pile cap and will come out of the pile cap about 1/16 inch.

The 11 inch cycles of horizontal displacement were followed by 4 cycles of 13.5 inches of horizontal and 0.45 inch of axial displacements. During the first cycle of this step, local buckling became more serious and the amplitude was about 1.5 inches. There was also some cracking of concrete cover on the pile cap. The concrete in the pile cap did not show spalling yet.

After completing four cycles of ±13.5 inch cycles of horizontal displacements, 20 cycles of ±16 inch horizontal displacements were applied. During the first cycle of 16 inch displacement, the local buckling became more pronounced and the tension yield lines increases. However, still there was no sign of yielding of web nor fracture. No more concrete spalling occurred. At the end of seventh cycle of 16 inches, some minor spalling of concrete occurred. Also at this time the bottom half of the north flange showed signs of local buckling just outside the pile cap.

At the end of 20 cycles of ±16 inch horizontal displacements the damage was in the form of local buckling of plastic hinge areas just outside the pile cap and at a location about 5 feet away from the pile cap. No sign of fracture could be seen.

Since specimen had not failed yet, the displacement history developed by researchers and approved by Sponsor was already exhausted. Therefore, to establish failure, it was decided to continue the test by extending displacement time history. The additional displacements were 4 cycles of ±16 inches of horizontal displacements and ±0.60 inches of vertical, 4 cycles of ±16 inches of horizontal displacements and ±0.75 inch of vertical, 4 cycles of ±16 inches of horizontal displacements and ±0.90 inches of vertical, 4 cycles of ±16 inch horizontal displacements and ±1.05 inch of vertical displacements.

After applying 4 cycles of ±16 inch horizontal displacements and ±0.6 axial, no new damage was observed. After applying 4 cycles of ±16 inch horizontal displacements and ±0.75 axial, there was still not much of new damage, the locally buckled areas continued to deform under compression and straighten slightly under tension.

After applying 4 cycles of ±16 inch horizontal displacements and ±0.90 axial, the overall buckling was very visible. The plastic hinge that had formed at a section about 5 feet from the pile cap continued to show sharper and sharper locally buckled areas. At the end of four cycles of 0.9 axial displacement, when specimen was under tension, a hair crack was noticed at the tip of locally buckled area of the plastic hinge that was about 5 feet below the pile cap. After applying 2 cycles of ±16 inch horizontal displacements and ±1.05 axial, there was another crack at the same location but along the web. The cracks joined and the entire south flange had fractured.

Fig 8 Fracture through locally buckled area at mid-length of Specimen B

Cyclic Behavior of Specimen C

The total number of inelastic cycles that Specimen C could tolerate prior to developing cracks was 69. The maximum cyclic horizontal displacement was ±16 inches. In the following a summary of behavior of Specimen C during the cyclic loading is provided.

Testing of this specimen started by applying two initial cycles of ±1.0 inch horizontal and ±0.15 inch vertical. At the end of these two cycles, there was no signs of yielding visible. However, it was realized that the stiffness of specimen when under tension is less than stiffness of specimen when under compression. This was investigated and it was felt that the flexibility is the result of elongation of bolts connecting the pile to the actuator.

The next step in loading history was applying two cycles of ±2.5 inches of horizontal and ±0.25 inch of vertical displacements. During this cycle, yield lines occurred on the south half of the bottom flange. There was no other yield lines visible. The yield lines occurred along the 11 feet of length of pile below the pile cap. This was an indicator of yielding being the result of axial effects.

The next step was applying two cycles of ±3.5 inches of horizontal and ±0.25 inch of vertical. During this cycle, yield lines occurred on the south half of the bottom flange of the pile as well as the south side of top flange. The amount of yielding of both flanges during this cycle became almost the same.

The next step was applying six cycles of ±6 inches of horizontal and 0.3 inch of vertical displacements. During the first cycle of this step, yield lines on the south half of both flanges became more visible and thicker. But, no yield lines could be seen on the web or on the north half of the flanges. There was no signs of local or overall bucking yet and other than flange yield lines as explained earlier, there was no other damage. The concrete in the pile cap was intact with no hair cracks.

The ±6 inches horizontal displacement cycles were followed by four cycles of ±8.5 inches of horizontal displacements. During first cycle of these series, when specimen was in compression, no additional yield lines could be seen. However, when axial force was reversed to tension, the specimen showed new yield lines over a length of about one foot just below the pile cap. There was no sign of local or overall buckling yet. Also, there was no sign of damage to pile cap concrete. There was no yield lines visible on the north side flanges or the web.

The next step was applying four cycles of 11 inches of horizontal and 0.45 inch of axial displacements. During the first cycle of this step, yield lines on the south half of both flanges became more visible and thicker. Also yield lines on the north side became more visible. But, no yield lines could be seen on the web. The north side yield lines had also propagated further. There was no signs of local or overall bucking yet and other than flange yield lines as explained earlier, there was no other damage. The concrete in the pile cap did not have any damage. It was clear now that the pile was yielding inside the pile cap since anytime tension was applied, the pile would stretch inside the pile cap and will come out of the pile cap about 1/8 inch.

The 11 inch cycles of horizontal displacement were followed by 4 cycles of 13.5 inches of horizontal and 0.45 inch of axial displacements. During the first cycle of this step, local buckling of both flanges started. The locally buckled area was about 20 inches long just below the pile cap. Yield lines on both flanges became more visible and thicker. Still, there was no yield lines visible on the web. There was no sign of overall buckling. The concrete in the pile cap showed spalling of cover at the tip of pile flanges. The spalled area was about 2x2 inches in the area and about 1/2 inch deep. Again, it was clear that the pile was yielding inside the pile cap since anytime tension was applied, the pile would stretch inside the pile cap and will come out of pile cap for about 1/8 inch.

After completing four cycles of ±13.5 inch cycles of horizontal displacements, 20 cycles of ±16 inch horizontal displacements were applied. During the first cycle of 16 inch displacement, the local buckling became more pronounced and the tension yield lines increases. However, still there was no sign of yielding of web nor fracture. No more concrete spalling occurred. At the end of these 20 cycles of ±16 inch horizontal displacements there was not much damage other than yielding of flanges, local buckling of both flanges and minor spalling of concrete in the pile cap. No sign of fracture could be seen.

After applying 4 cycles of ±16 inch horizontal displacements and ±0.6 inch vertical displacements followed by 4 cycles of ±16 inch horizontal displacements and ±0.75 inch vertical displacements there was still not much of new damage. The locally buckled areas continued to deform under compression and straighten slightly under tension.

After applying 4 cycles of ±16 inch horizontal displacements and ±0.90 axial, the overall buckling was very visible. A plastic hinge clearly had formed at mid-length. After applying 4 cycles of ±16 inch horizontal displacements and ±1.05 axial, there was still not much of new damage other than more and more overall buckling deformations and local buckling deformations at mid-length and just below the pile cap.

At this time specimen had not failed yet and the displacement history applied to previous specimens was already exhausted. Therefore, to establish failure, it was decided to continue the test by extending displacement time history. The additional displacements were 4 cycles of ±16 inches of horizontal displacements and ±1.8 inches of vertical displacements, 4 cycles of ±16 inches of horizontal displacements and ±3.6 inches of vertical displacements followed by 4 cycles of ±16 inches of horizontal displacements and ±4.50 inches of vertical displacements.

After applying 4 cycles of ±16 inch horizontal displacements and ±1.80 axial, there was still not much of new damage other than more and more overall buckling deformations and local buckling deformations of mid-length. At this time the tension side of mid-length hinge also formed local buckling.

The next cycles were ±16 inch horizontal displacements and ±3.6 inches vertical displacement. During these cycles, when specimen was under compression, the overall buckling was very pronounced and visible. When specimen was subjected to tension, 45-degree yield lines occurred throughout the length of specimen. During the last cycle of this series, a hair crack was observed at the tip of locally buckled area of flange at mid-length.

The next series of cyclic displacements were 4 cycles of ±16 inch horizontal displacements and ±4.5 inches vertical displacements. During the first tension part of these cycles, the tip of flange at locally buckles area of mid-length fractured. Fig. 9 shows mid-length plastic hinge area and crack at the tip of locally buckled area.

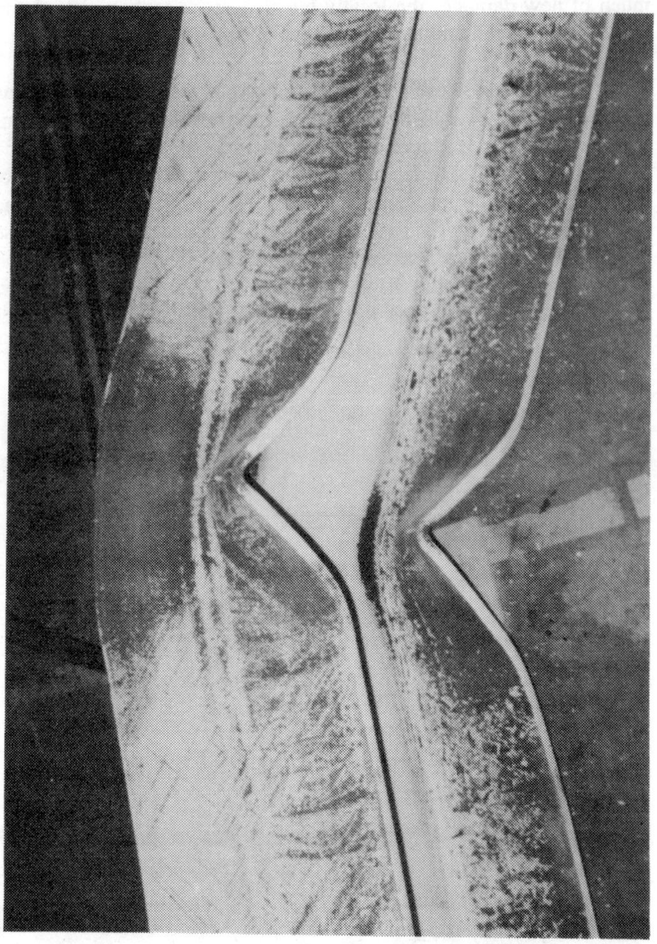

Fig 9 Fracture through locally buckled area at mid-length of Specimen C

CYCLIC TEST RESULTS

Fig. 10 shows variation of axial load versus bending moment for three specimens. The shaded regions indicate the areas within which the M-P combinations occurred during the tests. The figure also shows the M-P curves currently prescribed by the AASHTO Specifications (AASHTO, 1994) with dashed lines.

CONCLUSIONS

Based on analysis of test results, the following conclusions are reached:

1. The steel H-Piles that were tested were very ductile and capable of tolerating ±16 inches of cyclic horizontal displacements along with ±0.45 inches of vertical displacements for more than 25 cycles. The length of pile was 20 feet.

2. The final fracture for all three specimens occurred at the tip of locally buckled areas.

3. The plastic hinges that formed in the pile could tolerate cyclic rotations of at least 0.05 radians for 25 cycles without fracture.

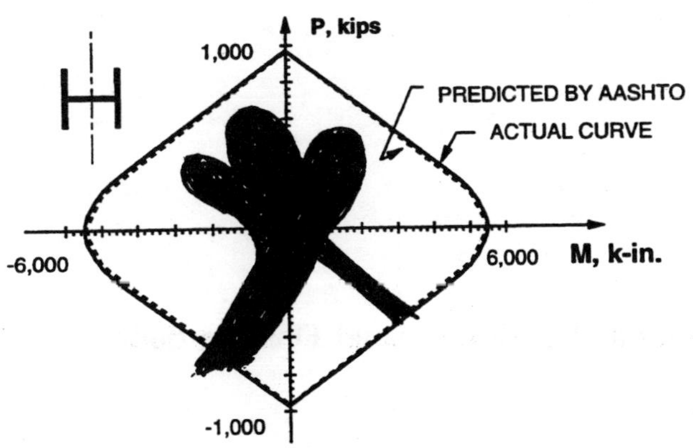

M-P Curves, Bending About Strong Axis

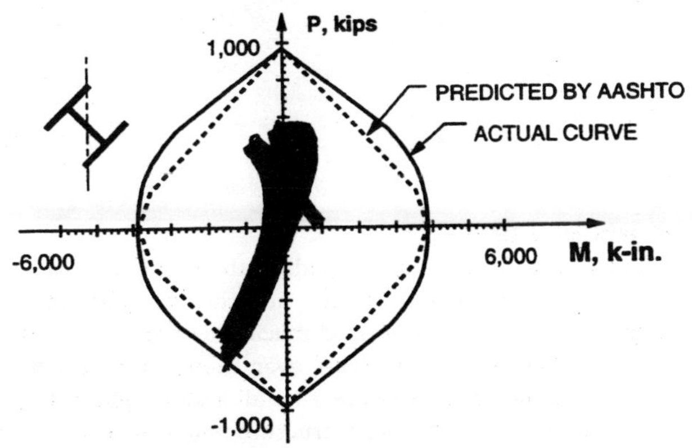

M-P Curves, Bending About 45-degree Axis

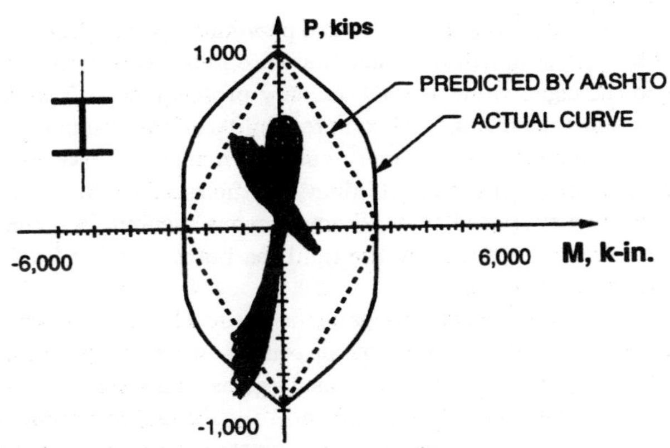

M-P Curves, Bending About Weak Axis

Fig. 10 Axial load-moment interaction for the cross section below the pile cap.

ACKNOWLEDGMENTS

The project summarized here was sponsored by the California Department of Transportation (Caltrans) and was conducted at the Department of Civil and Environmental Engineering of the University of California at Berkeley. The support and technical comments by Timothy Leahy, Reza Valizadeh, James Anderson, and James Roberts of Caltrans are sincerely appreciated. The assistance of Dr. Marcial Blondet, Dr. Lev Stepanov, Tod Merport, William Mac Cracken and Chris Moy of the Department of Civil Engineering was essential in conducting the tests. Graduate student research assistants Judy Liu and Mahmoud Hachem also were involved in design of test set-up and in conducting the tests. Undergraduate research assistant Alec Dawson conducted the material tests for steel and concrete for test specimnens.

REFERENCES

Astaneh-Asl, A, and Ravat, S (1997). "Cyclic Tests of Steel H-Piles," *Report No. UCB/CEE-Steel-97/01,* Dept. of Civil and Env. Engrg., Univ. of California, Berkeley.

AASHTO, (1994). "AASHTO-LRFD Specifications for Design of Highway Bridges", American Association of State Highway and Transportation Officials, Washington D.C..

Proceedings of the Seventh (1997) International Offshore and Polar Engineering Conference
Honolulu, USA, May 25-30, 1997

ISBN 1-880653-28-1 (Set); ISBN 1-880653-32-X (Vol. IV)

A Theoretical Study on the Effect of Seaquakes on a Two-Dimensional Floating Body

Yasushi Higo
Hiroshima University
Higashi-Hiroshima, Japan

ABSTRACT

This paper is presenting a reasonable numerical calculation method of seismic wave propagation in two-layered medium. The hydrodynamic pressure caused by seaquake is analyzed numerically and its dynamic effects on the floating structure is investigated.

In order to obtain the solution of the seismic wave field in two-layered medium, the eigen functions on the open boundary and the boundary integral equation are introduced. Furthermore, in order to understand the properties of the effect of the seaquakes on the two-dimensional floating structure, some typical calculations by using Boundary Integral Equation Method are carried out. Then the effect of seaquake forces acting on the two-dimensional floating structure is investigated from a view point of the structure safty.

Keywords: seaquake forces, two-dimensional floating body, two-layered medium, BIEM

1.INTRODUCTION

A very strong earthquake suddenly shook Kobe city in 1995. The damage caused by the earthquake was beyond expectations and this gave a great shock to us, even though we experience frequent earthquakes in Japan. Seizing the occasion, we are turning our eyes to the vast spaciousness of the ocean. That is to say, the conception which main facilities, e.g. airport, power plant etc., are made on marine structures for the sake of aseismatic design, is attracting the attention.

Marine structures had been generally considered to be aseismatic design. However, the effects of vertical motion of sea bed due to earthquake may not be small, so that it is considered that so-called seaquake which is caused by propagation of seismic motion of sea bed through seawater affects on a floating structure. There are some reports about the damage due to seaquakes(Sato, 1978).

There have been a lot of studies about the sea load due to water waves, wind and curent, but the study on effect of seaquakes has not been performed much(Hagiwara, 1986; Baba, 1987; Baba,1988; Kiyokawa, 1989), even though seaquake forces acting on the floating structure should not be ignored. The effect of seaquakes on a floating structure may be one of an item to be taken into further consideration from a view point of the structure safety.

The author made clear in the previous reports (Higo et al., 1996a; Higo et al., 1996b) that there are resonances corresponding to the eigen-value of the boundary problem on the seaquake forces and the effect of the compressibility of fluid cannot be ignored in the vicinity of these resonant frequencies. However, the seaquake forces approached infinity at the resonant frequencies, because we supposed that a homogeneous medium is bounded by the free surface and by the rigid sea bottom and there is no damping in this area.

The main aim of this paper is to develop the reasonable numerical calculation method of seismic wave propagation in two-layered medium which is considered as consisting of the seawater and the marine sediment and to investigate the property of seaquake forces acting on a floating structure systematically. Finally, the effect of the marine sediment on the seaquake forces is investigated.

In order to obtain the solution of the seismic wave fields, the Boundary Integral Equation Method (BIEM) is basically used. The eigen-functions on the open boundary are introduced and then the boundary integral equation is also introduced. Finally, the pressure distribution and the seaquake forces on a floating structure are obtained by solving the integral equation, and then the effect of seaquakes on the floating structure are discussed.

2.PROBLEM FORMULATION

We make the assumption that the fluid is inviscid, irrotational and compressible. Let us suppose that a two-layered medium is bounded by the free surface $y = 0$ and by the rigid sea bottom $y = h$ as in Fig.1. Each homogeneous layer is bounded by the horizontal surface $y = h_0$. The densities of upper (layer I) and lower (layer II) layers will be denoted by ρ_1 and ρ_2, respectively, and the sound velocity by c_1 and c_2 ($c_1 < c_2$). Furthermore, we make the assumption that one part of sea bottom is oscillating vertically and harmonically due to the P-wave of earthquake, and there is a rigid floating body having a breadth B and a draft d, right over the epicenter. The seismic waves propagate in seawater, and these give the effects on the floating body.

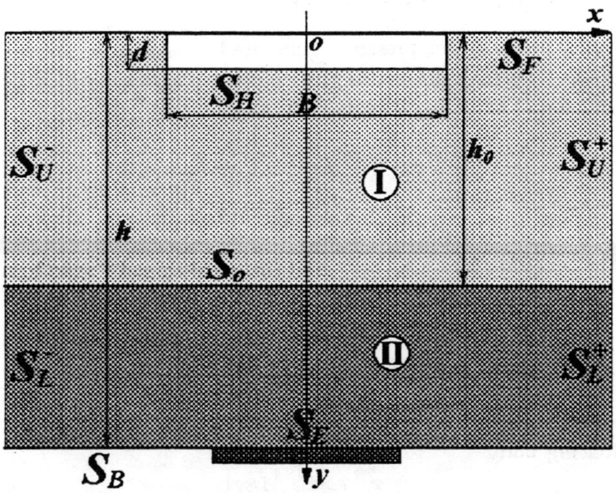

Fig.1 Coordinate system

2.1 EIGEN-FUNCTION

The seismic wave field will be specified by the velocity potential as follows:

$$\left.\begin{aligned} \Phi_1(x,y;t) &= \mathrm{Re}\left[\phi_1(x,y)e^{i\omega t}\right] \\ \Phi_2(x,y;t) &= \mathrm{Re}\left[\phi_2(x,y)e^{i\omega t}\right] \end{aligned}\right\}; \tag{1}$$

where $\Phi_1(x,y;t)$ and $\Phi_2(x,y;t)$ are the velocity potential in layer I and in layer II, respectively. The velocity potential $\phi_1(x,y)$ and $\phi(x,y)$ satisfy the following Helmholtz equations

$$\left.\begin{aligned} \nabla^2\phi_1 + k_1^2\phi_1 = 0; \quad k_1 = \omega/c_1 \\ \nabla^2\phi_2 + k_2^2\phi_2 = 0; \quad k_2 = \omega/c_2 \end{aligned}\right\}; \tag{2}$$

where k_1 and k_2 are wave numbers of layer I and layer II respectively. Therefore the velocity potentials describing a wave radiating from the source have the form

$$\left.\begin{aligned} \phi_1 &= \psi_1(y;\xi)e^{-i\xi x} \\ \phi_2 &= \psi_2(y;\xi)e^{-i\xi x} \end{aligned}\right\}. \tag{3}$$

Substituting (3) into (2), we obtain the following equations:

$$\left.\begin{aligned} \frac{d^2\psi_1}{dy^2} + \gamma_1^2(\xi)\psi_1 = 0 \\ \frac{d^2\psi_2}{dy^2} + \gamma_2^2(\xi)\psi_2 = 0 \end{aligned}\right\}; \tag{4}$$

where

$$\left.\begin{aligned} \gamma_1(\xi) &= \sqrt{k_1^2 - \xi^2} \\ \gamma_2(\xi) &= \begin{cases} \sqrt{k_2^2-\xi^2} & \text{for } \xi < k_2 \\ i\sqrt{\xi^2-k_2^2} & \text{for } \xi > k_2 \end{cases} \end{aligned}\right\}. \tag{5}$$

These velocity potentials must fulfill the following boundary conditions:

$$\left.\begin{aligned} &\psi_1 = 0 && \text{on } S_F \\ &\frac{\partial\psi_2}{\partial y} = 0 && \text{on } S_B \\ &\left.\begin{aligned} \rho_1\psi_1 = \rho_2\psi_2 \\ \frac{d\psi_1}{dy} = \frac{d\psi_2}{dy} \end{aligned}\right\} && \text{on } S_o \end{aligned}\right\}. \tag{6}$$

As a result, we can write the velocity potential of each layer as follows:

$$\left.\begin{aligned} \phi_1(x,y) &= \sum_{\ell=1}^{\infty} A_\ell \sin\gamma_{1\ell} y e^{-i\xi_\ell x} \\ \phi_2(x,y) &= \sum_{\ell=1}^{\infty} A_\ell \frac{\sin\gamma_{1\ell}h_0 \cos\gamma_{2\ell}(y-h)}{m\cos\gamma_{2\ell}(h_0-h)} e^{-i\xi_\ell x} \end{aligned}\right\}; \tag{7}$$

where

$$\left.\begin{aligned} m &= \rho_2/\rho_1 \\ \gamma_{1\ell} &= \sqrt{k_1^2 - \xi_\ell^2} \\ \gamma_{2\ell} &= \begin{cases} \sqrt{k_2^2-\xi_\ell^2} & \text{for } \xi_\ell < k_2 \\ i\sqrt{\xi_\ell^2-k_2^2} & \text{for } \xi_\ell > k_2 \end{cases} \end{aligned}\right\}, \tag{8}$$

and the eigenvalues $\gamma_{1\ell}$ and $\gamma_{2\ell}$ are the solutions of the following characteristic equation (Higo et al., 1994):

$$m\gamma_{1\ell} + \gamma_{2\ell}\tan\gamma_{1\ell}h_0 \tan\gamma_{2\ell}(h_0-h) = 0. \tag{9}$$

2.2 INTEGRAL EQUATION

As we have obtained the eigen-functions of two-layered medium, we discuss here the solution of the seismic field. As is well known, the principal solutions of the Helmholtz equation (2) can be represented by Hankel functions. We will introduce the boundary integral equation in which the kernel function consists of the principal solution of the Helmholtz equation, and find the solutions which satisfies the boundary conditions. The kernel functions of each layer are expressed by the following equations:

$$\left.\begin{aligned} G_1(P,Q) &= \frac{i}{4}\left[H_0^{(2)}(Kr) - H_0^{(2)}(Kr_1)\right] \\ G_2(P,Q) &= \frac{i}{4}\left[H_0^{(2)}(Kr) + H_0^{(2)}(Kr_2)\right] \end{aligned}\right\}, \tag{10}$$

where

$$\left.\begin{aligned} r &= \sqrt{(x-x')^2 + (y-y')^2} \\ r_1 &= \sqrt{(x-x')^2 + (y+y')^2} \\ r_2 &= \sqrt{(x-x')^2 + (y+y'-2h)^2} \end{aligned}\right\}. \tag{11}$$

The velocity potentials on each boundary are written by making use of Green's formula as follows:

$$\left.\begin{aligned} \frac{\phi_1(P_0)}{2} &= \int_{S_U}\left[G_1(P_0,Q)\frac{\partial\phi_1(Q)}{\partial n} - \frac{\partial G_1(P_0,Q)}{\partial n}\phi_1(Q)\right]dS \\ \frac{\phi_2(P_0)}{2} &= \int_{S_L}\left[G_2(P_0,Q)\frac{\partial\phi_2(Q)}{\partial n} - \frac{\partial G_2(P_0,Q)}{\partial n}\phi_2(Q)\right]dS \end{aligned}\right\}, \tag{12}$$

where $S_U = S_H + S_F + S_o + S_U^+ + S_U^-$ denotes the boundary of layer I and $S_L = S_E + S_B + S_o + S_L^+ + S_L^-$ denotes the boundary of layer II. Furthermore, S_E, S_H, S_F, S_B and S_o denote the seaquake center surface, the floating body surface, the free surface, the sea bottom surface and the boundary between two layers, respectively. S_U^+ and S_U^- denote the open boundary of layer I, and S_L^+ and S_L^- denote the open boundary of layer II. Considering the boundary conditions, we can rewrite equation (12) as follows:

$$\begin{aligned} \frac{1}{2}\phi_1(P_0) &= -\int_{S_H}\phi_1(Q)\frac{\partial G_1(P_0,Q)}{\partial n}dS \\ &+ \int_{S_o}\left[G_1(P_0,Q)\frac{\partial\phi_1(Q)}{\partial n} - \frac{\partial G_1(P_0,Q)}{\partial n}\phi_1(Q)\right]dS \\ &+ \int_{S_U^+ + S_U^-}\left[G_1(P_0,Q)\frac{\partial\phi_1(Q)}{\partial n} - \frac{\partial G_1(P_0,Q)}{\partial n}\phi_1(Q)\right]dS, \end{aligned} \tag{13}$$

and

$$\begin{aligned} \frac{1}{2}\phi_2(P_0) &= \int_{S_E}\left[v_n G_2(P_0,Q) - \frac{\partial G_2(P_0,Q)}{\partial n}\phi_2(Q)\right]dS \\ &+ \int_{S_o}\left[G_2(P_0,Q)\frac{\partial\phi_2(Q)}{\partial n} - \frac{\partial G_2(P_0,Q)}{\partial n}\phi_2(Q)\right]dS \\ &+ \int_{S_L^+ + S_L^-}\left[G_2(P_0,Q)\frac{\partial\phi_2(Q)}{\partial n} - \frac{\partial G_2(P_0,Q)}{\partial n}\phi_2(Q)\right]dS, \end{aligned} \tag{14}$$

where v_n is the normal velocity on the seaquake center surface. We also obtain the following relations on the boundary between layers:

$$\left.\begin{aligned} \rho_1\phi_1 &= \rho_2\phi_2 \\ \frac{d\phi_1}{dy} &= \frac{d\phi_2}{dy} \end{aligned}\right\} \quad \text{on } S_o. \tag{15}$$

As is evident from equations (7) and (8), the eigenfunction rapidly decreases with ℓ in the case of $\gamma_{2\ell} \geq k_2$, therefore we can ignore the eigenfunctions corresponding to a larger ℓ than ℓ_{max}. The velocity potential on the open boundary is represented by the sum of the first hundred eigenfunctions at most. Therefore if we divide the open boundary into a larger number than ℓ_{max}, we can determine the unique solution of the boundary integral equation by making use of the least squares method.

3.NUMERICAL RESULTS AND DISCUSSIONS

3.1 SEAQUAKE FORCES

The dynamic pressure generated in seawater can be expressed by making use of the velocity potential, which is the solution of the integral equation, as follows:

$$\left.\begin{aligned} p_1 &= -i\rho_1\omega\phi_1 \quad \text{in } I \\ p_2 &= -i\rho_2\omega\phi_2 \quad \text{in } II \end{aligned}\right\} \tag{16}$$

The seaquake forces acting on a floating body is written by equation (17).

$$f_{E2} = -i\rho_1\omega\int_{S_H}\phi_1(x,y)n dS \tag{17}$$

The non-dimensional seaquake forces written by the following equation is defined as the magnitude of the force acting on the floating body.

$$\widehat{f_{E2}}| = \frac{|f_{E2}|}{\rho_1\omega^2 aBd} \tag{18}$$

where a is amplitude of vibration of oscillating sea bottom. The non-dimensional frequency is denoted by k_2d.

3.2 VALIDITY OF PROPOSED METHOD

The main aim of this section is to verify the validity of the proposed method based on the formulation in the previous chapter. I carried out the calculations in the case of $n = c_1/c_2 = 1$ and $m = \rho_2/\rho_1 = 1$, and compared the results with the calculated results by the program corresponding to homogeneous one-layerd medium(Higo et al., 1996a).

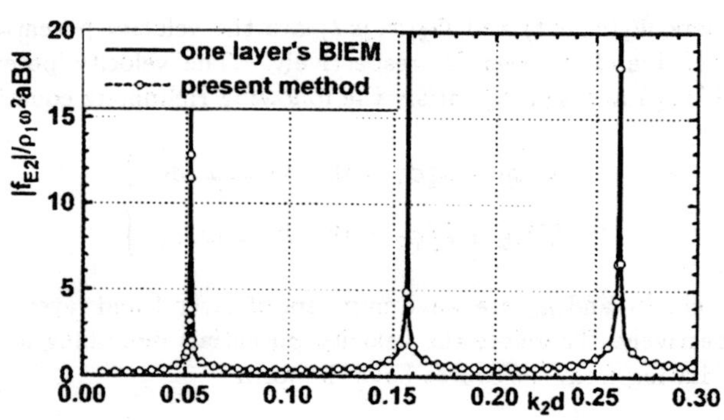

Fig.2 Comparison of numerical results of seaquake forces between one-layer's BIEM and present method

Fig.2 shows the calculated results of the seaquke forces in the case of $B/d = 2.0$, $h/d = 30.0$, $L/d = 5.0$ and $h_0/d = 15.0$. A solid line indicates the results calculated by the present program, and a solid line with circle indicates those by the program corresponding to homogeneous one-layerd medium in this figure.

It is clear that both results are identical, and it seems to be quite all right to consider that the comuting program is satisfactory.

It is clrear that there are sharp peaks at some frequencies as in Fig.2. These are caused by resonance in the seismic wave field which is bounded by free surface and rigid sea bottom. As we assume that the sea bottom is rigid and there is no absorption in the consisered field, the seismic wave never damp. Therefore, it is considered that the seaquake forces approach infinity at the resonant frequency. It means that the seaquake forces, which are calculated under these assumption, have quantitative problem in the vicinity of resonant frequency. On the other hand, it is considered that there may be energy damping on the interface between two layers in two-layer case. In order to investigate the effect of the marine sediment on the seaquake forces, the systematic calculations are carried out.

3.3 EFFECT OF INDEX OF REFRACTION

First of all, I investigated the effect of index of refraction 'c_1/c_2' on the seaquake forces. Fig.3 shows the calculated results in the case of $B/d = 2.0$, $h/d = 30.0$, $L/d = 5.0$, $h_0/d = 15.0$, $m = 2$ for various values of c_1/c_2.

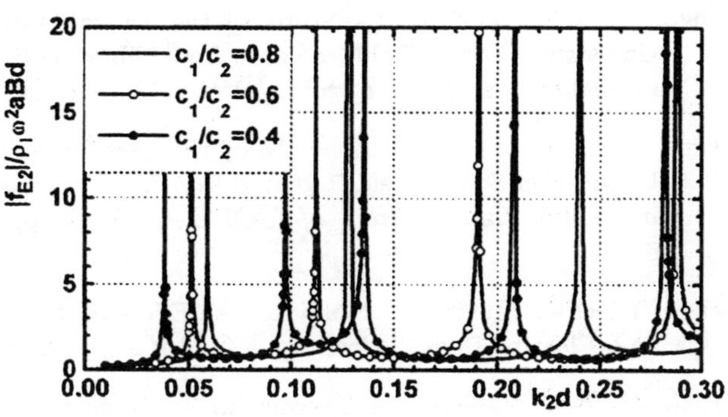

Fig.3 Effect of index of refraction on seaquake forces

It is clear that there are peaks at some frequencies of each index of refraction. These frequencies are corresponding to the frequencies which the wave number in the layer I 'k_1' approaches to eigen-value in the layer I 'γ_1' so that it is considered that these are caused by resonance of the seismic wave field in the layer I. Furthermore, it is clear that the interval of the resonant frequencies becomes wider with the increase of index of refraction. It means that the higher sound speed in the marine sediment becomes, the more resonant frequencies becomes. However, the magnitude of peak values of seaquake force doesn't become small compared to the case of homogeneous medium unexpectedly. It is considered that the marine sediment doesn't plays the great part as an acoustic damper.

3.4 EFFECT OF RELATIVE WATER DEPTH

Secondary, to investigate the effect of the relative water depth 'h_0/d' on seaquake forces, I carried out the calculations in the case of $B/d = 2.0$, $L/d = 5.0$, $h/d = 30.0$, $c_1/c_2 = 0.8$, $m = 2.0$ for various value of h_0/d.

Fig.4 shows the calculated results. It is clear that each resonant frequency is moving with the various relative water depth h_0, but the effect of relative water depth is small compared to the effect of index of refraction. It is relatively easy to expect the effect of water depth in the case of one-layered medium on the resonant frequency, but in the case of two-layered medium, we have to solve the characteristic equation (9). So we must carry out more detailed investigation, in order to make clear the effect of the rerative water depth on the seaquake.

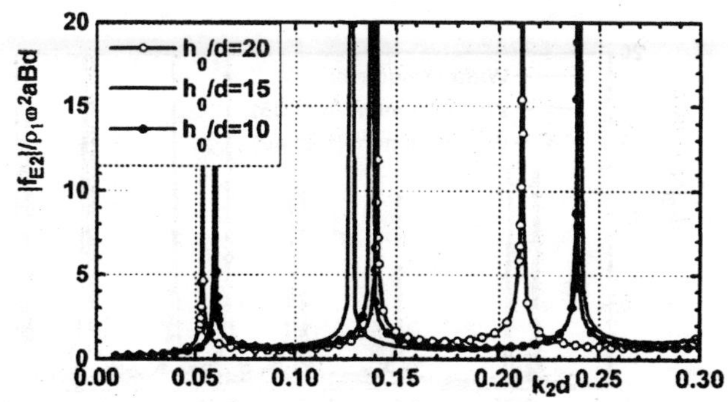

Fig.4 Effect of relative water depth on seaquake forces

3.5 EFFECT OF DENSITY RATIO

I carried out the calculations in the case of $B/d = 2.0$, $h/d = 30.0$, $L/d = 5.0$, $h_0/d = 15.0$, $c_1/c_2 = 0.8$ for the various value of m, in order to investigate the effect of the density ratio 'm' on seaquake forces. Fig.5 shows the calculated results.

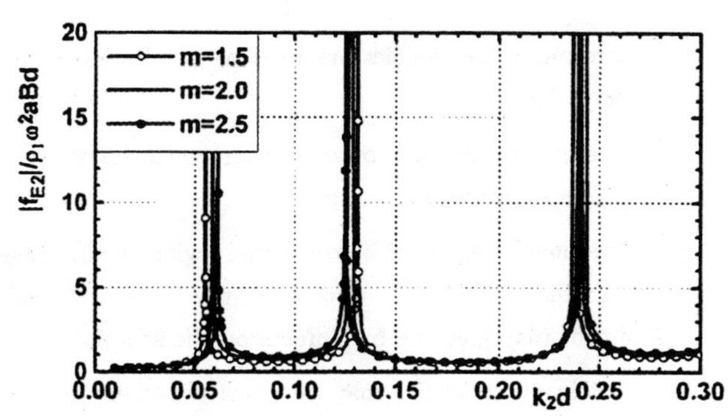

Fig.5 Effect of density ratio on seaquake forces

As is evident from Fig.5, it can be considered that there is very little effect of density ratio on the resonant frequency.

3.6 EFFECT OF SEDIMENT'S THICKNESS

Finally, inorder to investigate the effect of the marine sediment's thickness on seaquake forces, I carried out the calculations in the case of $B/d = 2.0$, $L/d = 5.0$, $h_0/d = 30$, $c_1/c_2 = 0.8$, $m = 2.0$ for various value of h/d. Fig.6 shows the calculated results.

In Fig.6, the solid line indicates the calculated results of the case of homogeneous medium, and this can be considered as the case of $h/d = 30.0$.

It is clear that the resonant frequency becomes more with the increase of the marine sediment's thickness. In this sense, the marine sediment may give the bad effects on the seaquake. Therefore it is considered that the floating structure may be constructed in the ocean area, where the marine sediment's thickness is relatively thin, from an aseismatic point of view.

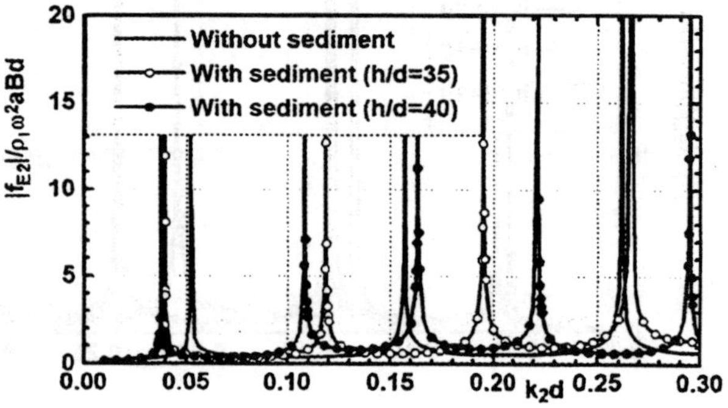

Fig.6 Effect of sediment's thickness on seaquake forces

4.CONCLUSIONS

The seaquake forces acting on a two-dimensional floating body in the two-layered seismic wave field were calculated numerically by making use of BIEM. The results may be summarized as follows:

1) The resonant frequencies are relating to the eigen-values of water field.
2) The resonant frequency becomes more with the increase of the index of refraction.
3) The resonant frequency becomes more with the increase of the marine sediment's thickness. In this sense, the marine sediment may give the bad effects on the seaquake.
4) The floating structure may be constructed in the ocean area, where the marine sediment's thickness is thin, from an aseismatic point of view.

Though I expected that the marine sediment plays the part as acoustic damper, the results which supports the expectation could not be obtained in the present study. We must solve this kind of problem by making use of not frequency domain solver, but time domain solver. Because the phenomenon of the problem is not absolutely permanent. So the following points are left as future problems:

1) Development of the solution method of time domain problem, in order to make clear the seaquake phenomenon.
2) Development of an experimental method at small tank.
3) Approach to a hydroelasticity problem, in order to investigate the dynamic behavior of structure.

REFERENCES

Baba, E. (1987). "A Study on the Effect of Seaquakes on a Floating Body", *J. of SNAJ*, Vol.162, pp.90-98

Baba, E. (1988). "Accelerations of a Floating Body Induced by Seaquaqke", *Proc. of 8th Ocean Engineering Symposium*, pp.305-311

Hagiwara, Y., et.al. (1986). "Seismic Response Property of Floating-type Structure for Nuclear Power Plant (NO.1)", *Report of the Central Research Institute of Electric Power Industry*, No.385028

Higo, Y., et.al. (1992). "A Study of the Two-dimensional Underwater Sound Field Radiated from Vibrating Bodies in Shallow Water", *J. of SNAJ*, Vol.172, pp.539-549

Higo, Y., et.al. (1994). "A Study of the Two-dimensional Underwater Sound Field Radiated from Vibrating Bodies in Shallow Water", *Proc. of ISOPE'94*, Vol.IV, pp.628-635

Higo, Y., et.al. (1996). "A Study on the Effect of Seaquakes on a Floating Body", *Proc. of KOJAM'96*, pp.269-276

Higo, Y., et.al. (1996). "A Theoretical Study on the Effect of Seaquakes on a Floating Body", *Transactions of The West-Japan Society of Naval Architects*, No.93

Kiyokawa, K. & Inada, Y. (1989). "Mechanism of Seaquake Generation, *Proc. of Coastral Engineering of Japan*, Japan Sosiety of Civil Engineering, Vol.36, pp.734-738

Sato, M. 1978, "Some Experiences concerning Seaquakes", *Kisho*, Vol.22, No.249, pp.26-28

Proceedings of the Seventh (1997) International Offshore and Polar Engineering Conference
Honolulu, USA, May 25-30, 1997

ISBN 1-880653-28-1 (Set); ISBN 1-880653-32-X (Vol. IV)

Dynamic Response Analysis of Semi Float Type Offshore Platform

Kenji Kawano, Katta Venkataramana and Tutomu Hashimoto
Kagoshima University
Kagoshima, Japan

Tomoyo Taniguchi
Kawasaki Heavy Industries, Ltd.
Kakogawa, Japan

Abstract

In order to develop the ocean space effectively, it is required to construct large scale offshore structures. Offshore platform with large deck area may provide great possibilities to develop ocean spaces which may be used as resident areas, airports, power stations, etc.. Since these structures have very heavy dead loads, reaction forces on the base foundation become severe and necessitate very firm foundation system. If these reaction forces can be reduced by some method such as large diameter members of buoyancy-type(semi float type). The concept of using large offshore platforms of semi float type can become popular in the offshore industry.

In this study, it is suggested that in order to implement the proper selection of buoyancy type member, it is important to clarify the relations between the effective loads, the external exciting forces, and allowable displacements of the soil-foundation system. Some parameters, which are used for the evaluation of wave force and the dynamic characteristics of the structure, give important effects on the response evaluation. It is shown that since these effects are evaluated with the perturbation method, the reliability index provides good indications of the dynamic response evaluation.

INTRODUCTION

Offshore platform with large deck area may provide great possibilities to develop ocean spaces which may be used as resident areas, airports, power stations, etc. Since these structures have very heavy dead loads, the reaction forces on the base foundation become severe and require very firm foundation system(Sigbjornsson et al (1978)). If these reaction forces on the base foundation system can be reduced using the buoyancy type large members, important contributions on developing the large ocean spaces can be provided by these offshore structures which experience more severe loads than those on land structures. In order to perform reliable design of such structures, it is important to clarify their dynamic characteristics. The evaluations of wave forces play one of the most important roles in the dynamic response analyses of the structure. Using the Airy wave theory, the wave forces on structural members with various diameters can be evaluated with the Morison's equation. It is suggested that if the reaction force from the offshore platform to foundation can be reduced with the buoyancy-type large members, this approach is very effective in developing structures with huge deck areas (Kawano et al (1996)). Applying stochastic properties on the motion of the water particles in the ocean wave, the wave forces are also represented as random processes. Applying the Bretschneider type power spectrum density functions for the sea surface motion, the dynamic response analyses can be performed using the random vibration approach (Jesien(1987)).

In this study, the dynamic response analyses of semi float type offshore platform subjected to wave loads are examined using random vibration approach. The equation of motion of the superstructure is represented with the finite element method, and that of foundation system is obtained with the impedance function model for the soil-pile foundation system. The governing equation of the total system is obtained using the substructure method(Yamada et al(1988)). Since the dynamic characteristics of input motions have important roles on the response evaluations of the structure, it is important, for reliable design of the structure, to perform the dynamic response evaluations to wave loads . The dynamic response analyses of the analytical models subjected to wave forces and current forces is performed with random vibration approach in order to examine the dynamic properties of the structure with various members of buoyancy types. Some parameters, which are used for the evaluation of wave force and the dynamic characteristics of the structure, may provide significant contributions on the response evaluation. Applying the perturbation method and the reliability index approach for the evaluations of the effects of these parameters, the reliable evaluation of the dynamic response can be carried out .

FORMULATION

Governing Equation of Motion

The analytical model of an offshore platform is shown in Fig.1. The sea wave has generally stochastic properties, and the wave force is usually expressed with the Morison's equation. The sea surface motion is represented with the power spectrum such as the Bretschneider type . Using the Bretschneider expression, power spectral density function of sea surface motion $S_{\eta\eta}(\omega)$ can be expressed with the mean wave period $\overline{T}$,and the mean wave height $\overline{H}$ as follows:

$$S_{\eta\eta}(\omega)=\frac{4210}{\omega^5}\left(\frac{\overline{H}}{\overline{T}^2}\right)^2\exp\left(-1052\left(\frac{1}{\omega\overline{T}}\right)^4\right) \tag{1}$$

Since these offshore structures may be subjected to current forces, these effects are examined in this analysis. The current force effects on the wave force may be expressed with the modified power spectrum of

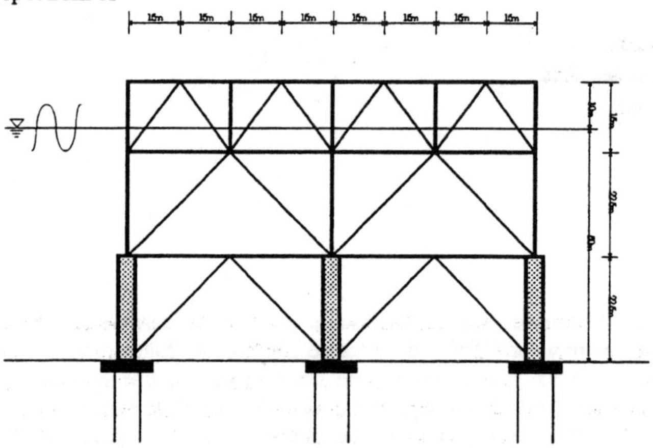

Fig.1 Analytical model of offshore structure with buoyancy type members

surface motion. The power spectrum $S^*_{\eta\eta}(\omega)$ for current in line with waves is represented as follows(Chakrabarti S. K. ,1987):

$$S^*_{\eta\eta}(\omega)=\left(\left[1+\left(1+\frac{u_0\omega}{g}\right)^{1/2}\right]\left[\left(1+\frac{u_0\omega}{g}\right)^{1/2}+\left(1+\frac{u_0\omega}{g}\right)\right]\right)^{-1}S_{\eta\eta}(\omega) \tag{2}$$

in which u_0 denotes the current velocity and g the gravitational acceleration.Since the water particle velocity and acceleration at any depth can be expressed with the sea surface motion using the Airy wave theory, the power spectral density function of these velocities for example can be expressed as follows:

$$S_{VV}(\omega)=\omega^2\left(\frac{\cosh(ks)}{\sinh^2(kh)}\right)^2 S^*_{\eta\eta}(\omega) \tag{3}$$

The superstructure is represented using the finite element method with beam elements and the dynamic characteristics of soil-foundation system are expressed with the impedance function model. The dynamic equation of motion is obtained by substructure method which is obtained by combining the equation of motion of superstructure and foundation and satisfying the compatibility and equilibrium equations at the interface. The equation motion of the superstructure is at first formulated for the wave force input which is represented by the modified Morison equation. Using the equivalent linearization method for the drag force term, the equation is linearized where the wave force is proportional to the relative velocity of water particle and structure(Penzien et al(1972)). The equation of motion of the superstructure with the fixed base conditions is given by

$$[\mathbf{M}+\mathbf{K}_m]\ddot{\mathbf{x}}+[\mathbf{C}+\overline{\mathbf{K}}_D]\dot{\mathbf{x}}+\mathbf{K}\mathbf{x}=\mathbf{K}_M\dot{\mathbf{v}}+\overline{\mathbf{K}}_D\mathbf{v} \tag{4}$$

in which

$$\mathbf{K}_m=\sum_e\int_e\rho(C_M-1)V_0\mathbf{N}(z)^T\mathbf{N}(z)dz$$

$$\overline{\mathbf{K}}_D=\sum_e\int_e\frac{\rho}{2}C_DA\sqrt{\frac{8}{\pi}}\sigma_v\mathbf{N}(z)^T\mathbf{N}(z)dz$$

$$\mathbf{K}_M=\sum_e\int_e\rho C_MV_0\mathbf{N}(z)^T\mathbf{N}(z)dz$$

The matrices, $\mathbf{M}$, $\mathbf{C}$ and $\mathbf{K}$ are the mass matrix, the damping matrix and the stiffness matrix of the superstructure, respectively, $\mathbf{v}$ and $\dot{\mathbf{v}}$ are the velocity and acceleration of the water particle , respectively, and V_0 and A are the corresponding volume and area of each structural member normal to these flow, and C_M and C_D are the inertia coefficient and the drag coefficient , respectively. The $N(z)$ is the shape function , ρ the density of water and σ_v denotes the *rms* relative velocity between water particle and structure. Since the several lower dominating vibration modes have significant contributions on the dynamic responses of the superstructure, the governing equation of motion of the total system can be expressed using these modes by the application of the substructure method(Yamada Y. et al,1988)). The governing equation of motion for the total system is then given by

$$\begin{bmatrix}\mathbf{I} & \tilde{\mathbf{M}}_{ap}\\ \tilde{\mathbf{M}}_{pa} & \tilde{\mathbf{M}}_{pp}\end{bmatrix}\begin{Bmatrix}\ddot{\mathbf{q}}\\ \ddot{\mathbf{x}}_p\end{Bmatrix}+\begin{bmatrix}[2\beta_j\omega_j] & \mathbf{0}\\ \mathbf{0} & \tilde{\mathbf{C}}_p\end{bmatrix}\begin{Bmatrix}\dot{\mathbf{q}}\\ \dot{\mathbf{x}}_p\end{Bmatrix}+\begin{bmatrix}[\omega_j^2] & \mathbf{0}\\ \mathbf{0} & \tilde{\mathbf{K}}_p\end{bmatrix}\begin{Bmatrix}\mathbf{q}\\ \mathbf{x}_p\end{Bmatrix}=\begin{Bmatrix}\mathbf{P}_a\\ \mathbf{P}_b\end{Bmatrix} \tag{5}$$

in which

$$\begin{Bmatrix}\mathbf{P}_a\\ \mathbf{P}_b\end{Bmatrix}=\begin{bmatrix}\Phi^T\mathbf{K}_M & \Phi^T\overline{\mathbf{K}}_D\\ \mathbf{G}^T\mathbf{L}^T\mathbf{K}_M & \mathbf{G}^T\mathbf{L}^T\overline{\mathbf{K}}_D\end{bmatrix}\begin{Bmatrix}\dot{\mathbf{v}}_a\\ \mathbf{v}_a\end{Bmatrix}=\mathbf{F}\ \mathbf{Z}$$

and the suffix a denotes the unrestrained nodal points of the superstructure and the suffix p denotes the pile soil foundation, the matrix $\mathbf{G}$ relates the displacements of the base nodal points to those of the soil-pile foundation, $\mathbf{x}_p$ denotes the displacements of the soil-pile foundation system, $\mathbf{\Phi}$ is the modal matrix of the superstructure, and $\mathbf{q}$ is the generalized displacement. Given the water particle velocity and acceleration at each nodal points, above equation can be solved by time integration methods such as the linear acceleration method. However, since these quantities have generally random properties, it is convenient to.express them with the power spectral density functions.

The dynamic response analysis of an offshore structure can be easily carried out applying the eigenvalue analysis to Eq.(5). Using the classical modal analysis and neglecting the modal coupling effects, Eq.(5) can be transformed into a general coordinate system. The governing equation of motion can be represented with the new eigenvector $\mathbf{\Psi}$ which is determined using the corresponding mass matrix and stiffness matrix of Eq.(5) as follows:

$$\ddot{\mathbf{y}}+[2\beta_j\omega_j]\dot{\mathbf{y}}+[\omega_j^2]\mathbf{y}=\mathbf{\Psi}^T\mathbf{F}\ \mathbf{Z} \tag{6}$$

in which

$$\begin{Bmatrix} \mathbf{q} \\ \mathbf{x}_p \end{Bmatrix} = \Psi \mathbf{y} , \quad \mathbf{Z} = \begin{Bmatrix} \dot{\mathbf{v}} \\ \mathbf{v} \end{Bmatrix}$$

Equation contains water particle velocity and acceleration terms due to the wave motion. As previously mentioned, since they have stochastic properties, it is reasonable to express them with the power spectral

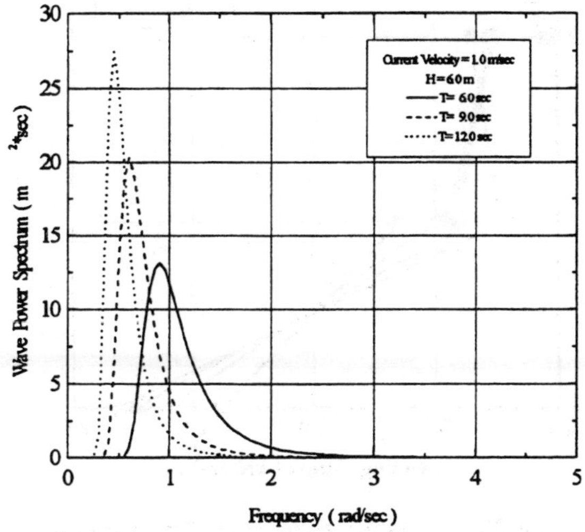

Fig.2(a) Bretschneider type power spectra

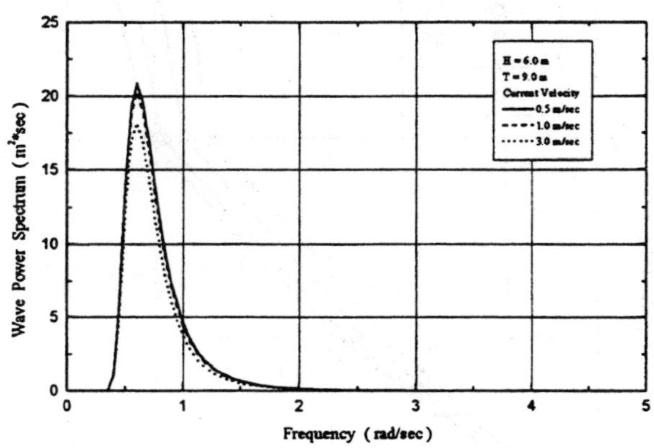

Fig.2(b) Bretschneide type power spectra

density function as shown in Eq.(1) and Eq.(2) in the frequency domain. Then, taking the Fourier transform of Eq.(7), the power spectral density function of the response vector $\mathbf{y}$ can be obtained as follows:

$$\mathbf{S}_{yy}(\omega) = \mathbf{H}(\omega)\,\mathbf{S}_{FF}(\omega)\,\mathbf{H}(\omega)^{T*} \qquad (7)$$

in which

$$\mathbf{H}(\omega) = [-\omega^2 \mathbf{I} + i\omega [\, 2\beta_j \omega_j \,] + [\, \omega_j^2 \,]\,]^{-1}$$

$$\mathbf{S}_{FF}(\omega) = \Psi\ \mathbf{F}\ \mathbf{S}_{ZZ}(\omega)\ \mathbf{F}^{*}\ \Psi^{T}$$

and * denotes the conjugate matrix. Assuming the stationary situation of the input wave forces, the responses can be easily expressed with the covariance matrix. Then, the covariance matrix of the vector $\mathbf{y}$ can be obtained using the inverse Fourier transform.

$$E[\mathbf{y}\mathbf{y}^T] = \int_{\infty}^{-\infty} \Psi\, \mathbf{S}_{yy}(\omega)\, \Psi^T\, d\omega \qquad (8)$$

Since the vector $\mathbf{y}$ includes the generalized displacements of the superstructure and displacements of the soil- foundation as expressed in Eq.(5) and Eq.(6), the covariance matrix of displacements of the superstructure can be obtained by the following transformation.

$$E[\mathbf{x}_a^c \mathbf{x}_a^c] = \Phi E[\mathbf{q}\mathbf{q}^T]\Phi^T \qquad (9)$$

in which x_a^c denotes the relative displacement of the superstructure. The root mean square (*rms*) responses can then be determined by taking the root of each diagonal term of the covariance matrix. Moreover, since the resultant force can be obtained with the displacement and the stiffness matrix, the covariance matrix of the resultant forces can be determined with the stiffness matrix of the structure as follows:

$$E[\mathbf{S}\mathbf{S}^T] = \mathbf{K}\, E[\mathbf{x}_a^c\, \mathbf{x}_a^{c\,T}]\, \mathbf{K}^T \qquad (10)$$

Finally, the *rms* responses of bending stress response and the shearing stress and the axial normal stress are evaluated with these resultant forces.

Evaluations of uncertain parameters

The governing equation of motion for offshore structures has many parameters with uncertain properties. The inertia coefficient and drag coefficient whose values must be chosen carefully, while evaluating wave force, have generally variations. Assuming these coefficients are random variables distributed statistically around their mean values, the coefficient matrices including these variables can be expressed with perturbation method.

$$\begin{aligned} \mathbf{K}_m &= \mathbf{K}_m^{(0)} + \varepsilon_1 \mathbf{K}_m^{(1)} \\ \mathbf{K}_M &= \mathbf{K}_M^{(0)} + \varepsilon_1 \mathbf{K}_M^{(1)} \\ \overline{\mathbf{K}}_D &= \overline{\mathbf{K}}_D^{(0)} + \varepsilon_2 \overline{\mathbf{K}}_D^{(1)} \end{aligned} \qquad (11)$$

in which ε_1 and ε_2 are random variables distributed statistically around their mean values for the inertia coefficients and the drag coefficient, respectively.

The surface motion of the random sea wave can be represented with the energy spectrum of the Bretschneider type as expressed in Eq.(1) and Eq.(2). This spectrum is a function of many variables such as wave number, mean wave height, mean wave period and so on. Since the mean wave height may provide significant effects on the dynamic response evaluation, it is also necessary to examine the variation of this parameter on the responses. The power spectrum of the wave energy of the surface motion can be expressed as the sum of the mean and perturbed values .

$$S^{*}_{\eta\eta} = S^{*(0)}_{\eta\eta}(\omega) + \varepsilon_3 S^{*(1)}_{\eta\eta}(\omega) \qquad (12)$$

in which ε_3 denotes the random variable distributed statistically around its mean value of the mean wave height.

The dynamic soil-structure interaction also have important effects on the dynamic response evaluation. The dynamic properties of the soil-foundation system generally can be evaluated with impedance functions which are functions of the shear wave velocity of soil. Assuming the shear wave velocity of a soil is random variable distributed statistically around its mean value, the corresponding matrix on the impedance function is given by

$$\mathbf{K}_p = \mathbf{K}_p^{(0)} + \varepsilon_4 \mathbf{K}_p^{(1)} \qquad (13)$$
$$\mathbf{C}_p = \mathbf{C}_p^{(0)} + \varepsilon_4 \mathbf{C}_p^{(1)}$$

in which ε_4 denotes the random variable distributed statistically around its mean value for the shear wave velocity of soil. Assuming the variations of coefficients of these parameters, the governing equation of motion on the offshore structure system can be expressed with the perturbation method. Then, the general coordinate system as expressed in Eq.(6) can be modified with these random variables as

$$\mathbf{S} = \mathbf{S}_0 + \sum_{j=1}^{4} \varepsilon_j \mathbf{S}_j \qquad (14)$$

in which $\mathbf{S}_j = \begin{bmatrix} \mathbf{q} & \mathbf{u}_p^{(j)} \end{bmatrix}^T$

Substituting Eq.(14) into Eq.(5) and neglecting the terms containing higher than the first orders in ε, the governing equation of motion for each random variable can be expressed as

$$\mathbf{A}_0 \ddot{\mathbf{S}}_j + \mathbf{B}_0 \dot{\mathbf{S}}_j + \mathbf{D}_0 \mathbf{S}_j = \mathbf{F}_j \qquad (15)$$

The coefficient matrices have the same matrices to those of Eq.(5). The vector $\mathbf{F}_j$ is represented with ones combining with the wave force and the corresponding responses. Since the left hand side of these equations have the same form, the dynamic response analyses can be easily carried out.

Therefore, the governing equation of motion including the random parameter effects can be solved in the frequency domain as expressed in Eq.(7). Since the coefficient matrices in Eq.(15) is also identical to those of Eq.(5), the dynamic response analyses can be similarly carried out using the classical modal assumption (Venkataramana et al (1996)).

In this study, emphasis is placed upon the maximum response evaluation of the dynamic responses and reliability on some reasonable response levels. Using the parameters of the response values obtained from the power spectral density function of the responses, the first probability on level crossing can be expressed as

$$L(\lambda) = \exp\left[-\frac{1}{\pi}\sqrt{\frac{\alpha_2}{\alpha_0}} t_0 \exp\left(-\frac{\gamma^2}{2} \right) C_1 \right] \qquad (16)$$

in which

$$C_1 = \frac{1 - \exp\left(-\sqrt{\frac{\pi}{2}} \cdot q_x \gamma \right)}{1 - \exp\left(-\frac{\gamma^2}{2} \right)}$$

$$q_x = \left(1 - \frac{\alpha_1^2}{\alpha_0 \alpha_2} \right)^{\frac{1}{2}}$$

$$\alpha_i = \int_0^{\infty} \omega^i G_x(\omega) d\omega \qquad (i = 0,1,2)$$

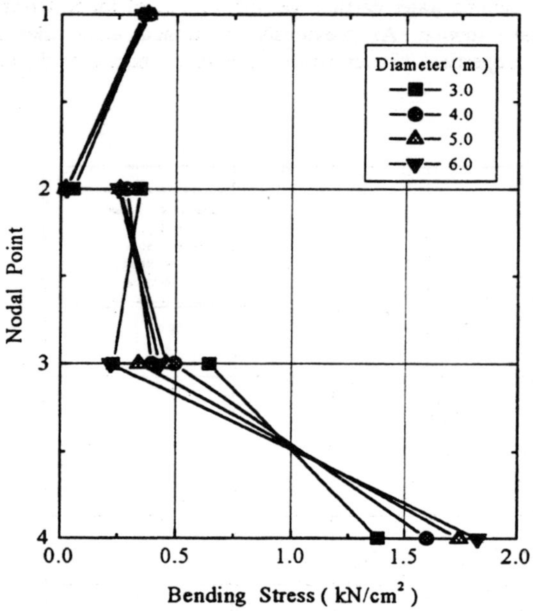

Fig.3(a) Dynamic responses of offshore structure with different buoyancy type members

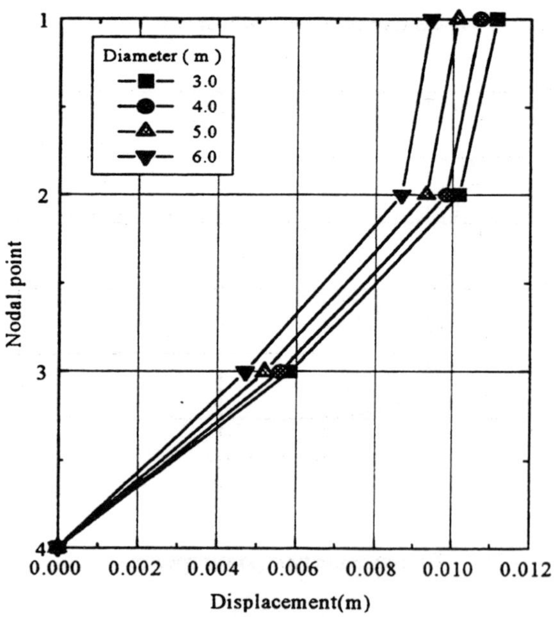

Fig.3(a) Dynamic responses of offshore structure with different buoyancy type members

and t_0 is the duration and γ is the ratio between the design level response and the *rms* response which includes the uncertain parameter effects. Generally, the design of the offshore structure is carried out such that the stresses due to the load conditions are within the allowable stress. In this case, the dynamic load effects are added to static load effects. Assuming that the static loads and the dynamic loads are applied to the offshore structure independently, the design stress is required to satisfy the following condition at any section of the member.

$$\sigma_0 + k\sigma_d \le \overline{S} \le \sigma_a \qquad (17)$$

in which σ_0 denotes the stress on the static load and σ_d the ones due to the dynamic loads, and σ_a denotes the allowable stress. The parameter k is determined taking into account the dynamic response properties. The allowable stress σ_a depends upon the strength of the member materials. Since these stresses due to the applied loads and the strength of the member have some variations, it is reasonable to evaluate with reliability index on these responses. The reliability index is expressed using the *rms* responses and the coefficient variation of each response as follows:

$$\beta = \frac{\overline{R} - \overline{S}}{\sqrt{\sigma_S^2 + \sigma_d^2}} \qquad (18)$$

in which

$$\sigma_s = \sqrt{\mathrm{var}(\sigma_0) + \mathrm{var}(k\sigma_d)}$$

$$\sigma_a = \overline{R}, \qquad \sigma_R = S_R * \overline{R}$$

and S_R denotes the coefficient variation of strength of member. Assuming that the stress $\overline{R}$ and its standard deviation σ_R are independent of time, they take constant values. On the other hand, the stress $\overline{S}$ and its standard deviation are function of time. The value of $k\sigma_d$, which is obtained by the maximum responses, and its standard deviation, are also function sof time.

NUMERICAL RESULTS AND DISCUSSIONS

Analytical model of an offshore structure is shown in Fig.1. This model has buoyancy-type member near the sea bed. The structure is *60m* height, and *120m* wide. The bouyancy member is *22.5m* height, and it has diameters between *3.0m* and *6.0m*. The foundation is supported by pile and soil, the shear wave velocity of soil *100m/s*. The displacement of each nodal point has horizontal, vertical and rotational components. The critical damping ratio of the first mode of the superstructure is *2%*. The wave force is represented with the modified Morison equation using the Airy wave theory. The response analysis is
performed with equivalent linearization of the drag force term.

Fig.2 (a),(b) show Bretschneider type power spectra with different mean wave periods and current velocity, respectively. The mean wave height is *6m* and the mean wave period is *9.0sec*. The peak value of the power spectrum decreases for increasing the current velocity as can be known with Eq.(4). Since the current flow causes significant drag force on the structure, the dynamic response is not affected by only decreasing the power spectrum of wave force.

Fig.3 shows the dynamic response of the offshore structure along the main member between node 1 and node 4. The structure is subjected to wave force of mean wave period *9sec* and mean wave height *9m*. The *rms* displacement and bending stress are slightly affected by the buoyancy type member with different diameters. Since the significant contribution of wave force is mainly near the water surface area, it

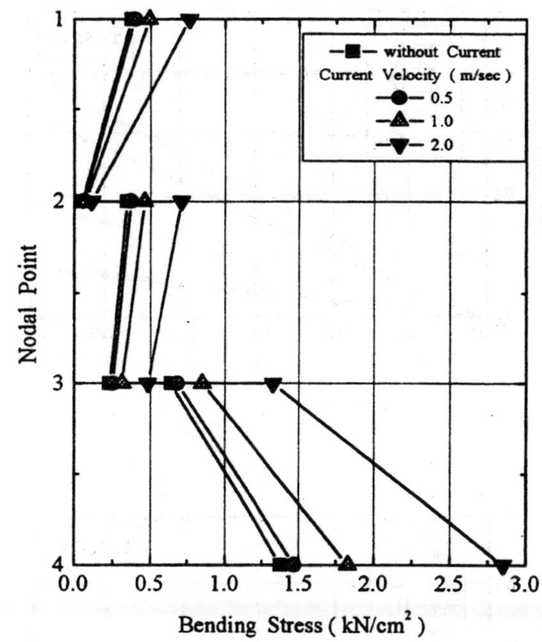

Fig. 4(a) Current force effects on dynamic rersponses

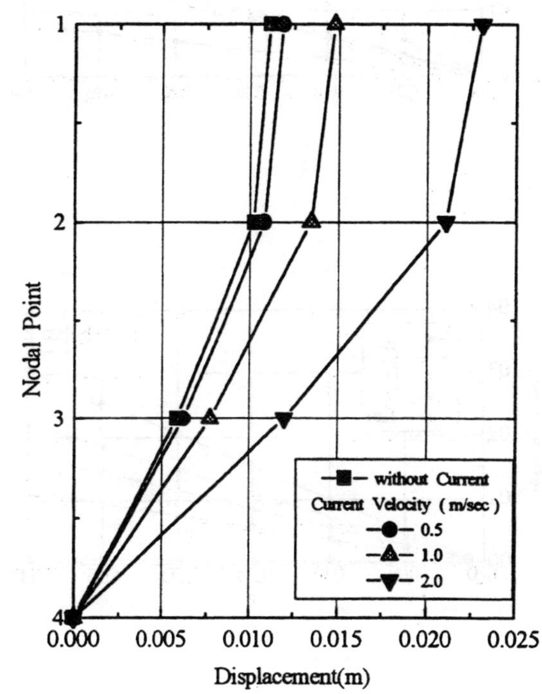

Fig. 4(a) Current force effects on dynamic rersponses

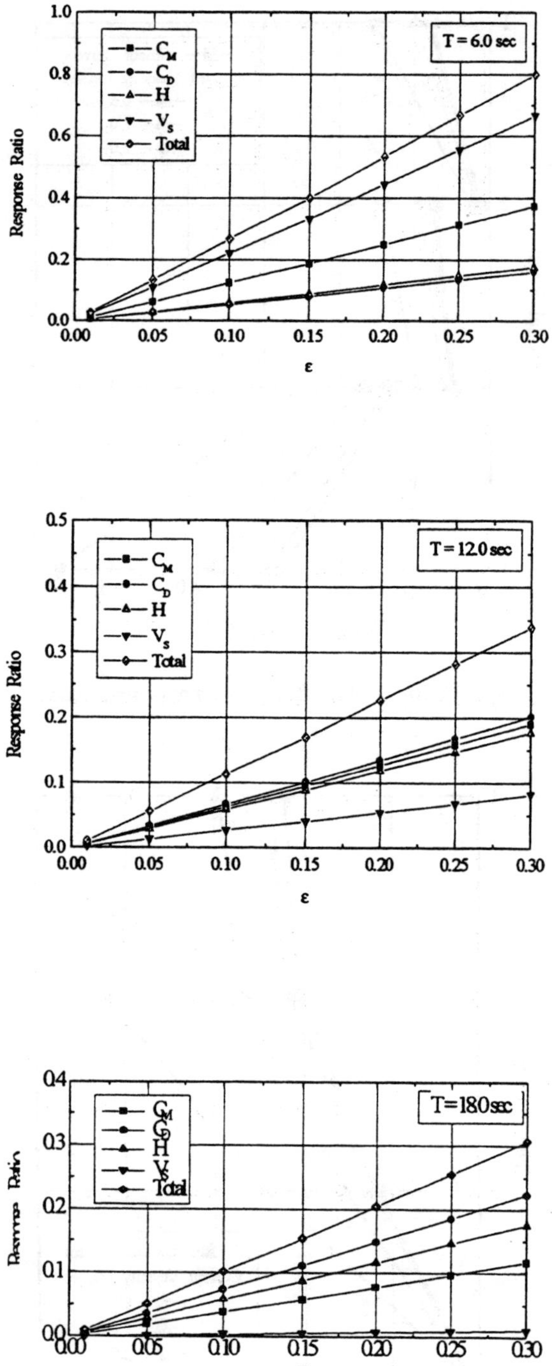

Fig.5 Random variable effects on the dynamic responses of the structure with buoyancy type member(D=3.0m)

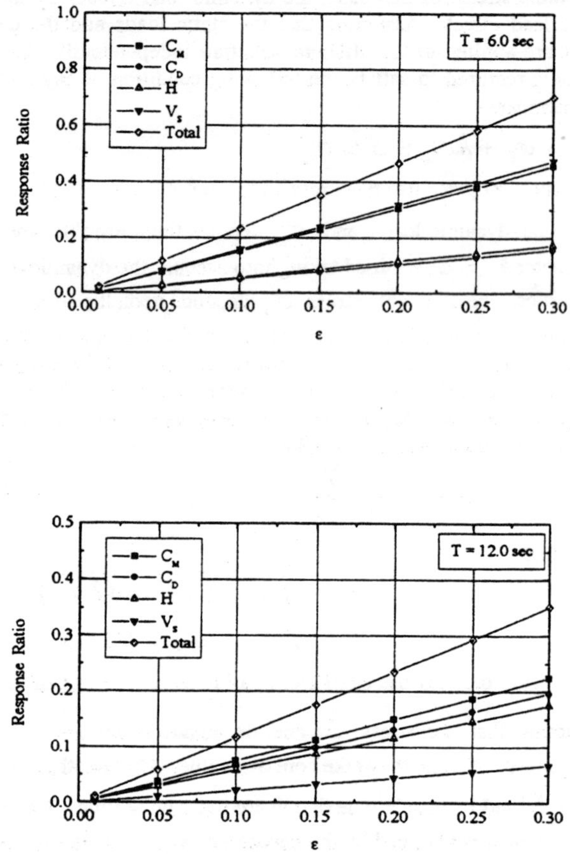

Fig.6 Random variable effects on the dynamic responses of structure with buoyancy type member(D=6m)

seems that these dynamic responses are not influenced by using the buoyancy type member with large diameter which is located near the sea bottom.

Fig.4 shows the current loading effects on the dynamic responses of the offshore structure model. The wave has the mean wave period *9.0*sec, and the mean wave height *6.0*m. The diameter of the buoyancy-type member is *3.0m* and the concentrated deck load *7840kN*. The results are given for current velocities, *0.5m/s,1.0m/s*,and *2.0m/s*, respectively. Comparing the responses without current loading with the ones to each current velocity, it is seen that current force provides significant effects on the dynamic responses. Since the current velocity takes almost the same value along the depth except near the sea bottom, the structural member is subjected to uniformly distributed current loading along the depth. The current loading near the water surface generally gives to larger effects on the foundation than the one acting near sea bottom. Therefore, it seems that it is important to examine and investigate the current loading effects on the dynamic response evaluations to the offshore structures with buoyancy type member of large diameters.

Fig.5 shows random variable effects on the dynamic responses of the offshore structure with buoyancy type member of diameter of

3m. The abscissa denotes the variation of coefficients on each random parameters, which C_M denotes the variation of coefficient of inertia coefficient, C_D denotes one of drag coefficient, H denotes one of the mean wave height, and V_s denotes one of the shear wave velocity of soil. The ordinate represents the response ratios to the responses which are obtained with the mean value of each parameter. The abscissa takes same values for each parameter denoting identical variations of parameters for simplicity. These results are represented in terms of the *rms* displacements at nodal point 1, and the mean wave height is *6m*.

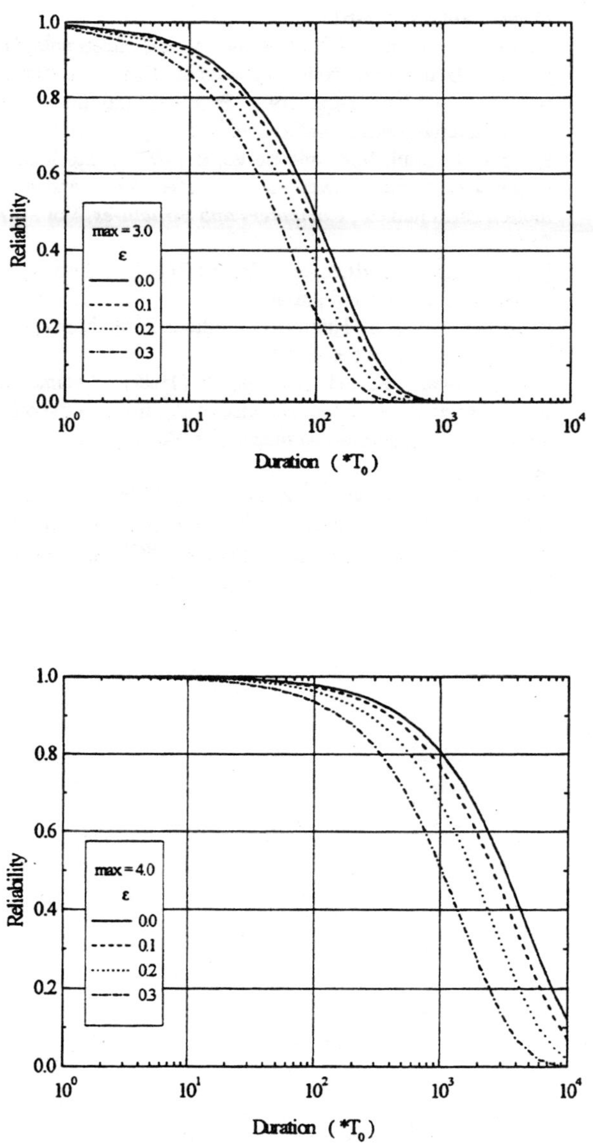

Fig.7 Uncertain parameter effects on level crossing probability

Fig.6 shows similarly random variable effects on the dynamic responses of the offshore structure with buoyancy type member of the diameter *6m*. The structure is subjected to wave forces which have the same situations as in the previous case. The random variables give the same tendency for the responses as obtained in the previous case.

These responses take large variations with increasing variations of coefficient on each parameter. Since the variations of each parameter provide important contributions on the dynamic response evaluations, it is important to examine these effects on the response of the structure to carry out reliable evaluations of the responses.

Fig.6 (a),(b) show the random variables parameters effects on level crossing probability of the offshore structure with buoyancy type member. The structure is subjected to wave force of the mean wave height 6m and mean wave height *9sec*. Taking the crossing level of the maximum responses at 3 times the *rms* response for the case when random variations are not considered, the probability of level crossing

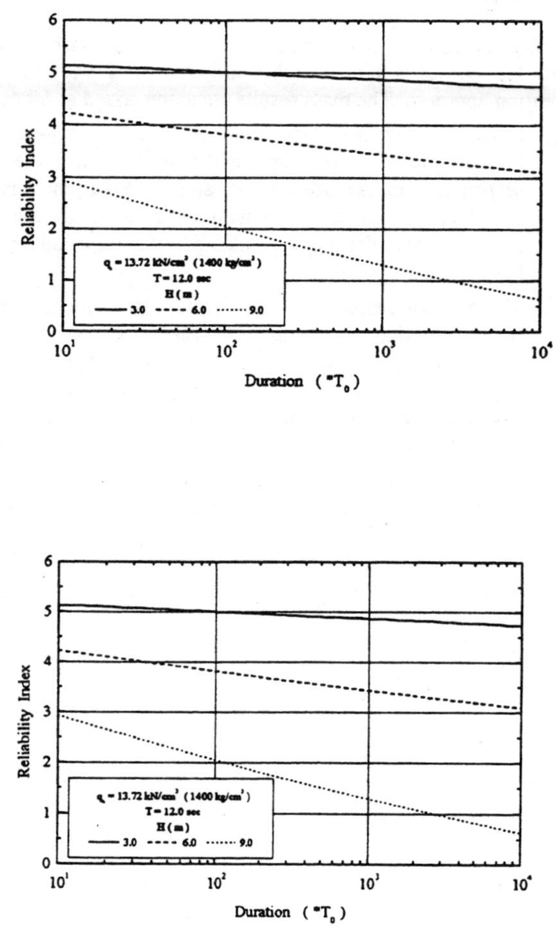

Fig.8 Reliability index on offshore structure (at nodal point 1)

has an increasing tendency. When the crossing level takes large values, the variations of random variables give important effects on the probability of level crossing. It is suggested that in order to enhance the reliability of the offshore structure to the dynamic forces, it is important to evaluate the random parameter effects on the dynamic responses.

Fig.8 (a),(b) show the reliability index on the offshore structure subjected to wave forces which have the mean wave period of *12sec* and the mean wave height varying from *3m* to *9m*. The reliability index is evaluated at nodal point 4 and is calculated with Eq.(18). The abscissa denotes the duration time and the allowable stresses of normal direction of the member are $13.7kN/cm^2$ and $18.6kN/cm^2$. The mean wave height gives significant contributions on the reliability evaluations. Using the member of nodal point 4 with the allowable stress $13.7kN/cm^2$, the reliability index decreases severely.

Conclusion

The dynamic response characteristics of the offshore structure with the buoyancy type member are examined with the random vibration approach. The main results are summarized as follows :

(1) When the buoyancy-type member is provided near the sea bottom , the wave force has slight contributions on the dynamic responses. On the other hand, since the current force gives significant effects on the response evaluations for the offshore structure with buoyancy-type member, it is important to evaluate these effects.

(2) The dynamic responses of the offshore structure are influenced by the variation of parameters which are related with the wave force evaluation and structural characteristics. Since these random parameters also influence the response values, it is necessary to examine their effects on the responses and the reliability evaluation.

(3) The reliability index can serve as indicator of the dynamic response evaluation of the offshore structure which has various parameters with randomly varying characteristics. The reliability index can also be used to quantify the random variable effects of the uncertain parameters of wave forces on the response evaluation. Application of the reliability approach to the structure of response evaluations can be useful to estimate not only the structural properties to the static loads but also the dynamic properties under dynamic loads with uncertain parameters.

REFERENCES

Chakrabarti S. K.(1987), Hydrodynamics of offshore structure, Computational Mechanics Publications

Jesien, W.(1987), 'Random vibrations of baltic drilling platform subjected to wind loads and water waves', *Earth. Eng. Struct. Dyna.* , Vol.15, 596-617

Kawano,K. , Komasa,K., Miyazaki, Y. and Hashimoto, H. (1996), ' Dynamic response analyses of offshore platform with buoyancy type large members', Proc. Of Sixth Inter.Offshore and polar engineering conference, Vol.I, pp.176-181

Penzien, J., Kaul, M.K. and Berge, B. (1972), 'Stochastic response of offshore towers to random sea waves and strong motion earthquakes', *Computers and Structures*, Vol.2, .733-756.

Sigbjornsson, R.,Bell,K., and Holand,I.(1978),' Dynamic response of framed and gravity structures to waves', Numerical methods in offshore eng., edited by O.C. Zienkiewicz et al, 245-280

Venkataramana, K. and Kawano, K.(1996) ,' Parametric uncertainties for dynamic response evaluation of offshore structures', Computational mechanics,Vol.18, No.6 ,pp.454-463

Yamada, Y., Kawano, K., Iemura, H. and Venkataramana, K. (1988), 'Wave and earthquake response of offshore structures with soil-structure interaction', Proc. of JSCE, *Structural Eng./Earthquake Eng.*, Vol.5, No.2, pp361s-370s.

Proceedings of the Seventh (1997) International Offshore and Polar Engineering Conference
Honolulu, USA, May 25-30, 1997

ISBN 1-880653-28-1 (Set); ISBN 1-880653-32-X (Vol. IV)

Deepwater Hyperbaric Welding - Initial Process Evaluation

Ian M. Richardson and John H. Nixon
Cranfield University
Bedfordshire, United Kingdom

ABSTRACT

Fusion welding technology, involving skilled welders has been used extensively offshore. As exploration and exploitation moves to deeper waters, a limit is reached beyond which direct manned intervention is no longer feasible and current welding technology ceases to be viable. This paper explores two alternative arc welding processes for operation in deep water, hyperbaric environments. Although modifications are required to the process control strategies, and the processes are inherently more complex than their shallow water counterparts, it is shown that good quality welds can be produced at pressures up to 100bar. Recent research has shown that welding processes can operate successfully at pressures at least up to 150bar and work is currently under way to extend these limits to 250bar, equivalent to 2,500m (8,200ft) water depth.

KEYWORDS

Underwater Welding, Hyperbaric Welding, Plasma, Gas Metal Arc, Flux Cored Arc, Deep water.

INTRODUCTION

There is a general consensus within the oil and gas industry that exploitation of hydrocarbon reserves in water depths significantly greater than those currently being developed will take place within the next decade. Many deepwater reserves world-wide are currently under assessment, including several areas in the Gulf of Mexico, offshore Brazil, in the Philippines, offshore West Africa and in the northern North Sea. Two major projects under way to develop technology for deep water exploitation include *'Deep Star'* led by Texaco, and *'Procap 2000'* led by Petrobras. These and others are examining the overall technological concepts needed to commercially extract oil and gas from deep water reserves.

In shallow waters, underwater (wet) or hyperbaric (dry) welding techniques are widely used for the joining of pipelines and structures. There is, however, comparatively little known about the capability of welding processes for operation in deep waters, particularly at depths inaccessible to divers. This paper reports on work carried out to assess the behaviour and suitability of two welding processes for deep water applications, and discusses the advantages, limitations and future development of the technology.

BACKGROUND

Welding is currently the standard technique for making high integrity connections underwater and although other techniques, such as mechanical connectors, are generally available, they are considered less reliable because of the increased possibility of leakage. Unfortunately, underwater welding becomes more difficult as the water depth increases. Moreover, since such operations currently require diver intervention, their range of application is limited to depths of no more than about 500m (~1,600ft). Significant developments in diverless technology will therefore be needed before welding technology can be applied in deeper waters. However, important advances have already been made and the technology for habitat deployment without diver assistance already exists. Due to the established nature of arc welding as a construction and repair technology, it is highly likely that remotely operated hyperbaric welding techniques will be developed provided the welding processes can be shown to be viable.

The Influence of Pressure on Welding Arcs

The influence of ambient pressure on welding process behaviour has been considered for most of the major arc welding processes at pressures equivalent to the first few hundred metres water depth.

Although the operation of electric arcs at pressures of several hundred atmospheres have been reported in the literature, these are generally long, stabilised, or transient in nature and unsuitable for welding applications (Bauder 1968). Studies of arcs suitable for welding operations, i.e., those which have the properties of a stable, controllable heat source, have been reported at pressures up to 135bar (Allum 1982a). With the exception of Cranfield's HyperWeld 250 installation (described below), the welding facility with the highest pressure rating known to be operational at the present time has a maximum working pressure of 110bar (Dos Santos et al 1990).

The influence of pressure on process performance and requirements is governed to a large extent by changes in arc properties. All arc processes involve the generation of an electrically conducting gaseous plasma; for welding purposes the arc must generally be self sustaining and provide a stable source of heat.

The structure of the arc is controlled by the material properties of the plasma, environmental conditions and the current which the arc must support. Equilibrium operating conditions occur when the rate of electrical energy supplied is balanced by the energy losses to the electrode, workpiece and surrounding atmosphere. Arc voltage may therefore be regarded as an indication of the total energy loss from the arc column (for a given welding current).

As the ambient pressure is raised above one atmosphere the physical properties of the plasma are altered and particle densities increase in proportion to the absolute pressure. Energy losses from the outer regions of the arc increase resulting in a rise in arc voltage. The arc contracts and adopts a smaller cross-section to minimise the additional energy losses. The gas entrained by the arc rapidly becomes turbulent in the cooler outer regions of the arc and the plasma velocity falls. This has the effect of increasing the arcs susceptibility to external disturbances such as air currents or stray magnetic fields. The contraction of the arc also results in an increased radiative energy loss which occurs as a result of a re-distribution of the arc energy (due to the columnar shape).

This general behaviour is common to all types of welding arcs presently used offshore. However, the results in terms of specific changes in arc behaviour depend on detailed environmental conditions and are therefore process dependent.

Current Welding Processes

Manual Metal Arc Welding

Manual metal arc or 'stick' welding is presently used for the majority of offshore welding applications. This process may operate either in the wet or in the a dry environment. In the former case, the arc is in direct contact with the water and relies on the flux constituents to provide the material for arc formation. Although wet welding is not favoured for operations in the North Sea, it is extensively used in the Gulf of Mexico and elsewhere where offshore structural design and loading requirements are not so severe.

During MMA welding, the flux decomposes on contact with the arc and performs a number of functions, including shielding the molten weld pool and metal droplets during transfer. The flux also supplies an excess quantity of easily ionisable material to the arc which substantially influences the energy balance behaviour. As a consequence, most authors (Lythal and Gibson 1976) report that the arc voltage is only slightly increased as pressure rises. However, voltage fluctuation, due to increased metal droplet detachment frequency and column electric field strength becomes more severe (Pandey and Swetnam 1985).

Process limitations associated with low productivity, low heat inputs at elevated pressures and depth sensitivity of the weld pool chemistry, together with the difficulty of adapting the process for diverless operations preclude serious consideration of this technique for automated deep water applications.

The Gas Tungsten Arc Process

The influence of pressure on the GTA process has been studied extensively and several comprehensive descriptions of arc and process behaviour at elevated pressures have been published (Allum 1982a, Knagenhjelm and Ovland 1982). The GTA arc may be described as free burning, i.e., the discharge is not subject to external constraints other than those imposed by the electrode geometry and environmental conditions. Arcs of 1mm length or greater tend to be dominated by the inert gas supplied as a shield for the electrode and molten weld pool, arc characteristics are therefore affected by the physical properties of the shielding gas (shorter arcs may have a significant metal vapour content and electrical behaviour can deviate from that predicted for a gas dominated plasma).

In practice, arc voltage is observed to increase in accordance with the square root of absolute pressure (Allum et al 1980). This is indicated by an increase in the electric field strength of the arc column. The total arc voltage may be approximated by the sum of the fall voltages and the plasma column voltage, the former remaining substantially independent of pressure.

As working pressure increases, the arc becomes increasingly less stable. Instabilities are thought to be generated in the region of gas entrainment close to the cathode. These may be due to buoyancy forces (Allum 1981) in the hot non-current carrying layer of the arc immediately adjacent to the arc core. Allum et al (1980) have shown that gas flow in the arc column is turbulent at pressures of only a few bar. This leads to unstable (stochastic and time dependent) energy losses and consequently voltage fluctuations. The amplitude of these oscillations increases with working depth due to the pressure dependence of the electric field.

Magnetic field interactions with the arc plasma also act to destabilise the process. Arc blow, due to residual magnetism in the workpiece or interactions occurring as a result of self induced fields in the arc and workpiece (associated with the current path through the workpiece) causes significant voltage disturbances (Allum 1982a) and oscillation amplitudes increase rapidly with increasing pressure.

The GTA process is presently qualified for repair operations for depths up to 500 - 600m. Beyond this limit it is unlikely in practice that the process stability can be sufficiently maintained to meet the stringent tolerance requirements for automated offshore repairs. Although test welds at greater depths have been successfully carried out in the laboratory, repeatability cannot be guaranteed.

Welding Processes for Deep Waters

Research to date has shown that the welding processes most likely to offer an acceptable degree of stability and control at pressures above 50 to 60bar are the plasma, solid wire gas metal arc (GMA) and flux cored arc (FCA) welding processes (Dos Santos et al 1989, Richardson 1989b, 1991, Hansen et al 1991). The GTA process, which is the currently preferred technique for semi-automated hyperbaric repairs, is unsuitable for very deep water operations due to the turbulent nature of the gas flow in the arc column and the consequent instability of the arc root. Instabilities are expected to limit the application of the GTA process to pressures below about 50bar (Nixon and Billingham 1987, Richardson 1989a). Although the flux cored arc process operating in the conventional mode is already used offshore at relatively modest depths, the mode of operation required for deep waters differs significantly and is discussed in detail below.

Figure 1 The 100bar chamber at Cranfield.

EQUIPMENT

Results are reported in the following section from experiments carried out in the 100bar chamber facility at Cranfield, and the HyperWeld 250 facility which has recently been commissioned.

The 100bar facility has been described in detail elsewhere (Richardson 1991). The pressure chamber is 1.25m long and 0.44m diameter and has a volume of approximately 1901 (figure 1). The internal work handling frame is equipped with a three axis motion control system with speed and position feedback control on each axis. The welding specimen is secured on a table which moves below the welding torch. The torch can be moved laterally or vertically with respect to the weld specimen. Arc power is obtained from a GEC H350sr series regulator power source with an open circuit voltage of 110V and a stable operating voltage up to approximately 100V. The output characteristic of the power source is continuously variable between constant current and constant voltage, with a dynamic response rate of 50 to 450A/ms. All parameters are computer controlled and may be modified on demand during the welding operation.

The HyperWeld 250 facility is rated to more than twice the pressure of any other current welding research facility. A detailed description is given by Nixon and Richardson (1995). Briefly, the pressure chamber consists of a main body, which can be traversed hydraulically away from a stationary closure plug, through which all supplies and control signals are fed to the internal systems (see figure 2). Gas supplies are stored at several pressures, and pumped into the vessel to achieve the required internal pressure and flow rate on two supply lines. On depressurisation, the gas is filtered and recovered into the gas storage banks automatically by the gas supply control system. Gas flow rate control is effected by a digital choked orifice system, in order to achieve the required degree of precision over the working pressure range of the vessel.

Currently, a four degree of freedom Cartesian welding manipulator is installed within the vessel, enabling flat plates to be welded in any position, using either the GMAW or plasma processes. As process knowledge is acquired, it is intended to replace this manipulator with an orbital system capable of welding complete pipe joints. The welding power supply is a compound unit, to be made up of five high frequency inverters capable of supplying about 250 volts at 450 amps.

Figure 2 The HyperWeld 250 chamber (upper) and internal systems (lower).

An auxiliary unit can supply about 700 volts at 50 amps, and is intended primarily for plasma keyhole welding.

To avoid the requirement for viewing ports, an image of the arc is conveyed from the viewing head to a small pressure window by a coherent fibre optic bundle. This enables the main viewing system, including neutral density filters and CCD television camera, to be operated at one atmosphere, improving accessibility.

The HyperWeld 250 is equipped with comprehensive data collection and analysis systems, utilising experience gained from previous hyperbaric and surface based welding projects. The various sub units are co-ordinated by a control system of similar type to that used on the THOR and PRS hyperbaric welding systems.

RESULTS

The GMA Process

Successful welds have been produced using the open arc pulsed current welding technique at pressures up to 80bar. At higher pressures, the increasing short circuit frequency and limitations in the available control system led to the generation of spatter in positions other than downhand which was considered excessive.

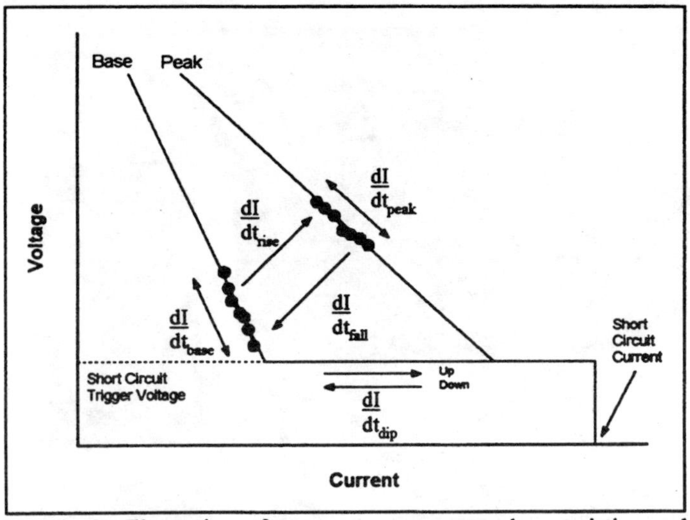

Figure 3 Illustration of power source output characteristics and dynamic response conditions for control of pulsed GMAW at high pressures.

The technique employed to control the arc is based on the self adjustment principle typical of conventional GMA at one atmosphere. At high pressures, the slope of the power source is adjusted such that a unit change in arc length leads to a measured change in arc current irrespective of the pressure. It is well known that the arc column electric field strength (E) is approximately proportional to the square root of absolute pressure (P) in bar; hence, the optimum slope is pressure dependent. The following expression is found to hold

$$k = \frac{E_1 \ P^{0.5}}{\text{Constant}} \qquad Eq\ 1$$

where E_1 is the one bar electric field strength in V/mm is of the order 0.8 for an argon gas an 1.8 for helium. k is the slope in VA^{-1}. The constant is dependent on the composition and diameter of the consumable and is found to be of the order 42 for a 0.8mm diameter A18 grade wire. i.e., to achieve self adjustment, a 42A change in current is required to correct for a 1mm change in arc length irrespective of pressure.

Typical power source output characteristics for a pulsed current welding operation are shown in figure 3. The position of a characteristics is defined by the specification of a set point and a slope. However, the operating point, i.e., the position of intersection between the arc characteristic and the power source characteristic is dependent on a number of factors, including the material, shielding gas, wire addition rate and the electrical extension of the wire.

In order to control the arc at high pressures, the process operates with a short arc length. The dynamic nature of the process is illustrated in figure 4, which shows a sequence of frames from a high speed cine taken at a framing rate of 4000/s. The rapid movement of the arc column and occasional short circuit are clearly evident. These pictures also show a significant level of fume formation, which was invariably encountered when welding at pressures in excess of a few bar.

Figure 5 shows a macrograph of a weld in 12.5mm thick steel made at 60bar in a helium shielding gas. The welding conditions are given in table 1 and the measured hardness values in figure 6. Although the bead placement could be improved, this figure clearly illustrates the degree of penetration achievable at high pressures.

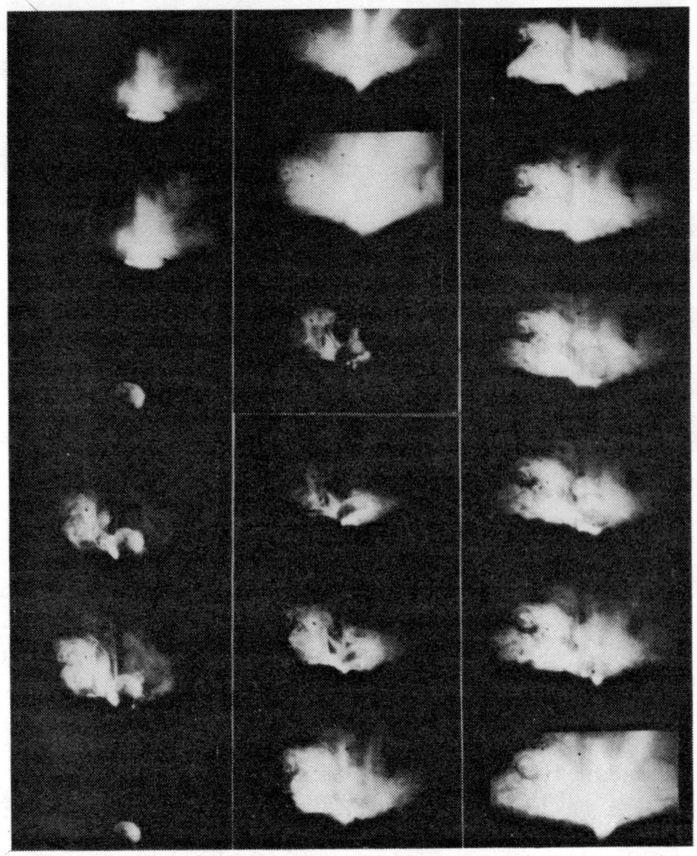

Figure 4 High speed cine film of the hyperbaric GMAW process. Frames (at 4000f/s) are consecutive looking down each column in turn from left to right.

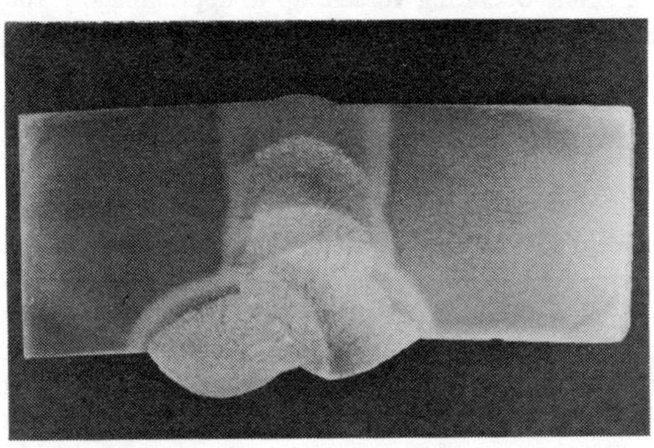

Figure 5 Macrograph of a welded joint made with the hyperbaric GMAW process at 60bar.

Welding trials with a range of materials and joint geometries have been undertaken and results indicate that good root welding conditions can be obtained provided a suitable joint geometry is employed. With a 60° included angle joint, 1mm nose, closed butt configuration, root welds were difficult to achieve. Current variations of more than 2A either side of the optimum value led to lack of penetration or burn through. Modification of the joint to a closed butt, 30° included angle geometry with a flat base 5mm wide and 4mm thick, improved the tolerance to current variation. Acceptable welds were obtained over a wide range of mean currents (160A ± 20A) without occurrence of lack of penetration, over penetration or lack of side wall fusion.

Peak Set Point	192A, 35.5V
Base Set Point	142A, 26.3V
Peak and Base Duration	4.0ms
Peak and Base Slope	0.33V/A
Travel Speed	4.0mm/s
Wire Feed Rate	10.0m/min
Mean Current	164 ± 2A
Mean Voltage	35 ± 1V

Table 1 Summary of GMA Welding conditions at 60bar.

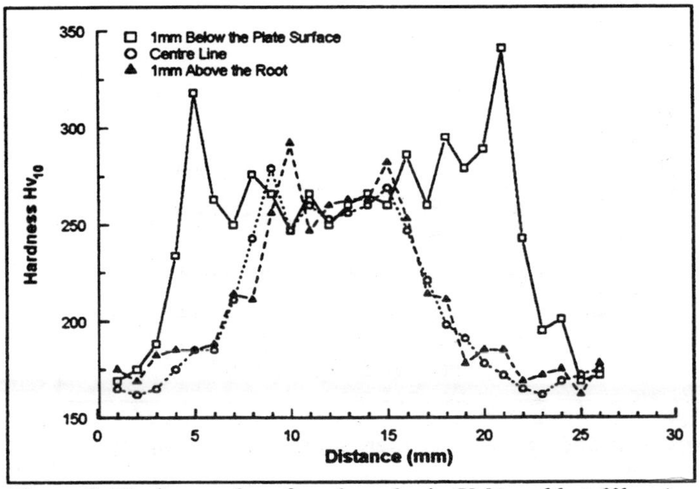

Figure 6 Hardness values for a hyperbaric GMA weld at 60bar (see figure 5).

With the exception of the slope of the power source output characteristics, optimum welding conditions remain relatively unaffected by pressure for ambient pressures in excess of about 20bar.

The Plasma Process

The plasma welding process has been studied in depth at Cranfield University over the past ten years. The majority of process changes occur over the first few bar of pressure where, for example, plasma velocity decreases, weld bead aspect ratio (width to depth) decreases and the arc energy distribution changes from electrode to column dominated; i.e., at high pressures, the majority of the power is generated in the column of the arc (Richardson 1991, Cave 1996).

The plasma process may operate in a melt-in or keyhole mode. Keyhole welding has been limited to pressures below about 40bar due to the high voltages required. However, analysis of the process behaviour suggests that the technique could be extended to higher pressures if suitable power sources become available (Richardson 1994).

The capability of the plasma process operating in the melt-in mode has been demonstrated by the production of test welds at 75 and 100bar (figure 7). The welds were made in 29mm thick X65 pipeline steel with an A18 grade filler wire.

Welding conditions similar to those employed in automated orbital TIG welding at pressures of the order of 20 to 40bar were selected and these were shown to produce acceptable weld bead profiles in positions between the downhand and the vertical. Operational envelopes were determined from initial bead on plate trials, to assess the influence of current, plasma gas flow rate, arc length, welding speed and wire addition rate on weld bead characteristics.

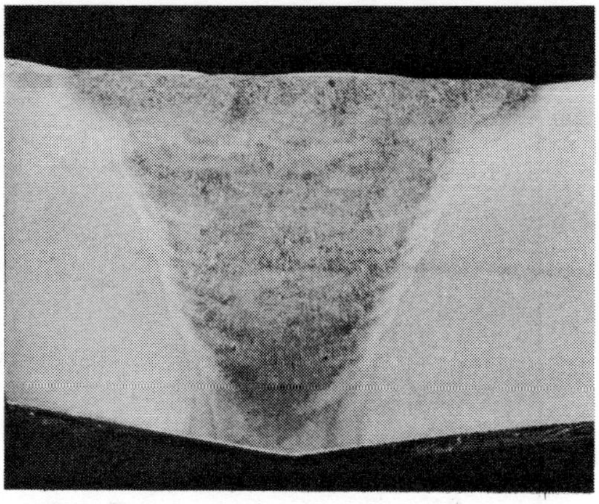

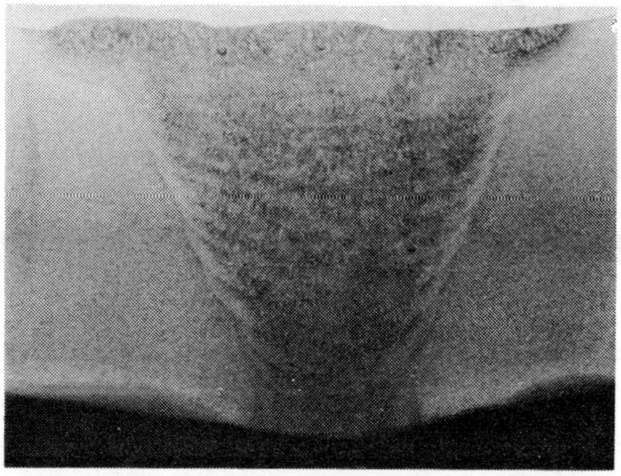

Figure 7 Macrographs of plasma welds made at 100bar (upper and 75bar (lower).

Hardness measurements from each weld were carried out according to AWS D3.6-93. Little difference was observed between the measurements for each weld, with the maximum hardness values less than 270 Hv_{10}. The mean values for the 75bar and 100bar welds are 210 and 216 Hv_{10} respectively, and the maximum values are 260, and 264 Hv_{10} both of which occurred in the weld metal capping passes.

Microhardness measurements in the heat affected zone were also taken using a 500g load. These were spaced at 250mm intervals along lines parallel to the upper plate surface at distances of 1mm, 14mm and 28mm; a summary of the results is presented in table 2. Again, the cap region is seen to have the higher hardness values, although the maximum hardness are less than 270 $Hv_{0.5}$, consistent with the general survey.

	75 bar			100 bar		
	Cap	Mid	Root	Cap	Mid	Root
Max.	236	220	220	261	220	202
Min.	192	189	178	203	196	183
Mean	212.1	206.3	197.6	220.1	206.8	193.5

Table 2 Summary of microhardness measurements from the HAZ of plasma welds made at 75 and 100bar. Measurements were made 250µm apart with a 500g load.

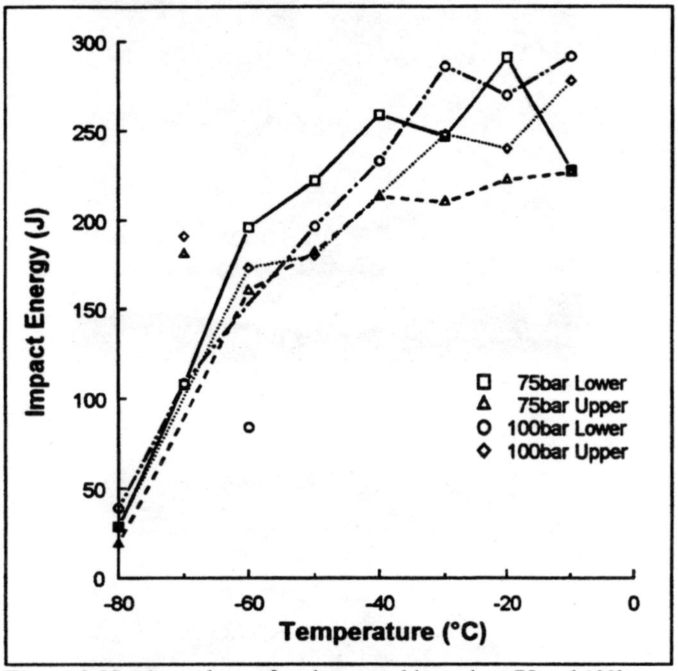

Figure 8 Notch toughness for plasma welds made at 75 and 100bar.

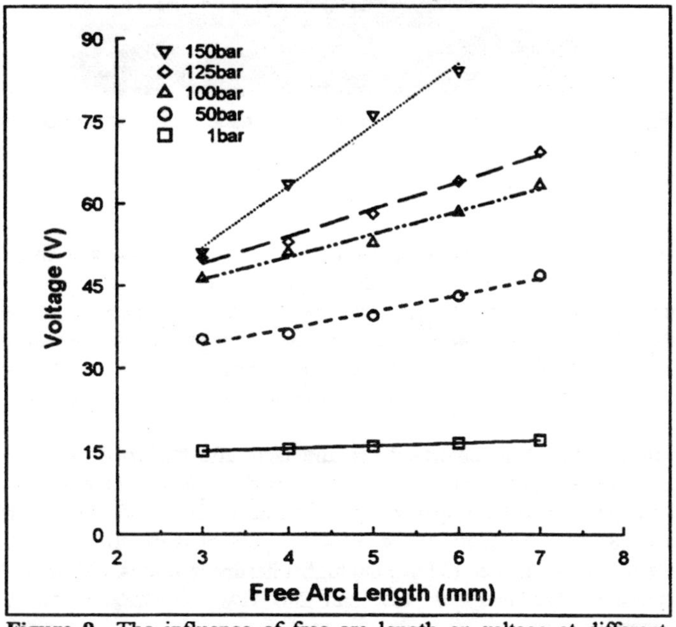

Figure 9 The influence of free arc length on voltage at different pressures for an argon plasma arc at 75A, 1.5mm constricted arc length and 6mm diameter orifice.

The measured hardnesses lie well within the normally accepted values for welds made in X65 pipeline steels and are considerably lower than the maximum of 325 H_v10 specified in AWS D3.6-93.

The results of the notch toughness tests are shown in figure 8. Upper shelf energies are above 180 Joules and transition temperatures below -50°C.

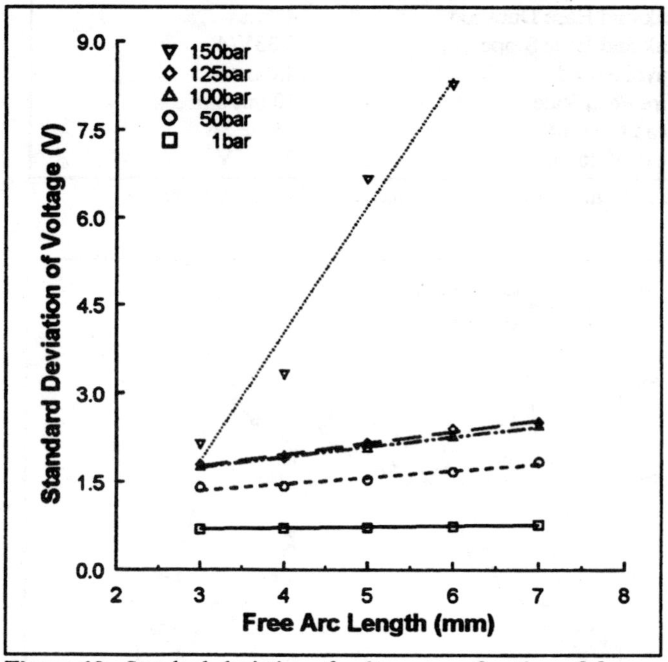

Figure 10 Standard deviation of voltage as a function of free arc length at different pressures for an argon plasma arc (see fig. 9)

Trends for samples taken from the upper parts of the weld are similar, and all measurements at temperatures above -60°C exceed the minimum value of 34J specified in AWS D3.6-93. These results should be treated with some caution because they represent only a single measurement at each temperature; nevertheless, consistent trends are evident and results indicate that welds with good toughness properties can be made at high ambient pressures.

Initial welding trials undertaken during the commissioning phase of the HyperWeld 250 facility have confirmed that plasma arc ignition and maintenance is readily achievable at pressures at least up to 200bar, albeit at short arc lengths due to voltage restrictions. Bead on plate welding trials have been carried out at pressures ranging up to 150bar at arc lengths similar to those used for production of the 75 and 100bar test joints. Voltage arc length curves are shown in figure 9. These indicate an increasing electric field strength in the free arc column, consistent with previous observations in the 1 to 100bar range (Cave 1996). Some indication of the stability of the arc may be obtained from measurement of the standard deviation of the arc voltage. Figure 10 shows results from data logged at a frequency of 10kHz for a period of 5s and indicates that the arc becomes less stable as the length of the free burning column increases and as pressure rises. The effect of increasing arc length on stability deterioration is shown by the gradient of the lines which increase with ambient pressure.

Pressure has a measurable effect on weld bead geometry, the bead width increasing with pressure when all other parameters are fixed (figure 11). For the series of trials reported here, positional instabilities were observed in the weld beads made at 150bar, the fusion boundary moving by 12.5% of the bead width for the 4mm long arc condition and by as much as 25% at 6mm. The positional instability of the anode root inferred from observations of the weld bead therefore mirror the rapid increase in standard deviation of the arc voltage at this pressure.

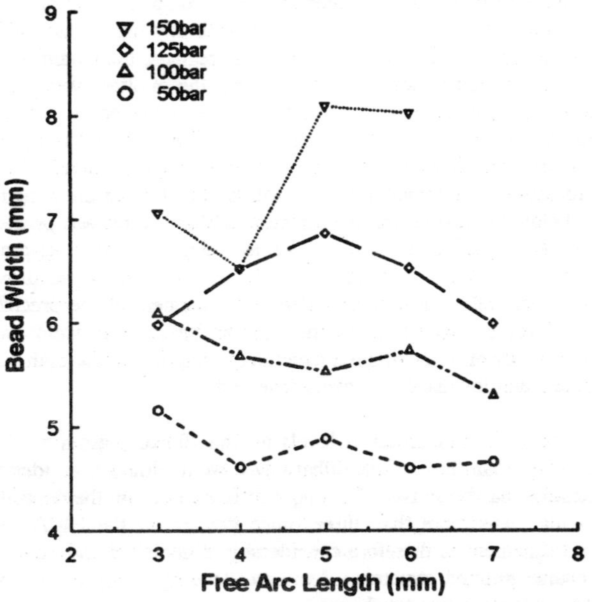

Figure 11 The influence of free arc length on weld bead width at different pressures for an argon plasma arc.

Discussion

Gas Metal Arc Welding Process

The mode of operation of the GMAW process reported here differs from the conventional approach in a number of important details. High ambient pressures lead to constriction of the arc column and arc roots. Unlike the one bar case, the arc roots are localised on the surface of the molten pool and wire tip, and there is no envelopment of the developing droplet by the arc plasma.

Root constriction and localisation results in the formation of strong plasma jets which destabilise the metal transfer process, causing ejection of droplets which then form spatter. The action of the plasma jets can be suppressed by minimising the arc length, typically to less than 1mm. Metal transfer then takes place either through a dip process or by droplet motion around the arc column. The droplet is presumably driven by the gas flow surrounding the plasma jets as metal transfer takes place in the overhead as well as the downhand position. However, from high speed cine films it is seen that the small current carrying core of the arc invariably acts to repel the droplet due to the reaction forces of the plasma jets.

It has been found that successful process control can only be obtained with small diameter consumables. Solid wires with diameters of 0.8 or 0.9mm and cored wires of 1.0mm diameter provide the most stable behaviour. This may be attributable to the resistivity of the consumable and the area of contact of the molten droplet with the unmelted wire. In order to transfer a droplet, sufficient force must be applied to overcome the surface tension retaining force which depends on the area of contact between the solid wire and molten droplet. The resistivity of the consumable influences ohmic heating in the solid wire; changes in the electrical extension (through variation of the torch to workpiece separation), can be used to control the current flow as the operating point moves along the power source characteristic (see figure 3). This is achieved because of the relatively high voltage drop in the wire extension.

When welding with large diameter consumables, the voltage difference afforded per unit change in electrical extension is small, and excessive movement of the welding torch is necessary to control the current, resulting in unacceptable changes in gas shielding to the molten weld pool and surrounding regions.

The use of a self adjusting characteristic appears to be necessary in order to control arc rooting behaviour. The reason for this is unclear, however, experimental trials have shown that constant current operation always leads to high spatter levels due to the formation of very strong plasma jets. It is reasonable to assume that the size of the cathode root will depend on the current flow. In constant current operation, the power source will force a fixed current to flow regardless of the availability of electron emission sites on the metal surface. This is thought to lead to excessive heating of a localised root area and consequently, the formation of the strong plasma jets observed. By contrast, the self adjusting characteristic will moderate the current flow, reducing emission (by virtue of the increased voltage) when insufficient emission sites are available. Hence some control over the plasma jet formation is maintained.

From table 1, it may be seen that the pulse parameters employed for hyperbaric welding at high pressures bear little relationship to those employed for pulsed GMAW at one bar. In the one bar case, the peak current is selected to provide spray transfer conditions and the time is selected to melt an appropriate quantity of material, often amounting to one drop per pulse. The base current and duration are chosen to minimise melting and reduce the overall mean current. Under high pressure conditions, the peak value (power source output characteristic set point) is also selected to yield a peak current above the spray transition current, and the peak time to melt the appropriate quantity of wire. However, it is not feasible to weld with a low base current due to frequent arc extinction, appreciable melting therefore occurs in the base portion of the cycle. For this reason, the length of the base duration is considerably shortened. The pulsing action helps to ensure droplet detachment. This is thought to be due to a physical shock mechanism as the plasma collapses or expands due to the current fall and rise. The response rate (rate of change of current with time) of the power source is critical to this process. Slow response gives rise to intermittent failure to detach, causing the formation of excessively large droplets on the end of the wire which are ultimately ejected to form spatter. Conversely, a very fast response rate leads to the formation of fine spatter droplets as the rapid expansion of the arc core shatters the molten droplet on the wire tip.

The self adjusting characteristic, together with the current control afforded by torch to workpiece variation for small diameter wires provides a unique mode of operation for high ambient pressures. For a given wire feed rate, the torch position can be altered to achieve the desired current. By reducing the wire feed rate and re-positioning the welding torch to achieve the same current, the balance of heating between the parent plate and the wire can be altered, affecting the plate dilution levels. At one bar, plate dilution is typically of the order 20% (i.e., 20% fused parent material to 80% deposited material). At high pressures this balance can be varied between 20% dilution and 60% dilution, enabling root and filler pass welding to be carried out with little risk of cold lapping or the lack of fusion defects typically associated with GMAW at one bar.

For the results reported here, an upper pressure limit of 80bar was encountered due to limitations inherent in the power source control system. Whilst control was acceptable in the arcing phase, the power source injected a current pulse of selected current and duration when a short circuit was detected. This pulse was always constant current and tended to destabilise the welding process as described above. The increasing short circuit frequency at pressures above 60bar led to a gradual deterioration of control. This may be overcome in the future by development of a more flexible power source control strategy, which will provide similar levels of static and dynamic control in both the dip and the open arc phases. Such a system is being designed as part of the HyperWeld 250 facility development at Cranfield University.

The Plasma Welding Process

The production of test joints at high pressures has demonstrated the feasibility of the *'soft'* plasma process for deep water hyperbaric applications. Process operation was comparable to that employed at lower pressures for GTA welding i.e., the arc was burning in the melt-in mode with the physical constriction acting to stabilise the arc column but without significantly increasing the force exerted by the arc on the weld pool.

It is well known that arc structure and behaviour at one atmosphere is influenced by welding torch design, welding parameters rarely being transferable between different makes of equipment (Richardson 1991). These differences may be attributable to the way in which the plasma forming gas is fed into the plenum chamber immediately upstream of the cathode tip. The subsequent arc structure will also depend on the shape of the plenum chamber, the relative position of the electrode tip to the constricting orifice, the throat length of the orifice, the degree of constriction and the internal orifice geometry. All of these factors influence the gas flow behaviour in the arc column and hence the convective heat dissipation.

The turbulent behaviour of the cold gas entering the plasma torch might be expected to reduce the influence of torch geometry on arc behaviour at elevated pressures. However, it has been shown that geometric considerations have a significant influence on both process stability and power dissipation, which in turn affect the position and size of the resultant weld pool (Cave 1996). For this reason, a welding torch was designed specifically to match the requirements of the hyperbaric environment, and was used for the production of the test joints reported above.

The welds were made with a mixed gas plasma, comprising argon and helium. The introduction of the gas mixture to replace pure argon led to improved stability of the arc column due to an increased plasma velocity, and a greater fluidity of the weld pool which improved the wetting characteristics at the pool boundary. The same approach has been adopted elsewhere, with different mixing ratios to produce hard plasmas for cutting and keyholing purposes at lower pressures (Smart 1993, Cave 1996). This behaviour is poorly understood at present, but it has been postulated that the phenomena is similar to that observed in high current plasmas used for materials processing purposes, and is related to gas demixing and the rapid diffusion of helium to the arc core.

One of the main difficulties encountered during production of the test joints was the limited visibility available due to refraction in the vicinity of the arc, caused by the turbulent motion of the cool gas surrounding the plasma column. The operator was effectively faced with multiple *'ghost'* arc images, and selection of the correct position for wire input proved problematic. The difficulties were reduced by filtering of the image, however, further development of viewing techniques will be required for deep water applications.

The quality of the welds produced at 100bar compare favourably with welds produced at one atmosphere. As with any multi-pass welded joint there are two distinct microstructural regions, the columnar (as-deposited) structure which forms when the molten weld metal solidifies, and the reheated structure which occurs when the columnar regions are modified by subsequent passes. The reheated region has both coarse and fine grained structures, former lying directly below any remaining columnar structure, whilst the fine grained regions occur below the coarse grained structure and are composed primarily of equi-axed grains of ferrite. The low deposition rate approach adopted (typically 0.7 to 0.8kg/hr) for the test welds led to small bead deposits per weld pass and considerable refinement of the preceding layers. With the exception of the capping passes, the welds were predominantly of a fine grained structure, giving rise to the high notch toughness and low hardness levels observed.

It is notable that the hardness levels for the 100bar plasma weld are much lower than those of the 60bar GMA weld, although an identical consumable has been used. The major difference lies in the deposition rate which was more than three times greater for the GMA weld. Grain refinement is therefore considerably diminished and columnar and coarse grained structures dominate, leading to higher hardness levels and lower notch toughness.

Welding trials undertaken in the HyperWeld 250 facility were carried out during commissioning trials for the gas pumping system. A modified Thermal arc PWH-7A cutting torch (see Richardson 1991) was employed and welds were carried out in a nitrogen atmosphere with argon shielding and plasma gases.

The results indicate a reasonable consistent trend with the arc column electric field strength increasing from approximately 3V/mm at 50bar to 4.9V/mm at 125bar. At 150bar the column field strength jumps to 11.1V/mm. The standard deviation of the voltage increases rapidly with increasing arc length, and does not follow the trend anticipated from the lower pressure trials. In fact, the 150bar data comes from a separate welding trial, and although the torch configuration did not change substantially between the two, a slight difference in the internal shape of the orifice was noted, the orifice for the 150bar trial having a throat length approximately 0.25mm greater than that of its predecessor. The effect of a small change in geometry in this critical area may account for the anomalous measurements shown in figures 9 and 10, and is in accord with observations made concerning the influence of torch and nozzle geometry at lower pressures (Cave 1996).

The measurements for weld bead width presented in figure 11 represent the average of five measurements for each arc length and pressure. The curves show no consistent trend with increasing arc length, which may be anticipated due to the columnar nature of the plasma arc. Bead widths, however, increase with increasing ambient pressure. This behaviour may be attributed to the increase in arc power generated as a result of the higher arc voltages at high pressures. For example, power dissipation for the 5mm long arc rises from 2.95kW at 50bar to 5.65kW at 150bar. Some of the additional power appears at the anode, the whilst the remainder is dissipated in the surrounding environment. Welding at higher pressures or longer arc lengths was not possible during this phase of the commissioning because only two of the five available power sources (for series operation) were operational. It is anticipated that results from trials extending process operation to 250bar will be published later in the year.

Conclusions

This paper provides an overview of the GMAW/FCAW and plasma welding processes in terms of their operation at high ambient pressures, together with some results to support process feasibility under these conditions. It is clear from the work undertaken to date, as well as results reported by other groups, that successful welding operations can be achieved at pressure well in excess of 50bar. Although laboratory conditions cannot match those offshore, confidence in both processes is such that outline equipment and operating specifications could be given for the welding of carbon manganese pipeline steels in the depth range 500m to 1,000m; although considerable work would be required to develop detailed procedures and robust equipment for operation under such conditions. In addition, it would be necessary to develop diverless installation and alignment capabilities to place the welding equipment at the installation or repair site.

It has been demonstrated that good welding properties are relatively easy to achieve at 100bar using the plasma process, and that the plasma arc can be easily ignited at pressures up to 200bar. Voltage limitations have so far prevented welding at pressures in excess of 150bar; however, based on observations to date, it is considered likely that process operations can be extended to greater pressures, provided suitable care is taken over the welding torch design.

The limitations on GMAW process control have so far prevented successful welds at pressures in excess of 80bar at Cranfield, although welds at higher pressures have been reported in the literature. The reasons for this limitation have been discussed and are again linked to power source deficiencies. It is believe that such problems can be overcome in the future and that successful process operation can be extended to significantly higher pressures.

Acknowledgements

The results reported in this paper are a compilation from a number of research programmes and the authors would like to thank the following people for significant contributions to the research: William Cave, Philip Hart, Helle Hansen, Anna Rassmussen and Neil Woodward. We are indebted to the following organisations for their financial and technical interest in hyperbaric welding research: Amoco, BP, Comex, Isotek, Norsk Hydro, Petrobras, Rockwater, Shell, Statoil and Texaco. Thanks are also due to the Marine Technology Directorate (acting on behalf of the UK's Engineering and Physical Sciences Research Council) for their continuing financial support.

References

Allum C J (1981) 'Gas Flow in the Column of a TIG Welding Arc' *J. Phys. D: Appl. Phys.*, Vol. 14, pp1041 to 1059, 1981.

Allum C J (1982a) 'Characteristics and Structure of High Pressure (1-42bars) Gas Tungsten Arcs', *Ph.d Thesis*, Cranfield Institute of Technology.

Allum C J, Pinfold B E and Nixon J H (1980) 'Some Effects of Shielding Gas Flow on Argon Tungsten Arcs Operating in High Pressure (1 to 14bar abs.) Environments' *Weld. J. Res. Sup.* 59, (7), pp199s - 207s July.

Bauder U H (1968) 'Radiation from High Pressure Plasmas' *J. Appl. Phys. Vol.* 39, No. 1, Jan.

Cave, W.R. (1996) 'Investigation of the Constricted Plasma Arc Process for Hyperbaric Welding at Pressures 1 to 100bar', *Ph.D Thesis*, Cranfield University, March.

Dos Santos J F, Szelagowski P and Schafstall H G (1989) 'Properties of Hyperbaric Flux Cored Arc Welds: Weld Metal Chemistry' *21st OTC Conf.*, Houston, Texas, USA.

Dos Santos J F, Szelagowski P and Schafstall H G (1990) 'Preliminary Investigations of the Effect of Short Circuit Variables on Metal Transfer above 60bar abs.' *22nd OTC Conf.*, pp127-134, Houston, Texas, USA.

Hansen H R, Rassmussen A and Richardson I M (1991) 'Hyperbaric GMA Process Control and Properties at Pressures 1 to 60bar' *Proc. ASM Conf. 'Offshore Mechanics and Arctic Engineering'*, Stavanger, Norway, June 23-28.

Knagenhjelm H O and Ovland S (1982) 'Deep Ex 81', *Norsk Hydro Report No. 47/82* Oct.

Lythall D J and Gibson D E (1976) 'Report on the Effect of Depth on Underwater Code Quality Gas Metal Arc *Welding' Offshore. Tech. Conf.*, Paper 2643 Dallas, Texas.

Nixon J H and Billingham J (1987) 'A Survey of Underwater Welding Techniques' *Endeavour, New Series*, Vol. 11, No. 3.

Nixon J H and Richardson I M (1995) 'The Design and Construction of a 250bar Hyperbaric Welding Research Facility' *Proc. 14th International Conf. 'Offshore Mechanics and Arctic Engineering'* Copenhagen, Denmark, Vol. 3, pp553-562, June.

Pandey R K and Swetnam D (1985) 'Dry Hyperbaric MMA Welding, The Effect of Water Depth on Welding of BS 4360 50E Structural Steel' *Weld. & Met. Fab.*, pp 232-236 Sept. & pp 278-281 Oct.

Richardson I M (1989a) 'Underwater Arc Welding - A Survey of Process Behaviour' *Weld. Rev.* pp11-20, Feb.

Richardson I M (1989b) 'Some Characteristics of the Constricted GTA (Plasma) Arc at Elevated Pressures' *Proc. IIW Int. Conf. 'Welding Under Extreme Conditions'* Helsinki, Finland, Sept. Publ. Pergamon Press.

Richardson I M (1991) 'Properties of the Constricted Gas Tungsten (Plasma) Welding Arc at Elevated Pressures', *Ph.D Thesis*, Cranfield Institute of Technology.

Richardson I M (1994) 'Hyperbaric Plasma Welding' *Offshore Research Focus*, No. 104, December.

Smart J (1993) 'Hyperbaric Plasma Cutting' *M.Sc Thesis*, Cranfield Institute of Technology.

Proceedings of the Seventh (1997) International Offshore and Polar Engineering Conference
Honolulu, USA, May 25-30, 1997

ISBN 1-880653-28-1 (Set); ISBN 1-880653-32-X (Vol. IV)

Monitoring of Molten Pool and Adaptive Control of Penetration in GTAW of Thin Mild Steel Plates

Y. Suga, M. Mukai and S. Usui
Keio University
Yokohama, Japan

K. Ogawa
Osaka Prefecture University
Sakai, Japan

ABSTRACT

A new welding robot system with visual sensors is developed in order to realize a full automatic welding of thin mild steel plates, such as automatic seam tracking and automatic control of welding conditions. Especially, in this study, a system to detect the dimension of molten pool using CCD camera and a penetration control system in gas tungsten arc welding (GTAW) of mild steel plates are investigated. In order to increase the robustness of the sensing system, some new methods, such as self-arc-lighting method, application of pulse current and etc., are proposed and the effectiveness of these methods is shown. As the results of the automatic welding experiments applied the control system with visual sensor, it is confirmed that the system constructed is effective for penetration control in automatic butt welding of thin mild steel plates.

KEY WORDS: *Visual sensor, Automatic welding, Welding robot, Molten pool, Penetration control*

INTRODUCTION

Recently, mechanization and robotization of welding has been investigated widely, and many types of intelligent welding robots which have some sensors, such as arc-sensor[1] and vision sensor[2-8], are developed. Especially, the adaptive control of welding conditions, in which the welding arc phenomena and the molten pool condition are observed during welding and furthermore the welding conditions are controlled in real-time, are remarked. For instance, a method to control the penetration of weld, in which the shape and dimensions of the molten pool are visually detected and welding conditions are controlled so as to keep them optimum, is remarked and investigated by some researchers[8].

On the other hand, in automatic butt welding of relatively thin plates, it is important to control welding conditions to obtain a soundfull penetration weld. Especially, it is difficult to obtain a full penetration weld, because the optimum welding condition changes with change of various welding situations, such as gap of groove, shape of groove welding speed and so on. Accordingly, estimation of molten pool condition by detecting the width of the molten pool using CCD camera was tried and furthermore the real-time control for full penetration welding was performed.

In this study, the principle of detection of penetration, detecting method of molten pool dimension and control system of welding condition according to the dimension of the molten pool are investigated.

HARDWARE SYSTEM OF WELDING ROBOT

Figure 1 shows the hardware system of the welding robot used in this study. The robot system is constructed by a three-axial Cartesian coordinates robot body, CCD cameras, image digitizers, personal computers (CPU: i486), equipments for gas

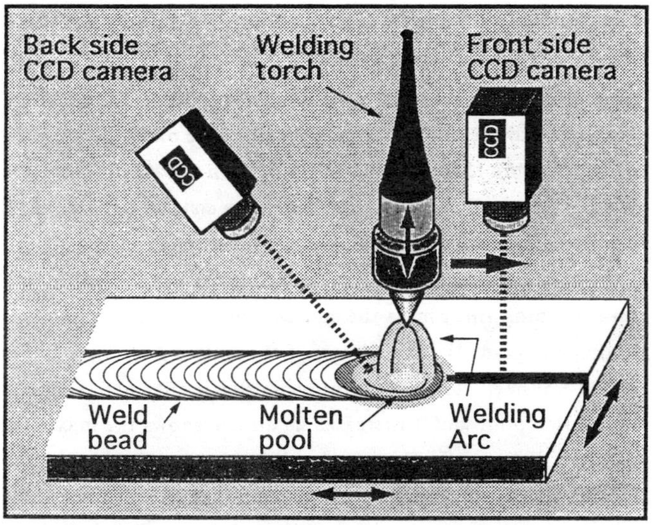

Fig.1 Arrangement of equipment

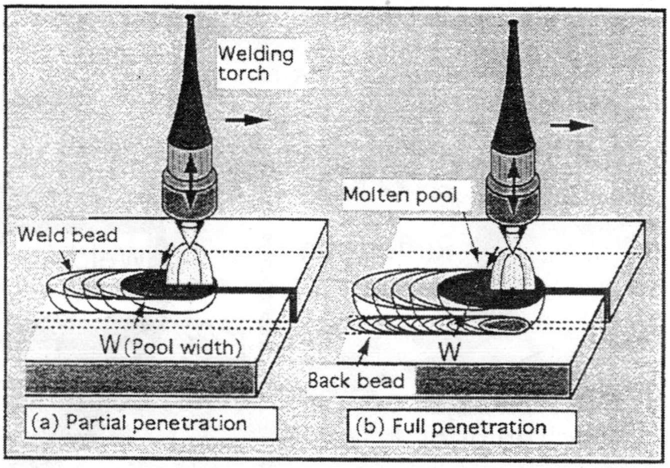

Fig.2 Principle to control penetration of weld

tungsten arc welding (GTAW) and so on. As the power source, a pulse current type D.C. electrical source with drooping characteristic was used. In automatic seam-tracking, a CCD camera equipped in front of the welding torch was used for detecting welding line of the joint. The image data taken by the camera was recorded in image digitizer (256 × 240 pixels, 6 bits) as the digital numerical value of 64 steps, and were taken into a personal computer. The time of 1/30s is required for photographing a single frame. Then the welding line position was detected by data processing and then the three axes of the robot were controlled in order to make the welding torch track the welding line automatically. For the adaptive control of welding condition in automatic penetration welding, a similar CCD camera equipped in back side of the welding torch was applied. The image taken was processed in another personal computer and welding conditions were controlled to obtain an optimum penetration weld bead. The view area of the camera was 10.2 × 9.6mm, and accordingly the resolution of the image was 0.04 mm/pixel.

As base metals for welding experiments, SPCC mild steel plates of 1.6 mm in thickness were used. In welding, gas tungsten arc welding (GTAW) process was applied and argon gas was used as the shielding gas.

DETECTION OF MOLTEN POOL WIDTH BY VISION SENSOR

Principle to estimate penetration of weld

Generally, it is important to obtain a penetrated weld bead which has an appropriate back bead width in arc welding of thin plates,. It is well known that the width of the back bead changes in proportion to bead width as shown in figure 2. In case the welding current is too small, the weld bead may be narrower and at the same time the penetration shallow. Consequently, the weld bead results in the partial penetration as shown in figure 2 (a). On the contrary, when the welding current is higher enough to penetrate the plates, a sound penetration weld bead with back bead was obtained as shown in figure (b). However, if the current is too high, the weld bead may melt down. As stated above, the back bead width changes approximately with surface bead width. Therefore, the control of the back bead width was tried by controlling the width of the weld bead in surface side during

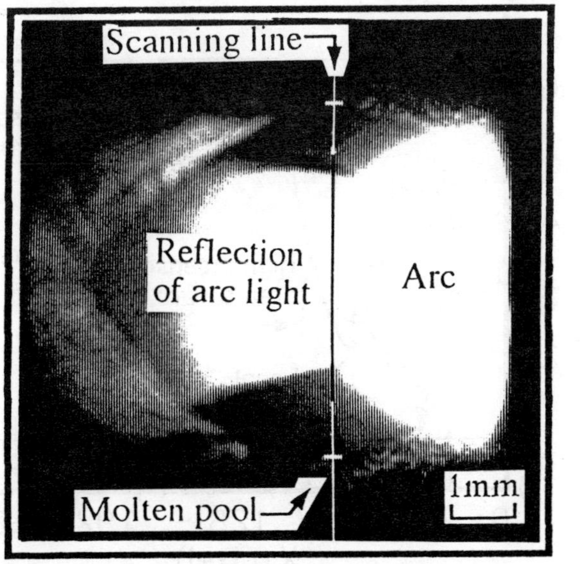

Fig.3 Appearance of molten pool

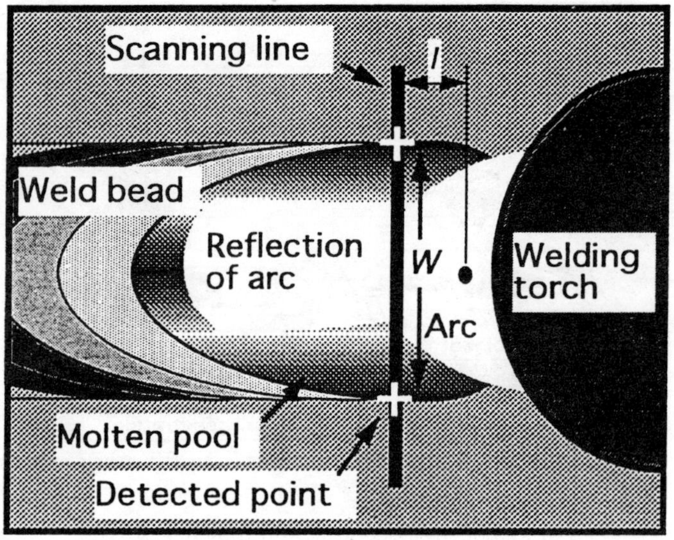

Fig.4 Schematic illustration of molten pool

welding. In order to realize the real-time measurement of the bead width in surface side, a method to detect the width of the molten pool by a CCD camera during welding was used.

Figure 3 shows an example of picture showing molten pool during welding of steel plates and figure 4 its schematic illustration which indicates a method to detect the molten pool width. The picture was taken by a CCD camera which was equipped at the back side of the welding torch and set at the angle of 45° . In the figure, a scanning line to detect the molten pool width, W, is shown by a straight line and detected points of edges between molten pool and base metal are indicated by cross cursors. In this picture, the central area of the molten pool is quite bright due to the reflection of high intensity arc light. However, the both sides of the pool is relatively dark. On the other hand, the base metal is somewhat brighter at the edge that is adjacent to the pool. Accordingly, the edge of the pool can be detected by the change of the brightness on a scanning line. The width of the pool, W, was measured as the distance between both edge points detected by the above mentioned method.

Figure 5 shows a typical example of brightness distribution on a scanning line across the molten pool. The brightness tends to decrease gradually from center of the molten pool toward base metal. However, it has the minimum value locally at the edge of the molten pool and it increases again and shows the maximum value in that area. In this system, accordingly, the edge of the molten pool was detected as a center point between the local maximum and minimum points of the brightness distribution on the scanning line.

Improved method detecting the molten pool edge

As the molten pool edge is detected by the increase of brightness at the edge of the base metal, it may be better to apply some high intensity lightings. However, generally, the lighting is not so bright as high intensity arc light in GTA welding. Accordingly, the light of the arc itself was applied for lighting the weld part using reflecting mirrors equipped around the torch as shown in figure 6. Because the arc light itself is used for lighting, the method was named self-arc-lighting method. Furthermore, in order to make use of the arc light effectively, the arc current was increased. However, it is impossible to increase the welding current carelessly, because the penetration of weld

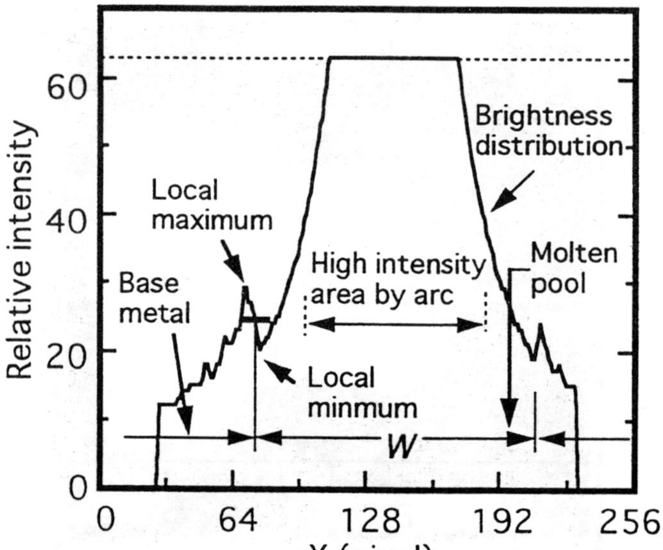

Fig.5 Brightness distribution

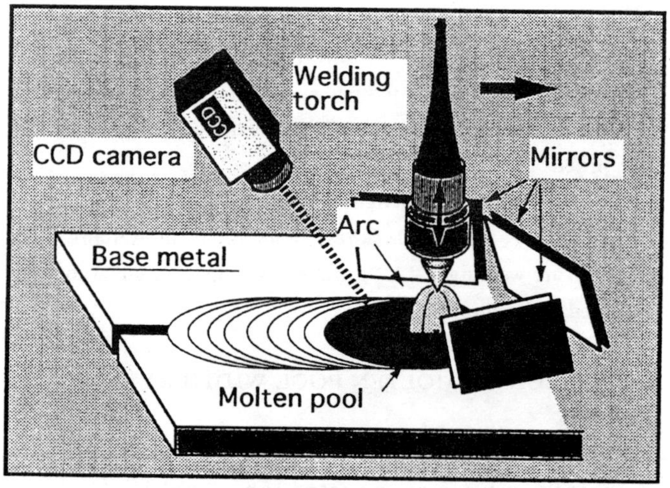

Fig.6 Arrangement of mirrors and torch

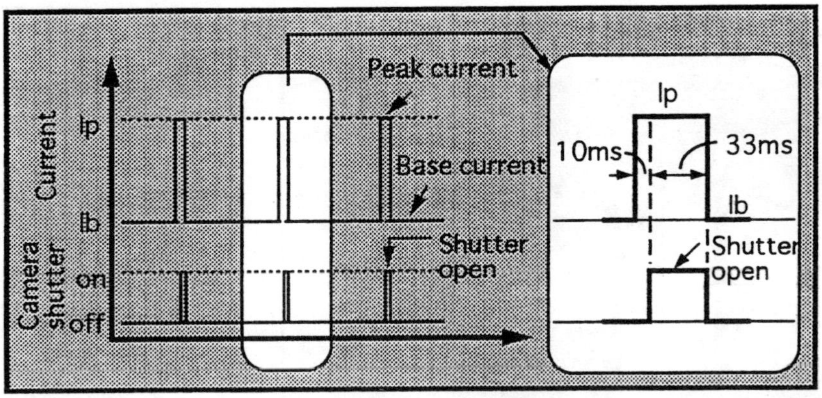

Fig.7 Timing chart of pulse current and camera shutter

changes sensitively with it. Therefore, a new system to use pulse current as the welding current was developed. As shown in figure 7, when an image is taken by CCD camera, the current is increased up to the peak current, Ip, at a short time. The precise timing chart of peak current and camera shutter is shown also in the figure. At first the computer turns on the current controller and accordingly the arc current increases up to the peak current. After 10ms, the computer make the shutter of the camera open. Since it takes 33ms to photograph the image, the peak current is kept during it. Then the current is decreased to normal value (base current). The intensity of peak current can be selected by an operator. Fig.8 (a) and (b) show the effect of the application of mirror and an increase of current. Figure (a) shows a brightness distribution on a scanning line, in case any improving method is not used. Figure (b) shows also the brightness distribution obtained by applying mirrors and pulse current. Comparing figure (a) and (b), it is seen that, applying the improved method, the brightness difference between molten pool and base metal increases and accordingly the edge of the molten pool can be detected more clearly.

Application of directional filter to detect molten pool edge

In some experiments to measure the molten pool width during actual welding, misjudgment occurred sometimes due to the high intensity arc light. Therefore, in order to improve the robustness of the system, a directional filter was applied to detect the edge point of the molten pool. That is to say, the direction of the edge of the molten pool may be parallel to welding direction in the area where dimension of the molten pool width is measured. Therefore, a linear smoothing filter (1 × 9 pixels) along the molten pool edge was applied as shown in fig.9. Because average data of nine points were used to judge the edge of molten pool, bad influence of noise is eliminated and relatively reasonable width of molten pool was obtained.

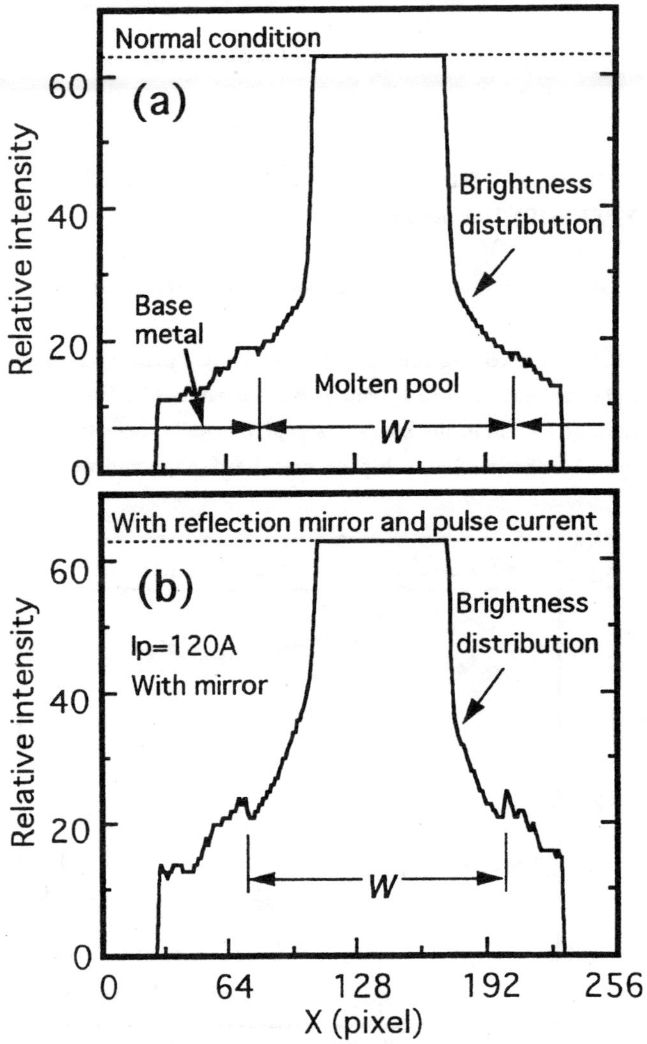

Fig.8 Effect of mirrors and pulse current on detection of molten pool width

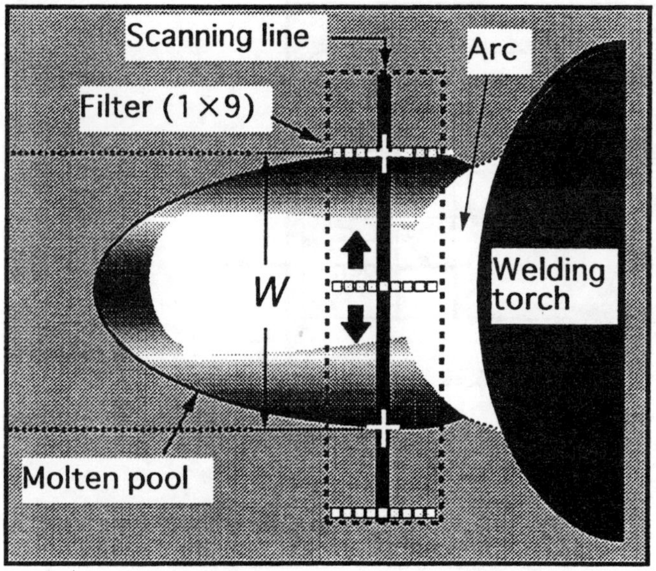

Fig.9 Application of linear filter for detecting molten pool width

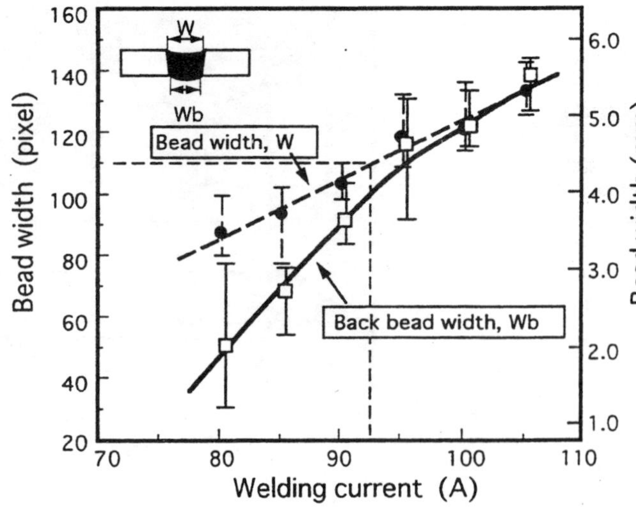

Fig.10 Relation between welding current and bead width

CONTROL OF THE PENETRATION WELDING AND EXPERIMENTAL RESULTS

Relation between weld bead width and penetration

Before the welding control experiments, relationship among bead width, back bead width and welding current in GTAW of steel plates of 1.6mm in thickness was experimentally investigated. Figure 10 shows relationship between bead shape and welding current. As the welding conditions 3.3mm/s (20cm/min) in welding speed, 2mm in arc length and 10 l/min in the flowrate of argon shielding gas were used. In the figure, the bead width and the back bead width increase with an increasing welding current. As the suitable back bead width in this welding condition seems to be in a range between 3.5 and 4.5mm approximately (about 90 or 110pixel), the optimum welding current seems to be about 92A. In this case, the surface bead width is 4 or 4.5mm. Therefore, it is seen that the target width of the molten pool should be set at about 4.5mm (110pixel) in penetration control.

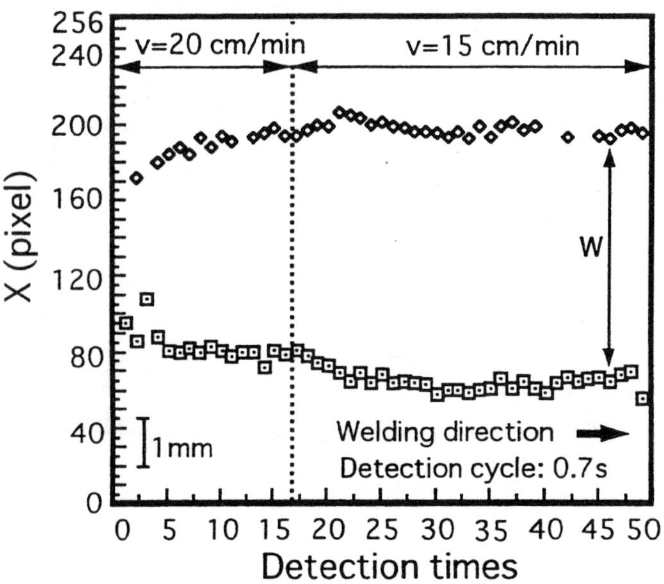

Fig.11 Detected result of molten pool width in TIG arc welding

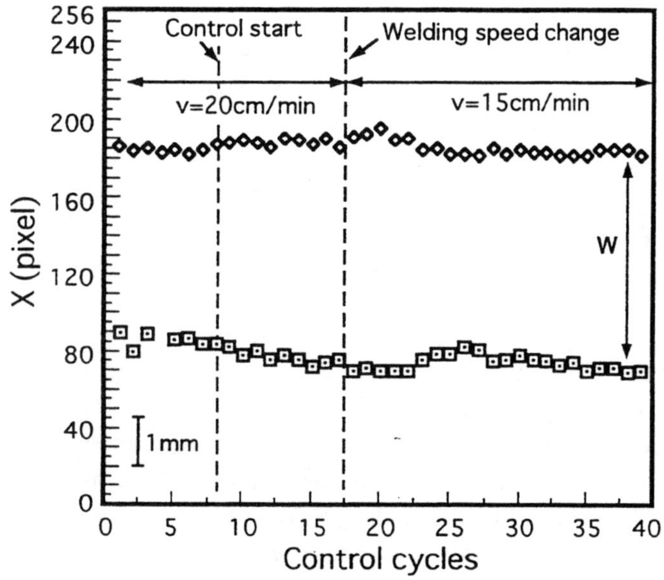

Fig.12 Result of penetration control in TIG arc welding

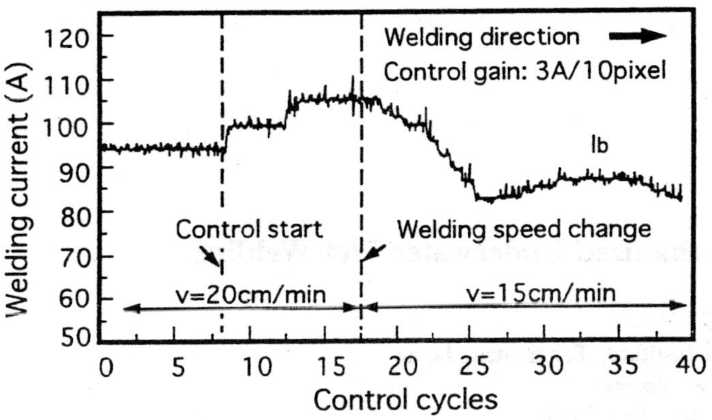

Fig.13 Change of welding current in automatic penetration welding

Results of welding experiments

In order to confirm the effectiveness of the control system, welding experiments of mild steel plates were performed. In welding, GTAW process was used. As the shielding gas, pure argon gas was used. The diameter of tungsten electrode was 2mm and the tip angle of the electrode 50°.

Figure 11 shows the measured results of change of molten pool width during welding without penetration control. In this case, as the welding speed is decreased suddenly from 20cm/min to 15cm/min, the molten pool width increases just after the decrease of the welding speed.

Figure 12 shows a penetration control result in welding of steel plates. The welding began at a welding speed of 20cm/min, but it decreased suddenly to 15cm/min as shown in figure 11. The penetration control was performed by controlling welding current and the gain was set at 3A/10pixel(7.5 A/mm). As the control method, a simple proportional control method was used. Fig.13 shows the change of the welding current during control experiment shown in figure 12. In the figures 12 and 13, although the bead width increases just after decreasing the welding speed, the welding current and the bead width decrease gradually and it is kept at a suitable bead width between 100 and 110 pixel.

CONCLUSIONS

A system to detect the dimension of molten pool by a vision sensor and a penetration control system in gas tungsten arc welding (GTAW) of thin mild steel plates are investigated. Main results obtained are summarized as follows:

1) A vision system to detect the width of a molten pool during gas tungsten arc welding (GTAW) of mild steel plates was constructed.

2) In order to detect the edge between molten pool and base metal, a method detecting the brightness change on a scanning line between them was proposed and its effectiveness was shown.

2) In order to detect the width of the molten pool without misjudgment, self-arc-lighting method, application of pulse current and application of linear filter were proposed, and the effectiveness of the methods was confirmed.

3) A penetration control system in automatic welding of mild steel plates was constructed. As the result of welding experiments, it is confirmed that the systems constructed are effective in the automatic penetration welding of mild steel plates.

REFERENCES

(1) H.Nomura, Y.Sugitani, M.Murayama: Development of Automatic Fillet Welding Process with High Speed Rotating Arc, Quat.J. Jap. Weld. Soc., Vol.4, No.3(1986)P.502 (in Japanese)

(2) E.Ide, H.Fujimura: The Present Technique in Welding Process, J. Jap. Weld. Soc., Vol.60, No.2 (1991)P.121 (in Japanese)

(3) T.Araya: Sensing Technology for In Process Control(1), J Jap. Weld. Soc., Vol.60, No.2 (1991)P.127 (in Japanese)

(4) Y.Suga, K.Takahara, M.Ikeda: Recognition of weld Line and Automatic Weld Line Tracking by Welding Robot with Visual and Arc Voltage Sensing System, J. Jap. Soc. Precision Eng., Vol.57, No.6 (1991)P.1060 (in Japanese)

(5) Y.Suga, M.Ikeda: On the Recognition of Weld Line by Visual Sensor in Automatic Welding of Machined Steel Plates, J. High Temp. Soc., Vol.18, No.3 (1992)P.138 (in Japanese)

(6) Y.Suga, M.Naruse, T.Tokiwa: Application of Neural Network to Visual Sensing of Weld Line and Automatic Tracking in Robot Welding, Proc. Int. Conf. 47th Int. Inst. Weld.'94, Beijing, (1994)P.269

(7) Y.Suga: Effect of Some Image Processing on the Seam Tracking by a Welding Robot with Visual Sensor, Proc. 5th ISOPE"95 vol.4 (1995)P.121

(8) K.Ohshima, S.Yamane: Fuzzy Control in Tracking of Welding Line, Proc.Int.Conf. on New Advance of Welding and Allied Processes, Beijing, Vol.1 (1991)P.138

Proceedings of the Seventh (1997) International Offshore and Polar Engineering Conference
Honolulu, USA, May 25-30, 1997

ISBN 1-880653-28-1 (Set); ISBN 1-880653-32-X (Vol. IV)

Consideration of Sensing Methods for Mechanized Underwater Wet Welding

M. Tanaka and T. Morita
Shikoku National Industrial Research Institute, Takamatsu, Japan
N. Kitamura and K. Tohno
NKK, Tsu Research Center, Tsu, Japan
T. Irie and H. Matsushita
Mitsui Engineering & Shipbuilding, Ichihara, Japan
Y. Ogawa and T. Sumitomo
Shikoku National Industrial Research Institute, Takamatsu, Japan

ABSTRACT

Mechanized underwater wet welding is desired for practical fabrication and maintenance technique of offshore structures. Wet welding with water curtain nozzle is a useful method except for detection of welding path. Water curtain assists the stability of local cavity around the weld pool. But the flow of water curtain on the weld plate contains fine bubbles of shield gas. These bubbles disturb the observation of weld pool. Accurate detection of real welding path and shape to be welded is the most important factor to obtain sound weld beads.

This paper describes a desired function of sensing method for mechanized underwater wet welding procedure. Three sensing methods to detect the welding path were considered. They were mechanical touch sensing, visual sensing and ultrasonic sensing. Advantages and disadvantages for these methods are summarized below:

The mechanical touch sensor is an established one for well-prepared groove face. It is reliable and durable where the groove of weld line is well prepared. Otherwise, this method requires exact observation on weld groove geometry before welding.

The visual sensing is an essential for intelligent robot, but it is sometimes very difficult to keep clear around the area to be observed in case of underwater welding. The water curtain also disturbs the visibility. The suitable way to guarantee the visibility is using a fiber scope technique, which is inlaid inside of the nozzle. A sophisticated technique is required for this method, but the accurate detection of groove face is possible. In addition, the observation of welding condition is also available.

The ultrasonic sensing is another good method to evaluate the groove geometry under water, especially in case of low visibility. The water curtain disturbs the accuracy of ultrasonic transmission, so the careful selection of the distance between the nozzle and the sensor is required.

KEY WORDS: Underwater welding, Wet welding, Seam tracking, Mechanized welding, Visual sensing, Ultrasonic sensing, Mechanical sensing

INTRODUCTION

Keen interest for utilization of offshore space and resources has been steadily growing up with improved offshore technology. Wet underwater welding technique is a key technology to improve its cost performance. However the usage in practice has been limited only for manual use. The results of mechanized wet underwater welding in laboratory test have shown reasonable quality. But, almost all fixed offshore structures are fabricated as pipe structures. The place to be proceeded by mechanized wet welding is limited. And the miscellaneous techniques to realize the mechanization have not reached to the stage for realistic practice, mainly by economical reason.

A Mega-Float project has being carried out in Japan since 1995. It is a technological research activity to realize a very large scale floating structure as a base for economical activities instead of land reclamation. Many box typed floating units are fabricated in the shipyards simultaneously, and they are joined at the offshore installation site. Underwater welding on a very long straight line is necessary in this case. The mechanized wet underwater welding technique is now closing up from an economical point of view. Fundamental research and development activities have being carried out to establish the mechanized wet welding methods for the Mega-Float(Ogawa, *et al*, 1996).

Considerations of suitable seam tracking method and evaluation system of groove geometry were carried out fundamentally. Following various sensing methods were considered. They are mechanical, visual, ultrasonic, and eddy current sensing systems. They must be worked under the severe condition. The circumstance is fully filled by seawater, and they are usually muddy by grinding or cleaning operation of groove face. Sometimes, tacked weld beads are existed inside the groove joint. This paper describes the advantages and disadvantages of them.

EXPERIMENTAL PROCEDURE

Structure of the Mega-Float

Fig.1 shows schematic illustration of the Mega-Float unit. The final dimension of the Mega-Float is 300m in length, 60m in width and 2m in draught. Nine floating units those dimensions are 100mL, 20mW and 2mD were joined at Yokosuka harbor. Inside of this unit is divided into many rooms by transverse and longitudinal bulkheads. It is relatively easy to obtain dry condition in offshore joining because of this structure. Then, almost all of them are welded by semi-automatic process in dry condition. Many methods to make dry condition were tried to consider the advantages and disadvantages of their procedure. In this point of view, mechanized wet welding was carried out in two blocks to consider practical problems in real practice for future application(Ogawa, *et al.*, 1996).

Fig.2 shows schematic of underwater wet welding process, which was carried out for offshore joining of the Mega-Float in July 1996. Water depth was at most 60cm in this case, then partially submerged type wet welding was carried out. An insert plate was used to compensate the setting accuracy of butt joint. Groove surface had been prepared in shipyards, and backing plate was also attached in surface. Then the joint groove geometry was almost the same for one weld line of a few meters. And the line was straight in this case. Plate thickness is 12mm and 45 degree V-groove butt joint was used.

Accurate positioning of weld line and accurate selection of weaving width are required for mechanized wet welding system, otherwise some defects such as under-cut and lack of fusion will be appeared. Several kinds of sensing systems were considered for this wet welding procedure.

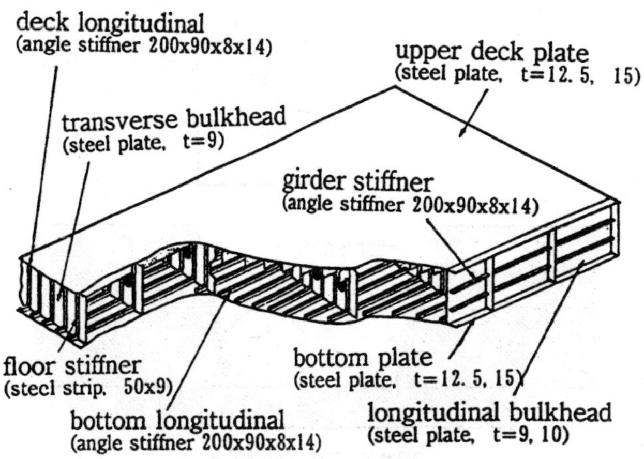

Dimensions of float unit : 100mL x 20mW x 2mD,
Weight : 680 – 770t,
Fabrication cost : about 130milion yen,
Material : rolled steel for ships equivalent to SM400

Fig.1 Cross-sectional view of the float unit.

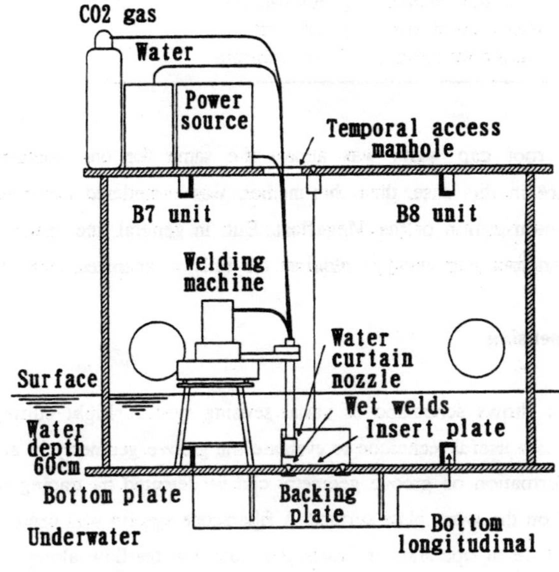

Fig.2 Schematic of underwater welding equipment.

Mechanical sensing

Fig.3 shows schematic illustration of mechanical sensing system. This system is established as mechanized welding system in air use. The sensing stylus touches on the root and the sidewall. The position of the stylus is controlled that the vertical and horizontal touching pressures are always kept within designed values. Table 1 shows the specification of used mechanical sensor. The radius of stylus is an important factor for the accuracy of seam tracking ability. The value was chosen by former experience.

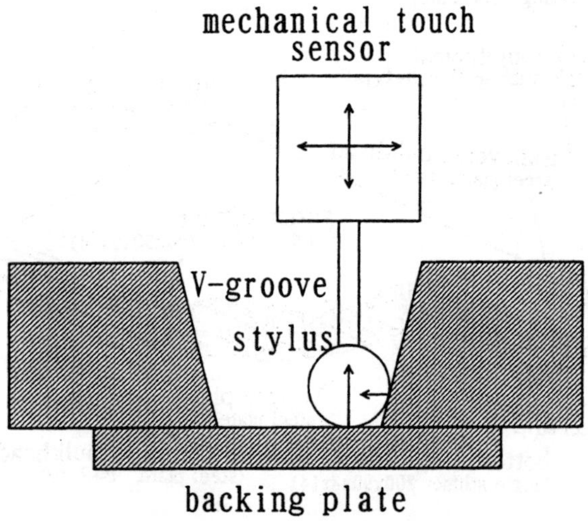

Fig.3 Schematic of mechanical touch sensor.

Table 1 Specification of mechanical sensor.

dimension	vertical and horizontal
accuracy	within 0.2mm
maximum weight	25 kg
maximum moment	500 kg·cm
maximum stroke	250 mm
maximum speed	200 mm/min

The root gap width was almost the same for one welding procedure in this case, then this method was considered for open field demonstration of the Megafloat. But, in general, the sensing ability on root gap width is required for fully mechanized process.

Visual sensing

Fig.4 shows schematic of visual sensing system. Light cutting method is a useful technique to evaluate the groove geometry in air. The information on groove geometry can be detected by tracing of slit line on the work plate surface. A fiberscope system was used in convenience of operation in water. Additional water flow along the axis of the video camera was used to avoid disturbance from following subjects, such as muddy water, high-speed flow of curtain water, and fine bubbles of shielding gas from the welding nozzle.

The quality of image is very important by visual sensing, but there are many disturbances to decrease the quality in underwater operation such as scattering and reduction of light by many fine obstacles in water. In the case of wet welding by water curtain technique, the weld portion is completely covered by high-speed curtain water flow. And fine bubbles of shielding gas are released at the out skirt of the curtain on the work plate. Then it is difficult to obtain clear image. In addition, some cleaning and grinding operations are carried out to remove rust and marine fouling on the groove face just before the welding. Then the visibility becomes quite low at the welding time.

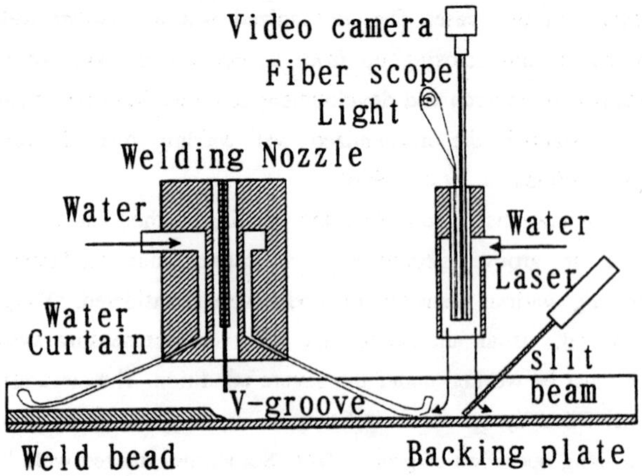

Fig.4 Schematic of visual sensing method.

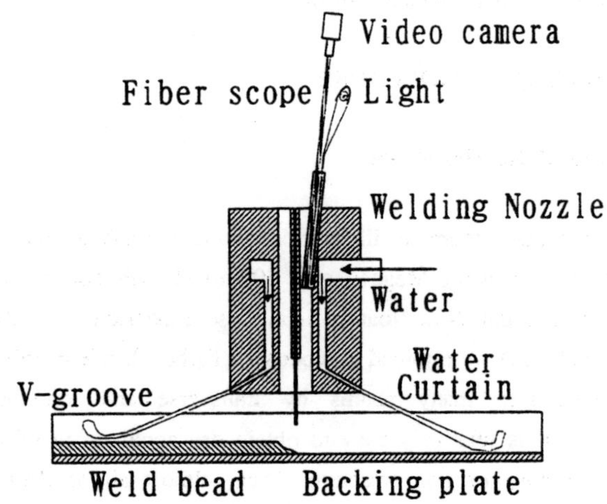

Fig.5 Schematic of internal visual sensing method.

Fig.5 shows another type of visual sensing system. A fiberscope image sensing system is installed inside the welding nozzle. It is dry inside the nozzle, then the problem due to in water operation is perfectly solved. But there are still many restrictions by structural limitation of this system. Active method such as light cut method is difficult to install, then passive method can only be adopted. There is a danger of damage caused by high temperature by welding arc, sputtering from weld pool and high intensity of arc.

Ultrasonic sensing

Fig.6 shows schematic of typical ultrasonic sensing. The distance between the probe/transducer and the work surface can be detected to check the time delay of ultrasonic reflection. The probe/transducer is scanned transversely, then groove geometry can be detected(Suga *et al.*, 1994). The reflection of ultrasonic beam on the sidewall of V-groove is hardly obtained in this case, but the surface of this wall is flat and uniform. Therefore, when the surfaces of work plates and root are detected precisely, the wall can be estimated. The purpose of this sensing is to detect the exact location of root edges. The diameter, focal length and frequency of used probe are 10mm, 32mm in water and 5MHz, respectively.

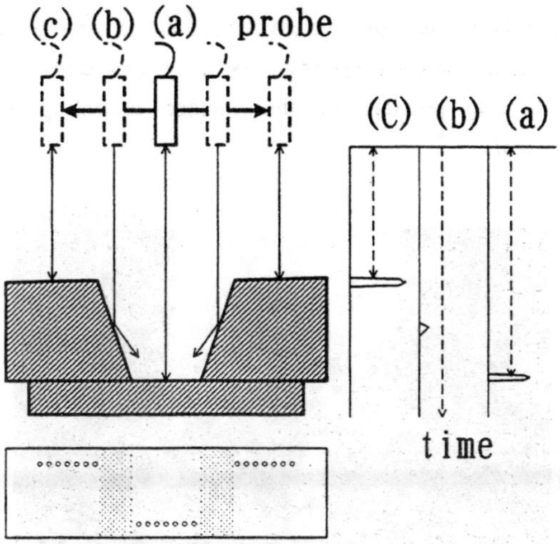

Fig.6 Schematic of ultrasonic sensing method.

A linear array sensor by electronic scanning is also considered its ability. This sensor has 90 elements in the array, and valid field size to observe is 50mm(Tanaka *et al.*, 1996).

Eddy current sensing

When an alternating current flows in a test coil near the work plate, it sets up eddy currents in the work plate. These eddy currents produce a magnetic field that interacts with the test coil by producing a back electromotive force and changing its impedance. A fluctuation of the distance between the probe and the work surface will change the magnitude of the eddy current in the work plate. And this will be shown as a change in the voltage of the test coil, due to its change of impedance. The diameter of the sensor head is 20mm, and the measurable distance is between 0 to 10mm (Tanaka *et al.*, 1996). Mechanical transverse scanning was proceeded to measure the groove geometry.

Measuring method

The characteristics of sensors above mentioned were measured against typical grooved butt joints. These probes were installed on the three-axis robot. The test plate was set in an acrylic water tank. The motion of the robot and the measuring procedure were controlled by a personal computer. Fig.7 shows schematic of experimental devices.

The ability of mechanical sensor was examined by using of the mechanized wet welding machine that was developed for the MegaFloat as shown in figure 8(Ogawa *et al*, 1996).

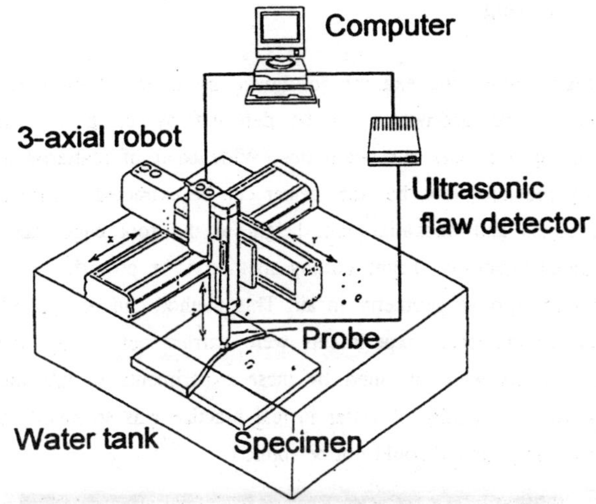

Fig.7 Schematic of experimental device.

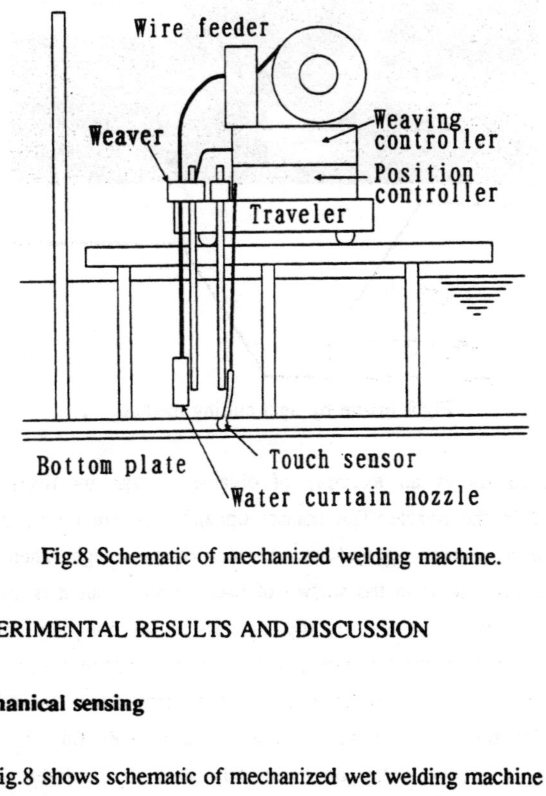

Fig.8 Schematic of mechanized welding machine.

EXPERIMENTAL RESULTS AND DISCUSSION

Mechanical sensing

Fig.8 shows schematic of mechanized wet welding machine with a mechanical touch sensor. Weaving width and relative torch position to the sensor were determined before the welding process. The weld line was straight and the root gap width was almost the same for one welding process, and there was no tacked weld bead in the groove, then successful welds were carried out for all trials.

A flexible shelter against sputtering was attached behind the sensor to avoid miss-operation of seam tracking by adhered sputters on the groove surface.

Visual sensing

Fig.9 shows the example of image by light cutting method. Exact groove geometry can be detected by using of image processing for laser slit beam line. Wide band of scattered light nearby the laser line, but real center can be extracted. Position of root edges and miss-alignment height at the root edge can be evaluated precisely. Light cutting method is a powerful tool to detect the groove geometry in air. This method can be applied in water. Fundamental experiments were carried out in tap water. Good results were obtained by these experiments in laboratory. However, the quality of water in real practice was so muddy that light cutting method could not be applied.

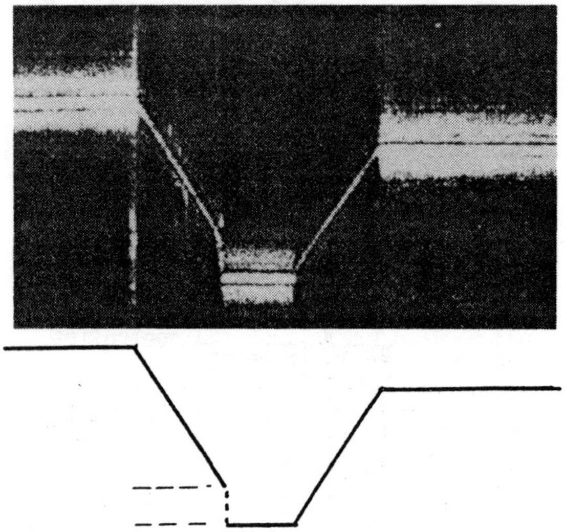

Fig.9 Image by light cutting method.

Fig.10 shows an example of obtained image by fiberscope inserted in the nozzle. The contact tip and wire are removed for easiness to understand the characteristics of the image. There are some uneven areas in the surface of backing plate, but it is almost flat. The fiberscope and its lighting axis is set obliquely to the nozzle axis, then the brightness of the image is different along the longitudinal position. The upper side of the image is the near side in actual location, and it is well lighted compared to the far side. The surface of groove walls are completely slant to the fiber scope system, then the area of sidewalls is dark. The contrast of brightness between the root surface and side wall is clear in the near side. The contrast in the far side is not so clear. The brightness along the traverse position is also different. Detecting of edge elements by unitary differential operator is not suitable in this case.

Fig.11 shows an example of detected edges by localized threshold. The suitable threshold condition was chosen by calculation of local brightness and difference from neighboring elements. The edge elements were detected neither more nor less. The exact edge position can be calculated by correction in distortion of image. One side of the edge line in the image is always in almost the same pixels, because the system is kept in the same position by mechanical touch sensing system. Visual sensing during welding operation is also possible by this method according to some preliminary tests.

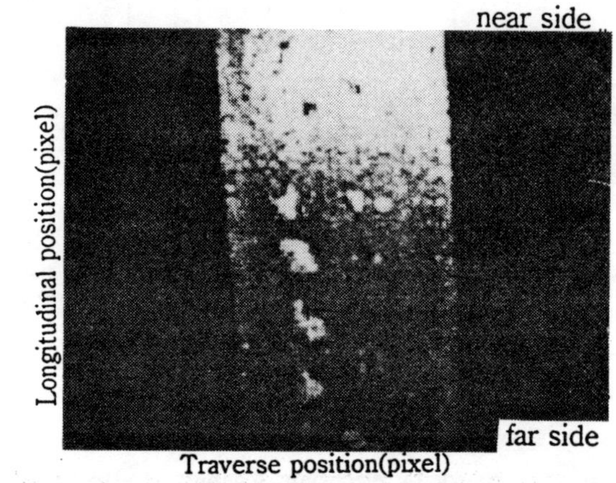

Fig.10 Example of obtained image data.

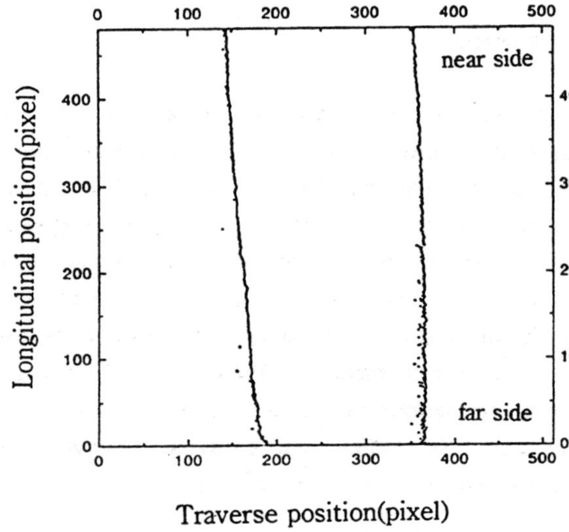

Fig.11 Detected edge by localized threshold.

Ultrasonic sensing and eddy current sensing

Fig.12 shows the example of edge detection for lap joint by ultrasonic sensing. Edge point of the lap joint was clearly detected. The focal spot of ultrasonic beam was small enough to detect the distance correctly. If the distance is measured by amplitude of reflected ultrasonic beam, it is difficult to obtain correct value. Because the reflected intensity is strongly influenced by the wall conditions such as surface roughness and surface angle. The fluctuation of intensity is especially high near the sharp edge.

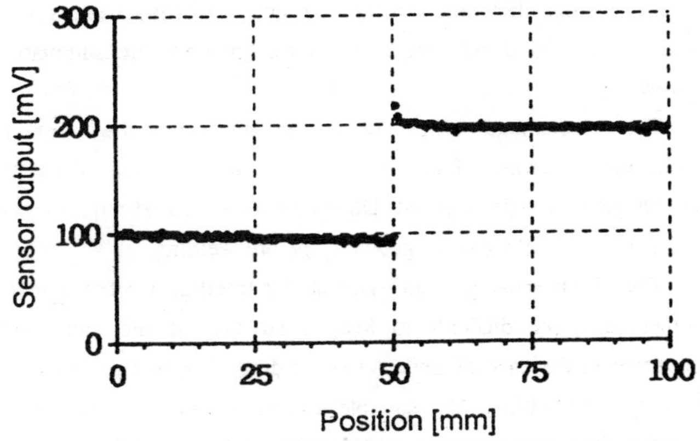

Fig.12 Detection of lap joint geometry by ultrasonic sensing.

Fig.13 shows the result by eddy current system for the same test piece. The focal spot diameter of it is about 20mm. It is too broad to detect the sharp edge, then the obtained result is not clear. Results applied for V- and I-grooved specimen show the same tendency. According to these rssults, it is difficult to estimate the correct groove geometry for butt joint by this eddy current method.

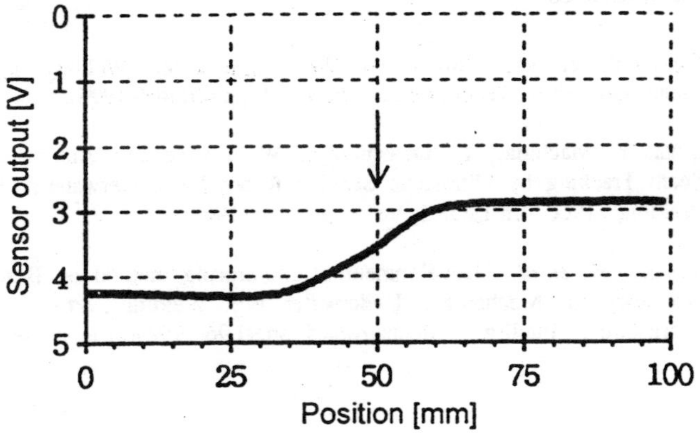

Fig.13 Detection of lap joint geometry by eddy crrent sensing.

Fig.14 shows the detected results by ultrasonic sensing on a butt joint, which was proceeded by CO2 welding to the second path. Thickness of this plate is 20mm, and the groove angle is 60 degree. Surface of the bead is not so rough, then the detection of the bead surface position is possible. Measured results of the wall surface position indicate wrong distances, then the process to correct these miss-indications is required.

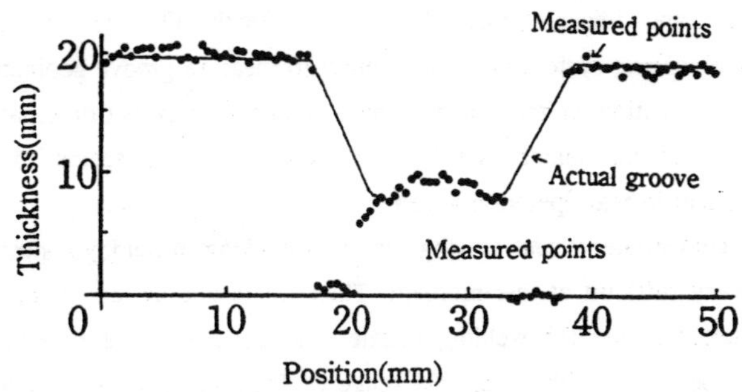

Fig.14 detection of but joint geometry by ultrasonic sensing.

Fig.15 shows an example of obtained image by the array sensor. Strips shaped ninety probe elements are implemented in one unit which effective field size is 50mm. The frequency of ultrasonic beam is 5MHz, and the beam is electronically scanned. Detected results are not so sharp, and there is some ghost image near surface. However, the position of the plates' surface and groove surface can be recognized by this image. One of the advantages of this sensing system is that the scanning is carried out electronically, then dynamic movement of the sensor is not necessary. And the groove geometry can be measured in a short duration. The wavelength is rather long compared to the light, then there is a possibility to measure in muddy condition.

Fig.15 Example of obtained image for V-groove butt joint by array sensor.

Combination of mechanical sensing and visual sensing

Mechanical sensing system is durable, and it is suitable for seam tracking. But the evaluation of groove geometry such as root gap width and miss-alignment height is difficult. On the contrary, visual-sensing system has much ability to measure groove geometry, if the position of the video camera is kept constant. Combination usage of mechanical sensing and visual sensing seems to be practical in real operation at sea.

Underwater visual sensing for seam tracking in muddy condition is very difficult at this moment. Then fiberscope system, which is installed inside the welding nozzle, was considered for measuring the root gap width. Available range of focal distance is limited in this fiberscope. Then the combination of this fiberscope and mechanical touch sensor is considered. The nozzle-work distance is kept constant in this case. Then the distance between the fiberscope head and the work surface is also kept constant. This means that the image is clear, and the position of the one side root edge is appeared almost the same pixel in the image. Therefore the evaluation of root edge position becomes easier. Analyzing procedures for obtained image data is also simplified.

The problem of miss-operation of mechanical sensor by tacked weld in the butt joint is also solved by this combination. Visual sensing system can detect the presence of tacked weld bead, then correction of miss-operation by mechanical sensing is applicable.

Ultrasonic sensing system has another possibility to join with mechanical sensing system. Array sensing system can be used instead of visual sensing system. This system has a possibility to measure in muddy condition. Further consideration of this ability should be researched.

Visual and ultrasonic sensing systems have good possibility, but they were too sophisticated method for the first trial in hot, humid and dirty condition. So, mechanical touch sensor was used for seam tracking, and the root gap width was manually evaluated before the welding trial. Mechanized wet welding trials for open field demonstration of the Mega-Float with mechanical touch sensor was carried out successfully.

CONCLUSION

Various kinds of sensing systems were considered for mechanization of underwater wet welding by water curtain nozzle. Several sensing methods to detect the welding path and groove geometry were considered. They were mechanical touch sensing, visual sensing, ultrasonic sensing, and eddy current sensing system. Advantages and disadvantages for these methods are summarized below:

The mechanical touch sensor is an established one for well-prepared groove face. It is reliable and durable where the groove joint is well prepared. Otherwise, this method requires exact evaluation on weld groove geometry before welding.

The visual sensing is an essential for intelligent robot, but it is sometimes very difficult to keep clear around the area to be observed in the case of underwater welding. The water curtain also disturbs the visibility. The suitable way to guarantee the visibility is using a fiber scope technique that is inlaid inside of the nozzle. A sophisticated technique is required for this method, but the accurate detection of groove face is possible. In addition, the observation of welding condition is also available.

The ultrasonic sensing is another good method to evaluate the groove geometry under water, especially in case of low visibility. The water curtain with many fine bubbles disturbs the accuracy of ultrasonic transmission, so the careful selection of the distance between the nozzle and the sensor is required.

REFERENCES

Ogawa,Y, *et al.*, "Study on Wet Underwater Welding for Megafloat", Proc. Techno-Ocean'96, Vol.2, pp.401-406(1996).

Suga, Y, Machida, A, "Detection of Weld Line and Automatic Seam Tracking by Ultrasonic Sensing Robot for Underwater Wet Welding", Proc. 4th ISOPE, Vol.3, pp.86-91(1994).

Tanaka, M, *et al.*, "Development of monitoring method on Bead Geometry for Mechanized Underwater Wet Welding", Proc. 1st Workshop on Intelligent Mechatronics, pp.91-96(1996)(in Japanese).

Proceedings of the Seventh (1997) International Offshore and Polar Engineering Conference
Honolulu, USA, May 25-30, 1997

ISBN 1-880653-28-1 (Set); ISBN 1-880653-32-X (Vol. IV)

Automation of Arc Welding for Large Scale Structures by a Visual Sensor

A. Ishii
University of Electro-Communications, Tokyo, Japan
N. Masaoka and T. Tsushima
Tomoe Corporation, Tokyo, Japan
Y. Ochi
University of Electro-Communications, Tokyo, Japan

ABSTRACT

Assuming that welding data among structural parts including positions and directions of their contact points and lines can be easily obtained from CAD data, we have proposed image processing algorithms to reliably detect only the terminal points for the welding, and to precisely detect the welding line for the torch position control. The proposed algorithms have been applied to a test piece model of large scale steel tube structures, and the continuous tracking along the welding line measuring its welding gap has been presented.

KEY WORDS: Welding, Arc Welding, Image Processing, Positioning, Weld Line, Hough Transform

INTRODUCTION

In the automation of arc welding, it is necessary to find the starting point for the welding and control the welding torch precisely along the welding line. But, in the actual welding, there are a lot of external variations including the change of the weld gap, the movement of workpiece due to local thermal expansion, and the existence of obstacles and weld tacks, so the applications of the automation of the arc welding must be limited. Recently visual sensors have been applied for the automation of arc welding in order to solve those problems[1]. They are mainly to detect a groove location with high precision by the slit lighting method[2], to detect a weld line with a small gap by an edge detection operator [3-4], to detect a weld line on a thin plate by considering lighting conditions [5-6], and to detect a weld line on a T-tube joint by the slit lighting method [7]. But, these weld line detection methods may be on the basis of local processing, and would fail to deal with partial occlusion of the weld line due to weld tacks and variation of lighting conditions. Moreover, there are few papers reported about the reliable detecting methods for the starting point for the welding.

In order to solve the above problems, we have investigated how a visual sensing system should be applied to the arc welding system. Under an important and practical assumption that some connection data among the structural parts, including their positions and directions of their contact points and lines, can be obtained from their CAD data, we have presented some image processing algorithms to reliably detect the start and end points of the welding, to measure the welding gap, and to find welding torch paths at the center of the welding gap. Finally, we have presented a result of the tracking on the weld line by a video camera placed on three-axis motor driven stage.

TEST PIECE AND IMAGE CAPTURE

A test piece is prepared in assuming the automation of the arc welding of steel tube transmission towers, and placed on turning rollers of the XYZ-θ motor driven stage as shown in Fig. 1. The test piece consists of, one steel tube whose diameter is ϕ216mm, three ring stiffener plates, and two rectangle gusset plates. The tube and plates are tacked with weld tacks whose lengths are from 30 to 60mm. The weld regions exist in straight and curved segments among the tube and the ring plates.

The test piece is rotated by the turning rollers as one of the gusset plates is directed in the horizontal position. The image is captured from 0.8m above the test piece by a CCD video camera. The image has 512 pixels in horizontal and 480 pixels in vertical, and each pixel has 256 gray levels. The actual size of the image corresponds to 180mm x 170mm. Ordinary fluorescent lamps on the ceiling are used as illumination.

RELIABLE DETECTION OF WELD LINES AND TERMINAL POINTS

There exists a fair amount of fluctuation of illumination in actual work sites, and there exist rust, assembly chalk markings, and bolt holes on the surface of weld parts. Also there exist weld tacks on the welding paths. Those external factors would make it difficult to reliably detect the weld lines and terminal points (as the starting and ending points) by local image processing methods. Let's consider applying Hough transform, which is a global processing and can detect specified lines in a noisy image [8-9].

One straight line detection

Let's consider reliable detection on one prescribed straight weld line. Fig. 2(a) shows one of example images on straight welding lines. The tube in the lower of the image is tacked with the gusset plate in the upper. Now, let's presume that an actual weld line (in horizontal direction) must exist inside a prescribed interested area (such as 200 x 512 pixels in Fig. 2(b)) no matter what, the actual positions will move away from the ones obtained by CAD data, due to error of machining, positioning and thermal strain during the welding. Then we show a simple method to reduce computing power in Hough transform. The method uses a window to confine the interested area as follows.

a) Edge detection

To find one straight line, its vertical edges are detected by directional derivative operators whose size is 3 x 3 pixels in Fig. 2(b) [10]. Fig. 2(b) shows a lot of edge pixels are detected in the welding gap boundaries, the weld tack and the surface of the tube.

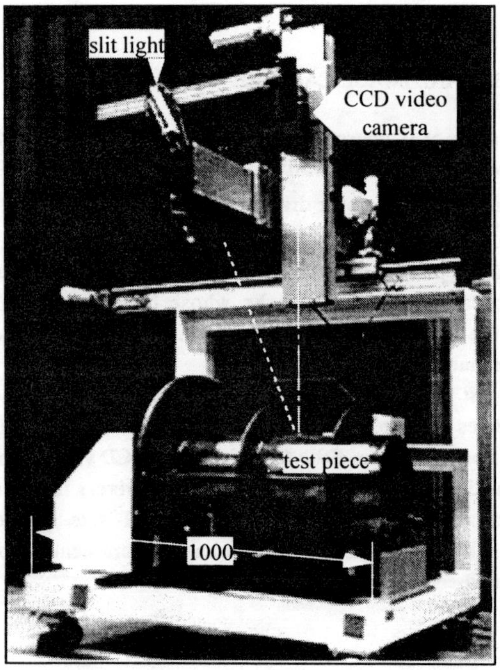

Fig. 1 Overview of experimental system.

The thresholding level L is chosen as follows, in order to avoid excess and deficiency of detected edge pixels.

$$\text{edge pixel } (i,j) = \{ (i,j) \; ; (A_{ij} \geq L) \}$$

$$L = \text{maximum}\left(\sum_{k=L}^{255} N_k \geq M_1 \right) \tag{1}$$

where A_{ij} is the intensity after the derivative processing, N_k is the number of pixels whose intensity is k, and M_1 is a specified value depending on the length of the detected straight line.

b) Local window

Detected edge pixels involve edge pixels in the weld tack and the rust on the tube as shown Fig. 2(b). Then, in order to detect only the area near the weld line, a histogram of edge pixels (corresponding to every row in Fig. 2(b)) is calculated and its peak Y-coordinate is obtained. Then, a local window, whose width is more narrow than that of the interested area (60 pixels in Fig. 2(c)), is set around the peak.

c) Hough transform

Hough transform must prepare a two-dimensional accumulator array for voting, and it requires an enormous amount of calculation and memory. However, since Hough transform could be executed only inside the local window, the computing power can be fairly

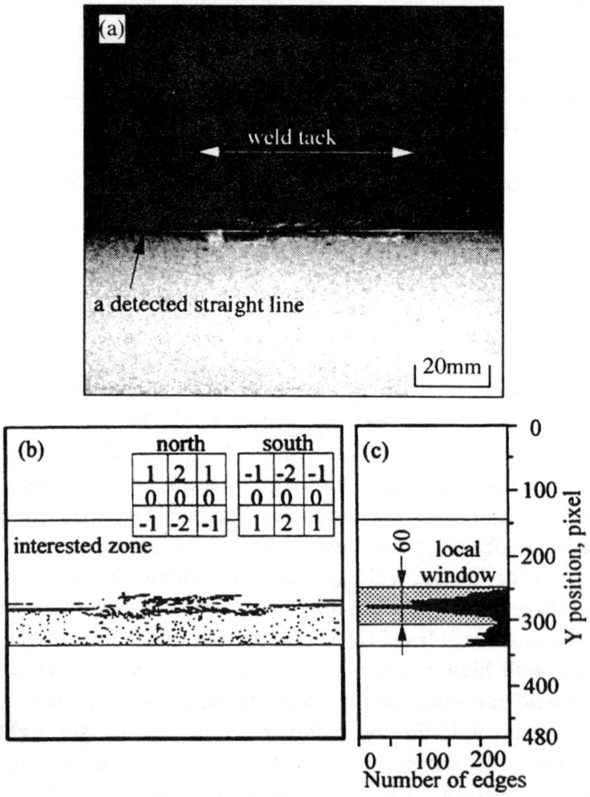

Fig. 2 Straight line detection by two-dimensional Hough transform. (a) an original image. (b) detected edge pixels within an interested area. (c) a peak detection in a histogram of edge pixels in Y position and a window setting.

reduced. The detected line is shown in Fig. 2(a) as a white solid line. It notes that the line is located on the weld gap boundary of the tube side. If the direction of the weld line doesn't direct towards the horizon or the perpendicular, Affine transform should be recommended in order to rotate the area interested, as the direction of the weld line coincides with the horizontal direction.

Terminal point detection

Let's consider the terminal point where weld lines intersect each other. Fig. 3(a) shows that a gusset plate and a ring stiffener are tacked to a main tube, and three weld lines (a-b, c-b, d-b) are intersected at the point 'b'. The weld line a-b is very distinct, but the other weld lines c-b and d-b are not so distinct. The point 'b' is the terminal point where the welding starts or ends, or the direction of the welding may change. At the same time, the point 'b' might be a standard point for checking whether the structural parts are be tacked correctly into the main tube before the welding. For these two purposes, let's consider reliable detection of the point 'b'.

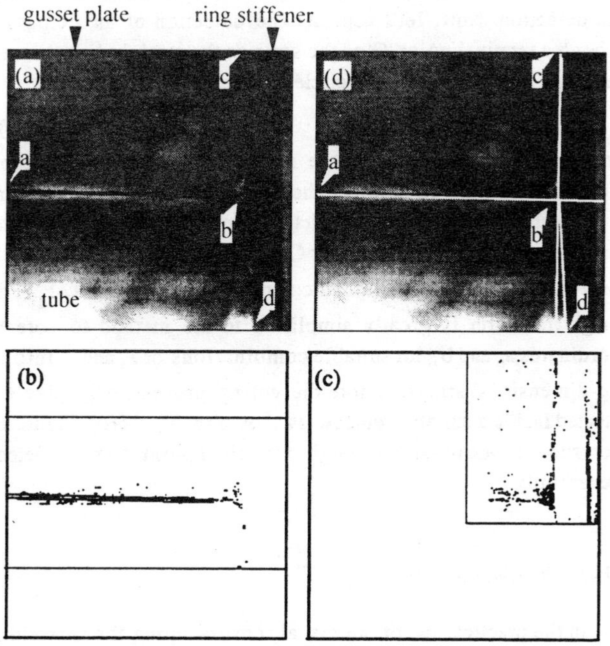

Fig. 3 Detection of terminal points. (a) an original image. (b) and (c) edge detection for 'ab' and 'cb' weld lines, respectively. (d) a detected terminal point 'b' and detection of a straight line 'db' through a point 'b'.

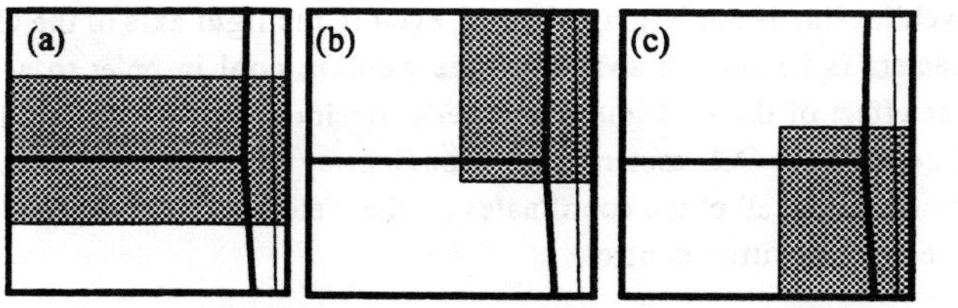

Fig. 4 Interested area for each weld zone.

Table 1 Coordinates of interested area for each weld zone.

weld line pattern	window area
tube - gusset plate	(0,115) - (511,365)
gusset plate - ring stiffener	(286,0) - (511,290)
tube - ring stiffener	(286,190) - (511,479)

First, on basis of the connection data among the tube and the plates, an interested area for each weld line to be detected can be prepared as shown in Fig. 4. Table 1 shows a connection pattern and coordinates for each weld line (an original point is the upper left in the image).

Second, straight lines a-b and c-b for their interested area are detected by using the same method of the previous section. The intersection of the detected straight line becomes the terminal point as shown in Fig. 3(d). However, there may appear several major histogram peaks for edge pixels as shown in Fig. 3(c). In that case, it is necessary to add information that the left major histogram peak is valid. By using this method, coordinates of all terminal points in the test piece can be checked according to the order of the welding.

Detection of a weld line passing through a given point

Now, let's consider to detect the weld line from the first terminal point to the next terminal point. If the two terminal points don't exist in one image, the detection area must be divided into several sections as shown in Fig. 5. The size of divided sections depends on the degree of the deviation from the straight line connected two

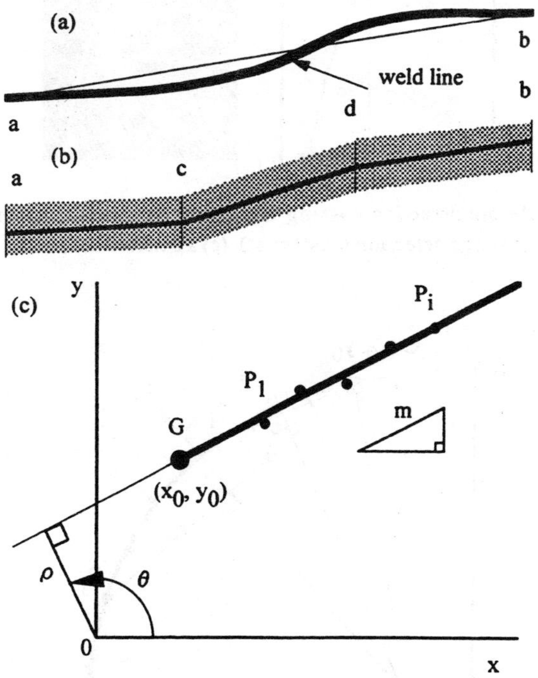

Fig. 5 A division of a weld line 'ab' and detection of a straight line passing through one point. (a) a weld line between terminal points 'a' and 'b'. (b) divided sections and detected straight lines. (c) a straight line passing through a given point 'G'.

points. If the divided weld line 'a-c' can be approximated into a straight line or an arc, the end point 'c' can be obtained as the end point of the straight line or the arc passing through the point 'a'. Also the point 'd' for the divided weld line 'c-d' can be obtained by using the obtained point 'c'. In order to detect a straight line and an arc passing through a given point, Hough transform has been greatly simplified as follows.

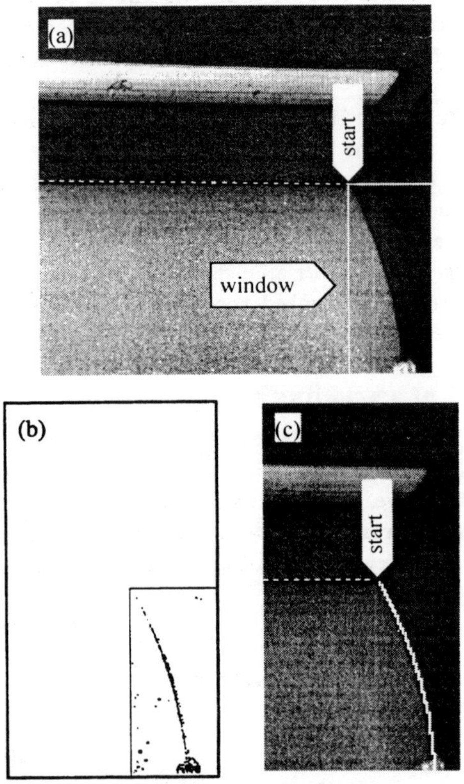

Fig. 6 An arc detection passing through one point. (a) an original image. (b) edge detection (east+west). (c) a detected arc (a white solid line).

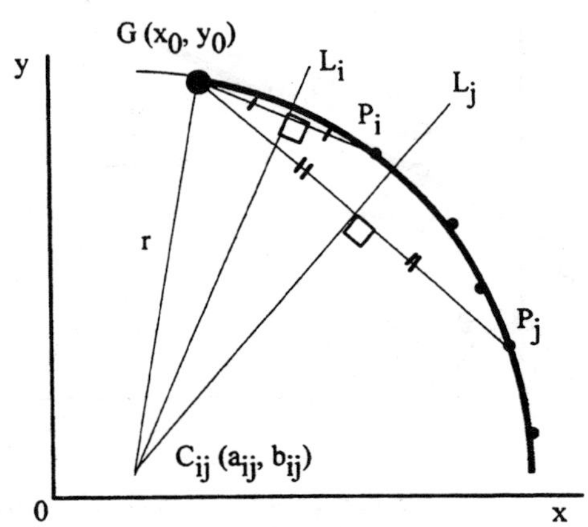

Fig. 7 Detection of an arc passing through a point G.

a) A straight line detection

In Fig. 5(c), a straight line P passing through a given point 'G (x_0, y_0)' with a slope 'm' is defined as follows,

$$(y - y_0) = m(x - x_0) \tag{2}$$

Let $P_1,\dots,P_n$ be a set of detected edge pixels. If the coordinates of each pixel P_i $(1 \le i \le n)$ are (x_i, y_i), each slope of the straight line passing through the points G and P_i is given as follows

$$m_i = (y_i - y_0) = (x_i - x_0) \tag{3}$$

Thus, the execution of Hough transform can be greatly simplified by using a one-dimensional accumulator array for the voting. A straight line 'd-b' in Fig. 3(d) is a result of detecting a straight line passing through the terminal point 'b'.

b) An arc detection

An intersection of the tube and the ring stiffener is observed as an arc with a large curvature with looking sideways as shown in Fig. 6(a). In general, the detection of an arc by Hough transform requires a three-dimensional accumulator array, and much larger storage memory and computer power are required than those of the straight line detection. Now, let's consider the detection of an arc passing through a terminal point 'G' as the same as the straight line detection (see Fig. 7). The formula of a circle is given as follows,

$$(x - a)^2 + (y - b)^2 = r^2 \tag{4}$$

Since the point 'G' is known, the number of the unknown becomes two. Moreover, the direction of the center and the rough diameter of an arc to be detected can be predicted by using the connection data. Also the coordinates of the center C (a_{ij}, b_{ij}) can be obtained from the point intersected by two perpendicular bisectors G-P_i and G-P_j. Then, Hough transform is greatly simplified to the method to vote two coordinates a_{ij} and b_{ij} for suitable combinations of P_i and P_j into two one-dimensional arrays. Before the voting process, edge pixels are detected inside a suitable window as shown in Fig. 6(b). Figure 6(c) shows that an arc passing through a terminal point 'start' is detected successfully.

WELD TRACKING

In the practical use of the visual sensors during the arc welding, it is very important how the effect of the arc light should be reduced. For the welding of large scale structures, it is considered that amount of the relative deformation among terminal points in a narrow section such as that of the present test piece must be small, although the structural parts can be deformed by thermal strain during the welding. This means that the precision of the tracking on the welding line is not largely affected, even if the light axis of the CCD camera is far enough away from the welding pool in order to avoid the effect of the arc light. Then, let's consider tracking the welding line when the CCD camera is set enough away from the welding pool, or measuring all of the coordinates on the welding line in advance for the torch position control.

Measurement of welding gap

It is necessary to change the welding conditions on the weld tack and on the wider weld gap. Now, let's consider checking the existence of the weld tack on the weld line, then, measuring the weld gap where the weld tack doesn't exist as follows.

a) Detection of the weld tack

Fig. 8 shows an intensity profile along the weld line shown in Fig. 2(a). The intensity on the weld line is estimated by the minimum intensity in the 7 pixels across the weld line, and the intensity on the weld gap is the intensity on the straight line jointing the both ends of the weld tack. It is found that the intensity

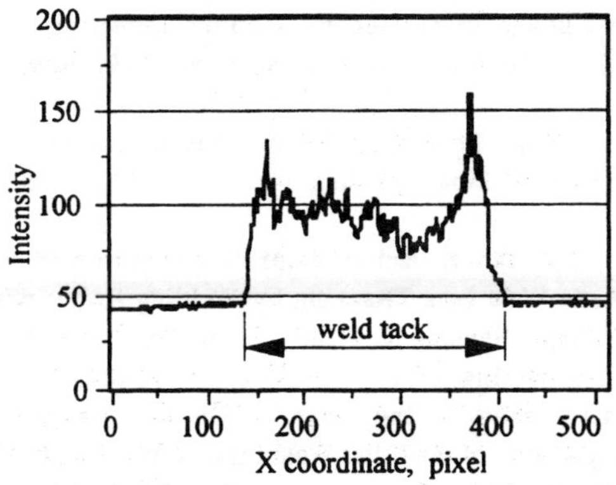

Fig. 8 An intensity profile on a weld tack (see Fig. 2(a)).

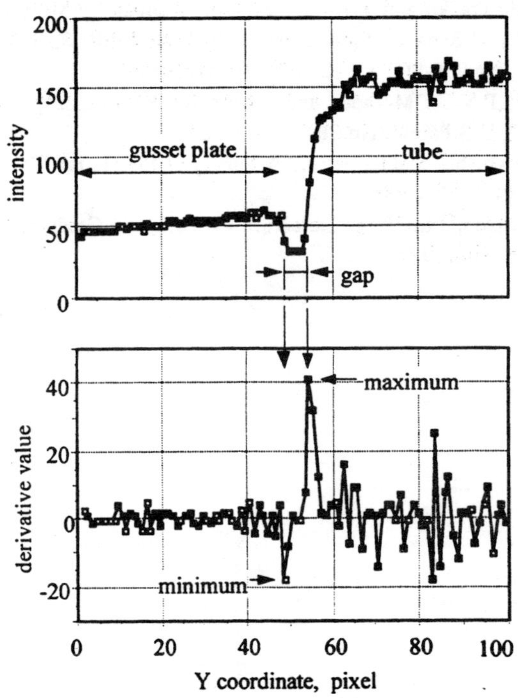

Fig. 9 Weld gap measurement. (a) an intensity profile around a weld gap. (b) detection of weld gap boundaries by using derivative operation.

on the weld tack changes violently, and the intensity on the weld tack is fairly larger than the one on the weld line. Then, the both ends of the weld tack are estimated as the points where the intensity exceeds a threshold level L_{th} along the weld line and the intensity change exceeds a threshold ΔL_{th}.

The coordinates on the straight line jointing the both ends are used for the torch control.

b) Measurement of the weld gap

Figs. (a), (b) show an intensity profile across the weld line and a profile of the derivative among neighborhood pixels. As the intensity decreases in the weld gap drastically, the weld gap is estimated as a distance between two points where the derivatives become the maximum and the minimum as searching from the minimum intensity point to the outside.

Tracking on the weld line

Although a line passing through the center of the measured weld gap seems to be a precise weld line, the both estimated ends of the weld gap can be affected considerably by the irregularity on the plate edge as shown in Fig. 10. The irregularity may produce an irregular metal reflection at the plate edge and cause a variation for the estimation of the ends of the weld gap. Then, in order to reduce such

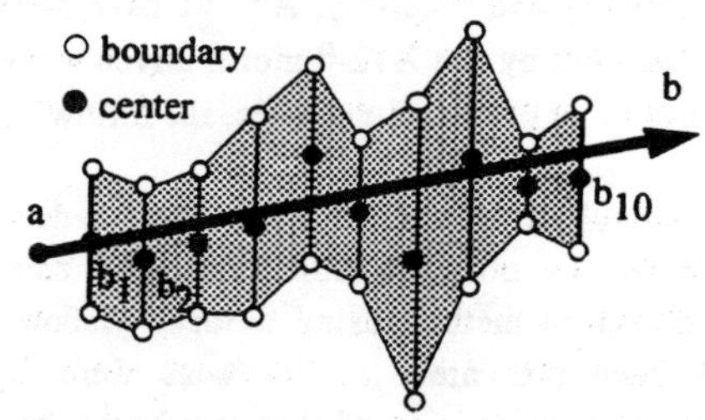

Fig. 10 Detection of a center weld line for a weld tack.

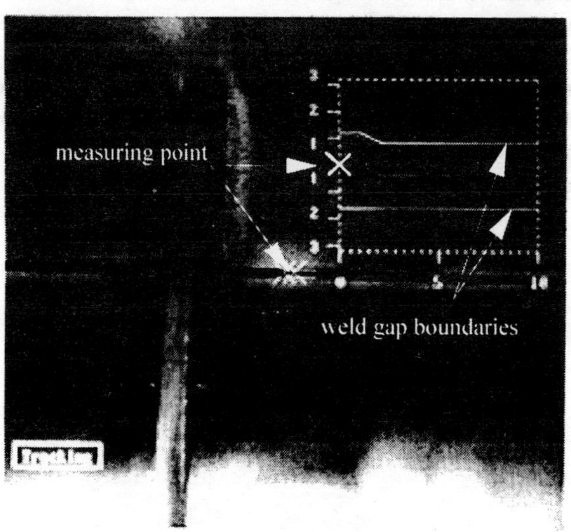

Fig. 11 A monitor display of the weld gap measurement and the weld line tracking.

a variation, the precise weld line for the torch control is estimated as follows. For the region of 10 pixels ahead of the present torch position 'a', the centers of the weld gap b_1,..., b10 are estimated. A weld line vector $\overline{ab}$ is obtained as a straight line passing through a given point in the previous section. As a result, the torch can be controlled smoothly by repeating this operation at proper intervals.

Fig. 11 shows one of images in the monitor display during a real time weld line tracking by the XYZ-θ motor driven stage in Fig. 1. In the figure a cross mark on the weld line shows the present torch position to be controlled. The superimposed figure shows the measurement of the weld gap. The vertical axis shows the distance from the center of the weld gap in unit of 'mm', and the horizontal shows the distance from the present torch position in unit of 'pixel'. One pixel corresponds to 0.35mm. From the superimposed figure, it is found that the boundaries and centers of the weld gap are precisely estimated, and a precise and reliable weld line tracking can be expected.

CONCLUSIONS

In the automation of arc welding for large scale structures, we have investigated that how visual sensors should be applied to the arc welding system in order to detect the weld line and to control the welding torch reliably and precisely. And we have presented a real time weld line tracking by the XYZ-θ motor driven stage in order to confirm the validity of proposed methods, the following results are obtained.

(1) Under a practical assumption that connection data among the structural parts can be obtained from their CAD data, a reliable straight line detection method using a local window and Hough transform has been presented. At the same time, by detecting terminal points reliably, it is found that coordinates of all terminal points can be checked before the arc welding.

(2) Methods to detect straight line and an arc passing through a terminal point given have been presented by simplifying Hough transform.

(3) A stable detection method of the center of the weld line for the torch control has been presented. By applying to the real time weld line tracking, the validity of the proposed methods has been confirmed.

ACKNOWLEDGEMENT

The authors would like to thank Dr. V. Lashkia at Okayama Science University, and Dr. M. Kikuchi at Tomoe Corporation who provided valuable assistance and discussion.

REFERENCES

[1]Edited by Japan Welding Society:"The Basis of Welding and Joining Engineering", Maruzen, 77, 1993 (in Japanese).

[2]Kawahara,M. and Taki,K.:"Tracking Control for Guiding Electrodes along Joints by Pattern Detection of Welding Groove", Transactions of the Society of Instrument and Control Engineers, 15-4, 492, 1979 (in Japanese).

[3]Inoue,K.:"Image Processing for On-Line Detection of Welding Process (I)", J. of Japan Welding Society, 49-9, 609, 1980 (in Japanese).

[4]Inoue,K.:"Image Processing for On-Line Detection of Welding Process (II)", J. of Japan Welding Society, 50-11, 1118, 1980 (in Japanese).

[5]Suga,Y., Takahara,K. and Ikeda,M.:"Recognition of Weld Line and Automatic Weld Line TRacking by Welding Robot with Visual and Arc Voltage Sensing System", J. of the Japan Society for Precision Engineering, 57-6, 1060, 1992 (in Japanese).

[6]Kitaoka,T., Sano,Y. and Suga,Y.:"On the Recognition and Automatic Tracking of the Fillet Weld Line by Welding Robot with a Visual Sensor", Proceeding of the 70th JSME Spring Annual Meeting, No.930-9(IV), 109, 1993 (in Japanese).

[7]Saito,K., Ishii,H. and Suga,Y.:"On the Recognition and Automatic Tracking of T-Joint Weld Line of pipes by Mobile Robot with a Visual Sensor", Proceeding of the 70th JSME Spring Annual Meeting, No.930-9(IV), 112, 1993 (in Japanese).

[8]Hough,P.V.C.:"Methods and Means for Recognizing Complex Patterns", U.S.Patent, No.3069654, 1962.

[9]Illingworth,J. and Kittler,J.:"A Survey of the Hough Transform", CVGIP, 44-1, 87, 1988.

[10]Robinson,G.S.:"Edge Detection by compass Gradient Masks", CVGIP, 6, 492,1977.

Proceedings of the Seventh (1997) International Offshore and Polar Engineering Conference
Honolulu, USA, May 25-30, 1997

ISBN 1-880653-28-1 (Set); ISBN 1-880653-32-X (Vol. IV)

Gas Removal and Separation for Underwater-Welding Applications with Shielding Gas Recycling

M. Creutz and D. Mewes
University of Hannover
Hannover, Germany

ABSTRACT

In this paper a combination of a shielding technique for underwater welding and a technique for shielding gas recycling is presented. Both of the techniques are based on centrifugal forces imposed to a two-phase fluid by a rotating impeller and are combined in one tool. Special emphasis is laid on the separation of the gas-water mixture that is removed from the welding area. Experimental and theoretical investigations on the separation quality and on the dynamic behaviour of the separator are presented. The shielding technique and the separation technique have been combined and the technical operation of the combined tool was proved.

KEY WORDS: Underwater welding, gas recycling, liquid-gas separation, gas removal, shielding technique, two-phase flow, electrical measurement techniques

NOMENCLATURE

A	m²	pipe cross section area
D	-	attenuation coefficient
h	m	liquid level inside the impeller
k	1/m⁴	pipe friction factor
K	-	dimensionless pipe friction factor
L	m	pipe length
p	N/m²	pressure
R	m	radius of the impeller
t	s	time
T	s	time constants
$\dot{V}$	m³/s	volumetric flow rate
x	m	height of the impeller
ρ	kg/m³	density
ω	1/s	angular velocity of the impeller

Subscripts

0, 1, 2	ambient, inlet and outlet conditions
L	liquid

INTRODUCTION

Underwater welding is encountered mainly in the maintenance of underwater structures such as pipelines and platform legs in the offshore industry. Starting out from manual metal arc welding processes in direct contact to the water or inside hyperbaric diving chambers (Ibara, et al. 1990, Sanchez-Osio, et al. 1993, Szelagowski et al. 1993), recent investigation focus on automated underwater welding techniques with shielding techniques that remove water from the welding zone. These shielding techniques allow for low cooling rates and thus improve the weld quality. Also the hydrogen formation and the resulting hydrogen cracks are minimised by avoiding a direct contact of water and the plasma arc. Recent investigations on different kinds of shielding techniques have been carried out by Arata et al. (1987), Hoffmeister et al ((1991), Küster et al. (1987), Hamasaki and Sakakibara (1984, 1978), Shinda et al. (1980, 1982), Coriatt and Roggen (1979), Draugelates et al. (1993, 1994), and Dos Santos et al. (1990, 1992). Most shielding techniques are based on water or/and gas impulse used for the removal of the water. The hydrostatic pressure corresponding to the water depth is not drastically reduced. Due to these facts, the mass flow of the plasma, shielding and focus gas is much higher in underwater welding than under atmospheric conditions. The fact that these gases are often costly inert gases (argon is commonly used) is one reason for the development of a gas removal, gas cleaning and recycling equipment.

Recently, underwater welding is also being in discussion for the maintenance of nuclear power plant structures such as pipes, liners and other metal structures. This application requires a complete removal of all kinds of contaminants within the gas and the surrounding water. Especially a free movement of hydrogen that is originated from water dissociation in the surrounding of the plasma arc has to be restricted.

In this article a concept is presented that combines a shielding technique, a local pressure reduction in the welding area and a gas removal system. The shielding technique and the local pressure reduction was developed by Steinkamp (1994) and Creutz et al. (1995). A centrifugal pump impeller is used to yield a pressure reduction and to prevent water from leaking into the welding area. The concept is shown schematically in Fig. 1. A rotating cylinder is placed around the welding torch. A disc is fixed to the underneath of the cylinder which rotates closely above the work surface. Using blades to increase the momentum transfer to the two-phase fluid, this disc is similar to the rotor of a radial pump, where the work surface corresponds to the stator. Inside the rotating cylinder, different shielding techniques (wire-brush, gas or liquid nozzles etc.) can be applied additionally. This system was shown to work well in combination with a plasma-MIG arc welding system. Weld qualities have been achieved that are close to those under atmospheric conditions (Creutz et al. 1995).

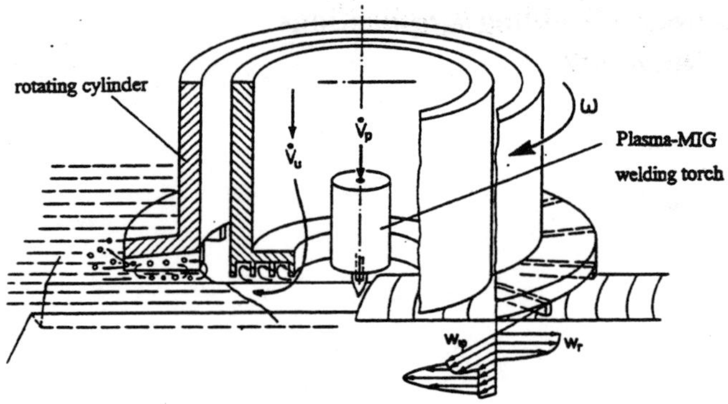

Fig. 1: Scheme of the water removal technique and the pressure reduction

The recycling concept is based on the above mentioned shielding and pressure reduction technique. At the outside of the rotating cylinder where the gas is dispergated in the surrounding water, the two-phase gas/water mixture has to be removed. The gas has to be separated from the liquid before it can be cleaned, split into the different gas components and recycled. The separation has to be carried out under highly transient conditions. Since the welding tool may operate in different working position with respect to gravity, a simple gravity separator would cause large problems if placed under water as well. For this reason, a separator is developed which is also based on centrifugal forces. The separator can be applied as an additional tool to the rotor of the shielding technique. This paper will focus on the separation technique and on the combination of the separation and the shielding technique.

TRANSIENT GAS-LIQUID SEPARATION AND THE COMBINATION WITH THE SHIELDING TECHNIQUE

In Fig. 2 the gas liquid separator is shown schematically. It consists of a stationary casing and a rotating shaft with an impeller. In contrast to the impeller of a centrifugal pump, this impeller is open on both sides. The gas-liquid mixture enters the separator in the axial direction. Typically, an intermittent two-phase flow pattern is observed at the inlet depending on the position of the separator with respect to gravity. Inside the impeller the mixture is set to rotate and the liquid is forced in the radial direction by means of centrifugal forces. The gas leaves the separator in the axial direction and may be further cleaned. The liquid is forced into a ring chamber from where it can be pumped to the water level or can simply flow out to the surrounding through a nozzle. Regarding the separation technique the reaction of the separator on transient mass flow rates at the inlet, i.e. the controlling of the system is of major interest:

Inside the impeller, a gas-liquid interface is developed. The position of that interface has a major influence on the pressure inside the ring chamber at the outer circumference of the impeller and vice versa. The inherent controlling of the separator system relies on the movement of this interface: Given a sudden rise in the liquid flow rate entering the separator, the interface will move radially towards the rotation axis. This movement enlarges the pressure in the ring chamber and thus leads to a higher water flow rate through the following pipe or nozzle. In other words, the radial position of the interface adopts itself to a given liquid flow rate and a given back pressure in the ring chamber.

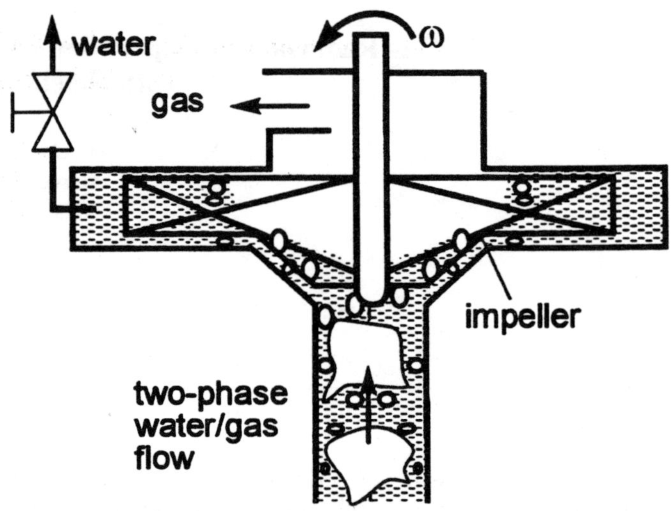

Fig. 2: Scheme of the gas-liquid separator

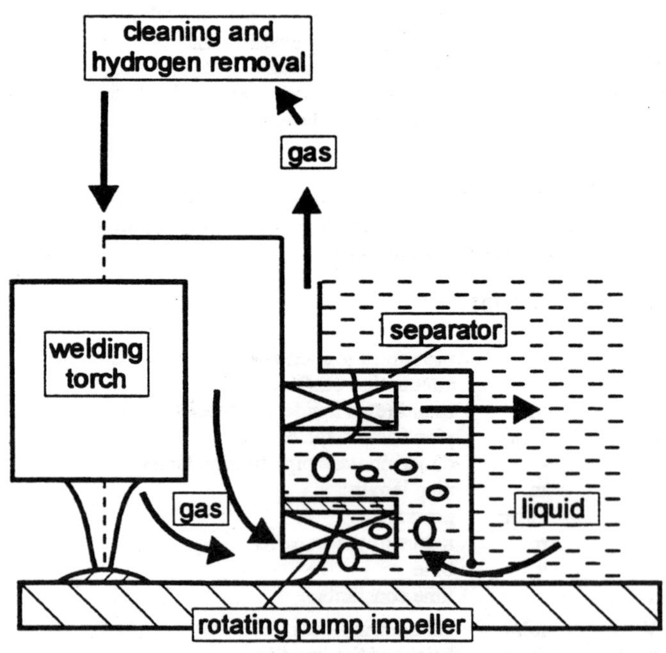

Fig. 3: Scheme of the combined shielding and separation technique

The concept of a combined tool including the rotating pump as a shielding technique a gas removal and the gas-liquid separator is schematically shown in Fig. 3. The shielding technique consist of a pump impeller and is mounted around the welding torch. A stationary shielding technique might be used between the welding torch and the impeller as shown in Fig. 1. Around the rotating cylinder, a stationary cylinder leads for the capture of the gas bubbles that are torn away from the impeller. The pressure inside this cylinder should be below the ambient pressure, leading to a liquid transport to the inside and preventing the gas from flowing out. The two phase mixture is lead to a second impeller which acts as the gas-liquid separator. The gas can then be led to further cleaning and hydrogen removal and recycled after being compressed.

DYNAMICS OF THE SEPARATOR

In Fig. 4 a scheme of the impeller, the ring chamber and a liquid pipe is plotted. For a given liquid flow rate in the separator entrance $\dot{V}_{L1}$, the radial position of the gas liquid interface h (liquid level), the back pressure of the separator p_2 and the flow rate at the outlet $\dot{V}_{L2}$ can be calculated. This calculation will be used to determine the mean, stationary liquid level as well as the dynamic behaviour of the system. The governing equations are the continuity equation of the liquid

$$2\pi x(R-h)\frac{\partial h}{\partial t} = \dot{V}_{L1} - \dot{V}_{L2} \quad , \qquad (1)$$

the Euler-equation of the impeller

$$p_2 - p_1 = \frac{\rho}{2}\omega^2\left(2hR - h^2\right) \quad , \qquad (2)$$

and the pressure drop in the liquid pipe

$$p_2 - p_0 = \frac{\rho}{2}k\dot{V}_{L2}^2 + \frac{\rho L_{pipe}}{A_{pipe}}\frac{\partial \dot{V}_{L2}}{\partial t} \qquad (3)$$

In these equations, k denotes the turbulent friction factor of the liquid pipe, ρ is the liquid density and $\omega = 2\pi f$ is the angular velocity of the impeller. The different volumetric flow rates, pressures and geometric quantities can be seen in Fig. 4. It is assumed that the gas is completely removed from the liquid, that there is no variation in the angular velocity and that the radial liquid velocity inside the impeller is small compared to the circumferential velocity and thus can be neglected.

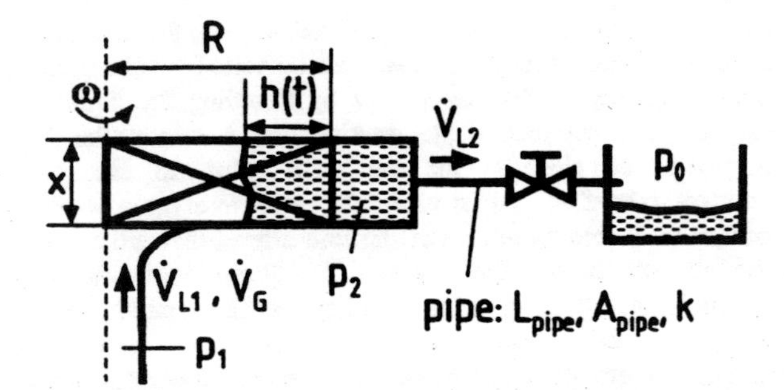

Fig. 4: Scheme of the dynamic system

Eq. 2 shows that the pressure difference p_2-p_1 depends on the liquid level h. The liquid level adapts itself to a given liquid flow rate and a given pressure difference and thus inherently controls the system. The liquid level reacts quickly and allows for separating intermittent inlet flow patterns such as slug and plug flows.

Eq. 1-3 are non linear and have to be solved numerically for a given time dependent liquid flow rate at the entrance with a Runge-Kutta method. The numerical solution are compared to experimental investigations on the transient behaviour of the separator.

If the deviations of the time dependent properties remain close to a stationary operational point, deeper insight in the controlling aspects of the separator can be gained by solving Eq. 1-3 analytically. By linearizing them at the operation point, a relation can be found between the dimensionless pressure p_2 and the dimensionless volumetric flow rates that enter and that leave the separator (see Appendix for the definition of the coefficients and the dimensionless quantities):

$$\dot{V}_{L1}^* = \dot{V}_{L2}^* + T_1\frac{\partial \dot{V}_{L2}^*}{\partial t} + T_2^2\frac{\partial^2 \dot{V}_{L2}^*}{\partial t^2}$$

$$K\left(\dot{V}_{L1}^* + T_D\frac{\partial \dot{V}_{L1}^*}{\partial t}\right) = p_2^* + T_1\frac{\partial p_2^*}{\partial t} + T_2^2\frac{\partial^2 p_2^*}{\partial t^2} \qquad (4)$$

The transient behaviour of the liquid flow rate which is described by the left hand side of the equations, depends on the flow pattern entering the separator. The right hand side describes the reaction of the separator. The time constants of the system T_1 and T_2 describe possible oscillations. The attenuation of the system is determined by the attenuation coefficient

$$D = \frac{T_1}{2T_2} = \frac{p_2 - p_1}{\omega \dot{V}_L}\sqrt{\frac{\pi x A_{pipe}}{2\rho^2 L_{pipe}}} \quad . \qquad (5)$$

If the flow entering the separator is a transient slug flow, a reduction of the angular velocity ω results in a steadier liquid outflow, lower variations in the liquid level and less gas remnants in the liquid.

EXPERIMENTAL INVESTIGATIONS ON THE SEPARATION QUALITY

Experimental investigations have been carried out on the quality of the gas-liquid separation. Of special interest is the rate of gas that leaves the separator through the liquid outlet and which is therefore lost for recycling purposes. These investigations have been carried out omitting the welding and the shielding technique. In Fig. 5 the experimental set up is shown. Water is pumped in a loop from a vessel through a thermostat. It is then mixed with gas from a supply line and led into a separator. The gas leaves the separator into the surrounding. By means of a nozzle at the gas outlet, the pressure of the whole system can be set to up to 5 bars. The liquid is lead back to the vessel where remnants of gas are separated by gravity. The liquid flow rate and the gas fraction inside the liquid flow is measured by means of a coriolis flowmeter. The accuracy of the determination of the gas fraction is below 0.1%. Pressures are measured at the inlet and at the outlet of the separator. The separator is made from stainless steel. Eight impeller blades are fixed at an outlet angle of 90 degrees. The outer diameter of the impeller is 120 mm, the impeller is closed at the sides except for the entrance region and the gas outlet. It can be set to a rotation of up to 50 Hz. The gas flow rate varies between 0.1 and 4 L_n/s.

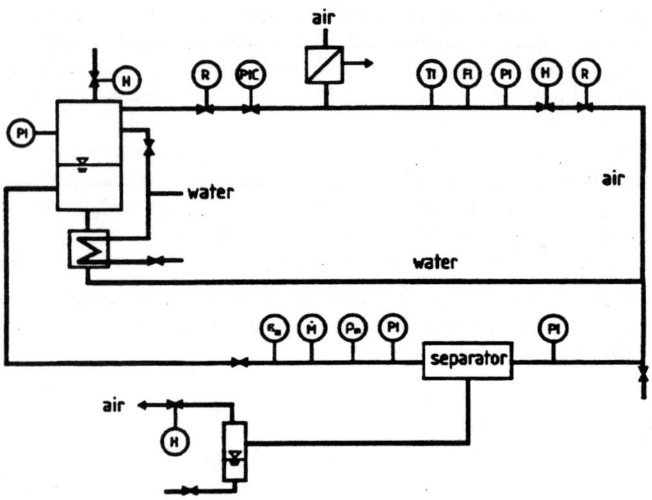

Fig. 5: Experimental set up for the investigations on the separation quality

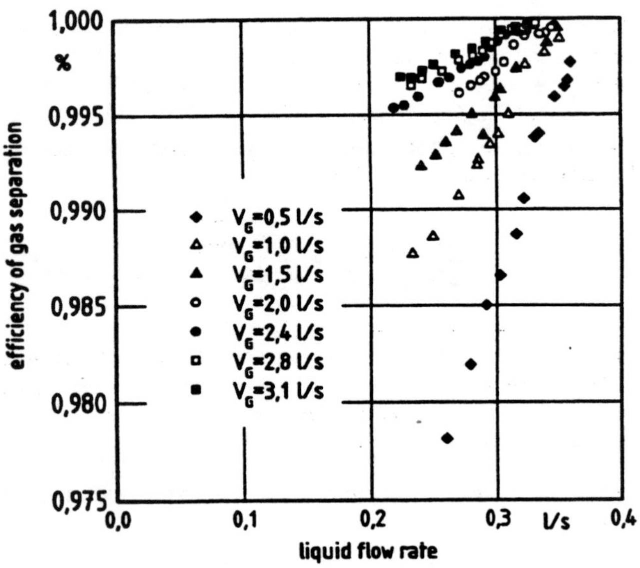

Fig. 6: Efficiency of the gas separation versus volumetric liquid flow rate

In Fig. 6 the efficiency of the gas separation is shown as a function of the liquid and the gas volumetric flow rate. The efficiency is defined as the fraction of the separated gas flow rate to the entering gas flow rate. The gas is separated at 98 ... 100 %. The void fraction in the liquid outlet is plotted versus the liquid flow rate for different gas flow rates and rotational velocities in Fig. 7. In contrast to the efficiency, the void fraction is shown to depend only little on the gas flow rate. The void fraction and thus the quality of the separation depends on the liquid flow rate and the rotational velocity of the impeller. By calculating the liquid level h inside the impeller, the two parameters can be reduced to one parameter. All measurement points shown in Fig. 7 are included in Fig. 8 as circles. Here, the void fraction is plotted versus the dimensionless liquid level $h^* = h/R$ in the impeller which is calculated for a measured pressure difference p_2-p_1 according to Eq. 2. The void fraction decreases with increasing liquid level. At the maximum liquid level of 0.58 where the radius of the gas outlet is reached, the separation efficiency equals 100%.

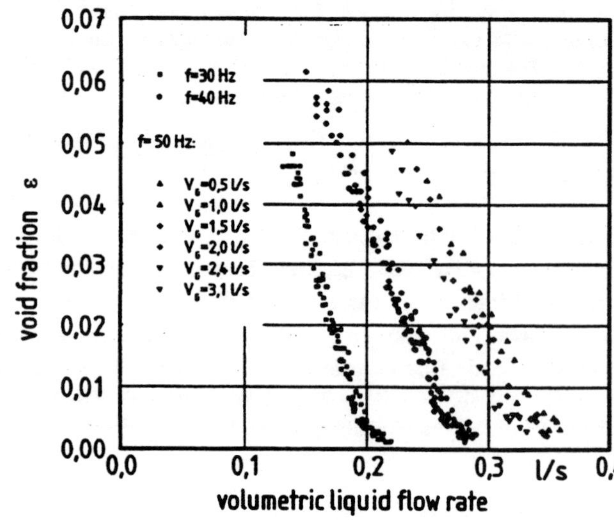

Fig. 7: Volumetric void fraction versus volumetric liquid flow rate

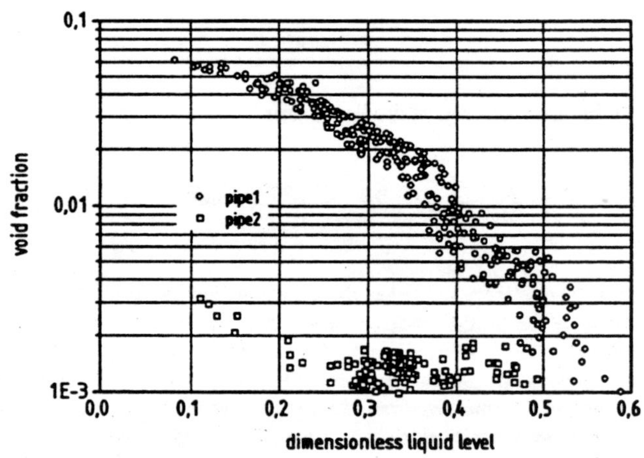

Fig. 8: Volumetric void fraction versus the calculated liquid level

The plot shown in Fig. 8 is a characteristic curve for the separator. In a technical process, the liquid level can be varied by changing the rotational velocity of the impeller or by throttling the liquid pipe behind the ring chamber. If the entering flow is steady, the liquid level should be close to the maximum value. In case of an intermittent inlet flow, the level should be set to a value below the maximum to enable variations of the liquid level that are necessary for the inherent control. The squares in Fig. 8 refer to measurements with a slight modification in the liquid outlet. The liquid has to pass through a cascade, where most remnants of gas are removed. By using this modification, a complete separation between liquid and gas is achieved for almost all liquid levels.

EXPERIMENTAL INVESTIGATIONS ON THE TRANSIENT BEHAVIOUR OF THE SEPARATOR

As mentioned before, the two-phase flow that enters the separator is not stationary. The gas flow changes abruptly on the beginning and the end of a welding process. The liquid flow rate is determined by the gap between the welding and shielding tool and the work surface. This geometry varies. Finally, intermittent flow patterns may occur depending on the position of the welding tool with respect to gravity, where a horizontal position is the worst case. For that reason, experimental investigations on the transient behaviour of the separator have been carried out with the separator mounted horizontally.

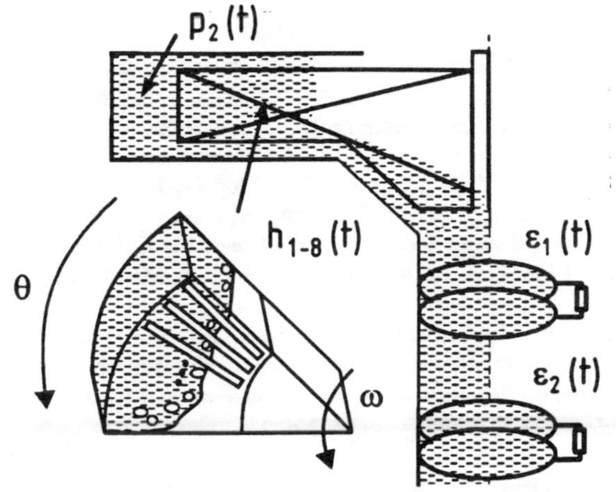

Fig. 9: Dynamic measurement system

The dynamic measurement set-up is shown in Fig. 9. The local liquid fraction inside the pipe at the inlet of the separator is measured with a frequency of up to 1 kHz via two pairs of ring electrodes that are mounted around the supply pipe. These two measurements allow for the calculation of the slug velocity of a two phase gas-liquid flow via a cross correlation. The time averaged liquid flow rate is determined by the coriolis flowmeters at the outlet pipe. The local liquid fraction, the slug velocity and the mean liquid flow rate lead to determining the dynamic liquid flow rate. The liquid level inside the impeller is measured by means of an electrical resistance sensor at eight circumferential positions. A picture of the sensor inside the impeller is shown in Fig. 10. Silver stripes are mounted at both of the insulated impeller covers in radial direction. Parts of theses pairs of stripes are electrically connected via a liquid bridge. The liquid level determines the height of the liquid bridge and thus the electrical resistance between the pair of electrodes. The eight pairs of electrodes are connected to a multiplex system. A three bit code controls the connection of one electrode pair to a Wheatstone resistance bridge. The displacement of the bridge is measured and corresponds to the electric resistance between the electrodes and thus to the liquid level inside the impeller. The electric signals leave and enter the rotating system via high-speed slip rings. The back pressure inside the ring chamber is measured.

In Fig. 11 the flow rate of the liquid that enters the separator and the pressure response at the outlet is plotted versus the time for an intermittent flow at the inlet. The inlet pipe has an inner diameter of 16 mm and is mounted horizontally. For relatively low liquid and gas flow rates, a plug flow exists with high temporal variations in the flow rates. While the inlet flow was kept constant, the damping coefficient was varied by changing the separator frequency. For a low damping coefficient (left diagram) the pressure reacts strongly on variations in the flow rate. By increasing the damping coefficient (right diagram) the back pressure of the separator as well as the flow rate in the outlet are much more constant.

Fig. 10: Photograph of the resistivity sensor and the impeller of the separator

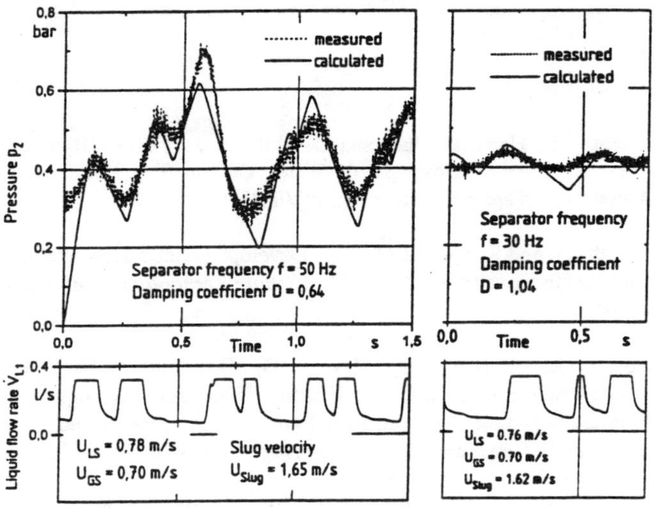

Fig. 11: Comparison of calculated and measured pressures at the outlet of the separator for two different damping coefficients

The non-linear differential equations 1-3 for the measured liquid flow rate at the inlet, the back pressure p_2 is calculated. The calculated values are shown in Fig. 11 and fit the measurements reasonably well.

EXPERIMENTAL INVESTIGATIONS ON SHIELDING GAS RECYCLING

As a preliminary study, the shielding technique and the separation technique have been combined in one test loop. The shielding technique is placed under water inside a tank. Gas is pumped to the inside of the rotating shielding impeller. At the outside of the impeller, a stationary casing captures the gas that is hurled out of the impeller. The two-phase gas/liquid mixture is then let through a pipeline into the separator. The separated liquid flows back to the water tank. The gas outlet is connected to a vacuum vessel leading to the driving pressure difference for the movement of the two-phase flow from the welding area towards the separator. Experimental investigations in this facility show that the combination of both of the techniques works well.

CONCLUSIONS

A separation technique for transient gas/water mixtures has been presented. Equations for the layout of the separator and for transient calculations have been presented. The calculations fit the experimental investigations reasonably well. Furthermore, it was shown that the separation technique can be adopted to a shielding technique for underwater-welding that is based on a rotating pump impeller. This leads to the ability of shielding gas recycling for underwater welding applications. Further investigations will be carried out on the removal of aerosols from the gas and on the separation of the different gas components, mainly on the removal of hydrogen.

ACKNOWLEDGEMENTS

The authors would like to thank the DFG (Deutsche Forschungsgemeinschaft) SFB (Sonderforschungsbereich) 264 for the generous financial support.

REFERENCES

Ararata, Y; Hamasaki, M, and Sakakibara, J (1981). "Hydrogen Content in Arc Atmosphere of Water Curtain Type Underwater Argon Arc," *Welding Transactions of JWRI*

Coriatt, G, and Roggen, R (1979). "Development of a New Welding Gun for Underwater Welding," *11. Offshore Technology Conference*, Houston.

Creutz, M, Bartzsch, J, Mewes, D, Draugelates, U (1995) "Underwater Plasma-MIG Arc Welding: Shielding Technique and Pressure reduction by a Centrifugal Pump, " *5. th Offshore and Polar Engineering Conference ISOPE*, The Hague, vol. 4, pp. 141-146

Dos Santos, JF; Manzenrieder, H; Cui, H., and Recoschewitz, J (1992). "Automatic and Diverless underwater welding: new systems and concepts," *Proc. of the 2nd Int. Offshore and Polar Engineering Conf.*, San Francisco, USA, Volume IV, 1992

Dos Santos, JF, Szelagowski, P; Schafstall, HG, and Dobernowsky, A (1990). "Preliminary investigations on the effect of short circuit variables on metal transfer above 60 bar abs.," *22nd Offshore Technology Conference*, Houston, pp. 127-134

Draugelates, U; Bouaifi, B; and Bartzsch, J (1993). "Underwater Welding using the Plasma MIG Method," *Proceedings of the 12th International Conference on Offshore Mechanics and Arctic Engineering*, Glasgow, pp. 175-181

Draugelates, U, Bouaifi, B, and Bartzsch, J (1994). "Investigations on Local Shielding Systems for Underwater Applications," *Proc. of the 4th Intern. Offshore and Polar Engineering Conference*, Osaka

Hamasaki, M and Sakakibara, J (1984). "Trockenes WIG/MIG-Unterwasserschweißen mit der "wire brush"-Düse," *Unterwasserschweißen und -schneiden, 1. Internationales Symposium Geesthacht.*

Hamasaki, M; Sakakibara, J; Minehisa, S; Sakahata, S, and Arata, Y (1978). "Automatic wet underwater welding at horizontal position of tubes using water curtain type GMA welding technique," *Trans. Jap. Weld. Soc.* 9, pp 69/74

Hoffmeister, H; Pfannkuche, H; Maas, U, and Huismann, G (1991). "Develpment and Testing of a Remote Controlled Local Dry Chamber Underwater GMAW Process Facility," *3. Intern. Symposium on Underwater Technology*, Geesthacht

Ibarra, S; Reed, RL; Smith, JK; O'Connor, PE, and C.E. Grubbs (1990). "Underwater Wet-Welding Techniques for Repair of Higher-Carbon-Equivalent Steels," *Proc. of the 22nd Offshore Technology Conf.*, Paper OTC 6214, S. 135-142, Houston, May 7-10

Küster, K; Hoffmeister, H, and Schafstall, HG (1987) "Hydrogen pick up during welding by using an advanced wet welding technique," *Proceedings of the 2. International GUSI-Symposion - Underwater Technology*, paper No. 9

Sanchez-Osio, A.; Liu, S; Olson MD , and Ibarra, S (1993). "Underwater Wet Welding Consumables For Offshore Applications," *Proc. of the 12th Int. Conf. on Offshore Mechanics and Arctic Engineering*, 1993, Volume III-A, Materials Engineering

Shinada, K, and Tamura, M (1982). "Trockenschweißen unter Wasser mit örtlichem Hohlraum - Entwicklung der Schweißmethode und Ergebnisse," *Schweißen und Schneiden* 34 Vol. 3

Shinada, K, Tamura, M, Nagatov, Y, Omae,T , Wada, H, Manabe, Y and Nabushige, T (1980). "Development of Underwater Welding Techniques," *Welding Research Abroad* 26.

Steinkamp, H(1994). "Rotierende zweiphasige Strömung zur lokalen Druckreduktion beim Schweißen unter Wasser," *Fortschritts-bericht VDI Reihe 7 Nr. 239, Düsseldorf, VDI-Verlag*

Szelagowski, P, VR Dos Santos und CJ Teixeira (1993). "Wet welding qualification trials at 35 msw.," *Proc. of the 12th Int. Conf. on Offshore Mechanics and Arctic Engineering*, Volume III-A, Materials Engineering, S. 133-146

APPENDIX

Dimensionless volumetric liquid flow rate and pressure

$$\dot{V}^*_{L1} = \frac{\dot{V}_{L1} - \dot{V}_{L1,0}}{\dot{V}_{L1,0}} \qquad p^*_2 = \frac{p_2 - p_{2,0}}{\frac{\rho}{2}\left(\frac{\dot{V}_{L1,0}}{A_{pipe}}\right)^2}.$$

Coefficients in Equation 4:

$$K = kA^2_{pipe},\ T_D = \frac{2L_{pipe}}{kA_{pipe}\dot{V}_{L1,0}},\ T_1 = \frac{\pi x k \dot{V}_{L1,0}}{\omega^2},\ T_2 = \sqrt{\frac{2\pi x L_{pipe}}{\omega^2 A_{pipe}}}$$

Proceedings of the Seventh (1997) International Offshore and Polar Engineering Conference
Honolulu, USA, May 25-30, 1997

ISBN 1-880653-28-1 (Set); ISBN 1-880653-32-X (Vol. IV)

Intelligent Image Processing for Abstraction and Discrimination of Defect Image in Radiographic Film

Kimiya Aoki and Yasuo Suga
Keio University
Yokohama, Japan

Abstract

The X-ray radiographic testing method is often used for detecting weld defects as a non-destructive testing method (NDT). Due to the difficulties in identifying small defects from the X-ray film, skilled laborers should be trained. However, recently it has been difficult to employ skilled laborers. Moreover, for the identification process, not only the laborers skill influences the testing result, but also it is difficult for skilled laborers to assess small flaws within a short time. In comparison, computer visual image processing system has some good characteristics, allowing objective assessment, high speed judgment, non-human's error etc.. Therefore, the image processing system allows weld defects to be detected using X-ray radiography in the presence of background noise. This paper deals with an image processing method for displaying defects by computer graphics. Furthermore an application of neural network to discriminate the type of defect is tried. As the result of the investigation, it is confirmed that the system constructed is effective to the abstraction and discrimination of small weld defects clearly.

Key words: Nondestructive inspection, Weld defect, X-ray film, Image processing,Characteristic value, Defect type, Neural network

1. Introduction

Several types of non-destructive testing method are used for detecting weld defects. Because the X-ray radiographic testing method is particularly useful in inspecting the inside of weld metal, it is often used in industries. However, since skilled inspectors for X-ray radiographic testing are gradually decreasing, recently several methods to detect weld defects from films automatically have been investigated to improve the processing efficiency and quantify the inspection results [1-5]. However, the image in a X-ray film shows a very low contrast and involves noise, so that it is difficult to automate the inspection of X-ray films. Therefore, in this study, an image processing system to abstract the indistinct weld defect images and display them clearly was constructed, and moreover an application of the neural network to judge their types from characteristic values of defect images was investigated.

2. Experimental equipment

Fig.1 shows the arrangement of equipment. The system is assembled by a CCD camera, an image digitizer (512×512 pixels, 8bit), a personal computer (CPU: i486) and etc.. Defect images in a X-ray film put on a lighting box are taken by a CCD camera and the image data are transmitted to the personal computer through the image digitizer as digital data. The image data are analyzed and processed in the computer, and the result is shown on a CRT or a monitor.

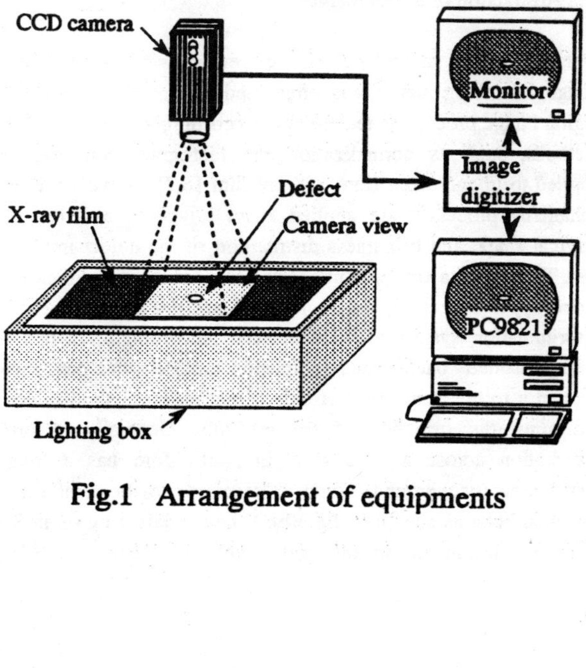

Fig.1 Arrangement of equipments

The films used in this study were acquired by X-ray inspection of welded joint of a steel pipe obtained by the submerged-arc welding process. The films include several defects, such as blowholes, slag inclusions, cracks or incomplete penetrations. An illustration of a typical X-ray film is shown in fig.2.

3. Outline of the system

The data processing system is constructed with two main parts. In the first part, an image of X-ray film is taken by a CCD camera and the defect images are abstracted clearly by some image processing (previous processing). In the second part, the defect type of the defect image abstracted by the previous processing is discriminated by a neural network .

In the previous processing, the best combination of the general image processing was discussed in order to abstract the defect images clearly. Then, the new methods named "special points connection method" and "background subtraction method" were constructed to abstract defect images from a noisy X-ray film. In the discrimination of the type of the defect, the neural network which is studied well in the domain of the pattern discrimination and so on were applied.

The previous processing will be discussed in next paragraph and the discrimination of the defect type in paragraph 5.

4. Previous processing

4.1 Acquisition of image

The area of the X-ray film put on a lighting box including defect images is taken by a CCD camera and the image data are transmitted to the personal computer as digital data (512×512 pixels, 8bit). Then, a window which is a small area in the acquired image for the image processing is set along the weld bead image. The shape of the window is square, and the size is limited to a small area of 128×128 pixels so as to decrease the calculation time for the image processing by a personal computer.

4.2 Abstraction of defect images

When an image taken by a CCD camera shows low contrast, several image enhancing methods is often used to enhance it in the field of digital image processing technology. Accordingly, the image enhancing methods, such as normalization and histogram equalization, were applied to image taken from a X-ray film [6,7]. However, when these enhancing processes are applied to an image of a X-ray film, the original shape and brightness distribution of the defect image may be lost. The lost data are sometimes important for the discrimination of the defect type. Therefore, these processes are not always suitable for a discrimination system of weld defects. Accordingly, in this study a method named "background subtraction method" was constructed and applied it to the abstraction of defect images. Fig.3 shows illustrations indicating the principle of the methods. Generally, a brightness distribution across a weld bead in X-ray film has a long range distribution (background) which corresponds to the reinforcement of the weld bead as shown in fig.3(b). When a defect exists in the weld bead, the brightness of the part which corresponds to the defect

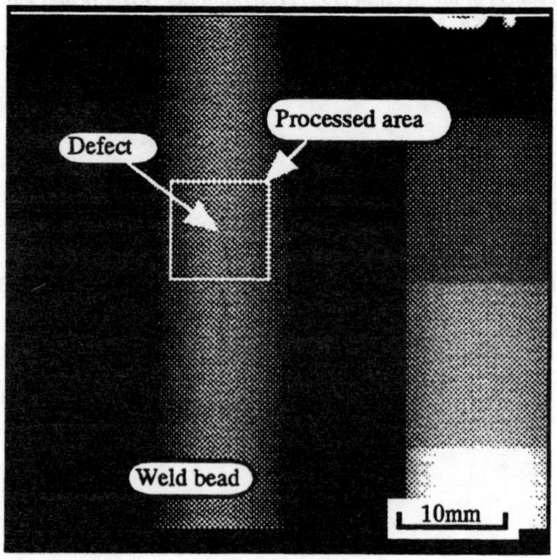

Fig.2 Original X-ray image
Defect ; crack

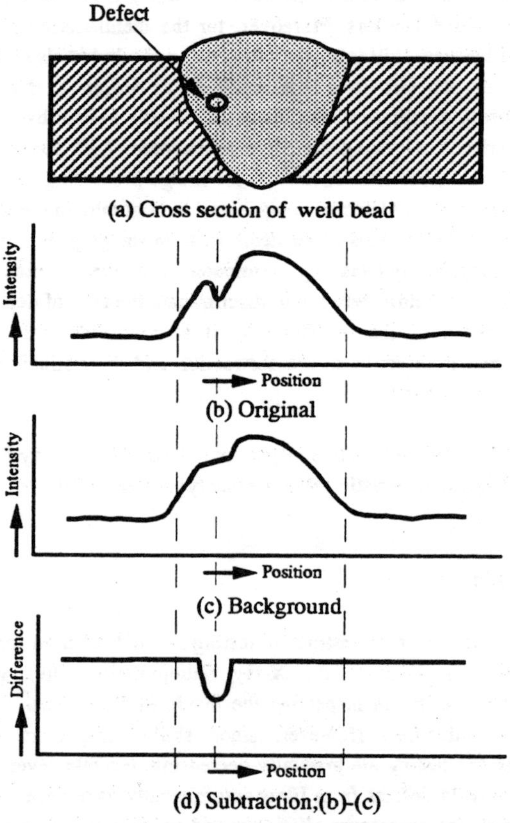

Fig.3 Background subtraction method

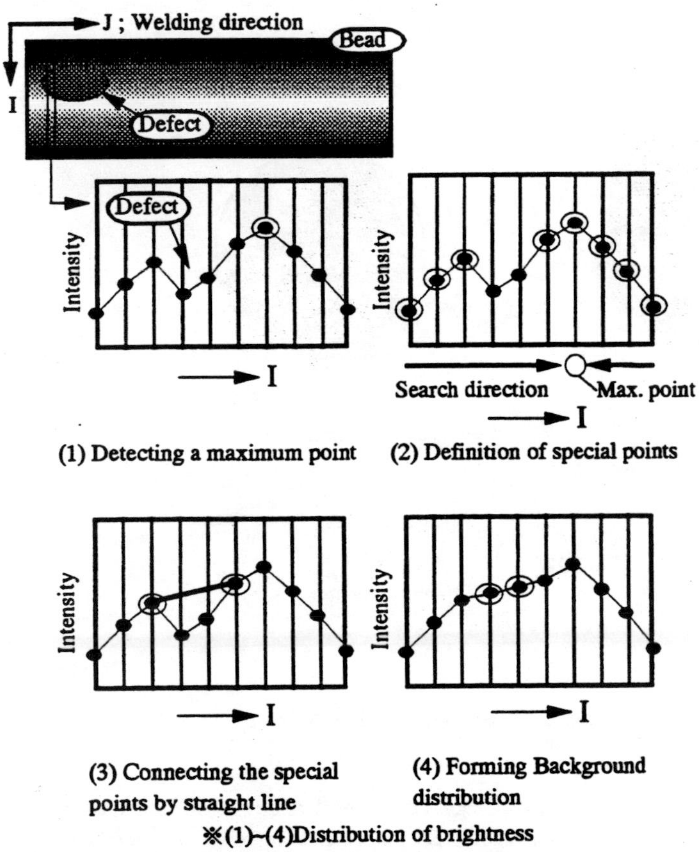

Fig.4 Forming Background distribution
"Special points connection method"

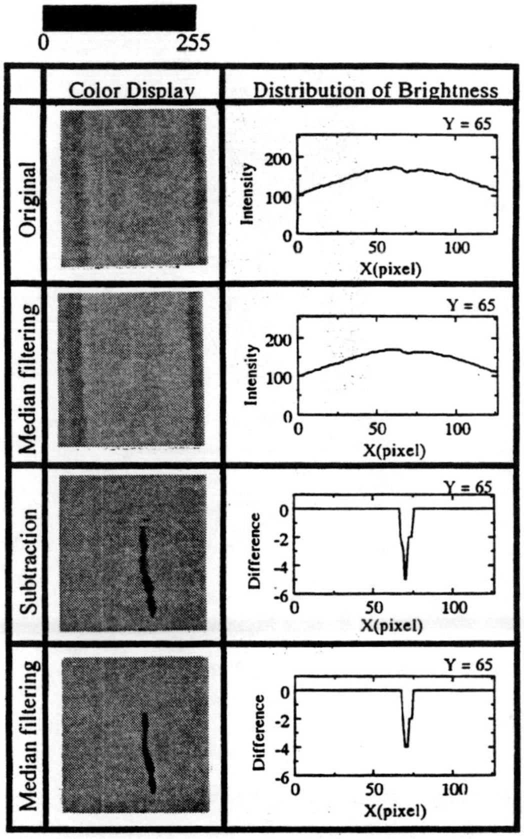

Fig.5 Effect of image processing

decreases as shown in fig.3(b). Accordingly, the low brightness area in the distribution is supplemented, and this distribution obtained is defined as the background line (fig.3(c)). This process is repeated all over the image, and a background image is obtained. Then, this background image is subtracted from the original image and accordingly only the low brightness area corresponding to the defect can be abstracted as shown in fig.3(d). In order to obtain the background distribution by which the defect signal of low brightness is not eliminated, a new method named "special points connection method" was proposed as shown in fig.4. That is to say, the brightness distribution was searched from both sides toward the maximum point of the remarked brightness distribution line, and in this process the point which is higher than previous searched point is defined as special point. Then these special points are connected by straight line. According to this process a relatively precise background distribution was obtained.

The image processing to abstract defect images without any enhancement processing was performed as follows. Fig.5 shows an example of the process in which a defect image shown in fig.2 was abstracted. That is to say, first of all, the median filter (3×3) was applied to eliminate noises (fig.5(b)), and then the background obtained by "special point connecting method" was subtracted (fig.5(c)). In this case, a threshold value was set and some residual noise due to small noise included in the background image is eliminated. Then a median filter (5×5) was applied again in order to eliminate noises surely as shown in fig.5(d). The abstracted defect image keeps information of not only shape and dimension, but also contrast of the defect image in the X-ray film. This contrast of the defect image is also important to discriminate the type of the defect. From the fig.5, it is confirmed that the system is effective to abstraction of defect images.

Last of the previous processing, the abstracted images were labeled in order to distinguish each other. The labeled defect images are subjected to the next process so as to discriminate type of defects one by one.

5. Discrimination of type of defects

5.1 Learning of neural network

In order to discriminate a type of defects, a neural network with three layer is applied. In this system, an error back-propagation algorithm was applied for network learning.

It may be easier to input the defect image abstracted by the previous processing to neural network directly. However, this method has some problems. That is to say, the size of image data (128×128 pixels) is too large to calculate in personal computer. Furthermore, since the system should cope with the change of the image position, rotation and so on. Therefore, if such system is realized, the the scale of the neural network system becomes huge and some hardware problems may occur. Accordingly, in this system, the quantitative characteristic values determined by the measurement of the abstracted image, such as defect position, shape, contrast and etc. were abstracted and applied to the neural network. Concretely, following eight characteristic values are used. (1) position of the defect image, (2) ratio of length, (3) defect

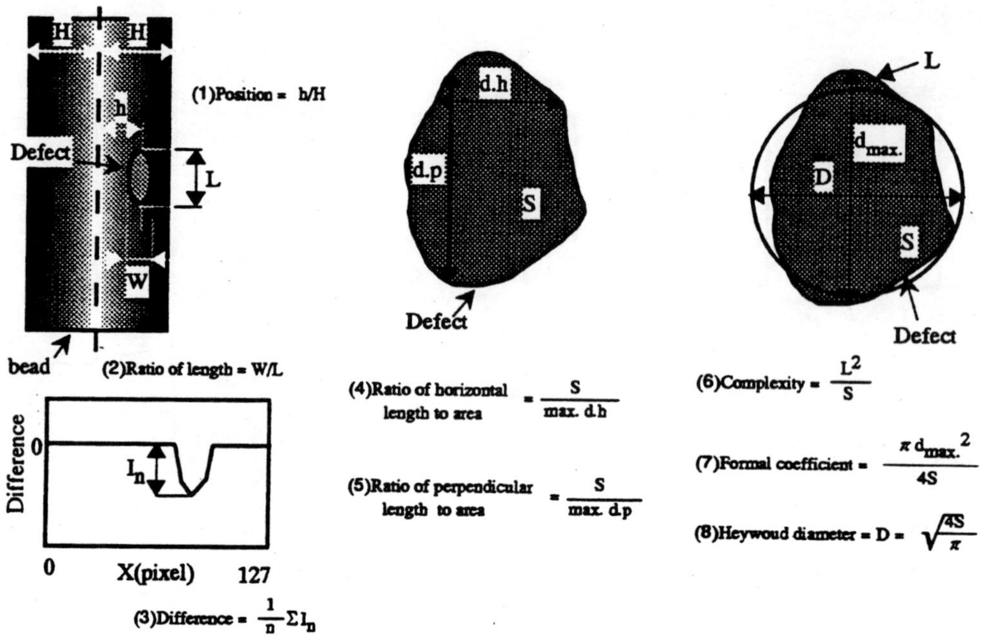

Fig.6 Characteristic values of a weld defect

contrast, (4) ratio of horizontal length to area, (5) ratio of perpendicular length to area, (6) Complexity, (7) Formal coefficient, (8) Heywoud diameter. These eight characteristic values are input into neural network as input parameters. Generally, three characteristic values from (1) to (3) may be enough to discriminate the type of normal size of defects. However, if these parameters take exceptional values because of imperfect abstraction of defect image in the previous processing, it is difficult for neural network to learn the defect pattern rightly. Therefore, in that case the other characteristic values of (4) to (8) concerning image shapes will be used to support it. A concrete definition of these parameters is given in fig.6.

In an actual learning of neural network, these characteristic values are transformed into relative values from 0 to 1 and then input to an input layer. Accordingly, the input layer of the neural network system has eight units. The output layer has five units, because each unit corresponds to five types of defects, slag inclusion, blowhole, longitudinal crack, incomplete penetration and undercut. In this system, typical teach data of twenty patterns were prepared for discrimination of defect type, and the neural network learned them.

5.2 Discrimination of defect type by neural network

The learned neural network discriminates the type of defect abstracted by the previous processing above mentioned. That is to say, the eight characteristic values are measured from an abstracted defect image automatically and input to neural network. Then the neural network calculates and outputs values to five units of the output layer. Because each output units corresponds to the type of defects, the output unit which has the highest value indicates the defect type.

An example of discrimination process of type of defect indicated in fig.2 is shown in fig.7. In the figure, the third unit in the output layer of the neural network shows the highest value of 0.987 and it is confirmed that the neural network has judged that the defect type of this defect is a crack. As the results of experiments of 45 defects, 41 defects are judged rightly. Therefore, the right discrimination ratio is 90% approximately. In this system, very small cracks less than 0.5mm length and small slag inclusions with less contrast in X-ray film were difficult to abstract and discriminate stably. Therefore, it will be necessary to develop more precise system, which can abstract smaller defects. However, it was confirmed that the system can be effective to abstract and discriminate weld defects larger than 0.5mm and may be applied to practical use.

6. Conclusions

An intelligent image processing for abstraction and discrimination of defect image in X-ray radiographic film was constructed. Main results obtained are summarized as follows:

1) A method to abstract defect image without enhancement processing was proposed to keep a contrast information of defect image in X-ray film.

2) In order to abstract the defect image, the background subtraction method was proposed and it is shown that the method is effective to abstract only defect image from image with brightness distribution due to a weld bead.

3) In order to obtain a precise background distribution, a new method named special points connection method was proposed, and it is confirmed that the method is effective to obtain a relative precise background image.

4) Some quantitative characteristic values determined by measuring shape and dimension of an abstracted image were proposed, and it was confirmed that the parameters are effective to discriminate the type of

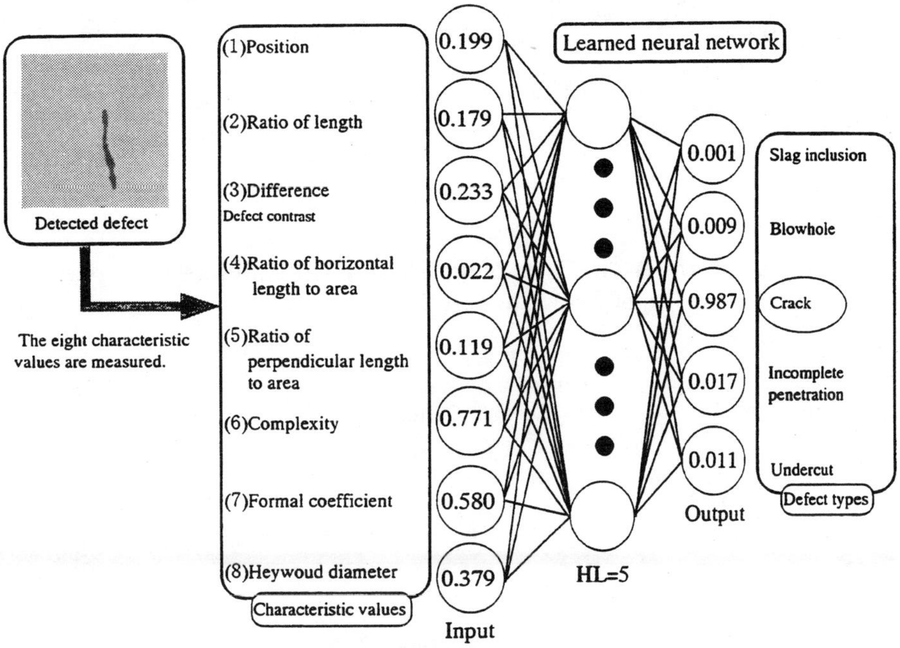

Fig.7 Result of discrimination of defect type

defects.

5) In order to discriminate the type of the defect abstracted by image processing, a neural network which has three layer was applied. As an input data, eight characteristic values of defect was used and output value indicates the defect type automatically. As the result of experiment, it was confirmed that the system is effective to abstract and discriminate the defects.

References

1)Daum, W, Rose, P, Heit, H, and Builtijes, JH, Automatic Recognition of Weld Defects in X-Ray Inspection, Brit J NDT, Vol.29, No.2, pp.79-82 (1987)

2)Inoue, K, and Kobayashi, M(1982). "Automatic Recognition of Weld Defects in Radio Graphic Test(Report 1), "Trans Jap Weld Res 1nst, Vol 11, No 2, pp123-132.

3)Itoga, K, Sugimoto, K, Michiba, K, Kato, Y, Sugita, Y, and Onda, Y(1991). "Development of Automatic Radiographic Inspection System Using Digital Image Processing and Artificial Intelligence," J Jap Weld Soc, Vol 60, No 1, pp68-73.

4)Shirai, Y(1967). "Automatic Inspection of X-Ray Photograph of Welding, "Pattern Recognition, Vol 1, pp257-261.

5)Wen, KL, Lin, JL, and Chang, SS(1993). "Computer-Aided X-Ray Radiography Image Processing Inspection for Welding," Proc 3rd Int Offshore and Polar Eng Conf, Singapore, ISOPE, Vol 4, pp462-466.

6)Suga, Y, Kojima, K, and Tominaga, T, "Detection of Weld Defects by Computer-Aided X-Ray Radiography Image Processing," Intn J Offshore Polar Eng, Vol 5, No.2, pp.142-146 (1995)

7)Aoki, K, Suga, Y, Higo, M, and Tanaka, H, "Abstraction and Discrimination of Weld Defects in X-Ray Radiography by Computer Aided Image Processing," The 6th International Symposium of the Japan Welding Society, Vol 2

Proceedings of the Seventh (1997) International Offshore and Polar Engineering Conference
Honolulu, USA, May 25-30, 1997

ISBN 1-880653-28-1 (Set); ISBN 1-880653-32-X (Vol. IV)

The Influence of the Main Factors on the Impact Toughness Properties of Low Carbon Steel Welds

T. Wegrzyn
University of Beira Interior
Covilhã, Portugal

ABSTRACT

Nitrogen, oxygen and metallographical structure are regarded as the main factors for impact properties low carbon steel weld deposits. Especially the influence of the variable amounts of nitrogen and oxygen on impact properties of different welds were tested. Furthermore, on the basis of the results of the influence of the variable amounts of nitrogen, oxygen and metallographical structure on impact properties, mathematical equations regarding those relationships have been given.

KEY WORDS: welding, nitrogen, oxygen, acicular ferrite, impact properties.

INTRODUCTION

The influence of nitrogen contents in weld metal deposit on impact properties was well analyzed in the last 15 years (1,2,3,4,5,6,7,8). Authors of the main publications present that the optimal content of nitrogen should not be greater than 100 ppm. The lowest amount of nitrogen in all weld metal gives the best impact results of the weld metal deposit. However also amount of oxygen and metallographical structure could be regarded as the important factors on impact properties and metallographical structure (3,4,5,6,7,8,9,10).

EXPERIMENTAL PROCEDURE

Welding Procedure

To asses the effect of nitrogen and oxygen on mechanical properties of deposited metals there were used basic electrodes prepared in experimental way. The electrode contained constant or variable proportions of the following components in powder form:

technical grade chalk	30%
fluorite	19%
rutile	4%
quartzite	3%
muscovite	1%
ferrosilicon (45%Si)	6%
ferromanganese (80%Mn)	4%
ferrotitanium (20%Ti)	2%
iron powder.	31%

This principal composition was modified by separate additions (at the expense of iron powder):

quartzite	up to 12%
magnetite	up to 10%
nitried ferromanganese	up to 10%
rutile	up to 12%
titanium.	up to 5%

The principal diameter of the electrodes was 4 mm. The standard current was 180A, and the voltage was 22V. A typical weld metal deposited had following chemical composition: 0.08% C, 0.9% Mn, up to 0.35% Si, 0.02%P, 0.02% S, up to 0.035% Al, up to 0,07 Ti and a variable composition of nitrogen (from 63 to 287 ppm) and oxygen (from 262 to 804 ppm).

In order to prepare weld metals containing different concentration of nitrogen , nitrided ferromanganese was added to the coatings of electrodes. A variation in the nitrogen amount in the deposited metals was analyzed from the extra low-nitrogen weld metals containing 63 ppm up to 287 ppm. In order to prepare weld metals containing higher concentration of oxygen, magnetite, quartzite and rutile was added to the coatings of electrodes. A variation in the oxygen amount in the deposited metals was also varied by adding titanium, magnetite, quartzite and rutile to the electrode coatings. As a result after welding the amount of oxygen in the weld metal was on the level from 262 to 804 ppm.

Weld Metal Testing

Only chemical analysis, micrograph tests, and Charpy V-notch impact toughness tests of the deposited metal were carried out. The Charpy tests were done mainly at +20°C using 5 specimens from each weld metal. The results of chemical analysis, micrographical structure and Charpy tests are given in Table 1.

Charpy V-notch impact toughness tests of the selected weld metal at lower temperatures were also done with 5 specimens. Results are given in table 2.

Results and Discussion

Based on the results from the table 1, the influence of nitrogen, oxygen and metallographical structure on the impact toughness properties of weld metal was established using the least square method.

Table 1. Relations between impact toughness properties of weld metal deposit at 20°C, and the metallographical structure and the amount of oxygen and nitrogen

Oxygen in weld metal deposits, ppm	Nitrogen in weld metal deposits, ppm	Acicular ferrite, %	Impact toughness of weld metal deposits, J	Theoretically counted (equation 1) impact properties, J	Error, 1) J
262	121	14	182	185	-3.39
303	99	20	198	195	2.73
354	102	18	181	182	-1.03
353	151	60	190	188	2.48
407	104	30	185	180	-4.69
448	98	29	175	175	-0.08
481	99	30	160	169	-8.91
606	150	70	135	147	-12.22
655	181	68	120	118	2.15
697	179	80	120	121	-0.65
804	182	79	105	97	7.66
346	63	18	220	215	5.26
348	97	42	210	206	4.37
351	147	47	175	180	-4.69
352	194	43	150	149	1.42
356	248	47	117	119	-2.28
350	287	55	100	98	2.47

1) error between measured and carculated impact values

The data from the table implies that impact toughness of weld metals depends on the amount of nitrogen and oxygen contents, and the micrographical structure as follows:

$$U_v = 295.98 - 0.19\,O_s - 0.59\,N_s + 0.81\,F_a \qquad (1)$$

where:

U_v - impact toughness of weld metal deposited at 20°C
O_s - oxygen contents of weld metal deposit, ppm
N_s - nitrogen contents of weld metal deposit, ppm
F_a - amount of the acicular ferrite, %

Maximal error is 12.21 J, while the least square error is 4.98 J.

According to the equation (1) the impact properties of metal weld deposits depends strongly on nitrogen amount. The coefficient for nitrogen is three times greater than the coefficient for oxygen. The influence of the factor $0.81\,F_a$ on the impact toughness of weld metal is not greater then factors such as: $0.19\,O_s$ and $0.59\,N_s$ because of its smaller value. The amount of oxygen and acicular ferrite in metal weld deposits are less important factors than the amount of nitrogen on the impact strength of weld metal (see fig.1).

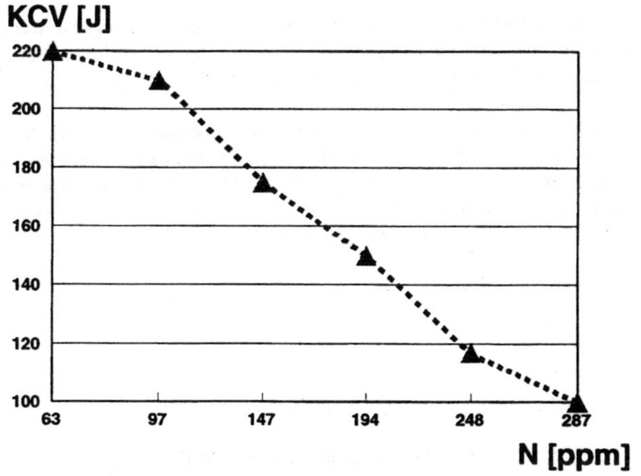

Figure 1. Impact strength of weld metal with various nitrogen content and relatively fixed amount of oxygen (350 ppm).

In order to analyze relations between impact toughness properties of weld metal in other temperatures, and the amount of oxygen, nitrogen and acicular ferrite in weld metal, there were compered data from Table 2.

Table 2. Relations between impact toughness properties of weld metal deposit at lower temperatures, and the metallographical structure and the amount of oxygen and nitrogen

O, N in weld metal deposited, ppm	Acicular ferrite, %	Temperature of impact tests, °C	Impact toughness of weld metal deposited, J	Theoretically counted (equation 2) impact properties, J	Error, J
353, 151	60	+20	190	193	3
353, 151	60	0	159	154	5
353, 151	60	-20	109	116	7
353, 151	60	-40	73	77	4
353, 151	60	-60	31	38	7
448, 98	29	+20	175	172	3
448, 98	29	0	130	138	8
448, 98	29	-20	97	103	6
448, 98	29	-40	71	69	2
448, 98	29	-60	28	34	6
655, 181	68	+20	120	117	3
655, 181	68	0	86	93	7
655, 181	68	-20	74	70	4
655, 181	68	-40	38	47	9
655, 181	68	-60	18	23	5

Impact properties of weld metal deposited are shown on figure 2.

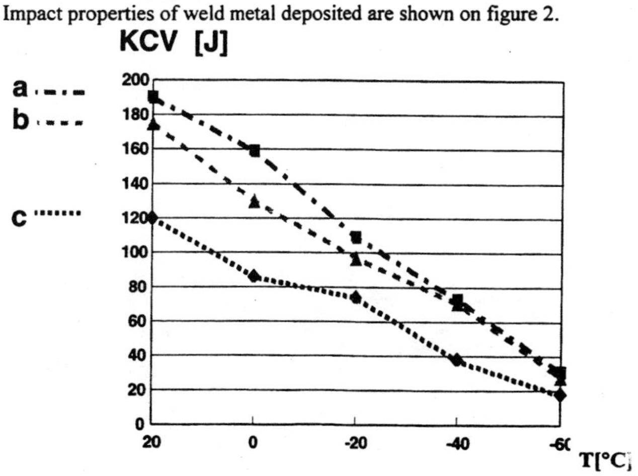

Figure 2. Impact strength of weld metal with various nitrogen and oxygen contents with data from table 2. (curve **a**: 353 ppm O, 151 ppm N; curve **b**: 448 ppm O, 98 ppm N; curve **c**: 655 ppm O, 181 ppm N)

Coefficients in the equation (1) simplified experimentally leads to simpler form of the equation (2) having the similar values of the maximal error (13 J) and the least square error (5,57 J). The experimental approximated temperature coefficient *P* describes changes of impact toughness of weld metal deposited in other tested temperatures:

$$U_v = (300 - 0{,}22 \cdot O_s - 0{,}6 \cdot N_s + F_a) \cdot P \quad (2)$$

where:

P - experimental approximated temperature coefficient is:

$P = 1$ for standard temperature +20°C
$P = 0{,}8$ for standard temperature 0°C
$P = 0{,}6$ for standard temperature -20°C
$P = 0{,}4$ for standard temperature -40°C
$P = 0{,}2$ for standard temperature -60°C

Impact toughness of weld metal deposits can be classified by equation (1) and (2) for weld metal of the following chemical composition only:
- less than 0,1% C
- 0,8 ± 0,2% Mn
- 0,40% Si
- (P+S) unless 0,045%
- from 63 to 287 ppm N
- from 262 to 804 ppm O

Impact toughness of weld metal deposited can be classified by equation (2) for weld metal of the similar chemical composition of C, Mn, Si, P, S. The percentage of O and N, are on the respective level :
350 - 650 ppm O
100 - 150 ppm N

According to the equation (2), the impact properties of metal weld deposit strongly depends on nitrogen amount and temperature of tests. The amount of oxygen and metallographical structure are not so important factors such as nitrogen and temperature of impact toughness tests in that case.

CONCLUSIONS

Examination of the influence of the variable amounts of nitrogen, oxygen and metallographical structure on impact properties of metal weld deposits with basic electrodes gave the following conclusions:

1. The toughness of the weld metals is much more strongly affected by the amount of nitrogen than by oxygen and the amount of acicular ferrite in weld metal.
2. Low-carbon basic electrode weld metals containing less than 100 ppm of nitrogen and the oxygen in range between 250 and 450 ppm can attain high impact strength of 200 J at +20°C according to the presented equation 1.
3. Low-carbon basic electrode weld metals containing less than 100 ppm of nitrogen and the oxygen in range between 250 and 450 ppm can attain good impact toughness (above 47 J) at temperatures not lower than -20°C according to the presented equation 2

REFERENCES

1. Węgrzyn T, *Oxygen in SMAW and GMAW process*, International Conference on Advances in Mechanical Engineering (1995)
2. L.van Nassau, van der Mee, *Effect of nitrogen on submerged arc metal properties,* IIW Doc II-A-661-85, 1985
3. Evans G M, *The effect of micro-alloying elements on the microstructure and properties of ferritic all-weld metal deposits*, IIW Doc II-A-855-92.
4. Węgrzyn T, *Oxygen and nitrogen in the low-carbon low-hydrogen electrode deposited metal* ,Przegląd Spawalnictwa 1992 (6)
5. Węgrzyn T, *Manganese, nickel and molybdenum in basic electrodes weld metal deposits*, Conference Materials of Częstochowa Polytechnic
6. Węgrzyn T, *Oxygen and nitrogen in low carbon basic electrode weld metal deposits,*
7. IIW-Comission II :Arc Welding, Sub-Commission II A:The Metallurgy of weld metal,1992,Doc.II-A-1181-92.
8. Węgrzyn J, *Oxygen in the submerged arc weld process*, Metal Construction 1985 **17**
9. Węgrzyn T, *The Impact properties of low alloyed gas shielded welds*, Materials Conference in Aveiro, Portugal 1995
10. Węgrzyn T, *Oxygen and nitrogen in SMAW and GMAW processes,* International Conference of Society of Offshore and Polar Engineers, Los Angeles 1996

Proceedings of the Seventh (1997) International Offshore and Polar Engineering Conference
Honolulu, USA, May 25-30, 1997

ISBN 1-880653-28-1 (Set); ISBN 1-880653-32-X (Vol. IV)

Development of Real-Time Quality Evaluation of Friction Welding by Acoustic Emission: 3rd Report. Effects of Initial AE Counts During Plastic Deformation in FRW

Sae-Kyoo Oh, Hyung-Dong Park and Bae-Sub Lee
Pukyong National University
Pusan, Korea

Hong-Keun Chang
Korea Institute of Machinery and Materials
Changwon, Korea

ABSTRACT

Both in-process quality control and reliability of the weld is one of the major concerns in applying friction welding. No reliable nondestructive monitoring method is available at present to determine the weld quality particularly in process of production.

So that, this paper presents an experimental examination and quantitative analysis for the effects of initial acoustic emission (AE) counts during plastic deformation of friction welding (FRW) for Zone A on the weld strength relating to the rotating speed as a new approach which attempts finally to develop an on-line quality monitoring system design for friction welds using AE techniques.

As one of the important results, it was well confirmed that the initial AE counts occurring during plastic deformation period of welding were quantitatively correlated with reliability at 95 % confidence level to the joint strength of welds of tube-to-tube (S20C to SUS 304) and then an AE technique using the initial AE counts can be reliably applied to in-process strength monitoring of the weld.

KEY WORDS : Quality evaluation, friction welding(FRW), initial acoustic emission(AE), welding condition, weld strength, cumulative total AE counts, joint efficiency

INTRODUCTION

In applying friction welding process, one of the imortant concerns is the reliable quality monitoring method for a good weld strength in process of production. However, no reliable nondestructive test method is available at present to monitor the weld strength quantitatively in process[1),2)] except the author, et al.'s reports[3),4),5)], even though studies on ultrasonic detection of weld strength were reported[6),7)] but no quantitative relationship was drawn because of considerable scattering of the data.

In this continued study since the former reports[8),9)], the relationship between the weld strength and the measurable initial AE counts was analyzed quantitatively. This was possible through relating the direct measurement and analysis of AE counts to the strength (tensile strength) of friction welds of stainless steel tubes to low carbon steel tubes using the initial AE counts for Zone A instead of the total AE counts for Zone A+B.

SPECIMENS AND PROCEDURES

The dimensions of welding workpieces and tension test specimens are the same as those in Figure 1. And the chemical compositions and tensile strength of base metals are listed in Table 1. The friction welding conditions (inertia type) used in this study are revealed in Table 2 including the diameters of workpieces.

Table 1 Chemical compositions and strength of base metals

Materials KS(AISI)	Chemical composition (wt %)									T.S*
	C	Mn	P	S	Pb	Cr	Mo	Ni	Si	σ_T
S20C(1020)	0.20	0.45	0.04	0.05	-	-	-	-	-	62.2
SUS304(304S.S.)	0.08	2.00	0.04	0.03	-	18.0	-	8.00	0.75	75.5
· Unit kgf/㎟, each specimen with 0.635㎜ R notch.										

Table 2 Welding conditions

Materials combination	diameter D, ㎜	Moment of inertia I, kgfm²	Initial rotation speed V, rpm	Axial pressure p, kgf/㎟	Initial energy E, kgfm
Tube-to-tube S20C-SUS304	OD 25.40 ID 19.05	0.472	796-2984	8.2	167-2351

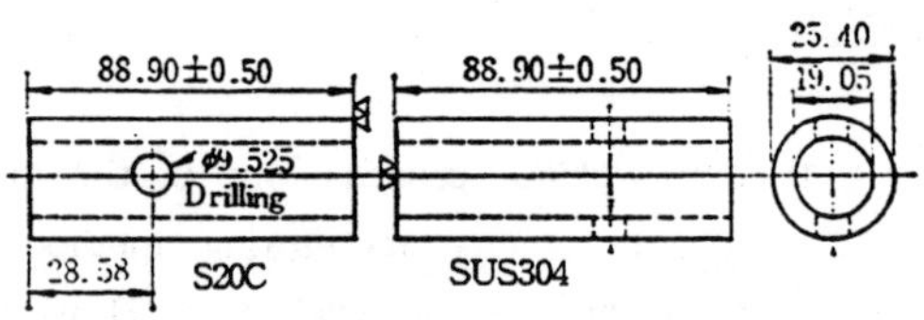

(a) Welding workpieces

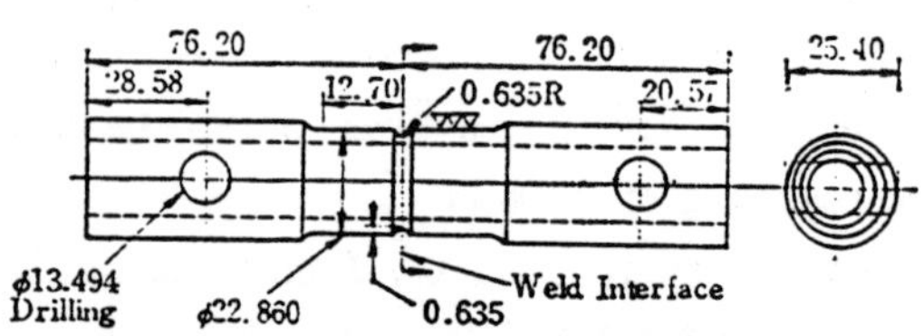

S20C-SUS304 tube welds

Each base metal with notch

(b) Tension test specimen

Fig. 1 Welding workpieces and tension test specimen

The AE transducer location diagram and the block diagram of electronic components for the AE monitoring equipment including the welding parameter-history measuring apparatus are given in Figure 2-(a) and (b)[3], respectively. A 22.225 ㎜ ϕ acoustic emission piezoelectric transducer (sensitivity-65dB) was tightly mounted with tape on the jaw, using high vacuum grease on the contacting surface ; At such a location as shown in Figure 2-(a), there was no thermal effect from the interface. The AE equipment was a standard commercial unit (Dunegan/ Endevco Model 3000). As shown in Figure 2-(b), the band pass filter was set to 100 through 300㎑ to remove the background noise, and AE signals were further amplified (0-60dB) after the preamplifier of 40 dB[3]. Therby, the total system gain setting was 75dB in this study, that is, counting each pulse that exceeded 75dB. About eighty workpieces for each metal pair for welding and measuring AE counts were prepared.

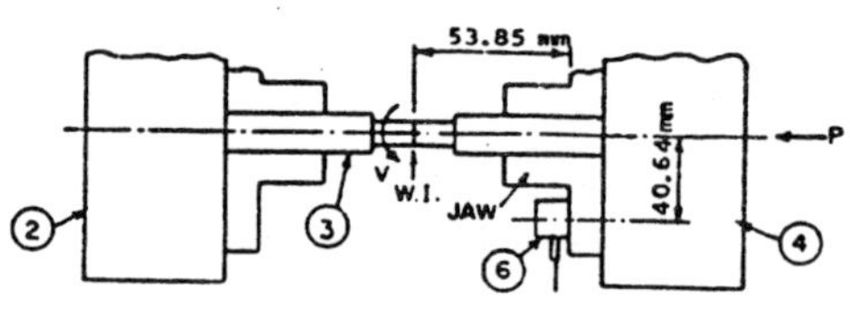

(a) Location of AE transducer

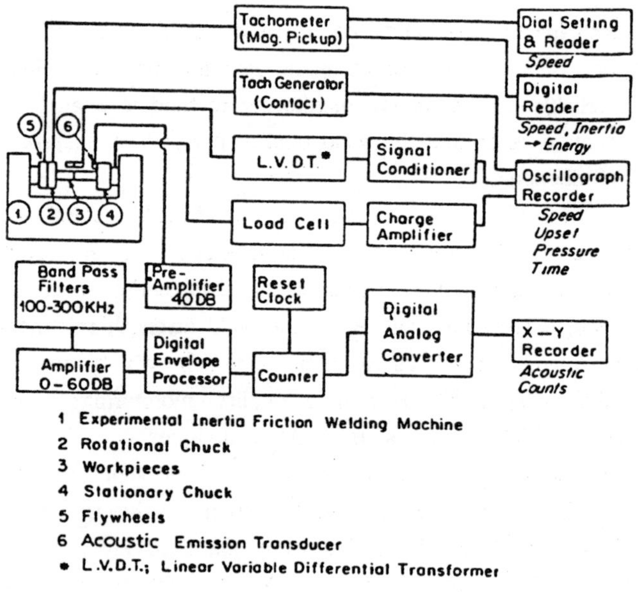

Fig. 2 (b) Block diagram of wleding-parameters measuring-apparatus and AE monitoring-apparatus

RESULTS AND DICUSSIONS

Initial AE Counts

Figure 3[3] shows the typical history of welding parameters and AE measurements in inertia friction welding. The initial AE counts (N_A, counts) at Zone A of plastic deformation in the welding period between welding start (W_S) and welding end (W_E) except Zone B of the cooling period between martensite formation start (M_S) and finish (M_F) were used for relating to the tensile strength (σ_T, kgf/㎟) of welded joints, because the initial AE counts at Zone A result primarily from the plastic deformation during welding and those at Zone B primarily from the martensitic phase transformation during cooling. The correlated history graphs as conditionent as those in Figure 3 were made according to every welding condition and used for analysis of each weldment and AE measurement.

Figure 4 shows a typical appearance of tube-to-tube friction welded joint of S20C to SUS304 steel.

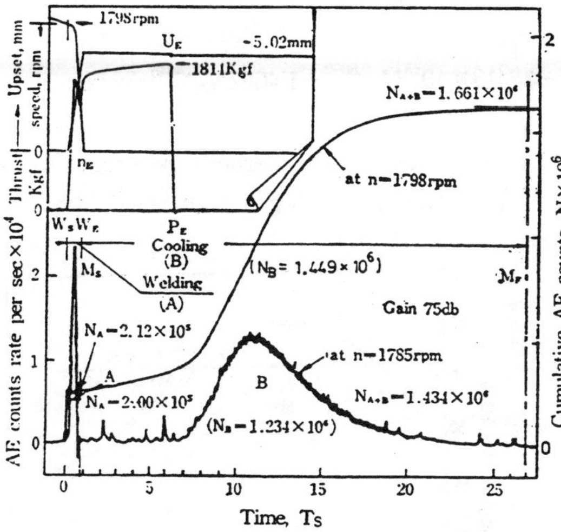

Fig. 3 Typical result of the histories of welding parameters and acoustic emission in inertia friction welding of S20C to SUS304 steel(tube-to-tube, OD 25.40㎜, ID 19.05㎜), (I=0.472kgfm², n=1798 and 1785 rpm, respectively)

Fig. 4 Typical appearance of tube-to-tube friction welded joint of S20C to SUS304 steel

Effects of Rotating Speed on Weld Strength and Initial AE Counts

The experimental results with the same materials and welding conditions as those in Figure 3 are shown in Figure 5. It is certain in Figure 5 that the weld strength (σ_T, tensile strength), initial AE counts (N_A) and welding energy (E) are all correlated cubically to rotating speed (n). So that, their experimental data points were fitted to each cubic curve, with the results that the reliability is kept at 95% confidence level, too. And then, the optim um rotating speed (OnZ) is obtained by the N_A-n curve as 1,600 through 2,400 rpm, at which the ON_AZ of 1.44×10^5 through 3.91×10^5 counts as well as the weld strength with the joint efficiency more than 100% can be obtained at the same time through the N_A-n and σ_T-n curves, respectively. And OEZ is 676 through 1521 kgfm, resulting in a good weld zone (GWZ).

Thus, it seems again that such initial AE counts within the ON_AZ do indicate the weld strength monitored with the joint efficiency more than 100% for friction welding of such same material pairs and same welding conditions, and that this non-destructive test method of friction weld quality monitoring using initial AE techniques can be more effectively applied in real time of friction welding.

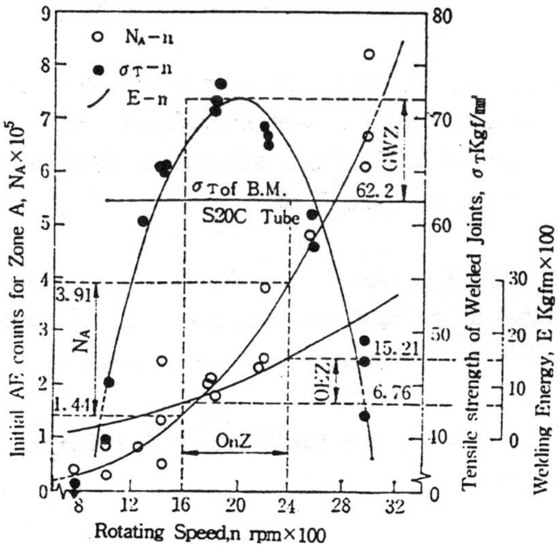

Fig. 5 Effect of rotating speed on initial AE counts and weld strength for 25.4㎜ tube-to-tube welds of S20C to SUS304 steels. AE gain; 75 dB

The relationship equations of N_A-n, σ_T-n and E-n in Figure 5 were calculated as follows :

$$E = In^2 / 1787.3 \qquad (1)^{1),2)}$$

$$N_A = 1.7310 \times 10^{-3} n^{2.4714} \qquad (2)$$

$$\sigma_T = -2.8214 \times 10^{-5} n^2 + 1.1208 \times 10^{-1} - 39.56 \qquad (3)$$

$$E = 2.6409 \times 10^{-4} n^2 \qquad (4)$$

where E is welding energy(kgfm), I is the selected inertia (0.472 kgfm²) of flywheel.

Effects of Initial AE Counts for Zone A on Weld Strength

Figure 6 shows the relationship between the inertia friction weld strength and initial AE counts for Zone A of friction welded joints of 25.4㎜ (3.175㎜ wall) tubes.

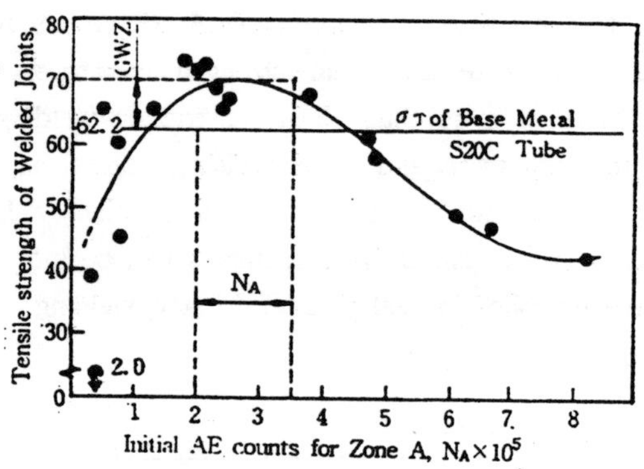

Fig. 6 Effects of initial AE counts for Zone A(plastic deformation) on weld strength of friction welded joint of 25.40(3.175㎜ wall) tubes, S20C to SUS304 steels. AE gain; 75 dB.

The empirical equation of σ_T-N_A for friction welding of S20C tube to SUS304 tube is given as the following, computed by the polynomial regression analysis :

$$\sigma_T = 4.2287 \times 10^{-16} N_A^3 - 6.5625 \times 10^{-10} N_A^2 + 2.5641 \times 10^{-4} N_A + 40.8 \qquad (5)$$

The σ_T-N_A Equation (5) shows the cubic which is very coincident with the σ_T-N_{A+B} equation in the former report[8),9)].

So, it is confirmed that the initial AE counts can be more effectively applied to real-time evaluation of weld strength.

Thus, by the above empirical equation of the σ_T-N_A relationship for the welded joint, the general empirical equation form can be modeled as the following cubic (a≠0) equation depending on materials:

$$\sigma_T = aN_A^3 + bN_A^2 + cN_A + d \qquad (6)$$

Comparison between Empirical and Calculated Equations and 95% Confidence Tests

Figure 7 shows the comparison between the calculated equation[8)] σ_{Tn}-N_A and the empirical equation σ_T-N_A : Such calculated equation derived from the combination of empirical σ_T-n Equation (3) and N_A-n Equation (2) are as follows[10)],

$$\sigma_{Tn} = -4.8418 \times 10^{-3} N_A^{0.8092} + 1.4682 N_A^{0.4046} - 39.56 \qquad (7)$$

The maximum mean % difference between Equation (5) and (7) is only 3.80%, showing a high reliability.

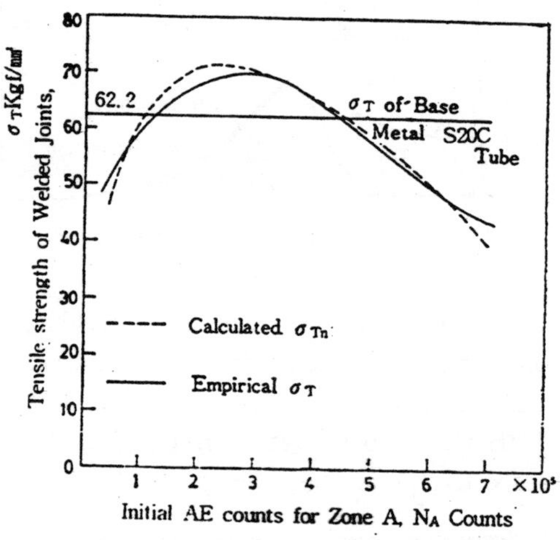

Fig. 7 Comparison between calculated and empirical equations for friction weld strength versus initial AE counts of 25.4㎜(3.175㎜ wall) tube-to-tube welds(S20C-SUS304). AE gain : 75 dB

Table 3 indicates the computation of residual sum of squares and pure error for initial AE counts during plastic deformation in friction welding on the empirical N_A-n equation of S20C-SUS304 tube welds in Figure 5 in order to investigate the 95% confidence of such empirical equations.

Table 3 Computation of residual sum of squares and pure error for initial AE counts during plastic deformation in friction welding on the empircal N_A-n equation of S20C-SUS304 tube welds

Trial No.	Initial AE Count, Counts Expected, N_A	Observed, N_A'	Trial NO.	Initial AE Count, Counts Expected, N_A	Observed, N_A'
1 a	189931	212000	6	47410	76000
b	189931	200000	7	79693	75000
2 a	317812	378000	8	110820	45000
b	317812	252000	9	107620	242000
3 a	459029	484000	10	110440	133000
b	459029	467000	11	195220	182000
4 a	661464	666000	12	307600	232000
b	661464	608000	13	669780	817000
5	47867	30000			

Residual sum of squares = $\Sigma(N_A'-N_A)^2 = 6.3730 \times 10^{10}$
pure error = $1/2\ \Sigma(N_A'b-N_A'a)^2 = 9.8365 \times 10^9$

Table 4 ANOVA table for testing the lack of fit on the empirical N_A-n equation (mean % error : 11.69) of S20C-SUS304 tube welds

Source	Sum of squares	Degree of freedom	Mean square	F-ratio
Residual	6.3730×10^{10}	16		
Pure error	9.8365×10^9	4	2.4591×10^9	
Lack of fit	5.3894×10^{10}	12	4.4912×10^9	1.83
Remarks	From F-table : 12, 4, 0.05 = 3.28, F14, 4, 0.05 = 3.23, then, F13, 4, 0.05 = 3.26			

For their 95% confidence examination, in the case of S20C-SUS304 tube welds, the residual sum of squares for the weld strength are computed statistically, and the ANOVA (analysis of variables) table for testing the lack of fit is calculated as in Table 4. The calculated F-ratio for the lack of fit for both the empirical equation at full range and N_A of initial AE and the calculated equation is 1.83 which is much smaller than the corresponding values of 3.26 from the F-table at 95 percent confidence level. This analysis suggests that not only there is no great danger of lack of fit between all the postulated models and the experimental data in this study, but also it seems possible to evaluate the weld strength quantitatively in real time during the process of friction welding production by the initial acoustic emission techniques[3),4),5),10)] more reliably than by the conventional methods[11)].

CONCLUSION

The results obtained in the study on development of Real-Time Quality Evaluation of Friction Welding by Acoustic Emission (Report 3)-Effects of Initial AE Counts are as follows :

(1) It was confirmed that the weld strength can be quantitatively and usefully evaluated or controlled in real time of friction welding process by using initial AE techniques because it was clarified experimentally and quantitatively that the friction weld strength and the initial acoustic emission counts for Zone A have the correlation between them, and the relationship model is expressed as $\sigma_T = aN^3 + bN^2 + cN + d$, which has a high reliability, and can be used for inprocess real-time evaluation or monitoring of friction weld strength as a nondestructive evaluating method using initial AE techniques.

(2) It was confirmed again statistically by error analysis and lack-of-fit testing that the empirical equations for the relationship between the weld strength and the initial AE counts computed by the regression analysis using the least squares method have also a reliability at 95% confidence level.

(3) The measurable initial AE counts are quantitatively related to the weld strength and the empirical equation is as follows :

$$\sigma_T = 4.2287 \times 10^{-6} N_A^3 - 6.5625 \times 10^{-10} N_A + 2.5641 \times 10^{-4} N_A + 40.8$$

ACKNOWLEDGEMENTS

This study has been undertaken as a part of the project financially supported by the 1993 Cooperation Research Fund of the Korea Science & Engineering Foundation (KOSEF No. 93-1-10-05-01-2, 1993-1995), sincerely much acknowledged by the authors.

REFERENCES

1) AWS ; Welding Handbook, Vol.3, pp.240, AWS, Miami, 1980.

2) JFWRC ; Friction Welding, pp.57, 139, Edited by A. Hasui. Corona, Japan, 1979.

3) S.K. Oh ; Friction Weld Strength Analysis by Acoustic Emission Techniques(1), J. of KSME, Vol.22, No.3, pp184-190, 202, 1982

4) K.K. Wang, S.K. Oh, G.R. Rief ; In-Process Quality Detection of Friction Welds Using Acoustic Emission Techniques, Proc. 63rd Annual AWS Convention, Session 5-a(Apr.27,1982). Welding Journal of AWS, Vol. 61, No. 9, September, Welding Research Supplement 312-316-s, 1982

5) S.K. Oh, A. Hasui, T. Kunio, K.K. Wang ; Effects of Initial Energy on Acoustic Emission Relating to Weld Strength in Friction Welding, Transaction of JWS, Vol. 13, No. 2, p.15-26. October, 1982, Proc. of 4th Int. Sym. of JWS, 4 JWS-V-8. Nov. 24-26, pp,713-718, 1982

6) K.K. Wang, S. Ahmed ; Ultrasonic Detection of Weld Strength for Dissimilar Metal Friction Welds, Proc. of 4th North American Metalworking Research Conf., pp.384, 1976

7) S. Ahmed ; Ultrasonic Detection of Weld Strength, A Technical Report, Cornell University, USA, 1975

8) S.K. Oh, J.H. Oh, T.E. Jeon, S.W. Oh ; Development of the Real-Time Quality Evaluation of Friction Welding by Acoustic Emission (Report 1)-Effects of Total AE counts for Zone A+B, Proc. 5th Int. Offshore and Polar Engineering Conference, ISOPE, Vol. 4-1, p.163-168, 1995.6

9) S.K. Oh, J.W. Oh, H.K. Chang ; Development of the Real-Time Quality Evaluation of Friction Welding by Acoustic Emission (2nd Report)-Effects of Welding Parameters, 6th Int. Offshore and Polar Engineering Conf., ISOPE, Vol. 4-2, p.177-184, 1996.5

10) S.K. Oh, K.K. Wang ; Effects of Welding Parameters on Weld Strength and Acoustic Emission in Friction Welding, Proc. of the Korean Society of Marine Engineers(Apr.23,1982), J. of KOSME, Vol. 7, No. 1, pp.21-31, 1983

11) S.K. Oh ; Study on Friction Welding of Valve Materials-On Improving the Friction Weld Quality for Exhaustive Valve Materials SUH3-SUH31, J. of KSME, Vol. 14-3, pp.221-232, 1974

Proceedings of the Seventh (1997) International Offshore and Polar Engineering Conference
Honolulu, USA, May 25-30, 1997

ISBN 1-880653-28-1 (Set); ISBN 1-880653-32-X (Vol. IV)

Friction Welding of Aluminum Alloy and Steel

H. Ochi
Osaka Institute of Technology, Osaka, Japan
K. Ogawa
Osaka Prefecture University, Sakai, Japan
Y. Yamamoto
Setsunan University, Neyagawa, Japan
Y. Suga
Keio University, Yokohama, Japan

ABSTRACT

Friction welding of 6061 aluminum alloy and steel was conducted. Sound welds were produced in all A6061/steel combinations, and tensile fracture occurred in thermal softened A6061 area. Joint strength depends on the width of this softened A6061 area. Tensile strength of T62 heat treated joint was high compared with that of the weld joint.

KEY WORD: Friction welding, Joint strength, Heat treatment, Hardness, 6061 aluminum alloy, Carbon steel, Stainless steel.

INTRODUCTION

Friction welding has been widely used for welding of similar and dissimilar metals. However, in friction welding of dissimilar metals, such as aluminum alloy and steel, brittle intermetallic compounds tend to form making it very difficult to obtain a sound weld (Elliott and Wallach, 1981). Presently, the demand for friction welding of such materials is increasing in many industrial fields. Thus, in order to produce a sound welded joint between such metals, friction welding was undertaken using an insert metal (Sassani and Neelam, 1988), (Ochi, et al, 1995). In such a case, the problem is to increase welding process. In this study, the direct friction welding of low-weldability aluminum alloy and steel was conducted, and suitable welding conditions for producing a sound weld were investigated. In particular, the effects of softened aluminum alloy on the joint strength were examined. Moreover, the effects of various heat treatments of the welded joint on the joint strength were examined.

EXPERIMENTAL

The materials used in this investigation included 6061-T6 aluminum alloy (A6061), S15C, S25C, S35C, S45C and S55C carbon steels and SUS304 stainless steel (AISI 304). The chemical compositions and mechanical properties of the base materials are shown in Table 1 and Table 2. The welding materials were machined to the shape and dimensions shown in

Table 1 Chemical compositions of the base materials (mass%).

Material	Si	Fe	Cu	Mn	Mg	Cr	Zn	Ti	Al
A6061-T6	0.59	0.16	0.20	0.01	0.95	0.06	0.01	0.01	Bal

Material	Si	Fe	Cu	Mn	Cr	C	P	S	Ni
S15C	0.20	Bal	0.01	0.42	0.09	0.17	0.017	0.018	0.02
S25C	0.22	Bal	0.08	0.40	0.14	0.22	0.017	0.014	0.04
S35C	0.18	Bal	0.01	0.72	0.06	0.34	0.013	0.017	0.01
S45C	0.26	Bal	0.02	0.78	0.06	0.46	0.019	0.016	0.03
S55C	0.25	Bal	0.01	0.79	0.13	0.54	0.027	0.014	0.02
SUS304	0.25	Bal	–	1.71	18.75	0.06	0.033	0.026	8.26

Table 2 Mechanical properties of the base materials.

Material	Yield strength (MPa)	Tensile strength (MPa)	Elongation (%)
A6061-T6	299	319	18
S15C	343	451	32
S25C	282	465	30
S35C	427	617	28
S45C	422	727	23
S55C	461	784	23
SUS304	579	765	39

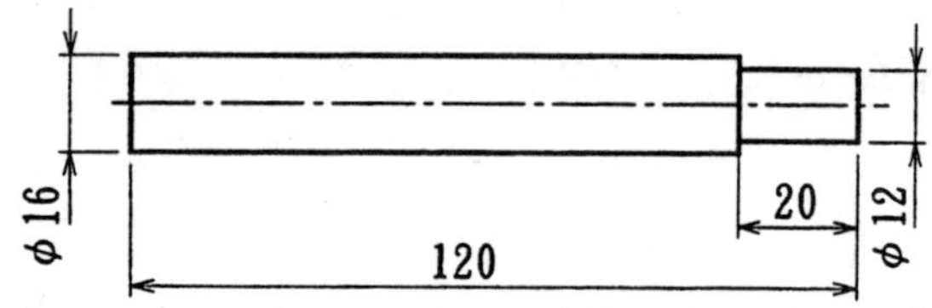

Fig.1 Shape and dimensions of the welding material (unit: mm).

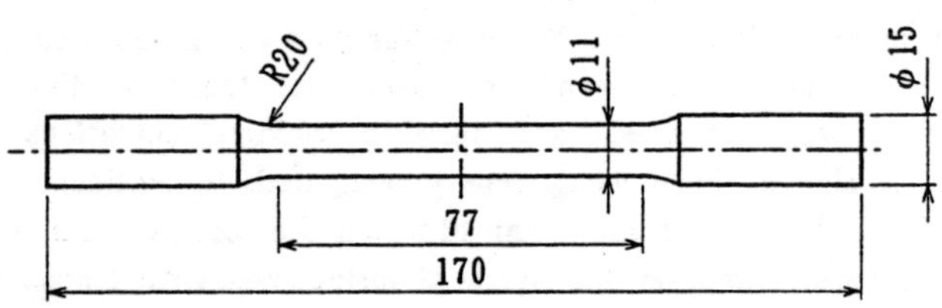

Fig.2 Shape and dimensions of tensile test specimen (nuit: mm).

Fig.1. Friction welding was carried out using a brake-type friction welding method. The effects of the welding conditions (friction pressure P_1, forge pressure P_2, friction time t_1 and rotation speed N) on the joint strength were examined. The joint strength was measured using a test specimen without a burr as shown in Fig.2 in tensile test. A hardness distribution of welds was measured using a vickers hardness tester.

Next, the welds were heat treated, then examined by both tensile and hardness tests. Heat treatments were T42 : at 525℃ for 1hr, T62 : at 175℃ for 8hr after T42 and T62' : at 175℃ for 8hr, respectively.

RESULTS AND DISCUSSION

The results show that, among these welding conditions, forge pressure had significant effects on the joint strength. Thus, all joints in A6061/steel combinations were welded under welding conditions of friction pressure 30MPa, forge pressure 40～280MPa, friction time 2s and rotation speed 4000rpm. The changes in total burn-off quantity and tensile strength with forge pressure are shown in Fig.3 and Fig.4, respectively. Total burn-off quantity and tensile strength were seen to increase when the forge pressure was increased. For the same welding conditions, the tensile strength of A6061/steel joints were much alike for all the steels used. Sound welds were produced in all A6061/steel combinations under forge pressure 200～280MPa, and the maximum tensile strength of these welds was about 280MPa.

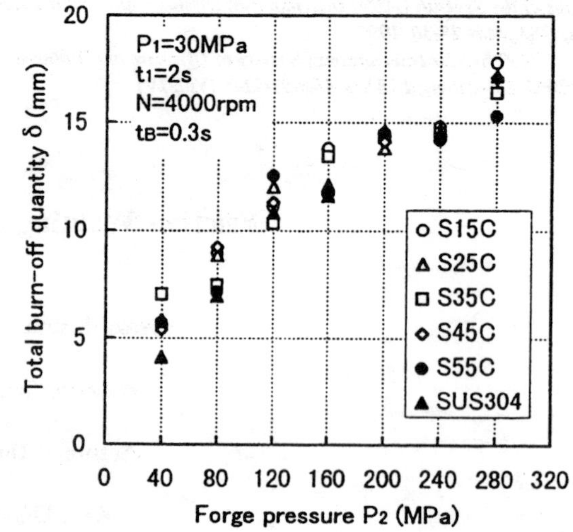

Fig.3 Changes in total burn-off quantity with forge pressure.

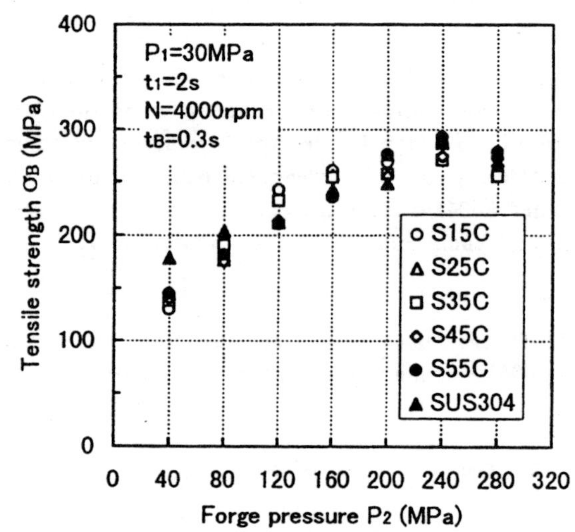

Fig.4 Changes in tensile strength with forge pressure.

The typical interfacial layers, appearance of tensile tested specimens and fractured surfaces in S45C side of A6061/S45C joints are shown in Fig.5. A burr was largely exhausted under the action of pressure when the forge pressure was raised. This result corresponded to the behavior of the total burn-off quantity. In the case of a 40MPa forge pressure, tensile fracture occurred at the weld interface. On the other hand, in the case of a 120MPa forge pressure, tensile fracture occurred in the A6061 base material, for

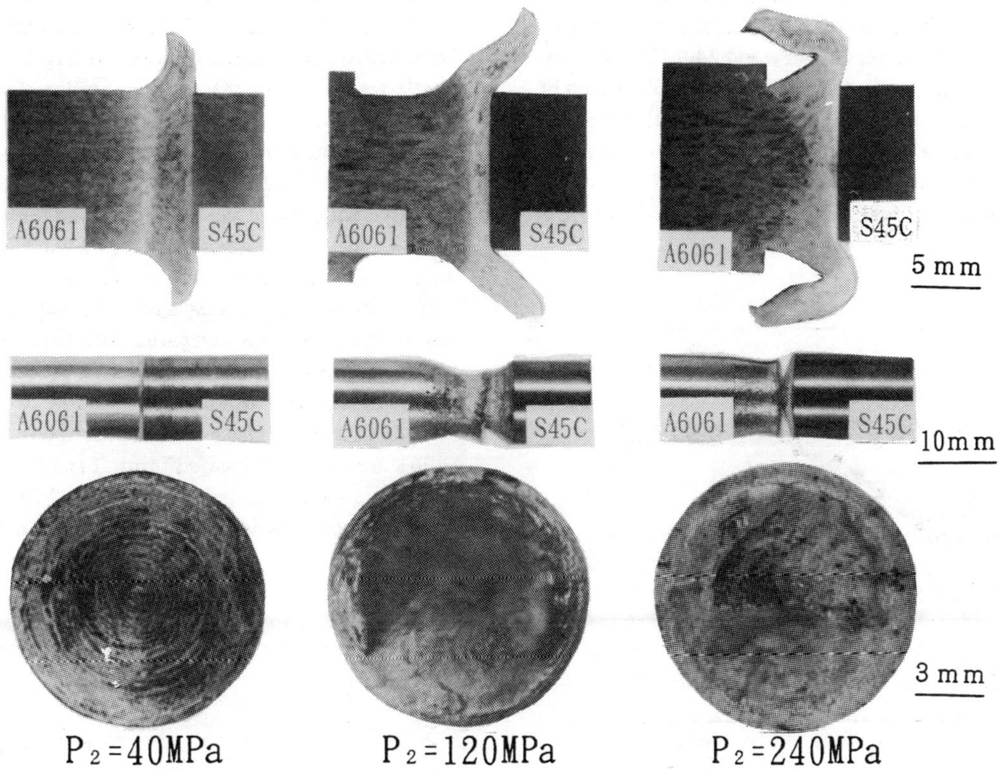

Fig.5 Appearance of interfacial layer, tensile tested specimen and fractured surface.

a 240MPa forge pressure case, tensile fracture almost occurred in A6061 base material and partially at the weld interface. It was observed that the tensile fracture surface approached closer to the weld interface according to increases in the forge pressure, for joints welded under forge pressure not less than 80MPa. Similar results were obtained for other A6061/steel joints.

Welded joints of heat treated aluminum alloy such as A6061-T6 often scften due to the welding heat (Hashimoto, 1980). In this investigation, tensile fracture of sound welds occurred in A6061 base material, however, these maximum tensile strengths are low compared with that of A6061 base material, which can be attributed to thermal softening caused by the friction heat. Hardness distribution in A6061 side of typical A6061/S45C and A6061/SUS304 joints are shown in Fig.6. A6061 of all welds were softened and the softened A6061 area of the joint welded under high forge pressure was narrow compared with the joint welded under low one. These facts can be ascribed that the softened A6061 was exhausted with burr under the action of forge pressure. It was concluded that the joint strength and fracture location depends strongly on the width of the softened A6061 area.

As mentioned above, aluminum alloy such as A6061-T6 often soften with welding heat, with the result that welds of these aluminum alloys are often heat treated after welding (Hashimoto, 1980). Thus A6061/S45C and A6061/SUS304 joints welded under 120MPa and 240MPa were heat treated. Hardness

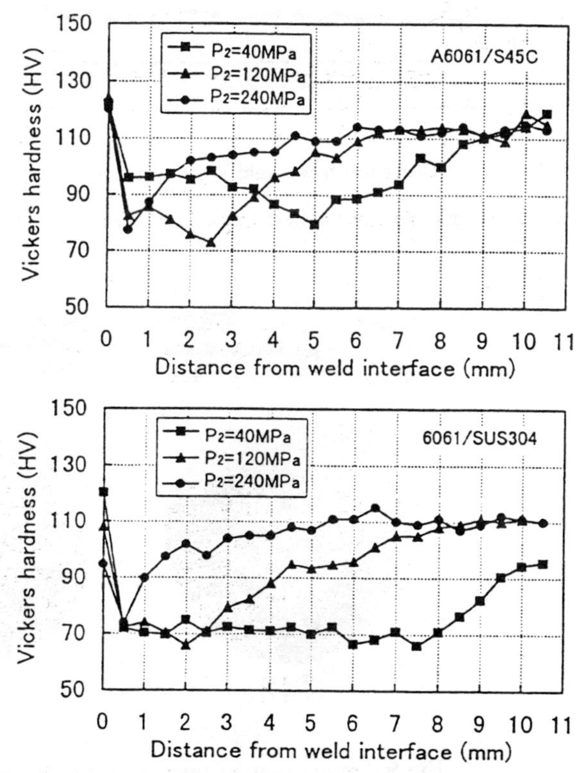

Fig.6 Hardness distributions of joints.

distribution in A6061 side of various heat treated A6061/S45C joints are shown in Fig.7. In the case of T42, the whole A6061 area of both joints welded under 120MPa and 240MPa softened. On the other hand, in the case of T62, the whole A6061 area of both joints hardened as same as the received material of A6061-T6. In the case of T62', although the softened A6061 area was narrow compared with that of the weld joint, the whole A6061 area had not hardened. Heat treated A6061/SUS304 joints showed the same hardness distribution such as A6061/S45C joints. Tensile strength of these heat treated A6061/S45C and A6061/SUS304 joints are shown in Fig.8. And appearance of tensile tested specimens and fractured surfaces in S45C side of heat treated A6061/S45C joints welded under 240MPa are shown in Fig.9. Fracture of other heat treated joints were similar to Fig.9. In the case of T42, tensile strengths of all joints were low compared with that of the weld joint, and tensile fracture occurred in the A6061 base material, due to the wide softened A6061 area. On the other hand, in the case of T62 and T62', tensile strength of both joints were high compared with that of the weld joint, and that of T62' were low compared with that of the T62 specimen. Tensile fracture of both T62 and T62' treated joints occurred at the weld interface, although that of T62' treated joint occurred partially in A6061. These facts are ascribed to remaining softened A6061 area of T62' treated joint. Tensile strength of T62 and T62' joints welded under 240MPa were high compared with that of joints welded under 120MPa, which can be attributed to the

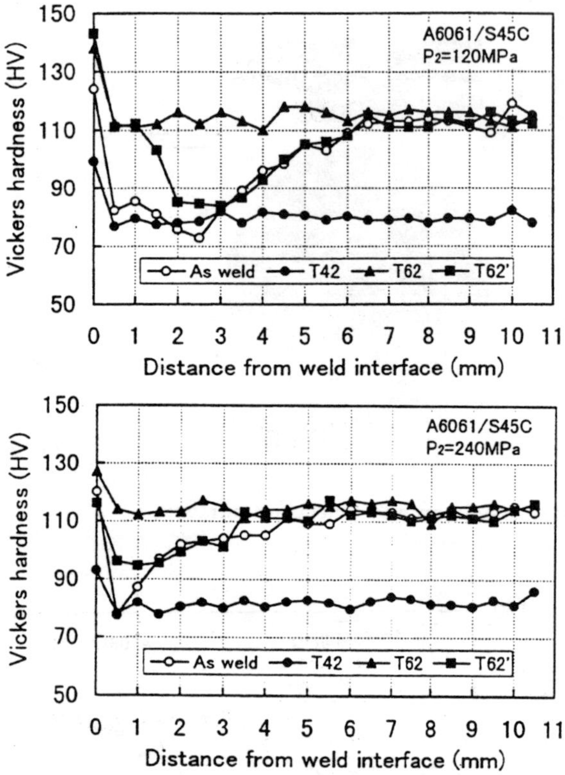

Fig.7 Hardness distributions of heat treated A6061/S45C joints.

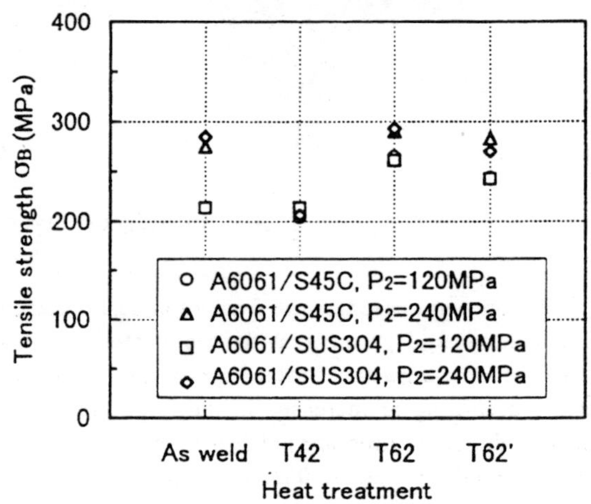

Fig.8 Tensile strengths of heat treatment joints.

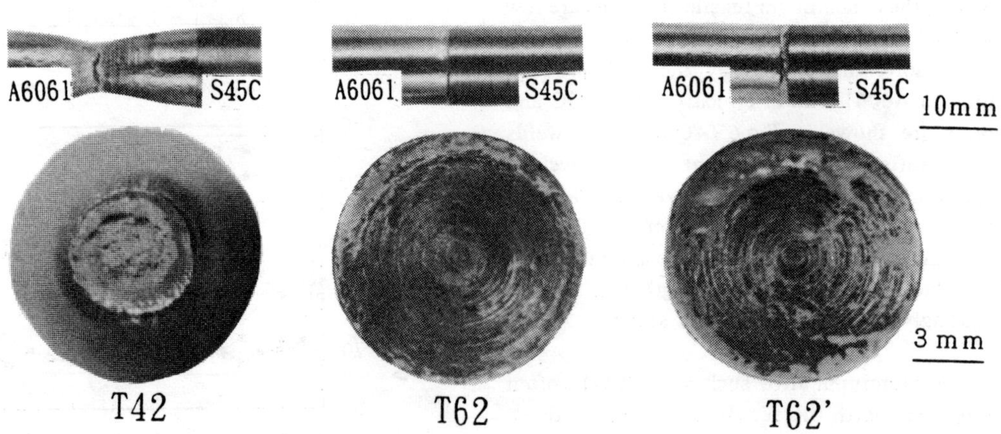

Fig.9 Appearance of tensile tested specimen and fractured surface of heat treated joints.

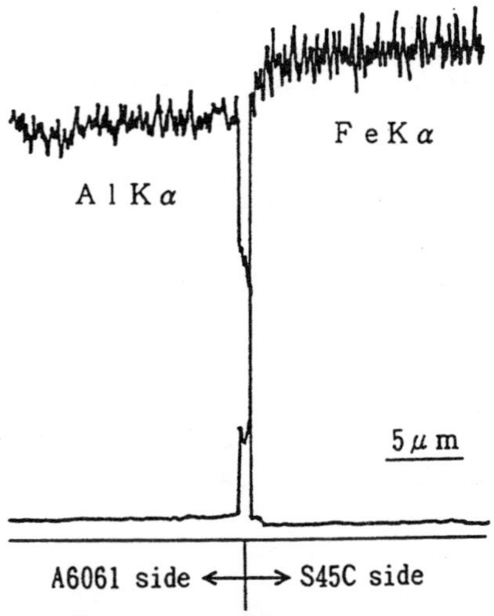

Fig.10 EPMA line scanning profile of interfacial layer of heat treated joint.

welding at weld interface of the weld joint welded under 240MPa was sounder than that of 120MPa.

Scott and Sequires (1966) investigated the effect of annealing the intermetallic layer of $FeAl_3$ with some Fe_2Al_5 formed at the weld interface between friction welded aluminum and stainless steel, and found appreciable growth after 4hr at 500℃, which was accompanied by a loss in strength. Thus, the intermetallic layer of heat treated welds was investigated using an electron probe micro analyser (EPMA). EPMA line scanning profile of T62 treated A6061/S45C joint welded under 240MPa are shown in Fig.10. It is not possible to clear the formation of Fe-Al intermetallic layer smaller than 1mm judging from a spatial resolving power of EPMA (1～2mm). Other welds show the same EPMA line scanning profile. In general, sound welds of aluminum and steel can be produced with an intermetallic thickness of 0.2～1mm (Elliott and Wallach, 1981). It was concluded that heat treatments in this investigation do not lead to a loss in strength.

CONCLUSIONS

Friction welding of 6061 aluminum alloy and steel was conducted, and the effects of the softened A6061 area and various heat treatments of the welded joint on the joint strength were examined. Under the same welding conditions, the tensile strength of A6061/steel joints were much alike for all the steels used. Sound welds were produced in all A6061/steel combinations, and the maximum tensile strength of these welds was about 280MPa. In this case, tensile fracture occurred in thermal softened A6061 area. Joint strength depends on the width of this softened A6061 area. The whole A6061 area of T62 heat treated joint hardened to the same extent as the received material of A6061-T6, and tensile strength of this joint was high compared with that of the weld joint.

REFERENCES

Elliott, S, and Wallach, ER (1981). "Joining Aluminum to Steel Part 2-Friction Welding," *Metal Construction*, Vol13, pp 221-225.

Hashimoto, K (1980). *The Welding Engineering*, Kogaku Tosyo, pp 78-281.

Ochi, O, Ogawa, K, Suga, Y and Yamamoto, Y (1995). "Friction Welding of Aluminum Alloy and Carbon Steel Using Insert Metal," *Proceedings of The 38th Japan Congress on Materials Research*, pp 80-85.

Sassani, F, and Neelam, JR (1988), "Friction Welding of Incompatible Materials," *Welding Research Supplement*, 264-270-s.

Scott, MH, and Squires, IF (1966). "Metallurgical Examination of Aluminum-Stainless Steel Friction Welds," *British Welding Journal*, No.3, pp 151-164.

Proceedings of the Seventh (1997) International Offshore and Polar Engineering Conference
Honolulu, USA, May 25-30, 1997

ISBN 1-880653-28-1 (Set); ISBN 1-880653-32-X (Vol. IV)

Friction Welding of 6061 Aluminum Alloy Pipe to S25C Carbon Steel Pipe

G. Kawai
Osaka Sangyo University, Daito, Japan
K. Ogawa
Osaka Prefecture University, Sakai, Japan
H. Tokisue
Nihon University, Narashino, Japan

ABSTRACT

Two kinds of tubular materials with different thermal and physical properties were friction welded with the following procedure and the subjects. A series of welding was carried out by a brake type friction welding method. Welded materials were JIS 6061 aluminum alloy pipe and JIS S25C carbon steel pipe. After welding, both outer burr and inner burr formed at welds during welding were measured. Also the welded joint quality was examined in tensile test.

key words: Friction welding, aluminum alloy, carbon steel, up-set burr, tensile strength, welding of dissimilar pipe, hollow degree in pipe.

INTRODUCTION

Not only a pipe is used frequently as a conduit which passes gas and liquid, but also for the purpose of a reduction in product weight and material cost. Up to now, some authours have reported the results of investigation on the friction welding of pipe consited of similar materials (Ogawa, 1988, Kobayashi, 1986, Shaitarov, 1979, Sutovskii, 1975, Tou, 1976, 1977). However, the study of friction welding on a dissimilar material pipe is not seen at all.

In the friction welding of pipe, the forming of inner burr at the inside of pipe differs from that of solid bar without any inner burr. The forming mechanism of inner burr varies with the welding condition, the sort of base material and specially the diameter and thickness of pipe. The outer burr formed at the outside of pipe can be confirmed and removed easily by a machining. However, it is so difficult to confirm and remove the inner burr. Thus, the inner burr produces the obstruction of flow in using of a friction welded pipe joint.

Thereupon in this paper, the friction welding of an aluminum alloy pipe to a carbon steel pipe having very different thermal and mechanical properties was carried out to discuss the forming mechanisum of both outer and inner burrs and the joint strength. That is, the influence of the welding condition and the shape of pipe on the deformation behaviour of welds and the tensile strength of pipe joints was examined. Especially in this study, to estimate the shape of inner burr from the shape of outer burr was tried.

EXPERMENTAL PROCEDURE

Materials used are Al-Mg-Si aluminum alloy solid bar (JIS-A6061-T6) and 0.45 % carbon steel solid bar (JIS-S45C) on the market widely. The chemical compositions and mechanical properties are shown in Table 1 and Table 2 respectively. The received materials of 20 mm in diameter were cut off to 80 mm for the aluminum alloy and 120 mm for the carbon steel in length by a saw and then 30 mm hole in depth was drilled on one end face in order to make a pipe. The pipes of 6 thicknesses, 1.0 mm, 1.5 mm, 2.0 mm, 2.5 mm, 3.0 mm and 3.5 mm were machined using drills of 18 mm, 17 mm, 16 mm, 15 mm, 14 mm and 13 mm in diameter respectively as shown in Fig.1. The form ratio (thickness / outside diameter) were 1.0/20, 1.5/20, 2/20, 2.5/20, 3.0/20 and

Table 1 Chemical compositions of base materials (mass%).

(a) Aluminum alloy

Material	Si	Fe	Cu	Mn	Mg	Cr	Zn	Ti	Bi	Pb	Al
A6061-T6	0.58	0.16	0.20	0.01	0.99	0.07	0.01	0.01	-	-	RE

(b) Carbon steel

Material	C	Si	Mn	P	S	Cu	Ni	Cr	Fe
S45C	0.45	0.27	0.72	0.022	0.016	0.040	0.045	0.090	RE

Table 2 Mechanical properties of base materials.

(a) Aluminum alloy

Material	Proof stress $\sigma_{0.2}$ (MPa)	Tensile strength σ_B(MPa)	Elongation ε (%)	Vickers hardness HV(micro)
A6061-T6	273	289	21.9	125

(b) Carbon steel

Material	Yield point σ_{ys} (MPa)	Tensile strength σ_B(MPa)	Elongation ε (%)	Vickers hardness HV(micro)
S45C	422	727	19	213

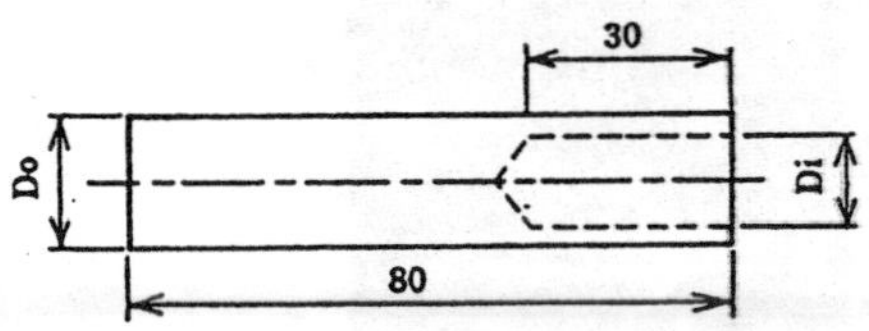

(a) Aluminum alloy

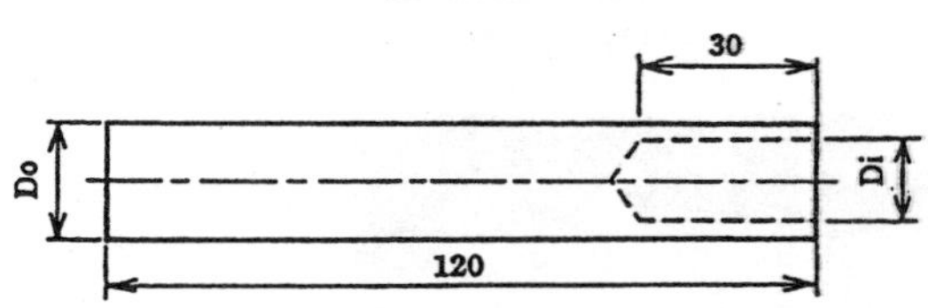

(b) Carbon steel

Fig.1 Shape and dimensions of base materials.

3.5/20 in turn. One end face of all materials to be welded was machined in even plan and then degreased by acetone before welding.

The equipment used was a brake type friction welding machine for experiment. The machine is capable of providing widely variable values of friction pressure P_1, forging pressure P_2, friction time t_1, rotating speed N and braking time t_B. Moreover the machine is instrumented to measure and record on a photocorder the speed of the rotating component, the resistance torque, the axial force and the axial shortening (up-setting) as they vary with time during the welding. Figure 2 is the cycle of the brake type friction welding process.

The friction welding conditions were consisted of each factors of the friction pressure P_1: 5, 10, 15 MPa, the forging pressure P_2: $2P_1$, $3P_1$ MPa, the friction time t_B: 1s, the braking time t_B: 0.3 s and the rotating speed N: 2.5, 3.0, 3.5 m/s. The constant- circumferential speed throughout each pipe was calculated at the point of $(Di/2+t)2/3$ on the end face to be welded as the rotating speed V, where Di and t are the inside diameter and the thickness of pipe respectively. These factors for the friction welding conditions are shown in Table 3.

The measuring positions on the up-set burr are shown in Fig.3. The measurement of burr shape was performed on a half-sectioned

Table 3 Factor for friction welding condition.

Inner diameter Di (mm)	Thickness of wall t (mm)	Rotating speed N(1/sec)	Friction pressure P_1 (MPa)	Forging pressure P_2 (MPa)
18.0	1.0	41.2 49.4 57.6		
17.0	1.5	41.9 50.3 58.7	5	
16.0	2.0	42.6 51.2 59.7	10	$2 \cdot P_1$
15.0	2.5	43.4 52.1 60.8		$4 \cdot P_1$
14.0	3.0	44.2 53.1 61.9	15	
13.0	3.5	45.1 54.1 63.1		

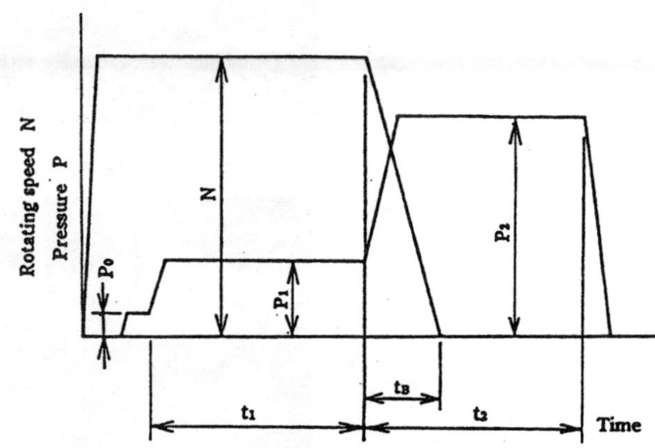

P_0:Pre-heating pressure
P_1:Friction pressure
P_2:Forging pressure
t_1:Friction time
t_2:Forging time
t_B:Braking time
N :Rotating speed

Fig.2 Cycle of the brake type friction welding process.

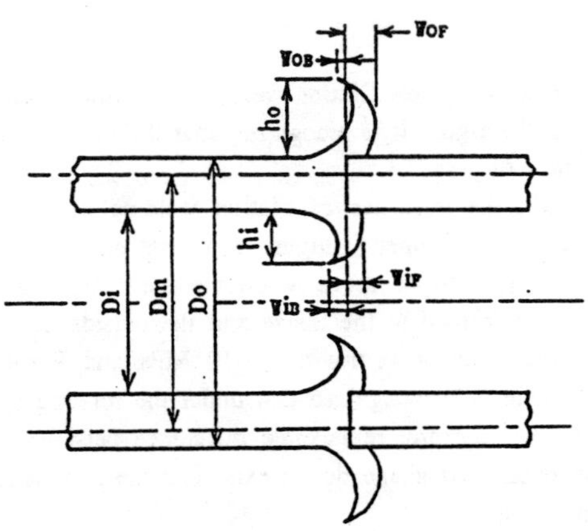

Fig.3 Shape of up-set burr.

area cut along axis center using a universal projector.

RESULTS AND DISCUSSION

Influence of friction welding condition on burr shape

In the friction welding of aluminum alloy to carbon steel, only the aluminum alloy side deforms forming up-set burr, while the carbon steeel side doesn't deform at all. Figure 4 shows the macro-view of the typical friction welds with various thicknesses of pipe. From the figure it is recognized that the burr shape changes with thickness of pipe.

First of all, the influence of friction welding condition on the burr shape was examined. Figure 5 shows an example of the influence of the rotating speed V on the burr shape h_o, h_i, W_{oF}, W_{oB}, W_{iF}, W_{iB} formed at the inside and the outside of D_i=17 mm pipe with the constant values of P_1=10 MPa and P_2=30 MPa. From the figure it is recognized that under the rotating speed from 2.5 m/s to 3.5 m/s any remarkable difference between inner burr shape and outer burr shape does't exist and the symmetric burr is formed in all cases.

INFLUENCE OF THICKNESS OF PIPE ON BURR SHAPE

Burr becomes complicated in accordance with its growth, that is, at the beginning of welding the edge of burr curls back and the edge progresses along the carbon steel keeping the curl back with the progress of welding. The burr shape varies with the welding conditions and the thickness of pipe. Figure 6 illustrates the change in burr size with the thickness of pipe welded under the

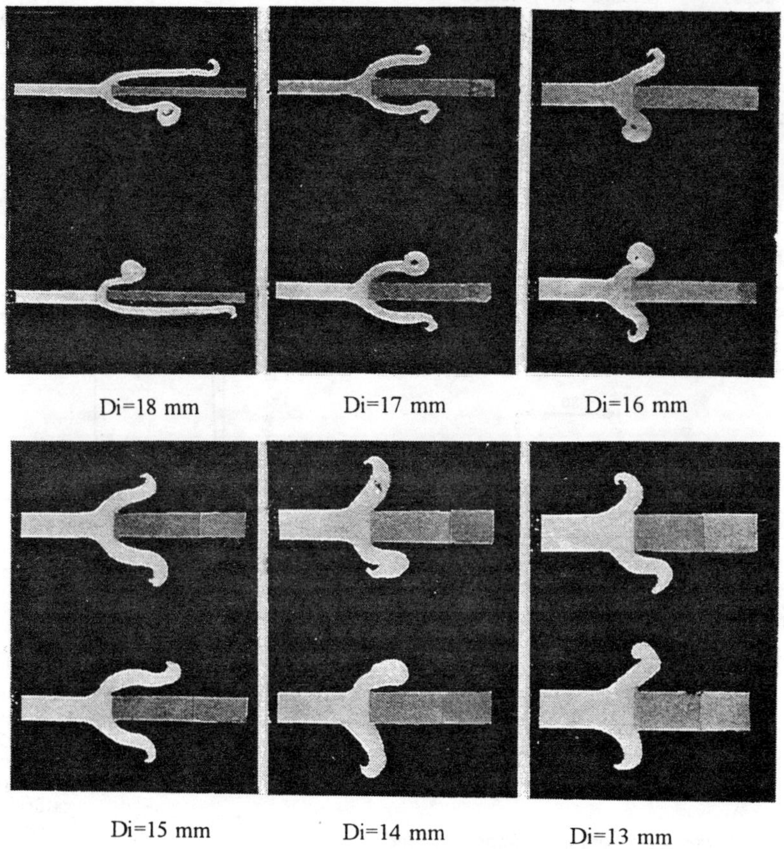

Fig.4 Macro-view of the friction welds (P_1=15MPa, P_2=30MPa and V=3.0m/s).

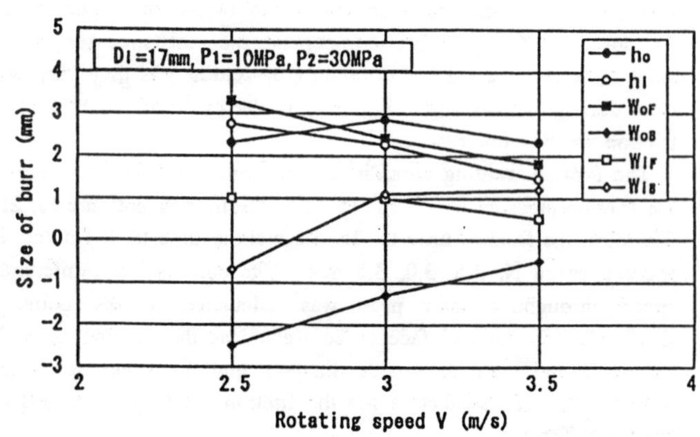

Fig.5 Relation between size of burr (burr height and burr width) and rotating speed.

condition of P_1=15 MPa, P_2=30 MPa, V=3.0 m/s. The height of outer burr ho doesn't change with increase of the thickness of pipe,while the height of inner burr hi changes with one. That is, the hollow degree of pipe decreases with increase of the thickness of pipe.

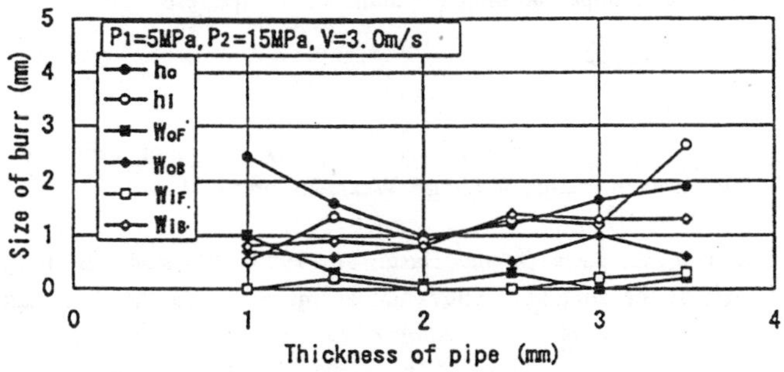

(a) Low pressure (P_1=5MPa, P_2=3 P_1)

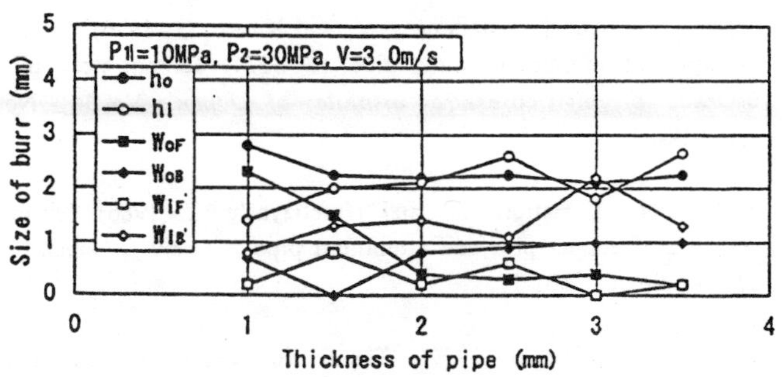

(b) Middle pressure (P_1=10MPa, P_2=3 P_1)

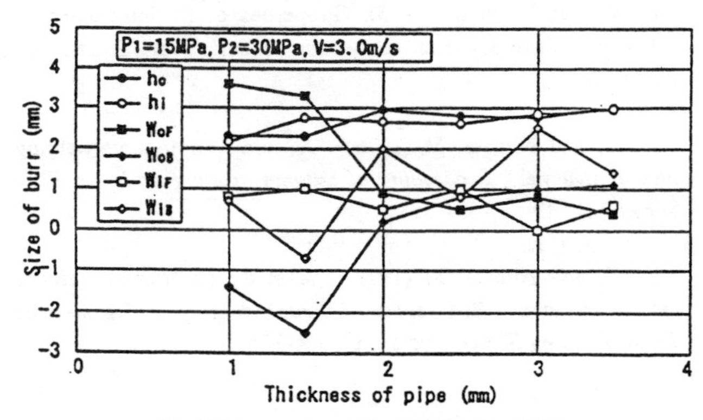

(c) High pressure (P_1=15MPa, P_2=2 P_1)

Fig. 6 Relation between size of burr and thickness of pipe.

INFLUENCE OF WELDING PRESSURE ON UP-SET QUANTITY AND TENSILE STRENGTH

Figure 7 shows the relation between the up-set quantity δ and the welding pressure P at the rotating speed V=3.0m/s. The up-set quantity increases in proportion to the welding pressure. Figure 8 shows the relation the tensile strength σ_B and the welding pressure P under the same condition with Fig.7. The tensile strength increases in propotion to the pressure except extremely thin pipe of 1 mm thickness (Ogawa, 1996). The pipes of 3 mm thickness show the maximam tensile strength of 170 MPa. Meanwhile the thin pipe of 1mm thickness shows the maximum tensile strength when the friction pressure P_1=10 MPa and the forging pressure P_2=20 MPa, but when the pressure over above value is adopted, the tensile strength decreases conversely.

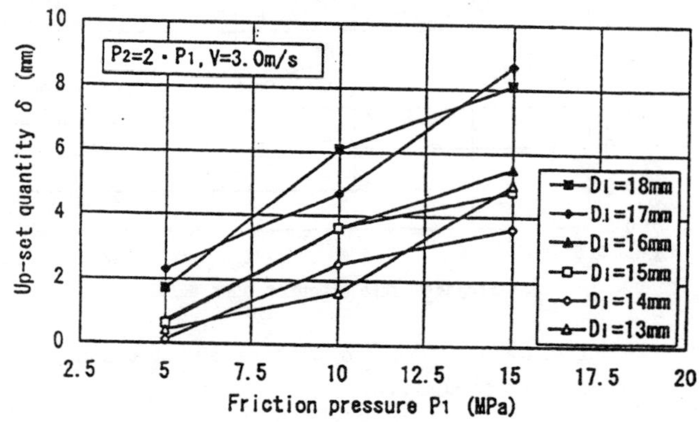

Fig.7 Relation between up-set quantity and friction pressure.

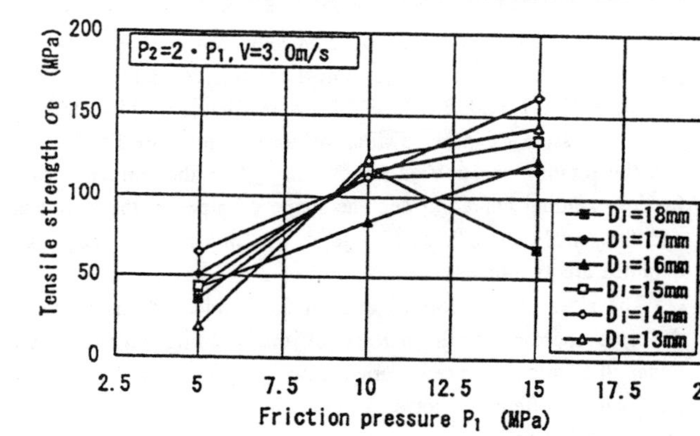

Fig.8 Relation between tensile strength and friction pressure.

RELATION BETWEEN SHAPE OF INNER BURR AND SHAPE OF OUTER BURR

The inner diameter of burr di and the outer diameter of burr do are obtained from results of measurment of the height of inner burr hi and the height of outer burr ho. Thus, the inner diameter of burr di is in proportion to the outer diameter of burr do except the pipe of 1 mm thickness and the inner diameter of burr di increases with decrease in thickness of pipe being independent of the rorating speed. Figure 9 shows the result obtained at the speed V=3.0 m/s in which the inner diameter of burr di becomes small and the inside space of pipe becomes narrow with increase of outer diameter of burr do.

From the fact, it was clear that in the pipe having an constant outside diameter of 20 mm and the inside diameters of 18 mm, 17 mm, 16 mm, 15 mm, 14 mm and 13 mm, the estimation of hollow degree of pipe was possible using the following equation. The equation to estimate the hollow degree (or inner diameter of burr di) of pipe was introduced as follows,

$$d_i=(4.152-1.105 d_o/D_o)D_i$$

where, d_o, D_i and D_o are the diameter of outer burr, the inner diameter of pipe and the outer diameter of pipe respectively.

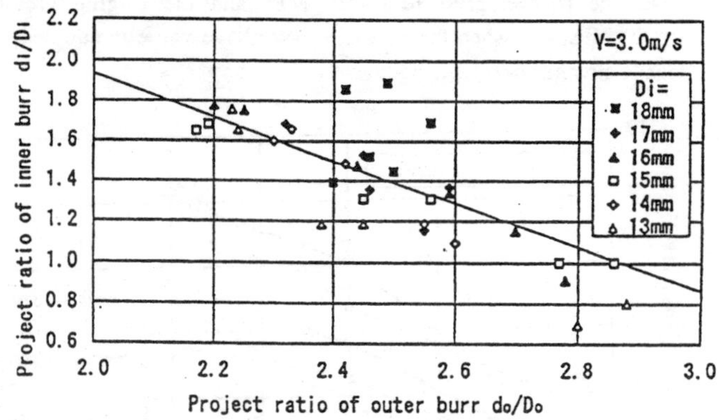

Fig.9 Relation between project ratio inner burr and project ratio of outer burr.

CONCLUSIONS

Aluminum alloys (JIS-A6061) pipe and carbon steel (JIS-S45C) pipe having various form ratio (thickness/outside diameter), 1.0/20, 1.5/20, 2.0/20, 2.5/20, 3.0/20, 3.5/20 were friction welded under the friction pressure P_1: 5 ~ 15MPa, the forging pressure P_2: $2P_1$, $4P_1$ and the rotating speed V: 2.5 ~ 3.5 m/s. And the influence of the friction welding condition and the shape of pipe on the burr shape and the tensile strength of pipe joint was investigated. The results obtained are as follows:

(1) The sizes of burr formed at both the inside and the outside of the pipe increases with increase of the welding pressure, while under the rotating speed varied from 2.5 m/s to 3.5 m/s any remarkable difference between inner burr shape and outer burr shape does't exist.

(2) The burr progresses complicately as the welding time elapses, that is, at first the tip of burr curls back and then the burr proceeds along the carbon steel in accordance with increase of the burr.

(3) The tensile strength of pipe joints increase in propotion to the welding pressure except extremely thin pipe of 1 mm thickness.

(4) It is possible to estimate the hollow degree of pipe by measuring the height of outer burr. The equation to estimate the hollow degree (or inner diameter of burr d_i) was introduced as follows,

$$d_i=(4.152-1.105 d_o/D_o)D_i$$

where, d_o, D_i and D_o are the diameter of outer burr, the inner diameter of pipe and the outer diameter of pipe respectively.

REFERENCES

Duffin, F, D, and Bahrani, A, S (1973). "Friction behaviour of mild steel in friction welding," Wear, 26, pp.53-74.

Ogawa, K, Kaga S, and Hasui, A (1988). "A study on friction welding of aluminum alloy tubular material -On the deformation behavior of weld-," *Proceeding of the 4th International Conference on Aluminum Weldments*, The Japan Light Metal Welding and Construction Association, pp.236-237.

Ogawa, K, Ochi, H, Yamamoto, Y, and Kawai, G (1996). "Friction welding of aluminum alloy pipe to S25C carbon steel pipe," *Journal of high pressure institute of Japan*, Vol.34, No.6, pp.381-387.

Kobayashi, H, Hattori, T, and Umebayashi, T (1986). "Friction welding of copper pipe and aluminum pipe," *Report for meeting of JFWA*, No.261, pp1-17.

Shaitarov, A, N, et al. (1979). "The friction welding of non-rotating joints in pipes using a cutting insert," *svar. Proiz.*, No.12, pp.16-17.

Sutovskii, P, M, et al. (1975). "Properties of friction-welded joints between drill pipes and connecting ends," Svar. Proiz., No.10, pp.25-27.

Tou, T, and Nimata, M (1976). "A study on friction welding of tubular material," *Proceeding of autumn meeting of Japan Welding Society*, No.19 , pp.354-355.

Tou, T, and Nimata, M (1977). "A study on friction welding of tubular material - 2nd report -," *Proceeding of spring meeting of Japan Welding Society*, No.19, pp.304-305.

Proceedings of the Seventh (1997) International Offshore and Polar Engineering Conference
Honolulu, USA, May 25-30, 1997

ISBN 1-880653-28-1 (Set); ISBN 1-880653-32-X (Vol. IV)

The New Eurocode 3 - Part 2 - Annex C: Approach for the Choice of Steel Material to Avoid Brittle Fracture

Natalie Stranghöner, Gerhard Sedlacek, Gero Stötzel, Winfried Dahl and Peter Langenberg
RWTH Aachen
Germany

Abstract

The appropriate choice of steel material is important for structures used in low temperature regions, in particular for structures susceptible to fatigue as bridges, pressure vessels, tanks and offshore structures, in order to avoid brittle fracture of parts of the structures. In the frame of the harmonization of the European design rules in the Eurocodes, a new approach, the Annex C of Eurocode 3, Part 2, has been developed, which gives rules to prevent brittle fracture of steel structures.

Keywords: Brittle Fracture, Choice of Steel Material, Fracture Mechanics, Eurocode 3

1 Introduction

The appropriate choice of steel material for steel structures is of great importance with regard to brittle failure of parts of the structures subjected to static or fatigue loading. With the introduction of welded steel structures instead of riveted joints in the early twenties of this century, by that time unknown cleavage failures of several steel structures occured, for instance: in the thirties the bridges of Rüdersdorf and Hardenbergstraße (Zoo), Berlin, both in Germany and the bridge in Hasselt, Belgium collapsed over night by brittle fracture and during World War II several liberty-ships of the United States of America failed with cleavage fracture. Several investigations in the thirties to fifties were carried out to determine the failure reasons. They came to the conclusion that the steel grade, the steel composition and the charpy energy of the steels as well as the plate thickness and the service temperature have a great impact on the susceptibility of steel structures to brittle or cleavage fracture (Stranghöner (1997)). Therefore, in the past each country developed its own national recommendation for the prevention of brittle fracture for various types of steel structures: steel bridges, pressure vessels, tanks and offshore structures.

The German guideline for the choice of steel material, the DASt-Ri 009, is based, for instance, on empirical results from large scale tests and engineering experiences coming out from research performed in these decades. Some rules like the French AFNOR are already based on fracture mechanic methods. In the frame of the harmonization of the European design codes, a concept for the choice of steel material for steel bridges has been developed, which is laid down in the Annex C of Eurocode 3, Part 2. In the following, firstly the safety concept of the Eurocodes is explained shortly. Then the Annex C concept is presented. Its applicability to typical parts of steel bridges is demonstrated in accordance to experimental large scale tests (Sedlacek et al (1996), Stranghöner et al (1996), Langenberg et al (1996b)).

2 Safety Concept of the Eurocodes

The safety concept of the Eurocodes and also of Eurocode 3, the European standard for the design of steel structures, is related to limit states which are:

- the serviceability limit states,
- the ultimate limit states and
- the fatigue limit states.

Herein, the *serviceability limit states correspond to states beyond which specified service criteria are not longer met.* The *ultimate limit states are those associated with collapse, or with other forms of structural failure which may endanger the safety of people* (Eurocode 3, Part 1-1). Fatigue limit states refer to the durability of structures.

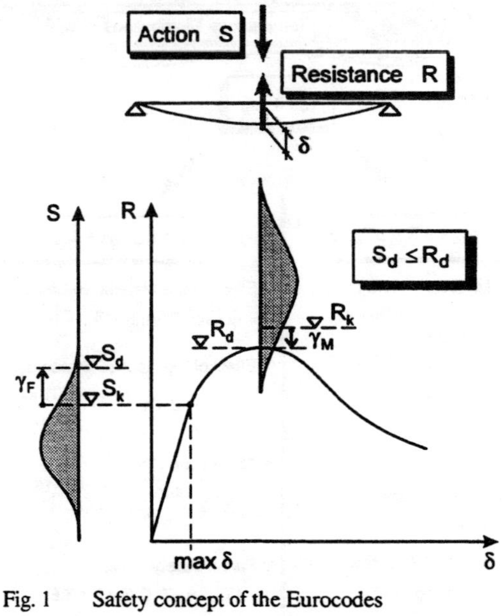

Fig. 1 Safety concept of the Eurocodes

The Eurocodes distinguish between actions S, e.g. loads, and resistances R, e.g. the bending resistance of a beam. Figure 1 illustrates the safety concept. In order to avoid failure of a structure, the design value of the action

S_d has to be equal or smaller than the design value of the resistance R_d: $R_d \geq S_d$.

Due to the fact that the action values and the values of the resistances scatter, partial safety factors have to be considered for both sides in the balance of the ultimate limit state. Herein, the actions and resistances are defined as characteristic values, e.g. as 5% fractiles. The partial safety factor for the action side γ_F leads to an increase of the action whereas the partial safety factor for the resistance side γ_M yields to a reduction of the resistance. Hereby, the partial safety factors take into account inaccuracies of the derived model describing the resistance, deviations between actual and nominal values, e.g. for the material behaviour, geometrical dimensions, etc. and the failure probability p_f, which corresponds to the probability, for which the structure fails when it is designed with the given concept. For normal failure consequences the failure probability is chosen to $p_f = 7 \cdot 10^{-5}$ which is adequate to the safety index $\beta = 3.8$ that is introduced instead of p_f for practical reasons.

According to this approach the design formulae for the determination of structural resistances are derived from experimental tests using a standardized test evaluation method. They must reflect the mechanical background in the limit states to describe the actual structural behaviour.

3 Annex C of Eurocode 3, Part 2

The mechanical behaviour of steel parts failing in brittle fracture can be best described by a fracture mechanic based concept which considers the main influencing parameters as the service temperature, the applied stresses in the tension members and the thicknesses of the structural details. The verification format of such a concept should be easy understandable for designers who most of all have no fracture mechanical knowledge.

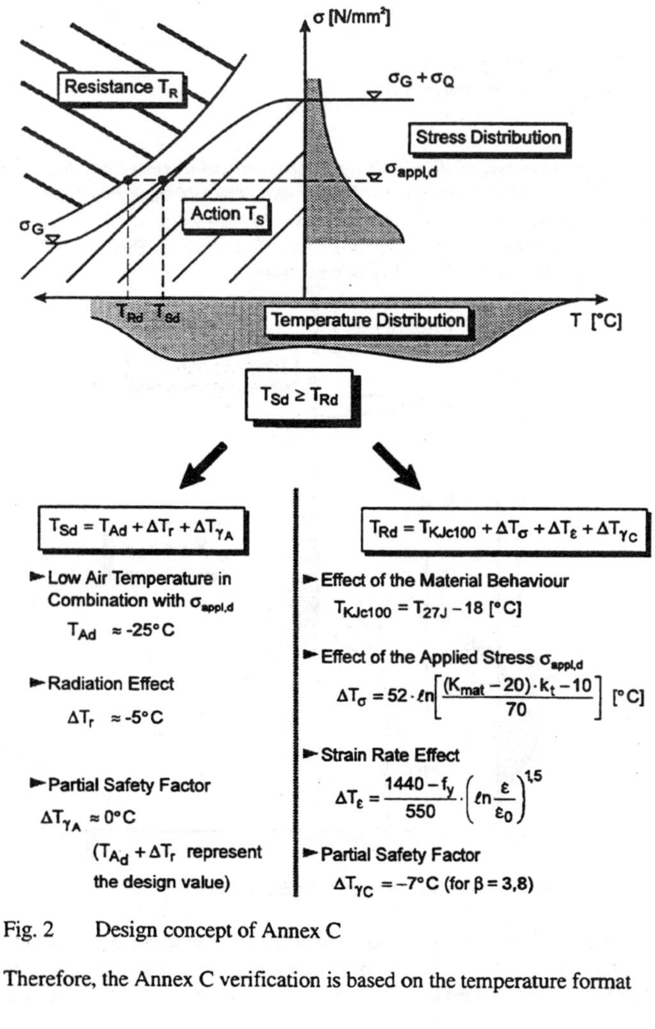

Fig. 2 Design concept of Annex C

Therefore, the Annex C verification is based on the temperature format

$$T_{Sd} \geq T_{Rd} \tag{1}$$

as illustrated in figure 2 in which

T_{Sd} represents the design value of the temperature action, that corresponds in principle to the service temperature and

T_{Rd} is the design value of the temperature resistance of the structural detail under consideration of the material toughness.

Dealing with temperatures instead of loads, see figure 1, the temperature action T_{Sd} must be higher than the temperature resistance T_{Rd} in order to fulfil the verification.

Due to the fact that brittle fracture of main parts of a structure may lead to a sudden collapse of the whole structure and there-fore, the safety of people may be endangered, the verification against brittle fracture is a ultimate limit state verification. Hence, partial safety elements have to be defined and considered. In the following, the determination of the design values of the action side and the resistance side is explained.

3.1 Temperature Action T_{Sd}

The temperature action T_{Sd}, as illustrated in figure 2, depends on the climatic conditions and consists of three elements:

- the most unfavourable air temperature T_{Ad},
- the temperature shift ΔT_r caused by radiation effects and
- the partial safety element ΔT_{γ_A}.

Herein, the two dimensional character of the action side is considered which results from the air temperature and stress effects from traffic. The most unfavourable air temperature has not to be the minimum air temperature statistically distributed over 50 years but the most disadvantageous low air temperature in combination with a high rush of traffic. The typical time history of the daily air temperature in winter months shows that the minimum air temperature on a winter day is at approximately 5 am whereas the highest traffic loads from heavy trucks, which contribute to the applied stresses $\sigma_{appl,d}$, may occur a few hours later.

However, for Germany T_{Ad} is taken as -25°C corresponding to the minimum air temperature at 5 am which is an admissible realistic but in some cases a conservative value (Mangerig (1995)). Of course, other countries as for example the Scandinavian countries or Canada have to define their own values. The temperature shift ΔT_r due to radiation effects is taken as -5°C (Mangerig (1995)). The partial safety element is set to $\Delta T_{\gamma_A} \sim 0$. The associated stresses $\sigma_{appl,d}$ are described in the next chapter.

3.2 Temperature Resistance T_{Rd}

The temperature resistance T_{Rd}, as given in figure 2, describes the resistance of a structural detail to brittle fracture at a specific temperature, i.e. the action temperature. T_{Rd} is mainly dependent on the toughness of the material, which can be described by means of a fracture mechanic concept. T_{Rd} consists of four additive elements:

- the temperature T_{KJc100} for which the material exhibits a fracture toughness of $K_{Jc} = 100 MPa\sqrt{m}$,
- the temperature shift ΔT_σ caused by fracture mechanical effects of the applied stresses $\sigma_{appl,d}$ in combination with the surface crack a_d,
- the temperature shift $\Delta T_{\dot{\varepsilon}}$ caused by strain rate effects and
- the partial safety element ΔT_{γ_c} that is calculated from a statistical evaluation of experimental full-scale tests.

The temperature T_{KJc100} is given by a correlation between the required temperature T_{27J} for a Charpy-V-energy of 27 Joules and the fracture toughness temperature at $K_{Jc} = 100 MPa\sqrt{m}$. The basic idea for this temperature correlation came from Sanz in the seventies (Marandet and Sanz (1976)) and was improved by Wallin (1990a) and Sedlacek et al (1993) for modern steels. The so called modified Sanz correlation reads:

$$T_{KJc100} = T_{27J} - 18 \quad [°C] \quad . \tag{2}$$

The temperature shift ΔT_σ takes account of fracture mechanical effects of the applied stresses $\sigma_{appl,d}$ and of the surface crack a_d. Wallin (1990a, 1990b, 1994) developed a master curve, as illustrated in figure 3, which describes the exponential increase of the toughness K_{Jc} over the normalized temperature $T - T_{KJc100}$ up to the initiation temperature T_i for stable crack propagation. This correlation is given by

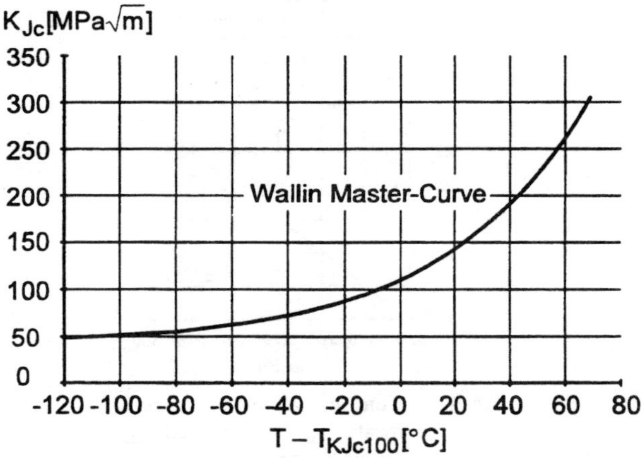

Fig. 3 Wallin Master-Curve

$$\Delta T_\sigma = T - T_{KJc100} = 52 \cdot \ell n \left(\frac{(K_{mat} - 20) \cdot k_t - 10}{70} \right) \quad [°C]. \tag{3}$$

k_t considers the influence of the plate thickness in the value of the material toughness K_{Jc} and is given by

$$k_t = \left(\frac{b_{eff}}{25\,mm} \right)^{\frac{1}{4}} \quad [-] \tag{4}$$

with $b_{eff} = 2 \cdot c$ [mm] for semielliptical surface cracks.

K_{mat} is determined according to the CEGB-R6-method, which gives a limit state curve for the calculated stress intensity factor K_I of the structural detail that is in the limit state equivalent to the material toughness K_{Jc}. Herewith, K_{mat} becomes

$$K_{mat} = K_{Jc} = \frac{\sigma_{appl.d} \cdot \sqrt{\pi \cdot a_d} \cdot Y}{k_{R6} \cdot \sqrt{1000}} \quad [MPa\sqrt{m}] \tag{5}$$

with

$\sigma_{appl.d}$	[MPa]	is the design value of the actual stresses in the structural detail due to primary and secondary stresses in combination with the service temperature T_{Ad}: $\sigma_{appl.d} = \sigma_p + \sigma_s$ [MPa],
a_d	[mm]	is the design crack size,
Y	[-]	is the stress intensity correction function for a semielliptical surface crack according to stress intensity handbooks,
$k_{R6} = \frac{1}{\sqrt{1 + 0.5 \cdot L_r^2}}$	[-]	is the option 2 limit curve from the CEGB-R6 failure assessment diagram for $L_r \le 1$ describing the plasticity correction factor and
$L_r = \frac{\sigma_p}{\sigma_{gy}(t)}$	[-]	with σ_p [MPa] as the primary stresses and $\sigma_{gy}(t)$ [MPa] as the state of general yielding related to the yield strength which is dependent on the plate thickness and is calculated by

$$f_y(t) = f_y - 0.25 \cdot \left(\frac{t}{t_0} \right) \quad [MPa]$$

with t [mm]= plate thickness, t_0 = 1 mm and f_y [MPa] = yield strength.

The primary stresses σ_p are stresses resulting from the applied loads, composed of the permanent actions G and variable loads Q. They can either be taken as the stresses resulting from design with actual loads or they may be taken as $\sigma_p = 0.5 \cdot f_y$ [MPa]. Herein, σ_p presents the most unfavourable loading case. It is justified by the following considerations:

- in consequence of the ultimate limit state, partial safety elements for the action side have to be taken into account,
- the partial safety element for permanent loads G corresponds to $\gamma_G = 1.35$,
- the partial safety element for variable loads Q is $\gamma_Q = 1.50$ and
- the sections used in bridges are class 3 -sections according to Eurocode 3. These are sections *in which the calculated stress in the extreme compression fibre of the steel member can reach its yield strength, but local buckling is liable to prevent development of the plastic moment resistance* (Eurocode 3, Part 1-1).

The following load combination has to be applied and fulfiled

$$\sigma(G_k \cdot \gamma_G) + \sigma(Q_k \cdot \gamma_Q) \le f_{yd} \tag{6}$$

with $f_{yd} = \frac{f_y}{\gamma_M}$ and $\frac{\sigma(G_k)}{\sigma(Q_k)} \approx 1.0$. Hence, equation (6) reads consequently

$$1.35 \cdot \sigma(G_k) + 1.5 \cdot \sigma(Q_k) \le \frac{f_y}{1.1}$$

$$\Leftrightarrow \quad 2.85 \cdot \sigma(G_k) \le 0.9 \cdot f_y \tag{7}$$

$$\Leftrightarrow \quad \sigma(G_k) \le 0.32 \cdot f_y \quad .$$

The primary stresses are defined as the actual stresses in the structural detail. Taking account of the frequent value $\psi_1 = 0.75$ for the dominant variable action, the primary stress can be derived to

$$\sigma_p = \sigma(G_k) + \psi_1 \cdot \sigma(Q_k) = 0.56 \cdot f_y \approx 0.50 \cdot f_y \quad . \tag{8}$$

The secondary stresses are approximated by $\sigma_s = 100$ MPa to take account of residual stresses resulting from welding. They are taken to be independent on the steel grade.

The crack size a_d is determined using an initial semielliptical surface crack with an initial depth

$$a_0 = 0.5 \cdot \ell n \left(\frac{t}{t_0} \right) \quad \text{with } t_0 = 1 \text{ mm} \tag{9}$$

dependent on the plate thickness and an initial crack width of $2c = 5 \cdot a_0$. The crack propagation from the initial crack a_0 to the design value of the crack a_d is calculated using the Paris equation.

The influence of the strain rate on fracture mechanic values is considered by the temperature shift $\Delta T_{\dot\epsilon}$ according to Falk (1993)

$$\Delta T_{\dot\epsilon} = \frac{1440\,MPa - f_y(t)}{550\,MPa} \cdot \left(\ell n \frac{\dot\epsilon}{\dot\epsilon_0} \right)^{1.5} \quad [°C] \tag{10}$$

with $\dot\epsilon_0 = 0.0001\ s^{-1}$.

The formula for T_{Rd} was restricted to T_{Mod}

$$T_{Mod} = T_{27J} - 18 + 52 \cdot \ell n \left[\frac{(K_{mat} - 20) \cdot k_t - 10}{70} \right] + \frac{1440 - f_y}{550} \cdot \left(\ell n \frac{\dot\epsilon}{\dot\epsilon_0} \right)^{1.5} \tag{11}$$

and was checked with 19 experimental large scale tests to determine the safety element ΔT_{γ_C}. These tests were performed with full-scale specimens with double edged through thickness cracks, as illustrated in figure 4, under static loading. Thus, strain rate effects could be neglected. Table 1 presents some experimental data including the experimental failure temperature T_{Exp} and the calculated failure temperature T_{Mod}.

From figure 4 it can be seen that the Annex C model describes the experimental behaviour quite satisfactory. The partial safety element ΔT_{γ_C}

was determined from a statistical analysis according to Annex Z of Eurocode 3, Part 1-1. This analysis resulted in the partial safety element $\Delta T_{\gamma_c} = -7$ °C. Statistical characteristics of T_{Mod} are the mean value deviation of $\bar{b} = -9.01$ °C and the standard deviation of $\sigma_D = 14.2$ °C. The negative value of the partial safety element is caused by the large deviation between the required nominal and the actual measured temperature T_{27J} at the Charpy-V-energy of 27 Joules.

Table 1 DECT-Full Scale Tests

Speci-men	Steel Grade	2a [mm]	a/w [-]	$\sigma_{fracture}$ [MPa]	T_{27J} [K]	T_{Exp} [K]	T_{Mod} [K]
1	S 355	60	0.2	350	248	238	303
2	S 355	60	0.2	349	248	253	303
3	S 355	60	0.2	377	213	254	273
4	S 690	60	0.2	503	186	232	259
5	S 690	60	0.2	494	203	223	274
6	S 690	60	0.2	477	203	233	271
7	S 885	60	0.2	784	203	243	304
8	S 885	60	0.2	824	203	258	308
9	S 885	60	0.2	822	203	263	308
10	S 855	60	0.2	760	205	251	304
11	S 355	37.36	0.126	362.3	263	232	305
12	S 355	39	0.13	332.9	173	163	202
13	S 690	36.36	0.121	354.9	233	232	262
14	S 690	36	0.12	521	233	248	292
15	S 690	37.2	0.124	395.2	203	204	240
16	S 890	38	0.127	703.2	233	273	313
17	S 890	38.4	0.128	651.4	233	278	307
18	S 890	40.9	0.136	662.9	198	232	276
19	S 890	39.9	0.133	606.9	198	245	269
For all specimens: t = 30 mm, 2W = 300 mm							

Furthermore, the Annex C model was checked with 69 full scale tests performed by Wallin (1990a, 1990b, 1994) at VTT, Finland. Wallin investigated □-profiles with surface flaws (SSCT/CCT-specimens) and ⊔-profiles with single edged notches. The ⊔-profiles were manufactured by welding so that residual stresses were induced. The specimens were tested statically in 3-point bending with a beam length of l = 600 mm. Figures 5 to 7 present the comparisons between experimental and calculated failure temperatures. It becomes obvious that also for these tests a good correlation was achieved between the experimental and calculated values. The mean value deviations and standard deviations were $\bar{b} = 0.45 - 7.07$ [°C] and $\sigma_D = 10.25 - 23.33$ [°C]. Partial saftey elements could not be evaluated because of the absence of nominal values for the geometrical dimensions and material data.

The afore described investigations were realized for base metals (BM). For the choice of steel material, fatigue cracks situated in the heat affected zones (HAZ) of welded structural details are also of great significance. First investigations carried out for 33 welded wide plates showed that Annex C is also applicable to these kinds of structural details. Figures 8 and 9 illustrate the results achieved for these plates (Langenberg et al (1996)).

The calculations to determine T_{Mod} were carried out with material parameters of the base metals, as illustrated in figure 8, and with material properties of the heat affected zones, as shown in figure 9. It becomes obvious that Annex C describes in both cases the tendency satisfactory. Nevertheless, the results using material data from the heat affected zones show a better mean value behaviour than the results considering material data from the base metal.

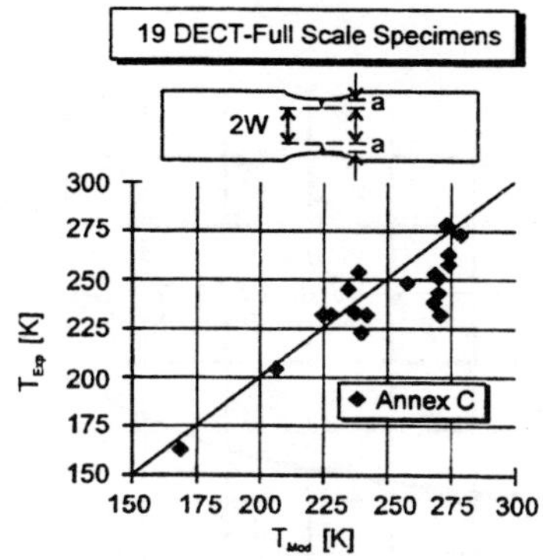

Fig. 4 Comparison between experimental T_{Exp} and calculated T_{Mod} failure temperatures

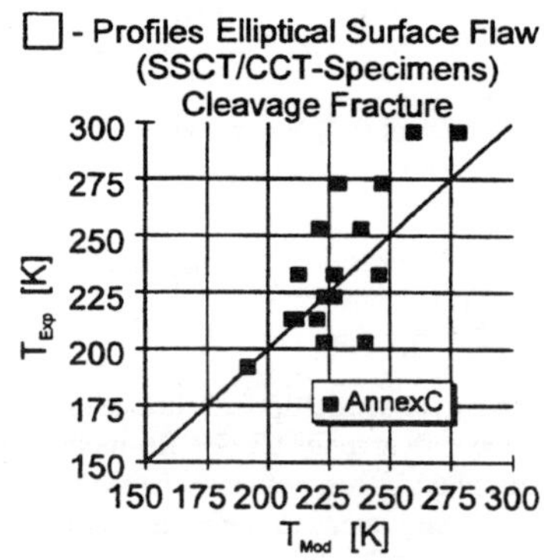

Fig. 5 Three-point-bending tests with □-profiles, SSCT/CCT-specimens, cleavage fracture, Wallin (1994), VTT, Finland

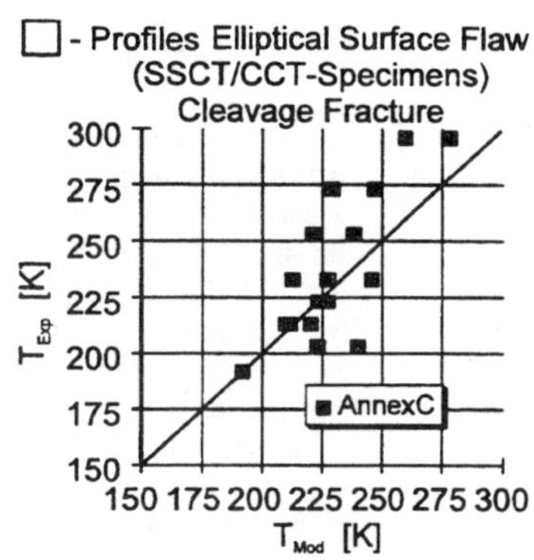

Fig. 6 Three-point-bending tests with □-profiles, SSCT/CCT-specimens, cleavage fracture, Wallin (1994), VTT, Finland

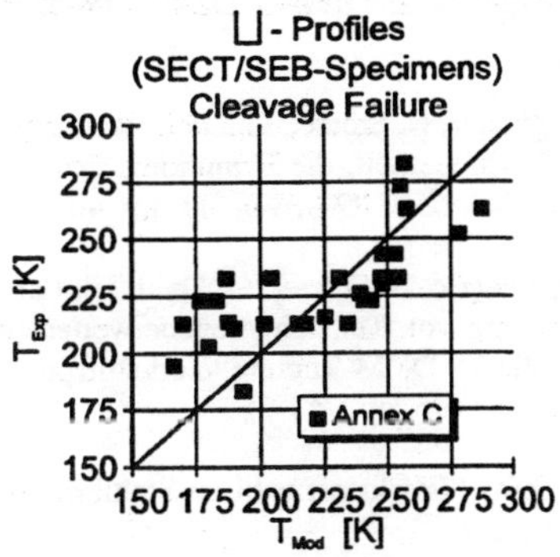

Fig. 7 Three-point-bending tests with ⊔-profiles, SECT/SEBT-specimens, cleavage failure, Wallin (1994), VTT, Finland

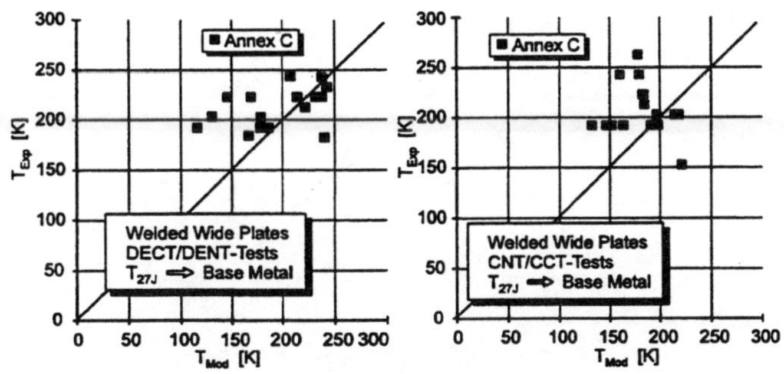

Fig. 8 Comparison of calculated and experimental failure temperatures for welded wide plates based on material parameters of the base metals

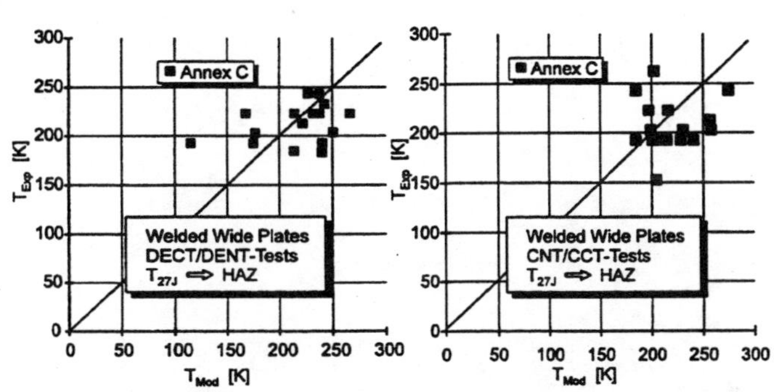

Fig. 9 Comparison of calculated and experimental failure temperatures for welded wide plates based on material parameters of the heat affected zones (HAZ)

However, the scatter is in both cases higher than those achieved for base metal specimens, see figures 4 to 7.

The larger scatter of the results from the welded wide plates in comparison to the base metal tests is caused by the inheterogenity of the material properties of the welded metal.

However, it is planned to investigate welded details by means of further tests. These tests will include wide plates with welded structural details as for instance longitudinal and transverse elements as they are common in steel bridges.

4 Tables for the Choice of Steel Material

With the explanations given afore it could be shown that the Annex C gives a reliable approach for the determination of the appropriate steel to avoid brittle fracture. In practical use, the presented method may take too much time, especially for those engineers who want to have a quick estimation for predesign. Therefore, it was decided to develop tables for the choice of steel material dependent on the steel grade and the plate thickness for some typical structural details. These tables may be considered as an user-friendly alternative to the temperature verification format given in Annex C.

For steel bridges such tables have been developed for structural details as illustrated in figure 10. The following input data were taken into account for the calculations:

- Eurocode 3 requires fatigue class 56 for this structural detail, which complies with $\Delta\sigma_c = 56$ MPa at $2\cdot10^6$ cycles.
- Primary stresses due to permanent and variable actions are determined to $\sigma_p = 0.50\cdot f_y$. Secondary stresses due to residual stresses are considered by $\sigma_s = 100$ MPa.
- The design value of the temperature action is $T_{Sd} = -30$ °C.
- The initial crack a_0 is a semielliptical surface crack at the welding seam.
- The fatigue loading is taken as one quarter of the total damage loading: $\frac{1}{4}\cdot\Delta\sigma_c^3\cdot2\cdot10^6$ which is equivalent to more than one quarter of the service life. This time period corresponds to the interval for the renewal of the corrosion protection, when inspections could be carried out.

The initial crack a_0 propagates due to fatigue loading to the design crack size a_d. With these input data and for a wide range of typical geometrical boundary conditions L/t and B/t, see also figure 10, the calculation of admissible plate thicknesses was carried out for different steel grades. For example, the Annex C method yields for the steel grade S 355 N/M and the given detail in figure 8 to an permissible plate thickness of t = 80 mm. Figure 10 presents some admissible plate thicknesses for various steel grades.

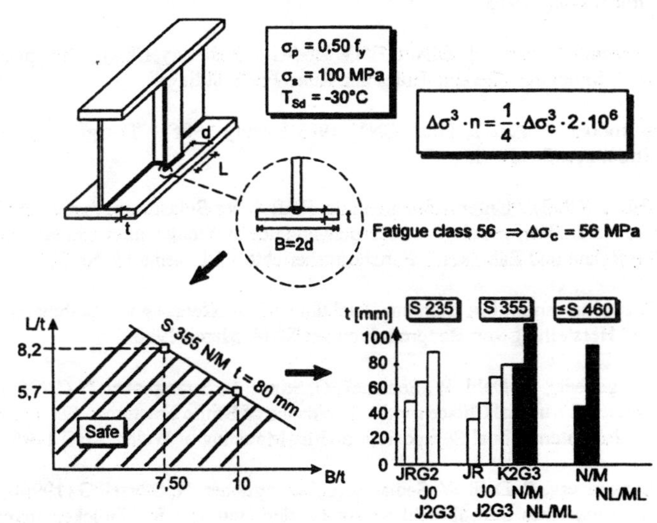

Fig. 10 Simplified choice of steel material according to Annex C

Due to the fact that the temperature action of -30°C may be just applicable to steel bridges in special regions, it is also necessary to give recommendations for the prediction of the maximum allowable plate thickensses for other temperature actions. Therefore, Eurocode 3, Part 2 considers this demand by a table for maximum plate thicknesses of steel plates and steel sections for different temperature actions from 0°C to -50°C.

Furthermore, the Annex C approach is not combined necessarily with the presented applied stresses $\sigma_{appl.d} = 0.5\ f_y$. The method is applicable for the actual stresses in the structural detail. This is important with regrad to the choice of steel for other steel structures than steel bridges.

5 Conclusion

The Annex C concept for Eurocode 3, Part 2 is a fracture mechanic based concept for the choice of steel material to avoid brittle fracture. Initially, it was just applied to steel bridges. Its extended applicability to other structures, as e.g. pressure vessels, tanks and offshore structures is desirable and has still to be developed. Such an extension of the applicability to other steel structures would reduce the number of different verification formats, which exist for all ranges of applications. For instance, the BSPD 6493 concept is a fracture mechanic based failure assessment concept that combines the CTOD-concept with failure assessment diagrams. It is a highly sophisticated method, but the application requires fracture mechanical knowledge by the user. Compared to this method the Annex C concept is based on the CEGB-R6 method combined with a temperature verification format. This method results in steel selection tables which allow a fast and economical steel selection in the daily practice. Herewith, the user is able to apply the Annex C approach with out any fracture mechanical knowledge. This is the great advantage of the Annex C concept. The discussion should be opened with regard to the question how far the Annex C concept can be applied to other fields of applications. *The steel does not know where it is used.*

6 References

AFNOR, A 36-010, "Choix des Qualités d´aciers pour construction métallique ou chaudronnée vis-à-vis du risque de rupture fragile", Association Française de Normalisation, May 1980

BSPD 6493. "Guidance on Methods for Accessing the Reliability of Flaws in Fusion Welded Structures", British Standard Institution Published Document, British Standard Institution 1991

DASt-Ri 009, "Empfehlungen zur Wahl der Stahlgütegruppe für geschweißte Stahlbauten", Deutscher Ausschuß für Stahlbau, Stahlbau Verlags GmbH Köln, 1973

Eurocode 3, Part 1-1, DIN V ENV 1993-1-1 (February 1992). "Design of Steel Structures, General Rules and Rules For Buildings"

Eurocode 3, Part 2, Draft ENV 1993-2 (July 1996). "Design of Steel Structures, Steel Bridges"

Falk, J (1993). "Untersuchungen zum Einfluß der Belastungsgeschwindigkeit auf das Verformungs- und Bruchverhalten an Stählen unterschiedlicher Festigkeit und Zähigkeit", Fortschrittsberichte VDI, Reihe 18, Nr. 117

Herstellungsrichtlinie Stahlbau, Richtlinie zur Ausführung von Stahlbauten und Herstellung von Bauprodukten aus Stahl, March 1996

Langenberg, P, Dahl, W, Sedlacek, G, Stözel, G, Stranghöner, N (1996a). "Annex C Material Choice for the Avoidance of Brittle Fracture in Eurocode 3", 2nd International Symposium on Mis-Matching of Welds, April 1996

Langenberg, P, Dahl W, Sedlacek, G, Stranghöner, N, Stötzel, G (1996b), "Europäisches Sprödbruchkonzept für den Bau von Stahlbrücken unter Berücksichtigung der Lebensdauer", DVM-Tagung Werkstoffprüfung, Bad Nauheim, Germany, December 1996, pp. 227-239

Mangerig I. (1995). Unpublished documents

Marandet, B, and Sanz, G (1976). "Ètude par la méchanique de las rupture de la ténacité d´aciers à resistance moyenne fournis en forte épaisseur", *Revué Métallurgie*, pp. 359-383

Sedlacek, G, Dahl, W, Stötzel, G, Liessem, A (1993). "Improvement of the Methods Given in Eurocode 3, Annex C for the Choice of Material to Avoid Brittle Fracture: Background and Revised Draft", IIW-Document X-1274-93, 1993

Sedlacek, G, Stranghöner, N, Stötzel, G, Dahl, W, Langenberg, P, Liessem, A (1996). "Die Tragsicherheit, die Ermüdungssicherheit und das Sprödbruchmodell", *Stahlbau 65 (1996)*, Vol. 11, pp. 407-414

Stranghöner, N, Sedlacek, G, Stötzel, G, Dahl, W, Langenberg, P (1996). "Statistische ermittlung von Teilsicherheitsbeiwerten für das Sprödbruchkonzpet im Eurocode 3", DVM-Tagung Werkstoffprüfung, Bad Nauheim, Germany, December 1996, pp. 241-252

Stranghöner, N (1997). "Werkstoffwahl im Stahlbrückenbau", (in preparation)

Wallin, K (1990a, 1990b, 1994). Methodology for Selecting Charpy Toughness Criteria for Thin High Strengths Steels, Part I, II, II, Jernkontorets Forskning, Nt. 4013/89, VTT Manufacturing Technology, Finland

Proceedings of the Seventh (1997) International Offshore and Polar Engineering Conference
Honolulu, USA, May 25-30, 1997

ISBN 1-880653-28-1 (Set); ISBN 1-880653-32-X (Vol. IV)

Study on Hot Spot Stress for Fatigue Strength Assessment of Fillet Welded Structure

Kanta Nihei, Fumihide Inamura and Shigeki Koe
Kawasaki Heavy Industries, Ltd.
Hyogo, Japan

Abstract

First, in order to investigate a universal method to evaluate hot spot stress (HSS) for various combinations of the plate thickness t and the weld leg length ℓ in fillet welded joints, stress analysis in the vicinity of weld toe has been carried out for 2-dimensional fillet welded joint models, by means of boundary element method (BEM). As a result, the authors propose the new equation to estimate HSS, σ_{hs}, for various fillet welded joints, $\sigma_{hs} = \sigma(X_{hss})$. This means that HSS is defined as stress at X_{hss}, the specified point distant from weld toe. And X_{hss}/t is the function of ℓ/t.

Second, the HSS values evaluated from the method proposed by authors and several other methods have been compared with one another to confirm the precision of those HSS values for representative fillet welded joints with structural stress concentration.

Third, in order to verify the validity of the proposed HSS evaluating method, the fatigue tests of fillet welded joints have been carried out. As a result, each S-N data based on HSS range evaluated from the new proposed method has gathered within a narrow band respectively, independent of structural stress concentration factors. And, it has been confirmed that the new proposed method to evaluate HSS is applicable to fatigue strength assessment of fillet welded structures.

1. Introduction

The fatigue strength of welded structure usually depends upon the structural stress concentration of the structure. As the fatigue crack initiates at the weld toe, hot spot stress (HSS) is one of the most useful parameters for fatigue strength assessment of the welded structures.

HSS means "structural stress", that is the stress at weld toe including only structural stress concentration but not including local stress concentration due to weld bead, as the definition is shown in Fig. 1.

There has been proposed several methods to evaluate HSS, which is an effective parameter to assess fatigue strength of various fillet welded structures. One is 2-points extrapolating method in which HSS is evaluated by linear extrapolation of stress at two specified points distant from weld toe (Niemi, 1995; Tamura, 1988; The Shipbuilding Research Association of Japan, 1991; Huther, 1991) and the other is 1-point representative method in which HSS is defined as stress at the specified point distant from weld toe (Nihei, 1993; Kawano, 1992). In these methods, specified point(s) to estimate HSS is defined as the function of plate thickness t. But, generally, the specified point(s) should be defined as the function of not only the plate thickness t but also the weld leg length ℓ. In this paper, we propose a new equation to estimate HSS for various fillet welded joints as the 1-point representative method considering t and ℓ.

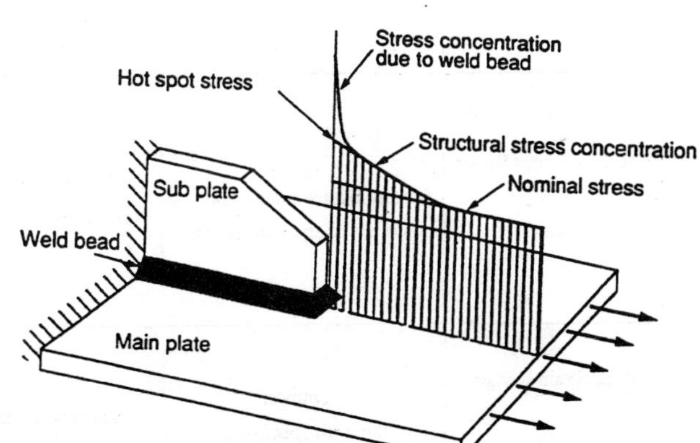

Fig.1 Definition of Hot Spot Stress

2. HSS for Fatigue Assessment of Welded Structures

2.1 Stress Categories and Definition of HSS

In fatigue assessment of welded structures, the following stress categories are used:

Nominal stress σ_n; HSS σ_{hs}; Peak stress ε_{pk}.

The comparison of these stresses is shown in Table 1. HSS has the advantages as follows.

1) The definition of HSS is clear even for complicated structure , but that of nominal stress is not clear for complicated structure. And the only one S-N curve is enough for fatigue assessment of the welded structures in the hot spot stress application, but the different S-N curve for each joint type is needed in the nominal stress application. The conventional nominal stress approach and the unified HSS approach for fatigue strength assessment of welded structures are shown in Fig.2, for reference.

2) It is relatively easy to obtain the value of HSS by means of FE analysis or strain measurment, but it requires much effort to obtain that of peak stress.

2.2 Definition of Fatigue Life

In HSS application, the definition of fatigue life for fillet welded joint should be the visual surface crack initiation life Nc. The reason using Nc as the fatigue life of the welded structure is as follows.

If Nf is defined as fatigue life in HSS concept, the propagated area of fatigue crack is beyond HSS area. Therefore, it is reasonable that Nc is defined as fatigue life under HSS concept.

2.3 Unified S-N diagram for weld toe failure of fillet welded joint

In HSS concept, the S-N diagram should be common one excluding the effect of structural stress concentration. Therefore, the S-N diagram of simple fillet welded joint is basicly the unified one which includes all the effects except structural stress concentration.

For the fatigue design of fillet welded joints, it is neccessary to consider the following parameters which cause the scattering of fatigue strength.

① Local stress concentration due to weld bead profile, K_w
② Welding residual stress at weld toe, σ_{res}
③ Thickness effect
④ Stress condition (membrane or bending)

3. Universal Evaluating Method of HSS

The category of fillet welded joints is basicly devided to T-joint and cruciform joint. The elastic stress distributions in the vicinity of weld toe has been analyzed for these two joint types under membrane and bending stress conditions, by means of boundary element method (Tsuta,1990). We have investigated the relation between the stress distribution and the weld leg length ratio (ℓ / t) , and formulated the equation for evaluating HSS.

3.1 Elastic Stress Analysis of T-joint and Cruciform Joint Model

HSS is defined as the stress at weld toe, including structural stress concentration but excluding local stress concentration due to weld bead. So, it is necessary to make clear the specified distance from weld toe, X, to which the effect of local stress concentration due to weld bead does not extend.

Table 1 Comparison of Stresses for Fatigue Strength Assessment of Fillet Welded Structure

Stress (Strain) for Fatigue Assessment	S-N Diagram	Definition of Fatigue Life	Calculation of Stress for Fatigue Assessment	Applicability for Fatigue Design
Nominal Stress Range $\Delta\sigma_n$ (elastic)	Different for each welded joint type	Visual crack initiation Nc, Specimen failure Nf, or Crack propagation through thickness Np	Elastic stress analysis · easy to define for simple structure, but difficult for complicated structure	[Merit] · easy to apply the fatigue design rule for simple structure [Demerit] · required to prepare different S-N diagram for each welded joint type · difficult to define "Nominal Stress" for complicated structure
Hot Spot Stress Range $\Delta\sigma_{hs}$ (elastic)	Common for any fillet welded joint type (simple fillet welded joint)	Visual crack initiation Nc at the weld toe (crack depth 1～2 mm)	Elastic stress analysis · easy to define for any fillet welded structure · possible to apply stress obtained from FE analysis or strain measurement effectively	[Merit] · The S-N Diagram is common. · "Hot Spot Stress" is easy to define for any fillet welded structure.
Peak Strain Range $\Delta\varepsilon_{pk}$ (elastic-plastic)	Common (smooth specimen)	Micro crack initiation Nc'	Elastic-plastic strain analysis · required to make much effort	[Demerit] · required to make much effort for peak strain calculation · not preccise enough to estimate the fatigue life of as-welded joint

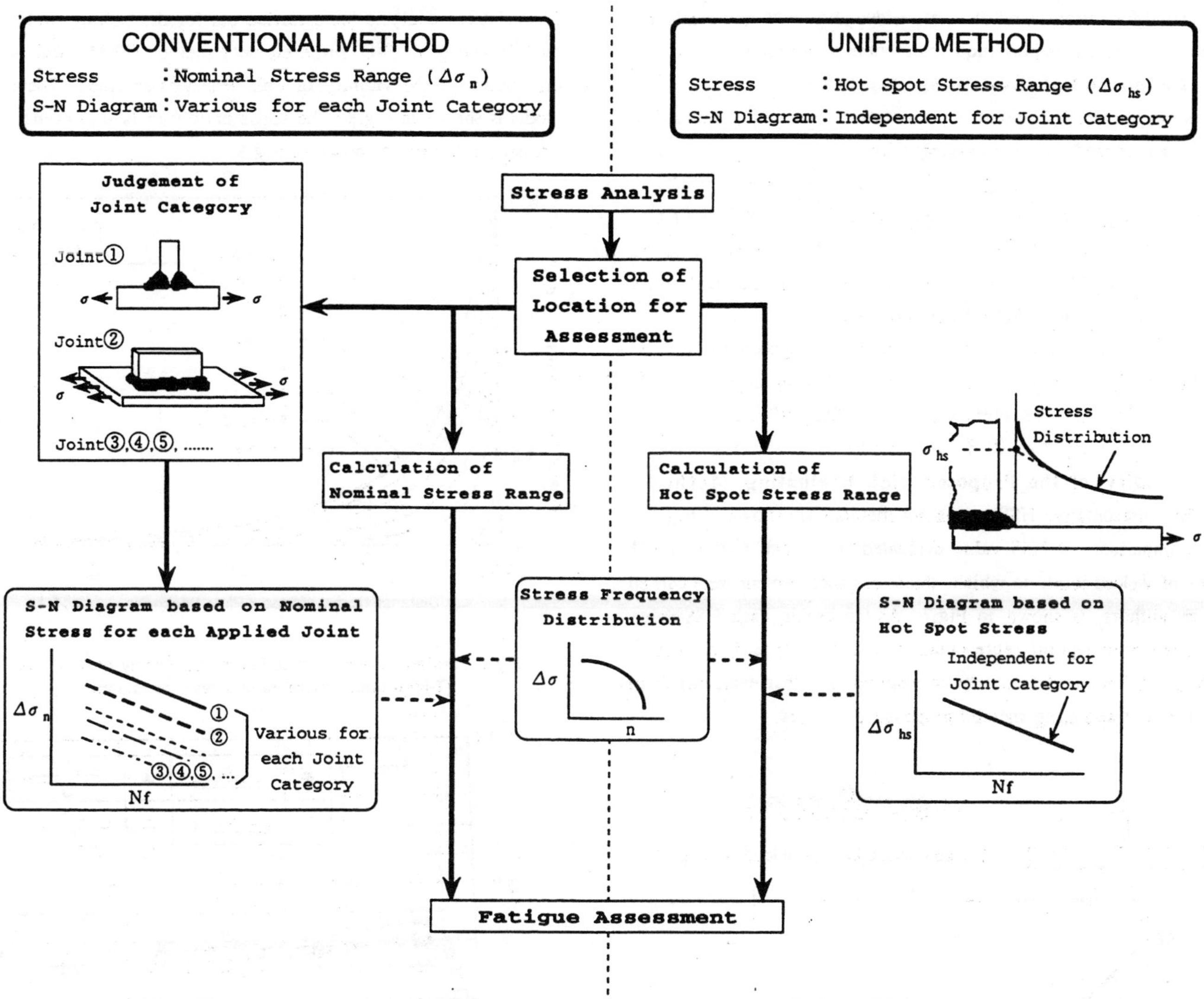

Fig.2 Comparison of Fatigue Strength Assessment Approaches for Welded Structure

The models for stress analysis by boundary element method (BEM) are shown in Fig.3. The models consist of main plate and weld bead to exclude the effects except for local stress concentration due to weld bead. The boundaries of each model consist of quadratic line elements, and the element mesh size near weld toe is sufficiently fine to calculate local stress precisely. The stress conditions are uniform membrane and uniform bending, and the elastic stress distributions in the vicinity of weld toe are analyzed in plain strain condition for various weld leg length ℓ and main plate thickness t, (ℓ / t = 0.01～3.0).

3.2 Results of Analysis and Discussion

The surface stress distributions for T-joint model under membrane stress condition, as an example, is shown in Fig.4. The stress in the vicinity of weld bead has the tendency to decrease rapidly when ℓ / t becomes higher. And, the same tendency was comfirmed for cruciform joint model and under bending stress condition.

The stress σ of 1.02 times nominal stress σ_n ($\sigma = 1.02\times\sigma_n$) was adopted as the stress not including the effect of local stress concentration due to weld bead. The distance, that the stress becomes 1.02 times nominal stress, was obtained and defined as X_{hss}. The relation between ℓ / t and X_{hss} / t is shown in Fig.5. X_{hss} / t decreases rapidly when ℓ / t becomes smaller than 0.2. On the other hand, X_{hss} / t converges to some constant value when ℓ / t becomes larger. It is found that X_{hss} / t is the function of ℓ and t. And the functions are different for joint type or stress condition.

So, the relation between ℓ / t and X_{hss} / t for each case has been formulated as follows.

$$X_{hss}/t = f(\ell/t) = \{a\cdot c\cdot(\ell/t) + b\cdot(\ell/t)^2\} / \{a + (\ell/t)^2\} \quad \cdots\cdots (1)$$

where a, b and c are the constants. Each constant value is shown in Table 2. From this equation, the distance X_{hss}, which is affected only by local stress concentration of weld bead, can be calculated for various kind of fillet welded joint with any leg length ratio ℓ / t and stress condition (membrane or bending).

3.3 Proposed HSS Evaluating Method

As the universal "1-point representative" evaluating method of HSS, we proposed the following equation.

$$\sigma_{hs} = \sigma (X_{hss}) \quad \cdots\cdots (2)$$

Here, $\sigma (X_{hss})$ is the stress at $X=X_{hss}$, where $X_{hss} = t \times f\ (\ell / t)$.

Eq. (2) means that HSS is approximated to the stress at the nearest position from weld toe where the effect of the local stress concentration due to weld bead does not extend.

From Fig.5, "0.3 t" method previously proposed by authors (Nihei, 1993) is found to be applicable for T-joint in $\ell / t \geqq 0.2$. And, equation (2) is expressed as follows ($X_{hss} = 0.3\,t$).

$$\sigma_{hs} = \sigma\ (0.3\,t) \quad \cdots\cdots (3)$$

3.4 Validity of the Proposed HSS Evaluating Method

The representative HSS evaluating methods are shown in Fig.6. The comparison of HSS value evaluated from each method for 3 types of welded joint, to which the stress measurment were carried out by authors, is shown in Fig.7. As the result, each HSS value evaluated from 1-point representative method proposed by authors ("$X_{hss} \cdot t$" method) is nearly corresponded to that evaluated from 2-points extrapolating method proposed by others.

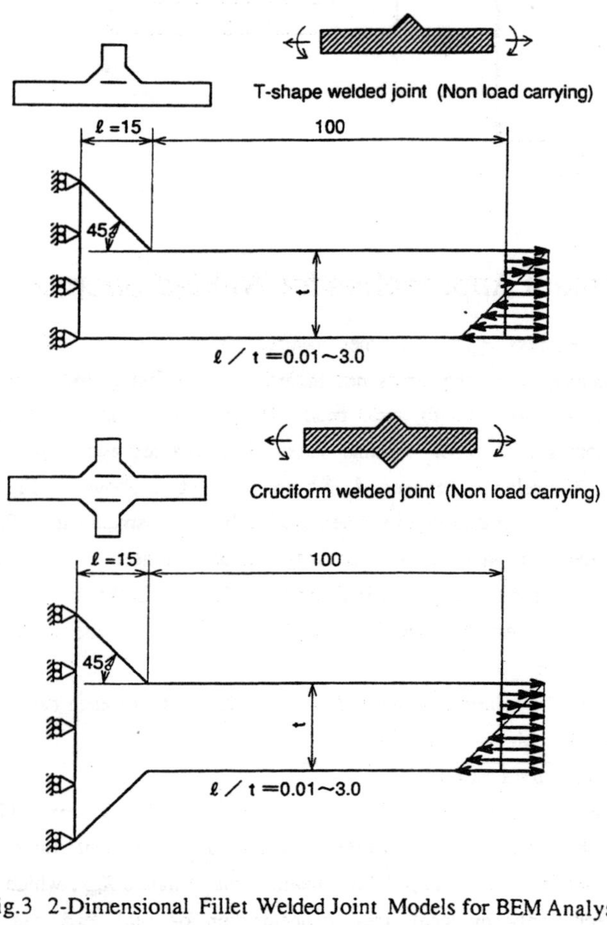

Fig.3 2-Dimensional Fillet Welded Joint Models for BEM Analysis

Elastic stress analysis were carried out for the boxing welded joint by 3 dimensional (3-D) finite element method (FEM), and the stress distribution in the vicinity of weld bead was obtained. The analysis model is shown in Fig.8. The stress condition is under tension. The stress distribution is shown in Fig.9.

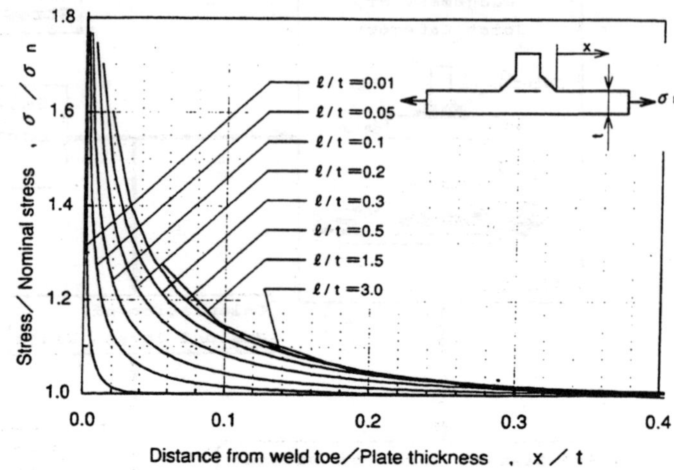

Fig.4 Surface stress distribution in the vicinity of weld toe (T-joint under membrane stress condition)

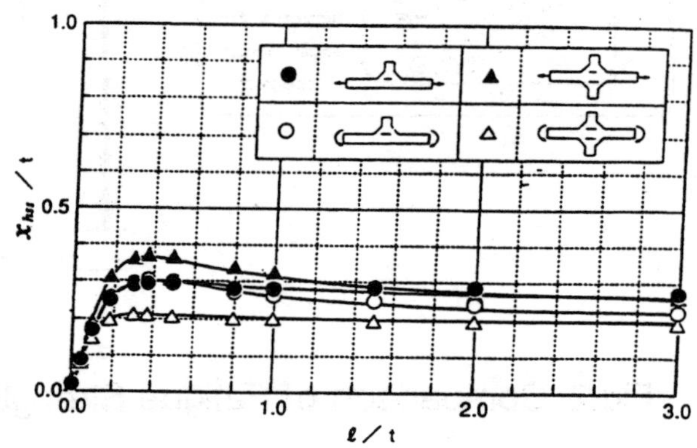

Fig.5 Relationship between ℓ / t and X_{hss} / t

Table 2 The coefficients of Eq. (1)

Fitting curves

$$\frac{x_{hss}}{t} = f\left(\ell/t\right) = \frac{c \cdot a \cdot \left(\ell/t\right) + b \cdot \left(\ell/t\right)^2}{a + \left(\ell/t\right)^2}$$

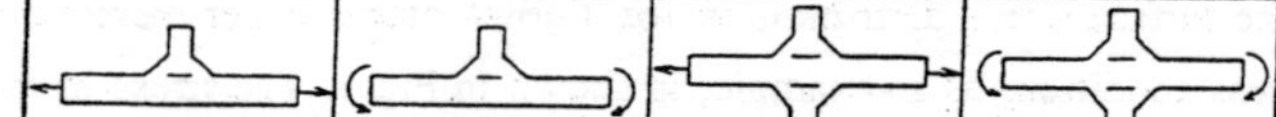

a=0.03	a=0.06	a=0.05	a=0.01
b=0.25	b=0.19	b=0.23	b=0.19
c=1.40	c=1.50	c=2.00	c=1.30

The stress at weld toe including only structural stress concentration was estimated from the shape of stress distribution beyond the position $X = X_{hss}$. The tangent line at $X = X_{hss}$ of stress distribution in the vicinity of weld bead was drawed, and the stress value extrapolated to weld toe position was regarded as "correct value" of HSS. Here, "correct value" of HSS is not equal to but very close to the real HSS value. And it is required to make much effort for evaluating "correct value" of HSS. For this joint, the comparison of HSS values evaluated from "$X_{hss} \cdot t$" method, other various methods ,and "correct value" are shown in Table 3. As the results, the value evaluated from "$X_{hss} \cdot t$" method is closed to "correct value". It is found that 1-point representative "$X_{hss} \cdot t$" method proposed by authors is applicable for practical use.

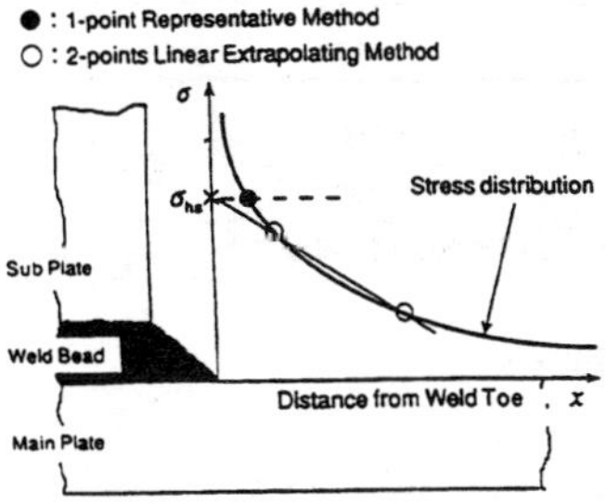

Evaluating method of HSS		Referred stress point(s) distant from weld toe
1-point Representative	$X_{hss} \cdot t$	$X_{hss} \cdot t$
	0. 3 t	0.3 t
	Kawano	5 mm
2-points Linear Extrapolating	Niemi	0.4 t, 1.0 t
	Tamura	0.5 t, 1.0 t
	SR202B	0.5 t, 1.5 t
	Huther	0.4 t, 2.0 t

Fig.6 Comparison of Evaluating Methods of HSS

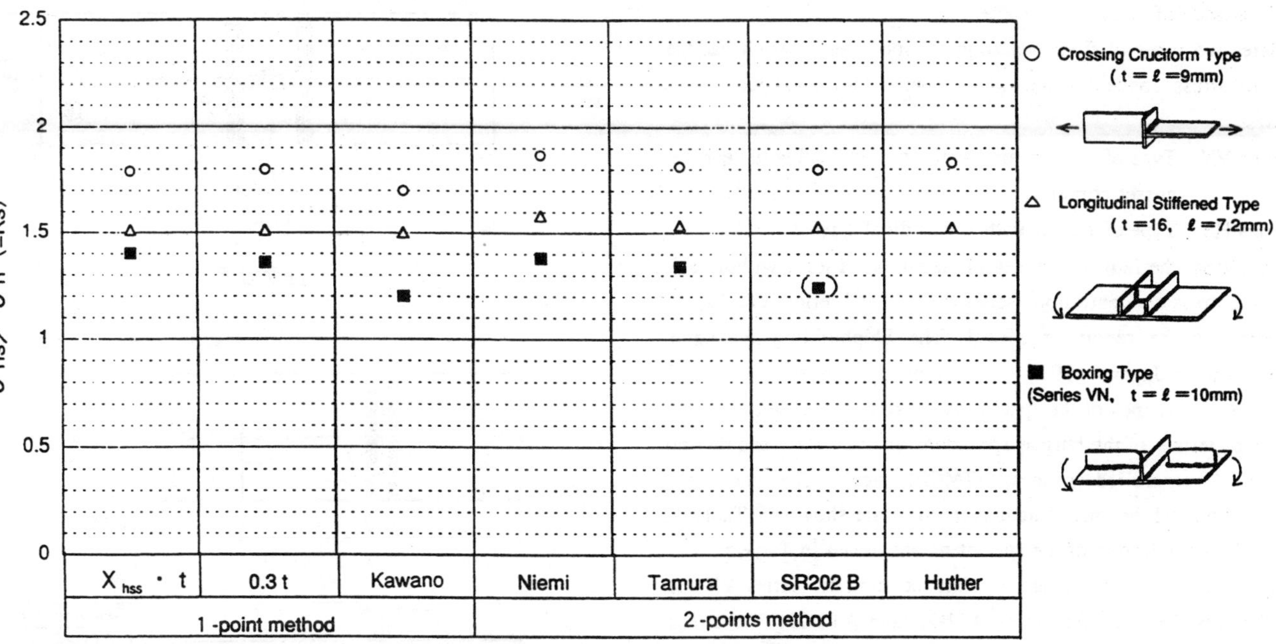

Fig.7 Comparison of Hot Spot Stress In Various Evaluating Methods

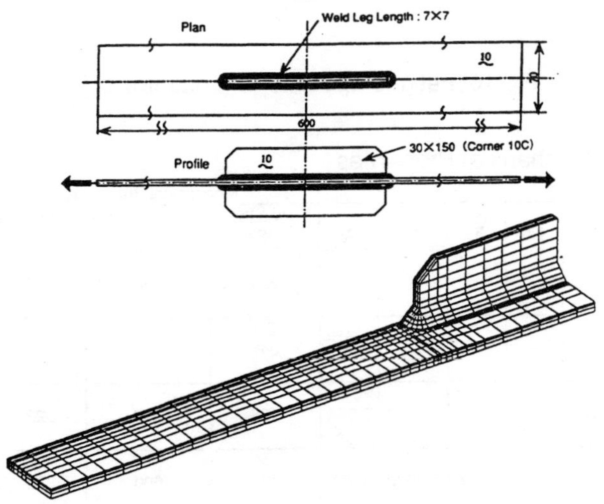

Fig.8 FEM Analysis Model

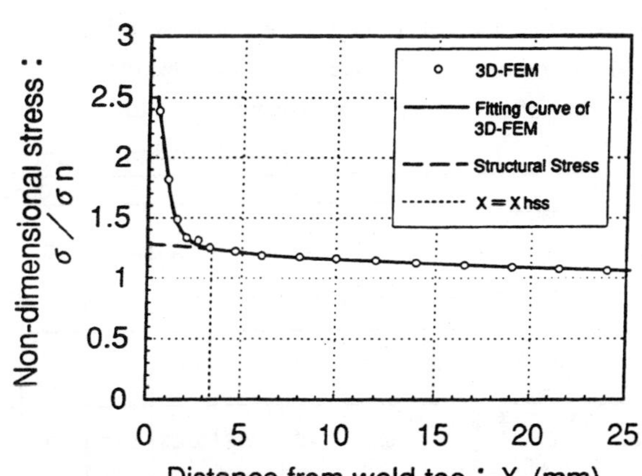

Fig.9 Stress Distribution in the Vicinity of Weld Toe (Boxing Type, Tension)

Table 3 Comparison of HSS Value Calculated in Various Methods (Boxing Welded joint, Tension, 3D-FEM)

Evaluating Method		Definition	σ hss/ σ n	Ratio to "Correct Value"
1-Point Representative	Xhss · t	Xhss · t	1.24	0.97
	0.3 t	0.3 · t	1.26	0.98
	Kawano	5 mm	1.22	0.95
2-Points Extrapolating	Niemi	0.4 t , 1.0 t	1.28	1.00
	Tamura	0.5 t , 1.0 t	1.27	0.99
	SR202 B	0.5 t , 1.5 t	1.26	0.98
	Huther	0.4 t , 2.0 t	1.27	0.99
"Correct Value"		Extrapolating Tangent Line at X = Xhss	1.28	1.00

4. Fatigue Test of Fillet Welded Joints

4.1 Method of Test

Here, two types of fillet welded joint specimens, which had the structural stress concentration, were prepared for the fatigue test as follows.

Series VN : Type of boxing joint with another weld bead near the gusset corner

Series HB : Type of T joint with longitudinal gusset

In addition, the fatigue test data of the simple T joint without any structural stress concentration (series AT1-22 and series AT2-22), referred to the report of SR202 (The Shipbuilding Research Association of Japan, 1991), was compared with series VN and series HB. The shapes of the specimens are shown in Fig.10.

The materials of the fatigue specimens are mild steel and thermo mechanical controlled process (TMCP) steel. The chemical composition and the mechanical properties are shown in Table 4. The welding conditions of the specimens are shown in Table 5.

The loading conditions are 4-points bending for series VN and HB, 3-points bending for series AT1-22 and AT2-22. The stress ratio R(= $\sigma_{min}/\sigma_{max}$) of all the fatigue tests is nearly equal to 0.1. The testing machines used were 100kN and 200kN servo-hydraulics with a sinusoidal cycle.

The inspecting methods of fatigue crack were as follows.

Visual observation : surface crack length at weld toe

Monitoring of strain range : crack depth at weld toe

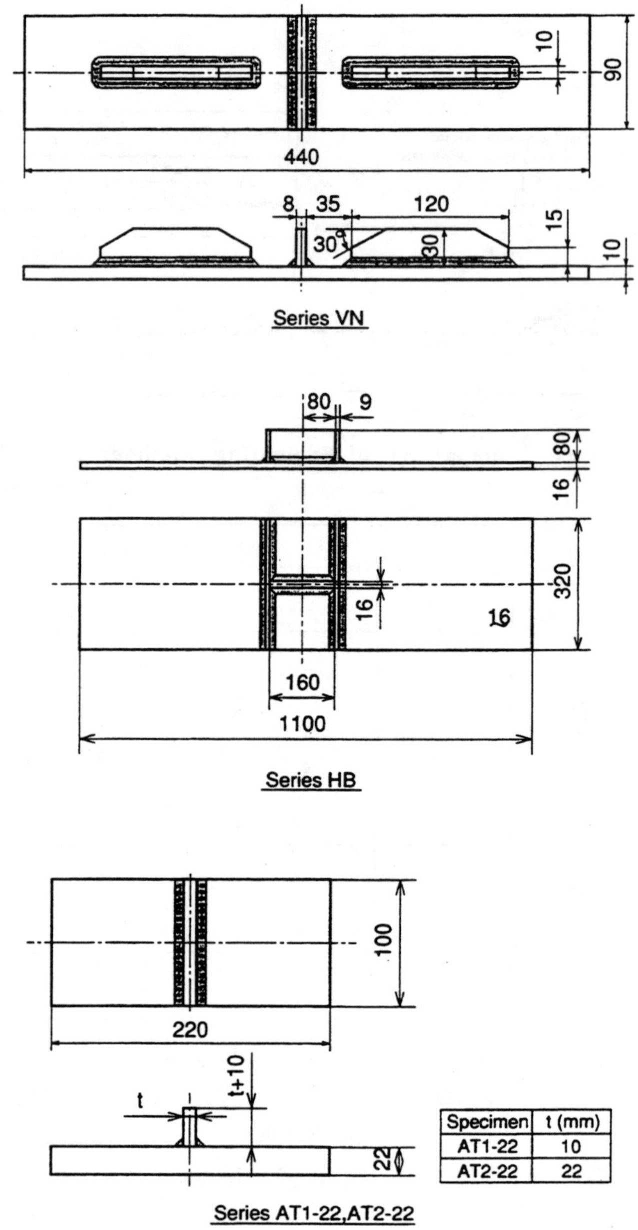

Fig.10 Fatigue Specimens (unit:mm)

Table 4 Chemical Composition and Mechanical Properties

Series of Specimen	Material	Chemical Composition (Wt%)														Mechanical Properties		
		C	Si	Mn	P	S	Cu	Ni	Cr	Mo	B	Nb	Ti	N	Al	Yield Stress (MPa)	Tensile Stress (MPa)	Elongation (%)
VN	LR GRADE A	0.18	0.07	0.66	0.021	0.005	-	-	-	-	-	-	-	-'	-	299	444	27
HB	JIS SM400B	0.15	0.18	0.67	0.013	0.008	-	-	0.05	-	-	-	-	-	-	346	475	30
AT1-22	Hi-tensile Steel (TMCP)	0.1	0.27	1.42	0.012	0.003	0.01	0.02	0.03	-	0.0001	0.01	-	-	0.044	438	547	22
AT2-22	Hi-tensile Steel (TMCP)	0.08	0.19	1.47	0.01	0.002	0.01	0.01	0.04	0.01	-	-	0.016	0.0032	0.33	428	500	22

Table 5 Welding Conditions

Item	VN	HB	AT1-22	AT2-22
Electrode Type	JIS Z3313 YFW-C50DR	JIS Z3312 YGW11	AWS E7016-G	AWS E7016-G
Electrode Diameter, mm	1.4	1.2	5	5
No. of Passes	1	1	2	8
Current, A	270	300	220	220
Voltage, V	29	31	25	25
Welding Speed, mm/min	690~710	350	190	180~190

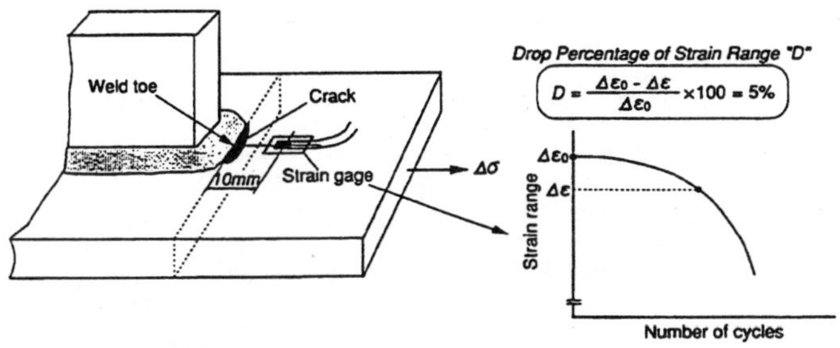

Fig.11 Concept of "5% Strain Drop" Method

Here, the crack initiation was defined as "5% Strain-Drop", which had been proposed by SR202 (The Shipbuilding Research Association of Japan, 1991). The concept of "5% Strain-Drop" method is shown in Fig.11. And in the fatigue test, the beach-marked test was carried out for a specimen to grasp the relation between the rate (%) of strain range dropping and crack depth.

4.2 Test Results and Discussion

4.2.1 Stress distribution

The longitudinal stress distributions along the center line of plate width are shown in Fig.12. The stress values are obtained from measured strain values in both series VN and HB. The stresses in Fig.12 are non-dimensional values devided by nominal stress.

The hot spot stress ranges were calculated by Eq.(1) to get the structural stress concentration factors Ks. As the results, Ks=1.40 for series VN and Ks=1.51 for series HB were calculated in average, respectively. In Fig.12, it is impossible to apply some 2-point linear extrapolating methods to calculate the reasonable HSS for series VN. In series VN, there exists another weld bead 16mm distant from weld toe at the gusset corner, and the referred stress points distant from the weld toe must be less than 16mm. In Huther's method, it is impossible to calculate HSS, because one of the referred stress points distant from the weld toe, 2.0 t , is 20mm and is over 16mm. And it is also impossible in SR202's method (The Shipbuilding Research Association of Japan, 1991). One of the referred stress points distant from the weld toe, 1.5 t , is 15mm and is less than 16mm, but the stress value at the position of 1.5 t is influenced by the local stress concentration at another weld bead, found in Fig.12. Previously shown in Fig.7, it is also possible to apply "0.3 t "method, which is one of the 1-point representative methods, to calculate the reasonable HSS for series VN.

4.2.2 Situation of Crack Initiation and Propagation

The beach-marked fracture surface of series VN is shown in Fig.13 as an example .The relations between the rate(%) of strain range dropping during the fatigue test and the crack depth at weld toe are shown in Fig.14. In Fig.14, the crack depth at 5% strain range dropping were 1 ~2mm in both series VN and HB. Here, the crack initiation life Nc is defined as the life at 5% strain range dropping.

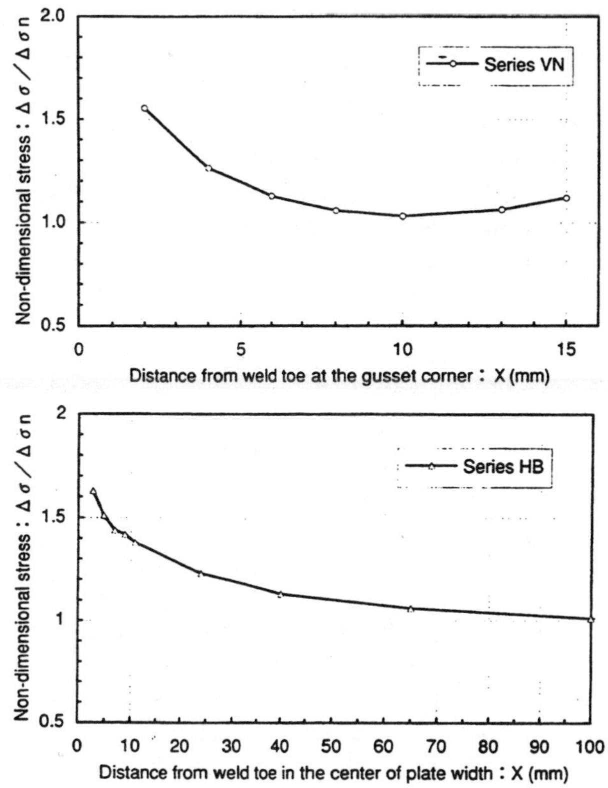

Fig.12 Stress Distribution of the Series VN and HB

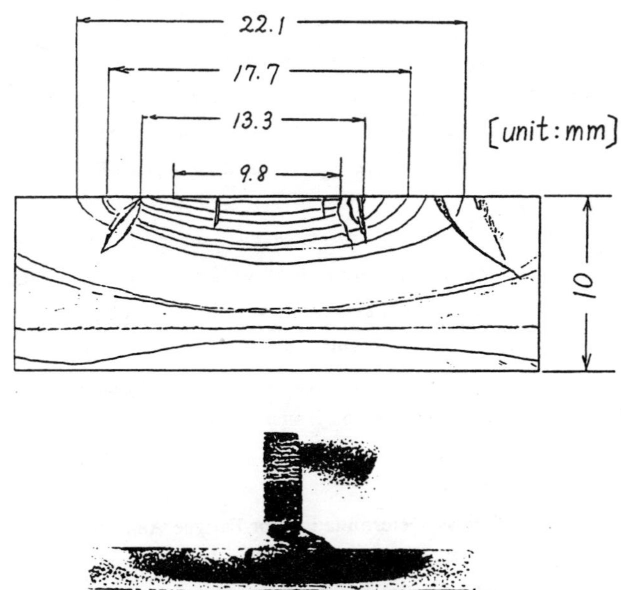

Fig.13 Beach-marked Fracture Surface (Series VN)

4.2.3 S-N Data

The relation between nominal stress range $\Delta\sigma_n$ and crack initiation life Nc in 4 series of fatigue specimens are shown in Fig.15. It is found that the fatigue strengths of the series VN and HB, with structural stress concentration, are lower than those of the series AT1-22 and AT2-22, without structural stress concentration. And the lines in Fig.15 are the average lines calculated as coefficient m=4 ($\Delta\sigma_n^m \times Nc$=Constant).

The relation between HSS range $\Delta\sigma_{hs}$, calculated by Eq.(1), and crack initiation life Nc are shown in Fig.16. It is found that the fatigue strength data of 4 series, with and without structural stress concentration, has gathered within a narrow band.

As a result, it has been confirmed that our proposed 1-point representative method to evaluate hot spot stress is applicable even for complicated welded structures, to which application of some 2-points extrapolating method is impossible to apply.

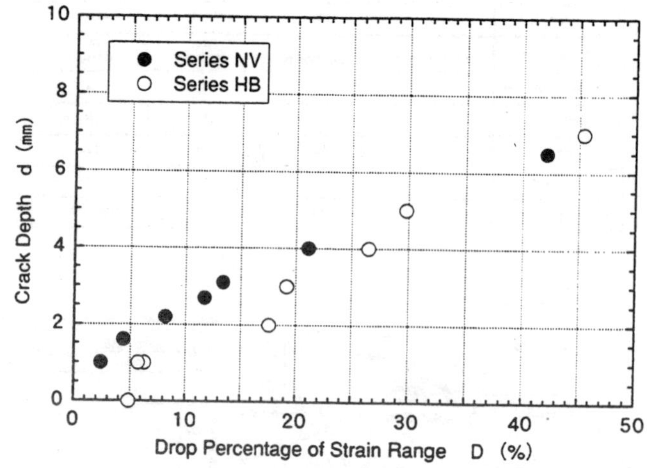

Fig.14 Crack Depth Plotted against Drop Percentage of Strain Range

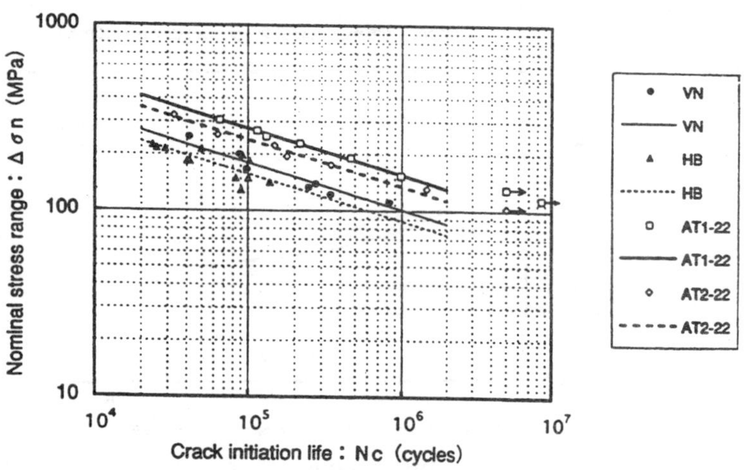

Fig.15 $\Delta\sigma$ n - Nc Diagram

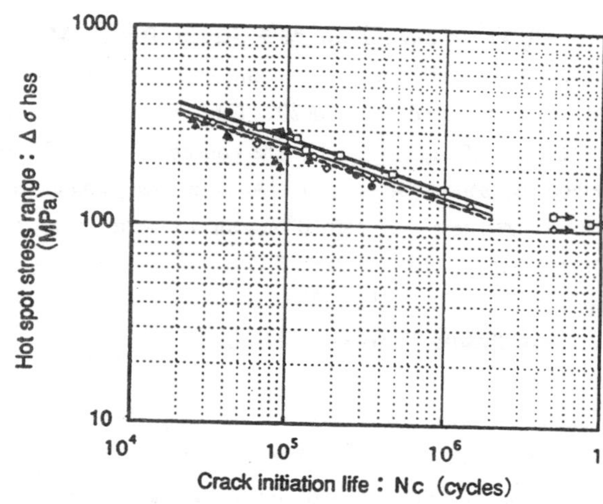

Fig.16 $\Delta\sigma$ hs - Nc Diagram

5. Conclusion

We have proposed the 1-point representative method ("$X_{hss}\cdot t$" method) to estimate hot spot stress for various fillet welded joints, that the hot spot stress is defined as stress at X_{hss}, the specified point distant from weld toe, where X_{hss}/t is the function of ℓ/t.

It has been confirmed that the hot spot stress value evaluated from "$X_{hss}\cdot t$" method proposed by authors is applicable for practical use for representative fillet welded joints with structural stress concentration, and also confirmed that our proposed 1-point representative method to evaluate hot spot stress is applicable even for complicated welded structures, to which application of some 2-points extrapolating method is impossible to apply.

References

- E.Niemi : "Stress Determination for Fatigue Analysis of Welded Components (IIS/IIW- 1221-93)",The International Institute of Welding,1995.
- H.Tamura : A Proposal of Practical Method for Analysis on Stress Consentration Due to Structural Discontinuity of Welded Joints., J. of JWS, Vol.2-No.2,1988
- K.Nihei, F.Inamura and S.Koe : Study on Unified Fatigue Strength Estimation Method for Fillet Welded PlateStructures — (1st Report) Simplified Calculating Method of Hot Spot Stress—. J. of Kansai Soc.N.A., Japan, No.220, September 1993
- H.Kawano et.al : Consideration of Referent Stress for Fatigue Strength Evaluation, J. of Seibu N.A.83,207/213,(1992).
- M.Huther and J.Henry : Recommendation for Hot-Spot Stress Definition in Welded Joints,IIW-IIS,Doc.XIII 1416-91,1991
- SR202 : Study on Fatigue Design of Offshore Structures and Qualification of Weldment, the shipbuilding research association of Japan, 1991
- T.Tsuta,S.Yamaji,et al : The Boundary Element Analysis System-SURFES, Proceedings of the 12th International Conference on Boundary Elements in Engineering, 1990.

Proceedings of the Seventh (1997) International Offshore and Polar Engineering Conference
Honolulu, USA, May 25-30, 1997

ISBN 1-880653-28-1 (Set); ISBN 1-880653-32-X (Vol. IV)

Effects of Static Load on Fatigue Strength of Ship Structure

W.S. Kim
Hyundai Heavy Industries Co.
Ulsan, Korea

Y. Tomita, K. Hashimoto and N. Osawa
Osaka University
Osaka, Japan

ABSTRACT

Boxing fillet weldments, which exist many parts of ship structure initially have residual stress approximately equal to yield stress of material around weld toe. Fatigue strength of the weldments has been evaluated by using experimental results of small as-welded specimens with initial residual stress. However, ship structural members undergo fairly large static loading, usually a tank test and ballast loading (ballasting) before going into service. This static pre-loading is expected to shake-down the initial residual stress.

Both the initial welding residual stress distribution on as-welded boxing fillet weldments and the shake-down residual stress distribution on statically pre-loaded weldments were examined by using FEM analysis and actual measurement. Load-controlled axial fatigue tests were performed on these weldments to examine the influence of the shake-down residual stress on fatigue strength.

The main points are as follows:

1) **When static tension load which can induced nominal stress of 50% to 80% of yield stress is applied onto a weldment, the initial welding residual stress of boxing fillet weldment reduces to almost zero, or compressive stress condition. This has been confirmed by FEM analysis and actual measurement.**

2) **The fatigue strength of boxing fillet weldment with shake-down residual stresses increases. This becomes more dramatic in the longer life range.**

3) **We propose S-N diagrams of boxing fillet weldments with shake-down residual stresses. Diagrams should be applied to fatigue strength evaluation of weldments in ship structure.**

KEY WORDS: Boxing fillet welded joint, Shake-down residual stress, FEM analysis, Sectioning method, Fatigue strength diagram

INTRODUCTION

Boxing fillet weldment is one of the most important ship structural components from the viewpoint of fatigue strength. One common type of fatigue failure is fatigue cracks on side longitudinals at their intersection with the transverse bulkhead and/or transverse member in VLCC /1/. Around welded joints there exists in general a residual stress, which after construction almost reaches to the yield point of the material. Fatigue strength of these weldments has been evaluated with the design curve developed from experimental results of small as-welded specimens with initial welding residual stress.

In general, static loading on these weldments induced by water pressure before service such as tank test and ballasting is relatively much greater than the cyclic loading induced by waves during service. With these static loading, the initial welding residual stress at the joints of welded structures (in most cases, where stress concentration occurs) is expected to be shake-down greatly by the elasto-plastic deformation behavior of material. After this, ship structural members are exposed to cyclic loading during her service life. In this regard, it seems more reasonable to assess the fatigue strength of welded joints with S-N curves which have taken into account the effect of shake-down welding residual stress (re-distributed welding residual stress) on fatigue strength.

In this paper, using the case of side londitudinals in the side cargo hold of a VLCC, the re-distribution of welding residual stress induced by static pre-loading is examined with FEM analysis and measured by an ordinary sectioning method for small specimens of non load-carrying boxing fillet weldments. Also, fatigue tests have been performed to evaluate the fatigue strength of both the as-welded and statically pre-loaded specimens.

STATIC PRE-LOADING APPLIED TO SIDE CARGO HOLD IN VLCC BEFORE SERVICE

Recently, a double hull structure has been mandated by IMO to protect against pollution of the sea, the outer space to be used as a ballast tank./2/ When sailing condition is ballast loaded, the static pre-loading applied to the side londitudinals in the side cargo hold consists only of water pressure. For this situation, the data on design water pressure on side londitudinals at the midship of a 300,000 ton class double hull VLCC is given by Classification Society rules./3/4/5/ The allowable nominal stress at side longitudinals around the water line where fatigue failure is concentrated can be calculated from the above rules, and reaches to about half of the yield point under ballast loaded conditions. Members at the intersection of the side londitudinals and transverse bulkhead and/or transverse member have structural stress concentration depending on the section shape of the longitudinals. As a result, the static pre-loading applied to the members under ballast loaded conditions reaches to about 50% to 80% of the yield point.

WELDING RESIDUAL STRESS DISTRIBUTION

The material used in this study is 360MPa class high strength steel with a nominal thickness of 10mm, Ship Classification Society (NK) standard KA36. The chemical composition and mechanical properties are listed in Table 1, while details of the specimen (non-load carrying type boxing fillet welded joint) are shown in Fig. 1. Attachments were welded to both sides of main plate (60mm width, in flat position) by CO_2 semi-automatic welding. Welding conditions and the welding sequence are listed in Table 2.

Table 1 Chemical composition (%) and mechanical properties for KA36

C	Si	Mn	P	S	Cu	Ni	Cr	T.Al
0.16	0.20	1.11	.016	.005	0.01	0.02	0.02	0.031

	Yield point σ_y (MPa)	Tensile strength σ_B (MPa)	Elongation (%)
Mill Seat	491	551	20
Tensile Test	356.2	540.6	-

Table 2 Welding conditions

Current	250A	②(④) Welding sequence
Voltage	26V	
Speed	30 cm/min	
Method	Semi-automatic CO_2	①(③) ( ): other side

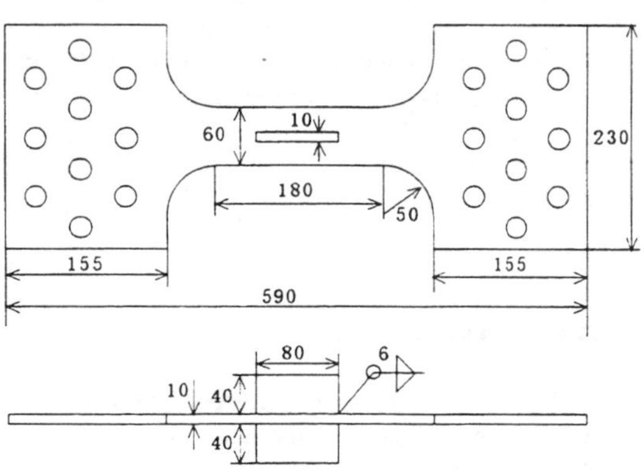

Fig. 1 Shape and size of specimen (unit; mm)

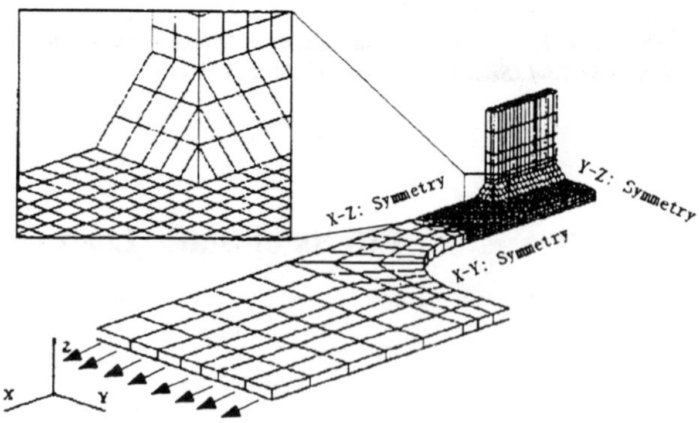

Fig. 2 FEM model

(1) FEM Analysis for Welding Residual Stress Distribution

FEM analysis was performed to calculate welding residual stress distributions of both the as-welded and statically pre-loaded specimens by thermal conduction analysis and thermal elasto-plastic analysis.

Set-up conditions for stress analysis were as follows.

1) **FEM Model**

Models are shown in Fig. 2. Because specimens used for residual stress analysis and fatigue tests are of a symmetrical shape with respect to X, Y and Z direction, only one-eighth of a specimen is taken into account in the model. The element is an eight-node solid one, and the element size around the weld toe is 2.0mm * 2.5mm * 1.6mm. The attachment is connected to the main plate only by means of the weld bead.

2) **Static Pre-load Induced by Ballast Water**

Various types of static pre-load conditions that we applied to as-welded specimens are summarized in Table 3. In each case, the force was applied uniformly to the free edge of the model. (See Fig. 2) At the boundaries, symmetrical condition was imposed. (See Fig. 2)

Table 3 Load case of static pre-load conditions

Load case	Pre-load	Pre-Load Pattern
1	as weld	0
2	$1.172\sigma_y$	0, $0.69\sigma_y$
3	$0.689\sigma_y$	0, $1.17\sigma_y$
4	$-0.689\sigma_y$	0, $-0.69\sigma_y$
5	$-1.172\sigma_y$	0, $-1.17\sigma_y$
6	$\pm 0.85\sigma_y$	0, 1, 2, 3, 4, 5, 6, $0.85\sigma_y$, $-0.85\sigma_y$

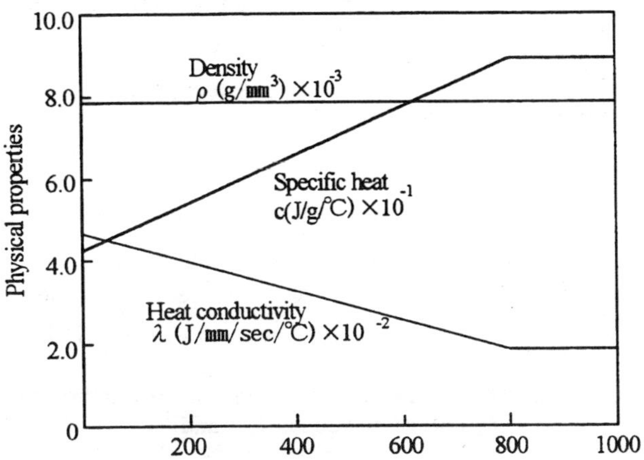

Fig. 3 Physical properties

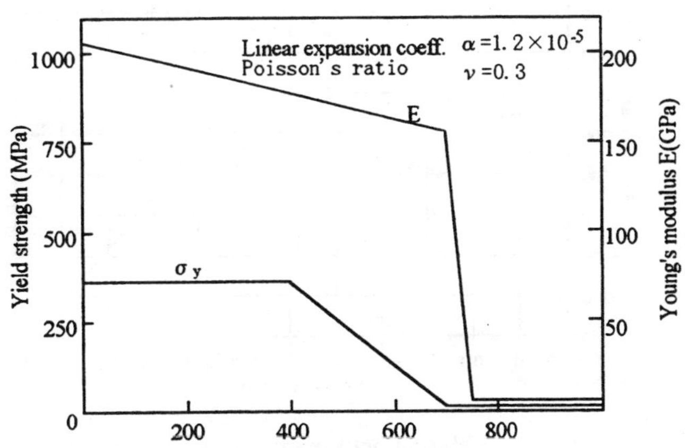

Fig. 4 Mechanical properties

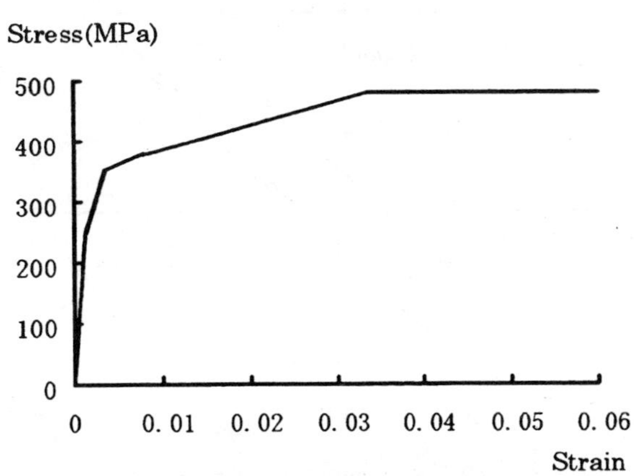

Fig. 5 Stress strain relation

3) Physical and Mechanical Properties of Material for FEM Analysis

Both the physical and mechanical properties of the material are shown in Figs. 3 and 4. Since the stress-strain relation of a material has an effect on the results of residual stress distribution, the stress-strain relation that we derived from a tensile test was used. This is approximated by a five-line graph, as shown in Fig. 5.

4) Heat Source Transfer and Heat Input

In order to simulate welding residual stress distribution by FEM analysis with high accuracy, both a heat source transfer and a heat input change accompanying the source transfer have to be taken into account in thermal conduction analysis. In this paper, however, because of the difficulty of the above-mentioned analysis, the following commonly used assumption has been adopted, because the heat source is fixed, the whole heat input calculated from the welding condition (See Table 2) is added to the whole weld bead all at once. Heat input duration is 0.5 sec, which is the time needed for the heat source to pass through each element at a given welding speed. (See Table 2) Then, the distribution of the temperature of each element is calculated by a non-stable thermal conduction analysis respect to time.

5) Welding Residual Stress Distribution

Examples of the result of the analysis are shown in Fig. 6 to Fig. 11. Fig. 6(a) shows the time history of temperature distribution, and 6(b) welding residual stress distribution $\sigma x,res$, which affects the fatigue strength of joints in the longitudinal direction of a plate surface. It is clear that just after finishing welding, the residual stress distribution varies much over time, and then becomes stable after about one minute has passed. Shake-down residual stress distributions $\sigma x,res$ are compared with the initial distribution in the longitudinal direction in Fig. 7 (cases 1 - 3) and Fig. 8 (cases 1, 4 and 5). (See Table 3) The shake-down residual stress at 1 mm away from the weld toe is summarized in Fig. 9. The initial welding residual stress near the weld toe, which reaches to almost the yield stress or more, decreases much with applied pre-tensile loading. When pre-compressive loading of about 70 % of yield stress is loaded, residual stress decreases to almost zero. In case of pre-compressive loading, the degree of reduction is much lower than the case with pre-tensile loading. When about 80 % of yield stress is loaded, residual stress decreases from its initial high to about half of yield stress, and then the shake-down residual stress remains nearly constant even if the pre-compressive loading increases.

The cyclic behavior of welding residual stress at 1 mm distance from the weld toe under reverse loading cycle conditions is described in Fig. 10. The numbers on the cyclic loops in Fig. 10 correspond with the numbers on the reverse loading cycle of Load case 6. The initial residual stress, 520 MPa, changes to compressive stress, -116 MPa after the first loading (1,2), and in the loading (3,4) that follows, the residual stress changes to the tensile stress of 89 MPa. Then, after the following loading (5,6) residual stress returns again to almost the same compressive stress (-98 MPa) as the value after the first loading. Based on this cyclic behavior, one can assume that the shake-down residual stress stays almost constant when the same reverse loading cycles are continuously loaded, and also in the case when the smaller reverse loading cycles are substituted.

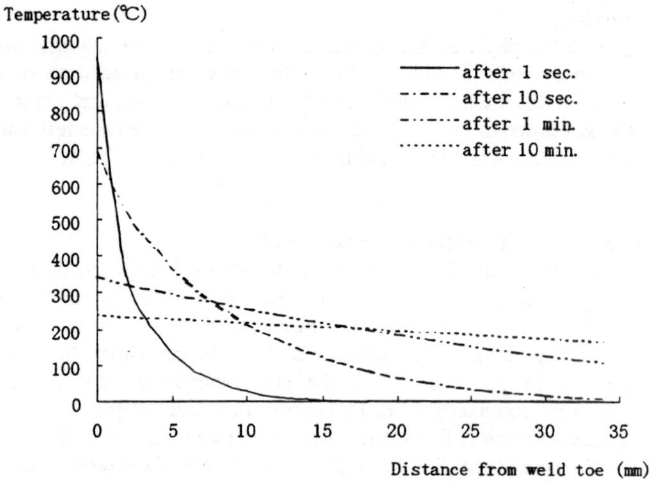

Fig. 6(a) Time history of temperature distribution

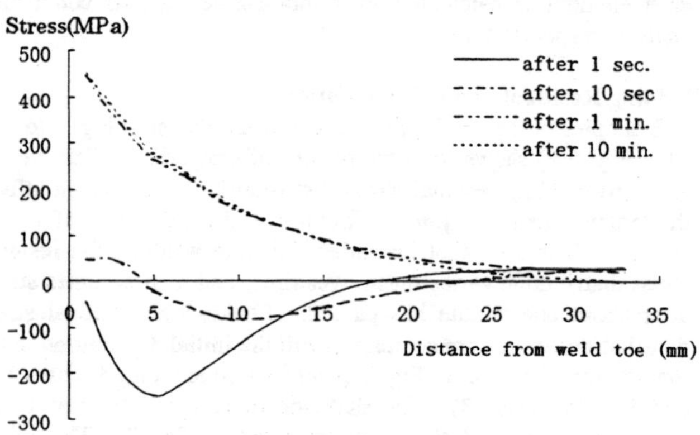

Fig. 6(b) Time history of residual stress sx,res distribution

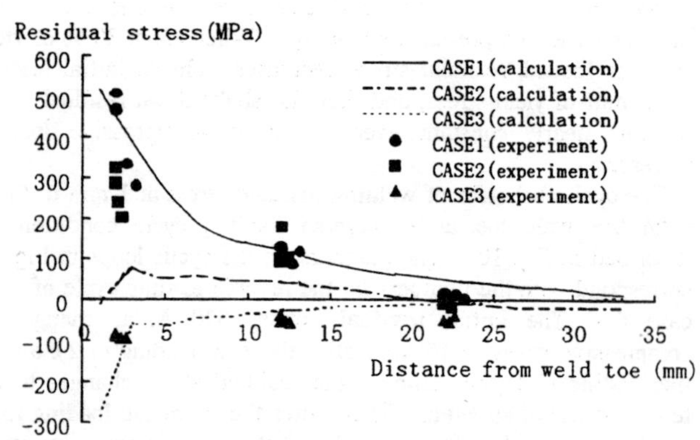

Fig. 7 Distribution of residual stress σx,res under Pre-tension (Cases 1-3)

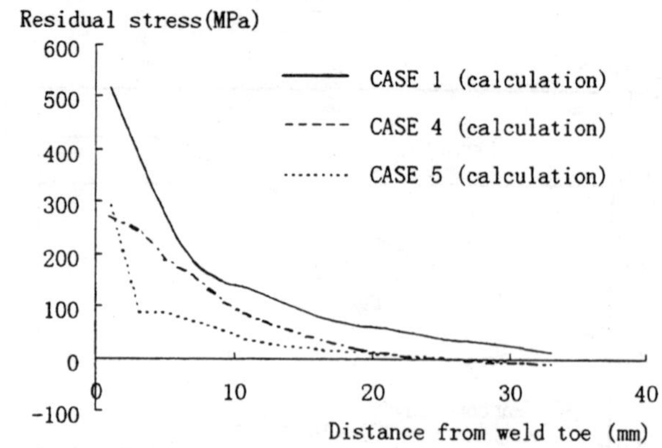

Fig. 8 Distribution of residual stress σx,res under Pre-compression (Cases 1, 4 and 5)

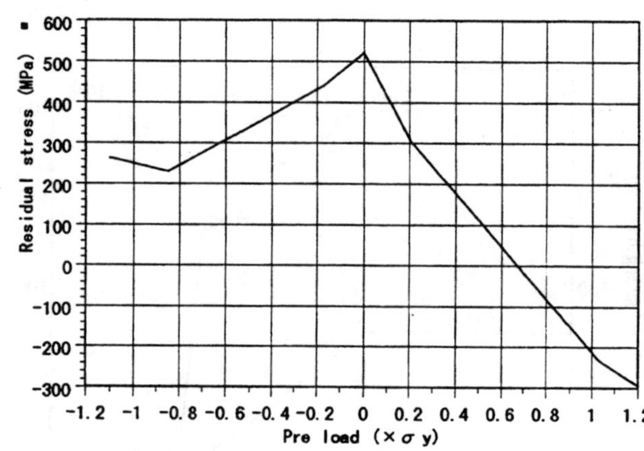

Fig. 9 Relation between σx,res and pre-load

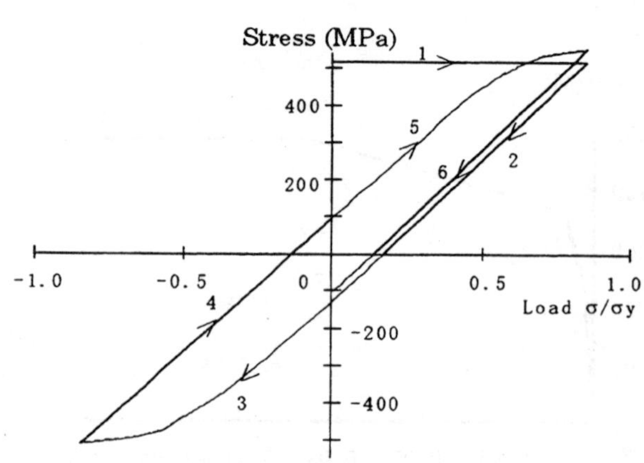

Fig. 10 Cyclic behavior of σx,res

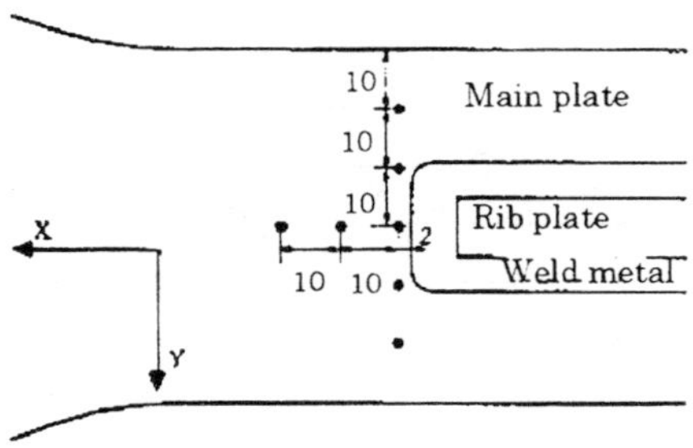

Fig. 11 Location of strain gages for measurement of residual stress distributions

Table 4 Fatigue test series

	Load Pattern		Load Pattern
Series 1 R=-1 Pre-load =0	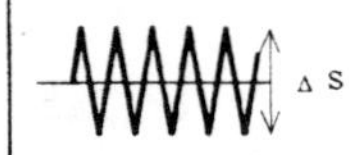	Series 3 R=-1 (mean stress =0) Pre-load = 1.17sy	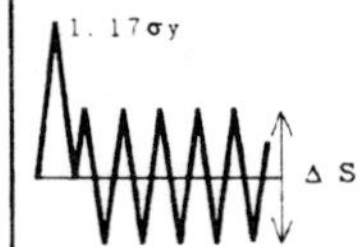
Series 2 R=-1 Pre-load =0.69sy	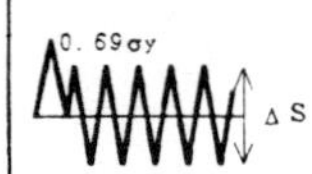		
Series 4 R=-1 Pre-load =0.97sy	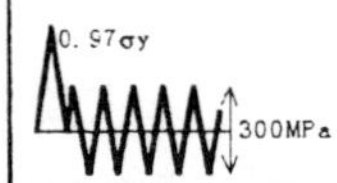	Series 7 Smax=0 Pre-load =0.69sy	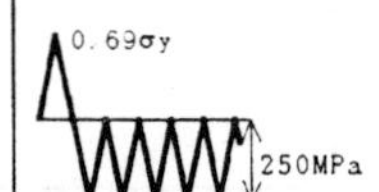
Series 5 R=-1 Pre-load =-0.69sy	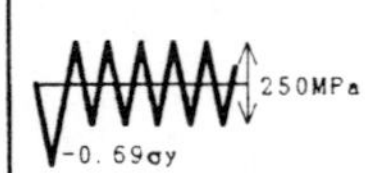	Series 8 R=0 Pre-load =0.9sy	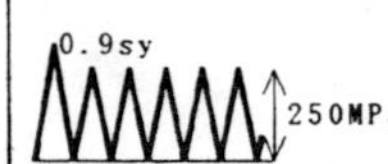
Series 6 R=-1 Pre-load =-0.9sy	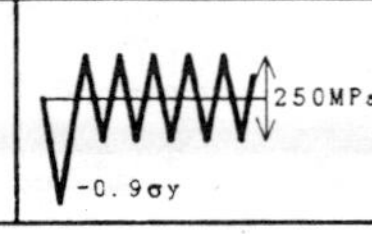	Series 9 S_{mean}=245 Pre-load =0.69sy	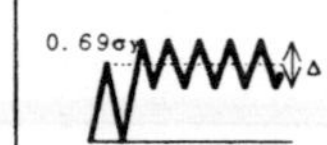

(2) Experimental Analysis for Welding Residual Stress Distribution

Initial and re-distributed welding residual stress distributions were measured by using an ordinary sectioning method. As shown in Fig. 11, two directional strain gages with a gage length of 1 mm were bonded on either side of the main plate. Then, by sectioning the main plate around the strain gages into small cubics, the relaxed strains were measured. Welding residual stresses were calculated from these measured strain. Examples of measured residual stress distribution are shown in Fig. 7. Results are in good agreement with those derived from the calculation.

FATIGUE STRENGTH OF BOXING FILLET WELDMENTS WITH INITIAL AND RE-DISTRIBUTED WELDING RESIDUAL STRESS

Fatigue tests carried out in this study were constant amplitude tests under load-controlled axial loading in room temperature and an air environment. Details of the fatigue tests performed on the specimens are listed in Table 4. Test frequency was approximately in the range of 6 to 12 Hz. Strain amplitude during fatigue tests was continuously recorded on a personal computer by using strain gages attached 5 mm from the weld toe. For the purpose of this study, crack initiation life was taken to be the point (the number of cycles) when the strain amplitude had decreased by 5 % from its initial value.

The place where fatigue cracks began was on the main plate next to the weld toe. Fatigue test results of Test Series 1, 2 and 3 are plotted in Figs. 12 and 13. Figure 12 represents the relation between nominal stress range ΔS and crack initiation life Nc, while Figure 13 shows failure life Nf. The S-N curves are represented by the following equations with a least squares regression analysis.

Test Series 1: $$\Delta S_{Nc} = 4863\ Nc^{-0.276} \quad (1)$$

$$\Delta S_{Nf} = 9135\ Nf^{-0.304} \quad (2)$$

Test Series 2: $$\Delta S_{Nc} = 2988\ Nc^{-0.213} \quad (3)$$

$$\Delta S_{Nf} = 3937\ Nf^{-0.224} \quad (4)$$

Test Series 3: $$\Delta S_{Nc} = 2449\ Nc^{-0.176} \quad (5)$$

$$\Delta S_{Nf} = 2895\ Nf^{-0.175} \quad (6)$$

As mentioned in the previous section (See Fig. 10), shake-down residual stress on pre-loaded specimens maintains the initial value that is assumed after pre-loading under reverse loading cycle conditions (mean load is zero). In other words, in our tests, the residual stress stayed constant at 520 MPa, -10 MPa and -280 MPa under reverse loading fatigue conditions. (Test Series 1, 2 and 3) We can estimate the fatigue strength of welded joints with an arbitrary residual stress as follows :

1) First the fatigue strength of welded joints with residual stress can be approximated by a straight line on a log-log scale ($\Delta S = C\ N^{m}$), as shown in Figs. 12 and 13.

2) Next graph the relation between constant C in equations 1 to 6 and residual stress $\sigma x,res$ (as in Fig. 14), and between constant n and $\sigma x,res$ (as in Fig. 15).

3) From these two graphs, the relations can be approximated by straight lines represented by following equations.

$$C_{Nc} = 3082 \times 10^{0.000374\ \sigma x,res} \quad (7)$$

$$C_{Nf} = 4218 \times 10^{0.000629\ \sigma x,res} \quad (8)$$

$$m_{Nc} = 0.000124\ \sigma x,res - 0.212 \quad (9)$$

$$m_{Nf} = 0.000158\ \sigma x,res - 0.222 \quad (10)$$

4) Inserting coefficients C and m in equations 7 to 10 into equation $\Delta S = C\ N^{m}$

5) Following two equations 11 and 12 generate the fatigue strength diagrams (S-N diagram) of welded joints with an arbitrary residual stress under reverse loading cycle conditions.

$$\Delta S = C_{Nc}\ Nc^{m_{Nc}} \quad (11)$$

$$\Delta S = C_{Nf}\ Nf^{m_{Nf}} \quad (12)$$

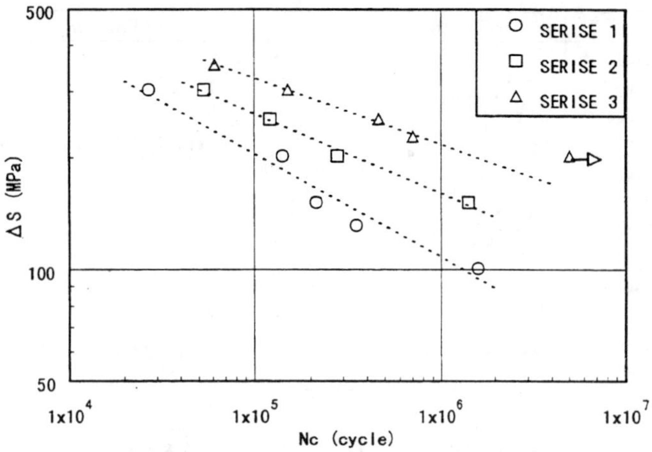

Fig. 12 ΔS-Nc curves

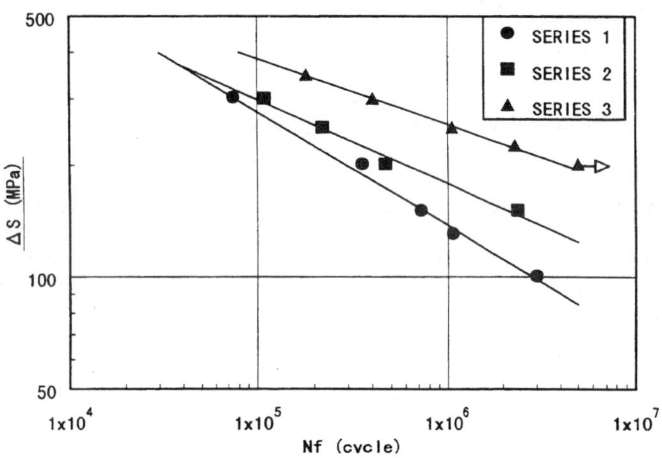

Fig. 13 ΔS-Nf curves

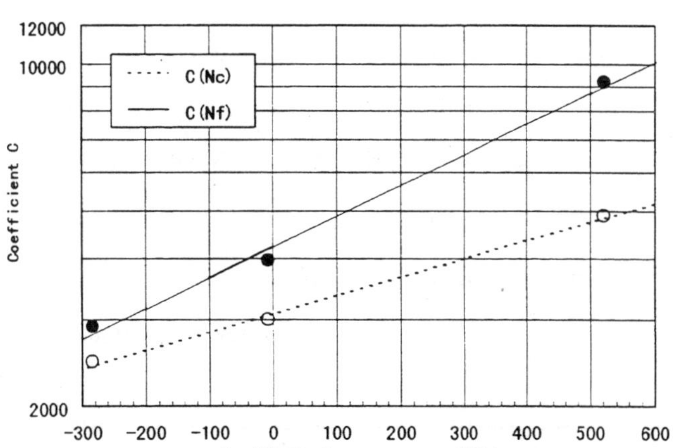

Fig. 14 Relation between σx,res and coefficient C

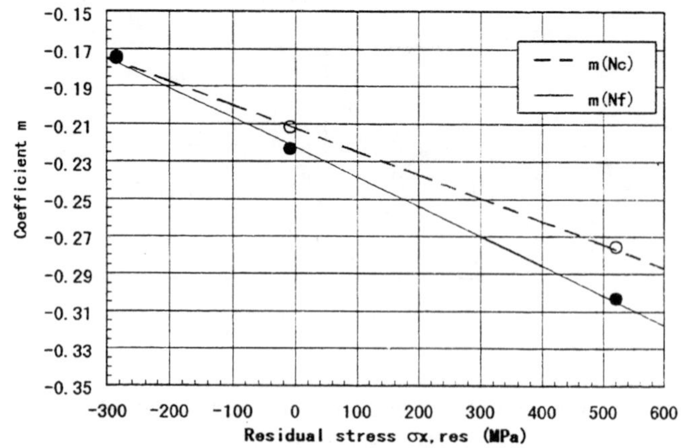

Fig. 15 Relation between σx,res and coefficient m

In order to verify the applicability of the derived equations, fatigue tests were performed. First, reverse loading cycles were loaded to the pre-loaded specimens. (Test Series 4, 5 and 6) Both calculated S-N diagrams and experimental for crack initiation life Nc in Fig. 16 and for failure life Nf in Fig. 17. It is clear from Figs. 16 and 17 that the derived equations can be used to evaluate the fatigue strength of welded joints with residual stress under reverse loading conditions.

Next, the applicability of the equations to other types of cyclic loading conditions (mean load is not zero) is examined. (Test Series 7, 8 and 9) When mean load is not zero, it is necessary to obtain the stress near the weld toe at the moment the mean load is loaded. This stress can be assumed to remain nearly constant during cyclic loadings, and is regarded as a calculable residual stress when applying the derived equations to fatigue strength evaluation. Inserting this calculable residual stress value into σx,res in Equations 11 and 12, S-N diagrams for Test Series 7, 8 and 9 are obtained. Both calculated and experimental results are shown in Figs. 16, 17, and it is clear that there is close agreement.

Up to now the fatigue strength of welded joints in ship structure has been evaluated based on data for as-welded small specimens which disregards the mean load effect. However, welded joints are subjected to various cyclic loading conditions and with a variety of welding residual stress states due to pre-loading before going into service. In order to evaluate the fatigue strength of welded joints more logically, the effect of both residual stress states and cyclic loading conditions on fatigue strength have to be taken into account for fatigue strength evaluation.

In this paper, the notion that the derived equations are appropriate for evaluating the fatigue strength of welded joints with a residual stress is verified. Applying these equations is thus desirable, but more experimentation is necessary to improve their reliability.

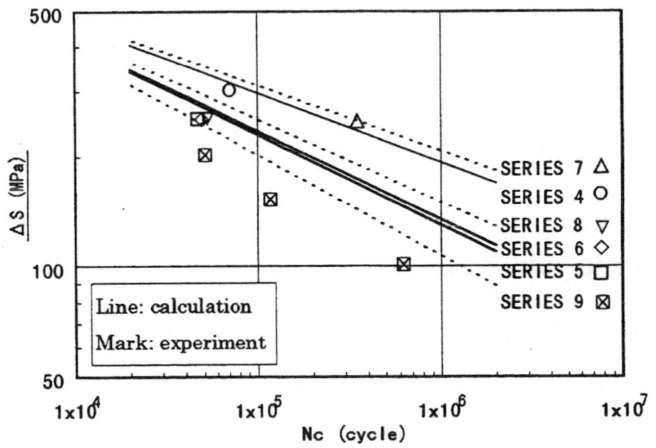

Fig. 16 Diagrams of Δ S versus Nc for both calculated life and experimental results

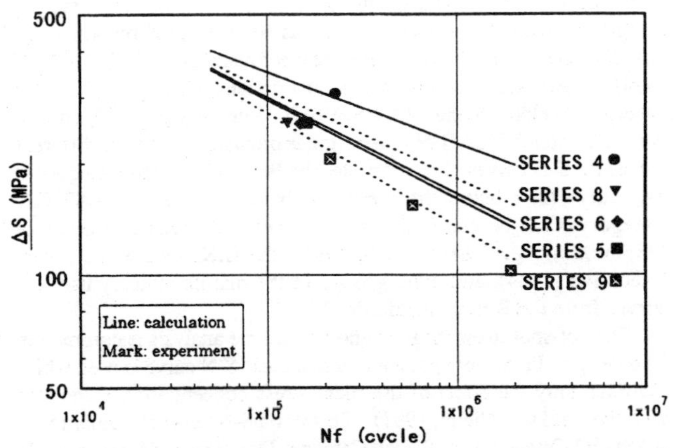

Fig. 17 Diagrams of ΔS versus Nf for both calculated life and experimental results

CONCLUSION

Fatigue strength of welded joints in ship structure has been evaluated by using experimental results of small as-welded specimens with the initial welding residual stress, disregarding the mean load effect./6/7/8/ However, residual stress in weldments changes with various static pre-loading which are applied to the structural members before going into service. Structural members are subjected to various cyclic loading conditions (mean load is zero, or not). It seems, therefore, more reasonable to assess the fatigue strength of weldments with S-N curves which have been taken into accout the effect of both shake-down residual stress states and mean load during cyclic loading on fatigue strength.

In this study, the shake-down residual stress was examined with FEM analysis and measured by an ordinary sectioning method for non load-carrying boxing fillet weldments. Fatigue tests also were performed to evaluate the fatigue strength of both as-welded and statically pre-loaded specimens.

The main results obtained in this study are as follows.

1) Initial welding residual stress at the weld toe decreases with static pre-loading which are applied to as-welded specimens.

2) In case of pre-compressive loading, the degree of reduction foi residual stress is much lower than the case with pre-tensile loading.

3) Since the fatigue strength of welded joints is affected by both residual stress and mean load under cyclic loading conditions, we propose equations which take into account above effects. The equations can be used to the fatigue strength evaluation of boxing fillet welded joints with any value of residual stress.

REFERENCES

/1/ 216 Research Committee of the Shipbuilding Research Association of Japan, Report of Research on the Strength of the Longitudinal foi Large Ships, No. 405 (1994)

/2/ International Marine Organization; 32nd MEPC March 1992, London

/3/ Lloyd Register of Shipping; Rules and Regulation for the Classification of Ships, Part 4, 1993

/4/ American Bureau of Shipping; Rules for Building and Classing. Steel Ships Part 5, 1993

/5/ Nippon Kaiji Kyokai; Rules and Regulations for the construction and Classification of ships, Part C, Chapter 29A, 1993

/6/ Lloyd Register of Shipping; Calculation Procedure for Fatigue Damage Assessment of Ship Structure Details, TPDD 88/3, 1989

/7/ Det Norske Veritas Classification AS; Fatigue Assessment of Ship, Report No.93-0432, 1993

/8/ America Bureau of Shipping; Guide for the Fatigue Strength Assessment of Tankers, 1992

Proceedings of the Seventh (1997) International Offshore and Polar Engineering Conference
Honolulu, USA, May 25-30, 1997

ISBN 1-880653-28-1 (Set); ISBN 1-880653-32-X (Vol. IV)

Restored Fatigue Life with Repair Welds

Barbara A. Kelly and Robert J. Dexter
Lehigh University
Bethlehem, PA, USA

ABSTRACT

Various types of repairs are discussed for fatigue cracks in marine structures, emphasizing the limited database on the fatigue performance of these repairs. Fatigue tests on full-scale-welded beams show that the original fatigue strength of various types of butt welds are all AWS Category D, regardless of the type of steel, whether they are two-sided or one-sided welds, with and without backing, and with and without the edges ground flush. Continued testing after vee-and-weld repairs of the fatigue cracks indicate that vee-and-weld repairs of through-thickness cracks have the same fatigue strength as the original new butt welds, even after repairing the same location up to four times. The fatigue strength of weld access holes can also be nearly restored with repair welds of cracks.

KEY WORDS: repair, weld, fatigue, ship

INTRODUCTION

Some commercial ships have exhibited extensive fatigue cracking at welded details and cutouts (Liu and Thayamballi, 1996). These fatigue cracks are often repaired by gouging a vee weld preparation along the length of the fatigue crack and welding (Huang and Dexter, 1996). The Tanker Structure Co-operative Forum (TSCF) has published several books which show pictorials of the types of cracks which commonly occur in tankers and the suggested repairs (TSCF, 1986; 1992; and 1995). The restored fatigue life after repair determines the adequacy of these repairs.

The few fatigue test data from full-scale specimens with repair welds indicate that the fatigue resistance of a vee-and-weld repair is equivalent to that of an ordinary butt weld in new construction. Therefore, the basic S-N curve that is used in initial fatigue design of welded ship details may also be used to predict the restored fatigue resistance of the vee-and-weld repairs. However, there is a need for additional test data to support this approach.

An S-N curve gives the fatigue strength in terms of an effective stress range as a function of the number of cycles of applied load. In the nominal-stress approach, there are many S-N curves which correspond to different detail categories. The effect of the global stress concentration associated with the geometry of the detail and the effect of the local stress concentration associated with weld toe are both reflected in the fatigue detail category. The nominal-stress fatigue design approach for welded details has been adopted by American Association of State Highway and Transportation Officials (AASHTO, 1994), American Institute of Steel Construction (AISC, 1994), and the American Welding Society (AWS, 1996). Eurocode 3 (1992) and the British Standard 7608 (1994) use similar provisions but have different names for S-N curves. For example, the British Standard Category D S-N curve for a butt weld is essentially the same as the AASHTO Category C S-N curve for a butt weld. American Bureau of Shipbuilding guidelines (ABS, 1992), the U.K. Health and Safety Executive (1984), and other groups in the marine industry use S-N curves from the British Standards.

The hot-spot stress range method of fatigue analysis is common in ship design. In the hot-spot approach, a basic S-N curve is used which includes only the effect of the local stress concentration associated with the weld toe (Niemi, 1991). This basic S-N curve is similar to the AASHTO Category C or British Category D S-N curves for butt welds. The global stress concentration factor (SCF) associated with the geometry of the detail, such as at the tip of a bracket, is reflected in the hot-spot stress range (Niemi, 1991).

Most studies of repair of fatigue cracks have focused on how to perform the repair, often under difficult conditions including fluctuating load. Such studies are beyond the scope of this paper. Underwater wet and wet-backed welds have been extensively studied but are not relevant to repair welds made in dry conditions (Dexter et. al, 1990; and Grubbs and Zanis, 1993).

Fatigue crack repair techniques can be categorized as 1) weld-toe surface crack repairs; 2) repair of through-thickness cracks; and, 3) modifying the connection to reduce the cause of the cracking (Fisher and Dexter, 1993; and 1994). Most previous studies of the residual fatigue strength of repaired fatigue cracks focused on weld-toe surface crack repairs such as hammer peening and weld grinding (Fisher and Dexter, 1993; 1994; Fisher et al, 1979; and Watkinson et al, 1971). The studies indicate that hammer peening is an effective repair technique for surface cracks less than 3 mm deep (Fisher and Dexter, 1993; 1994; and Fisher et al, 1979). The vee-and-weld repair for through-thickness cracks is an example of the second category and will be the subject of this paper.

Only two sources of data on the fatigue strength of repair welds for through-thickness cracks were found, both of these were studies performed at TWI. The first study involved the repair of details at attachment plates in full-scale tests of bridge girders (Wylde, 1983). The second study involved smaller axially loaded tests on fillet welded attachments (Gregory et al, 1989). Both of these studies concluded that the repair-welded details had a fatigue strength as good as the original details.

Uncertainty in the residual fatigue life of weld repairs prevails, particularly a mistaken belief that the steel itself has suffered fatigue damage and that replacing the weld does not restore the fatigue life back to the original expected life of a butt weld. Consequently, many types of weld repairs, such as "veeing and welding" a fatigue crack, are often only approved as temporary repairs which eventually require more substantial replacement of an entire panel. In many cases, this is unjustified because the weld repair will result in an adequate fatigue life, at least as long as the original detail. Therefore, fatigue tests were conducted to provide a basis for the acceptance of the vee-and-weld repair.

This paper describes results from fatigue tests on full-scale-welded beams with a variety of butt welds in the flanges and weld access holes in the webs. The tests establish the original fatigue strength of these details, i.e., the allowable range in the nominal stress, and the remaining fatigue strength after the cracked specimens are repaired in various ways. The lower bound of the test data for a particular detail is equated to an S-N curve. The S-N curves resulting from these tests provide a basis for calculations of the remaining fatigue life after repair which will enable more confident scheduling of inspection and repair. These calculations can be used to evaluate the fitness for purpose of weld repairs, e.g., to evaluate if the weld repairs qualify as "permanent".

DESCRIPTION OF EXPERIMENTS

Built-up welded I-section specimens were fabricated at Edison Welding Institute from 9mm thick steel plates as shown in Figure 1. The flange plates were either API-2W (TMCP) steel produced by Kawasaki Steel Corporation at Mizushima Works in Kurashi, Japan, or ASTM A131 (ABS) Grade AH-36 steel produced by Bethlehem Steel Corporation at the Burns Harbor, Indiana plant. The webs were ASTM A36 steel plate.

Each specimen contained six butt welds, including two-sided and one-sided welds with and without backing bars. In some of the welds, the runouts on the edges of the flange were ground flush with the flange edges, whereas in others the runouts were left un-ground. All original welds were made using shielded-metal arc welding (SMAW) process using E7010-G electrodes for the root passes, E7018 for the fill and E7028 for the cap. The electrodes were manufactured by Lincoln Electric Company.

The beams were designed to include the six butt welds within the constant moment region of the beam, three welds on each flange. The welds were spaced at 305 mm, to assure that a crack in one location would not affect the other welds. Also in order to increase the independence of these welds, the welds were staggered between the top and bottom flanges so that in any given cross section there would be no more than one butt welds.

The web of each beam was attached to both flanges by two 8 mm longitudinal fillet welds. Weld access holes 50 mm in diameter were thermally cut in the web of the beam at the location of each butt weld. For half of the weld access holes, the longitudinal fillet welds were terminated from 8 to 16 mm from the edge of the hole. In the other half, the longitudinal fillet weld was wrapped around the inside of the weld access hole.

The butt welds in the flanges of the I-sections are similar to splice welds in stiffener or girder flanges or hatch opening coaming. If the 152-mm wide flanges can be considered representative of continuous bottom plating, the specimens could also be considered representative of the details which are common on tanker bottoms at the weld access holes in way of the master butts.

The I-beams were tested in the four-point bending fixture illustrated in Figure 2. The load was applied from a 490 kN hydraulic actuator and was measured with a load cell. The actuator was run in load control at 3 Hz using a computer-controlled servohydraulic system. For ease in comparing test results, all tests were conducted with a zero mean load and a constant-amplitude cyclic load of plus and minus 100 kN. This load range of 200 kN created a nominal stress range of 120 MPa in the 9 mm thick flanges.

Strain gages were installed on each specimen to verify the actual stress range. The results from static tests showed good agreement with the calculated stress and good repeatability (no drift) over the several load excursions. The good agreement between measured and calculated stress showed that the beam and the loading fixtures were behaving as anticipated. Strain gage measurements made more than 100 mm away from any crack did not show any significant change as a result of the crack, therefore it was concluded that the cracking in one weld did not have any effect on the stress ranges in the other welds on board the specimen. Furthermore, it was also verified that the clamping of spice plates and hole drilling of web crack (temporary repairs) did not significantly affect the stress ranges at neighboring welds.

At first visual observation of cracking, the number of cycles, location and surface length of the crack were recorded. Loading was continued until the crack propagated to failure. Failure was defined as a through thickness crack which had propagated through at least half of the flange. Cracks were typically spliced with doubler plates on both sides of the flange which are clamped to the flanges with many C-clamps. Cracks in the web of the beams were drilled with a 38-mm diameter hole. These temporary measures enabled the experiment to continue until the specimen had at least four cracks. The specimen was then weld-repaired and returned to testing. Some details were cracked and repaired more than four times.

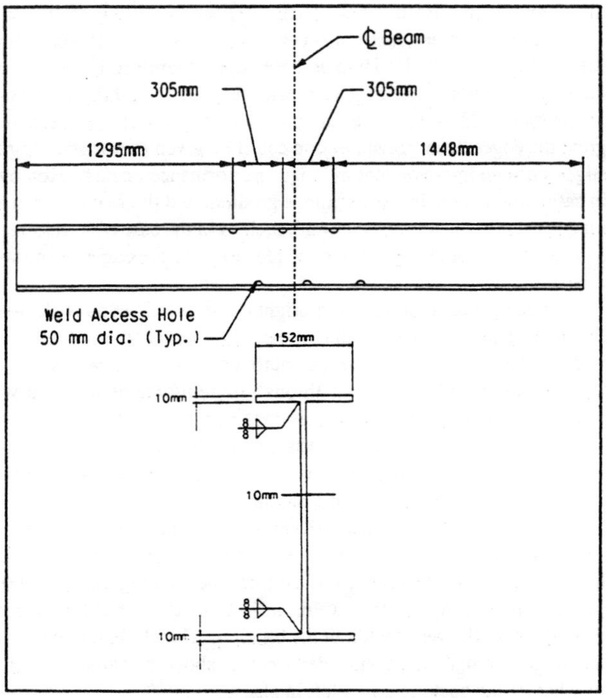

Figure 1. Welded I-Section Specimens.

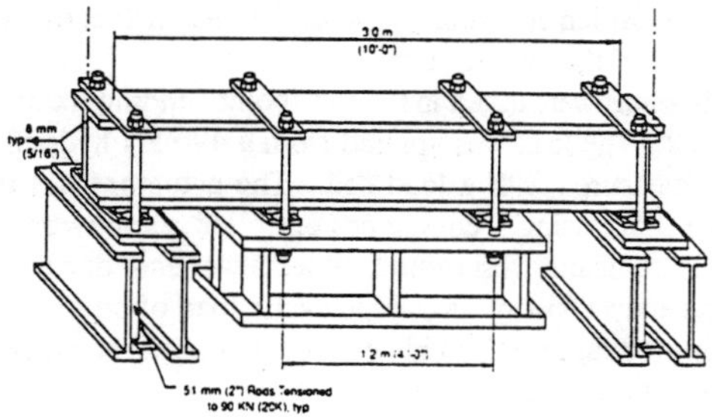

Figure 2. Loading Fixture Used for Testing.

RESULTS OF EXPERIMENTS

Butt Welds

The AWS Structural Welding Code - Steel classifies the fatigue strength of two-sided butt welds, i.e., welds that were gouged and welded from the second side, with reinforcement, with edges ground flush, and 100% ultrasonically-tested (UT), as Category C. The fatigue strengths are for two-sided welds that are not subjected to UT and one-sided welds are not given in AASHTO or AWS codes. BS 7608 and Eurocode 3 give the fatigue strength of one-sided welds with and without backing as the equivalent of AASHTO Category D, if complete penetration is assured.

Figure 3 compares all of the original butt welds grouped according to whether the edges were ground flush. If there were an effect of the edge condition, it would be expected that it would be favorable to grind the edges flush. Both types of data appear to belong to the same distribution, i.e., there does not appear to be any systematic trend in the data. Cracking rarely initiated at these edges, therefore it appears that the edge condition is not important with respect to the fatigue strength of the butt weld. Therefore, in applications where it is expensive to grind the edges flush, consideration could be given to not grinding the edges since no improvement in fatigue performance occurs. However, in cases where there is a stress range gradient and the maximum stress range occurs at the edge of the weld, such as in the case of a weld repair at a cutout, it is probably still a good idea to grind the edges of the weld flush.

The fatigue data for all of the original one-sided welds is shown in Figure 4. This S-N curve compares the welds with backing bars left in place to the one-sided welds made without backing. The welds without backing attempted to get melt-through or reinforcement at the weld root. The angle at the toe of this root reinforcement is difficult to control and in many cases it was steeper than allowable in AWS specifications. The cracks typically initiated at this toe reinforcement in the one-sided welds without backing.

In looking at the fracture surfaces of cracks in welds with backing bars it was common to find that the cracks initiated from the backing bar. Even though the cracks initiate from the backing bar, many data points representing original welds with the backing bar left in place exceed that of the welds without the backing. Therefore, it appears that the fatigue strength of all one-sided welds is about the same, regardless of whether the backing bar is left in place.

It should be noted that one crack of a one-sided weld with backing, along with two cracks from one-sided welds made with an open root are not included in this data. These welds cracked prematurely at unusually low cycles due to usually large weld defects and were clearly outliers from the rest of the data.

Figure 5 compares the fatigue data for all the original two-sided welds to all original one-sided welds. All of these data seem to have a similar distribution, therefore no significant difference between two-sided welds without UT and one-sided welds is observed. Presumably, if the two-sided welds are carefully inspected and weld defects are repaired, the fatigue strength could be taken as Category C compared to the original classification as Category D details. However, this adds considerable expense and may not be necessary if there are other more severe types of details which control the fatigue design. Ships, for example, are full of bracket-end details and other details where there is intersecting structural elements which are typically Category E details. These details will always crack before splice welds in plating and other butt welds without addition stress concentrations.

From the fatigue tests, it was found that all of the original butt welds have approximately the same Category D fatigue strength. The lack of sensitivity to the type of butt weld greatly simplified the rest of the test program since it is not necessary to test a variety of repair weld types. This finding also simplifies specification of repair welds and original welds because the weld type can be changed to suit individual circumstances without affecting the fatigue strength, provided good and complete penetration is assured.

Figure 6 shows that data from all the original butt welds compared to the repaired butt welds. All the repaired butt welds are two-sided welds with the edges ground flush. Although a few of the data from the repair welds were slightly below the Category D line, the majority of the data show a similar distribution of performance. Some of the repair welds were made under adverse conditions so the slightly poorer life of a few of the repairs can be rationalized. Therefore, it is concluded that there is no significant difference between the expected remaining fatigue life of a repair weld and the expected fatigue life of welds in new construction of equivalent quality. In other words, if it can be assured that the entire crack is removed, the fatigue life of a butt weld is completely restored by the gouging and re-welding repair.

Figures 7 shows the data for multiple repairs at one location. The repair data are separated according to the sequence of the times the detail was repaired. At this one location, the life of the third repair was on the low end at only 300,000 cycles in comparison to approximately 500,000 cycles on the first and second repair. However, this detail was repaired a fourth time and the life of this repair was over 1.1 million cycles. The low result for the third repair was anomalous and may have been due to incomplete removal of the crack or due to a severe weld defect. Since the crack was repaired, the fracture surface could not be examined.

Figure 8 also shows multiple repairs at one location and confirms that the repair welds perform as well, if not better than the original or previous repair weld. It is therefore concluded that multiple cycles of repair are not detrimental to the fatigue strength of TMCP steel. However, there are other practical limitations on the number of repairs, e.g., after three or four repairs at the same location the residual stress increases and weldability problems arise.

Beams with thicker flanges (25 mm) were also tested. In comparing the beams with 9 mm thick flanges to the beams with 25 mm thick flanges, it was found that the distribution of data is similar and therefore shows that there is no thickness effect in the range of typical ship plate thicknesses.

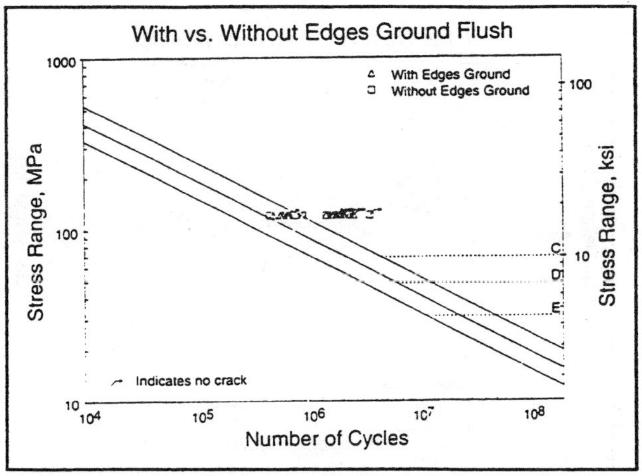

Figure 3. Fatigue Data for All Butt Welds With and Without Edges Ground Flush.

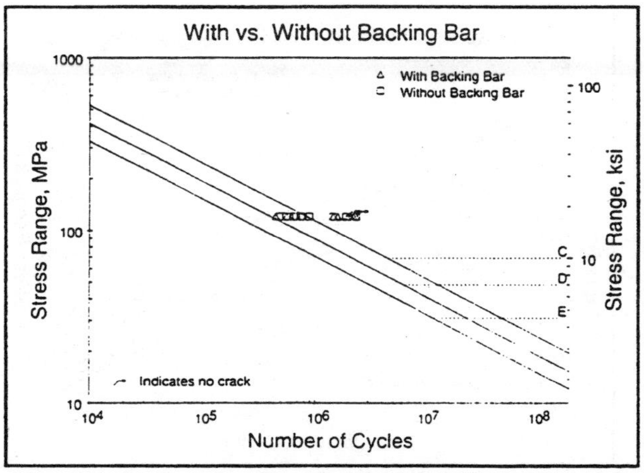

Figure 4. Fatigue Data for All One-Sided Welds With and Without Backing Bars.

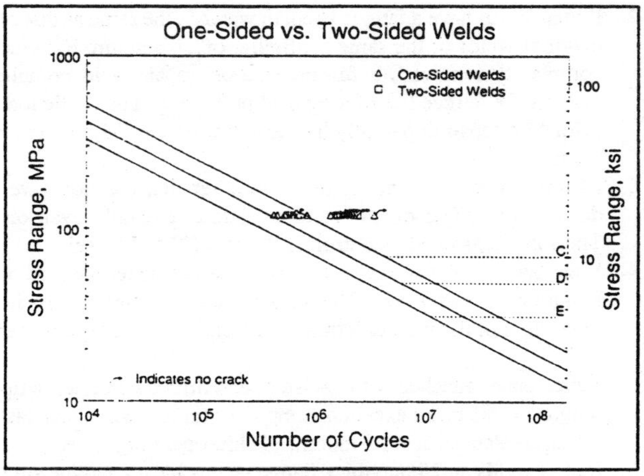

Figure 5. Fatigue Data for Original One-Sided and Two-Sided Welds.

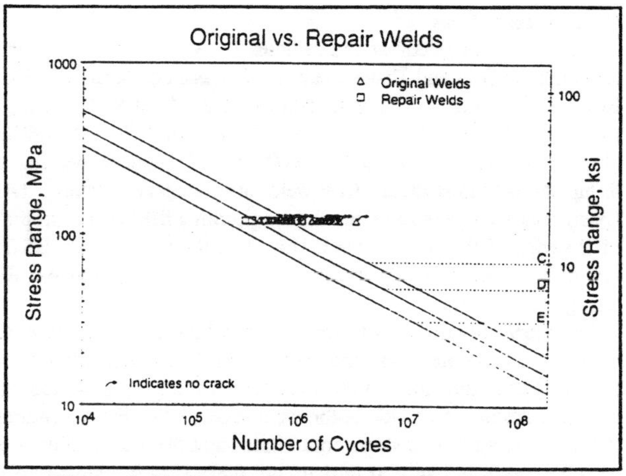

Figure 6. Fatigue Data for Original Butt Welds and Repaired Butt Welds.

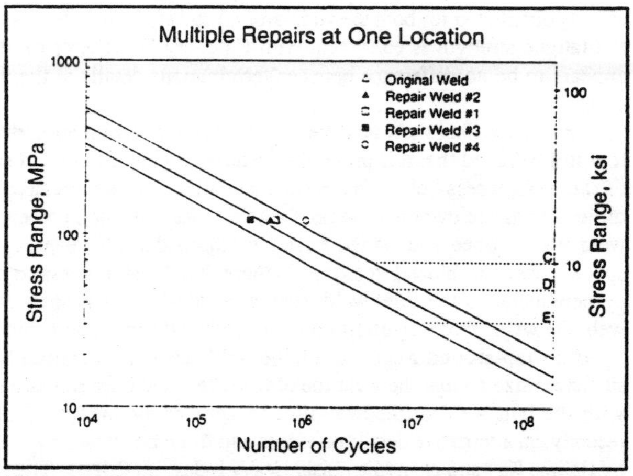

Figure 7. Fatigue Data for Original Butt Weld and Repaired Butt Welds at One Location Indicating Number of Repairs.

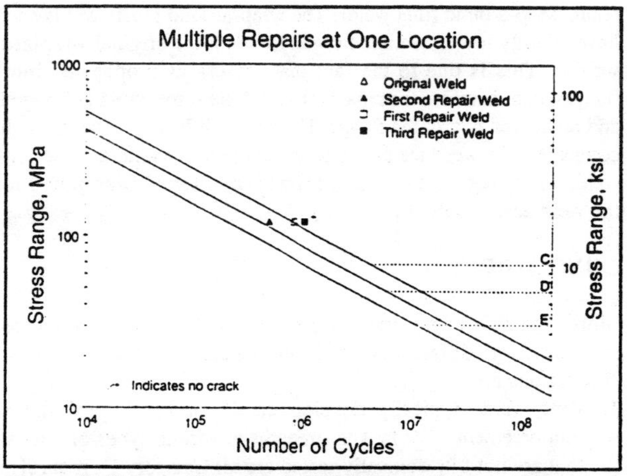

Figure 8. Fatigue Data for Original Butt Weld and Repaired Butt Welds at Another Single Location Indicating Number of Repairs.

Weld Access Holes

Cracks often occur at the ratholes or weld-access holes in the longitudinals (Liu and Thayamballi, 1996; Basar and Jovino, 1990; and Rolfe et al, 1993). The terminations of the fillet welds along the longitudinal at these holes are the critical detail. In the AASHTO (1994), AISC (1994), and AWS (1996) codes, details with terminations of longitudinal fillet welds are Category E details. (AISC specifications also require that the longitudinal fillet welds stop short of the edge of the access hole.) However, the British code BS 7608 (1994) indicated that weld access holes are the equivalent of AASHTO Category D details.

Figure 9 shows the results from cracks which occurred at the edges of weld access holes and compares the results for welds which were terminated 9.5 mm short of the weld access holes versus welds which wrapped around the inside of the weld access hole. Both termination details have a definite stress concentration at the toe as an area from which a crack is likely to initiate, but there is also a stop-start area where the welder made the fillet weld as far around the web at the access hole and then had to pick up the rod and start again from the other side of the beam.

However, contrary to expectations, Figure 9 shows that the data are equally distributed for both the wrap-around and stop-short details, and the fatigue strength is consistent with Category D. Therefore there appears to be no difference between terminations details at the weld access hole.

Though the data show no difference between the termination detail, it is still believed that it is preferable to terminate the fillet weld short of the weld access hole. This recommendation is made because it is difficult to assure that there are no discontinuities at the point where the weld was stopped and started from the other side. However, in the special case of biaxial loading if there are tensile stress ranges perpendicular to the fillet weld, such as vertical stress ranges in the web, the welds that stop-short create a notch and should be avoided.

If a wrap-around weld is used, the weld access hole should be of sufficient size so that the weld toe of the fillet weld does not coincide with the butt weld or the heat-affected zone of the butt weld. Just recently, an alternative detail for access and drain holes was explained by Hajime Kawano from Mitsubishi Heavy Industries (MHI). The MHI detail is continuous fillet welds which continue past the edge of the hole as just weld beads on the plate. The fillet welds run right over the groove weld and reinforcement.

Figure 10 compares the original wrap-around fillet welds to the repair wrap-around fillet welds. The wrap-around repair welds seem to have slightly lower average fatigue life than the original wrap-around welds. This is due to the fact that the cracks propagated into the flanges, and therefore the repairs also include some partially gouged out and re-welded part of the flange. Thus, it is difficult to establish a direct comparison between the performance of the repair weld and the original weld. In any case, the Category D line is a suitable lower bound for all the weld access hole details.

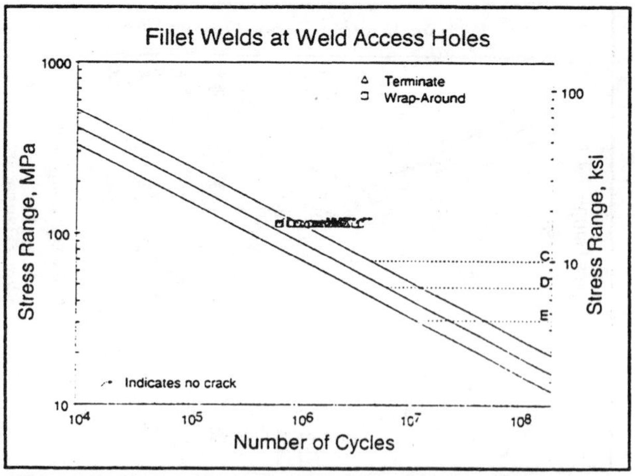

Figure 9. Fatigue Data for Cracks at Original Weld Access Holes Comparing the Stop-Short Detail to the Wrap-Around Detail.

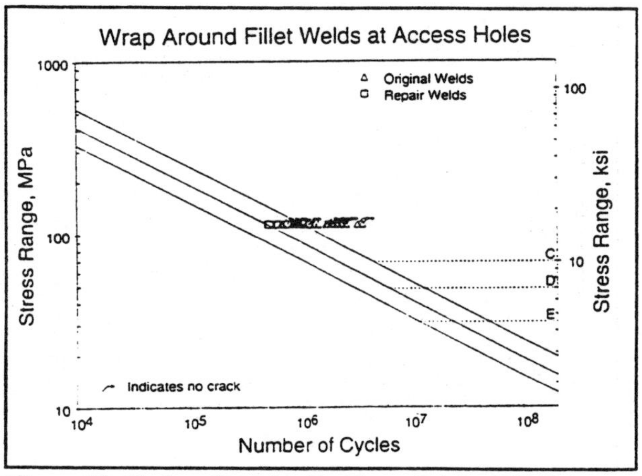

Figure 10. Fatigue Data for Cracks at Original Weld Access Holes Compared to Fatigue Data Cracks at Weld Repairs at Weld Access Holes.

CONCLUSIONS

More than 290 fatigue cracks were generated in welded full-scale test specimens with different butt weld and access hole configurations. The conclusions are:

1. Butt welds typically develop cracks at the toe of the weld reinforcement. The fatigue strengths of various types of butt welds that are not ultrasonically-tested are all Category D, regardless of the type of steel, whether they are two-sided or one-sided welds, with and without backing, and with and without the edges ground flush.
2. Repair welds have a fatigue strength which is the same as that of the original welds of the same configuration. Therefore if a crack is completely removed, a full-penetration repair weld completely restores the fatigue life of a cracked butt weld. The cyclic loading prior to a repair apparently has no effect on the fatigue life of the repair.
3. Multiple cycles of repair, up to four repairs, did not have any detrimental effect on the restored fatigue strength, even on the Thermomechanically Controlled Process (TMCP) steel.
4. Weld access holes typically develop cracks at the termination of the longitudinal fillet welds. The fatigue strength of weld access holes, regardless of the type of termination detail, is AASHTO Category D.
5. Splice plates attached with C-clamps and drilled holes at the tips of fatigue cracks make excellent temporary repairs which can last for the equivalent of several months of ship operating time.
6. Repair welds on 25 mm thick flange beams have a fatigue strength which is the same as that of the repair welds on 9 mm thick flange beams. Therefore, within the range of common ship plate thicknesses, there is no observed thickness effect on the fatigue strength of butt welds including repair welds.

ACKNOWLEDGMENT

This research was performed jointly by the Center for Advanced Technology for Large Structural Systems (ATLSS) at Lehigh University and the Navy Joining Center (NJC) at Edison Welding Institute. The project was sponsored by the Ship Structure Committee and the authors appreciate the patience and guidance of the Project Technical Committee. Professor John W. Fisher, Director of ATLSS Center, Rong Huang of Chevron shipping Company, and Bruce R. Somers of Concurrent Technologies corporation provided valuable guidance in defining the scope of the research project. Eric Kaufmann, Manager of the Welding Laboratory at ATLSS performed fractography and provided valuable advice. The authors would like to thank the technicians, welders, and other student at ATLSS for providing assistance throughout the laboratory testing. The authors are particularly grateful to all of those who responded to our survey, which helped to define the test program, who are too numerous to mention here.

REFERENCES

The American Association of State Highway Transportation Officials (1994). AASHTO LRFD Bridge Design Specifications, First Edition, Washington, D.C.

American Bureau of Shipbuilding (1992). Guide for Fatigue Assessment of Tankers, New York, NY.

American Institute of Steel Construction (1994). Load and Resistance Factor Design Specification for Structural Steel Buildings, Second Edition, Chicago.

American Welding Society (1996). ANSI/AWS D1.1, Structural Welding Code - Steel, Miami.

Basar, NS, and Jovino, VW (1990). Guide for Ship Structural Inspections, SSC-332, Ship Structure Committee.

British Standards Institute (1994). BS 7608, Code of Practice for Fatigue Design and Assessment of Steel Structures, London.

Dexter, R J, Norris, EB, Shick, WR, and Watson, PD (1990). Performance of Underwater Weldments, Report SSC-335, Ship Structure Committee.

European Committee for Standardization (CEN) (1992). ENV 1993-1-1, Eurocode 3: Design of Steel Structures - Part 1.1: General Rules and Rules for Buildings, Brussels.

Fisher, JW, and Dexter, RJ (1993). Weld Improvement and Repair for Fatigue Life Extension," Proceedings of the 12th International Conference on Offshore Mechanics and Arctic Engineering Conference (OMAE), 20-24 June 1993, Salama et al eds., ASME, Vol. III, Part B, Materials Engineering, pp. 875-882.

Fisher, JW, and Dexter, RJ (1994). Field Experience with Repair of Fatigue Cracks, AWS/WIC International Conference on Fatigue, Toronto, 9-10 May, pp. 45-52.

Fisher, JW, Hausammann, H, Sullivan, MD, and Pense, AW (1979). Detection and Repair of Fatigue Damage in Welded Highway Bridges, National Cooperative Highway Research Program Report 206, Transportation Research Board.

Gregory, EN, Slater, G, and Woodley, CC (1989). Welded Repair of Cracks in Steel Bridge Members, National Cooperative Highway Research Program Report 321, Transportation Research Board.

Grubbs, K, Zanis, C (1993). Underwater Repair Procedures for Ship Hulls (Fatigue and Ductility of Underwater Wet Welds), Report SSC-370, Ship Structure Committee.

Huang, RT, and Dexter, RJ (1996). Meeting minutes for Workshop Session 4: "Fabrication and Repair", Proceeding: Workshop and Symposium on the Prevention of Fracture in Ship Structure, March 30-31, 1995, Washington, D.C., Marine Board, Committee on Marine Structures, National Research Council.

Liu, D, and Thayamballi, A (1996). "Local Cracking in Ships-Causes, Consequences, and Control," Proceeding: Workshop and Symposium on the Prevention of Fracture in Ship Structure, March 30-31, 1995, Washington, D.C., Marine Board, Committee on Marine Structures, National Research Council.

Niemi, E (1991). Recommendations Concerning Stress Calculation for Fatigue Analysis of Welded Components, International Institute of Welding, IIW-XIII-1426-91.

Rolfe, ST, Hays, KT, and Henn, AE (1993). Fracture Mechanics Methodology for Fracture Control in VLCC's, Ship Structures Symposium '93, November 16-17, 1993, Arlington, VA, available from SNAME, Jersey City, NJ.

Tanker Structure Co-operative Forum (1986). Guidance Manual for the Inspection and Condition Assessment of Tanker Structures, issued by: International Chamber of Shipping and Oil Companies International Marine Forum, published by: Witherby & Co. Ltd., London, England.

Tanker Structure Co-operative Forum (1992). Condition Evaluation and Maintenance of Tanker Structures, published by Witherby & Co. Ltd., London, England.

Tanker Structure Co-operative Forum (1995). Guidelines for the Inspection and Maintenance of Double Hull Tanker Structures, issued by: International Chamber of Shipping and Oil Companies International Marine Forum, published by Witherby & Co. Ltd., London, England.

UK Health & Safety Executive (formerly the U.K. Dept. Of Energy) (1984). Fatigue Design Guidance for Steel-Welded Joints in Offshore Structures, H.M.S.O., London, England.

Watkinson, F, Bodger, PH, and Harrison, JD (1971). The Fatigue Strength of Welded Joints and Methods for Its Improvement, Proceedings, Conference on Fatigue of Welded Structures, The Welding Institute.

Wylde, JG (1983). The Fatigue Performance of Repaired Fillet Welds, The Welding Institute, Cambridge, England.

Proceedings of the Seventh (1997) International Offshore and Polar Engineering Conference
Honolulu, USA, May 25-30, 1997

ISBN 1-880653-28-1 (Set); ISBN 1-880653-32-X (Vol. IV)

Definition of Physical Wear-and-Tear of the Marine Transport Hydraulic Structures

Vladimir L. Menshikov
"Soyuzmorniiproject" Institute
Moscow, Russia

ABSTRACT

Technical exploitation of the marine transport hydraulic structures in Russia is made according to the Guides on service, inspection and repair of the structures. The repair of structures is made on the basis of the technical inspection results according to strict rules uniform for all the sea ports of Russia.

The treated information about the technical state of structures collected during their technical inspection is taken for the State Register by the head branch institute for technical inspection - SOYUZMORNIIPROJECT. The information collected in the Register is used for optimization of the investments for repair of the structures and definition of the conditions, which enable their effective operation with minimum expenses for the technical exploitation.

The main used parameter, which characterizes the technical state of the structure is its physical wear-and-tear. The method of the structure wear-and-tear definition standardizes: the list of quantitative and qualitative features of the technical state of the inspected objets; methods of determination of the technical state characteristics and its normative values; means of control, procedures for definition of the inspection volume and the form of the flow-chart control, requirements to the executors, the order of definition of the physical wear-and- tear parameters.

The defect list for the structural elements is obtained by using the results of the complex structure inspection. Three categories of the defects, which determine three types of the technical state of the inspected objects, have been established. The defect category is defined according to the table values of its characteristics. According to the defect category and its value determined is the physical wear-and-tear, which defines the type of the repair and reconstruction works, the composition and the volume of the repair.

KEY WORDS: Hydraulic structures, wear-and-tear, technical control, classifier, technical state, defect, workability of the structures.

INTRODUCTION

The marine transport enterprises in big ports of Russia have been transformed into joint-stock companies under the conditions of economic changes at the beginning of the 90ths. However, the hydraulic structures remained the State property and are used by the share-holders on a lease basis. To provide functions of the State control at the transport processes, the Sea Port Administrations have been created under direction of the Federal Service of Sea Fleet (Rosmorflot).

One of the problems of these administrations is the control of safety of the port hydraulic structures and creation of the conditions, which enable the most effective technical exploitation and reproduction.

To enable the most effective exploitation of the hydraulic structures in the modern conditions the Rosmorflot gave order to the State Design and Research Institute of Sea Transport (Soyuzmorniiproject) to develop and introduce the new system of technical control for marine structures (Menshikov, 1996).

TECHNICAL CONTROL OF STRUCTURES

The new monitoring system of the structures is entered into action in 1994 by two managing documents : " Rule about organization of the technical control..." РД 31.35.14-94 (1994) and "Manual on technical control..." РД 31.35.11-93 (1993). First of them is devoted to the questions of organization of the technical control work, tax, account (record-keeping) and analysis of the information on the technical condition of the structures, preparation and certification of the experts. The other is a methodical guide for the executors and contains dozens of techniques of the control operations, the recommendations for optimization of the work and evaluation of physical wear- and- tear of the structure elements.

The base of new inspection system is the unified and strictly formalized information gathering about technical state of structures, preparation of registry documents on different management levels and account of inspection results in State Register of sea transport hydraulic structures of Russia - computer formation system, which was developed by Soyuzmorniiproject . The information about technical state of structures is analysed and used in management decisions about investments into repair and reconstruction.

The management documents contain the following main elements of a system of the technical control of structures:

· The terms and definitions;

Circuit of the departmental technical control of the structures;

organization and interaction of its participants at the control fulfilment operations;

Procedure of registration and the account of the technical control results;

Nomenclature of the objects to be controlled, the controlled attributes, the norms of parameters of the technical state of the controlled objects;

Order of the control scooping and drawing up of its algorithm;

Means and methods of the control;

List of the control technical documents, their form, procedure of filling and storage;

Requirements to the executors.

METHODOLOGY OF THE CONTROL

The controlled objects are rationed by the classifier of hydraulic structures and their elements (Fig.1).

The technical control of the structures is reduced to the control of technical state of their elements, units of their connections, structures by revealing of defects (insignificant, considerable, critical defects) and valuation of their influence on the structure workability. Availability and the defect sizes are characterized by the established values of the parameters of state of the objects under controlled the qualitative attributes, for which the quantitative valuations are not applied.

The nomenclature of these parameters and attributes, as well as the limits of the allowable deviations are estabilieshed in РД 31.35.11-93 (1993).

The circuit of control (Fig.2) has three main stages:

. Scoping of the control, reception of the information on the state of the objects attributes, comparison of the received information to the normative values, revealing of conformity and definition of the type of the object technical state.

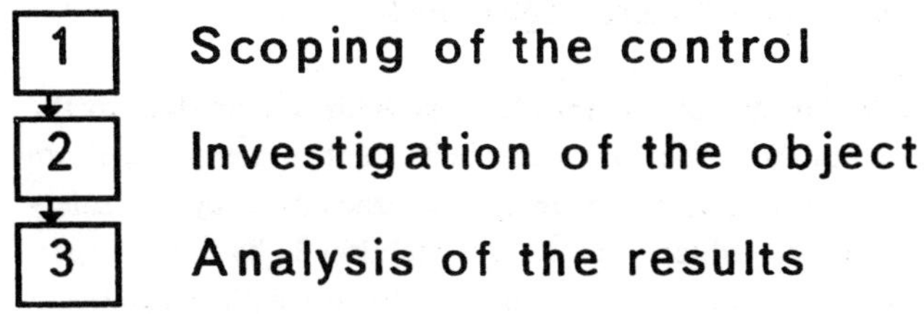

Fig.2. The circuit of the technical control

1. Choice of the controllable properties nomenclature, definition of the nomenclature and volume of quantitative and qualitative attributes, creation of the control algorithm.

2. Valuation of the object according to the qualitative attributes, analysis of the primary results, updating of the control algorithm. Fulfilment of the measuring, analysis of secondary results, updating of the control algorithm. Realization of Test (if necessary).

3.Processing and analysis of the results. Definition of defects of the elements. Drawing up of the fault register. Definition of the wear-and-tear of the main designs and the structure as a whole; drawing up of the conclusion about the technical state, recommendation for the exploitation mode, instructions on elimination of defects.

In accordance with ГОСТ 27.002-89 (1989) the following degrees of product availability of objects are allocated: good state, faulty state (but up state), down state, limiting state. The transition of the object from one state to another in connection with occurrence or elimination of defects is shown on Fig.3.

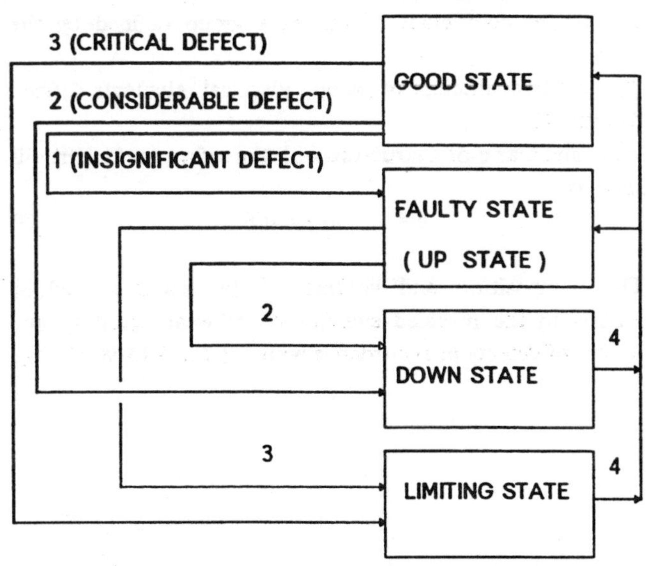

Fig.3 The circuit of main technical state and varying their events: 1- damage, 2 -failure, 3 - transition in limiting state with occurrence of critical defect, 4 - repair.

DEFINITION OF WEAR AND TEAR

The technical state of structures is defined according to the influence of the defects on workability of their elements. From the quantitative point of view it is characterized by the safety coefficient a, which value is established by the expert evaluation and makes 1>a> 0.8 for elements with insignificant defects, 0.8> a> 0.4 with considerable defects and d< 0.4 with critical defects.

The coefficient of safety for a group i of identical elements is defined as

$$a_i = \frac{\sum_{j=1}^{m} a_j}{m} , \quad (1)$$

where: a_f - particular significance of the coefficient of safety,

j=1,2,3...m - element number of the group i of identical elements,

m - quantity of elements in the group i.

The safety coefficient of the structure with n groups of identical elements is defined as

$$a_n = \frac{\sum_{i=1}^{n} a_i \ b_i}{\sum_{i=1}^{n} b_i} , \quad (2)$$

where: i=1,2,3....n - design circuit element ordinal number (identical elements group)

b - weight coefficient of elements group defined by the Table 1

a_i - safety coefficient of group i identical elements defined by formula (1).

Wear and tear c of a structure with the safety coefficient a is defined as

$$c= (1 - a) \cdot 100\% , \quad (3)$$

The composition and volume of the repair is defined according to the revealed significance of wear, quantity and character of defects in accordance with РД 31.35.13-90 (1990).

CONCLUSION

By the results of the technical control the documents, which confirm the validity of the structure to operation or forbidding operation and offering to remove revealed defects, are made out. These documents can be ordered in three operational regimens: the structure suitable to operation in a normal mode (conditional designation the "green mode"), suitable under condition when a normal mode is limited (restriction of loads, depth, use of separate elements of a design and other; a "yellow mode"), not suitable to operation ("red mode"). Under certain conditions the "yellow-green mode" can be appointed, at which the structure can be maintained in normal mode, but it is necessary to remove insignificant defects of separate elements, structures not presenting as a whole to work.

REFERENCES

Menshikov, VL (1996). "State System of Technical Control of Sea Transport Hydraulic Structures in Russia". Proceedings of the Fourth Pacific/Asian Offshore Mechanics Symposium, Pusan, Korea, pp.1-3.

ГОСТ 27.002.89 (1989). "Reliability in Engineering Main Concepts. The Terms and Definitions". State Standard, USSR, 31p.(in Russian).

РД 31.35.14-94 (1994). "Rules on Organization of the Technical Control Hydraulic Structures of the Marine Transport". Managing document of Federal Service of Sea Fleet, 92p. (in Russian).

РД 31.35.11-93 (1993). "Manual on Technical Control of Hydraulic Structures of the Marine Transport". Managing document of Federal Service of Sea Fleet, 248p. (in Russian).

РД 31.35.13-90 (1990)."Instruction on Repair of the Marine Transport Structures". Managing Document of Federal Service of Sea Fleet, 260p. (in Russian).

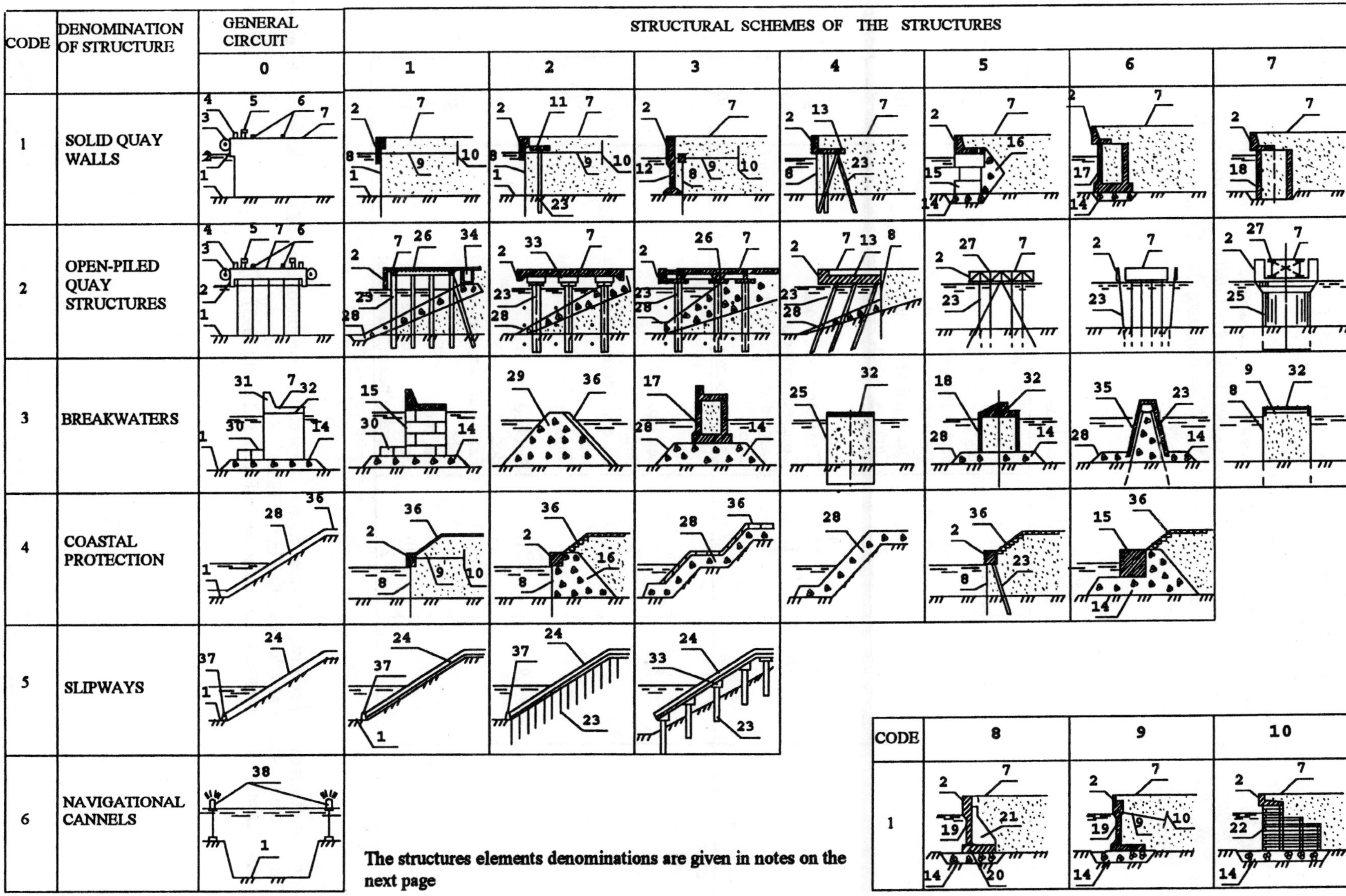

The structures elements denominations are given in notes on the next page

Fig.1. Classifier of hydraulic structures and their elements.

Notes to Fig. 1:

1. Sea bed; 2. Capping beam; 3. Fender; 4. Curb; 5. Mooring device; 6. Crane rail track; 7. Pavement; 8. Sheet-piled wall; 9. Tie rod; 10. Deadman anchorage; 11. Relieving platform; 12. Loading block; 13. Hight-level concrete deck; 14. Bed; 15. Concrete lockwork; 16. Rock fill; 17. Concrete caisson; 18 Large diameter shell; 19. Angle type precast wall unit face slab; 20. Foundation slab; 21. Counterfortunit; 22. Crib; 23. Pile; 24. Slipway rail track; 25. Sheet piled cell; 26. Cross-bar; 27. Span beam; 28. Rubble slope protection; 29. Rubble mound; 30. Apron block; 31. Parapet wall; 32. Breakwater deck; 33. Capital; 34. Retaining wall; 35. Precast element of breakwater; 36. Slope protection armour; 37. Foundationtion sill; 38. Navigation aids; 39. Foundation soil; 40. Fill material; 41. Drainage system; 42. Water supply system; 43. Electric supply system; 44. Communication system; 45. Railway tracks; 46. Special construction units Points; 39-46 are not shown.

Table 1

Weight coefficient (b) of the elements

Structure denomination	Circuit number	Element number	Element denomination	b,%
Sheet-piled wall	1.1	1	Sea bed	5
	1.2	8	Seet-piled wall	40
	1.4	9	Tie rod	25
		10	Deadman anchorage	10
		2,7	Capping beam, pavement	20
Solid quay wall	1.5	1	Sea bed	5
	1.6	14	Bed	10
	1.7	15	Concrete block work	65
	1.10	17	Concrete caisson	65
		18	Large diameter shell	65
		22	Crib	65
		2,7	Capping beam,pavement	20
Angle tipe precast wall: Counterfort wall				
	1.8	1	Sea bed	5
		14	Bed	10
		19,20	Face and foundation slabs	65
		2,7	Capping beam, pavement	20
Tied wall				
	1.9	1	Sea bed	5
		14	Bed	10
		19,20	Face and foundation slabs	30
		9	Tie rod	20
		10	deadman anchorage	15
		2,7	Capping beam, pavement	20

Notes: 1. If a constructive unit of a structure consists of several various elements, the weight of each of them is defined by expert valuation within the limits of general value specified in the Table.

2. Weight coefficients for other structures elements are given in РД 31.35.11-93 (1993).

Proceedings of the Seventh (1997) International Offshore and Polar Engineering Conference
Honolulu, USA, May 25-30, 1997

ISBN 1-880653-28-1 (Set); ISBN 1-880653-32-X (Vol. IV)

Recent Improvements in Automated Inspection and Fabrication Technology for Pipelines and Risers in Deep Water

J.C. Price
Brown and Root Energy Services
Houston, TX, USA

ABSTRACT

During the next decade, Oil and Gas developments will be from reservoirs in ocean waters down to 4000 feet and beyond. This will require the utilization of more enhanced installation techniques that will be based on welded fabrication of pipelines, risers, and bundles in the vertical (J-lay) rather than the normal horizontal position (S-lay). This paper reviews the recent improvements in Fabrication and Automated inspection technology that are taking place to optimize installation costs for J-Lay techniques.

Specifically, the recent improvements in the preferred method of Mechanized Gas Metal Arc Welding (GMAW), which has been used since 1991 in this application, are discussed, in addition to the latest developments in Radial Friction Welding. Correspondingly, the respective Inspection technology, that is currently in use on many Gulf of Mexico projects, where GMAW is utilized, is also detailed. References are made to several recent projects where these improvements in technology have been made to a high degree of success.

KEY WORDS: Deep Water, J-Lay, Fabrication, Welding, Automated, Mechanized, One Shot Process.

INTRODUCTION

The trend to develop Oil and Gas in deeper water is continuing with the installation of risers, pipelines and flowline bundles. In particular, Shell and its partners are proposing a network of pipelines in the Gulf of Mexico to link all proposed and current deep water production facilities. (Figure 1). With 7 billion cubic feet of natural gas per day expected out of the Gulf of Mexico by 2005, these facilities will be used to move gas to market as well as those of other producers. Beyond 800 feet of water there is virtually no infrastructure. In particular, the Mars field is considered the largest Oil and Gas discovery in the Gulf of Mexico in more than 20 years.

The greater use of subsea systems in the development of these world-wide Oil and Gas reserves, coupled with new contracting strategies, has created a highly competitive market place in which installation contractors must be able to offer a diverse range of technical capabilities. In particular, these developments have resulted in the requirement for more efficient J Lay techniques instead of conventional S Lay and Reeling techniques (Figures 2,3). To optimize the J-Lay technique it has been necessary to develop enhanced Inspection and Fabrication technology that can increase productivity and negate expensive and unnecessary repair situations. The use of this technology can then realize shorter installation times.

In the J-lay technique , the pipe is vertical as it is welded into the line and then lowered to the ocean bottom, creating a J configuration. The entire operation of fit-up, welding , inspection, repair and field coating takes place at a single station. As a result, joint to joint time is much longer than with S-lay techniques. It is therefore of prime importance to utilize Fabrication and complimentary Inspection technology that will enhance productivity , to the highest degree possible in one single station , and thus reduce installation times and cost as mentioned previously.

A previous paper has discussed candidate welding methods for J-lay installation concluding that Mechanized Gas Metal Arc welding and Radial Friction welding were the most promising for future development (Price et al, 1991).

The latest developments in these Fabrication techniques and the Automated Inspection technology, that is used in conjunction, will now be discussed in the following sections.

DEVELOPMENTS IN FABRICATION TECHNOLOGY

"S" lay construction has been used very successfully on thousands of miles of submarine pipelines. However, with few exceptions, this "S" lay technique has been limited to water depths of less than 1,000 feet. At depths of several thousand feet, the additional weight of the pipe exceeds the ability of the laybarge to maintain the necessary tension and prevent buckling. To overcome this problem the pipe must be laid in a vertical or near vertical position (J-lay), and the girth weld is deposited in the horizontal position.

This change in welding position drastically affects the girth welding operation. In S-lay, girth welding and other operations are spread over multiple stations and the joint to joint cycle time is determined by the slowest individual station. In the "J-lay" approach, all of the girth welding must be completed at a single station. Thus, the individual welding times per pass are additive, resulting in much longer joint-to-joint completion rates. The high cost of offshore construction places a premium on minimizing welding time. Therefore, the development of a rapid reliable single-station welding system, that can repeatedly deposit high-quality girth welds, is technically and economically required for deep water construction. (Randall et al ,1991).

For 27 years, the CRC-Evans Automatic Welding system for girth welding of pipelines has continued to progress technologically and grow in popularity, while welding more than 20,000 miles of cross-country and offshore pipelines around the world.(Randall, 1996). In Canada, Mechanized Gas Metal Arc Welding (MGMAW) with the CRC process, is currently used to weld 90 percent of all pipelines installed by Trans Canada pipelines Ltd (TCPL) and NOVA, an Alberta Corporation.

Much of the success of the system is attributable to the unique automatic internal welder line-up clamp. Recent changes to the power of the air drive motors increases the travel speed from 2.0- to 3.4 ft per second, thus reducing the joint-to-joint travel time and increasing the pipelay rate. For the J lay fabrication of risers, pipelines, etc, the internal line-up clamps are modified to align the weld joint and provide a copper backup bar for the root pass. During installation the internal lineup clamp is "nested" at the top of the derrick. The 160 foot length of pipe is then laid on the tilt-up "strongback" ramp which is rotated into the vertical position. The internal ,lineup clamp was then inserted into the end of the pipe and lowered into position at the weld joint. The clamp is provided with fail safe breaks and umbilicals to maximize reliability. The weld joint backup bar consists of a circumferential row of 1 inch by 2 inch copper segments. Each segment has a spring mounted on each end so that the segment can pivot and adjust to any high-low of the pipe I.D.

Welds are typically completed using the pulsed-gas-metal-arc-welding (PGMAW) process or conventional GMAW using the CRC-Evans M-220 and P-100 standard external welders, commonly known as "bugs" which travel around a preplaced circumferential band. The welding operator can vary wire-feed speed, travel speed, and oscillation width and frequency with fingertip controls located on the bug. The use of the CRC narrow bevel design has promoted the use of automated ultrasonics, since radiography can miss certain significant defects associated with this joint design. Further, improvements have been made in more precisely controlling welding parameters through the use of micro-processor technology, enabling faster travel speeds and increased production while still maintaining high weld quality. This has resulted in the commissioning of the P-200 Computer Controlled GMAW/PGMAW bug, which has an automatic motor oscillator to provide precise weld bead width in the joint. For small diameter pipelines CRC has developed the P-500 which has a remote pendant to control welding parameters. These bugs have a reduced number of working parts thus improving arc efficiency and downtime. The development of these new bugs has enabled the root, hot ,fill and cap passes to be completed with one bug rather than several different types, due to the micro-processor control capability.

CRC-Evans has been performing development work for J-lay applications since 1991. They have demonstrated that with their equipment using optimum welding procedures, welding arc times for 12-24 inch pipe for a wall thickness of 0.625 inch, vary between 4 1/2 to 5 ½ minutes. This translates to attainable total joint-to-joint cycle times of less than 30 minutes. Further, the rapid, sequential welding does not cause excessive thermal build - up. (Laing, 1991). The use of PGMAW compared to conventional GMAW provides 1) less spatter, which promotes less downtime on equipment, and 2) superior mechanical properties, particularly toughness, such that very high Crack Tip Opening Displacement (CTOD) values may be achieved. This results in the possibility of using Alternative Defect Acceptance Criteria, thus avoiding unnecessary repairs. The only disadvantage of using PGMAW is the appearance of very light porosity which can affect nick break performance. The maximum usable deposition rate for both processes is 11.5 pounds per hour.

Apart from the CRC-Evans equipment, other Contractors such as Allseas have used their "Phoenix" mechanized GMAW system succesfully on some J-Lay projects in the Gulf of Mexico.

Other welding methods, such as those in the "one shot" process category, are being developed as a fully integrated service capability with a multi-purpose pipe-lay vessel. Such a development has taken place on the "Seaway Falcon", with installation planned in the fall of 1996. It is anticipated that an 8 inch pipeline will be laid in the North Sea, with a "one shot" process, shortly afterwards. This development will include the installation of the latest friction welding technology, providing the potential for an increase in lay rate for J-Lay techniques and, most importantly, a capability for highly cost effective installation of specialized pipeline materials such as stainless steels and titanium. The process utilizes a heat resistant mandrel which is expanded inside the pipe bore at the weld location. The mandrel serves to align the abutting pipes and to react to the radial welding pressure. The close fit of the mandrel prevents flash ingress into the bore, leaving a smooth internal profile to the weld. (Journal Offshore Tech, 1996).

A solid ring of the same material as the pipe is machined on the ID to a sharper included angle than the included angle between the beveled ends of the abutting pipes. The ring is gripped with the jaws of the radial compression unit (RCU) and rotation and compression is started so that the ring is progressively forced into the V-shaped preparation. Frictional heat sufficient for welding is produced and after a suitable metal displacement, rotation is ceased and the radial pressure maintained to consolidate the weld.

Combining an on-deck firing line with a novel, adjustable J-Lay ramp, means that both shallow and deep water installations can be handled. The process proceeds with off line inspection before the pipe enters the firing line into station one for radial friction welding. The actual weld cycle time represents approximately 25-35 seconds of the 4.5 minutes required at station one to fully conduct the weld line-up, friction weld, internal and external NDT processes. At station 2 flash material is removed leaving a smooth cap profile for subsequent coating operations in station three. The final firing line 4 is reserved for re-instating thermal insulation at the field joint if required. Pipe from the firing-line then passes through the lay ramp, first undergoing a controlled plastic deformation to a maximum strain of 2%, followed by reverse bending and passing through a straightener. The variable inclination of the lay ramp allows the correct departure angle to be selected for the prevailing water depth, thus minimizing line tension and making deep water installation possible.

It is interesting to note that welding times for both Radial Friction welding and Mechanized GMAW are very similar at 5 minutes average arc/weld time

Additionally, there has been considerable progress in recent years in the development of Homopolar Welding (HPW), a single-shot solid -state resistance forge welding process utilizing a Homopolar Generator, that is also being developed as a single -station pipe welding method. HPW is an inherently controllable process and has a weld speed of three seconds which makes weld advance times of 12-20 minutes possible, as in the case of Mechanized GMAW and Radial Friction welding. (Pappas et al, 1996). Weld trials on API 5LX linepipe have produced good results, comparable to that achieved with GMAW. The goal of the current program is to produce a prototype system suitable for barge installation in 1998 for pipe diameters up to 12 inch diameter. Advances have also been made in automated Ultrasonic inspection of the welds which can be performed in a time of 30 seconds per weld.

DEVELOPMENTS IN INSPECTION TECHNOLOGY

At the forefront of the effort to develop more sophisticated inspection technology is automated Ultrasonic technology. Automated Inspection technology has been widely used in Canada since 1989 for inspecting Mechanized girth welds and is gradually replacing Radiographic examination on a worldwide scale.

The ability of radiography to detect voids or lack of fusion defects in mechanized GMAW girth welds depends on the orientation of the indication relative to the radiation beam. The inability of radiography to detect and display indications which do not lie in line with the beam is a physical limitation of the process and using current techniques, little can be done to rectify it. The quality characteristics of each radiograph are measured using an I.Q.I (Image Quality Indicator). Although the I.Q.I provides an overall indication of the quality of the Radiograph, in terms of sensitivity the plan view presentation cannot provide a depth assessment of the indication presented.

With sensitivity ranges achievable down to 1 percent of the wall thickness and no ability to determine depth it becomes easy for radiographers to become overly conservative in the interpretation of indications within the weld as rejectable. These disadvantages of radiography have provided the impetus to further develop ultrasonic techniques.

An audit carried out on a recent construction project of 55 km of 42 inch pipeline showed that 35 percent or repairs called for by radiography were grossly conservative. This conservatism has prompted continuous development of Automated Ultrasonic techniques for pipeline girth weld inspection.

Initially, the use of Ultrasonic inspection was in addition to Radiographic inspection due to difficulties in interpreting volumetric indications. Recent developments have been described that have taken place to provide coverage of indications such as porosity and root defects to further enhance interpretation (Ginzel et al,1993). Since 1991 the industry has been developing a technique to locate and quantify porosity in GMA welds since it is an intermittent reflector. Finite element models of infinitely long root porosity pores ,since the root variety is the most severe, were developed to provide a conservative assessment of the load bearing capabilities of a weld containing porosity.

The results of the study suggested that porosity up to ten percent projected area provides no structural deterioration to the weld. Hence an acceptance limit based on 5 percent projected area provides a suitable accept /reject criteria threshold notwithstanding the subjective nature of the porosity interpretation.

Ultrasonic detection of porosity was first used on a trial basis by TCPL followed by NOVA Corporation of Alberta in the summer of 1992. During the winter 1992/1993 construction season, porosity detection was first commercially utilized in the spread ultrasonic examination.

Repair statistics have been compiled to determine the efficiency of the Ultrasonic system in locating porosity by comparing ultrasonic and radiographic interpretations. The results demonstrated that an ultrasonic system is indeed capable of repeatedly detecting porosity and shows the superior performance in locating lack of fusion defects.

To date , the use of Ultrasonic examination , as a sole inspection method has been implemented by NOVA Gas Transmission LTD in Canada and applied to almost 1000 km of large diameter land pipeline construction since early 1994 (Hodgkinson et al, 1995). Much of the recent development work has been pioneered by Shaw Pipeline Services of Calgary, Alberta. The UT approach is a result of an extensive development effort by NOVA over the last 16 years, culminating in 1993, when they received regulatory approval to use UT as a stand alone system. The development of Mechanized Ultrasonic Weld inspection was started in mid 1990 and was completed in early 1992. All software and hardware was designed, developed and proven by Shaw Pipeline personnel based on Krautkramer ultrasonics with IBM support.

The technology developed by Shaw utilizes multiple specially designed focused shear, creeping wave, and tandem transducers that permit more accurate mapping of weld defects in a highly productive manner. The use of multiple focused ultrasonic probes requires sophisticated control apparatus to sequentially fire each probe, and to process and store the returning data. This allows for accurate depth, width, and length assessment of all critical weld defects. This is achieved with multi-channel instrumentation and computer data acquisition/manipulation and signal recording. Defects are then evaluated using alternative weld acceptance criteria from an Engineering Critical Assessment (ECA) based on Fracture Mechanics principles. This methodology has been adopted by TCPL since 1989 when ultrasonic inspection was first used commercially on pipeline girth welds. The current use of the technology by TCPL is the result of eight years of research and development in the technology.

Further , the rapid response time feature of the technology enables Ultrasonic personnel to relate weld discontinuities to welding operators, resulting in reduced repair rates and improved welder performance. In addition, there are no radiation hazards and no environmental concerns from processing chemicals. It also avoids inconsistent inspectors.

The Shaw Pipeline Mechanized Ultrasonic system has operated in the field under conditions ranging from + 40 degrees centigrade to - 40 degrees centigrade and has successfully inspected pipeline girth welds without a single day of production lost due to breakdown or inability to calibrate successfully. To date approximately 100,000 welds have been inspected with this system.

The most recent developments have concentrated on the technology required to accurately identify injurious defects in Shielded Metal Arc Welds (SMAW) in heavy wall applications. In the fall of 1996, Shell and Allseas assigned Shaw Pipeline Services to investigate this issue by comparing Radiography's and Ultrasonic's ability to fully identify lack of fusion (LOF) and interpass LOF or cold lap in pipe samples from the Shell Mensa II offshore project. The 8 inch diameter, 0.719 inch wall, pipe samples were sectioned to verify the type and size of defects. The conclusions from the test program were 1) That light slag or tight LOF are not picked up by radiography, but will definitely be shown by UT, 2) UT made no false calls or re-shoots. As a result of this investigation Shell and Allseas approved UT for manual weld inspection which will enable Shell to apply Alternative Acceptance Criteria from API 1104 for manual welds.

RECENT PROJECTS

The CRC welding equipment previously described has recently been used on a deep water project in the Gulf of Mexico. In particular, in the summer of 1996 two parallel pipelines were laid in 2,930 feet of water in Shell's Mississippi Canyon Block 807 (The Mars Field) using J Lay techniques. This was a record depth for the Gulf of Mexico. Each of the two parallel pipelines was about 26 miles long , one being a 14 inch (0.562 and 0.625 wall) X42 gas line , and the other an 18 inch (0.688 wall) X 42 oil line, both extending from Block 409 to a Tension Leg Platform (TLP). Construction was carried out in two 12 hour shifts per day with outstanding production. The best cycle times for adding a 160 foot section of 14 inch pipe was 14 minutes, 39 seconds. The highest joint production for a shift and a day were

46 and 96 respectively. The repair rate was 1.5 percent. (Teale et al, 1996). This production rate easily exceeds the 30 minute cycle time achieved several years ago. The other deep water J-Lay projects completely recently by Allseas, with Mechanized GMAW, have used their "Phoenix" system.

Table 1 details the 1996 deep water projects, in the Gulf of Mexico for which Shaw Pipeline has performed Automated Ultrasonics. All the projects have used Mechanized GMAW with the exception of Mensa II ,which used Shielded Metal Arc Welding.

CONCLUSIONS

1) The next decade of pipeline construction in the Gulf of Mexico will necessitate the development of a deep water infrastructure requiring enhanced J-Lay installation techniques.

2) Automated Ultrasonics has the potential to completely replace radiography on deep water J-Lay Projects.

3) The combination of Mechanized GMAW and Automated Ultrasonics provides the optimum method of reducing installation costs for deepwater J-lay Projects.

4) The recent installation of Radial Friction Welding on the "Seaway Falcon" and the advances made in Homopolar Welding has provided the possibility of a "one shot" welding process as a viable alternative to Mechanized GMAW.

5) The recent improvements in fabrication and inspection technology have produced shorter work station cycle times in J-Lay installations.

REFERENCES

Ginzel,E., Ginzel,R., Gross, B., Hoff. M., and Manual, P., "Developments in Ultrasonic Inspection for Total Inspection Of Girth Welds", Eighth Symposium on Line Pipe Research, AGA, Houston, Sept. 1996.

Hodgkinson, D., Dorling, D., and Glover, A., "Mechanized Ultrasonic Testing of Pipeline Girth Welds Produced by Gas Metal Arc Welding", 2nd Pipeline Technology Conference, Ostend , Belgium , Sept. 95.

Laing, B.S and Hoyt, D.S., " J-Lay Pipeline Welding ", Offshore Technology Technology Conference, Paper No. 6731, Houston, Texas, 1991.

Pappas,J., Harville,M.W., Weldon,J.M., Carnes,R.W., Hudson R.S., Nichols,S.P., " Homopolar Welding Development and Economics", ASME International, 1996.

Price, J.C. and Teale, R.A., "Welding Process Selection for the Installation of Deepwater Pipelines, 72 nd American Welding Society Annual Meeting, Detroit, 1991.

"Radial Friction Welding on the "Seaway Falcon", The Journal of Offshore Technology, August 1996.

Randall, M and Laing, B.S., " Development of Vertical Pipe (2G) Girth-Welding System for Tendons, Risers, and " J-Lay " Construction", 72 nd American Welding Society Annual Meeting, Detroit, 1991.

Randall, M, " Automatic Welding Technology Moves Forward ", Pipeline Digest, May 1996.

Teale, R.A., and Laing B.S., "J-Lay Welding on the Shell Mars Project", Australian Pipeline Industrial Association, Conference, Townsville, Australia, Oct., 1996.

TABLE 1 PROJECTS USING AUTOMATED ULTRASONICS

PROJECT	DEPTH FEET	CONTRACTOR	DIAMETER	GRADE	LENGTH
MARS J-LAY	2933	McDERMOTT	14/18	X-70	35 MILES
MENSA 1	5414	ALLSEAS	12.75	X-70	70 MILES
AUGER PHASE	3360	ALLSEAS	14/18	X-60/52	10 MILES
RAM/POWELL	3218	ALLSEAS	12.75	X42/52	26 MILES
RAM/POWELL	3200	ALLSEAS	14	X42-60	20 MILES
MENSA II	5414	ALLSEAS	6.625	X-52	17 MILES

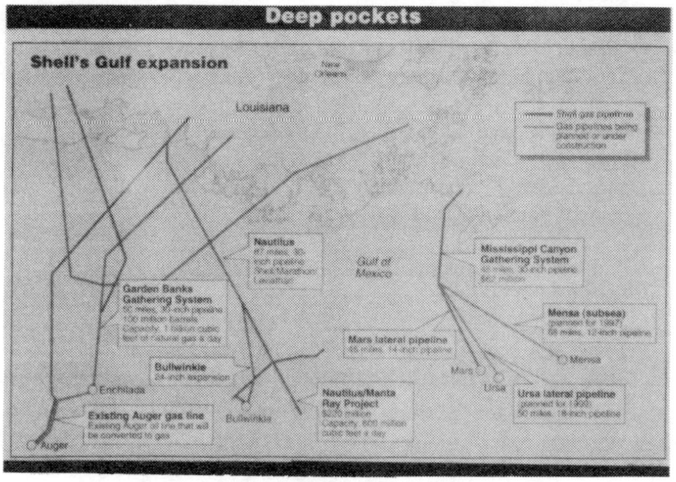

FIGURE 1 FUTURE GULF OF MEXICO PIPELINES.

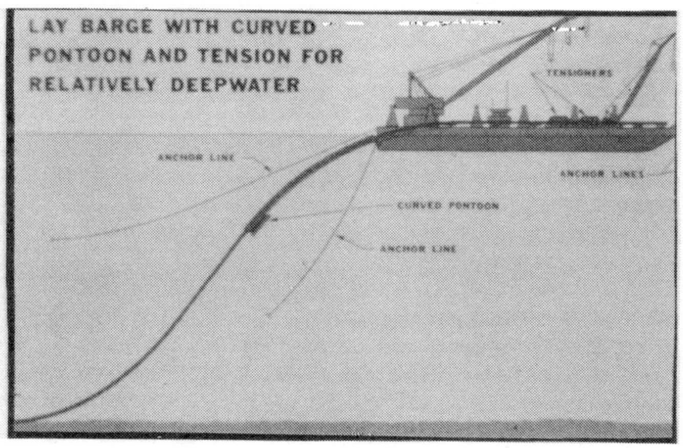

FIGURE 2 S LAY TECHNIQUE

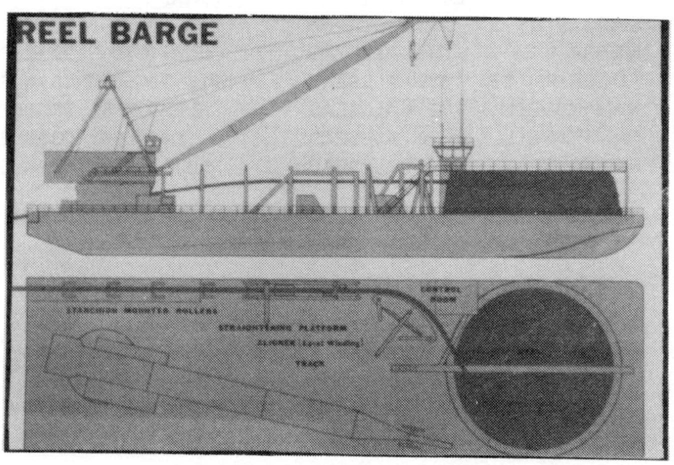

FIGURE 3 REELING TECHNIQUE

Proceedings of the Seventh (1997) International Offshore and Polar Engineering Conference
Honolulu, USA, May 25-30, 1997

ISBN 1-880653-28-1 (Set); ISBN 1-880653-32-X (Vol. IV)

Development of Long-Time Creep Safety Life Prediction of Steam Turbine Rotor Steel and AE Evaluation

Sae-Kyoo Oh
Pukyong National University, Pusan, Korea
Hong-Keun Chang
Korea Institute of Machinery and Materials, Changwon, Korea
Min-Hwa Chung and Sang-Guk Lee
Korea Electric Power Research Institute, Taejon, Korea

ABSTRACT

In this paper, a study on the prediction of long time creep rupture strength and safety life for 1.2% Cr-Mo-V and 11% Cr steam turbine rotor steels were performed by using round-bar type specimens under static load at 500~700℃ by the initial strain method (ISM) as a new approach, comparing with the results by other predicting methods such as time-temperature parameter (TTP) and minimum commitment method (MCM).

As an important result, the emirical equations for creep life and strength prediction of rotor steels were derived by using the ISM. And it was well clarified that the results by the ISM were safer than by the TTP for prediction creep strength, because not only 10^3~10^4 hours creep strengths obtained by ISM were maximum 5% lower than those by TTP but also 10^5 hours strengths by ISM were similar to those of the lower band by TTP, while 10^2 hours creep strengths by ISM were very similar to those by TTP or MCM.

KEY WORDS : Creep rupture strength, creep life prediction, initial strain method, rupture time, extrapolation, acoustic emission(AE)

INTRODUCTION

It is one of the fundamental necessities in designing, manufacturing, operating and maintaining the rotors to understand correctly the long time creep properties of steam turbine rotors used for power electric generation under high temperature and high pressure. However it is actually inevitable to predict the long time (10^5 hours) strength of practical necessity by extrapolation from the data of somewhat short time creep rupture test[1,2]. For this purpose, the time temperature parameter(TTP) method based on chemical reaction speed theory used for extrapolation, such as Larson-Miller parameter(LMP), Orr-Sherby-Dorn parameter (OSDP), Manson-Hafered parameter (MHP), Manson-Succop parameter (MSP), Manson-Brown parameter (MBP), etc., has been considered and used so far[2,3]. But these TTP's have some difficulties that the peculiar material constants must be determined for application and also the high degree functions must be used. Otherwise, the minimum commitment method(MCM) developed by the program based on TTP, is also very difficult to understand for actual application.

So, in this paper, as more simple method, the initial strain method(ISM)[4,5] applied to a study on the prediction of long term creep rupture strength and safety life for 1.2% Cr-Mo-V and 11% Cr steam turbine rotor steels under static load at 500~700℃, comparing with the results by TTP and MCM.

SPECIMENS AND EXPERIMENTAL PROCEDURES

Table 1 and Figure 1 show the chemical composition and mechanical properties of 1.2% Cr-Mo-V and 11% Cr steels, respectively, which have been used for steam turbine rotor steel under high temperature and high pressure for the power electric generation.

Figure 2 represents the profile of the specimens and applied specification used for creep test.

Table 1 Chemical composition of 1.2%Cr-Mo-V and 11% Cr steels

(a) 1.2% Cr-Mo-V

Comp.	C	Si	Mn	P	S	Ni	Cr	Mo	V	Al	Cu
Wt. %	0.25	0.03	0.65	0.007	0.005	0.64	1.21	1.12	0.29	0.004	0.03

(b) 11% Cr

Comp.	C	Si	Mn	P	S	Ni	Cr	Mo	V	Nb
Wt. %	0.16	0.24	0.67	0.005	0.003	0.58	11.01	0.92	0.23	0.05

BACKGROUND OF ISM

Oh[4] discovered that the creep initial strain (ε_0) can be quantitatively related to prediction of creep properties (fracture time t_r, creep rupture stress σ, etc.) in short term creep test of heat resisting steels (SUS303, SUH3) and the friction weld under static load at 700℃ and suggested the following models :

$$t_r = a(\varepsilon_0)^{-b}, \quad \varepsilon_0 = c(\sigma)^d \qquad (1)$$

where a, b, c and d are constants concerning materials.

It was also noted afterward by Oh et al.[5,6], Jeong[7], Kim[8] and Park[9] that the creep property prediction by ISM is more reliable than by other TTP methods.

It was also known by Mitra et al.[10], Maruyama et al.[11], and Ishikawa[12] that the initial strain in high temperature creep test for Al and Ni is similar to the strain under the same stress of the monotonic tensile-strain curve, the initial strain and stress have the linear relationship in high temperature creep test for 9Cr-Mo-V-Nb steel, and the initial strain affects on the transition creep rate in creep test for Cu under cyclic creep, respectively.

RESULTS AND DISCUSSIONS

Creep Life Prediction by ISM

Table 2 shows the typical creep test results and empirical and calculated data by ISM for 1.2% Cr-Mo-V steel under various creep stresses (13~806 MPa) at the high temperatures of 500, 600 and 700℃.

Figure 3 (a) and (b) indicates the relationship of creep rupture time(t_r, hr) versus steady state creep rate ($\dot{\varepsilon}_s$, hr^{-1}) (Monkman - Grant relation) for 1.2%Cr-Mo-V and 11%Cr steels under various creep stresses of 13~806MPa at high temperatures of 500, 600 and 700℃, respectively.

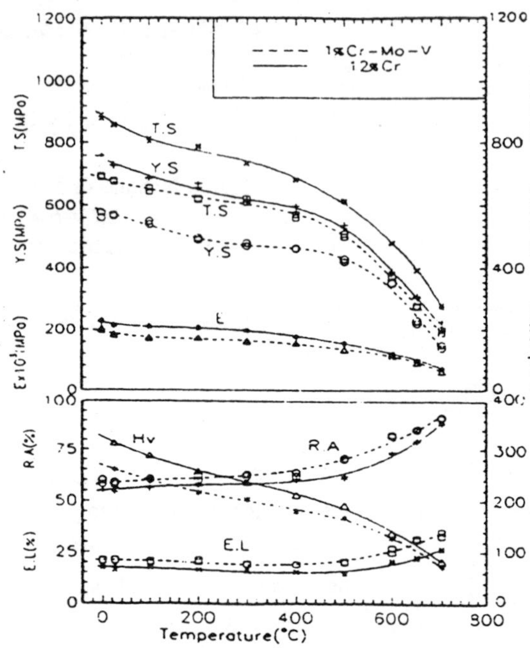

Fig. 1 Tensile test results of 1.2%Cr-Mo-V and 11%Cr steels (Y.S: yield strength, T.S : tensile strength, E: elastic modulus, R.A: reduction of area, E.L: elongation, H_v: Vickers hardness)

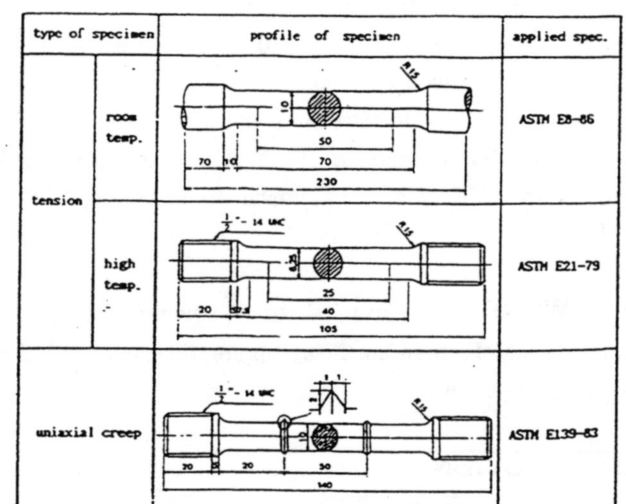

Fig. 2 Profile of specimens and applied specifications

The empirical equation was calculated by a regression analysis using the least squares method as the following :

for 1.2% Cr-Mo-V steel:

$t_r = 1.92 \times 10^{-1}\ \dot{\varepsilon}_s^{-0.71}\ (R^2=0.97)$

for 11% Cr steel:

$t_r = 5.45 \times 10^{-2}\ \dot{\varepsilon}_s^{-0.91}\ (R^2=0.98)$ (2)

As known in Figure 3, the t_r- $\dot{\varepsilon}_s$ relationship is fitted into straight line in a full-log coordinate system with no effects of temperature and can be modeled as follows by Equation (2) :

$$t_r(\dot{\varepsilon}_s)^m = K \quad (3)$$

where m and K are constants concerning materials. This equation model of the relationship between rupture time and steady state creep rate is very coincident with Monkman and Grant's[13].

Table 2 Typical creep test results and empirical and calculated data by ISM equations for 1.2%Cr-Mo-V steel

Temp-erat-ure (℃)	E (MPa)	σ (MPa)	ε_0 (%)	$\dot{\varepsilon}_s$ (%·hr-1)	t_r(hr) Cal.	t_r(hr) Emp.	Rem-ark
500	1.4×10^5	806	1.184	1.0×10^{-2}	5	6	
		503	0.584	3.8×10^{-3}	10	9.5	
		405	0.415	1.0×10^{-4}	10^2	1.1×10^2	
		327	0.301	5.3×10^{-6}	10^3	9.9×10^2	
		264	0.218	2.0×10^{-7}	10^4	-	Cal.
		213	0.158	7.7×10^{-9}	10^5	-	
600	1.17×10^5	385	0.512	9.9×10^{-3}	5	4	
		340	0.425	3.1×10^{-3}	10	8.5	
		235	0.255	1.0×10^{-4}	10^2	95	
		165	0.151	3.6×10^{-6}	10^3	1.01×10^3	
		116	0.089	1.4×10^{-7}	10^4	-	Cal.
		82	0.053	5.3×10^{-9}	10^5	-	
700	6.01×10^4	179	0.441	1.2×10^{-2}	5	5	
		148	0.331	4.0×10^{-3}	10	11	
		83	0.139	1.0×10^{-4}	10^2	105	
		47	0.059	5.3×10^{-6}	10^3	1.06×10^3	
		24	0	1.1×10^{-7}	10^4	-	Cal.
		13	0	3.0×10^{-9}	10^5	-	

E : elastic modulus, C, n : constants concerning materials at experimental temperatures, σ : stress, ε_0 : initial strain, $\dot{\varepsilon}_s$: steady state creep rate, t_r : creep rupture time(creep life)

Figure 4 shows the relationship of initial strain(ε_0, %) versus the normalized creep stress(σ/E), where σ was divided with elastic modulus(E, MPa) for reducing the effects of temperature and materials, was fitted into a straight line with high reliability under various static creep stress at the elevated temperatures of 500, 600 and 700℃ for 1.2%Cr-Mo-V and 11%Cr steels. The empirical equation was calculated as follows :

$$\varepsilon_0 = 2.71\times10^3(\sigma/E)^{1.50}$$

$(R^2=0.99,\ S=0.06)$ (4)

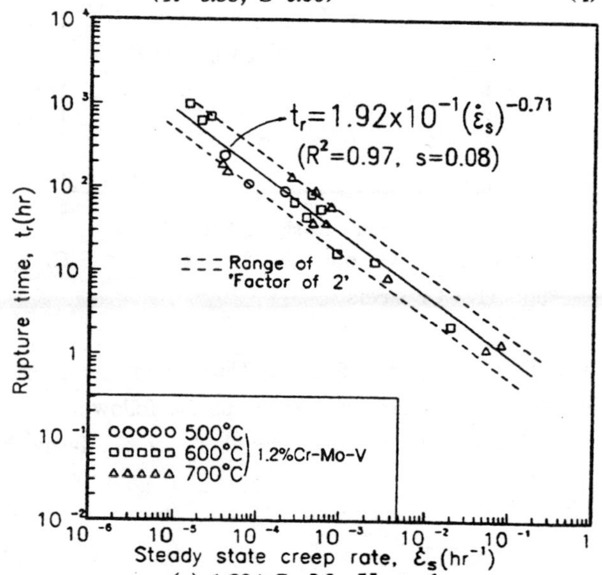

(a) 1.2% Cr-Mo-V steel

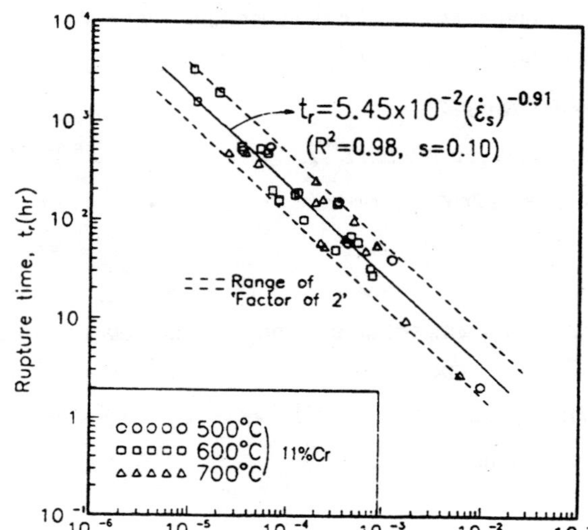

(b) 11% Cr steel

Fig. 3 Relationship of creep rupture time versus steady state creep rate (Monkman-Grant relation) for 1.2%Cr-Mo-V and 11%Cr steels (R^2 : coefficient of determination, S : standard deviation)

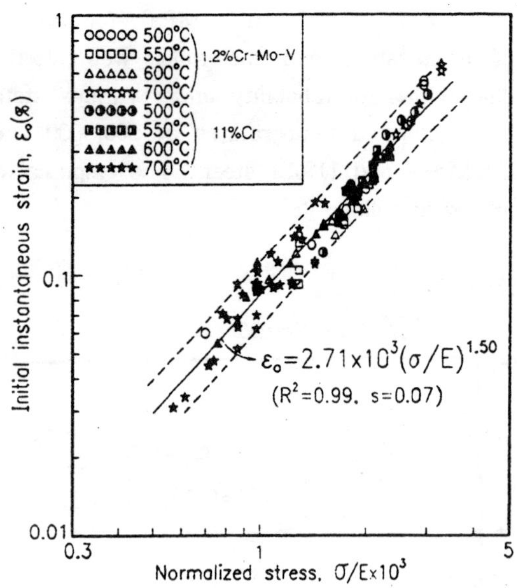

Fig. 4 Initial strain vs. normalized stress for 1.2%Cr-Mo-V and 11% Cr steels

where E is elastic modulus according to temperatures. This empirical equation can be modeled as the following, which is very coincident with Mitra et al.'s[10] and Maruyama et al.'s[11] :

$$\varepsilon_o = d(\sigma/E)^e \tag{5}$$

where d and e are constants concerning materials.

Meanwhile, $\dot{\varepsilon}_s-\sigma$ and $\dot{\varepsilon}_s-(\sigma/E)$ equations are expressed as the following :

$$\dot{\varepsilon}_s = A\sigma^n \text{(by Norton's[14])} \tag{6}$$

$$\dot{\varepsilon}_s = C(\sigma/E)^n \tag{7}$$

where A, C and n are constants concerning materials at experimental temperatures, and E is elastic modulus according to temperatures.

From Equation (5) and (7), ε_0 - $\dot{\varepsilon}_s$ equation can be derived as follows :

$$\varepsilon_0 = d(\dot{\varepsilon}_s / C)^{e/n} \tag{8}$$

where d, e, n and C are constants concerning materials. From Equations (3) and (8), the following ε_o - t_r equation can be derived :

$$\varepsilon_o = d(K/C^m t_r)^{e/mn} \tag{9}$$

And from ε_o - t_r and ε_o - σ equations according to each temperature variance of 500, 600 and 700℃ for 1.2%Cr-Mo-V and 11% Cr, $\sigma - T - \varepsilon_o - t_r$ equation can be derived as follows :

$$t_r/\sigma = \alpha(\varepsilon_o)^\beta \tag{10}$$

where α and β are constants concerning terms of temperature as the following :

for 1.2% Cr-Mo-V steel ;

$$\alpha = \frac{6.50\times10^{-5}T^2 - 6.54\times10^{-2}T + 16.54}{-6.68\times10^{-3}T^2 + 5.94T - 574.3}$$

$$\beta = -5.49\times10^{-5}T^2 + 8.80\times10^{-2}T - 38.1 \tag{11}$$

for 11% Cr steel ;

$$\alpha = \frac{6.16\times10^{-5}T^2 - 6.39\times10^{-2}T + 16.54}{-7.51\times10^{-3}T^2 + 6.59T - 576.6}$$

$$\beta = -1.84\times10^{-4}T^2 + 2.60\times10^{-1}T - 94.2 \tag{12}$$

where T is creep temperature(℃).

Reliablility of Creep Prediction by ISM

Figure 5 (a) and (b) show the comparison of creep rupture strength versus temperature relationship according to rupture time by ISM with TTP and MCM for 1.2% Cr-Mo-V and 11% Cr steels, respectively.

It is clarified in Figure 5 that not only the 10^3~10^4 hours creep rupture strengths obtained by ISM are maximum 5% lower than those by TTP but also 10^5 hours creep rupture strengths by ISM are similar to those of the lower band by TTP, while 10^2 hours creep rupture strengths by ISM are very similar to those by TTP and MCM.

Figure 6 indicates the comparison of the actual creep rupture time (t_r) with the calculated and predicted creep rupture time (t_r') from Equation (10), (11) and (12) by ISM for 1.2%Cr-Mo-V and 11%Cr steels under various static stresses at the temperatures of 500, 600 and 700℃, where both of them are very coincident each other, showing high reliability.

So that, it is confirmed that the long-time creep life prediction of steam turbine rotor steel(1.2%Cr-Mo-V and 11%Cr) by ISM could be accomplished at initial real-time in

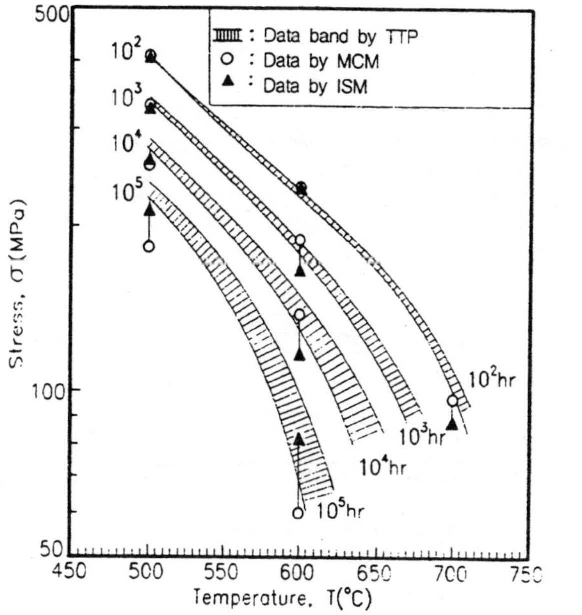

(a) 1.2% Cr-Mo-V

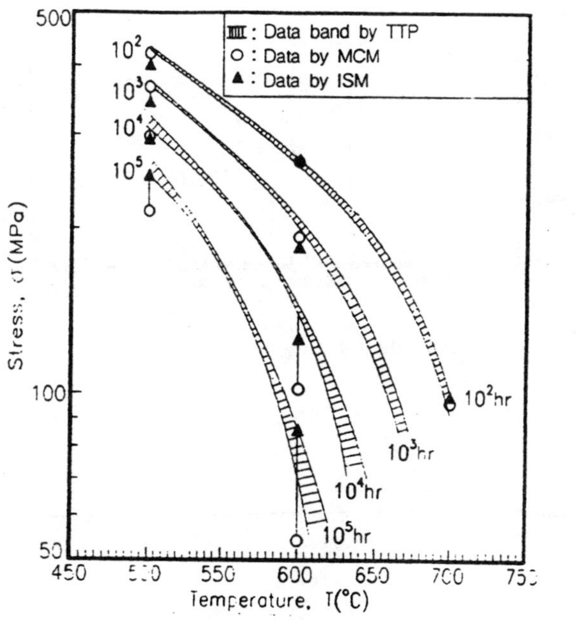

(b) 11% Cr

Fig. 5 Comparison of creep rupture strength-temperature relationship according to rupture time by ISM with by TTP and MCM for 1.2%Cr-Mo-V and 11%Cr steels

process by measuring the initial strain only, and utilized more reliably and safely for creep design, operation and maintenance of turbine rotors than by TTP or MCM.

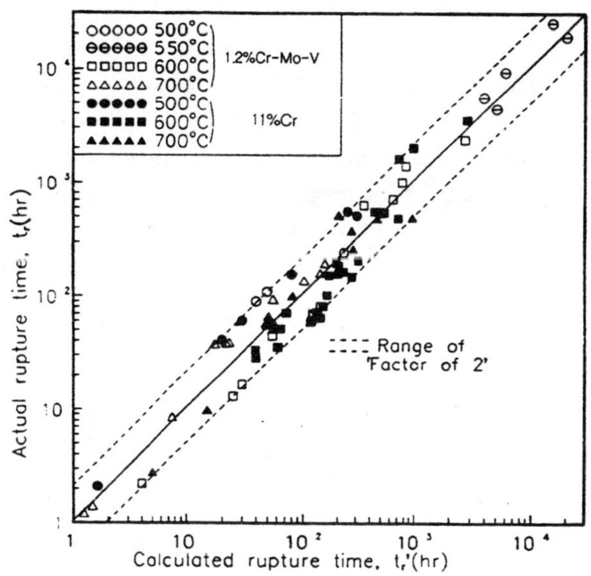

Fig. 6 Comparison of actual rupture time (t_r) with calculated and predicted rupture time (t_r') by ISM for 1.2%Cr-Mo-V and 11% Cr steels

Evaluation of High Temperature Creep by AE

Figure 7 shows the typical relationship between creep behavior and AE properties during creep test of 1.2%Cr-Mo-V steel under stress of 35~42 kg$_f$/mm^2 and 22~25 kg$_f$/mm^2 (t_r= 4~309 hours) at 700℃.

It is considered in Figure 7 that the various creep properties such as initial strain, transition creep, steady state creep, accelerated creep, crack and fracture in creep curve could be qualitatively related to AE properties such as counts vs. time, hits vs. time, event vs. time, etc.

CONCLUSION

The results obtained from the experimental and analytical study on the development of long-time creep life prediction of steam turbine rotor steels (1.2%Cr-Mo-V and 11%Cr) and AE evaluation are summarized as follows :

(1) The emirical equation for long-time creep life and creep rupture strength prediction of steam turbine rotor steel was derived by using the ISM.

(2) And it was well clarified that the result by the ISM was safer than by the TTP for predicting creep strength, because not only 10^3~10^4 hours creep strengths obtained by the ISM were maximum 5% lower than those by TTP but also 10^5 hours strengths

by the ISM were similar to those of the lower band by TTP, while 10^2 hours creep strengths by the ISM were very similar to those by TTP and MCM.

(3) It was confirmed that AE evaluation for monitoring high temperature creep properties in real time could be possible qualitatively at present.

Temperature (℃)	Load (kg_f)	Stress (kg_f/mm^2)	Initial strain (%)	Fractical time(hr)
700	600	7.64	0.025	34.25

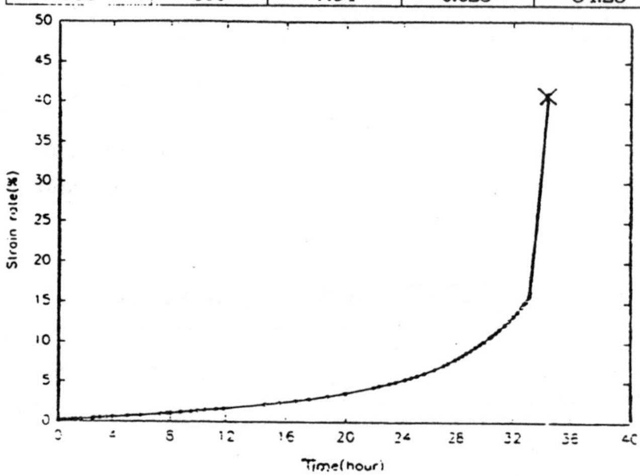

(a) Creep curve for 1.2%Cr-Mo-V steel

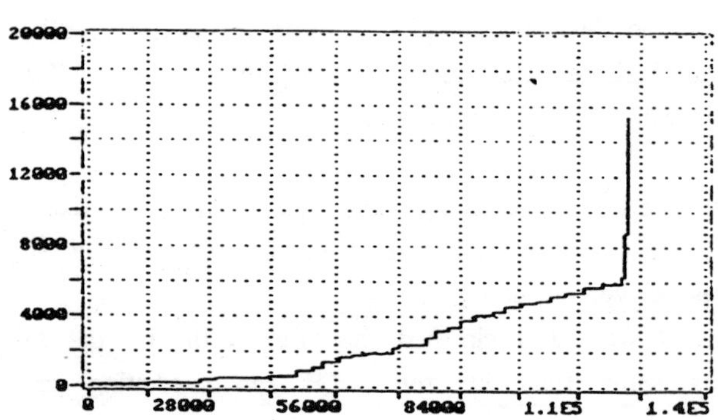

(b) AE comulative counts vs. time

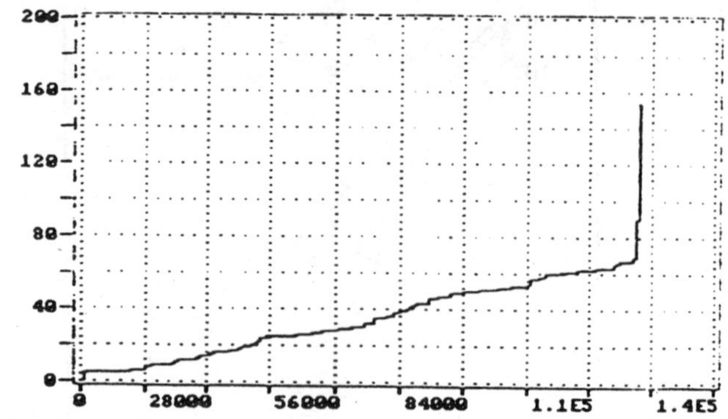

(c) AE comulative hits vs. time

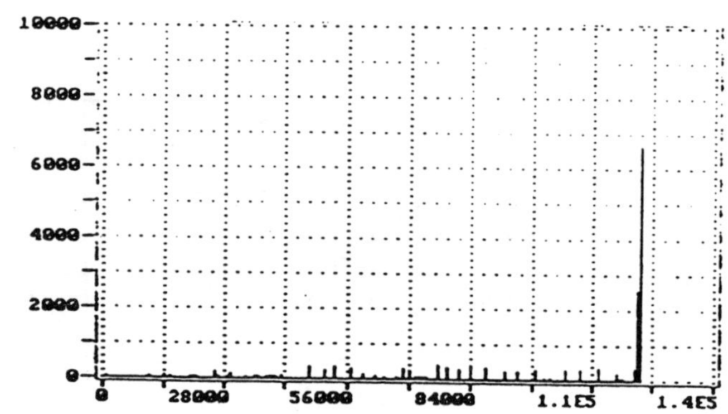

(d) AE count rate vs. time

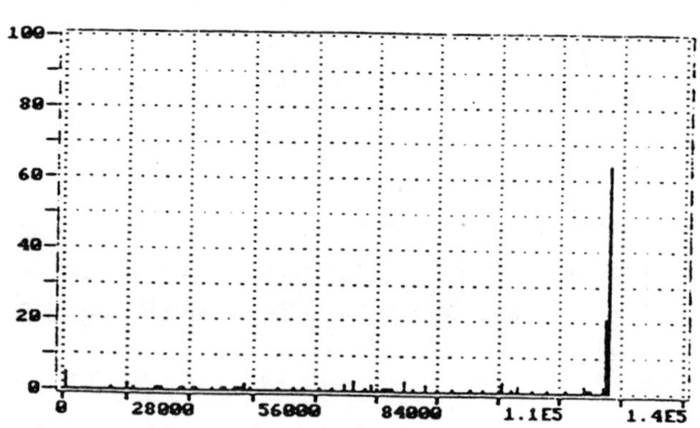

(e) AE hits vs. time

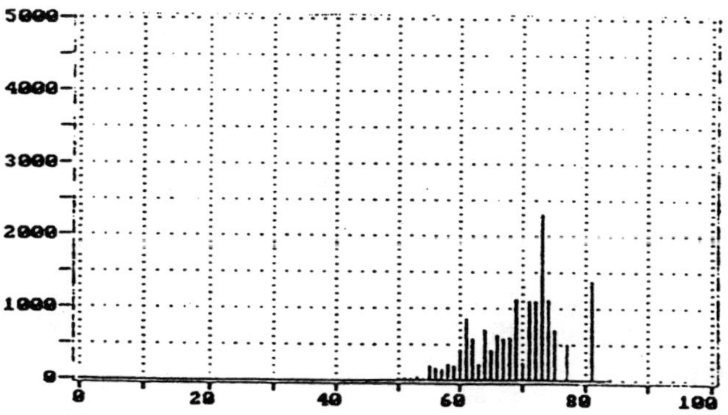

(f) AE counts vs. amplitude

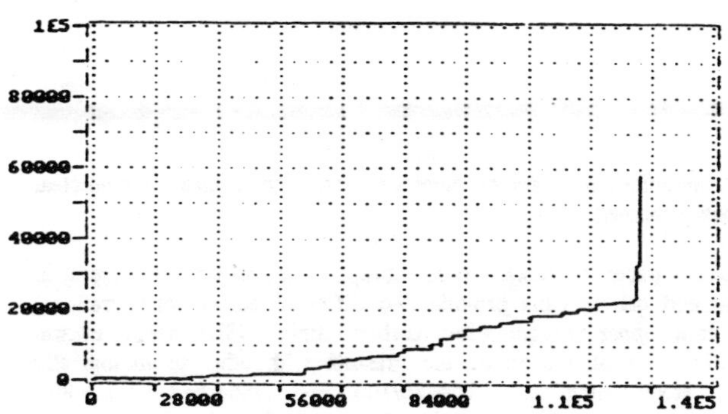

(g) AE energy vs. time

Fig. 7 Relation between creep behavior and AE properties during creep test of 1.2%Cr-Mo-V steel at 700℃

ACKNOWLEDGEMENT

This work is the project financially supported by the 1994 Research Fund of the Korea Electric Power Corporation and supervised by the Electrical Engineering & Science Research Institute, much acknowledged by the author.

REFERENCES

1) Araki, Toru(1979). "Data Sheets on the Elevated-Temperature Properties of 1Cr-1Mo-0.25 V Steel Forgings for Turbine Rotors and Shafts (ASTM A470-8)," NRIM Creep Data Sheet, No. 9A, pp.1-14.

2) Manson, S.S. and Ensign, C.R. (1971). "Interpolation and Extrapolation of Creep Rupture Data by the Minimum Commitment Method-Part I Focal-Point Convergence," pp.299-398.

3) Hizel, L.C. and Sherby, O.D.(1968). "Time-Temperature Parameters for Creep-Rupture Analysis," ASM Pub. No. D8-100, pp. 115.

4) Oh, S. K.(1982). "Study on Strength Analysis of Friction Welded Joints and In-Process Monitoring of the Welding Using Acoustic Emission Techniques," Ph.D. Thesis, Keio University, Japan, pp.85-146, 302.

5) Oh, S.K., Kunio, T. and Hasui, A.(1986). "Effect of Initial Strain on Evaluation of High Temperature Creep Properties of Friction Welded Joints," Proc. of Asian-Pacific Congress on Materials Strength Evaluation-86, pp.549-554.

6) Oh, Sae-Kyoo, Seon-Jin Kim and Tae-Eun Jeon(1996). "Development of Long-Time Creep Safety Life Prediction of Steam Turbine Rotor Steel and AE Evaluation(1st Report)-Possiblility of Creep Prediction by Initial Strain Method," Proceedings of 6th(1996) International Offshore and Polar Engineering Conference, ISOPE, Los Angels, pp.275-279.

7) Jeong, Soon-Uk(1987). "Study on Friction Welding of SUS304 to SCM4 and High Temperature Creep Properties of Friction Welded Joints under High Stress," M.S Thesis, National Fisheries University of Pusan, Korea, pp.1-38.

8) Kim, Seon-Jin(1989). "Study of Initial Real-Time Predictability for High Temperature Creep Life of Heat-Resisting Steels and the Friction Welded Joints," M.S Thesis, National Fisheries University of Pusan, Korea, pp.1-56.

9) Park, Joon-Yeong(1990). "Study on Predictability for High Temperature Creep Life of SUH3, SUS304 and the Friction Welded Joints by Initial Strain Method," M.S Thesis, National Fisheries University of Pusan, Korea, pp.1-46.

10) Mitra, S.K. and McLean, D.(1966). "Work Hardening and Recovery in Creep," Proc. R. Society, A295, pp.288-299.

11) Maruyama, K., Tamaka, C. and Oikawa, H.(1990). "Long-Time Creep Curve Prediction Based on the Modified θ Projection Concept," ASME, pp.92-97.

12) Ishikawa, H.(1987). "Relation Between Cyclic Creep and Pure Creep on Copper," Journal of Engineering Materials and Technology, Vol. 109, pp.221~225.

13) Monkman, F.C. and Grant, N.J., (1956). "Proc. ASTM," Vol. 56, pp.600-602.

14) Norton, F.H.(1929). "The Creep of Steel at High Temperature," Mcgraw-Hill, New York, pp.67.

Proceedings of the Seventh (1997) International Offshore and Polar Engineering Conference
Honolulu, USA, May 25-30, 1997

ISBN 1-880653-28-1 (Set); ISBN 1-880653-32-X (Vol. IV)

Some Aspects of Aging in a Low Carbon Bainitic Steel Containing Copper

G.J. Kulkarni and D.K. Biswas
Naval Materials Research Laboratory, Naval Dockyard
Mumbai, India

ABSTRACT

Microstructural and mechanical property evolution occurring as a function of aging are compared in two different microstructures namely martensite and bainite produced in the same steel. Transmission Electron Microscopy revealed that the kinetics of aging in martensite were faster than in bainite. This was reflected in the isochronal aging curves also. The martensitic specimens developed yield point after aging whereas the bainitic ones did not. These observations are interpreted as due to the inherent difference in the defect concentrations in martensite and bainite.

INTRODUCTION

The sequence of events occurring during the precipitation of face centered cubic (fcc) ε - copper from supersaturated solid solution of copper in body centered cubic (bcc) α - iron is identified and established using many advanced experimental techniques. First the coherent bcc copper clusters form which on attaining a critical size transform in-situ to fcc ε-copper precipitates that coarsen and elongate on further aging. Presence of bcc clusters in a certain size range is associated with the peak aged condition. Now, many authors have also confirmed the related phenomena such as orientation relationship between ε - copper and ferrite, precipitate morphologies and interphase precipitation of ε -copper. These studies have been performed mostly on copper containing pure iron or low alloy steels or in some cases on steels containing considerable alloy additions. Extensive references to previous work on precipitation of copper in steels can be found in many publications including LeMAy and Krishnadev, (1982) and Thompson and Krauss, (1996).

However, despite the advancements, a general agreement regarding the mechanisms involved in precipitation or in strengthening does not exist. The precipitate-matrix misfit strain and the difference in size of copper and iron atoms do not appear to strongly influence the transformation or the strengthening processes. The factors that more strongly play a role are the crystal defects such as vacancies and dislocations, their interaction with copper atoms and mobility of copper atoms. Most of the aging studies have been performed on martensitic (water quenched) specimens that do not offer much variation in their content and mix of defects. Some early researchers altered the state of ferrite (ferrite, cold worked ferrite, martensite and cold worked martensite, etc.), to study the effect of defect concentration on aging. While these studies point to the role of defects in precipitation of copper, there are contradicting results and no clear picture has emerged.

In the present paper we discuss some comparative aspects of aging in martensite and bainite (that provide two different defect concentrations) produced in a copper containing low carbon bainitic HSLA steel (Kulkarni et al., 1995). We also compare the results with other technologically prominent steels such as HSLA-80/100 and 'modified HSLA-80'. Development of the steel used in the present study was based on 'modified HSLA-80' (Hamburg and Wilson, (1988) ; Thompson,Colvin and Krauss, (1988,1990)), with incorporation of boron for better bainitic hardenability and improved grain refinement through controlled recrystallisation of austenite due to combined addition of Nb, B, Cu and Mo. As the development of this and similar class of steels has been recent, extensive studies on their microstructure and mechanical properties have not been available, and hence the present paper also attempts to fill this gap.

EXPERIMENTS

The steel used in this study (0.06% C, 1.5% Mn, 1% Ni, 1.2% Cu, 0.4% Mo, 0.05% Nb, 10- 20ppm B and 70 ppm S) was made in a 40 kg - vacuum induction melting furnace and cast into 120 mm diameter ingots. Proper care was taken to prevent the loss of boron from the melt. The ingots were processed to produce 12.5 mm plates as described elsewhere (Kulkarni et al., 1995).

Since the aim was to study aging in martensite and bainite produced in the same steel, two sets of specimen blocks (cut from the 12.5 mm plate) were subjected to austenitisation at 975 °C for 45 min.; one set was water quenched (WQ) while the other was air cooled (AC). Both sets of specimens were then isochronally aged at temperatures from 200 to 650 °C for 1 h. These heat-treated blocks were given the final machining to produce tensile test specimens (ASTM E8). Tensile tests were carried out on a screw driven INSTRON machine.

Undeformed ends of the expended test specimens were used for optical and transmission electron metallography. Thin-foil specimens for Transmission Electron Microscope (TEM) studies were prepared by twin jet electropolishing of 0.1 mm thin discs in a 90% acetic acid 10% perchloric acid electrolyte at 15 °C and 27 V. A JEOL 200 TEM was used for the purpose.

RESULTS AND DISCUSSIONS

MICROSTRUCTURE: Optical microscopy revealed fully martensitic structure in the WQ specimens while the WQ+Aged microstructure typically looked like tempered martensite. The AC and AC+Aged specimens were similar in appearance to the WQ+Aged specimens, in optical microscope.

TEM was used to follow the submicroscopic details. However, detailed electron diffraction studies were not pursued due to the following reasons: (1) Presence of oxides on the foils made the diffraction patterns quite complex and (2) Relatively low volume fraction of precipitates formed at the temperatures and times used in this study made the second phase spots quite weak in intensity. Nevertheless, the microstructural observations throw sufficient light on the intended investigations.

In TEM, the WQ specimens showed fully lath martensitic structure, figure 1. The laths were 0.05 to 0.25 μ thick and a few microns long. The laths showed a very high density of dislocations with no indication of twinning as expected for low carbon lath martensite. At some places very thin interlath phases were observed which are thought to be retained austenite, figure 2. Undissolved particles or precipitates were not observed commonly, which shows that the process of reaustenitisation had effectively dissolved the precipitates.

The WQ specimens aged upto 550 °C did not show much difference in overall microstructure except for a small decrease in dislocation density. No indication of recrystallisation was available and precipitates could not be seen very distinctly. However, a uniformly developed fine scale contrast was seen in specimens aged at 550°C which indicates the submicroscopic precipitation processes in the laths. The WQ+650°C aged specimens showed profuse intralath as well as some interlath precipitation of ε -Cu and a considerably reduced dislocation density. The copper precipitates both at intra- and interlath sites were elongated and clearly showed strain field contrast, figure 3 (centered dark field image). This shows that the precipitates are overaged and have crossed the spherical-shape stage. Only some of the copper precipitates inside the laths were associated with dislocations indicating that their nucleation took place at other sites too. All the elongated precipitates inside a given lath were parallel to each other which shows that they had the same variant of their orientation relationship (OR) with the matrix. Centered dark field images obtained from other spots in the diffraction patterns showed relatively fine and spherical microalloy carbides, figure 4. Occurrence of copper precipitates at lath boundaries shows that the latter can act as potential sites for nucleation and growth as they enhance the mobility of copper atoms due to the high defect content in them. The retained austenite at interlath sites seen in WQ specimens could also have formed some products of decomposition on aging, but this could not be confirmed. It is interesting to compare the above results with those for HSLA-100 (Foley and Fine, (1992); Mujahid et al., (1992); Mikalac and Vassilaros, (1992)). Most notably HSLA-100 steel shows formation of austenite along the lath boundaries (in the form of midribs) during aging at 640 °C. Also HSLA-100 showed recovery and even recrystallisation of laths. But the present steel showed more resistance to recrystallisation and did not show the presence of austenite even after aging at 650 °C. This is explained by the higher AC3 temperature of the present steel due to it's learner chemistry.

The AC specimens showed granular bainite, a mixture of lath ferrite and martensite-austenite (M-A) phases (Harbracken and Economopoulous, 1969), figure 5. In some areas the ferrite was of equiaxed nature. In general ferrite had a high dislocation density. No indication of undissolved particles or auto aging or interphase precipitation was observed. Aging upto 550 °C did not affect the microstructure noticeably and no clear visible indication of widespread precipitation was seen. However, specimens aged at 300 °C and examined elsewhere (Kulkarni and Biswas, 1993), showed Moire fringe patterns which indicate the presence of copper clusters. The M-A phases were somewhat opaque to the electron beam, but showed thick population of dislocations and no twinning could be observed. Detailed examination could not be done inside these M-A phases.

The AC+650 °C aged specimens clearly showed uniformly distributed precipitates inside the laths, figure 6, with no noticeable precipitation at interlath boundary sites. In general these ε - Cu precipitates were finer and spherical with no strain field contrast around them, figure 7 (centered dark field image). Only a few of these precipitates showed elongated morphology and strain field contrast. These specimens also did not show any clear indication of recrystallisation or formation of austenite. The absence of copper precipitates at the ferrite intra-lath sites could be due to the low angle nature of these boundaries which are known to be containing lesser number of defects compared to other high angle boundaries.

The above results provide an interesting comparison between aging in martensite and in lath ferrite. The copper- precipitates in martensite were larger, elongated, showed strain field contrast and also formed at interlath sites whereas those in ferrite were predominantly spherical and finer in comparison and formed only inside the laths. This shows that aging is relatively faster in martensite matrix than it is in ferrite matrix. This can be understood in terms of the higher overall defect concentration in the water quenched specimens. Thus, the nature of the matrix in which the precipitation occurs, clearly plays a role. Further experiments with systematic variation in dislocation concentration (cold work) and vacancy concentration (quenching temperature) can throw more light on the aging process as well as the strengthening mechanisms.

MECHANICAL PROPERTIES: All the tensile test results are shown in figure 8. The tensile strength (UTS) of WQ+Aged specimens remained somewhat the same upto about 500-550°C and then fell quite rapidly, where as the yield strength (YS) showed a small increase at 300°C before decreasing above 550°C. The effect of copper precipitation was not strongly felt in WQ+aged alloys. Earlier work on modified HSLA-80 also showed similar trends (Hamburg and Wilson, 1988). But the other copper containing steels such as HSLA-80/100 showed a pronounced peak in yield strength, though UTS was not much affected in them (Mikalac and Vassilaros, 1992). Absence of a strong peak in the present steels could be attributed to competition between softening of martensite and precipitation strengthening due to copper. In HSLA-100 the high amount of nickel(3-5%) could be preventing the softening of martensite thereby making the precipitation effect more apparent. The UTS however, is not strongly affected at small volume fraction of precipitates and hence remains steady in both steels. Consistent with the microstructural observations, the WQ+Aged steels did not show a significant drop in strength until 550°C. Above this temperature, loss of coherency and elongation led to overaging and concurrent drop in strength.

The AC+Aged specimens showed a constant YS and UTS upto about 400°C followed by a significant peak at about 450 °C and a gradual fall above 550°C. Occurrence of a peak (coherent BCC Cu phases) at a higher temperature and a lesser slope of the strength-aging temperature curve above 550 °C, in the AC+Aged alloys as compared to the WQ+Aged alloys also supports the earlier microstructural observation regarding the difference in the kinetics of aging in the two sets of specimens. The observation that the air cooled alloys can retain large amount of copper and produce substantial precipitation strengthening, could have important technological significance for this class of steels. The reason why UTS is affected by aging in AC+Aged alloys unlike in WQ+Aged is that UTS is sensitive to the presence of second phase particles which influence fracture. The AC+Aged specimens contain large volume fractions of second phase M-A particles which are also simultaneously undergoing precipitation strengthening thereby contributing to the UTS.

Aging at 400°C and above caused the appearance of yield point in WQ+Aged specimens, figure 9. This doesnot appear to be the simple quench aging effect. The AC+Aged alloys did not show any appearance of yield point even after aging at 650°C. This effect is probably due to the difference in the dislocation densities in the two sets of specimens, however, more careful microscopic studies are required.

CONCLUSIONS

1. The overall copper precipitation kinetics appear to be slower in bainite (air cooled) matrix than in martensite (water quenched) matrix.
2. This difference in kinetics is most likely brought about by the difference in defect concentration. Further experiments with controlled defect concentrations can throw more light.
3. The air cooled alloy retains considerable amount of copper in solution, thereby allowing precipitation strengthening.
4. The present alloys did not show recrystallisation on aging upto 650 °C and no austenite formed at this temperature.

ACKNOWLEDGMENTS We thank Dr. P C Deb, Director NMRL, for his encouragement and permission to publish this work

REFERENCES

Foley, RP, and Fine, ME (1992). Conf. Proc., *International Conf. on processing, microstructure and properties of microalloyed and other modern HSLA steels* , ed., A.J.DeArdo, pp 315-330.

Hamburg EG, and Wilson AD (1988). *Processing microstructure and properties of HSLA steels*, ed., A.J.DeArdo, TMS, pp 241-259.

Harbracken, LJ, and Economopolous, M (1967). *Transformations and Hardenability in steels*, Climax Molybdenum Co., p 69.

Kulkarni, GJ, Sarma DS, and Biswas, DK (1995). *Trans. IIM*, vol 48, No.3, pp 263-266.

Kulkarni, GJ, and Biswas, DK (1993). *Unpublished work*.

LeMay, I, and Krishnadev, MR (1982). *Copper in iron and steels*, John Wiley.

Mikalac, SJ, and Vassilaros, MG, (1992). Conf. Proc., *International Conf. on processing, microstructure and properties of microalloyed and other modern HSLA steels*, ed., A.J.DeArdo, pp 331-343.

Mujahid, M, Lis, AK, Garcia CI, and DeArdo, AJ, (1992). *ibid*, pp 345-356.

Thompson, SW, and Krauss, G (1996). *Metall. Mater. Trans.*, 27A, pp 1573-

Thompson, SW, Colvin, DJ, and Krauss, G (1990). *Metall. Trans.*, 21A, pp 1493-1507.

Thompson, SW, Colvin, DJ, and Krauss, G (1988). *Scripta Metall.*., 22, pp 1069-1074.

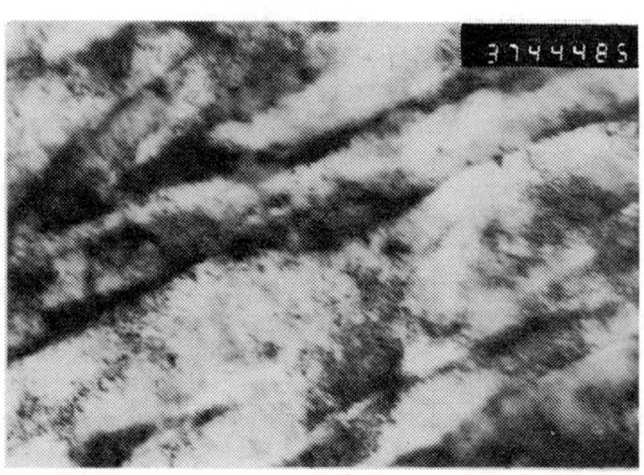

Figure 1. TEM image of WQ (lath martensite)

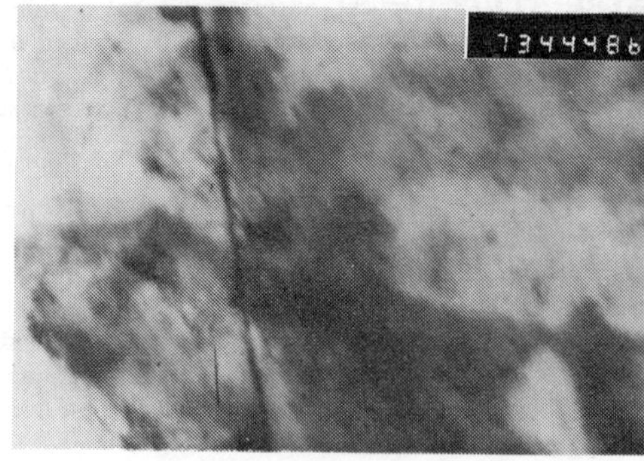

Figure2. TEM image of WQ(interlath phase)

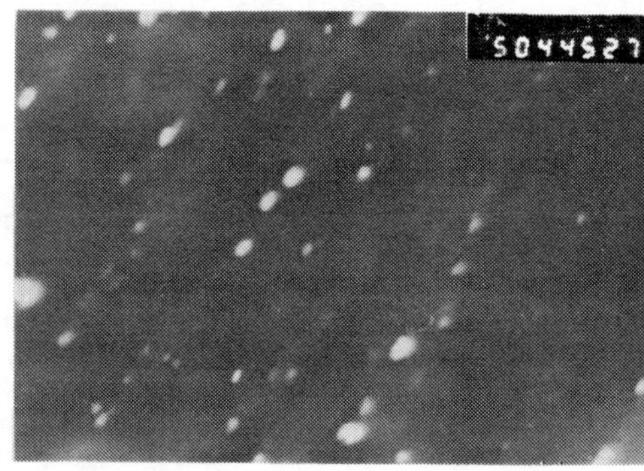

Figure3a. TEM image of WQ+650°C aged, dark field (interlath precipitates)

Figure 3b. TEM image of WQ+650°C - aged dark field (intralath precipitates)

Figure 4. TEM image WQ+650°C aged dark field (micro alloy carbides)

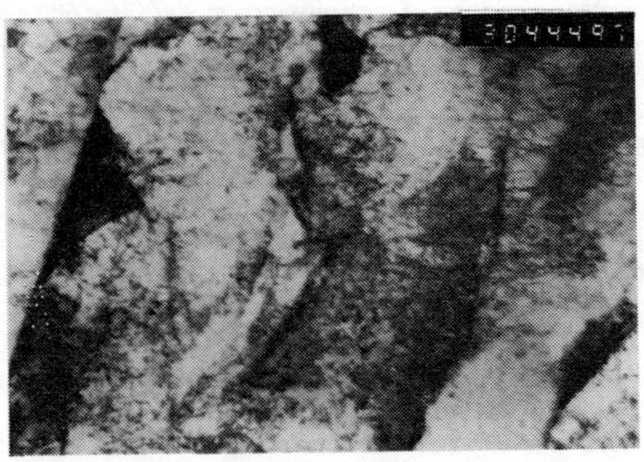

Figure 5. TEM image AC (granular bainite)

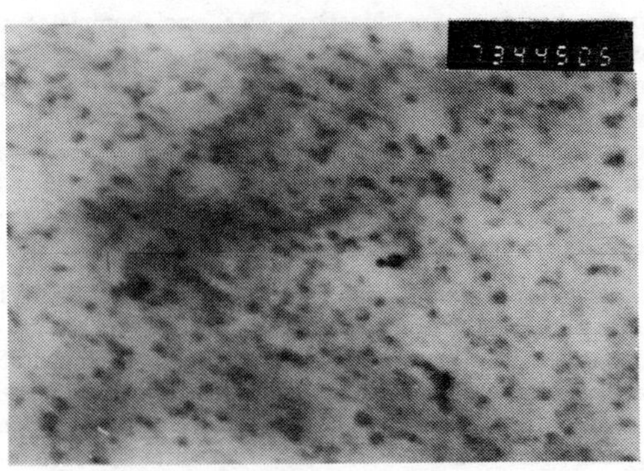

Figure 6. TEM image AC+650°C (general precipitation)

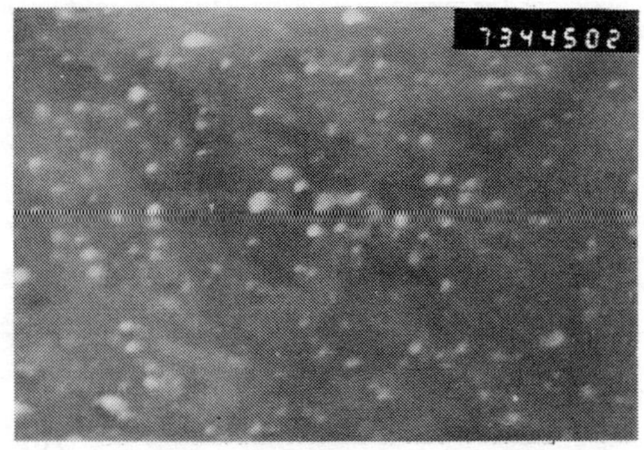

Figure 7. TEM age +650°C dark field (intralath precipitates)

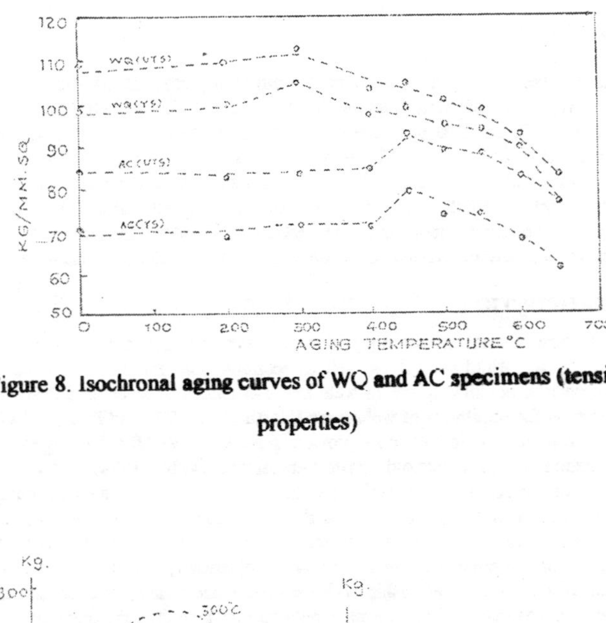

Figure 8. Isochronal aging curves of WQ and AC specimens (tensile properties)

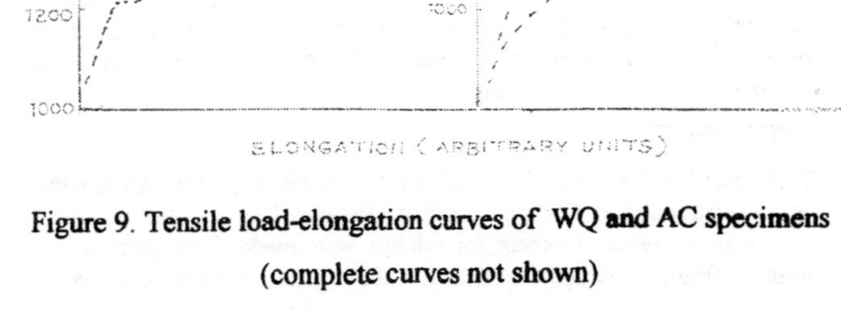

Figure 9. Tensile load-elongation curves of WQ and AC specimens (complete curves not shown)

Proceedings of the Seventh (1997) International Offshore and Polar Engineering Conference
Honolulu, USA, May 25-30, 1997

ISBN 1-880653-28-1 (Set); ISBN 1-880653-32-X (Vol. IV)

Nucleation of Proeutectoid Ferrite at Manganese-Sulphide Inclusion Sites in HY-80 Steel

D.K. Biswas and G.J. Kulkarni
Naval Materials Research Laboratory, Naval Dockyard
Mumbai, India

ABSTRACT

In this paper the results of isothermal simulation experiments on HY-80 steel with two different sulphur levels (i.e. different MnS inclusion contents) : 50 and 500 ppm, are presented. The steel with higher sulphur content showed considerably faster kinetics of austenite to ferrite transformation with multiple ferrite grains surrounding the inclusion sites. This kinetic effect was not observed in the steel with low sulphur content. The results indicate the nucleation of pro-eutectoid ferrite at MnS sites and a resultant increase in the kinetics of the overall transformation.

INTRODUCTION

Nucleation of transformation products of austenite at non metallic inclusion sites is of fundamental as well as technological interest. Presence of intragranular acicular ferrite in weld microstructures of steels has desirable effects on the toughness of weldments (North et al., 1979 and Evans, 1982). Inclusions in weldments can provide potential sites for heterogeneous nucleation of proeutectioid ferrite (Abson and Dolby, 1978) and can be used for improving the weld toughness. Hence this subject, though understood to a limited extent, has drawn considerable attention over the past two decades . Many modern techniques have been used to study the effect of inclusions, their surface chemistry, crystal structure, epitaxy/orientation relationship with the matrix and strain fields around the inclusions, as can be found in many publications including the ones listed in this paper (from Abson et al., 1978 to Ricks et al., 1982). While oxygen and titanium containing inclusions have been shown to assist the formation of intragranular acicular ferrite (Court and Pollard, 1989), the role of MnS has not been very clear. This paper attempts to study the role of MnS inclusions, in a more general context, using physical simulation techniques. This work was done as part of a larger programme to evaluate the effect of sulphur on the mechanical properties and weldability of HY-80 steel (Biswas, 1990). Though the present experiments do not replicate the complex conditions of weld transformations, they permit us to isolate and study the more fundamental case of isothermal nucleation of proeutectoid ferrite on second phase particles (MnS in this case).

EXPERIMENTS

HY-80 steel (C- 0.15, Mn- 0.3, Ni- 3, Cr-1, Mo-0.3, S-50ppm ; all in wt%) was resulphurised to 500 ppm sulphur (Biswas, 1990). Thus, two steels identical in all respects except for sulphur were made. Test specimens of 10mm × 10mm × 55mm, cut from the above steel plate were subjected to thermal cycles comprising of austenitisation at 900°C for 600 s followed by rapid cooling (≈100°C/s) to 650°C and holding for 900 s followed by quenching. A thermal cycle simulator (SMITWELD 1405) was used for this purpose. An in-situ high precision dilatometer was used to follow the transformations occurring during the thermal cycle. The cycle was repeated for steels with both the sulphur levels. Argon gas shield was used to prevent oxidation of the test specimens. After thermal cycling the specimens were metallographically examined using a JEOL-840 scanning electron microscope (SEM).

RESULTS AND DISCUSSION

Quantitative microscopic examination revealed that the volume fraction of MnS inclusions in the 50 ppm sulphur steel was 0.0013 and that in the 500ppm steel was 0.0024. With increasing sulphur the average length of inclusions remained constant ,10 to 12 μ, and the aspect ratio increased from 4 to 6.5. This leads to a higher number of thinner inclusions per unit volume and a correspondingly higher available surface area of MnS inclusions per unit volume. The average prior austenite grainsize for 50 and 500 ppm sulphur steels was 12 and 9 microns respectively. The thermal cycle imposed on the samples and the corresponding dilatometric traces are shown in figure 1. It can be seen that the steel with higher amount of sulphur (higher MnS) has a faster transformation kinetics. This steel completes majority of transformations in a relatively short time. The steel with 50 ppm sulphur transformed rather slowly. The dilatometric trace corresponding to quenching after isothermal hold also shows that the 500 ppm steel has a small kink whereas the 50 ppm steel has a larger kink. These kinks are due to formation of martensite from untransformed austenite and a larger kink indicates a higher amount of martensite. The two steels do not show much difference in the incubation period.

Figure 2 shows the corresponding SEM images of the two steels. Both the steels show polygonal ferrite (light areas) and martensite (the particles in relief). The steel with 50 ppm sulphur showed lesser amount of ferrite and a higher amount of martensite than the other steel. In the steel with 500 ppm sulphur, figure 2b, a MnS inclusion is seen to be completely surrounded by ferrite grains. This feature was very frequently observed at the MnS inclusion sites. Atleast three to four grains were seen associated with each MnS inclusion. The ferrite grains appear to be radiating outwards, thereby indicating their origin at the inclusion and growth away from the inclusion. The average dimension of such ferrite grains was higher along the length of the inclusion than along the normal direction . This indicates that the nuclei grew very quickly along the inclusion surface until stopped by an adjacent grain and then grew away from the inclusion. Thus the microstructural evidence strongly points to the nucleation of ferrite at 'MnS sites' and a possible faster transformation kinetics. It is assumed that only the nucleation rate is affected and not the growth rate. In weldments though acicular ferrite is of interest, in the present case however, the polygonal ferrite has formed due to the isothermal conditions. Nevertheless the present experiments show the role of MnS inclusions in general.

As seen in figure 2b, there is a space gap between the inclusion and the matrix. This is because of the difference in the thermal expansion coefficient between the two and it is expected to be present even at the temperature of transformation. Since all the MnS inclusions had this space gap , it follows that nucleation of ferrite occurred at the free surface of austenite surrounding the inclusions rather than on the surface of inclusions. It is interesting to note that there are many voids (without MnS) in the 50 ppm steel, figure 2a, and all of them are similarly surrounded by ferrite grains, indicating ferrite nucleation at voids also.

A theoretical model is available to compute the critical free energy of nucleation and the size of the critical nucleus (Liu 1995). Such models assume direct nucleation on inclusions and hence are not directly applicable to the present case, unless the conditions at the free surface are taken into account. Availability of free space which accommodates the transformation strains associated with ferrite formation (volume expansion) can provide additional driving force. Assuming no severe chemical gradients or thin compound layers, the potential of the inclusion sites for nucleation relative to that of austenite grain boundaries can also be computed.

The preliminary energy considerations and the experimental evidence indicate the nucleation of ferrite at MnS inclusion sites. However, the extent to which the overall kinetics is affected depends on the volume fraction of inclusions, the relative nucleation potential of inclusions and the magnitude of the surface area provided by the inclusions. In the present case the inclusions are expected to have played a strong role. The kinetics is also affected by the grain size and the chemistry. But these latter factors are known to induce a higher incubation period of transformation. In the present case the incubation period remained the same for both the sulphur levels and hence other factors may not have played a dominant role. Though a quantitative correlation between the observed increase in kinetics and the surface area of inclusions and the other factors is beyond the scope of the present paper, the results are indicative of nucleation at MnS sites and a consequent enhancement of transformation kinetics.

CONCLUSIONS

1. MnS inclusions aid in the heterogeneous nucleation of polygonal proeutectoid ferrite in isothermal transformations at least in HY-80 steel.

2. The nucleation activity at such sites can add to an observable increase in transformation kinetics.

3. The above considerations could be of significance in weldments where there is a high volume fraction of inclusions or in sintered components with a high amount of pores and voids.

ACKNOWLEDGMENTS We thank Dr. PC Deb, Director NMRL, for his encouragement and permission to publish this paper.

REFERENCES

Abson, DJ, and Dolby, RE (1978). *Weld. Inst.Res.Bull.*, vol.19, p 202.

Biswas, DK (1990). *PhD Thesis*, IIT Kharagur.

Court, SA, and Pollard, G (1989). *Metallography*, vol.22, no. 3, p 219.

Dowling, JM, Corbett, JM, and Kerr, HW (1986). *Metall. Trans.*, vol. 17A, p 1611.

Evans, GM (1982). *Welding J.*, vol. 61, p 125-s.

Farrar, RA, and Watson, MN (1979). *Met. Constr.*, Vol. 11, p 285.

Farrar, RA, and Harrison, PL (1987). *J. Mater. Sci.*, Vol. 22, p 3812.

Kayali, ES, Corbett, JM, and kerr, HW (1983). *J. Mater. Sci. Letts.*, Vol. 2, p 123.

Liu, WJ (1995). *Physical Metallurgy of Direct Quenched steels*, eds., Taylor, KA et al., TMS, p39.

Mills, AR, Thewlis, G, and Whiteman, JA (1987). *Mater. Sci. Technol.*, Vol. 3, p 1051.

North, TH, Bell, HB, Koukabi, A, and Craig, I (1979). *Welding J.*, Vol. 58, p 343.

Ricks, RA, Howell, PR, and Barrite GS (1982). *J. Mater. Sci.*, vol.17, p 732.

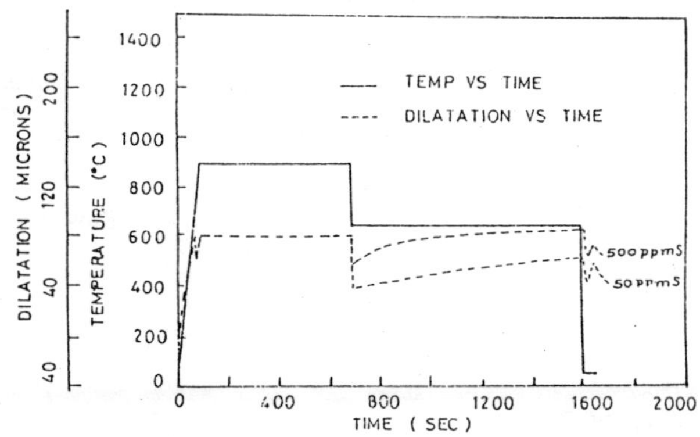

Figure 1. Isothermal simulation curves (Temperature vs. Time and Dilatation vs. Time)

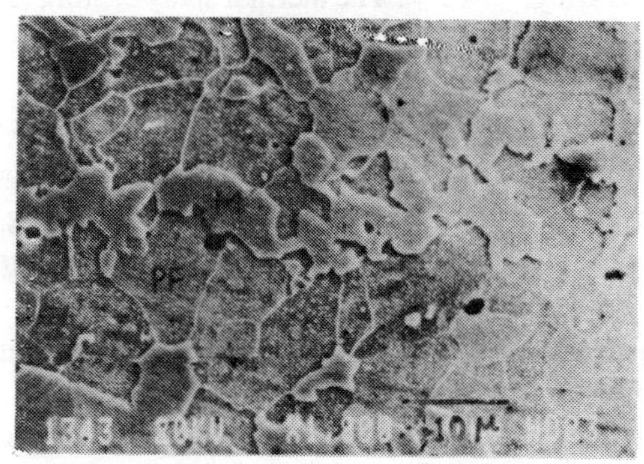

Figure 2a. SEM image of 50 ppm steel

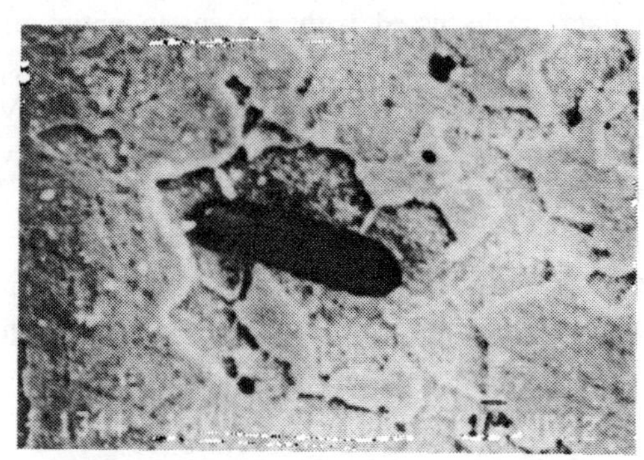

Figure 2b. SEM image of 500 ppm steel

Proceedings of the Seventh (1997) International Offshore and Polar Engineering Conference
Honolulu, USA, May 25-30, 1997

ISBN 1-880653-28-1 (Set); ISBN 1-880653-32-X (Vol. IV)

Fatigue Strength of Partially Thermal Sprayed Materials Under Corrosive Environment

H. Yara, T. Makishi and H. Arasaki
University of the Ryukyus
Okinawa, Japan

A. Ikuta
Kawasaki Heavy Industries, Ltd.
Chiba, Japan

ABSTRACT

Thermal spraying is one of the coating methods to improve the wear, heat and corrosion resistant properties of the metals and alloys [1,2]. The thermal spray coating technique is used to many kinds of thermal spray materials, and it becomes one of the objects of the surface treatment. In order to investigate the fatigue strength of the partially thermal sprayed specimens, the rotating bending fatigue test has been conducted under accelerated corrosive environment cycled among the salt water spray, the dry in the warm air and the exposure in the open air at room temperature.

Experimental studies on fatigue strength were performed under corrosive environment by using partially thermal sprayed Zn and Al powder spray materials to medium carbon steels (S45C). Some fundamental data on the mechanical and metallurgical properties of the fatigue strength under corrosive environment were obtained by using partially thermal sprayed specimens.

The data is discussed with a view to optimizing the methods, and also it is compared with those of the non-sprayed and painted specimens in terms of fatigue limit and Vickers hardness tests, and the macro and micro structural observations were also made with microscopes and SEM.

KEY WORDS: Partially thermal spray, Accelerated corrosive environment, Fatigue strength, Rotating bending fatigue test apparatus, Powder type flame spraying.

INTRODUCTION

Recently, various methods of surface treatment have been performed for improving the properties of metals and alloys. In many engineering applications, coated materials are subjected to service conditions where fatigue strength is critical. For non-coated materials, fatigue cracks initiate at the substrate surface.

Evaluation for extent of the loss of life of structural elements and industrial plant components by fatigue and corrosion is pragmatic importance for ensuring from safety of the system. Normally, industrial structural components are subjected to repeated load of varying intensity and direction under corrosive environment. Thus, even when the intensity of the individual impact of the load is below the static fracture strength, fatigue might be caused after a certain period to lead to undesired destruction. Progress of fatigue failure is strongly influenced by the corrosive environment. For example, in high-temperature high-humidity environments, coastal zones with a high-density of sea salt particles in the atmosphere and in industrial areas, a significant extent of fatigue life reduction would be caused [3].

Commonly, fatigue behavior corrosive environment is simulated by either fatigue test of corroded specimen, fatigue test under continuous application of droplets of salt water [4,5] or fatigue test under application of load being immersed in corrosive test solution [6]. In such methods of simulation, the actual situation of corrosion fatigue condition is not realistically reproduced.

In this work, the rotational bending fatigue test was undertaken for carbon steel specimens with thermal sprayed Zn and Al layers by gas flame thermal spraying under accelerated fatigue corrosion test cycles composed of salt water spraying, drying in the warm air and holding at room temperature. The same fatigue test was done for the as-machined specimen, as-blasted specimen and paint coated specimen in order to compare. Hardness distribution and fatigue strength were measured, and microstructure and fatigue surface were observed for these specimens to evaluate contributions of mechanical and metallurgical factors for fatigue corrosion behavior of the carbon steel component.

EXPERIMENT

Material description

Specimen material used for the fatigue test was a 32 mm Y rod of carbon steel S45C with about 0.45 carbon mass % of which chemical compositions and mechanical properties are shown in Table 1. The partially coated specimens with dimension as depicted in Figure 1 were machined out of the as-provided S45C rod. Diameter of the flat test part was set to 8 mm. The spray was done by using the powder type spray gun for thermal spray, and the gases used were oxygen and acetylene. Powder of Zn and Al were flame thermal sprayed to this part to thickness about 150 μ m. Paint coated specimen was prepared by twice applying white-epoxy resin to make the coating layer thickness 200~250 μ m. Table 2 summarized characteristics of these powder and Table 3 thermal spraying conditions.

Table 1 Chemical compositions and mechanical properties

a) Chemical compositions (mass %)

Material	C	Si	Mn	P	S	Ni	Cr	Cu
S45C	0. 45	0. 25	0. 79	0. 01	0. 01	0. 09	0. 03	0. 18

b) Mechanical properties

Lower yield point (MPa)	Ultimate tensile strength (MPa)	Elongation (%)	Reduction of area (%)
458	646	30	47. 3

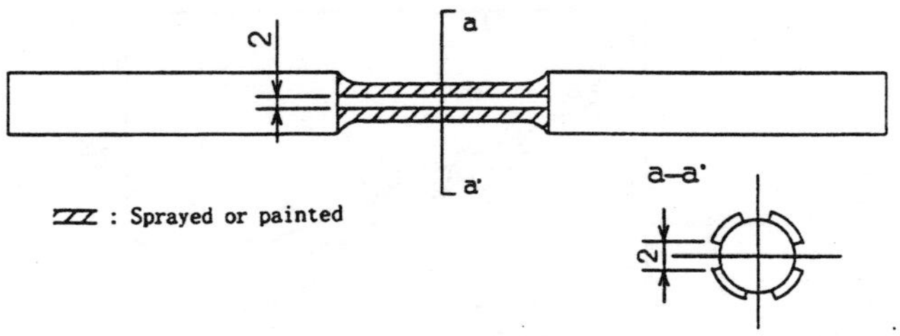

Fig. 1 Shape and size of fatigue specimen (mm)

Experiment apparatus and procedure

The fatigue tests were undertaken in Ono's rotating bending fatigue test apparatus (rate of revolution 3400 rpm, capacity 10 kgf · m) in air and under corrosive environment to derive relation between applied load and number of cycles. The rotating bending fatigue test was conducted under accelerated corrosive environment cycled of salt water spraying, drying in the warm air and holding at room temperature. The experimental apparatus used in this study was consisted of an

Table 2 Characteristics of spray materials

Characteristics	Zn	Al
Purity(%)	99.9	99.9
Grain size(μm)	50~200	50~100
Specific gravity	7.1	2.7
Melting point(K)	693	933

Table 3 Spraying conditions

Conditions		Zn	Al
Oxygen (O_2)	Pressure(MPa)	0.343	0.343
	Supply(m^3/hr)	1.875	2.5
Acetylene (C_2H_2)	Pressure(MPa)	0.0823	0.0823
	Supply(m^3/hr)	0.63	1.26
Air pressure(MPa)		0.276	0.392~0.412
Quantity of powder(kg/hr)		9.1	6.8
Distance(mm)		150	150
Thickness(μm)		150	150

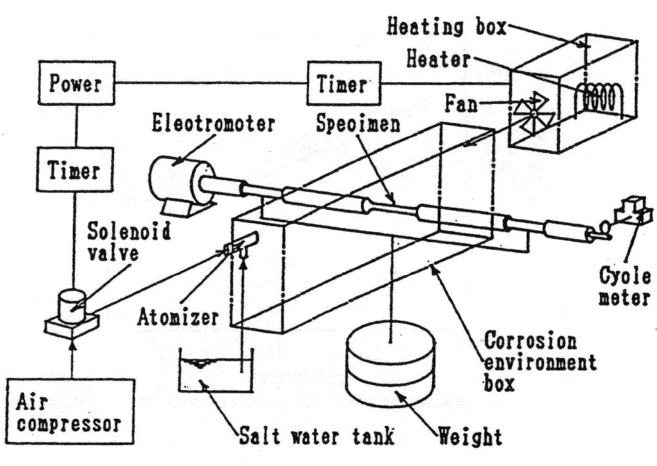

Fig.2 Schema of corrosion fatigue testing apparatus

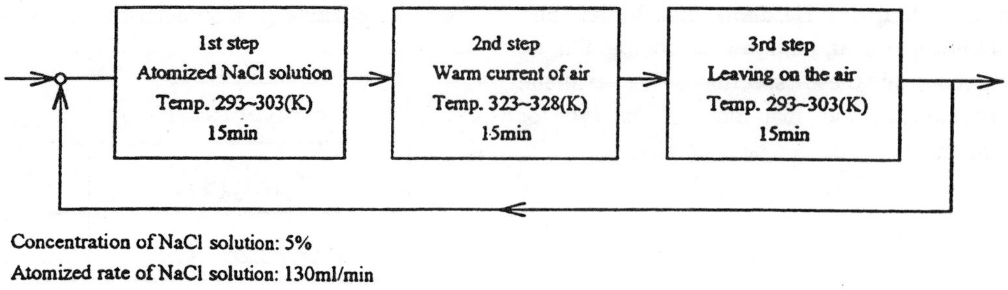

Fig.3 Flow chart of corrosive environment cycle

air compressor for salt water spray, and a heating box for blowing warm air. The experimental apparatus is shown in Figure 2. Figure 3 summarizes the sequence of the corrosion condition in unit cycle.

The blasting treatment with steel grit #100 was applied for the flat part of the S45C fatigue specimen using direct pressure type blasting apparatus prior to the partially Zn, Al and paint coated layer. The coating of Zn and Al were used thermal spray apparatus, and the coating of paint was manualed. When the specimen did not fail after 10^7 cycles of this sequence, the test was stopped at this moment. Applied load to the specimen was calculated using the formulas

$$M=WL/2=10W \quad (kg \cdot cm)$$
$$\sigma=32M/\pi d^3 = 320W/\pi d^3$$
$$=101.8W/d^3 \quad (kgf/cm^2)$$
$$=9.98W/d^3 \quad (MPa)$$

where L(=20 cm) refers to the distance between load supporting points, d(cm) the specimen diameter, M(kg · cm) uniform bending moment, W(kgf) mass of the hanging weight block and σ(kgf/cm^2) stress induced for the specimen.

RESULTS AND DISCUSSION

Fatigue strength

The measurement of the fatigue strength yields fundamental data on the mechanical properties of the partially coated specimens.

The S-N curves for various coated specimens under corrosive environment and in air are presented in Figure 4 by the rotating bending fatigue test. Under corrosive environment coated Zn, Al and paint specimens, the fatigue limit was about 280, 290 and 260 MPa which were almost same fatigue limit in air. Therefore, the coated specimens have anti-corrosive effect under corrosive environment. The fatigue limits of the partially sprayed Zn and Al specimens, 4 parts of 2 mm width axial direction at flat test part, were both about 270 MPa under corrosive environment. But the fatigue limit of partially paint coated specimen was about 215 MPa, that was about 40 MPa lower than the fatigue limit of paint coated specimen under corrosive environment. Thus, the thermal spray coating have lower galvanic position metal in comparison with the substrate, and acts on the chemical action of sacrificial anodes.

These results indicate that the application of thermal sprayed coating layer to existing industrial structure, in place of application of traditional paint coating, might be considered as practical measure for the service life elongation; and no reduction in fatigue strength is caused by the application of the thermal spraying. Where pore sealing processing is applied additionally to the thermal sprayed coating layer, further extent of the life elongation might be gained.

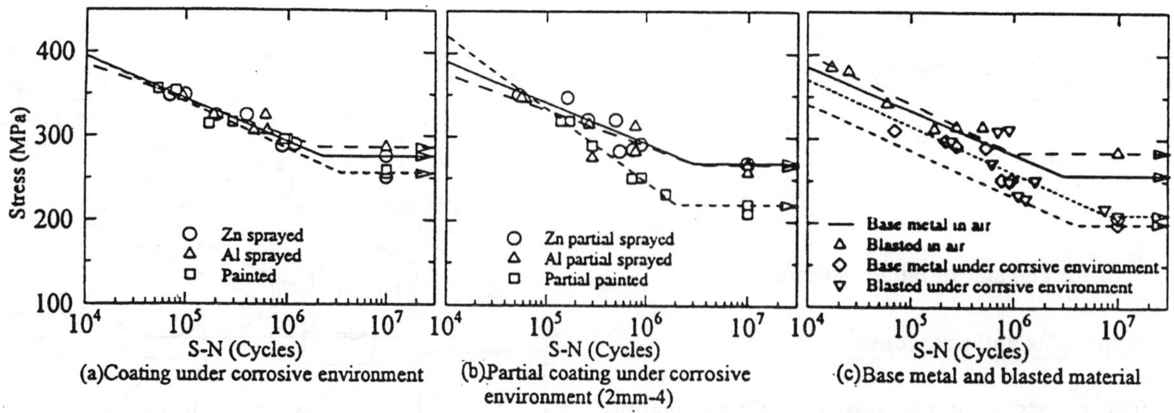

Fig.4 S-N curves for various coated specimens under corrosive environment and in air

The surface coarsening by blasting is indispensable to ensure good adhesion of the thermal sprayed layer to the substrate. There appeared to be no detrimental effect introduced by the blasting on the corrosion fatigue performance of any specimen. But, this conception seemed to be unjustified, judging from the fact that the fatigue limit values observed on the coated specimens under corrosive environment were compared to those on the non-coated specimens in air. Roughness Rmax of the specimen surface after the blasting was measured to be 40 μ m [7] . According to Nakazawa and Homma [8], roughness Rmax = 60 μ m was obtained by shot blasting, and the strength reduction induced by notch effect through generation of uneven surface profile seemed to be well compensated by the work hardening effect and by the compressive residual stress for the shot blasted surface. While the extent of notch introduction in the grit blasting and that in the shot blasting are not equal, reduction of surface strength caused by morphological modification introduced and by grit blasting in the present work might be also considered as being well compensated by physical surface hardening effect, and thus the loss of fatigue strength was avoided by the blasted thermal sprayed specimens.

Hardness distribution

Correlation is known to exist between the hardness and the fatigue strength. Thus, it is probable to estimate fatigue strength when the hardness is known. Actually, the vickers hardness is one of the mechanical properties measured, and also one of the factors closely related to the microstructure [1] .

The hardness near the surface of the fracture part was measured by using Micro Vickers Hardness Testing Machine, and the load for the each test was set at 0.98 N. The measured place is the axial direction near the surface and 0.5 mm intervals.

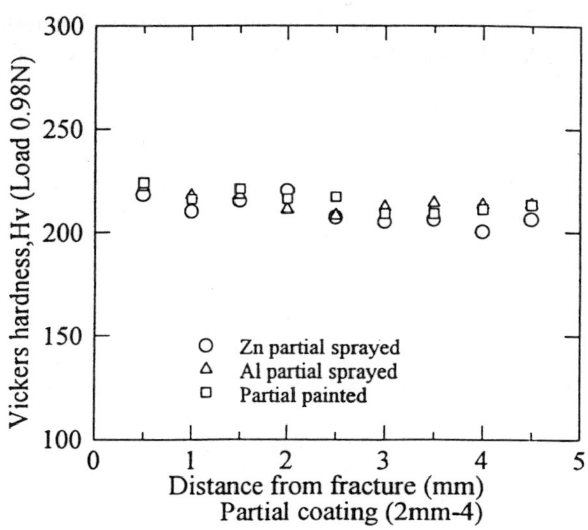

Fig.5 Distribution of Vickers hardness

Figure 5 summarizes the profiles of HV measured for specimens fractured near the fatigue limit. HV near the fracture surface was 222 for the Zn sprayed specimen, 217 for the Al sprayed specimen and 225 for the paint coated specimen. HV in these specimens tended to decrease with increasing distance from the fracture surface indicating that the hardening near the fracture surface was brought by stress concentration and plastic deformation induced around the crack which eventually led to fracturing failure. HV of Zn coating layer was around 30 and that of Al around 45 ; these low values of HV exhibited by the coating layers must be ascribable to the porous nature of the coating layers.

Macrostructure of near fatigue limit

Figure 6 shows general appearances of failed specimens near the fatigue limit. Significant extent of corrosion development was observed to the non-painted part of partially paint coated specimen. The observed features of corrosion for that specimen appeared to be similar to the corresponding structures in the field service. The white rust formed non-coated part of the partially Zn thermal sprayed specimen was $ZnCl_2$ and $Zn(OH)_2$, that of the partially Al thermal sprayed specimen was $AlCl_3$ and $Al(OH)_3$ for the reaction of salt water. That is, in the partially Zn and Al thermal sprayed specimens, the corrosion of non-coated substrate was effectively inhibited until the moment of fatigue failure.

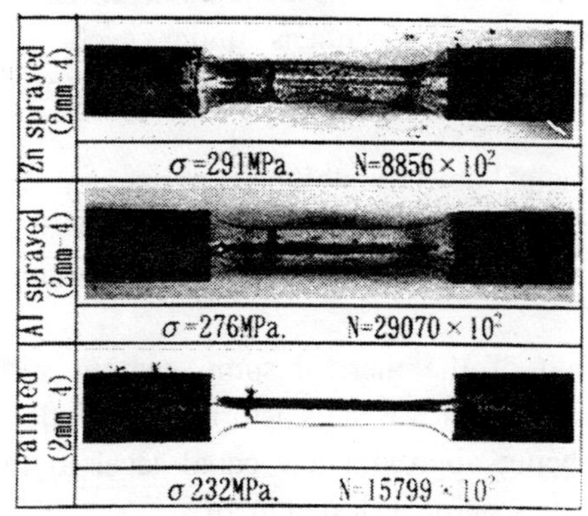

Fig.6 Example of fatigue fracture specimens

Fractograph of fatigue fracture surface

Figure 7 shows the examples of macrostructure by the optical microscope and microfractographs by scanning electron microscope (SEM) for the fatigue fracture specimens. The macrostructure of the fatigue fracture surface did not observe clam-shell mark nor ratchet mark. By detailed SEM observation of fracture surface, striation pattern characteristic of stably growing part of the fracture was seen for any specimen examined and greater part of the fracture surface was in the dimple pattern representative of ductile failure.

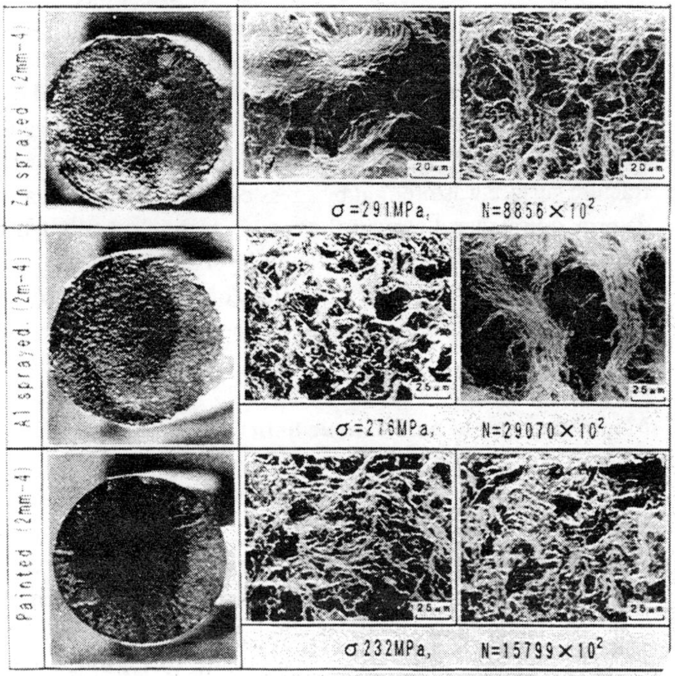

Fig.7 Typical fractographs of fatigue fracture surface

CONCLUSION

Fatigue corrosion test was undertaken for the partially Zn, Al and paint coated specimens under cyclic accelerated corrosion condition including salt water spraying and warm air blowing.

The principal objectives of the experimental study was to obtain the basic data on the mechanical and metallurgical properties. The experimental findings may be summarized as follows:

(1)Fatigue limit of the thermal spray coating specimen or paint coating specimen was satisfactorily high under corrosive environment being approximately equal level to the fatigue limit of the same specimens tested in air.

(2)Application of partially thermal sprayed coating layer was concluded to be effective for elongation of industrial structure in service as it would not cause undesired reduction of fatigue strength.

(3)The macrostructure of the fatigue fracture surface did not observe clam-shell mark nor ratchet mark.

(4)Striation pattern characteristic of fatigue fractured surface was observed by SEM observation of fractured surface of any specimen examined. Greater part of the fracture surface was in the dimple pattern representative of ductile fracture.

REFERENCES

[1] H.Shimura, Y.Enomoto, N.Ohtomo and T.Yamaguchi :"Wear - Resistant Metal Cladding by Low Pressure Laser Spraying", Japan Thermal Spraying Society.1988. pp.93.

[2] H.T.Steine, C.Wasserman and K.Kasai: "Hot Gas Corrosion Protective Coatings and Their Application", Japan Thermal Spraying Society, 1988, pp.405.

[3] H.Yara, A.Ikuta and K.Kotake: "Fatigue Characteristics of Thermal Spray - Coated Materials in Corrosive Environments", Corrosion Engineering 43.1994, pp.711.

[4] M.Nakajima,H.Kunieda and K.Tokaji:"Fatigue Corrosion Life Distribution Characteristics of S45C Carbon Steel in Atomosphere and in Salt Water", Collected Papers of Jpn Soc. Mech. Eng.,Series A, 57 [536] (1991).pp.727.

[5] M.Gotoh, S.Fukushima, H.Nishitani and A.Miura: "Fatigue Process of Tempered S45C Carbon Steel in 3% Salt Water", ibid., 57 [542] (1991), pp.2314.

[6] K.Siozawa, J.Mizutani, S.Nishino, S.Ebata, N.Yokoi and Y.Haruyama: "Corrosion Fatigue Crack Propagation Behavior of Forge-Casted Al Alloy Component", ibid., 57 [538] (1991), pp.1279.

[7] H.Yara and D.Fujiki: "Fatigue Strength of Thermal Sprayed Components", Collected Papers of Jpn Weld. Inst., 10 [2] (1992), pp. 287.

[8] H.Nakazawa and H.Homma: "Fatigue Strength of Metals". Yoh-ken-doh Publ., Tokyo, 1982, pp.51.

Proceedings of the Seventh (1997) International Offshore and Polar Engineering Conference
Honolulu, USA, May 25-30, 1997

ISBN 1-880653-28-1 (Set); ISBN 1-880653-32-X (Vol. IV)

Coalescence Conditions for Stress-Corrosion Cracking Based on Interacting Crack Pairs

B.N. Leis and R. Mohan
Battelle
Columbus, OH, USA

ABSTRACT

Crack coalescence by stress-corrosion cracking (SCC) can lead to cracks that eventually are large enough to be fracture-critical (unstable). Up to that point continuing coalescence produces ever longer cracks, which because of coalescence experience significant jumps in their average crack-growth rate. Therefore, coalescence can be an important aspect of the SCC failure process. This paper develops a criterion to determine when the conditions between two crack tips, expressed in terms of the linear-elastic fracture mechanics stress-intensity factor, K, will support the onset of coalescence. Whether the loading remains sufficient to support the continuation of the coalescence process can likewise be determined with this criterion. This criterion is based on satisfying the condition between the two tips that $K \geq K_{Coalesce}$, where $K_{Coalesce}$ is a measure of the material's resistance to cracking due to either SCC or fracture. This criterion is illustrated by its application to field data for the coalescence of SCC that resulted in a leak in a gas-transmission pipeline.

Key Words: Crack coalescence, crack spacing, criterion, stress-corrosion cracking (SCC), failure criteria

INTRODUCTION

Stress-corrosion cracking (SCC) occurs occasionally on the outside surface of gas-transmission pipelines. Where failure occurs, this stress-corrosion process produces an axial flaw that can involve the coalescence of many crack segments, which often have similar depths. Coalescence can lead to crack sizes that do not support ductile tearing but can continue to grow stably by SCC until the wall is penetrated (i.e., a leak occurs). This coalescence, which occurs at a rate determined by the kinetics of SCC, begins when the adjacent crack tips sense one another. As shown by Stoneseifer et al (1993), crack tips tend to grow toward each other for interacting colinear crack pairs and away from each other for coparallel pairs. Coalescence by SCC also can lead to crack sizes that either at coalescence or after additional SCC are large enough to support stable ductile tearing leading to a leak as the flaw grows stably through the wall. This coalescence occurs at a rate fixed by either the kinetics of SCC or stable tearing (the later depending on the crack size, the pipe pressure, and the pipe steel's tearing resistance (e.g., Leis et al (1993)). Finally, coalescence can lead to a crack whose overall length supports continued unstable axial growth at the current depth (i.e., a rupture). Such coalescence and the ensuing rupture occur very quickly. As discussed later, conditions for failure as a leak or rupture of the coalesced cracks can be assessed using existing failure criteria for an isolated crack (e.g., Leis et al (1993), Leis and Ghadiali (1994)), except in situations where deep coparallel cracks exist.

This paper develops a criterion that determines when crack-tips sense each other, which is a necessary condition for the onset of coalescence between adjacent crack tips. This criterion, when coupled with (1) crack growth kinetics and related analysis to track the time for coalescence and (2) a failure criterion to determine when the size of such coalesced cracks is critical at the current pressure, form the basis for analysis of stress-corrosion cracking in pipelines. Whereas criteria for coalescence that specify crack shape during and following the coalescence process could be pursued (e.g., anon. (1991,1994), Leek and Howard (1994)), the present work approximates the problem in terms of fracture mechanics analysis of the stress intensity field between adjacent crack tips. Suitability of this approximation is then judged by comparison with typical SCC data from field failures and laboratory experiments.

A COALESCENCE CRITERION BASED ON FRACTURE MECHANICS

Approach

Fracture mechanics enters the development of a coalescence criterion as the means to determine if and where the critical level of the linear elastic fracture mechanics stress intensity factor, K, for stress-corrosion cracking, is exceeded. While a rigorous analysis of the conditions for coalescence would consider the effects of the several mutually adjacent cracks that can occur with SCC, this paper approximates this complexity by considering the behavior of the mutually adjacent tips of a pair of cracks.

The approach is to determine under what conditions of crack size and tip spacing the K-field supports the growth of one crack-tip toward another, eventually culminating in their coalescence. Such a criterion for coalescence determines only if two cracks of known size, shape, and proximity can coalesce. Such a criterion does not determine if they have coalesced, because two tips that satisfy the condition for coalescence must still grow to meet each other. Thus, the merging of two cracks requires satisfying the coalescence criterion and thereafter tracking the growth of these cracks over the time required for them to actually join. Tracking this coalescence process requires evaluating the SCC kinetics over the course of this process. In this sense, the present work differs from other empirical studies, as for example that of Parkins and Singh (1990), which focus on crack positions with tips that have joined — as opposed to positions that could coalesce.

To determine if mutually adjacent flaws can coalesce requires knowledge of (1) the K field between the tips and (2) properties data for $K_{Critical}$ for SCC for the environment and pipe steel of concern. Because the value of K in a pipeline depends on the operating pressure, information on the operating conditions of the line is also needed. For the present work the value of K is found using the Finite Element Alternating Method (FEAM) as elaborated by Nishioka and Atluri (1983) and others. The FEAM as applied in this work is detailed by Stoneseifer et al, (1993). This method is particularly powerful and adaptable to problems such as the present. As outlined by Stoneseifer et al, (1993), this method provides for the evaluation of all three modes of cracking.

The Pipeline Situation

A coalescence criterion is developed for conditions representing the maximum operating pressure (MOP) in the pipeline. MOP for the pipeline is taken as the maximum allowable operating pressure (MAOP) for Class I conditions, which is 72 percent of the specified minimum yield stress (SMYS) in the United States and 80 percent of SMYS in Canada. A coalescence criterion covering hydrotest conditions, which here correspond specifically to a maximum wall stress at 110 percent SMYS, followed by a hold at any pressure less than 0.9 (110 percent SMYS), also has been developed. Because flat-plate solutions have been earlier shown to adequately determine the value of K in large diameter thin-wall pipelines (Stoneseifer et al (1992), this criterion has been developed without regard to pipe diameter.

Coalescence conditions have been developed by determining the relative positions of the adjacent tips of two cracks for which K over the path between the tips exceeds $K_{Critical}$. These relative positions have been developed for wall thicknesses that bracket many pipeline situations by considering wall thicknesses of 0.250, 0.300, and 0.375-inch (6.25, 7.62, and 9.53 mm) in pipe with properties representing steels typical of Grades X52 to X65. These absolute thicknesses are considered for crack depths, a, relative to the wall thickness, t, for depth/thickness (a/t) values given as 0.20, 0.50, and 0.75, in cracks with a half-length, c, leading to values of the crack aspect ratio, 2c/a, equal to 6. The value of $K_{Critical} = K_{SCC}$ has been taken as 15 ksi-inch$^{1/2}$ (16.34 MPa-m$^{1/2}$), which is typical of many pipeline steels in a carbonate-bicarbonate cracking environment (Fessler and Parkins (1986)).

Dealing with Mixed-Mode Cracking

Because coalescence is driven by opening mode cracking (i.e., K_I) and by the shear modes of cracking (i.e., K_{II} and K_{III}), conditions for coalescence must be determined in terms of a uniaxial "effective" K whose application may be viewed as analogous to the effective stress concept used in the analysis of other multiaxial problems. To this end, K is represented by an effective value, K_{eff}, defined as

$$K_{eff} = (G \cdot E / \beta)^{1/2} \quad , \qquad \text{(Eq. 1a)}$$

where the strain energy release rate, G, is

$$G = \frac{\beta}{E}(K_I^2 + K_{II}^2) + \frac{(1 + \nu)}{E} K_{III}^2 \quad , \qquad \text{(Eq. 1b)}$$

and $\beta = (1 - \nu^2)$ in plane strain or 1 in plane stress. The value of K_{eff} has been assessed in terms of K at $\phi = 0$, because the tips of adjacent cracks are closest at the surface.

Determining When Coalescence Controls Pipeline Failure

A pipeline section that locally contains several cracks could fail by growth of any one crack, or by their coalescence and subsequent growth through-wall, or by through-wall growth followed by incidental coalescence as the fracture encounters adjacent cracks as it runs down the pipe. Typically seven to nine shorter cracks can be found in the vicinity of an SCC origin, of which three to five contribute to the critical size and the remaining cracks are encountered as the rupture propagates. The last of these does not involve coalescence in the present context and so is not a concern hereafter. Which of the remaining two situations occurs depends on the pressure in the pipeline and the sizes, the shapes, and the locations of the cracks. Because failure for any situation occurs at the lowest pressure, which of these possibilities controls the behavior of the pipeline can be identified simply by comparing the failure pressures for each and then choosing that with the lowest failure pressure, as follows.

Consider first the case where a single crack grows to failure. In this case, the failure behavior can be predicted by a criterion like PAFFC (Leis and Ghadiali, 1994) for an isolated crack and there is no need to address coalescence. Thus, through-wall growth of individual cracks is not an issue in establishing the scope of failure processes that must be addressed to identify necessary conditions for coalescence. Therefore, the only possibility of interest here is the axial coalescence of two (or more) adjacent deep cracks. It is noteworthy that this situation can lead to failure without growth after coalescence at the deepest point along the individual crack segments. Fig. 1, based on analysis using PAFFC for typical pipeline parameters, is useful as the basis to demonstrate this possibility.

Fig. 1a plots the failure pressure of an axial external crack in a pipeline on the y-axis (as a fraction of SMYS), as a function of normalized crack length on the x-axis, for various crack depths given as fractions of the wall thickness for the various trends in the figure. Service at MAOP for a Class I location (i.e., at 72 percent of SMYS) is represented as a horizontal line at a value of 0.72 on the ordinate of this figure. With reference to the figure, for the pipeline represented by Fig 1a when operating at 0.72 (SMYS), failure of a single deep crack — with a depth say 0.9 a/t — is indicated to occur for normalized crack lengths greater than about 1.3. Consider now an identical pair of such cracks, shown schematically in Fig. 1b, spaced such that they just coalesce to form a crack whose total normalized length is 2.6, (i.e., equal to their combined lengths). Observe from Fig. 1a that the failure pressure for this coalesced pair is now only about 0.35 (SMYS). That is, simply doubling the crack length — as would occur by the coalescence of adjacent coaxial cracks — reduces the failure pressure by almost a factor of two, and without an increase in depth! Analysis similar to that just outlined at crack depths greater than about 0.6 a/t indicates that a critical crack size can likewise produced by axial coalescence — without need for any increase in crack depth.

It follows that the axial coalescence of two deep cracks can cause a failure without the need for additional through-wall growth. In such cases, the failure process is dictated by the stress field (i.e., K_{eff} field) at and between the adjacent crack tips. Likewise, for cracks shallower than 0.6 a/t, depth is a secondary consideration for coalescence, so that conditions for coalescence are controlled by the regions between crack-tips.[1] It follows that a coalescence criterion can be formulated without concern for differing effects of crack depth in terms of the surface stress fields between adjacent crack-tips.

Data to Define Coalescence Conditions

Solutions for the K-fields between the crack tips of pairs of geometrically identical adjacent cracks have been developed that could be used to plot contours of constant K between them ((Stoneseifer et al (1993), Leis (1996)). These results, which represent a variety of crack-tip positions for crack aspect ratios, 2c/a, equal to 6 and 1, are very comprehensive for 2c/a = 6 but somewhat limited for c/a = 1. Accordingly, the following sections deal with 2c/a = 6, which is common for SCC behavior in many crack colonies. While useful in a qualitative sense, contours of constant K do not provide a convenient, quantitative basis to determine if coalescence can occur or what path it might track between the adjacent tips. For this

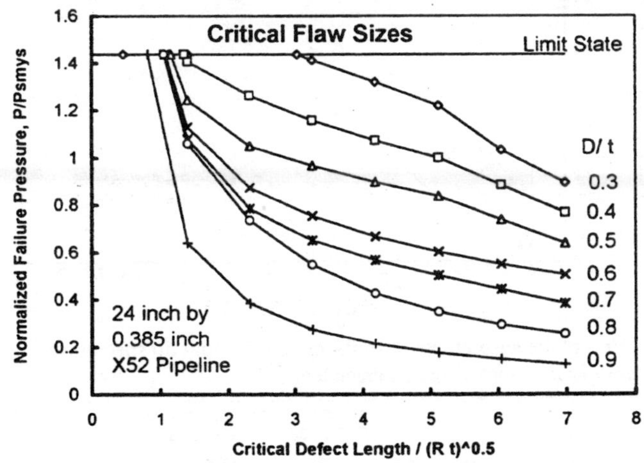

a) strong dependence od failure pressure for adjacent cracks

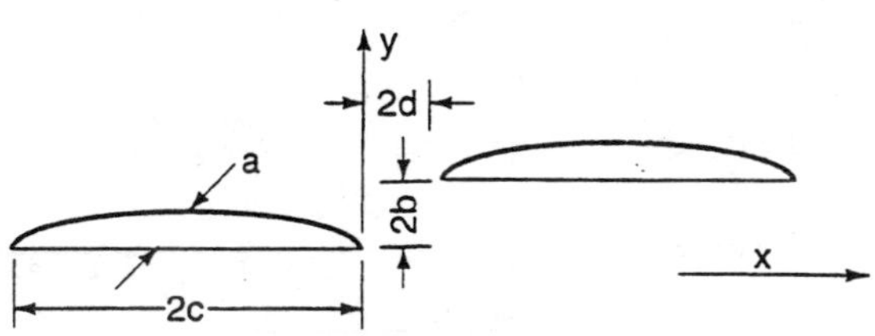

b) schematic illustrating to adjacent geometrically similar cracks whose interior tips are spaced to just coalesce

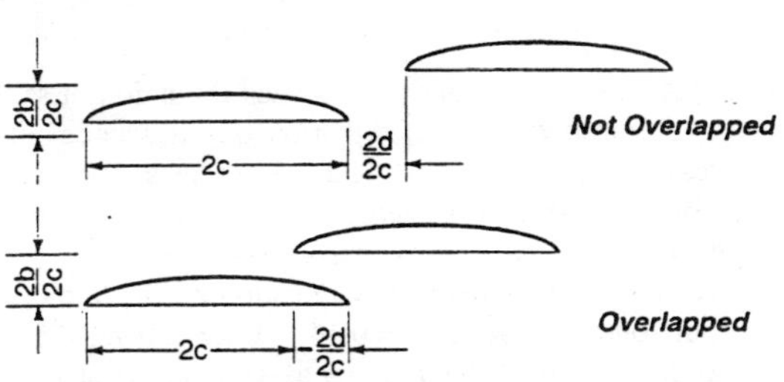

c) analysis notation for adjacent crack tips
Fig. 1 Aspects related to crack size and location

1 Leis, 1995 shows that the inside pipe surface (the so-called "back surface") does not strongly influence K for cracks shallower than 0.6 a/t).

reason, the results of Stoneseifer et al (1993) and Leis (1996) are recast into quantitative conditions for coalescence — specifically for 2c/a = 6. This work addresses coalescence in gas-transmission piping that, because of its high thickness to diameter ratio and loadings typically dominated by internal pressure, can be characterized as a membrane tension for most situations. Thus, the focus is on tension loaded crack pairs — the simplest situation in an array of cracks.

DEVELOPING CONDITIONS FOR COALESCENCE

Recent Literature

Various approaches can be found in the literature to deal with interacting cracks. Section XI of the ASME Boiler and Pressure Vessel Code (anon. 1994) and Section 8 of the British Standard PD6493 (anon. 1991) address interacting cracks in terms of a minimum spacing after which the approaching cracks are represented as an enveloping crack. Within the critical spacing, the ASME Code and the British Standard represent the behavior of adjacent crack tips by that of a single crack. Thus, a pair of approaching crack tips is characterized as a coalesced crack with crack-growth occurring along a common front. The critical spacing for interaction is represented in the ASME Code and the British Standard either as a fraction of the crack depth (the ASME Code) or as a fraction of the crack length (the British Standard). Others have suggested alternative approaches to that in the ASME Code and the British Standard, as for example the interaction factors due to Leek and Howard (1994).

Present Approach

This work presents an alternative approach to determine when interaction begins, at which point an "equivalent" enveloping crack could be used to represent the ensuing behavior. Crack tip spacings where $K_{eff} \geq K_{coalesce}$ have been determined in terms of the axial and circumferential distance between the adjacent tips. Each matched set of axial and circumferential distances represents an (x, y) pair of crack-tip spacings taken with respect to the interior tip of the left-hand crack, as shown in the schematic in Fig. 1c. The y-value, representing the circumferential spacing for a corresponding x-value in each (x, y) pair, is shown as a data point in Fig. 2c on coordinates of the circumferential spacing as a function of the crack length. Each of these points represents a combination satisfying the just stated condition for coalescence.

The choice of coordinates in Fig.1 follows from the empirical observation by Parkins and Singh (1990) that expressed the limiting value of the circumferential spacing for coalesced (completely merged) cracks in the form:

$$y = 0.14\,(LENGTH)\ , \qquad \text{(Eq. 2)}$$

which is represented in Fig. 2c as the solid line. Values of y less than that form Equation 2 could coalesce. This equation follows as limiting trend of a database of cracks whose lengths are primarily less than 0.2 inch (5 mm). Values of the circumferential spacing from the numerical analysis where (1) $K_{eff} \geq K_{coalesce}$ and (2) the crack lengths were on the order of or smaller than the length limit for data that underlie the empirical criterion all lie either on or below this limit, as can be seen in Fig. 2c. However, for longer cracks the numerical results diverge to values of y much larger than this limit.

Crack-Tip Positions for Coalescence — Normalized by the Crack's Length

Each of the just-discussed numerical results that determined a crack-tip spacing which support coalescence is replotted in Fig. 2b on axes that normalize the crack-tip spacings, which has been done by dividing by the corresponding crack length. Note that Fig. 2b includes both x and y spacings whereas Fig. 2c made use of only the circumferential spacing (the "y value" in Equation 2). Normalization by crack length in Fig. 2b is introduced to reduce the apparent dependence of the results on a/t. The very narrow band of results in Fig. 2b indicates this reduction in data spread is achieved.

In Fig. 2b, each (x, y) pair represents one of the crack-tip spacings given above in terms of (2d, 2b) or (axial-tip spacing, circumferential-tip spacing), which now becomes (2d/2c, 2b/2c) after dividing by the crack's length. Likewise, after dividing Equation 2b by the crack's length, it becomes y/length = 0.14, which in the format of Fig. 2b is a horizontal line at y = 2b/2c = 0.14. Note for this work that normalizing by crack length is analogous to normalizing by depth, because the ratio of the length to the depth is a constant. After this normalization, some small dependence on the applied stress remains, because the analysis for 72 and for 80 percent SMYS must produce different crack-tip spacings that satisfy the condition for coalescence: $K_{eff} \geq K_{coalesce}$. The trends formed by the data points in Figure 2b define combinations of crack length and the spacing of adjacent tips for which coalescence is possible according to $K_{eff} \geq K_{coalesce}$.

Note that a pair of axial cracks shows symmetry about an axis along the pipeline. For this reason the results in Fig. 2b are symmetric about the x-axis, 2b/2c = 0. However, such symmetry does not exist about the y-axis, 2d/2c = 0, because for crack spacings along the x-axis positive values indicate crack-tip positions that are axially separated whereas negative values indicate axially overlapped tips. With reference to Fig.2b, the reference crack lies along the negative x axis. Its right-most tip lies at (0, 0) in the coordinate system of Fig. 2b. That is, (0, 0) lies at the interior tip of the left-most crack in an adjacent pair of cracks (see the schematic in Fig. 1c). For positive values of x, crack tips approaching this crack that satisfy the condition that $K_{eff} \geq K_{coalesce}$ lie in the arc-shaped trend for positive x values (i.e., positive 2d/2c) in Fig. 2b. Coalescence (grow toward each other) begins when the inner tip of the right-most crack first reaches this arc-shaped trend defined by the analysis data points with positive values of x in Fig. 2b. In contrast, for negative values of x, overlapping crack tips whose spacings can still coalesce lie in the band of results that form the arc-shaped trend at negative x values. Values more negative than this band are past the possibility of coalescence. Spacings that lie below these arc-shaped bands or between them do not satisfy the requirement for coalescence. It is noted that once coalescence begins the process continues — unless the value of K decreases because of a decrease in the external loading. Arrest is possible in this situation.

There are several analytical results that satisfy $K_{eff} \geq K_{coalesce}$ that lie between the x axis and the limit defined by Equation 2, which in the format of Fig. 2b is a horizontal line at 2b/2c = 0.14. Observe from Fig. 2b that these numerical data points all lie up the y axis — that is at x = 0. For these smaller cracks the results from this numerical analysis are consistent with the empirical criterion in that they lie along the y-axis (i.e., 2d/2c = 0 for 2b/2c ≤ 0.14). Because 2d/2c = 0 for these empirical results the axial spacing between their tips is very small. The fact that this situation occurs for 2b/2c ≤ 0.14 means that their circumferential spacing around the pipe

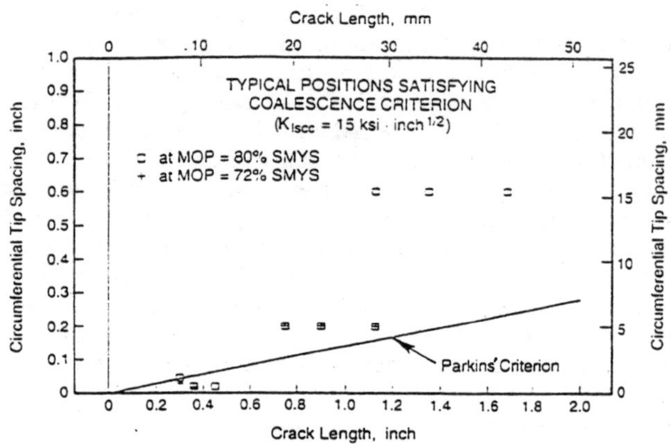

a) on coordinates of circumferential tip spacing vs. crack length

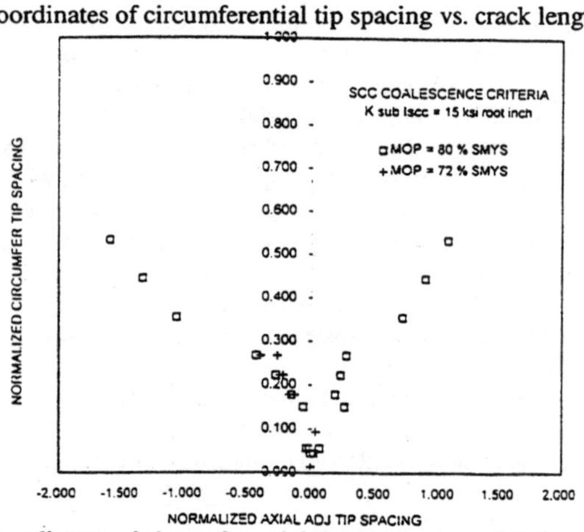

b) on coordinates of circumferential tip spacing vs. axial tip spacing, normalized by crack length

Fig. 2 Formats to examine coalescence criteria for adjacent cracks

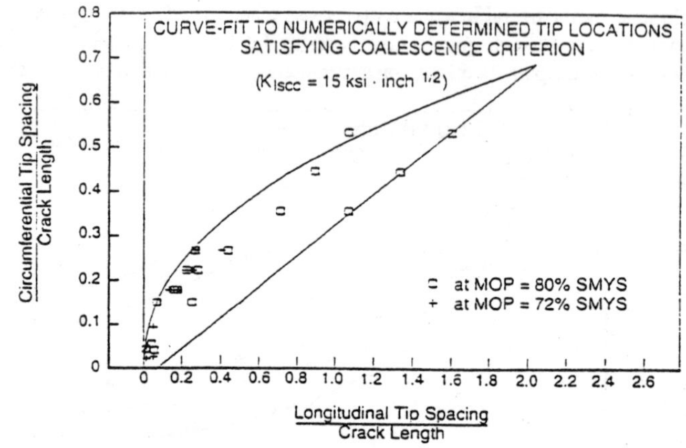

c) in terms of the absolute value of the circumferential tip spacing vs. axial tip spacing, normalized by crack length

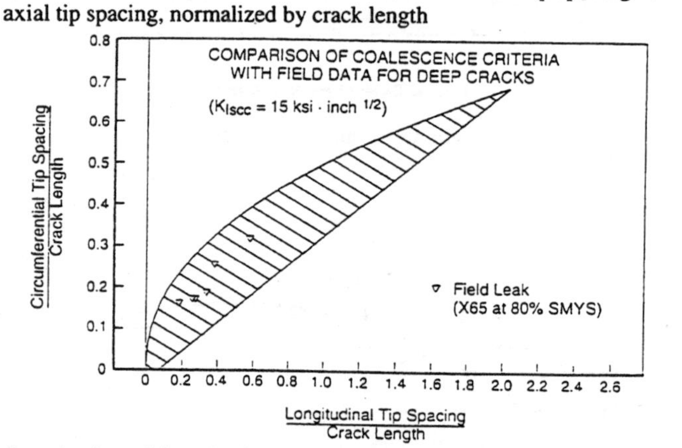

d) evaluation of the criterion by comparison with field data is a small fraction of the crack's length. In light of the close axial and circumferential crack-tip spacing, these empirical data represent a "dense" patch of stress-corrosion cracking. (The terminology "dense" applies to a specific spacing of cracks (Leis, 1995), which involves short and shallow cracking that is very closely spaced around the circumference of the pipe.) This has been confirmed by examining the data that underlie Equation 2 (Parkins and Singh (1990)). Thus, the present numerical results are consistent with the empirical data for smaller cracks. That the analytical results are consistent with the empirical criterion for physically smaller cracks occurs in spite of comparing analytical trends for the onset of coalescence with empirical trends for cracks that are coalesced. This somewhat surprising result occurs because the condition for the onset of coalescence based on $K_{eff} \geq K_{coalesce}$ is virtually coincident with the result for complete coalescence when the cracks are short — which is the case for the empirical criterion. But, when cracks become larger (equally their crack-tip spacings become larger in terms of Fig. 2b), the value of K is dominated by crack size. Thus, for longer cracks the values of K_{eff} and $K_{coalesce}$ will not be coincident in comparing situations based on the onset of coalescence versus its completion for larger cracks. In this context, the empirical and the analytical criteria must diverge. As will become apparent later other reasons also exist for such divergence. Fig. 2b indicates this divergence occurs above the line at 2b/2c = 0.14. This means that coalescence may be possible for crack tip spacings further apart than (y / length) = 2b/2c ≤ 0.14. Likewise, the axial spacing may be a factor in assessing if two cracks will coalesce. These data indicate that coalescence by SCC can occur in the region -1.6 ≤ 2d/2c ≤ 1.1 and 0 ≤ 2b/2c ≤ 0.55 for the range of crack size, wall thickness, and stress examined.

Crack-Tip Positions that Can Coalesce

The similar shapes of the trends in Fig.2b for possible coalescence at positive and negative values of x suggests that the simple expedient of characterizing the bands of data points that lie in these trends as absolute values could bring some simplicity to determining when coalescence can occur. Accordingly, the analytical results in Fig. 2b are examined in terms of absolute values, the results of which are shown in Fig. 2c. Fig. 2c indicates that there is only a small difference in the locations of positions that can coalesce for positive 2d/2c (i.e., axially separated cracks) and negative 2d/2c (i.e., axially overlapped cracks) — with positions for axially separated cracks being slightly closer to the tip than for overlapped cracks. The positions satisfying $K_{eff} \geq K_{Critical}$ evident in Fig. 2c are presented with respect to two limits on their (2d/2c, 2b/2c) locations. Whereas coalescence in regard to Fig. 2b could occur within the bands of data points that made up the two arc-shaped trends, coalescence under the expedient of using absolute values can occur within that same data spread as well as within the band that lies between the two arc-shaped trends. Because there is a small region where the data from these two trends does not overlap, as is evident in Fig. 2c considering the full region between these bands as crack spacings that satisfy $K_{eff} \geq K_{critical}$ introduces some limited amount of conservatism — the extent of which will be assessed later in Figure 2d. The bounds on the data points shown in Figure 2c have been obtained from a plot of the log of the axial and circumferential normalized spacings, which indicates that a power-law expression between 2d/2c and 2b/2c can serve as the upper bound and a linear expression between 2d/2c and 2b/2c is appropriate as the lower bound. This upper bound encompasses points that represent the larger 2b/2c tip spacings, while the lower bound trend reflects the smaller 2b/2c tip spacings for a given axial spacing, 2d/2c, that satisfy $K_{eff} \geq K_{Critical}$. The area between these bounds thus represents the crack tip spacings that satisfy the condition $K_{eff} \geq K_{Critical}$ for coalescence.

The upper bound trend shown in Fig. 2c, taken in conjunction with above noted lower bound, define a zone (enclave) at the crack-tip that satisfies the requirement $K_{eff} \geq K_{Critical} = 15$ ksi inch$^{1/2}$ (16.34 MPa). This enclave lies between:

$$\pm y_1 = c \left(\frac{\pm x}{2c}\right)^{0.40} \qquad \text{(Eq. 3a)}$$

and

$$\pm y_2 = 0.7\, c \left(\left(\frac{\pm x}{2c}\right) - 0.07\right). \qquad \text{(Eq. 3b)}$$

Because crack tip positions that lie in this enclave can coalesce, these equations can be used to determine if a pair of adjacent tips can coalesce by checking if their respective sites lie in this region.

Equations (3a) and (3b) represent the condition for coalescence by SCC, expressed with respect to a coordinate system located at the right tip of the longest crack in the pair of cracks considered for coalescence. The enclave of coalescence exists in all quadrants referenced to the tip, which is evident in the "±" symbols on both x and y. Therefore, this enclave lies ahead of and behind the crack-tip, as well as above and below the tip, which follows from the use of the absolute values in constructing Fig. 2c. In these equations, the symbol "2c" denotes the length of the longest crack in the pair of cracks considered for coalescence, which for the sake of simplicity given the reference zero for the above Equations is taken to be the left-most crack in the pair of adjacent cracks. The symbol y_1 represents the upper bound circumferential position whereas y_2 represents the lower bound position. If the value of $K_{Critical} \neq 15$ ksi inch$^{1/2}$ (16.34 MPa), which is embedded in the constants used here, then the constants in these equations change.

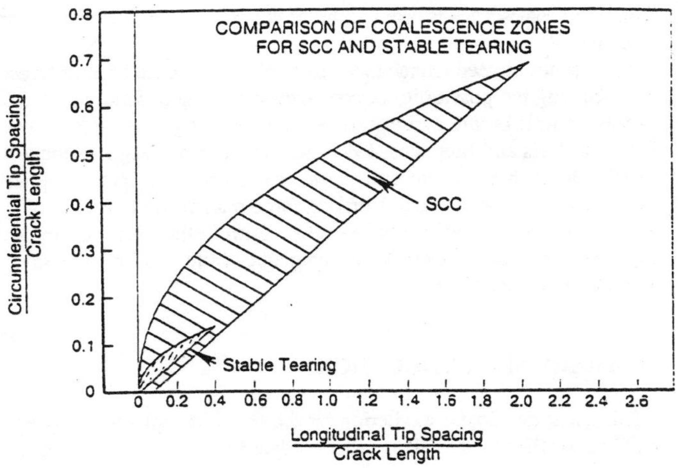

Fig. 3 The enclave of crack-tip spacings for coalescence for a 3-inch (36.2 mm) long axial surface crack, for both SCC and stable tearing

Coalescence by SCC versus Fracture

Crack-tip locations that satisfy the condition for the onset of coalescence by SCC (i.e., where $K_{eff} \geq K_{Critical}$ everywhere between tips) as expressed by Equation (3) are shown in Fig. 2d on coordinates similar to that in Fig. 2c. Whereas in Fig. 2c the symbols represented positions where $K_{eff} \geq K_{Critical}$ in the analyses, the symbols in Fig. 2d represent measurements made for cracks growing toward an adjacent crack, for deep cracks at a leak due to

SCC.[2] These field data range from the onset of coalescence through nearly complete coalescence. Note that the field data lie between the bounds developed analytically, which indicates that Equation (3) is suitable for practical applications. It follows that the expressions for the onset of coalescence by SCC developed through fracture mechanics can be used with some confidence as part of a model for SCC in pipelines.

The condition for coalescence by SCC prescribed by the above equations is illustrated in Fig. 3 as the larger of the two enclaves shown. This figure shows the enclave for coalescence in the (+x,+y) quadrant for a 3-inch (76 mm) long crack in a pipeline operating at a pressure that creates a wall stress equal to 0.72 SMYS. The size of the enclave can be determined by multiplying either of the x- or y-distances by 3 or 76, respectively, to get this size in inch or mm units. This leads to rather large tip spacings — even greater than the crack length! In turn this implies that coalescence by SCC can begin at quite large distances. Increases in $K_{Critical}$, as occur for more resistant steels can significantly reduce the size of this zone, which means cracks have to grow further or be more closely spaced circumferentially before coalescence can begin.

Role of Crack Depth and Coparallel Cracking

The above example did not include consideration of crack depth because the equations represent specifically $2c/a = 6$, which implicitly incorporates a prescribed crack depth. A significant depth dependence can be expected as $2c/a \rightarrow 0$. It follows that, while this format is useful to assess coalescence in a first generation model for pipeline SCC, a more general format may be needed for future generations of such models for the same reason that failure criteria for a single flaw include flaw depth. Once coalesced, the length of the coalesced pair can be defined as the tip-to-tip length of the two merging cracks as indicated in the preceding section. Unfortunately, too little work has been done to quantify cases where the depths of the coalescing cracks differ significantly. For this reason the depth after coalescence is taken as that of the deepest crack in the merging crack pair.

The presence of deep interacting coparallel cracks could become a factor in evaluating the possibility of coalescence leading to failure that is not addressed well by the earlier results. Analysis (e.g., Leis and Colwell, 1996 and Leis and Stephens, 1997) and experiments (e.g., Chouchaoui, 1993) indicate that such coparallel cracks can increase the local compliance which acts to reduce the crack driving force making analysis such as that presented earlier somewhat conservative. Until further work dealing with more general arrays of cracks is completed, little can be done save to accept this conservatism.

SUMMARY AND CONCLUSIONS

This paper developed a criterion for the onset of coalescence by either SCC or ductile tearing. This criterion determined only if two cracks of known size, shape, and proximity can coalesce. Therefore this criterion must be coupled with (1) a procedure that tracks the progress of adjacent crack-tips until they join and (2) a failure criterion that determines if the coalesced crack will grow stably through the wall or unstably along the pipe.

It was shown that failures involving coalescence by SCC or stable tearing are controlled primarily by the stress fields at or near the surface between the adjacent tips. A criterion for the onset of coalescence was based on exceeding a critical level of crack driving force between adjacent crack-tips. A consequence of this criterion was that SCC leaks and breaks form by axial coalescence *at the current depths* of the crack segments that merge to produce the failure — a tendency that was supported by field fracture features.

Conditions underlying coalescence were expressed in terms of the driving force for coalescence, K_{eff}, at and between adjacent crack tips. A criterion was formulated for the onset of coalescence that is applicable to the cracking patterns generated by SCC. For the shallow cracks, K_{eff} was much less than K_{Ic}, so that such coalescence could not initiate by stable tearing for such shallow cracks. But, because $K_{eff} \geq K_{Critical}$ coalescence of shallower cracks could initiate and continue by SCC. For longer and deeper cracks, K_{eff} can exceed both $K_{Critical}$ and K_{Ic}. Therefore, such cracks could coalesce by either of SCC and stable tearing, with SCC continuing until the value of K_{eff} exceeded K_{ic}.

Conditions for the onset of coalescence by either SCC or stable tearing were expressed as a set of crack tip positions in an enclave at either crack tip. The upper and lower bound circumferential crack-tip positions for coalescence, y_1 and y_2 respectively, were expressed as a function of the axial crack-tip position in a local coordinate system located at the right-hand tip of the longest crack in the pair of cracks in Equation 3. These conditions were illustrated for axial SCC in a pipeline under typical operating conditions and compared to field data for an in-service leak. These field results lay in the middle of the enclave defined by Equation 3, indicating that these equations are suitable for practical applications.

Conclusions that can be drawn from this work as it applies to crack coalescence in gas-transmission pipelines include:

- For smaller cracks $\leq$ 0.450 inch long (11.4 mm) with $a/t \sim 0.2$, operation at 80 percent of SMYS (as in Canada) just indicates the onset of coalescence and continued growth by SCC for cracks with small axial spacings whose tips lie circumferentially above each other.
- For smaller cracks $\leq$ 0.450 inch long (11.4 mm) with $a/t \sim 0.2$, operation at 72 percent of SMYS (as in the US) marginally supports the onset of coalescence or continued growth by SCC — Operation at much lower stresses is unlikely to support coalescence or continued macrocrack growth.
- Zones of coalescence at MAOP for cracks in dense patches have limited driving force for growth — so, long after nucleation the crack pattern is virtually unchanged.
- The fracture-mechanics-based coalescence criterion is simple — agrees with the existing empirical criterion for short and shallow cracks but diverges from it for longer (and deeper) cracks — and matches field observations over the range from short shallow cracks through longer cracks at depths approaching the wall thickness.
- Coalescence depends on crack length, applied pressure, pipeline steel properties, and implicitly on crack depth.
- This criterion represents the onset of coalescence and must be coupled with (1) analyses that track cracking beyond the onset of coalescence either by SCC kinetics or by stable tearing and (2) a failure criterion to determine critical conditions that lead to failure.
- Through-wall growth by SCC after coalescence is not necessary to induce failure — so ruptures can occur immediately after axial coalescence.

ACKNOWLEDGMENTS

This work has been supported in part by the Line Pipe Research Supervisory Committee of the Pipeline Research Committee, the Engineering Mechanics Department of Battelle Columbus, and TransCanada Pipelines Ltd.

2 This particular failure was selected because the coalescing cracks were free of the influence of adjacent deep coparallel cracks, which could cloud the suitability of this coalescence criterion. This issue is explored in the next section where array solutions are used to determine the effects of such adjacent cracks.

REFERENCES

Anon, (1994). ASME Boiler and Pressure Vessel Code, Section XI

Anon, (1991). British Standards Institution, PD 6493, Section 8

Chouchaoui, B (1993). "Evaluating the Remaining Strength of Corroded Pipelines", PhD Thesis, University of Waterloo, Canada

Fessler, R. R., and Parkins, R. N., "nace 320

Leek, TH, and Howard, IC (1994). "Rules for the Assessment of Interacting Surface Cracks under Mode I Load", Int. J. Pres. Ves. & Piping, Vol. 60, pp 323-339

Leis, BN, Brust, FW, and Scott, PM (1991). "Development and Validation of a Ductile Flaw Growth Analysis for Gas Transmission Line Pipe", Pipeline Research Committee, American Gas Association, AGA Catalog No. L51643, June 1991

Leis, BN, and Ghadali, ND (1994). "Pipe Axial Flaw Failure Criteria — PAFFC, Version 1.0 User's Manual and Software", Pipeline Research Committee, American Gas Association, AGA Catalog No. L51680

Leis, BN (1995). "Characterization of Axial Flaws in Pipelines, With a Focus on Stress-Corrosion Cracking", Pipeline Research Committee, American Gas Association, NG-18 Report 212

Leis, BN, and Colwell, JA (1996) "Initiation of Stress-Corrosion Cracking on Gas-Transmission Piping", ASTM STP 1298, Effects of Environment on the Initiation of Crack Growth, to appear

Leis, BN, and Stephens, DR (1997). "An Alternative Approach to Assess the Integrity of Corroded Line Pipe — Part I: Current Status", ISOPE '97

Nishioka, T, and Atluri, SN (1983). "Analytical Solutions For Embedded Elliptical Cracks, And The Finite Element Alternating Method For Elliptical Surface Cracks Subjected To Arbitrary Loadings", Engr. Frac. Mech., Vol. 17, No. 3, pp 247-268

Parkins, RN, and Singh, PM (1990). "Stress-Corrosion Crack Coalescence", Corrosion, 46, pp. 485

Stoneseifer, RB, Brust, FW, and Leis, BN (1992). "Stress-Intensity Factors for Long, Axial Outer Surface Cracks in Large R/t Pipes", 22nd National Fracture Mechanics Symposium, ASTM STP 1044, Volume II, pp 29-45

Stoneseifer, RB, Brust, FW, and Leis, BN (1993). "Mixed-Mode Stress-Intensity Factors for Interacting Semi-Elliptical Surface Cracks in a Plate", Engr Frac. Mech., Vol. 45, No. 3, pp 357-380

Proceedings of the Seventh (1997) International Offshore and Polar Engineering Conference
Honolulu, USA, May 25-30, 1997

ISBN 1-880653-28-1 (Set); ISBN 1-880653-32-X (Vol. IV)

Corrosion Behavior of Thermal Sprayed Metals on Weld of Galvanized Steel

Young-Gak Kweon and Jai-Hyun Park
Research Institute of Industrial Science and Technology
Pohang, Korea

ABSTRACT

Corrosion behavior of sprayed metal coating on weld of galvanized steel was studied with spraying materials of Zn, Al, and Zn/Al alloys. Observation of corrosion of sprayed metal and electrochemical analysis were carried out to verify preferential corrosion of sprayed zinc metal to plated zinc deposit. It was found that the preferential corrosion of sprayed zinc on weld due to higher driving potential for galvanic action between zinc and substrate. Among different spraying metals, Zn/Al(95/5)alloy showed the most desirable behavior in corrosion protection. In addition, top coating of epoxy acryle on sprayed zinc coating was found to be appropriate to extend the lifetime of coating.

KEY WORD : Galvanized steel, weld, thermal spraying, anodic metals, galvanic activity.

INTRODUCTION

Galvanized steel plates are frequently used for steel structures in corrosive environments. However, when the welding is used in fabrication process, zinc deposit disappears on weld of galvanized steel. Thermal spraying on weld with anodic metal of zinc is generally used to compensate the corrosion resistance of weld. Corrosion characteristics of sprayed and plated zinc would be different to emphasize the corrosion of a certain coating between two. In general, red rust was observed firstly on weld of galavnized steel structures. In this study, we analysed the corrosion bevavior of sprayed and plated zinc metal to verify the galvanic activity of metal coating relative to steel. Some kinds of spraying materials were studied to improve the corrosion resistance of weld. In addition, top coatings on sprayed metal with different kinds of polymer resins were also studied to prolong the the lifetime of metallic coating.

EXPERIMENTS

Galvanized steel pipes were used for corrosion test of sprayed and plated zinc coatings. Substrate steel was low carbon steel. In manufacturing process of pipes, galvanized steel band was formed to pipe with high frequency resistance welding. Zinc spraying was performed on weld just after machining of weld bead followed on welding. Corrosion of sprayed metal on weld was observed after long time field service in moist atmosphere.

Electrochemical analysis of sprayed and plated zinc were performed by measurement of corrosion potential and current on polarization curves.

For the improvement of protection capability of sprayed coating, pupe zinc, pure aluminium and Zn/Al alloys(85/15 and 95/5)were studied. Steel plates coated with these different materials were tested by salt spray test using 3% NaCl solution according to ASTM standard. For prolongation of coating lifetime, top coatings on sprayed metal were considered. Polymer resins, such as acryle modified alkid, stylene and maleic anhydride, epoxy acryle and ester urethane were tested after application on sprayed metal.

RESULTS AND DISCUSSIONS

A. Preferential corrosion of sprayed metal

Sprayed metal corroded preferentially to plated metal. Fig.1 shows the progressing stages of corrosion of zinc sprayed layer on weld of galvanized steel. In figure, plated layer means galvanized part non-affected by welding in the same specimen. In initial stage(Fig.1.a), thickness of sprayed layer was not uniform, in the range of 10 to 20 μm, while that of plated layer was relatively

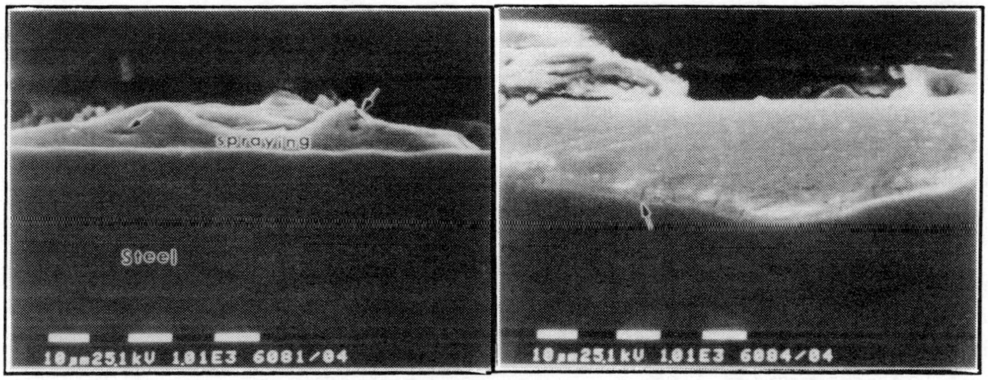

a) Initial stage

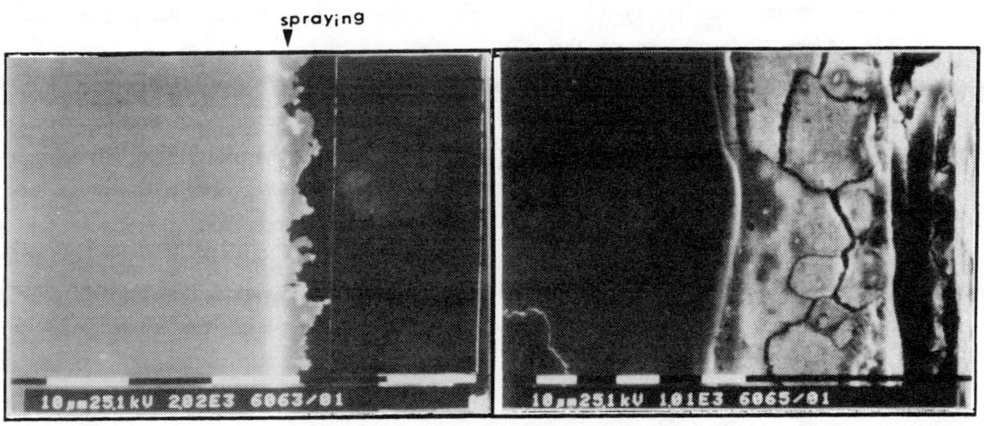

b) White rust stage

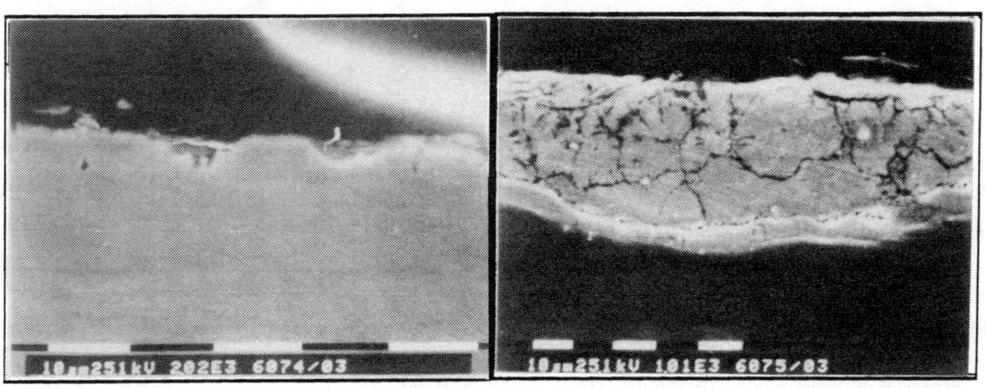

c) Red rust stage

Fig.1 Appearance of sprayed(left side) and plated(right side) layer on steel substrate after atmospheric corrosion in moist environment.

uniform in about 30 μm. After exposure to corrosive environment during certain period, white rust appeared on both sprayed and plated layer. However, the thickness of sprayed layer decreased significantly, while that of plated layer did not, as shown in Fig.1.b. When red rust was produced on sprayed part, that is, steel substrate began to corrode, most of sprayed metal existed no more, as shown in Fig.1.c, while plated layer preserved its initial thickness only with white rust of zinc.

B. Electrochemical analysis

Fig.2 shows the polarization curve for sprayed and plated zinc metal. Corrosion potential of plated zinc was -995mV/SCE, while that of sprayed zinc was -1035mV/SCE. The difference of corrosion potential was not so significant as to distinguish the galvanic activity between sprayed and plated metal. However, it is evident that the sprayed metal has the tendancy of rapid corrosion. Corrosion current density of sprayed metal was 29.63 $\mu A/cm^2$, while that of plated metal was 9.56 $\mu A/cm^2$. It is well known that the sprayed metal is porous. In general, zinc sprayed coating has about 5% of porosity in volume. It could be easily deduced that the sprayed metal has large surface to corrode per unit area than the plated metal.

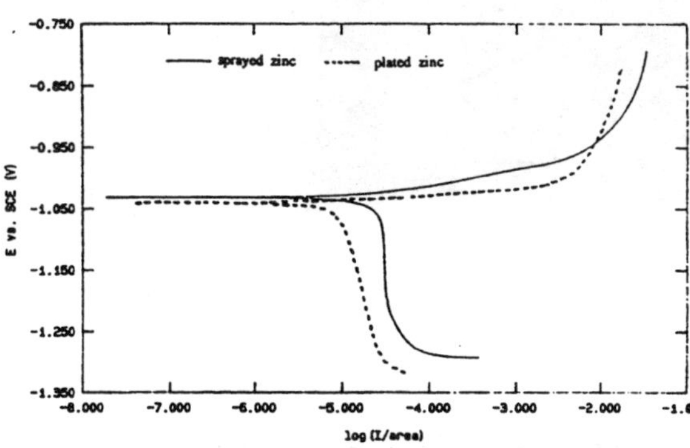

Fig.2 Polarization curves of sprayed and plated zinc

Preferential corrosion of sprayed metal can be explained also by galvanic activity of zinc to steel substrate. Fig.3 shows the polarization curves of each part of steel weld, such as weld metal, heat affected zone and base metal. Corrosion potential of weld metal was -445mV/SCE, while that of base metal and HAZ was about -540mV/SCE. It exist about 100mV difference in potential. In the case of steel weld without any coating, we can expect that weld metal is cathodic to base metal, that is, local corrosion of weld would not occur. However, in the case of zinc coating, galvanic action of zinc would be produced by potential difference between zinc and steel. The potential difference in a galvanic couple, called driving potential, can affect corrosion rate of anodic metal. Sprayed zinc on weld metal has higher driving potential than plated zinc on base metal of steel. Therefore, sprayed zinc exhibited higher corrosion rate than plated zinc. After exhaustion of sprayed zinc by corrosion, weld metal is exposed directly to corrosive environment, resulting in corrosion of weld.

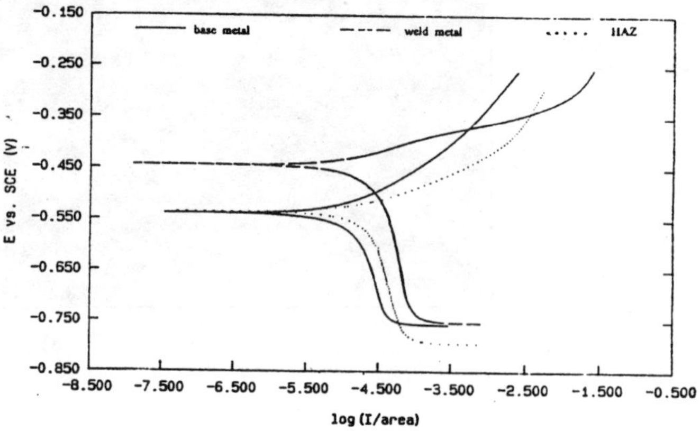

Fig.3 Polarization curves of each part of steel weld

C. Influence of spraying materials

In the thermal spraying for protection of steels from corrosion, various spraying metals are used, such as zinc, aluminium and Zn/Al alloys. Zinc is generally known as sacrificial anodic metal. Al is less anodic compared to zinc, but the corrosion rate of Al is relatively low to increase the lifetime of coating. Zn/Al alloy has intemediate characteristics. Fig.4 shows the results of salt spray test of steel plates spray-coated with different materials and thickness. We measured the time to appearance of red rust on the coating, that is, corrosion of substrate. Time to red rust increased

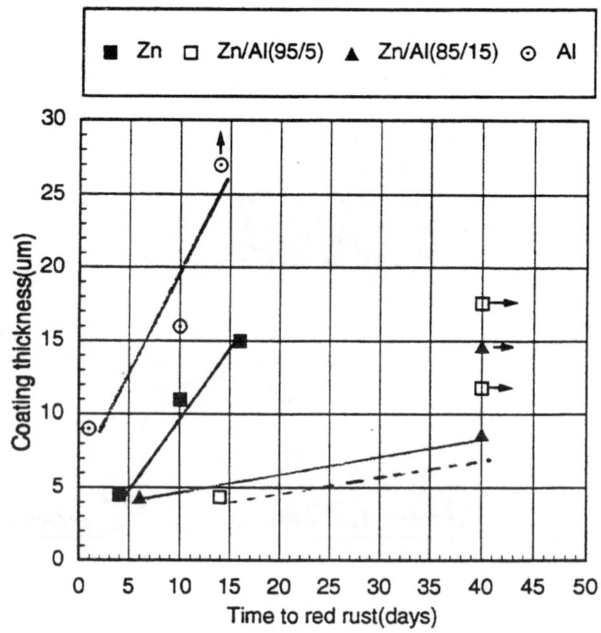

Fig.4 Results of salt spray test of steel plates spray-coated with different materials and thickness

almost proportionally with the increase of coating thickness. Al coating showed the rapidest appearance of substrate corrosion . It was not by the exhaustion of coating, but probably by insufficient anodic activity. Zn/Al alloy coating, especially Zn/Al(95/5) alloy, showed the most desirable behavior of retard of substrate corrosion. It can offer not only sufficient sacrificial anodic activity to steel but less corrosion rate of coating than pure zinc.

In the test with weld specimen, the same result was obtained. Zn/Al (95/5) alloy coating with 2 μm thickness exhibited red rust on its surface after 15days in salt spray test, while pure zinc coating with 10 μm thickness and pure Al coating with 7.6 μm thickness showed black spots of corrosion products only after 5 days and red rust after 12 days in the test.

D. Influence of top coatings

Sprayed metallic coating can extend its lifetime by application of top coating with polymer resin. Some kinds of resins were applied on zinc sprayed coating and tested in salt spray test. Table 1 summerized the results of salt spray test. Among different kinds of resins, epoxy acryle exhibited the best behavior to protect metallic coating. Corrosion resistance of sprayed metallic coating could improve by application of proper top coating.

Table 1. Result of salt spray test of sprayed Zn coating with top coating

Type of resin	Coating thickness(μm)		Time to appearance of red rust
	Spraying	Resin	
Acryle modify alkid	8	5	8 days
Stylene and maleic anhydride	8	5.7	6 days
Ester urethane	8	5.6	23 days
Epoxy acryle	8.6	5	$>$ 40 days

CONCLUSIONS

Corrosion of sprayed metallic coating on weld of galvanized steel was studied on corrosion and electrochemical behavior between sprayed and plated zinc, on different kinds of sprayed materials and on use of top coating with polymer resins. Following conclusions are drawn.

1) Sprayed metal corroded preferentially to plated metal. Sprayed zinc on steel weld has a higher driving potential to protect substrate than plated metal on base metal of steel. Corrosion rate of sprayed metal was also higher than that of plated metal probably because of its porosity.

2) Among different kinds of spraying materials such as Zn, Al and Zn/Al alloys, Zn/Al(95/5) alloy exhibited the best behavior to protect steel substrate from corrosion.

3) Application of polymer resins as a top coating on sprayed metal increased the lifetime of coating. The resin of acryle epoxy was found to be proper top coating on zinc sprayed metallic coating.

REFERENCES

1) Y.G. Kweon and C. Coddet(1992), "Behavior in seawater of zinc base coatings on AA 5086", Corrosion, vol.48, No.2, pp97-102.

2) Y.G. Kweon, R.W.Chang, Y.M.Yang and S.M.Lee(1992), "Blistering behavior of thermally sprayed metallic coatings in sea water", Proc. of ISOPE92, Vol.IV, pp19-21

Proceedings of the Seventh (1997) International Offshore and Polar Engineering Conference
Honolulu, USA, May 25-30, 1997

ISBN 1-880653-28-1 (Set); ISBN 1-880653-32-X (Vol. IV)

Prevention of Preferential Corrosion of Welded Joints of Carbon Steel in CO_2 Containing Environment

Shigeru Endo and Sakae Fujita
NKK Corporation, Fukuyama, Japan
Moriyasu Nagae
NKK America Inc., New York, NY, USA
Mitsumasa Iwata
Hiroshima University, Hiroshima, Japan

ABSTRACT

In order to mitigate the preferential corrosion of welded joints, the mechanism of the preferential corrosion needs to be clear. For this purpose, electrochemical tests and immersion tests of welded joints of line pipes in CO_2 saturated synthetic sea water were carried out. In addition to these tests, electrochemical potential distributions on the inner surface of two types of linepipes that have different chemistries were analyzed using a two dimensional boundary elements method. Corrosion rate distributions in the inner surface were also estimated by using the electrochemical potential distributions analyzed and polarization curves of the inner surfaces.

These test results show that the preferential corrosion of welded joints occurs in the manner of galvanic corrosion caused by the differences in electrochemical potential of weld metals and base metals. These differences are mainly caused by difference in their chemical compositions. Corrosion in the macro-cathod portion was also observed.

The addition of Cu, Ni and Mo to weld metals in amounts larger than that in the base metals is effective in preventing the preferential corrosion of the weld metals.

INTRODUCTION

Because of the buffering effect and its resulting larger capacity to provide H^+, higher corrosion rate is expected in CO_2 containing environment (Kodama, 1983). General corrosion and preferential corrosion are reported as typical corrosion types in this environment. As weld metal is generally different from base metal in chemical composition and microstructure, it is believed that these differences cause preferential corrosion of welded joints galvanically when these two dissimilar materials are coupled in corrosive environment. Furthermore, the area of weld metal is much smaller than that of the base metal, and as a result, enhanced corrosion of the weld metal may occur. Corrosion problems have occurred in pipelines and OCTG which are used in CO_2 containing fluids(Ikeda,1983, Ikeda, 1984, Videm, 1987). Severe weld metal corrosion in seam welded joints and girth welded joints were detected in a North sea sub-marine pipeline transporting oil that contains a small amount of water and CO_2. This preferential corrosion was mainly observed in girth welded joints (London, 1991, Steel, 1991). Some accidents related to the preferential corrosion of OCTG were also reported (Joosten, 1988). Meanwhile, erosion caused by turbulence of fluid due to the existence of welded beads is said to be one possible reason of uneven thinning around weldments (Kvernvold, 1992).

In this research, the likelihood and the mechanism of preferential corrosion of weldment in CO_2 containing environments were examined. The possibility of preferential corrosion of weldment was examined using an immersion test in synthetic sea water saturated with 0.1 MPa of CO_2. The mechanism of the preferential corrosion was investigated by comparing the electrochemical properties of base metals , heat affected zone (HAZ) and weld metals. Furthermore, the electrochemical potential distributions on the inner surface of two types of linepipes were analyzed using a two dimensional boundary element method (BEM). Corrosion current densities and the resulting corrosion rates on the inner surfaces were obtained using the calculated electrochemical potential and experimentally measured polarization curves. The experimentally obtained and the calculated corrosion rates were compared.

In order to develop the countermeasures to the preferential corrosion of welded joints, the effects of alloying elements, such as Ni and Mo in were investigated.

EXPERIMENTAL PROCEDURE

Immersion test of welded joints of linepipe

An immersion test of welded joints in a CO2 environment was carried out in order to determine the possibility of the preferential corrosion,

as well as to verify electrochemical experiment results. Gas metal arc welding (GMAW) simulating a practical girth welding of a pipeline was used to make welded joints. These welded joints were used in the immersion test. The chemical composition of a base metal and that of weld metals are shown in Table 1. The dimension of a typical immersion test specimen is shown in Fig. 1. Except for the top surface which includes a weld bead, the other five surfaces of the immersion test specimen were sealed by epoxy resin. The specimens were immersed in synthetic sea water saturated with 0.1 MPa-CO2 for two months. After two months, the weight loss, surface roughness and width of the weld metal were measured, from which the corrosion rates of both the base metal and the weld metal were calculated. For purpose of comparison, specimens made of base metal without weld bead were also immersed in the same environment.

Table 1 Chemical composition of the base metal and the weld metals for the immersion test

Material		C	Si	Mn	Cu	Ni	Mo
Base metal		0.04	0.28	1.34	0.25	0.16	-
GMAW	Min.	0.06	0.25	1.15	0.10	0.07	-
	Max.				0.15	2.97	0.5

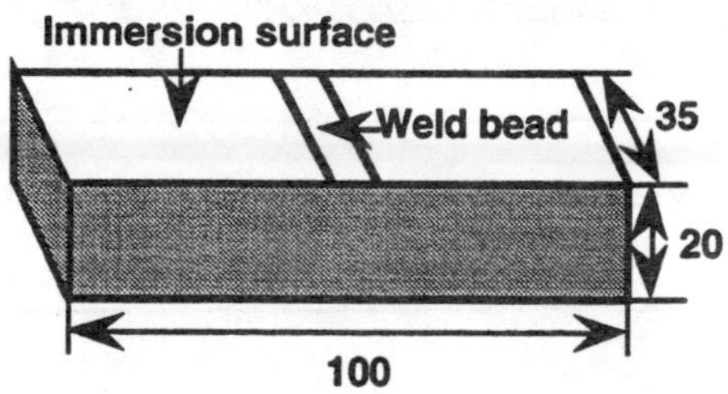

Fig. 1 Immersion test specimen

Electrochemical test

Electrochemical properties, such as immersion potential and anodic polarization curve, have to be different between a base metal and a weld metal in order to cause preferential corrosion of the welded joint in the manner of galvanic corrosion. Polarization curves of a base metal and two weld metals were measured in order to clarify the mechanism of preferential corrosion in a CO_2 containing environment. Chemical compositions of these metals are listed in Table 1. In addition to these measurements, polarization measurement of HAZ was also performed using a synthetic HAZ that has been reheated to a temperature of 1623K and had been cooled at a rate of 20K/sec. The polarization measurements were performed in the same type of solution as that used in the immersion test. The current was measured during potential scan from -900 mV to -600 mV (versus SCE) at a rate of 30 mV/hr. Furthermore, corrosion rates were determined from the measured polarization curves using Stern and Geary's method (Stern and A. L. Geary, 1957). An comparison of the immersion test results and the calculated corrosion rates was then made.

Numerical analysis of potential distributions and prediction of corrosion rates

A numerical analysis of the potential distributions on the inner surface of linepipes that were exposed to corrosive fluid was made using a two dimensional boundary element method (BEM) (Iwata). The main purpose of this analysis is to examine the effect of the area ratio of weld metal to base metal on the corrosion rate, and to clarify the mechanism of the preferential corrosion. Assuming linepipes with 965 mm outside diameter, 25.4 mm wall thickness and straight seam welding, 2 cases were analyzed. The linepipes are assumed to have the same base metal that has the chemical composition listed Table 1. The chemistry of the weld metals A and B are shown in Table 2.

Table 2 Chemical compositions of the weld metals used for BEM

	C	Mn	Cu	Ni	Mo
Weld Metal A	0.05	1.4	0.20	0.11	-
Weld Metal B	0.09	1.4	0.20	0.42	0.15

A schematic drawing of the numerical model is shown in Fig. 2. When the pipe is completely filled with electrolyte, the area ratio of base metal to weld metal is 44. This area ratio can be varied from 2 to 44 by changing the height of the electrolyte in the pipe. The fundamental equation of BEM consists of the first order assembled equation which has two unknown variables: the electrochemical potential, ϕ, and the current density, q, on the elemental nodes on the interface of electrolyte-electrode. The specific resistance of the electrolyte and the position of the elemental nodes determine the coefficient matrix of the assembled equation. The value of q is zero on the electrolyte-air interface and the symmetry axis. Furthermore, q at the inner surface is a polarization characteristics and can be expressed as a function of the potential, and as A result it can be eliminated from the assembled equation. The electrochemical potential can be determined by solving the assembled equation, when the specific resistance and the polarization characteristics of base metal, HAZ and weld metal are set as boundary conditions. The experimentally obtained polarization curves used are shown in Fig. 3. As applied polarization current, I_{appl} varies linearly with the potential, polarization resistance, $\Delta \phi / \Delta I_{appl}$, were used as the boundary conditions. The specific resistance of 20 Ω•cm is used for the electrolyte.

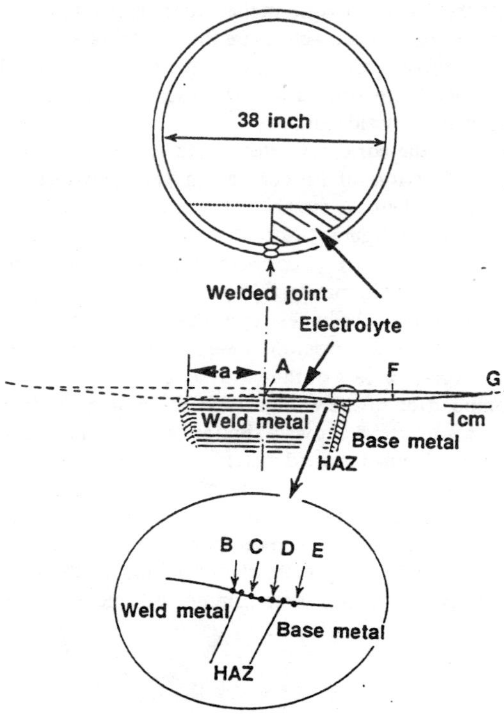

Fig. 2 Used element division method of internal surface of linepipe for BEM

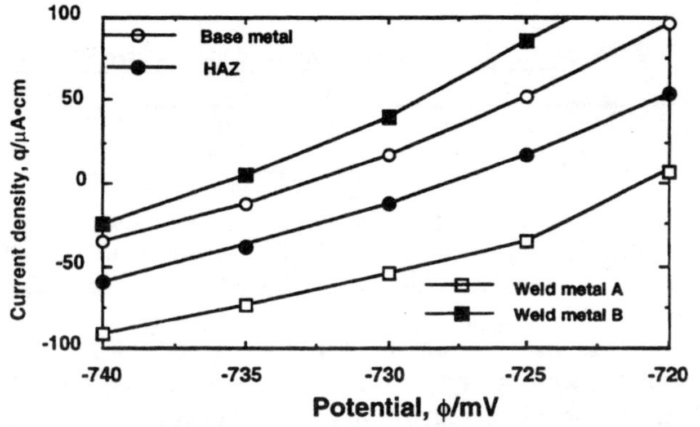

Fig. 3 Used boundary conditions for analysis of electrochemical potential distributions of internal surface of pipe by Boundary Element Method (BEM)

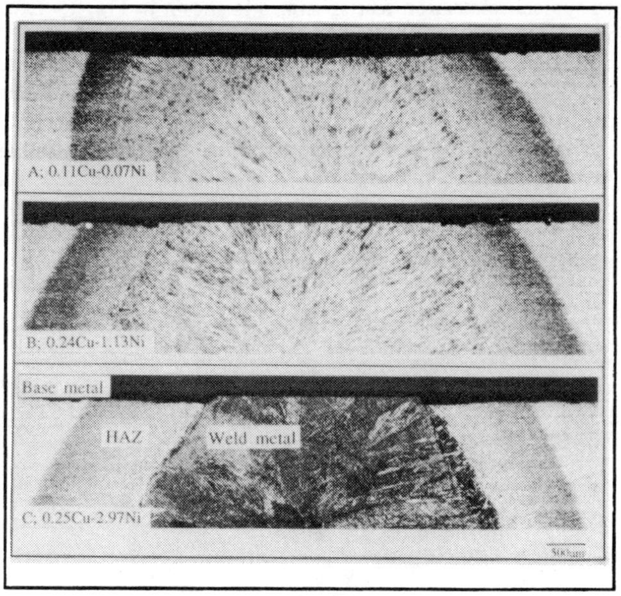

Fig. 4 Macrostructure of the immersion tested specimens

TEST RESULTS

Immersion test results

Fig. 4 shows the cross-section view on the macrostructure of three immersed specimens. Surface roughness these specimens is shown in the Fig. 5. The surface roughness was measured using non-contact type laser profile meter. The weld metal containing 0.11 % Cu and 0.07 % Ni is preferentially corroded Its corrosion rate is approximately 0.5 mm/year larger than that of the base metal. On the other hand, preferential corrosion is not observed in the weld metals which contain the larger amounts of Cu and Ni. Though, in the case of the weld metal that contains approximately 3 % Ni, slightly greater thickness reduction in the HAZ is observed. No knife line attack is observed regardless of the composition of the weld metals.

Fig. 6 shows the corrosion rates of the weld metal and the base metals as a function of Ni content in the weld metal. The corrosion rate of the base metals with and without a weld metal are also shown in this figure. Note that the Ni content in the base metals is a constant, here its corrosion rates are plotted versus the Ni content in the weld metals. The corrosion rate of the higher Ni-content weld metal becomes lower than that of the base metal, and hence preferential corrosion of weld metal was not observed in this case. Corrosion rate of the base metal varies little regardless of the composition of the weld metal. The base metal that is not coupled to a weld metal shows a similar corrosion rate as the base metals that are coupled to weld metals.

Fig.7 shows the corrosion rates of the weld metal and base metals as the function of Mo content in the weld metal. The weld metals that contain Mo show lower corrosion rate than that of the base metals, and hence, preferential corrosion of the weld metal was not observed in these Mo-containing weld metals.

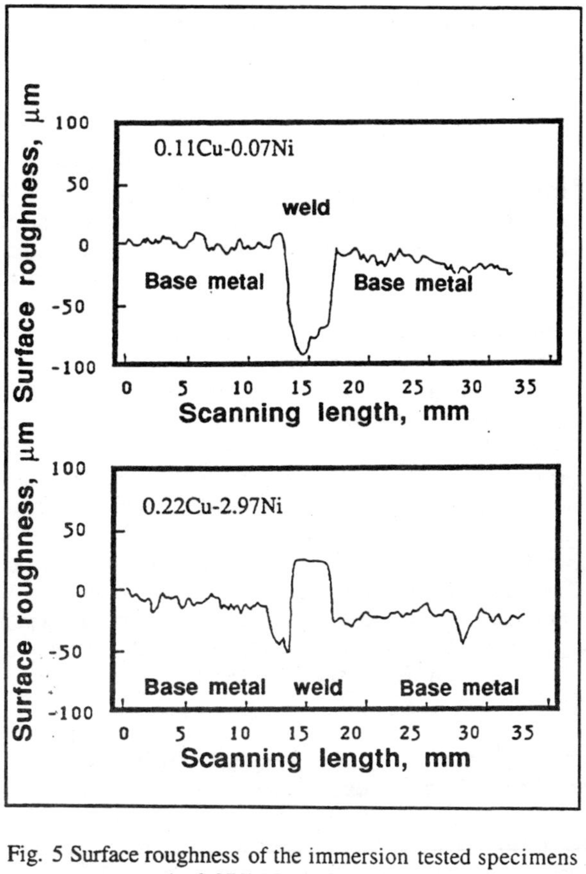

Fig. 5 Surface roughness of the immersion tested specimens (a; 0.07% Ni, b; 2.97% Ni)

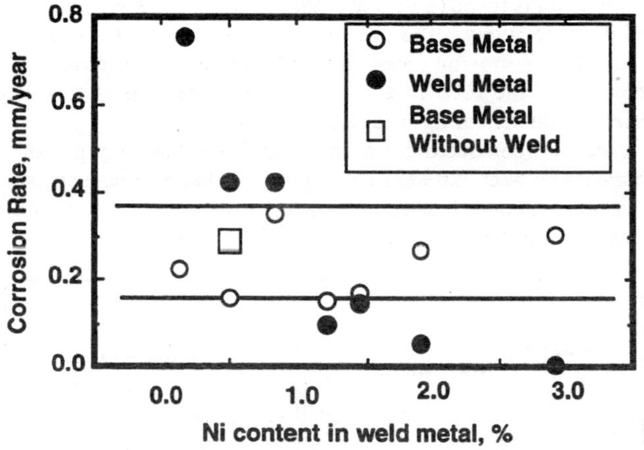

Fig. 6 The effect of Ni content in the weld metal on corrosion rate (303K-0.1 MPa CO2 saturated synthetic sea water)

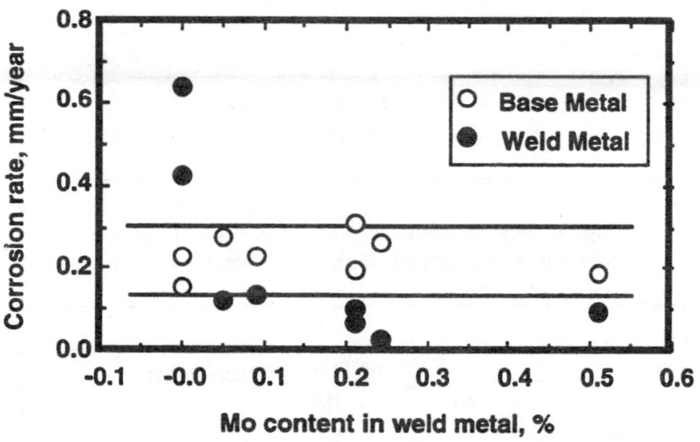

Fig. 7 The effect of Mo content in the weld metal on corrosion rate (303K-0.1 MPa CO2 saturated synthetic sea water)

The electrochemical characteristics of base metal, HAZ and weld metals

Polarization curves of the base metal, the HAZ and the weld metals with 0.07% Ni and 1.13% Ni in synthetic sea water saturated with 0.1 MPa CO_2 are shown in Fig. 8 through Fig. 10. Differences in the anode polarization curves and the corrosion potentials are observed. The corrosion potential of the weld metal with 0.07% Ni is 5 mV lower than that of the base metal, whereas the corrosion potential of the weld metal with 1.13% Ni is 10 mV nobler than that of the base metal. HAZ shows a nobler corrosion potential than the base metal. It is interesting to note that in the above described immersion tests, the higher Ni weld metal is not being preferentially corroded when coupled with the base metal. If a galvanic process is assumed, the immersion test results can be shown to be consistent with the observed relationships between the corrosion potentials and the metal chemistry.

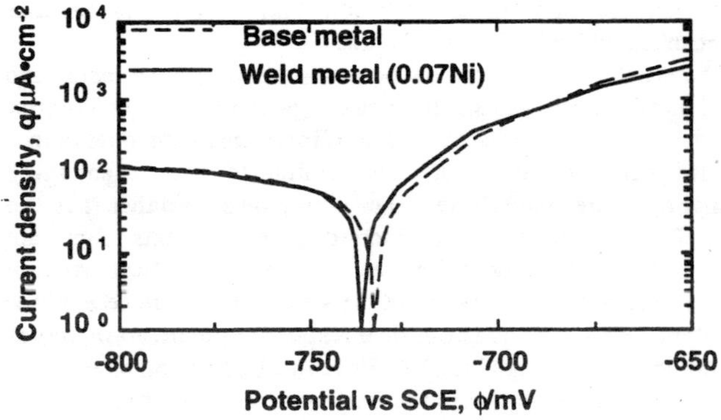

Fig. 8 Polarization curves of weld metal A and base metal (303K-0.1 MPa CO2 saturated synthetic sea water)

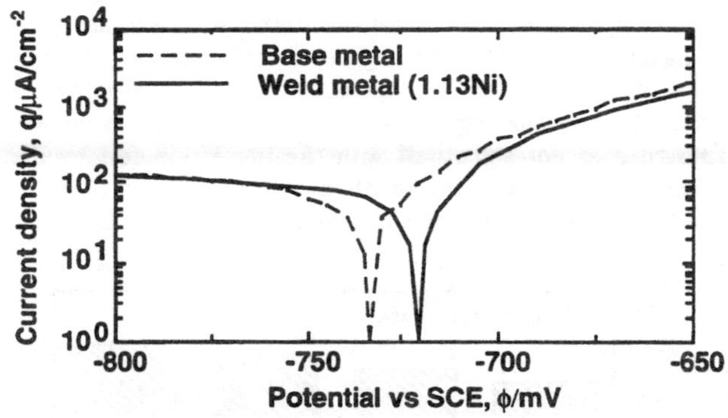

Fig. 9 Polarization curves of weld metal B and base metal (303K-0.1 MPa CO2 saturated synthetic sea water)

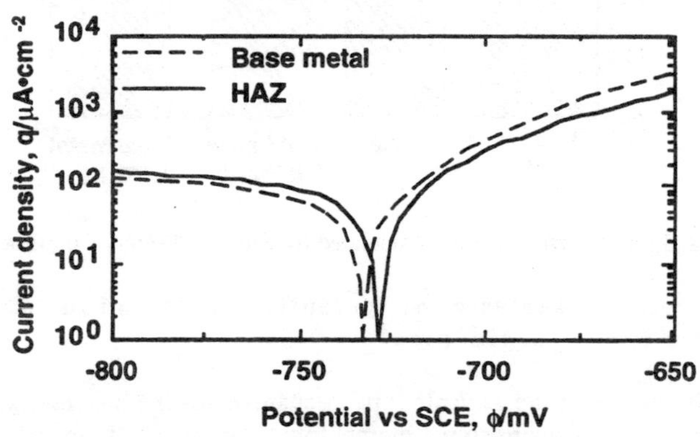

Fig. 10 Polarization curves of HAZ and base metal (303K-0.1 MPa CO2 saturated synthetic sea water)

DISCUSSION

Corrosion rate of the base metal and the weld metal

From the aforementioned test results, it is clear that the preferential corrosion occurs in the welded joints in the CO_2 environment and the chemical composition of the weld metal and the base metal affect resulting morphology in the weld region. The dependence of the corrosion potential on the chemical composition of the weld metals is found to describe well the tendency of preferential corrosion observed in the immersion test.

However, the preferential corrosion may occur via another mechanism which is discussed below:

The difference in corrosion potential between base metal and weld metals seems too small to cause galvanic type preferential corrosion. The observed corrosion of individual components in a welded joint progresses at rates similar to that when these components are stand alone. Thus it may be argued that it is the natural corrosion rates of individual components that are responsible for the observed preferential corrosion. The corrosion rates of the base metal, HAZ and the weld metals were determined by Stern-Geary's method from their respective polarization curves shown in Fig 8 through Fig. 10. Since these corrosion rates are in close agreement with that in the immersion tests, it is very likely that the preferential corrosion occurs in the manner of the aforementioned mechanism. The corrosion rates of the weld metals with 0.07% Ni and 1.13% Ni, the HAZ and the base metal calculated are shown in Fig. 11. Here the weld metals show greater corrosion rate than the HAZ and the base metal, regardless of their compositions. In comparison to the corrosion rate of the base metal and the HAZ, the higher corrosion rate of the weld metal with 0.07% Ni is in good agreement with the immersion test results. However, the larger corrosion rate of the weld metal with 1.13% Ni does not agree well with the immersion test results. Consequently, the aforementioned hypothesis can not explain the observed preferential corrosion behavior.

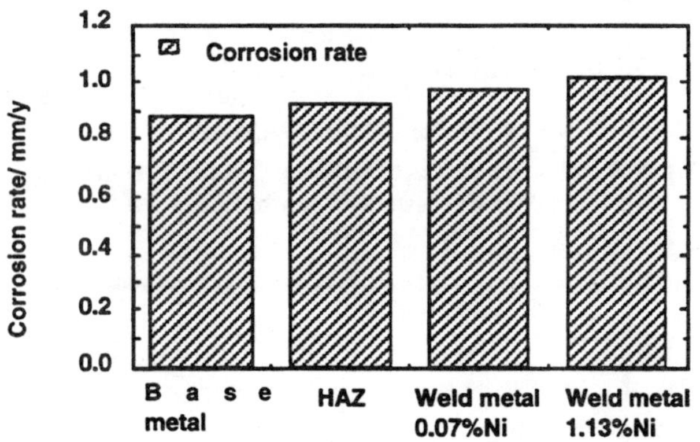

Fig. 11 The corrosion rates calculated by Stern and Geary's method

Numerical Analysis of potential distribution on inner surface of linepipe

By assuming the galvanic type mechanism for the preferential corrosion, a corrosion morphology predicted from the polarization characteristics and the immersion test results are in good agreement with each other. However, the differences in the measured potentials between the base metal and the weld metals seem to be too small to account for their dissimilar corrosion rate, which is believed to be the cause of preferential corrosion observed in the immersion test. In order to verify the galvanic type mechanism, numerical analysis of the potential distribution on the inner surface of the linepipe was carried out.

The calculated potential distributions in linepipes with weld metals A and B are shown in Fig. 12 and 13, respectively. In both cases, potentials of the weld metals approach that of base metal with increasing area ratio of base metal to weld metals, up to a value of 24. It is noted that up to a value of 8, the area ratio has strong effect on the polarization of the weld metal. The maximum polarization is found to be 3 mV and -10 mV for the weld metals A and B, respectively. The potentials of the base metals approach their natural corrosion potentials with increasing area ratio. The potential of the HAZ in the case of weld metal B is found to increase with increasing area ratio.

It can be concluded from these analysis that the potential difference between different components is large enough to cause polarization, and hence, galvanic corrosion is a possible mechanism for the observed preferential corrosion.

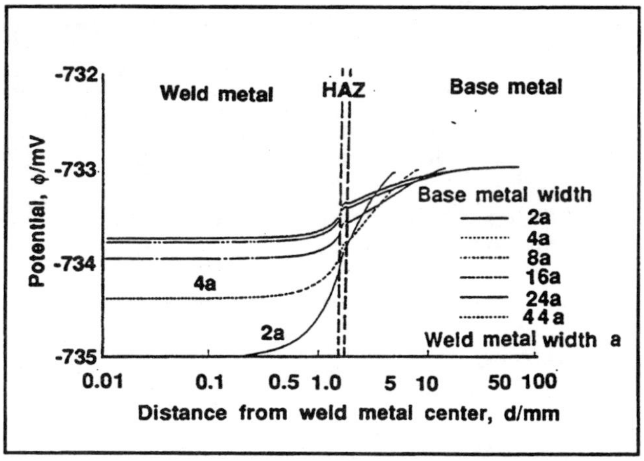

Fig. 12 Distributions of electrochemical potential on inner surface of linepipe (Weld metal A)

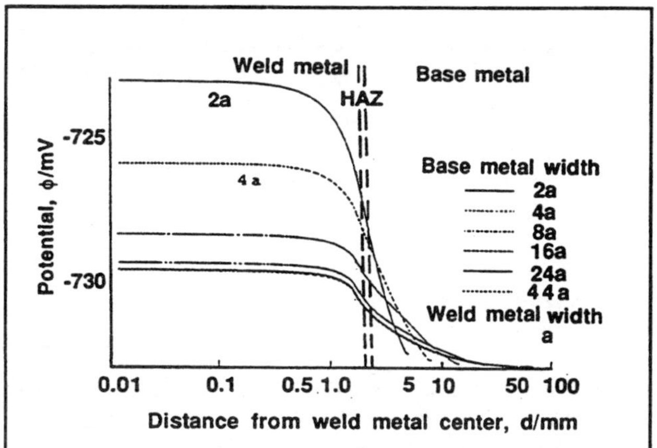

Fig. 13 Distributions of electrochemical potential on inner surface of linepipe (Weld metal B)

Anodic current density and corrosion rate

Since ($\phi-\phi_{corr}$) measured in this work is typically less than 50mV, I_{appl} does not have the same value as the anodic corrosion current density i_a which is directly related to corrosion rate. This is because the cathodic current density, i_c in equation (1) can not be neglected.

$$I_{appl}=i_a-i_c \quad (1)$$

Thus, the corrosion rate can not be directly determined from the equation, $q=F(\phi)$, used in the above numerical analysis.

Therefore, in this research the corrosion rate is determined using equation (2) which describes the anodic polarization curve.

$$\phi-\phi_{corr}=\beta_a \log(i_a/i_{corr}) \qquad (2)$$

where ϕ_{corr} and i_{corr} are the corrosion current density and corrosion potential. i_{corr} can be determined by Tafel extrapolation method or liner polarization method (Ed by M. G. Fontana and N. D. Greene). In this work , i_{corr} of the base metal, the HAZ and the weld metals had been obtained using liner polarization method. β_a is the Tafel gradient of the anode polarization curve, and had been determined to be 55 mV/decade (Gray, 1989).

The anodic corrosion current density, i_a, had been calculated for the segments A to G in Fig. 2 and the results are shown in Fig. 14. The first case considered deals with the weld metal A, which shows a less noble corrosion potential than that of the base metal. The anodic current density at each segment is by substituting the analyzed potentials into the equation (2). The segments A and B are in the weld metal, C and D are in the HAZ, and E to G are in the base metal, respectively. The weld metal has approximately 30 mA/cm2 higher current density than the base metal. This current density is equivalent to an extra corrosion rate of 0.3 mm/year, which is in quite good agreement with that observed in the immersion test. As the test condition does not consider flowing fluid and the resulting erosion, this calculated corrosion rate seems to be also in good agreement with the corrosion rate observed in a real pipeline.

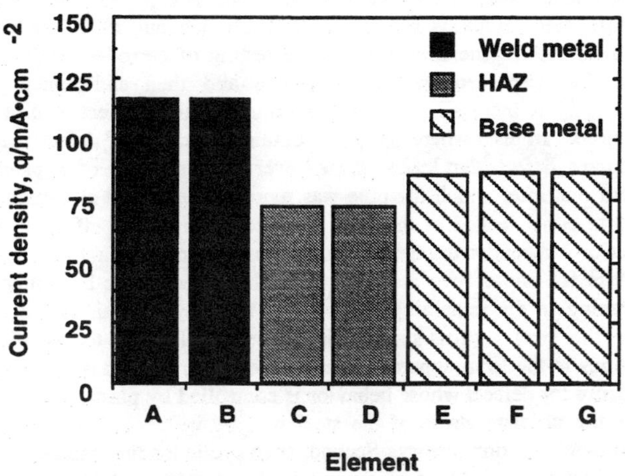

Fig. 14 Distributions of current density on internal surface of line pips (Weld metal A)

The second case considered deals with the weld metal B, which shows a more noble corrosion potential than that of the base metal. The results of the analysis are shown in Fig. 15. Here the weld metal shows a lower current density than the base metal adjacent to the HAZ. From the results shown in Figs 14 and Fig. 15, the base metals is polarized less than the weld metals, especially in the case of larger area ratio. The base metals have similar current density in the cases of both weld metals.

From the analyzed potential distributions and corrosion current density, the mechanism of the preferential corrosion is verified to be galvanic corrosion. However, this is a case in which the corrosion rate of the macro-cathod is very close to its natural corrosion rate, and the galvanic corrosion is driven by a small variation in the potential distribution.

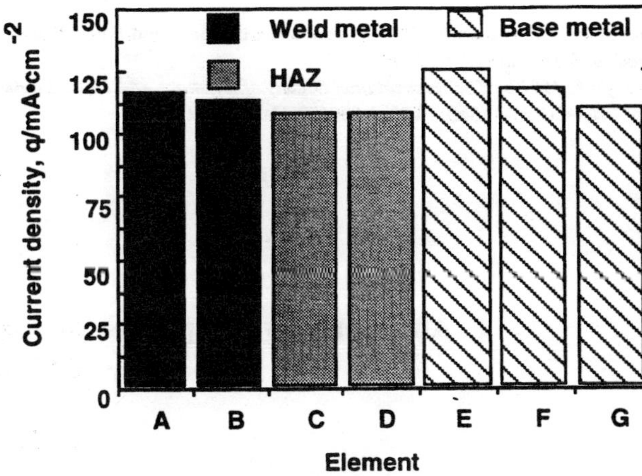

Fig. 15 Distributions of current density on internal surface of line pips (Weld metal B)

CONCLUSIONS

1) It is observed that preferential corrosion of welded joints occurs in a CO_2 environment, and the chemical compositions of weld metals and base metal affect the morphology of preferential corrosion.
2) The differences in the corrosion potential and the polarization curves are observed among different components in welded joint.
3) The corrosion rates calculated from the polarization curves can not reproduce the preferential corrosion of the welded joints observed in the immersion test.
4) From the analyzed potential distributions and corrosion current density, the mechanism of preferential corrosion is believed to be galvanic corrosion.

REFERENCES

Fontana G and Greene ND, Corrosion Engineering., McGraw-Hill, P342

Gray LGS et al., Corrosion 89, Paper No. 464, 1989

Ikeda A, et al., Corrosion 83 Paper No. 45, 1983

Ikeda A, et al., Corrosion 84 Paper No. 289, 1984

Iwata M and Huang Y, Proc. of the 12th Int. Conference on Boundary Elements in Engineering, Boundary Elements XII, Vol. 1, Applications in Stress Analysis, Potential and Diffusion, P 143-154

Joosten MW, et al., Corrosion 88 Paper No. 211

Kodama T, Fusyoku-Bousyoku Symposium, 1983, p 1-

Kvernvold O et al., Proc. Conf. OMAE"92, vol. V-B P 409-420

London CJ, Pipe & Pipeline International May-Jun. 1991, p7-13

Steel WJM, et al., May 6 1991, Oil & Gas Journal, P49-544) K 9)

Stern M and Geary AL, Journal of The Electrochemical Society, Vol. 104, No. 1, 1957, P56-63

Videm, et al., Corrosion 87 Paper No. 42, 1987

Proceedings of the Seventh (1997) International Offshore and Polar Engineering Conference
Honolulu, USA, May 25-30, 1997

ISBN 1-880653-28-1 (Set); ISBN 1-880653-32-X (Vol. IV)

An Alternative Approach to Assess the Integrity of Corroded Line Pipe - Part I: Current Status

B.N. Leis and D.R. Stephens
Battelle
Columbus, Oh, USA

ABSTRACT

This paper explores the basis for some of the current criteria for assessing the effect of metal-loss defects on the integrity of pipelines, with particular reference to important trends in the database of full-scale experiments on corroded pipe sections. Critical trends are identified as the foundation for an alternative to current integrity criteria for metal-loss defects. Finally, this paper outlines an approach to formulate an alternative criterion based on the limiting response of metal-loss defects observed in field and full-scale laboratory experiments. Part II of this paper develops this approach and presents numerically generated trends from parametric analysis that fix the functional form and the fitting-constants for this alternative criterion. From these trends a simple criteria for metal-loss defects is presented and demonstrated to be accurate based on comparisons to results from full-scale tests of end-capped pipe sections.

Conclusions from Part I of this paper relate to the formulation of the alternative criterion and its practical application, and include the following points. The use of the actual yield stress in place of SMYS in B31G and other similar or related criteria offers little improvement in either their accuracy or consistency. In contrast, the uniaxial ultimate tensile stress in place of SMYS in B31G and other similar criteria consistently correlates data from burst tests of defect-free end-capped pipes with little or no scatter. Circumferentially aligned (coparallel) adjacent defects cause a local increase in the compliance of the pipe wall, which explains their failure at pressures greater than that for a geometrically similar isolated defect. Axially aligned (colinear) interacting defects have an equivalent defect length greater than that of the longest of the individual defects, leading to a reduction in failure pressure. Test specimens free of significant axial stresses representing "open-cylinder" conditions, which are a useful "reference condition", failed at pressures less than specimens from corresponding end-capped tests. Thus, integrity criteria calibrated or validated using end-capped vessel test data could be nonconservative in applications to metal-loss defects in areas of bending-induced compression in any piping system or pipeline, earth-anchored, above-ground, or offshore.

Key Words: Corrosion, defect, failure, criterion, B31G, RSTRENG, alternative criterion, transmission pipeline

INTRODUCTION

The behavior of metal-loss defects due to corrosion has been studied for years using both experimental and analytical methods. Experimental work (e.g.,, Mok et al, 1990, Wang, 1991, Jones and Ritchie, 1992, Hopkins and Jones, 1992, Chouchaoui, 1993, Vieth and Kiefner, 1994) generally involved the testing of corroded sections of pipe that were removed from service and then end-capped and subjected to increasing internal pressure until a defect leaked or ruptured. In tests where the pipe contained a range of defect shapes and sizes, defects that leaked at the lower pressures were often patched or cut out, after which the pipe was repressurized, until all remaining defects either leaked or the pipe ruptured. While cost effective, this practice can render the data from failures beyond the first one nearly useless for one of two reasons. First, localized cyclic hardening or softening in the remaining ligament of the as-yet-to-fail defects can alter the steel's stress-strain behavior there, leading to significant changes in the local apparent ultimate tensile stress. Thus, the failure pressure for defects whose behavior is controlled by plastic collapse at near the ultimate stress of the steel will be influenced by the prior plastic-deformation history. Second, such cyclic loading can cause and also sharpen crack-like features within the metal-loss defects or cause stable tearing at already sharpened defects, requiring a significant post-mortem study to identify which data are valid in terms of a plastic collapse mechanism and which are not. The analytical work has ranged from "empirical models"(e.g., Shannon, 1974, anon, 1984, Kiefner and Vieth, 1989, Coulson and Worthingham, 1990a, Chouchaoui, 1993, Kiefner, et al, 1995) to "engineering analysis" (e.g., Stewart et al, 1995, Smith and Grigory, 1996), through three-dimensional numerical modeling using the finite-element method (e.g., Chouchaoui, 1993, Kanninen et al, 1993, British Gas, 1994), which has included material nonlinearity and occasionally geometric nonlinearity.

As the pipeline infrastructure ages, and maintaining the system to avoid costly leaks and ruptures due to corrosion becomes increasingly evident on the bottom line, pipeline operators and plant managers will want criteria to make integrity decisions free of significant embedded conservatism and high scatter. Criteria for integrity analysis must provide a consistent margin of safety to ensure the pipeline remains safe and serviceable on a cost-effective basis. Such criteria are critical to ensure that firm deliveries are met. They are also critical to avoid the significant direct and indirect costs related to lost product, replacement product, repairs, litigation, lost public confidence, increased regulatory pressure, and so on, all of which follow after a leak or a rupture. But, maintenance before it is needed to ensure safe serviceability comes with a cost penalty too. Such costs can run from hundreds of thousands of $US, depending on the situation, up into millions in the case of offshore failures (e.g., Keen, 1990, Sowerby, 1991). For these reasons, pipeline companies will need criteria that provide for consistent integrity analysis — with a known and reasonable factor of safety — to plan a timely, cost-effective response to integrity concerns. Eventually, criteria that facilitate a planned cost-effective response to integrity concerns will be a priority for all operating companies.

It follows that, while decision-making criteria for the continuing operation of pipelines in the presence of metal-loss defects exist today, as for example ASME B31G[1] and its modification (referred to as Modified B31G) that led to RSTRENG[2], operators eventually will move toward criteria that more consistently assess integrity in corroded plant piping and transmission pipelines — free of excessive conservatism such as the literature associates with these and other such criteria (e.g., Coulson and Worthingham, 1990b, Fu and Kirkwood, 1995).

This paper begins by evaluating the basis for some of the current metal-loss criteria, with particular reference to important trends in the database from full-scale experiments on corroded pipe sections. This evaluation leads to the identification of critical trends that provide the foundation for an alternative to the historical integrity criteria for metal-loss defects. Finally, this paper outlines an approach to formulate an alternative criterion based on the limiting response of metal-loss defects observed in field and full-scale laboratory experiments and presents the rudimentary and preliminary form of an alternative criterion that is presently being developed.

LESSONS AND TRENDS FROM FIELD AND FULL-SCALE TESTING

Pipe joints removed from service because they contain metal-loss defects often show a range of defect shapes and sizes. Some deep, isolated defects show a gradual wall thinning within a near-circular boundary on the pipe's outer surface, whose diameter ranges from as small as a few wall thicknesses up to many inches across. Other defects cover a very large surface area with rather uniform metal loss over the entire region. But, in general, the defects found on pipe removed from products and gas transmission service are neither isolated nor uniform in metal loss. Nevertheless the less, in most cases their shapes and sizes lie between or are a composite of the two just-noted extremes.

Two Limits on the Failure Pressure for Metal-Loss Defects

The results of proof tests done on joints of pipe removed from service because of metal-loss defects due to corrosion (e.g., Shannon, 1974, anon, 1984, Kiefner and Vieth, 1989, Coulson and Worthingham, 1990a, Chouchaoui, 1993, Kiefner, et al, 1995) are quite instructive in formulating an alternative approach to assess integrity in the presence of such defects. When proof tests are done on joints of pipe that show either of the just-noted extremes in defect shape, the results indicate two clear trends. In the case of the deep, smaller diameter pit-like defect, the wall stress at failure can approach that of the defect-free pipe. Indeed, failures in such tests occasionally have occurred in the full wall thickness of the pipe body — in areas remote to the metal-loss defects. Failure pressures in such tests tend to be comparable to the pressure calculated using the ultimate tensile stress of the pipe steel (in the usual formula for the hoop stress in a pressurized cylinder ($P = (\sigma_{UTS} \cdot t)/r)$) and the full-wall thickness. In contrast, in the case of the large areal defect with some appreciable loss in wall thickness, failure occurs consistently within the defect. Failures in such defects are observed at pressures comparable to that calculated using the ultimate tensile stress of the pipe steel in the same formula for hoop stress — but now using the net (minimum) wall thickness. That is, in the first case the pipe responds like there was no defect whereas in the second case the behavior is like that of a pipe with a wall thickness equal to the net thickness at the defect.

Fig. 1 schematically depicts the two extremes in the behavior of metal-loss defects. The upper limit indicated in the figure is representative of the case where the defects are short and the full-wall thickness acts to resist the pressure in spite of the presence of the defect. The lower limit reflects cases where the defect is so large that the presence of the surrounding pipe wall has no influence on the failure and the net (minimum) wall thickness controls the failure. This paper focuses on the failure behavior of these two extremes in metal-loss defect shape and size and to a lesser extent also addresses the range of behavior that lies between them, the latter being represented by the several curves drawn between the two extremes. Note in this context that the above noted occasional failures in the full-wall thickness containing non-failing short, deep pit-like defects could be viewed as part of the extreme associated with large areal defects, which in that case have zero depth.

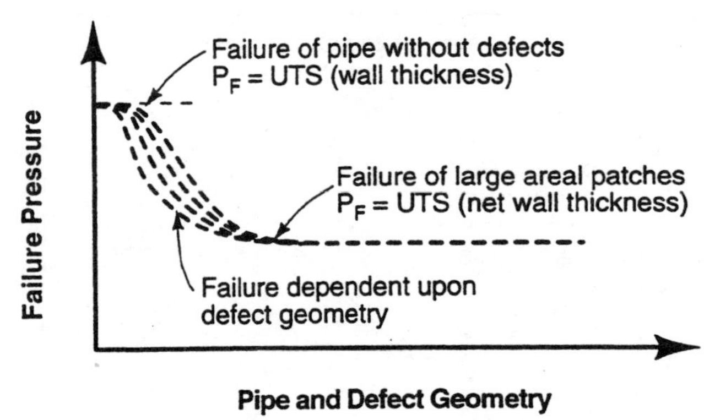

Fig. 1 Schematic depicting the failure conditions for the two extreme classes of metal-loss defect — short, deep pits, and large areal defects.

Defects Aligned Along and Around the Pipe

The results of full-scale tests on man-made metal-loss defects also can be instructive. Such work can define distances between defects where their interaction influences the failure behavior and leads to failures at a pressure much different than that of the largest defect in the group. Such work also is useful in understanding both the factors that control

1 The technical basis for the B31G criterion (anon., 1984) was developed under the auspices of the PRC, International (PRCI). The B31G criterion is under the jurisdiction of the ASME B31 Committee on Piping

2 The modifications to the B31G criterion and the related RSTRENG criterion (Kiefner and Vieth, 1989) and its continuing validation (Kiefner et al, 1995) have been developed under the auspices of the PRC, International (PRCI)

the failure behavior under more controlled conditions of defect size, shape, and position, and their relative significance. Of particular significance in regard to defect interaction are results such as that of Hopkins and Jones, 1992 and Chouchaoui, 1993. Such work shows that axially aligned defects reduce the failure pressure as compared to a single identical defect, because interacting axially aligned (colinear) defects act to increase the "equivalent length" of the defect, as compared to that of the longest defect in the group. Such results are shown in Fig. 2a, which plots the failure pressure for colinear slits and pits as a function of the failure pressure for a geometrically comparable but isolated single slit or pit. These data for colinear defects, which are shown as the +, x, and * symbols, all fail at pressures less than that for

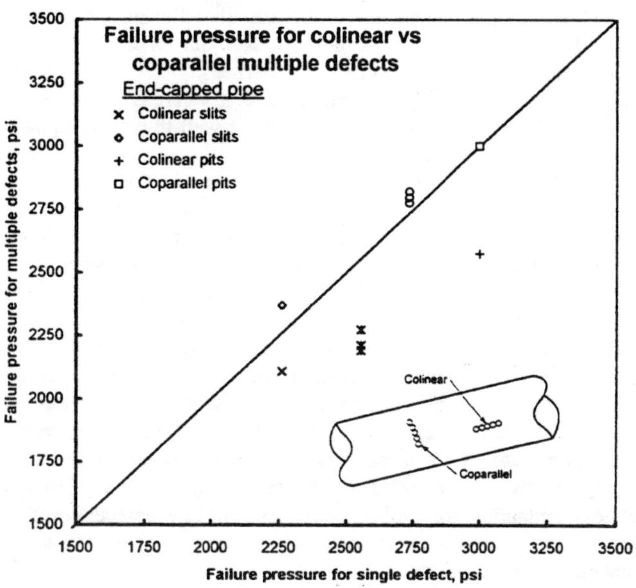

a) colinear and coparallel defects (data from Hopkins and Jones, 1992 and Chouchaoui, 1993)

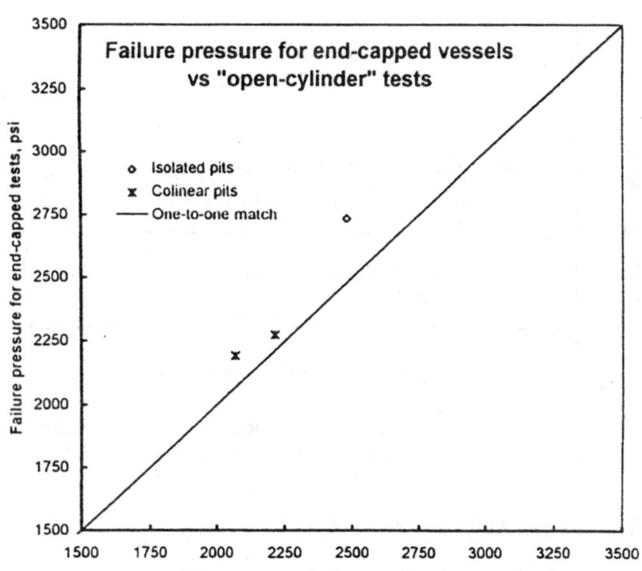

b) end-capped versus open-cylinder testing (data from Chouchaoui, 1993)

Fig. 2 Trends in failure pressure for some full-scale tests

the corresponding isolated single slit or pit.

That is they lie below the diagonal line in the format of Fig. 2a. Such defect interaction for colinear slits and pits is well known and is easily rationalized. In contrast to the lower failure pressure that can occur for defects aligned along the pipeline, circumferentially aligned (coparallel) defects can cause the failure pressure to increase as compared to an isolated, geometrically comparable, single defect. That this occurs is evident in the results for coparallel defects, like shown in Fig. 2a as the ○, □, and ◇ symbols, which lie above the diagonal line in the format of that figure.

It follows that one simple discriminating test for the validity of integrity analyses for crack-like as well as metal-loss defects is the ability of that analysis to correctly predict the distinctly different trends between colinear and coparallel defects. It is noted in this context that because coparallel defects fail at pressures in excess of that for an isolated, single defect, the failure behavior of this class of defects can lie outside the limits of the behavior developed at the earlier suggested extremes of an isolated pit and a large areal defect. To the authors' knowledge, this is the only class of defect to lie beyond those extremes or combinations formed by them.

Results for crack-like coparallel defects indicate that the apparent increase in failure pressure for this class of defects occurs because they act to increase the local compliance (e.g., Leis and Colwell, 1996), which diminishes locally the effect of the internal pressure on the defects. Note from Fig. 2a that the trend to fail at pressures greater than that for the corresponding isolated single defect is the same for both sharp and blunt coparallel defects. This trend must be common to both sharp and blunt coparallel defects if the analytically inferred local increase in compliance is the cause for the observed increase in failure pressure as compared to that for a comparably-sized isolated single defect. Likewise, the trend to fail at pressures less than that for the corresponding isolated defect is the same for both sharp and blunt colinear defects. Again this commonality between the trends for both sharp and blunt coparallel defects must occur if this postulated effect of nearly coaxial defects resulting in an increase in the equivalent defect length is to serve as a viable explanation. It is noteworthy in this context that the just noted analytical trends for the failure pressure of colinear and coparallel cracks also are observed in analytical results for blunt defects (e.g., Stephens et al, 1995). In light of these trends, one could expect that the behavior of defects with a diagonal alignment or with a diagonal component around the pipe could be quite complex, reflecting both the effects of an increase in local compliance and an increase in equivalent length!

End-Capped versus Open-Cylinder Conditions and Stress-State Effects

Finally, consider the difference in the behavior of defects under the end-capped testing conditions noted above and that under pipeline-service conditions. Chouchaoui (1993) reports on experiments (and analysis) of end-capped vessel tests of metal-loss defects supplemented by additional experiments wherein it is indicated that the testing fixture was such that the longitudinal stress, σ_L, was zero. Such a testing condition gives rise to an axial contraction due to the Poisson effect, so that $\epsilon_L = -(\nu / E) \cdot \sigma_H$, where ϵ_L is the axial strain, σ_H is the hoop stress, ν is Poisson's ratio, and E is the modulus of elasticity. Physically this testing condition was open ended, much like a long section of pipeline. However, buried transmission pipelines can have a virtual anchor along their length if they are buried in compacted and adherent soils, which resists axial displacement leading to a condition where $\epsilon_L \equiv 0$. The condition $\epsilon_L \equiv 0$ gives rise to a longitudinal stress $\sigma_L = \nu \cdot \sigma_H$, which indicates that the pipeline undergoes biaxial stressing. In biaxial principle stress space this case is represented as $\sigma_H / \sigma_L = 1/\nu$ (i.e., the slope is 1/v). As such, while physically similar to a pipeline, these open "cylinder" tests give rise to biaxial stressing whose effect could be appreciable because of the difference in ϵ_L as compared to a buried, earth-anchored pipeline. Nevertheless the less, this "open cylinder condition" serves as a useful reference state in terms of differences in axial straining that can occur in areas of compression, due to bending induced by piping system loads or by outside forces, such as a land slide.

As noted at the outset of this section, end-capped pipes historically were used in "vessel" tests to evaluate the failure conditions for metal-loss defects, and to calibrate empirical models for the integrity analysis of such defects. Physically, these end-capped vessels are much different than the open-ended pipeline because the pressure-load on the end caps creates a longitudinal stress $\sigma_L = \sigma_H / 2$. In biaxial principle stress space this stressing in these vessel tests occurs along a slope $\sigma_H / \sigma_L = 2$, which affects the failure pressure. Because failure of metal-loss defects generally occurs through a plastic-collapse mechanism (e.g., see Chouchaoui, 1993), the above mentioned value of Poisson's ratio within the defect is about 0.5 as failure approaches, consistent with the constant-volume requirement associated with metal plasticity. For deeper, longer defects for which failure can occur at pressures that cause only elastic response in the vessel, the value of v is 0.3 for the body of the vessel. In contrast, for shorter, even very deep defects, test pressures at failure cause inelastic response in the vessel, such that the value of v for the body of the vessel is 0.5. In cases where the value of v is 0.5, the above noted ratio $\sigma_H / \sigma_L = 1/v$ in a pipeline has a value of 2, which is exactly the case in end-capped vessel testing. In such cases end-capped results replicate in-service situations for buried earth-anchored pipelines and other straight piping runs that are constrained axially.

In spite of the physical differences between an end-capped vessel and the otherwise uncapped pipelines, there are other situations where end-capped vessel tests simulate directly the behavior of plant-piping systems or non-anchored transmission pipelines. For example, bends in plant piping create the an axial force equal to that of the end-capped vessel. Bends in transmission piping likewise create an axial load like that in an end-capped vessel, which is resisted by the pipeline unless it is anchored. For this reason, integrity criteria for metal-loss defects that were calibrated or validated using end-capped vessel results can be applied directly to axially constrained (e.g., earth-anchored buried) pipelines.

However, buried transmission pipelines also involve areas where low-angle bends are introduced to traverse hills or change directions. Some quite large-diameter pipelines also use formed elbows or fabricated miter-bends, which, like bends in plant piping, generate end-capped tension loads. In spite of these end-cap induced tension loads, areas where bending occurs in the piping create compression on one side of the piping or pipeline. Such compression can move the just-discussed stressing conditions into a quadrant other than that characteristic of the end-capped or the open-cylinder conditions. Fig. 2b helps illustrate this situation by comparing open cylinder data with corresponding results for an end-capped condition, based on results extracted from work by Chouchaoui (1993). In light of the tendency for failure pressure to be over-estimated by end-capped results in Fig. 2b for applications where there is minimal axial tension, it is reasonable to expect that axial compression would make the end-capped result appear even more nonconservative. Therefore, criteria developed for end-capped applications, such as B31G or RSTRENG, should be used with caution in applications other than earth-restrained buried pipelines. The influence of local compression deserves more attention — through both experiment and analysis.

It follows that criteria developed for earth-restrained pipelines or criteria calibrated or validated based on end-capped conditions may be nonconservative in some applications involving above-ground piping, like that around pumping and compressor stations and general plant piping, or offshore piping, or areas of bends in earth-anchored pipelines. Such situations may be more correctly represented by the open-cylinder condition or by results involving axial compression. Based on the data shown in Fig. 2b, such predictions can be nonconservative by about 6 percent on average, and by as much as 11 percent in the worst case. However, given the limited scope of this comparison, further work should be considered to better quantify this situation.

CURRENT CRITERIA FOR MAKING PIPELINE INTEGRITY DECISIONS

Because the two earlier discussed extremes in metal-loss defect shapes and sizes bound or form the basis for the behavior for most corrosion-induced defects, results developed at these two limits of pressure response can be expected to serve as bounds for the results of integrity analysis for most metal-loss defects. It follows that the successful prediction of the failure behavior of metal-loss defects at these two extremes is a good first-order test of the viability of related failure criteria. The range of failure pressures predicted by integrity analysis for metal-loss defects should reasonably lie between the results for these two bounds — with the exception of coparallel defects.

This section evaluates the range of failure pressures predicted by B31G and its modifications as embodied in RSTRENG, specifically in reference to the upper bound extreme of failure pressure — that is for cases where a large areal defect has zero depth. The underlying premise is that prior to being adapted to applications involving metal-loss defects such criteria should reasonably predict the failure pressure in the absence of defects. Thereafter, this section contrasts the predicted failure behavior of blunt defects to that for sharp defects. The premise here is that blunt defects, such as large area-loss defects, should fail at pressures in excess of that observed for sharp defects of otherwise similar length and depth.

Predicted Failure Pressure by Current Criteria as the Defect Size Tends to Zero

The ASME B31G criterion and its modifications as embodied in RSTRENG have their roots in the so-called NG-18 surface flaw equation (Maxey et al, 1972) which has the form:

$$S = S_o \left(\frac{1 - A/A_o}{1 - (A/A_o) \cdot M^{-1}} \right) \qquad \text{(Eq.1)}$$

In this equation S represents the nominal circumferential (hoop) stress at which failure occurs. In formulating Equation ,1 it was assumed that the material would fail in a ductile manner at a stress level represented by the symbol S_0. The effect of the metal-loss defect geometry is addressed by the term in the parenthesis. In the following sections S_0 as the "first term" in this equation whereas the symbols contained within the parenthesis are referred to as the "second term".

Equation 1 determines the failure pressure as the product of a term that defines the failure stress in the absence of defects and a term that reduces that stress to account for the defect size and shape, independent of its width. Because it is a product of two terms, when either is zero the failure pressure is zero. But, because the second term has a value of unity when the defect depth is zero, that is there is no metal loss, the failure pressure takes on the value of the first term. It follows that the first term is the material property that reflects the stress at which an unflawed pipe fails, which should be related to the failure behavior of an unflawed uniaxial sample of that pipe steel, provided that differences in stress and strain states are accounted for. Thus, the value of S_0 is not an arbitrary constant — rather, its value is prescribed by the stress level where failure occurs by plastic collapse in the pipe steel of interest. This value should be consistent with the failure pressure of defect-free pipe.

In spite of the S_0 having a prescribed value that should be a material property and also correlate the failure pressure of unflawed pipes observed in full-scale testing, the symbol S_0 was assigned various values over the years. Checking criteria for metal-loss defects indicates that the value used depends on the specific criterion or modification being considered. Initially, S_0 was termed a "flow stress" and defined in terms of a constant that multiplied SMYS. Subsequently, the value of S_0 was modified to a form where a constant was added to SMYS. But, regardless of its definition, its value of S_0 was defined in relation to the material's yield stress. The symbol A in Equation 1 was related to the area of the defect on the through-thickness longitudinal plane whereas the symbol A_0 had a value equal to the product of the defect length and the pipe's full wall thickness. Finally, the symbol M involved provision for pressure-driven bulging in the vicinity of the defect, which was expressed in various forms adapted from Folias' work (1965) on the behavior of through-wall slits.

Finally, Equation 1 and its modifications assume that the failure process develops in a ductile manner (i.e., net-section collapse), which, as just discussed, has been historically expressed in terms of some function of the yield stress. In this context, brittle fracture or some other form of cracking within the metal-loss defect in lieu of the presumed ductile fracture response could cause failure at a pressure less than that associated with the stress level substituted for S_0 (i.e., 1.1 x yield stress, yield plus 10 ksi, etc). This will be discussed more fully later.

Yield Stress as a Defect-Free Failure Criterion (Reference Failure Stress)

Figs. 3a through 3d present results for a range of full-scale "burst" tests on end-capped "defect-free" pipe sections, for line pipe grades ranging from X42 to EX100[3]. (Anon, 1996a, 1996b). Fig. 3a shows the failure pressures predicted for these defect-free burst tests based on Equation 1 as adapted for B31G and modified B31G (equally in this context RSTRENG). In such cases the second term in Equation 1 is unity so that the value of S_0 must correspond to the wall hoop stress for Equation 1 to correctly predict the failure pressure. The trends in this figure indicate that both the B31G and the modified B31G (equally the RSTRENG) criteria consistently underestimate the wall stress at which these various sections of line pipe failed. Because these tests were defect free, the full extent of this conservatism can be ascribed to the S_0 term in Equation 1. That is, the reference failure stress based on the yield stress is fully responsible for this conservatism. As expected from results that couple both terms in Equation 1, these trends indicate that the (1.1 x SMYS) definition of S_0 in the B31G criterion is most conservative, with the (SMYS + 10 ksi) definition of S_0 in the modified B31G criterion (equally the RSTRENG) being slightly less conservative for these vessel tests. In the worst case the B31G criterion is conservative by ~30,000 psi wall stress, which for that case reflects a 40 percent difference. The corresponding worst case prediction for the modified B31G criterion is conservative by ~22,000 psi, which is slightly better at about a 29 percent difference. That is, in spite of being "less conservative" due to changes to reduce the conservatism "designed into" B31G, the modified B31G (equally RSTRENG) criterion still underestimates the failure behavior for defect-free burst tests by as much as nearly 30 percent.

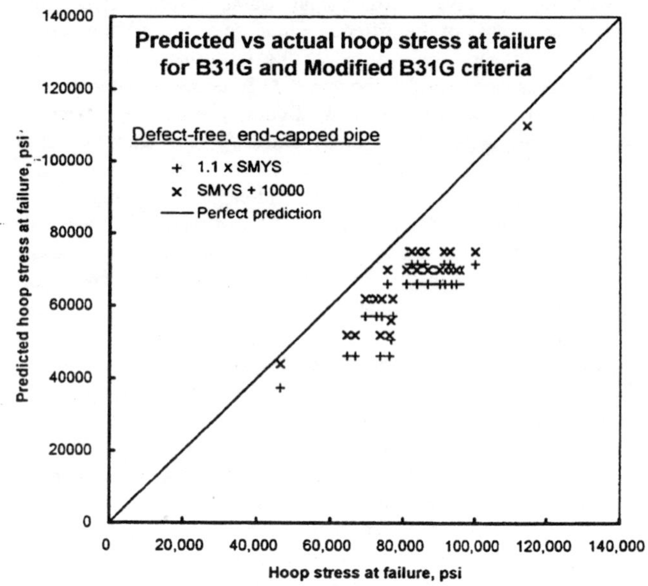

a) conservatism evident in predictions based on SMYS

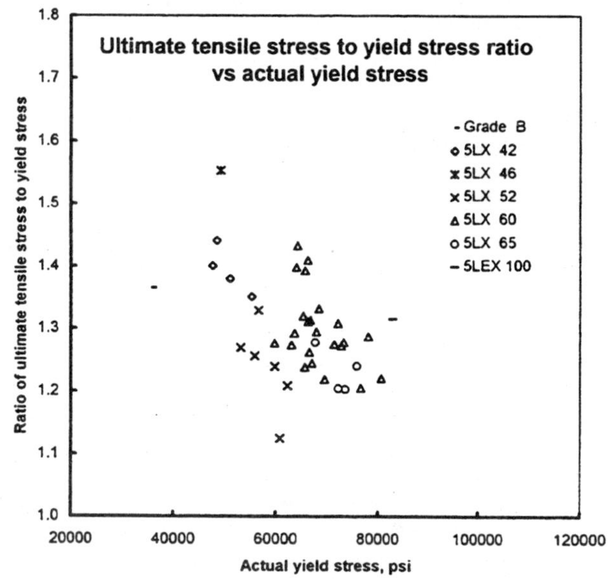

b) scatter inherent in UTS / AYS as a function of AYS

Fig. 3 Issues related to predicted versus actual failure pressure in defect-free, end-capped pipes

4 The grade notation used here is that from the API 5L specification for line pipe. The numbers following the letter X correspond to the specified minimum yield stress (SMYS) in ksi units. The designation EX 100 denotes an experimental X100 grade.

Actual versus Specified Minimum Yield Stress

The just noted analysis was repeated after replacing the value of SMYS in S_0 by the actual yield stress (AYS) in the respective definitions of S_0 in the B31G criterion and the modified B31G criterion (equally the RSTRENG) to determine the extent to which differences between SMYS and the AYS were responsible for the scatter in Fig. 3a. This analysis showed that the conservatism in Fig. 3a was only marginally reduced. It follows that the conservatism evident with respect to either the (1.1 x SMYS) definition of S_0 in the B31G criterion or the (SMYS + 10 ksi) definition of S_0 in the modified B31G criterion (equally the RSTRENG) is not due to the use of SMYS in lieu of the actual value.

Multiaxial Defect States versus a Uniaxial Reference Stress

Another possible explanation for this still significant conservatism evident in Fig. 3a may lie in the use of a uniaxial value of S_0 whereas the failure of these end-capped pipes includes a significant axial tension due to the pressure load on the end caps. An assessment of this effect was carried out by increasing S_0 in the B31G criterion and the modified B31G criterion (equally the RSTRENG criterion) by a factor of 1.155 to account for this difference in stress state. This multiplication merely shifted the data points in Fig. 3a up by a factor of 1.155, which obviously did nothing to decrease the scatter. Based on the results for this analysis, in the worst case the "multiaxial" versions of the B31G criterion and the modified B31G criterion are conservative by ~22,000 psi or ~19 percent. But, as implied above, this apparent improvement only results because this worst case difference now develops at a much larger failure stress — so there has been no net change in the scatter. Note that this "apparent" improvement in the accuracy leads to situations where the initial (1.1 x SMYS) definition of S_0 in the B31G criterion is now a less conservative measure of failure as compared to the (SMYS + 10 ksi) definition, which has been adopted in modifying the B31G criterion (equally RSTRENG) to reduce some of its "designed-in" scatter.

Why Scatter Occurs for a Yield-Based Reference Stress

Fig. 3b, which presents properties data for each of the steels represented in Fig. 3a, indicates that there can be significant scatter in the relationship between either SMYS or AYS and the ultimate tensile stress (UTS). Sources for the scatter evident in Fig. 3b lie in the variability of the UTS and the corresponding values of the AYS and the SMYS. It is well known that the UTS can be changed by variations in chemistry, but such changes also are manifest in the other mechanical properties. Since Fig. 3b matches corresponding values of these mechanical properties, this underlying dependence on chemistry becomes secondary. It is also well known the actual yield stress can vary appreciably as a result of steel making and processing and pipe making, without a corresponding effect on the UTS. Disparity between SMYS and the UTS can also exist in that line pipe steels have a specified minimum ultimate (tensile) stress (SMUS) that does not correlate strongly with AYS. Finally, variability in property correlations will exist for more recent line pipe because one grade higher can be sold as the lower grade, which means the AYS and SMYS associated with a given pipe joint or pipeline have little to do with the grade actually shipped and built into a pipeline. It follows that reference stress values based on SMYS (or a SMUS) can vary significantly without a corresponding variation in the limit state defined by the UTS.

It follows from the data shown in Fig. 3b that the major source of the scatter evident in Fig. 3a is use of the yield stress to predict limit-state failures. Related to this is the fact that the plastic collapse process that is presumed to underlie these failures is more closely characterized by the UTS than it is by the process of yielding. Thus, the scatter in Fig. 3a is due also to the use of a material property to define S_0 that has little to do with the physical process of plastic collapse. Because plastic collapse through the remaining ligament controls ductile failure at metal-loss defects (Chouchaoui, 1993), it may be better to adopt the UTS or other similar measures of a limit state[4] in lieu of functions of the yield stress.

A Limit State as a Failure Criterion — UTS as a Reference Failure Stress

Support for the use of the UTS as the value of S_0 is evident in Fig. 3c. This figure is comparable to Fig. 1 except that predictions for the same burst tests are now shown to be consistently and accurately predicted using the value of UTS from a uniaxial tension test. Not only does the UTS lead to a trend that scatters uniformly about the line of perfect predictions, it does so in a very tight band, which is as expected in light of the above discussion of Fig. 3b. This means that the quality of the predictions based on UTS are not only more accurate, they are also more precise, as compared to yield-stress-based values of S_0 as used in the B31G and the modified B31G (RSTRENG) criteria. While the UTS is viewed by some as more difficult to determine during the rehabilitation of older pipelines as compared to the value of SMYS, the UTS does correlate very well with the hardness of steel. Consequently, values of the UTS could be simply estimated in the ditch during rehab, on a joint-by-joint basis, using an appropriately calibrated portable-hardness tool. It follows that use of the UTS can be quite simple, even in cases where mill records of typical properties are not available.

While the UTS appears to be a simple and appropriate candidate to replace the reference stress in metal-loss criteria (e.g., S_0 in B31G), the data in Fig. 3d imply otherwise — at least for the present. Fig. 3d presents results for burst tests of end-capped sections of corroded X42 pipes made of a relatively low toughness line pipe steel. Predictions in this figure based on the UTS all lie well above the line that designates perfect predictions. Predictions in this figure based on the more usual use of a yield-based reference stress all lie well below the line that designates perfect predictions. Examination of the available records for these data leads to the speculation that fracture intervened in these failures before the pressure associated with plastic collapse and the UTS was achieved.

Fig. 3e illustrates one way that fracture could intervene causing failure before plastic collapse. The top view in Fig. 3e reflects a metal-loss defect in the surrounding full-wall thickness under a stress-free condition, which is considered as the initial condition for the subsequent discussion. To help illustrate how this initial condition evolves with increasing pressure, this figure enlarges the thickness in comparison to the length of the defect. The view to the right and below the initial condition in Fig. 3e illustrates the first step as the stress-free condition moves toward failure. This view represents the situation after the onset of local yielding, where related wall-thinning is progressing toward the limit state. The extent of this wall thinning that is associated with plastic collapse depends on the steel's mechanical properties. Very ductile steels will experience significant thinning prior to plastic collapse whereas less ductile steels will experience

5 The UTS may be viewed by some as an unacceptable choice in that it is currently not a design parameter. However, as limit-states design is accepted by the pipeline industry to the extent it has been by other industries, this perception would change. Current work in the pipeline industry suggests that this shift is plausible as well as underway (e.g., Zimmerman et al, 1996, Ellinas et al, 1989), at which point yield stress will have the status currently ascribed to the UTS.

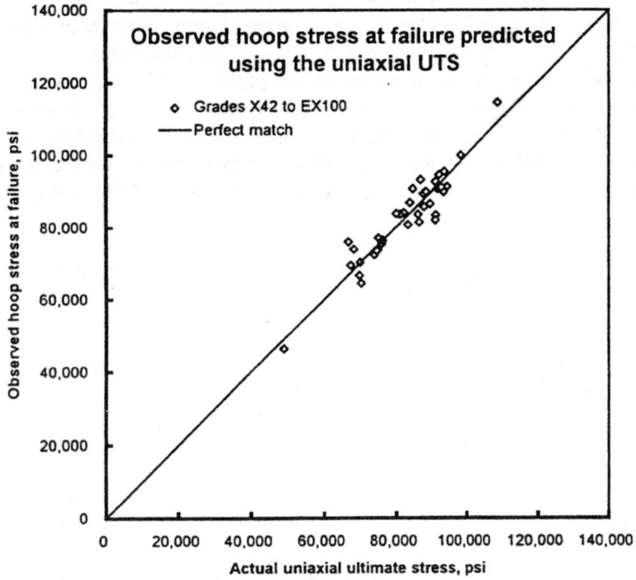

c) based on the ultimate tensile stress

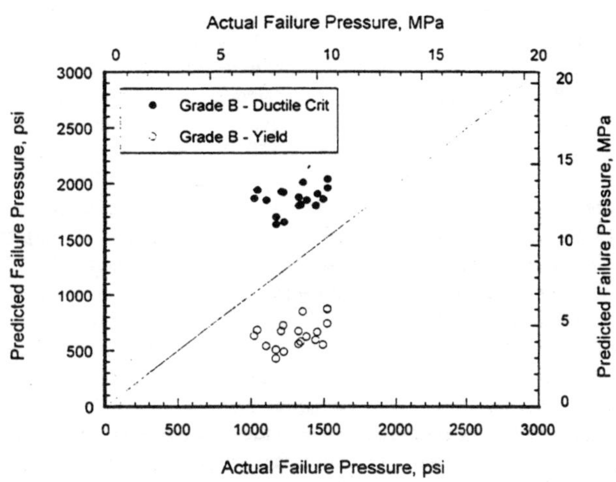

d) based on UTS in an application involving a lower toughness steel

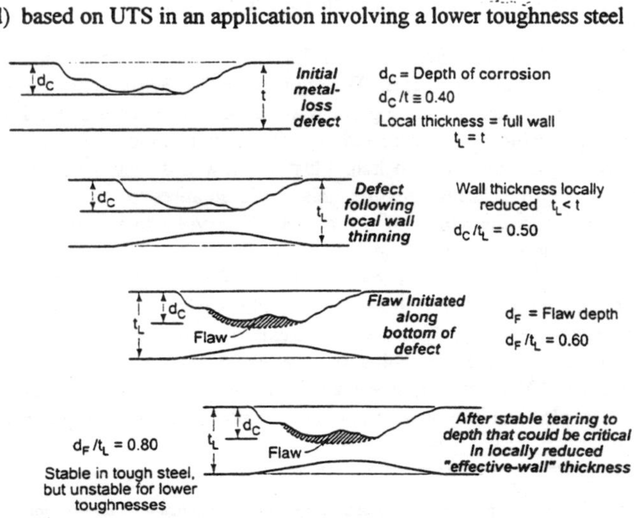

e) two failure scenarios depending on relative toughness and mechanical properties

Fig. 3 Issues related to predicted versus actual failure pressure in defect-free, end-capped pipes (continued)

proportionately less thinning. The next lower view to the right illustrates the failure process as the wall thinning progresses toward the limit state. It differs from the preceding view in that corrosion-induced micro-stress raisers that lie along the bottom of the defect have now initiated a flaw due to stable tearing, which is a void nucleation and growth process comparable to that involved with the ductile failure in a uniaxial tension test. This just initiated flaw is in this view considered to be of a sub-critical size. In this case, the stable tearing can continue along the tip of this part-through-wall crack, whose growth eventually could cause a leak or a rupture. Alternatively, depending on the toughness, a mechanism like quasi-cleavage could intervene leading to failure before the wall thinning leading prior to plastic collapse occurs. Considering that the cracking remains subcritical the wall thinning will continue, producing the situation in the bottom view in Fig. 3e. At this state a deepened crack now exists in a pipe wall that has suffered significant local thinning due to plastic flow, which is superimposed on the initial wall thinning due to the corrosion.

Failure can occur in one of two ways for the scenario played out in the sequence of views shown in Fig. 3e. Such failure could occur as the pressure in the burst test either is held at a high level or continues to increase. Which of these occurs depends on the relative ductility and toughness of the steel. For higher toughness steels, the initiation of the flaw is delayed and the progress of the tearing is slowed, which allows the wall to thin until the limit state is achieved (i.e., plastic collapse occurs). In contrast, for lower toughness steels the stable tearing initiates a flaw at relatively lower pressures. This flaw can then become fracture-critical at lower ratios of flaw depth to local-wall thickness. Alternatively, for the lower toughness steels failure could lead to unstable growth due to a quasi-cleavage mechanism. Because the defect can become critical at lower ratios of flaw depth to local-wall thickness, the amount of wall thinning that can occur before the flaw becomes critical is reduced in comparison to that needed for plastic collapse. That is, the developing flaw becomes unstable before the limit state associated with the UTS can be achieved. Which of these two possibilities controls failure depends on the steel's mechanical and fracture properties. Clearly, fracture intervening before thinning leading to plastic collapse is favored in lower toughness steels as compared to higher toughness line pipe steels.

Note that the reference failure stress for the two types of failure that arise from Fig. 3e is bounded above by the UTS for relatively tough steels and below by the AYS for lower toughness steels. The reference stress could be lower than the AYS in situations involving a quasi-cleavage mechanism. Coincidently, the predictions in Fig. 3d for the UTS and the AYS lie to either side of the line of perfect predictions, as they must for the speculative scenario played out in Fig. 3e to be viable in terms of a stable tearing fracture process. It follows from Figs. 3d and 3e that the failure criterion adopted as the reference stress in the present approach, or for criteria such as that reflected in Equation 1, probably must be more general than single-valued measures such as the UTS. It also follows that if stable tearing intervenes this reference stress will be bounded above and below by some measure of plastic collapse, like the UTS, and the AYS, respectively. In this context note that failure could occur at pressures well below that predicted by a lower bound for ductile response like the AYS in fracture-brittle materials, a topic that is beyond the scope of this paper.

Alternatives in Lieu of the UTS as the Reference Stress

Alternatives to the use of the UTS as a means to evaluate limit-state failure conditions for blunt defects have also been considered. Miller (1988) has reviewed the limit-states of test specimens and structures containing defects. Leis et al (1996) have developed an expression for the limit-state in a pipe wall as part of the upgrading of the PRC*I* failure criterion for sharp axial surface flaws, for conditions where the defects blunt prior to failing, such as can occur for shallower flaws and for very deep flaws in fracture-tough steels. This criterion, which has the form:

$$\sigma_f = (\sigma_u)^x \cdot (\sigma_y)^{1-x} , \quad (2a)$$

where σ_f, ϵ_f, σ_u, ϵ_u and σ_y and ϵ_y respectively denote the true stress and the true strain at fracture, ultimate stress, and yield stress conditions and

$$x = \log(\varepsilon_f / \varepsilon_y) \, / \, \log(\varepsilon_u / \varepsilon_y) \quad . \quad (2b)$$

The form of Equation 2 provides for values of the reference failure stress, $_f$, that range from the AYS to the UTS, depending on the nature of the material's stress-strain response beyond the AYS. That is, the failure stress from Equation 2 is not single-valued, but does include the UTS as an upper limit to the value of the reference failure stress. For the data presented in Fig. 3c, the use of Equation 2 in lieu of the UTS leads to virtually no difference in the quality of the predictions. The value of Equation 2 as a failure criterion can only be realized in situations where the toughness of the pipe steel precludes reaching the upper bound of Equation 2 (i.e., the UTS).

In light of Figs. 3a and 3c note that improvements to B31G and its current modified form may be possible through still further modification of reference stress, S_0, to a value based on the UTS. But, in light of Fig. 3d, the use of the UTS will not provide viable predictions in all cases. Figs. 3d and 3f suggest that the modified form of the reference stress S_0 needs to be somewhat more complicated than has been historically considered. And, because the calibration of the second term in Equation 1 depends implicitly on the value selected for S_0 during this calibration, it is likely that the second term would also evolve somewhat if the reference stress were changed in criteria like B31G. The next subsection explores issues related to defect size and shape, which are accounted by the second term in Equation 1.

Predictions of Failure Pressure for Blunt versus Sharp Defects

To evaluate the viability of the second term in Equation 1 the failure behavior of blunt (metal-loss) defects is compared to that for sharp (crack-like) defects on the premise that blunt flaws should fail at a pressure greater than a sharp flaw of the same length and depth. This evaluation can be made by comparing predictions from Equation 1 with experimental results from full-scale testing involving sharp flaws. But, because it is difficult to assemble such sharp flaw data covering a wide range of geometries within a given set of pipe steel properties, another approach must be used. For this reason this evaluation is made using numerical results to represent the sharp flaw situation. To facilitate comparison of the sharp and blunt flaw cases the sharp flaw results are superimposed onto a plot of failure stress versus flaw length and depth created using Equation 1 in the form of the B31G criterion and the modified B31G (equally here RSTRENG) criterion. While several combinations of values for S_0 and definitions of crack area, A, could be considered in such a comparison, only two such combinations involving the B31G and the modified B31G criteria are included herein.

Figs. 4a and 4b present results from B31G and the modified B31G criteria along with failure predictions for sharp flaws based on the PRC*I* ductile flaw growth model (Leis et al, 1991), as embedded in a software format known as the PRC*I* pipeline axial flaw failure criterion, PAFFC (Leis and Ghadiali, 1994). This pipeline axial flaw failure criterion has its roots in a nonlinear fracture mechanics (NLFM) based flaw growth model that is coupled with a net-section (plastic-collapse) criterion for the through-wall ligament[5]. The plastic-collapse criterion embedded in PAFFC is based on the form of Equation 2, which by virtue of Fig. 3c correctly predicts the behavior of blunt defects in the case where the defect depth approaches zero. A plastic-collapse criterion is needed in PAFFC to address the tendency for very shallow flaws, as well as very deep flaws in very tough materials, to blunt and thereafter fail by shear through the thinned wall. It follows that the analog to S_0 used in B31G and the modified B31G criteria that is embedded in the current (second) release of PAFFC provides a consistent and not overly conservative basis to predict the net-section collapse behavior in pipeline applications.

The problem in using results from one analysis to assess the viability of another, as was done in Figs. 4a and 4b, lies in deciding which of the two analyses is responsible for any observed disparity. To help resolve this problem the accuracy and precision of PAFFC is assessed in here in terms of Figs. 4c and 4d in cases dealing with sharp defects. In cases where the pipe is defect free, and because PAFFC shares the net-section collapse criterion outlined earlier in regard to Equation 2, the excellent predictions of failure pressure based on the use of the UTS evident in Fig. 3c are equally relevant to PAFFC. PAFFC is equally accurate and precise in dealing with sharp flaws. This is evident in the results are summarized in Fig. 4c, which deals with machined axial flaws (Leis et al, 1991), and Fig. 4d, which deals with "blind" predictions made for in-service and hydrotest failures in real pipelines (anon, 1996). These figures show very little scatter about the trend for an exact match with the burst tests (i.e., similar to Fig. 3c). It follows that PAFFC provides a viable means to estimate the failure pressure of sharp flaws and that it provides a reasonable basis for comparison with the blunt flaw trends from the B31G and the modified B31G criteria.

Comparisons of the failure behavior of a sharp flaw based on PAFFC with the blunt flaw response based on B31G and the modified B31G criteria are presented in Figs. 4a and 4b, respectively. It is evident from both of these figures that the blunt flaw predictions for shallower flaws based on the B31G and the modified B31G criteria can lie well below that for the sharper flaws, a trend that certainly is not what is expected. Cases where blunt flaws are predicted to fail at pressures less than the sharp flaws with identical depth and length are indicated by the dashed boundaries traced onto Figs. 4a and 4b. Note that in spite of the differences between B31G and the modified B31G criteria, the situations that produce these unexpected failure predictions are quite similar. The significant role of S_0 in this context is evident in the difference in the predicted failure stress for the defect-free case, that in this figure can be seen where the defect length is zero. The results compared in Figs. 4a and 4b indicate that the conservatism embedded in the use of S_0 based on yield stress or some function of it leads to a conservative error on the order of 15 to 20 percent. This same level of conservatism was evident in the earlier discussion related to S_0, in regard to the data trends discussed in regard to Fig. 3a.

---

6 Release 2 of PAFFC embeds the limit-states criterion in Equation 2 (see Leis et al,1996)

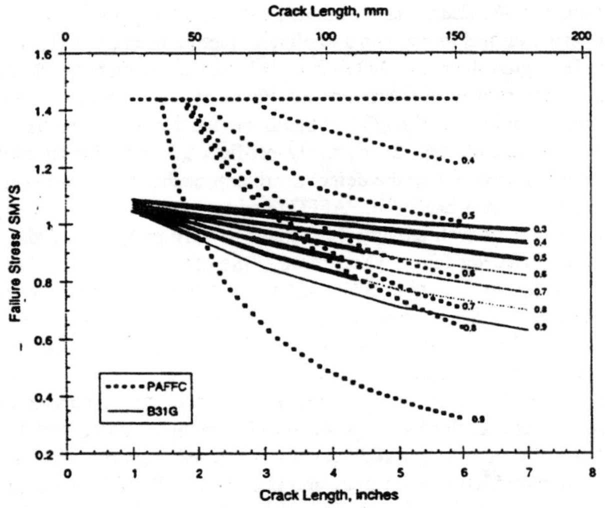

a) for blunt defects based on B31G (0.67 dL definition of flaw area)

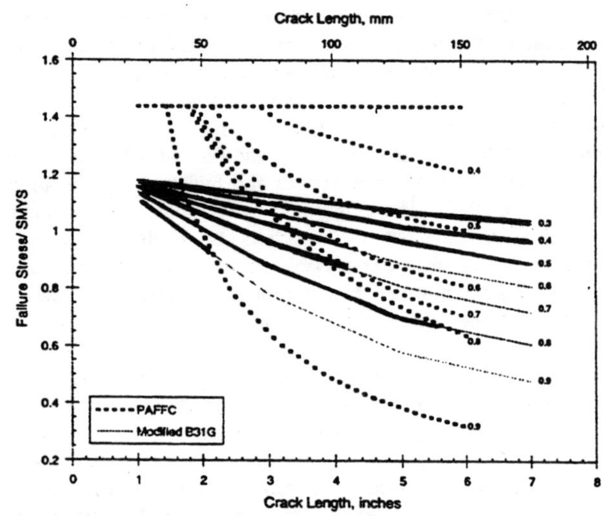

b) for blunt defects based on Modified B31G (0.85 dL definition of flaw area)

Figure 4 Predicted versus actual failure pressure in end-capped pipes with either blunt defects or sharp flaws

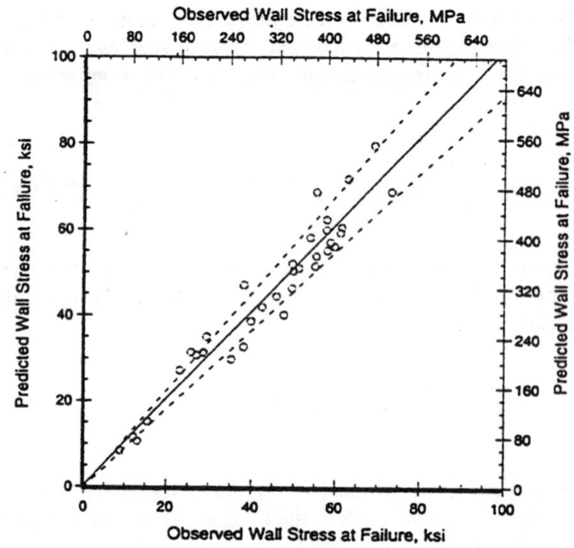

c) for machine-made axial surface flaws

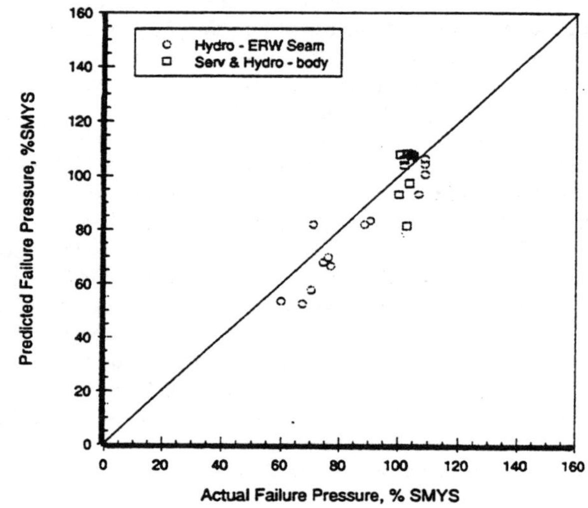

d) for naturally occurring stress-corrosion cracking

Figure 4 Predicted versus actual failure pressure in end-capped pipes with either blunt defects or sharp flaws (continued)

Consider now the role of the second term in Equation 1, which reduces the selected value of S_0 to account for defect size and shape, in cases where S_0 is used as defined in the B31G and the modified B31G criteria. Figs. 4a and 4b indicate that failure occurs at lower pressures for blunt defects as compared to sharp defects at defect depths less than about 60 percent of the wall thickness in such cases. Against the backdrop of Figs. 4c and 4d, which indicate the accuracy of PAFFC, these results suggest that the second term in Equation 1 is quite conservative for flaw depths less than ~60 percent of the wall thickness. However, this is an unfair comparison in that it includes the significant conservatism embedded in S_0 and thus overestimates the level of conservatism that can be attributed to the modeling of the defect size and shape. This bias due to S_0 can be offset simply by shifting the trends for the blunt flaws to match the trends with that due to PAFFC at zero flaw length. After this adjustment to S_0, the depth below which failure pressure for blunt defects is less than that for sharp flaws decreases to about ~40 percent of the wall thickness. While this is an improvement, it still indicates that the current form of the second term in Equation 1 embeds some significant amount of conservatism.

The just noted shift in S_0 to reduce the conservatism for shallower metal-loss defect predictions also leads to a corresponding upward shift of the blunt-defect trends for the longer, deeper defects. Thus, as alluded to earlier, a change in S_0 in Equation 1 to resolve difficulties with shallow, short-length defects will require a recalibration of the second term dealing with defect shape and size to offset this upward shift for longer, deeper defects. That is, a "fix" for the first term in Equation 1 will have a "ripple" effect on the second term. Furthermore, what "fixes" the first term to reduce the scatter and conservatism still associated with S_0 (and avoid the predicted failures for short, deep pits in one steel) may be confounded in a subsequent application to another data set. It follows that some significant difficulties can be encountered in the application of criteria based on Equation 1 to the extremes in defect geometry where the flaw size tends to zero and possibly also to a very large areal patch. For this reason, a form more general than that of Equation 1 appears to be necessary to implement an alternative approach for a metal-loss integrity criterion.

APPROACH TO DEVELOP ALTERNATIVE INTEGRITY CRITERIA

With reference to the schematic shown in Fig. 1, failure pressures associated with each of the bounds indicated there are, as suggested the basis to formulate an alternative criterion. A value near the UTS derived from Equation 2 serves in analogy to the value assigned to S_0 in B31G or its modifications in this setting, while the full wall thickness and the net wall thickness bound the form of the second term in analogy to Equation 1. The full wall thickness based upper bound relates to failures for short, deep defects, while the net (minimum) wall thickness based lower bound relates to failures in large areal defects. The approach to develop an alternative criterion involves determining the form of the functions that provide the transition between these two bounds. This is the subject of the second part of this paper, which explores the transition between the limits shown schematically in Fig. 1 as a function of defect length, depth, and shape, and thereafter comments on issues related to defect interaction.

SUMMARY AND CONCLUSIONS

Part I of this paper has examined the form of metal-loss criteria, focusing specifically on B31G and related criteria. This examination was done in the context of important trends in the database of full-scale experiments on corroded-pipe sections, which provided the basis to formulate an alternative integrity criterion for metal-loss defects. Part II of this paper continues the development of this alternative criterion based on the approach suggested here schematically in Fig. 1, in terms of the limiting response of short, deep and long, deep metal-loss defects that is consistent with the failure pressure behavior observed in field and full-scale laboratory experiments. Important conclusions reached en route to this alternative approach follow:

- Circumferentially aligned (coparallel) adjacent defects cause a local increase in the compliance of the pipe wall, which diminishes their effect as compared to an isolated, geometrically similar defect, which leads to failure pressures greater than that for the isolated, geometrically similar defect.
- Axially aligned (colinear) interacting defects create an equivalent defect length that is greater than the length of the longest of the individual interacting defects, which leads to failure pressures less than that for the longest of the individual interacting defects.
- The significant difference between the behavior of colinear and coparallel defects can serve as a good measure of the viability of a failure criterion, for blunt as well as sharp flaws.
- The stressing conditions in end-capped "vessel" tests are comparable to that of earth-anchored transmission pipelines, in spite of their being physically much different.
- Specimens free of significant axial stresses tested using "open-cylinder" conditions, which mimic the stressing conditions in some pipeline and plant piping conditions, failed at pressures less than that for the corresponding end-capped test specimens. These results point to the possibly significant influence of axial force on the validity of integrity criteria calibrated and/or validated by end-capped vessel test data in applications other than straight, earth-anchored pipelines. Compression stresses due to bending or other service loadings could alter the failure conditions in a nonconservative way as compared to that based on end-capped conditions. Until this aspect is studied further, care should be taken in applications of such results to unrestrained piping or pipelines or situations where the presence of compression stresses could render end-capped criteria somewhat nonconservative.
- The use of the actual yield stress in place of SMYS in B31G and other similar or related criteria offers little improvement in either the accuracy or the consistency of these criteria.
- The use of a multiaxial yield stress in place of the uniaxial value in B31G and other similar or related criteria offers little improvement in either the accuracy or the consistency of these criteria.
- The uniaxial ultimate tensile stress (UTS) consistently correlates data from burst tests of defect-free end-capped pipes with little or no scatter. The use of yield stress as a reference stress as in B31G and related criteria provides limited correlation for similar conditions and shows quite significant scatter. Therefore, improved predictions would result if the value of S_0 in B31G and other similar or related criteria were set at the UTS. However, in applications to steels with lower relative toughness, it is possible that this reference stress may fall below the UTS.
- Replacing the value of S_0 in B31G and other similar or related criteria offers the possibility of significant improvement in the quality and consistency of predicted failure pressures.
- An approach that offers the possibility for accurate and consistent predictions of metal-loss defects, which can be generalized to address complex loadings, complex defect shapes, sizes, and spacings, can be formulated to predict leak as well as rupture. It is developed in Part II of this paper.

REFERENCES

Anon., (1996). Battelle Archives on Line Pipe Steels — Gas Research Institute Projects

Anon., (1996). National Energy Board (of Canada) Report on Public Inquiry on Pipeline Stress-Corrosion Cracking, Proceeding MH-2-95, p. 137

Anon., (1996). Battelle Archives on Line Pipe Steels and Burst Pressure Response — Pipeline Research Committee, American Gas Association

Anon., (1984). ANSI/ASME B31G - Manual for Determining the Remaining Strength of Corroded Pipelines, ASME

British Gas (1994). "Line Pipe Corrosion Group Sponsored Project"

Chouchaoui, B (1993). "Evaluating the Remaining Strength of Corroded Pipelines", PhD Thesis, University of Waterloo, Canada

Coulson, KEW, and Worthingham, RG (1990a). "Pipe Corrosion - New Guidelines Promise More Accurate Damage Assessment", Oil &Gas J., Vol 88, No. 15

Coulson, KEW, and Worthingham, RG (1990b). "Pipe Corrosion - Standard Damage Assessment is Overly Conservative", Oil &Gas J., Vol 88, No. 16

Ellinas, CP, Walker, AC, Palmer, AC, and Howard, CR (1989). "Subsea Pipeline Cost Reductions Achieved Through the Use of Limit State and Reliability Methods", Offshore Petroleum Technology Conference, Amsterdam

Folias , ES, (1965). "An Axial Crack in a Pressurized Cylindrical Shell", I.J. of Fracture Mechanics, Vol. 1, pp 104 - 113

Fu, B, and Kirkwood, MG (1995). "Determination of Failure Pressure of Corroded Line Pipes using Nonlinear Finite-Element Method", 2nd Pipeline Technology Conference, Vol. II, pp 1 - 9

Hopkins, P, and Jones, DG (1992). "A Study of the Behavior of Long and Complex-Shaped Corrosion in Transmission Pipelines", 11th International Conference on Offshore Mechanics and Arctic Engineering, Canada, pp 211 - 217

Jones, DG, and Ritchie, D (1992). "Failure Behavior of Internally Corroded Line Pipe" 11th International Conference on Offshore Mechanics and Arctic Engineering, Canada

Kanninen, MF et. al. (1993). "Assessing the Reliability of Corroded Oil Transmission Pipelines under the Combined Loading Conditions Arising in Arctic Service", 12th Offshore Mechanics and Arctic Engineering Conference, Glasgow

Keen, J. (1990). "Corrosion Forces Repairs to Oil Pipelines", USA Today, February 5th issue

Kiefner, JF, Vieth, PH, and Roytman, I (1995). "Continuing validation of RSTRENG", Pipeline Research Supervisory Committee, American Gas Association

Kiefner, JF, and Vieth, PH (1989). "A Modified Criterion for Evaluating the Remaining Strength of Corroded Pipe", Pipeline Research Supervisory Committee, American Gas Association, Catalog No. L51609, see also Vieth, PH, and Kiefner, JF (1993). "RSTRENG2 Users Manual", Pipeline Research Supervisory Committee, American Gas Association

Leis, BN, Brust, FW, and Scott, PM (1991) "Development and Validation of a Ductile Flaw Growth Analysis for Gas Transmission Line Pipe", Pipeline Research Supervisory Committee, American Gas Association, Catalog No. L51543

Leis, BN, and Ghadiali, ND, (1994). "Pipe Axial Flaw Failure Criteria — PAFFC, Release 1.0 User's Manual and Software", Pipeline Research Supervisory Committee, American Gas Association Project, Catalog No. L51680

Leis, BN, Rudland, DL, and Eiber, RJ (1996). "Analysis of the Benefits of Hydrotesting in terms of Strength and Tightness Requirements for Onshore and Offshore Pipelines", Onshore, Offshore Design Applications Supervisory Committee, American Gas Association Project PR-3-9523

Leis, BN, and Colwell, JA (1996) "Initiation of Stress-Corrosion Cracking on Gas-Transmission Piping", ASTM STP 1298 Effects of Environment on the Initiation of Crack Growth, in press

Maxey, WA, Kiefner, JF, Eiber, RJ, and Duffy, AR (1972). "Ductile Fracture Initiation, Propagation, and Arrest in Cylindrical Vessels", ASTM STP 514 - Fracture Toughness, pp 70 - 81.

Miller, AG (1988). "Review of Limit Loads of Structures Containing Defects", I. J. Pressure Vessels & Piping, Vol 32, pp 197 - 327

Mok, DHB, Pick, RJ, Glover, AG (1990). "Behavior of Line Pipe with Long External Corrosion", *Materials Performance*, Vol. 29, No. 5, pp 75 - 79

Shannon, RWE (1974). "The Failure Behavior of Line Pipe Defects", I. J. Pressure Vessels & Piping, Vol. 2, pp 243 - 255

Smith, MQ, and Grigory, SC (1996). "New Procedures for the Residual Strength Assessment of Corroded Pipe Subjected to Combined Loads", International Pipeline Conference, Calgary, pp 387 - 400

Sowerby, T. (1991). "In-Line Inspection Prompts Forties Line Replacement", Oil &Gas J., June issue, pp 27 - 40

Stephens, DR, Bubenik, TA, and Francini, RB (1995). "Residual Strength of Pipeline Corrosion Defects Under Combined Pressure and Axial Loads", Pipeline Research Committee, American Gas Association, NG-18 Report No 216

Stephens, DR, and Leis, BN (1996). "Development and Validation of Failure Criteria for Residual Strength of Pipeline Corrosion Defects", Pipeline Research Supervisory Committee, American Gas Association Project PR-3-9509

Stewart, G, Klever, F, and Ritchie, D (1995). "An Analytical Model to Predict the Burst Capacity of Pipelines", 13th Offshore Mechanics and Arctic Engineering Conference, Houston

Vieth, PH, and Kiefner, JF (1994). "Database of Corroded Pipe Tests", Pipeline Research Supervisory Committee, American Gas Association

Wang, YS (1991). "An Elastic Limit Criterion for the Remaining Strength of Corroded Pipe", 10th International Conference on Offshore Mechanics and Arctic Engineering, Norway, p 179

Zimmerman, TJE, Chen, Q, and Pandey, MD (1996). "Target Reliability Levels for Pipeline Limit States Design", International Pipeline Conference, Calgary, pp 111 - 120

Proceedings of the Seventh (1997) International Offshore and Polar Engineering Conference
Honolulu, USA, May 25-30, 1997

ISBN 1-880653-28-1 (Set); ISBN 1-880653-32-X (Vol. IV)

An Alternative Approach to Assess the Integrity of Corroded Line Pipe - Part II: Alternative Criterion

B.N. Leis and D.R. Stephens
Battelle
Columbus, OH, USA

ABSTRACT

Part II of this paper builds on the review of the experimental database and the analysis of the related trends that defined the approach to develop an alternative criterion in Part I based on an upper limit on failure pressure for very small defects and a lower limit that reflects large areal defects. Results are developed for specific combinations of defect size and profile through parametric analyses. These results are used to: (1) compare this alternative approach with other criteria for metal-loss defects, and (2) derive the mathematical form of the alternative criterion. Results from these analyses are used to illustrate the effects of defect profile, size, and spacing, to define the transition between the two noted limits on failure pressure. This parametric analysis has been done using a special purpose numerical finite element model in conjunction with the failure criterion outlined in Part I. Trends in these data are used to identify the preliminary form of the function needed to modify the reference stress for defect-free pipe to account for the size, shape, and spacing of the defects. The results of this parametric analysis are cast into a mathematical format that is then used to predict the failure pressure response of the defect set that underlies the development of the ASME B31G criterion. These same data are used to infer defect sizes where the upper limit on failure pressure can be expected as well as defect sizes where the lower limit involving large areal defects will control failure.

Key Words: Corrosion, defect, failure, criterion, B31G, RSTRENG, alternative criterion, transmission pipeline

INTRODUCTION

Part I of this paper indicated the financial drivers and also identified the practical technical drivers that motivate this development of an alternative approach to assess the integrity of metal-loss defects in pipelines and piping systems. This second part builds on the earlier review of the experimental database and the analysis of the related trends. Results from the application of this alternative approach are compared with that from other predictive schemes for metal-loss defects as a first check of the viability of this approach. The premise is that, unless this comparison that focuses on a specific isolated defect geometry leads to results similar to those provided by other criteria for a case where they have enjoyed historical success, it is unlikely that the alternative approach will circumvent the problems with current criteria that were identified in Part I.

Given success in this first preliminary check, the alternative approach is then evaluated by its application to a broader set of defects, with current criteria again used as a benchmark. This second evaluation is made in view of the significant time and cost needed to develop even a limited parametric analysis to establish the functional transition between the two limiting failure pressures identified in Part I. As will become evident, these preliminary evaluations indicated the viability of the alternative approach evolved in Part I, so that trends that underlie the transition between the two limits of failure pressure are then developed and discussed. Trends in these data are used to identify the functional form needed to modify the reference stress for defect-free pipe to account for the effects of defect size, profile, and spacing. The result is presented as a mathematical expression that is then used to predict the failure pressure for much of the data that underlie the current empirically calibrated expressions, as for example ASME B31G (anon, 1984) criterion and the PRCI RSTRENG criterion (Kiefner et al, 1989).

DEVELOPING THE ALTERNATIVE CRITERION

A significant effort involving the parametric analysis of a broad range of defect profiles, sizes, and shapes is needed to generate the transition functions between the two limiting pressures identified in Part I. Because of the still uncertain success of this effort we began by comparing the trends of some current approaches to calculate failure pressure with the two bounds associated with the alternative criterion. This comparison is presented in Fig. 1 on coordinates of failure pressure, normalized by the failure pressure of the pipe remote to the defect, and defect length, L, normalized as $L / (R \cdot t)^{0.5}$, where R is the radius of the pipe and t is its wall thickness.

Preliminary Evaluation — Isolated Uniform Metal-Loss

Fig. 1 addresses failure predictions for uniform metal-loss defects with a depth equal to 50 percent of the wall thickness. By definition, the upper limit of the value of failure pressure in Fig. 1 is unity, which signifies the failure pressure of the full wall thickness, which is equal to the ultimate tensile (limit-state) stress. This upper limit lies across the upper edge of Fig. 1. As indicated in Part I, this upper bound relates to failures for short, deep defects. The lower limit for this alternative criterion is shown as the heavy, horizontal dashed line in Fig. 1. Part I indicates that this value is set by failures occurring in the net (minimum) wall thickness in the midst of a large areal defect. Because the example in Fig. 1 considers a defect whose depth is half of the wall thickness, this lower bound lies at a value of 0.5 on the y axis.

Fig. 1 also shows the predicted failure boundaries for each of three current criteria for metal-loss defects. Comparison of the predictions for these criteria with the upper limit shows the same degree of conservatism for short defects discussed in Part I of this paper in regard to ASME B31G and related criteria. This conservatism is evident in the difference between a value of one (i.e., a failure at the ultimate tensile stress (UTS) in the full-wall thickness) on the y-axis and the points where these three other criteria intersect this axis. Inspection of Fig. 1 shows that this difference is between 15 and 20 percent, which is of the same order discussed in regard to Fig. 3 in Part 1. Predictions for longer defects also show conservatism in comparison to the lower limit (the heavy dashed line). The conservatism for longer defects evident in Fig. 1 is less in comparison to that for the shorter length defects. Because of the step that occurs in the B31G criterion as the defect length increases, it is the most conservative of these criteria for long defects. The trend for RSTRENG matches the lower limit proposed in Part I almost exactly for longer defects, whereas the other two criteria are more conservative.

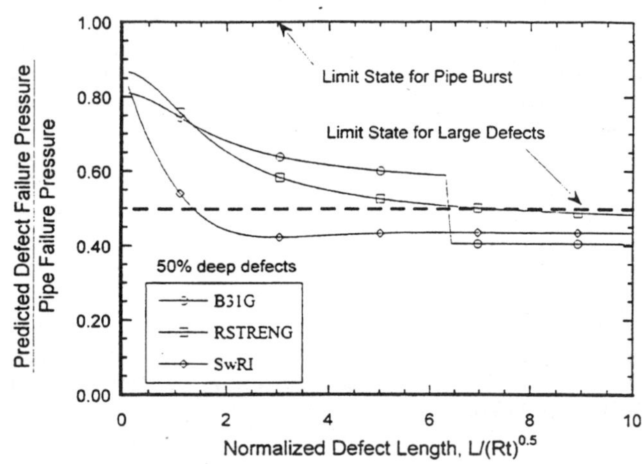

Fig. 1 Failure pressure as a function of defect length for uniform metal-loss to 50 percent of the wall thickness.

Fig. 1 confirms that shifting the value of S_0 to offset the conservatism evident for shorter, deeper defects (as discussed in Part I in regard to Figs. 4c and 4d) will lead to changes in the accuracy and precision of the predicted failure pressure for longer defects. In view of Fig. 1, such a shift for RSTRENG (equally here the Modified B31G criterion) to reduce the conservatism for short, deep defects will lead to nonconservative predictions for the longer defects — unless the second term in Equation 1 in Part I is altered. In contrast, Fig. 1 indicates that such a shift in S_0 will lead to more accurate predictions for all crack lengths for the SWRI criterion and for longer lengths only for the B31G criterion — at least for this specific defect depth. In the absence of difficulties in dealing with steels with lower toughness in terms of the UTS, as indicated in Part I, such a shift would be an attractive potential improvement to B31G or other such criteria.

Given that the trends in Fig. 1 do not indicate significant shortcomings in the limits adopted as the basis for the alternative approach, a more detailed evaluation was made. That evaluation made use of case-specific simulations for a range of isolated defect geometries using finite element analysis.

Application of the Alternative Approach to a Broader Defect Set

Broader evaluation of the alternative approach involves comparisons of the bounds postulated in Part I with results from predictions based on numerical analysis of the stresses and strains determined in the vicinity of the defect. Stresses and strains were calculated using a special purpose shell-based finite element code known as PCORR. Initially developed as a linear-elastic small-strain model (e.g., Stephens et al, 1995), in its current form PCORR has evolved significantly to include material nonlinearity, but the formulation remains geometrically linear (e.g., Stephens and Leis, 1997). Stresses and strains calculated in the ligament that either surrounds the metal-loss defect or defines the minimum wall thickness are compared with a failure criterion based on plastic collapse, as a function of increasing pressure, to determine the pressure at which failure is indicated. The limit state criterion outlined in Equation 2 in Part I of this paper is used to define failure. This criterion embeds material properties involving the yield stress, the tensile ultimate stress, and the fracture toughness, and for tough steels indicates a limit state at a stress that is comparable to the UTS (Leis et al, 1996). For fracture brittle steels, this failure criterion includes the possibility of failure at stresses even less than the yield stress.

PCORR coupled with the just noted limit state has been used to simulate the failure behavior of a large number of the metal-loss defects found in the corroded-pipe test literature. Its use here is to better determine the potential of this alternative approach in its most rudimentary form, without resorting to the use of complex descriptions of or modifications to the actual flaw shape and size. Defects included from the published literature (Veith and Kiefner, 1994, Chouchaoui, 1993, Kiefner et al, 1995) were those considered as "isolated" and obtained in testing free of prior test-pressure cycles, for pressure-only loading. A balanced mix of leaks and ruptures was sought. In this regard it is emphasized that whether a defect leaks or ruptures is determined by its length when it breaks through the wall — whereas the limit state that triggers initiation does not. Therefore, correctly predicting the failure pressure of both leaks and breaks, without bias for one or the other, will be used as one basis to judge the viability of this (or any other) alternative criterion. The last requirement imposed in selecting the data set was that the failure was due clearly to plastic collapse (shear instability through the wall, as opposed to fracture through the wall on a plane normal to the pipe wall).

Test data for the purpose of evaluating the alternative criteria were obtained from Veith and Kiefner, 1994, Chouchaoui, 1993, Kiefner et al, 1995, subject to the above-noted restrictions. All predictions were made assuming an elliptic defect profile. In all cases the maximum defect length and depth based on that reported in the original reference were used. The viability of the alternative criterion was evaluated in terms of both the accuracy and the consistency of the predicted failure pressure.

Fig. 2a presents the predicted failure pressures, which cover results for Grades from 5L X46 and above. The standard deviation for these predictions is only 13 percent. As noted above, these predictions are based on the raw defect sizes in conjunction with an elliptical profile. That is, no adjustments were made to the maximum length and depth of the defect, nor were any adjustments made in regard to their local depth profile. Yet, the predicted failure pressure lies within a scatter band that is uniform about the line denoting perfect predictions. The data for leaks are distributed among the data for breaks, without any bias beyond that evident for the breaks. Fig. 2b shows the RSTRENG predictions for the corresponding data analyzed in Fig. 2a. In contrast to the results in Fig. 2a, these predictions do not show uniform scatter about the line for perfect agreement. Moreover, they show different tendencies for failures that were leaks as compared to breaks. The standard deviation for these RSTRENG predictions, which included "fitting adjustments" to the defect size and shape, is 15 percent. Thus, the present approach provided better predictions and did so without resorting to a detailed defect profile or adjustments to the defect dimensions.

THE TRANSITION IN FAILURE PRESSURE BETWEEN THE TWO LIMITING DEFECT TYPES

Given the apparent success evident in Fig. 2, there is merit in further developing the alternative criterion, which eventually will require developing a functional transition between the two limits shown in Fig. 1. This alternative criterion could be implemented in terms of a numerical model that embeds PCORR and the limit-states criterion, as was just illustrated. However, because criteria that can be stated as a simple mathematical expression often are favored over detailed numerical methods, this alternative criterion will be implemented in this paper as an engineering model whose format is comparable to that of Equation 1 in Part I of this paper. Such an engineering model requires functions that characterize the effect of defect profile, size, and spacing on the failure pressure. To this end PCORR has been coupled with the limit-states failure criterion and used in a parametric analysis for a range of defect sizes and profiles to define the transition between the limits of failure pressure indicated in Fig. 1.

This section presents the results of the parametric analysis and discusses the dependence of the transition between the upper limit on failure pressure involving short defects and the lower limit involving large areal defects. These results reflect only a part of the scope of calculations needed to develop the functional transition between these limits for more general situations, such as interacting defects. This functional transition provides the basis for an engineering model that could be used for making integrity decisions in many applications, without resorting to case-by-case analysis using numerical schemes like PCORR. These results also are helpful in identifying the parameters controlling failure at metal-loss defects and in determining if a numerical analysis like PCORR adequately portrays the underlying mechanics. Of course, because such engineering models are most useful and accurate in applications to simpler defect geometries, the use of PCORR and the limit-state criterion is suggested in applications where a complete description of the complicated defect geometry is available.

The Parameters: the Pipeline, the Defect Sizes and Profiles, and the Steels

This parametric analysis reflects the behavior of a 24-inch (610 mm) diameter by 0.281-inch (7.1 mm) wall thickness line pipe, which unless indicated otherwise is made of an X52 steel. The results for this geometry and pipe steel are generalized in terms of the normalizing scheme introduced in Fig. 1. Results are presented for uniform metal-loss depths of 20, 50, and 80 percent of the wall thickness. The influence of changes in the strength of the line pipe steel are presented in terms of selected results for an X65 and an X70 steel. The yield strengths of the steels considered are 59.6, 70.1, and 76.1 ksi (411, 483, 524 Mpa) whereas their ultimate strengths are 73.6, 90.6, and 87.2 ksi (507, 624, 601 Mpa), for the X52, X65, and X70, respectively. Note that the UTS of the X65 steel is slightly higher than that of the X70 steel. Further details on the properties for these steels can be found in Leis et al (1991). For the sake of simplicity here, the limit-state for these analyses was taken as the UTS. The results of this parametric analysis are presented in Figs. 3a through 3d.

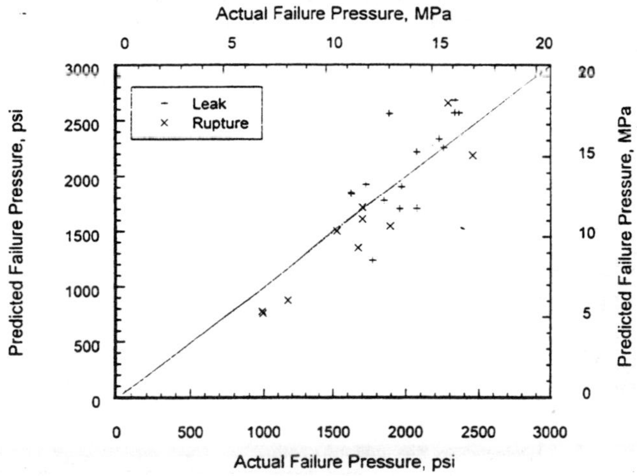

a) alternative approach, using maximum defect length and depth

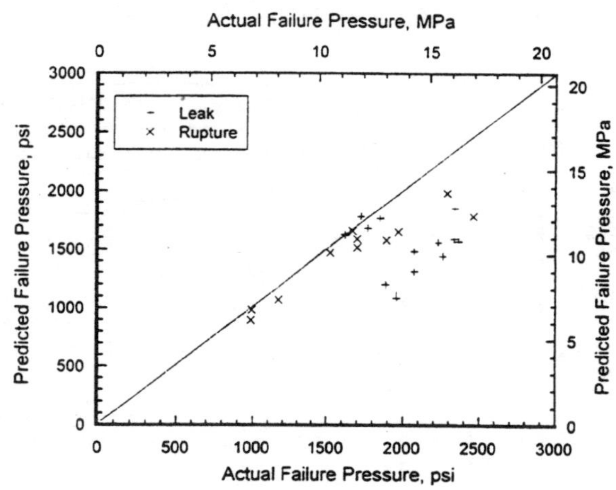

b) RSTRENG, using "adjusted" defect size

Fig. 2 Predicted failure pressure versus that observed for some full-scale end-capped tests

Effect of Defect Depth

Fig. 3a shows the predicted failure pressure for square, uniform depth metal-loss defects with depths of 20, 50, and 80 percent of the wall thickness, covering a range of defect lengths. The results show that for short, shallow defects the predicted failure pressure is approaching the upper bound value of unity, which is consistent with the limit represented for the alternative criterion as evident in Fig.1. For these short, shallow defects there is very little reduction in failure pressure, particularly for cases where $L / (r \cdot t)^{0.5} \leq \sim 0.5$. Even quite deep but still short defects show this tendency for little reduction in failure pressure, which is observed in end-capped tests (particularly in cases where the failure occurs in the pipe body remote to short, deep areas of corrosion). Beyond $L / (r \cdot t)^{0.5} = \sim 4$, the significance of defect length begins to diminish and a plateau of failure pressure is approached, after which the failure pressure becomes rather independent of defect length. Fig. 3a shows that this occurs for $L / (r \cdot t)^{0.5} > \sim 6$, although this value appears to depend on the defect depth, being largest for shallower defects.

Thus, based on Fig. 3a, where $L / (r \cdot t)^{0.5} \geq \sim 6$ pressure-loaded uniform-depth metal-loss defects behave as large areal defects. In such cases the failure pressure is proportional to the defect depth referenced

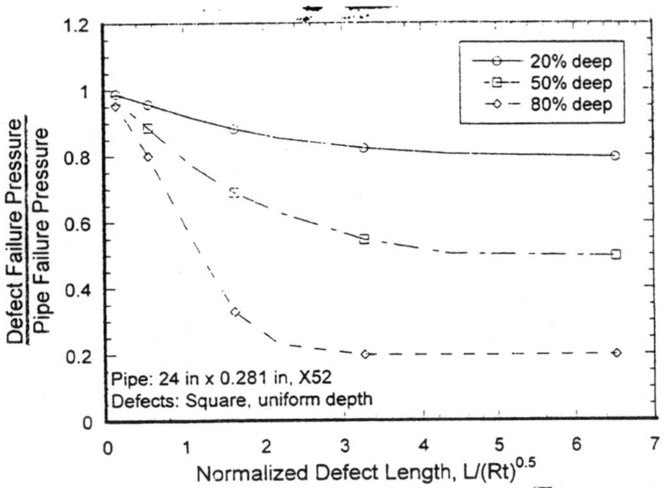

a) effect of defect depth for defects with uniform depth

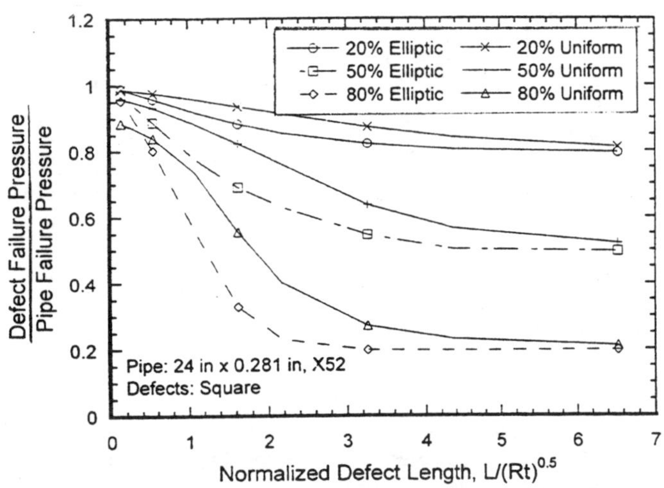

b) effect of defect shape for various defect depths

Fig. 3 Trends in predicted failure pressure as a function of defect shape and size for pressurized pipelines

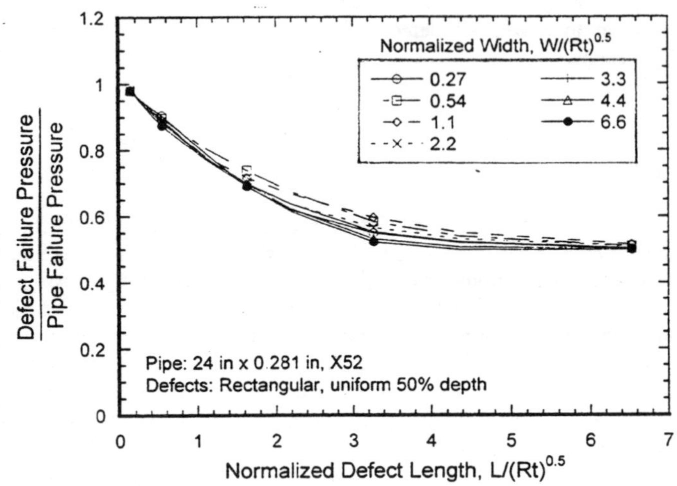

c) effect of defect length and width

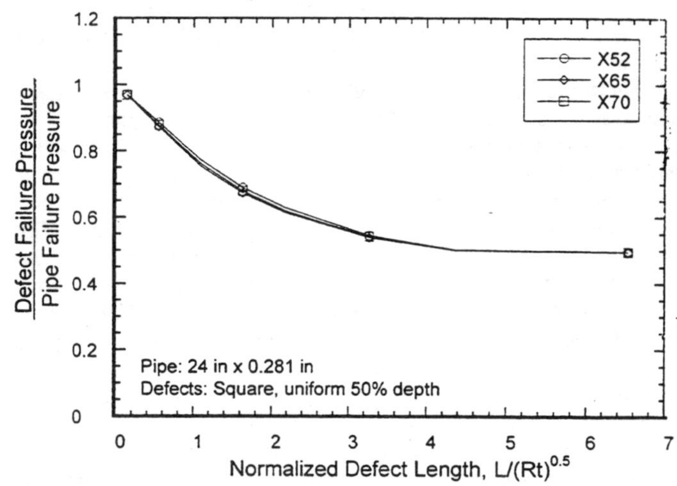

d) effect of material grade (strength)

to the UTS, as asserted in formulating the bounds shown in Fig. 1. Comparison of the failure pressures predicted numerically with that corresponding to the value referenced to the net wall thickness differ by less than one-half percent for the results in Fig. 3a. These results indicate clearly that the lower bound failure pressure is quite accurately set in terms of the net wall thickness and the UTS, as asserted in regard to Fig. 1. Because the lower bound failure pressure is observed in Fig. 3a to be proportional to the defect depth referenced to the UTS for large areal defects, the defect depth must be considered a primary geometry variable controlling failure pressure. The defect length also must be considered a primary variable for shorter defects. This follows from the data in Fig. 3a that show doubling the defect length can decrease the failure pressure by about a factor of three. These numerically predicted trends are consistent with experience in testing end-capped pipes, but they run contrary to other models that base failure on the yield stress, particularly in dealing with shorter defect lengths.

Effect of Defect Shape

Fig. 3b presents results for square, uniform metal-loss defects along with the corresponding results for defects whose profile is elliptical in both the axial and circumferential direction. Because there is more wall thickness remaining below the elliptical profile, this defect is "geometrically stiffer" than the uniform wall defect. For this reason it takes a higher pressure to develop the limit state in elliptic profile defects as compared to cases of uniform metal-loss. The trends in failure pressure for the stiffer elliptic profile defect thus lie somewhat above those for the uniform-depth defect. In the case of equilibrium-dominated large areal defects, the added stiffness limits the amount of bulging that can take place, resulting in lower stresses at a given pressure and higher failure pressures for a given maximum length, width, and depth. Fig. 3b indicates that $L / (r \cdot t)^{0.5} \geq \sim 6$ before the elliptic profile defects take on the behavior of large areal defects —

particularly for the shallower defects. In contrast, defects with a uniform depth profile show the behavior of large areal defects beyond $L / (r \cdot t)^{0.5} \geq \sim 3$.

The results in Fig. 3b suggest that shape is more important for deeper defects. These results for this pipe geometry and material indicate that the failure pressures for elliptical and uniform depth defects can differ as much as 100, 200, and 300 psi, respectively, for 20, 50, and 80 percent deep defects. These differences represent changes on the order of 5, 11, and 17 percent of the nominal pipe burst pressure, which are significant as compared to defect width, particularly for deep defects. This becomes more evident in comparing these results with those in the next subsection.

Effect of Defect Width

Fig. 3c presents the failure pressure predicted for a range of rectangular metal-loss defects with a depth equal to 50 percent deep of the wall thickness. The defect sizes considered span defect widths, W, from ~0.25 through ~6.5. Each line in this figure represents a different, constant defect width. Observe in comparing the trends in Figs. 3a, and 3b with those in 3c that, for pressure-only loading, the effect of significant changes in normalized defect width on the failure pressure is quite small. The overall behavior here is substantially the same as that for elastic results, with plasticity effects leading to some overlap of trends due to the modest effects of yielding and redistribution of stresses.

The numerically generated trends in Fig. 3c, as compared to that in Figs. 3a and 3b, clearly show defect length and depth as primary variables relative to defect width *for pressure-only loading*. This tendency is consistent with both experience and other empirical models for corrosion defects. For conditions of pressure-only loading this limited sensitivity indicates that defect width must be considered as a secondary variable in that varying the width by a factor of ~24 alters the failure pressure by 100 psi or less—which is less than 5 percent of the pipe burst pressure. In contrast to these results involving only internal pressure, results involving pressure combined with axial loading clearly show that width can be a primary variable (Stephens et al, 1995, Stephens and Leis, 1997, Chouchaoui, 1993).

Effect of Steel Strength (Grade)

Fig. 3d shows the results of a nonlinear plasticity analysis for square, uniform, 50 percent deep defects as the basis to evaluate the influence of pipe strength (grade). The yield strengths of the steels reported as X52, X65, and X70 are 59.6, 70.1, and 76.1 ksi, respectively, whereas their ultimate stresses are 73.6, 90.6, and 87.2 ksi, respectively. Recall that, as noted earlier, the ultimate strength of the X65 steel is higher than that of the X70 steel. In spite of this range of grades and the break in their pattern of ultimate strengths with grade, the numerically predicted failure behaviors are almost exactly coincident in the normalized format used for Fig. 3. However, were these results presented in terms of failure pressure, they would show a banding of the failure pressures in proportion to their respective UTS values. Because the UTS of a given grade can be above or below that of an adjacent grade, these results indicate that higher strength by grade will not necessarily afford an improved failure pressure for metal-loss defects. This tendency is comparable to that for sharp defects where, because toughness does not scale with grade, the resistance to crack-like defects also does not scale with grade.

FUNCTIONAL FORM FOR DEFECT SHAPE AND SIZE

Consider first the bounding values for this function that follow from Fig. 1. For short defects that are nearly as resistant to failure as the surrounding pipe wall — including the behavior of quite deep defects, particularly for lengths where $L / (r \cdot t)^{0.5} < \sim 0.5$, the function form of the alternative criterion must approach the value of the limit state given in Equation 2 in Part I of this paper. In terms of the failure pressure this upper limit, P_{FUL}, has the form:

$$P_{F_{LL}} = UTS \left(\frac{2t}{D} \right) \quad . \qquad \text{(Eq. 1)}$$

In this equation, D is the pipe diameter, t is the wall thickness.

Likewise, there is a lower limit in failure pressure, P_{FLL}, that reflects the behavior of the large areal patch where the failure pressure is proportional to the net-section wall thickness, t_N, which gives:

$$P_{F_{LL}} = UTS \left(\frac{2t}{D} \right) \left(1 - \left(\frac{d}{t} \right) \right) \quad . \qquad \text{(Eq. 2)}$$

In this equation, d is the depth of the defect and the rest of the symbols are as above. This second limit exists beyond large values of $L/(r \cdot t)^{0.5}$ on the order of ~6. Once this length is reached, the effect of the defect depth is proportional to the area loss for uniform metal-loss defects, with bulging developing over this full length. The functional dependence between these limits depends on the defect geometry and the steel's mechanical properties, and more generally also on the steel's fracture toughness.

For gradually changing profiles, such as the idealized elliptic situation considered herein, a term probably will be required to reflect their increased stiffness. While not addressed herein, the rate of strain hardening of the steel could also be important as it significantly influences the spread of plasticity due to stress redistribution that occurs as yielded areas begin to shed load into the surrounding pipe wall. It follows that the term dealing with the defect profile will also likely include provision for this effect — which can be particularly significant for the gradually deepening as well as irregularly shaped defects.

At present the alternative criterion for the failure pressure, P_F has the form of Equation 2, with a term included that reflects the effects of defect length as a function of the pipe's cross-section geometry, viz:

$$P_{F_{LL}} = UTS \left(\frac{2t}{D} \right) \left(1 - \left(\frac{d}{t} \right) f(\text{geometry}) \right) \quad . \qquad \text{(Eq. 3a)}$$

Based on the present results, the transition between these limits has the form:

$$f(\text{geometry}) = 1 - \exp C \left(\frac{L}{(R\, t_N)^{0.5}} \right) \quad . \qquad \text{(Eq. 3b)}$$

In this equation C is a constant whose value has been found by fitting the trends in Figs. 3a - 3d, subject to the bounds given in Equations 1 and 2. The value of C so determined is about is about -0.16, but is subject to evolve as additional numerical results are included.

This alternative criterion is termed "PCORRC", which is short for pipe corrosion failure criterion. The reference to PCORR signifies the roots of PCORRC in the numerical analysis of PCORR, which was introduced earlier as the genesis of Figs. 3a - 3d. The C signifies the use of PCORR as part of a failure criterion that is tied to the limit-state expression in Equation 2, which was introduced in Part I of this paper. It is emphasized that as written, PCORRC reflects failure by plastic collapse. If, for example, cracking due to either localized tearing or some brittle mechanism controlled failure as opposed to a shear instability, the use of PCORRC would lead to failure pressures that are nonconservative. But this is to be expected as a consequence of using a failure criterion in an application where the active failure mechanism is not built into the criterion.

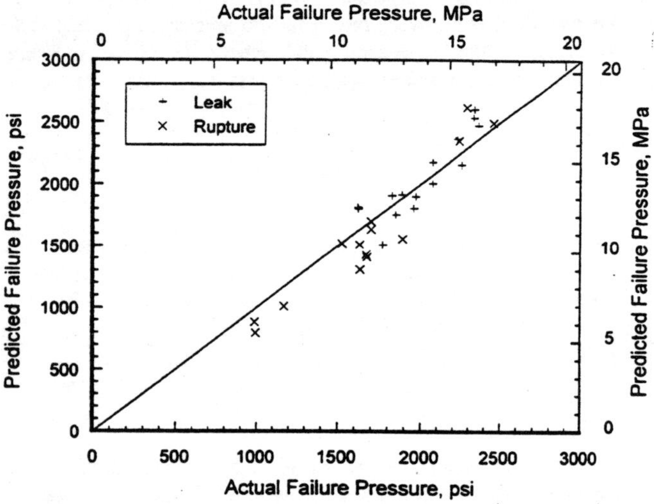

Fig. 4 Comparison of predicted failure pressure with the actual value

Equations 3a and 3b have been used to predict the failure pressure for a range of the data that underlies the empirical calibration of ASME B31G and RSTRENG and other data (Veith and Kiefner, 1994, Chouchaoui, 1993, Kiefner et al, 1995). In making these predictions the actual values of the mechanical properties have been used. Fig. 4 shows the results for these predictions. It is apparent that PCORRC provides outstanding predictions. The mean value of the ratio of predicted to actual failure pressure was 0.97, with a standard deviation of 0.105. In comparison, the means for the identical data set for ASME B31G and RSTRENG criteria were 0.80 and 0.81, respectively, with standard deviations of 0.145 and 0.142, respectively. It follows from this direct comparison that PCORRC provides a reduced factor of safety and more precise predictions. By virtue of this enhanced precision, engineers can make decisions that minimize rehabilitation and other such actions without jeopardizing safety or serviceability.

SUMMARY AND CONCLUSIONS

This second part of this paper developed the approach derived in Part I from the review of the trends underlying the experimental database, which identified two limits in the response of metal-loss defects — an upper limit in failure pressure involving short defects and a lower limit involving large areal defects. The key to implementing this alternative approach was developing functions of defect profile, size, and spacing that provide the transition between these upper and lower limits. Following verification of the alternative approach by its successful prediction of a range of defects, the functional dependence of failure pressure was demonstrated by its accurate and precise prediction of failure pressure for full-scale tests covering a range of defect and pipe geometries.

Because these preliminary studies and the form of these functions seem consistent with the available data for corroded-pipe tests, this work is expected to continue. Important conclusions reached in regard to developing this alternative approach follow:

- The alternative approach correctly predicted the failure behavior of the defect-free end-capped pipe tests in Part I, indicating that this alternative uses an appropriate reference stress that would be reduced to account for the size and shape of the metal-loss defect.
- The alternative approach ranks defect depth and length as first-order factors controlling the failure behavior of metal-loss defects — defect width is second order, at least for pressure-only loading.
- The alternative approach indicates that short defects are nearly as resistant to failure as the surrounding pipe wall — including the behavior of quite deep defects, particularly for defect lengths where $L / (r \cdot t)^{0.5} \leq \sim 0.5$. Even quite deep but still short defects show this tendency, which is observed in end-capped tests — particularly in cases where the failure occurs in the pipe body remote to short, deep areas of corrosion. This tendency is consistent with end-capped pipe testing, but generally not predicted by current models for metal-loss defects. Beyond $L / (r \cdot t)^{0.5} = \sim 6$, the significance of defect length begins to diminish and a plateau of failure pressure is reached thereafter that is independent of defect length.
- The alternative approach correctly predicts the tendency for large areal defects to fail at a pressure independent of size. Defect lengths must exceed $L / (r \cdot t)^{0.5} \geq \sim 3$ before the characteristics of a large areal defect develop for deeper defects, whereas $L / (r \cdot t)^{0.5} \geq \sim 6$ before this occurs for the shallower defects.
- The trends indicate that the functional form that accounts for the effects of defect shape, size, and spacing may be similar to that in several of the current models for large areal geometries as defined herein in terms of $L / (r \cdot t)^{0.5}$, but might differ appreciably at values of $L / (r \cdot t)^{0.5} \leq \sim 6$.
- The alternative failure criterion developed as a transition function of geometry between the upper and lower limits of the failure pressure, termed PCORRC, showed much improved accuracy and better precision than currently used criteria like ASME B31G or RSTRENG. Therefore, work continues to enhance its scope of applicability.

REFERENCES

Anon., (1984). ANSI/ASME B31G - Manual for Determining the Remaining Strength of Corroded Pipelines, ASME

Chouchaoui, B (1993). "Evaluating the Remaining Strength of Corroded Pipelines", PhD Thesis, University of Waterloo, Canada

Kiefner, JF, and Vieth, PH (1989). "A Modified Criterion for Evaluating the Remaining Strength of Corroded Pipe", Pipeline Research Supervisory Committee, American Gas Association, Catalog No. L51609, see also Vieth, PH, and Kiefner, JF (1993). "RSTRENG2 Users Manual", Pipeline Research Supervisory Committee, American Gas Association

Kiefner, JF, Vieth, PH, and Roytman, I (1995). "Continuing validation of RSTRENG", Pipeline Research Supervisory Committee, American Gas Association

Leis, BN, Brust, FW, and Scott, PM. (1991) "Development and Validation of a Ductile Flaw Growth Analysis for Gas Transmission Line Pipe", Pipeline Research Supervisory Committee, American Gas Association, Catalog No. L51543

Leis, BN, Rudland, DL, and Eiber, RJ (1996). Appendix 2 in "Analysis of the Benefits of Hydrotesting in terms of Strength and Tightness Requirements for Onshore and Offshore Pipelines", Pipeline Research Committee International, Project Report PR-3-9523

Stephens, DR, Bubenik, TA, and Francini, RB (1995). "Residual Strength of Pipeline Corrosion Defects Under Combined Pressure and Axial Loads", Pipeline Research Committee, American Gas Association, NG-18 Report No 216.

Stephens, DR, and Leis, BN (1997). "Development and Validation of a Failure Criteria for Residual Strength of Pipeline Corrosion Defects", Pipeline Research Committee International, Project Report PR-3-9509

Vieth, PH, and Kiefner, JF (1994). "Database of Corroded Pipe Tests", Pipeline Research Supervisory Committee, American Gas Association Catalog No. L51689.

Proceedings of the Seventh (1997) International Offshore and Polar Engineering Conference
Honolulu, USA, May 25-30, 1997

ISBN 1-880653-28-1 (Set); ISBN 1-880653-32-X (Vol. IV)

Transverse Fatigue Crack Growth in a Graphite/Epoxy Composite Subjected to Combined Hydrostatic and Axial Loading

S.T. Mear and H.G. Wheat
University of Texas at Austin
Austin, TX, USA

H.L. Marcus
University of Connecticut
Storrs, CT, USA

ABSTRACT

Transverse corrosion fatigue tests were performed on a carbon-fiber epoxy-matrix composite in air, simulated seawater, and distilled water environments. Crack growth rates were determined for ambient and elevated pressure (up to 4000 psi) fluid environments and were compared to baseline values for the composite tested in air. In addition, the effects of water chemistry and loading frequency on fatigue crack propagation were investigated. Finally, composite specimens saturated with simulated seawater were fatigued in order to determine the effect that prior moisture content has on crack growth rates of specimens tested in either air or immersed environments.

KEY WORDS: Composites, transverse fatigue crack growth, environmental effects, hydrostatic pressure effects

INTRODUCTION

In the coming years, exploration for offshore oil deposits will employ the use of tension-leg platforms (TLP) to drill at ocean depths of 10,000 feet where the hydrostatic pressure reaches 4400 psi. The high hydrostatic pressure, corrosive seawater, and cyclic loading induced by wave motion or ocean currents present a challenging environment for TLP components. A TLP is weight critical, and the potential for mass and cost savings afforded through the use of carbon or glass-reinforced epoxy composites make these composites ideal candidate materials for TLP components such as platform facilities, risers, and tethers (Kim, Hahn, and Williams, 1988). However, the effect of long term high pressure seawater exposure on these materials must be understood if they are to be used in structures with design lives of 20 or more years.

Previous research has identified both beneficial and deleterious effects of seawater and hydrostatic pressure on composite materials. These effects are dependent not only on the choice of the fiber and matrix composite system but also upon the composite laminate layup. In addition, the composite response to seawater and pressure exposure is highly dependent upon the mode of deformation imposed during testing. Beneficial effects of hydrostatic pressure may include the closure of flaws and microcracks (Shin and Pae, 1992a) and the reduction of the matrix free volume (Shin and Pae, 1992b), both of which are reported to improve the strength of the composite. In some cases large hydrostatic pressures are reported to suppress the debonding of fiber bundles, decreasing the strength of composite tensile specimens (Parry and Wronski, 1985). Finally, hydrostatic pressure has been reported to both inhibit (Kim, Hahn, and Williams, 1988) and increase (Sloan, 1991) the uptake of seawater in different composite systems.

The net effect of moisture absorption on the mechanical behavior of composites depends on the material's loading cycle and geometry-factors that determine the mechanism involved in the composite failure. Absorbed moisture lowers the glass transition temperature of an epoxy resin matrix, resulting in degradation of the creep rupture performance of the composite due to increased matrix plasticization (Saunders, Clark, and van Blaricum, 1990). However, increased matrix plasticity in aramid fiber composites improves fiber-dominated properties as tensile strength (Komai, Minoshima, and Shiroshita, 1991). Weakening of the fiber-matrix interface by moisture absorption can allow for increased fiber debonding and a corresponding rise in the interlaminar fracture resistance of graphite-epoxy composites (Sloan and Seymour, 1992) while decreasing the Mode I transverse fracture toughness of graphite-epoxy composites (Garg, 1986).

This project aims to determine the effect of seawater exposure on the growth of transverse fatigue cracks in polymer-matrix carbon-fiber-reinforced composite materials. During this study, fatigue testing of unidirectional graphite/epoxy composite compact tension specimens was performed to evaluate the growth of transverse cracks as a function of hydrostatic pressure, moisture content, loading frequency, and fluid environment. In addition, post-failure scanning electron microscopy was performed in order to characterize specimen failure surfaces.

EXPERIMENTAL PROCEDURES

Materials and Specimens

Compact tension (CT) specimens (W=1inch) were machined from 12 ply thick panels of AS4C/E719LT graphite epoxy unidirectional plates according to ASTM E647 (1991). The flat panels were fabricated using a filament winding and autoclave cure process by R-Cubed Composites from prepreg manufactured by Fiberite. The E719LT matrix material is a thermosetting, solution-based resin with proprietary elastomer additives for increased matrix toughness.

Specimens were notched parallel to the fiber direction using a 0.02-inch wide slitting saw blade. This configuration, referred to as $[90]_{12}$ in this paper, was chosen to ensure that transverse crack extension would remain perpendicular to the applied load. Figure 1 shows the geometry of the CT specimen as well as the orientation of fibers and plies relative to the machined notch. These "as-received" specimens were stored in a sealed container with desiccant to prevent moisture uptake prior to testing. The elastic properties of the test material were determined by tension and shear tests on coupon specimens and are given in Table 1 (Schapery, 1994).

Table 1. Material properties of as-received AS4C/E719LT.

E_1 (Msi)	E_2 (Msi)	G_{12} (Msi)	ν_{12}	V_f (%)
18.6	1.23	0.77	0.33	60

To evaluate the effects of moisture content on crack growth behavior, additional specimens were soaked in 106°F synthetic seawater until saturation of the epoxy matrix was reached. The elevated temperature was chosen to increase the moisture diffusivity and thus decrease the time required for complete saturation of the test specimens. The conditioning facilities consisted of large isothermal water baths maintained at the OTRC research facility located at Texas A&M University. Specimen weight gain as a function of time was determined by removing the specimens from the conditioning bath, drying them with a lint free towel, and then weighing them on an analytical balance. This procedure was performed on a weekly basis. The saturation limit was determined when the weight of the composite specimens no longer increased. These specimens are referred to as "conditioned" in this paper.

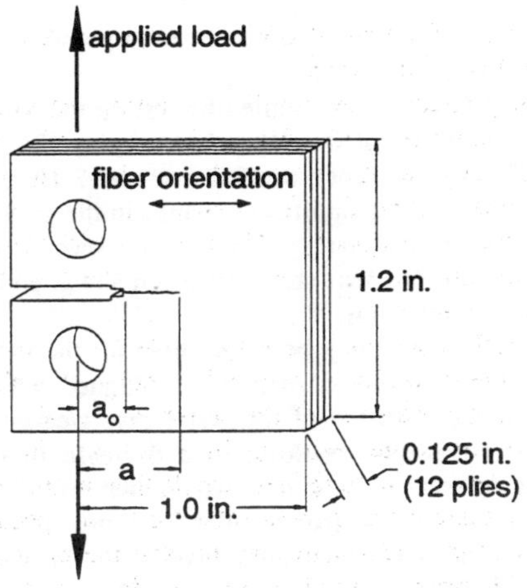

Figure 1. Compact tension specimen geometry showing orientation of fibers and plies relative to the machined notch.

Hydrostatic Pressure Test System

The experimental test system consists of an Inconel 600 hydrostatic pressure test chamber, system pump, fluid reservoir, fluid filter, and high pressure fittings, tubes and control valves (Fowkes, 1994). The dimensions of the test chamber accommodate the 1-inch wide CT specimen as well as grips and instrumentation. The test vessel is mounted in a 22 kip MTS servohydraulic fatigue test frame, which provides control of applied loads and displacements. Data are obtained via a PC-based digital acquisition program. Pressure is monitored using an optical sensor transducer with an operating range of 0-6000 psig and an accuracy of ±0.1% in that range.

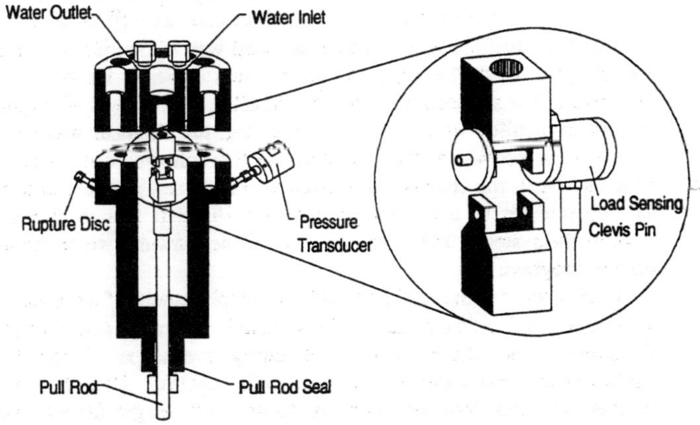

Figure 2. Schematic of high pressure test chamber and load sensing clevis pin.

The low load bearing capacity of the notched $[90]_{12}$ unidirectional specimens (approximately 50 pounds) coupled with friction loads at the pull rod seals and pressure fluctuations during chamber pressurization introduced an extremely difficult situation for monitoring the cyclic fatigue loads. A load sensing clevis pin with the capability of accurately monitoring loads of less than 1 pound and a maximum load capacity of 350 pounds at 5500 psi was used to monitor and control test loads. Since the load pin was mounted inside the test chamber, shown schematically in Figure 2, seal friction was eliminated as a possible source of error. In addition, the pin, which has an offset load due to the 4000 psi hydrostatic pressure of less than 2 pounds, minimized errors due to pressure fluctuations resulting from ram motion or from pumping of the pressurizing fluid.

Test Methods

Crack length in the enclosed chamber was monitored using the compliance method. Specimen compliance calibration was determined using both displacement and load control methods. Displacement control compliance testing was performed by loading the compact tension specimens at a crosshead rate of .005 in/min as outlined by Slepetz and Carlson (1975). Specimens were precracked in order to obtain a sharp crack away from the machined notch. Compliance was determined by loading the specimen until a visible increment of crack propagated from the precrack. During crack extension, the load decreased sharply and the crack was arrested. The new crack length was measured and the specimens were loaded and unloaded to determine the change in compliance. This method was chosen over one in which a machined notch is used because any fiber bridging in the composite is preserved when using a natural crack. Compliance values were also determined by load control cycling of specimens at the test loads and frequencies while observing the crack length through a monocular.

Both as-received and conditioned specimens were precracked in ambient air (70°F and 40-50% RH) prior to fatigue testing using the K_{max} decreasing method followed by additional precracking at the test loads. The final length of the precrack was measured under an optical microscope and was then compared with the value from the compliance calibration to ensure agreement between the actual and calculated crack lengths.

Load control tension-tension fatigue testing was performed in air, ambient seawater, 2000 psi and 4000 psi seawater as well as in ambient and 4000 psi distilled water. The mean load was 15 pounds with a ±10 pound cyclic load (R=0.2). Waveform was either 2 Hz or 0.2 Hz sinusoidal. The seawater solution used in all tests consisted of synthetic sea salt in distilled water. After precracking, the chamber was sealed, and the seawater was allowed to circulate through the chamber for ten minutes. Next, the chamber was taken up to the test pressure and held for two minutes to allow for pressure stabilization and to check for leaks in the system. There was no flow of the seawater while the tests were in progress.

During testing, the load and load line displacement of the specimen were monitored using a LabVIEW® based data acquisition routine. Compliance was determined by obtaining the slope of the load-displacement curves during load-increasing cycles. Compliance vs. number of cycles data were converted into crack length (a) vs. cycles (N) data using the compliance calibration curve. The seven-point incremental polynomial method outlined in ASTM E647 (1991) was then used to generate da/dN vs.ΔK data from the a vs. N data.

After testing, specimens were removed from the system, dried, and then stored in a vacuum desiccator for subsequent failure analysis. Specimen fracture surfaces were examined using both optical and scanning electron microscopy (SEM). To prevent surface charging, specimens were coated with a layer of Au-Pd prior to examination in either a JEOL JSM-35C or a JEOL JSM-840A scanning electron microscope. Micrographs were made of the representative fracture surface features.

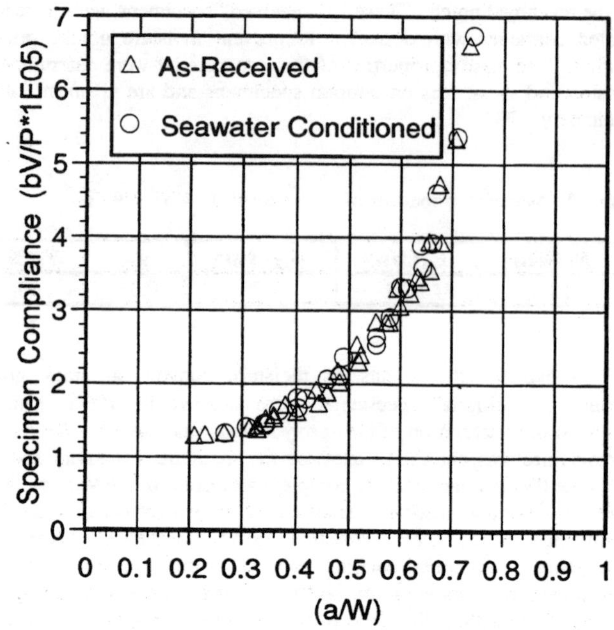

Figure 3. Specimen compliance vs. normalized crack length for as-received and seawater conditioned specimens (Mear, 1997).

EXPERIMENTAL RESULTS

Specimen compliance is plotted versus crack length in Figure 3. From the curve it was determined that the change in the compliance with crack length of both the as-received and conditioned specimens was large enough to use the compliance method for monitoring the crack length of the test specimen in the enclosed test chamber. A fifth order polynomial fit was used to fit the data points to a curve (Saxena and Hudak, 1978). No significant difference between the compliance of the as-received and conditioned specimens was seen. This could indicate that there was no substantial degradation of the interfacial bond between the fiber and the matrix as a result of moisture ingress into the specimen.

Fatigue crack growth data for the as-received AS4C/E719LT tested in air and simulated seawater are shown in Figure 4. Each curve represents three replicate tests for the given test condition. For all tests, the R-ratio was 0.2 and the frequency was 2 Hz. Crack growth rates for specimens tested in ambient and elevated pressure seawater are approximately ten times higher than growth rates in the specimen tested in ambient air. Growth rates in seawater do not appear to be related to the fluid pressure except possibly at crack velocities greater than $1\text{x}10^{-04}$ in./cycle.

To examine the effects of fluid chemistry, as-received specimens were tested in distilled water. Again, three replicate tests were performed for each test condition, the R-ratio was 0.2, and the frequency was 2 Hz. These data, superimposed over the data from Figure 4, are presented in Figure 5. As with the specimens tested in simulated seawater, the growth rates in those specimens tested in either ambient or pressurized distilled water were an order of magnitude higher than those tested in air. There was no apparent difference in growth rates due to the changes in water chemistry. Also, there did not appear to be a difference between the crack rates of specimens tested in ambient or pressurized distilled water except possibly at crack velocities greater than $1\text{x}10^{-04}$ in./cycle.

Frequency effects were studied by testing the as-received material in simulated seawater at 0.2 Hz while keeping the loading the same (R=0.2). These data, superimposed over the 2 Hz data, are given in Figure 6. There was no significant change in the growth rates resulting from the change in frequency. The rates for specimens tested in either ambient or pressurized seawater were again about ten times higher than those tested in ambient air.

Three mechanisms may be responsible for the increases in growth rates when the as-received material is fatigued while submerged in water. First, the presence of the water may weaken the fiber-matrix bond in the composite resulting in a decrease in toughness (Garg, 1986). This decrease in toughness would then result in a corresponding increase in fatigue crack growth rates for those specimens fatigued in water. A second mechanism may involve the presence of the nearly incompressible water in a fatigue-induced damage zone at the crack tip. If the water cannot be expelled during the down-loading interval of the fatigue cycle, it exerts pressure on the crack faces. This can change the global tension/tension loading into a local tension/compression stress field at the crack tip, again resulting in increased crack propagation rates (Kosuri and Weitsman, 1995). Finally, in the specimens tested in air, there may be fibrils of matrix material which extend across and join the crack faces behind the advancing crack tip. This bridging of the crack faces would lead to slower growth rates. In the specimens fatigue tested under water, these fibrils may plasticize and rupture sooner than they would in air, resulting in accelerated crack propagation.

To determine what effect long term exposure to seawater would have on fatigue propagation rates, fatigue tests with conditioned AS4C/E719LT specimens were conducted in ambient air, ambient seawater, and 4000 psi seawater. Two replicate tests were performed for each of the test conditions. The results are compared with the as-received data in Figure 7. Three results from this data stand out. First,

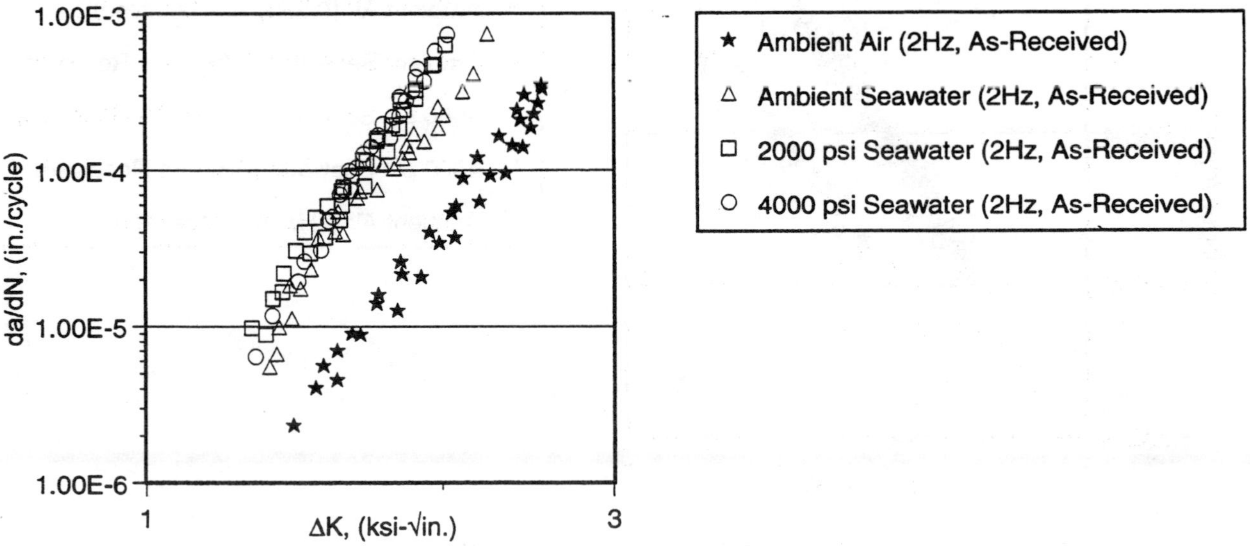

Figure 4. Transverse fatigue crack growth (2Hz) of as-received AS4C/E719LT in a seawater environment.

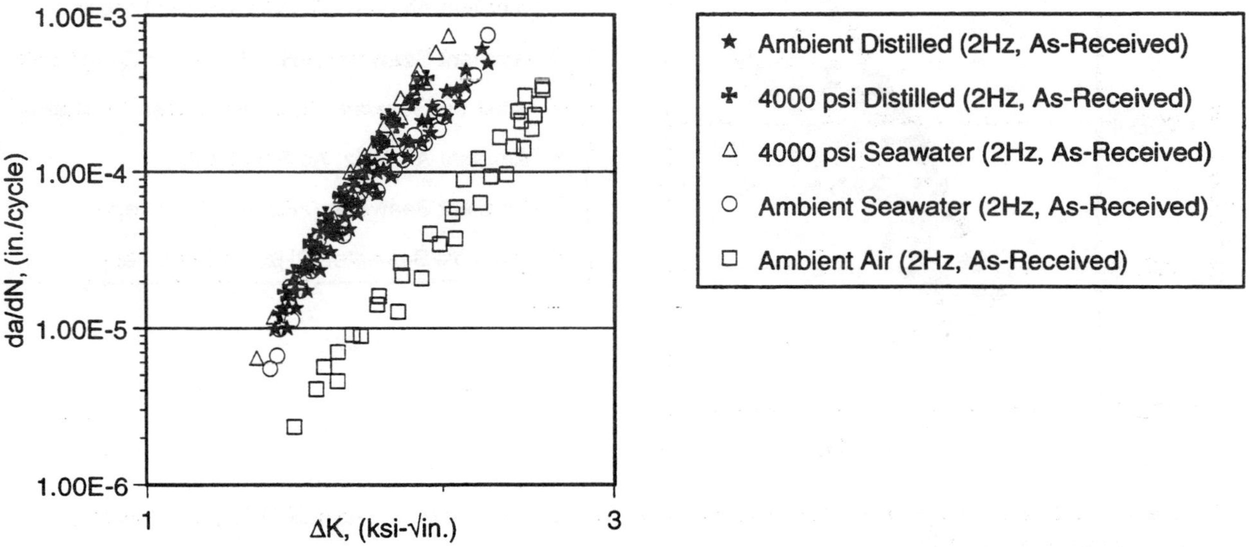

Figure 5. Transverse fatigue crack growth (2Hz) of as-received AS4C/E719LT in seawater and distilled water environments.

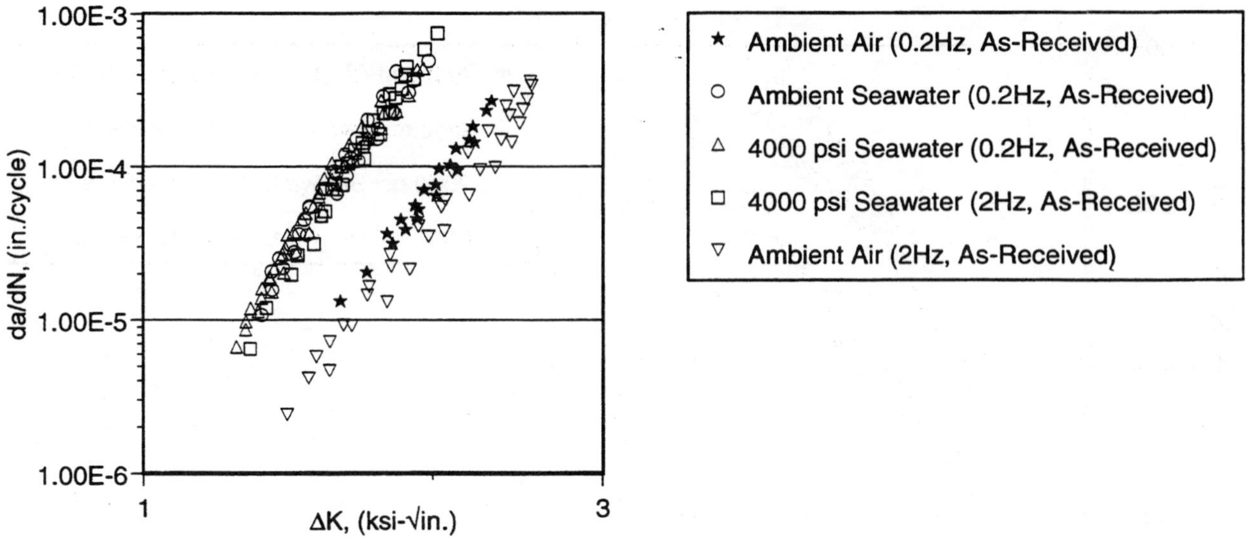

Figure 6. Transverse fatigue crack growth (2Hz and 0.2Hz) of as-received AS4C/E719LT in a seawater environment.

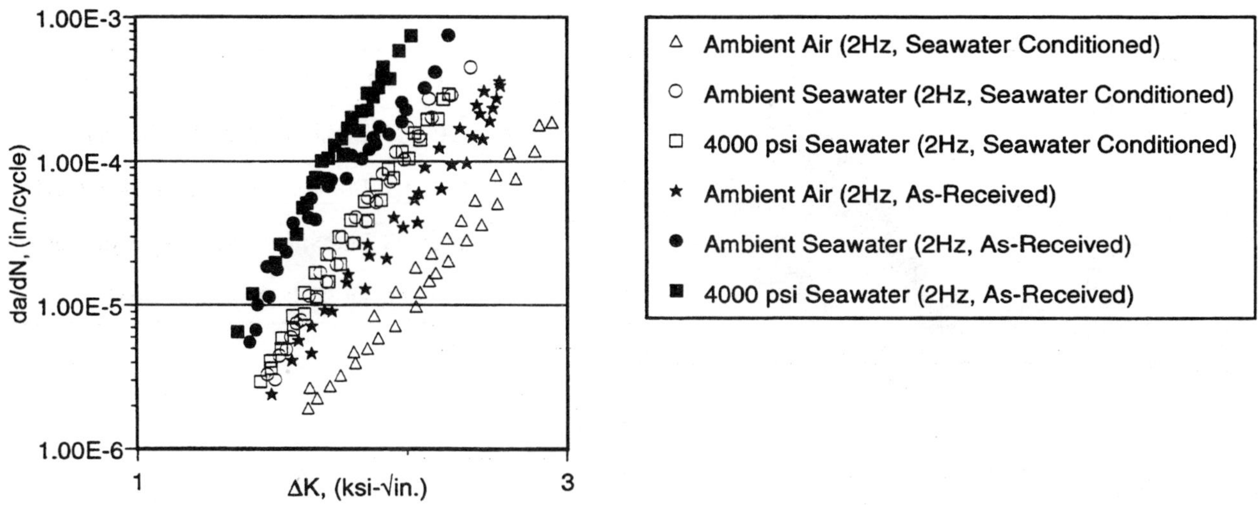

Figure 7. Transverse fatigue crack growth (2Hz) of as-received and seawater conditioned AS4C/E719LT in a seawater environment (Mear, 1997).

the crack propagation rates for the conditioned specimens tested in air are lower than the as-received air rates. In addition, the same may be said for the conditioned specimens tested in simulated seawater when they are compared to the as-received material tested while submerged. Finally, there is approximately a tenfold increase in growth rates for the conditioned specimens tested in water compared to conditioned specimens tested in air.

The decrease in growth rates for the conditioned material versus those found in the as-received material may be a result of two factors. First, while the failure stress for the two conditioned states is roughly the same, the strain to failure in the conditioned material is almost one third greater than that of the as-received composite (Schapery, 1995). The energy to propagate the crack in the conditioned specimens would go up accordingly. The second cause in the decrease may involve relief of curing strains that arise from the mismatch in thermal expansions of the matrix and fibers. The increase in moisture in the composite acts to relieve the strains in resin rich and lean regions as well as those found at voids and inclusions in the composite (Bradley, 1996).

Failure surfaces of specimens from each test condition were examined using optical and scanning electron microscopy. Representative micrographs of as-received and conditioned specimens are shown in Figure 8 and Figure 9 respectively. There were no significant differences between the failure surfaces for any of the test pressures, fluid environments, frequencies, or as a function of moisture content. In each case, the presence of features indicating fiber-matrix interface failure, matrix failure, and fiber failure were found. These include broken fibers free of matrix and deformed matrix material throughout the fatigue crack growth region. These features result from transverse cracking which is controlled by both fiber-matrix interface failure and matrix fracture (Garg and Ishai, 1985). Also present in all specimens were manufacturing flaws. Specimens tested in seawater contained salt crystals on their failure surfaces.

CONCLUSIONS

The transverse fatigue crack propagation rates of a unidirectional graphite-epoxy composite as a function of pressure, environment, frequency, and absorbed moisture content were investigated. This work was performed in an environmental test system that has a maximum working pressure of 5500 psi. The system includes an internally mounted load sensing clevis pin, which allows the low loads encountered during testing to be accurately monitored without error from seal friction and pressure fluctuation.

The crack growth rates of the as-received composite tested in water increased by a factor of ten over the composite tested in air. There were no noticeable differences in propagation rates between specimens tested at 2 Hz versus those tested at 0.2 Hz. In addition, the use of distilled water instead of simulated seawater did not seem to affect the growth rates.

Specimens that were saturated with seawater displayed a difference in growth rate when compared to the as-received material. Propagation rates for the composite tested in air were lower for the conditioned specimens when compared to the propagation rates in the as-received composite. This is also the case when the data for the conditioned test specimens fatigued in simulated seawater are compared with data for the as-received specimens tested under similar conditions. Finally, an increase in growth rates for the conditioned specimens tested under water is seen when they are compared with the same material tested in air. This difference between the propagation rates for the conditioned specimens tested in air or seawater may indicate that the entrapment of additional moisture in a crack front damage zone may be responsible for the increase in propagation rates when specimens are fatigued while submerged in water.

Optical and scanning electron microscopy of representative failure surfaces for all the test conditions did not show any significant differences in the failure morphologies as a function of test variables or specimen moisture content. Typical surface features including broken fibers, matrix failure, and processing defects were found on all specimens.

ACKNOWLEDGMENTS

This research was conducted with funding provide by the National Science Foundation and the Offshore Technology Research Center. In addition, the authors express their gratitude to Harold Kleeman for machining the test specimens and Jon Stolk for his help in preparing drawings used in this manuscript.

REFERENCES

ASTM E647 (1991). **"Standard Test Method for Measurement of Fatigue Crack Growth Rates,"** *Annual Book of ASTM Standards*, ASTM, Philadelphia, pp. 1-28.

Bradley, W.L. (1996). Personal Communication.

Fowkes, G.F. (1994). **"An Apparatus for the Characterization of the Corrosion Fatigue Behavior of Offshore Structural Materials Under Combined Hydrostatic and Axial Loading,"** Master's Thesis, The University of Texas at Austin.

Garg, A.C. (1986). **"Intralaminar and Interlaminar Fracture in Graphite/Epoxy Laminates,"** *Engineering Fracture Mechanics*; Vol. 23, No. 4, pp. 719-733.

Garg, Amar, and Ishai, Ori (1985). **"Hygrothermal Influence on Delamination Behavior of Graphite/Epoxy Laminates,"** *Engineering Fracture Mechanics*, Vol. 22, No. 3, pp. 413-427.

Kim, K.S., Hahn, H.T., and Williams, J.G.(1988). **"Application of Composites in TLP Tethers,"** *Conference on Offshore Mechanics and Arctic Engineering (OMAE) Proceedings*, Vol. 3, pp. 1-7.

Komai, K., Minoshima, K., and Shiroshita, S.(1991). **"Hygrothermal Degradation and Fracture Process of Advanced Fibre-Reinforced Plastics,"** *Material Science and Engineering*; Vol. A143, pp. 155-166.

Kosuri, R., and Weitsman, Y. (1995). **"Sorption Process and Immersed-Fatigue Response of GR/EP Composites in Sea Water,"** *Proceedings of the Tenth International Conference on Composite Materials, Volume VI: Microstructure, Degradation, and Design*, Anoush Poursartip and Ken Street, eds., Woodhead Publishing, pp. 177-184.

Mear, Steven T. (1997). **"Transverse Fatigue Crack Growth in a Graphite/Epoxy Composite."** *Ph.D. Dissertation*, the University of Texas at Austin, to be published May 1997.

Parry, T.V., and Wronski, A.S (1985). **"The Effect of Hydrostatic Pressure on the Tensile Properties of Pultruded CFRP,"** *Journal of Materials Science*, **20**, pp. 2141-2147.

Saunders, D.S., Clark, G., van Blaricum, T.J., and Preuss, T.E.(1980). **"Graphite/Epoxy Laminate Coupon Testing Programme,"** *Theoretical and Applied Fracture Mechanics*; Vol. 13, pp. 105-124.

Saxena, Ashok, and Hudak, S.J., Jr. (1978). **"Review and Extension of Compliance Information for Common Crack Growth Specimens,"** *International Journal of Fracture*, Vol. 14, No. 5, pp. 453-468.

Schapery, R.A. (1994). Personal Communication.

Schapery, R.A. (1995). Personal Communication.

Shin, E.S., and Pae, K.D (1992a). **"Effects of Hydrostatic Pressure on the Torsional Shear Behavior of Graphite/Epoxy Composites,"** *Journal of Composite Materials*, Vol. 26, No. 4, pp. 462-485.

Shin, E.G., and Pae, K.D (1992b). **"Effects of Hydrostatic Pressure on In-Plane Shear Properties of Graphite/Epoxy Composites,"** *Journal of Composite Materials*, Vol. 26, No. 6, pp. 828-868.

Slepetz, J.M., and Carlson, L. (1975). **"Fracture of Composite Compact Tension Specimens,"** *Fracture Mechanics of Composites, ASTM STP 593*, American Society for Testing and Materials, pp. 143-162.

Sloan, F.E. (1991). **"The Effects of Long-Term Seawater Exposure on Graphite/Epoxy Composite Materials,"** Dissertation, University of California, San Diego.

Sloan, F.E., and Seymour, R.J.(1992). **"The Effect of Seawater Exposure on Mode I Interlaminar Fracture and Crack Growth in Graphite/Epoxy,"** *Journal of Composite Materials*; Vol. 26, No. 18, pp. 2655-2673.

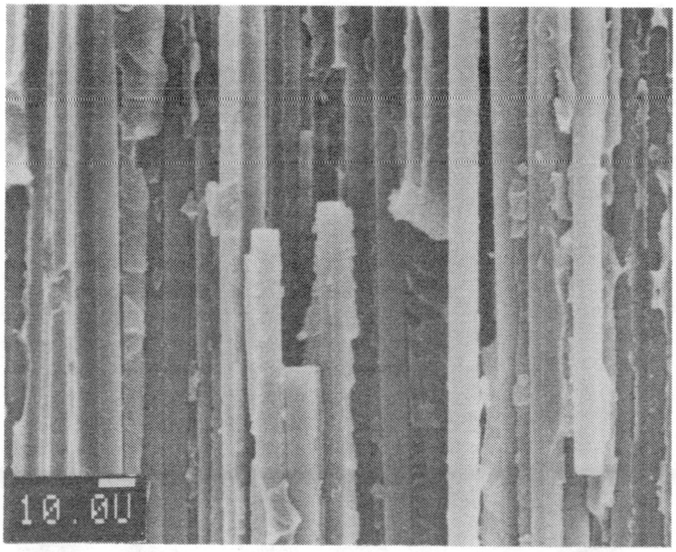

(a) Ambient Air

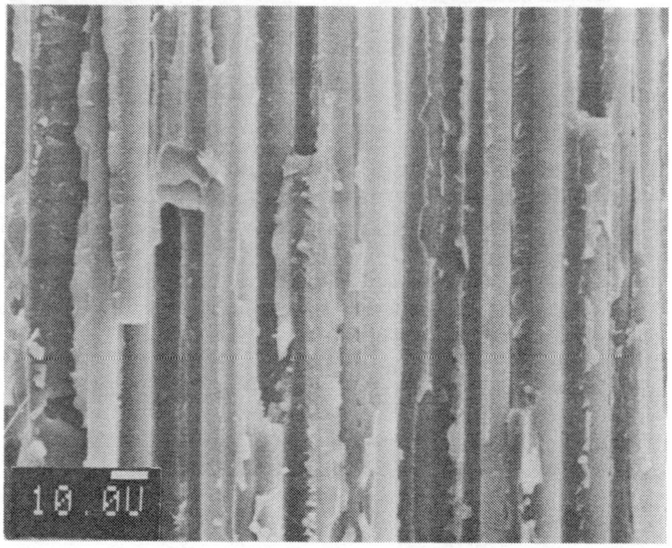

(b) 4000 psi Seawater

Figure 8. Fracture surface SEM micrographs of as-received AS4C/E719LT fatigued in (a) ambient air and (b) 4000 psi seawater.

(a) Ambient Air

(b) 4000 psi Seawater

Figure 9. Fracture surface SEM micrographs of seawater conditioned AS4C/E719LT fatigued in (a) ambient air and (b) 4000 psi seawater.

Proceedings of the Seventh (1997) International Offshore and Polar Engineering Conference
Honolulu, USA, May 25-30, 1997

ISBN 1-880653-28-1 (Set); ISBN 1-880653-32-X (Vol. IV)

Proposition of Polyhedrally-Stiffened Shells for Undersea Pressure Hulls

R.H. Knapp
University of Hawaii at Manoa
Honolulu, HI, USA

T.T. Le
Structural Solutions, Inc.
Aiea, HI, USA

ABSTRACT

The polyhedrally-stiffened cylindrical (PC) shell is a structural concept that offers design alternatives to the conventional ring-stiffened cylindrical shell for undersea pressure hulls. Both the defense and commercial ocean industries have a continuing requirement for pressure hulls in underwater systems. A significant percentage of these systems operate in water depths from near surface to 1,500 meters, depths well suited to PC pressure hulls. Some examples of applications for this structural concept include pressure housings for autonomous underwater vehicles, remotely operated vehicles, submersibles, buoys and sonabuoys.

In this paper, previous research on PC shells is expanded to include a broader range of materials suitable for undersea pressure hulls. Findings similar to those obtained for composite materials reported earlier have been obtained for aluminum, titanium and steel alloys. Parametric design studies using finite element analysis shows that the PC shell can be used with any of these materials advantageously. The undulated polyhedral geometry is shown to provide a substantially greater operating depth than a comparably sized circular cylinder with less control of fabrication tolerances needed. Thus, the PC hull behaves as a stiffened cylinder with the advantage of potentially lower fabrication costs.

KEY WORDS: Polyhedral cylinder, circular cylinder, ring-stiffened cylinder

STRUCTURAL CONCEPT

The PC shell concept has been discussed in prior publications (Miura, 1969, Knapp (1977), Knapp and Szilard (1978), Knapp and Dumlao, 1979, Knapp, Kudo and Auksmann, 1980, Knapp and Le, 1996a and 1996b), but is briefly reviewed here for completeness. The conventional view of a buckled cylinder is that it is a "failed" structure (Figure 1(b)); however, it has been proposed that the buckled cylinder can be treated as a new "unfailed" structural form described here as a polyhedrally-stiffened cylindrical (PC) shell. If designed initially in the ideal buckled pattern depicted in Figure 1(a), it has been shown that the PC shell possesses higher buckling resistance than a comparable circular cylinder (Knapp and Dumlao, 1979, Knapp and Le, 1996a and 1996b). It is proposed that such structures can perform well as undersea pressure hulls and may be cost-competitive with conventional circular cylindrical (CC) and ring-stiffened cylindrical (RC) pressure hulls.

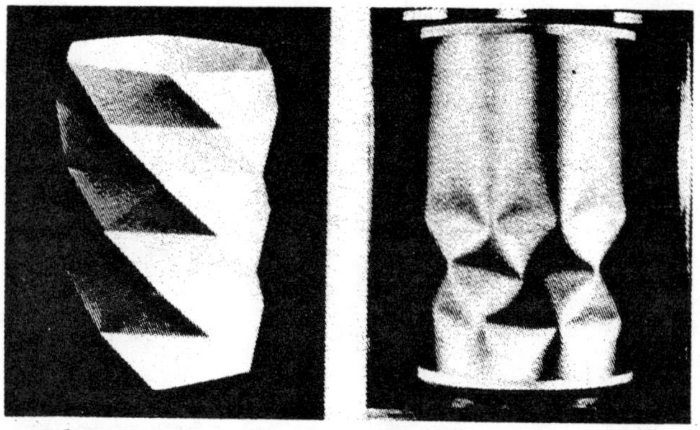

Fig. 1 (a) Ideal PC Shell (b) Buckled Cylinder

Figure 2 depicts this geometry with the amplitude of undulation given by the radial distance, A. This undulating surface, much like the structural action of corrugated sheet metal, adds structural stiffness to the shell. It has been confirmed by numerical analysis and experimental investigations that a hydrostatically loaded PC shell may possess several times the buckling resistance of a comparably sized circular cylinder having the same wall thickness and overall dimensions.

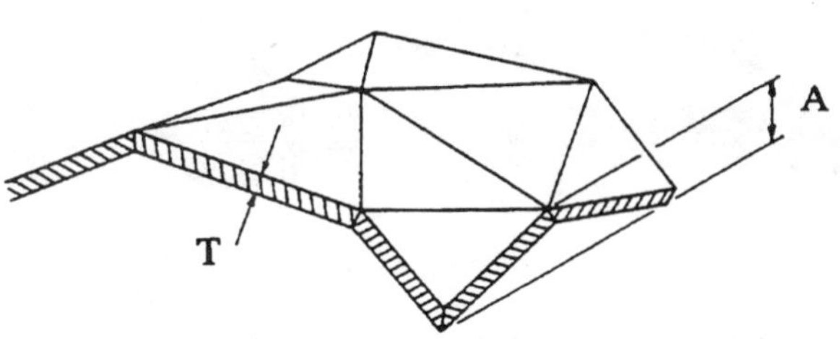

Figure 2. Undulated PC Shell Geometry

Since the PC shell is out-of-round by design, small variations in its shape do not affect buckling strength to the degree it would a circular cylinder. Therefore, it may be possible to relax fabrication tolerances and reduce manufacturing costs. Moreover, special forming operations necessary to make circular cylinders are unnecessary for PC shells. Whereas the circular cylinder must be formed accurately into a circular arc, the PC shell can be formed in a modular fashion from flat triangular plates or by laying up composite materials on a mold. Molding is simplified by taking advantage of the developable PC shell surface. Since it is a developable surface, composite materials can be laid up around a PC shell mold without stretching or distortion of the fabric. Furthermore, since the PC shell is an inherently stiffened cylinder, stiffening rings used in circular cylindrical hulls are unnecessary, and hence manufacturing costs are therefore relatively lower.

Recent efforts to develop ring-stiffened composite hulls have been disappointing due to the difficulties in joining or integrally molding rings to the cylindrical shell. Also, ring geometry must include fillets to reduce stress concentrations that add appreciably to the manufacturing cost. Elimination of the stiffening rings should advance the development of composite hulls by simplifying hull geometry and reducing fabrication costs (Knapp and Le, 1996b).

Most structural materials, including metals such as steel or aluminum and composites, can be used in the construction of PC hulls. Also, transparent materials, such as acrylic plastic, can be employed in the construction of PC hulls where visual or photographic observation through the hull is desired (Knapp, Kudo and Auksmann, 1980).

Undersea Pressure Hulls

With the addition of endcaps, the PC shell is well suited as an external pressure hull for undersea applications. Figure 3 shows one concept using hemispherical endcaps, but other shapes including flat plates also could be used.

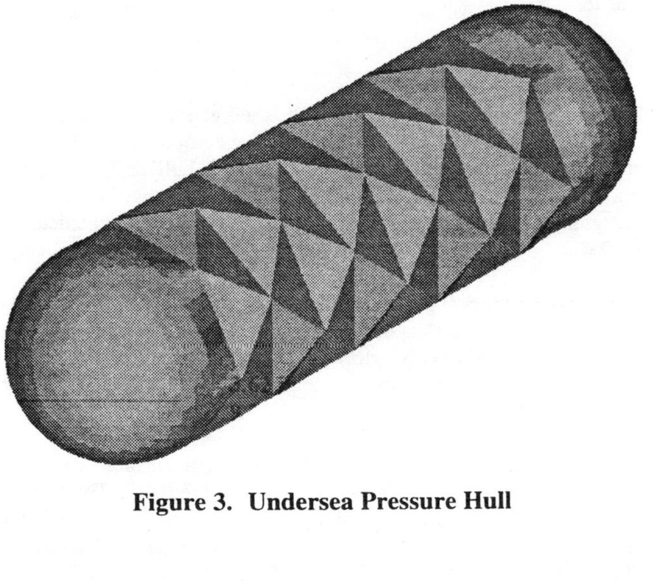

Figure 3. Undersea Pressure Hull

This study shows that such PC hulls have a structural advantage over circular cylinders for a range of materials to depths of 1,500 meters, and at depths to 600 meters this advantage is substantial.

Objectives

The primary objective of this research program has been to examine the potential application of PC hulls in the undersea environment. A technical program to demonstrate proof-of-concept for PC hulls included the following tasks.

First, laboratory test models were fabricated from an E-glass/epoxy composite material and tested to validate the design methodology. This effort consisted of designing models that could be buckled at atmospheric pressure. The test results were then compared with design predictions to confirm that modeling procedures are valid. These results are reported in Knapp and Le (1996a and 1996b).

Second, the validated design methodology was used to perform a parametrical design study where optimal PC hull geometries were found for each material investigated. This paper documents the parametrical design study.

PC HULL GEOMETRY

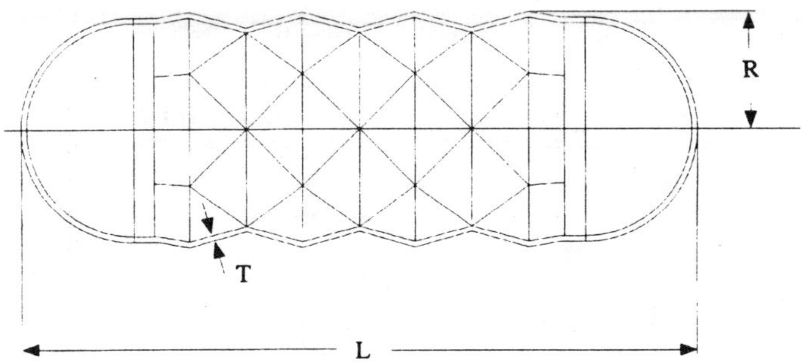

Figure 4. PC Hull Geometry

Candidate PC shell geometries for a complete structural system were investigated. Each component of the structural system, consisting of the PC shell, an end transition region and hemispherical end enclosures, were designed to achieve the greatest possible operating depth considering both material and buckling failure modes.

Figure 4 shows the fundamental PC shell geometrical parameters including the overall length, L, the radius, R, measured from the shell axis to the vertex of adjacent polyhedra on the outer surface, and a uniform wall thickness, T. In addition to these parameters, the number of polyhedra in the axial and circumferential (hoop) directions are given by M and N, respectively.

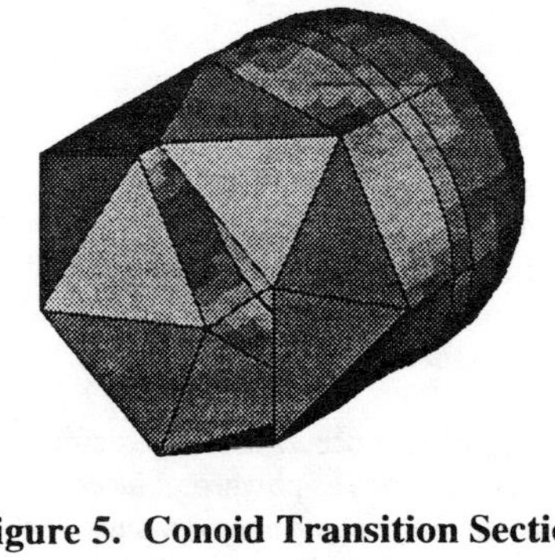

Figure 5. Conoid Transition Section

Conoid Transition Section and Endcap

In addition to optimizing the PC shell geometry, a means of joining endcaps to the shell is needed. For the purpose of comparing PC and CC shells, hemispherical endcaps were selected, although other shapes such as flat plates also could be used. To join the circular edge of the endcap to the straight polyhedral edges of the PC shell, a transition region is needed. For this study, a conoid section (Figure 5) was used. The diameter of the cylindrical ends was chosen to have the same circumferential length as the PC shell region. The axial length of the transition region was determined for each PC shell geometry such that failure depth was maximized.

MATERIAL PROPERTIES

The material properties used in the parametrical design study are presented in Tables 1 and 2. In these tables, S = material strength, E= elastic modulus, G= shear modulus, and ν = Poisson ratio. The subscripts, x and y, refer to in-plane coordinates and z is the transverse coordinate to describe the orthotropic properties of the composite materials.

Table 1. Composite Material Properties

E-glass/Epoxy [0/90]	Carbon/epoxy [0/90]
S = 550 MPa $E_x = E_y = 16$ Gpa $E_z = 6.9$ Gpa $G_{xy} = 4.1$ Gpa $G_{xz} = G_{yz} = 3.5$ Gpa $\nu_{xy} = 0.09$ $\nu_{xz} = \nu_{yz} = 0.18$	S = 590 MPa $E_x = E_y = 50$ GPa $E_z = 7.6$ GPa $G_{xy} = 48$ GPa $G_{xz} = G_{yz} = 1.7$ Gpa $\nu_{xy} = .09$ $\nu_{xz} = \nu_{yz} = .07$

Table 2. Metal Properties

Material	E (GPa)	ν	S (MPa)
MA-300 Steel	200	1/3	2070
HY-150 Steel	200	1/3	1030
7178-T6 Aluminum	70	1/3	540
6Al-4V Titanium	115	.34	1070

FINITE ELEMENT MODELING

A finite element modeling approach was developed to predict the buckling pressure and mode shape for both PC and CC hulls having hemispherical endcaps. The ANSYS program was used to obtain all finite element results reported in this paper (ANSYS, 1995).

For the PC hull, buckling occurs as a large deformation process involving snap-through deformations as polyhedral faces "fold" or bend during shell collapse. Thus, the finite element model must be able to account for large translations and rotations, orders of magnitude larger than the thickness of

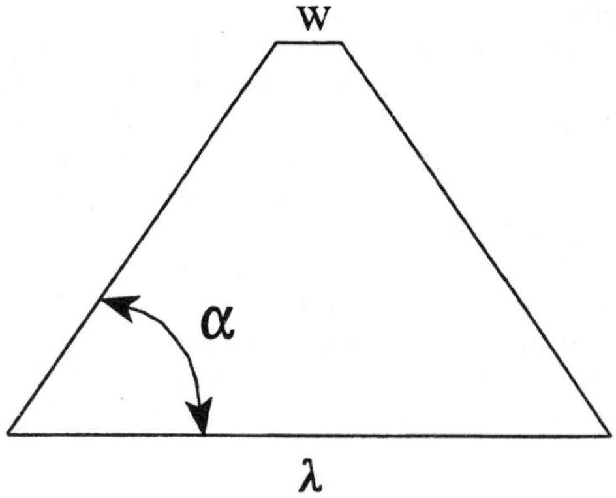

Figure 6. Polyhedral Geometry

the shell. Due to the discrete geometry of the PC hull, a number of buckling modes are possible, each yielding a different buckling pressure. To find the lowest and most probable buckling pressure, the geometry of the finite element model must be perturbed to induce the mode shape yielding the lowest buckling pressure. This was accomplished by finding several of the lowest mode shapes from a linear buckling analysis, then using an initial imperfect geometry affine to the linear modes for the geometrically nonlinear buckling analysis. The maximum amplitude of the linear mode (eigenvector) was specified as 0.5% of the theoretical shell radius for both the PC and CC hull finite element analyses. This value is consistent with the maximum error in the radius, R, encountered in fabricating the PC shell models (Knapp and Le, 1996b).

In order to make a direct comparison with a circular cylindrical (CC) hull, equivalent CC and PC hulls were analyzed using identical finite element modeling procedures. Equivalent here means that each hull had the same circumference and length, and the same hemispherical ends were used. Also, an equivalent PC hull would utilize the same amount of material in its construction as a CC hull. Hull materials were assumed to be linearly elastic.

OPTIMAL DESIGN

The design study seeks to determine PC shell geometrical parameters that optimize structural performance for a specified material. Optimal structural performance requires that the PC hull design will fail by buckling and material yield coincidently over a range of depths. The designs developed in this study were based on *nearly* coincident failure modes.

A range of polyhedral geometries was studied for each material. Variation of the radius-to-wall thickness ratio, R/T, and polyhedral discretization (number and shape of polyhedra) were considered. Polyhedral shape was varied according to the geometry depicted in Figure 6, given by the angle, α, and trapezoidal parameter, W/λ.

A more general tessellation of the PC shell surface would utilize trapezoidal polyhedra (W/λ>0) as shown in Figure 6. The trapezoidal tessellation has been found to reduce stress concentrations by eliminating the singular point shared by the vertices of adjacent triangular plates (Figure 3), but this shape also has been found to reduce buckling strength. Decreasing the angle, α, from 60° (equilateral triangular geometry) also reduces stress, but increases buckling resistance. As the number of polyhedra is increased, the more the PC shell resembles a circular cylinder with an attendant reduction of flat plate bending stresses and buckling

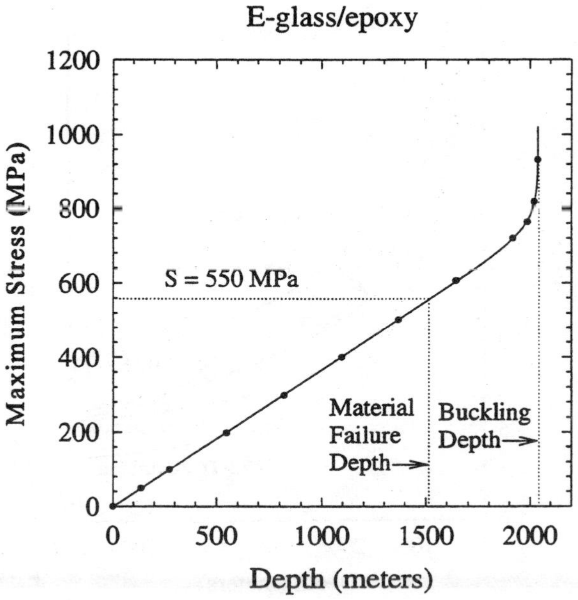

Figure 7. Typical Stress-Depth Curves

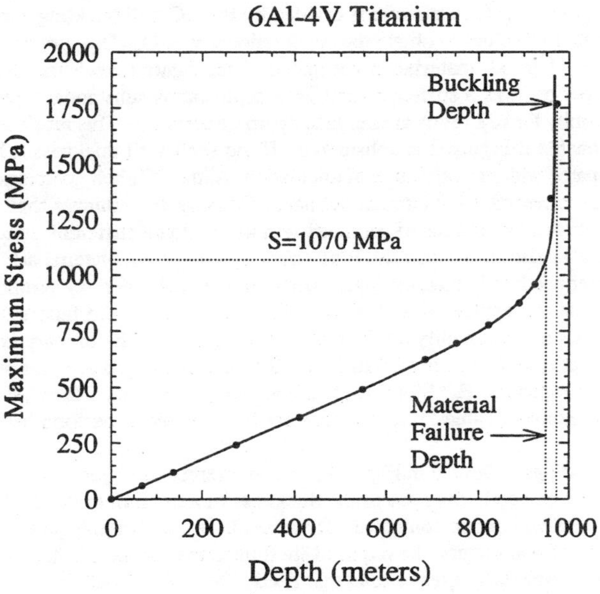

Figure 8. Typical Stress-Depth Curves

resistance. As the number of polyhedra is decreased, the larger will be the plate bending stresses and buckling resistance. Since an optimal design requires low bending stresses and high buckling resistance, the number of polyhedra actually selected will be a compromise for a particular operating depth. Structural analyses are required to reveal the most promising polyhedral geometries.

The aforementioned geometrical parameters were varied to achieve optimal PC shell performance for six candidate materials to depths of

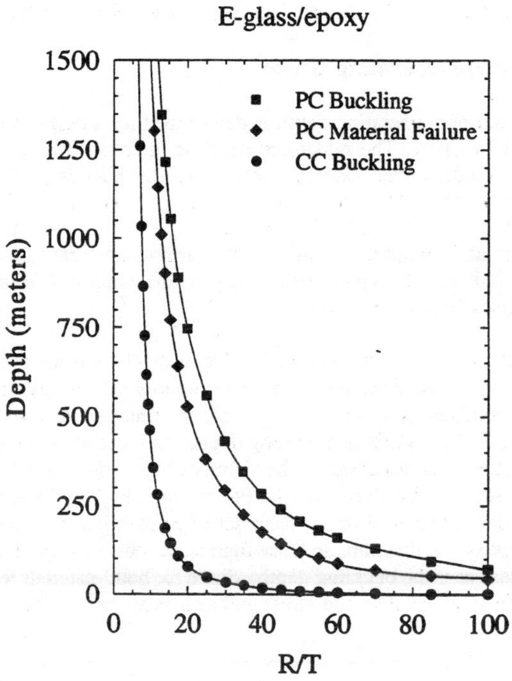

Figure 9. Typical Depth Curves

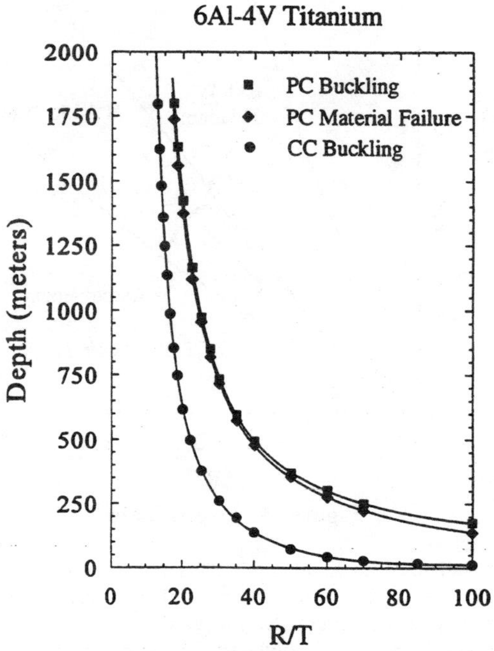

Figure 10. Typical Depth Curves

1,500 meters. The hull length-to-radius ratio, L/R ~ 7, was selected as a representative value for hulls of practical interest. Studies (Knapp and Dumlao, 1979) have shown that structural performance is not very sensitive to this parameter. The PC hull performs slightly better than the

CC hull with greater L/R ratios and slightly worse with smaller L/R ratios. In this study, the geometrical parameters, M, N, α and W/λ are varied in the ranges: $10 \le M \le 24$; $4 \le N \le 8$; $30° \le \alpha \le 60°$; $0 \le W/\lambda \le 0.2$.

Maximum Operating Depth

The maximum operating depth is determined for a range of R/T values from 10 to 100. The maximum depth is established by the dominant failure mode, either material yield or elastic buckling. For the two composite materials investigated, we assume a symmetrical [0/90] stacking sequence that yields identical elastic properties and strengths in both tension and compression in the two orthogonal directions. In this simplified case, it is appropriate to approximate material failure with the von Mises-Hencky criterion.

For each material and given R/T value, critical geometrical parameters including M, N, W/λ, and α were varied to yield the largest operating depth obtained by geometrically nonlinear finite element analysis. To determine the maximum operating depth, the location of the largest von Mises stress was identified (in the vicinity of the largest radial deflection) and plotted as a function of depth. For example, Figures 7 and 8 show the stress-depth curves obtained for a set of geometrical parameters for E-glass/epoxy and titanium. In these figures, the curves are seen to become nonlinear near the buckling depth. Since for both materials the material strength is lower than the stress induced near the buckling point, the

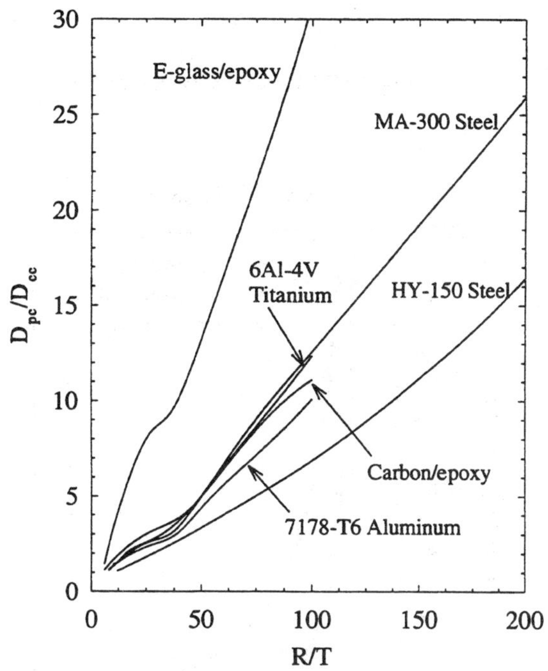

Figure 11. Depth Ratio

material failure depth governs this PC shell geometry. The analysis is repeated, varying M, N, W/λ, and α, until the greatest operating depth, either material or buckling failure, is found.

Having found the optimal polyhedral geometry for each R/T value, the results can be compiled as presented in Figures 9 and 10. Note that the material failure and buckling curves for the PC hull are close over the range of R/T. This result suggests that a nearly optimal design has been achieved. Each point on the PC shell buckling and failure stress curves represent unique values of the M, N, W/λ, and α parameters; however, for both materials, these parameters were constant over the selected range of R/T. The resulting optimal values for these parameters are presented in

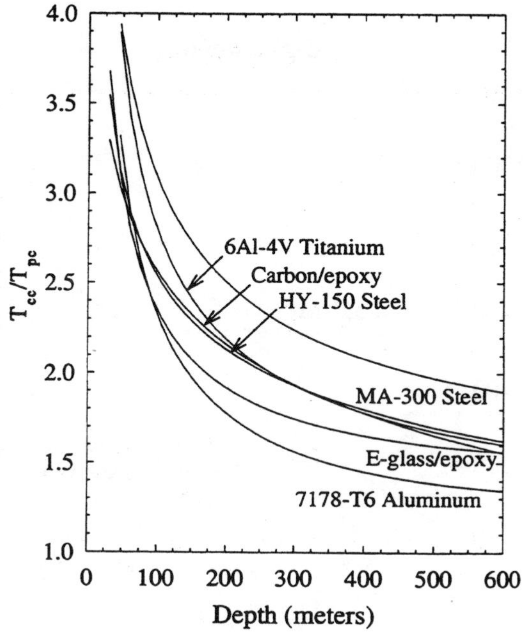

Figure 12. Thickness Ratio

Table 3. In addition to the PC shell curves, the buckling curve for the CC shell also is plotted.

To evaluate the performance of the PC hull relative to the CC hull, the PC hull governing failure curve is divided by the CC hull buckling curve at discrete R/T values to obtain the depth ratio curves, D_{PC}/D_{CC}, presented in Figure 11 for all materials investigated. This figure reveals that the PC hull performs as a stiffened shell with depth ratios substantially greater than unity for R/T ratios greater than approximately 10. This result seems reasonable if Figure 2 is considered. If the shell wall thickness is large compared with the amplitude of undulation, A, the PC shell geometrically begins to resemble the circular cylinder. Conversely, a thinner shell wall accentuates the undulated geometry and its attendant stiffening effect.

Figure 11 also shows that the E-glass/epoxy composite material achieves a better depth ratio than the higher performance carbon/epoxy composite. This can be explained as follows. For the PC hull, the failure depth depends predominantly on the material strength, S. The primary failure mode of the CC hull is buckling, and the buckling depth depends on the elastic modulus, E. Since E-glass/epoxy has a higher S/E value than carbon/epoxy (Table 1), it can therefore be expected to perform better.

The findings presented in Figure 11 can be interpreted in yet another way. For a given depth, the ratio of the thickness of a CC hull to the thickness of the PC hull can be computed. These results are shown in Figure 12. To depths of 600 meters, the required shell thickness of the CC hull is seen to be substantially greater than that needed for the PC hull for all of the materials studied.

Fabrication Error Sensitivity

To determine the sensitivity of PC shell buckling resistance to fabrication errors, an additional finite element study was conducted. In this case, the PC shell is assumed to be molded with E-glass/epoxy. Figure 13 is a plot of the Buckling Pressure Ratio, P_{PC}/P_{CC}, as a function of the Radial Imperfection for a particular PC shell geometry.

This curve is evidence that PC shells are more tolerant to fabrication errors than circular cylinders. First, it is noted that the P_{PC}/P_{CC} ratio increases

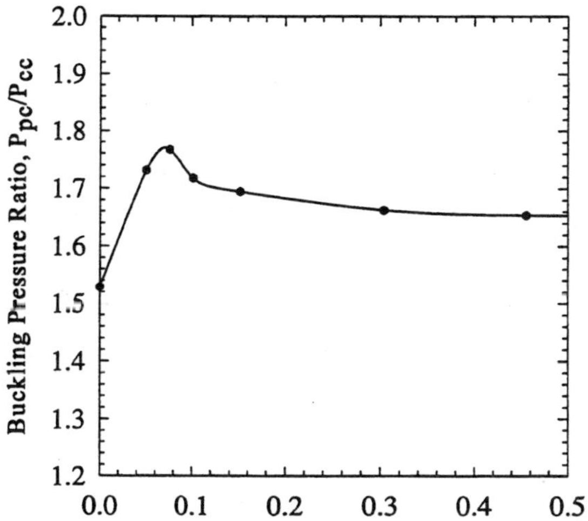

Figure 13. Error Sensitivity

with increasing geometrical imperfection. Second, at about 0.1% radial imperfection this ratio reaches a maximum and maintains a slowly diminishing value between 0.1% to 0.5%. As stated earlier, the PC shell is out-of-round by design and its buckling strength is less sensitive to variations in its shape than for a CC shell. Figure 14 confirms this hypothesis and suggest that fabrication tolerances might be relaxed to reduce manufacturing cost.

DESIGN DECISION MATRIX

The parametrical design studies have shown that optimal structural performance of the PC hull depends on the material yield stress and elastic modulus. Higher strength materials permit larger polyhedral surfaces to be used (bending stresses increase with increasing polyhedral edge length, λ). Materials with a higher elastic modulus create a stiffer structure to permit smaller polyhedral plates to be used (buckling strength decreases with decreasing edge length, λ).

After obtaining optimal designs for the six materials studied, it was found that certain geometries were best for a range of yield stress and elastic modulus values. Table 3 shows a design decision matrix of the best PC shell geometries as a function of these two material parameters. PC shell geometries for other materials might be estimated by use of this table. In this table, N = number of circumferential polyhedra; λ, W and α are as defined as in Figure 4.

CONCLUSIONS

Parametric design studies have shown that PC hulls have a structural advantage over CC hulls for a range of metallic and composite materials to depths of 1,500 meters. Geometrically nonlinear finite element analyses were used to determine the buckling loads of both CC and PC shells. These analyses considered also the effect of geometrical imperfections on buckling loads. The shape and size of the polyhedra were chosen to optimize structural performance for each material. Both triangular and trapezoidal surface tessellations were examined, and methods of joining the polyhedral cylinder with hemispherical ends were studied. For example, results for E-glass/epoxy PC hulls show potential increases in the operating depth from 3 to 30 times CC hull performance for $10 \leq R/T \leq 100$.

It is proposed that PC shells may be very well suited to molding composite pressure hulls. Molding techniques similar to those used for CC shells may be employed at a cost less than composite RC shells can now be produced. The potential of low-cost manufacturability of composite pressure hulls provides strong motivation to consider PC shells for undersea applications. In spite of the larger stresses occurring due to the bending of the flat polyhedral faces, it is shown that the PC hull can be designed to operate at a substantially greater depth than the CC hull for the same overall geometry and hull weight. Thus, the PC hull is recognized as a polyhedrally-stiffened shell, competitive with RC shells. In consideration of these findings, the paper concludes that the PC hull has the potential to advance development of composite pressure hulls.

Table 3. Design Decision Matrix

<table>
<tr><td></td><td colspan="6">Yield Stress (MPa)</td></tr>
<tr><td>Elastic Modulus (GPa)</td><td>AL 540</td><td>Gl 550</td><td>CA 590</td><td>HY 1030</td><td>TI 1070</td><td>MA 2070</td></tr>
<tr><td>Gl 16</td><td colspan="6">TRI
N = 4
α = 45°</td></tr>
<tr><td>CA 50</td><td colspan="6">TRI
N = 6
α = 35°</td></tr>
<tr><td>AL 70</td><td colspan="5" rowspan="3">TRA
N = 6
α = 40°
λ/W = 13</td><td rowspan="2">TRA
N = 6
α = 40°
λ/W = 13</td></tr>
<tr><td>TI 115</td></tr>
<tr><td>HY/MA 200</td><td>TRI
N = 5
α = 45°</td></tr>
</table>

Note: TRI = Triangular tessellation; TRA = Trapezoidal tessellation; AL = 7178-T6 aluminum; GL = E-glass/epoxy; CA = Carbon/epoxy; HY = HY-150 steel; TI = 6Al-4V titanium; MA = MA-300 steel.

ACKNOWLEDGEMENTS

This work was funded by the National Defense Center of Excellence for Research in Ocean Sciences (CEROS). CEROS is part of the Natural Energy Laboratory of Hawaii Authority (NELHA), an agency of the Department of Business, Economic Development & Tourism, State of Hawaii. CEROS is funded by the Advanced Research Projects Agency (ARPA) through grants to NELHA. This report does not necessarily reflect the position or policy of the Government, and no official endorsement should be inferred.

REFERENCES

ANSYS (1995). ANSYS User's Manual (version 5.2), Ansys, Inc., Houston, PA, USA, 1995

Knapp, RH (1977). "Pseudo-Cylindrical Shells: A New Concept for Undersea Structures," *Trans. ASME, J of Engineering for Industry*, Vol 99, No 2, pp 485-492.

Knapp RH, and Szilard, R (1978). "Nonlinear Stability Analysis of Pseudo-Cylindrical Shells," *ACTA Technica Academiae Scientiarum Hungaricae*, Vol 86, pp 9-41.

Knapp, RH, and Dumlao, C (1979). "Experimental Investigations of Prebuckled Cylinders Under External Pressure", *Trans. ASME, J of Engineering for Industry*, Vol 101, No 2, pp 178-184.

Knapp, RH, Kudo, R and Auksmann, B (1980), "Development of a New Concept for Undersea Pressure Hulls," *Trans. ASME, J of Engineering for Industry*, Vol 102, No 3, pp 253-257.

Knapp, RH, and Le TT (1996a). "Polyhedrally-Stiffened Composite Pressure Hulls," *Proc. 6th International Offshore and Polar Engineering Conference*, Vol IV, pp 292-298.

Knapp, RH, and Le TT (1966b). "Polyhedrally-Stiffened Cylindrical Shells for Undersea Composite Pressure Hulls," *Proc. Third International Conference in Computational Structures Technology*, Budapest, Hungary, August 21-23, 1996, Advances in Analysis and Design of Composites, Civil-Comp Press, Edinburgh, United Kingdom, pp 81-89.

Knapp, RH (1996). "Polyhedrally-Stiffened Cylindrical (PC) Pressure Hull," Patents Pending

Miura, K (1969). "Proposition of Pseudo-Cylindrical Concave Polyhedral Shells," Report No. 442, Institute of Space and Aeronautical Science, University of Tokyo.

Proceedings of the Seventh (1997) International Offshore and Polar Engineering Conference
Honolulu, USA, May 25-30, 1997

ISBN 1-880653-28-1 (Set); ISBN 1-880653-32-X (Vol. IV)

Construction Applications of Fiber Reinforced Polymer Composites: A Survey

T. Kant and V.P.V. Ramana
Indian Institute of Technology, Bombay, India
P.K. Dutta
US Army Cold Regions Research and Engineering Laboratory, Hanover, NH, USA
A. Mukherjee and Y. Desai
Indian Institute of Technology, Bombay, India

ABSTRACT

A review is made of the different applications of fiber reinforced polymer composite (FRPC) materials and carbon fiber reinforced plastic (CFRP) cables in construction. The potential FRPC application areas are classified into three groups, namely: (i) Repair and rehabilitation of structural elements using FRPC sheets, (ii) FRPC bars as a reinforcement material for concrete structural elements, and (iii) Structural elements made of FRPC. Literature relevant to FRPC material in construction is divided into the above three groups. This review focuses on experimental research only. No attempt has been made to identify different analytical studies available on FRPC. Sub-topics of the above three areas are mentioned along with partial references.

Key Words: FRPC, CFRP, rehabilitation, strengthening, fiber composite bar.

INTRODUCTION

FRPCs, hitherto used in aerospace, automobile, ship-building and chemical industries are now finding their application in construction industry. It is their resistance to corrosion, high strength-to-weight ratio and durability properties that make them potential construction materials along with conventional materials like steel and concrete. Erki and Rizkalla (1993) explained various types of FRPC reinforcement for concrete structures. The first application of composites was in the form of rebars and structural shapes. Later FRPC laminates were used to strengthen the concrete bridge girders by bonding them to the tension face of the girder.

Composites are the materials created through the combination of two material phases, one serving as the reinforcement (fibers) and the other as the matrix (resin). Different manufacturing processes and issues related to selection of materials can be found in Karbhari (1995). Meier (1992) discussed use of different CFRP materials in bridge construction. Construction details of prestressed bridge decks using CFRP tendons was presented by Rizkalla and Tadros (1994). The importance of using materials by combination (FRPC) in different construction areas (sewage pipes, concrete beam strengthening etc.) can be found in Meier (1988). Rizkalla and Abdelrahman (1995) explained different applications of FRPC material with practical examples. Tarricone (1995) and Victor (1995) outlined the present situation of composites in construction industry. A large volume of literature exists on various applications of FRPCs in construction. This paper mainly discusses and critically examines the available literature in the following three application areas, viz.,

- Repair and rehabilitation of structural elements.
- FRPC bars as reinforcement material for concrete elements.
- Structural elements made of FRPCs.

Partial list of literature relevant to the sub-topics of above areas is given in Table 1. Finally, current status of research on application FRPCs in construction industry in the form of concluding remarks is presented at the end.

REPAIR AND REHABILITATION OF STRUCTURAL ELEMENTS

Structural elements, namely, beams, columns and slabs are made of either steel or reinforced concrete. Until now most of the research work was on strengthening of reinforced concrete members. This review mainly emphasises on strengthening of reinforced concrete elements in general and reinforced concrete beams in particular. No effort was made to review strengthening of beams by steel plates. Review on strengthening by steel plates and composite laminates can be found in McKenna and Erki (1994). Few studies can also be found on strengthening structural steel elements. The initial work on strengthening steel elements stressed the need for assessment of durability of bond (Karbhari and Shulley, 1995) between composite and steel.

Epoxy bonded steel plates have been used effectively to strengthen concrete bridge girders. Later, it was found that corrosion of the steel plate affects the bond between the concrete and steel plate. To eliminate corrosion problem, the steel plate was replaced by a corrosion resistant FRPC plate. The first field application of this repair technique was carried out (Meier et al., 1992) on a prestressed concrete bridge in Lucerne, Switzerland. Nanni (1995) presented various case studies of strengthening concrete structures in Japan with external FRPC reinforcement.

Saadatmanesh along with his co-researchers, pioneered the work on strengthening beams with composite laminates. Saadatmanesh and Ehsani (1991a), Ritchie et al. (1991), and Meier and Kaiser (1991) proved beyond doubt the effectiveness of using composite laminates as an external reinforcement for strengthening. Most of the experiments conducted so far used glass fiber laminates. However, some works reported by Meier and Kaiser (1991), Meier et al. (1992), Shahawy et al. (1996 a,b), Takeda et al. (1996) used carbon fiber laminates. Ritchie et al. (1991) studied the effectiveness of strengthening using different types of FRPC laminates. Laminates made of glass, carbon and aramid fibers were used and the increase in ultimate strength was found to be ranging from 28 to 97% of that of unstrengthened beams for above types of laminates. Faza and Ganga Rao (1994) reported an increase of 200% in strength when CFRP laminates are wrapped around beams.

The adhesive material (epoxy) used to bond the FRPC laminates to concrete, plays an important role. Experiments with different epoxies were carried out by Saadatmanesh and Ehsani (1990a). They concluded that rubber toughened epoxies are best suited for strengthening. A better cracking behaviour was observed in all the strengthened beams. Meier and Kaiser (1991) in their experiments bonded a pre-cracked beam with a carbon fiber laminate which resulted in a uniform distribution of cracks through out the beam. The three observed general modes of failure are (i) Sudden tensile failure of the composite sheet, (ii) Failure of the concrete in compression region and (iii) Continuous peeling off of the composite laminate due to an uneven concrete surface, or sudden peel-off during loading due to development of shear cracks in the concrete. Triantafillou and Plevaris (1992) also identified four collapse mechanisms and obtained equations describing each failure mechanism using strain compatibility method.

Triantafillou et al. (1992) have advocated strengthening of beams by pretensioned FRPC laminates. Two non-linear equations (Triantafillou, 1990) were used to calculate the maximum achievable prestress and the same was confirmed by conducting experiments. Triantafillou and Deskovic (1992) carried out similar type of work on wood members. Plevris and Triantafillou (1992) studied the behaviour of wood flexural members reinforced with FRPC laminates on their tension faces. The work carried out by Meier et al. (1992) compares the difference between the strengthening of concrete beams with non-prestressed and prestressed glass fiber reinforced plastic laminates. Experimental results showed an increase of 33% in the ultimate load for beams with both prestressed and non-prestressed CFRP sheets, with the only difference between the two behaviors being the displacement at failure of the beams. Non-prestressed GFRP sheet beam showed displacement as that of non-strengthened beam. The deflection of prestressed CFRP sheet strengthened beam is about one half of that of non-strengthened beam. Saadatmanesh and Ehsani (1991a) provided pre tensioning in FRPC laminate by cambering the beam before bonding it to the beam. Improved cracking behaviour was observed in this cambered beam.

All the experimental studies discussed are backed by analytical studies at macro level. In all the experiments strain compatibility is used to predict the flexural behaviour. The study at micro level was reported by Kaliakan et al. (1996). Eight noded isoparametric element was used to model the concrete non-linear behaviour. This study was conducted by using ABAQUS program. The parametric study performed by Picard et al. (1995) highlights the importance of concrete compressive strength as the limiting factor in external strengthening. Also, strengthening will be more effective (Wei et al., 1991) for less longitudinal steel reinforcement ratios. The experimental results of Sharif et al.(1994) showed that the ductile behaviour of repaired beams is inversely proportional to laminate thickness. However, no explanation was offered by them for this kind of behavior.

The work related to shear strengthening of the beams with different fiber laminates is reported by Michael et al. (1995). Experimentally it was found that the shear strength increased by 60 to 150% of that of unreinforced beam. Orientation of fibers played an important role in influencing the shear strength contribution.

Long-term behaviour of the strengthened beams is also very important and the work in this direction is reported by Xie et al. (1995), Karbhari and Engineer (1996) and Plevris and Triantafillou (1994). The effect of environmental effects on adhesive bond joint between reinforced concrete and carbon fiber laminates was studied by Xie et al. (1995). Experimental results indicated that the water exposure tests did not show a significant effect on flexural strength. But the tests on temperature exposed samples showed a reduction in strength. A peel test to study the effect of environmental factors on the durability of bond between composite and concrete was developed by Karbhari and Engineer (1996). A methodology was also developed to understand the different mechanisms and modes of interfacial fracture. Karbhari and Shulley (1995) developed a wedge test to study the durability of the bond between steel and composite laminate. Creep and shrinkage of strengthened beams were studied by Plevris and Triantafillou (1994). Bank et al. (1995) presented different test methods to study the long term behaviour of FRPC structures. They also mentioned different theoretical models and experimental techniques to predict the effects of different environmental conditions.

An attempt to evolve design guide-lines for strengthened beams (with steel or composite laminate) was made by Ziraba et al. (1994) which are based on previous experimental studies and parametric study conducted by using a finite element program.

Fiber reinforced composite straps are also used to strengthen the concrete columns (Li et al., 1992) and unreinforced masonry walls (Saadatmanesh, 1994) against possible earthquake forces. An effective, economical method to increase the seismic resistance of reinforced concrete column in a structure is by lateral prestressing of the column by wrapping high strength composite straps around them. Analytical studies in this direction are reported by Saadatmanesh et al. (1994). A new concept of column in the form of circular FRPC encased concrete was put forwarded by Fardis and Homayoun (1981). Preliminary experimental work on this column system was reported by Iyer et al. (1995). Mirmiran and Shahawy (1996) developed analytical model to predict the ultimate strength of above column system.

Experimental studies on strengthening precracked solid and voided slabs using CFRP laminates was reported by Shahawy et al. (1996a). The difference in failure modes of strengthened and original slabs indicated the effectiveness of strengthening. The maximum deflection at ultimate loads for retrofitted and precracked voided slabs remain same but the ultimate load for retrofitted slab is about 1.78 times that of precracked slab.

FIBER REINFORCED COMPOSITE BAR AS A REINFORCEMENT FOR CONCRETE MEMBERS

The corrosion of steel bars in reinforced concrete structural members progresses rapidly due to the presence of initial cracks. This phenomenon drastically reduces their service life. To overcome this problem, researchers have started replacing corrosion prone steel bars with, chemically resistant, rust-free fiber reinforced composite bars.

The behaviour of FRPC bar reinforced concrete elements largely depends on the bond behaviour (Nanni et al., 1995) between concrete and composite bar. This paper reviews the behaviour of concrete elements with prestressed or non-prestressed FRPC bars as reinforcement. No effort was made to review the bond behaviour between composite bar and concrete.

The first experimental work using FRPC bars as reinforcement in concrete beams was reported by Rubinsky and Rubinsky (1954). Beams reinforced with FRPC bars behaved in the same way (Saadatmanesh and Ehsani, 1991b., Brown and Bartholomew, 1993) as those reinforced with conventional steel bars. Nawy and Neuwerth (1971) observed 3 to 5 times more cracks in beams reinforced with glass fiber bars when compared to those reinforced with steel. Similar kind of behavior was observed by Chaallal and Benmokrane (1996) at moderate to higher loading. In all the experiments conducted so far higher deflections were observed when compared to steel reinforced beams. This can be attributed to lower Young's modulus of the used FRPC bars when compared to steel. It was observed (Nanni et al., 1994) that the flexural behaviour of the beams can be predicted based on the knowledge of the tensile behaviour of FRPC bars and the assumptions used in conventional reinforced concrete analysis. The design example given by Nanni (1993) proved that there is not much saving of material when beams are designed with FRPC bars. Their use will be justified based on durability and magnetic permeability. The effective use of FRPC bars can be possible in prestressed concrete construction. The experimental work reported by McKay and Erki (1993) showed the improved ultimate strength of beams pretensioned with aramid fiber reinforced plastic tendons but the relaxation losses were high when compared to prestressed beams with steel bars. Also the load carrying capacity was not changed under cyclic loading. Square piles pretensioned with fiber glass and steel bars (Sen et al., 1993) showed a similar behaviour when driven into same soil.

The bond between fiber composite bar and concrete is an important factor in arriving at the design guide-lines. The review of literature related to this aspect is not presented here for the sake of brevity. However, a few important references are listed in Table 1.

FRPC bars are also used as reinforcement in slabs in the form of composite grids by Banthia et al. (1995) and Ahmad et al. (1994). Dutta and Baily (1995) outlined typical tests that have to be carried out on grid frame and grid reinforced concrete. Banthia et al. (1995) compared the behaviour of concrete slabs reinforced with FRPC grids to that of a slab reinforced with steel grid. The ultimate loads supported by slabs reinforced with FRPC were equal to or higher than that supported by the companion slab reinforced with steel. It was concluded that, no significant changes are needed to the various code equations when extending them to slabs reinforced with FRPC reinforcement. Ahmad et al. (1994) presented results of a study on the punching shear behaviour of concrete slabs reinforced with 3D continuous carbon fiber fabric under central concentrated load. Test results revealed that the carbon fiber reinforced concrete slabs exhibited a similar behaviour as that of steel reinforced slab but significant non- linear behaviour and reduction in stiffness was observed in the post cracking stage. The reason for non-linear behaviour after cracking was not explained.

CFRP bars known for their high tensile strength are used in bridges either for pretensioning deck slabs (Abdelrahman et al., 1995) or as cables (Kim and Meier, 1991) in cable stayed bridges. Abdelrahman et al. (1995) after experimentally investigating four concrete T-beams arrived at the design recommendations of beams prestressed with CFRP bars. They also presented different construction details. Different aspects relating to fabrication of prestressed beams using FRPC rods is explained in Sen et al. (1994).

It is again seen that most of the experimental studies are related to flexural behaviour of beams and slabs. No work is reported on behaviour of columns reinforced with FRPC bars.

FIBER REINFORCED COMPOSITES AS STRUCTURAL ELEMENTS

Growing maintenance and longevity problems in transportation structures have led to the use of FRPC material in bridges. Apart from bridge girder strengthening , FRPC structural shapes are used in bridge decks and CFRP cables in cable stayed bridges. Barbero and Ganga Rao (1991) presented an overview of the structural systems constructed with FRPC material and a review of the state of the art on the application of FRPC structural shapes. The use of CFRP cables would allow tripling of the limiting span of cable stayed bridges (Meier, 1986) in comparison to steel wires. Meier (1995) stressed the need to use CFRP cables in suspension bridges and reviews the structural behaviour of different FRPC structural shapes and use of CFRP cables in cable stayed bridges.

Bending stiffness of commercially produced unidirectional glass fiber reinforced plastic beams is low (Barbero et al., 1991., Ganga Rao and Nagaraj, 1995) when compared to that of steel sections of the same shape. Local buckling was responsible for total failure of the beam. Anisotropy properties increased the (Barbero and Raftoyiannis, 1993) buckling strength of column. Theoretical and experimental evaluation of structural behaviour of bridge decks with FRPC shapes was reported by Sotiropoulos et al. (1994). It was observed in this study that FRPC shapes performed satisfactorily as floor systems in terms of stiffness and structural efficiency. An application to design box and I-sections was put forwarded by Ganga Rao et al. (1995a). Lack of design criteria (Ballinger, 1990) and standard test methods are inhibiting the acceptance of FRPC shapes in the construction industry.

Triantafillou and Meier (1992) have presented a new idea of hybrid beam. The hybrid beam consists of GFRP box beams, combined with a layer of concrete in the compression zone and a unidirectional carbon laminate in the tension zone. Using non-linear optimization analysis Triantafillou et al. (1991) optimized the GFRP-reinforced thin-walled rectangular aluminium sections for a given strength and stiffness. The theoretical formulation is verified by conducting three point bending tests. After studying both short term and long term behaviour Deskovic et al. (1995 a,b) suggested design procedure for hybrid members.

TOPICS NOT COVERED IN THE SURVEY

Among the topics of current interest on composite materials and related areas which are not covered in the present survey are: strengthening of reinforced concrete beams with steel plates as external reinforcement and other related papers, bond behaviour between fiber reinforced plastic bars and concrete along with tests on FRPC rods, non-structural behaviour of composite materials (shapes) including their durability, section properties and analytical studies on fiber reinforced composite shapes.

CONCLUDING REMARKS

A review is made of the different applications of FRPC material in structural engineering. Discussion focuses on three application areas

Table-1. Some of the topics not discussed in the paper and a partial list of publications on these topics.

Topic	References
• Strengthening of RC beams with steel plates as external reinforcement and other related papers.	6,12,13,15,20,33,34,37,43,46, 47,48,49,51, 68,69,73,93,104.
• Bond behaviour between fiber reinforced plastic bars and concrete and tests on FRPC rods.	16,17,24,25,36,44,61,62,66, 95.
• Non structural behaviour of composite materials (shape) including durability and section properties.	4,22,21,31,32,88,85,86.
• Analytical studies on fiber reinforced composite shapes.	8.

namely 1. Repair and rehabilitation of structural elements, 2. FRPC bars as a reinforcement material for concrete elements and 3. Structural elements made of fiber reinforced composites. Considerable amount of literature was found on flexural and shear strengthening of concrete beams with FRPC laminates. Research has proved beyond doubt that FRPC laminates are an effective alternative to steel plates which were used earlier for strengthening. Even though their initial cost is higher, it is their light weight and better performance that make them preferable and superior to steel plates. Also their life-cycle cost is lower as compared to other conventional materials. Bond between composite and concrete, durability of bond, behaviour of bond under cyclic loading are some of the shadow areas that have to be addressed in detail. Most of the available literature pertains to strengthening reinforced concrete elements. Strengthening of steel structural elements is another potential area to be explored. Complete design guide-lines can be developed only after knowing the durability and long-term behaviour of the strengthened beams.

The literature on FRPC bars as reinforcement shows the importance of understanding bond between FRPC bar and concrete. Ductility is the main criteria while using FRPC bars in lieu of steel bars. Also there will be excessive deflection because of their lower Young's modulus as compared to steel. Using FRPC bars can be suggested at places where corrosion and magnetic permeability are the governing factors. These bars seem to be more promising in prestressed beam construction because of their high tensile strengths. Continuous FRPC reinforcement in the form of grids can also be an alternative to steel bars. But their application in the field should be backed by detailed research. No literature was found on columns with FRPC bars as reinforcement. Before arriving at the full fledged design guide-lines for concrete elements with FRPC bars, long-term behaviour of bond between FRPC bars and concrete has to be established.

ACKNOWLEDGEMENTS

This study was carried out under an USIF funded research project entitled, *Performance of fiber reinforced polymer composites as structural reinforcements in hot, cold and humid environment* (IND104) and partially supported by Department of Army of the US Army Research. The authors are also thankful to Dr. Jerry C. Comati, Chief of the Environmental Sciences Branch for his encouragement.

REFERENCES

1. Abdelrahaman, A, Tadros, G, and Rizkalla, SH (1995). "Test model for the first Canadian smart highway bridge," *ACI Structural Journal*, Vol 92, No 4, pp 451-458.
2. Ahmad, SH, Zia, P, Yu, TJ, and Xie, Y (1994). "Punching shear test of slabs reinforced with 3D carbon fiber fabric," *ACI Concrete International, Design & Construction*, Vol 16, No 6, pp 36-41.
3. Ballinger, CA (1990). "Structural FRP composites- Civil engineering materials of the future?," *Civil Engineering, ASCE*, Vol 60, No 7, pp 63-65.
4. Bank, LC (1990). "Shear properties of pultrated glass Fiber reinforced plastic materials," *ASCE Journal of Materials in Civil Engineering*, Vol 2, No 2, pp 118-122.
5. Bank, LC, Gentry, TR, and Barakatt, A (1995). "Accelerated test methods to determine the long-term behaviour of FRP composite structures: environmental effects," *Journal of Reinforced Plastics and Composites*, Vol 14, pp 559-586.
6. Banthia, N, and Dubeau, S (1994). "Carbon and steel micro fiber-reinforced cement based composites for thin repairs," *ASCE Journal of Materials in Civil Engineering*, Vol 6, No 1, pp 88-99.
7. Banthia, N, Al-Aslay, M, and Ma, S (1995). "Behaviour of concrete slabs reinforced with fiber reinforced plastic grid," *ASCE Journal of Materials in Civil Engineering*, Vol 7, No 4, pp 252-257.
8. Barbero, EJ, and Raftoyiannis, IG (1994). "Lateral and distortion buckling of pultrated I-beams," *Composites and Structures*, Vol 27, No 3, pp 261-268.
9. Barbero, EJ, and Raftoyiannis, IG (1993). "Local buckling of FRP beams and columns," *ASCE Journal of Materials in Civil Engineering*, Vol 5, No 3, pp 339-355.
10. Barbero, EJ, Shin-HamFu, and Raftoyiannis, I (1991). "Ultimate bending strength of composite beams," *ASCE Journal of Materials in Civil Engineering*, Vol 3, No 4, pp 292-306.
11. Barbero, EJ, and Ganga Rao, HVS (1991). "Structural applications of composites in infrastructure Part: I," *SAMPE Journal*, Vol 27, No 6, pp 9-16.
12. Basunbul, IA, Gubati, AA, Al-Sulaimani, GJ, and Baluch, MH (1990). "Repaired reinforced concrete beams, "*ACI Materials Journal*, Vol 87, No 4, pp348-354.
13. Berridge, PSA, and Donovan, LH (1956). "Prestressing restores weakened truss bridge," *ASCE Civil Engineering*, Vol 26, No 9, pp 48-49.
14. Brown, VL, and Bartholomew, CL (1993). "Fiber reinforced bars in reinforced concrete members," *ACI Materials Journal*, Vol 90, No 1, pp34-39.
15. Cairns, J (1996). "Alternative strategies for temporary support during structural repair of reinforced concrete beams," *ASCE Journal of Structural Engineering*, Vol 122, No 3, pp 238-246.
16. Chaallal, O, and Benmokrane (1993). "Physical and mechanical performance of an innovative glass-fiber-reinforced plastic rod for concrete and grouted anchorages," *Canadian Journal of Civil Engineering*, Vol 20, No 2, pp 254-268.
17. Daniali, S (1990). "Bond strength of fiber reinforced plastic bars in concrete," *Serviceability and Durability of Construction Materials*, Proc. of first materials engineering congress. Part-2. ASCE, pp 1182-1191.
18. Deskovic, N, Meier, U, and Triantafillou, TC (1995a). "Innovative design of FRP combined with concrete: long term behaviour," *ASCE Journal of Structural Engineering*, Vol 121, No 7, pp 1079-1089.
19. Deskovic, N, Triantafillou, TC, and Meier, U (1995b). "Innovative design of FRP combined with concrete: short-term behaviour," *ASCE Journal of Structural Engineering*, Vol 121, No 7, pp 1069-1078.

20. Dussek, IJ (1980). "Strengthening of bridge beams and similar structures by means of epoxy-resin-bonded external reinforcement," Transportation Research Record No. 785, Transportation Research Board, pp 21-24.
21. Dutta, PK, and Lampoo, RG (1993). "Behaviour of fiber-reinforced plastics as construction materials in extreme environments," *Proc. of the third International Offshore and Polar Engineering Conference.* Published by International society of offshore and polar engineers *(ISOPE)*, pp 339-344.
22. Dutta, PK (1995). "Durability of FRP composites," in *Fiber Reinforced Structural Plastics in Civil Engineering*, SL. Iyer and V. Kalyanaraman, eds. (Proceedings of the International Conference on Fiber Reinforced Structural Plastics in Civil Engineering, IIT-Madras, 18-20 December 1995), Tata McGraw-Hill, New Delhi. pp 360-370.
23. Dutta, PK, and Bailey, D (1995). "FRP composite grid/frame structures for reinforced concrete," in *Fiber Reinforced Structural Plastics in Civil Engineering*, SL. Iyer and V. Kalyanaraman, eds. (Proceedings of the International Conference on Reinforced Structural Plastics in Civil Engineering, IIT-Madras, 18-20 December 1995), Tata McGraw-Hill, New Delhi. pp 499-508.
24. Ehsani, MR, Saadatmanesh, H, and Tao, S (1995). "Bond of hooked glass fiber reinforced plastic (GFRP) reinforcing bars to concrete," *ACI Materials Journal*, Vol 92, No 4, pp 391-400.
25. Ehsani, MR, Saadatmanesh, H, and Tao, S (1996). "Design recommendations for bond of GFRP rebars to concrete," *ASCE Journal of Structural Engineering*, Vol 122, No 3, pp 247-254.
26. Erki, MA, and Rizkalla, SH (1993). "FRP reinforcement for concrete structures," *ACI Concrete International. Design and Construction*, Vol 15, No 6, pp 48-53.
27. Fardis, MN, and Homayoun, K (1981). "Concrete encased in fiberglass-reinforced plastic," *ACI Journal*, Vol 78, No 6, pp 440-446
28. Faza, SS, and Ganga Rao, HVS (1994). "Fiber composite wrap for rehabilitation of concrete structures," *Proceedings of the Materials Engineering Conference 804, ASCE*, pp 1135-1139.
29. Ganga Rao, HVS, Lopez-Anido, R, Dutta, PK, and Trovillion, JC (1995a). "Design of fiber reinforced plastic structural members," in *Fiber Reinforced Structural Plastics in Civil Engineering*, SL. Iyer and V. Kalyanaraman, eds. (Proceedings of the International Conference on Fiber Reinforced Structural Plastics in Civil Engineering, IIT-Madras, 18-20 December 1995), Tata McGraw-Hill, New Delhi. pp 481-489.
30. Ganga Rao, HVS, Nagaraj, V (1995b). "Static behaviour of pultrated GFRP sections. Infrastructure: New materials and methods of repair," in *Fiber Reinforced Structural Plastics in Civil Engineering*, SL. Iyer and V. Kalyanaraman, eds. (Proceedings of the International Conference on Fiber Reinforced Structural Plastics in Civil Engineering, IIT-Madras, 18-20 December 1995), Tata McGraw-Hill, New Delhi. pp 844-848.
31. Ganga Rao, HVS, Vijay, PV, and Altizer, SD (1995c). "Durability of glass composites under alkaline and prestress environment." in *Fiber Reinforced Structural Plastics in Civil Engineering*, SL. Iyer and V. Kalyanaraman, eds. (Proceedings of the International Conference on Fiber Reinforced Structural Plastics in Civil Engineering, IIT-Madras, 18-20 December 1995), Tata McGraw-Hill, New Delhi. pp 371-378.
32. Ghyslaine, M, and Mohammadi, Y (1995). "Compression creep of pultrated E-glass-reinforced-plastic angles, "*ASCE Journal of Materials in Civil Engineering*, Vol 7, No 4, pp 269-276.
33. Hamoush, SA, and Ahmad, SH (1990a). "Debonding of steel plate- strengthened concrete beams," *ASCE Journal of Structural Engineering*, Vol 116, No 2, pp 356-371.
34. Hamoush, SA, and Ahmad, SH (1990b). "Static strength tests of steel plate strengthened concrete beams," *Materials and Structures*. Vol 23, pp 116-125.
35. Iyer, SL, Kortikere, A, and Khubchandani, A (1995). "Confining concrete and sand with composite tubes" in *Fiber Reinforced Structural Plastics in Civil Engineering*, SL. Iyer and V. Kalyanaraman, eds. (Proceedings of the International Conference on Fiber Reinforced Structural Plastics in Civil Engineering, IIT-Madras, 18-20 December 1995), Tata McGraw-Hill, New Delhi. pp 188-202.
36. Javier Malver, L (1995). "Tensile and bond properties of GFRP reinforcing bars," *ACI Materials Journal*, Vol 92, No 3, pp 276-285.
37. Jones, R, Swamy, RS, Bloxham, J, and Bouderbalah, A (1980). "Composite behaviour of concrete beams with epoxy bonded external reinforcement," *International Journal of Cement Composites and Lightweight Concrete*, Vol 2, No 2, pp 91-107.
38. Kaliakan, VN, Chajes, MJ, and Januszka, J (1996). "Analysis of concrete beams reinforced with externally bonded woven composite fabrics," *Composites Part-B*, Vol 27B, No 3/4, pp 235-244.
39. Karbhari, KM, and Shulley, SB (1995). "Use of composites for rehabilitation of steel structures - Determination of bond durability," *ASCE Journal of Materials in Civil Engineering*, Vol 7, No 4, pp 239-245.
40. Karbhari, VM (1995). "Civil infrastructure and composite materials: Issues in materials and manufacturing," in *Fiber Reinforced Structural Plastics in Civil Engineering*, SL. Iyer and V. Kalyanaraman, eds. (Proceedings of the International Conference on Fiber Reinforced Structural Plastics in Civil Engineering, IIT-Madras, 18-20 December 1995), Tata McGraw-Hill, New Delhi. pp 17-26.
41. Karbhari, VM, and Engineer, M (1996). "Investigation of bond between concrete and composites: Use of a peel test," *Journal of Reinforced Plastics and Composites*, Vol 15, No 2, pp 208-227.
42. Kim, P, and Meier, U (1991). "CFRP cables for large structures," *Advanced Composites. Materials in Civil Engineering Structures Proceedings*. MT Div/ASCE/Las Vegas, pp 233-244.
43. Klaiber, FW, Dunker, KF, and Sanders, WW (1982). "Strengthening of single span steel-beam bridges," *ASCE Journal of Structural Engineering*, Vol 108, No 12, pp 2766-2780.
44. Larralde, J, and Silva-Rodriguez, R (1993). "Bond and slip of FRP rebars in concrete," *ASCE journal of Materials in Civil Engineering*, Vol 5, No 1, pp 30-40.
45. Li, MW, Saadatmanesh, H, and Ehsani, MR (1992). "Behaviour of externally confined concrete columns," *Materials engineering congress. ASCE*, pp 677-690.
46. Lloyd, GO, and Calder, AJJ (1982). "The micro structure of epoxy-bonded steel-to-concrete joints," Supp. Rept. No. 705, Transport and Road Research Laboratory, Crowthorne.
47. Maeda, Y, Matsui, S, Shimada, I, and Kato, H (1980). "Deterioration and repairing of reinforced concrete of highway bridges in Japan," Technology Reports, Osaka Univ., 30, pp 135-144.
48. Mancarti, GD (1984). "Strengthening of California steel bridges by prestressing," Transportation Research Record No. 950, Transportation Research Board, pp 183-187.
49. McDonald, MD, and Calder, AJJ (1982). "Bonded steel plating for strengthening concrete structures," *International Journal of Adhesion and Adhesives*, Vol 2, No 2, pp 119-127.
50. McKay, KS, and Erki, MA (1993). "Flexural behaviour of concrete beams pretensioned with aramid fiber reinforced plastic tendons," *Can. J. Civ. Engng, Vol* 20, No 4, pp 688-695.
51. McKenna, JK, and Erki, MA (1994). "Strengthening of reinforced concrete flexural members using externally applied steel plates and fiber composite sheets-a survey," *Canadian Journal of Civil Engineering*, Vol 21, No 1, pp 16-24.
52. Meier, U (1987). "Proposal for a carbon fiber reinforced composite bridge across the strait of Gibraltar at its narrowest

site," *Proc. Instn. Mech. Engrs*, Vol 201, No B2, pp 73-78.

53. Meier, U (1988). "Designing and producing materials by combination," *Materials and Structures*, Vol 21, pp 85-89.
54. Meier, U, and Kaiser, HP (1991). "Strengthening of structures with CFRP laminates," *Proc. Advanced Composite Materials in Civil Engineering Structures, MT. Div./ASCE/Las Vegas*, pp 224-232.
55. Meier, U, Deuring, M, Meier, H, and Schwegler, G (1992). "Strengthening of structures with CFRP laminates: Research and applications in Switzerland," *Advanced Composite Materials in Bridges and Structures* K. W. Neale and P. Labossiere, Editors; Canadian Society for Civil Engineering.
56. Meier, U (1992). "Carbon fiber-reinforced polymers: Modern materials in bridge engineering," *Structural Engineering International,* Vol 1, pp 7-12.
57. Meier, U (1995). "Evolution of suspension and cable stayed bridges through the use of CFRP cables," in *Fiber Reinforced Structural Plastics in Civil Engineering*, S.L. Iyer and V. Kalyanaraman, eds. (Proceedings of the International Conference on Fiber Reinforced Structural Plastics in Civil Engineering, IIT-Madras, 18-20 December 1995), Tata McGraw-Hill, New Delhi. pp 247-254.
58. Michael, JC, Ted, FJ, Deninnis, RM,Theodore, AT Jr, and William, WF Jr (1995). "Shear strengthening of reinforced concrete beams using externally applied composite fabrics," *ACI Structural Journal*, Vol 92, No 3, pp 295-303.
59. Mirmiran, A, and Shahawy, M (1996). "A new concrete filled hollow FRP composite column," *Composites Part-B*, Vol 27B, No3/4, pp 263-268.
60. Nanni, A, Bakis, CE, and Boothby, TE (1995). "Test methods for FRP-concrete systems subjected to mechanical loads: state of the art review," *Journal of Reinforced Plastics and Composites*, Vol 14, No 6, pp 524-558.
61. Nanni, A, Tanagaki, M, and Hasuo, K (1992). "Bond anchorage of pretensioned FRP tendon at force release," *ASCE Journal of Structural Engineering*, Vol 118, No 10, pp 2837-2854.
62. Nanni, A, Utsunomiya, T, Yonekura, H, and Tanigaki, M (1992). "Transmission of prestressing force to concrete by bonded fiber reinforced plastic tendons," *ACI Structural Journal.* Vol 89, No 3 pp 335-344.
63. Nanni, A (1993). "Flexural behaviour and design of RC members using FRP reinforcement," *ASCE Journal of Structural Engineering*, Vol 119, No 11, pp 3344-3359.
64. Nanni, A, Henneke, MJ, and Okomoto, T (1994). "Behaviour of concrete beams with hybrid reinforcement," *Construction and Building Materials*, Vol 8, No 2, pp 89-95.
65. Nanni, A (1995). "Concrete repair with externally bonded FRP reinforcement," *ACI Concrete International; Design and Construction*, Vol 17, No 6, pp 22-26.
66. Nanni, A, and Tanigaki, M (1992). "Pretensioned prestressed concrete members with bonded fiber reinforced plastic tendons: Development and flexural bond lengths (static)," *ACI Structural Journal*, Vol 89, No 4, pp 433-441.
67. Nawy, EG, Neuwerth, GE, and Phillips, CJ (1971). "Behaviour of fiber glass reinforced concrete beams," *Proceedings of the ASCE Journal of Structural Division*, Vol 97, No ST9, pp 2203-2215.
68. Oehlers, DJ, and Moran, JP (1990). "Premature failure of externally plated reinforced concrete beams," *ASCE Journal of Structural Engineering*, Vol 116, No 4, pp 978-995.
69. Oehlers, DJ (1992). Reinforced concrete beams with plates glued to their soffits, *ASCE Journal of Structural Engineering*, Vol 118, No 8, pp 2023-2038.
70. Picard, A, Bruno, M, and Eric, B (1995). "Strengthening of reinforced concrete beams with composite materials: Theoretical study," *Composite Structures*, Vol 33, No 2, pp 63-75.
71. Plevris, N, and Triantafillou, TC (1992). "FRP reinforced wood as structural material," *ASCE Journal of Materials in Civil Engineering*, Vol 4, No 3, pp 300-317.
72. Plevris, N, and Triantafillou, TC (1994). "Time-dependent behaviour of reinforced concrete members strengthened with FRP laminates," *ASCE Journal of Structural Engineering*, Vol 120, No 3, pp 1016-1042.
73. Raithby, KD (1980). "External strengthening of concrete bridges with bonded steel plates," Supp. Rept. No. 612, Transport and Road Research Laboratory, Crowthorne.
74. Ritchie, PA, David, AT, Le-Wulu and Guy, MC (1991). "External reinforcement of concrete beams using fiber reinforced plastics," *ACI Structural Journal*, Vol 88, No 4, pp 490-500.
75. Rizkalla, SH, and Tadros, G (1994). "A smart highway bridge in Canada," *ACI Concrete International; Design and Construction*, Vol 16, No 6, pp 42-44.
76. Rizkalla, SH, and Abdelrahman, AA (1995). "FRP for the 21st century," in *Fiber Reinforced Structural Plastics in Civil Engineering*, SL. Iyer and V. Kalyanaraman, eds. (Proceedings of the International Conference on Fiber Reinforced Structural Plastics in Civil Engineering, IIT-Madras, 18-20 December 1995), Tata McGraw-Hill, New Delhi. pp 3-16.
77. Rubinsky, IA, and Rubinsky, A (1954). "Preliminary investigation of the use of fiber glass for prestressed concrete," *Magazine of Concrete Research*, Vol 6, No 17, pp 71-78.
78. Saadatmanesh, H, and Ehsani, MR (1990a). "Flexural strength of externally reinforced concrete beams," *Serviceability and Durability of Construction Materials.* Proc. First ASCE Materials Engineering Congress. Part-2., ASCE, pp 1152-1161.
79. Saadatmanesh, H, and Ehsani, MR (1990b). "Fiber composite plates can strengthen beams," *ACI Concrete International; Design and Construction*, Vol 12, No 3, pp 65-71.
80. Saadatmanesh, H, and Ehsani, MR (1991a). "Reinforced concrete beams strengthened with GFRP plates-I: Experimental study," *ASCE Journal of Structural Engineering*, Vol 117, No 11, pp 3417-3433.
81. Saadatmanesh, H, and Ehsani, MR (1991b). "Fiber composite bar for reinforced concrete construction," *Journal of Composite Materials*, Vol 25, No 2, pp 188-203.
82. Saadatmanesh, H, Ehsani, MR, and Li, MW (1994). "Strength and ductility of concrete columns externally reinforced with fiber composite straps," *ACI Structural Journal*, Vol 91, No 4, pp 434-447.
83. Saadatmanesh, H (1994). "Fiber composites for new and existing structures," *ACI Structural Journal,* Vol 91, No 3, pp 346-354.
84. Sen, R, Issa, M, Wadsack, P, and Shahawy, M (1993). "Driving stresses in fiberglass pretensioned piles," *ACI Structural Journal,* Vol 90, No 6, pp 666-674.
85. Sen, R, Mariscal, D, and Shahawy, M (1993). "Durability of fiberglass pretensioned beams," *ACI Structural Journal*, Vol 90, No 5, pp 525-533.
86. Sen, R, Mariscal, D, Issa, M, and Shahawy, M (1993). "Durability and ductility of advanced composites," *Proc. of the Symposium on Structural Engineering in Natural Hazards Mitigation*, Published by ASCE, pp 1373-1378.
87. Sen, R, Spillett, K, and Shahawy, M (1994). "Fabrication of aramid and carbon fiber reinforced plastic pretensioned beams," *ACI Concrete International; Design and Construction*, Vol 16, No 6, pp 45-47.
88. Sen, R, Shahawy, M (1995). "Durability testing of fiber reinforced plastics," in *Fiber Reinforced Structural Plastics in Civil Engineering*, SL. Iyer and V. Kalyanaraman, eds. (Proceedings of the International Conference on Fiber Reinforced Structural Plastics in Civil Engineering, IIT-Madras, 18-20 December 1995), Tata McGraw-Hill, New Delhi. pp 353-359.

89. Shahawy, MA, Arockiasamy, M, Beitelman, T, and Sowrirajan, R (1996b). "Reinforced concrete rectangular beams strengthened with CFRP laminates," *Composites Part-B*, Vol 27B, No 3/4, pp 225-233.

90. Shahawy, MA, Beitelman, T, Arockiasamy, M, and Sowrirajan, R (1996a). "Experimental investigation on structural repair and strengthening of damaged prestressed concrete slabs utilizing externally bonded carbon laminates," *Composites Part-B*, Vol 27B, No 3/4, pp 217-224.

91. Sharif, A, Al-Sulaimani, GJ, Basunbal, IA, Baluch, MH, and Ghaleb, BN (1994). "Strengthening of initially loaded reinforced concrete beams using FRP plates," *ACI Structural Journal*, Vol 91, No 2, pp 160-168.

92. Sotiropoulos, SN, Ganga Rao, HVS, and Ahmed, NKM (1994). "Theoretical and experimental evaluation of FRP components and systems," *ASCE Journal of Structural Engineering*, Vol 120, No 2, pp 464-485.

93. Swamy, RN, Jones, R, and Bloxham, JW (1987). "Structural behaviour of reinforced concrete beams strengthened by epoxy bonded steel plates," *The Structural Engineer*, Vol 65, No 2, pp 59-68.

94. Takada, K, Mitsui, Y, and Murakami, K (1996). "Flexural strength of reinforced concrete beams strengthened with carbon fibre sheets," *Composites Part A*, Vol 27A, No 10, pp 981-987.

95. Tao, S, Ehsani, MR, and Saadatmanesh, H (1992). "Bond strength of straight GFRP bars," *Materials: Performance and Prevention of Deficiencies and Failures 92,* Materials Engineering Congress. ASCE, pp 598-605.

96. Tarricone, P (1995). "Composite sketch," *ASCE Civil Engineering*, Vol 65, No 5, pp 52-55

97. Triantafillou, TC (1990). "Mechanics of short term behaviour, *ASCE Journal of Engineering Mechanics*, Vol 117, No 7, pp 1652-1672.

98. Triantafillou, TC, Kim, P, and Meier, U (1991). "Optimization of hybrid aluminium/CFRP box beams," *International Journal of Mechanical Sciences*, Vol 33, No 9, pp 729-739.

99. Triantafillou, TC, and Meier, U (1992). "Innovative design of FRP combined with concrete," in *Advanced Composite Materials in Bridges and Structures* K.W Neale and P. Labossiere, Editors; Canadian Society for Civil Engineering. pp491-500.

100. Triantafillou, TC, and Plevaris, N (1992). "Strengthening of reinforced concrete beams with epoxy bonded fiber-composite materials," *Materials & Structures*, Vol 25, No 145, 201-211.

101. Triantafillou, TC, and Deskovic, N (1992). "Prestressed FRP sheets as external reinforcement of wood members," *ASCE Journal of Structural Engineering*, Vol 118, No 5, pp 1270-1284.

102. Triantafillou, TC, Deskovic, N, and Deuring, M (1992). "Strengthening of concrete structures with prestressed fiber reinforced plastic sheets," *ACI Structural Journal*, Vol 89, No 3, pp 235-244.

103. Victor, CLi (1995). "New construction materials proliferate in Japan," *ASCE Civil Engineering*, Vol 65, No 8, pp 38-41.

104. Wayne, FK, Kenneth, FD, and Wallace, WS (1982). "Strengthening of single span steel beam bridges," *ASCE Journal of Structural Engineering*, Vol 108, pp 2766-2780.

105. Wei, A, Saadatmanesh, H, and Ehsani, MR (1991). "Reinforced concrete beams strengthened with FRP plates II. Analytical and parametric study," *ASCE Journal of Structural Engineering*, Vol 117, No 11, pp 3434-3455.

106. Xie, M, Hoa, SV, and Xiao, XR (1995). "Bonding steel reinforced concrete with composites," *Journal of Reinforced Plastics and Composites*, Vol 14, No 9, pp 949-964.

107. Ziraba, YN, Baluch, MH, Basunbul, IA, Sharif, AM, Azad, AK, and Al- Sulaimani, GJ (1994). "Guide-lines toward the design of reinforced concrete beams with external plates," *ACI Structural Journal*, Vol 91, No 6, pp 639-646.

Proceedings of the Seventh (1997) International Offshore and Polar Engineering Conference
Honolulu, USA, May 25-30, 1997

ISBN 1-880653-28-1 (Set); ISBN 1-880653-32-X (Vol. IV)

Study of FEM Simulation of Cyclic Plastic Behavior of Face-Centered Crystal (f.c.c.) Polycrystalline Materials

Yasumitsu Tomita, Kiyoshi Hashimoto and Naoki Osawa
Osaka University
Osaka, Japan

Taira Ozaki
Komatsu Co., Ltd.
Japan

Abstract

In order to improve a prediction method of crack initiation and failure life, it is needed to acquire knowledge about microscopic deformation mechanisms of materials under cyclic loading conditions. For this reason, crystal-plastic FEM analysis for cyclic plastic deformation of f.c.c. polycrystalline aggregate is performed.

Some microscopic hardening rules which can derive actual macroscopic stress-strain relations under cyclic loading conditions are obtained based on the analysis. There is a tendency that the local plastic deformation on free surface is larger than that within the model.

KEY WORDS: Cyclic plasticity, Crystal plasticity, Finite element method, Fatigue strength

1. Introduction

The deformation behavior of metallic materials during fatigue process is described as cyclic plastic behavior. Micro mechanisms of the cyclic plastic behavior can be related to failure. Polycrystalline aggregate deforms non-uniformly, and the concentration of local plastic deformation is observed on free surface of aggregates. Since fatigue cracks usually start from free surfaces, it can be assumed that such concentration is related to crack initiation. Therefore, it is needed to acquire knowledge about the microscopic deformation mechanisms under cyclic loading conditions in order to improve a prediction method of the crack initiation and failure life.

Up to now, the models of dislocation movement under cyclic loading conditions have been proposed, but it is very difficult to decide the propriety of proposed models only by observing directly dislocation on free surface. For the judgment of propriety, it is useful method to compare the simulated macroscopic deformation behavior derived from the dislocation models with stress-strain relation of actual materials obtained from fatigue tests. For the simulation of both macroscopic and microscopic behaviors, crystalline FEM analysis is useful. However, there is no established way to determine the microscopic hardening rule based on dislocation model to analyze macroscopic cyclic plastic deformation.

In this paper, we propose two microscopic hardening rules which can be used for the analysis of cyclic plastic deformation. Then, we carry out FEM analysis of 2-dimensional and 3-dimensional deformation of f.c.c. polycrystalline using the proposed hardening rules, and investigate the characteristics of non-uniform local deformation. By using macroscopic stress-strain relation which can be calculated from the non-uniform local deformation, we discuss the propriety of proposed hardening rules under cyclic loading conditions.

2. Basic Equations

The main subject of this study is to analyze the behavior of cyclic plastic deformation during high cycle fatigue process in room temperature. The following is assumed.

1) Grain boundary slip is neglected, and the thickness of grain boundary is zero.
2) Inelastic deformation is caused only by crystal slip. Activation condition of the slip systems is given by Schmid's law.
3) Elastic deformation of crystal is specified by the generalized Hook's law.
4) Deformation is small so that infinitesimal deformation formulation of FEM analysis can be acceptable and lattice rotation of crystal can be neglected.

2.1 Constitutive equation

Based on the assumption 4) of infinitesimal deformation formulation, the 'stress' means Cauchy stress tensor which is equal to other type stress tensors, and 'strain' means the infinitesimal strain tensor.

The crystal slip system (n) consists of slip direction $\mathbf{s}^{(n)}$ lying in a slip plane whose normal is $\mathbf{m}^{(n)}$. Both $\mathbf{m}^{(n)}$ and $\mathbf{s}^{(n)}$ are taken to be unit vectors. We now define the second-rank tensor as follows

$$\mathbf{P}^{(n)} = \frac{1}{2}\left(\mathbf{s}^{(n)} \otimes \mathbf{m}^{(n)} + \mathbf{m}^{(n)} \otimes \mathbf{s}^{(n)}\right). \qquad (1)$$

Based on the assumption 4) of neglecting of the lattice rotation, $\mathbf{m}^{(n)}, \mathbf{s}^{(n)}$ and $\mathbf{P}^{(n)}$ remain constant during deformation.

Using the above tensor $\mathbf{P}^{(n)}$, for the slip system (n), the relationship between the increment of the resolved shear stress (R.S.S.) $\tau^{(n)}$ and the local stress tensor σ is given by

$$\dot{\tau}^{(m)} = P_{ij}^{(m)}\dot{\sigma}_{ij} \quad (2)$$

and the relationship between the increment rate of shear strain of the slip system $\gamma^{(n)}$ and the local plastic strain ε^{p} is given by

$$\dot{\tau}^{(m)} = P_{ij}^{(m)}\dot{\sigma}_{ij}. \quad (3)$$

The assumption 2) produces eq. (4) for the increment rate of total strain $\dot{\varepsilon}$

$$\dot{\varepsilon}_{ij} = \dot{\varepsilon}_{ij}^{e} + \dot{\varepsilon}_{ij}^{p} \quad (4)$$

where, $\dot{\varepsilon}_{e}$ is the increment rate of the elastic strain, and it is specified by the generalized Hook's law

$$\dot{\varepsilon}_{ij}^{e} = S_{ijkl}^{e}\dot{\sigma}_{kl} \quad (5)$$

where, S_{ijkl}^{e} denotes the elastic compliance tensor.

R.S.S. $\tau^{(n)}$ determines the behavior of plastic slip motion as follows. When $\tau^{(n)}$ reaches the initial critical resolved shear stress (C.R.S.S.) τ_0, which is initially taken equal to all slip systems, plastic slip initiates on the system (n). After initial yield occurs, plastic slip continues on system (n) if the R.S.S. keeps pace with the flow stress, which is increased by strain hardening. Thus, for active slip systems n and m, the increment of R.S.S. $\tau^{(n)}$ is described as

$$\dot{\tau}^{(n)} = \sum_{m=1}^{12} h^{(nm)}\dot{\gamma}^{(m)}, \quad \dot{\gamma}^{(n)} > 0 \quad (6)$$

while for inactive systems p

$$\dot{\tau}^{(p)} < \sum_{m=1}^{12} h^{(pm)}\dot{\gamma}^{(m)}, \quad \dot{\gamma}^{(p)} = 0 \quad (7)$$

where $h^{(nm)}$ denotes strain-hardening coefficient.

Based on equations (1) to (7), the constitutive equation for infinitesimal deformation can be derived as follows (Hill(1966)):

$$d\varepsilon_{ij} = \left(S_{ijkl}^{e} + \sum_{n}\sum_{m}\left[h^{(nm)}\right]^{-1} P_{ij}^{(n)} P_{kl}^{(m)} \right) d\sigma_{kl} \quad (8)$$

2.2 *Strain hardening*

Ohashi(1987) has developed the model for mean glide path distance of dislocations during both single slip and multiple slips under monotonous loading condition, and derived the approximate form of the strain-hardening coefficient $h^{(nm)}$ as

$$h^{(nm)} = \Omega^{(nm)} \frac{\partial F\left(f(\rho^{(m)})\right)}{\partial \rho^{(m)}} \frac{\partial \rho^{(m)}}{\partial \gamma^{(m)}}. \quad (9)$$

where,

$$F(\gamma^{(m)}) = F(f(\rho^{(m)})) = a\mu b\sqrt{\rho^{(m)}};$$
$$\tau_y = \tau_0 + \sum_{m=1}^{12} \Omega^{(nm)} F(\gamma^{(m)}) \quad (10)$$

$\rho^{(m)}, b, \mu, a, \tau_y$, and τ_0 denote the density of dislocations on slip system (m), the magnitude of burger's vector, the elastic shear modulus, a numerical factor of order 0.1, C.R.S.S., and the initial C.R.S.S, respectively. $\Omega^{(nm)}$ is the 'interaction matrix', which determines the latent hardening characteristics of the slip systems. Ohashi(1987) has determined this matrix by using the experimental results of Jackson-Basinski(1967) on a copper single crystal. The above hardening rule is developed for the case of monotonous tensile loading. In monotonous loading, the density of dislocations increases as the shear strain caused by plastic slips increases, and the right hand side of eq. (9) increases.

In this paper, we investigate the cyclic plastic deformation during fatigue process. It is well known that the density of dislocations does not change during cyclic deformation after the cyclic hardening is saturated. In such situations, it can be supposed that the hardening coefficient $h^{(nm)}$ does not vary during cyclic deformation. So, we can choose $h^{(nm)}$ to be

$$h^{(nm)} = \Omega^{(nm)} H_0. \quad (11)$$

Following Ohashi(1987), we also choose $\Omega^{(nm)}$ to be the values of Jackson-Basinski(1967).

2.3 *Evolution of C.R.S.S.*

In order to analyze cyclic plastic behavior, it is needed to define the way of determining C.R.S.S on the slip systems when the loading direction is reversed from tension or compression to compression or tension (the reversal of a loading).

It is considered that the sources of strain hardening are both the short and long range stress fields induced by accumulated dislocations during plastic slips. Next, we consider how these stress fields affect the plastic slip deformation when the loading direction is reversed. If these stress fields do not disappear during the reversal of a loading, the short range stress field becomes an obstacle to the plastic slip deformation, while the long one activates dislocation sources and decreases C.R.S.S. In such case, C.R.S.S. $\tau_y^{(m)}$ can be described in the form

$$\tau_y^{(m)} = R^{(m)} - \alpha^{(m)} \quad (12)$$

where, $R^{(m)}$ is the C.R.S.S due to the short range stress, and $\alpha^{(m)}$ is the back stress of the slip system (m). In order to define the evolution of $R^{(m)}$ and $\alpha^{(m)}$, we introduce a parameter C_r, which is the ratio of the contribution of short range stress to the increase of C.R.S.S. during plastic slips. Denoting $R_0^{(m)}$ and $\alpha_0^{(m)}$ are the values of $R^{(m)}, \alpha^{(m)}$ at the moment of the reversal of a previous loading, and $d\tau_y^{(m)}$ is the increment of C.R.S.S. from the moment of the reversal of a previous loading, $\tau_y^{(m)}$ can be described as

$$\tau_y^{(m)} = R^{(m)} - \alpha^{(m)}$$
$$;R^{(m)} = R^{(m)} + C_r d\tau_y^{(m)}, \quad \alpha^{(m)} = \alpha_0^{(m)} - (1-C_r)d\tau_y^{(m)}, \quad (13)$$
$$d\tau_y^{(m)} = \int \sum_{n=1}^{12} h_{mn}\dot{\gamma}^{(n)} dt.$$

The hardening is kinematic if $C_r = 0$, while it is isotropic if $C_r = 1$. Generally, C_r varies as the plastic strain varies. It is well known that annealed metals show cyclic hardening at the beginning of cyclic loading, then the cyclic hardening saturates as the number of loading cycles increase. The saturation becomes faster as the amplitude of stress or plastic strain becomes larger. Therefore, when C_r can be assumed to depend on the equivalent plastic strain $\kappa = \int_0^t \left[\frac{2}{3}\dot{\varepsilon}_{ij}^{p}\dot{\varepsilon}_{ij}^{p}\right]^{1/2} dt$, it can be written as

$$C_r(\kappa) = C_{r0}\exp(-k_r\kappa). \quad (14)$$

In this paper, we consider the steady cyclic deformation after the saturation of cyclic hardening, therefore, C_r is only the following case,

$$C_r \equiv 0. \quad (15)$$

On the other hand, some or all of the stress fields induced by plastic slips may disappear at the reversal of a loading, because the dislocation structures generated during plastic slips are quasi-stable. We also analyze the case where the increment of C.R.S.S. caused by plastic slips disappear during the reversal of a loading. For such case,

$$\tau_y^{(m)} = \tau_0 + \int \sum_{n=1}^{12} h_{mn}\dot{\gamma}^{(n)} dt \quad \text{when plastic slip occurs}$$
$$\tau_y^{(m)} = \tau_0 \quad \text{when unloading} \quad (16)$$

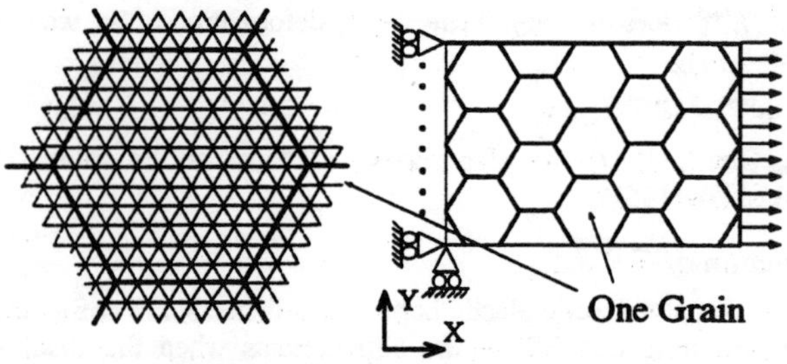

Fig. 1. 2-dimensional plane strain model of a polycrystalline aggregates

Table 1. Dimensions of finite element models for 3-dimensional polycrystalline aggregates

model name	n	i	j	k
2−8×4×4	2	8	4	4

Table 2. Material properties used for 2- and 3-dimensional FEM analysis

Young's modulus (MPa)	Poisson's ratio	Initial critical resolved shear stress (MPa)
70560	0.3	88.2

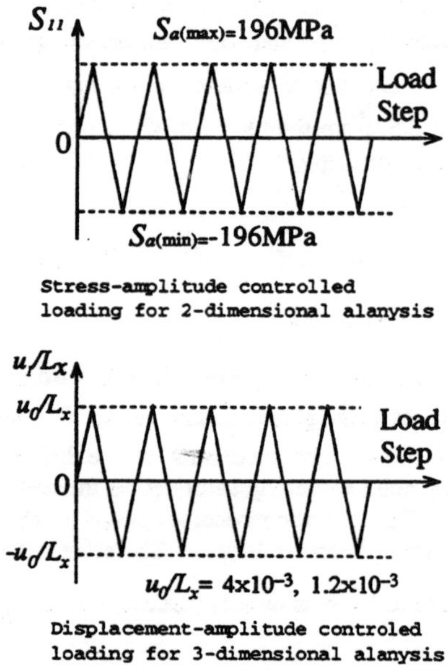

Fig. 3. Constant amplitude loading cyclels

3. Numerical Method

The problem studied here is to analyze the deformation behaviors of rectangular (in 2-dimensional analysis) and rectangular parallelepiped (in 3-dimentional analysis) f.c.c. polycrystalline aggregates under uniaxial cyclic tension-compression loading conditions.

The 2-dimensional rectangular model consists of 25 hexagonal crystal grains which consists of 216 plane strain triangular elements as shown in Fig.1. The 3-dimensional rectangular parallelepiped model consists of $i \times j \times k$ cubic crystal grain which consists of $n \times n \times n$ 8-node isoparametric solid elements as shown in Fig.2. The values of i, j, k, n used in calculations are listed in Table 1. The direction of each crystal grain is set randomly so that the whole model is nearly isotropic at the state of non-deformed condition.

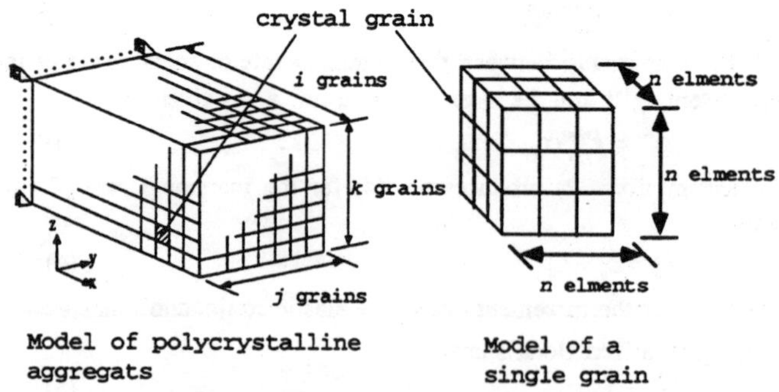

Fig. 2. 3-dimensional model of a polycrystalline aggregates

Material constants used in calculations are listed in Table 2. These values are used so as to simulate the plastic deformation behavior of aluminum alloy. The hardening rules and the evolution of C.R.S.S. during the reversal of a loading are as follows. Eq. (11) is used as the hardening rule, where H_0 is set to 2499MPa. Two models for determining C.R.S.S. during the reversal of a loading are used. One is the model defined from eqs. (13) and (15), which means that the slip systems have the fully kinematic hardening characteristics. The other is the model defined from eq. (16), which means that the stress fields induced by plastic slips disappear during the reversal of a loading.

In numerical calculations, cyclic loadings are performed through axial displacement rate are prescribed at the end of the models, while in both side the ends remain shear free (see Figs.1 and 2). In 2-dimensional calculations, the end displacement is controlled so that the average amplitude of the axial component of local stress is constant, while in 3-dimensional calculations it is controlled so that the displacement amplitude is constant. The loading conditions used in calculations are fully-reversed cyclic loading and the numbers of cycles are 5 to 10 as shown in Fig.3.

The time histories of deformation of the models are calculated in incremental fashion. Using the calculated time histories of local stress $\boldsymbol{\sigma}$ and local strain $\boldsymbol{\varepsilon}$, macroscopic stress $\mathbf{S}$ and macroscopic strain $\mathbf{E}$ are integrated as

$$S_{ij} = \frac{1}{V}\int_V \sigma_{ij} dV \quad , \quad E_{ij} = \frac{1}{V}\int_V \varepsilon_{ij} dV \qquad (17)$$

where V is the volume of the model. In this paper, the coordinate system is selected as the loading direction is parallel to x_1-axis. So, the relationship between S_{11} and E_{11} gives macroscopic stress-strain

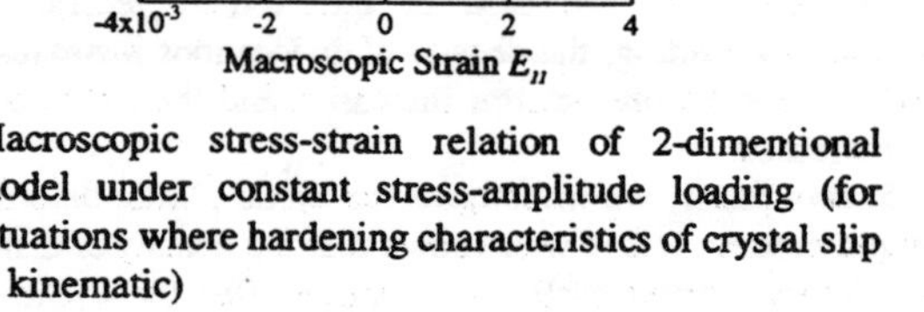

Fig. 4. Macroscopic stress-strain relation of 2-dimentional model under constant stress-amplitude loading (for situations where hardening characteristics of crystal slip is kinematic)

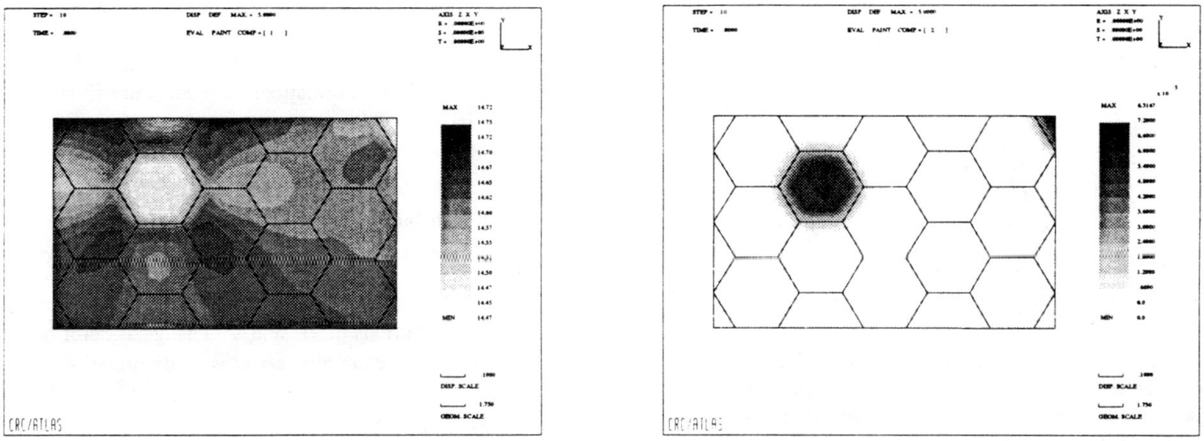

a) $\overline{\sigma}$ (left) and $\overline{\varepsilon}^p$ (right) at the moment of yielding during tensile loading of the 1st cycle

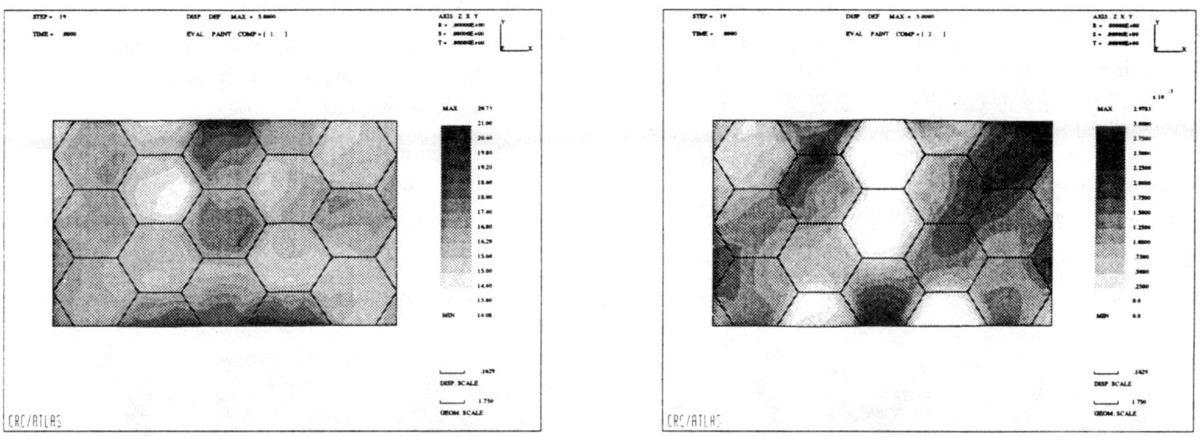

b) $\overline{\sigma}$ (left) and $\overline{\varepsilon}^p$ (right) at the moment of the maximum tensile loading of the 1st cycle

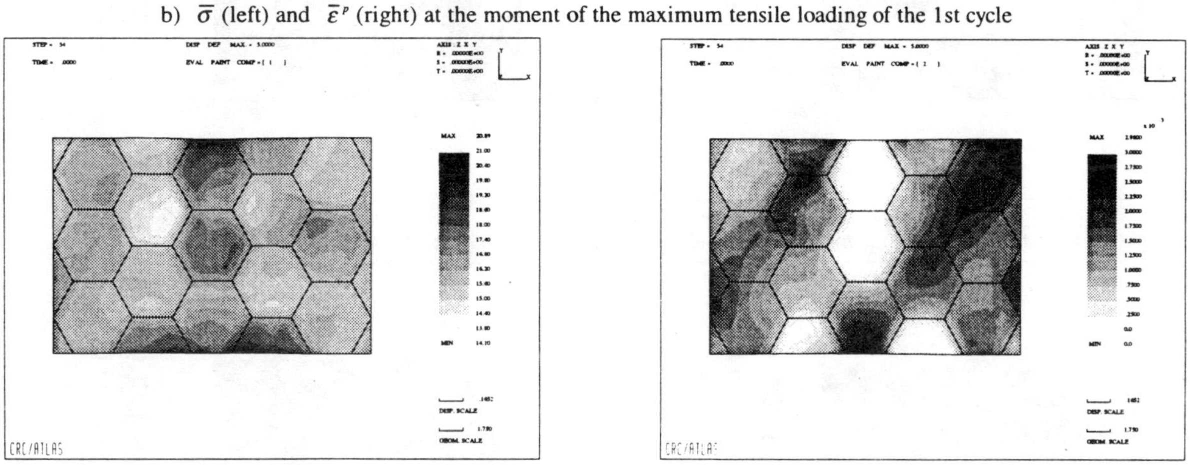

c) $\overline{\sigma}$ (left) and $\overline{\varepsilon}^p$ (right) at the moment of the maximum compressive loading of the 1st cycle

Fig. 5. Contours of equivalent stress and the 2nd invariant of plastic strain tensor for 2-dimentional model under constant stress-amplitude loading (for situations where hardening characteristics of crystal slip is kinematic)

curves, where suffix '11' denotes the components in loading direction of **S** and **E**.

4. Results and Discussion

4.1 Results of two-dimensional calculation (fully kinematic hardening slip)

Fig.4 shows the macroscopic stress-strain relation when the evolution of C.R.S.S. is given by eqs. (13) and (15). In this figure, except for the first 1/4 cycle, the stress-strain relation of each loading cycle retraces almost the same course. The shape of the hysteresis loop is symmetric with respect to the origin. It shows that the stable

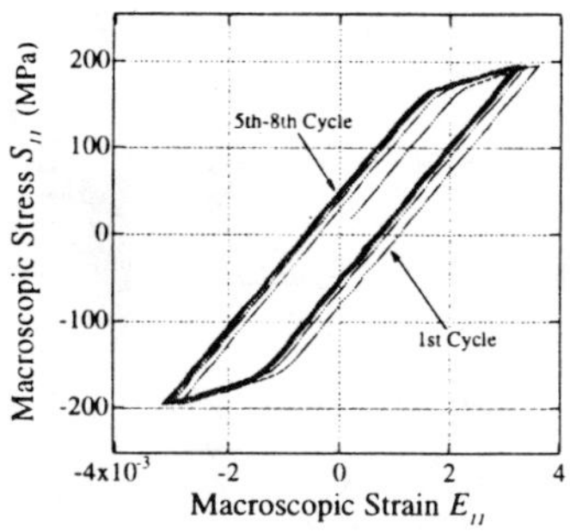

Fig. 6. Macroscopic stress-strain relation of 2-dimentional model under constant stress-amplitude loading (for situations where increase of C.R.S.S. disappears on the reversal of a loading)

hysteresis loop, which is the feature of cyclic plastic deformation of actual materials after cyclic hardening is saturated, can be generated when C.R.S.S. is increased only by long-range stress induced by accumulated dislocations, which works as back stress on slip systems.

Our target is to investigate non-uniform deformation of material from the microscopic view under cyclic loading conditions. Fig.5 shows the distribution of local equivalent stress $\bar{\sigma}$ and the 2nd invariant of local plastic strain $\bar{\varepsilon}^p$ at the point of yielding, the maximum tensile and compressive loading during the first cycle. $\bar{\varepsilon}^p$ is employed as a scalar parameter which represents the amount of local plastic deformation instead of the equivalent plastic strain κ. This is because κ increases monotonously even if the reversal of a loading occurs. The deformations of models are shown in Fig.5 with magnifying factor by 5.0.

Based on the results of the time history of $\bar{\sigma}$ and $\bar{\varepsilon}^p$, the following are revealed:

1) The distributions of $\bar{\sigma}$ and $\bar{\varepsilon}^p$ are non-uniform. The non-uniformness of $\bar{\varepsilon}^p$ is more severe than that of $\bar{\sigma}$. $\bar{\varepsilon}^p$ has a tendency to concentrate into grains on free surface of the model.

2) The distributions of $\bar{\sigma}$ and $\bar{\varepsilon}^p$ change as deformation becomes larger. For example, no plastic deformation occurs on the grain where $\bar{\varepsilon}^p$ becomes largest at the moment of the maximum loading.

3) There is a tendency that local stress becomes larger as local plastic strain becomes smaller.

4) At the moment of the maximum tensile and compressive loading, the distributions of $\bar{\sigma}$ and $\bar{\varepsilon}^p$ are almost the same for all loading cycles.

5) The relationship between $\bar{\sigma}$ or $\bar{\varepsilon}^p$ and the component in loading direction of macroscopic stress S_{11} of each loading cycle is identical, and also the shapes of the $\bar{\sigma} - S_{11}$ or

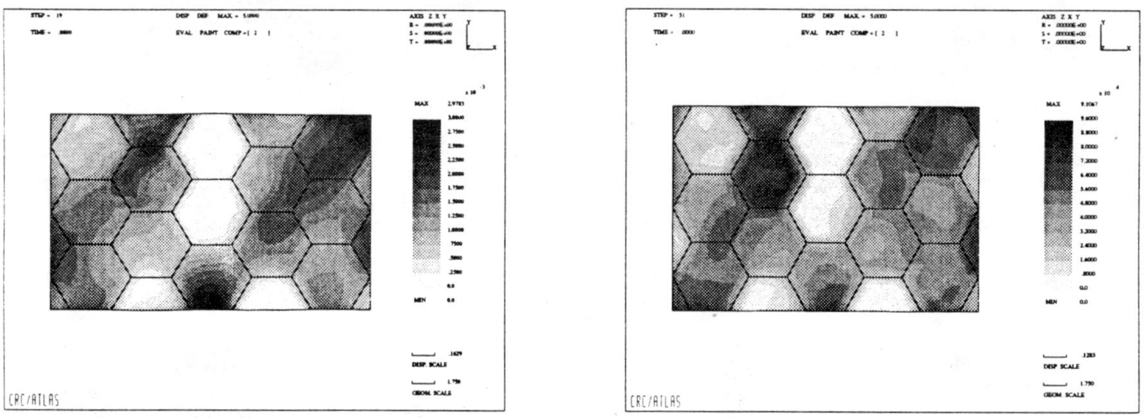

a) At the moment of the maximum tensile (left) and compressive (right) loading of the 1st cycle

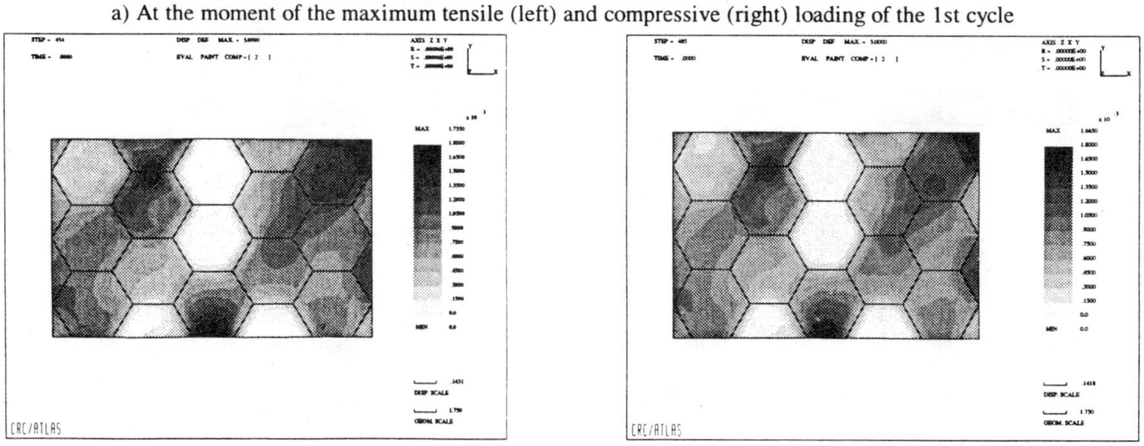

b) At the moment of the maximum tensile (left) and compressive (right) loading of the 5th cycle

Fig. 7. Contours of the 2nd invariant of plastic strain tensor for 2-dimentional model under constant stress-amplitude loading (for situations where increase of C.R.S.S. disappears on the reversal of a loading)

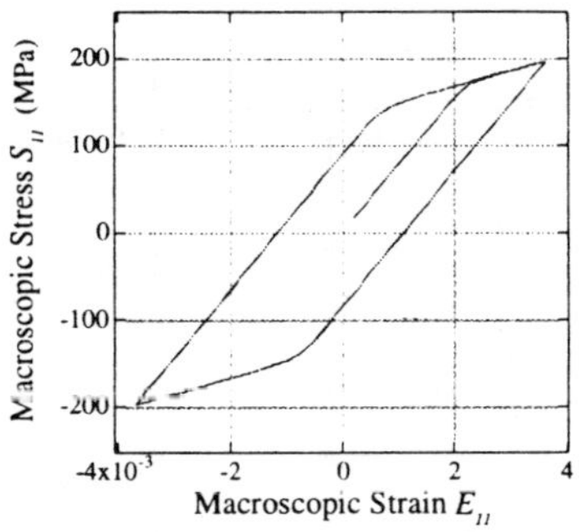

Fig. 8. Macroscopic stress-strain relation of 3-dimensional model under constant displacement-amplitude loading (for situations where hardening characteristics of crystal slip is kinematic)

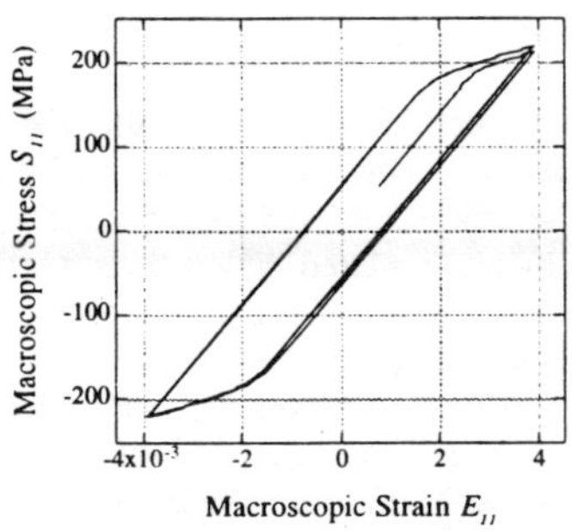

Fig. 9. Macroscopic stress-strain relation of 3-dimensional model under constant displacement-amplitude loading (for situations where increase of C.R.S.S. disappears on the reversal of a loading)

$\bar{\varepsilon}^p - S_{11}$ diagrams are symmetric with respect to the origin.

6) At all the places in the model, the activated slip systems and the amount of slip occurred during tensile loading are almost the same as those during compressive loading.

Based on the above characteristics, it is concluded that all places in the model deforms as plastic deformation caused only by 'reversible' slips.

4.2 Results of two-dimensional calculation (C.R.S.S. increment is assumed to disappear on the reversal of a loading)

Fig.6 shows the macroscopic stress-strain relations when the evolution of C.R.S.S. is given by eq. (16). It is apparent from the figure that the stress-strain curve does not retrace the same course at the beginning of cyclic loading cycle, but after about 5 cycles proceed the hysteresis loops converge to a certain stable state. The stable hysteresis loop can be generated when the increment of C.R.S.S disappears at the reversal of a loading.

Fig.7 shows the distribution of $\bar{\varepsilon}^p$ at the moments of the maximum tensile and compressive loading of 1st and 5th cycles.

Based on the results of the time history of $\bar{\sigma}$ and $\bar{\varepsilon}^p$, the following are revealed:

1) As shown in Fig.7 a), at the beginning of cyclic loading, the distribution of $\bar{\sigma}$ and $\bar{\varepsilon}^p$ in compressive loading are different from those in tensile loading.

2) As shown in Fig.7 b), after the hysteresis loops shown in Fig.6 converge, the distributions of $\bar{\sigma}$ and $\bar{\varepsilon}^p$ in compressive loading become almost the same as those in tensile loading.

3) There is a similarity between the distributions of $\bar{\sigma}$ and $\bar{\varepsilon}^p$ after the convergence of macroscopic stress-strain response and those calculated for situations where hardening characteristics of crystal slips is kinematic. That is, $\bar{\varepsilon}^p$ has a tendency to concentrate into grains on free surface of the model, and all the places in the model deforms as plastic deformation caused only by 'reversible' slips.

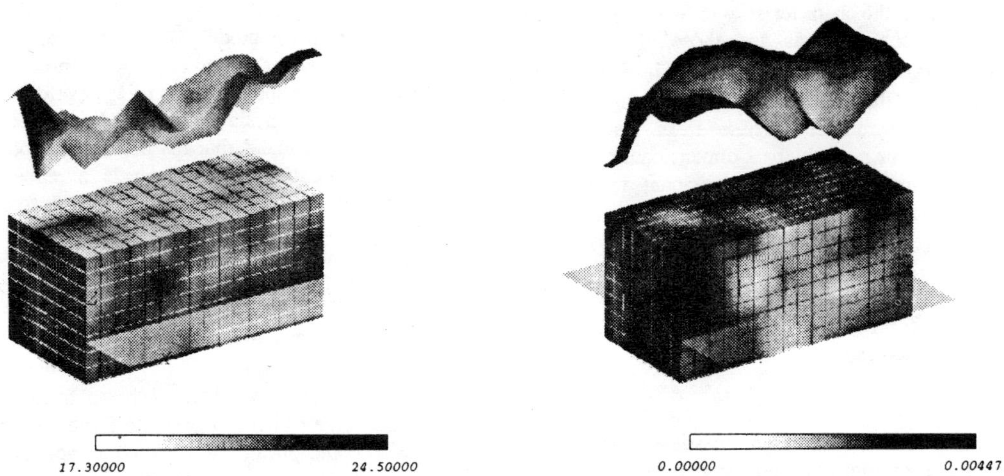

Fig. 10. Contours of equivalent stress (left) and the 2nd invariant of plastic strain tensor (right) for 3-dimensional model at the moment of the maximum tensile loading of the 1st cycle of the constant displacement-amplitude loading (for situations where hardening characteristics of crystal slip is kinematic)

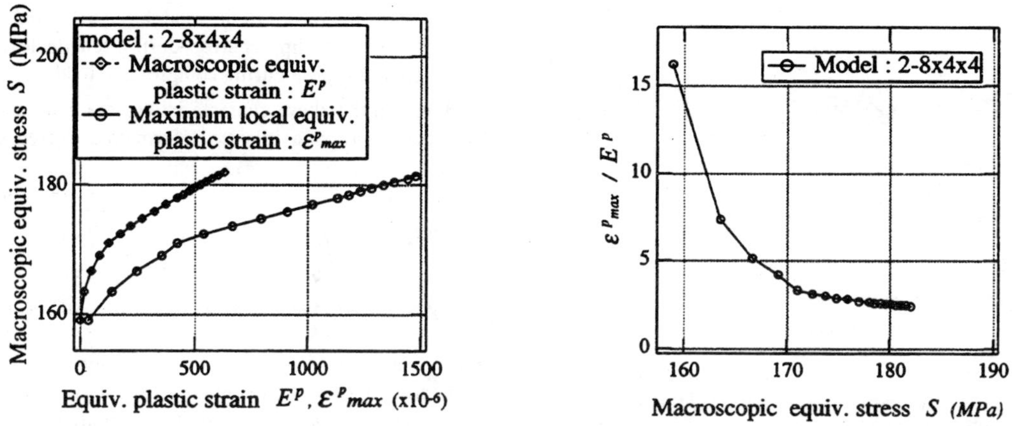

Fig. 11. The maximum and average (left) and the ratio of the maximum to average (right) of the 2nd invariant of local plastic strain in 3-dimensional model during tensile loading of the 1st loading cycle. (for situations where hardening characteristics of crystal slip is kinematic)

4.3 Results of three-dimensional calculation (macroscopic stress-strain response, distribution of local stress and strain)

Fig.8 shows the macroscopic stress-strain relation under displacement-controlled cyclic loading conditions when the evolution of C.R.S.S. is given by eqs.(13) and (15). Fig.9 shows the one when the evolution of C.R.S.S. is given by eq.(16). It is clear from these two figures that, the stable hysteresis loops can be generated when strain hardening of plastic slips is induced by kinematic hardening or the increment of C.R.S.S. disappears at the reversal of a loading. Such behavior occurs not only on 2-dimensional models under load-controlled loading conditions but also on 3-dimensional models under displacement-controlled loading conditions.

Fig.10 shows the distributions of $\bar{\sigma}$ and $\bar{\varepsilon}^p$ at the moment of the maximum tensile loading of 1st cycle when strain hardening of plastic slips is induced by kinematic hardening. The distributions of $\bar{\sigma}$ and $\bar{\varepsilon}^p$ at the moments of the maximum tensile and compressive loading of each cycle are almost the same as those in Fig.10.

Based on the results of time history of $\bar{\sigma}$ and $\bar{\varepsilon}^p$, it is clear that there is a similarity between the characteristics of local stress and strain distribution of 2-dimensional analysis and those of 3-dimensional analysis. That is, $\bar{\varepsilon}^p$ has a tendency to concentrate into grains on free surface of the model, and the local deformation during tensile and compressive loading are symmetric. These facts lead that only 'reversible' plastic slips occur not only for 2-dimensional models under load-controlled loading but also for 3-dimensional models under displacement-controlled loading.

Fig.11 shows the maximum value $\bar{\varepsilon}^p_{max}$, the average value $\bar{E}^p$ and the ratio of the maximum to the average $\bar{\varepsilon}^p_{max}/E^p$ of the 2nd invariant of plastic strain during tensile loading of the 1st cycle. Here, the average value $\bar{E}^p$ is defined as

$$E^p = \frac{1}{V}\int_V \bar{\varepsilon}^p dv\,, \tag{18}$$

and it represents the amount of macroscopic plastic strain. It is clear from Fig.11 that at the beginning of plastic deformation, the ratio $\bar{\varepsilon}^p_{max}/E^p$ is more than 10, then it converges to the value of 2 to 3 as plastic deformation increases.

4.4 Application to fatigue strength analysis

Based on the results of numerical calculations, we can employ either of two evolution rules for C.R.S.S. to examine the concentration of local plastic deformation during fatigue process by crystalline FEM analysis. One is the strain hardening mainly caused by long range stress field induced by accumulated dislocations which works as back stress of slip systems, the other is the disappearance of the increment of C.R.S.S. at the reversal of a loading.

Up to now, many dislocation movement models for analyzing cyclic plastic deformation after the saturation of cyclic hardening have been developed. Some examples of such models are;

1) Flip-flap motion of the edge segment of prismatic dislocation loops (Feltner(1965))

2) Bow-out of dislocation segments which does not work as dislocation sources. (e.g. Avery-Backofen(1963))

3) The dynamic recovery caused by cross slip cancels the dislocation multiplication during plastic slips. (e.g. Klesnil-Lukás(1980))

The mechanisms of models 1) and 2) do not increase the density of dislocation.

For these cases, the increase of C.R.S.S. is caused only by long range stress field induced by dislocation segments. On the other hand, the increase of C.R.S.S. caused by plastic slips disappear at the reversal of a loading when model 3) is assumed. The evolution of C.R.S.S. which is derived from the above three dislocation models can be represented by the evolution rules developed in this study.

From the analysis of the local deformation for 2- and 3- dimensional models, it is found that the distribution of local stress and strain are quite non-uniform, and there is a tendency that the local plastic deformations concentrate into grains on free surface. It is commonly supposed that PSB, which is considered to be the origin of fatigue crack, can be generated easily in such concentrated area of local deformation.

It should be emphasized that such non-uniformness is derived even if any initial inperfection is not considered in the FEM models used in this study. The non-uniformness seems to be due to the difference of crystal grain direction.

On the other hand, it is shown from Fig.11 that there exists the stress range which can generate finite local plastic deformation even if macroscopic plastic deformation is negligible. It can be considered that this stress range cause fatigue failure below yield point.

However, in this study we cannot simulate the generation of inreversible slip, which is the microscopic process for the generation of both PSB and intrusion, because of neglecting lattice rotation of grain. Our next effort will focus to performing the analysis which takes into account lattice rotation.

Conclusions

Two microscopic hardening rules which can be used for the analysis of cyclic plastic deformation are proposed. Based on the proposed hardening rules, FEM analysis of 2- and 3-dimensional deformation of f.c.c. polycrystalline are carried out. The characteristics of non-uniform local deformation under cyclic loading conditions are investigated.

The main results obtained are as follows.

1) The stable macroscopic stress strain hysteresis loops, which is the feature of cyclic plastic deformation of actual materials after cyclic hardening is saturated, can be generated when the strain hardening of plastic slips is induced by kinematic hardening or when the increment of C.R.S.S. disappears at the reversal of a loading.

2) The distribution of local stress and local plastic strain are quite non-uniform. There is a tendency that the local plastic strain concentrates into grains on free surface of the FEM model.

3) There exists the stress range which can generate finite local plastic deformation even if macroscopic plastic deformation is negligible.

References

Hill, R.(1966). *J. Mech. Phys. Solids,* **14**, p.95

Ohashi, T.(1987). *J. Japan Inst. Metals,* **51**, 1, p.37

Jackson, P.J. and Basinski, Z.S.(1967). *Canada. J. Phys.,* **28**, p.707

Feltner, C. E.(1965). *Phil. Mag.,* **12**, p.1229

Avery, D.H. and Backofen, W.A.(1963). *Acta Metal.,* **11**, p.653

Klesnil, M. and Lukás, P.(1980), *Fatigue of Metallic Materials,* Elsevier, New York

Proceedings of the Seventh (1997) International Offshore and Polar Engineering Conference
Honolulu, USA, May 25-30, 1997

ISBN 1-880653-28-1 (Set); ISBN 1-880653-32-X (Vol. IV)

A Micromechanical Study of the Freeze-Thaw Behavior of Polymer Composites

Piyush K. Dutta
U.S. Army Cold Region Research and Engineering Laboratory
Hanover, NH, USA

ABSTRACT

The essential quality of a good polymer composite is that the bond between the fiber and the matrix is well established and is continuous both around the fiber and along its length. When a load is applied in the direction of the fiber, the ratio of the load share depends on the relative elastic modulus of the fiber and the matrix. However, the elastic modulus of the polymer matrix is significantly influenced by the temperature. At low temperature the modulus of elasticity increases considerably, and thus load sharing changes between the fibers and the matrix. Also, because of the mismatch of coefficient of thermal expansion (CTE) of matrix and fiber, the matrix is usually stretched in the fiber direction during curing, and develops internal tensile stress, interfacial shear stress, hoop stress, and radial stress. On further cooling during the freezing process, the magnitude of all these induced stresses would usually increase, developing potential microcracks. The change in the radial or clamping stress which controls the crack development and propagation (fracture) both across and along the fiber would also change the composite's fracture behavior in the low temperature regime. More complex stresses are developed when the composites are constructed as laminates with each lamina (layers of fibers) having fiber orientations different from the adjacent ones. Reductions of strength and modulus of composites, following freeze-thaw cycling, as evident in experimental results, support this micromechanical theory of composites degradation.

Key Words: Composites, low temperature, thermal fatigue, thermal cracks, durability, cold regions, freeze-thaw, composite properties, composite design, polymer composites.

INTRODUCTION

In polymer composites high elastic modulus fibers are incorporated into a lower elastic modulus matrix to achieve structural reinforcement. Most common fibers are E-glass, carbon or graphite, and aramids (Kevlar). Typically the elastic modulus and strength of these fibers are of a magnitude higher than the polymer matrix in which these fibers are embedded. The essential quality of a good composite is that the bond between the fiber and the matrix is well-established and is continuous both around the fiber and along its length. Thus, a good composite's performance essentially depends on the interfacial bond quality. When a load is applied in the direction of the fiber orientation of the composite, the load is shared both by the fiber and the polymer matrix. The ratio of this load share depends on the relative elastic modulus of the fiber and the matrix. However, the elastic modulus of the polymer matrix is significantly influenced by the temperature. At lower temperature the modulus of elasticity increases considerably, and thus it is expected that load sharing between the fibers and the matrix would also change.

Another significant aspect of the temperature variation in polymer composites is related to the development of internal stresses. The typical fibers of polymer composites have extremely low coefficients of thermal expansion (CTE), whereas polymers have thermal expansion coefficients which are almost an order of magnitude higher. Thus, when the composites are cooled, the fibers do not shrink as much as the polymer matrix would tend to shrink, but because of the continuous interfacial bond with the fiber the matrix only shrinks as much as the fiber shrinks. Nevertheless, the matrix is stretched, and in the fiber direction, develops internal tensile stress and interfacial shear stress. Cooling also develops hoop stresses and radial stresses around the fiber and thus changes the clamping stress. The clamping stress controls the fracture behavior or crack development and propagation both across and along the fiber direction of unidirectional composites. More complex stresses are developed when the composites are constructed as laminates with each lamina (layers of fibers) having fiber orientations different from the adjacent ones. A large amount of experimental investigations on the effects of temperature on mechanical and fracture behavior of composites including strength, modulus, toughness, failure mode, as well as damage tolerance, have been reported in literature which have primarily aimed towards high temperature behavior. However, very little work has been done on mechanism and mechanics of behavior and failure at low temperature. The purpose of this paper will be to briefly discuss on the micromechanical aspects of the low temperature responses expected from the polymer composites and give some experimental results to prove that freeze-thaw cycling tend to degrade the composites.

MATERIAL MODEL

In its simplest form the polymer composite material can be visualized as a bundle of solid rods held together with a glue filling up all the void space around the fibers (Fig. 1). However, in the microscopic scale, these

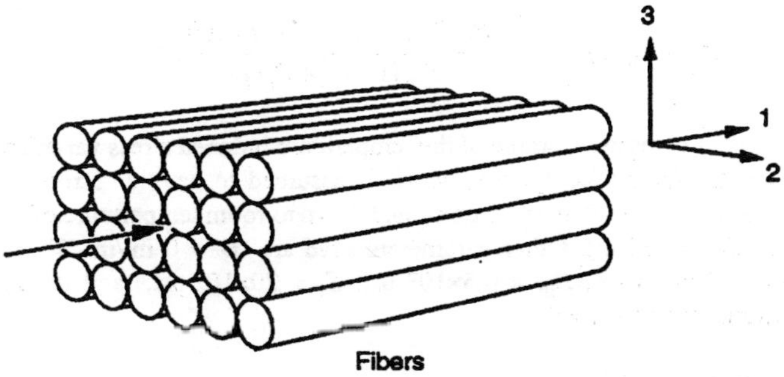

Fig. 1. Model of unidirectional fiber composite.

solid rods have diameters varying in the range of 5 μm to about 20 μm. It is obvious from Fig. 1 that the orientation of the fibers controls the overall behavior of the composite, and it is necessary to refer to these orientations and other directional properties with some designated axes, say 1, 2, and 3. Thus, we refer to a tensile or compressive stress σ, in fiber direction as σ_{11}, and the other two orthotropic directions as σ_{22} and σ_{33}. Other elastic parameters, such as strains (ε), modulii (E for Young's modulus, and G for shear modulus), or Poisson's ratios as well as the thermal and moisture expansion coefficients (α, β) are also referred to their directionality with appropriate subscripts.

Longitudinal stresses in unidirectional composite

The bundle of rods shown in Fig. 1 can be assumed to represent a model of the unidirectional composite in which fibers are oriented in 1–1 direction. If the dimensions of this bundle in 2–2 direction and 3–3 direction are extremely small, such that the block looks like a very long bar, then any changes in dimensions either in 2–2 direction or 3–3 direction due to Poisson's ratio effect will be negligible and can be ignored. Under this condition a load in 1–1 direction is referred as longitudinal load, and the stresses developed in this bar across the section in 2–3 plane due to this load is the longitudinal stress (σ_{11}). The strain (ε_{11}) from this stress will then be related by Hook's law as:

$$\sigma_{11} = E_{11}\varepsilon_{11}. \tag{1}$$

Although, in reality, changes in dimension in 2–2 and 3–3 directions are not to be ignored, we will reserve the discussions of the Poisson's ratio effects for later sections for simplicity and concentrate on the contributions of the fiber phase and the matrix phase in the elastic modulus.

We assume that materials of both the fiber and the matrix are homogeneous and isotropic. Using subscripts "f" for fiber, "m" for matrix and "c" for the composite, we can express the above composite strain as

$$\varepsilon_{11} = \varepsilon_m = \varepsilon_f. \tag{2}$$

On the unit length basis, and using the volume fractions of the fiber v_f, the composite stress

$$\sigma_{11} = \sigma_m (1 - v_f) + \sigma_f v_f. \tag{3}$$

Expressing these stresses in terms of strains and Young's modulus, Eq. 3 can be rewritten as

$$E_{11}\,\varepsilon_{11} = E_m\varepsilon_m (1 - v_f) + E_f\varepsilon_f v_f, \tag{4}$$

which in conjunction with Eq. 2 converts into:

$$E_{11} = E_m (1 - v_f) + E_f v_f. \tag{5}$$

Temperature effect

Composites shrink or expand according to the change of temperature. A change of temperature from T_0 to T_1 for a composite with the CTE = α_c in 1–1 direction develops the thermal strain ε_T, given by

$$\varepsilon_{11T} = (T_0 - T_1)\,\alpha_c \tag{6}$$

and the thermal stress, using Eq. 3:

$$\sigma_{11T} = E_{11}(T_0 - T_1)\,\alpha_c$$

$$= [E_m (1 - v_f)\,\alpha_m + E_f v_f \alpha_f]\,(T_0 - T_1), \tag{7}$$

which on simplification gives the CTE of the composites in 1–1 direction as

$$\alpha_c = \frac{E_m(1 - v_f)\alpha_m + E_f v_f \alpha_f}{E_{11}}. \tag{8}$$

While the change of temperature, say, to a lower temperature, would make the composite to shrink according to the value of α_c, the matrix usually having a higher TEC than the composite will develop a tensile stress in 1–1 direction given by

$$\sigma_{11T} = E_m(\alpha_m - \alpha_c)\Delta T \tag{9}$$

where $\Delta T = (T_0 - T_1)$. Using Eq. 5 and 8, Eq. 9 can be rewritten to give the thermally induced matrix stress as:

$$\sigma_{m11T} = \frac{E_m E_f(\alpha_m - \alpha_f)\Delta T}{E_m(1 - v_f) + E_f v_f} \tag{10}$$

Eq. 10 shows the thermal stresses induced in the matrix of the unidirectional composite as a function fiber volume fraction, temperature differential, and the difference of CTE between the fibers and the matrix.

Transverse and off-axis stresses

So far, we have considered the stresses only in 1–1 direction while the fibers too are also oriented in 1–1 direction. Now we will consider the fibers that are oriented at an angle to the direction of the loading. Referring to Fig. 1 for detail, note that if the stresses are applied in 2–2 direction, the tendency of this stress will be to pull the composite apart, by causing the interfacial bond between the fiber and the matrix to fail. It is obvious that the fibers will not share any load in this case. Thus, the maximum strength of the composite will be limited to the stress at which the fiber/matrix interfacial bond will fail, and the crack could propagate across the matrix to the neighboring fiber. It would be clear now that the composite will evidently have much lower strength in this direction than the longitudinal direction. Again the thermal strain in 2–2 direction can be expressed as

$$\varepsilon_{22T} = [v_f\alpha_f + (1 - v_f)\,\alpha_m]\Delta T \tag{11}$$

which on further simplification gives

$$\varepsilon_{22T} = [v_f(\alpha_f - \alpha_m) + \alpha_m]\Delta T. \tag{12}$$

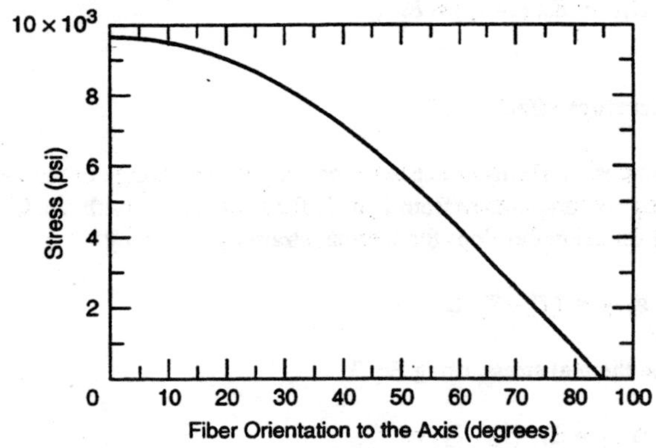

Fig. 2. Off-axis stress variationw ith fiber orientation.

In the above equations we have assumed that the CTE for the fiber, α_f, is the same in all directions; that is, the CTE is isotropic. Realistically the value of α_f is extremely small for most fibers (e.g., approximately −0.1 to 0.1 in./in.-°F for carbon fibers and Kevlar, and approximately 6×10^{-6} in./in.-°F for glass fibers, compared to 50 to 150 in./in.-°F for the polymer matrix. Thus, setting $\alpha_f = 0$, Eq. 12 can be rewritten as:

$$\varepsilon_{22T} = [(1-v_f)\ \alpha_m \Delta T. \qquad (13)$$

When the composite is free to contract in the transverse direction no stresses will be induced from cooling; therefore, $\sigma_{22T} = 0$. In general, it is practical to assume that the induced matrix thermal stress σ_θ in the axial direction when the fibers are oriented at an angle θ to the loading axis (refer to Fig. 2) is the cosine of the stress σ_{m11T}. Thus,

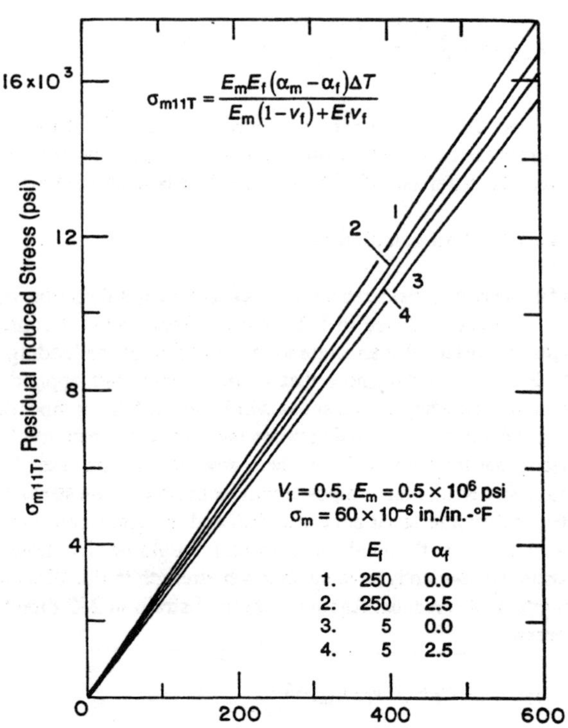

Fig. 3. Induced thermal stress in the matrix of unidirectional composites.

$$\sigma_\theta = \sigma_{m11T}\cos\theta = \frac{E_m E_f v_f(\alpha_m - \alpha_f)\Delta T\cos\theta}{E_m(1-v_f)+E_f v_f}. \qquad (14)$$

Fig. 3 gives an example of the temperature induced stress variation in a graphite/epoxy composite, which is assumed to have a "stress-free" temperature of 400°F and then cooled down to room temperature of 70°F (i.e., $\Delta T = 330$°F). Other parameters used are $\alpha_f = 0$ in./in.-°F, $\alpha_m = 60\times10^{-6}$ in./in.-°F, $E_m = 0.5\times10^6$ psi, $E_f = 20\times10^6$ psi, and the fiber volume fraction $v_f = 0.5$.

Laminate model

Laminates are composites which are made up of several unidirectional layers (laminae) of composites. The formulation of the thermal residual stresses in these composites can be obtained using classical laminated plate theory found in many texts (for example, Agarwal and Broutman, 1980), and is summarized here without any derivation. In a laminate of n-layers, the stress at any point in the kth layer is given by

$$\{\sigma\}_k = [Q]_k\{\varepsilon^0\} + Z[Q]_k\{k\} - [Q]_k\{\alpha\}_k\Delta T \qquad (15)$$

where $[Q]_k$ and $\{\alpha\}_k$ are the transformed stiffness matrix and thermal expansion coefficient vector of the layer, respectively; $\{\varepsilon^0\}$ and $\{k\}$ are the in-plane strain and curvature vectors of the mid-surface of the plate element, respectively: Z is the distance of the point from the midsurface, and $\Delta T = (T_0 - T_1)$. Eq. 15 is used to derive the stress resultants in the laminate as:

$$\{N\} = [A]\{\varepsilon^0\} + [B]\{k\} - \{N\}_T \qquad (16a)$$

$$\{M\} = [B]\{\varepsilon^0\} + [D]\{k\} - \{M\}_T \qquad (16b)$$

where $\{N\}$ and $\{M\}$ are the resultant forces and moments respectively per unit length acting on the element; and $[A]$, $[B]$, and $[D]$ are the respective extensional, coupling, and bending stiffness matrices of the plate element. The terms $\{N\}_T$ and $\{M\}_T$ are apparent thermal forces and moments defined by

$$\{N\}_T = \Delta T\Sigma\,[Q]_k\{\alpha\}_k t_k \qquad (17a)$$

$$\{M\}_T = \Delta T\Sigma\,[Q]_k\{\alpha\}_k z_k t_k \qquad (17b)$$

where t_k is the layer thickness and z_k is the distance of the kth layer from the midsurface.

Using Eq. 16 and 17 the thermal residual stresses in any layer of the laminate can be derived. In a symmetric laminate $[B] = 0$, and with no external loads applied, both $\{N\}$ and $\{M\}$ are also zero. Under these conditions Eq. 15 can be rewritten as

$$\{\sigma\}_k = [Q]_k[A]^{-1}\{N\}_T - [Q]_k\{\alpha\}_k\Delta T \qquad (18)$$

Eq. 18 can be used to calculate the residual stress distribution in the laminates due to a change in temperature. Because of the relative magnitude of the longitudinal and transverse modulii of elasticity and CTEs of each unidirectional lamina, a reduction in temperature will in general result in different stresses in different directions within the laminate. As a result high interlaminar shear stresses are induced.

Freeze-thaw effects

We can summarize that large residual stresses are developed in highly anisotropic composite materials due to large differences in CTEs, and because of extreme differences in the stiffness of the constituent elements. In high curing temperature environments these stresses become potentially damaging for polymer matrix composites. The damage usually begins with the formation of microscopic cracks in the matrix or the matrix fiber interface. When these cracks develop to a certain density and size, they coalesce to form macroscopic matrix cracks (Wang, 1986). Doner and Novak (1969) have shown that translamina cracks found in graphite-epoxy laminates were the result of thermal stresses induced as the composite cooled from a relatively strain-free curing temperature to room temperature. Further work by Spain (1971) indicated that the severity of cracking is directly related to the amount of resin shrinkage during curing. He also found that even if the composites were fabricated in a stress-free state, the subsequent thermal cycling usually produced microcracks. Transverse matrix cracking can affect laminate stiffness, strength, dimensional stability, and fatigue resistance. In aggressive environments because of the increased matrix crack density and resulting high permeability these materials may suffer reduced corrosion resistance.

Table 1. Influence of low temperature thermal cycling on the strength and elastic properties of graphite/epoxy composites.

Properties	*Fiber orientation*	*Uncycled*	*Cycled 10 times*	*Change (%)*
Tensile strength (psi × 10^3)	0°	247.9	263.7	+6.4
	45°	13.7	12.3	−10.2
	90°	8.8	8.3	−5.7
Modulus E_1 (psi × 10^6)	0°	18.30	17.84	−2.51
Modulus E_x (45°) (psi × 10^6)	45°	1.61	1.62	0.62
Modulus E_2 (psi × 10^6)	90°	1.41	1.37	−2.83
Shear modulus G_{12} (psi × 10^6)		0.57	0.58	1.75
Poisson's ratio, ν	0°	0.31	0.31	0
Poisson's ratio, ν	90°	0.02	0.02	0

EXPERIMENTAL RESULTS

Two groups of composite materials, the first unidirectionally reinforced 7-ply AS4/Hercules 3501 laminate specimens, and the second 6-ply $[90_20]_s$ S-2 glass/SP250 laminate specimens, were thermally cycled between 60° and −60°C in a thermal cycling chamber. Before thermal cycling, to remove moisture, the specimens were dried in an oven for 168 hours at 105°C.

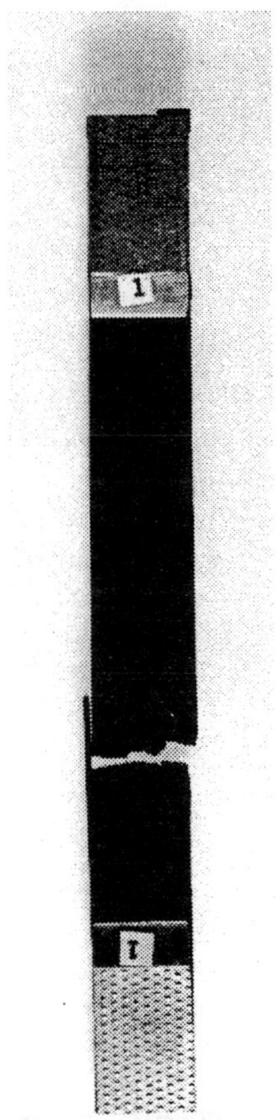

Fig. 4. Example of graphite-epoxy coupon test under ASTM 3039 procedure.

Graphite/epoxy Composites

These materials were cut into longitudinal tensile testing specimen sizes of 0.75 in. × 7.5 in. with fiber orientations at 0°, 45°, and 90° to the loading direction. From each batch 6 to 8 specimens were subjected to tensile testing as per the ASTM 3039 tensile test. Fig. 4 shows one of the test specimens after failure. The specimens were strain gaged and tensile tests were performed at room temperature; and on 6 to 8 more specimens after subjecting them to 10 thermal cycles between 60° and −60°C. The results of these tests are summarized in Table 1.

Besides the above tests additional thermal cycling of the 7-ply AS4/Hercules 3501 specimens of 90° fiber orientation were performed to see if the increase of thermal cycling degrades the matrix dominated failure behavior. Results from this test is shown graphically in Fig. 5.

Glass-epoxy composites

Fig. 6 shows the failure of one of a 6-ply $[90_20]_s$ S-2 glass/SP250 laminates which were thermally cycled between 60° and −60°C and then tested at room temperature as per the ASTM 3039 tensile test procedure. Six specimens were tested without any thermal cycling, and then a batch of six were tested after 10, 50, 100, and 150 cycles. The tensile strengths so determined are plotted in Fig. 7. There is a rapid degradation of strength following the first few thermal cycles, then the rate of degradation slows down with the increased number of thermal cycling. A curve fitting polynomial equation approximately represents the trend of degradation.

DISCUSSION

Low temperature environmental degradation is a major issue in the large-scale application of composites. In cold regions of the world repeated freeze-thaw of any outdoor structure made of composites cannot be avoided. The tests performed under this investigation clearly show the degradation of properties following freeze-thaw cycling. From micromechanical analysis, it is also evident that severe stresses are induced in laminates and in the microstructure of the unidirectional composite matrix from excursion to a low temperature state. The lower the temperature, greater will be the magnitude of the induced stresses and potential for microcrack formation and growth primarily in the matrix. As mentioned before, the repeated thermal cycling will cause coalescence of the of these microcracks and result in damage to the composite material. The total stress at any point in the composite material is the superposition of residual thermal stress and the applied mechanical stress. If the low temperature-induced stresses have already produced some damage, the specimen will then obviously fail at a lower applied load than the failure load of the uncycled specimens.

For the transversely (90°) loaded composites the failure is matrix dominated. Any microcracks in the matrix would have a stronger influence on the failure. It is evident from Table 1 and Fig. 5 that tensile strength degraded in 45° and 90° unidirectional composites. The E_1 and

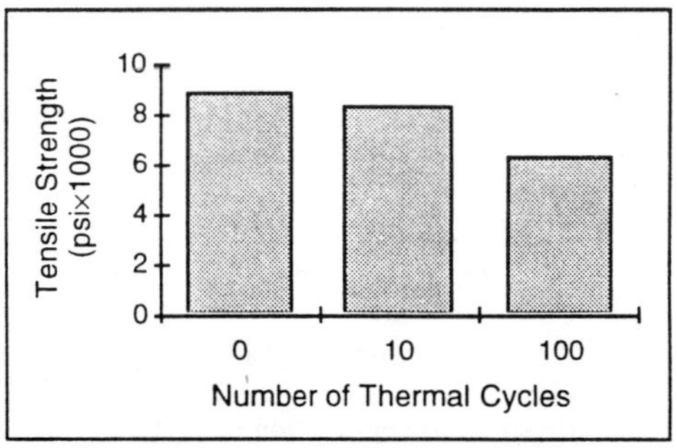

Fig. 5. Results of thermal cycling on the graphite-epoxy unidirectional (90° orientation) composite coupons.

Fig. 6. Example of a 6-ply $[90_2 0]_s$ S-2 glass/SP250 laminate tested after thermal cycling.

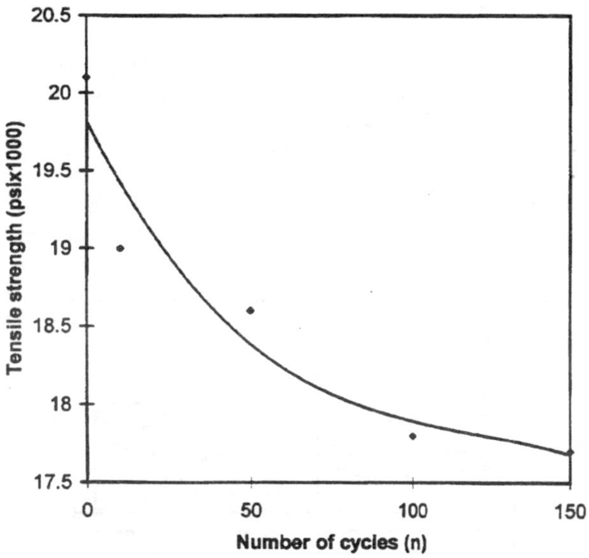

Fig. 7. Tensile strength degradation of 6-ply $[90_2 0]_s$ S-2 glass/SP250 laminates following thermal cycling. (Solid line represents the polynomial Eq. 19 with coefficients a, b, c *equal to –0.0423, 0.0003, and -9×10^{-7}, respectively).*

E_2 also have degraded with thermal cycling. It is to be noted that the tensile strength of 0° fiber composite has slightly increased. However, it has been postulated that the thermal cycling softens the matrix in some composites where fibers are not straight to begin with. This allows the fibers to become straight and distribute the stress more uniformly among them to allow more load to be applied before failure (Dutta, 1992).

For the glass-epoxy laminates of $[90_2 0]_s$ orientation, the degradation follows a trend which can be approximately represented by following polynomial equation:

$$\sigma_{11(n)} = \sigma_{11(0)} + an + bn^2 + cn^3 \tag{19}$$

where n = Number of thermal cycles
$\sigma_{11(0)}$ = Tensile strength without thermal cycling
$\sigma_{11(n)}$ = Tensile strength after n number of thermal cycles, and coefficients a, b, c are equal to –0.0423, 0.0003, and -9×10^{-7}, respectively.

It must be noted that the coefficients depend primarily on the temperature of excursion, the fiber and matrix type, and its layup, and must bear some relationship with the micromechanical processes of damage. The current amount of data are not sufficient to develop any quantitative relationship of the experimental results to the stresses developed, failure criteria and damage growth. For example, the data given in Fig. 5 for unidirectional graphite epoxy also follow a polynomial equation as Eq. 19, but the coefficients a, b, and c are now –0.055, 0.0003, and 0, respectively. While the coefficients b and c are almost the same, the coefficient 'a' varies with the composite type and layups.

CONCLUSIONS

Due to freeze-thaw thermal cycling, microstructural changes are predicted, and evidence of material property degradation have been found. There was an apparent reduction in transverse and shear strengths, indicating possible matrix degradation. The micromechanical analysis, showing development of a large amount of stresses with large temperature excursions, supports this conclusion only indirectly. A model predicting damage growth with low temperature thermal cycling is still not available. The elastic modulii also have degraded after thermal cycling. The degradation trend irrespective of the type of composites is similar with increasing number of thermal cycles. For a given type of composite, empirical formula for degradation of tensile strength with number of freeze-thaw cycles may be developed after a sufficient number of tests.

REFERENCES

Agarwal, B. and Broutman, LJ (1980) *Analysis and Performance of Fiber Compoiste*, John Wiley & Sons, New York.

Doner, DR and Novak, RC (1969) "Structural Behavior of Laminated Graphite Filamen Composites," *Proc. 24th Annual Technical Conference*, The Society of Plastic Industry Inc.

Dutta, PK (1992) "Tensile Strength of Unidirectional Fiber Composites at Low Temperature," *Proc. of the Sixth Japan-U.S. Conference on Composite Materials*, 22–24 June.

Spain, RG (1971) "Thermal Microcracking of Carbon/Resin Composites," *Composites*.

Wang, ASD (1986) "On Fracture Mechanics of Matrix Cracking in Composite Laminates," *Proc. International Symposium on Composite Materials and Structures*, Beijing, pp. 576–584.

Proceedings of the Seventh (1997) International Offshore and Polar Engineering Conference
Honolulu, USA, May 25-30, 1997

ISBN 1-880653-28-1 (Set); ISBN 1-880653-32-X (Vol. IV)

Design of Fast Ships for Minimal Motions

Alexander H. Day and Lawrence J. Doctors
Australian Maritime Engineering Cooperative Research Centre
Sydney, Australia

ABSTRACT

In previous studies, the hulls of ships have been optimized hydrodynamically in the sense of minimizing their resistance to forward motion with the intention of reducing fuel consumption of fuel and hence operating costs. A simple hull-generation model was used to define the vessel geometry based on a limited number of principal dimensions and form coefficients; the values of these parameters were optimized in order to minimize the total resistance. This approach provides a very practical form of output as it supplies the naval architect with a useful starting point from which refinements may be made to the hull in order to satisfy design constraints and goals not considered in the optimization.

In the current work, the hull-generation model has been extended to allow additional flexibility, and, in particular, to allow the vessel geometry to include a transom stern. The desired goal in this case is the improvement of the passenger comfort through the reduction of the vessel motions in sea waves, rather than the improvement of the resistance characteristics of the vessel.

The motions have been calculated using strip theory; the objective function is based on a prediction of the human response — in terms of motion sickness — to these vessel motions at a variety of points on the vessel in a selection of sea conditions. The optimization is carried out using a combination of a genetic algorithm with a hill-climbing technique. Results from some preliminary runs are given to illustrate the technique.

KEY WORDS: Ship motions, ship design, fast ships, optimization, genetic algorithm

INTRODUCTION

Much of the published work on the optimization of ship hydrodynamics has been concentrated on an objective of the reduction of the resistance, and hence fuel consumption and operating costs. The prediction of the resistance has been effected almost exclusively through the use of the landmark thin-ship theory of Michell (1898). Early work in this area using the Michell theory includes that of Havelock (1923), Lin, Webster, and Wehausen (1963), Kostyukov (1968), and Wehausen (1973).

More recently, Day and Doctors (1996) used a model based on a number of principal dimensions to define the vessel geometry, and optimized the values of these dimensions using a genetic algorithm (GA) in order to minimize the total resistance. This approach provides a very practical form of output as it can be used to supply the naval architect with a useful starting point from which refinements may be made to the hull in order to satisfy design constraints and goals not considered in the optimization. This study was based on earlier research, also utilizing the GA, and described by Doctors and Day (1995) and Day and Doctors (1996). Similar research aimed at minimizing ship resistance on the basis of overall ship parameters has been reported by Tuck and Lazauskas (1996).

Rather less work appears to have been reported in the area of minimization of ship motions. Hearn, Wright and Hills (1995), used a simple GA to perform some relatively minor modifications to an existing design, based on an NPL hullform. Two objective functions were examined, one involving the separation of the periods of pitch and roll, while the other involved the motion at a point located transversely at the maximal beam and longitudinally at the bow. A two-dimensional strip theory was used to compute the motions. Hearn, Wright, and Yaakob (1995) re-analyzed the hulls obtained using more sophisticated theories, and the rankings of the hulls examined were shown to be maintained for the zero-speed case, thus supporting the use of these simple and computationally inexpensive theories. They conclude that similar results are likely to be obtained in the case of non-zero forward speed.

The possibilities for considerable increases in seaworthiness of the vessel must be emphasized here. For example, Doctors (1995) investigated unusual hullforms with elongated bulbous bows on slender forms, created by manually combining features of promising basis hullforms. He found that reductions in the peak of the pitch response amplitude operator (RAO) in head seas of over 80% were achievable. Using a systematic procedure for searching for the best hulls is likely to be even more effective; it is this approach which is the subject of the present work.

The approach of Day and Doctors (1996) is extended here in that the vessel, represented by means of a small number of overall geometric parameters, may include a transom stern. As noted in the previous work on ship resistance, this type of approach leads to a practical result in that hullforms with wildly impractical sections and waterlines were avoided. In this preliminary study we will concentrate on an examination of the feasibility of the approach, and restrict our interest to ship motions, though it is intended that ultimately resistance and motions will be minimized simultaneously. We will specifically search for hullforms that minimize the susceptibility of the passengers to seasickness.

When the motions of the ship have to be solved many times over the course of an optimization, it is of course vital that the total computational resources do not become excessive. To this end, it has been decided to make use of the now well established linearized theory of Salvesen, Tuck, and Faltinsen (1970). This slender-ship or strip theory is usually assumed applicable to the case of Froude numbers up to about 0.6.

The method is founded on the assumption that the three-dimensional hydrodynamics can be approximated as simple sums of the two-dimensional hydrodynamics of the sections of the vessel. The saving in computational effort is several orders of magnitude in comparison to fully three-dimensional computations. Furthermore, since the approach is well behaved numerically, one might expect that the ranking of a selection of vessels, will be accurate, even if the predicted motions are less accurate than those obtained by a more elaborate and much more computationally intensive three-dimensional theory.

HULL PARAMETERIZATION

The catamaran demihull is essentially represented by four key lines, defining the section shape, the keel profile, the load waterline, and the profile chine line. These are illustrated in Figure 1. The section shape is a simple single-chine section, possessing vertical sides near the waterline together with a bottom consisting of a simple straight deadrise. Such a section is typically used in the design of modern high-speed catamarans.

The keel profile and the load waterline line are both constructed from a simple combination of a parabola, a straight section and another parabola, with the slopes continuous at the junctions. In the case of the profile chine line, the non-dimensional depth of the chine (i.e. the ratio of the depth of the chine to the draft of the section), rather than the dimensional depth of the chine, is constructed in this manner. It should be noted that negative values of this non-dimensional depth are allowed; this is interpreted physically as the chine line crossing the waterline and extending into the topsides of the hull. In this instance the chine line is ignored in the generation of the section shape, and a purely triangular shape is used.

Fifteen parameters are then required to define the demihull completely. Three are the principal dimensions: the waterline length and beam and the overall draft, L_{wl}, B_{wl}, and T. With these values chosen, three parameters are required to specify the waterline shape: the beam at the transom, B_t, the length of the non-parallel waterline aft, L_{npwla}, and the length of the parallel waterline, L_{pwl}. The keel profile can similarly be specified with four further parameters: the draft of the vessel at the transom, T_t, the length of the non-horizontal keel aft, L_{nhka}, the length of the horizontal keel profile, L_{hk}, and the draft at the bow T_b. The section beam, B_x, and draft, T_x, are then completely defined. Finally, the chine line requires five parameters: the non-dimensional depth at the transom, T_{ct}/T_t, the length over which the non-dimensional depth of the line is non-uniform aft, L_{cnua}, the non-dimensional depth in the uniform section, T_{cu}/T_x, the length over which the non-dimensional depth of the line is uniform, L_{cu}, and the non-dimensional value at the bow, T_{cb}/T_b. An additional parameter is required to specify the tunnel width, W_T, defined as the smallest transverse distance between the inner extremities of the demihulls' load waterlines.

This representation has two particularly desirable features. Firstly it allows a wide range of hullform geometry whilst requiring only a relatively small number of parameters. Secondly, the comparison of the motions of different vessels is only really meaningful if the displacements are comparable; in this representation it can be seen that only one parameter is required in order to define the beam of the demihull; thus, the demihull beam can conveniently be replaced by the volume displacement, ∇, as the independent variable.

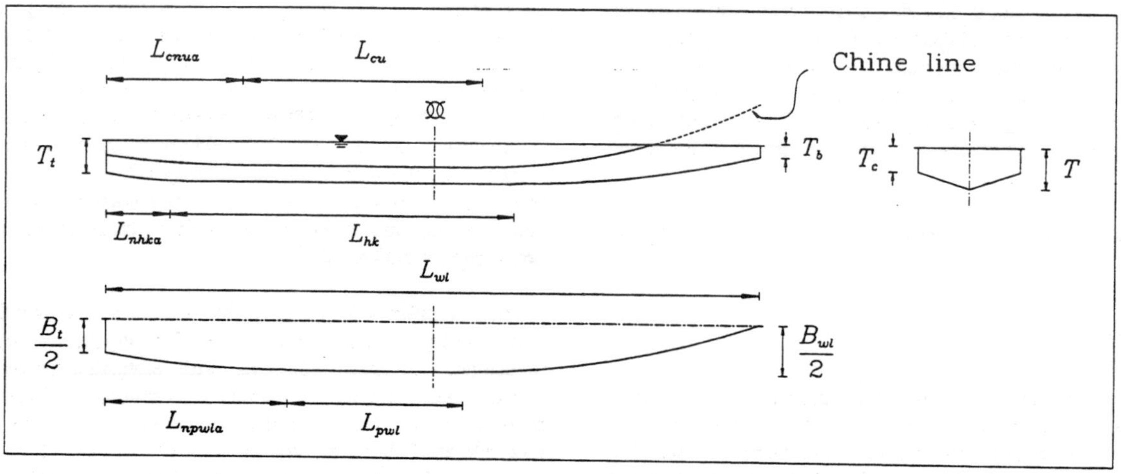

Figure 1: Hull parameterization

CALCULATION OF SHIP MOTIONS

As noted above, the ship motions were predicted here by the strip theory developed by Salvesen, Tuck, and Faltinsen (1970). The lion's share of the work in this theory is the computation of the hydrodynamics of the individual sections of the vessel, which are assumed to behave independently of each other. The hydrodynamics are represented by the following quantities: the added masses, the damping coefficients, and the exciting forces and moments. These quantities were computed using the boundary-element method (BEM), as developed by Doctors (1988). The principal advantage of the BEM (over, for example, conformal mapping techniques) is that arbitrary section shapes can be handled. A curious problem, however, is that of the so-called "irregular frequencies", at which the mathematical solution breaks down. This difficulty was overcome in the present context by distributing elements, or panels, not only on the submerged wetted portion of the section, but also on the internal free surface. A small additional computational effort is therefore required, but this approach has been found to eliminate the mathematical difficulty completely.

Despite the computational efficiency of this theory, it was decided that it was vital to increase the speed of the computation even further, because of the large number of test hulls that have to be processed in the search for the optimum vessel. It was therefore decided to constrain the vessel to a simple section type, That is, there are four straight sides in all, which becomes five, if one counts the internal free surface noted above.

The hydrodynamic coefficients for the motion of such a section can be expressed in terms of three parameters: the non-dimensional frequency of oscillation (i.e. the encounter frequency), the section beam–draft ratio, and the ratio of the chine depth to the section draft. One complicating factor was the fact that that the exciting forces are a function also of the (absolute) wave frequency, which would introduce a further parameter and increase the size of the hydrodynamic data base beyond a reasonable limit. For this reason, it was decided not to store the exciting forces in the data base but, instead, the velocity potentials on the surface of the section, from which the sectional exciting forces could be computed later as required. This less-than-ideal arrangement still led to a calculation of the hydrodynamics which was an around an order of magnitude faster for the numbers of elements per section adopted here than an approach based on direct calculation.

A data base was created consisting of the hydrodynamic coefficients and surface potentials calculated for a large number of combinations of these three parameters. The section beam–draft ratio was varied from zero to unity in twenty-one steps, and the draft–beam ratio varied from unity to zero in twenty-one steps. In the cases for which the beam–draft ratio was less than one, the encounter frequency was non-dimensionalized with respect to the draft of the vessel; conversely, when the draft–beam ratio was less than one, the beam was used as the non-dimensionalization factor.

The non-dimensional encounter frequency was split up into two regions, from the lower limit (taken here as 0.02) to unity, and from unity to the upper limit (taken here as 12.0). Twenty-three steps were used in the lower region, in which the step size was uniform in the encounter frequency. Twenty four steps were used in the upper region in which the step size was proportional to the inverse of the encounter frequency. Finally, the chine-depth–draft ratio was varied from 0.0 through 0.95 in twenty steps. The total number of records in the database was thus just under forty thousand.

For this preliminary study in ship motions, the decision was also made to restrict interest to catamarans. For this purpose, it was assumed that the separation between the hulls was sufficiently large so that the hydrodynamics of the vessel could be obtained by ignoring interactions between the hulls, and that the effect of viscous damping in roll was negligible. It was still a requirement, of course, to include the influence of the demihull separation by using the relevant version of the parallel-axis theorem, when transferring the hydrodynamic coefficients, originally computed about a local axis system centered on the demihull waterline, to the vessel axis system.

The effects of ship motions on passenger sickness are not well understood, and most of the approaches to the estimation of motion-sickness are based on the experiments of O'Hanlon and McCaulay (1974), who studied sinusoidal vertical acceleration only. The approach adopted in the first instance is the frequency-weighted root-mean-square vertical acceleration (FWRMSVA) approach as described by Stark (1982). In this method the RMS vertical acceleration (at a given point) is obtained over a range of frequencies; a weighted sum is then calculated in which the weights are proportioned to the inverse of the human response characteristic as derived from O'Hanlon and McCauley's data.

This is a convenient measure for these purposes as it may easily be calculated from the vertical acceleration response spectrum at a given point (which itself depends upon the combined heave, pitch and roll of the ship), and yields a single number which may be used directly as an objective for minimization. It is, however, unfortunate that the majority of available data on the *human response* to oscillatory motions considers only vertical accelerations, as the "corkscrewing" motions in which vertical accelerations are combined with lateral accelerations and pitch- and roll-induced angular accelerations are widely implicated by anecdotal evidence as being particularly undesirable. In terms of the calculations adopted here, it would be extremely easy to include further components of acceleration in an optimization objective, once their effects on passenger comfort were adequately understood. Nonetheless, it seems that the approach chosen is appropriate, at least until more complete criteria are available.

OPTIMIZATION

It is intended that the methods developed here can aid the naval architect in the decision making process in a number of different ways. One obvious way is to allow some "fine-tuning" of an existing design; in such a case it is likely that the principal dimensions are

Frequency (Hz)	Weight	Frequency (Hz)	Weight
0.050	0.200	0.315	0.620
0.063	0.420	0.400	0.438
0.078	0.520	0.500	0.260
0.100	0.745	0.630	0.152
0.125	0.924	0.800	0.060
0.160	1.030	1.000	0.025
0.200	1.000	1.250	0.009
0.250	0.860	1.600	0.004

Table 1: Weights for vertical accelerations

substantially constrained, and the aim is to examine optima in the locality of the existing design. If only one ore two variables are to be adjusted, and only one or two constraints imposed, this is essentially a simple "hill-climbing" type of task; however, with more variables and more constraints, the problem rapidly becomes much more difficult.

Alternatively, it is useful in some cases for the naval architect to have a tool for the exploration of highly unconstrained "blue-sky" problems such as the development of vessels to meet new markets or to adopt new technologies. This type of global optimization problem may expose the naval architect to large areas of unfamiliar terrain in the design space, so a tool for the systematic exploration of the space is potentially very powerful; however, the potentially complex geography of the terrain places very great demands on the optimization technique adopted.

Once the results have been obtained, they may be used by the naval architect in a variety of ways. With complex objectives such as those used here, it is likely that there will be a large number of solutions of almost equal merit. Since there will always be practical constraints which cannot feasibly be included in the model, it is desirable that many of the better solutions be examined as they may prove more useful than the one with the best value of the objective function. A comparison with the best solution will then allow the "price" paid for the additional constraints to be quantified — and thus aid the naval architect in the decision-making process.

It is vital, then, that the optimization technique is sufficiently flexible and powerful to meet the demands of both the local and global optimization tasks described in a computationally efficient manner, whilst ideally operating in some parallel manner such that solutions which are widely separated in the design space yet are of approximately equal merit can be compared and retained, if desired, for further study.

The specific aim of this study is to choose the decision variables — i.e. the geometric parameters described earlier — in such a way as to minimize the worst case FWRMSVA on the passenger deck, whilst simultaneously satisfying any external constraints on the ship. The motions will, in general, depend upon the geometry parameters in a rather complex manner; both the objective (here the motions), and any constraints imposed, are likely to be highly non-linear and probably highly multi-modal functions of the decision variables. Furthermore, the numerical calculation of the motions introduces some "noise" into the solution such that partial derivatives of the objective with respect to the decision variables may not be continuous. In order to overcome these difficulties a hybrid technique is adopted here in which a GA is coupled to a "hill-climbing" technique.

The GA is based upon a simplified analogue of Darwinian evolution, and is thus intrinsically parallel in nature, allowing the simultaneous exploration of widely separated regions of the design space. It has shown success in solving a wide range of complex engineering problems (see for example Day (1995), Gage (1995), Day and Doctors (1996)), though at the expense of a rather large number of computations. A simple GA is described in Michalewicz (1994).

The version adopted here is broadly similar to the "Structured Genetic Algorithm" (see Dasgupta and McGregor (1992)). In the version adopted here the variables are binary coded (using a Gray coding) within a specified range; the decision variable are thus intrinsically constrained. The algorithm employs tournament selection to choose the parents and the conventional binary operators of crossover and mutation to generate offspring. Genetic diversity is enforced throughout; offspring compete with each other and all parents. Finally, the population is re-initialized periodically (saving the single best individual) to enhance diversity amongst the relatively small populations.

In an attempt to improve the speed and efficiency of the GA in locating local optima, a conventional "hill-climbing" algorithm is periodically run, with the starting set of decision variables taken from the current best individual. On completion of the hill-climbing, the variables are re-encoded as binary, the individual is replaced with the (hopefully) improved version, and the GA continued. Once the GA is deemed to have converged — which is determined by examining the rate of change of the best and the average of the population — the hill-climber is run for a final time. Several multi-dimensional hill-climbing techniques were tried over a selection of test functions; whilst not the fastest algorithm, Powell's direction set approach (see Acton (1970)) was found to give good results in a variety of conditions.

Additional constraints are imposed on both the GA and the hill-climbing algorithm by the inclusion of penalty functions in the calculation of the objective function. This does not formally guarantee the satisfaction of the constraints; however, it has been found that the constraints have always been satisfied within a degree of accuracy consistent with the convergence of the hill-climbing algorithm.

There are no guarantees of global optimality when solving a non-linear problem with a technique such as this. In order to try and determine some measure of confidence in the quality of the converged results, a number of different runs should be carried out with different initial populations. If many or all of the results are closely spaced (in

Decision variable	Non-dimensional form	Minimal value	Maximal value
L_{wl}	$L_{wl}/\nabla^{1/3}$	4.00	10.00
T	T/L_{wl}	0.02	0.10
B_t	B_t/B_{wl}	0.00	1.00
L_{npwla}	$\frac{L_{npwla}}{(L_{wl}-L_{pwl})}$	0.00	0.95
L_{pwl}	L_{pwl}/L_{wl}	0.00	0.95
T_t	T_t/T	0.00	1.00
L_{nhka}	$\frac{L_{nhka}}{(L_{wl}-L_{hk})}$	0.00	1.00
L_{hk}	L_{hk}/L_{wl}	0.00	1.00
T_b	T_b/T	0.00	1.00
T_{ct}/T_t	–	-9.00	0.95
L_{cnua}	$\frac{L_{cnua}}{(L_{wl}-L_{cu})}$	0.00	1.00
T_{cu}/T_x	–	0.00	0.95
L_{cu}	L_{cu}/L_{wl}	0.00	1.00
T_{cb}/T_b	–	-9.00	0.95
W_T	W_T/L_{wl}	0.20	0.50

Table 2: Decision variables used in optimization

terms of the objective), then some reassurance may be taken that the results are reliable; conversely, if the results are widely scattered, then it is likely that the parameters controlling the algorithms need some adjustment, or that more runs are needed.

An additional degree of confidence can be gained in cases in which the problem is solved over a wide range of conditions by ensuring that the objective function varies smoothly and continuously. In this problem, for example, it is reasonable to expect that the FWRMSVA at a particular point of the optimal hull will vary smoothly and continuously with the speed of the vessel, even though the decision variables may vary very rapidly indeed with speed.

IMPLEMENTATION

The hull model, the motions technique, and the GA were implemented in a FORTRAN 77 program. The decision variables were essentially as described previously; however, they were all non-dimensionalized to enable the technique to be applied consistently to hulls of differing volume displacements. The particular non-dimensionalization of the variables employed, and the ranges allowed are shown in Table 2. In the cases of the length–displacement ratio and the length–draft ratio, the decision variables were taken as the logarithm of the non-dimensional ratio so that the wide ranges of the variables under consideration would be better covered.

The runs carried out for this preliminary study were based on a ship with volume displacement of 1000 m^3, a vertical center of gravity 3.2 m above the waterline, and a roll center 0.5 m above the waterline. Twenty-one equally spaced stations, each with twenty panels were used in the strip theory calculations. In addition to the intrinsic constraints on the decision variables, a number of additional constraints were imposed. The first pair of constraints related to the slopes of the waterline at the entry and at the transom which were restricted to 15 and 10 degrees respectively. A lower and an upper bound were placed on the demihull beam–draft ratio of 0.25 and 4.0. Finally, a lower and an upper bound were also placed on the product of the length and the beam, with the values set at 1500 m^2 and 3000 m^2.

It was assumed that the passenger deck had an area of 1500 m^3, and ran the full beam of the ship. The deck was located 7.7 m above the waterline and was situated as far aft as possible, on the assumption that the worst motions would be at the bow of the vessel. The longitudinal extent of the deck was governed by the beam of the ship. It was assumed that the worst location for passenger sickness within the rectangular deck area would be at the full beam of the vessel at either the extreme forward or the extreme aft end of the deck. At each point, the spectrum of vertical acceleration was calculated, at the frequencies given in Table 1; the FWRMSVA was then calculated from the spectra. The worse of the two values was taken as the objective. In cases in which the ship was examined at a variety of headings, the worst FWRMSVA of the location/heading combinations was chosen.

RESULTS

A small series of test cases was analyzed in order to examine the viability of the method. A single sea state was chosen, being a JONSWAP spectrum with significant wave height of 3.0 m and mean zero-crossing period of 6.0 s. Single headings of 90, 135 and 180 degrees were examined, corresponding to beam, bow-quartering and head seas, respectively. Speeds of 15.0 m/s and 20.0 m/s were considered. as typical of modern catamaran ferries, whilst yielding Froude numbers at which the strip theory yields results with reasonable accuracy.

In each case between five and eight runs were carried out with different initial populations in order to examine the scatter and hence gain some idea of the reliability of the optima. Rather more scatter then would be considered desirable was found, suggesting that the problem is highly multi-modal, and indicating that further runs, to explore the nature of the objective function, and some fine-tuning of the GA control parameters would give added confidence to the results.

As might reasonably be expected for a catamaran, the vessel performs well in beam seas at both speeds; at the slower speed the worst performance was found in head seas, whilst at the higher speed the bow-quartering seas yielded a higher FWRMSVA. The pitch component of the motions was clearly driving the geometry to maximize the longitudinal moment of waterplane area, as the vessels were extended almost to their length constraint, and right up to the entry angle constraint, with flat wide demihulls.

The results for head seas only are shown in Figure 2, while the results for the combination of all three sea directions are shown in Figure 3. The diagrams illustrate the profile, chine line and waterlines of the vessel plotted at the stations used for the strip theory. It can be seen that the hull forms are broadly similar in the two cases, in spite of the critical heading being 135 degrees in the faster case for all

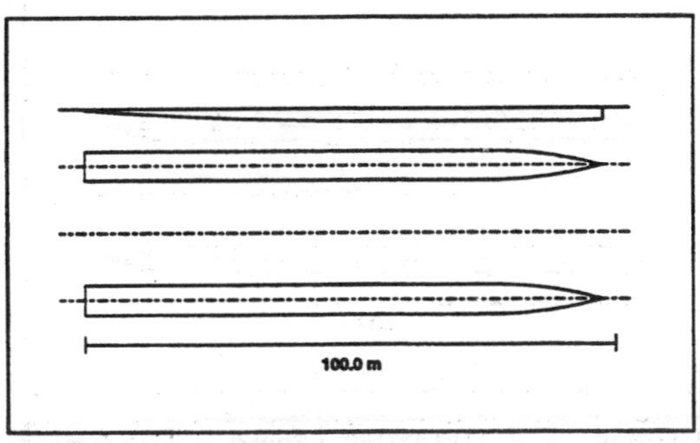

Heading	180 deg	Speed	15.0 m/s
Critical point	aft		
FWRMSVA	0.096g		

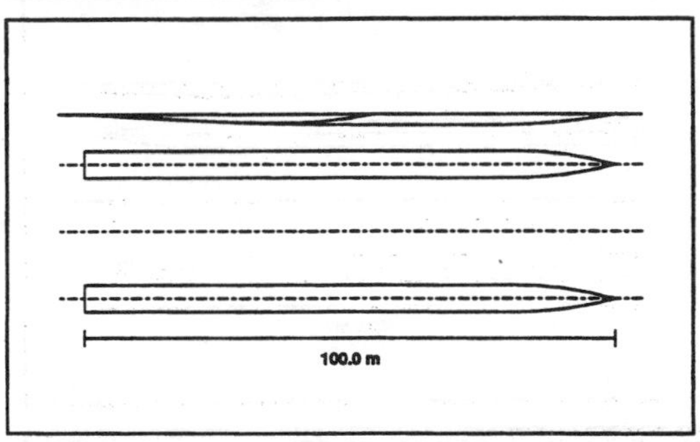

Heading	180 deg	Speed	20.0 m/s
Critical point	aft		
FWRMSVA	0.099g		

Figure 2: Hull geometry obtained for head seas

headings. It is interesting to note that the worse position in both cases is at the stern; this is presumably because the passenger deck (of fixed area) only extends forward around one-half of the waterline length with these total beams. The lower speed hulls have no submerged chine at all, while the higher speed hulls have a small chine in the aft half. A more sophisticated model might be developed in which the passenger deck was moved fore and aft in order to minimize the worst FWRMSVA — which would presumably occur when the values were equal at the forward and aft corners of the deck.

The demihulls are clearly somewhat impractical for a number of reasons, one obvious reason being the lack of room to fit water jets for propulsion. In order to examine the effects on the solution procedure of a more complex system of constraints, two further constraints were added, limiting the minimal waterline beam at the transom to 4.5 m, and the minimal chine depth at the transom to 2.0 m. When the problem was re-run, it was found that the chine stayed at the waterline almost all the way to the transom, and then dropped almost vertically over the aftmost station, thus satisfying the required transom area without adding any significant volume at the transom! In order to achieve a more realistic effect, the product of the uniform non-dimensional chine depth and the draft (i.e. $(T_{cu}/T_x)\times T$) was also constrained to be greater than 2.0 m. The results are shown in Figure 4.

It can be seen that a somewhat wedge-shaped demihull results, with triangular sections forward and almost rectangular sections aft. The length is reduced somewhat, as volume is forced into the stern, and the worse FWRMSVA is significantly worse, with head seas dominating at both speeds.

These hulls are still slightly unrealistic in some ways. It must be remembered that only one sea state was used in these calculations; further sea states could easily be incorporated, at a linear penalty in computational time. The runs incorporating all three headings each took around 8 to 10 hours on a DEC AlphaStation 500/266 computer, though this might be speeded up by fine tuning some of the GA control parameters; set against that, it would certainly be desirable to carry out more runs with different initial populations, as discussed earlier.

Notwithstanding, if the hulls obtained from a reasonable number of runs, with several sea-states considered, still proved slightly unrealistic, which seems likely, further constraints could be artificially imposed to force the hulls to conform to other practical requirements. It is felt, however, that a better approach, where possible, is to specify additional rational goals rather than additional essentially arbitrary constraints. For example, it is believed that adding the additional goal of minimizing the resistance is likely to produce more desirable results than constraining the hullform geometry. This dual goal optimization will be the subject of further study.

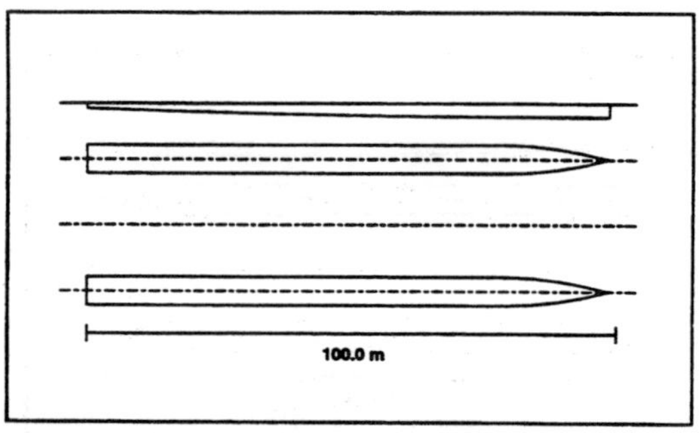

Heading	90/135/180	Speed	15.0 m/s
Critical point	aft	Critical heading	180
FWRMSVA	0.096g		

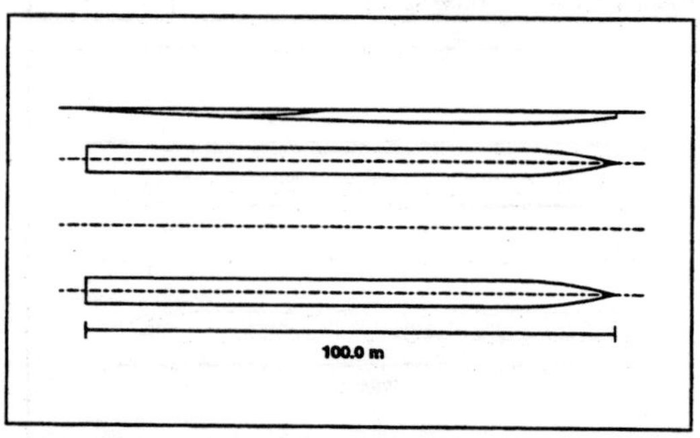

Heading	90/135/180	Speed	20.0 m/s
Critical point	aft	Critical heading	135
FWRMSVA	0.101g		

Figure 3: Hull geometry obtained for all seas

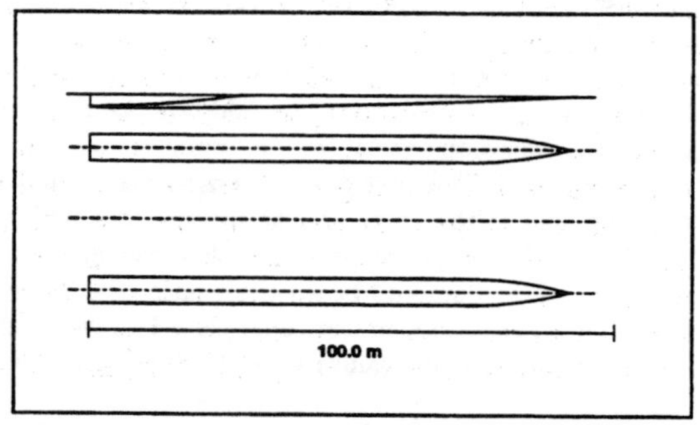

Heading	90/135/180	Speed	15.0 m/s
Critical point	aft	Critical heading	180
FWRMSVA	0.123g		

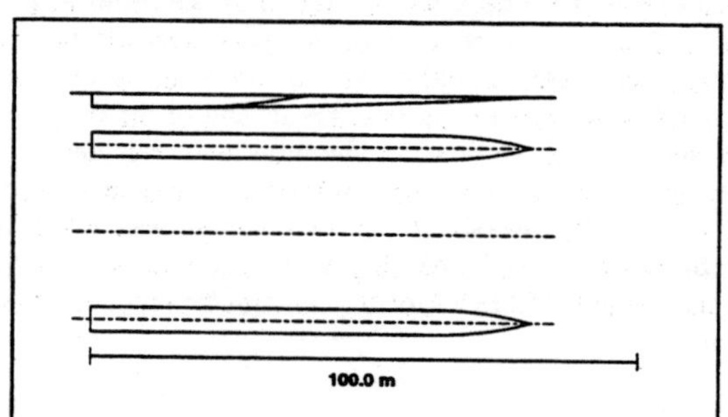

Heading	90/135/180	Speed	20.0 m/s
Critical point	aft	Critical heading	180
FWRMSVA	0.138g		

Figure 4: Hull geometry obtained for all seas — Transom dimensions constrained as described

CONCLUSIONS

A method has been developed by which the underwater geometry of catamarans may be optimized in the sense of minimizing the worst-case frequency-weighted root-mean-square vertical acceleration over the passenger deck, and thus maximizing passenger comfort. The method is suitable for the exploration of new concepts at the preliminary design stage as well as for the investigation of modification of existing designs.

Preliminary runs indicate that the method can be used within the scope of reasonable computational resources, and produces plausible results. However, it is likely that some fine tuning might yield improved results.

The hulls obtained by optimizing on motions only are somewhat impractical for a variety of reasons; however, it is shown by example that specific realistic constraints can successfully be imposed to enforce practical restrictions on hullform. Nonetheless it is believed that, where possible, adding additional rational goals, such as resistance, rather than essentially arbitrary constraints will yield better results.

ACKNOWLEDGEMENTS

The first author is a Senior Research Fellow with the Australian Maritime Engineering Cooperative Research Centre (AMECRC). The second author is on partial secondment from The University of New South Wales to AMECRC. They would like to thank Mr. L. Lazauskas of The University of Adelaide for his helpful discussions.

REFERENCES

ACTON, F.S. (1970). *Numerical Methods that work*, Harper & Row, Publishers, New York, 541 + xviii pp

DASGUPTA, D. AND MCGREGOR, D.R. (1992). "Engineering Optimization Using the Structured Genetic Algorithm", *Proc. European Conference on Artificial Intelligence (ECAI-92)*, Vienna, Austria

DAY, A.H. (1995). "The Design of Yacht Sailplans for Maximal Upwind Speed", *Proc. Twelfth Chesapeake Sailing Yacht Symposium*, US Naval Academy, Annapolis, Maryland, pp 97–116

DAY, A.H. AND DOCTORS, L.J. (1996). "Minimal-Resistance Hullforms for High-Speed Craft", *Trans. Royal Institution of Naval Architects*, London, England, 25+i pp

DAY, A.H., AND DOCTORS, L.J. "Resistance Optimization of Displacement Vessels on the Basis of Principal Parameters", *J. Ship Research*, submitted for publication

DOCTORS, L.J. (1988). "Application of the Boundary-Element Method to Bodies Oscillating near a Free Surface", *Computational Fluid Dynamics — Proc. International Symposium on Computational Fluid Dynamics ISCFD-Sydney*, Elsevier Science Publishers B.V., Amsterdam, pp 377–386

DOCTORS, L.J. (1995). "A Versatile Hull-Generator Program", *Proc. Twenty-First Century Shipping Symposium*, University of New South Wales, Sydney, New South Wales, pp 140–158, Discussion 158–159

DOCTORS, L.J. AND DAY, A.H. (1995). "Hydrodynamically Optimal Hull Forms for River Ferries", *Proc. International Symposium on High-Speed Vessels for Transport and Defence*, Royal Institution of Naval Architects, London, England, pp 5-1–5-15

GAGE, P.J. (1995). "New Approaches to Optimization in Aerospace Conceptual Design", NASA Contractor Report 196695, Ames Research Center, Moffett Field, California, 168+xix pp

HAVELOCK, T.H. (1923). "Studies in Wave Resistance: Influence of the Water-Plane Section of the Ship", *Proc. Royal Society of London*, Series A, Vol. 103, No. 723, pp 571–585

HEARN, G.H., WRIGHT, P.N.H., AND HILLS, W. (1995). "Sea-keeping for Design: Balancing the Vertical and Horizontal Motions of a Catamaran", *Proc. Third International Conference on Fast Sea Transportation (Fast '95)*, Lübeck-Travemünde, Germany, Volume 1, pp 205–220

HEARN, G.H., WRIGHT, P.N.H., AND YAAKOB, O. (1995). "Sea-keeping for Design: Identification of Hydrodynamically Optimal Hull Forms for Large High Speed Catamarans", *Proc. International Symposium on High-Speed Vessels for Transport and Defence, Royal Institution of Naval Architects*, London, England, pp 4-1–4-15

KOSTYUKOV, A.A. (1968). *Theory of Ship Waves and Wave Resistance*, Effective Communications Inc., Iowa City, Iowa, 400+vii pp

LIN, W.-C., WEBSTER, W.C., AND WEHAUSEN, J.V. (1963). "Ships of Minimum Total Resistance", University of California, Institute of Engineering Research, Berkeley, Report NA-63-7, 47+ii pp

MICHALEWICZ, Z.. (1994). *Genetic Algorithms + Data Structures = Evolution Programs*, Springer-Verlag Berlin, 340 + xvi pp

MICHELL, J.H. (1898). "The Wave Resistance of a Ship", *Philosophical Magazine*, London, Series 5, Vol. 45, pp 106–123

O'HANLON, J.F. AND MCCAULEY, M.E. (1974). "Motion Sick-ness Incidence as a Function of the Frequency and Acceleration of Vertical Sinusoidal Motion", *Aerospace Medicine*, Vol. 45, No. 4, pp 366–369

SALVESEN, N., TUCK, E.O., AND FALTINSEN, O. (1970). "Ship Motions and Sea Loads", *Trans. Society of Naval Architects and Marine Engineers*, Vol. 78, pp 250–279, Discussion: 279–287

STARK, D.R. (1982). "Marine Vehicle Ride Quality: A State-of-the-Art Assessment", *Transportation Research Record*, No. 894

TUCK, E.O. AND LAZAUSKAS, L. (1996). "Low-Drag Rowing Shells", *Proc. Third Conference on Mathematics and Computers in Sport*, School of Information Technology, Bond University, Queensland, Australia, pp 17–34

WEHAUSEN, J.V. (1973). "The Wave Resistance of Ships", *Advances in Applied Mechanics*, Academic Press, Inc., New York, Vol. 13, pp 93–245

Proceedings of the Seventh (1997) International Offshore and Polar Engineering Conference
Honolulu, USA, May 25-30, 1997

ISBN 1-880653-28-1 (Set); ISBN 1-880653-32-X (Vol. IV)

On Efficient Design of Swath Ships by Using Reliability-Based Optimisation

Yongchang Pu
Oxford Brookes University
Oxford, United Kingdom

ABSTRACT

In this paper the reliability-based optimisation technique was applied to a typical transverse frame in a built SWATH ship. The response amplitude operators of primary loads were evaluated by a three dimensional oscillating source distribution method in association with linearised potential theory. The extreme values were then calculated by spectral analysis. The structural weight was used as the objective function. It was found that a more efficient design of SWATH ships could be achieved by using reliability-based optimisation techniques.

KEY WORDS: Reliability analysis, reliability-based optimisation, structural analysis, component reliability

INTRODUCTION

SWATH is a novel concept. Because of its good seakeeping performance, large deck area, large transverse stability, high sustained speed at most headings to ocean wave, and the low vertical motion and acceleration amplitudes, which are important for deck operations, it is widely used in a variety of fields such as naval, ferry, strategic positioning, surveillance, oceanographic and hydrographic survey, personnel transport, offshore patrol, offshore range support, mine warfare, towing and vertically deployed aircraft platform, etc. On the other hand, SWATH has some disadvantages, namely, extreme weight sensitivity (structural weight reaches about 40% of the displacement), large draft and freeboard (to avoid heavy slamming in the wet-deck), potential for upper hull and lower deck impact, usually large beam ratio relatively slow speed in calm water, high construction cost. The SWATH concept is weight critical which suggests that the structural design cannot be too conservative.

In this paper reliability-based optimisation technique was applied to a typical transverse frame in a built SWATH ship in order to achieve a more efficient design. Firstly the transverse frame was idealised by a one-dimensional model by carrying out a series of finite element analysis. Then the primary loads on the ship were calculated by a three dimensional oscillating source distribution method, and their extreme values were evaluated by a spectral analysis. Finally the frame was optimised by considering the reliability constraints.

STRUCTURAL MODEL OF A TYPICAL TRANSVERSE FRAME

It is quite difficult to consider the whole SWATH ship. In this study the reliability-based optimisation technique was applied to a typical transverse frame in a built SWATH, which was idealised by a one-dimensional model shown in Fig. 1. The one-dimensional model was obtained from analyses of a series of finite element models using a general purpose finite element package LUSAS. Firstly, a coarse mesh of the whole ship was generated in MAESTRO, then the structure between two bulkheads near the midship section idealised by a three-dimensional finer model was analysed. Based on these results, the effective breadths for the two-dimensional model were determined using the following formula:

$$\sigma_{max} \times b_e = \int_{-b/2}^{b/2} \sigma \times dz$$

where σ_{max} and σ are the maximum and actual transverse stress respectively in the decks, b and b_e are the full breadth and effective breadth at transverse section. Finally, the one-dimensional model was created using the results in the two-dimensional model. Details of these analyses can be seen in Das and Pu (1993).

PRIMARY LOADS ON THE SWATH SHIP

The various loads that were considered are wave induced loads, buoyancy and structural weight and superimposed loads including acceleration effects.

The response amplitude operators of wave induced loads, vertical and horizontal acceleration on the ship in regular waves were evaluated by the program, MARCHS (Chan, 1991), in which a three dimensional oscillating source distribution method in association with linearised potential theory was used. The design extreme values with a value of 0.01 of the risk parameter in the lifetime were then calculated by a spectral analysis, in which the Pierson-Moskowitz spectrum and Ochi's formula (Ochi, 1973) was adopted. The significant wave height for ultimate strength design is 3 meters.

Five headings, namely 0°, 45°, 90°, 135°, 180° were chosen, the speed of the ship is zero because the maximum side force occurs when the ship is stationary in the beam seas, the maximum side force is 2180 kN. It is recognised that if the horizontal inertia force is in the same direction as hydrodynamic side force, the situation could be worse. For the same reason the vertical inertia force could increase the maximum shear force at the junction of cross-structure and haunch. Hence the horizontal and vertical accelerations were calculated by the spectral analysis. The significant values of the accelerations were estimated by the formula $2.0\sqrt{m_0}$, where m_0 is the area under the response spectrum. The significant values of the vertical and lateral accelerations at the point where the mid-frame to be optimised was taken are 1.37 (m/s^2) and 0.95 (m/s^2), only 14% and 9.7% of gravity acceleration respectively. Details of these calculations can be seen in Pu (1995).

The statistical properties of the loads were listed in Table 1. The uncertainties in the weight of the upper and lower parts of the structure (L_1, L_2, L_3) are assumed to be zero. The coefficient of variation of the other components of the loads (L_4, ..., L_{11}) are assumed to be 20%, based on past experience (Faulkner and Sadden, 1978).

Table 1: Loads used in the analysis

Type of the loads	Mean value (N)	COV
Side force (L_{10}, L_{11})	81636	0.2
Buoyancy (L_9)	56567.5	0.2
Weight of upper part (L_1, L_2)	18798.8	0
Weight of lower part (L_3)	18798.8	0
Vertical inertial force of upper part (L_4, L_5)	1642.3	0.2
Vertical inertial force of lower part (L_6)	1642.3	0.2
Lateral inertial force of upper part (L_7)	1149.8	0.2
Lateral inertial force of lower part (L_8)	1149.8	0.2

The way to apply the above loads was shown in Fig. 1. The boundary conditions are due to symmetry for central line node and restrained in vertical direction at ship side node to absorb unbalanced vertical force. All the loads were applied in the form of concentrated loads in order to reduce the total number of load cases, which is an important factor affecting the computational time.

RELIABILITY INDEX-BASED OPTIMISATION OF THE FRAME

There are two types of reliability-based optimisation formulations, one is the component reliability index-based optimisation, another is the system reliability index-based optimisation. Most work in reliability-based optimisation in recent decades is based on the elemental reliability (Murotsu, et al, 1983, Enevoldsen and Sorensen, 1994, Madsen and Hansen, 1991, Nikolaidis, E. and Burdisso, 1988, and others). In the analysis the reliability indices, which are usually evaluated by AFOSM (Advanced First Order Second Moment) method, of all the critical elements are restricted to an acceptable range, e.g. greater or equal to 3.0 (Faulkner, 1983). In practice the finite element method is usually needed to generate the coefficients of the safety margins of the critical elements.

The progress in system reliability index-based optimisation is relatively slow because of computational time. The structural system reliability analysis itself is time-consuming, and much time is spent in searching for significant failure modes and on multi-normal integration (if it is necessary). During the process of optimisation, the design variables are changed in each iteration, and the system reliability index of the structure needs to be re-evaluated. Because of the amount of calculation the method is only applied to relatively simple cases, although much effort has been made (Feng and Moses, 1986a, Thoft-Christensen and Murotsu, 1986, Enevoldsen and Sorensen, 1993, Feng and Moses, 1986b, Frangopol and Fu 1989, and others). The state-of-the-art in reliability-based optimisation was comprehensively reviewed by Frangopol and Moses (1994).

Used in the present study was the formulation for elemental reliability-based optimisation which was expressed as:

$$\min \quad C(\mathbf{d},\mathbf{X}) \tag{1}$$

$$\text{subject to} \quad \beta_{ci} \geq \beta_{cia} \qquad (i = 1,...,14) \tag{2}$$

$$d_j^L \leq d_j \leq d_j^u \qquad (j = 1,...,11) \tag{3}$$

where $C(\mathbf{d},\mathbf{X})$ is the objective function referring to the weight of the structure, **d** is the design variable vector, **X** is the random variable vector, β_{ci} is the reliability index of the ith critical section, β_{cia} , which is assumed to be 3.0, is the allowable reliability index for the ith critical section. The fourteen critical sections were marked in Fig. 1. d_j^L and d_j^u are the lower and upper bounds of jth design variable. Also shown in Fig. 1 were the eleven design variables d_1,...,d_{11}, whose lower and upper bounds in optimisation were listed in Table 2.

Table 2: The optimum structure

Design variables	Upper bound	Lower bound	Original	Optimal
d_1	3.5	6.0	4.85	3.5
d_2	3.5	6.0	4.85	3.5
d_3	800	1200	1000	999.99
d_4	3.5	6.0	4	3.5
d_5	3.5	6.0	4.85	3.5
d_6	3.5	6.0	4.85	3.5
d_7	3.5	6.0	4	3.5
d_8	65	85	75	75.28
d_9	8	12	10	12
d_{10}	150	300	200	200.26
d_{11}	4.0	8.0	6	5.61

The safety margins for the fourteen constraints are automatically generated. At present the plasticity condition was used as the failure criteria of critical sections, and the interaction of bending moment and axial force, as well as the buckling effect, was considered in the analysis. Only the loads and material strength were treated as random variables, the others were deterministic. So the safety margins were expressed as:

$$M_i = \sigma_{yi} w_i(\mathbf{d}) - C_i^T x_t \qquad (i=1,...,14) \tag{4}$$

where:

M_i is the safety margin for the ith critical section

$w_i(\mathbf{d})$ is section modulus of the ith critical section

σ_{yi} is yield stress of the ith critical section

$$C_i^T = \left(\gamma \frac{w_i(\mathbf{d})}{A_i(\mathbf{d})} \text{sign}(F_{xi}) \quad 0 \quad \text{sign}(M_{zi}) \right) \tag{5}$$

$A_i(\mathbf{d})$ is cross-section area of the ith critical section

γ is the factor to consider the effect of buckling.

$$\gamma = \begin{cases} 1 & \text{for tension} \\ \sigma_{yi} / \sigma_{ei} & \text{for compression} \end{cases} \tag{6}$$

σ_{ei} is the buckling stress of the member, is estimated by Perry-Robertson approach as follows:

$$\frac{\sigma_{ei}}{\sigma_{yi}} = \frac{1}{2}\left[1 + (1+\eta)\frac{\sigma_E}{\sigma_{yi}}\right] - \sqrt{\frac{1}{4}\left[1 + (1+\eta)\frac{\sigma_E}{\sigma_{yi}}\right]^2 - \frac{\sigma_E}{\sigma_{yi}}}$$

where $\sigma_E = \pi^2 E / (L/r)^2$ is Euler stress, $r = \sqrt{I/A}$ is radius of gyration, L is the effective length of the column, $\eta = 0.3(L/100r)^2$

$x_t = (\, F_{xi} \quad F_{yi} \quad M_{zi} \,)$ is the nodal forces in the member. (7)

The nodal forces can be expressed by external loads as follows:

$$F_{xi} = \sum_{j=1}^{ml} a_{ij}(\mathbf{d}) L_j \tag{8}$$

$$F_{yi} = \sum_{j=1}^{ml} c_{ij}(\mathbf{d}) L_j \tag{9}$$

$$M_{zi} = \sum_{j=1}^{ml} b_{ij}(\mathbf{d}) L_j \tag{10}$$

where ml is the number of external loads, L_j (j = 1,...,ml) is external load. a_{ij}, b_{ij} and c_{ij} are coefficients.

so the safety margin in Eq.(4) is re-arranged as:

$$M_i = \sigma_{yi} w_i(\mathbf{d}) - \sum_{j=1}^{ml} \left\{ \gamma \frac{w_i(\mathbf{d})}{A_i(\mathbf{d})} \text{sign}(F_{xi}) a_{ij}(\mathbf{d}) + \text{sign}(M_{zi}) b_{ij}(\mathbf{d}) \right\} L_j \tag{11}$$

The reliability indices are then calculated by the AFSOM method. Because the first derivatives of the constraints are needed in the optimisation they are evaluated by the method suggested by Bjerager and Krenk (1989).

$$\frac{d\beta}{d\theta} = \frac{1}{|\nabla M(u^*)|} \frac{\partial M}{\partial \theta} \tag{12}$$

where θ is a deterministic parameter, at present it is a design variable d_k, k=1,...,11. u^* is the design point in standard normalised space.

$|\nabla M(u^*)|$ is already calculated in AFOSM. From Eq.(4), $\frac{\partial M}{\partial \theta}$ is equal to

$$\frac{\partial M}{\partial \theta} = \sigma_{yi} \frac{\partial w_i(\mathbf{d})}{\partial \theta} - \left\{ \frac{\partial C_i^T}{\partial \theta} x_t + C_i^T \frac{\partial x_t}{\partial \theta} \right\} \tag{13}$$

where:

$$\frac{\partial C_i^T}{\partial \theta} = \left\{ \gamma \text{sign}(F_{xi}) \left[\frac{\partial w_i(\mathbf{d})}{\partial \theta} A_i(\mathbf{d}) - w_i(\mathbf{d}) \frac{\partial A_i(\mathbf{d})}{\partial \theta} \right] / A_i^2(\mathbf{d}) \quad 0 \quad 0 \right\} \tag{14}$$

$$\frac{\partial x_t}{\partial \theta} = \left\{ \frac{\partial F_{xi}}{\partial \theta} \quad \frac{\partial F_{yi}}{\partial \theta} \quad \frac{\partial M_{zi}}{\partial \theta} \right\} \tag{15}$$

The derivatives $\frac{\partial w_i(\mathbf{d})}{\partial \theta}$, $\frac{\partial A_i(\mathbf{d})}{\partial \theta}$, $\frac{\partial F_{xi}}{\partial \theta}$, $\frac{\partial F_{yi}}{\partial \theta}$, and $\frac{\partial M_{zi}}{\partial \theta}$ in Eqs.(13)-(15) are evaluated by the numerical difference method.

RESULTS AND DISCUSSIONS

A sequential quadratic programming method developed by Schittkowski (1985) was used in the optimisation. The optimum values for the design variables were listed in Table 2. It was found that the design variables d_1, d_2, d_4, d_5, d_6, d_7 and d_{11} were at their lower bounds, while the remaining variables were hardly changed. There is potential for reducing the structural weight. The weight of the structure was 312.4 and 243.7 kg for original and optimum structures respectively. Hence the optimum structure is 28% lighter, and at the same time has better safety than the original one.

Other important information obtained from the analysis is the sensitivity factors (e.g. the first derivatives of constraints to design variables, $\frac{d\beta_j}{dd_k}$, j=1,...,14, k=1,..., 11), which clearly show how the reliability index of each critical section varies with the change of each design variable. The sensitivity factors of the optimum structure were listed in Table 3. It was shown that the design variables had only local effect on the reliability. For instance, the design variables in the upper box area have little influence on the reliability indices in the haunch area, and the design variables in haunch area hardly affect the reliability of the upper box. The sensitivity factors corresponding to d_3, d_8 and d_{10} are very small for all constraints. This is why these variables were nearly kept unchanged during the optimisation process, which sounds suggest that these variables could be excluded from the analysis.

CONCLUSIONS

The objective of this paper was to apply some sophisticated techniques in primary loads prediction and optimum design to a typical frame of a built SWATH ship to achieve a more efficient design. This is very important for SWATH ships because they are weight critical. The response amplitude operators of primary loads on the ship in regular waves were evaluated by the program, MARCHS, in which a three dimensional oscillating source distribution method in association with linearised potential theory was used. The extreme values were then calculated by a spectral analysis.

The reliability-based optimisation technique was then applied to optimise the structure. The linearised plasticity condition was used as the failure criteria of the critical sections, and the interaction of bending moment and axial force including the buckling effect was considered in the analysis.

It is shown that the optimum structure is 28% lighter, and at the same time has better safety than the original one. It may be said that a more efficient design could be achieved.

ACKNOWLEDGEMENT

This work was developed based on the author's PhD study in the University of Glasgow. The author would like to take this opportunity to express his gratitude to Professor D. Faulkner and Dr. P.K. Das for their supervision and friendly co-operation. Thanks are due to Professor K. Schittkowski and Dr. H.S. Chan for their generous help.

REFERENCES

Bjerager, P. and Krenk, S.(1989): 'Parametric Sensitivity in First Order Reliability Theory', *Journal of Engineering Mechanics*, ASCE, Vol. 115, No. 7, July, pp 1577-1582.

Chan, H.S.(1991): 'Prediction of Motion and Structural Responses of Mono- and Twin-Hull Ships Advancing in Waves', Report NAOE-91-16, Department of Naval Architecture and Ocean Engineering, University of Glasgow.

Das, P.K. and Pu, Y. (1993): 'Finite Element Based Structural Analysis of SWATH', Report NAOE-93-20, Department of Naval Architecture and Ocean Engineering, University of Glasgow.

Enevoldsen, I. and Sorensen, J.D.(1993): 'Reliability-Based Optimization of Series Systems of Parallel Systems', *J. of Structural Engineering*, ASCE, Vol. 119, No. 4, April, pp 1069-1084.

Enevoldsen, I. and Sorensen, J.D.(1994): 'Reliability-Based Optimization in Structural Engineering', *Structural Safety*, **15**, pp 169-196.

Faulkner, D and Sadden, J.A. (1978): 'Toward a Unified Approach to Ship Structural Safety', RINA, pp 1-15.

Faulkner, D.(1983): 'On Selecting a Target Reliability for Deep Water Tension Leg Platforms', *Proc. 11th IFIP Conference on System Modelling and Optimisation*, Copenhagen, Denmark, July 25-29, pp 490-513.

Feng, Y.S. and Moses, F. (1986a): 'Optimum Design, Redundancy and Reliability of Structural Systems', *Computers and Structures*, 1986, Vol. 24, No. 2, pp 239-251.

Feng, Y.S. and Moses, F.(1986b): 'A Method of Structural Optimization Based on Structural System Reliability', *J. Struct. Mech.*, **14**(4), pp 437-453.

Frangopol, D.M. and Fu, G.(1989): 'Limit States Reliability Interaction in Optimum Design of Structural Systems', *ICOSSAR'89*, pp 1879-1886.

Frangopol, D.M. and Moses, F.(1994): 'Reliability-Based Structural Optimization', Chapter 13 in *Advances in Design Optimization*, ed. by H. Adeli, Chapman & Hall, pp. 492-570.

Madsen, H.O. and Hansen, P.F.(1991): 'A Comparison of Some Algorithms for Reliability Based Structural Optimization and Sensitivity Analysis', *Proc. of the 4th IFIP WG7.5 Conference, 'Reliability and Optimization of Structural Systems'*, ed by R. Rackwitz and P. Thoft-Christensen, Munich, Germany, Sept. 11-13, pp 443-451.

Murotsu, Y., Kishi, M., Okada, H., Yonezawa, M. and Taguchi, K.(1983): 'Probabilistically Optimum Design of Frame Structure', *Proc. of 11th IFIP Conference on System Modelling and Optimization*, Copenhagen, Denmark, July 25-29, pp 545-554.

Nikolaidis, E. and Burdisso, R.(1988): 'Reliability Based Optimization: A Safety Index Approach', *Computers and Structures*, Vol.28, No.6, pp 781-788.

Ochi, M.K. (1973): 'On Prediction of Extreme Values', *Journal of Ship Research*, March, pp. 29-37.

Pu, Y. (1995): 'Reliability Analysis and Reliability-Based Optimisation Design of SWATH Ships', PhD Thesis, Department of Naval Architecture and Ocean Engineering, University of Glasgow, June.

Schittkowski, K.(1985): 'NLPQL: A FORTRAN Subroutine Solving Constrained Nonlinear Programming Problems', *Annals of Operations Research*, **5**, pp 485-500.

Thoft-Christensen, P. and Murotsu, Y.(1986): 'Application of Structural Systems Reliability Theory', Berlin, Heidelberg, New York, Springer.

Table 3: Sensitivity factors of the optimum structure

	constraint 1	constraint 2	constraint 3	constraint 4	constraint 5	constraint 6	constraint 7	constraint 8	constraint 9	constraint 10	constraint 11	constraint 12	constraint 13	constraint 14
d_1	0.99	0.99	0.99	0.99	-0.20e-11	-0.95e-11	-0.53e-11	-0.20e-11	-0.22e-11	0.10e-11	0.11e-11	0.88e-11	0.15e-10	0.17e-10
d_2	0.99	0.99	0.99	0.99	-0.20e-11	-0.95e-11	-0.53e-11	-0.20e-11	-0.22e-11	0.10e-11	0.11e-11	0.88e-11	0.15e-10	0.17e-10
d_3	0.13e-1	0.13e-1	0.13e-1	0.13e-1	0.13e-1	0.13e-1	0.13e-1	0.14e-1	0.14e-1	0.14e-1	-0.41e-2	-0.14e-2	0.22e-2	0.15e-2
d_4	0.41	0.41	0.41	0.41	0.16e-10	0.13e-10	0.11e-10	0.70e-10	0.72e-11	0.88e-12	-0.11e-10	-0.29e-11	0.12e-10	0.96e-11
d_5	0.36e-10	0.35e-10	0.36e-10	0.38e-10	0.99	1.02	1.02	1.06	1.06	1.11	-0.18	-0.65e-1	0.99e-1	0.70e-1
d_6	0.36e-10	0.35e-10	0.36e-10	0.38e-10	0.99	1.02	1.02	1.06	1.06	1.11	-0.19	-0.65e-1	0.99e-1	0.70e-1
d_7	0.28e-10	0.30e-10	0.30e-10	0.31e-10	0.40	0.41	0.41	0.43	0.43	0.44	-0.50e-1	-0.18e-1	0.27e-1	0.19e-1
d_8	0.29e-11	0.29e-10	0.29e-11	0.29e-11	0.11e-2	0.11e-2	0.11e-2	0.11e-2	0.11e-2	-0.62e-3	0.85e-1	0.66e-1	0.49e-1	0.51e-1
d_9	-0.41e-10	-0.41e-10	-0.41e-10	-0.41e-10	0.85e-2	0.87e-2	0.87e-2	0.89e-2	0.89e-2	-0.49e-2	0.66	0.51	0.37	0.39
d_{10}	0.29e-11	0.29e-11	0.29e-11	0.29e-11	0.16e-2	0.16e-2	0.16e-2	0.16e-2	0.16e-2	-0.90e-3	0.87e-1	0.64e-1	0.42e-1	0.47e-1
d_{11}	-0.22e-10	-0.22e-10	-0.22e-10	-0.21e-10	0.61e-2	0.65e-2	0.65e-2	0.72e-2	0.72e-2	-0.49e-2	0.32	0.25	0.98e-1	0.14

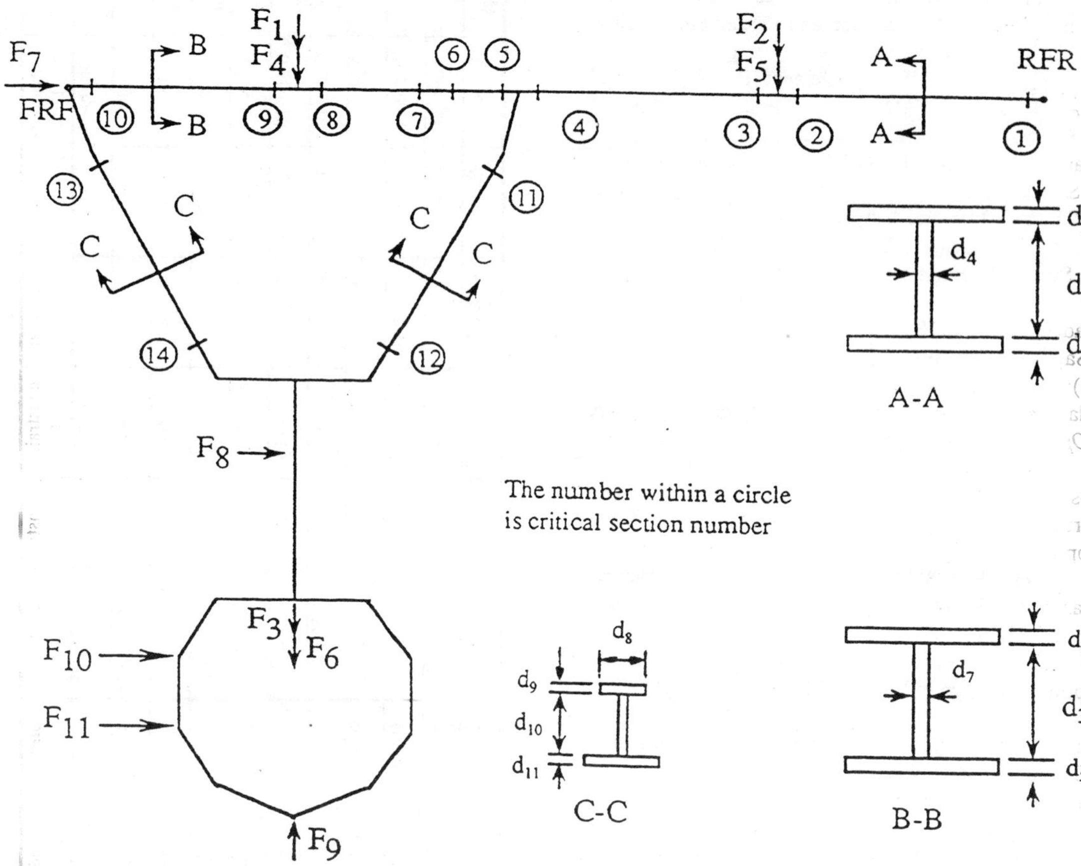

Fig 1 The typical frame in a built SWATH ship

Proceedings of the Seventh (1997) International Offshore and Polar Engineering Conference
Honolulu, USA, May 25-30, 1997

ISBN 1-880653-28-1 (Set); ISBN 1-880653-32-X (Vol. IV)

Comparison of Structural Analysis Methods for Swath Ship Design

L.H. Seidl and K.F. Cheung
University of Hawaii at Manoa
Honolulu, HI, USA

S. Wang
American Bureau of Shipping
Houston, TX, USA

ABSTRACT

The structure of a Swath ship can be analyzed by the hydro-elastic, rigid-dynamic and quasi-static approaches. These are standard methods used in the industry and academic research. Through a numerical model of an actual Swath ship design, the present paper provides a comparative study of these methods and discusses their applicability. The deflections and stresses evaluated from the three methods are compared. Hydro-elastic effects are shown to be more important in certain areas where the structure is more flexible.

KEYWORDS: hydro-elastic, quasi-static, rigid-dynamic, ship design, ship structure, structural analysis, Swath ship.

INTRODUCTION

Accurate predictions of structural stress and deformation in Swath ships assumes greater importance as recent designs involve more advanced applications and lighter construction. The inherent geometry of Swath ships and the need for a lightweight construction yield a more flexible structure than generally found in conventional ships. When subject to oscillatory loading, a Swath ship might exhibit structural vibrations in addition to the rigid-body response. Excessive vibrations can be objectionable to passengers, but most importantly can lead to fatigue failure of the structure. The design and operational experience of a Swath ship is presented by Seidl et al. (1993). In general, the structural stress and deformation in a Swath ship can be evaluated by the hydro-elastic, rigid-dynamic and quasi-static approaches.

The concept of hydro-elasticity was introduced in the late 1950's by analogy to aero-elasticity (e.g., Flax, 1960; and Heller, 1964). The method received more attention in marine dynamics after the work of Bishop and Price (1979) and Wu (1984), and has been used in a number of recent research projects involving flexible floating structures (e.g., Reane et al., 1991; Ertekin et al., 1993; and Newman, 1994). In the hydro-elastic approach, both the structural and rigid-body responses are included in the evaluation of the hydrodynamic and inertia forces. The solution is obtained by a modal superposition of the ship's dry mode shapes, where the hydrodynamic pressure associated with each mode is treated as an external load.

When a relatively rigid structure is considered, the structural vibration and its associated hydrodynamic effects are negligible. This reduces the hydro-elastic approach to the rigid-dynamic approach with only the hydrodynamic and inertia forces associated with the rigid-body response used in the calculation. This dynamically based approach is specified for design and evaluation of ship structures in the American Bureau of Shipping Safehull System and Rules 2000 (ABS, 1994 and 1996). The simplest approach is the quasi-static, which ignores both the rigid-body and structural responses of the ship (e.g., Reilly et al., 1988). This approach is commonly used in the early stages of design, as it does not require input of the ship's mass properties.

The quasi-static and rigid-dynamic approaches are commonly used in the industry and by certifying agencies, while the hydro-elastic approach which is supposedly more accurate is mainly used in academic research. However, the accuracy and performance of the quasi-static and rigid-dynamic approaches in relation to the hydro-elastic approach has not been studied adequately. Previous academic studies were mostly carried out for idealized structures in order to demonstrate the method of hydro-elasticity. The present study emphasizes the application to a realistic Swath ship. Through a numerical model of an actual Swath ship design, the deflections and stress distributions calculated from the three approaches are evaluated and compared. The selection of a structural analysis method for Swath ship design and evaluation is discussed.

STRUCTURAL ANALYSIS OF SWATH SHIPS

Hydro-Elastic Approach

For Swath ships, the primary hydrodynamic loading is dominated by wave scattering effects, which in some situations may be several times larger than the Froude-Krylov forces. To evaluate the wave scattering pressure, the hydrodynamic analysis must take into account the three-dimensional flow around the ship. Two-dimensional methods based on strip theory are unable to account for the complicated flow pattern near the ends of the struts. Empirical procedures based on model tests are available to evaluate the primary wave loading on Swath ships (e.g., Sikora and Dinsenbacher, 1990).

However, these empirical methods only cover a limited range of design parameters and their applications are somewhat restricted. The three-dimensional source distribution method based on potential theory is commonly used to evaluate the primary wave loading on ships and offshore structures (e.g., Faltinsen and Michelsen, 1974; and Seidl, 1978). Let t denote time, ω the angular frequency and ρ the density of water. In the hydro-elastic approach, the hydrodynamic pressure can be expressed as:

$$p = i\omega\rho[\phi_I + \phi_D + \sum_{j=1}^{N} \xi_j \phi_{Rj}]e^{i\omega t} \quad (1)$$

where ϕ_I is the incident potential; ϕ_D is the scattered potential; and ξ_j and ϕ_{Rj} are respectively the response and radiation potential of the j-th mode. For j = 1 to 6, ξ_j refer to the response amplitude operators (RAO) of the rigid body modes; and for j = 7 to N, ξ_j refer to the principle coordinates of the elastic modes.

In the numerical procedure, the wetted surface is discretized into a number of plane panels, in which a uniform distribution of the source strength is assumed. The flow potentials ϕ_D and ϕ_{Rj} (j = 1 to 6) are evaluated using the corresponding boundary conditions for the rigid-body motions, while the potentials ϕ_{Rj} (j = 7 to N) are evaluated from the ship's dry mode shapes, which in turn can be evaluated from a finite element model of the ship structure. With the potentials defined, the equations of motion for the N degrees of freedom $\xi = (\xi_1, \xi_2, ..., \xi_N)$ can be assembled:

$$\mathbf{M}\ddot{\xi} + \mathbf{B}\dot{\xi} + \mathbf{C}\xi = \mathbf{F} \quad (2)$$

in which **M**, **B** and **C** are respectively the generalized mass, damping and stiffness matrices and **F** is the generalized hydrodynamic force vector given by:

$$\mathbf{F} = \int_S p\mathbf{n}\,dS = i\omega\rho\int_S [\phi_I + \phi_D + \sum_{j=1}^{N} \xi_j \phi_{Rj}]e^{i\omega t}\mathbf{n}\,dS \quad (3)$$

where **n** is the generalized normal vector. Since the radiation potentials ϕ_{Rj} are associated with the responses of the ship, they can be treated more conveniently as added mass and damping and moved to the left-hand side of the equation (2). With the matrices and excitation force vectors defined, the rigid-body and elastic responses of the ship can be evaluated. The resulting deformation and stress distribution are then calculated from the mode shapes and principle coordinates.

Rigid-Dynamic Approach

In the rigid-dynamic approach, the ship is assumed to be rigid, so that the structural deformation is ignored in the evaluation of the hydrodynamic and inertia loading. Only the fluid and inertia forces associated with the rigid-body response are used in the calculation. The hydrodynamic pressure includes the radiation pressure as well as the excitation due to the incident and scattered waves

$$p = i\omega\rho[\phi_I + \phi_D + \sum_{j=1}^{6} \xi_j \phi_{Rj}]e^{i\omega t} \quad (4)$$

With the potentials associated with the rigid-body motions defined, the equations of motion for the six degrees of freedom can be assembled. The response amplitude operators ξ_j (j = 1 to 6) for the rigid-body motions and the resulting hydrodynamic pressure distribution can be evaluated using the same procedure as in the hydro-elastic analysis.

The structural inertial force is calculated from the response amplitude operators and the total mass of the ship, which includes the structural and non-structural components such as the main propulsion engine, fuel tank, ballast tanks and deck house. The structural inertia force vector is given by:

$$\mathbf{f} = -m\omega^2\mathbf{u}e^{i\omega t} \quad (5)$$

in which m is the distribution of the ship mass and the displacement vector **u** is given by

$$\mathbf{u} = \begin{Bmatrix} \xi_1 + \xi_5 z - \xi_6 y \\ \xi_2 + \xi_6 x - \xi_4 z \\ \xi_3 + \xi_4 y - \xi_5 x \end{Bmatrix} \quad (6)$$

where (x, y, z) are the Cartesian coordinates of a point on the ship.

The hydrodynamic pressure and structural inertia force are applied statically in the finite element model of the ship to evaluate the corresponding deformation and stress. All the applied loads and the calculated deformation and stress are presented in complex form with real and imaginary parts. The amplitudes of the stress and deformation, which are needed in the design and evaluation of a Swath ship can be calculated from the finite element output. Although the structural analysis described here is static, the applied loads account for the motion of the ship and the system is statically balanced.

Quasi-static Approach

The quasi-static approach assumes the ship is rigid and fixed in the evaluation of the hydrodynamic loading, and therefore only the pressure distribution due to the incident and scattered waves is considered. Since the ship is assumed to be rigid and fixed, the response amplitude operators of the ship need not be evaluated. The analysis begins with the calculation of the hydrodynamic pressure distribution on the wetted surface of the hull using only the incident and scattered potentials:

$$p = i\omega\rho[\phi_I + \phi_D]e^{i\omega t} \quad (7)$$

The real and imaginary parts of the hydrodynamic pressure can be evaluated using the same procedure as for the hydro-elastic analysis. A finite element model of the ship structure is used to evaluate the real and imaginary parts of the resulting deformation and stress, from which the corresponding amplitudes can be evaluated.

The quasi-static approach is very useful for preliminary design, in which the structural details and mass distribution on the ship are yet to be determined. Because of its simplicity, this approach has been widely used in the industry. However, the fluid and structural inertial forces are not included in the analysis, and as a result, the system is not statically balanced. A 'soft-spring' technique is used in the finite element analysis to prevent the formation of a singular stiffness matrix. The limitations of the quasi-static approach are not very well understood, especially when the dynamic response of the ship is significant or when a relatively flexible structure is considered.

Port-Starboard Symmetry

Because of the port-starboard symmetry of Swath Ships, as for most ships, only one half of the structure is required in the hydrodynamic and structural models (e.g., Wu, 1984, Wu et al., 1993). This represents a substantial reduction in computing time and storage, and gives rise to more manageable database files. Although the pressure distribution is not always symmetric, they can be separated into the symmetric and anti-symmetric components evaluated using only one-half of the ship geometry. The symmetric and anti-symmetric components of the hydrodynamic pressure can be calculated by

$$p_S = \frac{1}{2}(p^P + p^s) \quad (8)$$

$$p_A = \frac{1}{2}(p^P - p^S) \quad (9)$$

where p^P and p^S are the pressure distributions calculated respectively on the port and starboard sides of the ship.

The deformations and stresses are evaluated separately for the symmetric and anti-symmetric cases using the corresponding loading and boundary conditions. For the symmetric modes, all anti-symmetric displacements (sway, roll and yaw) at the center plane of the finite element model are set to zero. For the anti-symmetric modes, all symmetric displacements (surge, heave, pitch) at the center plane are set to zero. The resulting deformations and stresses on the port and starboard sides of the ship are obtained by combinations of the symmetric and anti-symmetric results. In the quasi-static and rigid-dynamic analyses, the deflections or stresses on the port and starboard sides are obtained by

$$R^P = R_S + R_A \quad (10)$$

$$R^S = R_S - R_A \quad (11)$$

where R denotes the deflection or stress, and the subscripts S and A indicate respectively the symmetric and anti-symmetric components.

In the hydro-elastic analysis, each mode shape plays the role of one load case. Hence, the deformations or stresses on the port and starboard sides of the ship can be calculated by a summation of the symmetric and anti-symmetric modal responses

$$R^P = \Sigma\xi_i(R_S)_i + \Sigma\xi_j(R_A)_j \quad (12)$$

$$R^S = \Sigma\xi_i(R_S)_i - \Sigma\xi_j(R_A)_j \quad (13)$$

where ξ_i and ξ_j are the principle coordinates for the symmetric and anti-symmetric modes respectively.

RESULTS AND DISCUSSION

Swath Ship S20T10

The Swath ship S20T10 is selected for this comparative study. The ship structure has been designed to a rather detailed extent so that a meaningful structural analysis can be carried out and the weight distribution be calculated. The ship has a displacement of 62 tonnes and a design speed of 22 knots. A lightweight construction using aluminum is employed and the structural rigidity is comparable to other Swath ship designs in its class. The ship is designed as a multi-mission platform, adaptable to such use as an offshore work vessel, a patrol craft, a sport fishing vessel and a small passenger excursion vessel.

The deck and profile scantling drawings and a typical cross-section of the ship are shown in figures 1, 2 and 3 respectively. The length, molded beam and draft of the ship are 20 m, 12 m and 2 m respectively. Each pontoon has a length of 19 m and is connected to the upper hull by two struts. The transverse bulkheads or deep frames are placed at 1 m spacing, while the longitudinal bulkheads or deep frames are placed at 1.5 m spacing. Stiffeners are placed at 0.5 m spacing in both the transverse and longitudinal directions. Both the hull and deck plate are 6 mm, while the thickness of the bulkheads ranges from 6 to 8 mm as indicated in figure 1. The ship structure is designed according to the "carrier principle", in which the passenger accommodations are located in the deckhouse mounted on the carrier vessel. Since the deckhouse is mounted on resilient mounts, it does not contribute to the stiffness of the ship structure.

Finite Element Model

Due to symmetry, only one half of the Swath ship is modeled. The finite element model of the Swath ship is shown in figure 4. The model is composed of 5562 shell, beam and mass elements with 2877 nodes. As shown in the figure, a local fine mesh is developed over a portion of the ship, where detailed stress distributions are required and the structural details are modeled more accurately. In the fine mesh, each of the stiffeners and frames is modeled as a combination of shell and beam elements to correctly reproduce its connection to the hull. The rest of the ship is modeled with largest possible elements to adequately reproduce the locations of the bulkheads and deep frames of the structure. In the coarse mesh, the stiffeners and deep frames are modeled as beam elements with the moment of inertia augmented to account for the combined effects with the plating.

The structural mass of the Swath ship constitutes about 38% of the total ship mass. In the hydro-elastic and rigid-dynamic analyses, the effect of non-structural mass (e.g. main propulsion engine, fuel tank, ballast tanks and deck house) would have a major influence on the inertia force and response of the ship, and hence the stress distribution. In the finite element model, the non-structural ship mass is modeled by lumped mass elements placed at existing nodes, at which concentrated loads are expected to transmit to the ship structure. To prevent the lumped masses from affecting the local vibration modes of the bulkheads and decks, they are carefully placed at the intersection between the deck and bulkheads or deep frames. In this model, approximately one-third of the non-structural ship mass is located above the waterline, and two-third of the non-structural ship mass is in the pontoons.

Hydrodynamic Analysis

One half of the ship is modeled in the hydrodynamic analysis. The struts and pontoon of the ship below the waterline are discretized into 256 facets with 268 nodes. The hydrodynamic forces and responses are obtained using the source distribution method as described by Seidl (1978). It should be noted that the same mesh under the waterline is used in the hydrodynamic and finite element models. Although this is not a requirement of the procedures, it effectively facilitates the hydrodynamic to structural coupling.

The primary force acting on the structure of a Swath ship is the so-called 'prying force' in beam seas. The entire transverse structure of the ship is determined by this force and the associated moment. In beam seas, the encounter frequency is in dependent of the vessel's speed, and therefore a zero-forward speed hydrodynamic solution is appropriate. To illustrate the loading conditions, figure 5 shows the prying force, the sway excitation force and response amplitude operator of the Swath ship. The results presented here correspond to unit amplitude beam waves in deep water. The response amplitude operator is computed for a rigid structure and will be slightly modified if hydro-elasticity is considered.

For the 3.5 second incident waves, the distance between the pontoon centerlines is slightly greater than one half of the wavelength and a standing wave is formed between the pontoons with the anti-node located at the ship centerline. The transverse forces on the two hulls are nearly equal, but in the opposite directions giving rise to a large prying force, and a small excitation force and response amplitude operator in sway. Standing waves are also formed under the ship for the 2.2 second incident waves, in which the wavelength is approximately equal to the clear distance between the pontoons. Although the prying force is a maximum at 2.2 seconds, the wave height is limited by the wavelength. The most critical wave period for the design of this ship is expected to be around 3.5 second and slightly higher. For longer wave periods, the ship follows the motion of the water particles and the associated structural stress is expected to be small.

Free Vibration Analysis

In the hydro-elastic analysis, a free vibration test of the finite element model is first performed to evaluate the mode shapes and natural frequencies of the 'dry' vessel. Since only one-half of the vessel is modeled, the symmetric and anti-symmetric modes of vibrations are evaluated separately using the corresponding boundary conditions of the centerplane. For each case (symmetric and anti-symmetric), the first 7 elastic modes of vibrations are evaluated. Table 1 summarizes the results of the free vibration analysis. The natural frequencies of the Swath ship are found to be well above the wave excitation frequency.

Figure 6 shows the first symmetric mode shape corresponding to the prying motion of the pontoon, while figure 7 shows the first anti-symmetry mode shape corresponding to the twisting motion of the deck. The undeformed shape of the ship is also shown to provide a reference for comparison. Usually, the first couple of modes are most important in the evaluation of the hydro-elastic response. For this model, the 7-th symmetric mode and the 8-th anti-symmetric mode or higher correspond to vibrations of the stiffened decks and bulkheads with frequencies greater than 27.9 Hz. As there is no driving hydrodynamic force for these modes, these vibrations will not contribute to the resulting stress distribution. It should be noted that no local vibration associated with the lumped masses is observed in the finite element results.

Table 1 Natural frequencies of symmetric and anti-symmetric modes

Mode	Frequency (Hz)	
	Symmetric	Anti-symmetric
1	6.2668	7.2743
2	11.2595	11.0082
3	16.3103	14.0080
4	23.3819	16.5314
5	24.9822	20.0654
6	25.9439	25.1927
7	27.9439	27.2162

Structural Analysis

The deformation and structural stress in the Swath ship are evaluated using the hydro-elastic, rigid-dynamic and quasi-static approaches for the wave periods of 3.5, 4.0 and 4.5 seconds. All three analyses are based on the same computer program package and the same hydrodynamic and finite element models are used. The results presented here correspond to unit amplitude beam waves in deep water. In the quasi-static approach, only the pressure associated with the incident and scattered waves is used, while the rigid-dynamic approach includes the hydrodynamic pressure due to all six rigid-body modes and the structural inertia force as well. In the hydro-elastic analysis, the first 14 elastic modes are used.

Figure 8 shows the amplitudes of the horizontal pontoon deflections along the keel. In the quasi-static and rigid-dynamic approaches, only the symmetric component is used in the calculation of the structural deformation giving rise to identical results on the port and starboard sides of the ship. In the hydro-elastic analysis, both the symmetric and anti-symmetric modes are used and the pontoon deflections on the port and starboard sides are therefore different. In should be noted that only the elastic modes are used in the calculation of the structural deformation. The deflections of the pontoons shown in figure 8 corresponds primarily to the flexure bending of the pontoons and struts. The rigid-dynamic approach gives slightly higher predictions of the deflection compared to the quasi-static approach, while the hydro-elastic approach gives consistently higher predictions. The pontoons which account for approximately 48% of the total ship mass has a significant contribution to the dynamic response. The discrepancy between the hydro-elastic prediction and the other two is more pronounced near the stern of the ship, where the pontoon is more flexible and hydro-elastic effects are more important.

Figures 9 and 10 show the distributions of the von Mises stress at bulkheads 30 and 34 (see figure 1 for bulkhead locations). Bulkhead 30 is located at the haunch area and is the most critically loaded structural component on the ship. The stresses shown are for the elements on the lower hull and inboard side of the strut along the bulkheads. In all three analyses, both the symmetric and anti-symmetric stress components are used in the calculation of the stress distributions. Since the anti-symmetric stress components are about an order of magnitude less than the symmetric counterparts, the calculated stresses on the port and starboard sides of the ship are very similar. The quasi-static and rigid-dynamic approaches give very similar stress distributions indicating the dominance of the scattered wave pressure, while the hydro-elastic approach gives slightly higher or lower predictions compared to the other two approaches. Since bulkheads 30 and 34 are located at the fore strut of the ship, where the pontoon is relatively rigid and hydro-elastic effects are not as important, the good agreement of the stresses obtained from the three method is expected.

CONCLUSIONS

The procedures and applications of the hydro-elastic, rigid-dynamic and quasi-static approaches have been summarized. The hydrodynamic pressure is evaluated by the source distribution method, while the ship structure is modeled by the finite element method. All three analysis methods can be performed by the same computer program package, whereby for the quasi-static and rigid-dynamic analyses only the rigid-body modes are used. The pressure distributions and responses are separated into symmetric and anti-symmetric components, which in turn can be evaluated more efficiently using only one half of the ship.

Through a numerical model of an actual Swath ship design, the structural deformations and stresses obtained from the three approaches are evaluated and compared. The quasi-static and rigid-dynamic approaches give very similar results, which are also reasonably close the hydro-elastic predictions in areas where the structure is relatively rigid. For more flexible components near the stern of the ship, the hydro-elastic approach is expected to give more accurate predictions of the deflection and stress.

REFERENCES

ABS 1994. Technical Information. American Bureau of Shipping and Affiliated Companies.

ABS 1996. Guide for Dynamic Based Design and Evaluation of Container Carrier Structures. American Bureau of Shipping and Affiliated Companies.

Bishop, R.E.D. and Price, W.G. 1979. *Hydroelasticity of Ships*. Cambridge University Press, Cambridge, U.K. 423 pp.

Ertekin, R.C., Riggs, H.R., Che, X.L. and Du, S.X. 1993. Efficient methods for hydroelastic analysis of very large floating structures. *Journal of Ship Research*, **31**(1), 58-76.

Faltinsen, O.M. and Michelsen, F.C. 1974. Motions of large structures in waves at zero Froude number. *Proceedings of the International Symposium on the Dynamics of Marine Vehicles and Structures in Waves*, London, pp. 3-18.

Flax, A.H. 1960. Aero and Hydroelasticity, in *Structural Mechanics*, edited by J.N. Goodier and N.J. Hoff, Pergamon Press, New York, pp. 94-159.

Heller, S.R. 1964. Hydroelasticity, in *Advances in Hydroscience*, edited by V.T. Chow, Academic Press, New York, Vol. 1, 94-159.

Newman, N.J. 1994. Wave effects on deformable bodies. *Applied Ocean Research*, **16**, 47-59.

Reilly, E.T., Shin, Y.S. and Kotte, E.H. 1988. A prediction of structural load and response of a Swath ship in waves. *Naval Engineers Journal*, 251-264.

Reane, A.J., Temarel, P., Wu, X.J. and Wu, Y. 1991. Hydroelasticity of nonbeamlike structures. *Transactions of the Royal Society of London*, **A334**, 339-355.

Seidl, L.H. 1978. Theoretical Background - Part I: Documentation for Program Package VESDYN. Ocean Engineering Consultants, Inc., Honolulu, Hawaii.

Seidl, L.H., Clifford, W.F. and Cummings, J.P. 1993. Design and operational experience of the SWATH ship Navatek I. *Marine Technology*, **30**(3), 153-171.

Sikora, J. and Dinsenbacher, A. 1990. Swath structure: Navy research and development application. *Marine Technology*, **27**(4).

Wu, Y.S. 1984. Hydroelasticity of Floating Bodies. Ph.D. Dissertation, Brunel University, U.K.

Wu, Y.S., Wang, D., Riggs, H.R. and Ertekin, R.C. 1993. Composite singularity distribution method with application to hydroelasticity. *Marine Structures*, **6**, 143-163.

ACKNOWLEDGMENT

This study was supported in part by the National Defense Center of Excellence for Research in Ocean Sciences (CEROS) under Grant No. MDA972-94-1-0010/CEROS-40464.

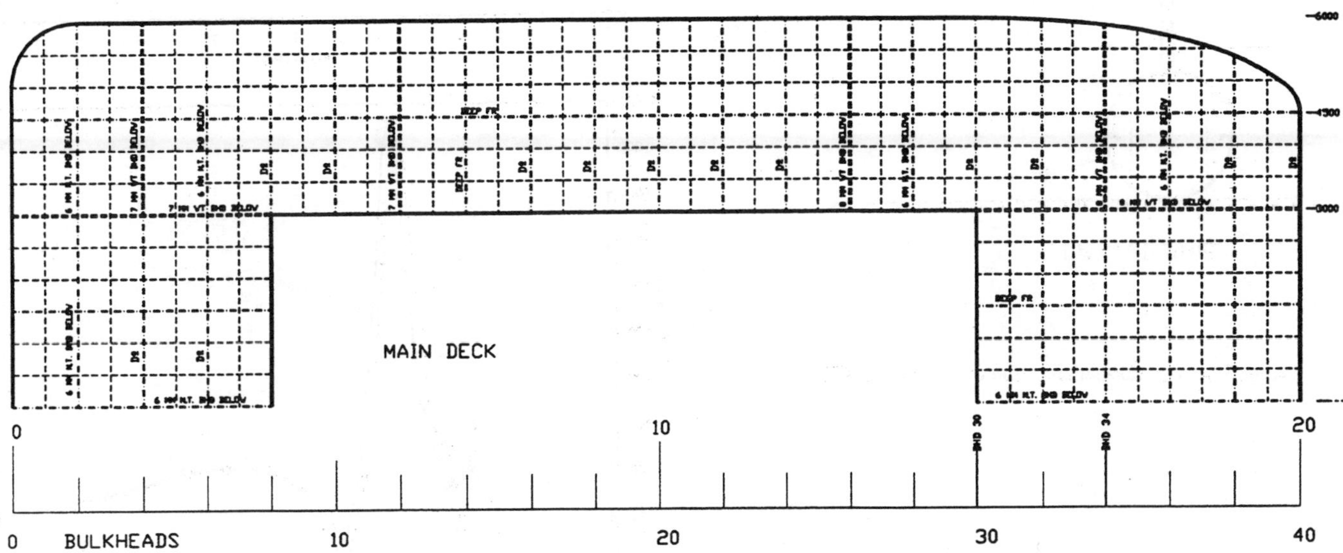

Figure 1 Deck scantlings of the Swath ship S20T10.

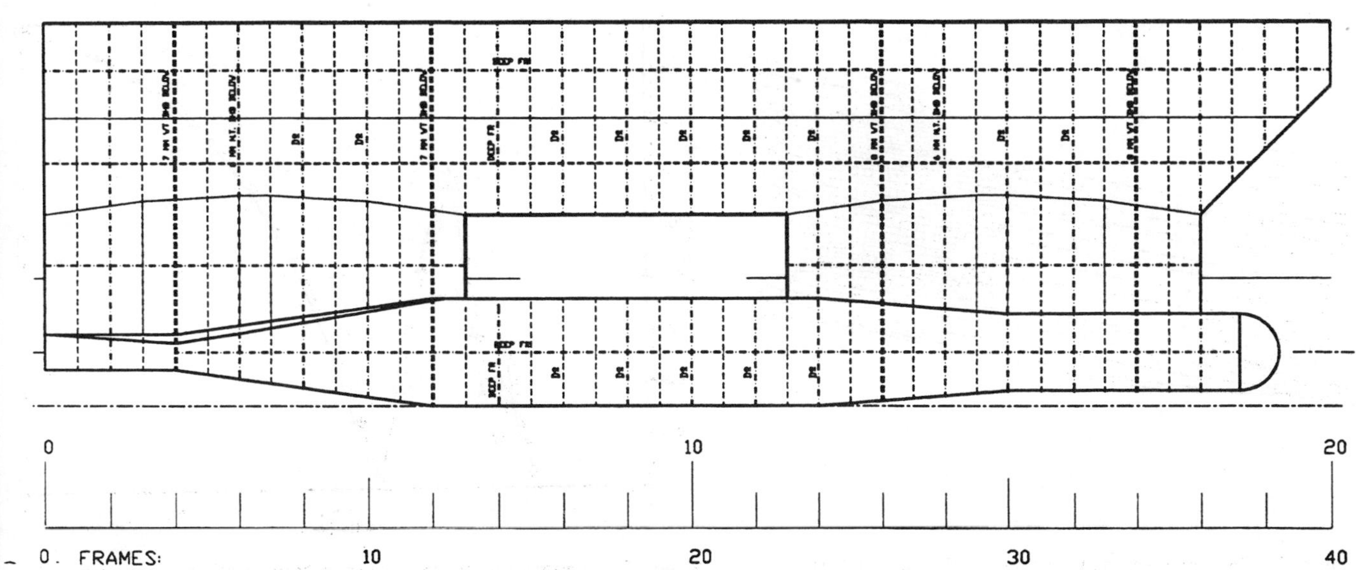

Figure 2 Profile scantlings of the Swath ship S20T10.

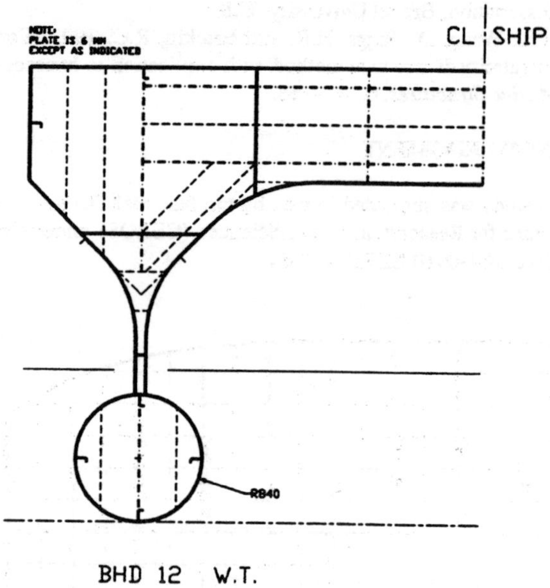

Figure 3 A typical section of the Swath ship S20T10.

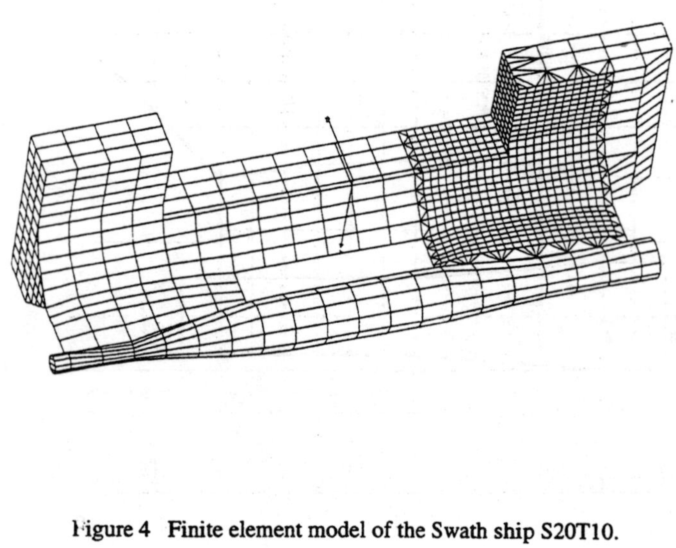

Figure 4 Finite element model of the Swath ship S20T10.

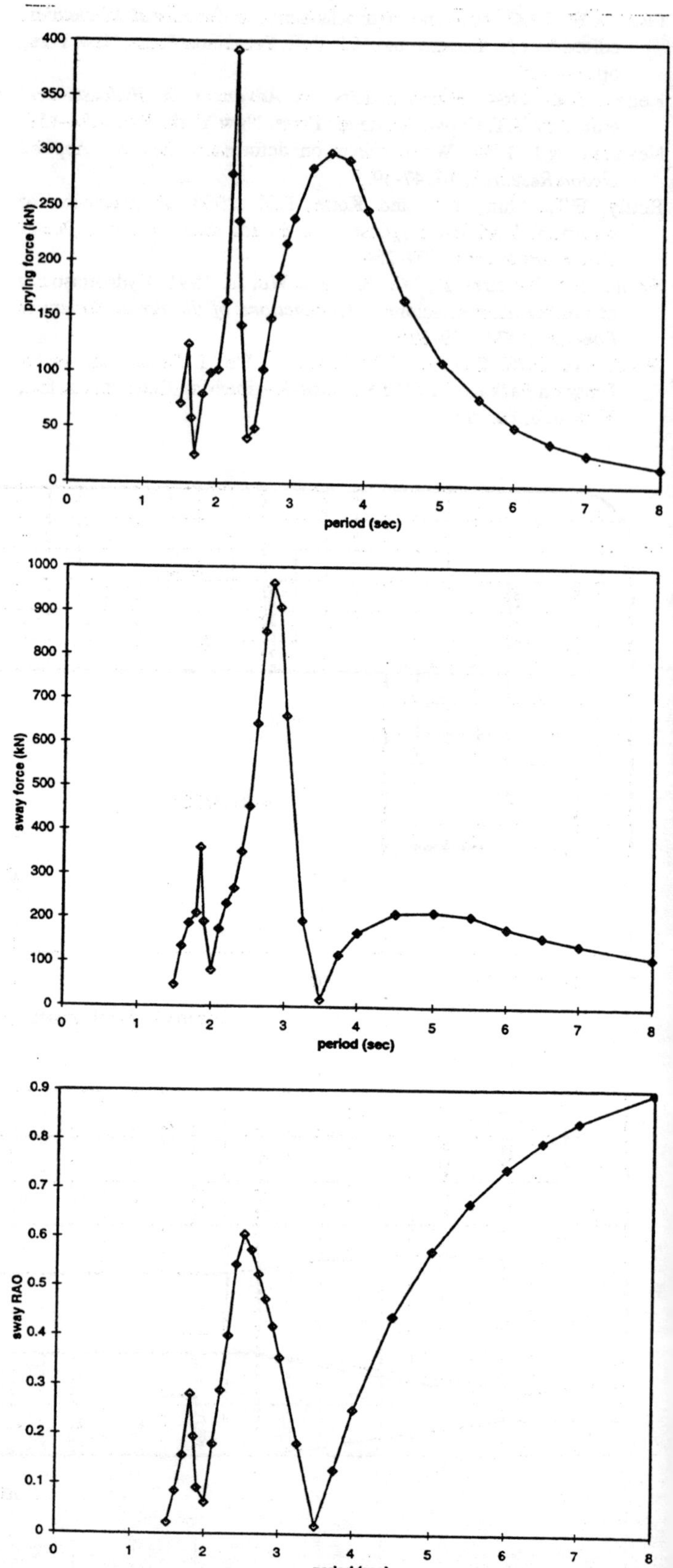

Figure 5 Prying force, wave excitation force and response amplitude operator in sway for regular beam waves of unit amplitude.

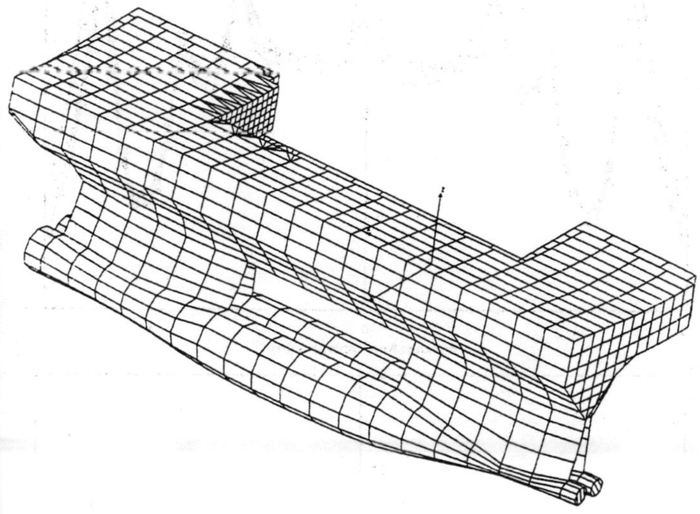

Figure 6 First symmetric mode shape.

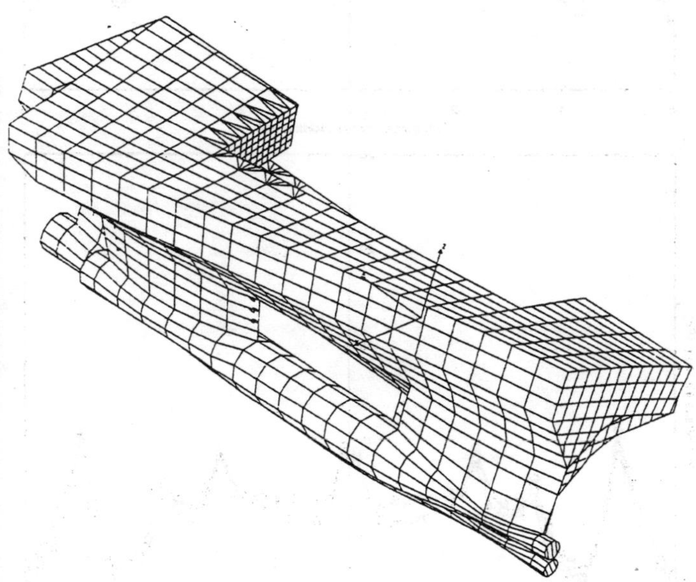

Figure 7 First anti-symmetric model shape.

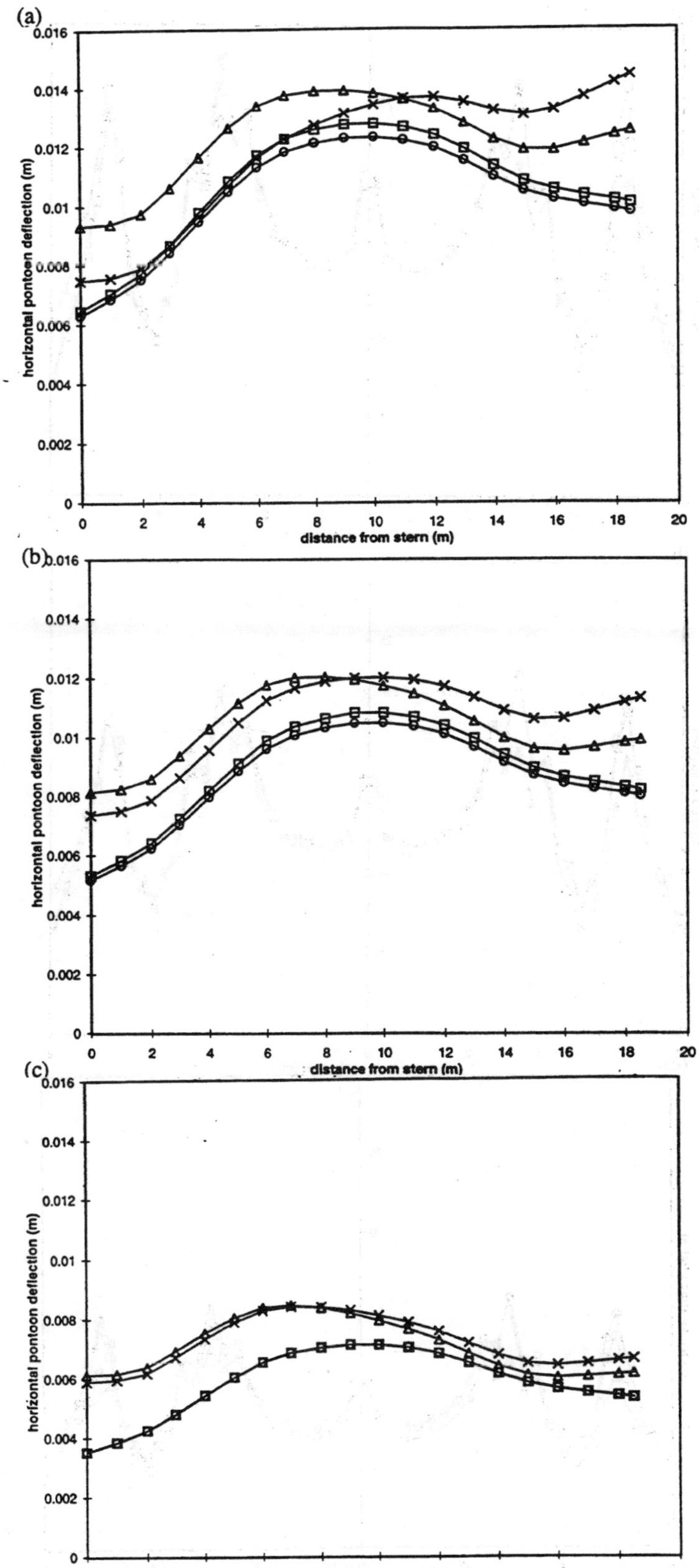

Figure 8 Horizontal pontoon deflections for regular beam waves of unit amplitude. (a) 3.5 s. (b) 4.0 s. (c) 4.5 s. Δ, hydro-elastic (port); ×, hydro-elastic (starboard); □, rigid-dynamic; O, quasi-static.

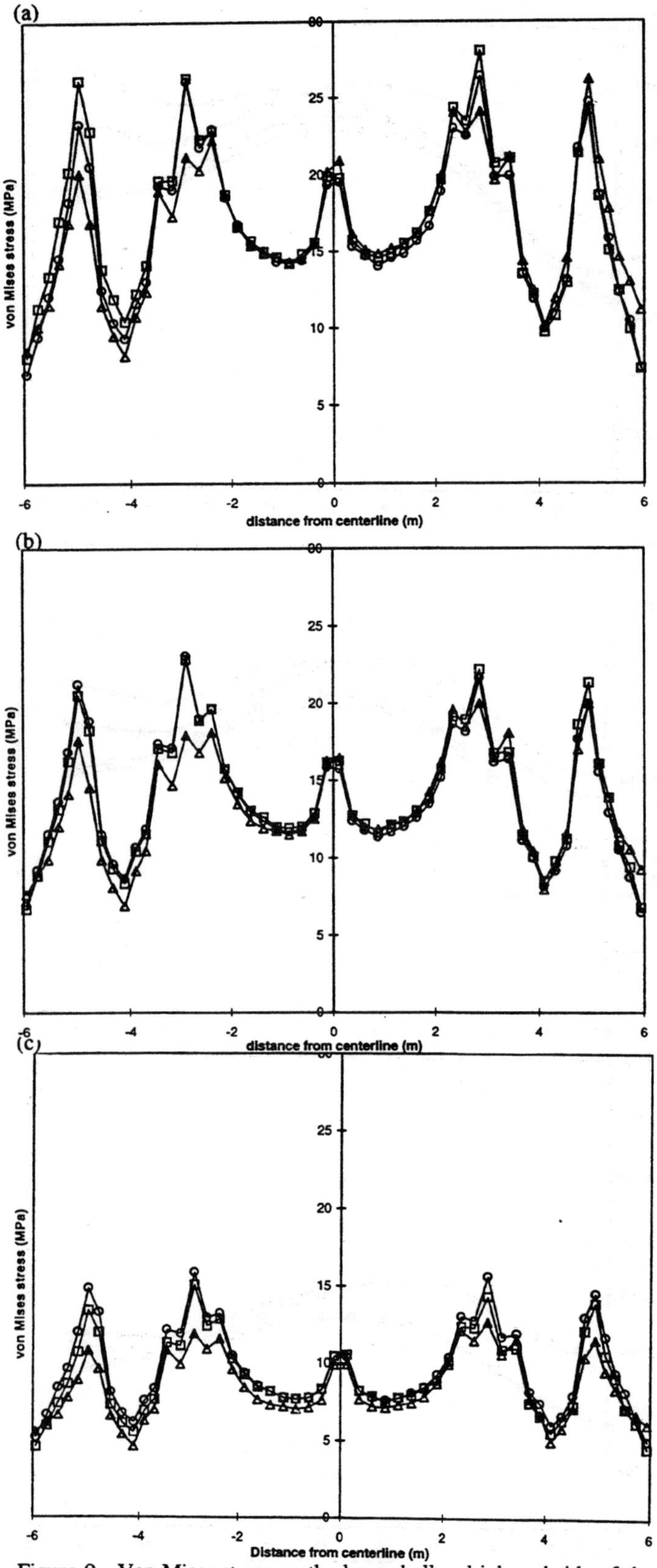

Figure 9 Von Mises stress on the lower hull and inboard side of the strut along bulkhead 30 for regular beam waves of unit amplitude. (a) 3.5 s. (b) 4.0 s. (c) 4.5 s. Δ, hydro-elastic; □, rigid-dynamic; O, quasi-static.

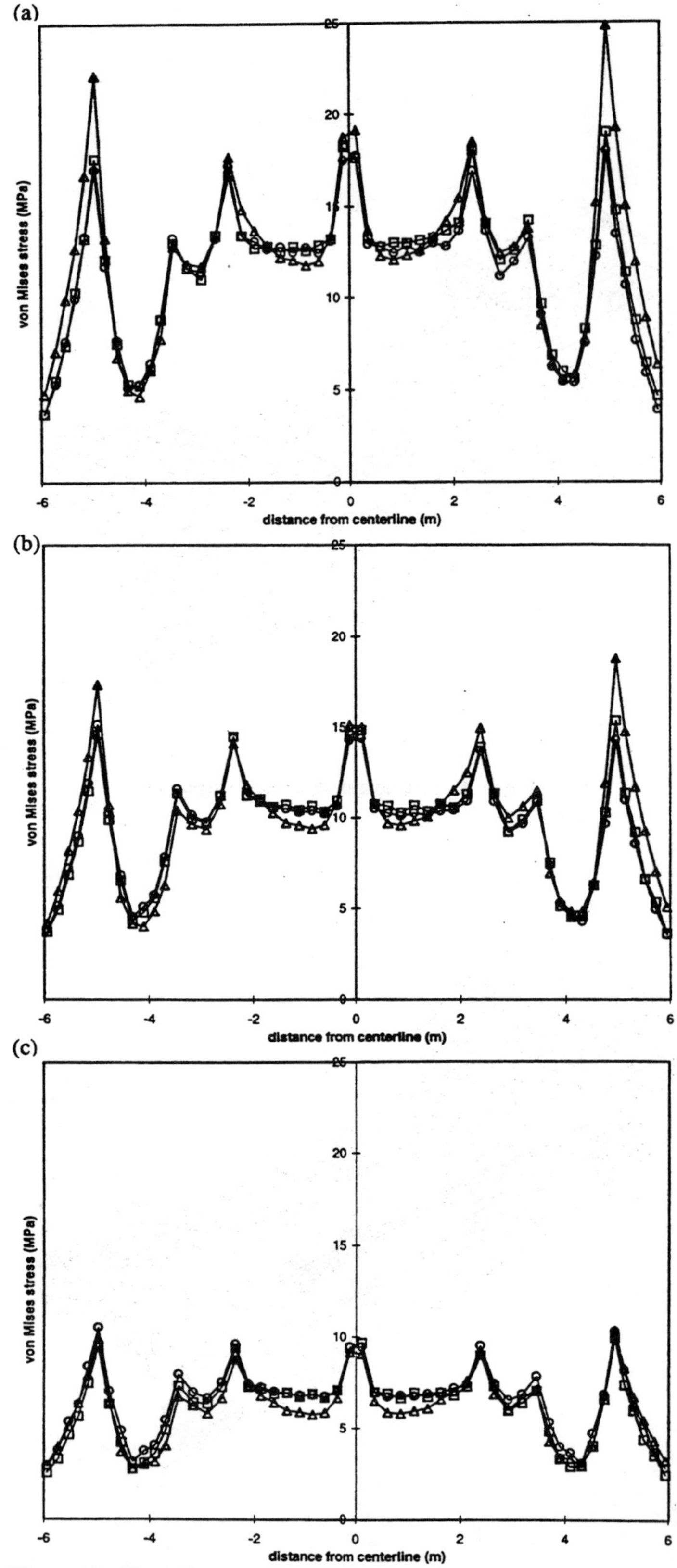

Figure 10 Von Mises stress on the lower hull and inboard side of the strut along bulkhead 34 for regular beam waves of unit amplitude. (a) 3.5 s. (b) 4.0 s. (c) 4.5 s. Δ, hydro-elastic; □, rigid-dynamic; O, quasi-static.

Proceedings of the Seventh (1997) International Offshore and Polar Engineering Conference
Honolulu, USA, May 25-30, 1997

ISBN 1-880653-28-1 (Set); ISBN 1-880653-32-X (Vol. IV)

An Experimental Study on the Structural Behavior and Vibrational Characteristics of Sandwich Plate with Aluminium Honeycomb Core

C.Y. Son
Inha University, Inchon, Korea
I.T. Kim
Hong-Ik University, Chochiwon, Korea
J.S. Paik
Halla Heavy Industries Co., Seoul, Korea

ABSTRACT

Aluminium honeycomb sandwich elements not only have high flexural rigidity and strength per density but also excellence in anti-vibration and anti-noise properties. Their properties are very useful for building airplane and high speed ship which need lighter-weighted and more strengthened elements. In this paper, the Reissner-Mindlin's shear deformation and rotary inertia plate theory was considered the rectangular sandwich plate which expanded from Hamilton's principle. This paper is that results of the equation of motion given by Hamilton's principle compared with that of Rayleigh-Ritz method given by energy relation. The applied sandwich models have isotropic and symmetrical aluminium faces and aluminium honeycomb cores. It is considered of four plates of different thickness faces and different honeycomb cell size cores. They were used seawater survey ship (SWSS) in Japan in 1993[1][10]. The applied boundary conditions are simple support, fixed support and free support.

KEYWORDS

Aluminium Honeycomb Core, Sandwich Plate, Vibrational Characteristics, Cell Size, Hamilton's Principle

1. Introduction

To develop large sized high speed marine craft, We must possess various essential technology including the newest ship hull research, applying light material, constitution of a large output propulsive system and development of effective seakeeping system etc. As ship hull element type of large sized high speed marine craft, when aluminium sandwich structure[2] is compared with a single plate, it obtain lighter weight as four times as than single plate at same stiffness. Because honeycomb has many merit of strong energy absorption,radio frequency shielding, light diffusion etc., if they were connected into strong stiff sandwich structure, large sized high speed marine craft can obtain lighter weight which essential condition of large sized high speed marine craft.

When Aluminium Honeycomb Sandwich Plate (AHSP) that inadequate systematic research about various structural behavior is applied to real structure design, many questions are occur. Therefore, This research is studied usefulness of AHSP Structure about existed experimental formulas and analytic formulas which based on elasticity as experimental research. To research linear elastic behavior and directional characteristics of cell about AHSP, static bending test is carried by MIL-STD 401B and To analyze excellent crushing and energy absorption of honeycomb core, compression test is carried. Especially, structural assessment is carried with virtual single element using equivalent stiffness theory.

Also, This paper is made flat plate which based on dimension of aluminium sandwich element applying for 13m Sea Water Survey Ship (SWSS) in Japan in 1993 and carried vibrational experimental research.

In this paper, the Reissner-Mindlin's shear deformation and rotary inertia plate theory was considered the rectangular sandwich plate which expandedfrom Hamilton's principle. This paper is that results of the equation of motion given by Hamilton's principle compared with that of Rayleigh-Ritz method given by energy relation. The applied sandwich models have isotropic and symmetrical aluminium faces and aluminium honeycomb cores. Natural frequency of experimental analysis using modal analysis package program, CADA-PC is compared with natural frequency using Rayleigh-Ritz method. And this research is compared with analytic results and finite element program, NISA.[5]

2. Vibration Analysis Theory of Sandwich Plate

As sandwich model at plate theory of Mindlin, bending strain energy of upper and lower face, shear strain energy of core and kinetic energy of total system are applied. Because core which feeble density and strength has characteristics which not resist bending of strong face, shear coefficient of sandwich plate is approximate 1. Fig. 2.1 show Analytic model of Aluminium Honeycomb Sandwich Plate.

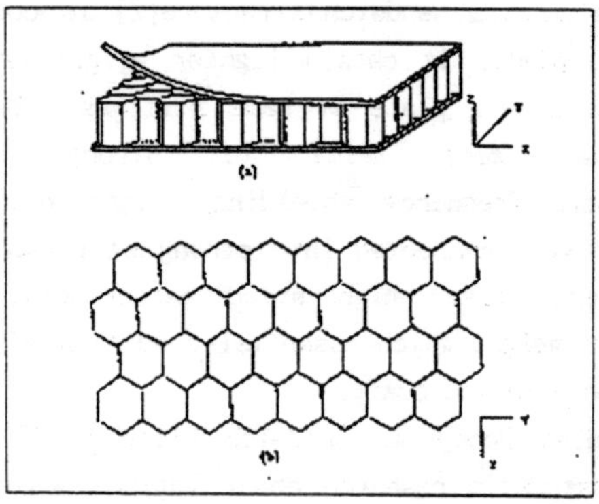

Fig. 2.1 Dimensions of Aluminium Sandwich Plate with Aluminium Honeycomb Core

2.1 Assumption[6]

(i) Bonding of face and core is perfect.
(ii) Face and core are isotropic, especially core is antiplane core which has constant shear distribution regardless distance of neutral axis.
(iii) Because stiffness of xy plane of core is smaller than face stiffness, It is ignored.
(iv) z-directional stress of face, core is ignored.
(v) Shear Deformation of Face is ignored.

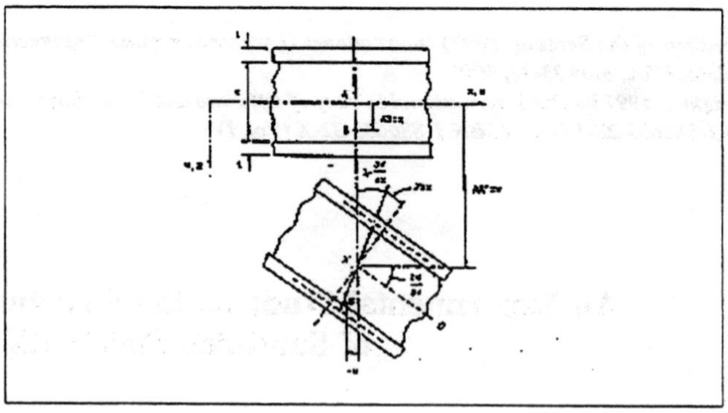

Fig. 2.2 Section through Deflected Sandwich Plate in zx-Plane

2.2 Equation of Motion[11][4]

2.2.1 Equation of Motion of Face and Core When it is considered vertical inertia and angle inertia, kinetic energy of isotropic, homogeneous plate

$$T = \frac{1}{2}\int\int_R \int_{-t_{uf}-t_{uc}}^{t_{lf}+t_{lc}} \rho \dot{u}_i \dot{u}_i \, dz \, dA$$
$$(i = x, y, z)$$
$$= \frac{1}{2}\int\int_R \int_{-t_{uf}-t_{uc}}^{t_{lf}+t_{lc}} \rho \, [z^2(\frac{\partial \psi_x}{\partial t})^2 + z^2(\frac{\partial \psi_y}{\partial t})^2 + (\frac{\partial w}{\partial t})^2] dz \, dA$$

Because sandwich plate is different density of upper and lower face, core, according thickness, Above equation about sandwich plate is as followed :

$$T = \frac{1}{2}\,[\rho_{uf} I_{uf} + \rho_c I_c + \rho_{lf} I_{lf}\,] \int\int_R \{(\frac{\partial \psi_x}{\partial t})^2 + (\frac{\partial \psi_y}{\partial t})^2\}\, dA + \frac{1}{2}\,[\rho_{uf} t_{uf} + \rho_c t_c + \rho_{lf} t_{lf}\,] \int\int_R (\frac{\partial w}{\partial t})^2 \, dA \quad Eq.(2.1)$$

Then, I_{uf}, I_c, I_{lf} is Second Inertia Moment of Upper Face, Core, Lower Face.

2.2.2 Shear Strain Energy of Cores

From above, By assumption (iii), Because deformation by plane stress are $\varepsilon_{xy}=0$, $\varepsilon_{xx}=\varepsilon_{yy}=0$, only is considered energy by shear deformation.

$$U_c = \frac{1}{2}\int\int_R\int_{-t_c}^{t_c}(2\sigma_{yz}\varepsilon_{yz} + 2\sigma_{xz}\varepsilon_{xz})\,dz\,dA$$
$$= \frac{1}{2}\,k\,t_c\,G_c\int\int_R\{(\frac{\partial w}{\partial y} - \psi_y)^2$$
$$+ (\frac{\partial w}{\partial x} - \psi_x)^2\}\,dA \quad Eq.(2.2)$$

2.2.3 Bending Strain Energy of Face

Bending strain energy of Face is

$$U_f = U_{lf} + U_{uf}$$

At equation of bending strain energy U, Face is considered only plane stress and $\sigma_z = 0$, $\gamma_{xz} = \gamma_{yz} = 0$ is, Bending strain energy of upper and lower face is as followed Bending Strain Energy of Face is

$$U_f = \frac{1}{2}\int\int_R\{\frac{E}{1-\nu^2}(I_{uf}+I_{lf})[(\frac{\partial\psi_x}{\partial x})^2+(\frac{\partial\psi_y}{\partial y})^2$$
$$+2\nu(\frac{\partial\psi_x}{\partial x}\frac{\partial\psi_y}{\partial y})]+\frac{E}{2(1+\nu)}(I_{uf}+I_{lf})(\frac{\partial\psi_x}{\partial y}$$
$$+\frac{\partial\psi_y}{\partial x})^2\}dzdA \quad Eq(2.3)$$

2.2.4 Substitution of Equivalence Relationship

If Lagrangian which calculated above energy relationship is compared with Lagrangian of indlin theory,

① $\rho_{uf}I_{uf} + \rho_c I_c + \rho_{lf}I_{lf} = (\rho I)^*$

② $\rho_{uf}t_{uf} + \rho_c t_c + \rho_{lf}t_{lf} = (\rho h)^*$

③ $k\,t_c\,G_c = (k\,h\,G)^*$

④ $D_{uf} + D_{lf} = D^*$

⑤ $G(I_{uf}+I_{lf}) = \frac{E}{2(1+\nu)}(I_{uf}+I_{lf}) = T^*$

Therefore, If equivalence relationship of ①~⑤ is used, Equation of motion of sandwich plate model can obtain.

$$\{\nabla^2 - \frac{(\rho h)^*}{(khG)^*}\frac{\partial^2}{\partial t^2}\}\{(D)^*\nabla^2 - (\rho I)^*\frac{\partial^2}{\partial t^2}\}\,w$$
$$+ (\rho h)^*\frac{\partial^2 w}{\partial t^2} = 0 \quad Eq(2.4)$$

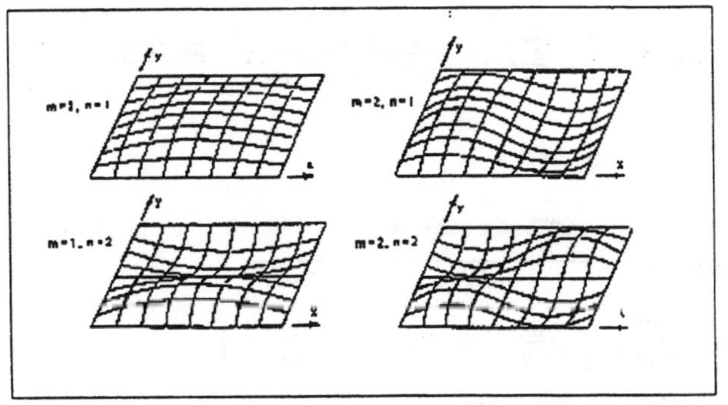

Fig. 2.3 Examples of Mode Shapes

2.4 Applying for Rayleigh-ritz Method of Sandwich Plate

To apply for Rayleigh-Ritz method, Shape functions of w, ψ_x, ψ_y are as followed.

$$w(x,y,t) = \sum_{i=1}^{\infty}\sum_{j=1}^{\infty}a_{mn}\sin\frac{m\pi x}{a}\sin\frac{n\pi y}{b}\sin\omega t \quad Eq.(2.5)$$

$$\psi_x(x,y,t) = \sum_{i=1}^{\infty}\sum_{j=1}^{\infty}b_{mn}\cos\frac{m\pi x}{a}\sin\frac{n\pi y}{b}\sin\omega t \quad Eq.(2.6)$$

$$\psi_y(x,y,t) = \sum_{i=1}^{\infty}\sum_{j=1}^{\infty}c_{mn}\sin\frac{m\pi x}{a}\cos\frac{n\pi y}{b}\sin\omega t \quad Eq(2.7)$$

where, ψ_x, ψ_y are shape functions which are described deflected angle. In case of isotropic materials, b_{mn}, c_{mn} of coefficients of shape function are same. Therefore, Kinetic energy, strain energy of face and core are as followed as above shape functions :

$$T_{max} = \frac{1}{2}(\rho I)^*\omega^2\frac{ab}{4}\sum_{m=1}^{\infty}\sum_{n=1}^{\infty}2b_{mn}^2 + \frac{1}{2}(\rho h)^*\omega^2\frac{ab}{4}\sum_{m=1}^{\infty}\sum_{n=1}^{\infty}a_{mn}^2$$
$$Eq.(2.8)$$

$$U_{c,max} = \frac{1}{2}(khG)^*\frac{ab}{4}\sum_{m=1}^{\infty}\sum_{n=1}^{\infty}\{(\frac{m\pi}{a}a_{mn} - b_{mn})^2$$
$$+(\frac{n\pi}{b}a_{mn} - b_{mn})^2\} \quad Eq.(2.9)$$

$$U_{f,max} = \frac{1}{2}\{D^*\sum_{m=1}^{\infty}\sum_{n=1}^{\infty}b_{mn}^2\{(\frac{m\pi}{a})^2+(\frac{n\pi}{b})^2$$
$$+2\nu\frac{mn\pi^2}{ab}\} + T^*\sum_{m=1}^{\infty}\sum_{n=1}^{\infty}b_{mn}^2(\frac{n\pi}{b}+\frac{m\pi}{a})^2\}\frac{ab}{4}$$
$$Eq.(2.10)$$

If Lagrangian which calculated from Eq. (2.8~10) is differential by coefficient of shape function, a_{mn}, b_{mn} to obtain minimum value,

$$\frac{\partial(T-U_c-U_f)}{\partial a_{mn}} = \frac{\partial(T-U_c-U_f)}{\partial b_{mn}} = 0$$

From this relationship, natural frequency are obtained at (m,n) mode, that is

$$\omega_{mn} = \frac{1}{2}\sqrt{\sum_{i=1}^{3} f_i + \frac{\sqrt{\sum_{i=4}^{16} f_i}}{a^2b^2\,(\rho h)^*\,(\rho I)^*}} \quad Eq.(2.11)$$

where,

$$f_1 = 2(khG)^* \left\{ \frac{(\frac{m\pi}{a})^2+(\frac{n\pi}{b})^2}{(\rho h)^*} + \frac{1}{(\rho I)^*} \right\}$$

$$f_2 = \frac{D^*}{(\rho I)^*} \left\{ (\frac{m\pi}{a})^2+(\frac{n\pi}{b})^2+\frac{2mn\nu\pi^2}{ab} \right\}$$

$$f_3 = \frac{T^*}{(\rho I)^*} (\frac{m\pi}{a}+\frac{n\pi}{b})^2$$

$$f_4 = 4\ a^4b^4\,((khG)^*)^2\,((\rho h)^*)^2$$

$$f_5 = 4\ a^2\ b^2\ D^*\ (khG)^*\pi^2\ ((\rho h)^*)^2\ (\ b^2m^2+a^2n^2 + 2abmn\nu\ )$$

$$f_6 = (D^*)^2\ \pi^4\ ((\rho h)^*)^2\ (\ b^2m^2+a^2n^2\ )^2$$

$$f_7 = 4\ (D^*)^2\ \nu\ \pi^4\ ((\rho h)^*)^2\ \{\ a^2n^2+b^2m^{2+}abmn\ \nu\ \}\ abmn$$

$$f_8 = 16\ a^3\ b^3\ m\ n\ ((khG)^*)^2\ \pi^2\ (\rho h)^*\ (\rho I)^*$$

$$f_9 = -4\ D^*\ (khG)^*\ \pi^4\ (\rho h)^*\ (\rho I)^*\ (\ b^2m^2+a^2n^2\ )^2$$

$$f_{10} = -8\ D^*\ (khG)^*\ \nu\ \pi^4\ (\rho h)^*\ (\rho I)^*\ (\ b^2m^2+a^2n^2\ )\ abmn$$

$$f_{11} = 4\ ((khG)^*)^2\ \pi^4\ ((\rho I)^*)^2\ (\ b^2m^2+a^2n^2\ )^2$$

$$f_{12} = 4\ a^2b^2(khG)^*\ \pi^2\ ((\rho h)^*)^2\ T^*\ (bm+an)^2$$

$$f_{13} = 2\ D^*\pi^4((\rho h)^*)^2\ T^*\ \{\ (b^2m^2+a^2n^2)^2+2(b^2m^2 + a^2n^2)\ abmn\ \}$$

$$f_{14} = 4\ D^*\ \nu\ \pi^4\ ((\rho h)^*)^2\ T^*\ (bm+an)^2\ abmn$$

$$f_{15} = -4(khG)^*\ \pi^4\ (\rho h)^*\ (\rho I)^*\ T^*\ (b^2m^{2+}a^2n^2)^3$$

$$f_{16} = ((\rho h)^*)^2\ (T^*)^2\ \pi^4\ \{\ (b^2m^2+a^2n^2)\ +\ 2\ abmn\ \}^2$$

3. Sandwich Specimen test Methods[8]

3.1 The Properties and Dimension of Test Specimen

Aluminium Honeycomb Sandwich Specimens of this paper are consist of combination of face and core which show next tables.

TABLE 1. Mechanical Properties of faces

Material	Young Mod. E (kg/mm²)	Ult. Tens. Strength σ_U (kg/mm²)	Yield Strength σ_y (kg/mm²)	Elongation (%)
A5083H321	7,245	37.4	27.3	13
A6061T651	7,031	29.5	24.6	12

TABLE 2. Mechanical Properties of A3003H19 Honeycomb Core (mm)

Items	Cell Size	Density	Cell Size	Density
	12.7	41	19.2	28
Foil Thickness	0.0635		0.0635	
0.2% Yield Stress	19.4		19.4	
Ult. Tens. Streng.	20.9		20.9	
Elongation, %	4		4	
Comp. Streng.	0.11		0.066	
Shear Streng., L	0.109		0.064	
Shear Streng., W	0.071		0.027	
Shear Mod., L	20.999		18.247	
Shear Mod., W	13. 456		4.077	
Young Mod.	7,040		7,040	

* Unit of Density : kg/m³

Unit of Stress, Strength, Modulus : kg/mm²

TABLE 3. Dimensions of Vibration Test Specimens Edhesive (Epoxy)

Faces	Core
A5083H321(6t)	A3003H19 (20t), Cell Size 12.7 mm
A6061T651(4t)	A3003H19 (20t), Cell Size 19.2 mm

3.2 Sandwich specimen Flexural Test

(80 * 600), Span length (500)(mm)

3.2.1 Significance and Use

Flexural tests on flat sandwich construction may be conducted to determine flexural and shear stiffness of the construction, shear modulus and shear strength of the core, or compressive or tensile strength of the facings.

3.2.2 Procedure

This experiment was performed by ASTM.

Fig. 3.1 Experimental Example of 3 point bending

3.2.3 Test Result

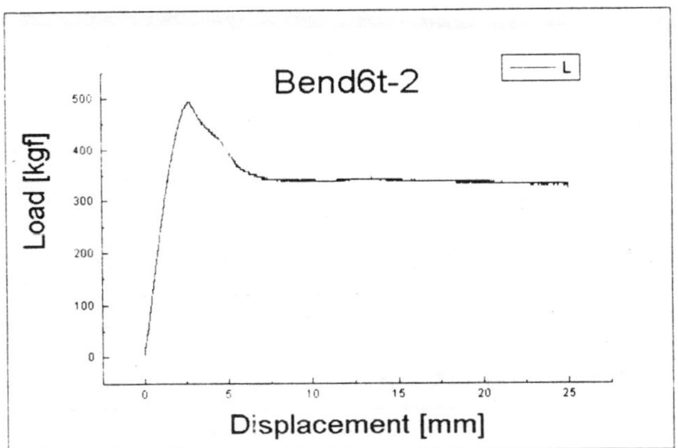

Fig 3.2 Load deflection curve for 3-point bending test

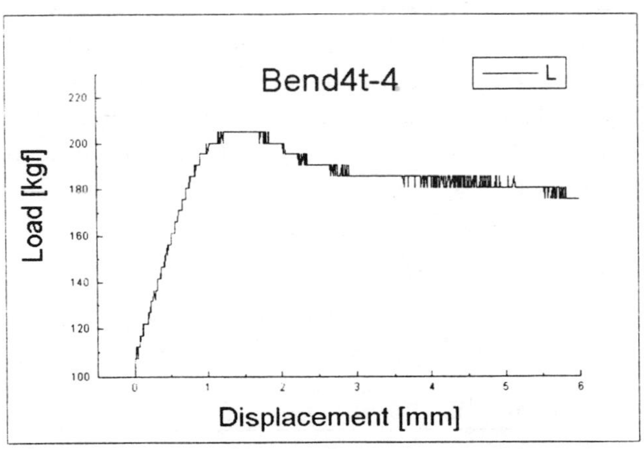

Fig 3.3 Load-deflection curve for 3-point bending test

Table. 4 Test result of 3 point bending(4t)

A6061T651 face(4t), A3003H19 (20t) Cell size 19.2	
Load (kgf)	Displacement (mm)
214.84375	0.885714285
209.96094	0.828571429
214.84375	1.028571429
209.96094	1.428571429
229.42919	0.314285714

Table. 5 Test result of 3 point bending(6t)

A5083H321 face(6t), A3003H19 (20t) Cell size 12.7	
Load (kgf)	Displacement (mm)
332.03125	2.971428571
493.16406	2.585714288
473.63281	2.085714288
419.92188	2.157142857
371.09375	2.142857143

3.3 Edgewise Compression Test (80 * 300), (mm)

3.3.1 Procedure

This experiment was performed by ASTM.

Fig. 3.4 Experimental example of edgewise strength test of length direction[7]

3.3.2 Test Result

Table. 6 Test result of edge wise compression

Specimen	Max load (kgf)	Specimen	Max load (kgf)
4t-1, 19.2	12480.46875	6t-1, 12.7	14721.679688
4t-2, 19.2	11816.40625	6t-2, 12.7	14672.851563
4t-3, 19.2	12294.92188	6t-3, 12.7	14414.062500
4t-4, 19.2	10029.29688	6t-4, 12.7	14687.500000
4t-5, 19.2	11943.35938	6t-5, 12.7	15185.546875

3.4 Sandwich Flatwise Crushing Test[9] (100 * 100), (mm)

3.4.1 Procedure

This experiment was performed by ASTM.

Fig. 3.5 Experimental example of flat wise compression test

3.4.2 Test Result

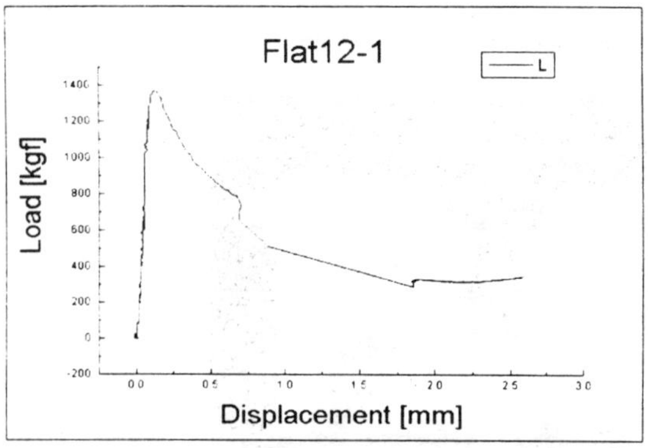

Fig 3.6 Load-deflection curve flat with test

Table 7. Crushing strength by dividing the max load by area of the ram

A6061T651 face(4t), A3003H19 (20t) Cell size 19.2	
Max load (kgf)	Crushing strength (kgf/mm^2)
634.2968	0.0885
620.1171	0.0839
668.9453	0.0905
659.1796	0.0890
507.8125	0.0687

Table 8. Crushing strength by dividing the max load by area of the ram

A5083H321 face(6t), A3003H19 (20t) Cell size 12.7	
Max load (kgf)	Crushing strength (kgf/mm^2)
1215.8203	0.1645
1186.5234	0.1606
1181.6406	0.1599
1337.8906	0.1539
1240.2343	0.1678

4. Result and Considerations

4.1 The Properties and Dimension of Test Specimen

Aluminium Honeycomb Sandwich Specimens of this paper are consist of combination of face and core which show table 1,2

TABLE 9. Dimensions of Vibration Test Specimens (1200*1000), (mm) Edhesive (Epoxy)

Specimens	Faces	Core
ALAV	A5083H321(6t)	A3003H19 (20t), Cell Size 12.7 mm
ALBV	A6061T651(8t)	A3003H19 (20t), Cell Size 19.2 mm
ALCV	A6061T651(4t)	A3003H19 (20t), Cell Size 19.2 mm
ALDV	A6061T651(8t)	A3003H19 (20t), Cell Size 12.7 mm

(Example) ALAV : Aluminium A Vibration Plate

4.2 Comparison of Test Results, Analytic calculation and NISA Results

Above proposed 1200*1000(mm) vibrational specimens of four which combinate face and core are divided into elements of 6*7 (size 20*20 cm) respectively as show Fig. 5.1. Frequency analysis data are imported at 2 Channel FFT 3550. And it is carried on mode analysis with Modal Analysis Package Program, CADA-PC.

1	7	13	19	25	31	37
2	8	14	20	26	32	38
3	9	15	21	27	33	39
4	10	16	22	28	34	40
5	11	17	23	29	35	41
6	12	18	24	30	36	42

Fig. 4.1 FEM Vibrational Analysis Modal Elements of 1200*1000 (mm^2) Specimens.

From Fig. 4.2 to Fig4.3 respective graphs are compared with natural frequencys of test results, analytic calculation and NISA results.

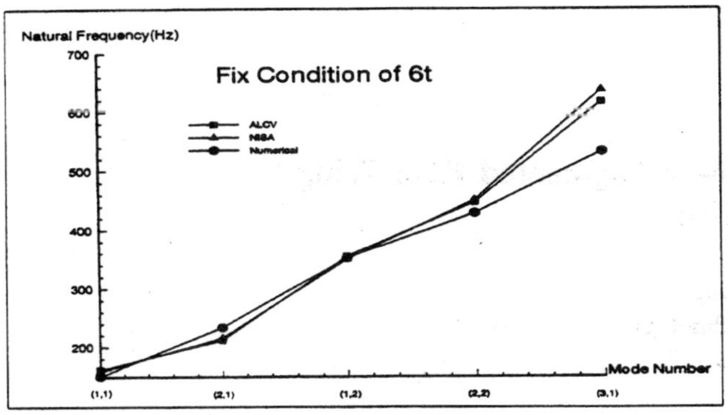

Fig. 4.2 Comparison of Natural Frequency of NISA and Numerical Solution

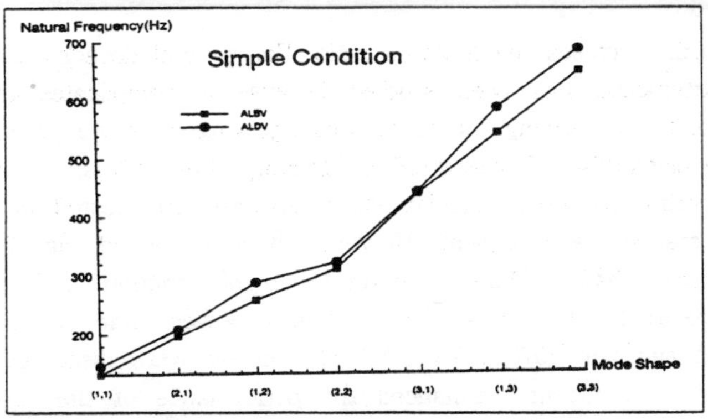

Fig. 4.3 Comparison of Natural Frequency of Different cell size

5. Conclusion

1. Less decreasing cell size, that is less increasing density, more increasing natural frequency.
2. More increasing face thickness, more increasing natural frequency.
3. When they are compared with natural frequencies of test results, analytic calculation and NISA results, Natural frequency is the largest than any other and following is natural frequency ofISA results.

The theoretical analysis is considered rotary inertia and shear Deformation, But NISA is not.

The reason of increasing test results is thought of imperfection of boundary conditions.

In my opinion, it must consider shear deformation about face axis in case of increasing face thickness

REFERENCE

[1]Paik, J.K., and Lee, Y. W., "On Materials and Structures of Large High Speed Ship Hulls",Proc.of SNAK, November 1995, pp. 513-518

[2]Corden, J., "Honeycomb Structure",Engineered Materials Handbook, Vol. 1,'Composites',ASM International, Ohio,1987, pp. 721-728

[4]S.M. Jun and C.S. Hong, "Finite Element Stress Analysis of Clamped Honeycomb Sandwich Plate under Uniformly Distributed Load", Journal of The Aerospace Sciences,Vol. 11, No. 1, pp.45-54, June, 1983

[5]Hu-Nan Chu, "Influence of Transverse Shear on Nonlinear Vibrations of Sandwich Beams with Honeycomb Cores", Journal of The Aerospace Sciences, pp. 405-410, May,1961

[6]H.G. Allen, "Analysis of Structural Sandwich Panels", Chapter 1, Chapter 12

[7]W.W. Troxell and H.C. Engel,"Column Characteristics of Sandwich Panels Having Honeycomb Cores", Journal of The Aerospace Science, pp. 413-421, 1947

[8] MIL-STD-401B "Sandwich Constructions and Core Materials : General Test Methods", 26, Sept, 1967

[9]H. Smallen and W.F. Roberts, "Properties of Stainless Steel Sandwich Using Low-Density Honeycomb Cores", Welding Research Supplement,pp.90s-96s, Feb,1961

[10]Y. Kaneko, K. Takeuchi and K. Aokage, "Design and Construction of a Seawater Survey Ship Built Using Aluminium Honeycomb Panels", FAST'93,pp.449-4606

[11]M.P.Bieniek and A.M. Freudenthal, Frequency-Response Functions of Orthotropic Sandwich Plates", Journal of The Aerospace Science, pp. 732-736, Sept, 1961

Proceedings of the Seventh (1997) International Offshore and Polar Engineering Conference
Honolulu, USA, May 25-30, 1997

ISBN 1-880653-28-1 (Set); ISBN 1-880653-32-X (Vol. IV)

Numerical Study on 3-Dimensional Power-Augmented Ram Wing in Ground Effect

Seung-Hyun Kwag
Halla Institute of Technology
Wonju, Korea

ABSTRACT

A Power-Augmented Ram Wing in Ground (PAR-WIG) is a wing in ground effect, whose lift is enhanced by the airflow of a thruster placed in the upstream of the wing, to make a high ram pressure region under the wing. In the present paper, 3-dimensional PAR effects were numerically studied by solving the Navier-Stokes equations. Pressure distributions and velocity vectors were simulated around the pressure and suction sides of the wing surface. Through the numerical simulation, Cp values and lift/drag ratio are carefully reviewed by changing the height/chord; 0.05, 0.1, 0.3 and 0.8, and the angle of attack; 4,6,10 degrees. The shape of model is NACA 0012 and span/chord is 3.0. According to the numerical results, the relationship between lift/drag and height/ chord is fairly reasonable.

KEY WORDS: PAR(power-augmented ram), ground effect, wing, Navier-Stokes, NACA 0012, lift, drag, Cp values.

INTRODUCTION

The WIG (Wing In Ground effect) ship combines the features of both the airplane and ship, attracting the attention as the next generation of high speed craft, because it gives us the newest type of highly efficient vessel operating over 100 knots. A wing operating in close proximity to the ground exhibits a reduction in induced drag, which increases the lift/drag ratio. For several decades, this phenomenon has been studied because it complicates the takeoff and landing for the advancing wing near the ground (Lippisch,1964, Lockeed,1962). During the 1970s, small experimental WIG vehicles were designed and tested in a coastal sea environment. However, it was noticed that the vehicles had a high structural weight fraction and an undesirably high takeoff-power/cruise-power ratio. In the mid and late 1970s, the PAR phenomena were discovered, which significantly enhanced the performance of the WIG concept(Smithey,1977). At low speeds, a static pressure rise occurs under the wings, which lifts the vehicle out of the water. The development of WIG/PARWIG crafts had been made mainly for the military purpose. However, after the collapse of the Russia communism, it is known to western countries that the Russian technology in these crafts is the most advanced one. Recently, an interest in the development of super-high speed ships is rapidly growing worldwide. Among several types of crafts, it is expected that a PARWIG craft is a potential candidate among the fast sea transport means for the next generation; military naval ship, passenger ship, cargo/container vessel and marine leisure craft, etc. In this respect, the development of the commercial WIG has been undergoing in several countries such as Japan, Germany, Australia, Korea. Recently an international workshop on WIG was held in Sidney in 1995.

In the present study, 3-dimensional PAR was numerically reviewed by solving the Navier-Stokes equations (Masuda, 1991, Panel method). Pressure distributions and velocity vectors are simulated around the pressure and suction sides of the wing. Through the numerical simulation, Cp values and lift/drag ratio are carefully reviewed by changing the height/chord and the angle of attack. The shape of model is NACA 0012 and span/chord is 3.0. With numerical results, the relationship between lift/drag and height/chord is closely discussed.

NUMERICAL SCHEME

Governing Equation

The Navier-Stokes and continuity equations are employed in the numerical simulation. They can be written in the nondimensionalized form as follows :

$$u_t+uu_x+vu_y+wu_z = -p_x+\frac{1}{R_n}\nabla^2 u$$
$$v_t+uv_x+vv_y+wv_z = -p_y+\frac{1}{R_n}\nabla^2 v$$
$$w_t+uw_x+vw_y+ww_z = -p_z+\frac{1}{R_n}\nabla^2 w \tag{1}$$
$$u_x + v_y + w_z = 0$$

where u, v and w are the velocity components in the uniform flow, lateral and vertical direction, respectively ; p is the pressure ; Rn is the Reynold number. Here, x, y and z represent coordinates in a Cartesian system ; x in the uniform flow direction, y lateral direction and z normal to x and y. All the variables are nondimensionalized by the chord length L and uniform velocity Uo. Subscripts denote partial derivatives with respect to the referred variables. The transformation links the partial derivatives expressed on the physical and computational domains through the following formula :

$$\xi = \xi(x,y,z),\ \eta = \eta(x,y,z),\ \zeta = \zeta(x,y,z) \tag{2}$$

Substituting (2) into (1), the followings can be obtained.

$$u_t+Uu_\xi+Vu_\eta+Wu_\zeta = -(\xi_x P_\xi+\eta_x P_\eta+\zeta_x P_\zeta)+\frac{1}{R_n}\nabla^2 u$$
$$v_t+Uv_\xi+Vv_\eta+Wv_\zeta = -(\xi_y P_\xi+\eta_y P_\eta+\zeta_y P_\zeta)+\frac{1}{R_n}\nabla^2 v \tag{3}$$
$$w_t+Uw_\xi+Vw_\eta+Ww_\zeta = -(\xi_z P_\xi+\eta_z P_\eta+\zeta_z P_\zeta)+\frac{1}{R_n}\nabla^2 w$$

$$\xi_x u_\xi+\eta_x u_\eta+\zeta_x u_\zeta+\xi_y v_\xi+\eta_y v_\eta+\zeta_y v_\zeta+\xi_z w_\xi+\eta_z w_\eta+\zeta_z w_\zeta = 0 \tag{4}$$

where U, V and W are the unscaled contravariant components of velocity defined as follows :

$$U = \xi_x u + \xi_y v + \xi_z w$$
$$V = \eta_x u + \eta_y v + \eta_z w \tag{5}$$
$$W = \zeta_x u + \zeta_y v + \zeta_z w$$

The Laplacian equation can be expressed as,

$$\nabla^2 q=(\xi_x^2+\xi_y^2+\xi_z^2)q_{\xi\xi}+(\eta_x^2+\eta_y^2+\eta_z^2)q_{\eta\eta}+(\zeta_x^2+\zeta_y^2+\zeta_z^2)q_{\zeta\zeta}+2(\xi_x\eta_x+\xi_y\eta_y+\xi_z\eta_z)q_{\xi\eta}$$
$$+2(\eta_x\zeta_x+\eta_y\zeta_y+\eta_z\zeta_z)q_{\eta\zeta} \tag{6}$$
$$+2(\zeta_x\xi_x+\zeta_y\xi_y+\zeta_z\xi_z)q_{\zeta\xi}+(\xi_{xx}+\xi_{yy}+\xi_{zz})q_\xi$$
$$+(\eta_{xx}+\eta_{yy}+\eta_{zz})q_\eta+(\zeta_{xx}+\zeta_{yy}+\zeta_{zz})q_\zeta$$

Computational Strategy and Boundary Condition

The calculation proceeds through a sequence of loops each advancing the entire flow configuration through sufficiently small finite time increment Δt. The output of each loop is taken as an initial condition for the next and the computation is performed until the steady state. An Euler explicit scheme is used for the time marching procedure. Pressures are obtained throughout the fluid domain by solving the Poisson equation. Iterations are stopped when the pressure difference between two consecutive approximations is smaller than a certain quantity ε, chosen a priori. The new pressure field generates a new velocity field. The velocity component updating is achieved by using the time-forward difference form of the momentum equations.

The marker particles are moved to their new positions by using either Lagrange or Euler techniques. The still state condition is used as an initial condition and the flow is accelerated within several time steps, up to the given constant velocity, for the numerical stability. All the spatial derivatives employ the second order central difference, while for convection terms the third order upstream difference is used.

The boundary condition for off PAR is as follows. At the upstream the flow starts from zero and is accelerated up to the predefined speed. Thus, each horizontal component of velocity has the same constant value depending on the time step. The vertical component is equal to zero in each point of the upstream boundary and remains the same during the pressure computation. At downstream, the wavy motion influence is so gentle that the zero gradient extrapolation is used for both components of velocity and pressure. On the body surface, the no slip and neumann condition are used for the velocity and pressure respectively.

RESULTS AND DISCUSSION

Computational Conditions

The present work represents the numerical calculation of three dimensional wing. The ratio of span/chord is 3.0 with the shape of NACA 0012 wing section in the spanwise direction at 4, 6, 10°, angle of attack. The height from the ground is 0.05, 0.1, 0.3 0.8. The computing domain is 3.5 times the chord length in the streamwise direction. All parameters are non-dimensionalized based on the wing chord L and uniform velocity Uo. The minimum grid spacing in z direction is 0.002. The oncoming flow has been accelerated during the first 500 steps. The relaxation factor is given 0.8 to solve the Poisson equation. The maximum number of iteration per step is 10 to 50, which depends on the pressure difference between the present and next calculations. The grid is made H-H topology to easily treat the bottom as a free surface in the future research, but relatively a large number of iterations are requested near the leading edges. Fig.1 shows the coordinate system for the computation; x, y and z represent coordinates in a Cartesian system, x in the uniform flow direction, y in the lateral and z normal to the x-y plane. Fig.2 shows the sectional grid view. The number of grid used here is extremely insufficient to simulate the flows in all interesting regions, but some places are relatively clustered, for example, near the body surface, around tip and trailing edges. The number of grid is 101x32x44, and the computation domain is x(-1.0, 3.5), y(0.0, 0.8) and z(0.0,-1.5). Here the Reynolds number is 10^3.

Computational Results and Discussion

The computation is carried out at 4, 6, 10°, angle of attack. Fig.3 is the computation result at 4, 6° and some comparison is made at 6° as shown in Table 1 (Riegels,1961, Abbott,1958). The lift and drag coefficients are compared with those by Hirata and BIEM, in which the lift shows a reasonable result in its trend. Here the total drag is made up of the pressure and frictional drag. In low Reynolds number calculation, the frictional drag shows a large value.

Table 2,3,4,5 show the PAR effect and lift & drag coefficients at 4, 6, 10°. According to the results, when the body approaches to the ground, the velocity becomes small below the wing surface due to the PAR effect and the lift becomes large due to the pressure increase. In the simulation, the thrust is applied in the form of uniform flow for the PAR 1 and -45° inclined flow for the PAR 2. The location is in front of the leading edge, off half the chord length. To investigate the effect of PAR, the propulsor is put in the upstream of wing and the flow is accelerated through the wing with 10° angle of attack for the analysis. The thrust gives 30% increase in the velocity from the momentum theory. Fig.4 shows the calculated results for the height of 0.8 and angle of attack of 10 degree; (a) pressure contour at midspan, (b) pressure on the upper surface, (c) pressure around the tip, (d) velocity below the wing, (e) velocity at midspan. Due to the clustering of grid near the tip and leading edge, the iteration number for the poisson solver is flexible, from 10 to 50. The separation with a back flow is well simulated in this case.

Fig.5 shows the results for the height of 0.3 and angle of attack of 10 degree; (a) to (d) for off PAR, (e) for PAR 1, (f) for PAR 2. The fluctuation of pressure is seen in the upstream of the wing for PAR 2, which is due to an added thrust in an application of boundary condition.

Fig.6 shows the results for the height of 0.1 and angle of attack of 10 degree. The effect of PAR is well seen when the body is near ground. The pressure difference is a little large between the lower surface of the wing and the ground.

Fig.7 shows the results for the height of 0.05 and angle of attack of 10 degree. The effect of PAR is noticeably well shown because it gives the venturi action near the ground and shows some influences on the pressure coefficients. Fig.8 gives pressure coefficients on the wing surface; (a) is for the height of 0.1, (b) for the height of 0.3, (c) is for off PAR,

For this study, all numerical simulations were carried out by the SUN Sparc center 2000 at Halla Institute of Technology.

CONCLUDING REMARKS

In the present study, 3-dimensional PAR was numerically studied by solving the Navier-Stokes equations. Pressure distributions and velocity vectors are simulated around the wing, and then Cp values and lift/drag ratio are carefully reviewed by changing the height/chord; 0.05, 0.1, 0.3, 0.8, and the angle of attack; 4,6,10 degrees. Through the simulation, the followings are realized.

(1) The N-S solver is confirmed to be useful and valid

for three dimensional PAR analysis. It is through the comparison of lift/drag ratio around the 3-D wing. The lift increases due to the increase of ram pressure below the wing because it gives the venturi action near the ground.

(2) The PAR is confirmed to be promising because the lift can be aerodynamically controlled by the location and thrust magnitude of PAR. The ground effect also gives a reasonable reliability.

ACKNOWLEDGMENT

This paper was supported by NON DIRECTED RESEARCH FUND, Korea Research Research Foundation, 1996.

REFERENCES

Abbott,I.H. and Von Doenhoff,A.E.(1958), "Theory of Wing Sections", Dover Publication, New York

Chung,K.H., Chun,H.H.,et al (1995), "Experiment Studies on WIG Effect and Design and Construction of a PARWIG Craft", Proc. of Autumn Conf. of Soc. of Nav. Arch., Korea

Hirata, N.(1993), "Simulation on Viscous Flow around Two-Dimensional Power-Augmented Ram Wing in Ground Effect", Jour., Soc., Naval Arch. of Japan, Vol. 174, pp.47-54

"Janes Surface Skimmers", Janes Publishing Company Limited, London, pp.26-28, England

Kataoka,K, Ando,J.(1992), "Free Surface Effect on Characteristics of Two-Dimensional Wing", Trans. of West-Japan Soc. of Naval Arch, Vol.83, pp.21-30

Lippisch,A.M.(1964), "The Aerodynamic Ground Effect and Development of the Aerofoil Boat", Luftfahrtechnik-Raumfahrteknik 10, pp. 261-269,

Lockeed California Company Report (1962) LA/ME/2148, Sep., "Preliminary Economic Analysis of Winged Hull Vehicles",

Smithey,W.J.H.(1977), Papadales,B.S.,Jr., Chaplin, H.R., "Effect of Turbulent Jet Mixing on the Static Lift Performance of a Power-Augmented-Ram Wing", David Taylor Naval Ship Research and Development Centre ASED-389, Sep.

Masuda,K., Suzuki,K. (1991), "Simulation of Hydrodynamic Effects of 2-Dimensional WIG moving near the Free Surface, Jour. of Soc. of Naval Arch of Japan, Vol.170, pp.83-92

Riegels, F.W.(1961), "Aerofoil Sections", Butter worths, London

Table 1 Comparison of lift and drag at α=6°

	Present solver	Hirata	BIEM	Abbott
C_L	0.6136	0.65	0.71	0.64
C_D	0.032	0.0117	-	0.0084

Table 2. Effect of PAR at h/c=0.05 and α=10°

	C_L	Cf	Cp	C_D (total)
off PAR	1.3099	0.0143	0.083	0.097
PAR 1	1.5722	0.0150	0.084	0.099
PAR 2	1.5961	0.0153	0.085	0.100

Table 3. Effect of PAR at h/c=0.10 and α=10°

	C_L	Cf	Cp	C_D (total)
off PAR	1.0743	0.0106	0.081	0.092
PAR 1	1.2918	0.0109	0.082	0.093
PAR 2	1.4778	0.0113	0.082	0.093

Table 4. Effect of PAR at h/c=0.30 and α=10°

	C_L	Cf	Cp	C_D (total)
off PAR	0.7875	0.0101	0.078	0.088
PAR 1	1.2262	0.0104	0.079	0.089
PAR2	1.4523	0.0106	0.079	0.090

Table 5. Effect of Angle-of-Attack(α) at h/c=0.8

	C_L	Cf	Cp	C_D (total)
α=10°	0.7295	0.0101	0.074	0.084
α=6°	0.6136	0.0099	0.032	0.042
α=4°	0.5616	0.0104	0.028	0.038

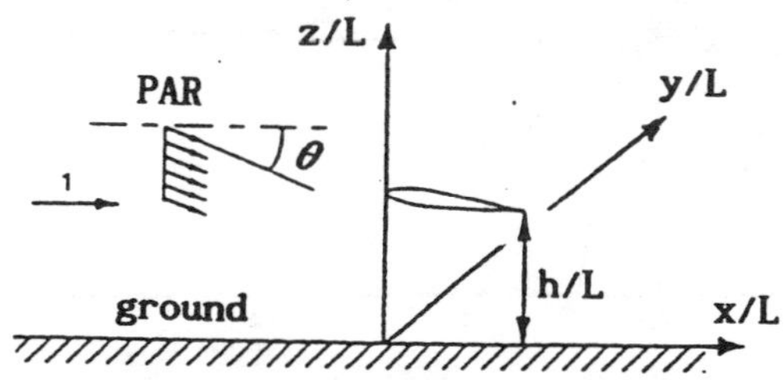

Fig. 1 Coordinate definition

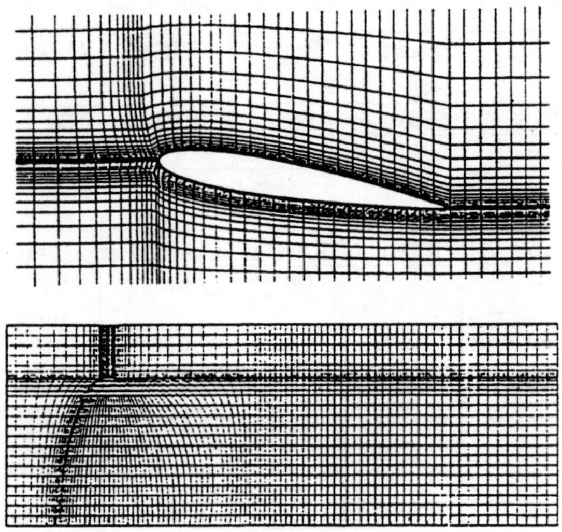

Fig. 2 Sectional grid view

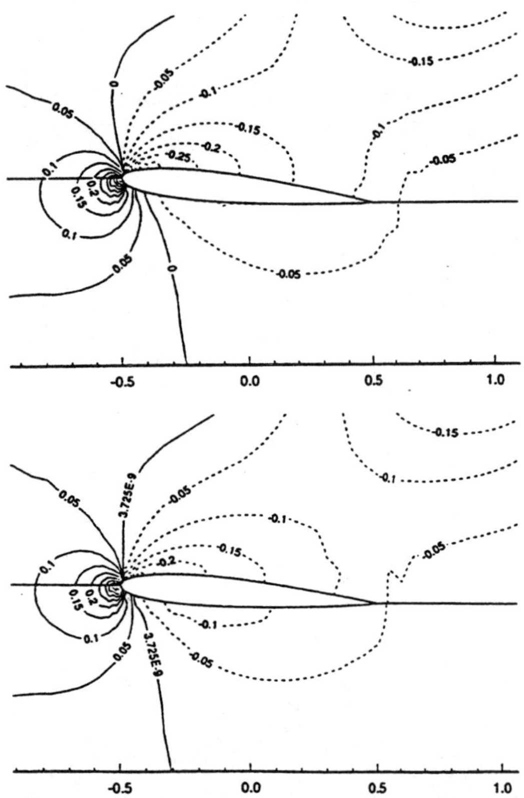

Fig. 3 Computed results (off PAR)
(above: α=6 °, below: α=4 °)

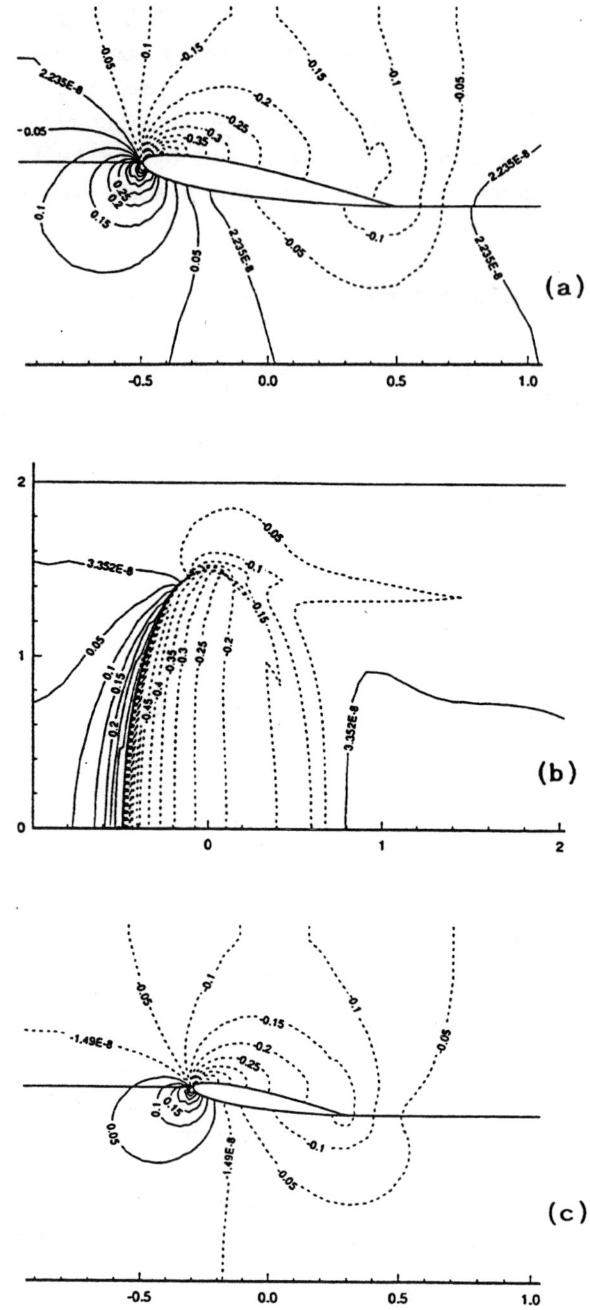

Fig. 4 Computed results at h/c=0.8 (off PAR)
(a) Pressure contour at midspan(j=3)
(b) Pressure contour on upper surface
(c) Pressure contour around tip(j=14)

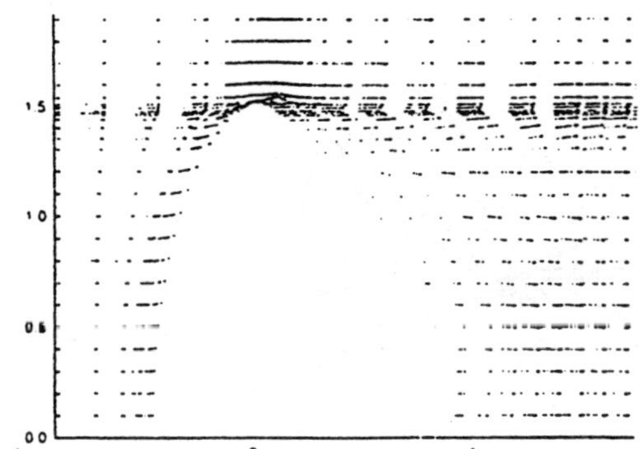

(d)

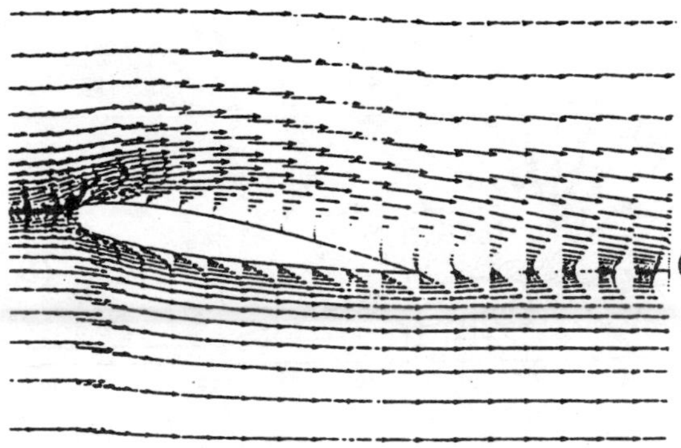

(e)

Fig. 4 (Continued)
(d) Velocity vector at k=23
(e) Velocity vector at midspan(j=3)

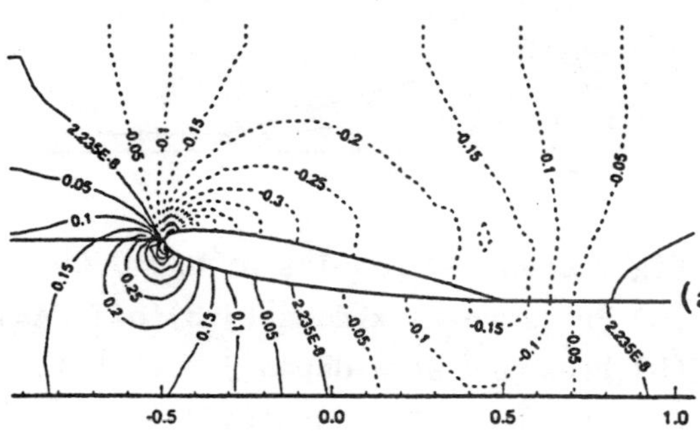

(a)

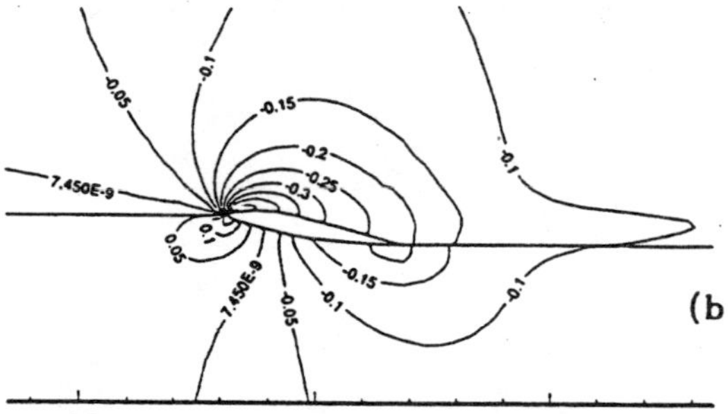

(b)

Fig. 5 Computed results at h/c=0.3
(a) Pressure at midspan(j=3)(off PAR)
(b) Pressure around tip(j=14)(off PAR)

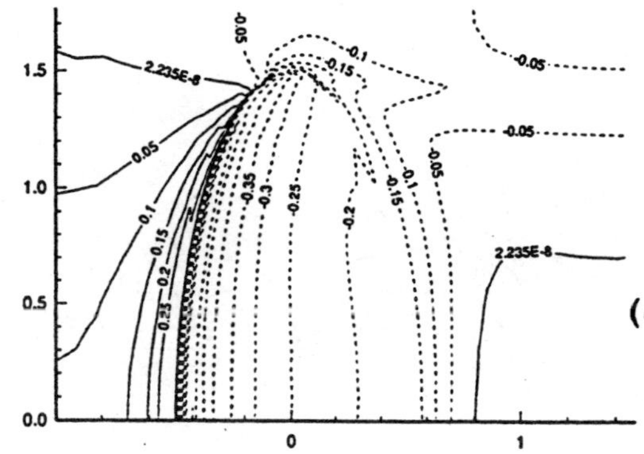

(c)

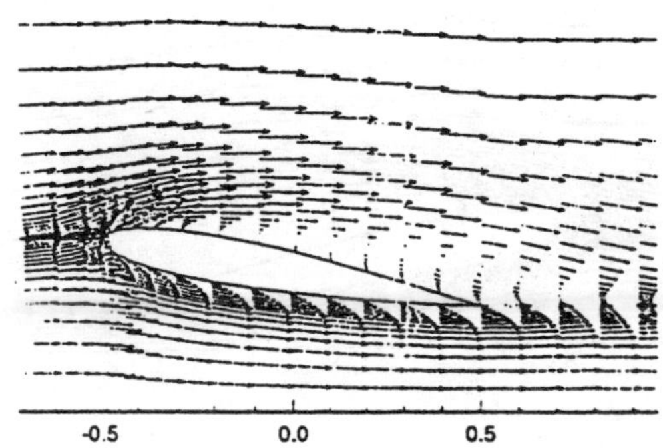

(d)

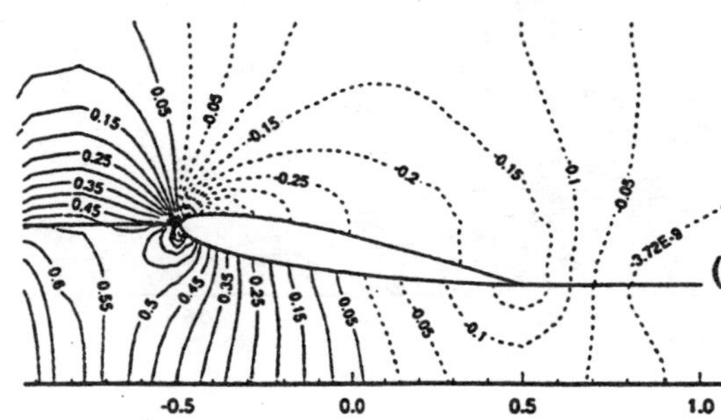

(e)

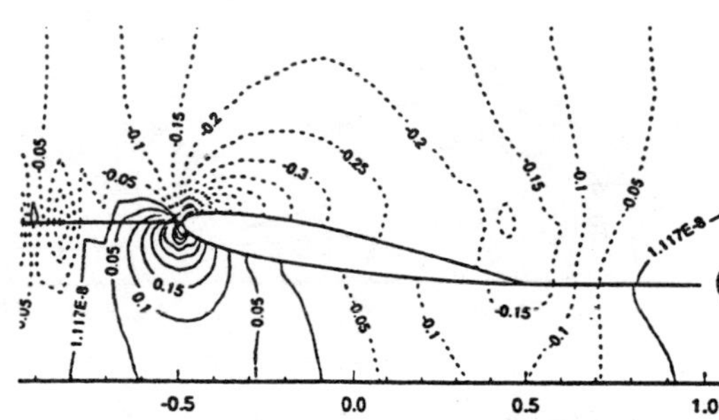

(f)

Fig. 5 (Continued)
(c) Pressure at upper surface(off PAR)
(d) Velocity at k=23(off PAR)
(e) Pressure at midspan(j=3)(PAR 1)
(f) Pressure at midspan(j=3)(PAR 2)

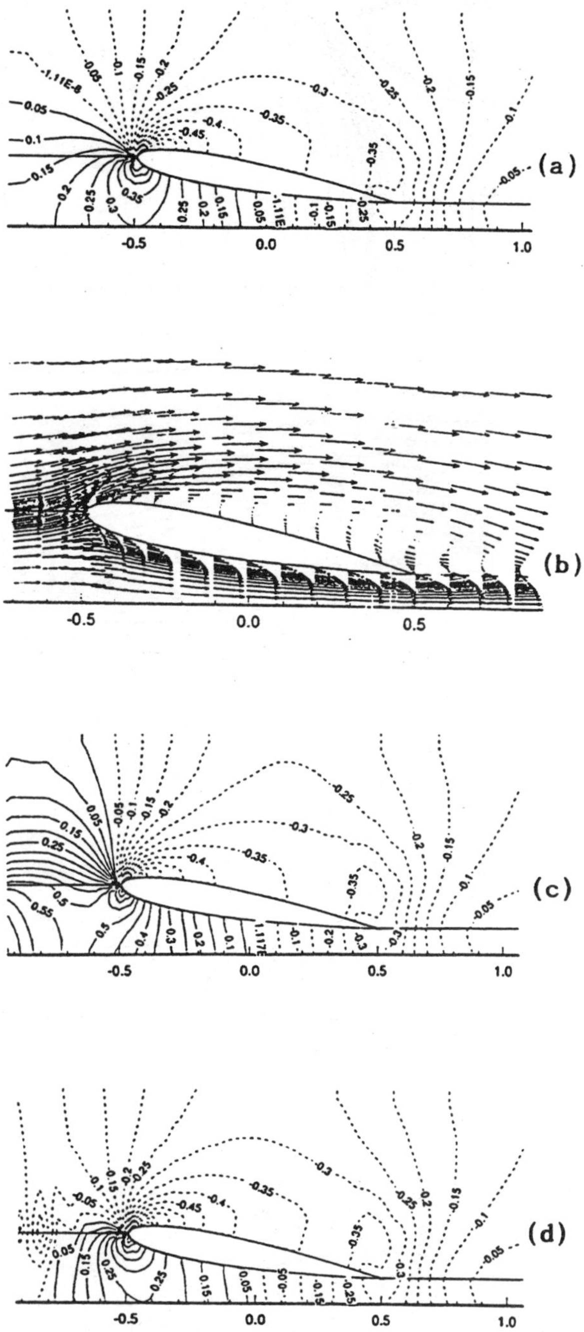

Fig.6 Computed results at h/c=0.1

(a) Pressure at midspan(j=3)(off PAR)

(b) Velocity at midspan(off PAR)

(c) Pressure at midspan(j=3)(PAR 1)

(d) Pressure at midspan(j=3)(PAR 2)

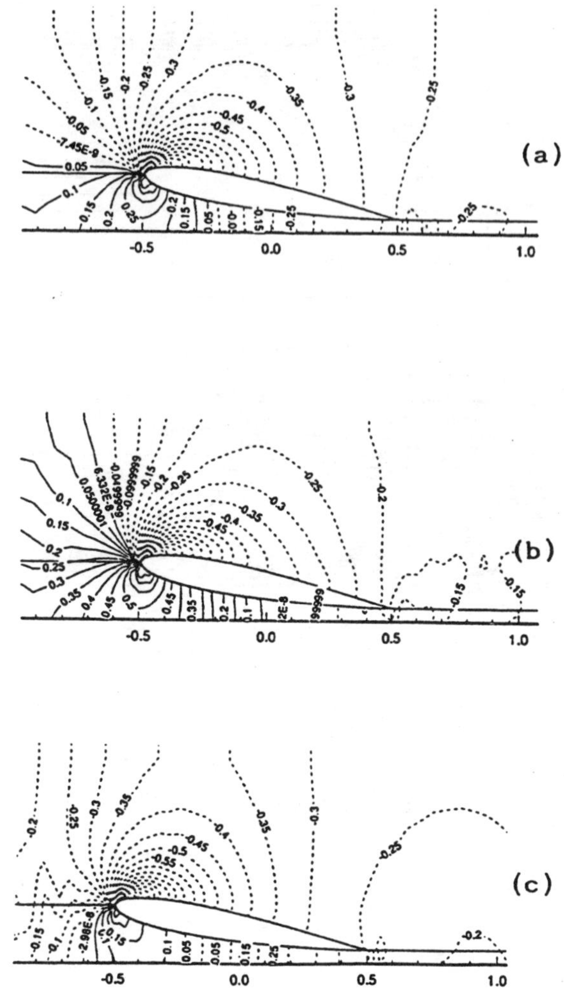

Fig.7 Computed results at h/c=0.05

(a) Pressure at midspan(j=3)(off PAR)

(b) Pressure at midspan(j=3)(PAR 1)

(c) Pressure at midspan(j=3)(PAR 2)

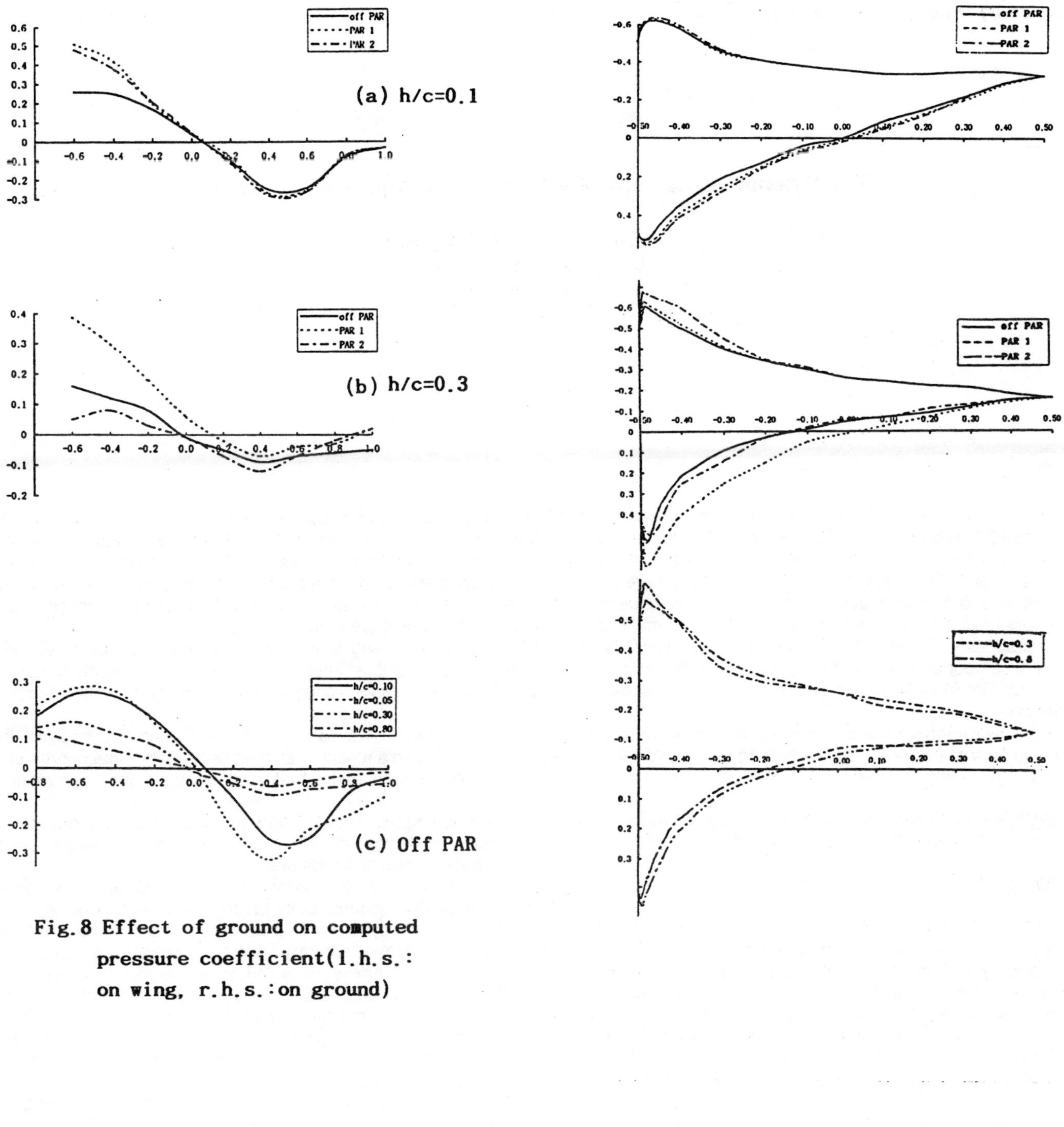

Fig. 8 Effect of ground on computed pressure coefficient(l.h.s.: on wing, r.h.s.:on ground)

Proceedings of the Seventh (1997) International Offshore and Polar Engineering Conference
Honolulu, USA, May 25-30, 1997

ISBN 1-880653-28-1 (Set); ISBN 1-880653-32-X (Vol. IV)

The Dynamic Behaviour of a Fast Sailing Monohull in Waves

F. Louarn, W.G. Price and P. Temarel
University of Southampton
United Kingdom

ABSTRACT:

Three dimensional hydroelasticity theory is applied to investigate the dynamic behaviour of a WOR 60 hull in regular waves encountered at arbitrary heading. This is an interesting application of hydroelasticity as the hull is relatively short with an L_{oa}/B_{max} ratio of 3.6. In addition, unlike previous applications of hydroelasticity, the hull is fitted with keel and bulb, as well as, mast and standing rigging. The objectives of the investigation are twofold, namely understanding the global structural dynamics (i.e. natural frequencies and mode shapes) of a fast sailing monohull in order to create a suitable dynamic model, and obtaining global structural responses (e.g. stresses) in realistic operational conditions in order to identify critical areas of the structure.

A reasonably detailed F.E. model of the WOR 60 is constructed, using shell and beam elements, which contains all main structural components, such as longitudinals, bulkheads etc.. To this model (i) the ballast keel and (ii) the mast and standing rigging are added. The modular nature of the model allows assessment of the influence of various components on the dynamic behaviour of the vessel.

In the *wet* analysis the dynamic behaviour of the complete structure in regular waves is examined. During this analysis the wetted surface of the WOR 60 is idealised with panels containing a source (Green's function) at their centre. Principal coordinates for the rigid body motion and distortions are obtained. Stresses are subsequently evaluated using modal summation.

KEYWORDS: hydroelasticity, yacht dynamics, wave loading, hull / keel connection, fluid-structure interaction

INTRODUCTION

Traditionally naval architects and yacht designers have restricted themselves to static or quasi static considerations when dealing with the structural design of a ship moving in waves. Accordingly, the primary stress level which relates to the hull girder bending is often based upon a scenario whereby the craft is balanced on a trochoidal wave form so that crests appear at the bow and stern. Global equilibrium exists between gravity and buoyancy forces. Nevertheless the local net forces calculated at, say, discrete stations result in approximate shear force and bending moment distributions. In the case of sailing yachts, in addition, the compression of the mast around amidships normally has a major effect upon the aforementioned distributions. Secondary and tertiary levels of stress are generally treated in a simple manner based upon the local hydrostatic pressure.

The structural problems experienced by the WOR 60s in the last 1994 Whitbread round the world race, e.g. loosening of the keel to hull connection and local delamination of the sandwich shell, highlight the limitations of the current design methodologies. The solution to such problems requires a good prediction of wave-induced dynamic loads under realistic operating conditions.

The aim of this paper is to investigate how three dimensional hydroelasticity theory may be applied to study the structural dynamics of a sailing yacht progressing through a wave system at arbitrary speed and heading. As suggested earlier the example of a WOR 60 is relevant and has, therefore, been selected. The geometry of this particular craft is obviously far from "beamlike". The use of a three dimensional finite element model consisting of beam and shell elements for the "dry hull" analysis (Price et al, 1994), allows for overcoming the restriction featured in earlier work where slender structures were investigated using a Timoshenko or Vlasov (Bishop and Price, 1979) beam analogy.

The first part of this paper focuses on the finite element model with special attention devoted to the influence of important items such as the ballast keel and rig upon the dynamic characteristics of the structure.

In the analysis of the "wet hull" a panel method involving a pulsating source distribution over the wetted surface of the hull is used in order to obtain hydrodynamic coefficients and principal coordinates. These describe the coupled rigid and flexible responses of the structure in regular waves for given incident wave amplitude and frequency and for selected ship speed and heading. Stresses are calculated using the summation of the modal stresses multiplied by the corresponding principal coordinates.

I THEORETICAL BACKGROUND:

Three dimensional hydroelasticity theory was developed to predict the dynamic responses of flexible structures in a seaway. This is carried out in two stages, namely "dry" and "wet" analyses whose main features are discussed in this section (Bishop et al, 1986).
The equation of motion for a discretised system can be expressed in matrix form as:

$$M\ddot{U}+D\dot{U}+KU=Z \qquad (1)$$

where $\ddot{U},\dot{U},U$, are vectors of nodal accelerations, velocities and displacements respectively and M, D and K correspond, respectively, to the mass, damping and stiffness matrices of the "dry" structure. Z is the external force vector representing, for instance, the fluid loadings.

The first part of the analysis considers the free undamped vibrations of the system in vacuo. Equation (1) then reduces to:

$$M\ddot{U}+KU=0 \qquad (2)$$

A finite element approach allows for solutions to be obtained for eqn. (2) in the form of natural frequencies, ω_r, and principal mode shapes as well as other dynamic characteristics such as generalised masses, a_{rr}, and modal stresses, σ_r.

Applying Rayleigh's theorem, and, thereby, expressing the total displacement of the structure as a summation of distortions in the principal modes, equation (1) can be generalised as:

$$a\,\ddot{p}(t)+b\,\dot{p}(t)+cp(t)=Z(t) \qquad (3)$$

where, a, b (b_{rr} = $2\nu_r.\ \omega_r.a_{rr}$) and c (c_{rr} = $\omega_r^2.a_{rr}$) are the diagonal generalised mass, structural damping and stiffness matrices respectively, with ν_r representing a modal damping factor. p(t) denotes the vector of principal coordinates which contains n terms as, for all practical purposes, only n mode shapes are admitted in the summation, (r = 1,...,6 denote the rigid body modes: surge, sway, heave, roll, pitch and yaw and r ≥ 7 denotes the flexible mode shapes). The generalised excitation in regular sinusoidal waves is represented by the vector Z(t) whose elements are of the form:

$$Z_r(t)=\left[\Xi_{or}+\Xi_{dr}+H_r+R_r\right]e^{i\omega_e t} \qquad (4)$$

The vectors Ξ_o and Ξ_d are the generalised excitations due to incident waves (Froude-Krilov) and diffracted waves. They depend upon wave amplitude a, frequency ω, forward ship speed U, and wave heading χ.

$$H_r=\sum_{k=1}^{n}p_k\left(\omega_e^2.A_{rk}-i.\omega_e.B_{rk}\right)$$ is the rth generalised radiation force generated by the motions and distortions of the structure. It is therefore a function of the principal coordinates p_k. The terms A_{rk} and B_{rk} refer to the generalised added mass and hydrodynamic damping respectively. Note that they account for coupling between modes r and k and are both encounter frequency (ω_e) dependent.

$$R_r=-\sum_{k=1}^{n}p_k.C_{rk}$$ corresponds to the rth generalised restoring force.

Z(t) is represented by a velocity potential of the following form:

$$\Phi(x,y,z,t)=\Phi_o(x,y,z,t)+\Phi_d(x,y,z,t)+\sum_{r=1}^{n}\Phi_r(x,y,z).p_r(t) \qquad (5)$$

Subscripts o, d and r refer to the incident, diffracted and radiation potentials. On linearising the boundary conditions a composite source distribution technique provides a numerical solution for Z(t) based upon the above velocity potential (Bishop et al, 1986).

In regular waves the principal coordinates vary with time as $p(t)=p.e^{i.\omega_e.t}$. Thus eqn. (3) becomes:

$$\left[-\omega_e^2.(a+A)+i\omega_e.(b+B)+(c+C)\right].p=\Xi_o+\Xi_d \qquad (6)$$

providing the solution for the complex principal coordinate amplitudes p in regular sinusoidal waves.
The corresponding responses of the structure can be obtained, using modal summation. For example the stress component σ_x is given as:

$$\sigma_x(x,y,z,t)=e^{i.\omega_e.t}.\sum_{r=7}^{n}\sigma_{x_r}(x,y,z).p_r \qquad (7)$$

where the summation commences from r=7 as the modal stresses are zero for the first six rigid body modes.

II ANALYSIS OF THE DRY STRUCTURE

The only method currently available to establish the dynamic characteristics of a non beam-like vessel, such as a WOR 60, is F.E.A.. Use is made of a commercial package, Ansys (1994).

II.1 Structural Finite Element model

The data for a preliminary design study of a Whitbread 60 served as a basis for the development of a FE model in six modules quoted below:

WOR1: shell + transverse members (bulkheads and frames)
WOR2: WOR1 + longitudinal beams
WOR3: WOR2 + ballast keel
WOR4: WOR3 + mast and standing rigging
WOR5: WOR4 + mast compression due to fore and aft stay tension
WOR6: WOR5 + mast compression due to shroud pretension (athwartship)

Linear shell elements (shell63) adequately represent the essential structural properties of the shell. They not only offer membrane stiffness but also bending stiffness capabilities. This is useful as the plating, whilst acting primarily as a membrane, could exhibit some degree of bending due to the relatively fat configuration of the WOR 60. The same type of elements is used for the bulkheads and the structure securing the keel to the hull. The estimation of average material properties and scantlings is based upon the use of the laminate constitutive equations (Smith,1990). It is interesting to note the quasi-isotropic behaviour of the materials, (balsa core, S-glass/kevlar epoxy skins or monolithic laminate for the keel area) which is an outcome of the extensive implementation of multiaxial clothes. Hence, for convenience, the material is treated as isotropic in the F.E. models.

As for the transverse frames and longitudinals, a hybrid formulation (Hughes, 1988) of a 3.D. beam element (beam4) has been adopted whereby the axial stiffnesses of the beam and attached plating are treated independently whilst an effective breadth of plating is accounted for in the calculation of the beam flexural rigidity. Bonding is only possible at nodal positions so that the correct stiffening effect of the beams depends upon sufficient mesh refinement.

The ballast keel comprises a cast iron fin connecting a lead alloy bulb to the hull. They are conveniently idealised as a plate of variable thickness (2 shell elements across the chord) and as a non uniform beam, the important factor for the bulb being the mass distribution.

Beam elements are used in order to model the aluminium mast whereas the steel shrouds and stays require tension only spars (link8) for they do not possess compressive stiffness. The base of the mast is simply hinged at the node coinciding with the top of the fin leading edge. The connection with the deck must allow the mast to slide through in a vertical plane whilst restraining its displacements in the deck plane. So called "contact elements" (contact52) are found suitable. These are spar elements with compressive and out of plane frictional (neglected in this case) stiffness properties. Note that their use may lead to convergence difficulties and it may be necessary to decrease the nominal compressive stiffness in order to obtain a solution (Ansys, 1994).

The correct mass distribution is obtained by adjusting densities across the mesh.

Local distortions may arise in a dynamic analysis as some nodes, insufficiently supported in the idealised structure, start behaving as mechanisms. They may corrupt the mode shapes and lead to erroneous generalised masses. It can be observed that the optional bending stiffness selected for the plating readily helps in preventing some of these. For the same purpose, massless plate elements have also been added locally linking orthogonal elements between the shell and the bulkheads, thereby, also improving the modelling of the joints.

The complete model is shown in figure 1.

II.2 Characteristics of the dry hull

Because of its large size (6930 degrees of freedom for WOR6) the eigenvalue problem is reduced to one of equivalent master degrees of freedom.

A Guyan's reduction method (Ansys, 1994) is implemented based upon 650 automatically selected masters which exclude rotations. According to the authors' experience better mode shapes, i.e. less local distortions, can be obtained from Ansys if translations alone are considered whilst selecting the masters.

In the case of models WOR5 and WOR6 the structure is pre-stressed under the action of rigging loads. For example, for the complete model, WOR6, a two step static analysis was initially carried out first applying stay then shroud loading. A total static force of 325000 N is applied at the base of the mast, externally balanced by reactions set up at the rigging lines attachment points (chainplates), where pin end support boundary conditions are assumed. Evaluated pre-stresses are allowed for, subsequently, in the modal "dry" hull analysis by superimposing a stress stiffness matrix onto the structure stiffness matrix (Ansys, 1994).

The mode shapes are normalised with respect to a unit maximum translation. Note that, since we are interested mainly in the responses of the hull, the degrees of freedom associated with the rig are deliberately omitted from the normalisation procedure.

When normalising a mode shape with respect to a unit maximum translation the corresponding generalised mass is obtained as the inverse of the square of this displacement.

Examples of flexible mode shapes corresponding to WOR6, in the lower band of the frequency range, are shown in fig.2. Comparative results for the principal modes of main significance are presented in tables 1 and 2. Note that, for models WOR1 and WOR2, $r = 11$ corresponds to the first distortion mode. The high frequencies associated with the flexible mode shapes, e.g. 2.2 Hz for the lowest mode ($r = 7$), suggests that the structure is very stiff. Note that for models WOR4, WOR5 and WOR6 flexible modes associated with the distortions of the mast are not included as the hull does not distort. Some of these mast related distortions occur at frequencies lower than 2.2 Hz.

III ANALYSIS OF THE WET HULL

III.1 Hydrodynamic Model

Four and, occasionally, three cornered panels are used in order to discretise the wetted surface of the hull. For convenience rather than necessity, the underwater part of the finite element mesh is transposed into a hydrodynamic mesh except for the bulb where more refinement is required.

Port-starboard symmetry considerations allow for increased computational efficiency. The resultant half model possesses 203 panels, each containing a pulsating source at its centre. A view of the full grid is presented in fig.3.

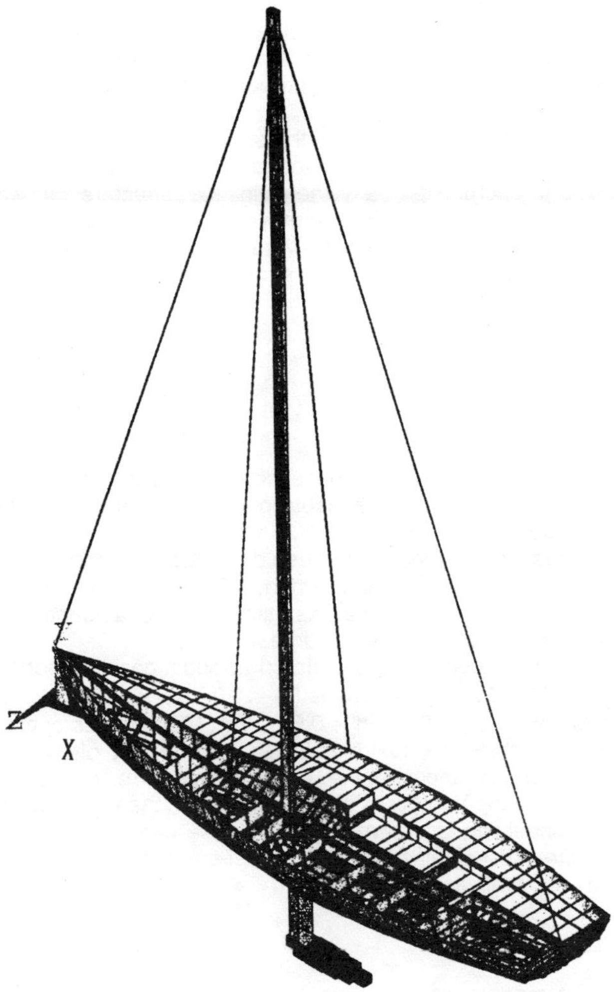

Fig.1. Complete* F.E. mesh for model WOR6
* some deck elements are not shown to allow for viewing of the internal structure

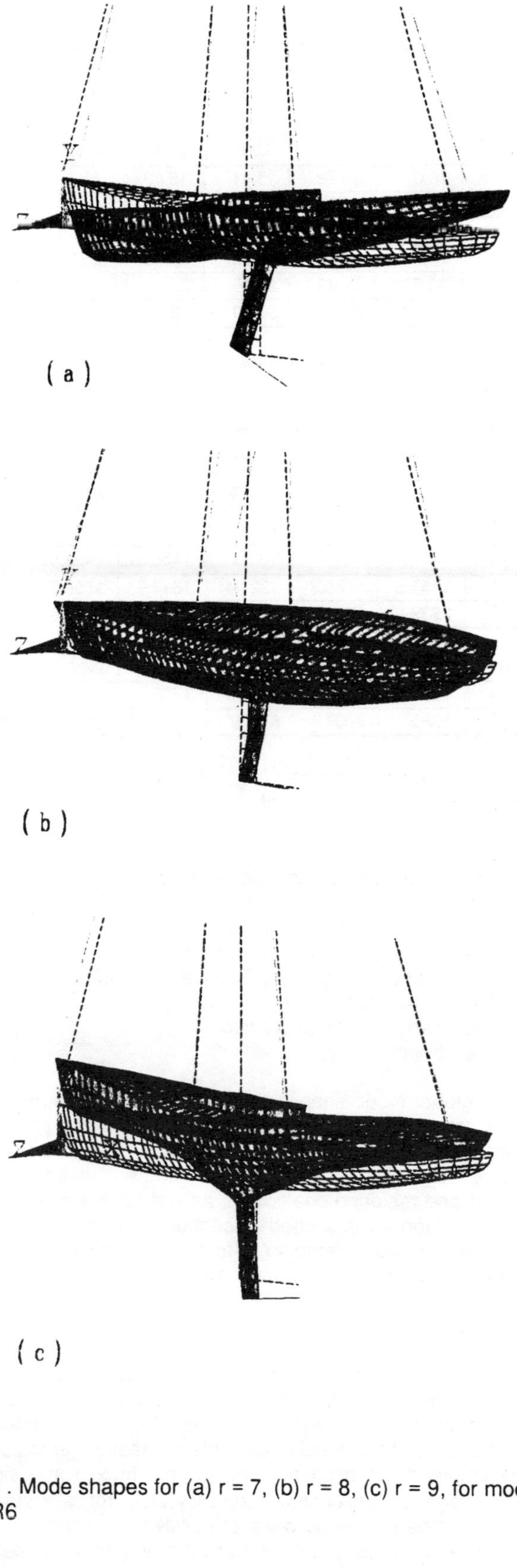

Fig.2 . Mode shapes for (a) r = 7, (b) r = 8, (c) r = 9, for model WOR6

r *	wor1	wor2	wor3	wor4	wor5	wor6
7 (S)			2.10	2.14	2.16	2.16
8 (A)			3.89	5.85	4.36	4.40
9 (S)			8.18	8.78	7.85	7.90
10 (A)			8.05	8.16	8.15	8.15
11 (A)	15.07	15.22	15.41	15.83	15.54	15.54
12 (S)	17.04	17.32	18.77	19.10	18.72	18.74
13 (S)	23.29	23.75	23.97	24.24	23.75	23.98
14 (A)	25.91	26.18	25.98	25.76	25.72	25.72
15 (A)	30.53	31.46	27.40	27.24	27.26	27.26
16 (S)	31.42	33.53	33.60	33.80	33.50	33.49

table 1. Dry natural frequencies (Hz) for various models of the WOR 60 hull.

r *	wor1	wor2	wor3	wor4	wor5	wor6
7 (S)			5648.1	5649.5	4466.7	4378.5
8 (A)			3115.8	3032.7	3750.1	3742.7
9 (S)			3410.3	3293.3	3317.1	3299.0
10 (A)			1369.3	1455.3	1423.8	1424.6
11 (A)	1143.5	1278.3	1321.4	1614.9	1556.9	1561.4
12 (S)	1539.4	1720.4	1190.7	1171.9	1198.0	1189.9
13 (S)	1592.6	1694.3	1929.4	1925.8	2070.6	2066.9
14 (A)	1782.7	2288.9	2440.6	2932.3	2882.1	2899.4
15 (A)	1281.5	1160.2	1776.9	1736.0	1762.7	1779.5
16 (S)	459.2	1057.4	840.9	846.5	864.2	855.7

table 2. Generalised masses ($kg.m^2$) associated with the dry structure for various models of the WOR 60 hull.

* Modal index r:
(S) and (A) denote symmetric and antisymmetric modes respectively.
7: hull / keel (x,y) - 8: hull / keel (y,z) - 9: bottom sag
10 : fin twist (oy) - 11:1 node torsion
12 : 2 node bending (x,y) - 13: 3 node bending (x,y)
14: 2 node torsion & bending (x,z)
15: 2 node bending (x,z) and torsion
16: 4 node bending (x,y)

Ten global mode shapes (r = 7 to 16) corresponding to WOR6, (see section II.2), have been selected and classified according to symmetry criteria (see fig.5). They are used to evaluate the generalised added mass, hydrodynamic damping and restoring coefficients as well as the generalised wave excitation as described in section I. Note that, at this stage, no viscous damping effects are included in the analysis. The principal coordinates are evaluated from equation (6).

III.2 Rigid Body Responses

The rigid body displacements are computed accounting for coupling effects with flexible distortions. Experimental results obtained from towing tank tests on a parent model are used as a means of validation. Plots of heave (amplitude / wave amplitude) and pitch (amplitude / max. wave slope) R.A.O.s against encounter frequency, ω_e, for a range of Froude numbers in head seas are shown in figure 4. Note that, unlike the experimental set-up, the numerical procedure is not restricted to head seas.

III.3 Flexible Responses

The responses of the WOR 60 (model WOR6) were evaluated when sailing upwind at a realistic heading angle of 135 degrees. The incident wave amplitude and frequency are 1.0m and 1.15 rad/s respectively, corresponding to rough sea conditions in the Northern Atlantic (Rawson and Tupper, 1983). It has to be pointed out on one hand, that the speed of 16 knots, Fn=0.65, may be somewhat optimistic in these conditions and on the other, that the assumed upright sailing attitude is a limitation of the current numerical model. The variations of the symmetric (r = 7,9,12,13,16) and antisymmetric (r = 8,10,11,14,15) principal coordinate amplitudes with frequency of encounter are shown in fig. 5. It should be remembered that during the evaluation of the responses using modal summation, not only the amplitude but also the phase of the principal coordinate together with the magnitudes of modal characteristics, e.g. distortion, stress etc., determine the contribution from each principal mode.

Equation (7) indicates how the six stress components may be evaluated from the knowledge of modal stresses and principal coordinates, anywhere in the structure. It was decided that the most convenient way to present these results was through the use of a Von Mises equivalent stress which is based upon the total shear strain energy in the material and expressed as

$$\sigma_{vm} = \begin{bmatrix} \left[(\sigma_x - \sigma_y)^2 + (\sigma_y - \sigma_z)^2 + (\sigma_z - \sigma_x)^2 \right] .0.5 \\ +3.\left(\tau_{xy}^2 + \tau_{yz}^2 + \tau_{zx}^2 \right) \end{bmatrix}^{0.5} \qquad (8)$$

This scheme is only valid because, as discussed in section II.1 the laminates have isotropic equivalent properties. Orthotropic
materials normally require the use of an Azzi-Tsai formulation of the equivalent stress (Smith, 1990). Moreover, In the context of a dynamic analysis, stresses are a function of spatial coordinates as well as time.
Stress magnitudes were calculated at nodal positions for the model WOR6 in regular waves (a = 1.0 m, ω = 1.15 rad/s, ω_e = 2.88 rad/s, χ = 135°, Fn = 0.65). Von Mises stress contour plots over the entire hull indicate that the vicinity of the keel connection experiences the largest stress magnitudes. A close-up of the Von Mises stress contours in the vicinity of the keel connection is shown in fig.6. These can be compared with stresses obtained from a quasi-static grounding case study shown in fig.7.

IV DISCUSSION

The modular nature of the finite element model allows for an appreciation of the impact of various structural items upon the dynamic characteristics of the WOR 60 (see tables 1 and 2). The effect of longitudinal beams (comparing models WOR1 and WOR2) is an increase of natural frequency for all modes of interest. As for bending, (mode 12, 13, 16 and 14, 15 to an extent), a simple beam analogy can explain this behaviour. Natural frequencies for a simple beam in bending are known to be proportional to ($EI / \mu L^4)^{0.5}$. The inclusion of longitudinals implies a significant increase in EI, (E_{shell}=25 Gpa, $E_{long.}$=57.5 Gpa), with a corresponding small increase (+ 9%) in μ, resulting in higher natural frequencies.

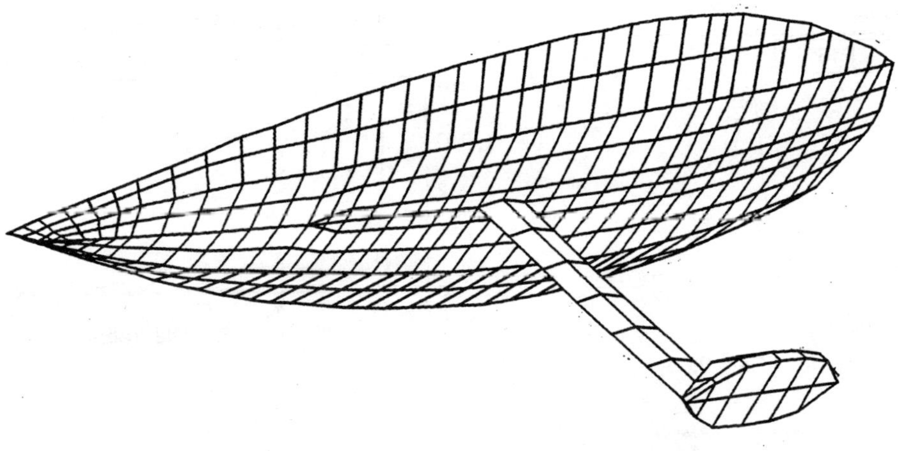

Fig. 3. Hydrodynamic panel idealisation of the wetted hull surface of the WOR 60.

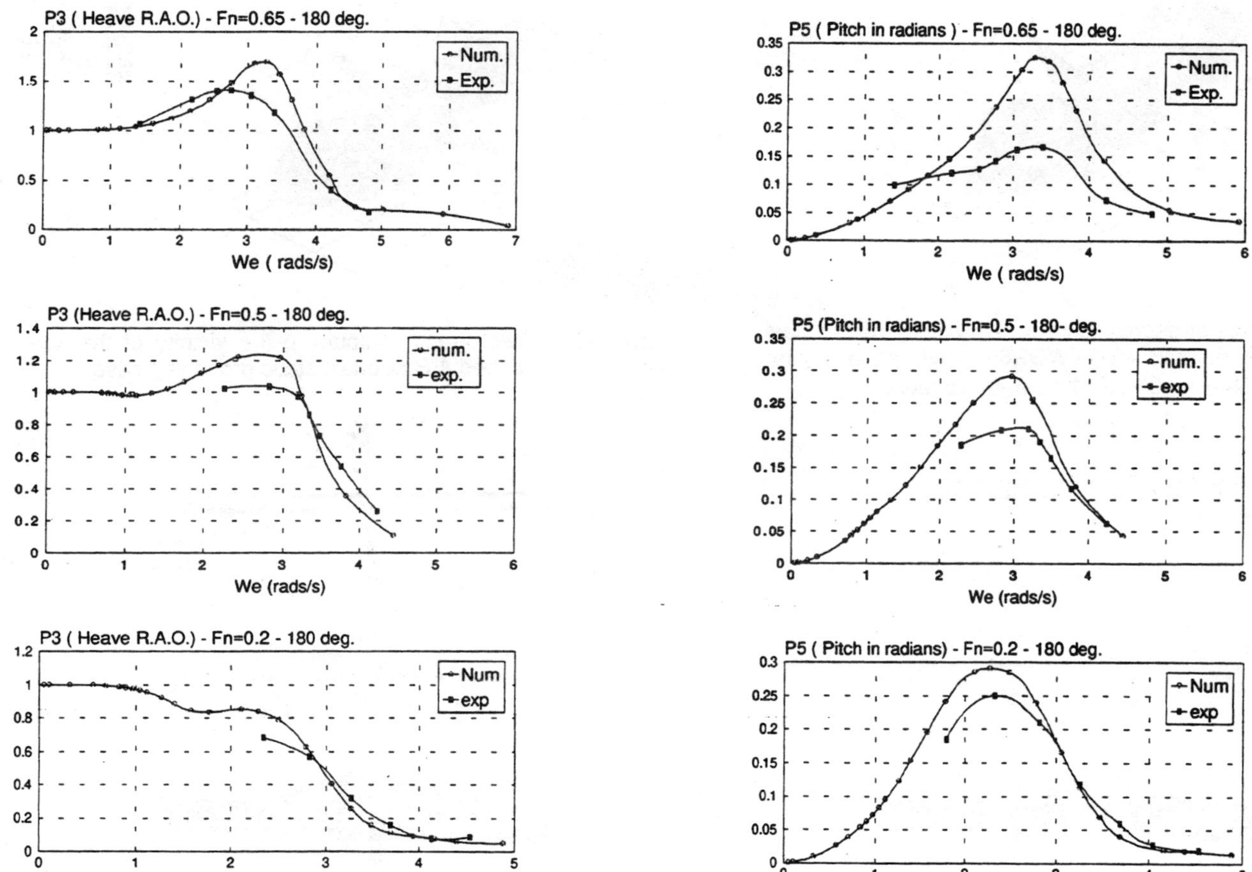

Fig.4. Rigid body motion (heave and pitch) R.A.O.s of the WOR 60 obtained from numerical and experimental analyses for various Froude Numbers in head seas.

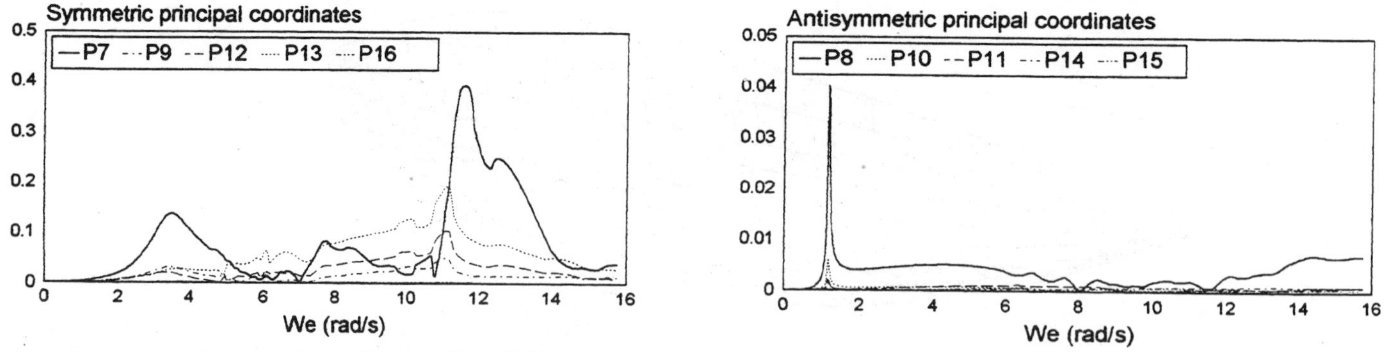

Fig.5. Amplitudes of principal coordinates (Fn = 0.65, χ = 135°, a = 1.0 m)

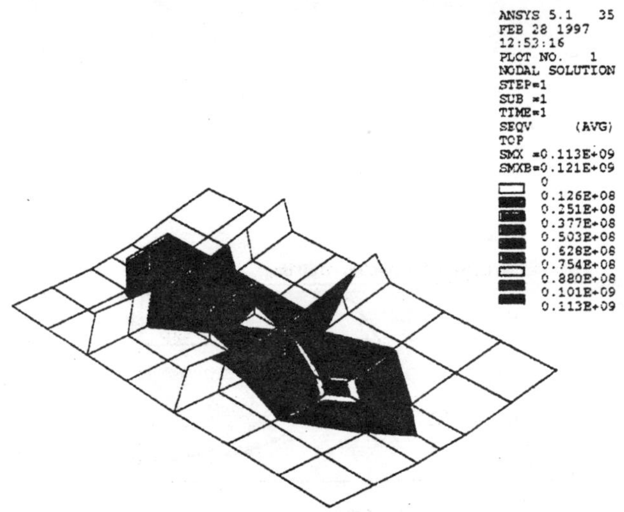

Fig.6. Von Mises stress contours in the vicinity of the keel (looking inside the hull). Model WOR6 is travelling in regular waves (Fn=0.65, χ = 135°, a =1 m, ω = 1.15 rad/s)

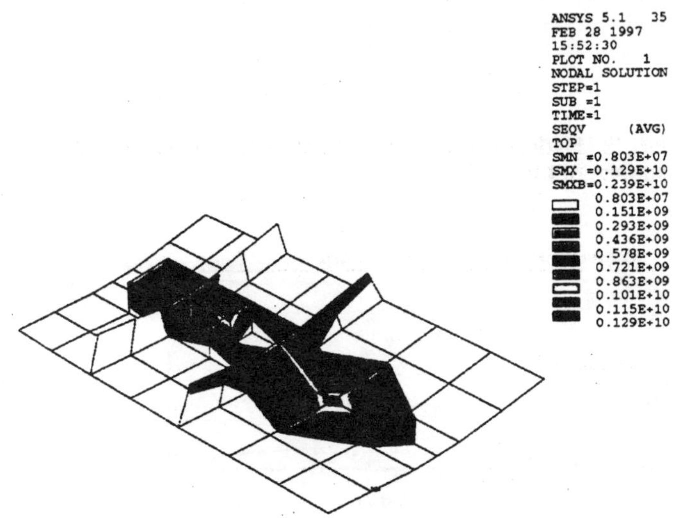

Fig.7. Von Mises stress contours in the vicinity of the keel based on prescribed A.B.S. quasi static grounding case

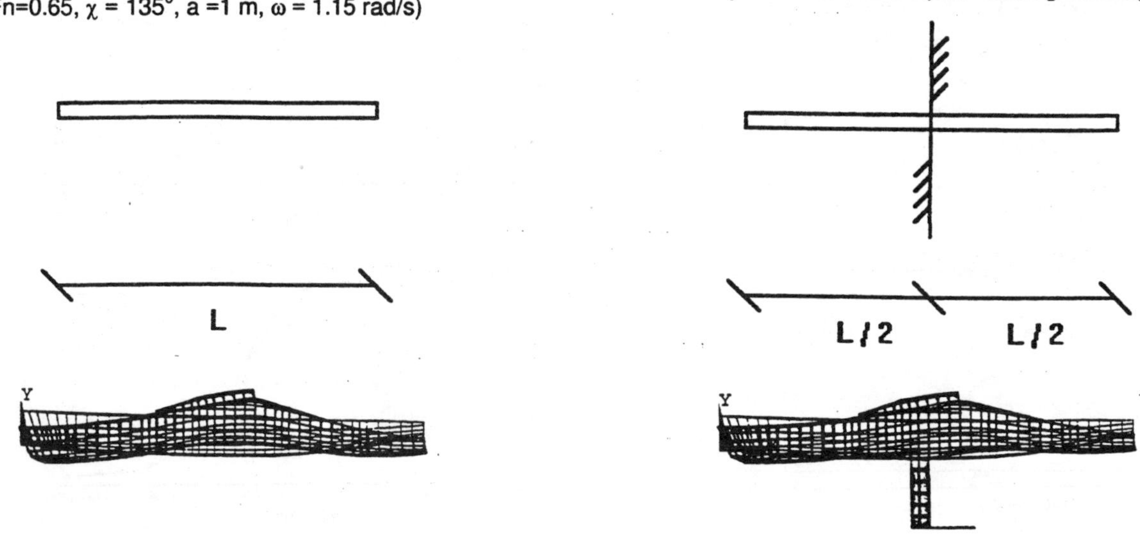

Fig.8. Simple beam analogy for comparison between models WOR2 and WOR3 for mode 12.

This stiffening effect does not affect torsional distortions to the same degree since the shear moduli of rigidity of the beams and shells are very similar, 0.34 Gpa and 0.40 Gpa respectively.

Generalised masses can , broadly, be described as scaling factors applying to mode shapes normalised for a unit maximum displacement. They are inversely proportional to the square of the maximum translation corresponding to the unit generalised mass normalisation scheme. As one would expect smaller distortions are obtained from WOR2 resulting in larger generalised masses. It must be pointed out,however, that, because the lack of longitudinals induces local distortions in WOR1, thereby corrupting generalised masses, direct comparison with WOR2 in this respect requires caution. This may explain some deviations from the trend observed in generalised masses (e.g. for mode 15). Note that despite the limitations highlighted in section II.1, regarding to connection of beam to shell elements, the expected global stiffening effect is present.

As we further upgrade from WOR2 to WOR3, new mode shapes, where keel and hull oscillate about one another, are introduced (r = 7, 8, 9 and 10), the first three of which are shown in fig.2. Their relatively low natural frequencies suggest that they are likely to be significant with regards to the response of the structure to wave loading. Again a simple beam model offers a reliable means of explaining the changes in natural frequencies for the bending distortions (e.g. r = 12). It can be shown that the addition of the keel, in the vicinity of amidships, actually modifies the boundary conditions locally as illustrated in fig. 8 for the 2 noded bending case. That is to say, a free-free beam is turned into two half cantilevers. Simple beam analogy relationships lead to an estimated end fixity of 0.75 amidships, which explains the 8 % increase in natural frequency observed in mode 12. The same approximation can also show that the three node bending (r = 13) is largely unaffected, a node being "naturally" present around amidships. The same can be said for the four node bending (r = 16). The behaviour of the generalised masses cannot entirely be explained using the same arguments, reminding us of the non beamlike character of this hull. Animation of the mode shapes clearly shows that the the hull girder flexes about the keel inducing large transverse distortions of the midship sections, at deck level in particular. As a result, despite a smaller deflection of the hull girder, the generalised masses are smaller (mode 12 and 16).The case for mode 13 is not similar as the existence of a node around amidships (WOR2) inhibits most of the orthogonal distortions (i.e. hull girder bending and sagging of deck plating) which would be incurred by the clamping effect. Antisymmetric modes (r = 11, 14 and 15) consistently show increased generalised masses (or smaller distortions) as a result of the stiffening of the structure.

The introduction of the rig, (WOR4), mainly has a marked effect on the natural frequencies associated with mode 8, (51% increase), where transverse oscillations of the largely undistorted hull relative to the keel are experienced (see fig.2). The immediate consequences of bringing the rig into the analysis are an increase in mass (10%), in transverse moment of inertia (260%) and, we believe, midship section stiffness. The mechanics by which these factors contribute to the global dynamic behaviour of this hull require further investigation. No significant difference can be depicted between the dynamic characteristics of models WOR5 and WOR6 suggesting that shroud pre-tension is not a major parameter as far as the global dynamic behaviour of the structure is concerned. Loading of the structure from the stays (i.e. models WOR4 and WOR5) affects mode 7 and 8 in terms of generalised masses. It is strongly felt that this is related to the fact that the mast compression is applied directly at the leading edge of the fin. Nevertheless, further investigations are required for this to be ascertained.

The predicted amplitudes of heave and pitch motions in regular head waves, together with experimental measurements, were shown in fig.4. The agreement between experimental and analytically predicted results becomes worse as the Froude number increases from 0.2 to 0.65. The heave and, in particular, the pitch responses become largely overpredicted by the potential flow analysis approach. The source of the discrepancy relates to the typical limitations of a linear hydrodynamic theory whereby above water flares and overhangs are not considered (Kuhn and Schlageter, 1994). In the case of the WOR 60 hull the presence of a pronounced stern overhang, undoubtedly, causes a lot of damping in pitch. As Fn increases and this yacht moves into a semi-displacement mode, she undergoes a bow-up trim which further enhances the influence of the stern overhang. The frequencies at which the motion amplitudes peak are, generally, very consistent except for the heave response at Fn=0.65. Note that the analytical predictions indicate a greater degree of coupling between heave and pitch compared to experimental measurements. On the other hand, the reliability and suitability of the experimental results carries certain elements of doubt. That is to say the relatively small size of the tank (60m x 4m x 1.5m) makes it difficult to damp out the transverse wave system accompanying the generation of the longest waves. Furthermore, the experimental model is similar but not identical to that used in the numerical analysis, (neither rig or keel were fitted but correct pitch gyradius, displacement and CG location were achieved).

The principal coordinates are plotted against encounter frequency, ω_e , in fig.5. The symmetric principal coordinates display a peak at the heave / pitch resonance, around 3.5 rad/s, due to coupling. They also have a peak around 11.5 rad/s which is believed to be the wet resonance frequency corresponding to mode 7. The antisymmetric principal coordinates have a peack at 1.2 rad/s as a result of coupling with roll.

Stresses have been calculated for the entire structure travelling at Fn = 0.65 in regular quartering waves of 1.0 m amplitude encountered at a frequency ω_e= 2.88 rad/s. The stress contours shown in fig.6 confirm an important comment made earlier on the dynamic behaviour of such a yacht, namely that of a rigid body perturbed by distortions set up at the keel/hull connection. The structure responds in a similar fashion to a quasi-static grounding case prescribed by A.B.S. which exclude mast compression effects (A.B.S., 1994). The magnitude of these stresses (see fig.7), however, are much higher, compared to the dynamic stresses in waves shown in fig.6. This is to be expected since the former case refers to "normal", even though rough, sailing conditions whereas the latter describes an extreme load case. On the other hand, the static stresses being near the U.T.S. of the material suggests a weakness of the design studied, in the keel/hull connection, in complying with the prescribed structural requirements.

Note, finally, that the use of nodal stresses or element stresses to describe the stress distribution over the hull may produce different results particularly in regions where elements with vastly differing material properties share the same nodes.

CONCLUSIONS

The three-dimensional hydroelasticity theory has been applied to investigate the dynamic behaviour of a WOR 60 yacht form in waves.

The modular nature of the FE model provided an insight into the influence of various structural components, the mast and rigging and, in particular, the keel, on the dynamic behaviour of this yacht form.

The dominance of mode 7 is confirmed qualitatively by the final predictions of flexible responses to wave excitation (fig.6 and 7). It is felt that the moment inherently present in the dry structure and created by the offset of the bulb centre of gravity off the fin axis, is responsible for the depicted behaviour. The analysis in waves has confirmed the importance of the connection of the keel to the hull in terms of dynamic wave loads. These findings were also confirmed by the application of an extreme grounding load case prescribed by A.B.S.. Nevertheless, a better modelling of this region should be studied in view of the stress magnitudes produced.

Useful conclusions have also been drawn regarding the influence of the mast and rigging on the global dynamic behaviour of this yacht form (e.g. shroud tension does not have an important effect). Nevertheless, further work is necessary in this area.

Comparisons of heave and pitch motion predictions with experimental measurements have shown good agreement at low to medium Froude Numbers.

A time domain simulation in realistic seaways, including the effects of slamming loads and responses, will provide further evidence of the effects of dynamic loads.

This application has illustrated the versatility of hydroelasticity analysis in investigating the dynamic behaviour of a fast sailing monohull.

REFERENCES

A.B.S. (1994)."Guide for Building and Classing Offshore Racing Yachts"

Ansys 5.1 (1994) manuals." Procedures", "Elements", "Theory"

Bishop, RED, Price, WG and Wu, Y (1986)." A general hydroelasticity theory of floating structures moving in a seaway". Phil. Tran. R. Soc. Lond.

Bishop, RED and Price, WG (1979). "Hydroelasticity of Ships". Cambridge University Press

Hughes, OF (1988)."Ship structural design A rationally-Based, Computer-Aided, Optimisation approach". S.N.A.M.E.

Kuhn,JC and Schlageter,EC (1994)."The effects of flare and overhangs on the motions of a yacht in head seas". 11th Chesapeake Sailing Yacht Symposium

Price, WG, Temarel, P and Keane, AJ (1994) "Hydroelastic analysis of a SWATH in waves". Proc. Int. Conf. Hydr. Mar. Tec., Trondheim

Rawson, KJ and Tupper, EC (1983)."Basic Ship Theory", vol.1. Longman Scientific

Smith, CS (1990)."Design of Marine Structures in Composite Materials". Elsevier Applied Science

Proceedings of the Seventh (1997) International Offshore and Polar Engineering Conference
Honolulu, USA, May 25-30, 1997

ISBN 1-880653-28-1 (Set); ISBN 1-880653-32-X (Vol. IV)

Prediction of Hydrodynamic Forces on High Speed Vessels

Mikio Takaki and Xin Lin
Hiroshima University
Japan

ABSTRACT

We discuss the validity of the existing theories for estimating the seakeeping qualities of high speed vessels. They are strip theory, high speed slender body theory, three-dimensional Green function method and Rankine panel method. The boundary-value problems of them are solved under the same linearization of uniform flow. Numerical results concerning the hydrodynamic forces and motion amplitudes in head sea conditions are compared with the experimental data. We have gotten the conclusion from comparative studies that, at present stage, the high speed slender body theory is the most useful and most accurate method for estimating the seakeeping qualities of a high speed boat.

KEY WORDS: high speed boat, seakeeping quality

INTRODUCTION

Recently many kinds of high speed vessels have been developed and investigated. The performances of high speed vessels are strongly affected by wave loads. Therefore reliable estimation of seakeeping qualities is one of the most important tasks with respect to the development of high speed vessels. So far strip theory has been used as the main tool for estimating the performance of a ship in waves (Tasai and Takagi (1969), Salvessen *et al.*(1970), Kobayashi *et al.*(1973)). However, its validity and limitation for high speed cases have not yet been investigated clearly.

In this paper, we apply four existing theories, the strip theory, the high speed strip theory (Chapman (1975), Saito *et al.*(1979), Yamasaki *et al.*(1985), Faltinsen *et al.*(1991), Ohkusu *et al.*(1990, 1991, 1994)), the three-dimensional Green function method (Iwashita *et al.*(1989)) and the Rankine panel/source method(Yasukawa (1990), Nakos *et al.*(1990), Takagi (1990), Lin *et al.*(1993)), to a high speed boat. We evaluate the hydrodynamic forces and the motion amplitudes in head sea condition. We have also carried out three kinds of experiment, the forced oscillation tests, measurement of wave exciting forces and ship motion amplitudes to examine the accuracy of each theory. We discuss the validity of all the methods above mentioned by comparing the calculated results with the experimental data (SHH Report (1994)). In particular, the limitation of strip theory for the high speed boat is made clear from the comparative studies. Finally, we have gotten the conclusion that, at present stage, the high speed strip theory is the most useful and the most accurate method for estimating the seakeeping qualities of the high speed boat.

THEORY

Grouping roughly the theories for estimating the seakeeping qualities of a ship in waves, we have four theories, the strip theory, the high speed slender-body theory (HSST or 2-1/2 dimensional theory), the three-dimensional Green function method and the Rankine panel/source method. These theories have the common parts to formulate the boundary conditions. First we, therefore, show the boundary conditions which the velocity potential must satisfy.

Let us consider a ship at high speed in regular incident waves on deep water. A Cartesian coordinates system *o-xyz* is used in this paper which is shown in Fig. 1. S_H represents the wetted hull surface, S_F the free surface, and $\boldsymbol{n}$ the normal vector points inward the fluid domain.

Velocity potential describing the flow around an oscillating ship with forward speed U can be expressed in the form of

$$\Phi(\boldsymbol{r};t) = U[-x + \phi_s(\boldsymbol{r})] + \Re[\phi(\boldsymbol{r})e^{i\omega_e t}], \quad (1)$$

$$\phi = \frac{g\zeta_a}{\omega_0}(\phi_0 + \phi_7) + i\omega_e \sum_{j=1}^{6} \xi_j \phi_j, \quad (2)$$

$$\phi_0 = ie^{Kz - iK(x\cos\chi + y\sin\chi)}, \quad (3)$$

where $K = \omega_0^2/g$ is the incident wave number, χ the directional angle of the incident wave, $\omega_e = \omega_0 - KU\cos\chi$ the encounter frequency, ϕ_s the steady flow velocity potential, and ϕ the unsteady wave flow velocity potential. In the unsteady component, ϕ_0 is the incident wave potential and ϕ_7 the diffraction wave potential.

The potential ϕ_j ($j = 1 \sim 7$) must satisfy the Laplace equation and the following boundary conditions;

Hull surface condition [H]:

(for radiation potential)

$$\frac{\partial \phi_j}{\partial n} = n_j + \frac{U}{i\omega_e} m_j, \quad (j = 1 \sim 6) \quad \text{on } S_H \tag{4}$$

$$\begin{aligned} (n_1, n_2, n_3) &= \boldsymbol{n}, & (m_1, m_2, m_3) &= -(\boldsymbol{n} \cdot \nabla)\boldsymbol{V}, \\ (n_4, n_5, n_6) &= \boldsymbol{r} \times \boldsymbol{n}, & (m_4, m_5, m_6) &= -(\boldsymbol{n} \cdot \nabla)(\boldsymbol{r} \times \boldsymbol{V}), \\ \boldsymbol{r} &= (x, y, z), & \boldsymbol{V} &= \nabla(-x + \phi_s), \end{aligned}$$

(for diffraction potential)

$$\frac{\partial \phi_7}{\partial n} = -\frac{\partial \phi_0}{\partial n}. \quad \text{on } S_H \tag{5}$$

Linear free surface condition [F]:

$$\left[\left(i\omega_e - U\frac{\partial}{\partial x}\right)^2 + g\frac{\partial}{\partial z}\right] \phi_j = 0, \quad \text{on } z = 0 \tag{6}$$

as well as the radiation condition [R].

Strip Theory

Strip theory is the most popular method, and there have been many studies on it. We will skip most parts of its formulation. Strip theory accounts for the effect of forward speed in a simple way. The linear free surface condition is simplified as follows;

$$\frac{\omega_e^2}{g}\phi_j - \frac{\partial \phi_j}{\partial z} = 0. \quad \text{on } z = 0 \tag{8}$$

The unsteady waves propagate towards the direction perpendicular to the ship length. This theory does not take into account the three-dimensional effect.

2.2 High Speed Slende-body Theory

In high speed slender-body theory, the velocity potential satisfies the two-dimensional Laplace equation and the hull surface condition the same as the strip theory, while on free surface it satisfies the three-dimensional free surface condition like 3-D Green function method and Rankine panel/source method. Boundary value problems that $\phi_j\,(j = 2 \sim 6)$ have to satisfy are given by;

$$[\,\mathrm{L}\,] \quad \frac{\partial^2 \phi_j}{\partial y^2} + \frac{\partial^2 \phi_j}{\partial z^2} = 0, \quad \text{in } \Omega \tag{9}$$

$$[\,\mathrm{F}\,] \quad \left(i\frac{\omega_e}{U} - \frac{\partial}{\partial x}\right)^2 \phi_j + K_0\frac{\partial \phi_j}{\partial z} = 0, \quad \text{on } z = 0 \tag{10}$$

$$[\,\mathrm{H}\,] \quad \frac{\partial \phi_j}{\partial N} = n_j + \frac{U}{i\omega_e} m_j, \quad \text{on } S_H \tag{11}$$

$$[\,\mathrm{R}\,] \quad \phi_j = \frac{\partial \phi_j}{\partial x} = 0, \quad \text{at } x = x_0 \tag{12}$$

where $K_0 = g/U^2$, $\boldsymbol{N}$ denotes the normal vector of 2D-section the same as that in strip theory.

In order to solve these boundary value problems, we introduce new potentials ψ_j;

$$\psi_j(x, y, z) = \phi_j(x, y, z) e^{-i\frac{\omega_e}{U}x} \tag{13}$$

$$[\,\mathrm{L}\,] \quad \frac{\partial^2 \psi_j}{\partial y^2} + \frac{\partial^2 \psi_j}{\partial z^2} = 0, \quad \text{in } \Omega \tag{14}$$

$$[\,\mathrm{F}\,] \quad \frac{\partial^2 \psi_j}{\partial x^2} + K_0\frac{\partial \psi_j}{\partial z} = 0, \quad \text{on } z = 0 \tag{15}$$

$$[\,\mathrm{H}\,] \quad \frac{\partial \psi_j}{\partial N} = \left(n_j + \frac{1}{i\omega_e} m_j\right) e^{-i\frac{\omega_e}{U}x}. \quad \text{on } S_H \tag{16}$$

The normal derivative term in [F] can be expressed in terms of ψ on S_F by virtue of finite difference scheme (FDS), like

$$\frac{\partial \psi}{\partial n} = -\frac{\partial \psi}{\partial z} = \kappa\left\{\psi - d(x; y, 0)\right\}, \quad \text{on } z = 0 \tag{17}$$

where $\kappa = U^2/g(\Delta x)^2$, $d(x, y, 0)$ is determined by FDS. In this paper, we have used the first order FDS and got

$$d(x; y, 0) = 2\psi_{i-1,j} - \psi_{i-2,j}. \tag{18}$$

Applying Green's second identity, we can get the boundary integral equation for determining ψ;

$$\begin{aligned} \frac{1}{2}\psi(P) = &-\iint_{S_F}\left[\kappa\left\{\psi(Q) - d(Q)\right\} - \psi(Q)\frac{\partial}{\partial n}\right] G dS \\ &-\iint_{S_H}\left[\frac{\partial \psi(Q)}{\partial n} - \psi(Q)\frac{\partial}{\partial n}\right] G dS, \end{aligned} \tag{19}$$

where

$$G = -\frac{1}{2\pi}\log R, \quad R = \sqrt{(y - \eta)^2 + (z - \zeta)^2}. \tag{20}$$

Radiation pressure can be expressed as

$$P_j(x, y, z) = (i\omega_e)^2 \rho U \xi_j e^{i\frac{\omega_e}{U}x}\frac{\partial \psi_j}{\partial x}. \tag{21}$$

and the force F_{ij} in the i–th direction due to the motion in j–th direction as follows;

$$F_{ij} = -\iint_{S_H} P_j n_i dS, \quad (i, j = 2 \sim 6). \tag{22}$$

Furthermore, the wave pressure can be expressed as follows;

$$P(x, y, z) = \frac{\rho g \zeta_a}{\omega_0}\left[U e^{i\frac{\omega_e}{U}x}\frac{\partial \psi_7}{\partial x} - i\omega_0\phi_0\right], \tag{23}$$

and we have the wave exciting forces in the i–th direction as follows;

$$E_i = -\iint_{S_H} P n_i dS, \quad (i = 2 \sim 6). \tag{24}$$

2.3 3-D Green Function Method

Applying Green's second identity to the velocity potential ϕ_j with 3-D wave Green function G, we can express velocity potential at point P on S_H as

$$\frac{1}{2}\phi_j(P) = -\iint_{S_H}\left[\frac{\partial \phi_j(Q)}{\partial n} - \phi_j(Q)\frac{\partial}{\partial n}\right] G(P, Q) dS. \tag{25}$$

Here the line integral term has been neglected, $P = (x, y, z)$ and $Q = (\xi, \eta, \zeta)$ represent the field point and source point respectively.

Solving the above integral equation by virtue of hull surface condition (4), we can determine the unstedy flow velocity potential ϕ_j, and then we can evaluate the linear unsteady hydrodynamic force F_{ij} by

$$\begin{aligned} F_{ij} &= (i\omega_e)^2 \rho \xi_j \iint_{S_H} \left(n_i - \frac{U}{i\omega_e} m_i \right) \phi_j \, dS \\ &\equiv -(i\omega_e)^2 \xi_j \left[A_{ij} + \frac{B_{ij}}{i\omega_e} \right], \quad (i,j = 1 \sim 6). \end{aligned} \tag{26}$$

The wave exciting force in the i-th direction, E_i, can be given by

$$\left(-i\frac{\omega_0}{\omega_e} \right) \frac{E_i}{\rho g \zeta_a} = \iint_{S_H} \left(n_i - \frac{U}{i\omega_e} m_i \right) (\phi_0 + \phi_7) \, dS. \tag{27}$$

2.4 Rankine Panel Method

Applying Green's second identity with free space Green function, we get the boundary integral equation of velocity potential ϕ_j

$$\begin{aligned} &\alpha(P)\phi_j(P) - \iint_{S_H + S_F} \phi_j(Q) \frac{\partial}{\partial n} G(P,Q) dS \\ &+ \iint_{S_F} \frac{\partial \phi_j}{\partial n} G(P,Q) dS = - \iint_{S_H} \frac{\partial \phi_j}{\partial n} G(P,Q) dS, \end{aligned} \tag{28}$$

where $\alpha(P)$ is the interior solid angle at point P, and $G(P,Q)$ is of

$$G(P,Q) = \frac{1}{\sqrt{(x-\xi)^2 + (y-\eta)^2 + (z-\zeta)^2}}. \tag{29}$$

Similar to HSST, the normal derivative term of ϕ_j in the integration over S_F can be expressed in terms of the potential on S_F by virtue of the free surface condition. There are many kinds of constant and high order Rankine panel/source methods. The merit of this theory is that we can apply arbitrarily any free surface condition to the above equation. Many successful applications have been reported to steady, unsteady and time domain potential wave flow problems. To make the comparative study we have, however in this paper, performed the numerical calculations by applying the linear free surface condition based on the uniform flow and constant panels with upstream FDS. Once we determine the velocity potential ϕ_j, we can evaluate the hydrodynamic force F_{ij} and the wave exciting force E_i similarly to 3-D Green function method discribed in section 2.3.

EXPERIMENT

The model ship used in the experiment is a high speed fishing boat. The principal particulars and the body plan of the ship are shown in Table 1 and Fig. 2 respectively. We have carried out three kinds of experiment, which are measuring the radiation forces, the wave exciting forces and motion amplitudes for heave and pitch. Experimental conditions are shown in Table 2.

COMPARISON AND DISCUSSION

We have focused the longitudinal motions of heave and pitch in the comparative studies. The coupling equations between heave and pitch motions can be expressed as follows;

$$\left. \begin{aligned} (M + A_{33})\ddot{z} + B_{33}\dot{z} + C_{33}z + A_{35}\ddot{\theta} + B_{35}\dot{\theta} + C_{35}\theta \\ = F_{zc}\cos\omega_e t - F_{zs}\sin\omega_e t \\ (J_{55} + A_{55})\ddot{\theta} + B_{55}\dot{\theta} + C_{55}\theta + A_{53}\ddot{z} + B_{53}\dot{z} + C_{53}z \\ = M_{\theta c}\cos\omega_e t - M_{\theta s}\sin\omega_e t \end{aligned} \right\}. \tag{30}$$

4.1 Strip Method

Fig. 3 shows the comparisons of the motion amplitudes between the experimental results (marks of black circle) and the numerical ones (solid lines) due to strip theory at $Fn = 0.4$, 0.6 and 0.8. The calculated results at $Fn = 0.4$ are in a good agreement with the experimental ones. The discrepancies between both results, however, become larger as the ship speed increases. On the other hand, the marks of triangle, which are estimated by using the experimental results of added mass and damping coefficients in motion equations (30), agree well with the measured ship motions. It indicates that the above discrepancies are caused by the bad estimation of hydrodynamic coefficients. From these figures, we can say that the strip theory is valid for estimations of seakeeping qualities at ship speed less than $Fn = 0.4$ at most. In the next section, We will, therefore, carry out the comparative studies among the theories to estimate the seakeeping qualities at ship speed of $Fn = 0.6$ where strip theory fails to estimate accurately.

4.2 Other Theories

We have also performed the numerical calculations by applying the other three theories to examine their merits and demerits. They are the high-speed slender-body theory (HSST), 3-D Green function method and Rankine panel method. These calculations have been performed under the same assumptions of the uniform flow and the linear free surface condition.

The comparisons of the numerical results and experimental ones for the diagonal terms and coupling terms of hydrodynamic coefficients, the wave exciting forces and the motion amplitudes are shown in Fig. 4 to Fig. 7 respectively. HSST has the best agreement with the experimental results for both the radiation force coefficients and the wave exciting forces among the three theories. HSST, therefore, estimates well the motion amplitudes.

From theoretical point of view, 3-D Green function method is more rational than HSST. However, the estimation accuracy of the hydrodynamic forces is not as good as that of HSST. In particular, the estimation accuracy becomes worse as the oscillating frequency becomes higher, as shown in Fig. 4. These numerical results are estimated with 292 panels on half wetted body. Even if the panel number increased to 1239, the improvement of accuracy is very small. It seems that these discrepancies are due to other factors, such as the neglect of the line integral term. Further studies should be carried out to make clear these problems from now on.

In the comparison, Rankine panel method fails to get the reasonable prediction. In particular, the resultant damping coefficients B_{33} and B_{55} are only half of those estimated by HSST or

3-D Green function method. The resultant motion amplitudes become much larger than that of other theories around the wave length ratio λ/L of 2.2, as shown in Fig. 7. We think that this error may mainly originate to the neglect of the effect of Kelvin wave flow. The high-speed fishing boat used in this study has a transom stern. Wave surface will leave the bottom tangentially there. The draft at the stern is of order $O(d)$. While in our calculation, the constant free surface panels are located on the undisturbed free surface. It seems that this may cause large error. This method also should be improved to get a reasonable prediction of seakeeping performance of the ship with a transom stern.

Generally speaking, we should involve the effect of steady wave flow in the estimation of unsteady wave problem in high speed cases. Rankine panel/source method is the most hopeful theory to do that. Application of high order boundary element in Rankine panel method will enable us to treat even nonlinear steady wave flow, and provides a much reliable fundation to the prediction of seakeeping performance of high speed ships. Our group of Hiroshima University is now working in this direction. Fig. 8, at the low right corner of this page, shows an example of the application of 4-th order B-spline Rankine panel method in steady wave problem which is linearized on double-body flow. We hope that we can show some results in near future.

CONCLUDING REMARKS

The main conclusions obtained in this study can be summarized as follows:

(1) Strip theory is valid for ship speed less than $Fn = 0.4$, and the estimating accuracy will become worse when ship speed is higher than $Fn = 0.4$.

(2) High speed slender-body theory is the most useful method for estimation of seakeeping qualities of high speed vessels at present stage.

(3) 3D Green function method has some problems such as the line integral, free surface condition, the number of discretized panels, and consuming CPU time that should be improvedin future.

(4) Rankine panel/source method should be the most hopeful theory. However, the involvement of the effect of Kelvin wave flow is indispensable especially for a high speed ship with transom stern.

REFERENCES

Chapman, R. B., 1975, "Free-Surface Effects for Yawed Surface-Piercing Plates", Journal of Ship Research, Vol. 20, No. 3, pp. 125–136.

Faltinsen, O. and Zhao, R., 1991, "Numerical predictions of ship motions at high forward speed", Phi. Trans R. Soc. Lond., No. 334, pp. 241–252.

Iwashita, H. and Ohkusu, M., 1989, "Hydrodyna-mic Forces on a Ship Moving with Forward Speed in Waves", JSNAJ, Vol. 166, pp. 187–216.

Kobayashi, M. *et al.*, 1973, "A Computer Program for Theoretical Calculation of Sea-keeping Quality of Ships (Part 1 Method of Theoretical Calculation)", MITSUI ZOSEN TECHNICAL REVIEW, No. 82, pp. 1–34.

Lin, X., *et al.*, 1993, "Effect of the Steady Disturbance on the Unsteady Flow around a Ship in Waves", JSNAJ, Vol. 174, pp. 151–162.

Nakos, D. and Sclavounos, P., 1990, " Ship Motions by Three-Dimensional Rankine Panel Method", Proc. of 18th Symposium on Naval Hydrodynamics, Michigan, pp. 21–40.

Ohkusu, M. and Faltinsen, O., 1990, "Prediction of Radiation Forces on a Catamaran at high Froude Number", Proc. of 18th Symposium on Naval Hydrodynamics, Michigan, pp. 5–20.

Ohkusu, M., *et al.*, 1991, "Prediction of Hydrodynamic Forces on High-Speed Catamarans in a seaway ", JSNAJ, Vol 170, pp 153–161.

Ohkusu, M. and Wen, G., 1994, "PREDICTION OF SHIP MOTION OF HIGH-SPEED SHIP IN WAVES", Rept of Study on Seakeeping Quality of High-Speed Vessels, pp. 80–95.

Saito, K. and Takagi, M., 1978, "A Solution for the Oscillating Slender-Body with Forward-Speed", JSNAJ, Vol. 144, pp. 96–105.

Salvesen, N., Tuck, E. O. and Faltinsen, O. M., 1970, "Ship Motions and Sea Load", TSNAME, Vol. 78.

Takagi, K., 1990, "Calculation of Unsteady Pressure by Rankine Source Method", J. Kansai Soc. N.A., No. 219, pp. 47–56.

Takaki, M. and Iwashita, H., 1994, "On the Estimation Methods of the Seakeeping Qualities for the High Speed Vessel in Waves", 11th Marine Dynamic Symposium, SNAJ, pp. 389–432.

Tasai, F. and Takagi, M., 1969, "Theoretical Prediction of Motion Response and Calculating Method in Regular Waves", 1st Symposium on Seakeeping Qualities, SNAJ, pp. 1–52.

Yasukawa, H., 1990, "A Rankine Panel Method to Calculate Unsteady Ship Hydrodynamic Forces", JSNAJ, Vol. 168, pp. 131–140.

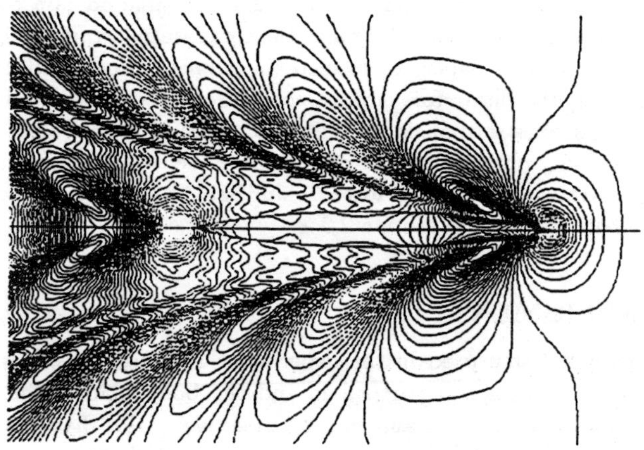

Fig. 8 Contours of Kelvin wave potential around a modified Wigley hull at $Fn = 0.3$

Table 1 Principal particulars of high-speed fishing boat

Overall Length	L_{all} (m)	2.543
Length bt Perpendiculars	L_{pp} (m)	1.840
Breadth	B (m)	0.428
Depth	D (m)	0.158
Draught at AP	d_A (m)	0.118
Draught at Midship	d_M (m)	0.088
Draught at FP	d_F (m)	0.059
Displacement	Δ (kg)	52.3
Position of C.G.	x_G (m)	0.163
Radius of Gyration	K_l/L_{pp}	0.275

Table 2 Experimental Cconditions

	Radiation Force	Wave Exciting Force	Motion Amplitude
Ship Speed	Fn=0.4, 0.6	Fn=0.4, 0.6	Fn=0.4, 0.6, 0.8
Wave Direction		Head Sea Cond.	Head Sea Cond.
Wave Height		$H/\lambda = 1/50$	4.5 (cm)
Wave Length		$\lambda/L = 0.5 - 4.0$	$\lambda/L = 1.0 - 4.0$
F.O. Amplitude	$z_a = 2$ (cm)		
F.O. Frequency	$KL = 1.0 - 25$		

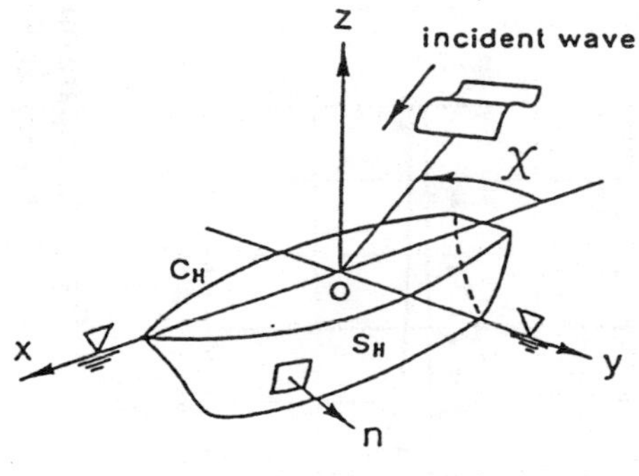

Fig. 1 Coordinate system

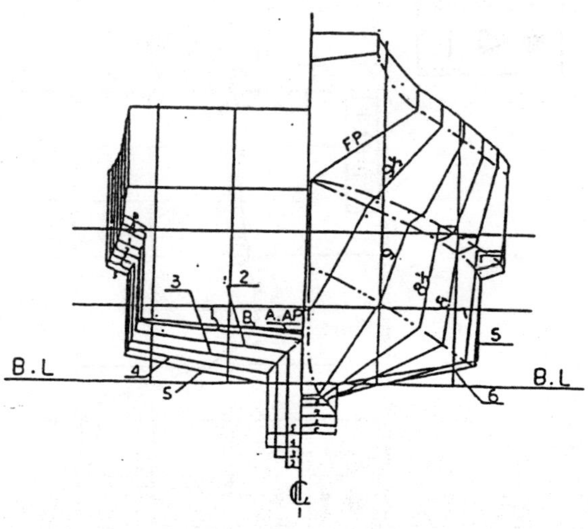

Fig. 2 Body plan of high-speed fishing boat

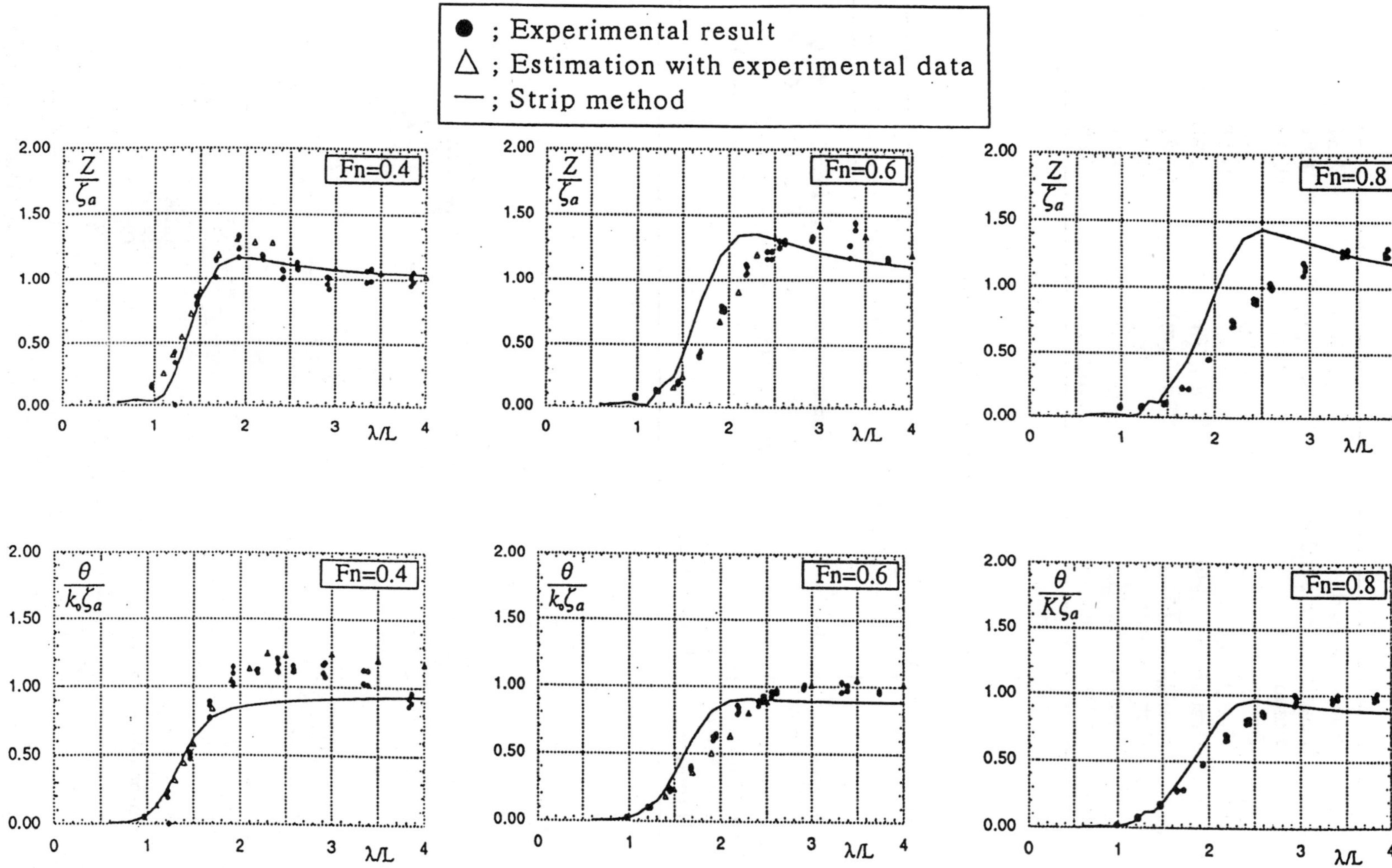

Fig. 3 Comparison between experimental results and numerical ones due to strip theory at Fn = 0.4, 0.6, 0.8

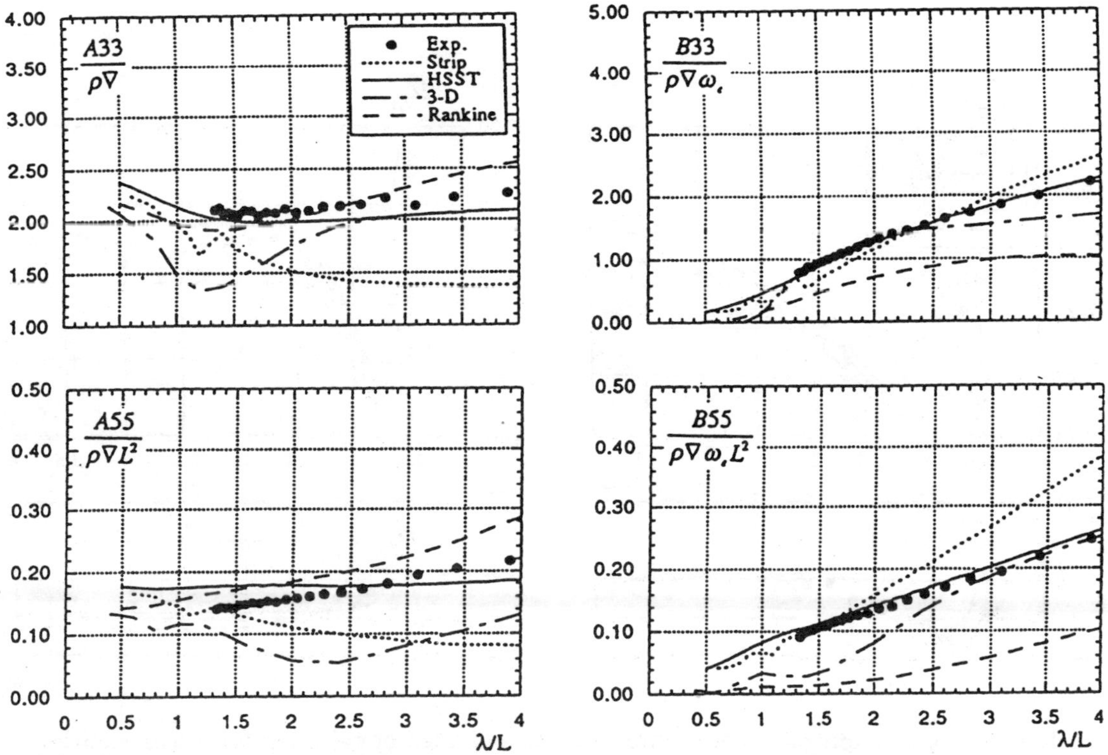

Fig. 4 Comparison of experimental results and theoretical ones for main terms at Fn = 0.6

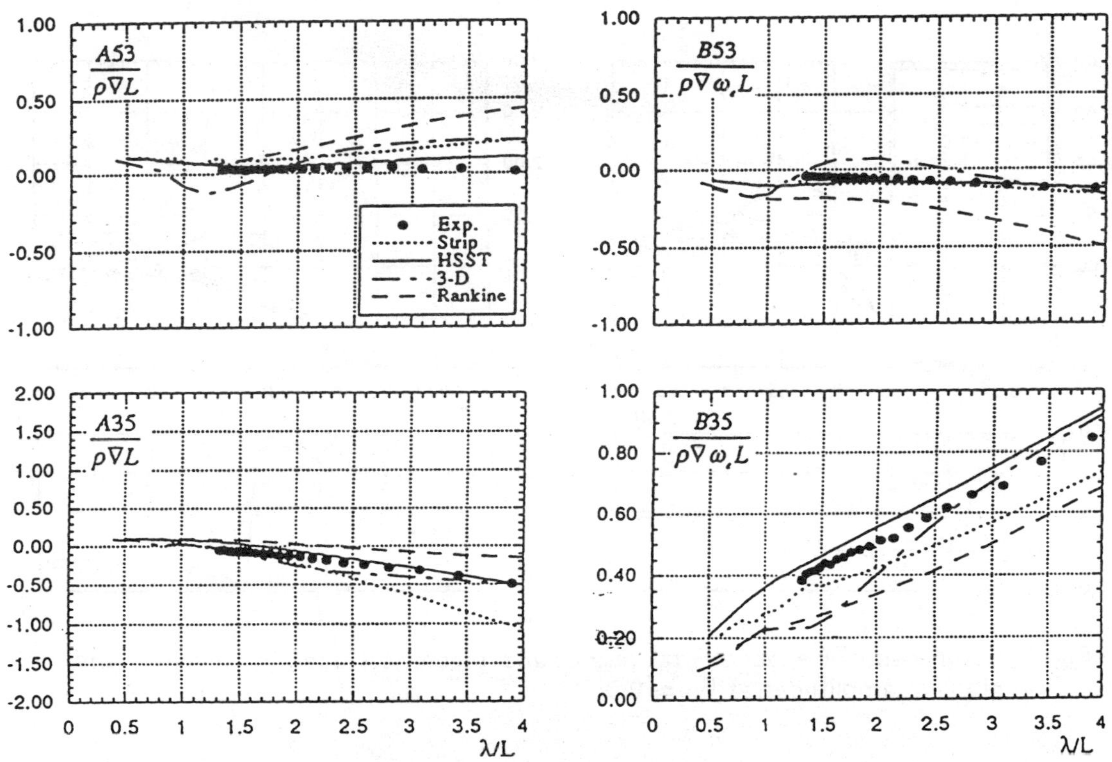

Fig. 5 Comparison of experimental results and theoretical ones for coupling terms at Fn = 0.6

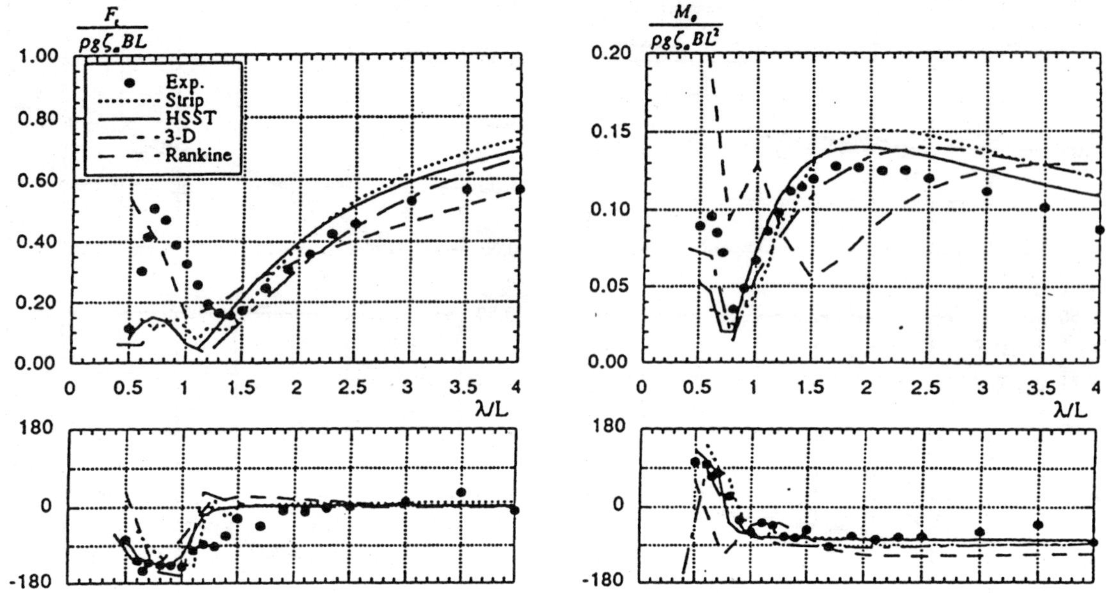

Fig. 6 Comparison of experimental results and theoretical ones for wave exciting forces and moments at Fn = 0.6

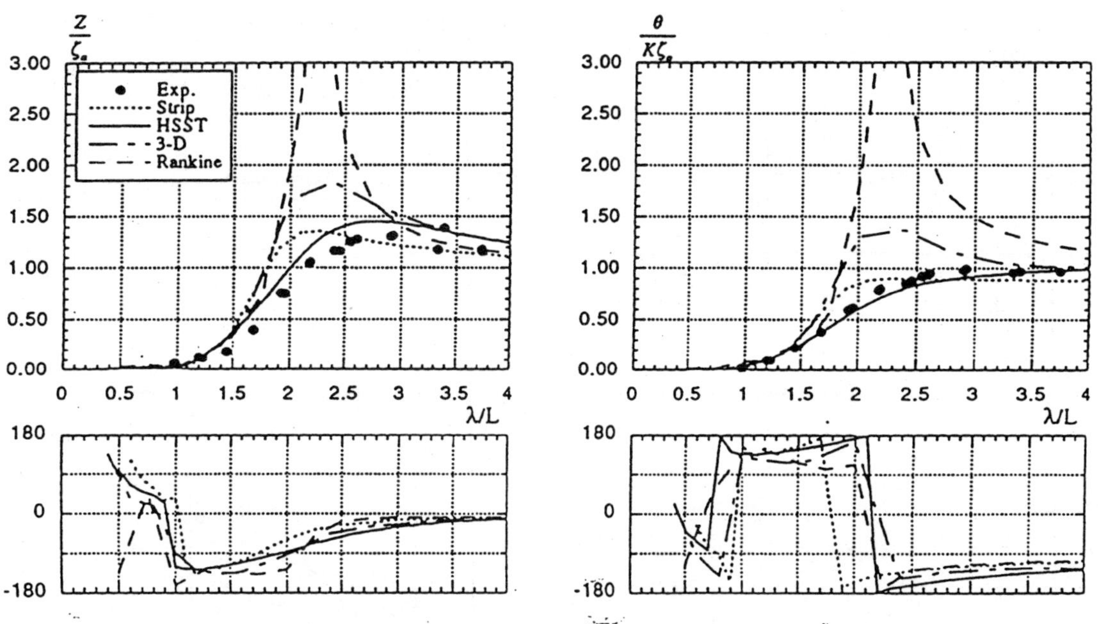

Fig. 7 Comparison of experimental results and theoretical ones for heaving and pitching amplitudes at Fn = 0.6

Proceedings of the Seventh (1997) International Offshore and Polar Engineering Conference
Honolulu, USA, May 25-30, 1997

ISBN 1-880653-28-1 (Set); ISBN 1-880653-32-X (Vol. IV)

Prediction of the Effects of Control Surfaces on the Motion of a Catamaran

Woo-Joon Yoo, Jae-Hong Park, Tae-Young Lee and Il-Geun Oh
Samsung Heavy Industries Co., Ltd.
Daejon, Korea

ABSTRACT

A car-ferry catamaran of 80m class has a large demand in the global market of high-speed ships and SHI(Samsung Heavy Industries) has recently developed its own design. The seakeeping characteristics or ride quality is an important factor to consider necessarily in designing such vessels. A high-speed ship is inevitably accompanied by the excessive motion which causes passengers to experience extremely unpleasant discomfort, such as seasickness, nausea, and vomit. The present paper addresses an effort to improve the seakeeping property by developing the active ride control technology using control surfaces. An adequate mathematical model describing the motion of a vessel subjected to wave excitations is established and used to analyze the dynamic property of the catamaran. Heave and pitch motions are only considered in this paper. Then the effects of control surfaces are investigated by comparing the controlled responses with the uncontrolled ones. The fins or foils demonstrate the capability of reducing the motion significantly.

KEY WORDS : ride quality, high-speed catamaran, active ride control system, seakeeping, seasickness

INTRODUCTION

The worldwide market of car-ferry is growing rapidly. Its tendency of demand is characterized by increasing the speed up to 50 knots and enlarging the size up to 100m. The high speed is achieved by designing a light-weight ship. Designing such a large-size high-speed vessel requires an accurate understanding of wave loads imposed on the vessel in waves.

The operation of a high-speed catamaran in waves produces greater emphasis on the aspects of vessel motions and resultant structural loads. This is due to the larger encounter frequencies arising from the higher speed, with attendant increase in accelerations, as well as the need to reduce motion magnitudes relative to the free surface.

A car-ferry catamaran of 80m class was recently developed by SHI in its own design. The design particulars of the developed catamaran are listed in Table 1 and its profile view is shown in Figure 1.

Table 1. Design Particulars of the Car-Ferry Catamaran

Length overall (m)	81.3
Length at design WL (m)	77.8
Beam overall (m)	19.5
Draft at full displacement (m)	2.9
Service speed at 90% MCR	37.5
No. of passengers (person)	600
No. of vehicles	50 Cars + 4 Buses
Propulsor	Waterjet × 4 sets

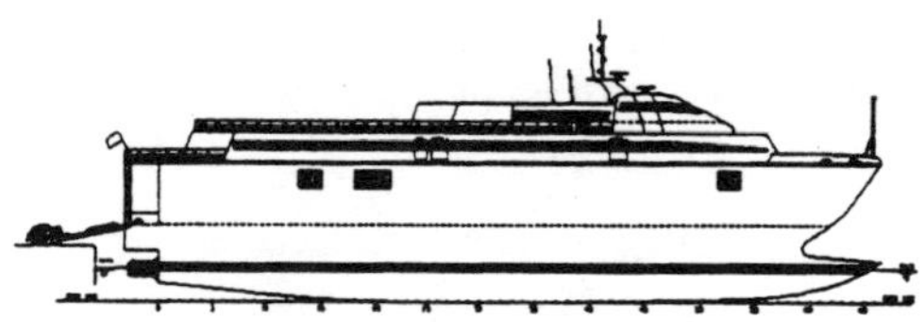

Figure 1. The Profile View

Since the seakeeping characteristics or ride quality is an important factor to consider necessarily in designing such a vessel, it is absolutely required to develop a method for determining and reducing motions and loads. The need for increased habitability requires consideration of adequate types of Ride Control System (RCS). For the newly-developed catamaran case the primary RCS is in the form of lifting surfaces, such as foils and trim tabs.

The present paper describes frequency and time domain models for determining motions of newly-developed high-speed catamaran in waves. It also describes the modeling procedure of motion controls and demonstrates its effects. The general basis for use of such time domain treatments are described in the following sections of this paper.

ESTIMATION OF SEAKEEPING PERFORMANCE

Equations of Motion

In order to calculate and predict the motion of a high-speed catamaran in waves, the six degrees of freedom should be considered. However, in this paper the longitudinal equations of motion which are primary motions of a catamaran are only analysed and the equation for the translational mode of x-axis is neglected. The advancing speed V is assumed constant even if the draft changes significantly. Then the equations of motion are formulated relative to an axis system whose origin is located at the vessel CG.

Figure 2 shows the coordinate system.

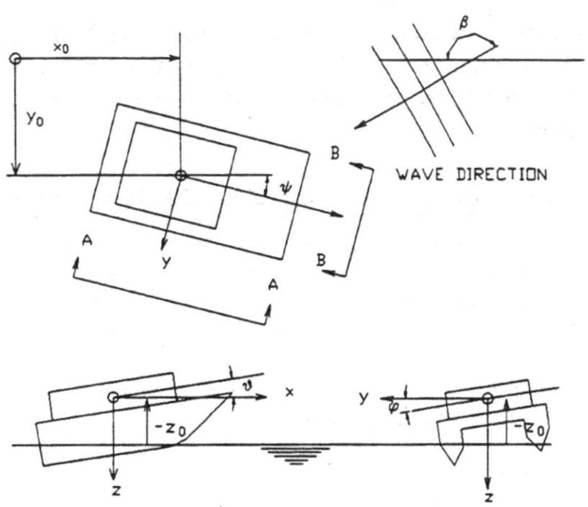

Figure 2. The Coordinate System

The heading angle β is defined as 180° for head sea and 0° for following sea. The incident waves on the free surface are assumed as unidirectional sinusoidal waves. The surface wave elevation relative to the calm water equilibrium plane is represented by the expression

$$\eta = A\sin\left[\omega_e t - k\left(x\cos\beta + y\sin\beta\right)\right] \quad (1)$$

where

A = wave amplitude

k = wave number $= 2\pi / \lambda = \omega^2 / g$

λ = wave length

ω = circular wave frequency

ω_e = encounter wave frequency $= \omega - kV\cos\beta$ (2)

The equations of motion are simplified for engineering efficiency. Then the linearized equations of motion for vertical motions are as follows (Bhattacharyya,1978; Lewis,1987).

$$Z_{\ddot{z}}\ddot{z} - Z_{\dot{z}}\dot{z} - Z_z z - Z_{\ddot{\theta}}\ddot{\theta} - Z_{\dot{\theta}}\dot{\theta} - Z_\theta \theta = Z_{waves} \quad (3)$$

$$M_{\ddot{\theta}}\ddot{\theta} - M_{\dot{\theta}}\dot{\theta} - M_\theta \theta - M_{\ddot{z}}\ddot{z} - M_{\dot{z}}\dot{z} - M_z z = M_{waves} \quad (4)$$

where $Z_{\ddot{z}}$ is the sum of mass and added mass. $Z_{\dot{z}}$ is damping coefficient and Z_z is restoring force coefficient which depends on the waterline area. Other coefficients are defined similarly.

Methods of Calculation

A suitable wave spectrum is first chosen for the particular seaway in which the vessel is to operate. In this study an irregular wave system of 2.5 m significant wave height using ITTC spectrum shown in Figure 3 is considered to estimate the seakeeping performance. ITTC spectrum is defined as follows (Lewis,1987).

$$S(\omega) = \frac{A}{\omega^5}\exp\left(\frac{-B}{\omega^4}\right) \quad (5)$$

$$A = \frac{173H_{1/3}^2}{T_1^4}, B = \frac{691}{T_1^4} \quad (6)$$

where T_1 is averaged period and $H_{1/3}$ is significant height of waves. As illustrated in Figure 3, the wave energy is concentrated in the frequency range between 0.6 rad/sec and 0.9 rad/sec. This implies the possibility of an excessive magnitude of motion when the heave RAO (Response Amplitude Operator) function lies in the condition of resonance.

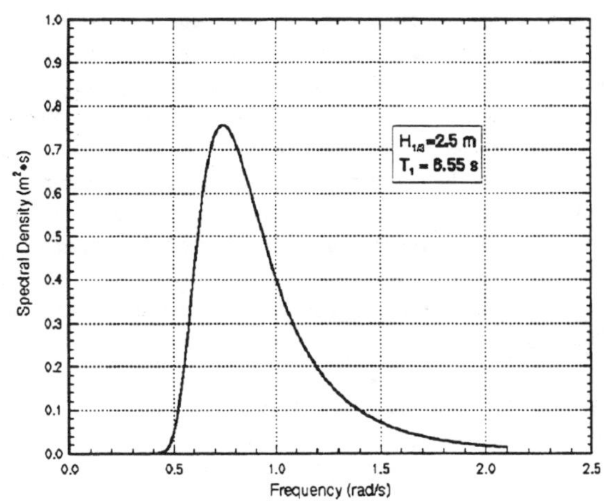

Figure 3. The ITTC Wave Spectrum

Next a plot is obtained in which the ordinates represent the amplitude of motion. In this study it was obtained analytically in regular waves and extended to irregular waves (Salvensen,et al.,1970). To get the amplitude of motion analytically the external force due to waves and the hydrodynamic coefficients (added masses and damping coefficients corresponding to the encounter frequencies) are required. The diagram of the amplitude of motion is modified so that the ordinate represents the ratio or the square of the motion amplitude divided by the square of the wave amplitude. This diagram is termed the response amplitude operators (RAO) or simply the transform spectrum. The RAO diagram is transformed into a base of absolute wave frequency to be integrated with the wave spectrum. Figure 4 shows such a diagram as obtained in the present study.

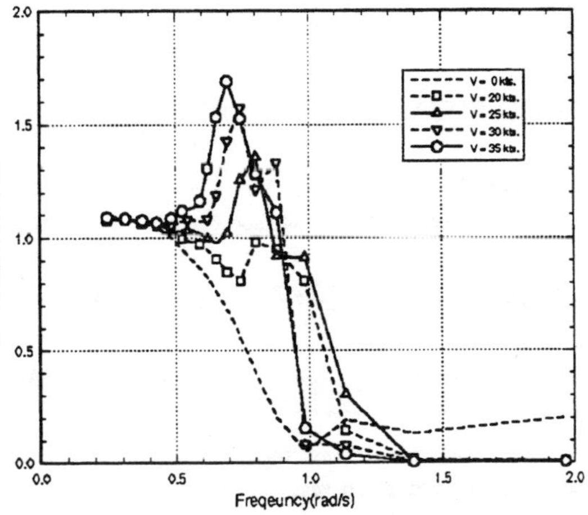

Figure 4. Heave RAO

By integration of multiplying the ordinates of wave spectrum by the ordinates of RAO for the corresponding wave frequencies, the motion in irregular sea is obtained. This is formulated by the following equation describing the RMS (Root Mean Squared) magnitudes of heave (Bhattacharyya,1978; Lewis, 1987).

$$\sigma_z = \sqrt{\int_0^\infty |T_z(\omega)|^2 S(\omega) d\omega} \tag{7}$$

where σ_z is RMS value of heave and $T_z(\omega)$ is heave RAO. For vertical acceleration the encounter frequency should be multiplied as follows

$$\sigma_{\ddot{z}} = \sqrt{\int_0^\infty |T_z(\omega)|^2 S(\omega) \omega_e^4 d\omega} \tag{8}$$

where $\sigma_{\ddot{z}}$ is RMS value of vertical acceleration.

Calculation Results and Discussion

It is widely known that the seakeeping performance becomes worse as the speed of a vessel increases(Faltinsen,1993). Although this is due to various complicated reasons, the most important two factors can be pointed out as follows. First, the increase of vertical acceleration, which is one of the major measurement of seakeeping performance, is proportional to the square of encounter frequency as equation (8). Thus in head sea condition the encounter period decreases or the encounter frequency increases as vessel speed increases. Secondly, the frequency range of the peak magnitude of Heave RAO moves closer to the frequency range of concentrated energy.

Figure 5 shows the vertical acceleration with varying the ship speed for the cases of head sea, beam sea, and following sea. In the case of head sea the calculated result is compared to the experimental ones. The tests were performed at KRISO (Korea Research Institute of Ships and Ocean Engineering) for the catamaran (Shin, 1995). They show considerably good agreement.

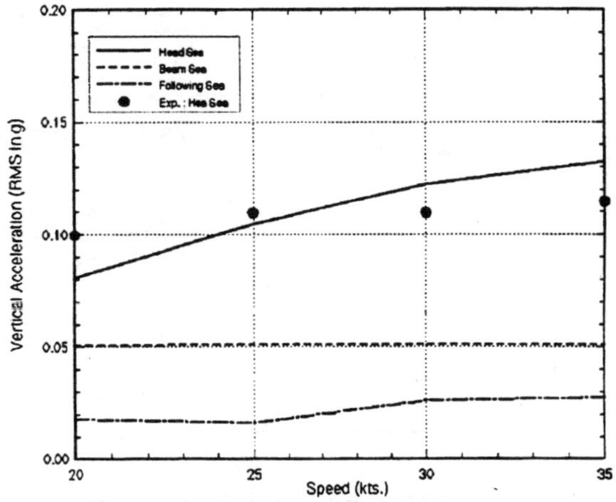

Figure 5. Vertical Acceleration at Midship with Varying Ship Speed

It is also found from Figure 5 that the increase of vertical acceleration with ship speed is not so large as the case of conventional low-speed ships. In the case of high-speed ships, the vertical acceleration does not increase noticeably but even decreases sometimes as the ship speed increases (Shin,1995). This seems to be a contradictory result to the previous explanation of the present section. However, a close inspection of Figure 3, wave spectrum, and Figure 4, heave RAO, unveils the reason. When the ship speed is 25 knots, the resonance region of heave RAO of the catamaran is found in Figure 4 around the frequency range between 0.6 rad/sec and 0.9 rad/sec. This frequency range corresponds very closely to the frequency range of peak magnitude of wave energy in Figure 3 of wave spectrum. So if the speed is increased further or greater than 2 5 knots, the resonance region of heave RAO moves to the even lower frequency side, which is now beside the frequency range of peak wave energy. Then this matching condition of heave RAO and wave spectrum result in a decrease of heave.

A simple method to reduce the ship motion is to decrease the speed of vessel (Faltinsen,1993). It may still be a good method for much higher waves than 2.5m (Shin,1995). However, it cannot be a good solution when the significant wave height is around 2.5 m, at which severe motion response occurs most frequently for high-speed ships. Therefore, to produce a high-class ship with good seakeeping performance, the active ride control system is not an option but a necessity.

Vertical acceleration is a measure of seakeeping performance and is estimated by an analysis in the frequency domain. ISO requires the results of 1/3 octave band analysis of vertical acceleration and proposes a standard by specifying the articles 2631/3.

Figure 6 shows the 1/3 octave band data of RMS vertical acceleration recommended by ISO 2631/3 seasickness criteria for 30 minutes and 2 hours. It also shows the RMS vertical accelerations at midship obtained by calculations and experiments for ship speeds 25 knots and 35 knots at head sea condition. Circles are used to show the experimental results clearly (Shin,1995). The calculation results and experimental results show good agreement qualitatively except that calculation yields larger peak magnitudes and lower corresponding frequencies than experiments.

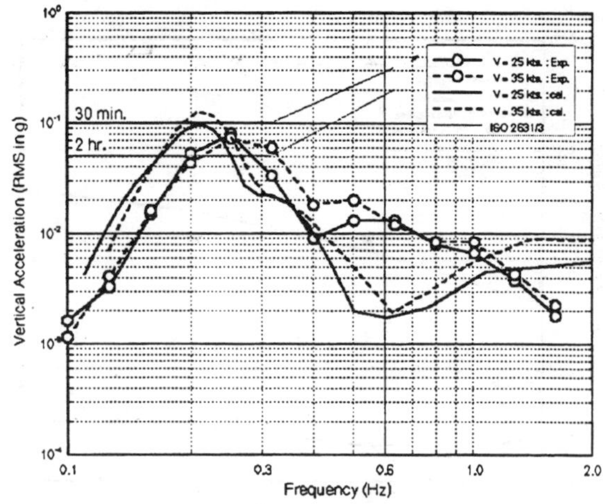

Figure 6 1/3 Octave Band Data of Vertical Acceleration at Midship

Assessment of seakeeping performance with the present method reaches a conclusion that the newly-developed catamaran does not possesses a very good seakeeping capability for a long voyage.

Figure 7 illustrates the RMS values of vertical acceleration in the unit of g (gravity) with varying the heading angle as a parameter.

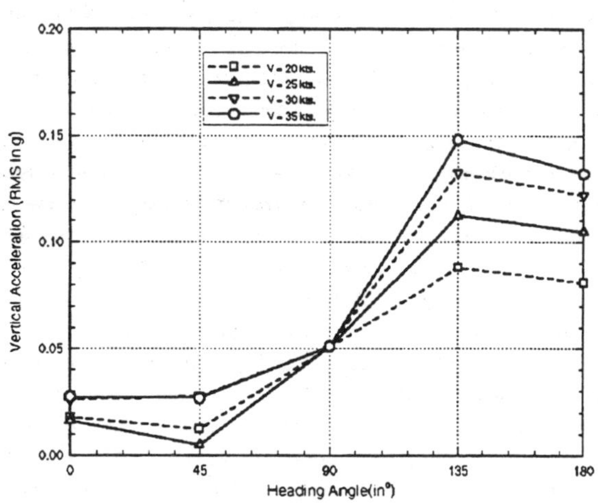

Figure 7. Vertical Acceleration at Midship with Varying Heading Angle

It shows that the largest vertical acceleration occurs in the case of bow sea and the smallest in the case of quartering sea. This is again explained by the relation between speed and encounter frequency, and the relation between heave RAO and wave spectrum.

According to the last relation of equation (2), the encounter frequency ω_e for following sea decreases as ship speed V increases. The encounter frequency of following sea is lower than that of head sea after all, and lower than the frequency range of concentrated wave energy. Since the vertical acceleration is proportional to the square of encounter frequency, it is smaller for following sea than head sea.

When large head waves are upcoming, the ship is steered to avoid head sea condition resulting in possibly bow sea condition. Figure 7, however, shows that the vertical acceleration for bow sea is even larger than that for head sea. This means the change of heading angle for improving bad ride condition in head sea does not guarantee an effective result or drives the ship into an even worse situation. Therefore, an active Ride Control System is absolutely required in order to secure a good ride quality.

MODIFIED EQUATIONS OF MOTION

The equations of motion for heave and pitch of a catamaran, including the effect of lifting surface controls, are given by

$$(m + A_{33})\ddot{z} - Z_{\dot{z}}\dot{z} - Z_z z - Z_{\ddot{\theta}}\ddot{\theta} - Z_{\dot{\theta}}\dot{\theta} - Z_\theta \theta = Z_w + Z_c \tag{9}$$

$$(I_y + A_{55})\ddot{\theta} - M_{\dot{\theta}}\dot{\theta} - M_\theta \theta - M_{\ddot{z}}\ddot{z} - M_{\dot{z}}\dot{z} - M_z z = M_w + M_c \tag{10}$$

where Z_c and M_c are lifting force and moment due to control surface in heave and pitch respectively.

The vertical control force Z_c, acting on a lifting-surface fin and trim tab is given by

$$Z_c = Z_{fc} + Z_{tc} \tag{11}$$

where Z_{fc} is the lift force due to the forward fin and Z_{tc} is that due to the trim tab located at transom. It is assumed that the sizes and angles of forward-fin and trim-tab system are the same for both port and starboard to control the total vertical force acting on the newly-developed catamaran. The expressions of vertical forces for the forward fin are given by

$$Z_{fc} = -C_{L_a}(h_{fc})\rho S_{fc} V^2 \delta_{fc} \tag{12}$$

$$Z_{tc} = -C_{L_a}(h_{tc})\rho S_{tc} V^2 \delta_{tc} \tag{13}$$

where $C_{L_a}(h_{fc})$ and $C_{L_a}(h_{tc})$ are the lift coefficient obtained from following equations (14) and (15), S_{fc} and S_{tc} are areas of control surfaces, and δ_{fc} and δ_{tc} are the commanded deflection angles of forward fin and trim tab respectively, ρ is water density, and V is ship's speed (Kaplan,1988).

The lift coefficient for the forward fin is given by

$$C_{L_a}(h_{fc}) = \frac{2\pi A}{2 + (A^2 + 4)^{1/2}} \cdot \frac{h_{fc}^2 + \frac{c^2}{16}}{h_{fc}^2 + \frac{c^2}{8}} \tag{14}$$

where A is the aspect ratio, h_{fc} is the fin's submergence depth below the free surface, and c is the fin's chord. The lift coefficient for a trim tab is given by

$$C_{L_a}(h_{tc}) = 0.046 \tag{15}$$

and is found in Savitsky et al.(1976). The resulting pitch moment due to lift forces is then given by

$$M_c = -x_{fc} Z_{fc} - x_{tc} Z_{tc} \tag{16}$$

where x_{fc} is the distance from the CG of a ship to the location of forward fin and x_{tc} is that of trim tab.

CONTROL SYSTEM DESIGN

Forward fins and trim tabs are used for controlling the coupled modes of heave and pitch given by equations (9) and (10). The equations of motion including the effects of control surfaces are represented by state equations as follows :

$$\begin{bmatrix} Z_{\ddot z} & -Z_{\dot z} & -Z_{\ddot\theta} & -Z_{\dot\theta} \\ 0 & 1 & 0 & 0 \\ -M_{\ddot z} & -M_{\dot z} & M_{\ddot\theta} & -M_{\dot\theta} \\ 0 & 0 & 0 & 1 \end{bmatrix}\begin{bmatrix} \ddot z \\ \dot z \\ \ddot\theta \\ \dot\theta \end{bmatrix} =$$

$$\begin{bmatrix} 0 & Z_z & 0 & Z_\theta \\ 1 & 0 & 0 & 0 \\ 0 & M_z & 0 & M_\theta \\ 0 & 0 & 1 & 0 \end{bmatrix}\begin{bmatrix} \dot z \\ z \\ \dot\theta \\ \theta \end{bmatrix} + \begin{bmatrix} Z_w \\ 0 \\ M_w \\ 0 \end{bmatrix} +$$

$$\begin{bmatrix} -C_{L_\alpha}(h_{fc})\rho S_{fc}V^2 & -C_{L_\alpha}(h_{tc})\rho S_{tc}V^2 \\ 0 & 0 \\ x_{fc}C_{L_\alpha}(h_{fc})\rho S_{fc}V^2 & x_{tc}C_{L_\alpha}(h_{tc})\rho S_{tc}V^2 \\ 0 & 0 \end{bmatrix}\begin{bmatrix} \delta_{fc} \\ \delta_{tc} \end{bmatrix} \quad (17)$$

or

$$\dot x = Ax + Dw + Bu$$
$$y = Cx \quad (18)$$

where $x = [\dot z \;\; z \;\; \dot\theta \;\; \theta]$ is the state vector, A is a 4×4 matrix of system coefficients, B is a 4×2 coefficient matrix for a 4×1 input vector, D is a 4×2 coefficient matrix for a 4×1 disturbance vector, y is a 2×1 matrix for output vector and C is a 2×4 coefficient matrix for output vector (Kuo,1987). The conventional optimal controller as shown in Figure 8 is applied to the present motion control system.

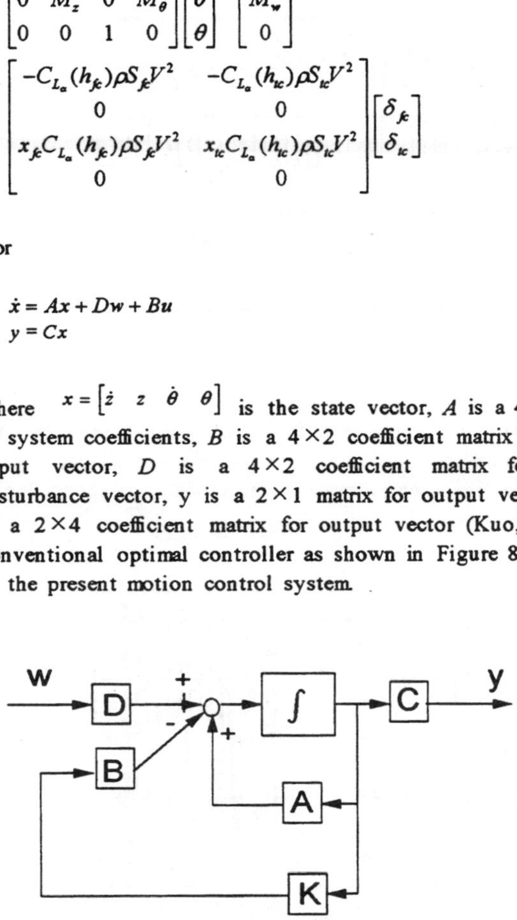

Figure 8. Optimal Feedback Controller

Optimal control problem is to find an admissible control u that causes the system of equation (18) to follow an admissible trajectory x. The trajectory x is determined by minimizing the performance measure

$$J = \lim_{t\to\infty}\int_0^t (x^T Q x + u^T R u)dt \quad (19)$$

where Q and R are weighting matrices, Q is a real symmetric positive semi-definite matrix and R is a real symmetric positive definite matrix (Kirk,1970). Equation (19) can be minimized for u to give

$$u = -R^{-1}B^T P x = -Kx \quad (20)$$

where the existence of R^{-1} is assured since R is a positive definite matrix. P of equation (20) is given by the following Riccati equation.

$$A^T P + PA + Q - PBR^{-1}B^T P = 0 \quad (21)$$

REAL TIME SIMULATION AND RESULTS

Forward fins and trim tabs of simple shape are applied to reduce the motion response of a newly-developed catamaran. The size and configuration of forward fins and trim tabs are shown in Figure 9.

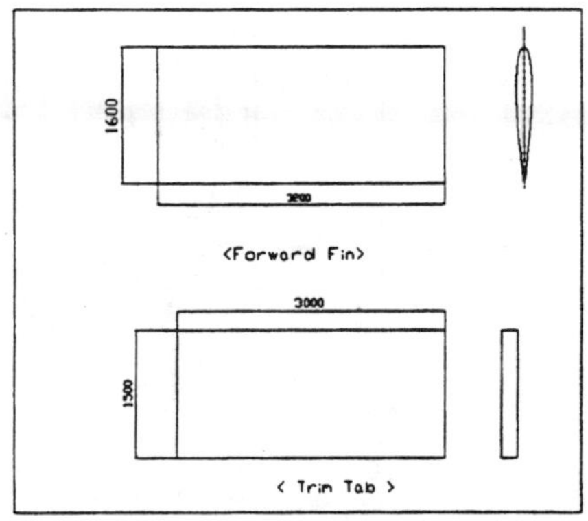

Figure 9. The Configuration of Control Surfaces

The simulation of motion response is performed for the newly-developed catamaran equipped with this system of forward fin and trim tab at both hulls. The simulation conditions are 35 knots ship speed, regular plane waves with amplitude 3m and length 2 Lpp. For the design of optimal controller, the weighting matrices Q and R are selected by trial and error.

The time histories of the response pitch and heave are shown in Figures 10 and 11 respectively.

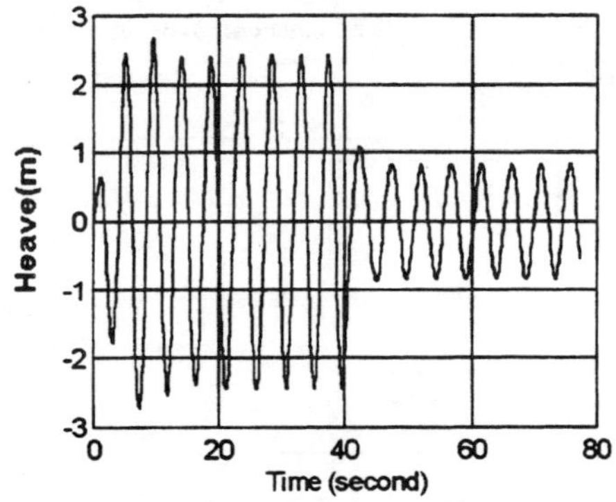

Figure 10. The Response of Heave Motion

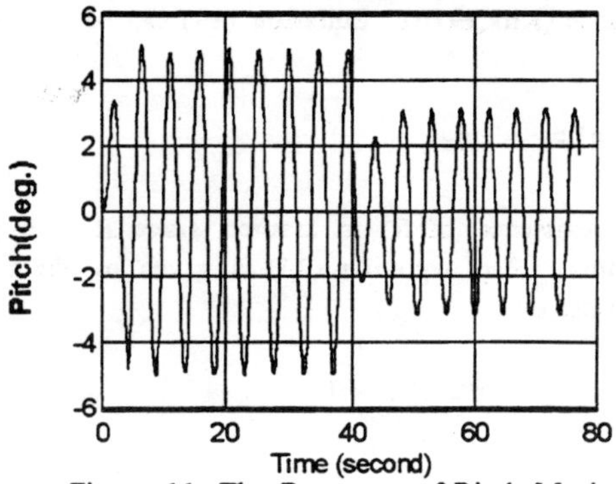

Figure 11. The Response of Pitch Motion

The effects of control are clearly observed. The pitch motion is reduced up to 30 % in control-on condition while the heave motion is reduced dramatically up to 70 %. The variations of angle of attack of forward fin and trim tab are shown in Figure 12. The variation of forward fin is bounded to move within ±20° and trim tab within ±10°.

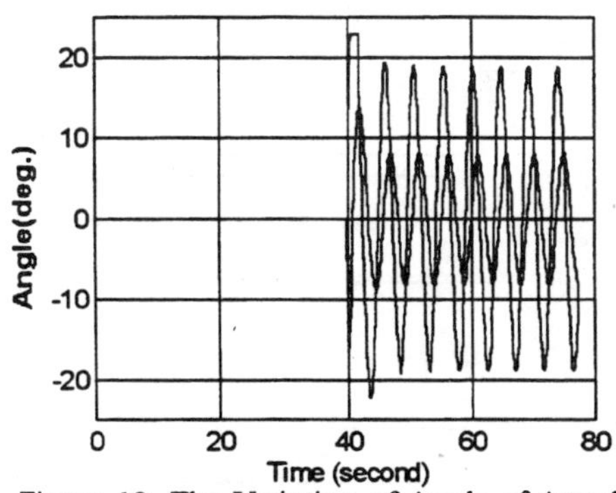

Figure 12. The Variation of Angle of Attack of Forward Fin and Trim Tab

Unfortunately, this type of forward fin is not suitable for practical operation in spite of such an excellent control effect. The forward fin should be located at the farthest position from center of gravity of the ship to maximize the moment arm. Then, in rough seas, the pitch motion increases, the forward fin easily comes out of the surface of water, and almost no control effect results in. The control surfaces should be immersed deep enough to avoid this situation. Therefore, T-foil having a flap as shown in Figure 13 is newly proposed. The use of the trailing-edge flap is for the convenience in the operation mechanism of T-foil.

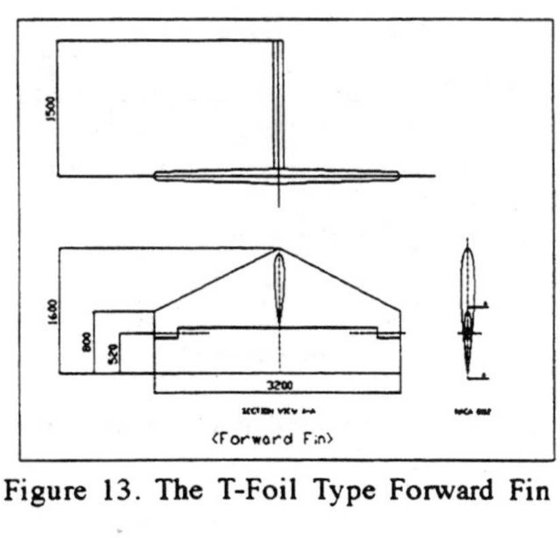

Figure 13. The T-Foil Type Forward Fin

The flap type cannot give enough lift force compared with the all-movable type. The simulation results of flap-type forward fin are shown in Figures 14 and 15.

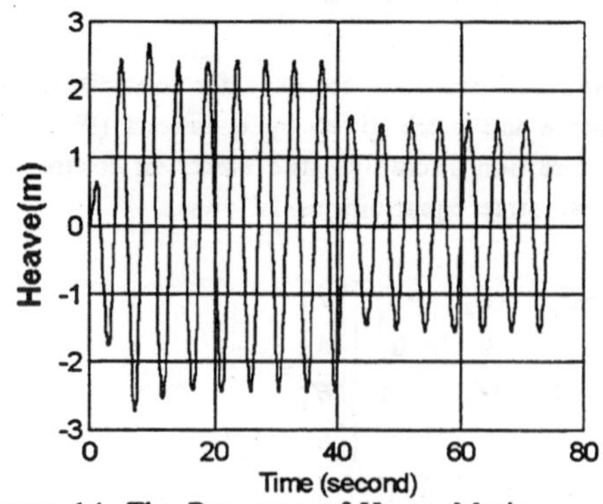

Figure 14. The Response of Heave Motion

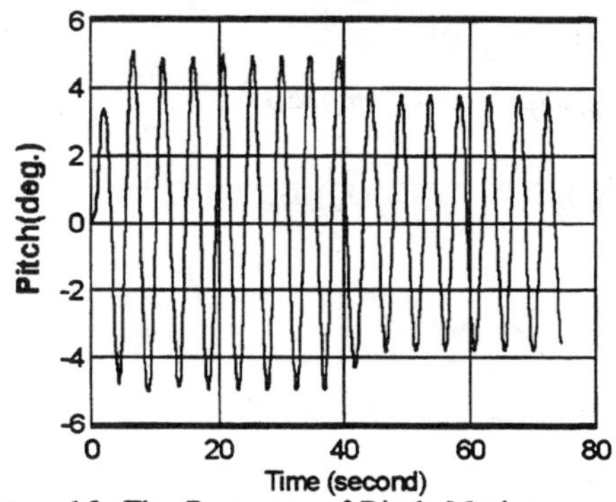

Figure 15. The Response of Pitch Motion

The control is not very effective in comparison with the previous simulation and the saturation of angle of attack occurs due to the limitation of movable angle of flap as shown Figure 16.

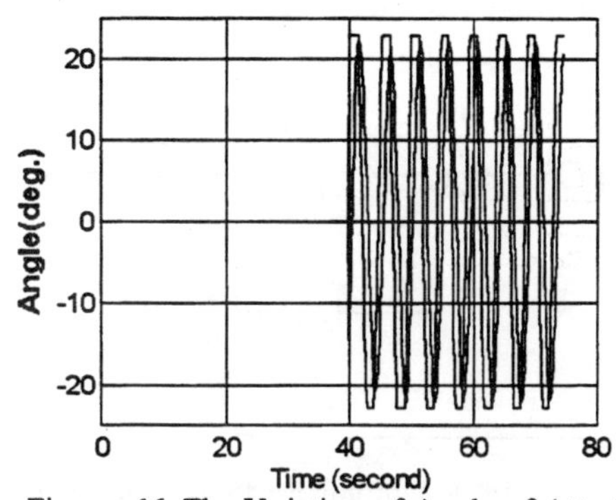

Figure. 16 The Variation of Angle of Attack of T-Foil Type Forward Fin and Trim Tab

The limitation angle, ±25° for both T-foil and trim tab, was set to prevent the lifting surface from stall.

Figure 17 shows the amplitude of heave variation with varying wave heights. As wave height increases, heave response also increases. This seems to be because hydrodynamic external forces increase almost linearly with wave height. Control effect declines from wave height of 3.0 m and above. As shown in Figure 16, the saturation occurs at wave height of 3.0 m. Thus the control system does no good or cannot overcome the external disturbance, i.e., wave, for higher waves or a very rough sea condition.

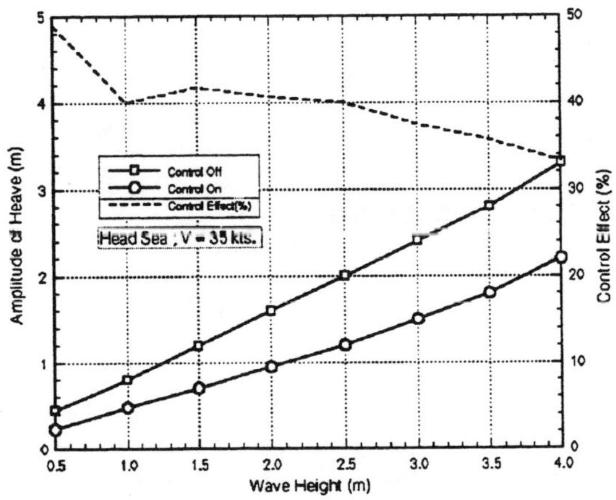

Figure 17. The Control Effect for Heave with Varying Wave Height

The lift is proportional to the square of velocity of the incoming fluid. So the control surface cannot produce an effective result when ship speed is lowered. It is, however, very effective when ship speed is high enough. Figure 18 illustrates this fact clearly. The ride control system is useless for speeds lower than 25 knots, and is effective for speeds higher than 25 knots.

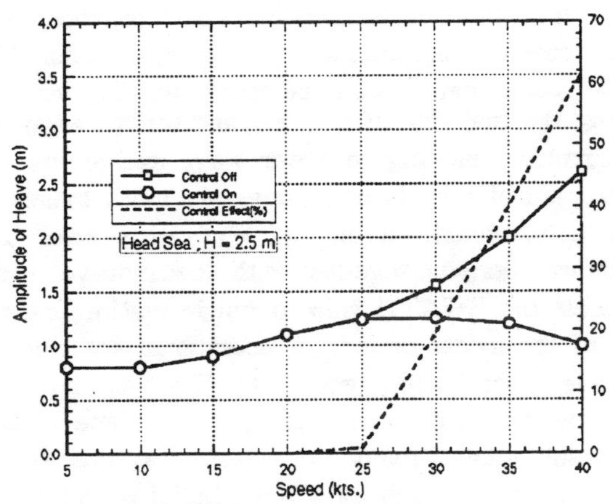

Figure 18 The Control Effect for Heave with Varying Ship Speed

SUMMARY AND CONCLUSIONS

In this paper we assessed, the seakeeping quality of a newly-developed high-speed catamaran in irregular waves mainly by the numerical calculation and comparison with experimental results. Especially we concentrated into the effect of vertical acceleration on ship motion when a ship is operated following an imposed criteria. The Ride Control System of the developed ship has been designed and validated by the simulation in a simple condition. Based on the present study for inexperienced high-speed catamaran the following points are pointed out conclusively.

- As the speed of a ship increases, the seakeeping performance becomes worse.
- The poor seakeeping quality of a high- speed ship is inevitable for the case of head sea because heave RAO and wave spectrum results in the resonance condition around the range of operating speed.
- To improve the seakeeping performance an active Ride Control System is absolutely required. The present study presents a design of such a system composed of T-foil type forward fins and trim tabs as control devices.
- The motion of a high-speed catamaran is reduced to an acceptable level when control surfaces are activated. This supports the necessity of a motion control system in high-speed ships.

For further study, an advanced algorithm to effectively control the motion should be developed and the dynamic characteristics of control surface should be investigated sufficiently.

REFERENCES

Bhattacharyya, R. (1978), "Dynamics of Marine Vehicles", John Wiley & sons, Inc.

Faltinsen, O., et al. (1993), "Report of the High-Speed Marine Vehicles Committee", Proceedings of 20th ITTC

Kang, C. G., Hong, S. Y., Suh, S. H., Lee, C. M., Kim, Y. G., and Gong, I. Y. (1993), "Attitude Control System for a High-Speed Catamaran with Hydrofoils in Waves", FAST '93, Yokohama, Japan

Kaplan, P. (1988), "Frequency Domain Models for Calculation of Motions and Loads of SES and Catamarans in Waves", Technical Report, Hydromechanics, Inc.

Kirk, D. E. (1970), Optimal Control Theory; An Introduction, Prentice Hall, Inc.

Kuo, B. C. (1987), Automatic Control Systems, Prentice Hall, Inc., Fifth Edition

Lewis, E. V. (1987), "Principles of Naval Architecture", Second Revision, SNAME

Salvensen, N., Tuck, E. O., and Faltinsen, O. (1970), "Ship Motions and Sea Loads", Trans. SNAME, Vol. 78

Savitsky, P. and Brown, P. W. (1976), "Procedures for Hydrodynamic Evaluation of Planing Hulls in Smooth and Rough Water", Marine Technology

Shin, M. S., et al. (1995), "A Study on Hull Form of 600P High-Speed Catamaran for SHI"", Technical Report, Korea Research Institute of Ships and Ocean Engineering

Yoo, W. J. and Oh, I. G. (1996), "A Study of Motion Characteristics of a High Speed Catamaran Ferry in waves", Proc. of Society of Naval Architecture of Korea

Proceedings of the Seventh (1997) International Offshore and Polar Engineering Conference
Honolulu, USA, May 25-30, 1997

ISBN 1-880653-28-1 (Set); ISBN 1-880653-32-X (Vol. IV)

Seakeeping Analysis of a 30-Knot Coastal Passenger Swath Ship

H.H. Chun
Pusan National University
Pusan, Korea

M.S. Kim and Y.R. Joo
Samsung Heavy Industry R&D Center
Daejon, Korea

ABSTRACT

A coastal SWATH ship carrying 350 passengers with a cruising speed of 30 knots in a sea state of 4 has been jointly developed by Samsung and Pusan National University. The operating route is between Pohang and Wooleng island in the east sea of the Korean peninsular which is known to be a harsh sea environment. The present paper is to describe the seakeeping analysis of the design ship by means of computations and experiments. A potential based 3-D panel method is used to develop a computational method which can predict the 6-degree of freedom motion responses of a SWATH ship travelling in waves at an arbitrarily heading angle. An empirical cross flow effect of the submerged bodies is taken into account together with the stabilizing fins effect. The motion responses measured in regular head waves with a 1/18 scale model are compared with the computational results. Based on the motion responses in regular waves, the seakeeping qualities of the design SWATH in the operating seas are analysed by the spectrum analysis.

KEY WORDS: SWATH ship, 3-dim. panel method, 6-degree motion responses, viscous damping, fins, spectrum analysis

1. INTRODUCTION

The need to develop high speed marine vehicles is growing due to the increase of the world sea trade and passenger transit and also due to its fast transportation demand. Due to the limit in increasing speed beyond a certain level by the conventional hull concept, new types of high speed hull forms have been developed worldwide. A good seakeeping quality seems to be of prime importance for high speed ships operating, in particular, in open seas with a long endurance.

It is generally known that a catamaran is a good hull form for high speed in view of stability, seakeeping and speed performance. The increased number of building high speed catamarans in these days seems to reflect this claim. Among them, a SWATH (Small Waterplane Area Twin Hull) ship has superior seakeeping characteristics and a minimum speed degradation in severe seas. This remarkable performance can be achieved by virtue of submerging the majority of the buoyant volume away from the sea surface, resulting in lower wave excitations. As a result, the natural periods of motions are much longer than those of a monohull having an equivalent size. These longer natural periods together with lower wave exciting forces enable the SWATH ship to run in platforming mode for most of the modal periods of operational sea conditions except some cases in following seas. However, due to the increased wetted area and small waterplane area, a SWATH ship has many drawbacks compared with equivalent monohulls such as a large initial power and low TPC(tonnes per centimetre immersion) etc. The advantages and disadvantages of the SWATH concept are well described by MacGregor (1989).

Samsung and Pusan National University have been jointly developing a coastal SWATH passenger whose operating route is open seas between Pohang and Wooleng island in the east sea of the Korean peninsular. The desired cruising speed is 30 knots in sea state of 4 with an acceptable passenger ride quality. By increasing the waterplane area more or less compared with that of the conventional SWATH ship, an optimum SWATH design in view of speed and seakeeping has been completed and the results of the calm water resistance tests showed to be promising by Joo et al (1996). The present paper is to describe the seakeeping analysis of the design ship by means of computations and experiments.

By developing a 3-dim. panel method to compute 6 degree motion responses of a SWATH ship in stationary and travelling in waves at an arbitrarily heading angle, Chun et al (1990) compared the computational results with a wide ranging of experimental results in head, quartering, beam and following seas, showing good correlations between two results. Chun et al's method is extended in this paper by adding an empirical viscous cross flow effect suggested by Lee & Curphey (1977) used for the motion calculation of a SWATH ship with a strip method. Also, the stabilizing fins effect on the motion responses is considered by following Lee & Curphey's suggestion (1977). The motion responses measured in regular head waves with a 1/18 scale model are compared with the computational results. The seakeeping qualities of the design SWATH in the operating seas are analysed by the spectrum analysis.

2. FORMULATION OF THE PROBLEM

A right handed reference 0-xyz system which travels in the positive x direction with speed U at a wave heading angle β (β=180° for head seas) is used as shown in Fig. 1. The origin is in the plane of the undisturbed free surface above the centre of gravity.

2.1 Nonlinear Boundary Value Problem

By assuming an inviscid, irrotational and incompressible fluid, the problem is reduced to solving a governing equation (Laplace's equation) with the following boundary conditions:

$$\nabla^2\Phi=0$$

$$\Phi_{tt}+g\Phi_z+2\nabla\Phi\cdot\nabla\Phi_t+\frac{1}{2}\nabla\Phi\cdot\nabla(\nabla\Phi\cdot\nabla\Phi)=0 \quad \text{on } z=\zeta$$

$$(\vec{V_s}-\vec{V})\cdot\vec{n}=0 \quad \text{on S}$$

$$\frac{\partial\Phi}{\partial z}=0 \text{ at } z\rightarrow-\infty \tag{1}$$

and an appropriate radiation condition which makes the solution unique. In the above equations, S is the wetted surface, $\vec{V_s}$ and $\vec{V}$ are the surface and fluid velocities , $\vec{n}$ being an unit normal vector. Since the boundary conditions above are highly nonlinear and furthermore, the free surface (ζ) is not known beforehand, it is nearly impossible to solve the problem directly. Therefore, a linearization procedure is necessary.

2.2 Linearised Boundary Value Problem

If assuming that a slender body travels in waves whose amplitudes are small, the water disturbances due to the body may be also small and then, the total velocity potential can be linearly superposed by the steady potential and the unsteady potential:

$$\Phi=[\ -Ux+U\Psi]\ +\phi \tag{2}$$

By the same principle, the unsteady potential can divide into incident, diffraction and radiation components as follows:

$$\phi=[\ \xi_o(\phi_I+\phi_D)+\sum_{j=1}^{6}\xi_{j}\phi_j]\ e^{-i\omega_e t} \tag{3}$$

where j=1, 2, 3 indicate surge, sway, heave and j=4, 5, 6 roll, pitch, yaw components. If the forward speed is assumed to be small, the steady potential may be small in magnitude compared to the unsteady potential and then, the boundary conditions above can be reduced to a linearised free surface boundary condition with simplified body boundary conditions, see Salvensen et al (1970) and Faltinsen et al (1974) :

$$(iw_e+U\frac{\partial}{\partial x})^2\phi+g\frac{\partial}{\partial z}\phi=0 \quad \text{on } z=0 \tag{4}$$

$$\frac{\partial}{\partial n}(\phi_I+\phi_D)=0 \quad \text{on S} \tag{5a}$$

$$\frac{\partial\phi_j}{\partial n}=-iw_e n_j+Um_j \quad \text{on S} \tag{5b}$$

where g is the gravitational acceleration constant with the following notations:

$$(n_1,n_2,n_3)=\vec{n},$$

$$(n_4,n_5,n_6)=\vec{r}\times\vec{n}$$

$$m_j=0 \text{ for } i=1,2,3,4$$

$$m_5=n_3, \text{ and } m_6=-n_2 \tag{6}$$

where r is a position vector with respect to the origin. Further, assuming that the frequency of oscillation is high together with a small forward motion (ie, $\omega_e > U\partial/\partial x$), eq'n (4) can be reduced to the following, see Salvensen et al (1970) and Inglis (1980) :

$$-w_e^2\phi+g\frac{\partial\phi}{\partial z}=0 \quad \text{at } z=0 \tag{7}$$

An efficient algorithm to calculate the green function which satisfies the above eq'n, the infinite sea bottom condition and the radiation condition was developed by Telste & Noblesse (1986) and this is used for the present study.

2.3 3-D Panel Source Distribution Method

By the virtue of Green's theorem, the velocity potential can be expressed as follows:

$$\int\int_S \sigma(Q) G(P,Q) ds(Q) = 4\pi\phi(P) \quad \text{for p inside fluid} \quad (8)$$

where σ(Q) is the source strength distributed on the mean wetted surface and G(P,Q) is the Green's function describing the influence at a point P(x,y,z) due to an unit source strength at a point Q(ξ,η,ζ). The unknown source density can be found by imposing the body boundary condition on the above eq'n and results in the following equation:

$$-\frac{1}{2}\sigma(P) + \frac{1}{4\pi}\int\int_S \sigma(Q)\frac{\partial G(P,Q)}{\partial n} ds(Q) = V_n$$

for p body surface (9)

where V_n is the normal water velocity on the body surface and is given in eq'n (5) for the radiation and diffraction problems. The surface integral in the above eq'n can be written as the sum of the integrals over the N panels of area ΔS_j and the source strength is assumed to be a constant over each panel. Having found N unknown source strengths, the fluid velocity in a given direction at a point can be obtained by the spatial derivatives of the velocity potential, eq'n (8), in that direction.

2.4 Pressures and Forces

The fluid pressure is obtained by the following linearised unsteady Bernoulli's eq'n:

$$p = -\rho[-iw_e - U\frac{\partial}{\partial x}]\,\phi e^{-iw_e t} \quad (10)$$

The forces and moments acting on the hull can be calculated by the following:

$$H_i = -\int\int_S p n_i ds = F_i + E_i \quad \text{for } i=1,2,\cdots 6 \quad (11)$$

The wave exciting forces due to the incident and diffracted waves are obtained by the following:

$$F_i = F_i^I + F_i^D$$
$$= -\rho i w_0 \int\int_S n_i \phi_I ds + \rho\int\int_S (\phi_j^0 + \frac{U}{iw_e}\phi_j^U)\frac{\partial \phi_I}{\partial n} ds$$
$$= -\rho\int\int_S \{iw_0 n_i \phi_I - \frac{\partial \phi_I}{\partial n}\phi_j^0\}\, ds$$
$$\pm[\frac{\rho U}{iw_e}\int\int_S \frac{\partial \phi_I}{\partial n}\phi_{3,2}^0 ds]_{j=5,6} \quad (12)$$

where + for j=5 and - for j=6. And the forces and moments due to the radiation potentials are given in the followings:

$$E_i = -\rho\int\int_S (iw_e n_i + U m_i)\sum_{j=1}^{6}\xi_{j0}\phi_j ds = \sum_{j=1}^{6} T_{ij}\xi_{j0} \quad (13)$$

$$T_{ij} = -\rho\int\int_S (iw_e n_i + U m_i)\phi_j ds$$
$$= -(A_{ij}\ddot{\xi}_j + B_{ij}\dot{\xi}_j)$$
$$= w_e^2 A_{ij} + iw_e B_{ij} \quad (14)$$

T_{ij} indicates the i-th directional forces and moments by the unit motion displacement in the j-th direction. A_{ij} and B_{ij} indicate the added mass and damping coefficients, respectively, as given in the followings:

$$A_{ij} = \frac{\rho}{w_e} Im[\int\int_S n_i\phi_j ds] - \frac{\rho U}{w_e^2} Re[\int\int_S m_i\phi_j ds] \quad (15a)$$

$$B_{ij} = \rho Re[\int\int_S n_i\phi_j ds] + \frac{\rho U}{w_e} Im[\int\int_S m_i\phi_j ds] \quad (15b)$$

2.5 Viscous Damping and Fin Effects

Using an empirical formulae given by Thwaites(1960) for the force subject to a slender body at a small incident angle of attack in a uniform stream, Lee & Curphey(1977) considered the viscous damping effect of the submerged bodies of a SWATH ship, which travels at a uniform speed in regular waves, on its motion calculations. The vertical force of the submerged body of the SWATH ship at a small trim angle of α was given by Lee & Curphey(1977) as follows:

$$F_V = \frac{1}{2}\rho A_p'(U^2 a_0\alpha + C_D w|w|) \quad (16)$$

The first term is due to the viscous lift and the second term is a drag due to the cross flow. A_p' is the projected area of the submerged body into the xy-plane, a_o viscous lift coefficient, C_D the cross flow drag coefficient. The value of a_o found from the experiments on airship models with circular or polygonal cross sections is known to be about 0.07 and C_D is between 0.4 and 0.7, increasing with the increase of the slenderness ratio, Allen et al (1951). a_o of 0.07 and C_D of 0.5 are used in this paper. w is the vertical component of the relative water particle velocity w.r.t. the hull and is given at the center of the submerged body (x, $\pm$b(x), $-d_1$(x)) as follows:

$$w = \dot{\xi}_3 + y\dot{\xi}_4 - x\dot{\xi}_5 - \frac{\partial \phi_I}{\partial z}(x, \pm b(x), -d_1(x)) \quad (17)$$

where the dot point on the head means the time derivative. The trim angle of the body or the incidence angle of the oncoming local water flow at a section x is given by

$$\alpha(x) = \xi_5 + \frac{w}{U} \quad (18)$$

With the incidence angle of the oncoming local water flow and relative water velocity at a section x being known, the vertical force F_V at that section can be calculated and then the total vertical forces can be obtained by integrating all along the body length. Similarly, horizontal and moments can be calculated using the above equations and detailed derivations can be found in Lee & Curphey (1977) and Kim (1997).

The hydrodynamic effects of the fins on the motions are from several aspects such as the interference between the body and fins, the unsteady effect, boundary layer effect and the free surface effect etc. Following the method suggested by Lee & Curphey (1977), the interference effect between the body and fins is only considered in this paper. The force generated by the fins oscillating with a forward speed is the sum of the lift, cross flow drag and inertia force. The lift force L(t) of a fin at a small angle of incidence is given by

$$L(t)=\frac{1}{2}\rho U^2 A^{(f)} C_{L\alpha}\,\alpha(t) \tag{19}$$

$A^{(f)}$ is the planform are of the fin, $C_{L\alpha}$ the lift coefficient curve which is obtained empirically by taking into account the body-fin interference, as given by Lee & Curphey (1977) :

$$D(t)=\frac{1}{2}\rho A^{(f)} C_D^{(f)} w|w| \tag{20}$$

where $C_D^{(f)}$ is the cross flow drag coefficient of the fin.

The inertia force of the fin is given by

$$I(t)=(m^{(f)}+a_{33}^{(f)})(\ddot{\xi}_3-l\,\ddot{\xi}_5+b_1\,\ddot{\xi}_4) \tag{21}$$

where l and b_1 are x and y coordinates of the centre pressure point of the fin, respectively. $m^{(f)}$ is the mass of the fin, $a_{33}^{(f)}$ is given by $\rho s c^2 \pi/4$ and is the added mass of the fin (s is the span and c chord).

The total force subject to a fin is the sum of the above three components and the whole forces of N fins can be simply superposed. The moments can be also calculated by multiplying the distance to the centre of the pressure on the fins and the forces. Also, detailed derivations are given in Lee & Curphey (1977) and Kim (1997).

3. MODEL TESTS

Using a 1/18 scaled model of the design SWATH whose general arrangement and principal dimensions are given in Fig.2 and Table 1, respectively, the tests were conducted in the towing tank of Pusan National University. The length, breadth and water depth of the towing tank are 90m, 5m and 3m, respectively. Using the 6 component motion test device, the motions of the model in stationary were measured in regular waves with various heading angles. In addition, motions were measured with two forward speeds, 1.213m/s (full scale 10 knots) and 2.43 m/s(full scale 20 knots), in regular head waves. The surge motion is restricted for the forward speed cases and all the motion tests were conducted without stabilizing fins.

The radius gyrations and GM_T and GM_L of the model without fins measured by the pendulum and the incline test, respectively, are given in Table 2 together with its natural periods of motions measured. Since the 6 component motion test device can not be attached at the centre of gravity of the model and is attached at a point on the cross deck 0.2m (full scale 3.58m) above the centre of gravity, it was suspected that the system could change the natural periods of the motions of the model. As seen in Table 2, the natural periods of motion for the model attached with the device become, in general, longer than those of the free model. In particular, the roll natural period becomes much longer for the model with the device. This fact influenced the motion responses measured, in particular, for the quartering and beam seas, resulting in the motion peaks occurred not at the model natural periods but at some other periods. Therefore, the motion responses measured in these two waves can not be compared with the computational results. The motion responses measured in head seas are only compared with the computational results.

4. COMPUTATIONAL RESULTS AND DISCUSSIONS

Using the symmetry property, the model is represented by a number of quadrilateral panels for the demihull and it was found that around 200 panels or some more for the demihull give a converged solution. Fig. 1 shows the panel arrangement with 245 panel numbers for the demihull which is used for the present calculations.

Figs.3, 4 and 5 shows the computational and experimental surge, heave and pitch RAOs vs the non-dimensional encountering frequency ($\omega_e\sqrt{L/g}$, L = length of the submerged body). The exp. 1, 2 and 3 in the figures indicate the repeat tests whose wave heights are slightly changed to check the linearity of the motion responses. As seen in all the figures, the computed results with only the potential damping give two high values at the frequencies near the natural frequencies whereas the results with the viscous damping shows much reduced peaks. The maximum peak value at the natural frequency could slightly vary with the C_D values which is in the range of 0.4 - 0.7, increasing with the increase of the slenderness ratio, as mentioned in section 2.5. For the present calculations, C_D is fixed to be 0.5.

As the wave length becomes longer, it is usual that the surge response divided by the wave height becomes unity or slightly larger than the unity. However, the present experimental results did not show this trend. The heave RAO in Fig.4 shows two peaks. The first one for the experiment is the non-dim. encountering frequency of 2.0 which is the pitch natural frequency of the model attached with the motion test device. The second peak is for the heave natural period and the computational and experimental RAO peaks are slightly different as each other, being 2.4 for the experiment and 2.6 for the computation. This results seems to be due to the presence of the motion test device as aforementioned. The non-dimensional encountering frequency of 2.4 is for the heave natural period of the model with the device and the frequency of 2.6 at which the computed peak occurs is for the one of the free model. Apart from these facts, the computed results with the viscous damping give fairly good correlations with the experimental results.

Figs.6 and 7 shows the computational and experimental heave and pitch RAOs at a full scale of 10 knots. When the viscous damping effect is taken into account, the two results give good correlations. At a speed of 20 knots, the heave reponses measured agree well with the computed results with the viscous damping whereas the pitch responses measured near the resonant frequency are much larger than the computational results with the viscous damping. The large pitch peak measured is due to the fact that the surge motion is restricted for the forward speed tests and hence this surge restriction can exaggerate the vertical motions, in particular, the pitch motions at higher speeds. Therefore, when the computed motion responses at the forward speed case are compared with the experimental results, it is recommended that the motion responses should be measured with the model free running without any restriction.

The effects of the fins on the motion responses are shown in Figs.10, 11 and 12 at the zero forward speed, in Figs.13 and 14 at 10 knots, and in Figs.15 and 16 at 20 knots. The measured results are for the model without the fins and the computed results are with the viscous damping taken into account. As seen in the results, the fin effects on the motions is negligible for the stationary case and as the speed increases, the large responses over the resonant frequency range are much damped down.

Figs.17, 18 and 19 shows the iso-lines of the heave, pitch, roll responses of the design SWATH with two pair of fins as functions of wave frequency and wave headi at 30 knots. For the heave responses as seen in Fig.17, t large motions occur in head seas with wave period around 9 seconds. Also, the large pitch motions occur head seas whose wave period is around 8 seconds, as se in Fig.18. The operating route of the design ship between Pohang and Wooleng island in the east sea of t Korean peninsular. The annual average wave height of tl sea is known to be around 1.2m with the modal period of ~4 seconds. Therefore, it can be seen from Figs.17 and that the heave and pitch motions are very small in wave heading angles at this modal period range. T maximum roll motions can be occurred in the ste quartering seas at this range as seen in Fig.19 but magnitude seems to be not significant.

Based on these regular motion responses, the followi ITTC sea spectrum is used to analyse the seakeepi qualities of the design ship in the operating route.

$$S(\omega)=\frac{124(H_{W_{1/3}})^2}{\omega^5 T_{02}^4}\ \exp\left(-\frac{496}{\omega^4 T_{02}^4}\right)$$

$T_{02}=2\pi\sqrt{\dfrac{m_0}{m_2}}$: Average zero-crossing period

$H_{W_{1/3}}$: Significant wave height

ω : Circular frequency in radious per second

$$m_0=\int_0^\infty S(\omega)d\omega$$

$$m_2=\int_0^\infty \omega^2 S(\omega)d\omega$$

Two average zero crossing periods of 3.5 and 5.0 seco are used. It is understood that these sea states are mo similar to those of the operating seas. Table 3 shows motion responses in several wave heading angles with two wave periods together with the heave accelerations the centre of gravity. 3° pitch, 5° roll and 0.1G's total R accelerations at C.G. are commonly indicated comf limits. The motion responses and heave accelerations the design ship are much lower than those of the li criterions. It can be understood that the heave accelera is more or less larger than that of the usual SWATH s This is due to the fact that as mentioned in introduction, the waterplane area of the design SWA ship is more or less increased compared with that of conventional SWATH ship in order to increase the sp performance at the cast of the motion responses in wav If the fins are active in operating, the motions respon can be much reduced.

5. CONCLUSIONS

Based on the present study, the following conclusions can be drawn:

- The present computational results where the viscous damping effect is taken into account give good correlations with experimental results even at higher speeds except few cases which have experimental uncertainties with the present motion test device system for the SWATH model with a cross deck well above the water level.
- In order to obtain more reliable data for forward speed cases, the motion responses should be measured with the model free running without any restriction in motion using such as LVDT (Linear Variable Displacement Transformer) system or LED(Light Emitting Diaodes) system and then, the experimental results should be compared with the computational results.
- The seakeeping qualities of the design SWATH ship with fixed two pair of fins in the operating seas at 30 knots is much lower than those of the motion limit criterions. In addition, the fins are active in operating, the seakeeping performance of the ship will be much enhanced.

REFERENCES.

Allen, HJ et al (1951). "A Study of Effects of Viscosity on flow over Slender Inclined Bodies of Revolution," *NACA Report 1048*

Chun, HH et al. (1990). "A Wide Ranging Study on the Motions of SWATH Ships with and without Forward Speeds," *Proc. of ASME 9th Int. Conference on OMAE*, Houston

Faltinsen, OM, and Michelsen, FC (1974). "Motions of Large Structures in Waves at Zero Froude Number," *Proc. of Int. Symp. on Dynamics of Marine Vehicle*

Inglis, RB, (1980). "A Three-Dimensoinal Analysis of the Motion of a Rigid Ship in Waves," *Ph.D thesis, University College London*, U.K.

Joo, YR et al. (1996). "Development of a High Speed, Coastal Passenger Catamaran with a Superior Seakeeping Quality," *Proc. of the Spring Meeting, SNAK*, Koje, Korea

Kim, MS (1997). "A study on the Prediction of Motion and Wave Loads of High Speed Twin-Hull Ships," *M.Sc. thesis, Pusan National University*, Korea

Lee, CM, and Curphey, RM (1977). "Prediction of Motion, Stability and Wave Load of Small Waterplane Area, Twin-Hull Ships," *Transaction SNAME*, Vol. 85

MacGregor, JR (1989). "A Computer Aided Method for Preliminary Design of SWATH Ships," *Ph.D Thesis, University of Glasgow*, U.K.

Salvensen, N, Tuck, EO, and Faltinsen, OM (1970). "Ship Motions and Sea Loads," *Transaction SNAME*, Vol. 78

Telste, JG, and Noblesse, F (1986). "Numerical Evaluation of the Green Function of Water-Wave Radiation and Diffraction," *J Ship Res*, Vol 30.

Thwaites, B (1960). "Incompressible Aerodynamics," *Oxford University Press*

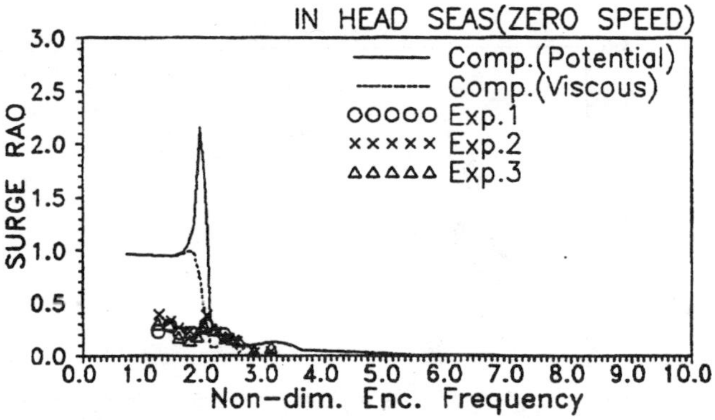

Fig. 3 Surge response vs. $\omega_e\sqrt{L/g}$ of SWATH-S1 model (without fin)

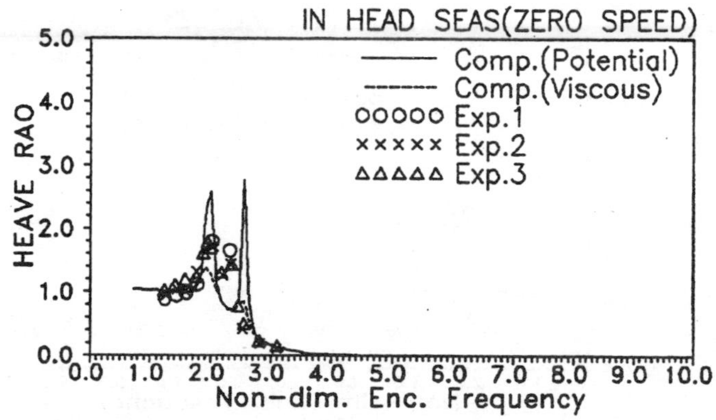

Fig. 4 Heave response vs. $\omega_e\sqrt{L/g}$ of SWATH-S1 model (without fin)

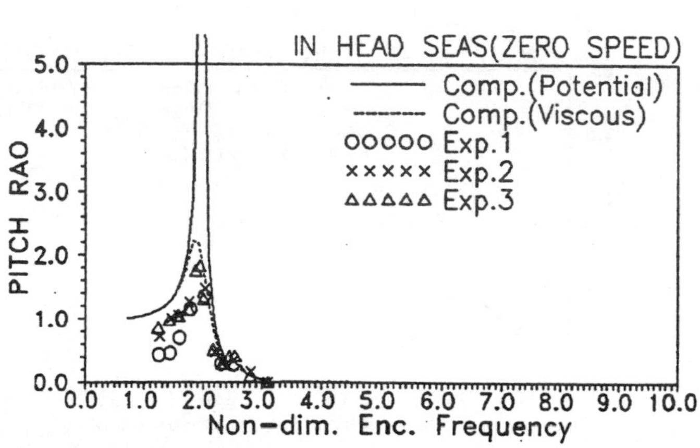

Fig. 5 Pitch response vs. $\omega_e\sqrt{L/g}$ of SWATH-S1 model (without fin)

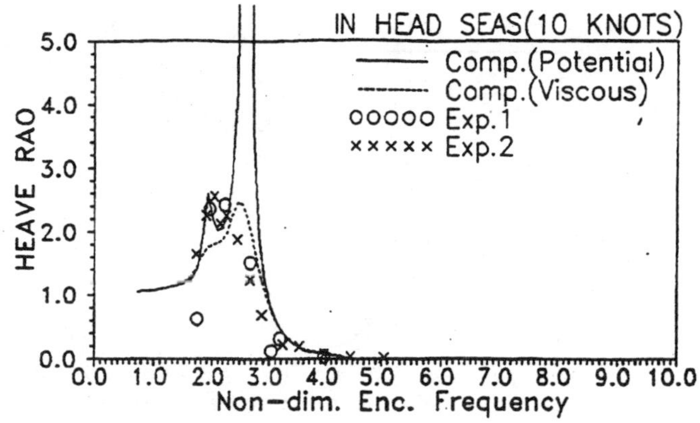

Fig. 6 Heave response vs. $\omega_e\sqrt{L/g}$ of SWATH-S1 model (without fin)

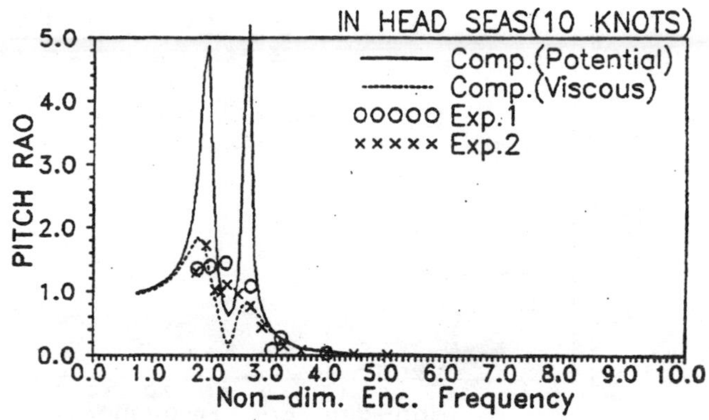

Fig. 7 Pitch response vs. $\omega_e\sqrt{L/g}$ of SWATH-S1 model (without fin)

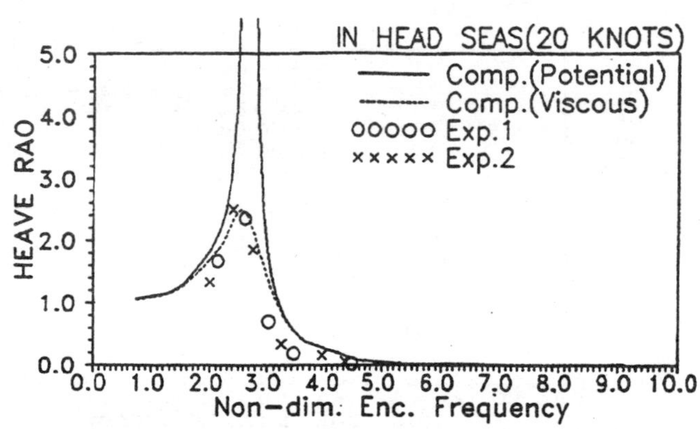

Fig. 8 Heave response vs. $\omega_e\sqrt{L/g}$ of SWATH-S1 model (without fin)

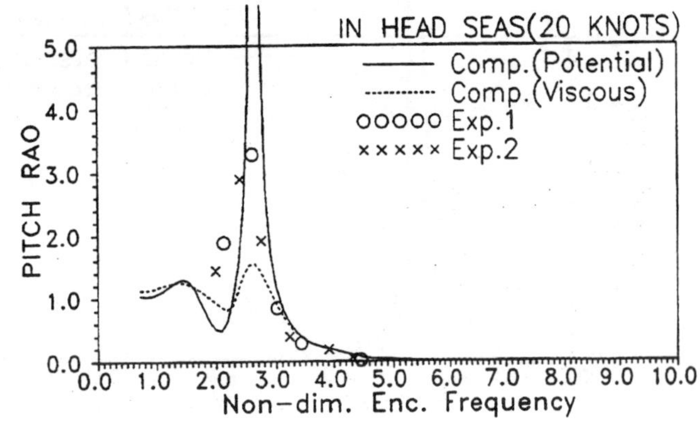

Fig. 9 Pitch response vs. $\omega_e\sqrt{L/g}$ of SWATH-S1 model (without fin)

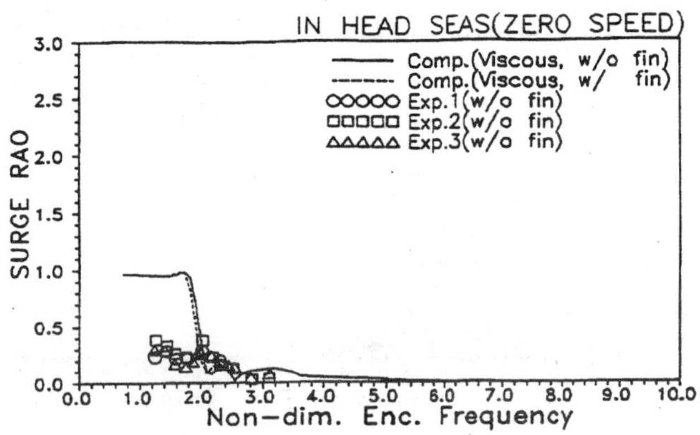

Fig. 10 Surge response vs. $\omega_e\sqrt{L/g}$ of SWATH-S1 model (with and without fin)

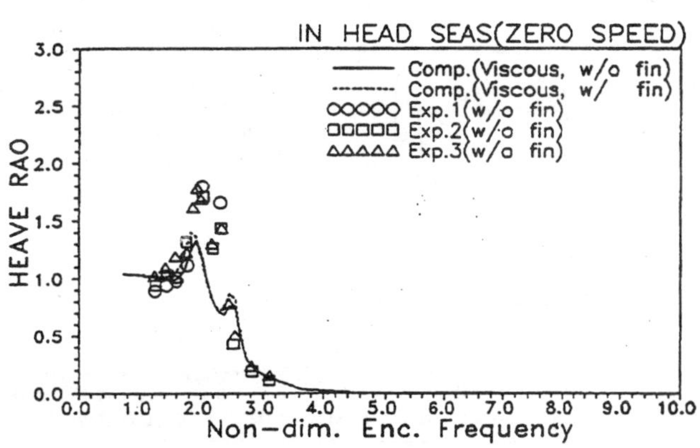

Fig. 11 Heave response vs. $\omega_e\sqrt{L/g}$ of SWATH-S1 model (with and without fin)

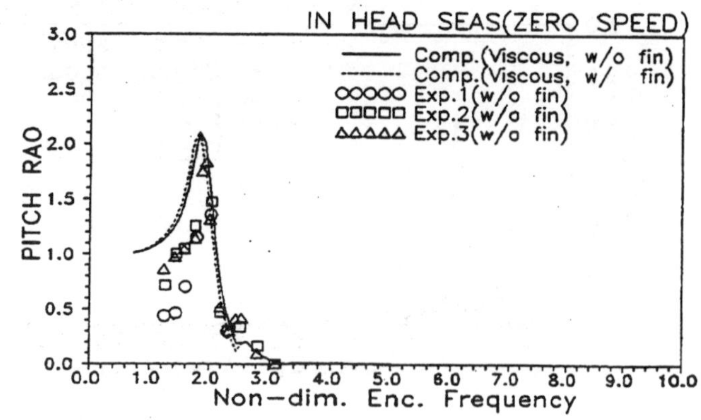

Fig. 12 Pitch response vs. $\omega_e\sqrt{L/g}$ of SWATH-S1 model (with and without fin)

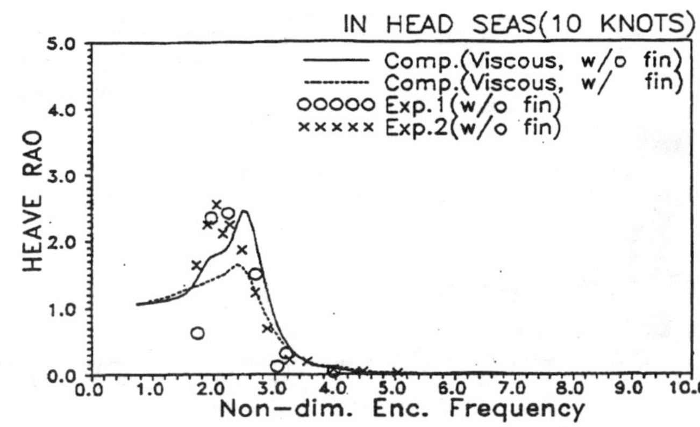

Fig. 13 Heave response vs. $\omega_e\sqrt{L/g}$ of SWATH-S1 model (with and without fin)

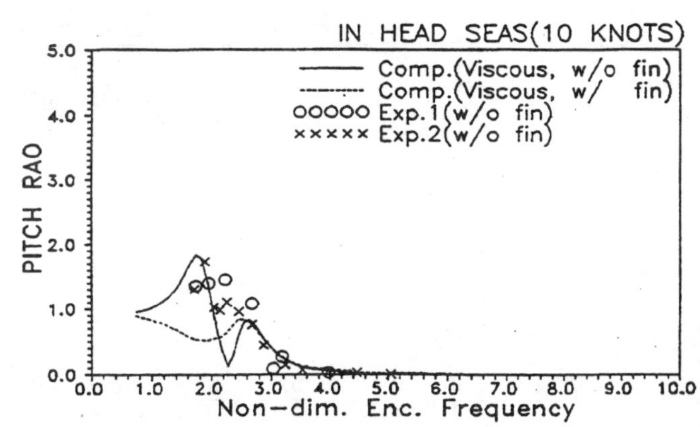

Fig. 14 Pitch response vs. $\omega_e\sqrt{L/g}$ of SWATH-S1 model (with and without fin)

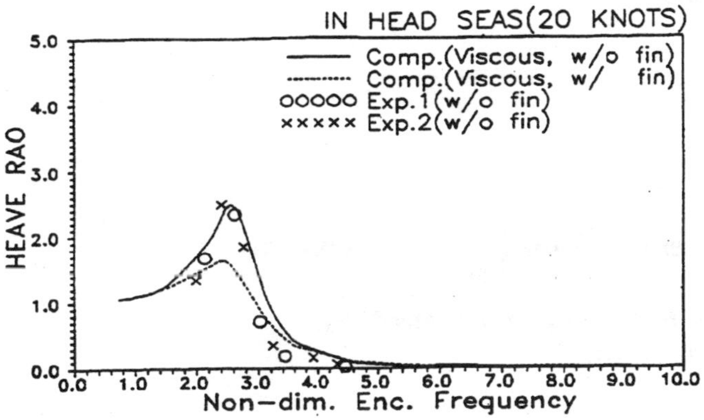

Fig. 15 Heave response vs. $\omega_e\sqrt{L/g}$ of SWATH-S1 model (with and without fin)

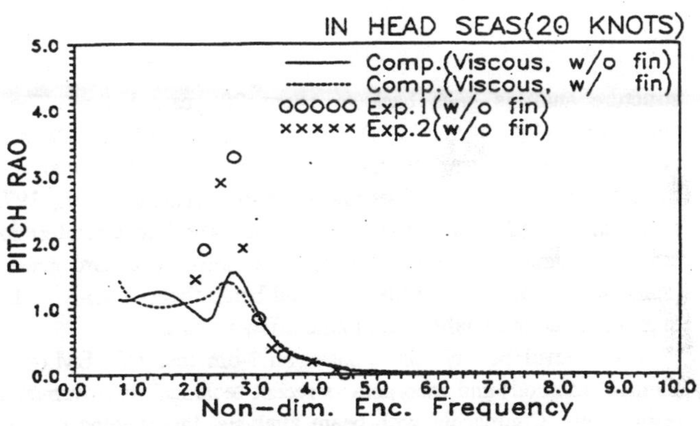

Fig. 16 Pitch response vs. $\omega_e\sqrt{L/g}$ of SWATH-S1 model (with and without fin)

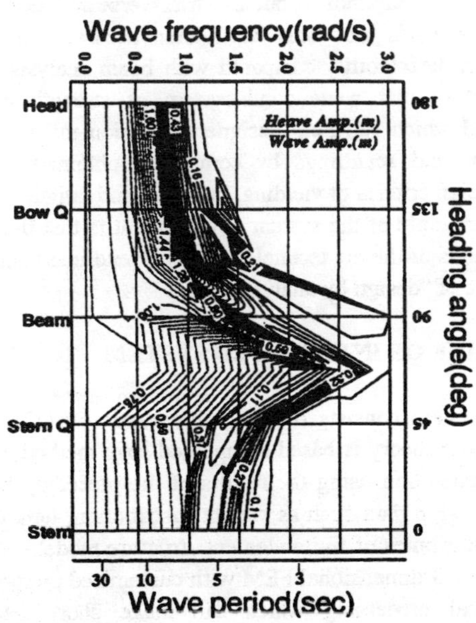

Fig. 17 Heave response of SWATH-S1 with fins as a function of heading and wave frequency at 30 knots

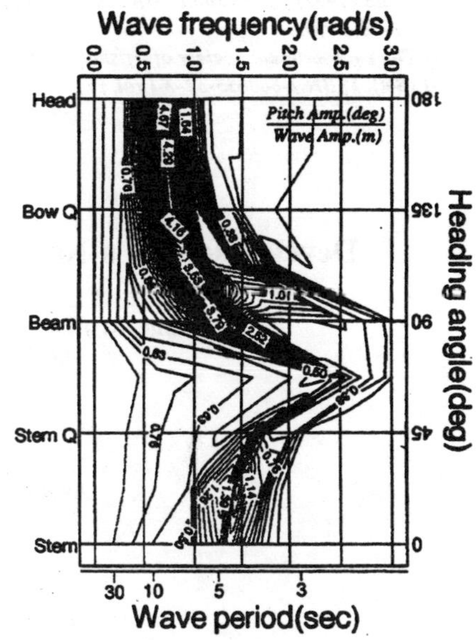

Fig. 18 Pitch response of SWATH-S1 with fins as a function of heading and wave frequency at 30 knots

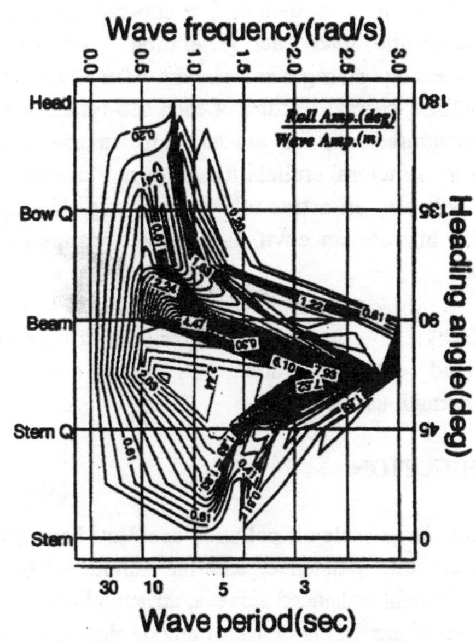

Fig. 19 Roll response of SWATH-S1 with fins as a function of heading and wave frequency at 30 knots

Proceedings of the Seventh (1997) International Offshore and Polar Engineering Conference
Honolulu, USA, May 25-30, 1997

ISBN 1-880653-28-1 (Set); ISBN 1-880653-32-X (Vol. IV)

Development of a New Integrated System of Ship Structure Analysis

O. Niho, K. Nishimura, T. Tsubokawa, H. Kaminaga and T. Maeda
Mitsui Engineering & Shipbuilding Co.
Chiba, Japan

K. Shimada
Akishima Laboratories (Mitsui Zosen) Inc.
Tokyo, Japan

ABSTRACT

A new integrated system of ship structure analysis has been developed. This system consists of five sub-systems of hydrodynamic analysis / automatic load data generation / structural analysis / statistical processing / strength assessment. After a brief introduction of the system outline, special features of each sub-system are described.

This integrated system assists ship structural designers in enhancing the structural reliability through "design by analysis" method and will play an active role in designing ship structures with high reliability in accordance with increasing demands among shipping communities.

KEY WORDS : design by analysis, hydrodynamic analysis, automatic load data generation, structural analysis, statistical processing, strength assessment

1. INTRODUCTION

The recent large-scale oil pollution accidents have brought double hull oil tankers into existence, and the serious casualties in bulk carriers have brought enhanced surveys, structural reinforcements, etc. Today, it is one of increasing demands among shipping communities to enhance the degree of safety and reliability of ship structures.

Strutcural design of conventional ships such as oil tankers and bulk carriers has been, so far, based on "design by rule" method, where nominal strength is assessed in comparison with past experiences and records. Recent remarkable advancement of computer and engineering technology, however, makes it possible to perform "design by analysis", where absolute strength is assessed by simulating actual structural behaviors as close as possible.

The newly developed system called MISTRAL (= Mitsui Integrated system of STRuctural And Load analysis) can simulate the structural response by integrated analyses, i.e., direct load analysis of ships in operation, structural response through interface program from the obtained load, statistical processing and strength assessment.

History of this kind of integrated computer systems for ship structure analysis goes back to about 20 years before. The first generation (Akita et al., 1974) is direct longitudinal strength analysis system based on beam theory integrating ship motion analysis and statistical processing. The second generation (Nakata et al., 1975) is so-called refined beam theory which can take into consideration of torsional characteristics such as container ships. The third generation (Kanetsuna et al., 1978) utilizes spatial beam theory mainly dedicated to global structure analysis of mobile offshore units.

The developed system is based on 3-dimensional FEM (= Finite Element Method) and incorporates recent technical improvements and refinements. Comparing with beam analysis, this method requires to deal with huge amount of data such as spatial distribution of load data, degrees of freedom of FEM model and so on. This system, however, is well practical for usual design work owing to remarkably advanced data processing capacity and user interface. 3-dimensional FEM analysis can analyze not only longitudinal but also transverse and local strength as a whole. This leads not only to higher accuracy but also better phenomenon detectability compared with beam analysis. It is another special feature that a new sub-system of strength assessment is incorporated which can help designer's decision making in structural arrangement and scantlings by comparing obtained responses and strength design criteria of yielding, buckling and fatigue.

The principle of the system development is that the system shall have the-states-of-the-art technology, user friendliness and an efficient performance of "design by analysis".

2. OUTLINE OF INTEGRATED SYSTEM

This system consists of five sub-systems of: (1) hydrodynamic analysis whose theory is based on updated strip method, (2) automatic load data generation using the results of the preceding hydrodynamic analysis and ship data such as weight distribution, tank configuration defined as attributes of finite element structure model, (3) structural analysis using 3-dimensional FEM with customized pre/post processor, (4) statistical processing which can make short- and long-term predictions as to expected design parameters, and (5) strength assessment in the light of yield, buckling and fatigue criteria which assists designers to make prompt decisions. These sub-systems have been given new practical functions to deal with a mass of analysis data easily.

2.1 Hydrodynamic Analysis

The "hydrodynamic analysis" sub-system gives responses of motion and load in a regular wave based on updated strip method.

SCS (= Ship Calculation System) (Kobayashi, M. et al., 1973) was developed based on NSM (= New Strip Method) by MES (= Mitsui Engineering & Shipbuilding Co., Ltd.). NSM has been utilized for some decades for ship motion analysis because it gives good predictions in spite of a simplified method. But NSM has some shortcomings that wave forces in the vertical direction diverge at $\omega_e = 0$, which occurs when the ship runs in the following or quartering waves at relatively high speed. The present system is so developed as to be free from the shortcomings of NSM.

(1) Brief description of analysis method

i) Solution of 2-dimensional boundary value problem

Fluctuating pressures and hydrodynamic forces on the hull surface can be described using velocity potentials. Total velocity potential is described as a sum of component potentials :

$$\Phi_T = \Phi_R + \Phi_D + \Phi_0 \qquad \cdots\cdots(1)$$

where Φ_R, Φ_D and Φ_0 are radiation, diffraction and incident-wave potentials, respectively. Unknown potentials, Φ_R and Φ_D should be solved by Laplace equation, boundary conditions on the hull and free surface and radiation condition.

Major differences between the present method and SCS lie in the analysis of diffraction problem as explained below.

3-dimensional free surface condition is simplified in SCS, ignoring the ship-lengthwise derivative, to the 2-dimensional free surface condition

$$\omega_e^2 \Phi_D - g\partial \Phi_D / \partial z = 0 \qquad \cdots\cdots(2)$$

, where

g	=	gravity acceleration
z	=	coordinate in vertical direction
ω_e	=	circular frequency of encounter ($=\omega$-KVcos χ)
ω	=	circular frequency of incident wave
K	=	wave number
χ	=	direction of incident wave

This is the same as for the radiation problem. Ignoring the ship-lengthwise derivative, however, is not adequate for diffraction problem because the incident wave varies with wave number K. Another problem is that vertical shearing force and moment diverge at $\omega_e = 0$ due to the divergence of 2-dimensional added mass.

The present system has overcome this problem as mentioned below.

Assuming Φ_D varies with K such that

$$\Phi_D = \phi_D \exp(-iK_x \cos\chi + i\,\omega_e t) \qquad \cdots\cdots(3)$$

an alternative 2-dimensional free surface condition for diffraction problem

$$\omega^2 \Phi_D - g\partial \Phi_D / \partial z = 0 \qquad \cdots\cdots(4)$$

is obtained, which gives no singularity at $\omega_e = 0$.

Another insufficient point of SCS(NSM) is that prediction of fluctuating pressures for short waves does not have good accuracy. This can be attributed to the simplified procedure in NSM for diffraction problem that boundary condition is imposed only at one representative point instead of all over the hull surface. The present system adopts the exact boundary condition .

Furthermore, the present system uses 2-dimensional singularity distribution method, which makes the system applicable for arbitrary-shaped cross section like bulbous form, while SCS uses multi-pole expansion method and has less applicability.

ii) Improvement of computational method of wave force on hull section

Gravitational force and buoyancy are not necessarily balanced on each cross section, which yields static components of vertical shearing force and bending moment. When the ship rolls, this imbalance acts as dynamic components of horizontal shearing force and bending moment, which become significant at roll resonance.

The system takes this effect into consideration. An example is shown for horizontal shearing force as follows :

$$\int dF_{IY}/dx\,dx = \omega^2/g \int w(x)(y_G + x\,\psi)\,dx + \underline{(\int B(X)\,dx - \int w(x)\,dx)\,\phi} \qquad \cdots\cdots(5)$$

, where

F_{IY}	=	horizontal shearing force
w(x)	=	weight distribution
B(x)	=	buoyancy distribution
y_G	=	sway motion
ϕ	=	roll motion
ψ	=	yaw motion

Eq.(5) shows an inertial component and the underlined is added for gravitational components.

Next example is influence of GM distribution on torsional moment. Integration of restoring roll moment along the ship length

$$\rho\,\phi \int \nabla(x)\,GM(x)\,dx \qquad \cdots\cdots(6)$$

, where

ρ	=	density of sea water
∇(x)	=	displacement volume distribution
GM(x)	=	metacentric height distribution

should be estimated precisely. In the present system, GM distribution is calculated, assuming that gravity center on each section lies in the constant height.

(2) Constitution of the system

The present system is composed of following five STEPs. Users can re-start analysis from any STEP, because input and output data are stored in common files and used in each STEP.

STEP1 for generation and check of input data
STEP2 for 2-dimensional hydrodynamic analysis
STEP3 for motion analysis
STEP4 for analysis for wave force on hull section, fluctuating pressure and acceleration
STEP5 for output of results of analysis

(3) Example of analysis

Let us highlight conditions and items on which the present system and SCS show significant differences.

Firstly, computed results of vertical shearing force and bending moment at midship of SR108 container ship in the following waves are shown in Fig. 1 and Fig. 2, respectively, together with experimental results (Kobayashi, K. et al., 1981). Results of SCS diverge at $\omega_e = 0$ (λ/L=0.48; λ: Wave Length, L:Ship Length), while results of the present system show good agreement with experimental ones without divergence.

Secondly, computed results of torsional moment at midship of SR108 container ship are shown in Fig. 3 comparing with experimental results (Kobayashi, K. et al.,1981). Differences of computed results of torsional moment become marked due to the consideration of GM distribution. The computed result of torsional moment considering GM distribution has a distinct peak at the roll resonance period.

Thirdly, the present system can give a realistic fluctuating pressure distribution: on weather-side hull of a VLCC, the pressures are higher than those on lee-side hull as shown in Fig. 4. The result of the present system shows good agreement with experimental one because of strict boundary condition on hull.

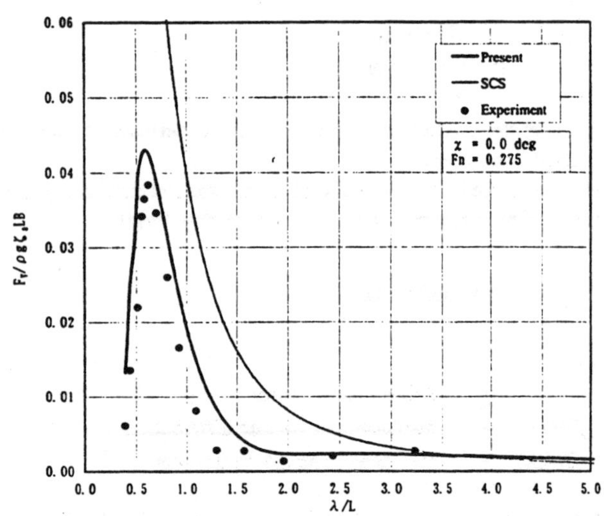

Fig.1 Vertical Shearing Force at Midship (Container Ship)

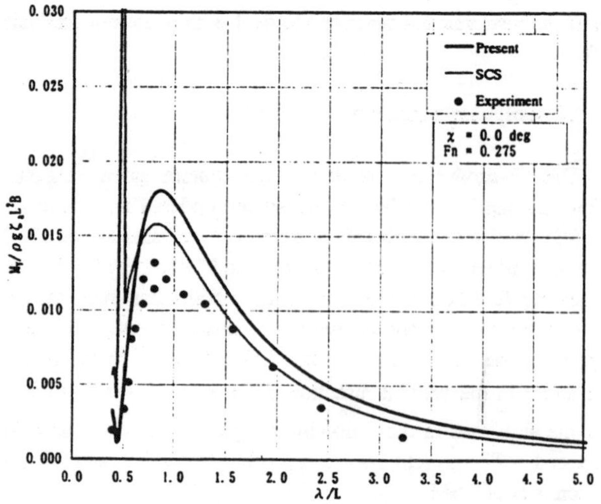

Fig.2 Vertical Bending Moment at Midship (Container Ship)

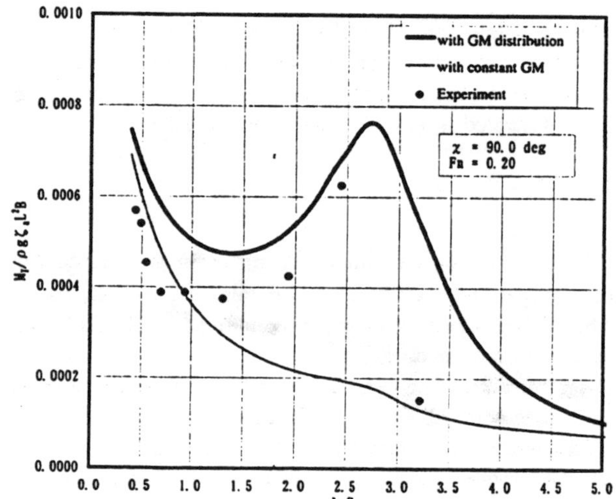

Fig.3 Torsional Moment at Midship (Container Ship)

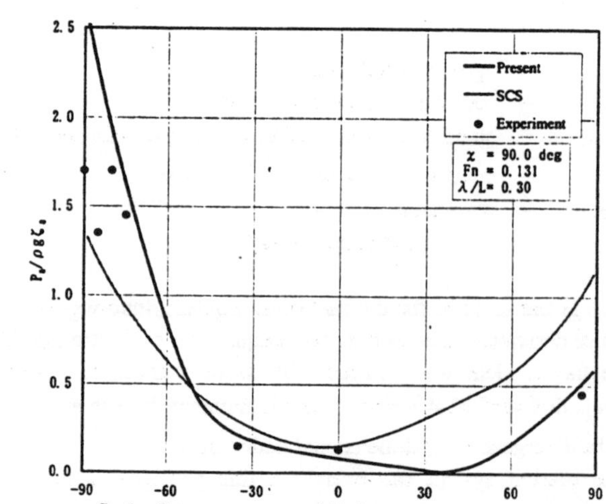

Fig.4 Hydrodynamic Pressure along Girth at Sq. St. 4.0 (VLCC)

2.2 Automatic Load Data Generation

The "automatic load data generation" sub-system generates load data for FEM analysis corresponding to various loads acting on ship in wave, where the results of hydrodynamic analysis, such as ship motion, acceleration and dynamic sea pressure are used. Following loads are generated by this sub-system.

a) sea pressure (static, dynamic)
b) internal tank pressure (static, dynamic)
c) inertia force (gravity, dynamic inertia force due to ship motion)

Necessary operations for users are only to define basic information of the ship (e.g. outer shell configuration, tank arrangement, weight distribution) as attributes of FEM model, using advanced graphical user interface. By making use of these informations and the results of hydrodynamic analysis, load data for FEM analysis is generated by the system automatically.

The following methods of load data generation are prepared ;

1) Momentary Load Direct Generating Method
2) Discrete Analysis Method

The detail procedure of generating loads data using above two methods is described below.

(1) Momentary load direct generating method

When starting the sub-system, a user is requested to input following data.

- Principal particulars of the ship (Length, Beam, Depth, ...)
- Load condition and wave condition in which load data is generated

These conditions must be selected among those analyzed by hydrodynamic analysis.

- Time subdivision within the wave period for which load data is generated

Then, load data is generated according to following procedures.

i) Sea pressure

In the first place, a user needs to define ship's outer shell by selecting FEM elements on CRT. Then, static sea pressures corresponding to the draft and the trim of each load condition are generated for elements below waterline automatically.

Amplitudes and phases of dynamic sea pressures in wave are separately calculated by hydrodynamic analysis at several calculation points in several sections. By 3-dimensional interpolation from these results, momentary dynamic sea pressures at each node of outer shell of FEM model are calculated and load data of FEM analysis is generated.

Generated load data of sea pressures can be checked by various graphical outputs on CRT such as vector plot, color contour plot and numerical value plot. For example, Fig. 5 shows a contour plot of dynamic sea pressure on VLCC in diagonal sea.

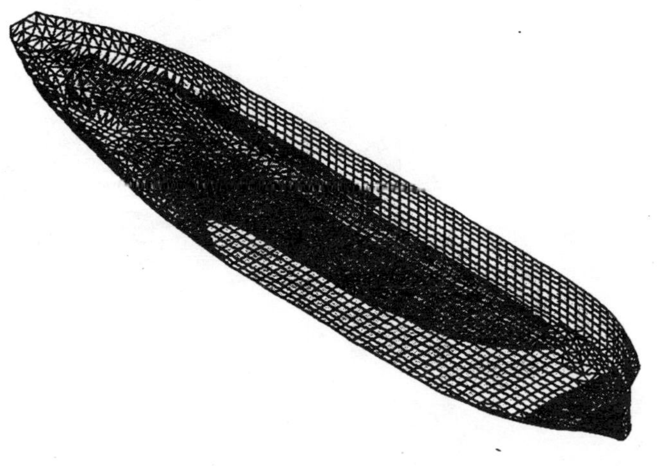

Fig.5 Contour Plot of Dynamic Sea Pressure on VLCC in Diagonal Sea (χ=150 deg, λ/L=0.6, Hw=8.0m)

ii) Internal tank pressure

Firstly, tank configurations are defined on CRT by selecting FEM elements which compose the boundary of each tank, then the name of the tank is given.

Secondly, loading pattern of each load condition is defined by specifying filling ratio (or liquid level/weight) of each tank.

Finally, static and dynamic tank pressures are calculated considering dynamic accelerations at the center of each tank, roll angle and pitch angle due to ship motion obtained by hydrodynamic analysis.

Similar to sea pressure, generated load data can be checked by various graphical outputs on CRT. For example, a contour plot of dynamic tank pressure is shown in Fig. 6.

iii) Inertia force

Weight distribution of the ship (light weight, deadweight) except liquids in tanks is defined as density of FEM elements or concentrated mass at nodes.

Inertia forces acting on each node can be calculated by multiplying the mass by dynamic acceleration at the node.

Other additional functions of this sub-system are as follows :

- automatic generation of symmetric and anti-symmetric load components for half breadth model
- graphical plot of ship motion and wave surface (Fig. 7)

(2) Discrete analysis method

In case when the number of loading cases of FEM analysis become huge due to many load and wave conditions, discrete analysis method as described below is an efficient way. In discrete analysis method, FEM analysis is carried out for each discrete load pattern, and structural response in each wave condition is obtained as a combination of the results for unit discrete load.

i) Definition of discrete load patterns

The key factor to conduct such an analysis with accuracy under the limited number of unit loads is to define discrete load patterns properly

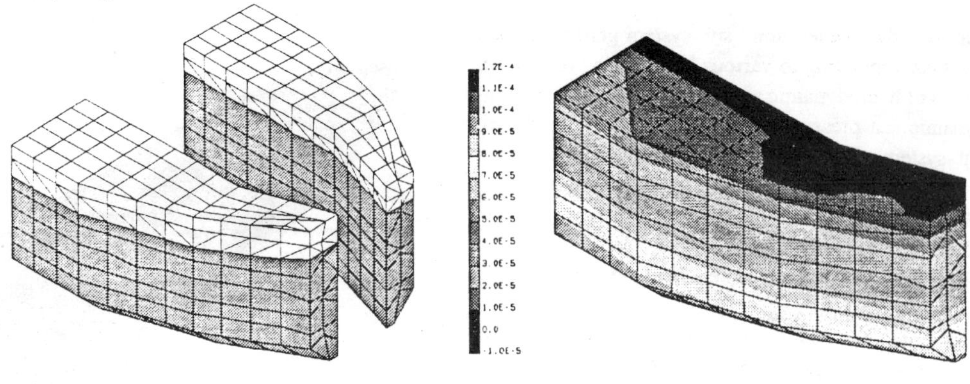

(a)Loading Level Diagram (b)Pressure Contour Plot

Fig.6 Generation of Internal Tank Pressure

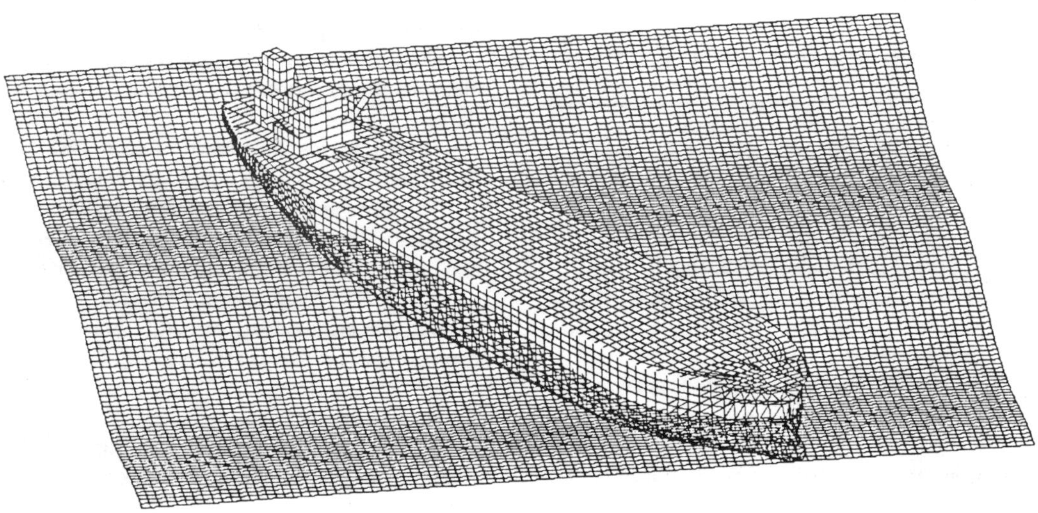

Fig.7 Graphical Display of Ship Motion in Regular Wave(χ =150deg, λ /L=0.6, Hw=8.0m)

taking account for the characteristics of each load component. This sub-system can generate following discrete load patterns easily.

Since internal tank pressure and inertia forces are dependent on ship motion and acceleration, it is possible to express any distribution of these loads in various wave conditions as linear combination of discrete load patterns corresponding to 6 components of ship motion and inclination due to pitch and roll.

On the other hand, it is not so easy to express dynamic sea pressure distribution by combination of a few discrete load patterns. This sub-system provides following 3 methods for discrete patterns of sea pressure.

a) linear approximation (Fig. 8-a)
b) bi-quadratic approximation (Fig. 8-b)
c) 2-dimensional Fourier series expansion (Fig. 8-c)

Users shall select one of these methods and define grids on hull surface, then the discrete load patterns for FEM analysis are generated.

ii) Calculation of load magnification factor

Load magnification factors are defined as coefficients of linear combination of each discrete load pattern which compose momentary load distribution in each wave condition. These factors are calculated from the results of hydrodynamic analysis and saved in a file.

In "structural analysis" sub-system and "strength assessment" sub-system, combined results of discrete load patterns with load magnification factors are used. The combination of the results of discrete load patterns can be achieved using a function of "structural analysis" sub-system.

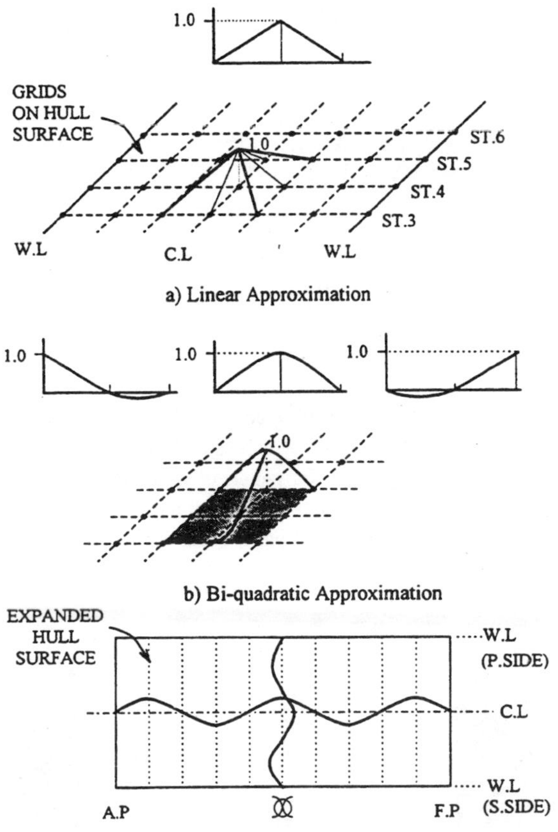

a) Linear Approximation

b) Bi-quadratic Approximation

c) 2-dimensional Fourier Series Expansion

Fig.8 3 Discrete Patterns of Sea Pressure

2.3 Structural Analysis

The "structural analysis" sub-system consists of a customized pre/post-processor of KSWAD and a solver of MSC/NASTRAN for FEM analysis. Following functions are specially added to deal with a large mass of data quicker and more practicably.

(1) Geometric data interface

This sub-system can convert in-house hull lines system SILC (= Surface Image Line Creator, Arase et al., 1991) data and 2-dimensional CADAM data into the pre/post processor as geometrical shape data. When using CADAM data, the data is converted via standard format file.

(2) Anisotropic-plate element

The sub-system generates anisotropic elements from sectional properties and configurations of both stiffened plate and corrugated plate. Instead of anisotropic element stresses, the sub-system calculates "plate" stresses so as to be directly used for the strength assessment.

Using the anisotropic plate elements with both in-plate and out-plate stiffness of those plates, a whole ship model analysis can be performed under a limited number of elements.

(3) Mesh generation for zooming analysis

This sub-system can generate fine meshes from coarse mesh and create boundary displacements at the fine mesh nodes by interpolation.

2.4 Statistical Processing

The "statistical processing" sub-system can give expected design parameters by performing the frequency domain short- and long-term predictions.

In addition to the ordinary functions of statistical processing, two functions, i.e., the long-term prediction considering non-linearity against wave height and the ship operational profile, have been introduced. Also, databases for statistical prediction and the function of graphical/tabulated output on CRT enable a user to perform an efficient and speedy design work.

(1) Flow of statistical processing

Input data for this sub-system is RAOs (= Response Amplitude Operator) of ship motion, acceleration and dynamic sea pressure, which is calculated by preceding hydrodynamic analysis, and taken into this sub-system through the standard file. If necessary, a user can define RAOs directly as input data of this sub-system. This sub-system has another function to combine the several RAOs considering their phase lag, and the user can generate new RAOs with the combined effect of each load component easily.

The databases of this sub-system are as follows :

- Wave spectrum ; ISSC spectrum, JONSWAP spectrum
- Wave directional distribution ; $\cos^n \theta$ distribution, triangular distribution
- Wave scatter data ; BMT, WALDEN, S&O, SR163

The wave scatter data for a certain sea route can be obtained easily by combining the data of each area referring to the chart on CRT. The additional data defined by the user can be added to the above database.

(2) New statistical processing

The following new functions are introduced in this sub-system.

i) Long-term prediction considering non-linearity

The linearity against wave height is the basic assumption for the frequency domain statistical prediction. However it is well known that some load components such as rolling motion and dynamic sea pressure around waterline have significant non-linearity against wave height. In order to perform rational design, the ship designers are required to consider this effect if it is dominant. For this purpose, this sub-system make it possible to define several RAOs for each group of significant wave height, and the user can obtain the expected value considering the effect of non-linearity approximately. An example of calculation results is shown in Fig. 9. In this figure, the expected dynamic sea pressure considering non-linearity against wave height is considerably reduced from that of linear assumption.

ii) Long-term prediction considering the ship operational profile

In general, structural design is performed under the expected load which is obtained from the assumption that the number of encounter waves in her life is 10^8 and she serves worldwide trading. Each ship has individual operational profile such as the number of days in operation per year and the design speed corresponding to sea state etc.

In this sub-system, the new function has been added, which can count the number of encounter waves based on her operation profile considering the encounter period to wave and can define RAOs for each sea state. Accordingly, structural design could be performed under the expected parameter based on individual operational profile. This function may become especially useful for high speed craft which has an operational restriction of sea state.

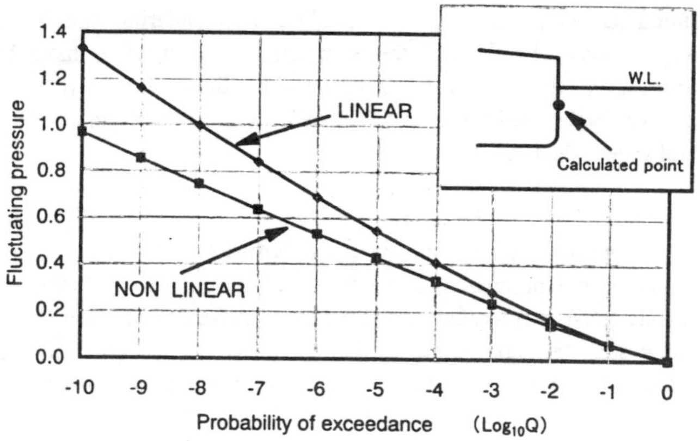

Fig.9 Long-term Prediction of Dynamic Sea Pressure considering Non-linearity against Wave Height

2.5 Strength Assessment

The "strength assessment" sub-system is used for assessment of the stress obtained by structural analysis in the light of yielding, buckling and fatigue criteria.

The sub-system can efficiently extract necessary information for the determination of scantlings from massive analysis results. Such system is indispensable when carrying out integrated analysis of load and structure using a whole ship model, since both the number of load conditions and the number of structural members subject to strength assessment are huge.

(1) Yielding assessment

To begin with, users shall define the stress components to be assessed and the allowable stress values for each loading conditions on an input panel, and define relevant elements to be assessed.

Then, the sub-system calculates the usage factor of each element for specified load conditions. The calculated usage factor can be displayed on CRT as color plot or value plot.

The sub-system has a function to show the most critical usage factor among examined loading conditions. The most critical usage factor can be displayed as color plot and the corresponding loading condition number is displayed on the element at the same time as shown in Fig. 10.

Finally, the system estimates necessary reinforcement (increase of plate thickness/sectional area) and display them on CRT.

(2) Buckling strength assessment

Recently, the buckling strength assessment has become decisive due to extensive usage of higher tensile strength steel, so that it requires considerable amount of design work.

MES developed a computer system for buckling strength assessment which extract working stress for plate panel to be examined from the result of FEM analysis and conduct buckling strength assessment according to buckling strength criteria given by Classification Rules. The system has been used for the design of many merchant ships.

This sub-system has been added some new functions for the purpose of efficient assessment of large amount of analysis results obtained by whole ship FEM analysis under many loading conditions. The new functions are as follows :

a) efficient definition of assessment panels

The assessment panels are defined using the advanced functions of FEM preprocessor for mesh generation. If necessary, CAD data can be utilized.

b) automatic calculation of reinforcement measures

It has been one of the most time consuming work to determine the best reinforcement measures (increase of plate thickness and/or panel breaker arrangement) for panels of which buckling strength is found not sufficient. The sub-system estimates the required plate thickness and effective zone of panel breakers.

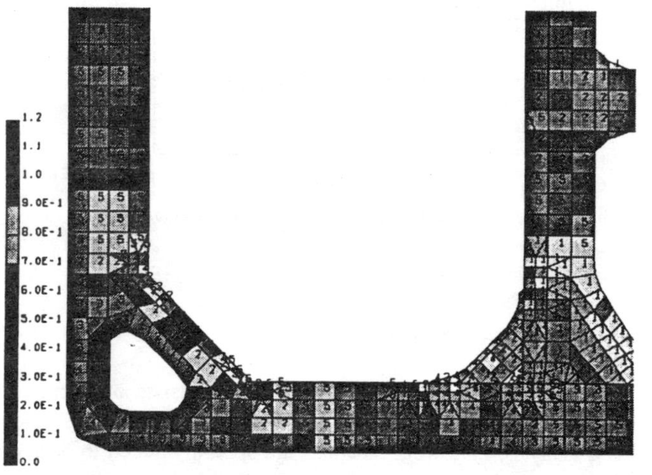

Fig.10 Severest Usage Factor(Figures on elemonts denote the loading condition number which gives the severest usage factor.)

(3) Fatigue strength assessment

This sub-system can be applied to following two methods of fatigue assessment.

<u>Simplified Analysis</u> - Fatigue strength is assessed by comparing stress amplitude in design wave and allowable stress amplitude calculated from assumed Weibull parameter and specified S-N curve.

<u>Detail Analysis</u> - Fatigue strength is assessed by comparing damage factor and/or fatigue life calculated from long-term distribution of stress obtained by "statistical processing" sub-system with allowable damage factor or specified life time of the ship.

The functions of the system is described below, which are common for the 2 methods except specified.

i) Definition of Locations for Fatigue Assessment and Stress

Components to be considered

Assessment point is defined as center of elements, node, or any points specified by coordinates. The stress components subject to fatigue assessment is to be also specified for each assessment point.

ii) Selection of S-N Curve

S-N curve can be defined by users and can be stored in database. Users shall specify an S-N curve used for each location for fatigue assessment.

iii) Calculation of Stress used for Fatigue Assessment

Simplified Analysis - Calculation of Working Stress Amplitude

Stress amplitude at the locations in each loading condition is calculated according to the interpolation method specified for the location, and combined according to the characteristics of loading conditions as follows :

- difference of maximum and minimum values within one cycle of wave period
- square summation of SINE and COSINE components
- combination using coefficient of correlation
- any combination with linear combination and square summation

Detail Analysis - Generation of Stress RAO and Results of Statistical Processing

Stress amplitude in various regular waves are calculated from the results of FEM analysis and they are saved as a response function of stress in the input file of "statistical processing". Then, long-term prediction considering wave scatter diagram of specified area is carried out by "statistical processing" sub-system. The results of long-term prediction stored in output file are taken into this system.

iv) Fatigue Assessment Calculation

Using the data prepared as above, fatigue assessment calculation according to each method is carried out.

3. CONCLUDING REMARKS

A new integrated system of ship structure analysis is developed aiming at the advancement and rationalization of structural design. It is expected to be a powerful tool for "design by analysis". We are going to do our best efforts to enhance the structural reliability and rational structural designs by making use of this system.

REFERENCES

Akita, Y. et al. (1974). "Total System of Analysis on the Longitudinal Strength of Ships", *J. Soc. of Naval Arch. of Japan*, Vol.135 (in Japanese)

Arase, S. et al. (1991). "CAD/CAM System for Ship Hull Structure", *4th International Marine Systems Design Conference*

Kanetsuna, M. et al. (1978). "A Design and Analysis Program for Offshore Structures", *Mitsui Zosen Technical Review*, No.103 (in Japanese)

Kobayashi, M. et al. (1973). "A Computer Program for Theoretical Calculation of Sea-Keeping Quality of Ships (Part1)", *Mitsui Zosen Technical Review*,No.82 (in Japanese)

Kobayashi, K. et al. (1981). "Model Tests on Ship Motions and Wave Loads of a Container Ship in Oblique Waves", *Technical Review of Sumitomo Heavy Industries, Ltd.*, Vol.29, No.86 (in Japanese)

Nakata, Y. et al. (1975). "Statistical Prediction of Normal Stress of Container Ship", *J. Soc. of Naval Arch. of Japan*, Vol.137 (in Japanese)

Proceedings of the Seventh (1997) International Offshore and Polar Engineering Conference
Honolulu, USA, May 25-30, 1997

ISBN 1-880653-28-1 (Set); ISBN 1-880653-32-X (Vol. IV)

The Influence of a Ship's Rudder on the Scouring Action of a Propeller Wash

G.A. Hamill and J.A. McGarvey
Queen's University of Belfast
Belfast, United Kingdom

ABSTRACT

The problems which have become associated with the scouring action of propeller jets is common knowledge to those working in the design or maintenance of harbours and inland waterways. Excessive movements of sediment can result in dangerous undermining of quay structures and in the release of pollutants from within the sea bed. The need to predict the maximum depth of scour is essential to the design of quay structures. Based on an experimental investigation, the influence of the ship's rudder on the scouring process is discussed.

KEY WORDS

Propeller scour, harbour, sediment movement, jets, quay structures.

INTRODUCTION

One consequence of the increases in ship size, installed engine power, and volume of marine transport is an increased attack by propeller slipstreams on the bed and banks of harbour basins and navigation channels. A ship's propeller produces thrust by drawing in water, accelerating it and discharging it in the form of a wash. This wash is a highly turbulent, submerged jet, the velocities within which depend upon the operating characteristics of the propeller and the speed of advance of the ship. The wash entrains the surrounding water and so the velocities within it decay with distance from the propeller.

In this way the jet expands and dissipates its energy by diffusion. If the ship is moving in a confined area, for example a waterway or in shallow waters, the diffusion process within the wash will be restricted and the energy remaining within the jet may cause damage to the bed and banks. The presence of the ship's rudder splits the wash into two streams, Hamill & McGarvey (1996), with one stream being directed towards the sea bed.

RELATED RESEARCH

A survey of harbours in Sweden, Bergh (1981) has shown that 16 out of the 18 ports surveyed have suffered propeller induced damage in recent years. Similar problems have been encountered in several United Kingdom ports and the British Ports Association commissioned a review, Prosser (1986) of existing knowledge.

Limited work is available on the influence of the rudder on the wash produced by the rotating propeller and its scouring capabilities. The majority of the work available deals with methods for predicting velocity distributions within the wash. The work of Fuehrer (1985), Robakiewicz (1987) and Verhey (1983) provide limited details of the processes involved in the diffusion of such a wash around a ships propeller. Qurrain (1994) proposed an equation that would allow the maximum depth of scour, at any time, to be calculated for a propeller operating in the absence of a rudder. This equation took the form

$$\varepsilon_u = \Omega\left[\ln(t)\right]^{\Gamma} \qquad 1$$

where ε_u is the maximum scour depth in mm at any time t, in seconds. The other coefficients in the equation depend on the propeller size and sediment characteristics and are given by the equations

$$\Omega = \left[\frac{C}{d_{50}}\right]^{-4.758}\left[\frac{D_p}{d_{50}}\right]^{2.657}\left[F_0\right]^{3.517} \qquad 2$$

and

$$\Gamma = \left[\frac{C}{d_{50}}\right]^{0.768}\left[\frac{D_p}{d_{50}}\right]^{-0.339}\left[F_0\right]^{-0.479} \qquad 3$$

where C is the clearance between the bed and the propeller, d_{50} is the average sediment size, D_p is the propeller diameter and F_0 is the densimetric Froude number which is given by

$$F_0 = \frac{V_0}{\sqrt{g\, d_{50} \dfrac{\Delta\rho}{\rho}}}$$

where V_0 is termed the efflux velocity of the propeller wash.

It was not until the recent work of McGarvey (1996) however, that a full understanding of the diffusion process and subsequent scouring has become available. McGarvey showed that the wash was split into two streams, one which was directed towards the sea bed while the other was directed towards the water surface. As the rudder was turned into each of these streams the diffusion pattern of the wash changed. This paper forms part of the findings of this work.

EXPERIMENTAL SET-UP

In order to investigate the influence of the ships rudder the washes from two different propellers (131mm and 76 mm in diameter), each rotating at three speeds, acting over two beds of different sediment size, were investigated in a laboratory tank 7.5m x 4.4m x 1m deep. The sediment was well sorted so that the threshold of movement was similar for each grain. The experimental rudders were manufactured in accordance with the Ship Design Manual. According to a literature survey, an angle of 35 degrees was found to be the practical limit within which a rudder is an effective turning device. It was decided to conduct tests with the rudders at angles that varied between 35 degrees on either side of the propeller axis centreline. The development of the scour pit was monitored with time, until an asymptotic state was achieved.

GEOMETRY OF THE SCOUR PROFILE

The geometry of the scour profile produced by a propeller jet depends on the water velocity induced by the propeller, and the clearance between the bed and the propeller. The final eroded profiles obtained for the 131 mm diameter propeller without the rudder present was observed and plotted as shown in Figure 1. These results were obtained for the jet acting on a fine sediment and the geometry is typical for that obtained for all speeds for a jet without a rudder. It was observed that the maximum depth of scour lies on the propeller centreline with the erosion and the subsequent deposition approximately symmetrical about the propeller axis. The scour depths along the propeller centreline were plotted as shown in Figure 2. There is an increase in erosion up to a distance of approximately 6 D_p and as the velocities decay with distance, there is a subsequent reduction in the scouring to 10 D_p, and thereafter deposition of the eroded material. The eroded profiles are approximately symmetrical about the propeller axis, and this is due to the symmetrical diffusion of the propeller jet without a rudder.

Figure 1 Isometric View of Final Eroded Profile Without the Rudder Present

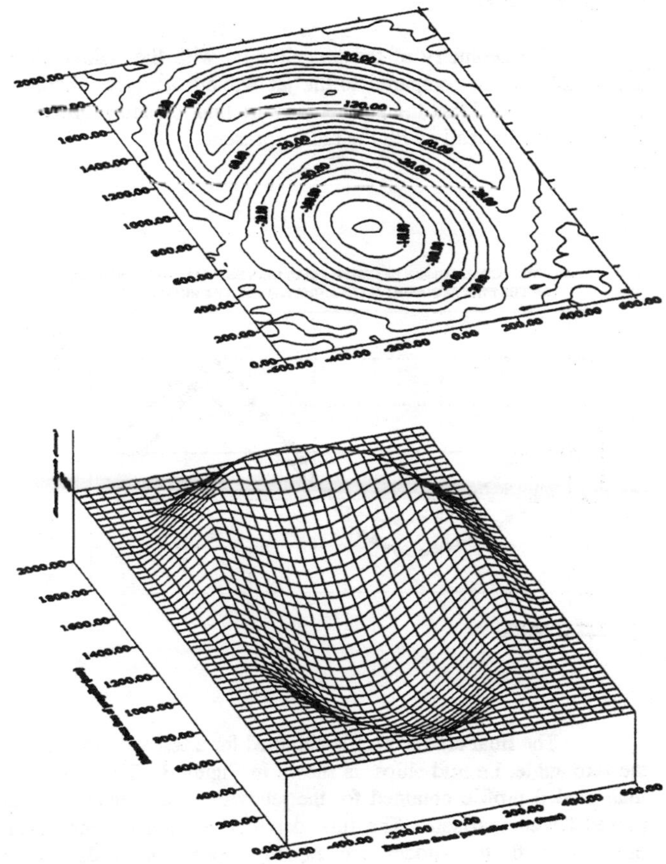

131 mm Diameter Propeller
Propeller Speed = 500 rev/min

FIGURE 2 FINAL ERODED PROFILE MEASURED ON FINE SEDIMENT WITHOUT THE RUDDER PRESENT

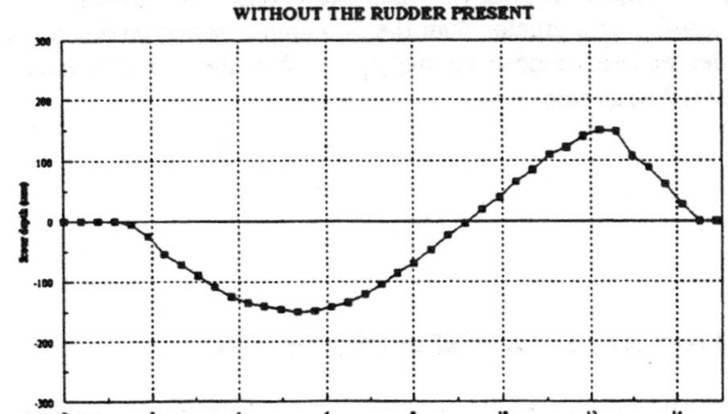

McGarvey (1996) showed that the rudder dramatically changes the diffusion characteristics of a propeller wash. The rudder splits the flow into two, separate, high velocity streams, one directed towards the surface, and the other directed downwards towards the bottom. The angle of diffusion of the jet was greater and the magnitudes of the velocities within the propeller jet were greater than that expected without the rudder.

The geometry of the propeller jet with the rudder changed dramatically when compared to the jet without a rudder therefore it would be expected that the geometry of the final eroded profiles would be altered dramatically with the rudder present.

FIGURE 3 COMPARISON BETWEEN FINAL DEPTHS OF EROSION MEASURED ON THE FINE SEDIMENT WITH AND WITHOUT THE RUDDER PRESENT

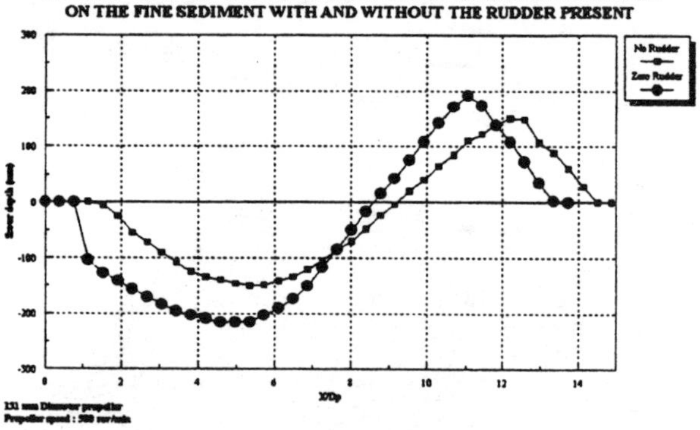

The final eroded profile obtained for a test with the rudder at the zero angle, i.e mid ships, is shown in Figure 3. The depth of the final eroded profile obtained for the jet without the rudder was also plotted for comparison. The final depth of erosion has significantly increased with the rudder present, by approximately 25 %, with greater depths of erosion experienced much closer to the propeller. This is due to the greater diffusion angle of the jet, with the rudder present, indicating that within this region the magnitude of the velocities are greater. The maximum depth of erosion was found to be greater with the rudder positioned at zero degrees in all of the experiments conducted, when compared to the scouring for the jet without a rudder.

The final eroded profiles of the scour pits produced at different rudder angles were investigated. The maximum scour depth for negative rudder angles, or when the rudder was turned into the bottom stream, was greater than the maximum scour produced when the rudder was set at positive angles, or when the rudder was turned into the surface stream.

LOCATION OF MAXIMUM SCOUR DEPTH

The location of the maximum eroded depth has been shown to be independent of the propeller rotational speed, Stewart (1992). It was decided to compare the location of the maximum velocity within the jet with the location of maximum erosion, as it was expected that they would correspond to approximately the same location. Figure 4 shows the final eroded contours obtained for one of the tests carried out, and is typical of all the results obtained.

The location of the maximum velocity, based on measurements made by the authors, transversely within the jet is superimposed on this figure for comparison. It can be seen that the location of the maximum eroded depth corresponds quite well to the location of the maximum velocity.

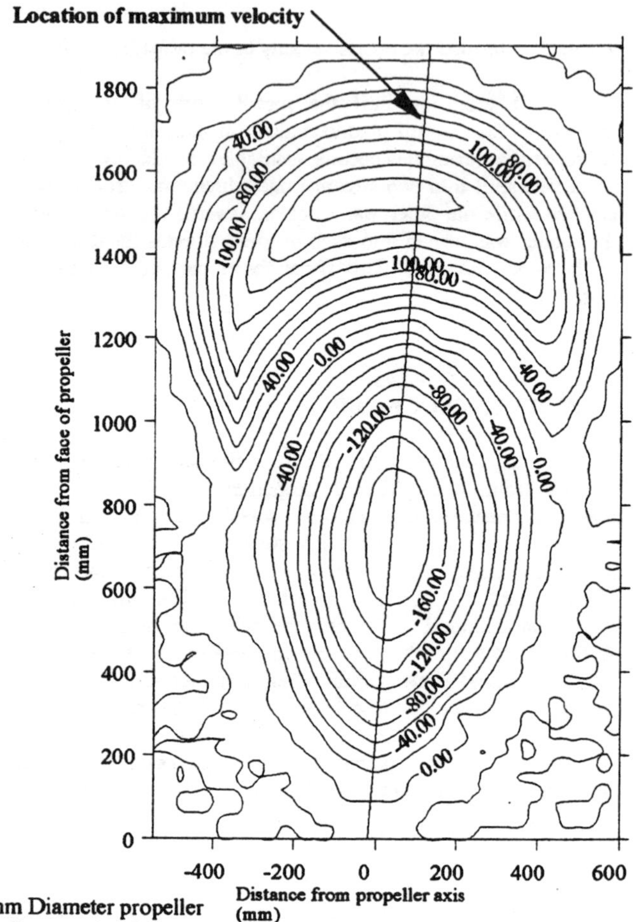

76 mm Diameter propeller
Propeller speed : 1500 rev/min

Figure 4 Location of Maximum Scour compared to Locus of Maximum Velocity

This trend was observed for all rudder angles tested with only a slight variation observed between the locations of the maximum velocity and the location of the maximum eroded depth. Therefore, the equations developed to locate the locus of the maximum velocities, Hamill & McGarvey (1996), may be used to locate the region within which the maximum eroded depth behind the propeller will occur.

PREDICTION OF MAXIMUM SCOUR DEPTH

The presence of the rudder has been shown to change the geometry of the final eroded profiles. There was a significant increase in the final depth of erosion when compared to the scour depths observed for the jet without a rudder, and in some tests by a factor of 2 or greater. However at the 35 degree rudder position it was observed that some of the final eroded depths were approximately 80 to 90% of the maximum depth observed for the jet without a rudder.

It was therefore decided to express the maximum eroded depth with the rudder present, ε_{mr}, as a ratio of the maximum depth observed for the jet without a rudder, ε_u. Analysis revealed that an arithmetic model provided the best correlation, with an R^2 of 0.932 and a standard error of 0.128, for the measured data as follows,

$$\frac{\varepsilon_{mr}}{\varepsilon_u} = 0.75 - 0.07\,F_0 + 0.02\left(\frac{D_p}{d_{50}}\right) - 0.15(1+\alpha) \qquad 4$$

where α is the rudder angle, measured from the propeller axis of rotation, and all other terms are as defined earlier.

CONCLUSION

The rudder configuration had a significant effect on the geometry of the final eroded profile and location of the maximum depth of erosion. The location of the maximum eroded depth on the bed, was found to be dependent on the magnitude and placement of the rudder angle. It was found that the location of the maximum eroded depth corresponded well to the location of the maximum velocity within the jet. The equations presented by Hamill &McGarvey (1996) to determine the locus of the maximum velocities within the jet may also be applied to the region at which the position of the maximum eroded depth will occur.

It was found that with an increase in rudder angle there was a reduction in the length of the final eroded profile, however there was more erosion at distances closer to the propeller.

There was a significant increase in the final depth of erosion with the rudder present, when compared to the jet without a rudder. The investigation revealed that the final eroded depth could be increased by up to a factor of 3 with the rudder present.

The maximum eroded depth with the rudder present was established as a function of the maximum depth for the jet without a rudder, and the various experimental parameters including the rudder configuration.

REFERENCES

BERGH H., CEDERWALL K., (1981)" Propeller erosion in harbours", Bulletin No TRITA-VBI-107, Hydraulics Laboratory, Royal Institute of Technology, Stockholm, Sweden.

FUEHRER M., POHL, H., ROMISCH K., (1987), " Propeller jet erosion and stability criteria for bottom protection of various constructions.", P.I.A.N.C., Bulletin No. 58, 1987

HAMILL, G.A., McGARVEY, J.A., (1996) "Designing for propeller action on harbours" ICCE'96, September 1996, Orlando, Florida, USA.

McGARVEY, J.A, (1996) "The influence of the rudder on the hydrodynamics, and the resulting bed scour, of a ship's screw wash" Thesis submitted to the Queens University of Belfast for the degree of Doctor of Philosophy.

PROSSER, M.J., (1986) " Propeller induced scour", British ports association report No. RR2570, February 1986.

QURRAIN, R., (1994)" Influence of the sea bed geometry and berth geometry on the hydrodynamics of the wash from a ships propeller ", Thesis submitted to the Queens University of Belfast for the degree of Doctor of Philosophy.

ROBAKIEWICZ, W., (1987), " Bottom Erosion as an Effect of ship propeller action near the harbour quays ", PIANC, Bulletin No.58, 1987, pp 89-106.

STEWART, D.P.J.,(1992), " Characteristics of a ships screw wash and the influence of quay wall proximity ", Thesis submitted to the Queens University of Belfast for the Degree of Doctor of Philosophy.

VERHEY, H.J.,(1983), " The stability of bottom and banks subjected to velocities in the propeller jet behind ships ", Delft Publication No.303, April 1983, Delft Hydraulics Laboratory, Netherlands.

Proceedings of the Seventh (1997) International Offshore and Polar Engineering Conference
Honolulu, USA, May 25-30, 1997

ISBN 1-880653-28-1 (Set); ISBN 1-880653-32-X (Vol. IV)

A Finite-Volume Method for Three Dimensional Viscous Flow Around Ship Stern

Tieli Li, Yan Lin and Yanying Wang
Dalian University of Technology
Dalian, China

ABSTRACT

In this paper, a numerical method (FVM) for solving the Reynolds-averaged Navier-Stokes equations of three-dimensional viscous flow around ship stern in nonorthogonal body-fitted coordinate(BFC) has been developed. The Navier-Stokes and the continuity equations in Cartesian coordinates are transformed to a generalized curvilinear coordinates using finite difference approximations in the transformed domain. Solutions of the finite difference equations are obtained iteratively by using a pressure-velocity correction algorithm (SIMPLE-C). Finally, calculating example of SSPA-720 arė given by this method.

INTRODUCTION

With the currently increasing computer capability, numerical simulations of 3D incompressible flow problems using Reynolds-average Navier-Stokes equations (RANSE) are now becoming more feasible in Naval Architecture and Marine Engineering. Numerical methods include finite difference method, finite element method ,finite volume method, etc. A finite volume method including SIMPLE, SIMPLEC, SIMPLER SIMPLEN has been applied for computation of ship stern flow, such as Patel V.C. (1990) in polar coordinates by using SIMPLER, Zhou L.D.(1994) in polar coordinates by using SIMPLEC, Cura Hochbaum A.(1994) in body-fitted coordinates by using SIMPLE. In this paper, SIMPLEC is applied for computation of viscous flow around ship stern in nonothogonal body-fitted coordinates.

MATHEMATICS MODEL

We consider the flow around a ship moving steadily ahead. The Navier-Stokes equations and the continuity equation are applied to describe incompressible flow for Newtonian fluid (like water) and can be written in Cartesian coordinates as:

$$\frac{\partial u}{\partial t}+\frac{\partial E}{\partial x}+\frac{\partial F}{\partial y}+\frac{\partial G}{\partial z}=S \qquad (1)$$

where (x, y, z) represent the Cartesian coordinates, and

$$u=[\rho,\ \ \rho u,\ \ \rho v,\ \ \rho w]^T$$

$$E=[\rho u,\ \ \rho uu-\mu u_x,\ \ \rho uv-\mu v_x,\ \ \rho uw-\mu w_x]^T$$

$$F=[\rho v,\ \ \rho vu-\mu u_x,\ \ \rho vv-\mu v_y,\ \ \rho vw-\mu w_z]^T$$

$$G=[\rho w,\ \ \rho wu-\mu u_x,\ \ \rho wv-\mu v_y,\ \ \rho ww-\mu w_z]^T$$

$S=$

$$\left[0,(\mu u_x)_x+(\mu v_x)_x+(\mu w_x)_x-P_x,(\mu u_y)_y+(\mu v_y)_y+(\mu w_y)_y-P_y,(\mu u_z)_z+(\mu v_z)_z+(\mu w_z)_z-P_z\right]$$

Equation (1) is transformed into a general curvilinear coordinate system $(\xi,\ \ \eta,\ \ \zeta)$, as follows:

$$\frac{\partial u}{\partial t}+\frac{\partial E}{\partial \xi}\frac{\partial \xi}{\partial x}+\frac{\partial E}{\partial \eta}\frac{\partial \eta}{\partial x}+\frac{\partial E}{\partial \zeta}\frac{\partial \zeta}{\partial x}+\frac{\partial F}{\partial \xi}\frac{\partial \xi}{\partial y}+\frac{\partial F}{\partial \eta}\frac{\partial \eta}{\partial y}+\frac{\partial F}{\partial \zeta}\frac{\partial \zeta}{\partial y}+\frac{\partial G}{\partial \xi}\frac{\partial \xi}{\partial z}+\frac{\partial G}{\partial \eta}\frac{\partial \eta}{\partial z}+\frac{\partial G}{\partial \zeta}\frac{\partial \zeta}{\partial z}=s \qquad (2)$$

where

$$\xi_X=J*(y_\eta z_\zeta-y_\zeta z_\eta)\quad \xi_y=-J*(x_\eta z_\zeta-x_\zeta z_\eta)\quad \xi_z=J*(x_\eta y_\zeta-x_\zeta y_\eta)$$

$$\eta_X=-J*(y_\xi z_\zeta-y_\zeta z_\xi)\quad \eta_y=J*(x_\xi z_\zeta-x_\zeta z_\xi)\quad \eta_z=-J*(x_\xi y_\zeta-x_\zeta y_\xi)$$

$$\zeta_X=J*(y_\xi z_\eta-y_\eta z_\xi)\quad \zeta_y=-J*(x_\xi z_\eta-x_\eta z_\xi)\quad \zeta_z=J*(x_\xi y_\eta-x_\eta y_\xi)$$

$$J=\left[x_\xi(y_\eta z_\zeta-y_\zeta z_\eta)-x_\eta(y_\xi z_\zeta-y_\zeta z_\xi)+x_\zeta(y_\xi z_\eta-y_\eta z_\xi)\right]^{-1}$$

The $k-\varepsilon$ turbulence model has been employed for turbulent flow computations,

$$(\rho k)_t+\left(\rho u_i k-\frac{\mu_{eff}}{\sigma_k}k_{xi}\right)_{xi}=\rho(G-\varepsilon) \qquad (3)$$

$$(\rho\varepsilon)_t+\left(\rho u_i\varepsilon-\frac{\mu_{eff}}{\sigma_\varepsilon}\varepsilon_{xi}\right)_{xi}=\rho\frac{\varepsilon}{k}(C_1 G-C_2\varepsilon) \qquad (4)$$

where the effective viscosity μ_{eff} is calculated from:

$$\mu_{eff}=\mu+\mu_t=\mu+\rho C_\mu k^2/\varepsilon$$

and the turbulent kinetic production term, G is defined as:

$$G = C_\mu \frac{k^2}{\varepsilon}\left[(u_y + v_x)^2 + (v_z + w_y)^2 + (w_x + u_z)^2 + 2(u_x^2 + v_y^2 + w_z^2)\right] \quad (5)$$

and the turbulence model constants can be taken from:

$C_\mu = 0.09,\quad \sigma_k = 1.0 \quad \sigma_\varepsilon = 1.3 \quad C_1 = 1.44 \quad C_2 = 1.92$

DISCRETIZATION OF CONSERVATION EQUATIONS

The discretization of the conservation equations for the Finite-volume approach is given by Dimirdzic et al. (1980) and Spalding(1972). Finite volume methods integrate over a single cell before dependent variables are approximated by the cell-centered nodal values. The RANSE can be described by the general transport equation:

$$\frac{\partial}{\partial t}(\rho\Phi) + \nabla\cdot(\rho\vec{V}\phi - \Gamma_\Phi \nabla\Phi) = S_\Phi \quad (6)$$

where Φ is one of the dependent variables, $u_i,\ k,\ and\ \varepsilon$ Γ_Φ are the corresponding diffusion terms and S_Φ is the source term. The complete discretization of (2) includes approximations of the convective, diffusive, and pressure terms at the cell sides and the mass term.

$$a_p \upsilon_p - \sum_{nb} a_{nb}\upsilon_{inb} = S_i + S_i^{(p)} \qquad nb = e,w,n,s,t,b \quad (7)$$

The non-negative coefficients a_{nb} of the unknown velocities in the neighbouring cell centers stem from convective fluxes through the side nb and the opposite side of cell P and diffusive flux through the side nb. The coefficient a_p of the considered cell is the sum of the coefficients a_{nb} of unknowns in the six adjacent cells. S_i covers the optional external force and explicitly treated parts of convective and diffusive terms. The pressure terms $S_i^{(P)}$ are treated explicitly within the used SIMPLEC method. The velocity components, u, v and w, are solved and stored at the grid nodes and the pressure P is located at the centers of control volume of u, v and w. Mass conservation equation is used later:

$$\sum_{nb} Fluss_{nb} = 0\ , \qquad nb = e,w,n,s,t,b \quad (8)$$

BOUNDARY CONDITIONS

Boundary conditions (BC) on all boundaries of the computational domain are :

The no-slip condition on the ship hull and a symmetry condition on symmetry planes.

The inlet boundary condiction: The x-component of velocity is set to negative ship speed. The other components are zero. The dissipation rate ε follows from

$$\varepsilon = C_u^{3/4} k^{3/2} / l$$

where the turbulent length $l = E * y$, $E = 0.42$, y is the distance from the wall.

The outlet boundary condiction: The gradients in x-direction of all variables are set to be zero . The downstream boundary plane, outer boundary, the center plane and the water surface are treated as planes of symmetry. On the ship hull, the wall function is used for the no-slip condition.

GRID GENERATION FOR 3D SHIP HULL

The ship hull geometry is given by β spline function, see Fig. 1. Grid is generated based on the solution of Poisson equations, given by Lin Yan (1996) etc.

SOLUTION PROCEDURES

First, the finite difference momentum equations (for u, v and w) can be written as:

$$a_p v_{ip}^* = \sum a_{nb} v_{inb}^* + S_i + S_i^{(p^*)} \quad (9)$$

where v_{inb}^*, p^* represent the solutions of equations (9) . To satisfy the continuity equation , the velocitiy componets and pressure are corrected according to the following relations:

$$p^{**} = p^* + p^1$$

$$v_i^{**} = v_i^* + v_i^1$$

$$Flux_{nb}^{**} = Flux_{nb}^* + Flux_{nb}^1$$

The following solution steps are contained for The SIMPLEC method.

(1) guess initial velocity and pressure field.
(2) solve for the velocity field using equations (9).
(3) solve for $k-\varepsilon$ equations.
(4) solve the pressure correction equation.
(5) correct the velocity and pressure fields.
(6) go back to step(2) until solution converges.

APPLICATION TO SHIP FLOWS

The flow around stem for a steadily advancing ship (speed u_0) is computed neglecting the free surface (double-body flow). Computation are limited to a certain Reynolds numbers (Re=5*10^6) for model SSPA 720. See Fig. 2.

REFERENCES

1 Broberg L. Calculation of ship stern flows using a body-fitted co-ordinate system, Division of Mechanics, Chalmers University of Technology, Gothenburg , Sweden, 1984

2 Patel V C, Chen H C. Ju S. Ship Stern and wake flows: Solution of the Fully-Elleptic Reynolds-Averaged Navier-Stokes Equations and Comparions with experiments. IIHR Report No.323,Iowa Institute of Hydraulic Research, University of Iowa, Iowa City , 1988

3 Cura Hochbaum A. A Finite-Volume Method for Turbulent Ship Flows. Ship Technology Research, Germany, 1994, Vol41, 135~148

4 Gao Q X, Zhou L d. Ship stern solver by 3D Unsteady RANS Equation. Journal of Hydrodynamics, 1994,9(4). p487~497.

5 Lin Y, Li T L. Three dimensional auto numerical grid generation around ship hulls by PDE method. Journal of Dalian University of Technology. 1996,36(4)

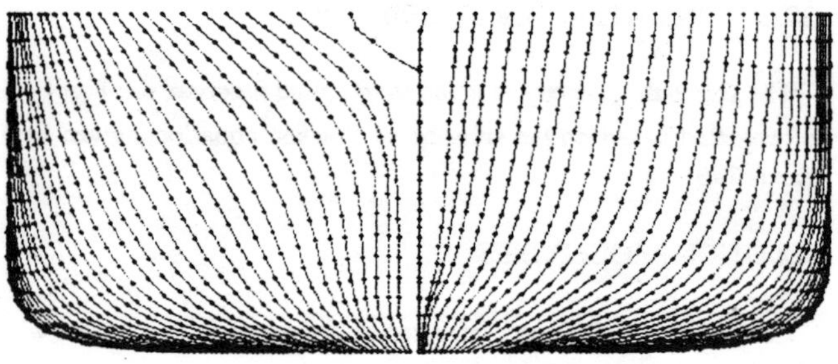

Fig 1. Grid generation for 3D ship hull

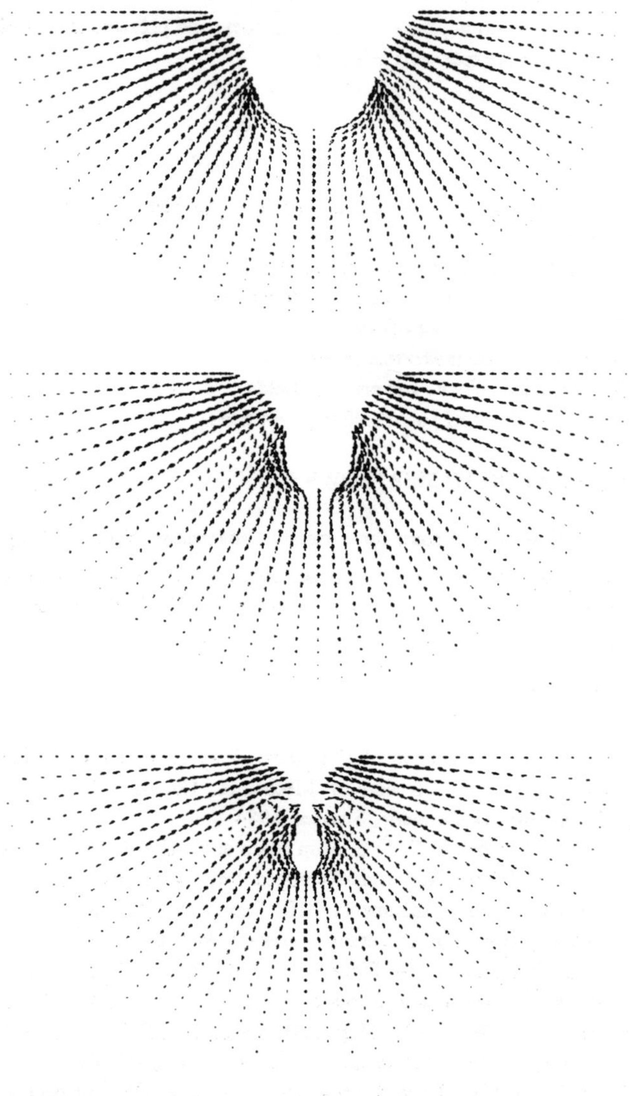

Fig. 2 The vector distribution of transverse velocities at x/L=0.7,0.8,0.9

Proceedings of the Seventh (1997) International Offshore and Polar Engineering Conference
Honolulu, USA, May 25-30, 1997

ISBN 1-880653-28-1 (Set); ISBN 1-880653-32-X (Vol. IV)

Ship Form Description with Function Parameter

Yan Lin, Zhaohui Zhu, Zhuoshang Ji and Yinsheng Dai
Dalian University of Technology
Dalian, China

ABSTRACT

A new style of mathematics description, which is a kind of function with function parameters, is derived from the boundary conditions of the side flat end line, the bottom flat end line and so on, and from some geometric information of the body plan. The major advantages of the function are presented in the paper. The function with better pliable and tough behavior; and with fewer undetermined coefficients can approach form line easily under the control of some body plan parameters. The practical value of the new function is proved by the three different ship form examples given in this paper.

KEY WORDS: Ship hull, form design, mathematics description

INSTRUCTION

There are mainly two approaches to ship hull form design: One is the offset-point design[1~5], the other is the function description design (mathematics form)[6,7]. The latter approach uses a function $F(x,y,z)$ to describe the form, so compared with the spline description form, it has major advantages in the control of the number and the position of the inflection points, of the uniform variance of curvature and of the variance of the UV form and so on. However, the problem of how to make up a function description of forms hasn't been solved in a satisfactory way yet, while the makeup of the function description of form is the key to mathematics form. Usually, high order functions are used in order to describe forms more properly, while on the other hand, the higher the power of the function gets, the more undetermined coefficients are needed. That is to say, at the beginning of the design process, the designer can't go on with his form design until he obtains a lot of geometric information of forms, and thus can't meet the demands of practical production designs. Aiming at the problems mentioned above, this paper puts forward a mathematics description with function parameters. This description not only exhibits better pliable and tough behavior but also contains fewer undetermined coefficients. Even when only a few geometric parameters are available, the new function can describe ship hull forms better and even go one step further to design forms[7].

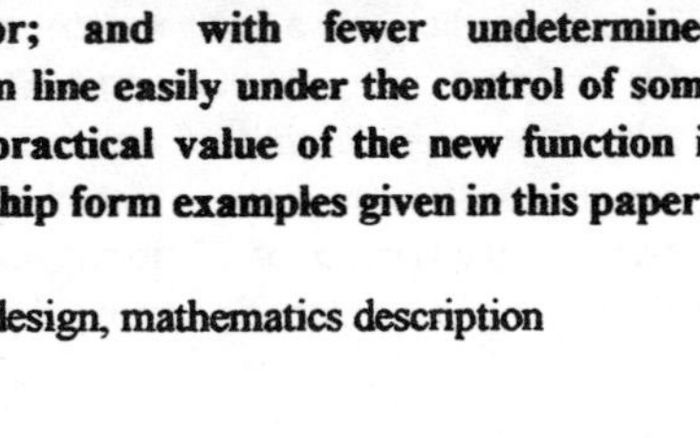

Fig. 1 Hull coordinate system

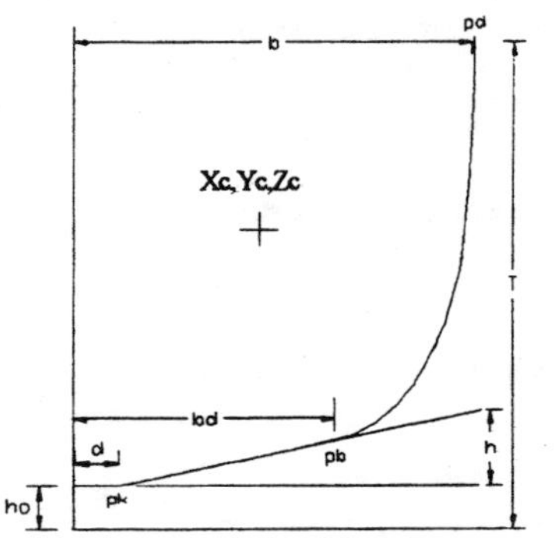

b-half breadth
T- draft in side flat end line
h-deadrise, d-half width of keel
h_0-from bottom to baseline
X_c,Y_c,Z_c-geomestric center
A-half area of body under T
M_y,M_z-moment of area
K_0-slope in p_b and p_k
K_1-slope in p_d

Fig. 2 Typical body plan

FUNCTION DESCRIPTION MAKEUP

Suppose the function description to be :

$$z=\begin{cases} h_0 & 0\le y\le d \\ h_0+k_0(y-d) & d\le y\le b_d \\ f(y) & b_d\le y\le b \end{cases} \tag{1}$$

In the following part, we'll only discuss the function $z=f(y)$ $(b_d\le y\le b)$,

let

$$f(y)=T-(b-y)g(y) \tag{2}$$

where $g(y)$ is an undetermined function, $f(y)$ satisfise

$$f(y)_{y=b}=T;\quad f(y)_{y=b_d}=T-(b-b_d)g(b_d)=k_0(b_d-d) \tag{3}$$

from Equation (3), we get

$$g(y)=C_1+(b_d-y)h(y) \tag{4}$$

where $C_1=k_0+T/(b-b_d)$; $h(y)$ is an undetermined function;

Substitute Equation (4) into (2), we get

$$f(y) = T - (b-y)[C_1 + (b_d - y)h(y)] \quad (5)$$

Derive Equation(5), we get

$$f'(y) = C_1 + (b_d + b - 2y)h(y) - (b-y)(b_d - y)h'(y) \quad (6)$$

and Equation (6) satisfies

$$f'(y)_{y=b} = k_1 \quad (7)$$

so

$$h(y) = C_2 - (b-y)p(y) \quad (8)$$

where $C_2 = (k_1 - C_1)/(b - b_d)$, substitute Equation (4) and (8) into (2), we get

$$f(y) = T - (b-y)\{C_1 + (b_d - y)[C_2 - (b-y)p(y)]\} \quad (b_d \le y \le b) \quad (9)$$

where $p(y)$ can be a higher order polynomial or a expression of some other kinds. If $p(y)$ takes the n order polynomial, then $f(y)$ will possess n+3 order polynomial. What's more, $f(y)$ should also satisfy

$$f'(y)_{y=d} = k_0 \quad (10)$$

$$A = \int_{h_o}^{T} y dz \quad (11)$$

$$M_y = \int_{h_o}^{T} yz dz \quad (12)$$

$$M_z = \int_{h_o}^{T} \frac{1}{2} y^2 dz \quad (13)$$

If k_0、A、M_y and M_z are known terms, then taken

$$p(y) = \sum_{i=0}^{3} a_i y^i \quad (14)$$

From Equation(10) to (13), we could get the undetermined coefficients $a_i(k_0, A, M_y, M_z)$ $(i = 0,1,2,3)$.

FUNCTION DESCRIPTION SPREAD

In order to approach the body plan curve with a large curvature K, we could piecewise the curve. In most cases, we can come to better results when a curve is divided into two parts. However, it needs careful study to decide the position of the inflection point of a curve. From the fairing perspective, the place at $\min(K^2)$ is usually considered as a quite proper inflection point.

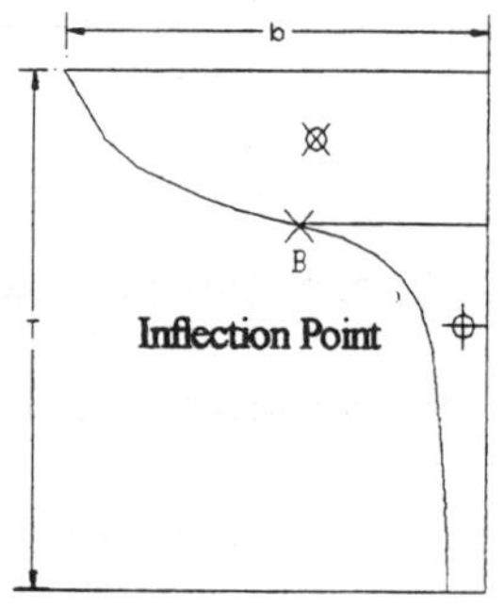

Fig. 3 piecewise in curve

Suppose the point $B(y_i, z_i)$ to be the inflection point of the curve, the functions $f_d(y)$ $(b_d \le y \le y_i)$ and $f_u(y)$ $(y_i < y \le b)$ to represent the body plan curve respectively, and the point B to divide each of A, M_y and M_z into upper and lower parts. (see Figure 3), then on the condition that C^2 is a continuous derivation at the point B, we can derive the functions $f_d(y)$ $(b_d \le y \le y_i)$ and $f_u(y)$ $(y_i < y \le b)$ using the same method as above. No repetition is needed here.

FORM DESCRIPTION SAMPLES

The form description of an oil tanker, see Figure 4.

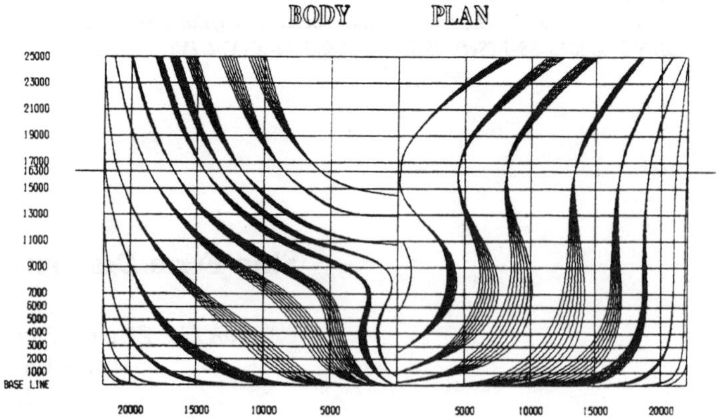

Fig. 4 The form description of an oil tanker

CONCLUSIONS

(1) To use the function $F(x, y, z)$ to describe the ship form has great advantages in the control of the number and the position of the inflection points, of the uniform variance of curvature, and of the variance of UV form.

(2) This mathematics description with function parameters not only exhibits the pliable and tough behavior but contains less undetermined coefficients, which means that we could have an approach to the ship hull form even when only a few geometric parameters are available.

(3) The great success of the approach to ship hull forms achieved by using this mathematics description with function parameters builds a solid basis for ship hull forms design with this mathematics function. As to the design of the variance of UV form under the limited constraint conditions, please refer to the author's continued papers.

REFERENCES

1 Lin Yan, Ji Zhuoshang, Dai Yinsheng . Mathematics description and computer method for B spline hull surface. Shipbuilding of China. 1996(3) (in Chinese)

2 Lin Yan, Ji Zhuoshang, Dai Yinsheng . Ship hull geometric modeling design and software system. Journal of Dalian University of Technology. 1996,36(2):212~214 (in Chinese)

3 Lin Yan, Ji Zhuoshang, Dai Yinsheng . Software system for hull geometric design. Software Development and Application. 1995(6):1~15

4 Lin Yan, Ji Zhuoshang. Software system for ship form design. Ship Engineering. 1995(2):10~13

5 Lin Yan, Ji Zhuoshang. Study for ship form design. Journal of Shanghai Jiao Tong University. 1994(1):16~23

6 Ji Zhuoshang, Lin Yan. A practical method for ship form design by changing UV form. Shipbuilding of China. 1995(3):10~13

7 Zhu Zhaohui. Mathematics ship form description with function parameters and design of the UV form changing under limited constraint condition :[Paper for Bachelor of science], Dalian: Dalian University of Technology, 1996

Proceedings of the Seventh (1997) International Offshore and Polar Engineering Conference
Honolulu, USA, May 25-30, 1997

ISBN 1-880653-28-1 (Set); ISBN 1-880653-32-X (Vol. IV)

A CFD Model of a Marine Propeller Wash

P.B. Brewster, G.A. Hamill, S. Raghunathan, D.J. Robinson and H.T. Johnston
Queen's University of Belfast
Belfast, United Kingdom

ABSTRACT: Computational fluid dynamic techniques are used as a tool to predict the velocities produced in a propeller wash, which may be used to predict the scour produced downstream of a ship's propeller. A finite volume approach is used to solve the governing differential equations, incorporating a moving mesh to simulate the rotating propeller blades. The computational results are in close agreement with the experimental data available. Further investigations are ongoing.

KEYWORDS
Propeller wash, propeller scour, Computational fluid dynamics, propeller slipstream.

1. INTRODUCTION

As a consequence of the increase in the size and installed engine capacity of ships, in particular roll-on roll-off ferries, the jets produced by the vessel's propellers frequently result in the erosion of the seabed. The manoeuvring of ships in harbours, where keel clearances are small, enhance the amount of erosion leading to quay structures being undermined and possibly failing. Cases of such propeller induced damage have been well documented, Qurrain (1994). Prevention of such failures requires a detailed knowledge of the fluid mechanics of the turbulent jets responsible for the resulting scouring action.

A ship's propeller, as it rotates, draws in water which is accelerated and discharged as a jet. This jet is complex in nature, with its formation being dependant on the propeller blade geometry, and its diffusion dependant on the relative magnitude of the velocity components it contains. A propeller jet has been shown to have three components. The main component is termed the axial component, with a direction parallel to the axis of rotation of the propeller. Two other components exist, termed radial and tangential. Investigations in this field of work have involved a semi-empirical approach based on the measurement of velocities in the wash produced by model propellers. Such modelling has many limitations.

In practice a typical ships' propellers may reach 3m in diameter and physically measuring velocities in full scale is not practical. Accurate predictions made from the scale model tests are limited by scale effects, and by the narrow range in which predictions may be made. Producing a three dimensional picture of the propeller wash, from physical measurements, is time consuming and this limits the amount of data that can be collected in order to form accurate predictive equations. A mathematical model of the problem could possibly reduce the limitations imposed by the physical model. Therefore, an approach using modern Computational Fluid Dynamic (CFD) techniques has been introduced.

In the Civil Engineering application for which the CFD approach is being utilised the distribution of the axial velocity, up to eight propeller diameters downstream, in the resultant wash is required. Investigations, Qurrain (1994), showed that it was in this range that the jet would impinge on the sea bed and cause scouring. Studies in the field of Naval Architecture have involved providing a description of the flow near to the propeller and are not particularly relevant to the present investigation.

The physical aspects of any fluid flow are governed by three fundamental principles: The conservation of mass, momentum and energy. These fundamental principles can be expressed in the form of partial differential equations. CFD is the method used to replace the governing equations of fluid flow with algebraic equations; finite volume or finite difference and solving these equations in space and/or time to obtain a numerical description of the complete flow field of interest, Anderson (1992).

The problem of modelling the complexities of the propeller jet has been overcome using a modern CFD software package called STAR CD. STAR CD employs finite volume techniques to solve the governing equations through out the calculation mesh. The mesh is unstructured allowing the complex geometry involved to be incorporated into the model. As it is the complex blade geometry which influences the formation of the wash it has been decided to model the situations whereby the surface of the propeller blades are physically modelled using baffle cells (zero thickness solid boundaries) surrounded by a mesh arrangement which incorporate moving mesh and sliding mesh facilities in a cylindrical polar co-ordinate system.

This paper describes how, using the baffle cell/moving mesh technique, along with a time marching method, a steady state computation of the wash downstream of the a propeller can be obtained. The results of these investigations are presented in this paper.

The effects of grid refinement, domain size and boundary conditions were investigated in order to optimise the mesh settings for solving the equations of the mathematical model. Standardised mesh settings were incorporated in studying the effect the turbulence model has on the CFD predictions, and these predictions are compared to experimental results.

2. GOVERNING EQUATIONS WITHIN CFD CODE

The mass and momentum conservation equations solved by STAR-CD are the Navier-Stokes equations for general incompressible and compressible fluid flow, and a moving co-ordinate frame. For turbulent flow these equations were time averaged and the appropriate turbulence model was adopted. STAR-CD employs mathematical models of turbulence to determine the Reynolds stresses and turbulent scalar fluxes. The option selected was the *k-e* model (Launder 1974) comprising of differential equations for the turbulence energy *k*, its dissipation rate *e*. The differential equations governing the conservation of mass, momentum, energy etc. within the fluid were discretised by the finite volume (FV) method, Patankar (1980). They were first integrated over the individual computational cells and over a finite time increment and then approximated in terms of the cell-centred nodal values of the dependant variables. This approach has the merit of ensuring that the discretised forms preserve the conservation properties of the parent differential equations.

The FV equation applies over an arbitrary time increment, spanning the 'old' and 'new' time levels. The fluxes and sources prevailing over this interval are taken to have their new time level values i.e. a fully implicit formulation is used. This practice avoids a stability related time step and in principle, the fully implicit formulation allows any magnitude of time step to be used. In practice other considerations impose limits.

The Upwind Differencing scheme was used to express the convective and diffusive fluxes in terms of the nodal values. The scheme was first-order accurate and selected the nearest upwind neighbour value for the dependant variable involved. This form of interpolation preserved the correct physical bounds on the dependant variable under all conditions, but the grid had to be refined to minimise the truncation error.

2.1 General mesh motion

Mesh motion involving the 'arbitrary' (in reality within distortion limits) motion of the cell vertices was catered for from the outset by working with the general forms of the governing differential equations for an arbitrary moving co-ordinate frame. The development and solution of the discretised forms of these was straightforward within the finite volume framework, provided that appropriate measures were taken to ensure obedience of the space conservation law, Demirdzic (1988), which was allowed for in STAR-CD as part of the extended PISO algorithm employed, Issa (1990).

2.2 Internal sliding mesh

To allow the modelled blades to move within the general mesh the sliding mesh feature of STAR-CD was utilised. The feature involves the general mesh shearing and then sliding at a pre-defined interface such that two adjacent regions of mesh can move relative to each other without limit. The shearing movement was accommodated by the mesh motion treatment described above. Sliding involves the detachment of facing cells on either side of the interface and reconnection with their nearest neighbours. In the treatment employed by STAR-CD the two sides remain implicitly coupled during this process and care is taken, via appropriate interpolation in time and space, to preserve flux continuity across the interface and avoid introducing spurious perturbations to the flow field. The cell connectivities change during sliding, but this was easily accommodated by the indirect addressing methodology used in STAR-CD.

3 CFD MODELLING OF PROPELLER WASH.

The solution of the mathematical model of the propeller jet was based on physically modelling the surface of the blades using zero thickness baffle cells and incorporating this into the moving mesh features of the CFD code.

3.1 Grid generation

In setting up the CFD model the problem being investigated was described in spatial terms. This involved defining the overall size and shape of the flow domain and subdividing the domain into a mesh of discrete, finite volume cells.

The unstructured grid was generated through the extrusion of two dimensional surfaces. A collection of shells were used as the starting point and this surface was extruded in space to create a three dimensional block.

The solution domain, shown in figure 1, consisted of an inner cylinder with a diameter 1.5 times the propeller diameter. The cylinder contained the baffle cells representing the surface of the propeller blades. The tank containing the rotating propeller reached up to two propeller diameters upstream and eight propeller diameters downstream of the propeller.

The complex geometry of the blades meant that the initial two dimensional shell was created by building up the mesh from its constituent cells. The parameters describing the geometry of the propeller blades i.e. pitch, chord, rake angle etc., were used to define each point on the blade surfaces in cylindrical polar co-ordinates. A system of vertices were created from these co-ordinates and cells were attached to these vertices to define the volume elements constituting the model of the blade, as shown in Figure 2. The resulting two dimensional was extruded to produce a cylinder, in which the baffles were then added. The faces of the cylinder were then extruded to produce the mesh for the solution domain.

The internal characteristics of the flow regime were catered for by careful control of the mesh spacing within the solution domain. By using cylindrical polar co-ordinates radiating from the propeller axis, the mesh was finer where the flow characteristics were changing most rapidly.

FIGURE 1. GEOMETRY OF PROBLEM DOMAIN

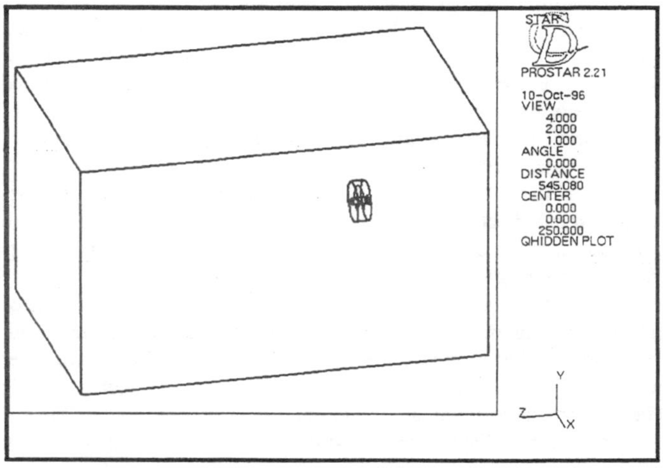

FIGURE 2. TWO DIMENSIONAL SHELL INCORPORATING BLADE GEOMETRY

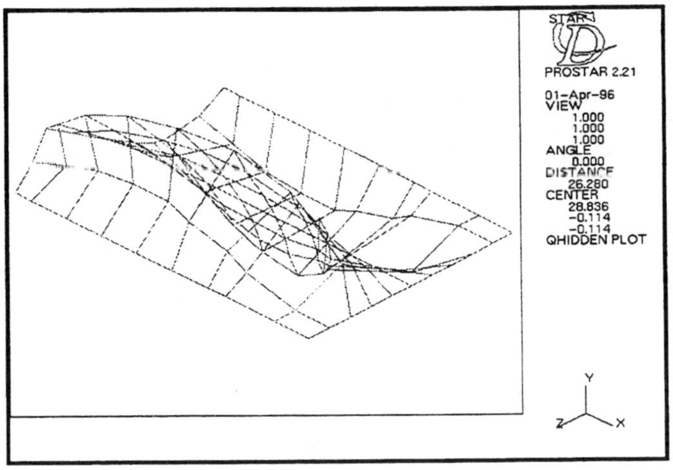

3.2 Boundary conditions

The propeller blades were modelled using impermeable baffles. Both surfaces of the baffles are treated as walls by the code during the flow analysis stage. The rotating hub of the propeller is represented by a wall boundary condition where a rotational velocity, equal to the revolving speed of the propeller, is specified. Attach boundaries were specified at the interface between the rotating cylinder and the rest of the mesh domain to be used with the internal sliding mesh in the CFD code. A slip surface boundary was specified at the location of the free surface. Inlet and outlet boundaries were specified at appropriate locations, upstream and downstream of the propeller.

3.3 Moving mesh features

The cylinder containing the baffle cells was to move at a rotational speed equal to the propeller speed being investigated. This was catered for through the mesh motion and internal sliding mesh treatment employed. The treatment involved specifying the grid connections at the sliding interface and the vertex movements during every time step, to simulate the motion of a rotating propeller.

A time marching technique was employed to iterate to the steady-state. The time step was calculated by taking into account the speed of rotation of the inner cylinder and the number of cells making up one revolution on the face of the cylinder

3.4 Grid independence

Different grid set-ups were investigated in order to establish a standard problem specification that was independent of the mesh structure. Varying the size of the computational domain, the type of boundary conditions and cell refinement were employed and changes were continued until the solution obtained was independent of these grid parameters. The standard grid set-up was then employed to investigate the effect of the turbulence model on the CFD predictions.

A graph of iteration number against velocity, from a selected cell location, was plotted in order to determine the point at which convergence of the solution occurred as shown in figure 3.

FIGURE 3. CONVERGENCE DATA FOR SOLUTION

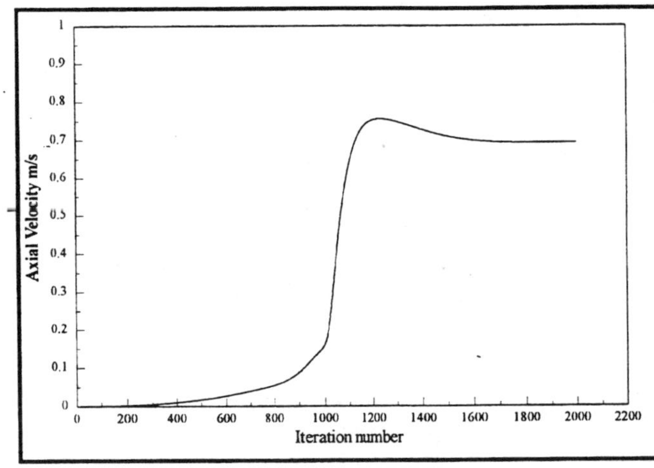

3.5 Turbulence Model

The turbulence model selected for the CFD analysis was the *k-e* model (Launder 1974). When choosing this option in STAR-CD the user must specify a characteristic length of the solution domain. After investigating the sensitivity of the CFD results to this parameter it was decided to specify a characteristic length of 1Dp for further investigations.

4. RESULTS AND DISCUSSION

The vector plot as shown in Fig. 4 gives a pictorial representation of the diffusing jet obtained from the CFD model. This represents closely the findings of previous experimental work, McGarvey (1996). The two high velocity regions either side of the propeller axis are prominent in the 'wash' as is the initial low velocity core immediately downstream of the hub. As the jet diffuses downstream of the propeller the velocities in the central core increase, the width of the jet expands and the magnitude of the maximum velocity in each section perpendicular to the propeller axis decreases. These trends can be extracted from both the CFD and experimental results and they are similar in both cases.

FIGURE 4. VECTOR PLOT OF THE AXIAL COMPONENT OF VELOCITY

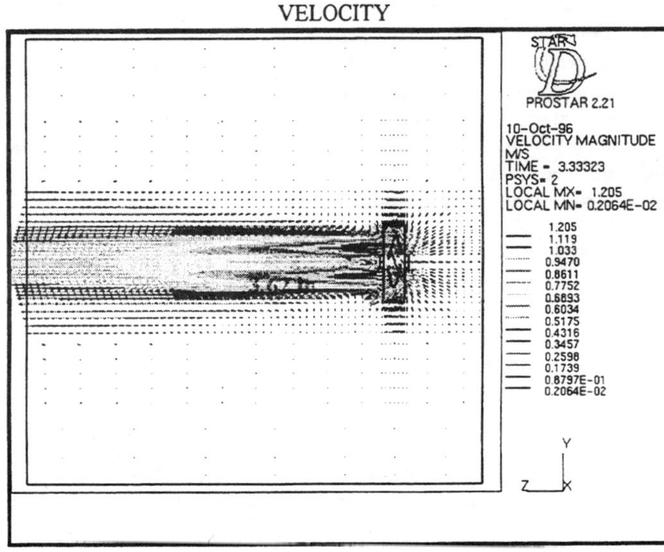

4.1 Velocity distribution at face of propeller

The action of the propeller causes water to be drawn into the blades and accelerated downstream due to the thrust provided by the moving blades. The velocity distribution at the cutting edge of the propeller is dependant on the size, speed of rotation and geometry of the propeller, (McGarvey 1996). This initial distribution diffuses downstream to produce the propeller wash. The efflux distribution from the CFD analysis was compared to the experimental data collected (McGarvey 1996) using laser Doppler Anemometry. The experimental data was obtained by taking 1000 readings and the average, maximum and minimum values were recorded at each point in the flow. The experimental data was plotted with the CFD predictions as shown in Figure 5.

FIGURE 5. AXIAL VELOCITY DISTRIBUTION AT THE FACE OF THE PROPELLER

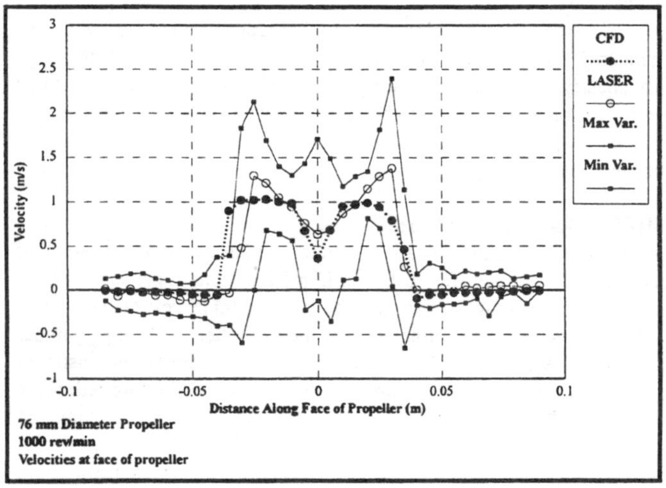

The CFD and experimental profiles are similar and show two high velocity regions either side of a low velocity core. The experimental data indicates two peak values of velocity either side of the propeller hub which are not present in the CFD predictions. The flow in the propeller wash is highly turbulent at this location and the turbulence model employed in the CFD analysis may not sufficiently represent the turbulent fluctuations and as a result fails to accurately model the flow at this point. However, the CFD values do fall between the maximum and minimum limits from the experiments and this would suggest that maybe the averaging process employed in the experiments was insufficient and that the CFD predictions are better approximations for the flow in this region

4.2 Velocity distribution in propeller wash

Figures 6 and 7 graphically illustrate the relationship between the distribution of the axial component of velocity at different sections downstream of the propeller. The velocity distributions obtained from the CFD approach are compared to the experimental data obtained by McGarvey 1996. Figure 6 is an example of the distribution of velocity in the zone of flow establishment. At 1.32Dp downstream from the propeller both the CFD predictions and the experimental results are in close agreement and lie in the middle of the regions defined by the maximum and minimum variations obtained from the experimental investigation. The magnitude of the velocity in the two peak regions has decreased while the velocity downstream of the hub has increased due to the mixing in the jet.

At 2.37Dp downstream, as shown in Figure 7, the CFD analysis slightly over-predicts the experimental data. This would suggest that there is more mixing present in the experiments and the model of turbulence used in the CFD analysis does not simulate this. The difference is small and the CFD predictions still lie in the middle of the region defined by the maximum and minimum experimental values which suggests that the CFD model provides a good prediction method for the phenomena involved.

FIGURE 6. AXIAL VELOCITY DISTRIBUTION IN PROPELLER WASH

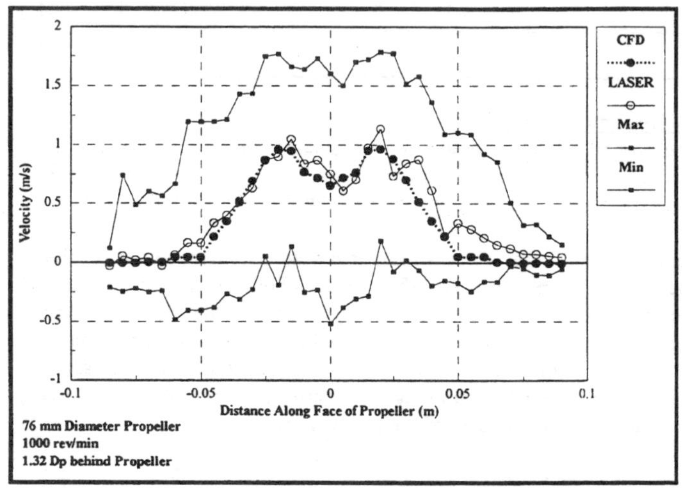

FIGURE 7. AXIAL VELOCITY DISTRIBUTION IN PROPELLER WASH

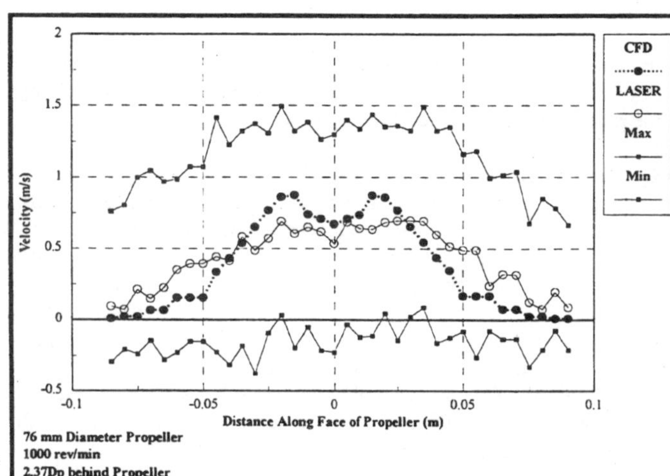

5. CONCLUSION

A mathematical model of a propeller jet has been developed whereby the blades of the propeller are physically modelled and by incorporating the moving mesh features in a modern CFD code, the resulting velocities in the propeller wash can be obtained. Preliminary results for the velocities downstream of the propeller indicate that the CFD predictions agree reasonably well with the experimental data. The grid set-up used can be non-dimensionalised and employed in further investigations relating to a sensitivity analysis of the CFD analysis and the scale effects. The CFD approach can be used as a step towards predicting downwash scour. Further investigations are in progress.

REFERENCES:

Anderson, J.D. (1992). *Computational Fluid Dynamics An Introduction,* Springer-Verlag, Berlin, Heidleberg and New York.

Demirdzic, I., & Peric, M. (1988). "Space conservation law in finite volume calculations of fluid flow," *Int. J. Numer. Methods in Fluids,* Vol. 8, 1037-1050.

Issa, R.J., Ahmadi Befrui, B., Beshay, K., & Gosman, A.D. (1990). "Solution of the implicitly descretised reacting flow equations by operator-splitting," *J. Comp. Phys.*

Launder, B.E., & Spalding, D.B. (1974) "The Numerical Computation of Turbulent Flow," *Comp. Meth. in Appl. Mech. & Eng,* Vol. 3, 269-270.

McGarvey, J.A. (1996) *The influence of the rudder on the hydrodynamics, and the resulting bed scour, of a ship's screw wash,* Thesis Submitted to the Queens University of Belfast for the Degree of Doctor of Philosophy.

O'Brien, T.P. (1962). *The Design of Marine Screw Propellers, Chapter 1, Some Initial Considerations,* pp 25-38, National Physical Laboratory, Teddington, England.

Patankar, S.V. (1980). *Numerical Heat Transfer and Fluid Flow,* Hemisphere.

Qurrain, R. (1994). *Ship Propeller Wash Characteristics and Induced Bed Movements,* Thesis Submitted to the Queens University of Belfast for the Degree of Doctor of Philosophy

Proceedings of the Seventh (1997) International Offshore and Polar Engineering Conference
Honolulu, USA, May 25-30, 1997

ISBN 1-880653-28-1 (Set); ISBN 1-880653-32-X (Vol. IV)

Navy Total Asset Visibility: A System Component Review

Robert F. Paguio and Ramon Flores
Naval Facilities Engineering Service Center
Port Hueneme, CA, USA

ABSTRACT

The Naval Facilities Engineering Service Center (NFESC) is developing a logistics management system to support the shifting operational paradigm of the Unites States Navy. The shift from shore-based to sea-based facilities will require a more efficient use of operating space and will not allow for the storage of an oversupply of goods. NFESC is using Radio Frequency Identification technologies to develop a refined asset tracking system that will allow the Navy to keep a much more accurate accounting of its inventory status. Users at all points in the supply chain will have the capability to access autonomous, real-time container manifest, and asset location information as items are sent from warehousing facilities in the continental United States to off shore bases throughout the world.

Key Words: logistics, tracking, wireless, radio frequency, inventory, tags, assets

INTRODUCTION

The Naval Facilities Engineering Service Center is tasked to study the logistical issues involved in shifting the United States Navy (USN) and Marine Corps (USMC) away from operations and exercises in foreign territories that require a large land dependency. One issue being addressed is the accurate tracking of supplies so that operating forces are neither overwhelmed or left short-handed. During the chaos of a military conflict, the movement of supplies is far too rapid to keep accurate records using manual methods. NFESC proposed to address this problem using an automated asset tracking system based on radio frequency identification (RFID) technology. Initial efforts were begun under a project entitled Recording and Tracking Technologies (RTT) sponsored by the United States Marine Corps. Using the RTT system as a baseline, a more refined system is being developed to adapt to U.S. Navy specific requirements under a project entitled Navy Total Asset Visibility (NTAV) sponsored by the Office of Naval Research. This paper presents the results of new NTAV system development and test results of RTT equipment tests.

BACKGROUND

The United States Navy is moving further away from land-based operations in support of the Department of Defense initiative to enhance the Nations' strategic mobility and rapid deployment capability. Additionally, the growing economy and increasing independence of foreign nations is creating a situation in which U.S. presence is no longer being supported as the recent examples of the closing of U.S. bases overseas.

The movement away from a land-based supply structure will require a more accurate accounting of supplies. Distribution channels must operate from a "pull" system in which end-users initiate shipment rather than a "push" system in which a "dumping" of supplies occurs. The new paradigm will leave little room for the storage of excess supplies. A crucial requirement to fulfill this shift is to provide commanders and supply managers with a system that can give them inventory status within minutes. Early solutions to this problem were paper based or required too much human intervention. Recent military conflicts such as Desert Storm/Shield clearly indicate the need for a more automated approach.

During Desert Shield/Storm over 22,000 of the 45,000 containers sent to the gulf had to be opened and completely unstuffed to determine their contents. This was a clear indication that the existing Logistics Marking and Reading Systems (LOGMARS) were not reliable in an expansive and chaotic logistics environment. LOGMARS utilize bar codes and legacy databases to record and track military assets. Micro-Circuit Technology in Logistics Applications (MITLA) incorporated by NTAV moves beyond the barcode into radio frequency communication. MITLA has been tested on a limited scale by both the USMC and US Army. During these exercises, MITLA has proven to reduce manpower and time requirements while increasing reliability of asset tracking (Paguio, 1996; Angel, 1994). Initial efforts have produced a very well planned baseline system. The NTAV effort attempts to refine this system with critical enhancements that have not been considered by other services (Paguio, 1996)

Existing System

Recording and tracking of assets will take place in many environments, and will provide item information to many command and force elements. The Marine Corps has identified a need for total asset visibility, in-transit visibility, in-the-box visibility, and real time status information for improved distribution management. Total asset visibility represents a need for global information regarding shipment and item movement. In-transit visibility represents a need for visibility throughout a battlefield area. In-the-box visibility provides unit level identification of assets that are packaged for various means of transport. Status information and data distribution show a need for logistics information throughout the command elements as well as at a working level for the warfighter.

The baseline RTT system consists of a high capacity radio frequency (RF) tag and a family of interrogator configurations. The RF tags are 128Kbyte receiver/transmitters called, "SealTags" produced by Savi Technology, Incorporated. The SealTag is designed to be somewhat weather resistant and is powered by a 6 volt battery. Self-contained within the tag is a beeper the size of thumbtack that can be activated to give an audible position signal. Operating at 433mhz, it can contain up to 256 Bytes of Standard Memory and 128K Bytes of Extended Memory. The increased memory in the SealTag allows it to incorporate what is called a Tag Database Engine (DBE).

The DBE incorporates both a database and a database operator. This allows information to be stored in the tags in a database format. The importance of the DBE is that it allows users to conduct searches by specified fields. For example, the user could send out a query for "size 10 combat boots". All tags within the interrogation area would then search their databases and only the tags with the size 10 combat boots would respond. The DBE also provides for "manifesting-on-the-fly" which allows the user to bar-code scan items into the database or simply add and delete individual records without the need to obliterate existing information and rewrite the entire database to the tag. This allow for rapid reconfiguration and recording of supplies at the Seabase, port, or landing zone. Search commands in the DBE are initiated via a hand-held or fixed interrogator.

There are essentially two methods of accessing RF tagging information. The first is to use a fixed interrogator. Fixed interrogators resembles a smoke-detector and is about half the size of a bowling ball. A single fixed interrogator attached to the central computer acts as a dissemination/collection point to tags and other fixed interrogators. Other fixed interrogators are usually set-up in a matrix pattern creating an interrogator network. In a warehousing area, these interrogators could be hardwired together, but this is impractical in a field environment. A more robust configuration is to attach the interrogators to RF links which act as RF modems to relay information from one interrogator to another. Field deployed interrogators must also be self-powered and self-mounting. In this context, a field deployed interrogator is a set of equipment referred to as a Rapidly Deployed Interrogator (RDI).

RDIs attach to batteries with solar recharge, mount on tripods (telescoping to about 7.6 m), and utilize wireless communication through RF Links. Individual interrogator ranges start at a minimum of 91.4 m. Item location using current hardware could be identified by which interrogator/s pick up specific tags. Tag location would then be specified within a grid point of the interrogator matrix.

The network is operated by a PC-based system which utilizes a user friendly Graphical-User-Interface or GUI access. System commands are initiated using point and click methods in a Windows environment significantly reducing the learning curve needed for implementation into the field. Once completed, a more robust interface with existing logistics systems will allow users to utilize systems they are already familiar with as the front-ends to the programming of the tags.

The second way to communicate with the tags is with a Hand-Held Interrogator (HHI). HHIs allow communication to the RF tags but at limited ranges (30.5 - 45.7 m). They allow users all of the functionality of the fixed interrogator/central computer at reduced ranges and reduced viewability (no large computer screen to view , sort, and process data; no mouse or trackball to move around displays; etc.). Like a fixed interrogator, an HHI can search for individual items through the tag database, read/write to tag memory, turn on tag beeper to help identify its location, incorporate a bar codes scanner, etc. As an added feature the HHI also has a "Find Tag" mode which allows the user to use it like a divining rod. In this mode a single thermometer bar appears on the screen and increases or decreases in size depending on how close the user is to the targeted tag.

The basic components of the RTT system provide the foundation for further efforts designed to increase the capability and address application issues which the current system cannot fulfill. One such effort is a low capacity RF tag called a "Radel". This tag is designed to act as an active Radio Frequency tag with read/write capability that behaves as a transmitting barcode label. It is intended for use on pallets, multi-pack items, small high-value assets, and other critical items that are packed into an ISO Container. The Radel consists of several integrated components packaged to the size of a ¼ deck of cards. Each Radel is manufactured with a unique 8 character permanent ID, a 2.4 GHz wake-up receiver, 2.4 GHz slot antenna design, and a 433Mhz transmitter. Early test and evaluation efforts showed this tag system to have a 100% identification rate when queried in a metal container environment. Evaluation of the automating manifesting system are still underway (MCA, 1996).

Planned System

The Navy Total Asset Visibility project is an effort to leverage the millions of research and development dollars and several years of work being conducted in the RFID area. As much as it will drive the development of new product designs, it will also find a means of providing a platform for the integration of other technologies. NTAV seeks to take the Radel development out of the box and into the field. The initial design of the Radel hardware under the USMC program was to provide a means of autonomous manifesting. This meant that the Radel would only function as part of a system which would only work with a container. The Radel was accompanied with a specialized reader which would transfer data to a SealTag.

The USMC is organized in a much more unitized and containerized fashion than the USN. Preliminary NTAV studies show that the Navy possesses much more break-bulk medium value assets than the USMC. It was also found that asset tracking on a MOBS or a Seabase would require the ability to identify individual items without the benefit of containerization.

The NTAV system will require a much more choke point oriented approach to logistics tracking. NTAV will take advantage of the newly enhanced SealTag design which will allow for limited range portal reading as well as the wide area acquisition provided in earlier versions. This new SealTag design requires a portal interrogator as well. Basically, communication to tags passing through the portal is conducted within a limited window at 2.4GHz. Only tags within the portal window will receive the signal and respond back at 433Mhz. This same type of system is what was being used for the Radel Container system. Unfortunately, the new portal interrogator is not designed to read Radels.

The development target for the NTAV system is to adapt the existing system to read Radel-type tags just as it reads the new SealTags. This would allow NTAV to capitalize on the development of the In-the-Box Radel system, while still having the ability to track break-bulk items as they pass specified choke points. In essence, break-bulk items would be tracked as they were moved from point to point. If one of these tagged items is then placed into a container or holding bin equipped with a SealTag, its information is fed up to the SealTag to create a manifest of that site. NTAV is looking to provide this capability and seeks to have an integrated interrogator which would provide dual portal and batch collection. The baseline system is depicted in figure 1.

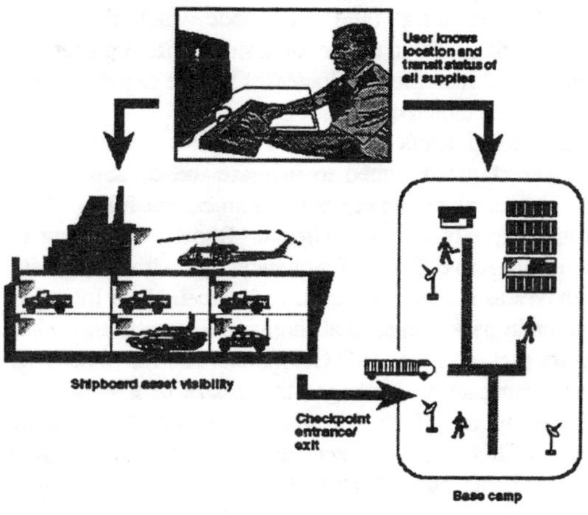

Figure 1 Baseline System

Component Testing

The SAVI hardware components underwent component and system level testing at the Naval Construction Battalion Center (NCBC), Port Hueneme, CA, during the period of 30 July through 3 September. Weather throughout the testing period was generally clear and temperate but somewhat humid with temperatures varying between 18.3 and 29.4 C, except for a verifying RF-linked interrogator test that was conducted in a light rain (MCA, 1996)

The emphasis of component level testing was to empirically determine data transfer profiles between tags and interrogators and amongst interrogators using the RF links. These profiles were generated for upload and download procedures while varying amount of data stored in the tags as shown in table 1.

Planned Data Fill (%)	Actual Data Fill (%)	Database Size (KB)	No. of Records
25	24.2	31	337
50	53.9	69	775
75	77.3	99	1,112
100	94.5	121	1,359

Table 1 Standard Databases Used for tag testing.

The configuration for this part of the testing consisted of an RF link between the host computer and a Fixed Interrogator with RF interrogation of the SealTags. The RF link was later replaced by a direct connection. Upload and download times for both cases are shown in Figure 2.

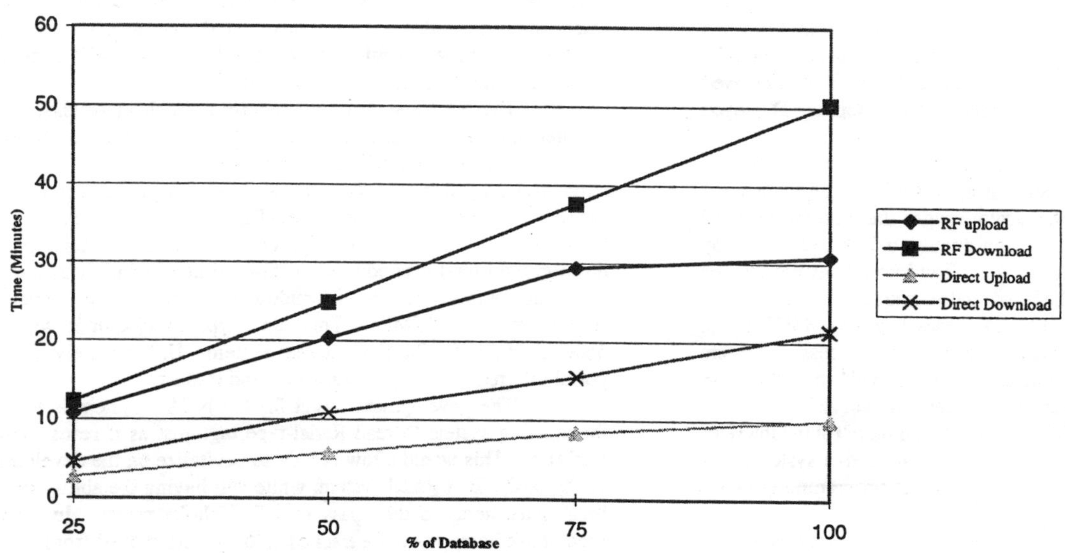

Figure 2 Load Times for RF and Direct Connection

For system level testing, a 457.2 X 487.7 meter simulated marshaling area was used. The layout was 9 Fixed Interrogators, each with an RF link, in a square grid. SealTags were strategically distributed within the grid and were collected in two configurations: RF-linked interrogators and direct connected interrogators. These configurations are compared in Figure 3.

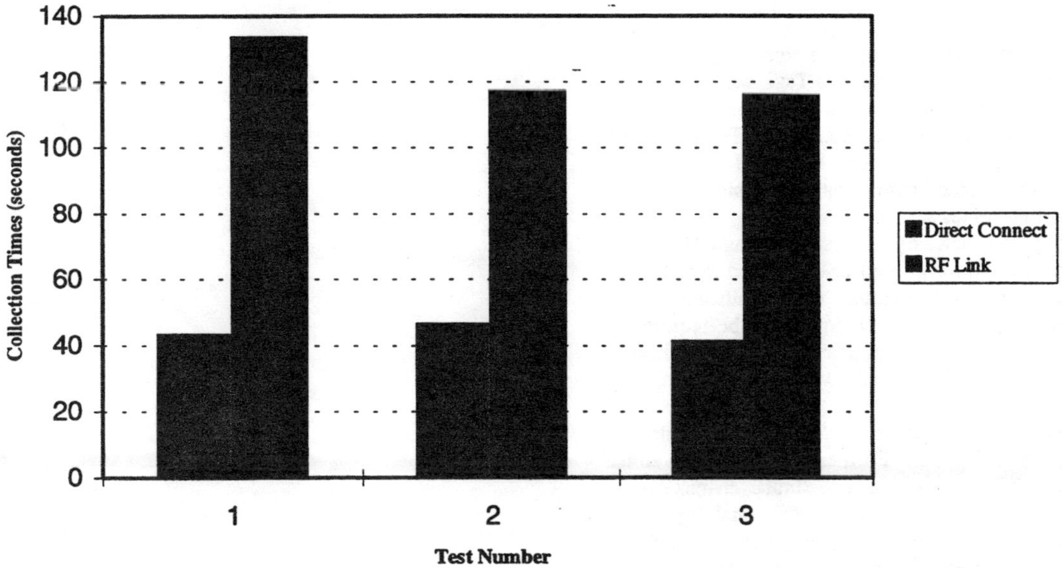

Figure 3. Direct Connect Vs. RF-Linked Collections

Another test dealt with Electromagnetic Interference (EMI) Testing. EMI testing was performed on the NTAV components to assess the system's vulnerability to radiated electric fields over the frequency range of 14 kHz to 10 GHz. The field strength levels used were 10 V/m over 14kHz to 2 GHz, and 40 V/m from 2GHz to 10 GHz. The modulation format used during the testing was a 1 kHz, 50% duty cycle, pulse modulation. The operating frequency of the tags and interrogators (433 Mhz) was the main are of interest. The system was deemed susceptible if loss of communications between an of the system components resulted from the application of the test levels. The system incurred no permanent failures as the result of these test levels. None of the sub-systems were found to posses any vulnerabilities below 200 MHz. The sub-systems would respond to low frequency signals below 200 MHz, but by bypass filtering the transmitter out, responses in the sub-systems were eliminated.

The final test was ongoing as of the publication of this document. This involved a test of the Radel System. A demonstration conducted in April of 1996 indicated the Radel system hand 100% reliability within a containerized environment. Five to ten tags were oriented in a metal container and an collection was initiated when the container door was shut. The data was then transferred to a SealTag on the outside of the container. The Sealtag essentially acted as a dynamic manifest. Interrogation of the SealTag revealed that all tags were collected. As tags were removed and new tags placed in the container, the manifest updates remained consistent. Preliminary results of tests on the same system (started November 1996) in an open environment indicate a variable reliability rate. In this case tags are simply place at varying distances from the Reader (Interrogator) device and a collection routine is initiated. The placement of all components is no longer in a container, but in the hallway of a building or an open field. Preliminary results indicate a reliability rate from 45 to 85 percent.

DISCUSSION

During RF uploading and downloading, there appeared to be considerable variation in the times required to upload the various databases into the SealTags. This is most likely due to the line-of-sight susceptibility of the RF links. The profiles in figure 2 show that the RF upload and download are significantly greater than the upload and download times of the direct connect collections. When uploading and downloading SealTags using direct connections, the data flow appeared to be very steady with only minor variations in the loading times. Downloading times were about twice as long as uploading times for the largest databases (full 128 Kbytes), but they were comparable for the low end (31 Kbytes).

Comparing the RF-linked interrogator system to the direct connect interrogator system shows significant differences in SealTag collection times. The RF-linked configuration required about three more times longer to collect the same information as did the direct connect configuration. This difference is significant when planning the system layouts for ship storage areas, piers, marshaling yards, and warehouses.

The EMI tests showed that the Savi components will not be vulnerable to interference from most ship based RF emitters. The system was tested over the frequency range of 14 kHz to 10 GHz. The field strength levels used were 10 V/m over 14kHz to 2 GHz, and 40 V/m from 2GHz to 10 GHz.

Radel testing shows that the initial design actually utilized the characteristics of the metal container to obtain its reliability.

Reflections of the RF signal within the container provided the Reader several opportunities to obtain the tags signal. In an open environment such reflections may not occur and so varying degrees of reliability are achieved.

CONCLUSIONS

During component level testing, the communications link between the fixed interrogators and the SealTags was established using two different configurations from a host computer. Both configurations showed good reliable communications. The RF-linked configuration showed longer upload and download times when compared to the direct connect configuration. This trend was continued during system level tests.

The level of performance of this system during EMI testing was found to be very suitable for ground or ship based installations that are not in close proximity to emitters with significant energy levels. Some frequency management techniques will be required if HF and UHF emitters, possessing significant power at applicable harmonics, are to be used in close proximity to a system utilizing these components.

The current Radel design is inadequate to provide the capability desired by the U.S. Navy. Further development to either refine this system or to create a system with similar characteristics needs to be conducted. Anything less than a 100% reliability would result in an unacceptable accounting variance which would lower confidence in the system and result in an intentional over-ordering of supplies.

ACKNOWLEDGEMENTS

This work was funded by the Marine Corps Systems Command Amphibious Warfare Technology Directorate and the Office of Naval Research.

REFERENCES

Angel, R., (1994). "Agile Sword 1994 Exercise Reconstruction and MITLA Evaluation", a field memorandum. Center for Naval Analysis.

MCA Engineers (1996). "Independent Field Test of Savi Technology; Radio Frequency Transponders for Seabase Transfer of CSS Assets".

Paguio, R.F. , (1996). "*Recording and Tracking Technologies: Opening a Window to Totall Asset Visibility*", International Society of Offshore and Polar Engineers", Los Angeles CA.

Proceedings of the Seventh (1997) International Offshore and Polar Engineering Conference
Honolulu, USA, May 25-30, 1997

ISBN 1-880653-28-1 (Set); ISBN 1-880653-32-X (Vol. IV)

Developing and Testing a Rough-Water Flexible Connector System for Pontoon Barges

William G. Hatch, Erick T. Huang and Joseph L. Barthelemy
U.S. Naval Facilities Engineering Service Center
Port Hueneme, CA, USA

ABSTRACT

This paper presents the test results from a series of sea trials involving a new flexible connection system for joining multiple pontoon barges in elevated seaways. The tests were conducted to verify the performance of proposed rigging systems and operational procedures, and establish the credibility of the new flexible connector concept. The new connector and supporting methods of rigging and joining are designed to reduce and relieve the dynamic loads that occur as a result of relative motions and tug-induced forces. Information that is presented includes a description of the test the flexible connector system, description of test conditions and procedures, details of calibration, summary of results, and list of lessons learned.

KEYWORDS: Navy-Lightered (NL) pontoon, barge, causeway section, constant tension winch, flexible connector, separation force, pretension, in-line configuration, hip-tow configuration and modified in-line configuration.

BACKGROUND

Amphibious operations have been an integral part of the Navy's mission since World War II when pontoons played a significant role during invasions in both the Pacific and Mediterranean theatres. The basic pontoon structure has changed little during the past 50 years, and is comprised principally as a floating array of watertight, steel "cans" that are bolted together to form a completed barge structure as described in the Pontoon System Manual (1994). Amphibious structures made from Navy-Lightered (NL) pontoons include barge ferries, causeways and other floating platforms. However, size and construction limit the utility of the present NL system. Individual causeway sections, sized 27.4 m in length, 6.4 m in width, and 1.5 m in depth, must be transported aboard dedicated ships - ships that are being retired but not replaced by the Navy. The 1.5 m depth dimension limits cargo capacity to just 600 KN, while the effective cross-sectional area does not provide the inherent dynamic stability required for operating in waters above sea state 2 conditions.

The vision of the future amphibious operation sees a self-sustaining commercial containership transporting families of "building blocks" to an operational areas where these individual modules may be offloaded by existing container handling equipment and then assembled on the water as barges. The current research and development program at the Naval Facilities Engineering Service Center (NFESC) recommends the development of standard pontoon modules that are sized 12.2 m in length, 7.3 m in width, and 2.4 m in depth, a tradeoff between the requirements for shipping within standard container cells and the limitations of cargo handling and on-site assembly (Kane, 1996). Two types of connector are required to assemble floating structures at sea. A unit barge consists of three modules, joined end-to-end, using rigid-type connectors to fix articulation between mating faces, thereby producing a continuous-deck surface. After a number of such barges are assembled, they may be linked using flexible-type connectors to produce any of a number of useful floating platforms. Although NFESC is also developing the technology required to build the rigid module connector (Huang 1996 and 1997), it is the flexible connector used to join multiple barges that is the subject of this paper.

The primary operational objective in designing the flexible connector is to develop the capability to link multiple barges in elevated seaways through sea state 3 conditions, and to maneuver strings of barges in waves and swell on the open ocean up to sea state 5 conditions.

PROBLEM STATEMENT

A major deficiency typical of NL assets and modular pontoon structures in general is collision damage sustained by barges and the tugs that are supposed to be controlling them. Most connector related damage occurs as floating structures are joined on the water. After a proper connection is completed, the operational platform is generally stable against failure or damage under normal operating conditions. Making a flexible connection in calm seas is generally not hazardous because mating sections are easily controlled. However, in conditions of sea state 2 and above, surge and sway motions left unchecked can literally throw neighboring craft together. Existing standard operating procedures limit the execution of end-to-end connections to calmer waters whenever an unassisted marriage is attempted. The unassisted marriage is the method of choice by the NL crews because it is completed without the need for extraneous marriage bridle rigging. In the most basic sequence, two NL barges are maneuvered in close proximity head-to-head, and at the opportune moment, protruding male and female pipe shear connectors that link the two are "eased" into engagement. Variations on the basic theme include the use of overlapping barge hulls, ships or other physical means to help index and align the free ends of the sections as they are brought together. After the initial link is completed, and while under continued control by tender craft, Navy flexors are passed between barges and locked into position to secure the connections. Although the ability to complete an unassisted marriage has been proven on calmer waters, continued execution has resulted in repeated damage to hulls and connectors. The reinforced corners of the barges and the leading edges of the protruding pipe shear connectors are damaging and can easily open a hole in the thin skin of a pontoon. Both the execution time and safety hazards associated with this operation increase dramatically with building sea state, making the unassisted marriage an extreme liability in sea state 3 conditions.

NFESC CONCEPT

The NFESC concept is a connector system that provides the means to maintain proper alignment between mating barges as well as the control required to subdue erratic random motions as barges are drawn together. The system concept consists of a flexible connector proper, plus all hardware and rigging required to complete the connection. The primary benefits of the new connector are application in higher sea states, increased safety to personnel, reduced connecting times and less damage to the operational assets. The baseline concept is described by Huang (1995)

Aligning and Controlling

The concept connector system uses constant tension winches, marriage bridles, and elastomeric transitions that combine to bring mating barges together while progressively synchronizing their motions. Relative motions and snap loads are contained through effective tensioning, aligning and damping prior to final connector engagement, and dynamic stresses and moments are relieved as external loads are transferred to the connector body. In the sequence of events leading up to the linking of two barges, each section is taken under control by a tug. Each leg of a split-leg bridle harness is passed through an empty tunnel aboard the receiving barge and is connected to a flexible connector housed within the matching tunnel of the donating barge. As the winch-controlled bridle is drawn in to reduce the distance between barges, the two tender craft use opposing thrust to maintain a steady tension on the wire rope bridle legs. At the proper level of pretension, the approaching barges are held in proper alignment and the relative motions between sections begin to come into phase. Once the flexible alignment section of the connector enters the receiver tunnel, the ends of the barges are synchronized. In the event of sudden or dramatic surge motion, hazardous snap loads may be a threat. To minimize the possibility of this danger, NFESC is testing a constant tension winch that can overhaul the wire at a pre-determined load.

Flexible Connector Design

The flexible connector proper is comprised of linked leading and trailing heads, and two guillotine collars. When two causeway sections are connected, the leading head is locked by its guillotine within the tunnel of the receiving section, while the trailing head is locked by its guillotine within the tunnel of the donating section. A hinge pin joining the two heads is the link that allows articulation and translates all tensile and compressive loads into shear loads.

The flexible connector assembly consists of the flexible connector components just described, plus a guiding alignment member. The alignment member is the transitional element that guides the leading head into its receiving tunnel. The alignment member consists of a leader of high-strength chain encased within a compliant urethane sleeve of large cross-sectional area. The chain provides the strength required to resist the high tensile forces while the tapered urethane body allows smooth entry of the chain and prevents kinking as the connector is drawn into the receiving tunnel. The flexible connector system includes the flexible connector assembly plus the split-leg bridle harness and all other required rigging, wire rope, and hardware.

Figure 1 is a drawing of the flexible connector assembly (shown without the locking guillotines) that depicts the two heads, the shear pin, and the alignment member. Figure 2 pictures the closing moments of a marriage sequence as the flexible alignment member is drawn into its mating receiver tunnel.

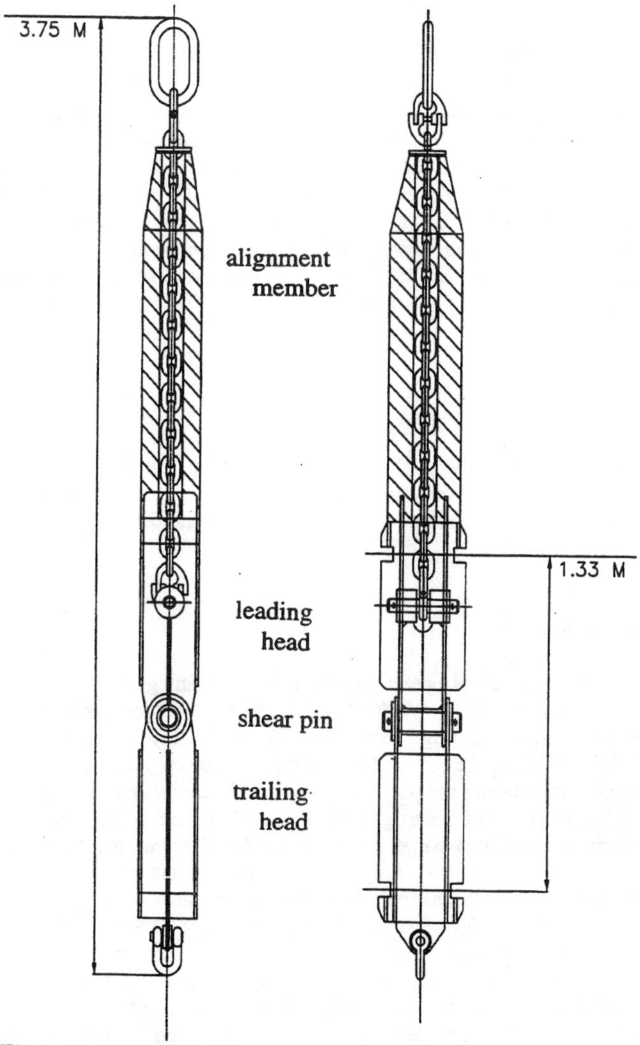

Figure 1. NFESC flexible connector assembly.

Figure 2. Entry of the alignment member during a marriage.

CONNECTOR TESTS

A series of demonstration tests was conducted using NL causeway sections as the test platforms, and prototypes of the NFESC flexible connector as the linking members. The tests were conducted at the Naval Amphibious Base, Coronado, California, using tugs and crews provided by Amphibious Construction Battalion One (ACB-1). The primary objectives were to verify performance of the connector and establish the most effective methods and equipment for operating under sea state 3 conditions.

Hardware and Setup

The test platforms consisted of two NL causeway sections that had the protruding NL pipe shear connectors at the ends removed. The platforms were reconfigured to accept the prototype connectors and upgraded to meet the requirements of the test plan. The ends were fitted with two different types of fender material to absorb impact energy during testing on the open seas.

Prototype connector. The prototype flexible connectors were fitted with special load indicating pins in place of the usual main hinge pin joining the two heads. The pin was designed to hinge the platforms while absorbing all tensile, compressive and shear forces to a maximum of 222.4 KN (50,000 lbs) per connector. The bi-axial gage configuration enabled the recording of the applied load vectors, and post processing produced the magnitude and direction of the resultant.

Rigging and winch. The split bridle legs were made from 1.9 cm wire rope. Two sizes of leg were used, 19.8 m and 27.4 m, to model, respectively, the connecting of two barges and multiple barges (with simulated cargo on deck). The primary operating winch used to draw lines was a pneumatic "air tugger", configured as a constant tension winch that provided line tensions as great as 44.5 KN (10,000 lbs) while operating in either the normal mode or the constant tension mode. In the later, the danger of snap loads was greatly reduced because the constant tension winch could overhaul or pay out when excessive forces were applied, or take up slack in the lines following sudden surge. The level of tension was controlled by air pressure regulated at the control console by the winch operator.

Instrumentation. In addition to the two load pins that replaced the connector hinge pins, load cells were installed at selected hard points and in-line in order to record the magnitude of applied line loads. The load cells were also used to monitor the marriage bridle loads and the bollard pull witnessed by the tugs. The analog output was recorded on a datalogger that displayed real time readings as it digitized and stored the information for subsequent downloading to a computer. The motions of the barges were monitored using a Humphreys SA07-0904-1 gyro-stabilized package that provided output data on roll, heave, surge, pitch, sway and yaw. A Datawell Wave Rider was deployed at the test site to provide local information on wave height and period that was recorded on the datalogger.

Test Parameters

The tests were conducted according to a matrix of physical and environmental variables. The parameters that were used to make up the test matrix were as follow:

Arrangement of barge and tug. Several different arrangements of barge and controlling tug were tried in order to evaluate the effects of attachment point and position. In the "in-line" configuration shown in Figure 3, each tug was attached behind its respective barge (or string) to form a continuous line. In the "hip-tow" approach shown in Figure 4, each tug was tied alongside its respective barge. A modified in-line arrangement used a barge/tug combination on a hip-tow configuration followed by a barge/tug in-line. The modified in-line configuration is shown in Figure 5. A final configuration required the use of a single tug tied alongside the lead barge.

Level of pretension. The level of pretension - that is, the force of separation maintained on the lines between approaching barges, was another variable. In addition to changing the magnitude of pretension, several methods of developing the force of separation were tried, including two tugs engaged in opposing thrust, a single tug pulling against an anchor, and the exploitation of hydrodynamic drag that occurs naturally as a barge is taken under tow.

Rate of closure. The speed of the winch was varied to the maximum level allowed by the separation force. The overhaul point of the constant tension winch was adjusted as the separation force varied.

Sea Conditions. Although the physical state of the seas was not a factor that could be controlled, a cross-section of local conditions was achieved by testing in areas ranging from dockside in protected harbors, to the open ocean 2 miles offshore, and by testing during winter and early spring when the desired sea state 3 conditions could be approached gradually by changing the location of the test site. Connector tests were conducted with the array of barges and tugs headed into the seaway, traveling parallel to the waves, and rotating through 180 degrees.

Calibration and Pilot Tests

The initial sequence of tests was conducted in the protected bay to establish procedures and train the crew. All efforts to prepare instrumentation were conducted under calm conditions, including the verification of deck fixture integrity and winch capability, calibration and comparison of load cell data, and measuring of tug thrust capability. Each of the test options was completed first in calm water areas to identify potential problems or hazards before journeying into the open ocean.

Thrust and load cell calibration. Initial tests were conducted to establish the level of thrust force that could be produced by the supporting tugs, and at the same time calibrate in place the connector load pins to at least 44.5 KN. In order to acquire useful data for the power curve, engine speed was stepped up for short intervals to 2,000 RPM, even though normal operating procedures do not allow speeds in excess of 1800 RPM. Data values were used to establish bollard pull as a function of engine speed, with thrust values ranging from 10.7 KN at idle to 48 KN with both engines revved to 2,000 RPM.

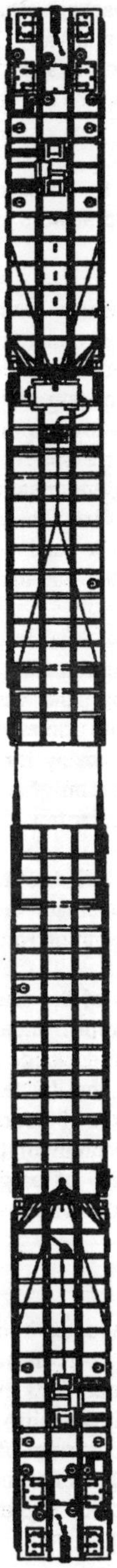

Figure 3. In-line Configuration.

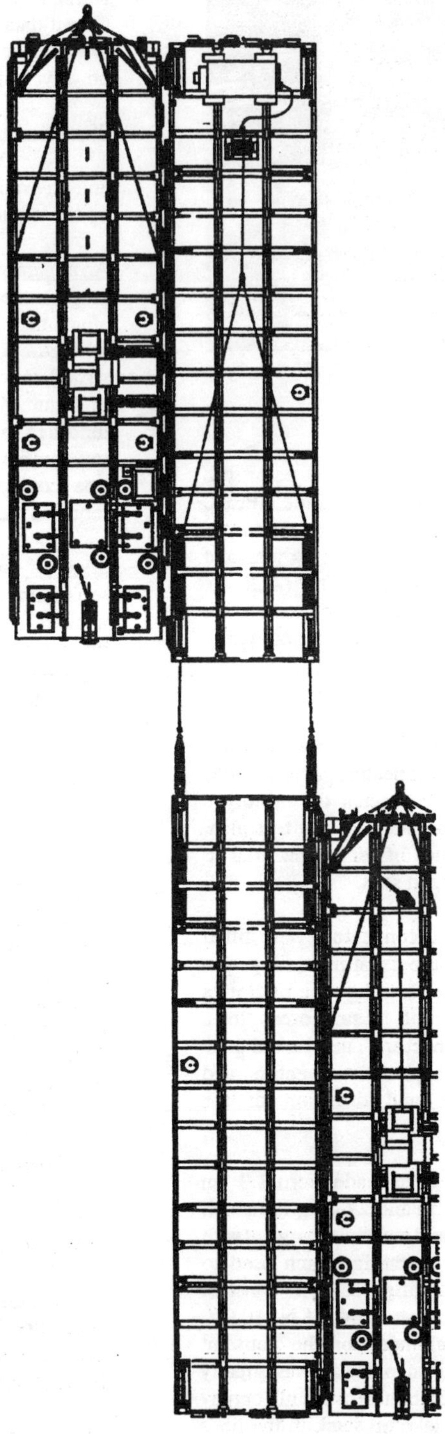

Figure 4. Hip-tow configuration.

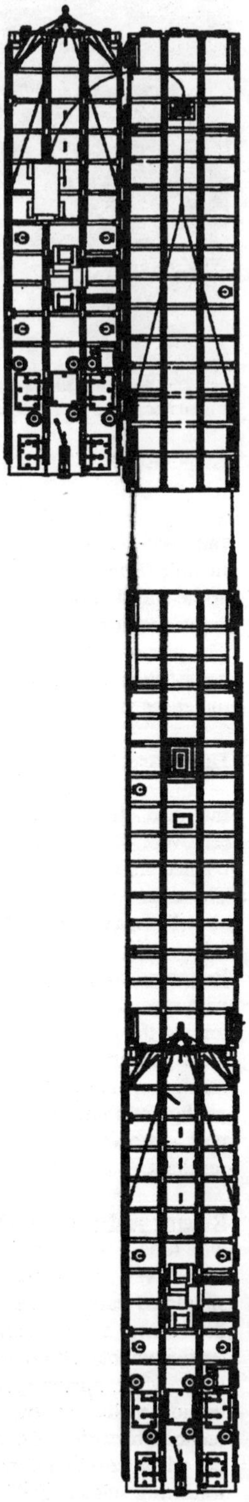

Figure 5. Modified in-line.

Winch pull. Tests were conducted to ensure that the pull of the winch could not exceed the safe load limits of the deck fittings. Load cells were placed between each snatch block and padeye, and a third load cell was positioned at the junction of the winch cable and the two split-leg marriage bridles. The tests demonstrated that the 84.5 KN pull of the winch did not produce more than 133.4 KN of force or exceed a 30 degree angle on the fittings. Levels were within the 4:1 safety factor applied to wire and rigging gear. Figure 6 is a typical data plot of load and pressure recorded during operation of the constant tension winch.

Ocean Tests

Sea conditions encountered during the periods of testing were unseasonably calm. Calm water tests benefited from still air and glassy water found in the protected areas of San Diego Bay, but the open ocean two miles west of Point Loma produced only a 1.3 m, 10 second swell with decaying waves. The ambient sea state did produce some significant load spikes in the connector pins when barges were connected as shown in Figure 7, but wave conditions were not sufficiently dynamic to properly test the constant tension winch during actual marriages of barges. The forces witnessed by the connector pins were of very short duration, whereas the constant tension winch is designed to overhaul and payout in response to loads applied over a longer interval of time. Open ocean tests were conducted with barge/ tugs in the hip-tow and in-line configurations.

Hip-tow configuration. Several connector tests were completed with barge and tug attached in hip tow configuration, the method favored by experienced NL crews to complete unassisted marriages on calm water. When this technique is used, there is a marked tendency for the trailing section to yaw and sway while being drawn toward the leading section, especially in calm waters and when working under low levels of line pretension. When connectors do not enter receivers smoothly on the first attempt, there is a tendency to overshoot, with momentum of the trailing barge being transferred to the lead barge through contact. Forces are intermittently upset by the application and removal of forces from the constant tension winch as the operator draws the line. The hip-tow configuration effectively produces a moment couple on the towed barges as the natural result of eccentric thrust and drag forces. Trials conducted in both calm water and open seas verified that connections could be made more quickly and with greater ease at higher levels of pretension force, even though a higher rotational speed was produced by the eccentricity of drag and thrust. Release of the guillotine mechanism used to lock and unlock connectors, causeway separation and subsequent marriage became routine accomplishments, with connections completed at higher levels of pretension being more positive and controllable than those made at lower levels. At 1200 - 1500 RPM, tugs produced the favorable 26.7 KN of pretension required for marriage. Although rotation caused by eccentric thrust and drag was unavoidable, it did not present a problem and was ignored whenever there was adequate room to maneuver and favorable seas were encountered. When sea or swell became dangerous at certain angles of attack, the deck operation was suspended until the window had passed.

In-line configuration. Barges and tugs were configured for in-line testing. It was anticipated that increased control would be afforded by keeping forces due to thrust and drag in line. Tugs were operated using the opposing thrust produced at 1200 - 1800 RPM engine speed to maintain pretension forces while the winch drew sections together. The in-line configuration proved to be a more controllable arrangement of barge and tug as it was possible to attain high separation forces with winch loads confined to the middle of the efficiency band of the constant tension winch where control sensitivity is good. There were no observable lateral movements between converging sections the tendency to overshoot receivers was dramatically reduced. During some of the repetitions, the operation was actually predictable and uneventful.

Modified in-line configuration. Although the in-line configuration does provide the most positive control, it requires more time to set up (two additional end-to-end matings) and potentially exposes the stern of one of the tugs to the seas. The compromise shown in Figure 5 places one tug in a hip tow with the lead barge while the trailing barge is connected in-line with a tug. This places the most maneuverable craft in the stern with good visibility for alignment control and allows both tugs to be steaming in the same direction.

LESSONS LEARNED

The primary objective of the sea trials conducted by NFESC was to demonstrate a new flexible connector technology for linking multiple barges on the open seas - a technology that included a prototype flexible connector, a constant tension winch, and novel operational procedures. The test matrix was designed to: (1) establish the effectiveness of various techniques in aligning and controlling barges as they are married in sea state 3 conditions. (2) verify the performance characteristics of a prototype flexible connector that retains the strength and integrity of the existing Navy flexor, but eliminates the dangers of protruding male and female pipe shear connectors at the end face of a causeway section; (3) demonstrate the ability of a constant tension winch to reduce the level of loads in marriage bridle lines caused by surge, sway and other random barge motions. Although weather conditions and sea states prevailing during periods of testing seldom matched the ideal environment called out in the test plan, the experience gained and information collected have been highly insightful. The following are observations and recommendations stemming from the experience.

Flexible Connector

The prototype flexible connector performs well and the concept shows great promise as a sea state 3 connecting system. A general recommendation is that the leading and trailing heads should be made from a quality cast material rather than as a welded fabrication. The hinge pin should be designed to provide durable, low maintenance service. The flexible connector system requires additional development to dampen the high dynamic shock loads that are transferred through the hinge pin, and to compensate for loads imposed by yaw and lateral shear. The benefits of improved safety and greater damage control derived from the smooth fendered ends of barges using the new connector system offer payoffs that make the continued effort well worthwhile.

Constant Tension Winch

The constant tension winch performs well, although the overhaul and constant tension features were not fully demonstrated because the lower-than-expected sea states did not offer an adequate operational challenge to the system. The 44.5 KN capacity of the winch is adequate for connections completed on seaways up to sea state 3. Operator intervention is necessary during the final moments of connection when it is often necessary to stop and synchronize the pull of the winch with the motion of the mating platforms. The calm water and at-sea tests did reinforce the belief that a constant tension winch could be a valuable asset in controlling the magnitude of loads imposed on the wire bridle system. A properly balanced winch/wire combination will provide safer, more controllable connections in significantly less time than current procedures permit.

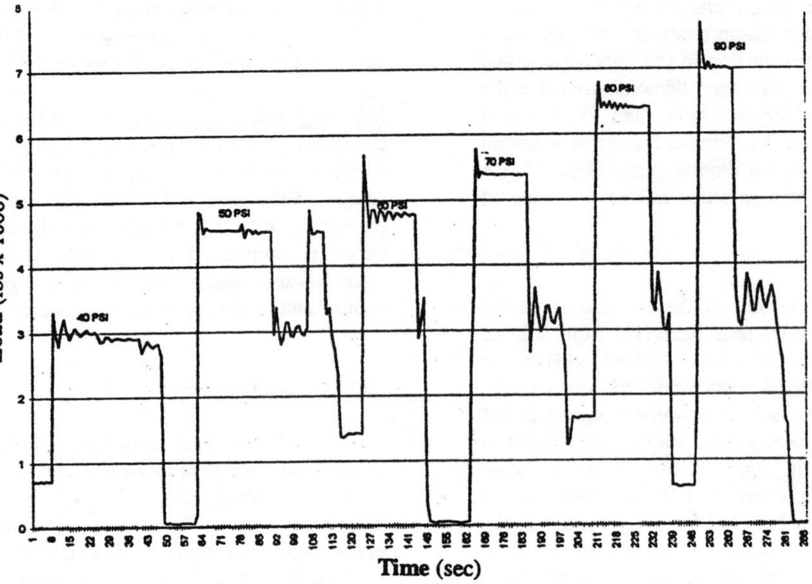

Figure 6. Typical plot of constant tension winch load as a function of control pressure.

multiply	by	to obtain
psi	6.8947	kilopascals
lbs	4.4482	newtons

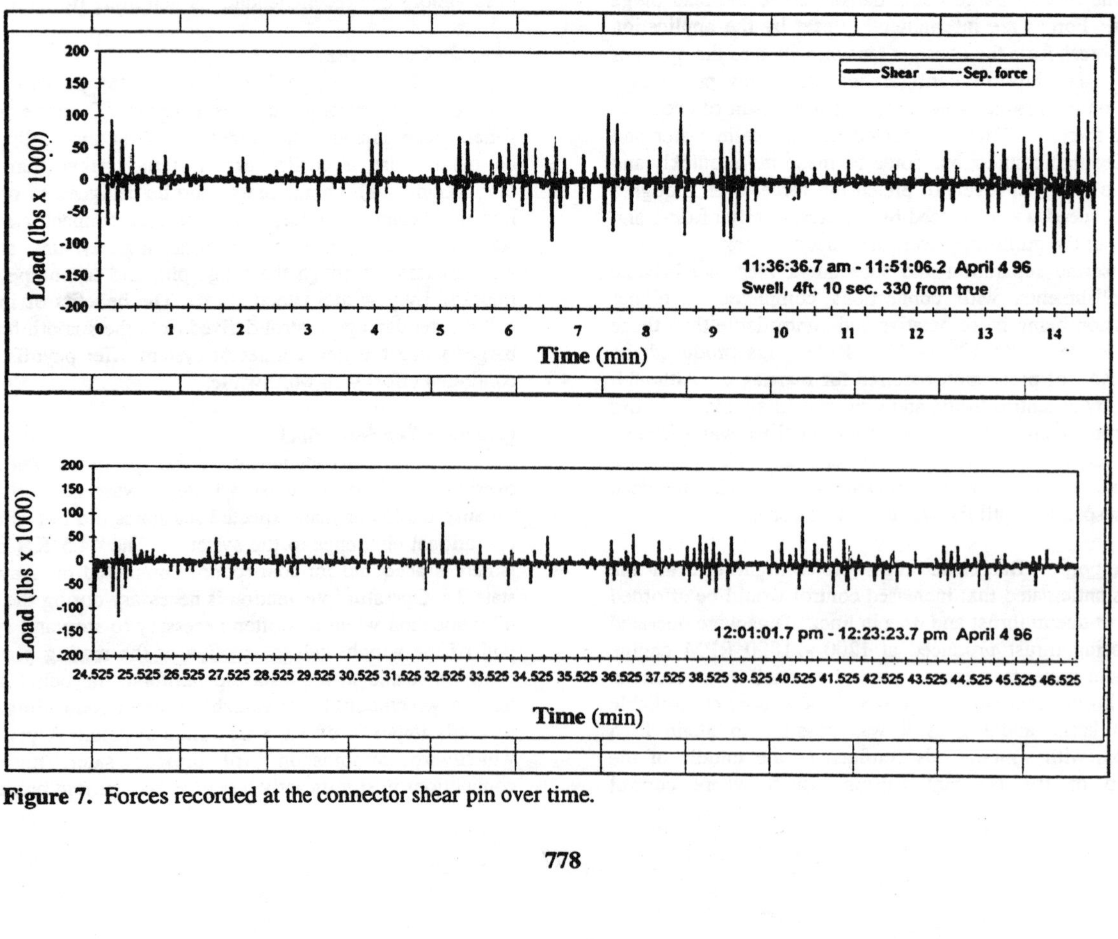

Figure 7. Forces recorded at the connector shear pin over time.

Rigging and Configuration

When one or both of the escort tugs is tied alongside in a hip tow configuration, the matrix of tugs and barges will rotate when taken under tow. The amount of rotation is a function of the magnitude of thrust and the level of line tension maintained by the winch. Operators should allow the tow to rotate, completing connections as expeditiously as possible. Keeping tugs in line with the barges, however, provides superior control when using marriage bridles between sections. The in-line method is not always practical, however, because many extraneous factors can affect the availability of connecting points between the tugs and the barges that they are trying to connect. The modified in-line configuration is a useful compromise that allows both tugs to maintain the same heading, but produces less of a rotational moment than occurs in the hip-tow method. The modified configuration permits reasonably rapid connection of successive barges to produce a causeway string and proved to be the method of choice by operating personnel. In all marriages, the level of pretension on the bridle lines must be maintained high enough to maintain alignment of the barges. Low separation forces result in excessive yaw and sway motions as the barges are drawn toward marriage. Deck fittings, sheaves and fairleads need to be certifiable and placed in locations that minimize bridle wear or entanglement. Bridles are suitable for multiple connection operations, but must be inspected and maintained regularly.

CLOSING

The market potential for a novel at-sea flexible connector that is safe, effective and easy to use extends far beyond the military sector. Navy lighters offer but one configuration of pontoon assets world wide. The spin-off of technology to maritime commerce is more than a likely consequence since there are many types of barge systems and floating applications within the commercial market. Currently, commercially used flexible connecting systems are limited almost exclusively to inland waters where surface wave conditions are subdued. The NFESC alternative for linking barges at sea opens a potential window of opportunity to over-water commerce and transportation.

ACKNOWLEDGEMENTS

This work has been sponsored by the Office of Naval Research. The demonstration tests were conducted at the Naval Amphibious Base, Coronado, California, using tugs and crews provided by Amphibious Construction Battalion One (ACB-1).

REFERENCES

Huang, E.T. (1995). "Conceptual Development of Open Sea Module Connection Techniques", Technical Memorandum TM-2067-AMP, Naval Facilities Engineering Service Center, Port Hueneme CA.

Huang, E.T. (1996). "An Open Sea Modular Construction Method: Rigid Pontoon Connectors", *Sixth International Offshore and Polar Engineering Conference*, Los Angeles, California.

Huang, E.T. (1997). "Modular Dynamics of Coupled Pontoons in Seaways", *Seventh International Offshore and Polar Engineering Conference*, Honolulu, Hawaii.

Kane, P., (1996). "LOTS Development", Military Engineer, Volume 88, Number 579, August-September, pp 42-43.

Navy Pontoon Manual (1994), NAVFAC P-401, Naval Facilities Engineering Command, Alexandria, Virginia.

Proceedings of the Seventh (1997) International Offshore and Polar Engineering Conference
Honolulu, USA, May 25-30, 1997

ISBN 1-880653-28-1 (Set); ISBN 1-880653-32-X (Vol. IV)

Rapid Steel Pile Cutting Technology

Sheng S. Lin
Naval Facilities Engineering Service Center
Port Hueneme, CA, USA

ABSTRACT

This paper presents the results of the technological assessment of steel pipe pile cutting technologies, and the preliminary tests of an automated plasma arc cutter.

A rapid pipe cutting technology is needed in support of the Navy Modular Elevated Causeway System, which transfers cargo from ships moored offshore to an unimproved beach staging area. The developed cutter will be used to improve the efficiency of the causeway system installation by reducing the pile cutting time from 20 minutes with the current practice to less than 3 minutes.

KEY WORDS: Elevated causeway, pile cutting, amphibious cargo transfer, steel cutting technologies, explosives and shaped charges, abrasive water jet

INTRODUCTION

The U.S. Navy is developing a rapid pipe steel pile cutting technology to improve the installation of the Navy Modular Elevated Causeway System (ELCAS(M)).

The elevated causeway system is used to transfer logistics cargo from ships moored offshore to any unimproved beach staging area during an amphibious operation. The causeway system consists of a ramp, roadway, pierhead, and driven pile foundations. A part of the elevated causeway during construction is shown in Fig. 1.

Fig. 1. ELCAS(M) during construction.

The modules of causeway sections, construction equipment, and supporting vehicles are unloaded from ships, ferried to the beach staging area, and assembled. The installation of the modular causeway sections begins from the beach toward the ocean by a cantilevering process. The installation sequence is from ramp, roadway, to pierhead.

The causeway sections are elevated to 6 meters above the water surface at high tide to eliminate any environmental interference from waves, the tide, or current during installation and operation.

The causeway system adopted the modular concept that each structural module is prefabricated to a standard dimension. The modules require minimal preparation or modification in the field. The modules and their parts are interchangeable to keep the inventory simple.

The entire causeway system is supported by driven piles. After the piles are driven into the soil to develop the required bearing capacity, the modular causeway section is attached to the driven piles. Piles extending 1.5 meters or more above the causeway must be quickly cut off to allow a forklift to transport the piles and causeway sections for the next installation. Fig. 2 shows a forklift carrying a 12-meter causeway section over the cutoff piles at a transverse spacing of 7.2 meters. A quick pile cutting method will improve the completion of the ELCAS(M) installation. The cut off piles are hauled back to the marshaling yard and welded to form a pile section.

Fig. 2 Forklift carrying a causeway section over the ramp.

The objective of this paper is to document the results of the technical assessment of steel pipe pile cutting technology and present the cutting test results with an automated plasma arc cutter.

ASSESSMENT OF STEEL CUTTING TECHNOLOGIES

Current Navy Practice

The current cutting method is to use a hand-held oxyacetylene torch to cut a steel pile. This requires a welder and takes about 20 minutes per cut. The pile's cutting surface quality is rough and requires resurfacing before using.

Cutting Technologies

There are many steel cutting technologies that are commercially available. These cutting technologies include, but are not limited to: mechanical and abrasive cutting, explosives and shaped charges, laser cutting, abrasive water jet, thermal flame cutting, plasma arc cutting, chemical combustion, and low temperature cutting. A brief summary of each of these cutting technologies follows.

Mechanical and Abrasive Cutting. Power tools such as a pipe cutter, a hacksaw, an abrasive cutting disk, and a band saw are commercially available. The mechanical and abrasive cutting process involves: turning, sharpening, milling, broaching, drilling, and grinding (Marks, 1987).

In general, the cutting time is longer than torch cutting. The advantage of this cutting process is that these tools are simple, well proven, and suitable for small diameter steel pipe pile. The disadvantages are that these tools require a longer cutting time than the oxyacetylene torch and require periodical maintenance.

Explosives and Shaped Charges. Explosives and shaped charges have been used for cutting during structural demolition and clearing during a combat scenario. There are many explosives and shaped charges available. One flexible shaped charge called explosive cutting tape has been tested on steel plates, pile, and structural steel (Scott and Saccone, 1989). The explosive cutting tape is quick and effective. It takes about 2 to 3 minutes to set up (wrapping, tapping, and wiring), and a few more microseconds to cut off a 25.4 millimeter steel plate. The advantage is the explosive cutting tape is quick and easy to use. The disadvantages are that the fragments might reach a distance of about 300 meters. This will force the installer to evacuate during the detonation period resulting in temporarily shutting down the ELCAS(M) installation. Cracks caused by the explosion are difficult to detect and the steel piles cannot be reused.

Laser Cutting. Laser cutting has been used to cut plastic, rubber, hard metals, and diamonds. The laser beam micro-fractures the target materials with a precision stream of shock waves (Hecht, 1989). The cutting speed in general is slower than the thermal flame. The advantages are that the laser cutting yields a superior cut surface, generates minimal material waste, and can reach a narrow space via a flexible optical fiber cable. The disadvantages are that the overall laser cutting system is bulky, heavy, and expensive. The level of personnel technical training, and maintenance and repair in a combat situation would be difficult to achieve.

Abrasive Water Jet. Pressurized water jetting of up to 3,500 kgf/cm^2 has been used to cut plastic, ceramics, fiber, etc. (Tikhomirov, et al., 1992). Often, suspended abrasive solids, such as garnet, silica, and silicon carbide are added to increase the micro-cutting process. Carbon steel can be cut at a rate of 150 to 450 millimeters per minute for 6.5-millimeter plate and 100 to 200 millimeter per minute for a 19-millimeter steel plate. The advantages of using the abrasive water jet are that it yields an excellent cut surface, generates low heat, and the cutting fluid can be recycled. The disadvantage is that the pressurized jetting system is subjected to high pressures and has a high leakage potential that creates a maintenance and repair problem.

Thermal Flame Cutting. Thermal flame cutting applies a quick heat source to a steel surface 1.5 times the steel's melting temperature, which range from 1,250°C to 1,450°C (Sacks, 1976). An oxyacetylene flame can generate a surface temperature of about 3,500°C that melts the steel pipe. Thermal flame cutting is the current technology used in the ELCAS(M) installation. The thermal flame cutting is simple, cheap, and easy to maintain. Thermal flame cutting forms the basis to compare other cutting technologies.

Plasma Arc Cutting. Plasma arc cutting transfers the gas (air, nitrogen, etc.) into an ionized gas state (plasma). When the plasma strikes a target, the plasma arc releases heat in excess of 28,000°C (Marks, 1987). This temperature can melt any known metal. For a carbon steel 12.7 millimeters thick, the cutting rate is 1,500 to 2,000 millimeters per minute. The advantages of the plasma arc cutter are: it is portable, quick, and easy to operate. The disadvantage is it has potential electric shock due to the use of high voltage electricity if the system is not properly grounded.

Chemical Combustion. Chemicals such as aluminum, magnesium, hafnium, niobium, titanium, zirconium, and a mixture of these chemicals with other chemicals such as carbon, can generate high temperatures in a short time when combusted (CRC Handbook of Chemistry and Physics, 1989-1990 and Encyclopedia of Chemical Technology, 1978). For example, a mixture of powdered aluminum and iron oxide (called thermit) can generate a temperature exceeding 2,200°C and has been used in welding railroad tracks. Magnesium can generate a temperature of 2,000°C that can melt steel pipe. The cutting rate by chemical combustion is instantaneous. The advantage of using the chemical combustion method is that it is quick. The disadvantages are that these chemicals are difficult to handle since many can spontaneously self-combust in air and generate smoke and fume after combustion. The smoke and fumes are toxic and harmful to the operator.

Low Temperature Cutting. Metals become brittle in low temperatures and can be cut off with less effort. In theory, low temperature cutting seems to be easy and environmentally sound since the liquefied gas evaporates into the air after it is applied. The disadvantage is that freezing induces thermal contracting cracks in both longitudinal and transverse directions. These thermal cracks are difficult to control and the breaking surface may not be at the area of interest (Zarate, 1996). Also, these cracked pipe piles can not be reused.

Assessment of Cutting Speed

The cutting speed is controlled by the type of technology, material, and thickness of a work piece. Table 1 lists the relationship of cutting speed with other cutting technologies including plasma arc, abrasive water jetting, laser, and oxyacetylene gas (McGeough 1988). For a thickness of 12.7 millimeters, the cutting speed is in the decent order of plasma arc, oxyacetylene, and abrasive water jet cutting.

Table 1. Relationship of Cutting Speed with Other Technologies

Cutting Method	Approximate Cutting Speed[a] (m/min)
Plasma Arc[b]	1.4 - 1.5
Thermal Gas[b]	0.6 - 0.7
Laser, CO^2, 1 kW[b]	0.25 - 0.40
Abrasive Water Jetting	0.12 - 0.25

[a]Work Piece = 12.7 millimeters thick
[b]After J. A. McGeough, 1988

Selection of a Cutting Technology

The selection criteria were developed prior to choosing a candidate cutting method. The criteria were:

1. Cutting speed must be faster than current practice.
2. Only unskilled labor must be used.
3. Operational safety must be improved.
4. Shipping volume and weight must be minimized.

Based on the selection criteria, the plasma arc cutting method was selected as the first candidate for further evaluation.

The Naval Facilities Engineering Service Center (NFESC) retained a contractor to develop a rapid automated, pile cutter (Cardinal Scientific Inc., 1996). Fig. 3 shows the model automated, plasma arc cutter. A series of cutting tests was performed and is presented.

Fig. 3. Model of the automated plasma arc cutter.

TEST SETUP

A series of tests to assess cutting performance parameters were performed at the NFESC from July to September 1996. The test piles, test equipment, and power requirement are briefly summarized as follows.

Test Piles

The steel pipe piles used in the cutting tests were A252 steel. The steel pipe has a 610 millimeters outside diameter, and 12.7 millimeters thick. The steel pipe pile has the same grade as the steel pile used in ELCAS(M).

Test Frames

The cutter consists of a plasma arc unit, an automated control switch box, a clamshell, a steel frame, and an automated driver unit. The steel frame serves as a working platform that supports the plasma arc unit, the clamshell, and driver unit. The steel rollers allow the steel frame to move around. The clamshell has a latch that opens to allow the cutter to hold the pipe pile. The clamshell can accommodate a pipe pile of outside diameters up to 760 milllimeters. The switch box controls the piercing time, cutting speed, automation or manual operation, and on-off power switch.

Power and Air Requirements

The plasma arc cutter is an ESAB, PCM-1000I series A (ESAB, 1995). The input power was 3-phase, 208/230 VAC, 50/60 Hz, 40-ampere per phase, while the output was 108 DC and 85 amperes. The air pressure ranges from 4.5 to 5.5 kgf/cm^2 at a rate of 0.00252 m^3/s.

TEST RESULTS

Standoff Distance

Standoff distance is measured between the plasma arc tip and steel pipe pile. Four standoff distances were used: 1.6, 3.2, 4.8, and 6.4 millimeters. At the distances of 1.6 and 3.2 millimeters, the molten materials struck back from the steel surface and frequently damaged the tip of the plasma arc. At the distance of 6.4 millimeters, the cutting rate was slower than 4.8 millimeters, therefore, the standoff distance of 4.8 millimeters was used during the tests.

Effect of Current (Amperes)

The effect of the cutting current ranging from 50 to 80 amperes on the cutting time was tested. The results of the tests are shown in Fig. 4. At a cutting current of 80 amperes, the average cutting time was 2.7 minutes. While at a current of 50 amperes, the average cutting time was 6.6 minutes. At the current of 80 amperes, the pipe piles can be cut off within 3 minutes, which is significantly less than the current practice of 20 minutes. The cutting surface is also much better.

Beveling Angles

A series of cutting tests was performed with beveling angles ranging from 0 (normal to pile) to 45 degrees. The results are shown in Fig. 5. The cutting time averaged 2.7 minutes for 0-degree beveling angle and 5.8 minutes for a 45-degree beveling angle. The increase in cutting time is due to the thickness increase with beveling angles.

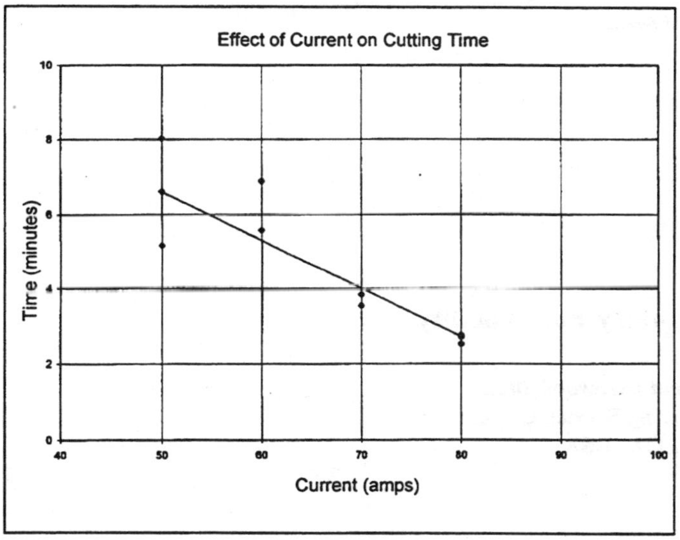

Fig. 4. The effect of current versus cutting time.

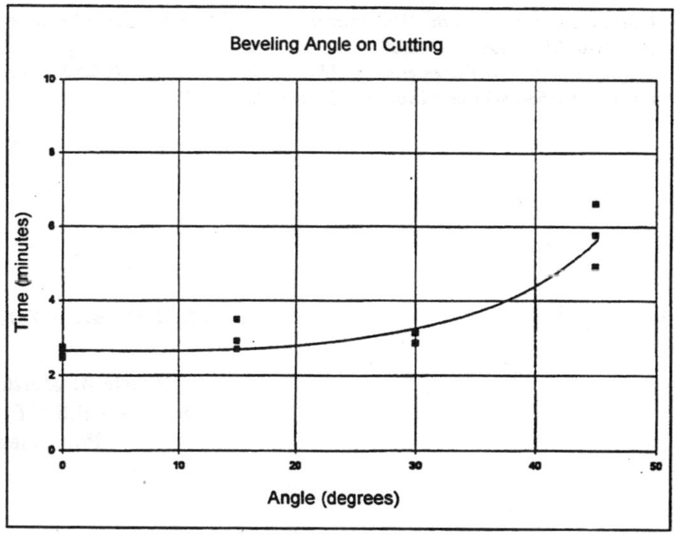

Fig. 5. The effect of beveling angle on cutting time.

SUMMARY

An assessment of steel cutting technologies was made. Based on factors such as cutting speed, labor requirement, operational safety, maintenance required, and shipping volume and weight, the automated cutter using plasma arc was selected. Increasing the cutting current will reduce the cutting time. At a current of 80 amperes, the cutting time was less than 3 minutes. The cutter nozzle tip can be adjusted to perform beveling, which will be used in the welding process. Increasing the beveling angle requires cutting through more steel, which will increase the cutting time.

ACKNOWLEDGMENT

The author wishes to thank the Office of Naval Research for funding this project, Cardinal Scientific Inc. for designing and fabricating the plasma arc cutter under a Navy contract, and Messrs. Monte Faust and Samuel Oppedisano for assisting in the cutting tests.

REFERENCES

Cardinal Scientific, Inc. (1996). "Portable Rapid Pipe Pile Cutter System," Naval Facilities Engineering Service Center contract report, Port Hueneme, California.

CRC Handbook of Chemistry and Physics (1989-1990). CRC Press, Inc., Boca Raton, Florida.

Encyclopedia of Chemical Technology (1978), by Kirk-Othmer, 3rd edition, Vol 5, John Wiley & Sons, New York, New York.

ESAB (1995). "Instructional Manual for PCM-10000, Plasma Arc Cutting Package," Florence, South Carolina.

Hecht, E (1989). "Optics," 2nd edition, Addison-Wesley Publishing Co., Reading, Massachusetts.

"Marks Standard Handbook for Mechanical Engineers" (1987). McGraw-Hill Book Co., New York, New York.

McGeough, JA (1988). "Advanced Methods of Machining," Chapman & Hall, New York, New York.

Sacks, R (1976). "Welding: Principles and Practices," Chas. A. Bennett, Co., Peoria, Illinois.

Scott, S and J Saccone (1989). "The Linear Shaped Charge Comparison Report," U.S. Army Engineer Waterways Experiment Station, Letter Report, Vicksburg, Mississippi.

Tikhomirov, R, V Babanin, E Petukhov, I Starikov, and V Kovalev (1992). "High Pressure Jet Cutting," ASME Press, New York, New York.

Zarate, D (1996). NFESC personal communication, Port Hueneme, California.

Proceedings of the Seventh (1997) International Offshore and Polar Engineering Conference
Honolulu, USA, May 25-30, 1997

ISBN 1-880653-28-1 (Set); ISBN 1-880653-32-X (Vol. IV)

Design of an Expeditionary Port Facility

Michele A. Murdoch and Glenwood Bretz
Naval Facilities Engineering Service Center
Port Hueneme, CA, USA

ABSTRACT

The Naval Facilities Engineering Service Center (NFESC) was tasked to develop a conceptual design package for an expeditionary port facility capable of rapid installation, simultaneous offload of two deep draft cargo ships during operations, and located up to 3 km offshore. The system design for the Rapidly Deployed Pier (RDP) includes a floating multi-deck pierhead moored by a single point mooring at the bow and a compliant mooring system aft, allowing the pierhead to be rotated about the SPM pivot point. The floating pierhead is connected to shore by a fixed, elevated approachway constructed using a cantilevered construction method.

KEY WORDS: Piers, ports, expeditionary, deployable, causeways, RO/RO, single point moorings.

INTRODUCTION

In October 1995, the Defense Advanced Research Projects Agency (DARPA) tasked the Naval Facilities Engineering Service Center (NFESC) to develop a system concept and design package for an expeditionary port facility capable of offloading deep draft cargo ships across unimproved shorelines. The work was led by NFESC, with contracted support provided by M. Rosenblatt & Son, Inc. in association with Elliott Bay Design Group, IMODCO, Inc., MAR, Inc., MCA Engineers, Inc., The Logistics Management Institute, Chesapeake Marine Design, Inc., and USA Models. This paper presents the results of the design and development effort.

SYSTEM MISSION

The mission of the RDP is to expeditiously provide port facilities for deep draft cargo ships in areas where adequate port facilities do not exist or have been damaged. Current methods for providing logistics support over unimproved shorelines require in-stream offload of ships to lighters, which transport the cargo either directly to the beach or to a shallow water expeditionary pier facility. While these methods are adequate for delivering relatively small quantities of cargo, they are not suitable for larger scale operations that require frequent offloading of modern containerships, Roll-On/Roll-Off (RO/RO) ships, and other deep draft ships bringing cargo to the operating area.

ENVIRONMENTAL CRITERIA

The RDP pierhead is designed for installation in 15 meters water depth. The approachway connecting the pierhead to the shore can be installed at sites with seafloor slopes ranging from 1:200 to 1:10, and in seafloor soils ranging from soft clays to medium dense sands. Table 1 shows the environmental design criteria for installation, operation and survival of the RDP system. The pierhead and approachway are designed to withstand the survival conditions shown in the table with 24 hours advanced warning of a major storm, with the system operational within 48 hours of the storm's passing. If weather and seas exceeding the survival conditions are anticipated, the pierhead can be released from its moorings and relocated to deeper water or safe haven until conditions are safe enough for return to the installation site for reconnection.

Table 1: Environmental Design Criteria

	Installation	**Operation**	**Survival**
Sea State	2-3	3-4	6
Sig. Wave Ht.	1.2 m	1.8 m	4.6 m
Max. Wave Ht.	2.3 m	3.4 m	8.5 m
Max. Energy Per.	6.5 sec	7.2 sec	11.3 sec
Period Range	5.1 - 15.4 sec	6.1 - 16.2 sec	9.9 - 17.4 sec
Sust, Wind Speed	8.2 mps	12.9 mps	25.7 mps
Current Speed	1.0 mps	1.0 mps	2.1 mps
Tidal Range	2.4 m	2.4 m	2.4 m
Temp. Range	- 28° to + 65° C		
Rain Conditions	Yes	Yes	Monsoon
Sea Ice	None		
Earthquake Load	None		
Sea state data from Bales, et al (1984)			

SYSTEM OVERVIEW

The RDP consists of a 290 m long floating pierhead connected to the shore by a fixed elevated approachway up to 3 km in length, depending on nearshore slope. An artist's conception of the system is shown in Figure 1. The pierhead is moored by a single point mooring (SPM) located at the bow (shoreward end) of the pierhead, and by four anchor legs at the stern. This mooring configuration ensures minimal movement of the pierhead at the bow where it interfaces with the fixed approachway, but allows the pierhead to be rotated over an arc of ninety degrees about the rotation point defined by the SPM. Figure 2 shows another view of the system concept, with the pierhead at an angle to the approachway.

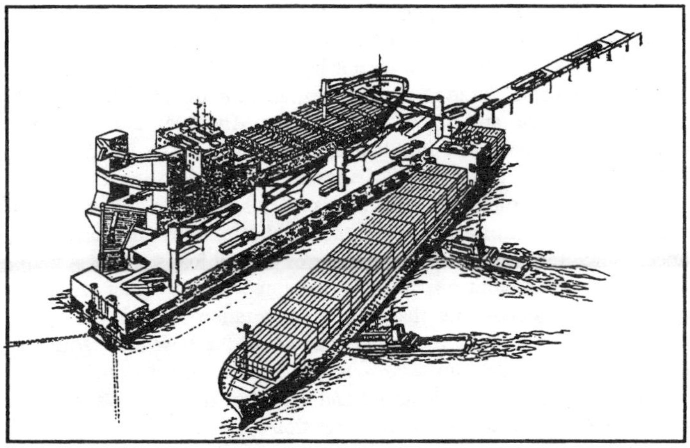

Figure 1: RDP System Concept

Figure 2: RDP System with Angled Pierhead Orientation

The approachway system provides a fixed, elevated roadway connecting the pierhead to the shore. The approachway is built of spans roughly 19 m long, using a cantilevered construction method. Each span is supported by driven steel pipe piles. The width of the roadway allows for normal two-way traffic of RO/RO vehicles and tractor-trailers carrying containerized or palletized cargo.

The RDP will most likely be a pre-positioned asset. A typical RDP installation will remain in place for approximately one year. Once the mission is complete, the RDP system will be retrieved, refurbished and prepared for re-deployment.

PIERHEAD SYSTEM

The pierhead, designed by M. Rosenblatt & Son, Inc., in association with Elliott Bay Design Group, is a multi-deck barge structure capable of transporting all approachway components and construction equipment required for system installation. A summary of vessel characteristics is shown in Table 2, and the outboard profile in shown in Figure 3. The vessel length and beam were limited to facilitate passage through the Panama Canal.

Table 2: RDP Pierhead/Barge Vessel Principal Characteristics	
Length Overall	289.6 m
Length at Waterline	274.4 m
Length of Transit Waterline	273.0 m
Maximum Beam	32.3 m
Depth, Molded (Weather Deck at side)	21.6 m
Full Load Draft (Deployed)	11.6 m
Draft to Transit Waterline	8.9 m
Light Ship Weight (Including Permanent Ballast)	41,762 tonnes
Displacement to Transit	70,201 tonnes
Full Load Displacement (Deployed)	93,359 tonnes

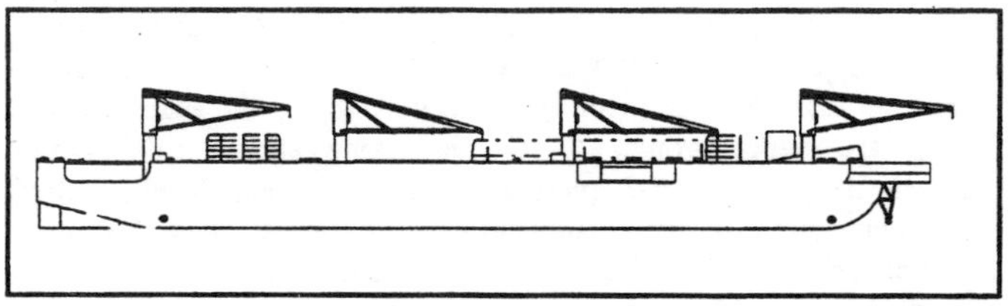

Figure 3: Outboard Profile of RDP Pierhead/Barge

Weather Deck

The weather deck is configured for pierhead crane LO/LO operations when installed at the operations site, and is the primary cargo deck for approachway spans and piles during transit. The bow of the weather deck is cantilevered 15.2 m over the end of the hull, extending 13.4 m beyond the pivot point of the single point mooring. The shape of the bow extension is circular, with its center located above the pivot point, so that the pierhead can rotate under the pierhead bridge span along a rail built into the periphery of the extension. The main extent of the weather deck ends 36.6 m from the stern of the vessel, and resumes for the last 9.1 m of the vessel. The two sections are connected by a walkway along the centerline of the pierhead. The two "cut-away" sections on either side of the pierhead, each 27.4 m long and 14.6 m wide, provide clear horizontal and vertical access for the deployment of stern ramps from RO/RO cargo vessels offloading at the RDP. The aft canopy provides needed weather deck level area for cargo ship and pierhead mooring hardware.

Main Deck and RO/RO Throughput Paths

During throughput operations, the main deck provides the thoroughfare for RO/RO vehicles entering the pierhead at the stern ramp access sites and two side ports. Freeboard at the main deck is 3.7 m, providing ramp access for all RO/RO capable ships expected at the

RDP. Freeboard of the main deck can be adjusted higher to suit specific cargo ships using the salt water ballast system. During transit, the deck is used for cargo stowage, including approachway spans. At the stern, the main deck is configured to provide a clear area 36.6 m long by the full 32.3 m width of the barge for receiving either fixed or slewing RO/RO ramps and vehicle traffic flow. Two side ports with clear deck areas, one each located port and starboard near amidships, provide access to RO/RO capable cargo ships with side port ramps. Each side port provides a 15.2 m wide by 5.2 m high clear opening. The side ports are secured by weather-tight doors during transit. To provide the maximum area available for vehicle transit and maneuvering, the main deck is clear of stanchions, providing clear access across the full interior width of the pierhead. The clear deck height of the main deck is 4.1 m, and permits RO/RO vehicle transit of UH-60A utility helicopters when configured for transport by sealift. An internal ramp leads from the main deck to the weather deck where the traffic path continues over the bow extension and onto the approachway. The ramp is sheltered and secured by weather tight doors on the weather deck.

Lower Decks

The second deck, third deck and tank top are used for cargo stowage during transit and operation, and are divided into cargo holds by the watertight bulkheads. The forward compartment of the main deck also provides room for the single point mooring hardware. Lockers for the stern mooring chain are provided in the aft peak tank. Fourteen salt water ballast wing tanks (seven port, seven starboard) extend through the third deck from the tank top to the second deck. Hatches are aligned through all decks for moving cargo to and from the weather deck using the pierhead cranes. Once RDP system components and installation equipment have been offloaded, the lower decks of the pierhead are free for other use, and could be configured to provide support services for expanded pierhead operations, including maintenance and repair of RO/RO cargo, transient storage for offloaded cargo, or configured with modular suiting to provide a variety of service support functions.

Pierhead Cranes

Four pedestal cranes are built into the pierhead structure to provide LO/LO offloading capability. The cranes are supported by the pierhead structure between the main and weather decks. Each crane is capable of lifting 75 LT at a radius of 14.0 m, representative of a combat-loaded M1A1 tank, and 42 LT at a radius of 46.0 m, representative of a fully loaded 12.2 m ISO container plus handling gear. The crane requirements selected for conceptual design of the RDP were met by MacGregor-Hägglunds crane model GL7514/4246-35. Pedestal extensions place the base of each crane boom 9.1 m above the weather deck, allow direct line-of-site observation of lifts from vessels with greater freeboards, but do not prevent the RDP from transiting under major bridges.

Transverse Thrusters

Integral to the pierhead/barge are two tunnel-type 3000 HP transverse thrusters, one forward and one aft, to provide maneuvering assistance. The thrust provided is sufficient for the pierhead/barge to withstand a 0.25 mps beam current.

Cargo Ship Berthing and Mooring Provisions

Twenty winches with a rated pull of 249 kN are arranged port and starboard, ten on each side of the weather deck. Bitts, chocks, sheaves and fairleads are arranged along the perimeter of the weather deck to support cargo ship mooring. Two separate fendering systems are required for the RDP. For RO/RO cargo vessels, floating fenders (camels) are required to provide sufficient standoff from the pierhead. For LO/LO operations, an expediently installable hull-mounted fendering system is available on the starboard side of the pierhead.

PIERHEAD MOORING SYSTEM

The mooring system for the pierhead was designed by IMODCO, Inc., and consists of an external chain turret mooring at the bow (Figure 4) and four stern anchor legs secured by drag embedment anchors. This configuration allows the orientation of the pierhead to be adjusted over an arc of 90 degrees (forty-five degrees in either direction from the initial installation orientation). The capability to rotate the pierhead about the SPM allows the pierhead orientation to be set such that environmental loads from currents, sea state and winds are minimized. This in turn increases the number of sites and conditions under which the RDP can be used, reduces overall loads on the system, and allows throughput operations to continue over a wider range of environmental conditions than if the orientation was fixed. The mooring system reduces pierhead excursion at the bow to a radius of 0.6 m during seastate 4 operations, and to a radius of 6.1 m in survival conditions. Both the chain turret and the stern mooring systems are configured with quick-release disconnection systems to allow rapid site evacuation in the event that conditions exceeding seastate 6 are anticipated.

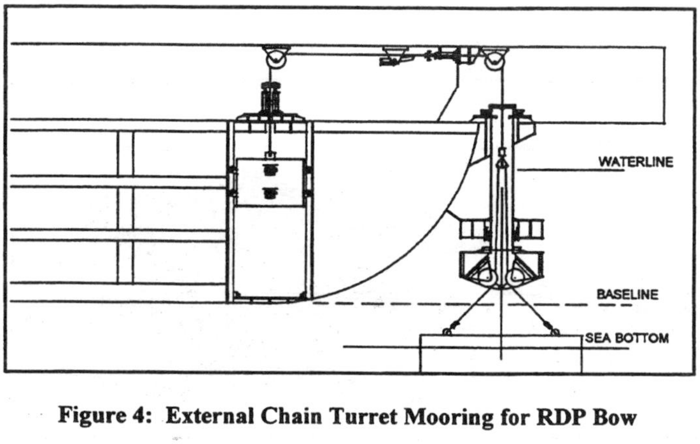

Figure 4: External Chain Turret Mooring for RDP Bow

Single Point Mooring at Bow

One end of the SPM anchor chain is connected to the seafloor by a piled base, and the other is connected to a counterweight internal to the hull of the RDP. The counterweight provides an instant reaction force to environmental loads, minimizing pierhead motions at the bow. External to the hull of the RDP, the anchor chain is contained in the 1.9 m diameter turret column that remains stationary over the anchor base. The turret column is suspended from the RDP main deck by three rows of roller bearings that allow the RDP to rotate about the turret column while transferring both horizontal and vertical mooring loads to the hull. The column is supported horizontally by a bottom collar cantilevered from the bow of the RDP about 6.7 m above baseline, containing a spherical bearing that allows the collar to rotate about the column with the pierhead.

A single chain length connects the system between the internal counterweight and a swivel positioned within the turret column. The swivel connects to a flounder plate where three lengths of anchor chain are connected. These three chain legs are spaced at 120 degrees, reeved through a sheave table located at the base of the turret, and connected to the seafloor anchor base at pad-eyes. Within the hull of the RDP, the chain path between the counterweight and turret column includes chain jacks and a chain stopper for assisting mooring installation operations, and two sheaves for guiding the chain. Each sheave is configured with a 445 kN brake. The brakes help absorb the mooring loads, and provided for significant reductions in mooring hardware size as reflected in the current design.

The anchor base is hexagonal in shape and is secured to the seafloor by six 1.2 m diameter, 1.9 cm thick steel pipe piles. Three pad-eyes spaces at 120 degrees are used to connect the anchor legs to the chain base. The length of driven pile required is dependent upon soil type, ranging from 17 m to 40 m for the range of soils considered in the conceptual design.

The counterweight consists of a 5.5 m diameter 3.7 m deep bucket filled with fine iron ore, and will weight approximately 340 tonnes in the dry moon pool. The counterweight has a vertical range of motion of 10.4 m within the moonpool. The moonpool contains four equally spaced vertical rails with self-adjusting bogey wheel assemblies that guide the counterweight within the moonpool. The bottom of the moonpool is watertight and is capable of holding the static load of the counterweight and dynamic loads incurred during transit.

Stern Mooring System

The layout for the stern mooring system is shown in Figure 5. The stern mooring system consists of four anchor legs, each roughly 950 m in length. The four legs are made from two continuous anchor lines that pass through the rigging at the RDP stern. The two anchor line assemblies are identical and are installed in parallel at the stern of the pierhead. When one end of the chain is under tension, the other is slack. Each of the four legs is secured with two 20 tonne MAG type anchors connected to an 550 m long loop of wire rope, which in turn is connected to 400 m of anchor chain using a underwater fairleader at the connection. These fairleaders distribute the load on each mooring leg evenly between the two drag embedment anchors securing each leg. At the transom of the pierhead/barge, each of the four legs is guided by sheaves and controlled by two chain jacks.

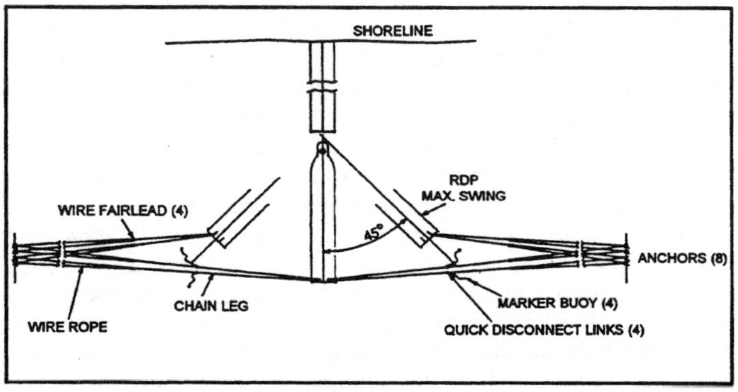

Figure 5: Site Layout for Stern Mooring System

The chain jacks that control each leg are connected to the port operations center and can be remotely activated, allowing the operator to swing the stern to a position where the mooring load is distributed evenly between the two legs on the side resisting the load. Each jack is also used to stopper off the chain to hold the desired pierhead orientation. Four underwater fairleaders mounted below the baseline of the hull, all free to rotate in the horizontal plane, keep the anchor chain beneath the baseline of the deepest draft (11.6 m) cargo ship at berth.

APPROACHWAY SYSTEM

The approachway system was designed in house at NFESC, and is comprised primarily of standard spans each nominally 19.0 m long, 7.6 m wide and 2.0 m deep. The span width provides for two-way traffic over the length of the system. All spans are supported by 0.6 m diameter, 1.3 cm thick driven steel pipe pile. The approachway deck is elevated 10.4 m over mean lower low water (MLLW), providing an air gap sufficient to prevent uplift loads at the maximum design wave at maximum tide. The system is connected to the pierhead with a special one-way pierhead bridge span that accommodates the six degrees of motion of the pierhead. At the beach, a seven degree ramp makes the transition from full elevation to the shore. Near mid-span, a traffic by-pass roughly 55 m long and an additional 7.6 m wide provides a pull-out area for disabled vehicles, or to allow extra-wide vehicles to pass. The by-pass is made up of standard components.

The approachway system is installed using a cantilevered construction method (Figure 6), and can be installed simultaneously from both the beach and pierhead to speed construction. Mobile construction cranes are used for all construction except for the three most seaward spans installed offshore. Pierhead crane support is used to install these sections, after which one of the mobile construction cranes can be lifted onto the newly installed spans. Construction proceeds from both ends until the gap between them is roughly 36 m or less. A combination of up to four special intermediate and short length spans are used to close the gap to a final offset of 3.2 m maximum. A simply supported bridge span is used to close the final gap.

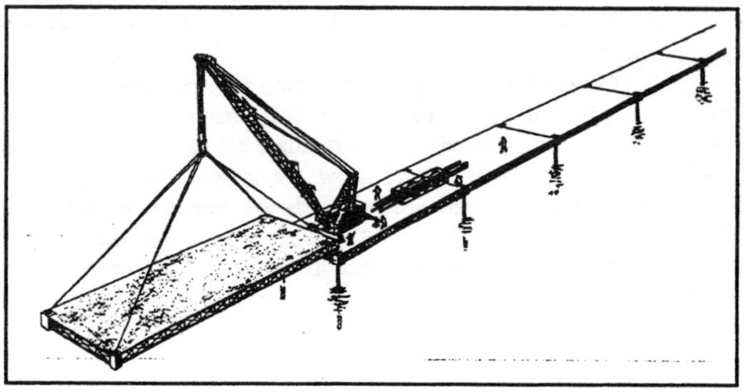

Figure 6: Cantilevered Construction Method for Approachway

Approachway Spans

All approachway spans are constructed of four longitudinal trusses that are tied laterally and diagonally for stabilization. Trusses are constructed of structural steel tubing. Span decks are made of high-strength steel grating, providing the wear surface on the spans and distributing vehicle and construction loads to the stringers and trusses. All grating is removable to provide access below, if needed, and to facilitate repair or replacement if damaged. Special hardened load points are provided to support construction crane loads during installation.

The standard spans have two internal spudwells located at the end of the span with the female connectors. Four four-pile spans are used during construction in locations where fully supported and stable platforms are required for initial construction efforts. The four-pile spans are nearly identical in design to the standard spans, with two internal spudwells added at the male connector end of the span. The intermediate and short gap-closing spans (12 m and 7.6 m long, respectively), also have the same width, depth and basic design of the standard spans. Only the truss elements are sized differently in these spans to account for the different load distributions. Internal spudwells, pile grippers and connection mechanisms are identical on all span types.

The connection mechanism consists of four pin-type male connectors on one end of the spans, and four matching hoop female connectors on the other end. The female end of the span points in the direction of approachway construction. The connectors are used during construction to cantilever each span as it is installed, and also provide the structural continuity between adjacent spans once supported by piles.

Spans are secured to the foundation piles by double wedge-action type pile grippers. The gripper is made up of 8 pairs of wedges spaced symmetrically around the pile. Torque applied to actuating bolts running through the outer wedge bears the outer wedge against the inner wedge, which in turn grips the side of the pile. The inner wedge face is serrated to improve gripping performance. The commercially available pile gripper assembly selected in the RDP design is manufactured by Hunting Marine Contractor Services, Inc.

Beach Ramp

Special gripper adapters are installed between the span and pile gripper assembly on the three spans that make up the beach ramp, providing a seven degree offset to the grippers. This allows the piles to be driven perpendicular to grade while maintaining the proper ramp angle. These adapters allow standard two and four-pile spans to be used for beach ramp construction, instead of requiring specialized components.

A special seven degree beach ramp adapter assembly is used to make the final transition between the beach ramp and the first level approachway span. The adapter has standard connector assemblies on each end, and is configured to allow the piece to be used during construction either from the beach towards the pierhead, or vice versa. The assembly is approximately 1.9 m long, and matches the spans in width and depth. The adapter does not require pile support and fully transfers the load of a cantilevered span to the pile-supported span it is hung from.

Approachway Bridge Span

The bridge span used to close the final gap in the approachway, created during simultaneous construction operations from offshore and beach-side, is a simply supported span nominally 4.9 m long, 7.3 m wide, and 0.3 m deep. The span is secured to the two supporting approachway spans by bolts at each corner. The span can accommodate horizontal offset between the two spans up to 1.2 m, and vertical misalignment up to 0.3 m.

Pierhead Bridge Span

The span connecting the floating pierhead to the fixed approachway is roughly 15.2 m long, 6.4 m wide, and 1.5 m high. The design gap between the approachway and the pierhead, when over its mooring center, is 9.1 m, based on the expected watch circle of the pierhead, and on bridge length requirements to maintain a reasonable angle. The structural design of the span is similar to that of the other approachway components, except that the road surface of the span is located at the base of the truss to minimize vertical offset between the bridge and the pierhead and approachway. The bridge provides a clear width of 5.5 m suitable for one-way traffic of all design vehicles.

Rollers support the approachway end of the bridge span, and ride in tracks installed on the first offshore span of the approachway. This connection primarily accommodates surge motion and tidal fluctuations, although some sway and yaw are accommodated within the track and roller system. At the pierhead, a curved beam at the base of the bridge span fits into a curved track along the periphery of the pierhead bow extension. The pierhead connection accommodates the roll, heave, pitch, sway, and yaw expected of the pierhead during seastate 4 operating conditions. The bridge span maintains horizontal alignment with the approachway, allowing the pierhead to rotate under the bridge span when pierhead orientation is adjusted. A come-along or winch will be used to help shift the bridge span along the curved track during pierhead orientation adjustment operations.

DEPLOYMENT

All RDP system components and equipment are self-contained on the pierhead/barge during deployment. A minimum of two ocean-going commercial tugs will be used for deployment, installation, and operational support of the system. The support vessel requirements for RDP transit, installation and operations can be fulfilled by either commercial combination tugs or Anchor Handling Supply Tugs (AH/TS) with minor bow modifications, depending upon the sea conditions across the transit route. A cost comparison of charter and purchase arrangements indicates that a peacetime availability contract with a commercial tug charter organization would be more advantageous for providing system support.

INSTALLATION SEQUENCE

The principal installation sequence for a system with a 915 m long approachway is shown in Figure 7. The estimated construction time required for installation of an RDP system with a 915 m long approachway is 207 hours (9 days), and 406 hours (17 days) for a system with a 3,000 m long approachway. These estimates were based on simultaneous construction operations and installation in a medium dense sandy seafloor, typical of sandy nearshore areas. Installation in weaker soils will require additional installation time.

Once the RDP arrives on site, the pierhead is installed in a temporary mooring consisting of a gravity-based SPM situated to support initial offshore approachway construction, and the stern mooring system installed in its semi-permanent location and configuration. With the pierhead secured in its temporary mooring, offshore construction of the approachway begins, concurrent with lighterage offload of construction equipment and system components to the beach for beach-end construction, and installation of the semi-permanent SPM. Beach-end construction commences with the arrival of the first lighterage load. Simultaneous approachway construction continues until the final gap is 36.5 m or less. The gap-closing spans

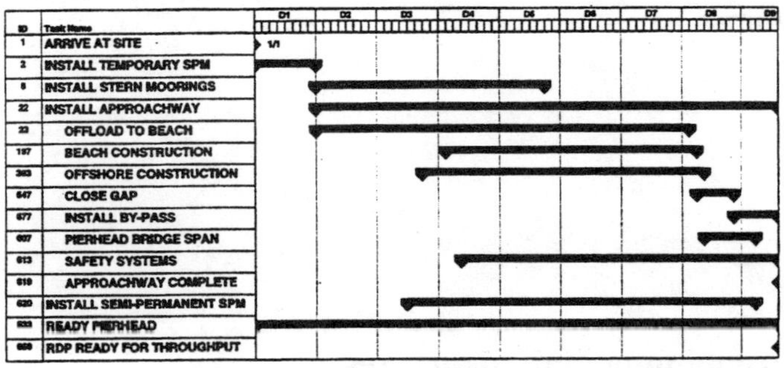

Figure 7: Installation Sequence for RDP System with 915 m Approachway

are installed, and in the sequence illustrated in Figure 7, the traffic by-pass is installed from the beach end after the gap is closed, although the by-pass spans can be installed at any time.

The pierhead bridge span is deployed once the pierhead has been moved to its semi-permanent mooring location. Figure 7 shows this occurring after the conclusion of approachway construction, but can take place at any time after the first three seaward spans of the approachway are installed. During approachway construction, the semi-permanent single point mooring anchor is installed from the pierhead, and the pierhead is prepared for throughput operations,. Installation of the temporary SPM mooring system, stern mooring and approachway construction are the primary critical path items for RDP installation.

CONCLUSIONS

A conceptual design for the RDP system, consisting of a floating pierhead and fixed elevated approachway, has been developed to provide cargo offloading facilities for deep draft cargo ships in situations where port facilities are damaged, inadequate, or non-existent. The unique adjustable orientation feature of the pierhead increases the number of sites and conditions under which the RDP can be used, and allows throughput operations to continue over a wider range of environmental conditions compared to fixed orientation systems. The RDP is self-deployed to the installation site, rapidly installed, and can be retrieved at the conclusion of each mission for refurbishment and re-use. This system will provide an efficient offload capability for cargo ships bringing RO/RO, container, and breakbulk cargo in support of U.S. military and humanitarian operations in areas where adequate port facilities do not exist.

ACKNOWLEDGEMENTS

The authors would like to acknowledge M. Rosenblatt & Son, Inc., Elliott Bay Design Group, IMOCDO, Inc., MAR, Inc., The Logistics Management Institute, MCA Engineers, Inc., Chesapeake Marine Design, Inc., USA models, Carderock Detachment Naval Surface Weapons Center (CD NSWC), and the NFESC in-house concept development and design teams for their contributions to the successful development of the RDP conceptual design. In addition, we wish to acknowledge the Defense Advanced Research Projects Agency (DARPA) and their Technical Agent at CD NSWC for sponsoring the effort.

REFERENCES

Bales, S.L., W.T. Lee, and J.M. Voelker (1984), *Standardized Wave and Wind Environments for NATO Operational Waters*, David Taylor Naval Ship Research and Development Center Report DTNSRDC/SPD-0919-01, July 1981, Revised March 1984.

Proceedings of the Seventh (1997) International Offshore and Polar Engineering Conference
Honolulu, USA, May 25-30, 1997

ISBN 1-880653-28-1 (Set); ISBN 1-880653-32-X (Vol. IV)

Buckling Strength of Lightened Aluminium Hull Structures

Yoshiteru Tanaka and Kazuyoshi Matsuoka
Ship Research Institute, Ministry of Transport
Japan

ABSTRACT

The product of large scale high speed aluminum vessels has increased. For those vessels, reduction of structural weight becomes an important subject, and structural elements of light weight but high strength are sought for.

For hull structures of aluminum ships, extrusions from A5083S-H112 alloy, pi shaped section, had been popular in Japan. However, from the view point of the lightening above mentioned, extruded hollow sections from A6N01S-T5 alloy and pre-ribbed panels of A5083P-H321/A5083S-H112 were developed and are going to be applied to hull structures.

In this paper, the buckling strength of hull structures by these extrusions and pre-ribbed panels is dealt with. Using these elements, test models were manufactured. Axial compression tests were performed on the models. On those models, FEM analyses were carried out, and simplified calculation methods were examined. Based on the results, the estimate methods on the buckling strength of aluminum hull structures were proposed. Furthermore, the efficiency of the elements for hull structures, in saving structural weight, was studied.

1. INTRODUCTION

Recently, the design and product of large scale high speed vessels have increased. For those vessels, high strength aluminum elements such as pi shaped extrusions from A5083S-H112 alloy, extruded hollow sections from A6N01S-T5 alloy and pre-ribbed panels of A5083P-H321/A5083S-H112 have been used. For the rational use of these elements and to design lighter hulls, it is necessary to recognize not only the mechanical properties of the materials but the structural strength of the elements and the members with them.

Authors have been studied the fatigue strength of welded joints [1)-4)], the strength of heat affected softened zone by welding [5)] and the structural strength of extrusions and pre-ribbed panels of aluminum alloys [6)-8)], with the aim of the establishment of evaluation methods on the structural strength of lightened aluminum hull structures. In this paper, the results of the studies about the buckling strength of outer shell under axial compressive loads [9)-11)] are reported.

2. COMPRESSION TESTS

2.1 Materials

The elements used for the test models are pi shaped sections, pre-ribbed panels and hollow sections. The cross section shapes and dimensions are presented in Fig.1 and the mechanical properties are shown in Table 1.

2.2 Test models

The models imitated hull structures were fabricated by pi shaped sections (A5083S-H112), pre-ribbed panels (A5083P-H321/A5083S-H112) and hollow sections (A6N01S-T5) (see Fig.1). The shapes and the dimensions of the models are presented in Figs. 2 to 4. These models consisted of skin plates, longitudinal stiffeners and 4 pieces of transverse girders.

Though the models $\Pi-1$ and $\Pi-2$ shown in Fig.2 were made of pi shaped sections, the shapes of the cross sections were different (see Fig.1). The model $P-1$ shown in Fig. 3 were made of pre-ribbed panel, which is a stiffened panel of skin plates of A5083P-H321 alloy and bulb plates of A5083S-H112 alloy. The shape of the stiffeners (bulb plates) is the same as those of the model $\Pi-2$. On the other hand, the models $H-1$ and $H-2$ were made of the same hollow section, however, the arrangement of the brackets between the face plates and the transverse girders and the existence of the tripping

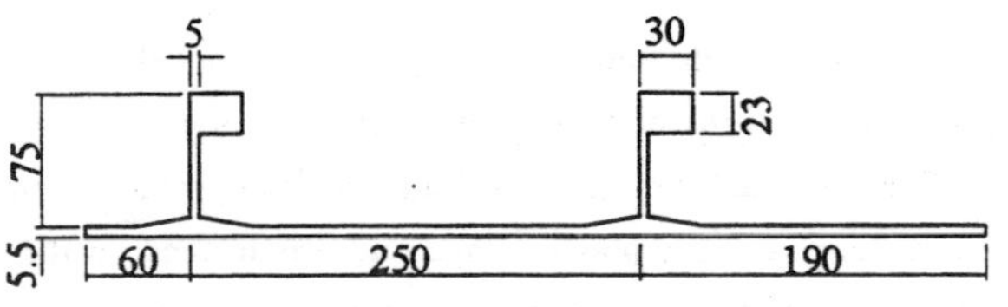

Pi shaped section (Test model Π − 1)

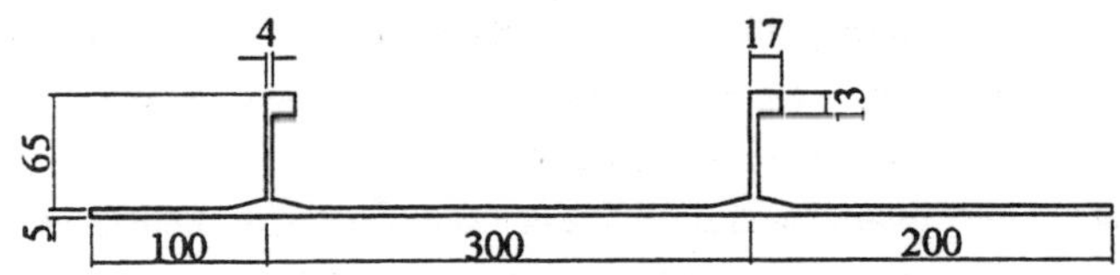

Pi shaped section (Test model Π − 2)

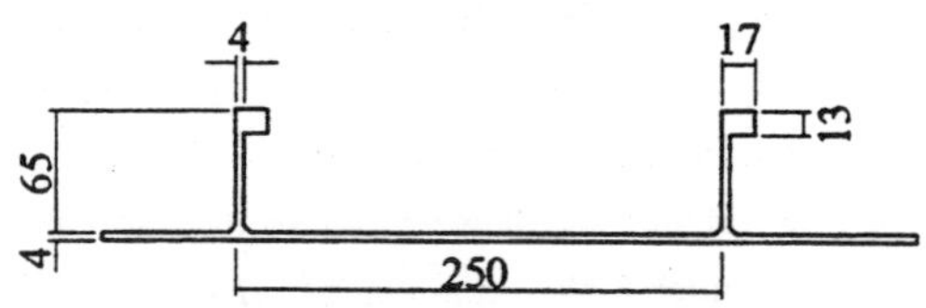

Pre-ribbed panel (Test model $P-1$)

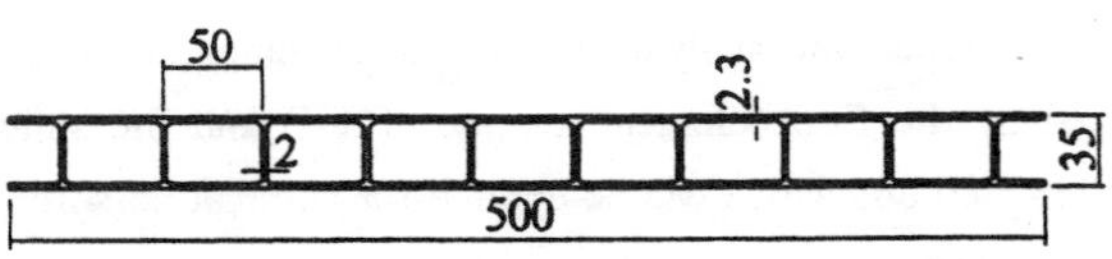

Hollow section (Test models $H-1$ and $H-2$)

Fig.1 Cross section shapes of elements (unit : mm)

Table 1 Mechanical properties of aluminum alloys

Test model	Material	0.2% offset strength (MPa)	Tensile Strength (MPa)	Elongation (%)
Pi-section (Π-1)	A5083S-H112	201	314	21
Pi-section (Π-2)	A5083S-H112	192	344	19
Pre-ribbed panel (P-1)	A5083P-H321	232	348	14
	A5083S-H112	200	303	29
Hollow section (H-1)	A6N01S-T5	237	269	12.4
Hollow section (H-2)	A6N01S-T5	247	274	14.8

Specimen Π-1

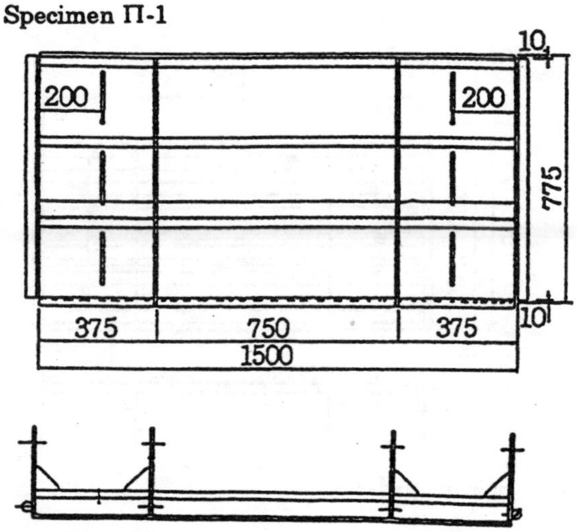

Specimen Π-2

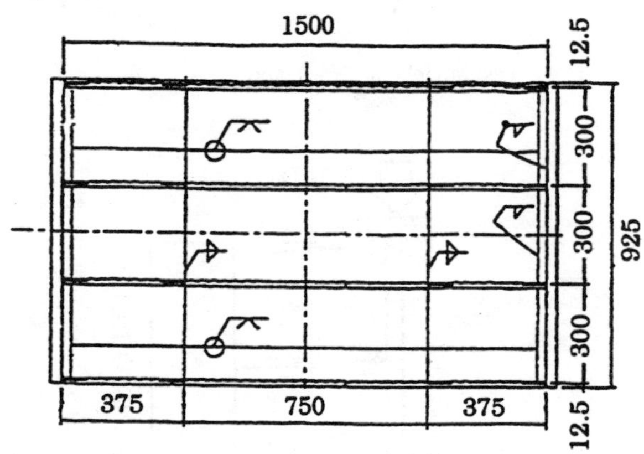

Fig. 2 Test models of Π − 1 and Π − 2 (unit : mm)

brackets on the both end spans were different (see Fig. 4).

2.3 Test setup

The setup of the compression tests is illustrated in Fig. 5. Four transverse girders of each model were supported elastically, and the axial compressive loads were applied on the both ends of the models. Since the central span is the longest among the three spans divided by the transverse girders, the first buckling occurs in it.

2.4 Test results

The results of the experiments are presented as the ultimate load in Table 2. The ultimate load of 0.2% O.S. in the table means multiple of cross section area and 0.2% offset strength. And σ_{eff} represents the effective stress (average compressive stress), that is, the quotient of ultimate load divided by cross section area .

The collapsing process of both the pi shaped section (test models Π − 1 and Π − 2) and the pre-ribbed panel (test model $P-1$) was the following way. The first elastic buckling occurred in the skin plates, and the lateral buckling of the longitudinal stiffeners (bulb plates) followed, then the applied compressive load reached the ultimate load. In the case of the pi shaped section, there was a slight difference between the skin plates (the part of panel) and the bulb plates in the wave length of buckling. This difference was more remarkable on the model Π − 1 than Π − 2 . On the contrary, in the case of the pre-ribbed panel, the wave length of the skin plates in buckling almost corresponded to the wave length of the stiffeners in lateral buckling.

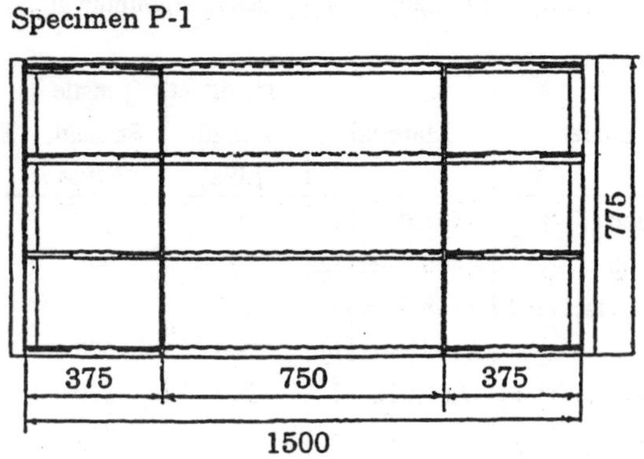

Fig. 3 Test model of $P-1$ (unit : mm)

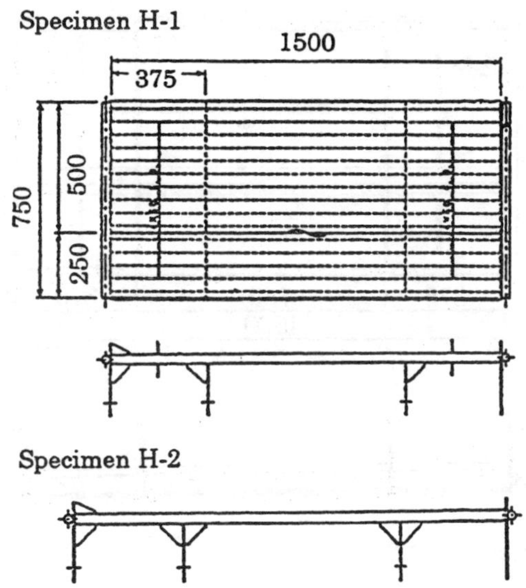

Fig. 4 Test models of $H-1$ and $H-2$ (unit : mm)

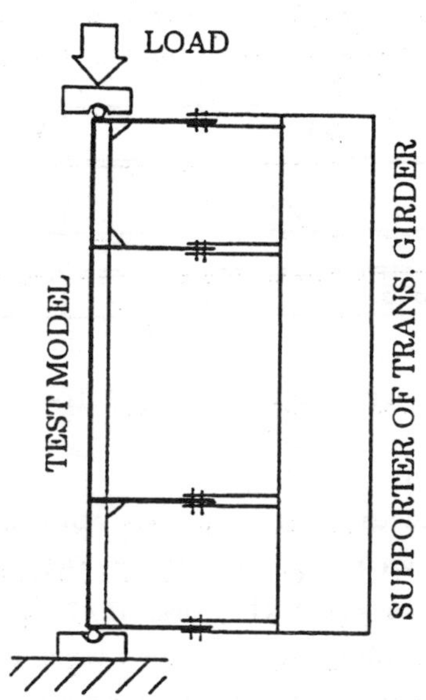

Fig. 5 Test setup

Two types of the hollow sections (test models $H-1$ and $H-2$) represented the different collapsing processes. On the model $H-2$, firstly the deformation like Euler buckling mode occurred in the central span, and the plastic hinge line was formed on the center line of the span, then the applied compressive load reached the maximum load.

On the contrary, the behavior of the model H-1 was the same as that of $H-2$ until the Euler buckling mode occurred, however, the plastic hinge lines were produced obviously at the part of the fillet weld of the tripping brackets on the end spans and at the both sides of the central span in the condition of the ultimate deformation.

This definite difference in the effective stress between the models $H-1$ and $H-2$ depended on the existence of these plastic hinge lines. In the model $H-2$, the fillet weld lines with the transverse girders were divided and stiffened by the brackets and the plastic hinge lines of full width were not formed and the collapse load of $H-2$ was much higher than that of $H-1$.

3. ANALYTICAL METHODS

In this paper, the analyses on the axial compression tests are performed by the finite element method (FEM) and the simplified calculation methods which can be applied for application software of personal computers.

3.1 FEM analyses

The elasto-plastic calculations for the stiffened panels such as the models $\Pi-1$, $\Pi-2$ and $P-1$ are performed by supposing the condition of the continuous stiffened plate. The thick shell element is used for both the skin plates and the bulb plates. The transverse girders are not modeled but the lateral deflection of the nodal points on the skin plate corresponded to the intersection is constrained. And the lateral deflection of the skin plates is given as initial deflection in a sine wave which is at most one tenth of the plate thickness.

Moreover, in carrying out the elasto-plastic calculations, the relationships between the stress and the plastic strain of the parent metal and the heat affected softened zones are utilized. This relationships are obtained from the tensile tests of the specimen containing the heat affected softened zones.

The collapse mechanisms of the models H-1 and H-2 (hollow section) were different from those of the stiffened panels such as the pi shaped sections or the pre-ribbed panels. Every local panel of the hollow sections carried the applied load effectively without buckling, until the general collapse occurred. In this case, the hollow section is supposed to be a plate whose area and rigidity is equal to it, and whose lateral deflection at the intersection with four transverse girders is constrained. This model is analyzed by the FEM elasto-plastic calculations according to the relationships between the stress and the plastic strain .

Table 3 Results of compression tests and calculations

Test model	Ultimate load (kN)		Exp.	σ_{eff}(MPa)	FEM	(1)	(2)	(3)
	Experiment	0.2% O. S.	/0.2% O. S.	Experiment				
Π-1	1050, 1070	1678	0.625, 0.638	126, 128	157.1	(173.6)	153.6	-
Π-2	918	1330	0.690	132	129.0	137.6	(144.9)	-
P-1 (S = 300 mm)	-	-	-	-	118.5	117.8	(146.2)	-
(S = 250 mm)	581	1042	0.558	123	134.9	136.5	(147.3)	-
(S = 200 mm)	-	-	-	-	145.9	(157.4)	148.6	-
(S = 150 mm)	-	-	-	-	179.2	(180.7)	150.2	-
H-1	631, 594	1098	0.575, 0.541	136, 128	184.1	-	-	137.7
H-2	800, 802	1144	0.699, 0.701	173, 173	184.1	-	-	180.0

(1) Calculation after 3.2
(2) Calculation after 3.3.1
(3) Calculation after 3.3.2

3.2 Simplified calculation by effective width of stiffened panels

After the buckling of the skin plates, the stress contribution of the stiffened panels at the moment the stiffeners buckle laterally is simplified as shown in Fig. 6. Then the collapse load and the effective stress are calculated according to the following way. The stress σ_1 in the center part of the skin plate except for the effective width, the stress σ_2 in the effective width of the skin plates containing the stiffeners and the effective width W are calculated from the following equations.

$$\sigma_1 = 3.6E\left(\frac{t}{S}\right)^2 \tag{1}$$

$$\sigma_B^{cr} = \left(\frac{\pi^2}{12}\right)E\left(\frac{b}{S}\right)^2 \tag{2}$$

$$\sigma_2 = \sigma_B^{cr} \qquad \left(\sigma_B^{cr} \le \frac{\sigma_Y}{2}\right) \tag{3-1}$$

$$= \sigma_Y\left(1 - \frac{\sigma_Y}{4\sigma_B^{cr}}\right) \qquad \left(\sigma_B^{cr} > \frac{\sigma_Y}{2}\right) \tag{3-2}$$

$$W = 0.60t\sqrt{\frac{E}{\sigma_1}} \tag{4}$$

where, E : Young's modulus
t : the thickness of skin plate
S : the distance between stiffeners (stiffener space)
b : the bulb width
σ_Y : yield stress (0.2% offset strength of parent metal)

Equation (1) represents the elastic buckling stress of the skin plate when the wave length of the buckling is equal to the stiffener space. σ_B^{cr} shown in Eq. (2) represents the elastic lateral buckling stress of the bulb plate when the wave length of the buckling is equal to the stiffener space. The stress of stiffeners denoted σ_2 is the modification of Eq. (2) by Eq. (3), that is, Johnson's formulation. If necessary, Johnson's modification is also applied on the stress σ_1. According to Fig. 6, the collapse load is obtained by the integration of the stress given in Eqs. (1) through (4) about the cross section area.

3.3 Simplified calculation methods considered softened zones

In calculating the collapse load by the effective width of the stiffened panels, the effect of the softened zone is not considered. When the elastic buckling of the skin plates occurs with the same wave length as the stiffener space and the stiffeners collapse by lateral buckling with the same mode, the supporting condition of the joints between the transverse girders and the skin plates are independent from the collapse mechanism, therefore, the effect of the heat affected softened zone is negligible.

In the case of the over all buckling as a column, the softened zone at the intersection between the skin plates and the transverse girders influence the collapse mechanisms.

Here, the calculations of the collapse load considering the softened zone are performed by the simplified methods (Simplified Two-flange Column Model [12]), that is, the outer shell is supposed to be a column with two flanges.

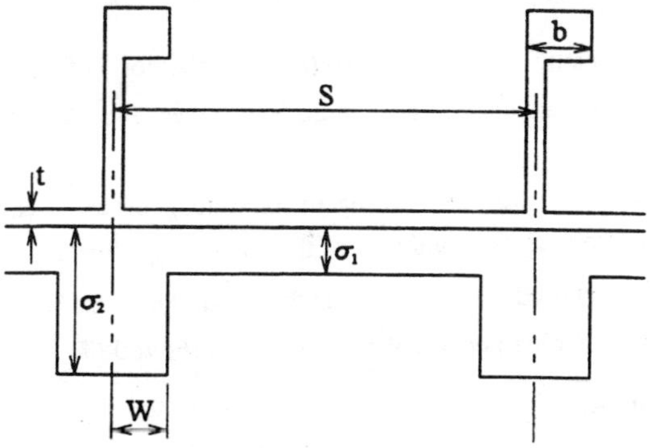

Fig. 6 Simplified stress distribution at failure

3.3.1 Stiffened panels

The buckling as a column of one span divided by two transverse girders in the outer shell is considered for the models $\Pi-1$, $\Pi-2$ and $P-1$. The stiffeners fall down when the over all buckling occurs to the stiffened panels. Since the stress of the lateral buckling in the stiffener is extremely small in the condition that the both ends are supported, the bending moment of the central span is negligible in the condition of the over all buckling. Then, as shown in Fig. 7, this mechanism is replaced to the problem of the elasto-plastic buckling of a column, that is, the both ends of the column are simply supported and the softened zone exists at the central part of the span. Here, the bulb of the stiffener is supposed to be one face plate of A_1 in the cross section area and the skin plate is supposed to be the other face plate of A_2, by neglecting the web of the bulb plates of the pi shaped sections and the pre-ribbed panels.

When the uniform axial compressive stress σ reaches σ_{eff}, the lateral deflection is assumed to occur without increase of the axial load. Then the relationships between the increase of the compressive stress of the bulb denoted $\delta\sigma_1$ and that of the skin plate denoted $\delta\sigma_2$ become Eq. (5), because the axial load dose not change.

$$A_1\delta\sigma_1 + A_2\delta\sigma_2 = 0 \tag{5}$$

From Eq. (5), both the bulb and the skin plate can not keep the condition of the plastic loading, therefore, either one returns to the unloading (elastic) condition.

The change of the moment derived from this change of the stress is expressed in Eq. (6), defining the increase of the compressive stress of the bulb for plus.

$$\delta M = DA_1\delta\sigma_1 = -DA_2\delta\sigma_2 \tag{6}$$

Where, D is the distance between the centers of gravity of both the bulb and the skin plate. Neglecting the elastic strain, the increase of the curvature δK corresponding to the moment of Eq. (6) is presented in Eq. (7), where $\delta\varepsilon_1^P$ is the increase of the plastic strain of the bulb and $\delta\varepsilon_2^P$ is that of the skin plate.

$$\delta K = \left(\delta\varepsilon_1^P - \delta\varepsilon_2^P\right) / D$$

$$= \delta\varepsilon_1^P / D \quad \text{or} \quad -\delta\varepsilon_2^P / D \tag{7}$$

Where, because of the unloading (elastic behavior) of either the bulb or the skin plate as presented in Eq. (5), either $\delta\varepsilon_1^P$ or $\delta\varepsilon_2^P$ is equal to zero.

Defining $d\sigma / d\varepsilon^P$ (tangent modulus) in the heat affected softened zone at the condition of the stress σ_{eff} for E_t, the tangent modulus denoted $(EI)_t$, which is the rigidity against bending in this condition, is obtained as follows from Eqs. (6) and (7).

$$\delta M = (EI)_t \delta K \tag{8}$$

$$(EI)_t = D^2 A_1 E_t \quad \text{or} \quad D^2 A_2 E_t \tag{9}$$

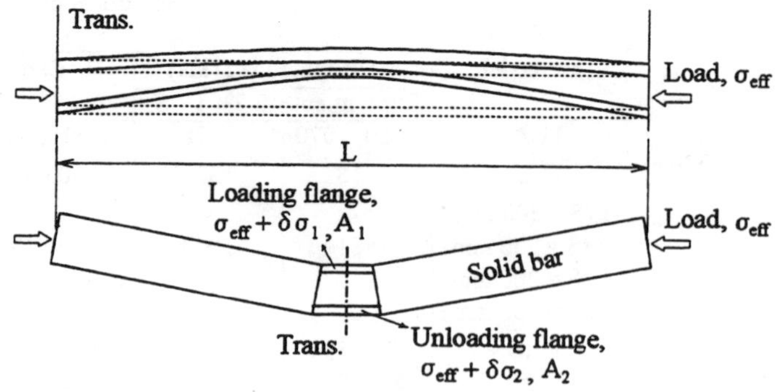

Fig. 7 Simplified two-flange column

The collapse load P_{cr} is obtained by substituting the tangent modulus of the rigidity against bending for Euler's buckling formula Eq. (10). Where L is the length of the span. The smaller value of A_1E_t and A_2E_t in Eq. (9) becomes the collapse load shown in Eq.(10). Then the effective stress is obtained as presented in Eq.(11).

$$P_{cr} = \pi^2 (EI)_t / L^2 \tag{10}$$

$$\sigma_{eff} = P_{cr} / (A_1 + A_2) \tag{11}$$

In the case of the stiffened panels treated in this paper, the models $\Pi-1$, $\Pi-2$ and $P-1$, the tangent modulus $(EI)_t$ of the bulb shown in Eq. (9) is smaller than that of the skin plates. Therefore, the cross section area of the bulb (A_1) and the relationships between the stress and the plastic strain in the softened zone of A5083S-H112, that is, the differential of Eq. (12) are used as the tangent modulus (E_t) of the relationships between the stress and the strain.

$$\sigma = 108 + 140\left(\varepsilon^P\right)^{0.173} \tag{12}$$

The effective stress is obtained after the iterative calculations of $\sigma = \sigma_{eff}$ on the assumption that Eqs. (10) through (12) simultaneously stand. As a result, the smaller one of the values derived from this subsection (3.3.1) and the previous section (3.2) is adopted for the collapse load (or effective stress) of the stiffened panels.

3.3.2 Hollow section

Both sides of the face plates of the hollow sections under compression work effectively till the final stage of the loading, therefore, the equilibrium of the moment needs to be considered. Figure 8 shows the deformed condition of the model $H-1$.

The softened zone was created only in the one side (inner side) of the face plates, which was the weld part with the transverse girders in the model $H-1$. In Fig. 8, B and D represent the width of the softened zone and the distance between the both centers of gravity of the inner and the outer face plates respectively. A_1 and A_2 represent

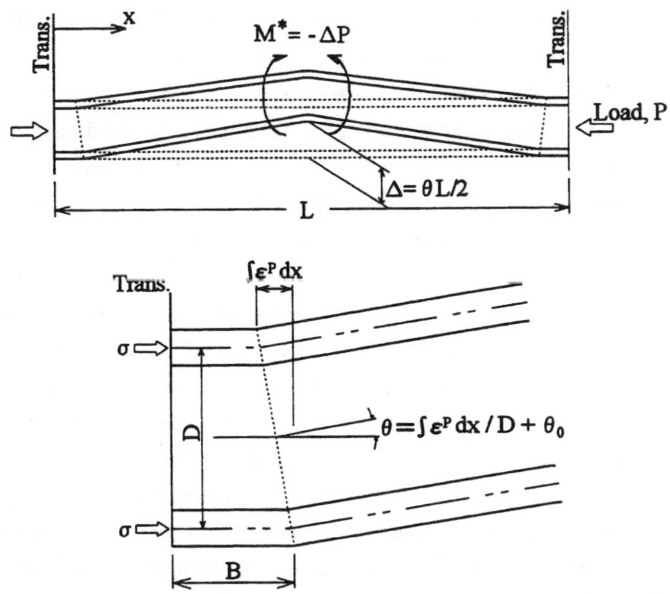

Fig. 8 Failure mechanism of hollow section

the both cross section area. The area A_1 is equal to A_2 for this hollow section. Assuming that the equal compressive stress σ is acting in the both face plates at the welded part with the transverse girders. Then the inner plate shrinks longitudinally because the plastic strain ε^P is produced in the softened zone. In this way, the angular distortion of the weld part with the transverse girders shown in Eq. (13) is created. Neglecting the elastic strain, the lateral deformation at the center of the span is expressed in Eq. (14).

$$\theta = \int_0^B \varepsilon_p dx / D + \theta_0 \qquad (13)$$

$$\Delta = \theta L / 2 \qquad (14)$$

Where, θ_0: the initial angular distortion by welding the transverse girders

L: the length between the transverse girders

In addition, θ_0 is assumed to be given in the following Eq. (15), which is derived from the transverse shrinkage by the inherent stress method [13].

$$\theta_0 = 0.16\alpha Q / EtD \qquad (15)$$

where, Q: the heat input per unit length for the weld of a transverse girder

E: Young's modulus

t: the thickness of inner face plate

α: the sharing rate of heat between the transverse girder and the face plate (assumed to be 0.5 here)

From this lateral deformation and the axial force $P = (A_1 + A_2)\sigma$, the moment M^* shown in Eq. (16) is created at the center of the span.

$$M^* = -\Delta P \qquad (16)$$

The moment M^* is modified in Eq. (17) as well as Eq. (6).

$$M^* = DA_1(\sigma_1 - \sigma) = -DA_2(\sigma_2 - \sigma) \qquad (17)$$

where, σ_1: the compressive stress of the inner face plate at the center of the span

σ_2: the compressive stress of the outer face plate at the center of the span

The compressive stress σ_2 become larger than the stress σ_1 in this deformation mechanism. When the stress σ_2 reaches the elastic limit (σ_e) of the parent metal, this mechanism lose the stability.

The effective stress σ_{eff} is obtained as the stress σ, that is the stress at the moment σ_2 reaches the elastic limit (σ_e), by using the relationships between the stress and the plastic strain and by equalizing both Eqs. (16) and (17).

3.4 RESULTS OF CALCULATIONS AND THE EVALUATION

Both the effective stress calculated by the simplified method and the results of experiments are shown in Table 2. The smaller value of the results of the simplified methods (1) and (2) in Table 2 is adopted by the reason mentioned in the subsection 3.3.1. The values which were not adopted are bracketed. In addition, the calculated effective stress of the model $P-1$ whose stiffener space is assumed to be 300, 200 and 150 mm is shown in the table.

In the models $\Pi-1$ and $H-1$, the results of the experiments are quite different from those of the FEM calculations. The results of the simplified methods also differ from the test results for the model $\Pi-1$. Comparing the results of the FEM with those of the simplified methods, the FEM results of the models $H-1$ and $P-1$ (when the stiffener space is assumed to be 150 mm) give much larger the effective stress than the other results.

The model $\Pi-1$, in which the results of both the FEM and the simplified calculations exceeded the test results, has much bigger bulbs of the bulb plates than those of the models $\Pi-2$ and $P-1$. Therefore, the share rate between the bulb and the web of the bulb plate is not appropriate and the part of the bulb did not work effectively. That is, since the transverse rotation of the bulb was produced at the joint between the bulb and the transverse girders, it is presumed that E_t shown in Eq. (9) was reduced.

The FEM analyses for the models $H-1$ and $H-2$ correspond to the test results of the model $H-1$, because the effect of the softened zones are considered and the existence of the brackets was not considered. However, the results of the FEM analyses for the models $H-1$ and $H-2$ are rather close to the test results of the model $H-2$ as shown in Table 2, therefore it is not to be said that the FEM analyses are successful. Even if the hollow section material is replaced to the unhomogeneous plate whose section rigidity is equal in the FEM modeling, its applicability for the plastic collapse accompanied with the lateral deformation is bad and the 'Simplified Two-flange Column Model' in the section 3.3 is more applicable.

When the stiffener space is assumed to be 150 mm for the model $P-1$, the result of the FEM analysis, 179.2 MPa corresponds to the buckling stress of the skin plate by Eq. (1).

And considering that the use of A5083S-H112 bulb plate whose 0.2% offset strength and the elastic limit are as much as 200 MPa and 110 MPa respectively, this stress is too large. Since the collapse mode assumed in the FEM analysis corresponds to the simplified calculation method using the effective width (see (1) in Table 2), it seems to be the cause of the excessive effective stress that over all collapse as a column cannot be evaluated.

4. DESIGN STRENGTH OF LIGHTENED HULL STRUCTURES AND EFFICIENCY OF SAVING WEIGHT

4.1 Calculation methods of buckling strength

When the FEM analyses are used for the calculations of buckling strength, the appropriate modeling for the object is required. In addition, there is much complexity such as the definition of mesh, the boundary conditions, the initial imperfections and the incremental quantities in the nonlinear analyses. Though it is reasonable that the FEM is utilized in the design stages, the appropriate initial imperfections, the relationships between the stress and the strain and the iterative quantities need to be defined, for the models which can express as many the collapse behaviors as possible.

On the other hand, the simplified calculation methods stated in the sections 3.2 and 3.3 are based on the experimental fact and the experiences, however, these methods give the sufficient estimate value in the extent of the practical use, for the pi shaped sections, the pre-ribbed panels and the hollow sections.

4.2 DESIGN STRESS

Among the pi shaped sections, the pre-ribbed panels and the hollow sections treated in this paper, the pi shaped sections are the only outer shell element exactly corresponding to the conventional design methods.

At present, the stress of 110 MPa is adopted for the standard of the design stress, which is the standard 0.2% offset strength of A5083S-H112 and 85% of that of A5083-O. The standard value is quite small compared to the elastic limit or the 0.2% offset strength of A5083S-H112 used here (they are about 170 and 200 MPa in the parent metal, 110 and 150 MPa in the heat affected zone respectively). Nevertheless, seeing the test results of the models $\Pi-1$ and $\Pi-2$ made of the pi shaped sections (the effective stress is about 130 MPa), this standard value is quite appropriate for the structural elements so far. That is, as far as the pi shaped sections are used, the buckling strength of them (compressive collapse strength) is need not to be evaluate .

In order to reduce structural weight, it is necessary to use high strength material and to establish high design stress. On the contrary, in the present condition, the 0.2% offset strength in the softened zone obtained by the experiments is often adopted as the standard value, for the material whose strength is raised by work hardening or heat treatment. However, concerning the buckling strength, the effect of the softened zones varies with the collapse mechanisms. When the structures of stiffened panels reach the ultimate strength by preceding of the buckling of the skin plates and by the lateral buckling of the stiffeners (see Fig. 6), there is the rare effect of the softened zones. The effect of the softened zone appears when the outer shell elements collapse by the over all buckling as a column (see Figs. 7 and 8). Even in this case, the buckling strength cannot be replaced directly with the strength of the softened zone. Because even the effective stress (130 MPa) of specimen H-1, which is considered to be affected most seriously by softening, is large enough compared to the elastic limit 78 MPa or the 0.2% offset strength 105 MPa of most seriously softened zone.

As stated above, regarding the buckling strength of the lightened structures, the strength of the softened zones should not be the upper limit easily, but should be evaluated by the method shown in this paper.

4.3 EFFECT OF SAVING WEIGHT

The effect of saving weight on the pre-ribbed panels and the extruded hollow sections treated in this paper is investigated by the following way.

There is no effect of saving weight for the model $P-1$ fabricated by the pre-ribbed panel, seen from the test results that the effective stress was 123 MPa (see Table 2). Because the stiffener space (250 mm) was wide for the thickness of the skin plate (4 mm) and the elastic buckling stress of the panel was extremely small (64.5 MPa).

According as the stiffener space is reduced, the effective stress of the model $P-1$ rises as seen in Table 2. That is, in order to make the most of the high strength of A5083P-H321 plate in pre-ribbed panels, it is necessary to reduce stiffener space and to raise the buckling stress of the skin plate. However, the collapse mechanism of the pre-ribbed panels becomes the mode of the over all buckling as a column, when the stiffener space is smaller than 200 mm for the thickness of the skin plate and the bulb plate treated in this paper. At this time, the effective stress is subject to the relationships between the stress and the plastic strain in the softened zone of the bulb plate shown in Eq. (12). Therefore, if the stiffener space is smaller than 200 mm, the effective stress of 150 MPa becomes the upper bound. It is the considerable effect of saving weight that the effective stress rises up to 150 MPa, however, improving the relationships between the stress and the plastic strain of the softened zone in the stiffeners, in order to raise the effective stress further.

The high effective stress is obtained for the hollow sections, if the softened zones are stiffened appropriately as seen in the model $H-2$ (see Table 2). The brackets which divide the softened zones of the model $H-2$ is effective for the restraint of the initial imperfections (θ_0 in Eq.(13)). However, since the softened zones caused by the corner boxing weld around the end of the brackets influences the ultimate strength, the excessive installation of brackets reduces the effective stress. Therefore, the bracket space should be determined with considering the stress concentration at the end of the brackets and the fatigue strength of the structure.

5. CONCLUSIONS

On the buckling strength of the aluminum hull structures by the pi shaped sections, the pre-ribbed panels and the excluded hollow sections, the experimental and analytical studies were carried out. And also, the simplified calculation methods of the buckling strength was proposed. Conclusions from this study were presented as follows.

(1) Comparing the results by the simplified calculation methods with the experimental results and the results of the FEM calculations, the effectiveness of the simplified calculation methods was confirmed.

(2) If the structure by pi shaped extrusions from A5083S-H112 is designed with the limit strength of 110 MPa, the examination of buckling can be omitted.

(3) When the high strength material such as A5083-H321 or A6N01S-T5 are used, it is not rational and practical that regarding the strength of the softened zones as the upper limit of the buckling strength.

(4) To make the best use of high yield strength of A5083P-H321 plate on pre-ribbed panel, the stiffener space needs to be considered. Moreover, the improvement of the buckling strength can be expected by the exchange of the material of the bulb plate.

(5) The buckling strength of the extruded hollow sections is seriously influenced by the heat affected softened zones by welding. Therefore, in order to raise the buckling strength of them, it is required to prevent from forming long plastic hinge lines by dividing the softened zones with brackets etc.

REFERENCES

1) Matsuoka, K., Uemura, T., Kamata, K., Chiaki, S.: Fatigue strength of welded joints of 6N01 aluminum alloy extrusions, J. of JLMWC, Vol.32, (1994), No.5, 195.

2) Matsuoka, K., Uemura, T., Kamata, K.: Fatigue strength of as welded and improved joints of 6N01 aluminum alloy extrusions, J. of SNAJ, Vol. 176, (1994), 475.

3) Matsuoka, K., Uemura, T., Fujita, Y.: Fatigue strength of welded joints of Al-Mg-Si alloy (6N01) extrusions, Proc. 6th Int. Conf. on Aluminum Weldments, (1995), 503.

4) Sakuma, M., Matsuoka, K.: Fatigue strength of welded joints of thin-wall aluminum extrusion, Proc. of 65th general meeting of Ship Research Institute, (1995), 189.

5) Kitamura, S., Sakuma, M., Matsuoka, K.: Strength of heat affected softened zone of aluminum alloy due to welding, Proc. of 65th general meeting of Ship Research Institute, (1995), 193.

6) Tanaka, Y., Matsuoka, K., Chiaki, S.: Structural strength of extruded aluminum hollow sections, Proc. of 63rd general meeting of Ship Research Institute, (1994), 147.

7) Matsuoka, K., Tanaka, Y., Chiaki, S., Fujita, Y.: Structural strength of extruded aluminum pi and hollow sections, Proc. 6th Int. Conf. on Aluminum Weldment, (1995), 225.

8) Matsuoka, K., Tomita, Y., Tanaka, Y.: Buckling strength of extruded aluminum hull structures, Proc. Of 26th Annual Meeting of JLMWC, (1996), 41.

9) Tanaka, Y., et al.: Research projects on strength of aluminum high-speed vessels, Proc. 2nd Int. Forum on Aluminum Ships, (1995).

10) Tomita, Y., Matsuoka, K., Minoda, K., Hotta, S.: Study on weight critical structure of aluminium alloy ships in Japan, Proc. 2nd Int. Forum on Aluminium Ships, (1995).

11) Tanaka, Y., et al.: Study on strength of aluminum alloy structures, Proc. of 67th general meeting of Ship Research Institute, (1996), 141.

12) Shanley, F. R.: Inelastic column theory, J. Aeronautical Science, Vol. 14 (1947), No. 6, 261.

13) Matsuoka, K.: Inherent stress approach on residual stresses in weldments, Proc. of 67th General Meeting of Ship Research Institute, (1996), 197.

Proceedings of the Seventh (1997) International Offshore and Polar Engineering Conference
Honolulu, USA, May 25-30, 1997

ISBN 1-880653-28-1 (Set); ISBN 1-880653-32-X (Vol. IV)

A Study on the Characteristics of BTA Deep Hole Drilling for Marine Part Materials

Sung-Bo Sim
Pukyong National University
Pusan, Korea

Tae-Ok Jun
Dong-A University
Pusan, Korea

ABSTRACT

The deep hole drilling such as BTA(Boring and Trepanning Association)drilling and gun drilling is capable of machining a hole with a large length to diameter ratio in single pass. The main feature of the deep hole drilling stabilizes cutting force resultants necessary for self guiding of the drill head, also the cutting tool edges are unsymmetrically located on the drill head. This provides an increasing demands for the prediction of hole geometry and other dynamic stability of the deep hole drilling guidance. In this study, the effects of BTA tool vibration on hole profile in deep hole drilling are analyzed by the method of domain analysis, also the mechanism and causes of multi lobe shape in hole profile are discussed. The profile roundness is influenced with cutting speed and feed rate, chip flow, tool wear and etc. Also in this study, it is investigated why the wear of BTA drill tips is caused mainly by the Ti, W atoms at high temperature during BTA drilling.

KEY WORDS : BTA (Boring and Trepanning Association), Lobe, Guidepad, Single tube system

INTRODUCTION

The deep hole drilling is defined as that the depth to diameter ratio is over 5 and can be classified largely as BTA (Boring and Trepanning Association) drilling and gun drilling.[1)]

In tip configuration, fluid induction and chip removal, BTA drilling differs from the conventional gun drill. However supporting bearing arrangement is similar to the conventional gun drill and the basic guidance is identical. The tool cuts in a true circular pattern and the tip follows to the direction of its own axis and is mounted to a tubular steel shank. The tooling is designed for deep hole drilling machine or specially equipped horizontal lathes, both of them must include a large volume and high pressure fluid system. Such a system can result in high production rates, accurate deep or shallow holes with good surface finish.

The BTA drilling was studied by Beisner, Pearson, Griffiths, Evans, Pfleghr, Rao, Osman, Sakuma, Taguchi, etc. Osman and etc. have studied about the optimum cutting edge design of BTA drill which can give minimum cutting force, also Sakuma and Taguchi studied about the profile of deep hole and its generating mechanism, but in these study, the lobe characteristics were not investigated for complete theoretical or experimental results. Accordingly, This study has tried to investigate more complete charateristics of multi lobe form and wear of cutting edges by performing CNC BTA drilling with tool rotating head.[2)]

THEORETICAL BACK GROUND

Commonly, BTA drilling is similar with a reamer with 4 cutting edges or a twist drill and also it is known as simple form of multi point edge tool.

Between multi edges and workpiece, formation of multi lobes is given in Fig.1 by tool rotating.(In this case, pentagonal type is selected for convenience)

In the case of multi edge tool of BTA drill or twist drill and reamer etc, Sakuma reported that polygonal holes with S lobes most frequently was formed, where S=nZ+1 (S:number of lobes, n: integer, Z: number of cutting edges).

Usually corresponding to n=1, the cutting edges of BTA Drill and its guide pads are equally spaced with the number of multi edge tool with 90 degree interval such as reamer Z=4. Consequently the number of lobes can be 3 or 5 usually.

The BTA drill axis center rotates with self-excited vibration and forced vibration, the formation of machining hole surface is jagged repeatedly and the cutting edge makes jaggy and then next cutting edge (next is guide pad in BTA drill) moves along the jagged surface supported. Among tool diameter D_t, tool radius R_t, locus of tool axis center with harmonic function and the tool axis rotating radius R_a, revolution diameter D_a, angular velocity ω_a, the vector locus of circumferencial coordinate of optional cutting edge can be summarized as $\vec{R}$.

$$\vec{R} = |\vec{R}|\, e^{j\theta} = |\vec{R_a}|\, e^{j(\omega_a t+\phi_o)} + |\vec{R_t}|\, e^{j(\omega_t t+\theta_o)} = x\vec{i} + y\vec{j} \quad (1)$$

Accordingly, each coordinate component x, y on the locus of optional maxmium circumferencial cutting edge A can be given by following formula[3)]

$$x_a = \frac{D_a}{2}\cos(\phi_o + \omega_a t) + \frac{D_t}{2}\cos(\theta_o + \omega_t t) \quad (2)$$

$$y_a = \frac{D_a}{2}\sin(\phi_o + \omega_a t) + \frac{D_t}{2}\sin(\theta_o + \omega_t t) \quad (3)$$

where $2\pi/Z$ is interval angle of cutting edge and Z is the number of cutting edges. The following edges lag in $2\pi/Z$, $2(2\pi/Z)$, ..., $(Z-1)(2\pi/Z)$ interval and each edge contacts on the workpiece respectively.

The locus of following cutting edge or guide pad, B can be given as follows.

$$x_b = \frac{D_a}{2}\cos(\phi_o + \omega_a t) + \frac{D_t}{2}\cos(\theta_o - \frac{2\pi}{Z} + \omega_t t) \quad (4)$$

$$y_b = \frac{D_a}{2}\sin(\phi_o + \omega_a t) + \frac{D_t}{2}\cos(\theta_o - \frac{2\pi}{Z} + \omega_t t) \quad (5)$$

Moreover, the condition that the moving locus of cutting edge or guide pad B corresponding to cutting edge A must be same is also as follows.

$$x_b(t = t_o + 2\frac{\pi}{(Z\omega t)}) = x_a(t = t_o) \quad (6)$$

$$y_b(t = t_o + 2\frac{\pi}{(Z\omega t)}) = y_a(t = t_o) \quad (7)$$

The formula (6) and (7) must satisfy as this condition $\omega_a/\omega_t = nZ$ (where n: integer, same direction of ω_a and ω_t : +, Opposite direction of ω_a and ω_t : -).

Namely, if the BTA drill axis revolves at an equal speed for drill rotation of nZ times on its own axis, each cutting edge moves along the same identical paths. The locus of cutting edge can be given as follows.

$$|\vec{R}| = \sqrt{x^2+y^2} = \frac{1}{2}\sqrt{[D_a^2 + D_t^2 + 2D_aD_t\cos((nZ-1)\omega_t t + (\phi_o - \theta_o))]} \quad (8)$$

Formula (8) means that the |nZ-1| lobes are formed.

EXPERIMENTAL METHOD

Experimental Apparatus

Table 1 is the list of experimental machine and instruments. Fig. 2 shows the photograph of experimental equipment.

In this experiment, deep hole drilling machine used is the type of workpiece fixed on the table.

Fig. 3 shows the flow of experimental apparatus. BTA drill is connected to single type tube. In this case, the coolant is machining oil (S-CUT-7E) under 2.5 MPa fluid pressure conditions.

Kistler Piezo type tool dynamometer was used for measuring force and torque for the analysis of the frequency of axis vibration with charge amplifier (type 5007) and A/D converter (TEAC MR-30).

For an analyzing FFT signal treatment and power spectrum, the IBM-AT PC was used with the internal organs PCL-818 lab card.

Workpiece and Tool

In this experiment, the workpiece and tool used appeares on Table 2 and chemical compositions, mechanical properties on Table 3 and Table 4. For the

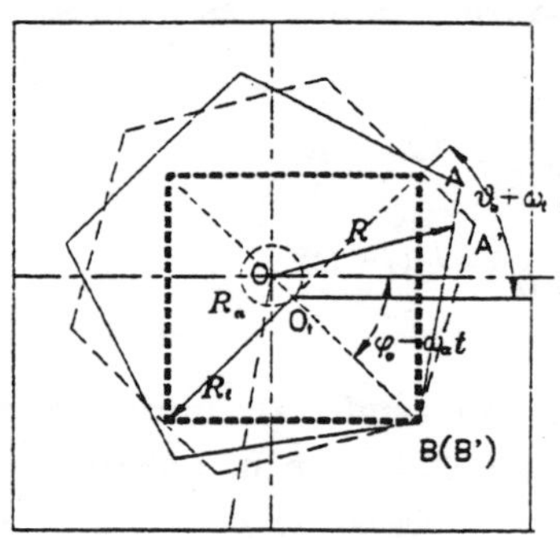

(a) For BTA tool rotating

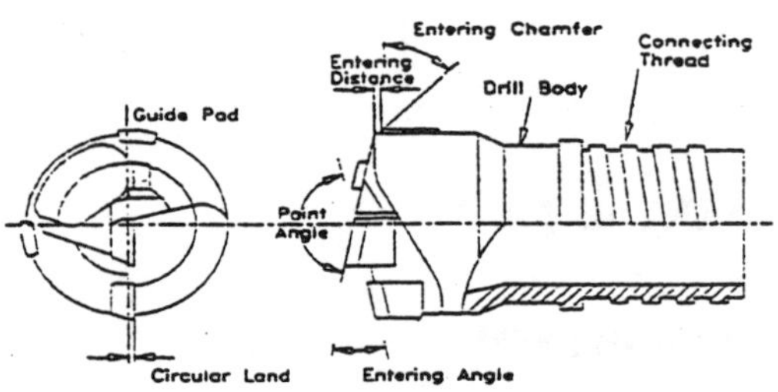

(b) BTA drill

Fig. 1 Generation of workpiece lobe for the rotation of tool and BTA drill shape

removal of internal stress, heat treatment was taken. The lathe working was performed suitably to the experimental conditions.

Fig. 1(b) shows the BTA drill shape.

Experimental Method

In BTA drilling experiments in Fig. 2, angular type fixture was installed on the CNC deep hole machine table, and then the workpiece loaded on the tool dynamometer chuck with tool alignment of ±0.025mm was adjusted. At the sametime, BTA drill head and drill tube fixed were assembled with the collet chuck, After all, the tool should be penetrated through guide bush, and then perform the deep hole machining according to macro-program operation. The tool dynamometer measured the cutting force and then the measured signal was amplified by charge amplifier, recorded on the data recorder. FFT signal treatment and power spectrum analysis were performed by IBM-AT486PC and PC-818 lab card from data recorder.

In additional experiments, deep hole profile measurement was performed. Profile measuring instrument of Taylor Hobson Ltd., roughness tester and SEM were used and 3-dimensional measuring machine was used for a few kinds of measurements.

The basic format of CNC deep hole drilling, Macro program for Fig. 4 might be used in this experiment.

```
N0010 G92  X0 Y0 Z0
N0020 G186 X_ Y_ Z_ K-60 B-5 D-5 T1
               F_ S_
```

Upper format was used in following experimental basic machining process in Fig. 4.

Table 5 shows the experimental cutting conditions.

EXPERIMENTAL RESULTS AND CONSIDER-RATION

Cutting Resistance and Lobe

Fig. 5 shows the torque with ϕ19.25 diameter BTA deep hole drilling, in the condition of S45C steel workpiece, V=71 m/min, f=0.165 mm/rev. Cutting force is sum of pure cutting torque, friction torque and burnishing torque. The main difference against common drilling was the results caused from burnishing of guide pad in reference. [3][4]

After coolant is flushed until 0.5 second, no loading torque appears, the reason of which is that cutting edge is located naturally on true position which is the arrangement of NC pecking cycle of macro-program.

The purpose of arrangement is to protect eccentricity of tool position, chipping and fracture caused by cutting force. In the middle section with no loading, after measuring for 2 seconds, the measured profile was influenced by guide pad.

In Fig. 5 cutting torque increases during first 2 seconds and then gradually stabilized.

After no loading during the sencondary deep hole drilling, cutting force was risen faster than inital drilling that has been large amplitude of torque in first action. In 0~2 seconds, its reason is considered

Fig.2 Experimental apparatus

Table. 1 List of experimental apparatus

No.	Name	Model	Maker
1	Deep hole drilling machine	STGB -1500CNC	Shinil machine Co. Ltd.
2	Surface roughness tester	M20 -0003	Taylor Hobson
3	3D measuring machine	EC -3200 PA 600A	Kosak Seimitsu
4	Roundness tester	RTH TR150	Taylor Hobson
5	Optical microscope	RM -TH -10	Meiji techno Co. Ltd.
6	Tool dynamometer	9271A	Kistler
7	Charge amp.	5007	Kistler
8	Recorder	MR -30	TEAC
9	P C	IBM-AT 486	Sam Bo
10	Labcard	PCL -818	Advantech Co.
11	Oscilloscope	5504	Hung Chang
12	SEM	S -2700	Hitachi
13	SEM -EDX	840A	JEOL

Table. 2 List of tool and workpiece

Tool diameter	Tool material	Work piece size	Work piece material
Ø19.25	P70	Ø40x200 Ø50x60	S45C

Table. 3 Chemical compositions of S45C(%)

C	Si	Mn	P	S
0.47	0.31	0.65	0.013	0.028

Table. 4 Mechanical properties of S45C

Yielding strength MPa	Tensile strength MPa	Hardness H_B	Elongation (%)
365	596	200	19.8

Table 5 Cutting conditions for S45C

Workpiece number	Revolution rpm	Cutting speed m/min.	Feedrate mm/rev
1			0.15
2	1072	65	0.1
3			0.17
4			0.18
5			0.125
6			0.137
7	1170	71	0.147
8			0.155
9			0.165
10			0.175
11			0.132
12	1220	74	0.141
13			0.149
14			0.158
15			0.12
16	1340	81	0.128
17			0.136
18			0.144
19			0.168
20	960	58	0.179
21			0.19
22			0.201
23			0.199
24	810	52.8	0.213
25			0.225
26			0.18
27			0.165
28	1072	64.8	0.158
29			0.144
30			0.201

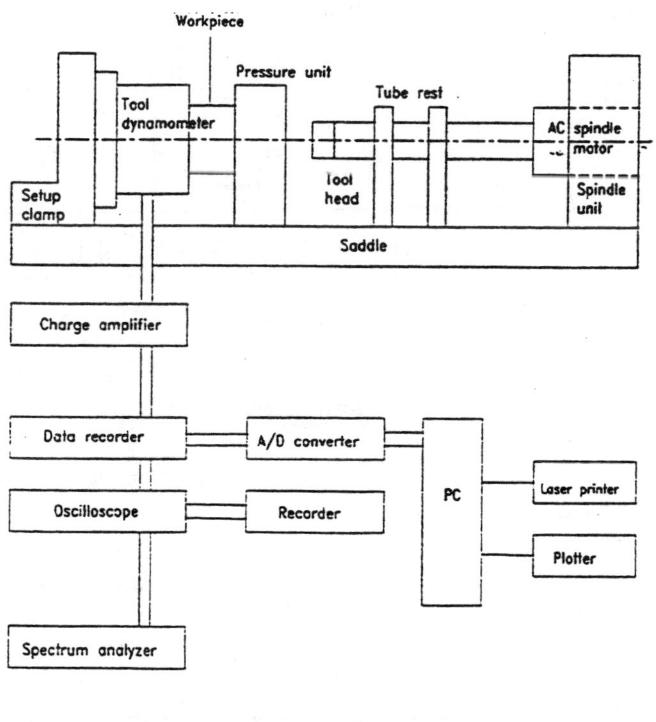

Fig.3 Configuration of experimental setup for deep hole drilling

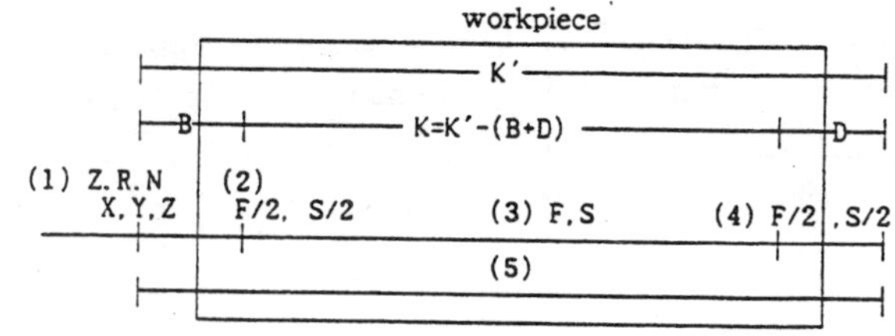

K: Normal cutting B: Initial reduced cutting speed
D: Period reduced cutting speed T: Number of repeat
(1) Locating by rapid feeding
(2) B section for initial reduced cutting speed
(3) Section for normal cutting speed
(4) D section for period reduced cutting speed
(5) Rapid return after finish

Fig.4 Deep hole drilling process

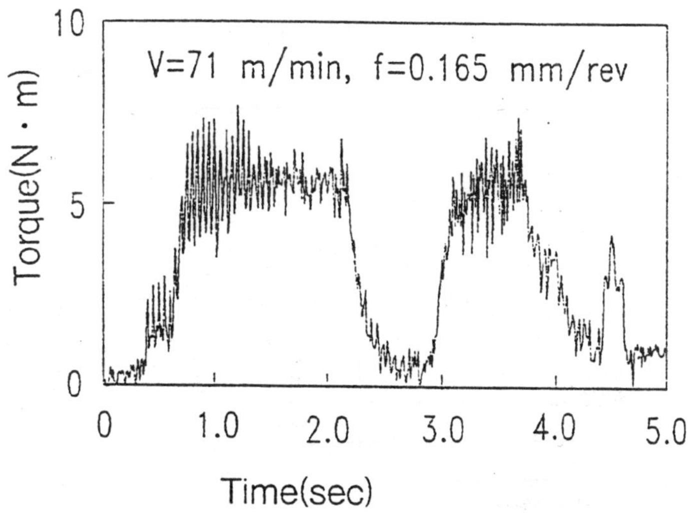

Fig.5 Configuration of torque signal

that the tool is finding the true position for initial drilling.

Fig. 6 is the result of power specturm in frequency domain after FFT treatment of cutting torque singnal for Fig. 5 in the condition of ϕ19.25 BTA drill, S45C steel, V=71m/min cutting speed and f=0.165 mm/rev feeding constantly.

In Fig. 6(a) represents power spectrum in linear scale, (b) in log scale.
The peak frequency is indicated as f_s=39, 58.6, 78.1, 97.6, 117.15 etc. and may be shown in the figure.

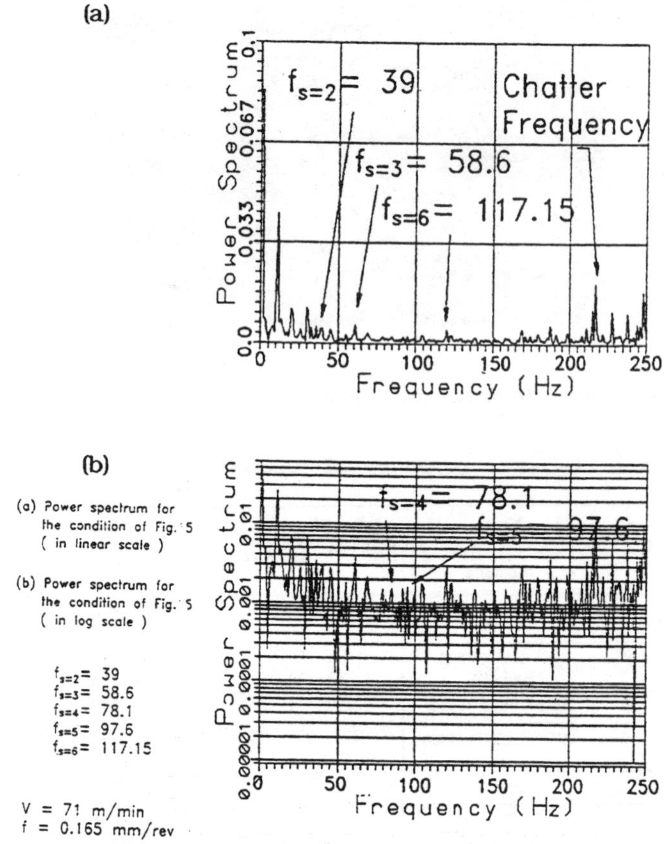

Fig.6 Power spectrum of torque signal for the condition of FIg.5

Fig. 7 is the locus of lobe(S=2~5) computed and theoretical formula(8) was used.

Fig. 8(a),(b),(c),(d) shows the formed lobe after deep hole drilling on the condition of S45C steel, ϕ19.25 tool diameter, V=71m/min(N=1170rpm), f=0.165 mm/rev. We can observe that actually similar shape may be shown in theoretical results in Fig. 7.

In this experiment, the three lobe was mostly frequent. We can suppose that 6 lobes workpiece may be 60° turning phase of 3 lobes and 2 or 3 combinations may be in this case. It appeared that the characteristic of lobes was to be piled from elliptic base 2 lobes to 3 lobes. The number of lobe S appears in resonance frequency domain in Fig. 8 very well.

Within frequencey domain, high power frequency is related to workpiece lobe frequency and this frequency may be f_s=SN/60(Hz) which shows the mapping into frequencey domain and also involving frequency of each lobe. In condition of V=71 m/min(N=1170rpm), the lobe frequency coincided well to $f_{s=2}$=39, $f_{s=3}$=58.6,

f_{s-4}=78.1, f_{s-5}=97.6, f_{s-6}=117.15Hz.

If the lobe of deep drilled hole was dominant, the feature would be clearly within frequency domain. Inversely, we can calculate the number of lobe with formula $S=60f_s/N$, and the weak mode could be seen in log coordinates.

As well, abnormal frequency such as chatter, etc. was detected. The feature of chatter could be confirmed distinctively at frequency 217Hz.

As mentioned above, the number of lobe can be investigated according to deep hole shape. It appears that the deep hole of 3 lobes coulde be found widely.

The reason of this one can be considered as the highest energy at frequency of f_s=58.6Hz.

Fig. 9 is the magnified photo. of deep hole drilled workpiece under the condition of V=71 m/min, f=0.165 m/rev for S45C steel with ϕ19.25 BTA drill.

The part A is the boundray line of cutting surface by cutting edge and burnishing surface by guide pad. The part of lower boundary line is cutting surface and the part of upper boundary line is burnishing surface. The burnishing surface by guide pad reveals better excellent machining surface than tool edge surface, and the state of white brilliant mirror surface.

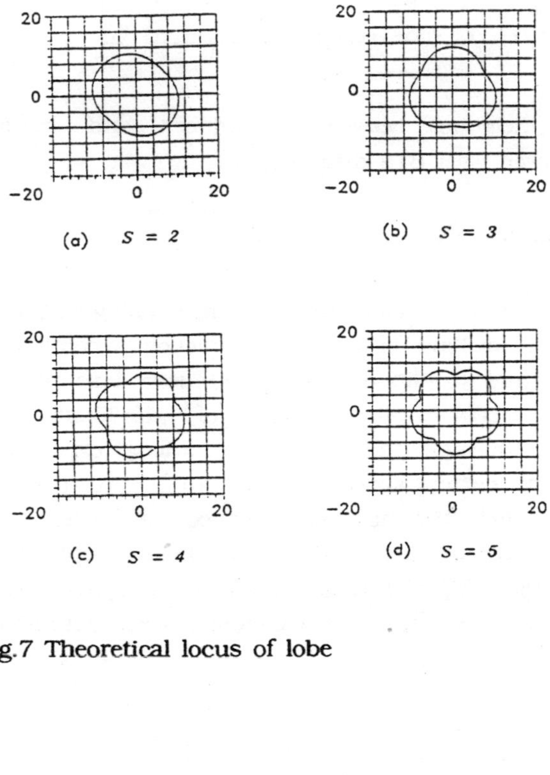

Fig.7 Theoretical locus of lobe

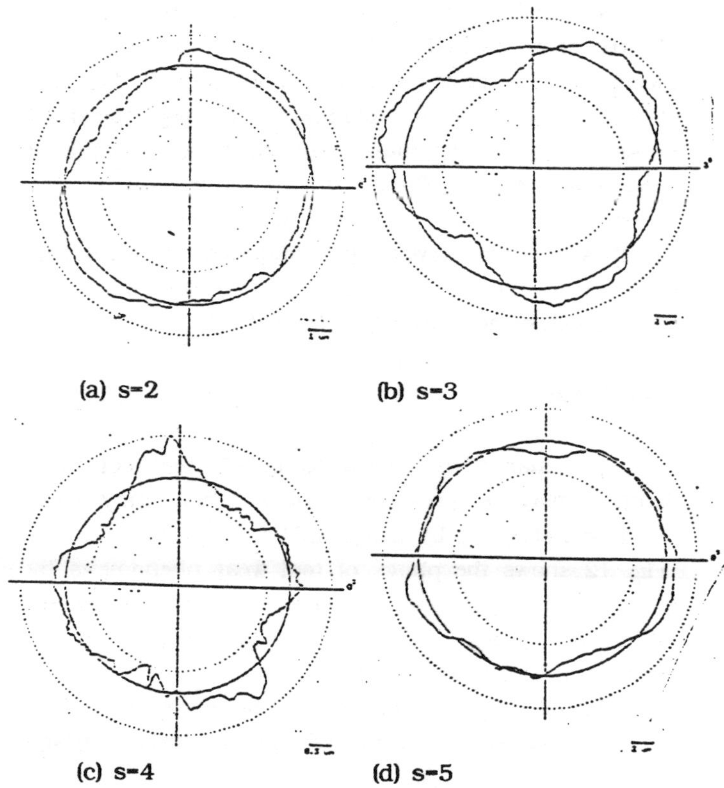

Fig.8 Experimental locus of lobe profile for BTA drilled hole (ϕ19.25 V=71m/min, f=0.165 mm/rev)

At part B and C, it can be known that the sharpened V groove in deep hole circumference must be considered as chattering. Also cutting edge was dug into the cutting surface projectively, and the reason is considered as the influence of chatter.

Fig.9 Photo. of hole profile (ϕ19.25, V=71m/min, f=0.165 mm/rev)

In cutting condition of V-65m/min, f-0.15mm/rev and ϕ19.25 BTA, deep hole drilled workpiece profile was measured by 3D measuring machine in Fig. 10.

As the same method in Fig. 11, the cutting condition was fixed on workpiece material, such as tool diameter of ϕ19.25, feed of f-0.02mm/rev, cutting speed of V-60m/min to 120m/min, according to axial depth of deep hole of 0~200mm, the roundness was also measured and the average value was pointed.

In these results, mostly the roundness appears less than 9㎛, in the over cutting speed, V-110m/min over, the roundness decreased less than 4㎛.

Tool Wear

Using Uehara's theoretical formula[5], the tool wear quantity may be presumed but in this study wear trend was observed by using SEM .

Fig. 12 shows the photo. of tool wear phenomena by SEM on cutting tool edge after BTA deep hole drilling under the cutting condition of V-120 m/min, f-0.08 m/rev and ϕ19.25 BTA drill.

At this time, Fig. 12(a) shows the outer tip crater wear surface which is viewed on the rake surface edge.

Fig. 12(b) is the outer tip flank wear surface, which is located on the relief surface. Here in Fig. 12(a), part A is one extreme crater wear point which is part B in Fig. 13 point of rake surface.

For this part, in Fig. 14 the Ti and W seperating conditions was observed by SEM-EDX on B point of Fig.13. Part A (Fig. 12(a)) is applicable to point B in Fig.13. Here it can be considered that the bending moment of chip can become largest one.

At the point B, it is considered that the temperature may be 900~1200℃ on the carbide tool edge by Geoffrey Boothroyd[6]

Also, it could be considered why the BTA tool tip is weared during the deep hole cutting such as SEM-EDX photograph in Fig. 14. Ti and W decreased from new one(beforce cutting) to the used one(after cutting), and the reason upon this is because the Ti and W atoms is made to TiC, WC at high temperature. TiC, WC is helpful to deep hole cutting, but V-71m/min, f-0.165 cutting condition can decrease Ti and W atoms about 20~25% by macrospect observing.

Hence, it can be known as the seperating development of Ti and W. The seperating development is considered as tool wear accelerating that it might reduce the tool life.

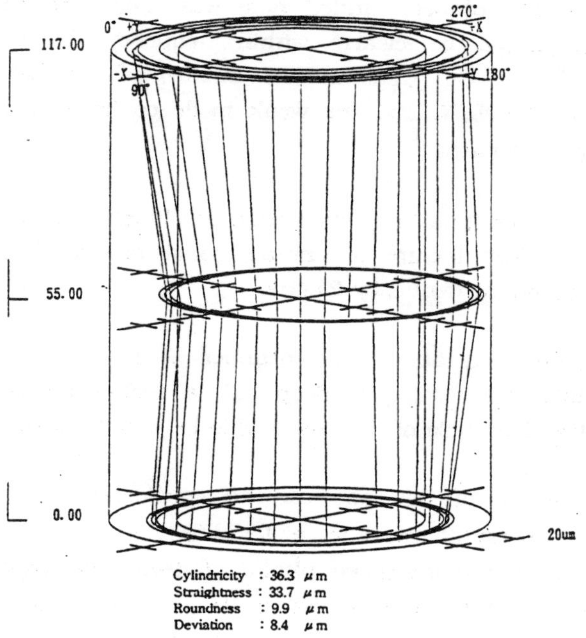

Fig.10 Geometric accuracy output for BTA drilled hole (ϕ19.25 V-65m/min, f-0.15 mm/rev)

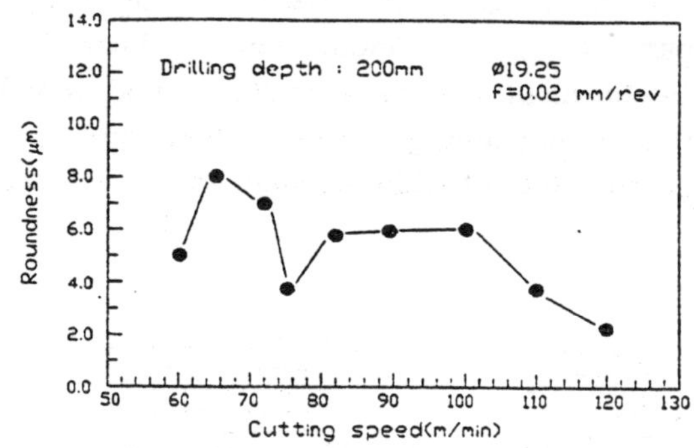

Fig.11 Relation between roundness and cutting speed with BTA drill (S45C)

CONCLUSION

This experiment has been performed using P70 BTA deep hole drilling to tool axis direction according to experimental conditions with CNC deep hole drilling machine. BTA drill diameter of ϕ19.25mm was selected, through the deep hole drilling experiment along with cutting speed and feed, then cutting torque was detected. The next performance was survey and analysis consideration for deep hole shape after machining, chip forming, roughness and tool wear conditions. Upon this, the conclusion was obtained as follows.

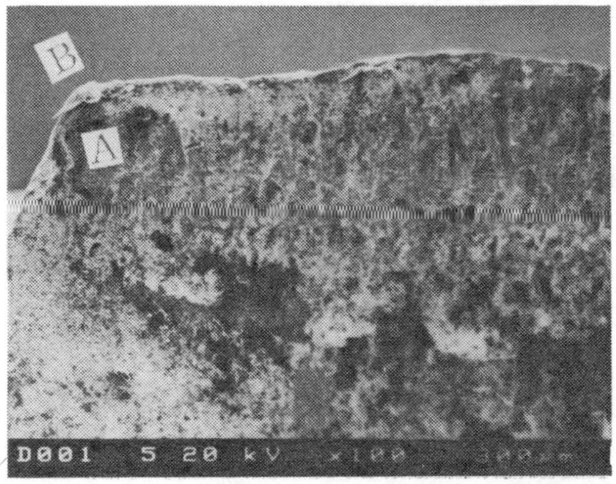

(a) crater wear of outer tip

(b) Flank wear of outer tip

Fig.12 Wear profile of cutting edge by SEM
(ϕ19.25 V=71m/min, f=0.165 mm/rev)

1. The Results of pad action and the pure cutting torque, friction torque and burnishing torque was generated and causes to the number of lobe generated. The number of lobe was analyzed by the method of frequency domain analysis.

2. When the BTA drilling was performed, the lobe with 2-5 pieces was generated according to the rotating speed and tool axis revolving speed.

3. The roundness of drilled workpiece with BTA becomes worse as increasing feed, drilling depth, tool wear and hign speed, etc.

4. The wear of BTA drill could appear in the crater wear, flank wear and chipping. As the composition of Ti and W decreases, the TiC, WC are separated at high temperature by friction and wear.

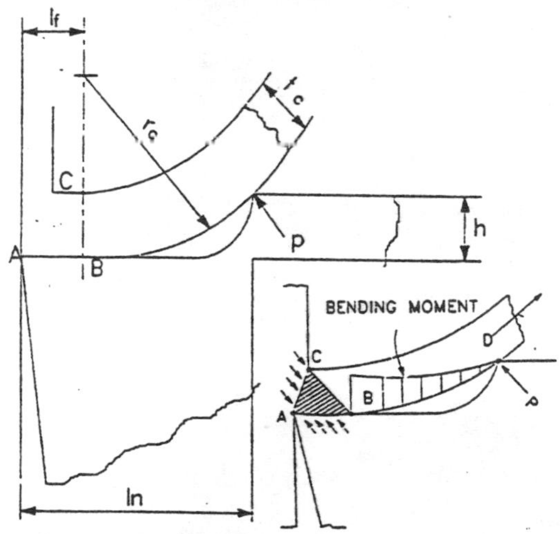

l_f : contact length between the chip and tool
l_n : chip-breaker distance
h : chip-breaker height
t_c : chip thickness
r_c : chip-curvature radius

Fig.13 Chip curling mechanism

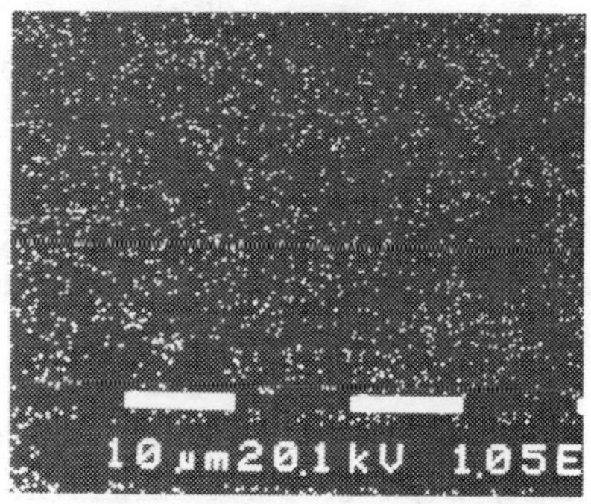

(a) Ti (before drilling)

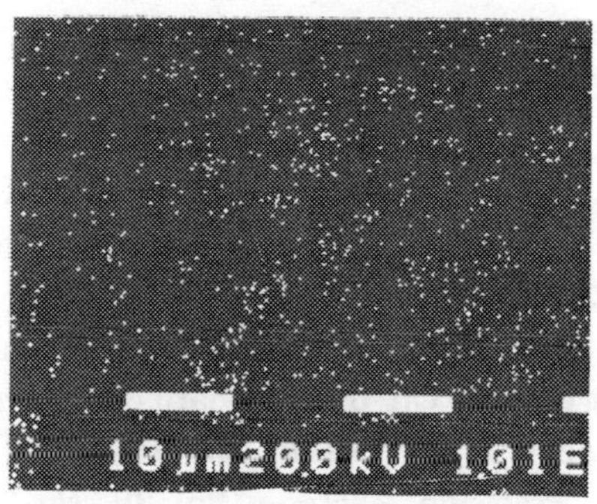

(b) Ti (after drilling)

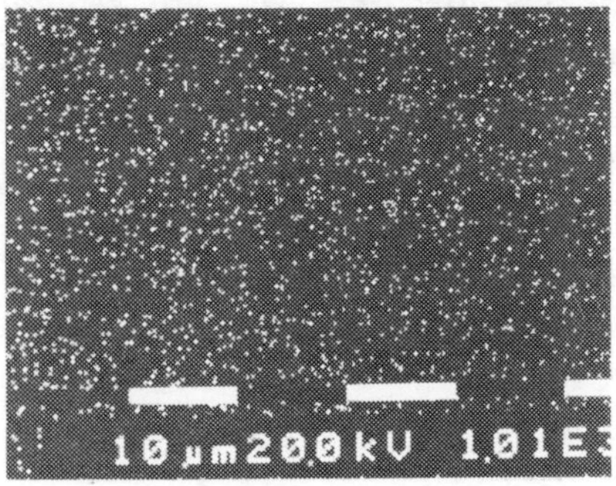

(c) W (before drilling)

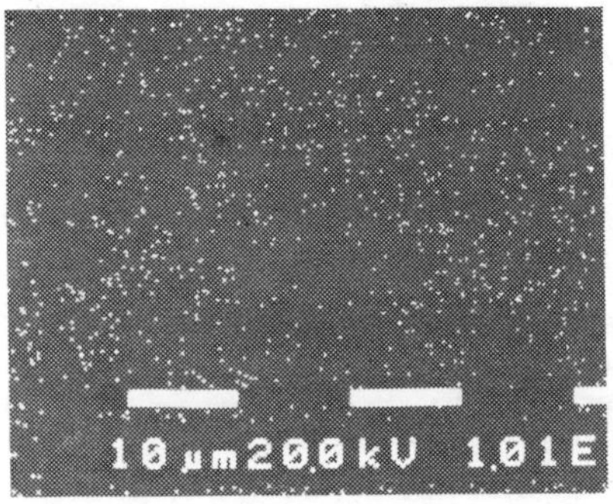

(d) W (after drilling)

Fig.14 Atomic distribution for BTA drill tip by SEM-EDX (ϕ19.25 V=71m/min, f=0.165 mm/rev)

REFERENCES

1. H. J. Swinehart (1967) "Gun Drilling, Trepanning, & Deephole Machining", ASTME
2. Sung Bo Sim, (1995) "A Study on the Characteristics of Deep Hole Drilling for Carbon Steel", PhD thesis, Dept. of Mech. Eng. Graduate School, Dona-A University, Pusan, Korea, Feb. pp 1-170
3. Keizo Sakuma, Koichi Taguchi, Akio Katsuki (1978) " Study on Deep - Hole Boring by BTA System Solid Boring Tool", Precision Machine, Vol 44, No 9, pp 69-74.
4. K. Sakuma, K. Taguchi, A. Katsuki (1981) "Self-Guiding Action of Deep-Hole-Drilling Tools", Annals of the CIRP, Vol 30/1
5. Kunio Uehara, Keji Nishina, Hideo Takeshita, Kazuya Uchida(1972) "On the Mechanism of Crater Wear of Carbide Cutting tool in the Regionof Low Cutting Speed", Precision Machine, Vol 38, No 3. pp 261-267
6. Geoffrey Boothroyd (1975) "Fundamentals of Metal Machining and Machine Tools", McGraw-hill International Book Company
7. M.O.M. Osman, V. Latinvic, B. Greuner (1981) "On the Performance of Cutting Fluids for BTA Deep-Hole Machining", Int. J. Prod. Res. Vol 19, No 5, pp 491-503

Proceedings of the Seventh (1997) International Offshore and Polar Engineering Conference
Honolulu, USA, May 25-30, 1997

ISBN 1-880653-28-1 (Set); ISBN 1-880653-32-X (Vol. IV)

Recent Developments in Evaluating and Designing Against Offshore Fire and Explosion

G.A. Chamberlain
Shell International Oil Products
Chester, United Kingdom

ABSTRACT

Ignition of an accidental release of flammable fluid on an offshore structure may lead to an explosion and fire hazard to surrounding personnel and plant. Over recent years considerable progress has been made in quantifying these hazards and in applying the knowledge to plant design and layout, and effective protection systems. Fire hazard severity in terms of thermal loading to engulfed equipment and to the surroundings, as well as the carbon monoxide and smoke hazards to personnel can now be evaluated for a whole range of offshore scenarios to a level sufficient for the design of effective protection. Similarly the heat flux loading to fire-engulfed equipment by natural gas, propane, butane and kerosene/methane jet fires has been measured, so that response calculations and can be put on a firmer basis and the efficacy of fire protection coatings, water sprays and blowdown strategies can be evaluated. Currently there is more uncertainty in the model predictions for the severity of vapour cloud explosion although the global mechanisms are understood. Large and small scale experiments have been carried out recently to improve knowledge of the complex interactions between combustion, induced fluid flow, and turbulence. Computer models of vented explosions are generally able to predict cases where low flame speeds and overpressures are generated, but become increasingly less reliable in geometries where strong combustion/turbulence feedback is created. Since high internal overpressures are associated with the latter it is important to ensure adequate protection so that escalation can be avoided. Thus the strength of the surrounding walls must be appropriately designed, and/or explosion mitigation systems should be effective. Water sprays, for example, can reduce the severity of gas explosions but there are limitations to their use. This paper discusses the current state of knowledge of the hazard loadings from fire and explosion and the applicability of predictive models to safety engineering.

KEY WORDS: Offshore platforms, fire hazards, explosion hazards, hazard mitigation, safety engineering.

INTRODUCTION

In today's regulatory environment the onus is placed increasingly on the plant operator to demonstrate that safety risks are tolerable and as low as reasonably practicable. Physical effects such as explosion overpressure and heat radiation have to be calculated to assess the threat to people and whether escalation and further threat are real possibilities. This paper reviews recent developments in the quantification of these effects. Accurate assessments based on realistic accident scenarios can avoid costly overdesign in terms of protective measures and ensure that all hazards and effects are correctly identified.

The response of plant and structure to accidental loadings and the probability of such events are equally important, but are not considered here. However, the cost effective solution to tolerable risk-taking must ultimately involve combination of all three aspects.

EVENT TREES - OVERVIEW OF HAZARD CONSEQUENCES

An accidental release of flammable hydrocarbon results in a variety of consequences, dependent on the type and initial state of the hydrocarbon and whether or not ignition occurs. These consequences are conveniently displayed by means of an event tree such as shown in Figure 1. By means of gross simplification this tree covers ALL INITIAL outcomes - fire, explosion or safe dispersion.

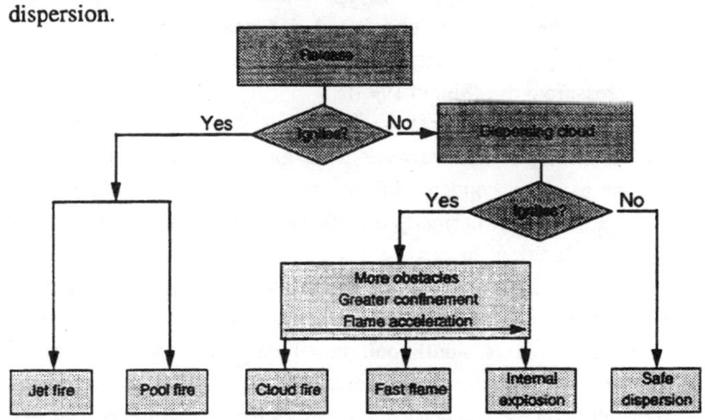

Fig. 1 Event tree showing all consequences of flammable release.

Further analysis reveals that just three event trees, determined simply by reference to the state of the fluid before release, are required to cover all the different physical effects from the possible range of fire types and explosion scenarios. The three fluid pre-release conditions determining the "source term" for the event trees are:

1. Liquid at ambient pressure, e.g. diesel, gasoline or oil tank,
2. Liquid at pressure above ambient, e.g. LPG, pumped crude oil,
3. Gas or vapour above ambient pressure, e.g. gas pipeline, vapour space in separator, gas blowout.

Complete definition of the source term then requires calculation of the outflow rate, and the physical form and dimensions of the release.

For example, a liquid release from a gasoline tank will always result firstly in the formation of a liquid pool, immediate ignition of which results in a transient fireball which burns back to a spray jet fire and burning pool. If ignition is delayed, the pool first spreads and partially evaporates. The vapour cloud disperses. If the surroundings are confined, such as for a release inside a building, then a vented explosion could occur on ignition. A cloud dispersing in the open produces a flash fire with no blast. An ignited release in a congested but unconfined plant environment could lead to the generation of fast flames with associated blast. The combination of confinement and congestion, such as in a typical offshore module, can lead to significant blast.

A liquid release from pressurised storage initially leads to a spray jet. The spray would consist of both vapour and liquid phases, in the form of droplets, if the liquid is volatile e.g. condensate, or simply droplets for non-volatile releases e.g. stabilised crude oil. Mechanical and thermodynamic forces both play a role in causing the jet to break up in to droplets. Just 1 or 2 bar is required for atomisation of the jet (Bowen and Shirvill, 1994) independent of any flashing processes that may be occurring. The effect this has on subsequent dispersion, degree of liquid rainout and fire behaviour is critical to an assessment of the hazardous consequences, and hence the extent and type of protection system required. Jet and pool fires can occur on immediate ignition after an initial transient fireball, but liquid rainout is unlikely if a released liquid, such as LPG, readily flashes to vapour in the atmosphere. The fate of a cloud undergoing delayed ignition depends on the degree of confinement and congestion in the surroundings, as before. Catastrophic loss of containment and ignition of pressurised liquid stored above its atmospheric vapour pressure, such as LPG, results in a type of fireball known as a BLEVE (boiling liquid expanding vapour explosion). The underlying physical processes are now well enough understood for model formulation and prediction of the transient fireball heat flux and explosion overpressure pulse (Shield 1993).

A pressurised gas release normally results in a dispersing gas jet which may contain aerosols if the gas cools sufficiently during release to allow condensation of some or all of the components. Entrainment of air normally is sufficient to re-evaporate the aerosol droplets. Ignition creates the classic jet fire, typified by refinery flares. In extraordinary cases the release of hot gas containing heavy components could lead to condensation and agglomeration of droplets and a rain-out pool. In all cases however, as before, delayed ignition results in either a cloud fire or explosion dependent on the geometry and layout of the surroundings in which the release occurs.

Methodologies, models, rules and guidance now exist with varying degrees of sophistication to allow engineers to quantify ALL of the above physical effects. In most cases the models, usually in the form of computer tools, have been validated over large ranges of conditions. The remainder of this paper is aimed at a state-of-the-art review of the appropriate and available knowledge.

RELEASES AND SOURCE TERMS

Gas/vapour release

Outflow from a vessel or small hole in pipe

The efflux of gas through a hole (<full bore) in a pipe or process vessel can be modelled using the standard expressions for isentropic nozzle flow (TNO 1992). Losses are accounted for by the introduction of a discharge coefficient (or contraction coefficient), C_D - this incorporates both velocity losses and losses due to the real flow area being less than the area of the nozzle. Typical values of C_D will vary between 0.65 and 1.0 (Coulson et al 1990). In general the higher the Reynold's number of the gas, the closer C_D will be to unity, however the shape of the orifice/nozzle will also be important. In the absence of a better value, C_D =0.85 represents a general value for hazard assessment work. However, for a conservative calculation of jet fire hazards, one might choose a C_D value higher than 0.85, and for conservative design of an emergency depressurisation system, one might choose a C_D value of lower than 0.85.

It is normally assumed that the pressure, specific volume (i.e. the reciprocal of the density) and temperature of the gas are related by the ideal gas law and this is suitable for most purposes. The exit temperature may in practice differ from that predicted from the above ideal gas relations, because of the Joule-Thomson effect, whereby the energy associated with the intermolecular forces in a gas will vary with pressure. The effects of low temperatures on the structural integrity of materials should be considered in designing a process, e.g. emergency depressurisation systems and related pipework.

Full bore rupture of pipework leading from vessel

This case is different from the full bore rupture of a pipeline, which is outside the scope of this paper. It may be assumed that the inventory in the pipe is negligible as compared to the inventory of the vessel. Shortly after rupture has occurred, a pressure profile will be established characterised by a monotonic decay between the vessel and the pipe rupture. There will be frictional losses along the length of the pipe and so the isentropic flow equations will not be valid. For small lengths of pipe (<5m), the equations do, however, provide a conservative upper bound. A set of equations can be formulated to describe the situation where the flow along the pipe is adiabatic but not isentropic. This type of flow is called Fanno flow and is discussed in standard texts (Imrie 1973).

Condensation in Vessel and Emerging Jet

The exit temperature of a gas may be considerably less then the stagnation temperature. In some cases, the isentropic relationships predict state conditions that are no longer consistent with the fluid being in a gaseous state. This circumstance frequently arises if the release from the gas phase of a storage vessel is at saturation pressure. Whether condensation does occur will depend on the timescale for the attainment of thermodynamic equilibrium. In the case of steam, droplet formation is quite slow and so material is likely to exit an orifice as a metastable gas, with droplet condensation occurring downstream of the exit plane. For the case of hydrocarbons, droplet formation is quite a fast process and condensation may occur in the nozzle. The presence of droplets is significant since it will lead to an increased mass flow rate, because the two-phase fluid will have a higher density than the gas.

During blowdown of a pressure vessel liquid droplets can also form in the vapour space because if the pressure decays adiabatically, the temperature will drop. The Imperial College BLOWDOWN program (Haque et al 1992) can predict this phenomenon but the general level of understanding of this problem is far from complete.

Jet expansion downstream of orifice

If there is a discontinuity in pressure between the exit point p_{exit} and atmospheric pressure, p_{atm}, then the gas will accelerate as it emerges from the orifice. At the atmospheric plane, the expanding gas will have velocity u_j There are two ways of quantifying this expansion. In the first, the orifice is considered to be part of an ideal converging-diverging nozzle. In such a case, the flow through the orifice can be treated as reversible. For an adiabatic and reversible (hence isentropic) process, it can be shown that,

$$\frac{1}{2}u_j^2 = \frac{1}{2}u_{exit}^2 + p_{exit}^{\frac{1}{\gamma}} v_{exit} \frac{\gamma}{\gamma - 1}\left[p_{exit}^{\frac{\gamma-1}{\gamma}} - p_{atm}^{\frac{\gamma-1}{\gamma}} \right] \quad (1)$$

where γ is at p_{exit}. The specific volume v of the gas at the atmospheric plane can be determined from the isentropic relation

$$p_{exit} v_{exit}^{\gamma} = p_j v_j^{\gamma} \quad (2)$$

and the ideal gas relation may then be used to calculate the temperature. The Joule-Thomson effect may also influence the temperature. In reality the presence of various shock structures downstream of the orifice will mean that the flow is not isentropic. The value of u_j calculated in this way has proved convenient for developing correlations for gaseous jet fires, but the values are unrealistically large.

In the second method the flow from the exit point to the atmospheric plane is considered to be adiabatic and irreversible. The change in fluid momentum is equated to the force caused by the pressure drop. This yields the following expression.

$$A(p_{exit} - p_{atm}) = \dot{m}(u_j - C_D u_{exit}). \quad (3)$$

The temperature change of the fluid can be calculated from the change in enthalpy of the fluid, where the c_p is the heat capacity of the gas at constant pressure

$$\frac{1}{2}[u_j^2 - u_{exit}^2] = c_p (T_{exit} - T_j) \quad (4)$$

Joule-Thomson cooling should be considered also. There is evidence, however, that the temperature rise in the immediate post shock region is such that the fluid rapidly regains its original stagnation temperature. The specific volume of the gas at the atmospheric plane can be calculated from the pressure and temperature using the ideal gas law. For both the methods for calculating the expanded jet velocity, mass conversation may be used to determine the area of the expanded jet.

Two-phase and liquid releases

Two-phase flow occurs when:

- The stagnation condition at the inlet to the pipe or orifice is two-phase.
- The stagnation condition at the inlet to the pipe or orifice is liquid and the liquid flashes whilst in the pipe/orifice.

It is a vitally important consideration in emergency relief systems as well as providing the source terms for fire and dispersion models and enables prediction of realistic release rate and duration. In the first condition, if the gas and liquid phases are travelling at equal velocities through the pipe/orifice, then the flow can be described as homogeneous. This assumption is reasonable for short pipes and pipework around process vessels. For full bore pipeline ruptures, homogeneous flow is less likely and complexities introduced by the decompression wave render the subject outside the scope of this paper. Another assumption is that the flow is at true thermodynamic equilibrium. The combined model (Leung, 1990) is termed the Homogeneous Equilibrium Model (HEM). Non-equilibrium effects only become important where the fluid at the inlet to the pipe/orifice is all-liquid and the liquid can flash whilst in the pipe/orifice, as in the second condition given above, but the extent to which it does so is controlled by the time to reach true thermodynamic equilibrium. If the liquid reaches equilibrium it can be modelled using the HEM. If it does not reach equilibrium it is most appropriately treated as a liquid in its passage through the pipe/orifice and it will flash once it is clear of the orifice. In general the transition from non-equilibrium to equilibrium flow is not well understood, but the following rule conforms to the best present understanding, obtained by comparing the equilibrium calculation with experimental releases of pressurised propane (Hirst 1986):

Residence time<0.0026s: Non-equilibrium calculation (liquid flow).

Residence time≥0.0026s: Equilibrium calculation (two-phase flow).

FIRE

Jet Fires

A jet fire is a turbulent diffusion flame resulting from the combustion of a fuel continuously released with some significant momentum in a particular direction. In principle any pressurised fuel has the potential to generate a jet fire.

Flame stability

Not all releases give rise to stable flames. A correlation, validated by extensive experimental data (Kalghatgi 1981, Birch et al 1988, McCaffrey et al 1986), shows that for unobstructed releases of natural gas from orifices less than about 30mm diameter, certain drive pressures will not sustain a stable flame, see Figure 2.

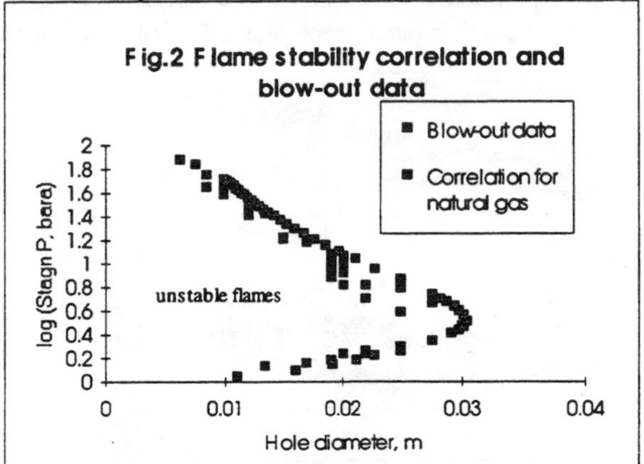

The blow-out velocity u_{jb} derived from equation (4) can be described by the empirical relationship (also shown in Figure 2),

$$u_{jb}/S_u = 0.0028\, R_H^{1.1} \left(\rho_j/\rho_{air} \right)^{-1.5} \quad (5)$$

where ρ is the fluid density, and subscripts j and "air" refer to the expanded jet and to ambient conditions respectively. H, the distance to the stoichiometric concentration in the Reynolds number R_H. Thus accidental damage to small bore high pressure fittings might reasonably be expected not to result in a stable flame, except that on an offshore platform the likelihood of flame stabilisation by impact on adjacent surfaces is high. Figure 2 refers only to natural gas. The increased burning velocity S_u associated with higher hydrocarbon gases results in greater stability and smaller critical diameters. On the other hand jet releases of liquid kerosene up to 2.5 kg/s are unable to sustain stable flames when released as a horizontal jet because of insufficient droplet evaporation (Davenport 1994b).

Jet flame size and heat flux

A jet fire can be divided into three zones:

1. The jet source and expansion down to the flame lift-off point;
2. The main body of flame, which can be divided into a zone of continuous turbulent combustion where large eddies entrain air and smaller eddies mix air with fuel, and a region where flame appears intermittently and;
3. The buoyant thermal plume where cold soot and combustion products escape as smoke.

On an offshore platform free jet fires could be considered unlikely but, to date, most predictive models have been developed and validated for such fires, and it is worthwhile reviewing the progress made. Impacting and confined fire behaviour is the subject of ongoing research, discussed later. The main hazard from a free jet fire is thermal radiation to the surroundings. The simplest expression for the radiation flux received by an object q is:

$$q = \tau V S \tag{6}$$

where τ is the atmospheric transmissivity, V is the view factor of the flame surface and S is the flame surface emissive power. For most calculations the Shell jet flame model (Chamberlain 1987, Johnson et al 1994) which represents the flame as a solid body - the frustum of a cone radiating uniformly over its surface - adequately predicts the incident flux for safety engineering purposes, and is considerably more accurate than single point source representations (e.g. in API RP521) for objects close to the flame. The view factor is derived from the flame location and size, which is based on Froude number correlations. The surface emissive power is set equal to FQ/A_f where F is the fraction of heat radiated from flame surface area A_f and Q is the net heat release. F is correlated with gas velocity u_j defined previously.

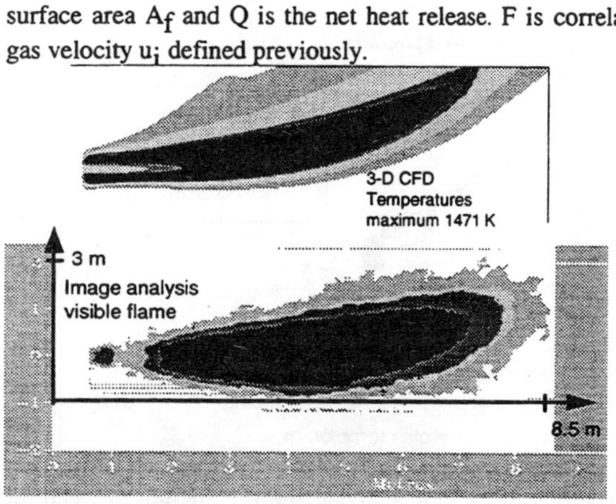

Fig. 3 Comparison of image analysed propane flame (contours at 50, 10, 5, 1% occurrence) with CFD predicted temperatures.

The applicability of empirical models is however limited by the range of experimental data on which they are based. Computational fluid dynamics (CFD) models have been improved in recent years but require considerable computer power, computer time and expertise. Some results however are remarkably good. As an example, Figure 3 shows the predicted 3-D flame temperatures (max 1471K, outer edge 490K) compared with a visible image analysis of a 0.3 kg/s sonic propane flame (Johnson et al 1997). Additional output from the model includes such important information as convective and radiative components of heat flux within the flame and external radiation field, which have been validated by experiment for natural gas as well as propane gaseous flames.

The intermediate path of 1-D integral or phenomenological modelling has been exploited by British Gas in their THRAIN model (Caulfield et al, 1993).

With partial funding from the CEC, Shell and British Gas (Sekulin et al 1995, Davenport 1994a) carried out many tests with propane, butane and mixtures of natural gas and butane at flow rates up to 28 kg/s for the single component fuels and 2.5 kg/s for the mixtures. In general, the flames were more radiative than natural gas and produced more smoke. They were also more buoyant and wind affected. For butane concentrations up to 40% the flame properties were similar to those of natural gas. With funding from the API, Davenport (1994b) carried out similar tests with kerosene/natural gas jet fires. The fires were even more radiative than with butane mixtures, but pure kerosene releases could not sustain flames under the test conditions. Measured initial heat fluxes to the rear of a 0.9m diameter steel pipe are shown in Figure 4.

Fig. 4 Heat flux at rear of target

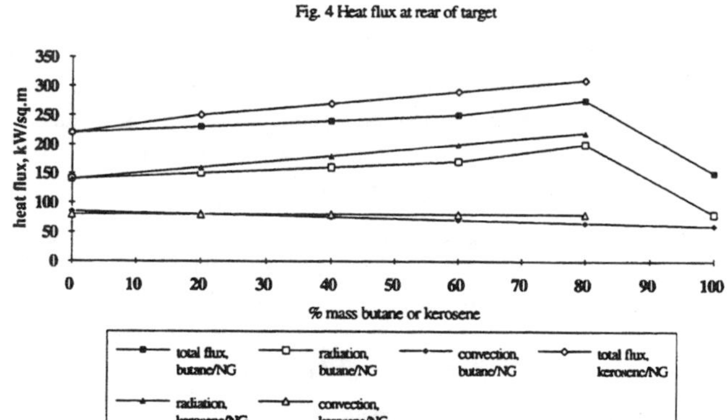

In an offshore structure an accidental ignited release is likely to impact on adjacent objects. Thus the thermal loading is required to assess the risk of escalation by failure of pipework, or vessels, or structure. The object surface is subjected to convection from the high velocity hot combustion products and to radiation emitted by the flame. Not all the thermal load is absorbed and some is re-radiated to the flame or to the atmosphere. Assuming that the flame and surface are diffuse grey bodies the absorbed flux q_a can be written,

$$q_a = \varepsilon_s \sigma\left(\varepsilon_f T_f^4 - T_s^4\right) + \varepsilon_s\left(1-\varepsilon_f\right)\sigma T_o^4 + h\left(T_f - T_s\right) \tag{7}$$

where ε is emissivity, σ is the Stefan Boltzmann constant, T is absolute temperature and h is the convective heat transfer coefficient. Subscripts f, s and o refer to the flame, surface and ambient respectively. h and T_f can be derived experimentally and combined in physical models (Cracknell et al 1995) to predict the thermal load to fire engulfed objects, for example see Figure 5. Advances in application of CFD models now also show the correct trends (Johnson 1997). An alternative approach is to use the direct measurements obtained from large scale experiments, such as the result shown in Figure 6, taking care not to extrapolate too far from the experimental conditions.

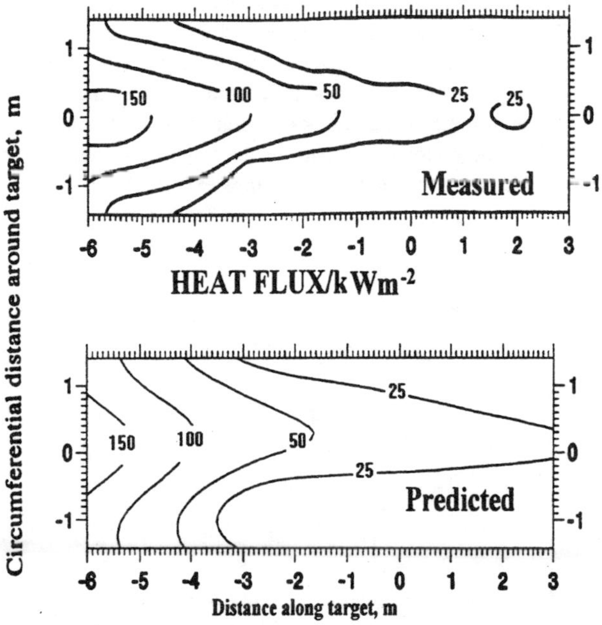

Fig.5 Measured and predicted contours of flux density for a 8.5 kg/s sonic natural gas flame 28m from the release point.

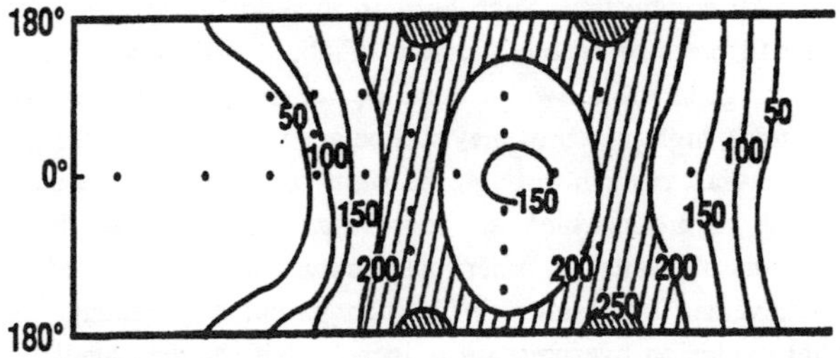

Fig. 6 Spatial distribution of total flame impingement heat flux on a 900 mm diam. pipe placed 9m downstream of a 8-10 kg/s sonic natural gas release. The cylindrical target has been unwrapped so that 0° corresponds to the part of the target directly facing the release (Cowley et al, 1991).

Pool Fires

A pool fire is a turbulent diffusion fire burning above a pool of vaporising liquid fuel where the fuel vapour has very low initial momentum. The vaporisation of high boiling point fuels is controlled mainly by radiative feedback to the pool surface. The hazard effects are controlled largely by the size of the pool and the fuel type. These determine the mass burning rate, usually expressed as kg/s/m^2 of pool area, and the size of the visible flame above the pool. Wind tilt drags the flame beyond a confining wall boundary and can bring the fire closer to downwind objects. Two distinct regions are created in the flame zone - a region of continuous burning and a more smoky region where flame appears intermittently. Empirical models (Considine 1984, Johnson 1992, Pritchard 1992) usually represent the flame as a tilted cylinder which take into account these two regions by assigning different factors for smoke obscuration, although the Shell version makes no such distinction and hence is less accurate for received radiation within one diameter of the flame. The external radiation field is calculated utilising a view factor and flame surface emissive power given by equation (6).

Organisation	Pool size, m	Fuel	Total heat flux, kW/m^2
AEA Winfrith	0.5 x 9.5	kerosene	150
US Dept of Transportation	-	kerosene	138
US Coast Guard	-	kerosene	110-142
US Dept of Transportation	-	kerosene	136-159
Sandia	-	kerosene	113-150
HSE Buxton		kerosene	130
Shell Research	4 x 7	kerosene	94-112
Hagglund		JP-4	120-150
Sandia	9 x 18	JP-4	90-164
Sandia	9 x 18	JP-4	80-120
Shell Research	35	LNG	250-300
Shell Research	12 x 15	Propane. Butane	230-270

TABLE 1 Experimentally measured heat fluxes in pool fires

Relatively little information is available on the thermal loading from pool fires to engulfed objects. However, it is known that radiative heat transfer dominates (about 80%) and that flame temperatures reach about 1000^oC. A convective heat transfer coefficient of 39 W/m^2/K was derived from data on aviation fuel pool fires (Schneider et al 1989). Other studies have shown that as the molecular weight of the fuel decreases the gases and soot become much hotter. The soot concentration in lighter molecular weight hydrocarbon flames is also much lower than in heavier hydrocarbon flames, thus they achieve their maximum radiative transfer at much longer emitting path lengths than for heavier hydrocarbon fuels. Therefore, the heat transfer to objects engulfed in natural gas fires for example will continue to increase with the size of the fire long after the heat transfer in kerosene fires for example has reached its maximum. Table 1 shows some experimentally measured heat fluxes.

Confined fires

The behaviour of jet and pool fires can be significantly modified when the fire is confined by walls and ceilings, such as the boundaries of offshore compartments. If the compartment openings are small or the release rate of fuel is high, the fire may not be able to entrain enough air for complete combustion inside the compartment. The fire is then said to be ventilation controlled. As well as the hazards already discussed, additional ones exist (Wighus et al 1991). These include external flaming from compartment openings, impaired visibility along escape routes, increased CO hazard, explosion hazard from unburnt fuel if the fire terminates due to lack of oxygen, and possible increased thermal loading to objects due to greater amounts of hot soot. Recent large scale studies (Chamberlain 1994, 1996) have confirmed that fire global stoichiometry is a useful correlating parameter. Jet fires burning in slightly fuel rich conditions represent worst case because they combine high heat release with enhanced soot production. Pool fires, on the other hand, are more severe in over-ventilated conditions. When ventilation controlled the pool fire burning rate is governed by the size of the ventilation opening and the evaporation rate tends to drop to match the lower burning rate. Thus the pool fire limits itself to the slightly under-ventilated regime. Persaud et al (1997) show that the effect of insulation on the compartment walls is to raise the average smoke temperature by 200-400°C.

Mitigation

Water spray systems are routinely installed on offshore platforms for mitigation of fire hazards, but their effectiveness has only been assessed in recent years (Wighus 1991). A study by Shell and British Gas (Shirvill et al 1994) has shown at realistic scale that a dedicated deluge system (10.2 litres/m^2/min) cannot be relied upon to prevent excessive temperature rise when a vessel (2m diam, 8m long) is subjected to a high velocity gas jet fire (3 kg/s). The high velocity jet readily penetrated the water film. High pressure water from a monitor was however more effective. Area deluge typical of the arrangement used offshore has been shown (Chamberlain 1995) to extinguish compartment jet fires in the test conditions used. This however may be undesirable because of the risk of further ignition and explosion. The main mechanism is thought to be one of inerting the entrained air by steam. The extent of hot surroundings play an important part in determining the steam loading and further work is necessary to develop safety engineering guidance. Pool fires do not extinguish in the same deluge conditions, but these fires can be successfully controlled.

Passive fire protection coatings can guarantee effective protection against offshore fires, if they are applied correctly and tested in realistic fire conditions (Shirvill 1992, Hunt et al 1997).

EXPLOSION

Overpressure generation

The burning of a vapour cloud in the absence of confinement or congestion does not generate significant overpressure. In order to generate what may be regarded as an explosion, there must be either congestion, i.e. obstacles in the path of the flame, or confinement of the gas cloud. In a typical offshore module, for example, both of these will be significant. But in discussing explosion potential, it is useful to make a distinction between these two factors: congestion and confinement.

In the absence of turbulence, the (laminar) burning velocity of a hydrocarbon flame is quite low, around 0.5 m/s. As the flame ball grows this can increase owing to instabilities, but only by a factor of two to three. However, the expansion pushes gas ahead of the flame. If the resulting flow passes obstacles (e.g. vessels, pipework), turbulence will be generated. The flame speed in a turbulent flow is much greater than the laminar flame speed, principally because the turbulence wrinkles the flame, creating a much greater area for reaction. Therefore, when the flame reaches the turbulent region downstream of the obstacles, it burns faster. This faster burning in turn creates faster flow and so a higher level of turbulence downstream of the next group of obstacles. The more intense turbulence results in even faster burning, and so on. This process is known as the Shchelkin mechanism (Figure 7). Thus, even where there is little confinement, high flame speeds may be generated by congestion. The high velocities associated with the rapid burning are sufficient to generate high pressure, i.e. an explosion. Understanding of the mechanism described above allows characterisation of the types of congested but unconfined plant likely to cause high overpressures even without confinement.. These are typically chemical or refinery plant with large amounts of pipework or other obstructions. Locations where obstacles block more than 40% of the path a flame may have to travel through, or where there are closely repeated rows of obstacles, are particularly bad. On the other hand, lightly congested plant no longer needs to be treated as if it were as hazardous as the most congested regions. Furthermore, since the explosion is generated only by the congested region, the volume of the explosion source is related not to the whole flammable cloud, but only to the volume of the congested region. This means that the overpressures experienced at a distance are lower than they would be if derived from a larger dispersing cloud.

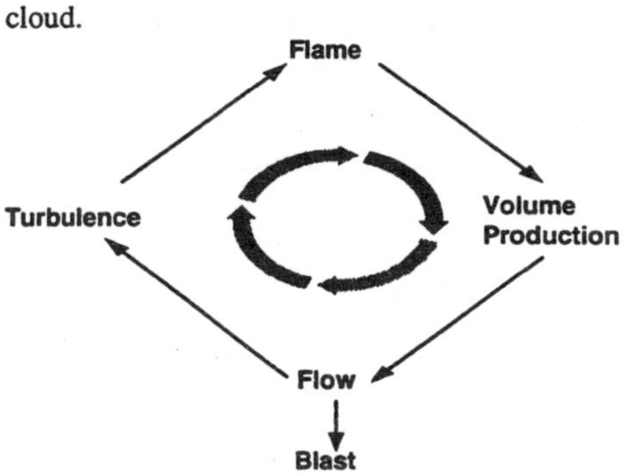

Fig. 7 The Shchelkin mechanism of flame acceleration.

An explosion in a building is confined by the walls of the building, which restrict the escape of the extra volume of gas generated by combustion. Such burning in a completely enclosed volume will produce an overpressure of about 8 bars because of this volume generation. Even if there are vents in the walls (partial confinement), high pressures may still be generated because the gas cannot flow fast enough through the vents to relieve the pressure. Explosions in areas which are more than 60% enclosed by a combination of walls or other obstructions may be treated as confined, vented explosions. With such partial confinement, the potential explosion overpressure is increased if the area available for venting is small, if the obstructions to the flow provide large blockage, or if the path from a possible ignition location to the main venting involves passing many obstacles.

Another feature of such incidents is the "external explosion" (Harrison et al 1987). As the flame burns through a vented enclosure, most of the gas is pushed out of the vent ahead of the flame. This is a consequence of the 8:1 volume expansion. Outside the vent it forms a mushroom-shaped, highly turbulent "starting jet". When the flame emerges through the vent, it burns rapidly through this gas, creating an external explosion. The external explosion may cause damage at some distance from its source, and also has influence on the pressure inside the enclosure, so that the maximum internal pressure is usually reached after the external explosion.

Various methods which have been used for the prediction of overpressures associated with vapour cloud explosions have been discussed by Cates (1991). Historically the first method, tried and used for many years, was the analogy with TNT explosions. However, it is important to emphasise that the TNT equivalence method is not appropriate for vapour cloud explosions. The "congestion assessment method" is one example of the new approaches currently being adopted for gas explosions in unconfined, congested regions. For confined, vented explosions essentially two types of models are used:

1. CFD models have the advantage that they can represent the detailed geometry of plant or offshore module. However, such models require long run times (tens of hours) on large computers and, as yet, they are not necessarily more accurate than simpler models. There are several CFD explosion models in existence; examples are EXSIM from Telemark, Norway, FLACS from Christian Michelsen Research, Norway (Van Wingerden et al 1993), and AutoReagas from TNO in the Netherlands.

2. The second category comprises models which idealise the geometry in order to simplify and speed up the calculations, but which are based an understanding of the physics underlying the various processes involved in the explosion. Because such "phenomenological" models are based on physical understanding, they are better able to extrapolate outside of the range of existing data than earlier correlation-based models. A prime example is SCOPE (Shell Code for Overpressure Prediction in gas Explosions) by Puttock et al (1995).

In some circumstances, the best way to predict a potential explosion in a congested area may be to perform scaled experiments. A theory for scaling explosion experiments was developed by Taylor and Hirst (1988), based on Gouldin's (1987) fractal model of turbulent combustion. The fractal scaling theory allows explosion assessments to be performed on scale models by using more reactive fuel (Samuels 1993), such as a 1/12 scale model and ethene in place of methane as the flammable gas. An alternative approach which can also be used is oxygen enrichment of the fuel/air mixture (Catlin 1991).

Mitigation

A number of techniques exist, such as explosion relief, blast containment, release of halons or dusts, or optimised layout, which can mitigate the effects of an explosion. All techniques can be traced back to a weakening of the feedback loop in the Shchelkin mechanism. Recent work (Johnson 1995) has shown that water sprays may also suppress gas explosions and an optimised system could potentially offer an elegant solution for protection of existing plant. Unfortunately current knowledge is not sufficiently advanced. Whilst the flame speed and overpressure generated by congestion can be considerably reduced by conventional water sprays, the overpressure can become greater in highly confined geometries. In the former case the relatively large water droplets are shattered into smaller droplets by the fast moving flames, whereas in the latter the turbulence generated by the spray increases the flame speed. Fine water mists generate less turbulence but it is difficult to achieve sufficient loading of the flame front unless the droplet density is, so far impractically, high.

CONCLUDING REMARKS

This paper cannot do justice to the rapid advances of recent years in the assessment of the combustion hazards posed by accidental releases of hydrocarbons. The interested reader is strongly urged to read the references cited herein to appreciate the detailed approaches taken. While research continues to address remaining areas of uncertainty, such as explosion prediction, chemistry of compartment fires, general aspects of turbulent combustion, and water droplet/flame interactions, many validated methods are now available to the engineer for cost effective design.

ACKNOWLEDGEMENTS

The author is indebted to the members of the Fire and Explosion Hazard Groups of Shell's Thornton Research Centre whose continuing efforts make significant impacts to knowledge in this field.

REFERENCES

Birch, AD, Brown, DR, Cook, DK, and Hargrave, GK (1988). "Flame stability in under expanded natural gas jets", Comb. Sci. and Tech., Vol. 58.

Bowen, PJ, and Shirvill, LC (1994). "Combustion hazards posed by the pressurised atomisation of high-flashpoint liquids", J.Loss Prev. Process Ind., Vol. 7, No. 3.

Catlin, CA (1991). "Scale effects on the external combustion of offshore modules caused by venting of a confined explosion", Comb. Flame, Vol. 83.

Chamberlain, GA (1987). "Developments in design methods for predicting thermal radiation from flares", Chem.Eng.Res.Des. Vol. 65.

Chamberlain, GA (1994). "An Experimental study of large scale compartment fires", Trans. IChemE, Vol. 72, Part B.

Chamberlain, GA (1995). "An Experimental study of water deluge on compartment fires", International Conference on Modelling and Mitigating the Consequences of Accidental Releases of Hazardous Materials, AIChE, New Orleans.

Chamberlain, GA (1996). "The hazards posed by pool fires in offshore platforms", Trans. IChemE., Vol. 74, Part B.

Cates, AT (1991). "Fuel gas explosion guidelines", Intl. Conf. on Fire and Explosion Hazards, Moreton-in-the-Marsh, Inst. Energy.

Caulfield, M, Cook, DK, Docherty, P, and Fairweather, M (1993). "An integral model of turbulent jets in a cross flow. Part 2 - fires", Trans. IChemE, Vol. 71.

Cracknell, RF, Davenport, N, and Carsley, AJ (1995). "A model for heat flux on a cylindrical target due to the impingement of a large-scale natural gas flame", 2nd European Conf. on Major Hazards Onshore and Offshore, IChemE, Oct.

Considine, M (1984). "Thermal radiation hazard ranges from large hydrocarbon pool fires", Safety and Reliability Directorate Report SRD R297, UKAEA.

Coulson, JM, Richardson, JF, Backhurst, JR, and Harker, JH (1990). "Chemical Engineering, Vol. 1 Fluid Flow, Heat Transfer and Mass Transfer", 4th edition, Pergamon, Oxford.

Cowley, LT, and Pritchard, MJ (1990). "Large scale natural gas and LPG jet fires and thermal impact on structures", Gastech 90, 14th International LNG/LPG Conference, Amsterdam.

Davenport, N (1994a). "Large scale natural gas/butane mixed fuel jet fires. Final report to the European Commission", Shell Research report no. TNER.94.030.

Davenport, N (1994b). " "Large scale natural gas/kerosene mixed fuel jet fires. Final report to the American Petroleum Inst.", Shell Research report no. TNER.94.061.

Gouldin, PC (1987). "An application of fractals to modelling premixed turbulent flames", Comb. and Flame, Vol. 68.

Harrison, AJ, and Eyre, JA (1987). "External explosions as a result of explosion venting", Comb. Sci. and Tech., Vol. 52.

Haque, MA, Richardson, SM, and Saville, G (1992). "Blowdown of pressure vessels - 1. Computer Model", Trans. IChemE, part B, Proc. Safety Environ. Protection, Vol. 70.

Hirst, WJS (1986). "Combustion of large scale jet releases of pressurised liquid propane", Proceedings 3rd Symposium on Heavy Gas Risk Assessment III, D.Reidel Publishing Co., Netherlands.

Hunt, RJ, Shirvill, LC, and Earles, GA (1997) "Design and application of passive fire protection for the deck of the Mars tension leg platform", Offshore Technology Conference, Houston.

Imrie, BW (1973). "Compressible Fluid Flow", Butterworths, London.

Johnson, AD, Brightwell, HM, and Carsley, AJ (1994), "A model for predicting the thermal radiation hazard from large scale horizontally released natural gas jet fires", Trans. IChemE., Vol. 72, Part B.

Johnson, AD, Ebbinghaus, A, Imanari, T, Lennon, SP, and Marie, N (1997). "Large-scale free and impinging turbulent jet flames - numerical modelling and experiments", IChemE Symp. Series 141, Hazards XIII, UMIST, Manchester.

Johnson, DM, and Shale, GA (1995). "Research on the mitigation of explosions on offshore oil and gas Platforms", International Conference on Modelling and Mitigating the Consequences of Accidental Releases of Hazardous Materials, AIChE, New Orleans.

Kalghatgi, GT (1981). "Blow-out stability of gaseous jet diffusion flames. Part 1: Still air", Comb. Sci. and Tech., Vol. 26.

Leung, JC (1990). "Two-phase flow discharge in nozzles and pipes - a unified approach", J. Loss Prev. Proc. Ind., Vol. 3, 27.

McCaffrey, BJ, and Evans, DD (1986). "Very large methane jet diffusion flames", 21st Symposium (International) on Combustion, The Combustion Institute.

Persaud, MA, Chamberlain, GA, and Cuinier, C (1997). "A model for predicting the hazards from large scale compartment jet fires", IChemE Symp. Series 141, Hazards XIII, UMIST, Manchester.

Puttock, JS, Cresswell, TM, Marks, PR, Samuels, B, and Prothero, A (1995). "Explosion assessment in confined vented geometries. SOLVEX large-scale explosion tests and SCOPE model development", Shell Research report to UKHSE, OTO 96 004.

Puttock, JS (1995). "Fuel gas explosion guidelines - The Congestion Assessment Method", 2nd European Conf. on Major Hazards Onshore and Offshore, IChemE, Manchester.

Pritchard, MJ, and Binding, TM (1992). "FIRE2: A new approach for predicting thermal radiation levels from hydrocarbon pool fires", IChemE Symposium Series No. 130.

Samuels, B (1993). "Explosion hazard assessment of offshore modules using 1/12 scale models", Trans. IChemE, Vol. 71, Part B.

Sekulin, AJ, and Acton, MR (1995). "Large scale experiments to study horizontal jet fires of mixtures of natural gas and butane", British Gas report to CEC, GRC R0367.

Shield, SR (1993). "A model to predict the radiant heat transfer and blast hazards from LPG BLEVEs", AIChE Symp. Series, Vol. 89.

Shirvill, LC, and White, GC (1994). "Effectiveness of deluge systems in protecting plant and equipment impacted by high velocity natural gas jet fires", International Symposium on Heat and Mass Transfer in Chemical Process Industry Accidents, Rome.

Schneider, ME, and Kent, LA (1989). "Measurements of gas velocities and temperatures in a large open pool fire", Fire Technology, Feb.

Taylor, PH, and Hirst, WJS (1988). "The scaling of vapour cloud explosions: a fractal model for size and fuel type", 22nd Intl. Symp. on Combustion, Poster paper.

TNO (1992). "TNO Yellow Book", 2nd edition, Chapter 2.

Van Wingerden, K, Storvik, I, Arntzen, B, Teigland, R, Bakke, JR, Sand, IO, and Sorheim, HR, (1993) "FLACS-93 - A new explosion simulator", 2nd International Conference and Exhibition on Offshore Structural Design Against Extreme loads, ERA Report 93-0843.

Wighus, R, Meland, O, and Vembe, B (1991). "Smoke hazard in offshore platform fires", SINTEF Report STF25 A91007.

Wighus, R (1991). "Active fire protection: extinction of enclosed gas fires with water sprays", SINTEF Report STF25 A91028.

Proceedings of the Seventh (1997) International Offshore and Polar Engineering Conference
Honolulu, USA, May 25-30, 1997

ISBN 1-880653-28-1 (Set); ISBN 1-880653-32-X (Vol. IV)

Risk Based Decisions for Entrance Channel Operation and Design

Andrew L. Silver and John F. Dalzell
Naval Surface Warfare Center
Bethesda, MD, USA

Abstract

Economic and environmental pressures have required the Navy to minimize dredging when determining the depth of entrance channels for deep draft ships. To maintain the capability of these deep draft ships transiting shallow entrance channels safely, the Navy has developed a risk-based methodology to aid in grounding avoidance and minimizing required channel depth.

This methodology generates a prediction of the underkeel clearance of the ship during transit of a channel of a defined depth and calculates the associated risk. The risk of the ship touching the channel bottom is calculated based upon the uncertainties in each of the major parameters contributing to the underkeel clearance prediction. A summary of the details of the underkeel clearance prediction, the risk model, and its application in an operational system and channel depth determination is provided.

Key Words

Channels, Ports, Waves, Ship Motions, Dynamic Drafts, Tide, Operator Guidance.

Introduction

Some of the U.S. Navy's large deep draft ships have home ports in harbors that have entrance channels that are both shallow and exposed to waves. Economic and environmental pressures have required minimizing the amount of dredging of the channels. This has resulted in less than fully accessible channels for these ships. To respond to these constraints, the Navy has developed an operational system, known as the Environmental Monitoring and Operator Guidance System (EMOGS), to predict the net effective clearance between the keel of the deep draft Navy vessels and the channel bottom before they transit the shallow entrance channel. A risk analysis has been developed for the system and applied to aid in assessing the amount of risk of grounding for the calculated underkeel clearance. Additionally, the methodology has been used to determine the optimum channel depth for a deep draft ship. The optimum channel depth is defined as the depth which allows the maximum accessibility with the least amount of dredging. This paper will provide a full description of both EMOGS, as it currently is configured, and the channel depth determination methodology along with a synopsis of the risk model provided in the appendix.

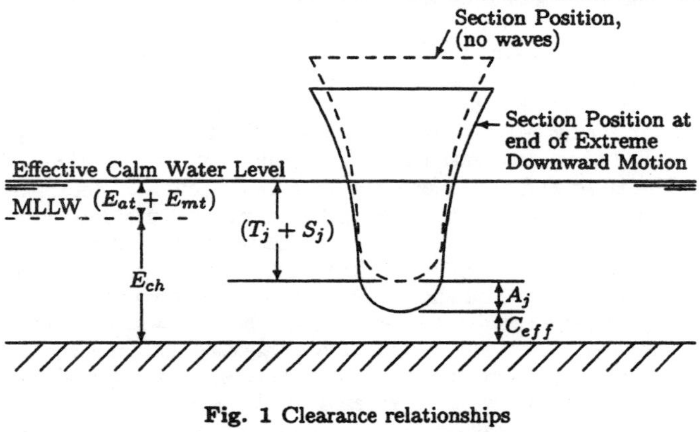

Fig. 1 Clearance relationships

EMOGS Description

The operational system to warn ships of dangerous transit conditions of shallow entrance channels is EMOGS. EMOGS predicts the estimated underkeel clearance during a given transit of a shallow water channel for a specific ship. This underkeel clearance can then be used to evaluate the risk of the ship's keel impacting the channel bottom. The effective underkeel clearance, C_{eff}, is calculated by taking the difference between the effective channel depth and the effective dynamic draft of the ship. The major parameters used in determining the effective channel depth and dynamic draft are illustrated in Fig. 1, and identified in Eq. 1.

$$C_{eff} = (E_{ch} + E_{at} + E_{mt}) - (T_j + S_j + A_j) \qquad (1)$$

In Eq. 1, E_{ch} is the estimated channel depth at mean lower low water (MLLW), E_{at} is the predicted change in the water level from MLLW due to astronomic tides, E_{mt} is the change in water level from MLLW from extreme wind speed from an offshore or onshore direction and/or extremes in barometric pressure. The variables T_j, S_j, and A_j represent the static draft, sinkage (or squat) and motions allowance at either the bow or stern, j, respectively. As shown in Fig. 1 and described above, the effective water depth in the channel is dependent upon three major parameters. The first, and primary factor, is the depth of the water at MLLW during the time of transit. The value for this parameter is input by the EMOGS user. The value can be either obtained from results of the Corps of Engineers surveys of the channel or set as the project depth, or guaranteed depth of the channel at MLLW. If the matrix of depths from the Corps of Engineers surveys are used, they are statistically analyzed to determine the "controlling depth" for different sections of the channel. These channel sections are determined from changes in the channel orientation and sedimentary regimes. The result of the statistical analysis of the channel survey yields the 99th percentile shallowest depth for each section. The statistical analysis also identifies all recorded depths and corresponding locations that are shallower than the "controlling depth" and notes them as "hot spots". Because the frequency of the surveys was not set, the chance of using a deeper water depth from an old survey than actually existed at the time of transit was high. Therefore, the preferred method of calculating effective channel clearance is by using the project depth. The effect of different sediment types on the effective channel depth was not taken into account because, for the channels for which EMOGS was designed, the channel bottoms are composed of coarse sand.

The second major factor in determining the effective water depth of the channel is the variation in water level due to astronomic tides. The value of this factor is calculated by a program, fully integrated in the EMOGS software, which predicts the change in the astronomic tide levels from MLLW for the time and date of the estimated time of arrival of the ship to the channel entrance. The program also tracks the change in water level from MLLW through the entire transit. This program is based on accepted astronomic prediction theory (Shureman, 1940) by accounting for the most significant of the 37 major tidal constituents in the area of the channel. The predicted tide levels were calibrated, for St. Mary's Inlet, with tide data obtained from a gage installed in an area near the channel (Lillycrop, et al, 1991).

The third and final parameter used by EMOGS to determine the effective water depth is the change in water depth due to meteorological effects. The predominant meteorological factors that effect the water level are extremes in barometric pressure and wind speed both on and off shore. This parameter is obtained primarily by real-time water level gages near the sites where EMOGS is installed. The water level data are collected continuously and averaged every 15 minutes as the deviation of the water level from MLLW. Since the water level responds slowly to extreme meteorological events, EMOGS uses a six hour average of the 15 minute interval data. As shown in Eq. 1, channel depth, astronomical and meteorological tide variations are summed to yield the effective water depth in the channel before the ship transits.

The second set of parameters on the right hand of Eq.1 describe the predicted dynamic draft of the ship during the transit. As illustrated in Fig. 1, this draft is determined by estimating the static draft in sea water, T_j, the speed dependant sinkage and trim which leads to an estimate of the squat at bow or stern, S_j, and an allowance for the wave-induced vertical displacement at the bow or stern of the ship, A_j.

The default static draft of the ship within EMOGS is the fully loaded design draft. The user can override this value if more information is known. The sinkage and trim conditions represent changes in the draft due to the ship speed in shallow water. These parameters are calculated within EMOGS based on shallow water sinkage and trim theory put forth by Beck, Newman and Tuck (1975). The sinkage values from the program were calibrated by the results of model tests of the ship conducted in the shallow water basin at the Waterways Experiment Station.

The final contribution to ship dynamic draft in the EMOGS calculation is the wave-induced vertical displacement. This parameter is calculated by combining the measured wave spectrum from near the channel and the motion transfer functions of the ship. The measured directional wave spectrum is obtained by an array of pressure sensors mounted on the sea bottom near the channel at a range mark. This wave array collects data continuously. Every three hours, the data are analyzed and the resulting (non-parameterized) directional spectrum is available to the EMOGS user in the form an average spectral density for each of the cells of a frequency–direction matrix. The wave array was installed, maintained, and analyzed by the Scripps Institution of Oceanography (Seymour and Higgins, 1977). The wave spectrum is then combined with the database of ship motion transfer functions. Ship motion transfer functions are defined as the response of the ship per unit wave amplitude. These motion transfer functions are the results from a strip theory ship motions program which is calibrated by means of the results of model tests. There are six motions in the database which vary according to the ship's encounter frequency and direction with the wave and ship speed. The motions most important in determining the dynamic draft are the vertical up and down motion of heave, and the angular up and down motion of pitch. These motion transfer functions are coupled to produce the vertical displacement on the centerline at the bow and stern of the ship through the following equation:

$$H_j = z + X_j\theta \qquad (2)$$

where H_j is the vertical displacement transfer function, the subscript j identifies the location on the ship (bow or stern), Z is the heave motion transfer function, X_j is the distance from the ship's center of gravity to the bow or the stern and θ is the pitch transfer function. The variance of the motions, σ_j^2 , is calculated by integrating the product of the square of the modulus of the vertical displacement transfer function, $|H_j|$, referred to as the Response Amplitude Operator (RAO), with the directional wave spectrum, $S(\omega, \eta)$, over frequency, ω, and direction, η, according to the following equation:

$$\sigma_j^2 = \int_0^{2\pi} \int_0^{\infty} S(\omega, \eta)\, |H_j(\omega, \eta)|^2\, d\omega\, d\eta \qquad (3)$$

Because the wave data collected by the slope array represents only a "snapshot" of the actual wave field, and because the wave field is a random process, there is a high likelihood that the waves that would produce the greatest amplitude of motion would not be included in the measured data. Therefore, EMOGS processes the square root of the motion variance, rms, through the following statistical equation after Ochi (1973) to yield a motions allowance, A_j:

$$A_j = \sigma_j \sqrt{2 \ln \left[\frac{L_r \sigma_{vj}}{2\pi\alpha U \sigma_j} \right]} \qquad (4)$$

where σ_j is the rms motion at the bow or stern of the ship, σ_{vj} is the rms time derivative of the motion, or velocity, at the bow or stern, L_r, is the length of the reach or succession of reaches of the channel during the transit, U is the transit speed of the ship, and α is the risk parameter which determines the frequency of exceeding the motions allowance. The risk parameter within EMOGS is set to be 1 in 100 or that in 100 downward excursions of the ship during a transit, one would greater than the calculated allowance. This motions allowance is therefore defined as the predicted extreme vertical displacement of the ship at the bow or the stern during the transit of the channel. The EMOGS motion calculations, as it is installed, are carried out only for major changes in direction of the channel to account for the change in ship-to-wave direction. Different conditions can be adapted for motions calculations for other installations of EMOGS.

Because of the prevailing conditions at the ports where EMOGS has been installed, current has not been accounted for in the calculations of any of the components of net effective clearance.

In addition to the near real-time mode, EMOGS has the capability to obtain forecasted underkeel clearances for up to 48 hours in advance of the transit. This is done by obtaining the 24 and 48 hour forecast wave data from the Fleet Numerical Meteorology

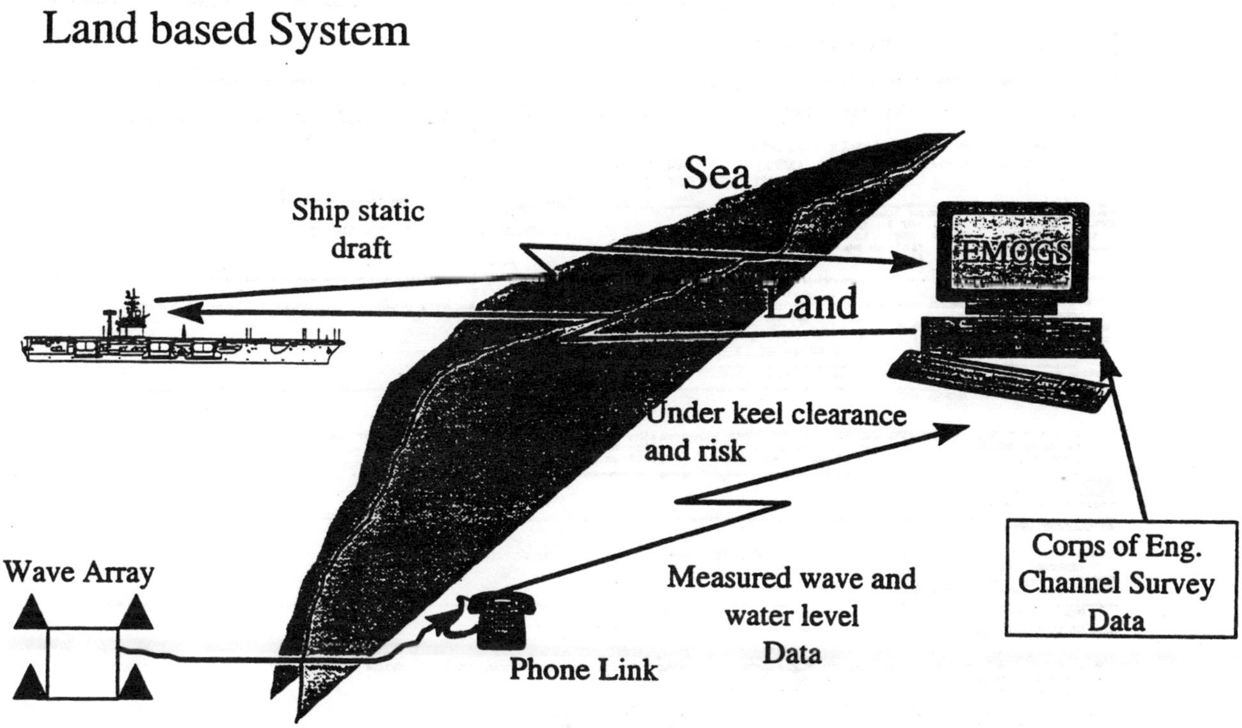

Fig. 2 EMOGS data flow

and Oceanography Center's Wave Model (FNMOC WAM). The WAM calculates forecasted wave data throughout the world in a grid spaced at 0.5 degrees apart in longitude and latitude. The directional wave spectral output of the model conforms with the format set in Pierson (1982). The EMOGS forecast mode is used by the Navy Base to establish trends of safe passage through the channel.

Figure 2 depicts the flow of data and information within EMOGS as it has been installed. Briefly, the ship motion data base is combined with wave information from either the slope wave array, or the FNMOC WAM using the user supplied ship operating conditions for the predicted transit. This yields the extreme vertical motions which are added to the user supplied static draft, and the calculated sinkage and trim to yield the predicted effective dynamic draft of the ship. This dynamic draft is subtracted from the effective channel depth to yield the effective underkeel clearance predicted for the transit.

The output for EMOGS is in an electronic worksheet form and shown in Fig. 3. This worksheet contains the values for each of the major parameters contributing to the ship's predicted underkeel clearance for the estimated time of arrival (ETA) at the channel entrance, the predicted underkeel clearance and associated risk of grounding for three different speeds and one and two hours before and after ETA. The risk model, explained in detail in the Appendix, accounts for the uncertainties inherent in the values for each of the major parameters and calibrates the risk of grounding during the transit for each underkeel clearance prediction.

As it is used operationally, the EMOGS system provides the information and the user, in this case the operational Navy, decides the acceptable risk level.

Uncertainty in the Underkeel Prediction

To calibrate the accuracy of the underkeel clearance prediction, and the relative risk of the ship touching the channel bottom, the uncertainties in each of the six constituent parameters in Eq. 1 are estimated. These uncertainties are inherent due to measuring techniques and assumptions made in the calculations. All error bands reported below represent 2 standard deviations of a normal distribution.

The controlling depth of the channel at MLLW is input by the user. As previously mentioned, this value can be obtained either through a statistically derived shallowest depth from recent Corps of Engineers surveys, or input the standard project depth. For purposes of this uncertainty analysis, it is assumed that the user will be inputting the project depth rather than the statistically derived shallowest depth from the Corps survey of the channel. This assumption was made because it was not known how often the channel would be surveyed, and the project depth represented the shallowest depth the Corps of Engineers would allow before maintenance dredging. By using the project depth in EMOGS, the effective clearance is then related to the risk of touching the bottom of a hypothetical channel of exactly constant depth. By this definition, the channel depth is a constant to which no uncertainties apply.

The atmospheric tide level for the channel is calculated with the values of the 37 harmonic constituents of the tidal cycle. The error introduced in the calculation on tide level stems from the calibration of these constituents with measured data. The measured data came from the water-level gage installed by the U.S. Army Corps of Engineers. According to Lillycrop et al (1991), this gage had a nominal error of 0.03 feet (1 cm). The data available for calibrating spanned several non-continuous months. This precluded accurate calibration of the tidal constants that had return cycles of several months to a year. Therefore, the error band assumed

EMOGS Advisory Report

Report Printed at 12:13:56 (ST) Fri Dec 27, 1996 [17:13:56 (GMT) Fri Dec 27, 1996]

Approved by (sign and date):

Prepared by: EMOGS default values
Preparation Time/Date: 1213 (ST) 27-DEC-1996
1713 (GMT) 27-DEC-1996

Transit Direction: Inbound (Entering Reach 1)
ETA Channel Entry: 1800 (ST) 13-JUN-1996
(at Reach 1) 2300 (GMT) 13-JUN-1996
Static Drafts: (ft) at bow
(ft) at rudder

[OBSERVED WAVE DATA USED]

* * * * * * * CHANNEL DEPTH COMPUTATIONS (** Indicates Sedimentation Hot Spot Exists, separate analysis follows) * * * * * * *

CHANNEL SEGMENT:	Reach 1			Reach 2		
Channel Depth (at MLW, Ft.)						
Meteorological Tide (Ft.)						
Mean Speeds	8 Kt	10 Kt	12 Kt	8 Kt	10 Kt	12 Kt
Astronomical Tide Stage (Ft.)	3.7	3.8	3.8	3.7	3.8	3.8
EFFECTIVE CHANNEL DEPTH (Ft.)						

* * * * * * SHIP DRAFT AND MOTION COMPUTATIONS (Controlling values: Suffix "r" for Rudder, "b" for Bow Hydrophone) * * * * * *

CHANNEL SEGMENT:	Reach 1			Reach 2		
Mean Speeds	8 Kt	10 Kt	12 Kt	8 Kt	10 Kt	12 Kt
Static Drafts (Ft.)						
Underway Sinkage (Ft.)						
Extreme Motion Estimate (Ft.)						
EFFECTIVE DRAFTS (Ft.)						

* SEDIMENTATION HOT SPOT COMPUTATION *

CHANNEL SEGMENT:	Reach 2		
Hot Spot Channel Depth (Ft.)	40.0		
Correction to Clearance (Ft.)	-5.0		
Mean Speeds	8 Kt	10 Kt	12 Kt
HOT SPOT CLEARANCE (Ft.)	.5r	1.6r	2.1r

* * * * * * * * * * * * * EFFECTIVE CLEARANCES FOR CHANGE IN ETA (ASTRONOMICAL TIDE VARIATION ONLY) * * * * * * * * * * * * *

CHANNEL SEGMENT:	Reach 1			Reach 2		
Mean Speeds	8 Kt	10 Kt	12 Kt	8 Kt	10 Kt	12 Kt
Clearance (Ft.) ETA 2200 GMT	3.0r	3.8r	4.2r	5.6r	6.6r	7.0r
Clearance (Ft.) ETA 2300 GMT	3.1r	4.0r	4.6r	5.8r	6.6r	7.1r
Clearance (Ft.) ETA 0000 GMT	2.8r	3.6r	4.1r	5.1r	6.2r	6.7r
Clearance (Ft.) ETA 0100 GMT	2.0r	2.9r	3.6r	4.4r	5.5r	6.1r

CHANNEL SEGMENT:	Reach 1			Reach 2		
Mean Speeds	8 Kt	10 Kt	12 Kt	8 Kt	10 Kt	12 Kt
NET EFFECTIVE CLEARANCES (Ft.)	3.1r	4.0r	4.4r	5.8r	6.6r	7.1r

EMOGS Advisory Report - Risk Analysis Results

Report Printed at 12:13:56 (ST) Fri Dec 27, 1996 [17:13:56 (GMT) Fri Dec 27, 1996]

Prepared by: EMOGS default values
Preperation Time/Date: 1213 (ST) 27-DEC-1996
1713 (GMT) 27-DEC-1996

* * * * RISK ESTIMATES, NOMINAL ETA AND CHANNEL DEPTH * * * * * * *

CHANNEL SEGMENT:	Reach 1			Reach 2		
Mean Speeds	8 Kt	10 Kt	12 Kt	8 Kt	10 Kt	12 Kt
RISK ESTIMATES:	1/9500	1/10000	1/10000	1/10000	1/10000	1/10000

* * * * RISK ESTIMATES, SEDIMENTATION HOT SPOTS * * * * * * * * * *

CHANNEL SEGMENT:	Reach 2		
Mean Speeds	8 Kt	10 Kt	12 Kt
RISK ESTIMATES:	1/260	1/3800	1/10000

* * * * RISK ESTIMATES, ETA VARIATION * * * * * * * * * * * * * * *

CHANNEL SEGMENT:	Reach 1			Reach 2		
Mean Speeds	8 Kt	10 Kt	12 Kt	8 Kt	10 Kt	12 Kt
RISK, ETA 2200 GMT	1/8100	1/10000	1/10000	1/10000	1/10000	1/10000
RISK, ETA 2300 GMT	1/9500	1/10000	1/10000	1/10000	1/10000	1/10000
RISK, ETA 0000 GMT	1/5800	1/10000	1/10000	1/10000	1/10000	1/10000
RISK, ETA 0100 GMT	1/1600	1/10000	1/10000	1/10000	1/10000	1/10000

Fig. 3 Sample page of EMOGS advisory report

for astronomic tide for the whole channel was 0.5 feet (0.15m).

The change in water level due to local meteorological conditions was calibrated by measured data. The typical magnitude of this parameter is on the order of 0.3 feet but can be as high as 2.0 feet. An error band of 0.5 feet was chosen to cover the uncertainty in even the most extreme cases.

The uncertainties in the parameters composing effective dynamic draft, static draft, sinkage and trim, and extreme vertical motion were determined in the following way. The exact static draft of the vessel as it is entering the channel on an inbound transit may be hard to determine because the personnel cannot see the draft marks while underway. The draft taken dockside may have some error because the water may be brackish and not representative of the water in the channel. Because of these factors, an error band of 0.5 feet (0.15m) is assumed for the static draft. Sinkage and trim calculated estimates were calibrated with model test data in shallow water. The resulting error band for this parameter was also 0.5 feet (0.15m) based on the scatter between the individual runs in the model test data. Combining the error bands of these two parameters yields an error band of approximately 0.7 feet (0.2 m).

The vertical motions of the ship in the channel are the product of the RAOs of the ship and the measured directional wave spectrum. The motion RAOs are linearized analytical estimates. However, most ship systems are weakly non-linear, or for a limited range of excitation frequencies, the ratio of response and wave excitation amplitudes vary weakly. Therefore there is some error introduced by this assumption. The motion RAOs used within EMOGS have been calibrated by model test results.

The wave data used in the EMOGS calculation are a representative sample of the wave conditions in the channel based on data collected prior to the time of transit. These data are not the actual waves that the ship encounters during the transit, but are used as a representation of the wave climate in the channel. This is regarded as the major source of error in the estimation of dynamic draft prediction during the transit. The uncertainty in the wave data for both the risk model and the full scale data comparison was determined using measured waves with buoys. The current configuration of EMOGS has the Scripps slope array of bottom mounted pressure sensors. According to Seymour and Higgins (1977), there is no bias in the waves measured with the slope array and little variability. However, because there is a time difference between the wave measurement and the transit time, variability in the wave estimate will probably be close to the variability of the waves measured by the buoys, but the bias would be eliminated.

To quantify the error in the wave-induced ship vertical motion estimate, full scale motion data, measured as the ship transited the channel, was compared with the predicted motions using the measured wave data from buoys. The resulting ratio between the measured and predicted motions yielded 0.8 with a scatter of 30%. This result assumed that there were no serious biases in either the motion transfer function file or the measured wave data. Further investigation revealed that the bias is introduced primarily from the measured directional wave spectrum from the buoys.

Channel Depth Determination

The prediction of underkeel clearance was also used to help determine the "optimum" channel depth for a class of deep draft surface vessels. The "optimum" channel depth is defined as the minimum dredge depth which provides the maximum amount of safe access to the harbor. The two ports studied to determine the "optimum" channel depth were San Diego and Mayport. The methodology used to determine the depth used a wave climatology, or the persistence of a range of significant wave heights and modal periods as the wave input, and the underkeel clearance predictions for each hypothetical transit.

The wave climatologies used for both ports were modifications of the result of a compilation of wave data by the Corps of Engineers in their Wave Information Study (Jensen et al, 1992, and Hubertz, et al, 1993). These data were compiled by grid points in tables showing the percent of occurrences of each wave height and modal period combination for each major wave direction. Each of the wave height, period and direction combinations were then spread in a point spectrum using a simplified version of the JONSWAP spectrum (Goda, 1985, and Hasselman, 1973). The shape of each spectrum was weighted according to the modal period. In San Diego, the closest grid point to the channel was approximately six miles away. The wave climate was transferred to the seaward channel entrance through a wave transformation computer model developed by O'Reilly and Guza (1993) for the southern California bight. At Mayport, the grid point was five miles from the seaward entrance of the channel. For this case, the waves were transferred to the channel entrance by the RCPWAVE transformation model (Ebersole et al., 1986).

The uncertainty used in the channel depth determination was essentially the same as used in EMOGS except for the wave input. In EMOGS the waves were measured by a device. The wave climatology used for the channel depth was generated from FNOC using the Global Spectral Wave Model (GSOWM). GSOWM wave data were compiled and compared with data from buoys deployed by the National Buoy Data Center by Wittmann and Clancy (1993). It was found that the bias of these forecasted GSOWM wave data is approximately 1.3 feet (0.4m) with a scatter index of approximately 0.4. The scatter index is defined as the ratio of the standard deviation of the error to the mean of measured observations. This implies that there is a potential two-sigma "error" of ±80% in the forecast wave predictions.

The primary goal of the risk computation was to aid in determining the overall channel accessibility. Risk analysis using the same model as EMOGS, explained in the Appendix, was performed on 174 representative wave conditions at San Diego and 54 wave conditions for Mayport with the energy spread in a JONSWAP form. The ship operating conditions used in the risk analysis were inbound and outbound transits at ship speeds of 6, 10, and 14 knots. A range of project depths was considered for each port. The risk of touching bottom was computed for each of the assumed ship operating conditions and project depths in each of the representative wave conditions. For each of these situations the channel was considered accessible if the resulting risk levels were less than 0.01 (1 in 100). The percent of occurrence of each wave condition was weighted by zero or one according to whether the channel was considered inaccessible or accessible according to the risk computation. The resulting weighted percent occurrence was then multiplied by 365 to obtain the number of days per year that the channel was safe to transit for each wave condition. The days of channel access was then summed over all the wave conditions to yield the total number of days of accessibility per year for a given project depth. The end result was a chart showing the days of accessibility for each project depth at MLLW and different tide levels above MLLW.

Summary

A system of predicting underkeel clearance and the associated risk of grounding for deep draft vessels transiting shallow entrance channels has been developed and used both operationally and for channel design. The operational use provides near real-time predictions of the underkeel clearance and the relative safety of the transit for the prevailing environmental conditions. This system has been installed at a Navy base and operational since 1989. The methodology has been used to determine optimum channel depth at two locations for Navy aircraft carriers. The result has saved the Navy a considerable amount in dredging costs and reduced environmental impacts.

References

Beck, RF, Newman, JN and Tuck, EO (1975) Hydrodynamic Forces on Ships in Dredged Channels. *Journal of Ship Research*, **19**, 3.

Ebersole, BA, Cialone, MA and Prater, MD (1986) Regional Coastal Processes Numerical Modeling System, Report 1, RCP-WAVE – A Linear Wave Propagation Model for Engineering Use. Technical Report CERC-86-4. US Army Corps of Engineers, Waterways Experiment Station, Coastal Research Engineering Center, Vicksburg, MS.

Goda, Y (1985) *Random Seas and Design of Maritime Structures.* University of Tokyo Press.

Gumbel, EJ (1958) *Statistics of Extremes.* Columbia University Press, New York.

Hasselmann, K, et al (1973) Measurement of Wind-wave Growth and Swell Decay During the Joint North Sea Wave Project (JONSWAP). *Deutsche Hydrographischen Zeitschrift*, **A8**, 12.

Hubertz, JM, Brooks, RM, Brandon, WA and Tracy, BA (1993) Hindcast Wave Information for the US Atlantic Coast. Wave Information Studies of US Coastlines, WIS Report No. 30. Coastal Engineering Research Center, Waterways Experiment Station, US Army Corps of Engineers, Vicksburg, MS.

Jensen, RE, Hubertz, JM, Thompson, EF, Reinhard, RD, Borup, BJ, Brandon, WA, Payne, JB, Brooks, RM and McAneny, DS (1992) Southern California Hindcast Wave Information. Wave Information Studies of US Coastlines, WIS Report No. 20. Coastal Engineering Research Center, Waterways Experiment Station, US Army Corps of Engineers, Vicksburg, MS.

Lillycrop, WJ, Howell, GL, Garcia, AW, Grogg, WE and Andrew, ME (1991) Automated Real Time Elevation System. In: *Advances in Tidal Hydrodynamics*, B. Parker, Ed. John Wiley & Sons.

Ochi, MK (1973) On Prediction of Extreme Values. *Journal of Ship Research*, **17**, 1.

O'Reilly, WC and Guza, RT (1993) A Comparison of Spectral Wave Models in the Southern California Bight. *Coastal Engineering*, **19**, 263–282.

Papoulis, A. (1965) *Probability, Random Variables and Stochastic Processes.* McGraw-Hill, New York.

Pierson, WJ (1982) The Spectral Ocean Wave Model, a Northern Hemisphere Computer Model for Specifying and Forecasting Ocean Wave Spectra. DTNSRDC Report 82/011.

Seymour, RJ and Higgins, AL (1977) A Slope Array for Measuring Wave Direction. *Proceedings of a Workshop on Coastal Processes Instrumentation,* University of San Diego.

Seymour, RJ, Castel, JD, McGehee, DD, Thomas, JO and Reilly, WC (1993) New Technology in Coastal Wave Monitoring. *Proceedings of Waves '93,* New Orleans, Louisiana.

Shureman, P (1940) Manual of Harmonic Analysis and Prediction of Tides. Special Publication 98. U.S. Department of Commerce, Coast and Geodetic Survey. (Reprinted in 1958 with corrections).

Appendix: Summary of the risk model

The risk model was developed in the operational guidance context; that is, the assessment, before the transit begins, of the minimum under-keel clearance which might exist during the transit. It is reasonable to consider risk in the context of all the possible values of the minimum instantaneous channel clearance which might be experienced during each of a long (hypothetical) series of transits under statistically constant conditions. The event of interest, touching bottom, will almost certainly occur if the minimum clearance is negative. Thus for present purposes;

> Risk is defined as that proportion of all possible transits under statistically constant conditions in which the minimum channel clearance would be negative.

To estimate risk under this definition it is necessary to compute the probability density of the minimum clearance, and find its area up to a minimum clearance of zero. For present purposes denote the minimum clearance during a transit by C_M, and the probability density of this minimum clearance by $f_{CM}(C_M)$. Then the risk is mathematically defined as

$$\text{Risk} = \int_{-\infty}^{0} f_{CM}(C_M)\, dC_M \tag{5}$$

In general, the instantaneous minimum clearance, C_M, is considered to be the algebraic sum of six more-or-less random variables which parallel those described in the net effective clearance defined in Eq. 1.

$$C_M = D_{ch} + D_{at} + D_{mt} - T - S + Z \tag{6}$$

where D_{ch} is the local channel depth, D_{at} is the local astronomical tide, D_{mt} is the local meteorological tide, T is the local ship draft, S is the local ship sinkage, and Z is the local maximum downward motion of the ship. (Ship motion is taken to be positive if the ship moves upward, and accordingly, the sign of Z for interesting events will be negative.)

The use of the word "local" in conjunction with the ship draft, sinkage and motion is meant to call attention to the fact that these quantities vary with position on the ship. In practice, clearance and risk is evaluated for a number of positions on the ship, though, for the present exposition, one location is assumed. Similarly, the use of the word local in conjunction with the channel depth and tides means that the values pertain to conditions where ever the ship happens to be when the minimum clearance occurs.

There are two kinds of risk of interest; the risk of "touching project depth" and the risk of "touching bottom". The difference is in the treatment of the channel depth. In the risk of touching project depth, the channel bottom is assumed to be perfectly flat with depth equal to project depth or other controlling depth; that is, the channel depth is considered to be deterministic. In computing the risk of touching bottom, an approximate description of the statistics of the depth of the channel is taken into account. Though more realistic, the "touching bottom" case requires data which are ordinarily impossible to obtain in a timely fashion.

Extreme motion

It is assumed that, within the duration of a single transit, the time history of the motions of the ship is a realization of a stationary zero mean Gaussian process. The current interest is only in the minima of the process. It is generally reasonable to assume that ship motions are narrow banded, though the treatment by Ochi (1973) of the extreme minima of the general wide band Gaussian process is used to estimate a motions allowance in the main text. This treatment involves a truncation so that only negative minima are considered, and a restriction that the process not be unusually broad. The net effect is that the derived extreme values are essentially those which would have resulted from a initial narrow band assumption. The narrow band model is thus assumed as an initial step in the risk model.

To make the assumption concrete, a randomly selected minimum of the motions process will be denoted as Z_R, where the negative sense of Z_R is downward, and positive values are impossible by the definition of "downward". If it is assumed that the standard deviation of the motions process, σ, is given, the (Rayleigh) conditional probability density of Z_R may be written:

$$f_{ZR}(Z_R|\sigma) = \frac{|Z_R|}{\sigma^2} \exp\left[-\frac{Z_R^2}{2\sigma^2}\right] \quad \text{for } Z_R \leq 0 \tag{7}$$

Now denote the number of statistically independent motion minima during a transit by K. The smallest minimum expected in a sample of K statistically independent minima is the quantity finally desired, and will be denoted by Z. Applying the theory of extremes (Gumbel 1958) with Eq. 7 considered the density of the initial process, the conditional probability density of Z may be written

$$f_{ZK}(Z|\sigma,K) = K\left(1-\exp\left[-\frac{Z^2}{2\sigma^2}\right]\right)^{K-1} \cdot \frac{|Z|}{\sigma^2}\exp\left[-\frac{Z^2}{2\sigma^2}\right] \quad \text{for } Z \le 0 \qquad (8)$$

If the motions process is observed for time T_{ex}, the expected total number of observed minima may be estimated by $T_{ex}/T_2 = T_{ex}\sigma_V/2\pi\sigma$, where T_2 is the expected zero crossing period and σ_V is the standard deviation of the motions velocity. Now the total number of motions minima is not the same as the number of statistically independent motion minima in the sample because successive minima in a relatively narrow band process are ordinarily correlated. However, vertical ship motions are ordinarily broad enough that the difference is less than a factor of two. Thus, if K is taken to be the total expected number of minima, the final result for extreme values tends to be conservative, but not outrageously so. Making the assumption, $K \approx T_{ex}\sigma_V/2\pi\sigma$, and substituting in Eq. 8, there results the conditional probability density of the extreme minimum, Z, during a transit; given values of the standard deviations of motion and velocity (σ, σ_V) and the exposure time, T_{ex}:

$$f_Z(Z|\sigma,\sigma_V) = \frac{T_{ex}\sigma_V}{2\pi\sigma}\left(1-\exp\left[-\frac{Z^2}{2\sigma^2}\right]\right)^{\frac{T_{ex}\sigma_V}{2\pi\sigma}-1} \frac{|Z|}{\sigma^2}\exp\left[-\frac{Z^2}{2\sigma^2}\right] \quad \text{for } Z \le 0 \qquad (9)$$

It may be noted that the exposure time is implicitly taken to be deterministic in Eq. 9. This exposure time is estimated by the product of the channel length and the ship speed. Though the channel length is a fixed constant, the ship speed during the transit is not generally exactly constant. Accordingly, the effect of the neglect of a minor amount of variability in speed is routinely checked in the guidance algorithm by a speed sensitivity check.

The next step in the development of the risk model is to remove the conditionality from Eq. 9 in the standard way (Papoulis 1965). This operation may be expressed as follows:

$$f_{ZM}(Z) = \int_{-\infty}^{\infty}\int_{-\infty}^{\infty} f_Z(Z|\sigma,\sigma_V)\, f_{\sigma,\sigma_V}(\sigma,\sigma_V)\, d\sigma\, d\sigma_V \qquad (10)$$

where Eq. 9 is understood to be substituted in the integrand, $f_{\sigma,\sigma_V}(\sigma,\sigma_V)$ is the joint density of σ and σ_V, and the result, $f_{ZM}(Z)$, represents the deconditioned probability density of the largest downward excursion in the transit.

The values of the motion and velocity standard deviations for a given transit have been found to have a substantial variability. We do not have sufficient data to establish the form of the joint density of the standard deviations, and in order to produce a working risk model have fallen back on the bivariate Normal density model for this purpose. To establish notation the joint density will be written:

$$f_{\sigma\sigma_V}(\sigma,\sigma_V) = \frac{1}{2\pi s_\sigma s_{\sigma_V}\sqrt{1-\rho^2}} \cdot \exp\left[-\frac{1}{2(1-\rho^2)}\left\{\left(\frac{\sigma-\sigma^*}{s_\sigma}\right)^2 -2\rho\left(\frac{\sigma-\sigma^*}{s_\sigma}\right)\left(\frac{\sigma_V-\sigma_V^*}{s_{\sigma_V}}\right)+\left(\frac{\sigma_V-\sigma_V^*}{s_{\sigma_V}}\right)^2\right\}\right] \qquad (11)$$

Given the form of Eqs. 9 and 11, it may be next to impossible to make an analytical evaluation of Eq. 10. The writers have not tried. The evaluation of Eq. 10 was carried out numerically. In the process it was necessary to truncate the infinite limits of $f_{\sigma,\sigma_V}(\sigma,\sigma_V)$ within the range $\pm5\sigma, \pm5\sigma_V$. The range of Z returned was programmed to be such that the error on the probability distribution at the lower limit of Z was less than 0.0001

Since there are a half dozen variables to contend with, a considerable sensitivity analysis was carried out. The results of the sensitivity analysis may be summarized best by listing the parameters in the order of influence on the computed density.

1. σ^*, the expected value of the motion standard deviation. The value is the estimated value for a given transit, corrected for any systematic biases which may have been determined from an uncertainty analysis. Since this is a scaling parameter, it is the most important to estimate correctly
2. s_σ is the standard deviation of the estimated random error in the vertical motion standard deviation. It has a strong effect on the density function.
3. T_{ex} and σ_V are the exposure time and the standard deviation of the motion velocity. These parameters occur only as a product which, in part, determines the sample size. It is important to get the magnitude of the product correctly, but small differences produce small effects on the density.
4. ρ, the correlation coefficient between σ and σ_V, is capable of producing visible effects on the density, but small errors again produce very small effects on the density.
5. s_{σ_V}, the standard deviation of the estimated random error in the motion velocity standard deviation appears to be a very weak parameter. It appears only necessary to determine the correct order of magnitude.

Calm water channel clearance

If the ship motion is negligible, the remaining five variables in Eq. 6 determine what may be termed the "Calm Water Clearance", C_C. Dropping Z from Eq. 6,

$$C_C = D_{ch} + D_{at} + D_{mt} - T - S \qquad (12)$$

Under a combination of assumptions, each of the variables on the right hand side of Eq. 12 may be treated as a random variable. The principal assumptions are

1. The variability is defined in the context of a hypothetical long series of transits under statistically identical conditions.
2. The calm water clearance of interest in each transit is evaluated at a randomly chosen position in the channel. Normally, this position will be determined by the location of the ship at the time of the extreme motion minimum. Since the time of the extreme is essentially unpredictable. it is assumed that the location of interest may be anywhere within the extent of the channel.

Both fundamental assumptions are necessary for the interpretation of the channel depth as a random variable. The soundings from a channel survey may be summarized in the form of a frequency function, and from the frequency function a mean depth, D^*_{ch} and a standard deviation, s_{ch}, may be approximated. Though the soundings (the actual shape of the channel bottom) are deterministic, the assumption that locations will be picked at random for a hypothetically long series of transits allows the channel depth to be represented as a random variable. For lack of systematic information to the contrary the probability density of the channel depth is assumed to be Normal with mean D^*_{ch} and standard deviation s_{ch}. As noted in the first part of this appendix, when risk of touching project depth is of interest, D^*_{ch} is set equal to project depth and s_{ch} is zeroed.

Astronomical tide computations are quite accurate for tides at tide gages. In the present case it is necessary to estimate the instantaneous tide at any location in a long channel which is usually somewhat remote from the tide gage. Errors in this prediction are inevitable. It is assumed that the astronomical tide variable is Normally distributed with mean value D^*_{at} and standard deviation s_{at}.

Similarly, the prediction of meteorological tide is usually highly error prone. The meteorological tide variable is modeled as a

Normal variable with mean value D^*_{mt} and standard deviation s_{mt}.

The accuracy with which ship drafts, T, may be determined before the transit varies with the situation, but the errors are not always negligible. For risk model purposes draft is assumed to be a Normal variable with mean value T^* and standard deviation, s_T. Similarly, the sinkage determination is not always of high quality, especially since sinkage is sensitive to actual channel depth which is not always known exactly. Again for risk model purposes, the sinkage variable is assumed to be Normal with mean value S^* and standard deviation s_S.

Because of the divers natures of the various sources of uncertainty in the five components of calm water clearance, it is within reason to assume that they are all mutually independent. Thus, since all components have been assumed to be Normal, the calm water clearance becomes another Normal variable. Denoting the probability density of C_C as $f_{CC}(C_C)$,

$$f_{CC}(C_C) = \frac{1}{\sqrt{2\pi}s_C} \exp\left[-\frac{1}{2}\left(\frac{C_C - C_C^*}{s_C}\right)^2\right] \qquad (13)$$

where s_C is the standard deviation and C_C^* is the mean value of the calm water clearance. The relations between the calm water standard deviation and mean, and the individual parameters which define the mean and variability of each of the five constituents are as follows:

$$\begin{aligned} s_C &= \sqrt{s_{ch}^2 + s_{at}^2 + s_{mt}^2 + s_T^2 + s_S^2} \\ C_C^* &= D_{ch}^* + D_{at}^* + D_{mt}^* - T^* - S^* \end{aligned} \qquad (14)$$

With the discussion above, the random variable, C_C is interpreted as the instantaneous calm water clearance at a randomly chosen channel location in a hypothetical long series of transits under statistically identical conditions.

The density of the minimum clearance

Comparing Eqs. 6 and 12, the minimum under-keel clearance random variable, C_M may be written:

$$C_M = C_C + Z \qquad (15)$$

Equations 10 and 13 are advanced as models for the probability densities of each of the two components on the right side of Eq. 15 Because the physical origins of the two components are quite different, is is reasonable to assume that they are statistically independent. With the independence assumption the probability density of C_M may be expressed as a convolution (Papoulis 1965):

$$f_{CM}(C_M) = \int_{-\infty}^{\infty} f_{CC}(C_M - Z)\, f_{ZM}(Z)\, dZ \qquad (16)$$

where the functions in the integrand are the densities defined in Eqs. 10 and 13. Because the computation of Eq. 10 had to be numerical, that of Eq. 16 had to be also. The infinite limits of the Normal function, f_{CC} were truncated at $\pm 5 s_C$. The net effect of the various numerical approximations which had to be made along the way was that no great confidence was thought justified in those values of computed risk, Eq. 5, which turn out to be less than 0.0001.

The Proceedings of the Seventh (1997) International OFFSHORE AND POLAR ENGINEERING CONFERENCE

VOLUME III, 1997

Numerical Waves, Nonlinear Waves, Wave Statistics and Measurements, Internal Waves, Hydrodynamic Forces, Dynamic Responses, Airgap and Metocean, Vortex Shedding and Flow-Induced Vibrations, Coastal Hydrodynamics

edited by:

Jin S. Chung, Colorado School of Mines, Golden, Colorado, USA
C.H. Kim, Texas A & M University, College Station, Texas, USA
Michel Olagnon, Ifremer, Plouzane, France
Shigeru Naito, Osaka University, Osaka, Japan

presented at:

The Seventh (1997) International Offshore and Polar Engineering Conference held in Honolulu, USA, May 25-30, 1997

organized by:

International Society of Offshore and Polar Engineers

sponsored by:

International Society of Offshore and Polar Engineers (ISOPE)
with cooperating societies and associations

International Society of Offshore and Polar Engineers (ISOPE)
P.O. Box 1107 Golden, Colorado 80402-1107 U.S.A.

International Standard Book Number: ISBN 1-880653-28-1 (Set)
ISBN 1-880653-31-1 (Vol. III)
Library of Congress Catalog Card Number: 97-70292

Cooperating Organizations:

Canadian Association of Petroleum Producers (CAPP)
American Society of Civil Engineers (ASCE) - Engineering Mechanics Division
Korea Committee for Ocean Resources and Engineering (KCORE)
Canadian Society of Civil Engineers (CSCE) - Engineering Mechanics Division
Chinese Society of Ocean Engineers (CSOE)
Chinese Society of Naval Architects and Marine Engineers (CSNAME)
Chinese Society of Theoretical and Applied Mechanics (CSTAM)
Russian Academy of Sciences
Singapore Structural Steel Society (SSSS)
Norwegian Petroleum Society (NPF)
The Institution of Engineers Australia (IE Australia)
Kansai Society of Naval Architects, Japan (KSNAJ)
IRO (The Netherlands)
Technical Research Centre of Finland (VTT)
The Society of Materials Science, Japan (JSMS)
The Offshore Engineering Society (OES), United Kingdom
Ukraine Society of Mechanical Engineers (USME)
IFREMER, France
Scott Polar Research Institute (SPRI), UK
The Institute of Engineers Indonesia (PII)

Printed and bound in USA

FOREWORD

We greatly appreciate the excellent responses and help we have received from colleagues around the world in the successful organization of the 7th International Offshore and Polar Engineering Conference (ISOPE-97), Honolulu, May 25 -30, 1997. The Conference features **91** sessions **of 517** refereed papers and **10** plenary sessions from 40 countries.

The purposes of the ISOPE conference are to:

- promote technological progress and activities, international technological transfer and cooperation, and opportunities for engineers to maintain and improve technical competence; and
- provide a timely international forum for technical activities, cooperation, opportunity and fellowship among researchers and engineers,

by developing focused session topics with papers of high quality in both originality and significance of the results which were accepted after rigorous review. The conference promotes interdisciplinary interaction between academia and industry.

The International Society of Offshore and Polar Engineers (ISOPE) has already held 12 successful international meetings:

- the 1st (1990) European Offshore Mechanics Symposium (ISOPE EUROMS-90) in Trondheim, 1990;
- the 1st (1990) Pacific/Asia Offshore Mechanics Symposium (ISOPE-PACOMS-90) Seoul; PACOMS-94 (SOSC-94) Beijing; and PACOMS-96 Pusan.
- the Annual ISOPE conferences, starting in Edinburgh, 1991 were held in San Francisco, Singapore, Osaka, The Hague and Los Angeles. Since 1992, the annual ISOPE Conference has been held with the world's largest technical program of its kind with refereed papers.
- the 1st (1995) ISOPE Ocean Mining Symposium (ISOPE OMS-95), Tsukuba; the 2nd ISOPE OMS-97 will be held in Seoul, November 24-25, 1997.
- the 1st (1996) International Deep-Ocean Technology (IDOT-96) Symposium and Workshop, Los Angeles.

This year the ISOPE-97 TPC expanded greatly the number of papers on recent worldwide developments in environment, advanced marine technology, coastal engineering, Sakhalin oil and gas. We shall continue to add other emerging topics to the future conferences.

On behalf of all the individual TPC members (listed in this volume), we would like to thank the reviewers (listed in this volume) for their help in maintaining the technical accuracy and quality of papers. We greatly appreciate the help from colleagues around the world, particularly from the authors and reviewers in meeting the deadlines. The efficient handling of the conference organization by the ISOPE staff is greatly appreciated. TPC received 517 mats, examined individual papers, page by page, paginated, and retyped all individual titles according to the ISOPE format, all in 3 weeks, ready for the printer. Only about 5% of the papers were redone by the authors, following the ISOPE mat format and font size. It was possible with the help of fax, international express mail and wordprocessor.

It is our pleasure to welcome participants from all over the world to the ISOPE-97 Conference in Honolulu. We hope that the participants will also join us at the ISOPE-98 Conference to be held in Montréal, Canada, May 24-29, 1998.

Co-chairmen of the ISOPE-97 Conference:

Jin Chung, Colorado School of Mines, USA
Bob Frederking, National Research Council, Canada
Ron Knapp, University of Hawaii, USA
Michel Olagnon, Ifremer, France
Yukio Ueda, Kinki University, Japan

Acknowledgement

The success of the 7th International Offshore and Polar Engineering Conference (ISOPE-97), USA, May 25-30, 1997, is credited to the teamwork of the members of the Technical Program Committee (TPC), the conference cosponsors, the cooperating organizations and the reviewers (see the list).

The TPC would like to express its sincere appreciation to the TPC members' organizations for their support of the TPC members, and University of Hawaii TPC colleagues for their support toward the successful conduct of the conference events, and to the ISOPE staff for the efficient handling of the conference organization and proceedings.

ISOPE-97 Technical Program Committee (TPC)

Prof. T. Yao, Hiroshima Univ., Higashi-Hiroshima City, Japan
Prof. R.W. Yeung, Univ. of California, Berkeley, CA, USA
Prof. S.C.S. Yim, Oregon State Univ., Corvallis, OR, USA

ISOPE-97 Executive Committee

Prof. Jin S. Chung (Chairman), Colorado School of Mines, USA
Dr. Robert M.W. Frederking, National Research Council, Canada
Prof. Ronald H. Knapp, University of Hawaii, USA
Prof.-Dr. Ram S. Puthli, Karlsruhe University, Germany

Conference Co-chairmen

Jin S. Chung, Colorado School of Mines, USA
Robert M.W. Frederking, National Research Council, Canada
Ron H. Knapp, University of Hawaii, Honolulu, USA
Michel Olagnon, Ifremer, France
Yukio Ueda, Kinki University, Wakayama, Japan

Hawaii Committee

Prof. R.H. Knapp (Chairman), University of Hawaii at Manoa, Honolulu, HI, USA
Prof. H.-J. Krock, University of Hawaii at Manoa, Honolulu, HI, USA
Prof. H.R. Riggs, University of Hawaii at Manoa, Honolulu, HI, USA

Partial List of Reviewers

Abe, N
Abel, A
Akyarli, A
Albert, D G
Alessandrini, B
Ali, P
Allyn, N
Andronicou, L
Aoki, S
Aoki, S I
Arai, M
Asai, K
Aso, K
Athanassios, T
Athanassoulis, G A
Bahlavouni, A
Bai, Y
Baier, V
Balasubramanian, S
Bangaru, N V
Barthelemy, J
Bea, R G
Bekker, A T
Berhault, C
Bertsimas, D J
Bitner-Gregersen, E
Bjørnsen, T
Blackmore, R
Boo, S Y
Borekci, O
Boswell, L F
Bouchard, G
Bowen, A
Brandes, H G
Bretz, G
Brown, A J
Bryden, I G
Burcharth, H F
Burgess, J J
Busch, W
Capanoglu, C
Caridis, P A
Carter, D J
Cavanié, A
Chan, D B
Chan, E S
Chandler, B
Chaplin, J R
Che, X L
Chen, H C
Cheung, K F
Chiang, E
Chien, L K
Chin, Y P
Cho, S R
Choi, H S
Choo, Y S
Chow, K W
Christensen, J
Chun, H H
Chung, J S
Chwang, A T
Clément, A
Cole, I S
Contento, G
Cornett, A M
Cox, D T
Croasdale, K R
Crookshank, N E
Dalane, J I
Damy, G
Das, B M
Deguchi, I
Demirbilek, Z
Den Haag, T
Dier, A
Dijkstra, G J
Dijkstra, O D
Dillon, W P
Doi, Y
Dos Santos, J F
Dursthoff, W
Dutta, P K
Duvat, G
Easson, W J
Engebretsen, T
Ertekin, R C
Fairchild, D P
Falzarano, J M
Farzaneh, M
Ferrant, P
Ferrante, A J
Fish, A M
Fithen, R
Fithen, R M
Ford, S F
Forristall, G Z
Francescutto, A
Frederking, R
Fujikubo, M
Fujino, M
Fujisaku, H
Fukumoto, M
Fuwa, T
Fylling, I
Garza-Rios, L
Goda, Y
Goel, A P
Gokce, T
Gokkus, U
Gouleilmos, A M
Grass, A J
Gresnigt, A M
Grilli, S
Grundy, P
Guerin, G
Gunbak, A R
Guo, T
Gutfraind, R
Handa, K
Hardy, C
Haritos, N
Harrison, G E
Haver, S
Herion, S
Hessler, R R
Higo, Y
Hiraishi, T
Hobbs, R E
Holder, G D
Hollebone, B
Hong, S W
Hong, S Y
Honji, H
Hopkins, M
Hosoyamada, T
Hover, F
Huang, T
Huang, T L
Huang, X L
Hyodo, M
Hyun, B S
Ifuku, M
Ikeda, Y
Isaacson, M
Iwata, K
Jacobsen, V
Jakobsen, J B
Jang, Y S
Jiang Q
Jin, D
Jo, C H
Johnson, B H
Jones, S J
Ju, F
Juhl, J
Kagawa, K
Kajiwara, H
Kaneko, K
Kao, C C
Kapdasli, S
Karahan, E
Karr, L A
Kashiwagi, M
Kawamura, Y

Kawano, K
Kershenbaum, N Y
Kielland, P
Kim, C H
Kim, D H
Kim, M H
Kim, S B
Kitamura, M
Kitamura, R
Kite-Powell, H L
Kiyokawa, T
Knapp, R H
Kioka, W
Kobayashi, A
Kobayashi, T
Kokawa, H
Kolle, J J
Koo, J Y
Koterayama, W
Krock, H J
Krogstad, H E
Kubo, M
Kujala, P
Kuriyama, Y
Kuroiwa, T
Kvamme, B
Kvenvolden, K A
Kyozuka, Y
Labeyrie, J
Laflamme, J
Laforte, J L
Lampela, K
Landrini, M
Langen, I
Le Boulluec, M
Le Goff, J P
Lee, S J
Lee, S K
Lee, Y G
Leira, B
Leonard, J
Li, J C
Li, Y C
Lian, D
Liu, Y H
Lin, G
Lin, M C
Lin, S S
Ling, S
Lipp, S
Liu, Y L
Liukkonen, S
Lozowski, E P
Luton, M J
Ma, Z L
Maeda, H
Magda, W
Makkonen, L
Manabe, Y
Mansard, E P D
Maple, J A
Marcus, H S
Martin, A J
Mase, H
Masson, D
Masuda, K
Masuda, Y
Matsui, T
McClimans, T
McClintock, F
McClure, G
McComber, P
Mebarkia, S A
Meon, G
Meylan, M H
Miao, G P
Mimura, N
Mitsuyasu, H
Miura, K
Mizutani, N
Modi, V J
Moe, G
Mogridge, G R
Mohr, W C
Molin, B
Moo-Young, H
Mori, K
Moshagen, H
Muggeridge, D B
Murakawa, H
Nabeshima, Y
Nadaoka, K
Naito, S
Nakai, T
Nakamura, M
Nam, D
Narayanan, R
Natvig, B J
Nerzic, R
Nestegard, A
Niedzwecki, J M
Niho, O
Nomura, S
Nwogu, N
O'Donoghue, T
Ogawa, K
Ogawa, Y
Oh, S K
Ohtaka, K
Olagnon, M
Onitsuka, K
Onoufriou, T
Oppenheim, B
Orisamolu, I R
Oshiwa, K
Otsuka, K
Oumeraci, H
Paik, J K
Pamukcu, S
Papanikolaou, A
Park, I K
Parsey, M R
Paté-Cornell, M E
Patrikalakis, N M
Pedersen, P T
Penna, R
Perkins, N C
Pervizpour, M
Pesce, C P
Powers, E J
Pradhan, B S
Prakash, S
Prasad, S
Prevosto, M
Puri, V K
Puthli, R S
Randall, R E
Raschke, S
Rigaud, V
Riggs, H R
Rijken, O
Riska, K
Roby, K
Rodd, J L
Roesset, J M
Ronold, K
Ryerson, C C
Saeki, H
Sakai, T
Sakakiyama, T
Sakamoto, Y
San, K
Sanni, S
Sarmento, A
Savenije, R Ph A C
Sayed, M
Sayer, P G
Scarsi, G
Schjølberg, I
Schriever, G
Senjanovic, I
Sergy, G
Setoguchi, T
Shen, F
Shen, H T
Shibayama, T
Shigematsu, T
Shin, J G
Shin, M S
Shinoda, T
Shogaki, T
Sibetheros, I
Sigbjörnsson, R
Silver, A L
Skjong, R
Sklenicka, V
Skop, R A
Skourup, J
Snijder, H H
Sodhi, D S
Sørensen, J D
Spencer, P
Squire, V A
Stansberg, C T
Stear, J
Strømsem, K
Su, M Y
Suga, Y
Sumer, B M
Sun, Z C
Suzuki, H
Suzuki, K
Szilder, K
Tabeta, S
Takagi, K
Takahashi, H
Takaki, M
Talley, W K
Tanaka, H
Tanaka, Y
Tanizawa, K
Tedeschi, R
Temperville, A
Teng, B
Thayamballi, A
Thiel, H
Timco, G W
Tomita, H
Tomita, Y
Toyosada, M
Triantafyllou, M S
Tsai, C P
Tsai, T K
Tsuboi, H
Tsunekawa, Y
Tsurusaki, K
Tuhkuri, J
Tulin, M
Uchida, K
Uchida, M
Ueda, Y
Utsunomiya, T
van der Vegte, G J
Vassalos, D
Venkataramana, K
Verley, R
Wadhams, P
Wang, B P
Wang, J
Wang, K H
Wang, Y S
Wang, Y X
Wang, Z Z
Watabe, T
Weave, T J
White, H B
Williams, A N
Williams, F M
Winterstein, S
Wright, B D
Wu, K Q
Xu, T
Yajima, H
Yamaguchi, S
Yamazaki, T
Yao, T
Yashima, A
Yasuzawa, Y
Yeung, R W
Yim, J Z
Yoneyama, H
Yoon, H K
Yoon, J H
Zdravkovich, M M
Zhang, D H
Zhang, H X
Zhang, J
Zhao, Y N
Zhou, L D

ISOPE Publication Order Information

International Journal of Offshore and Polar Engineers - Quarterly (March, June, September, December) (ISSN 1053-5381).

Editors: Jin S. Chung, J. Wardenier, R.M.W. Frederking, W. Koterayama. **Topics**: *offshore engineering, ocean engineering, environment, offshore mechanics, polar (Arctic and Antarctic) engineering, advanced marine technology, materials and technology and energy*: **Yearly subscription rates**: US$120. (Member, complimentary) and US$100 per a set of annual back issues.

Proceedings of the 1st (1990) European Offshore Mechanics Symposium (EUROMS-90), Trondheim, Norway, 20-22 August 1990. A single volume: (ISBN 0-9626104-4-5; LCCN 90-084691); approx. 590 pp., US$80. (Member, US$60.)

Proceedings of the 1st (1990) Pacific/Asia Offshore Mechanics Symposium (PACOMS '90), Seoul, Korea, 24-28 June 1990. For a set of 3 volumes, (ISBN 0-9626104-0- 2; LCCN 90-082039): approx. 1260 pp., US$150 per set and US$70 (Member, US$50) per volume.

Proceedings of Special Offshore Symposium China (SOSC-94) / 3rd (1994) Pacific/Asia Offshore Mechanics Symposium (PACOMS-94), Beijing, China, 16-18 April 1994. A single volume (ISBN 1-880653-15-X); 797 pp., US$100.

Proceedings of 4th (1996) Pacific/Asia Offshore Mechanics Symposium (PACOMS-96), Pusan, Korea, October 31-November 1, 1996. A single volume (ISBN 1-880653-27-3); 325 pp., US$60 (Member, US$50.)

Proceedings of 1st (1995) SOPE Ocean Mining Symposium, Tsukuba, Japan, November 21-22, 1995. A single volume (ISBN 1-880653-21-4, LCCN 95-77708): 238 pp., US$100 (Member, US$80.)

Proceedings of 2nd (1997) ISOPE Ocean Mining Symposium, Seoul, Korea, November 24-25, 1997. A single volume (ISBN 1-880653-33-8): 250 pp. est., US$100 (Member, US$80.)

Proceedings of the 1st (1991) International Offshore and Polar Engineering Conference (ISOPE-91), Edinburgh, U.K., 11-16 August 1991. For a set of 4 volumes, (ISBN 0-9626104-5-3; LCCN 91-071635): approx. 2190 pp., $320 (US$240; member) and $100 ($80; member) for a single volume.

Proceedings of the 2nd (1992) International Offshore and Polar Engineering Conference (ISOPE-92), San Francisco, USA, June 14-19, 1992. For a set of 4 volumes, (ISBN 1-880653-00-1; LCCN 91-78280): 2,899 pp. est., $340 (Member, $260) and $100 ($80; member) for a single volume. See below for postage and handling.

Proceedings of the 3rd (1993) International Offshore and Polar Engineering Conference (ISOPE-93), Singapore, 6-11 June 1993. For a set of 4 volumes, (ISBN 1-880653-05-2; LCCN 92-76219): 2,965 pp., $340 (ISOPE Member, $260) and $100 ($80; member) for a single volume. See below for postage and handling.

Proceedings of the 4th (1994) International Offshore and Polar Engineering Conference (ISOPE-94), Osaka, Japan, April 10-15, 1994. For a set of 4 volumes, (ISBN 1-880653-10-9; LCCN 93-80555): 2,700 pp. est., $340 (ISOPE Member, $260) and $100 ($80; member) for a single volume. See below for postage and handling.

Proceedings of the 5th (1995) International Offshore and Polar Engineering Conference (ISOPE-95), The Hague, June 11-16, 1995. For a set of 4 volumes, (ISBN 1-880653-16-8; LCCN 94-73796): 2,543 pp., $340 (ISOPE Member, $260). 2,589 pp., $340 (ISOPE Member, $260) and $100 ($80; member) for a single volume. See below for postage and handling.

Proceedings of the 6th (1996) International Offshore and Polar Engineering Conference (ISOPE-96), Los Angeles, May 26-31, 1996. For a set of 4 volumes, (ISBN 1-880653-22-2): 2,326 pp., $360 (ISOPE Member, $280). For a single volume, $110 (ISOPE Member, $80). See below for postage and handling.

Proceedings of the 7th (1997) International Offshore and Polar Engineering Conference (ISOPE-97), Honolulu, May 25-30, 1997. For a set of 4 volumes, (ISBN 1-880653-28-1): 3,548 pp., $380 (ISOPE Member, $300). For a single volume, $120 (ISOPE Member, $80). See below for postage and handling.

Postage (surface mail) and handling: add for each of ISOPE-92 through ISOPE-97 Conference Proceedings orders

A set	USA $15	Canada $20	Other countries $35
A volume	USA $8	Canada $10	Other countries $15

How To Order

To order, check or tick ☑ the appropriate boxes, fill in the information below, enclose the correct amount of payment or purchase order, and mail to ISOPE Order Department, P.O. Box 1107, Golden, Colorado 80402-1107 USA

❑ Member rate ISOPE Member No. _______ ❑ Non-member rate

❑ **International Journal of Offshore and Polar Engineering**
❑ **Proceedings of the 1st (1990) Pacific/Asia Offshore Mechanics Symposium**
❑ Vol. I ❑ Vol. II ❑ Vol. III ❑ A Set
❑ **Proceedings of the 3rd (1994) Pacific/Asia Offshore Mechanics Symposium (Special Offshore Symposium China)**
❑ **Proceedings of the 4th (1996) Pacific/Asia Offshore Mechanics Symposium**
❑ **Proceedings of the 1st (1990) European Offshore Mechanics Symposium**
❑ **Proceedings of the 1st (1995) ISOPE Ocean Mining Symposium**
❑ **Proceedings of the 2nd (1997) ISOPE Ocean Mining Symposium**
❑ **Proceedings of the 1st (1991) International Offshore and Polar Engineering Conference**
❑ **Proceedings of the 2nd (1992) International Offshore and Polar Engineering Conference**
❑ **Proceedings of the 3rd (1993) International Offshore and Polar Engineering Conference**
❑ **Proceedings of the 4th (1994) International Offshore and Polar Engineering Conference**
❑ **Proceedings of the 5th (1995) International Offshore and Polar Engineering Conference**
❑ **Proceedings of the 6th (1996) International Offshore and Polar Engineering Conference**
❑ **Proceedings of the 7th (1997) International Offshore and Polar Engineering Conference**

Proceedings of ISOPE-91~97 (*check*):
❑ Vol. I ❑ Vol. II ❑ Vol. III ❑ Vol. IV ❑ A Set

Method of Payment: Advance payment must accompany order (including the postage and handling charge above for 1992-97 Proceedings.) International money order in U.S. dollars, bank checks in U.S. dollars drawn on a U.S. bank, or credit card (Visa or MasterCard only) (*payable to ISOPE*) will be accepted. Remit with the mailing address including the postal (or zip) code for shipping.

❑ international money order in U.S. dollars is enclosed.
❑ bank check in U.S. dollars through a U.S. bank is enclosed.

❑ credit card no. ________________________________ *Circle one*: Visa MasterCard American Express

Exp. Date ________ Cardholder's name ____________________________

Signature ______________________ Date __________________

Ship To Name ______________________________

Company ____ Dept. ___________________________________

Address ______________________________________

City ____________________ State (Province) _______________

Country _______________________ Postal (or Zip) Code _____________

Prospective authors are invited to submit abstract to

ISOPE-98 Montréal, Canada, May 24-29

The 8th (1998) International

Offshore and Polar Engineering Conference

Montréal, Canada, May 24-29, 1998

First

Call For Papers

Abstract Deadline	**August 1, 1997**
Manuscript (for review) Deadline	**November 30, 1997**

The **International Society of Offshore and Polar Engineers (ISOPE)** is organizing about 60 sessions of refereed papers in cooperation with international technical program committee (TPC) members and cooperating organizations. Delegates from more than 35 countries are expected. All manuscripts will be peer-reviewed prior to the final acceptance for the conference proceedings. Papers of archival value may be further reviewed for the *International Journal of Offshore and Polar Engineering*.

Main fields of interest are:

Offshore Technology	**Environment**	**Materials & Welding**
Offshore Mechanics	**Geotechnics**	**Tubular Structures**
Ocean Engineering	**Energy, Resources**	**Risers, Pipelines, Cables**
Polar Engineering	**TLP, SPAR, FPS**	**Comp Methods/CAD**
Deep-Ocean Tech	**Adv Marine Tech**	**Safety, Risk , Reliability**
Underwater Systems	**Hydrodynamics**	**Coastal Engineering**
Measurements	**Underwater Eng**	**Emerging Topics**

The abstract should stress the originality and the significance of the results, and include the complete names, addresses and fax numbers of all authors.

Send 2 copies of your abstract in 300-400 words to: (1) One of the TPC members (session organizers) or (2) **ISOPE-97 TPC**, ISOPE, P.O. Box 1107, Golden, Colorado 80402-1107 USA: FAX 1-303-420-3760.

Sponsors: *ISOPE.* **Cooperating Organizations and Societies**: *ASCE (USA), CAPP (Canada), CSCE (Canada), CSOE (China), CSNAME (China), CSTAM (China), IE Australia, IFREMER (France), IRO (The Netherlands), JSMS (Japan), KCORE (Korea), KSNAJ (Japan), NPF (Norway), OES (UK), PII (Indonesia), RAS (Russia), Scott Polar Res Inst (UK), SSSS (Singapore), and VTT (Finland).*

The Proceedings of
The Seventh (1997) International
OFFSHORE AND POLAR ENGINEERING CONFERENCE

Honolulu, Hawaii, USA, May 25-30, 1997

VOLUME I

OFFSHORE RESOURCE & TECHNOLOGY
SAKHALIN OIL & GAS
RESOURCES & ENERGY (WAVE ENERGY, GAS HYDRATES, DEEP-OCEAN MINING)
TLP, VERY LARGE FLOATING PLATFORMS (VLFS) AND FPSO
OFFSHORE SYSTEMS
ENVIRONMENT (DEEP-SEA IMPACT, WASTE AND TRANSPORT, POLLUTANT, CONTAMINATION, ECOLOGY)
GEOTECHNICAL ENGINEERING

VOLUME II

OFFSHORE AND ARCTIC PIPELINES
MECHANICS OF RISERS, CABLES AND MOORING
UNDERWATER VEHICLES, CONTROL AND SYSTEMS
POLAR AND ICE ENGINEERING
JOIA PROJECTS
ARCTIC AND ANTARCTIC SYSTEMS
ARCTIC OIL POLLUTION
ICE REMOTE SENSING

VOLUME III

NUMERICAL WAVES
NONLINEAR WAVES
WAVE STATISTICS AND MEASUREMENTS
INTERNAL WAVES
HYDRODYNAMIC FORCES
DYNAMIC RESPONSES
AIRGAP AND METOCEAN
VORTEX SHEDDING AND FLOW-INDUCED VIBRATIONS
COASTAL HYDRODYNAMICS

VOLUME IV

TUBULAR STRUCTURES
WELDING AND ROBOTICS
FATIGUE AND FRACTURE
STEEL, CORROSION AND NDT
COMPOSITE MATERIALS
MECHANICS AND STRUCTURES
RELIABILITY, RISK AND SAFETY
COLLISION/IMPACT/DAMAGE
MARINE & UNDERWATER SYSTEMS
EARTHQUAKE ENGINEERING

CONTENT

HYDRODYNAMICS: Forces

COASTAL HYDRODYNAMICS

ADDITIONAL PAPERS

Proceedings of the Seventh (1997) International Offshore and Polar Engineering Conference
Honolulu, USA, May 25-30, 1997

ISBN 1-880653-28-1 (Set); ISBN 1-880653-31-1 (Vol. III)

On the Contribution of Swell to Sea Surface Phenomena

Hisashi Mitsuyasu
Hiroshima Institute of Technology
Hiroshima, Japan

ABSTRACT

Swell can be observed very frequently in the ocean. But little attention has been paid to its contribution to other sea surface phenomena. On the basis of our recent studies, various important phenomena relating to the swell are discussed in this article. Firstly, the change of the swell by turbulent wind is considered. It is shown that the magnitude of the decay rate of the swell by adverse wind is almost the same with that of the growth rate of the swell by following wind. Secondly, the effects of the swell on various sea surface phenomena are discussed. As will be expected, the effects of the swell are negligible unless the swell steepness is large. However, with the increase of the swell steepness, the swell shows interesting effect on sea surface phenomena. The swell propagating against the wind intensifies the growth of wind waves, if the swell steepness is large. This is quite contradictory to the well-known phenomena that the steep swell propagating in the direction of the wind suppresses wind waves. The former phenomenon cannot be explained by the mechanism proposed by Phillips & Banner(1974); it explains only the latter case, even though the mechanism should be applicable to the both phenomena. The similar contradictory effects of the swell are seen in the effects of the swell upon wind-induced drift current. The swell propagating against the wind intensifies the drift current, while the swell propagating in the direction of the wind shows little effect upon the drift current. In spite of the drastic changes of wind waves by the swell, the microwave intensity backscattered from the sea surface is not much affected by the swell. This is probably due to the fact that the swell changes only a dominant part of the wind wave spectrum and it does not affect the high frequency region of the spectrum, which mainly governs the Bragg scattering mechanism of the microwaves at wavy surface.

KEY WORDS: Ocean waves, wind waves, swell, coexistence of wind waves and swell

INTRODUCTION

It is an ubiquitous phenomenon in the ocean that wind waves in some particular wind area coexist with swell. Figure 1 is an example of the wave spectra observed in the North Atlantic Ocean, which shows the transition of the spectrum from coexistent system of the swell (low frequency peak) and wind waves (high frequency peak) in an initial stage to wind waves dominating over the swell in a later stage. Such swell is generated in other wind areas and not necessarily propagating in the direction of the local wind (Fig.2). However, there are few systematic studies on the interactions among wind, wind waves and swell, except for several pioneering studies (Mitsuyasu 1966, Phillips & Banner 1974, Hatori et al. 1981, Donelan 1987, Masson 1993). Reflecting such situations, many numerical wave models do not consider the interactions between the swell and wind waves.

Recently we made a series of laboratory experiments which are concerned, in a wide sense, with the interaction between wind waves and swell (Mitsuyasu & Honda1982, Mitsuyasu & Yosida 1989, 1991, Cheng & Mitsuyasu 1992, Mitsuyasu & Kusaba 1992). The results of these studies are combined in the present paper to show the important contribution of the swell to the sea surface phenomena.

EXPERIMENTAL DATA

Wind and wave data from four sources are used in the present study; (1) The first data are from Mitsuyasu & Honda(1982) which studied the wind-induced growth of mechanically generated waves in a laboratory tank, (2) the second data are from Mitsuyasu & Yoshida(1989) which studied the attenuation of the swell by adverse wind and the change of wind waves by opposing swell, (3) the third data are from Cheng & Mitsuyasu (1992) which studied the surface drift current induced by the wind and swell and (4) the fourth data are from Mitsuyasu & Kusaba (1992) which studied the effect of swell on the microwave back scattering from wind wave surface.

The first three studies, (1), (2) and (3), were done in the same wind-wave flume 0.8m high, 0.6m wide, and with a commonly used test section 15m long, though the experimental arrangements are different one another. The water depth was kept at 0.335m for the first one and 0.353m for the second and third ones. The detailed experimental arrangements and procedures for these three studies are referred to the individual papers. A large wind-wave flume 2m high, 1.5m wide and 54m long (Kusaba & Masuda 1988) was used in the fourth experiment. In the fourth experiment microwave backscattering from the wavy water surface was measured at the fetch 20m in the flume. Wind profile over water surface and waves were also measured at the same fetch. The measurements were done in the following surface conditions of the water; Swell without wind, swell under the action of wind, and pure wind waves. Fundamental parameters in the experiment are as follows; Swell period T: 1.0, 1.5 (sec), initial swell steepness H_0/L_0: 0.04, wind speed Ur: 3.0, 6.0, 10.0 (m/s), microwave scatterometer used in the experiment: 9.6 GHz and HH polarization, and the range of incident angles of the microwave θ : $-35^{\circ} \sim +35^{\circ}$.

THE CHANGES OF SWELL BY THE WIND

When the swell comes into a wind area, wind-induced growth or decay of the swell appears depending on whether or not the wind direction is the same as the propagation direction of the swell. In comparison with many studies on the growth of water waves (swell) by the following wind, studies are relatively few on the attenuation of the water waves (swell) by the adverse wind.

Figure 3 shows a comparison of the normalized growth rate of the swell β/f for the following wind (the broken line) and the normalized decay rate of the swell β/f for the adverse wind (the solid line and the experimental data), which is reproduced from Mitsuyasu & Yoshida (1989). The ordinate U_*/C is a normalized friction velocity of the wind.

The empirical relation for the normalized growth rate (the broken line) is given by

$$\beta/f = 0.34(U_*/C)^2 , \qquad (1)$$

and that for the normalized decay rate (the solid line) is given by

$$\beta/f = 0.52(U_*/C)^{2.37} , \qquad (2)$$

where f and C are frequency and phase velocity of the swell respectively. Although the expressions, (1) and (2) do not appear so similar, these relations give quite similar values of β/f in the range, $0.1< U_*/C <1.0$. That is, the magnitude of the decay rate of the swell for the adverse wind is almost the same as that of the growth rate of the swell for the following wind in this range of U_*/C .

Although the experimental data agree fairly well, in the above range of U_*/C, with the predictions of other formula of the decay rate (eg. Al-Zanaidi & Hui 1984, Tsuruya 1988), the trend in the change of our measured β/f is slightly different from those given by these formula.

Furthermore, in our experiment, the wind-induced growth of the swell was measured under the coexistence of wind waves (ordinary case), and under the suppression of wind waves by adding a surfactant to the water. It is interesting that almost the same relation (1) is satisfied for the both cases, if the measured friction velocity of the wind is used as a wind parameter. It should be mentioned, however, that the result does not mean the wind-induced growth of the swell being insensitive to the wind waves. The growth of the swell itself is much intensified by the coexisting wind waves, but the friction velocity (momentum flux) also increases by the coexistence of wind waves. As the results, almost the same relation (1) holds for the both cases. According to the study of Masson (1993), nonlinear energy transfer between swell and wind waves is negligible unless the frequencies of the swell and wind waves are very close. Since the above experiment satisfies this condition, we can consider that the increase of the growth rate of the swell by the coexistence of wind waves can be attributed largely to the increase of the direct wind input to the swell, and the increase of the wind input to the swell is caused by the change of the surface roughness of the swell, which is caused by the superposition of wind waves.

THE EFFECT OF SWELL ON THE GROWTH OF WIND WAVES

As already studied by many authors, e.g., Mitsuyasu (1966), Phillips & Banner(1974) and Donelan (1987), the swell propagating in the direction of the wind suppresses the wind waves when the swell steepness increases. The phenomena can be explained fairly well by the mechanism proposed by Phillips & Banner(1974).

According to Mitsuyasu & Yoshida(1989,1991), however, the swell propagating against the wind waves shows a quite different and interesting phenomenon that the swell amplifies the wind waves with the increase of the swell steepness (Fig.4). The latter phenomenon is quite contradictory to the prediction of the Phillips & Banner's mechanism, because the mechanism gives an attenuation of wind waves even when the swell is propagating against the wind. The amplification of the wind waves by opposed swell is very important for practical applications such as numerical wave predictions.

Although detailed discussion of this problem are left for future studies, a simple and possible explanation is given as follows; At the crest of the opposing swell, where the wind stress is supposed to be large, propagation speed of the wind waves decreases due to the effect of orbital motion of the opposing swell. Therefore, the rate of the momentum transfer from the wind to the wind waves increases at the swell crest. On the contrary at the trough of the swell, where the wind stress is supposed to be small, propagation speed of the wind waves increases. Therefore the rate of the momentum transfer from the wind to the wind waves decreases at the swell trough. Now if we consider the difference in the time of exposure to the wind at the crest and that at the trough, it will be natural to consider that total momentum transferred from the wind to the wind waves increases by the effects of the opposing swell.

However, further studies are needed to clarify not only the new mechanism for the change of wind waves by the opposing swell but also the effect of the angle between the propagation direction of the swell and that of wind waves.

THE EFFECT OF SWELL ON THE SURFACE DRIFT CURRENT

Sea surface wind generates surface drift current simultaneously with wind waves. Cheng & Mitsuyasu(1992) made a comprehensive laboratory experiment to clarify the effect of swell on the surface drift current generated by the wind. In their study, the surface drift current was measured under the following three different conditions;

(1)without swell, (2) with swell propagating in the direction of the wind, and (3)with swell propagating against the wind.

The drift current generated by the wind without swell showed the similar property with that reported previously. The surface drift velocity U_S is proportional to the friction velocity of the wind U_*, and their relation is given by

$$U_S/U_* = 0.52, \qquad (3)$$

which is quite similar to the following relation obtained by Wu (1973),

$$U_S/U_* = 0.55. \qquad (4)$$

When the swell is propagating in the direction of the wind, the surface drift current is not much affected by the swell ((a) in Fig.5). On the other hand, when the swell is propagating against the wind, the surface drift current is intensified considerably with the increase of the swell steepness ((b) in Fig.5). The increase of the measured surface drift current is much larger than the theoretical prediction of the Phillips & Banner's mechanism(Cheng & Mitsuyasu 1992). This is very interesting but confusing result. Because one of the key points of the Phillips and Banner's mechanism is that wind waves attenuate by wave breaking owing to the enhanced surface drift current by the swell. In spite of the large increase of the surface drift current by the opposing swell as shown in this section, however, wind waves are amplified as discussed in the section 4. The result is quite contradictory to the Phillips & Banner's mechanism. Further studies are needed to clarify the problem.

THE EFFECT OF SWELL ON THE MICROWAVE BACKSCATTERING FROM WIND WAVE SURFACE

Since the successful experiment of the SEASAT-A in 1978, the microwave scatterometer has become a powerful tool for observing sea surface wind in a global scale. The principle of the measurement of sea surface wind is based on the intimate relation between backscattered intensity of microwaves and high frequency components of wind waves (Bragg waves), which strongly depend on the wind speed (Stewart 1985).

As shown in section 4, wind waves are affected by the swell when the swell steepness is large. From this fact we will expect some effects of the swell on the microwave backscattering from wind wave surface. In order to clarify the problem, Mitsuyasu & Kusaba(1992) made a laboratory experiment. The experimental conditions are briefly shown in the section 2. Figure 6 is a typical example of the result for the wind speed U=10m/s, in which the intensity of microwaves backscattered from wind waves coexisting with swell is compared with that from pure wind waves. Two swells with periods T= 1.0 sec and T= 1.4 sec but with the same steepness H_0/L_0 = 0.04 are used in the measurements. The results are shown as a function of incident angle θ . It can be seen that the effect of the swell is negligible for the microwave backscattering from the wind wave surface.

The result can be well understood if we look at the wave spectra shown in Fig.7, in which the spectrum of wind waves coexisting with the swell propagating in the direction of wind waves, is compared with that of pure wind waves. As shown in Fig.7, the high frequency region of the wind wave spectrum, which is very important for the Bragg scattering of microwaves, is not much affected by the coexisting swell, though the dominant region of the wind wave spectrum is much affected by the swell.

CONCLUDING REMARKS

The magnitude of the decay rate of the waves by adverse wind is almost the same to that of the growth rate of the waves by the following wind. The result can be applicable not only to the decay of swell but also to the decay of wind waves by the adverse wind.

The contribution of the swell to the sea surface phenomena is negligible, if the swell steepness is small. However with the increase of the swell steepness, the swell shows various interesting effects on the sea surface phenomena depending on the propagation direction of the swell relative to the wind direction. The swell propagating in the direction of the wind suppresses the wind waves as usually observed, while the swell propagating against the wind amplifies the wind waves. That is, the effect of the swell on the wind waves is quite different depending on the propagation direction of the swell relative to the direction of the wind waves. The amplification of wind waves by the swell propagating against the wind will contribute to generate abnormally huge waves in the ocean, though the mechanism is still not clear.

The wind-induced surface drift current is also affected by the swell if the swell steepness is large. The effect depends on the direction of the swell relative to the wind. A large increase of the surface drift current is caused when the steep swell is propagating against the wind, while the change of the surface drift current is negligible when the swell is propagating in the direction of the wind. Although the contribution of this phenomenon to the practical problems are not clear now, the result is very interesting dynamically. Because this result, in association with the amplification of wind waves by opposing swell, gives rise serious questions on our previous explanation of the phenomena, and they require further studies to obtain new mechanisms.

The backscattering of microwaves from the wind wave surface is insensitive to the coexisting swell. This is due to the fact that the high frequency region of the wind wave spectrum is not much affected by the swell. In practical applications, the result is fortunate for us. Because it requires no modification of the algorithm of the scatterometer presently used to measure the sea surface wind.

REFERENCES

Al'Zanaidi, MA and Hui, HW (1984). "Turbulent air flow over water waves -A numerical study", J. Fluid Mech, Vol 48, pp 225-246

Cheng, Z and Mitsuyasu, H (1992). "Laboratory studies of surface drift current induced by wind and swell", J. Fluid Mech, Vol 243, pp 247-259

Donelan, M (1987). "The effect of swell on the growth of wind waves", Johns Hopkins APL Tech. Dig., 8(1), pp 18-23.

Hatori, M, Tokuda, M and Toba,Y (1981). "Experimental study on strong interaction between regular waves and wind waves - I ", J. Oceanogr. Soc. Japan, Vol 37, pp 111-119.

Masson, D (1993). "On the nonlinear coupling between swell and wind waves", J. Phys. Oceanogr. Vol 23, pp 1249-1258.

Mitsuyasu, H and Honda, T (1982). " Wind-induced growth of water waves", J. Fluid Mech, Vol 123, pp 425-442.

Mitsuyasu, H and Yoshida, Y (1989). "Air-sea interactions under the existence of swell propagating against the wind", Bulletin of the Res. Inst. Appl. Mech., Kyushu Univ. (in Japanese), No.68, pp 47-71.

Mitsuyasu, H and Yoshida, Y (1991), "The effect of swell on the growth of wind waves", In Oceanography of Marginal Seas (ed. K. Takano), Elsevier, pp 381-391

Mitsuyasu, H and Kusaba, T (1992. "Microwave backscattering from wind waves coexisting with swell", Proc. Annual Meeting of the Oceanographic Society of Japan, (in Japanese), pp224-225.

Phillips, OM and Banner, M L (1974). " Wave breaking in the presence of wind drift and swell", J. Fluid Mech, Vol 66, pp625-640.

Stewart, RH (1985). "Method of Satellite Oceanography", Univ. Calif. Press, pp360

Tsuruya, H(1988). "Experimental study on the wave decay in an opposing wind", Coastal Engineering in Japan Vol 30, pp25-43.

Wu, J (1973). "Prediction of near-surface drift currents from wind velocity", J. Hydraulic Div. ASCE , Vol 99, pp1291-1302.

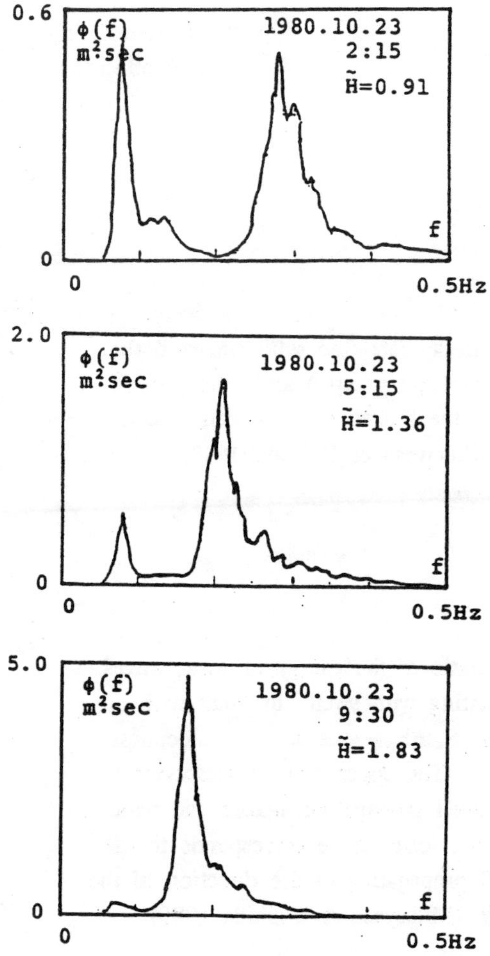

Fig.1 Growth of the spectrum of ocean waves coexisting with swell. The second spectrum was measured three hours later than the first one and the third one was measured seven hours later than the first one.

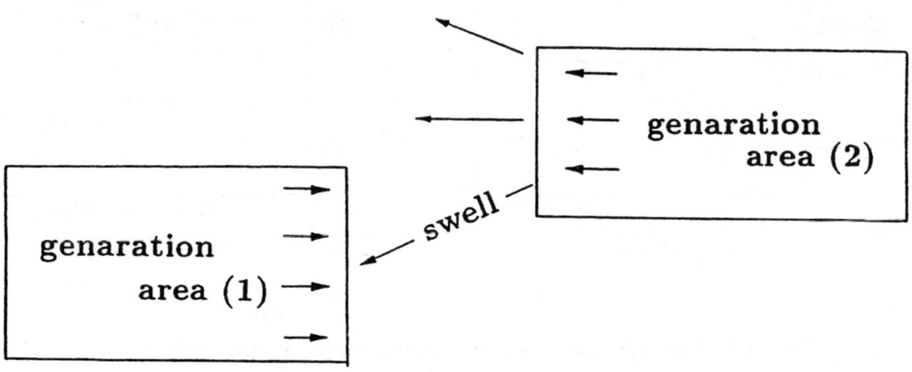

Fig.2 A schematic figure showing a coexistence of wind waves and swell.

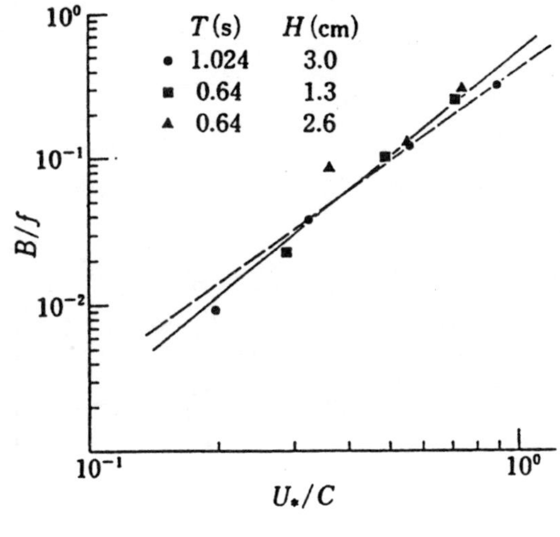

Fig.3 The decay rate (the solid line and the experimental points) and the growth rate (the broken line) versus U_*/C. (Mitsuyasu & Yoshida 1989)

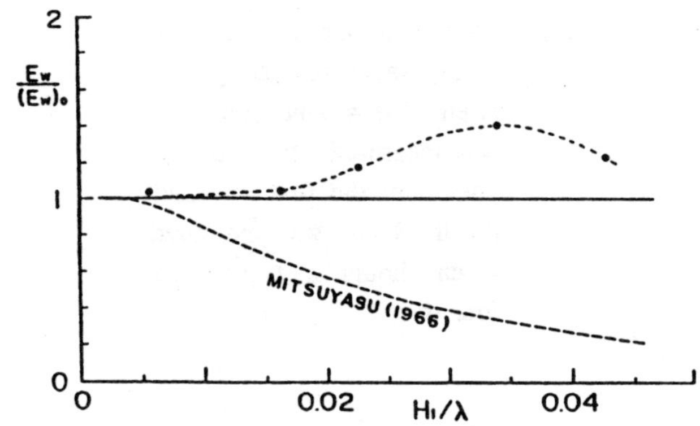

Fig.4 The ratio of the energy of wind waves coexisting with swell to that without swell, $E_w/(E_w)_0$, versus swell steepness, H_1/λ. The upper curve corresponds to the swell propagating against the wind and the lower curve corresponds to the swell propagating in the direction of the wind. (Mitsuyasu & Yoshida 1989)

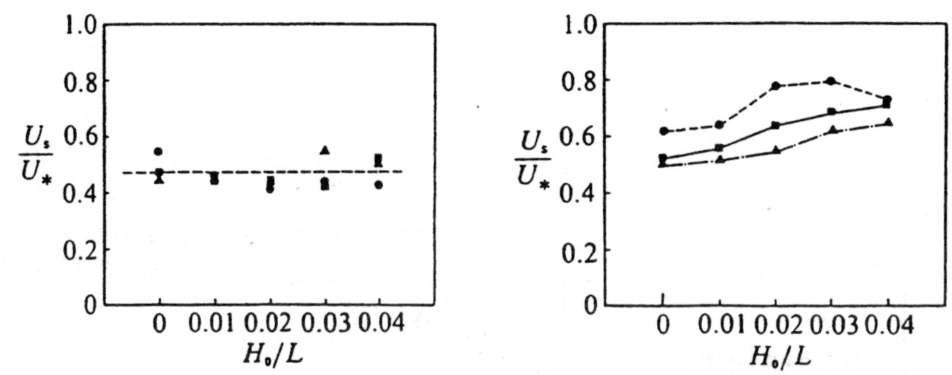

(a) (b)

Fig.5 The dimensionless surface drift velocity U_s/U_* versus swell steepness H_0/L.
(a) The swell is propagating in the direction of the wind.
(b) The swell is propagating against the wind.
Wind speed U_r: ● 3 m/s, ■ 5 m/s, ▲ 7 m/s
(Cheng & Mitsuyasu 1992)

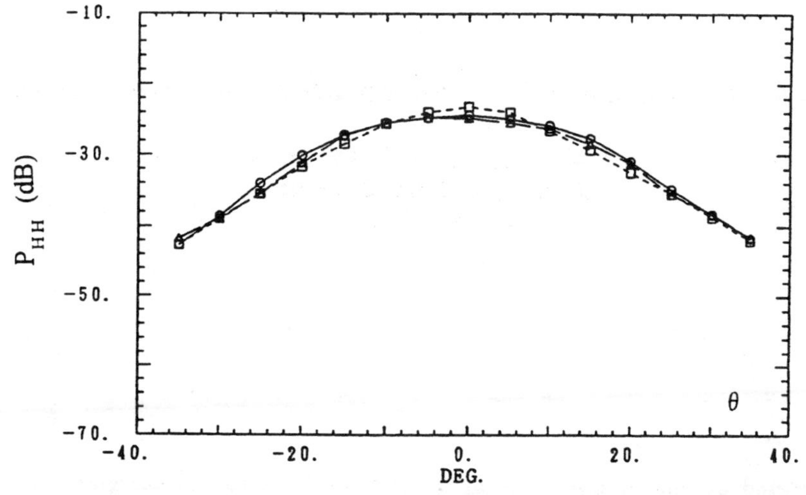

Fig.6 Microwave backscattering from pure wind waves and wind waves coexisting with swell
○: pure wind waves, △, □: wind wave coexisting with swell
(steepness: H_0/L_0=0.04, periods: △ T=1 sec, □ T=1.5 sec)

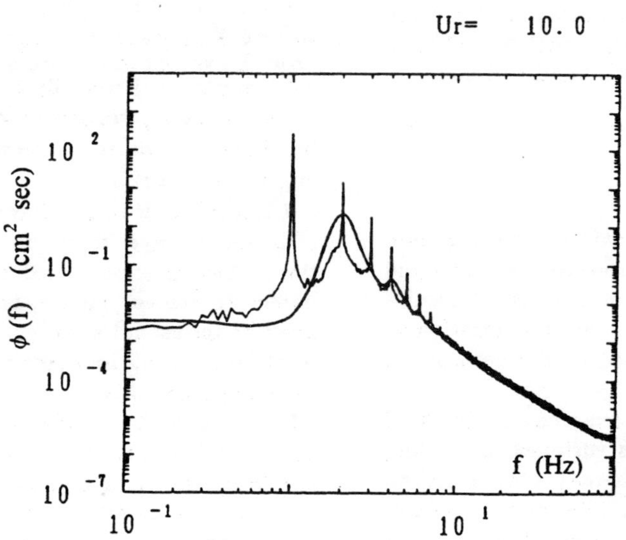

Fig.7 Frequency spectrum of pure wind waves (smooth curve) and
that of wind waves coexisting with swell (spiky curve).
wind speed: U=10 m/s, swell: H_0/L_0=0.04, T=1 sec

Proceedings of the Seventh (1997) International Offshore and Polar Engineering Conference
Honolulu, USA, May 25-30, 1997

ISBN 1-880653-28-1 (Set); ISBN 1-880653-31-1 (Vol. III)

On the Dynamics and Control of Fluid-Structure Interaction Instabilities

V.J. Modi, M.L. Seto and S.R. Munshi
University of British Columbia
Vancouver, Canada

ABSTRACT

Vibration problems encountered by bluff bodies in ocean environment present considerable similarity to vortex resonance and galloping type of instabilities, extensively studied by industrial aerodynamicists. The paper briefly reviews control of this class of vibration problems using both passive as well as active procedures involving nutation dampers and the Moving Surface Boundary-layer Control (MSBC), respectively. It is suggested that both the procedures present considerable promise in resolving vibration problems of bluff offshore structures caused by ocean waves and currents.

KEY WORDS: Vibration Suppression, Nutation Dampers, MSBC.

INTRODUCTION

Vibration of mechanical systems is a universal phenomenon extending to outer space, continuum atmostphere, earth and ocean based structures. Suppression of mechanical oscillations, forced or self-excited, or at least their minimization to an acceptable level, is a challenge faced by engineers in all disciplines.

Perhaps the earliest vibration problems caused by fluid flow, to receive significant as well as sustained attention, pertain to flutter and divergence of aerodynamic surfaces. Advent of the space age brought to light vibration problems of highly flexible large space structures [Modi, 1996], such as the proposed International Space Station Alpha, operating in the free molecular and microgravity environment. However, here the attention is focused on the earth and ocean based structures, usually bluff in geometry, subjected to wind, waves and currents.

The spectacular collapse of the Tacoma Narrows Bridge in 1940 provided considerable impetus to the understanding of wind effects on civil engineering structures. Ever since, a vast body of literature aimed at fluid dynamics and dynamics of bluff bodies has evolved which has been reviewed quite effectively by Cermak (1975), Welt (1988), Modi et al. (1995), Seto (1996), Munshi (1996) and others. The fluid-structure interaction induced instabilities of concern have been the vortex resonance and self-excited oscillations of the galloping type. Primarily the interest has been on vibrations of smokestacks, transmission lines, suspension bridges, tall buildings, submarine detection systems, offshore platforms, etc. (Modi, 1982).

It is important to emphasize that the basic problem is that of boundary-layer formation on a bluff body and its separation, which is experienced by both ground as well as ocean based structures. Hence, the vast body of literature developed by industrial aerodynamicist can be used to advantage in understanding and resolving vibration problems of structures exposed to the ocean environment.

Over the years, a wide variety of devices have been developed which aim to control this form of instabilities. Broadly speaking, they can be classified as: (a) passive; and (b) active; procedures (Figure 1). The passive techniques may be further divided into two categories: (a_1) spoilers, which tend to change fluid dynamic characteristics of structures in such a way as to interfere with and weaken the excitation force; (a_2) dampers, which provide a mechanism for dissipation of energy or its transfer from one part of the system to the

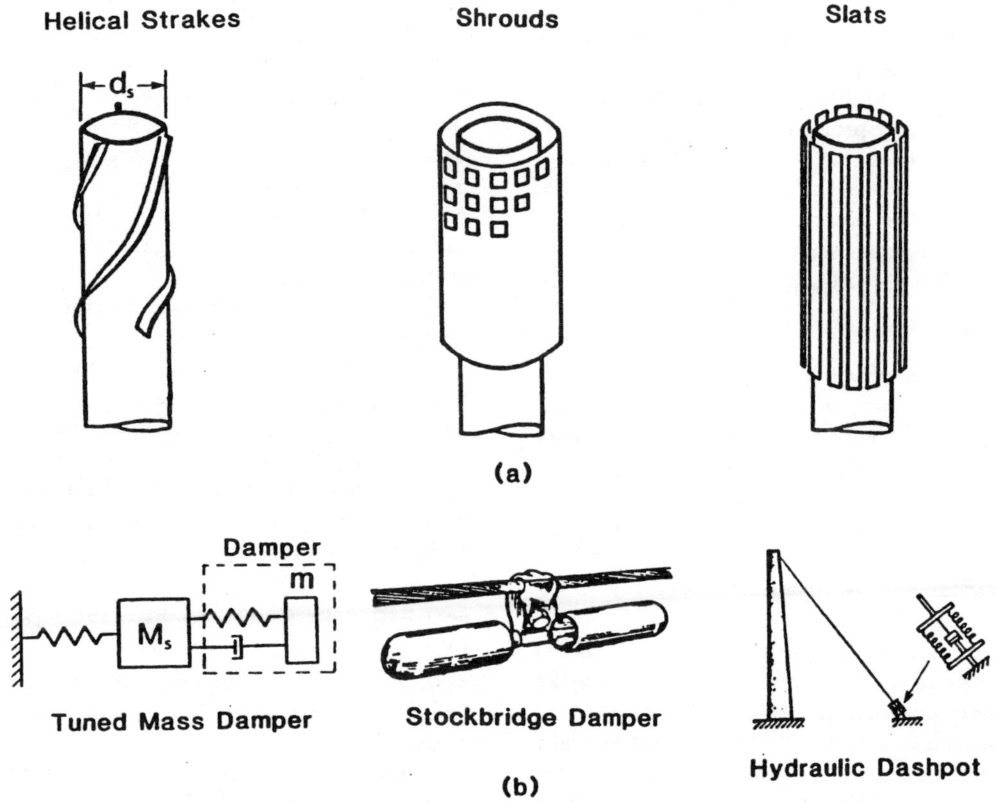

Fig. 1 Typical conventional passive devices used to: (a) modify character of the flow; (b) promote energy dissipation.

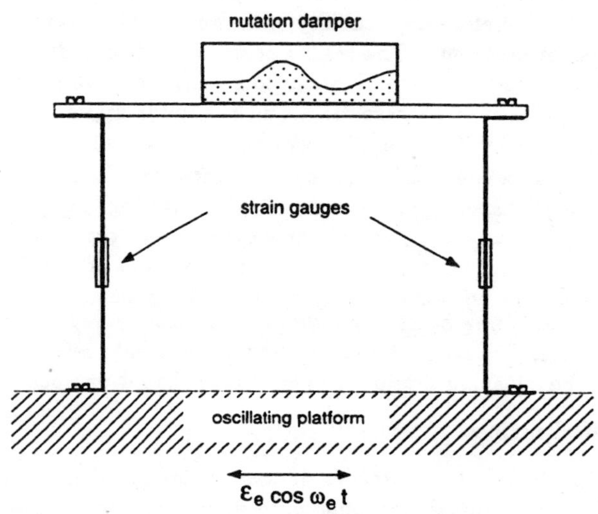

Fig. 3 A force balance used to measure the reduced damping of nutation dampers under steady state excitation.

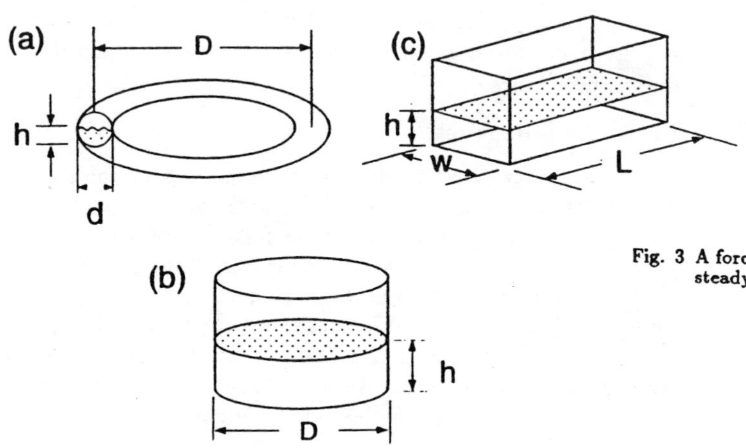

Fig. 2 Several configurations of sloshing liquid nutation dampers: (a) toroidal; (b) circular cylindrical; (c) rectangular.

other. Zdravkovich (1981, 1984) as well as Every et al. (1982) have presented indepth reviews of contribution in these categories. Among the passive devices, dissipation of energy through sloshing liquid, referred to as the nutation damper, appears particularly promising. On the other hand, active procedures normally aim at regulating the boundary-layer separation, as in the case of the Moving Surface Boundary-layer Control (MSBC). This paper briefly reviews control of flow-induced vibrations of bluff structures using passive nutation dampers and active MSBC.

NUTATION DAMPER

This passive mechanism of energy dissipation, through a liquid undergoing sloshing motion at the characteristic resonance frequency, can have a wide variety of geometries, from toroidal ring to rectangular or circular cross-section cylinders (Figure 2). Furthermore, it may carry internal devices in the form of baffles, screens, moving spheres, particle suspensions, etc. interacting with the sloshing motion. Such dampers are ideally suited for suppressing relatively low frequency vibration. They have been in use for quite some time in controlling the librational motion of satellites (Anzai et al., 1981) where the period may range from around 90 minutes to 24 hours! The space environment dictates nutation dampers to be completely filled (no free surface). The dissipative wake of a sphere provides a mechanism for damping of the nutational motion. As the frequency of oscillations encountered in the wind-induced instability of bluff bodies, earthquake response of buildings, and wave-excited vibrations of offshore structures is relatively small, the authors thought it appropriate to explore applicability of the nutation dampers to this class of problems. The results show the nutation damper to be particularly attractive due to its efficiency, simplicity in design, ease of construction, and negligible maintenance as discussed by Tamura, Kousaka and Modi (1992).

A word about the designation 'nutation damper' would be appropriate. The classical term 'nutation' represents a component of the general three-dimensional rotation in the Eulerian description. Pitch and roll librations of satellites as well as the fundamental oscillations of cantilever structures such as smokestacks, buildings, etc. closely represent the Eulerian nutation. Lately, several authors have referred to it as the 'sloshing liquid damper', 'tuned liquid damper', etc.

Dissipation

Efficiency of a nutation damper is measured in terms of the conventional damping ratio η or the reduced damping parameter $\eta_{r,l}$ defined as

$$\eta_{r,l} = \frac{F_s}{2M_l\epsilon_e{\omega_e}^2}\sin\phi .$$

Here F_s = sloshing force; M_l= liquid mass; ϵ_e, ω_e = excitation amplitude and frequency, respectively; and ϕ = phase between the sloshing force and displacement. The reduced damping parameter can be measured quite readily using an oscillating platform, driven by a Scotch-Yoke mechanism, with a strain gage balance (Figure 3). Both η and $\eta_{r,l}$ can be expressed in terms of energy dissipated per unit mass of liquid per oscillation cycle as shown by Welt (1983, 1988). The energy dissipation depends on a large number of variables associated with the damper geometry, liquid parameters and excitation characteristics. Figure 4 shows variation of η with the liquid height h and the system frequency f for a toroidal damper with a circular cross-section (diameter=d). D is the mean diameter of the toroid and a is the initial displacement. The presence of an optimum liquid height is apparent. Note, the damper performance improves significantly at a lower system frequency.

The next logical step was to explore the possibility of improving the dissipation rate further. To that end, introduction of floating particles appeared attractive. The premise is that they may promote dissipation of energy through an increase in exposed area, higher particle drag and inelastic collisions among themselves as well as with the wall. In general, for a given aspect ratio of particles, there was an optimum value of the seeding density (ρ_p) leading to the peak reduced damping. Figure 5 shows the effect of seeding density on the performance of a circular cylindrical damper at an excitation frequency coinciding with the liquid natural frequency ($\hat{\omega}_e = 1$). For the particle aspect ratio of 1, the peak $\eta_{r,l}$ of ≈ 2.2 represents an increase of around 30%. Note, a broad peak suggests relative insensitivity to the particle seeding density.

A comment concerning application of the results to design of a prototype damper would be appropriate. Design of the structure, where the damper is to be installed, is of course available; so is the wind profile from meteorological data. Furthermore, building code provides information as to the peak amplitude (stress) that can be acceptable. Thus the amount of energy to be dissipated per cycle of oscillation to limit the amplitude can be calculated. For example, during vortex excitation, the resonance condition would be critical. One can now select the liquid height and geometry (say diameter of the circular damper) so that the peak sloshing condition is estabilished at the structural resonance. The reduced damping data can now be used to calculate the energy dissipated per cycle. This would then establish the total number of dampers needed to meet the building code requirement. Modi et al. (1988) have explained in detail the design procedure.

Flow Visualization

In order to get better physical appreciation of the energy dissipation through sloshing liquid, a visual study of the liquid modes was carried out. The forced vibration facility shown in Figure 3 was used to that end. The fundamental as well as subharmonics and higher harmonics were recorded on a S-VHS video. The video captured the evolution of travelling waves, their collision among themselves and with the wall, disintegration, dissolution and regeneration. It helped understand the conditions leading to higher dissipation.

Wind Tunnel Tests

The effectiveness of the dampers in controlling vortex res-

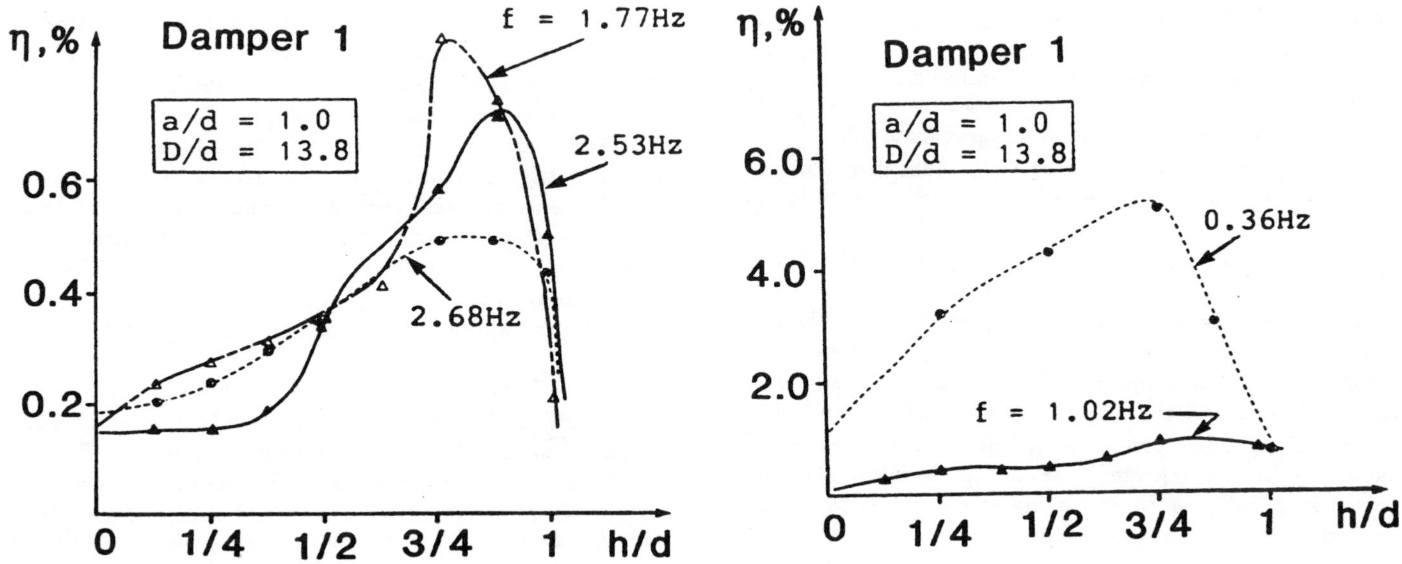

Fig. 4 Effect of liquid height and system frequency on the performance of a toroidal damper.

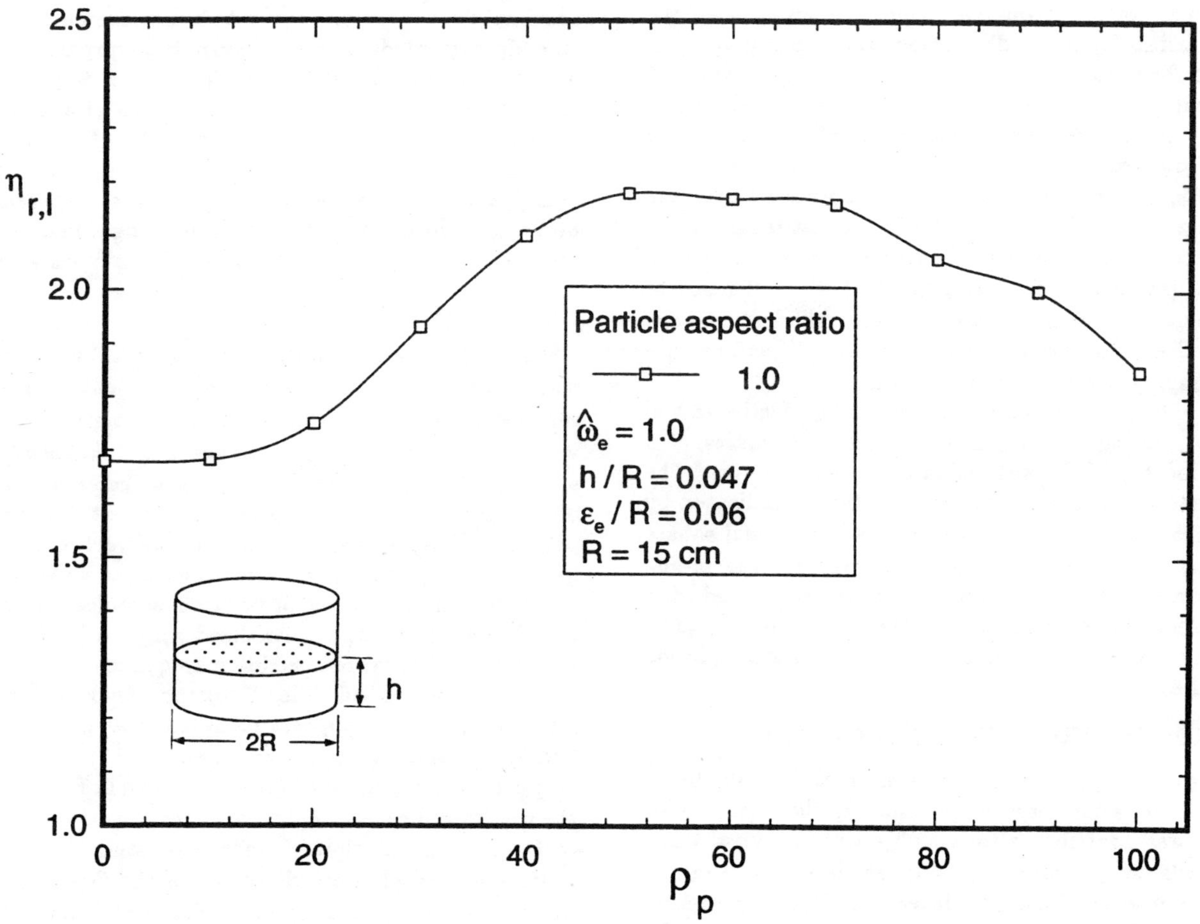

Fig. 5 A representative plot showing the effect of particle concentration on reduced damping for a circular damper. Note, the liquid height is held fixed at $h/R = 0.047$.

onance and galloping instabilities for a family of two and three-dimensional models was studied in both laminar and turbulent flows. A closed circuit laminar flow wind tunnel with a test-section of 0.69 × 0.91 × 2.44 m is able to produce a stable flow with velocity ranging from 0.3 - 30 m/s and a turbulence level of less than 0.1%. Turbulent flow tests were conducted in a boundary-layer tunnel (1.58 × 2.44 × 24.4 m) with 20.74 m of roughness board installed in the test-section upstream of the model to produce the desired boundary-layer thickness and turbulence intensity

The dampers were mounted on a variety of bluff body aerodynamic models during the tests. The models had either circular or square cross-section. The two-dimensional models essentially spanned the height of the tunnel (0.69 m) while the models of finite span simulated three-dimensional conditions. As can be expected, the amount of information obtained is literally enormous. Only a few typical results are presented here to indicate trends.

Figure 6 shows a typical set of results for a toroidal nutation damper mounted on a circular cylinder model undergoing vortex resonance. Dramatic reduction in plunging response was observed even with a small amount of water. The vibratory response reduced further by as much as 30% with an optimal value of the particle density.

Effectiveness of a rectangular damper in suppressing the galloping instability is vividly demonstrated in Figure 7. In absence of the nutation damper, the square prism exhibited large amplitude oscillations, almost approaching the physical limit of the test-apparatus, at a reduced velocity of $U \approx 16$. However, with the nutation damper having a mass ratio of 3%, the galloping is virtually suppressed over the entire range of the wind speed used in the test-program.

Figure 8 compares relative performance of the three damper geometries (square, toroidal and circular) with reference to the galloping of a two-dimensional square cylinder. The circular cylindrical damper shows the best overall performance with a damping of over 90% in amplitude.

Experiments were also conducted to assess effectiveness of the damper in suppressing fluid induced instabilities of a structure located in the wake of an upstream structure. The problem has received considerable attention as indicated by Zdravkovich (1981, 1984). A pair of two-dimensional square prisms and circular cylinders were selected for the in-line wake response study. Even with the critical spacing and wind speed leading to the peak response, the damper proved to be remarkably successful in reducing the vibration amplitude by around 85%.

VIBRATION SUPPRESSION THROUGH MSBC

An alternate approach to vibration control would be to modify the excitation forces responsible for these vibrations. A moving wall contributes to this objective in several ways: (a) it retards the growth of the boundary-layer; (b) it injects momentum into the boundary-layer and delays its separation. Using rotating cylinders as momentum injecting elements, Modi et al. (1991, 1992) have shown effectiveness of the MSBC for lift augmentation and stall delay of airfoils, as well as drag reduction of bluff bodies. More recently, Munshi (1996) assessed the effectiveness of the MSBC concept in controlling vortex resonance and galloping type of instabilities. A comprehensive study based on static and dynamics experiments in the wind tunnel, complemented by water channel flow visualization, were carried out on a family of bluff bodies (flat plate, D-section, rectangular and square prisms). Here, only a small sample of the results, relevant to the discussion, is presented. It conclusively shows that the MSBC is quite successful in suppressing the flow induced instabilities.

Test Procedures

The bluff body models, fitted with end plates and rotating cylinders, used in the present study are shown in Figure 9. The static tests were carried out in a closed circuit low speed, low turbulence wind tunnel mentioned earlier. Pressure distribution around the body was measured using a Scanivalve transducer (sampling rate, 100Hz; sampling period, 5s). The aerodynamic forces were obtained by integrating the mean pressure distribution around the body. Vortex shedding frequency was obtained through the spectral analysis of the pressure versus time signal. For dynamic experiments, a test-rig was specially designed, which held a set of four air bearings for the model support, a displacement transducer, an eddy current damper, and springs to impart desired stiffness to the model. A light weight balsa wood model, fitted with rotating cylinders, was free to vibrate transverse to the wind tunnel flow direction. Also, flow visualization tests were carried out using a closed circuit water channel facilty and Plexiglas models, employing a suspension of fine polyvinyl chloride powder and slit lighting, to observe streaklines. Both still photographs and video movies were taken.

Results and Discussion

Figure 10 shows the variation of the Strouhal number (St) with the angle of attack (α) and the momentum injection parameter U_c/U (U_c = cylinder surface velocity, U=freestream velocity). Classical results of Fage and Johansen (1927) are also included for comparison. A remarkable increase in the Strouhal number with cylinder rotation at a given α is apparent. It suggests an increase in the vortex shedding frequency coressponding to narrowing of the wake. Obviously, this will delay the onset of vortex resonance. In the case of the D-section, a large normalized amplitude of ≈ 5, corresponding to the vortex shedding frequency, was observed in absence of the MSBC. This frequency peak becomes gradually smaller and disappears completely as momentum injection is increased to $U_c/U = 2$ (Figure 11). The plot of St versus α for a rectangular prism is shown in Figure 12. Again the results indicate complete suppression of vortex shedding at $U_c/U = 2$. A flow visualization study, reported later, confirmed this physical character of the flow. For small α, the criterion for galloping instability is $dC_y/d\alpha > 0$ (Parkinson and Brooks, 1961). Here C_y is the lateral force coefficient. The results of the C_y variation with α for the square

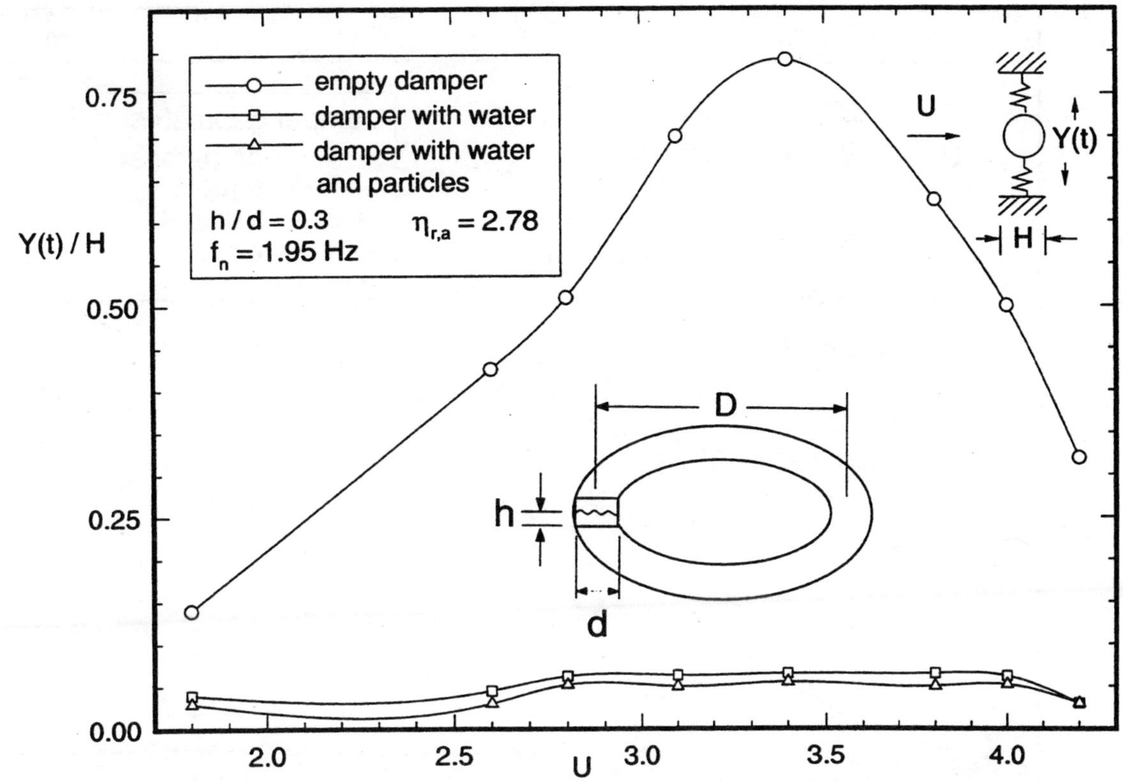

Fig. 6 Wind tunnel test results for a two-dimensional circular section model in presence of a toroidal damper.

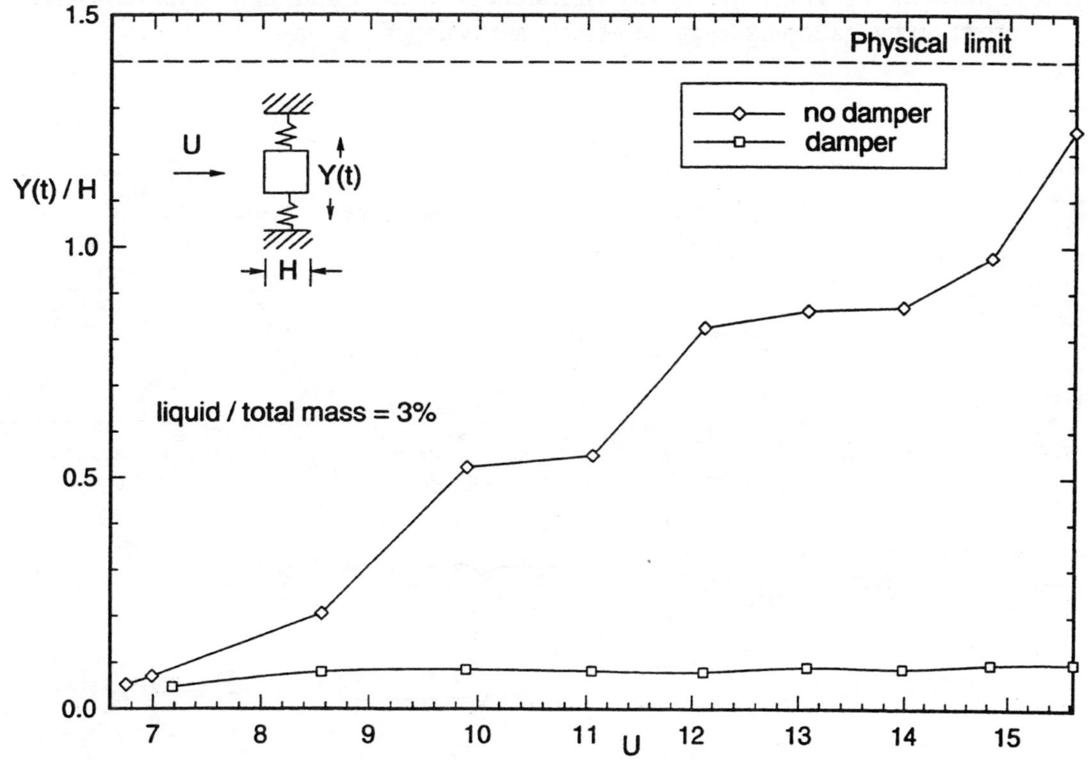

Fig. 7 Galloping response of a square prism showing remarkable effectiveness of a rectangular damper.

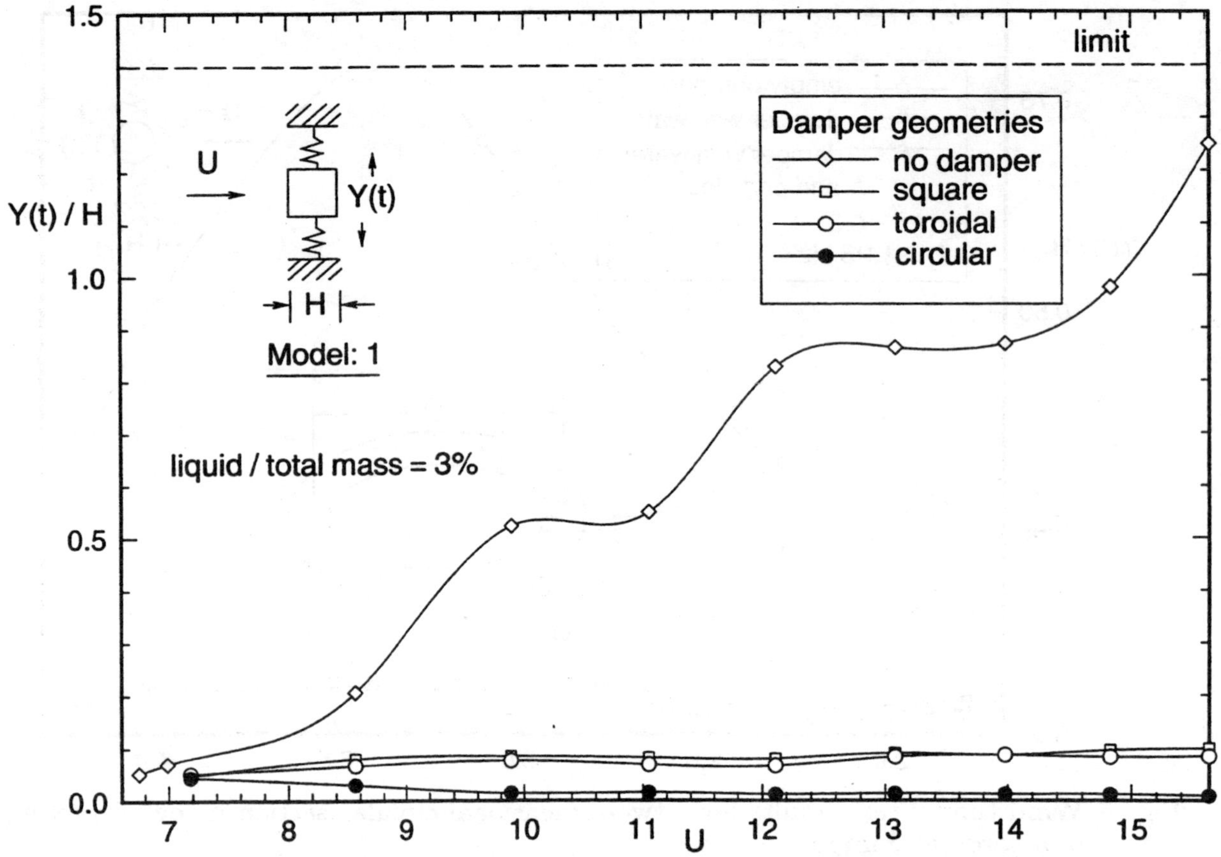

Fig. 8 Comparative performance of three different geometry nutation dampers showing their effect on the galloping response of a square cylinder.

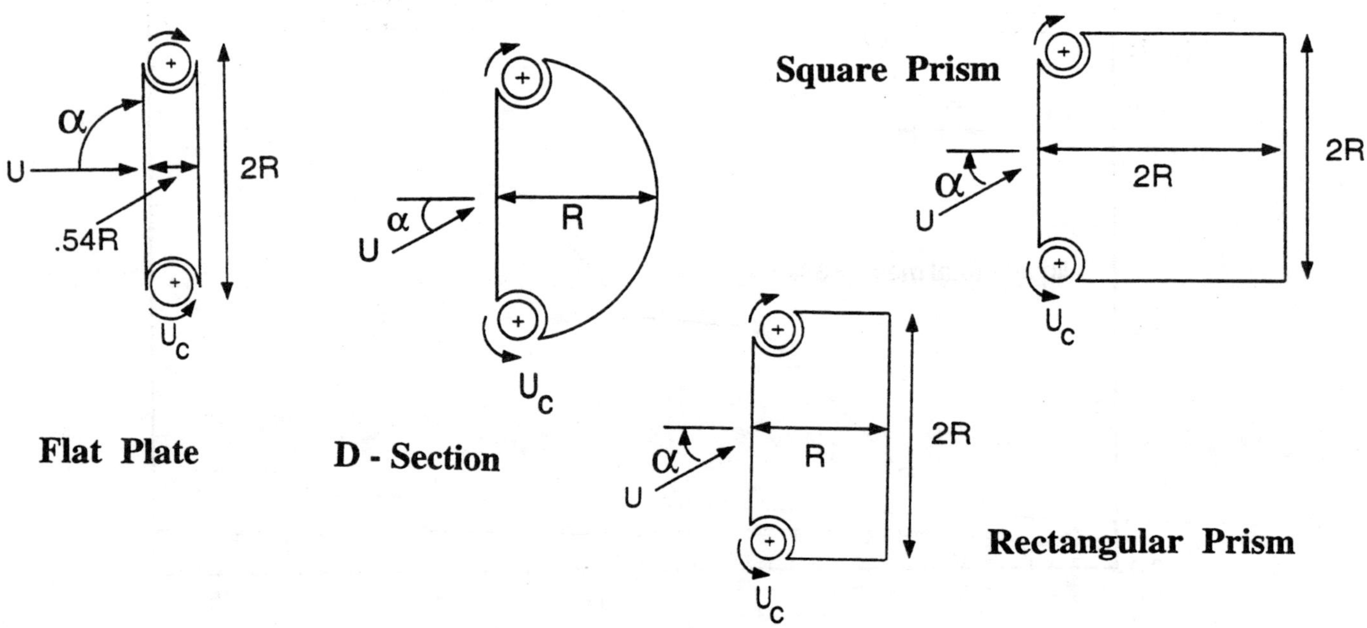

Fig. 9 Schematic diagrams of four bluff geometries used to assess effectiveness of the MSBC in suppressing flow induced vibrations.

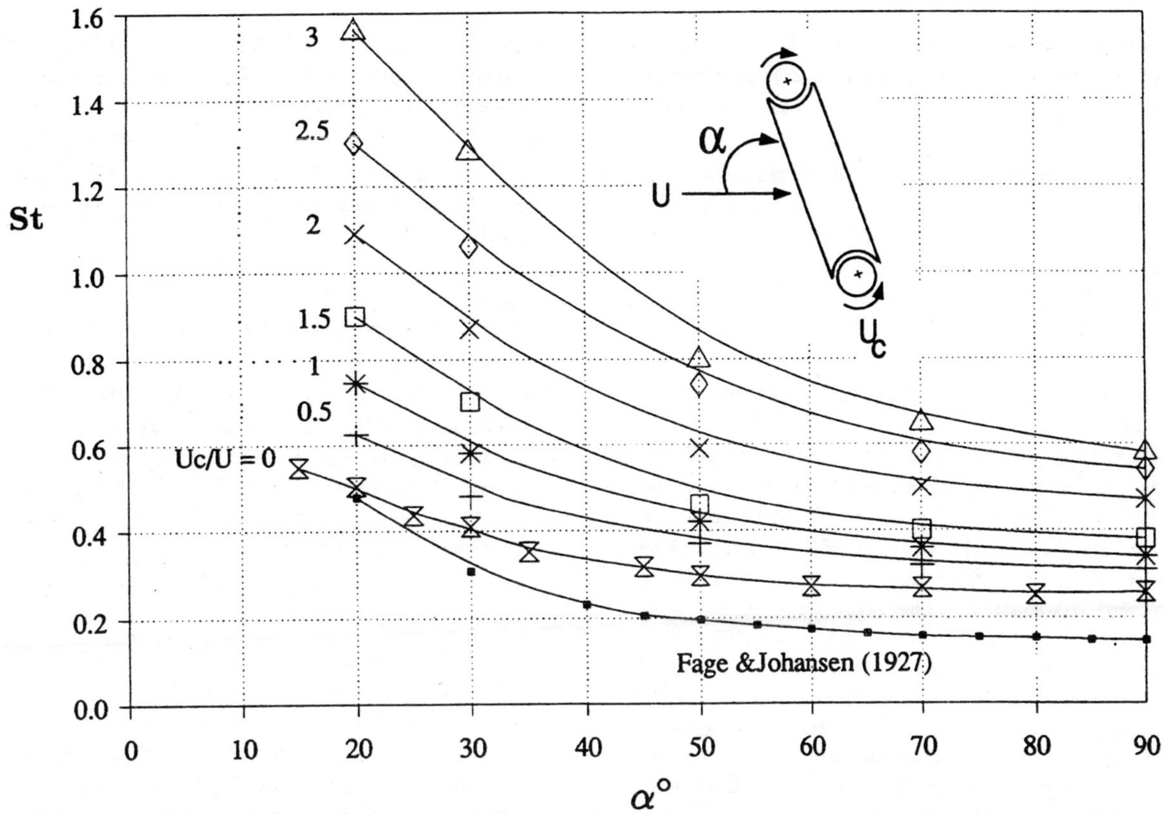

Fig. 10 Effect of momentum injection and angle of attack on the Strouhal number associated with a flat plate.

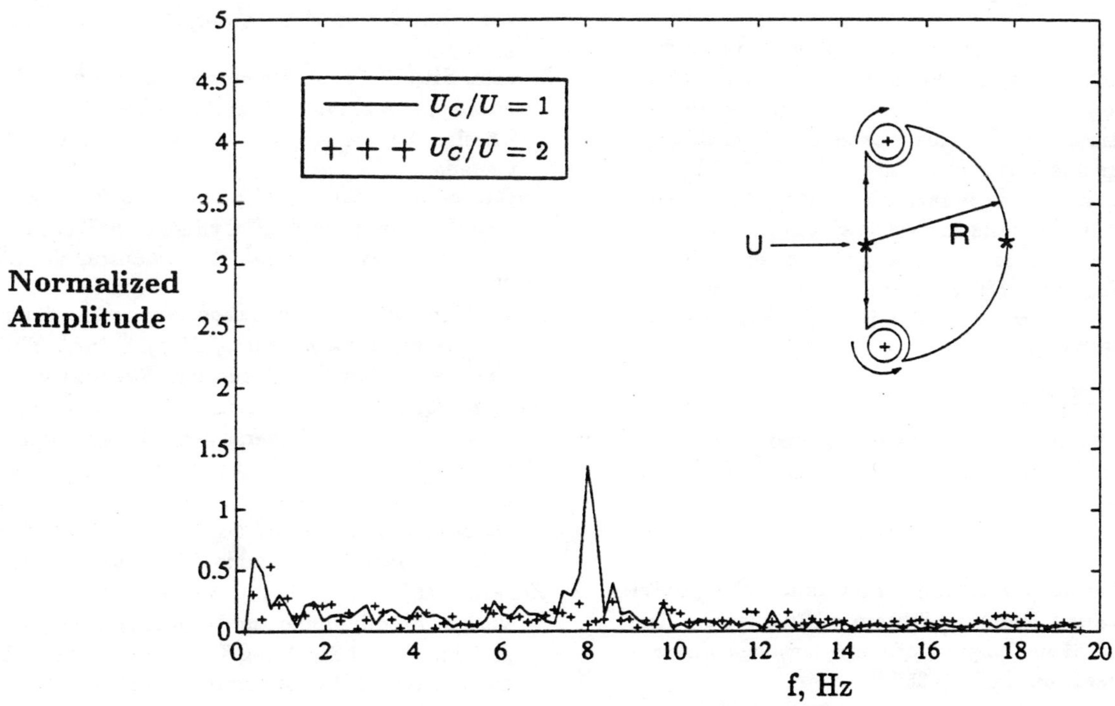

Fig. 11 Power spectral density representation of the wake pressure showing reduction in the strength of the shed vortex in presence of the momentum injection. Note, for $U_c/U = 2$, the vortex shedding is virtually eliminated.

prism are shown in Figure 13. For $0 < \alpha < 5^\circ$, the square prism shows strong instability ($dC_y/d\alpha \approx +2.6$) in absence of the momentum injection. As the momentum injection is increased to U_c/U =4, $dC_y/d\alpha$ becomes 4.6 ($\alpha \leq 5^\circ$) and the stability is restored. The above prediction was confirmed during the dynamic experiments in the wind tunnel. The amplitude response of the square prism at α =0, as affected by the MSBC, is shown in Figure 14. As the wind speed is increased from zero, the square prism experiences vortex resonance ($0.9 \leq V \leq 1.25$) and later galloping instability ($V > 20$). In contrast, both the instabilities are suppressed even with a small amount of momentum injection ($U_c/U = 1$).

The flow visualization study was carried out for all the four bluff body models to check the absence of vortex shedding (Figure 15). In absence of the momentum injection, a large vortex can be seen in the process of formation. The wake vortices displayed considerable turbulence, with associated mixing and dissipation. The separating shear layer from the top has a sinusoidal character and divides the wake into well-defined pockets of diffused vorticity, generated from the viscous dissipation of shed vortices. The overall configuration of the wake reverses intself in an alternate manner in phase with the vortex shedding process. The video revealed considerable unsteadiness of the wake, oscillating from side to side due to the low pressure region associated with the nascent vortex. However, with the momentum injection, the wake gradually becomes narrower and the vortex formation is suppressed.

CONCLUDING REMARKS

A vast body of literature aimed at control of fluid-structure interaction problems is available. Depending on the situation, appropriate procedure can be used to suppress the undesirable oscillations. In particular, passive control through nutation dampers and active procedure of the MSBC appear quite attractive. Extensive parametric studies, flow visualization and wind tunnel tests suggest both the procedures to be remarkably successful in suppressing the vortex resonance as well as galloping type of instabilities. These elegantly simple designs are cost-effective, easy to implement and require virtually no maintenance.

ACKNOWLEDGEMENTS

The investigation reported here was supported by the Natural Sciences and Engineering Research Council of Canada, Grant No. A-2181.

REFERENCES

Anzai, T, Ikeuchi, M, Igarashi, K, and Okanuma, T (1981). "Active Nutation Damping System of Engineering Test Satellite-IV," *22nd Congress of the International Astronautical Federation*, IAF 81-350.

Cermak, JE (1975). "Application of Fluid Mechanics to Wind Engineering," *Journal of Wind Engineering and Industrial Aerodynamics*, Vol 97, No 1, pp 9–38.

Every, MJ, King, R, and Weaver, DS (1982). "Vortex Excited Vibrations of Cylinders and Cables and their Suppression," *Ocean Engineering*, Vol 9, pp 135–157.

Fage, A, and Johansen, FC (1927). "On the Flow of Air Behind an Inclined Flat Plate of Infinite Span," *Proceedings of the Royal Society of London*, Series A, Vol 116, pp 170–197.

Modi, VJ (1982). "Dynamics of an Inflatable Structure Suitable for use in Submarine Detection: A Brief Review," *Proceedings of the Indian Academy of Sciences*, Vol 5, Part 1, pp 51–73.

Modi, VJ, Welt, F, and Irani, MV (1988). "On the Suppression of Vibrations Using Nutation Dampers," *Proceedings of the ASME Pressure Vessels and Piping Conference*, Pittsburgh, PA, U.S.A., PVP Vol 133, pp 17–25.

Modi, VJ, Mokhtarian, F, Fernando, MSUK, and Yokomizo, T (1991). "Moving Surface Boundary-Layer Control as Applied to Two-Dimensional Airfoils," *Journal of Aircraft*, Vol 28, pp 104–112.

Modi, VJ, Shih, E, Ying, B, and Yokomizo, T (1992). "On the Drag Reduction of Bluff Bodies Through Momentum Injection," *Journal of Aircraft*, Vol 29, pp 429–436.

Modi, VJ, Welt, F, and Seto, ML (1995). "Control of Wind-Induced Instabilities through Application of Nutation Dampers: A Brief Review," *Engineering Structures*, Vol 19, No 9, pp. 626–638.

Modi, VJ (1996). "Dynamics and Control of Large Flexible Space Structures," Invited Plenary Address, *36th Israel Annual Conference on Aerospace Engineering*, Tel Aviv/Haifa; also *Proceedings of the Conference*, pp A-9 to A-36.

Munshi, SR (1996). *Aerodynamics of Bluff Bodies in Presence of the Moving Surface Boundary-Layer Control*, Ph.D. Thesis, University of British Columbia, Vancouver, Canada.

Parkinson, GV, and Brooks, NPH (1961). "On the Aeroelastic Instability of Bluff Cylinders," *Transactions of the ASME, Journal of Applied Mechanics*, Vol 28, pp 252–258.

Seto, M (1996). *An Investigation on the Suppression of Flow-Induced Vibrations of Bluff Bodies*, Ph.D. Thesis, University of British Columbia, Vancouver, Canada.

Welt, F (1982). *A Parametric Study of Nutation Dampers*, M.A.Sc. Thesis, University of British Columbia, Vancouver, Canada.

Welt, F (1988). *A Study of Nutation Dampers with Application to Wind-Induced Oscillations*, Ph.D. Thesis, University of British Columbia, Vancouver, Canada.

Zdravkovich, MM (1981). "Review and Classification of Various Aerodynamic and Hydrodynamic Means for Suppressing Vortex Shedding," *Journal of Wind Engineering and Industrial Aerodynamics*, Vol 7, pp 145–189.

Zdravkovich, MM (1984). "Reduction of Effectiveness of Means for Suppressing Wind-Induced Oscillations," *Engineering Structures*, Vol 6, pp 344–350.

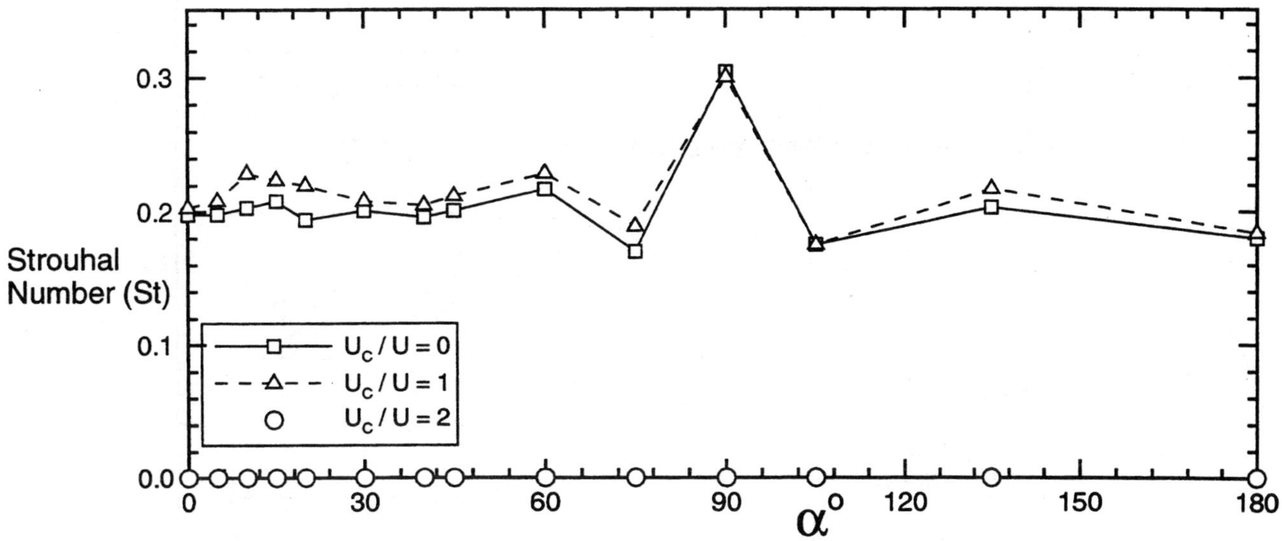

Fig. 12 Effect of momentum injection on the Strouhal number of a rectangular prism.

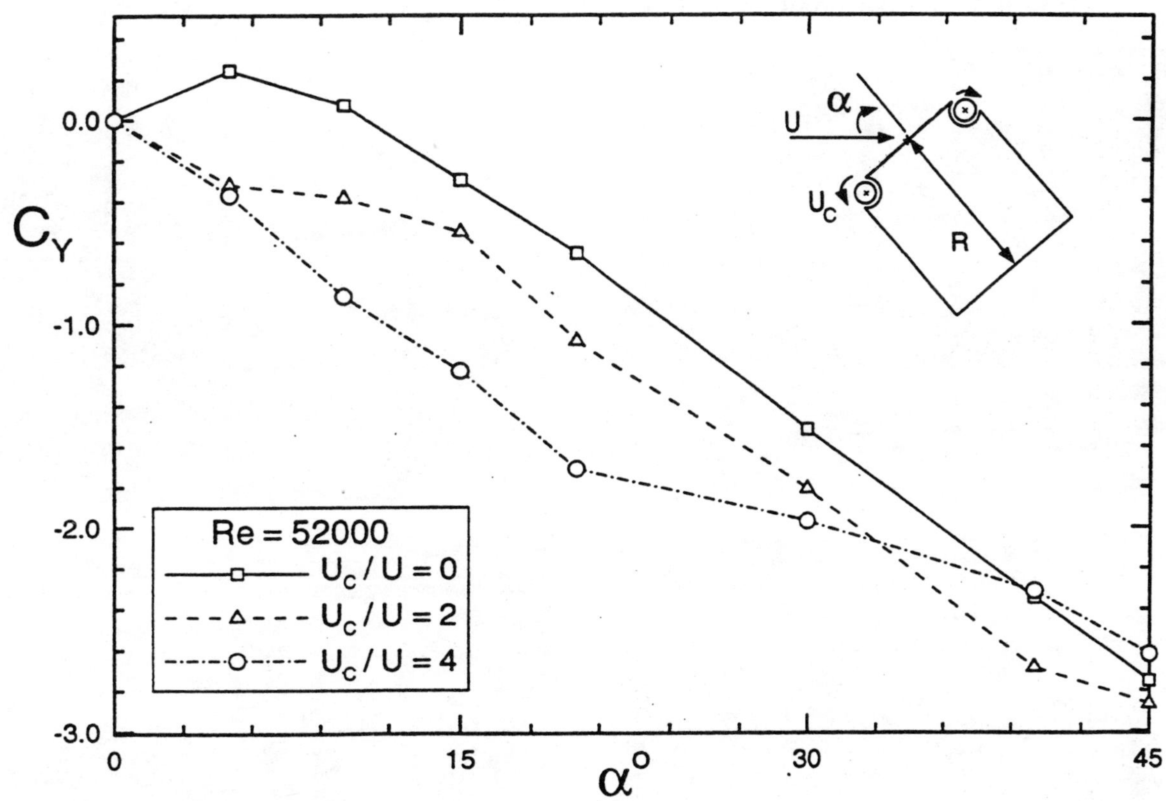

Fig. 13 The square prism is unstable ($dC_y/d\alpha > 0$) for $\alpha \leq 5°$ in absence of momentum injection ($U_c/U = 0$). With the cylinder rotation ($U_c/U = 2, 4$), the stability is restored completely ($dC_y/d\alpha < 0$).

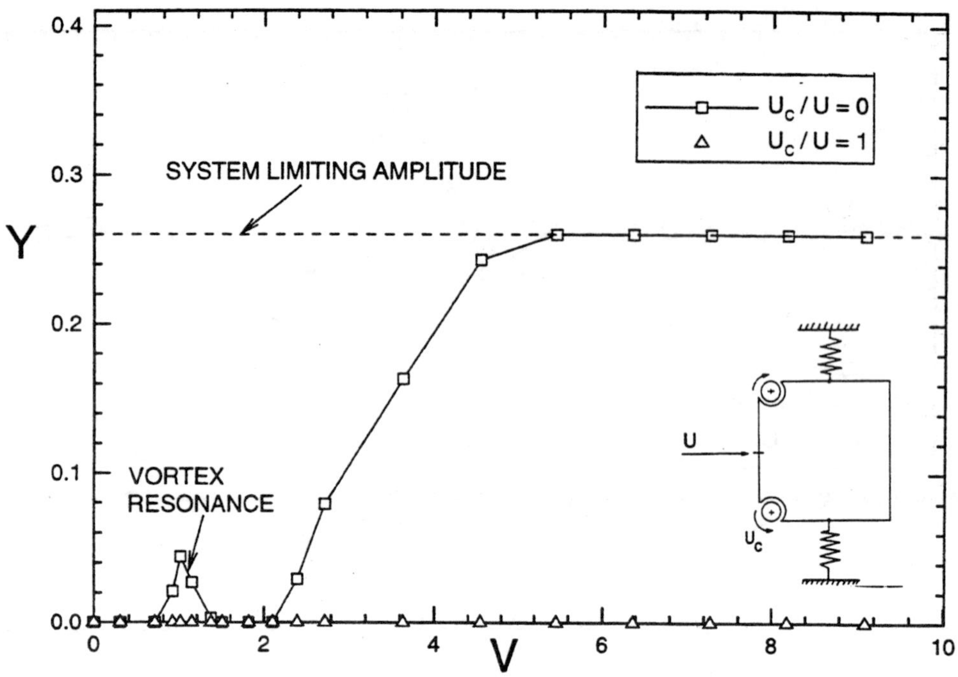

Fig. 14 Dynamic response of the square prism without and with momentum injection. Note the presence of both vortex resonance and galloping in absence of the MSBC ($U_c/U = 0$). The square prism becomes completely stable even with the momentum injection as small as $U_c/U = 1$.

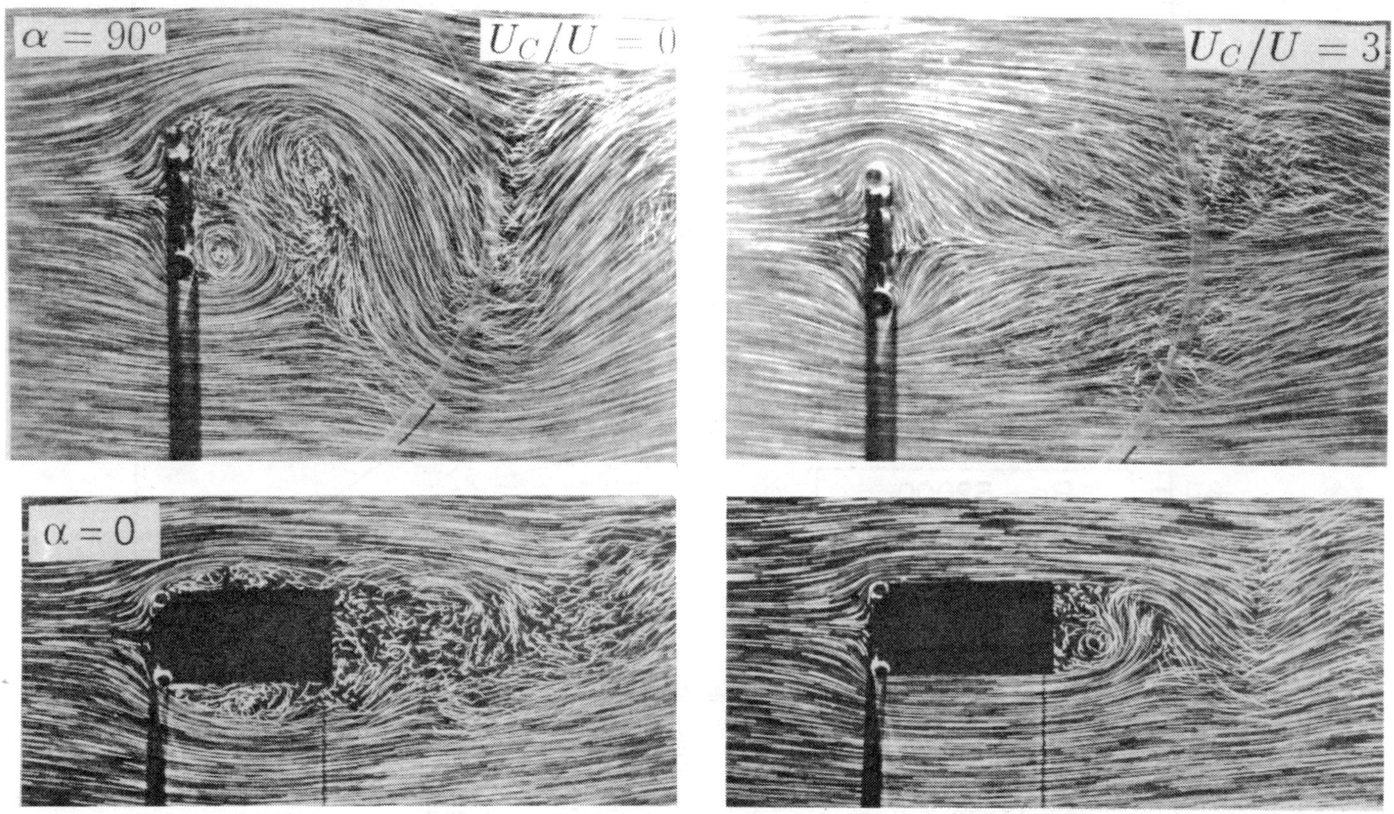

Fig. 15 Typical flow visualization photographs showing, rather dramatically, effectiveness of the MSBC in suppressing the vortex shedding from bluff bodies.

Proceedings of the Seventh (1997) International Offshore and Polar Engineering Conference
Honolulu, USA, May 25-30, 1997

ISBN 1-880653-28-1 (Set); ISBN 1-880653-31-1 (Vol. III)

Wave-in-Deck Forces

Timothy D. Finnigan
Chevron Petroleum Technology Company
San Ramon, CA, USA

Chuck Petrauskas
Chevron Petroleum Technology Company
La Habra, CA, USA

ABSTRACT

Under the current competitive situation, Chevron needs to make risk-based decisions with regard to continued use of existing platform assets. Chevron's Gulf of Mexico fleet is dominated by aging platforms. About 70% of the platforms (344 platforms, excluding caissons) are older than 20 years. In addition, a large percentage of the platforms offshore Nigeria are also old, and regularly being considered for extended use. Continued use of existing assets will save millions of dollars in Capital Expenses that can be redirected to other developments. Platform failure probabilities and consequences are required input to the risk analysis. Because many of the older platforms do not meet present day standards for deck height, failure probability calculations require accurate estimates of (a) deck-inundation wave/current forces and (b) statistical characterization of wave crest heights.

KEY WORDS: Offshore, platforms, deck, forces, crest, heights, waves

INTRODUCTION

In order to develop a hydrodynamic model for deck forces, Chevron conducted a series of wave tests on a 1:28 scale model of a 141-ft. water depth steel jacket platform during the winter of 1988-89. The model included an instrumented deck structure representative of older Gulf of Mexico platforms which consisted of tubular and wide flange members. The deck had three levels: a sub-cellar deck without equipment; a cellar deck with equipment modeled; and a main deck without equipment. The test program consisted of a series of unidirectional regular and irregular waves, and multi-directional irregular waves consistent with Gulf of Mexico hurricanes and winter storms. There is presently no other comprehensive model wave force data of this nature available in the literature. To characterize the statistics of crest heights Chevron has over the years maintained a data base of measured crest height and wave height data and has continually evaluated its mathematical models.

Two basic methods for predicting deck loads are considered: a "component" approach in which wave loads are computed on individual structural members and equipment as the wave passes through the structure; and a "silhouette" approach in which the load is computed on the projected area of the deck. The component approach can account for varying amounts of deck equipment and structural members, and member shielding and blockage can be taken into consideration. However, the method requires a comprehensive description of the deck and equipment, which is not always available for older structures. The silhouette approach, which was adopted for API 20th edition, can be used with calibration factors that take into consideration the relative density of the structure based on the amount of structural members and equipment.

MODEL TEST PROGRAM

Models

A model similar in size, water depth, and framing to a Chevron platform recently removed from service was built to a scale of 1:28. The model included both the jacket and the deck, upon which one of three basic deck equipment configurations were used: a "bare" deck, a "moderate" deck, and a "full" deck. The density of deck equipment increased respectively for these three models. The different deck configurations were installed on removable trays which were made out of wire mesh carefully chosen to scale the flow properties expected through standard platform steel grating. Boat landings, piping, stairs, handrails, and other appurtenances were not modeled to simplify data analysis.

Conductors were modeled in some of the tests to determine whether or not they influenced the flow through and hence the loads on the deck. The deck model was free to "float" on the jacket legs, and did not have direct contact with the conductors. The loads on the conductors were transmitted to the jacket structure as they would be on prototype structures. Early numerical simulations indicated deck grating had a significant effect on the wave loads. This was because loads on individual elements were computed without considering the shielding effect of leading members. However, it was believed that flow shielding would significantly reduce the loads. Therefore, tests were conducted with and without the grating at the scaffold level to determine its influence on the wave loads.

Figure 1 shows a "typical" older Gulf of Mexico platform which guided the development of the model used in these tests. Figures 2 and 3 show the full jacket model with deck, and a close up of the deck model respectively. The framing seen at the top of the model was used to support instrumentation, and was not part of the deck structure. The schematic in Figure 4 shows the deck elevation views from the side and end-on.

Figure 1: Prototype "Aging" Platform Used As Basis For Models

Figure 2: Jacket + Deck Model

The scaffold deck, comprised mainly of WF4x13 framing members and hung from the cellar deck with 12 inch members, was located 29 ft. above MLLW. The cellar deck, which contained the deck equipment, was comprised of 8, 12, and 14 inch members and was located 34 ft. above MLLW. The main deck was comprised of 14, 21, 24, and 33 inch members and was located 43 ft. above MLLW.

Figure 3: Close-Up of Deck Model

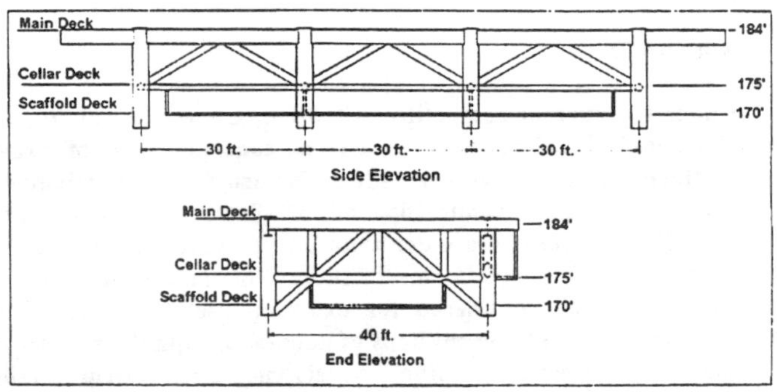

Figure 4. Elevation schematic of deck model

Figures 5 and 6 show the bare deck and full deck trays used in the wave tests. The moderate deck tray had about half as much equipment as the full deck. Six conductors were used in some of the tests with the moderate deck configuration to assess the effect conductors would have on the flow through the deck. Also, tests were conducted on the moderate deck setup with the grating removed from the scaffold deck to assess the impact deck grating has on the loads.

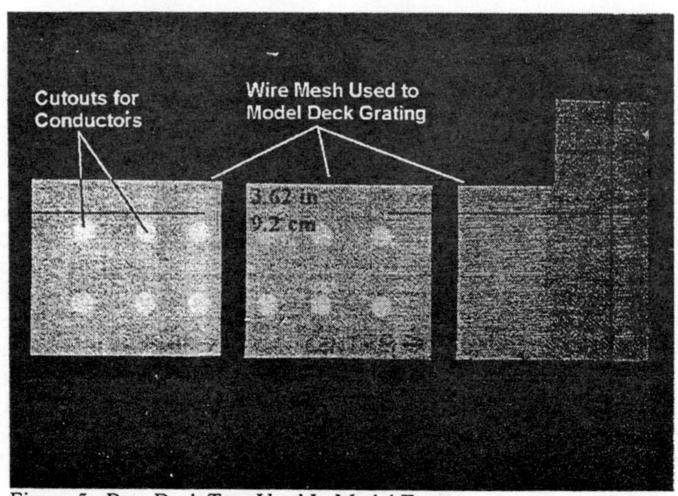

Figure 5: Bare Deck Tray Used In Model Tests

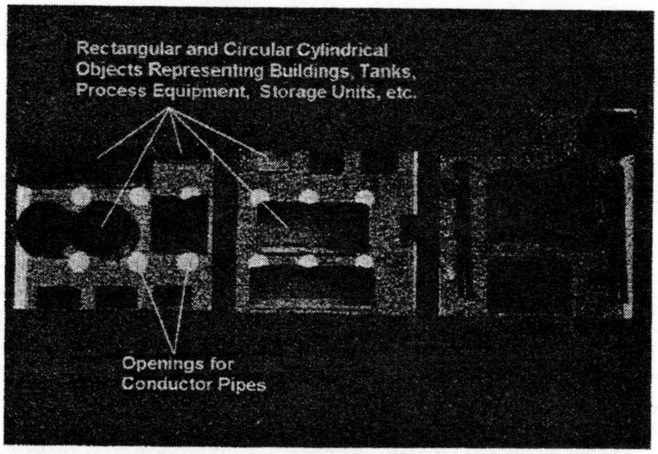

Figure 6: Full Deck Tray Used In Model Tests

Test Program

Wave tests were conducted at NRC-Canada, in Ottawa in the directional sea basin. The model was tested at three headings: 0-degrees or end-on; 45-degree or diagonal; and 90-degrees or broadside. Regular and irregular uni-directional waves and irregular multi-directional wave tests were conducted. Due to the possibility of reflections from the side walls and beach for long simulations, and the interest in only selecting wave crests which could impact the deck, wave "snapshots" were used in the tests. A wave selection process was then established to get a distribution of crest heights that ranged from just below the scaffold deck to just above the main deck level.

Snapshot Selection - Multiple wave time histories 3 hours in length (prototype scale) were numerically simulated from a typical Gulf of Mexico design wave spectrum. These time histories were screened to identify wave segments 3 minutes in length (prototype scale) with the following criteria: 1) the segments contain 2 to 3 waves with a crest height between the bottom of the scaffold and top of the main deck; and 2) at least 2 to 3 waves separated these large crests for statistical independence. Due to wave nonlinearities and other idiosyncrasies of wave making, it was not possible to exactly recreate the numerically simulated wave traces. Therefore, a calibration phase was conducted in which the selected waves were fed into the drive signal of the wave maker, and waves were created in the wave basin. Only those wave segments which satisfied the above criteria and were not breaking were used in the model tests.

Conditional Simulations - In order to make a direct comparison of wave forces in unidirectional and multidirectional irregular waves, the wave record at the centerline of the model in unidirectional waves was required to match the wave record in multidirectional waves. This was accomplished in the wave basin by first creating a multidirectional irregular wave from the simulated wave traces. After waves with the desired characteristics were identified, a unidirectional wave trace was generated given or "conditioned upon" the multidirectional wave measured at the center of the test section.

Figure 7 shows a segment of a long simulated wave record and a wave episode selected for testing with two crests high enough to impact the deck. Figure 8 shows a comparison of a measured unidirectional wave segment and its corresponding measured multidirectional wave segment.

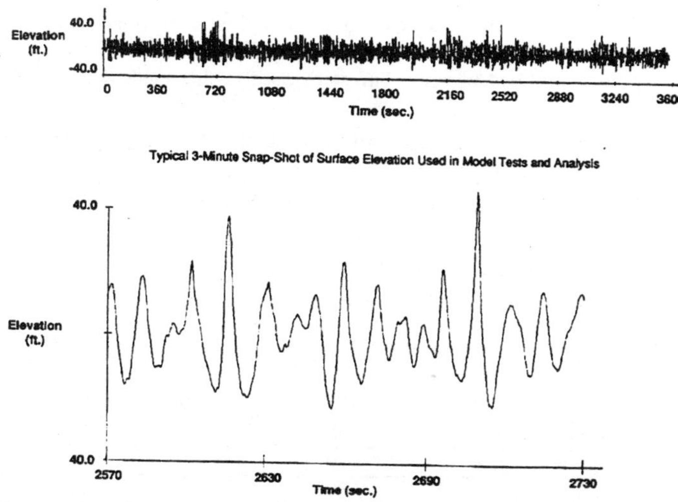

Figure 7: Segment Of Simulated Wave Record And A Wave "Episode"

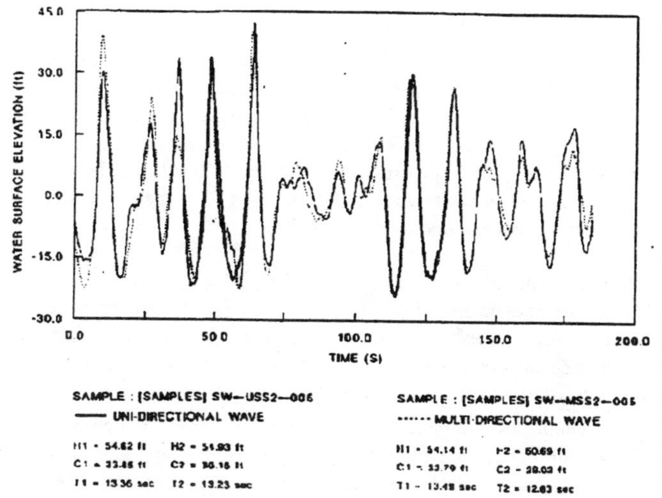

Figure 8: Measured Unidirectional Vs. Multidirectional Wave Segment

DECK FORCE PROCEDURES

Two deck force procedures are evaluated in this paper. One is a detailed component approach which relies upon the quantification of the various structural members and deck equipment impacted by the wave. The other is a "silhouette" approach which requires only the gross dimensions of the deck. The first procedure is more accurate, and is appropriate when a numerical deck model has already been built for structural analysis purposes. The second procedure is specified in Supplement 1 of the 20th Edition of the API RP2A (1996). It is appropriate when details about the deck structure are lacking, or to create general guidelines for wave in deck loads for risk assessments.

Component Model

A space frame computer model is built, including deck legs; vertical, horizontal, and diagonal support tubulars, plate girders, truss beams, deck beams, deck grating, and deck equipment. Forces are computed for a range of wave heights for a specific period (peak spectral period) whose crest heights range from just below the lowest deck (scaffold deck for these tests) to just above the highest deck (main deck). This method is chosen in lieu of computing the deck forces for a specific wave height and period. The steps used in this method are outlined below:

1. Create a computer model of the deck including all wide flange and tubular members,
2. Model deck equipment as rectangular or circular cylinders,
3. Model deck grating as lumped areas throughout the deck, using only 35% of the actual projected area in the lateral direction,
4. Choose a drag coefficient according to the table below:

Heading	with equipment		without equipment	
	without conductors	with conductors	without conductors	with conductors
end-on	0.6	0.55	0.67	0.6
diagonal	1.0	1.10	1.0	1.10
broadside	0.85	0.80	0.85	0.80

5. Use Regular Stream Function Wave Theory for wave kinematics,
6. Use a wave spreading factor of 0.88 (multiply the fluid velocities) for hurricanes and 0.95 for winter storms to account for directional spreading effects.

The values of 0.88 and 0.95 are derived from the wave loads measured on the jacket alone from an earlier part of the same model test program, and is supported by numerical computations with cosine squared spreading. The drag coefficients used in the computations are based on a rudimentary blockage procedure (Finnigan, 1992; Lambrokos, 1989; Taylor, 1991), in which the flow is assumed to be steady relative to the duration of wave impact.

7. The wave forces are computed using Morison's formulation. An inertia coefficient of 1.5 is used, but deck forces are found to be relatively insensitive to inertia forces due to dominance by the horizontal crest velocities (horizontal crest accelerations are near zero when crests are high enough to impact the deck).
8. Compute a wave crest height equal to 1.07 times that computed with Stream Function Theory for the given wave height and period. The 7% increase is based on Chevron in-house studies from calibrations of Stream Function Wave Theory with measured crest heights.
9. Choose the wave force vs. crest height distribution for a given wave heading; and
10. Find the wave force from the table or plot that corresponds to the computed crest height.

Silhouette Model

The Silhouette Method presented below was developed by a sub-group of Work Group 3, 'Environmental Loading', of the API Task Group on Assessment of Existing Platforms (Petrauskas, 1996). Participants were Tim Finnigan (Chevron), Rick Mercier (Shell), John Heideman (Exxon), and Gene Berek (Mobil). Guidelines are described in Supplement 1 of the Wave Force Commentary of the 20th Edition of the API RP2A. The steps for this method are summarized below:

1. Compute the crest height using an appropriate non-linear wave theory (no 7% correction);
2. Compute the gross wetted area, A, normal to the wave heading as in Figure 9. In this computation, the legs and bracing above the cellar deck are considered part of the wetted area, while the legs and bracing below the cellar deck are not;
3. Compute the maximum horizontal water particle velocity, u, at the lower of the crest or top of the main deck level. Use equation 2.3.1b.2 of the Supplement;
4. Compute the wave force according to the formula below:

$$F_{deck} = \tfrac{1}{2}\,\rho\, C_D\, A\,(\alpha_{WKF} * u + \alpha_{CBF} * U)^2 \;; \qquad (1)$$

where ρ is the water density, C_D is the drag coefficient which accounts for the heading and "solidity" of the structure, A is the wetted area, α_{WKF} is the wave kinematics factor (0.88 for hurricanes, 1.0 for winter storms), u is the wave kinematic velocity, α_{CBF} is the current blockage factor, and U is the current velocity. The recommended drag coefficients for three deck types: "heavy" or very dense equipment; "moderate" with half the density as heavy; and "bare" are noted below. The "Full" deck in the present model tests is considered only a "Moderate" deck in terms of solidity in the API guidelines. This is because modern offshore platforms have much denser decks than that modeled in the test program. The selection of the density is subjective, and left to the judgment of the engineer. The guidelines contained in API RP2A are based on a "Bare" deck similar to the Bare deck used in these tests, a "Moderate" deck similar to the Full deck used in these tests, and a "Full" deck which is essentially a solid block.

Drag Coefficients used in Silhouette Approach

Type	end-on or broadside	diagonal
heavy	2.5	1.9
moderate	2.0	1.5
bare	1.6	1.2

These drag coefficient values are based on a combination of results from these model tests, theoretical considerations of momentum flux (Mercier), and experience from platform damage during hurricane Andrew. A coefficient of variation of 35% is applied to the loads predicted with this method when used to develop extreme load distributions for risk assessment purposes.

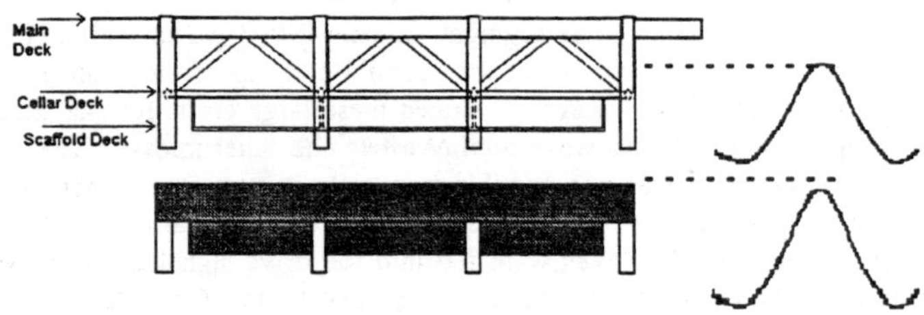

Figure 9. Shaded Area Used In Computing Loads For Silhouette Method.

ANALYSIS AND RESULTS

Analysis

Only multidirectional wave snapshots are used to evaluate the deck force procedures. There are several reasons for this. 1) The multidirectional waves more accurately depict the design wave properties in the Gulf of Mexico; 2) Due to the steepness of the waves representing the design wave heights for the Gulf of Mexico location studied in this paper (141 ft. water depths), the unidirectional waves were relatively unstable, and had a much higher incidence of breaking which cannot be adequately modeled in the wave basin; 3) The repetitive nature of the regular waves in which the turbulent trough of a preceding wave influenced the crest kinematics of the following wave, led to unsteady wave force statistics (the deck forces steadily decreased in value for each passing wave).

Due to the variability of crest height vs. wave height, and the sensitivity of the deck forces to how much of the deck is wetted by the wave, the measured wave forces are compared to the predicted wave forces as a function of the crest height rather than the wave height.

Component Method Results

The deck forces predicted using the "component" procedure are compared to the measured deck forces for the various deck configurations in the figures below; however, it is not possible in the present paper to show results for all headings and deck configurations. Therefore, comparison of predicted and measured forces are shown for a few cases.

In Figure 10, the predicted forces are shown by the dotted line, and the measured forces as open circles. The scaffold deck is first impacted when the crest height reaches 29 ft. There is a jump in wave forces at the 34 ft. crest height, reflecting when waves first impact the cellar deck. There's another jump at 43 ft. when the main deck is impacted.

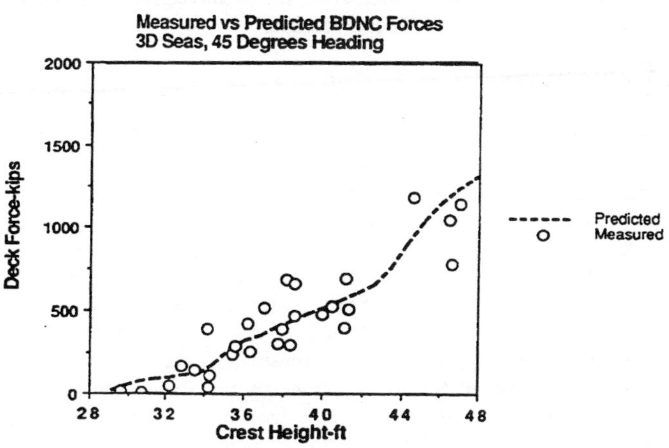

Figure 10. Measured And Predicted Bare Deck Forces, Diagonal Heading, Component Approach

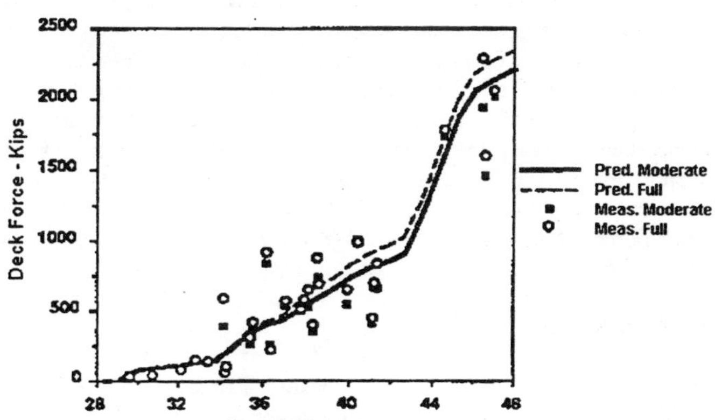

Figure 11. Measured And Predicted Moderate Deck Forces, Broadside, Component Approach

Figure 11 shows predicted moderate deck and full deck forces vs. measured moderate deck and full deck forces for the broadside wave heading. This figure more dramatically shows the large increases in deck force as the crest impacts successive deck levels.

Another way to assess the importance of basing deck force calculations on crest height is shown in Figure 12.

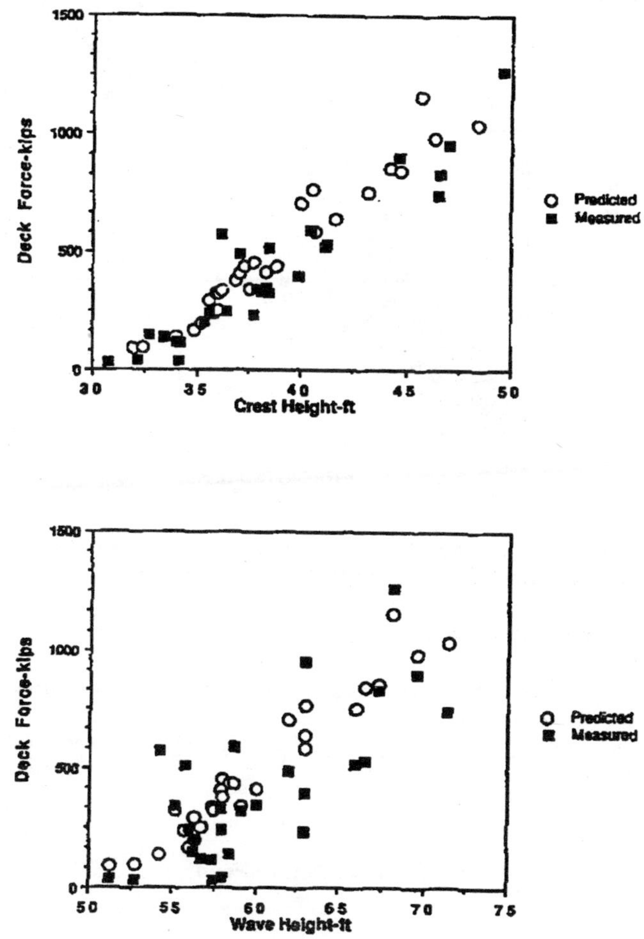

Figure 12. a (top) - Measured And Predicted Forces For Bare Deck, End-On, Based On Crest Heights; b (bottom) Measured And Predicted Forces Based On Measured Wave Heights.

Figure 12a shows the scatter in predicted (open circles) and measured (filled squares) deck forces for the bare deck, end-on condition when basing forces on measured crest heights. Figure 12b shows the scatter when forces are based on the measured wave heights.

These results demonstrate how important it is to have an accurate estimate of crest height to predict the deck forces. Often waves with the same trough to crest "wave height" will have significantly different crest heights. Figure 11 indicates that a variation in crest height of only 4 feet (from 42 to 46 feet) could result in up to a 300% difference in deck forces (from 700 kips to 2000 kips). This 1000 kip + difference in jacket base shear would also result in significant differences in overturning moment of the jacket due to the high force center.

A plot of all measured vs. predicted deck forces based on the component procedure is shown in Figure 13, which indicates a bias of 0.5% and a coefficient of variation of 33%.

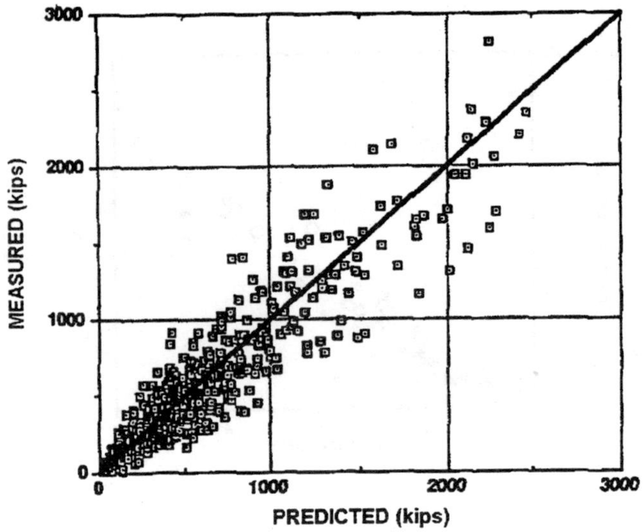

Figure 13. Measured Vs. Predicted Deck Force Data Based On Measured Crest Height, (all 2D & 3D tests, 532 impacts).

Silhouette Method Results

The deck forces predicted with the Silhouette method as outlined in API-RP2A are compared to measured forces for two conditions: Bare Deck, 45 degree heading, and Full Deck, 90 degree heading. Recall that the "Full" deck in the model tests represents only a "Moderate" deck in the API guidelines.

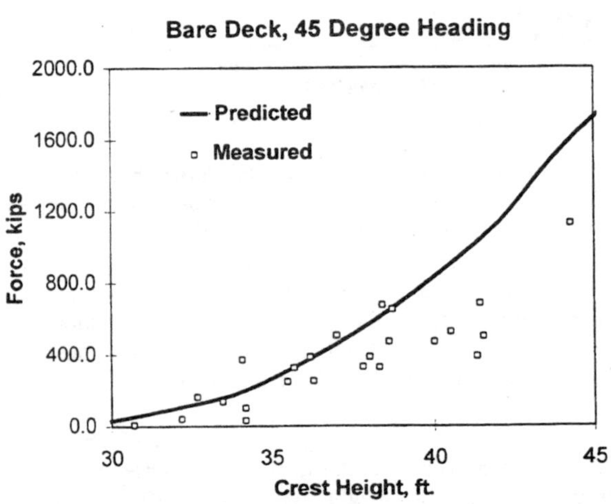

Figure 14. Comparison Of Measured And Predicted (API Guidelines) Bare Deck Forces, 45 Degree Heading

Figure 14 shows the predicted force (line) for the Bare Deck, 45 degree heading case plotted against the measured (open squares) forces. The API guidelines more or less defines an envelope to the measured forces. The same is observed for the Moderate Deck, 90 degree heading in Figure 15

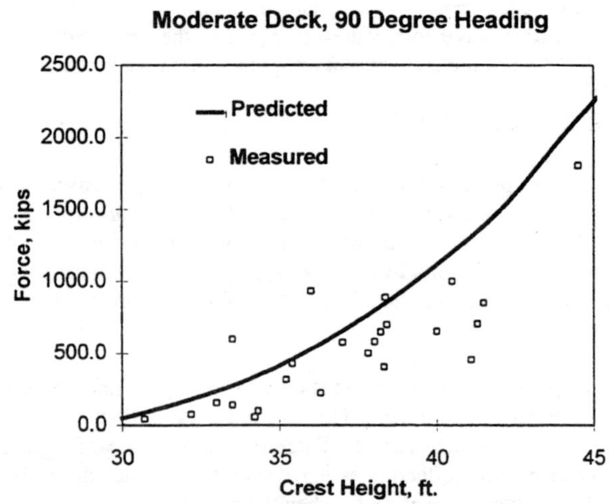

Figure 15. Comparison Of Measured And Predicted (API Guidelines) Moderate Deck Forces, 90 Degree Heading

CONCLUSIONS

Model tests were conducted in 1988-89 to assess the hydrodynamic loads on decks of aging platforms in the Gulf of Mexico. The test results formed the basis of a "component-based" deck force calculation procedure, which is shown to accurately predict the wave-in-deck forces for typical older Gulf of Mexico platforms. The data also formed the basis for a more generic "silhouette" approach for wave-in-deck force calculations that has been incorporated into in Supplement 1 of the 20th Edition of the API RP2A. Wave-in-Deck force calculations are shown to be highly sensitive to crest heights. More research is required to develop a crest height distribution which can be used in conjunction with wave-in-deck force calculation procedures to more accurately assess offshore platform failure probabilities.

ACKNOWLEDGEMENT

The authors would like to thank the following organizations and individuals for support in the research that led to the publication of these results. Chevron Petroleum Technology Company, for supporting our efforts to evaluate this problem; the National Research Council, Canada Wave Facility for conducting a thorough and comprehensive wave test program; Jim Allender (Chevron), for his support during the model tests; and John Heideman (Exxon), Rick Mercier (Shell), and Gene Berek (Mobil) for their involvement in the Deck Force sub-group; and Majid Atef (Chevron) for his help with the data analysis.

REFERENCES

Finnigan, TD (1992). "Current Blockage Effects On Model-Scale Offshore Platform," Civil Engineering In The Oceans V.

Lambrokos, KF et al (1989). "Wake And Shielding Effects On Hydrodynamic Loading," E & P Forum Workshop On Wave And Current Kinematics And Loading.

Petrauskas, C et al (1996). "Supplement 1 to RP 2A-WSD, 20th Ed."

Taylor, PH (1991). "Current Blockage: Reduced Forces on Offshore space-frame structures," OTC 6519.

Proceedings of the Seventh (1997) International Offshore and Polar Engineering Conference
Honolulu, USA, May 25-30, 1997

ISBN 1-880653-28-1 (Set); ISBN 1-880653-31-1 (Vol. III)

Assessment of Mean Water Levels and Tides from Satellite Data

C. Le Provost and F. Rabilloud
IMG/LEGI
Grenoble, France

M. Olagnon
Ifremer
Brest, France

ABSTRACT

Extreme SWLs (Still Water Levels) are used at the design stage of many fixed offshore structures, for instance to prevent the risk of an insufficient deck clearance, or to avoid tether detensionning for TLPs. With respect to the LAT (Lowest Astronomical Tide) bathymetric level, the SWL needs to take into account tidal amplitude and "storm surge", which are of metric order of magnitude in the North Sea.

Altimetric satellite measurements have been extensively used in the last years to improve tidal amplitude prediction models and to monitor long term changes in the average sea level. These very long term trends are extensively studied and the results published within the scientific community. However, the offshore industry is also interested in accurate prediction of tides and SWL at specific dates and times, in order to retrieve absolute bathymetry and/or instantaneous storm surge values.

The present paper discusses the processing of altimetric satellite measurements, in order to identify the tidal components and the mean water level at a given location.

It is shown that even in the absence of a calibrated tidal model, tides can be predicted to an accuracy of a few centimeters, and that the storm surge residual may thus be estimated with a similar precision.

Case studies are performed for several locations in the English Channel and the North Sea.

Conclusions are drawn as to the extraction of North Sea storm surges from satellite measurements, and possibilities of extension to other areas than the North Sea are discussed.

KEYWORDS: Storm surge, water level, tides, satellite altimetry.

INTRODUCTION

The design of sufficient deck clearance for fixed offshore structures was recently identified as one of the areas requesting special research efforts[5]. Difficulties in the estimation of the maximum level that "green water" may reach with respect to the lower decks of platforms stem mainly from lack of knowledge in the statistics of crest heights, but also, especially for intermediate or shallow water depths, from the large variations of the still water level. These variations are composed of a deterministic part, the astonomical tide, and of a stochastic part that reflects the static effect of barometric pressure and wind shear, the solitary waves that may be generated by the displacements of pressure lows, currents effects, and many other unknown sources. The still water level may thus significantly depart from the predicted, deterministic, astonomical tide. The existence of these deviations is well known, and the corresponding "storm surges" have been studied with respect to the risk of coastal land flooding, for instance in countries such as the Netherlands.

In order to extract the anomalies time-history, one needs to have access to precise measurements of the actual still water level on one hand, and to a high quality tidal prediction on the other hand.

Long term time-series of level measurements are available for a number of coastal locations where tide gauges have been operating over long durations, and thus provide through harmonic analysis the tidal components of the astronomical tide, and the residual anomalies after removal of this deterministic tidal history. For locations farther offshore, and especially oil fields where human activity began only recently, it is much more difficult to estimate the distributions and extreme values of the anomalies: measurements are difficult to perform, and numerical tidal models require a precise description of the bathymetry and estimation of some empirical coefficients such as dissipation terms.

With the launch of the high precision altimetry satellite Topex/Poseidon, it became possible to calibrate tidal models of the world's oceans to an accuracy of a few centimeters[9, 10]. These improvements were however until now mostly limited to the deep ocean, because of the increased complexity of continental shelf tides, comprising a large number of individual components that may be aliased together by the repeat period of the satellite, and of the distance before which measurements can be relied upon for satellite tracks going off the coast.

The present paper deals with the estimation of tidal (deterministic) components of the still water level by the use of satellite altimetry. In the following, we show that harmonic analysis of Topex/Poseidon altimeter measurements leads to identification of the tidal components' phases and amplitudes with an accuracy similar to that of the best available models, even for complex tide areas.

As a result, the non-deterministic part of the variation of the mean level may also be estimated to an accuracy sufficient for most engineering applications. Another paper in the same conference ([6]) deals with the statistical analysis of these extracted surges and set-downs.

The tidal results of satellite altimetry may however also be used for themselves at locations where accurate calibrated tidal models are not available.

POSITION OF THE PROBLEM

Water levels measurements

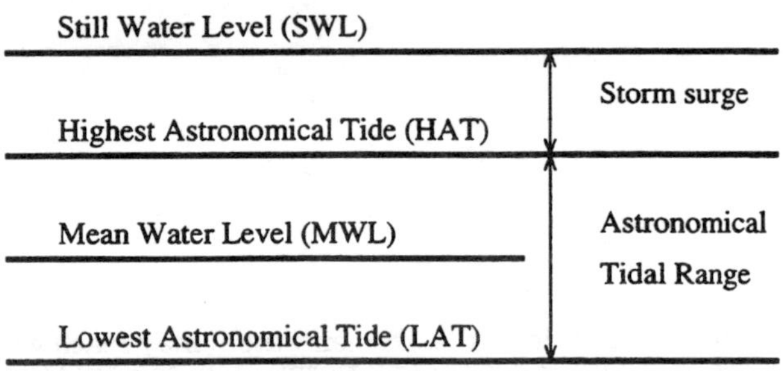

Figure 1. Definitions of design water levels

Figure 1 presents the common practice nomenclature for design water levels. These levels may be used for the definition of the part of a structure submitted to hydrodynamic loads, for the design of the deck clearance, for the implementation of docking equipment, for the definition of the splashing zone, or for the tension of TLP tendons.

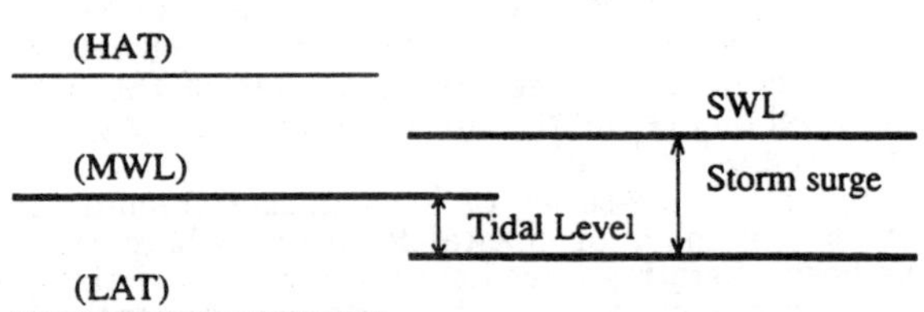

Figure 2. Common case water levels

A more common case is shown as figure 2. The level difference with respect to the LAT (usually known as chart datum) is split into a predictable, deterministic astronomical tide part, and a stochastic part, that includes the static effect of barometric pressure and of wind shear, the solitary waves that may be generated by the displacement of pressure lows, current effects, and many other unknown sources. One should note that although the storm surge is represented as positive in figure 2, it might as well be negative, and thus the still water level SWL may be observed in some instances lower than the LAT level. A conclusion of the analysis of the surges extracted by the present study is that this case may occur for high sea-states as well as for good weather, and should be cared for in the design of TLPs for instance [6].

Satellite-borne altimeters measure the instant still water level, averaged over an area roughly 7 km in length by 4 km in width. The long term mean of the instant still water level, which may be obtained by harmonic analysis and/or averaging of a great number of satellite measurements at a given location, provides the so-called mean water level (MWL). The Topex-Poseidon French-US satellite was specially designed for high precision altimetry and orbitography in order to study ocean topography, and its measurements were thus used in the present study.

The altimetric measurement is obtained from the time it takes a sharp radar pulse to travel back and forth between the satellite and the ocean surface. A number of corrections are performed to account for the influence of some parameters such as significant wave height and atmospheric effects. The most important point is then to know with high precision the exact position of the satellite with respect to the earth, and this is achieved for Topex/Poseidon by a sophisticated tracking system. The resulting precision on the satellite position is 2 cm, and on the sea-satellite distance about 3 cm.

The Topex/Poseidon orbit has a repeat cycle of 9.9156 days (127 revolutions), and the orbital pathes are 315 km apart at the equator, with an inclination angle of 39.5° with respect to the equator.

Due to a failure of its tracking system, the ERS1 satellite could not achieve nominal precision specifications. Cross-calibration with Topex/Poseidon enabled later on to recover a good amount of data, but as ERS1 is in phase with the sun, it is not very adequate for the purpose of identifying tidal harmonic components, some of which are also in phase with the sun.

Tidal components identification

Tides are generated by the newtonian attraction of the moon and the sun on the water of the oceans. The respective positions of the moon, and the sun with respect to the earth may be characterized by three parameters: distance, longitude, latitude. These parameters are periodic functions of 5 fundamental angles, summarized in table I.

Angle	Associated period
Mean Moon Longitude	27.32 days
Mean Sun Longitude	365.24 days
Lunar Orbit Perigee	8.85 years
Lunar Orbit Ascent	18.6 years
Solar Orbit Perigee	20 900 years

Table I
Fundamental tidal periods

Doodson ([4]) proposed a development of the generating potential into a sum of constant phase and amplitude sine functions of the local mean lunar time of the form (for the lunar potential):

$$V_L = \frac{3}{4} K M_L \frac{\rho^2}{D_L^3} \sum_i \lambda_i C_i cos(Arg_i)$$

where λ_i is a latitude factor, C_i is a constant amplitude, and Arg_i is a linear combination of the local mean lunar time and of the 5 above angles. He gave the amplitude values of the most important 400 components. This leads to a very complex decomposition, and the earlier development of Darwin ([2]) is usually prefered. This other development was fully detailed by Schureman ([11]).

The main differences are that moon longitude and latitude are referenced to the lunar orbit, and that mean solar time is used. Within each group of periods (long-term, diurnal, semi-diurnal, ...), only the most energetic ones are kept, and the other ones are accounted for by a so-called "nodal correction", *i.e.* a long term modulation f_i of the

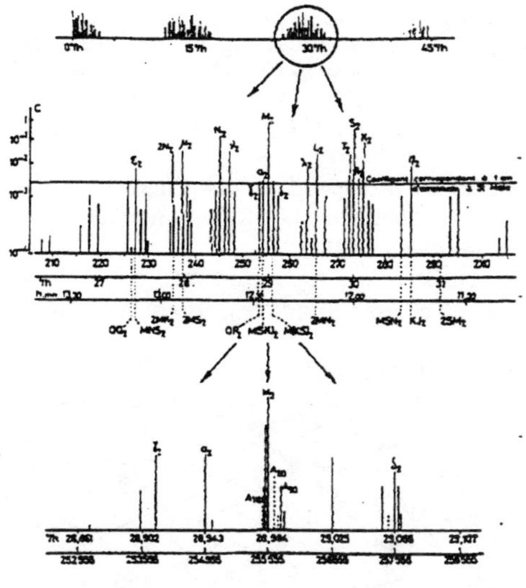

Figure 3. Complexity of the generating potential spectrum

amplitude and u_i of the phase:

$$V_L = \frac{3}{4} K M_L \frac{\rho^2}{D_L^3} \sum_i f_i C_i cos(V_i + u_i)$$

where V_i are linear combinations of the mean horary angle of the sun and of the 3 first above angles, C_i are the average amplitudes, and the f_i factors depend on the lunar orbit excentricity, slant, and period ratio to that of the sun. This generating potential is represented at figure 3.

Despite the simplification introduced by the modulation factors, the resulting spectrum retains frequency bands very close together, for instance the measured semi-diurnal ones summarized in Table II.

Component	Period
M_2	12h 25mn
N_2	12h 39mn
ν_2	12h 37mn
S_2	12h 00mn
K_2	11h 59mn

Table II
Semi-diurnal periods

Indexing refers to the group of periods which the component belongs to, with a standing for yearly, m for monthly, f for half-monthly, 1 for diurnal, 2 for semi-diurnal, ...

For coastal waters, finite depth effects create harmonic compositions of the above components. For instance, the interaction wave M_4 is the second harmonic of M_2, and $2MS_6$ is the sixth-diurnal component resulting from the interaction of M_2 and S_2, with angular frequency $\omega_{2MS_6} = 2\omega_{M_2} + \omega_{S_2}$.

Some of these combinations happen to have exactly the same frequency as components of the generating potential, and the corresponding components are called mixed. For instance, $2MS_2$ is mixed with μ_2, $2MN_2$ with L_2, $3MSN_2$ with ν_2.

The resulting sea level can thus be approximated by the following formula:

$$lS_{ap}(x,y,t) = S_0(x,y) + \sum_{i=1}^{N} f_i(t) S_i(x,y) cos\,(\omega_i t + (V_0 + u)_i(t) - g_i(x,y))$$

Values for V_{0i}, $f_i(t)$ and $u_i(t)$ were explicited by Schureman ([11]) for components of astronomical origin, and can be derived from the former using the theory of tides propagation in coastal waters developed by Le Provost ([7]) for interaction components.

The problem is thus the identification of the set of S_i and g_i pairs from the level measurements $S_{ob}(x,y,t)$ coming from the satellite altimeters. The difficulties stem from the fact that some frequencies are very close together, and that, although the development of Darwin avoids very long periods, the orbit period of the satellite may alias shorter ones to very long ones.

HARMONIC ANALYSIS

Least squares fit

In order to facilitate the fit, the sea level is rewritten as:

$$S_{ap}(t) = S_0 + \sum_{i=1}^{N} f_i(t) \; [a_i cos\,(\omega_i t + (V_0+u)_i(t)) + b_i sin\,(\omega_i t + (V_0+u)_i(t))]$$

The fit consists in minimizing the quadratic difference between S_{ap} and S_{ob},

$$E = \sum_k (S_{ob}(t_k) - S_{ap}(t_k))^2$$

where t_k are the measurement times.

One may note that the partial derivatives of E with respect to a_i and b_i may easily be expressed in closed form.

Handling aliasing

It is well-known that the frequency contents of a signal that stands beyond the Nyquist frequency ω_N (half the sampling frequency) gets folded back, modulo $2\omega_N$, to the interval $[-\omega_N, \omega_N]$ according to a phenomenon called aliasing and commonly experienced in the form of stroboscopic effects. Further, positive and negative frequencies can not be distinguished.

For instance, the $M2$ component, the actual period of which is 12h 25, is aliased by the Topex/Poseidon orbiting frequency to 62.1 days, *i.e.* 6.26 satellite cycles.

This gives raise to two problems that would not be encountered with conventional tide-gauges:

1. different components may get aliased to very close frequencies ;
2. some components may get aliased to very low frequencies, or even to a constant (infinite period).

In order to distinguish between two frequencies ω_i and ω_j, the duration of the measurement that is processed needs to represent an integer or a large number of times $\frac{1}{\omega_i - \omega_j}$. This would lead, for instance, to an integer or large number of times 650.4 cycles of the Topex/Poseidon satellite (17.66 years) to discriminate M_6 from L_2. The requirement is the same if one of the frequencies is aliased close to zero, *i.e.* for components aliased to infinite or almost infinite periods. This is the case for the solar component with sun-synchronous satellites such as ERS1 and ERS2.

Only 120 cycles of the Topex/Poseidon satellite were available at the time of the study, thus, in order to overcome the problem of very close frequencies, some pairs of components are merged into a "carrier" and a "secondary" wave, using information from a model or a nearby tide gauge as to their amplitude ratio and phase difference. Physical considerations are used to govern the pairing of components: for a given area, some components of similar origin or of similar frequencies do appear to have also similar amplitude and phase patterns.

It should be noted that, when the aim of the study is to remove the predicted tide from the altimeter measurements in order to extract the storm-surge, prediction is only needed at the times and places of the actual measurements. The results of the procedure are thus correct where it matters, even in the case of wrong amplitude ratios and phase differences, guessed for pairing a secondary wave with its carrier.

If the tidal coefficients were to be used for accurate tidal prediction at other times and locations, it would be advisable to proceed to further verification of the paired amplitudes and phases, either with models or field measurements.

Procedure qualification

Noise	residual Std.Dev.	components perturbation
None	2×10^{-12}	0
White 1 cm	0.3 cm	0.5 cm
White 10 cm	4.0 cm	4.6 cm
White 100 cm	45.1 cm	45.2 cm
Red 1 cm	0.3 cm	0.5 cm
Red 10 cm	5.0 cm	2.8 cm
Red 100 cm	25.7 cm	32.3 cm

Table III
Effect of measurement noise

The identification procedure was qualified by using simulated tides for an English Channel location: Saint Servan. Of the 61 components that were determined by various studies and summarized by Desnoës and Simon [3], only the 36 for which the amplitude is larger than 1 cm were considered.

The resulting (synthetic) signal was then sampled according to the Topex/Poseidon overpasses, for the 105 cycles of October 12th, 1992 to August 10th, 1995. As the chosen coastal location is subject to complex tides and the measurement duration is short, 10 component-pairings had to be introduced. Three of the remaining components, namely $2MN_6$, S_{sa} and M_6 could not be paired, and had to be imposed. Their amplitudes are respectively 1.9, 6.4 and 2.5 centimeters, and should be set in perspective with the 6+ meters of high spring tide overall amplitude in Saint Servan.

A first check on this synthetic signal verified that the least squares fit brought the residual to zero within machine accuracy. The synthetic signal was then perturbed with white or red noise. Table III summarizes the effect of this pertubation.

APPLICATION TO NORTH SEA LOCATIONS

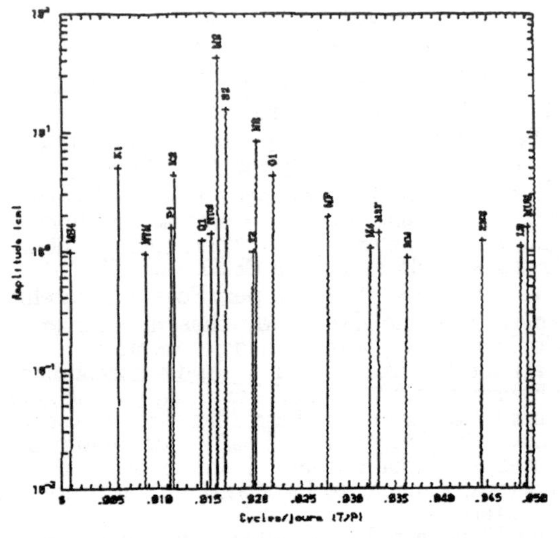

Figure 4. Spectrum of aliased components at Frigg location

The intent of this section is to apply the previously described procedure to some locations in the North Sea where *in situ* measurements are also available for comparison, and extract the residual elevation (storm surge or set-down), in order to provide input for the distribution comparisons reported in [6].

The signal to be analysed is obtained by substracting from the satellite altitude (*Orbit*) of the 120 cycles available at the time of the study:

1. the raw altimeter distance measurement (*Range*),
2. environment corrections (*Dry, Wet, Iono, Bem*),
3. the geoid,
4. terrestrial and load tides.

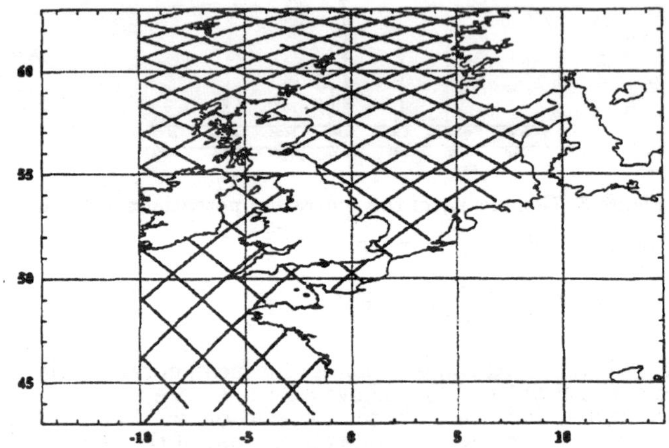

Figure 5. Orbital tracks for Topex/Poseidon

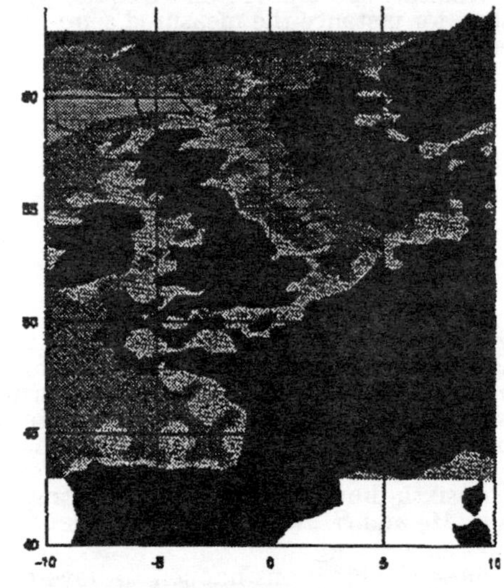

Figure 6. M_2 component amplitude

Reference measurements were obtained from field measurements on the oil platforms at Frigg, K13a and Euro.

From analysis of tides at North Sea coastal locations, about 60 components can be identified. For the sake of simplicity, only those with amplitudes higher than 1 cm were isolated from the field measurements for the present study and retained. However, the annual and semi-annual components S_a and S_{sa} were eliminated from the analysis, because of strong interference with climate seasonal variations. Using time-frequency analysis, it was verified that these components could definitely not be considered as deterministic enough for the duration of the available signal to enable to keep them in the fitting procedure. Especially, these two components appeared far from being constant, and removal from the field measured signal of their supposed amplitudes did not reduce the residual energy.

For the 19 remaining components, shown at figure 4, 6 pairings were performed, according to table IV.

Carrier	K_1	S_2	M_2	N_2	M_2	S_2
Carried	P_1	K_2	μ_2	ν_2	L_2	T_2

Table IV
North Sea pairings

As every component that might have caused a problem could be paired, it was not necessary to further impose any other component value.

The components extracted from the Topex/Poseidon measurements were compared to the Flather model for North Sea tides. The vector difference, defined as:

$$d = \sqrt{(a_1 cos(p_1) - a_2 cos(p_2))^2 + (a_1 sin(p_1) - a_2 sin(p_2))^2}$$

is less than 5 cm for the M_2 component along the satellite tracks, see figures 5 and 6, and increases close to the coasts, where the satellite altimeter fails to measure on outbound orbits.

Frigg field

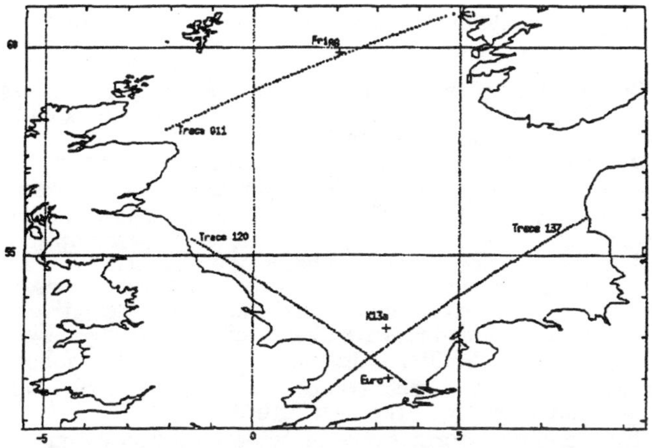

Figure 7. Locations of the reference field measurements

The measurement point at Frigg Field is located in central North Sea, see figure 7,at a water depth of 99.4 m, and only 7 km away from track 011 of Topex/Poseidon. Reference residual is computed from harmonic analysis of the hourly level measurements performed on-board the quarters platform of the field. The characteristics of these on-board measurements are such that one may truly consider that tidal components are fully eliminated, and that the reference residual is only the stochastic part of the water level elevation, related with meteorological effects.

Four "satellite" residuals were computed, and compared to this reference residual, namely:

1. T/P level minus tide obtained from T/P harmonic analysis.
2. T/P level minus tide of model FES94.1 [8].
3. T/P level minus tide of Cartwright and Ray model [1].
4. T/P level minus tide of Schwiderski model [12].

Comparison is shown on figure 8 and summarized, by decreasing order of concordance, in table V. It is clear from this table that the T/P Analysis reaches the quality of the best model.

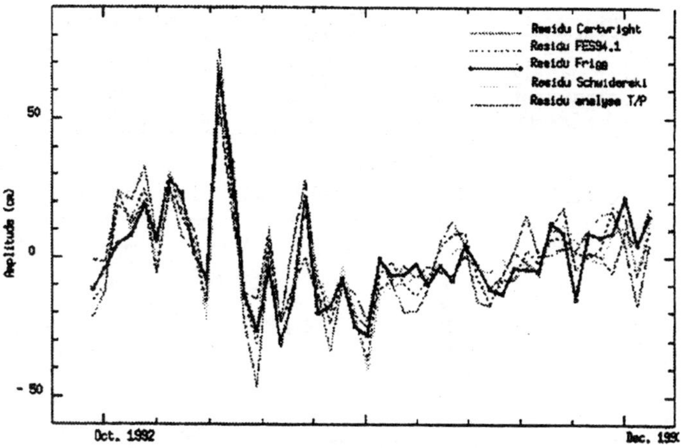

Figure 8. Residual comparison at Frigg location

Tide model	RMS diff. (cm)	Correlation
FES94.1	7.8	0.91
T/P Analysis	7.9	0.89
Schwiderski	8.4	0.88
Cartwright & Ray	11.9	0.81

Table V
Frigg T/P residuals compared to field measurements

K13a platform

The same comparisons were performed at the K13a Platform location. Water depth is only 22 m, and the closest satellite track (137) is nearly 49 km away. Because of the shallow water and of the distance between the satellite track and the field measurement, one can expect the high gradient to significantly penalize the comparison.

Results are summarized in table VI.

Euro platform

Water depth at Euro Platform is about the same as at K13a, but the closest satellite track (120) is closer (less than 19 km away). From the results presented in table VII, it is clear that tide models experience some problems at this location. It should be noted however that relatively small phase errors may lead to large decorrelation in the comparison, especially for locations where tidal amplitudes and rates of variation are large.

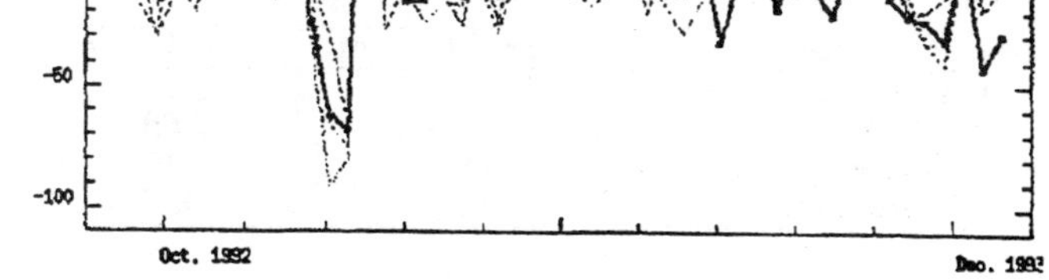

Figure 9. Residual comparison at K13a location

Tide model	RMS diff. (cm)	Correlation
FES94.1	18.3	0.82
T/P Analysis	20.8	0.74
Cartwright & Ray	21.2	0.80
Schwiderski	44.1	0.53

Table VI
K13a T/P residuals compared to field measurements

Location Location	Distance to track	T/P Analysis	Field Analysis
Frigg	7 km	-84/+81	-84/+77
K13a	49 km	-84/+98	-114/+117
Euro	19 km	-117/+145	-118/+134

Table VIII
Astronomical Tidal Ranges

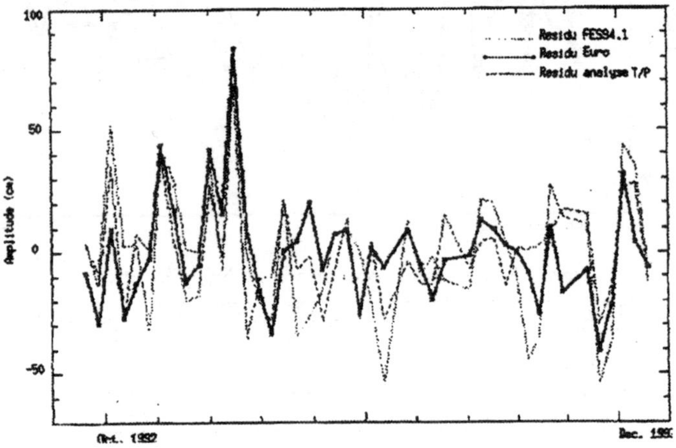

Figure 10. Residual comparison at Euro location

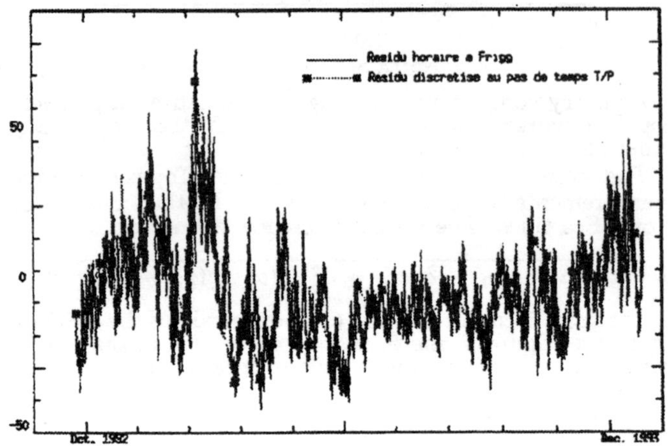

Figure 11. Residual sampled hourly and at T/P period

Further analysis

An attempt was made to utilize track crossing points to improve the sampling, and thus the harmonic analysis. It appears that no significant improvement is observed on the analysis, and that the main "quality" parameter governing the extracted residual is the distance to the nearest track. Analysis, as it is performed, is of good accuracy and difficult to improve.

This is confirmed by comparison of the prediction of the astronomical tidal range from T/P analysis and from field measurements analysis, summarized in table VIII.

Because of high gradients, the tide components amplitudes can thus only be assumed to be correctly identified along the satellite tracks. It is very likely however that the residual (storm surge or set-down) is estimated along these tracks with an accuracy such as that for Frigg, and that the larger differences observed at the other experimental locations can be fully explained by their distance to the nearest track.

One should note that only 120 samples of the storm surge are available at a given location. One can see on figure 11 how few values this represents with respect to hourly sampling. In order to estimate distributions and extreme values, this number is clearly too low, and a study of spatial variability and coherence, in order to gather a significant population over a larger area is then necessary.

CONCLUDING REMARKS

Tidal components were identified from the Topex/Poseidon altimetric measurements. Along the satellite tracks, and in the North Sea case, these components are of similar accuracy as those of the best numerical models available to date.

Tide model	RMS diff. (cm)	Correlation
T/P Analysis	16.5	0.70
FES94.1	20.7	0.69
Cartwright & Ray	50.3	0.11
Schwiderski	85.8	-0.09

Table VII
Euro T/P residuals compared to field measurements

Qualification of the procedure also shows that errors less than 10 cm can reasonnably be expected along satellite tracks for any tidal configuration.

Due to the relatively short duration of operation of the Topex/Poseidon satellite, some components can however not be distinguished one from the other. A procedure was devised that solves the problem by the use of assumptions with respect to their relative phases and amplitudes. This procedure of pairing components is such that if it is reasonably applied, the extracted residual should be insensitive to errors in the assumed relative phases and amplitudes. With the on-going of the T/P mission, and the foreseen follow-up missions, most of these components pairing should even become unnecessary in the near future.

The residual (storm surge or set-down) was estimated from satellite alone with an accuracy better than 8 cm for a North-Sea location very close to a satellite track. As long as coastal proximity does not impair the satellite measurements, distance to the nearest track seems to be the only significant factor that may increase the error in that estimation of the residual.

Due to the low number of samples at a given location, it would be necessary to utilize measurements over some sizeable geographic area in order to estimate distributions and predict extreme values. Discussion of this problem can be found in Hajji & Olagnon [6].

ACKNOWLEDGEMENTS

The present study was performed within the SCHOS (Surges and Wave Characteristics by Satellite Observation) project of the Clarom, co-funded by the French Fonds de Soutien aux Hydrocarbures. The participants and contributors were Ifremer, IMG/LEGI, Météomer, Optimer, STN-MTE, Elf Aquitaine Production, and EPSHOM. Their permission to publish is gratefully acknowledged here.

REFERENCES

[1] **Cartwright, D.E. & Ray, R.D.** "Oceanic tides from Geosat altimetry", *J. of Geophysical Research*, Vol. 95,

(1990), pp. 3069–3090.

[2] **Darwin, G.H.** "Report on harmonic analysis of tidal observations", Brit. Ass. for Adv. of Sciences, (1883), pp. 48–118

[3] **Desnoës, Y. & Simon, B.** "Analyse et prédiction de la marée - Application aux marées du Havre et de Brest", EPSHOM Report No 441, Brest, (1975).

[4] **Doodson, A.T.** "The harmonic development of the tide generating potential", Proceedings of the Royal Society, A - vol.100 (1921), pp. 305–328

[5] **E&P Forum** "Overview summaries" *Workshop on uncertainties in the design process* Report No 3.15/219 London, 1995.

[6] **Hajji, H. & Olagnon, M.** "Contribution of satellite data to storm surge climatology", proc. 7th ISOPE (1997).

[7] **Le Provost, C.** "Contribution à l'étude des marées dans les mers littorales. Application à la Manche", *Thèse d'Etat*, Grenoble, (1974).

[8] **Le Provost, C., Genco, M.-L. & Lyard, F.** "Spectroscopy of the world ocean tides from a finite element hydrodynamic model", *J. of Geophysical Research*, Vol. 99 C12, (1994), pp. 24777–24797.

[9] **Le Provost, C., Bennett, A.F. & Cartwright, D.E.** "Ocean Tides for and from TOPEX/POSEIDON", *Science*, Vol. 267, (1995), pp. 639–642.

[10] **Le Provost, C., Genco, M.-L. & Lyard, F.** "Modeling and predicting tides over the World Ocean", *Coastal and Estuarine Studies*, Vol. 47, (1995), pp. 175–201.

[11] **Schureman, P.** "Manual of harmonic analysis and prediction of tides", Special Pub. N°98, US Dept. of Commerce - Coast. and Geoditic Survey, (1958).

[12] **Schwiderski, E.N.** "On charting global ocean tides", *Rev. of Geophys. and Space Phys.*, 18, (1980), pp. 243–268.

Proceedings of the Seventh (1997) International Offshore and Polar Engineering Conference
Honolulu, USA, May 25-30, 1997

ISBN 1-880653-28-1 (Set); ISBN 1-880653-31-1 (Vol. III)

Contribution of Satellite Data to Storm-Surge Climatology

Hafedh Hajji
Météomer
Puget sur Argens, France

Michel Olagnon
Ifremer
Brest, France

ABSTRACT

Accurate knowledge of extreme SWLs (Still Water Levels) is required at the design stage of many fixed offshore structures, either to prevent the risk of an insufficient deck clearance for extreme crest heights, or to avoid tether detensionning for TLPS. With respect to the LAT (Lowest Astronomical Tide) bathymetric level, the SWL needs to take into account tidal amplitude and "storm-surge", which are both of metric order of magnitude in the North Sea.

Altimetric satellite measurements have been extensively used in the last years to improve tidal amplitude prediction. Since the accuracy of both level measurements and tidal predictions is now reaching the centimeter order, it becomes possible to extract from satellite measurements the storm surge residual, and to study its characteristics as those of any other metocean parameter.

The present paper discusses briefly the extraction of the storm surge residual from satellite measurements. It then studies the statistical distribution of storm surges for several locations in the North Sea with special emphasis on the preparation of extrapolation to extremes: separate analysis of pressure and other components of the storm surge identification of independent samples, definition of climate consistent areas.

Down-surges are also investigated.

Conclusions are drawn as to the definition of a design storm-surge from satellite measurements alone, and possibilities of extension to other areas than the North Sea are discussed.

KEYWORDS: Storm surge, water levels, satellite altimetry.

INTRODUCTION

Accurate knowledge of extreme SWLs (Still Water Levels) is required at the design stage of many fixed offshore structures, either to prevent the risk of an insufficient deck clearance for extreme crest heights, or to avoid tether detensionning for TLPS. With respect to the LAT (Lowest Astronomical Tide) bathymetric level, the SWL needs to take into account tidal amplitude and "storm-surge", which are both of metric order of magnitude in the North Sea.

Altimetric satellites measurements have been extensively used in the last years to improve tidal amplitude prediction. The accuracy of both level measurements and tidal predictions is now reaching the centimeter order, and it becomes possible to extract from satellite measurements the so-called "storm surge" residual, and to study its characteristics as those of any other metocean parameter.

In the first section, we introduce briefly the extraction of the storm-surge residual from satellite measurements. Details of the processing of altimetric satellite measurements in order to identify the tidal components and the mean water level will be found in another paper presented in the same conference ([7]).

As reported by Le Provost *et al.*, only one measurement every ten days can be obtained with the Topex/Poseidon satellite. When considering a given location, this means that only about 120 measurements were available at the time of the study, and that this number is not increasing quickly.

In order to assess storm surge distributions with sufficient accuracy and confidence to derive extremes, one might bring in data from other satellite tracks if they are close enough, and use longer portions of each satellite track than the only closest point. To this aim, a first step is to define how far and how long are acceptable, how many statistically independent samples come out of the process, and what is the correspondence between these geographical extents and durations at a single point, in order to determine the ratio of extrapolation to a given project life.

We study in a second part the identification of independent surge-states defined in terms of SWLs. It is followed, in a third part, by the statistical analysis of extracted surges and set-downs distributions.

An attempt is made to reduce the statistical variability of the storm surge by extracting the barometric pressure dependent part. As long-term databases exist for this part, it would then lessen the consequences of the uncertainty on extreme storm surge created by the small size of the database used for extrapolation.

STORM SURGE RESIDUAL DERIVATION

Still Water Level (SWL) is the instantaneous average of the water surface level, removing waves. It can also be

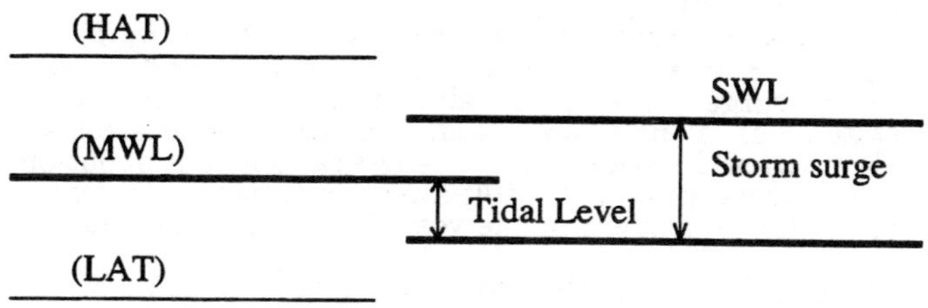

Figure 1. Water levels

referenced to the long-term average level, Mean Water Level (MWL). The difference between SWL and MWL consists of the tidal amplitude and the storm surge (or set-down), as shown in Figure 1.

The derivation of SWLs from altimetric satellite measurements is discussed in reference ([7]). The intent of the present section is to introduce briefly the extraction of the storm surge residual from satellite measurements.

Satellite measurements can provide the mean water level (MWL) by long-term averaging. The Topex-Poseidon satellite was specially designed for high precision altimetry and orbitography in order to study topography, and its measurements could thus be used for the present study.

The principle of the water level measurements is as follows:

- each altimetric measurement provides, after environmental corrections (tropospheric, earth tide,...) the distance between the ocean surface and the satellite every second,
- considering the knowledge of the satellite position at each measurement instant, one can compute the sea level height with respect to a fixed referencial,
- the difference between the level obtained and its long term average can be decomposed into oceanic tide and storm-surge.

"*Storm surge*" accounts for the static effect of barometric pressure and wind shear, for the solitary waves that may be generated by the displacements of pressure lows, for currents effects, and for many other unknown sources.

To analyse the mean water level at a given location, one has to assess and remove the tidal contribution. In [7], a method to assess the mean water levels and the tides from satellite data is described. This method was applied to Topex-Poseidon measurements as well as to in-situ measurements, giving the storm-surge residual data bases used in this study.

DEFINITION OF SURGE-STATES

Storm surge is a continuous function $\delta(x, y, t)$. One may define a surge-state as a convex domain in x, y, t where all δ values belong to the same distribution, that is taken as a normal one $\mathcal{N}(\Delta_0, \sigma)$. We can measure the δ function along a line (satellite track), and for some single time value: $\delta_\tau(x) = \delta(x, y = f(x), t = \tau)$, or at a single point (tide gauge), during some time duration: $\delta_\chi(t) = \delta(x_0, y_0, t)$. In both cases, a surge-state appears as a segment of constant ordinate distribution in the measurements.

On either $\delta_\tau(x)$ or $\delta_\chi(t)$, we may apply the identification techniques for stationary segments proposed by Labeyrie *et al.* [3, 4, 5] on the basis of an Hinkley [1] test. The procedure progresses along the function history, searching for transitions from one stationary constant mean distribution to another one. Within a surge-state, values are of the form $\delta = \Delta + \theta$, where Δ stands for the current estimate of the surge-state average, and θ is a random variable with normal distribution $\mathcal{N}(0, \sigma)$. In case of a transition to another surge-state, $\theta = \delta - \Delta$ would belong to $\mathcal{N}(\mu, \sigma)$, with $\mu \neq 0$.

At each new sample, the detector thus tests the two hypotheses $\theta \in \mathcal{N}(0, \sigma)$ and $\theta \in \mathcal{N}(\mu, \sigma)$. In that way, the signal is split into a sequence of segments: surge-states. Considering these individual surge-states, and retaining only one value within each of them (maximum, average or minimum, depending on the application), considerably improves the independence of the statistical samples, though complete independence is only asymptotically ensured.

The durations of the surge states are a key factor for extrapolation in the case of discontinuous measurements such as satellite ones. The equivalent duration for the database is approximated by the number of surge states times their average duration. With the above definition, it is possible to use surge states identified along satellites tracks, and to affect to them average durations coming from field measurements, since both identification procedures rely on a consistent definition.

Stationarity of in situ measurements

The three same locations as in reference [7] were used. Their locations are recalled in table I.

The transition detector uses five parameters. Two of them, α and β, characterize the intrinsic statistical variability of the variable of interest (surge) within a stationary segment. The variability is expressed as a linear function of the variable: $\sigma = \alpha\Delta + \beta$. It can be computed by regression of the absolute value of the sample to sample increments over the samples values.

Two other parameters are used to characterize threshold uncertainty on the variable: $\mu_{min} = A\Delta + B$. A and B are chosen according to one's knowledge of the problem. In our case, the results of [7] lead to prefer a small proportional dependence, 5%, as error is mostly independent of surge, and an additional constant of 5 cm, as this is the order of magnitude of the differences between different models and field data.

The fifth parameter, s, is an adimensionalized detection threshold [3].

The chosen set of values for time histories of the surge at Frigg is thus: $\alpha = 0.013$, $\beta = 1.5$ cm, $A = 0.05$, $B = 5$ cm, $s = 7$. This leads to an average stationary duration of 6.5 hours for the Frigg location. The sets of parameters for the other locations are slightly different in α and β, and give a little less than 4 hours for the two other locations.

Because of the importance of heuristics in this choice, a sensitivity study was performed around these values. Around the retained values, each of the A, B, and s parameters was allowed to vary.

Stationarity along satellite tracks

The same transition detector as for time histories may be used for space lines along the satellite tracks. The set of retained parameters at Frigg in this case is the same as for time histories, apart from a small change of α to 0.012.

The average stationary length along satellite tracks is then about 60 km for Frigg location, and 40 km for the other ones.

One might infer from the above results that stationary lengths and durations are linked by a kind of relationship, close to one hour per 10 km, as can be verified on figure 2. This would however require further confirmation for other locations, and the experimental conditions did not allow a comparison of individual geographical and temporal surge states. If such a relationship could be further assessed, field measurements would not be necessary anymore and

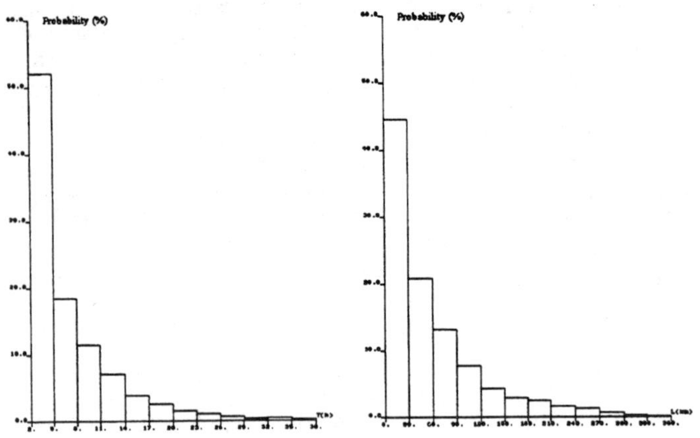

Figure 2. Frigg - Surge states durations and lengths

the equivalent database duration could be determined from satellite measurements alone.

Surge states characteristics

Statistics made with satellite data pertaining to a single location would suffer from an insufficient number of surge states: the 120 satellite cycle values available would be equivalent to about 600 hours at Frigg, *i.e.* 25 days, which clearly do not enable extrapolation to design values. However, if the surge climate can be shown to be consistent over some area around the location of interest, one may bring in several values from each track, and even values from other nearby satellite tracks to approximate the long-term distribution.

In order to define the size of surge climate consistent areas, circular areas of increasing diameter around Frigg station were studied. The statistical characteristics of the distribution depart from the in-situ ones when the radius exceeds 200 km.

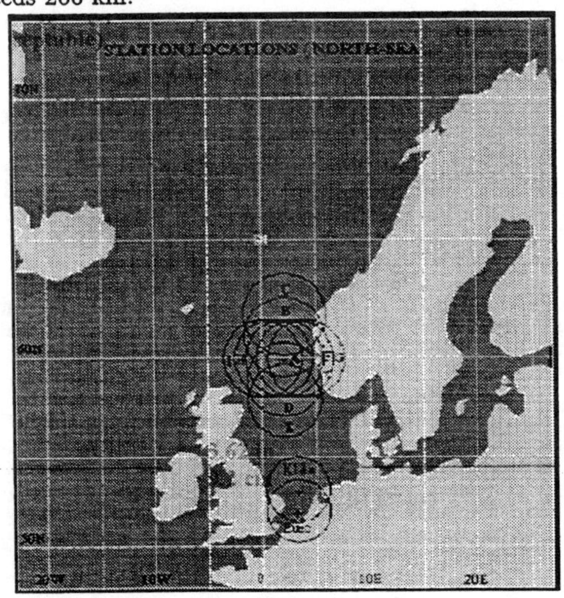

Figure 3. North Sea locations

In order to refine these areas, nine shifted areas around Frigg station were chosen (figure 3). From Table II, it appears that the gradient in the distributions is significant in the N-S direction, as can be seen from the differences in fitted 3.3 year extremes between areas located in the north (B and C) and areas located in the south (D and E), and much less in the E-W direction.

It thus appears that an elliptic area of 600 km by 300 km with an E-W major axis would enable to consider a consistent surge climate, and allow to use about 2000 surge states, i.e. a little more than one year of equivalent single location duration. The coverage ratio is thus about one third: three years of satellite measurements are necessary to get the equivalent of one year of field measurements.

COMPARISON OF DATABASES - SUITABILITY TO EXTRAPOLATION

General statistics

Table III compares the surge statistics performed on 15 months of data for the tide gauges of the three stations, and the Topex/Poseidon statistics for the same period and the 200 km radius area around the locations. The tidal components were estimated from the measurements using the techniques detailed in [7], and the computed tidal SWL removed from the measured one, yielding the surge residual. The sampling period is 10 minutes for the field data.

Figures 4 and 5 show the scatter-diagrams of extracted surge as a function of significant wave-height for Frigg location (For T/P surges, the wave height is the satellite altimeter measured significant wave height).

Representativity

When considering the aforementioned 15-month period, it is remarquable that for the Frigg location, the numbers of identified surge states are similar for field and satellite measurements. The estimation of extreme storm surges can in that case be considered to have a similar base for extrapolation as would field measurements for the duration of operation of the Topex/Poseidon satellite.

In comparison, the Euro and K13a locations are disadvantaged, mostly because proximity to the coast reduces the area over which surges may be considered.

Extremes

Analysis and prediction of design extreme values of surge are based on the works of Labeyrie [2]. This approach allows to:

- take into account the different time-scales (stationary states, overall measurement period, planned life length of the project) involved,
- take care of the dynamics of the large values [4],
- give estimates of uncertainties and confidence intervals.

The prediction is implemented in the STEXOM package [2], and provides, for a given planned life of the structure, the statistical distribution of the expected maximum of a parameter over this life span. One obtains thus exceedance probabilities, as well as the associated confidence intervals.

It proceeds through three steps:

1. fit a parametric model to the observed distribution, with special emphasis on the goodness of fit on tail part of the distribution,
2. find the type of Fisher-Tippett law to which the distribution of the maximum of the observed set is the most likely to converge, and its parameters,
3. extrapolate from the duration of the measurements to that of the projected life of the structure.

" Bounding " model parameters, corresponding to a given confidence ratio, are also computed in STEXOM at the fitting phase. They are also extrapolated and provide " bounding " distribution for extreme values.

Extreme values are computed separately for up-surge and for down-surge. Table IV shows the 2- and 10-year values for the field and satellite measurements.

Agreement is better for down-surges than for up-surges, and of course for the Frigg location where the database represents a larger sample set than for the other locations.

When examining these results, one should also take note that a very exceptional single event occurred in February 1993, and that extrapolation is very sensitive to the measurement of this single event, that is only partially taken into account by Topex/Poseidon.

EFFECT OF BAROMETRIC PRESSURE

The static effect of barometric pressure on surge is well-defined: 1.01 cm/hPa. Surge can thus be decomposed into a static pressure component and a residual, in the expectation that prediction of the static pressure component can be made with great efficiency due to the long history of available measurements and to the availability of weather numerical simulation programs. The residual should be of reduced variability and lead to less extrapolation uncertainties if one could consider surge as "static barometric pressure effect + noise".

Frigg location

Barometric measurements were performed on the Frigg location, and it is thus possible to remove the actual pressure effect from the surge measurements. This was done for the tide-gauge data.

For the satellite data, the pressure information issued by the European Centre for Medium range Weather Forecast (ECMWF) numerical model was used. This procedure accounts for the assumption that field information should not be expected to be available for satellite measurements. The statistics are summarized in table V.

It appears that variability is only reduced by about one fourth, but that the distribution of surges is brought closer to a gaussian distribution.

It should be noted that pressure and H_S are very correlated variables. The above results can be used to infer that surge and H_S are only weakly dependent, and this can be verified on figures 4 and 5.

A very important consequence for design is that large set-downs may occur even for high H_S values, and have to be specially accounted for in the design of TLPs. For instance, two of the largest observed H_S at Frigg, close to 10 m, occurred jointly with a set-down of 20 to 30 cm of the SWL.

Euro and K13 locations

As no pressure measurement is available for these locations, only removal of the ECMWF pressure from the satellite measurements was attempted. The order of magnitude of the variability reduction is the same as for Frigg ($\frac{1}{4}$), and even less for Euro.

CONCLUSIONS

Surge and set-down values were obtained for 3 locations of the North Sea by extraction from altimetric measurements of satellite Topex/Poseidon.

Surge states of stationary surge value were defined, and their durations (field measurements) and lengths (satellite measurements) extracted. These durations and lengths are dependent on the location, but might be related by a proportionality factor.

Although the satellite operates on a ten-day cycle, it is possible, with the help of the above definition, to make up for the small number of measurements at a single location by analysing data over some larger geographic area. Distributions could be estimated in that way, that can be assumed of equivalent representativity as field measurements during one third of the satellite flight duration.

Extremes may thus be estimated from the now 5+ years of Topex/Poseidon measurements with similar confidence as from 2 years of field data. With the continued mission of Topex/Poseidon, this confidence is continuously increasing.

It appears from the present study that extreme set-downs may occur for high H_S values, and that this risk should not be underestimated for TLP design.

It should be possible to extend most of the steps of the study to areas where no field measurements are available, especially as more and more altimeter measurements become available. To this aim, it would be necessary to verify on other locations the relationship between stationary surge states lengths and durations.

One should take care however that these results are limited to the area of operation of the satellite, i.e. that nearshore coastal locations cannot be studied that way without further research.

Lastly, removal of the effect of barometric pressure alone cannot significantly reduce the variability of the surge level.

ACKNOWLEDGEMENTS

The present study was performed within the SCHOS (Surges and Wave Characteristics by Satellite Observation) project of the Clarom, co-funded by the French Fonds de Soutien aux Hydrocarbures. The participants and contributors were Ifremer, IMG/LEGI, Météomer, Optimer, STN-MTE, Elf Aquitaine Production, and EPSHOM. Their permission to publish is gratefully acknowledged here.

REFERENCES

[1] **Hinkley, D.V.**, "Inference about the change-point from cumulative sum tests", *Biometrika*, **58**, pp.509-523

[2] **Labeyrie J.**, "STEXOM: A statistical package for the prediction of extreme values", *Proc, 4th Int. Conf. On Engineering Software*, Springer Verlag, Berlin, ch.15 (1985), pp.17-32

[3] **Labeyrie J.**, "Stationary and Transient States of Random Seas", *Marine Structures* 3(1990), pp.43-58.

[4] **Labeyrie J.**,"Time Scales and Statistical Uncertainties in the Prediction of Extreme Environmental Conditions", *Reliability Engeering and Safety*, 32 (1991), pp.243-266

[5] **Lavielle M. & Rio E.**, "Identification d'états de mer stationnaires ou de transitions par détection de ruptures d'un modèle", *Rapport de recherche, Mathématiques appliquées*, Laboratoire d'études mathématiques des phénomènes aléatoires de Brest (1987).

[6] **Tournadre J. & Robin A.**, "Analyse des échelles de temps et d'espace de la hauteur significative des vagues à la Plateforme QP du champ de Frigg", DRO-EO,DITI, Ifremer-Brest, 27 pp.

[7] **Le Provost, C., Rabilloud, Olagnon M.**, 1997, "Assessment of mean water levels and tides from satellite data", proc. 7th ISOPE (1997)

Station	Latitude	Longitude	Depth
Frigg Field	02° 03' 53".825 E	59° 52' 42".421 N	99.4 m
Euro Platform	03° 16' 35".000 E	51° 59' 55".000 N	20-30 m
K13a Platform	03° 13' 13".000 E	53° 13' 04".000 N	22 m

Table I
Locations of the in situ measurements

Area	Samples	min	max	std. dev.	3.3yr min	3.3yr max
A	37414	-0.48 m	0.64 m	0.13 m	-0.41 m	0.37 m
B	31068	-0.48 m	0.64 m	0.13 m	-0.43 m	0.34 m
C	20804	-0.48 m	0.64 m	0.13 m	-0.38 m	0.31 m
D	38804	-0.50 m	0.69 m	0.13 m	-0.43 m	0.51 m
E	39087	-0.58 m	0.69 m	0.14 m	-0.58 m	0.56 m
F	33313	-0.48 m	0.64 m	0.13 m	-0.41 m	0.37 m
G	27611	-0.48 m	0.64 m	0.13 m	-0.45 m	0.36 m
H	36425	-0.48 m	0.64 m	0.13 m	-0.41 m	0.37 m
I	28957	-0.48 m	0.73 m	0.13 m	-0.41 m	0.35 m

Table II
Statistics for 9 areas around Frigg station

	T.Gauge Frigg	T.Gauge Euro	T.Gauge K13a	T/P Frigg	T/P Euro	T/P K13a
Nbr. states	1747	2990	3093	1548	418	1081
Std. dev.	0.15 m	0.24 m	0.25 m	0.13 m	0.20 m	0.21 m
Min	-0.39 m	-0.99 m	-1.16 m	-0.48 m	-0.79 m	-0.79 m
Max	0.78 m	2.08 m	1.81 m	0.64 m	1.06 m	1.24 m
Skewness	0.9	1.1	0.9	0.5	0.6	1.2
Kurtosis	4.6	9.0	6.1	3.5	5.0	7.7

Table III
Surge statistics comparison (1992/10/1 to 1993/12/31)

	T.Gauge Frigg	T.Gauge Euro	T.Gauge K13a	T/P Frigg	T/P Euro	T/P K13a
equiv. duration	3 yr	3 yr	3 yr	3 yr	0.5 yr	1.2 yr
2-years up	0.80	1.7	1.6	0.62	1.4	1.5
10-years up	0.93	2.0	1.9	0.72	*1.6*	*1.8*
2-years down	-0.39	-1.1	-1.2	-0.37	-1.1	-0.91
10-years down	-0.44	-1.3	-1.3	-0.45	*-1.2*	*-1.1*

Table IV
Surge extremes comparison

	Tide-Gauge SS - pe	Tide-Gauge raw SS	T/P SS - ECMWF pe	T/P raw SS
Mean value	- 0.02 m	0.00 m	- 0.04 m	0.00 m
Standard deviation	0.11 m	0.14 m	0.12 m	0.14 m
Min	- 0.44 m	- 0.39 m	- 0.52 m	- 0.48 m
Max	0.51 m	0.78 m	0.37 m	0.64 m
Skewness	0.37	0.87	-0.13	0.37
Kurtosis	3.67	4.65	3.25	3.26

Table V
Effect of pressure removal on statistics (1992/1/1 to 1993/12/31)

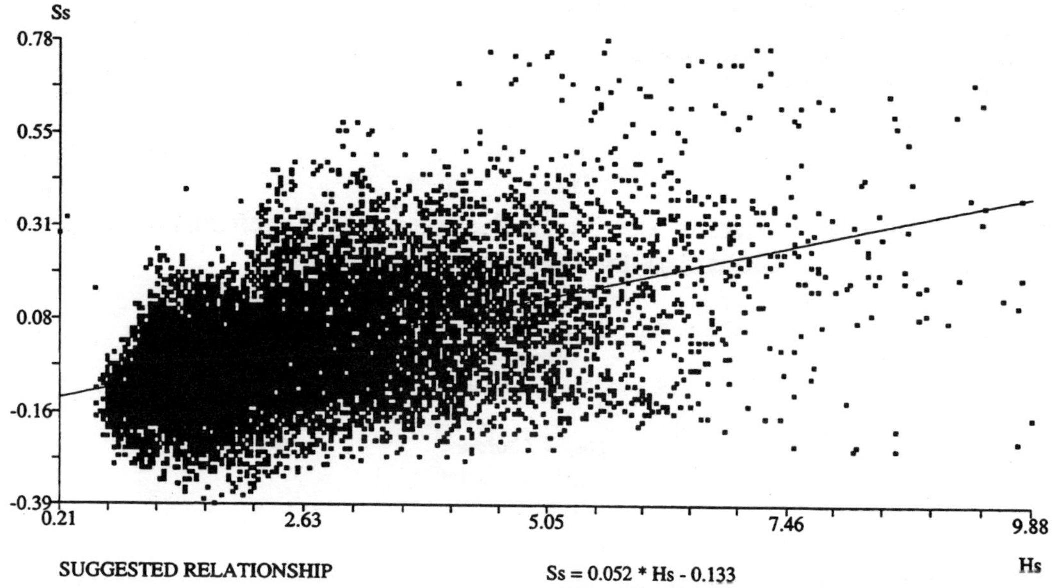

Figure 4. Frigg - In-situ Surge vs. H_S

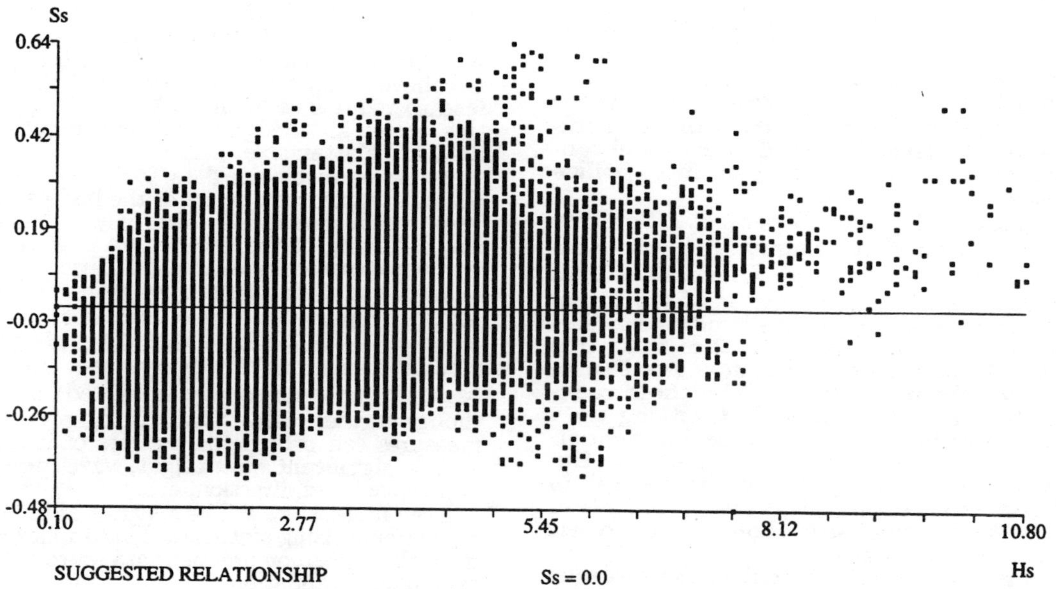

Figure 5. Frigg - T/P Satellite Surge vs. H_S

Proceedings of the Seventh (1997) International Offshore and Polar Engineering Conference
Honolulu, USA, May 25-30, 1997

ISBN 1-880653-28-1 (Set); ISBN 1-880653-31-1 (Vol. III)

SUCCESS: A Joint Project to Use Novel Techniques to Teach Future Engineers About Satellite Metocean Data Sources

Michel Olagnon
Ifremer
Brest, France

Georg Lindgren
Lund University
Lund, Sweden

ABSTRACT

Data sources have always been a common concern for engineers requiring metocean parameters for their studies. The Ocean Engineering and Naval Architecture professions have been asking for improvement and extension of the available data sources in all their committees and workgroups [2, 13]. However, the main bulk of engineers are still not sufficiently aware of the specificities of satellite obtention of metocean data to be in a position to utilize them efficiently.

The EU sponsored success project (Satellite Uses: a Common Course for Engineering and Sea-conditions Studies) gathers some Universities and Engineering Schools of eight countries of the European Union, Ifremer and two small metocean services companies. It stems from the fact that, due to the novelty of the subject, professors need some material to help them include it into their regular teaching. The project is aimed at setting up a teaching package to be used in Engineering Schools and Universities of Technology throughout the European Union. It will provide all the "hardware" in the form of textbook, excercises, videos, and other graphical aids, that a professor should wish to use in order to deliver the course.

The following of this paper describes the foreseen usage of videos in this package, in order to:

- **Provide a "portable animated blackboard".**
- **Relate the actual metocean phenomena to their mathematical and statistical descriptions and measurements.**
- **Show how metocean parameters are used in industrial applications.**

It should however be noted that the paper does not present significant scientific results but rather focusses on an initiative to integrate swiftly advances in measuring and processing techniques into the current practice by addressing the training of future practitioners.

KEYWORDS: Metocean data sources, teaching, satellite measurements.

INTRODUCTION

The design, sizing and operation of ships and offshore structures exhibit a strong sensitiveness to metocean parameters (wind, wave, and current), sometimes a linear dependency, but sometimes quadratic or even with higher exponents. It is believed that better knowledge of these parameters can lead to reductions in the risks and uncertainties, and that this better knowledge would only require more measured data to be attainable.

As far as metocean data sources are concerned, one of the most significant points of the last years is that satellite data, which were mostly used only for research purposes at the beginning of the decade, are now getting widely used for operational engineering applications. A review of available references on this subject will soon be available in the ISSC 97 report [3]. Either used as such, or with the help of numerical and statistical models, satellite remote sensing measurements may shortly represent the primary source of "actual" (not numerically simulated) metocean data. Although further research work is still needed, remote sensing measures can provide parameters of interest to engineers, such as significant wave height, waves period and direction, wind speed and direction, etc.

Even more, they will be a source of full data sets and not only summarizing statistics. This implies that they require, more than common practice data sources, sophisticated processing and validation.

Due to the novelty of these techniques, their potential users are faced with two problems:

- they have to be made aware of the availability of satellite measurements, and of the relative differences between the miscelleanous data sources ;
- they need to understand the characteristics of these data and of their statistical representations to a sufficient extent, not only to use them, but also to discuss with certification authorities which have not yet issued "off the shelf" rules of usage.

A group of academia, research organizations and metocean services providers thus set up a training project: SUCCESS (Satellite Uses: a Common Course for Engineering and Sea-conditions Studies) to address mainly the academic curriculum of the engineers, before they become active. It tries to fulfil the need that professors have of some material to help them include the subject into their regular teaching.

The aim is to set up a teaching package, to be used in Engineering Schools and Universities of Technology throughout the European Union. The partners will make available all the "hardware" that a professor should wish to use in order to deliver the course. The project outcome thus consists of a textbook, of overhead transparencies, especially graphs and computer outputs, of exercises, and commented solutions as separate sets, and of some introductory and motivating video presentations mixing the supporting scientific paradigms with actual relevant industrial problems and their treatment.

The course is split into 3 modules, each aiming at the presentation of one step of Earth Observation (EO) data applications: sources, dedicated processing, practical use. It will be prepared in several of the EU languages, and be made available, at no more cost than handling charges, to any School or University of the EU. Professors might then choose to deliver the course themselves, or to request the assistance of a correspondent within the group involved in the development of the course.

It is considered that at least 30 educational institutions in the EU might include the course into their regular cycle.

A student having received this teaching should then be able to understand, specify and use metocean parameters coming from satellite measurements with no more pain than if they came from any other more commonplace origin. She/He should also be aware of the relevance of some related research problems with regards to practical industrial applications.

IDEAS TO BE CONVEYED

In some instances, engineering for the offshore industry might seem to be reduced to a blind application of standard, agreed upon, recipies. Indeed, the profession has issued standards and recommendations, and is constantly updating them, see Johansen *et al.*[4].

However, we believe that in teaching future engineers, one is faced with three challenges that one may not ignore:

1. give them the "engineering sense" that enables them to verify, even though slide-rules are not used anymore, that the output of the application of a standard procedure matches the expected order of magnitude, and that the governing assumptions were not obviously violated.
2. have them recognize the progress of the tools and information that they use, and let them find ways to introduce these novelties in the design process as soon as they are reliable, without waiting, in a chicken and egg issue, for general integration in the State of the Art.
3. get them ready to contribute to the emergence of new concepts, that will require somewhat different techniques for their design, but that will bring increased exploitation safety and profit.

The present project deals mainly with the second challenge, but should also try to contribute to the other two. Above all, we should not assume that students already have any prior knowledge of this engineering sense.

Due to the large contribution of waves to the definition of extreme and operational loads on offshore structures, most ideas to convey should be related to sea-state characterization in terms of waves. By the words extreme and operational, we mean rare events and those more common conditions that influence fatigue of materials and the logistics of normal operation. This implies that the characterizations all have to be made by means of statistical distributions and probabilities.

Significant wave height

Significant wave height is the basis of most descriptions of sea-states. However, its definitions leave a large range of uncertainty as to what it exactly represents. To an engineer, significant wave height can mean:

- the output of some measurement of the sea-state intensity.
- the energy contents of a sea-state.
- the basis for the prediction of the maximum wave within a sea-state.

Each of these definitions requires further analysis, especially in order to be used in connection with satellite measurements. Reference [1] gives three possibilities for significant wave height estimation from a time-history of the water surface elevation:

- the actual value $H_{\frac{1}{3},d}$, average of the highest one third of the ordered down-zero-crossing waves set (prefered), or $H_{\frac{1}{3},u}$ with the up-zero-crossing waves.
- the estimated value H_{m_0}, set to 4 times the square root of the zeroth order spectral moment of the water surface elevation signal.
- the estimated value H_σ, set to 4 times the standard deviation of the water surface elevation. This estimate, obtained from Fast Fourier Transform of the time signal, should be the same as the previous estimate (from the properties of FFT), although the values to be estimated may differ (from the effects of non linearities and finite bandwidth, Longuet-Higgins 1980 [9]) In practice, those differences also appear on the estimates. The first of these two definitions, more convenient since wave measurement records are usually stored as spectral densities, is commonly prefered.

It is noteworthy that common practice puts the estimates ahead of the actual value, to such extent that significant wave height is often opposed to $H_{\frac{1}{3}}$ in order to emphasize the "energetic" definition. However, we agree with the International Association for Hydraulic Research that the most useful value is the "statistical" $H_{\frac{1}{3}}$ definition, and that it is the one that should be taught.

It is clear that statistical characteristics of a sea-state are better grounds for extreme wave prediction than an intensity measurement that is biased by filtering and limited bandwidth. Rychlik ([12]) showed that, with no more assumptions than gaussian elevation, significant wave height could even be estimated from standard field measurements by simple level crossings counting.

In the case of satellite altimeter measurements, the sensor output is indeed related to the factual water surface elevation distribution rather than to the energetic contents:

$$m(t) = \int^t \int e(z)p(z-\eta)\, d\eta\, dz$$

where $e(z)$ is the square pulse emitted by the radar and $p(\eta)d\eta$ the water surface elevation p.d.f.

There is thus no reason to use an intermediate step of energetic parameters for extreme estimation, and we believe that the idea to be conveyed is that significant wave height is a statistical characteristic of a sea-state.

Directional spectra

Directional spectra provide a representation of the joint frequency-amplitude distribution of the wave components within a sea-state. Confusion is very easy between the spectrum itself, a mathematical entity implying the Airy model of linear superimposition of independent components, and its estimates from measurements.

When dealing with directional energy spectra one is faced with the distinction between the $(f,\ \theta)$ frequency-direction representation and the $(k_x,\ k_y)$ cartesian wave-number representation. The first respresentation lends itself to intuitive representation of theoretical models, especially in relation with the non-directional frequency spectra that may be obtained by analysis of single-point records of water surface elevation time-history. It is also the one that is commonly estimated from directional buoys measurements, thanks to the well-known 1963 Longuet-Higgins method [8], which is consistent with the Airy model.

The latter, cartesian, representation is most easily estimated by a 2-D Fourier transform of an image of the water surface elevation, such as can be obtained from airborne measurements. Satellite carried Synthetic Aperture Radars (SAR) provide such a Fourier transform, but for surface slopes instead of elevation, and modified by several complex non-linear "transfer functions". A description of these measurements can be found in [11]. The properties of the estimator are somewhat different of those of the results of directional buoys measurements. The non-linear part of the spectrum is also accessed simultaneously to the linear part by the method.

It is thus necessary to explain both representations. The $(f,\ \theta)$ one because it is the one of the model and of many measurements, the $(k_x,\ k_y)$ one because it is that of satellite measurements and of non-linear computations. However, the transformation of actual wave characteristic distributions between the two representations is not straightforward. Traditionally, the $(f,\ \theta)$ representation can easily be used for description of quantities like wave period in single point records, while the $(k_x,\ k_y)$ representation gives the statistical wave length properties; for details, see Lindgren *et al.* [7] in this conference.

Sea-state definition and time scales

As long as only single point time history measurements are considered, some very approximative notions of sea-states and time scales may be sufficient, as common practices do not induce significant deviations with respect to continuous measurements, and as results are also needed most of the time for single point time history usage.

Stationarity with respect to H_S can be used as the criterion to define sea-states [5, 6]. Due to the mechanics of wave generation, it is very unlikely in practice that the spectral shape can change without significant wave height changing simultaneously. Such a definition is thus fully consistent with the intended use of significant wave height as the conditionning variable for the prediction of the extreme wave that a fixed structure will be likely to be submitted to.

Satellite measurements represent "frozen time" views of the wave and wind fields. Even though one may apply the same techniques to identify stationary pieces in the space "slices" of significant wave height measurements along satellites tracks, relationships need to be established between time stationarity and space stationarity, because the applications remain prediction of extreme waves at single locations.

One also has to convey the notion that point to point comparisons may be impossible, but that distributions may be estimated from satellite measurements over some climate consistent area [10], and successfully compared [11].

MAKING USE OF VIDEO

It is more common to find video presentations in a commercial context than in a teaching one. Moreover, most "scientific" videos are only used to show experiments requiring special facilities to audiences in conferences where these facilities are not available.

In our case, videos were prefered to computer demonstrations because they are more portable and because it is easier to include shots of actual applications.

Yet, we did not want the course to be a stand-alone video package. The need for a textbook or teaching authority is by no means reduced because of the use of videos, but it is expected that this authority sees his(her) work made much easier from the support of such tools. During discussions with the partners, it was also decided that the textbook should not request the videos to be understood, and should stand by itself. Videos should be used as motivating introductions on one hand, and illustrations of results, and of orders of magnitude on the other hand. However, if the video is shown before and after the course to a student, (s)he should be able to understand more on the second viewing.

We now describe the intended purposes for the videos.

To give some initial motivations to the students

We did not want to take the risk to assume that working with satellite data could be sufficiently attractive to catch and keep the interest of the audience. Also, we wanted to state a number of problems in a way such that students would understand that these are actual cases to which they may be faced at any time in their future job. The intent of this part of the videos is thus not to explain or demonstrate anything, but to set up the context and provide an introduction for the explainations of the professor.

This objective can be illustrated by the first video made for the project. Although some parts of the synopsis are also meant to give orders of magnitude and to present actual applications, the most spectacular sequences were chosen to arouse interest in the audience. For instance, structures or ships are shown in extreme weather conditions.

Show the use of metocean parameters in industrial applications

Especially, this should allow the students to find by themselves the questions raised by satellite data, with the correct orders of magnitude as well for the stakes as for the information. It should also convince them that daily forecasts, whatever accurate and exact they may be, are only a small part of the requirements.

It should stress the importance of joint probabilities, such as wave-height and wave period or wave and wind occurrence, that can be accessed by the use of SAR products or by combinations of several sensors such as SAR and scatterometer. The derivation of extreme conditions from measurements should also receive emphasis.

A possible video theme under this heading would be to report, from the satellite metocean statistics viewpoint, on an actual on-going building and installation of an offshore structure. One might also consider, on a given location to location transport, the choice of routes and their effect on the chances of running into high waves.

Provide a "portable animated blackboard"

A large part of the course should consist of computer animations and graphs, in order to present some scientific concepts that are difficult to explain without them. The images will be generated after the needs are identified from the preparation of the textbook. The first part of this paper gives an idea of the notions to convey.

Some possible subjects falling into this part can be:

- Animate, on a map, a storm, having it grow and move realisticly, show a ship and barge convoy and its speed, and how the satellite moves with respect to these objects.
- Show the distance between such a convoy and the closest satellite measurement in time and in space, and how it evolves with time.
- Present geographical boxes, and data coming into them as the satellite spins around the world.

Relate metocean phenomena to mathematical and statistical descriptions and measurements

It is surprising that many students do not link correctly mathematical descriptions, statistical processes, or even sensors' outputs to the "empirical" view of the metocean conditions or of their effects on structures or the environment. The images in this part should remind them that models and measures are only incomplete approximations of a physical reality, and explain what is approximated, and how.

A possible illustration for this part could be obtained by going and taking shots of the sea, with some fixed structure to set scale, at a location close to a satellite track every time the satellite goes by. The satellite measurements could then be shown in parallel to the actual scenes.

CONCLUDING REMARKS

Satellite measurements of metocean conditions represent a dramatic step forward for the availability of sea state statistics. However, full utilization of these statistics may ask for special training of engineers that need to use them, in order to make them conscious of the essence of the data, of their differences with previously commonly used values, and of the special processing techniques that are required.

A course is being written, that makes full usage of novel techniques such as video presentations, in order to provide Professors with the material to update their teaching. It will also include a textbook, exercises and overheads.

The ideas to convey are linked both to precise mathematical or statistical definitions and to actual physical phenomena. In this context, the advantage of videos is that they can offer means of motivating the students, that they can show actual industry applications, that they provide a sort of portable animated blackboard, and that they can give a notion of orders of magnitude by relating the actual metocean phenomena and their mathematical/statistical descriptions and measurements.

The conventional material, and expecially the textbook, will be an opportunity to clarify the accuracy limits of each kind of measurement, and to discuss the intrinsic nature of the metocean parameters of interest and the uncertainties introduced by the estimation methods, in a context of engineering applications.

The need for a precise and common vocabulary for metocean applications was clearly felt in the preparation of this course, and we would recommend that such an effort should be undertaken in an international collaboration.

ACKNOWLEDGEMENTS

The SUCCESS project benefits from support and funding of the DG XII (Science, Research and Development) of the European Commission and of the European Space Agency. The partners in this project are the Technical University of Denmark, Ecole Centrale de Paris, Ifremer, University of Lund, Instituto Superior Tecnico Lisboa, Météomer, National Technical University of Athens, Satellite Observing Systems, University of Genova, and Delft University of Technology. Their contributions are gratefully acknowledged here. Our thanks go also to those who helped us learn about the use of video to convey ideas, and who were not able to participate in this project.

REFERENCES

[1] **International Association for Hydraulic Research** "List of Sea State Parameters" *Supplement to bulletin* No 52 (1986).

[2] **ISSC** "Report of the technical committee I.1 – Environmental conditions", *Proc. 12th ISSC*, Vol. 1, Elsevier ed., 1994.

[3] **ISSC** "Report of the technical committee I.1 – Environmental conditions", *Proc. 13th ISSC*, Vol. 1, Pergamon ed., 1997.

[4] **Johansen, A.R., Loppinet, A., Reeve, P.T.N., Thomas, G.A.N., Thorp, G. & Vanzini, R.** "International Standards for the Exploration & Production Industry - Their Development and Use", *Proc. Offshore Technology Conference*, OTC 7759, 1995.

[5] **Labeyrie, J.** "Stationary and transient states of random seas", *Marine Structures*, 3-1, 1990, pp. 43-58.

[6] **Labeyrie, J.** "Time Scales and Statistical Uncertainties in the Prediction of Extreme Environmental Conditions", *Reliability Engineering and Systems Safety*, 32, 1991, pp. 243-266.

[7] **Lindgren, G., Rychlik, I., & Prevosto, M.** "The relation between wave length and wave period distributions in random Gaussian waves", *Proc. 7th ISOPE*, 1997.

[8] **Longuet-Higgins, M.S., Cartwright, D.E., & Smith, N.D.** "Observations of the directional spectrum of sea waves using the motions of a floating buoy", *Ocean Wave Spectra*, Prentice Hall, 1963.

[9] **Longuet-Higgins, M.S.** "On the distribution of the heights of sea waves: some effects of nonlinearity and finite bandwidth", *J. Geophys. Res.*, vol. 85, no. C3, pp. 1519–1523

[10] **Olagnon, M., Chen, Q.-L., Charriez, P., & Lasnier, P.** "Validity of satellite data for design wave estimation in Chinese offshore areas", *Proc. Special Offshore Symposium China*, Beijing (1994) pp. 599–612

[11] **Olagnon, M.** "Use of satellite data for metocean design parameters estimation", *Proc. Oceanology International*, Brighton, 1996.

[12] **Rychlik, I.** "A note on significant wave height" *J. Ocean Engineering* vol 23 No.6, 1996, pp. 447-454.

[13] **E&P Forum** "Overview summaries" *Workshop on uncertainties in the design process* Report No 3.15/219 London, 1995.

Proceedings of the Seventh (1997) International Offshore and Polar Engineering Conference
Honolulu, USA, May 25-30, 1997

ISBN 1-880653-28-1 (Set); ISBN 1-880653-31-1 (Vol. III)

Satellite Measurements and Their Application to the Pechora Sea

Hafedh Hajji
MétéoMer
Puget sur Argens, France

Richard A. Sproson
PARAS
Isle of Wight, United Kingdom

Abstract

This paper intends to show the contribution of satellite data towards studying the extreme wave climate in the Pechora Sea in northern Russia. Paras Ltd. used satellite measurements, output from the WINCH model and output from numerical wave models developed by MétéoMer to determine design wave criteria for the proposed platform site in the Prirazlomnoye Field in the Pechora Sea. The MétéoMer models made use of the European Centre for Medium Range Weather Forecasts (ECMWF) wind fields to determine the wind forcing as well as satellite measurements to calibrate the wind field and the sea-state parameters as well as to define the sea ice coverage. The results illustrate the considerable benefit of having used these different databases for estimating the design 100-year significant wave height for the area of interest.

I/ INTRODUCTION

Nowadays, marine operators prospect all over the world and increasingly in polar regions. Whatever the location, the design and requalification of marine structures or the preparation of marine operations first require the construction of a reliable sea-state database.

Satellites - combined with wave models - bring new and powerful means for performing such metocean studies (Charriez and Olagnon, 1995 and Charriez et al, 1995). Firstly, remote sensed information allows for the well-known calibration of wave model input (winds) and output (waves). Secondly, the ERS scatterometer measurements, available in all-weather conditions, may be implemented in a procedure that allows the acquisition of a reliable knowledge of temporal and spatial sea-ice coverage and variability, thus dramatically improving wave model calculations.

In the spring of 1996, BHP Petroleum Ltd. commissioned Paras Ltd. to study the extreme wave climate within the Prirazlomnoye Field which lies in the shallow water region in the southern part of the Pechora Sea (see Figure 1). As there was no in-situ measured wave data Paras used the output from a procedure which combined satellite data and numerical wave models developed by MétéoMer together with output from the Norwegian Met. Institute's WINCH model to determine design wave criteria for the site of the proposed platform.
This paper briefly presents satellite measurements, introducing their principle and the derivation of the associated operational parameters. Some details will be given for the scatterometer data, including the method developed by the Centre ERS d'Archivage et de Traitement (CERSAT) the French Processing and Archiving Facility (PAF) to produce wind vector and ice measurements. Then, we will present the different numerical wave models as well as the procedure developed to insure satellite measurements. The implementation of this procedure will be illustrated by some results we have reached to optimize structural design parameters for the Prirazlomnoye Field in the Pechora Sea.

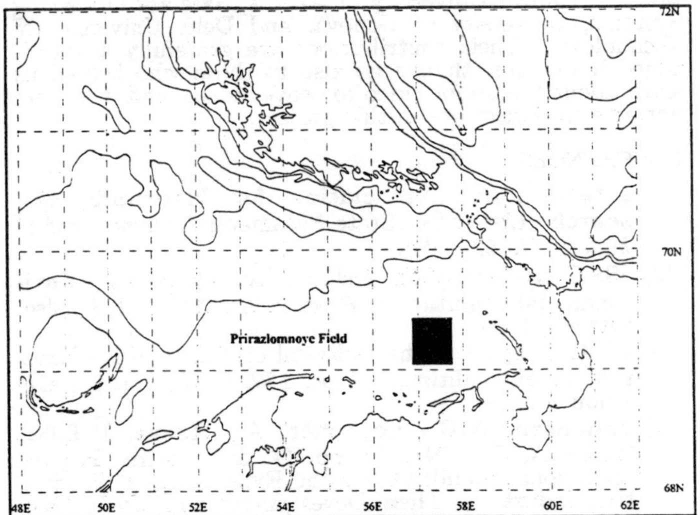

Fig. 1 - Chart of Pechora Sea showing Location of Prirazlomnoye Field

II/ SATELLITE MEASUREMENTS

Satellites have been observing the oceans for several years and have profoundly changed our insight into oceanography and the means of investigation of the marine sciences. Following the successful experimental mission of Seasat in 1978, the Geosat satellite returned reliable sea-state data during the periods March 1985 to September 1986 and November 1986 to December 1989. Since July 1991, the ERS Radar Altimeter and Synthetic Aperture Radar (SAR) satellites have significantly supplied us with information on wave heights, periods and directions. In this section, we briefly present SAR in wave mode and give a detailed description of the ERS scatterometer including ice detection.

II.1) Radar Altimeter

The Radar Altimeter is a nadir-pointing pulse radar which is designed to make measurements of the echoes from ocean and land surfaces. In the ocean mode, the echo characteristics of interest are:

- the time delay with respect to the transmitted pulse - this provides the measure of altitude;
- the slope of the echo leading edge, which is related to the width of the height distribution of the reflecting facets, and thus to wave height parameters; and
- the power level of the echoed signal, which depends on small scale roughness and thus to the nadir wind speed.

The radar echoes over ice sheets, particularly the rough surfaces at the continental margins, show much greater variance in shape than oceanic echoes. In order to maximize the data return in these areas, this mode includes features to improve its 'robustness'.

II.2) SAR in Wave Mode

The objective of this mode is to derive the ocean wave direction and wave length from small SAR images (5 km × 5 km), since the ocean surface waves modulate the radar backscattering of the sea surface. To retrieve the ocean spectrum, an ocean-SAR inversion is used. It consists in finding an ocean wave spectrum in accordance with the observed SAR-image spectrum (Bitner-Gregersen et al., 1995).

II.3) Scatterometer and Ice Detection

The ERS Wind Scatterometer Mode

The ERS scatterometer operating at the C-band frequency of 5.3 GHz (wavelength 5.7 cm) is an active microwave radar. It uses three sideways looking antennae; one pointing normal to the satellite flight path, one pointing 45° forward and the third one pointing 45° backward. The radar illuminates a 500 km wide swath with incidence angles ranging from 18° to 57°. Radar measurements are sampled on a 25 km squared grid. The result is three independent backscatter measurements relating to cell centre nodes, which have been obtained using the three viewing directions and are separated by only a very short time delay. The measurement triplet is then processed in order to compute the wind vector (speed and direction). The processing technique uses a mathematical model which defines the relationship between backscatter, wind speed, wind direction and the incidence angle of the observation.

The principal backscattering mechanism for the scatterometer measurements is the Bragg mechanism. The Normalized Radar Cross Section (NRCS) is related to the sea surface wave density spectrum in the wavelength range $L = \lambda/(2\sin\theta)$ where λ is the radar wavelength and θ is the incidence angle. The higher the wind speed, the higher the wave density in that range and the higher the NRCS values. The NRCS dependency on the wind speed is modulated with the wind direction with respect to the beam azimuth angle.

The CERSAT Algorithm and Wind Products

A mathematical model has been developed which defines the relationship between backscatter, wind speed, wind direction and the incidence angle of the observation. This model is used for all radar-wind scatterometers and is based on empirical calibration. The experience with Seasat, of relating radar backscatter to surface winds and wind stress, showed the necessity for human intervention in calibrating the retrieval algorithm.

The ERS C-Band model:
Results of campaigns led by the European Space Agency (ESA) have been analysed by Long (1985). The empirical model relating σ_0 to the wind vector is described by the equation:

$$\sigma_0 = 10^{b0}(1+b_1\cos 2\varphi + b_2\cos 2\varphi)$$

where:

$$b_0 = a+b(V-\delta)^{1/2}$$
$$\theta = (\theta_i-36)/19$$
$$\varphi = \phi-\Psi$$

a, b and δ are arrangements of polynomials of order up to three in θ, b_1 and b_2 are arrangements of polynomials of order up to two in θ and V. θ_i is the incidence angle, V is the wind speed, ϕ is the wind direction and Ψ is the antenna azimuth angle.

The CERSAT algorithm:
The CERSAT has spent considerable effort in developing efficient software to process the scatterometer data and evaluate the influence of different factors (e.g. noise, bias, model function errors) on the accuracy of the results. This is achieved particularly by validating the ERS C-Band model against an independent dataset of NOAA (National Oceanic and Atmospheric Administration) buoy data (Quilfen and Cavanié, 1991; Quilfen and Bentamy, 1994).

The fundamental concept of the C-Band algorithm is to determine all points in the space of wind speed and direction which represent local minima of a specified distance. The CERSAT has chosen the maximum likelihood estimator (MLE) instead of the Euclidean or least squares distance. The MLE gives a measure of the probability that the point lies at a distance from the model solution.

The wind scatterometer processing is based on three steps:

- find the different wind solutions fitting the σ_0;
- build up an ice mask directly related to the error between the measurements and the backscattering model predictions; and
- choose one wind solution and discard the ice points using either of two general methods - a process which is called the ambiguity removal and which is done either by comparison with meteorological wind fields or by assessing the probability of each measurement resulting from comparison with the theoretical results of the backscattering model.

Ice detection:

The backscatter is a function of the incidence angle and it is different for sea water and for ice water. Moreover, Gohin (1995) and Ezraty and al. (1995) showed that the behavior of the backscatter is essentially linked to the ice type, including marginal ice, smooth first-year ice of the consolidated pack or multi-year ice. For instance, the NRCS at high incidence angle attains its highest values for the marginal ice which is broken and rafted.

Considering these differences, the MLE is used to detect sea ice. Thus, the longest deviations between the σ_0 observed and the σ_0 obtained from the wind model are attained over sea ice for low and high incidence angles and not for intermediate incidence angles. To get homogeneous information, ice masks are constructed on the basis of interpolation through a kriging algorithm and for each measurement point an ice flag number is assessed.

III/ NUMERICAL WAVE MODELS

III.1) WINCH Model

The WINCH model is a deep water, discrete wave prediction model which is a modified version of the SAIL model which was one of the models included in the SWAMP study. The model runs on a rectangular grid (transverse Mercator projection) with a spacing of 75 km for the North Sea, the Norwegian Sea and the Barents Sea and 150 km for the eastern North Atlantic. The time step is 2 hours for propagation and 1 hour for growth. The propagation scheme employs downstream interpolation based on 21 grid points and includes great circle effects. The wave spectrum is divided into 24 direction bands with 15° bandwidth and 15 frequency bands ranging from 0.04-0.24 Hz. Each spectral bin is propagated and grown individually and the high frequency tail is restored to a f^5 form. Each bin is grown by the largest of three factors: zero growth, exponential growth of the Miles type or parametric growth of the JONSWAP type. After growth of each wave component, non-linear interactions and dissipation are achieved by use of reference spectrum of the JONSWAP type and directional relaxation. For that part of the spectrum with frequencies greater than the Pierson-Moskowitz peak frequency, f_{PM}, the spectrum is restored to the reference spectrum by removing excess energy, while bands with less energy than reference spectrum are left unchanged. Energy at frequencies below f_{PM} are treated as well and are subject to advection and geometric damping only. Wave parameters are archived at 6-hour intervals. H_{m0}, T_p and θ_{pp} (for total wave conditions) were extracted by the Norwegian Met. Institute for grid point 191 (69.4°N, 57.6°E) adjacent to the Prirazlomnoye Field. Although the model does not take into account the effects of the shallow water bathymetry of the Pechora Sea it does incorporate the effects of the seasonal and inter-annual variation in the ice sheet in the Barents and the Pechora Sea. The sea ice charts are prepared by the Norwegian Met. Institute using NOAA/AVHRR satellite data and observations from ships and aircraft.

III.2) MétéoMer Models

MétéoMer perform operational analysis and prediction of ocean waves using two versions of the Integrated Swell Forecasting System (ISFS) model. The deep water version (ISFS-DW) is a discrete spectral wave model in which the directional spectrum is represented by a discrete number of finite bandwidth spectral components travelling in a specified number of directions. The generation and dissipation terms of the model are empirically linked to the observed fetch and duration. The model equation is solved by standard finite difference techniques over a series of time-steps on a bathymetric grid along orthodromic lines converging to the location of interest. The coastal model (ISFS-SW) allows for effective transfer of the offshore deep water directional wave spectrum into shallow water taking into account shoaling, refraction, wind sea generation by local winds, non-linear wave-wave interactions and dissipation by bottom friction and wave breaking. This model also operates using a bathymetric grid and retropropagated rays. For the Pechora Sea study (Ciambelli, 1996) the ISFS models were operated using wind fields from the European Centre for Medium Range Weather Forecasts (ECMWF) and monthly ice charts to hindcast directional wave height spectra at 3-hour intervals for the period from May 1992 to October 1995. The wind fields were calibrated against altimeter and scatterometer wind data and model heights against altimeter wave data (Lasnier et al., 1992, 1993). The deep water directional spectra (72 direction bands with 5° bandwidth and 23 frequency bands ranging from 0.04-0.33 Hz) were first created for the Barents Sea and the coastal model was then used to transfer offshore wind-sea and swell and to hindcast locally generated wind-sea for the site (69.25°N, 57.30°E) in the Prirazlomnove Field. The coastal model used a bathymetric grid produced from charts published by the UK Admiralty and by the Russian Ministry of Defence. The bathymetric grid in the southern area of the Pechora Sea was referenced to mean tidal level using information supplied by Russian oceanographic institutions via BHP Petroleum Ltd. H_{m0}, T_2, T_p, θ_{pp} and θ_m were then derived from the total directional spectra. Both models take into account the seasonal and inter-annual variation in the ice sheet in the Barents and the Pechora Sea.

IV/ APPLICATION TO THE PRIRAZLOMNOYE FIELD

For the Pechora Sea project in northern Russia, Paras required the following information:

- ice field coverage over the Barents Sea and, more precisely, the Pechora Sea;
- ECMWF wind fields for the period 1992-1996;
- Geosat and ERS1 satellite *Hs* measurements for the period 1986-1989 and 1992-1995;
- ERS1 SAR in wave mode sea-state spectra over the Barents Sea for the period 1992-1995; and
- coastal hindcast database (H_{m0} with associated wave periods and directions) for the period 1992-1995.

In order to meet these requirements, operational procedures have been applied which combine sea-state satellite measurements with metocean deterministic models for the calculation of directional wave height spectra and tools for the analysis of environmental databases (Charriez et al., 1995).

IV.1) Ice Coverage Maps

The aim of this stage was to implement an operational procedure for the determination of the monthly ice layer on the Barents Sea and particularly on the Pechora Sea. This procedure is based upon the use of the CERSAT ERS scatterometer data. The results comprise the ice location together with the associated ice concentration for each 0.5° × 0.5° box in the area of interest. The monthly ice concentration maps (for example see Figure 2) are then used for the numerical wave model computations. Thus, orthodromic lines converging to the location of interest may be computed using this information.

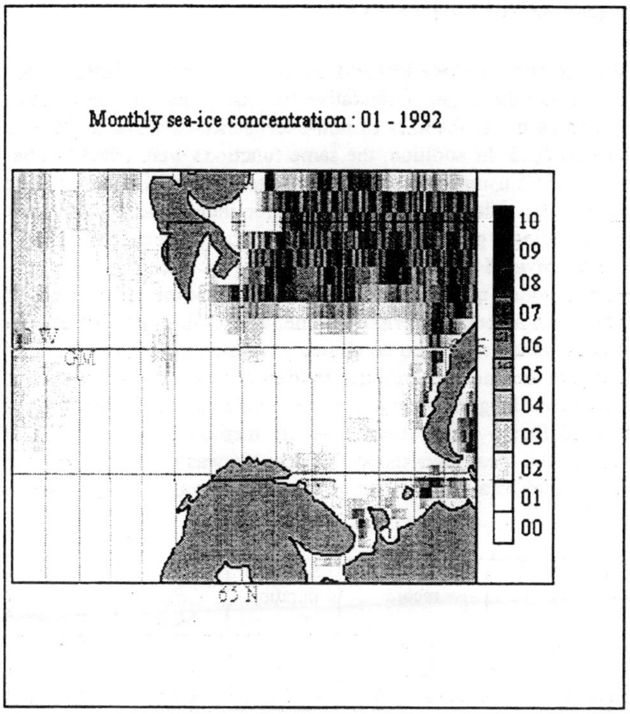

Fig. 2 - Example of ice map

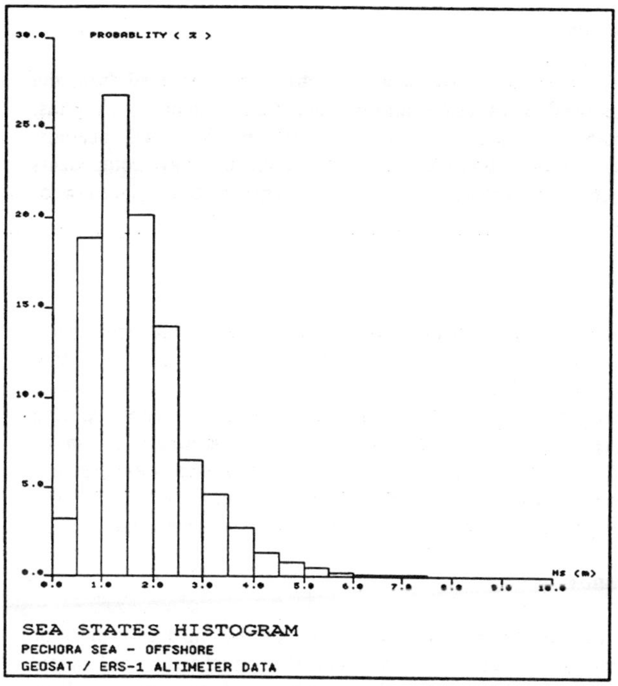

Fig. 3 - Satellite Altimeter *Hs* Histogram for the Local Offshore Climate Coherent Area 69.0°-71.5°N x 50°-59°E

IV.2) Satellite Offshore Sea-state Database

Within this stage, a satellite offshore sea-state database is built. This database is based mainly on 7 years of H_{m0} altimetric measurements (Geosat and ERS1) and upon 3 years of sea-state spectra issued from the ERS1 SAR in wave mode.

Altimetric H_{m0} database:

Altimeters from Geosat (11/1986-9/1989) and ERS1 (from 12/1991) provide offshore H_{m0} and wind intensity (*ff*) parameters every second (or every 7 km) along their ground tracks. This sampling regime is similar to the current engineering practice of estimating H_{m0} from a sample of 20 minutes duration.

Satellite measurements are spatial ones. Therefore, in order to build a statistically significant database, a climate consistent offshore area is defined around the location of interest, by the use of meteorological considerations. *Hs* altimeter measurements are then selected from within this area and routine qualifications are applied to each satellite dataset. As a result, an initial H_{m0} database, (32,069 samples in a period of 7 years, 1987-1989 and 1992-1995) was obtained for the local offshore climate coherent area 69.0°-71.5°N x 50°-59°E. Figure 3 presents an histogram of the significant wave heights within the Pechora Sea.

Sea-state spectra from ERS1 SAR in wave mode:

The area chosen for data selection is 69.0°-71.5°N x 50°-59°E. SAR in wave mode data were then selected from this area and SAR-inversion applied to each satellite dataset. Because the low spatial (or temporal) sampling of SAR in wave mode measurements (every 200 km) did not offer enough data for the elaboration of consistent statistics in a local area, these measurements were only used to derive qualitative information about sea-state directional structures inside the Barents Sea.

IV.3) Hindcast Offshore Sea-state Database

For this stage, the present lack of SAR-in-wave-mode data led to the development of an offshore directional wave height spectra database through the use of a metocean model. This database is mainly based on the implementation of the metocean model ISFS-DW for the calculation of directional wave height spectra from wind fields. This 3 hour step database allowed the characterisation of the coastal transfer onto the Pechora location, thus relying upon a complete description of sea-states through their directional spectra, and the construction of the final database, after the coastal transfer of the whole directional wave height spectra database, i.e. the reduced parameters (H_{m0} with associated periods and directions).

Inputs:

Inputs to this model consisted of wind fields originating from several meteorological models. These wind fields have been submitted to pre-processing for time (from 12 to 3 hours) and space interpolations (from 1.5° x 1.5° to 0.5° x 0.5°).

Computation:

Offshore directional wave height spectra were computed from the previous wind fields, taking into consideration the monthly ice grids. The model is a punctual one, using spectro-directional methods: generation, propagation and dissipation of spectral wave components computed along orthodromic lines converging to the location of interest.

Outputs:

The result of this metocean model is an offshore directional wave height spectrum, i.e., the energy density of wave components within each sea-state, per period (here, from 3 s to 23 s, with a 1 s step) and per direction (with a 5° step). The typical sea-state reduced spectral parameters can then be computed from each spectrum. It can be pointed out here that each spectrum can include several wave systems, for instance those corresponding with a wind-sea and a swell component, simultaneously generated by corresponding winds.

Calibration:

In addition to the previous description of this model, it must be outlined here that the input (meteorological wind fields) and output (sea-states) of this model have been calibrated through the use of the available satellite measurements, that is statistical comparisons of ECMWF wind fields with altimeter and scatterometer wind measurements. From these ECMWF calibrated wind fields, systematic calculations of directional wave height spectra were achieved on available *Hs* altimeter measurements in the local offshore climate coherent area. As a result, temporal and spatial simultaneous couples (satellite H_{m0}, model H_{m0}) were built, allowing the final calibration of the offshore model.

IV.4) Coastal Sea-state Database

The determination of coastal wave spectra incorporates both the transfer of offshore swell and the modelisation of wind-sea on the location of interest.

The aim of the first step of this present stage was to implement an operational procedure for the coastal transfer of the above offshore databases. The coastal model used allows for the effective transfer of every offshore sea-state, fully described by its complete directional wave height spectrum. The implemented model can take into account refraction, shoaling, interaction with local winds and currents, bottom friction and wave breaking. This model uses the same approach as the offshore one, solving the wave action conservation equation on retropropagated rays.

The second step of this stage was to implement a coastal sea-state wind generation model for the determination of wind-sea on the coastal location. Coastal directional wave height spectra are a combination of swell and wind-sea. Each offshore directional wave height spectrum is then transferred to the coastal location by applying to each of its offshore wave components the action conservation equation along the corresponding refracted orthogonal. Local wind is also taken into account for the generation of wind-sea. Thus, for every offshore directional wave height spectrum, the corresponding complete coastal one is determined.

V/ EXTREME WAVE CLIMATE

The Weibull and the three Fisher-Tippett functions were fitted to that part of the non-directional cumulative frequency distribution (cfd) of H_{m0} for which the probability of non-exceedence of wave height was greater than 0.05. In addition, the same functions were fitted to non-directional cfd's using only data from the tail of the distribution. The time series H_{m0} data were also sampled for independent storm peak events exceeding a given threshold. The number of peaks exceeding a given level, divided by the number of years of record, gives the rate of exceedence which can then be used to find the expected number of occurrences in a specific period of time. The probability distribution of the peak values was then combined with the rate of occurrence of peaks to give the unconditional distribution of peak values from which extreme wave heights were estimated using the Exponential and Generalised Pareto distributions. The cfd method was applied to all datasets but the 'peak over threshold' analysis was used only with the hindcast model data. The results of the extreme value analysis are as follows:

Data Source	Period of record	Sea-state duration	100-year H_{m0} Mean	100-year H_{m0} Range
WINCH model	55-95	2-hr	8.7 m	8.0-9.6 m
MétéoMer Model	92-95	3-hr	7.4 m	5.8-8.5 m
Geosat/ERS1	86-89,91-96	≈ 7.5-min	8.0 m	5.9-8.8 m

The WINCH hindcast model assumes that the Pechora Sea has an homogenous deep-water bathymetry. However, the MétéoMer model takes into account the actual bathymetry of the Pechora Sea and specifically the shallow-water area around the Prirazlomnoye Field. Therefore, the MétéoMer model is more likely to give an accurate representation of the wave climate in the study area. However, the MétéoMer model hindcast only covers the period May 1992 to October 1995. Analysis of the WINCH model hindcast database shows that 100-year extreme wave heights estimated using the period 1992-1995 overestimate by 10% the extreme value derived from the entire period July 1955 to June 1995. Therefore, a correction factor of 0.9 was applied to the mean 100-year significant wave height derived from the MétéoMer model giving a value of 6.7 m for the Prirazlomnoye Field.

VI/ CONCLUSION

The results illustrate the considerable benefit of using both the MétéoMer specific site database and the long term 40-year WINCH database which led to estimating with confidence the design 100-year significant wave height for the Prirazlomnoye Field.

ACKNOWLEDGMENTS

MétéoMer wish to express their thanks to the European Space Agency (ESA), The Centre National d'Etudes Spatiales (CNES) and to the CERSAT, for all their kind support to these services.
Paras would like to thank BHP Petroleum Ltd. for their kind permission to present a summary of the results of the investigation into the extreme wave climate within the Prirazlomnoye Field.

REFERENCES

Bitner-Gregersen E., Bonicel D., Hajji H., Olagnon M., Parmentier G., 1995:
World-wide characteristics of *Hs* and *Tz* for long-term load responses of ships and offshore structures.
Proc. Int. Offshore and Polar Engineering Conf., ISOPE, 1996.

Charriez P., Olagnon M., 1995:
Relevance of satellite data for a reliability approach of design wave estimation.
Proc. Offshore Mechanics and Arctic Engineering, OMAE, 1995.

Charriez P. , Hajji H., Thouvenin V., 1995:
Combined use of satellite data, metocean and prediction models for a site study.
Proc. Int. Offshore and Polar Engineering Conf., ISOPE, 1995.

Ciambelli, P., 1996:
Sea state and ice characterisation: Pechora Sea - Northern Russia.
MétéoMer report No. PP1192/96, August 1996.

Ezraty R., Gohin F., Bryère Ph., Maroni Ch., Cavanié A., 1995.
Determination of sea ice extent and ice types using the ami-wind Antarctica, 1992-1995.
Proc of the 2d ERS Applications Workshop, ESA-SP-383, London, 1995.

Gohin F., 1995:
Some active and passive microwave signatures of Antarctica sea ice from mid-winter to sprint.
Int. J. Remote Seinsing, 16, 2031-2054, 1995.

Lasnier P., Charriez P., Hajji H., Cauneau F., 1992:
Use of SAR-wave, altimeter and wind scatterometer data in an operational swell forecasting system.
Proc. Space and Sea Colloquium, E.S.A.Cannes, 1992.

Lasnier P., Hajji H., Loeul S., Bonicel D., Chapron B., 1993:
Use of SAR-Wave mode, altimeter and wind scatterometer data in operational wave forecasting system.
Proc. European Space Agency 2nd ERS1 Symposium, Hamburg, 1993.

Long A.E, 1986:
Toward a C-Band radar sea echo model for the ERS-1 scatterometer.
Proc, first Int. Coll. on Spectral signatures, ESA, Les arcs, Held in December 1985.

Sproson R.A, 1996:
Design wind and wave criteria for the platform site in the Prirazlomnoye Field.
Paras Ltd Report No, 0086, Prepared Paras Ltd for BHP Petroleum Ltd, 28th October 1996.

Quilfen Y., A. Cavanié, 1991:
A high precision wind algorithm for ERS-1 scatterometer and its validation.
Proc. IGARSS 1991, Espoo, Finland, 873-876, 1991.

Quilfen Y., A.Bentamy, 1994:
Calibration/Validation of ERS-1 scatterometer precision products..
Proc. IGARSS 1994, Pasadena, United States, 945-947, 1994.

Proceedings of the Seventh (1997) International Offshore and Polar Engineering Conference
Honolulu, USA, May 25-30, 1997

ISBN 1-880653-28-1 (Set); ISBN 1-880653-31-1 (Vol. III)

Runup on a Structure Due to Waves and Current

Bjarne Büchmann
Technical University of Denmark, Lyngby, Denmark
Jesper Skourup
Danish Hydraulic Institute, Hørsholm, Denmark
Kwok Fai Cheung
University of Hawaii at Manoa, Honolulu, HI, USA

ABSTRACT

A numerical wave tank is considered in which the interaction between waves, current and a structure is simulated by a 3D Boundary Element Model in the time domain. Through a perturbation expansion the model is formulated to second order in wave steepness and to first order in current speed. The boundary value problem is separated into a known incident wave field and an unknown scattered wave field. Active wave absorbers are applied numerically at the radiation boundaries. The present paper concern the wave runup on a bottom mounted vertical circular cylinder. The numerical results show good agreement with previous analytical and numerical solutions for second-order wave diffraction without a current and first-order wave diffraction with a colinear current. The second order wave runup with a current is illustrated by numerical examples. It is shown that the second order terms can be of significant importance when calculating wave runup on a body in a current.

Key words: Boundary Element Method, Numerical Wave Tank, Runup, Three-dimensional, Wave-Current Interaction.

INTRODUCTION

Numerical modeling of the interaction of waves and a current with fixed or floating bodies is of significant interest in engineering applications. These include the computation of added resistance on ships, slow drift oscillations of floating production platforms, runup on fixed off-shore structures and high frequency forces in the mooring lines of floating oil production systems. Although linear models are commonly used in the design and analysis of these structures many of the important effects are of higher order, which can only be accounted for by nonlinear models.

Within the framework of potential theory a numerical method widely used for simulating non-linear wave-structure interaction is the Boundary Integral Equation Method (BIEM) which is applied numerically as a Boundary Element Model (BEM). Romate (1989) followed by Broeze (1993) and van Daalen (1993), developed a 3D BEM for simulation of waves in a numerical wave tank. Skourup *et al.* (1992) used a 3D BEM to study the interaction between waves and a fixed vertical cylinder. Also Ferrant (1996) developed a 3D BEM in order to study higher order effects such as ringing. However, these 3D boundary element models for nonlinear problems are quite demanding in computational time and storage, and are usually applicable to a relatively small domain and cover a short simulation time.

In order to improve the computational efficiency of nonlinear simulations, finite order models based on Stokes' theory have been proposed. The second-order wave diffraction problem has recieved particular attention and numerous solutions have been developed in the frequency domain (e.g Eatock-Taylor and Hung, 1987, Kim and Yue, 1990, and Kriebel, 1990). Of equal importance is the wave-current interaction with a structure. Zhao and Faltinsen (1988) and Nossen *et al.* (1991), respectively, developed a 2D and a 3D linear BEM to study wave diffraction and radiation by arbitrary bodies with a small forward speed in the frequency domain. As an alternative to simulations in the frequency domain, Isaacson and Cheung (1992) developed a time domain BEM to study to second order wave diffraction around a three dimensional body. Cheung *et al.* (1996) later included a weak current in a 3D time domain BEM correct to first order in both wave steepness and current speed and showed that a current can significantly affect the runup on the body. These finite order models in the time domain require considerably less computation time than the fully non-linear models, but their application require an accurate treatment of the radiation condition at each order.

Recently, Skourup (1996) studied the use of an active wave absorption technique to improve the radiation condition in time-domain simulations of a numerical wave tank using a linear BEM. In the present paper, the formulations by Cheung *et al.* (1996) and by Skourup (1996) are adopted and extended to second order in wave steepness and to first order in current speed. Results of the first and second order runup around a bottom-mounted vertical circular cylinder in waves and a current are given.

Governing equations and boundary conditions

Consider a three-dimensional irrotational flow in a homogeneous, incompressible and inviscid fluid. Let t denote time and $\mathbf{x} = (x, y, z)$ the position vector. A velocity potential $\phi(\mathbf{x}, t)$, can be defined such that the velocity $\mathbf{u}$ in the three spatial dimensions can be written as

$$\mathbf{u} = (u, v, w) = \nabla\phi = (\phi_x, \phi_y, \phi_z) \tag{1}$$

where ∇ is the gradient operator and the subscripts x, y and z denote the partial derivatives in the respective directions. The continuity equation combined with the condition of an irrotational flow yields the well known Laplace equation

$$\nabla^2\phi = 0 \tag{2}$$

The Laplace equation is defined in a domain Ω bounded by a horizontal sea bed Γ_b, a free surface Γ_f, a body surface Γ_c and a set of vertical boundaries Γ_{r1}, Γ_{r2}, Γ_{r3} and Γ_{r4} truncating the infinite domain as shown in Figure 1.

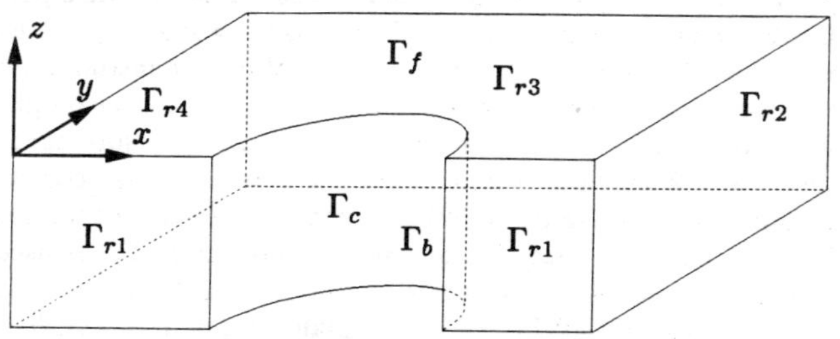

Figure 1: Definition of fluid domain and boundaries

Two non-linear boundary conditions are applied at the free surface Γ_f. The kinematic free-surface boundary condition states that a particle on the surface remains there

$$\eta_t - \phi_z + \nabla'\phi \cdot \nabla'\eta = 0 \ , \ z = \eta \tag{3}$$

where $\nabla' = (\frac{\partial}{\partial x}, \frac{\partial}{\partial y}, 0)$ is the gradient operator in the horizontal plane and '$\cdot$' denotes a scalar product. The dynamic free-surface boundary condition is given by the Bernoulli equation

$$\phi_t + g\eta + \tfrac{1}{2}\left|\nabla\phi\right|^2 = 0 \ , \ z = \eta \tag{4}$$

in which the pressure is assumed to be constant on the free surface and thus can be omitted. The sea bed Γ_b is considered to be impermeable and thus the velocity component perpendicular to the sea bed is zero

$$\phi_n = 0 \ , \ \text{at } \Gamma_b \tag{5}$$

The subscript n denotes partial differentiation along the unit normal vector $\mathbf{n}$ pointing outwards of the fluid domain.

Flux conditions are applied on the control surfaces Γ_{r1}, Γ_{r2}, Γ_{r3} and Γ_{r4}

$$\phi_n = f(\mathbf{x}, t) \tag{6}$$

Such conditions can be applied to simulate wave generation, wave radiation, wave reflection and to define a symmetry plane.

In time domain models wave absorption on a control surface is often performed by a radiation conditions or using a sponge layer on the free surface or both. The use of a sponge layer often requires an extention of the computational domain, and this results in an increase of the computing time and memory requirements. Skourup (1996) and Skourup and Bingham (1996) used active absorption of the outgoing waves in which the radiation condition is simulated by a number of wave paddles along one or more lateral boundaries of the domain. This technique is well established in 2D wave flume experiments. The Active Wave Absorption Control System (AWACS) technique which is used here has been developed at the Danish Hydraulic Institute for simultaneous wave generation and absorption in flumes (Schäffer *et al.*, 1994).

In the present numerical model both a sponge layer and active wave absorption using the AWACS have been implemented. The two methods can be used separately or in combination as done by Clément (1996) in his 2D BEM. However, only the active absorption is considered in the present paper.

Second order wave-current formulation

The major difficulty in solving free surface potential flows is the treatment of the free surface boundary conditions (3) and (4), which are both non-linear and defined on a free surface not known prior to solving the problem. Since neither very steep waves nor waves in very shallow water are studied here, a Stokes procedure can be used to expand the free surface about a known steady solution. In this paper the waves are expanded around a steady current solution as opposed to a perturbation about the still water solution $\phi = 0$, $\eta = 0$. Thus, the potentials and free surface elevations are expanded by perturbation series as

$$\phi = \phi^{(0)} + \epsilon\phi^{(1)} + \epsilon^2\phi^{(2)} + \cdots \tag{7}$$
$$\eta = \eta^{(0)} + \epsilon\eta^{(1)} + \epsilon^2\eta^{(2)} + \cdots \tag{8}$$

where ϵ is a small perturbation parameter relating to the wave steepness and the superscripts $^{(0)}$, $^{(1)}$ and $^{(2)}$ denote the order of the solution. Thus, in general the free surface boundary conditons would be expanded in the vertical direction from the free surface elevation corresponding to the steady solution $\eta^{(0)}$ up to the actual free surface η. A two-parameter perturbation procedure may at first glance seem to be a more appropriate choice than using (7) and (8) (i.e., a perturbation parameter ϵ related to the wave steepness and a parameter δ related to the Froude number). However, it can be shown that (7) and (8) include all terms to second order in ϵ and to first order in δ. In the case of the fully separated problem, one has to first solve the linear wave diffraction problem without a current and then solve the linear wave/current coupling problem. The same applies for the second order problem. As a result of using the one parameter perturbation expansion, only one solution is needed at each order, and the mathematical formulation is simpler. The solution at each order of ϵ might include some higher order terms with respect to the Froude number, but the total solution is still correct to first order in δ.

A weak current is considered in the present application, and thus only the linear part of the steady current solution is used. The steady current potential $\phi^{(0)}$ can be taken as the solution to the double body flow with no steady waves introduced, since these would be of second or higher order in the current speed. Therefore $\eta^{(0)} = 0$ and the Taylor series expansions are taken about the still water level $z = 0$. However, the steady current potential $\phi^{(0)}$ will be non-zero and will influence the higher order components of the wave field.

Applying a Taylor series expansion to (3) and (4) about $z = 0$ gives

$$\begin{aligned}(\eta_t - \phi_z + \nabla'\phi \cdot \nabla'\eta)_{z=0} & \\ +\eta\frac{\partial}{\partial z}(\eta_t - \phi_z + \nabla'\phi\cdot\nabla'\eta)_{z=0} + \cdots &= 0 \quad (9)\\ (\phi_t + g\eta + 1/2\ |\nabla\phi|^2)_{z=0} & \\ +\eta\frac{\partial}{\partial z}(\phi_t + g\eta + 1/2\ |\nabla\phi|^2)_{z=0} + \cdots &= 0 \quad (10)\end{aligned}$$

Introducing the perturbation expansions (7) and (8) into the free surface boundary conditions (9) and (10) and collecting terms of order ϵ^k ($k = 0$, 1 and 2) leads to

$$\phi_z^{(k)} - \eta_t^{(k)} = F_1^{(k)} \quad (11)$$

$$\phi_t^{(k)} + g\eta^{(k)} = F_2^{(k)} \quad (12)$$

in which the forcing terms $F_1^{(k)}$ and $F_2^{(k)}$ are given as

$$F_1^{(0)} = 0 \quad (13)$$

$$F_2^{(0)} = 0 \quad (14)$$

$$F_1^{(1)} = \nabla'\phi^{(0)}\cdot\nabla'\eta^{(1)} + \eta^{(1)}(\phi_{xx}^{(0)} + \phi_{yy}^{(0)}) \quad (15)$$

$$F_2^{(1)} = -\nabla'\phi^{(0)}\cdot\nabla'\phi^{(1)} \quad (16)$$

$$\begin{aligned}F_1^{(2)} &= \nabla'\phi^{(0)}\cdot\nabla'\eta^{(2)} + \nabla'\phi^{(1)}\cdot\nabla'\eta^{(1)}\\ &+\eta^{(2)}(\phi_{xx}^{(0)} + \phi_{yy}^{(0)}) + \eta^{(1)}(\phi_{xx}^{(1)} + \phi_{yy}^{(1)}) \quad (17)\end{aligned}$$

$$\begin{aligned}F_2^{(2)} &= -\nabla'\phi^{(0)}\cdot\nabla'\phi^{(2)} - \frac{1}{2}\left|\nabla\phi^{(1)}\right|^2 - \eta^{(1)}\phi_{zt}^{(1)}\\ &- \eta^{(1)}\nabla'\phi^{(0)}\cdot\nabla'\phi_z^{(1)} + \eta^{(1)}\phi_z^{(1)}(\phi_{xx}^{(0)} + \phi_{yy}^{(0)}) \quad (18)\end{aligned}$$

At zeroth order (i.e. $k = 0$) $\eta^{(0)} = 0$ and $\phi_t^{(0)} = 0$. Contrary to the case with no current, it is noted that the k'th order forcing terms ($k \geqslant 1$) also depends on $\phi^{(k)}$ and $\eta^{(k)}$.

A physical wave tank is usually bounded by vertical walls, where waves are generated along one of the sides. In the present formulation, however, an analytical solution for the incident wave field is used, and the lateral boundaries Γ_{r2}, Γ_{r3} and Γ_{r4} are used for absorption of the scattered waves. The wave potentials and elevations are therefore separated into the incident and scattered wave components.

$$\phi^{(k)} = \phi_w^{(k)} + \phi_s^{(k)} \quad (19)$$

$$\eta^{(k)} = \eta_w^{(k)} + \eta_s^{(k)} \quad (20)$$

where subscripts w and s denote respectively the incident and the scattered components. For $k \geqslant 1$, (11) and (12) can be rearranged into

$$\frac{\partial\eta_s^{(k)}}{\partial t} = \frac{\partial\phi_s^{(k)}}{\partial z} + \left(\frac{\partial\phi_w^{(k)}}{\partial z} - \frac{\partial\eta_w^{(k)}}{\partial t}\right) - F_1^{(k)} \quad (21)$$

$$\frac{\partial\phi_s^{(k)}}{\partial t} = -g\eta_s^{(k)} - \left(\frac{\partial\phi_w^{(k)}}{\partial t} + g\eta_w^{(k)}\right) + F_2^{(k)} \quad (22)$$

The analytic solution for long-crested second order Stokes waves riding on a uniform current over a horizontal sea bed is used to define the incident waves. The solution can be obtained from the second order solution without a current by taking the wave and current directions and phase shift into account (see e.g. Jonsson and Wang, 1980). For waves and a current at angles β_w and β_c with the x-axis, respectively, the surface elevations and potentials can be written as

$$\eta_w^{(1)} = \frac{H}{2}\cos\theta \quad (23)$$

$$\phi_w^{(1)} = -\frac{\omega H}{2k}\frac{\cosh k(z+h)}{\sinh kh}\sin\theta \quad (24)$$

$$\eta_w^{(2)} = \frac{1}{16}kH^2(3\coth^3 kh - \coth kh)\cos 2\theta \quad (25)$$

$$\begin{aligned}\phi_w^{(2)} &= -\frac{3}{32}\omega H^2\frac{\cosh 2k(z+h)}{\sinh^4 kh}\sin 2\theta\\ &-\frac{1}{8}\frac{gkH^2}{\omega h}(x\cos\beta_w + y\sin\beta_w - Ut\cos(\beta_c - \beta_w))\\ &-\frac{1}{16}\omega^2H^2\frac{t}{\sinh^2 kh} \quad (26)\end{aligned}$$

$$\begin{aligned}\theta &= (Uk\cos(\beta_c - \beta_w) + \omega)\,t\\ &- k\,(x\cos(\beta_w) + y\sin(\beta_w)) \quad (27)\end{aligned}$$

where $\omega = 2\pi/T$ is the angular frequency of the incident wave relative to the current (i.e., the frequency corresponding to the given wavelength, L, without the current). T is the wave period, $k = 2\pi/L$ is the scalar wave number, H is the wave height, h is the water depth and U is the current speed. With the introduction of the current, the absolute frequency (i.e., the encounter frequency with the body), is generally not equal to $\omega/2\pi$. Note that the term $-U^2/2$ has been omitted from the potential when compared to Jonsson and Wang (1980), since the present formulation is linear in current speed. In the present application Γ_{r1} is used as a symmetry plane to improve the computational efficiency. Since Γ_{r1} is situated parallel to the (x, z)-plane at $y = 0$, β_c and β_w will be either zero or π (180°). Thus, only incident waves in following or opposing current is considered here, even though this is not a restriction of the formulation.

With the incident waves and the current defined, the remaining unknowns in the problem are the scattered wave components $\phi_s^{(1)}$, $\phi_s^{(2)}$, $\eta_s^{(1)}$ and $\eta_s^{(2)}$. The initial conditions corresponds to an undisturbed second-order wave train on a uniform current (i.e., the body has no effect on the flow). The scattered waves are generated by imposing the body boundary condition. In order to avoid an abrupt initialization and the following instabilities near the structure a ramp function is multiplied to the body boundary condition. The ramp function is defined such that its first and second derivatives are zero at time $t = 0$. After a few wave periods the ramp function is unity while its derivatives tend to zero.

ACTIVE WAVE ABSORPTION CONTROL SYSTEM

In the present application the body is fixed, and thus the Taylor series expansion about Γ_c is not needed. At the lateral boundaries the AWACS is used to absorb the first and second order scattered waves (see Skourup, 1996; Skourup and Bingham, 1996).

The AWACS for 2D wave flume experiments is based on a frequency domain expression for the wavemaker paddle position which may be expressed as

$$X_a = (2A_I - A_0)F \quad , \quad F = \frac{1}{i\sum_{j=0}^{\infty} c_j^*} \quad (28)$$

in which ' * ' denotes complex conjugation and i is the imaginary unit. Here A_I and A_0 are the complex amplitudes of the prescribed incident wave elevation and the elevation measured on the wave board. The complex transfer function, F, is based on a real transfer function c_0 and some imaginary transfer functions c_j. Here c_0 (the Biésel transfer function) relates the amplitudes of the generated wave to the amplitude of the wave paddle motion in the case of pure wave generation. The rest of the series in (28) is imaginary, and relates the amplitude of the evanescent modes at the wave board to the amplitude of the paddle motion. For a piston type wavemaker we have

$$c_j = \frac{4\sinh k_j h}{2k_j h + \sinh 2k_j h} \tag{29}$$

where k_j is the j'th solution to the linear dispersion relation in complex form

$$\omega_a^2 = g k_j \tanh k_j h \tag{30}$$

and ω_a is the absolute wave frequency. It has one real solution, k_0, and an infinite series of imaginary solutions k_j ($j \geqslant 1$, where $ik_j > 0$). Practical application in the time domain is made by use of the recursive digital filter

$$v_n = \sum_{k=0}^{M} a_k u_{n-k} + \sum_{k=1}^{N} b_k v_{n-k} \tag{31}$$

where the input u_n is $2\eta_I - \eta_0$ evaluated at time $t = n\Delta t$ and the output v_n is the wavemaker position, $X(n\Delta t)$. The transfer function of this recursive filter is fitted to match the transfer function F over a wide frequency range by optimization of the filter coefficients (a_k, b_k) under stability constraints. With these coefficients, (31) provides the instantaneous paddle position for irregular as well as regular waves. The system only needs the instantaneous elevations of the prescribed incident-wave elevation, η_I, and the elevation measured at the wave board, η_0. No advanced knowledge on the waves to be generated or absorbed is required. For pure wave absorption $\eta_I = 0$. Further details of this formulation can be found in Schäffer *et al.* (1994) and in Schäffer and Skourup (1996).

The AWACS technique described above may also be used in a 3D numerical wave tank by considering a number of 2D wave absorbers placed next to each other and working independently (i.e. a quasi-3D absorber). Each absorber is then governed by the digital recursive filter determined by the AWACS theory and by the local time histories of elevation and position of the wave absorber. This method was also used by Skourup (1996) and Skourup and Bingham (1996). An extension to a fully 3D active wave absorption method was given by Schäffer and Skourup (1996), but this has not yet been implemented in the present version of the programme.

In the present application the AWACS is used only on the scattered wave components and reacts to all second order scattered waves as if they were free waves. Thus the boundary conditions used on the absorbing boundaries are

$$\phi_n^{(k)} = \delta_t^{(k)} \tag{32}$$

where $\delta_t^{(k)}$ is the piston velocity found by the AWACS.

Numerical solution

The wave tank considered here is bounded by a horizontal sea bed, a free surface, the body surface and a set of vertical boundaries truncating the domain. The free surface boundary conditions have been expanded by a Taylor series, so that the geometry of the computational domain is invariant in time.

In order to solve the boundary value problem a Boundary Integral Equation Method based on Green's 2nd identity is used

$$\alpha(\mathbf{x})\,\phi(\mathbf{x},t) = \int_\Gamma \left(\phi(\boldsymbol{\xi},t)\frac{\partial G(\mathbf{x},\boldsymbol{\xi})}{\partial n} - G(\mathbf{x},\boldsymbol{\xi})\frac{\partial \phi(\boldsymbol{\xi},t)}{\partial n}\right) d\Gamma \tag{33}$$

in which $\boldsymbol{\xi} = (\xi_1, \xi_2, \xi_3)$ is a point on the integration path, while $\mathbf{x}$ is an observation point, i.e. $\boldsymbol{\xi}$ is situated at the boundary Γ. The term α depends only on the position of the observation point $\mathbf{x}$, and is equal to 4π if $\mathbf{x}$ is inside the fluid domain Ω and zero if $\mathbf{x}$ is outside Ω. If $\mathbf{x}$ is on the boundary Γ then α is the interior angle of the boundary at the location $\mathbf{x}$. The kernel function $G(\mathbf{x},\boldsymbol{\xi})$ is the free space Green's function

$$G(\mathbf{x},\boldsymbol{\xi}) = \frac{1}{r} = \frac{1}{|\mathbf{r}|} = \frac{1}{|\boldsymbol{\xi} - \mathbf{x}|} \tag{34}$$

where r is defined as the length of the vector $\mathbf{r} = \boldsymbol{\xi} - \mathbf{x}$.

The normal derivative of the free space Green's function is easily obtained using the chain rule for differentiation

$$\frac{\partial G(\mathbf{x},\boldsymbol{\xi})}{\partial n} = \frac{\partial G(\mathbf{x},\boldsymbol{\xi})}{\partial r}\frac{\partial r}{\partial n} = \frac{-1}{r^2}\frac{\mathbf{r}\cdot\mathbf{n}}{r} = \frac{-\mathbf{r}\cdot\mathbf{n}}{r^3} \tag{35}$$

Specific harmonic functions, $G_j = 1/r_j$, are added to the Green's function (34) to account for the boundary conditions at the horizontal bottom and the symmetry plane Γ_{r1} (the (x,z)-plane). Choosing r_j carefully the resulting Green's function satisfies the boundary conditions on Γ_b and Γ_{r1}, and numerical integrations need not be performed over these boundaries. The resulting Green's function, taking into account the horizontal bottom and the symmetry about the (x,z)-plane, is

$$G(\mathbf{x},\boldsymbol{\xi}) = \sum_{j=1}^{4}\frac{1}{r_j} = \sum_{j=1}^{4}\frac{1}{|\boldsymbol{\xi}_j - \mathbf{x}|} \tag{36}$$

where $\boldsymbol{\xi}_1 = (\xi_1, \xi_2, \xi_3)$, $\boldsymbol{\xi}_2 = (\xi_1, -\xi_2, \xi_3)$, $\boldsymbol{\xi}_3 = (\xi_1, -\xi_2, -\xi_3 - 2h)$ and $\boldsymbol{\xi}_4 = (\xi_1, \xi_2, -\xi_3 - 2h)$. If a symmetry plane is not used the $j = 2$ and $j = 3$ terms vanish. Similarly, if the bottom is not horizontal the $j = 3$ and $j = 4$ terms are omitted. Thus, if no reflective planes are used G reduces to (34).

The problem is solved using the collocation method, applying the governing integral equation (33) to discrete points (nodes) at the boundary. To perform the integrations numerically, the boundary is divided into NELEM quadrilateral elements, where nodes are located at each corner of the elements. A total of N collocation nodes are located on the entire boundary.

Without any approximation the governing integral equation (33) can be written as

$$\alpha(\mathbf{x}_i)\,\phi(\mathbf{x}_i,t) = \sum_{i=1}^{\text{NELEM}} \int_{\Gamma_i} \phi(\boldsymbol{\xi},t) G_n(\mathbf{x}_i,\boldsymbol{\xi})\, d\Gamma - \sum_{i=1}^{\text{NELEM}} \int_{\Gamma_i} G(\mathbf{x}_i,\boldsymbol{\xi})\phi_n(\boldsymbol{\xi},t)\, d\Gamma \tag{37}$$

In the numerical integrations bi-linear shape functions are used to describe the shape of the boundary elements and the boundary functions. Hence, the integrals can be expressed by the values of the integrands at the nodal points. Since $G(\mathbf{x},\boldsymbol{\xi})$ and $G_n(\mathbf{x},\boldsymbol{\xi})$ are singular when $\mathbf{x}=\boldsymbol{\xi}$, the integrals in (37) will be singular when integrating over an element where $\mathbf{x}_i$ coinsides with one of the nodes of the element. Thus, special care is taken for these 'singular' elements. A Gauss-Legendre quadrature formula is used for all regular elements, whereas for singular elements a similar method using orthogonal polynomials is used to achieve an accurate integration (cf. Cristescu and Loubignac, 1978).

Using these shape functions and integration methods, (37) can be reduced to the following linear system

$$\sum_{j=1}^{N}(c_{ij}+A_{ij})\Phi_j^{(k)} = \sum_{j=1}^{N}B_{ij}\left(\Phi_n^{(k)}\right)_j \quad , \quad i=1,\cdots,N \ , \ k=1,2 \qquad (38)$$

where $\Phi_i^{(k)}=\phi^{(k)}(\mathbf{x}_i,t)$, $\left(\Phi_n^{(k)}\right)_i=\phi_n^{(k)}(\mathbf{x}_i,t)$ and

$$c_{ij} = \begin{cases} 0 & \text{, if } \ i\neq j \\ \alpha(x_i) & \text{, if } \ i=j \end{cases} \qquad (39)$$

Alternatively, the diagonal elements of $(c_{ij}+A_{ij})$ can be determined using the so-called rigid mode technique, (i.e. $c_{ii}+A_{ii}=-\sum_{j=1,\ j\neq i}^{N}A_{ij}$). It is emphasized that the matrices A_{ij} and B_{ij} only contain information regarding the Green's function and the geometry of the problem, and these are not functions of time or the order of the problem.

At intersecting boundaries where two or more elements meet at an angle, the double and triple node techniques are used to account for the discontinuity of the normal vector. Since applying (37) at the double and triple nodes will result in two or three identical equations, special care must be taken at these nodes in order to avoid a singular system. The redundant equations are replaced with equations stating that the potential is continuous at the double and triple nodes

$$\phi(\mathbf{x}_i,t)=\phi(\mathbf{x}_j,t) \quad \text{,if } \mathbf{x}_i=\mathbf{x}_j \qquad (40)$$

In general, the steady current potential can be solved numerically, for instance by solving the double body flow using a boundary integral formulation. However, the analytical solution to the flow around a bottom mounted vertical circular cylinder with a rigid lid condition at the free surface is known and applied directly in the numerical model as

$$\phi^{(0)}(\mathbf{x})=U(x\cos\beta_c+y\sin\beta_c)\left(1+\frac{a^2}{x^2+y^2}\right) \qquad (41)$$

where a is the radius of the cylinder.

Time stepping procedure

The free surface boundary conditions are satisfied by a time-integration procedure. At each time step, the procedure consists primarily of two distinct operations performed for each of the first and second order problems. First $\Phi_i^{(k)}$ and $\eta^{(k)}(\mathbf{x}_i,t)$ are updated on the free surface and $(\Phi_n)_i$ on the remaining boundaries using the corresponding boundary conditions, and then the governing integral equations (38) are solved to obtain the rest of the unknowns.

The time stepping procedure used on the free surface is an Adams-Bashforth-Moulton predictor-corrector method. The free surface boundary conditions (21) and (22) are used to calculate the time derivatives of $\phi_s^{(k)}$ and $\eta_s^{(k)}$ at the previous time step. This in turn requires the calculation of spatial derivatives on the free surface using finite difference approximations. The finite difference approximations are obtained by satisfying 2D Taylor expansions to second order at the node of interest and at five neighbouring nodes.

$$\begin{aligned} f(x_i,y_i) &= f(a,b)+(x_i-a)f_x(a,b)+(y_i-b)f_y(a,b) \\ &+\frac{1}{2}(x_i-a)^2f_{xx}(a,b)+(x_i-a)(y_i-b)f_{xy}(a,b) \\ &+\frac{1}{2}(y_i-b)^2f_{yy}(a,b) \end{aligned} \qquad (42)$$

In (42) (a,b) are the coordinates of the node in which spatial derivatives of the function f are to be determined and (x_i,y_i) $(i=1,\cdots,5)$ are the coordinates of the other five nodes. The function values $f(x_i,y_i)$ and $f(a,b)$ are known and thus $f_x(a,b)$, $f_y(a,b)$, $f_{xx}(a,b)$, $f_{xy}(a,b)$ and $f_{yy}(a,b)$ are the unknowns. The system of equations is solved for all five unknowns, but in order to get as accurate results as possible only the first derivatives $f_x(a,b)$ and $f_y(a,b)$ are used. Second deriviatives are found by repeating the procedure with the first derivatives as input. The five nodes are chosen carefully to make sure they are both close to the central node and distributed evenly in space so that the resulting system will be well conditioned.

A 4th order explicit Adams-Bashforth method is used to update $\phi_s^{(k)}$ and $\eta_s^{(k)}$ to the subsequent time step $(t+\Delta t)$.

$$f(t+\Delta t)=f(t)+\Delta t\sum_{j=0}^{3}a_j\frac{\partial f(t-j\Delta t)}{\partial t} \qquad (43)$$

where $(a_0,a_1,a_2,a_3)=(55/24,-59/24,37/24,-9/24)$.

Now an approximation to $\frac{\partial\phi_s^{(1)}}{\partial t}$ at the new time step can be evaluated using (22), and a 4th order implicit Adams-Moulton procedure can be applied to get an improved estimate of $\phi_s^{(k)}$ at the new time step.

$$f(t+\Delta t)=f(t)+\Delta t\sum_{j=0}^{3}b_j\frac{\partial f(t+\Delta t-j\Delta t)}{\partial t} \qquad (44)$$

where $(b_0,b_1,b_2,b_3)=(9/24,19/24,-5/24,1/24)$. The implicit updating is repeated until the solution has converged. In this formulation of the problem it is not feasible to make an implicit updating of η using (21), since Φ_n is not known on the free surface at the new time step prior to solving the boundary integral equations. Values of $\Phi_n^{(k)}(t+\Delta t)$ can be found using the output from the AWACS on the absorbing boundaries and a no-flux condition on the body. At this point the boundary integral equations (38) are solved to find the remaining unknowns in the problem.

Numerical results

The numerical experiments presented here are carried out using a domain bounded by the free surface Γ_f, the sea bed Γ_b, the surface of a body Γ_c and the lateral boundaries Γ_{r1}, Γ_{r2}, Γ_{r3} and Γ_{r4} (see Fig.1). In the present application the body is a bottom mounted vertical circular cylinder, and thus the analytical solution (41) can be applied. Using the symmetry conditions no boundary elements are needed on Γ_b and Γ_{r1}. About 3400 nodes are used in the model, most of which are placed on the free surface and on the cylinder. The free surface grid is generated such that the smaller elements are placed in the vicinity of the cylinder (see Fig.2).

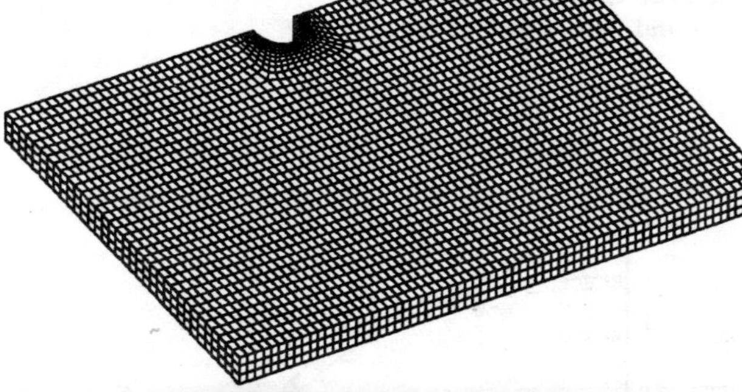

Figure 2: Grid used for comparison with results by Cheung *et al.* (1996). In this particular case the length of the wave tank is $26.1h$ corresponding to 4.1 wavelengths of the incident wave.

The length of the domain ranges from three to four wavelengths of the incident waves and the width is half of the length, so in this case a square physical wave tank is simulated. In order to suppress instabilities near the cylinder a weak low-pass filter has been used on the surface elevations at the free surface nodes adjacent to the cylinder.

In the experiments around 400 time steps per wave period were used to ensure accurate numerical results. Since relatively small elements are used near the cylinder a large number of time steps per wave period is needed in order to satisfy the local Courant number criterion. In a numerical testing it was possible to use 100 time steps per period corresponding to a local Courant number of approximately 1.8 based on the smallest element size and the absolute celerity of the incident waves. The limiting factor on the simulation is the memory rather than the CPU time. The largest jobs require about 220Mb of memory and the corresponding computation time was around 70 minutes CPU on an IBM RS6000 work station.

A series of numerical experiments were carried out to investigate the runup on a bottom mounted vertical circular cylinder in a current and to examine the performance of the model. In each experiment the parts of the time series in which initial conditions or reflections from the lateral boundaries prevailed were not used in the estimation of the runup. The resulting time series cover at least a few wave periods. In Figure 3 a time series plot of the elevations at the front of the cylinder is shown to illustrate the performance of the absorbing boundaries. The time series have been normalized with the relative period of the incident waves. It is noted that the encouter frequency with the cylinder is slightly higher than the relative frequency of the incident wave due to the weak following current. Some reflection at second order is noted after approximately six periods, and in this case the time interval $4T < t < 6T$ is used to determine the runup on the cylinder (i.e. more than two *absolute* (encounter) wave periods).

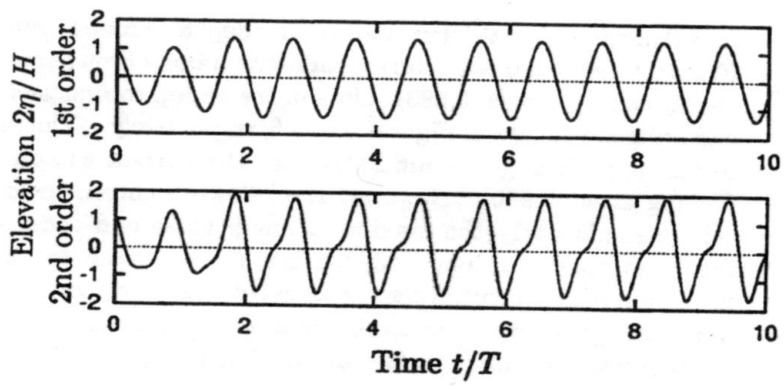

Figure 3: Time series of elevations to first and second order at the front of the cylinder ($\beta = \pi$) for $ka = 0.374$, $kh = 1.036$, $kH = 0.402$ and a following current of depth Froude number $Fr_d = 0.050$.

Non-dimensional wave numbers kh of near unity are chosen here since the AWACS works well in this range of frequencies (see Schäffer *et al.*, 1994). The parameters in the computation are non-dimensionalized by the radius of the cylinder a and the acceleration due to gravity g. The wave steepness, and thus the nonlinearity of the problem, can be specified in the post processing of the data. A range of wave steepnesses kH can thus be examined using the numerical output from one simulation.

A solution for the interaction between second order Stokes' waves and a large bottom mounted circular cylinder is given by Kriebel (1990). Wave runup results were given in a subsequent paper (Kriebel, 1992). These results have been reproduced using the present numerical model, and a comparison is given in Figure 4 for the parameters $ka = 0.374$, $kh = 1.036$ and $kH = 0.402$. In the figure, β denotes the angle of the position on the cylinder with respect to the x-axis. The first order runup is the well known solution by MacCamy and Fuchs (1954). The agreement between the results of the present numerical model (found by averaging over a few wave periods) and the theoretical runup results of Kriebel (1992) is very good. The small difference is attributed to truncation errors in the numerical scheme.

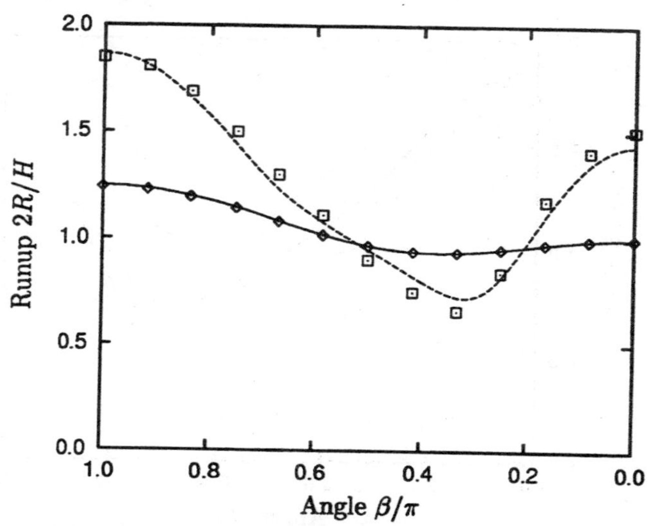

Figure 4: Runup on a cylinder for $ka = 0.374$, $kh = 1.036$ and $kH = 0.402$. No current in this case. Results to first order (—) and second order (- -). Results from Fig. 7(e) in Kriebel (1992) to first order (◇) and second order (□).

Introducing a weak current of a depth Froude number $Fr_d = U/\sqrt{gh} = \pm 0.050$ (corresponding to a body Froude number of $Fr_a = U/\sqrt{ga} = \pm\ 0.083$), plots of the runup to first and second order are shown in Figures 5 and 6, respectively. The runup to first order changes significantly with the current as noted by Cheung *et al.* (1996). However, the following current does *not* change significantly the maximum runup to second order. On the other hand, the opposing current *does* decrease the maximum runup to second order. It is noted that the second order terms change the wave runup significantly. Both the maxima and the shape of the runup profiles are modified.

The runup on a cylinder was calculated by Cheung *et al.* (1996) using first order perturbations for waves and current. A particular numerical experiment utilized $ka = 1$ and $kh = 1$. The runup was investigated for $Fr = U/\sqrt{gh} = U/\sqrt{ga} = \pm 0.10$. Repeating their experiment the first order runup is plotted in Figure 7. It is noted that the result compares well with the result by Cheung *et al.* (1996). Choosing $kH = 0.20$ the runup to second order is plotted in Figure 8. It is noted that the second order terms increase the maximum runup and that in this case the following current does increase the maximum runup to second order compared to the case without current. For the opposing current the maximum runup in the linear case (Fig.7) is found near $\beta/\pi = 0.66$ as opposed to $\beta/\pi = 1.0$ in the case with no current. The maximum runup to *second* order with the opposing current is found at around $\beta/\pi = 0.56$ (Fig.8).

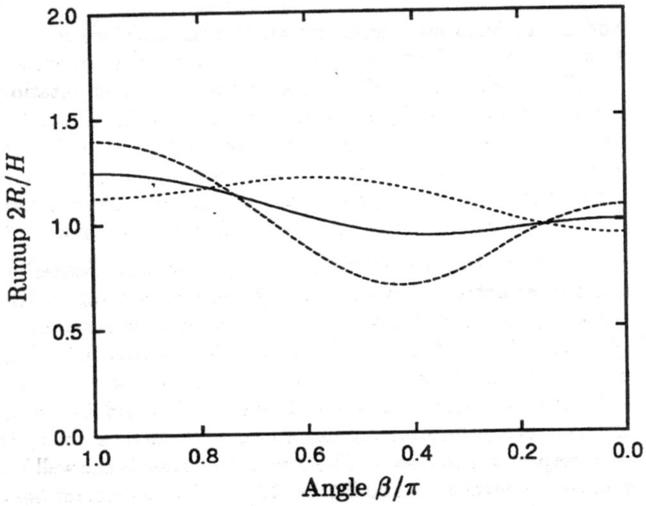

Figure 5: Runup on a cylinder for $ka = 0.374$ and $kh = 1.036$. Results to first order without current (—), in following current of depth Froude number $Fr_d = 0.050$ (--) and in opposing current of $Fr_d = -0.050$ (···).

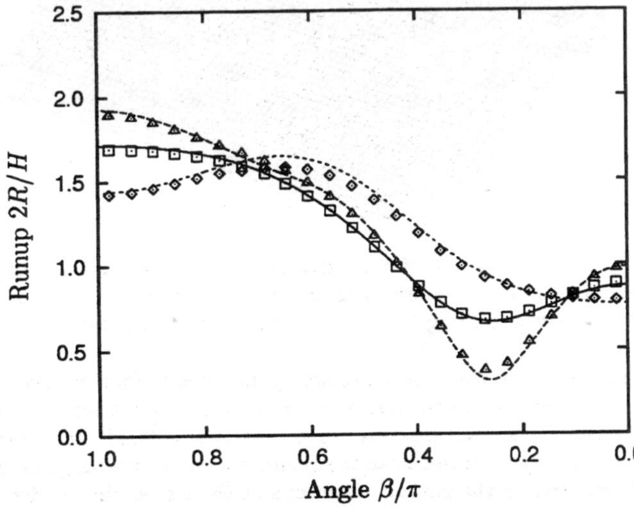

Figure 7: Runup on a cylinder for $ka = 1.0$ and $kh = 1.0$. Results to first order without current (—), in following current of Froude number $Fr = 0.10$ (--) and in opposing current of $Fr = -0.10$ (···). Results from Cheung *et al.* (1996) without current (□), in following current of Froude number $Fr = 0.10$ (△) and in opposing current of $Fr = -0.10$ (◇).

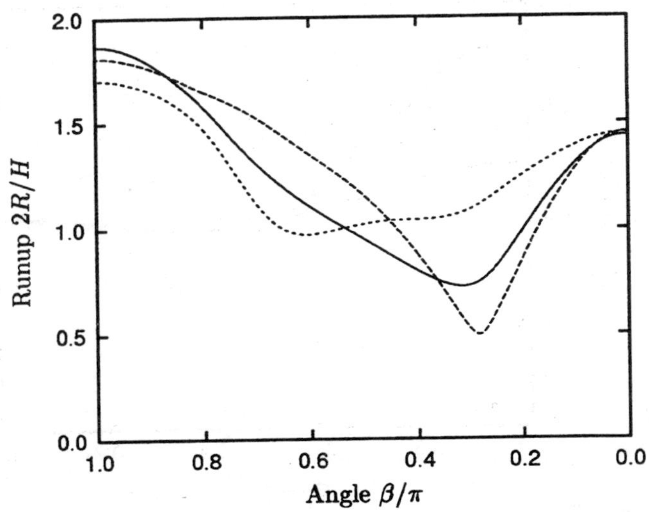

Figure 6: Runup on a cylinder for $ka = 0.374$, $kh = 1.036$ and $kH = 0.402$. Results to second order without current (—), in following current of depth Froude number $Fr_d = 0.050$ (--) and in opposing current of $Fr_d = -0.050$ (···).

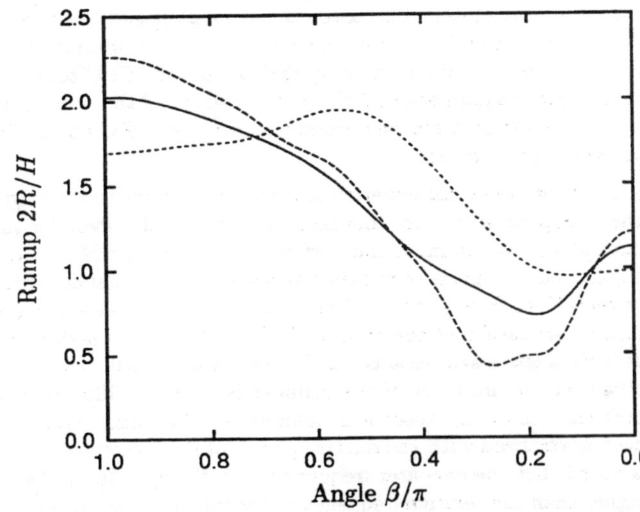

Figure 8: Runup on a cylinder for $ka = 1.0$, $kh = 1.0$ and $kH = 0.20$. Results to second order without current (—), in following current of depth Froude number $Fr = 0.10$ (--) and in opposing current of $Fr = -0.10$ (···).

Conclusions

A time-domain second order boundary element model for wave-structure interaction in a weak current has been developed. The model has been applied to calculate the runup on a bottom mounted vertical circular cylinder for two different radius to depth ratios. The model has been verified by comparison with numerical results for second order wave runup on a cylinder and results for first order wave runup on a cylinder in a current.

It is shown that the second order terms can be of significant importance for estimating the wave runup on a body in a current. Future applications will include other body geometry and waves and current at angles other than 0° or 180°. Applications to floating structures and use of other numerical methods to improve the accuracy and reduce the computation time will be considered.

Acknowledgements

This work was funded in part by the Danish National Research Foundation. Their support is greatly appreciated. David Kriebel is thanked for sending us his original data for second order wave runup on a bottom mounted vertical circular cylinder.

References

Broeze, J. (1993). "Numerical Modelling of Nonlinear Free Surface Waves with a 3D Panel Method," Ph.D. dissertation, University of Twente, The Netherlands

Cheung, K.F., Isaacson, M. and Lee, J.W. (1996). "Wave diffraction around a three-dimensional body in a current," *ASME Journal of Offshore Mechanics and Arctic Engineering*, Vol.118, No.4, 247–252.

Clément, A. (1996). "Coupling of two absorbing boundary conditions for 2D time-domain simulations of free surface gravity waves," *Journal of Computational Physics*, Vol.126, 139–151.

Cristescu, M. and Loubignac, G. (1978). "Gaussian quadrature formulas for functions with singularities in 1/r over triangles and quadrangles," from *Recent Advances in B.E.M.*, Editor: Brebbia, C.A., Pentech Press, London, 375–390.

van Daalen, E.F.G. (1993). "Numerical and Theoretical Studies of Water Waves and Floating Bodies," Ph.D. dissertation, University of Twente, The Netherlands

Eatock-Taylor, R. and Hung, S.M. (1987). "Second order diffraction forces on a vertical cylinder in regular waves," *Applied Ocean Research*, Vol.9, No.1, 19–30.

Ferrant, P. (1996). "Computation of higher order diffraction effects using a fully nonlinear simulation method," *Proc. of the 11th International Workshop on Water Waves and Floating Bodies*, Hamburg, Germany.

Isaacson, M. and Cheung, K.F. (1992). "Time-domain second-order wave diffraction in three dimensions," *Journal of Waterway, Port, Coastal, and Ocean Engineering*, Vol.118, No.5, 496–516.

Jonsson, I.G. and Wang, J.D. (1980). "Current-depth refraction of water waves," *Ocean Engineering*, Vol.7, No.1, 153–171.

Kim, M.H. and Yue, D.K.P. (1990). "The complete second-order diffraction solution for an axisymetric body. Part 2. bichromatic incident waves and body motions," *Journal of Fluid Mechanics*, Vol.211, 557–593.

Kriebel, D.L. (1990). "Nonlinear wave interaction with a vertical circular cylinder. Part I: Diffraction theory," *Ocean Engineering*, Vol.17, No.4, 345–377.

Kriebel, D.L. (1992). "Nonlinear wave interaction with a vertical circular cylinder. Part II: Wave run-up," *Ocean Engineering*, Vol.19, No.1, 75–99.

MacCamy, R.C. and Fuchs, R.A. (1954). "Wave forces on piles: a diffraction theory," U.S. Army Corps of Engineers, *Beach Erosion Board*, Tech. Memo. No. 69, Washington D.C.

Nossen, J., Grue, J. and Palm, E. (1991). "Wave forces on three-dimensional floating bodies with small forward speed," *Journal of Fluid Mechanics*, Vol.227, 135–160.

Romate, J.E. (1989). "The Numerical Simulation of Nonlinear Gravity Waves in Three Dimensions using a Higher Order Panel Method," Ph.D. dissertation, University of Twente, The Netherlands.

Schäffer, H.A. and Skourup, J. (1996). "Active absorption of multidirectional waves," *Proc. 25th Int. Conf. on Coastal Engineering*, Orlando, Florida, ASCE, New York.

Schäffer, H.A., Stolborg, T. and Hyllested, P. (1994). "Simultaneous generation and active absorption of waves in flumes," *Proc. Int. Symposium: WAVES - Physical and Numerical Modelling*, University of British Columbia, Vancouver, Canada, 90–99.

Skourup, J. (1996). "Active absorption in a numerical wave tank," *Proc. of the Sixth International Offshore and Polar Engineering Conference*, Los Angeles, USA, Vol.3, 31–38.

Skourup, J. and Bingham, H.B. (1996). "Active absorption of radiated waves in a 3D boundary element model," *Proc. of the 11th International Workshop on Water Waves and Floating Bodies*, Hamburg, Germany.

Skourup, J., Sterndorff, M.J. and Hansen, E.A. (1992). "Numerical modelling of wave-structure interaction by a three-dimensional non-linear boundary element model: a step towards the numerical wave tank," *Ocean Engineering*, Vol.19, No.5, 437–460.

Zhao, R. and Faltinsen, O.M. (1988) "Interaction between waves and current on a two-dimensional body in the free surface," *Applied Ocean Research*, Vol.10 No.2, 87–99.

Proceedings of the Seventh (1997) International Offshore and Polar Engineering Conference
Honolulu, USA, May 25-30, 1997

ISBN 1-880653-28-1 (Set); ISBN 1-880653-31-1 (Vol. III)

A Rational Approach to Free Surface Flow by Using Higher-Order Boundary Elements

Hong G. Sung
Seoul National University, Seoul, Korea
Sa Y. Hong
Korea Research Institute of Ships and Ocean Engineering, Daejon, Korea
Hang S. Choi
Seoul National University, Seoul, Korea

Abstract

In this paper, an improvement of the boundary element method is attempted by using discontinuous elements for effective treatment of the free surface problem with intersecting vertical boundaries. The GMRES algorithm is used for solving large dense nonsymmetric linear systems of equations, while the 4th-order Runge-Kutta algorithm is implemented for time integration. Advantages of the present numerical scheme are demonstrated in terms of accuracy and convergence for typical boundary value problems with free surface.

Keywords : Free surface waves, Boundary elements, Discontinuous elements, GMRES algorithm, Runge-Kutta method

1. Introduction

Recently the "numerical wave tank technique" is getting its weight for the design analysis of offshore structures, partly because of physical restrictions of experimental wave tanks(Kim, 1995). Since the pioneering work of Longuett-Higgins & Cokelet(1976) for periodic nonlinear waves in two dimensions, there have been a lot of studies on this problem. Most studies are based on the boundary element method, which is known to be very effective with respect to the operation count for large systems(Romate, 1989). Dommermuth et al.(1988) successfully applied this method in order to predict the kinematics of plunger-type breaking waves within the framework of potential theory. Meanwhile essential extensions to three dimensions are made by Xü(1992), Boo(1993) and Lee et al.(1994).

The assumption of the irrotational motion of ideal fluid renders the introduction of a velocity potential, which is the solution of the Laplace equation. By using Green's function, the governing equation can be converted into a boundary integral equation as follows :

$$\alpha(\mathbf{x})\phi(\mathbf{x}) = \int_{\partial\Omega} [\, \phi(\xi)\frac{\partial}{\partial n_\xi} G(\xi, \mathbf{x}) - \frac{\partial}{\partial n_\xi}\phi(\xi) G(\xi, \mathbf{x})] dS(\xi), \quad \mathbf{x} \in \partial\Omega, \tag{1}$$

where $\alpha(\mathbf{x})$ is the interior solid angle, ϕ the velocity potential. Green's function is given by

$$G(\xi, \mathbf{x}) = -\frac{1}{4\pi}\frac{1}{|\xi - \mathbf{x}|}. \tag{2}$$

In order to evaluate the solid angle correctly, the following analytic expression can be utilized.

$$\alpha(\mathbf{x}) = \int_{\partial\Omega} \frac{\partial}{\partial n_\xi} G(\xi, \mathbf{x})\, dS. \tag{3}$$

The solvability of the Laplace equation requires appropriate boundary conditions over the whole boundary surface($\partial\Omega \equiv S$) enclosing the domain(Ω). The fluid domain is bounded by the free surface(S_f), the bottom(S_h), the body surface(S_b) and the radiation boundaries(S_r). Necessary conditions for each boundary are stated below :

$$\frac{D}{Dt}\phi(\mathbf{x},t)=\frac{1}{2}|\nabla\phi|^2 - z - p_f, \quad \mathbf{x}\in S_f, \tag{4}$$

$$\frac{D\,\mathbf{x}_f}{Dt} = \nabla\phi, \quad \mathbf{x}\in S_f, \tag{5}$$

$$\mathbf{n}\cdot\nabla\phi(\mathbf{x},t) = 0, \quad \mathbf{x}\in S_h, \tag{6}$$

$$\mathbf{n}\cdot\nabla\phi(\mathbf{x},t)=\mathbf{n}\cdot\mathbf{V}(\mathbf{x},t), \quad \mathbf{x}\in S_b, \tag{7}$$

where all the variables have been normalized by corresponding characteristic quantities. In the above, D/Dt is the material derivative, p_f the atmospheric pressure, $\mathbf{x}_f$ material coordinates of fluid particle on the free surface, $\mathbf{n}$ the outward normal vector and $\mathbf{V}$ the velocity of the body. Although no unique numerical way of handling the open boundary condition has been set up, the Sommerfeld radiation condition seems to be practical.

$$\frac{\partial\phi}{\partial t} + C\frac{\partial\phi}{\partial n} = 0, \quad x\in S_r \tag{8}$$

In the above, C denotes the phase speed. It is well known that the success of the numerical scheme depends strongly on the correct estimation of the phase speed on the open boundary. In this paper, the phase speed is estimated approximately by combining the above equation and the dynamic free surface condition on the downstream. This method was successfully used by Boo(1993) and Kim(1996).

2. Procedure of Boundary Element Analysis

It is well known that an initial boundary value problem can be solved by time integrating of its corresponding boundary-value problem. Therefore, the performance of the method is affected by the boundary-value problem solver and the time stepping scheme. In this section, the former is briefly discussed under numerical aspects. The free surface and the radiation boundary have Dirichlet-type conditions and other boundaries have Neumann-type conditions. Thus we have "a mixed Dirichlet-Neumann problem". Since the constant panel method is not a proper choice for this mixed problem(Xü, 1992), "bi-quadratic boundary elements" are taken. Furthermore, it is generally true that higher-order elements cost much less computing time than constant panels. The treatment of the intersection line between two Dirichlet-type boundaries, i.e., the free surface and the radiation boundary is of great importance to higher-order boundary element methods. It is pointed out that the use of discontinuous elements may be relevant(Brebbia, 1984). Herein single-edge discontinuous elements are used together with double-edge discontinuous ones. Throughout the calculations presented herein, freedom nodes are located at $d=2/3$ for both kinds of discontinuous elements.

The boundary integral equation prescribed above is discretized by introducing geometric boundary elements and another elements for physical variables. In doing so, discontinuous elements are included for physical variables along the intersecting lines. Typical geometrical and variable elements are shown in Figure 1. In the figure, dotted lines correspond to the geometrical elements and solid lines to the variable ones. With the discontinuous elements, distinct basis function sets may be utilized for geometry and variables

$$\Psi = \sum_{m=0}^{8} N_m(s,t)\,\Psi^j_m \quad \text{over } E_j, \tag{10}$$

where N_m is the product Lagrangian basis function, Ψ can be either velocity potential or its normal derivative or the position vector of the boundary surface, and E_j denotes the j-th element. In this study, nine nodes bi-quadratic elements are used. Its parametric space is sketched in Figure 2. The basis function N_m is determined according as which type of element E_j is used and whether Ψ is position vector or variables. The discretized boundary integral equation is solved usually by "the collocation method" in most engineering literatures. It is because the global method such as the Galerkin method may be cumbersome for complex and nonsmooth geometries. The discretized and collocated boundary integral equation results in the following linear system of equations.

$$\alpha(\mathbf{x}_i)\phi(\mathbf{x}_i)=\sum_j\sum_m[\phi^j_m D^j_m(\mathbf{x}_i)-\phi^j_{nm}S^j_m(\mathbf{x}_i)], \tag{11}$$

$$\text{where } D^j_m(\mathbf{x}_i) = \int_{E_j} N_m \frac{\partial}{\partial n_\xi} G(\xi,\mathbf{x})dS,$$

$$S^j_m(\mathbf{x}_i) = \int_{E_j} N_m G(\xi,\mathbf{x})dS, \tag{12}$$

$$\alpha(\mathbf{x}_i) = \sum_j\sum_m D^j_m(\mathbf{x}_i).$$

Finally this set of equations is rearranged into the matrix form by keeping it in mind that the unknown variable is uniquely defined at each collocation point, whether it is velocity potential or its normal derivative. The resulting system of linear equations has its dimension of the total number of collocation points.

$$A\begin{bmatrix}\phi\\ \phi_n\end{bmatrix} = \mathbf{b} \tag{13}$$

The matrix is generally dense, nonsymmetrical and large,

especially in three dimensional cases. In addition to the large amount of arithmetic operations, the solution of the system is required in every step or substep of time integration for the updated system matrix and the righthand-side vector. Normally a boundary element analysis code pays most portion of computing time for the composition of influence matrices and the solution of linear equations. These two problems are discussed in the following sections.

3. Numerical Integration

3.1 For collocation points outside the element

When the singularity is distributed constantly over a planar element, an exact integration may be possible. Otherwise, there is no analytic solution and a numerical integration is required. The product Gauss-Legendre quadrature is widely used. It is to note that the critical issue of the Gauss quadrature is the order of quadrature. Most works, however, seem to be done without any rational approach to this problem. In this paper, a practical and effective estimation of the order of quadrature is presented.

First, geometric parameters are defined from the relation between element and collocation point as follows :

$$p_0 = r_{min} / l_c,$$
$$p_1 = \max\{\angle Q_0PQ_2, \angle Q_7PQ_3, \angle Q_6PQ_4\}/\pi, \qquad (14)$$
$$p_2 = \max\{\angle Q_0PQ_6, \angle Q_1PQ_5, \angle Q_2PQ_4\}/\pi,$$

where r_{min} is the minimum distance between the element and the collocation point, and l_c is the characteristic length of the element. Roughly speaking, p_0 is the normalized distance, while p_1 and p_2 are the maximum perspective angles along each parametric direction. Definitions of parameters are depicted in Figure 3. Next, the order of quadrature for each parameter is found from the predetermined function :

$$\nu_s = [f(p_0, p_1)],$$
$$\nu_t = [f(p_0, p_2)], \qquad (15)$$
$$\text{where } [f] = \nu \quad \text{if } \nu - \frac{1}{2} \le f < \nu + \frac{1}{2}.$$

In order to determine the function $f(p, q)$, almost exact evaluation of elemental integral can be obtained from adaptive numerical integration or Romberg integration. The functional relation of the set of orders and parameters is fitted by combining appropriate polynomials and piecewise step functions.

3.2 For collocation points on the element : self influence

If the singular integrand of the self influence is not evaluated carefully, impractically large number of integration points are required in order to keep a certain degree of accuracy. An effective method for this problem is to transform the parameter space (s, t) into another one (σ, τ), and to apply the Gauss quadrature over this space. A variety of numerical tests shows that the so-called "triangle polar coordinate mapping"(Li et al., 1985) is one of most efficient transformations. For example, for a singular point at (s_0, t_0), the triangle polar coordinate mapping leads to

$$(s,t) = (1-\sigma)(s_0, t_0) + \sigma(1-\tau)(s_1, t_1) + \sigma\tau(s_2, t_2). \qquad (16)$$

Figure 4 shows that a triangle is mapped into a square and the singular point into the vertical line. Due to this beneficial degeneracy, the order of singularity is reduced. Since Xü(1992) clearly described the detailed procedure for continuous elements, only the case of discontinuous elements is discussed in this paper.

Let the collocation point be $\mathbf{x}_k$ in the local node number. The parametric space (s, t) is triangulated into subtriangles with the help of the triangle polar coordinate mapping. The triangulation depends upon the collocation point, and an example for continuous elements is shown in Figure 5. In fact, similar triangulations occur for discontinuous elements. Under the mapping, the basis function set is transformed

$$N^g_m(s,t) = N^g_m(s_k, t_k) + \sigma H^g_m(\sigma, \tau), \qquad (17)$$

where N^g_m is the geometric basis function and H^g_m is the appropriate polynomial function of σ and τ.

Due to this relation, the relative position vector takes the form

$$\begin{aligned} \xi - \mathbf{x}_k &= \sum N^g_m(s,t)\,\mathbf{x}_m - \sum N^g_m(s_k,t_k)\,\mathbf{x}_m \\ &= \sigma \sum H^g_m(\sigma,\tau)\,\mathbf{x}_m. \end{aligned} \qquad (18)$$

The differential area on the parametric space (s, t) is written by

$$ds\,dt = 2\sigma A_\nu\, d\sigma\, d\tau, \qquad (19)$$

where A_ν is the area of the corresponding subtriangle.

The basis function set for variables is now converted into the triangle polar coordinates. Unlike the case of geometric basis functions, the basis functions for variables are mapped

$$N^v_m(s,t) = \delta_{mk} + \sigma H^v_m(\sigma, \tau), \qquad (20)$$

where N^v_m and H^v_m are interpreted as the counterparts of N^g_m and H^g_m to the case of variable basis functions. Equations (17) - (20) are substituted into (12) in order that the self influence may become completely regular in the triangle polar coordinates, except $m = k$ for double-layer distribution. Fortunately enough, the case of $m = k$ can be exactly cancelled out by the analytic expression of the solid angle, which is the so-called row sum formula.

The single disadvantage of the use of the triangle polar coordinate mapping in evaluating the self influence is to determine the transformed form of basis functions H_m^g and H_m^v in (σ, τ), which is lengthy, but straightforward.

4. GMRES Algorithm

As the size of systems for simulating three dimensional nonlinear waves becomes larger and larger, a strong demand for a fast iterative solver is unavoidable. The conjugate gradient method(Golub & Van Loan, 1989) with preconditioning is widely used for well-structured and sparse systems of linear equations, which are reduced from these numerical methods such as finite difference and finite element method. Saad & Shultz (1986) developed the Generalized Minimal Residual algorithm, which requires neither symmetry nor positive definiteness. The algorithm is generalized from the Minimal Residual algorithm and based on the Arnoldi process, which is analogous to the Lanczos algorithm for nonsymmetric matrixes. The GMRES algorithm is known to be economic in operation counts and robust in numerical aspects. The algorithm can be briefly summarized as below :

The generalized minimal residual(GMRES) algorithm

$$Ax = f$$

1. Start : choose x_0, compute $r_0 = f - Ax_0$ and $v_1 = r_0 / \|r_0\|$

2. Iterate : For $j = 1, 2, \cdots, k$,

$$\begin{aligned} h_{ij} &= (Av_j, v_i) \quad i = 1, 2, \cdots, j, \\ \hat{v}_{j+1} &= Av_j - \sum_{i=1}^{j} h_{ij} v_i, \\ h_{j+1,j} &= \| \hat{v}_{j+1} \|, \\ v_{j+1} &= \hat{v}_{j+1} / h_{j+1,j}. \end{aligned} \tag{21}$$

3. Form the approximate solution

$x_k = x_0 + V_k y_k$, which minimizes

$\|\beta e_1 - \overline{H}_m y\|$, $y \in R^m$ and $\beta = \|r_0\|$,

$$\overline{H}_m = \begin{bmatrix} h_{1,1} & h_{1,2} & h_{1,3} & \cdot & \cdot & \cdot & h_{1,m-1} & h_{1,m} \\ h_{2,1} & h_{2,2} & h_{2,3} & \cdot & \cdot & \cdot & h_{2,m-1} & h_{2,m} \\ 0 & h_{3,2} & h_{3,3} & \cdot & \cdot & \cdot & h_{3,m-1} & h_{3,m} \\ 0 & 0 & h_{4,3} & \cdot & \cdot & \cdot & h_{4,m-1} & h_{4,m} \\ \cdot & \cdot & \cdot & \cdot & \cdot & \cdot & \cdot & \cdot \\ \cdot & \cdot & \cdot & \cdot & \cdot & \cdot & \cdot & \cdot \\ \cdot & \cdot & \cdot & \cdot & \cdot & \cdot & \cdot & \cdot \\ 0 & 0 & 0 & \cdot & \cdot & \cdot & h_{m,m-1} & h_{m,m} \\ 0 & 0 & 0 & \cdot & \cdot & \cdot & 0 & h_{m+1,m} \end{bmatrix},$$

$V_m = [v_1 | v_2 | v_3 | \cdots | v_m]$ and

$e_1 = [1 \ 0 \ \cdots \ 0]^T$; 1st standard basis of R^{m+1} .

In the practical implementation of the GMRES algorithm, minimization over Krylov subspace should be programmed with a great care. The successive QR factorization can be effectively utilized in the minimization procedure. It is also recommended that a restarting version is used in the case of extremely large system because of the quadratic growth of computing time for $\overline{H}_m$.

5. Time Integration Scheme

The 4th-order Adams-Bashforth-Moulton method was used in the work of Longuett-Higgins & Cokelet (1976), where the classical 4th-order Runge-Kutta method initiated the starting procedure. Stability consideration, however, shows that the former is not well suited(Romate, 1989). Furthermore, the Runge-Kutta method has many variants, and maximum stability or minimum truncation error can be achieved by adjusting coefficients(Gear, 1971). In this paper, the 4th-order Runge-Kutta method with minimum truncation error is adopted. The truncation error of this method is about 54% of that of the classical methods, while the computation procedure is not complex and it requires no extra memory.

6. Numerical Results and Discussion

Test calculations for linear regular waves have been carried out in order to examine the accuracy, convergence and stability of the boundary element method together with the performance of the GMRES algorithm and also the time integration scheme. A rectangular region is taken as computation domain and linear wave potentials are given on the free surface and radiation boundary, while normal velocities are imposed on other boundaries.

The accuracy and convergence is checked as functions of the grid size. Figure 6 shows the performance of the present boundary element analysis. The horizontal axis is the characteristic element length normalized by wave length and the vertical axis represents the normalized maximum error. Circles designate the maximum error on Dirichlet boundaries and diamonds those on Neumann boundaries. It is concluded that at least 8 elements per wave length are necessary for keeping the error below 1%. The convergence property can be approximated by the following equation.

$$\max|\phi_n(\underline{x}_i) - \phi_{n\,i}| \le C_1 h^{2.10}$$
$$\max|\phi(\underline{x}_i) - \phi_i| \le C_2 h^{2.77} \qquad (22)$$

Convergence estimation may depends on the definition of the error. Normally maximum error at collocation points is sufficient. Though the accuracy and convergence on Dirichlet boundary is a little bit poor in comparison with those on Neumann boundary, the error norm on Dirichlet boundary converges to zero fast enough to acquire reasonable accuracy. Thus the present boundary element analysis is believed to have favorable accuracy and convergence properties.

The classical Gauss elimination requires $O(N^3)$ arithmetic operations, where N is the system size. Since the total number of unknowns ranges normally in order of thousand for practical solutions for three dimensional nonlinear waves, the Gauss elimination may be quite burdensome even for high-speed computers. Hence it is generally accepted that boundary element methods should be supported by a fast iterative solver. The GMRES algorithm has been referred to as the most efficient algorithm for solving linear system of equations, which requires $O(N^2)$ operations. The performance of the GMRES algorithm is illustrated in Figure 7. Test calculations have been carried out on SGI workstation of 43 MFLOPS. Triangles, diamonds and circles correspond to Gauss elimination, the GMRES algorithm with no preconditioner and that with diagonal preconditioner, respectively. The GMRES algorithm with diagonal preconditioner requires the minimum computing time, which is proportional to N^2. For the GMRES algorithm with no preconditioner, however, computing time amounts to $O(N^{2.5})$. The required dimensions of the Krylov subspace are shown in Figure 8 for these two algorithms. For $N = 1602$, for example, the GMRES algorithm with diagonal preconditioner proves that the dimension of the Krylov subspace is less than one quarter of that for the GMRES algorithm with no preconditioner. It is directly related to the significant reduction of computing time. Thus the GMRES algorithm is recommended to use with an appropriate preconditioner, which may depend upon the boundary geometry.

To investigate the accuracy and stability of the time stepping scheme, the propagation of linear waves is simulated with the Runge-Kutta method(Figure 9). Exact potentials of steady waves are fed from the upstream boundary. Initial values of potential and surface elevation are specified on the free surface, and potentials are on the downstream boundary. At this stage, the exact value of the phase speed is given on the downstream radiation boundary. Time integration is made by the 4th-order Runge-Kutta method with the minimum truncation error. Since the absolute stability requirement of the present method is equal to that of the classical 4th-order Runge-Kutta method, the minimum time interval can be easily obtained by considering the evolution equations and maximum wave number of error components. In this calculation, time step is taken as one tenth of the stability requirement. From the figure, it is observed that the scheme is quite accurate and stable.

Finally the so-called ideal wave maker problem is considered in order to investigate the efficacy of the Sommerfeld radiation condition and the approximating technique for the phase speed(Figure 10). The normal velocity of the linear monochromatic wave is fed on with modulation during first several periods in order to avoid undesirable impulsive responses. After initial modulation phase, the calculated wave tends to the expected steady solution. However, the wave amplitude seems to be slightly decreased as indicated in the figure. Just as in the case of Figure 9, no active smoothing or filtering technique has been applied. It is believed in the present stage that this is mainly caused by somewhat inaccurate evaluation of the phase speed. Namely the numerically calculated phase speed shows oscillating behavior around the exact value. Thus additional efforts should be made for the evaluation of the phase speed.

7. Summary

An accurate and efficient boundary element method has been developed for numerical analysis of three dimensional wave problems. The GMRES algorithm has been adopted in order to accelerate the solution procedure for a large dense nonsymmetric system of equations. Accuracy, efficiency and stability of the present boundary element method have been proved through example calculations. Accurate evaluation of the phase speed may improve the numerical result. In addition to this, some special techniques such as smoothing and regridding can be included for simulating the nonlinear free surface flow.

Acknowledgement

This study has been supported financially by Korea Research Institute of Ships and Ocean Engineering. Authors deeply appreciate its support.

References

Boo, S.Y. 1993, Application of Higher Order Boundary Element Method to Steady Ship Wave Problem and Time Domain Simulation of Nonlinear Gravity Waves, PhD Dissertation,

Texas A&M

Brebbia, C.A. 1984, *Topics in Boundary Element Research. Vol. 1 : Basic Principle and Applications*, Springer-Verlag, Berlin

Dommermuth, D.G., Yue, D.K.P., Lin,W.M., Rapp, R.J., Chan,E.S. & Melville, W.K. 1988, Deep-water Plunging Breakers : A Comparison Between Potential Theory and Experiments, *J. Fluid Mech.*, 184: 267-288.

Gear, C.W. 1971, *Numerical Initial Value Problems in Ordinary Differential Equations*, Prentice-Hall, Inc.

Golub, G.H. & Van Loan, C.F. 1989. *Matrix Computations*. Johns Hopkins University Press, 2nd ed.

Kim, C.H. 1995, Recent Progress in Numerical Wave Tank Research, Proceedings of ISOPE-95, The Hague. Vol.3 pp 1-9

Kim, D. 1996, Interaction of a Large Three-Dimensional Body with Waves and Currents by THOBEM, PhD Dissertation, Texas A&M

Lee, C.C., Liu, Y.H. and Kim, C.H. 1994, Simulation of Nonlinear Waves and Forces Due to Transient and Steady Motion of Submerged Sphere, International Journal of Offshore and Polar Engineering, Vol.4, No.3, pp 174-182

Li, Hong-Bao, Han, Guo-Ming & Mang, Herbert A. 1985, A New Method for Evaluating Singular Integrals in Stress Analysis of Solids by the Direct Boundary Element Method, International Journal for Numerical Methods in Engineering, Vol. 21, pp 2071-2098

Longuett-Higgins, M.S. & Cokelet, E,D. 1976, The Deformation of Steep Surface Waves on Water, I. A Numerical Method of Computations, Proc. Roy. Soc. London, A 350: 1-26.

Romate, J.E. 1989, The Numerical Simulation of Nonlinear Gravity Waves in Three Dimensions Using a Higher Order Panel Method, PhD Thesis, University of Twente

Saad, Y. & Schultz, M.H. 1986, GMRES: A Generalized Minimal Residual Algorithm for Solving Nonsymmetric Linear Systems, SIAM J. Sci. Stat. Comput., 7: 856-869.

Xü, H. 1992, Numerical Study of Fully Nonlinear Water Waves in Three Dimensions, PhD Dissertation, MIT

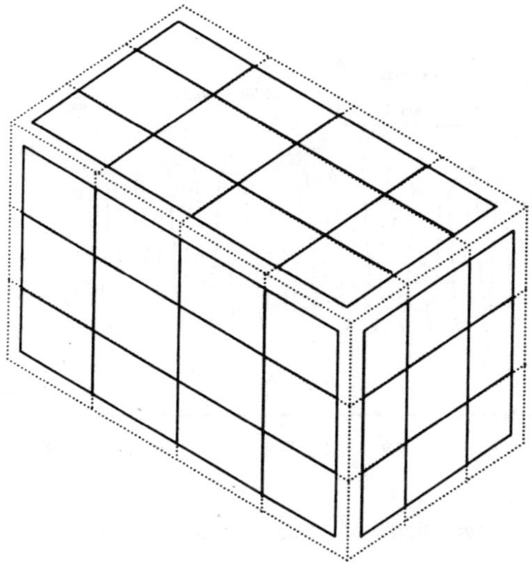

Figure 1 Typical element distribution with discontinuous elements

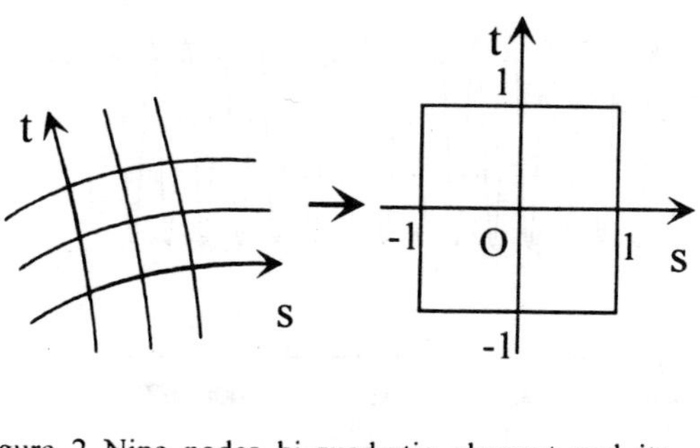

Figure 2 Nine nodes bi-quadratic element and its parametric space

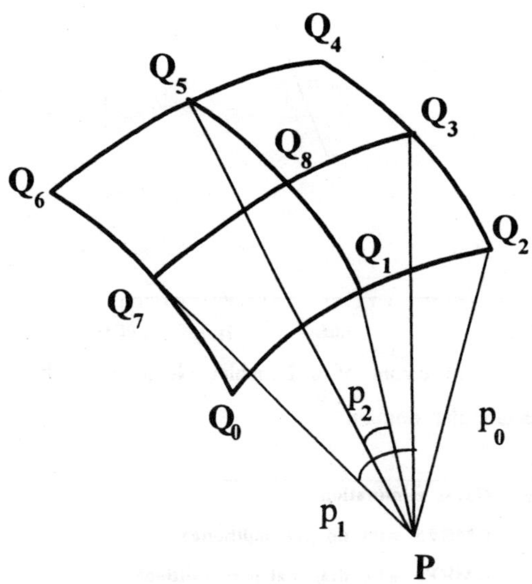

Figure 3 Geometric parameters to determine the order of quadrature

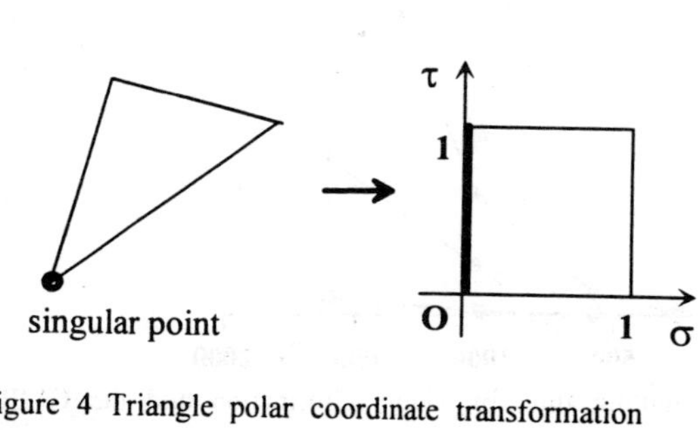

Figure 4 Triangle polar coordinate transformation

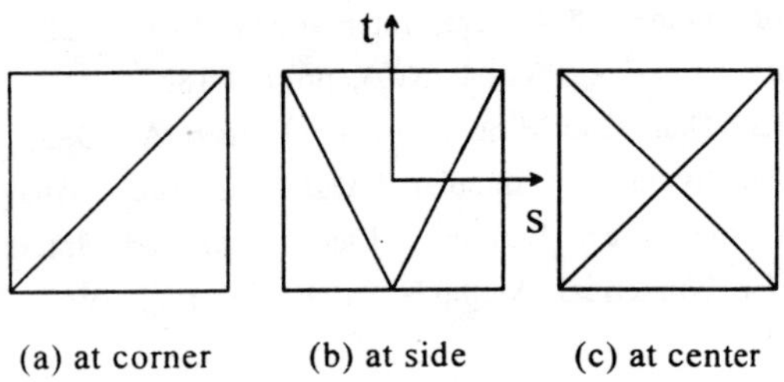

Figure 5 Triangulations in continuous elements

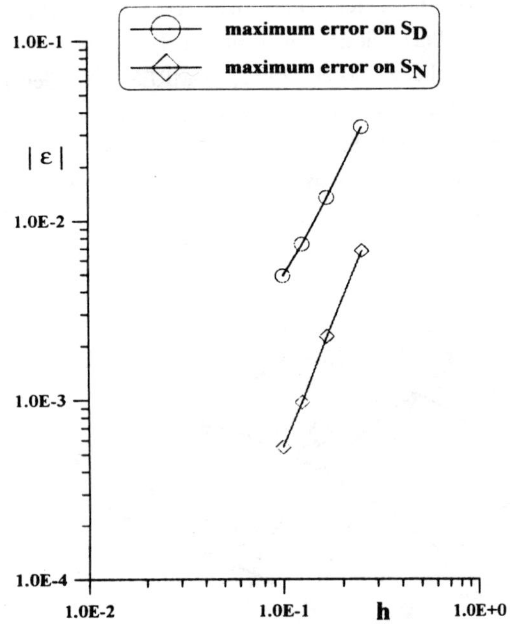

Figure 6 Maximum errors of a Dirichlet Neumann problem in a rectangular domain

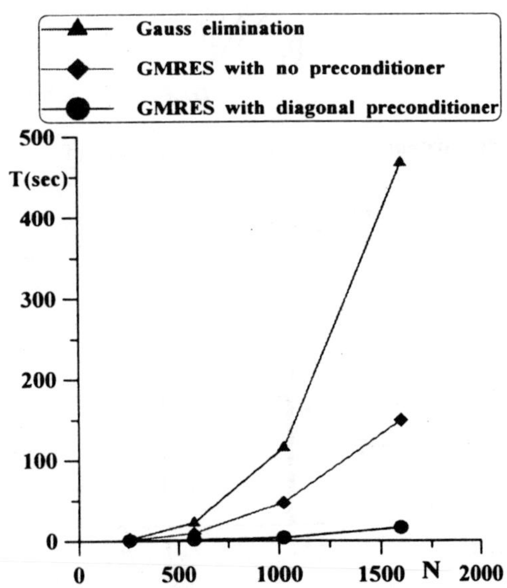

Figure 7 Computing time by Gauss elimination and the GMRES algorithm

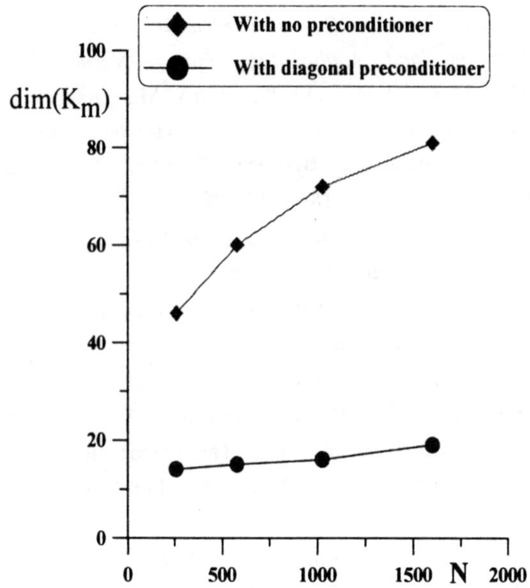

Figure 8 Dimensions of Krylov subspace

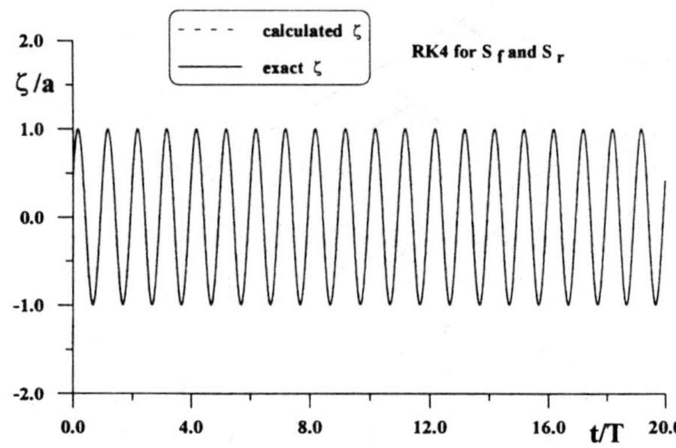

Figure 9 Time evolution of free surface elevation at far downstream

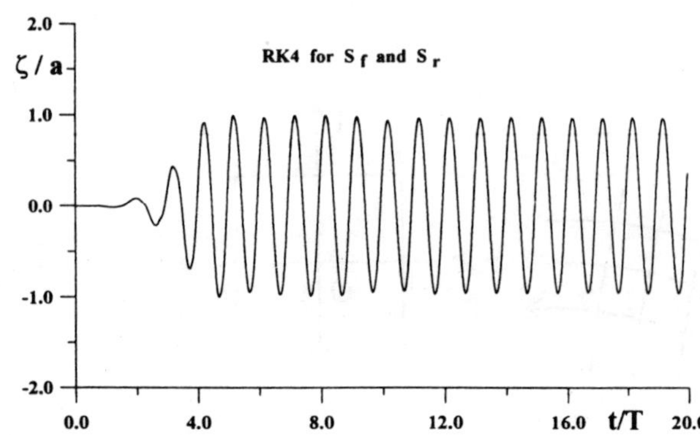

Figure 10 Time evolution of free surface elevation at far downstream for ideal wave maker problem

Proceedings of the Seventh (1997) International Offshore and Polar Engineering Conference
Honolulu, USA, May 25-30, 1997

ISBN 1-880653-28-1 (Set); ISBN 1-880653-31-1 (Vol. III)

Open Boundary for Horizontally 2-D Wave Analysis with Boundary Element Method

Ken-ichiro Hamanaka
Hokkaido Tokai University
Sapporo, Japan

Yukio Sato
Kitami Institute of Technology
Kitami, Japan

ABSTRACT

Open boundary condition used in the analysis of wave fields with the boundary element method is discussed. Hamanaka (1995) proposed a new open boundary conditon which is ideally appropriate for any wave condition. But this boundary condition is formulated under an assumption that an opening and two boundaries on both sides of the opening make a straight line. In actual harbors, most of the boundaries on both sides of the opening have a finite angle. In the present paper, the validity of this condition is examined in detail. It is confirmed that this open boundary condition is still valid in such configurations of boundaries.

KEY WORDS: Open boundary condition, boundary element method, inclined boundaries, Helmholtz equation.

1. INTRODUCTION

It is of considerable importance to analyze the interior and exterior wave fields of the harbors, in particular to analyze the resonance of waves for harbor designs. To analyze the wave fields in actual harbors, numerical methods are used instead of a purely analytical method because of the complex geometry of actual harbors.

In many well maintained harbors, the water depth is assumed to be constant, then the wave fields are governed by 2-D Helmholtz equation. In this case, the boundary element method can be utilized implemented by Green's integral equation. This method has several advantages over other numerical methods, (1) it is easy to treat the arbitrary geometry of harbors, (2) necessary unknown variables are collocated only on the boundaries and (3) only one numerical procedure is needed without any kind of iteration.

To apply this method to actual harbors, several types of boundary conditions should be established. In particular, the open boundary condition is most important because it is used not only on openings but also on incident boundaries to transmit the waves reflected from other boundaries. Furthermore, the open boundary condition is extended to the partial reflection boundary.

Usually a transmitting condition for normal incident waves has been conventionally used as an open boundary condition; Issacson and Qu (1990) and Chou and Han (1993) for example. Hamanaka (1995) showed that with this condition, numerical reflected waves are generated at the opening, and he proposed a new open boundary condition which is ideally appropriate to any kind of wave condition. He also showed the validity of this condition through numerical experiments for several wave conditions.

Although this open boundary condition is ideally appropriate to any kind of wave condition, this boundary condition is formulated under the assumption that the opening and the two boundaries on both sides make a straight line. Obviously, in actual harbors, most boundaries on both sides of the opening are diagonally opposed to each other. In the present paper, the validity of this open boundary condition is examined in detail for these types of boundaries.

To do this examination, a potential matching method is utilized without use of any special open boundary condition on the opening. We install one more region outside the opening and apply the integral equation in both regions independently. On the opening, matching conditions are applied to the potentials and their normal derivatives; that is, the potentials and their normal derivatives of the both regions should be identical on the opening. This condition is physically natural, therefore it can be expected to obtain an exact solution. The results with this method are compared with that of the analysis using the proposed open boundary condition. The angle of two boundaries on both sides of the opening is varied both positively and negatively. No significant discrepancy is found in the comparisons. This means that the proposed open boundary condition is still valid in the cases of diagonal boundaries.

2. GOVERNING EQUATION AND INTEGRAL EQUATION

Periodic and irrotational fluid motion is assumed in an incompressible and non-viscous fluid. Then the velocity potential $\hat{\Phi}$ exists and is described as,

$$\hat{\Phi} = \mathrm{Re}[\hat{\phi}\exp(-i\hat{\omega}\hat{t})] \tag{1}$$

where the symbol Re represents the real part of a complex value, $\hat{\phi}$

is the reduced velocity potential, $\hat{\omega}$ is the angular frequency and $\hat{t}$ is time.

Each variable is non-dimensionalized by $\hat{\omega}$ and the gravitational acceleration $\hat{g}$.

$(x,y,z)=(\hat{x},\hat{y},\hat{z})\hat{\omega}^2/\hat{g},\ t=\hat{\omega}\hat{t},\ \ \overline{\phi}=\hat{\phi}\hat{\omega}^3/\hat{g}^2$ etc.

where the sign ^ denotes the dimensional variables.

The velocity potential is governed by the Laplace equation with the bottom condition on $z=-h$ and the linearized free surface condition on $z=0$.

$$\begin{cases}\Delta_3\overline{\phi}=0 & \\ \partial\overline{\phi}/\partial z=0 & on\quad z=-h \\ \partial\overline{\phi}/\partial z=\overline{\phi} & on\quad z=0\end{cases}$$

where $\Delta_3=\partial^2/\partial x^2+\partial^2/\partial y^2+\partial^2/\partial z^2$, and h denotes the water depth.

The usual manner of separation of variables leads to

$$\overline{\phi}=\phi(x,y)Z(z)$$

where $\quad Z(z)=\cosh k(z+h)$.

Then ϕ is governed by the Helmholtz equation.

$$\Delta\phi+k^2\phi=0 \tag{2}$$

with the dispersion relation

$$k\tanh kh=1$$

where $\Delta=\partial^2/\partial x^2+\partial^2/\partial y^2$, and k denotes the wave number.

Under a suitable differentiability, two functions ϕ and G satisfy Green's formula for the Helmholtz operator.

$$\int_V[\phi(\Delta G+k^2G)-G(\Delta\phi+k^2\phi)]dv=\int_S(\phi\frac{\partial G}{\partial n}-G\frac{\partial\phi}{\partial n})ds \tag{3}$$

where n denotes the normal unit on the boundary S of the region V, taken outward.

The Hankel function of the first kind and zero-th order is applied in Eq.(3) as a function G.

$$G=H_0^{(1)}(kr)\quad,\quad r=\sqrt{(x-x_0)^2+(y-y_0)^2} \tag{4}$$

which satisfies the requirement of singularity at the point (x_0,y_0).

Substituting Eq.(4) into Eq.(3) and using Eq.(2) leads to the integral equation.

$$c\phi(x_0,y_0)=\int_S(\phi\frac{\partial G}{\partial n}-G\frac{\partial\phi}{\partial n})ds \tag{5}$$

The constant c in Eq.(5) takes the following values corresponding to the location of singular point.

$c=4i$ inside the region

$c=2i$ on the boundary.

Eq.(5) states that once the potential and its normal derivative on the boundary are obtained, the potential at any point in the region can be estimated by this equation.

3.NUMERICAL PROCEDURE WITH BOUNDARY ELEMENT METHOD

To obtain numerically the potential and its normal derivative on the boundary, Eq.(5) is used in a discretized form. The boundary in Eq.(5) is divided into small segments and the singular point is located on the i-th segment. The integral in Eq.(5) is approximated by the simple trapezoidal rule.

$$2i\phi_i=\sum_j^S(\phi_j\overline{E}_{i,j}-E_{i,j}\overline{\phi}_j)\quad,\quad\overline{\phi}_j=\frac{\partial}{\partial n}\phi_j \tag{6}$$

where on $i\neq j$

$$E_{i,j}=\int_{\Delta S_j}Gds=G(x_j,y_j:x_i,y_i)\Delta S_j$$

$$\overline{E}_{i,j}=\int_{\Delta S_j}\frac{\partial G}{\partial n}ds=\frac{\partial}{\partial n}G(x_j,y_j:x_i,y_i)\Delta S_j$$

on $i=j$

$$E_{i,j}=\Delta S_j+\frac{2i}{\pi}\Delta S_j(\log\frac{k\Delta S_j}{4}-1+\gamma)$$

$$\overline{E}_{i,j}=0$$

where γ is Euler's constant.

Applying the boundary conditions, the potential or its normal derivative can be eliminated and the equation may be solved using a standard matrix method.

4. OPEN BOUNDARY CONDITION

In this section, the open boundary condition proposed by Hamanaka (1995) is briefly introduced.

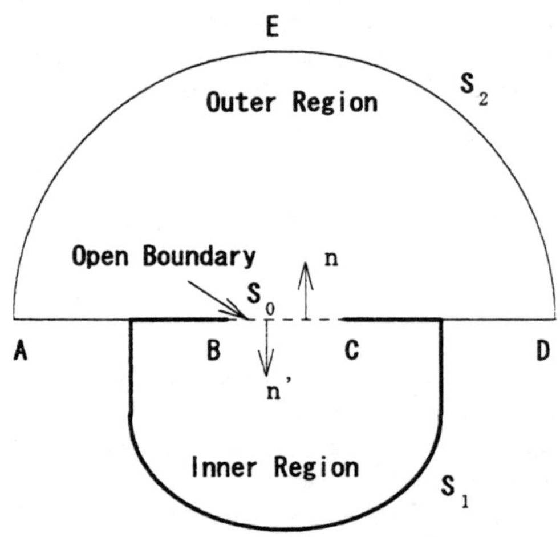

Fig. 1 Schematic diagram of open boundary

The region under consideration is referred to as an inner region, surrounded by S_0 and S_1 (see Fig.1). S_0 denotes an open boundary. To formulate a condition on it, we consider a semi-infinite space outside the open boundary. A far boundary AED in Fig.1 is installed on this space at infinity from the open boundary. The region surrounded by the far boundary and the straight line AD is referred to as an outer region. The integral equation Eq.(5) is applied to the outer region. The singular point $P:(x_0,y_0)$ is located on the open boundary, then Eq.(5) is described as follows.

$$2i\phi(P) = \int_{S_0+S_2} \left(\phi\frac{\partial G}{\partial n'} - G\frac{\partial \phi}{\partial n'}\right)ds \quad (7)$$

where n' denotes normal unit on the boundary of the outer region. Because the singular point P is located on the straight line AD,

$\partial G / \partial n' = 0$ on AD

$\partial \phi / \partial n' = 0$ on AB and CD.

The functions G satisfies the radiation condition on the far boundary AED and ϕ also satisfies the radiation condition for all incident wave conditions. Therefore only the integral on the open boundary is needed in the right hand side of Eq. (7). Then considering that the directions of n and n' are opposite, the following expression is obtained.

$$2i\phi(P) = \int_{S_0} G\frac{\partial \phi}{\partial n}ds \quad (8)$$

Eq.(8) gives a new open boundary condition. This condition does not depend on the incident angle and the number of components of waves. It is valid not only for plane waves but also for cylindrical waves.

This condition Eq.(8) is discretized in the same manner of Eq.(6) and used with Eq.(6) to solve the problem with the boundary element method. For the partial reflection and incident boundaries, Eq.(8) can be used extensibly (see Hamanaka, 1995).

5. POTENTIAL MATCHING METHOD

To examine the validity of the open boundary condition Eq.(8) with inclined boundaries on both sides of an opening, we perform an another type of analysis with a potential matching method.

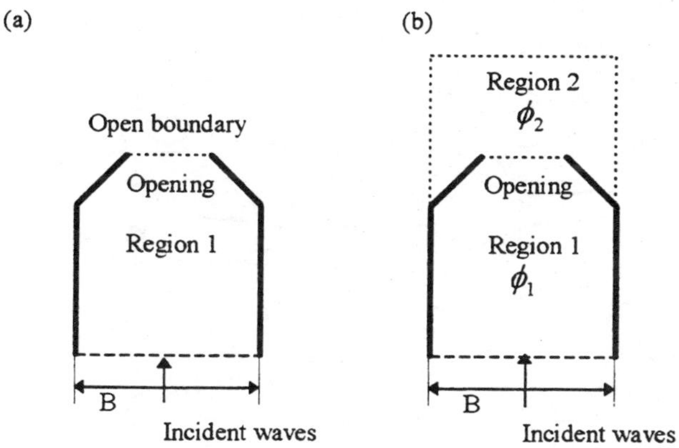

Fig. 2 Schematic diagram of boundaries for two methods

Region 1 in (a) and (b) of Fig. 2 is the region under consideration. The bold lines denote reflection boundaries, the dotted line, an opening, and the broken line, an incident boundary. In (b) of Fig.2, Region 2 is installed outside the opening for use of the potential matching method. The integral equation Eq.(5) is applied to Regions 1 and 2 independently. On the opening, matching conditions are applied to the potentials and their normal derivatives.

$\phi_1 = \phi_2$ on the opening

$\partial \phi_1 / \partial n_1 = -\partial \phi_2 / \partial n_2$ on the opening

where n_1 and n_2 denote the normal unit on the opening in Regions 1 and 2.

Three boundaries are installed in Region 2 surrounding the opening. The open boundary condition Eq.(8) is applied to these boundaries. There exist diffracted waves form the opening in Region 2. It has been confirmed that the diffracted waves are sufficiently transmitted through these boundaries with Eq.(8) by Hamanaka (1995).

The integral equations in the two regions and the matching conditions are discretized in a similar manner to Eq.(6), and solved simultaneously.

On the opening in (a) of Fig.2, the open boundary condition Eq.(8) is applied directly. Results with the two methods will be compared numerically in the next section.

6. NUMERICAL RESULTS

The non-dimensional water depth is taken as, $h = 1$. The non-dimensional wave length is about 2π on this depth. The non-dimensional width of the region is taken as, $B = 30$. And the width of segment is about 0.5. The incident waves approach normally to the incident boundary. The amplitude of the incident waves is unit 1.

Fig.3 shows an example of water surface displacement in a case when the opening is placed at an outward diagonally opposed angle from the region. (a) is the result with the direct use of the open boundary condition on the opening. (b) is the result on Region 1 with the potential matching method. No significant discrepancy is found between two results. Fig.4 shows contour lines of the amplitude of the same results of Fig.3. Only very small differences are found in the contour lines. It can be seen that the accuracy is sufficient for practical uses.

Fig.5 and Fig.6 show the case when the opening is placed further outward. Only very small differences are found in the contour lines in this case, too.

Fig.7 and fig.8 show the contour lines of two cases when the opening is placed at an inward diagonally opposed angle inside the region. The water surface displacements are omitted because no significant discrepancy is found. The differences in the contour lines are sufficiently small in these cases too.

7. CONCLUSION

The validity of the open boundary condition proposed by Hamanaka (1995) is examined in detail in the cases when the opening is placed at an inward or outward opposed angle. Numerical experiments are performed in the both cases. A potential matching method is used for comparison with the direct use of the proposed open boundary condition. No significant discrepancies are found in all cases. Therefore, it is confirmed that the proposed open boundary condition is still valid for the opening with diagonal boundaries.

REFERENCES

Chou, C. R. and W. Y. Han, 1993: Wave-induced oscillations in harbours with dissipating quays, Coastal Engg. Japan, Vol. 36, No. 1, pp. 1-23

Isaacson, M. and S. Qu, 1990: Waves in a harbour with partially reflecting boundaries, Coastal Engg., Vol. 14, pp. 193-214

Hamanaka, K., 1995: Fundamental solution and boundary conditions on oscillations in habors of arbitrary geometry, 12th Australasian Coastal and Ocean Engg. Conf., pp.163-168

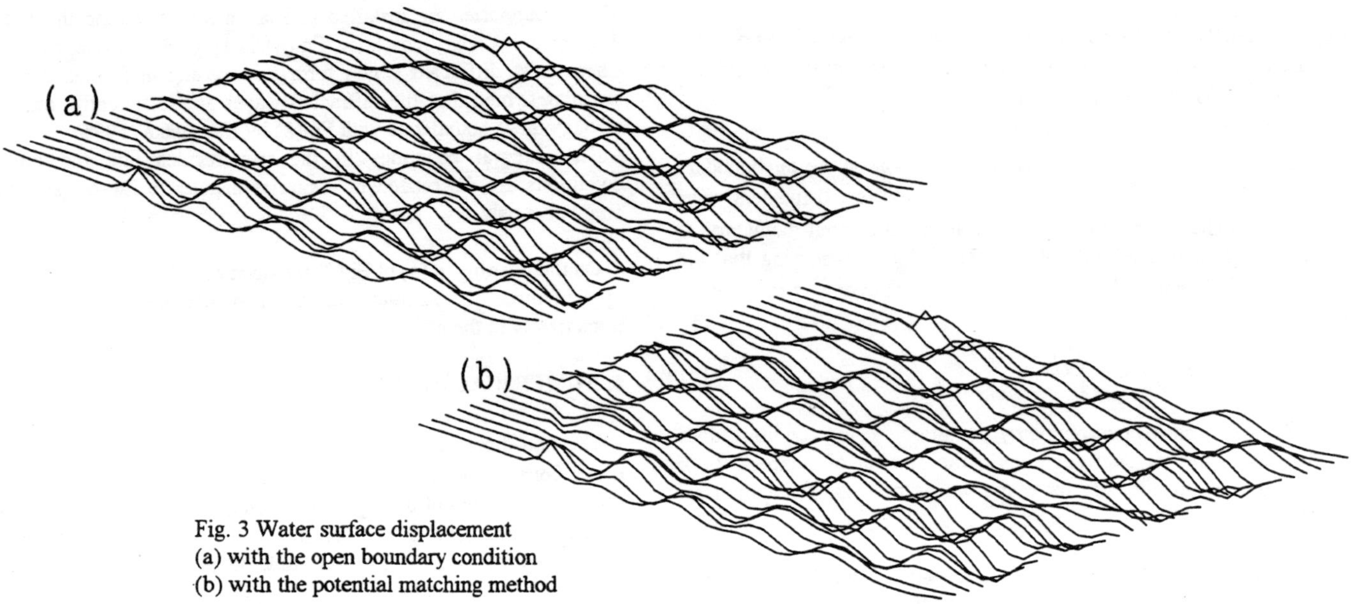

Fig. 3 Water surface displacement
(a) with the open boundary condition
(b) with the potential matching method

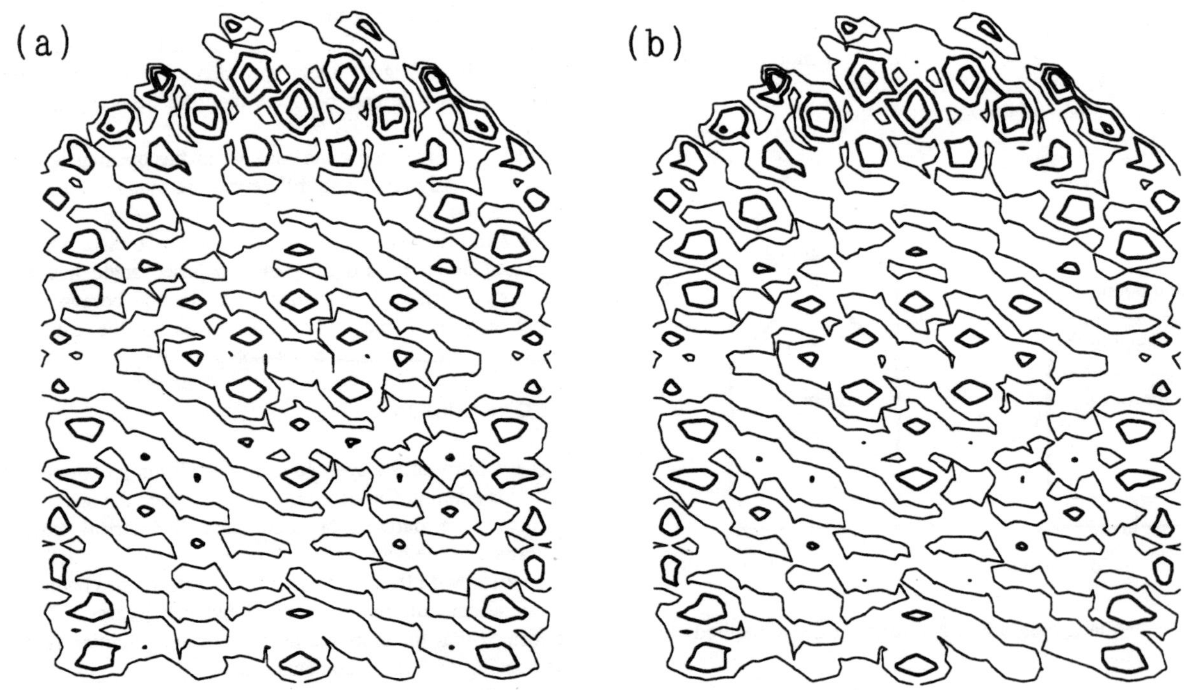

Fig. 4 Contours of amplitudes of 1 (thin line), 2 and 3 (bold lines)
(a) with the open boundary condition, (b) with the potential matching method

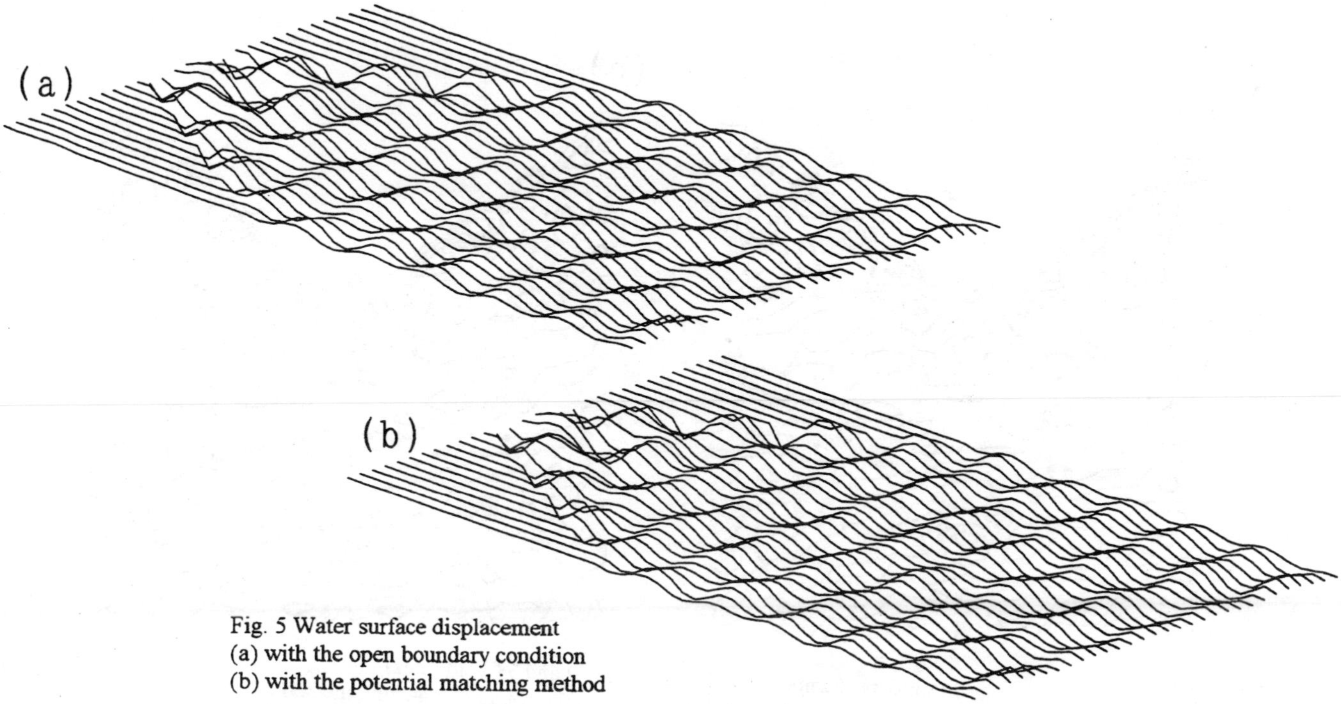

Fig. 5 Water surface displacement
(a) with the open boundary condition
(b) with the potential matching method

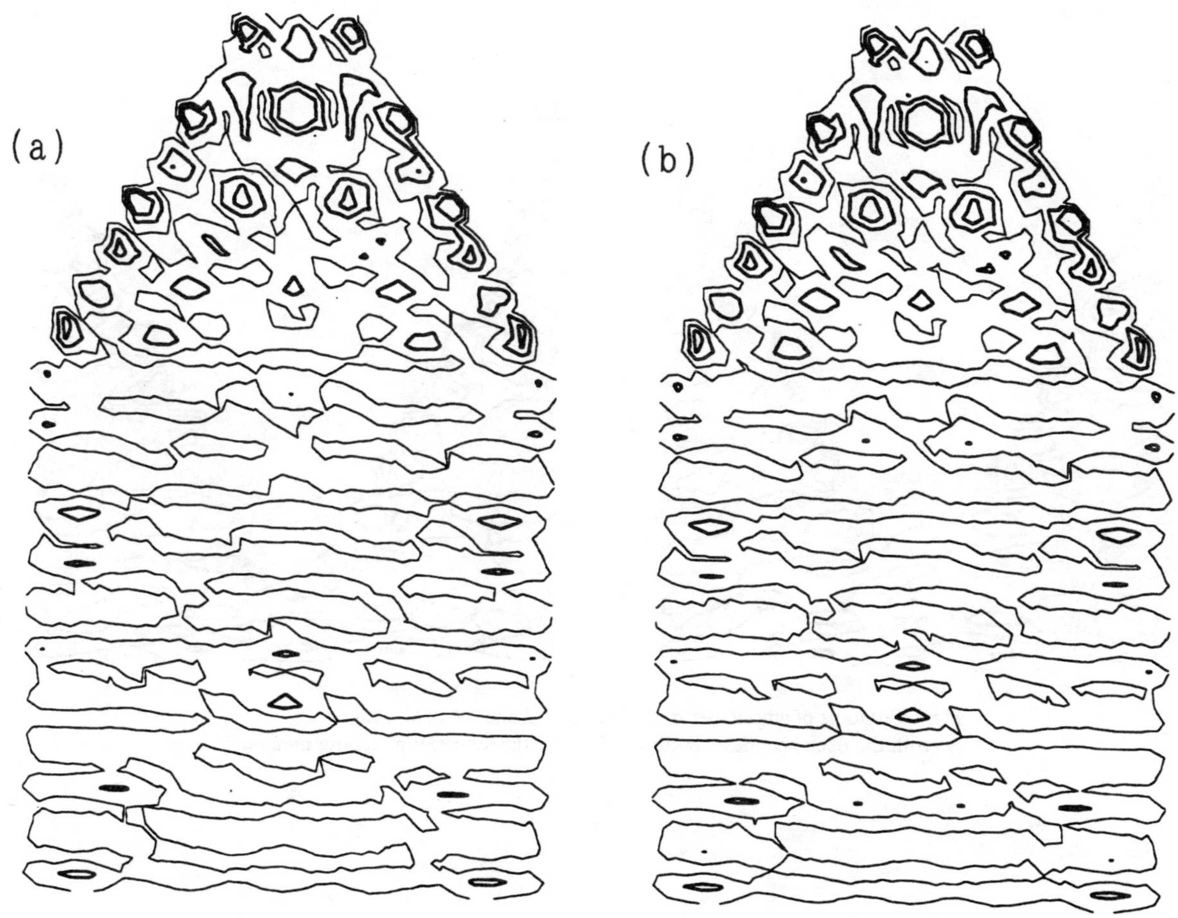

Fig. 6 Contours of amplitudes of 1 (thin line), 2 and 3 (bold lines)
(a) with the open boundary condition, (b) with the potential matching method

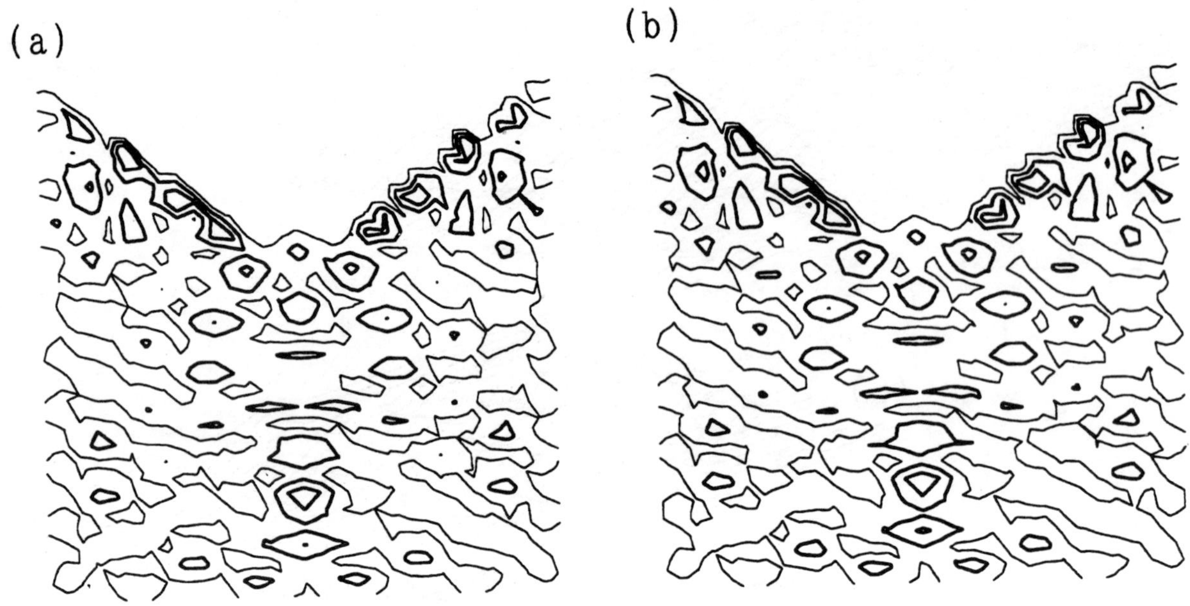

Fig. 7 Contours of amplitudes of 1 (thin line), 2 and 3 (bold lines)
(a) with the open boundary condition, (b) with the potential matching method

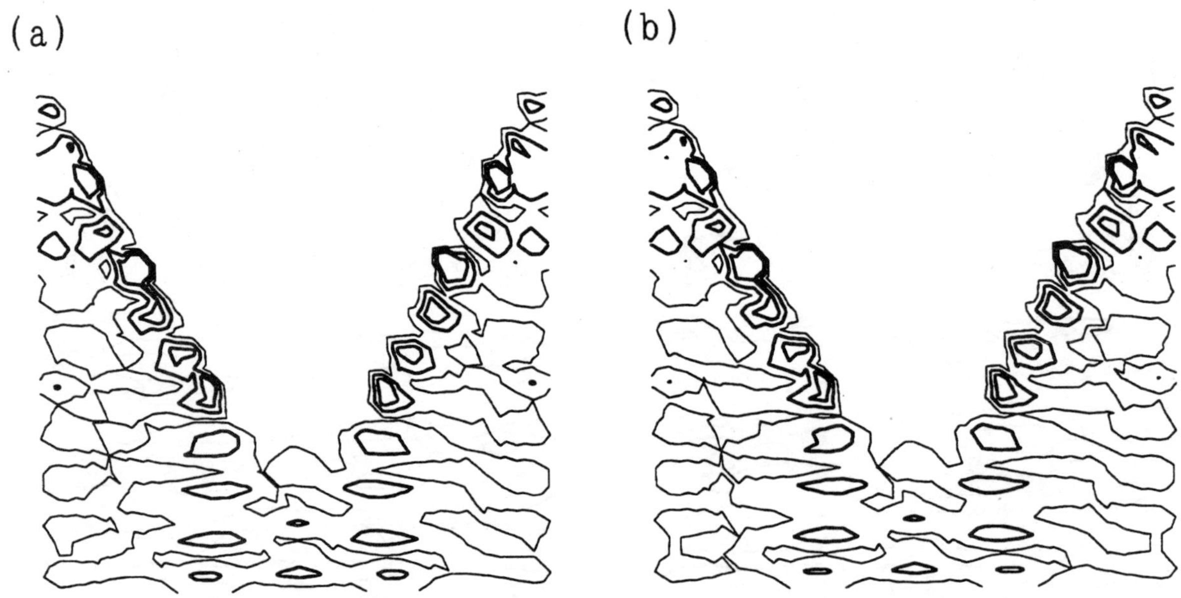

Fig. 8 Contours of amplitudes of 1 (thin line), 2 and 3 (bold lines)
(a) with the open boundary condition, (b) with the potential matching method

Proceedings of the Seventh (1997) International Offshore and Polar Engineering Conference
Honolulu, USA, May 25-30, 1997

ISBN 1-880653-28-1 (Set); ISBN 1-880653-31-1 (Vol. III)

A Study on Parametric Roll Motions by Fully Nonlinear Numerical Wave Tank

Katsuji Tanizawa
Ship Research Institute
Tokyo, Japan

Shigeru Naito
Osaka University
Osaka, Japan

ABSTRACT

Parametric roll motion of a two-dimensional floating body is studied numerically, theoretically and experimentally. For numerical study, a fully nonlinear numerical wave tank is applied. This is a time domain simulation program which solves simultaneous equations of ideal fluid motion and floating body motions. To satisfy the radiation condition at the tank ends, an artificial damping technique is applied. Using this numerical wave tank, motions of a floating body in regular waves are simulated and the critical wave height, which excite the parametric roll oscillation, is estimated. Theoretical estimation of this criteria is also given from the stability analysis of coupled Mathieu type equation of heave and roll motions. To validate these numerical and theoretical results, an experiment is performed. Comparison among numerical, theoretical and experimental results show that the simulated motions well agree with the measured motions in both harmonic and parametric oscillations, and as a result, the criteria estimated by the numerical simulations agrees with the measured criteria qualitatively and quantitatively, meanwhile the theoretical criteria agrees qualitatively.

KEY WARDS : Parametric oscillation, Numerical wave tank, Fluid-body interaction, Fully nonlinear simulation, Acceleration potential, Implicit body surface boundary condition

INTRODUCTION

Responses of floating bodies such as ships or ocean structures to incident waves are one of the main concern in ocean engineering. The responses are usually treated as harmonic assuming small amplitude wave and body motions. Under the assumption, the frequency responses have been investigated by linear or perturbation theories. But, the body motions are not always harmonic in real ocean. When amplitude of wave and body motions are large in rough seas, nonlinear effects become dominant. Capsizing in the plunging breaker is the extreme example. Even if the amplitudes are small, nonlinearity due to body shape, mooring force, free water on the deck etc. affect to the body motions and chaotic or parametric motions may be resulted. In this paper, harmonic and parametric roll motions of a two-dimensional body are studied numerically, theoretically and experimentally.

Parametric roll motions are arising from the time dependent variation of the metacenter height of the body due to relative heave motions to the free-surface. If the body has sensitive shape to parametric excitation, the roll angle diverges exponentially and results in capsizing. In particular, when the wave period is nearly equal to the half of the natural roll period, parametric motion is very easy to be excited even in a small amplitude wave.

For the analysis of parametric roll motions, time domain fully nonlinear simulation can be a powerful tool. Time domain fully nonlinear simulation methods are studied by Vinje [5)], Cointe [7)], Tanizawa [8, 14)], Kang [9)], Van Daalen [10)], Sen [11)], Cao [12)] and others in the past decade and fully nonlinear numerical wave tanks have been developed. In this study, using the numerical wave tank [15)], the parametric motion together with the harmonic motion are simulated seamlessly in time domain and the critical wave height, which excite the parametric motion, is estimated. For the validation of the numerical results, an experiment is performed.

The parametric oscillation is a classical problem and has been studied using Mathieu type model equations in a number of fields of science. So, also in this paper, Mathieu type simultaneous equations of heave and roll motions are composed and the criteria is estimated from the stability analysis of this model equations.

Through these studies, the applicability of the numerical wave basin to the parametric motions are discussed.

FULLY NONLINEAR NUMERICAL WAVE TANK

Mathematical formulation

Motions of a floating body inside a two-dimensional wave basin is considered. As Fig.1 shows, fluid domain Ω is bounded by free-surfaces S_f , a piston wave maker S_p , bottom and rigid wall S_b and a floating body S_s . Here, gravitational acceleration g , density of fluid ρ and width of floating body B are chosen as units to nondimensionalize the problem. An space-fixed Cartesian coordinate system $o-xz$ is used with x coincident with the calm free-surface and z positive upward. The fluid is assumed to be homogeneous, incompressible, inviscid and its motion irrotational. The fluid motion can be described by a velocity potential ϕ and its time derivative ϕ_t . In the fluid domain Ω , ϕ and ϕ_t satisfies Laplace's equation

$$\nabla^2\phi = \nabla^2\phi_t = 0 . \tag{1}$$

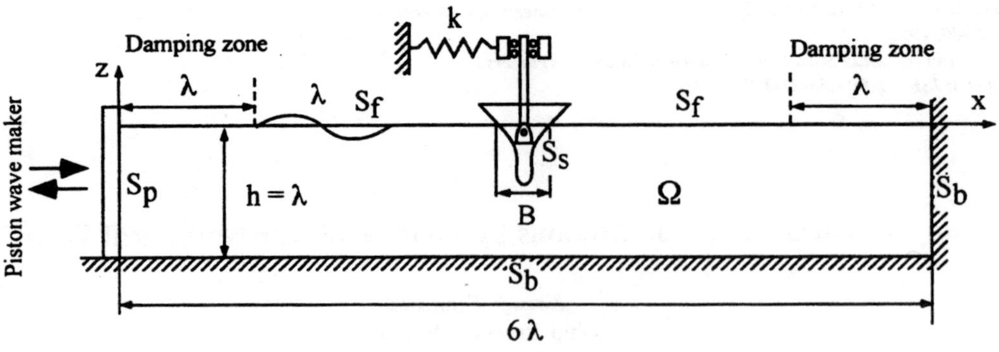

Fig.1 Simulation with damping zones

Green's second identity can be applied on both ϕ and ϕ_t

$$c(\mathcal{Q})\left\{\begin{array}{c}\phi(\mathcal{Q})\\ \phi_t(\mathcal{Q})\end{array}\right\} = \int_S \left\{\begin{array}{c}\phi(\mathcal{P})\\ \phi_t(\mathcal{P})\end{array}\right\}\frac{\partial}{\partial n}\ln r(\mathcal{P},\mathcal{Q}) - \ln r(\mathcal{P},\mathcal{Q})\left\{\begin{array}{c}\dfrac{\partial\phi(\mathcal{P})}{\partial n}\\ \dfrac{\partial\phi_t(\mathcal{P})}{\partial n}\end{array}\right\} dS\ , \qquad (2)$$

where $\mathcal{P},\mathcal{Q}$ are points on the boundary, n is outward normal direction of the boundary, $r(\mathcal{P},\mathcal{Q})$ is distance between $\mathcal{P}$ and $\mathcal{Q}$, $c(\mathcal{Q})$ represents the angle subtended at $\mathcal{Q}$ by boundaries.

On the free-surface, kinematic boundary condition and dynamic boundary condition for zero atmospheric pressure are applied as

$$\frac{D\phi}{Dt} = -z + \frac{1}{2}(\nabla\phi)^2 \qquad (3)$$

$$\frac{D\boldsymbol{x}}{Dt} = \nabla\phi\ , \qquad (4)$$

where $\boldsymbol{x} = (x, z)$. On the body surface, impermeability condition with respect to ϕ is expressed as

$$\frac{\partial\phi}{\partial n} = V_n\ , \qquad (5)$$

where V_n denotes the normal velocity of the body surface. Denoting translating and angular velocities of the body as $\boldsymbol{v}_0$ and $\boldsymbol{\omega}$ respectively, V_n is written as

$$V_n = \boldsymbol{n}\cdot(\boldsymbol{v}_0 + \boldsymbol{\omega}\times\boldsymbol{r})\ . \qquad (6)$$

Impermeability condition on the body with respect to ϕ_t [14] can be written as

$$\frac{\partial\phi_t}{\partial n} = -k_n\,(\nabla\phi - \boldsymbol{v}_o - \boldsymbol{\omega}\times\boldsymbol{r})^2 + \boldsymbol{n}\cdot(\dot{\boldsymbol{v}}_o + \dot{\boldsymbol{\omega}}\times\boldsymbol{r}) + \boldsymbol{n}\cdot\boldsymbol{\omega}\times(\boldsymbol{\omega}\times\boldsymbol{r}) + \boldsymbol{n}\cdot 2\boldsymbol{\omega}\times(\nabla\phi - \boldsymbol{v}_o - \boldsymbol{\omega}\times\boldsymbol{r}) - \frac{\partial}{\partial n}\left(\frac{1}{2}(\nabla\phi)^2\right)\ , \qquad (7)$$

where k_n is curvature of body, $\dot{\boldsymbol{v}}_o, \dot{\boldsymbol{\omega}}$ are translating and angular accelerations of the body respectively.

On the floating body surface, $\dot{\boldsymbol{v}}_o, \dot{\boldsymbol{\omega}}$ can not be specified explicitly and implicit boundary condition should be applied. Denoting the inertia tensor of the floating body as $\mathcal{M}$ and generalized normal vector of body surface as $\boldsymbol{N} = (\boldsymbol{n}, \boldsymbol{n}\times\boldsymbol{r})$, the implicit boundary condition is written as

$$\frac{\partial\phi_t}{\partial n} = \boldsymbol{N}\mathcal{M}^{-1}\left\{\int_{S_s} -\phi_t \boldsymbol{N} ds\right\} + \boldsymbol{N}\mathcal{M}^{-1}\left\{\int_{S_s}\left(-z - \frac{1}{2}(\nabla\phi)^2\right)\boldsymbol{N} ds + \boldsymbol{F}_g\right\} + q - \frac{\partial}{\partial n}\left(\frac{1}{2}(\nabla\phi)^2\right)\ , \qquad (8)$$

where $\boldsymbol{F}_g$ is sum of gravity, mooring force and other external forces acts to the body and q is the term which cab be explicitly evaluated from the solution of velocity field as

$$q = -k_n\,(\nabla\phi - \boldsymbol{v}_o - \boldsymbol{\omega}\times\boldsymbol{r})^2 + \boldsymbol{n}\cdot\boldsymbol{\omega}\times(\boldsymbol{\omega}\times\boldsymbol{r}) + \boldsymbol{n}\cdot 2\boldsymbol{\omega}\times(\nabla\phi - \boldsymbol{v}_o - \boldsymbol{\omega}\times\boldsymbol{r})\ . \qquad (9)$$

With these boundary conditions and Green's second identity with respect to ϕ and ϕ_t, both velocity and acceleration fields can be solved numerically by BEM. The solutions are integrated with respect to time by 4th order Runge-Kutta method, then fluid and the body motions are simulated in time domain. The free-surface is traced by MEL [4].

Artificial damping zone

Following a preceding work of Cointe et al. [7], damping terms are added to dynamic and kinematic free-surface boundary conditions to give artificial damping effect to free-surface. The free-surface boundary conditions inside a damping zone are given as

$$\frac{D\phi}{Dt} = -z + \frac{1}{2}(\nabla\phi)^2 - \nu(x_e)(\phi - \phi_e) \qquad (10)$$

$$\frac{D\boldsymbol{x}}{dt} = \nabla\phi - \nu(x_e)(\boldsymbol{x} - \boldsymbol{x}_e)\ . \qquad (11)$$

where $\nu(x_e)$ is the damping coefficient

$$\nu(x) = \left\{\begin{array}{ll}\alpha\omega\left(\dfrac{x - x_0}{\lambda}\right)^2, & for \quad x_0 \le x \le x_1 = x_0 + \beta\lambda\\ 0, & for \quad x < x_0 \ \ or\ \ x > x_1\end{array}\right. . \qquad (12)$$

In the definition of $\nu(x)$, ω and λ are angular frequency and wave length of the incident wave respectively. The performance of this damping zone is controlled by two nondimensional parameter α and β. α is used to control the strength of damping and β is used to control the length of damping zone. $\phi_e, \boldsymbol{x}_e$ are reference values. This damping zone damps down differences $\phi - \phi_e$ and $\boldsymbol{x} - \boldsymbol{x}_e$. When the damping zone is applied in front of a rigid wall and works as a simple absorber, the reference values are set to $\phi_e = 0, \boldsymbol{x}_e = (x_e, 0)$. And when the damping zone is applied in front of a wave maker and works as an absorbing wave maker, the reference values are set to the solution of the wave generated by the wave maker. This solution can be computed by numerical simulation of the wave without bodies in the tank. For practical purpose, linear analytical solution can be a good substitution.

Linear propagating wave generated by a piston wave maker is described as

$$\phi(x,z,t) = \frac{4s\tanh kh \sinh kh}{\omega(2kh+\sinh 2kh)}\cosh k(z+h)\cos(kx-\omega t) \quad (13)$$

$$\eta(x,t) = -\left.\frac{\partial\phi}{\partial t}\right|_{z=0} = \frac{4s\sinh^2 kh}{2kh+\sinh 2kh}\sin(kx-\omega t) \quad (14)$$

where s is stroke of the wave maker and k is wave number of the generated wave. Wave reflection coefficient of this damping zone is less than 2% , when the tuning parameter is appropriately set to $\alpha=\beta=1$ for a regular wave. [7, 15].

SIMULATION WITH EXPERIMENTAL VALIDATION

Experiment

The experiment was performed in the sea-keeping test basin of Ship Research Institute. Fig.2 shows the plan and side views of the test basin and Fig.3 shows the section shape of the test model used in the experiment. To keep two-dimensionality, the test model was set between the rigid current plates installed in the center of the test basin. Fig.4 shows the measuring equipment of the motions. The swaying carriage is moored by weak and linear springs in horizontal direction to restrict free drifting. The principal dimensions of the model and measuring equipment are give in Table 1 and Table 2.

Table 1 Principal dimensions of the bow section body

Item		M-K-S system	
Breadth	B	0.600	m
Breadth at W.L.	$B_{W.L.}$	0.304	m
Depth	D	0.532	m
Length	L	0.997	m
Draft	d	0.370	m
Displacement	W	48.55	kg
Center of inertia	KG	0.242	m
Radius of inertia	R_I	0.193	m
Metacenter height	GM	0.032	m
Natural period of heave	T_h	1.07	sec
Natural period of roll	T_r	2.66	sec

Table 2 Principal dimensions of the measuring equipment

Swaying carriage	Weight	5875 g
	Measuring span	±40 cm
Heaving rod	Weight	453 g
	Measuring span	±20 cm
Gimbal	Weight	525 g
	Measuring angle	±30°
Load cell	Weight	137 g
	Capacity	±5 kgf
Mooring system	Balance weight	500 g
	Spring constant	51.07 N/m

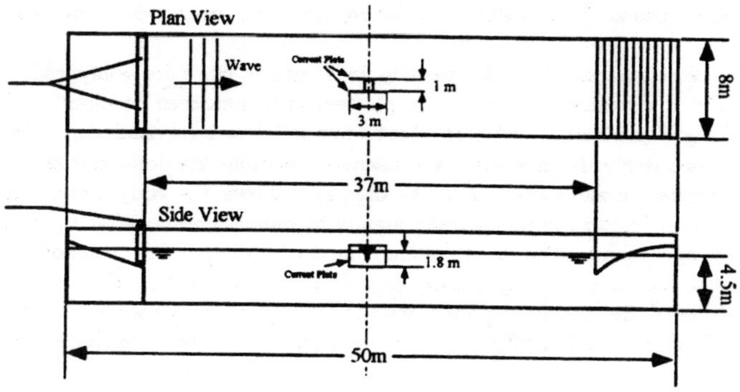

Fig.2 Test basin used for the experiment

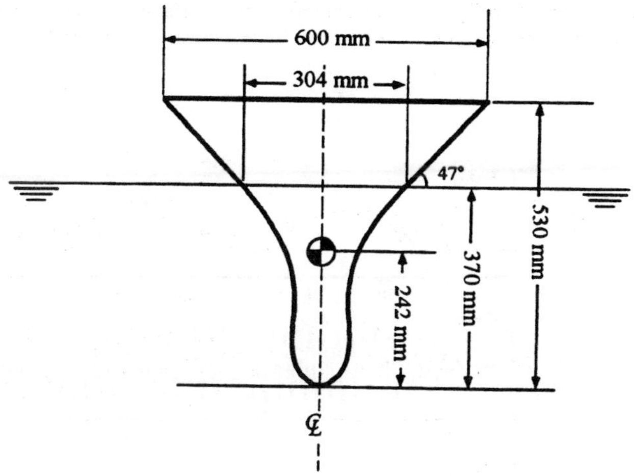

Fig.3 Body shape of the bow section model

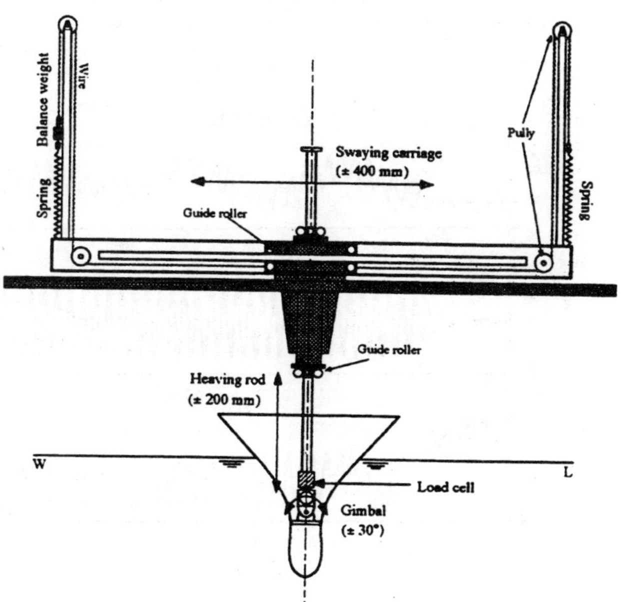

Fig.4 Meauring equipment

Comparison between simulated and measured body motions

First, harmonic body motions in a regular wave are compared. In Fig.5, simulated motions are plotted with measured motions. The length and height of the incident wave are $\lambda = 1.4m$ and $h_w = 5cm$ respectively. In this wave, the harmonic motions are dominant in both simulation and experiment. As the figure shows, the body motions are converging to the periodically steady harmonic motions after the transient motions. The low-frequency horizontal motion due to mooring is also damping down gradually around the position where the restoring force of the mooring balances with the steady wave drift force. The simulated harmonic motions of the body agree with the measured motion in overall.

Next, body motions in a different wave are compared in Fig.6. The length and height of the incident wave are $\lambda = 3.6m$ and $h_w = 8cm$ respectively. In this wave, roll motions are not harmonic and significant double period oscillations are observed both in simulation and experiment. This double period oscillation is the typical characteristic of the parametric oscillation. When damping is linear, the parametric motion does not have a limit cycle and diverges exponentially. As this figure shows, the simulated motions well describe the exponential growth of the parametric oscillation and agreement with the measured motions are very good. The growth speed of roll angle in the simulation is faster than that of measured motion because the viscous effect is not considered in the simulation.

To visualize this parametric roll motion, simulated instantaneous positions of the body at $t/T_w = 21.25, 21.50, \ldots, 25.00$ are plotted in Fig.7, where T_w is the period of the incident wave. Horizontal thin lines are calm water surface. The positions arranged in the same row are distant $2T_w$ in time. We can observe that the parametric roll motion is excited by the heave motion like a swing pumping and its growth is very natural. The amplitude of roll motion is large but wave radiation is not significant and no wave breaking is occurred near the body. The same natural and smooth parametric roll motions are observed also in the experiment.

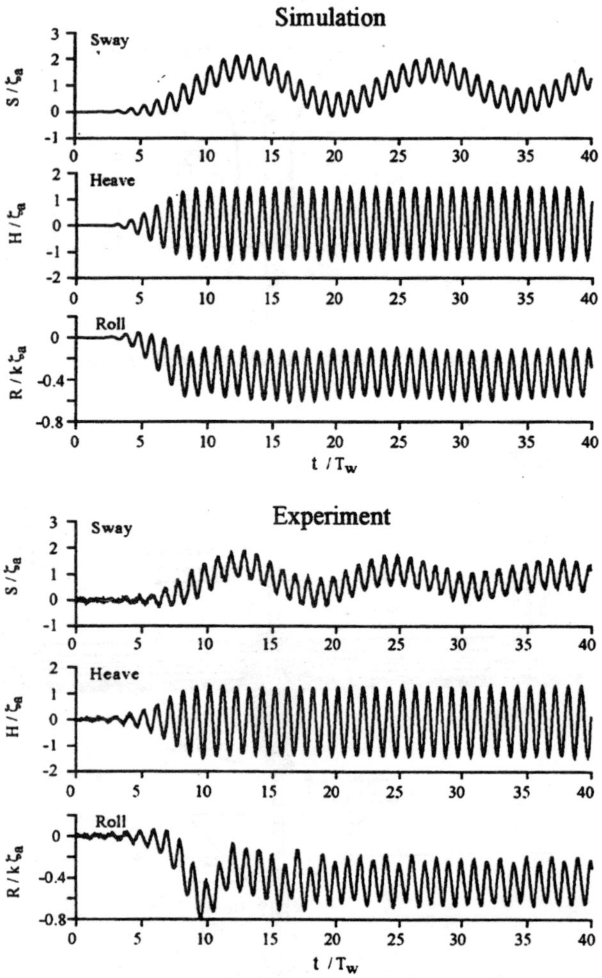

Wave Length $\lambda = 1.4$ m
Wave Height $h_w = 2\zeta_a = 5$ cm

Fig. 5 Comparison of body motions between simulation and experiment

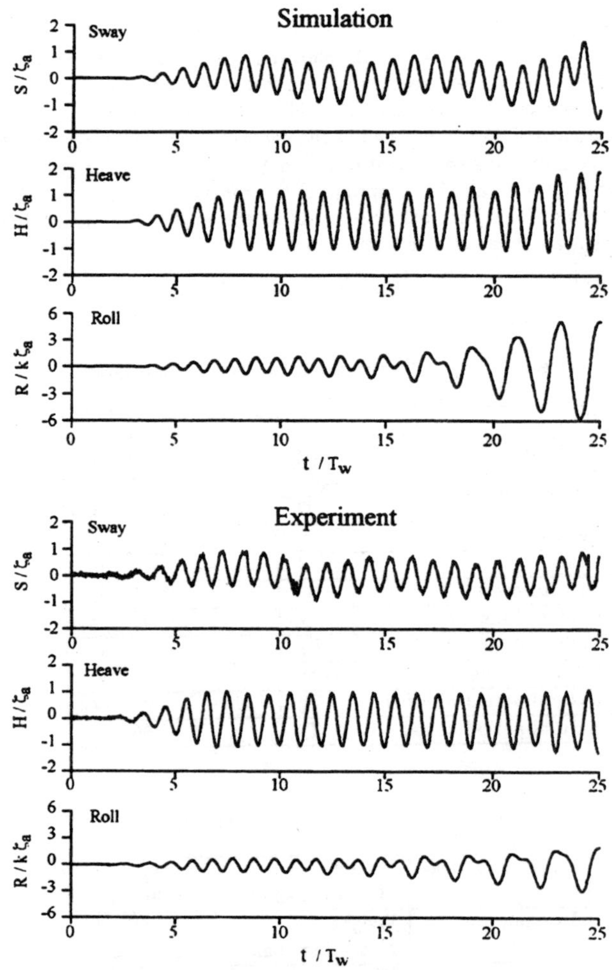

Wave Length $\lambda = 3.6$ m
Wave Height $hw = 2\zeta_a = 8.0$cm

Fig.6 Comparison of body motions between simulation and experiment

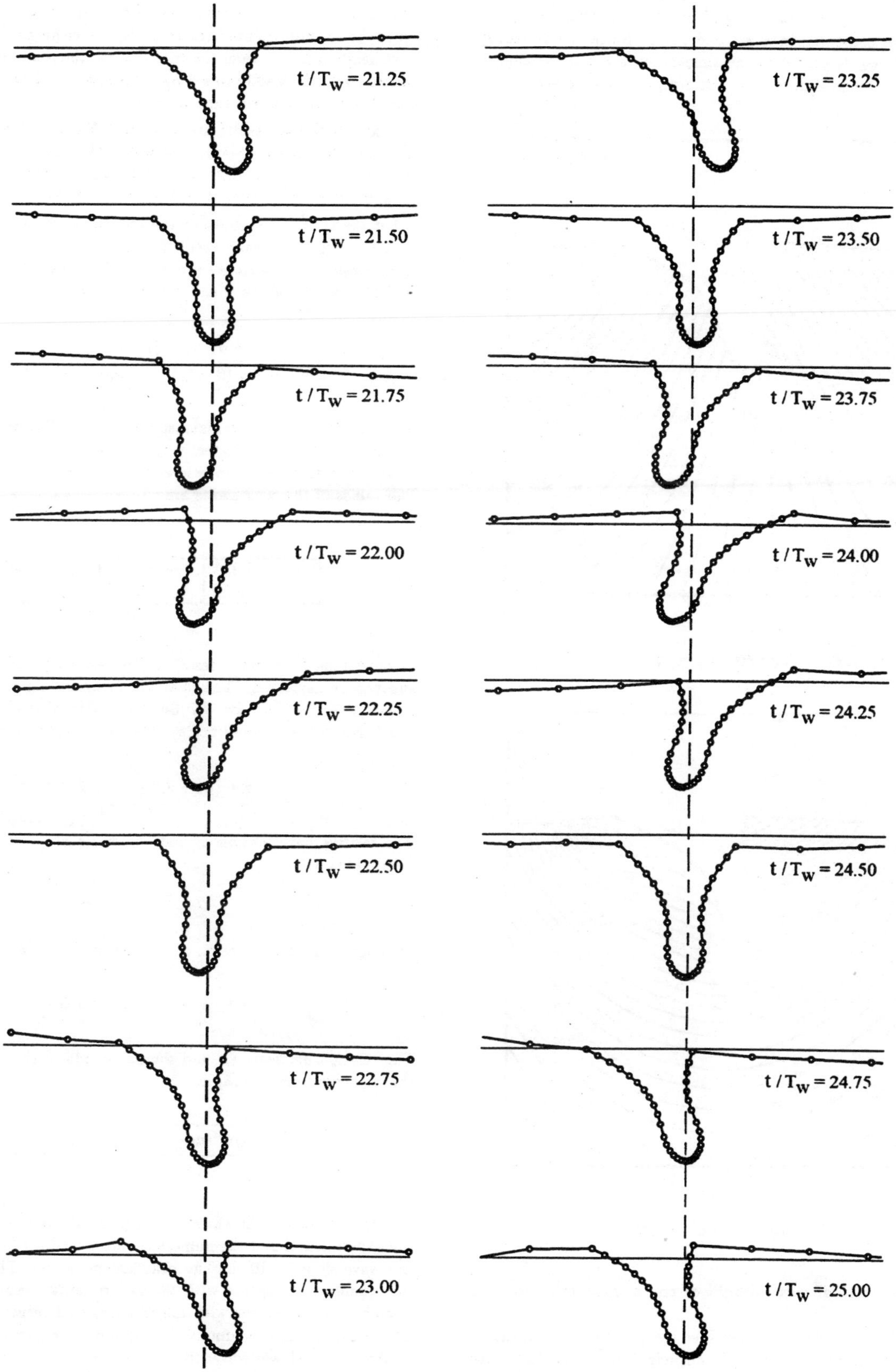

Fig.7 Simulated parametric motions of the bow section body

To observe the velocity and hydrodynamic pressure field, contour plots of the velocity potential ϕ and its time derivative ϕ_t at t/T_w = 24.00 are presented in Fig.8 and Fig.9. Since the boundary value problem on ϕ_t is solved in the simulation, the computation of ϕ_t distribution in the fluid domain is also simple and accurate.

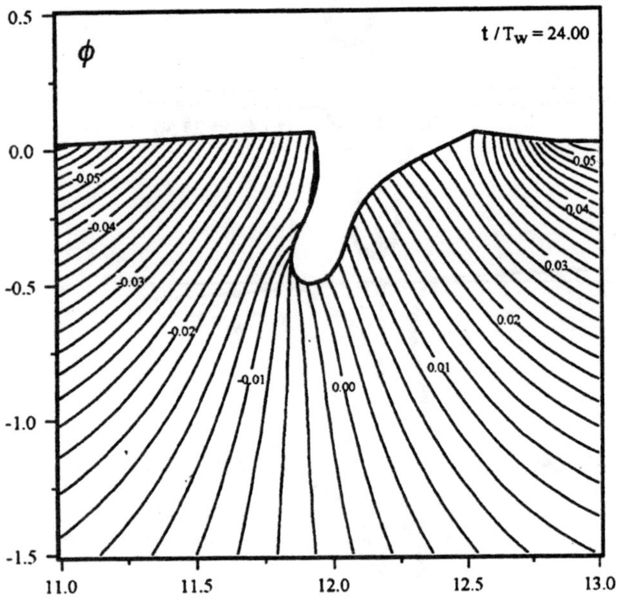

Fig.8 Contour plot of ϕ

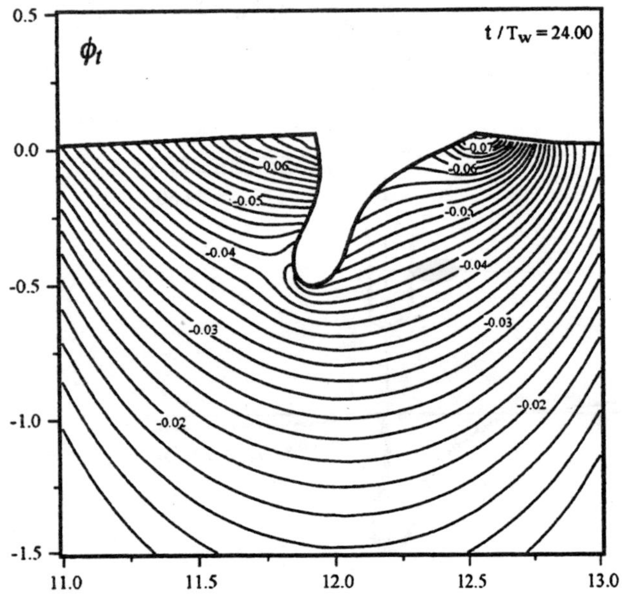

Fig.9 Contour plot of $\partial\phi/\partial t$

Analysis of the critical wave height of the parametric motions

First, the critical wave height determined by the experiment is shown in Fig.10. Combinations of the length and height of the incident wave used in the experiment are plotted. The △ marks indicate that the body motions are harmonic in that wave and the ▼ marks indicate that they are parametric. When the body motions are parametrically excited, the significant double period oscillations are observed. As this figure shows, motions are harmonic when wave height is low. And when wave height is higher than the critical level, the parametric oscillation is excited. The thick solid line drawn between △ and ▼ is the critical wave height determined experimentally.

Next, numerical results are plotted in Fig.11. The △ marks indicate that the simulated body motions are harmonic and the ▼ marks indicate that they are parametric. The thick broken line drawn between △ and ▼ is the critical wave height determined numerically.

For the theoretical study of the critical wave height, Mathieu type model equation is usually used. The critical wave height is given as the eigenvalues of the Mathieu equation. The Mathieu type simultaneous model equations of heave and roll motions can be written as

$$\ddot{z}_G + c_1\dot{z}_G + a_1(z_G - \xi) = 0 \tag{15}$$

$$\ddot{\theta} + c_2\dot{\theta} + a_2(\theta - \xi_x) + b_2(z_G - \xi)(\theta - \xi_x) = 0\ , \tag{16}$$

where z_G, θ, ξ, ξ_x are nondimensional heave, roll, free-surface elevation and free surface slope respectively and $a_1 \sim c_2$ are coefficients determined from the hydrodynamic and hydrostatic property of the body. The values of the coefficients are

a_1	2.839	——		c_1	0.7308
a_2	0.3495	b_2	9.783	c_2	0.02541

To determine the eigenvalues of eq.(16), exciting force is not important and can be neglected. The damping term can be removed by substituting $\vartheta e^{-c_2(z_G-\xi)/2}$ for θ. But, since damping coefficient c_2 is so small that we can simply neglect the damping term to have

$$\ddot{\theta} + \{a_2 + b_2(z_G - \xi)\}\theta = 0\ . \tag{17}$$

The relative heve motion of body to free-surface $z_G - \xi$ can be approximated as

$$z_G - \xi = \xi_r \cos\omega t\ , \tag{18}$$

then equation of roll motion is reduced to be a Mathieu equation

$$\ddot{\theta} + \{a_2 + b_2\xi_r \cos\omega t\}\theta = 0\ . \tag{19}$$

The relation between ξ_r and the amplitude of the incident wave ξ_a can be given from eq.(15) as

$$\xi_r = \frac{\sqrt{\omega^4(a_1 - \omega^2 + c_1^2)^2 + a_1^2 c_1^2 \omega^2}}{(a_1 - \omega^2)^2 + c_1^2\omega^2}\xi_a\ . \tag{20}$$

The critical wave height given from eq.(19) is plotted together with the experimental and the numerical values in Fig.12. The region above the wave slope 1/10 has no practical meaning. This theoretical result qualitatively agrees with the experimental result, meanwhile the numerical result agrees with the experimental result qualitatively and quantitatively. In the numerical simulation, the parametric motions are excited in little lower waves than those of in experiment. This difference is considered to result from the viscous effect.

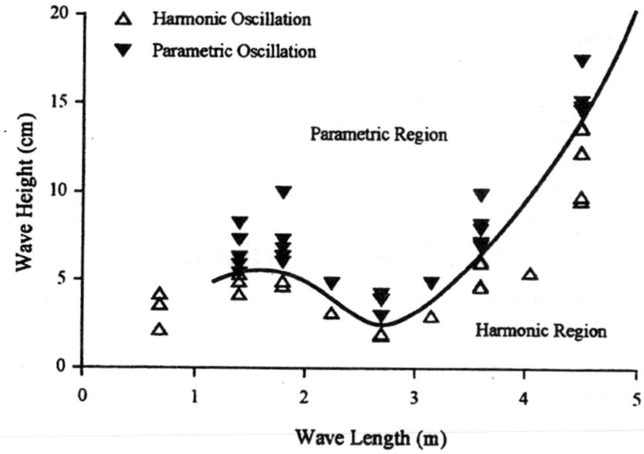

Fig.10 Criteria of parametric and harmonic regions given by experimentation

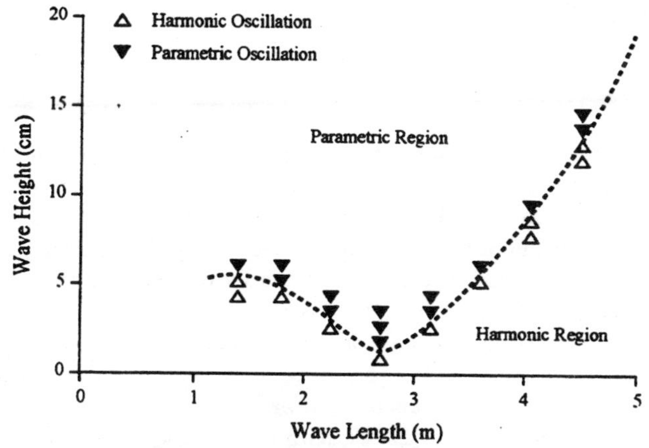

Fig.11 Criteria of parametric and harmonic regions given by the nonlinear simulation

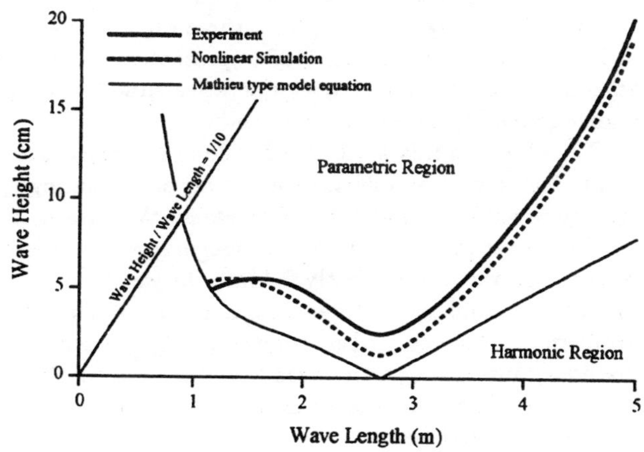

Fig.12 Criteria of parametric and harmonic regions. Comparison among experiment, the nonlinear simulation and the Mathieu type model equations

CONCLUSION

In this study, the fully nonlinear numerical wave tank is applied to simulate parametric roll motions of a two-dimensional floating body and the critical wave height is determined. The results are validated by the experiment. The following items are main conclusions.

1. The simulated harmonic body motions well agree with the measured harmonic motions in overall.
2. The simulated parametric motions also agree with the measured parametric motions.
3. The critical wave height analyzed by Mathieu type model equation agree with the experimental results qualitatively.
4. The critical wave height determined by the simulation well agree with the experimental results quantitatively.

In conventional analysis of the harmonic, parametric and the other nonlinear motions of floating bodies, the interaction between wave and floating bodies is represented as hydrodynamic coefficients. These coefficients are given from linear or perturbation theories in general. But now, we have fully nonlinear numerical wave tank. With this new technology, the nonlinear interaction can be directly simulated in time domain. The parametric roll motion is only one application of the numerical wave tank. This powerful and rational tool can be applied to a variety of nonlinear problems.

REFERENCES

1) Hsu,C.S. (1977), "On nonlinear parametric excitation problems", *Advances in Applied Mechanics, Vol.17*, pp245-298
2) Skomedal,N.G. (1982), "Parametric exitation of roll motion and its influence on stability", *Proc. of 2nd Int. Conf. on Stability of Ships and Ocean Vihecles, Tokyo*, pp113-125
3) Kyozuka,Y. (1977), "A study n the unstable roll motions of moored floating bodies in regular waves.", *Master thesis, Osaka Univ.*, pp1-82
4) Longuet-Higgins,M.S. and Cokelet,E. (1976), "The deformation of steep surface waves on water", *Proc. Roy. Soc. ser.A350*, pp1-26
5) Vinje, T. and Brevig, P. (1981), "Nonlinear Ship Motions", *Proc. of the 3rd. Int. Conf. on Num. Ship Hydro.*, ppIV3-1-IV3-10
6) Sen, D., Pawlowski, J.S., Lever, J. and Hinchey, M.J., (1989), "Two-dimensional numerical modeling of large motions of floating bodies in waves", *Proc. 5th Int. Conf. Num. Ship Hydro., part1*, pp257-277
7) Cointe,R.,Geyer,P.,King,B.,Molin,B. and Tramoni,M. (1990), "Nonlinear and linear motions of a rectangular barge in perfect fluid", *Proc. of the 18th Symp. on Naval Hydro., AnnArbor, Michigan*, pp85-98
8) Tanizawa, K. (1990), "A numerical method for nonlinear simulation of 2-D body motions in waves by means of B.E.M.", *Journal of SNAJ*, Vol.168, pp223-228
9) Kang,C.G., and Gong,I.Y. (1990), "A numerical solution method for three-dimensional nonlinear free-surface problems", *Proc. of the 18th Symp. on Naval Hydro., Annbor, Michigan.*, pp427-438
10) Van Daalen, E.F.G. (1993), "Numerical and Theoretical Studies of Water Waves and Floating Bodies", *Ph.D. thesis, University of Twente, The Netherlands*
11) Sen,D. (1993), "Numerical simulation of motions of two-dimensional floating bodies", *Journal of Ship Research, vol.37*, pp307-330
12) Cao, Y., Beck, R. and Schultz, W.W. (1994), "Nonlinear motions of floating bodies in incident waves", *9th Workshop on Water Waves and Floating Bodies, Kuju, Oita*, pp33-37
13) Tanizawa, K. (1995a) "A Nonlinear Simulation Method of 3-D body Motions in Waves", *10th Workshop on Water Waves and Floating Bodies, Oxford*, pp235-239
14) Tanizawa, K. (1995b) "A Nonlinear Simulation Method of 3-D Body Motions in Waves", *Journal of SNAJ*, Vol.178, pp179-191
15) Tanizawa,K. (1996) "Long time fully nonlinear simulation of floating body motions with artificial damping zone", *Journal of SNAJ*, Vol.180, pp311-319

Proceedings of the Seventh (1997) International Offshore and Polar Engineering Conference
Honolulu, USA, May 25-30, 1997

ISBN 1-880653-28-1 (Set); ISBN 1-880653-31-1 (Vol. III)

Nonlinear Irregular Waves and Forces on Truncated Vertical Cylinder in a Numerical Wave Tank

S.Y. Boo
Korea Naval Academy
Chinhae, Korea

C.H. Kim
Texas A&M University
College Station, TX, USA

ABSTRACT

A numerical scheme was developed to predict nonlinear diffractions acting on a floating structure by weakly nonlinear irregular waves. The Stokes weakly nonlinear irregular waves were firstly fed into the wave tank through the inflow boundary and then the scattered wave fields due to the structure were computed by solving the boundary integral equation in the time stepping procedure. The integral equation was approximated using a quadratic-order boundary element method. The instantaneous free surfaces was integrated in time using the fourth-order multi-step method. Open boundary condition was treated by combining two schemes of absorbing beach and stretching. The proposed numerical scheme was verified by comparing linear results with theory and then applied to compute the diffractions acting on a truncated vertical circular cylinder by inputting the linear and nonlinear irregular waves. Additionally, the numerical scheme was examined by the balance check of the momentum-impulse.

KEY WORDS: Nonlinear diffraction, irregular waves, numerical wave tank, higher order boundary element method, open boundary condition.

INTRODUCTION

An improved prediction technique of the extreme wave load gives a reliable data for design of the offshore structures. Linear irregular waves are commonly used in statistically computing the wave load. A method for estimation of the kinematics of the irregular waves was developed by a kinematic boundary condition fit (Forristall 1985). However, natural ocean waves are known to be nonlinear in the storm seas so that their crest heights are bigger than trough depths. Thus, weakly nonlinear irregular waves may exist in the severe seas due to the interaction of the waves. Longuet-Higgins (1963) and Dean&Sharma (1981) developed the Stokes second-order irregular wave model for deep and finite water depth, respectively. Pierson (1993) proposed a theory for the Stokes third-order nonlinear irregular waves. The third order model was developed using third-order perturbation scheme and can treat the long-crest seas in the deep water.

Numerical schemes for the computation of nonlinear wave loads in the time domain have been developed by many researchers, for instance, Boo&Kim (1996), Ferrant (1995), Isaacson&Cheung (1992), Kim *et al* (1994), Lin&Yue (1989), Yang & Ertekin (1992). The literature for the simulation of diffractions by nonlinear irregular waves are very rare. Xu *et al* (1993) reported on the nonlinear irregular wave simulation by Green-Naghdi theory in a three dimensional numerical wave tank. Boo&Kim (1994) presented the numerical scheme for the nonlinear irregular wave simulation using a higher order boundary element method.

The object of this work is to develope a numerical scheme to predict nonlinear diffractions acting on a floating structure by weakly nonlinear irregular waves. We have employed the Stokes second-order nonlinear irregular waves as incident waves. The waves were firstly fed into the wave tank through the inflow boundary and then fully nonlinear boundary value problem was solved in the time-stepping procedure. The boundary integral equation was formulated using the Rankine source and approximated using a quadratic-order boundary element method.

The instantaneous free surfaces were integrated in time using the 4th order predictor and corrector method with Eulerian approach. Open boundary condition was treated by combining two schemes of absorbing beach and stretching (Boo&Kim 1996). The proposed numerical scheme was applied to simulate the diffractions by the linear and weakly nonlinear irregular waves. Accuracy was investigated by checking the balance of the momentum-impulse.

MATHEMATICAL FORMULATION

The origin of a Cartesian coordinate system is in the plane of the undisturbed free surface with the x–axis positive in the direction of incoming wave propagation and positive z–axis in the opposite direction of the gravity as shown in Fig. 1. It is assumed that the fluid is ideal, incompressible, and its motion is irrotational. The free surface tensions are also ignored and water depth is finite.

The fluid motion can be described by the total velocity potential of $\phi(\vec{x}, t)$,

$$\phi(\vec{x}, t) = \phi_i(\vec{x}, t) + \phi_s(\vec{x}, t) \quad (1)$$

where $\phi_i(\vec{x}, t)$ is the incident potential, $\phi_s(\vec{x}, t)$ the disturbed potential, $\vec{x}$ the coordinates and t the time. The potential $\phi(\vec{x}, t)$ satisfies the Laplace equation

$$\nabla^2\phi(\vec{x}, t) = 0 \quad (2)$$

in the space occupied by the fluid. The kinematic and dynamic free surface conditions can be derived from the material derivatives and Bernoulli equation. The resulting equations on Γ_F are

$$\frac{\partial \zeta}{\partial t} + \frac{\partial \phi}{\partial x}\frac{\partial \zeta}{\partial x} + \frac{\partial \phi}{\partial y}\frac{\partial \zeta}{\partial y} - \frac{\partial \phi}{\partial z} = 0 \quad (3)$$

$$\frac{\partial \phi}{\partial t} + \frac{1}{2}\nabla\phi \cdot \nabla\phi + g\zeta = 0 \quad (4)$$

The body boundary condition is imposed under the instantaneous wetted body surface by requiring no normal flux. On the surface of body Γ_S, it can be expressed as :

$$\frac{\partial \phi}{\partial n} = 0 \quad (5)$$

The unit normal vector is denotes by $n = (n_x, n_y, n_z)$ and it is pointing out of the fluid domain. On flat bottom boundary Γ_B, the following nonpenetrating condition is used,

$$\frac{\partial \phi}{\partial n} = 0 \quad (6)$$

For the remaining boundaries which include the side boundary Γ_W and downstream boundary Γ_R, open boundary condition is necessary. It will be explained in detail.

Since the proposed problem is solved in the time domain, the initial condition also must be satisfied. The condition is

$$\begin{aligned} \phi(\vec{x}, t) &= 0 \quad t \leq 0 \quad \text{in the fluid domain} \\ \eta(\vec{x}, t) &= 0 \quad t \leq 0 \quad \text{on the free surface} \end{aligned} \quad (7)$$

FORMULATION OF INTEGRAL EQUATION

The direct boundary integral equation for solving the prescribed boundary value problems is derived using Green's theorem. Fredholm integral equation of the second kind can be written as:

$$\begin{aligned} \hat{C}\phi \quad &+ \int_{\Gamma_{S+W}} \phi\frac{\partial G}{\partial n}d\Gamma - \int_{\Gamma_{F+R}} G\frac{\partial \phi}{\partial n}d\Gamma \\ &= \int_{\Gamma_{S+W}} G\frac{\partial \phi}{\partial n}d\Gamma - \int_{\Gamma_{F+R}} \phi\frac{\partial G}{\partial n}d\Gamma \end{aligned} \quad (8)$$

The solid angle $\hat{C}$ is 0 or 4π when the field point is outside the boundary or in the fluid domain. When the field point is located on the nonplanar boundary, the solid angle is computed using special manner. Two symmetry planes are utilized to model the half fluid domain and exclude the bottom boundary. The resulting Green's function is

$$G = \frac{1}{R_1} + \frac{1}{R_2} + \frac{1}{R_3} + \frac{1}{R_4} \quad (9)$$

The distances between two points are given as:

$$\begin{aligned} R_1 &= \sqrt{(x - x_i)^2 + (y - y_i)^2 + (z - z_i)^2} \\ R_2 &= \sqrt{(x - x_i)^2 + (y + y_i)^2 + (z - z_i)^2} \\ R_3 &= \sqrt{(x - x_i)^2 + (y + y_i)^2 + (z + z_i + 2d)^2} \\ R_4 &= \sqrt{(x - x_i)^2 + (y - y_i)^2 + (z + z_i + 2d)^2} \end{aligned}$$

where d is the water depth. The (x, y, z) and (x_i, y_i, z_i) are the field and source point, respectively.

The total pressure on the instantaneous wetted surface of body can be computed from the Bernoulli's equation

$$p = -\rho\left(\frac{\partial \phi}{\partial t} + \frac{1}{2}\nabla\phi \cdot \nabla\phi + gz\right) \quad (10)$$

The forces and moment acting on the cylinder can be obtained by

$$\vec{F} = \int_{\Gamma_S} p\,\vec{n}\,d\Gamma \quad (11)$$

$$\vec{M} = \int_{\Gamma_S} p\vec{n} \times \vec{r}\,d\Gamma \quad (12)$$

where $\vec{r}$ is the position vector.

BOUNDARY ELEMENT FORMULATION

The geometries of each element are represented by polynomial functions and thus the entire curved boundary can be approximated by a number of higher order elements. Within the boundary elements, physical variables are interpolated by the higher order polynomials. Thus, the resulting approximations are written as:

$$\begin{aligned} (x, y, z) &= \sum_j N_j\,(x_j, y_j, z_j), \quad \phi = \sum_j N_j\phi_j \\ \frac{\partial \phi}{\partial n} &= \sum_j N_j\left(\frac{\partial \phi}{\partial n}\right)_j \end{aligned} \quad (13)$$

where N_j is the shape function at the j-th node. The ϕ_j and $(\frac{\partial\phi}{\partial n})_j$ are the velocity potential and normal velocity at the j-th node. Typical shape functions of 9-node quadrilateral (Fig. 2) are written as:

$$\begin{aligned}
N_j &= \frac{1}{4}\xi_j\xi(1+\xi_j\xi)\eta_j\eta(1+\eta_j\eta) && j=1,3,5,7 \\
N_j &= \frac{1}{2}(1-\xi^2)\eta_j\eta(1+\eta_j\eta) && j=2,6 \quad (14)\\
N_j &= \frac{1}{2}(1-\eta^2)\xi_j\xi(1+\xi_j\xi) && j=4,8 \\
N_j &= (1-\xi^2)(1-\eta^2) && j=9
\end{aligned}$$

A discontinuous elements is also used on the boundary with edges or corners, especially on the truncation boundary of the down stream. A disconticontiuity parameter for 9-node quadrilateral (Fig. 3) is chosen as $\delta = 2/3$ and the shape functions with single-edge discontinuity are derived as:

$$\begin{aligned}
N_j &= \frac{1}{10}\xi(3\xi-2)\eta_j\eta(1+\eta_j\eta) && j=1,7 \\
N_j &= \frac{1}{4}(1+\xi)(2-3\xi)\eta_j\eta(1+\eta_j\eta) && j=2,6 \\
N_j &= \frac{9}{20}\xi(1+\xi)\eta(\eta_j+\eta) && j=3,5 \\
N_j &= \frac{9}{10}\xi(1+\xi)(1-\eta)(1+\eta) && j=4 \\
N_j &= \frac{1}{5}\xi(3\xi-2)(1-\eta)(1+\eta) && j=8 \quad (15)\\
N_j &= \frac{1}{2}(1+\xi)(2-3\xi)(1-\eta)(1+\eta) && j=9
\end{aligned}$$

By substituting equation (13) and (14) into equation (8), the integral equation is represented in the discretized form as

$$\sum_{k=1}^{NN}\delta_{ik}\hat{C}_i\phi_k + \sum_{k=1}^{N_{S+W}}\hat{H}_{ik}^{S+W}\phi_k + \sum_{k=N_{S+W}+1}^{NN}\hat{G}_{ik}^{FR}\frac{\partial\phi}{\partial n}_k = \sum_{k=1}^{N_{S+W}}\hat{G}_{ik}^{S+W}\frac{\partial\phi}{\partial n}_k + \sum_{k=N_{S+W}+1}^{NN}\hat{H}_{ik}^{FR}\phi_k \qquad (16)$$

where $i = 1,2,\cdots,NN$, $NN =$ the total number of nodes, N_S=number of nodes on Γ_S, N_W =number of nodes on Γ_W and

$$\begin{aligned}
\hat{H}_{ik}^{S+W} &= \sum_e\sum_j \delta_{kr}\int_{\Gamma_{S+W}^e}\frac{\partial G}{\partial n}N_j|J|d\xi d\eta \\
\hat{H}_{ik}^{FR} &= \sum_e\sum_j \delta_{kr}\int_{\Gamma_{F+R}^e}\frac{\partial G}{\partial n}N_j|J|d\xi d\eta \qquad (17)\\
\hat{G}_{ik}^{S+W} &= \sum_e\sum_j \delta_{kr}\int_{\Gamma_{S+W}^e}GN_j|J|d\xi d\eta \\
\hat{G}_{ik}^{FR} &= \sum_e\sum_j \delta_{kr}\int_{\Gamma_{F+R}^e}GN_jd\xi|J|d\eta \\
\delta_{ik},\delta_{kr} &= \text{Kronecker delta function}
\end{aligned}$$

where r is the node number at the j-th node on the e-th element, e the element number and $|J|$ the Jacobian.

Since the discretized integral equation is variant in time, all the boundary surfaces are regridded at each time step. The influence coefficients are also recomputed using the updated grids.

If the normal velocity on an arbitrary boundary surface is known, the fluid velocity on the boundary can be computed using the following equation,

$$\begin{bmatrix}\frac{\partial\phi}{\partial x}\\ \frac{\partial\phi}{\partial y}\\ \frac{\partial\phi}{\partial z}\end{bmatrix} = \begin{vmatrix}\frac{\partial x}{\partial\xi} & \frac{\partial y}{\partial\xi} & \frac{\partial z}{\partial\xi}\\ \frac{\partial x}{\partial\eta} & \frac{\partial y}{\partial\eta} & \frac{\partial z}{\partial\eta}\\ n_x & n_y & n_z\end{vmatrix}^{-1}\begin{vmatrix}\frac{\partial\phi}{\partial\xi}\\ \frac{\partial\phi}{\partial\eta}\\ \frac{\partial\phi}{\partial n}\end{vmatrix}$$

The first order derivatives of wave elevation are also evaluated as:

$$\begin{bmatrix}\frac{\partial\zeta}{\partial x}\\ \frac{\partial\zeta}{\partial y}\end{bmatrix} = \begin{vmatrix}\frac{\partial x}{\partial\xi} & \frac{\partial y}{\partial\xi}\\ \frac{\partial x}{\partial\eta} & \frac{\partial y}{\partial\eta}\end{vmatrix}^{-1}\begin{vmatrix}\frac{\partial\zeta}{\partial\xi}\\ \frac{\partial\zeta}{\partial\eta}\end{vmatrix}$$

OPEN BOUNDARY CONDITION

Open boundary condition by introducing the absorbing beach and stretching was well implemented in simulating the fully nonlinear diffractions(Boo&Kim 1996) for monochromatic waves. The scheme is introduced for the present works. Lateral boundary of the tank was treated as rigid wall.

Firstly, the absorbing beach is placed in the longitudinal as well as transverse direction of the tank in order to partially absorb the waves. The dynamic free surface condition on the beaches can be written as:

$$\frac{\partial\phi}{\partial t} + \frac{1}{2}\nabla\phi\cdot\nabla\phi + g\zeta + \mu\phi = 0 \qquad (18)$$

where the tuning factor μ is determined in the following equations,

$$\mu = \begin{cases}\mu_x & : \bar{x} > x_b,\ \bar{y} < y_b\\ 0 & \text{otherwise}\end{cases}$$

where

$$\mu_x = 2\left(\frac{\bar{x}-x_b}{L_x}\right)^2$$

Here x_b is the distance from the inflow boundary to the longitudinal beach front. The coordinate $\bar{x}$ is measured from the inflow boundary.

Secondly, a stretching method is applied to the outflow boundary (Fig. 4). The resulting equations of the stretching for the velocity potential and velocity are

$$\phi(\bar{x},t) = \phi_o(\bar{x},t)\frac{cosh\ k(z_e+d)}{sinh\ kd} \qquad (19)$$

$$\frac{\partial\phi(\bar{x},t)}{\partial n} = \frac{\partial\phi_o(\bar{x},t)}{\partial n}\frac{cosh\ k(z_e+d)}{sinh\ kd} \qquad (20)$$

where $\phi_o(\vec{x}, t)$ is the velocity potential on the instantaneous free surface at the upper edge of the open boundary and k the wave number. The mapped z-coordinate z_e is defined as

$$z_e = \frac{(z + d) \times d}{\zeta_o + d} - d \qquad (21)$$

where d is the water depth and ζ_o the location of the instantaneous free surface at the upper edge of the open boundary.

Lastly, in treating the boundary condition on the inflow boundary, stretching scheme which is anlogous to one on the outflow boundary is employed. The mapped coordinate on the boundary is

$$z_e = \frac{(z + d)}{\zeta_o + d} \times (d + \zeta_i) - d \qquad (22)$$

where ζ_i is ζ of the incident wave. The velocity potentials and velocities on the boundary are computed using the coordinates and utilized for the input wave generation.

CHECKS OF ACCURACY

A global check of the time stepping accuracy is provided by the volume error relative to the initial volume (Tanizawa 1995). The total volume (V_x, V_y, V_z) of the fluid in the three directions is computed using

$$(V_x, V_y, V_z) = \sum_e \int_e (x\ n_x, y\ n_y, z\ n_z) d\Gamma \qquad (23)$$

The total energy E_t and workdone W should balance in the fluid where E_t is the sum of the potential energy E_p and kinetic energy E_k.

$$\begin{aligned} E_t &= E_p + E_k \\ &= \frac{1}{2} \sum_e \int_e z^2\ n_z d\Gamma + \sum_e \int_e \phi \frac{\partial \phi}{\partial n} d\Gamma \qquad (24) \\ W &= \sum_e \int_0^t \int_e p\ \frac{\partial \phi}{\partial n} d\Gamma dt \qquad (25) \end{aligned}$$

The conservation of the total momentum (V_x, V_y, V_z) and impulse (I_x, I_y, I_z) in the three directions is computed using

$$\begin{aligned} (M_x, M_y, M_z) &= \sum_e \int_e (\phi\ n_x, \phi\ n_y, \phi\ n_z) d\Gamma \qquad (26) \\ (I_x, I_y, I_z) &= \sum_e \int_0^t \int_e (p\ n_x, p\ n_y, p\ n_z) d\Gamma \qquad (27) \end{aligned}$$

NUMERICAL RESULTS

The kinematic and dynamic free surface conditions was integrated in time using multi-step method. The fourth-order Adams-Bashforth multi-step method is firstly applied to the free surface conditions and then the Adams-Moulton scheme is used to correct the dynamic condition once and thus the velocity potential on the free surface can be computed. In order to let the 4th order predictor-corrector method start initially, the initial values for the three steps were computed from the incident wave.

The wave elevation, velocity potential, normal derivative of potential, and velocities were filtered numerically using five-point filtering scheme. Additional smoothing in the transverse direction was also performed by alternating the directions.

The waves at the inflow boundary increased gradually using a modulation function which satisfies the initial calm water condition. The modulated potential is given by $\phi_m(\vec{x}, t) = f_m(t)\phi(\vec{x}, t)$, where $f_m(t)$ is a modulation function.

$$f_m = \begin{cases} \frac{1}{2}\left[1 - cos(\frac{\pi t}{T_m})\right] & : t < T_m \\ 1 & : t \geq T_m \end{cases}$$

where T_m is the modulation time which is chosen as one wave period. A numerical instability was avoided by choosing the appropriate time increment. The longitudinal absorbing beach was placed at the 3/4 tank length from the inflow boundary.

The continuous quadratic element was used to discrete the enclosed boundary except the intersection region near the downtank boundary, where the partially discontinuous element with parameter $\delta = 2/3$ was distributed.

The computed diffraction force and moment were normalized as:

$$CF_x' = \frac{F_x}{\rho g R^3}, \quad CF_z = \frac{F_z}{\rho g R^3}, \quad CM = \frac{M}{\rho g R^4}$$

Simulation of linear regular wave

The present numerical scheme was firstly verified by simulating linear regular wave. The wave length, L and period, T are $1.68m$ and $1.04\ sec$, respectively. The wave slope was chosen as 0.07. The simulation was performed for $130T$, where T is the wave period and its partial record was compared with theoretical input wave in Fig. 4. The numerical results are in good agreement with the theory.

Simulation of linear irregular waves

The time series of the linear irregular waves can be generated using the four Airy wave components. The input parameters for the irregular waves are given in the following table.

Table 1. Parameters for the irregular waves

Airy wave frequency (rad/sec)	ω_1	5.25
	ω_2	5.55
	ω_3	6.73
	ω_4	7.46
frequency averaged (rad/sec)	ω	6.25
wave length averaged (m)	L	1.68
repetition period (sec)	T_r	128

The principal dimensions of the wave tank were length × width × depth = $2L \times 0.5L \times 0.5L$. The wave slope of the incident waves were 0.035. Eight elements of quadratic order were distributed on the one wave length. The simulation was performed by inputting the theoretical waves through the inflow boundary and linear boundary value problem was solved every time step. The proposed open boundary condition needs one wave frequency in order to implement the stretching scheme. The irregular waves, however, have several different wave components and thus the averaged frequency given in Table 1 was used for the present simulation. The irregular waves were simulated for the repetition period of 128 seconds in the HOBEM numerical wave tank with a time increment of $T/50$. The partial records of the waves measured at the upstream and tank center are shown in Fig. 5. The waves at the upstream are the theoretical values. The simulation waves fairly agree with the theoretical ones. The amplitude spectrum of the waves were compared in Fig. 6. Linear properties are shown to be maintained during the simulation. Additional simulations were carried out to investigate the energy for different time step of $T/80$ and mesh density of 12 elements per wave length. The energy difference were within 1.2 percents for both cases. It can be stated that the present open boundary modeling can be applied to simulate multi-component waves.

Diffractions by linear irregular waves

The linear irregular waves given in Table 1 were again utilized to simulate the diffractions due to a truncated vertical circular cylinder while the wave slope was chosen as $ka = 0.035$. The wave tank dimensions were length × width × depth = $11R \times 6R \times 6R$ where R denotes the radius of the circular cylinder. Radius and draft of the cylinder were 0.082 and 0.3 meters, respectively. The cylinder dimensions were chosen as 1/100 model scale of the ISSC TLP column. The number of quadratic elements on the cylinder and free surface were 16 and 78, respectively. The total number of nodes were 600.

Figure 7 is partial records of the horizontal and vertical forces as well as moment acting on the cylinder in which the hydrostatic force was not included. Magnitude of the horizontal forces is shown to be bigger than one of the vertical forces. The amplitude spectrum of the diffraction forces and moment are shown in Fig. 8. The frequency components of the forces and moment are agree well with the ones of the input waves. Figure 9 shows the total wave fields around the cylinder at $t = 30$ seconds. Some effects due to the modeling of the open boundary are weakly presented around the downstream region.

Diffractions by nonlinear irregular waves

The tank size, cylinder dimensions and meshes were the same of those which were used in the linear diffraction analysis. The Stokes second-order irregular waves was fed through the inflow boundary of the wave tank and then the fully nonlinear boundary value problem defined in equation (1)-(7), was solved in the time marching scheme. The Stokes irregular waves are represented by the sums and differences of combinations of frequencies (Pierson 1993). The theory for the second-order irregular waves produces higher crest height and shallow trough depth. In order to generate the Stokes second-order irregular waves, the four wave frequencies in Table 1 were again utilized. The constant steepness of each wave was chosen to be 0.035. The average wave length and other characteristics of the waves are referred to Table 1. The same mesh density as the linear diffraction was used while the time step was reduced as $T/70$.

Figure 10 shows the time histories of the diffraction forces and moment acting on the truncated vertical circular cylinder. Vertical forces are much smaller than horizontal forces. In Fig. 11, the fully nonlinear horizontal forces were compared with the linear forces which were already presented in Fig. 7. In the relatively bigger wave packet, the nonlinear forces are generally bigger than the linear forces. However, a qualitative comparison of nonlinear results with linear ones can be done using spectral analysis. This will be studied in the future.

The accuracy of the present time-stepping scheme was investigated in the Fig. 12 and Fig. 13. Figure 12 displays the percent error of the fluid volume relative to the initial one. The error oscillates during the simulation but is within about 1.5 percents. The balance of the momentum-impulse in the horizontal direction are drawn in Fig. 13. Those values are shown to be well balanced.

SUMMARY AND CONCLUSIONS

In the present research, a numerical scheme was developed to predict fully nonlinear diffractions acting on a floating structure by nonlinear irregular waves. The Stokes second-order nonlinear irregular waves was chosen as incident waves. The waves were firstly fed through the inflow boundary and the boundary integral equation which was approximated using a quadratic-order boundary element method, was then solved in the time-stepping procedure. The instantaneous free surfaces were located using the 4th order predictor and corrector method. Open boundary condition was treated by combining two schemes of absorbing beach and stretching.

The proposed numerical scheme was verified by simulating linear regular and irregular waves and applied to compute the diffractions due to a truncated vertical circular cylinder by the linear and nonlinear irregular waves. The simulation was performed for the repetition period of 128 seconds of the linear irregular waves without being interrupted and thus spectral analysis was done. The harmonic components of the simulated linear forces was shown to be agree well with frequencies of the incident waves.

The magnitude of the nonlinear forces were relatively bigger than one of the linear forces. The overall numerical scheme was examined by the balance check of the momentum-impulse. They were proven to be well conserved during the simulation.

REFERENCES

Boo,S.Y. & Kim,C.H. (1994) Simulation of Fully Nonlinear Irregular waves in a 3-D Numerical Wave Tank. *Proc. 4th International Offshore and Polar Engineering Conference*, Vol. III, Osaka, Japan, pp. 17-23.

Boo,S.Y. & Kim,C.H. (1996) Fully Nonlinear Diffraction Due to a Vertical Circular Cylinder in a 3-D HOBEM Numerical Wave Tank. *Proc. 6th International Offshore and Polar Engineering Conference*, Vol. III, L.A., U.S.A., pp. 23-30.

Dean,R.G. & Sharmar,J.N. (1981) Simulation of Wave Systems Due to Nonlinear Directional Spectra, *Proc. of the Intl. Symposium on Hydrodynamics in Ocean Engineering*, Norwegian Institute of Technology, August, Vol.2, pp. 1211-1222.

Ferrant, P. (1995) Time Domain Computation of Nonlinear Diffraction Loads Upon Three-dimensional Floating Bodies. *Proc. 5th ISOPE Conference*, Vol. III, Hague, Netherlands, pp.280-288.

Forristall, G.Z. (1985) Irregular Wave Kinematics from a Kinematic Boundary Condition Fit (KBCF). *Applied Ocean Research*, Vol.7, pp. 202-212.

Isaacson,M. & Cheung,K.F. (1992) Time-Domain Second-Order Wave Diffraction in Three Dimensions. *Journal of Waterway, Port, Coastal & Ocean Engineering*, ASCE, Vol. 118, No. 5, pp. 495-516.

Kim,J.W., Bai,K.J & Yoon,K.S. (1994) A Numerical Computation of Nonlinear Wave Exciting Forces on a Three-dimensional Vertical Circular Cylinders. KSNA *Spring Conference*, Korea, pp. 358-363.

Lin,W.-M. & Yue,D.K.P. (1991) Numerical Solutions for Large-Amplitude Ship Motions in the Time Domain. *18th Symposium on Naval Hydrodynamics*, U. Michigan, Ann Arbor, pp. 41-66.

Longuet-Higgins,M.S. (1963) The Effect of Nonlinearities on the Statistical Distributions in the Theory of Sea Waves. *Journal of Fluid Mechanics*, Vol.17, pp.459-480.

Pierson,W.J. (1993) A Third Order Oscillatory Perturbation Expansion for Sums of Interacting Long Crested Stokes Waves. *Journal of Ship Research*, Vol.37, No.4, pp.345-383.

Tanizawa, K. (1995) A Nonlinear Simulation Method of 3-D Body Motions in Waves. *JSNA*, Vol. 178, pp. 180-191.

Xu,Q., Pawlowski,J.S. & Baddour,R.E. (1993) Simulation of Nonlinear Irregular waves by Green-NaghdiTheory. *Intl. Workshop on Water Waves and Floating Bodies*, Newfoundland, Canada.

Yang,C. & Ertekin,R.C. (1992) Numerical Simulation of Nonlinear Wave Diffraction by a Vertical Cylinder. *J. Offshore Mech. and Arctic Eng.*, Vol. 114, pp. 36-44.

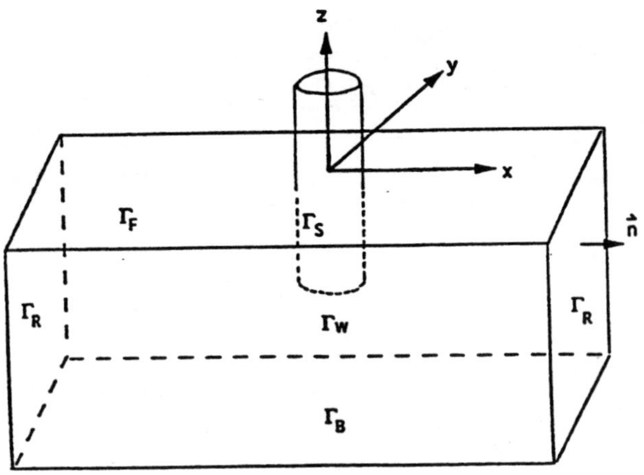

Fig. 1 Coordinate systems and boundary Surfaces

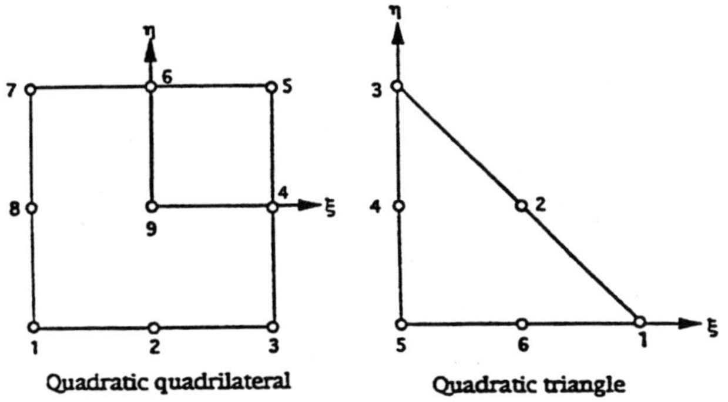

Fig. 2 Continuous quadrilateral and triangular elements of quadratic order

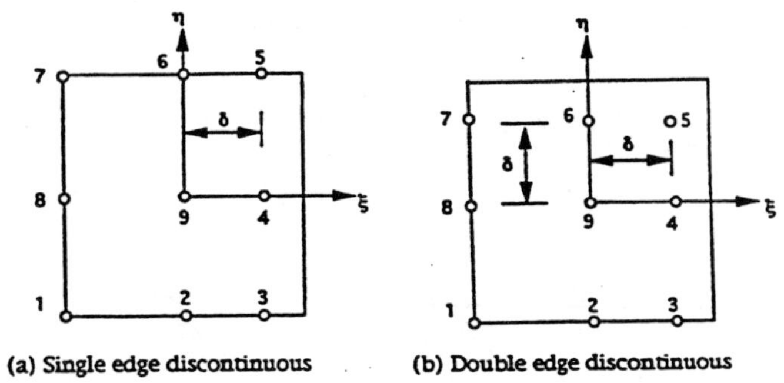

Fig. 3 Partially discontinuous quadrilaterals of quadratic order

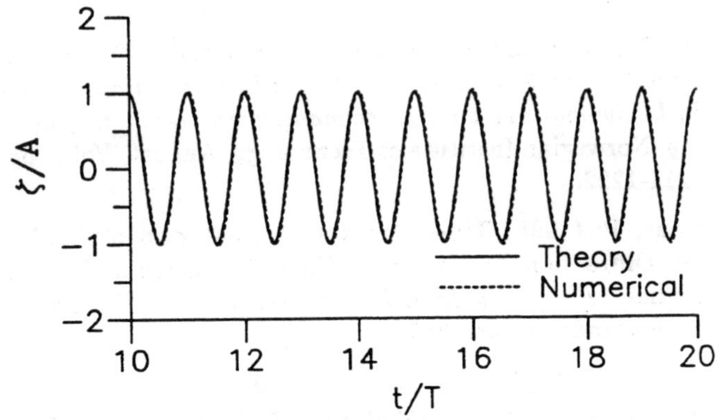

Fig. 4 Comparison of simulated linear regular wave with theory

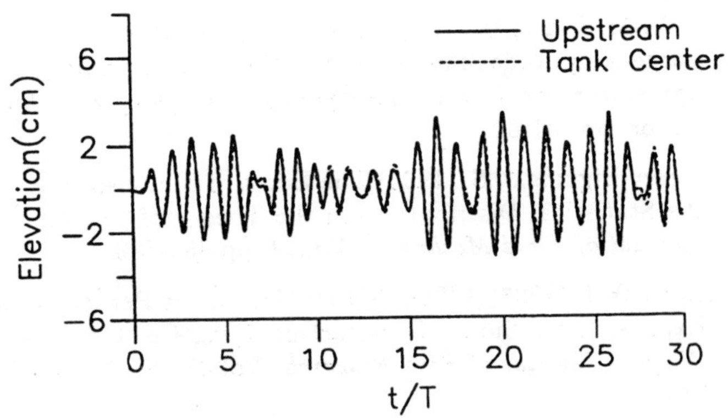

Fig. 5 Comparison of simulated linear irregular waves

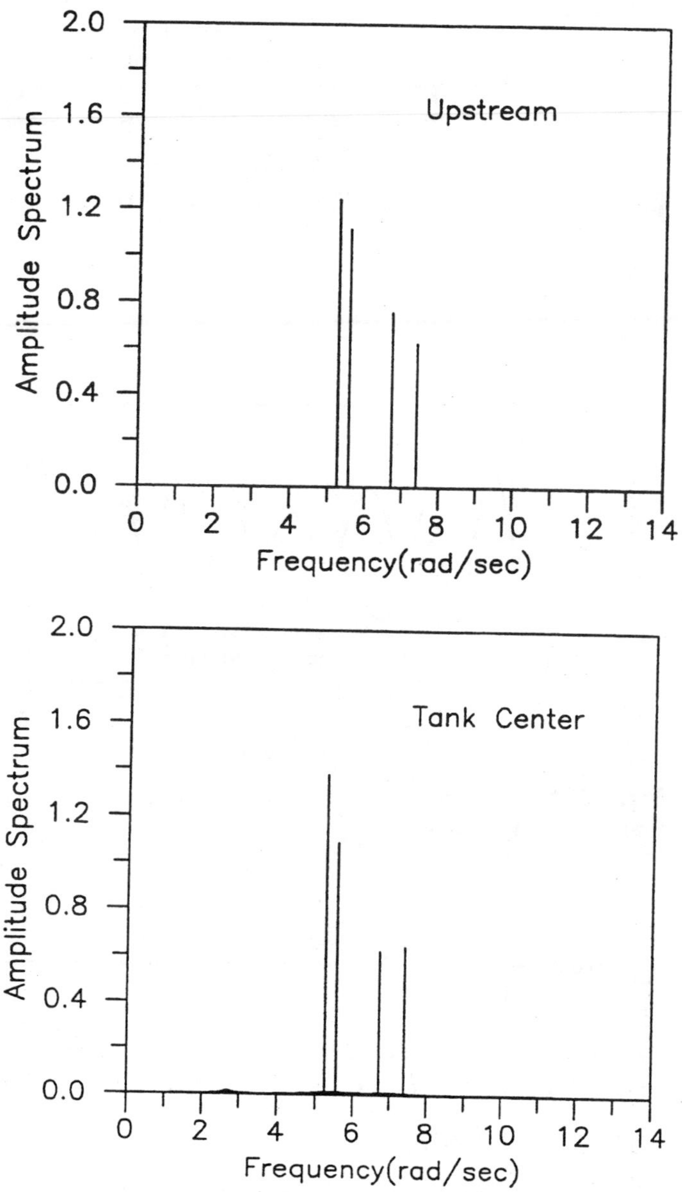

Fig. 6 Amplitude spectra of the linear irregular waves measured at upstream and tank center

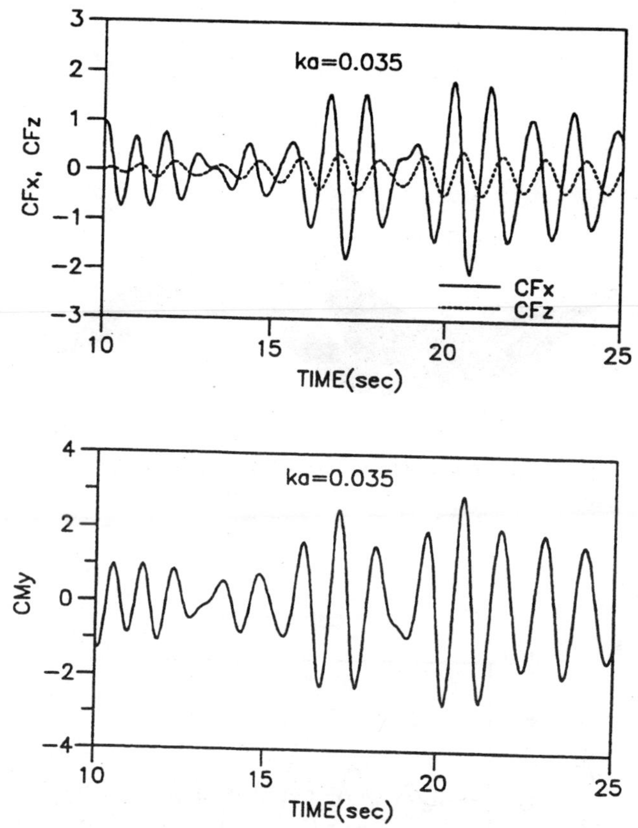

Fig. 7 Time histories of the linear diffraction forces and moment acting on the cylinder by the linear irregular waves

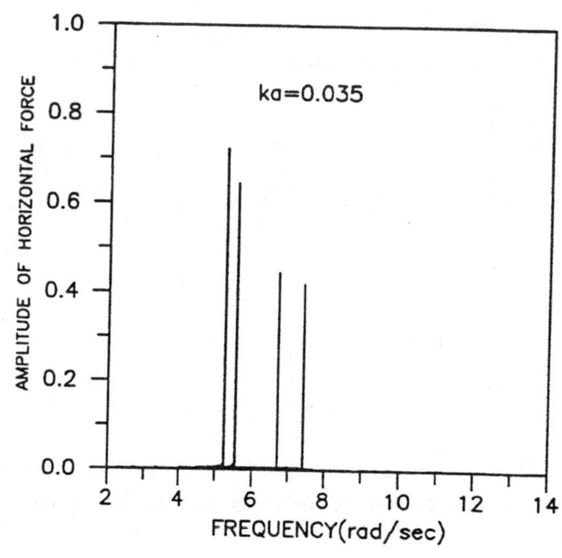

Fig. 8 Amplitude spectrum of the linear diffraction force

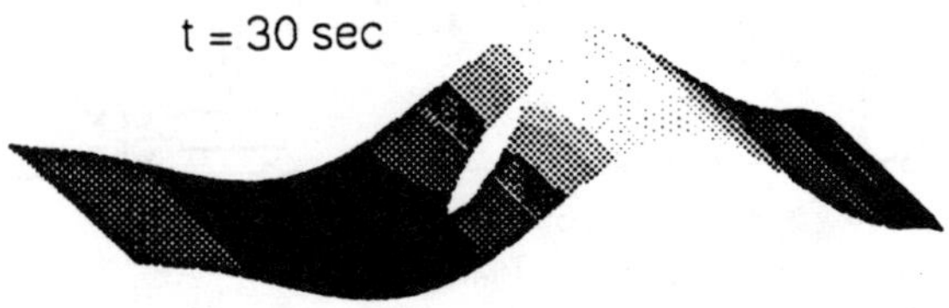

Fig. 9 Total wave field around cylinder due to linear irregular waves

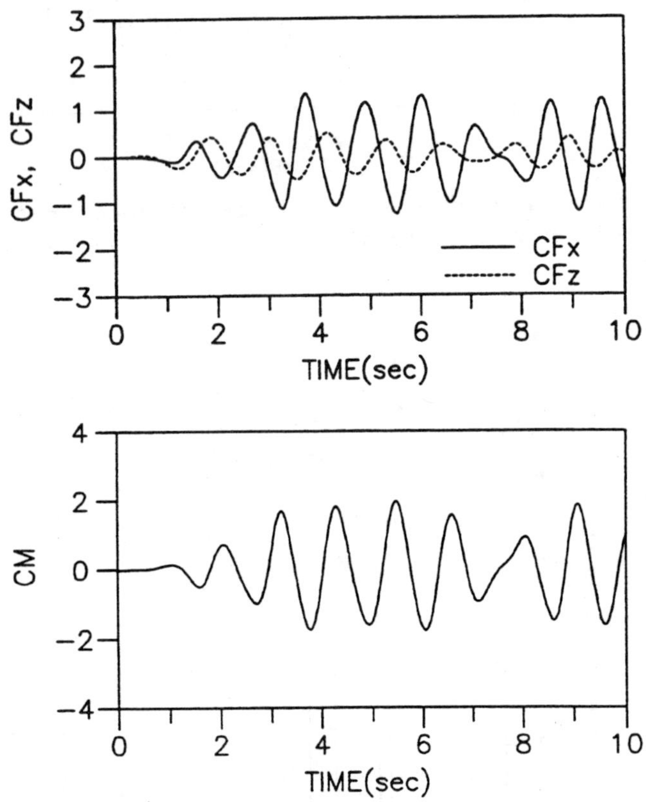

Fig. 10 Time histories of the nonlinear diffraction forces and moment acting on the cylinder by the nonlinear irregular waves

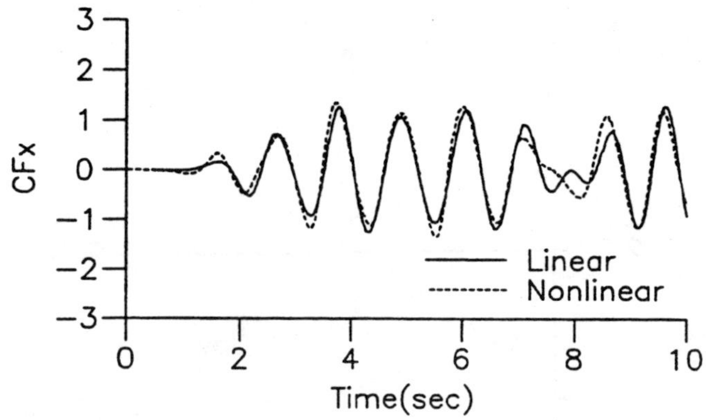

Fig. 11 Comparison of the linear forces with the fully nonlinear forces

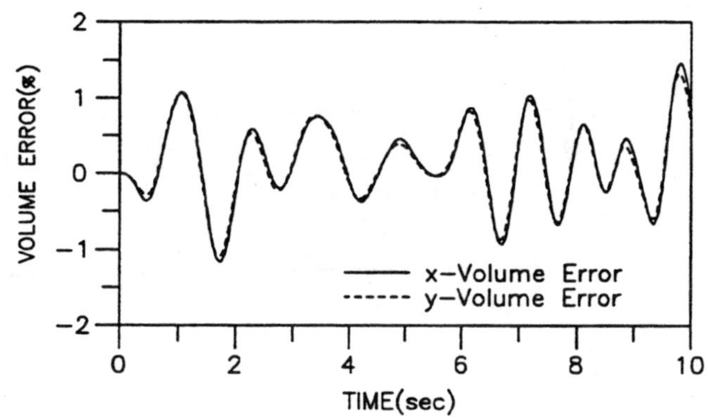

Fig. 12 Percent error of the volume for the nonlinear irregular wave simulation

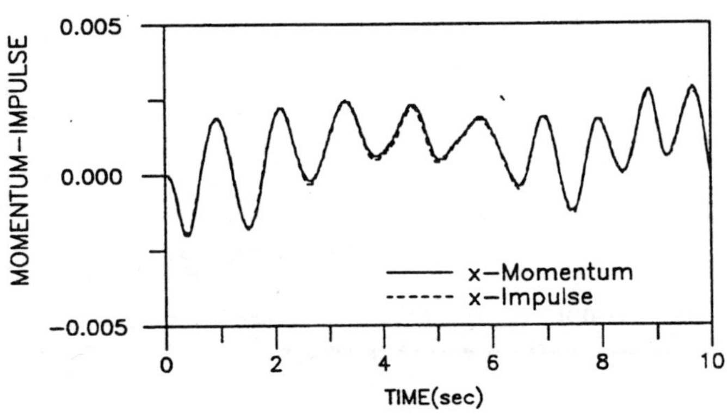

Fig. 13 Balance check of the momentum and impulse for the nonlinear irregular wave simulation

Proceedings of the Seventh (1997) International Offshore and Polar Engineering Conference
Honolulu, USA, May 25-30, 1997

ISBN 1-880653-28-1 (Set); ISBN 1-880653-31-1 (Vol. III)

Wave Generation and Active Absorption in a Numerical Wave Flume

Jesper Skourup and Hemming A. Schäffer
Danish Hydraulic Institute
Hørsholm, Denmark

ABSTRACT

A numerical wave flume is simulated using a Boundary Element Method (BEM) model combined with a time stepping procedure. The boundary conditions in the BEM are similar to those in a physical wave flume, e.g. with a wavemaker at one end boundary and an absorbing condition at the other. The free water surface is described by two non-linear boundary conditions, and the sea bed by a no-flux condition.

Active absorption by a wave paddle is used for eliminating outgoing waves at the two ends of the wave flume. These waves are: 1) the progressive waves propagating towards the outflow boundary, and 2) the waves coming back to the generation boundary due to reflection from the outflow boundary. The method used at the Danish Hydraulic Institute for simultaneous generation and active absorption of waves in flumes is implemented as boundary conditions and tested for regular as well as irregular waves. Comparison with theoretical prediction of absorption coefficients shows good agreement. Although the active absorption system has previously been successfully tested in a physical wave flume, the numerical flume provides an attractive tool for further development.

Key Words: Boundary Element Method, Numerical Wave Flume, Active Absorption

1 INTRODUCTION

Wave flume experiments (physical as well as numerical) are normally performed by generating and absorbing the waves at the respective end boundaries. The quality of an experiment is highly dependent on the accuracy by which the waves are generated and how well they are absorbed. The waves are typically generated by the motion of a wave paddle. This motion is governed by a transfer function which for a given paddle type, water depth and wave period relates the paddle stroke to the wave height. The transfer function is determined by solving a boundary value problem governed by the Laplace equation for the velocity potential with boundary conditions at the free water surface, at the wave board and at the sea bed. Since the boundary conditions at the free water surface are non-linear and the position of the surface is unknown (i.e. part of the solution), some assumptions must be made. Using the classical method of perturbation theory in combination with Taylor series expansions of the boundary conditions at the free surface and at the wave board leads to a boundary value problem for the contribution at each order to the paddle movement. The first order wavemaker theory corresponding to linear Stokes wave theory has long been well established (see e.g. Svendsen, 1985, for a review) and this theory is still being widely used for wave generation in most physical facilities.

When the ratio of the wave height to the wave length (i.e. the wave steepness) becomes large (typically more than 20% of the maximum wave steepness) the use of first order wavemaker theory produces waves that are not of permanent shape as they propagate down the flume. This is because the higher harmonics of the waves become significant when compared to the first order waves. Using second order wavemaker theory normally solves this problem. Schäffer (1996) re-derived the second order wavemaker theory in a unifying and compact form, and some inconsistencies of previous derivations were identified and corrected. Skourup (1995) used a BEM model for simulating a numerical wave flume using first and second order control signals for a piston wavemaker according to the theory by Schäffer (1996). The effects of including second order terms in the wavemaker control signal were confirmed to be crucial when a constant shape of the waves was desired. In the present paper, however, only waves with small steepness are considered and thus only first order wavemaker theory is used.

The absorption of waves in flume experiments is just as important as having an efficient and accurate wave generation. The waves to be absorbed are the progressive waves that propagate towards the end of the flume, and the reflected waves (from a structure or from the end of the flume) which propagate towards the wavemaker. The first type of waves may be absorbed using passive absorbers such as a beach in a physical model or a sponge layer in a numerical model. An active wave absorber (e.g. a wave paddle) could also be used in both physical and numerical models for absorbing the incident waves. A combination of a passive and an active absorber (as by e.g. Clément, 1996) is another alternative, but in the present paper only an active absorber is considered. At the wavemaker the reflected incident waves will be re-reflected back into the flume unless they are compensated for in the wavemaker control signal. Wavemakers equipped with a control system for simultaneous generation and active absorption of waves are often referred to as absorbing wavemakers. In addition to the control signal needed for wave generation, an absorbing wavemaker system also needs an on-line response of some measurable quantity sensitive to the appearance of reflected waves approaching the wavemaker. The method used at the Danish Hydraulic Institute (DHI) is to measure the surface elevation at the wave board. Then, by using a technique based on a digital recursive filter, the wave board position at the next time level is determined so that the reflected waves are absorbed at the paddle while the desired progressive waves leave the wave board unaffected. The system is called Active Wave Absorption Control System (AWACS), and it supports both piston and hinged type wavemakers. Further details can be found in Schäffer et al. (1994).

In the present paper the active absorption technique described above is used in a numerical wave flume which is based on a BEM model (see Grilli et al., 1989, and Skourup, 1995). The AWACS will be used at both ends of the flume acting either as an active wave absorber or as an absorbing wavemaker. Examples of applications are given with regular waves and with irregular waves.

2 MATHEMATICAL FORMULATION

The irrotational flow of an incompressible fluid with a free surface is considered. Hence, potential theory can be used and the flow can be described by a velocity potential $\phi(\bar{x},t)$ so that the fluid velocity is given by $\bar{u} = \nabla\phi$ where $\bar{x} = (x,z)$ is a position vector of an "observation point" and t is the time. Using the mass conservation equation gives that ϕ must satisfy the Laplace equation

$$\nabla^2\phi = 0 \tag{1}$$

throughout the fluid domain $\Omega(t)$ which is shown in Fig. 1.

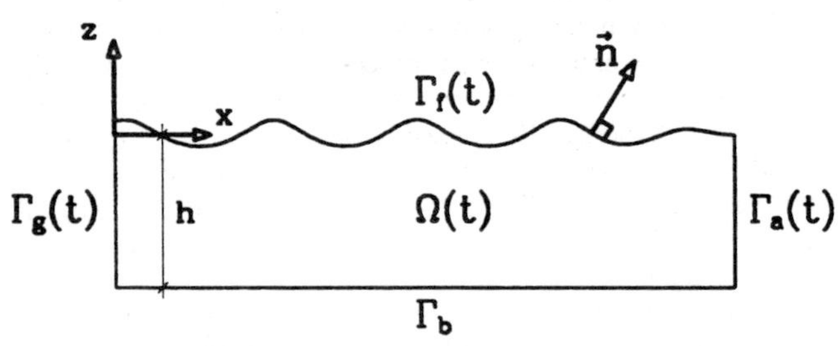

Figure 1. Fluid domain and definition of boundaries

The non-linear boundary conditions for ϕ on the free surface $\Gamma_f(t)$ are the kinematic condition

$$\frac{D\bar{r}}{Dt} = \left(\frac{\partial}{\partial t} + \bar{u}\cdot\nabla\right)\bar{r} \;=\; \bar{u} = \nabla\phi, \quad z = \eta \tag{2}$$

where $\bar{r}$ is a position vector of a water particle at the free surface, η, and the dynamic condition (the Bernoulli equation)

$$\frac{D\phi}{Dt} = -gz + \frac{|\nabla\phi|^2}{2}, \quad z = \eta \tag{3}$$

since $D\phi/Dt = \phi_t + |\nabla\phi|^2$. D/Dt is a particle following (Lagrangian) operator, and ϕ_t is an abbreviation of $\partial\phi/\partial t$. In (3), g is the acceleration due to gravity. The pressure term in the Bernoulli equation is taken as zero.

On the bottom, Γ_b, (which in the present application is horizontal) the boundary condition is

$$\phi_n = 0, \quad z = -h \tag{4}$$

where n is the co-ordinate along the unit normal vector $\vec{n}$ pointing outwards from the fluid domain and ϕ_n is an abbreviation of $\partial\phi/\partial n$. The water depth is denoted h.

At the end boundaries $\Gamma_g(t)$ and $\Gamma_a(t)$ conditions accounting for the generation and absorption of waves must be imposed. The motion of a piston type absorbing wavemaker is simulated at the boundary $\Gamma_g(t)$ and the velocity of the wavemaker is specified as a Neumann boundary condition. For an absorbing wavemaker and using linear theory, the complex amplitude X_a of the position X of the wave board at still water level is given in the frequency domain as (cf. Schäffer et al., 1994)

$$X_a = \frac{2a_I - a_o}{iA - D} \tag{5}$$

where a_I is the complex amplitude of the desired incident waves at the wave board, a_o is the complex amplitude at the front of the wave board and i is the imaginary unit representing a 90° phase shift. The ratio A between the wave height H and the stroke S of the wavemaker is given by the Biésel (1951) transfer function

$$A = \frac{4\sinh^2 kh}{2kh + \sinh 2kh} \tag{6}$$

where k is the wave number (found as a solution to the dispersion relation $\omega^2 = gk \tanh kh$, where ω is the angular wave frequency). D is the sum of all the evanescent-mode transfer functions present due to the mismatch between the shape of the wave board and the horizontal velocity profile of the progressive wave

$$D = \sum_{j=1}^{\infty} \frac{4\sin^2 k_j h}{2k_j h + \sin 2k_j h} \tag{7}$$

where k_j is an evanescent-mode wave number (the j'th solution to $\omega^2 = -gk_j \tan k_j h$).

The expression (5) is formulated in the time-domain using a digital, recursive, causal filter (see Schäffer et al., 1994). With this filter, the time series of a desired incident surface elevation η_I, an on-line measurement of surface elevation η_o on the front of the wave board and previous wave board positions provide the new position of the wave board, i.e.

$$X(t) \;=\; \sum_{j=1}^{M} d_j X(t - j\Delta t) + \sum_{j=0}^{M} c_j (2\eta_I(t - j\Delta t) - \eta_o(t - j\Delta t)) \tag{8}$$

where M is the order of the recursive filter, Δt is the time increment and d_j and c_j are filter coefficients designed to match the amplitude and phase characteristics of $1/(iA\text{-}D)$ over a broad range of frequencies.

The boundary condition at the wave generator (oscillating about the mean position x=0) is given as the time derivative of the wave board position X_{gen}, i.e.

$$\phi_n = -\frac{dX_{gen}}{dt}, \quad x = 0 \tag{9}$$

where the minus sign appears because of the opposite directions of the x-axis and the outward normal vector at the wave board.

At the other end boundary $\Gamma_a(t)$ an active wave absorber is situated. The motion of the wave absorber (oscillating about the mean position $x=l$, where l is the length of the flume) is given by (8) where $\eta_l=0$ (since no wave generation takes place) and the (Neumann) boundary condition is given as the time derivative of the wave board position X_{abs}, i.e.

$$\phi_n = \frac{dX_{abs}}{dt}, \quad x = l \tag{10}$$

The boundary conditions (9) and (10) are found by numerical differentiation of (8). When the end boundary $\Gamma_a(t)$ is a reflective wall the boundary condition there is $\phi_n = 0$.

3 INTEGRAL EQUATION SOLUTION

The boundary value problem described above is solved by using an integral equation method based on Green's second identity

$$\alpha(\bar{x})\phi(\bar{x},t) = \int_{\Gamma(t)} \phi(\bar{\xi},t)G_n(\bar{x},\bar{\xi}) - G(\bar{x},\bar{\xi})\phi_n(\bar{\xi},t) \; d\Gamma \tag{11}$$

In (11) $\bar{\xi}$ is the position vector of an "integration point" at the boundary $\Gamma(t)$ and $\bar{x}$ is the position of an "observation point" which can be situated anywhere. The Green's function is $G(\bar{x},\bar{\xi}) = \ln|\bar{\xi} - \bar{x}| + \ln|\bar{\xi}^* - \bar{x}|$, where $\bar{\xi}^*$ is the reflection of $\bar{\xi}$ about the horizontal sea bed, and G_n is the normal derivative of G. By use of the "reflected" Green's function integration over the horizontal sea bed is avoided. The factor $\alpha(\bar{x})$ depends on the boundary geometry ($\alpha(\bar{x}) = \pi$ for the observation point situated at a smooth part of the boundary).

For the numerical solution of (11) the boundary of the computational domain is subdivided into a finite number of small segments, each segment connecting two adjacent discretization points situated at the boundary curve. By representing the geometry as well as the boundary functions ϕ and ϕ_n at each segment by a linear variation, the governing integral equation can be formulated in terms of the values of the variables ϕ and ϕ_n in the discretization points, and all integrals are determined analytically. Hence, a linear algebraic system of equations is formed, in which each element is collocated by integrations related to the discretization points. The matrices are fully populated and the L-U factorisation and direct solution of the system of equations is found by using standard library routines.

Instead of determining the factor $\alpha(\bar{x})$ in each discretization point the so-called "rigid-mode-technique" (see e.g. Grilli et al., 1989) is used. Using this technique the diagonal elements of the matrix containing the $\alpha(\bar{x})$-values in the discretization points and the integrals related to G_n can be expressed as minus the sum of the off-diagonal elements in the same row of the matrix.

Continuity conditions are used at the intersection points between different parts of the boundary (i.e. at the "corners") by prescribing the velocity potential ϕ as a boundary condition at both boundaries adjacent to the intersection point. To facilitate these requirements numerically, two discretization points (double nodes) are situated there. Compatibility conditions (cf. Grilli et al., 1989) are also used at the corners.

4 UPDATING OF FLUID DOMAIN AND BOUNDARY CONDITIONS

After the solution to the Laplace problem for ϕ has been found, the fluid domain and the related boundary conditions are updated to a subsequent time level. The non-linear Lagrangian (particle following) approach by Dold and Peregrine (1984) is used. Their approach is based on truncated Taylor series in time for the position vector of a free surface particle and for the velocity potential at the free surface. The time increment is denoted Δt, and the Lagrangian (particle following) operator D/Dt is defined in (2). The term $O(...)$ is the Landau symbol denoting order of magnitude.

$$\bar{r}(t+\Delta t) = \bar{r}(t) + \sum_{k=1}^{n} \frac{(\Delta t)^k}{k!}\frac{D^k\bar{r}(t)}{Dt^k} + O((\Delta t)^{n+1}) \tag{12}$$

$$\phi(\bar{r}(t+\Delta t), t+\Delta t) = \phi(\bar{r}(t),t) + \sum_{k=1}^{n} \frac{(\Delta t)^k}{k!}\frac{D^k\phi(\bar{r}(t),t)}{Dt^k} + O((\Delta t)^{n+1}) \tag{13}$$

The fully non-linear expansion coefficients $D^k\bar{r}/Dt^k$ and $D^k\phi/Dt^k$ are obtained by successive solutions of the Laplace equation for the velocity potential ϕ and its time derivatives, where the solution of a Laplace problem corresponding to the expansion order k provides the non-linear boundary conditions for the Laplace problem at order $k+1$. The solution of the Laplace problem for ϕ by use of Green's second identity (11) gives ϕ and ϕ_n at the whole boundary of the computational domain. Hence, the gradient of the velocity potential, and thereby the first order expansion coefficients in (12) and (13) can be evaluated using the free surface boundary conditions (2) and (3). The tangential derivatives at the free surface are found using a Hermite cubic spline interpolation of the elevation and of the velocity potential. The boundary condition at the free surface for the Laplace problem with the first order time derivative is obtained from the Eulerian form of the dynamic condition

$$\phi_t = -gz - \frac{|\nabla\phi|^2}{2}, \quad z = \eta \tag{14}$$

The conditions at the remaining parts of the boundary are obtained by differentiation of the boundary conditions with respect to t. A simple numerical differentiation of (8) is used at the lateral boundaries.

Just as the governing differential equation (1) and the integral equation (11) are valid for ϕ, these may also be written for time derivatives of any order of ϕ. Hence, a mathematical problem is formulated in the same geometry as before, but now with ϕ_t and ϕ_{tn} as solution at the boundary of the computational domain. The numerical solutions of the second (and of subsequent) Laplace problems are computationally very fast compared to the solution of the first Laplace problem. This is because the kernel functions G and G_n of the governing integral equation are functions of the boundary geometry alone, and this is unchanged, since all derivatives are computed at the same time level. In the present work the terms up to $n=2$ are retained in the series (12) and (13).

The practical implementation of the Neumann condition at the absorbing wavemaker is accomplished by first updating the free surface and thereafter updating the wavemaker. The small gap that may appear due to a difference between the horizontal updating of a free surface particle at the wavemaker and the updating of the wavemaker according to (8), is taken into account simply by substituting the horizontal co-ordinate of the free surface particle with the position of the wavemaker. For the computations presented in this paper, the numerical errors due to this operation are negligible.

5 NUMERICAL EXAMPLES

A number of tests were made with different conditions at the boundaries. In the first series of tests a setup was used in which regular as well as irregular, progressive waves were generated by using $\Gamma_g(t)$ as a conventional wavemaker while $\Gamma_a(t)$ was used as a pure absorber (i.e. an absorbing wavemaker with η_I=0). In the second series of tests a setup was used with $\Gamma_g(t)$ as an absorbing wavemaker, while $\phi_n \equiv 0$ was imposed at $\Gamma_a(t)$ to simulate a fully reflective wall. The latter setup was used for studying transient regular waves running forth and back in a long flume as well as standing waves in a short flume. The wave periods, T, flume lengths, l, and discretizations are summarized in Table 1, and the water depth was h=1.0m throughout.

Table 1. Wave periods, flume lengths and discretization for the numerical wave flume tests. The water depth was h=1.0m throughout.

T (s)	l (m)	Δx (m)	Δt (s)	Fig.
Regular progressive waves				
0.7 0.8	2.5	0.05	0.0125	2
1.0 1.1 1.3 1.6 2.0 2.5 4.0	5.0	0.10	0.025	2
10.0	10.0	0.20	0.025	2
Irregular progressive waves				
1.0 - 2.0	10.0	0.10	0.025	3-4
Transient regular waves				
1.25	46.3	0.15	0.025	5
Regular standing waves				
2.5	8.75	0.20	0.025	6
2.5	7.0	0.20	0.025	7

For the regular, progressive wave tests the wave lengths L ranged from 0.77m to 31m. Hence, the longest waves were much longer than the flume. The largest number of computational nodes was 93 and the computing time was around one CPU minute for simulating 1200 time steps on an IBM RS6000 computer. The wave height was chosen to be small so that linear theory was adequate for wave generation and absorption. Based on time series of free surface elevations measured at two locations in the flume a reflection analysis (cf. Goda and Suzuki, 1976) for each test was made in order to obtain an absorption coefficient (equal to one minus the reflection coefficient). The reflection coefficient is defined as the ratio between the amplitude of the reflected wave and the amplitude of the incident wave. Figure 2 shows the result for each frequency in comparison with a curve giving the theoretical result as can be obtained from the theory behind the AWACS and the recursive filters involved. The purpose of the computations leading to the results shown in Fig. 2 is to depict that the AWACS works as expected in the numerical wave flume. The steepness of the computed waves was chosen to be small so that second and higher order components were insignificant. Thus an optimal condition for the active absorber was given. It is emphasized that the model is fully nonlinear, and that it is indeed able to simulate the propagation of steep waves (see e.g. Skourup, 1995). The purpose of the present computations is, however, not to show how the model works for waves of increasing steepness, but to verify the implementation of the AWACS and the theory behind it. Future publications will include simulations of waves of higher steepness. The results in Fig. 2 shows that the agreement with theory is very good and somewhat better than for the physical flume experiments given by Schäffer et al. (1994). Although better physical flume experiments have been made since 1994, they still do not give quite as good results as those obtained from the present numerical tests.

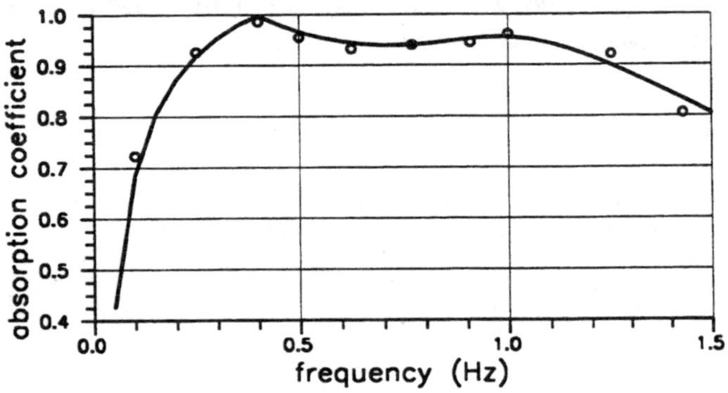

Figure 2. Absorption coefficient versus wave frequency. (o = numerical , ___ = theoretical)

A progressive wave test was made for irregular waves synthesized from a band-limited ($0.5Hz < f < 1.0Hz$) white noise spectrum with significant wave height H_s = 0.01m, using random phases. Figure 3 shows the surface elevation time series during the time range from 25s to 35s at a distance of 3.0m from the wavemaker. The theoretical prediction based on linear wavemaker theory relating paddle motion and wave elevation is shown for comparison. This theory is a good approximation for the present values of the wave height and period. The slight deviation is mainly due to imperfections in the active absorption. A more consistent verification could be made by comparing the simulation results with a time series determined from a test using a very long wave flume or from a test using a perfect wave absorber. However, the first test demands a long computing time and large storage requirements, and it might not be possible with the present formulation (the LONGTANK formulation by Wang et al., 1995, might be a possible method for such a simulation). The second type of test will not be possible since no perfect absorber is available.

The absorption coefficients are found using a reflection analysis based on measurements with two wave gauges. The reflection coefficients are computed as the square-root of the ratio between the energy of the reflected wave spectrum and the energy of the incident wave spectrum for each wavelet in the spectrum. Comparison with the theoretical curve as shown in Fig. 4 is quite satisfactory.

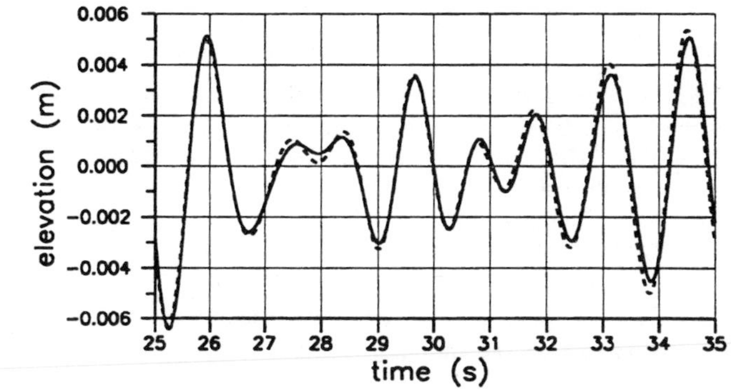

Figure 3. Time series plots of expected and measured elevation 3.0m from wavemaker using a white noise spectrum.

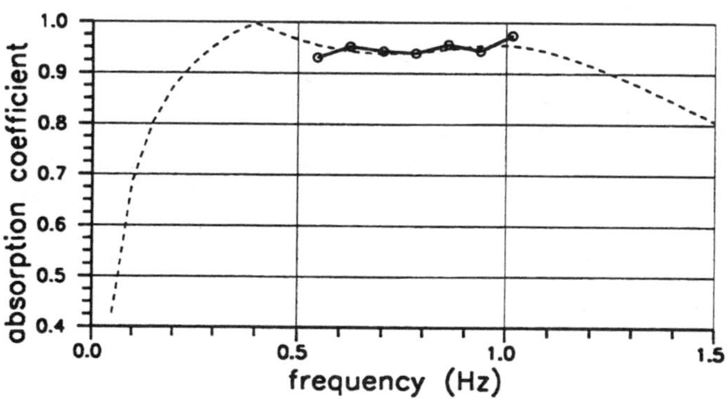

Figure 4. Theoretical and measured absorption coefficient for irregular waves based on a white noise time series.

The flume setup is now changed to have an absorbing wavemaker at one end and a fully reflective wall at the other. A transient made up by a group of regular waves was generated, propageted to the reflective wall, and returned to the absorbing wavemaker, where most of it was absorbed. Time series of the elevation measured in the middle of the flume (i.e. 23.15*m* from the wavemaker) is shown in Fig. 5. This test was inspired by the physical flume tests by Schäffer et al. (1994), who did not have the possibility to put a generator in both ends of the flume. Due to dispersion some overlap exists between the three time intervals for which the incident waves, the reflected waves and the (small) re-reflected waves show up in the elevation. Trying to separate these time intervals anyway, the square root of the ratio of the total variances from the third and second time interval provides an estimate of the reflection coefficient for the absorbing wavemaker. For the present case this corresponds to an absorption coefficient of 0.86, which is not quite as good as the values obtained above. This is probably due to the overlap mentioned above. Note that during this test no such thing as a switching from generation to absorption mode was used. Only the specified η_I was given as a sequence of regular waves followed by zeroes.

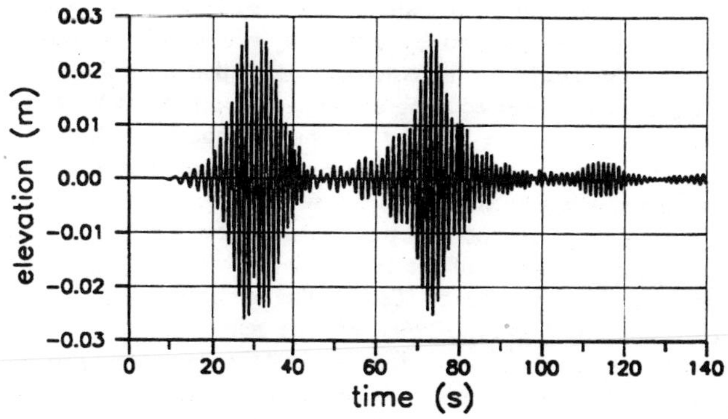

Figure 5. Surface elevation time series half way down the flume. Incident, reflected and re-reflected waves can clearly be identified.

Still with an absorbing wavemaker at one end of the flume and a fully reflective wall at the other, two standing wave cases were studied. In the first case a flume length of l=5/4 L (where L is the wave length) was chosen. Thus, the reflected wave becomes in anti-phase with the incident wave at $\Gamma_g(t)$. Figure 6 shows time series of: a) the surface elevation η_o measured at $\Gamma_g(t)$, and b) the movement of the absorbing wavemaker, $\Gamma_g(t)$.

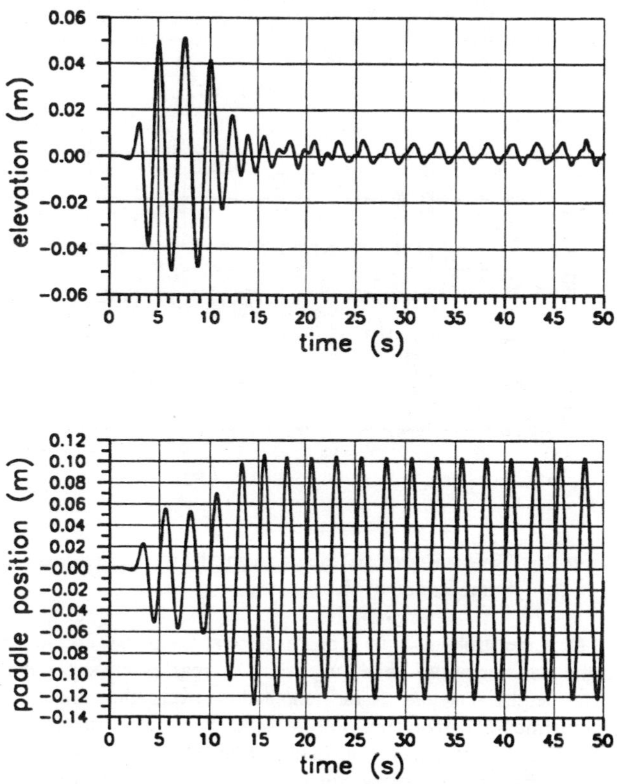

Figure 6. Time series of : a) elevation and b) wavemaker position for flume length $l = 5/4\ L$.

Neglecting the evanescent modes, linear theory would predict a node to develop in the surface elevation and $\Gamma_g(t)$ would be required to double its movement as compared to the case of plain wave generation. This is easily appreciated by setting $a_o = 0$ in (5). Including the evanescent modes in the discussion, it is noted that while the elevations of the opposite directed progressive waves cancel each other, the evanescent modes are doubled so that η_o does not quite vanish. Furthermore, it may still be shown theoretically that (5) requires exactly twice the paddle movement in this standing-wave case as that required for generation of the incident wave (using linear theory). To confirm this a Fourier analysis of the time series of wavemaker motion was made over the period 30*s* to 50*s*, and the first order component of the stroke was found to be 0.226*m*. For the generation of a progressive wave with height H=0.1*m*, period T=2.5*s* at the water depth h=1.0*m* the stroke of a piston wavemaker is determined using the Biésel transfer function (6) to be 0.113*m*. Thus the stroke of the wavemaker is doubled for this case as expected.

In the second case the length of the flume was chosen as $l=L$ to get an antinode at $\Gamma_g(t)$. In Fig. 7 time series of: a) η_o, and b) the position of $\Gamma_g(t)$ are given.

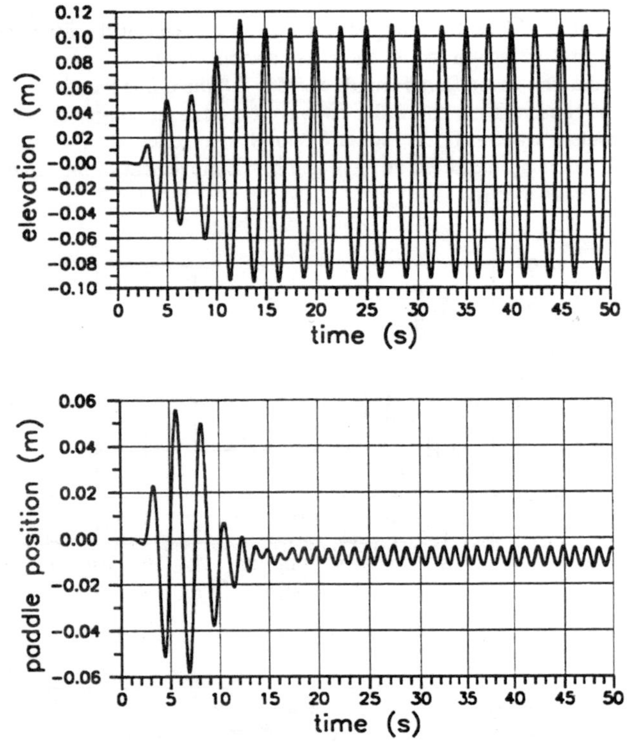

Figure 7. Time series of : a) elevation and b) wavemaker position for flume length $l = L$.

Now the stroke of $\Gamma_g(t)$ drops to zero (at the primary wave period) as the standing wave builds up, while η_o reaches twice the incident wave height. Due to finite amplitude effects the stroke has a superharmonic component. Furthermore, the mean position of the wavemaker is not at $x=0$. The actual mean position may be explained by looking at the theory for a second order standing wave in conjunction with the characteristics of the AWACS filter as shown in the following.

The time averaged mean water surface at a fully reflecting wall for a second order standing wave is given by (cf. Svendsen and Jonsson, 1980):

$$\bar{\eta} = k\frac{H^2}{4}\coth 2kh \tag{15}$$

The mean water surface $\bar{\bar{\eta}}$ in the flume (i.e. the spatial and temporal average) is different from zero when the mean position of the wavemaker differs from zero. Hence, the time averaged value of the elevation at the wavemaker is given by:

$$\overline{\eta_o} = \bar{\bar{\eta}} + \bar{\eta} \tag{16}$$

Furthermore, it can be shown that the response of the AWACS filter to zero wave frequency (i.e. infinitely long wave period) is:

$$\bar{X} = -6.62\overline{\eta_o} \tag{17}$$

(Note that this finite response at zero frequency is due to a high pass filter inferred in the AWACS to prevent slow drift of the paddle). Finally, continuity requires (omitting small terms):

$$l\bar{\bar{\eta}} - \bar{X}h = 0 \tag{18}$$

For the present case (15) yields $\bar{\eta}$ =0.00237*m* and solving (16)-(18) gives $\bar{X} = -0.00807m$ and $\bar{\bar{\eta}} = -0.00115m$. These values agree very well with the following mean values obtained from the numerical flume tests (averaging over the time range from 30*s* to 50*s*): $\bar{X} = -0.00810m$ and $\bar{\bar{\eta}} = -0.00113m$.

6 CONCLUSIONS

A method for simultaneous generation and active absorption of waves developed for a physical wave flume is implemented into a numerical model based on a Boundary Element formulation. For a numerical wave flume equipped with a generator in one end and an active wave absorber in the other, the absorption coefficient is found to correspond well with the theoretical one over a broad range of frequencies for regular waves and for irregular waves. Simultaneous wave generation and absorption also works satisfactorily.

The present numerical model is a good tool for detailed investigation and verification of methods used for active absorption. The numerical model can also provide results in some cases when measurements cannot be made in a physical wave flume due to instrument or model facility constraints. Changes of wave absorber type and characteristics are easily implemented into the numerical model as compared with a physical model.

The technique for simultaneous generation and absorption of waves is concurrently being extended to a 3D numerical model (a numerical wave tank) where waves are generated and absorbed along the side walls of the tank. Some examples of 3D results are given in Skourup (1996), Schäffer and Skourup (1996) and Skourup and Bingham (1996). Future extensions will cover other types of wave paddles as well as floating structures in the fluid domain.

ACKNOWLEDGEMENT

This work was funded by the Danish National Research Foundation. Their support is greatly appreciated.

REFERENCES

Biésel, F (1951). Etude théorique d'un type d'appareil à houle. *La Houille Blanche*, **6**(2), 152-165.

Clément, A (1996). Coupling of two absorbing boundary conditions for 2D time-domain simulations of free surface gravity waves. *Journal of Computational Physics*, **126**, 139-151.

Dold, JW and Peregrine, DH (1984). Steep unsteady water waves. An efficient computational scheme. *Proc. 19th Int. Conf. of Coastal Engineering*, Houston, Texas, **1**, 955-967. ASCE, New York.

Goda, Y and Suzuki, Y (1976). Estimation of incident and reflected waves in random wave experiments. *Proc. 15th Int. Conf. of Coastal Engineering*, Honolulu, Hawaii, **1**, 828-845. ASCE, New York.

Grilli, ST, Skourup, J and Svendsen, IA (1989). An efficient boundary element method for non-linear water waves. *Engineering Analysis with Boundary Elements*, **6**(2), 97-107.

Schäffer, HA (1996). Second order wavemaker theory for irregular waves. *Ocean Engineering*, **23**(1), 47-88.

Schäffer, HA, Stolborg, T and Hyllested, P (1994). Simultaneous generation and active absorption of waves in flumes. *Proc. Int. Symposium: WAVES - Physical and Numerical Modelling*, University of British Columbia, Vancouver, Canada, 90-99.

Schäffer, HA and Skourup, J (1996). Active absorption of multi-directional waves. *Proc. 25th Int. Conf. of Coastal Engineering*, Orlando, Florida. ASCE, New York.

Skourup, J (1995). Analytical second order wavemaker theory verified in a numerical wave flume. *Proc. 2nd Int. Conf. on Computer Modelling of Seas and Coastal Regions (COASTAL'95)*. Cancún, Mexico, 167-174.

Skourup, J (1996). Active absorption in a numerical wave tank. *Proc. 6th. Int. Offshore and Polar Engineering Conference (ISOPE-96)*, Los Angeles, **3**, 31-38.

Skourup, J and Schäffer, HA (1995). Active absorption in boundary element modelling of non-linear water waves. *Proc. of the third Int. Congress on Industrial and Applied Mathematics, (ICIAM 95)*, Hamburg, Germany.

Skourup, J and Bingham, HB (1996). Active absorption of radiated waves in a 3D boundary element model. *Proc. 11th Int. Workshop of Water Waves and Floating Bodies*, Hamburg, Germany.

Svendsen, IA (1985). Physical modelling of water waves. *Physical Modelling in Ocean Engineering*, Dalrymple, RA, ed., A.A. Balkema, Rotterdam.

Svendsen, IA and Jonsson, IG (1980). Hydrodynamics of Coastal Regions. Published and supported by Den Private Ingeniørfond. Technical University of Denmark.

Wang, P, Yao, Y and Tulin, MP (1995). An efficient numerical tank for non-linear water waves, based on the multi-subdomain approach with BEM. *Int. Journal for Numerical Methods in Fluids*, **20**, 1315-1336.

Proceedings of the Seventh (1997) International Offshore and Polar Engineering Conference
Honolulu, USA, May 25-30, 1997

ISBN 1-880653-28-1 (Set); ISBN 1-880653-31-1 (Vol. III)

Numerical Simulation of Short Wave-Wave Interaction

A. Clément
Ecole Centrale de Nantes
Nantes, France

L. Gil
F.C.T. - U.N.L.
Monte de Caparica, Portugal

Abstract : The interaction between two colliding monochromatic wave packets in water of finite depth is investigated through numerical experiments. A two dimensional numerical wave tank is used for this purpose. Each end of this flume is a piston acting as a wavemaker, and can be switched to be a wave-absorber when the wave packet emitted from the other end reaches it ; doing so, waves are not reflected in the basin. The main effect of the nonlinear interaction during head-on collision consists in a phase lag of the wave train due to a decrease of the phase velocity. In the present paper, the influence of the frequency, amplitude and wave number of the two incident waves on this phase lag is investigated by numerical experiments. Results established analytically by perturbation methods for infinite water depth are recovered here numerically in finite depth conditions.

Key. Words: hydrodynamics, free surface, nonlinear waves, numerical wave tank, wave interaction.

1 - Introduction.

The non-linear interaction between two or more different water wave trains has been widely studied using perturbation theories (e g :Longuet-Higgins 1962, Phillips 1960, Su & Mirie 1980), Fourier (Fenton & Rienecker 1982) or spectral methods (Dommermuth & Yue 1987), or numerical simulations (e g : Grilli et al. 1989, Olmez & Milgram 1995a, 1995b). Being exempt of assumption, except inviscid fluid and being limited only by (numerical) wave breaking, the latter approach has proved to be very useful in investigating this kind of nonlinear free surface phenomenon. In the present work, it is used to investigate the effect of head-on collision of short nonlinear gravity waves in water of finite depth. They are short in that their wavelength is not long compared to the water depth ; the latter is then finite but not shallow.

In the case of two long (solitary) waves, the nonlinear interaction results in a time shift and the emission of a dispersive wave train in the soliton tail (Su & Mirie 1980). When a long wave interacts with short dispersive waves, both of them experience a Doppler effect and a shift in phase (Clamond 1994).
We shall see in this paper that for two short waves, the same effect is observed on the phase and that it may be calculated, in some finite water depth cases, by a formula proposed by Longuet-Higgins & Phillips (1962) for infinite depth.

The problem is stated in two dimensions in the general framework of free surface potential flows. No other assumptions than inviscid fluid are made and the nonlinear initial boundary value problem is solved in the time-domain using a direct Mixed Euler-Lagrange BEM method (Clément 1995, 1996b). Each end of the numerical wave tank acts as a piston wavemaker generating monochromatic wave packets modulated by an envelope function.

2 - The numerical wave tank

In all the subsequent equations the physical variables are non-dimensionalized with respect to : the water depth h for the length variables, and $(h/g)^{1/2}$ for the time variable. The fluid density is set to unity.

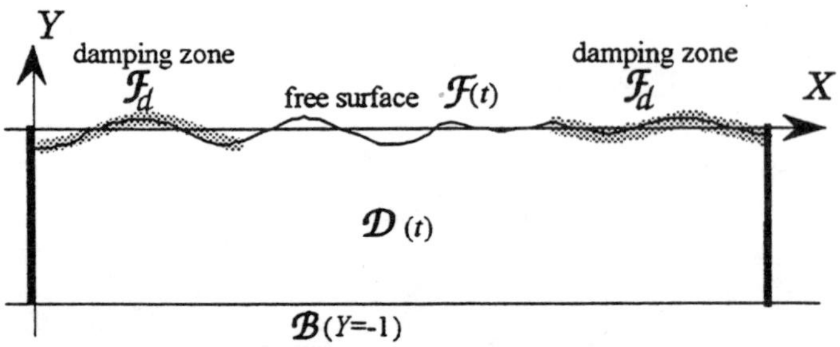

Fig.1. The computational domain

The usual assumptions of free surface potential flow theory are made : the fluid is incompressible and inviscid, surface tension is neglected, the atmospheric pressure above the free surface is constant and chosen as the pressure reference , the flow is irrotational. The fluid is at rest at time t=0.

Owing to the above assumptions, the fluid velocity $\bar{V}$ derives from a scalar potential $\Phi(M,t)$ function of time and space:

$$\bar{V}(x,y,t) = \bar{\nabla}\Phi(x,y,t) \qquad (1)$$

The potential Φ is easily shown to be the solution of the following nonlinear initial boundary value problem:

$$\Delta\Phi(x,y,t) = 0 \quad \text{in the fluid domain } \mathcal{D}(t) \qquad (2)$$

$$\frac{\partial\Phi}{\partial y}(x,-1,t) = 0 \quad \text{on the sea bottom } \mathcal{B}(t) \qquad (3)$$

$$\left.\begin{aligned} \frac{Dx}{Dt} &= \frac{\partial\Phi}{\partial x} \\ \frac{Dy}{Dt} &= \frac{\partial\Phi}{\partial y} \\ \frac{D\Phi}{Dt} &= -y + \frac{1}{2}(\nabla\Phi)^2 \end{aligned}\right\} \text{on the free- surface } \mathcal{F}(t) \qquad (4)$$

where the operator D/Dt is a Lagrangian time derivative.

The vertical ends of the basin can be moved horizontally, as pistons. They may act either as wavemakers or as active wave absorbers. In both cases, the no flux condition leads to a Neumann condition on these surfaces.

- <u>wave generation</u> : the boundary condition simply reads:

$$\frac{\partial\Phi}{\partial x}(x,y,t) = \begin{cases} \dot{f}_1(t) & M \in C1 \\ \dot{f}_2(t) & M \in C2 \end{cases} \qquad (5)$$

where f_1 and f_2 denote the prescribed wavemaker(s) motion, and the overdot represents differentiation with respect to time. In the present study, the wave packets were generated by sinusoidal piston velocity sequences, beginning and ending with a tapering *tanh* modulation function (see 6). A time delay δ_1 (resp. δ_2) before starting was also allowed in order to locate the interaction area in the central zone of the basin (see Fig.2). Let T_1 (resp T_2) denote the period, ω_1 (resp. ω_2) the angular frequency of the motion, α_1 (resp α_2) the stroke of the piston motion, and n_1 (resp. n_2) the number of periods in the wave packet. Then we have:

$$f_1 = \alpha_1 \tanh(t-\delta_1)\sin(\omega_1(t-\delta_1))\tanh(n_1T_1-(t-\delta_1)) \qquad (6)$$

for $\delta_1 \le t \le \delta_1 + n_1T_1$

, and idem, with subscript 2, for the right boundary $C2$.

-<u>wave absorption</u> : The vertical end walls acts as wavemakers in the manner described above during the first part of the simulation. Once the two wave packets has been generated and propagate towards each other in the basin, a flag is modified in the computer code in order to switch the boundaries $C1$ and $C2$ to the wave absorption mode. The wave absorption is achieved by coupling a "numerical beach" method on the terminal part $\mathcal{F}_d$ of the free surface adjacent to the vertical boundary C (Fig.1), to a piston wave-absorber condition on C itself (Clément 1996a).

- The *"numerical beach"* condition consists in adding a dissipative term to the dynamic boundary condition (4c) which then becomes:

$$\frac{D\Phi}{DT} = -y + \frac{1}{2}\left(\bar{\nabla}\Phi\right)^2 - \nu(x)\frac{\partial\Phi}{\partial n} \qquad M \in \mathcal{F}_d \qquad (7)$$

where $\nu(x)$ is a cubic function of the abscissa starting smoothly from zero at the beach entrance (see Clément 1996b).

- The *piston wave-absorber* condition stipulates that the wall moves as a piston reacting to (the first order component of) the hydrodynamic force upon it, with a velocity equal to the force itself, that is :

$$\frac{\partial\Phi}{\partial x}(x,y,t) = \int_{-1}^{\eta(t)} \frac{\partial\Phi}{\partial t}(x,y,t)dy \qquad M \in C \qquad (8)$$

Owing to the complementarity of the methods in terms of bandwidth, the coupling was shown to give excellent results (amplitude absorption coefficient > 93% whatever the wave frequency).

Conditions (2) to (8) above, and initial rest conditions at t=0 constitutes an initial boundary value problem (IBVP) which is solved, in our computer code CANAL, by the mixed Euler-Lagrange (MEL) method first introduced by Longuet-Higgins & Cokelet (1976), then further developed by a lot of authors.

The updating of the fluid domain geometry and of the Dirichlet condition on the free-surface from one time step to the next one is achieved by the integration of the ODEs (4) by a fourth order Runge-Kutta method. The tangential velocity needed in the evaluation of the right hand side of (4) is computed by means of a weighted *Arctang* method (Hasley - 1977). Then, at each time step, a mixed Neumann-Dirichlet boundary value problem has to be solved in the new fluid domain, with an updated Dirichlet condition on the free surface, and Neumann conditions (5)-(8) on the vertical end walls.

This problem is solved by a direct BEM using Rankine sources and normal dipoles linearly distributed over the segments of the polygon approximating the fluid domain boundary. A double-node technique is used at the corners where potential continuity is enforced.

3 - a generic case

In this section, the numerical simulation of a typical interaction sequence is described and commented. The main parameters of the interaction are schemed on Fig.2. Waterfall views of the free-surface history can be seen on Fig.3 an Fig.4. The former one presents the reference data set (*i.e* : no interaction) obtained with the following parameters : $[\omega_2 = 2. \ , a_2 = 0.04, \ \delta_2 = 0.]$. The total length of the basin in both runs is L=26.

The left going wave packet (P2, Q2, R2, S2) propagating at the group velocity (C_g=0.244) is clearly identified on Fig.3. One can notice the development of two systems of diverging waves ahead (S2,P2,s2) and behind (R2, Q2, r2) the main wave packet. This spreading phenomenon is due to the spectral content of the ramp functions in (6) which are broad-banded, and which thus generate these transient dispersive waves.

On Fig.4, the same wave group now collide with another one (P1, Q1, R1, S1) of longer wavelength, travelling in the opposite direction. This second group is defined by : $[\omega_1 = 1.\ , a_1 = 0.12]$. A delay, $\delta_1 = 30.$ (=$|OP1|$), is left before starting the wavemaker in such a way that the interaction area (A, B, C, D) is located approximately in the center of the tank. Once the two wave groups have been generated, the mode of motion of the end walls is switched from wavemaker to wave-absorber simultaneously at $t_a = 58$. A smooth transition is obtained, provided t_a satisfies : $|LQ2| < t_a < |LS1|$, and $|OQ1| < t_a < |OS2|$. One can easily see that the main effect of the interaction lies in a decrease of the shorter wave phase velocity, as it was shown by Longuet-Higgins (1962) in the infinite water depth case. On the contrary, the group velocity seems to remain unaffected by the interaction, at least at first order. The same effect occurs for the longest wave group, but is difficult to observe at first glance.

Let us now focus our attention on the shorter wave of angular frequency ω_2. We define the interaction duration as : $\Theta = t_A - t_D$, where t_A and t_D may be computed from the group velocities, according to the linear wave theory. Because the change in phase velocity, Δc_2 , is effective only during this finite interval of time, it results in a spatial shift, say χ, given by :

$$\chi = \Delta c_2 \Theta \qquad (9)$$

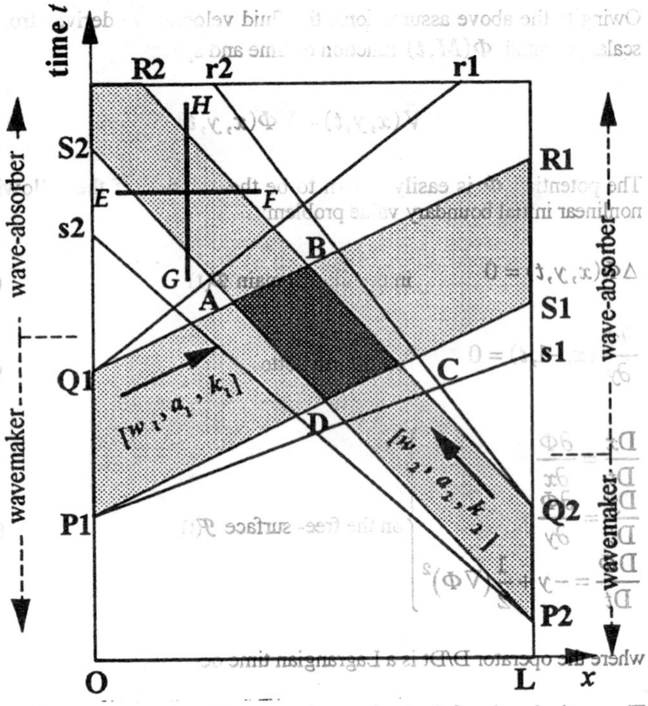

Fig. 2 : Sketch of a wave/wave interaction run.

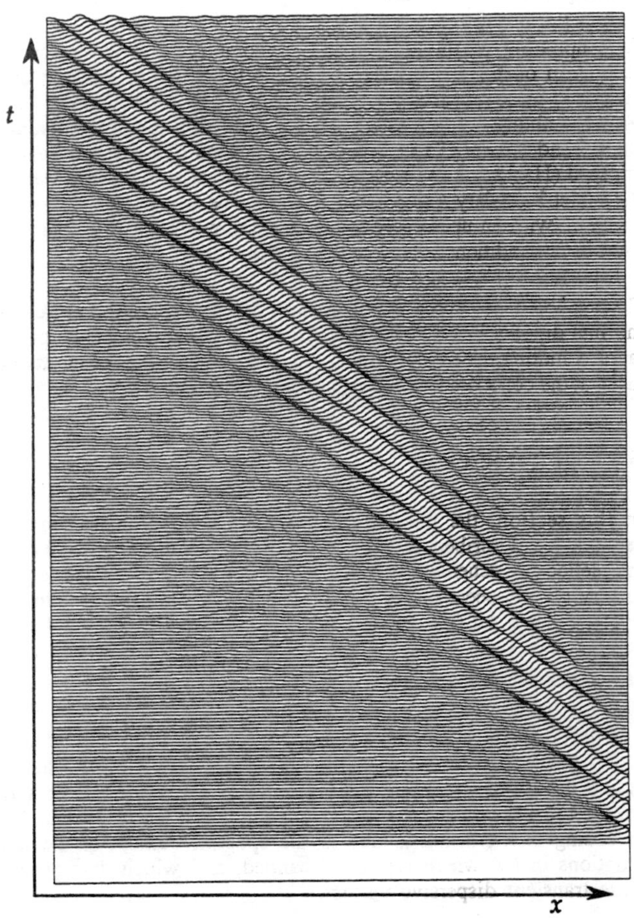

Fig.3 : Reference data ; ω_2=2.

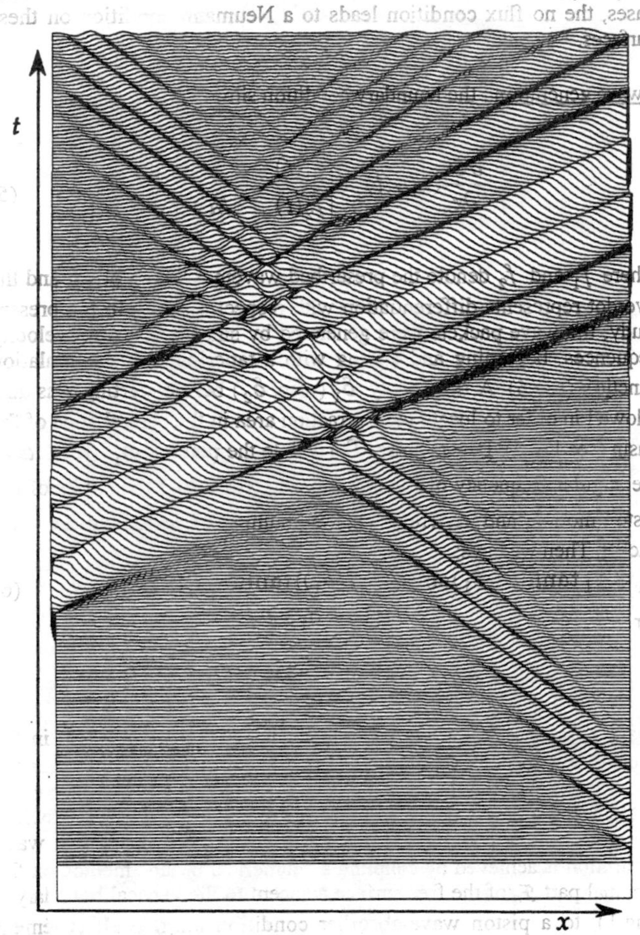

Fig.4 : waterfall plot of an interaction sequence ; ω_2=2, ω_1=1.

This shift may be observed and measured by comparing free-surface elevations in both cases : in the presence of the secondary wave train (Fig.4) or without it (Fig.3). This analyse may be performed either along a cut at fixed time (like EF on Fig.2), or at a fixed probe location (GH). Let us first consider an example of such a pair of signals on Fig.5. It is a cut at fixed time : t=89. The continuous line is the record of the undisturbed left going wave packet, whereas the dashed line shows the same wave group after interaction with the rightgoing one.

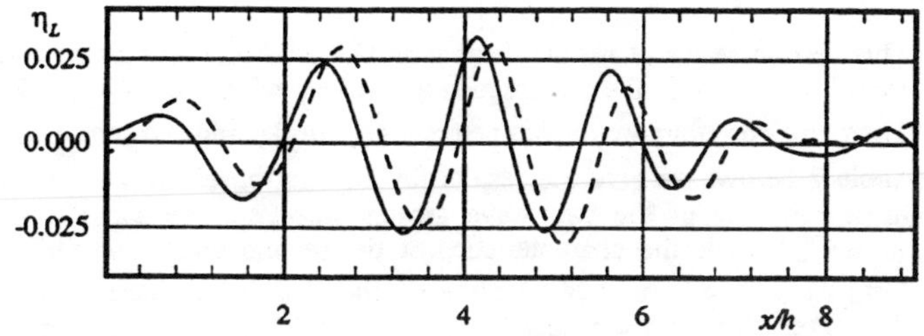

Fig.5 : spatial shift of the leftgoing short wave (ω_2=2.) resulting from interaction with the longer (ω_1=1.) rigthgoing wave (see Fig.4). (———): reference case without interaction ; (-----) : after interaction. t=89.

On the contrary, Fig.6 below shows a cut at a fixed abcissa of the same pair of signals. Here, the shift between the signals results in a time lag which will be denoted by τ in the sequel.

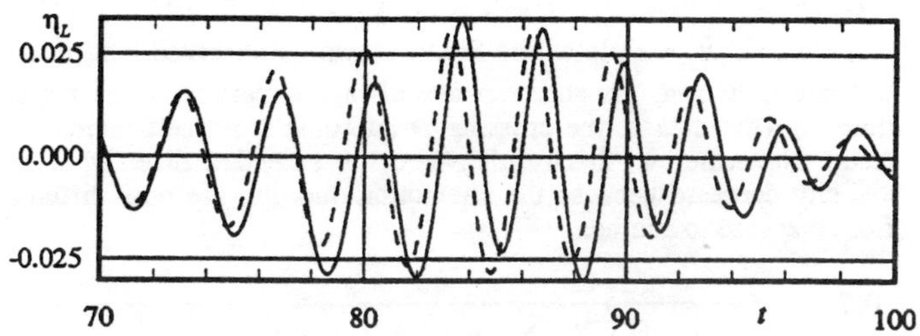

Fig.6 : time lag between the right-going waves, with (-----) and witout (———) interaction. Probe location (X=4.)

In these simulations, the wave packets are relatively short, and a steady state cannot be reached within the group. Thus the shift, in space or time, must be measured in an average sense rather than in a strict deterministic fashion. Here, we choose to define τ and χ by computing the cross-correlation between the two free-surface elevation signals (Fig.5-6).

Let us denote by $\tilde{\eta}(x,t)$ the undisturbed free surface, which will be the reference, and $\eta(x,t)$ the actual free-surface at the same place and time, but after interaction. Two coefficients may be defined from these signals : a time cross-correlation coefficient $r_t(x,\tau)$:

$$r_t(x,\tau)=\frac{1}{(t_H-t_G)}\int_G^H \tilde{\eta}(x,t)\eta(x,t+\tau)dt \qquad (10)$$

and a space cross-correlation coefficient $r_x(\chi,t)$:

$$r_x(\chi,t)=\frac{1}{(x_F-x_E)}\int_E^F \tilde{\eta}(x,t)\eta(x+\chi,t)dx \qquad (11)$$

The time lag between the signals is defined as the value of the variable τ where $r_t(x,\tau)$ is maximum, while the spatial shift χ corresponds to the maximum of $r_x(\chi,t)$. The strategy adopted here to evaluate the change in phase velocity is therefore a "best fit" in a statistical sense, rather than in a deterministic one. It will work as well for the analysis of the experimental data which are currently being collected.

When applied to the present generic case (Fig.6), this analysis gives the time cross-correlation coefficient curve plotted on Fig.7 below.

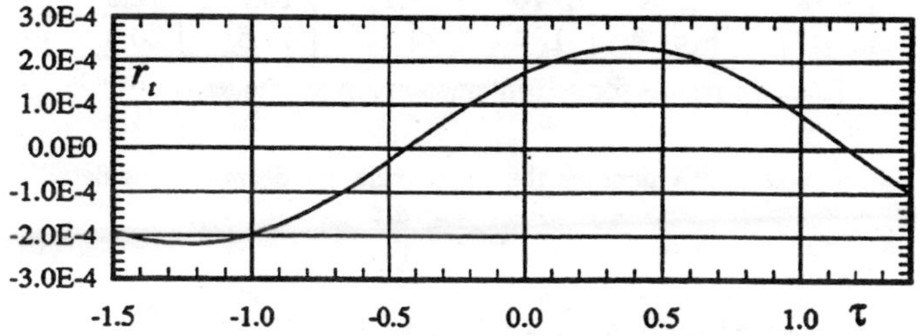

Fig.7 : time cross-correlation coefficient $r_t(x,\tau)$ for the data set on Fig.6

From this curve, the precise value of τ maximizing r_t is found to be equal to $\tau=0.383$. The same analysis preformed in the spatial domain leads to : $\chi=0.201$.

Would the monochromatic wave packet be long enough for a steady regime of standing waves to take place, the linear theory should lead to the relation :

$$\omega\tau=k\chi \qquad (12)$$

between the time lag and the spatial shift, where k derives from the dispersion relation :

$$k\tanh k=\omega^2 \qquad (13)$$

Both approaches were used in the analysis of the numerical runs presented in the forthcoming sections.

4 - Results

4-1 Self interaction

The present section deals with a series of run where the two monochromatic wave packets were strictly identical. These cases are therefore referred to as : *self interaction*. The total wave group duration is six periods ($n_1=n_2=6$). The nondimensional circular frequency is set to : $\omega_1=\omega_2=2.$, resulting in a linear wave number of : $k=4.0026$ and a wavelength of : $\lambda=2\pi/k=1.5698.$, which denotes deep water depth conditions. The tank length is L=10, and the interaction time $\theta=9.42$.

In this series, all the above parameters were kept constant, except the wave amplitude which was varied from 1.57×10^{-2}, to 5.04×10^{-2}. Results of the analysis in both space and time are presented in Table 1.

$a_1(=a_2)$ (*100)	τ	χ	Δc_τ (*1000)	Δc_χ (*1000)	Δc_L (*1000)
1.568	.034	.017	1.78	1.75	1.96
2.079	.061	.031	3.24	3.30	3.46
2.500	.094	.047	5.00	4.99	5.00
2.649	.108	.050	5.73	5.28	5.62
2.880	.124	.065	6.58	6.98	6.64
3.293	.162	.080	8.59	8.49	8.68
3.477	.183	.092	9.72	9.79	9.68
3.687	.211	.101	11.2	10.7	10.9
4.062	.254	.124	13.5	13.2	13.2
4.275	.290	.136	15.4	14.4	14.6
4.424	.313	.145	16.6	15.4	15.7
4.784	.362	.170	19.2	18.1	18.3
5.037	.417	.184	22.1	19.5	20.3

Table 1 : results for self-interaction run at frequency : $\omega = 2.$

In this table, Δc_τ denotes the phase velocity decrease calculated in the time domain, as :

$$\Delta c_\tau = \frac{\omega\tau}{k\Theta} \tag{14}$$

, whereas we have in the space domain :

$$\Delta c_\chi = \frac{\chi}{\Theta} \tag{15}$$

Together with these two measures, we give a third estimation of the shift based on the mode-coupling approach of Longuet-Higgins and Phillips (1962). According to these authors, the leading order component of the change in phase velocity in such a self interaction case should be equal to :

$$\Delta c_L = a^2\omega k \tag{16}$$

where a is the wave amplitude. This relation was derived for two colliding steady monochromatic primary wavetrains in water of infinite depth. Here, we are practically in deep water conditions ($\lambda \cong 1.56h$), and furthermore we assume that it holds also when the water depth is finite, provided the corresponding dispersion relation (13) is satisfied by the wave number k.

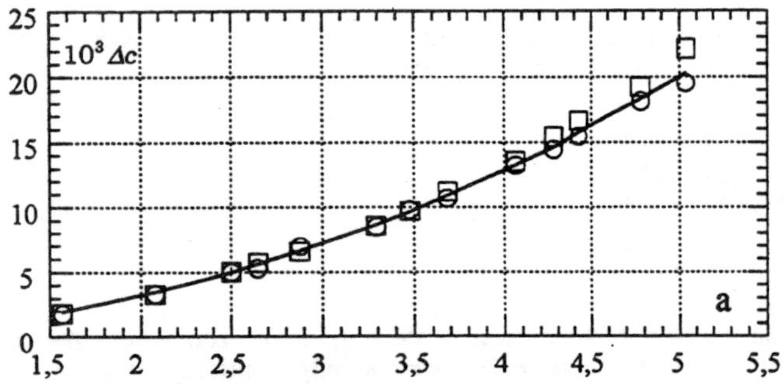

Fig.8 : phase velocity decrease versus wave amplitude (sqare : Δc_τ), (circle : Δc_χ), (—— : Δc_L)

Numerical results of Table 1 are plotted on Fig.8 above. The dots mark the estimation using either the time cut method (10)-(14), or the space cut (11)-(15), while the line shows the results obtained by the mode-coupling method (16). The very close agreement between the two approaches, although very different in nature, made us confident in the present method of analysis of our numerical experiments. A weak discrepancy appears between time domain and space domain methods for the largest amplitudes. It gives a measure of the order of magnitude of the error which was observed to be always less than 10% in this series of self interaction cases. The source of this error lies essentially in the limited length (*i.e* duration) of the wave packet which was choosen as short as possible is order to reduce the computation time. Due to it, a larger error is introduced in estimating the actual interaction duration by Θ as we do, and in the calculation of the wave amplitude.

4-2 Dual interaction

This second series of results deals with the head-on collision of two wave packets of different frequency, wavenumber and amplitude. There are now four wave parameters : $\{a_1, \omega_1\}$, and $\{a_2, \omega_2\}$. In Table 2 below, we give the results for the first cases we get. In all these run, one of the two wave groups was kept the same (with $\omega_2 = 2.$) , while the characteristics of the second one were varied. Here again, we focus our attention on the change in phase shift of the waves after their crossing.

ω_1	k_1	a_1 (*100)	ω_2	k_2	a_2 (*100)	τ_1	τ_2
1.4	2.029	8.89	2.0	4.003	4.45	.118	.610
1.7	2.907	6.16	2.0	4.003	4.45	.205	.420
2.3	5.290	3.07	2.0	4.003	4.45	.350	.150
2.6	6.760	2.07	2.0	4.003	4.45	.392	.072
2.9	8.410	1.83	2.0	4.003	4.45	.448	.064

Table 2 : results of dual-interaction

The curves showing the evolution of the shifts τ_1 and τ_2 versus wavenumber k_1 are plotted on Fig.9 . It can be observed that, in our test configuration, the shorter wave always experiences the higher time lag (the data at the crossing point come from self-interaction tests). Nevertheless, this result cannot be extended directly to the velocity decrease because the interaction duration are now different from one case to another.

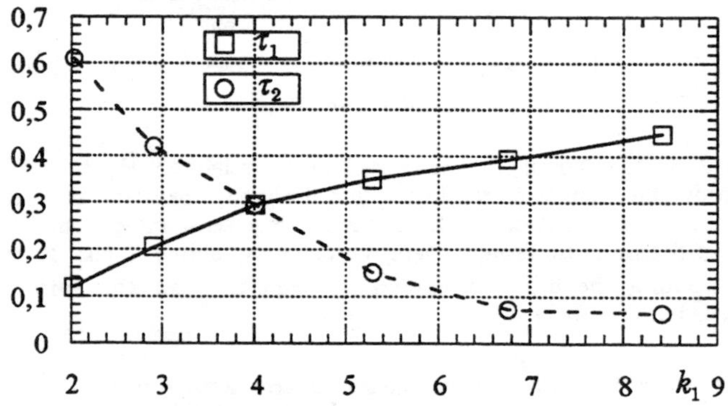

Fig.9 : phase shift of the first wave packet.
second wave packet : ($\omega_2 = 2.$, $a_2 = .0445$, $n_2 = 6$)

According to Longuet-Higgins & Phillips (1962), the velocity deviation for two different colliding wavetrains should be given by :

$$\left.\begin{aligned} \Delta c_2 &= a_1^2 k_1 \omega_1 \\ \Delta c_1 &= a_2^2 k_1 \omega_2 \end{aligned}\right\} \quad k_1 < k_2 \tag{17}$$

These results are valid for steady regular incident waves, but not for wave packets. Thus, a comparison of our results with (17) requires an estimation of an equivalent regular wave amplitude, and of the interaction time interval. This evaluation was found far less easy for the present dual-interaction cases than it was for self-interaction reported in the previous section. Then, we prefer to wait for forthcoming experimental results to compare with. This comparison will appear in a future paper.

5 - Conclusion

A numerical wave tank was used to model the head-on collision of two wave packets in water of finite depth. The nonlinear interaction results in a change in the phase velocity of both waves. In the case of two identical wavetrains, the resulting phase lag obtained by the present numerical method compare closely with those of the classical mode-coupling approach. The same effect on wave velocities is observed when the interacting wave packets are different, but cannot be compared so easily to the analytical formula which was derived for steady regular wavetrains in infinite water depth.

6 - Acknowledgement

The first author is indebted to the *Direction des Recherches et Études Techniques* of the French Ministry of Defence which supported the part of the work related to software development. The second author's research is financially supported by JNICT and FCT-UNL portuguese organizations.

7 - References

Clamond D. (1994). *"Amplitudes et phases dans la théorie des ondes de gravité"*. Th. Université J.Fourier. Grenoble.

Clément A., Mas S.: (1995):" Hydrodynamic Forces Induced by a Solitary Wave on a Submerged Cylinder."; *Proc. 5th Int Offshore and Polar Engineering Conf.* ISOPE95, The Hague, Vol.III, pp339/347.

Clément A. (1996a):" Coupling of two absorbing boundary conditions for 2D time-domain simulations of free surface gravity waves"; *J. Comp. Physics.*, 126, pp139-151.

Clément A. (1996b):" Dynamic Non-Linear Response of OWC Wave Energy Devices"; *Proc. 6th ISOPE*, Los Angeles, Vol. I, pp91-96.

Dommermuth D.G., Yue D.K.P. (1987) :"A high-order spectral method for the study of nonlinear gravity waves"; *J. Fluid Mech.*, 184, pp267-288.

Fenton J.D., Rienecker M.M. (1982) : "A Fourier Method for solving nonlinear water-wave problem : Application to solitary wave interactions". *J. Fluid Mech.*, 118, pp4117-443.

Grilli S.T., Skourup J., Svendsen I.A. (1989) : "An efficient boundary element method for nonlinear water waves". *Engng. Anal. Bound. Elem.*, 6 (2), pp97-107.

Halsey N.D. (1977) : *" Potential flow analysis of multiple bodies using conformal mapping "*; Thesis. California State University.

Longuet-Higgins M.S. (1962) :"Resonant interaction between two trains of gravity waves"; *J. Fluid Mech.*, 12, pp321-332.

Longuet-Higgins M.S., Phillips O.M. (1962) :"Phase velocity effects in tertiary wave interactions"; *J. Fluid Mech.*, 12, pp333-336.

Longuet-Higgins M.S., Cokelet E.D. (1976) : "The deformation of steep surface waves on water: 1- A numerical method of computation."; *Proc Roy. Soc.*, Londres A350 .

Olmez H.S., Milgram J.H. (1995a) :"On the dynamics of unsteady gravity waves of finite amplitude". *J. Fluid Mech.*, 289, pp199-226.

Olmez H.S., Milgram J.H. (1995b) :"Numerical methods for nonlinear interactions between water waves". *J. Comp. Phys.*, 118, pp62-72.

Phillips O.M. (1960) :"On the dynamics of unsteady gravity waves of finite amplitude". *J. Fluid Mech.*, 9, pp193-217.

Su C.H., Mirie R.M. (1980) : "On head-on collision between two solitary waves". *J. Fluid Mech.*, 98 (3), pp509-525.

Proceedings of the Seventh (1997) International Offshore and Polar Engineering Conference
Honolulu, USA, May 25-30, 1997

ISBN 1-880653-28-1 (Set); ISBN 1-880653-31-1 (Vol. III)

Development of a Viscous Numerical Wave Tank and Numerical Evaluation of Dynamic Loads on Submerged Bodies at Small KC Numbers

V. Armenio and M. Favretto
DINMA, University of Trieste
Trieste, Italy

SUMMARY

In this paper, a preliminary study on the interaction between water waves and a fixed submerged body was performed, using the primitive variables formulation of the Navier-Stokes Equations (NSE). The governing equations, together with the appropriate non-linear boundary conditions over the free surface, were solved with an improved MAC method (SIMAC), recently developed at the University of Trieste.

In the first part of the work, a Numerical Wave Tank (NWT) was developed by means of a pneumatic wave-maker located at the left hand side of the tank. Then, the dynamic loads over submerged rectangular and square cylinders were evaluated in the range of small Keulegan-Carpenter numbers.

KEY WORDS : numerical wave tank, dynamic loads in a viscous regime, low KC numbers

INTRODUCTION

Over the last few years, numerical simulations of free surface viscous flows have been successfully attempted. NSE, written in the primitive variables formulation, were solved by finite difference or finite volume techniques using both structured (Yeung et al., 1992a) and unstructured grids (Hino et al., 1993). The simulations were carried out in a wide area of applied hydrodynamics, including liquid sloshing (Armenio et al., 1996), ship viscous resistance (Alessandrini et al., 1994), and the evaluation of dynamic loads over fixed or floating structures (Gentaz et al., 1996; Yeung et al. 1995a). Two main different ideas are widely in use to describe the free surface movement, the moving grid technique and the MAC-type method respectively.

The first one is effective when small or moderate non overturning waves are studied. In this case, the governing equations are solved in a generalised co-ordinates system within a time-dependent physical domain mapped onto a fixed rectangular computational domain. Despite its own intrinsic accuracy, the moving grid technique fails when high distorted free surface patterns are obtained, mainly due to numerical troubles which occur in the evaluation of the metric tensors.

The MAC-type method is based on the solution of the equations in a fixed-in-time domain of integration, which is split into two different sub-domains: one full of liquid and the other empty. At each time step, the equations are solved into the full of liquid sub-domain. According to the local velocity field, the free surface position is then updated either by the use of the kinematic free surface boundary condition or by the use of a Lagrangian approach. The cells are finally re-flagged depending on their position as compared to the new free surface profile. This treatment can easily handle very distorted free surface patterns, including jets and sprays at the solid boundaries and breaking waves.

An improved MAC method (SIMAC), that can solve high Reynolds numbers free surface flows, was used in this work. The algorithm (Armenio, 1997) was first developed with the aim to obtain an accurate evaluation of large amplitude sloshing loads for partially filled unbaffled and baffled tanks. The typical problem of the MAC method, which consists in the need to use coarse grids, was overcome by merging well-established techniques such as the 'fractional step approximate factorisation' and the 'multigrid acceleration technique' together with a standard MAC method.

Over the last decade only few attempts have been made to approach the theoretical and numerical study of the wave propagation and interaction with fixed or floating bodies using a NSE solver. In 1993 Arai et al. studied the diffraction problem over a two-dimensional rectangular free surface piercing body. After neglecting the viscous stress, the authors solved the Euler's equation by means of a MAC-type method. In 1995 Yeung et al. (1995a) solved the three-dimensional viscous flow around a fixed vertical circular cylinder using a moving grid solver, focusing the attention on Reynolds number values up to 10000.

In this paper SIMAC was used to develop a two dimensional NWT and to evaluate the dynamic loads on submerged fixed bodies. The results are presented in section 3. As it will be thoroughly shown in the next sections, the work is still at a development stage. However, preliminary results obtained for deeply submerged bodies, looks to be very encouraging.

GOVERNING EQUATIONS AND NUMERICAL TECHNIQUE

Let us consider a Numerical Wave Tank as shown in Fig. 1. The fluid is Newtonian and incompressible and the governing equations are written in the frame of reference fixed to the basin (Fig. 1) . The primitive variables formulation of the unsteady laminar NSE reads as:

$$\frac{\partial u}{\partial x} + \frac{\partial v}{\partial y} = 0 \qquad (1)$$

$$\frac{\partial u}{\partial t} + \frac{\partial u^2}{\partial x} + \frac{\partial uv}{\partial y} + \frac{\partial P}{\partial x} = \nu \left[\frac{\partial^2 u}{\partial x^2} + \frac{\partial^2 u}{\partial y^2} \right] + Bx \qquad (2)$$

$$\frac{\partial v}{\partial t} + \frac{\partial uv}{\partial x} + \frac{\partial v^2}{\partial y} + \frac{\partial P}{\partial y} = \nu \left[\frac{\partial^2 v}{\partial x^2} + \frac{\partial^2 v}{\partial y^2} \right] + By \qquad (3)$$

being u and v the velocity components referred to the above-mentioned frame of reference, $P = \frac{p - p_{atm}}{\rho}$ the relative pressure, with p and ρ the physical pressure and the fluid density. B_x and B_y stand for the components of the body forces which act over an unitary volume of fluid. In the case herein studied $B_x = 0$ and $B_y = -g$, being g the gravitational acceleration.

The following kinematic and dynamic boundary conditions are applied at the instantaneous free surface position $\eta = \eta(x,t)$:

$$\frac{\partial \eta}{\partial t} + u \frac{\partial \eta}{\partial x} - v = 0 \qquad (4)$$

$$P = 0 \qquad (5)$$

The no-slip and the no-flow conditions are finally applied at the rigid walls.

SIMAC will be briefly described hereafter. Equations 1-3 are solved with the following two steps procedure:

$$\frac{\hat{\mathbf{u}} - \mathbf{u}^n}{\Delta t} = -\frac{1}{2}\left[3\,\mathrm{Conv}(\mathbf{u})^n - \mathrm{Conv}(\mathbf{u})^{n-1}\right] + \frac{1}{2}\nu\,\nabla^2\left[\hat{\mathbf{u}} + \mathbf{u}^n\right] - \nabla\Phi^n + \mathbf{B}^n \qquad (6)$$

$$\frac{\mathbf{u}^{n+1} - \hat{\mathbf{u}}}{\Delta t} = -\left[\nabla\Phi^{n+1} - \nabla\Phi^n\right] \qquad (7)$$

being Φ a computational pressure related to the physical pressure P as follows:

$$P^{n+1} = \Phi^{n+1} + \frac{\nu\,\Delta t}{2}\left[\nabla^2\Phi^{n+1} - \nabla^2\Phi^n\right] \qquad (8)$$

As in a standard fractional step method, the continuity equation is re-written as a Poisson's equation for the (computational) pressure:

$$\nabla^2\Phi^{n+1} = \frac{1}{\Delta t}\nabla\hat{\mathbf{u}} + \nabla^2\Phi^n \qquad (9)$$

As shown in Eq. 6, the Crank-Nicholson and the Adams-Bashforth schemes are respectively used for the time discretisation of the diffusive and convective terms, thus the algorithm is formally second order in time.

The algorithm works as follows: at first, the intermediate velocity field $\hat{\mathbf{u}}$ is obtained by Eq. 6; the RHS of Eq. 9 is then calculated in order to solve the Poisson's equation. After that the pressure distribution is evaluated, the new free divergence velocity field is calculated with Eq. 7. The position of the free surface is then updated by the use the kinematic free surface condition (Eq. 4), according to the local velocity field. The velocity and pressure points are re-flagged depending on their own position vis-à-vis the free surface profile. A velocity field is finally extrapolated beyond the free surface and the cycle can start again for the new time-iteration.

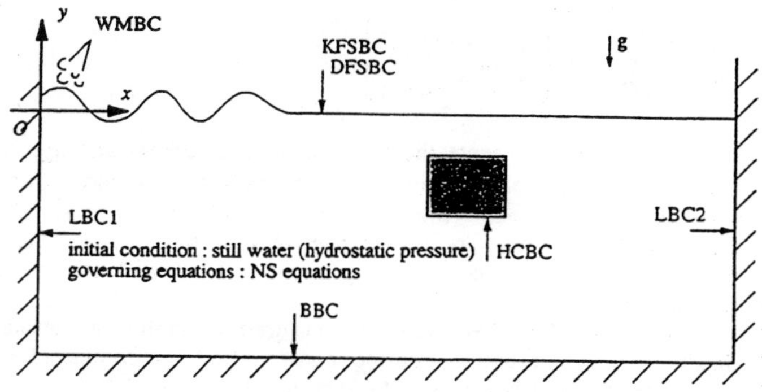

Fig. 1: Schematic representation of the NWT with the frame of reference.

The above-mentioned extrapolation is required for the calculation of the convective and diffusive terms and to calculate the velocities of the markers which identify the free surface location. The problem of the extrapolation of the velocity field beyond the free surface has not been fully solved in the MAC community as yet. In this work, the extrapolation of the velocity field above the free surface is performed as follows (Fig. 2). We first express the slope of the vertical distribution of a velocity component by the use of a second order backward formula, then the same slope is evaluated with a second order centred formula, which involves the unknown u_0. Finally, after equating the two formulas we obtain:

$$u_0 = (\Delta y_0 + \Delta y_1)\left[a_1\, u_1 + a_2\, u_2 + a_3\, u_3\right] \qquad (10)$$

being

$$a_1 = \frac{2\Delta y_1 + \Delta y_0}{\Delta y_1(\Delta y_0 + \Delta y_1)}, \quad a_2 = \frac{1}{(\Delta y_0 + \Delta y_1)} - \frac{\Delta y_1 + \Delta y_2}{\Delta y_1\, \Delta y_2}$$

and $a_3 = \dfrac{\Delta y_1}{\Delta y_1\, \Delta y_2 + \Delta y_2^2}$.

The markers, which individuate the position of the free surface, are located at the same abscissa of the pressure points . The markers are moved by the use of Eq. 4 as follows:

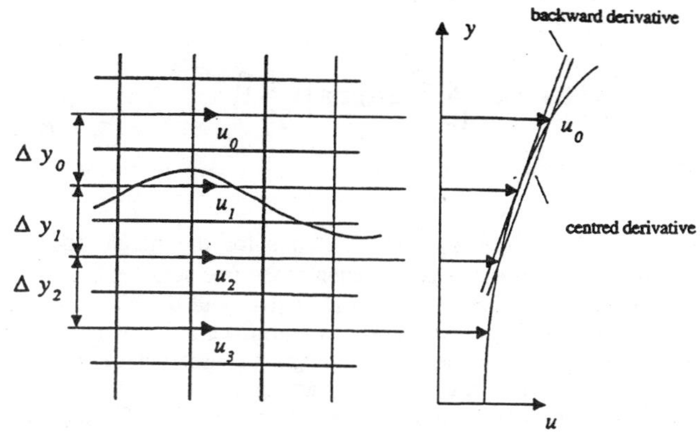

Fig. 2: Geometrical representation of the technique used for the extrapolation of the velocity components above the free surface.

$$\eta_k^{t+1} = \eta_k^t + \Delta t \left[v_k^t - u_k^t \left\langle \frac{\partial \eta}{\partial x} \right\rangle_k^t \right] \quad (11)$$

In Eq. 11 η represents the position of the markers and u_k, v_k stand for the actual velocities of the markers. u_k and v_k are evaluated by bilinear and linear interpolation among the neighbouring velocity nodes respectively. Finally, the space-derivative in Eq. 11 is computed by means of a Schweikert's spline in tension.

The above described treatment ensured a smaller numerical dissipation and a better volume conservation as compared to techniques previously employed (Armenio, 1997).

In SIMAC (Armenio et al., 1996; Armenio 1997) the pressure equation (Eq. 9) was solved with the Additive Correction Multigrid (ACM) technique. Nevertheless, as regards the NWT problem, the implementation of the wave-maker (numerically treated as a periodic time variation of the pressure at the free surface) caused some troubles to the multigrid procedure. For such reason Eq. 9 was here solved by means of the SOR technique. However, in the next future the ACM technique will be implemented for the NWT problem as well.

NUMERICAL RESULTS

This section is divided into two main sub-sections. In the first one the viscous numerical wave tank is described, in the second one numerical computations of dynamic loads over submerged square and rectangular cylinders are presented.

The viscous Numerical Wave Tank

Over the last few years, numerical simulations provided an effective tool for the study of water waves dynamics. An excellent review on the techniques currently in use for the inviscid NWT is in (Kim, 1995). The generation and propagation of water waves into a two-dimensional wave tank, was herein investigated taking into account the liquid viscosity. This circumstance made it possible to evaluate dynamic loads which derive from vortex generation and shear stresses too. A pneumatic wave-maker was located at the left side wall. It operates by means of an imposed periodic variation of the pressure over the free surface of the water, and it acts over a narrow area of the free surface near the left wall of the tank. On the opposite side, a rigid vertical wall was considered. Obviously, this matter introduced the possibility to obtain wave profiles affected by reflection phenomena. In order to overcome such problem, the implementation of an open boundary (radiation) condition as outlined in (Clement, 1996) is at present in progress.

The following formula was applied for the wave generation near the left side wall:

$$P(x,t) = \frac{P_p}{2} \left[1 + \cos \frac{2 x \pi}{L_w} \right] S(t) \sin \left(\frac{2 \pi}{T} t \right) \quad \text{for} \quad 0 < x < \frac{L_w}{2} \quad (12)$$

In Eq. 12 one has: P_p, which stands for the height (double amplitude) of the pressure perturbation at the free surface; L_w and T which respectively stand for the wave length and the wave period; the term inside the square brackets, term which distributes the pressure along the first half wave length using a cosine law; the term S(t) which ensures a gentle starting of the wave-maker; a sinusoidal term with period T, which is responsible for the time variation of the pressure at the free surface. The term S(t) takes up the following form:

$$S(t) = \frac{1}{2} \left[1 - \cos \frac{2 \pi t}{\tau} \right] \quad \text{for} \quad 0 < t < \tau$$
$$S(t) = 1 \quad \text{for } t > \tau \quad (13)$$

In the computations hereafter presented, τ was chosen equal to twice the wave period.

Grid sensitivity tests were first carried out using four different uniform grids (constant aspect ratio $\Delta x/\Delta y$ =2.75) and a kinematic viscosity equal to $1.0*10^{-6}$ m^2/s. The wave period was chosen equal to 1.57 s, so that the waves to be generated were 3.85 m long. The liquid depth inside the tank (hereafter referred as h) was chosen equal to the wave length (h = 3.85 m) and a tank long six times the wave length was used. Finally 1/21 steep waves (wave steepness = $2\zeta_a/L_w$ being ζ_a the wave amplitude) were generated.

Fig. 3 shows the comparison of the wave profiles obtained using four different uniform grids (respectively 15, 20, 25 and 30 cells per wave length) at the final time t = 12 T. The volume variation versus time is shown in Fig. 4. In short, it is clearly shown that the algorithm works satisfactory, even when steep waves are to be simulated. The volume variation in time is very small, less than 0.002 % for the coarsest grid.

With the aim to check for steadiness and for the presence of high order harmonic components in the wave pattern, some more numerical tests were performed. The same tank and wave length as previously seen were investigated. At first, a small amplitude wave (wave steepness = 1/191) was considered. Fig. 5a and 5b show the discrete spectra of the wave pattern along the tank (as shown in Yeung et al., 1992b) for two different time intervals, respectively $8 < t/T < 10$ and $10 < t/T < 12$. A steady state in the wave propagation is reached in the left part of the tank . The energy is at a large extent confined in the first harmonic component, as to be expected due to the small amplitude of the wave. The amplitudes of the other harmonic components are significant in the right part of the tank, where a steady state condition has still to be reached. The analysis of Fig. 5b clearly shows the presence of wave reflections, so that steady conditions can be reached only in a small section of the tank for a small time interval. In order to overcome this problem, a computational domain longer than that herein shown will be used in the next sub-section.

In Fig. 6, the discrete spectrum obtained for $8 < t/T < 10$ for a large wave amplitude (wave steepness = 1/16) is reported. It is evidenced that the second harmonic component has a significant value ($a_2 = 5.0\ 10^{-3}$ m) comparable with that of the equivalent Stokes wave ($a_{2(Stokes)} = 5.01\ 10^{-3}$ m).

Fig. 5, 6 also show the presence of a sub-harmonic component ($\omega_w/2$), which has to be ascribed to some resonance phenomena which always occur inside the closed tank. The implementation of a radiation condition will be effective against this phenomenon.

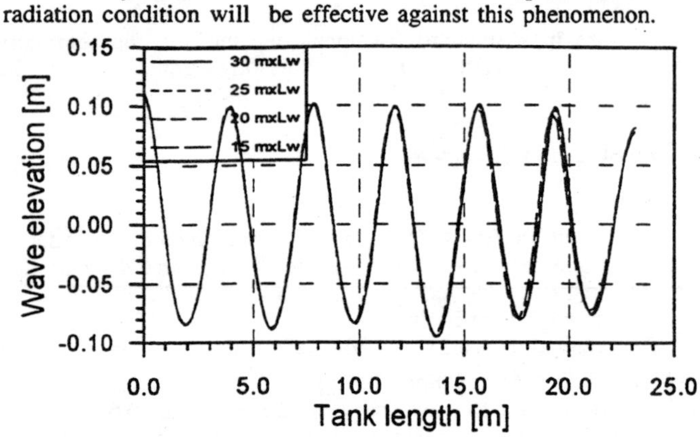

Fig. 3: Wave pattern in the tank at t = 12 T for four different uniform grids (wave steepness = 1/21).

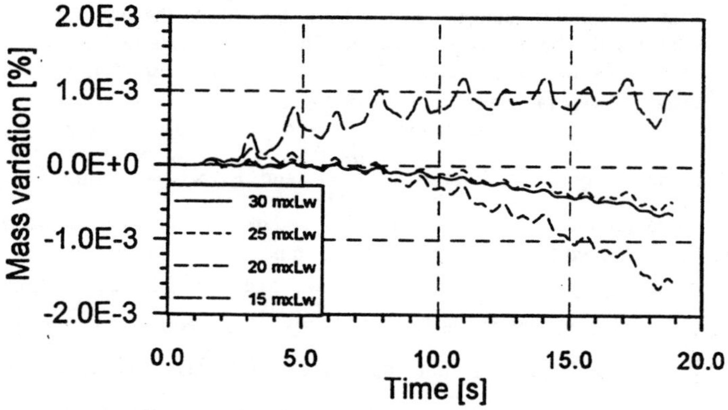

Fig. 4: Percentage of mass variation versus time for the computation of Fig. 3.

The calibration of the wave-maker was then performed. Unfortunately, we were unable to find analytical results for the wave-maker herein developed, so that the comparison between our numerical results and well established data in literature was not possible.

However, the complete wave-maker linear theory for plane waves produced by a paddle (Dean et al., 1992) shows a linear relationship between the stroke of the paddle and the wave height for each value of kh, being $k = \frac{2\pi}{L_w}$ the wave number.

Based on these results, we investigated the relationship between the pressure P_p and the wave height by means of numerical experiments. In our case too, a linear relationship was found. The results were then collected into a single graphic which contains all the characteristics of the pneumatic wave-maker herein described. This graphic is shown in Fig. 7. As well recognised in the literature, it was observed that the function $H = \frac{2\zeta_a}{P_p} = f(kh)$ is linear in the range of shallow water $\left(k\,h < \frac{\pi}{10}\right)$, whereas an asymptotic behaviour is noticed in the case of deep water $(k\,h > \pi)$

Evaluation of dynamic loads

When viscous effects are significant, the Morison equation is often employed for the evaluation of dynamic loads over submerged structures. If we consider an horizontal cylinder parallel to the crest line of the waves, the Morison equation can be written both for the horizontal and for the vertical components of the force per unit of length as follows:

$$F_x(t) = \frac{\rho A_x}{2} |u|\, u(t)\, C_{D_x} + \rho\, V\, \dot{u}(t)\, C_{M_x} \tag{14}$$

$$F_y(t) = \frac{\rho A_y}{2} |v|\, v(t)\, C_{D_y} + \rho\, V\, \dot{v}(t)\, C_{M_y} \tag{15}$$

In Eq. 14 and 15 A_x and A_y denote the projected area of the body per unit of length in the x and y directions respectively, u(t), v(t), $\dot{u}(t)$ and $\dot{v}(t)$ the time-dependent velocity and acceleration components of the undisturbed wavy field at the position of the body, $|u|$ and $|v|$ the modules of velocity components and V the body volume per unit of length. C_{M_x}, C_{M_y} are the coefficients of the inertial force, which represents the force in-phase with the fluid acceleration, whereas C_{D_x}, C_{D_y} are the coefficients of the forces in-phase with the fluid velocity (drag forces). With the aim to save space, the procedure herein used for the evaluation of the force coefficients is not reported. However, this procedure is well illustrated in (Faltinsen, 1990; Arai, 1995; Koterayama et al., 1995; Yeung et al., 1995b).

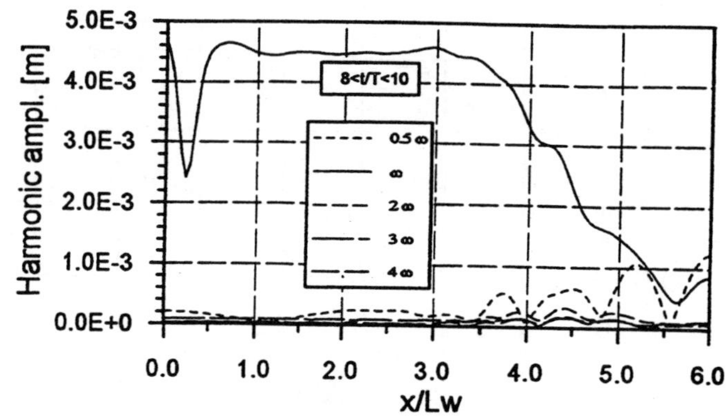

a: time window 8 < t/T < 10

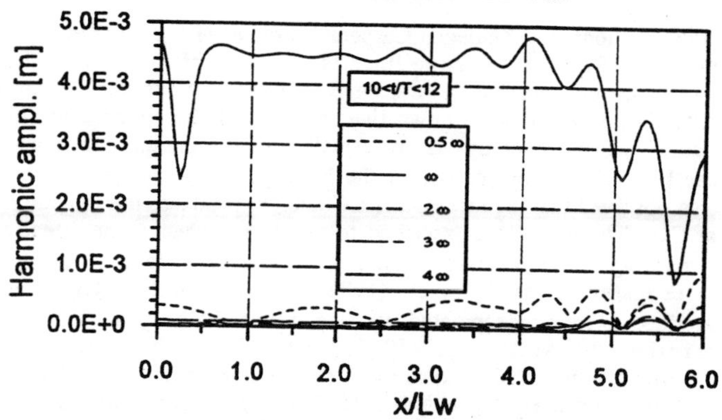

b: time window 10 < t/T < 12

Fig. 5 : Harmonic components of the wave system for a wave steepness = 1/191:

In order to evaluate the coefficients, the time records of the forces which act over the body are required. At each time step, such forces were computed by integration of the pressure and shear forces over the body.

In previous investigations, it was well established that a submerged body in an oscillating flow experiences inertial and drag forces, whose magnitude is strictly related to several factors: the vertical position and the geometry of the body; the relative motion between the neighbouring fluid and the body itself (i. e. planar or

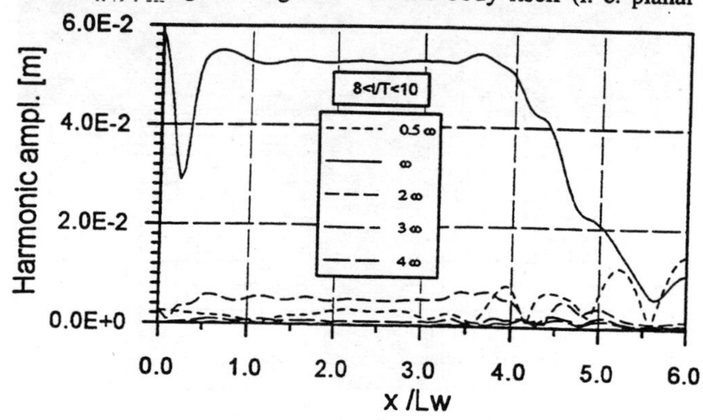

Fig. 6 : Harmonic components of the wave system in the time window 8 < t/T < 10 (wave steepness = 1/16).

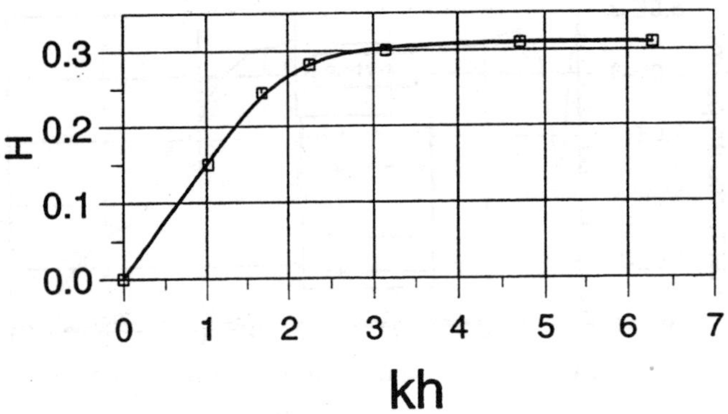

Fig. 7 : Calibration curve of the pneumatic wave-maker.

wavy motion); the Keulegan-Carpenter number (KC); the Reynolds number (Re). In an experimental work, which regards the interaction between water waves and a submerged circular cylinder (Chaplin, 1984), it was shown that for KC < 1 the contribution of liquid viscosity over C_M is very small, thus the inertial coefficient can be evaluated by the use of the linear potential theory. It was also observed that the increase of KC number produces the reduction of C_M. The matter was ascribed to non-linear phenomena (steady streaming) which arise inside the boundary layer. In the investigation of Ikeda et al. (1988), the reduction of C_M was ascribed to rotating separated vortices around the body. In a recent experimental work, Arai (1995) analysed the dynamic loads experienced by circular, square and rectangular horizontal cylinders in wavy flows. By the analysis of the dynamic loads and of the velocity field around the body, the author concluded that the reduction of the inertial forces can be ascribed to a rotating force in opposition of phase with the fluid acceleration. This force is basically due to the circulation of the velocity field around the body.

In the present work we numerically evaluated the wave loads over a square and two rectangular cylinders . In order to cut down wave reflections over the body, the length of the tank was chosen equal to 8 wave lengths, and 5 wave lengths were left downstream the body. The same experimental configuration as in (Arai, 1995) was re-produced. In particular, the wave period was chosen T=1.05 s, thus L_W = 1.7212 m. The water depth was h = 0.78 m and the centre of buoyancy of the bodies was put 0.093 m below the still water level. This matter made the effect of the Froude number over the induced loads negligible.

A square and a rectangular cylinder as in (Arai, 1995) were tested. In addition, the rectangular section, but rotated by an angle of 90°, was studied. The square section had breadth equal to 0.045 m, the first rectangular section had the same breadth as the previous one whereas the height was 0.024 m. The second rectangular section was 0.024 m breadth and 0.045 m height. The liquid kinematic viscosity was put equal to $1.0 \cdot 10^{-6}$ m^2/s.

In such investigations KC and Reynolds numbers are respectively defined as $KC = \frac{|\mathbf{u}|\,T}{D}$ and $Re = \frac{|\mathbf{u}|\,D}{\nu}$ being D a characteristic dimension of the body and $|\mathbf{u}|$ the module of velocity (as evaluated by means of the linear potential theory) at the centre of the body itself. In our cases we had $|\mathbf{u}| = 4.26\,\xi_a$ so that $KC = 99.4\,\xi_a$ and $Re = 1.917 \cdot 10^5\,\xi_a$. The frequency parameter $\beta = \frac{Re}{KC} = 1928.57$ is not dependent on the wave amplitude. Our simulations were carried out for wave steepness ranging from 1/191 up to 1/24 ($0.447 \le KC \le 3.58$ and $862 \le Re \le 6904$). Results obtained at $KC \ge 3.58$ were considered of poor quality, mainly due to wave reflections at the end of the wave tank. Due to the values of Re, turbulence was neglected.

In Fig. 8 a view near the body of the stretched grid used for the square cylinder is shown . In all the cases studied, the maximum and minimum aspect ratios of the cells ($\Delta x/\Delta y$) were 14 and 0.5 respectively. The computational grid was generated in such a way that the first velocity point was $1 \cdot 10^{-3}$ m far from the body surface. The maximum allowable time step was chosen equal to 0.005, and during the computation it could be reduced according to Courant number = 0.4. The simulations were carried out for about 20 wave periods (t = 21 s), using the CRAY J916/8-1024 supercomputer of the University of Trieste.

Hereafter we refer to the square section as "case 1", to the horizontal elongated rectangular section as "case 2" and to the vertical elongated section as "case 3".

Fig. 9 shows the time records of the shear and pressure forces computed at KC=0.447 for case 1. It is well evidenced that the shear force is very small compared to the other, so that drag is at a large extent due to viscous pressure. Fig. 10-12 show the polar diagram of the forces (F_y versus F_x) for the three cases, considering both small and moderate wave amplitudes. Fig. 10 (case 1) shows that the curve is practically circular when the wave steepness is very small and the problem is at a large extent linear and potential. The increase of the wave steepness produces a less regular polar diagram, basically due to viscous effects over the body. Fig 11 (case 2) shows the polar diagrams for the same wave amplitudes as in Fig. 10. The diagram is now elliptical, due to the elongation of the

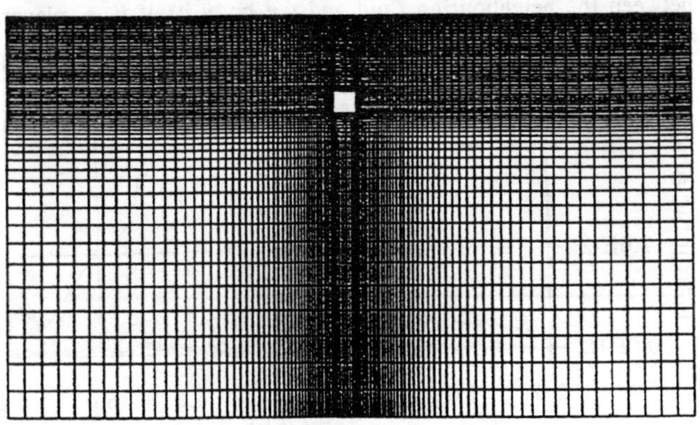

Fig. 8 : Sketch neighbouring the body of the computational grid used for the square cylinder.

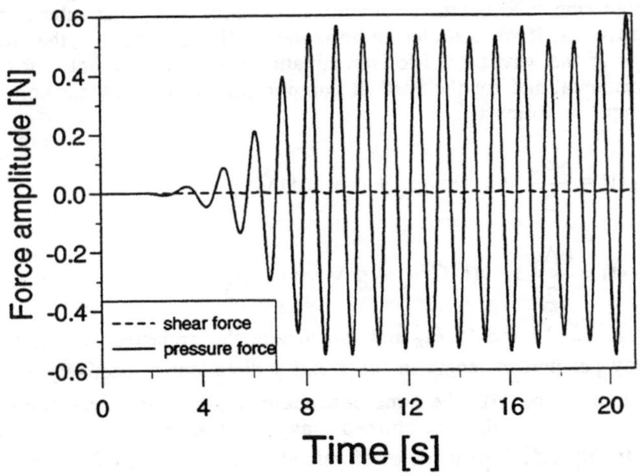

Fig. 9 : Time record of the shear and pressure forces for a small amplitude wave (wave steepness = 1/191; KC = 0.0447) in the case of the square section (case 1).

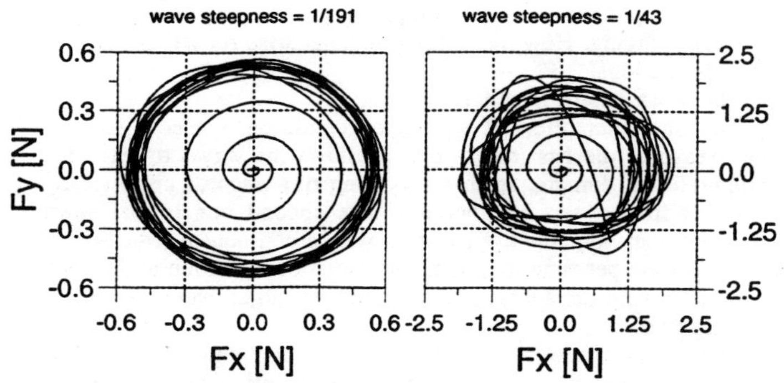

Fig. 10 : Rotating force (case 1) for two different wave amplitudes.

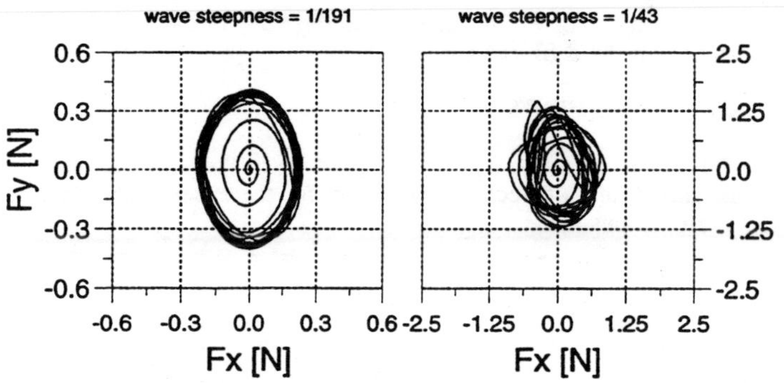

Fig. 11 : Rotating force (case 2) for two different wave amplitudes.

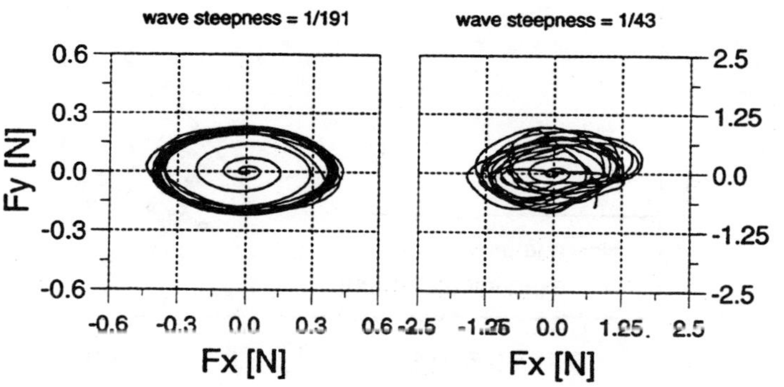

Fig. 12 : Rotating force (case 3) for two different wave amplitudes.

section, and irregularities are evidenced for the large wave steepness.

The comparison between Fig. 10 and Fig. 11 shows that the overall force which acts over the rectangular section is about 25 % lesser than that experienced by the square section. Fig. 12 (case 3) shows that, according to the change of geometry, the forces are now rotated by 90° if compared to those reported in Fig. 11.

Finally, the inertial and drag coefficients were calculated. The comparison between numerical and experimental results for case 1 and case 2 are in Fig. 13 - 16. Fig. 13 and Fig. 14 report the C_{Mx} and C_{My} values for case 1, whereas Fig. 15 and Fig. 16 contain C_{Mx} and C_{My} for case 2. The coefficients were obtained after averaging over 8 wave periods the maximum values of the forces. In addition, the variances were calculated in order to get information about the presence of unsteadiness in the time record. For comparisons, the experimental values obtained by Arai were also reported.

Small values of the variance are indicative of steady state conditions. Fig. 13 - 16 show that the results were weakly affected by wave reflection for KC < 3. When the KC number increases (i.e. large waves are generated), large transients are expected in the time records of the forces, due to vortex shedding from the body. In these flow conditions, the time records of the forces can be troubled by wave reflections before that a steady motion is reached. A sketch of the time record of the vertical force for case 2 at KC=3.58 (Fig. 17) shows the above described situation.

According to the experimental analysis, a reduction of the inertial coefficients with the increase of KC is observed . The minimum values of C_M are observed for values of KC neighbouring 3. After that the C_M values increase again. The square section experiences very similar values of the vertical and horizontal coefficients (according to the traces of the polar diagram of Fig. 10), whereas the horizontal elongated rectangular section (case 2) experiences C_{My} values larger than C_{Mx}. The agreement between

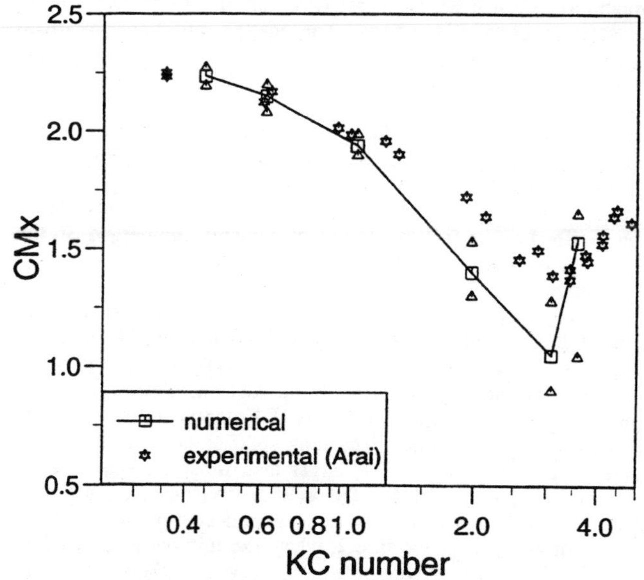

Fig. 13 : Square section (case 1) : numerical and experimental (Arai) inertial coefficients for the x-direction versus KC.

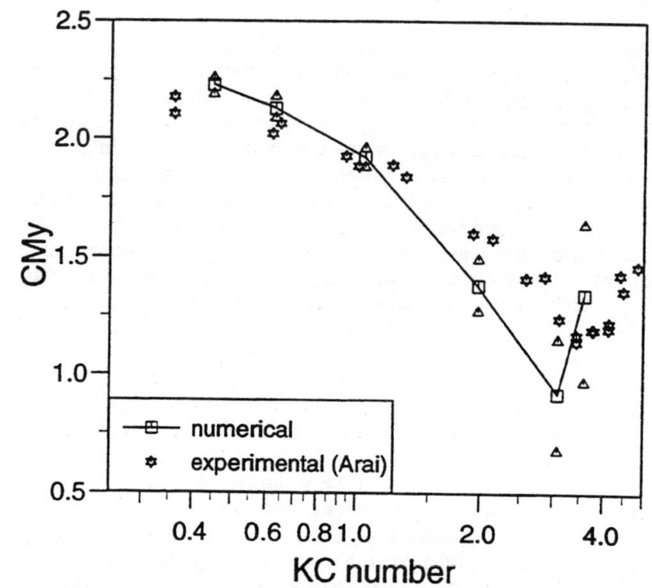

Fig. 14 : Square section (case 1) : numerical and experimental (Arai) inertial coefficients for the y-direction versus KC.

numerical and experimental results is in general satisfactory. Some disagreements are observed for KC larger than 2 due to the above mentioned wave reflection.

As shown in previous experimental observations, the decrease of the inertial coefficients when KC increases is ascribed to a rotating force due to fluid circulation around the body. This force is in opposition of phase with the inertial one. When KC is larger than 3, vortex shedding is observed from the body. This phenomenon is responsible for the break down of the circulation around the body, thus producing the increase of the C_M coefficients. Fig. 18 (case 1) is representative of such situation. When KC is very small (Fig. 18a and Fig. 18b) the liquid motion is at a large extent potential, apart some small re-circulation zones around the corners. The C_M coefficients are the same as in potential computations. When KC increases (Fig. 18c and 18d), due to liquid viscosity, a circulation rises. The matter produces a strong reduction of the inertial coefficients. Finally for KC ≥ 3 (Fig. 18e and Fig. 18f) vortex shedding is well evidenced in the velocity field.

The inertial coefficients for case 3 are reported in Fig. 19. The comparison with the coefficients evaluated for case 2 (Fig. 15 and Fig. 16) shows that the rotation of the body by 90° produces a corresponding rotation of the inertial forces. This circumstance is to be ascribed to the fact that the body is deeply submerged, so that the influence of the free surface (thus of the Froude number) is not appreciable.

The drag coefficients for case 1 are in Fig. 20, whereas those for the rectangular cylinders are in Fig. 21. The analysis of the results suggests that the relative importance of C_{Dx} over C_{Dy} is not dependent on the aspect ratio of the section. This conclusion, supported by the experimental results of Koterayama et al. (1995), needs further investigations. The drag coefficients derived for the square section are about 25 % larger than those obtained for the rectangular sections. As for the inertial coefficients, minimum values of the drag coefficients are experienced for values of KC neighbouring 3, except for case 2, where the minimum is found for KC = 2. The comparison of numerically calculated C_D with experimental values was not possible, because in Arai's experiments only the C_M values were evaluated.

CONCLUDING REMARKS

In this paper, the two-dimensional laminar NSE together with the non linear free surface boundary conditions were solved for the numerical evaluation of dynamic loads over submerged bodies at low KC numbers. NSE were solved using an improved MAC method (SIMAC) recently developed at the University of Trieste. In the first stage of the work a viscous numerical wave tank was implemented. A pneumatic wave-maker was considered at the left hand side of the tank, whereas a rigid wall was considered at the right hand side. This did not allow the waves to leave the computational domain since they were fully reflected. Grid sensitivity tests and checks on mass conservation showed a satisfactory behaviour of the algorithm.

In the second part of the work, the numerical evaluation of inertial and drag forces over submerged horizontal cylinders was carried out. In particular, some of the experiments of Arai were reproduced. According to previous experimental investigations it was observed that the increasing of KC from 0.5 to about 3.0 produces a large reduction of the inertial coefficients. The analysis of the velocity field nearby the body clearly showed that the reduction of C_M is strictly related to a rotating force in opposition of phase with the particles acceleration, force which is generated by the circulation of the liquid around the body. A further increasing of KC produces vortex shedding away from the body, and the break down of the circulation. As regards the drag coefficients, a behaviour similar to that obtained in experimental investigations carried out by other authors was obtained.

The comparison between numerical and experimental results is very encouraging. However some problems arise for KC ≥ 3 due to wave reflections.

The work is still in progress.

The implementation of a radiation condition is needed for two main reasons: the first one is that it allows the waves to leave the computational domain, so that very long time records up to steady state conditions can be obtained. The second one is that short computational domains can be used, with a consequent reduction of the CPU-time per time iteration. The radiation condition at the right wall will permit the analysis at KC numbers larger than those herein investigated.

The ACM multigrid technique, as implemented in the first version of SIMAC, will be used for the solution of the pressure equation. This improvement will cut down the computational efforts dramatically.

Finally, the work will be extended to the study of the wave-body interaction in the diffraction regime.

ACKNOWLEDGEMENTS

The author wishes to express his gratitude to Prof. S. Arai from Ashikaga Institute of Technology, Japan, for the experimental data and useful discussions.

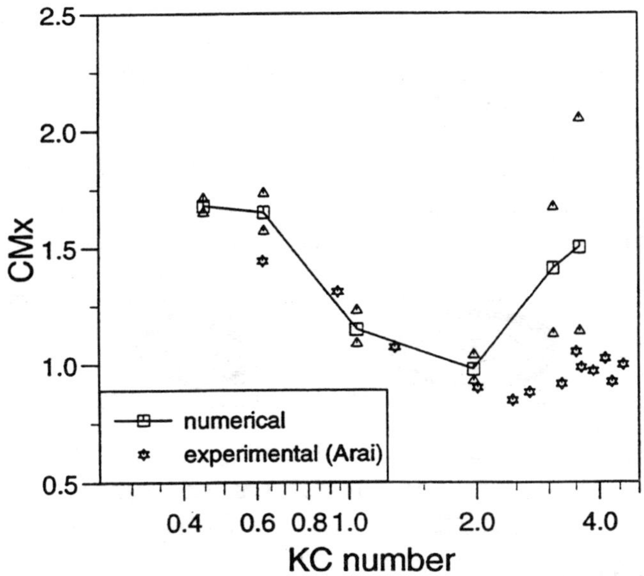

Fig. 15 : Horizontal elongated rectangular section (case 2) : numerical and experimental (Arai) inertial coefficients for the x-direction versus KC.

REFERENCES

Alessandrini, B., Delhommeau, G., 1994. Numerical calculation of three-dimensional viscous free surface flow around a serie 60 CB = 0.6 ship model. *Proc. Int. Workshop for Improvement of Hull Form Designs*, Vol. 1, pp. 95-104, Tokio.

Arai, M., Paul, U. K., Cheng L. Y., Inoue, Y., 1993. A technique for open boundary treatment in numerical wave tanks. *Jou. of the Soc. of Naval Architects of Japan*, Vol. 173, pp. 45-50.

Arai, S., 1995. Forces and circulation of horizontal cylinders submerged in regular waves. *Proc. of the Fifth International Offshore and Polar Engineering Conference, The Hague*, Vol. 3, pp. 348 - 355.

Armenio, V., 1997. An improved MAC method (SIMAC) for free surface unsteady high Reynolds flows. *Int. Jou. For Numerical Methods in Fluids,* Vol. 24, pp. 185-214.

Armenio, V., La Rocca, M., 1996. On the analysis of sloshing of water in rectangular containers: numerical study and experimental validation. *Ocean Engineering,* Vol. 23, pp. 705-739.

Chaplin, J. R., 1984. Nonlinear forces on a horizontal cylinder beneath waves. *Jou. Fluid Mechanics,* Vol. 147, pp. 449-464.

Clement, A., 1996. Coupling of two absorbing boundary conditions for 2D time-domain simulations of free surface gravity waves. *Jou. of Computational Physics,* Vol. 126, pp. 139-151.

Dean, R. G., Dalrymple, R. A., 1992. Water wave mechanics for engineers and scientists. Advanced series on Ocean Engineering, Vol. 2, World Scientific.

Faltinsen, O. M., 1990. Sea loads on ships and offshore structures. Ocean Technology Series, Cambridge University Press.

Gentaz, L., Alessandrini, B., Delhommeau, G., 1996. Motion simulation of a cylinder at the surface of a viscous fluid. *Schiffstechnik/Ship Technology Research,* Vol. 43, pp. 3-18.

Hino, H., Martinelli, L., Jameson, A., 1993. A finite-volume method with unstructured grid for free surface flow simulation. *Proc. 6th Int. Conf. on Numerical Ship Hydrodynamics.*

Kim, C. H., 1995. Recent progress in numerical wave tank research: a review. *Proc. of the Fifth International Offshore and Polar Engineering Conference, The Hague,* Vol. 3, pp. 1-9.

Koterayama, W., Hu, C., 1995. Wave forces on horizontal cylinders at low Keulegan-Carpenter and Reynolds numbers. *Proc. of the Fifth International Offshore and Polar Engineering Conference, The Hague,* Vol. 3, pp. 189 - 195.

Ikeda, Y., Otsuka, K., Tanaka, N., 1988. Wave forces acting on a horizontal submerged cylinder in regular waves at low KC numbers. *Jou. of the Society of Naval Architects of Japan,* Vol. 163, pp. 214 - 221.

Yeung, R. W., Ananthakrishnan, P., 1992a. Oscillation of a floating body in a viscous fluid. *Jou. of Engineering Mathematics,* Vol. 26, pp. 211-230.

Yeung, R. W., Vaidhyanathan, M., 1992b. Non-linear interaction of water waves with submerged obstacles.*Int. Jou. For Numerical Methods in Fluids,* Vol. 14, pp. 1111-1130.

Yeung, R. W., Yu, X., 1995a. Unsteady waves near a vertical circular cylinder in a viscous fluid. *Proc. 10th Int. Workshop on Water Waves and Floating Bodies,* Oxford.

Yeung, R. W., Yu, X., 1995b. Wave structure interaction in a viscous fluid. *Proc. 14th Int. Conference on Offshore Mechanics & Artic Engineering, OMAE, pp. 383-394,* Copenaghen.

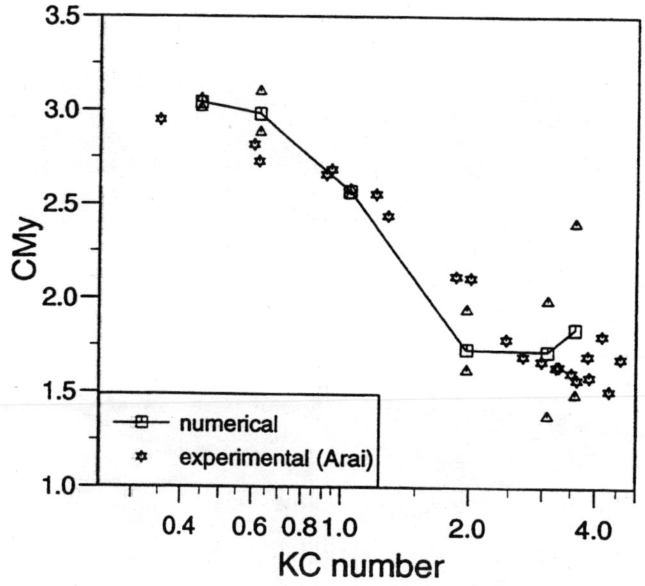

Fig. 16 : Horizontal elongated rectangular section (case 2) : numerical and experimental (Arai) inertial coefficients for the y-direction versus KC.

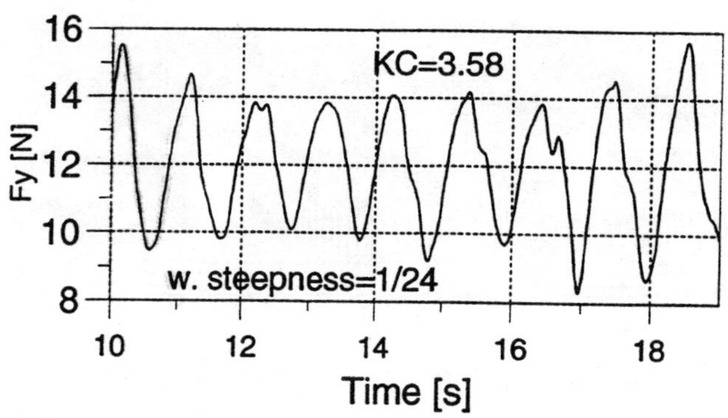

Fig. 17 : Horizontal elongated rectangular section (case 2) : time record of the vertical force for wave steepness = 1/24 (KC = 3.58).

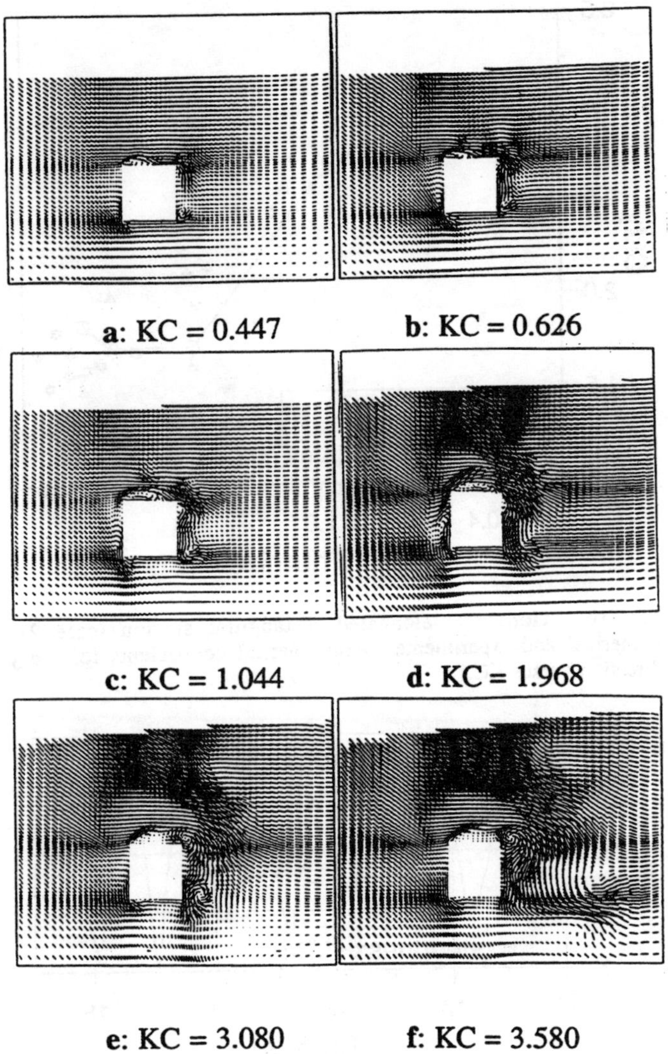

Fig. 18 : Square section: velocity field nearby the body at t=12.7 s, for 6 KC.

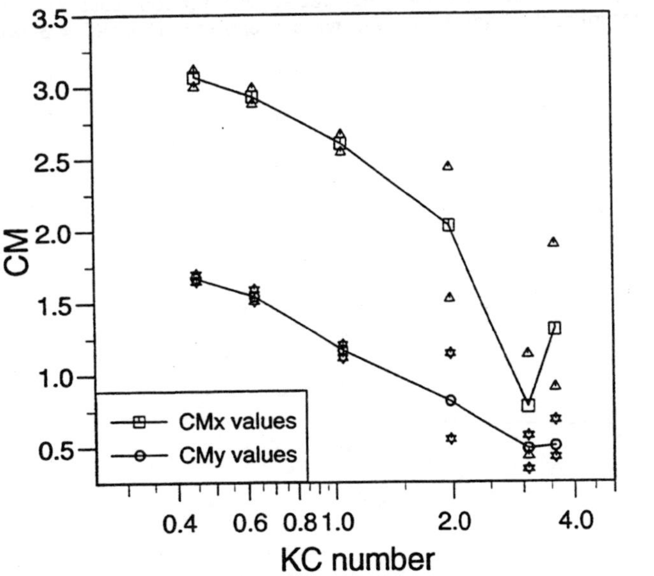

Fig. 19 : Vertical elongated rectangular cylinder: inertial coefficients versus KC.

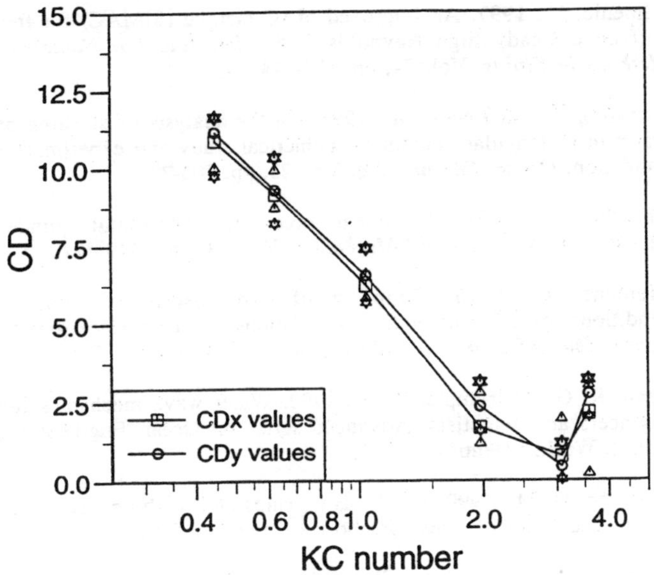

Fig. 20 : Square section: drag coefficients versus KC.

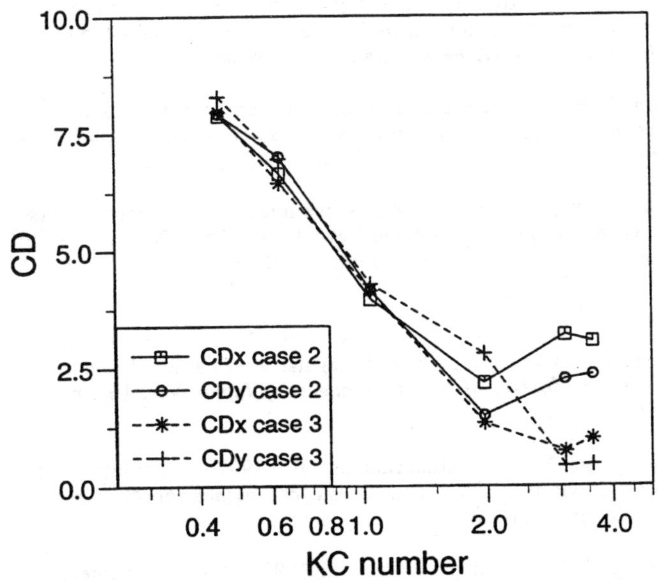

Fig. 21 : Rectangular cylinders : drag coefficients versus KC for cases 2 and 3.

Proceedings of the Seventh (1997) International Offshore and Polar Engineering Conference
Honolulu, USA, May 25-30, 1997

ISBN 1-880653-28-1 (Set); ISBN 1-880653-31-1 (Vol. III)

Wave-Current-Body Interaction by a Time-Domain High-Order Boundary Element Method

D.J. Kim and M.H. Kim
Texas A&M University
College Station, TX, USA

Abstract

The effects of uniform steady currents (or small forward velocity) on the interaction of a large three-dimensional body with waves are investigated by a time-domain higher order boundary element method (**THOBEM**). Using regular perturbation with two small parameters ϵ and δ associated with wave slope and current velocity, respectively, the boundary value problem is decomposed into the zeroth-order steady double-body-flow problem at $O(\delta)$ with a rigid-wall free-surface condition and the first-order unsteady wave problem with the modified free-surface and body-boundary conditions expanded up to $O(\epsilon\delta)$. Higher-order boundary integral equation methods are then used to solve the respective problems with the Rankine sources distributed over the entire boundary. The free surface is integrated at each time step. The Sommerfeld/Orlanski radiation condition is numerically implemented to absorb all the wave energy at the open boundary. Using the developed numerical method, wave forces, wave field and run-up, mean drift forces and wave drift damping are calculated.

Introduction

Reliable predictions of wave forces and run-up on large offshore structures in combined waves and currents are of practical importance in the design and operation of these structures in the actual sea conditions. Recently, several numerical methods to solve the wave-current-body interaction problem have been developed mostly in the frequency domain (e.g. Zhao et al, 1988; Wu & Eatock-Taylor, 1990; Emmerhoff & Sclavounus, 1992; Nossen et al, 1991; Teng & Eatock Taylor; 1995). The frequency-domain analysis is in general mathematically complicated and cannot easily be extended to fully nonlinear problems. Isaacson & Cheung (1992) and Prins (1995) recently employed alternative time-domain analyses to solve the same problem using constant panel methods. In addition to simplifying the mathematical treatment, the time-domain approach can also include transient responses.

In this paper, we developed a time-domain high-order boundary element method (**THOBEM**) to solve the interaction of a three-dimensional body with regular waves and uniform steady currents by a perturbation method. The entire boundary surface was discretized into a set of curved surface elements and higher-order interpolation functions were used to describe the velocity potential on each element. The use of the higher-order boundary element method (**HOBEM**) is thought to have certain benefit compared to conventional constant panel methods (CPM). **HOBEM** is particularly advantageous for the calculation of wave run-up and second-order forces, where the velocity potential and its derivatives are needed exactly on the intersection of free surface and body surface (Chau, 1989; Liu et al, 1995). In addition, using **HOBEM**, the accurate computation of the second-order spatial derivatives of the velocity potential on the integral surface is possible, which precludes any unnecessary modification of the so-called m_j terms (e.g. Zhao & Faltinsen, 1990).

In the present paper, the effects of uniform steady currents (or small forward velocity) on wave-frequency and mean drift forces, and run-up on large three-dimensional structures are investigated. From these results, wave drift damping can also be calculated. If current speed (or forward velocity) is small, the effects of flow separation are unimportant and the problem can be formulated within the framework of potential theory. Short ship-wave systems are disregarded in the current approach. As an indication of the applicability of the potential theory, it may be assumed that flow separation effects need not be considered when the relevant Keulegan-Carpenter (KC) number is less than about 3 (Isaacson & Cheung, 1992). Zhao et al. (1988) experimentally proved that flow separation around a hemisphere did not occur when KC<1 and current speed is less than maximum wave-induced velocity. Under such condition, potential theory reasonably matched with experimental results.

In our perturbation approach, both wave slope parameter ϵ and current speed parameter δ are assumed to be small so that the original boundary value problem can be divided into the zeroth-order time-independent problem at $O(\delta)$ and the first-order unsteady problem including $O(\epsilon)$ and $O(\epsilon\delta)$ (e.g. Isaacson & Cheung, 1992). To solve these problems, Rankine sources are distributed over the entire boundary surface. The resulting boundary integral equation is numerically solved by a higher-order boundary element method in the time domain. At the open boundary, the Sommerfeld/Orlanski radiation condition(Orlanski, 1976) was numerically applied.

Mathematical Formulation

We consider the interaction of a large three dimensional body with regular waves and uniform currents. It is assumed that regular incoming waves propagate to the positive x-direction and uniform currents flow in the opposite direction with velocity U. The current may also be regarded as the forward velocity of a floating body in the positive x direction. The relevant boundary value problem can be constructed using two Cartesian coordinate systems; the space fixed coordinates $\vec{x}_o = (x_o, y_o, z_o)$, and the body fixed coordinates moving with the mean forward speed $\vec{x} = (x, y, z)$, as can be seen in Figure 1. The z-axis is positive upward, and the origin of the coordinate system is located on the undisturbed mean free surface.

Assuming ideal fluid and neglecting surface tension, the fluid velocity vector $\vec{V}(\vec{x}, t)$ can be defined by $\nabla\Phi$, with the velocity potential $\Phi(x, y, z, t)$ satisfying the following Laplace equation and boundary conditions:

$$\nabla^2\Phi = 0, \quad \text{in the fluid domain} \tag{1}$$

$$\frac{\partial\Phi}{\partial t} - U\frac{\partial\Phi}{\partial x} + \frac{1}{2}\nabla\Phi\cdot\nabla\Phi + g\eta = 0, \quad \text{on the free surface} \tag{2}$$

$$\frac{\partial\eta}{\partial t} - U\frac{\partial\eta}{\partial x} + \nabla\Phi\cdot\nabla\eta = \frac{\partial\Phi}{\partial z}, \quad \text{on the free surface} \tag{3}$$

$$\vec{V}_s\cdot\vec{n} - \frac{\partial\Phi}{\partial n} = 0, \quad \text{on the body} \tag{4}$$

$$\vec{n}\cdot\nabla\Phi = 0, \quad \text{on the bottom} \tag{5}$$

where g =gravitational acceleration and $\vec{V}_s\cdot\vec{n}$ =normal body velocity . The first and second conditions on the free surface $\eta(x, y, t)$ are fully non-linear dynamic and kinematic free surface conditions, respectively. The symbol $\vec{n}$ represents the unit normal vector pointing out of the fluid domain.

To make the solution of the above boundary value problem unique, we have to impose an extra condition, called the radiation condition at far field. In this paper, the Sommerfeld/Orlanski method (Orlanski, 1976) is used as a radiation condition:

$$\frac{\partial Q}{\partial t} + C(t)\frac{\partial Q}{\partial n} = 0, \text{on the open boundary} \tag{6}$$

where Q is either scattered or radiated velocity potentials and $C(t)$ is the time-dependent phase velocity.

Solving this fully non-linear problem directly in the time domain requires tremendous computing time since the influence matrix should be set up and inverted at each time step as the free surface moves to a new position. Assuming small wave amplitudes and body motions, a perturbation procedure may be introduced to make this problem more manageable. Then, the calculation of influence matrix coefficients is required only once and it can be used repeatedly for all time steps and different wave conditions, which makes the current approach efficient and computationally more attractive.

By introducing two independent small parameters i.e. the wave steepness parameter $\epsilon = H/\lambda$ (wave height/wave length) and the current speed parameter(or Froude number F_n) $\delta = U/\sqrt{gl}$, where l=characteristic body length, the velocity potential and wave elevation can be expressed by the following perturbation series:

$$\Phi(\vec{x}, t) = \underbrace{\bar{\phi}(\vec{x})}_{O(\delta)} + \underbrace{\phi^{(1)}(\vec{x}, t)}_{O(\epsilon,\epsilon\delta)} + \underbrace{\phi^{(2)}(\vec{x}, t)}_{O(\epsilon^2,\ldots)} + \cdots, \tag{7}$$

and

$$\eta(\vec{x}, t) = \underbrace{\bar{\eta}(\vec{x})}_{O(\delta^2)} + \underbrace{\eta^{(1)}(\vec{x}, t)}_{O(\epsilon,\epsilon\delta)} + \underbrace{\eta^{(2)}(\vec{x}, t)}_{O(\epsilon^2,\ldots)} + \cdots, \tag{8}$$

where $\bar{\phi}$ is the zeroth-order steady disturbance potential and $\phi^{(1)}$ and $\phi^{(2)}$ are the time-dependent first-order and second-order potentials. The steady ship wave system $\bar{\eta}$ at $O(\delta^2)$ is neglected, as pointed out earlier.

Using Taylor expansion about the mean free and body surfaces and substituting eq.(7-8) into the boundary value problem (1)-(6), a series of new boundary value problems at each order can be obtained. As mentioned before, the terms higher than $O(\epsilon, \epsilon\delta)$ and $O(\delta)$ are neglected under the assumption that both parameters are small. The second-order wave-body interaction problem occurring at $O(\epsilon^2)$, for example, can be studied independently, as described in detail in Kim & Yue (1989).

The boundary value problem for the steady disturbance potential $\bar{\phi}$ can be obtained by collecting terms of order $O(\delta)$. All these terms are time-independent and linear functions of the current velocity U. Since the wave field associated with $\bar{\eta}$ has the order of magnitude $O(\delta^2)$, the free surface boundary can be considered as wall boundary on $z = 0$, which is frequently called double-body flow problem. Then, the disturbance potential $\bar{\phi}$ satisfies the Laplace equation, no-flux body and bottom conditions, and the free-surface-wall boundary condition. The resulting integral equation can be solved by the integral equation method (Kim, 1996).

The boundary value problem for the first-order unsteady wave potential $\phi^{(1)}$ including forward speed effects may be established by following the aforementioned perturbation procedure and collecting terms of order $O(\epsilon)$ and $O(\epsilon\delta)$. Then, the kinematic and dynamic free-surface conditions are given on $z = 0$ by

$$\frac{\partial\phi_i^{(1)}}{\partial z} - \frac{\partial\eta_i^{(1)}}{\partial t} + U\frac{\partial\eta_i^{(1)}}{\partial x} = \frac{\partial\bar{\phi}}{\partial x}\frac{\partial\eta_i^{(1)}}{\partial x} + \frac{\partial\bar{\phi}}{\partial y}\frac{\partial\eta_i^{(1)}}{\partial y} + \eta_i^{(1)}\frac{\partial^2\bar{\phi}}{\partial z^2} \quad on \quad S_f, \tag{9}$$

$$\frac{\partial \phi_i^{(1)}}{\partial t} + g\eta_i^{(1)} - U\frac{\partial \phi_i^{(1)}}{\partial x}$$
$$= -\frac{\partial \bar{\phi}}{\partial x}\frac{\partial \phi_i^{(1)}}{\partial x} - \frac{\partial \bar{\phi}}{\partial y}\frac{\partial \phi_i^{(1)}}{\partial y} \quad on \quad S_f. \tag{10}$$

If there is no current (or no forward velocity), the terms associated with U and the double-body potential $\bar{\phi}$ should vanish and the above equations are reduced to classical linear free-surface boundary conditions. The boundary value problem for the velocity potential $\phi^{(1)}$ can be decomposed into diffraction and radiation problems. For convenience, we denote incident and diffraction potentials by $\phi_0^{(1)}$ and $\phi_7^{(1)}$, and the radiation potential by $\phi_i^{(1)}$, where $i = 1 - 6$ represents each mode of six degree-of-freedom motions.

The regular incident wave can be written with respect to the moving reference frame using the encounter frequency ω given by

$$\omega = |\omega_o - Uk\cos\beta|, \tag{11}$$

where k is the wave number i.e. $\lambda = 2\pi/k$, β=wave heading with respect to the x axis, and ω_o is the original circular frequency of the incident wave. Then, the incident wave potential and its free surface elevation can be written in the form

$$\phi_0^{(1)} = \frac{gA}{\omega_o}\frac{\cosh k(z+h)}{\cosh kh}\sin\left[(kx\cos\beta + ky\sin\beta - \omega t)\right], \tag{12}$$

where A =wave amplitude, and h =water depth.

The body boundary condition can also be formulated using the same perturbation procedure. If the unsteady oscillatory motion of the body is small, the local displacement can be written in the form

$$\vec{\alpha} = \vec{X} + \vec{\Omega} \times \vec{r}, \tag{13}$$

where $\vec{r}$ denotes the position vector of the wetted body surface, and $\vec{X}$ and $\vec{\Omega}$ are the unsteady translational and rotational displacements of the body. For the unsteady potential, $\phi^{(1)}$, Timman & Newman (1962) derived the body boundary condition to account in a consistent manner for the interaction between the steady and oscillatory flow fields. This boundary condition including the so-called m_j term (the second right-hand side term) can be written in the form

$$\frac{\partial \phi^{(1)}}{\partial n} = \frac{\partial \vec{\alpha}}{\partial t}\cdot\vec{n} + \left[\left(\vec{W}\cdot\nabla\right)\vec{\alpha} - (\vec{\alpha}\cdot\nabla)\vec{W}\right]\cdot\vec{n} \quad on \quad S_b, \tag{14}$$

where

$$\vec{W} = \nabla\left(\bar{\phi} - Ux\right). \tag{15}$$

For the diffraction problem, the right hand side of eq.(14) becomes zero since there is no body motion . Therefore, the body boundary condition for the diffraction problem is simply given by

$$\frac{\partial \phi_0^{(1)}}{\partial n} + \frac{\partial \phi_7^{(1)}}{\partial n} = 0 \quad on \quad S_b. \tag{16}$$

Finally, the open (or radiation) boundary condition for the unsteady potential is given by the same Sommerfeld/Orlanski condition as (6).

Having obtained respective velocity potentials in the preceding sections, we are able to calculate the forces and moments acting on the body by waves and currents. The resulting pressure on the wetted body surface is given by the following unsteady Bernoulli equation

$$p = -\rho\left(\frac{\partial \Phi}{\partial t} - U\frac{\partial \Phi}{\partial x} + \frac{1}{2}\nabla\Phi\cdot\nabla\Phi + gz\right), \tag{17}$$

where ρ =fluid density. The forces and moments with respect to the center of gravity can then be calculated from the integration of the pressure over the instantaneous wetted body surface. After carrying out the integration and collecting respective terms, we obtain the first-order forces and moments in terms of $\bar{\phi}$ and $\phi^{(1)}$ as follows:

$$\vec{F}^{(1)} = -\rho\int_{S_b}\left(\frac{\partial \phi^{(1)}}{\partial t} - U\frac{\partial \phi^{(1)}}{\partial x} + \nabla\phi^{(1)}\cdot\nabla\bar{\phi}\right)\vec{n}\,dS - \vec{F}_{HS}^{(1)} \tag{18}$$

$$\vec{M}^{(1)} = -\rho\int_{S_b}\left(\frac{\partial \phi^{(1)}}{\partial t} - U\frac{\partial \phi^{(1)}}{\partial x} + \nabla\phi^{(1)}\cdot\nabla\bar{\phi}\right)(\vec{r}\times\vec{n})\,dS - \vec{M}_{HS}^{(1)} \tag{19}$$

In the above equations, $\vec{F}_{HS}^{(1)}$ and $\vec{M}_{HS}^{(1)}$ are hydrostatic restoring forces and the expressions are given, for example, in Liu et al. (1995).

The mean drift forces can be obtained from the time-average of the second-order forces:

$$\left\langle\vec{F}\right\rangle^{(2)} = \langle -\rho\int_{S_b}(\vec{\alpha}\cdot\nabla)\left(\frac{\partial \phi^{(1)}}{\partial t} - U\frac{\partial \phi^{(1)}}{\partial x} + \nabla\phi^{(1)}\cdot\nabla\bar{\phi}\right)\vec{n}dS + \frac{1}{2}\rho g\int_{W_o}\left(\eta^{(1)} - \alpha_3\right)^2\vec{N}dW + \vec{\Omega}\times\vec{F}^{(1)} - \frac{1}{2}\rho\int_{S_b}\left(\nabla\phi^{(1)}\cdot\nabla\phi^{(1)}\right)\vec{n}dS - \vec{F}_{HS}^{(2)}\rangle \tag{20}$$

where $\vec{\alpha} = (\alpha_1, \alpha_2, \alpha_3, \alpha_4, \alpha_5, \alpha_6)$ is the first-order body displacement, as defined in eq.(13), $\vec{N} = \vec{n}/\sqrt{1 - n_3^2}$, and W_o is the mean water line contour of a body. $\vec{F}_{HS}^{(2)}$ is the second-order hydrostatic-pressure-force contribution (Liu et al., 1995). If a body is restrained, all the motion related terms vanish.

Numerical Method

A boundary integral equation based on the distribution of simple sources is formulated for the numerical solution. Using basic singularities over the entire boundary surface Γ, the method can be used for any bottom topography or lateral boundaries. The boundary integral equation for the velocity potential is given by

$$4\pi c(p)\phi(p) + \int_\Gamma \frac{\partial G(p,q)}{\partial n(q)}\phi(q)\,d\Gamma = \int_\Gamma G(p,q)\frac{\partial \phi(q)}{\partial n(q)}\,d\Gamma, \tag{21}$$

where G is the Rankine Green function, $c(p)$ is the solid angle, $q = (x', y', z')$ is the source point, and $p = (x, y, z)$ is the field point. For convenience, it is assumed that the water depth is constant and the body is symmetric with respect to xz plane. Then we can include in G the image sources with respect to the bottom and the symmetry plane, and the integration surfaces can be reduced to the body surface, free surface, and open boundary surface of half fluid domain.

On the free surface and open boundary, the velocity potential is to be given, while on the body boundary, the normal velocity is known, so the boundary integral equation (21) can be rewritten in the form:

$$
\begin{aligned}
&4\pi c(p)\phi(p) - \\
&\int_{S_f \cup S_r} G(p,q)\frac{\partial \phi(q)}{\partial n(q)}\, d\Gamma + \int_{S_b} \frac{\partial G(p,q)}{\partial n(q)}\phi(q)\, d\Gamma \\
&= \int_{S_b} G(p,q)\frac{\partial \phi(q)}{\partial n(q)}\, d\Gamma - \int_{S_f \cup S_r} \frac{\partial G(p,q)}{\partial n(q)}\phi(q)\, d\Gamma \quad (22)
\end{aligned}
$$

To solve the above integral equation in an accurate and efficient manner, a higher-order boundary element method was developed. The body geometry and other surfaces are discretized by quadratic isoparametric elements and the physical variables on each element are described by quadratic interpolation (or shape) functions. The spatial derivatives of the velocity potential was then obtained from the differentiation of the shape functions using Jacobian transformation (Kim, 1996). To resolve the so-called Dirichlet-Dirichlet corner problem, discontinuous elements are used at the intersection of free-surface and radiation boundaries. The details can be found in Kim (1996).

The algebraic equation for the unknowns ϕ and ϕ_n can simply be written in the form

$$[A]\{X\} = [B]\{Y\}. \quad (23)$$

The matrices on both sides are only geometry dependent and invariant in time since the calculation domain is fixed. Therefore, the inversion of the left-hand side matrix is required only once so that it can be repeatedly used at each time step. In a fully non-linear time domain simulation, the influence matrix must be set up and solved at every time step because the computational domain continuously changes.

On the body surface, the normal derivative of the potential is explicitly known from the body boundary condition, while on the open boundary surface and still water surface, the potential needs to be calculated at each time step using the respective boundary conditions. For the time-integration of the free surface, Adams-Bashforth and Adams-Moulton methods are used. First, wave elevation was calculated using the fourth-order Adams-Bashforth method applied to the kinematic free-surface condition. The corresponding potential was obtained using the fourth-order Adams-Moulton method applied to the dynamic free-surface condition.

At the open boundary, potential values are calculated using Sommerfeld/Orlanski condition. For this, the time-dependent phase velocity $C(t)$ should first be calculated before the integration is performed. In this paper, instead of employing numerical differentiation for ϕ_t, we used the dynamic free-surface boundary condition as follows:

$$
\begin{aligned}
C(t) &= \left(-\frac{\partial \phi^{(1)}}{\partial t} / \frac{\partial \phi^{(1)}}{\partial n} \right) \\
&= \left(g\eta^{(1)} - U\frac{\partial \phi^{(1)}}{\partial x} + \frac{\partial \bar{\phi}}{\partial x}\frac{\partial \phi^{(1)}}{\partial x} + \frac{\partial \bar{\phi}}{\partial y}\frac{\partial \phi^{(1)}}{\partial y} \right) \\
&\quad / \frac{\partial \phi^{(1)}}{\partial n}. \quad (24)
\end{aligned}
$$

The resulting phase velocity was used along the vertical open-boundary surface. The time-dependent phase velocity calculation could be failed when $\phi_n \to 0$. To avoid this and make the scheme more robust, a numerical scanner is implemented. This scanner computed $C(t)$ at several points close to the open boundary and then they are averaged. In this procedure, the points where $\phi_n \to 0$ were excluded. The averaged phase velocity $\bar{C}(t)$ was then used to update the velocity potential at the open boundary. A similar numerical scanner was also used in the two-dimensional wave simulation by Jagannathan (1988), in which he used numerical differentiation for ϕ_n; whereas we obtained ϕ_n directly from the field solution, which makes the present phase velocity calculation more efficient. Through extensive numerical tests, it is proved that the present numerical open boundary condition works well even for very long time simulation. To avoid an abrupt initial condition, a cosine ramp (or modulation) function was used over two wave periods both in diffraction and radiation problems.

Results and Discussions

For illustration, a series of numerical simulations were performed for the diffraction of a bottom-mounted vertical cylinder and a square barge in uniform coplanar or adverse currents. Figures 2 and 3 show the computational domains and discretizations used in the present study. Each quadratic element has nine nodal points. The convergence test was carried out and the first-order force converged more rapidly than the second-order mean force.

To test the performance of the present open boundary condition, a numerical wave-tank (20m-10m-3m) simulation was conducted. Figures 4a,4b, and 4c show the free-surface-profile time series at three different locations (upstream, midstream, and downstream) and a three-dimensional snap-shot can be seen in Figure 4d. For this simulation, 60 quadratic elements (322 nodal points) in half fluid domain, wavelength=12.57m, and time step=1/80 of wave period were used. The linear wave was continuously supplied at the upstream boundary. It can be observed that there is no reflection from the downstream open boundary. To study the diffraction of a three-dimensional body in the open sea, the open boundary condition was implemented at all sides of the truncation boundaries. Then, the body-boundary condition was gradually imposed to generate scattered outgoing waves. Figure 5 shows the time history of the wave force on a bottom mounted vertical cylinder ($a = h$) for three different Froude numbers ($F_n = -0.1, 0, 0.1$). These results clearly show that the Orlanski open boundary condition performs well even when weak currents exist. The total wave energy inside the computational domain was also checked and it was satisfactorily conserved for all different wave and current conditions when the present Orlanski condition was used.

We next calculated the wave run-up, wave exciting forces, and the second-order mean drift forces on the same bottom-mounted vertical cylinder for three different current speeds. Figure 6a shows the wave run-up around the vertical cylinder for $ka = 0.5$. The positive and negative Froude numbers correspond to the adverse and coplanar currents, respectively. The wave run-up is increased near $\theta = 0$ and π and decreased near $\theta = \pi/2$ in the presence of coplanar currents. The opposite trend holds true for adverse currents. Figure 6b shows the wave run-up at the upwave point ($\theta = \pi$) as function of ka for three different current speeds. When the incident wave and the current propagate in the same direction, the run-up at the upwave point is significantly increased for all wave frequencies considered. On the other hand, the run-up is decreased by about the same amount in adverse currents.

The wave exciting forces and mean drift forces with and without currents are presented in Figures 7 and 8. The correctness of the no-current result was checked against the analytic solutions given in Kim & Yue (1989). The wave-frequency force is slightly increased or decreased due to currents depending on pertinent incident wave frequencies. However, the mean drift force is always increased in coplanar currents and always decreased in adverse currents regardless of incident wave frequencies. The modification of mean drift forces by currents is more pronounced than that of wave-frequency forces. The wave drift damping was subsequently obtained using 3-point central differentiation with the computed drift-force results for $F_n = 0$ and ± 0.04. The present result shows good agreement with Grue & Biberg's (1993) frequency-domain computation, as shown in Figure 9.

Figures 10, 11, and 12 show three-dimensional snap-shots and contour plots of diffracted wave fields around a square barge for $ka = 1$ and $kD = 0.444$, where a is the half length of each side and D =draft. To observe the variation of free surface near the barge more closely, the wave run-up is plotted in Figure 13 along the circumference of the barge starting from the lee-side center point to the weather-side center point. The maximum run-up occurs at the weather side and it is increased in coplanar currents and decreased in adverse currents.

Concluding Remarks

A time-domain higher-order boundary element method has been developed based on the potential theory and perturbation procedure to study wave-current interactions with a large three dimensional body of arbitrary shape. To solve the boundary value problems for the zeroth-order double-body potential and the first-order unsteady potential, Rankine sources were distributed over the entire boundary surface and the resulting matrix equations were solved repeatedly after updating the boundary values at each time step. A special Orlanski open boundary condition has been devised and turned out to perform well. Discontinuous high-order elements were used to circumvent the Dirichlet-Dirichlet corner problem. The present numerical method was validated through convergence tests as well as comparison with published data or analytic solutions if available. The developed computer program was then used to investigate the change of wave field, wave-frequency exciting forces, and mean drift forces in uniform coplanar or adverse currents. It is found that the current effects are in general more significant on the wave run-up and second-order mean drift forces than on the first-order wave exciting forces.

The THOBEM developed in this paper can be straightforwardly extended to fully non-linear free-surface problems. The computer program is directly applicable to various wave-current-body interaction problems including arbitrary bottom and lateral boundaries.

References

Chau, F.P. (1989). "The second-order velocity potential for diffraction of waves by fixed offshore structures" *Ph.D. Thesis, Univ. of College, London*

Emmerhoff,O.J. and Sclavounos,P.D.,(1992)."The slow drift motion of arrays of vertical cylinders". *Journal of Fluid Mechanics*, Vol.242, 31-50.

Grue, J. and Biberg, D. (1993). "Wave forces on marine structures with small speed in water of restricted depth" *J. of Applied Ocean Research*, Vol. 15, 121-135.

Isaacson,M. and Cheung,K.F.,(1992)."Wave-current interaction with a large structure". *Proc. of International Conference on Civil Eng. in Oceans*, ASCE, College Station,Texas, Nov

Jagannathan,S.,(1988)."Non-linear free surface flows and an application of the Orlanski boundary condition". *International Journal for Numerical Methods in Fluids*, Vol.8, 1051- 1070.

Kim, M.H. and Yue, D.K.P.,(1989)."The complete second-order diffraction solution for an axisymetric body. Part 1. Monochromatic incident waves". *Journal of fluid Mechanics*, Vol.200, 235-264.

Kim, D.J.,(1990)."Interaction of a large 3D body with waves and currents by THOBEM". *Ph.D. Thesis, Texas A&M Univ.*

Liu, Y.H., Kim, M.H., and Kim, C.H. (1995). "The computation of second-order mean and double-frequency wave loads on compliant TLP by HOBEM" *Journal of Offshore and Polar Engineering*, Vol. 5, No. 2, 111-119.

Nossen,J. ,Grue,J. and Palm,E.(1991)."Wave forces on three dimensional floating bodies with small forward speed". *Journal of Fluid Mechanics*, Vol.227, 135-160.

Orlanski,J.E.(1976)."A simple boundary condition for unbounded hyperbolic flows". *Journal of Computational Physics*, Vol.21, 251-269.

Prins, H.J.,(1995)."Time-domain calculations of drift forces and moments". *Ph.D Dissertation*,T.U.Delft.

Teng,B. and Eatock Taylor,R.(1994)."Application of a higher order boundary elemnet method in the calculation of wave run-up in a weak current". *Proc. of the Fourth International Offshore and Polar Eng. Conference*, Osaka, Japan, April.

Timman, R. and Newman, J.N. (1962). "The coupled damping coefficients of symmetric ships" *Journal of Ship Research,* Vol. 5, No.4, 34-55

Wu, G.X. and Eatock-Taylor, R. (1990). "The hydrodynamic force on an oscillating ship with low forward speed" *Journal of fluid Mechanics,* Vol.211, 333-353

Zhao, R., Faltinsen, O., Krokstad, J.R., and Aanesland, V. (1988). "Wave current interaction effects on large volume structures" *Proc. Behavior of Offshore Structures, BOSS'88,* Vol. 2, 623-638.

Zhao, R., Faltinsen, O. (1990) "A discussion of the m_j terms in the wave-current-body interaction problem" *Proceedings of the 5th International Workshop on Water Waves and Floating Bodies.*

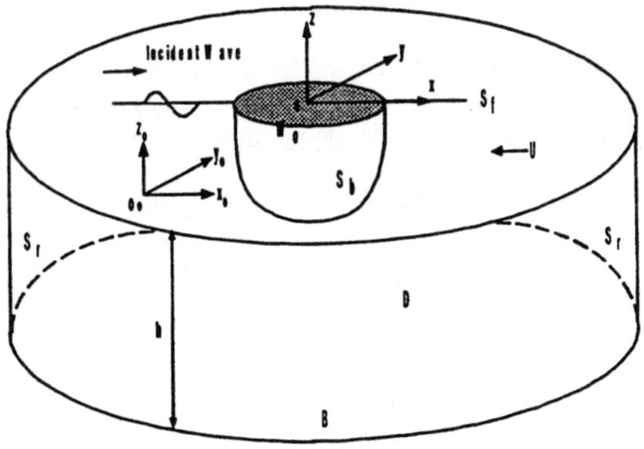

Figure 1: Coordinate system and definition of boundary surface

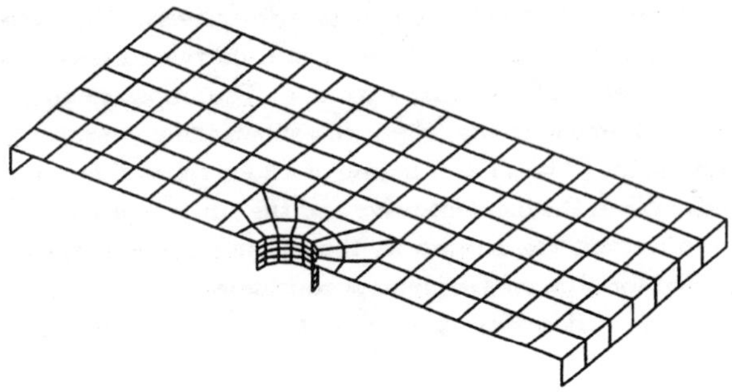

Figure 2: Grid system for a bottom-mounted vertical cylinder($h/a = 1$)

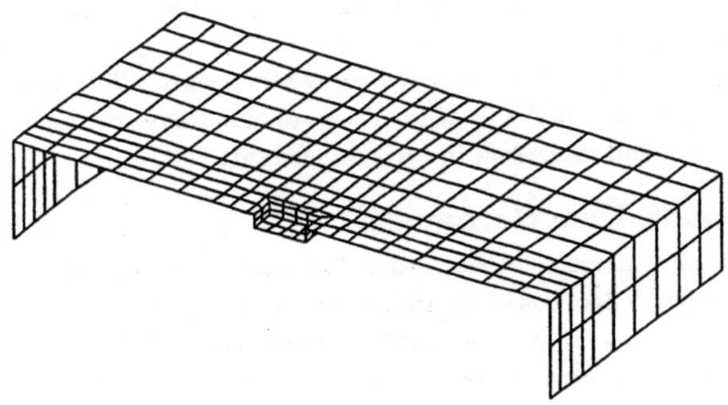

Figure 3: Grid system for a rectangular barge ($LxBxD =$ 90x90x20$m, h/a = 3.333$)

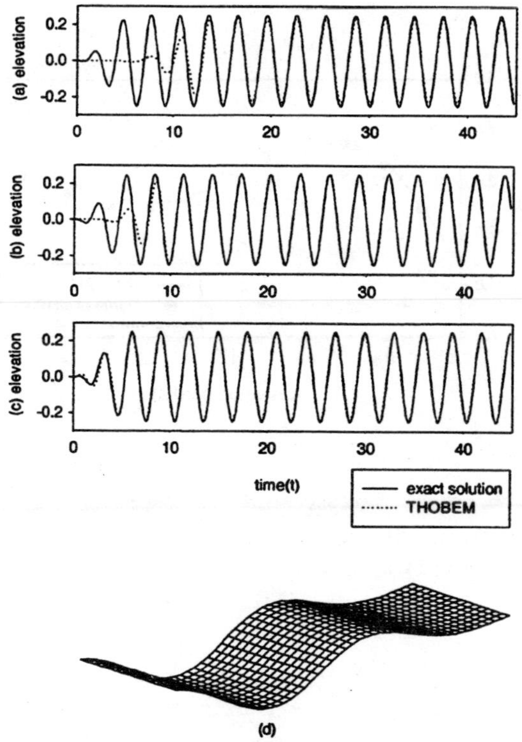

Figure 4: Time history of wave elevations for a single wave($k = 0.5$); (a)at downstream (b) midstream (c) upstream (d) snapshot at $t = 3T$

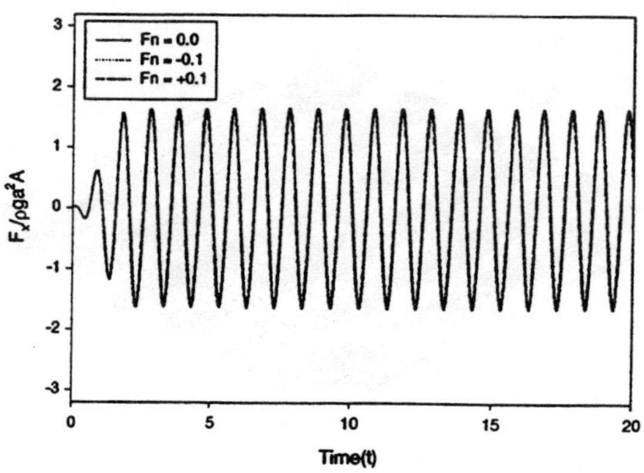

Figure 5: Time history of first-order forces for a bottom-mounted vertical cylinder($h/a = 1$, $ka = 1$)

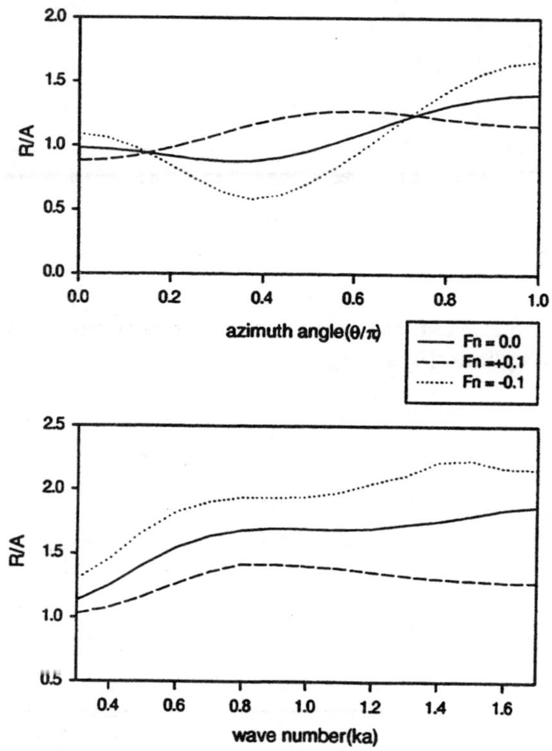

Figure 6: Maximum wave run-up around bottom-mounted vertical cylinder($h/a = 1$); (a) as function of azimuth angle($ka = 0.5$) (b) run-up at upwave point($\theta = \pi$) as function of wave number

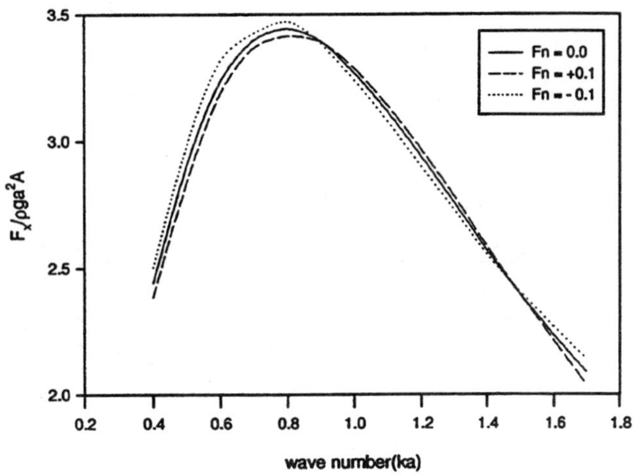

Figure 7: First-order wave forces on a bottom-mounted vertical cylinder($h/a = 1$)

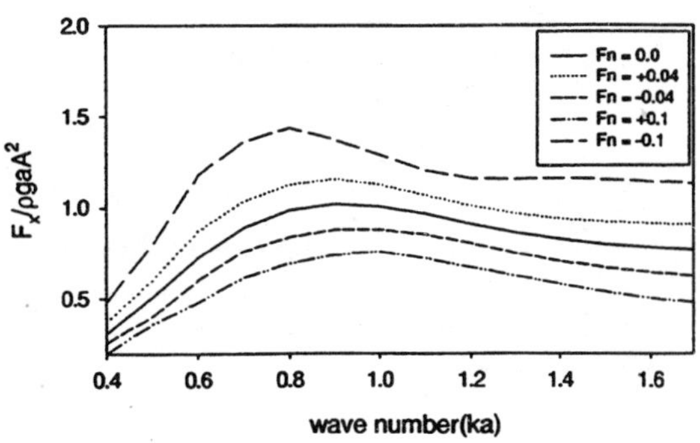

Figure 8: Mean drift forces on a bottom-mounted vertical cylinder($h/a = 1$)

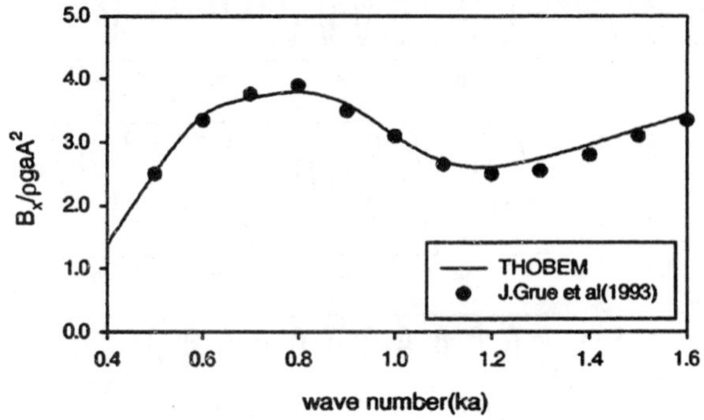

Figure 9: Wave drift damping on a bottom-mounted vertical cylinder($h/a = 1$)

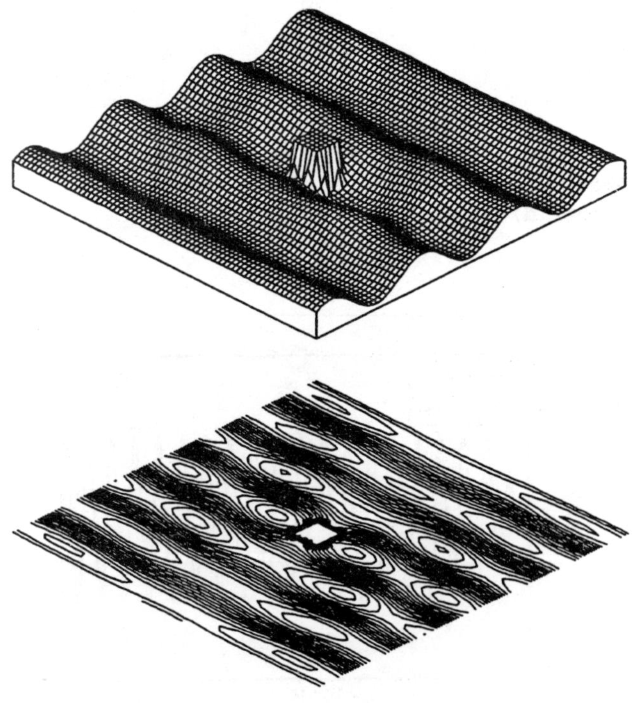

Figure 10: Perspective view and contour plot of the total wave fields around the rectangular barge($Fn = 0.0, ka = 1.0$)

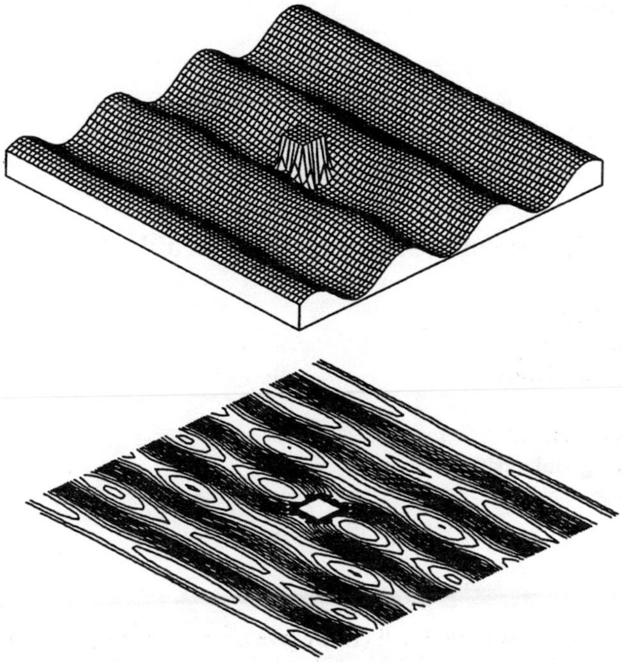

Figure 11: Perspective view and contour plot of the total wave fields around the rectangular barge($Fn = 0.05, ka = 1.0$)

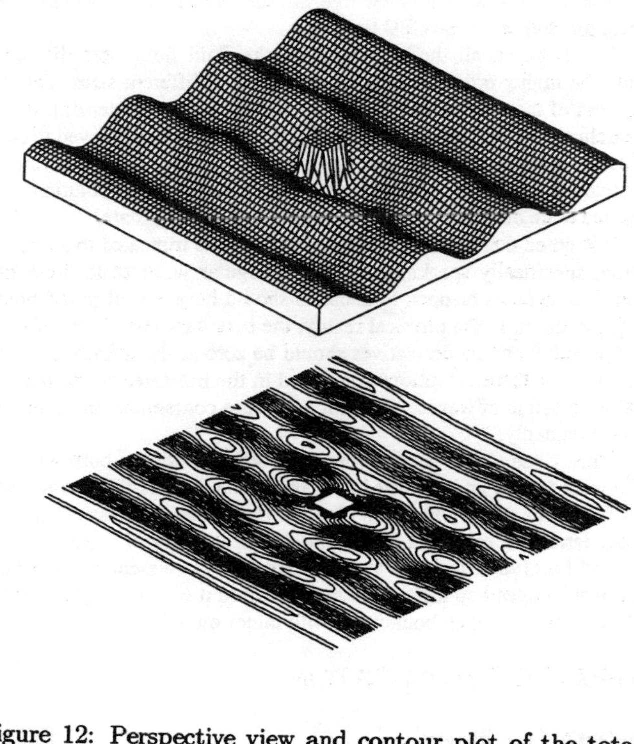

Figure 12: Perspective view and contour plot of the total wave fields around the rectangular barge($Fn = -0.05, ka = 1.0$)

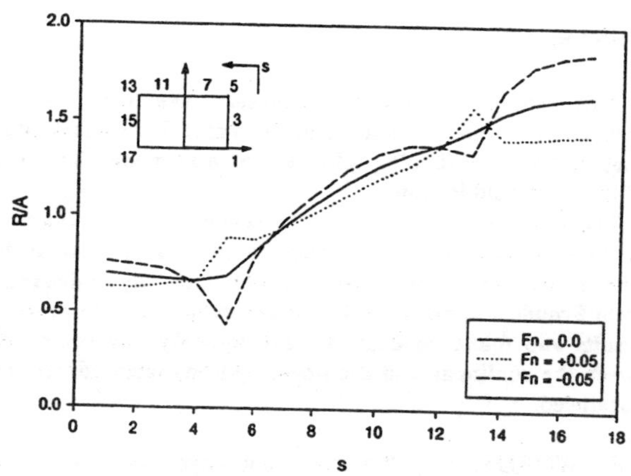

Figure 13: Maximu wave run-up around rectangular barge along the body contour($h/a = 3.333$,$ka = 1$)

Proceedings of the Seventh (1997) International Offshore and Polar Engineering Conference
Honolulu, USA, May 25-30, 1997

ISBN 1-880653-28-1 (Set); ISBN 1-880653-31-1 (Vol. III)

Numerical Simulation for the Wave Resistance of a Submerged Spheroid

Chung-Chyi Lee
Chinese Naval Academy
Kaohsiung, Taiwan, China

Ming-Chung Fang
National Cheng Kung University
Tainan, Taiwan, China

ABSTRACT

This paper describes the nonlinear wave field produced by a translating submerged spheroid. The exact free surface condition is used and a time-stepping scheme is used to model the development of the wave field in time.

The spheroid immersed in a three-dimensional wave tank starts abruptly with a constant velocity in an axial horizontal direction. The present solutions of the induced linear hydrodynamic forces from Froude's number 0.4 to 0.8 are compared with other authors' results and found to be close, subsequently, the comparisons between the nonlinear and the linear hydrodynamic forces are rather paralleled.

KEY WORDS: Wave Resistance, Submerged Spheroid, Higher Order Boundary Element Method

INTRODUCTION

With the increasing power of computational capability of computers, the evaluations of the wave-making resistance for a ship's motion in waves are not necessarily remained in the Neumann-Kelvin type of problem. Time domain analysis can solve the fully nonlinear free surface more easily than before.

Ever since Hess and Smith (1962) used the panel method to solve the motion of ships in potential flow, the boundary element method has been extensively applied by many researchers. In the early development, the source was assumed to be a uniform distribution over the element. Later, the linear distribution and thereafter the higher order of quadratic distribution of the source were developed. In this paper, the nine-node isoparametric quadrilateral element, which is used in the finite element method, was adopted as the boundary element. This higher order boundary element method (HOBEM) can be referred to Liu et al. (1991), Lee et al. (1994), and Boo et al. (1994). The advantages of HOBEM are fast convergence and none gap problems.

There are two reasons that we investigate the wave-making resistance of a submerged spheroid. Firstly, it resembles the submarine or some other kind of underwater vehicle; secondly, it avoids the treatment of the complicated intersections between the free surface and the ship. There are many papers discussing this, but the solutions in some cases are still uncertain. Farell (1973) gave a distribution of constant source over the surface of the spheroid and obtained a steady state solution related to the linear free surface waves. Doctors and Beck (1987) chose quadrilateral panels and used the combination of three different methods, that is, monopole, panel, and Galerkin methods, to integrate the Green function. The spheroid was also discretized into a different number of elements. They used Richardson's extrapolation to predict a convergent solution as the discretized panels approached infinity. Ferrant (1991) used a tabulation-interpolation procedure to evaluate a body-nonlinear spheroid in a linear free surface with the time-domain Green function and saved the CPU time.

In this paper, all the boundaries in the fluid field were discretized into the higher order boundary elements with different sizes. The integration of the Green function was performed on every element. In time marching, the fourth-order Runge-Kutta method was employed to calculate the velocity potential and wave height by way of the dynamic and kinematic boundary conditions on the free surface. The solution is expected to be obtained from the transient to the steady state.

It is noted that when the infinite fluid field is truncated to a finite region, specifically speaking, a 3-D rectangular wave tank, these truncated boundaries or open boundaries should have a well-posed boundary condition. In the physical reality, the boundary conditions of velocity potential and its derivatives should be zero at the infinity positions, however, if these conditions were used in the truncated boundaries, the false reflection of waves will occur and cause contamination of the solution eventually.

Thus, Chan (1977) used an open boundary condition borrowed from Orlanski (1976) and modified it in the calculation of the wave-making resistance of a ship-like body. The relevant papers discussing the open boundaries are Yang and Ertekin (1992), Isaacson and Cheung (1991), Li and Lu (1987), and Lee and Leonard (1987). We chose the original Orlanski's open boundary condition and used it on the upstream, downstream, and side open boundaries with minor modifications.

THEROTICAL FORMULATION

Potential Flow Theory

Considering the spheroid to move horizontally below the free surface with an abruptly constant velocity U in the opposite direction of positive x -axis, the water region is truncated into a rectangular wave tank. According to Cartesian's right-handed coordinate system, the positive x -axis points to the downstream, y -axis to the side wall, and z -axis to the upwards position. The origin o is located at the intersection of the still water surface and the z -axis which passes through the centroid of the spheroid as shown in Fig. 1. All the Physical quantities are nondimensionl. The characteristic length L is the major axis of the spheroid and the characteristic velocity U is the moving velocity. The Froude number is defined as $F_r = U/\sqrt{gL}$, where g is the gravitational acceleration.

Assuming the fluid is to be irrotational, incompressible, and inviscid, the total velocity potential Φ is

$$\Phi(x,y,z;t) = u(t)x + \phi(x,y,z;t) \tag{1}$$

where ϕ is a perturbation velocity potential and $u(t)$ signifies moving velocity of the spheroid in time domain. The Laplace equation is

$$\nabla^2\phi(x,y,z;t) = 0 \tag{2}$$

The kinematic and dynamic boundary conditions on the free surface are

$$\eta_t = -u(t)\eta_x - \phi_x\eta_x - \phi_y\eta_y + \phi_z \tag{3}$$

$$\phi_t = -u(t)\phi_x - \frac{1}{2}\nabla\phi\cdot\nabla\phi - \frac{\eta}{F_r^2} \tag{4}$$

where (ϕ_x, ϕ_y, ϕ_z) are perturbation velocities in the (x,y,z) directions. If the infinitesimal wave theory is assumed, the second-order terms can be neglected, then the waves are linear.

The Sommerfeld radiation conditions for the downstream and upstream boundaries are

$$\alpha_t \pm c\alpha_x = 0 \tag{5}$$

respectively, and for the side open boundary is

$$\alpha_t + c\alpha_y = 0 \tag{6}$$

where c is the outgoing phase velocity, i.e., wave celerity of perturbation wave; α is any variable, it is designated here as velocity potential ϕ , wave height η and derivatives of velocity potential ϕ_x and ϕ_y . It is also noted that ϕ_x and ϕ_y are only substituted into equation (5) and (6) respectively.

The kinematic boundary condition on the surface of the spheroid can be given by

$$\phi_n = -u(t)n_x \tag{7}$$

The dynamic pressure of the fluid from the unsteady Bernoulli's equation can be determined by

$$p = -\phi_t - u(t)\phi_x - \frac{1}{2}\nabla\phi\cdot\nabla\phi \tag{8}$$

The wave resistance, lift force, and pitching moment are computed by

$$F_x = \int p n_x ds \tag{9}$$

$$F_z = \int p n_z ds \tag{10}$$

$$M_y = \int p(z_c n_x - x_c n_z) ds \tag{11}$$

where (n_x, n_y, n_z) are unit normal vectors in the direction of coordinates (x,y,z) . The x_c and z_c are the coordinates located at the centroid of the spheroid and equal to $(x, z+d)$, where d is the submergence measured from the centroid of the spheroid to the still free surface.

The coefficients of wave resistance, lift, and pitching moment are defined as

$$C_w = F_x \Big/ \tfrac{1}{2}\rho U^2 S \tag{12}$$

$$C_l = F_z \Big/ \tfrac{1}{2}\rho U^2 S \tag{13}$$

$$C_m = M_y \Big/ \tfrac{1}{2}\rho U^2 S \tag{14}$$

where S is the area of wetted surface of the spheroid and can be calculated analytically by

$$S = 0.5\pi(B/L)^2 L^2 + 0.5\pi(B/L)L\left(\sin^{-1} e/e\right) \tag{15}$$

B/L is beam-length ratio and e can be denoted as

$$e = \sqrt{1-(B/L)^2}$$

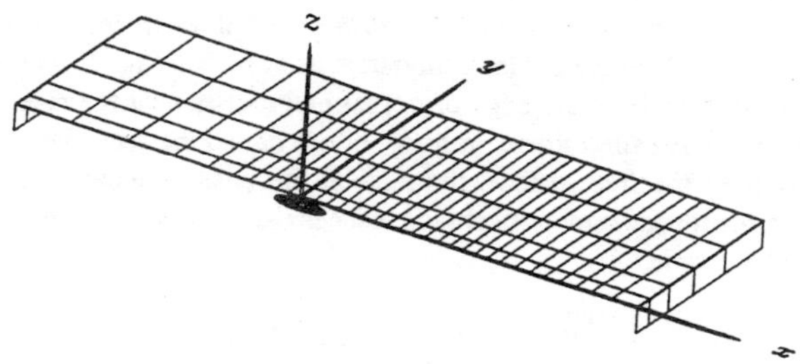

Fig 1 The sketch of coordinate system for the moving spheroid.

Boundary Element Method

The boundary integral equation of three-dimensional potential flow is

$$C(p)\phi(p) + \frac{1}{4\pi}\int_{\Gamma_q} W(\cdot) d\Gamma_q = 0 \tag{16}$$

and

$$W(\cdot) = \frac{\partial G(p,q)}{\partial n}\phi(q) - G(p,q)\frac{\partial \phi(q)}{\partial n}$$

where $C(p)$ is the solid angle of the field point. The integral surfaces of the source point q include upstream, downstream, side, bottom and the spheroid surfaces. The nine-node isoparametric quadratic quadrlateral elements are used here. For the purpose of saving the CPU time, the method of images are used on $x-z$ plane and the flat bottom, thus the Green function can be expressed as

$$G = \sum_{j=1}^{4} \frac{1}{R_j} \tag{17}$$

where R_j is the distance between the field point and the integrating source point. It can be expressed individually as

$$R_1 = \left[(x_p - x_q)^2 + (y_p - y_q)^2 + (z_p - z_q)^2\right]^{\frac{1}{2}}$$

$$R_2 = \left[(x_p - x_q)^2 + (y_p + y_q)^2 + (z_p - z_q)^2\right]^{\frac{1}{2}}$$

$$R_3 = \left[(x_p - x_q)^2 + (y_p - y_q)^2 + (z_p - z_q + 2h)^2\right]^{\frac{1}{2}}$$

$$R_4 = \left[(x_p - x_q)^2 + (y_p + y_q)^2 + (z_p - z_q + 2h)^2\right]^{\frac{1}{2}}$$

where h is the depth measured from the still free surface to the flat bottom.

Radiation Boundary Condition

The treatment of the body and free surface conditions is standard, therefore, the main difficulty with carrying out this simulation is the proper treatment of the radiation boundary condition (RBC). Since the multiple nodes technique is applied on the all edges and corners of the rectangular fluid domain, the treatment of the RBC is also closely related to the operation of the multiple nodes technique. It is called the edge boundary, when two faces (e.g., downstream/upstream/side vertical face and free surface) are intersected each other; it is called the corner boundary, when three faces (e.g., downstream vertical face, side vertical face and free surface) are intersected. There are two nodes at each selected position on the edge boundaries and three nodes at each selected position on the corner boundaries, as shown in Fig. 2. On the free surface, the node at the edge or corner boundaries near the side of the free surface is called the main node, others called the secondary nodes. Away from the free surface, the node at the edge boundaries near the downstream/upstream vertical face is the main node, and the other is the secondary node.

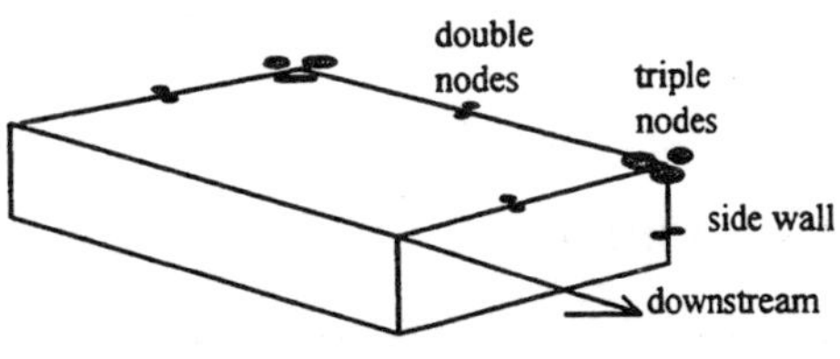

Fig. 2 Sketch of multiple nodes technique.

The Dirichlet boundary condition ϕ is prescribed on the free surface. We used the one-dimensional RBC in the eq. (5) first, and then used the eq. (6) to determine the new ϕ on each main node for the next time step. However, there is a redundant value on each main node at the corner boundaries. In order to consist the value of the main nodes at the corners with other neighboring main nodes at the edge, We used the values just obtained from the eq. (5) to substitute into the eq. (6) and obtained another new value at the corners. The wave elevation η on the main nodes at the radiation boundaries is also determined in a similar way.

The Neumann boundary condition ϕ_n is prescribed on all vertical open boundaries. We assumed ϕ_x as ϕ_n and used eq. (5) to determine the new ϕ_n on each secondary node at the edges and corners for the downstream and upstream vertical open boundaries. The ϕ_y is assumed as ϕ_n and the eq. (6) is used to determine the new ϕ_n on each secondary node at the edges and corners for the side vertical open boundaries. There is no redundant value at each corner as described above, since there should have two different ϕ_n values on each secondary node at the corners for two different vertical open boundaries.

The numerical scheme of the one-dimensional RBC in the finite difference form was adopted from the papers of Orlanski (1976), Isaacson et. al., (1991) and Lee et. al. (1994). The effective distance of the radiation boundaries was chosen by the numerical test and experience. Generally, the computational domain in our case is

x: $-2.8\lambda \sim 3.3\lambda$; y: $-2\lambda \sim 2\lambda$; z: $\eta \sim -0.25\lambda$

The dominant wave length λ is $2\pi / k$, where k is the wave number and equals to g / U^2 or $1/(LF_r^2)$.

Time Marching Procedures

At $t = 0$, the initial condition for the velocity of the spheroid is $u(t) = -1$. The velocity potential and wave height on the free surface are $\phi = 0$ and $\eta = 0$, respectively. The normal derivatives on all the open boundaries are $\phi_n = 0$. The procedures for the computations are as follows: Firstly, The HOBEM is used to solve the boundary integral equation to get the values of ϕ_n^t on the free surface and ϕ^t on others boundaries. Secondly, By time marching , the fourth-order Runge-Kutta method gives the velocity potential ϕ^{t+1} and wave height η^{t+1} at the nodes on the free surface for the next time step $t+1$. At this time Orlanski's open boundary conditions also give ϕ^{t+1}, η^{t+1} and ϕ_n^{t+1} on the multiple nodes at the intersections of the free surface and the vertical open boundaries. Finally, The magnitude of ϕ_n^{t+1} on the nodes below the still free surface is assumed to be attenuated by a factor e^{kz}, since the linear wave theory satisfied that $\phi_n = \phi_n|_{z=0}\, e^{kz}$, where k is the wave number. The procedures above will be repeated again, until the expected solutions of hydrodynamic force coefficients and wave formation can be obtained from the transient to the steady state.

RESULTS AND DISSCUSION

In order to compare the solutions with most relevant papers, the beam-length ratio B/L of the spheroid is taken as 1:5, and the submergence is $0.16L$. There are 20 elements (summing up 93 nodes) on the surface of half the spheroid in the longitudinal direction as shown in Fig. 2. The elements divided on the surface of the rectangular wave tank are generally $35 \times 5 \times 2$ and the time step size is around $1/40$ of the dominate wave period.

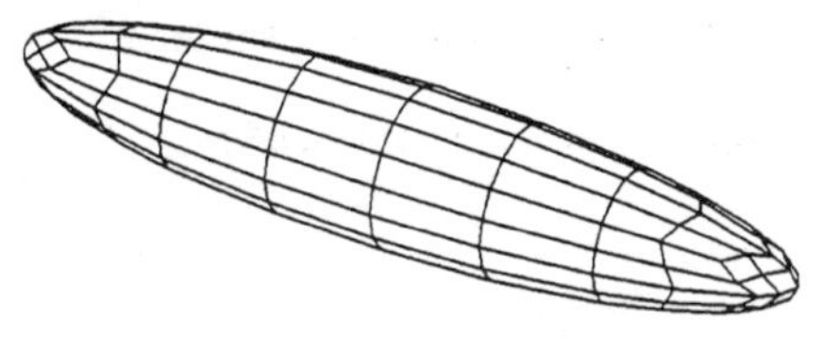

Fig. 3 Subdivision diagram of the spheroid.

In the linear wave motion, the present solutions of the curves of the hydrodynamic force coefficients from the Froude number 0.4 to 0.8 are compared with Doctors' solutions in the frequency domain and Ferrant's solutions in time domain as shown in Fig. 4 to 8 respectively. The force F_i on the ordinate is designated as the wave resistance F_x, lift force F_y, and pitching moment M_y. The scale of force coefficients $F_i / \rho g L^3$ can be obtained by multiplying a conversion factor $F_r^2 / 3.98466$ to the equation (12) to (14). The time scale on the abscissa is a value of the nondimensionl time divided by F_r and expressed as $t'\sqrt{g/L}$. Observing that the solution is still in oscillation,

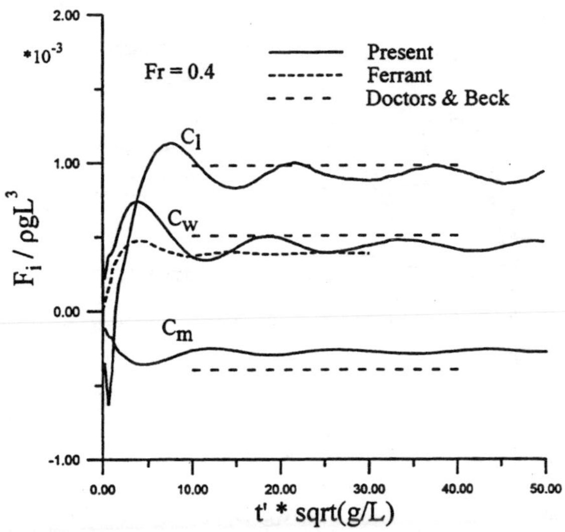

Fig. 4 Linear force coefficients for Froude's number 0.4.

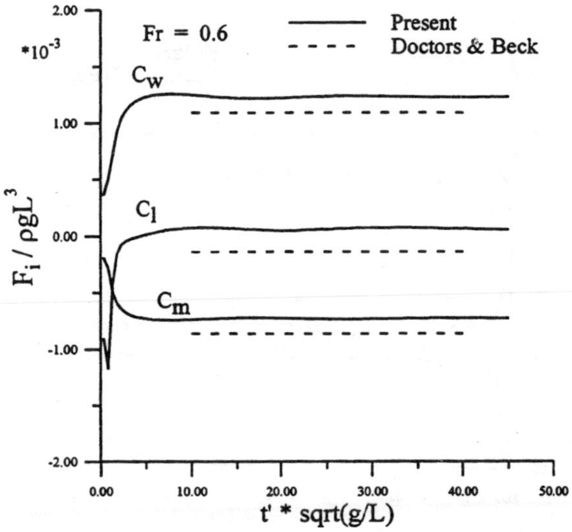

Fig. 6 Lnear force coefficients for Froude's number 0.6.

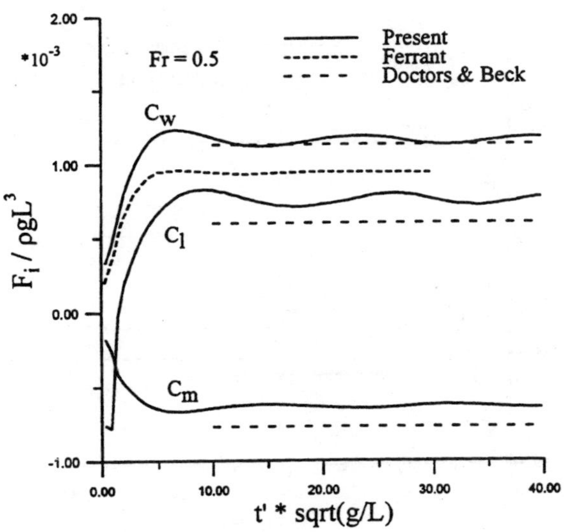

Fig. 5 Linear force coefficients for Froude's number 0.5.

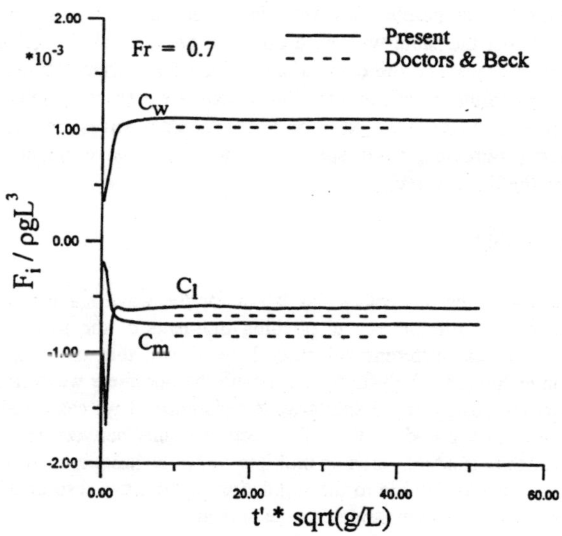

Fig. 7 Linear force coefficients for Froude's number 0.7.

when $F_r = 0.4$, however, when the Froude number goes higher and higher, the solution becomes nearly steady in quite a short time. The comparisons also shown that the present solutions are in a tendency approach to the solutions from those authors' works. In order to know the extent of the accuracy, the last peak values of the wave resistance in Fig. 4 and 5 were collected and compared with those Ferrant's ultimate solution in time domain, Farell's analytic, and Doctors' convergent solutions in the frequency domain as shown in table 1.

In the nonlinear wave motion, the computational cost is much higher. Considering about the stability of the solution, the smaller time step size is usually taken. For example, we took time step size 0.2 in the nonlinear wave motion and 0.25 in the linear wave motion for the case of the Froude number 0.4. The numerical filtering was also used at each

F_r	C_w (Present)	C_w (Ferrant)	C_w (Doctors)	C_w (Farell)
0.4	0.01172	0.01267	0.0131	0.01132
0.5	0.01826	0.01671	0.018	0.01822

Table 1

time step, whereas it was used at every three time step in the linear wave motion. We chose two cases, the Froude number 0.4 and 0.8, to compare their nonlinear solutions with linear one as shown in Fig. 9 and 10. Those two curves are almost in parallel with a definite difference. It is also seen that the lift coefficients are lower in the nonlinear waves than in the linear waves. For the Froude number 0.4, the wave

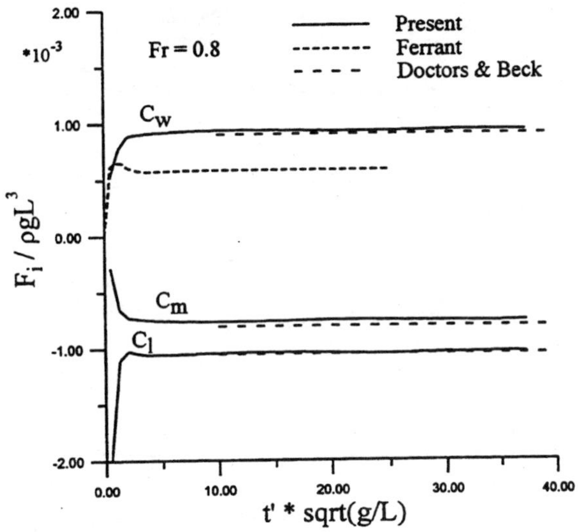

Fig. 8 Linear force coefficients for Froude's number 0.8.

resistance is higher in the nonlinear wave than in the linear wave, however, for the Froude number 0.8, both nonlinear and linear wave resistance are almost the same with each other. In order to get insight into the nonlinear behaviors, the comparisons of the free surface elevations between the nonlinear wave and the linear wave are presented in Fig. 11 and 12. These results indicate that the nonlinear stern waves are slightly steep, their progressing wave speeds are faster and wave heights are lower than the linear waves.

CONCLUSION

The results of the numerical simulation for the wave making resistance and related lift forces and pitching moments of the submerged spheroid at several different velocities have shown that time domain simulation technique is helpful to understand the nonlinear wave effects to the accurate design of the ship's wave resistance. If we can consider the treatment of the nodes at the intersection points between the free surface and the structure, such as multiple nodes technique, then these methods are also applicable to the simulation of the wave resistance for the surface ships and catamarans in time domain.

ACKNOWLEDGEMENTS

This research was supported by the National Science Council of Taiwan, ROC, Grant Number NSC 85-2611-E-012-001.

REFERENCES

Boo, SY, Kim, CH and Kim, MH (1994). "A Numerical Wave Tank for Nonlinear Irregular Waves by 3-D Higher Order Boundary Element Method," Int J Offshore and Polar Eng, ISOPE, Vol 4, No 4, pp 265-272.

Chan, RK-C (1977). "Finite Difference Simulation of the Planar Motion of a Ship," Proc 2nd Intl. Conf Num Ship Hydro, Univ California, Berkely, CA, pp 39-52.

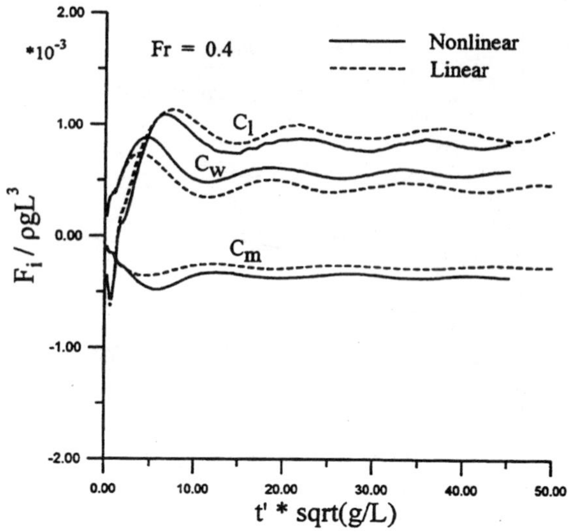

Fig. 9 Nonlinear and linear force coefficients for Froude's number 0.4.

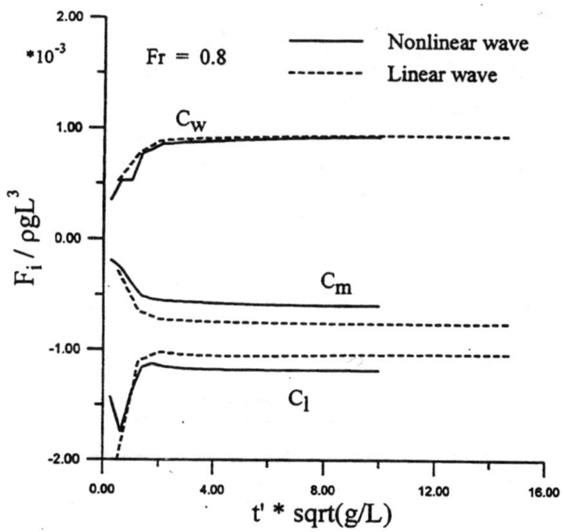

Fig. 10 Nonlinear and linear force coefficients for Froude's number 0.8.

Doctors, LJ and Beck, FB (1987). "Convergence Properties of the Neumann-Kelvin Problem for a Submerged Body," J of Ship Research, Vol 31, No 4, pp 227-234.

Farell, C (1973). "On the Wave Resistance of a Submerged Spheroid," J of Ship Research, Vol 17, No 1, pp 1-11.

Ferrant, P (1991). "A Coupled Time and Frequency Approach for Nonlinear Wave Radiation," 18th symposium on Naval Hydro, The Univ of Michigan, Ann Arbor, National Academy Press, Washington, DC.

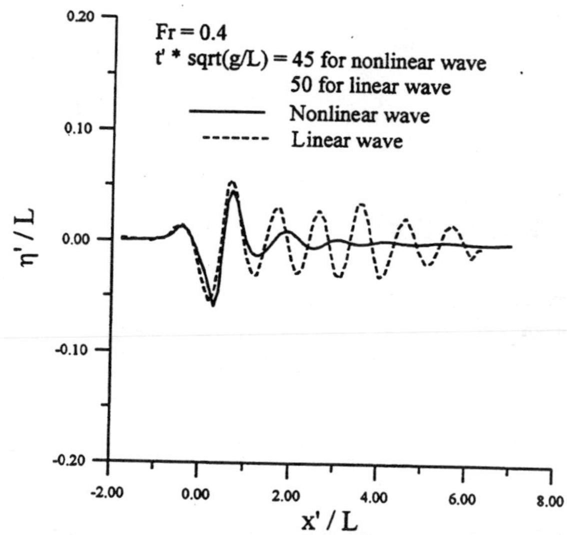

Fig. 11 Nonlinear and linear wave elevations for Froude's number 0.4.

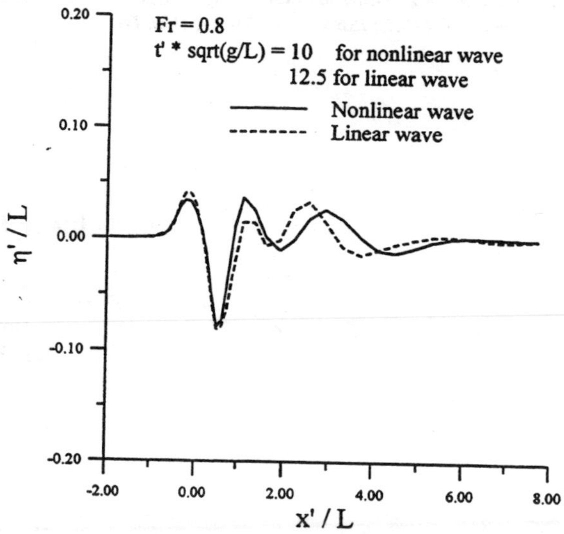

Fig. 12 Nonlinear and linear wave elevations for Froude's number 0.8.

Hess, JL and Smith, AMO (1962). "Calculation of Non-lifting Potential Flow About Arbitrary Three Dimensional Bodies," Douglas Report No ES 40622, Long Beach, CA.

Isaacson, M and Cheung, KF (1991)."Time Domain Simulation of Second-Order Runup and Wave Forces," Proc 1st Int Offshore and Polar Eng Conf, Edinburgh, United Kingdom, pp 468-475.

Lee, CC, Liu, YH and Kim, CH (1994). "Simulation of Nonlinear Waves and Forces Due to Transient and Steady Motion of Submerged Sphere," Int J Offshore and Polar Eng, ISOPE, Vol 4, No 3, pp 174-182.

Lee, JF and Leonard, JW (1987). "A Time- Dependent Radiation Condition for Transient Wave-Structure Interactions," Ocean Eng, Vol 14, pp 469-488.

Li, BY and Lu, YL (1987). "A Numerical Model for Non-linear Wave Diffraction Around Large Offshore Structures," Int J for Num Method in Fluids, Vol 7, pp 1343-1355.

Liu, YH, Kim, CH and Lu, XS (1991). "Comparison of Higher-Order Boundary Element and Constant Panel Methods for Hydrodynamic Loadings," Int J Offshore and Polar Eng, ISOPE, Vol 1, No 1,pp 8-16.

Orlanski, I (1976). "A Simple Boundary Condition for Unbounded Hyperbolic Flows," J Comp Physics, Vol 21, pp 251-269.

Yang, C and Ertekin, RC (1992). "Numerical Simulation of Nonlinear Wave Diffraction by a Vertical Cylinder," Trans of the ASME, Vol 114, pp 36-44.

Proceedings of the Seventh (1997) International Offshore and Polar Engineering Conference
Honolulu, USA, May 25-30, 1997

ISBN 1-880653-28-1 (Set); ISBN 1-880653-31-1 (Vol. III)

Non-linear Internal Wave Kinematics

Alastair J. Martin and William J. Easson
University of Edinburgh
Edinburgh, United Kingdom

Abstract

Large amplitude regular internal waves are modelled in a wave tank using water and saline solution to create a density profile. The parameter range covers long and short waves in a quasi two-layer stratification and also on a broader interface. The wave kinematics are measured using Particle Image Velocimetry and compared to weakly non-linear theory.

It is found that the second order Stokes theory underpredicts the velocities in the extremes of the short waves by about 40%. Both linear and solitary wave theory fail to adequately predict the kinematics of long waves, especially the vertical velocity components.

Keywords: Internal waves, particle image velocimetry, wave kinematics.

Introduction

Internal waves are known to occur in most ocean areas, especially the deep ocean where stable stratification is often found. New petroleum discoveries have prompted a move into areas (e.g. west of Shetland) where internal waves are more likely to be a problem. The implications of internal waves for offshore operations were first raised by Osborne, Burch and Scarlett (1978) and later by Bole, Ebbesmeyer and Romea (1994). The presence of internal waves in sea areas of current interest to the UK offshore industry has been confirmed by Sherwin (1991).

While many aspects of internal wave behaviour have been extensively studied (e.g., Thorpe, 1968 & 1975), few measurements of detailed wave kinematics have been performed. This paper presents results of a programme of experiments which aimed to measure wave kinematics in large amplitude waves, and comparisons with theory.

Theoretical description

Stokes waves

The following analysis is based on that of Thorpe (1968), though the Boussinesq approximation is not made. It can be shown that the streamfunction ψ in a stratified flow satisfies the vorticity equation

$$\left.\begin{aligned}
\nabla^2\psi_{tt} - \frac{g\bar{\rho}'}{\rho}\psi_{xx} &= Q_v + Q_b + Q_d + Q_\nu \\
Q_v &= \mathrm{J}(\nabla^2\psi, \psi)_t \\
Q_b &= -\frac{g}{\rho}\mathrm{J}(\tilde{\rho}, \psi)_x \\
Q_d &= \frac{1}{\rho}\Big\{ \tilde{\rho}_t[\mathrm{J}(\nabla^2\psi, \psi) - \nabla^2\psi_t] \\
&\quad + \Big(\tilde{\rho}_x[\mathrm{J}(\psi_x, \psi) - \psi_{xt}] \\
&\quad + \rho_z[\mathrm{J}(\psi_z, \psi) - \psi_{zt}] \Big)_t \Big\} \\
Q_\nu &\simeq \nu[\nabla^2\psi_{xx} - \nabla^2\psi_{zz}]_t
\end{aligned}\right\} \quad (1)$$

Here the sign convention for the streamfunction is

$$u = -\psi_z, \quad w = \psi_x \quad (2)$$

with horizontal and vertical co-ordinates x and z, velocity components u, w, density represented by ρ, gravity by g and time by t.

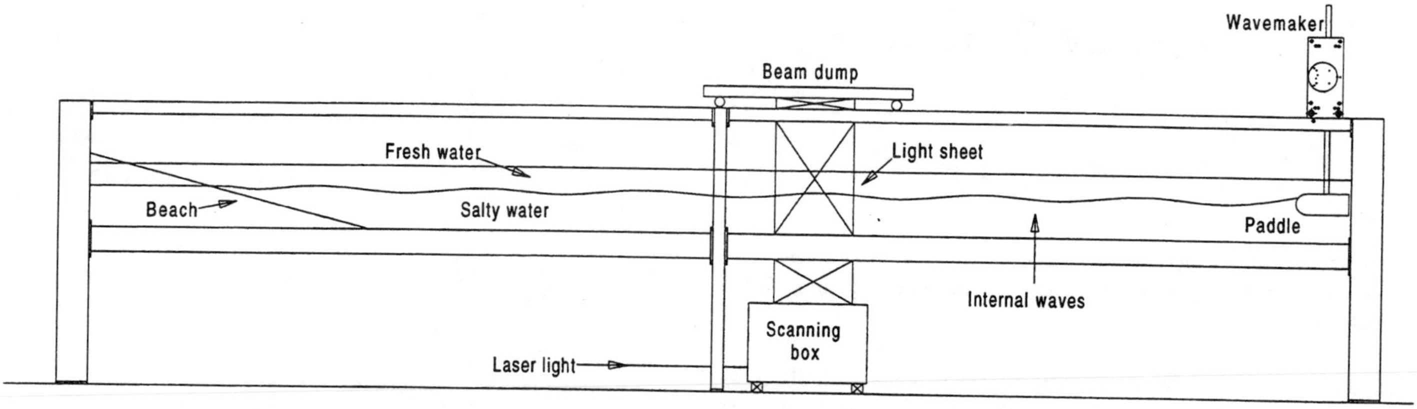

Figure 1: Wave tank, wavemaker and illumination system

Subscripts denote differentiation, and J represents the Jacobian derivative. The density ρ is split into a mean static profile with depth $\overline{\rho}$ plus a fluctuation due to the wave motion $\tilde{\rho}$.

The static density profile for the experiments was represented by

$$\overline{\rho} = \rho_0 \left(1 - \alpha \tanh\left(\frac{z}{\delta}\right)\right) \tag{3}$$

where ρ_0 is the median density, so that the range of density variation is $\Delta\rho = 2\alpha\rho_0$, and δ describes the thickness of the interface.

The terms on the right hand side of (1) represent non-linearity, large density differences and viscosity. The term Q_v collects the non-linear velocity terms. The non-linear terms which result from the buoyancy fluctuation are grouped in Q_b. The terms which disappear under the Boussinesq approximation are grouped in Q_d, while the viscous terms are represented by Q_ν. The left hand side is all that remains after linearisation, the application of the Boussinesq approximation and the assumption of an inviscid flow.

The Boussinesq approximation rejects the variation in density where it appears in the combination $\overline{\rho}+\tilde{\rho}$ on the assumption that $\overline{\rho} \gg \tilde{\rho}$. These terms are of order α. In the laboratory this is comparable to the non-linear velocity terms, and so use of the Boussinesq approximation would be inconsistent with consideration of non-linearity. In the ocean, $\alpha O(10^{-3})$ and the Boussinesq approximation can be made.

Scaling considerations show that for the experiments the viscous terms are significant in the boundary layers, less so in the interface between the fresh water and saline solution, and not at all in the rest of the fluid. When a two-layer system is used, viscosity becomes far more significant in the interface because of the intensity of the shear there. For the relatively diffuse interfaces used, the viscous contribution is expected to be comparable to the third-order velocity component. In the ocean, viscosity is even less significant, even if an eddy viscosity is used.

The density fluctuation $\tilde{\rho}$ is found from

$$\tilde{\rho}_t + \overline{\rho}'\psi_x - \mathrm{J}(\tilde{\rho}, \psi) = 0 \tag{4}$$

and the wave shape $\eta(x, z, t)$ is found from an equation similar to the kinematic free surface boundary condition for surface waves

$$\eta_t - \psi_z\eta_x - \psi_x = 0 \tag{5}$$

so that η defines surfaces of constant density.

Equation (1), if Q_d and Q_ν are discarded, is equivalent to that obtained by Thorpe (1968). Following Thorpe, solutions were obtained by means of a perturbation expansion in $\epsilon = ak$, where a is the amplitude and k is the wavenumber. For example, for the streamfunction

$$\begin{aligned}\psi(x, z, t) \\ &= \epsilon\psi_1(x, z, t) + \epsilon^2\psi_2(x, z, t) + \ldots \\ &= \epsilon\Psi_1(z)e^{jk(x-ct)} + \epsilon^2\Psi_2(z)e^{2jk(x-ct)} + \ldots\end{aligned} \tag{6}$$

where c is the phase speed. Similar expansions were used for the density fluctuation $\tilde{\rho}(x, z, t)$ and the wave shape $\eta(x, z, t)$.

The results are, for the first order component

$$\Psi_1'' + \frac{\overline{\rho}'}{\overline{\rho}}\Psi_1' + \left(-\frac{g\overline{\rho}'}{c_0^2\overline{\rho}} - k^2\right)\Psi_1 = 0 \tag{7}$$

and for the second order

$$\begin{aligned}\Psi_2'' + \frac{\overline{\rho}'}{\overline{\rho}}\Psi_2' + \left(-\frac{g\overline{\rho}'}{c^2\overline{\rho}} - 4k^2\right)\Psi_2 = \\ \frac{g\overline{\rho}'^2}{2c^3\overline{\rho}^2}\Psi_1^2 + \frac{\overline{\rho}'}{2c\overline{\rho}}\left(\Psi_1'^2 - \Psi_1\Psi_1''\right) \\ + \frac{g\overline{\rho}'}{c^3\overline{\rho}}\Psi_1^2 + \frac{2}{c}\Psi_1'^2 - \frac{\overline{\rho}'^2}{2c\overline{\rho}^2}\Psi_1\Psi_1' - \frac{2k^2}{c}\Psi_1^2\end{aligned} \tag{8}$$

To second order there is no correction to the phase speed. An equation for the third order component has been obtained but is not used here.

Equation (7) is an eigenvalue problem for which an infinite number of wave modes are possible. Here only the first mode is sought. These equations were solved using the "rigid lid" (flat upper surface) and flat bottom boundary conditions. A Runga-Kutta scheme was used for the first order component, while for the second order component a finite difference scheme had to be employed.

Korteweg-de Vries theory

The Stokes expansion above does not perform well when the waves are long compared to one or other layer depths. Benjamin (1966) adapted the shallow water surface wave theory of Korteweg & de Vries (1895) to the case of internal waves in a continuously stratified fluid. Benjamin did not apply the Boussinesq approximation and so his analysis can be used directly. The solitary wave result, in a frame of reference moving with the wave, is

$$\begin{aligned}\eta(x,z) &= \frac{\epsilon J\phi}{K}\operatorname{sech}^2\left(\frac{(\epsilon J)^{1/2}x}{2I^{1/2}}\right) \\ I &= c^2\int_{z_1}^{z_2}\rho\phi^2 dz \\ J &= \frac{(c^2 - c_0^2)}{\epsilon}\int_{z_1}^{z_2}\rho\left(\phi'\right)^2 dz \\ K &= c_0^2\int_{z_1}^{z_2}\rho\left(\phi'\right)^3 dz\end{aligned} \tag{9}$$

where c is the phase speed, z_1 and z_2 are the top and bottom boundaries and ϵ is a small parameter. The mode shape of the wave is given by $\phi(z)$, which is the solution to

$$\phi'' + \frac{\overline{\rho}'}{\rho}\phi' - \frac{g\rho'}{\rho c_0^2}\phi = 0 \tag{10}$$

This equation is identical to (7) for an infinite wavelength. The phase speed of an infinitessimal wave of infinite wavelength is c_0. The amplitude of the wave defines the actual phase speed c.

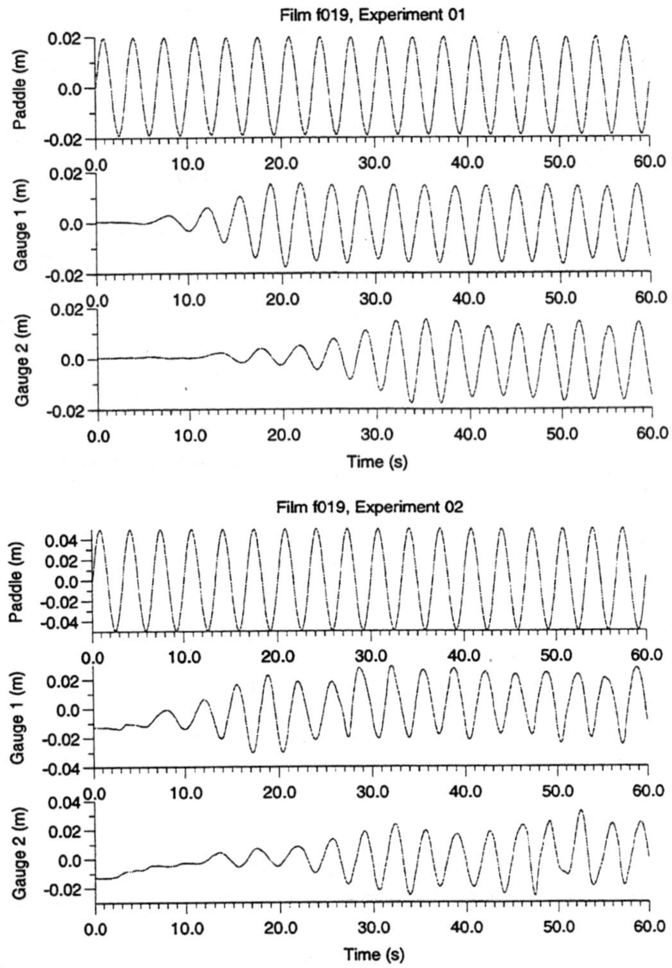

Figure 2: Wave gauge traces with $f = 0.3Hz$, $\Delta\rho/\overline{\rho} = 9.4\%$. Top picture: $H_{1/3} = 0.031m$; bottom picture: $H_{1/3} = 0.038m$

Experimental equipment

The experiments were conducted in a wave flume of length 6m, width 0.4m and total depth 0.6m (figure 1). The wavemaker was a vertically oscillating paddle controlled by a computer. Wave shapes were recorded using wire wave gauges, which measured the resistance of the water column at a fixed location. Since the conductivity of the saline solution dominated over the conductivity of the fresh water, the gauges measured the height of the interface between the two. This was adequate when the thickness of the interface was not large compared to the depth of one or other layer. All of the experiments described here used regular waves.

For most of the experiments a tanh density profile was set up in the tank, that is, one in the form described by equation (3). The parameters ρ_0 and α were measured with hygrometers before the tank was stratified. The interface thickness δ and its elevation were obtained using an aspirating density probe mounted on a vertical traverse. In some of the experiments a linearly stratified upper layer was used; the wave gauges produced greatly attenuated signals in this case and actual wave heights could not be measured.

Velocities were measured using an implementation of Particle Image Velocimetry described by Gray & Bruce (1995). A flow field of width 0.5m was illuminated by a 15W Argon-Ion laser using the scanning beam technique. The flow was seeded with conifer pollen. Images were recorded on 120 format photographic film. A rotating mirror image shifting system was used to resolve directional ambiguity. The negatives were analysed automatically using the Young's Fringe auto-correlation technique to obtain velocity vector maps measuring up to 40 $\times$ 40 vectors.

The velocities being measured were towards the lower end of the dynamic range of the PIV system. For this reason small amplitude waves could not be measured. All of the cases presented here had moderate to high steepnesses. Wave height gauges were placed just upwave and just downwave of the measurement zone. Physical constraints made it difficult to place the gauges directly in the measurement zone.

Results

Short waves

As stated above, all of the waves presented here were of moderate to large amplitude. The range of steepness for the short waves was $ak \simeq 0.1$–0.25.

For the cases where $\lambda < h_{1,2}/3$ approximately, where λ is the wavelength and $h_{1,2}$ are the upper and lower layer depths respectively, the measured velocities were compared to second order Stokes calculations. In order to do this, the significant wave height for the wave train ($H_{1/3}$) was calculated from the wave gauge trace. The significant wave height was used because the variability in the wave train made it difficult to predict the exact shape of the wave in the measurement zone. Typical wave gauge traces illustrating this are shown in figure 2.

Once the wave height was known, a theoretical velocity map could be calculated for the parameters of interest. The phase of the predicted wave relative to the measured wave was found by adjusting it until the error (the difference between measured and predicted velocities) was minimised. The match between measured and predicted velocity maps could then be evaluated by various means.

Typical vector maps are shown in figure 3. In this figure, two measured vector maps from the same experimental run are seen; these maps were separated by one wave period. There is a high degree of agreement between the measured maps. Also shown is a theoretical vector map phase matched to the second measured map. This shows second order Stokes calculations. The theoretical map is qualitatively very similar to the measured maps.

Figure 4 illustrates one means of evaluating the degree of agreement between measurements and theory. Plotting measured velocity components against theoretical components shows a linear

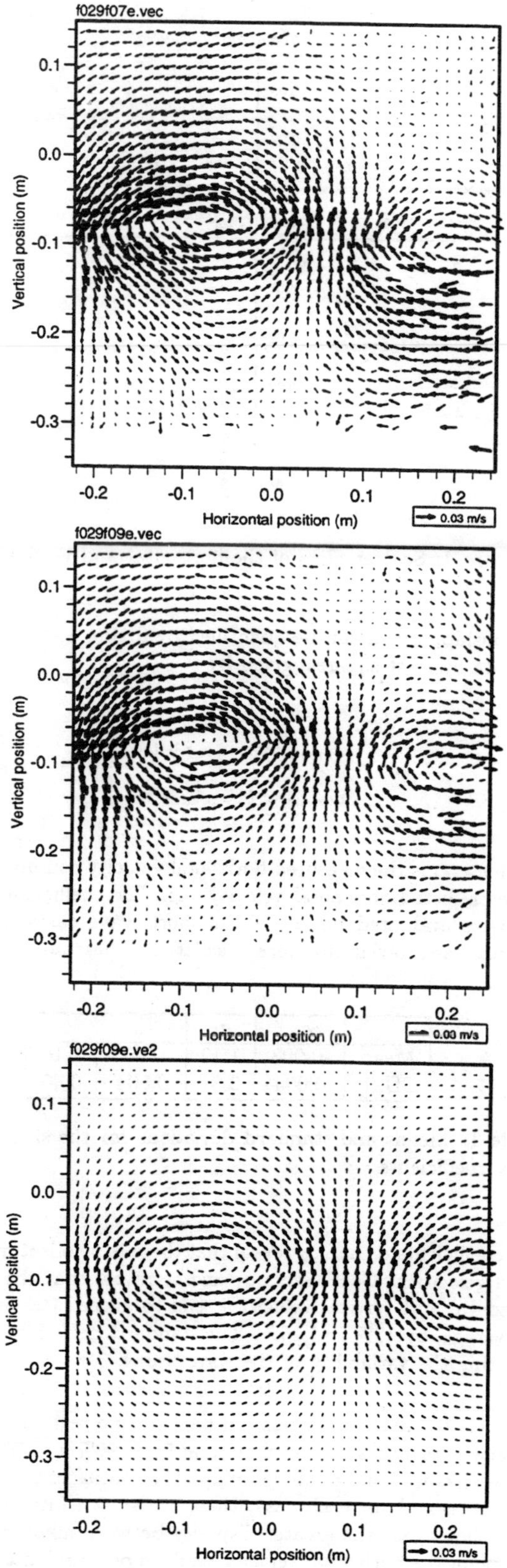

Figure 3: Successive views of a short wave crest and trough with $f = 0.3Hz$, $\Delta\rho/\overline{\rho} = 8.8\%$, $H_{1/3} = 0.030m$; with theoretical vector map (bottom).

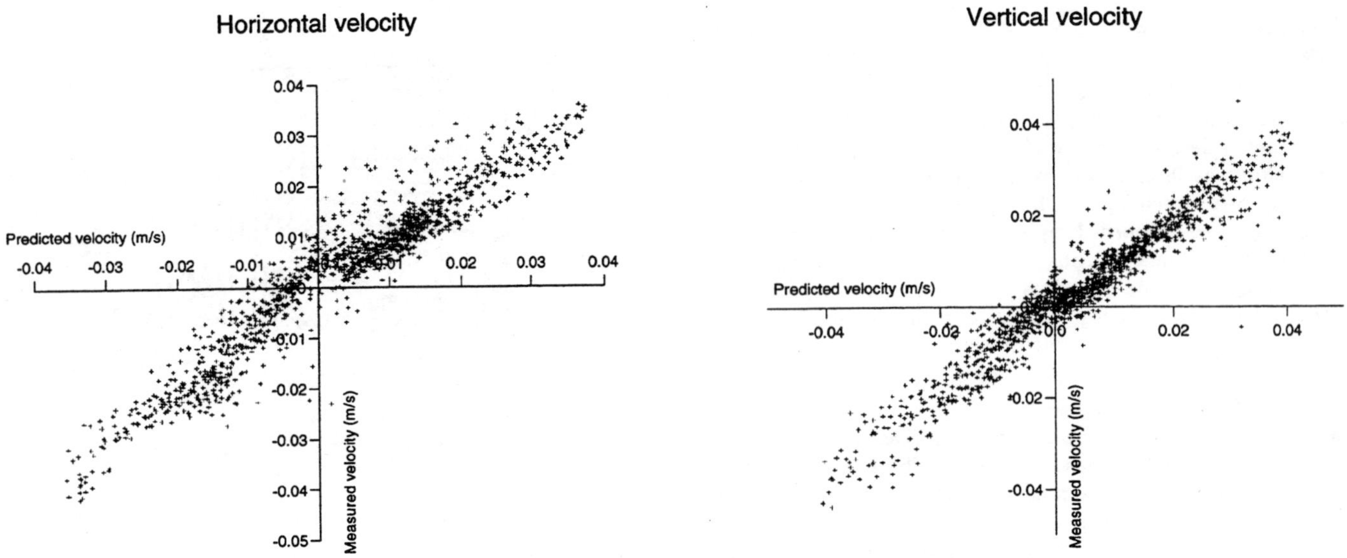

Figure 4: Scatter plots of the horizontal and vertical velocities of one vector map against second order predictions for short waves

relationship, which can be quantified by fitting a straight line to the data by least squares, given by

$$\left.\begin{aligned} u_{meas} &= a_1 u_{pred} + a_0 \\ w_{meas} &= b_1 w_{pred} + b_0 \end{aligned}\right\} \qquad (11)$$

where u_{meas} is the measured horizontal velocity component, u_{pred} is the predicted horizontal velocity, etc. The coefficients $a_{0,1}$ and $b_{0,1}$ were evaluated for all of the short wave cases and means and standard deviations were calculated. These are presented in table 1.

	a_0	a_1	b_0	b_1
Mean	-0.003	1.12	-0.002	1.16
SD	0.004	0.31	0.004	0.40

Table 1: Means and standard deviations for the short wave velocity coefficients

The coefficients a_0 and b_0 reveal that there is no offset in the measured velocities. Coefficients a_1 and b_1 reveal that the measured horizontal and vertical velocities tend to exceed the predictions, on average, by 12% and 16%. The standard deviations are large, reflecting the variability of the wave trains.

In order to make comparisons in the extreme regions of the flow, velocity percentiles were evaluated at 95%, 90% and 70% for each of the measured and predicted maps. The ratios of the percentiles were measured for each case; these are shown in table 2. This table shows that the greater disparity between measurement and theory for the horizontal velocity tends to be found in the extreme regions of the flow. This would be expected since the higher order corrections to the Stokes theory exert most influence near the interface, where the horizontal velocity maxima are found.

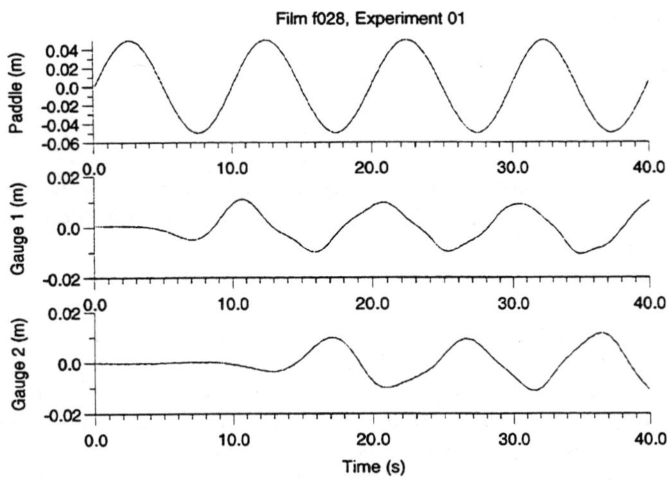

Figure 5: Long waves with equal layer depths, $f = 0.1Hz$, $H_{1/3} \simeq 0.020m$, $h_{1,2} \simeq 0.25m$, $\Delta\rho/\rho = 8.8\%$.

Long waves

Long wave experiments were subdivided into those where the depths of the upper and lower layers were equal, and cases where the upper layer was much thinner than the lower. The wavelength was in the range $\lambda > 10h_{1,2}$ at least, and $\lambda > 16h_{1,2}$ most of the time. The steepness range expressed in terms appropriate to the long waves was $H_{1/3}/h_1 \simeq$ 0.03–0.4.

Figure 5 shows the wave gauge traces of an experiment where the layer depths were equal. The wave shape is not sinusoidal. The symmetry in the flow favours odd harmonics over even ones, tending to make the wave shape more triangular. In contrast, the strong asymmetry in the flow in the case of unequal layer depths

	95%		90%		70%	
	u	w	u	w	u	w
Mean	1.41	1.26	1.37	1.23	1.36	1.26
SD	0.33	0.33	0.32	0.33	0.33	0.35

Table 2: Ratios of measured to predicted velocity percentiles for the short wave cases

	95%		90%		70%	
All cases	u	w	u	w	u	w
Mean	1.72	3.99	1.67	3.82	1.62	3.73
SD	0.53	1.68	0.49	1.69	0.42	1.84
Equal layer depths	u	w	u	w	u	w
Mean	2.07	4.12	1.97	4.05	1.78	4.09
SD	0.38	1.83	0.37	1.87	0.39	2.05
Unequal layer depths	u	w	u	w	u	w
Mean	1.20	3.80	1.22	3.48	1.38	3.21
SD	0.21	1.49	0.22	1.41	0.35	1.41

Table 4: Ratios of measured to predicted velocity percentiles for the long wave cases

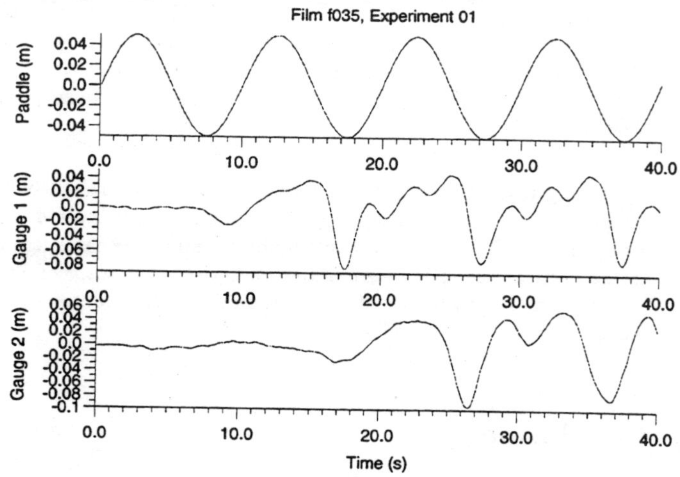

Figure 6: Long waves with unequal layer depths, $f = 0.1Hz$, $\Delta\rho/\rho = 9.7\%$, $h_1 = 0.065m$, $h_2 = 0.186m$, $H_{1/3} = 0.071m$.

(figure 6) favours even harmonics, sharpening the troughs and flattening the crests. This wave train looks more like a series of solitary or cnoidal waves than a packet of sinusoidal waves.

Both of these figures suggest that the paddle motion was not particularly well matched to the wave motion. For the second figure, the smaller parasitic bumps on the wave train may be free harmonics which arose from this. Alternatively, they could be considered to be part of a sequence of solitary waves which was evolving from the disturbance created by the paddle (Osborne & Burch, 1980).

It was not possible to run the Stokes expansion past first order for these waves as the second order term would not converge when λ was large. In any case, it can be shown that a small order Stokes

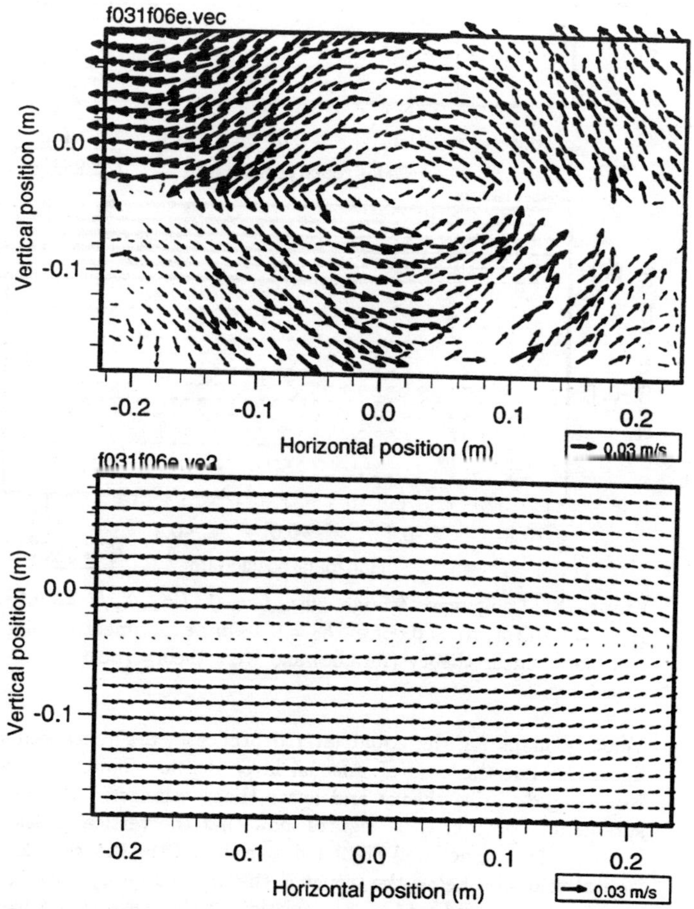

Figure 7: Long wave trough with $f = 0.1Hz$, $\Delta\rho/\overline{\rho} = 9.7\%$, $H_{1/3} = 0.022m$, $h_{1,2} = 0.15m$, experimental (top) and theoretical velocities

expansion produces anomalous bumps in the wave shape for even moderate amplitude long waves (Dean & Dalrymple, 1991).

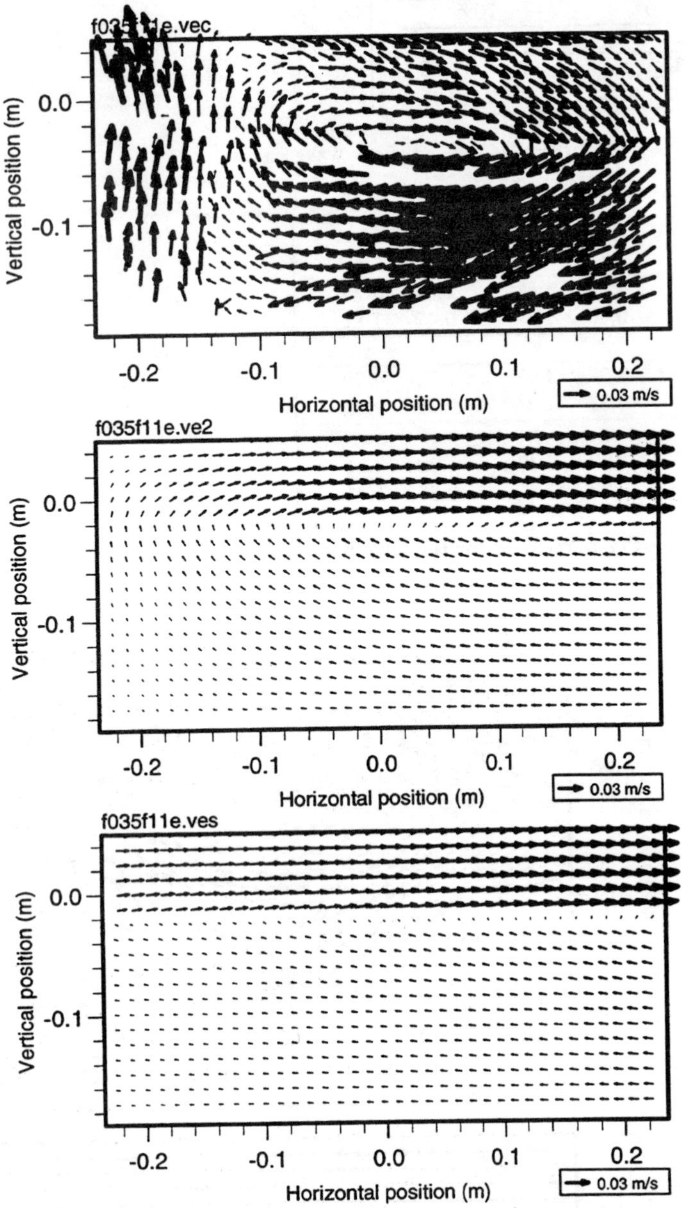

Figure 8: Long wave trough with $f = 0.1Hz$, $\Delta\rho/\overline{\rho} = 9.7\%$, $H_{1/3} = 0.071m$, $h_1 = 0.065m$, $h_2 = 0.186m$; experimental velocities (top), linear theory (middle) and KdV theory (bottom)

Velocity maps for the equal layer depth cases are presented in figure 7. For the case of unequal layer depths, figure 8 shows Korteweg-de Vries theory and linear theory alongside the measurements. Both of these figures show that the vertical motions in particular are not well predicted by either of the theories. Horizontal velocities below the trough in the unequal layer depth case are also badly predicted. The deviation of the wave shape from sinusoidal (or sech^2) is probably responsible for this in both cases.

The long wave data was analysed using the velocity coefficients and percentiles calculations detailed above. Only linear theory was evaluated in this way, because solitary wave theory would only produce meaningful results when compared to a wave trough

	All cases			
	a_0	a_1	b_0	b_1
Mean	-0.009	1.08	-0.005	1.24
SD	0.005	0.49	0.006	3.08
	Equal layer depths			
	a_0	a_1	b_0	b_1
Mean	-0.006	1.08	-0.004	1.47
SD	0.003	0.56	0.005	3.20
	Unequal layer depths			
	a_0	a_1	b_0	b_1
Mean	-0.012	1.07	-0.006	0.89
SD	0.006	0.39	0.008	3.02

Table 3: Means and standard deviations for the long wave velocity coefficients

in unequal layer depths, leaving too few cases for statistics to be meaningful.

Tables 3 and 4 show the results for the velocity coefficients and percentiles respectively. Most revealing in the former are the standard deviations in the vertical velocities, which show that the correlation between measurement and prediction was very poor. This would be expected where the wave shape had been poorly predicted.

Surprisingly, the percentiles show that velocities were slightly better predicted in unequal layer depths than in equal layer depths, especially for the horizontal component. In all of the cases, the extreme vertical velocities were underpredicted by a factor of about 4. Horizontal velocities exceeded predictions by a factor of 2 at most.

Conclusions

It has been shown that two-dimensional short waves travelling along the interface between two miscible fluids produce velocities which can be predicted to within about 16% on average, and within about 40% in the extremes, using second order Stokes theory. The extremes are found near the interface, and it is there that higher orders of expansion would make corrections. These higher orders would not be valid if viscosity were not included, however.

Long waves were compared with linear theory and Korteweg-de Vries theory, with statistical comparisons for the former. It was found that assumptions made in the theory about the wave shape appeared to prevent the theory from accurately predicting the kinematics. The extreme horizontal velocities exceed linear predictions by a factor of up to 2, while extreme vertical velocities exceed predictions by a factor of 4.

Both horizontal and vertical velocities are significant in deep water offshore engineering. The latter is important especially when catenary risers and tethers are used, and also where internal waves occur at diving depth. As pointed out by Evans (1996), the vertical motions associated with the large amplitude waves observed by Osborne & Burch (1980) could cause a diver to undergo large depth changes in a short period of time.

It would be advantageous to measure the kinematics of smaller amplitude waves, in the hope of observing the domains of validity of the linear and weakly non-linear theories. It has not proved possible to measure smaller waves with the existing PIV system.

Acknowledgments

This work is part of a joint project funded by the UK Health and Safety Executive, in collaboration with the Department of Civil Engineering, University of Dundee, and the Department of Physics and Astronomy, University of Edinburgh. The authors acknowledge helpful conversations with Dr Don Smith (HSE), Dr Colin Grant (BP Exploration) and Dr Alan Evans (University of Kent at Canterbury)

References

Benjamin, T. Brooke (1966).
Internal waves of finite amplitude and permanent form.
J. Fluid Mech., **25** (2), 241–270.

Bole, J.B., C.C. Ebbesmeyer & R.D. Romea (1994).
Soliton Currents in the South China Sea: Measurements and Theoretical Modeling.
Proc. 26th Ann. Ocean Tech. Conf., Houston, Texas, USA, 2–5 May (OTC 7417).

Dean, Robert G. & Dalrymple, Robert A. (1991).
Water wave mechanics for engineers and scientists,
Singapore: World Scientific.

Evans, W.A.B. (1996).
Exact Numerical Solutions to *Large Amplitude* Internal Solitary Waves.
Proc. 6th Int. Conf. Offshore & Polar Eng. (ISOPE 1996), 124–130. Los Angeles, USA, 26–31 May. Golden, Colorado: International Society of Offshore and Polar Engineers.

Gray, C. & T. Bruce (1995).
The Application of Particle Image Velocimetry (PIV) to Offshore Engineering.
Proc. 5th Int. Offshore & Polar Eng. Conf. (ISOPE '95), **3**, 701–708

Korteweg, D.J. & De Vries, G. (1895).
On the change of form of long waves advancing in a rectangular canal, and on a new type of long stationary waves.
Phil. Mag., Ser. 5, **39**, 422–443.

Osborne, A.R., T.L. Burch, & R.I. Scarlet (1978).
The influence of Internal Waves on Deep-Water Drilling.
J. Petroleum Tech., **30** 1497–1504.

Osborne, A.R. & Burch, T.L. (1980).
Internal Solitons in the Andaman Sea.
Science, **208** (4443), 451–460

Sherwin, T.J. (1991).
Evidence of a Deep Internal Tide in the Færoe-Shetland Channel,
In: Parker, Bruce B. (ed.). *Tidal Hydrodynamics*, 469–488, New York: Wiley

Thorpe, S.A. (1968).
On the shape of progressive internal waves.
Phil. Trans. Roy. Soc. A, **263**, 563–614.

Thorpe, S.A. (1975).
The Excitation, Dissipation and Interaction of Internal Waves in the Deep Ocean,
J. Geophys. Res., **80** (3), 328–338

Proceedings of the Seventh (1997) International Offshore and Polar Engineering Conference
Honolulu, USA, May 25-30, 1997

ISBN 1-880653-28-1 (Set); ISBN 1-880653-31-1 (Vol. III)

Nonlinear Models of Internal Tide and Internal Solitary Wave Evolution Over a Continental Slope

Peter E. Holloway
University of New South Wales, Australian Defence Force Academy, Canberra, Australia
Efim Pelinovsky and Tatyana Talipova
Institute of Applied Physics, Russian Academy of Sciences, Nizhny Novgorod, Russia
Belinda Barnes
University of New South Wales, Australia (Presently at Australian National University)

ABSTRACT
A numerical solution to the generalized Korteweg-de Vries (K-dV) equation, including horizontal variability and dissipation, is used to model the evolution of an initially sinusoidal long internal wave, representing an internal tide. The model shows the development of shocks and solitons as it propagates shorewards over the continental slope and shelf. Model results are compared to current meter and thermistor observations from the Australian North West Shelf and demonstrate good agreement with wave amplitude and waveform. It is found from observations that the coefficient of quadratic nonlinearity in the K-dV equation changes sign from negative in deep water to positive in shallow water and this plays a major role in determining the form of the internal tide transformation. On the shelf there is strong temporal variability in the nonlinear coefficient due to both background shear flow and the large amplitude of the internal tide which distorts the density profile over a wave period. Both the model and observations show the formation of an initial shock on the leading face of the internal tide. In shallow water, the change in sign of the coefficient of nonlinearity causes the shock to evolve into a tail of short period sinusoidal waves. After further propagation a second shock forms, on the back face of the wave, followed by a packet of solitons. Friction is found to be important in limiting the amplitudes of the evolving waves. Modelling of the generation of the internal tide, using a primitive equation model, is briefly discussed and shows strong distortion of the internal tide from a sinusoidal form as well as the formation of bores and short period waves.

KEY WORDS:
Internal solitary waves; Internal tides; nonlinear modelling; internal wave observations

INTRODUCTION
The action of the barotropic tide advecting density stratified water up and down sloping topography can result in the generation of tidal period internal waves, known as internal tides. These waves are a common feature of many continental slope and shelf regions, see for example the review by Huthnance (1989). These internal tides are most commonly seen at semi- diurnal periods (12.42 hr for M_2) and have wavelengths of 10's of km. In many instances these long internal waves are observed to form into nonlinear waves as they propagate. These nonlinearities can be in the form of bores (internal hydraulic jumps) and internal solitary waves with soliton-like character including cnoidal wave-trains. Observations from the Australian North West Shelf (NWS) show a long internal tide (wavelength $\sim 20km$), propagating onshore, that evolves into bores and short period oscillatory waves in the region of the shelf break (Holloway, 1987, Smyth and Holloway 1988).

Internal waves can be energetic phenomena, often associated with bottom intensification of the currents (e.g. Holloway, 1988), that can scour sediment from around seabed structures as well as providing significant loadings on structures in the ocean.

This paper presents a number of observations from the NWS of nonlinear internal tides showing bores and internal solitary waves. Secondly, a review is presented of modelling carried out by Holloway et.al (1997) where a Korteweg-de Vries (K-dV) model is used in an attempt to explain some of the observed features of the evolution of the long internal tide into nonlinear waves. Finally, a primitive equation (hydrostatic) model is used to simulate the generation of the internal tide, forced by the surface tidal oscillations, and some nonlinearities in the resulting internal tide are examined.

OBSERVATIONS OF INTERNAL WAVES ON THE AUSTRALIAN NORTH WEST SHELF
Observations of internal tides and nonlinear internal waves from the NWS are presented in this section. The observations are obtained from moored current meter data including velocities and vertical displacements of isotherms calculated from temperature records from the current meters. Fig. 1 shows the location of the three moorings, named *Slope*, *Break* and *Shelf*, and Fig. 2 shows a cross section with mooring and instrument positions. Data were

recorded at 2 min intervals using vector averaging acoustic current meters or at 5 min intervals using InterOcean S4 electrogagnetic current meters, as shown in Fig. 2.

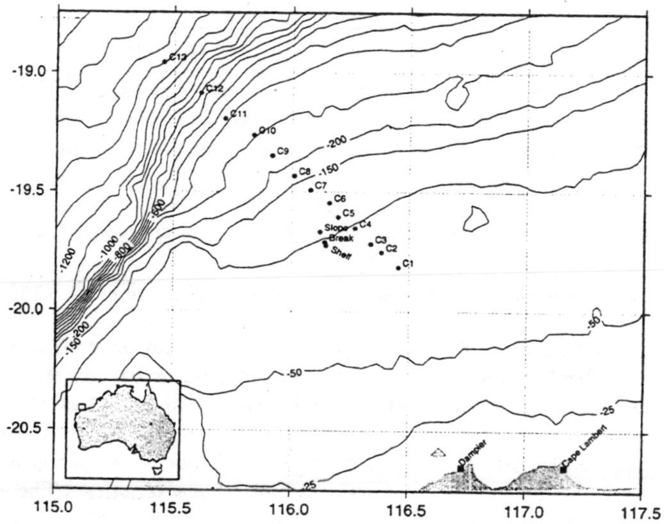

Fig. 1. Map of the Australian North West Shelf showing bathymetry, CTD station locations from 1995 (C1 to C13) and mooring locations *Slope*, *Break* and *Shelf* from 1992.

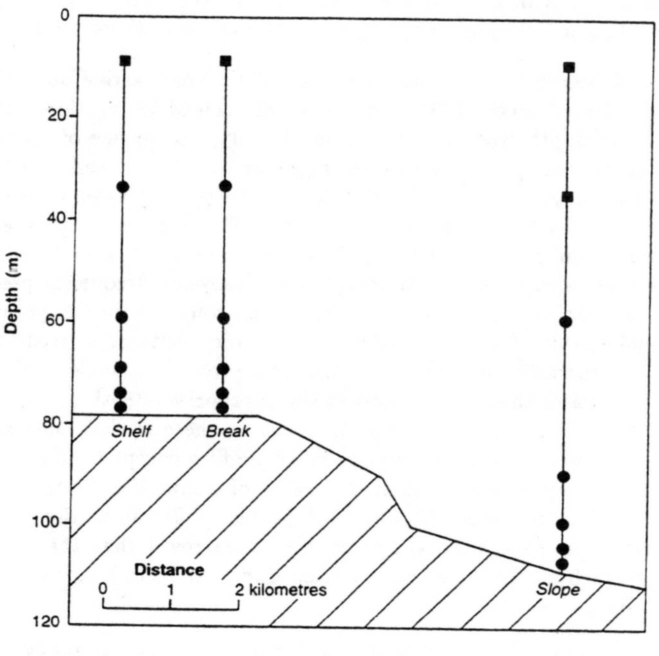

Fig. 2. Bathymetric cross section showing mooring locations and current meter positions for the 1992 data set. Current meters were Steedman Acoustic current meters (•) or InterOcean S4 electromagnetic current meters (□).

Fig. 3 presents time series of isotherm displacements from the 3 moorings over a period of 21 days that show strongly oscillatory signals. Spectra of each series show peaks at approximately 12 hr period reveallling a large amplitude internal tide. The height of the oscillations are up to 50 m indicating very large amplitudes compared to water depth (e.g. $a/H \sim 0.25$). Coherence and phase analysis gives a shoreward propagation of the oscillations with an average lag of 1.72 hr between *Slope* and *Break* and a lag of 1.09 hr between *Break* and *Shelf*. If the internal waves are propagating in a direction orthogonal to the mooring line (directly onshore) then these time lags correspond to phase speeds of 0.82 and 0.38 ms^{-1}, between each mooring pair. These values are consistent with expected values strongly supporting the existance of shoreward propagating internal tides. Further examples of internal tide observations from this region are given by Holloway (1984,1994).

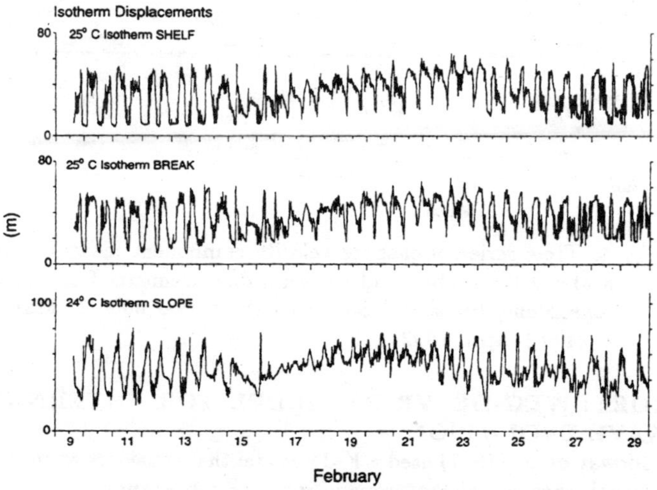

Fig. 3. Time series of isotherm displacements, plotted as height above the seabed, at locations *Slope*, *Break* and *Shelf*, over a 21 day period. Locations are shown in Fig. 1.

The oscillations seen in Fig. 3 are strongly distored from sinusoidal, indicating the waves to be nonlinear. A more detailed example of the waves is given in Fig. 4 (from Holloway et.al (1997). This plot shows isotherm displacements and onshore currents at 10 m above the sea bed from the 3 mooring locations. The currents associated with the internal waves are generally strongest near the seabed. The isotherm displacements are characterised by steep faced waves (shocks) at the leading face and sometimes at the trailing face of the wave. Short period oscillatory waves (internal solitary waves) are are often seen following the shocks. At times the short oscillatory waves occur throughout the waveform (e.g between B_4 and B_5 at *Shelf*). The currents at *Slope* are very intense (0.80 ms^{-1}) and show strong internal solitary waves. The onshore propagation of the waves is also seen by tracking events from *Slope* through to *Shelf*. Considerable change in the waveform is also evident as the waves propagate, particularly from *Slope* to *Break*, a distance of 5070 m. For events B_2 and B_4 the shocks strengthen between *Slope* and *Break* whereas B_3 is seen to delelop into a shock on the trailing face of the wave. There is a strong similarity between the isotherm and current signals, as would be expected, except that the current also contains a signal from the semi-diurnal barotropic tide. Further examples from the NWS are given by Holloway et al. (1997).

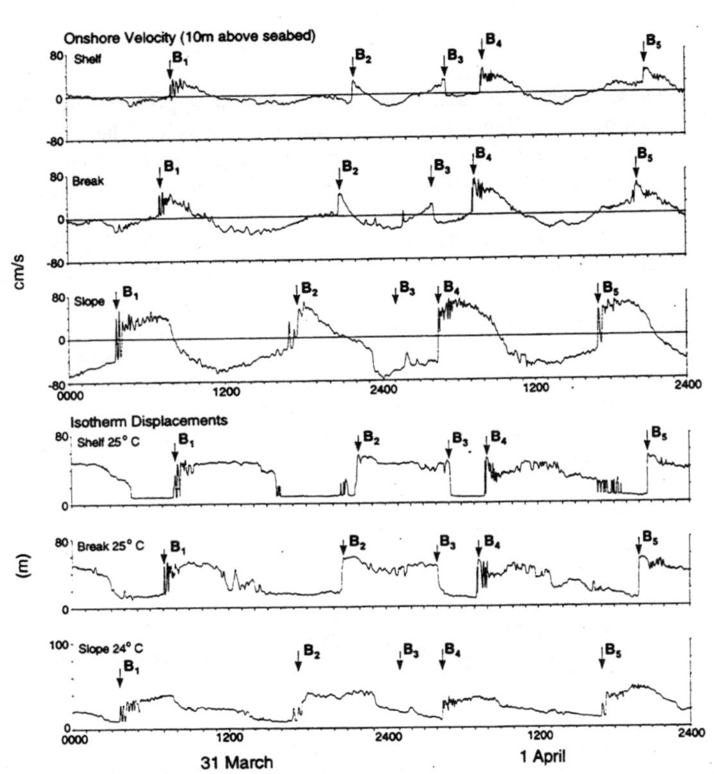

Fig. 4. Time series of onshore velocity component measured 10 m above the seabed and isotherm displacements from locations *Slope*, *Break* and *Shelf* over the period 0000 31 March to 2400 1 April 1992.

KORTEWEG-DE VRIES MODEL FOR INTERNAL WAVE EVOLUTION

Holloway et el. (1997) used a K-dV model that considers an initial internal-wave waveform propagating over topography with variable depth ($H(x)$) in the presence of a background density profile, which produces a background buoyancy frequency profile ($N(z)$), and with a background shear current ($U(z)$). The initial waveform can be arbitrary and the K-dV model predicts the evolution of the wave. The modified K-dV equation is employed to describe the nonlinear internal-wave evolution assuming the waves are long (wavelength exceeds the water depth), are small amplitude (amplitude is small compared to the water depth), and that the topography and environmental parameters are slowly varying in space. This is a weakly nonlinear theory.

The K-dV equation, including quadratic dissipation, is written as (eg. Zhou and Grimshaw (1989)),

$$\frac{\partial \eta}{\partial t} + c\frac{\partial \eta}{\partial x} + \alpha\eta\frac{\partial \eta}{\partial x} + \beta\frac{\partial^3 \eta}{\partial x^3} + \frac{c}{2Q}\frac{dQ}{dx}\eta + \frac{kc^2}{\beta}\eta|\eta| = 0 \quad (1)$$

where $\eta(x,t)$ is the wave displacement,

$$Q = \frac{c^2 \int_{-H}^{0} (c-U)\,(d\Phi/dz)^2\,dz}{c_0^2 \int_{-H}^{0} (c_0-U_0)\,(d\Phi_0/dz)^2\,dz} \quad (2)$$

and comes from the variable depth where values with index "0" are the values at some origin x_0 and k is the friction coefficient. The parameters α, β and c are coefficients of nonlinearity, dispersion, and phase speed of long internal waves, respectively. They are determined by the background density and horizontal velocity profiles and are defined as (in the Boussinesq approximation)

$$\alpha = \frac{3}{2}\,\frac{\int_{-H}^{0} (c-U)^2\,(d\Phi/dz)^3\,dz}{\int_{-H}^{0} (c-U)\,(d\Phi/dz)^2\,dz} \quad (3)$$

$$\beta = \frac{1}{2}\,\frac{\int_{-H}^{0} (c-U)^2\,\Phi^2\,dz}{\int_{-H}^{0} (c-U)\,(d\Phi/dz)^2\,dz} \quad (4)$$

where z is a vertical coordinate, positive upwards. The phase speed and vertical structure of the vertical-displacement amplitude of a wave mode $\Phi(z)$ are determined by the solution of the eigenvalue problem

$$\frac{d}{dz}[(c-U)^2\frac{d\Phi}{dz}] + N^2(z)\Phi = 0\,,\ \Phi(-H) = \Phi(0) = 0 \quad (5)$$

The numerical scheme is described by Holloway et al. (1997).

Numerical solutions of (1) to (5) are found once stratification and background velocity are defined over a bathymetric cross-section and an initial waveform is given. The model assumes that the internal wave can be described by a modal function and the waves are first mode. This neglects effects of steep topography and the propagation of waves along characteristics. The coefficients in (1) are also assumed constant but are allowed to vary slowly with distance such that at each value of x, local values are used.

COEFFICIENTS IN THE K-dV EQUATION

In studying the transformation of the internal tide on the NWS it is necessary at first to investigate the spatial variability of the coefficients of the K-dV equation (1) as the nature of the wave evolution is strongly dependant on the values of these coefficients.

Observations are used from a CTD survey carried out on the NWS in January 1995 (summer). At each of 13 locations, ranging in depth from 65 to 1382 m (Fig. 1), a sequence of repeated profiles were measured every 30, 60 or 90 min, depending on the water depth, over a 13 hr cycle. These profiles have been averaged at each location to remove variability induced from internal waves and the resulting CTD cross-section is used to define the coefficients α, β and c. For comparison, buoyancy frequency profiles from climatic data are calculated from profiles of the temperature and salinity for this region in a summer - autumn period taken from the atlas of Levitus (1982). This provides a single profile for deep water that is truncated at the appropriate depth in shallower water. For depths greater than 200 m there is close agreement between the measured and climatic profiles except for differences in the depth of the maximum value of Brunt-Vaisala frequency. For shallow water the observed profiles differ from the Levitus values, in particular, the pycnocline is narrower than the climatic pycnocline. This can be expected due to bottom boundary layer mixing over the shelf.

Coefficients of the K-dV equation are calculated for the 1995 data and Levitus data without the effects of background shear flow using (2), (3) and (4) with $U = 0$. The results are plotted in Fig. 5 as a function of distance across the slope/shelf region. It is seen that the estimates of phase speed and the dispersion coefficient from climatic data and those from the observations are very similar for all water depths. Both c and β depend primarily on the depth with a weaker dependance on the shape of the vertical profiles of buoyancy frequency. However, the nonlinear coefficients differ considerably, particularly in shallower water, reflecting the strong dependance of α on the shape of the vertical

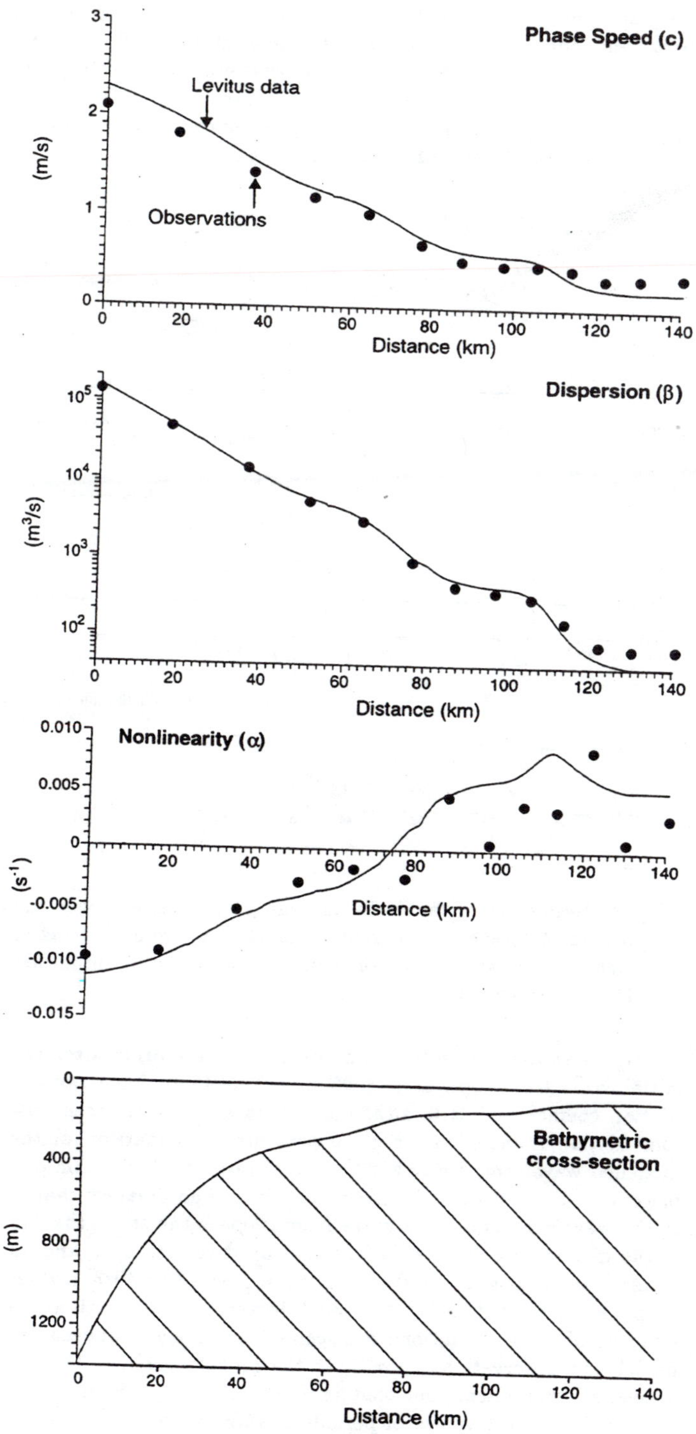

Fig. 5. Coefficients of the K-dV equation calculated from CTD data of 1995 (points) and from Levitus climatic data (solid line) as a function of distance across the shelf-break region. The bathymetric cross-section is also shown.

density profile. The sign of α is seen to change from negative in deep water to positive in shallow water, the zero value occuring at about 180 m water depth. The change in sign indicates an expected change in polarity of any solitons from downward in deep water to upward in the shallow water.

K-dV MODEL SIMULATIONS OF INTERNAL WAVE EVOLUTION

In this section results are presented from model runs using a variety of initial waveforms and K-dV coefficients from the 1995 data and Levitus data sets. Firstly, sinusoidal waves with period of 12hr and various amplitudes originating at a depth of 1382m (corresponding to the most offshore CTD data location), both with and without frictional dissipation, are considered. Results are insensitive to the position of this offshore origin as little evolution of the waveform occurs in deep water. Secondly, an observed waveform from the observations at *Slope* is used as the initial waveform and the evolution compared to observations at *Break* after the wave has propagated approximately 5 km.

Fig. 6 shows modelled waveforms at different distances from the origin with c, α and β calculated from the 1995 data. The run is for an initial wave amplitude of 10 m and without frictional dissipation ($k = 0$). In this case the wave propagates over 100 km (to water 150 m deep) before the wave steepens sufficiently to form a shock on the leading face of the wave. A group of solitons then form and grow in amplitude as the wave continues to move shoreward. However, the height of the solitons exceeds 100m, a value that is unrealistically high for these wave depths. It is clear that some dissipation is needed to reduce these wave heights.

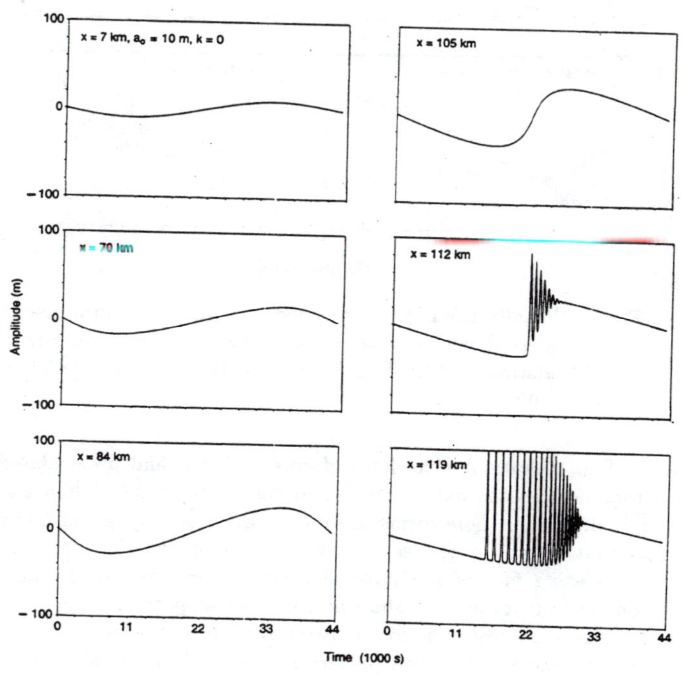

Fig. 6. Results from the K-dV model. The waveform over one period is shown at different distances (x) from the origin. Calculations are for data from 1995. Initial amplitude 10 m, $k = 0$.

Fig. 7 shows modelled waveforms with K-dV coefficients calculated from the 1995 data, initial amplitude 30 m and frictional dissipation with $k = 0.0013$. With the larger initial wave amplitude than the example above, the wave form steepens more rapidly and forms a shock on the trailing face of the wave after propagating approximately 70 km (to water 220 m deep). This is in the region where the nonlinear coefficient is negative and the solitons that form from the shock have negative polarity. These waves then propagate through the region where α changes sign from negative to positive. The effect on the solitons is to transform these waves into oscillatory cnoidal waves that are superimposed on the long internal tide. After further propagation a second shock forms and a new group of positive solitons form. Remnants of the cnoidal waves are still present producing a "noisy" waveform. The height of the shocks, solitons and cnoidal waves are up to 50m, consistent with values seen in the observations.

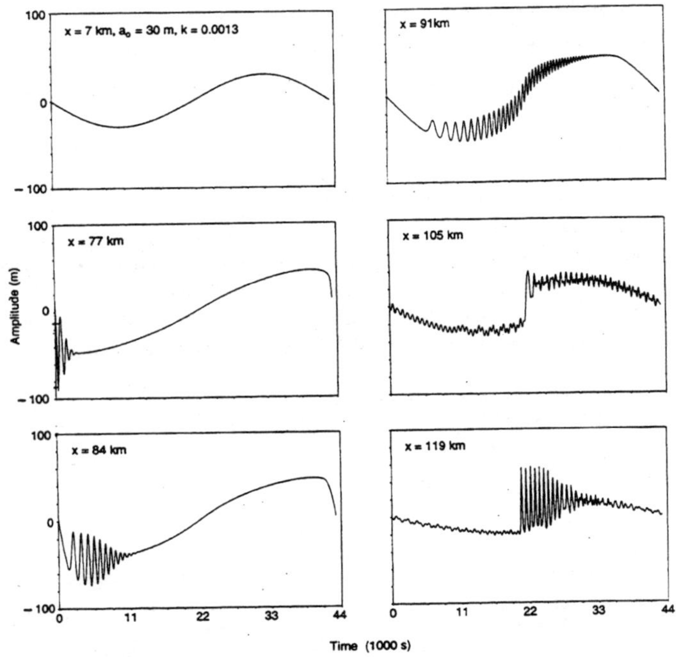

Fig. 7. Results from the K-dV model. The waveform over one period is shown at different distances (x) from the origin. Calculations are for data from 1995. Initial amplitude 30 m, $k = 0.0013$.

Fig. 8 shows modelled waveforms with c, α and β calculated from the Levitus data, with initial wave amplitude of 30m and $k = 0.0026$. A bore forms after the wave has propagated 90km so that it is in a region of positive α. Solitons then form on the leading face of the wave and grow in number as the wave continues to propagate shorewards. The amplitude of the soliton envelope decays almost linearly away from the bore. A very smooth and regular wave pattern is seen with maximum soliton height of about 60m.

For a more detailed comparison of the numerical simulations with observed data, use is made of the simultaneous measurements of isotherm displacements and currents at the locations *Slope*, *Break* and *Shelf*. Isotherm displacements, over a 12 h period, observed at the *Slope* location are used as the initial waveform and the K-dV model is used to predict the wave transformation as it propagates to *Break* and then to *Shelf* locations. Isotherm displacements from 0000 to 1200 on 13 April are used in the example presented here. Background vertical profiles of the density and shear flow fields and the corresponding coefficients of the K-dV equation are obtained by averaging the observed data from *Slope* over the 12 hr period 0000 to 1200. The averaging for *Break* and *Shelf* is carried out for the time interval 0300 - 1500, allowing for the propagation time of internal waves. The values of α at *Slope, Break* and *Shelf* are 0.00344, 0.00161 and $-0.00018s^{-1}$ respectively. The spatial variation of the coefficients between the locations are approximated by a cubic spline.

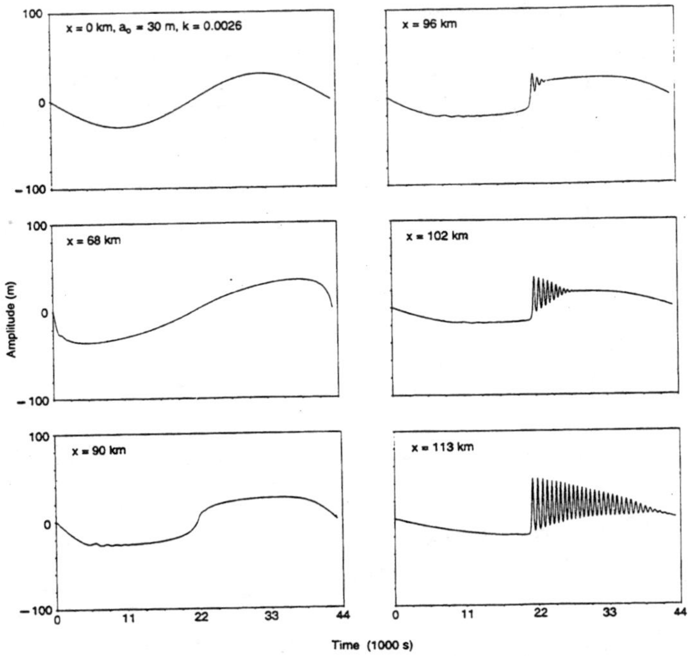

Fig. 8. Results from the K-dV model. The waveform over one period is shown at different distances (x) from the origin. Calculations are for Levitus climatic data. Initial amplitude 30 m, $k = 0.0026$.

Fig. 9 shows the initial and predicted waveforms after the initial wave has propagated 4920 m and 5248 m for a bottom friction coefficient $k = 0$, although results are similar when bottom friction is included. Fig. 10 presents a comparison of the modelled waveform at $x = 4920m$ and the observed isotherm displacement time series from *Break* location. The greatest change in the waveform as it propagates from *Slope* is the strengthening of the forward shock, formed from the positive sign of α, and an increase in amplitude of the short-period waves. With further propagation towards *Shelf*, the shock increases in strength and a large number of oscillations are seen. Comparison between the model and observations show that the model describes the low frequency component and leading shock correctly. Simulations predict the correct values of periods of short waves and their locations. There are differences between predicted and observed amplitudes of the short wave tail most likely due to limitations of the theory, which is valid for relatively small amplitude waves and for constant values of coefficients of K-d V equation (over 12 hr). However, the agreement between theory and observations is reasonable.

EFFECTS OF TEMPORAL VARIABILITY AND SHEAR ON THE K-dV COEFFICIENTS

The solutions above show the strong dependance of the wave evolution on the distribution of α. Equally it is found that for shallow water around the shelf break, background shear has a large influence on the modal structure and hence on the value of α. This is demonstrated by the results plotted in Fig. 11. Modal functions are plotted for two different times, either side of a bore, with temperature (giving the density) and velocity profiles as shown. If the background flow is neglected, the modal functions show that the maximum wave amplitude would occur below mid-depth for the depressed stratification (0200 hr) and above mid-depth for the raised stratification (0600 hr). Values of α are positive and negative respectively, as could be anticipated. However, the shear strongly distorts the modal functions, significantly changing the depths of maximum wave amplitude and reversing the signs of α, hence changing the nature of solitary waves that might occur under these conditions. This not only demonstrates the importance of background shear but also shows that these waves are strongly nonlinear. The assumption of constant coefficients is violated in the presence of these large amplitude waves.

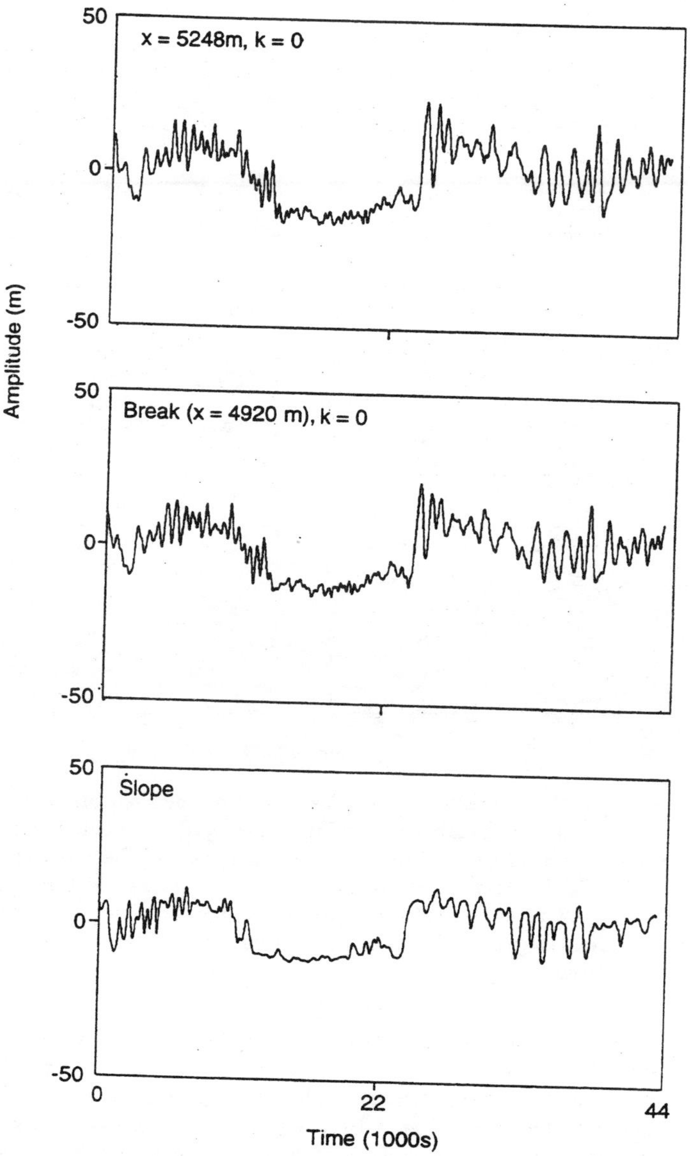

Fig. 9. Numerical simulation, using the K-dV model, of wave transformation using the waveform at *Slope* as the inital waveform. Predicted waveforms at $x = 4920m$ (near *Break*) and at $x = 5248m$ are shown for $k = 0$. Data are for a 12 hr period on 13 April 1992.

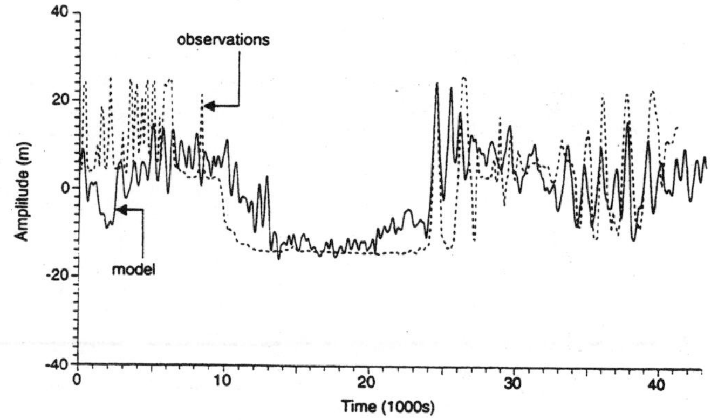

Fig. 10. Comparison of predicted waveform at ($x = 4920m$) with observed isotherm displacements at *Break* on 13 April 1992. Results are calculated with $k = 0$.

PRIMITIVE EQUATION MODEL OF INTERNAL TIDE GENERATION

Modelling of the generation and propagation of the internal tide for the NWS has been carried out by Holloway (1996). Forced by the sea-surface displacement associated with the barotropic tide, the model producess a barotropic tidal flow across a section of the continental shelf which in turn produces the internal tide. The model is a nonlinear primitive equation model using sigma corodinates and the Mellor Yamada (1982) turbulence closure scheme to model vertical mixing of momentum, heat and salt (Blumberg and Mellor, 1987). The model employs the hydrostatic assumption which is not strictly valid for high frequency internal waves when the wave frequency is similar to the buoyancy frequency. However, model runs using a high spatial resolution do show nonlinear, short period waves forming from the internal tide as it propagates over the shelf break region. Note that Lamb (1994) has used a non-hydrostatic primitive equation model to investigate the formation of tidally forced internal solitary waves.

The results of Holloway (1996) show several regions over the continental slope where the internal wave characteristics are approximately parallel to the seabed i.e. the seabed slope is critical. Near these regions, at approximately 350m and 125m depths, strong generation of the internal tide is seen in the model with bottom intensification of the currents. The waves propagate onshore with wave energy going both onshore and offshore, following characteristic paths. Strong dissipation of the signal occurs over the outer continental shelf, consistent with the observed character of the internal tide.

In this paper we present results from the internal tide generation model, forced by M_2 period sea surface oscillations with an amplitude at the offshore boundary of 0.95m representing Spring Tide values. Stratification is from the CTD profile at location

C13 (Fig. 1) and bathymetry from a combination of bathymetric charts and the ships cruise track. The model domain is from the coast out to a depth of 1600 m at a distance of 300 km. There is no variability in the alongshore direction. Imbedded in the model is a fine resolution grid with grid spacing of 100m over a distance of 35 km around the shelf break and there are 61 vertical grid levels.

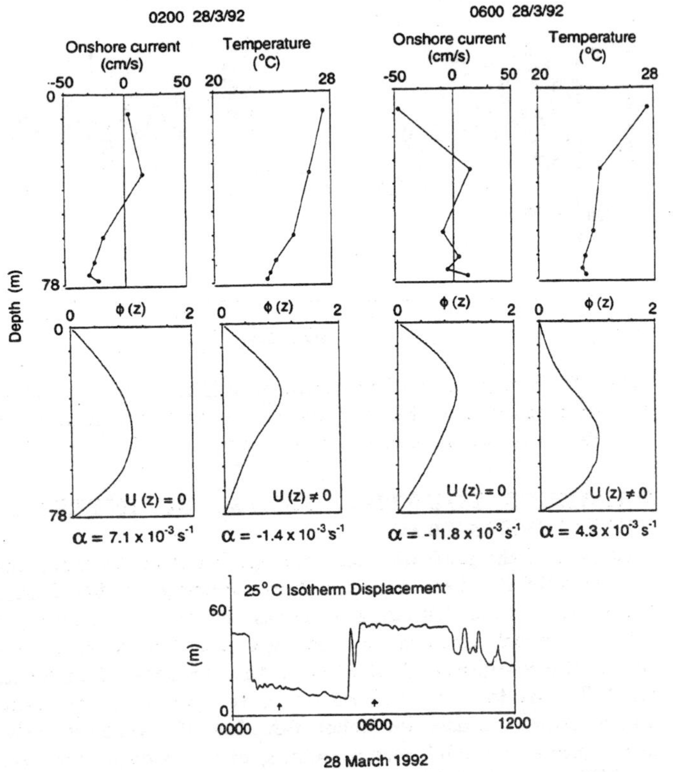

Fig. 11. Velocity and temperature profiles for times ahead and behind a bore along with vertical modal functions calculated with and without the influence of background current, and corresponding values of the nonlinear coefficient (α).

Snapshots of the density field around the shelf break, at 2 hr intervals over a wave cycle, are plotted in Fig. 12. The first panel shows a deep trough in the internal tide propagating onshore. This progresses in the same direction as the barotropic tidal flow and the leading face of the internal tide is seen to steepen and to form a bore with a packet of short period oscillatory waves. These waves dissipate after propagating for approximately 4 hr. The leading face of the internal tide then slows as the barotropic tide turns to flow offshore. However, the internal tide continues propagating onshore indicating that the phase speed is faster than the barotropic tidal flow. At this time the trailing face of the internal tide steepens and another deep trough begins to form. The model demonstrates the nonlinear steepening of the internal tide and the formation of a bore or shock on the leading wave-face, followed by a packet of short period waves, consistent with observations from the region, e.g. Fig. 4.

DISCUSSION

This paper has presented observations and modelling of the generation and evolution of internal tides and nonlinear internal solitary waves on the Australian North West Shelf. A review of modelling on the NWS based on the solution of the K-dV equation, incorporating slowly varying topography, background shear flow, stratification and frictional dissipation is presented. This model is valid to first order in wave amplitude. It is seen that the model explains a number of features of the observed short period waves and their formation from the internal tide. The results also demonstrate the sensitivity of the wave evolution on the coefficient of nonlinearity and hence on the background stratification and shear. In particular, α changes sign at a depth of approximately 180 m on the NWS and solitons passing through this region are transformed to oscillatory waves.

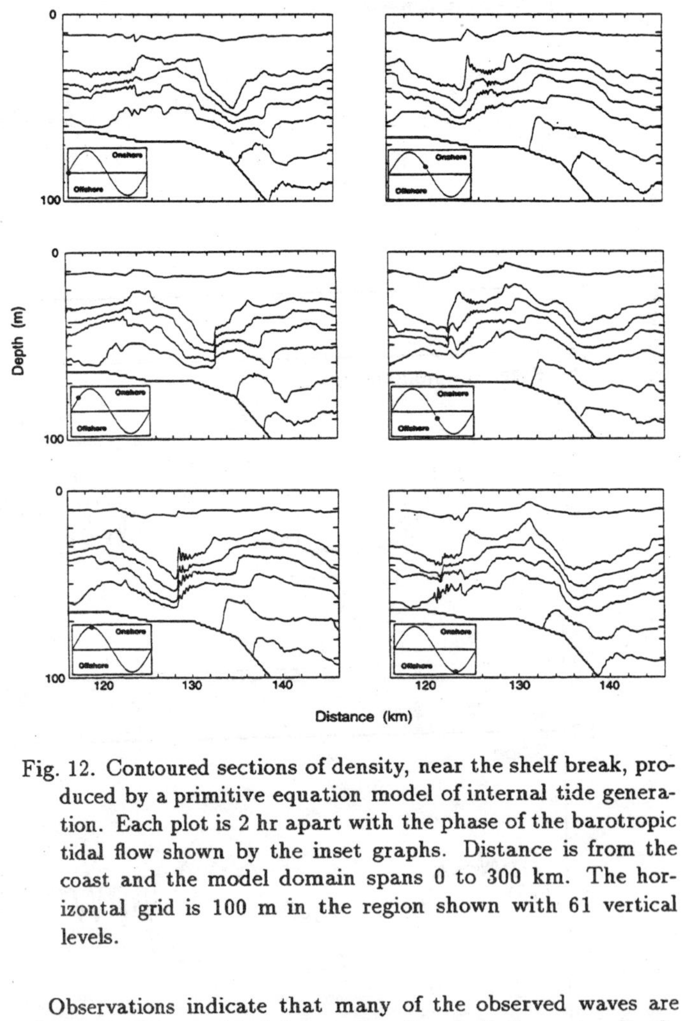

Fig. 12. Contoured sections of density, near the shelf break, produced by a primitive equation model of internal tide generation. Each plot is 2 hr apart with the phase of the barotropic tidal flow shown by the inset graphs. Distance is from the coast and the model domain spans 0 to 300 km. The horizontal grid is 100 m in the region shown with 61 vertical levels.

Observations indicate that many of the observed waves are strongly nonlinear with heights exceeding half the water depth. In these cases weakly nonlinear theory is not strictly valid. The nonlinear coefficient is seen to change sign at different phases of the wave and the assumption of constant coefficients is invalid. Several alternative approaches are available to account for the strong nonlinearities. The K-dV model can be extended to second order and cubic nonlinearity included. This work is in progress by the authors. Results are also presented from a primitive equation model. The advantage of such a model is that it does not employ many of the simplifying assumptions of the K-dV model. For example, slowly varying depth is assumed in the K-dV model whereas in places the depth is critical compared with the internal wave characteristics. Also, there is no assumption of small amplitude waves in the primitive equation model, although results presented in this paper are from a hydrostatic model. The disadvantage of primitive equation models is that it is more difficult to elucidate the essential physics of the proceesses involved, and they can be very demanding on computer resources.

To further our understanding of the generation and evolution of tidally forced internal waves, propagating over varying topography and through an environment with varying stratification and background shear flow, it is clear that a variety of modelling approaches are needed along with good quality observations.

ACKNOWLEDGMENTS

Financial support for TT and EP to work in Australia was provided by a Bilateral Science and Technology Grant. Support for BB has been provided by an ARC Small Grant. The data from 1992 was collected under an ARC Large Grant in collaboration with PH, Ray Steedman and Chris Fandry. The 1995 data was collected from R.V. Franklin by PH in collaboration with Peter Craig and Miles Furnas.

REFERENCES

Blumberg, A.F. and G.L. Mellor (1987) A description of a three-dimensional coastal ocean model, p1-16, In: Three-dimensional coastal ocean models, Vol. 4, Ed. N. Heaps, pp208. American Geophys. Union, Washington, DC.

Lamb, K.G. (1994) Numerical experiments of internal wave generation by strong tidal flow across a finite amplitude bank edge. *J. Geophys. res.*, 99, 843-864.

Holloway P.E. (1984) On the semi-diurnal internal tide at a shelf-break location on the Australian North West Shelf. *J. Phys. Oceanogr.*, v. 14, 1778 - 1790.

Holloway P.E. (1985) A comparison of semi-diurnal internal tides from different bathymertic locations on the Australian North West Shelf. *J.Phys. Oceanogr.*, v. 15, 240-251.

Holloway P.E. (1987) Internal hydraulic jumps and solitons at a shelf break region on the Australian North West Shelf. *J. Geophys. Res.*, v. C95, 5405 - 5416.

Holloway P.E. (1994) Observations of internal tide propagation on the Australian North West Shelf. *J. Phys. Oceanogr.*, v. 24, 1706 - 1716.

Holloway, P.E., E. Pelinovsky, T. Talipova, B. Barnes, (1997), "A nonlinear model of internal tide transformation on the Australian North West Shelf". *J. Phys. Oceanogr.* (In Press).

Huthnance J.M. (1989) Internal tides and waves near the continental shelf edge. *Geophys. Fluid Dyn.*, 48, 81-106.

Levitus S. (1982) "Climatological atlas of the World Ocean". NOAA Professional Paper 13.

Mellor, G.L. and T. Yamada (1982) Development of a turbulence closure model for geophysical fluid problems. *Rev. Geophys. Space Phys.* 20, p851-875.

Smyth N.F., Holloway P. (1988), "Hydraulic jump and undular bore formation on a Shelf Break". *J. Phys. Oceanogr.* v. 18, 947 - 962.

Zhou X., Grimshaw R. (1989), "The effect of variable currents on internal solitary waves". *Dyn. Atm. Oceans*, v. 14, 17 - 39.

Proceedings of the Seventh (1997) International Offshore and Polar Engineering Conference
Honolulu, USA, May 25-30, 1997

ISBN 1-880653-28-1 (Set); ISBN 1-880653-31-1 (Vol. III)

Temporal Behaviour of a Mixed Layer in a Stratified Flow

S.A. Walker
University of Edinburgh, Edinburgh, United Kingdom
G.A. Hamill
Queen's University of Belfast, Belfast, United Kingdom
H.A. Barnes
Unilever Research, Liverpool, United Kingdom
H.T. Johnston
Queen's University of Belfast, Belfast, United Kingdom

ABSTRACT

The erosion of a salt water impoundment subjected to a fresh water flow is presented. Velocity and density profiles were obtained at various time intervals, from which the temporal behaviour of the interface was observed. Mixing across the interface was found to result in the formation of a third layer, the characteristics of which changed with time as the impoundment eroded. Two distinct mechanisms of erosion have been identified based upon the temporal behaviour of the interfacial layer.

KEYWORDS: Stratified flows, Interfacial Mixing, Barrages

BACKGROUND

The discharge of land runoff to the sea results in complex processes due to the interaction of two chemically and physically different water masses in the estuary. The estuary often acts as a sink, not only to sediment moving alongshore or downstream, but also for nutrients and pollutants. Because of this there has been an upsurge of interest in estuaries as 'marine lungs', acting as nurseries and refuges for many organisms and providing resources for industrial and urban growth.

When the river flow is high, the salt water and fresh water mix. When the river flow is low, a stratification forms as the less dense fresh water from the river overrides the salt water. Due to the frictional forces developed along the interface between the fluids, the lower saline mass will develop into a wedge. The salt wedge remains quasi-stationary, advancing or retreating due to variations in river flow rates and tidal cycles. Density stratification in fluids can also occur as a result of differences in temperature and concentration of suspended sediment. Thermally stratified flows may be generated by solar heating or by the discharge of power plant or industrial plant cooling water. A special case of highly stratified flow is an entrapped wedge, which occurs when a pond of saline water becomes retained behind a barrier of some form. The saline water is isolated from the ocean by the barrage and is dependent upon the transfer of oxygen from the fresh water. When vertical mixing is predominately of a diffusive nature, there is little transfer of oxygen between the fresh and salt water layers, and a stagnant layer with low dissolved oxygen content results.

The entrapped wedge has become more common as attempts are made to improve the estuarine environment. New barrages have been constructed for rehabilitation of the estuary, examples of which are the Tees, Cardiff Bay and Lagan barrages, or their principle objective may be to prevent flooding, as in the Thames, the Dutch North Sea Defence and the Venice Barrages, as discussed by Lewin (1996). It is possible to remove the saline water by drawing it off at low level in order to alleviate the problems associated with entrapped wedge. Alternatively mechanical methods, such as stirrers and aeration systems, are sometimes used to help break up the stratification.

The behaviour of stratified flows has been studied experimentally in laboratory flume tests by many previous researchers such as Keulegan (1966), Christodoulou (1986), Dermissis (1990), Grubert (1989) and Foo et al (1995). From these experiments a number of dimensionless parameters which influence the mixing processes and entrainment rates have been found. There is little agreement on the actual equations to describe the rate of entrainment. Considerable differences exist when differing experimental apparatus are used, such variations include oscillating grids, density underflows and buoyant shear flows. The lack of work conducted with a temporal theme is also evident with the most recent and relevant having been conducted by Grubert (1989).

The most commonly used dimensionless number is the Richardson number, which is a measure of the ratio of gravity forces to inertial forces in a shear flow having a vertical density gradient. It is widely used as a measure of the stability of the flow with regards to vertical mixing and has two forms; the gradient Richardson number defined as:

$$Ri_g = -g/\rho \frac{\partial\rho / \partial z}{\{\partial v / \partial z\}^2} \tag{1}$$

where $\partial\rho / \partial z$ is the vertical density gradient, $\partial v / \partial z$ is the vertical gradient of horizontal velocity, g is the acceleration due to gravity and ρ is the fresh water density; and the bulk Richardson number defined as:

$$Ri_b = g\frac{\Delta\rho H}{\rho V^2} \tag{2}$$

where $\Delta\rho$ is the density difference, H is the depth of fresh water and V is the average fresh water velocity. It is also known as the inverse square of the densimetric Froude number.

Other dimensionless parameters such as the Reynolds number (used by Grubert (1980) and Curran (1993) to define the degree of turbulence) and the inverse of the Keulegan number, used by Dermissis and Parthenaides (1985) and Arita and Jirka (1987) to investigate interfacial shear stress, are also relevant to entrainment.

A large number of entrainment experiments have been performed, most of which concentrate on the entrainment laws. The type of turbulence produced, the experimental set-up and the *Ri* range have been shown to influence the equations produced. Little work has been carried out on the temporal erosion of the saline layer and the parameters which influence it. Reynolds number has been shown by Grubert (1989) to have an influence on the entrainment rate but the main dimensionless parameter, used to describe entrainment, is the bulk Richardson number (Ri_b). The main aim of this study was to develop a more complete understanding of the behaviour of the entrapped wedge with particular attention to the behaviour of the interfacial or mixed layer.

APPARATUS AND PROCEDURE

Experiments were conducted in a laboratory flume, at The Queen's University of Belfast, which was 20m long, 0.75m wide and 0.75m deep. A schematic of the experimental set-up is shown in Figure 1. The salt water was retained between two baffles which were 17.5 m apart. The salt water depth was kept constant at 0.25m whilst the fresh water depths were varied by means of an adjustable weir. The fresh water was pumped over the top of the stationary salt water impoundment and various velocities and depths, combined with differing salt water densities were chosen for the investigation.

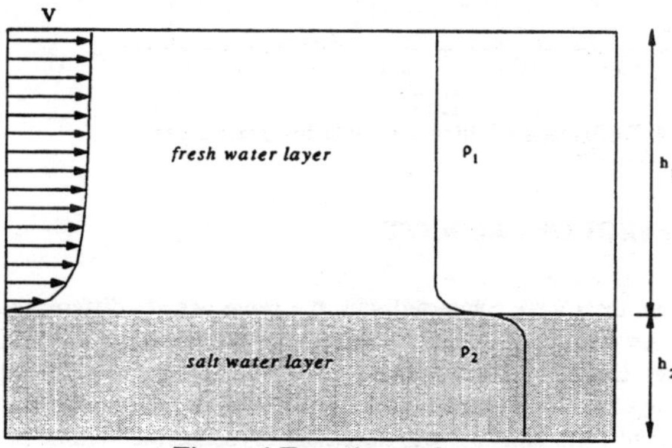

Figure 1 Experimental set-up

Fresh water was mixed with dried salt and dye in storage tanks and then pumped into the salt water impoundment to a depth of 0.25 m. Fresh water was then filled slowly on top, over the course of 12 to 15 hours (natural diffusion between the layers was found not to change the density profile significantly). Flow was initiated slowly in the upper layer by starting the fresh water pump. Once the flow had reached the required rate the water in the storage tanks was changed slowly. This was necessary due to the contamination of the fresh water which occurred as a result of mixing. The degree of contamination was small, but large enough to change the salinity of the fresh water over a substantial time period. The scanning process was commenced after flow was initiated. Velocity and density profiles were obtained at the centre of the channel. Velocity measurements were obtained using LDA with a fibre optic attachment to limit the effect of differing refractive idices. Density measurements were obtained using a conductivity probe with a small measuring volume. The probe consisted of two wires 5 mm in diameter which were completely covered except for 2 rings, 2 mm wide, near the bottom. Both probes were mounted on a linear actuator which was computer automated. The probes were placed in the salt water, usually about 20 to 30 mm below the interface in order to obtain a reading of maximum salinity. An average was taken over a period of 30 seconds and the probes were then moved up 5 mm and another point measurement was obtained. Due to the temporal nature of the experimental investigation it was necessary to obtain as many profiles per hour as possible. The profile obtained was therefore not a full vertical traverse but was limited to 120 mm around the interface. With time the interface moved downwards, as salt water was removed by entrainment from the impoundment, and it was necessary to incorporate movement of the base profile position into the scanning program.

DENSITY PROFILES

Velocity and density profiles were obtained at regular intervals during each test. A number of typical density profiles obtained during a test are plotted in Figure 2. The profiles are plotted in a dimensionless form, for comparison with differing tests, by plotting Z_0/H against $S/Smax$. where Z_0 is the position from the original interface, H is the initial depth of fresh water, S is the local salinity and $Smax$ is the maximum salinity within the profile. On the vertical scale 0 represents the original interface and 1 represents the water surface. On the horizontal scale 0 is the fresh water and 1 is fully saline water. The interface is clearly defined although the initial profile is not square due to a combination of mixing during filling and chemical diffusion.

The profile obtained after 25 minutes was similar to that reported by Arita and Jirka (1987) and Curran (1993) with a more sudden change to fully fresh than to fully saline. When flow was initiated, large interfacial shear waves were observed along the interface. After 1 hour 30 minutes another profile was taken from which a third layer is visible. Figure 2a-c presents a number profiles obtained after various time intervals. All the profiles show similar trends with a relatively small region, in relation to the depth of fresh water, over which the change from fully fresh to saline occurs. In between, is a region which is continually changing in density and thickness, known as the mixed or interfacial layer. This region is marked on the figures. To allow comparisons to be made of each test the relative density of the mixed layer was plotted against time. This was defined as the average density within the layer, a schematic of this definition is given in Figure 3. The linear portion of the density profile was used to define the density. In this region the density gradients were minimum, indicating the presence of a stable layer. To aid in further comparisons, the thickness of the interfacial layer was plotted with time. Visual observations of the thickness could have been used but this method was deemed less accurate as the layer was sometimes difficult to observe due to very small density differences. From the density profile shown in Figure 4, the thickness of the layer was defined as occurring between the points of maximum salinity gradient.

The behaviour of the layer due to differing fresh and salt water conditions was assessed. This was achieved by varying the three input parameters (V, H and $\Delta\rho$) individually as variations in the three parameters led to variations in Re and Ri_b which were the dimensionless terms of interest. It was therefore possible to achieve the same value of a dimensionless term with several combinations of input parameters. Although all three parameters were varied, only the effect of a variation of the fresh water velocity is presented.

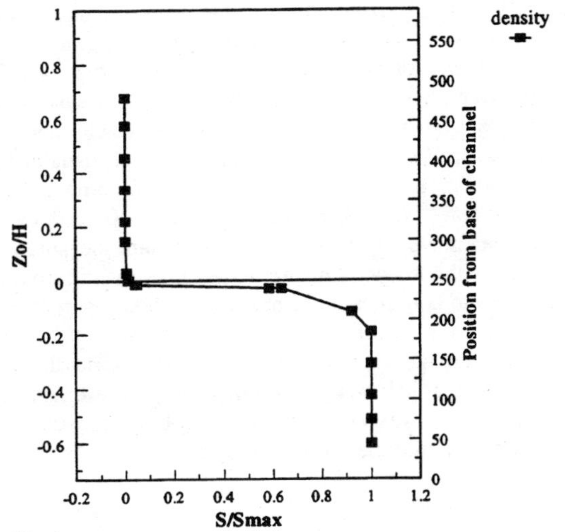

25 minutes

(a)

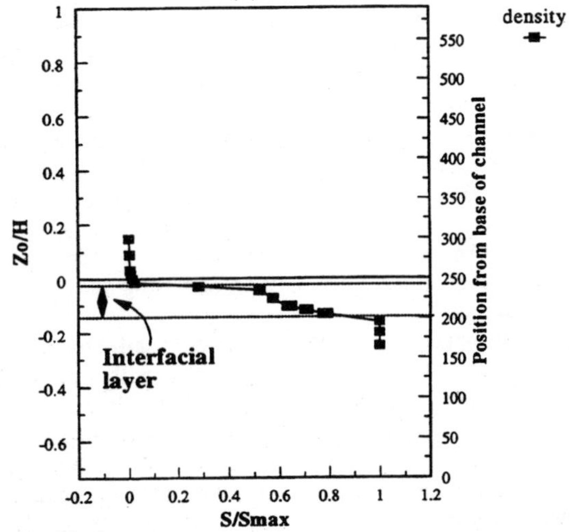

1 hr 30mins

(b)

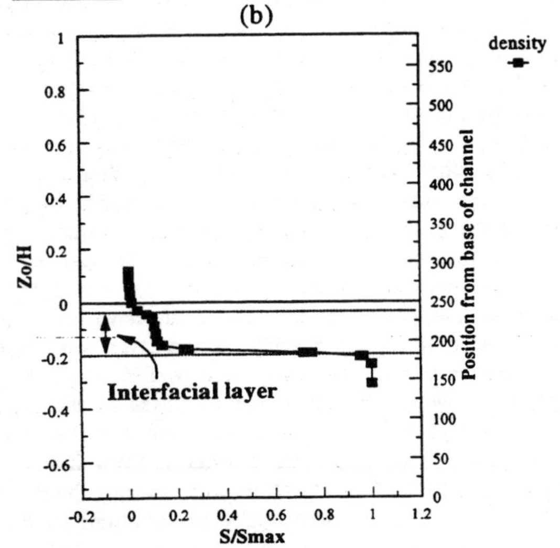

5 hours

(c)

Figure 2 Density profile at differing time intervals (a) after 25 minutes (b) after 1hour 30minutes (c) after 5 hours

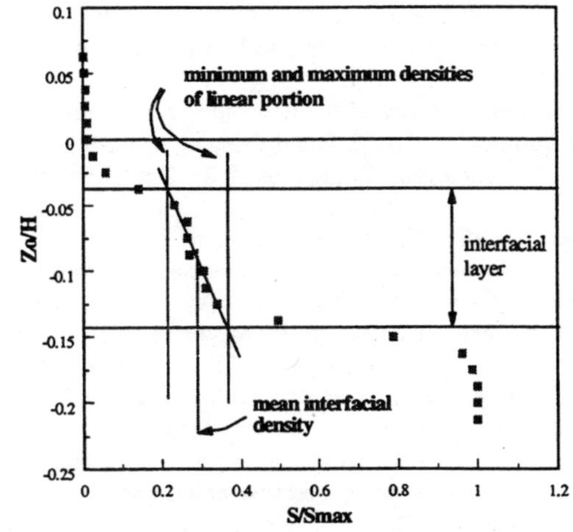

Figure 3 Definition of density of interfacial layer

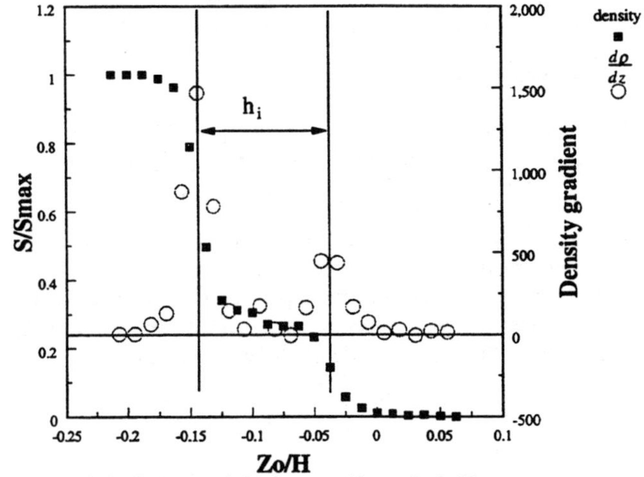

Figure 4 Definition of thickness of interfacial layer

THE EFFECT OF VELOCITY

Eight tests were conducted with the same density difference and fresh water depth. The depth chosen was 400 mm and the density difference chosen was 25 kg/m^3. A density difference of 32 kg/m^3 would have been preferred for better simulation of field conditions but due to project limitations it was necessary to reduce the volume of salt required for each test. The velocities were varied from 6 mm/s up to 70 mm/s. A velocity of 70 mm/s was the maximum possible for the chosen depth of fresh water to allow profiles to be obtained without major variations during the vertical traverse. The relative density and thickness of the interfacial layer for each test are presented for comparison. A summary of the tests conducted is given in Table 1.

Figure 5a shows a density profile obtained 1 hour and 35 minutes after the start of the test, for the lowest velocity of 6 mm/s. The Reynolds number for this test was 1095, which according to Curran (1993) is in the laminar region. With lack of any interfacial waves there was no mixing and even after 20 hours no significant change in the density profile was observed, as shown in Figure 5b.

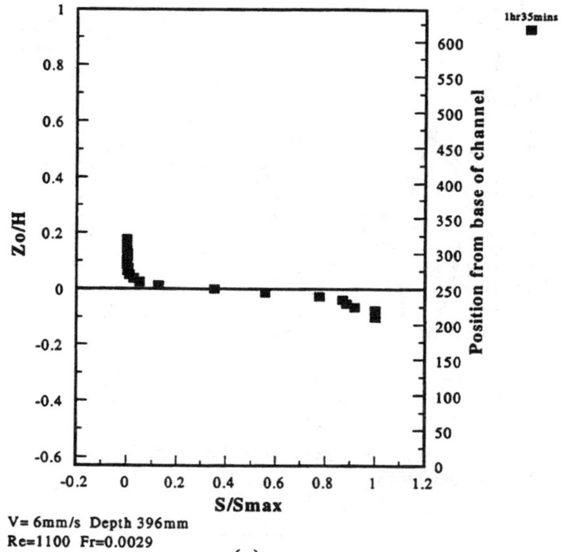

(a)

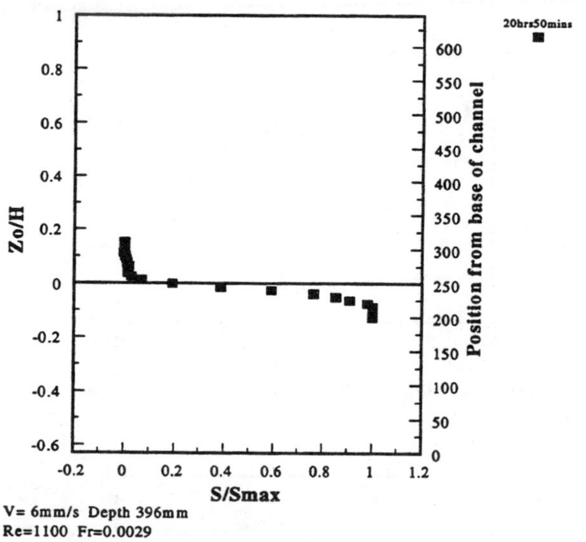

(b)

Figure 5 Density profiles for test 6/396/25 (a) after 1 hr 35 mins (b) after 20 hrs 50 mins

Table 1 Summary of tests conducted with varying velocity

Test	Velocity mm/s	Depth mm	Density kg/m^3	Ri$_b$ ($g\Delta\rho H / \rho V^2$)
6/400/25	6	400	1025	2964
20/400/25	20	400	1025	245
25/400/25	25	400	1025	157
30/400/25	30	400	1025	104
40/400/25	40	400	1025	61
50/400/25	50	400	1025	39
60/400/25	60	400	1025	27
70/400/25	70	400	1025	20

When the velocity was increased to 20 mm/s flow was in the turbulent region with a Reynolds number of 3850. Small interfacial waves were observed along the interface and with time an interfacial or mixed layer formed. The history of the layer is plotted in Figure 6.

The test was allowed to run for 100 hours after which it was stopped even though it had not completely eroded the impoundment. For the first eighteen hours there was a rapid decrease in the density after which a quasi-equilibrium condition was reached with a very slow rate of dilution.

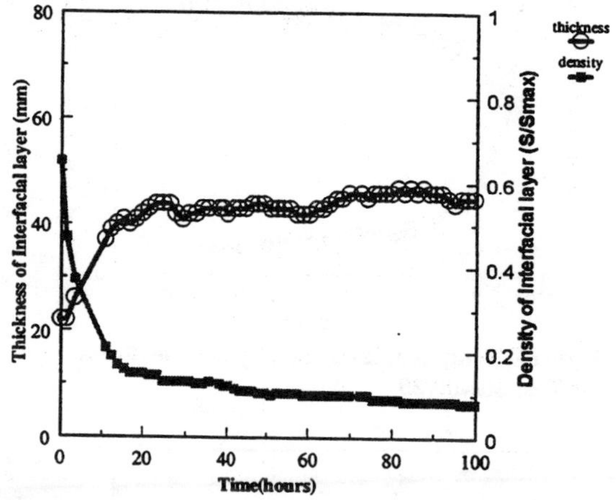

Figure 6 Temporal layer density and thickness for Test 20/400/25

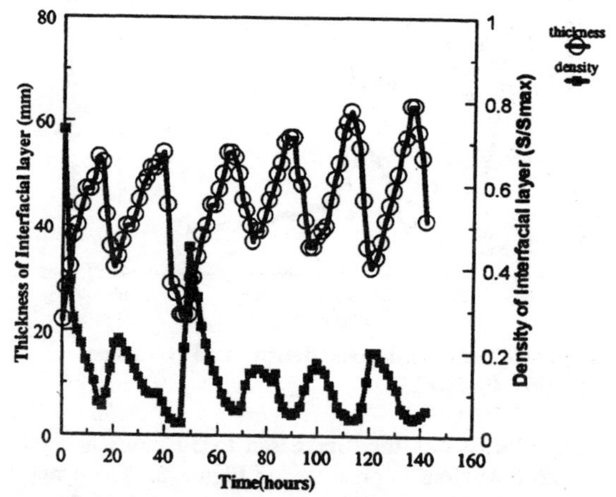

Figure 7 Temporal layer density and thickness for Test 25/400/25

The corresponding thickness of the layer shows a rapid increase in thickness over the first eighteen hours after which a constant value of 45 mm was reached. Slight fluctuations in the thickness were observed which were attributed to the method of defining the layer. The 5mm spacing between measurement points in the profile also limited the accuracy but a lower spacing would have taken too long to obtain a profile.

Increasing the velocity to 25 mm/s and keeping all other parameters the same gave the result shown in Figure 7. Again for the first eighteen hours there was a rapid decrease in density as the layer established. After this an increase in density was observed which corresponded to the upper part of the layer being removed. This resulted in a thinner denser layer and the figure shows a pattern of maximum densities corresponding with minimum thicknessses. At eighteen hours the density of the layer had dropped to 0.06 S/Smax which is equivalent to 1001.5 kg/m^3. The density then increased to approximately 0.2 S/Smax before decreasing and continuing the cycle. The thickness of the layer followed the same basic cyclic pattern with a range of 35 mm to 55 mm at the measurement point.

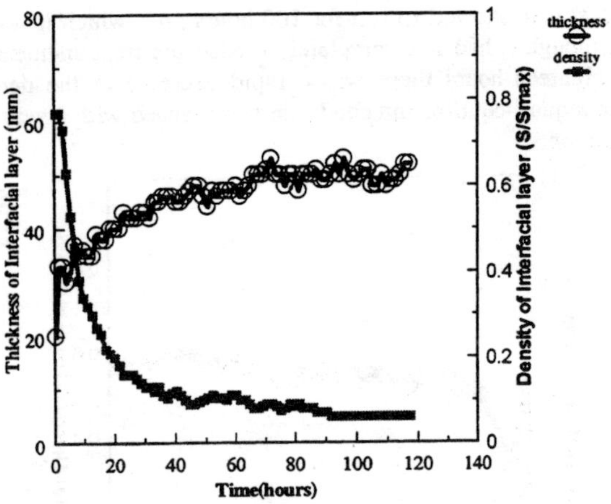

Figure 8 Temporal layer density and thickness for Test 30/400/25

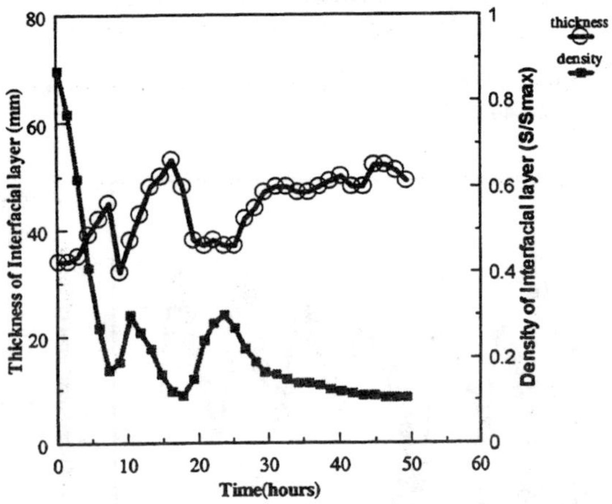

Figure 9 Temporal layer density and thickness for Test 40/400/25

The velocity was then increased to 30 mm/s and the resulting mixed layer behaviour is presented in Figure 8. The cyclic behaviour has been replaced by a stable pattern similar to that with a velocity of 20 mm/s. In this case the layer took over 25 hours to establish before settling into the quasi-equilibrium condition. At 100 hours the density of the layer showed no variation with small fluctuations being observed due to the measurement process. The temporal behaviour of the layer for a velocity of 40 mm/s is plotted in Figure 9. This time a more rapid change in density was observed and the impoundment had eroded after 50 hours. (The time to erode the impoundment is the time taken for the interface to fall below the lowest measuring point). The stable behaviour of the previous velocity has been replaced by the cyclic behaviour again.

When the velocity was increased to 50 mm/s the cyclic behaviour of the layer was again visible as shown in Figure 10. The increase in velocity had the effect of reducing the duration of the test to 25 hours. After 5 hours the layer had established and a minimum density of 0.12 S/Smax was reached before an increase in density was observed and the cyclic behaviour commenced. Only two cycles were observed during the test with a time interval of 10 hours. The thickness and density of the layer for a velocity of 60 mm/s is plotted in Figure 11. The increase in velocity reduced the duration of the test to 11 hours.

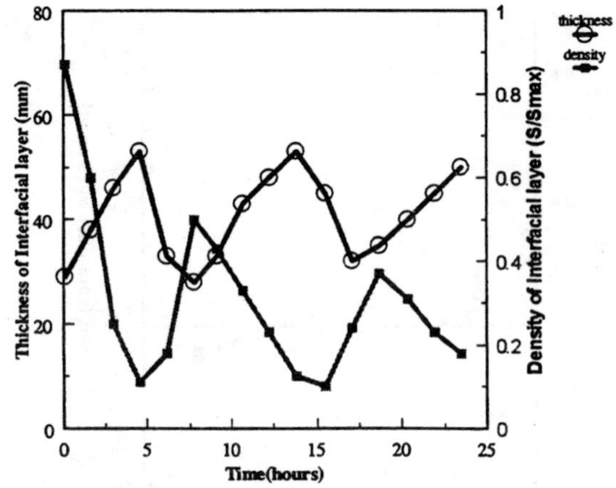

Figure 10 Temporal layer density and thickness for Test 50/400/25

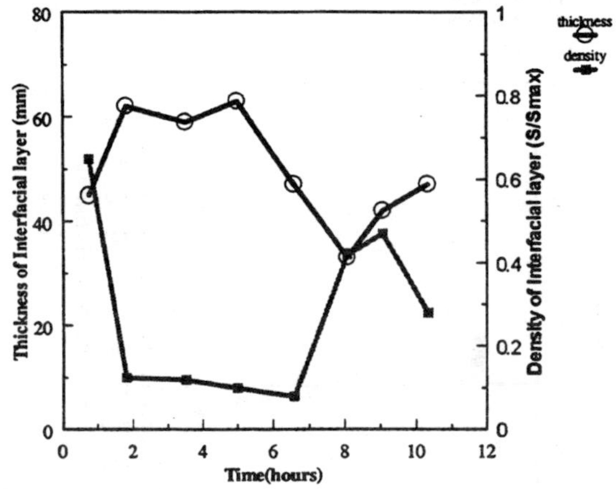

Figure 11 Temporal layer density and thickness for Test 60/400/25

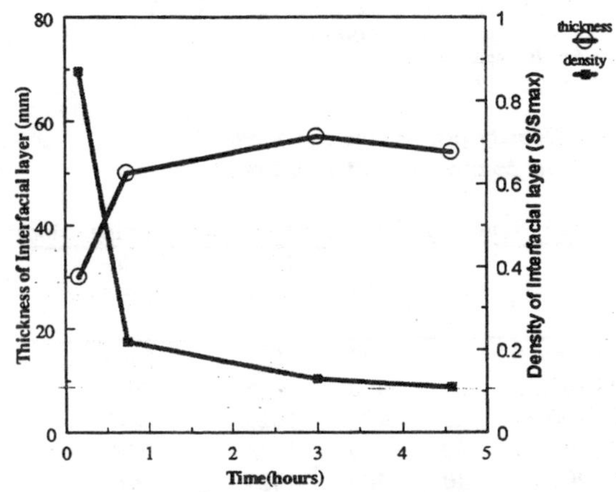

Figure 12 Temporal layer density and thickness for Test 70/400/25

The layer was considerably thicker during the first cycle, compared to the previous tests, at 60mm after which it reduced to 30mm and repeated the cycle. Only one full cycle was observed during the test before the impoundment had eroded.

When the velocity was increased still further to 70 mm/s the impoundment eroded very quickly as shown in Figure 12. This velocity was beyond the acceptable value in order to allow profiles to be measured before major variation had occurred from the time at which measurement started. The increased rate also reduced the duration of the test and it is not clear whether the cyclic behaviour would have been observed.

CONCLUSIONS

The mixing across a sheared density interface has been studied in a laboratory facility. Mixing across the interface has been found to result in a third layer which has a thickness range between 10 to 60 mm. The characteristics of the layer were monitored with time in order to establish the mechanism of salt water erosion from a saline impoundment. Below a threshold value of Reynolds number no mixing was found.

The tests conducted to monitor the effect of velocity have shown that the layer can behave in two different and distinct ways. In the first a cyclic variation in density and thickness of the layer was observed. In the second, the density reduces while the thickness increases and constant values are reached. At this stage a steady state condition is achieved. The results show that the time taken to remove the saline water generally reduces with increasing velocity. There appears to be no simple relationship, however, between velocity and which of the two mechanisms form. The results of this series also show that no obvious pattern exists between which of the mechanisms form and the value of Ri_b.

The results presented are part of an ongoing study of interfacial layer behaviour, in which the interfacial layer is monitored under differing fresh and salt water conditions. Using this it is intended to obtain the trigger between the two mechanisms presented.

ACKNOWLEDGEMENTS

Funding from The Department of Education, Northern Ireland (DENI) for this work is gratefully acknowledged. Helpful conversations with Prof. N.W.H. Allsop (HR Wallingford) and Dr. A Taylor (Imperial College, Department of Mechanical Engineering) are also acknowledged.

REFERENCES

Arita, M. and Jirka, G.H. (1987). Two-layer model of saline wedge. I: Entrainment and interfacial friction. *Journal of Hydraulic Engineering*, Vol.113, pp 1229-1248.

Christodoulou (1986). Interfacial mixing in stratified flows. *Journal of Hydraulic Research.* Vol. 24, pp 77-92

Curran, G. (1993). The effect of interfacial layer development on fresh water flows over an entrapped saline intrusion. *PhD thesis submitted to The Queen's University of Belfast, N. Ireland.*

Dermissis, V. (1990). Velocity distribution in arrested saline wedges. *Journal of Waterway, Port, Coastal and Ocean Engineering*, Vol. 116, pp 21-42.

Dermissis, V. and Parthenaides, E. (1985). Dominant shear stresses in arrested saline wedges. *Journal of Waterway, Port, Coastal and Ocean Engineering*, Vol.111, pp 733-752.

Foo, M.H., Shuy, E.B. and Chen, C.N. (1995). Entrainment across a density interface inside a flume compartment, *Journal of Hydraulic Research*, Vol. 33, pp 181-196.

Grubert, J.P. (1980). Experiments on arrested saline wedge, *Journal of Hydraulic Engineering*, Vol. 106, pp 945-960.

Grubert, J.P. (1989). Interfacial mixing in stratified channel flows, *Journal of Hydraulic Engineering*, Vol. 115, pp 887-905.

Keulegan, G.H. (1966). The mechanism of an arrested saline wedge. "Hydrodynamics of Estuaries and Fjords", *Proc. 9th International Liege Colloquium on Ocean Hydrodynamics*, Elsiever Scientific Publishing Company, pp 546-574.

Lewin, J. (1996). Mechanical aspects of water control structures. *Proc. Institution of Civil Engineers, Water Maritime and Energy*, Vol. 118, pp 29-38.

Proceedings of the Seventh (1997) International Offshore and Polar Engineering Conference
Honolulu, USA, May 25-30, 1997

ISBN 1-880653-28-1 (Set); ISBN 1-880653-31-1 (Vol. III)

Internal Tide Driving of Estuarine-Shelf Water Exchange

Tateki Fujiwara and Tetsuya Takahashi
Kyoto University
Kyoto, Japan

Katsuyuki Abo
National Research Institute of Aquaculture
Mie Prefecture, Japan

ABSTRACT

A role of internal tide in estuary-shelf water exchange is studied in Gokasho Bay, which is a drowned river valley estuary facing continental shelf of the Pacific Ocean. Thermistor chains and current meters were moored in and out of the bay for 15 days in the four seasons. The internal tide is predominant on the shelf throughout the stratified period. This internal tide propagates into Gokasho Bay as a progressive internal wave and acts as a major source of the fluctuations in temperature, salinity and velocity in the bay. Water exchange between the estuary and shelf waters is driven by the internal tide rather than the surface tide.

KEY WORDS: Internal tide, estuarine-shelf interaction, drowned river valley estuary, assessment of water exchange

INTRODUCTION

There are a lot of semi-enclosed and small (several kilometers in length) drowned river valley estuaries along the southern coast of Japan. Because these bays have small tidal prisms (the volume between high and low water levels), it is presumed that the water exchange between the bay and the shelf waters is inert. Nevertheless the water quality in these bays have been kept relatively well until now, despite prosperous aquaculture.

Hagino (1993) studied water exchange in Nomi Bay and indicated that the water exchange was not driven by the surface tide but mainly by the internal tide during the stratification period. The internal tide having a period of 12 hours is prevalent and causes the two-layer flow structure during the stratification period in the bay. When the stratification disappears in autumn, the internal tide also ceases and the magnitude of current velocity reduces remarkably, which results in hypoxia in the benthic layer.

The internal tide is a common phenomenon on the continental shelf waters (Nagashima, 1989; Pineda, 1995; Rosenfeld, 1990; Summers and Emery, 1963; Winant, 1974), and it may control estuarine-shelf water exchange. However, the internal tide in a small estuary has not been studied well. In this paper, we will study the internal tide in Gokasho Bay, which has similar hydrographic condition as Nomo Bay, and assess its role in water exchange between the bay and the shelf waters.

OBSERVATIONS

Gokasho bay is located in central Japan (Fig. 1). The bay has a length of 7.5 km and a mean depth of 20 m. The mouth of the bay has a width of 2.5 km and a maximum depth of 30m. The surface tide in the bay is dominated by a semi-diurnal tide and the amplitudes of M_2 and S_2 tide are 44.1 cm and 19.8 cm, respectively. At the head of the bay, there are two small rivers.

In this study we took the bay axis in the direction of 20

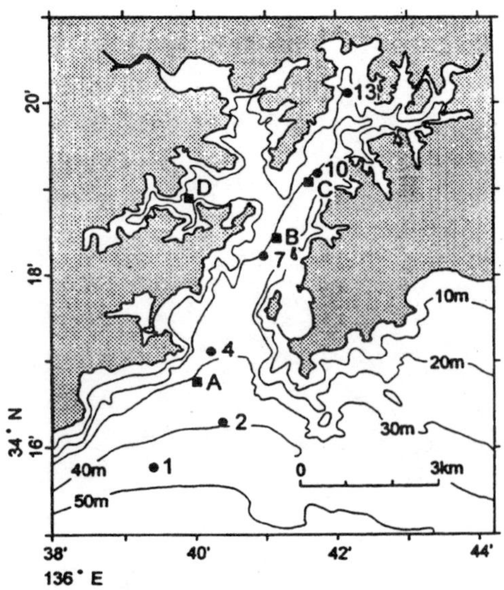

Fig. 1 Map of Gokasho Bay.

degrees north. "Mie prefecture" moored current meters at 14 stations in and out of the bay for 15 days in the four seasons (May, June, September, November in 1993). At these stations, temperature, salinity, and current were measured at 10 minutes intervals at depths of 2m below the sea surface, the half depth, and 2m above the bottom. In addition to this observation, we conducted a temperature profile measurement with thermistor chains at stations A, B, C, and D (black squares in Fig. 1) at 10 minutes intervals in September, 1994.

OBSERVED RESULTS

The isotherm depths are shown in Figure 2. The uppermost panel shows the predicted variations of the tide level. The level fluctuates with the semi-diurnal period. The first and second halves of the observation periods correspond to the spring and neap tides, respectively. In the lower panels, the isotherm also fluctuates regularly with the semi-diurnal period, indicating persistent existence of the internal tide on the shelf. Looking the isotherm of 24°C, we can see that the range of the internal tide exceeds 20m at station A (September 8, 9). The internal tide is more obvious at station B and its range amounts to 10 m. Although the height of the internal tide at station D, which is located in a branch away from the bay axis, is smaller than those at the other stations, the fluctuation with the semi-diurnal period is still obvious.

In order to indicate the propagation of the internal tide into the bay a magnified version of part of Figure 2 is shown in Figure 3. The arrows indicate the times of high waters (H.W.). At station A, the isotherms attain their highest level at 6.6 hours after the high water of the surface tide. The internal tide propagates from the bay mouth to station B, to station C, and to station D with 3.0, 4.7, 5.9 hours time lag, respectively. As the internal tide propagates into the bay, the wave becomes steeper at the front and gentler at the backward.

Figure 4 shows the currents (positive landward) in the upper and lower layers at the mouth of the bay. The currents in the upper and lower layers oscillate with the semi-diurnal period. However, the current directions of the surface and the bottom layers oppose each other.

The longitudinal cross-sectional variations of the current and the temperature on 17 September are shown in Figure 5. The shadows indicate a water temperature below 23 °C, which corresponds to the upper limit of the thermocline. This isotherm oscillates vertically off the bay mouth. When the

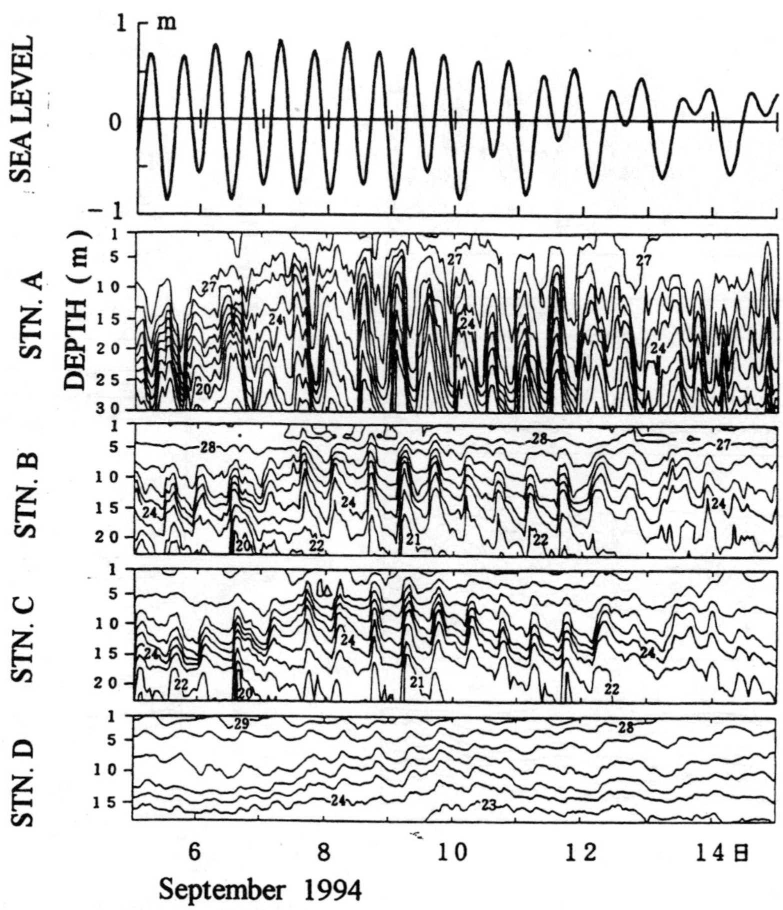

Fig. 2 Oscillation of the sea level (a) and isotherms (b - e).

isotherm rises to the highest level off the bay mouth, the water in the lower layer flows into the bay, while the water in the upper layer flows out (Fig. 5a). During the descending of the isotherm, retreating flows on the sloping bottom are prominent at stations 2 and 4 (Fig. 5 b and c). After this down-slope flow, the isotherm rises again off the bay mouth and the water in the lower layer flows into the bay. This cycle of oscillation is repeated with a 12 hour time period.

The variation of the bottom currents along the longitudinal axis are indicated in Figure 6. Outflow begins at the mouth and propagates into the bay with the speed of 0.37 ms^{-1}; inclination of chain lines in the figure represents this speed. The progress of the internal wave is one directional and no sign of reflection appears even in the vicinity of the bay head. This implies that the internal tide propagates into the bay as a progressive wave and the kinetic energy of the wave is lost within the bay. Figure 6 also shows that the current is intensive at "a backwash" of the internal tide (receding internal tide) at the bottom especially at station 2 and 4.

Figure 7 shows the residual current profile. At the surface the currents tend to be outflowing, at the middle inflowing, and at the bottom outflowing. The bottom outflow is the result of the strong backwash flow on the bottom, while the inflow at the middle depth is presumed to be the result of the Stokes drift of the progressive internal tide (cf. Fig. 7.09 in Mann and Lazier, 1991).

WATER EXCHANGE

To evaluate a role of internal tide in estuary-shelf exchange, the dispersive salt flux was separated into four components (Dyer, 1973; Fischer et al, 1979). These were a salt fluxes due to the: 1) shear of residual flow, 2) surface tide, 3) internal tide and 4) turbulence. The data used in this study were taken at the stations indicated in Figure 7. At first, we separated the

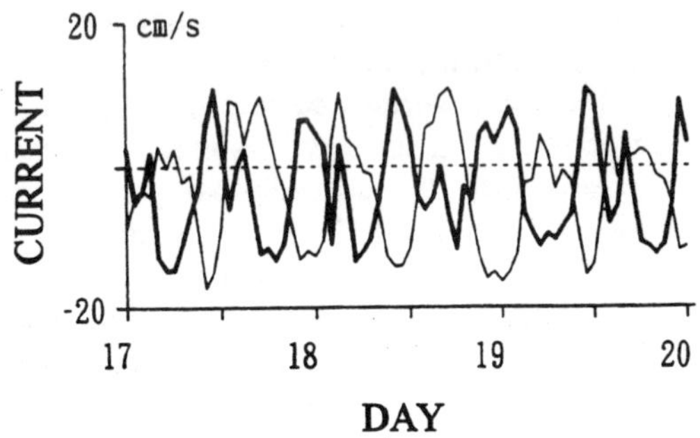

Fig. 4 Longitudinal current at station 4. Thick and thin lines represent bottom and surface currents, respectively.

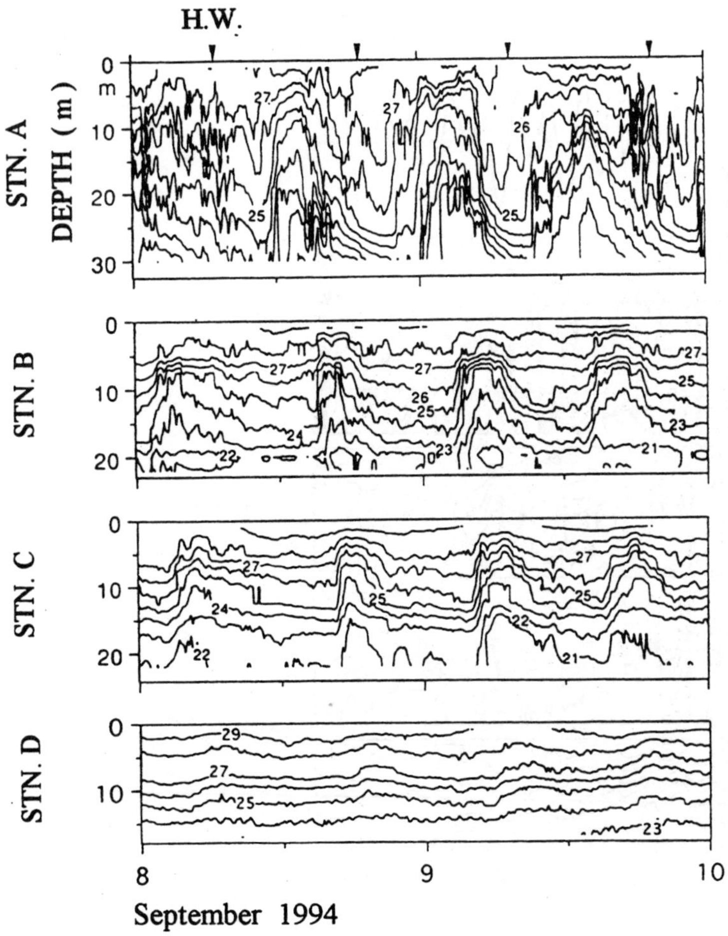

Fig. 3 Oscillation of isotherms.

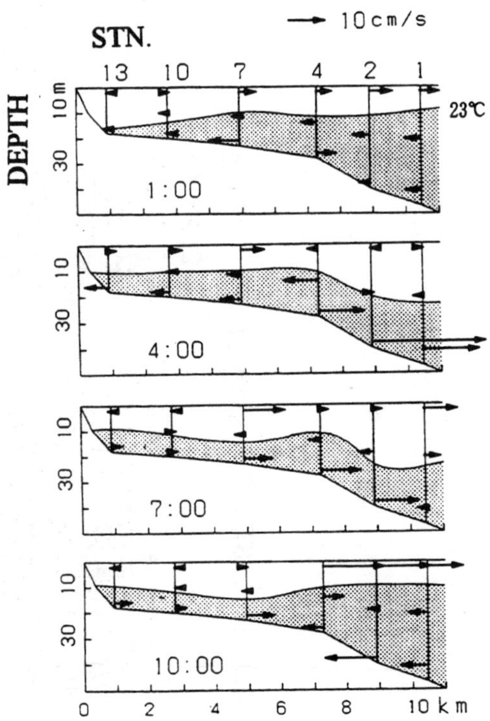

Fig. 5 Variation of currents and thermocline on the longitudinal cross section on September 17, 1993.

longitudinal velocity component (u ; positive landward) into three terms depending on the temporal period:

$$u = \bar{u} + U + u' \quad (1)$$

where first, second and third terms in the right-hand side of equation (1) are the residual current, tidal current (in this case consisting of a semi-diurnal and a diurnal component) and turbulent current, respectively. Then, each term is furthermore separated into two terms: spatial average (denoted by the bracket) and deviation from the spatial average (denoted by a subscript d). Thus, the current velocity u is separated into 6 independent terms.

$$\begin{aligned} \bar{u} &= \langle \bar{u} \rangle + \bar{u}_d \\ U &= \langle \bar{U} \rangle + U_d \\ u' &= \langle u' \rangle + u'_d \end{aligned} \quad (2)$$

Temporal and spatial variation of the salinity, s, is also separated into 6 terms in the same way. Since both the velocity of the current and salinity are separated into 6 independent terms, time and spatial average of salinity flux $\langle \bar{F} \rangle = \langle \overline{us} \rangle$ is given as:

$$\langle \bar{F} \rangle = \underset{1}{\langle \bar{u} \rangle \langle \bar{s} \rangle} + \underset{2}{\langle \overline{u_d} \rangle \langle \overline{s_d} \rangle} + \underset{3}{\overline{\langle U \rangle \langle S \rangle}} + \underset{4}{\langle \overline{U_d S_d} \rangle} + \underset{5}{\overline{\langle u' \rangle \langle s' \rangle}} + \underset{6}{\langle \overline{u'_d s'_d} \rangle} \quad (3)$$

On the right-hand side of the equation (3), the term 1 indicates the advection flux and the terms 2-6 indicate collectively the dispersion flux. The dispersion flux consists of the fluxes due to the shear of the residual current (term 2), surface-tide (term 3), internal tide (term 4), and turbulence (terms 5 and 6).

In our analysis, the temporal decomposition represented by the equation (1) was conducted by a harmonic analysis method for one-day-data, and the spatial decomposition was done by the average over three depths at each station: surface, middle, and bottom; except for station 13 where we averaged only surface

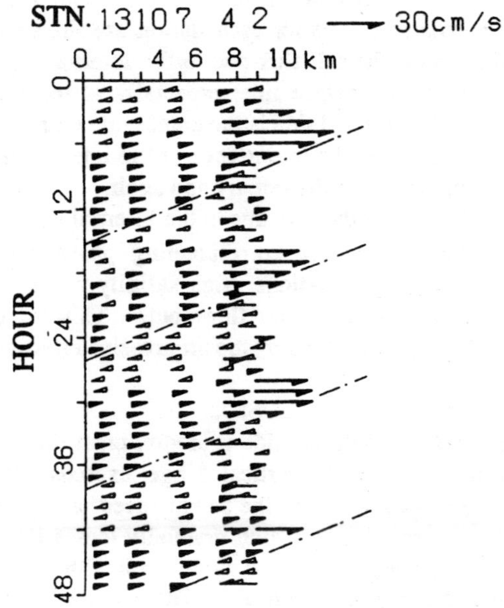

Fig. 6 Temporal variation of the bottom currents on the axis line of the bay on September 17, 1993 .

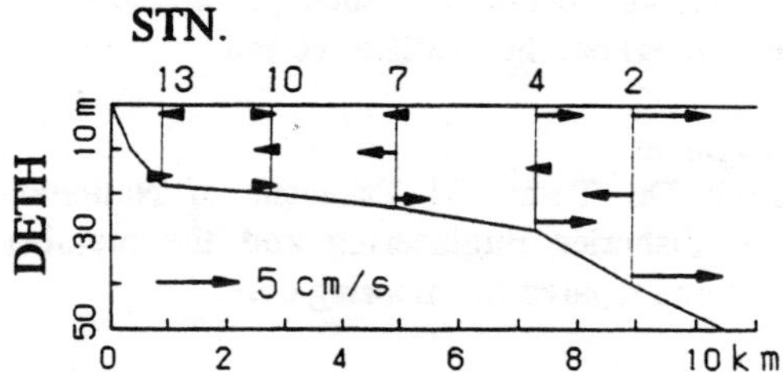

Fig. 7 Residual currents.

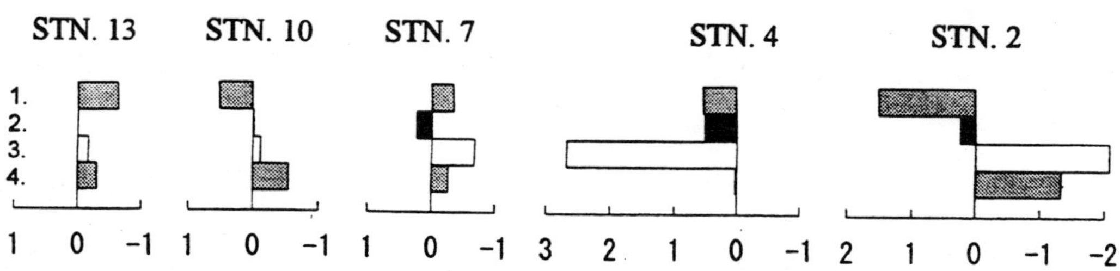

Fig. 8 Components of dispersive salt flux at each station. Unit: psu cm s^{-1}. Bars in figure represent the salt flux due to the shear of residual current (1), surface tide (2), internal tide (3) and turbulence (4). Landward flux is positive.

and bottom positions. We calculated the values of terms 2 - 6 daily from the 17 September to 19 September, 1993 and obtained values for each term averaged over the three days.

The calculated results for each station are shown in Figure 8. At all stations, the salt flux due to the internal tide exceeds the flux due to the surface tide; especially at the bay mouth (station 4). The internal tide is the dominant contributor to the salt flux. This implies the importance of the internal tide in the water exchange between the estuary and shelf.

The high value of the flux due to the internal tide at stations 4 and 2 is caused by the corresponding phase between the velocity and salinity variations; high salinity water flows into the bay and low salinity water flows out. This oscillation of salinity and current is driven by the progressive internal tide.

DISCUSSION AND CONCLUSION

On the shelf, the internal tide is predominant throughout the stratified period, and its vertical range exceeds 20 m. The internal tide propagates into the bay at speed of 0.37 ms^{-1} as a progressive internal wave, which gradually looses its energy in the bay. This propagation speed agrees well with the calculated speed from the two layer model. The wave of internal tide beats on the inclined bottom and drives a strong down-slope flow on the bottom, when it recedes. This backwash flow can transport suspended sediment near the bay mouth to the inner shelf.

When we conduct water quality assessment of estuaries, the effects of the internal tide have been ignored. However, the internal tide plays an important role in water quality variations in estuaries facing a continental shelf, where the internal tide is prominent throughout the stratified period.

Acknowledgment

We thank Dr. Tomoyuki Takeuchi of National research Institute of Fisheries Engineering and the member of Mie Prefecture who prepared the mooring data.

REFERENCES

Dyer, K. R. (1973). "Estuaries: A Physical Introduction", John Willy and Sons, London, pp. 140.

Fischer, H. B., List, E. J., Koh, R. C. Y., Imberger, J., and Brooks, N. H. (1979). "Mixing in Inland and Coastal Waters", Academic Press, London, pp. 483.

Hagino, S. (1993). "Hydrography in stratified aquaculture field", Report of Fisheries Hydrography Research Institute, No. 14, pp. 1-49 (in Japanese).

Mann, K. H. and Lazier, J. R. N. (1991). "Dynamics of Marine Ecosystems, - Biological-Physical Interactions in the Ocean", Blackwell, Boston, pp. 466.

Nagashima, H. (1989). "Review of the study on internal tide", Kaiyo (Marine) Monthly, Vol. 21, No. 9 (special issue on Internal Tide), pp. 509-512 (in Japanese).

Pineda, J. (1995). "An internal tidal bore regime at nearshore stations along western U.S.A.: Predictable upwelling within the lunar cycle", Continental Shelf Research, Vol. 15, No. 8, pp. 1023 - 1041.

Rosenfeld, L. J. (1990). "Baroclinic semidiurnal tidal currents over the continental shelf northern California", J. Geophysical research, Vol. 95, pp. 22,153-22,172.

Summers, H. J. and Emery, K. O. (1963). "Internal waves of tidal period off Southern California", J. Geophysical Research, Vol. 68, pp. 827-839.

Winant, C. D. (1974). "Internal surges in coastal waters", J. Geophysical Research, Vol. 79, pp. 4523-4526.

Proceedings of the Seventh (1997) International Offshore and Polar Engineering Conference
Honolulu, USA, May 25-30, 1997

ISBN 1-880653-28-1 (Set); ISBN 1-880653-31-1 (Vol. III)

A Flow Field Over Upwelling Producing Structure in an Oscillatory Flow with or without Stratification

G. Tsujimoto
Kobe City College of Tech., Japan
K. Michioku
Kobe University, Japan
H. Nishide
Hanshin Express Corporation, Japan

ABSTRACT

Numerical calculation of the flow field over the upwelling producing structure in an oscillatory flow is conducted using the k- ε turbulence model. Particular emphasis is on the flow field with or without stratification. The characteristic phenomenon such as the vortex formation was simulated. Irrespective of the value of Ri, the density interface will be destroyed when the height of the upwelling producing structure is higher than that of the density interface. Also with decreasing KC number, the separated vortex with intense vorticity is formed and a fast residual flow is seen near the upwelling producing structure through the experiment.

KEY WORDS :Turbulence model, upwelling flow, stratification, oscillatory flow

1.Introduction

An upwelling flow has the major effect on creating fishing grounds, for it carries the water mass which contains the nutrition from the lower layer to the upper layer. Therefore, it is important to study the development of an upwelling producing structure. The previous research, however, has been focused on the flow field without stratification, and there are only a few approaches for the stratified flow field. The direction of the upwelling flow may change from vertical direction to horizontal direction near the density interface due to the stratified layer, and the destruction of the density interface does not happen easily. So it is necessary to study the flow field with stratification.

In this paper the flow field over the upwelling producing structure settled at the sea bottom is examined experimentally and theoretically. The experiment was conducted in a two dimensional water tank where several kinds of upwelling producing structure are settled and the velocity profile is measured by using an electromagnetic current meter. The numerical simulation is also carried out by using the k- ε turbulence model under an oscillatory flow with or without stratification.

2.Previous studies

Ohno et al.(1979) measured the location from the sea bottom where the vortex is formed over the upwelling producing structure, Akeda et al.(1989) observed the movement of the density interface under uni-directional flow and divided the stratification profile into three types by using the density Froude number. Also Matsui and Aritomi (1988) conducted a numerical simulation in three dimensional water body. One of the authors Tsujimoto(1992) calculated a laminar flow field with stratification. However, there are only a few studies on an oscillatory flow with or without stratification.

3.Experimental Method

The experiment was performed in a water flume , which is 0.88m long,0.88m wide and 0.45m deep. As shown in Fig.1, six types of upwelling producing structure are settled at a depth of 0.25m. In Fig.1-(c), settling spaces S are set to be 5cm,8cm and 10cm,respectively. These structures are oscillated horizontally in a water flume, and the flow field is produced. An electromagnetic current meter is used for the measurement of velocity and a potential meter is used to measure the velocity phase. Table 1 gives the experimental conditions and the experimental apparatus is shown in Figure 2.

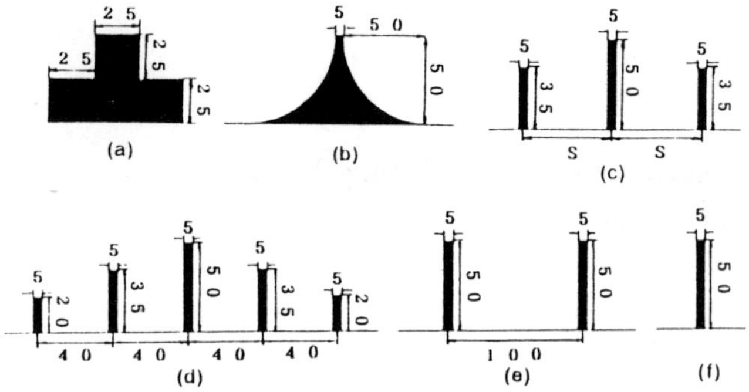

Figure 1 Upwelling producing structure(unit :mm)

Table 1 Experimental conditions

Run	Amplitude (cm)	Period (sec)	KC	Re
1	5	15	6.3	1047
2	5	5	6.3	3142
3	8	24	10	1047
4	11	33	14	1047
5	14	42	18	1047
6	17	51	21	1047

4.Numerical approach

4.1 Characteristic scales

The basic equations describing the fluid motion consist of the continuity equation, momentum equation, conservation equations for heat T, turbulent kinetic energy k and its dissipation rate ε. In order to normalize the governing equations, the following characteristic scales are chosen.

a) Velocity : $U_0 = 2 \cdot \pi \cdot S \cdot f$

b) Length : $L_0 = d$

c) Temperature : $T_0 = T_1 - T_2$

d) Time : $t_0 = L_0 / U_0$

where d is the height of the upwelling producing structure, S is the amplitude of oscillation ,f is the frequency of oscillation,T_1 and T_2 are the initial temperature of the upper-layer and the lower-layer fluid, respectively. And the dimensionless terms are given as follows.

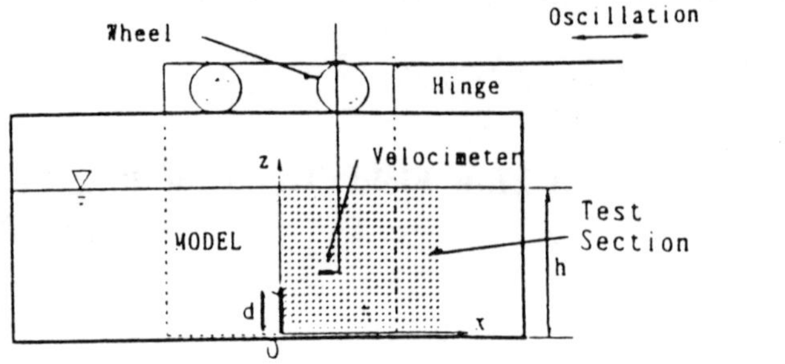

Figure 2 A schematic figure of the experimental apparatus

$$\tilde{T} = (T_1 - T_2)/T_0, \tilde{t} = t/t_0, (\tilde{x}, \tilde{z}) = (x, z)/L_0,$$
$$(\tilde{u}, \tilde{w}) = (u, w)/U_0, \tilde{k} = k/U_0^2, \tilde{\varepsilon} = \varepsilon L_0/U_0^3$$

where x, z are the Cartesian coordinates(x: horizontal direction, z: vertical direction), u and w are the velocity in the x and z direction, respectively, in which the tilde stands for the dimensionless terms. Figure 3 shows a schematic illustration of the two dimensional oscillatory flow field with stratification.

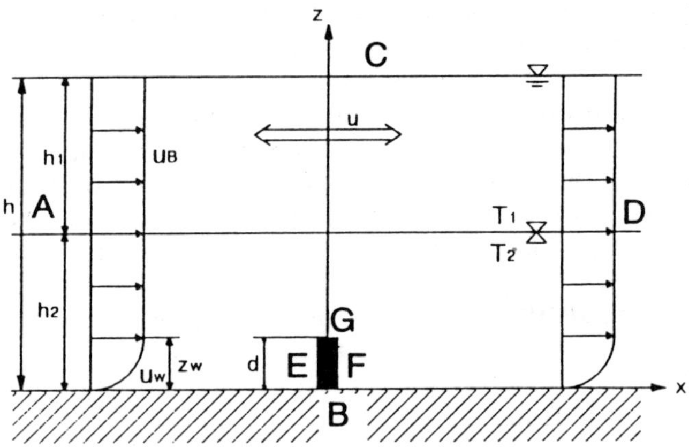

Figure 3 A schematic figure of a two dimensional oscillatory flow field

4.2 Basic equations

The basic equations normalized without the tilde are as follows:

a) Continuity equation

$$\frac{\partial u}{\partial x} + \frac{\partial w}{\partial z} = 0 \qquad (1)$$

b) Momentum equation

$$\frac{\partial u}{\partial t}+\frac{\partial u^2}{\partial x}+\frac{\partial uw}{\partial z}=-\frac{\partial p}{\partial x}+\frac{\partial}{\partial x}\left\{\left(\frac{1}{R_e}+2\nu_t\right)\frac{\partial u}{\partial x}\right\}$$

$$+\frac{\partial}{\partial z}\left\{\left(\frac{1}{R_e}+\nu_t\right)\frac{\partial u}{\partial z}+\nu_t\frac{\partial w}{\partial x}\right\} \tag{2}$$

$$\frac{\partial w}{\partial t}+\frac{\partial uw}{\partial x}+\frac{\partial w^2}{\partial z}=-\frac{\partial p}{\partial z}+\frac{\partial}{\partial z}\left\{\left(\frac{1}{R_e}+2\nu_t\right)\frac{\partial w}{\partial z}\right\}$$

$$+\frac{\partial}{\partial x}\left\{\left(\frac{1}{R_e}+\nu_t\right)\frac{\partial w}{\partial x}+\nu_t\frac{\partial u}{\partial z}\right\}+R_i\cdot T \tag{3}$$

c)Heat conservation

$$\frac{\partial T}{\partial t}+\frac{\partial uT}{\partial x}+\frac{\partial wT}{\partial z}=\frac{\partial}{\partial x}\left\{\left(\frac{1}{R_e}\cdot P_r+\frac{\nu_t}{\sigma_t}\right)\frac{\partial T}{\partial x}\right\}$$

$$+\frac{\partial}{\partial z}\left\{\left(\frac{1}{R_e}\cdot P_r+\frac{\nu_t}{\sigma_t}\right)\frac{\partial T}{\partial z}\right\} \tag{4}$$

d)Conservation of turbulent kinetic energy

$$\frac{\partial k}{\partial t}+\frac{\partial uk}{\partial x}+\frac{\partial wk}{\partial z}=\frac{\partial}{\partial x}\left\{\left(\frac{1}{R_e}+\frac{\nu_t}{\sigma_k}\right)\frac{\partial k}{\partial x}\right\}$$

$$+\frac{\partial}{\partial z}\left\{\left(\frac{1}{R_e}+\frac{\nu_t}{\sigma_k}\right)\frac{\partial k}{\partial z}\right\}+P+G-\varepsilon \tag{5}$$

e)Conservation of dissipation rate

$$\frac{\partial \varepsilon}{\partial t}+\frac{\partial u\varepsilon}{\partial x}+\frac{\partial w\varepsilon}{\partial z}=\frac{\partial}{\partial x}\left\{\left(\frac{1}{R_e}+\frac{\nu_t}{\sigma_\varepsilon}\right)\frac{\partial \varepsilon}{\partial x}\right\}-C_{2\varepsilon}\frac{\varepsilon^2}{k}$$

$$+\frac{\partial}{\partial z}\left\{\left(\frac{1}{R_e}+\frac{\nu_t}{\sigma_\varepsilon}\right)\frac{\partial \varepsilon}{\partial z}\right\}+C_{1\varepsilon}\frac{\varepsilon}{k}(P+G)\cdot\left(1+C_{3\varepsilon}R_f\right) \tag{6}$$

$$P=2\nu_t\left\{(\partial u/\partial x)^2+(\partial w/\partial z)^2+(\partial u/\partial z+\partial w/\partial x)^2\right\} \tag{7}$$

$$G=-R_f\cdot\nu_t/\sigma_t\cdot\partial T/\partial z \tag{8}$$

$$\nu_t=C_\mu\frac{k^2}{\varepsilon} \tag{9}$$

where p is the pressure, k is the turbulent kinetic energy, ε is the energy dissipation rate, Pr(=7.0) is the Prandtl number, σ_t (=1.0) is the turbulent Prandtl number, R_f (=-G/(P+G)) is the flux-type Richardson number, R_e(=UoLo/ν) is the Reynolds number, Ri(=α gToLo/Uo^2) is the Overall Richardson number. Other parameters are given as follows:

h=h/d, $h_1=h_1/d$, $h_2=h_2/d$, KC=2 π S/d

where h is the total water depth, h_1 and h_2 are the initial depth of the upper layer and the lower layer. The value of h is set to be 5.0 according to the field data. The constants in these equations are as proposed by Rodi as given as follows:

$$C_\mu=0.09,\sigma_k=1.0,\sigma_\varepsilon=1.3,C_{1\varepsilon}=1.44,C_{2\varepsilon}=1.92,C_{3\varepsilon}=1.0$$

4.3 Solution procedure

In Tabel-2 the boundary conditions are shown according to Figure 3. By using he open channel flow resistance formula, the horizontal velocity is written as follows:

$$u_w=u_*\{5.5+5.75\log_{10}(u_*zR_e)\}\cos\theta \qquad (0<z<z_w)$$

$$u_B=\cos\theta \qquad (z_w<z<h) \tag{10}$$

$$u_*=\sqrt{f/8}$$

where u_* is the friction velocity and f is the friction coefficient . The equations from (1) to (6) are discretized by the method of Patankar, and the SIMPLE algorithm is used to solve the equations. The value θ (=0) indicates the velocity phase when the horizontal velocity reaches maximum.

Table 2 Boundary conditions

	u	w	k	ε	T
A	eq.(10)	0	$\partial^2k/\partial x^2=0$	$\partial^2\varepsilon/\partial x^2=0$	$T_1=1,\ T_2=0$
B	0	0	0	$2/\mathrm{Re}(\partial k^{0.5}/\partial z)^2$	$\partial T/\partial z=0$
C	$\partial u/\partial z=0$	0	0	$2/\mathrm{Re}(\partial k^{0.5}/\partial z)^2$	$\partial T/\partial z=0$
D	eq.(10)	0	$\partial^2k/\partial x^2=0$	$\partial^2\varepsilon/\partial x^2=0$	$T_1=1,\ T_2=0$
E	0	0	0	$2/\mathrm{Re}(\partial k^{0.5}/\partial z)^2$	$\partial T/\partial x=0$
F	0	0	0	$2/\mathrm{Re}(\partial k^{0.5}/\partial z)^2$	$\partial T/\partial x=0$
G	0	0	0	$2/\mathrm{Re}(\partial k^{0.5}/\partial z)^2$	$\partial T/\partial z=0$

5. Results and discussion

5.1 Flow field over upwelling structure without stratification

It is thought that the mechanism of vortex formation is related to the settling space of the upwelling producing structure. In Fig.4 the velocity profiles over each structure are shown at the velocity phase of π /4. The cases of Fig.4(a),(c1),(c2) and (d) have similar velocity profiles. When the settling space is set to be more than two times longer than the height of the upwelling producing structure for cases (c), the scale of separated vortex produced over the upwelling producing structure has the same size. Also there seems to be little difference in velocity profiles between case (b) and case (f).

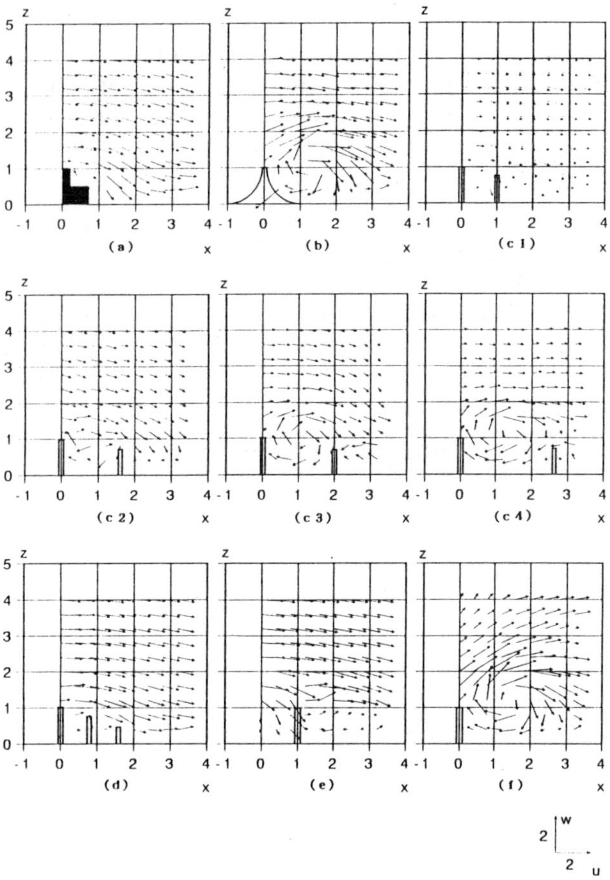

Figure 4 Velocity profiles over upwelling producing structures

Figure 5 shows the distribution of vorticity ω (= ∂ u/ ∂ z- ∂ w/ ∂ x) at z=1.0 in the (x, θ)-coordinates. Comparing case(b) with case(f), the region of concentrated vorticity are formed over the upwelling producing structure for case(f) due to its diffusion. Therefore, after the flow direction changes($\theta = \pi /2$), the positive vorticity strongly interacts with the negative vorticity and the development of the negative vorticity is prevented.

As shown in Fig.4, the development of separated vortex which has the positive vorticity is prevented in case(c), because the settling space between the upwelling producing structure is narrow. After the flow direction changes from the right to the left, the separated vortex disappeared rapidly and the negative vorticity developed. With increasing settling space, as in case(c2),case(c3) and case(c4), the negative vorticity developed for the velocity phase of the right flow or the positive vorticity for that of the left flow. The profile of vorticity distribution has a few similarities between case(a) and case(c). In this paper, six types of upwelling producing structure have been studied experimentally. Only case(f) is handled in the numerical approach because its shape is simpler than the other types.

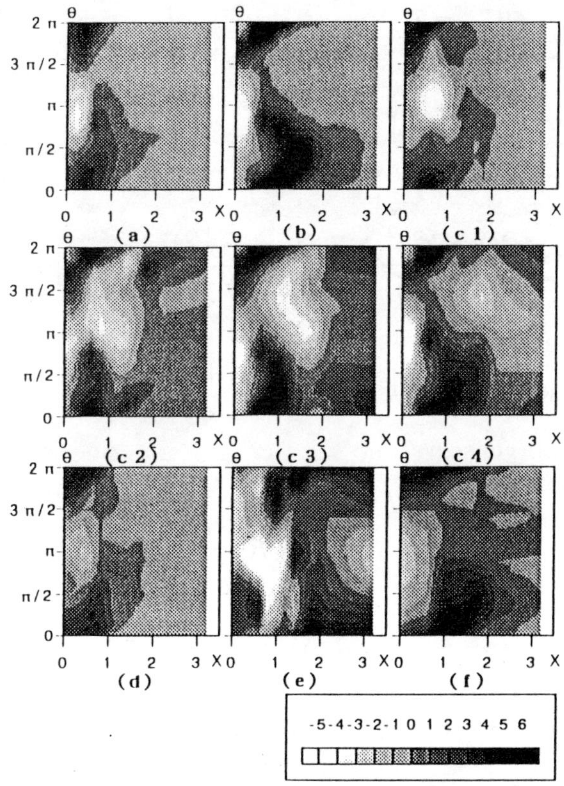

Figure 5 Vorticity distribution in the (x, θ)-coordinate

Figure 6 shows the comparison between calculated results and experimental data on velocity profile and vorticity for run 1 at the velocity phase of 0 π . A little difference on the vertical profile of velocity is seen, a good agreement, however, on the distribution of vorticity is observed.

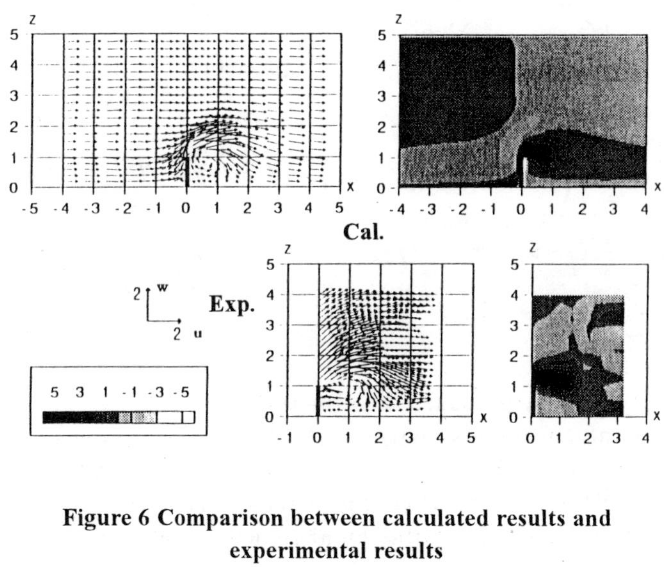

Figure 6 Comparison between calculated results and experimental results

Figure 7 shows the calculated turbulent kinetic energy at the velocity phase of $\pi/8$. With increasing KC number, the separated zone is developing and the separated vortex tends to be formed at some distance from the upwelling producing structure. Thus the distribution region of turbulent kinetic energy can be extended.

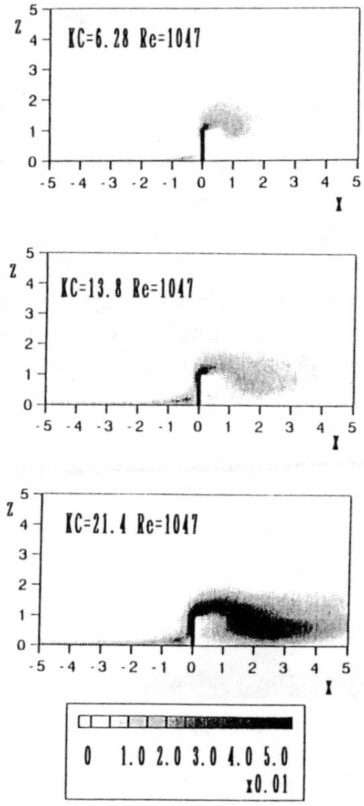

Figure 7 Turbulent kinetic energy

5.2 Relationship between vorticity and KC number

The upwelling flow carries the water body from the lower layer to the upper layer, making the mixing of both layer possible. As shown in the experimental results, the separated vortex has a major effect on producing the upwelling flow. Both vorticity and circulation are often used as an index of intensity of vortex; therefore the vorticity is used herein. Nishide et al.(1994) shows experimentally that the vorticity is constant at the initial stage of its development, irrespective of KC number, while depending on KC number at the final stage. Figure 8 shows the relationship between the maximum value of vorticity and KC number during one period. With decreasing KC number, the vorticity tends to increase. Thus is similar to the calculated results.

5.3 Distribution of residual flow

The mean velocity averaged over one period is named as a residual flow, which is produced locally by separated vortex. The residual flow is a uni-directional flow carrying such substance as a nutrient salt, having the major effect on upwelling flows. Figure 9 shows the experimental values of residual velocity for various values of KC number. With decreasing KC number, as the separated vortex with intense vorticity is generated near the upwelling producing structure, as shown in Fig.7, a fast residual flow is generated. On the contrary, with increasing KC number, a fast residual flow is not produced due to the diffusion of vorticity, and the region of high turbulence intensity may extend as shown in Fig.7.

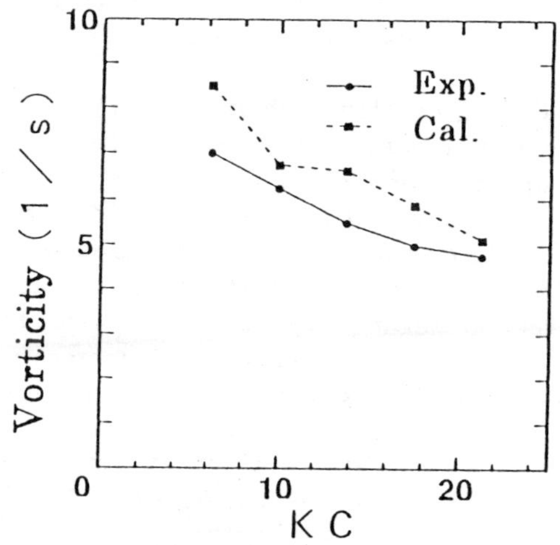

Figure 8 Relationship between vorticity and KC number

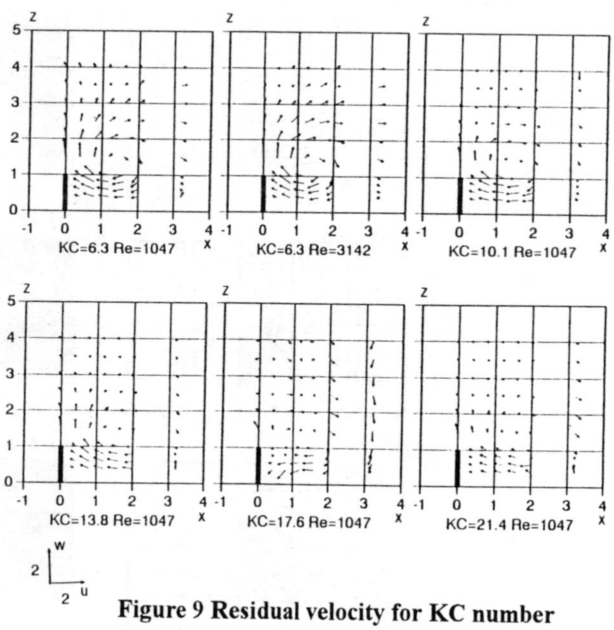

Figure 9 Residual velocity for KC number

5.4 Density profile under uni-directional flow

Figure10 shows the comparison of the Density profile under uni-directional calculated density interface with the experimental results by Akeda et al.(1989),who investigated the profile of density interface under uni-directional flow. In Fig.10 the solid line represents the experimental results. The height of density interface(T=0.05) starts to fall behind the upwelling producing structure, i.e. x=1.0, and it becomes minimum at x=4.0. Entrainment of the bottom fluid is also prominent here. It is difficult to compare the calculated interface profile with the measured one because of the limitation in measurement by flow visualization using dyes. The agreement between the calculation and experiment is, however, satisfactory.

5.5 Oscillatory flow with stratification

Figure 11 shows the calculated results at the velocity phase of $3\pi/2$ for run 1. The separated vortex formed at the top of the upwelling producing structure near the phase of the maximum velocity, which is similar to the calculated results without stratification (no figure). However, the negative vorticity is not extended rapidly due to the density interface . With increasing Ri, the positive vorticity is also formed near the density

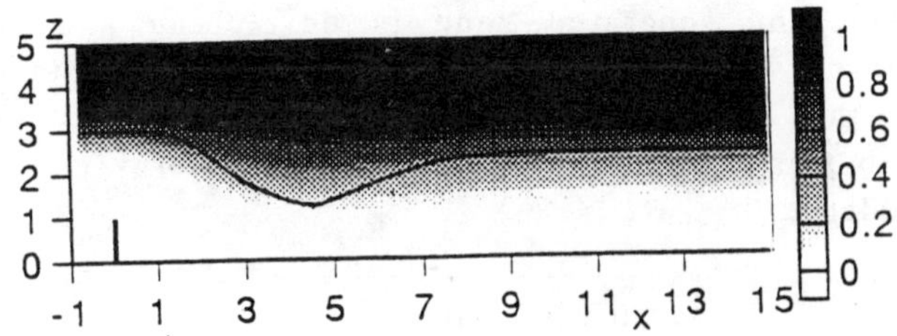

Figure 10 Profile of the density interface

interface at the phase of flow direction changes and it consumes the negative vorticity. But the destruction of vertical stratification of the density interface is progressive irrespective of Ri. On the other hand, when the height of the initial density interface is higher than that of the upwelling producing structure for h_2=2.5, the scale of separated vortex is the same as that without stratification, and the negative intense vorticity is seen, the intense vorticity is formed only near the upwelling producing structure under the present calculation condition. Therefore, the vorticity formed near the

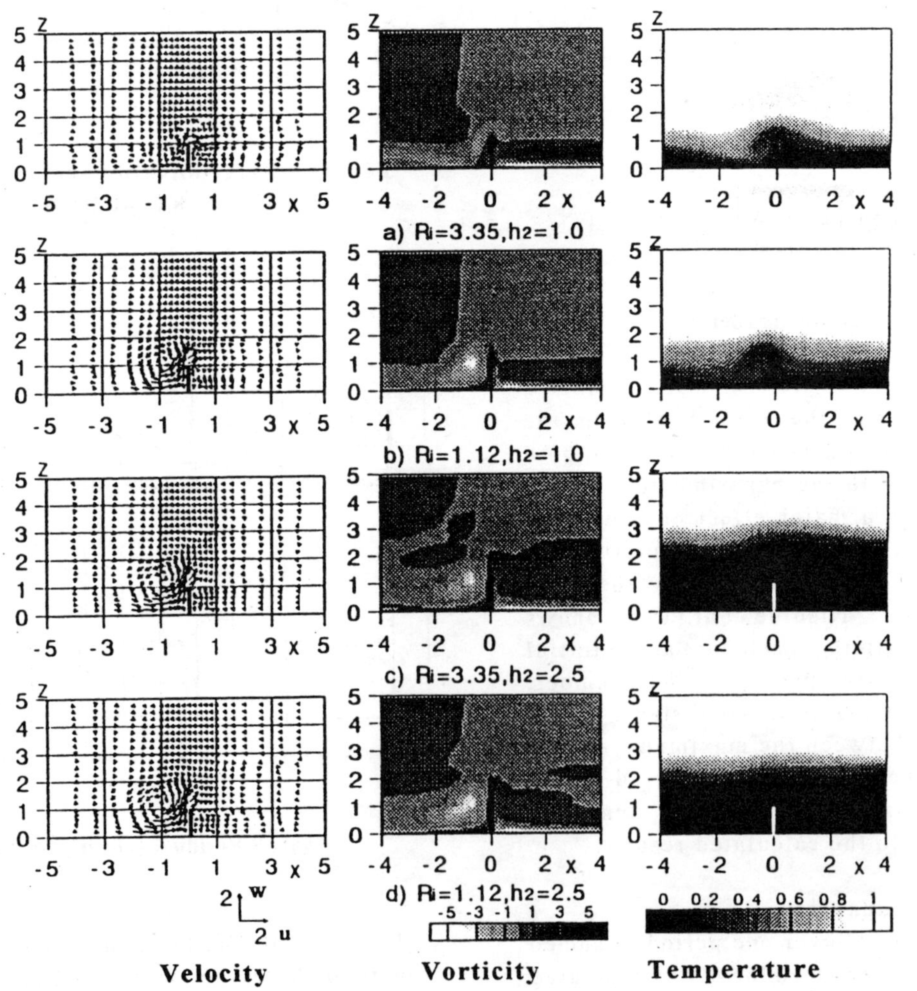

Figure 11 Calculated results under oscillatory flow with stratification

density interface has not enough intensity to destroy the density interface, and for larger Ri, the strong mixing of the density interface does not happen easily.

6. Conclusions

It is shown that the separated vortex over an upwelling producing structure is a very important factor to promote the upwelling flow. With decreasing of KC number, the separated vortexes with large of vorticity formed near the upwelling producing structure , so that a fast residual flow is formed nearby. On the other hand, with increasing KC number the scale of the separated vortex is larger than that of the above-mentioned one and the value of the residual flow is smaller, too.

The k- ε turbulence model is applied to a stratified flow and the profile of the density interface could be simulated for an uni-directional flow. Furthermore it is shown that the destruction of the density interface took place for smaller Ri in an oscillatory flow . Also when the value of h_2 is under 1.0, i.e., the height of upwelling producing structure is higher than the density interface, the upper layer and the lower layer will be mixed irrespective of the value of Ri.

It was found that the parameters; KC number, Ri and h_2, are important to promote the upwelling flow and especially h_2 is very significant through the present study. The present study includes no real field conditions. For example, the vale of KC number could be over 1000 in the field. The next step is to extend the numerical conditions to the real field and study a movement of bottom water by using Euler-Lagrange's method. Because the quantitative estimation on the outflow rate of the upwelling flow is needed.

References

Akeda M., H. Imamura, M. Ohtake and N. Aritomi(1989): Experiment on the development of upwelling producing structure, Proc.36th Japan Conf. on Coastal Eng.,pp.849 - 853 ,(in Japanese).

Ohno K., Y. Masuda, T. Miyazaki, K. Takahashi, M. Ukita and S. Seki(1991) : Experiment on the upwelling flow, Technical Memo. of Coastal Science Center,Vol.50-57,(in Japanese).

Tsujimoto G.(1992): Flow structure and density interface over coastal structure in a stratified oscillatory flow, 47th Abstract of Annual Convention,pp.1216-1217.

Nishide H., K. Michioku and G.Tsujimoto (1988): Local flow over coastal structure under an oscillatory flow, 49th Abstract of Annual Convention, pp.298-299.

Matsui M. and N. Aritomi(1988):Simulation of flow by using upwelling producing structure; "Maritex", Technical Memo. on Goyo Construction Comp.,Vol.17.pp.49-54.

Proceedings of the Seventh (1997) International Offshore and Polar Engineering Conference
Honolulu, USA, May 25-30, 1997

ISBN 1-880653-28-1 (Set); ISBN 1-880653-31-1 (Vol. III)

Shallow Water Equilibrium Range Waves

R.-Q. Lin
David Taylor Model Basin, Carderock Division
Bethesda, MD, USA

W. Perrie
Bedford Institute of Oceanography
Dartmouth, Canada

ABSTRACT

Zakharov and Filonenko (1966) found the equilibrium energy spectrum, $E_e = \alpha\omega^{-4}$ in deep water, by applying a conformal transformation method to the steady state energy balance equation. Following their approach, we derive the equilibrium energy in finite water depth. For the linear case in the deep water limit, our formulation agrees with the ω^{-4} solution by Zakharov and Filonenko (1966). In the shallow water, our formulation is approximately proportional to ω^{-2} and can be shown to be consistent with observations.

To study the wave breaking dissipation energy, we define a single wave breaking condition for finite depth water, motivated by the deep water formulation by Banner and Phillips (1974), using a normalized wave amplitude concept for finite water depth. Wave breaking dissipation energy was evaluated as the difference between the equilibrium energy spectrum, E_e, and corresponding maximum energy spectrum, using an appropriate coupling joint probability, following Yuen (1982).

KEY WORDS: Equilibrium range, wave breaking, finite water depth, conformal transformation method

INTRODUCTION

Wave breaking dissipation of wind-generated waves occurs all the time in the real ocean, but is not well understood. Traditionally, wave dissipation is calculated using the difference between the equilibrium energy spectrum of wind-generated waves and the real maximum energy spectrum (Longuet-Higgins, 1969) and a *coupling* joint probability. Therefore, to understand and estimate wave dissipation we have to understand two concepts: equilibrium energy spectrum and the joint probability for wave breaking conditions.

For deep water, Phillips (1958) used similarity arguments to suggest that the equilibrium spectrum E_e is proportional to ω^{-5}, where ω is wave frequency. This is called the ω^{-5} Phillips' equilibrium range. About a decate later, Zakharov and Filonenko (1966) analytically determined that E_e is proportional to ω^{-4} instead of ω^{-5}, using a conformal transformation to obtain Kolmogorov solutions. This was confirmed observationally by Toba (1973) and later by a number of oceanographers, for example Donelan et al (1985). Recently, Perrie and Toulany (1989) found that E_e is approximately proportional to ω^{-2}, by analysing the shallow water wave data from the Canadian Atlantic Storm Program (CASP) off Nova Scotia.

One of the goals of the present study is to first obtain the equilibrium energy spectrum in finite water theoretically and to present a verification in terms of observed spectra. As yet, we do not investigate the genuine three-dimensional energy distribution as suggested by some recent field observations presented by Su (1982a and 1982b). Secondly, we want to explore the relation which must exist between the finite depth equilibrium energy variation and wave breaking dissipation. However, it must be admitted that progress on clarifying the parameterization of wave breaking has been slow. Models for wave breaking conditions have hardly advanced since Longuet-Higgins (1969) first obtained a wave breaking condition for a narrow band spectrum. He assumed the single wave breaks when the vertical acceleration at the crest approaches $-0.5g$, where g is the acceleration due to gravity. Phillips and Banner (1974) and Banner and Phillips (1974) suggested a single wave breaking formulation by considering the difference between wave phase speed and surface drift. Wright (1976) improved the wave breaking formulation by including the impact of current. Yuen (1982) used statistical analysis to study wave breaking. He introduced a second-order joint probability for wave breaking, dependent on wave amplitude a and frequency ω, for a wide band spectrum. Huang et al (1986) adopted Yuen's theory to study the Wallop's wave breaking spectrum.

Alternately, Hasselmann et al (1988) used a simple heuristic formula for wave breaking, dependent on frequency ω alone. This functional, originally derived for deep water, was tuned to give appropriate wave energy growth with fetch, compared to observed fetch-relations for growing, evolving deep water waves, as reported by Hasselmann et al (1973), for example. These two formulations are different, which shows that there is little understanding of the wave breaking phenomenon.

In this paper, we analytically study the equilibrium energy spectrum in finite water depth by using a conformal transformation for the steady state energy transfer equation, in Section 2. Basic equations and boundary conditions are used to obtain the normalized wave amplitude in finite water depth and to define wave breaking conditions in Section 3. We also provide a calculation of the dissipation due to wave breaking.

EQUILIBRIUM ENERGY SPECTRUM

We are concerned with obtaining the equilibrium energy spectrum in finite water depth and in verifying it against observed data. Our formulation should agree with that of Zakharov and Filonenko (1966) in the limit of deep water.

From Lin and Perrie (1996), the nonlinear transfer equation for action A is:

$$\begin{aligned}\frac{\partial A_{(K_i)}}{\partial \tau} = 4\pi \int\int\int_{-\infty}^{\infty} & T^2_{i,1,2,3}\,\delta(\vec{K}_i + \vec{K}_1 - \vec{K}_2 - \vec{K}_3) \\ & \delta[\omega_{(\vec{K})} + \omega_{(\vec{K}_1)} - \omega_{(\vec{K}_2)} - \omega_{(\vec{K}_3)}] \\ \{A_{(K_3)}A_{(K_2)}[A_{(K_1)} + A_{(K_i)}] - & \\ A_{(K_1)}A_{(K_i)}[A_{(K_3)} + A_{(K_2)}]\} & d\vec{K}_1 d\vec{K}_2 d\vec{K}_3,\end{aligned} \tag{1}$$

where the coupling coefficient $T_{0,1,2,3}$ is given by

$$T_{0,1,2,3} = P_1 + P_2 \tag{2}$$

where

$$\begin{aligned}P_1 = & \\ -V^{(-)}_{(\vec{K}_3,\vec{K}_3-\vec{K}_1,\vec{K}_1)} V^{(-)}_{(\vec{K}_i,\vec{K}_2,\vec{K}_i-\vec{K}_2)} & \{\frac{1}{\omega_{(\vec{K}_1-\vec{K}_3)} - \omega_{(\vec{K}_3)} + \omega_{(\vec{K}_1)}} \\ & + \frac{1}{\omega_{(\vec{K}_2)} + \omega_{(\vec{K}_i-\vec{K}_2)} - \omega_{(\vec{K}_i)}}\} \\ -V^{(-)}_{(\vec{K}_2,\vec{K}_i,\vec{K}_2-\vec{K}_i)} V^{(-)}_{(\vec{K}_1,\vec{K}_1-\vec{K}_3,\vec{K}_3)} & \{\frac{1}{\omega_{(\vec{K}_1-\vec{K}_3)} - \omega_{(\vec{K}_1)} + \omega_{(\vec{K}_3)}} \\ & + \frac{1}{\omega_{(\vec{K}_i)} + \omega_{(\vec{K}_2-\vec{K}_i)} - \omega_{(\vec{K}_2)}}\} \\ -V^{(-)}_{(\vec{K}_2,\vec{K}_2-\vec{K}_1,\vec{K}_1)} V^{(-)}_{(\vec{K}_i,\vec{K}_3,\vec{K}_i-\vec{K}_3)} & \{\frac{1}{\omega_{(\vec{K}_1-\vec{K}_2)} - \omega_{(\vec{K}_2)} + \omega_{(\vec{K}_1)}} \\ & + \frac{1}{\omega_{(\vec{K}_3)} + \omega_{(\vec{K}_3-\vec{K}_i)} - \omega_{(\vec{K}_i)}}\},\end{aligned}$$

and

$$\begin{aligned}P_2 = & -V^{(-)}_{(\vec{K}_3,\vec{K}_i,\vec{K}_3-\vec{K}_i)} V^{(-)}_{(\vec{K}_1,\vec{K}_1-\vec{K}_2,\vec{K}_2)} \times \\ & \{\frac{1}{\omega_{(\vec{K}_1-\vec{K}_2)} - \omega_{(\vec{K}_1)} + \omega_{(\vec{K}_2)}} + \frac{1}{\omega_{(\vec{K}_i)} + \omega_{(\vec{K}_3-\vec{K}_i)} - \omega_{(\vec{K}_3)}}\} \\ & -V^{(-)}_{(\vec{K}_i+\vec{K}_1,\vec{K}_i,\vec{K}_1)} V^{(-)}_{(\vec{K}_2+\vec{K}_3,\vec{K}_2,\vec{K}_3)} \{\frac{1}{\omega_{(\vec{K}_i+\vec{K}_1)} - \omega_{(\vec{K}_i)} - \omega_{(\vec{K}_1)}} \\ & \qquad + \frac{1}{\omega_{(\vec{K}_2+\vec{K}_3)} - \omega_{(\vec{K}_2)} - \omega_{(\vec{K}_3)}}\} \\ & -V^{(+)}_{(-\vec{K}_2-\vec{K}_3,\vec{K}_2,\vec{K}_3)} V^{(+)}_{(\vec{K}_i,\vec{K}_1,-\vec{K}_i-\vec{K}_1)} \{\frac{1}{\omega_{(\vec{K}_2+\vec{K}_3)} + \omega_{(\vec{K}_2)} + \omega_{(\vec{K}_3)}} \\ & \qquad + \frac{1}{\omega_{(\vec{K}_i+\vec{K}_1)} + \omega_{(\vec{K}_i)} + \omega_{(\vec{K}_1)}}\} \\ & +W_{(\vec{K}_i,\vec{K}_1,\vec{K}_2,\vec{K}_3)},\end{aligned}$$

with coefficient $V^{(\pm)}_{(\vec{K}_i,\vec{K}_1,\vec{K}_2)}$ given by

$$\begin{aligned}V^{(\pm)}_{(\vec{K}_i,\vec{K}_1,\vec{K}_2)} = & \\ \frac{1}{8\pi\sqrt{2}}\{ & [\vec{K}_i \cdot \vec{K}_1 \pm K_i K_1 \tanh(K_i h)\tanh(K_1 h)][\frac{\omega_{(K_i)}\omega_{(K_1)}K_2}{\omega_{(K_2)}K_i K_1}]^{\frac{1}{2}} \\ & +[\vec{K}_i \cdot \vec{K}_2 \pm K_i K_2 \tanh(K_i h)\tanh(K_2 h)][\frac{\omega_{(K_i)}\omega_{(K_2)}K_1}{\omega_{(K_1)}K_i K_2}]^{\frac{1}{2}} \\ & +[\vec{K}_1 \cdot \vec{K}_2 + K_1 K_2 \tanh(K_1 h)\tanh(K_2 h)][\frac{\omega_{(K_1)}\omega_{(K_2)}K_i}{\omega_{(K_i)}K_1 K_2}]^{\frac{1}{2}}\},\end{aligned} \tag{3}$$

and coefficient $W_{(\vec{K}_i,\vec{K}_1,\vec{K}_2,\vec{K}_3)}$,

$$\begin{aligned}W_{(\vec{K}_i,\vec{K}_1,\vec{K}_2,\vec{K}_3)} = & \\ \bar{W}_{(-\vec{K}_i,-\vec{K}_1,\vec{K}_2,\vec{K}_3)} + & \bar{W}_{(\vec{K}_2,\vec{K}_3,-\vec{K}_i,-\vec{K}_1)} - \bar{W}_{(\vec{K}_2,-\vec{K}_1,-\vec{K}_i,\vec{K}_3)} \\ -\bar{W}_{(-\vec{K}_i,\vec{K}_2,-\vec{K}_1,\vec{K}_3)} - & \bar{W}_{(-\vec{K}_i,\vec{K}_3,\vec{K}_2,-\vec{K}_1)} - \bar{W}_{(\vec{K}_3,-\vec{K}_1,\vec{K}_2,-\vec{K}_i)},\end{aligned} \tag{4}$$

where

$$\begin{aligned}\bar{W}_{(\vec{K}_i,\vec{K}_1,\vec{K}_2,\vec{K}_3)} = & \frac{1}{64\pi^2}[\frac{\omega_{(K_i)}\omega_{(K_1)}}{\omega_{(K_2)}\omega_{(K_3)}}K_i K_1 K_2 K_3 \\ & \tanh K_i h \tanh K_1 h \tanh K_2 h \tanh K_3 h]^{1/2}[2(\frac{K_i}{\tanh K_i h} \\ & + \frac{K_1}{\tanh K_1 h}) - |\vec{K}_1 + \vec{K}_3|\tanh(|\vec{K}_1 + \vec{K}_3|h) - \\ & |\vec{K}_1 + \vec{K}_2|\tanh(|\vec{K}_1 + \vec{K}_2|h) \\ & -|\vec{K}_i + \vec{K}_3|\tanh(|\vec{K}_i + \vec{K}_3|h) - \\ & |\vec{K}_i + \vec{K}_2|\tanh(|\vec{K}_i + \vec{K}_2|h)].\end{aligned} \tag{5}$$

Other variables in Equations (1)-(5) are time τ, wavenumber $\vec{K}_i$ and depth h. For steady state solutions, $\frac{\partial A}{\partial \tau} = 0$. Angular averaging of Equation (1) over the plane, assuming the linear case, implies:

1) $\omega = (K \tanh Kh)^\alpha$,

2) $V^{(\pm)}_{(\vec{K}_i,\vec{K}_1,\vec{K}_2)} = (K \tanh Kh)^\beta \tanh^{-3/2} Kh(1 + c_1 \tanh^2 Kh)$;

3) $W_{(\vec{K}_i,\vec{K}_1,\vec{K}_2,\vec{K}_3)} = (K \tanh Kh)^\gamma \tanh^{-1} Kh(1 + c_2 \tanh^2 Kh)$

4) $T_{i,1,2,3} = (K \tanh Kh)^\gamma (\tanh^{-1} Kh)(c_3 \tanh^2 Kh + c_4 + c_5 \tanh^{-2} Kh)$

where α, β, γ, c_1, c_2, c_3, c_4 and c_5 are constants. Subscripts on the right side of the equations have been neglected because ω, $V^{(\pm)}_{(\vec{K}_i,\vec{K}_1,\vec{K}_2)}$, $W_{(\vec{K}_i,\vec{K}_1,\vec{K}_2,\vec{K}_3)}$ and $T_{i,1,2,3}$ are homogenous functions of $\vec{k}_i$, $\vec{k}_1$, $\vec{k}_2$, and $\vec{k}_3$.

Following Zakharov and Filonenko (1966), we obtain:

$$\int Q_o d\omega_1 d\omega_2 d\omega_3 = 0, \tag{6}$$

where

$$Q_o = D_{\omega_i,\omega_1,\omega_2,\omega_3} - D_{\omega_1,\omega,\omega_2,\omega_3} - D_{\omega_2,\omega_1,\omega_i,\omega_3} - D_{\omega_3,\omega_2,\omega_1,\omega_i}$$

with

$$D_{\omega_i,\omega_1,\omega_2,\omega_3} = P_{\omega_i,\omega_1,\omega_2,\omega_3} \delta_{\omega_i+\omega_1-\omega_2-\omega_3} \{A_{(\omega_3)} A_{(\omega_2)} [A_{(\omega_1)} + A_{(\omega_i)}] - A_{(\omega_1)} A_{(\omega_i)} [A_{(\omega_3)} + A_{(\omega_2)}]\} \tag{7a}$$

and

$$P_{\omega_i,\omega_1,\omega_2,\omega_3} = (k_i k_1 k_2 k_3 \tanh k_i h \tanh k_1 h \tanh k_2 h \tanh k_3 h)^{\alpha-1} \frac{dk_i dk_1 dk_2 dk_3}{d\omega_i d\omega_1 d\omega_2 d\omega_3} (|T_{k_i,k_1,k_2,k_3}|^2 \delta_{\vec{k}_i+\vec{k}_1-\vec{k}_2-\vec{k}_3}). \tag{7b}$$

Here, variable $P_{\omega_i,\omega_1,\omega_2,\omega_3}$ is a homogenous function of order $\omega^\Gamma \tanh^{-2} kh[c_3 \tanh^2 kh + c_4 + c_5 \tanh^{-2} kh]^2$, where $\Gamma = \frac{3(\gamma+d)}{\alpha} - 4$ and γ, d and α are constants to be determined later. We apply conformal transformations to the terms of Equation (6): the conformal transformation for the second term is:

$$\tilde{\omega_1} = \frac{\omega_i^2}{\omega_1}, \quad \tilde{\omega_2} = \frac{\omega_i \omega_2}{\omega_1}, \quad \tilde{\omega_3} = \frac{\omega_i \omega_3}{\omega_1}, \quad \tilde{\omega_i} = \omega_i, \tag{8}$$

for the third term:

$$\tilde{\omega_1} = \frac{\omega_i \omega_1}{\omega_2}, \quad \tilde{\omega_2} = \frac{\omega_i^2}{\omega_2}, \quad \tilde{\omega_3} = \frac{\omega_i \omega_3}{\omega_2}, \quad \tilde{\omega_i} = \omega_i, \tag{9}$$

and for the fourth term:

$$\tilde{\omega_1} = \frac{\omega_i \omega_1}{\omega_3}, \quad \tilde{\omega_2} = \frac{\omega_i \omega_2}{\omega_3}, \quad \tilde{\omega_3} = \frac{\omega_i^2}{\omega_3}, \quad \tilde{\omega_i} = \omega_i. \tag{10}$$

Equation (6) is thus reduced to:

$$\int D_{\omega_i,\omega_1,\omega_2,\omega_3} [1 + (\frac{\omega_i}{\omega_1})^y - (\frac{\omega_i}{\omega_2})^y - (\frac{\omega_i}{\omega_3})^y] d\omega_1 d\omega_2 d\omega_3 = 0 \tag{11}$$

and

$$y = \frac{3(\gamma+d)}{\alpha} - 1 - 3x. \tag{12}$$

The integration in Equation (11) becomes zero at $y = -1$ and $x = 0$. Alternately, another solution is $x = (\gamma+d)/\alpha$ and the distributions:

$$A_\omega = q\omega^{-\frac{\gamma+d}{\alpha}} \frac{\tanh^{2/3} kh}{(c_3 \tanh^2 kh + c_4 + c_5 \tanh^{-2} kh)^{2/3}}, \tag{13a}$$

and

$$E_\omega = 4\pi g^2 q \omega^{-(\gamma+d)} \frac{\tanh^{2/3} kh}{(c_3 \tanh^2 kh + c_4 + c_5 \tanh^{-2} kh)^{2/3}}, \tag{13b}$$

where $\gamma = 3$, $d = 1$, $\alpha = 1/2$, $\beta = 4/7$, q is constant and $E_\omega \equiv E(\omega)$ is the frequency spectrum of wave energy.

Therefore, Equation (13b) is

$$E_\omega = b_1 P^{1/3} \omega^{-4} \frac{\tanh^{2/3} kh}{(c_3 \tanh^2 kh + c_4 + c_5 \tanh^{-2} kh)^{2/3}}$$
$$E_\omega = b_2 Q^{1/3} \omega^{-11/3} \frac{\tanh^{2/3} kh}{(c_3 \tanh^2 kh + c_4 + c_5 \tanh^{-2} kh)^{2/3}}, \tag{14}$$

where b_1 and b_2 are constants, P and Q are the fluxes of energy to large k and small k, respectively, and where $Q = P/\omega$, following the notation of Zakharov (1991). Therefore, in the limit of shallow water waves, as $kh \to 0$, $E_\omega \tilde{\to} \omega^{-2}$.

OBSERVATIONS

It is difficult to measure the equilibrium range of the wave spectrum in finite depth water. Although the instruments may be placed in finite depth water, obtaining good data requires sufficiently high wind conditions that the waves *feel* the bottom. This requires a nontrivial fetch for growing and developing waves. Instruments placed in $25m$ depth water and $20km$ from land, for example, will almost always report deep water wave spectra *if* the winds are offshore, from land to ocean. To report finite depth waves which touch the bottom, the wind must be onshore, from ocean to land, and be of sufficient magnitude and duration that associated waves feel the bottom.

During the winter 1986, directional wave buoys were deployed off Nova Scotia as part of the Canadian Atlantic Storms Program (CASP). A full discussion of the wave component of CASP is given in Dobson et al (1989), and some of the difficulties in analysis and interpretation of the wave data are given in Perrie and Toulany (1990). The details of these analyses will not be recounted here. We only note that while the wind was offshore most of the time, so that much of the wave spectral data was deep water, some finite depth wave measurements were obtained, when the wind was onshore.

Assuming the energy spectrum has a simple one-dimensional representation,

$$E(\omega) = \begin{cases} 0 & \omega < \omega_m \\ Y\omega^{-q} & \omega > \omega_m \end{cases}$$

where Y and q are constants associated with Phillips' α coefficient, ω_m is the spectral peak frequency, the total energy E_o of the spectrum has the variation

$$E_o = \frac{Y}{-q+1}\omega_m^{-q+1}. \qquad (15a)$$

This implies that a deep water ω^{-4} spectrum would be associated with a total energy $E_o \sim \omega_m^{-3}$, whereas for a shallow water ω^{-2} spectrum, $E_o \sim \omega_m^{-1}$. In terms of wavenumber K, which in deep water, may be approximately related to ω through the dispersion relation $\omega^2 = gK$, this implies a total energy variation

$$E_o \sim K_m^{-3/2}. \qquad (15b)$$

In shallow water, K is related to ω through the approximate dispersion relation $\omega^2 = gK \tanh Kh \approx gK^2h$, which implies a total energy variation

$$E_o \sim K_m^{-1}. \qquad (15c)$$

Figures 1-2 present wave spectra from onshore wind conditions, as measured during CASP. Data are stratified according to dimensionless frequency ω_h, scaled with depth h, where $\omega_h \equiv \omega\sqrt{\frac{h}{g}}$. Manipulation of the approximate shallow water dispersion relation implies to first order, $\omega_h = Kh$. Thus, when $\omega_h \tilde{<} 2.0$ the wave can be said to touch the bottom. When $\omega_h \tilde{>} 2.0$ the wave can be said *not* to touch the bottom. Figure 1 shows that, for the shallow water waves from CASP, in terms of dimensionless variables, $\tilde{E}_o \sim \tilde{K}_m^{-1.08}$. For deep water waves, Figure 2 shows that $\tilde{E}_o \sim \tilde{K}_m^{-1.54}$. These results are in good agreement with Equations (15b) and (15c).

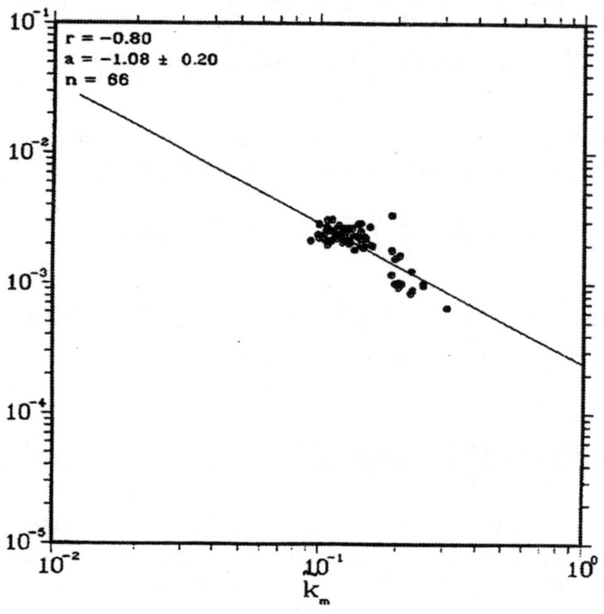

Fig. 1 Wave data from finite depth water in CASP, showing an energy variation of the form $E_o \sim K_m^{-1.08}$ when $1.5 < \omega_h < 2.0$. The correlation coefficient is -0.80 and there are 66 data points. Dimensionless total energy, $\tilde{E}_o$ is defined in terms of wind speed at 10m height, U_{10} as $\tilde{E}_o = E_o g^2/U_{10}^4$. Dimensionless wavenumber at the spectral peak is $\tilde{K}_m = K_m U_{10}^2/g$.

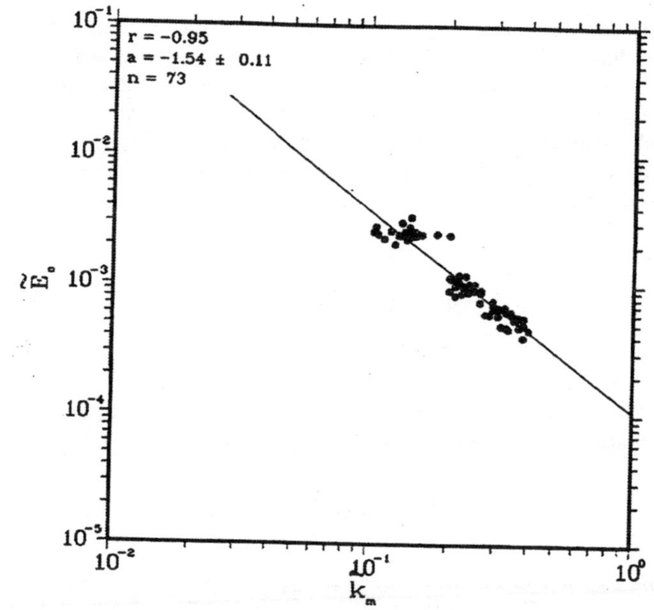

Fig. 2 Deep water wave data from CASP showing an energy variation of the form $E_o \sim K_m^{-1.54}$ when $2.0 < \omega_h < 4.7$. The correlation coefficient is -0.95 and there are 73 data.

FINITE DEPTH DISSIPATION

In this section, we use the wave amplitude in finite water depth in order to re-define the wave breaking conditions of Phillips and Banner (1974), and Banner and Phillips (1974). This leads to a calculation of the dissipative energy by wave breaking.

The usual basic equations and boundary conditions for potential incompressible flow are the following:

1. Continuity Equation:

$$\nabla_H^2\Phi + \frac{\partial^2\Phi}{\partial Z^2} = 0, \qquad for \qquad -h \le Z \le \eta \qquad (16)$$

where

$$\nabla_H = \vec{i}\frac{\partial}{\partial x} + \vec{j}\frac{\partial}{\partial y}$$

denotes the horizontal gradient.

2. Boundary Conditions:

At the free surface: $Z = \eta$

(a) Kinematics:

$$\frac{\partial\eta}{\partial t} + \nabla_H\Phi\cdot\nabla_H\eta = \frac{\partial\Phi}{\partial Z}, \qquad (17a)$$

(b) Dynamics:

$$\frac{\partial\Phi}{\partial t} + \frac{1}{2}(\nabla_H\Phi)^2 + \frac{1}{2}(\frac{\partial\Phi}{\partial Z})^2 + g\eta = 0. \qquad (17b)$$

At the bottom:

$$\frac{\partial\Phi}{\partial Z} + \nabla_H h\cdot\nabla_H\Phi = 0. \qquad (18)$$

In the above equations t, H, h, g, η, and Φ are time, the horizontal coordinate, the depth of the ocean, the gravity acceleration, the free surface elevation, and the potential function. We define Φ^o as the potential function at $z = 0$, which implies $\Phi^o_{(r,t)} = \Phi_{(r,z,t)}|_{z=0}$, and define ω as the frequency. Following Lin and Perrie (1996) and expanding all functions as power series in ϵ, which is a small parameter, we obtain,

$$\begin{aligned} \Phi &= \epsilon\Phi_1 + \epsilon^2\Phi_2 + \cdots \\ \eta &= \epsilon\eta_1 + \epsilon^2\eta_2 + \cdots \\ \omega &= \epsilon\omega_1 + \epsilon^2\omega_2 + \cdots \end{aligned} \tag{19}$$

First-order linear results are obtained by substituting the perturbation expansions of Equation (19) into Equations (17a) - (17b), implying

$$\frac{\partial^2\Phi_1}{\partial t^2} + g\frac{\partial\Phi_1}{\partial Z} = 0. \tag{20a}$$

The solution to Equation (20a) and corresponding linearizations of equations (15)-(18) is

$$\begin{aligned} \Phi_1 &= \Lambda_1 \exp\{i(\omega t + \vec{K}\cdot\vec{r})\}\cosh[|K|(Z+h)] \\ \eta_1 &= a\exp\{i(\omega t + \vec{K}\cdot\vec{r})\} \end{aligned} \tag{20b}$$

where K is the wave number, and K $= |\vec{K}|$; $\Lambda_1 = \dfrac{iag}{\omega\cosh Kh}$ and a is the wave amplitude. Substituting (20b) into (20a), we obtain the linear dispersion relationship,

$$\omega_1^2 = gK\tanh(Kh). \tag{21}$$

To second order in the perturbation expansions at $z = 0$, Equations (17a) - (17b) give,

$$\frac{\partial\eta_2}{\partial t} + \nabla_H\Phi_1^o\cdot\nabla_H\eta_1 = 0, \tag{22}$$

$$\frac{\partial\Phi_2^o}{\partial t} + \frac{1}{2}(\nabla_H\Phi_1^o)^2 + g\eta_2 = 0. \tag{23}$$

Combining Equations (22) and $\dfrac{\partial(23)}{\partial t}$ and substituting Equation (20b), we obtain

$$\frac{\partial^2\Phi_2^o}{\partial t^2} = 2i\omega K^2\Lambda_1^2\cosh^2 Kh\exp\{2i(\omega t + \vec{K}\cdot\vec{r})\}; \tag{24}$$

which implies that η_2 and Φ_2^o assume the form,

$$\begin{aligned} \eta_2 &= a_2\exp\{2i(\omega t + \vec{K}\cdot\vec{r})\} \\ \Phi_2^o &= \Lambda_2\cosh Kh\exp\{2i(\omega t + \vec{K}\cdot\vec{r})\}, \end{aligned} \tag{25}$$

where we have

$$a_2 = \frac{a^2K(3+\tanh^2 Kh)}{4\tanh^3 Kh}, \qquad \Lambda_2 = \frac{3aK}{8\sinh^3 Kh}\Lambda_1. \tag{26}$$

The second-order nonlinear dispersion term, ω_2, is equal to zero.

Given Equation (26), the obvious choice for the expansion coefficient ϵ, following Lin and Perrie (1996), is twice the ratio of a_2 and a,

$$\epsilon = 2\frac{a_2}{a} = \frac{aK(3+\tanh^2 Kh)}{4\tanh^3 Kh}.$$

This provides us with the motivation to define the normalized finite depth wave amplitude as $aK(3+\tanh^2 Kh)/4\tanh^3 kh$.

FINITE DEPTH WAVE BREAKING

Phillips and Banner (1974) define single wave breaking conditions in deep water as,

$$a_{max} = \frac{c^2}{2g}(1 - \frac{q+u}{c})^2 \tag{27a}$$

Applying the normalized finite depth wave amplitude concept, as defined in above section, we immediately obtain the corresponding single wave breaking conditions, for finite water depth:

$$a_{max} = \frac{c^2}{2g}(1 - \frac{q+u}{c})^2\frac{4\tanh^3 Kh}{3+\tanh^2 kh} \tag{27b}$$

where c is the wave phase speed, q is the surface drift and u is the current.

Yuen (1982) suggested a second - order joint probability for wave breaking for wide band wave spectra, as a function of amplitude, a, and frequency, ω:

$$\begin{aligned} P(a,\omega) = &\frac{2a^2\omega^3}{(2\pi)^{3/2}\nu_2^{1/2}\,\Delta_1^{1/2}\,\alpha^2 R} \\ &\exp[-\frac{a^2}{2\nu_o}]\exp[-\frac{a^2}{2\nu_o\Delta_1}(\nu_2 - \nu_o\omega^2/\alpha)^2], \end{aligned} \tag{28}$$

where $\nu_o = 0.0153$, $\nu_2 = 0.4438$, $\nu_4 = 16.5020$, $\Delta_1 = (\nu_o\nu_4 - \nu_2^2)$, α is positive constant, and R is constant equal to:

$$R = \int\int_S P(a, a' = 0, a'')|a''|da da'',$$

where $a' = \dfrac{\partial a}{\partial x}$ and $a'' = \dfrac{\partial^2 a}{\partial x^2}$ and S is the appropriate region of integration in (a, a'')-space.

DISSIPATIVE ENERGY

Using the Yuen (1982) joint probability for wave breaking, of the previous section, the dissipative wave energy may be parameterized in terms of the difference between the equilibrium energy spectrum and the maximum of the energy spectrum. In Section 2, we discussed the equilibrium energy spectrum in finite depth water. We now express wave breaking conditions in finite depth water, in terms of the equilibrium energy spectrum and the maximum energy spectrum.

Assuming a Gaussian distribution for the wave spectrum, the total dissipative energy is:

$$\int_{-\infty}^{\infty}\int_{a_{max}}^{\infty}\frac{1}{2}\rho g(a^2 - a_{max}^2)P(a,\omega)dad\omega = 1.2E_o \exp(-E_{max}/E_o), \quad (29)$$

where E_{max} is the energy associated with crests in a single wave train at the point of breaking. That is $E_{max} = \frac{1}{2}\rho g a_{max}^2$ is the energy when wave breaking occurs. The 1.2 factor is generated by considering wide band spectra, as compared to the narrow band spectrum of Yuen (1982).

SUMMARY

We analytically derived the equilibrium energy spectrum in finite depth, following the conformal transformation method of Zakharov and Filonenko (1966), water as: $E_w = \alpha_2\omega^{-4} \frac{\tanh^{2/3} kh}{[c_3 \tanh^2 kh + c_4 + c_5 \tanh^{-2} kh]^{2/3}}$. Of course this formulation agrees completely with Zakharov and Filonenko's (1966) formulation, in the limit as water depth becomes deep ($h \rightarrow \infty$). However, in the shallow water, the equilibrium energy spectrum is entirely different with the equilibrium energy spectrum in deep water. Moreover, we have shown that our formulation also agrees well with the observational data presented by Perrie and Toulany (1989) for shallow water waves.

We derived a normalized wave amplitude for finite depth water in order to extend the Phillips-Banner single wave breaking conditions. In finte water depth, our derived single wave breaking condition, for wave amplitude a, is: $a_{max} = \frac{c^2}{2g}(1 - \frac{q+u}{c})^2 \frac{4\tanh^3 kh}{3+\tanh^2 Kh}$ where c is the wave phase speed, q is the surface drift and u is the current. This leads to an estimate of the dissipative energy due to wave breaking in finite water depth, using the formulation from Longuet-Higgins (1969), $\int_{a_{max}}^{\infty}\int_{-\infty}^{\infty}(E_e - E_{max})P(a,\omega)d\omega da$. where $P(a,\omega)$ is the joint probability for wave breaking from Yuen (1982). However, this wave breaking condition in finite water depth has not yet to test with observational data.

ACKNOWLEDGEMENTS

We would like to thank Professors O. M. Phillips of the Johns Hopkins University and V. E. Zakharov of the University of Arizona for many useful suggestions and comments. This work has been supported by grants from the Office of Naval Research under the Coastal and Physical Oceanography Programs. W. Perrie was supported by the Federal (Canada) Panel on Energy Research and Development.

References

[1] Banner, ML, and Phillips, OM (1974). "On the incipient breaking of small scale waves," *J. Fluid Mech.*, Vol 65, part 4, pp 647-656.

[2] Dobson, F, Perrie, W, and Toulany, B, (1989). "On the deep-water fetch laws for wind-generated surface gravity waves," *Atmosphere-Ocean* Vol 27 (1), pp 210-236.

[3] Donelan, MA, Hamilton, J, and Hui, WH, (1985). "Directional spectra of wind-generated waves," *Phil. Trans. R. Soc. Lond. A*, Vol 315, pp 509-562.

[4] Hasselmann, K, Barnett, TP, Bouws, E, Carlson, H, Cartwright, DE, Enke, K, Ewing, JA, Gienapp, H, Hasselmann, DE, Kruseman, P, Meerburg, A, Miller, P, Olbers, DJ, Richter, K, Sell, W, Walden, H, (1973). "Measurements of wind-wave growth and swell decay during the Joint North Sea Wave Project (JONSWAP)," *Deutsche Hydrograph. Zeit., Erganzungsheft*, Reihe A (8), Nr. 12, pp 1-95.

[5] Hasselmann, S, Hasselmann, K, Komen, GK, Janssen, P, Ewing, JA, and Cardone, V, (1988). "The WAM model-A third generation ocean wave prediction model," *J. Phys. Oceanogr.*, Vol 18 , pp 1775-1810.

[6] Huang, NE, Bliven, LF, Long, SR, and Tung, CC, (1986). "An Analytical model for oceanic whitecap coverage," *J. Phys. Oceanogr.* Vol 16, no. 10, pp 1597-1604.

[7] Lin, RQ, and Perrie, W, (1996). "A New Coast Wave Model. Part III: Nonlinear wave-wave interaction for wave spectral evolution," In press in *J. Phys. Oceanogr.*

[8] Longuet-Higgins, MS, (1969). "On wave breaking and the equilibrium spectrum of wind-generated waves," *Proc. Roy. Soc. Lond. A.* Vol 310, pp 151-159.

[9] Perrie, W, and Toulany, BJ, (1989). "Shallow water wave data from the CASP experiment," WAM meeting in Valencia.

[10] Perrie, W, and Toulany, BJ, (1990). "Fetch relations for wind-generated waves as a function of wind-stress scaling," *J. Phys. Oceanogr.*, Vol 20, pp 1666-1681.

[11] Phillips, OM, (1958). "The equilibrium range in the spectrum of wind-generated waves," *J. Fluid Mech.*, Vol 4, pp 426-434.

[12] Phillips, OM, and Banner, ML, (1974). "Wave breaking in presence of wind drift and swell," *J. Fluid Mech.*, Vol 66, part 4, pp 625-640.

[13] Su, MY, (1982a). "Three dimensional deep-water waves, Part I, Experimental measurement of skew and symmetric wave patterns," *J. Fluid Mech.*, Vol 124, pp 73-108.

[14] Su, MY, (1982b). "Evolution of groups of gravity waves with moderate to high steepness," *Phys. Fluids*, Vol 25, pp 2167-2174.

[15] Toba, Y, (1973). "Local balance in air-sea boundary processes. III. On the spectrum of wind waves," *J. Oceanogr. Soc. Japan*, Vol 29, pp 209-220.

[16] Wright, JW, (1976). "The wind drift and wave breaking," *J. Phys. Oceanogr.*, Vol 6, pp 402-405.

[17] Zakharov, VE, and Filonenko, NN, (1966). "The energy spectrum for stochastic oscillation of a fluid's surface," *Doklady Akad. Nauk.*, Vol 170, pp 1292-1295.

[18] Zakharov, VE, (1991). "Inverse and direct cascade in the wind-driven surface wave turbulence and wave-breaking," ML Banner, RH Grimshaw Eds., *Breaking Waves IUTAM Symposium*, Sydney/Australia.

[19] Yuen, Y, (1982). On the statistical properties of sea waves, PhD Thesis in Department of Marine, Earth, and Atmospheric Sciences of North Carolina University, 80pp.

Proceedings of the Seventh (1997) International Offshore and Polar Engineering Conference
Honolulu, USA, May 25-30, 1997

ISBN 1-880653-28-1 (Set); ISBN 1-880653-31-1 (Vol. III)

On Some Characteristics of Breaking Waves Generated by a Submerged Circular Cylinder

Beom-Soo Hyun and Yong Heon Shin
Korea Maritime University
Pusan, Korea

ABSTRACT

The present study aims to understand the physical mechanism of breaking waves. Experiments have been carried out at circulating water channel to investigate the non-breaking and breaking waves generated by a submerged circular cylinder steadily moving under the free surface. Free surface profiles observed at various submerged depths of cylinder are complemented by the measurements of the pressure distribution on cylinder surface, the head losses occurring at the wakes of body and breaker, and finally the turbulent intensity. Their mutual correlation is also investigated. It is found that the head loss profile is an excellent indicator of the strength and extent of breaker. Very high turbulent intensities measured at and just downstream of the breaker indicate the consequence of energy transfer of wave breaking into turbulence.

KEY WORDS : Breaking wave, circular cylinder, depth of submergence, surface pressure, total head loss, turbulent intensity, wave/wake interaction

INTRODUCTION

The breaking wave emerges when the free surface meets certain critical conditions imposed by many factors. High speed marine craft with hydrofoils is a typical example when the lifting surface approaches to the free surface, inducing various kinds of waves and resulting in the change of surface pressure and drag. It has been reported that the appearance of breaking wave results in the increase of ship resistance as much as 15% (Baba, 1969). Similar problem is encountered when the wave load on ocean structures needs to be properly estimated, while the consideration of wave breaking is limited to empirical corrections in design procedure.

Breaking of water surface waves is related to many different mechanisms and physical phenomena such as instability, shear flow and vorticity transfer, turbulence generation and so on, even when the two dimensional aspects of breaking waves are considered. Lee (1995) pointed out, from his literature review on breaking waves, that the breaking waves generated by a submerged body moving under the free surface is phenomenally similar to the "spilling breaker" occurred on shallow-water coastal area. It is known that the spilling breaker, which is characterized by the generation of the spilling of water droplet on wave crest and little influence at other parts of wave, occurs when the slope of sea bottom is relatively mild. Once the wave breaking starts, the body experiences the additional drag so that the momentum deficit and energy dissipation into turbulence are in main concern.

Recently several experimental studies reveal that the breaking-wave phenomena around a submerged body are governed by Reynolds number as well as Froude number and the depth of submergence, meaning the viscosity plays a certain role in the generation of breaking wave. The notable features of the flow near breaking region are summarized to the following two findings (Banerjee, 1995; Shin and Mori, 1989; Wu, 1995); (1) the vertical distribution of velocity profiles near the free surface resembles that of turbulent boundary layer and wake, and (2) turbulent stresses can not be ignored at some locations of free surface. It is therefore the recent research trend in CFD that the treatments of turbulent stress are necessary for the simulation of breaking wave to impose the appropriate free-surface boundary condition. However, little progress is made theoretically yet. The theoretical background will not be discussed more in this paper, and interested readers may refer Lee (1995) and Hyun and Shin (1996).

Among many experimental studies including Parkin et al. (1956), honorably named as a benchmark test, probably the most thorough and prominent result in this area may be the study made by Duncan (1983). He developed a theory predicting the wave-breaking resistance using measured wave height, velocity and total head loss at the downstream of NACA0012 foil section. The relation between the critical Froude number and the depth of submergence was investigated as the criteria of wave breaking. Momentum deficit, corresponding to the increase of drag, was found to be proportional to the size of breaking region and the inclination angle of wave. It was also claimed that the wave breaking resistance reached up to 3.1 times higher than the maximum possible wave resistance obtained by nonlinear potential wave theory calculated at non-breaking wave condition.

The present study aims to the understanding of the wave-breaking phenomena as well as the interaction of free surface and wake of body. This paper consists of measurements and analyses of turbulent flow around breaking waves generated by a submerged cylinder installed at the circulating water channel. The analysis of the interaction between wake and free-surface is a main task. The key contents and methods included in this paper are (1) measurements of pressure, mean velocities and total head losses and their analyses to study the effects of submergence depth on breaking-wave profile, and (2) measurements of turbulent intensity at and vicinity of the breaker to find out the relationship between breaking wave and turbulence.

FACILITY, MODEL AND INSTRUMENTATION

Experiment on a two-dimensional cylinder was performed in circulating water channel (CWC) with its dimensions of 5m in length, 1.8m in width and 1.2m in height. The maximum flow speed at test section is 2 m/s, being achieved by two 0.8m-diameter impellers with maximum power of 22 kW each. Flow uniformity at the test section is about 2-3% when the speed of surface flow accelerator is properly adjusted. The rotational speeds of impeller and surface flow accelerator are calibrated against mean flow velocity at the test section using a Prandtl tube.

A circular cylinder of 7 cm in diameter made of acryl has its length of 1 m, which can be extended up to 1.5 m in case when the two dimensionality needs to be ensured in some cases. One side of cylinder is stuck to the wall of CWC, while a thin flat plate is attached the other side to avoid the interferences by supporting strut and other stuffs. The 36 pressure taps of 4 mm in diameter were holed with a spacing of 10 degree in azimuthal direction for the measurement of pressure distribution on cylinder surface. The wave profile was observed by eye and also recorded on video tape for post-analysis. Distributions of velocity and total head loss were measured at the downstream of cylinder using standard Prandtl tubes, located both inside the wake and the undisturbed flow region. The one-dimensional TSI hot-film prove was employed for the measurement of turbulent intensity at and around a breaker. Anemometer called "Intelligence system" by TSI and its associated software, "Flowpoint", were used for the data acquisition and signal processing finally by personal computer.

EXPERIMENTS

Coordinates and the schematics of breaking wave and following wavetrain are shown in Fig. 1. Here the black region denotes the breaking-wave region, and the grey the region where its effect is sustained. For the experiments, Froude number ($Fr = V/\sqrt{gD}$) was set to Fr=1.0 where V and D denote the free-stream velocity and the diameter of cylinder, respectively. In this case, the corresponding flow velocity was V=0.83m/s, and the Reynolds number ($Re = VD/\nu$) was $Re = 5.8\times10^4$.

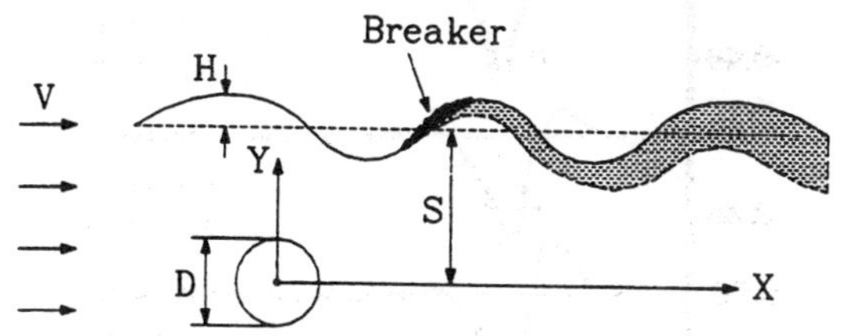

Fig. 1 Coordinate System

Wave profiles as well as surface pressure distributions on cylinder were measured at various submergence depths, ranging from 7cm to 45cm (S/D = 1.0 ~ 6.43). The measurements of velocity and total head loss distributions were made by wake survey at various X locations in front of, on and mostly downstream of breaker to see the generation, development and decay of the breaker. Three different depths of submergence, S/D = 4.29, 2.86, 2.14, were selected; a non-breaking case, a case of moderate breaking but steady regular wave sustained at further downstream, and a case of wave-breaking dominant so that no visible steady wave exists. Finally, turbulence intensity was measured at same condition made for wake survey.

RESULTS AND DISCUSSIONS

Wave profiles

Wave profiles at various depths of submergence (S/D) are shown in Fig. 2, where the cylinder axial location and the location of wave-breaking observed are marked also. The depth-based Froude number, $Fr_s = \frac{V}{\sqrt{gS}}$, is also shown in this figure as a reference for the readers who are interested in Froude number dependency of breaking wave.

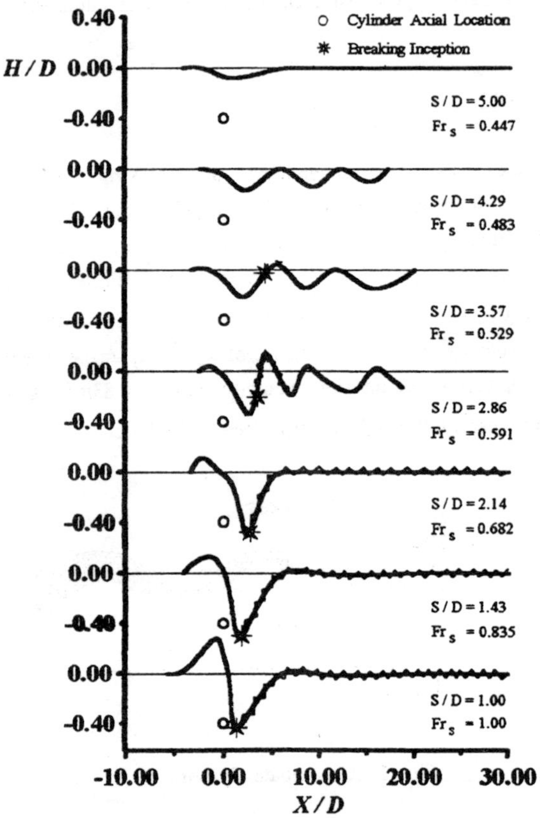

Fig. 2 Wave Profiles at Various Submergence Depths
(a) S/D = 5 (b) 4.29 (c) 3.57 (d) 2.86 (e) 2.14 (f) 1.43

It shows well, in sequence, the nearly calm free surface at S/D=6.43, the appearance of steady regular wave at S/D=5, the development and growth of wave, and the incipience of breaking wave (or free-surface instability) characterized by many ripples in first wave around S/D=3.57. This incipience occurred sporadically showing the lack of repeatability, and was observed to be very sensitive to surface flow condition in CWC, creating various kinds of disturbances such as free-stream turbulence and surface flow roughness. This argument is consistent with the fact that the incipient wave-breaking in the present experiment begins earlier than Duncan (1983), whose measurements were made at wave flume. The angle of wave slope was also very sensitive to flow condition and somewhat undeterministic, varying 7 to 12 degrees.

This instability propagated further downstream to eventually become a stable breaking wave when a submerged depth is decreased to S/D=2.86. The wave breaking appeared clearly and steadily. But it was still confined only at the first wave and the non-breaking waves were sustained downstream of it. Between S/D=3.57 and S/D=2.86, the angle of wave slope varies drastically from around 7 to 30 degrees at the location of the initial wave-breaking. Unfortunately, Froude number dependency of wave-breaking is not clearly found yet. More detailed measurements should be made between these two S/D's later to extract more information on wave- breaking phenomena.

With S/D further decreasing, the region of breaking wave grows larger in extent, and consequently the region of regular steady wave gradually diminishes. As S/D reaches less than 2.14, the breaking wave dictates all the free surface. The non-breaking waves are no longer visible. The growth of breaking wave results in the weakening of non-breaking wave, and generates the rough turbulent flow field on free surface. This observation is also consistent with Duncan (1983), where the increase of breaking wave resistance and the decrease of steady inviscid potential wave resistance are obtained as the depth of submergence decreases. With extremely shallow water (S/D=1), the wave-like surface profile over the body is nothing but the overflow of water over the body as like the flow over hydraulic dam. It causes the strong breaking region just behind a body, resembling a hydraulic jump.

Surface pressure distributions

Pressure distribution ($C_p = \frac{p - p_o}{(1/2)\rho V^2}$) on the cylinder surface at S/D=6.43 is shown in Fig.3.

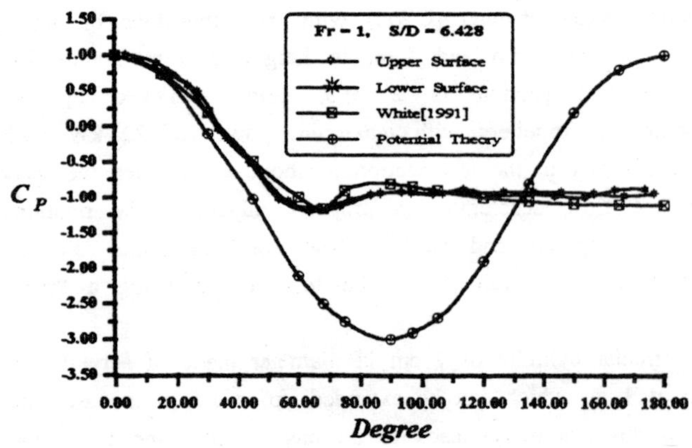

Fig. 3 Surface Pressure Distribution at Infinite Depth
(S/D = 6.43)

Since the pressure distributions on upper surface and lower surface give almost same results, this data can be regarded as a reference equivalent to the results at infinite depth. It agrees well with the measured data without free surface at $Re=1.86\times10^5$ given in White (1991). Comparing with the potential flow result also shown here, the discrepancies are evident even in front of the separation point observed at 65 degrees approximately.

Fig. 4(a) shows the pressure distribution on upper surface at various S/D. Up to 70 degrees in azimuthal direction, the pressure increases as S/D is decreasing. Separation point is delayed and moves to downstream from 60 deg at S/D=6.429 to 110 deg at S/D=1. Afterward, almost invariant pressure distribution is observed as expected, its value being about $C_p=-1$ to -1.2 depending on S/D. On the lower surface shown in Fig. 4(b), the pressure distributions at various S/D give almost same values except the vicinity of separation points, which are around 60 degree to 70 degree depending on S/D, implying that the influence of free surface on lower surface is less sensitive to S/D compared to that on upper surface.

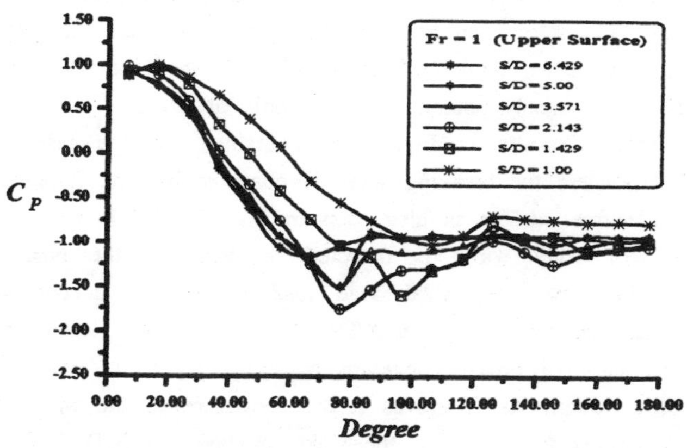

(a) Upper Surface

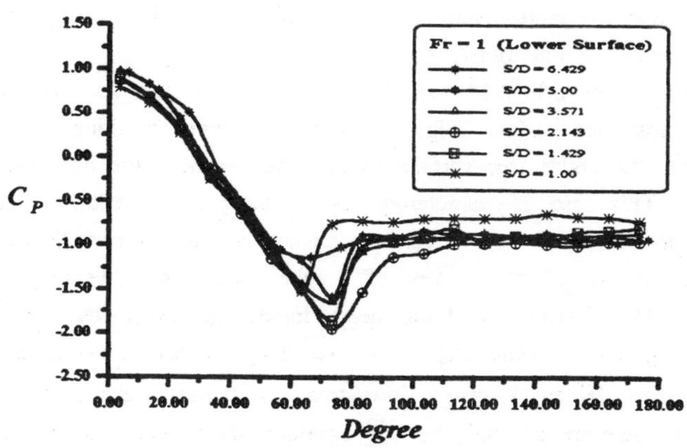

(b) Lower Surface

Fig. 4 C_P at Various Submergence Depths

Fig. 5 is presented to see the effect of submerged depth on cylinder surface more clearly. Here the pressure difference between upper and lower surfaces is noticeable beginning from S/D=3.57. Higher pressure on upper surface means the negative lift on cylinder. The reason of the delay of separation on upper surface with decreasing S/D seems to be the increase of kinetic energy due to the acceleration on flow near free surface, by supplying enough energy on the detarded flow near body surface. It can be concluded that there exists a certain relationship between wave profile and pressure distribution, although any indication of incipient wave-breaking is not clearly seen in pressure distribution.

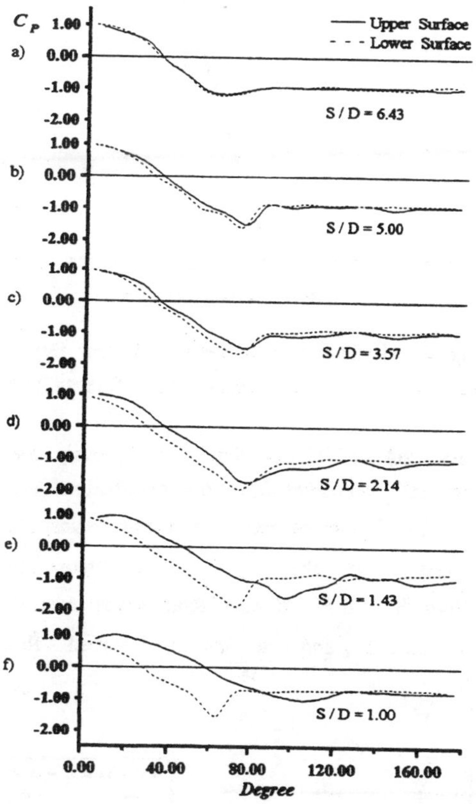

Fig. 5 C_P on Upper and Lower Surface

(a) S/D = 6.43 (b) 5 (c) 3.57 (d) 2.14 (e) 1.43 (f) 1

Total head loss distributions

Before presenting the distributions of measured flow velocity and total head loss at and downstream of submerged body, it is worth noting the result shown in Fig. 6, borrowed from Duncan [8]. This vertical velocity profile in case of wave breaking clearly shows the velocity defect near the free surface (consequently momentum deficit) at second trough after the breaker. Also shown is the calculated velocity profile of linear Stokes wave, showing the gradual increase of flow velocity toward free surface. This figure is a definite clue to justify the boundary layer and wake (or mixing layer) approximation for a certain confined region on free surface flow after wave breaking. Similar trend is seen in Fig. 7 in the present study.

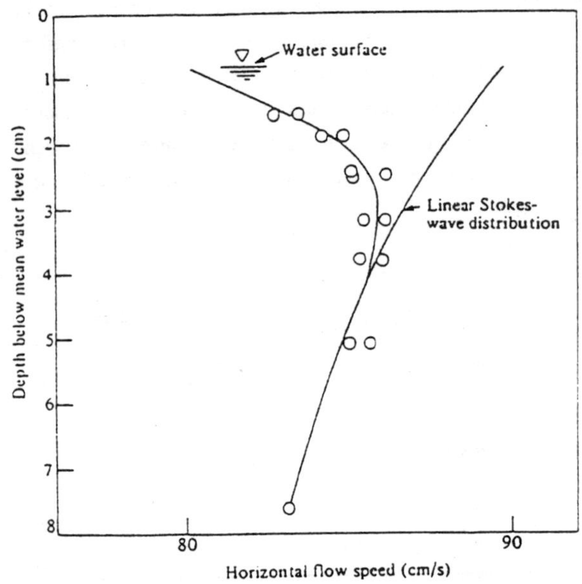

Fig. 6 Horizontal Flow Velocity Versus Depth (2nd Trough After the Breaker, from Duncan (1983))

Since this kind of velocity defect is obviously the effect of viscosity, especially turbulent flow, the distributions of total head loss ΔH ($= H - H_0$) at several X locations downstream of body are investigated to see the way how the viscous flow region develops. Here, H and H_0 denote total heads measured at the location of interest and at the undisturbed flow region, respectively.

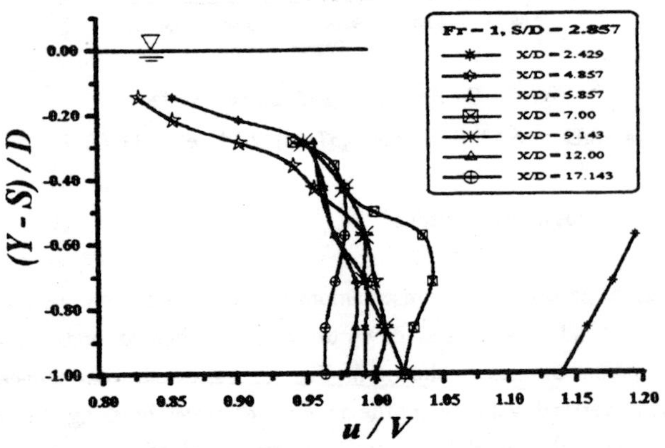

Fig. 7 Horizontal Flow Velocity Versus Depth (S/D = 2.857)

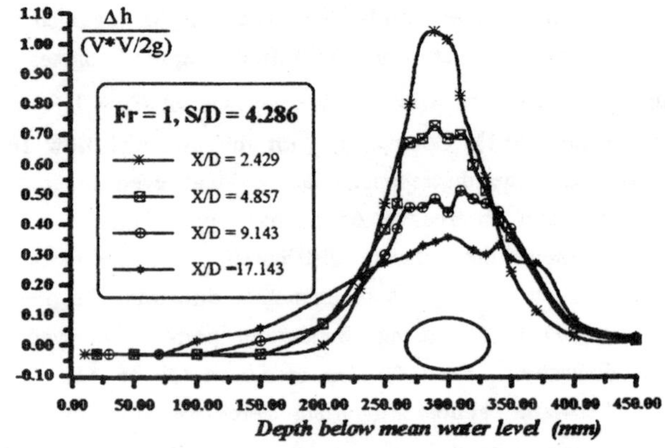

Fig. 8 Head Loss Distributions Behind Cylinder at Various Axial Locations (S/D = 4.29)

Figure 8 shows the distribution of total head loss in non-breaking condition of S/D=4.29. The development of body wake is only seen, which shows a typical two-dimensional wake behavior. Of course, the total head losses near free surface are not observed, no indication of wave breaking throughout all X locations. Any interaction between the free surface and the body wake is not shown up to X/D = 17. This is the case when the well-developed steady potential wave is only observed.

Interesting results in Fig. 9 are obtained for the case of S/D=2.86, where the breaking wave is observed by eye. Enlarged view near free surface is also seen in Fig. 10. While no head loss except body wake is detected in front of the breaker (X/D=2.43), very steep total head loss is seen in a confined region near the free surface at X/D=4.86, which is the location of breaker. Wake of breaker develops by extending its influence to vertical direction. As it moves further downstream, the wake of breaker begins to merge to the wake of body at X/D=7. Free surface disturbed by body wake shows very turbulent behavior and results in the steady regular wave diminishing. It is also likely that the mutual interaction of body and breaker distorts the body wake asymmetrically.

Results at S/D=2.14 are shown in Fig. 11 and its enlarged view near free surface (Fig. 12) where the wave breaking totally dictates the entire free surface so that the regular wave no longer exists. This case is characterized by a stronger interaction of the breaker and the body in overall compared to Figs. 9 and 10, and earlier merge of the wakes of breaker and body (merging at X/D=3.43). Profile of total head losses generated by wave-breaking shows some trends of extending its region wider and decreasing ΔH as it moves downstream, a typical wake behavior. It can therefore be said that the boundary layer and wake model of breaker may be physically correct.

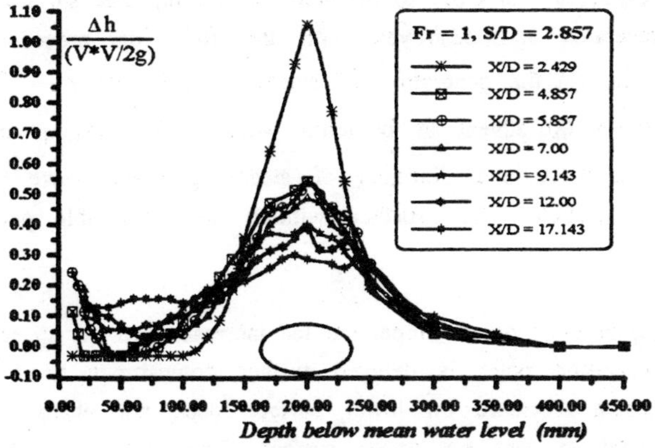

Fig. 9 Head Loss Distributions Behind Cylinder at Various Axial Locations (S/D = 2.86)

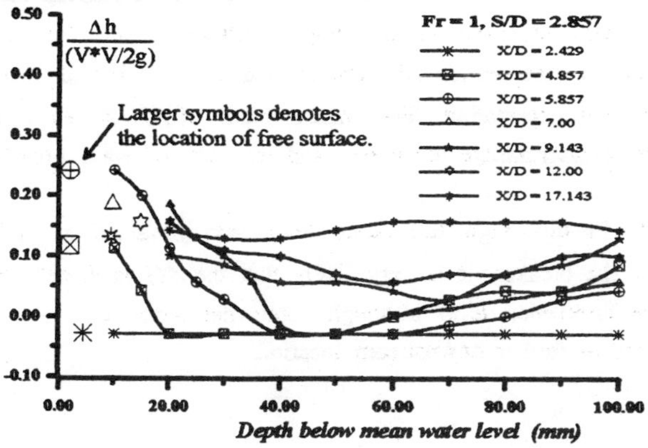

Fig. 10 Enlarged View of Fig. 9 in Breaking Region

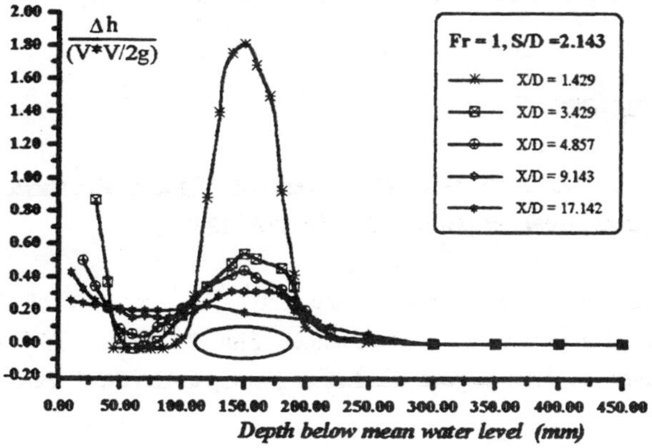

Fig. 11 Head Loss Distributions Behind Cylinder at Various Axial Locations (S/D = 2.14)

From the above arguments, it is concluded that the measurement of ΔH is an effective means to define the extent and intensity of the breaker. Its extent turns to be much larger than the area of air entrainment into water, what we can observe by eye. More detailed measurements are necessary to obtain the wave breaking resistance accurately by integrating ΔH at each X station.

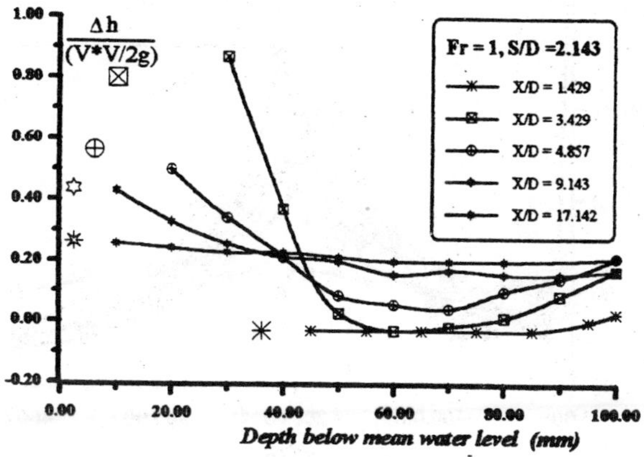

Fig. 12 Enlarged View of Fig. 11 in Breaking Region

Turbulence intensity

Turbulent intensity was measured to clarify the direct correlation of turbulence to total head loss. Here, only one component turbulent intensity is measured using one-dimentional hot-film probe for its simplicity. It should be noted that one component of turbulence does not mean the velocity fluctuating in one direction exclusively. This is just a fluctuation of velocity in X-Y plane since one-dimensional hot-film sensor can not resolve any desired component of instantaneous velocity.

Measured turbulent intensity distributions are shown in Fig. 13 for S/D=2.857 and Fig. 14 for S/D=2.143. Although experimental results do not shows the relation between breaking wave and turbulence level clearly, due to the difficulties of using hot-film on water, very high turbulent intensity (strong fluctuating velocity) is observed at the locations just downstream of the breaker near free surface. Somewhat suspiciously, some peak values reach over 50% of freestream velocity, indicating the direct consequence of energy transfer of wave-breaking into turbulence. While this result is consistent with ΔH distribution up to X/D=4.857, turbulence level at futher downstream was suprisingly low. It is very likely, but not perfectly clear, in the present stage that the amount of momentum deficit is proportional to the turbulence level in some sense.

For the conclusion, the forced wave-breaking by a submerged body causes the free surface flow being turbulent. Those results implicitly confirm the old findings of Baba (1969), who applied wake survey result to estimate the wave-breaking resistance stem from the bow of ship.

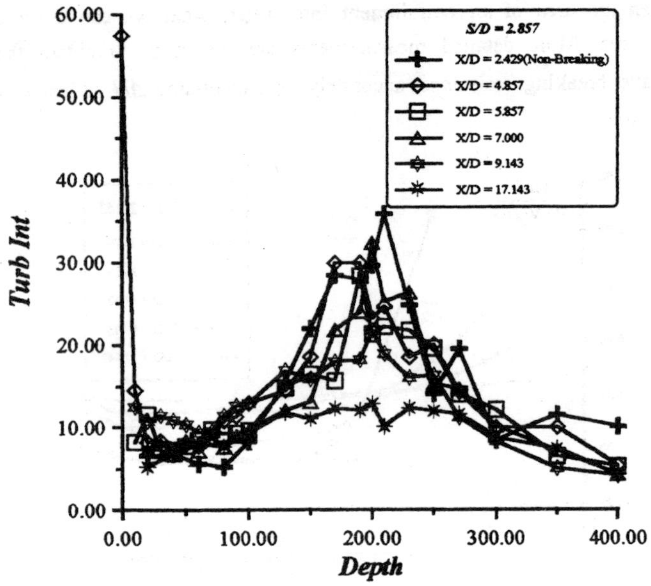

Fig. 13 Turbulent Intensity Distribution at Various Axial Locations (S/D=2.86)

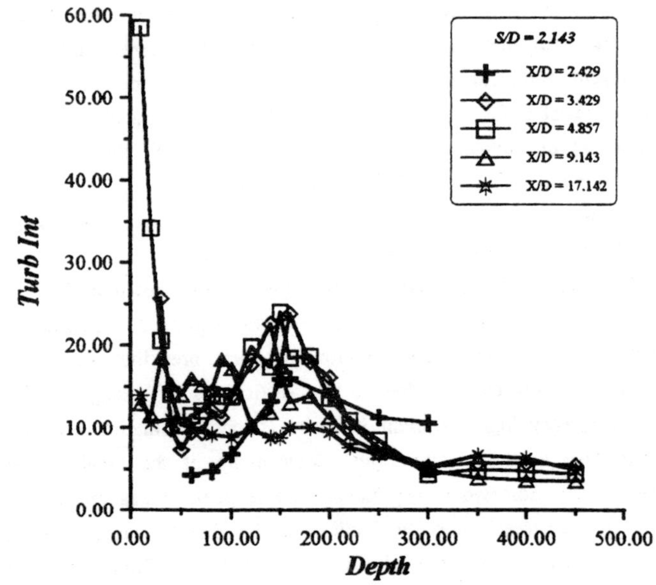

Fig. 14 Turbulent Intensity Distribution at Various Axial Locations (S/D=2.14)

CONCLUSIONS AND RECOMMENDATIONS

The present study on surface waves generated by a circular cylinder moving steadily under the free surface can be summarized as follows.

(1) As the depth of submergence decreases, starting from fully submerged case to close to the free surface, the free surface flow is generated and destroyed with the following procedure in sequence; ① the generation of steady potential wave → ② its growth → ③ advent of breaking wave → ④ development of breaking wave and diminish of steady potential wave → ⑤ entirely turbulent free surface without any noticeable potential wave.

(2) Pressure on the upper surface increases and the appearance of separation point is delayed further downstream as S/D is decreasing. Pressure on lower surface does not change much compared to that on upper surface. This result implies the production of the negative lift on the body.

(3) The interaction between the wave breaking and the body wake is investigated by measuring the total head loss distribution downstream of the breaker and the body. The very steep increase of ΔH was observed in the region influenced by the breaker. Merging of the wakes of breaker and body occurred relatively faster than expectation. The measurement of ΔH is an effective means to discriminate the extent and intensity of wave breaking.

(4) Finally, high turbulence level was obtained at and just behind the breaker. It is very likely that the ΔH is closely related to the turbulence level although somewhat suspicious results are obtained at further downstream location.

ACKNOWLEGMENTS

This research was supported by the Korea Science and Engineering Foundation (KOSEF) under Contract 96-0200-14-01-3.

REFERENCES

Baba, E. (1969), "A New Component of Viscous Resistance", *J. Soc. Naval Archi. of Japan* (SNAJ), Vol. 125.

Banerjee, S. et al (1995), "Structural Characteristics of Free-surface Turbulence in Open-channel Flows", *Proc. 1995 ONR Workshop on Free-Surface Turbulent Flows*, CALTECH, Feb. 27 - Mar. 1.

Duncan, J.H (1983), "The Breaking and Non-breaking Wave Resistance of a Two-dimensional Hydrofoil", *J. Fluid Mech.*, Vol. 126.

Hyun, B.S. and Shin, Y.H. (1996), "Study on Non-breaking and Breaking Waves Generated by a Submerged Cylinder (I)", *Proceedins of the Fall Meeting, Soc. Naval Archi. of Korea,* Seoul Nat'l Univ., Seoul, Nov. 7-8. (in Korean).

Lee, S.J. (1995), "Review on the Models of Breaking Waves", *J. Research Inst. of Industrial Technology,* Chungnam Nat'l Univ., Taejon, Vol. 10, No.1, June. (in Korean)

Parkin, B.R., Perry, B. and Wu, T.Y. (1956), "Pressure Distribution on a Hydrofoil Running Near the Water Surface", *J. Applied Physics*, Vol. 27, No. 3.

Shin, M.S. and Mori, K.H. (1989), "On Turbulent Characteristics and Numerical Simulation of Two- dimensional Sub-breaking Waves", *J. Soc. Naval Archi. of Japan,* Vol. 165.

White, F.M. (1991), "Viscous Fluid Flow", 2nd ed., *McGraw Hill.*

Wu, J.Z. (1995), "Interfacial Vorticity Dynamics and its Relevance to Free-surface Turbulence", *Proc. 1995 ONR Workshop on Free-Surface Turbulent Flows*, CALTECH, Feb. 27-Mar.1.

Proceedings of the Seventh (1997) International Offshore and Polar Engineering Conference
Honolulu, USA, May 25-30, 1997

ISBN 1-880653-28-1 (Set); ISBN 1-880653-31-1 (Vol. III)

Analysis of Near-Crest Pressure Gradient of Irregular Water Waves as a Dynamic Criterion of Breaking

K. Nadaoka, O. Ono and H. Kurihara
Tokyo Institute of Technology
Tokyo, Japan

ABSTRACT

To elucidate the physical mechanism of breaking the pressure field of irregular waves approaching breaking was analyzed in detail through laboratory experiments. Video images of irregular waves approaching the break, which were taken in a unidirectional wave channel with a bottom slope of 1:20, were analyzed to obtain data sets of the fluctuation of the free surface elevation both in space and time. A new technique has been developed to estimate the velocity and pressure field of irregular nonlinear waves from these data. From the analysis of the experimental results with this technique, it was found that when a wave breaks the vertical pressure gradient decreases rapidly and becomes nearly zero in the vicinity of the wave crest. By incorporating this fact as a breaking criterion into a wave equation, a numerical simulation to predict the breaker depth was attempted, and the computational result was found to give close agreement with an empirical breaking criterion.

KEY WORDS: Wave breaking, Breaking criterion, Irregular wave, Nonlinear wave, Pressure gradient

1. INTRODUCTION

Wave breaking is one of the most significant phenomena in coastal dynamics. It causes wave damping, nearshore currents and sediment transport and others. Although recent progress in water wave modeling provides efficient computational tools to precisely describe a wave field, but their applicability is limited to non-breaking waves. For the description of breaking waves including the prediction of breaking points, one must employ a more or less empirical way. Therefore the modeling of the wave breaking is one of the most significant subjects to be solved to provide wave models with more general applicability. On this line, in the present study, an attempt has been made to provide a breaking criterion which has a physically sound background and applicability even to irregular waves.

As one of the famous breaking criteria, Rankine (1886) suggested that when a wave breaks the horizontal particle velocity at the wave crest becomes greater than the wave speed C. However it is difficult to define the wave speed C for irregular waves. Using the ratio of the particle velocity at the crest to the wave speed, Stokes (1847) indicated the limiting angle at the crest of the highest wave in case of a deep water wave train with a symmetric profile. But in general the actual waves are not symmetric nor periodic. Some empirical criteria, on the other hand, have been suggested but they can be applied only to periodic regular waves. Besides the empirical methods are not able to apply to waves over arbitrary bottom topography. Recently, Schäffer (1992) proposed a breaking wave model in which waves are assumed to break when the wave front slope becomes steeper beyond a prescribed value. But according to our experimental result, as will be shown later, irregular waves do not break with a unique slope.

First in this paper we consider the vertical pressure gradient at the free surface, which may be used as a physical breaking criterion under general situation including nonlinear irregular waves. Second a new technique is developed to estimate the pressure field of nonlinear irregular waves from the free surface fluctuations both in space and time. Next, we show the pressure gradient fluctuation analyzed with this technique, and indicate that when a wave breaks the vertical pressure gradient decreases rapidly and becomes nearly zero in the vicinity of the wave crest. Finally we incorporate this fact as a breaking criterion into a wave equation developed recently by Nadaoka et al. (1994) and show that the computed breaker depth gives close agreement with the empirical breaking criterion by Goda (1970).

2. VERTICAL PRESSURE GRADIENT AT THE FREE SURFACE

For deep water waves, "Stokes' 120 degrees angle" is one of the famous breaking criteria, as mentioned before. According to this theory, the crest angle becomes 120 degrees at the wave breaking. The corresponding vertical acceleration and hence the vertical pressure gradient divided by the density ρ is 0.5 g (Longuet-Higgins,1963). In these theories waves are assumed to be symmetric in their profiles around the crest even at the breaking. However this assumption is unrealistic because actual breaking waves show appreciable deformation with highly skewed and asymmetric profile.

In irrotational fluid, the lowest pressure can exist only on the boundary (Lamb,1932). In the water waves, the pressure should be lowest at the free surface. Therefore the vertical pressure gradient at the free surface never can become negative. In this study the vertical pressure gradient is assumed to be a breaking criterion and a laboratory experiment was performed to examine this assumption and to find its critical value.

It is arduous to measure the pressure gradient at the free surface, thus the vertical momentum equation was employed to evaluate the pressure gradient. The vertical momentum equation at the free surface is

$$-\frac{1}{\rho}\frac{\partial p}{\partial z}\bigg|_{z=\eta} = g + \frac{\partial w}{\partial t}\bigg|_{z=\eta} + u_s\frac{\partial w}{\partial x}\bigg|_{z=\eta} + w_s\frac{\partial w}{\partial z}\bigg|_{z=\eta}, \tag{1}$$

where ρ is the water density, p the pressure, t the time, g the gravitational acceleration, x and z are the horizontal and vertical coordinates, u and w are the velocity components in the x and z direction, η the free surface displacement, and the suffix s indicates the variables at $z=\eta$. Because of the presence of the derivative of w with respect to z, Eq.(1) is complicated in its form to apply. However the derivatives of w with respect to x and t at the free surface are transformed as,

$$\frac{\partial w}{\partial t}\bigg|_{z=\eta} = \frac{\partial w_s}{\partial t} - \frac{\partial \eta}{\partial t}\cdot\frac{\partial w}{\partial z}\bigg|_{z=\eta}, \tag{2}$$

$$\frac{\partial w}{\partial x}\bigg|_{z=\eta} = \frac{\partial w_s}{\partial x} - \frac{\partial \eta}{\partial x}\cdot\frac{\partial w}{\partial z}\bigg|_{z=\eta}. \tag{3}$$

Substituting the above relations and the boundary condition at the free surface,

$$w_s = \frac{\partial \eta}{\partial t} + u_s\frac{\partial \eta}{\partial x} \tag{4}$$

into Eq. (1), we obtain

$$\begin{aligned}-\frac{1}{\rho}\frac{\partial p}{\partial z}\bigg|_{z=\eta} &= g + \frac{\partial^2\eta}{\partial t^2} + u_s{}^2\frac{\partial^2\eta}{\partial x^2} + 2u_s\frac{\partial^2\eta}{\partial t\partial x} \\ &+ \frac{\partial u_s}{\partial t}\frac{\partial \eta}{\partial x} + u_s\frac{\partial u_s}{\partial x}\frac{\partial \eta}{\partial x}\end{aligned} \tag{5}$$

To evaluate the vertical pressure gradient at the free surface from Eq.(5), the free surface fluctuation and the horizontal velocity at the free surface must be known. The free surface fluctuation η can be easily measured in laboratory experiment, while the horizontal velocity at the free surface is not easy to be measured. Therefore the latter was estimated with a newly developed method described below.

3. A NEW TECHNIQUE TO ESTIMATE VELOCITY FIELD UNDER IRREGULAR WAVES

Although Dean's stream function method (1965) is a well-known technique to estimate a velocity field from the free surface displacement of nonlinear waves, it can not be applied to irregular waves. Therefore Dean's method can not be employed for our purpose to evaluate the horizontal velocity at the free surface u_s in eq.(5). Hence an alternative new technique was developed in this study to estimate the velocity field of irregular and nonlinear waves from the free surface fluctuations both in space and time . If the variation in the water depth is so small that the depth may be assumed to be locally uniform, the general solution of the Laplace equation of the velocity potential ϕ can be given as

$$\phi(x,z,t) = \int_{-\infty}^{\infty} A(k,t)\frac{\cosh k(h+z)}{\cosh kh}\exp(ikx)dk, \tag{6}$$

where A is a time-varying wave number spectrum, k the wave-number, h the local depth, and i the imaginary unit. The kinematic boundary condition at the free surface is,

$$\frac{\partial \phi}{\partial z}\bigg|_{z=\eta} = \frac{\partial \eta}{\partial x}\cdot\frac{\partial \phi}{\partial x}\bigg|_{z=\eta} + \frac{\partial \eta}{\partial t}. \tag{7}$$

Substituting Eq.(6) into Eq.(7) and discretizing the resultant equation, we have

$$\begin{aligned}&\left[1+\left(\frac{\partial\eta}{\partial x}\bigg|_m\right)^2\right]\sum_{n=1}^{N}A_n(t)k_n\frac{\sinh k_n(h_m+\eta_m)}{\cosh kh}\exp(ik_nx_m)\Delta k|_n \\ &= \frac{\partial\eta}{\partial x}\bigg|_m\sum_{n=1}^{N}A_n(t)\left\{\left[\frac{\partial}{\partial x}\left(\frac{\cosh k_n(h+\eta)}{\cosh k_nh}\right)\right]_m \right.\\ &\left.+\frac{\cosh k_n(h_m+\eta_m)}{\cosh k_nh_m}ik_n\right\}\exp(ik_nx_m)\Delta k|_n + \frac{\partial\eta}{\partial t}\bigg|_m.\end{aligned} \tag{8}$$

The suffix m and n indicate respectively the discrete points on the space x and the wave-number k, and m runs from 1 to M. In the above equation, the unknown variable is only A_n.

Substituting $A_n=a_n+ib_n$ into Eq.(8) and taking the real parts, we finally obtain

$$\sum_{n=1}^{N} a_n\alpha_{mn} + b_n\beta_{mn} = \gamma_m, \qquad (m=1,\cdots,M) \tag{9}$$

where,

$$\begin{aligned}\alpha_{mn} = &\left\{-\left[1+\left(\frac{\partial\eta}{\partial x}\bigg|_m\right)^2\right]k_n\frac{\sinh k_n(h_m+\eta_m)}{\cosh k_nh_m}\sin k_nx_m\right.\\ &+\frac{\partial\eta}{\partial x}\bigg|_m k_n\frac{\cosh k_n(h_m+\eta_m)}{\cosh k_nh_m}\cos k_nx_m \\ &\left.+\frac{\partial\eta}{\partial x}\bigg|_m\left[\frac{\partial}{\partial x}\left(\frac{\cosh k_n(h+\eta)}{\cosh k_nh}\right)\right]_m\sin k_nx_m\right\}\Delta k|_n,\end{aligned}$$

$$\begin{aligned}\beta_{mn} = &\left\{\left[1+\left(\frac{\partial\eta}{\partial x}\bigg|_m\right)^2\right]k_n\frac{\sinh k_n(h_m+\eta_m)}{\cosh k_nh_m}\cos k_nx_m\right.\\ &+\frac{\partial\eta}{\partial x}\bigg|_m k_n\frac{\cosh k_n(h_m+\eta_m)}{\cosh k_nh_m}\sin k_nx_m \\ &\left.-\frac{\partial\eta}{\partial x}\bigg|_m\left[\frac{\partial}{\partial x}\left(\frac{\cosh k_n(h+\eta)}{\cosh k_nh}\right)\right]_m\cos k_nx_m\right\}\Delta k|_n,\end{aligned}$$

$$\gamma_m = \frac{\partial\eta}{\partial t}\bigg|_m$$

a_n and b_n in Eq.(9) may be determined from the following conditions

$$\frac{\partial S}{\partial a_n} = 0, \quad \frac{\partial S}{\partial b_n} = 0, \quad (n=1,\cdots,N) \tag{10}$$

where S is defined as

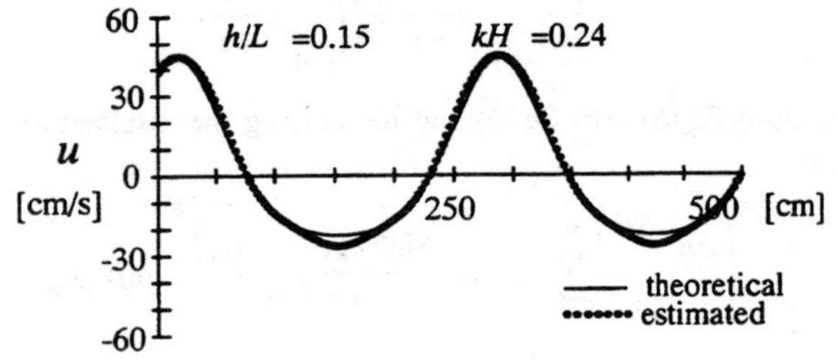

a) Stokes wave

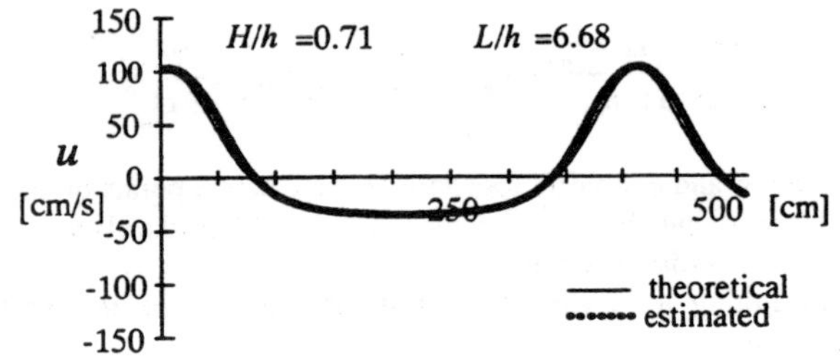

b) cnoidal wave

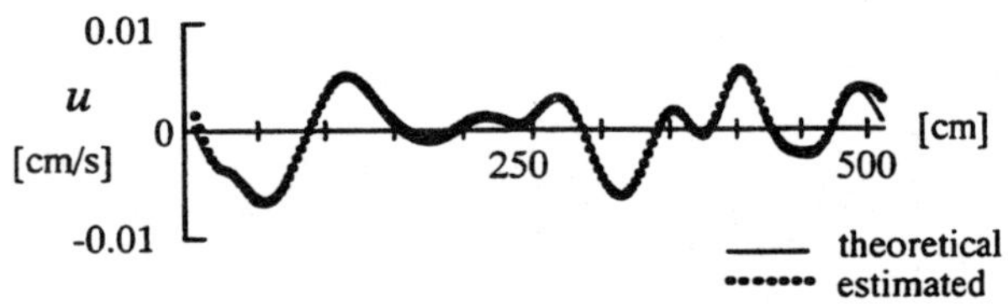

c) linear irregular wave

(h: water depth, H: wave height, L: wave length)

Fig. 1 Theoretical and estimated velocities at the free surface.

$$S=\sum_{m=1}^{M}\left[\sum_{n=1}^{N}\left(a_n\alpha_{mn}+b_n\beta_{mn}\right)-\gamma_m\right]^2 . \tag{11}$$

The resultant equations to determine a_n and b_n are

$$\sum_{m=1}^{M}\alpha_{lm}\left[\sum_{n=1}^{N}\left(a_n\alpha_{mn}+b_n\beta_{mn}\right)\right]=\sum_{m=1}^{N}\alpha_{lm}\gamma_m , \tag{12}$$

$$\sum_{m=1}^{M}\beta_{lm}\left[\sum_{n=1}^{N}\left(a_n\alpha_{mn}+b_n\beta_{mn}\right)\right]=\sum_{m=1}^{N}\beta_{lm}\gamma_m . \tag{13}$$

Although the wave field to which we want to apply the present method is nonlinear irregular waves, we have examined the validity of this method by applying it to the third-order Stokes and the second-order cnoidal waves and a linear irregular wave, because these waves have their theoretical expressions to be compared. Figure 1 shows the results of these comparison and indicates the good performance of the present method to estimate velocity fields under Stokes and cnoidal waves as typical nonlinear but regular waves and linear irregular waves. From these results one can expect that this method is applicable also to nonlinear irregular waves.

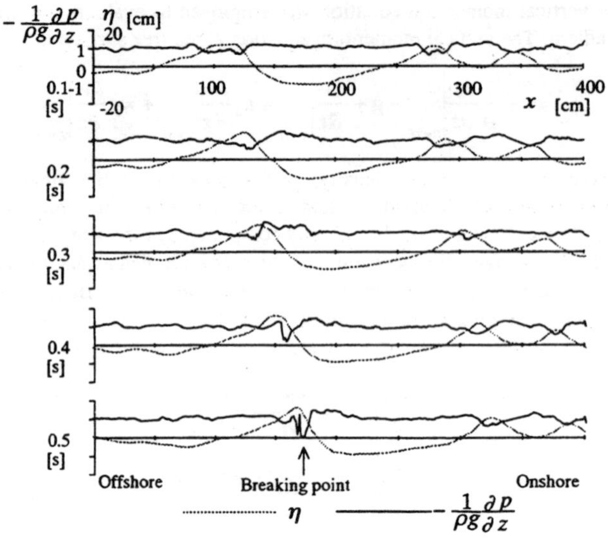

Fig.2 Time sequence of the free surface deformation and the corresponding vertical pressure gradient.

4. LABORATORY EXPERIMENT

The wave channel used for the experiment is equipped with a flap-type wave generator at one end and a slope of 1:20 at another end. The width of the channel is 40 cm, and the still water depth was kept to be 40 cm. Irregular waves generated in the channel were recorded by video camera at the interval of 1/30 s from the side of the channel, and the free surface elevation was detected by an image analysis of the video data. The generated incident irregular waves have a Bretschneider type spectrum with the mean period of 1.0 second and the mean height of 5.3 cm. The instant of the wave breaking is defined as that when a wave just starts to overturn.

Figure 2 shows a time sequence of the on-offshore distribution of the normalized pressure gradient (solid lines) and the surface displacement (dotted lines) in which the breaking occurs at the location indicated. The waves propagate from the left to the right. As the wave approaches the break, the pressure gradient rapidly decreases and becomes nearly zero when the wave breaks. The similar behavior of the pressure gradient was observed in almost all other waves approaching the break.

Figure 3 represents the variations in the smallest vertical pressure gradient near the wave crest and the corresponding change in η_{max}/h with the wave propagation, where η_{max} denotes the crest height. The waves broken within the observed section are classified as "breaking", and others are classified as "non-breaking". The doted lines represent the "non-breaking" wave, while the solid lines show the "breaking" wave. In the "non-breaking" wave, the vertical pressure gradient does not show appreciable decrease even when η_{max}/h becomes quite large. In contrast, the "breaking" waves indicate the abrupt decrease as the wave approaches the break and the vertical pressure gradient becomes nearly zero at the instant of the wave breaking. These results suggest that the vertical pressure gradient is more effective than the nonlinear parameter η_{max}/h as the breaking criteria and that the critical value of the pressure gradient is zero.

Figures 4 and 5 show respectively the minimum vertical pressure gradient within a wave and the minimum (steepest) foreside surface slope at the instant of the wave breaking. The horizontal axis in each figure is represented with the ratio of the surface elevation to the water depth at the location where the pressure gradient or the surface slope is

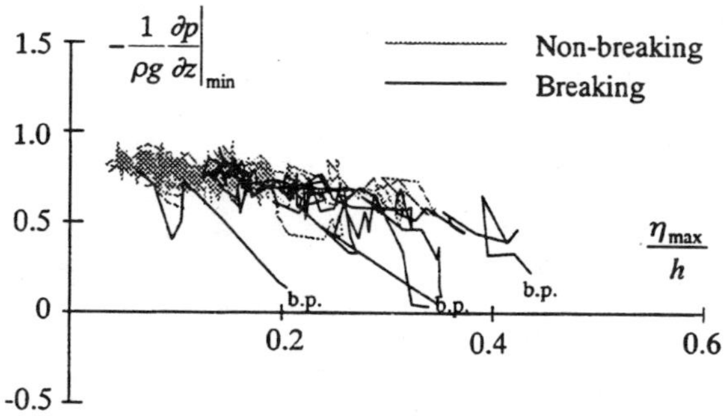

Fig.3 Variation in the normalized vertical pressure gradient and the corresponding value of η_{max}/h for 12 progressive waves; η_{max} denotes the crest height.

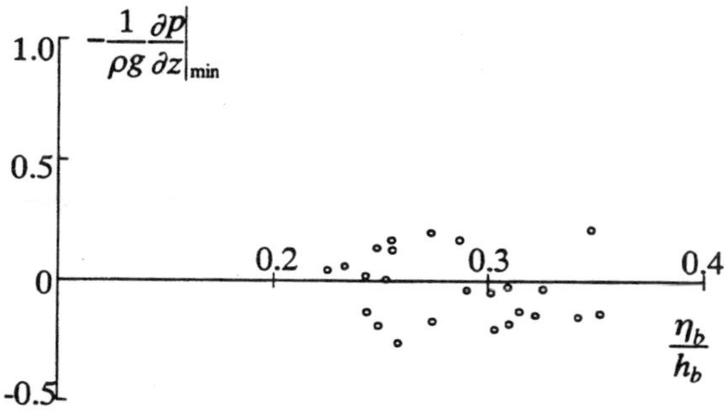

Fig.4 The minimum vertical pressure gradient at the instant of the wave breaking.

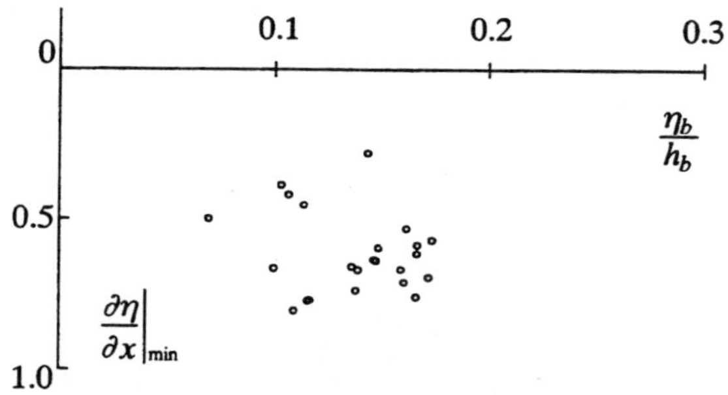

Fig.5 The minimum (steepest) foreside surface slope at the instant of the wave breaking.

minimum, respectively. When the waves break the pressure gradients become nearly zero, whereas the foreside surface slopes show appreciable scatter. These results give another demonstration to show that the vertical pressure gradient can be used as an effective breaking criterion, which is applicable even to irregular waves.

5. A WAVE MODEL COMPUTATION OF BREAKER DEPTH

By incorporating the zero pressure gradient condition as a breaking criterion into Nadaoka et al.'s (1994,1997) wave equation, which is a newly developed equation for nonlinear-dispersive waves on a bottom with arbitrary depth, a numerical simulation to predict the breaker depth was attempted. The numerical simulation was performed for a two

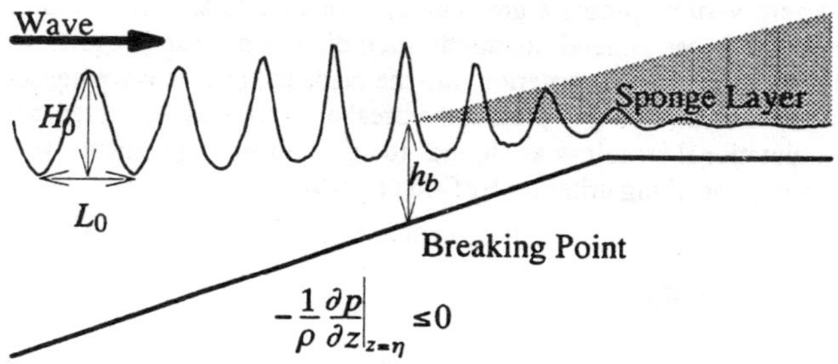

Fig.6 Sketch of the computational domain.

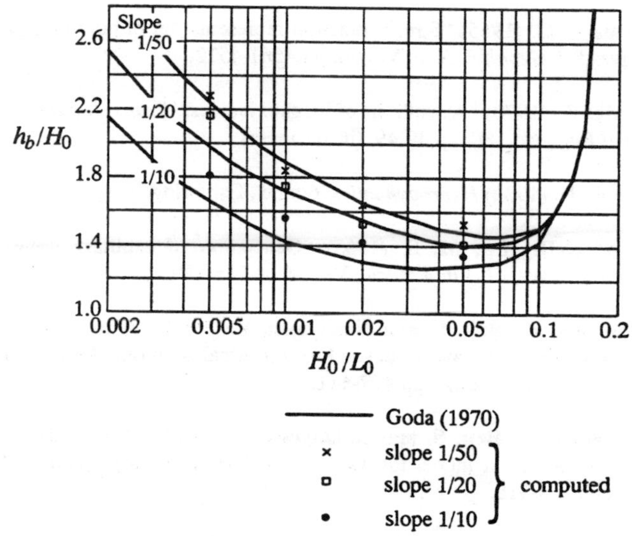

Fig.7 Computed breaker depth compared with the Goda's(1970) empirical criterion.

dimensional waves on slopes of 1:50, 1:20 and 1:10 . As shown in Fig.6, a sponge layer with a linearly increasing damping coefficient was installed at the region after the breaking to get rid of possible wave reflections otherwise the reflected waves may contaminate the computational wave field concerned. The computation was done for regular waves because existing breaking criterion are all for regular waves. We adopted the Goda's (1970) empirical breaking criterion for the comparison, which is widely used in Japan.

Figure.7 shows the breaker depth h_b relative to the offshore wave height H_0, taking the offshore wave steepness H_0/L_0 as the horizontal axis. The crosses indicate the computed breaker depth, while the solid line represents the Goda's empirical criterion. The good agreement between them demonstrates that the zero vertical pressure gradient condition may be effectively used with a wave equation to predict the breaking point.

6. CONCLUSION

A new technique was developed to estimate the pressure gradient of nonlinear irregular waves from the free surface fluctuations in both space and time. With this technique the vertical pressure gradient of irregular waves approaching the breaks was analyzed in detail through laboratory experiments. The experimental results show the vertical pressure gradient near the crest rapidly decreases and becomes nearly zero at the instant of the wave breaking, indicating that the condition of the zero vertical pressure gradient may be used as an effective breaking criterion under general situations including nonlinear irregular waves. By incorporating this criterion into the Nadaoka *et al.*'s wave equation, a numerical simulation to predict the breaker depth was attempted, and the computational result was found to give close agreement with the empirical breaking criterion by Goda (1970).

REFERENCES

Beji, S. and Nadaoka, K. (1997) "A time-depended nonlinear mild-slope equation for water waves," *Proc. R. Soc. Lond. A*, Vol. 453 pp.319-332.

Dean, R. G. (1965) "Stream function representation of nonlinear ocean waves," *J. Geophys. Res.*, Vol.70, pp.4561-4572.

Goda, Y . (1970) "A synthesis of breaker indices," *Trans. Japanese. Soc. Civil Eng.,* No.180, pp.39-49. (in Japanese)

Lamb, H. (1932) *Hydrodynamics,* 6th ed., Cambridge Press.

Longuet-Higgins, M. S. (1963) "Generation of capillary waves by gravity waves," *J. Fluid Mech.,* Vol.16 pp.138-159.

Nadaoka, K., Beji, S. and Nakagawa, Y. (1994) "A fully-dispersive weakly nonlinear wave model and its numerical solution," *Proc. 24th Int. Conf. on Coastal Eng.,* pp.427-441.

Nadaoka, K., Beji, S. and Nakagawa, Y. (1997) "A fully-dispersive weakly nonlinear model for water waves," *Proc. R. Soc. Lond. A*, Vol. 453 pp.303-318.

Rankine, W. J. (1864) "Summary of properties of certain stream line," *Phil. Mag., Ser.* pp.4-29.

Schäffer, H. A., Deigaard, R. and Madsen, P. A. (1992) "A two-dimensional surf zone model based on the Boussinesq equations," *Proc. 24th Int. Conf. Coastal Eng.*, pp.576-589.

Stokes, G. G. (1847) "On the theory of oscillatory waves," *Trans. Camb. Phil. Soc.* Vol.8, pp.441-455.

Proceedings of the Seventh (1997) International Offshore and Polar Engineering Conference
Honolulu, USA, May 25-30, 1997

ISBN 1-880653-28-1 (Set); ISBN 1-880653-31-1 (Vol. III)

Shallow-Water Nonlinear Wave Instabilities and Energy Transfer

Ming-Yang Su
Naval Research Laboratory
Stennis Space Center, MS, USA

ABSTRACT

There are essentially two types of three-dimensional instabilities of nonlinear surface gravity waves (Types I and II). Their relative importance depend on their wave steepness as well as on the water depth on which waves travel. In shallow water, Type II instability is one-order stronger than Type I instability and has the strongest instability in the directions away from the main wave propagation direction. The current theories about nonlinear wave-wave interactions in shallow water include three types: The first is unidirectional (Freilich and Guza, 1984). The second is three dimensional in general and strongest in the primary wave direction, but is suppressed in very shallow water (Hasselman, 1962). The third is also 3-dimensional but is enhanced in shallow water (McLean, 1982, Su et al., 1982b, Liu and Huang, 1996). This paper shall discuss the physical meaning and the relationship among the above nonlinear wave instabilities and the nonlinear wave-wave interactions.

INTRODUCTION

Wave breaking of the surface gravity waves, either generated by natural wind over sea, or generated in the laboratory wave tank, is the most nonlinear process for the surface wave dynamics either in deep water or over shallow water along the beach.

Surface wave breaking is not due to the fact that the associated wave height reaches the theoretical highest limit (the Stokes limit), but rather due to the intrinsic instability of the surface gravity wave with its wave steepness much lower than that of the Stokes-limited wave. The wave instability is actually a form of nonlinear resonance of the primary wave with other waves of different frequencies traveling either in the same direction of the primary wave, or in other directions. These nonlinear instabilities of surface waves are also responsible for the nonlinear energy exchanges among sets of waves of different frequency and directions.

The two most significant controlling wave parameters of these nonlinear instabilities and energy transfers are wave steepness (ak) and water depth (h), where a is the wave amplitude (½ of wave height) and k is the wave number. First, we shall review the existing theories of the effects of these two parameters on various kinds of wave instabilities. Next, we shall review the laboratory experimental results of these wave instabilities which tend to confirm the theoretical predictions. Afterward, we shall review a few field measurements available that tend to demonstrate the significance of these nonlinear wave characteristics in real wind-generated ocean waves. Finally, a concluding remark shall point out a few new efforts in field experiments that shall be carried out to give a more detailed comparison between the theoretical prediction and actual field measurements.

REVIEW OF THEORIES

As pointed out in the introduction, two main types (I and II) of wave instabilities for gravity waves on finite water depth have been recognized. The type I instabilities may be further divided into two sub-classes for easy discussions: (a) two-dimensional instabilities and (b) three-dimensional instabilities. We shall review main characteristics of these different types separately below.

TYPE I TWO-DIMENSIONAL INSTABILITIES

These types of instabilities have perturbations propagating in the same direction as the fundamental unperturbed waves. Based on a small perturbation method, Benjamin (1967) shows that the gravity waves of small, but finite amplitude, are unstable as a form of resonant interactions on water depth kh > 1.363, and stable for kh < 1.363. His analysis is valid for small perturbations, but arbitrary wave lengths. In parallel, but using the averaged Lagrangian method, Whitham (1967) proves that the inviscid, irrotational, water wave equation is of hyperbolic or elliptic type, depending on hk being less than or greater than 1.363. For the hyperbolic case, any changes in the wave trains propagate at the characteristic velocities. For the elliptic case, modulations in the wave trains grow exponentially and thus are unstable. In Whitham's analysis, the frequency of perturbations should be close to the fundamental frequency, but the perturbation amplitude is allowed to be large in comparison with the unperturbed waves.

More recently, Johnson (1977) makes a detailed analysis of the two-dimensional modulation of water waves in the neighborhood of the critical depth $(hk)_c$ = 1.363 by employing a modified nonlinear Schrodinger equation derived from the multiple-scale technique. He finds that the wave instabilities depend also on wave steepness and that the envelope solitons near $(hk)_c$ are flatter than those in deep water. Larson (1978), using the Benney and Roskes equations, show that the envelope solitons on finite water depth will force a long gravity wave, that the wave length of the soliton becomes infinite as hk approaches $(hk)_c$ and that the wave packet cannot exist for $hk < (hk)_c$.

TYPE I THREE-DIMENSIONAL INSTABILITIES

In this type of wave instabilities the perturbations propagate obliquely to the fundamental unperturbed waves. Benney and Roskes (1969) show that the wave trains are unstable to three-dimensional perturbations for both $hk > 0.38$ and $hk < 0.38$, and neutrally stable only for $hk = 0.38$. Furthermore, the most unstable modes in the range $hk > 0.38$ lie more obliquely to the fundamental wave direction than in the range $he < 0.38$. A different analysis using Whitham's averaged Lagrangian technique by Hayes (1973), and another using the multiple-scale technique by Davey and Stewartsen (1974), have both arrived at similar criteria for wave instabilities due to three-dimensional perturbations. More recently, Bryant (1978) using computer aid computes numerically the dependence of wave steepness (*ak*) on the oblique instability of the wave train.

So far we have discussed the instabilities of continuous wave trains. For the case of wave packets which contain only a finite number of waves in each group, Larson (1979) shows that envelope solitons are unstable to three-dimensional disturbances for $hk > 1.363$, and that envelope solitons are more unstable in water of moderate depth than in deep water. He suggests that steep swell may be more unstable to transverse disburbances when over continental shelves than in deep oceans.

Most recently, McLean (1981b) has made more comprehensive numerical computations using the full water-wave equation for this type of instabilities on water depths both above and below $(hk)_c$, and on wave steepness up to about 90% of the theoretically maximum steepness.

TYPE II INSTABILITIES

These types of instabilities are associated with higher order resonance from the linear dispersion than the lowest order resonance associated withType I instabilities. It was first discovered for deep-water gravity waves theoretically by McLean (1982a) and experimentally by Su et al. (1982a). McLean (1982b) has extended his deep-water analysis (McLean 1982a) to the case of finite depth.

The Type II instability is weaker than Type I instability for small steepness and becomes dominant for sufficiently steep waves. The most unstable (i.e., fastest growing) modes of Type II are always three-dimensional, having a perturbation wave length equal to two fundamental wave lengths. These most unstable modes up to about 90% of their theoretical maximum steepness are calculated by McLean (1981b). Two features are evident: (1) For a given depth (hk), the crestwise component of the perturbation wave number decreases as *ak* increases; and (2) For a given wave steepness (*ak*), the crestwise component of the perturbation wavenumber decreases with the water depth.

One important basic difference between the Type I and Type II instabilities is their perturbation velocities. For Type I, it is equal to the group velocity of the unperturbed waves. On the other hand, for Type II, it is equal to the phase velocity of the unperturbed waves; the perturbation co-propagates, or is phase-locked with the basic waves.

Another type of shallow-water instability was proposed theoretically by Freilich and Guza (1984). This type of instability involves three co-linear waves and may cause slow energy transfer between interesting waves and cause slow phase shift. The most important difference from the above Type I and II instabilities is that the triad interaction cannot induce directional energy transfer.

Some more recent theoretical computations of nonlinear wave-wave energy transfer have been carried out by Zakharov (1991), Zakharov and Shabat (1971) and Lin, Perrie and Huang (1996).

REVIEW OF EXPERIMENTAL MEASUREMENT

In Su (1982a) and Su et al. (1982a), some extensive experiments were conducted to investigate the quantitative nonlinear evolution of both wave trains and wave packets with moderate to high wave steepness ($ak = 0.1$ to 0.3) in a laboratory long wave tunnel and in a large open basin in deep water condition. As the wave steepness increases from around 0.10 to 0.25, they find that the type of wave instability switches from the Type I two-dimensional Benjamin-Feir instability to the three-dimensional Type II instability. For the initial $ak \approx 0.20$ to 0.25, three-dimensional cusp-shaped wave form are observed to occur from the original two-dimensional waves. These cusp-shaped waves are very similar, in appearance, to the short-crested breaking waves under the strong wind forcing. This observations seem to suggest that breaking of naturally generated wind waves are caused by Type II instability, rather than the often mistaken conception that surface wave breaks because it reaches the Stokes limiting highest wave. A few waves after the formation of these cusp-shaped waves, the wave train produces some waves that propagate away from their original direction. Hence, the Type II instability not only generate waves of different frequencies, but also waves of different directions. In other words, the Type II instability produce directional spreading of wave energy in addition to the Benjamin-Feir type up-shifting and down-shifting in wave frequencies.

Another sets of experiments were conducted by Su et al. (1982b) in which the effects of water depth on wave instability were investigated for both the evolution of the wave trains and wave packets. It was found that as the water depth decreases from the deep water condition to shallow-water condition, the Type I instability is greatly suppressed while the Type II instability enhanced. In other words, the Type II instability occurs at much lower initial wave steepness when the water depth become sufficiently small. These experimental observations and measurements are in complete agreement with the theoretical computations described above.

REVIEW OF FIELD MEASUREMENTS

Other than short-crested breaking waves often observed in the open ocean under strong wind forcing as mentioned in the last section, the clear evidence of difference between the Type I and Type II instabilities were much harder to be observed directly and / or measured statistically. Even so, in the last few years, several carefully executed field experiments did provide some definite confirmation of the important effects of the Type II instability. (Evidence of the Type I instability in the ocean wave evolution have been conformed for at least 20 years. See, for example, Hasselman et al., 1978). Long et al. (1994), using some aircraft radar for sea surface wave measurement, found that the fetch-limited directional wind wave spectra under steady offshore wind conditions show a very dominant direction at an angle of 30° to the mean wind direction. These authors suggest that the Phillips resonance wind wave generation (Phillips, 1977) may provide a physical mechanism to explain such directional wave spectra. However, it is equally possible to be caused by the Type II three-dimensional instability. Elgar et al. (private communication, 1996) also measured some directional wave spectra, using a bottom-mounted array of pressure gauges in the shallow-water close to the beach, that show distinctive two-prong feature of the resultant wave spectra at the final stage of steep nonlinear wave evolution under the influence of shallow-water Type II instability and / or other type of instabilities as yet unknown. Fortunately, in the forthcoming large-scale shallow-water coastal experiment (Sandy Duck 97 Experiment) at Duck, NC this summer, more detailed directional wave spectra under various wind / wave conditions shall be measured. These directional wave spectra may help to resolve some uncertainty we have at the present.

CONCLUDING REMARKS

This paper presents a concise review of the nonlinear wave instabilities and their effects on the wave breaking, energy transfer and directional wave spectra from the viewpoints of theory, laboratory controlled experiments and natural field observations and measurements. In particular, the attention is paid to the difference between the Type I and Type II instabilities, and difference due to the water depth on these two types of wave instabilities. From all these three viewpoints, it is fairly clear that more attention is needed to put to the Type II instability in order to explain some distinctive three-dimensional (or directional) features of ocean waves under strong wind forcing that is relatively less understood in comparison with moderate wind-wave generation and evolution.

REFERENCES

Benjamin, T.B., 1967: Instability of Periodic Wave-Trains in Nonlinear Dispersive Systems. Proc. Roy. Soc. London, A., 299, p. 59.

Benney, D.J. and G.J. Roskes, 1969: Wave Instabilities. Studies Appl. Math., 48, p. 377.

Bryant, P.J., 1978: Oblique Instability of Periodic Waves in Shallow Water. *J. Fluid. Mech.*, 86, p. 783.

Davey, A. and K. Stewartson, 1974: On Three-Dimensional Packets of Surface Waves. Proc. R. Soc. London, A., 338, p. 101.

Freilich, M.H. and R.T. Guza, 1984. Nonlinear Effects on Shoaling Surface Waves. Phil. Trans. Roy. Soc. London, A. 31. 1-41.

Hasselmann, S., Hasselmann, K., Komen, G.K., Janssen, P., Ewing, J.A. and Cardone, V., 1988: The WAM model-A third generation ocean wave prediction model. *J. Phys. Oceanogr.*, Vol. 18, 1775-1810.

Hayes, W.D., 1973: Group Velocity and Nonlinear Dispersive Wave Propagation. Proc. R. Soc. London, A., 332, p. 199.

Larsen, L. H., 1979: An Instability of Packets of Short Gravity Waves in Waters of Finite Depth. *J. Phys. Ocean.*, 9, p. 1140.

Lin, R.Q., W. Perrie, and N.E. Huang, 1996: A New Coast Wave Model. Part III: Non-Linear wave-wave interaction for wave spectral evolution. Submitted to *J. Phys. Oceanogr.*

Long, S.R., N.E. Huang, E. Mallo-Christensen, F.C. Jackson, and G.L. Geernaert, 1994. Directional wind wave development. Geo. Res. Letters. V. 21, No. 23, p. 2503-2506.

Longuet-Higgins, M., 1969: On wave breaking and the equilibrium spectrum of wind-generated waves. Proc. Roy. Soc. Lond. A. Vol. 310, 151-159.

McLean, J.W., 1982a: Instabilities of Finite-Amplitude Water Waves. J.F.M. 114, p. 315-330.

McLean, J.W., 1982b: Instabilities of Finite-Amplitude Waves on Water of Finite Depth. J.F.M. 114, p. 331-341.

Phillips, O.M., 1958: The equilibrium range in the spectrum of wind-generated waves. *J. Fluid Mech.*, Vol. 4, 426-434.

Phillips, O.M. and M.L. Banner, 1974: Wave breaking in presence of wind drift and swell. *J. Fluid Mech.*, Vol. 66, Part 4, 625-640.

Phillips, O.M., 1977. The Dynamics of the Upper Ocean Cambridge University Press.

Su, M.Y., 1982a: Three dimensional deep-water waves, Part I, Experimental measurement of skew and symmetric wave patterns. J. Fluid Mech., Vol. 124, 73-108.

Su, M.Y., 1982b: Evolution of groups of gravity waves with moderate to high steepness. Phys. Fluids, Vol. 25, 2167-2174.

Su, M.Y., M. Bergin, P. Marler, R. Myrick, 1982a: Experiments on Nonlinear Instabilities and Evolution of Steep Water Wave Trains. J.F.M. 124, p. 54-72.

Su, M.Y., M. Bergin, R. Myrick, and J. Roberts, 1982b. Experiments on shallow-water wave groups and breaking. Proc. First Int. Conf on Meteorology and Air / Sea Interaction of the coastal zone. The Hauge, Netherlands. p. 107-112.

Zakharov, V.E. and Shabat, A.B., 1971. Exact Theory of Two-Dimensional Self-Focusing and One-Dimensional Self-Modulating Waves in Non-Linear Media. Zh. Eksp. Teor. Fiz., 61: 118-134. Translated in Sov. Phys. JETP 34: 62-69 (1972).

Zakharov, V.E., 1991: Inverse and direct cascade in the wind-driven surface wave turbulence and wave-breaking. *M.L. Banner, R.H. Grimshaw Eds.* Breaking Waves IUTAM Symposium Sydney / Australia.

Proceedings of the Seventh (1997) International Offshore and Polar Engineering Conference
Honolulu, USA, May 25-30, 1997

ISBN 1-880653-28-1 (Set); ISBN 1-880653-31-1 (Vol. III)

Experimental Studies on Energy Dissipation in Breaking Waves

Jun Zhang and Eustorgio Meza Conde
Texas A&M University
College Station, TX, USA

Richard J. Seymour
Offshore Technology Research Center
College Station, TX, USA

ABSTRACT

A plunging wave breaking in a transient wave train was generated in a two-dimensional wave flume using a wave energy focussing technique. Surface elevation measurements of the transient wave train were made at several locations before and after the isolated plunging breaker. Applying a nonlinear decomposition approach to the measured elevation (Zhang et al. 1996), the free-wave components comprising the transient wave train were derived. By excluding the contribution from bound waves, a comparison of the spectra of free-wave components before and after wave breaking will accurately reveal the energy loss due to the breaking as a function of frequency. It was found that the energy loss is exclusively from wave components at frequencies higher than the spectral peak, especially from those wave components whose energy levels exceed the corresponding saturated level in a fully developed wave spectrum.

KEY WORDS: Wave breaking, energy dissipation, irregular waves.

INTRODUCTION

Wave breaking in deep or intermediate-depth water is the dominant sink in the ocean surface wave energy budget, yet the understanding of ocean wave breaking and estimates of the energy dissipation due to breaking are inadequate (Phillips 1977,1984, WAMDI 1988). Because energy dissipation due to wave breaking is extremely difficult to measure in the field, as an initial step it is desirable to measure the dissipation in well controlled laboratory experiments (Rapp & Melville 1990). This study focuses on energy dissipation due to wave breaking as a function of frequency in the absence of wind.

Ocean waves are almost always irregular, that is, an ocean wave field consists of many wave components with different frequencies and advancing in different directions. These wave components may result from a variety of mechanisms, such as resonant wave-wave interaction (instability), air-sea interaction (wind energy input) and transport from other parts of the ocean (swell). Regardless of their sources, when crests of the wave components with significant amplitudes are coincident, the local resultant surface wave can become very steep and breaking can be initiated. This is the direct cause of wave breaking and of limiting wave heights in deep or intermediate water depth.

When wave components travel together, strong wave-wave interaction inevitably occurs. Contrary to the weak wave interactions (resonant wave interactions), the effects of the strong wave interactions on wave characteristics such as elevation and pressures can be observed immediately, within one period or one wavelength, which is of the same time or length scale as wave breaking. The effects of the strong wave-wave interactions, such as bound waves, exist only temporarily. That is, when the free-wave components whose interaction results in the bound waves no longer travel together, the corresponding bound waves disappear (Yuen & Lake 1982). On the other hand, the resonant wave-wave interactions may permanently transfer energy from one free-wave component to others. Their effects, however, are negligible in a short distance, say in a scale comparable to wave breaking. Thus in quantifying the energy loss due to wave breaking, the energy transfer resulting from resonant wave-wave interaction can be neglected.

Bound-wave components may change significantly, depending on whether or not the free-wave components are in the position to have strong wave-wave interaction. In a transient wave train generated in the laboratory by focussing wave energy at a particular point in the wave flume, the strong wave-wave interaction becomes significant when the longer (lower frequency) wave components starts to overtake the shorter (higher frequency) wave components. Consequently, the energy of the bound-wave components in a transient wave train generated for the study of wave breaking may change significantly in a time or length scale comparable to the wave breaking. Thus, to accurately quantify the energy loss due to wave breaking, it is desirable to filter out the energy involved in the bound-wave components. The separation of the bound-wave components from the free-wave components comprising a transient wave train can be accomplished through a nonlinear decomposition of the wave train (Zhang et al. 1996), in which the bound-wave components are filtered out from the measured (resultant) wave spectrum. The free-wave components are assumed to be steady in a short duration and distance and in the absence of breaking. Therefore, the differences in the amplitudes of free-wave components before and after an isolated breaking wave may accurately reveal quantitative information about the energy loss during the wave breaking. This information may have important implication in modeling energy dissipation due to wave breaking in an ocean wave field.

Experiment Set-Up and Procedure

The experiment was conducted in a two-dimensional wave flume at the Hydromechanics Laboratory of the Civil Engineering Department at Texas A&M University. The flume is 36.1 m in length overall, 0.91 m wide and 1.22 m deep. The water depth during the experiment was 0.91 m. The downstream absorbing beach, located 31.5 m from the wavemaker, is constructed of fibrous mats 5 cm thick, supported by perforated metallic sheet on a 1 : 3.3 slope. Wave generation is provided by a dry-back, hinge-flap wavemaker powered by a brushless synchronous AC-servo motor. A very steep transient wave train was formed by sequentially generating a series of waves from high to low frequencies that superpose at a downstream location. As the wave train travels along the flume, it becomes very steep and eventually breaks. The mechanism of generating a transient wave train was similar to that used by Rapp & Melville (1990), but our transient wave spectrum is broad banded and has a sharp peak while theirs is relatively narrow banded and has "top hat" shape. Two resistance-type surface-piercing wave gauges were used to measure the wave elevation before and after the wave breaking. Because the wave-maker is computer-controlled, the transient wave train can be almost identically repeated. Therefore, the wave gages could be moved in different wave runs to record wave elevations at slightly different locations with respect to the wave breaking. The positions of the wave gages in the wave flume during two different wave runs are shown in Figure 1. For

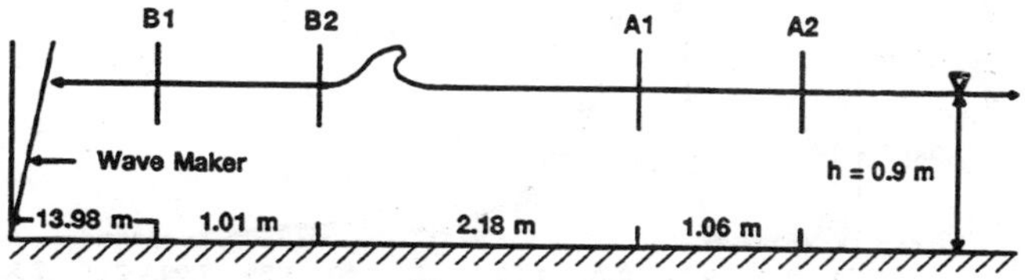

Figure 1. Sketch of wave gauges for the measurment of transient waves

reference, the location where wave-breaking occurred is also marked. The wave gages were calibrated with a linear curve fit. The standard deviations of the calibration were between plus and minus 0.2 mm. The sampling rate was 50 Hz.

Free-wave Components and Their Energy

A general view of wave elevations measured at fixed locations before and after the wave breaking are plotted as a function of time in Figure 2. The (resultant) amplitude spec-

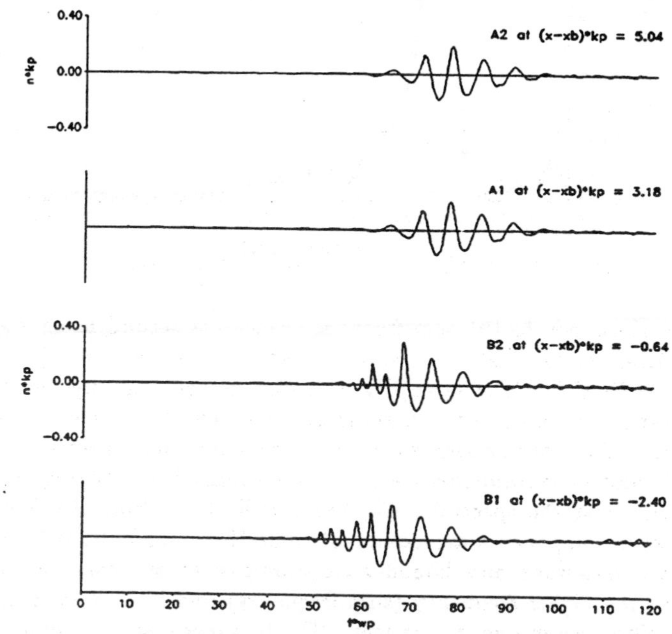

Figure 2. Time history of surface displacement at various distances from the observed breaking point, xb. Wave peak parameters fp = 0.63 Hz and kp = 1.74 m−1. A single plunging wave was observed between (x−xb)*kp = −0.64 and 3.18.

tra before and after the wave breaking were obtained using the Fast Fourier Transform (FFT), of 1024 point (about 20 s) of elevation measurements, and are given respectively in Figures 3 and 4. The frequency at the spectral peak is about 0.65

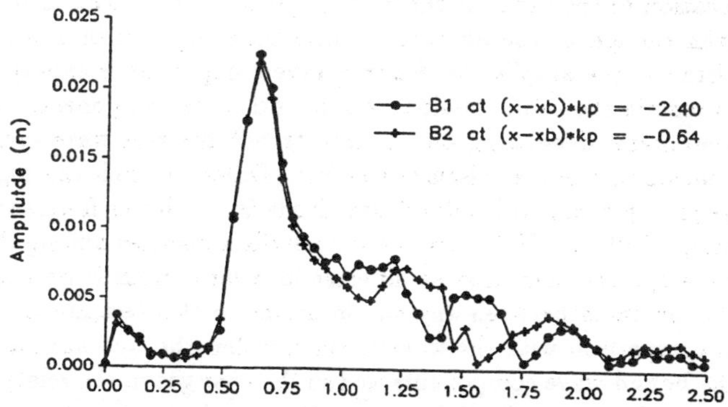

Figure 3. Resultant wave elevation spectra before breaking.

Hz and the wave energy beyond the frequency of 2.5 Hz is insignificant so the spectra are truncated here. Figure 3 shows that overall the spectra before the wave breaking are similar and particularly the amplitudes nearby the peak are almost identical. However, there are significant differences within a frequency range (1.00 - 1.50 Hz) higher than the peak frequency. Similar features are also observed in Figure 4 for the spectra measured after breaking. This high frequency range

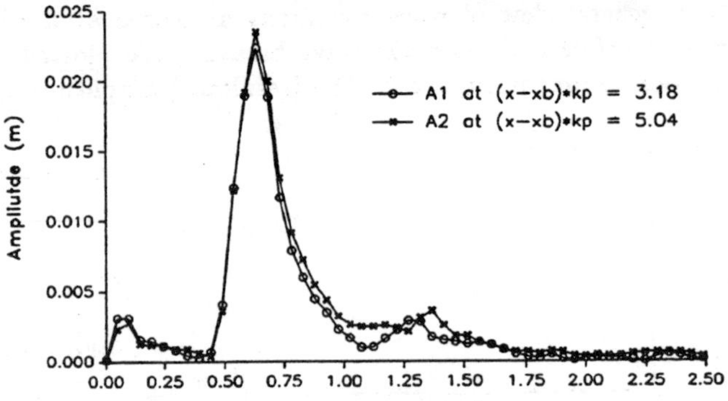

Figure 4. Resultant wave elevation spectra after breaking.

corresponds to the sum-frequency range (or second harmonic band) of the wave components within the peak range (0.50 - 0.75 Hz). The large discrepancies in the high frecuency range are expected. A resultant wave (FFT) amplitude is essentially the superposition of a free-wave component and a bound-wave component of the same frequency. At the frequencies near the spectral peak, the contributions from the free-wave components are dominant while the contributions from the free-wave and bound-wave components are comparable at the higher frequency (sum-frequency) and lower frequency (difference-frequency) ranges. The frequency and wavenumber of a free-wave component (also known as a linear wave component) obey the dispersion relation while the frequency and wavenumber of a bound-wave component generally do not. Therefore, a free-wave and bound-wave component, though of the same frequency, travel at quite different phase velocities. A resultant wave amplitude, the superposition of a free-wave and bound-wave component, depends on the relative phases between them and hence may change significantly from one location to the other at the high frequency range. This is true even though in the absence of wave breaking and in a short distance, the amplitude of a free-wave component is expected to remain the same. Besides, the bound wave components are temporary, depending on whether or not the free-wave components interact as discussed earlier. Different from the high frequency range, a resultant amplitude in the lower frequency range (0.05-0.20 Hz) shows very small discrepancies among the wave spectra measured at different locations. This is because the distances between any two locations in this test are small in comparison with the wavelength of either the free-wave and the bound-wave components and so the changes in the relative phases from one location to the other are insignificant. As a result, changes in the resultant amplitude in this low frequency range are small.

To separate the free-wave components from the bound-wave components, a numerical scheme based on the hybrid wave model is employed. The model was developed recently by Zhang et al. (1996), using a deterministic approach to the strong wave-wave interactions in irregular waves. The hybrid wave model satisfies basic hydrodynamic principles such as the Laplace equation, the free-surface and bottom boundary conditions, at least to second order in wave steepness. For fast convergence, two different perturbation approaches are employed for modeling wave-wave interaction in an irregular wave field (Zhang et al. 1993). In the case of a unidirectional irregular wave train, the input to the model is a time series of surface elevation measurement at a fixed point. The output of the model are the amplitudes and initial phases of the free-wave components of the wave train. Knowing the free-wave components, the resultant wave elevation and kinematics nearby the measurements can be computed. The comparisons between the computations and the corresponding measurements showed that the hybrid wave model is reliable and much more accurate than linear wave theory and its modifications (Spell et al. 1996, Randall et al. 1994). For simplicity, the details of the hybrid wave model are omitted.

Using the surface elevation measurements at a fixed point, the amplitudes and initial phases of free-wave components consisting of a transient wave train can be computed. The elevation amplitude spectra of the free-wave components at the locations B1 and B2 (before the wave breaking) are shown in Figure 5 and those at locations A1 and A2 (after the wave breaking) in Figure 6. Comparing Figures 5 and 6 with Figures 3 and 4, three interesting features are observed.

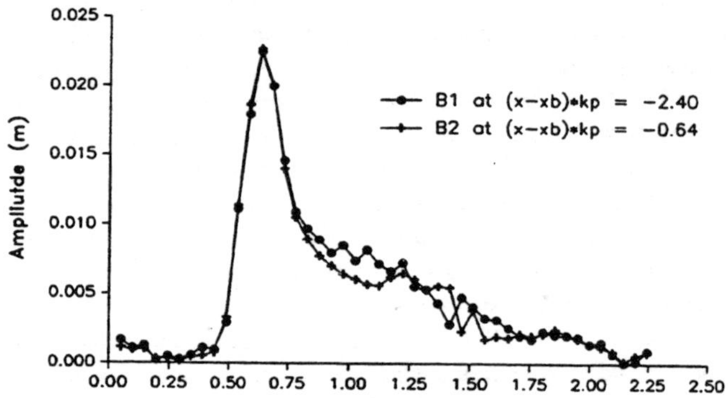

Figure 5. Free wave elevation spectra before breaking.

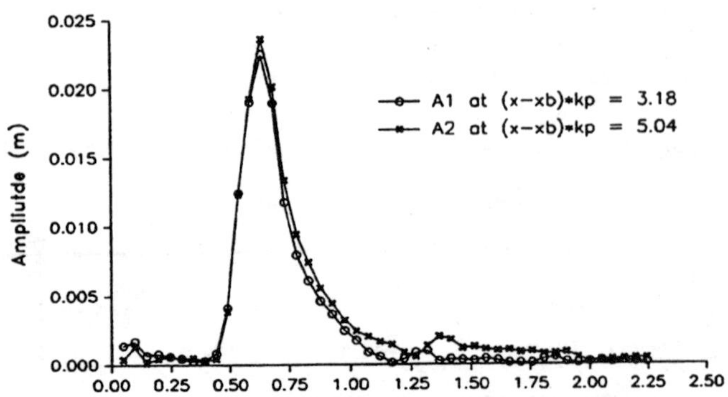

Figure 6. Free wave elevation spectra after breaking.

Figure 5 shows that the free-wave spectra measured before the wave breaking are nearly identical over the entire frequency range, and a similar observation is made for the free-wave spectra after the breaking in Figure 6. However, the corresponding resultant wave spectra are the same only close to the spectral peak (0.65 Hz), having substantial differences in the frequency range higher than the peak as shown in Figures 3 and 4. As mentioned previously, the differences in the high frequency range of the resultant wave spectra are due to the contributions of bound-wave components resulting from the sum-frequency interactions. In the free-wave spectra derived using the hybrid wave model, the contribution of bound-wave components is excluded and the spectra are hence very similar in the entire frequency domain.

Comparing Figure 3 with Figure 5, it can be seen that the amplitudes of the resultant wave spectra before the wave breaking in the sum-frequency and difference-frequency ranges are in general greater than those in the corresponding free-wave spectra. Similar trends are also observed in the comparison of Figure 4 with Figure 6. These differences result from the exclusion of bound-wave components in the free-wave spectra.

The above observations indicate the free-wave spectra remain approximately the same over a short distance and in the absence of wave breaking. Therefore, the differences between the free-wave spectra before and after the wave breaking may be attributed to the wave breaking. That is, the breaking accounts for the changes in the amplitudes of the free-wave components. Comparing the free-wave spectra before and after wave breaking, it is found that the amplitudes of the free-wave components near the spectral peak remain virtually the same. That is, there is no significant energy loss near the spectral peak due to wave breaking although wave components there are most energetic. On the other hand, in the frequency range higher than the spectral peak (1.0-2.0 Hz), the free-wave amplitudes before the breaking are much greater than those after the breaking, indicating that the energy loss due to wave breaking is mainly from the frequency range higher than the spectral peak.

Quantification of Breaking Energy Loss

The comparisons between the spectra before and after wave breaking indicate that the free-wave components near the spectral peak (0.5-0.75 Hz) remain almost the same. The free-wave components in the frequency range higher than the spectral peak are considerably smaller after the wave breaking than those before the wave breaking, indicating that the wave energy loss in this breaking event is dominantly contributed by the free-wave components of frequencies higher than the peak frequency in spite of the fact that the free-wave components near the spectral peak are the most energetic. The observation of Rapp & Melville (1990) suggested that the energy loss came mainly from the high end of the first-harmonic band (near the spectral peak) and very little from the second-harmonic band (sum-frequency range). Both observations have in common that the energy loss comes from the wave components of frequencies higher than the peak frequency. However, our observation indicates that the major energy loss is from a relatively high frequency range. The differences between these two studies may result from the differences in the wave spectra. The spectra in their experiments were of "top hat" shape which has excessive energy (above the saturated level) near the high frequency end of the spectral peak and very little free-wave energy in the high frequency range, although they only gave the resultant wave spectra. On the other hand, the spectra in this study are of broad banded and have a relatively sharp spectral peak. In other words, our spectra have much more free-wave energy in the second harmonic band and less excessive energy near the high frequency end of the spectral peak than those of Rapp & Melville. The above argument indicates that the energy loss due to wave breaking is dominantly from, if not solely, the wave components whose frequencies are higher than the peak frequency and whose energy is excessive with respect to the saturated energy level although a critical level of saturated energy as a function of frequency is not well defined yet.

Energy loss from high frequency wave components due to wave breaking can be anticipated based on intuitive and causal observations of the strong wave-wave interactions in ocean waves. In a transient or irregular wave field, when free-wave components of different frequencies travel together, short-wave (higher frequency) components riding on the surface of long-wave (lower frequency) components become shorter and steeper at the crests and longer and flatter at the troughs of long wave components, this is known as short wave modulation (Phillips 1981, Longuet-Higgins 1987, Zhang and Melville 1990). When the crests of many free-wave components align, local resultant wave elevation becomes extremely steep and may lead to wave breaking. Wave breaking, of both spilling and plunging type, seems to occur at steep wave crests where short-wave components are located, this is why the energy loss due to the breaking is at the expense of the short-wave components.

The differences in the free-wave amplitudes normalized by the peak amplitude as a function of frequency are plotted in Figure 7 for waves before breaking and Figure 8 for waves after

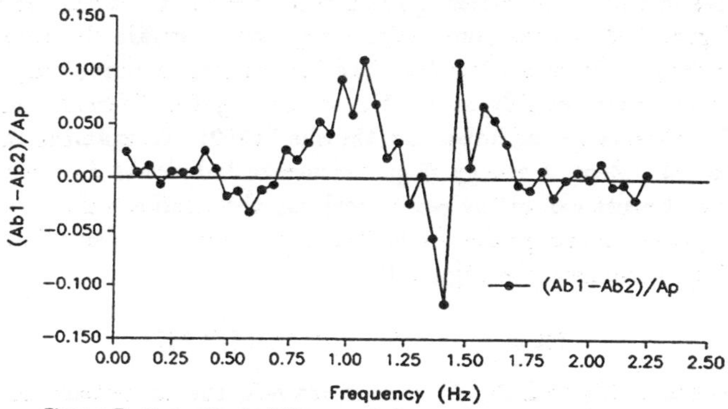

Figure 7. Normalized difference between free wave elevation spectra at B1 and B2. Ap = 0.022 m at B1.

breaking. They confirm our previous observation that the differences are small and may be attributed to high-order effects which were not included in the hybrid wave model, measurements errors and viscous loss at the side walls of the flume. To quantify the energy loss due to the breaking, first the free-wave spectra before and after wave breaking were respectively averaged and normalized by the average spectral peak amplitude. Next the differences between the normalized average free-wave amplitudes before and after breaking were computed as a function of frequency, this result is shown in Figure 9. To provide

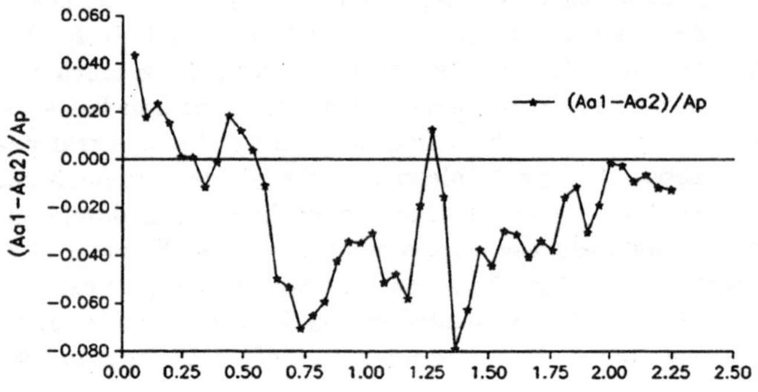

Figure 8. Normalized difference between free wave elevation spectra at A1 and A2. Ap = 0.022 m at A1.

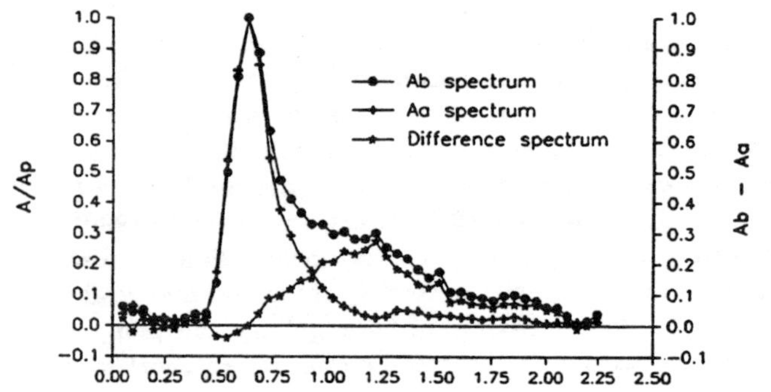

Figure 9. Normalized averaged free wave elevation spectra, before and after breaking, and their difference spectrum. Ap = 0.0228 m.

a reference to the normalized free-wave amplitude difference as a function of frequency, the normalized average spectra before and after the breaking are also plotted in the background. Figure 9 shows that the major energy loss is within the range of frequencies 0.9 - 2.0 Hz. The largest loss in amplitude is about 0.275 roughly at 1.2 Hz, which is quite different from the observation of Rapp and Melville (1990). To examine the loss of potential energy in percentage of individual free-wave components before the wave breaking, the relative differences in potential energy defined in Equation (1) are given as a function of frequency in Figure 10.

$$Rloss_{PE} = 1.0 - A_a^2(f)/A_b^2(f), (1)$$

where $A_b(f)$ and $A_a(f)$ are respectively the amplitude of a free-wave component before and after the breaking at frequency f. It is shown that the relative potential energy loss in the frequency range of 1.0 - 2.0 Hz is between 0.8 - 1.0 while near the spectral peak it less than 0.05. It is also shown that there are relative gains of potential energy after the breaking at frequencies smaller than the peak frequency. Considering that the amount of absolute gain in potential energy is small in comparison with possible errors made in the measurements or assumptions, a definite conclusion can not be given at this moment.

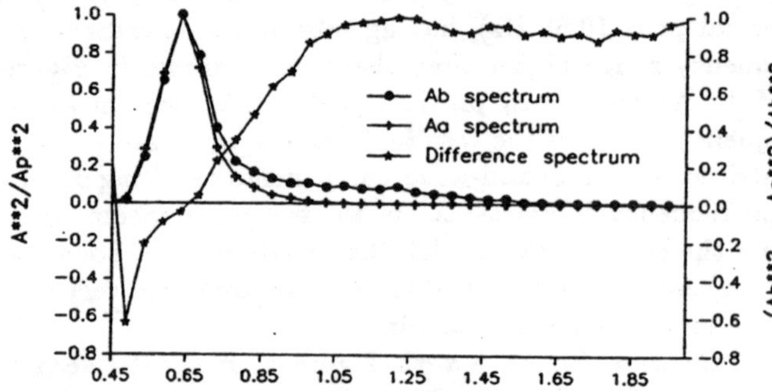

Figure 10. Normalized averaged free wave elevation spectra, before and after breaking, and their normalized difference spectrum. Ap = 5.2E−04 m.

This study suggests that the energy loss during wave breaking is at the expense of free-wave components which satisfy two conditions: 1) their frequencies are higher than the spectral peak and 2) their energy levels are great in comparison to the saturated energy level. It seems that the energy dissipation of a high frequency free-wave component in a breaking event depends on how great its energy is in comparison with its saturated level. This argument is consistent with the assumption that the dissipation rate becomes much greater once the energy at certain frequency is above its saturated level (Phillips 1984). Although the findings made here are based on the wave breaking of a transient wave train, they indicate the necessity to examine the empirical formula which assumes that the relative energy loss in ocean waves is proportional to the square of the frequency (Komen 1984).

Future Work

Further investigation of the energy loss due to different types of wave breaking, spilling and plunging, and in a wave field of a spectrum close to JONSWAP or Pierson- MosKwitz should be carried out. In addition, wave pressure and kinematics before and after wave breaking should also be measured and analysed.

Acknowledgement

This research is supported by the Offshore Technology Research Center which is sponsored in part by the National Science Foundation Engineering Research Centers Program Grant Number CDR-8721512.

REFERENCES

Komen, G.J., Hasselmann, S. and Hasselmann, K. 1984 On the existence of a fully developed wind-sea spectra. *J. Phys. Oceanogr.* **14**, 1271-1285.

Longuet-Higgins, M. S. 1987 The propagation of short surface waves on longer gravity waves. *J. Fluid Mech.* **177**, 293-306.

Phillips, O. M. 1977 The dynamics of the upper ocean, Cam-

bridge University Press.
Phillips, O. M. 1981 The dispersion of short wavelets in presence of a dominant long wave. *J. Fluid Mech.* **107**, 465-485.
Phillips, O. M. 1984 On the response of short ocean wave components at a fixed wavenumber to ocean current variations. *J. Phys. Oceanogr.* **14**, 465-485.
Randall, R.E., Zhang, J. and Longridge, J.K. 1993 Laser Doppl anemometer measurements of irregular water wave kinematics, *Ocean Engng.*, **20**, 541-554.
Rapp, R.J. and Melville, W.K. 1990 Laboratory measurements of deep-water breaking waves, *Philos. Trans. R. Soc. Lond. A*, **331**, 735-800.
WAMDI Group 1988 The WAM model-A third generation ocean wave prediction model, *J. Phys. Oceanogr.*, **18**, 1775-1810.
Spell, C.A., Zhang, J. and Randall, R.E. 1996 Hybrid wave model for unidirectional irregular waves, Part II. Comparison with laboratory measurements, *Applied Ocean Research*, **18** 93-110.
Yuen, H.C. and Lake, B.M. 1982 Nonlinear dynamics of deep-water gravity waves, *Adv. Appl. Mech.*, **22**, 67-229.
Zhang, J, and Melville, W.K. 1990 Evolution of weakly non-linear short waves riding on long gravity waves, *J Fluid Mech*, **214**, 321-346.
Zhang, J, Hong, K and Yue, DKP. 1993 Effects of wavelength ratio on wave modelling, *J Fluid Mech*, **248**, 107-127.
Zhang, J., Chen, L., Ye, M. and Randall, R.E. 1996 Hybrid wave model for unidirectional irregular waves," Part I. Theory and numerical scheme, *Applied Ocean Research*, **18** 77-92.

Proceedings of the Seventh (1997) International Offshore and Polar Engineering Conference
Honolulu, USA, May 25-30, 1997

ISBN 1-880653-28-1 (Set); ISBN 1-880653-31-1 (Vol. III)

Generation of Second-Order Long Waves at Depth Discontinuity

Hiroaki Toita and Masaru Mizuguchi
Chuo University
Tokyo, Japan

ABSTRACT

This paper describes theoretical analysis and experiment for second-order long waves generation both due to perfect reflection at vertical wall and due to depth discontinuity when a wave group is incident. We theoretically analyze these problems using the long wave equations with a forcing term in time domain. In experiment, we apply a method to suppress free long wave induced by wave group generation at a wave maker.

For vertical wall, theoretical analysis agrees well with experimental result. For depth discontinuity, quantitative comparison shows that the energy loss, frictional dissipation and modulation due to nonlinearity may be important.

KEYWORD

long wave generation, wave group, depth discontinuity, vertical wall

1. INTRODUCTION

Molin(1982) theoretically argued long wave generation over a sloping bottom and at depth discontinuity. He applied the second-order nonlinear theory to the two primary waves of different frequencies and obtained semi-analytical solution of generated long waves. On the other hand, Liu(1988) applied WKB method. He considers this problem on arbitrary bottom profile.

Here we study long wave generation at depth discontinuity from another point of view, applying the long wave equations with a forcing term given by Longuet-Higgins and Stewart(1962) and analyze the generation problem in time domain. Incident wave groups are split to reflected and transmitted wave groups at depth discontinuity. Thus, the forcing term changes at the discontinuity. Free long waves may be generated by this change of the forcing term in addition to the usual reflection, transmission phenomena of the bounded long waves to the incident wave group.

First, we deal with a preliminary case of perfect reflection to investigate the effect of the interaction between incident and reflected grouping waves on the long waves. Validity of the theoretical treatment is checked with laboratory experiments. In order to simplify the phenomenon, we use a single wave packet both in the analysis and in experiment. We also suppress long waves induced by wave group generation in experiments using Mizuguchi's method(1995).

2. PERFECT REFLECTION OF A WAVE GROUPS AND LONG WAVES

2.1 THEORITICAL APPROACH

We assume that the incident and reflected wave groups propagate at the group velocity without any modulation. The surface elevation of the incident wave group may be given by

$$\eta_i = A_i(x - C_g t)\cos(kx - \omega t) \tag{1}$$

Here, A_i is the slowly varying amplitude of incident wave group, k, ω and C_g are the wave number, angular frequency and the group velocity of primary waves. As the phase of primary waves does not shift by reflection, the surface elevation of the reflected wave group can be written as

$$\eta_r = A_r(x + C_g t)\cos(kx + \omega t) = K_r A_i(x + C_g t)\cos(kx + \omega t) \tag{2}$$

A_r is the amplitude of reflected wave group, K_r is the reflection coefficient defined by wave amplitude ratio. Total displacement of the water surface, η, is given by $\eta_i + \eta_r$. Total envelope of wave groups can be written as

$$A_p^2 = A_i^2(x - C_g t) + A_r^2(x + C_g t) + 2A_i(x - C_g t)A_r(x + C_g t)\cos(2kx) \tag{3}$$

Horizontal and vertical velocities of incident and reflected waves are given by

$$u_i = \frac{A_i\omega}{\sinh kh}\cosh k(z+h)\cos(kx-\omega t)$$

$$u_r = \frac{-A_r\omega}{\sinh kh}\cosh k(z+h)\cos(kx+\omega t)$$

$$w_i = \frac{A_i\omega}{\sinh kh}\sinh k(z+h)\sin(kx-\omega t)$$

$$w_r = \frac{-A_r\omega}{\sinh kh}\sinh k(z+h)\sin(kx+\omega t) \quad (4)$$

Total of the velocities are

$$u = u_i + u_r,\; w = w_i + w_r \quad (5)$$

Water depth h is assumed to be constant. Radiation stress is defined by

$$S_{xx} = \int_{-h}^{0}\rho(\overline{u^2 - w^2})dz + \frac{1}{2}\rho g\overline{\eta^2} \quad (6)$$

(Longuet-Higgins and Stewart 1962). ρ is the fluid density, g is the gravitational acceleration. Substituting Eqs.(1), (2), (4) and (5) into Eq.(6), S_{xx} can be calculated.

$$S_{xx} = S_{xxi} + S_{xxr} + S_{xxir} \quad (7)$$

$$S_{xxi} = (n - \frac{1}{4})\rho g A_i^2(x - C_g t) \quad (8)$$

$$S_{xxr} = (n - \frac{1}{4})\rho g A_r^2(x + C_g t) \quad (9)$$

$$S_{xxir} = -\frac{1}{2}A_i(x - C_g t)A_r(x + C_g t)\cos 2kx \quad (10)$$

S_{xxi}, S_{xxr} and S_{xxir} denote the radiation stress of incident, reflected waves and of their interaction, and n is the ratio of the group velocity to the phase speed of primary waves.

Next we seek a solution of the second-order long waves. Governing equations are given by Longuet-Higgins and Stewart (1962) as

$$\frac{\partial\bar\eta}{\partial t} + h\frac{\partial\bar u}{\partial x} = 0 \quad (11)$$

$$\frac{\partial\bar u}{\partial t} + g\frac{\partial\bar\eta}{\partial x} = -\frac{1}{h}\frac{\partial}{\partial x}(\frac{S_{xx}}{\rho}) \quad (12)$$

$\bar\eta$, $\bar u$ are the surface elevation and horizontal velocity of second-order long waves. Eliminating $\bar u$ from Eqs.(11) and (12), we obtain the following wave equation with a forcing term which has a form of spatial curvature of radiation stress.

$$\frac{\partial^2\bar\eta}{\partial t^2} - gh\frac{\partial^2\bar\eta}{\partial x^2} = \frac{\partial^2}{\partial x^2}(\frac{S_{xx}}{\rho}) \quad (13)$$

When the wave group propagates at group velocity without modulation, Eq.(13) has the bounded solution as follows (Longuet-Higgins and Stewart 1962).

$$\eta_* = -\frac{S_{xx}(x - C_g t)}{\rho(gh - C_g^2)} \quad (14)$$

For the incident and reflected wave groups, solutions of long waves bounded to them are obtained by substituting Eqs.(8), (9) into Eq.(14) as

$$\eta_{*i} = -(n - \frac{1}{4})g\frac{A_i^2(x - C_g t)}{gh - C_g^2} \quad (15)$$

$$\eta_{*r} = -(n - \frac{1}{4})g\frac{A_r^2(x + C_g t)}{gh - C_g^2} \quad (16)$$

The interaction term may be assumed to be quasi-steady, as the interacting part is not propagating. Eliminating the term of time derivative in Eq.(13), quasi-steady solution is given by

$$\eta_{*ir} = -\frac{S_{xxir}(x,t)}{\rho gh} \quad (17)$$

Substituting Eq.(10) into Eq.(17), we obtain

$$\eta_{*ir} = \frac{1}{2h}A_i(x - C_g t)A_r(x + C_g t)\cos 2kx \quad (18)$$

This result indicates that interaction term, whose amplitude becomes maximum at the vertical wall, fluctuates with long period in time but with a half-length of primary waves, or short length, in space. The total bounded long waves are linear superposition of Eqs.(15), (16) and (18).

$$\eta_* = \eta_{*i} + \eta_{*r} + \eta_{*ir} \quad (19)$$

When Kr=1, no free long waves are generated, as the Eq. (19) satisfies the boundary condition u=0 at vertical wall. When $Kr \neq 1$, free waves might be generated.

2.2 EXPERIMENT

Experiments were carried out in a 40.0m long, 0.3m wide flume with a piston-type wave maker. Figure 1 shows the laboratory flume and measuring points of surface elevation. Water depth is constant. Vertical wall is installed at a distance of 15m from the wave paddle. We generated a single wave packet as follows.

$$\eta(t) = \begin{cases} A(t)\sin(\frac{2\pi}{T}t) & 0 < t < T_g \\ 0 & t < 0,\; t > T_g \end{cases} \quad (20)$$

where

$$A(t) = (\frac{a_{max}}{2})\left\{1 - \cos(\frac{2\pi}{T_g}t)\right\}$$

a_{max} is the maximum amplitude of envelope, T is the period of primary waves and T_g is the duration of wave group in time. In generating a wave group, we suppress the generation of free long wave at the wave maker by Mizuguchi's method(1995), or by adding the following velocity of the bounded long wave to that of primary waves.

$$u_L(0,t) = \frac{C_g}{h}\eta_*(t) \quad (21)$$

We assume that the displacement of the wave paddle is very small, so that the total velocity is converted to the displacement of the wave paddle by following relationship.

$$\frac{d\xi}{dt} \fallingdotseq u\;(0,t) = u_p(0,t) + u_L(0,t) \quad (22)$$

Here, ξ is the displacement of the wave paddle, $u_p(=\sqrt{g/h}\eta)$ is the velocity of primary wave. Table 1 shows experimental cases. We measured surface elevation by new capacitance-type wave gages of high quality. Sampling frequency is 50 Hz.

2.3 EXPERIMENTAL RESULT

Figures 2 and 3 show time series data of surface profiles for Cases 1 and 2. Dotted line denotes theoretical envelope $|A_p|$. Thick solid line denotes experimental result. Incident wave group propagates with the group velocity and reflects at vertical wall, forming standing wave field in front of the wall. Wave group amplitude shows a little decrease with propagation. The decrease of the wave group amplitude may have a few causes. The frictional dissipation appear to be dominant in this experiment. Modulation due to linear dispersion and nonlinearity may be negligibly small.

Figures 4 and 5 show time series data of long wave components obtained by low-pass filtering the data of Figs.2 and 3. Dotted lines denote linear superposition of the bounded long waves Eqs. (15) and (16), the interaction component being ignored. Thin solid lines denote Eq. (19) the interaction term included. Agreement between experimental and theoretical results of the incident bounded long wave shows that only the bounded long wave is generated (Mizuguchi and Toita, 1996).

Next we examine the behavior of long waves near the wall by plotting space profiles. Figure 6 shows space profiles of long period component at t=25s, when the wave of maximum amplitude just hits the vertical wall. Not only the component of long length but also the component of short length exist. Agreement between experimental result and theoretical one given by Eq.(19) shows the component of short length is really generated by quasi-steady interaction of incident and reflected wave groups. However, its amplitude is negligibly small compared with incident and reflected bounded waves. No free wave is generated at the vertical wall in this experiment, as predicted by the theory.

3. GENERATION OF LONG WAVES AT DEPTH DISCONTINUITY

3.1 THEORITICAL APPROACH

We divide the region into two at depth discontinuity (see Fig. 7). We assume that wave group length is much longer than water depth and wave group propagates without modulating in both regions. Incident wave group is split into reflected and transmitted wave groups at depth discontinuity. In region 1, surface elevation may be given by

$$\eta_1 = A_i(x - Cg_1 t)\cos(k_1 x - \omega t) + A_r(x + Cg_1 t)\cos(k_1 x + \omega t) \tag{23}$$

where

$$A_r = K_r A_i$$

Subscripts in number denote the region. In region 2, surface elevation can be written as

$$\eta_2 = A_t(x - Cg_2 t)\cos(k_2 x - \omega t) \tag{24}$$

where

$$A_t = K_t A_i$$

Here, A_t is the transmitted wave group amplitude. K_t is the transmission coefficient defined by amplitude ratio of transmitted and incident wave group.

Here we consider the situation where, even for the primary waves, linear long wave theory can be applied. Matching conditions for the primary waves are continuity of both surface elevation and volume flux at depth discontinuity.

$$\lim_{x \to 0-} \eta_1 = \lim_{x \to 0+} \eta_2 \tag{25}$$

$$\lim_{x \to 0-} h_1 u_1 = \lim_{x \to 0+} h_2 u_2 \tag{26}$$

Then K_r and K_t are given by

$$K_r = \left(1 - \sqrt{h_2/h_1}\right) / \left(1 + \sqrt{h_2/h_1}\right)$$

$$K_t = 1 + K_r \tag{27}$$

(see, for example, Mei 1983)

Next, we calculate second-order long waves. Here, the interaction term of incident and reflected wave groups is ignored, because the amplitude of the short-length component is much smaller than that of incident and reflected bounded long waves, as shown in Section 2. The interaction term becomes smaller when K_r is less than unity. In region 1, surface elevation can be written as follows.

$$\overline{\eta_1} = \eta_{*i}(x - Cg_1 t) + \eta_{*r}(x + Cg_1 t) + G(x + C_1 t) \tag{28}$$

where

$$C_1 = \sqrt{gh_1}$$

and G is the unknown function for the free waves propagating offshore. Similarly, in region 2, surface elevation can be written as follows.

$$\overline{\eta_2} = \eta_{*t}(x - Cg_2 t) + F(x - C_2 t) \tag{29}$$

η_{*t} is the bounded long wave for the transmitted wave group, which is given by

$$\eta_{*t} = -\frac{S_{xxt}}{gh_2 - C_{g2}^2} \tag{30}$$

where

$$C_2 = \sqrt{gh_2}$$

S_{xxt} is radiation stress of transmitted wave group. Function F stands for the unknown free long waves propagating on shore. Matching condition is given by

$$h_1 \overline{u_1} = h_2 \overline{u_2} \tag{31}$$

for continuity equation. For the momentum equation with the forcing term, matching condition may be given by

$$\rho g h_2 \overline{\eta_1} + S_{xx1} = \rho g h_2 \overline{\eta_2} + S_{xx2} \tag{32}$$

where the pressure at the vertical face is taken into account as illustrated in Fig 8. Substituting Eqs.(28) and (29) into Eqs.(31) and (32), we obtain

$$\begin{aligned} G = \Big\{ & \left[C_1 + C_{g1} - C_1(C_1^2 - C_{g1}^2)/C_2^2\right]\eta_{*i} \\ & + \left[C_1 - C_{g1} - C_1(C_1^2 - C_{g1}^2)/C_2^2\right]\eta_{*r} \\ & - \left[C_{g2} - C_{g2}^2/C_2^2)\right]\eta_{*t} \Big\} / (C_1 + C_2) \end{aligned} \tag{33}$$

$$\begin{aligned} F = \Big\{ & \left[C_{g1} - C_2 + (C_1^2 - C_{g1}^2)/C_2\right]\eta_{*i} \\ & - \left[C_{g1} + C_2 - (C_1^2 - C_{g1}^2)/C_2\right]\eta_{*r} \\ & - \left[C_{g2} - C_{g2}^2/C_2\right]\eta_{*t} \Big\} / (C_1 + C_2) \end{aligned} \tag{34}$$

Again, substitution of Eqs.(33) and (34) into Eqs.(28) and (29) gives the solution.

3.2 EXPERIMENT

Experiments were carried out in the same wave flume, in which 0.1m height shelf is installed x=18.47m beyond. Figure 7 shows experimental setup and measuring points of surface elevation. Incident wave group is again given by Eq.(20). We also suppress free long wave generation by using Eqs.(21) and (22). Table 2 shows experimental cases. Reflected coefficient Kr and transmitted coefficient Kt, calculated from Eq(27), are 0.17 and 1.17 for both cases. No wave breaks at the depth continuing in these cases.

3.3 EXPERIMENTAL RESULT

Figures 9 and 10 show time series data of surface profile for Cases3 and 4. Solid lines denote experimental result. Dotted lines denote envelope of wave groups calculated by Eqs.(23) and (24). In experiment, Kr=0.13, evaluated at x=5.0m, and Kt=0.92, evaluated x=19.47m. Both values are smaller than theoretical ones.

In region 1, the incident wave group propagates without significant modulation and small reflected group is present. In region 2, the primary waves become sharp-crested, producing the higher harmonic waves, whose frequency is two times of primary waves and is traveling slower. They slowly separate by different propagating speeds. In region 2, the wave group becomes flat with propagation.

Figures11 and 12 show time series data of long wave components obtained by low-pass filtering the data in Figs. 9 and 10. Solid lines denote experimental result. Dotted lines denote Eqs.(28) and (29). It is noted that Eq.(28) does not take into account of reflection at the wave paddle. In region 2, a long wave, whose surface is slightly positive in front and negative in rear, is generated. This positive part is newly generated free long wave toward onshore. This long wave appears to be gradually growing ,as negative bounded long wave lags behind the positive free long wave. In region 1, negative free long wave and bounded long wave toward offshore are observed in the experiment.

Qualitatively, the agreement between the analysis and the experiment is satisfactory. Quantitatively, experimental result requires further improvement in region 2. Energy loss around depth, frictional dissipation with flume wall and modulation due to nonlinearity may have to be taken into account analysis for the long waves.

4. CONCLUSION

We studied theoretically long wave generation both at depth discontinuity and at vertical wall. When a bounded wave with a wave group is incident at vertical wall, not only incident and bounded waves but also the interaction component, whose amplitude is much smaller than them, are newly generated . Solution (18) agrees well experimental result in time and space. At depth discontinuity, two free waves, one of which propagates toward offshore and the other onshore, are generated in addition to the bounded long waves to reflected and transmitted wave groups. Experimental result shows qualitative agreement with theoretical analysis. Quantitative comparison indicates that energy loss around depth discontinuity, frictional dissipation with flume and modulation due to nonlinearity may be important.

REFERENCES

Liu, P. L-F. (1988). "A note on long waves induced by short-wave groups over a shelf " J Fluid Mech, Vol 205, pp.163-170.

Longuet-Higgins, M.S. and R.W. Stewart (1962). "Radiation stress and mass transport in gravity waves, with applications to 'surf beats' " J Fluid Mech, Vol 13, pp.481-504.

Mei, C. C. (1983) "The Applied Dynamics of Ocean Surface Waves " JOHN WILEY & SONS, 740p.

Mizuguchi, M. (1995). "Radiation stress approach to generation of long waves by a wave group in a flume " Coastal Engineering in Japan Vol 38, pp.1-4.

Mizuguchi, M. and H, Toita (1996). "Generation of second-order long waves by a wave group in a laboratory flume and its control " Proc.25th ICCE, Orlando (in press)

Molin, B. (1982). "On the generation of long-period second-order free-waves due to changes in the bottom profile " Ship. Res. Inst. Papers, Vol 68, Tokyo

Table. 1 Experimental case (vertical wall)

	a_{max}(cm)	T(s)	T_g(s)	h(cm)
case1	0.5	1.0	12.0	10.0
case2	0.5	1.2	14.4	10.0

Table.2 Experimental case (depth discontinuity)

	a_{max}(cm)	T(s)	T_g(s)	h_1(cm)	h_2 (cm)
case3	1.0	1.4	16.8	20.0	10.0
case4	1.5	1.4	16.8	20.0	10.0

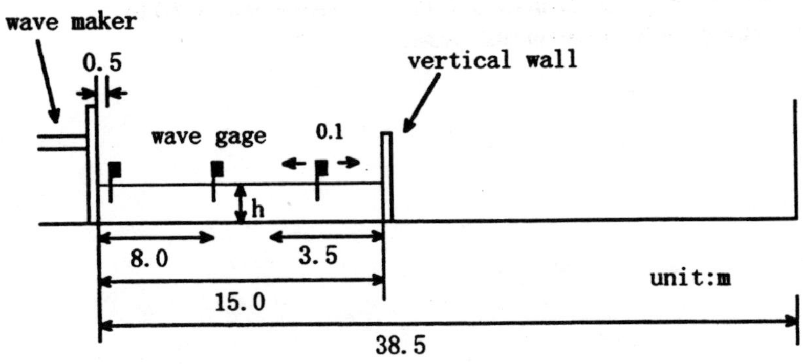

Fig.1 Laboratory flume(vertical wall). Positions of wave gages are shown.

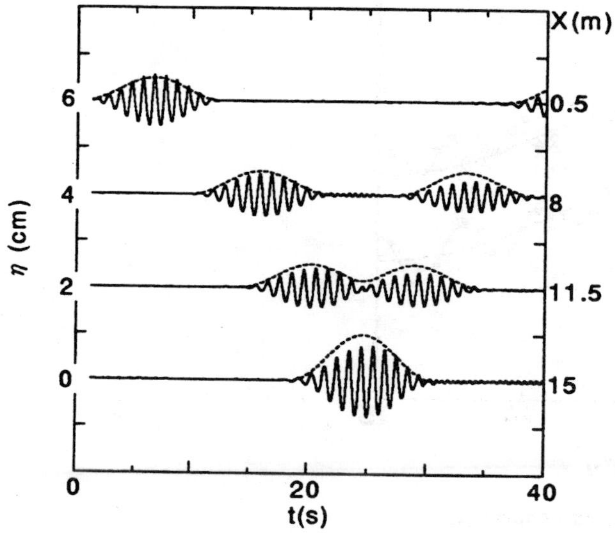

Fig.2 Surface profile of wave group (case1)

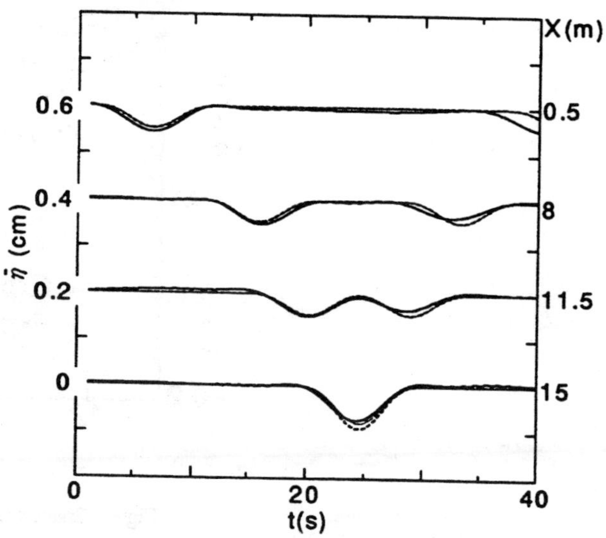

Fig.4 Long waves obtained by low-pass filtering the data in Fig.2 (case1)

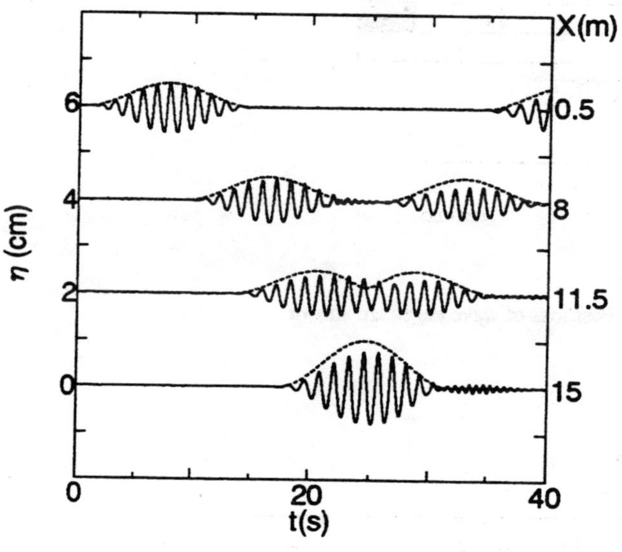

Fig.3 Surface profile of wave group (case2)

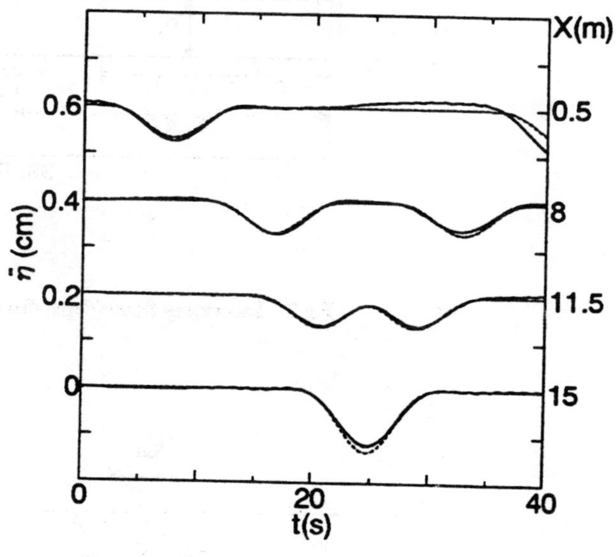

Fig.5 Long waves obtained by low-pass filtering the data in Fig.3 (case2)

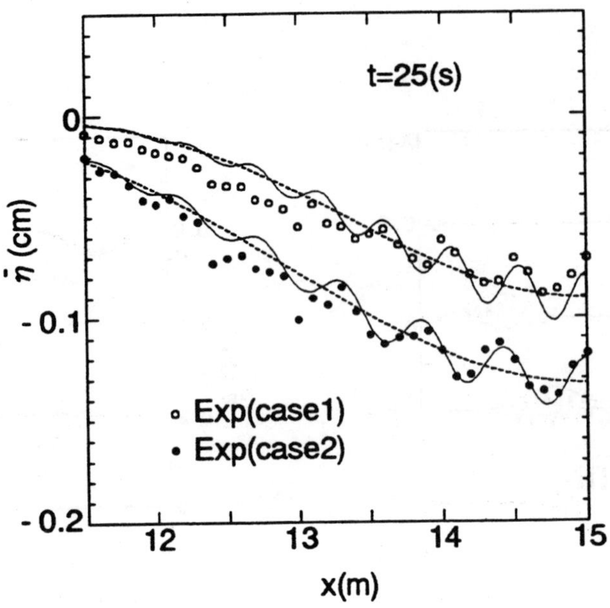

Fig.6 Space profiles of long waves near wall

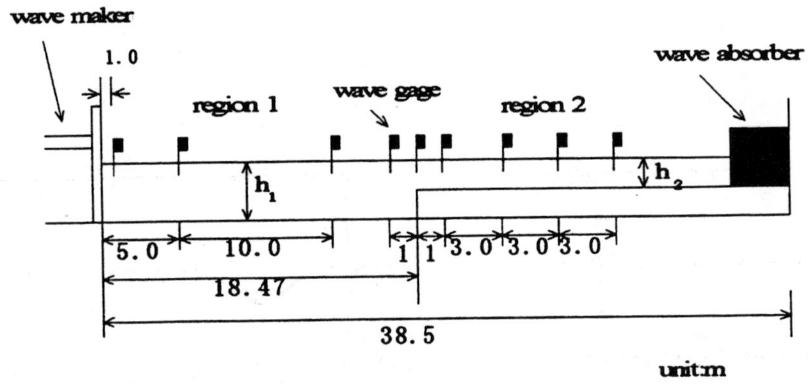

Fig.7 Laboratory flume(depth discontinuity). Positions of wave gages are shown.

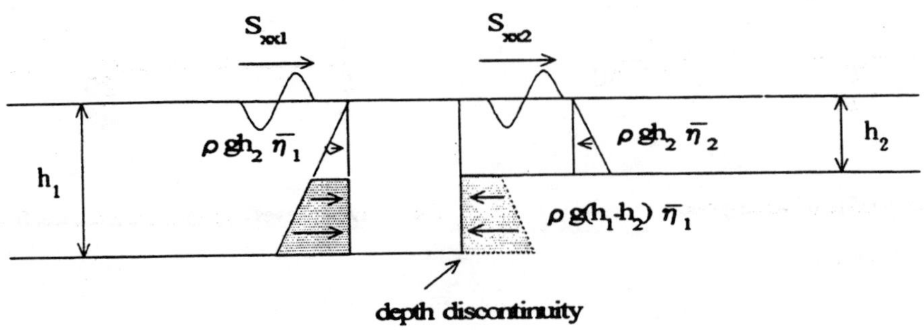

Fig.8 Momentum balance at depth discontinuity

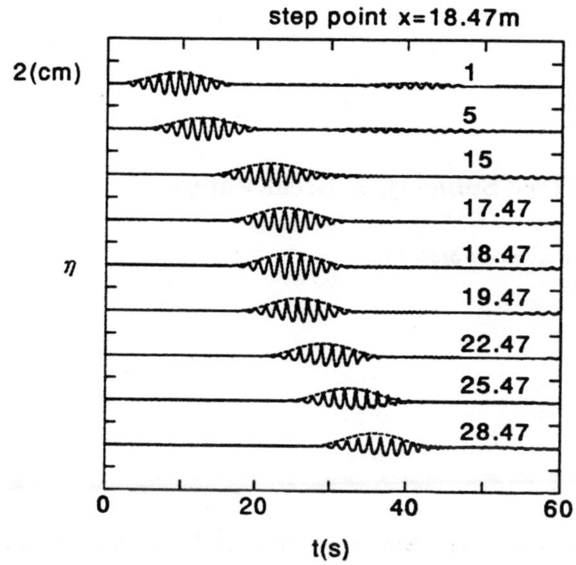

Fig.9 Surface profile of wave group

(case3)

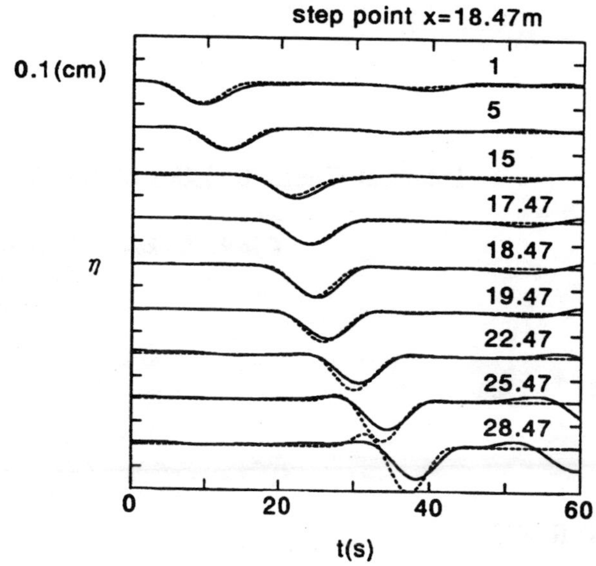

Fig.11 Long waves obtained by low-pass filtering the data in Fig.2

(case3)

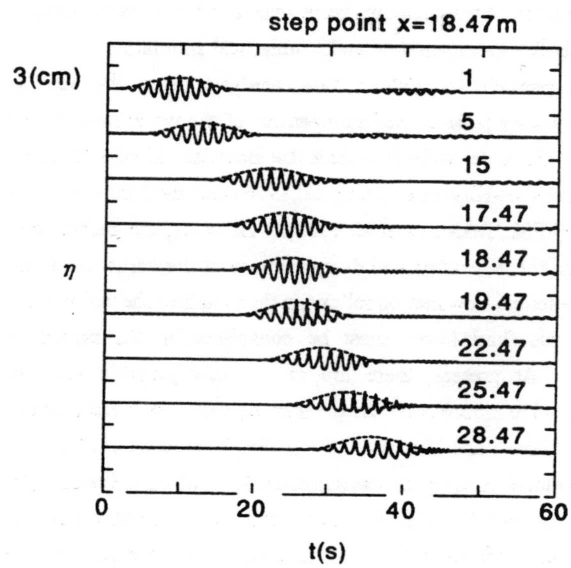

Fig.10 Surface profile of wave group

(case4)

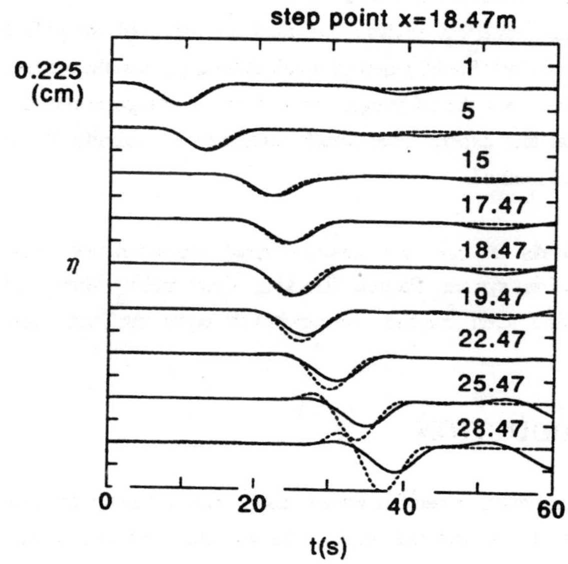

Fig.12 Long waves obtained by low-pass filtering the data in Fig.3

(case4)

Proceedings of the Seventh (1997) International Offshore and Polar Engineering Conference
Honolulu, USA, May 25-30, 1997

ISBN 1-880653-28-1 (Set); ISBN 1-880653-31-1 (Vol. III)

Internal Characteristics of Breaking Waves Over Submerged Breakwaters

Kiyoshi Takikawa, Fumihiko Yamada and Kensaku Matsumoto
Kumamoto University
Kumamoto, Japan

ABSTRACT

Two-dimensional laboratory experiments and numerical simulations were performed to clarify changes that occur in a breaking wave as it passes over a submerged breakwater. Spatial distributions of the flow pattern, velocity, and vorticity were measured using video images and a Particle Tracking Velocimetry (PTV) flow visualization system

In the numerical analysis, we used an improved Simplified Maker and Cell (SMAC) method to calculate changes in the breaker as it passed over the submerged breakwater. A comparison of the numerical and experimental results confirmed the validity of the numerical method.

Keywords : Breaking wave, internal characteristics, flow visualization system, Particle Tracking Velocimetry, Simplified Maker and Cell method, Navier-Stokes equations, numerical method

INTRODUCTION

Wave motion, especially wave breaking plays an important role in sediment transport and beach profile change. Impact breaking wave pressures can transform coastal structures. As a result, an important consideration in coastal engineering is understanding the internal characteristics of breaking waves. To date, many theoretical and experimental studies have been carried out to learn more about breaker transformation. In recent years, due to advances in measuring techniques such as Laser-Doppler Velocimetry (LDV) and Video Tape Recording (VTR), new information about the turbulence structure and velocity fields in surf zones has been available (Stive, 1980; Nadaoka and Kondo,1982; Sakai et al.1982,1984; Hattori and Aono, 1985; Okayasu et al., 1986). However, because wave breaking is a violent phenomenon with many entrained air bubbles, the extent to which the breaker transformation can be clarified through experimentation is limited. Therefore, in this paper we attempt to clarify the internal characteristics of the breaking wave over a submerged breakwater through both experimentation and numerical simulation.

It is possible to perform flow analysis using the velocity potential theory prior to the impingement of the water mass by the breaker. Possible methods include the Boundary Element Method (BEM, Longuet-Higgins and Cokelet,1976) and the Finite Element Method (FEM, Takikawa et al.,1983). However, the deformation process in the generation and disappearance of the vortex is caused by non-potential flow motion following the inrush of the water mass, and viscous fluid forces must be considered in the numerical analysis. At present, there are only a few possible ways to calculate wave motion after the breaker (Miyata, 1986; Sakai et al., 1986; Takikawa and Yamada, 1991,1996; Petit et al.,1994). We have developed method for numerical analysis of wave breaking by using FEM and Simplified Marker and Cell (SMAC) methods (Amsden and Harlow, 1970). With regard to wave breaking on a slope, we have been able to improve measurement accuracy beyond that achieved with prior experimental methods (Takikawa and Yamada, 1991,1996).

In this study, as the breaker type is different from that of the slope, we considered the wave breaking over a submerged breakwater where the water depth changes sharply. In order to clarify the internal characteristics of the breaking wave over the submerged breakwater, we measured velocity, vorticity ($\omega = \partial u/\partial y - \partial v/\partial x$), and skewness ($\gamma = \partial u/\partial y + \partial v/\partial x$) using Particle Tracking Velocimetry (PTV) with video image processing, and performed the numerical simulations using the methods described above.

EXPERIMENTAL EQUIPMENTS AND PROCEDURES

The experiments were conducted in a two-dimensional wave channel at the Department of Civil and Environmental Engineering at Kumamoto University. The wave channel was 30m long, 0.5m wide, and 0.8m deep. A submerged breakwater 1.0m in length and 0.4m high was installed in the wave channel. Regular waves were generated at a depth of 0.5m using a piston-type wave generator. Water surface elevations were measured using a capacitance-type wave gage along the centerline of the wave channel. Also, pictures of wave motion and flow patterns around the submerged breakwater were recorded through the glass side wall with a charge coupled device (CCD) camera (CCD-V90, SONY Co. Ltd, shutter speed 0.001 sec). A schematic diagram of the experimental equipment and coordinate system is shown in Figure 1.

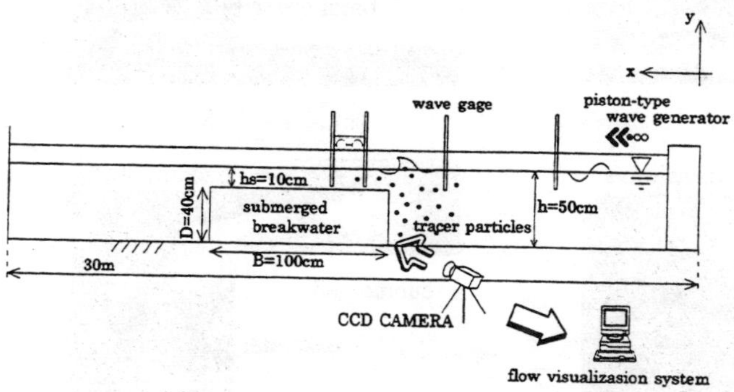

Fig. 1 Experimental equipment

Table-1 Experimental conditions

	T(sec)	h(cm)	hs(cm)	Hi(cm)	H_0/L_0	D(cm)	B(cm)	Breaker type
Case-1	1.27	50.0	10.0	10.0	0.040	40.0	100.0	spilling
Case-2	0.90	50.0	10.0	12.0	0.097	40.0	100.0	two-step
Case-3	1.70	50.0	10.0	4.5	0.011	40.0	100.0	drop-type

Experimental conditions are listed in Table-1. In the table, T is the wave period, h is the water depth in offshore uniform depth, hs is the water depth over the submerged breakwater, Hi is the incident wave height, H_0/L_0 is the calculated deepwater wave steepness, D is the height of the submerged breakwater, and B is the length of the submerged breakwater. The subscripts 0, i, and s correspond to values in deepwater waves, incident waves, and over the submerged breakwater, respectively. Only regular waves were used in this experiment.

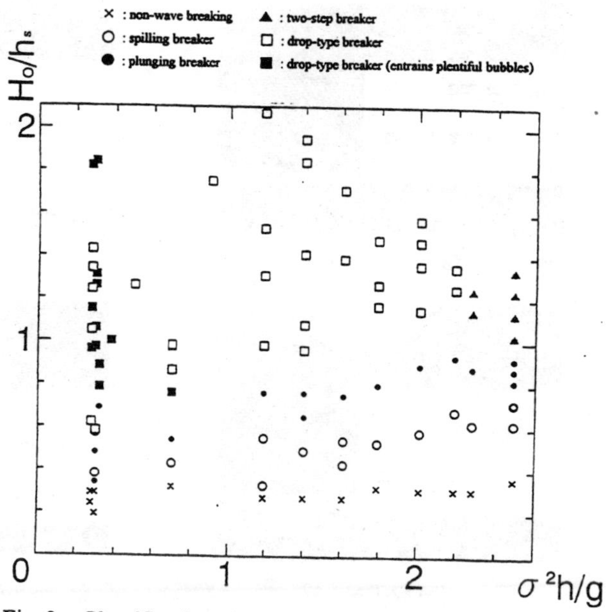

Fig. 2 Classification of the breaker types over the submerged breakwater

Breaker types over the submerged breakwater

The wave motion and flow patterns recorded on video tape were observed carefully. Photos 1～3 are stills of video images presented in time-lapse format and referred to as Case-1, Case-2, and Case-3, respectively.

In Photo 1, an incident wave collides with a return flow at the corner of the submerged breakwater (in (b)). The return flow then propagates off-shore, disturbing the water surface (in (d)). The breaker type in Case-1 is similar to the spilling breaker on a slope.

In Photo 2, just as in Photo 1, the incident wave collides with the return flow at the corner of the submerged breakwater (in (b)). Following that, the water mass from the crest strikes the trough surface, and a splash occurs (in (c)). Next, two splashes of equal scale are observed simultaneously (in (d)). This breaker type has been referred to as a two-step breaker by Katano et al. (1992). Furthermore, from this observation, the 1st splash is caused by the collision of the incident wave and return flow, and the 2nd splash emerges from the original incident wave.

In Photo-3, the return flow recedes from the corner of the submerged breakwater before colliding with the incident wave (in (a)). After that the return flow collides with the incident wave, and flow pattern is disturbed violently, producing plentiful bubbles. Also, splashes are observed after the incident wave impinges on the water surface over the submerged breakwater. Although not as clear as in Photo 2 and separated by a time lag, two splashes, the second larger than the first, are seen to exist. We conclude that these phenomena are based on the same mechanism observed in Photo 2. After all, the 1st splash caused by the return flow is

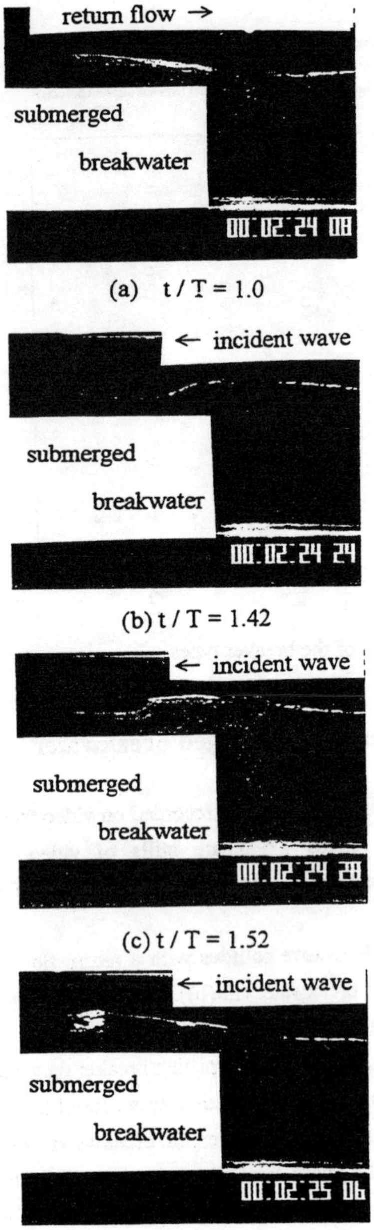

(a) t / T = 1.0

(b) t / T = 1.42

(c) t / T = 1.52

(d) t / T = 1.73

Photo 1 Case-1 (spilling breaker)

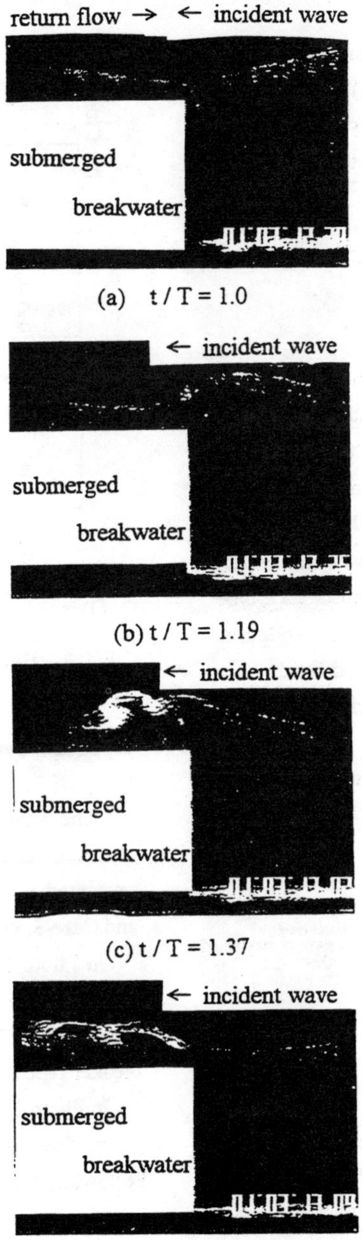

(a) t / T = 1.0

(b) t / T = 1.19

(c) t / T = 1.37

(d) t / T = 1.63

Photo 2 Case-2 (two-step breaker)

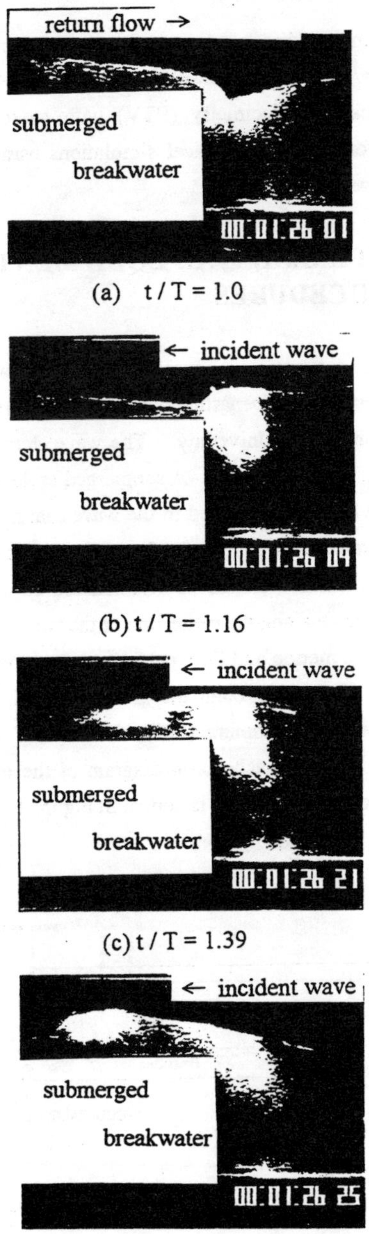

(a) t / T = 1.0

(b) t / T = 1.16

(c) t / T = 1.39

(d) t / T = 1.47

Photo 3 Case-3 (drop-type breaker)

pushed back by the incident wave, and 2nd splash is caused by the original incident wave colliding with the water surface above the submerged breakwater. These phenomena produce a large region of turbulence including plentiful bubbles at both the front and rear of the wave crest, and around the corners of the submerged breakwater. In this paper, we will call this breaker type the drop-type breaker.

Classification of breaker types over the submerged breakwater

We attempted a rough classification of these breaker types. Ninety-five sets of laboratory experiments were performed changing only incident wave height (H_0) and wave period (T). Water depth (h) remained constant at 50cm throughout these experiments.

Figure 2 shows the classification results of the breaker types over the submerged breakwater, in which the axis of abscissa is the non-dimensional period $\sigma^2 h/g$ ($\sigma = 2\pi/T$: angular frequency, g is the gravitational acceleration) and the ordinate axis is the ratio of equivalent deep water height (H_0) to water depth over the submerged breakwater (hs). In figure 2, a spilling breaker and plunging breaker mean that the breaker types are closely similar to those on a slope.

In the range of 0 to 1 on the ordinate axis, the breaker type changes non-wave breaking , spilling breaker, plunging breaker as H_0/ hs increases. Their distribution is uniform along abscissa axis $\sigma^2 h/g$. Also, the two-step breaker and drop-type breaker which are not observed on a slope appear when the value of H_0/ hs exceeds 1. These unique breaker types tend to be classified by $\sigma^2 h/g$, and two-step breaker is observed only in cases where the incident wave period is relatively short.

Furthermore, these experiments were carried out by changing only the wave height and wave period. The scale of the submerged breakwater and water depth were not changed. In the future, it will be necessary to conduct experiments under various conditions.

Flow analysis using a flow visualization system

An image analysis of the flow pattern was performed using a flow visualization system (Current、KANOMAX Co. Ltd.). The image analysis has the advantage of measuring the spatial distribution of the flow pattern relative to LDV. A Flow visualization system based on digital image processing techniques has been developed for analyzing two-dimensional flow fields (Kobayashi, 1992). Velocities are measured by a Particle Tracking Velocimetry (PTV) which traces the movement of polystyrene tracer particles (diameter ≒ 1 mm) in the video image.

Figures 3 and 4 show the image analysis results of Case-2 (two-step breaker). Figure 3 (a) and (b) show the velocity distribution around the corner of the submerged breakwater and correspond to Photo 2 (c) and (d). Arrow length is proportional to velocity. Figure 4 shows the vorticity ($\omega = \partial u/\partial y - \partial v/\partial x$) distributions corresponding to figure 3. Figure 3 (a) shows the results of splashing over the submerged breakwater. The largest upward flows occurred at the front side of the submerged breakwater. In figure 4 (a), vorticities of 140～170 (1/s) are distributed around the corner of the submerged breakwater and the water surface near the bore. Here, a clockwise direction signifies a positive vorticity value. In figure 3 (b), complicated flow patterns are observed around the front side of the submerged breakwater, and vorticities of 224 (1/s) are seen there (fig. 4 (b)).

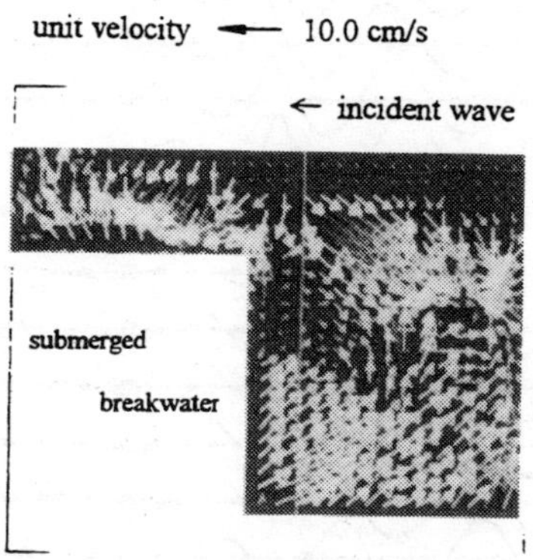

(a) t / T = 1.33

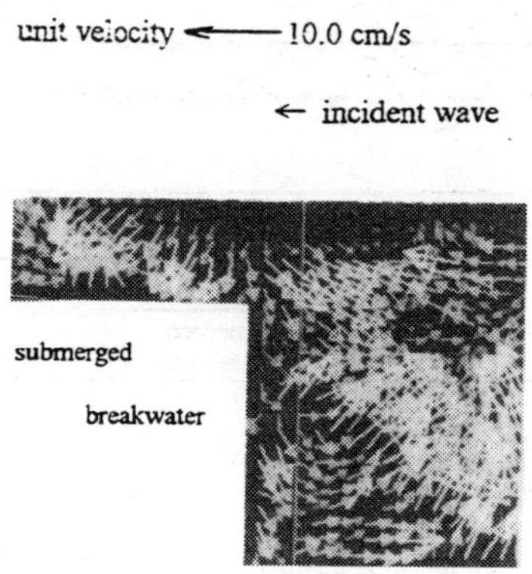

(b) t / T = 1.67

Fig. 3 Velocity distribution by PTV

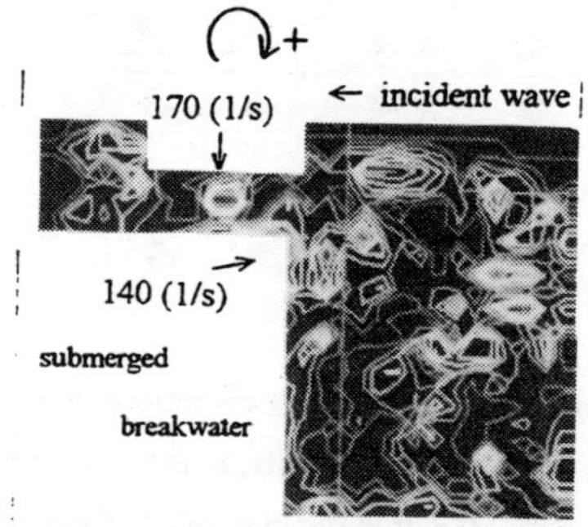

(a) t / T = 1.33

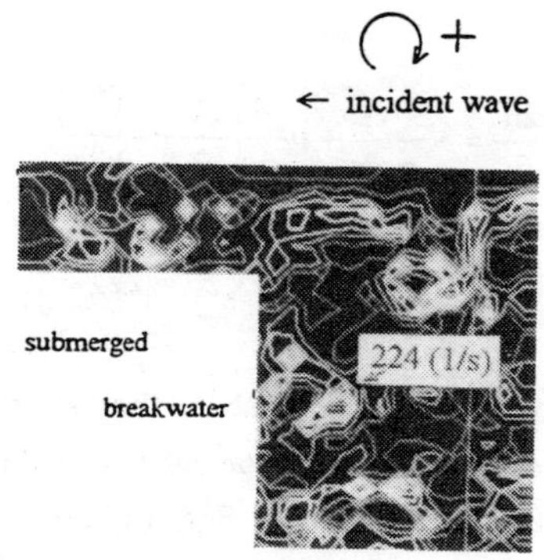

(b) t / T = 1.67

Fig. 4 Vorticity distribution by PTV

NUMERICAL ANALYSIS OF WAVE BREAKING OVER THE SUBMERGED BREAKWATER

Wave breaking is a violent phenomenon characterized by plentiful bubbles and the occurrence and dissipation of the vortex. As a result, an analysis of the finite amplitude wave based on the potential theory is possible prior to wave breaking, but fluid viscosity must be taken into account after the breaker. Here, like Sakai et al. (1986), the SMAC method was used. We also improved the SMAC method processing after the inrush of the water mass that had been a problem previously. (Takikawa and Yamada, 1991,1996)

Governing equations and computational conditions

The governing equations for an incompressible two-dimensional free-surface flow are as follows :

Continuity equation

$$D=\frac{\partial u}{\partial x}+\frac{\partial v}{\partial y}=0 \qquad (1)$$

Momentum equations (Navier-Stokes equations)

$$\frac{\partial u}{\partial t}+u\frac{\partial u}{\partial x}+v\frac{\partial u}{\partial y}=-\frac{1}{\rho}\frac{\partial p}{\partial x}+\upsilon\left(\frac{\partial^2 u}{\partial x^2}+\frac{\partial^2 u}{\partial y^2}\right)$$

$$\frac{\partial v}{\partial t}+u\frac{\partial v}{\partial x}+v\frac{\partial v}{\partial y}=-\frac{1}{\rho}\frac{\partial p}{\partial y}+\upsilon\left(\frac{\partial^2 v}{\partial x^2}+\frac{\partial^2 v}{\partial y^2}\right)+g \qquad (2)$$

where , u is horizontal velocity (x-direction) and v is vertical velocity (y-direction). ρ is the density, ν is the kinematic viscosity of water, p is pressure , and g is the gravitational acceleration.

When the diffusive terms in Eqs. (2) are transformed to vorticity (ω) using the Continuity equation (1), the momentum equations are as follows:

$$\frac{\partial u}{\partial t}+\frac{\partial u^2}{\partial x}+\frac{\partial uv}{\partial y}=-\frac{1}{\rho}\frac{\partial p}{\partial x}+\upsilon\frac{\partial}{\partial y}\left(\frac{\partial u}{\partial y}-\frac{\partial v}{\partial x}\right)$$

$$\frac{\partial v}{\partial t}+\frac{\partial uv}{\partial x}+\frac{\partial v^2}{\partial y}=-\frac{1}{\rho}\frac{\partial p}{\partial y}-\upsilon\frac{\partial}{\partial x}\left(\frac{\partial u}{\partial y}-\frac{\partial v}{\partial x}\right)+g \qquad (3)$$

Here, the convective terms are expressed in the conservative form.

A staggered mesh system with rectangular cells of equal spacing is employed for the discretization. Eqs. (3) are represented in finite-difference form by forward differencing in time and centered differencing in space for the diffusive terms. Also, differencing of the convective terms is employed via the second-order upwind difference method (Doner-Cell method) (Takikawa and Yamada,1996).

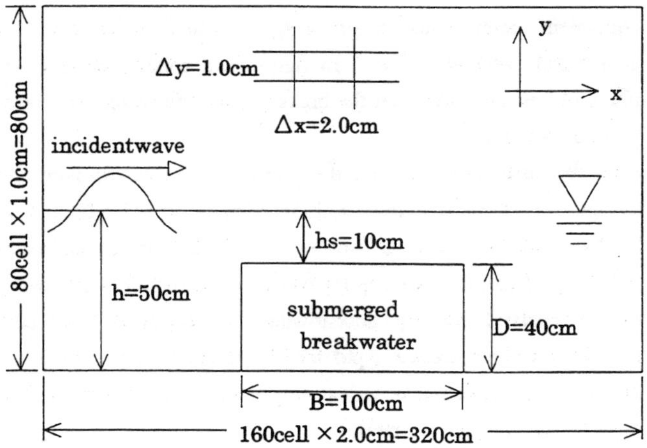

Fig. 5 Computational region

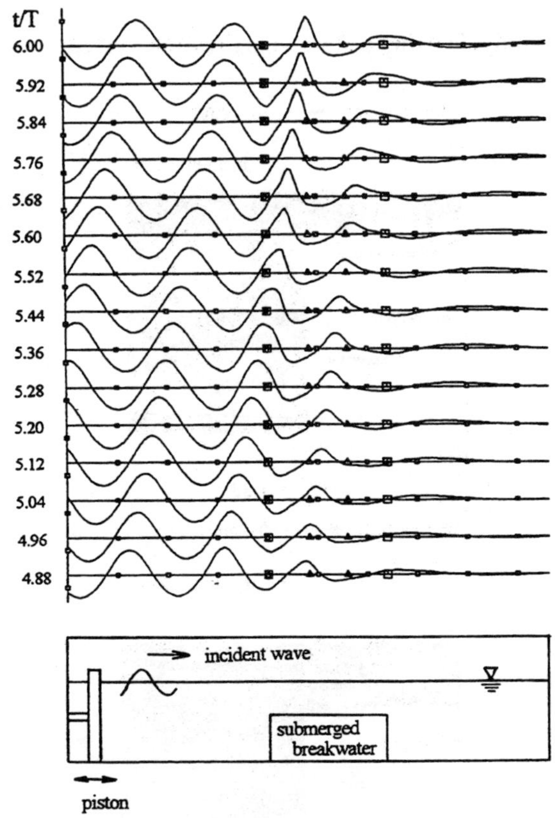

Fig. 6 Time-variation of wave profiles (FEM)

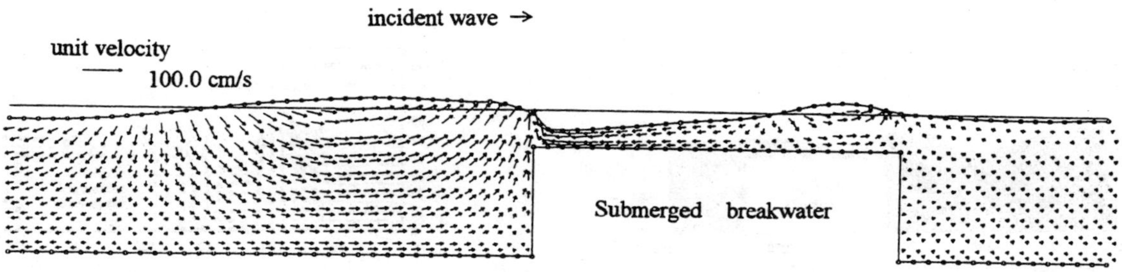

Fig. 7 Initial condition for the SMAC calculation (velocity)

The x-axis and y-axis were aligned facing shoreward and vertically upward, respectively (fig. 5). Here, in the calculation, the incident wave propagates in the opposite direction relative to the experiment The computational grid cells that were used in the SMAC method had dimensions of 2.0cm in the x-direction and 1.0cm in the y-direction . The computational region was divided into 160 meshes in the x-direction and 80 meshes in the y-direction. A time-step of 0.003 sec., determined from stability conditions, was used in the calculations.

Because the return flow over the submerged breakwater subsequent to wave breaking may have a strong effect on successive incident waves, we prepared two-wavelength wave for initial conditions in front of the submerged breakwater and performed numerical calculations during a two-wave period. Here, we used the FEM results immediately before the breaker as the initial conditions for the SMAC calculation (Takikawa et al., 1983).

Figure 6 shows FEM results based on the potential theory and time-variation of wave profiles over the submerged breakwater. Comparison with laboratory data shows good agreement. In this case, FEM results at non-dimensional time $t / T=5.52$ was used as the initial condition for the SMAC calculation (figure 7). However, because the computational mesh of the FEM method differed from that of the SMAC method, we interpolated the initial data for the SMAC method. This series of operations can be input directly into a computer.

Results and discussions of the numerical analysis

As a result of the calculations having been performed in two dimensions, three dimensional phenomena such as an obliquely descending eddy could not be reproduced.

Figure 8 shows the results of the numerical simulations for the same conditions as those created for the Case-2 (two-step breaker), and depicts marker particles for each time step. Figure 9 shows the velocity vectors corresponding to figure 8. Figure 8 (a) and 9 (a) show the calculation result of the 1st incident wave approaching the submerged breakwater. After that, the water mass emerged rom the wave crest, impinging on the water surface above the submerged breakwater. In figure 8 (b) and 9 (b), the return flow over the submerged breakwater receded from the edge of the submerged breakwater after the 1st incident wave broke . Next in figure 8 (c) and 9 (c), the return flow collides with the 2nd incidentwave at the edge of the submerged breakwater and the flow pattern around the edge is disturbed. Although figure 9 (a) and (c) display the velocity vectors when the 1st and 2nd incident waves approach the submerged breakwater, the transformation of the water mass beyond the breakwater edge were different in each case. Accordingly, figure 9 (c) shows the jumping-type breaker produced by the collision of the return flow and incident wave as witnessed in the experiments.

Figure 10 (a) and (b) show the vorticity distributions corresponding to figure 8 (b) and (c). Here, a counterclockwise direction signifies a positive vorticity value. In figure 10 (a), the water mass impinges on the water surface over the submerged breakwater and vorticities of about 60 (1/s) occur at the water surface. At this time, the distribution of vorticity is observed at the edge of the submerged breakwater, and maximum vorticities of 120 (1/s) occur there. Although at variance with the times recorded in the experiment, phase recording in figure 3 (a) coincides approximately with the calculation results with maximum value of 170 (1/s). In figure 10 (b), the return flow and 2nd incident wave collide at the corner of the submerged breakwater, and maximum vorticities about 200 (1/s) are measured there.. In the experiment, the maximum value of vorticities is 224 (1/s) (fig. 3 (b)). Consequently, we determine that the numerical analysis to be quantitatively accurate.

However, although the maximum vorticity value in the numerical analysis occur at the corner of the submerged breakwater, in the experiment are located a little below the corner. This may be attributed to not considering the influence of bubbles and fluid compressibility in the numerical analysis. Moreover, it is difficult to monitor the tracer particles in the video images in the region of turbulence.

Figure 11 shows the distribution of skewness ($\gamma = \partial u / \partial y + \partial v / \partial x$) corresponding to figure 10. Here, in the case of skewness, shear in a shoreward direction signifies a positive value. In figure

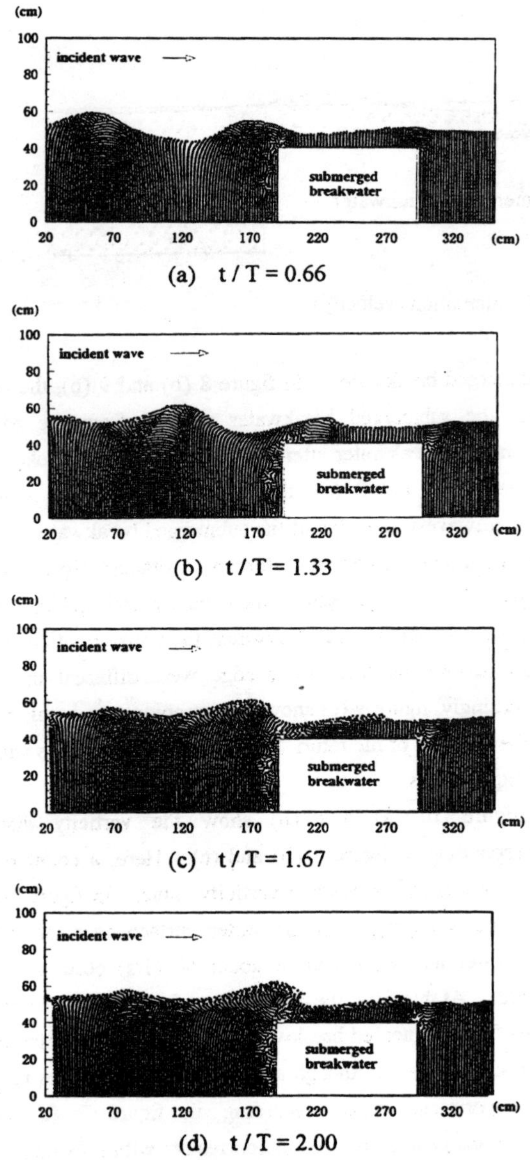

Fig. 8 The calculation results of breaker transformation (Case-2, marker particles)

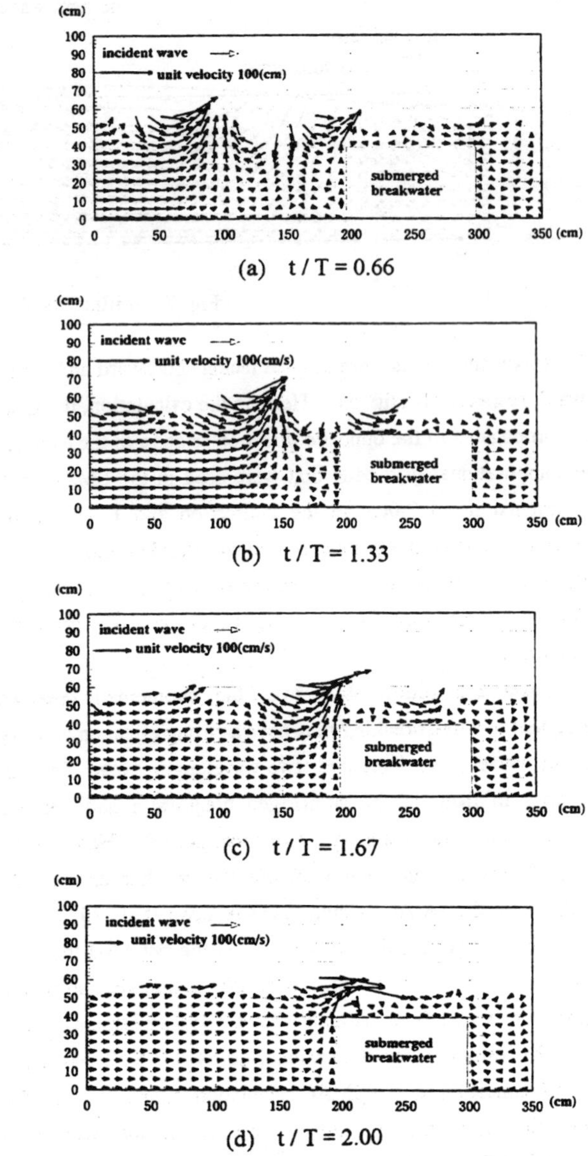

Fig. 9 The calculation results of velocity distribution (Case-2)

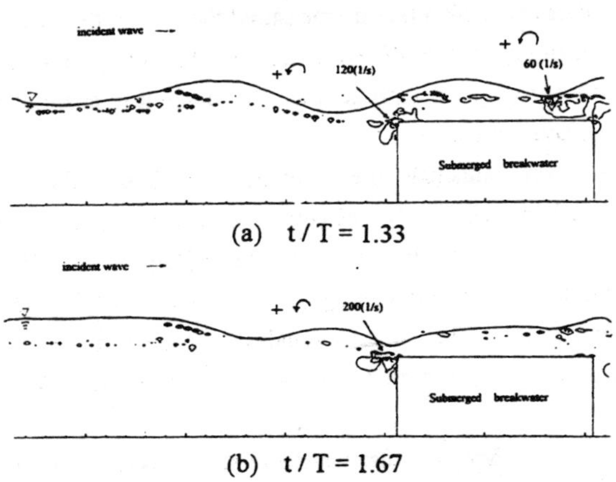

Fig. 10 Vorticity distribution (Case-2)

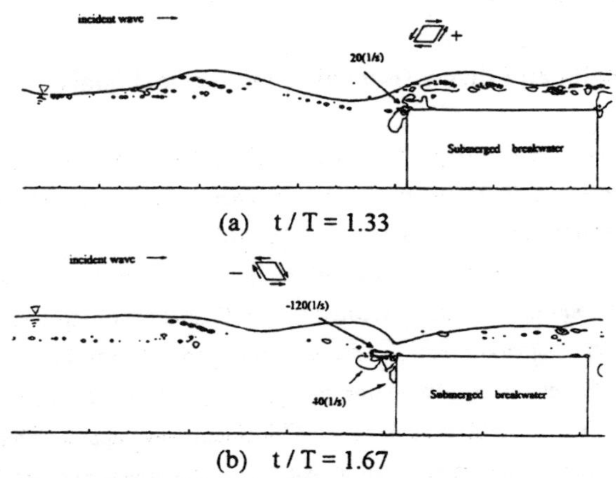

Fig. 11 Skewness distribution (Case-2)

11 (a), extreme distributions of skewness are not observed, and their maximum value is about 20 (1/s) and occurs at the corner of the submerged breakwater. A skewness value of -120 (1/s) is observed at the corner of the submerged breakwater in figure-11 (b). Also, a value of 40 (1/s) is observed in the same region. These reveal the violent disturbances caused by the return flow which drops at the corner of the submerged breakwater and collides with the incident wave.

CONCLUSION

In this study, we attempt to clarify the internal mechanism of the breaking waves over a submerged breakwater through experimentation and numerical analysis. In the experiment, velocities in the surf zone were measured using the flow visualization system, and the space distribution characteristics of vorticity (ω) and skewness (γ) were discussed. Also, breaker types were generally classified. Accordingly, the vorticity and skewness were most pronounced when the return flow collided with the incident wave at the corner of the submerged breakwater. Their scale was clarified quantitatively. Also, the two-step and drop-type breaker, not observed in wave breaking on slopes were classified by H_0/ hs and $\sigma^2 h / g$.

In numerical analysis, we used a modified SMAC method and compared the results with those of the experiments. Because the results compared reasonably well, a detailed numerical examination regarding wave deformation and the breaking water mechanism was possible. We will obtain much knowledge through the application of these calculations. However, the problem of air mixing and three-dimensional turbulence remains and must be resolved. Also, breaking water mechanisms on an irregular sea bed are also important. At present, we are working to develop a numerical simulation and flow visualization method for these as well.

ACKNOWLEDGEMENT

Part of this study was supported by Grant-in-Aid for Encouragement of Young Scientists (representative : F. Yamada) of The Ministry of Education, Science and Culture of Japan.

REFERRENCES

Amsden, A.A and F.H.Harlow (1970). "The SMAC method: a numerical technique for calculating incompressible fluid flows," Los Alamos Scientific Lab., N. Mex. (USA), LA-4370

Hattori,S. and Aono, T. (1985). "Experimental study on turbulence structure under breaking waves," Coastal Eng. Jpn.,JSCE, Vol 28, pp 6-10

Katano, A., Murakami, S. and Hattori, S. (1992). "Display system of wave dissipation properties over the submerged breakwater with wide crown," Proc. Coastal Eng., JSCE, Vol 39, pp 646-650 (in Japanese)

Kobayasi, T. (1992). "Advance in Computer-Aided-Flow Visualization," Proc. 6th Int. symp. on flow visualization, pp 5-13

Longuet-Higgins, M.S, and C.D.CokElet.(1976). "The deformation of steep surface waves on water," Proc. Roy. Soc.London, A364, pp 1-28

Miyata, H. (1986). "Finite-difference simulation of breaking waves," J.Comput.Phys Vol 65, pp 179-214

Nadaoka,K. and Kondoh, T. (1983). "Laboratory measurements of velocity field structure in the surf zone by LDV," Coastal Eng. Jpn., JSCE, Vol 25, pp 125-146

Okayasu, A., Shibayama, T. and Mimura, N. (1986). "Velocity field under plunging waves," Proc. 20th Int. Coastal Eng. Conf., Taipei. ASCE, Vol 1, pp 660-674

Petit, H. A. H., Tonjes, P., Van Gent ,M.R.A., Van den Bosch, P. (1994). "Numerical simulation and validation of plunging breakers using a 2D Navier-Stokes model," Proc.24th Int Coastal Eng. Conf., Kobe ASCE, Vol 1, pp 511-524

Sakai, T., Inada, Y. and Sandanbata, I. (1982). "Turblence generated by wave breaking on beach," Proc. 18th Int. Coastal Eng. Conf., Cape Town. ASCE, Vol 1, pp 3-21

Sakai, T., Sandanbata, I. And Uchida, M. (1984). "Reynolds stress in surfzone," Proc. 19th Int. Coastal Eng. Conf., Houston, TX. ASCE, Vol 1, pp 42-53

Sakai, T., Mizutani, T., Tanaka, H., and Tada, Y. (1986). "Vortex formation in plunging breaker," Proc. 20th Int. Coastal Eng. Conf., Taipei. ASCE, Vol 1, pp 711-723

Takikawa, K., Iwagaki, Y. and Nakagawa, M., (1983). "The analysis of the breaker transformation and inside mechanism of the wave on a slope by finite element method," Proc.Coastal Eng., JSCE, Vol 30, pp 20-24 (in Japanese)

Takikawa, K. and Yamada, F.(1991). "Development of numerical analysis system of deformation and breaking wave on sloping bottom," Proc.Coastal Eng., JSCE, Vol 38, pp 61-65 (in Japanese)

Takikawa, K. and Yamada, F.(1996). "Internal characteristics and numerical analysis of plunging breaker on a slope," Int. J Coastal Eng. (in press)

Proceedings of the Seventh (1997) International Offshore and Polar Engineering Conference
Honolulu, USA, May 25-30, 1997

ISBN 1-880653-28-1 (Set); ISBN 1-880653-31-1 (Vol. III)

Breaking Waves and Offshore Design

William J. Easson
University of Edinburgh
Edinburgh, United Kingdom

SUMMARY

Literature on breaking waves is reviewed in the context of offshore structure design. Particular attention is paid to deep and intermediate water. The effects of three-dimensionality and randomness on the water particle kinematics are considered. Recommendations are made for alteration of the design wave elevation/height ratio and deep water steepness limits.

KEYWORDS: Breaking waves, large waves, wave steepness, crest asymmetry, wave kinematics, air gap, loading, offshore structures

INTRODUCTION

The study of breaking waves has proceeded on a wide range of fronts: analytical, numerical, experimental, and full scale research from shore and ship-based to satellite photography. The investigators approach the problem from a wide range of backgrounds: mathematicians, engineers, physicists and oceanographers have all contributed to furthering our knowledge of wave breaking. Additionally, these investigators have often been looking at particular aspects of breaking with a specific purpose in mind. The transfer of knowledge between, say, the oceanographer and the engineer is limited, and often negligible, as the list of references appended to learned papers in these two fields demonstrates.

To bring together all of the research on wave breaking has become an almost impossible task, so this study will concentrate on breaking wave probability and breaking wave limits and wave kinematics in the context of offshore wave loading. A fuller review of forces and shallow water effects is published in Easson (1996). Banner and Peregrine (1993) provide a more general introduction to deep water breaking waves, covering experimental and theoretical developments for probability and kinematics.

The breaking wave is only one example of the limiting wave and for this reason both large non-breaking and breaking waves are considered. Small breaking waves are of limited interest to a design engineer calculating extreme loading and, as the following sections demonstrate, large non-breaking waves are more frequent than large breaking waves.

BREAKING CONDITIONS

Regular wave limits

In deep water, regular, 2-D waves the maximum possible wave height, H, of the Stokes wave is about 1/7 of the wavelength, L, and the limit may be expressed in terms of the wave period, T, as:

$$H_{max} = sgT^2 \qquad 1$$

where s=0.027. This wave also has a limiting downward acceleration in the crest of a=0.5g and a horizontal water particle velocity at the crest of u=c, the wave celerity (Longuet-Higgins, 1969).

This theoretical wave omits some of the key attributes of an extreme water wave. The main difference is that it does not show the observed asymmetry between the forward and rear crests. However, it does demonstrate the asymmetry between the amplitude of the crest, η, and the depth of the trough which is observed for large waves. The values for the limits of the Stokes regular wave have been taken to be indicators above which breaking occurs. Therefore we may state that, for a regular wave, breaking will occur if one of the following conditions is satisfied:

$$s > 0.027;\ a > 0.5g;\ u > c \qquad 2$$

As well as giving us the specific conditions of Equation 2, this approach has also given the main methods by which we analyse wave records to determine predictors of breaking. In other words, even in a fully 3-dimensional random sea, concepts such as the wave steepness, particle velocity and acceleration, and other geometrical parameters associated with individual crests are used as predictors of breaking.

Kjeldsen (1981) has summarised the significant geometrical parameters for a wave as shown in figure 1.

The most significant of these is the horizontal asymmetry factor, μ, representing the degree to which the height of the wave is due to the elevation of the crest, η,

$$\mu = \eta / H \quad (3)$$

For a small amplitude wave μ=0.5; for a Stokes limiting wave μ=0.67. This gives a further condition to add to those of Equation 2 for the regular breaking wave, namely that a wave will be breaking if

$$\mu > 0.67 \quad (4)$$

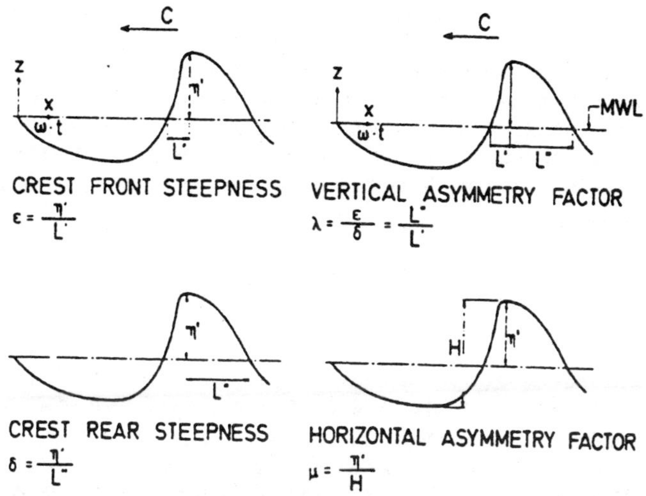

Figure 1: Attributes of an extreme wave.

The above conditions on wave steepness and wave asymmetry should be regarded as limiting cases for waves which are about to break where the crest fronts are almost parallel, the depth to wavelength ratio is near the deep-water condition, and the group lengths are long. She (1994) shows that, for a regular, focused wave in a 3-D tank the crest asymmetry, μ, is not sensitive to angle and has a mean value of 0.65. The wave steepness was close to the Stokes limit for the smallest spreading angle investigated (30 degrees), with s=0.026 (figure 2). Similarly, although asymmetry is a function of depth, for the deepest

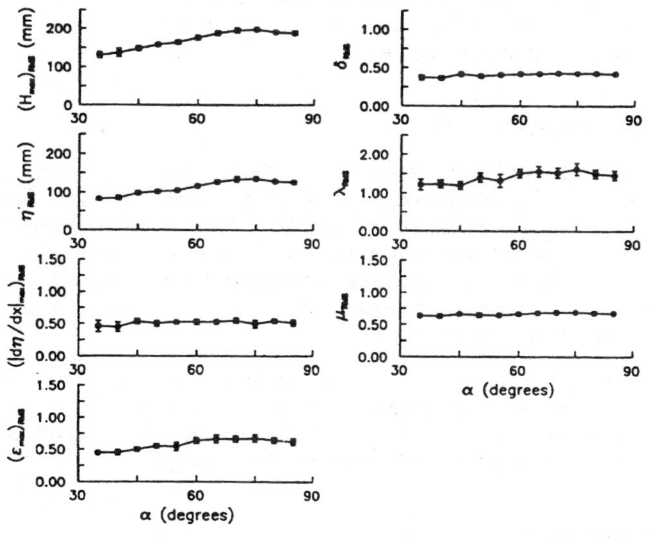

Figure 2: Observed limits of wave breaking parameters and ratios for single frequency waves focused over a range of angles

shoaling wave experiments, Griffiths (1992) finds μ=0.68. However, his greatest wave steepness is s~0.020 which is significantly less than the Stokes value.

Random waves

The most commonly accepted limit for wave breaking in random seas is that proposed by Ochi and Tsai (1983), based on the wave steepness,

$$H = 0.020 g T^2 \quad (5)$$

The value of s=0.020 is significantly lower than the regular wave condition (Equation 2) and was derived from wave flume tests with representative sea spectra. Ochi (1990) also reports values of s=0.019 and s=0.021 obtained by independent researchers. Ochi demonstrates how this relationship can be used to predict the probability of breaking in two dimensions based on the joint probability of period and wave height with the cut-off relationship above.

In his thesis on the kinematics of wave groups, Sutherland (1992) found that waves in a shorter group broke at a lower height. The study concentrated on the particle velocities in two component waves with the same mean period where the number of waves in the group varied from 3 to 12 according to the separation of the two wave components. Comparisons were made with various theories and with measured regular wave kinematics. It was found that, in the shortest groups, the maximum height which could be obtained was 65% of the regular wave limit for the mean period.

Srokosz (1986), has related the acceleration condition for breaking with the steepness relationship and found that s=0.020 is equivalent to a=0.4g, which is smaller than the result for regular waves. Srokosz also pointed out that the theory is based on Gaussian statistics and would break down if the sea surface was non-Gaussian. Nevertheless, he produced an equivalent formulation for breaking probability to that of Ochi, in terms of acceleration, with

$$B = \exp\left(-\frac{\alpha^2 \gamma^2}{2 m_4}\right) \quad (6)$$

where m_4 is the fourth spectral moment, and α is the coefficient of acceleration in a=αg.

Snyder and Kennedy (1983) use a different approach, still based on the acceleration criterion, to find the proportion of the sea surface covered by breakers, where

$$\beta = 1 - \Phi\left(\alpha g m_4^{-1/2}\right) \quad (7)$$

Both of the above methods rely on the fourth moment of the spectrum and the cut-off frequency of the calculation or the measuring system, if applied to experimental or field measurements.

The asymmetry value (equation 4) is not so well researched for random waves. Bonmarin (1989) watched regular waves in a long tank modulate due to their instability and measured the characteristics of breaking. All waves had s< 0.027, and he divided the results for asymmetry into spilling and plunging breakers,

Plunging	$0.62 < \mu < 0.93$	$\bar{\mu} = 0.77$	8
Spilling	$0.59 < \mu < 0.8$	$\bar{\mu} = 0.69$	9

This indicates that plunging waves generally have a higher crest-to-trough ratio than spillers, and that a breaking criterion of μ~0.6 is appropriate for random waves.

Three-dimensional waves

Very little work has been done on the effects of 3-dimensionality on wave breaking. However, the results which do exist are particularly significant, in that they question the whole basis of the above (2D) formalism.

Mather (1988) carried out studies on the joint probability of period and wave height for waves generated in the 3-D facility at the University of Edinburgh (figure 3). The purpose of the work was to investigate the Ochi and Tsai steepness limit for three dimensions. The results were completely unexpected. Although $H=0.020gT^2$ did indeed form an upper limit for the probability density function, it did not provide a breaking criterion. In fact, the breaking waves were evenly

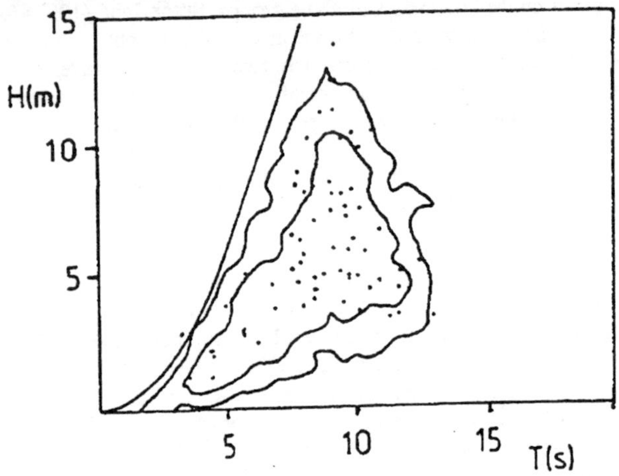

Figure 3: Probability density function of height and period for a random sea state. Also plotted are breaking events (dots) and the curve from equation 5

scattered throughout the H,T plane, indicating that, in a random, 3-dimensional sea, waves of any steepness were likely to break. Conversely, the probability of a large wave being a non-breaking wave is far more likely than it breaking.

Furthermore, in a similar, offshore study (Holthuijsen, 1986), the breakers were again marked on the wave record and separate probability density functions plotted. The histograms of the joint probability density functions were remarkably similar again, although the mean height of breakers was about 30% greater than for all waves. They also plotted the likelihood that any wave group would contain a breaking wave, and, in confirmation of Sutherland's result above, found that a wave in a shorter group was more likely to break. Further findings were that the average position of a breaker was forward in the group, and that the elevation of breaking waves was 60% greater than the elevation of the non-breakers. They also state (but, unfortunately do not show) that other geometric criteria, such as the asymmetry, μ, are not sufficient to discriminate breaking waves from non-breaking waves. The Mather study did not generate enough breaking waves to plot accurate mean values of breaking wave height, and the Holthuijsen investigation was conducted in intermediate water depth, so it is not possible to comment on the greater mean height of breaking waves for the deep water case.

In a study to investigate purely the 3-D effects, without the random nature of the wave spectrum, She (1994) looked at the focusing of a regular wave in a 3-D tank (figure 2). The spread of the waves was increased from 30 to 80 degrees and it was found that the steepness at which the wave broke increased from s=0.026 to s=0.040, for waves with components focusing from almost a full half-circle. The asymmetry remained at around μ=0.65 for all angles, in agreement with the value for a regular Stokes wave.

In tests performed in a flume, but with a circular paddle, short-crested regular waves were generated (Kolain, 1993). This had the effect of producing waves with three-dimensional characteristics in the flume which were found to modulate into groups. Values of $0.015<s<0.034$ were observed, in common with the Sutherland 2-D group results ($s>0.014$) and the She 3-D results ($s<0.040$). Kolain proposes an explanation for the high upper limit based on the theory that, as the wave is composed of components from different angles, the forward fluid particle velocity (added vectorially) will be smaller than its equivalent long crested wave of the same height. A shorthand calculation based on the down-wave (η_x) and cross-wave (η_y) slope is proposed,

$$s=0.027(1+\varepsilon^2) \qquad \text{where} \quad \varepsilon=\frac{\sqrt{\eta_y^2}}{\sqrt{\eta_x^2}} \qquad 10$$

This reduces to the standard case for a long crested wave $\varepsilon = 0$, and for a wave which is circular, gives $\varepsilon = 1$, and an upper limit of steepness s=0.054, which exceeds the maximum value obtained by She. However, it is worth noting here that the largest wave ever produced in the 3-D basin at Edinburgh was a bulls-eye with 180 degree components and frequency focusing which had an estimated steepness (H/L) of around 1 in 4, equivalent to s~0.050. Kolain concluded that wave steepness was not a good predictor for breaking.

It would appear that the simple, geometric formulation of the breaking criteria break down when applied to very wide band seas and fully 3-dimensional seas. In his paper on velocity measurements at the free surface, Melville (1988) anticipated that "simple 'breaking criteria' based on local wave properties may prove to be incorrect or at least ambiguous." Studies in which the proportion of waves observed to break and the proportion predicted to break are in agreement may well be fortuitous if the predictions are derived from the limiting (2-D) Stokes wave, as randomness and 3-dimensionality act in opposite senses on the steepness criterion.

WAVE KINEMATICS

The kinematics of water particles in a breaking wave of given form are now fairly well understood. New programming techniques have enabled detailed solutions to be calculated. New optical measurement techniques such as laser anemometry (LDA), and more recently particle image velocimetry (PIV) have enabled kinematics to be measured in the crest to trough region of the wave. This is the area in which the wave theories tend to diverge rapidly.

However, the application of the most advanced methods is not trivial and the form which the breaker may take is not always clear. It is therefore worth reviewing some of the key concepts before examining the detailed application of available theories to design.

Regular waves

There seems to be general agreement that, for regular waves, the traditional solutions provide a good fit. Swan (1990) measured the kinematics in the crest of regular waves up to s=0.010 and found good agreement with Stokes V. Sutherland (1992) compared measurements of regular waves with steepnesses between s=0.010 and s=0.017 with a variety of theories and found that the fourier method of Reinecker and

Fenton (1981) and the Dean's stream function method (Dean, 1965) gave the best fit to the measured crest kinematics. The other comparisons - linear theory, Wheeler & Chakrabarti stretching, delta stretching and extrapolation all performed badly, unless high order wave components were added in to assist with the non-linearity. For the Stokes wave, the stream function theory and the fourier approximation, the crest velocity in a maximum steepness wave always tends to the phase velocity.

Random Waves

Random, non-breaking waves may be described by the application of high order theories (Dean's stream function may be applied to an asymmetric wave profile), or by stretching techniques. However, Sutherland (1992) found that for two component wave groups, stretching theories out-performed the fourier approximation, which could not model the rapid change in horizontal velocity at the wave crest. Similarly, Gutmestad (1993) proposes that Wheeler stretching provides a good engineering approximation to the crest velocities, based on measurements made at Trondheim. Thus it would appear that the kinematics of random waves in 2-D may be described by some simple superposition theory.

Maximum velocity in a breaking wave

Using the regular wave breaking limit, it seems natural to assume that the maximum velocity in a breaking wave will be slightly greater than the wave celerity. This must be allied to the observation that the wave celerity is dependent on the wave steepness, and for shallow water waves, on the water depth. Easson (1984) measured the crest water particle velocity at the spout of a deep water breaking wave using laser anemometry, and the phase velocity by measuring the progress of the zero down-crossing of the wave crest and found that u~c in the spout.

Griffiths (1992) tabulated the findings of a number of investigators and found that with one exception, maximum velocities for deep and shallow plunging breakers were around the phase velocity. For the anomalous result (Kjeldsen, 1980), where u=2.8c, it was proposed that the anomaly was due to the method of calculation of the phase velocity, which is non-trivial for a summation of component frequencies and could result in an under-estimate of c by a factor of 3.

Interestingly, more recent work by Kjeldsen (1989) indicates maximum velocities in the region of $1.3c_o$ - $1.5c_o$ which, given that the phase velocity increases by around 20% for large waves ($c \sim 1.2c_o$). suggests that the crest kinematics are slightly greater than the local phase velocity, as expected.

Chaplin (1992), measured breaking wave kinematics into the crest at two scales, in the City University flume, with H~0.5m, and in the Delta flume at de Voorst, in the Netherlands, with H~3m. He concluded that the maximum velocity always lay in the range $u=c_o$ to $u=1.3c_o$.

She (1994) gives velocity measurements in 3-D breaking waves. These are symmetric single frequency breakers produced at the glass wall of a 3-D wave tank. Measurements were made in the plane of symmetry of the wave, which had spreading angles of +/-30° to +/-60°. The severity of breaking was dependent on the spreading angle, with 30° spreading only resulting in a spilling breaker. The plungers all had maxima between $u=c_o$ and $u=1.5c_o$; the spiller had a maximum at $u=c_o$. She showed that, for his measurements, the velocities under the crest in the plunging breakers did, in fact, tend to the celerity at the free surface. But the maxima were ahead of the crest in the area of the plunging spout.

Velocity field of a breaking wave

Skyner (1992,1996) has provided the first exact comparison of a numerical breaking wave and an experimental wave in a 2-D flume. Using the code of Dold & Peregrine (1986) he produced a numerical breaking wave by frequency superposition. Modelling the exact spectrum in the wave flume he was able to show that the breaking position, time of breaking, height, wave profile and internal kinematics were in agreement to within a few percent. It is therefore possible to predict the kinematics of a 2-D breaking wave exactly if enough information about the composition of the wave components is known.

New (1985), using the Dold & Peregrine code, demonstrated that breaking waves had velocities slightly in excess of the celerity at the crest, and that unusually large accelerations (~5g) were present in the crest front region, below the plunging spout. It was also found that the acceleration/velocity relationship was no longer orthogonal.

Chaplin (1992) found that a 17th order stream function solution gave a reasonable fit to his velocity profiles, and that Stokes V was inadequate. Easson (1988), in breaking waves on beaches work, found that the measured kinematics were in agreement with the predictions of Peregrine, and that high order stream function theory also gave an adequate approximation. Again, low order stream function and Stokes V methods underpredicted the crest velocities. The cnoidal theory, commonly used for shallow water waves was found to be insufficiently accurate for breaking waves in shallow water tests.

Over the past decade, the position regarding wave kinematics has become much clearer. It is now widely accepted that high order theories are required for steep waves. Although Peregrine has proposed the use of the non-linear programs for the prediction of breaking waves it is not yet clear that they exist in a suitable form for general use.

DESIGN IMPLICATIONS

The above findings will now be reviewed in the context of offshore design. Two main areas will be addressed. Choice of wave parameters and choice of wave kinematic theory. For a full treatment of the effects of large breaking and non-breaking waves on offshore structures, it would also be necessary to consider the methods of loading calculation. For example, are the coefficients of Morison's equation significantly altered by the 3-dimensionality of the wave kinematics?

At the present time, a single parameter wave design is in common use. The design wave is calculated from the statistics of the wave records in the area of interest. In this formulation, the significant wave height is used to determine the extreme wave height, the maximum wave elevation, and the wave period. It is possible to examine the derived ratios of steepness and crest asymmetry directly.

The extreme wave height may be calculated from the significant wave height where

$$H_{50} = 1.86 H_{S50} \qquad 11$$

and the extreme wave elevation (the maximum extent of the wave above mwl) is

$$C_{50} = 1.03 H_{S50} \qquad 12$$

The ratio of these is

$$\mu_{50} = 0.55 \qquad 13$$

The zero-crossing period and wave period associated with the 50 year design wave are derived as follows:

$$3.2(H_{S50})^{1/2} < T_z < 3.6(H_{S50})^{1/2} \quad 14$$

$$1.05T_z < T_{ass} < 1.4T_z \quad 15$$

Substituting the range of associated wave periods (Equation 15) and the design wave height (Equation 11) into Equation 14 gives

$$0.007 < \frac{H_{50}}{gT_{ass}} < 0.017 \quad 16$$

This calculation gives a maximum steepness for the design wave which is below the breaking limit for a regular wave (s=0.027), or even the commonly accepted limit (s=0.020) for a random 2-D wave. However, the upper limit is within the range of breaking wave steepness found for waves which belong to groups of around 3, which had a frequent occurrence according to Holthuijsen(1986). The wave asymmetry (equation 13) is also below that found for breaking waves. It follows that the use of these equations necessarily leads the designer into a regime where the waves may be treated as non-breaking.

A number of studies have shown that breaking waves have been found not to be limited to maximum steepness in 3D sea states. This suggests that a more reasonable approach may be to include breaking waves in the design process, but within the present extreme wave height. This would allow for the findings that the breaking events are randomly distributed amongst the range of wave steepnesses. Of course, the probability of encountering a breaking wave of any steepness is much less than that of a non-breaking wave, so the fifty-year breaking wave might have a smaller height.

All of the studies reviewed have shown that the maximum crest elevation is of the order of 65% of the wave height. The definition of maximum crest elevation in terms of the significant wave height leads to a value of 55% of the wave height being in the crest. Given the weight of evidence for the 65% figure it is suggested that the equation

$$C_{50} = 0.65H_{50} \quad 17$$

be used for the calculation of the crest elevation. This will have implications for the design air gap of a structure, as well as increasing the local kinematics of the wave field.

Once the form of the wave has been determined, the kinematic theory should be capable of modelling the full range of velocities which are appropriate to the wave steepness. If the above procedure is followed, it is to be expected that the largest wave will be steep, and a stream function method will be the most appropriate for modelling the wave kinematics.

SUMMARY

Limiting parameters for large and breaking waves have been investigated with respect to randomness and three-dimensionality. It is found that wave steepness is a useful predictor of breaking in 2-D waves, having a value of s=0.027 in regular waves and s=0.020 in random waves. However, if the wave groups in the random sea are short, the breaking limit may be reduced as far down as s=0.014, and the largest wave in the group may not break. Wave steepness has also been found to act as an upper limit on the H-T pdf of 3-D random seas, but was not a good indicator of breaking. Breaking waves are apparently randomly distributed amongst the non-breaking waves in the H-T pdf. The wave asymmetry, expressed in terms of the maximum elevation to wave height ratio seems to be a good indicator of breaking, though further work is required in three dimensions.

Velocity fields in 2-D breaking waves are now well predicted by fully non-linear simulations, and have been compared to detailed experimental results. Where fully non-linear solutions are not available stretching theories give velocity profiles which are good approximations to the velocity profile under the wave crest. Further work is required in the prediction of velocities in 3-D waves.

ACKNOWLEDGEMENTS

I wish to gratefully acknowledge funding from the UK Health and Safety Executive (HSE), and the assistance of Don Smith in pursuing the review which has led to this report. However, the conclusions and suggestions made here are my own, and do not necessarily reflect those of the HSE.

REFERENCES

1. Banner, M.L. and Peregrine, D.H., 1993, "Wave breaking in deep water", Ann Rev Fluid Mech, **25**, 373-97.
2. Chaplin, J., 1992, "Breaking wave forces on a vertical cylinder", OTH 90 324.
3. Dean, R.G., 1965, "Stream function representation of non-linear ocean waves", J Geophysical Res, (18), 4561-72.
4. Dold, J.W. and Peregrine, D.H., 1986, "An efficient boundary integral method for steep unsteady water waves", in Numerical Methods for Fluid Dynamics II, ed Morton, K.W. and Baines, M.J., 671-9, OUP.
5. Easson, W.J. and Greated, C.A., 1984, "Breaking wave forces and velocity fields", Coastal Engineering, pp 233-241
6. Easson, W.J., Griffiths, M.W. and Greated, C.A., 1988, "Kinematics of breaking waves in coastal regions", Proc. 21st. Int. Conf. on Coastal Engineering, A.S.C.E., New York.
7. Griffiths, M., Greated, C.A. and Easson, W.J., 1992, "Measured internal kinematics for shoaling waves with theoretical comparisons", J. Waterway, Port, Coastal and Ocean Engineering, **118**, 280-99, ASCE.
8. Gutmestad, O.T. and Haver, S., 1993, "Uncertainties in the prediction of wave kinematics in irregular waves", in Wave Kinematics and Environmental Forces, SUT **29**, 75-79.
9. Holthuijsen, L.H. and Herbers, T.H.C., 1986, "Statistics of breaking waves observed as whitecaps in the open sea", J. Phys Oceanography, **16**, 290-297.
10. Kjeldsen, S.P., 1981, "Determination of severe wave conditions for ocean systems in a 3-dimensional ocean seaway", VIII Congress of the Pan-Am Inst of Naval Eng.
11. Kjeldsen, S.P., 1989, "Breaking waves", in Water Wave Kinematics, ed Torum and Gutmestad, Kluwer, 453-74.
12. Kjeldsen, S.P and Myrhaug, D., 1980, "Wave-wave interactions,current-wave interactions, and resulting extreme waves and breaking waves", Coastal Engineering, 2277-2303.
13. Kolain, A. and Tulin, M., 1993, "Laboratory measurement of breaking inception and post-breaking dynamics of steep, short-crested waves", ISOPE 93, 45-51.

14. Longuet Higgins, M.S., 1969, "On wave breaking and the equilibrium spectrum of wind-generated waves", Proc. Roy. Soc. Lond. Ser A, **310**, 151-59.

15. Mather, D., Snook, M., Easson, W.J., Greated, C.A. and Bryden, I.G., 1988, "Probability of wave breaking in three-dimensional seas", Proc. 21st Int. Conf. on Coastal Engineering, A.S.C.E., New York.

16. Melville, W.K. and Rapp, R.J., 1988, "Surface velocity field in steep and breaking waves", JFM, **189**, 1-22.

17. New, A.L., McIver, P. and Peregrine, D.H., 1985, "Computations of overturning waves", JFM, **150**, 233-251.

18. Ochi, M.K., 1990, "Stochastic description of offshore environment", in Water Wave Kinematics, ed Torum and Gutmestad, Kluwer, 31-56.

19. Ochi, M.K. and Tsai, C-H, 1983, "Prediction of occurrence of breaking waves in deep water", J. Phys Oceanography, **13**, 2008-19.

20. "Offshore Installations: Guidance on design, construction and certification", 4th edition, 1990, HMSO.

21. Reinecker, M.M. and Fenton, J.D., 1981, "A fourier approximation method for steady water waves", J Fluid Mechanics, **104**, 119-37.

22. She, K-M, Greated, C.A. and Easson, W.J., 1994, "Experimental study of three-dimensional wave breaking", J. Waterway, Port, Coastal and Ocean Engineering, **120**, 20-36, ASCE.

23. Skyner, D.J., 1992, The Mechanics of Extreme Water Waves,Thesis, The University of Edinburgh.

24. Skyner, D.J., 1996, A comparison of numerical predictions and experimental measurements of the internal kinematics of a deep-water plunging wave, J Fluid Mech., to be published.

25. Snyder, R.L. and Kennedy, 1983, "On the formation of whitecaps by a threshold mechanism. Part I: Basic formalism", J. Phys Oceanography, **13**, 1505-18.

26. Sutherland, J., 1992, The Dynamics of Non-Linear Wave Groups, Thesis, The University of Edinburgh.

27. Swan, C., 1990, "Wave kinematics within the crest to trough region", in Environmental Forces on Offshore Structures and Their Prediction, SUT **26**, 45-60.

28. Srokosz, M.A., 1986, "On the probability of wave breaking in deep water", J. Phys Oceanography, **16**, 382-85.

Proceedings of the Seventh (1997) International Offshore and Polar Engineering Conference
Honolulu, USA, May 25-30, 1997

ISBN 1-880653-28-1 (Set); ISBN 1-880653-31-1 (Vol. III)

Analysis of Ocean Waves by Crossing- and Oscillation-Intensities

I. Rychlik
University of Lund
Sweden

M.R. Leadbetter
University of North Carolina
Chapel Hill, NC, USA

ABSTRACT

In this paper, relationships between wave characteristics and upcrossing- and oscillation-intensities are used to construct conservative bounds for significant wave heights and amplitudes and for moments of waves crests and troughs. A concept of oscillation filtering is introduced for removing small waves. Results are illustrated by five examples in which both Gaussian and non-Gaussian models for a sea or a load are considered.

KEY WORDS: Crossings, extrema of Morison's force, ocean-waves, fatigue, oscillations, rainflow cycles, significant crest height, significant amplitude, slams.

BASIC DEFINITIONS AND RESULTS

Let $x(t)$ be the height of the sea level at a fixed point as a function of time t. In oceanographic applications $x(t)$ is often seen as a sequence of waves where each wave can be described by means of its highest and lowest values (crest, trough), or by means of its amplitude (= crest - trough) and wave period, describing the duration of a single wave. There is no general agreement about the formal definition of a wave. Often one uses the so called mean downcrossing wave, where the wave is considered as a part of a function between the consecutive downcrossings of the mean sea level, (see the following definition).

Definition 1 *Assume $x(t)$ crosses a fixed reference level u_* finitely many times, (here the level most frequently crossed by x). Denote by t_i , $0 < t_1 < t_2 < \ldots$, the times of downcrossings of u_* . The trough and crest m_i^* , M_i^* , say, of the i th wave are the global minimum and the global maximum of $x(t)$, $t_i < t < t_{i+1}$, respectively. The so called crest front amplitude H^* is defined as the difference between the crest and the trough, i.e.*

$$H_i^* = M_i^* - m_i^* = \max_{t_i<t<t_{i+1}} x(t) - \min_{t_i<t<t_{i+1}} x(t). \qquad (1)$$

(In the following the reference level $u_* = 0$.)

A simple measure of severity of waves are the significant wave amplitude and the significant crest height, denoted by H^s , M^s , averages of the highest one-third of amplitudes H_i^* , crests heights M_i^* , respectively. A more complete statistical characteristic of waves are (crest, trough)- and (wave period, amplitude)-distributions.

Two major uses of wave characteristic distributions are to predict extreme waves and fatigue lifetimes of marine structures. Clearly, using the mean crossing waves one neglects small oscillations superimposed on major waves. However, for fatigue accumulation in marine structures, it is well known that even small oscillations can contribute to the damage and hence have to be analyzed. A simple form of cycle, which is useful in fatigue analysis and also for approximations is the so called 'min-max' cycle.

Definition 2 *Denote the i th local maximum in $x(t)$ by M_i and the previous local minimum by m_i . Then the wave form between m_i and M_i will be termed the 'i th min-max cycle' and referred for brevity as the cycle (m_i, M_i) . The sequence of local extremes in $x(t)$, denoted by $TP(x)$, i.e.*

$$TP(x) = [m_1, M_1, m_2, M_2, m_3, \ldots] \qquad (2)$$

will be called a sequence of turning points. (For simplicity of presentation the following convention is used; if 0 is the point of local minimum, e.g. the right derivative at zero is positive then $m_1 = x(0)$, otherwise the part of the function x before the first local minimum is omitted.)

It is generally agreed that the fatigue is a "rate independent" process, which means that only the sequence of turning points is essential for fatigue life prediction and not the exact path of x between the local extremes. However, it takes at least 10^4 local maxima before fatigue failure, and often more than 10^7 . Consequently in practice it is not possible to use the whole structure of the sequence $TP(x)$ to predict the fatigue lifetime. Instead, the distribution of min-max cycles are sometimes used. However the predictors are often biased, giving longer lifetimes than observed in experiments. Much better agreement is achieved by using the distribution of the so called 'rainflow cycles', which are local maxima and minima of x, paired using the hysteresis properties of the material, see Brokate and Sprekels (1996). The rainflow method was introduced by Endo, the first paper in English being Matsuishi and Endo (1967). Here the definition given in Rychlik (1987) will be used.

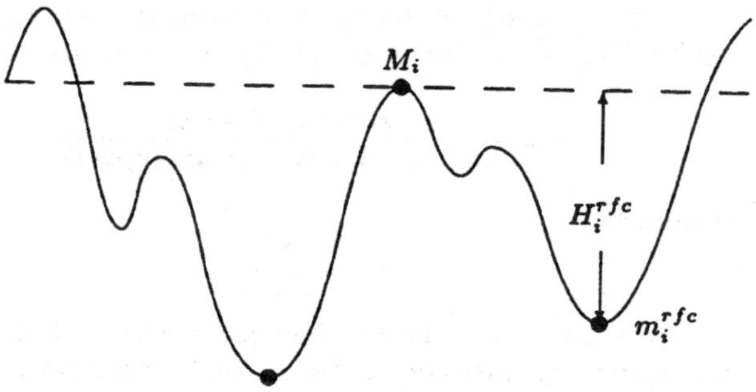

FIG. 1. Definition of rainflow cycle.

Definition 3 *To define the rainflow cycles, each local maximum in the load sequence is paired with one particular local minimum, determined as follows: from the i th local maximum (value M_i) one determines at the lowest values in forward and backward directions between M_i and the nearest points at which the waveform exceeds M_i. The larger (less negative) of those two values, denoted by m_i^{rfc}, is the rainflow minimum paired with M_i, i.e. m_i^{rfc} is the least drop before reaching the value M_i again on either side. Thus the i th rainflow pair is (m_i^{rfc}, M_i) and the rainflow amplitude is $H_i^{rfc} = M_i - m_i^{rfc}$; see Figure 1.*

As is evident the rainflow minimum does not necessarily immediately follow or precede the corresponding maximum so that the "cycle" (m_i^{rfc}, M_i) can include a number of oscillations. However, it may be shown that each minimum corresponds to only one maximum in forming a rainflow pair.

Another important characteristics of sea state (or a load process), which contains fundamental information about waves and cycles, are *upcrossing-* and *oscillation-intensities*, defined as follows.

Definition 4 *The upcrossing intensity $\mu(u)$ of the level u by $x(t)$, and oscillation intensity $\mu^{osc}(u,v)$ for the interval $[u,v]$ are defined by*

$$\mu(u) = \lim_{T\to\infty} \frac{1}{T}\left(\begin{matrix}\text{number of upcrossings of } u\\ \text{by } x(t),\ t\in[0,T]\end{matrix}\right), \tag{3}$$

$$\mu^{osc}(u,v) = \lim_{T\to\infty} \frac{1}{T}\left(\begin{matrix}\text{number of upcrossings of } [u,v]\\ \text{by } x(t),\ t\in[0,T]\end{matrix}\right) \tag{4}$$

assuming that the limits exist, the number of upcrossings of $[u,v]$ being the largest index n such that there are times $0 \le s_1 < t_1 < s_2 < \ldots < s_n < t_n$, satisfying $x(s_i) < u \le v < x(t_i)$ and $s_n < T$.

Obviously $\mu^{osc}(u,v)$ is equal to the average rate of oscillations between sets $(-\infty, u)$ and $(v, +\infty)$ by $x(t)$, which motivates the name. The upcrossing and oscillation intensities are related by the following formulae

$$\mu^{osc}(u,u) = \mu(u), \qquad \mu^{osc}(u,v) \le \min_{u\le z\le v} \mu(z). \tag{5}$$

Note that, in order to extract sequences of trough and crests (m_i^*, M_i^*), rainflow cycles (m_i^{rfc}, M_i), min-max cycles (m_i, M_i) or estimate upcrossing and oscillation intensities only the sequence of turning points $TP(x)$ is needed. If the signal $x(t)$ is available it is a simple task to extract $TP(x)$ and then to compute e.g. H^s, M^s or to fit arbitrary parametric distributions to the observed waves or cycle characteristics. However it is also important to find a statistical model for sea state based on more mathematical principles and it is natural to model $x(t)$ as a stochastic process. In particular it is important to compute the significant wave amplitude or crest height and the distributions of waves (or cycles) characteristics from the properties of a stochastic process $x(t)$. This is a difficult mathematical problem and explicit results are known only for some special cases. By contrast, the upcrossing intensity $\mu(u)$ is readily computed using "Rice's formula", see (12). In the paper the relationships between upcrossing-, oscillation-intensities and statistical distributions of wave (cycles) characteristics will be used to provide conservative bounds for the distributions.

In fatigue analysis one often needs to compute moments of the rainflow amplitude $H^{rfc} = M - m^{rfc}$, see Figure 1. Most commonly, it is assumed that the mean fatigue life is proportional to the inverse of a damage intensity D, say, defined by

$$D = (\text{intensity of cycles}) \cdot cE[(H^{rfc})^\beta], \quad \beta > 1, \tag{6}$$

where β and c are material constants, see Rychlik (1996a) for more detailed discussion. Obviously, conservative estimates of fatigue life can be obtained by computing upper bounds for the moments.

The relevant bounds for significant values and moments of the rainflow amplitude $E[(H^{rfc})^\beta]$ will be derived using the following elementary lemma.

Lemma 1 *Let X be a positive random variable and f a function with a positive derivative f'. If $f(0) = 0$ then*

$$E[f(X)] = \int_0^\infty f'(x)P(X > x)\,dx, \tag{7}$$

and, with X^s denoting the significant value of X, defined as the conditional expectation of X given that $X \ge x_{2/3}$, where $P(X > x_{2/3}) = 1/3$,

$$X^s = 3\int_0^\infty \min(P(X > x), 1/3)\,dx. \tag{8}$$

Further, let $X \le Y$ be random variables. Consider a positive function $f(u,v)$ such that $f(u,u) = 0$, for all u, and denote

$$f_2(u,v) = \frac{\partial f(u,v)}{\partial v}, \qquad f_{12}(u,v) = \frac{\partial^2 f(u,v)}{\partial u \partial v}. \tag{9}$$

If $f_2(v,v) = 0$, $f_{12}(u,v) \le 0$, $u < v$ and for all $y < z$, $\int_y^z \int_y^v |f_{12}(u,v)|\,du\,dv$ is finite then

$$E[f(X,Y)] = -\int_{u<v} f_{12}(u,v)P(X < u, Y > v)\,du\,dv. \tag{10}$$

Clearly, in order to bound the significant crest height M^s it is sufficient to find a bound for $P(M^* > u)$ and, since the function $f(u,v) = (v-u)^\beta$, $\beta > 1$, satisfies the assumptions of Lemma 1, the moments $E[(H^{rfc})^\beta]$ can be conservatively estimated if $P(m^{rfc} < u, M > v)$ can be bounded. In the following sections, those bounds will be derived, using upcrossing intensities. More precisely, first a stochastic model for a sea level is given, together with Rice's formula for upcrossing intensities. Next a bound for the wave crest height distribution is derived and related to the well known Rayleigh approximation. The bound is used to conservatively compute the significant wave height and amplitude and three example are also given. Then, the joint distributions for trough and crest height for mean crossing waves, rainflow and min-max cycles, respectively, are presented and results are applied to bound moments of the rainflow amplitude for the so called Morison force. Finally, in the last section the sequence of waves is obtained as a result of oscillation filtering of x.

STOCHASTIC MODELING OF OCEAN DATA

The standard assumptions for sea state under stationary conditions is that the model $x(t)$ is a stationary and ergodic stochastic process with mean $E[x(t)]$ assumed to be zero and spectral density $S(\omega)$, where ω is the angular frequency. Standard classes of spectral densities $S(\omega)$ which are suitable to describe sea state data are well established from experimental studies. It is generally accepted that a stationary fully-developed sea state at a deep water location can be satisfactorily modeled by a Gaussian process and hence the spectral density function fully defines the statistical properties of $x(t)$.

Real data $x(t)$ seldom perfectly support the Gaussian assumptions for the process $x(t)$. However, since the Gaussian case is well understood, the sea state is often modeled using Gaussian processes. Obviously, if the wave characteristics derived from $S(\omega)$, using Gaussian assumption, do not agree with observations, then a broader class of models is needed. The simplest (but widely effective) model for a non-Gaussian sea is where $x(t)$ is a function of a stationary Gaussian process $\tilde{x}(t)$ with variance one ($V[\tilde{x}(0)] = 1$), i.e.

$$x(t) = G(\tilde{x}(t)), \qquad \frac{dG}{dx} > 0 \quad \text{and} \quad G(0) = 0. \tag{11}$$

In order to use the model one needs to estimate the spectral density of $\tilde{x}(t)$ as well as the transformation G. There are some parametric formulae for G proposed in the literature, e.g. by Ochi and Ann (1994), while in Rychlik et al (1997) the upcrossing intensity and formulae (18) were used to estimate G nonparametrically.

UPCROSSING INTENSITY

The upcrossing intensity $\mu(u)$ is given by the celebrated Rice's formula, see Rice (1944, 1945), and Leadbetter et al (1983)

$$\mu(u) = \int_0^\infty z\, f_{x'(0),x(0)}(z,u)\, dz, \tag{12}$$

where $f_{x'(0),x(0)}$ is the joint density of the derivative $x'(0)$ and the value $x(0)$. In a special case when $x(t)$ is a zero mean Gaussian process $\mu(u)$ can be written in the following closed form

$$\mu(u) = \frac{1}{2\pi}\sqrt{\frac{\lambda_2}{\lambda_0}}e^{-\frac{u^2}{2\lambda_0}}, \tag{13}$$

where $\lambda_0, \lambda_2, \ldots, \lambda_{2k}$, are the so called spectral moments defined by

$$\lambda_{2k} = \int_{-\infty}^{+\infty} \omega^{2k} S(\omega)\, d\omega, \qquad k = 0, 1, \ldots. \tag{14}$$

The intensity of waves in a Gaussian sea is $\mu(0) = (2\pi)^{-1}\sqrt{\lambda_2/\lambda_0}$, while the irregularity factor, defined by

$$\alpha = \frac{\text{intensity of waves}}{\text{intensity of local maxima (cycles)}}, \tag{15}$$

is given by

$$\alpha = \frac{\lambda_2}{\sqrt{\lambda_0 \lambda_4}}. \tag{16}$$

We turn now to the non-Gaussian models and assume that the upcrossing intensity μ has a single maximum at the level $u_*(=0)$. Then there exists unique transformation G, $G(0) = 0$, such that the simple transformed Gaussian process $x(t) = G(\tilde{x}(t))$, see (11), has the upcrossing intensity $\mu(u)$. The following equation then holds

$$\frac{\mu(u)}{\mu(0)} = e^{-\frac{(G^{-1}(u))^2}{2}}, \tag{17}$$

where G^{-1} is the inverse function of G. If the upcrossing intensity is known, the last equation can be used to obtain the inverse transformation G^{-1} by

$$G^{-1}(u) = \begin{cases} \sqrt{-2\log \mu(u)/\mu(0)} & \text{if } u \geq 0, \\ -\sqrt{-2\log \mu(u)/\mu(0)} & \text{if } u < 0. \end{cases} \tag{18}$$

CREST HEIGHT DISTRIBUTION

In this section the relation between upcrossing intensity $\mu(u)$ and the distribution of crest height M^* is discussed. First, it is necessary to be precise how the distribution of M^* is defined.

Often, in practice, only one sample of a process $x(t)$ is observed. All waves (or cycles) in $x(t)$ are then found and, using a suitable statistical technique, the distribution of wave characteristics are estimated. Consequently, by assumed ergodicity of $x(t)$,

$$\begin{aligned} 1 - F_{M^*}(u) &= \lim_{T\to\infty} \frac{\begin{pmatrix}\text{number of waves in } [0,T] \text{ with crest} \\ \text{higher than } u\end{pmatrix}}{\text{number of waves in } [0,T]} \\ &= \frac{\text{intensity of waves with crest higher than } u}{\text{intensity of waves}} \\ &= P(M^* > u), \end{aligned} \tag{19}$$

where P is the so called Palm probability, see Leadbetter et al (1983) and Lindgren and Rychlik (1991) for more detailed discussion of Palm distributions.

Now, it is easy to see that, with $u_* = 0$, the intensity of waves is equal to $\mu(0)$ and that

$$(\text{intensity of waves with crest } M^* > u) = \mu^{osc}(0,u). \tag{20}$$

Consequently, using the inequality (5) the distribution of M^* can be bounded as follows

$$P(M^* > u) \leq \min_{0 \leq z \leq u} \frac{\mu(z)}{\mu(0)}. \tag{21}$$

By symmetry, one also has that

$$P(m^* < v) \leq \min_{v \leq z \leq 0} \frac{\mu(z)}{\mu(0)}. \tag{22}$$

Since in oceanography the most important case is when $x(t)$ has upcrossing intensity $\mu(u)$ with only one maximum at $u = u_*(= 0)$, then the inequality (21) can be written in more explicitly as

$$1 - F_{M^*}(u) \leq e^{-\frac{(G^{-1}(u))^2}{2}} = P(G(R) > u), \tag{23}$$

for all $u \geq 0$, where R has a Rayleigh distribution with density $f_R(r) = r\,\exp(-r^2/2)$ and G^{-1} is defined by (18).

For a Gaussian sea the equation (23) clarifies the relation between crest height distribution and the so called *Rayleigh approximation*, postulating that M^* is approximatively Rayleigh, commonly used in practice. Since $\mu(u)/\mu(0) = \exp(-u^2/2\lambda_0)$, i.e. $G(x) = \sqrt{\lambda_0}x$, then, for all $u \geq 0$,

$$1 - F_{M^*}(u) \leq P(\sqrt{\lambda_0}R > u). \tag{24}$$

Finally, from inequalities (23) and (22) it follows that

$$E[f(M^*)] \leq E[f(G(R))], \tag{25}$$
$$E[f(-m^*)] \leq E[f(-G(-R))]. \tag{26}$$

for any f satisfying the assumptions of Lemma 1.

SIGNIFICANT WAVE HEIGHT AND AMPLITUDE

Since $1 - F_{M^*}(v) \leq P(G(R) > v)$ then, by (8), the significant crest height is bounded by

$$M^s \leq 3\int_0^{+\infty} \min(e^{-\frac{(G^{-1}(v))^2}{2}}, 1/3)\,dv. \tag{27}$$

Similarly, the significant trough height is bounded as follows

$$-m^s \leq 3\int_{-\infty}^{0} \min(e^{-\frac{(G^{-1}(v))^2}{2}}, 1/3)\,dv. \tag{28}$$

Now, by combining (27) and (28) the significant amplitude

$$H^s \leq M^s - m^s \leq 3\int_{-\infty}^{+\infty} \min(e^{-\frac{(G^{-1}(v))^2}{2}}, 1/3)\,dv. \tag{29}$$

Finally, for a Gaussian sea the bounds can be written explicitly

$$M^s < 2.003\sqrt{\lambda_0} \quad \text{and} \quad H^s = 2 * M^s < 4.006\sqrt{\lambda_0}. \tag{30}$$

Example 1: North Sea wave data. The North Sea data, supplied by M. Olagnon, IFREMER, France, consist of approximately 18 minutes of measurements with sampling frequency 2Hz. The standard deviation is estimated to be 3.12 meters and the irregularity factor to be 0.69. A more extensive statistical analysis of the data can be found in Rychlik et al (1997).

From the data the significant amplitude and crest height H^s, M^s, have been evaluated, having values 12.58, 7.26 meters, respectively. Note that $2 * M^s = 14.52$ is much higher than the observed H^s value, clearly indicating that the Gaussian model is not appropriate for the data.

Next, the upcrossing intensity $\mu(u)$ has been estimated by the observed upcrossing intensity. Using (18) and (27), the bound for M^s is computed giving the value 7.37. Similarly, the bound for H^s is found to be 12.92. Clearly, the upper bounds for the significant crest height and amplitude are very close to the observed values.

Finally, if the Gaussian model for $x(t)$ were assumed, than the 'Rayleigh' bound (30) gives $M^s < 2.002 * 3.12 = 6.25$ and $H^s < 12.5$ meters. One can see that the Gaussian assumption (Rayleigh approximation) would lead to a non-conservative estimate of a significant crest height, the difference being one meter.

However the significant amplitude is very close to that observed. This seems to be contradictory. A closer look at the data shows that both crests and troughs are higher than predicted by a Gaussian model. Since amplitudes are differences between heights of crests and troughs, the prediction errors tend to cancel.

Example 2: Extremes of bending moment under slamming. When a ship proceeds at certain speeds in rough seas, slamming take place. That is to say, the front part of hull bottom sustains large forces resulting from impact with the sea surface. One of important design parameters is a maximal bending moment, denoted in the following by $x(t)$. A part of a typical measurement of a bending moment is shown in Figure 2. In this one can easily recognize the slowly varying wave pattern (the so called *wave induced* bending moment), and the superimposed high frequency oscillations due to slamming. The local minima are much bigger (in the absolute value) than the local maxima and hence only the distribution of minima will be considered in this example

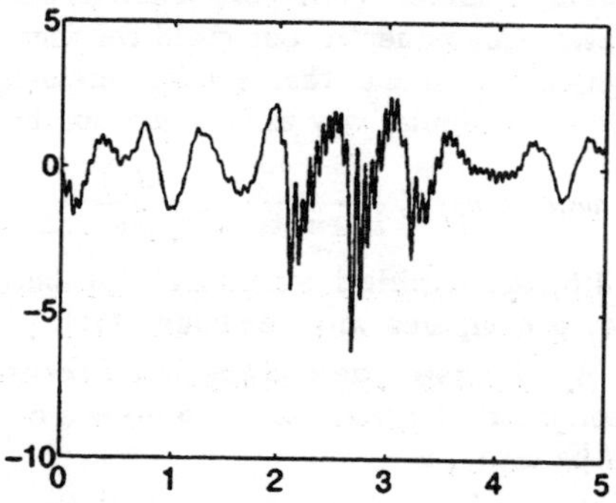

FIG. 2 A part of a measured bending moment $x(t)$.

Clearly the bending moment is a result of the superposition of two different physical processes; the wave train acting on a ship, and the slamming events. Consequently the statistical properties of small and high minima could be entirely different. Since only the low minima are of interest and should be adequately modeled it seems to be reasonable to consider a conditional approach, by omitting some of the minima in the analysis. There are several possibilities; One could select all local minima lower than a given threshold. However, the component of bending moment due to slamming has a high frequency and causes clustering of the low minima which has to be considered in the analysis. Another approach is to select all first local minima after the slam occurrences. For this one needs to define a slamming event more precisely and then to identify slams in $x(t)$. Here the concept of mean downcrossing waves, $u_* = 0$, will be employed and the distribution of trough height m^* used to model the low minima.

After extracting the sequence of troughs m_i^* from $x(t)$ one can fit a suitable distribution to the data. Such a fit will not be presented here, but instead the accuracy of the bound based on the upcrossing intensity will be checked. The intensity of upcrossings $\mu(u)$ is estimated by the observed intensity of upcrossings by $x(t)$. In Figure 3(a) an empirical distribution of m^* is compared with the conservative bound $\mu(u)/\mu(0)$. One can see that the bound is sharp for low minima. Further, the observed significant trough height is -3.35 while the conservative bound (28) is -3.87.

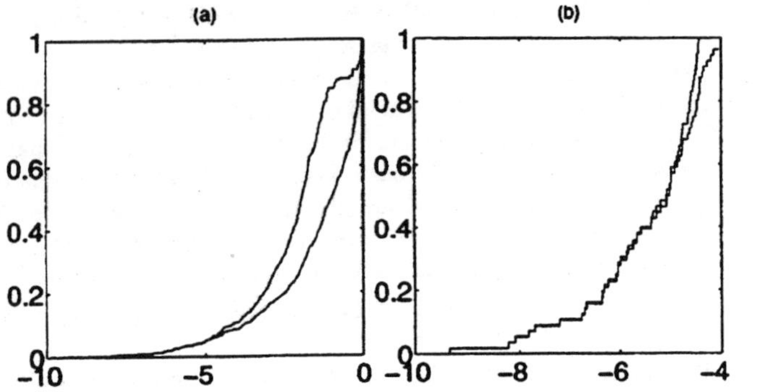

FIG. 3 (a) The empirical distribution of bending moment minima m^* compared with the bound $\mu(v)/\mu(0)$; (b) The conditional empirical distribution of bending moment minima lower than -4 compared with the bound (29).

Clearly, the description of the extremal properties of $x(t)$ using the sequence of crests and troughs of waves is a 'filtering operation' that removes small oscillations superimposed on a major wave pattern. However there are still some waves with small amplitudes which are less important. Some of the small waves in the example are artifacts of the method and do not corresponds to real sea waves. Consequently, one could consider only the waves with trough heights smaller than a given threshold v_0, say, here taken to be -4. In similar way as (22) one can then show that

$$P(m^* < v | m^* < v_0) \leq \frac{\mu(v)}{\text{intensity of waves with } m^* < v_0}. \quad (31)$$

In Figure 3(b) the empirical conditional distribution of $P(m^* < v | m^* < -4)$ is compared with the bound (31).

Example 3: Transformed Gaussian process. In this example a parametric formula for G proposed in Ochi and Ahn (1994) will be used, i.e.

$$G^{-1}(u) = \frac{1}{\sigma\gamma}(1 - e^{-\gamma u}), \qquad \gamma = \begin{cases} \gamma_a & \text{if } u \geq 0, \\ \gamma_b & \text{otherwise,} \end{cases} \quad (32)$$

and $\sigma > 0$. The parameter values are $\sigma = 1$, $\gamma_a = 0.1$ and $\gamma_b = 0.15$. The spectrum $S(\omega)$ is of Pierson-Moskowitz type, i.e.

$$S(\omega) = \frac{c}{(\omega)^5} \exp\left(-\beta(\frac{\lambda_m}{\omega})^4\right), \quad (33)$$

with $\lambda_m = 0.075$, $\beta = 0.09$ and c chosen so that the first spectral moment of $S(\omega)$ is equal to one, i.e. $V[\tilde{x}(0)] = 1$. Observe that the spectrum has an infinite fourth spectral moment, which means that the Gaussian process has an infinite number of local extremes in a finite interval.

Consequently, high frequencies of the spectrum are cut off leading to the value 0.582 for the irregularity factor α, see (15).

By employing numerical integration to compute (27) with G^{-1} given by (32) the bound for the significant crest height $M^s \leq 2.27$ for the process $x(t) = G(\tilde{x}(t))$ is obtained.

Finally, in order to check the accuracy of the bound five sample paths of the transformed process $x(t)$ have been simulated and the significant crest heights $\hat{M}^s$ estimated. Those values are 2.21, 2.27, 2.26, 2.31, 2.28 and should be compared to the average bound equal to 2.27. Note that the 'Rayleigh' bound for the significant crest height for the untransformed Gaussian process $\tilde{x}(t)$ is $M^s \leq 2.003$ and underestimates the simulated significant crest heights.

THE JOINT (TROUGH, CREST)-DISTRIBUTION

As in the previous section, a precise definition of the (crest, trough)-distribution is first presented

$$P(m^* < u, M^* > v) = \frac{\begin{pmatrix}\text{intensity of waves with crest} \\ M^* > v \text{ and trough } m^* < u\end{pmatrix}}{\text{intensity of waves}}. \quad (34)$$

(Here, again, P is the Palm probability.) In the previous section it was shown that $P(M^* > v) \leq \mu(v)/\mu(0)$ and $P(m^* < u) \leq \mu(u)/\mu(0)$, consequently,

$$P(m^* < u, M^* > v) \leq \min(\frac{\mu(u)}{\mu(0)}, \frac{\mu(v)}{\mu(0)}), \quad (35)$$

$u < 0 < v$. In fact somewhat stronger result also holds

$$P(m^* < u, M^* > v) \leq \min_{u \leq z \leq v} \frac{\mu(z)}{\mu(0)}, \quad (36)$$

$u < 0 < v$, since

$$\begin{pmatrix}\text{intensity of waves with crest} \\ M^* > v \text{ and trough } m^* < u\end{pmatrix} \leq \mu^{osc}(u, v). \quad (37)$$

For unimodal upcrossing intensities, equations (36) and (35) coincide and can be written in the more explicit way

$$P(m^* < u, M^* > v) \leq P(G(-R) < u, GR) > v), \quad (38)$$

$u < 0 < v$. Since the function $(v - u)^\beta$, $\beta > 1$, satisfies the assumptions of Lemma 1, one has, for the wave amplitude H^*, the following bounds for moments

$$E[(H^*)^\beta] \leq E[(G(R) + G(-R))^\beta]. \quad (39)$$

SOME DISTRIBUTIONAL RESULTS FOR RAIN-FLOW CYCLES

By extracting the sequence of troughs and crests from $x(t)$ some local extremes have been omitted. The average proportion of wave crests to local maxima (cycles) is the irregularity factor α. For a Gaussian sea, the distribution of crest height M^* is unknown, except in the degenerate case of a pure sine process, but it has been shown that $M^* \leq \sqrt{\lambda_0}R$, where R is a standard Rayleigh variable. (This holds if the second spectral moment is finite, i.e. $\lambda_2 < +\infty$, since, otherwise, the intensity of waves is infinite.) For the distribution of the height M of a local maximum a more exact result is well known, viz

$$M = \sqrt{\lambda_0}(\alpha R + \sqrt{1 - \alpha^2}X),$$

where R, X are independent, standard Rayleigh and Gaussian variables. (The formula is valid when the fourth spectral moment is finite, $\lambda_4 < +\infty$, otherwise one has infinitely many local extremes.) For a transformed Gaussian process $x(t) = G(\tilde{x}(t))$, defined by (11) one has $M^* \le G(R)$ and $M = G(\alpha R + \sqrt{1-\alpha^2}X)$.

We turn now to the joint distribution of a maximum and its rainflow minimum. In Rychlik (1993) it was proved that

$$\begin{pmatrix}\text{intensity of rainflow cycles such that} \\ M > v \text{ and } m^{rfc} < u\end{pmatrix} = \mu^{osc}(u,v). \quad (40)$$

This implies that

$$P(m^{rfc} < u, M > v) \le \alpha \min_{v \le z \le u} \frac{\mu(z)}{\mu(0)} \le \alpha \min(\frac{\mu(u)}{\mu(0)}, \frac{\mu(v)}{\mu(0)}), \quad (41)$$

$u < v$, where α is the irregularity factor defined by (15).

Now if a function f satisfies the assumptions of Lemma 1 then

$$E[f(m^*, M^*)] \le E[f(-G(-R), G(R))],$$
$$E[f(m^{rfc}, M)] \le \alpha E[f(-G(-R), G(R))]. \quad (42)$$

In particular, moments of rainflow amplitude, used in prediction of fatigue life, can be bounded as follows

$$E[(H^{rfc})^\beta] \le \alpha E[(G(R) + G(-R))^\beta], \quad \beta > 1. \quad (43)$$

Example 4: Morison force. Let $y(t)$ be the velocity of a fluid around a circular pile, and $F(t)$ be the force acting on the pile. By Morison's equation the force F is the following function of the process $y(t)$

$$F(t) = K_M y'(t) + K_D y(t)|y(t)|, \quad (44)$$

where K_M, K_D are positive constants. The properties of $F(t)$ are often discussed in ocean engineering literature, (see e.g. Naess (1983)), where the velocity is usually assumed to be a stationary Gaussian process.

Let $y(t)$ be a stationary Gaussian process with nonnegative mean $E[y(t)] = v \ge 0$. For simplicity only, assume that $V[y(t)] = V[y'(t)] = 1$ ($\lambda_0 = \lambda_2 = 1$) and $V[y''(t)] = \lambda_4 < \infty$. For a positive constant K define

$$x(t) = y'(t) + Ky(t)|y(t)|. \quad (45)$$

The force F considered in equation (44) can always be transformed into this form by suitable choice of scales.

The upcrossing intensity $\mu(u)$ of $x(t)$ can be expressed in the form of a single integral which has to be computed numerically, namely

$$\mu(u) = \frac{1}{2\pi}\sqrt{\frac{\lambda_4}{\lambda_2}(1-\alpha^2)} \int_{-\infty}^{+\infty} e^{-\frac{1}{2}(z-v)^2 - \frac{1}{2}(u-Kz|z|)^2} \cdot \Psi\left(\frac{\alpha}{\sqrt{1-\alpha^2}}(z - v - 2K|z|u + 2K^2z^3)\right) dz, \quad (46)$$

where $\Psi(z) = \phi(z) + z\Phi(z)$, ϕ and Φ being the standardized normal density and distribution functions and α is the irregularity factor of the $y(t)$ process, i.e. $\alpha = \frac{\lambda_2}{\sqrt{\lambda_4 \lambda_0}}$.

In this example, the bounds (43) for moments of rainflow amplitude $E[(H^{rfc})^\beta]$, with $\beta = 2,3,4,5$, are used to bound the damage intensity D, defined by (6). Using (43) the damage intensity is bounded by

$$D \le c\beta(\beta-1)\int_{u<v} (v-u)^{\beta-2} \min(\mu(u), \mu(v))\, du dv. \quad (47)$$

Let $y(t)$ be a stationary Gaussian process with covariance function

$$r(\tau) = \frac{\sin(\sqrt{3}\tau)}{\sqrt{3}\tau},$$

and mean v, ($y(t)$ is so called low frequency white noise). The y-process has $\lambda_0 = \lambda_2 = 1$ and $\lambda_4 = 1.8$.

For $K = 1$, by numerical integration of (47), the damage intensities, with $\beta = 2,3,4,5$ and $c = 1$, are bounded by, for $v = 0$; 6.27, 71.96, 1130.0, 22342.7, respectively, and for $v = 1$; 10.34, 126.81, 1963.9, 36828.9, respectively.

In order to check the accuracy of the bound three sample paths of $x(t)$ have been simulated and the damage intensities estimated. The results are for $v = 0$: (5.87, 5.83, 6.04), (62.8, 64.1, 65.5), (931.3, 1026.4, 958.1), (17408.4, 21816.2, 17153.3), and for $v = 1$: (9.78, 9.73, 10.1), (117.5 116.1, 125.7), (1820.5 1817.2, 1995.3), (34537.0, 36126.4, 38438.1).

Min-max cycles

The joint distribution of maximum and the preceding minimum is not known, even for Gaussian processes (except the pure sine process). However there are very accurate numerical algorithms to compute the joint distribution of (m, M), see Rychlik and Lindgren (1993) and (1995). Further a bound for $P(m < u, M > v)$ can be derived from the observation that

$$\begin{pmatrix}\text{intensity of min-max cycles} \\ \text{such that } M > v \text{ and } m < u\end{pmatrix} \le \mu^{osc}(u,v). \quad (48)$$

and hence $P(m < u, M > v) \le P(m^{rfc} < u, M > v)$. Consequently,

$$E[H^\beta] \le E[(H^{rfc})^\beta], \quad (49)$$

$\beta > 1$, where $H = M - m$. Obviously, from the first inequality it follows that longer fatigue lifetimes are always predicted when using the min-max cycles than the rainflow cycles.

OSCILLATION FILTER

Consider a continuous function $x(t)$, $t \in [0,T]$, and let $N^{osc}_{(u,v)}(x)$ denote the number of its upcrossings of interval $[u, v]$. Note that, if $u < v$ then $N^{osc}_{(u,v)}(x)$ is always finite, regardless how irregular $x(t)$ may be, e.g. infinitely many waves or cycles are allowed.

In this section the sequence of turning points $TP(x)$ will be represented by a continuous polygonal function obtained by drawing straight lines between points in $TP(x)$. Consequently $N^{osc}_{(u,v)}(TP(x))$ is well defined. The same is assumed for the sequence of waves $[m_1^*, M_1^*, m_2^*, \ldots]$.

Definition 5 *Let UV be a set of pairs (u, v), $u < v$. Any operation of filtering (or smoothing) of x such that the resulting function y, say, has the same number of upcrossings of intervals $[u, v]$ for all pairs $(u,v) \in UV$ as $x(t)$ has, e.g. $N^{osc}_{(u,v)}(y) = N^{osc}_{(u,v)}(x)$ will be called an oscillation filter.*

Examples of oscillation filters are:

(i) Let UV_1 to be the set of all pairs (u,v) such that $u < v$. Then the operation of extracting sequence of turning points

$$x(t) \to [m_1, M_1, m_2, \ldots] = TP(x) \tag{50}$$

is an oscillation filter. This is evident since as previously noted only the sequence of turning points is needed to compute $N^{osc}_{(u,v)}(x)$ as well as the sequences of troughs and crests (m_i^*, M_i^*), rainflow cycles (m_i^{rfc}, M_i), min-max cycles (m_i, M_i) or the upcrossing intensity, if such exist.

(ii) The procedure of extracting mean downcrossing waves, i.e.

$$x(t) \to [m_1^*, M_1^*, m_2^*, \ldots], \tag{51}$$

is also an oscillation filter defined by the set $UV_2 = \{(u,v) : u < u_* < v\}$

(iii) This example is taken from fatigue analysis where, in order to accelerate fatigue tests, one eliminates rainflow cycles with small amplitudes from x, see Rychlik (1996d). Such a filter, called also a rainflow filter, is defined by a set $UV_3 = \{(u,v) : v - u > v_0\}$, where v_0 is a fixed threshold.

(iv) In Example 2 all troughs m_i^* that are lower than a fixed level v_0 were extracted. This can be done by an oscillation filter defined by $UV_4 = \{(u,v) : u < v_0 < u_* < v\}$. Similarly, the intersection of sets UV_2 and UV_3, i.e. $UV_5 = \{(u,v) : u < u_* < v \text{ and } v - u > v_0\}$, will give the sequence of waves that has no amplitudes smaller than v_0. Note that a pair (m_i^*, M_i^*) such that $M_i^* - m_{i+1}^* \le v_0$, but both $H_i^* > v_0$ and $H_{i+1}^* > v_0$ will also be removed.

The advantage of using the oscillation filters is that the distributions of crest or trough heights, in the filtered signal, are always proportional to the intensity $\mu^{osc}(u,v)$, $(u,v) \in UV$, and are true conditional distributions. This is not the case if one is changing spectral density function by means of some linear filters. Another, perhaps a more theoretical advantage, is that one can handle stochastic processes with infinite number of cycles or waves in a systematic way. Processes with infinite numbers of cycles or waves are sometimes used as models for loads or sea states. For example, if the fourth spectral moment λ_4 is infinite, e.g. if $x(t)$ is a Gaussian process with a Pierson Moskovitz type of spectrum (33), then, $x(t)$, $t \in [0,T]$, has infinitely many local extremes (cycles). In the case of Morison's force in Example 4, if the $y(t)$ process has a Pierson's Moskowitz spectrum (33), then the upcrossing intensity $\mu(u) = \infty$, see (46). See also Example 5.

In order to explicitly construct an oscillation filter for a general set UV one can use the following property

$$N^{osc}_{(u,v)}(x) = \begin{pmatrix} \text{number of rainflow cycles such that} \\ M > v \text{ and } m^{rfc} < u \end{pmatrix} \tag{52}$$

proved in Rychlik (1993). Consequently, for any fixed (u,v), $N^{osc}(u,v)$ will be unchanged if all pairs of extremes (M_i, m_i^{rfc}), such that $u \le m_i^{rfc} < M_i \le v$, are removed from $TP(x)$. In order to provide a working algorithm for this approach one needs a more elaborate definition of rainflow cycles than given in Definiton 2. The following definition, which consider all possible anomalies in x, is given for completeness.

Definition 6 *Let $x(t)$, $t \in [0,T]$, be a continuous function. For any fixed $t \in [0,T]$ let t^-, t^+ be the following time points*

$$t^- = \begin{cases} \sup\{s \in [0,t) : x(s) > x(t)\}, \\ 0 \quad \text{if} \quad x(s) \le x(t) \text{ for all } s \in [0,t), \end{cases}$$

$$t^+ = \begin{cases} \inf\{s \in (t,T) : x(s) \ge x(t)\}, \\ T \quad \text{if} \quad x(s) < x(t) \text{ for all } s \in (t,T) \end{cases} \tag{53}$$

and let x_t^-, x_t^+ be the lowest points in the intervals $[t^-, t]$, $[t, t^+]$, respectively, i.e.

$$x_t^- = \inf\{x(s) : t^- \le s \le t\},$$
$$x_t^+ = \inf\{x(s) : t \le s \le t^+\}, \tag{54}$$

then the time for rainflow minimum $s^{rfc}(t)$ is given by

$$s^{rfc}(t) = \begin{cases} \sup\{s \in [t^-, t] : x(s) = x_t^-\} & \text{if } x_t^+ \le x_t^- \text{ or } t = T \\ \inf\{s \in [t, t^+] : x(s) = x_t^+\} & \text{otherwise.} \end{cases} \tag{55}$$

A function $y(t) = x(s^{rfc}(t)), t \in [0,T])$, will be called a rainflow minima function of x. and a pair $(y(t), x(t))$ such that $y(t) < x(t)$ will likewise be called a rainflow cycle originating at t, with top $x(t)$, bottom $y(t)$.

Example 5: Continuous process $x(t)$ with infinite intensity of waves. A zero mean stationary Gaussian process with spectral density

$$S(\omega) = \frac{1}{2\pi} \frac{\sigma^2}{(\omega)^2 + \beta^2}$$

is called an Ornstein-Uhlenbeck diffusion. By choosing $\sigma = \sqrt{2\beta}$ one has that $V[x(0)] = 1$. However, the second spectral moment is infinite ($\lambda_2 = \infty$), and hence the number of zerocrossings by x in $[0,T]$ is infinite. However, the intensity $\mu^{osc}(0,v)$ exists and is given in an explicit form

$$\mu^{osc}(0,v) = \frac{2\beta}{\sqrt{2\pi} \int_0^v \exp(\frac{z^2}{2})\, dz},$$

see Rychlik (1996c) and Scheutzow (1994) for more details. For the process with finite intensity of waves, the density of crest height is given by $f_{M^*}(v) = -\frac{d}{dv}\mu^{osc}(0,v)/\mu(0)$, i.e. $\mu(0)$ is a normalizing constant. For the Ornstein-Uhlenbeck process the derivative $-\frac{d}{dv}\mu^{osc}(0,v)$ has a singularity at zero and cannot be renormalized to a probability density function. However, one can define the conditional probability of a crest given that it is higher than some threshold v_0, say, by

$$P(M^* > v | M^* > v_0) = \frac{\int_0^{v_0} \exp(\frac{z^2}{2})\, dz}{\int_0^v \exp(\frac{z^2}{2})\, dz}, \quad v > v_0. \tag{56}$$

Finally, one can show that there is a constant v_1, $v_1 \ge v_0$, such that

$$P(M^* > v | M^* > v_0) \le P(R > v), \quad \text{for all} \quad v \ge v_1,$$

where R is a standard Rayleigh variable. This is a similar bound to (24) which was shown for Gausian processes with finite second spectral moment.

ACKNOWLEDGEMENT

This work was supported in part by the Office of Naval Research under Grant N00014-93-1-0841.

References

Brokate M. and Sprekels J. (1996) *Hysteresis and Phase Transitions*, Springer Verlag, New York.

Leadbetter, M. R., Lindgren, G. and Rootzén, H. (1983): *Extremes and Related Properties of Random Sequences and Processes.* Springer Verlag, New York.

Lindgren, G. and Rychlik, I. (1991): Slepian Models and Regression Approximations in Crossings and Extreme Value Theory. *International Statistical Review*, **59**, pp. 195-225.

Matsuishi, M. and Endo, T. (1968) *Fatigue of metals subject to varying stress.* Paper presented to Japan Soc. Mech. Engrs, Jukvoka, Japan.

A. Naess (1983). Prediction of extremes of Morison-type loading - an example of a general method. *Ocean Engng.*, **10**, pp. 313-324.

Ochi, M.K. and Ahn, K. (1994): Probability distribution applicable to non-Gaussian random processes. *Probabilistic Engineering Mechanics*, **9**, 255-264.

Rice, S. (1944). The mathematical analysis of random noise. *Bell Syst. Techn. J.*, **23**, pp. 282-332.

Rice, S. (1945). The mathematical analysis of random noise. *Bell Syst. Techn. J.*, **24**, pp. 46-156.

Rychlik I. (1987) A new definition of the rainflow cycle counting method, *Int. J. Fatigue*, **9**, pp. 119-121.

Rychlik I. (1993): Note on Cycle Counts in Irregular Loads. *Fatigue Fract. Eng. Mater. and Struc.*, **16**, pp. 377-390.

Rychlik I. (1996a) Fatigue and stochastic loads, *Scandinavian Journal of Statistics*, **23**, pp. 387-404.

Rychlik I. (1996b) A note on significant wave height, *Ocean Engineering*, **23** pp. 447-454.

Rychlik I. (1996c) Extremes, rainflow cycles and damage functionals in continuous random processes, *Stochastic Process. Appl.*, **63**, pp. 97-116.

Rychlik I. (1996d) Simulation of load sequences from Rainflow matrices: Markov method, *Int. J. Fatigue*, **18** , pp. 429-438.

Rychlik, I. and Lindgren, G. (1993): CROSSREG - A Computer Package for Extreme Value and Wave Analysis, with Reliability Applications. *Probability in the Engineering and Informational Sciences*, **7**, pp. 125-148.

Rychlik, I. and Lindgren, G. (1995): *WAVE analysis toolbox - a tutorial.* Dept of Mathematical Statistics, Lund University. ISSN 0281-1994.

Rychlik, I., Johannesson P. and Leadbetter, M. R. (1997) Modelling and Statistical Analysis of Ocean-Wave Data Using Transformed Gaussian Processes, *Marine Structures*, **10**, 13-47.

Scheutzow (1994) A law of large numbers for upcrossing measures, *Stochastic Process. Appl.*, **53**, pp. 285-305.

Proceedings of the Seventh (1997) International Offshore and Polar Engineering Conference
Honolulu, USA, May 25-30, 1997

ISBN 1-880653-28-1 (Set); ISBN 1-880653-31-1 (Vol. III)

Long Waves Accompanied by the Reflection of Short-Wave Groups

Wataru Kioka, Satoru Yamane and Kenji Kashihara
Nagoya Institute of Technology
Nagoya, Japan

Shin-ichi Aoki
Toyohashi University of Technology
Toyohashi, Japan

ABSTRACT

Behavior of long waves associated with the reflection of short-wave groups due to an impermeable wall is numerically investigated using the Boussinesq equations. The sea bottom is assumed to be horizontal and of variable depth. Two sinusoidal waves of slightly different frequency and their sub-harmonic component are given in various water depths at the incident boundary. The numerical results indicate that the temporally long wave components of the bottom velocity, bottom pressure and surface elevation differ from the spatial ones in their amplitudes. The wave number-frequency spectra are also calculated to examine nonlinear interaction terms between the incident and reflected short-wave components. It will be shown that the temporal components become smaller than the spatial ones due to the nonlinear interaction terms with the increase of relative water depth.

KEY WORDS: Long-period wave, Wave reflection, Boussinesq equation, Nonlinear interaction

INTRODUCTION

Progressive short waves with a narrow frequency band are known to be accompanied by long set-down waves traveling with the groups. The reflection of wave groups by coastal and offshore structures may generate the long waves locked to the envelopes of the incident and reflected short waves. The long waves induced by the reflection of short-wave groups are, however, not fully understood, although these are closely connected with the problems of wave run-up and slow-drift oscillation of a moored body near reflective boundaries such as seawalls and breakwaters. In this study, the behavior of long waves associated with the reflection of short-wave groups due to an impermeable wall is numerically investigated using the Boussinesq equations. The temporal and spatial long-wave components are computed for the surface elevation, the bottom velocity and the bottom pressure. The standard Boussinesq equations in terms of the depth-averaged velocity is used as a numerical model, which can incorporate the fully reflective boundary condition in the numerical scheme with accuracy.

LONG WAVES IN TERMS OF THE RADIATION STRESS

Long-period components accompanied with the reflection of short-wave groups can be expressed by applying multiple-scale perturbation method up to the leading order as follows.

The envelope wave profile of group waves propagating in the positive x-direction over constant water depth h may be defined as $A(x_1 - C_g t_1)$ with C_g being the group velocity of short-waves, and (x_1, t_1) the slowly varying independent variables. The envelope wave of reflected group waves may be then expressed as $K_R A(x_1 + C_g t_1)$ with the reflection coefficient K_R when phase interference influences can be disregarded. The long-period waves $\tilde{\eta}$ for the second-order (x_1, t_1) are derived by the multiple-scale perturbation method (e.g., Kioka et al., 1993):

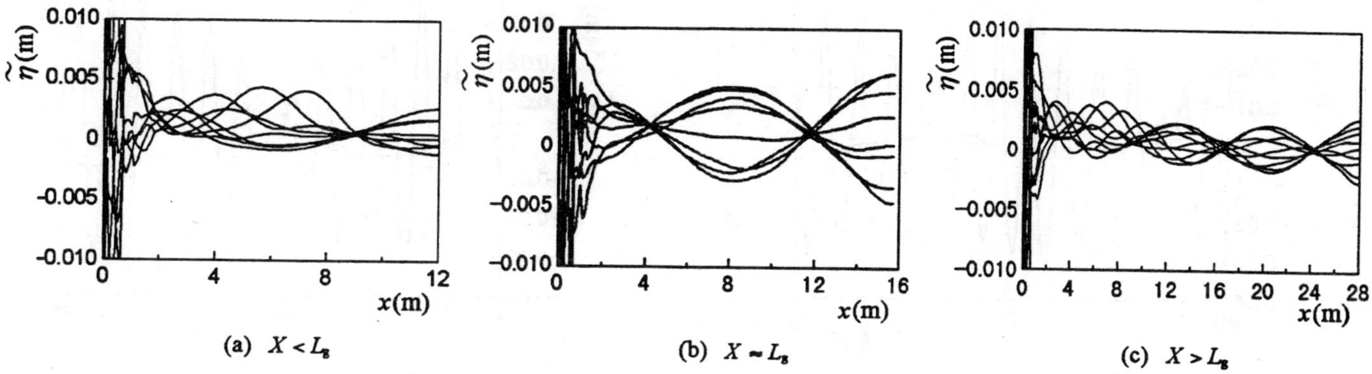

Fig. 1 Motion of the spatially long waves for different lengths of wave channel

$$\tilde{\eta} = -\frac{1}{\rho}\frac{1}{gh - C_g^2}\left[S^I(x_1 - C_g t_1) + S^R(x_1 + C_g t_1)\right] \tag{1}$$

where S^I and S^R are the radiation stresses in the direction of incident and reflected waves respectively as in the forms

$$S^I = \rho g\left(\frac{C_g}{C} - \frac{1}{4}\right)A^2(x_1 - C_g t_1) \tag{2}$$

$$S^R = \rho g\left(\frac{C_g}{C} - \frac{1}{4}\right)K_R^{\,2}A^2(x_1 + C_g t_1) \tag{3}$$

with C being the short-wave celerity. The above equations indicate that long-period wave $\tilde{\eta}$ can be described simply as the superposition of the incident and reflected set-down waves. The reflection coefficient of the long-period wave is given as $K_R^{\,2}$. If we consider the special case where the incident wave is sinusoidally modulated with the maximum amplitude A_0, i.e., $A = A_0\cos(Kx_1 - \Omega t_1)$, and where K_R equals 1 , Eqs. (1)-(3) reduce to

$$\tilde{\eta} = -\frac{gA_0^{\,2}}{gh - C_g^2}\left(\frac{C_g}{C} - \frac{1}{4}\right)\cos 2Kx_1 \cos 2\Omega t_1 \tag{4}$$

where K and Ω denote respectively the wave number and the angular frequency of the wave envelope.

NUMERICAL MODEL

The standard Boussinesq equations in terms of the water surface elevation η and the depth-averaged horizontal velocity u (Peregrine, 1967) are used as the numerical model :

$$\frac{\partial \eta}{\partial t} + \frac{\partial}{\partial x}[(h+\eta)u] = 0 \tag{5}$$

$$\frac{\partial u}{\partial t} + u\frac{\partial u}{\partial x} + g\frac{\partial \eta}{\partial x} = \frac{h}{2}\frac{\partial^3(hu)}{\partial x^2 \partial t} - \frac{h^2}{6}\frac{\partial^3 u}{\partial x^2 \partial t} \tag{6}$$

On the upright wall, the reflection coefficient K_R is assumed to be 1 . The fully reflective boundary condition is satisfied by imposing

$$u = 0, \quad \frac{\partial \eta}{\partial x} = 0 \qquad \text{(on wall)} \tag{7}$$

Using the velocity at an arbitrary depth as the representative velocity instead of the depth-averaged one, the dispersion properties of Boussinesq equations can be improved for intermediate water depth (see Nwogu,1993). On the reflective boundary, however, the simulation with the alternative velocity requires some iterative calculations because the reflective boundary condition cannot be specified by Eq. (7). For the sake of simplicity of numerical simulations, the depth-averaged velocity is adopted here for the model equations.

At the incident boundary, time series of the water surface elevation are given using two wave components with amplitudes a_1 and a_2, wave numbers k_1 and k_2 and angular frequencies ω_1 and ω_2 :

$$\eta = a_1\cos(k_1 x - \omega_1 t) + a_2\cos(k_2 x - \omega_2 t) \tag{8}$$

The radiation condition at the incident boundary is satisfied by applying the linear Boussinesq equations to absorb the reflected waves. The free long-period waves of the phase lagged from the set-down waves in 180° inevitably occur, because the radiation condition does not exactly satisfy the Boussinesq equations. To avoid the generation of spurious free waves, the sub-harmonic components η_- through the quadratic transfer functions $D_\pm$ based on the Boussinesq equations are added to the short-wave components expressed by Eq. (8) :

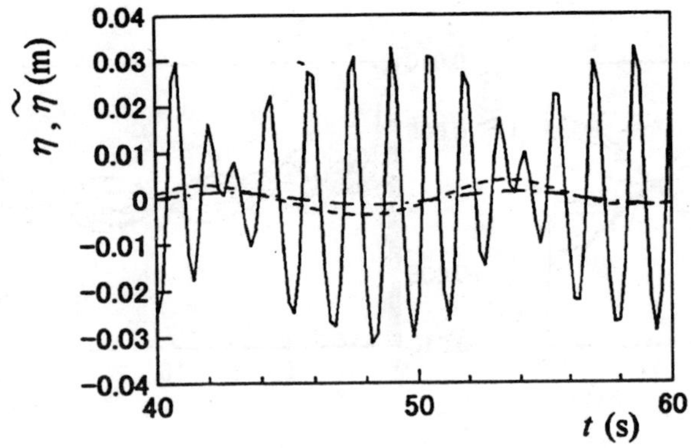

(a) Temporal profiles

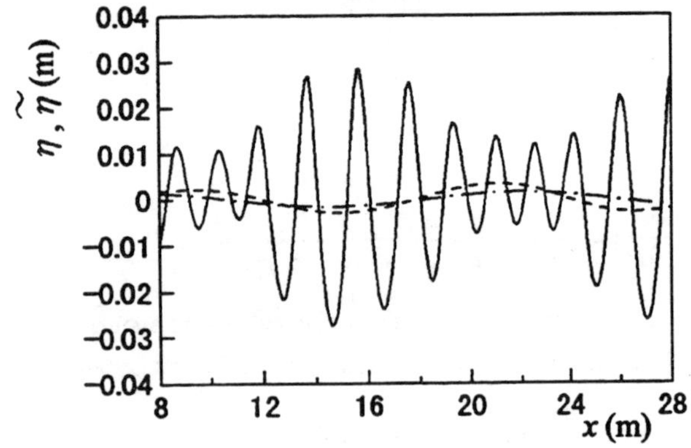

(b) Spatial profiles

Fig. 2 Comparison of the surface elevations ($h = 0.2$ m) (- - - - - - : long waves filtered from the computed wave profiles ; — · — · : long waves predicted from Eq. (4))

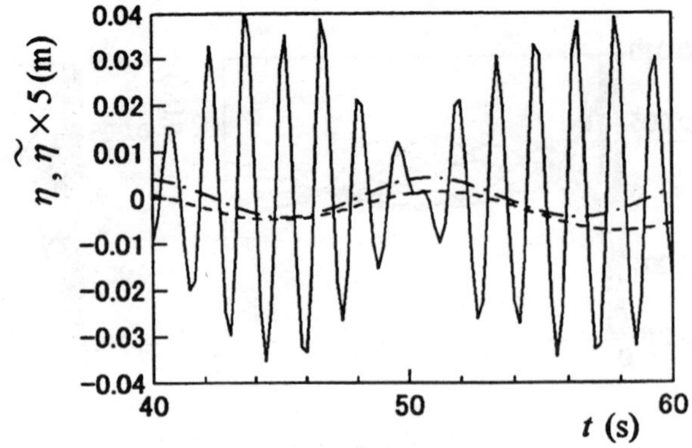

(a) Temporal profiles

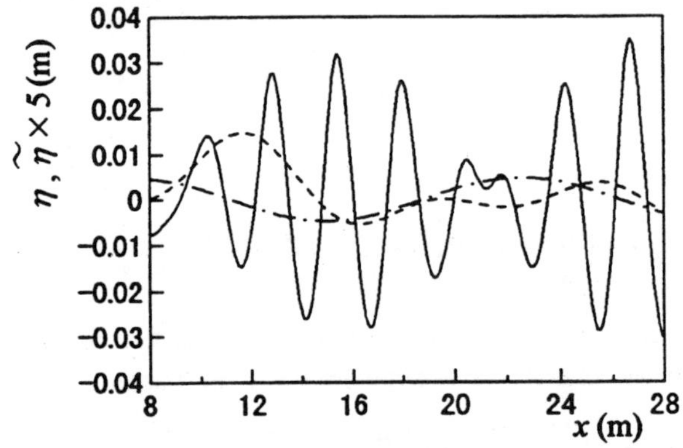

(b) Spatial profiles

Fig. 3 Comparison of the surface elevations ($h = 0.4$ m) (- - - - - - : long waves filtered from the computed wave profiles ; — · — · : long waves predicted from Eq. (4))

$$\eta_- = a_1 a_2 D_- \cos(k_- x - \omega_- t) \qquad (9)$$

where $k_\pm = k_1 \pm k_2$, $\omega_\pm = \omega_1 \pm \omega_2$, and D_- is the negative part of the quadratic transfer function

$$D_\pm(\omega_1, \omega_2) = \frac{\omega_\pm k_\pm h\left(1 + \frac{1}{3}(k_\pm h)^2\right)(\omega_1 k_2 h + \omega_2 k_1 h) + \omega_1 \omega_2 (k_\pm h)^2}{2 k_1 k_2 h^3 \left[\omega_\pm^2\left(1 + \frac{1}{3}(k_\pm h)^2\right) - g k_\pm^2 h\right]} \qquad (10)$$

The finite difference method is used to solve the equations with 5 grid-point centered differences for space derivatives and forward differences in time. For all cases, the computations are carried out using the spatial grid size $\Delta x = 0.1$ m, and the time increment $\Delta t = 0.02$ s.

COMPUTED RESULTS IN CONSTANT WATER DEPTHS

The sea bottom is assumed to be horizontal. Two wave components of amplitudes $a_1 = a_2 = 2$ cm and periods $T_1 = 1.4$ s and $T_2 = 1.6$ s are given at the incident boundary in constant water depths $h = 20$, 30, 40 and 50 cm.

Appropriate Lengths of Wave Channel

Before proceeding the numerical simulations, the effective lengths of the numerical wave flume required to produce proper characteristics of the standing long waves are examined through the comparisons of the computed surface elevations for various flume

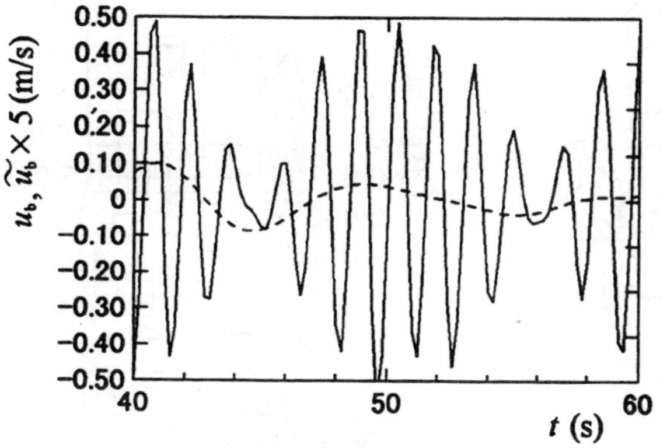

(a) Temporal profiles

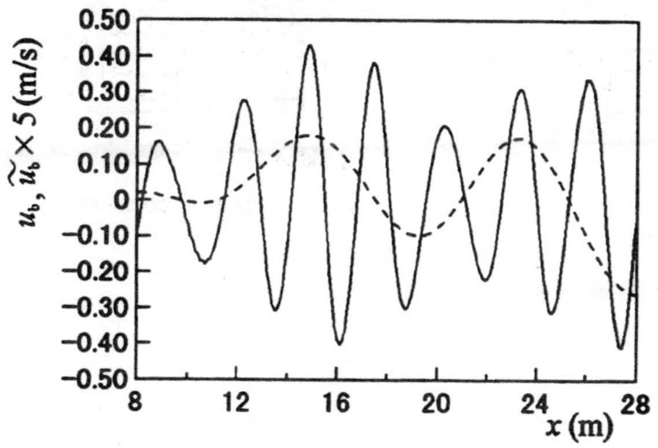

(b) Spatial profiles

Fig. 4 Comparison of the bottom velocities ($h = 0.4$ m)
(· · · · · : filtered long wave components)

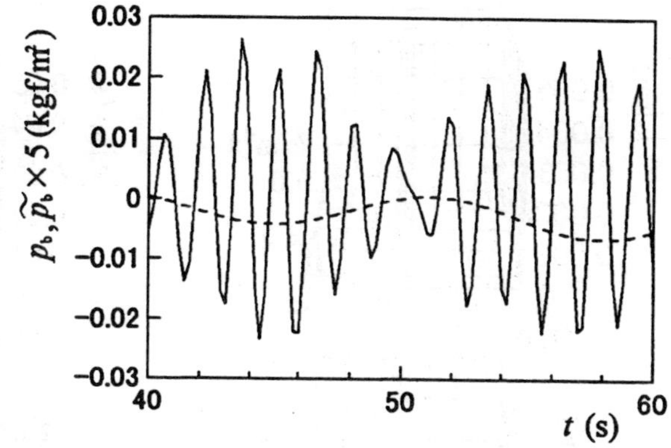

(a) Temporal profiles

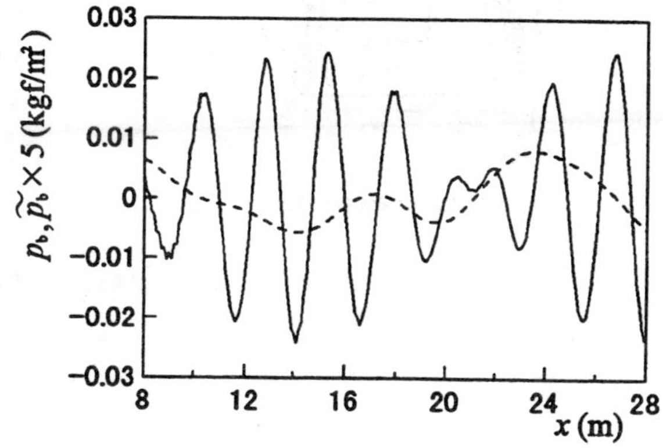

(b) Spatial profiles

Fig. 5 Comparison of the bottom pressures ($h = 0.4$ m)
(· · · · · : filtered long wave components)

lengths X.

Fig. 1 shows the behaviors of spatially long-period waves for the different channel lengths at the water depth of $h = 0.2$ m. The wave length L_g obtained from the group period T_g and the group velocity C_g of the incident waves is 15.7 m. In the case of the channel length $X < L_g$ (Fig. 1(a)), the phase interference effects are observed for the generated long-period waves. When $X = L_g$, the resonance between the incident and reflective boundary occurs as shown in Fig. 1(b). The reflective waves are not completely absorbed at the incident boundary for the numerical results because the radiation condition for the reflective waves is imposed according to the linear theory. Therefore, the wave profiles of long-period components near $X = 0$ are disturbed in all cases. Fig. 3(c) indicates that the generated standing waves are stable when the channel lengths X are longer than the long-period wave length L_g.

That is, in the case of $h = 0.2$ m, the channel length must be at least about 20 m.

Comparison between Time and Spatial Profiles

Figs. 2 and 3 show the comparisons between the temporal surface elevations at the wall boundary and the spatial surface profiles over the channel in constant depths $h = 0.2$ m and 0.4 m. The spatial profiles of long wave components $\widetilde{\eta}$ are obtained through a low-pass filter at the phase of maximum wave amplitudes. The profiles of long-period components are exaggerated by multiplying by 5 for the cases of $h = 0.4$ m. It can be seen from Fig. 2 that the amplitudes of long-period waves are nearly equivalent for time series and spatial profiles, and that both long-period waves almost correspond to the solution given in Eq. (4). Fig. 3 indicates, however, that the time series of the

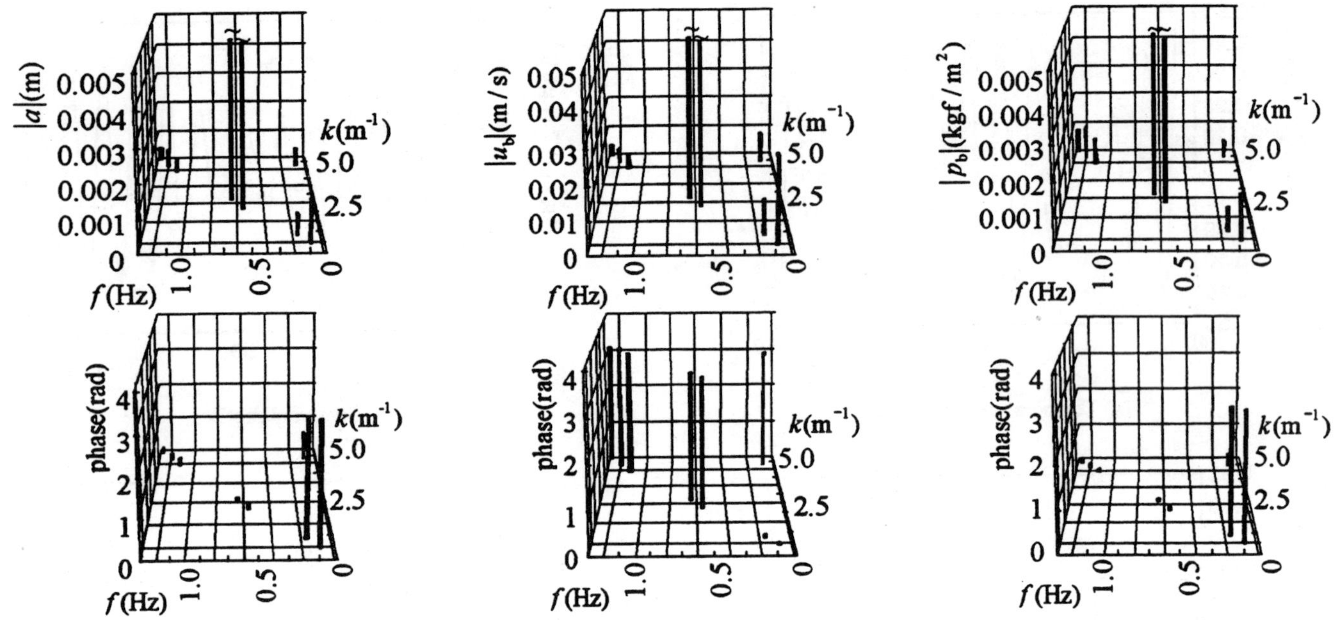

(a) Surface elevation (b) Bottom velocity (c) Bottom pressure

Fig. 6 Wave number-frequency spectra

long-period waves are small as compared with the spatial profiles. In this deeper water depth, the differences between the time series and the spatial profiles cannot be predicted from Eq. (4).

For the long-period components of the bottom velocities $\tilde{u}_b$ at the water depth $h = 0.2$ m, the temporal profiles nearly coincide with the spatial ones at the phases of a maximum amplitude. The temporal profiles become smaller than the spatial ones in the deeper water depth ($h = 0.4$ m) as shown in Fig. 4. The similar tendency is observed for the bottom pressures p_b as shown in Fig. 5.

Wave Number-Frequency Spectra

As shown in Figs. 3 to 5, in the case of relatively deep depth ($h = 0.4$ m), there exists a significant discrepancy between the temporal and spatial profiles. It may be instructive for explaining there differences to investigate the properties of nonlinear interaction between incident and reflective components. The wave number-frequency spectra (Hosoi and Ishida, 1970) are calculated from the temporal and spatial wave profiles through complex Fourier transform. The quadratic interaction terms associated with the long waves, i.e., the amplitudes of the spectra and phase of the components with $k_+ = k_1 + k_2$, $f_- = f_1 - f_2$ and $k_- = k_1 - k_2$, $f_+ = f_1 + f_2$ are investigated from the wave number-frequency spectra.

The results for the surface elevations η, the bottom velocities u_b, and the bottom pressures p_b are respectively shown in Fig. 6 (a)-(c). The long waves (f_-, k_+) with the frequency $f_- = 0.089$ Hz but with the spatially short-wave number $k_+ = 4.58$ m^{-1} are observed for all wave profiles of the surface elevations, the bottom velocities and the bottom pressures. These components are lagged by 180° from the phase of the long-period component (f_-, k_-). Thus the components (f_-, k_+) reduce the amplitudes of the temporal profiles. The components (f_+, k_-), differences between the incident component (f_1, k_1) and the reflective one ($-f_2, k_2$), are very small, and do not contribute to the spatially long wave profiles.

The differences between the temporal and spatial profiles of the long waves appear to be greater with the increase of the water depth. From the wave number-frequency spectra, the interaction terms (f_-, k_+) are found to have large effects on the temporally long waves. The interaction components (f_-, k_+) and (f_+, k_-) are further calculated through the quadratic transfer function given in Eq. (10). The results are plotted in Fig. 7 as a function of the relative water depth $k_0 h$ (k_0 : wave number of deepwater short wave). The frequency difference $\omega_- = \omega_1 - \omega_2$ is assumed to be $\omega_- = 0.1(\omega_1 + \omega_2)/2$. It is seen that the frequency difference $D_-(\omega_1, -\omega_2)$ between the incident component ω_1 and the reflected one $-\omega_2$ is always small, while the interaction terms $D_+(\omega_1, -\omega_2)$

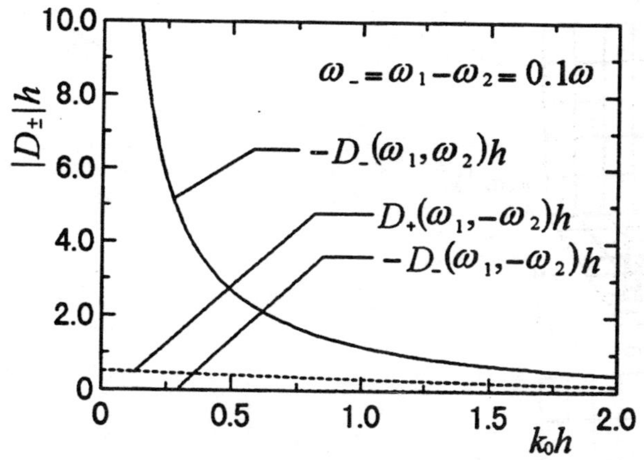

Fig. 7 Quadratic transfer function for the sub-harmonic components

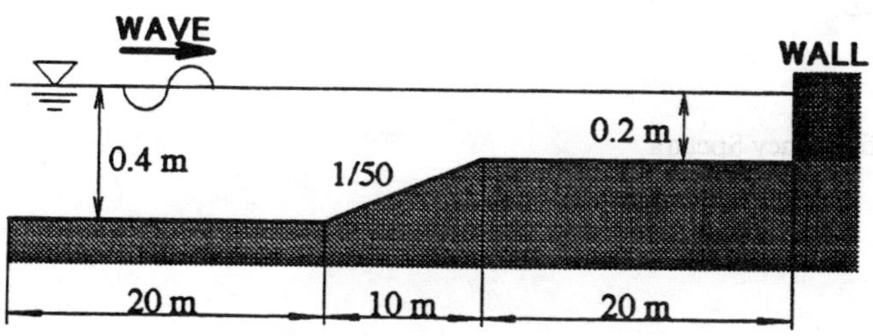

Fig. 8 Numerical setup

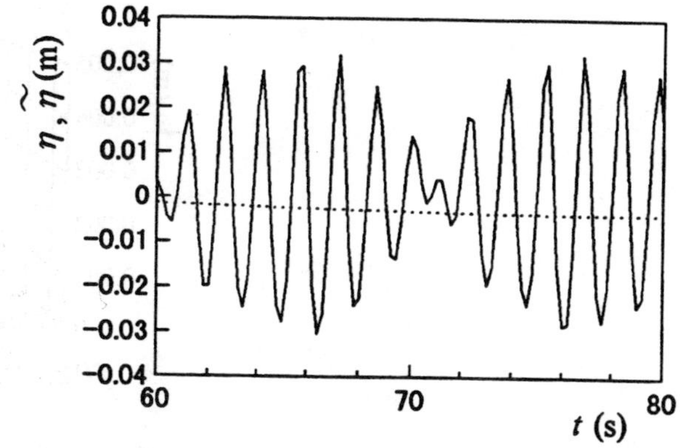

(a) Temporal profiles

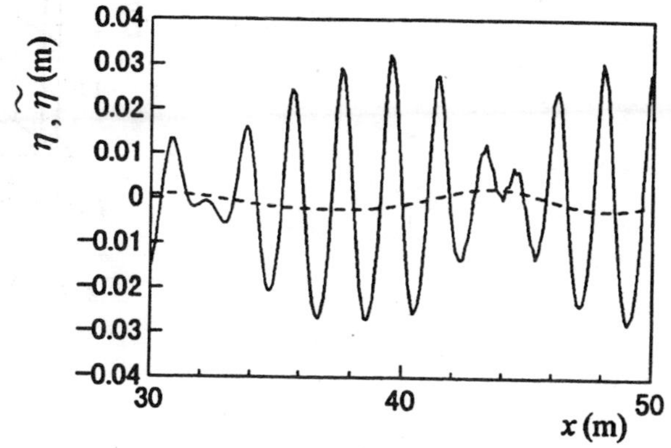

(b) Spatial profiles

Fig. 9 Comparison of the surface elevations ($h = 0.2$ m)
(· · · · · : filtered long wave components)

from the sum frequency cannot be disregarded from the incident set-down wave $D_-(\omega_1, \omega_2)$ for the relatively deep waters. The results shown in Fig. 7 are not exactly equal to the numerical solutions of the Boussinesq equations (3) and (4), as the quadratic transfer function given by Eq. (10) is an approximate solution of the Boussinesq equations. The difference in the amplitude between the temporal and spatial profiles is expected about 20% for $k_0h = 1.0$, according to Fig. 7. The calculations using the quadratic transfer function based on Stokes theory (e.g., Hamada, 1964) lead to about 80% difference for $k_0h = 1.0$.

COMPUTED RESULTS IN VARIABLE DEPTHS

The properties of long waves in variable depth are expected to differ from the previous results for constant depths due to the generation of free waves (Kioka and Ishihara, 1993). A slope of 1/50 is placed 20 m away from the vertical wall. The numerical layout is shown in Fig. 8. The long-period surface elevations $\tilde{\eta}$ at the wall boundary appear to be again smaller than the spatial profiles, as shown in Fig. 9. The amplitudes of the spatial and temporal long wave profiles are, however, smaller than those for constant water depth of $h = 0.2$ m (Fig. 2).

The wave number-frequency spectra over the water depth $h = 0.2$ m are shown in Fig. 10. The long-period waves with approximately 1.2 times faster than the speed of the set-down wave components (f_-, k_-) are detected, which indicates that the free waves unlocked to the wave groups are generated on the slope. The long-period free waves behave to reduce the amplitudes of the long-period waves in the present case.

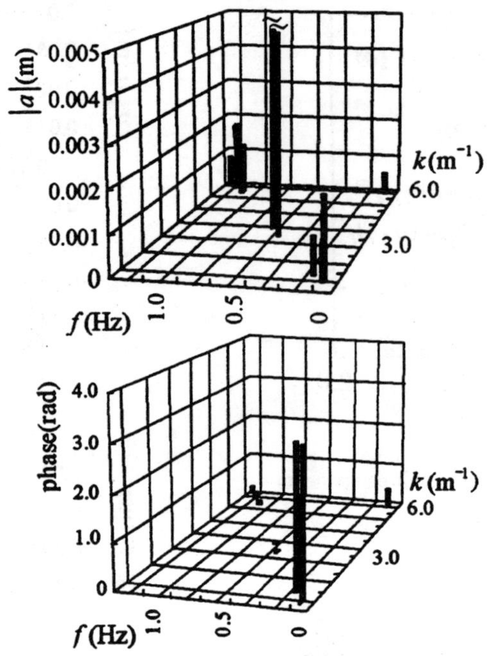

Fig. 10 Wave number-frequency Spectra

CONCLUSIONS

Behavior of long waves associated with the reflection of short-wave groups due to the upright wall is investigated using the Boussinesq equations. The following conclusions can be drawn :

1) In order to generate standing waves of long-period waves, the length of the wave channel must be longer than the wave length of incident long-period waves.
2) When the water depths are relatively small, the long-period waves, both spatial and temporal waves, can be described as the superposition of the incident and reflected set-down waves, ignoring the nonlinear interaction terms between the incident and reflected short waves.
3) As the relative water depth increases, the nonlinear interaction terms cannot be no longer ignored. The temporal long-wave components of surface elevation, bottom velocity and bottom pressure become smaller than the spatial ones due to the nonlinear interaction.
4) When a slope is constructed in front of the vertical wall, the free long waves occur at the slope, so that the long waves in front of the wall cannot be described as the superposition of the incident and reflected set-down waves.

REFERENCES

Hamada, T. (1964), "Secondary interactions of water surface waves," Proc. Coastal Eng., JSCE, Vol.11, pp.12-18 (in Japanese).

Hosoi, M. and Ishida, A. (1970), "The study of nonlinearity in the secondary wave phenomenon," Proc. Coastal Eng., JSCE, Vol.17, pp.47-52 (in Japanese).

Kioka, W. and Ishihara, K. (1993), "Second-order long waves induced by the refraction of short-wave groups," J. Hydraulic, Coastal and Environmental Eng., No.467, pp.75-82 (in Japanese).

Kioka, W., Kashihara, K. and Iwagaki, Y. (1993), "Resonance of long waves in a harbour by groups of short waves," J. Hydraulic, Coastal and Environmental Eng., No.473, pp.55-64 (in Japanese).

Nwogu, O. K. (1993), "Alternative form of Boussinesq equations for nearshore wave propagation," J. Waterway, Port, Coastal and Ocean Eng., Vol.119, No.6, pp.618-638.

Peregrine, D. H. (1967), "Long waves on a beach," J. Fluid Mech., Vol.27, pp.815-826.

Proceedings of the Seventh (1997) International Offshore and Polar Engineering Conference
Honolulu, USA, May 25-30, 1997

ISBN 1-880653-28-1 (Set); ISBN 1-880653-31-1 (Vol. III)

Performance of Active Wave Absorption Systems: Comparison of Wave Gauge and Velocity Meter Based Systems

Tue Hald and Peter Frigaard
Aalborg University
Aalborg, Denmark

Abstract

A comparison of wave gauge based and velocity meter based active absorption systems is presented discussing advantages and disadvantages of the systems. In detail one system based on two surface elevations, one system based on a surface elevation and a horizontal velocity and one system based on a horizontal and a vertical velocity are treated. All three systems are based on digital FIR-filters. For numerical comparison a performance function combining the frequency response of the set of filters for each system is derived enabling discussion on optimal filter design and system setup.

Irregular wave tests with a highly reflective structure with the purely wave gauge based system and the wave gauge velocity meter based system are performed. The wave tests depict the differences between the systems.

Key words: Active wave absorption, frequency response, digital filters, filter design, system performance, numerical and physical model tests.

Introduction

Coastal engineering problems are often solved by means of physical and numerical models. Physical and numerical modelling of coastal engineering phenomena require the capability of reproducing natural conditions. One of the problems associated with the modelling of waves in both numerical and laboratory wave flumes is the presence of rereflected waves altering the characteristics of the wave train incident to the model structure repeatedly.

Consequently, an effective absorption of the waves propagating towards the wave generator is necessary. By making the wave generator simultaneously generating the incident wave field and absorbing the reflected wave field the problems caused by rereflection can be reduced significantly.

The principle in constructing a combined wave generator and active wave absorber requires, Gilbert (1978):

1. A means of detecting reflected waves approaching the wave paddle.
2. A means of making the paddle generate waves that are, in effect, equal and opposite to the reflected waves so that the reflected waves are cancelled out reaching the paddle.

Milgram (1970) presented a system in which waves in a channel were absorbed by means of a moving termination at the end of the channel. The motion of the termination needed for absorption was determined by analog filtering of a surface elevation signal measured in front of the termination. This active wave absorption system was not used in a combined generation and absorption mode.

The common approach to detect the reflected wave when performing simultaneous wave generation and active wave absorption is by measuring the surface elevation with a wave gauge positioned on the face of the paddle. The reflected wave signal is then determined as the difference between the measured surface elevation signal and the generated surface elevation signal estimated from the wave generator control signal. In consequence, the actual reflected wave is not being estimated with the possibility of accumulating errors. Several systems based on this approach have been presented in the litterature, e.g. Bullock and Murton (1989), Hirakuchi et al. (1990) and Schäffer et al. (1994).

Recently, an active wave absorption system based on real time separation of the wave field in front of the paddle in incident and reflected waves has been installed in a wave flume at Hydraulics & Coastal Engineering Laboratory, Aalborg University (Frigaard (1993)). This open loop system is entirely different from the above described and does not accumulate errors because the actual reflected wave is determined continuously. The system shows excellent absorption characteristics, Frigaard and Christensen (1994).

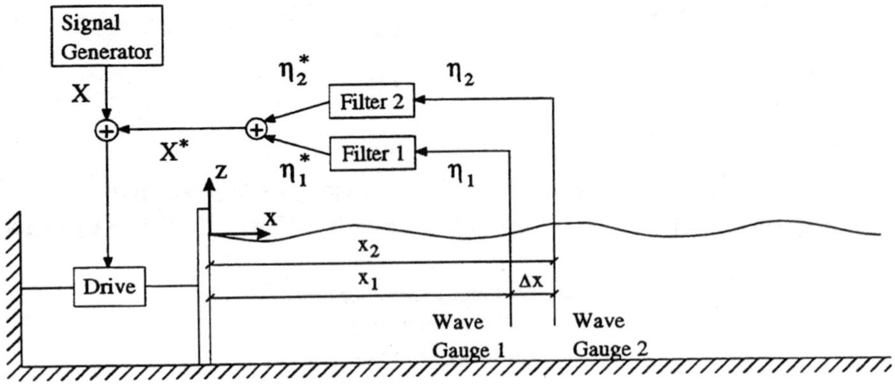

Fig. 1 Principle of active absorption system.

In the following three absorption systems based on the theoretical considerations by Frigaard and Brorsen (1995) are described: one based on two surface elevation signals, one based on a surface elevation and a horizontal velocity signal and one based on a horizontal and a vertical velocity signal. The systems are termed (η, η), (η, u) and (u, w). Based on both numerical and physical tests the systems are compared discussing the differences between the three systems as well as optimal filter design.

Principle of Active Absorption System

The active absorption system is operated by means of on-line signals from digital FIR-filters. In essence, a digital filter relates N input η^{k-i} with an output x^k by convolving the input data with N precalculated time domain terms h^i called filter coefficients. The input/output relation of the FIR-filter is given by the discrete convolution integral

$$x^k = \sum_{i=-M}^{i=M} h^i \eta^{k-i} \quad , \quad M = \frac{N-1}{2} \tag{1}$$

with the filter coefficients determining the impulse response of the filter. Given a desired frequency response the corresponding filter impulse response is designed by computing the inverse Fourier transform of the complex frequency response function. Notice that the filter output is delayed $\frac{N-1}{2}$ time steps relative to the input. For systems operating in real time this time delay must be removed.

Calculation of the paddle displacement correction signal needed for absorption of the reflected waves is done by digital filtering and subsequent superposition of two surface elevation or velocity signals measured in front of the wave generator, see principle in fig. 1.

When active absorption is applied the paddle displacement correction signal is addedto the input paddle displacement signal read from the signal generator causing the wave generator to operate in a combined generation/absorption mode.

Having outlined the principle of the system the only remaining problem is to specify the frequency response of the applied filters.

Frequency Response of Digital Filters

Offspring is taken in the system illustrated in fig. 1 where the wave flume is equipped with two wave gauges.

The surface elevation signal at a position x may be regarded as a sum of harmonic components since the effect of the near field local disturbances disapear when $x > 3d$ (Biesel (1951)). Considering an isolated component of frequency f the surface elevation arising from this component may be written as the sum of the corresponding incident and reflected wave components

$$\begin{aligned} \eta(x,t) &= \eta_I(x,t) + \eta_R(x,t) \\ &= a_I \cos(\omega t - kx + \phi_I) + \\ &\quad a_R \cos(\omega t + kx + \phi_R) \end{aligned} \tag{2}$$

where

f	:	frequency
$a = a(f)$	:	wave amlitude
$k = k(f)$	:	wave number
$\phi = \phi(f)$	:	phase

and indices I and R denote incident and reflected, respectively.

Provided a linear relation exists between a given paddle displacement signal and its corresponding surface elevation signal the paddle displacement correction signal $X^*(t)$ which cancels out the reflected component without disturbing the incident component is given by

$$X^*(t) = B a_R \cos(\omega t + \phi_R + \phi_B + \pi) \tag{3}$$

where

B	:	piston stroke/wave height relation
ϕ_B	:	phase shift between paddle displacement and surface elevation on the face of the paddle

In the following it is shown that it is possible to amplify and phase shift the surface elevation signals from the two wave gauges in such a way that their sum is identical to the paddle correction signal corresponding to absorption of the reflected component as given by eq. (3).

At the two wave gauges (fig. 1) we have:

$$\eta(x_1,t) = a_I\cos(\omega t - kx_1 + \phi_I) + a_R\cos(\omega t + kx_1 + \phi_R) \quad (4)$$

$$\begin{aligned}\eta(x_2,t) &= a_I\cos(\omega t - kx_2 + \phi_I) + a_R\cos(\omega t + kx_2 + \phi_R) \\ &= a_I\cos(\omega t - kx_1 - k\Delta x + \phi_I) + a_R\cos(\omega t + kx_1 + k\Delta x + \phi_R) \quad (5)\end{aligned}$$

where $x_2 = x_1 + \Delta x$ has been substituted into eq. (5).

An amplification of G^η and a theoretical phase shift ϕ^η are introduced into the expressions for $\eta(x,t)$. The modified signal is denoted η^*. For the i'th wave gauge signal the modified signal is defined as

$$\eta^*(x_i,t) = G^\eta a_I\cos(\omega t - kx_i + \phi_I + \phi^{\eta_i}) + G^\eta a_R\cos(\omega t + kx_i + \phi_R + \phi^{\eta_i}) \quad (6)$$

This gives at wave gauges 1 and 2

$$\eta^*(x_1,t) = G^\eta a_I\cos(\omega t - kx_1 + \phi_I + \phi^{\eta_1}) + G^\eta a_R\cos(\omega t + kx_1 + \phi_R + \phi^{\eta_1}) \quad (7)$$

$$\eta^*(x_2,t) = G^\eta a_I\cos(\omega t - kx_1 - k\Delta x + \phi_I + \phi^{\eta_2}) + G^\eta a_R\cos(\omega t + kx_1 + k\Delta x + \phi_R + \phi^{\eta_2}) \quad (8)$$

The sum of $\eta^*(x_1,t)$ and $\eta^*(x_2,t)$ which is termed $\eta^{calc}(t)$ is

$$\begin{aligned}\eta^{calc}(t) &= \eta^*(x_1,t) + \eta^*(x_2,t) \\ &= 2G^\eta a_I\cos(\frac{k\Delta x + \phi^{\eta_1} - \phi^{\eta_2}}{2}) \\ &\quad \cos(\omega t - kx_1 + \phi_I + \frac{-k\Delta x + \phi^{\eta_1} + \phi^{\eta_2}}{2}) + \\ &\quad 2G^\eta a_R\cos(\frac{-k\Delta x + \phi^{\eta_1} - \phi^{\eta_2}}{2}) \\ &\quad \cos(\omega t + kx_1 + \phi_R + \frac{k\Delta x + \phi^{\eta_1} + \phi^{\eta_2}}{2}) \quad (9)\end{aligned}$$

It is seen that $\eta^{calc}(t)$ and $X^*(t) = Ba_R\cos(\omega t + \phi_R + \phi_B + \pi)$ are identical signals in case

$$2G^\eta\cos(\frac{k\Delta x - \phi^{\eta_1} + \phi^{\eta_2}}{2}) = B \quad (10)$$

$$kx_1 + \frac{k\Delta x + \phi^{\eta_1} + \phi^{\eta_2}}{2} = \phi_B + \pi + n2\pi \quad (11)$$

$$\frac{k\Delta x + \phi^{\eta_1} - \phi^{\eta_2}}{2} = \frac{\pi}{2} + m\pi \quad (12)$$

where $n, m \in (0, \pm1, \pm2, ..)$.

Solving eqs. (10)-(12) with respect to ϕ^{η_1}, ϕ^{η_2} and G^η with $n = m = 0$ gives

$$\phi^{\eta_1} = \phi_B - k\Delta x - kx_1 + 3\pi/2 \quad (13)$$

$$\phi^{\eta_2} = \phi_B - kx_1 + \pi/2 \quad (14)$$

$$G^\eta = \frac{B}{2\cos(-k\Delta x + \pi/2)} \quad (15)$$

Eqs. (13)-(15) specify the frequency responses, i.e. the amplification factors and phase shifts of filters 1 and 2 in fig 1.

Considering a situation with a spatially co-located wave gauge and velocity meter measuring the horizontal velocity this gives for the two input signals

$$\eta(x_1,t) = a_I\cos(\omega t - kx_1 + \phi_I) + a_R\cos(\omega t + kx_1 + \phi_R) \quad (16)$$

$$u(x_1,z,t) = a_I\omega\frac{\cosh(k(z+d))}{\sinh(kd)}\cos(\omega t - kx_1 + \phi_I) - a_R\omega\frac{\cosh(k(z+d))}{\sinh(kd)}\cos(\omega t + kx_1 + \phi_R) \quad (17)$$

By similar calculations as for the (η,η)-system the amplification G^η of the surface elevation signal, the amplification G^u of the velocity signal and the theoretical phase shift $\phi^\eta = \phi^u$ of both the surface elevation and the velocity signal are found for the (η,u)-system

$$G^\eta = \frac{B}{2} \quad (18)$$

$$G^u = -\frac{B}{2\omega}\frac{\sinh(kd)}{\cosh(k(z+d))} \quad (19)$$

$$\phi^\eta = \phi_B + \pi - kx_1 \quad (20)$$

$$\phi^u = \phi_B + \pi - kx_1 \quad (21)$$

Finally, for spatially co-located horizontal and vertical velocity signals the two input signals are

$$w(x_1,z,t) = -a_I\omega\frac{\sinh(k(z+d))}{\sinh(kd)}\sin(\omega t - kx_1 + \phi_I) - a_R\omega\frac{\sinh(k(z+d))}{\sinh(kd)}\sin(\omega t + kx_1 + \phi_R) \quad (22)$$

$$u(x_1,z,t) = a_I\omega\frac{\cosh(k(z+d))}{\sinh(kd)}\cos(\omega t - kx_1 + \phi_I) - a_R\omega\frac{\cosh(k(z+d))}{\sinh(kd)}\cos(\omega t + kx_1 + \phi_R) \quad (23)$$

The (u,w)-system is very similar to the (η,u)-system since an amplification of the vertical velocity signal by $-\frac{\sinh(kd)}{\omega\sinh(k(z+d))}$ and a phase shift of $\frac{\pi}{2}$ simply give the surface elevation (c.f. eq. (22)). Considering the frequency responses of the two filters this gives

$$G^u = -\frac{B}{2\omega}\frac{\sinh(kd)}{\cosh(k(z+d))} \quad (24)$$

$$G^w = -\frac{B}{2\omega}\frac{\sinh(kd)}{\sinh(k(z+d))} \quad (25)$$

$$\phi^u = \phi_B + \pi - kx_1 \quad (26)$$

$$\phi^w = \phi_B + \pi - kx_1 - \frac{\pi}{2} \quad (27)$$

Even though the theoretical frequency response of the different filters easily can be calculated from the eqs. (13)-(15),

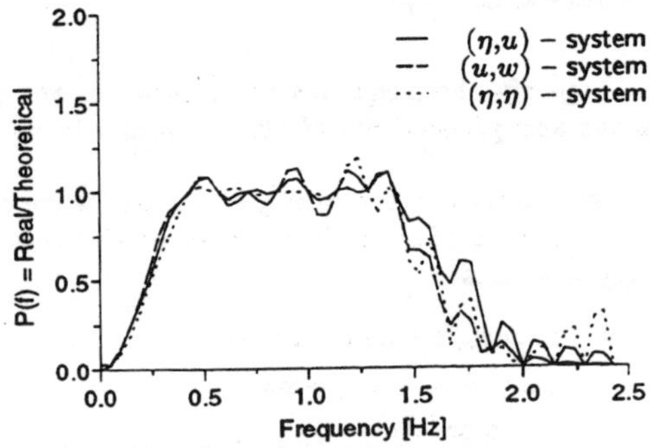

Fig. 2 Performance of absorption systems.

eqs. (18)-(21) and eqs. (24)-(27) actual realization or fitting of the theoretical frequency responses in FIR-filters is rather difficult. Any fitting of the filter coefficients to the above derived theoretical frequency responses obviously involves some error to be minimized in order to obtain the best possible performance. The fitting error is strongly dependent on: Water depth d, location of wave gauges x_1, x_2 and/or velocity meters x_1, z, number of filter coefficients N and the sample frequency of the filter f_s. This means that the actual performance of the systems is very depending on the setup of the filters.

Optimal Filter Design

To evaluate the efficiency and to enable optimization of an active absorption system the effect of the fitting error over some frequency area has to be quantified for the filter sets. The combined frequency response of the two filters can be determined by considering a simple cosine input

$$\eta(t) = \cos(\omega t) \tag{28}$$

and the frequency response functions of the two filters

$$\mathcal{H}_1(\omega) = G_1(\cos(\phi_1) + i\sin(\phi_1)) \tag{29}$$

$$\mathcal{H}_2(\omega) = G_2(\cos(\phi_2) + i\sin(\phi_2)) \tag{30}$$

By convolution of the input cosine with the two frequency response functions and subsequent addition of the outputs the following output is obtained

$$\begin{aligned} X(t) &= G_1\cos(\omega t + \phi_1) + G_2\cos(\omega t + \phi_2) \\ &= \{G_1\cos(\phi_1) + G_2\cos(\phi_2)\}\cos(\omega t) - \\ &\quad \{G_1\sin(\phi_1) + G_2\sin(\phi_2)\}\sin(\omega t) \end{aligned} \tag{31}$$

The gain corresponding to this combined input/output relation is

$$\begin{aligned} G_X &= \{(G_1\cos(\phi_1) + G_2\cos(\phi_2))^2 + \\ &\quad (G_1\sin(\phi_1) + G_2\sin(\phi_2))^2\}^{\frac{1}{2}} \end{aligned} \tag{32}$$

Calculating G_X for both the theoretical frequency response functions corresponding to eqs. (13)-(15), eqs. (18)-(21) or eqs. (24)-(27) as well as for the realized impulse response functions corresponding to the filter coefficients, the ratio $P(f)$

system		(η,η)	(η,u)	(u,w)
Δf	[Hz]	0.143	0.143	0.143
x_1	[m]	3.1	3.1	3.1
x_2	[m]	3.3	-	-
z	[m]	-	-0.15	-0.15
d	[m]	0.55	0.55	0.55

Table 1 Setup parameters.

signifies the performance of the actual system

$$P(f) = \frac{G_{X,real}}{G_{X,theo}} \tag{33}$$

This ratio is termed the performance function and gives the actual degree of absorption on every frequency desired. In case of 100% absorption $P(f) = 1$.

By calculating the error ϵ defined as

$$\epsilon = \int_{f_{low}}^{f_{high}} (1 - P(f))^2 df \tag{34}$$

setup parameters for each system can be evaluated. f_{low} and f_{high} are two prespecified low and high cutoff frequencies between which the filters are fitted.

In fig. 2 the performance function is plotted for the setup parameters given in table 1. All systems show excellent absorption characteristics within the frequency area [0.3;1.5] and are basically very similar in performance.

Generally, to obtain the best possible fit Δf must be small. On the other hand a small Δf results in a large filter delay that must be compensated for in the filters. In order to compensate for this delay the wave gauges or velocity meters have to be positioned far away from the paddle with the risk that accumulated phase errors due to the dispersion relation (correct to 2nd order) effect the system more. Both limitations must be considered when selecting Δf.

The main difference between the systems is not seen in fig. 2, but is found in the sensitivity to the internal gauge spacing $x_2 - x_1$ for the (η,η)-system. The frequency response of the (η,η)-system is very sensitive to changes in the internal gauge spacing since a singularity occurs because of the phase difference between the two input signals. Opposite, the (η,u)-system and the (u,w)-system are based on colocated input signals with no singularities occuring making the filters easier to realize.

A disadvantage of the two velocity meter based systems compared to the (η,η)-system is the higher sensitivity to non-linear waves.

For the tested setup the absorption frequency range covers the frequency range in which the main part of the energy is concentrated in the wave spectrum - only low frequency or long periodic waves are not covered.

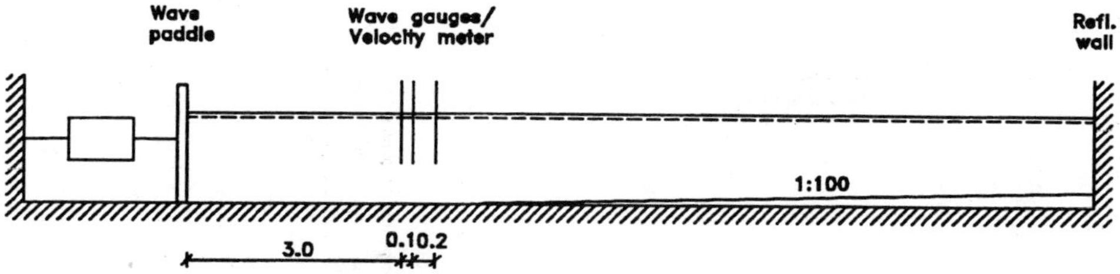

Fig. 3 Wave flume and model test setup.

In case another sea state with another peak wave length (or peak frequency) than those covered by the system described in table 1 is desired. The system setup parameters in table 1 are simply Froude scaled with a length scale determined by the ratio between the new and the old peak wave length (or peak frequency). The more elaborate procedure calculating ϵ and $P(f)$ for the different parameters is of course still applicable.

It should be noted that the gauge spacing from the wave paddle must be large than $3d$ for the gauges to be placed in the far field, since the elevation model in eq. (2) neglects the near field local disturbances.

Physical Model Test Comparisons

In order to determine the performances of the different active absorption methods described above, the (η,η)-system and the (η,u)-system were implemented in the control system of a piston-type wave generator in Hydraulics & Coastal Engineering Laboratory, Aalborg University.

The geometry and test setup of the wave flume are given in fig. 3 and the setup parameters for the two systems are those listed in table 1.

The Biesel phase ϕ_B and gain B (see eq. (3)) are determined using the linear transfer functions derived by Biesel (1951).

When active absorption was applied, the surface elevation time series were recorded and digitized by means of a PC equipped with an A/D-D/A-card, digital filtering and superposition were performed, and the resulting paddle displacement correction signal was added to the input signal read from the signal generator.

At the far end of the flume a vertical fully reflecting wall was mounted. The flume is equipped with three pairs of wave gauges mounted on a beam at distances of 3.0 m, 3.1 m and 3.3 m from the wave paddle. These gauges are used for reflection measurements as well as absorption. A water depth of d = 0.55 m was maintained throughout the test series.

All tests were performed with exactly the same input from the signal generator: a wave paddle displacement signal corresponding to a JONSWAP-spectrum with significant wave height H_s = 0.08 m, peak period T_p = 1.2 sec and peak enhancement factor $\gamma = 3.3$, sampled at a frequency of f_s = 40 Hz, and generated by means of digital filtering of Gaussian white noise in the time domain. All tests were run for 10 min.

The 0.08 m significant wave height is fairly high resulting in non-linear waves with some breaking in form of white capping. Furthermore, severe cross mode wave action was observed. These severe wave conditions should accentuate eventual differences between the systems.

In each test the incident and reflected spectra were resolved as described by Mansard and Funke (1980) and for the systems (η,η) and (η,u) the incident wave spectra are given in fig. 4.

Comparing the significant wave height for the two systems these are in the same order of magnitude though slightly higher than the target wave height. The differences between the two systems must be derived from the energy spectra showing more long periodic energy for the (η,η)-system than the (η,u)-system signifying that the purely wave gauge based system tends to generate long periodic waves. Also when comparing with the case of no absorption more long periodic energy is observed. Regarding the (η,u)-system the wave energy spectrum is far more peaked.

The tendency to generate long periodic energy can be explained by the high sensitivity of the (η,η)-system to uncorrelated pink noise on the two input signals, see fig. 5a and fig. 5b. Uncorrelated pink noise could be electrical noise or surface fluctuations due to breaking.

Since severe cross mode action occured during testing, the sensitivity to cross mode action is simulated for both systems by adding correlated pink noise to the two input signals. Because of the surface elevation model in eq. (2) the two sensors (surface elevation or velocity) will recogn cross mode action as correlated noise. The amplitude amplification for the two systems are shown in fig. 5c and fig. 5d.

Surprisingly, the sensitivity of the two systems is in the same order of magnitude. For the wave flume used, the cross mode of 1st order corresponds to a frequency of 1 Hz and as noted from fig. 4a and fig. 4b this corresponds very well with the frequency at the energy peak.

A matter that is not depicted in fig. 5d is that the actual separation of incident and reflected waves for this system is erroneous because the cross mode action only influences the wave gauge. This is not the case for the (η,η)-system and might explain the peaked incident wave spectrum for the (η,u)-system.

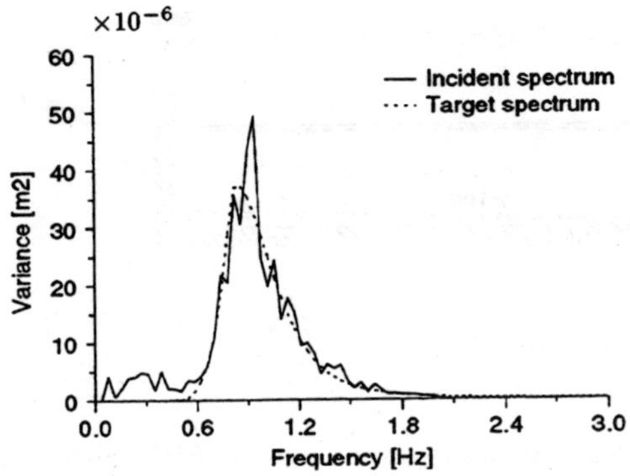

Fig. 4a Incident spectra for (η, η)-system.

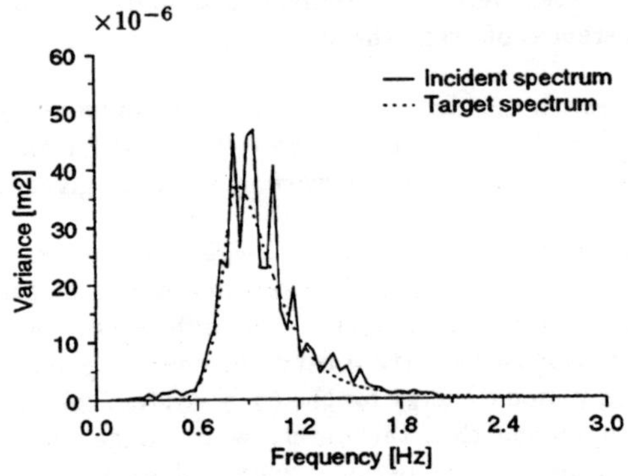

Fig. 4b Incident spectra for (η, u)-system.

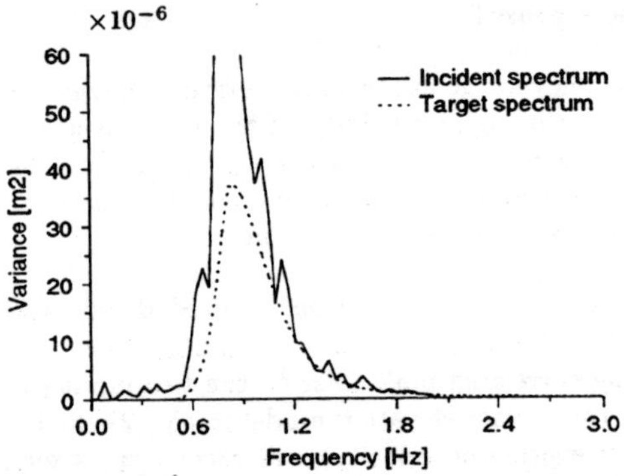

Fig. 4c Incident spectra with no absorption.

system	$H_{s,target}$ [cm]	$H_{s,inci}$ [cm]
(η, η)	8.4	8.5
(η, u)	8.4	8.6
no abs.	8.4	11.4

Table 2 Spectral wave characteristics.

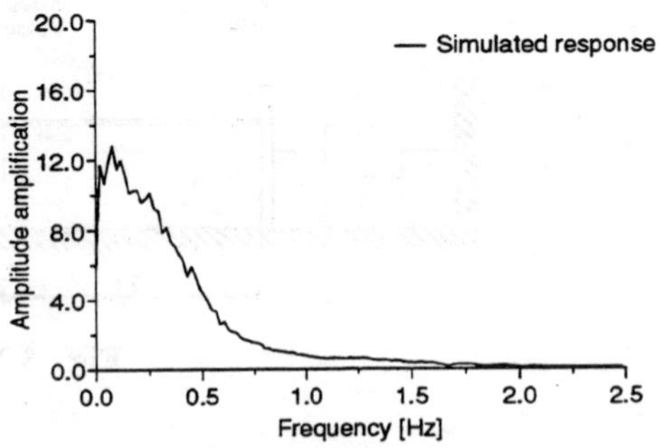

Fig. 5a Amplitude amplification for uncorrelated pink noise for (η, η)-system.

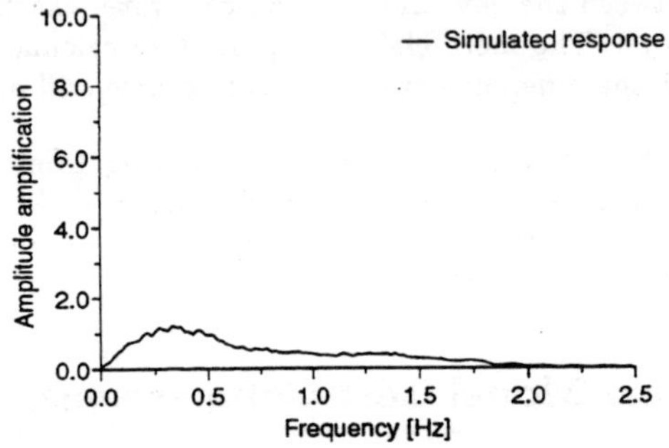

Fig. 5b Amplitude amplification for uncorrelated pink noise for (η, u)-system.

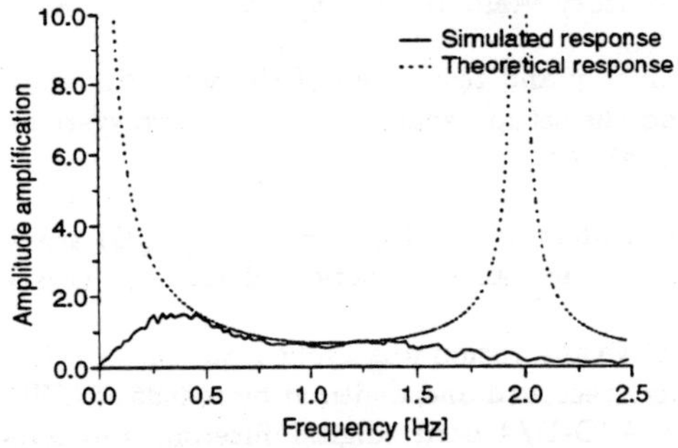

Fig. 5c Amplitude amplification for sim. cross mode action for (η, η)-system.

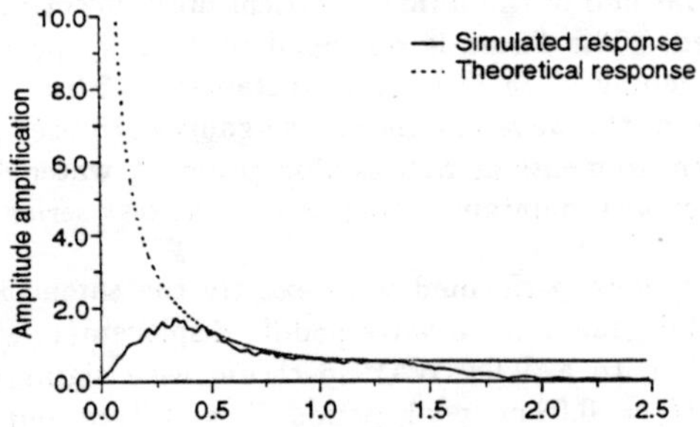

Fig. 5d Amplitude amplification for sim. cross mode action for (η, u)-system.

Conclusion

Three types of active absorption systems are described. A purely wave gauge based system termed (η, η), a combined wave gauge, velocity meter based system termed (η, u) and a purely velocity meter based system termed (u, w). For each system the theoretical frequency response of the filter systems is presented.

To optimize the frequency response of the filters a performance function has been derived. Numerical tests with the (η, η)-system, the (η, u)-system and the (u, w)-system demonstrate that all three filter systems easily are realized with excellent frequency response in the relevant frequency area. All systems show similar absorption characteristics

Tests performed with both the (η, η)-system and the (η, u)-system imply good absorption characteristics even at very high levels of reflection. Furthermore, both systems appear to be very stable.

Regarding the (η, u)-system the incident wave spectrum is more peaked than that of the (η, η)-system. Regarding the (η, η)-system some long periodic waves tend to be generated. A possible solution to these problems might be by demanding zero gain at $f = 0.0$ and averaging measured surface elevations over the width of the flume.

Converting a conventional wave generator to an absorbing wave generator based on the principle presented above is relatively inexpensive considering the improvements achieved: the only requirement in its most simple form is two conventional wave gauges and a PC equipped with an A/D-D/A-card. These facilities will normally be available in most laboratory environments. If a PC equipped with an A/D-D/A-card is used as signal generator for the wave generator, the wave gauges can simply be connected to this computer, allowing to perform signal generation and correction signal calculation simultaneously.

Acknowledgements

The work carried out is sponsored by the Danish Technical Research Council under the frame programme Marin Teknik 2.

References

Biesel, F., 1951,
Les Apparails Generateurs de Houle en Laboratoire, La Houille Blanche, Vol. 6, nos. 2, 4 and 5.

Bullock, G.N. and Murton, G.J., 1989,
Performance of a Wedge Type Absorbing Wave Maker, Journal of Waterway, Port, Coastal and Ocean Engineering, Vol. 115, No. 1.

Frigaard, P., 1993,
AWASYS - Active Wave Absorption System - Users Guide, Hydraulics & Coastal Engineering Laboratory, Department of Civil Engineering, Aalborg University.

Frigaard, P., Brorsen, M., 1995,
A Time-Domain Method for Separating Incident and Reflected Irregular Waves, Coastal Engineering, Vol. 24, Nos. 3 and 4.

Frigaard, F., Christensen, M., 1994,
An Absorbing Wave-Maker Based on Digital Filters, Proceedings of Int. Symp.: Waves - Physical and Numerical Modelling, Vol. 1, Vancouver.

Gilbert, G., 1978,
Absorbing Wave Generators, Hydr. Res. Station notes, Hydr. Res. Station, Wallingford, Oxon, United Kingdom, Vol. 20, Nos. 3-4.

Hirakuchi, H., Kajima, R., Kawaguchi, T., 1990,
Application of a Piston-Type Absorbing Wavemaker to Irregular Wave Experiments, Coastal Eng. Japan, Vol. 33, No. 1.

Mansard, E. and Funke, E., 1980,
The Measurement of Incident and Reflected Spectra Using a Least Squares Method, Proceedings 17th International Conference on Coastal Engineering, Vol. 1, Sydney, Australia.

Milgram, J.S., 1970,
Active Water-Wave Absorbers, J. Fluid Mech., Vol. 43, No. 4.

Schäffer, H.A., Stolborg, T., Hyllested, P., 1994,
Simultaneous Generation and Active Absorption of Waves in Flumes, Proceedings of Int. Symp.: Waves - Physical and Numerical Modelling, Vol. 1, Vancouver.

Proceedings of the Seventh (1997) International Offshore and Polar Engineering Conference
Honolulu, USA, May 25-30, 1997

ISBN 1-880653-28-1 (Set); ISBN 1-880653-31-1 (Vol. III)

Simulation of Design Storm Wave Conditions with Tailored Wave Groups

Günther F. Clauss and Walter L. Kühnlein
Berlin University of Technology
Berlin, Germany

ABSTRACT

Extreme wave conditions arise from the most unfavourable superposition of the related severe sea spectrum. When generating irregular seas at model scale this crucial condition related to the most critical storm spectra may occur if the random phase shift of all superimposing waves is zero at a certain position. Alternatively a project oriented design spectrum can also be generated in such a sequence that all waves superimpose without phase shift at a given position. The analysis of these nonlinear transient wave trains is based on the definition of a complex wave information which characterizes the (linear) information of a wave group. Combined with a modified wave celerity (information speed) the propagation of high wave groups can be calculated. As a result, the nonlinear registrations and associated wave group data are derived from the respective wave information.

Model tests with breaking transient wave packets prove that the obtained maximum wave force on a structure is much higher as compared to model tests with irregular waves, for a given spectrum. As a consequence, it must be stated that high irregular wave tests are not sufficient for evaluating design wave loads. Only tests series with deterministic transient wave groups promise reliable results.

KEY WORDS

Extreme wave conditions; nonlinear transient wave excitation; nonlinear transient wave packets; wave information; linear superposition of wave information; "freak" wave.

INTRODUCTION

Extreme wave conditions in a 100-years design storm arise from the most unfavourable superposition of component waves of the related severe sea spectrum. When generating irregular seas at model scale this crucial condition related to the most critical storm spectra eventually may occur if the random phase shift of all superimposing waves is zero at a certain position — a coincidence which requires a very extended test duration. It will take around 50 hours/$\sqrt{\text{model scale}}$ to simulate a storm with a duration of two days.

As an alternative the component waves of a project oriented design spectrum are tentatively generated in such a sequence that all waves superimpose without phase shift at a given position (see Fig. 1). This option is based on spectral analysis which describes the apparently chaotic wave field of a natural seaway as the superposition of an infinite number of independent harmonic waves with amplitude ζ_{an} and random phase γ_n

$$\zeta(t) = \sum_{n=1}^{\infty} \zeta_{an} \cos(\omega_n t - \gamma_n). \quad (1)$$

Each component wave contributes an amount of energy to the seaway proportional to its squared wave amplitude (Clauss et al., 1992). In general, superposition with random phase yields the well-known long-term registration of irregular waves with accidental high peaks and low troughs (Fig. 1). The most unfavourable event in storm wave conditions is the in-phase superposition of component waves in the seaway. As these "freak" waves are possible events during the life-cycle of a marine structure it may be selected as the relevant design wave. Of course, we expect high non-linearities. Therefore the scope of this paper covers the following topics:

- Design waves are the in-phase superposition of all component waves of the most severe sea spectrum: this rare event can be intentionally simulated in seakeeping tests to investigate maximum loads.
- The genesis of design waves is a highly nonlinear process. However, as these short and high wave groups evolve from long and low (linear) wave groups, the nonlinear characteristics can be evaluated from linear principles. Consequently, extremely high "freak" waves have a linear past.

The simulation technique is based on a computer controlled nonlinear procedure which governs the generation of a short, specifically tailored wave train with a "design" spectrum. As the wave

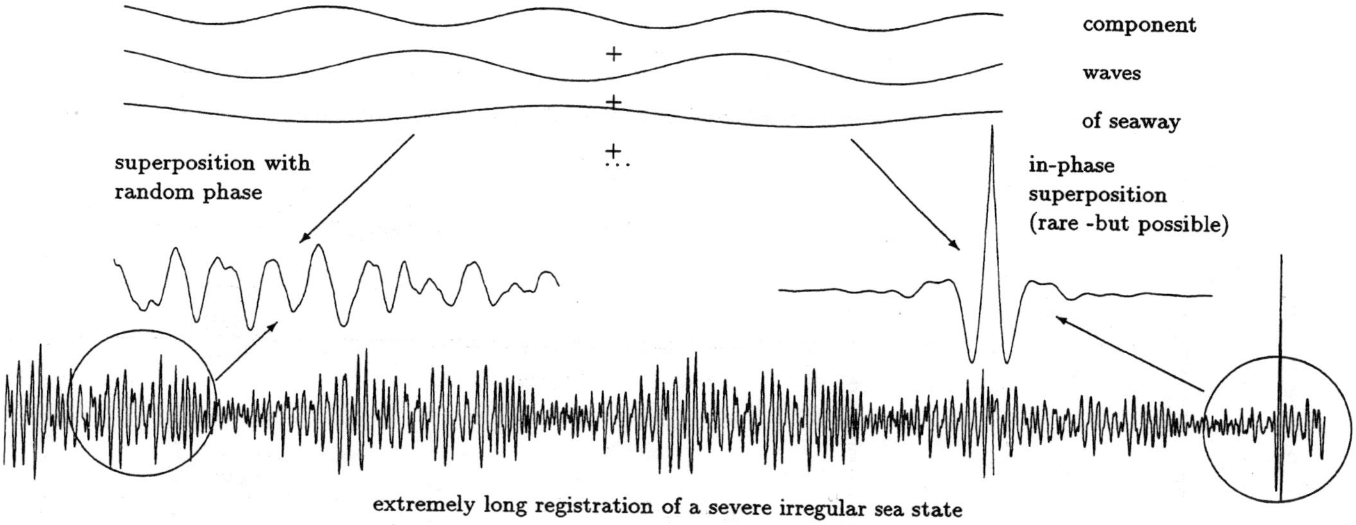

Fig. 1: Design wave as rare event of component wave superposition

train is exactly defined in space and time it is easy to transpose the registration to any position along the tank.

This paper proposes a new method for the description of the shape variation of nonlinear transient wave packets during its propagation. Short and high wave groups with strong nonlinear characteristics evolve from long and low wave groups, which are described by linear principles. Based on this fact, the idea was born that the linear complex Fourier transform (surface velocity) of the long and low wave group could be used as the characteristic information of the wave train. Consequently, all nonlinear wave elevations, velocities and accelerations should be derivable from this linear wave information. Also the transformation of the wave train registration from one position to any other position is feasible by using the linear wave information. During its metamorphosis the total energy of the transient wave is invariant, if breaking phenomena are excluded.

- At first, the paper describes the generation of linear transient wave trains and its linear transformation to selected positions. This is an important prerequisite as the same procedure will be used for the transformation of the linear wave information to establish a nonlinear wave train. In this case the linear wave celerity as a function of the wave frequency is replaced by a nonlinear "information" celerity which depends on frequency, space, and time.

- The next section introduces the development of the numerical nonlinear description of the entire wave field of the transient wave train by integrating the mutually dependent particle motion equations in time domain. The wave information is the primordial cell in which all nonlinearities are hatched.

- Finally, the paper presents model test results with a permeable wall, to illustrate the efficiency of the wave packet technique. The tests have been performed at the "Großer Wellenkanal Hannover" with irregular sea states, as well as with breaking and non-breaking transient wave trains. Tests with breaking transient wave trains prove that this is the worst condition for a given wave spectrum as compared to irregular seas. Even in one hour model tests with very high irregular waves we never have succeeded to achieve similar dramatic wave loads for the given spectrum. In particular the total maximum force on the permeable wall with the breaking transient wave train is twice as compared to an irregular sea state test and the maximum force on the top element of the wave absorber is even twenty times higher.

TRANSIENT WAVE TECHNIQUE FOR SEAKEEPING TESTS

Transient waves for model excitation were originally proposed by Davis and Zarnick in 1964, and further developed by Takezawa and Hirayama (1976). Clauss and Bergmann (1986) recommended a special type of transient waves, i.e. Gaussian wave packets, which have the advantage that their propagation behaviour can be predicted analytically (Chakrabarti and Libby, 1988).

With increasing efficiency and capacity of computers the restriction to a Gaussian distribution of wave amplitudes has been abandoned, and the entire process is now performed numerically (Clauss and Kühnlein, 1995). The shape and width of the wave spectrum can be selected individually for providing sufficient energy in the relevant frequency range. As a result the wave train is predictable at any instant and at any stationary or moving location. In addition, the wave orbital motions as well as the pressure distribution and the vector fields of velocity and acceleration can be calculated. According to its high accuracy the technique is capable of generating special purpose transient waves:

- defined wave groups for seakeeping tests;
- very high "freak" waves;
- counter-propagating wave groups with predictable superposition at selected positions. If both wave groups are identical, their in-phase superposition in the center plane leads to the doubling of vertical and cancellation of horizontal motions, velocities and accelerations (or vice versa if phase shift is 180°).

Any experiment performed with this technique is exactly repeatable, its duration is short and the energy flux is predictable.

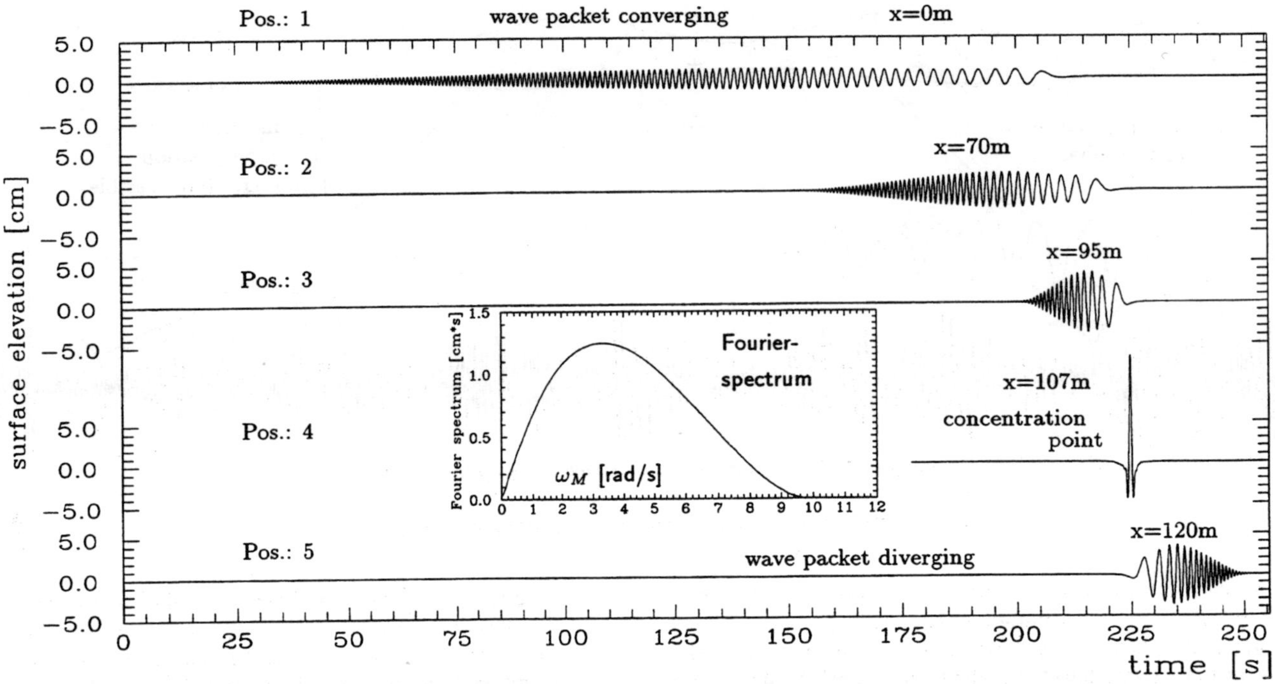

Fig. 2: Wave packet at different positions (water depth d=4.20m) and amplitude spectrum of the related complex Fourier transform

Even complicated interactions of wave group and structure are easy to analyze, as the reflected and transmitted wave train can be separated from the initial wave packet.

Linear Transient Wave Description

The synthesis and up-stream transformation of wave packets is developed from its so-called concentration point. At this position all waves are superimposed without phase shift resulting in a single high wave peak. Starting from this concentration point, the Fourier transform is transposed to the selected interaction position of structure and wave train, and finally to the position of the wave generator.

On this basis, the wave spectrum is defined by the complex Fourier transform

$$a(x,t) = \sum_{j=-(n/2-1)}^{n/2-1} |A(x,\omega_j)|\, e^{i\,(\omega_j\,(t-t_c)-k_j(x-x_c))}\,\Delta\omega, \qquad (2)$$

with $\omega_j = j\Delta\omega$, $\Delta\omega = \frac{2\pi}{T} = \frac{2\pi}{n\Delta t}$ (T=duration of registration), and $k_j = f(\omega_j, d)$. It is characterized by the amplitude spectrum and the related phase distribution. During propagation the amplitude spectrum remains invariant, however, the phase distribution and the related shape of the wave train varies with its position. At the concentration point, all waves are superimposed without phase shift resulting in a single high wave peak. To center the resulting wave train in its time frame, the Fourier transform is shifted by $T/2 = \frac{n}{2}\Delta t$

$$\begin{aligned} a(x{=}x_c,t) &= \sum_{(j)} |A(x_c,\omega_j)| e^{i\,\omega_j\,\frac{T}{2}} \cdot e^{i\,\omega_j(t-t_c-\frac{T}{2})}\,\Delta\omega \\ &= \sum_{(j)} |A(x_c,\omega_j)|\,(-1)^j\, e^{i\,\omega_j(t-t_c-\frac{T}{2})}\,\Delta\omega. \qquad (3) \end{aligned}$$

Consequently, the Fourier transform at the concentration point ($t = t_c$; $x = x_c$) is real, but the sign of the Fourier coefficients $A(x_c,\omega_j) = (-1)^j |A(x_c,\omega_j)|$ is alternating.

From its concentration point, the Fourier transform of the wave train is transformed to the upstream position at the wave board by introducing an average group velocity $\frac{1}{2}[c_{gr}(\omega_{\min}) + c_{gr}(\omega_{\max})]$ and replacing ωt by $\omega t - k(\omega)(x - x_c)$:

$$a(x,t) = \sum_{(j)} (-1)^j |A(x_c,\omega_j)|\, e^{i\left(\omega_j(t-t_c-\frac{T}{2})-k_j(x-x_c)\right)}\,\Delta\omega. \qquad (4)$$

With the new time shift

$$t_s = \frac{1}{2}\left(\frac{x-x_c}{c_{gr}(\omega_{min})} + \frac{x-x_c}{c_{gr}(\omega_{max})}\right) \qquad (5)$$

this results in

$$a(x,t)=\sum_{(j)}(-1)^j |A(x_c,\omega_j)| e^{i\,\omega_j\,t_s} e^{i\left(\omega_j\cdot(t-t_c-\frac{T}{2}-t_s)-k_j(x-x_c)\right)}\,\Delta\omega. \qquad (6)$$

Fig. 2 shows a wave train as a function of time at different positions (water depth d=4.20m) and the related amplitude spectrum of the complex Fourier transform.

The wave train can be transformed from a function of time $a(t, x_0)$ at a fixed location x_0 to a function of space $a(x, t_0)$ at a given time t_0 (see Fig. 3).

$$\left\{\begin{array}{c} a(x_0, t{=}t_{min}) \\ a(x_0, t{=}t_{min}+\Delta t) \\ a(x_0, t{=}t_{min}+2\,\Delta t) \\ a(x_0, t{=}t_{min}+3\,\Delta t) \\ \cdots \\ a(x_0, t{=}t_{min}+ \\ (n-1)\Delta t) \end{array}\right\} \rightleftharpoons \left\{\begin{array}{c} a(x{=}x_{min}, t_0) \\ a(x{=}x_{min}+\Delta x, t_0) \\ a(x{=}x_{min}+2\,\Delta x, t_0) \\ a(x{=}x_{min}+3\,\Delta x, t_0) \\ \cdots \\ a(x{=}x_{min}+ \\ (n-1)\Delta x, t_0) \end{array}\right\} \qquad (7)$$

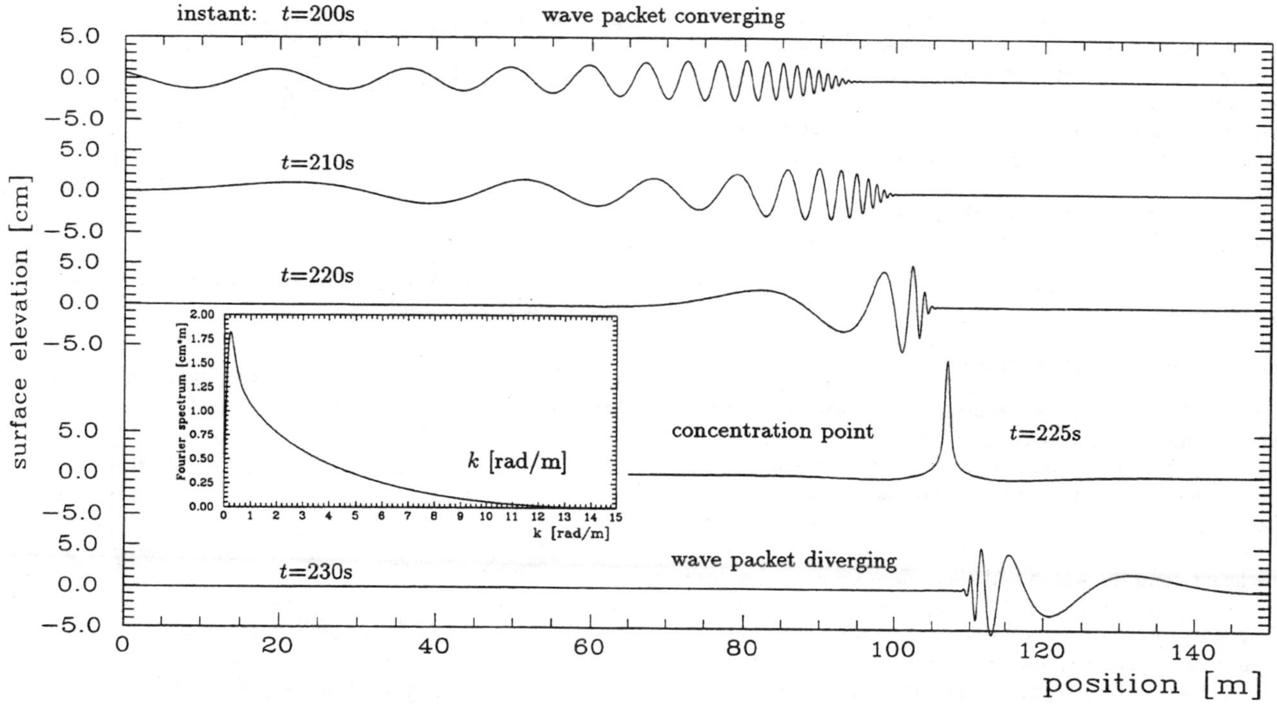

Fig. 3: Transformation of wave packet from time domain to a representation of instantaneous wave profiles at selected instants and its related Fourier spectrum as a function of the wave number k (water depth d=4.20m)

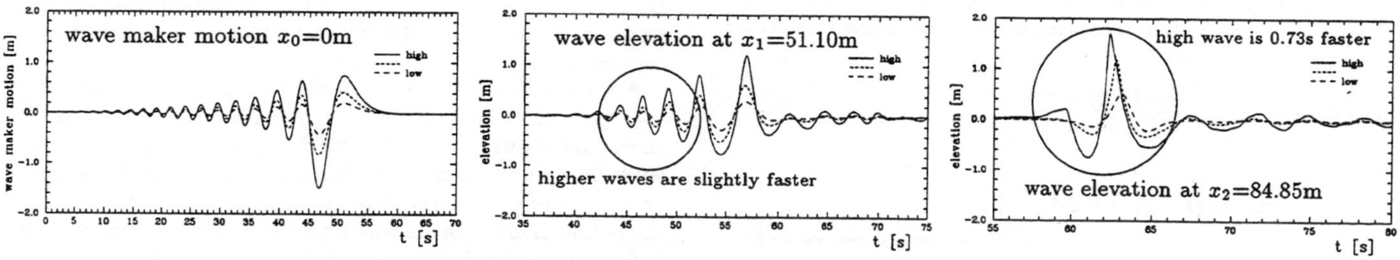

Fig. 4: Comparison of measured water elevations of low and high wave packets at different positions (water depth d=4.41m)

Note that this transformation can be done without loosing any information of the transient wave train.

For wave packets with high elevations this procedure cannot be performed by using linear theory. Fig. 4 shows the comparison of low and high wave trains, measured at two positions. It illustrates that a high wave train is slightly faster when propagating. Note that all three wave trains are generated by synchronous wave maker motions, which have been calculated by linear theory. The diagram at the right hand at the position x_2=84.95m illustrates that the entire high wave packet is about 0.73s faster. The illustration also visualizes the existence of lower waves following the wave train. This phenomenon is a consequence of incorrect wave maker motions which have been calculated by linear theory. To avoid this error an improved procedure for the generation of nonlinear wave groups has been developed.

Nonlinear Transient Wave Description

For regular waves nonlinear solutions have been introduced by Gerstner (1802), Stokes (1880), and Crapper (1957). Recent solutions for irregular seas have been proposed by Sand (1982), Mansard et. al. (1987), and Stansberg, (1995). None of these methods is applicable to the analysis of converging wave groups which become shorter, higher, and therefore **faster** (see Fig. 4 and 8) during its way through the channel.

The new method proposed in this paper is based on the fact that short and high wave groups with strong nonlinear characteristics evolve from long and low wave groups, which are basically characterized by linear principles. As the total energy of the transient wave is invariant during its metamorphosis, the initial linear Fourier transform of the surface velocity is selected as the backbone of "wave information". Based on this "wave information" the shape variation of the linear transient wave train during propagation is calculated. At selected positions a nonlinear expansion is accomplished by integrating the mutually dependent particle motion equations in time domain, resulting in a numerical nonlinear description of the transient wave train as a function of time or space at any fixed or moving reference point. After the nonlinear wave train passed the concentration point, the primordial linear Fourier spectrum of the long, low wave group can be found again, which can be used to monitor the accuracy of the transformation.

The numerical method confirms the experimental observation that nonlinear high-amplitude waves

- are slightly faster than small waves, and therefore also slightly longer;

- generate a wave induced current.

Continuity of the information content of the wave group is the basic idea of the following derivation. Also the total energy of the wave group is invariant during propagation although its local amplitudes, phase relations, and total length are changing. For characterizing the propagation of the nonlinear wave group we introduce the new parameter "wave information" I — and assign it to the linear complex water particle surface velocity of a deep water wave (real part = horizontal, imaginary part = vertical).

At first, the procedure is applied to analyze the propagation of high regular waves, and will be then expanded to nonlinear wave packets. In the case of a regular (high) wave the complex wave information I is simply given by the harmonic oscillation

$$I(t, x, w_a, \omega) = w_a \cdot e^{-i(kx-\omega t)}, \tag{8}$$

which corresponds to the orbital surface velocity $w_a = \zeta_a \omega$ of a linear deep water wave. The nonlinear particle motions are developed from the iterative integration of the coupled equations

$$\begin{aligned} \xi(t,x,z) &= \int_0^{\tau=t} \Bigg[\left(\frac{\tanh[k(\zeta(\tau)+d]}{\tanh(kd)} \right)^{\frac{1}{kd}} \\ &\quad \cdot \frac{\cosh[k(z+\zeta(\tau)+d)]}{\sinh(kd)} \cdot \underbrace{w_a \cos[-k(x+\overline{x}(\tau))+\omega\tau]}_{\text{Re } I(t,x,w_a,\omega)} \Bigg] d\tau; \\ \zeta(t,x,z) &= \int_0^{\tau=t} \Bigg[\left(\frac{\tanh[k(\zeta(\tau)+d]}{\tanh(kd)} \right)^{\frac{2}{kd}} \\ &\quad \cdot \frac{\sinh[k(z+\zeta(\tau)+d)]}{\sinh(kd)} \cdot \underbrace{w_a \sin[-k(x+\overline{x}(\tau))+\omega\tau]}_{\text{Im } I(t,x,w_a,\omega)} \Bigg] d\tau. \end{aligned} \tag{9}$$

k is the wave number, x and z are starting positions ($\xi(t = 0, x, z) = 0, \zeta(t = 0, x, z) = z$) of the water particle before the wave has arrived, and $x+\xi(t_{\text{end}}, x, z), z+\zeta(t_{\text{end}}, x, z)$ are the end positions of the particles after the wave has passed (t_{end}=51.2s, see Fig. 6). The first part of Eq. 9 are correcting functions due to the wave induced variation of water depth d (shallow water effects). Note that for deep water waves as well as for very small amplitude waves (linear) in shallow water the functions are 1 and may be neglected without affecting the results in Eq. 9 and 13. For higher regular waves in shallow water the correction by using these functions is confirmed by experimental (measured) results.

As shown in Eq. 9 and in Fig. 5, the particle position follows from the orbital motion and convection. The wave information

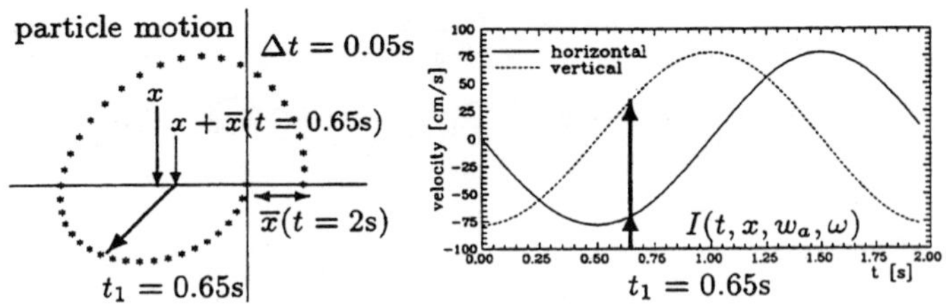

Fig. 5: Integration of particle motions from the linear wave information $I(t, x, w_a, \omega)$

is related to the position $x + \overline{x}(\tau)$, and $\overline{x}(\tau)$ is the mean value of the excursion due to its corrected convection:

$$\overline{x}(t) = \left(\xi(d, k, t_{\text{end}}, z = \zeta) - \frac{\int_0^{t_{\text{end}}} \int_0^{\zeta} \xi \; dz \, dt}{d} \right) \cdot \frac{t}{t_{\text{end}}} \tag{10}$$

At the first iteration step $\overline{x}(t)$ is set to 0. The celerity c of the regular wave follows from the linear celerity

$$c_{\text{linear}}(d, k) = \sqrt{\frac{g}{k} \tanh kd} \tag{11}$$

and the wave induced convection velocity, i.e.

$$c(d, k) = c_{\text{linear}}(d, k) + \frac{\xi(d, k, t_{\text{end}}, z = 0)}{t_{\text{end}}}. \tag{12}$$

During iteration the entire procedure is started again at Eq. 8. Generally, only one or two iteration loops are necessary to get convergent results.

Fig. 6 shows a 0.45m regular wave (period T=2s, water depth d=1.5m). Note that the wave is asymmetric, and the particles at the water surface are traveling downstream whereas at deeper levels an opposite convective current is observed. These results are confirmed by experimental investigations. Due to convection, the period of the moving water particle at the surface is slightly higher, and at the bottom is slightly lower as related to the wave period of T=2s. (At the surface a particle has moved downstream by about $\overline{x} \approx$0.12m after finishing one orbital track: therefore the resulting period of the water particle motion is $T = T_0 + \frac{\overline{x}}{c} \approx$2.04s)

If switching from this Lagrangian description of orbital motions of individual particles for regular (high) waves to Eulerian description, which gives the motion characteristics at a specified location $P(x_0, z_0)$ (with different particles passing by) we obtain from Eqn. 8-12

$$\begin{aligned} \xi(t, x_0, z_0) &= \int_0^{\tau=t} \Bigg[\left(\frac{\tanh[k(\zeta(\tau)+d]}{\tanh(kd)} \right)^{\frac{1}{kd}} \cdot \frac{\cosh[k(z_0+\zeta(\tau)+d)]}{\sinh(kd)} \\ &\quad \cdot w_a \cos[-kx_0 + \omega(\tau - \frac{\xi(\tau)}{c})] \Bigg] d\tau; \\ \zeta(t, x_0, z_0) &= \int_0^{\tau=t} \Bigg[\left(\frac{\tanh[k(\zeta(\tau)+d]}{\tanh(kd)} \right)^{\frac{2}{kd}} \cdot \frac{\sinh[k(z_0+\zeta(\tau)+d)]}{\sinh(kd)} \\ &\quad \cdot w_a \sin[-kx_0 + \omega(\tau - \frac{\xi(t)}{c})] \Bigg] d\tau. \end{aligned} \tag{13}$$

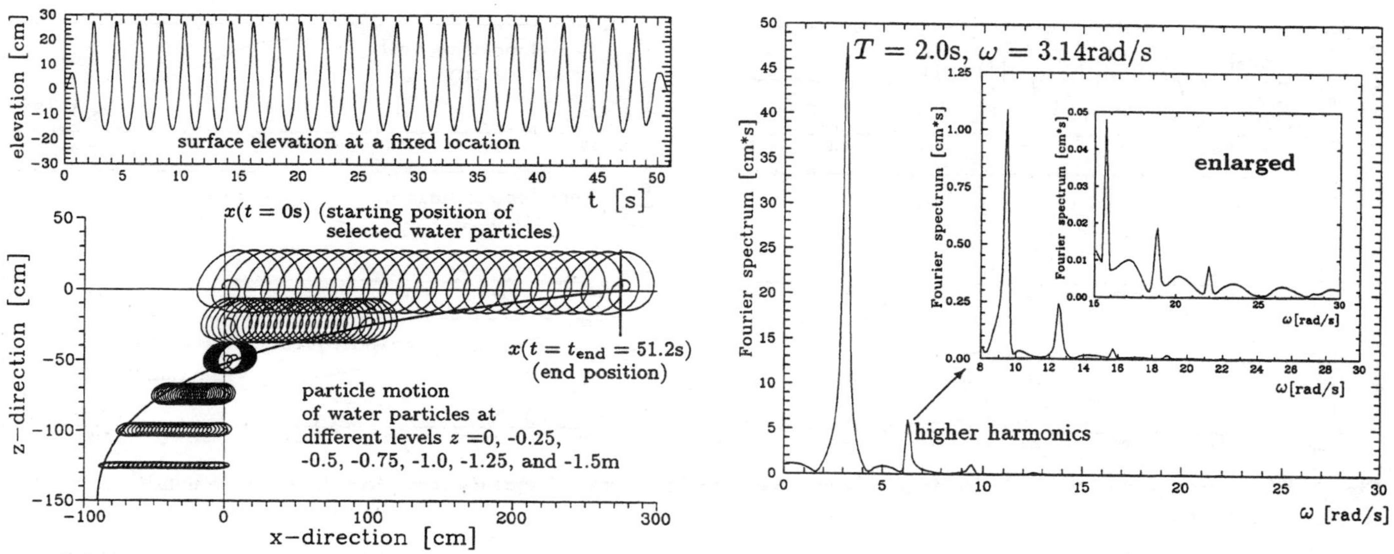

Fig. 6: Calculated surface elevation, water particle motion, and Fourier spectrum of a regular wave (T=2s, d=1.5m)

$\xi(t)$ and $\zeta(t)$ are mutually dependent. Eq. 13 is iteratively solved in time domain. As a result, Fig. 6 shows the surface elevation at a fixed location.

So far the application of the "wave information" technique for analyzing the propagation of high regular waves has been illustrated. To extend our model to transient wave trains we express wave frequency ω and wave number k as functions of time $f(t, x_0)$. As the first derivation of a harmonic function $f(t) = ae^{i(\omega t)}$ results in $\dot{f}(t) = i\omega ae^{i(\omega t)}$, the frequency of harmonic functions is introduced as:

$$\omega(t) = \frac{|\dot{f}(t)|}{|f(t)|}. \tag{14}$$

At a fixed location x_0 the wave frequency $\omega(t, x_0)$ simply follows from the ratio of the vertical velocity amplitude $w_a(t, x_0, z = 0)$ and the motion amplitude $\zeta_a(t, x_0, z = 0)$ (linear theory), i.e.

$$\omega(t, x_0) = \frac{w_a(t, x_0, z = 0)}{\zeta_a(t, x_0, z = 0)}, \tag{15}$$

with

$$w_a(t, x_0, z = 0) = \left| \sum_{(j)} W(x_0, \omega_j, z = 0)\Delta\omega + i \sum_{(j)} W(x_0, \omega_j, z = 0)e^{-i\frac{\pi}{2}}\Delta\omega \right| \tag{16}$$

and

$$\zeta_a(t, x_0, z = 0) = \left| \sum_{(j)} A(x_0, \omega_j, z = 0)\Delta\omega + i \sum_{(j)} A(x_0, \omega_j, z = 0)e^{-i\frac{\pi}{2}}\Delta\omega \right|. \tag{17}$$

$W(x_0, \omega, z = 0)$ and $A(x_0, \omega, z = 0)$ are the complex Fourier transforms of the vertical wave surface velocity, and of the wave elevation. As an illustration, Fig. 7 shows a wave group with its associated vertical wave velocity, and the accompanying envelopes. According to Eq. 15 we calculate the wave frequency from the ratio of the wave velocity envelope and the surface elevation envelope. This can be done very accurately, as the calculation of the nonlinear wave field always starts from the linear information. Finally the wave number follows from

$$k(t, x_0) \cdot d \cdot \tanh(k(t, x_0) \cdot d) = \frac{\omega^2(t, x_0)}{g} d \tag{18}$$

(Fig. 7). The Fourier transform $W(x_0, \omega_j)$ of the fictive velocities (derived from linear theory) has been selected as the significant quantity which controls the shape of a wave group during propagation. In analogy to Eq. 8 we introduce the complex wave information:

$$I(t, x_0) = \sum_{(j)} W(x_0, \omega_j)\Delta\omega + i \sum_{(j)} W(x_0, \omega_j)e^{-i\frac{\pi}{2}}\Delta\omega. \tag{19}$$

and calculate the wave particle motions according to the modified Eq. 9.

$$\begin{aligned}
\xi(t, x, z) &= \int_0^{\tau=t} \left[\left(\frac{\tanh[k(\zeta(\tau) + d]}{\tanh(kd)} \right)^{\frac{1}{kd}} \cdot \frac{\cosh[k(z + \zeta + d)]}{\sinh(kd)} \cdot \mathrm{Re}[I(\tau, x + \overline{x})] \right] d\tau; \\
\zeta(t, x, z) &= \int_0^{\tau=t} \left[\left(\frac{\tanh[k(\zeta(\tau) + d]}{\tanh(kd)} \right)^{\frac{2}{kd}} \cdot \frac{\sinh[k(z + \zeta + d)]}{\sinh(kd)} \cdot \mathrm{Im}[I(\tau, x + \overline{x})] \right] d\tau,
\end{aligned} \tag{20}$$

with $k = f(t, x + \overline{x})$, and the correcting functions due to shallow water effects.

The celerity, i.e. the propagation velocity of information follows from linear celerity and an additional term due to nonlinear convection effects (see Eq. 12)

$$c_I(t, x) = c_{\text{linear}}(t, x) + c_{\text{add}}(t, x) \tag{21}$$

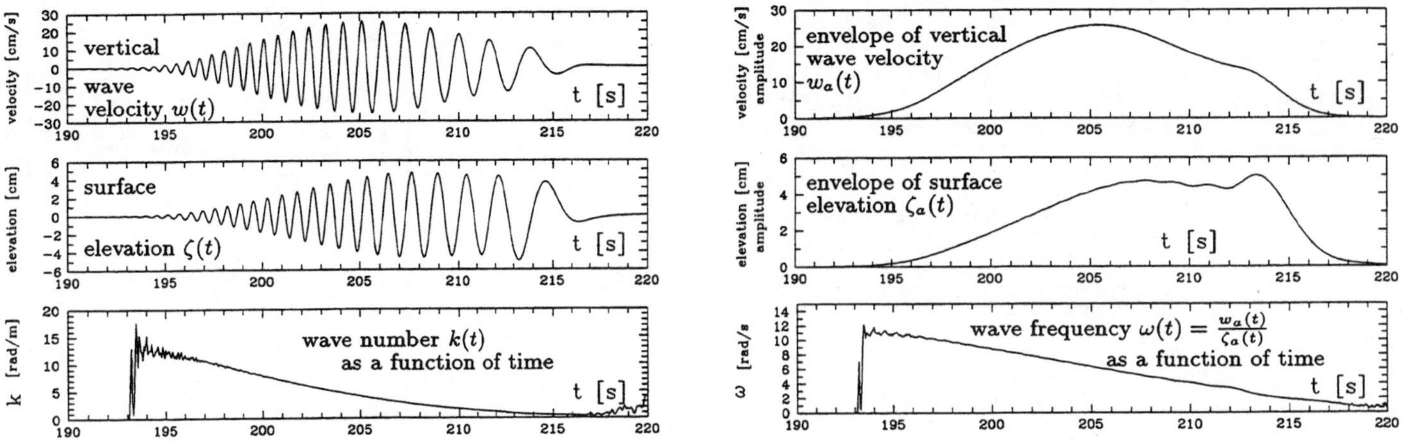

Fig. 7: Wave velocity, elevation, velocity amplitude and wave amplitude, wave frequency $\omega = f(t, x_0)$, and wave number $k = f(t, x_0)$ as a function of time

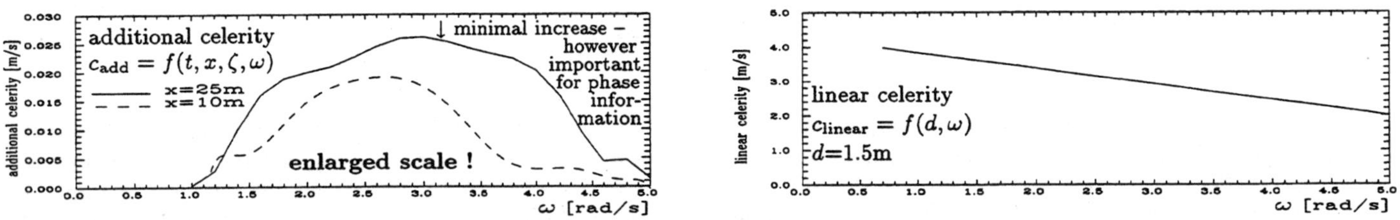

Fig. 8: Linear celerity c_{linear} and additional celerity for the wave packet (Fig. 10, x_1=10m, x_2 =25m), d=1.5m

($c_{linear} = f(\omega(t), k(t))$, $c_{add} = \dot{\bar{x}}(t, x, z = 0)$)

The surface motions at a specified location (Eulerian description) follow from

$$\xi(t, x_0, z_0) = \int_0^{\tau=t} \left[\left(\frac{\tanh[k(\zeta(\tau) + d]}{\tanh(kd)} \right)^{\frac{1}{kd}} \cdot \frac{\cosh[k(z_0 + \zeta + d)]}{\sinh(kd)} \cdot \mathrm{Re}[I(\tau - \frac{\xi}{c_I}, x_0)] \right] \mathrm{d}\tau;$$

$$\zeta(t, x_0, z_0) = \int_0^{\tau=t} \left[\left(\frac{\tanh[k(\zeta(\tau) + d]}{\tanh(kd)} \right)^{\frac{2}{kd}} \cdot \frac{\sinh[k(z_0 + \zeta + d)]}{\sinh(kd)} \cdot \mathrm{Im}[I(\tau - \frac{\xi}{c_I}, x_0)] \right] \mathrm{d}\tau, \quad (22)$$

with $k = f(t - \frac{\xi}{c_I})$. On this basis, all nonlinear elevations, motions, velocities, and accelerations of nonlinear transient wave trains can be predicted for any time and location. As shown in Fig. 8 the additional celerity due to wave height (see Eq. 21) depends on wave frequency and position of the wave group. Even if it is only 5% or less of the value for the linear celerity c_{linear}, it influences the phase shift which turns out to be a very sensitive quantity.

The transformation of the linear wave information $I(t, x_0)$ from one position to any other location is similar to the transformation of a linear wave packet: only the linear wave celerity has to be replaced by the nonlinear "information" speed $c_I(t, x)$ (see Eq. 21).

Fig. 9 shows the nonlinear water particle motions of a wave packet at four different positions. If compared to linear analysis it has been confirmed that the effective particle motions at the concentration point (right hand diagram) calculated by nonlinear theory are almost 50% higher. The illustration also shows a wave induced convective velocity in the direction of wave propagation at and near the water surface and a flow in the opposite direction at the bottom of the wave tank.

Fig. 10 demonstrates the transformation of a transient wave train, registered at position x_1=10m (**A**), to position x_2=25m (**D**), and x_3=39.1m (**G**). The registered wave train **A** is high nonlinear with steep crests and flat troughs. Its associated Fourier spectrum **AF** shows characteristic irregularities. By iterative procedures the linear information **B** (surface velocity) with its associated spectrum **BF** is filtered from wave train **A** (Note: the first derivation of the reduced (low pass filtered) registration **A** is used as a take-off function for the linear wave information **B**). The nonlinear wave elevations are calculated from this initial wave information and compared to registration **A**. The comparison **A-X** between calculation **X** and registration **A** is used to improve the wave information (the nonlinear elevations **X** are recalculated and compared again to registration **A**, and the procedure will be repeated until the comparison is satisfying). As the numerical procedure always starts from the linear wave information the wave frequency as a function of time can be calculated by using linear methods.

If the linear wave information (based on the linearized Fourier spectrum **BF**) is sufficiently satisfactory it is transferred to x_2=25m and x_3=39.1m. The resulting linear wave information **C**, **F** is transformed into the nonlinear wave trains **D**, **G** with its associated spectra **DF**, **GF** by using Eq. 22 (and the related Eq. 15-18). Note that this numerical result is confirmed by measured wave trains **E, H** as well as the difference between measured and calculated wave trains **E-D**, **H-G**, and the comparison of the associated Fourier spectra **EF/DF** and **HF/GF**).

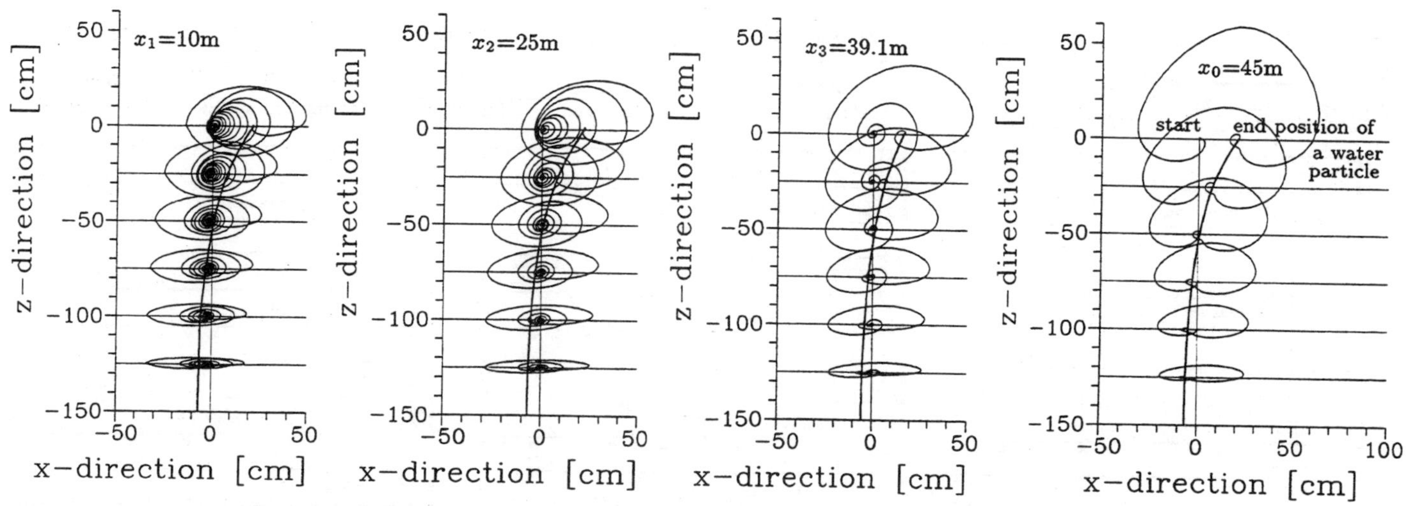

Fig. 9: Calculated water particle motions of a wave packet at x_1=10m (left hand), x_2=25m, x_3=39.1m,and at x_0=45m (concentration point, at right hand), d=1.5m (for t_{start} and t_{end} see Fig. 10)

With a Pentium computer (150MHz) the entire process of transferring a wave train (8192 points) from position x_1=10m to position x_2=25m, or x_3=39.1m takes about 4s. Four intermediate positions (x=15, 20, 30, 35m) are used to improve the information celerity c_I (see Eq. 21 and Fig. 8).

It is a tremendous advantage of the proposed method, that the nonlinear wave field is only calculated at positions, where it is needed. The transformation of the wave train to different positions is done by transposing the linear wave information. The information of different wave frequencies is superimposed as it is known from linear wave theory. Finally, at selected points, the nonlinear wave elevations are developed from the superimposed wave information.

MODEL TESTS

To illustrate the potential of the transient wave technique in generating design wave loads, the interaction of waves and a transparent vertical structure is presented. Extreme wave conditions in a 100-years design storm arise from the most unfavorable superposition of component waves of the related severe sea spectrum (Kjeldsen et al., 1979). As these "freak" waves are possible events during the life-cycle of a marine structure it may be selected as the relevant design wave. This rare event can be intentionally simulated in seakeeping tests to investigate maximum loads. Pioneering contributions on freak waves which include probability density distributions for steepness and asymmetry have been published by Myrhaug and Kjeldsen (1984). In this context it should be noted that the presented wave packet technique is based on physical reality with a continuously changing envelope of the propagating wave train (see Fig. 2 and 3). It is not comparable to the Newwave-theory which is using an artificial standard wave packet (Rozario et al., 1993).

As shown in Fig. 12, the investigated wave filter consists of 24 horizontal beams (□ 180 × 180 × 7mm, gap between beams 65mm, i.e. porosity p=26.6%). Their wave forces are measured individually, as each beam is connected to the vertical test structure by an integrated force probe. The entire research project, which is aiming for the optimization of multi-element wave filters and absorbers, is a joint venture of the Technical Universities Braunschweig and Berlin, funded by the Federal Ministry of Education, Research, and Development. While this complex project is underway, only one single preliminary test result is presented in this paper to illustrate the difference of experiments in irregular seas and transient wave trains. This test has been performed at the Large Wave Research Facility "Großer Wellenkanal" Hannover (length 330m, width 5m, water depth d=4.01m) which is jointly run by the Universities of Braunschweig and Hannover.

Fig. 11 presents two realizations of the sea spectrum H_s=1.25m and T_p=8s, both measured 3.6m behind the wave maker, i.e. - irregular waves with random phase and a specified tailored wave train which converges to a 3.2m "freak" wave at its concentration point (see Fig. 12). This transient wave train was generated in such a way that it breaks exactly into the permeable wall. By using irregular waves the probability to get a wave sequence which breaks exactly into the structure is next to nothing.

Note that the maximum wave height of the 1000-seconds irregular wave registration is about H_{max}=2.6m, which appears dramatic as compared to the harmless transient wave train at the position x=3.6m. However, 93 meters and 15 seconds later this transient wave train rears up and breaks into the vertical wave filter with terrifying consequences. As shown in Fig. 12 more than 125kN are measured on the top beam of the filter, a blow which only lasts for a few milliseconds. Extremely high loads are also measured at all other beams. If compared to irregular wave tests these loads are at least twice as high, and at the top beam even 20 times as high. A more detailed analysis reveals that the irregular waves reach the top beam of the transparent wall only a few times. The measured force gives a maximum value of about 6kN for the irregular sea state with a peak period T_p=8s and a significant wave height of H_s=1.25m (Fig. 13). As a consequence it seams inevitable, that the evaluation of reliable design loads requires model test series with tailored wave packets as they represent the most unfavourable superposition of all spectral component waves.

CONCLUSIONS

In this paper a new approach for the analysis of nonlinear transient wave trains is proposed. It is based on the definition of a complex wave information which characterizes the (linear) information of a wave group. Combined with a modified wave

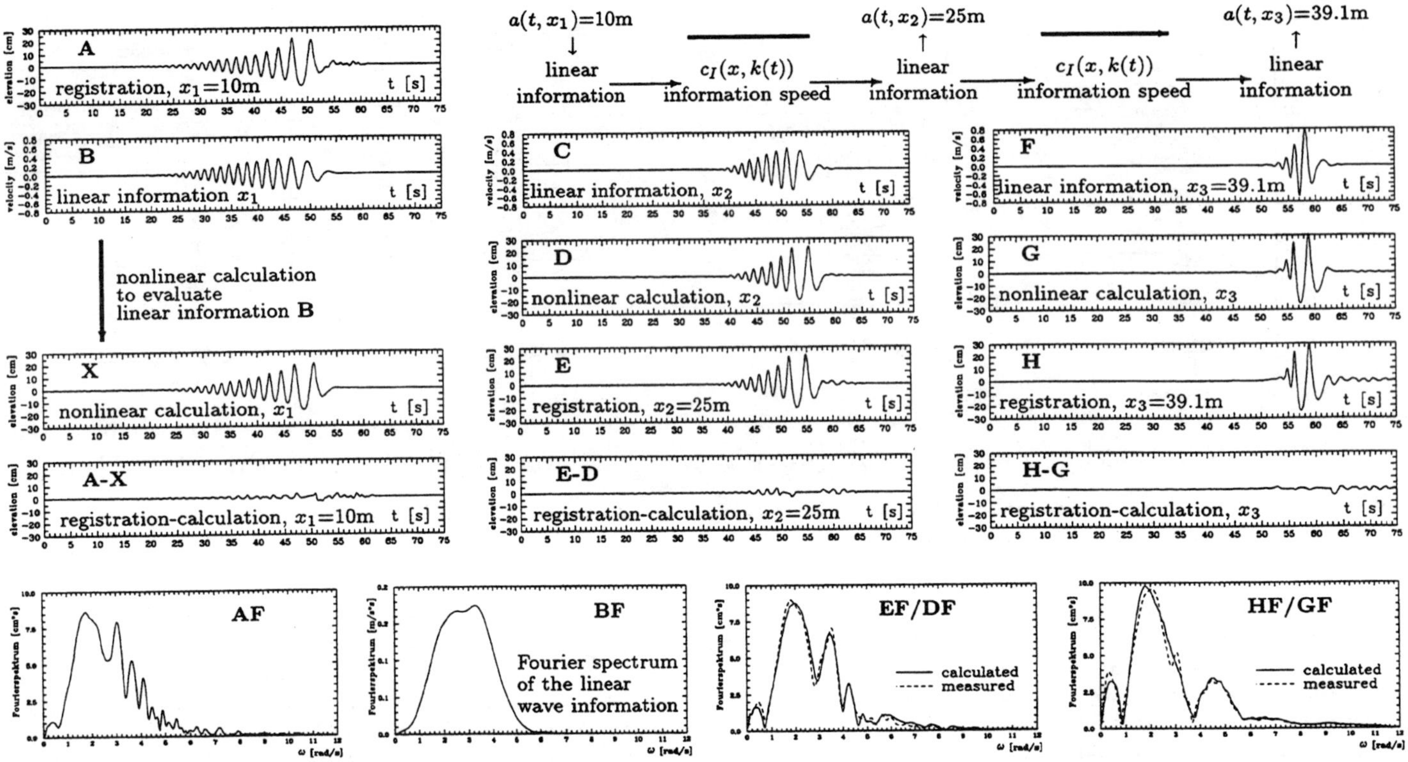

Fig. 10: Transformation of a wave packet, measured at x_1=10m, to the position x_2=25m, x_3=39.1m, and comparison between calculated and measured water surface elevations, water depth d=1.5m

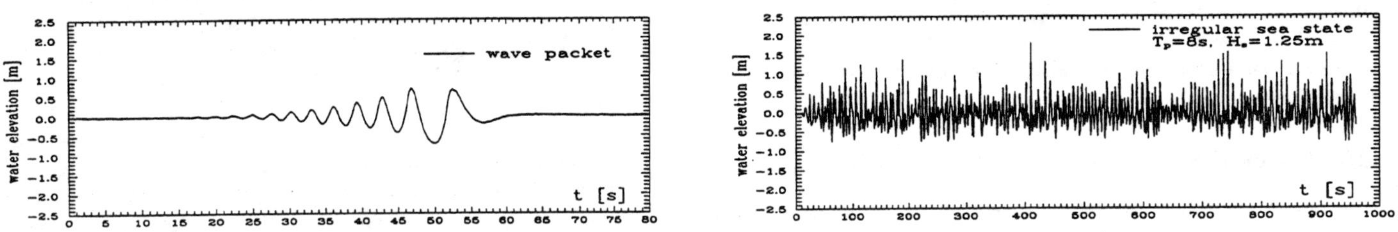

Fig. 11: Irregular sea state and transient wave train, measured at x=3.6m behind wave maker position

celerity (information speed) the propagation of high wave groups can be calculated. As a result, the nonlinear registrations and associated wave group data are derived from the respective wave information. The procedure allows the evaluation of all relevant wave parameters (motion, pressure distribution, velocities, accelerations), and serves as a sound basis for the analysis of offshore structures.

Model tests with breaking transient wave packets prove that the obtained maximum wave force on a structure is much higher as compared to model tests with irregular waves, for a given spectrum. As a consequence, it must be stated that high irregular wave tests are not sufficient for evaluating design wave loads. Only tests series with deterministic transient wave groups promise reliable results.

ACKNOWLEDGMENTS

The development of the transient wave technique has been funded by the Federal Ministry of Research and Development -BMFT- for more than ten years. The related projects dealt with analytical and numerical methods for generating and optimizing "tailor-made" wave trains and its application to seakeeping tests. The authors wish to thank the sponsor for the long-standing support which allowed us to develop and improve the technique and expand it to include nonlinear effects of very high wave groups. Based on this technique the Ocean Engineering Section (Head: Dr. Randl) of the above Ministry (Now: Federal Ministry of Education, Research and Technology - BMBF) recently granted a new project on the development of wave absorbers for harbour and coastal protection (MTK 0578). In this project our Institute of Naval Architecture and Ocean Engineering is jointly cooperating with the Leichtweiß-Institute of the Technical University Braunschweig -Prof. Dr. Oumeraci- and the Large Wave Tank "Großer Wellenkanal Hannover" (GWK: a joint coastal engineering research facility of the Universities of Braunschweig and Hannover). Just one preliminary test of this program is presented in this paper to illustrate the potential of the transient wave technique. The authors are indebted to the research members (Dipl.-Ing. Bergmann and Dipl.-Ing. Kuhlmann) as well as to the GWK-Crew Dr. Tautenhain and Dipl.-Ing. Grüne for their valuable cooperation.

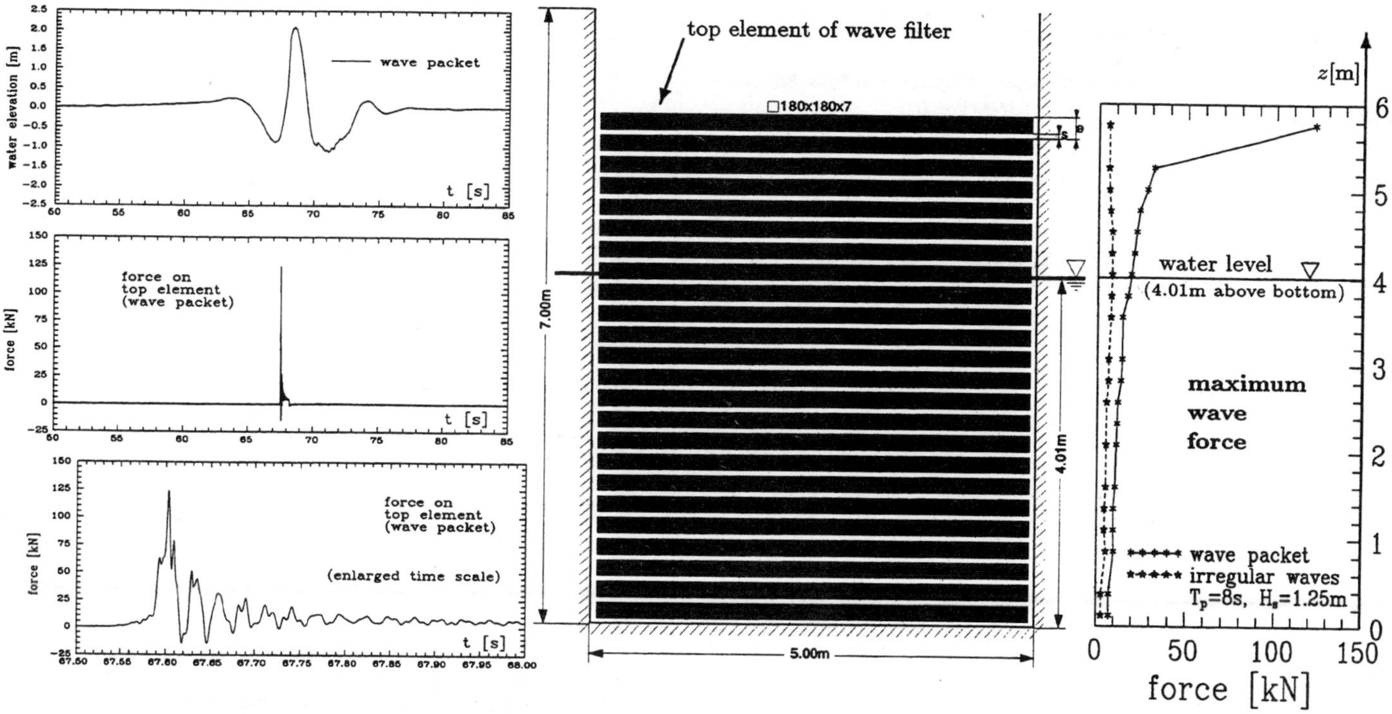

Fig. 12: Breaking transient wave packet ($H_{\max} \approx 3.2$m) with its resulting wave force on the top element of a permeable wall used as a vertical wave filter (porosity p = 26.6%, water depth d=4.01m), as well as the comparison of the maximum horizontal wave force as a function of water level, as measured with irregular waves or with the breaking transient wave train (H_s=1.25m, T_p=8s)

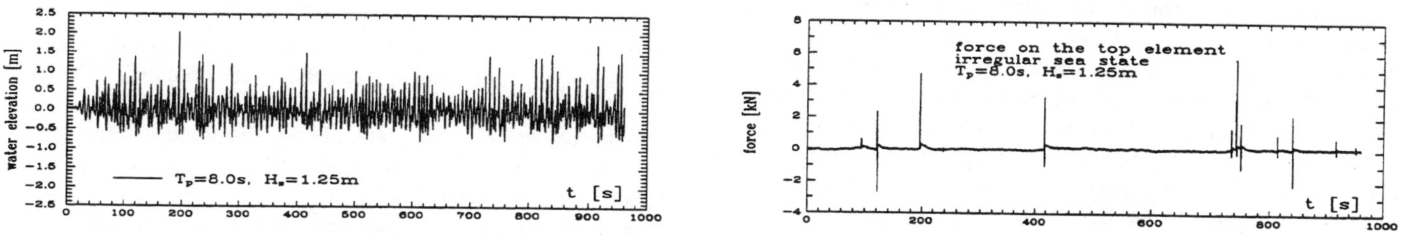

Fig. 13: Irregular sea state with its resulting horizontal wave force on the top beam of a transparent wall (H_s=1.25m, T_p=8s, water depth d=4.01m)

REFERENCES

S.K. Chakrabarti and A.R. Libby. Further verification of Gaussian wave packets. *Applied Ocean Research*, Vol.10(No.2), 1988.

G. Clauss and J. Bergmann. Gaussian wave packets — a new approach to seakeeping tests of ocean structures. *Applied Ocean Research*, Vol.8(No.4), 1986.

G.F. Clauss and W.L. Kühnlein. Transient wave packets - an efficient technique for seakeeping tests of self-propelled models in oblique waves. In *Third International Conference on Fast Sea Transportation*, pages 1193–1204, Vol. 2, Lübeck-Travemünde, Germany, September 1995.

G.F. Clauss, E. Lehmann, and C. Östergaard. *Offshore Structures.* Volume 1:Conceptual Design and Hydrodynamics, Springer Verlag London, 1992.

M.C. Davis and E.E. Zarnick. Testing ship models in transient waves. In 5th *Symposium on Naval Hydrodynamics*, 1964.

S.P. Kjeldsen and D. Myrhaug. Parametric modeling of joint probability density distributions for steepness and asymmetry in deep water waves. *Applied Ocean Research*, Vol. 6, No. 4, pages 207-220, 1984.

S.P. Kjeldsen, T. Vinje, and P. Brevig. Kinematics of deep water breaking waves. In *OTC 3714*, pages 317-321, Houston, 1979.

E.P.D. Mansard, S.E. Sand, and P. Klinting. Sub- and superharmonics in natural waves. In *Proceedings of the 6th OMAE Symposium*, Houston, June 1987.

J.B. Rozario, P.S. Tromans, P.H. Taylor, and M. Efthymiou. Comparison of loads predicted using 'NEWWAVE' and other wave models with measurements on the Tern structure. In *Wave Kinematics and Environmental Forces*, pages 143–159, Society for Underwater Technology. Printed in the Netherlands, 1993.

S.E. Sand. Wave grouping described by bounded long waves. *Ocean Engineering*, 9(6):567–580, July 1982.

C.T. Stansberg. Second-order random wave modelling in space and time, with comparison to experiments. In *Proceedings of the 14th Int. Conference on Offshore Mechanics and Arctic Engineering (OMAE)*, Copenhagen, June 1995.

S. Takezawa and T. Hirayama. Advanced experiment technique for testing ship models in transient water waves. In *11th Symp. on Naval Hydrodynamics*, 1976.

Proceedings of the Seventh (1997) International Offshore and Polar Engineering Conference
Honolulu, USA, May 25-30, 1997

ISBN 1-880653-28-1 (Set); ISBN 1-880653-31-1 (Vol. III)

Upstream Soliton Generation by a Slender, Vertical Strut and Ship: Boussinesq Equations

R.C. Ertekin
University of Hawaii at Manoa, Honolulu, HI, USA
Z.M. Qian
Aker Engineering Inc., Houston, TX, USA
J.V. Wehausen
University of California, Berkeley, CA, USA

ABSTRACT

It is well known that a "forcing" moving at a constant speed in a shallow-water channel generates two-dimensional precursor solitons in the upstream region. Such a "forcing" can be on the surface, e.g., a ship, or be submerged, e.g., a moving bump on the bottom. In this work, we use the generalized Boussinesq equations to study a slender, vertical strut and also a ship moving at a constant speed near the critical Froude number (based on water depth). By assuming that the ship is wall-sided, we obtain the vertical-strut case by letting the draft be equal to the water depth. Therefore, it suffices to formulate the ship case. To do that, we consider two regions in the fluid domain: 1) the region below the keel of the ship, and 2) the remaining fluid domain. Separate equations, specialized to these two regions, are supplemented by a matching condition at the interface. Furthermore, the boundary condition on the ship is applied along the centerplane of the ship, based upon an approximation, assuming that the ship is slender. This way, we avoid the additional complexity of numerical grid generation. The Boussinesq equations are then solved in the time domain, using a finite-difference method, to generate the three-dimensional nonlinear waves in the downstream region and the two-dimensional solitons in the upstream region. The results predicted for the soliton properties and wave resistance are compared with the available experimental data.

1. INTRODUCTION

Since the rediscovery by Huang et al. (1982a,b) of two-dimensional upstream solitons generated by a three-dimensional ship model moving with a constant forward speed in a model tank, there have been many studies of shallow-water soliton generation during the past decade and a half. Systematic experiments were conducted by Ertekin (1984) and Ertekin et al. (1984). This research first showed that soliton generation is directly controlled by the blockage coefficient, which is defined as the quotient of the maximum cross-sectional area of the ship and the cross-sectional area of the channel, and by the speed of the ship.

These kinds of nonlinear waves can also be generated by several types of two-dimensional or three-dimensional disturbances other than a ship. The pioneering research carried out by Wu and Wu (1982) showed that this phenomenon can also take place in the case of a two-dimensional pressure disturbance on the free surface when numerical solutions of the shallow-water wave equations of Boussinesq type, derived by Wu (1981), are obtained. The cases of a two-dimensional moving surface-pressure disturbance and two-dimensional moving bottom-bump disturbance were studied by Ertekin (1984) and Ertekin et al. (1984), respectively, using the equations of the Green-Naghdi restricted theory as well as the Boussinesq equations. Calculations for similar problems were carried out by Akylas (1984) and Cole (1985) based on the Korteweg-de Vries (KdV) equation. Lee (1985) analyzed numerically a two-dimensional moving surface-pressure disturbance and a two-dimensional moving bottom-bump disturbance using the KdV and Boussinesq equations, and conducted experiments for a moving bottom-bump. Numerical results for a three-dimensional moving pressure disturbance were obtained by Ertekin et al. (1986) based on the Green-Naghdi equations, by Wu and Wu (1982) based on the Boussinesq equations, and by Katsis and Akylas (1987) based on the Kadomtsev-Petviashvili equations (see also Wu, 1987). The generation of solitons for different types of channel cross-sections was studied both numerically and experimentally by Teng (1990).

Three-dimensional calculations for the case of ships have been carried out by different researchers in order to make direct comparisons with the experimental work of Ertekin (1984). The treatment of the three-dimensional ship boundary condition is complicated because of the difficulties in introducing a ship hull into the solution of the shallow-water wave equations, which themselves are two dimensional. An asymptotic perturbation technique was used by Mei (1986) in conjunction with the KdV equations to initiate a study on ship solitons, but since the ship was modeled as a thin strut, only two-dimensional downstream waves have resulted. This method was extended to ships by Choi and Mei (1989). A strut with an arbitrary beam was studied by Ertekin and Qian (1988) and Qian (1988) based on Boussinesq-type equations. Numerical results showed quantitative agreement with experimentally obtained solitons upstream and cnoidal waves downstream. Later, additional results on ship wave resistance were presented by Ertekin and Qian (1988,1989) and by Ertekin et al. (1990), and compared with the experimental data of Ertekin (1984). A wall-sided ship of small blockage was modeled by Ertekin and Qian (1990), which unified the strut and the ship cases through the body-boundary condition. The case of a strut was also independently simulated both numerically and experimentally by Cui and Wu (1988). Bai et al. (1989) used the finite-element method to model a strut moving in a shallow-water channel, which was the first alternative approach to using shallow-water wave equations. Choi et al. (1990) continued this work for moving strut and ship, based on both the finite-element method and the asymptotic shallow-water wave equations. Similar work that uses various shallow-water wave equations can be found, for example, in Tomasson and Melville (1991) and in Chen and Sharma (1992).

In this paper, a wall-sided ship of flat keel is considered and the generalized Boussinesq equations (gB) obtained by Wu (1981) are used as the governing equations. A simple condition representing the ship boundary is derived as a unified boundary condition that can be easily extended to the strut case used in the study by Qian (1988). To simplify the solution, a thin-ship case is solved numerically so that the ship boundary condition can be imposed directly on the centerplane of the channel. Thus, the computational domain becomes rectangular and the generation of a boundary-fitted coordinate system is not required. Although the ship model used here is not the Series 60 one that was used by Ertekin (1984), Choi et al. (1990) and Chen and Sharma (1992), the blockage effect, which is considered the most important parameter (see Ertekin, 1984), is confirmed here for different drafts of the ship.

2. ANALYTICAL MODEL

2.1 Governing Equations

The generalized Boussinesq (gB) equations obtained by Wu (1981) will be adopted. The fluid is assumed to be incompressible and inviscid, and of constant mass density ρ. The assumption of irrotationality of flow is made so that a velocity potential can be introduced. Two parameters are defined: $\alpha = A/h, \varepsilon = h/L$, where A is a typical wave amplitude, h the constant waterdepth, and L the length of the ship. These two parameters, α and ε, are introduced to represent the nonlinear and dispersive behaviors of waves, although they are assumed to be small for long waves in shallow water. The gB equations are simplified by maintaining both nonlinearity and dispersion in case the Ursell number is unity, i.e., $\alpha = \varepsilon^2$, after the functions are averaged along the vertical direction and expanded, so that the final equations have an error of order $(\alpha\varepsilon^4, \alpha^2\varepsilon^2)$.

The gB equations for constant water depth, as obtained by Wu (1981), are:

$$\zeta_t + \nabla[(\zeta + h)\nabla\phi] = 0\,, \tag{1}$$

$$\phi_t + \frac{1}{2}|\nabla\phi|^2 + g\zeta = \frac{h^2}{3}\nabla^2\phi_t\,, \tag{2}$$

where ζ is the surface elevation, ϕ the layer-mean potential of horizontal velocity, g the gravitational acceleration, t the time, and $\nabla = (\partial/\partial x, \partial/\partial y)$, where x and y are the horizontal coordinates. The bottom no-flux condition and the kinematic and dynamic free-surface conditions are satisfied within the approximation that led to derivation of these equations. The Eqs. (1) and (2) are then in the two-dimensional horizontal plane and are used to determine the three-dimensional behavior, i.e., $\zeta = \zeta(x,y,t)$, $\phi = \phi(x,y,t)$

2.2 Treatment of the Ship Boundary Condition

Although the governing equations, Eqs. (1) and (2), are two-dimensional, the ship hull, which does not extend to the bottom of the channel in general, is three-dimensional as shown in Fig. 1. Hence, this three-dimensional boundary can not be directly considered in the solution of the shallow-water wave equations. In the works of Choi and Mei (1989) and Chen and Sharma (1992), the perturbation solution near the ship follows the slender-body theory and matches the solution to the one in the region far from the ship. In this paper, a simplified ship boundary condition is derived based on the continuity equation and is then imposed directly on the solution without use of a perturbation expansion. A wall-sided ship with a flat keel, as shown in Fig. 1, is chosen to simplify the problem.

In the inertial coordinate system, used in Eqs. (1)-(2), the body boundary condition on $y=f(x)$ and $-d \le z \le \zeta$ is written as:

$$\mathbf{n}\cdot\nabla\phi_P = -U\,n_x\,, \tag{3}$$

where ϕ_P is the velocity potential of the three-dimensional flow, d the draft and U the constant forward speed of the wall-sided ship, moving from right to left in the negative Ox direction, and the unit normal vector of the body, pointing into the fluid, is given by

$$\mathbf{n} = (n_x, n_y, 0)\,,\quad n_x = \frac{-f_x}{\sqrt{1+f_x^2}}\,,\quad n_y = \frac{1}{\sqrt{1+f_x^2}}\,. \tag{4}$$

The Eq. (3) may also be obtained of course by using the relation between the velocity potentials in the inertial and moving coordinates, i.e., $\phi_P = \bar{\phi}_P - U\bar{x}$, where $\bar{\phi}_P$ is the velocity potential in the moving coordinates, $\overline{Oxyz}$, fixed in the body.

We now define the depth-averaged velocities for the two regions, i.e., Regions I and II in Fig. 1(b), by

$$\nabla\phi^{I} = \frac{1}{(h+\zeta)}\int_{-h}^{\zeta}\nabla\phi_P\,dz\,,\qquad \nabla\phi^{II} = \frac{1}{(h-d)}\int_{-h}^{-d}\nabla\phi_P\,dz\,. \tag{5}$$

These depth-averaged velocities can be shown to lead to the same order of error that is contained in Eqs. (1)-(2) when $\nabla h = 0$

[see Wu (1981), Eq. (40)]. In other words, ϕ^I is the same as ϕ_P in Eqs. (1) and (2).

In order to relate the two functions ϕ^I and ϕ^{II}, we shall exploit the continuity of ϕ_P across the boundary between Regions I and II. In particular, if ϕ_P^- is the value of ϕ_P on the I-side of the boundary and if ϕ_P^+ is the value on the II-side, then it follows that $\nabla\phi_P^- = \nabla\phi_P^+$ at this interface. In order to make use of the continuity, consider the integral of $\mathbf{n}\cdot\nabla\phi_P^-$ over the I-side of the interface boundary plus the ship boundary:

$$\begin{aligned}\int_{-h}^{\zeta}\mathbf{n}\cdot\nabla\phi_P^-\,dz &= \int_{-h}^{-d}\mathbf{n}\cdot\nabla\phi_P^-\,dz + \int_{-d}^{\zeta}\mathbf{n}\cdot\nabla\phi_P^-\,dz\\ &= \int_{-h}^{-d}\mathbf{n}\cdot\nabla\phi_P^+\,dz + \int_{-d}^{\zeta}\mathbf{n}\cdot\nabla\phi_P^-\,dz,\end{aligned}$$

or, using Eq. (3) and keeping in mind that $\mathbf{n}$ is independent of the z coordinate,

$$(h+\zeta)\mathbf{n}\cdot\nabla\phi^I(x,f(x),t) = (h-d)\mathbf{n}\cdot\nabla\phi^{II}(x,f(x),t) - (\zeta+d)Un_x \tag{6}$$

We now introduce Eq. (4) into Eq. (6) to obtain the interfacial matching condition:

$$(h+\zeta)\left[-f_x\phi_x^I + \phi_y^I - f_xU\right] = (h-d)\left[-f_x\phi_x^{II} + \phi_y^{II} - f_xU\right] \tag{7}$$

which is to be enforced on y=f(x).

The variables will now be made dimensionless according to Eq. (8):

$$x = Lx',\ y = By',\ z = hz',\ z = Az',\ \phi^{I,II} = \frac{gAL}{U}\phi'^{I,II},$$
$$Fr = \frac{U}{\sqrt{gh}},\ \alpha = \frac{A}{h}, \tag{8}$$

where B is the beam of the ship and Fr is the depth Froude number.

Upon substitution of the dimensionless variables given by Eq. (8) into Eq. (7), and removal of the primes for convenience, one has:

$$\begin{aligned}&(1+\alpha\zeta)\left[-\alpha\left(\frac{B}{L}\right)^2 f_x\phi_x^I + \alpha\phi_y^I - \left(\frac{B}{L}\right)^2 f_x Fr^2\right]\\ &= \left(1-\frac{d}{h}\right)\left[-\alpha\left(\frac{B}{L}\right)^2 f_x\phi_x^{II} + \alpha\phi_y^{II} - \left(\frac{B}{L}\right)^2 f_x Fr^2\right].\end{aligned} \tag{9}$$

The Eq. (9) evidently relates the solutions in Regions I and II. To evaluate the terms containing ϕ^{II}, consider the three-dimensional potential flow, ϕ_P, in Region II. Upon integration of the Laplace equation in the z direction and application of the no-flux condition at the keel and at the bottom of the channel, we obtain

$$\phi_{xx}^{II}\left(\frac{B}{L}\right)^2 + \phi_{yy}^{II} = 0. \tag{10}$$

Integration of Eq. (10) in the y direction yields the following equation for $\phi_y^{II}(x,f(x),t)$:

$$\phi_y^{II} = -\left(\frac{B}{L}\right)^2\int_0^{f(x)}\phi_{xx}^{II}\,dy, \tag{11}$$

where the integration constant is zero because ϕ_y^{II} vanishes on the symmetry boundary.

Upon substituting Eq. (11) for ϕ_y^{II} in Eq. (9), one observes that the terms containing the coefficient $\alpha(B/L)^2$ are smaller than the terms containing only α or $(B/L)^2$, since both α and $(B/L)^2$ are small by assumption, and Fr=O(1). Neglecting such higher-order terms, the interfacial matching condition, Eq. (9), becomes:

$$(1+\alpha\zeta)\left[\alpha\phi_y^I - \left(\frac{B}{L}\right)^2 f_x Fr^2\right] = \left(\frac{d}{h}-1\right)\left[\left(\frac{B}{L}\right)^2 f_x Fr^2\right], \tag{12}$$

which is to be enforced on y=0 (centerplane) due to the thin-ship assumption. Hence, after returning to physical variables,

$$\phi_y^I = Uf_x\frac{(\zeta+d)}{(\zeta+h)}. \tag{13}$$

Since the vertically averaged ϕ^I is the same as ϕ used in Section 2.1, Eq. (13) becomes

$$\phi_y(x,f(x),t) = Uf_x\frac{(\zeta+d)}{(\zeta+h)}. \tag{14}$$

This equation is to be satisfied along the horizontal boundary line of the ship, which describes the combined effect from both the moving ship and the flow underneath it. When d=h, the ship is reduced to a strut, and thus, Eq. (14) becomes $\phi_y = Uf_x$, derived by Mei (1986). Moreover, the form of Eq. (14) remains the same in the moving coordinate system fixed in the body since Eq. (14) does not involve the x derivative of the velocity potential. Finally, therefore, the three-dimensional ship boundary is simplified so that the two-dimensional shallow-water wave equations can be used directly.

2.3 Other Boundary Conditions and Wave Resistance

The ship is assumed to start moving abruptly in the initially quiescent water. The initial conditions for the gB equations are therefore given by

$$\zeta(x,y,0) = 0, \quad \phi(x,y,0) = 0. \tag{15}$$

On the two fictitious boundaries (see Fig. 1), namely on EO, far in front of, and on CD, far behind the ship, Orlanski's open-boundary condition with constant phase speed, i.e., $c/\sqrt{gh} = \pm 1.0$, is used to avoid significant reflections (see Orlanski, 1976). This condition was successfully used by Wu and Wu (1982), Ertekin (1984) and Qian (1988) among others. The open boundary conditions are as follows:

$$\zeta_t \pm c\,\zeta_x = 0, \quad \phi_t \pm c\,\phi_x = 0, \tag{16}$$

where the + and - signs refer to the downstream and upstream open boundaries, respectively.

The no-flux boundary condition on the wall ED, and on the symmetry boundaries OA and BC, is given by

$$\phi_y = 0. \tag{17}$$

The wave resistance R is obtained by integrating the pressure on the wetted ship-boundary. Since the bottom of the wall-sided ship is flat by assumption, the integration will be carried out only along the vertical part of the ship hull. The pressure at each point can be derived in the following form for the gB equations:

$$p = -\rho g(z-\zeta) + \rho(zh + \frac{z^2}{2})\nabla^2\phi_t. \tag{18}$$

Therefore, the wave resistance becomes

$$R = 2\int_0^L p_1 f_x dx,$$

$$p_1 = \frac{1}{2}\rho g(\zeta+d)^2 + \frac{1}{6}\rho(\zeta+d)\left[\zeta^2 + (\zeta-d)(3h-d)\right]\nabla^2\phi_t. \tag{19}$$

3. Numerical Results and Discussion

The boundary-value problem is solved at each time step. It should be emphasized that, in general, Eq. (14) is satisfied on the exact interface y=f(x). The ship here is assumed to be thin, so that Eq. (14) can be imposed directly on the centerplane of the channel. Therefore, the computational domain becomes rectangular. The final calculations are done in the moving coordinate system fixed in the body. The governing equations, Eqs. (1) and (2), and the open boundary conditions, Eq. (16), are transformed into the moving coordinate system, given by $x = \bar{x} - Ut$, $y = \bar{y}$, $z = \bar{z}$, and then solved by the modified Euler method for time integration. During each time step, the velocity potential, ϕ, is solved iteratively by the Successive-Over- Relaxation (SOR) method subjected to the boundary conditions, Eqs. (14) and (17) in moving coordinates. The central-difference formulas are used to maintain the accuracy in spatial discretization.

The geometric parameters of these problems are selected to correspond to the experiments of Ertekin (1984). These experiments, which were conducted in the Ship Model Towing Tank of the University of California at Berkeley, are also described by Ertekin et al. (1984). The ship model employed was the Series 60 model of 0.8 block coefficient. The model length on the waterline was 1.52m and the beam was 0.23m.

The blockage coefficient S_B is defined as the quotient of the cross-sectional under-water area of the strut or ship and the cross-sectional area of the channel below the still-water level. The boundary curve in the horizontal plane of the strut or ship is assumed parabolic: $y = 2Bx(1-x/L)/L$, where L is the length and B the beam of the ship. The blockage coefficients used in the computations match those of the experimental model for both the strut and ship. The ship is discretized by 101 points along the ship boundary. The grid size $\Delta x = \Delta y$ varies between 0.1 and 0.15, and the time interval Δt varies between 0.05 and 0.08. The number of nodes varies from 801x41 to 801x81. The number of time steps in the calculations ranges from 2000 to 4500. The tolerance of the convergence for solving the SOR matrix is set at 10^{-8}. The surface elevation and velocity potential in the gB model are filtered by a five-point formula at each time step in both the x and y directions (see Shapiro, 1975) to eliminate the saw-tooth behavior of the waves. The amplitude and speed of the first soliton are computed after this soliton is completely developed, while the period of generation (the period between the first and second solitons) is computed from the moving gauges the farthest from the ship. See Ertekin (1984), Fig. 97, for the locations of the moving gauges attached to the carriage.

Conservation of mass is checked to verify the accuracy of the numerical calculations. The dimensionless water mass, $m/(\rho h^3)$, is given by

$$\int_{A_c} (1+\zeta)dA_c + \int_0^t \left[\int_{l_R} (1+\zeta)udx - \int_{l_L} (1+\zeta)udx - \right] dt, \tag{20}$$

where l_L, l_R represent the open boundaries at the left and right ends of the channel, respectively, and A_c denotes the area of the computational domain in the horizontal plane. A change of mass of less than 5% is achieved in most of the calculations.

The perspective and contour plots, surface elevations at numerical moving wave gauges, and wave resistance are shown in Figs. 2 and 3 for $S_B = 3.14\%$ and $U/\sqrt{gh} = 1.0$ and for the thin strut and the ship, respectively. The exact strut results, i.e., without the use of the simplified boundary condition developed here, were presented by Qian (1988). Two-dimensional solitons are generated ahead of the thin strut or thin ship as seen in Figs. 2 and 3. The downstream waves are three-dimensional and do not seem to have reached steady state yet. Also, the wave patterns both upstream and downstream are similar for both the strut and ship cases, although upstream solitons are generated earlier in the case of the ship. Cross-channel waves also form following the diverging waves behind the stern (see the perspective plots in Figs. 2(a) and 4(a)). The soliton generation is delayed in the case of the thin strut or ship, compared with the case of the exact strut results presented in Qian (1988).

The calculated results for the upstream soliton and wave resistance are listed in Table 1 for $S_B = 2.32\%$, 3.14% and 7.53%, for the exact strut (Qian, 1988), thin strut and thin ship cases, and are compared with the experimental data (Ertekin, 1984). In Table 1, A_s is the upstream soliton amplitude, U_s the soliton speed, and T_g the period of generation of solitons. The soliton amplitude, speed and period of generation, corresponding to the thin strut and the thin ship show the same tendencies as those computed for the exact strut and ship, when compared with the experiments. The average wave-resistance is close to the total wave-resistance measured in the experiments. The wave profile at different locations and the wave resistance are plotted as functions of time in Figs. 2(c) and 3(c), and compared with the experimental records shown in Fig. 4. The comparison is generally acceptable.

Table 1 Comparison of dimensionless properties of solitons and mean resistance for the exact strut (Qian, 1988), thin strut, thin ship and experiments (Ertekin, 1984).

Case	S_B (%)	$\frac{d}{h}$	$\frac{U}{\sqrt{gh}}$	$\frac{A_s}{h}$	$\frac{U_s}{\sqrt{gh}}$	$\frac{UT_g}{h}$	R(N)
Exact Strut	2.32	N/A	N/A	N/A	N/A	N/A	N/A
Thin Strut	2.32	1.00	1.00	0.30	1.09	106	1.1
Thin Ship	2.32	0.25	1.00	0.31	1.15	116	1.2
Experiment	2.32	0.25	1.00	0.28	1.13	98	1.9
Exact Strut	3.14	1.00	1.00	0.33	1.16	69	4.5
Thin Strut	3.14	1.00	1.00	0.34	1.14	69	4.3
Thin Ship	3.14	0.33	1.00	0.41	1.18	55	6.6
Experiment	3.14	0.33	1.00	0.38	1.16	67	6.6
Exact Strut	7.53	1.00	1.00	0.53	1.22	38	4.8
Thin Strut	7.53	1.00	1.00	0.52	1.22	39	4.6
Thin Ship	7.53	0.33	1.00	0.62	1.26	32	8.2
Experiment	7.53	0.33	1.00	0.49	1.20	47	6.0

The results demonstrate the validity of the simplified boundary condition for a thin strut and a thin ship in shallow-water soliton generation. In general, both the thin-strut and thin-ship models can give the same results as the exact model when the blockage coefficient is small. The thin-strut results are almost identical to the exact-strut results within the assumptions used. Similar conclusions can be reached when one abandons the slender-body assumption and uses numerical grid-generation to enforce the exact body-boundary condition, provided that the blockage coefficient is small (see Qian, 1994).

Acknowledgment: This material is based upon work partially supported by the U.S. National Science Foundation under grant numbers MSM-8706910 and BCS-8958346. The computer time was provided by San Diego Super Computer Center and by IBM Corporation. SOEST Contribution No. 4373.

REFERENCES

Akylas, T.R. (1984), "On the Excitation of Long Nonlinear Water Waves by a Moving Pressure Distribution", *J. Fluid Mechanics*, Vol. 141, pp. 455-466.

Bai, K.J., Kim, J.W. and Kim Y.H. (1989), "Numerical Computations for a Nonlinear Free Surface Flow Problem", *Proc. 5th International Conference on Numerical ship Hydrodynamics*, Hiroshima, September, pp. 305-320, (Ed. by K.-H. Mori, Nat. Acad. Press, Washington, D.C., 1990).

Chen, X.N. and Sharma, S.D. (1992), "A Slender Ship Moving at a Near-Critical Speed in a Shallow Channel", *Proc. 18th IUTAM Congress*, Israel.

Choi, H.S., Bai, K.J., Kim, J.W. and Cho, H.H. (1990), "Nonlinear Free Surface Waves due to a Ship Moving Near the Critical Speed in a Shallow Water", *Proc. 18th Symposium on Naval Hydrodynamics*, Ann Arbor, Michigan, pp. 173-189, (National Academy Press, Washington, D.C., 1991).

Choi, H.S. and Mei, C.C. (1989), "Wave Resistance and Squat of a Slender Ship Moving near Critical Speed in Restricted Water", *Proc. 5th International Conference on Numerical Ship Hydrodynamics*, Hiroshima, September, pp. 337-350, (Ed. by K.-H. Mori, Nat. Acad. Press, Washington, D.C., 1990).

Cole, S.J. (1985), "Transient Wave Produced by Flow Past a Bump", *Wave Motion*, Vol. 7, pp. 579-587.

Cui, X.Y. and Wu, D.M. (1988), "Waves Caused by a Slender Ship Advancing near Critical Speed in a Shallow Channel", *Shipbuilding of China*, No. 103, pp. 1-11.

Ertekin, R.C. (1984), "Soliton Generation by Moving Disturbances in Shallow Water: Theory, Computation and Experiment", Ph.D. Dissertation, University of California, Berkeley, California, v+352 pp.

Ertekin, R.C., Webster, W.C. & Wehausen, J.V. (1984), "Ship-Generated Solitons", *Proc. 15th. Symp. on Naval Hydrodynamics*, Hamburg, September, pp. 347-364 (Ed. by C.M. Lee, National Academy Press, Washington, D.C., 1985).

Ertekin, R.C., Webster, W.C. & Wehausen, J.V. (1986), "Waves Caused by a Moving Disturbance in a Shallow Channel of Finite Width", *J. Fluid Mech.*, Vol. 169, pp. 275-292.

Ertekin, R.C. and Qian, Z.M. (1988), "Soliton Generation and Ship Wave Resistance in Shallow Waters. Part I: Vertical Strut", Univ. of Hawaii, Dept. of Ocean Engng., Report No. UHMOE-88101, Honolulu, December, ii+50pp.

Ertekin, R.C. and Qian, Z.M. (1989), "Numerical Grid Generation and Upstream Waves for Ships Moving in Restricted Waters", *Proc. 5th International Conference on Numerical ship Hydrodynamics*, Hiroshima, September, pp. 321-336, (Ed. by K.-H. Mori, Nat. Acad. Press, Washington, D.C., 1990).

Ertekin, R.C. and Qian, Z.M. (1990), "Soliton Generation and Ship Wave Resistance in Shallow Waters. Part II: Vertical Strut and Ship", Univ. of Hawaii, Dept. of Ocean Engng., Report No. UHMOE-90107, Honolulu, October, 44 pp.

Ertekin, R.C., Qian, Z.M. and Wehausen, J.V. (1990), "Upstream Solitons and Wave Resistance", in Engineering Science, Fluid Dynamics, Ed. by G.T. Yates, World Scientific Publishing Co., New Jersey, pp. 29-43.

Huang, De-Bo, Sibul, O.J. & Wehausen, J.V. (1982a), "Ships in Very Shallow Water", *Festkolloquium zur Emeritierung von Karl Wieghardt*, März 1982, Institut für Schiffbau der Universität, Hamburg, Bericht Nr. 427 (April 1983), pp. 29-49.

Huang, De-Bo, Sibul, O.J., Webster, W.C., Wehausen, J.V., Wu De-Ming & Wu T.Y. (1982b), "Ships Moving in the Transcritical Range", *Proc. Conf. on Behavior of Ships in Restricted Waters,* November, 1982, Varna, Bulgaria, Vol. 2, pp. 26/1-26/10.

Katsis, C. and Akylas, T.R. (1987), "On the Excitation of Long Nonlinear Water Waves by a Moving Pressure Distribution. Part 2. Three-Dimensional Effects", *J. Fluid Mech.*, Vol. 177, pp. 49-65.

Lee, S.J. (1985), "Generation of Long Water Waves by Moving Disturbances", Ph.D. Dissertation, California Institute of Technology, Pasadena.

Mei, C.C. (1986), "Radiation of Solitons by Slender Bodies Advancing in a Shallow Channel", *J. Fluid Mech.*, Vol. 162, pp. 53-67.

Orlanski, I. (1976), "A Simple Boundary Condition for Unbounded Hyperbolic Flows", *J. Comput. Phys.*, Vol. 21, pp. 251-269.

Qian, Z.M. (1988), "Numerical Grid Generation and Nonlinear Waves Generated by a Ship in a Shallow-Water Channel", M.S. Thesis, Univ. of Hawaii, Dept. of Ocean Engng., Honolulu, Hawaii.

Qian, Z.M. (1994), "Calculations of Three-Dimensional Nonlinear Ship Waves and Ship Resistance in a Shallow Water Channel", Ph.D. Dissertation, Univ. of Hawaii, Dept. of Ocean Engng., Honolulu, Hawaii.

Shapiro, R. (1975), "Linear Filtering", *Maths. Comput.* Vol. 29, pp. 1094-1097.

Teng, M.H. (1990), "Forced Emissions of Nonlinear Water Waves in Channels of Arbitrary Shape", Ph.D. Thesis, California Institute of Technology, Pasadena, CA.

Tomasson, G.G. and Melville, W.K. (1991), "Flow Past a Constriction in a Channel: a Model Description", *J. Fluid Mech.*, Vol. 232, pp. 21-45.

Wu, T.Y. (1981), "Long Waves in Ocean and Coastal Waters", *J. Engng. Mech.*, ASCE, Vol. 107, pp. 501-522.

Wu, D.M. and Wu, T.Y. (1982), "Three-Dimensional Nonlinear Long Waves due to Moving Surface Pressure", *Proc. 14th. Symp. Naval Hydrodynamics*, (National Academy Press, Washington, D.C., 1983), pp. 103-125.

Wu, T.Y. (1987), "Generation of Upstream Advancing Solitons by Moving Disturbances", *J. Fluid Mech.* Vol. 184, pp. 75-99.

Wu, D.M. and Wu, T.Y. (1987), "Precursor Solitons Generated by Three-Dimensional Disturbances Moving in a Channel", *Proc. IUTAM Symp. on Non-Linear Water Waves*, Tokyo, Japan, pp. 18-19.

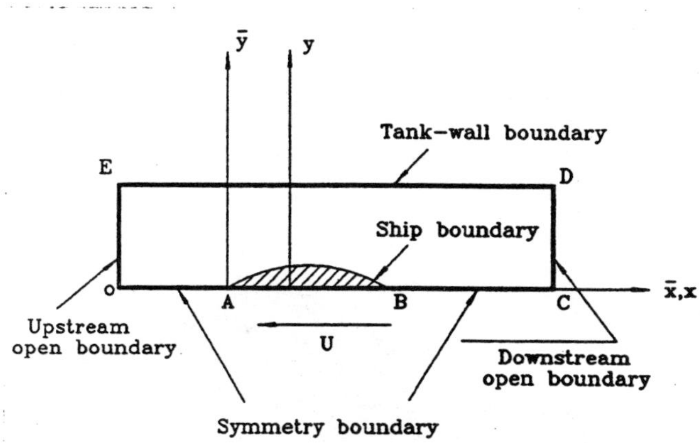

(a)

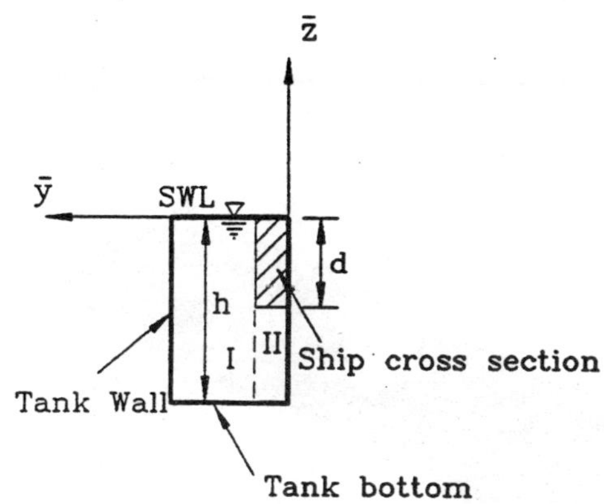

(b)

Figure 1 Configuration of the physical region: (a) plan view; (b) cross-sectional view.

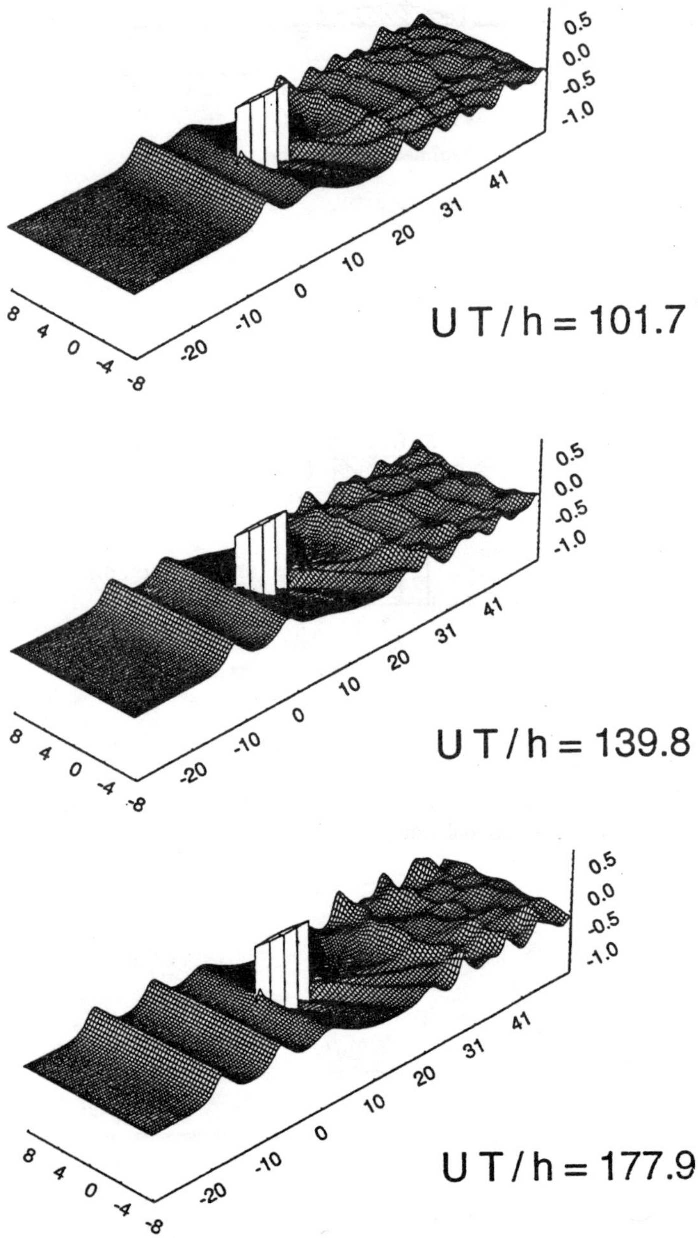

Figure 2(a) Perspective plots of computed surface elevation for thin strut. S_B =3.14%, Fr=1.0, and d/h=1.0.

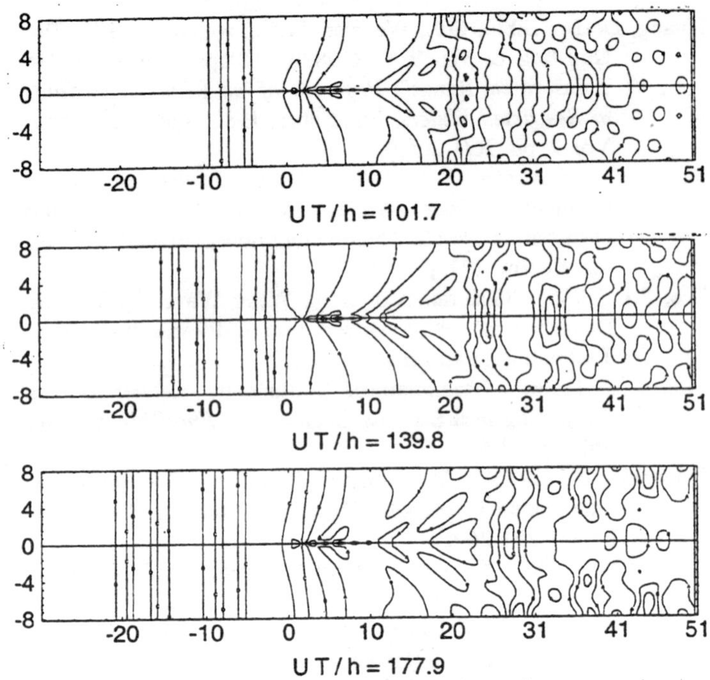

Figure 2(b) Contour plots of computed surface elevation for thin strut. S_B =3.14%, Fr=1.0, and d/h=1.0.

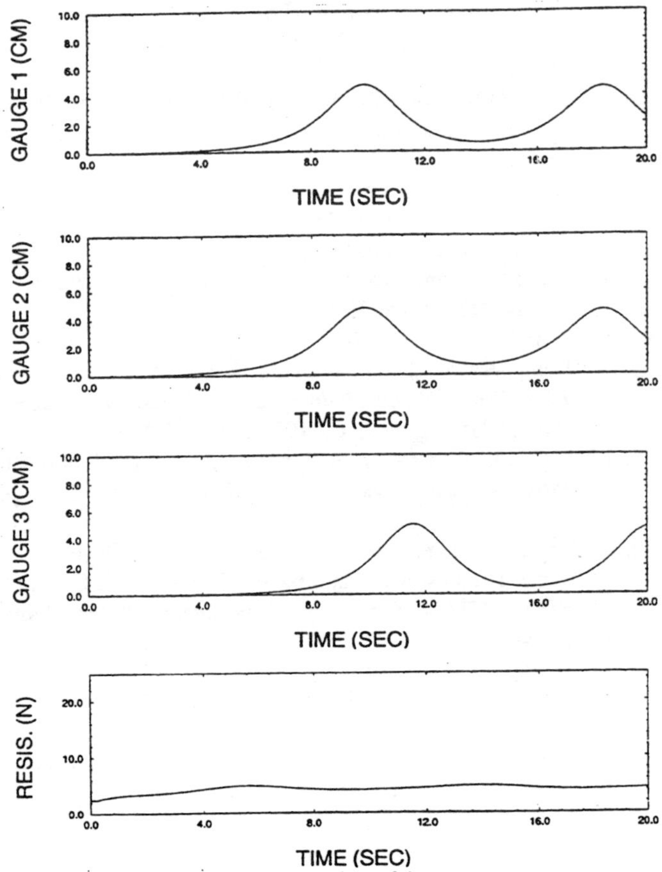

Figure 2(c) Computed surface elevation at four numerical, moving wave-gauges and wave resistance for thin strut. S_B =3.14%, Fr=1.0, and d/h=1.0, corresponding to Test No. 735 of Ertekin (1984).

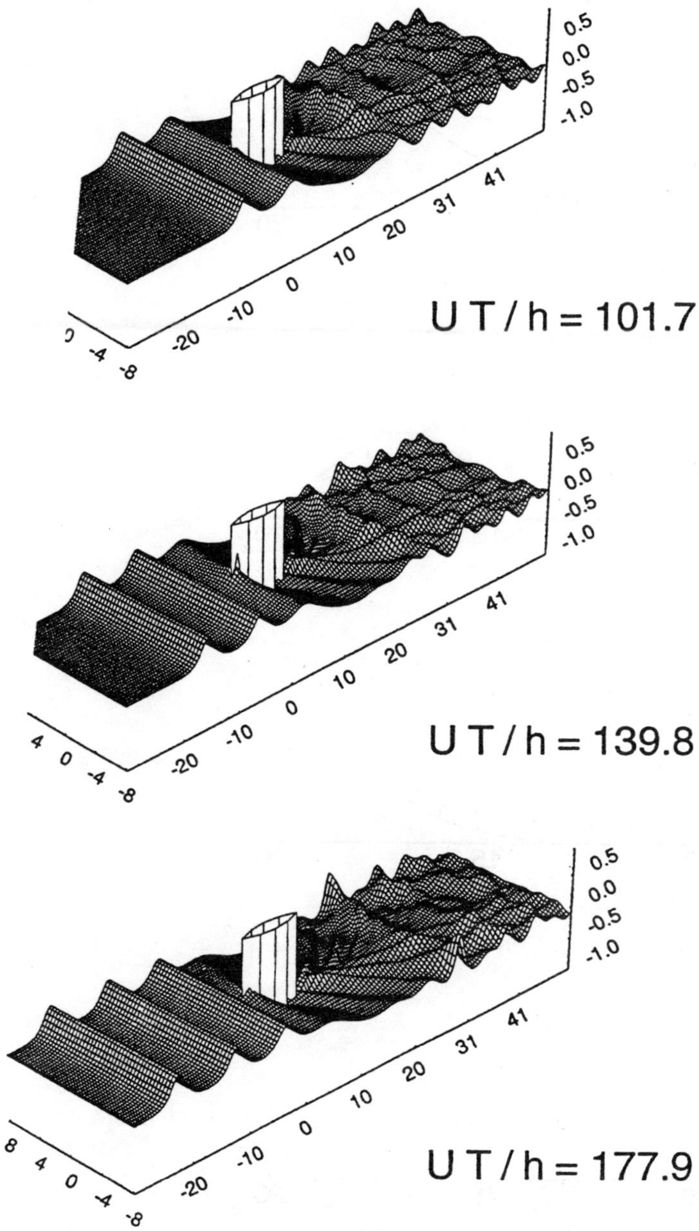

Figure 3(a) Perspective plots of computed surface elevation for thin ship. S_B =3.14%, Fr=1.0, and d/h=0.33.

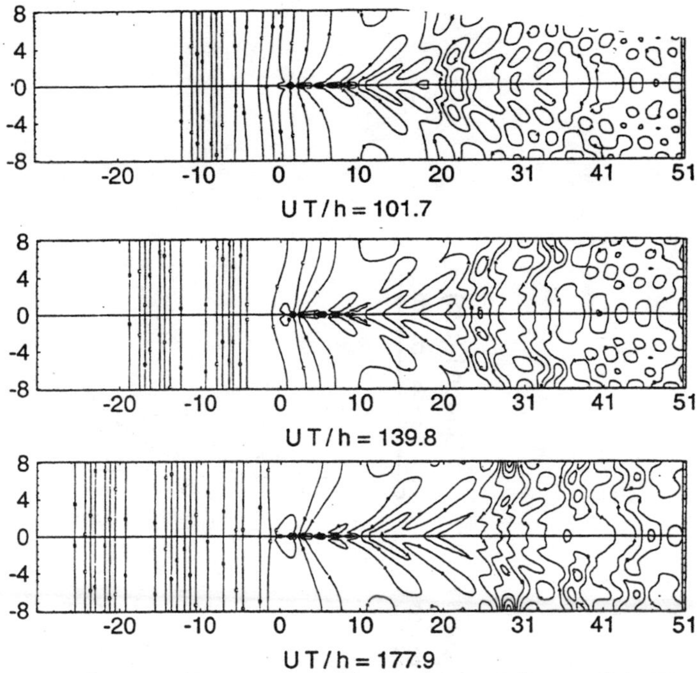

Figure 3(b) Contour plots of computed surface elevation for thin ship. S_B =3.14%, Fr=1.0, and d/h=0.33.

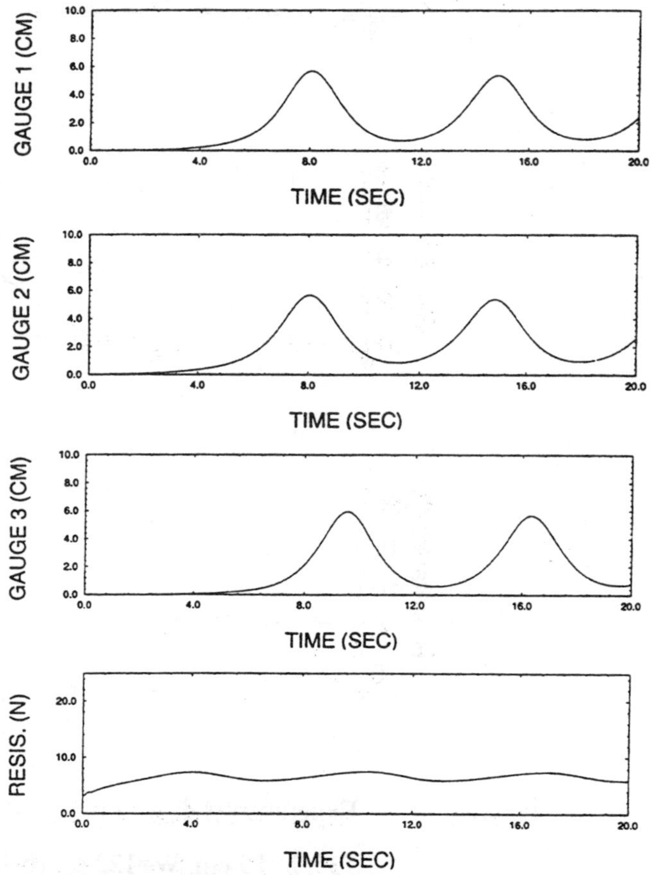

Figure 3(c) Computed surface elevation at four numerical, moving wave-gauges and wave resistance for thin ship. S_B =3.14%, Fr=1.0, and d/h=0.33, corresponding to Test No. 735 of Ertekin (1984).

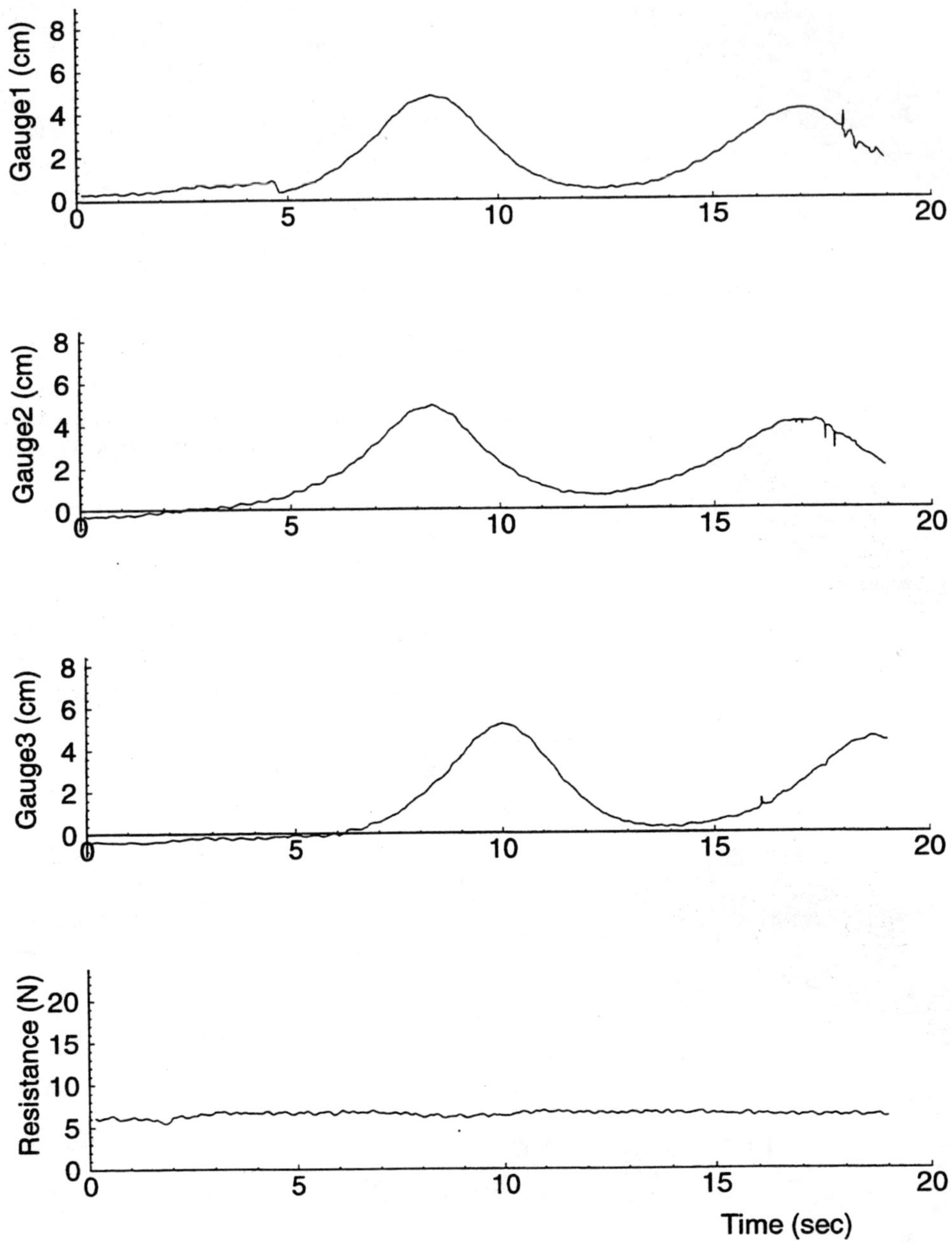

Figure 4 Experimental data (Ertekin, 1984) of surface elevation and total resistance, Test No. 735, h=15 cm, W=122 cm (half-width of the tank), d=5 cm, S_B =3.14%, Fr=1.0.

Proceedings of the Seventh (1997) International Offshore and Polar Engineering Conference
Honolulu, USA, May 25-30, 1997

ISBN 1-880653-28-1 (Set); ISBN 1-880653-31-1 (Vol. III)

Three-Dimensional Characteristics of Velocity Field and Effects of Undertow in a Surf Zone

Y. Watanabe and N. Mori
Hokkaido University
Sapporo, Japan

ABSTRACT

The velocity field in a surf zone was experimentally investigated, and the effects of an undertow after wave breaking on turbulence and wave motion are discussed. The mechanisms of production and transportation of three-dimensional turbulence in a surf zone are also discussed.

It is important to understand the reciprocal relationship between turbulence caused by wave breaking and a cross-shore steady flow, such as a rip current or an undertow, as both of these are closely related to sediment rolling up and cross-shore sediment transport. On the other hand, high-strain turbulent flows that have three-dimensionally coherent eddy structures with oblique vortexes are generated in a surf zone, and the steady flow is strongly influenced by the three-dimensional vortexes generated at this time. However, these have been few investigations of this problem. Therefore, this study was designed to investigate the three-dimensional characteristics of the velocity field and bed erosion in a surf zone, and the effects of undertow on them, in order to clarify the mechanisms of fluid motion and the cross-shore sediment transport in this region.

KEY WORDS: wave breaking, undertow, three-dimensional fluid velocity, particle image velocimetry, bed erosion

INTRODUCTION

In a surf zone, there is a three-dimensionally complex velocity field with various scales that affect each other in a surf zone. As a typical example, the reciprocal action between a steady flow, such as a rip current or an undertow, and turbulence produced by wave breaking contributes greatly to the suspension and transport of bottom sediments. Therefore, it is important to investigate the mechanisms by which turbulence caused by wave breaking influences the cross-shore steady flow, which is directly relevant to sediment transport. In this paper, we focus on the undertow as the cross-shore steady flow, which is a result of mass transport by waves. A wave field with an undertow in a two-dimensional wave flume may be regarded as one with a rip current in the field. In this study, the wave breaking mechanism in the case where the undertow is restrained was compared to the mechanism in the case where the undertow is uncontrolled in order to understand the effects of the nearshore current on the wave breaking process.

In coastal region, nearshore currents and complex bottom topography generally form a three-dimensional velocity field. On the other hand, in a surf zone, high-strain turbulent flows that have three-dimensionally coherent eddy structures with oblique vortexes are generated, and the steady current is strongly influenced by the three-dimensional vortexes generated at this time. Since a two-dimensional wave flume is a three-dimensional space with side walls, the fluid motion in experiment would be different from that in the field, due to restriction of fluid motion by the side walls in the case where three-dimensional fluid motion is generated by wave breaking. Although these factors are closely related to each other, there has been little study of their relationship.

Therefore, this study was designed to investigate the three-dimensional characteristics of the velocity field and bed erosion in a surf zone, and the effects of undertow on them, in order to clarify the mechanisms of fluid motion and the cross-shore sediment transport in this region.

This paper is composed of four main sections. First, the effects of undertow on breaking conditions, such as breaking wave height and breaking water depth, are explained. Second, the contribution of undertow to the instantaneous velocity field after wave breaking is discussed, based on the distributions of fluid velocity obtained by particle image velocimetry. Third, it is shown that three-dimensionality of velocity field in a surf zone is closely related to the existence of an undertow. Finally, the contributions of undertow to sandy bed erosion velocity and the three-dimensional bed topography formed by local erosion are discussed.

EXPERIMENTAL FACILITY

Experiments were conducted in a wave flume that was 24 m long × 1 m deep × 0.6 m wide. The wave flume was equipped with a wave generator that had an absorption function to remove reflected waves, and a wave absorber was set at the end of the wave flume (See Fig. 1). Circulation pipes, which were able to control the wave setup and undertow by opening and shutting of their rids, were installed under the bottom in the wave flume.

Wave gages were positioned as shown in Figure 1. Fluid velocity was horizontally and vertically measured at 2-cm intervals in the surf zone by a three-dimensional acoustic Doppler velocimeter. Video images of the particle image velocimetry were taken using a high-speed video camera. The main experiments, except for the first section of results, were carried out under the wave conditions shown in Table 1.

PARTICLE IMAGE VELOCIMETRY

To clarify the contributions of vortexes caused by wave breaking to the wave field, it is necessary to measure the spatial distribution of fluid velocity in a surf zone, including in the plunging jet and the bore front, as vortexes are transported to the inner fluid region and affect the velocity field under the trough level of waves.

However, it is very difficult to measure fluid velocity in a plunging jet and a bore front after breaking, due to mixing with air bubbles. (To Avoid this problem, the measurement of fluid velocity by an acoustic Doppler velocimeter was conducted only in the region under the trough level of waves and only in the case where the breaker was a spilling-type breaker.)

Thus, we investigated fluid velocity in this region by particle image velocimetry (PIV). This method estimates fluid velocities from video images that record the movement of markers mixed in the wave flume. In this method, the following spatiotemporal correlation of some frames are first calculated.

$$C = \frac{\int \left(f(\mathrm{x},t) - \overline{f(\mathrm{x})}\right)\left(f(\mathrm{x}+\Delta\mathrm{x},t+t') - \overline{f(\mathrm{x})}\right)dt}{\left[\int \left(f(\mathrm{x},t) - \overline{f(\mathrm{x})}\right)^2 dt \int \left(f(\mathrm{x}+\Delta\mathrm{x},t+t') - \overline{f(\mathrm{x})}\right)^2 dt\right]^{1/2}} \qquad (1)$$

x and t' denote the distance between the pixels of the particle image and lag time, respectively. $f(x,t)$ denotes the function of brightness and a bar means the average of $f(x,t)$.

Assuming the velocity field is locally stationary, the fluid velocities can be estimated by the following fundamental definition.

$$u = \Delta x / t'_{\max}, \qquad (2)$$

where t'_{max} is the lag time between the frames having the maximum correlation value.

In the experiments, we used white markers with a specific gravity of 1.005 and a diameter of 2 mm, and a high-speed video camera that can take 250 frames per sec. The brightness of the particle images were digitized into a 256-gray scale with a resolution of 10 pixels/mm. The noise in the image data was removed by a Gaussian filter. It was found that the results measured by PIV had an accuracy of about 80% if the pixel had a correlation of greater than 0.8.

RESULTS

Contribution of undertow to breaker index (breaking conditions)

The dependence of the breaking condition, such as breaking wave height and breaking water depth, on the existence of an undertow is discussed in the section.

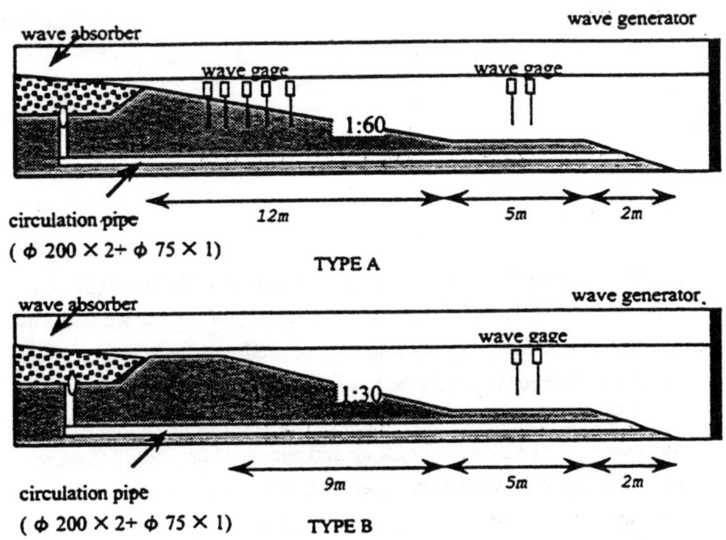

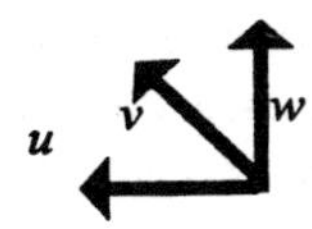

Fig. 1 sketch of the wave flume

case no.	wave conditions	breaking type	circulation pipes
case 1 (a)	T= 1.2 sec, Hb= 10 cm, hb=15 cm, bottom slope; 1:60	spilling breaker	closing rids (uncontrolled undertow)
case 1 (b)	T= 1.2 sec, Hb= 10 cm, hb=15 cm, bottom slope; 1:60	spilling breaker	opening rids (restrained undertow)
case 2 (a)	T= 2.4 sec, Hb= 12 cm, hb=15 cm, bottom slope; 1:30	plunging breaker	closing rids (uncontrolled undertow)
case 2 (b)	T= 2.4 sec, Hb= 12 cm, hb=15 cm, bottom slope; 1:30	plunging breaker	opening rids (restrained undertow)
case 3 (a)	T= 1.2 sec, Hb= 10 cm, hb=15 cm, bottom slope; 1:30	spilling breaker	closing rids (uncontrolled undertow)
case 3 (b)	T= 1.2 sec, Hb= 10 cm, hb=15 cm, bottom slope; 1:30	spilling breaker	opening rids (restrained undertow)

Table 1 wave conditions and control for c culation pipes

Fig. 2 shows the ratio of breaking wave height for deep water wave height (H_b/H_0) to the deep water wave steepness (H_0/L_0). Fig. 2 (A) shows the experimental results obtain by other researchers and the breaker index curve. The present experimental results are plotted in Fig. 2 (B). ○ and ● represent the results in the case of restrained undertow and uncontrolled undertow, respectively. Slight differences in the relative breaking wave height between each case can be seen in this figure. Since a development of radiation stress is suppressed before wave breaking, the breaking point is shoreward located when an undertow is restrained. However, this difference can not be regarded as the quantitative consequence of an undertow, because there are dispersed results caused by experimental errors, such as a disunion of the definition of wave breaking and a difference in the scale of wave flume, in the experiment on wave breaking (See Fig. 2 (A)). The contributions of undertow to the wave field are examined in following sections from another viewpoint.

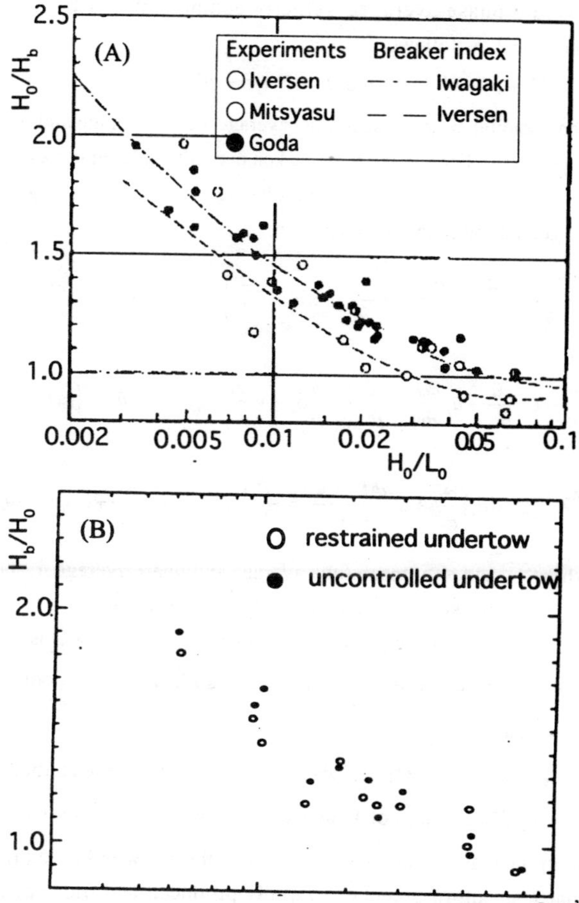

Fig. 2 Comparion of the breaker index (A) to the present results (B)

Distribution of fluid velocity afer wave breaking

Fig. 3 shows an example of velocity vector plot at the plunging point in case 2 (a) , which is obtained by PIV. There are large-scaled rotating motions in different directions in the inner fluid region as well as fluid motion in the rebounded jet with an upward high velocity shown near the arrow. PIV is advantageous for measuring the fluid velocity in a surf zone, which is a spatially complex structure.

Fig. 4 shows the horizontal velocity distributions at the breaking point, obtained by PIV, in cases 1 (a) and 1 (b) (See Table 1). In Fig. 4 and 5, (a) and (b) represent uncontrolled undertow and restrained undertow, respectively. In Fig. 4 (a), a higher fluid velocity is concentrated near the forward part of the wave crest, compared to that in 4 (b). This indicates that the offshoreward transport of the kinematic energy by undertow causes a concentration of energy near the breaking point, and intensifies the unsymmetrical velocity distribution.

Horizontal velocity distributions around the bore front at the point where 2 m shoreward from the breaking point in cases 2 (a) and 2 (b) are presented in Fig. 5. The distance l_a of the bore front precedently propagating over the inner fluid region in Fig. 5 (a) is longer than the distance l_b in 5 (b), and the vertical gradient of horizontal velocity in 5 (a) is steeper than that in 5 (b).

It is likely that the undertow contributes to the transportation of vortexes and the motion of the bore front through wave breaking process. This will be discussed in the next section.

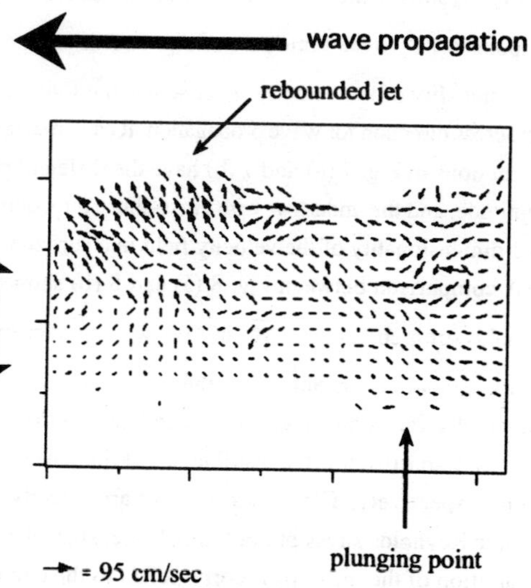

Fig. 3 Velocity vector at the plunging point in case 2 (a)

Three-dimensional velocity field after wave breaking

The three-dimensional characteristics of fluid velocity are discussed in this section. Fluid velocity was horizontally and vertically measured at 2-cm intervals in a surf zone by a three-dimensional acoustic Doppler velocimeter. To investigate the contributions of Reynolds stress to each velocity component, it is convenient to decompose the measured fluid velocity into the time-averaged velocity (steady flow) component, the periodic phase-averaged velocity component and the turbulence component, as follows.

$$u_i = U_i + \tilde{u}_i + u_i' \qquad \text{and} \quad p = P + \tilde{p} + p' \quad , \qquad (3)$$

where capital, ~ and ' mean time average, phase average and turbulence, respectively. Each average was taken over 30 waves in our experiment.

Substituting (3) into Navier-Stokes equation and averaging each term, we get the following three equations.

$$U_j \frac{\partial U_i}{\partial x_j} = \frac{\partial}{\partial x_j}\left(-\frac{P}{\rho}\delta_{ij} + \nu \frac{\partial U_i}{\partial x_j} - \overline{u_i' u_j'} - \overline{\tilde{u}_i \tilde{u}_j}\right) \qquad (4)$$

$$\frac{\partial \tilde{u}_i}{\partial t} + U_j \frac{\partial \tilde{u}_i}{\partial x_j} + \tilde{u}_j \frac{\partial U_i}{\partial x_j} = \frac{\partial}{\partial x_j}\left(-\frac{\tilde{p}}{\rho}\delta_{ij} + \nu \frac{\partial \tilde{u}_i}{\partial x_j} + \left(\overline{\tilde{u}_i \tilde{u}_j} - \tilde{u}_i \tilde{u}_j\right) - \left(\langle u_i' u_j' \rangle - \overline{u_i' u_j'}\right)\right) \qquad (5)$$

$$\frac{\partial u_i'}{\partial t} + \left(U_j + \tilde{u}_j\right)\frac{\partial u_i'}{\partial x_j} + u_j' \frac{\partial (U_i + \tilde{u}_i)}{\partial x_j} = \frac{\partial}{\partial x_j}\left(-\frac{p'}{\rho}\delta_{ij} + \nu \frac{\partial u_i'}{\partial x_j} + \left(\langle u_i' u_j' \rangle - u_i' u_j'\right)\right), \qquad (6)$$

where $\overline{}$ and < > mean time average and phase average, respectively. δ_{ij} is Kronecker's delta and ν is kinematic viscosity.

As expressed in equation (4), time-averaged velocity U_i is affected by spatial distributions of the Reynolds stress $\overline{u_i' u_j'}$ and the periodic Reynolds stress $\overline{\tilde{u}_i \tilde{u}_j}$.

Fig. 6 shows the spatial distributions of time-averaged turbulent energy $k\left(=\overline{u_i' u_i'}/2\right)$ at depths of 4 cm, 6 cm and 8 cm in cases 1 (a) and 1 (b). The sharp increase in turbulent energy near the arrow in Fig. 6 (a) suggests that a strong turbulence is rapidly produced and transported to the measuring point. In Fig. 6 (b), the turbulent energy gradually increases without the peak seen in 6 (a). The difference of the intensity of turbulent energy between 6 (a) and 6 (b) is thought to be caused by the advection effects shown in the left-side terms of equation (6).

Fig. 7 shows the distributions of the periodic Reynolds stress component $\overline{\tilde{v}\tilde{v}}$ in the same case of Fig. 6. This component means three-dimensionality of the velocity field, so v is the fluid velocity component in the cross direction for wave propagation. It can be seen that the horizontal distributions in Fig. 7 (a) and 7 (b) have the different characteristics and magnitude, and the undertow with cross-shore direction contributes to the three-dimensionality of the velocity field in a surf zone.

A comparison between Fig. 8 (a) and 8 (b) shows that the periodic Reynolds stress component $\overline{\tilde{u}\tilde{v}}$ in 8 (a), which is concerned with radiation stress and contribute greatly to the three-dimensionality of the steady flow, horizontally varies on a lager scale than that in 8 (b).

Figs. 9 and 10 show the distributions of time-averaged u and v at 4 cm depth, respectively. These steady flows are directly influenced by the periodic Reynolds stress but not turbulence, since the onsetting position of variation of the mean flow corresponds to the one of variation of the periodic Reynolds stress. The fact that the time-averaged v in Fig. 10 (a) is spatially undulated suggests that there is a three-dimensionally circulating mean flow in this region.

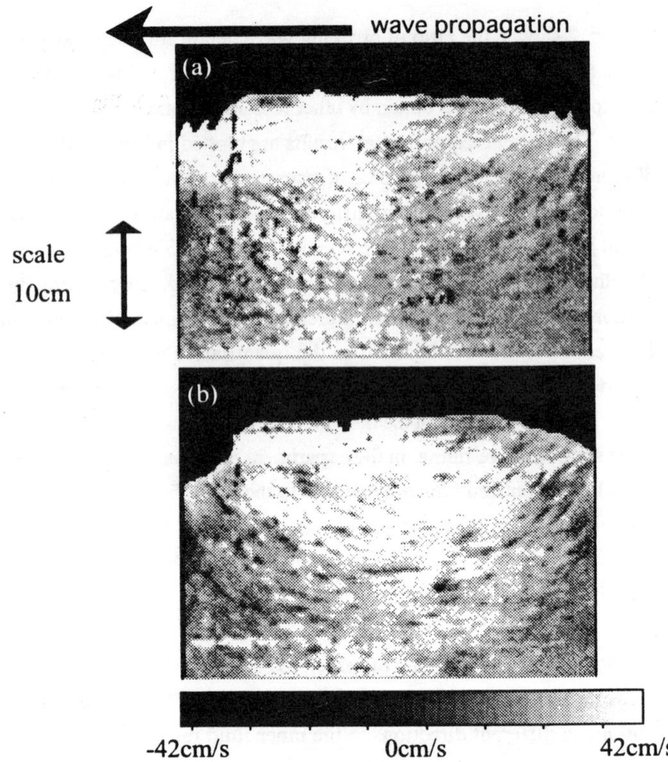

Fig. 4 Horizontal velocity distribution at the breaking point in case 1 ((a) uncontrolled undertow, (b) restrained undertow)

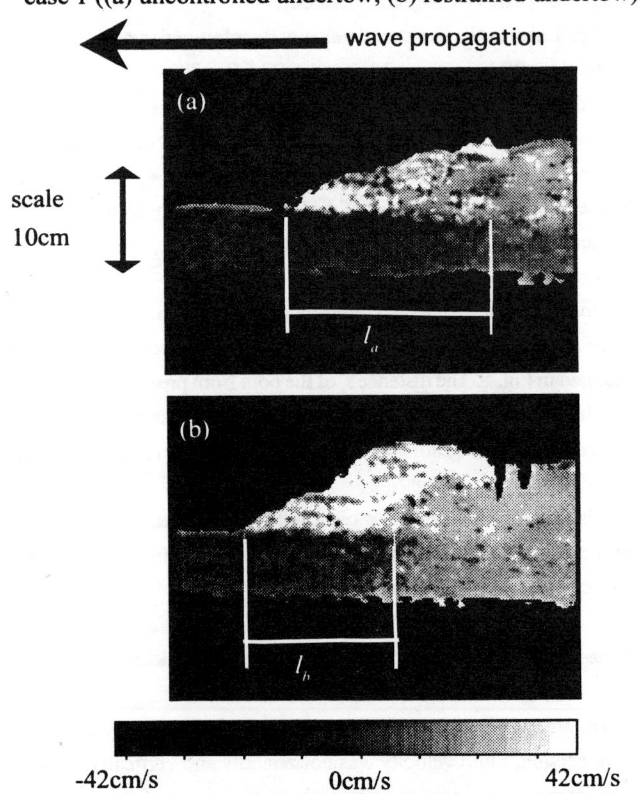

Fig.5 Horizontal velocity distribution at the point 2 m shoreward from the breaking point in case 2 ((a) uncontrolled undertow, (b) restrained undertow)

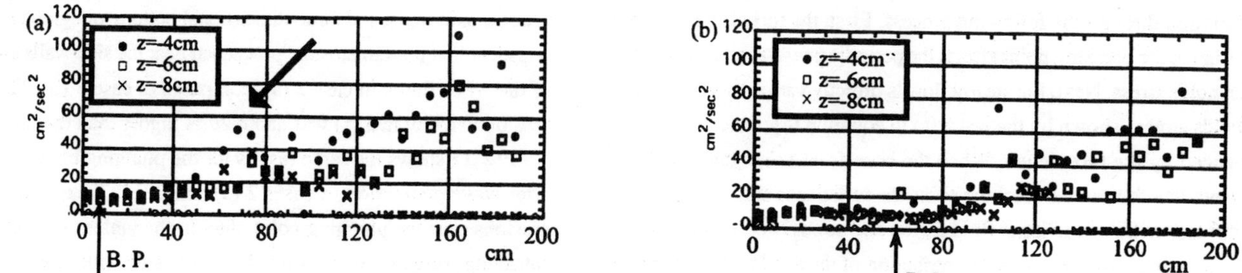

Fig. 6 Distribution of turbulent energy k in case 1
((a) uncontrolled undertow, (b) restrained undertow)

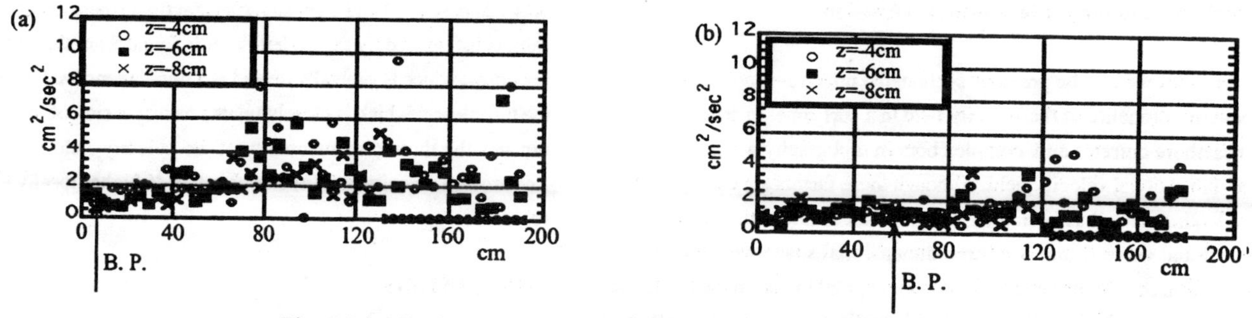

Fig. 7 Distribution of periodic Reynolds stress component $\overline{\tilde{v}\tilde{v}}$
in case 1 ((a) uncontrolled undertow, (b) restrained undertow)

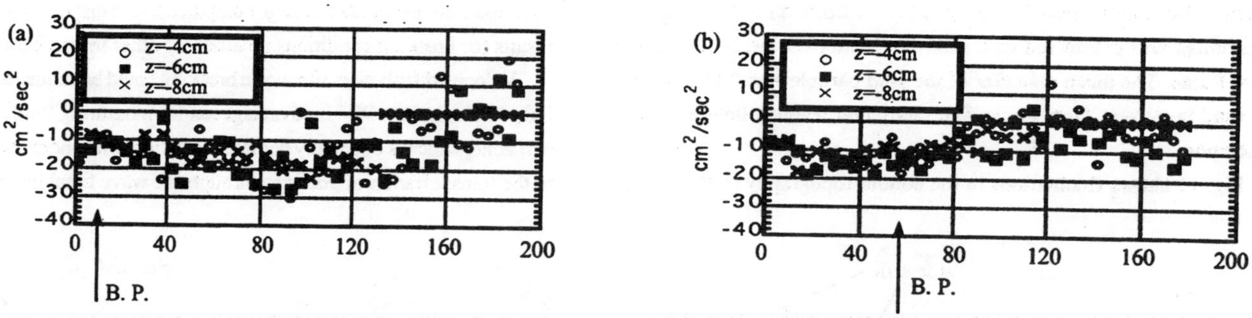

Fig. 8 Distribution of periodic Reynolds stress component $\overline{\tilde{u}\tilde{v}}$
in case 1 ((a) uncontrolled undertow, (b) restrained undertow)

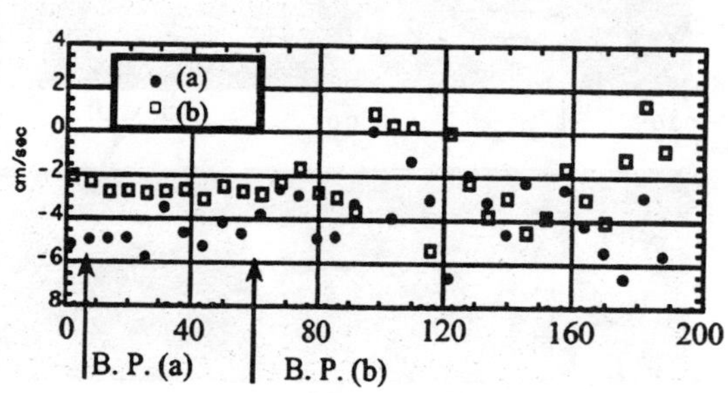

Fig. 9 Distribution of time-averaged u in the case 1
((a) uncontrolled undertow, (b) restrained undertow)

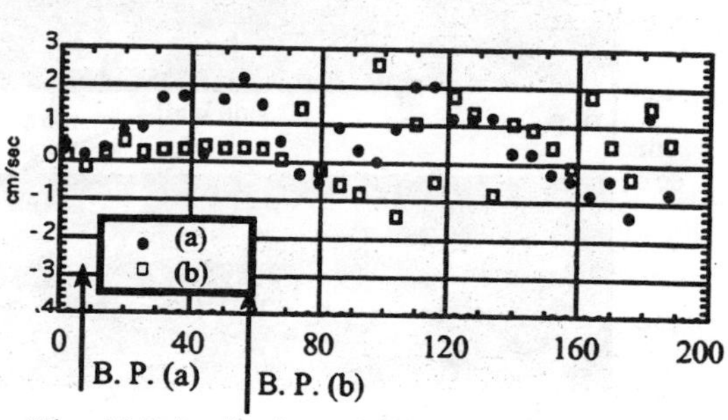

Fig. 10 Distribution of time-averaged v in case 1
((a) uncontrolled undertow, (b) restrained undertow)

These results indicate that small-scale turbulence affects large-scale fluid motion through the following process. First, the turbulence near the bore front at the breaking point affects the periodic phase-averaged velocity as Reynolds stress. Next, the steady flow is directly varied by the periodic Reynolds stress, shown by the last term in equation (4), which have been influenced by the turbulence. When the bore front propagates to about one wave-length from the breaking point, the turbulence sufficiently grows and diffuses to the entire inner fluid region and strongly contributes to the steady flow in a surf zone, so the contribution of the third term in equation (4) is greater in this region.

Contribution of the Undertow to Bed Erosion

As mentioned in the previous section, the undertow contributes to the three-dimensionality of the velocity field in a surf zone. In coastal region, the nearshore currents and complex bottom topography generally form a three-dimensional velocity field. Although these factors are closely related to each other, there has been little study of their relationship. Since a two-dimensional wave flume is a three-dimensional space with side walls, fluid motion in an experiment will not correspond to that in the field, due to restriction of fluid motion by the side walls in the case where three-dimensional fluid motion is produced by wave breaking.

In this section, we examine the preceding argument and study on the effects of three-dimensional velocity in a wave flume on bed erosion. The experiment was conducted on a sandy bed laid over the bottom of the wave flume. The mean diameter of the sand particle was 0.15 mm, and the sand layer was 5-cm thick. The sandy bed levelled before a wave generation.

Fig. 11 shows fluctuations in the bottom topography at 30 minutes after wave generation in cases 3 (a) and 3 (b). Erosions were concentrated near the side walls, and three-dimensional bottom topography was formed. It is clear that bottom erosion proceeded near the side walls and the center of the wave flume in case 3 (a), compared to case 3 (b). The maximum erosion velocity in 3 (a) was about 20 % higher than that in 3 (b).

Fig. 12 shows the same results for the plunging breaker at 6 minutes after wave generation in cases 2 (a) and 2 (b). There were large-scale erosions near the plunging point, duet to the violent vortexes caused by splashing. Erosions concentrated near the side walls were intensified in the case of a plunging breaker, compared to the case of a spilling breaker in Fig. 11. The fact that the sandy bottom topography is obviously different between case 2 (a) and 2 (b) suggests that the suspension and transport of sand is influenced by the undertow. It is found from this results that the undertow, which is generally treated as a two-dimensional flow, intensifies the turbulence, which is closely related to suspension of the sediment and induces the three-dimensionality of the velocity field near the bottom. The erosion velocity in 2 (a) was about 70 % higher locally than that in 2 (b).

CONCLUSIONS

The breaking wave height and the breaking depth are slightly smaller in the case of restrained undertow than in the case of uncontrolled undertow, but these differences are not large compared to the dispersive experimental results for breaking conditions obtained by other researchers.

Velocity distribution after wave breaking could be estimated by particle image velocimetry. PIV is advantageous for measuring fluid velocity in a surf zone, which is a spatially complex structure. The undertow contributes to the transportation of vortexes through the wave breaking process.

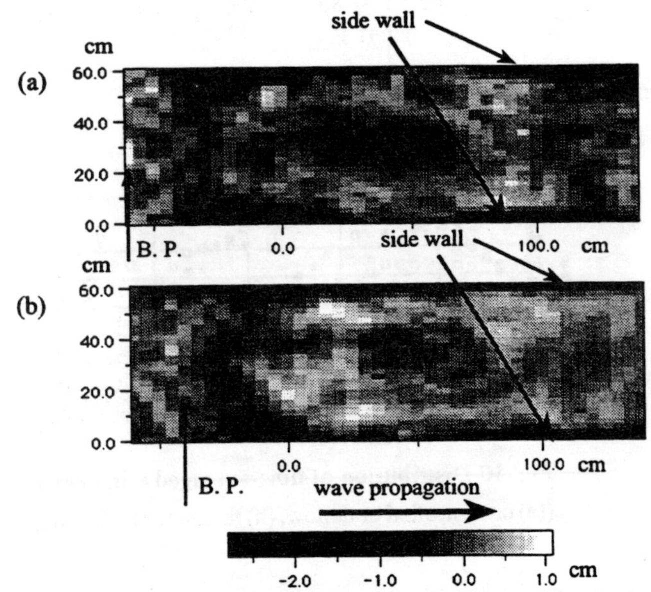

Fig. 11 Fluctuation in bottom topography at 30 minutes after wave generation in case 3 ((a) uncontroled undertow, (b) restrained undertow)

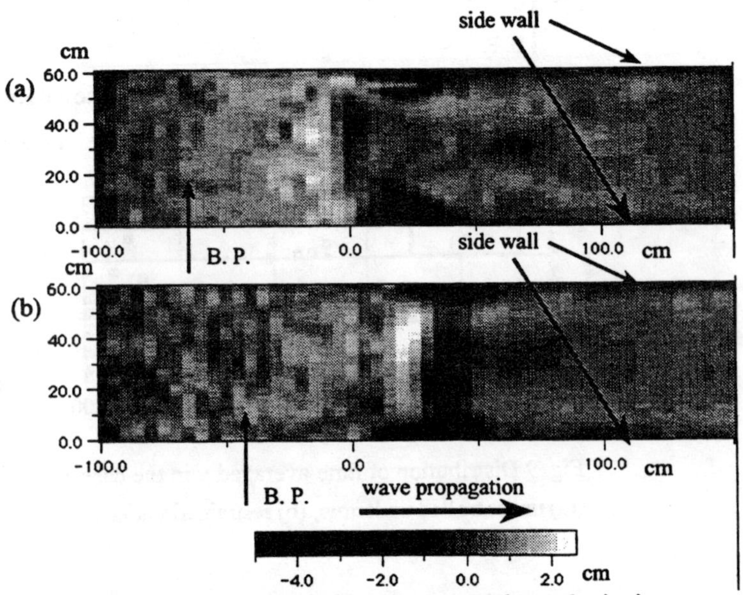

Fig. 12 Fluctuation in bottom topography at 6 minutes after wave generation in case 2 ((a) uncontroled undertow, (b) restrained undertow)

It was found that the undertow, which is generally considered to be two-dimensional, contributed to the three-dimensionality of the velocity field in a surf zone, and that the time-averaged velocity component v with a cross direction for wave propagation was spatially undulated. It is likely that there is a three-dimensionally circulating steady flow in this region.

The bottom erosion is notably accelerated by the existence of an undertow. The maximum erosion velocities in the case of an uncontrolled undertow were about 20 % higher for the spilling breaker and about 70 % higher for the plunging breaker than those in the case of an restrained undertow, respectively.

The undertow intensifies the turbulence, which is closely related to suspension of the sediment and to the three-dimensionality of the velocity field in a surf zone.

References

Akio Okayasu, Tomoya Shibayama and Nobuo Mimura, Velocity Field under Plunging Waves, 1986, Proc. Coastal Eng., pp. 660 - 674.

F. C. K. Ting, J. T. Kirby, Observation of undertow and turbulence in a laboratory surf zone, Coastal Engineering, pp.51-80, 1994

F. C. K. Ting, J. T. Kirby, Dynamics of surf-zone turbulence in a strong plunging breaker, Coastal Engineering, pp. 177-204, 1995

Mikio Hino, Kazuo Nadaoka, Atsusi Omata, 1984, Coherent eddy structure of turbulence in a surf zone and fluid motion, 36th Proc. of Coastal Eng. in JSCE, pp. 1-5, (in Japanese)

Proceedings of the Seventh (1997) International Offshore and Polar Engineering Conference
Honolulu, USA, May 25-30, 1997

ISBN 1-880653-28-1 (Set); ISBN 1-880653-31-1 (Vol. III)

Characteristics of Long-Period Waves Observed in a Port

Tetsuya Hiraishi
Port and Harbour Research Institute, Ministry of Transport
Yokosuka, Japan

ABSTRACT

Long-period waves with periods of 1 ~ 5 minutes sometimes terminate the cargo handling in Japan's several major ports by causing the breaking of mooring ropes. To compare the characteristics of long-period waves in such ports to those unaffected, field observations of long-period waves were made in two ports influenced by such waves. Their observations show that the long-period waves are mainly induced by the set-down waves bounded in the grouping waves and that the amplitude of long-period waves can be reduced inside an harbor by using breakwaters.

KEY WORDS : Field observations, Long-period waves, Port planning

INTRODUCTION

In several Japanese major ports, the termination of loading and unloading is often caused by the large surge motion of the moored vessels even when the wave heights in a harbor are quite small. The large surge motion results from the resonance of mooring system due to long-period waves(Suganuma et.al.,1995). The termination of cargo handling by long-period waves becomes significant in ports like Tomakomai, Kashima, Noshiro, Shibushi, etc.

Figure 1 shows the location of ports investigated in this study. In Noshiro-port located in the northern part of Japanese main island, the mooring ropes have been often broken by the effects of large surge motions. **Table 1** shows the reported damages in the port of Noshiro. Similar damages are common in the other Japanese major ports mentioned above. A field observation study at Tomakomai-port located in the northern part of Japan facing to the Pacific ocean, has revealed the occurunce of long-period waves which cause large ship motions(Suganuma et.al.,1995).

The mooring line damages have been reported from other major ports along the Japanese coasts. On the other hand, at Naha-port, the largest terminal in Okinawa-island, mooring line damages and interruptions of loading and unloading have not been observed. The Okinawa district, located in the southern area, is surrounded by several coral reefs. The long- period waves become significant when wave groups pagate in the shallow depths. The

Figure 1 Location of investigated ports

Table 1 Damage in Noshiro-port(Sep.'92-Mar.'93)

date	D. W. T.	number of broken ropes	offshore $H_{1/3}$ (m)
Sep. 30, '92	60, 000	4	2. 97
Nov. 4, '92	60, 000	22	-
Dec. 12 , 92	60, 000	18	5. 26
Mar. 1 , 93	60, 000	22	5. 27

amplitude of long–period waves may significantly be increased in shallow water depths and over the reefs. A wide reef expands around the Naha–port. High surge motions of ships can be in the Naha–port if long–period waves could cause damages to the mooring lines. We expect some reduction in amplification of the long–period waves due to the topography of this port. A preventive plan for long–period waves would be proposed if a reduction in the amplification of long–period waves is verified by our field investigation. We have conducted field observations of long–period waves at the offshore and in the harbors of Noshiro– and Naha–ports. We present these observations and investigate the differences in the characteristics of long–period waves in both ports.

OBSERVATION CONDITION

Figure 2 shows the location of measurement points in the Noshiro–port. Port engineers have often reported long–period surge motions of the moored vessels in this port. The points **P1** and **P2** indicate the location of wave gages installed on the sea bed outside(−50m) and inside(−12m) of the port, respectively. A continuous measurement program was carried out during 1993 and 1994. For the periods before 1993, analog records corresponding to the time when the mooring lines were broken, were converted to the digital data, and analyzed. In the analysis, the measured water surface elevation was divided into the three profiles, each with different frequency band; 1) **wind–wave component** (frequency f > 1/30Hz), 2) **long–period wave 1** (1/30>f>1/600Hz), and 3) **long–period wave 2** (1/600 Hz>f). For each component, wave heights and periods were calculated. The frequency spectrum was

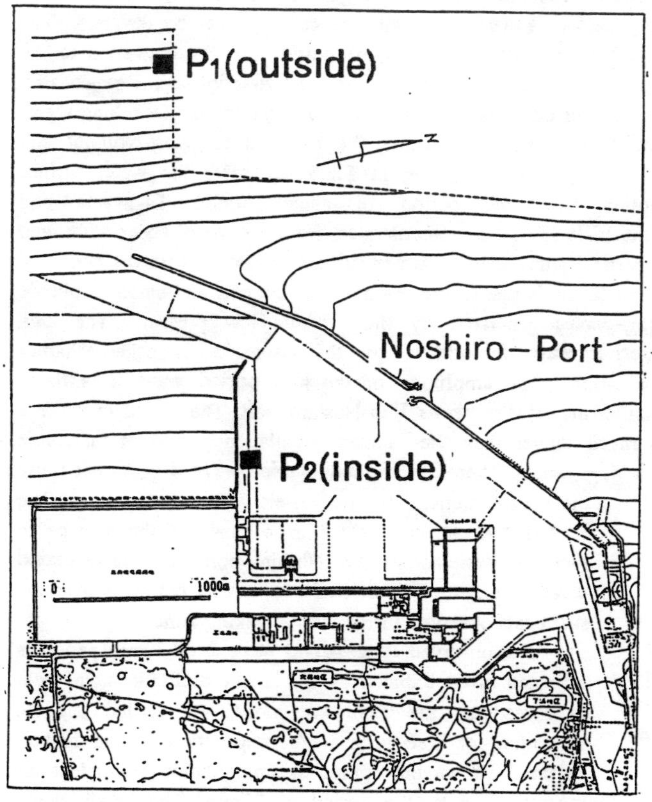

Figure 2 Location of measurement point in Noshiro–port

also analyzed using 20–min long continuous wave records.

Figure 3 shows the locations of measurement points for the Naha–port. Waves were measured offshore (**P0**, water depth=50m), in the reef area located at the northern side of the port, and inside (**P1**). Observations were made in February and March of 1995. The port is divided into the two parts; one is the outer area behind the offshore breakwater and another inner harbor protected by the breakwaters **B**. The inner harbor is used for the loading and un–loading of cargo vesssels. The point **P2** is the measurerment point 18m deep in the outer region and the point **P1** 17m deep in the inner harbor. Over the reef, the points **St.1**(−2m), **St.2**(−5m) and **St.3**(−12m) were used to measure the water profile and two horizontal components of currents. The measure data was analysed every hour for 20 minutes, and was divided into the wind–wave component(f>1/20Hz) and the long–period component (f<1/20Hz).

CHARACTERISTIC OF OBSERVED LONG WAVE

(1) Observations in Noshiro–port

Figure 4 shows an analyzed frequency spectrum at the offshore point **P1** in Noshiro–port. The analysis date corresponds to the time when the mooring lines of vessels were broken inside the port. In the figure, two clear peaks with periods of 11.1s and 2.9min represent the spectrum of wind–wave and **long–period wave 1**, respectively. The energy of wind–waves entering the port is controlled by the breakwaters designed to block these waves. The **long–period wave component** 1 may easily penetrate into the port. Therefore, the observed results demonstrate that the large surge motion of the moored vessels are mainly induced to the penetration of the **long period wave 1.**

Longuet–Higgins and Stewart(1962) has theoretically derived the generation of long–period oscillation of averaged mean water level bounded in the grouping waves. The bounded long–period water surface oscillation is called "set–down wave" as the resulting

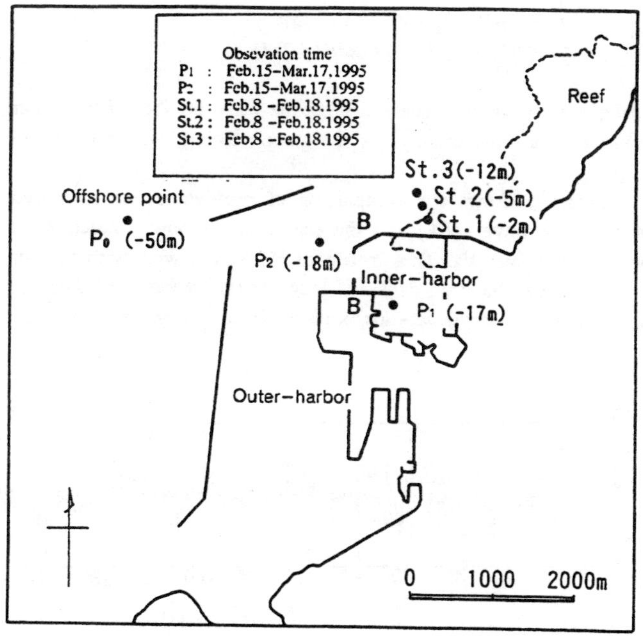

Figure 3 Location of measurement point in Naha–port

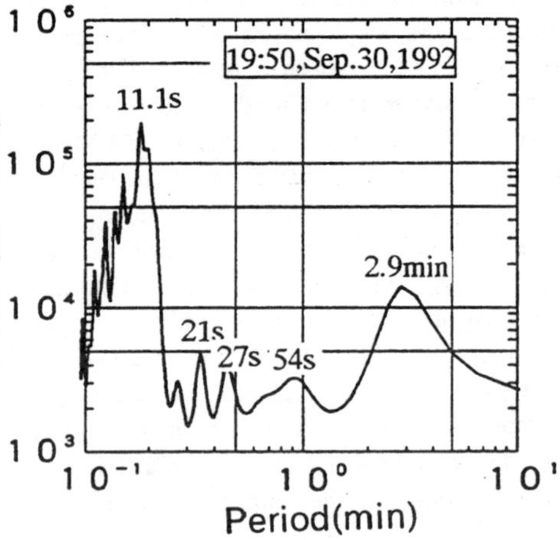

Figure 4 Wave frequency spectrum at P1 (Noshiro–port)

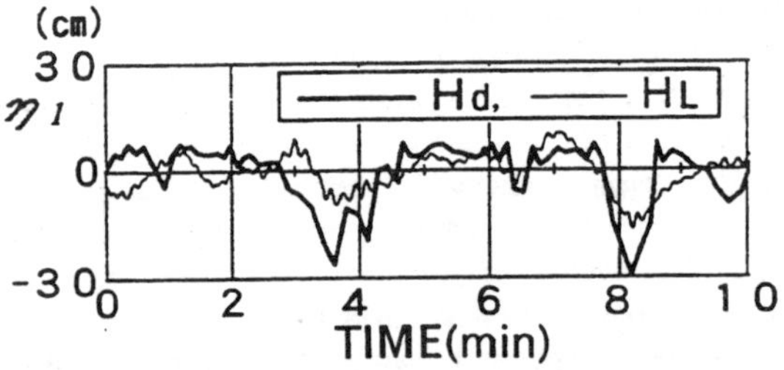

Figure 5 Comparison of measured and estimated long–period wave profiles

water surface motion remains under the initial mean level. The water surface elevation, η of set–down wave is estimated by the following equations;

$$\eta = -Sx/[\rho(gd - Cg^2)] + \text{const.} \qquad (1)$$

$$Sx = (1/8)\rho g H^2 (1 + 2kd/\sinh 2kd) \qquad (2)$$

where, g is the gravity acceleration, k the wave number, d the water depth, ρ the water density, Cg the group velocity, and H the wave height.

Figure 5 shows the comparison of profiles of the measured long–period wave 1(H_L) and the estimated set–down wave(H_d) at the offshore point **P1**. The profile of H_d agrees well with that of H_L. Therefore, the main cause of long –period wave 1 offshore of the Noshiro port is the set–down wave due to the grouping waves.

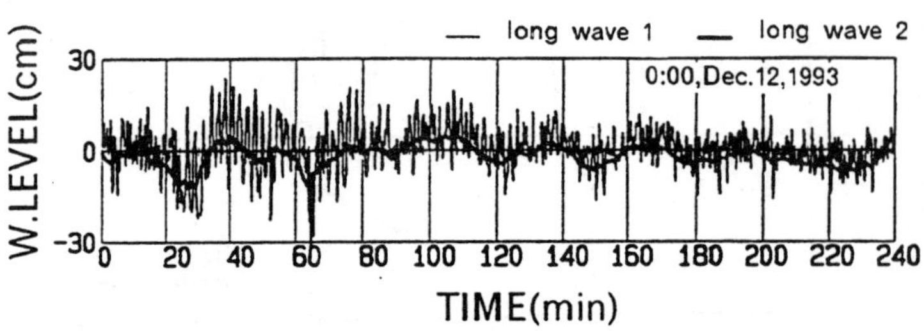

Figure 6 Observed long period wave profiles at P2

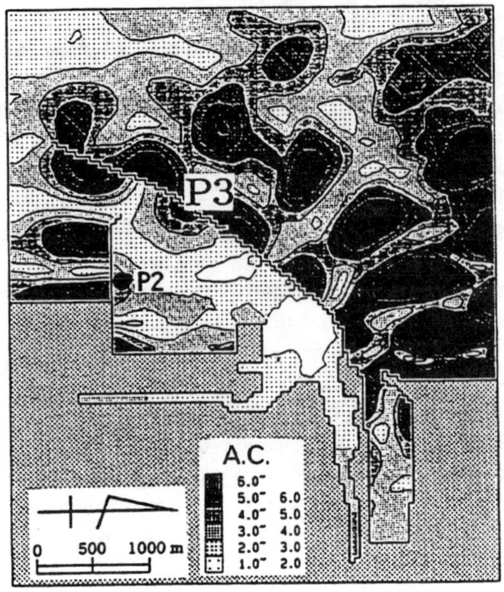

Figure 7 Computed amplification coefficient for period of 32min

The set–down wave may become smaller as the wind wave energy is reduced by the breakwater if the bounding phenomena is not broken by the wave shoaling, refraction and diffraction.

Figure 6 shows the profiles of **long–period wave 1** and 2 obtained at the **P2** inside the port. The amplitudes of **long period wave 1** is 10 to 50cm which almost similar to the amplitude of outside measured waves in Fig.5. Therefore, the amplitude of **long–period wave 1** is not reduced by the breakwaters. The bounded wave component in the outside may be converted to the free–long wave by the effects of wave diffraction at the breakwater. The shelter effect of breakwater is not large for the free–long wave. The free–long wave reaches the berth in the port without any decreasing. The thick line in Fig.6 represents the water surface motion with the period of about 30min. Takayama and Hiraishi(1988) have demonstrated from the field observation and numerical study that the water surface oscillation with the period of 30 to 60min is due to the resonance of the port and small amplitude long–waves generated by the offshore low pressure. The **long –period wave 2** induced by the resonance is called "harbor oscillation". The amplitude of the **long–period wave 2** is much smaller than that of **wave 1** in Noshiro–port. The period of **wave 2** is much longer than the natural oscillation period of the ship mooring system. Therefore, the **long–period wave 2** could not cause the severe damage to the vessels and mooring lines. The **long–period wave 1** is the main cause of the damage to mooring systems in the Noshiro–port. The **long–period wave 1** is named "port surf beat" as its period is near the surf beats observed in the wave breaking zone.

The observed **long–period wave 2** of 30min period is expected to be induced by the harbor oscillations. In order to investigate the resonance mode, we carried out the simulation of free–long wave propagation in the harbor. **Figure 7** shows the computed amplification coefficient for long–wave with the period of 32min. The amplification coefficient represents the ratio of computed long–period wave height in the harbor to the incident wave height at the offshore boundary. The amplification factor becomes large, as much as 5 to 6, at the inner most part. The amplification coefficient is small at the mouth of the port, and it ranges from 1 to 3. Therefore, the large harbor oscillations occur in the harbor. The oscillation with the period of about 30min corresponds to the resonance with first mode whose node is located at the harbor entrance and anti-node at the inner most part.

(2) Observations in Naha-port

Figure 8 shows the variation of significant heights of long period component waves $H_{l1/3}$ observed at the offshore site and on the reef in the Naha-port. During observation, the heights at **Po** are not so significant and the maximum significant height is less than 10cm. The long period wave height becomes larger at **St.2** and at **St.1** as the water depth becomes shallower. **Figure 9** shows the directional spectrum derived from the long period component of water elevation and currents at **St.2**. The main energy of long-waves is propagating from the north-west direction. A small peak of energy distribution appears in the **SSE** direction. The peak represents the long- period waves reflected at the shore line.

Figure 10 shows the comparison of long-period wave height obtained in field observation ($H_{l1/3}$) and that computed from Longuet-Higgins and Stewart(1962) theory ($H_{d1/3}$). At the field

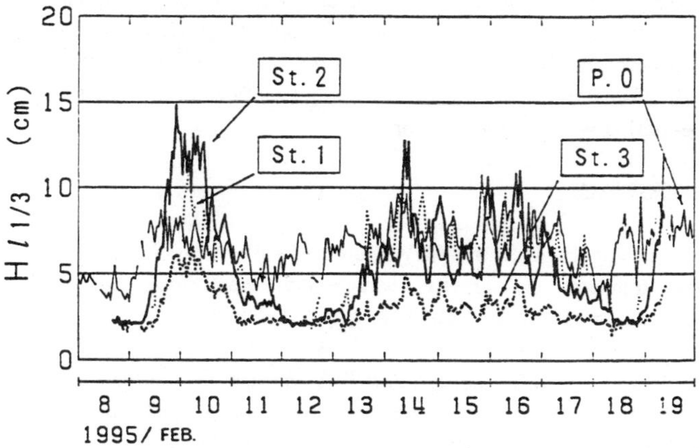

Figure 8 Variation of observed long-period wave height (Naha-port)

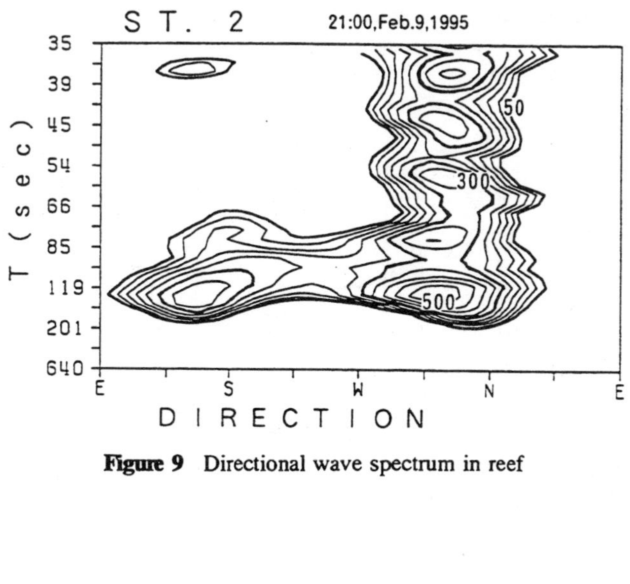

Figure 9 Directional wave spectrum in reef

observation point on the reef **St.3**, the order of $H_{l1/3}$ is equal to that of $H_{d1/3}$. Therefore, the bounded wave component forms the long-period wave components. The total wave energy, however, consists of the incident and reflected long- wave components at **St.2** as indicated in Figure 9. The height of bounded long-wave is much smaller than the observed long wave height at **Po**. Therefore, the long period wave at the offshore point **Po** is composed of not only the bounded long wave, but also the reflected free waves from coast lines.

We ccompared the wave energy spectrum at the offshore point **Po** and at the observation points in the port to determine the infuluences of long-period waves in the inner ports of the harbor.

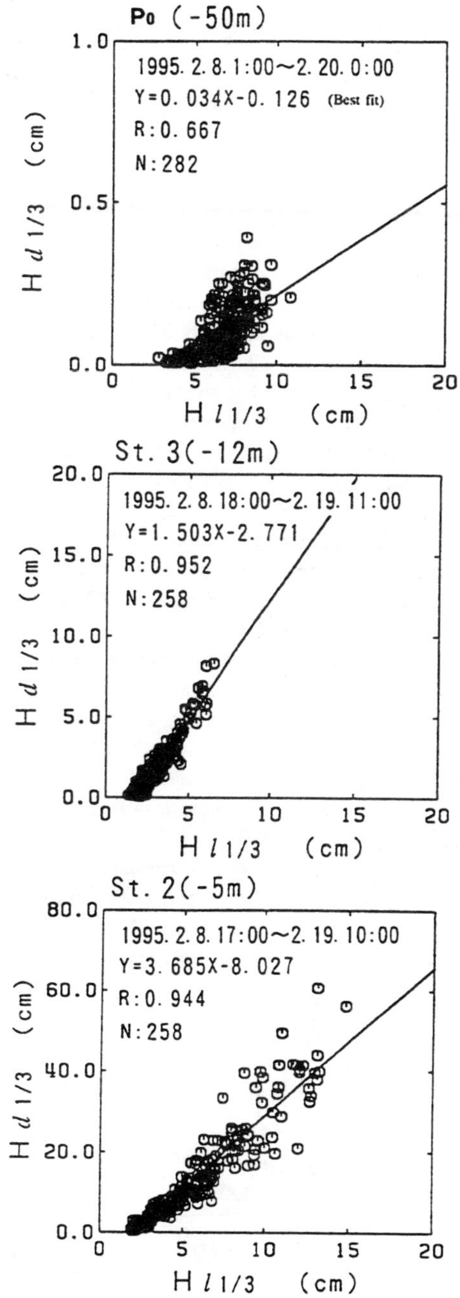

Figure 10 Height comparison of observed long-period wave and estimated set-down wave

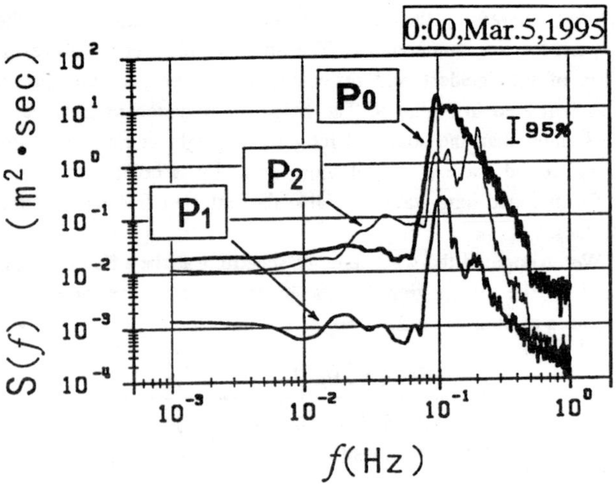

Figure 11 Variation of wave power spectrum obtained in Naha–port

Figure 11 shows the variation of wave power spectrum obtained in the Naha–port. The peak frequency of wind–wave component is 10^{-1} Hz. The peak value of wind–wave component at the point **Po** is becomming much smaller in the port. The frequency range of long–period wave is under 5×10^{-2} Hz. The spectral density of long–period wave component at **P2** is almost identical to that at **Po**. The energy frequency density of long–period component at point **P1** becomes much smaller than that measured at Po and P2. This demonstrates that the long–period wave energy may be reduced in the inner harbor at **P1**.

Figure 12a–c shows the amplification coefficients caluculated as the route of the ratio of spectrum density. Figures 12a, 12b and 12c represents the amplification coefficient of **P2/Po**, **P1/Po** and **P1/P2**, respectively. Figure 12a shows that the long– period wave induced by the group–bounded waves and the free–waves reflected from the shore line are not reduced in the outer part of the Naha–port. In the comparison of the energy at **P1** and **Po** shown in Fig.12b, the amplification coefficient is less than 0.5 in the whole frequency range. In Fig.12c, the amplitudes of long– period wave components at the inner harbor is reduced less than 50% compared to the amplitude outside.

The long–period wave energy, therefore, is not reduced outside, but it is significantly reduced inside the harbor area. The reason of such rapid decrease is the effects of inner–breakwater **B** and other small breakwaters protecting the inner harbor. The long–period waves causing the mooring line damages are induced in the offshore area of the Naha–port. These, however, do not penetrate into the inner harbor mainly because of the breakwaters protecting this harbor.

The long–period waves induced in the offshore area penetrate into the inner most part of the Noshiro port without any reduction. The loading and un–loading of vessels in the Noshiro–port is sometimes terminated by the propagation of these long–period waves. Therefore, extension of breakwaters is necessary to prevent these undesirable effects of the long –period waves.

CONCLUSIONS

The following main conclusions are derived from the observation results:

(1)Long–period waves with periods of 1 ~ 5min and harbor oscillations with periods of 30min are observed in the inner part of the Noshiro–port.

(2)Long–period wave profiles at the offshore side of Noshiro–port are almost identical to the wave profiles derived theoretically from the set–down wave bound grouping wave. Therefore, long–period waves which cause the cargo handling accidents, are mainly induced by the offshore grouping waves.

(3)In the Naha–port, the amplitudes of long–period waves are significant at the offshore area. The long–period wave amplitudes in the outer region behind offshore breakwaters are almost equal to those offshore. These become very small in the inner part of Naha–port, and do not cause any damage to the mooring facilities. Small groins and inner breakwaters will be effective to reduce the long–period waves in this harbor.

REFERENCES

Longuet–Higgins,M.S. and R.W.Stewart(1962):Radiation stress and mass transport in gravity wave with application to 'surf beats',*J.Fluid Mechanics*,Vol.13,No.4,pp.481–504.

Takayama,T. and T. Hiraishi(1988):Amplification mechanism of harbor oscillation derived from field observation and numerical simulation, *Technical Note of The Port and Harbour Research Institute*, No.636, 70p.

Suganuma,F.,M.Kamiya,Y.Atsumi and N.Koizumi (1995):Field survey of occurence probability of long period wave and vessel motion, *Proc. of Japanese Conf. of Coastal Engi.*, Vol.42,pp.951–955.

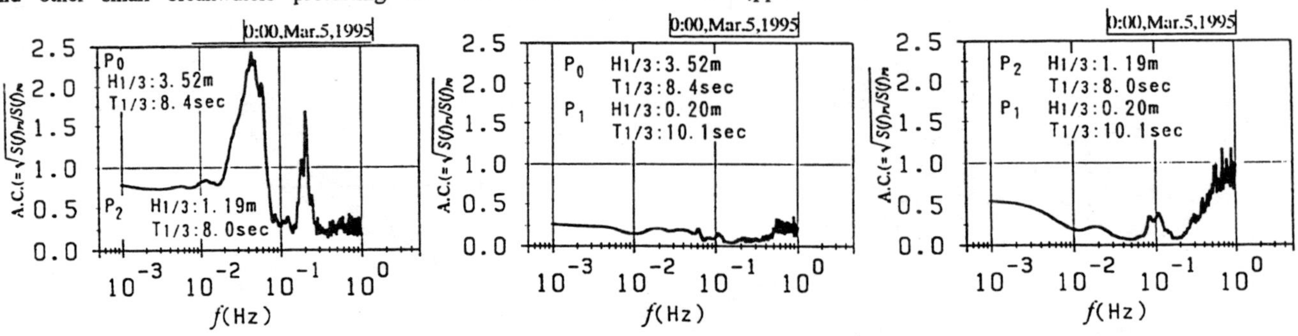

Figure 12 Amplification coefficient in Naha–port

Proceedings of the Seventh (1997) International Offshore and Polar Engineering Conference
Honolulu, USA, May 25-30, 1997

ISBN 1-880653-28-1 (Set); ISBN 1-880653-31-1 (Vol. III)

Accuracy Estimates of Multidecade Return Period Wind Velocity Values Over Shelf

Vladimir N. Kryjov
EPS Eco-System
Moscow, Russia

ABSTRACT

Estimates of the accuracy of up to 100 year return period wind velocity calculation via Weibull probability distribution approximation are analysed. Contribution of possible observation, probability, and approximation errors are shown. Also, dependence of the accuracy upon histogram partition is analysed and recommendations for decreasing errors are suggested.

KEY WORDS: Wind velocity, accuracy estimates, Weibull distribution.

1. INTRODUCTION

Multidecade return period wind velocity values are of great interest for shelf engineering project implementations. It is one of the main environmental characteristics which determines both construction design and material selection. As a rule wind velocity values are calculated for return period of up to 100 years, meanwhile the basis series are some one to three decades. Russian polar and offshore hydrometeorological stations feature especially short observation data series, moreover, computer available data cover only the latest 15 to 20 years. Also, newly founded site located stations have short time series of observations.

The shortness of data series does not make those data unuseful. They could be used in environment studies, but it is necessary to realise the degree of confidence of those data. So that, all multidecade return period wind velocity values should be accompanied with error bars.

Possible errors which arise during real procedure of long return period wind velocity calculations are analysed in the presented paper.

The technique to get multidecade return period wind velocity value is standard. Empirical data probability distribution is simulated with any appropriate theoretical probability distribution model, and required wind velocity values are simulated using probability values corresponding to a required return period. For the wind velocity probability distribution simulation, the most appropriate and the most usable model is the two parameter Weibull distribution. So, the error bars can be expressed through the error bars of distribution parameters.

Two notes should preface the following considerations: (1) the difference between an empirical distribution and a Weibull one is not statistically significant; (2) the term "error" should be realised as "possible error" so that error bars are, essentially, a confidence interval.

2. METHOD DESCRIPTION

For the wind velocity probability distribution simulation the most appropriate and the most usable model is the two parameter Weibull distribution. According to it, a random wind velocity observed value will exceed the certain wind velocity value v with the probability $F(v)$ which is given as

$$F(v) = 1-P(v) = \exp\left(-\left(\frac{v}{b}\right)^{a}\right) \tag{1}$$

where $P(v)$ is a cumulative distribution function, b and a are parameters entailing the scale and shape of the distribution. In the case of wind velocity $b > 0$, $a > 0$.

Eq.(1) can be rewritten in logarithmic form:

$$\ln(-\ln F) = a\,(\ln v - \ln b) \tag{2}$$

It should be noted that for a certain return period of L years $F = F_L$ is a constant, since

$$F_L = 1/(L^*l) \quad (3)$$

where l is an annual number of observations. $l = 365.25*8$, 8 being the daily number of observations.

For a certain return period of L years, the goal L year return period wind velocity value V_L can be obtained as

$$V_L = \exp(a^{-1} \ln(-\ln F_L) + \ln b) \quad (4)$$

So, V_L could be considered to be a function of two variables, those are a and b parameters. The V_L estimation error is determined with a and b estimation errors δa and δb. The approximate upper bound of V_L could be expressed as

$$\delta V_L = \left|\frac{\partial V_L}{\partial a}\right|\delta a + \left|\frac{\partial V_L}{\partial b}\right|\delta b \quad (5)$$

Taking into account Eq.(4)

$$\frac{\partial V_L}{\partial a} = -\frac{1}{a^2}\ln(-\ln F_L)\exp(a^{-1}\ln(-\ln F_L)+\ln b) \quad (6)$$

$$\frac{\partial V_L}{\partial b} = \frac{1}{b}\exp(a^{-1}\ln(-\ln F_L)+\ln b) \quad (7)$$

Substitution of Eq.(6) and Eq.(7) into Eq.(5) gives the V_L relative error dependence on a and b estimation errors:

$$\frac{\delta V_L}{V_L} = \frac{\ln(-\ln F_L)\delta a}{a^2} + \frac{\delta b}{b} \quad (8)$$

To figure out δa and δb one should return to Eq. (2). It is an equation of a straight line:

$$y = Ax + B \quad (9)$$

where variables y and x represent ln(-ln$F$) and ln$v$ respectively; parameters A and B represent a and $-a\ln b$, respectively. And, obviously, δa and δb can be expressed through δA and δB as

$$\delta a = \delta A \quad (10)$$

$$\delta b = \delta\left(\exp\left(-\frac{B}{a}\right)\right) = \left|\frac{\partial\left(\exp\left(-\frac{B}{a}\right)\right)}{\partial B}\right|\delta B = \frac{b}{a}\delta B \quad (11)$$

The technique to get A and B is standard. The analysed range of v variation should be divided into M classes and empirical $F_i = n_i/n$ is to be calculated for each v_i, n being sample series size, v_i being the lower boundary of i class, n_i being a number of all wind velocity values which exceed v_i, i.e. upper class boundary being open, and i being the number of class (i=1,...,M). The analysed v range does not coincide with the whole v range because for v = 0, i.e. $F(v)$ = 1, the value of ln(-ln$F$) is not defined. Also, according to practice, the Weibull simulation usually gives the best result for the v range limited with the smallest v values of 3 to 5 m/s. Then, after rearranging Eq.(1) terms into Eq.(9) terms and obtaining a corresponding series of x_i and y_i from the empirical probability distribution, one can use the linear least squares estimation. Equations for A and B calculation are well known; but because they will be useful in further considerations, they are to be written here as

$$A = \frac{M\sum_{i=1}^{M}x_i y_i - \left(\sum_{i=1}^{M}x_i\right)\left(\sum_{i=1}^{M}y_i\right)}{\Delta} \quad (12)$$

$$B = \frac{\left(\sum_{i=1}^{M}x_i^2\right)\left(\sum_{i=1}^{M}y_i\right) - \left(\sum_{i=1}^{M}x_i\right)\left(\sum_{i=1}^{M}x_i y_i\right)}{\Delta} \quad (13)$$

where

$$\Delta = M\sum_{i=1}^{M}x_i^2 - \left(\sum_{i=1}^{M}x_i\right)^2 \quad (14)$$

On the assumption that δy_i are random and independent of each other, the equations for δA and δB can be written as

$$\delta A = \sqrt{\sum_{i=1}^{M}\left(\frac{\partial A}{\partial y_i}\delta y_i\right)^2} \quad (15)$$

$$\delta B = \sqrt{\sum_{i=1}^{M}\left(\frac{\partial B}{\partial y_i}\delta y_i\right)^2} \quad (16)$$

Substituting of Eqs.(12) and (13) into Eqs. (15) and (16) leads to the explicit connection between possible errors in the A and B calculation and the deviation between the Weibull distribution and the empirical one:

$$\delta A = z\sigma_y\sqrt{\frac{M}{\Delta}} \quad (17)$$

$$\delta B = z\sigma_y\sqrt{\frac{\sum_{i=1}^{M}x_i^2}{\Delta}} \quad (18)$$

Further still, the previous manipulation gave the quantitative statistical measure of δA and δB, it is the standard deviation σ. Stating z = 1.96 in Eqs. (17) and (18) one gets the 95% confidence interval of the A and B estimation.

Substitution of Eqs. (17) and (18) into Eqs. (10) and (11), respectively, and then into Eq.(8) stating z = 1.96, gives the formula for the V_L relative error corresponding to the 95% confidence interval, i.e. the ratio of the V_L 95% confidence interval to the V_L value:

$$\frac{\delta V_L}{V_L} = z\sigma_y\left(\frac{\ln(-\ln F_L)\sqrt{\frac{M}{\Delta}}}{a^2} + \frac{\sqrt{\frac{\sum_{i=1}^{M}x_i^2}{\Delta}}}{a}\right) \quad (19)$$

The expressions under square root signs depend upon the histogram partition parameters, mainly, upon M and n_M. The effect of their variation is to be discussed in the sixth section.

σ_y was suggested to have three sources. One of them is obvious and visible: it is the difference of the Weibull distribution vs. the empirical one - the approximation error σ_{ya} :

$$\sigma_{ya} = \sqrt{\frac{\sum_{i=1}^{M}(y_i - Ax_i - B)^2}{M-2}} \quad (20)$$

If the Weibull model were the exact theoretical representation of natural probability distributions, this difference may be believed to be the only error which includes all other possible errors. But as it is only a model, it is more or less accurate but not exact. Meanwhile, observations are carried out with some error and a basic data series is not general population but a randomly selected sample. It becomes evident if σ_{ya} is close to zero: it shows only that an empirical distribution is close to Weibull one, but it testifies neither that observations have been more accurate nor that confidence interval of F_i has decreased.

So, the observation error σ_{yo} and the probability error σ_{yp} should be analysed separately from the approximation error. These errors are latent ones, they could be considered to be possible deviations of the empirical distribution from the Weibull one.

Each certain class observation error could affect F_i via n_i, i.e. some number of v values could fall into a certain class instead of the neighbouring one by mistake because the wind measurements have been made with errors. This effect can be simulated with a shift of the class boundary σ_{vi} which depends upon the individual wind observation accuracy δv and n_i.

$$\sigma_{v_i} = \frac{\delta v}{\sqrt{n_i}} \qquad (21)$$

The following comparison of F_i and $F(v_i + \sigma_{vi})$ gives the evaluation of this certain class observation error.

Each class probability error or, more accurately, confidence interval of F_i depends upon n_i and the sample series size n. The term "probability error" was taken in analogue to the term "probable error" which is usually applied to a confidence interval of a series mean

Such an independence among the observation, probability, and approximation errors allows the specification for the combined error: it should be not the simple sum of them but the square root of the sum of their squares:

$$\sigma_y = \sqrt{\sigma_{yo}^2 + \sigma_{yp}^2 + \sigma_{ya}^2} \qquad (22)$$

3. CALCULATION

To be absolutely strict, both observation and probability errors should be estimated through the empirical distribution and, in particular, it is necessary to take empirical n_i. However, assuming that the difference between the Weibull and empirical distribution is not significant, one could replace empirical n_i with the corresponding Weibull n_i and get the confidence intervals caused by the observation errors and the probability errors of the theoretical Weibull distribution.

Taking into account the above consideration, observation errors can be estimated knowing only the Weibull distribution parameters and the accuracy of the wind velocity measurement. There is no strict evaluation of an individual wind velocity measurement accuracy. The special experiment was carried out by the World Meteorological Organisation in 1988 - 1990, and it showed that the mean difference between ship observations and reanalysed data is some 1.5 m/s (Kent et al., 1991; Kent and Taylor, 1991). That figure was used in the presented paper. The measurement error was believed to be random since systematic instrumental errors have to be eliminated during instrument checking.

The observation error σ_{yo} was calculated as the deviation from the theoretical Weibull distribution and was given as

$$\sigma_{yo} = \sqrt{\frac{\sum_{i=1}^{M}(\ln(-\ln F(v_i + \sigma_{v_i})) - \ln(-\ln F_i))^2}{M-2}} \qquad (23)$$

where σ_{v_i} (Eq.21) is one standard deviation or a 68% confidence interval of v_i.

Also, the probability error σ_{yp} was calculated using the theoretical Weibull distribution and was given as

$$\sigma_{yp} = \sqrt{\frac{\sum_{i=1}^{M}(\ln(-\ln(F_i + \sigma_{F_i})) - \ln(-\ln F_i))^2}{M-2}} \qquad (24)$$

where σ_{F_i} is a 68% confidence interval of F_i. For $i < M$, σ_{F_i} was computed using a normal distribution. For $i = M$, σ_{F_i} was computed using a binomial distribution. As confidence intervals of the latter are asymmetric, the greatest one was used in calculating σ_{yp}.

Wind velocity time series are strongly correlated. The coefficient of autocorrelation with lag=1 r_i is some 0.7 to 0.75. Alekseev (1975) showed that in the case of correlated series for more accurate evaluation of dispersion, series size n should be reduced with the factor $(1-r_i^2)$. Meanwhile, German (1971) and German and Levikov (1988) showed that extreme values could be considered to be independent and noncorrelated. The M class in the presented paper was selected to include ten extreme wind velocity values. So, for all classes but M class standard deviation of v_i and F_i were calculated using $n_{ie} = n_i(1-r_i^2) = 0.5\, n_i$ instead of n_i.

There is no possibility to evaluate approximation errors without an empirical distribution. Nevertheless, the rough evaluation of it was used. This evaluation was obtained using Amderma, Marre-Sale and Ust-Kara polar station 18 year observation data and is given as

$$\sigma_{ya} = \frac{0.07}{\sqrt{N}} \qquad (25)$$

where N is the sample series size in years.

Certainly, it is very rough evaluation, in particular, it does not take into consideration that σ_{ya} depends upon a. Furthermore, the dependence of N is not so simple, especially at N greater than a couple of decades, but nevertheless it reflects the order of σ_{ya}.

The joint standard deviation of all three errors was calculated via substitution of Eqs. (23), (24), and (25) into Eq.(22). This procedure is not entirely correct because all three errors should be analysed simultaneously using equation of Eq.(21) type, i.e.

$$\sigma_y = \sqrt{\frac{\sum_{i=1}^{M}(\ln(-\ln(F_i + \sqrt{(\delta F_o)_i^2 + (\delta F_p)_i^2 + (\delta F_a)_i^2})) - \ln(-\ln F_i))^2}{M-2}} \qquad (26)$$

The value y in Eq.(21) represents the bilogarithm of possible empirical probability including all possible errors. However, unlike observation and probability δF_i, there is no possibility to estimate each δF_i for approximation, because the set of them strictly depends upon empirical distribution and histogram partition. Use of Eq.(22) allows the application of some approximate estimation of σ_{ya}, i.e. Eq.(25).

A comparison of results obtained using Eq.(22) vs. using Eq.(26) was carried out for the observation and probability errors (Table 1). The value σ_y decreases slightly for Eq.(22) but the difference was within 10% for long sample series and within 5% for short ones. So, using of Eq.(22) instead of Eq.(26) is justified.

Table 1: Comparison of $\sigma_{y\ observation+probability}$'s obtained using Eq.(22) vs. using Eq(26).

		$\sigma_{y\ observation+probability}$					
N (year)	***a***	1.0	1.2	1.4	1.6	1.8	2.0
5	Eq.(22)	0.031	0.032	0.034	0.037	0.040	0.045
	Eq.(26)	0.031	0.032	0.035	0.038	0.042	0.047
	Difference	0.000	0.000	-0.001	-0.001	-0.002	-0.002
10	Eq.(22)	0.027	0.028	0.029	0.032	0.036	0.040
	Eq.(26)	0.027	0.028	0.030	0.033	0.037	0.042
	Difference	0.000	0.000	-0.001	-0.001	-0.001	-0.002
20	Eq.(22)	0.024	0.025	0.027	0.029	0.032	0.035
	Eq.(26)	0.024	0.025	0.027	0.030	0.033	0.037
	Difference	0.000	0.000	0.000	-0.001	-0.001	-0.002
50	Eq.(22)	0.021	0.021	0.023	0.025	0.028	0.031
	Eq.(26)	0.021	0.021	0.023	0.026	0.029	0.033
	Difference	0.000	0.000	0.000	-0.001	-0.001	-0.002
100	Eq.(22)	0.018	0.019	0.021	0.022	0.026	0.028
	Eq.(26)	0.018	0.019	0.021	0.023	0.027	0.030
	Difference	0.000	0.000	0.000	-0.001	-0.001	-0.002

4. RESULTS

The main accuracy estimation results are shown in Table 2. These are the possible relative error values for various return periods L shown as the ratio of 95% confidence interval δV_L to L year return period wind velocity value V_L. They were obtained using $z = 1.96$, $b = 6$, $v_l = 5$, $M = 10$, $n_M = 10$. The explanation for $b = 6$ is given in the comments to Table 3. The explanation for suggested histogram partition is given in the fifth section.

Since an accuracy of multidecade return wind velocity values highly depends upon basic series size N, the figures for a wide range of N are shown. The dependence upon N is obvious: the longer the basic series, the more accurate result can be obtained. It is typical for all statistical evaluations. Also, the entire reasonable range of a parameter is covered. In accordance with Eq.(8) and Eq.(19) increasing a decreases the error. Return period lengths were selected to meet the requirements of Russian standards.

The figures performed in Table 2 could be applied for the accuracy evaluation of multidecade return period wind velocity values obtained using empirical distributions with similar parameters and histogram partition.

The confidence interval can be obtained using the relative error from Table 2. For example, at reasonable conditions such as N is 20 to 30 years, a is 1.4 to 1.6, b is 6 to 7, the relative error $\delta V_{100}/V_{100}$ is within 15%, that means that for V_{100} = 35 - 40 m/s the 95% confidence interval δV_{100} is some 5 to 6 m/s. Similar conditions are typical, for example, for a number of North Western Siberian polar stations.

Table 2: Relative error $\delta V_L/V_L$(%) corresponding to 95% confidence interval.

	$\delta V_L/V_L$(%)					
a	1.0	1.2	1.4	1.6	1.8	2.0
N (year)	**L** = 1 year					
V_L(m/s)	47	33	26	21	19	16
5	29.5	24.6	21.8	20.2	19.3	18.9
10	23.1	19.2	17.1	16.0	15.5	15.3
20	18.6	15.5	13.8	13.1	12.8	12.8
50	14.8	12.2	11.0	10.5	10.4	10.6
100	12.8	10.6	9.5	9.1	9.2	9.4
	L = 5 years					
V_L(m/s)	57	39	30	24	21	18
5	30.5	25.4	22.5	20.8	19.9	19.4
10	23.8	19.8	17.6	16.4	15.9	15.7
20	19.3	16.0	14.3	13.4	13.1	13.1
50	15.3	12.6	11.3	10.8	10.7	10.9
100	13.2	10.9	9.8	9.4	9.4	9.6
	L = 10 years					
V_L(m/s)	61	41	31	25	21	19
5	30.9	25.7	22.7	21.0	20.1	19.6
10	24.1	20.1	17.8	16.6	16.1	15.9
20	19.5	16.2	14.4	13.6	13.3	13.3
50	15.4	12.8	11.4	10.9	10.8	11.0
100	13.4	11.0	9.9	9.5	9.5	9.7
	L = 20 year					
V_L(m/s)	65	44	33	26	22	19
5	31.2	26.0	23.0	21.3	20.3	19.8
10	24.4	20.3	18.0	16.8	16.2	16.0
20	19.7	16.3	14.6	13.7	13.4	13.4
50	15.6	12.9	11.6	11.0	10.9	11.1
100	13.5	11.1	10.0	9.6	9.6	9.8
	L = 50 years					
V_L(m/s)	71	47	35	28	23	20
5	31.7	26.4	23.3	21.5	20.5	20.0
10	24.8	20.6	18.2	17.0	16.4	16.2
20	20.0	16.6	14.8	13.9	13.5	13.5
50	15.8	13.1	11.7	11.1	11.0	11.2
100	13.7	11.3	10.1	9.7	9.7	9.9
	L = 100 years					
V_L(m/s)	75	49	36	29	24	21
5	32.0	26.6	23.5	21.7	20.7	20.2
10	25.0	20.7	18.4	17.1	16.5	16.3
20	20.2	16.7	14.9	14.0	13.6	13.6
50	16.0	13.2	11.8	11.2	11.1	11.3
100	13.8	11.4	10.2	9.8	9.8	10.0

Table 2 illustrates the error dependence upon a and N, but errors depend upon b also. This dependence is shown in Table 3. Relative error $\delta V_{100}/V_{100}$(%) corresponding to 95% confidence interval was obtained using the same histogram partition parameters as in the case of Table 2, but $b = 8$ and $b = 10$.

Table 3: Relative error $\delta V_{100}/V_{100}$(%) corresponding to 95% confidence interval for b = 8; b = 10.

	b = 8					
a	1.0	1.2	1.4	1.6	1.8	2.0
V_{100} (m/s)	100	66	48	38	32	28
N (year)						
5	30.6	24.6	21.0	18.6	17.0	15.9
10	24.0	19.2	16.4	14.6	13.4	12.7
20	19.4	15.5	13.2	11.8	11.0	10.4
50	15.4	12.2	10.5	9.4	8.8	8.5
100	13.4	10.6	9.0	8.2	7.7	7.5
	b = 10					
a	1.0	1.2	1.4	1.6	1.8	2.0
V_{100} (m/s)	125	82	61	48	40	35
N (year)						
5	30.0	23.7	19.8	17.2	15.4	14.1
10	23.5	18.5	15.4	13.4	12.1	11.1
20	19.1	14.9	12.4	10.8	9.8	9.1
50	15.2	11.8	9.8	8.6	7.8	7.3
100	13.2	10.2	8.5	7.4	6.8	6.4

Comparison of the relative error $\delta V_{100}/V_{100}$ values obtained using various values of the parameter b shows that increasing b slightly decreases the error. b = 6 is the lower bound of empirical b evaluations which polar stations feature (Koshinsky, 1978), so Table 2 essentially gives the upper bound of possible errors. That is the reason for b = 6 was taken for the computation of the Table 2 figures.

Both Table 2 and Table 3 illustrate dependence of possible overall errors upon values of distribution parameters but not show the contribution of each parameter error into the overall error.

Table 4: Relative error $\delta V_{100}/V_{100}$(%) corresponding to 95% confidence interval caused with a estimation errors only.

	$\delta V_{100}/V_{100} = \ln(-\ln F_{100})\delta a/a^2$					
a	1.0	1.2	1.4	1.6	1.8	2.0
N (year)	Observation Error					
5	2.7	3.2	3.6	4.0	4.2	4.4
10	2.2	2.7	3.1	3.4	3.6	3.8
20	1.9	2.3	2.6	2.9	3.2	3.4
50	1.5	1.9	2.2	2.5	2.7	2.9
100	1.3	1.7	2.0	2.2	2.5	2.6
	Probability Error					
5	9.7	7.3	5.9	4.9	4.1	3.6
10	8.2	6.2	4.9	4.0	3.4	3.0
20	7.0	5.3	4.2	3.4	2.9	2.5
50	5.9	4.4	3.5	2.9	2.4	2.1
100	5.2	3.9	3.1	2.5	2.1	1.8
	Approximation Error					
5	9.9	7.7	6.3	5.3	4.5	4.0
10	6.8	5.3	4.3	3.6	3.1	2.7
20	4.7	3.6	2.9	2.5	2.1	1.8
50	2.9	2.2	1.8	1.5	1.3	1.1
100	2.0	1.5	1.2	1.0	0.9	0.7
	Combined Error					
5	14.1	11.1	9.4	8.2	7.5	7
10	10.9	8.6	7.2	6.4	5.9	5.6
20	8.7	6.8	5.8	5.2	4.8	4.6
50	6.7	5.3	4.5	4.1	3.9	3.8
100	5.8	4.5	3.9	3.5	3.4	3.3

Table 4 and Table 5 illustrate the contribution of a and b estimation errors into the overall errors, and the contribution of errors of various types into a and b estimation errors.

Table 5: Relative error $\delta V_{100}/V_{100}$(%) corresponding to 95% confidence interval caused with b estimation errors only.

	$\delta V_{100}/V_{100} = \delta b/b$					
a	1.0	1.2	1.4	1.6	1.8	2.0
V_{100}(m/s)	75	49	36	29	24	21
N (year)	Observation Error					
5	3.4	4.4	5.5	6.5	7.4	8.4
10	2.9	3.8	4.7	5.6	6.5	7.4
20	2.5	3.3	4.2	5.0	5.8	6.6
50	2.1	2.8	3.6	4.3	5.0	5.8
100	1.8	2.5	3.2	3.9	4.6	5.3
	Probability Error					
5	12.2	10.2	8.8	7.9	7.3	6.8
10	10.6	8.7	7.6	6.7	6.2	5.7
20	9.3	7.6	6.6	5.8	5.3	4.9
50	8.0	6.5	5.6	4.9	4.5	4.1
100	7.3	5.9	5.0	4.4	4.0	3.7
	Approximation Error					
5	12.5	10.7	9.5	8.6	8.0	7.5
10	8.8	7.5	6.6	6.0	5.5	5.2
20	6.2	5.2	4.6	4.2	3.8	3.6
50	3.9	3.3	2.9	2.6	2.4	2.2
100	2.8	2.3	2.0	1.8	1.6	1.5
	Combined Error					
5	17.8	15.4	14.1	13.4	13.2	13.2
10	14.1	12.1	11.1	10.7	10.6	10.7
20	11.5	9.8	9.1	8.8	8.8	9.0
50	9.2	7.8	7.2	7.1	7.2	7.5
100	8.0	6.8	6.3	6.2	6.3	6.6

The approximation and probability errors are similar, observation error is a little bit less. All errors depend upon sample series size N, they decrease with its increasing. With a increasing possibility and approximation errors decrease, observation error increases in accordance with Eq.(23).

5. RECOMMENDATIONS

As it ensues from Eq.(19), the error estimation depends upon histogram partition. It is controlled with M and n_M or v_M, the lowermost boundary v_1 = 5 m/s having been fixed (Section 2). Variation of histogram partition leads not only to variation of the expression in parenthesis in Eq.(19) but to a variation of the standard deviation as well and, finally, to a variation of the accuracy. Variation of histogram partition dependent Eq.(19) components and of observation and probability errors with variation of histogram partition is shown in Table 6.

Dependence upon M is the strongest. Increasing M reduces both σ_y and the Eq.(19) expression in parenthesis. However, M increasing is limited with natural v_l and v_M values as well as reasonable class width. The class width of some 2 m/s seems to be the most optimal. M reduction considerably reduces the accuracy of the estimation.

Table 6. Dependence of Eq.(19) components upon histogram partition. $N = 20$ years, $a = 1.4$, $b = 6$, $V_{100} = 36$ m/s.

	$n_M = 10$			$n_M = 10$
	$M = 5$	$M = 10$	$M = 15$	$M = 11$
σ_{yo}	0.017	0.014	0.013	0.013
σ_{yp}	0.029	0.022	0.019	0.021
$(M/\Delta)^{0.5}$	1.181	0.752	0.594	0.710
$(\Sigma x^2/\Delta)^{0.5}$	3.419	2.124	1.662	2.002
$(\delta V_{100}/V_{100})_{obs.}$	13.3	6.8	4.7	6.2
$(\delta V_{100}/V_{100})_{prob.}$	23.1	10.8	7.0	9.6
$v_M (m/s)$	28	28	28	28
	$M = 10$			
	$n_M = 5$	$n_M = 10$	$n_M = 15$	$n_M = 20$
σ_{yo}	0.017	0.014	0.012	0.011
σ_{yp}	0.029	0.022	0.019	0.017
$(M/\Delta)^{0.5}$	0.735	0.752	0.764	0.773
$(\Sigma x^2/\Delta)^{0.5}$	2.108	2.124	2.135	2.144
$(\delta V_{100}/V_{100})_{obs.}$	8.5	6.8	5.9	5.5
$(\delta V_{100}/V_{100})_{prob.}$	14.0	10.8	9.4	8.6
$v_M (m/s)$	29	28	27	26

On the other hand, decreasing n_M which means increasing v_M and, therefore, more possibilities of M increasing leads to σ_y increasing, although the Eq.(19) expression in parenthesis slightly reduces simultaneously. So that, some accuracy reduction is the result of n_M decreasing. However, that could be partly compensated with M increasing within the same velocity limits, it is illustrated with the rightmost column in the upper pert of Table 6.

Then, the above considerations have concerned only observation and probability σ_y components, they have not touched upon approximation component as it was defined with Eq.(25) which did not allow such an analysis. Although, analyses of polar station natural data showed that a v_M shift to the lower wind velocity values led to increase of σ_{ya}.

So, the v_M shift to greater v values rather increases the accuracy then reduces it in spite of its decreasing due to the n_M reduction. Certainly, the exact valuation depends upon each certain distribution; but since involvement of the uppermost distribution edge does not give essential reduction of σ_{ya}, it is better to select v_M such that n_M being some 10 to 20 values.

6. CONCLUDING REMARKS

The histogram partition being rational, the presented estimates are the upper bound of possible errors of multidecade return period wind velocity values, caused with possible observation errors, possible errors due to general population replacement with sample series, and due to approximation of an empirical probability distribution with a Weibull one. They depend upon distribution parameters and sample series size. This dependence is natural, and these possible errors can not be reduced. Also, possible errors depend upon histogram partition.

Taking into account conclusions from the fifth section there is a possibility to select optimal histogram partition parameters and improve the accuracy of wind velocity estimation.

REFERENCES

Alekseev, GA (1975) "Methods of Estimation of Hydrometeorological Data Random Errors", 96 p., Hydrometeoizdat, Leningrad (In Russian).

German, VH (1971) "Research and Calculation of Probability Characteristics of Extreme Sea Level", *Proceedings of State Oceanographic Institute*, V107, 152 p., Hydrometeoizdat, Moscow (In Russian).

German, VH and Levikov SP (1988) "Probability Analysis and Simulation of Sea Level Variations", 232 p., Hydrometeoizdat, Leningrad, (In Russian).

Kent, EC, Truscott, BC, Taylor, PK and Hopkins, JS (1991) "The Accuracy of Ship's Meteorological Observations Results of the VSOP-NA", *Marine Meteorology and Related Oceanographic Activities*, Report No.26, 86 p., WMO/TD-No.455, WMO Geneva.

Kent, EC and Taylor, PK (1991) "Ships Observing Marine Climate. A Catalogue of the Voluntary Observing Ships Participating in the VSOP-NA", *Marine Meteorology and Related Oceanographic Activities*, Report No.25, 124 p., WMO/TD-No.456, WMO Geneva.

Koshinsky, CD (1978) "Regime Characteristics of Strong Winds over the Seas of Soviet Union", Part 2, 392 p., Hydrometeoizdat, Leningrad (In Russian).

Proceedings of the Seventh (1997) International Offshore and Polar Engineering Conference
Honolulu, USA, May 25-30, 1997

ISBN 1-880653-28-1 (Set); ISBN 1-880653-31-1 (Vol. III)

Statistic Comparison Between Wave Height Time Histories from Different Simulation Techniques for a Campeche Bay Sea State

Jorge Silva-Ballesteros and Federico Barranco-Cicilia
Instituto Mexicano del Petróleo
Mexico, D.F.

ABSTRACT

The objective of this work is to compare three simulation methods for a Gaussian unidimensional stationary random process, which represent irregular sea waves. The methods under comparison are the Deterministic Spectral Amplitude method, the Non-deterministic Spectral Amplitude method; and the Convolution Integral method. The stochastic process to simulate is the sea surface elevation $\eta(t)$ for a 100 yr. return-period storm in the Campeche Bay, Gulf of Mexico. In all methods such processes will be obtained from a Pierson-Moskowitz type spectrum. Parameters considered for comparison are the statistical characteristics of time histories (irregular sea waves) produced from several samples.

KEY WORDS: Wave spectral density, transfer function, random process, variance, Pierson-Moskowitz type spectrum, ergodicity, Campeche Bay.

INTRODUCTION

When analyzing the structural behavior of offshore structures an important aspect to deal with is the generation of synthetic sea states (irregular waves histories) which represent, with a high confidence level, the most critical conditions presented during a 100 yr. return period storm. Since irregular wave histories are random processes, the best way to perform the structural analysis is using a probabilistic approach through the random vibration theory, where a wave spectral density function $S_{\eta\eta}(\omega)$ is required. Such density characterizes, from a statistical point of view, the corresponding sea state.

The use of this wave spectral density along with a specific transfer function of the structure allows the analyst to obtain the statistical characteristics of the structural response. However, for several cases a time domain analysis is required, then irregular wave histories have to be generated from wave spectral densities. In this work, three methods to generate such histories are presented, and statistically compared: the Deterministic Spectral Amplitude method (DSA), the Non-deterministic Spectral Amplitude method (NSA), and the Convolution Integral method (CI). The main difference between these three methods is the type of process they produce: while the DSA produces ergodic processes, the NSA produces non-ergodic ones, and the CI method can produce either ergodic or non-ergodic processes, depending upon the white noise generator. A stationary process is called "ergodic" if the averages taken along any single sample are the same as the ensemble averages, each sample function is then completely representative of the ensemble that constitutes the random process. This fact could have much effect on the variance of the processes generated, and when these histories are applied to the structure, in order to study its behavior, significant errors are expected to be present due to the great variation of variance and higher order moments. It should be pointed out that structural response depends mainly on such moments.

SEA SURFACE ELEVATION, $\eta(t)$, BY SPECTRAL REPRESENTATION (DSA) AND FAST FOURIER TRANSFORM METHOD (NSA)

Let $\eta(t)$ be a stochastic process at least weakly stationary, therefore the mean value has to be constant and the autocorrelation function should not depend on time t, but on τ. Then:

$$E[\eta(t)] = \mu_\eta = 0 \tag{1}$$

$$E[\eta(t+\tau)\eta(t)] = R_{\eta\eta}(\tau) \tag{2}$$

This stochastic process has a spectral density function, which is, according to the Wiener theorem, defined as:

$$S_{\eta\eta}(\omega) = \frac{1}{2\pi}\int_{-\infty}^{\infty} R_{\eta\eta}(\tau)e^{-i\omega\tau}\,d\tau \tag{3}$$

Taking the following theorem: to every real value one-directional, one-variate stationary stochastic process $\eta(t)$ with mean value equal to zero and two sided power spectral density function $S_{\eta\eta}(\omega)$, two mutually orthogonal real processes $u(\omega)$ and $v(\omega)$ with orthogonal increments can be assigned such that:

$$\eta(t) = \int_0^{\infty}\left[\cos(\omega t)du(\omega) + \sin(\omega t)dv(\omega)\right] \qquad (4)$$

It is easy to demonstrate (Shinozuka, 1991) that $\eta(t)$ can be represented as a wave superposition with deterministic amplitude and random phase angle, this is the called "Spectral Representation method" (DSA):

$$\eta(t) = \sqrt{2}\sum_{n=1}^{N} A_n \cos(\omega_n t + \phi_n) \qquad (5)$$

where:

$A_n = \left\{2S_{\eta\eta}(\omega_n)\Delta\omega\right\}^{1/2}$ is the amplitude of wave component having frequency ω_n,

$\omega_n = n\Delta\omega$, and

ϕ_n = random phase angle uniformly distributed over 0 and 2π.

For NSA method, the sea surface elevation is represented by Discrete Fourier Series (Shinozuka, 1991):

$$\eta(t) = \sum_{n=1}^{N}\left(a_n \cos\omega_n t + b_n \sin\omega_n t\right) \qquad (6)$$

where the coefficients a_n and b_n are independent zero-mean random variables which variances can shown to be (Vanmarcke et al, 1993):

$$E\left[a_n^2\right] = \begin{cases} 2\Delta\omega S_{\eta\eta}(\omega_n), & n = 1, \frac{N}{2}+1 \\ \frac{\Delta\omega}{2}\left[S_{\eta\eta}(\omega_n) + S_{\eta\eta}(\omega_{N-n+2})\right], & n = 2,\ldots,\frac{N}{2} \end{cases} \qquad (7)$$

$$E\left[b_n^2\right] = \begin{cases} 0, & n = 1, \frac{N}{2}+1 \\ E\left[a_n^2\right], & n = 2,\ldots,\frac{N}{2} \end{cases} \qquad (8)$$

The expected values for such coefficients are calculated from:

$$Z_{a_n} = \Gamma_n \sqrt{E\left[a_n^2\right]} \qquad (9)$$

$$Z_{b_n} = \Lambda_n \sqrt{E\left[b_n^2\right]} \qquad (10)$$

In equations (9) and (10) Γ_n and Λ_n are normal standard random variables.

Z_{a_n} and Z_{b_n} are obtained for the all range of frequencies applying symmetry conditions from the Nyquist frequency, $\omega_{\frac{N}{2}+1}$,

Finally, a Fast Fourier Transform (FFT) routine is applied to perform the inverse discrete transform on equations (9) and (10) to produce the sea surface elevation, $\eta(t)$.

It is easy to see that equation (5) produces an ergodic random process while equation (6) generates a non-ergodic random process, although both are at least weakly stationary. Since $\eta(t)$ is produced by the sum of N Gaussian processes we can take only one contribution from the sum in equation (5):

$$\eta_j(t) = \sqrt{2}A_j \cos(\omega_j t + \phi_j) \qquad (11)$$

A_j is not a random variable.

The mean value of this component is:

$$E\left[\eta_j(t)\right] = E\left[\sqrt{2}A_j \cos(\omega_j t + \phi_j)\right]$$

$$E\left[\eta_j(t)\right] = \sqrt{2}A_j E\left[\cos(\omega_j t + \phi_j)\right]$$

$$E\left[\eta_j(t)\right] = \mu_{\eta j} = 0 \qquad (12)$$

and the variance is:

$$Var\left(\eta_j(t)\right) = E\left[\eta_j^2(t)\right] - \mu_{\eta j}^2$$

$$Var\left(\eta_j(t)\right) = \sigma^2_{\eta j} = E\left[\eta_j^2(t)\right]$$

$$\sigma_{\eta j}^2 = A_j^2 \qquad (13)$$

Since A_j is not random, the variance of $\eta_j(t)$ for the DSA method does not variate from one sample to another, then $\eta_j(t)$ is ergodic as well as $\eta(t)$.

Rewriting equation (6), and using the same procedure as in equation (5), the following result is achieved:

$$\eta_j(t) = C_j \cos(\omega_j t - \theta_j) \qquad (14)$$

where:

$$C_j = \left(a_j^2 + b_j^2\right)^{1/2} \quad \text{; random variable}$$

$$\theta_j = tg^{-1}\left(\frac{b_j}{a_j}\right) \quad \text{; random phase angle.} \tag{15}$$

For this case, the obtained mean value is:

$$\mu_{\eta j} = 0$$

and the variance:

$$\sigma_{\eta j}^2 = \frac{\sigma_{cj}^2}{2} \tag{16}$$

The variance of $\eta_j(t)$ depends on the variance of C_j, then for each sample different statistics are produced therefore $\eta_j(t)$ and $\eta(t)$ are non-ergodic for this case.

SEA SURFACE ELEVATION BY CONVOLUTION INTEGRAL

There exists a unique relationship between the input and its corresponding output for linear systems, that relationship can be expressed by the transfer function $H(\omega)$ in the frequency domain or by the impulse response function $h(t)$ in time domain. Both functions are related each other by the Fourier pair transform:

$$h(t) = \frac{1}{2\pi}\int_{-\infty}^{\infty} H(\omega)e^{i\omega t}d\omega \tag{17}$$

$$H(\omega) = \int_{-\infty}^{\infty} h(t)e^{-i\omega t}dt \tag{18}$$

Naito et al (1995) obtained the following $h(t)$ from a Pierson-Moskowitz type spectrum and a white noise spectrum:

$$h(t) = \begin{cases} -\sqrt{AB^{-3/4}} \cdot I\left(\frac{5}{2}\right) & t \geq 0 \\ 0 & t < 0 \end{cases} \tag{19}$$

where:

$$A = \frac{173H_S^2}{T_0^4} \quad ; \quad B = \frac{691}{T_0^4}$$

H_S: Significant wave height
T_0: mean wave period

$$I(\gamma) = \sum_{j=1}^{2}\left\{\sum_{k=1}^{n-2}\sum_{\iota=1}^{k} Q_1(\gamma)\tau_0^{\iota-1}x_j^k\left(x_j - x_{3-j}\right)^{-k+\iota-1} + \sum_{k=1}^{n-1} Q_2(\gamma)\tau_0^{k-1}x_j^{k-1} + x_j^{n-1}\sum_{\iota=1}^{n-1} Q_3\tau_0^{\iota-1}\left(x_j - x_{3-j}\right)^{-n+\iota} + \frac{\tau_0^{n-1}}{(n-1)!}x_j^{n-1}\right\}\left(x_j - x_{3-j}\right)^{-n} x_j^{n-\gamma+1} e^{\tau_0 x_j}$$

$$Q_1(\gamma) = \frac{(2n-\gamma)(2n-\gamma-1)\cdots(2n-\gamma-(n-k-2)) \cdot (-n)(-n-1)\cdots(-n-(k-\iota))}{(\iota-1)!(n-k-1)!(k-\iota+1)!}$$

$$Q_2(\gamma) = \frac{(2n-\gamma)(2n-\gamma-1)\cdots(2n-\gamma-(n-k-1))}{(k-1)!(n-k)!}$$

$$Q_3(\gamma) = \frac{(-n)(-n-1)\cdots(-n-(n-\iota-1))}{(\iota-1)!(n-\iota)!}$$

with:

$$\tau_0 = i\sqrt[4]{B}\text{; } x_1 = \exp(i\pi/4)/\sqrt[4]{n}\text{; } x_2 = \exp(i3\pi/4)/\sqrt[4]{n}$$

Then, $\eta(t)$ can be obtained using the convolution integral:

$$\eta(t) = \int_{-\infty}^{\infty} h(t)N(t-\tau)d\tau, \tag{20}$$

where $N(t)$ is the time history from white noise spectrum.

NUMERICAL APPLICATION

Different computer algorithms were written to deal with equation (5), (6) and (20) in order to generate irregular sea waves. The sea state for Campeche Bay was represented by a Pierson-Moskowitz type spectrum as shown in figure 1, with H_S=10.4 m and T_0= 10 seg, mean value (μ_η) and variance (σ_η^2) are zero and 6.77, respectively.

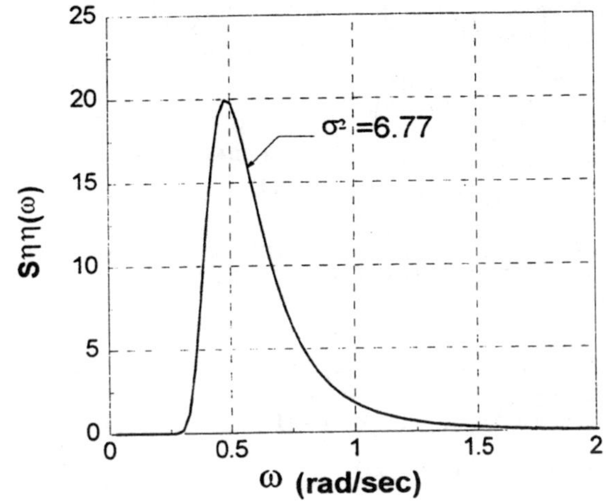

Figure 1. Pierson-Moskowitz type sea wave spectrum for Campeche Bay. H_S=10.4 m and T_0=10 sec.

Wave Spectra Estimation.- In order to compare the characteristics of the produced samples wave spectra estimation $\bar{S}_{\eta\eta}(f)$ are required, in this work a discrete Fourier transform is applied on $\eta(t)$, therefore:

$$\bar{S}_{\eta\eta}(f) = \frac{1}{T_0}\left\| \Delta t \sum_{k=0}^{N-1} \eta(k\Delta t) e^{-i2\pi k f \Delta t} \right\|^2 \qquad (21)$$

where T_0 is the total time length of $\eta(t)$ and $\| \|$ denotes norm of the argument.

Deterministic Spectral Amplitude Method.- Several samples were produced for a time period of 250 sec. For each sample, a spectral density estimation $\bar{S}_{\eta\eta}(f)$ was calculated having the same variance and higher order moments than those from the original spectrum. It follows from the fact that this method generates ergodic processes.

Figures 2 and 3 show two samples and their corresponding spectral densities, it is easy to see that no dispersion in spectral amplitudes is present.

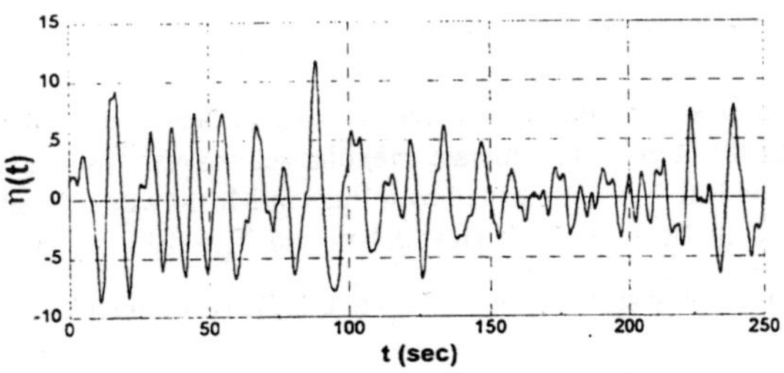

a) Irregular wave time history.

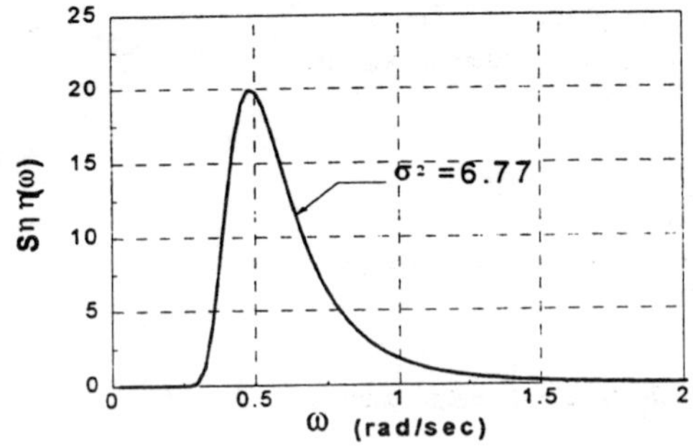

b) Sea wave spectrum.

Figure 2. Deterministic spectral amplitude method (sample 1).

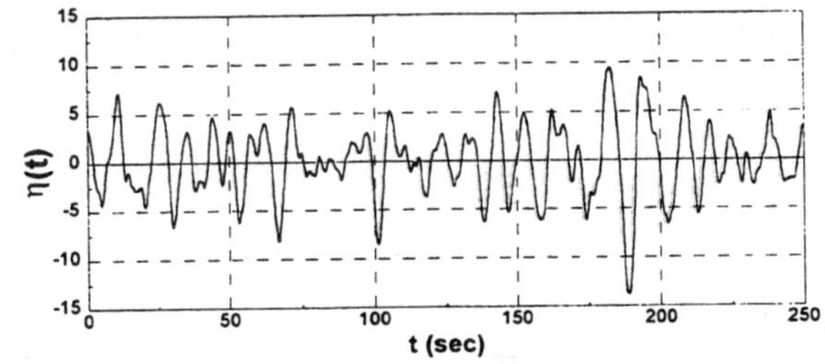

a) Irregular wave time history.

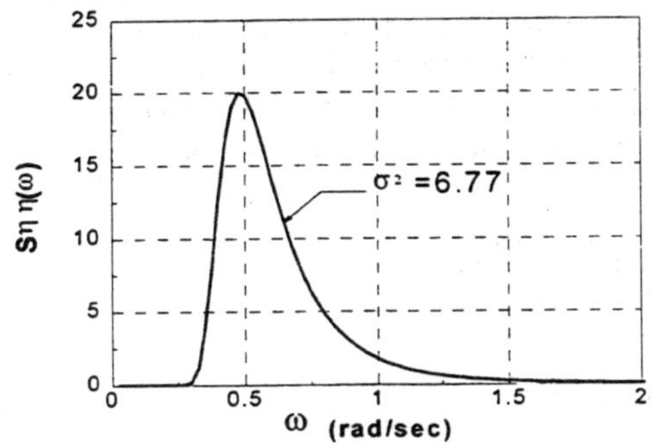

b) Sea wave spectrum.

Figure 3. Deterministic spectral amplitude method (sample 2).

Non-Deterministic Spectral Amplitude Method.- Since this method produces non-ergodic wave time histories, a very high number of samples had to be generated because the variance differs from one sample to another, as it can be seen in figures 4 and 5. After 1000 samples the mean value of their corresponding spectra shows a shape that adequately fits the original P-M type spectrum (see figure 6). However, if a mean time history from the 1000 samples is taken, there is not guarantee that its spectral shape will fit the P-M type spectrum.

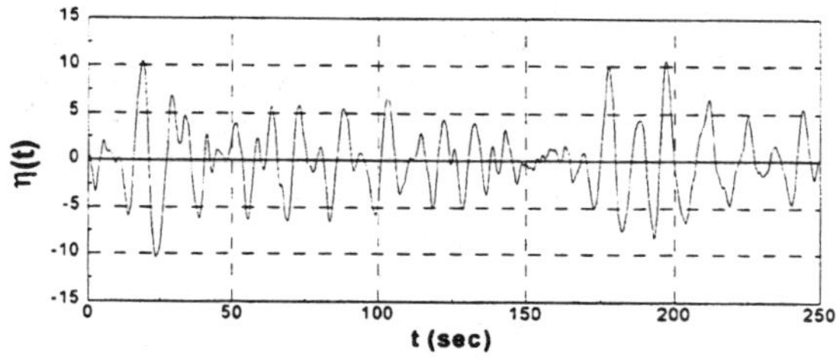

a) Irregular wave time history.

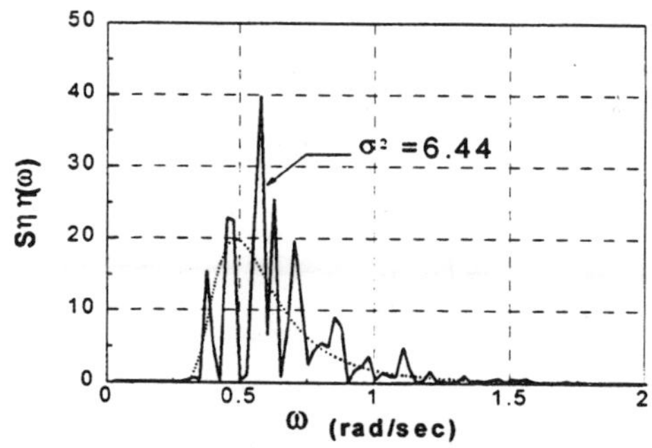

b) Sea wave spectrum.

Figure 4. Non-Deterministic spectral amplitude method (sample 1).

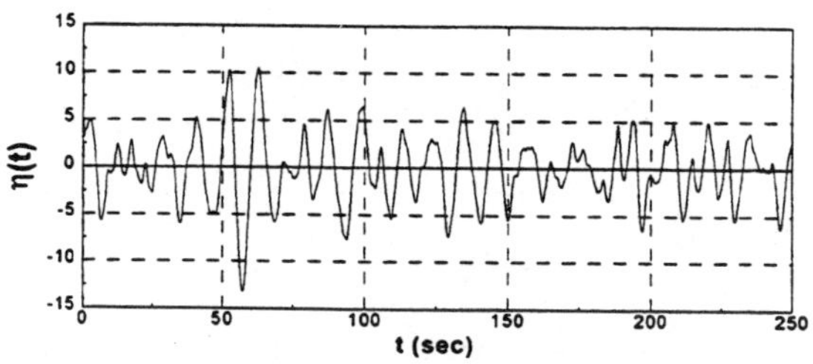

a) Irregular wave time history.

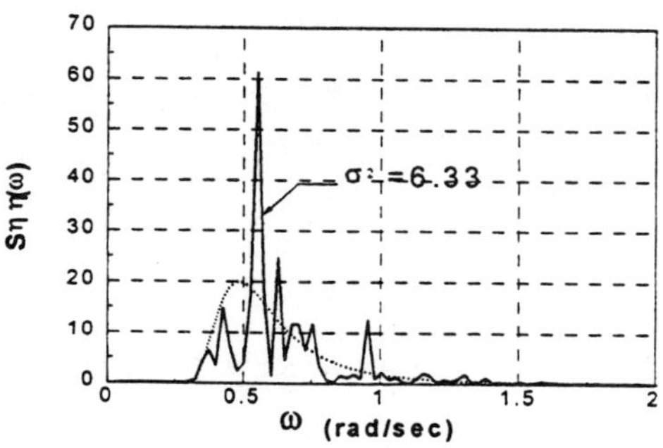

b) Sea wave spectrum.

Figure 5. Non-Deterministic spectral amplitude method (sample 2).

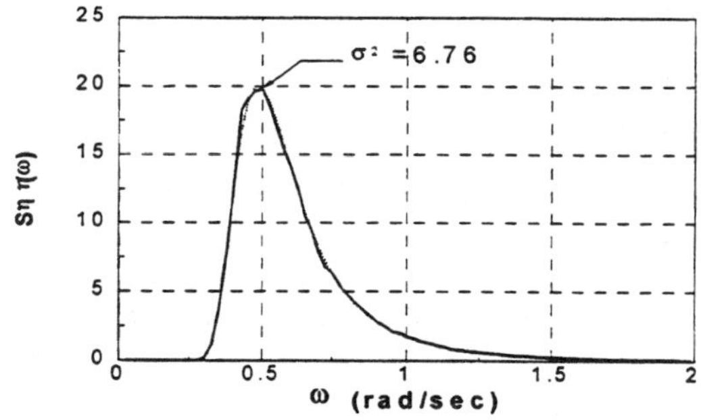

Figure 6. Mean spectrum after 1000 samples. NSA method.

Figures 4.b and 5.b show the wave spectra generated from two samples. Due to the fact that spectral amplitudes are scaled by two random variables Γ_n and Λ_n as can be seen in equations (9) and (10), many spikes are shown at different frequencies. This fact results in considerable differences in the moments for each sample spectrum, as it can be seen in Table 1. Figure 7 shows the high dispersion of the variance for the 1000 samples generated, and figure 8 shows the normalized variance of the mean spectrum when different number of samples were taken. From these figures along with the fact of variation in moments, it is easy to see the great uncertainty in selecting a time wave history to be applied for structural analysis.

	1000 simulations		P-M
Spectral moments	Maximum	Minimum	"Real"
M_0	13.57	5.14	6.77
M_2	5.14	1.807	3.15
M_4	7.02	4.88	7.06

Table 1. Values of spectral moments

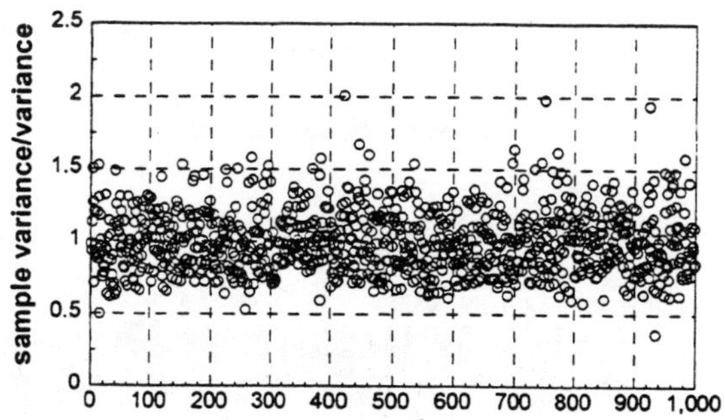

Figure 7. Dispersion of the variance for the 1000 samples. NSA method.

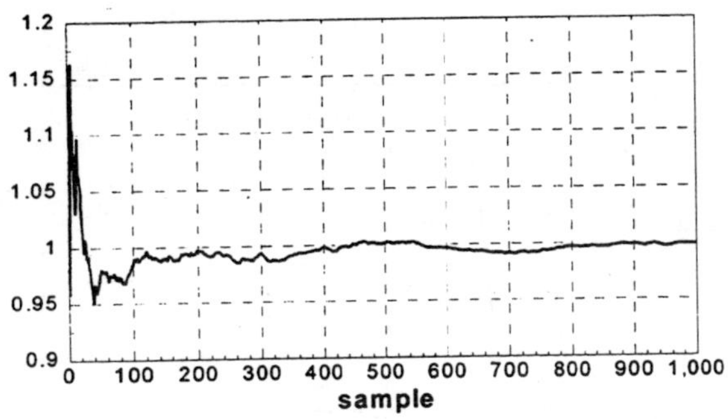

Figure 8. Normalized variance vs. samples. NSA method.

Convolution Integral Method.- As it is mentioned in the introduction of this paper, the ergodicity of the process depends on the time history N(t) (from white noise spectrum) which should be included in equation (20). Since in our case N(t) was a non-ergodic time series, then $\eta(t)$ is non-ergodic even though h(t) is deterministic.

Figure 9.a shows the impulse response function h(t), obtained for a P-M type spectrum which does not change from each sample to another, while figure 9.b shows the time history from white noise spectrum which is statistically different from each one of the thousand generated. Due to this fact the wave spectrum for each sample shows high dispersion as it is shown in figure 9.d.

After 1000 samples the mean spectral density fits the P-M spectrum as can be seen in figure 10.

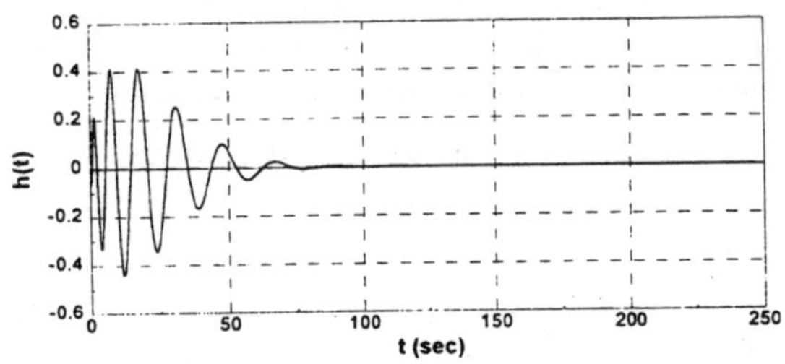

a) Impulse response function.

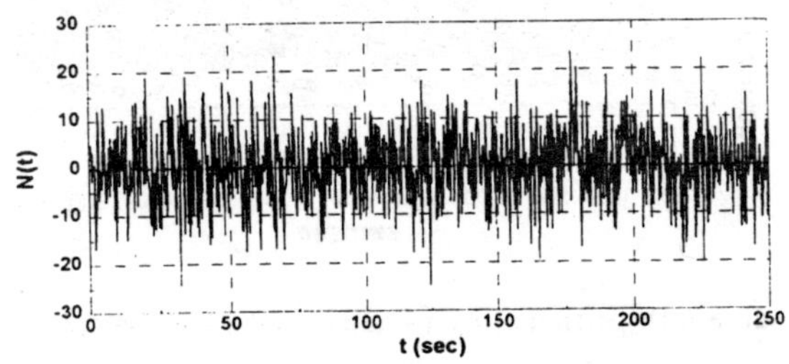

b) Time history from white noise spectrum.

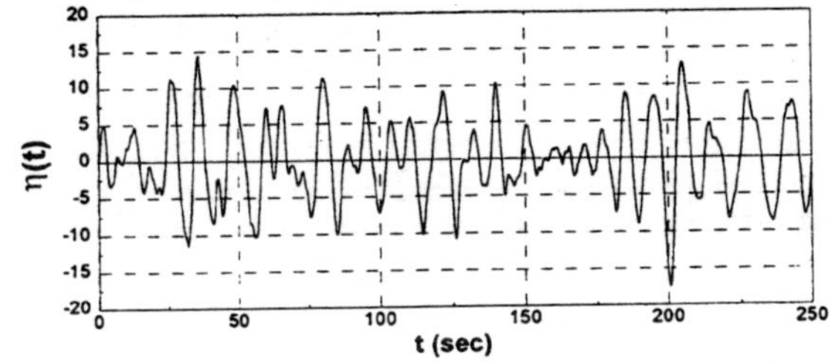

c) Wave time history.

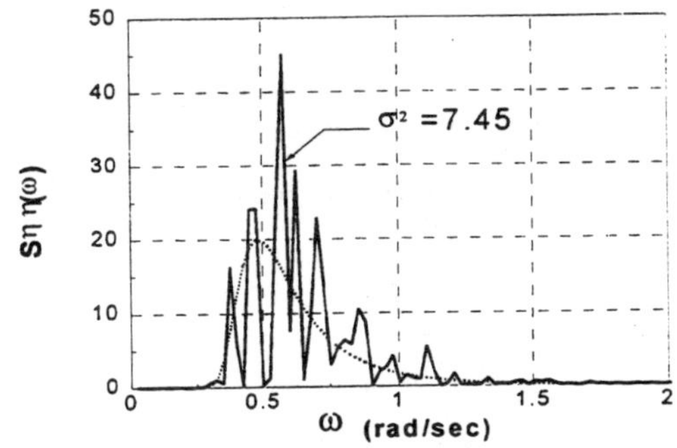

d) Sea wave spectrum.

Figure 9. Convolution integral method.

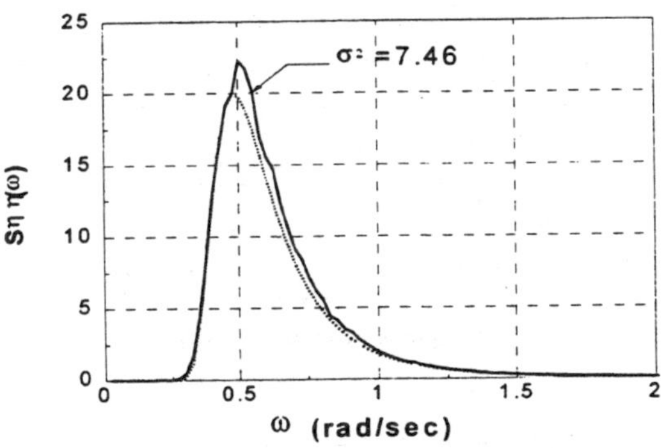

Figure 10. Mean spectrum after 1000 samples. CI method.

CONCLUSIONS

When subjected to a irregular waves (random) the response of a structure can be obtained through its transfer function as:

$$S_{\mu\mu}(\omega) = \left|H_1(\omega)\right|^2 S_{\eta\eta}(\omega) \quad (22)$$

where:

$S_{\mu\mu}(\omega)$ =spectral response density function

$H_1(\omega)$ =structural transfer function.

Since the process is narrow band, then the most probable value of the maximum response, or in other words, the mode of μ_{max} is:

$$(\mu_{max})_{mode} = \sqrt{2\ln\{2\upsilon^{+}D\}}\ \sigma_{\mu} \qquad (23)$$

where

σ_{μ} =standard deviation of the response.
D =expected time of the sea state will be present.
υ^{+} =zero crossing frequency of the response.

$$\upsilon^{+} = \frac{1}{2\pi}\left[\frac{M_2}{M_o}\right]^{1/2}$$

$$M_o = \int_{-\infty}^{\infty} S_{\mu\mu}(\omega)d\omega$$

$$M_2 = \int_{-\infty}^{\infty} \omega^2 S_{\mu\mu}(\omega)d\omega$$

Since $(\mu_{max})_{mode}$ strongly depends on the moments of the spectral response density, and considering $S_{\mu\mu}(\omega)$ is obtained from $S_{\eta\eta}(\omega)$ (equation 22), dispersion of the sea wave spectral density may result in a wrong estimation of $(\mu_{max})_{mode}$.

Due to the fact that NSA and CI (when using non-ergodic white noise generator) methods produce irregular sea waves which spectral density present high dispersion, the authors recommend to use carefully the corresponding time histories when carrying out structural response analysis.

However the authors recommend to use the Deterministic Spectral Amplitude since it produces ergodic samples and taking advantage of current computer capacity, it is not very high time consuming to get these time series.

REFERENCES

1. Brebbia, C.A., and Walter, S., "Dynamic Analysis of Offshore Structures", Newnes-Butterworths, 1979.
2. Hallan, M. G., et. al., "Dynamics of Marine Structures: Methods of calculating the dynamic response of fixed structures subjected to wave and current action", CIRIA Underwater Engineering Group, 1978.
3. Julius, S. B., "Random Data Analysis and Measurement Procedures", John Wiley and Sons, 1986.
4. Lee, Y.W., "Statistical Theory of Communication", John Wiley and Sons, 1967.
5. Morooka, C. K., and Yokoo, I. H., "Numerical Simulation and Spectral Analysis of Irregular Sea Waves", Proceedings of the Fifth International Offshore and Polar Engineering Conference, June 1995.
6. Naito, S. , and Minouva, M, "Representation of Pierson-Moskowitz Type Spectrum in Time Domain", Proceedings of the Fifth International Offshore and Polar Engineering Conference, June 1995.
7. Shinozuka, M., and Deodatis, G., "Simulation of Stochastic Processes by Spectral Representation", Applied Mechanics Reviews, Vol. 44, No. 4, April 1991.
8. Vanmarcke, E.H., Heredia-Zavoni, E. and Fenton, G.A., "Conditional Simulation of Spatially Correlated Earthquake Ground Motion", Journal of Engineering Mechanics, ASCE, 1993.

Proceedings of the Seventh (1997) International Offshore and Polar Engineering Conference
Honolulu, USA, May 25-30, 1997

ISBN 1-880653-28-1 (Set); ISBN 1-880653-31-1 (Vol. III)

A Wave Climate Study for an LNG Terminal Near Marmara Ereglisi, Turkey

Adnan Akyarli
Dokuz Eylul University
Izmir, Turkey

Ahmet C. Yalciner
Middle East Technical University
Ankara, Turkey

ABSTRACT
A liquefied natural gas (LNG) terminal has been constructed near Marmara Ereglisi - *a coastal town located at the northern coast of the Sea of Marmara in Turkey* - to unload imported LNG from Algeria. Regarding the wave related issues in the design of the terminal and offshore operations, a wave climate study - *comprising direct wave measurements carried out by a directional type wave buoy during one year period and hindcasting studies based on ten years of hourly wind data recorded at Tekirdag meteorological station* - have been performed.

KEY WORDS
Wave climate, directional wave measurements, wave hindcasting, LNG terminals, Sea of Marmara.

INTRODUCTION
A liquefied natural gas (LNG) plant - *to receive LNG imported from Algeria* - has been constructed near Marmara Ereglisi which is a coastal town located at the northern coast of the Sea of Marmara in Turkey. As a part of the works, an LNG terminal - *for tankers between 40.000 - 125.000* m^3 *category* - and intake and outfall structures - *to supply and discharge sea water to be used in the vaporisers* - have been designed (Fig. 1).

Wave conditions are likely to be critical with regard to LNG vessels moored at berths when the period of waves are within the approximate range of 6 to 12 sec. The limiting wave height of waves is dependent upon the size of the vessel. The following significant wave heights are the maxima which can be tolerated by vessels at the berth, whilst continuing to discharge cargo (Table 1) (L.G. Moucehl & Partners, 1988).

If the discharge of cargo is discontinued, the vessel can remain at the berth and be expected to tolerate waves having periods exceeding 6 sec, with significant heights of 1.5 to 2.0 m.

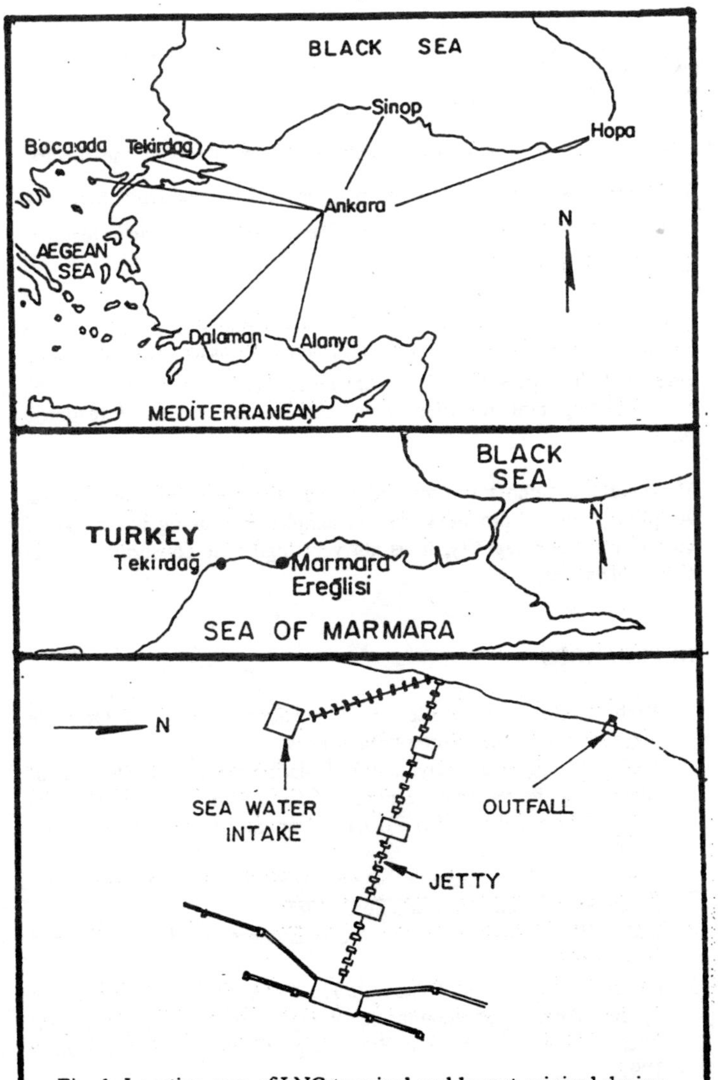

Fig. 1: Location map of LNG terminal and layout original design

Table 1: Limiting significant wave heights for LNG vessels at the berth

Vessel Capacity (m^3)	Wave orientation	Limiting significant wave height (m)
40.000	Head/Stern	0.8
40.000	Beam	0.6
125.000	Head/Stern	1.5
125.000	Beam	1.0

Attendant tugs and mooring boats will unlikely be able to manoeuvre to assist the ship when significant wave heights reach 1.5 to 2.0 m. Therefore, if critical wave conditions are likely to occur to the extent that vessel will need to leave the berth, this should be done before wave heights exceed those acceptable for the operation of tugs.

As noted above, if waves are to occur with periods exceeding 6 sec and heights greater than 1.5 to 2.0 m, the vessels will unlikely be able to remain at the berth. This introduces a further constraint in that it may not be possible to remove part loaded tankers from a berth to ride out a storm.

Safety guide of the Oil Companies International Maritime Forum for LNG terminals (OCIMF, 1978) notes the following operating limitations for "Refrigerated Gas Carriers":

- ***Ship stability***: The design of refrigerated gas carriers is such that ballast water and cargo must be handled simultaneously. This generally means that care must be taken to maintain the ship in a condition of sufficient positive stability. At a critical period in the operation, it may not be possible to move vessel off the berth at short notice. This should be taken into consideration when planning emergency procedures.

- ***Slack tanks***: A refrigerated gas carrier, because of its design, will at any time have a relatively low margin of positive stability. For this reason, it is not possible for some ships to put to sea with one or more partially filled tanks since this will further reduce the margin of stability. Some designs of cargo containment systems have the additional problem of the "sloshing" effect caused by virtual movement of the liquid in a partially-filled tank which can cause structural damage to the unit. This also means that the ship may not be able to vacate a berth a short notice during cargo operations.

Regarding the wave related issues in the design of terminal and offshore operations, a wave climate study has been performed :
i) to prepare the navigation procedure for LNG tankers,
ii) to confirm the design of LNG marine terminal,
iii) to modify the manual of the terminal operations.

This paper introduces the findings of the wave climate study which comprises both direct wave measurements carried out by a directional type wave buoy at the terminal site during one year by the supervision of the first author (Temel Surveying & IMST, 1991), and hindcasting studies of Delft Hydraulics (1990) which are based on ten years of wind data recorded at Tekirdag meteorological station (Fig. 1).

WAVE MEASUREMENTS

Methodology

Wave measurements were conducted by an ENDECO 1156 Directional WAVE-TRACK Buoy deployed at the vicinity of the proposed LNG terminal to be about 1 km offshore where the water depth is 17 meters. This system accurately measures wave direction and heave using a wave orbital following buoy designed as an "inverted pendulum" (Bainard, 1989). Simultaneous wind measurements were also performed by an anometer attached to the buoy at 1.5 m height above the still-water level.

The system was adjusted to make wave measurements during the first 17 minutes of every three hours (0, 3, .., 21 GMT), and sampling rate was adopted as 1/2 second (i.e. two hertz). However, the interval has automatically been reduced to one hour after the significant wave height has become greater than a pre-determined value.

Data were recorded in digital forms after the on-line transmission to the PC of the onshore station, and then analysed by zero-crossing and digital bandpass filter methods to obtain pertinent wave characteristics (Temel Surveying & IMST, 1991). Table 2 shows the wave characteristics obtained after the analysis of wave data. Here, while the first ten values are determined from zero-crossing analysis, last figure corresponds to wave direction having the maximum wave energy which is given by digital bandpass filter method (Bainard, 1983).

Table 2: Wave characteristics obtained after the analysis of measured wave data

DATE	HOUR	(1)	(2)	(3)	(4)	(5)	(6)	(7)	(8)	(9)	(10)	(11)
1 DEC	0:00	256	7.5	5.0	1.71	4.87	0.94	3.99	0.60	4.78	1.18	178
1 DEC	1:00	248	7.0	5.0	1.77	4.93	0.82	4.09	0.51	5.08	1.10	184
1 DEC	2:00	254	7.0	5.0	1.37	4.67	0.82	4.02	0.51	4.56	1.04	178
1 DEC	3:00	261	6.5	4.5	1.82	4.70	0.86	3.91	0.53	4.46	1.15	168
1 DEC	4:00	259	7.5	5.0	1.40	4.73	0.82	3.93	0.53	4.54	1.04	171
1 DEC	5:00	267	6.5	4.5	1.29	4.50	0.76	3.82	0.48	4.42	0.97	170

(1) N: NUMBER OF INDIVIDUAL WAVES IN THE RECORD
(2) T_{max}: MAXIMUM WAVE PERIOD (SEC)
(3) T_{Hmax}: PERIOD OF THE HIGHEST INDIVIDUAL WAVE (SEC)
(4) H_{max}: MAXIMUM WAVE HEIGHT (M)
(5) T_s: SIGNIFICANT WAVE PERIOD (SEC)
(6) H_s: SIGNIFICANT WAVE HEIGHT (M)
(7) T_{mean}: MEAN WAVE PERIOD (SEC)
(8) H_{mean}: MEAN WAVE HEIGHT (M)
(9) $T_{1/10}$: AVERAGE PERIOD OF HIGHEST 1/10 INDIVIDUAL WAVES (SEC)
(10) $H_{1/10}$ AVERAGE HEIGHT OF HIGHEST 1/10 INDIVIDUAL WAVES (M)
(11) WAVE DIRECTION IN DEGREE (0 INDICATES THE NORTH AND INCREASES IN THE CLOCKWISE DIRECTION)

Wave measurements were carried out between 10 February 1990 - 12 February 1991, and 3964 recordings were made. Wave-height data is presented by cumulative probability distribution which shows the percentage exceedence of significant and maximum wave heights (WMO, 1988) (Fig. 2).

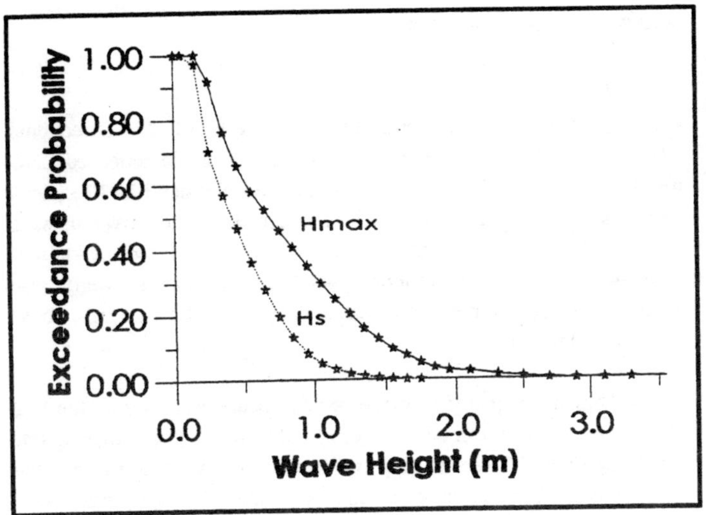

Fig. 2: Percentage exceedence of significant and maximum wave heights

According to this figure:

- Measured significant wave heights are less than 2.0 m, and exceedence probability of maximum wave height of 2.0 m is 2.6%,
- The probabilities of exceedence for 1.5 m significant and maximum wave heights are 0.5% and 10%, respectively,
- The probabilities of exceedence for 1.0 m significant and maximum wave heights are 5% and 29.5%, respectively.

These findings verify that wave conditions have been moderate at Marmara Ereglisi during the measurements.

Distribution of Individual Wave Heights

It is generally accepted that marginal probability distribution of individual wave heights in a wave record follows the Rayleigh distribution. Rayleigh distribution yields the following relationships between significant and other characteristic wave heights (Goda, 1979):

$$H_{mean} = 0.625\, H_s$$
$$H_{1/10} = H_s/\sqrt{2}$$
$$H_{max} = (\ln N/2)^{1/2}\, H_s$$

In order to examine the validity of these relationships, the best fit lines ($H_* = \alpha_i H_s + \beta_i$) between H_s and H_* - *which correspond respectively to* H_{mean}, $H_{1/10}$ *and* H_{max} *(most probable maximum value) of totally 3964 records*- were obtained by linear regression (Fig. 3). The coefficients of the best fit lines are presented in Table 5. Here, theoretical values of Rayleigh distribution and empirical values obtained by Turhan (1996) and Ozhan and Abdalla (1993b) after the analysis of wave data collected within NATO TU-WAVES project at the Mediterranean Sea *(Alanya and Dalaman)*, Aegean Sea *(Bozcaada)* and the Black Sea *(Sinop)* (Fig. 1) are also provided.

Referring to Fig. 3, it can be concluded that individual wave heights measured at the Sea of Marmara satisfactorily follows the Rayleigh distribution similar to the waves observed at other seas surrounding Turkey (Turhan, 1996).

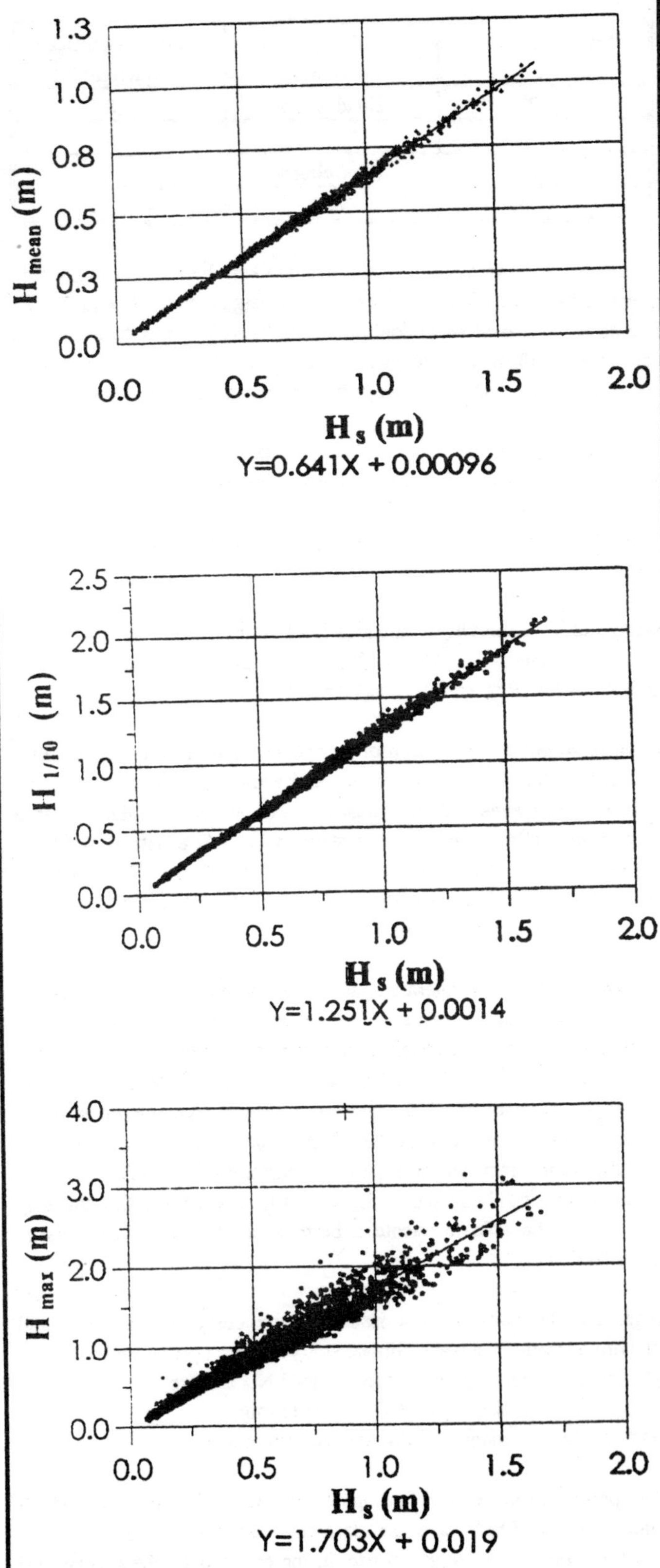

Fig. 3: Best fit lines between H_s and H_* (H_{mean}, $H_{1/10}$, H_{max})

Table 3: Relationships between characteristic wave heights

Parameter (H_*) →	H_{mean}		$H_{1/10}$		H_{max} (3)	
Station ↓	α_1	β_1	α_2	β_2	α_3	β_3
Alanya (1)	0.642	0.000	1.244	0.001	1.663	0.016
Dalaman (1)	0.636	0.001	1.249	0.003	1.658	0.031
Bozcaada (1)	0.642	-0.001	1.247	0.005	1.779	0.011
Sinop (1)	0.637	-0.001	1.244	0.005	1.631	0.063
M.Ereglisi	0.641	0.001	1.251	0.001	1.703	0.019
Overall (2)	0.638	-0.001	1.249	0.000	1.658	0.003
Theoretical	0.625	0.000	1.270	0.000	1.646	0.000

(1) average of monthly values given by Turhan (1996).

(2) average of the figures given for Alanya, Dalaman, Bozcaada and Sinop by Turhan (1996).

(3) Theoretical value of α_3 has been calculated assuming N = 225 individual waves.

Distribution of Individual Wave Periods

The best fit lines ($T_* = \alpha_i T_s + \beta_i$) between T_s and T_* - *which correspond T_{mean}, $T_{1/10}$ and T_{max} values of totally 3964 records, respectively* - were also obtained by linear regression (Fig. 4). The coefficients of best fit lines are given in Table 4 together with the values obtained for the other neighbouring seas of Turkey from the wave measurements performed within NATO TU-WAVES project (Turhan, 1996). Here, empirical ranges - *suggested by Goda and Nagai (1974) from the analysis of field wave data* - are also provided.

Table 4: Relationships between characteristic wave periods

Parameter (T_*) →	T_{mean}		$T_{1/10}$		T_{max}	
Station ↓	α_1	β_1	α_2	β_2	α_3	β_3
Alanya (1)	0.853	0.005	1.007	-0.009	1.012	0.041
Dalaman (1)	0.829	0.031	1.006	0.013	0.999	-0.060
Bozcaada (1)	0.852	0.020	1.009	0.007	1.011	-0.086
Sinop (1)	0.816	0.034	1.003	0.059	1.072	-.0145
M.Ereglisi	0.781	0.241	1.009	0.040	1.026	0.056
Overall (2)	0.832	0.032	1.008	0.036	1.015	0.084
Goda & Nagai	0.7-1.1	–	0.9-1.1	–	0.6-1.3	0.000

(1) average of monthly values given by Turhan (1996).

(2) average of the figures given for Alanya, Dalaman, Bozcaada and Sinop by Turhan (1996).

Similar to other stations, α_i values calculated for Marmara Ereglisi are between the limits given by Goda and Nagai (1974). Here, the most significant deviation occurs for T_{mean} since the wave periods are relatively small (T_s < 6 sec).

After the analysis of wave data collected within NATO TU-WAVES project, Turhan (1996) states that observed probability distribution of individual wave periods does not follow any of three widely-used theoretical distributions (namely: Bretschneider, Longuet-Higgins and Cavanier-Arhan-Ezraty distributions) which is also valid in case of Marmara Ereglisi.

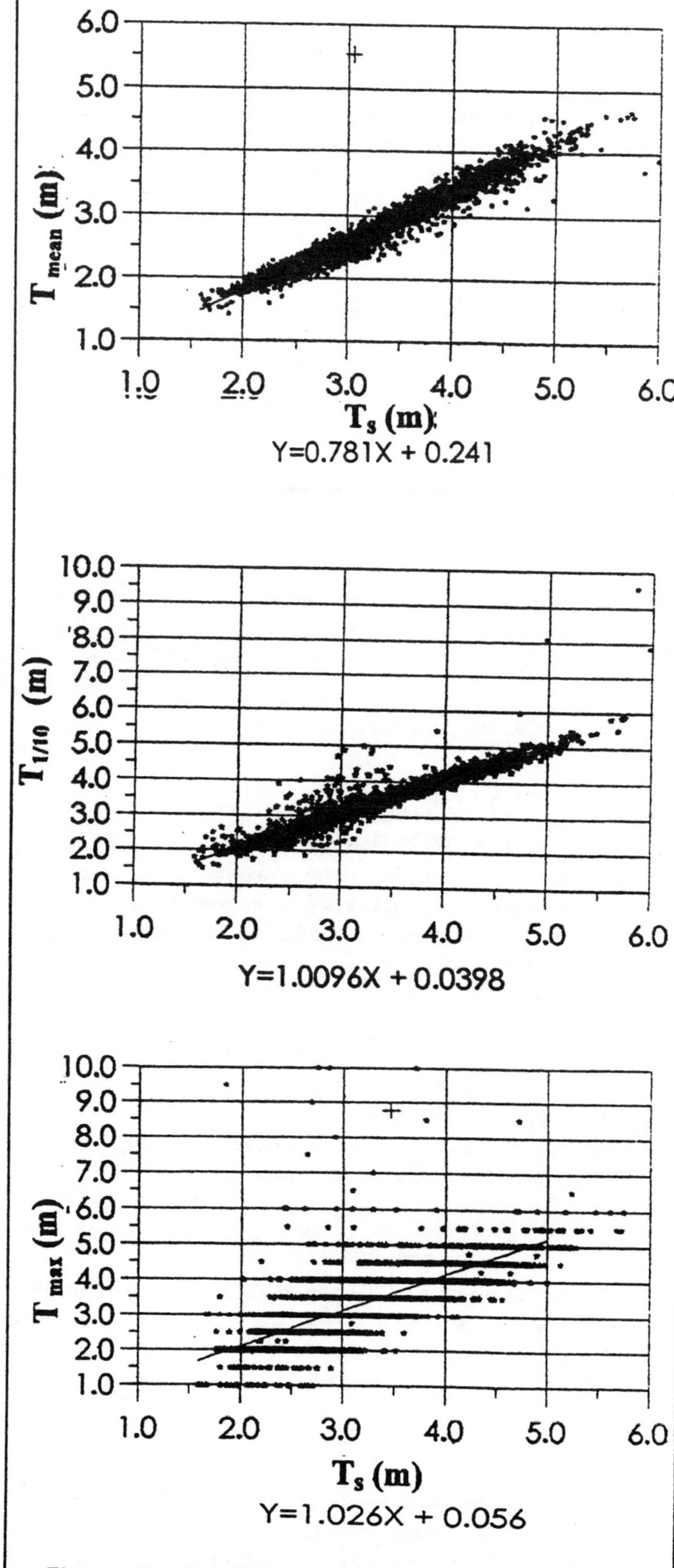

Figure 4: Best fit lines between T_s and T_* (T_{mean}, $T_{1/10}$, T_{max})

Joint Distribution of Significant Wave Heights and Zero-Upcrossing Periods

A particularly useful way of presenting wave climate data is an estimate of the joint distribution of H_S and T_Z - *often called "the scatter plot" combining both wave height and period data in one figure-* (WMO, 1988). Fig. 5 shows the frequencies counted in the specified joint intervals of wave heights and periods. Here, curves corresponding pre-selected wave steepness values - *respect to deep-water wave length (L_O)* - are also presented.

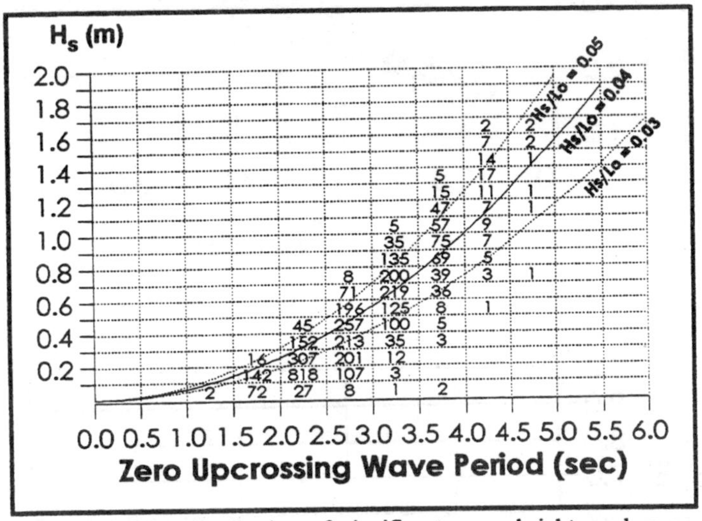

Figure 5: Joint distribution of significant wave heights and zero-upcrossing periods

After the correlation of conditional averages of T_z - *which is calculated as (7x4.25 + 2x4.75) /(7+2) = 4.36 sec for the interval of H_s = 1.5 - 1.6 m* - with corresponding H_s - *which is accepted as the mean value of the same interval (H_s = 1.55 m)* -, it has been found that $H_s/L_0 = 0.05$ if $H_s \geq 1.2$ m. Furthermore, considering the well-known relation between L_0 and T_z ($L_0 = gT_z^2 / 2\pi$), we can derive the following equation to determine T_z in case of H_s is known:

$$T_z = 11.21 \times (H_s / g)^{1/2}$$

T_z versus $T_{Hs} = 10 \times (H_s / g)^{1/2}$

According to Hansen (1981), the following relation exists between H_s and T_{Hmax} (period of heights individual wave):

$$T_{Hmax} \approx (H_s / g)^{1/2}$$

Considering the equations $T_{Hmax} = 0.9\ T_p$ (spectral peak period) - *given by Goda (1978)* - and T_p/T_z = 1.408 or 1.414 - *valid in case of Pierson-Moskowitz-Kitaigorodski and Neumann spectrums, respectively* -, it can be concluded that

$$T_z \approx (H_s / g)^{1/2}$$

which indicates the linear proportionality between T_z and $(H_s / g)^{1/2}$. Following the methodology proposed by Gouloumis and Moutzouris (1995) for the evaluation of wave data collected at southern Aegean Sea, a total of 3964 wave records were used to plot T_z versus $T_{Hs} = 10 \times (H_s / g)^{1/2}$ (Fig. 6).

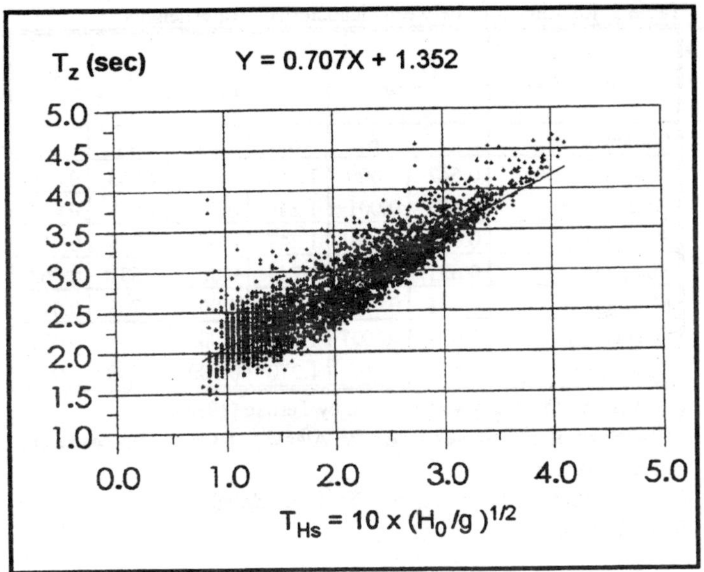

Figure 6: Plot of T_z versus $T_{Hs} = 10 \times (H_0 / g)^{1/2}$

Since the quantity of T_{Hs} seems generally to be lower than T_z - *particularly in case of $T_{Hs} \geq 3.5$ sec (i.e. $H_s \geq 1.2$ m)* -, it may be safer to use a wave hindcasting model - *like SMB model* - to predict T_z and then, to estimate H_s by the inverse equation:

$$H_s = gT_z^2 / 100$$

Here, the similarity with $T_z = 11.21 \times (H_0 / g)^{1/2}$ should be mentioned also stating that estimation of H_s is more conservative if the second equation is used. However, if the calculation of T_z corresponding to a pre-determined H_s is requested, previous relation should be applied to obtain higher values.

WAVE HINDCASTING

Wind Climate

The wave climate of Marmara Ereglisi - *affecting the LNG terminal* - were also determined by Delft Hydraulics (1990) using the hourly averaged wind data recorded at Tekirdag meteorological station for 10 years during 1975-1984. Frequency distribution of wind speeds from 16 directions - *each representing a 22.5 degree sector* - were obtained (Table 5).

With respect to the frequency of occurrence of wind, the directions NNE, N, NNW, WNW and SSE were found to be more important in decreasing order. However, according to the occurrence of higher wind speeds, the order of importance of directions are NE, SSE, S and SSE, respectively. From these important directions, only winds from SSE can generate waves affecting the LNG terminal (Delft Hydraulics, 1990).

Hindcasting Studies

Wave hindcasting calculations were carried out by using PM spectrum method for 11 directions (from NE to W). The fetch lengths in these directions vary from 11.0 km to 90.1 km (Table 5).

Table 5: Number of occurrences of wind velocities in various classes (1975-1984)

Direc-tion	Fetch length	Wind velocity (m/sec)				
—	km	3.0-5.0	5.1-8.0	8.1-13.5	13.6-18.0	≥18.1
NNE	–	3276	3992	935	2	0
NE	11.0	355	529	144	0	0
ENE	23.6	615	466	43	0	0
E	63.3	341	122	4	0	0
ESE	90.1	1050	218	22	0	0
SE	85.2	670	104	20	0	0
SSE	73.5	2080	873	85	3	0
S	55.8	680	679	107	3	0
SSW	47.0	689	68	129	4	0
SW	65.7	514	422	17	0	0
WSW	64.6	189	24	2	0	0
W	25.2	284	50	1	0	0
WNW	—	3384	542	28	0	0
NW	—	1721	346	27	0	0
NNW	—	2416	1313	219	2	0
N	—	2190	1737	294	2	0

As the result of hindcasting calculations, hourly time series of significant wave height and period in deep water for over ten years were obtained. Log-linear cumulative exceedence probability distribution and extreme value distribution of significant wave heights were derived. It was found that large waves occur in SSW to ESE sector SSE being the dominant direction (Table 6).

Both the log-linear and extreme value distributions indicate that the deep water significant wave heights with a return period of 50 years is about 4.0 m. The most likely direction of this wave is SSE (Table 6).

This result is also supported by the extreme value distribution of significant wave heights hindcasted by using wind data derived from synoptic weather maps (Delft Hydraulics, 1990).

Table 6: Significant wave heights in meters for various return periods

Direction	Return period (year)				
—	1	5	10	25	50
ENE	1.09	1.29	1.38	1.50	1.58
E	1.34	1.61	1.73	1.88	2.00
ESE	1.77	2.15	2.32	2.53	2.76
SE	1.63	2.00	2.16	2.38	2.54
SSE	2.45	2.97	3.19	3.49	3.71
S	1.92	2.27	2.42	2.62	2.77
SSW	1.92	2.32	2.49	2.72	2.89
SW	1.67	2.02	2.18	2.38	2.53
WSW	0.87	1.07	1.16	1.27	1.36
W	0.81	0.98	1.05	1.14	1.22
Extreme	—	2.70	3.05	3.84	4.17

Wave Refraction

Wave refraction diagrams were prepared for 10 directions (ENE to W). It was shown that waves from ENE to SSE are not affected by refraction until they reach -17 m water depth in front of the LNG pier. Waves from S decrease in height slightly, those from SSW moderately, and waves from SW, WSW and W very significantly due to wave refraction before they arrive at the pier (Table 7) (Delft Hydraulics, 1990).

Table 7: Refraction coefficients and wave directions (***increasing clock-wise respect to N***) at d = -17 m water depth

Wave period	Deep water wave direction									
sec (1)	ENE	E	ESE	SE	SSE	S	SSW	SW	WSW	W
3.0	–	–	–	–	–	–	-	0.20	–	–
								180*		
4.2	–	–	–	–	–	–	0.86	0.22	–	–
							198	***180****		
5.3	0.98	1.00	1.0	1.0	1.0	0.91	0.78	0.31	0.19	0.15
	66	***90***	***112.5***	***135***	***157.5***	***175***	***189***	***180****		
6.5	–	1.0		1.0	0.97	0.86	0.65	0.27	–	–
		90		***135***	***153***	***170***	***181***	***180****		
7.6	–	–		–	0.94	0.81	0.58	0.28	–	–
					150	***164****	***175****	***180****		

(1) for periods smaller than 4.7 sec, d = -17 m is deep water.
(*) change of 22.5 degrees direction sector.

Near-shore Wave Statistics

By using wave height change due to refraction and shoaling, and wave direction change due to refraction, the hourly series of Hs were transferred from deep water to -17 m location in front of the pier (Delft Hydraulics, 1990).

It was concluded that the probability distribution of waves from S and SSW are modified, whereas the deep water distributions for waves from NE to SSE can be used for the pier location without transformation. The pier is not exposed to waves from SW, WSW and W directions.

Since it occurs no significant transformation of waves from the dominant direction of SSE while moving from deep water to the pier location, the design wave characteristics (corresponding to various return periods) determined for deep water may also be used for the pier location (Table 8).

Table 8: Significant wave heights in meters for various return periods at d = -17 m water depth

Direction	Return period (year)				
—	1	5	10	25	50
ENE	1.09	1.29	1.38	1.50	1.58
E	1.34	1.61	1.73	1.88	2.00
ESE	1.77	2.15	2.32	2.53	2.76
SE	1.63	2.00	2.16	2.38	2.54
SSE	2.45	2.97	3.19	3.49	3.71
S	1.91	2.13	2.22	2.34	2.44
SSW	1.52	1.65	1.70	1.78	1.83

CONCLUSION

This paper evaluates the first systematically directional wave measurements in Turkey, and outlines the hindcasting studies performed to determine wave climate of the Sea of Marmara similar to previous papers of Akyarli (1995) and Ozhan and Abdalla (1993a) second of which present the verification of these hindcasting studies.

Since the Sea of Marmara is a favourable region for the construction of marine structures, this information will assist the engineers to make more reliable designs during their future works. Here, it is also recommended to refer Besiktepe et al (1995) and Yalciner et al (1993) for the physical oceanography of and tsunami generation in the Sea of Marmara.

There are some efforts to review the features of the wave climatology of the seas surrounding Turkey (Ergin and Ozhan, 1986; Akyarli, 1993; Abdalla, et al, 1995; Turhan, 1996). Furthermore, NATO TU-WAVES project has been started to make systematic wave measurements at these seas (Ozhan and Abdalla, 1993b).

In the content of this program, however, only non-directional measurements are conducted at the Sea of Marmara, and consequently, the wave data evaluated above seems to remain to be the only existing source on the directional properties of the wind waves generated in this basin.

ACKNOWLEDGEMENT

Authors express their appreciation to BOTAS (Turkish National Pipeline Co-operation), Sezai Turkes - Fevzi Akkaya Construction Company and Temel Surveying Company for their various supports during the implementation of the studies.

REFERENCES

Abdalla, et al (1995). "Measurements and Modelling of Wind Waves Along the Turkish Mediterranean Coasts and the Black Sea", *Proceedings of The Second International Conference on the Mediterranean Coastal Environment -MEDCOAST95*, Taragona-Spain, pp 1899-1910.

Akyarli, A (1993). "Wave Climate Studies Along the Turkish Coast of the Black Sea", *Coastline of the Black Sea - Special Volume of Proceedings of Coastal Zone 93*, New Orleans-USA, pp 556-570.

Akyarli, A (1995). "Wave Measurements for the Design of the Hamidabad Natural Gas Pipeline System", *Proceedings of the Fifth International Offshore and Polar Engineering Conference*, The Hague-Holland, Vol III, pp 105-111.

Bainard, EC (1983). "High Resolution Directional Wave Spectra Using the FFT Digital Band Pass Filter Method", *Proceedings of the Buoy Technology*, Marine Technology Society Gulf Coast Section and NOAA Data Buoy Center, pp 55-61.

Bainard, EC (1989). "Type 1156 Directional WAVE-TRACK Buoy", OCEAN '89, Seattle, Washington, 7 p.

Besiktepe, et al (1994). "The Circulation and Hydrography of the Marmara Sea", *Progress in Oceanography*, Vol 34, pp 285-334.

Delft Hydraulics (1990). "*Hydraulic Studies for LNG-Terminal near Marmara Ereglisi, Turkey (Vol I, Report on Hydraulic Studies)*", prepared for Sezai Turkes - Fevzi Akkaya Construction Company, 37 p + Appendix.

Ergin, A and Ozhan E (1986). "*Wave Hindcasting and Determination of Design Wave Characteristics for 15 Sites*", Middle East Technical University - METU KLARE Technical Report No 35, Ankara-Turkey, 137 p (in Turkish).

Goda, Y, and Nagai, K (1974). "Investigation of Statistical Properties of Sea Waves with Field and Simulation Data", *Report Port and Harbour Research Institute*, Ministry of Transport, Japan, Vol 13, No 1, pp 3-37.

Goda, Y (1978). "The Observed Joint Distribution of Periods and Heights of Sea Waves", *Proceedings of 16th Conference on Coastal Engineering*, Hamburg-Germany, pp 227-246.

Goda, Y (1979). "A Review on Statistical Interpretation of Wave Data", *Report Port and Harbour Research Institute*, Ministry of Transport, Japan, Vol 18, No 1.

Gouloumis, SM and Moutzouris, CI (1995). "Wave Measurements Under Fetch and Duration Limited Conditions in the South Aegean Sea", Fourth International Conference on Coastal & Port Engineering - COPEDEC IV, Brazil, Vol III, pp 2036-2047.

Hansen, NEO (1981). "Determination of Design Waves for Steel Platforms", Institute of Hydrodynamic and Hydraulic Engineering (ISVA), Technical University of Denmark, *Progress Report 55*, pp 31-37.

LG Mouceht & Partners, 1988). "*BOTAS LNG Import Terminal, Sea of Marmara - Turkey, Wave Persistence Study*", prepared for MW Kellogg Ltd, 21 p + Appendix.

OCIMF (1978). "*Guidelines and Recommendations for the Safe Mooring of large Ships at Piers and Sea Islands*", Oil Companies International Maritime Forum (OCIMF).

Ozhan, E and Abdalla, S (1993a). "Engineering Applications of the Simplified Third generation Wave Model (METUS3)", *Proceedings of The First International Conference on the Mediterranean Coastal Environment -MEDCOAST93*, Antalya-Turkey, pp 1181-1195.

Ozhan, E and Abdalla, S (1993b). "Wave Climatology of the Turkish Coast: NATO TU-WAVES", *Proceedings of The First International Conference on the Mediterranean Coastal Environment -MEDCOAST93*, Antalya-Turkey, pp 1221-1227.

Temel Surveying & IMST (1991). "*Wave, Current and Meteorological Measurements for LNG Terminal Near Marmara Ereglisi, Turkey*", prepared by Temel Surveying & Institute of Marine Science and Technology of Dokuz Eylul University, Izmir-Turkey for Consortium of SN Technigaz & Sezai Turkes - Fevzi Akkaya Construction Company, Istanbul-Turkey.

Turhan, MM (1996). "*Short-term Statistical Analysis of Wind-Wave Characteristics Measured Along the Turkish Coast*", an unpublished MSc thesis submitted to Middle East Technical University, Ankara-Turkey, 165 p.

WMO (1988). "*Guide to Wave Analysis and Forecasting*", World Meteorological Organisation, No. 702, Geneva, Switzerland.

Yalciner, AC, Kuran, U, Akyarli, A, Imamura, F (1993). "Generation and Propagation Process of Tsunamis in the Sea of Marmara", presented at International Society for the Prevention and Mitigation of Natural Hazards - HAZARDS '93, China.

Proceedings of the Seventh (1997) International Offshore and Polar Engineering Conference
Honolulu, USA, May 25-30, 1997

ISBN 1-880653-28-1 (Set); ISBN 1-880653-31-1 (Vol. III)

A Wave/Wind Climatology for the Gulf of Mexico

W.J. Teague and P.A. Hwang
Naval Research Laboratory, Stennis Space Center, MS, USA
D.W.-C. Wang
Computer Science Corporation, Stennis Space Center, MS, USA
E.F. Thompson
Coastal Engineering Research Center, Vicksburg, MS, USA
G.A. Jacobs
Naval Research Laboratory, Stennis Space Center, MS, USA

ABSTRACT

Significant wave heights, wave period, and wind speed over the Gulf of Mexico have been derived from the NASA altimeter on the TOPEX / POSEIDON satellite for 1992–1995. TOPEX / POSEIDON ground tracks in this region are approximately 300 km apart and have an along track resolution of about 7 km. Climatological maps of the Gulf of Mexico for these parameters have been generated for annual and seasonal time periods. Strong seasonality is clearly shown from these maps. Wave heights and wind speeds are found to increase from east to west across the Gulf. The fall and winter seasons are found to be most intense, with the largest significant wave heights and highest wind speeds observed during the fall season of 1995. Summer seasonal values were typically very low. The satellite data compare well with the National Data Buoy Center (NDBC) in situ buoy data and Army Corps of Engineers Wave Information Studies (WIS) Gulf of Mexico wave hindcast model statistics. Statistics of wave parameters derived from TOPEX / POSEIDON are improved over earlier satellite missions due to the improvements in altimeter accuracy and longer satellite lifetime.

Key Words – waves, winds, climatology, satellite, buoy

INTRODUCTION

Spaceborne sensors such as satellite altimeters provide long-term and continuous coverage of wave and wind fields of the world oceans. It has been shown convincingly that spaceborne altimeter measurements of significant wave height, H (henceforth referred to as wave height), agree with in situ buoy measurements to within 0.15 m (*Gower, 1996*) and measurements of wind speed to about 1.8 m/s (*Freilich and Challenor, 1994*). Recent study of tilting effects on radar backscatter indicates that the altimeter wind speed (U) measurements are accurate to better than 1 m/s (*Hwang et al., 1996*), and are within the accuracy of buoy measurements. The combination of wind speed and wave height further yields the information of wave period (T). It is therefore quite feasible that we can obtain the three critical parameters of a wave field (H, T and U) from satellite altimeters and can provide a long-term monitoring of the wave climate of the world oceans.

Measurements of H, T, and U are important for assimilating into numerical models, for forecasting of the weather, and for ocean circulation studies. Moored surface buoys are used to provide such data which have high resolution in time but are very limited in spatial coverage. In addition, buoy measurements are subject to a harsh environment that can introduce errors and loss of data. Satellites provide a broader coverage and have high resolution along track but have a lower time resolution as defined by the satellite's orbit. However, satellites have the advantage of using just a single instrument, in lieu of dozens of buoys, and operate in a relatively stable environment. Since the satellite coverage is global and the expected lifetime is better than five years, long term climate conditions can be studied. Currently, there are almost four full years of TOPEX / POSEIDON data. These data are suitable for averaging and gridding in order to produce climatological maps of waves and winds. For an initial effort the Gulf of Mexico is used as a testbed for studying the regional wave and wind properties.

This paper presents climatological maps of wave heights, wave periods, and wind speeds for the Gulf of Mexico. The maps are spot-checked with buoy measurements for validation and are compared to Army Corps of Engineers Wave Information Studies (WIS) wave hindcast model statistics. Seasonal and yearly averages are presented.

SATELLITE DATA

Satellite data used in this study are from TOPEX / POSEIDON. Henceforth, the satellite will be referred to simply as TOPEX. TOPEX was launched August 10, 1992 as a joint effort by the United States National Aeronautics and Space Administration (NASA) and the French Space Agency, Centre National d'Etudes Spatiales (CNES), for studying global ocean circulation (*Fu et al., 1994*). TOPEX data utilized in this study (Figure 1) are obtained from the Geophysical Data Records (GDRs) processed at the Physical Oceanography Distributed Active Archive Center (PODAAC) at the Jet Propulsion Laboratory. TOPEX has a repeat period of 9.92 days and an equatorial cross-track separation of 315 km. Along track resolution in the Gulf of Mexico is about 7 km. TOPEX measures the range from the satellite to the sea surface from which sea-surface height is obtained, significant wave height (the average height of the highest third of the waves), and the normalized radar backscatter cross section (σ_0) derived from the K_u band from which wind speed is derived. The footprint over which the wave and wind measurements are made ranges from about 10 to 70 km^2, depending on the sea state (*Chelton et al., 1989*). The wave height data and σ_0 are not interpolated along track for this analysis.

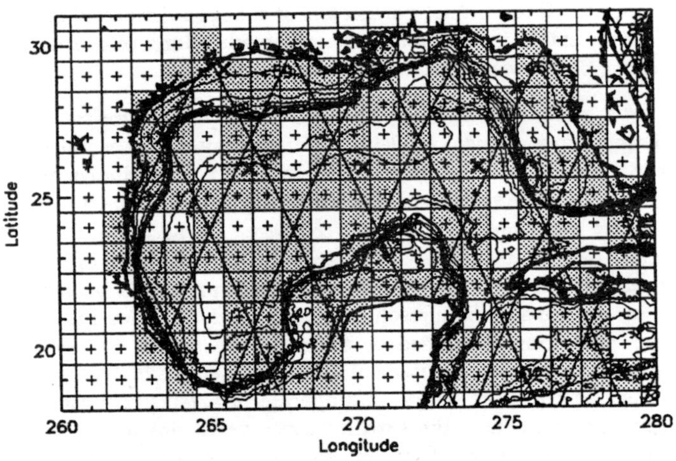

Figure 1. TOPEX tracks (heavy lines), NDBC buoy locations (x), bathymetry, and grid (+) are shown for the Gulf of Mexico. Grid squares are shaded where there are TOPEX data.

Equations for wave height, wind speed, wave period, and wave steepness are given below. See *Hwang et al. (1997)* for more details in the equation development. Wind speed at 10 m above the sea surface approximated with a fifth order polynomial fit by *Witter and Chelton* (1991, Table 1) is given by

$$U = \sum_{k=0}^{5} a_k \sigma_0^k, \tag{1}$$

where $a_0 = -2.2188 \times 10^2$, $a_1 = 1.0649 \times 10^2$, $a_2 = -1.7309 \times 10^1$, $a_3 = 1.2955$, $a_4 = -4.6079 \times 10^{-2}$, and $a_5 = 6.3270 \times 10^{-4}$. Empirically, the peak period of the wave field, T, is related to wind speed, U, and wave height, H, (*Hasselmann et al., 1976*), and is given by

$$U/(gT) = 0.048(U^2/(gH))^{0.67}, \tag{2}$$

where g is the gravitational constant. Using the TOPEX derived U and H, the period calculated from (2) is found to be slightly less (by 6%) than the buoy measured peak period (*Hwang et al., 1997*). The wave number, K, corresponding to T is calculated from the deep water dispersion relationship (*Kinsman, 1965*), given by

$$K = (4\pi^2)/(gT^2). \tag{3}$$

Defining the wave steepness as the product of wave number and wave amplitude (one-half of the wave height), the wave steepness can also be obtained from TOPEX. Comparisons of TOPEX and buoy measurements in the Gulf of Mexico indicate that the wave height rms difference is 0.15 m and that the wind speed rms difference is 1.2 m/s. The wave periods derived from TOPEX are within 10% of the peak period measured by buoys. For a detailed comparison of buoy measurements of H, T, and U with TOPEX measurements, see *Hwang et al. (1997)*.

In forming the climatological maps, the TOPEX data are averaged over one-degree squares. TOPEX data from one or more tracks that fall within any part of a square are used in the average for that square. The TOPEX data have high spatial resolution of about 7 km along track but have temporal resolution of only about 10 days for any particular track. In Figure 1, TOPEX tracks and grid are shown for the Gulf of Mexico along with buoy locations. The one-degree grid squares where there are TOPEX data are shaded. Although this averaging technique is primitive, over 75% of the grid squares are supported by data. An inverse distance averaging technique is used to interpolate grid values where there are no data. The data are then further interpolated to a 0.2 degree grid for smoothing purposes.

In this study, significant wave heights and wind speeds for the Gulf of Mexico are discussed for the 1993–1995 period. The analysis periods include annual means and the four seasons. Winter is defined as January through March, spring as April through June, summer as July through September, and fall as October through December.

Mean wave height and wind speed are shown for 1995 in Figure 2. Wave period and wave steepness are shown for 1995 in Figure 3. Corresponding maps were also computed for 1993 and 1994, but are not shown here. Average wave heights are found to be larger in 1995 than in 1993 and 1994. Interannual variability is definitely present. For each year, mean wave height increases from east to west and ranges in amplitude from about 0.5 m off the coast of Florida to about 1.5 meters off the coast of Texas. Similarly, mean wind speed ranges from about 4 m/s to nearly 8 m/s. Wave periods range from about 3 to 6 seconds and wave steepness ranges from about 0.08 to 0.095. Wave heights are largest, particularly in the western Gulf of Mexico, in 1995 and smallest in 1994. Low wave heights are found along the coasts at water depths less than 100 m except for the north-south oriented coasts of Texas and Mexico.

Striking seasonal differences are found for wave height and wind speed. Maps of wave height and wind speed for

Figure 2a. Wave height from TOPEX.

Figure 2b. Wind speed from TOPEX

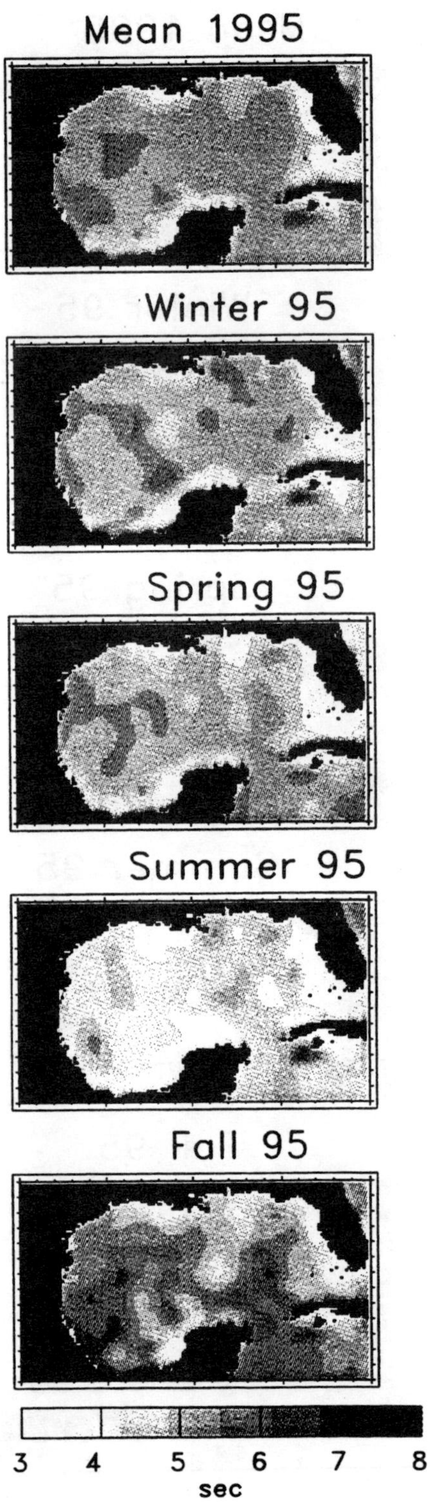

Figure 3a. Wave period from TOPEX

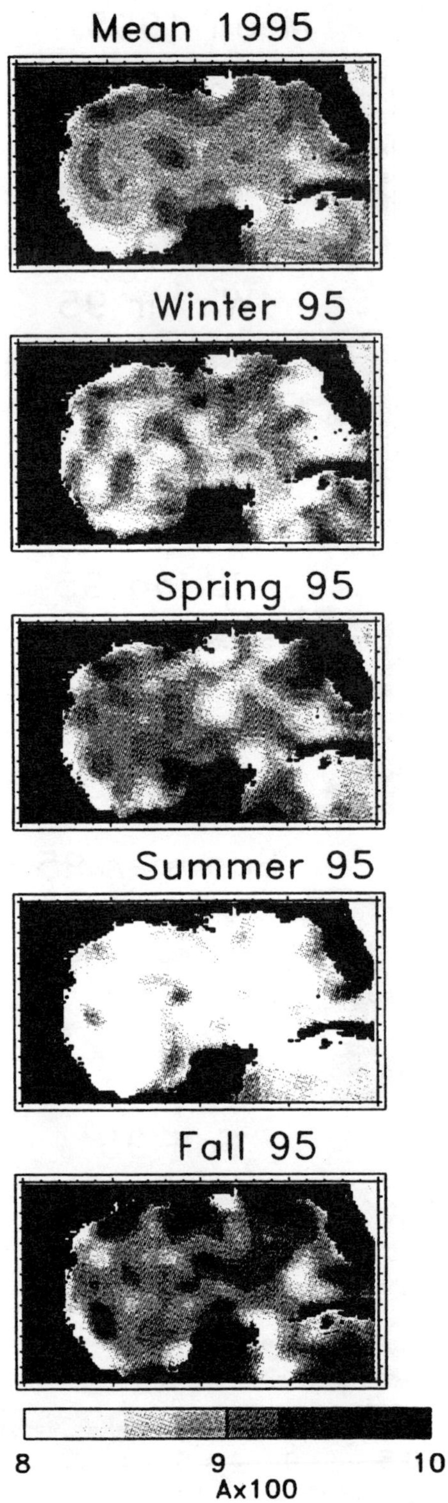

Figure 3b. Wave steepness from TOPEX

individual seasons for 1995 are presented in Figure 2. Lowest wind speeds and wave heights are found in the summer period. Highest wind speeds and wave heights are found in fall and winter. Spring appears to be a transition period. Similar conditions are found in 1993 and 1994. Interestingly, lowest wind speeds are found for summer, 1995 and highest wind speeds are found for fall, 1995. Correspondingly, largest overall wave heights are found in fall, 1995.

Wave period and wave steepness for individual seasons for 1995 are presented in Figures 3. Waves are generally steepest during the fall and winter, and least steep during summer, corresponding to the longer wave periods and shorter wave periods, respectively. However, summer, 1993 is an exception since steepness is relatively large compared to steepness in summer, 1995.

BUOY DATA

National Data Buoy Center (NDBC, part of the National Oceanic and Atmospheric Administration) buoys located in the Gulf of Mexico (see Figure 1) at water depths which ranged from about 20 m to depths deeper than 3000 m provide data for comparison with the satellite data. The buoys provide measurements of significant wave height, wind speed, atmospheric pressure, and air and sea temperature. Buoy wave height and wind speed estimates are available once per hour and are given to 0.1 m and 0.1 m/s, respectively. These measurements are not quite the same as the altimeter measurements since they are point measurements in time and space and are subject to their inherent instrumental problems (*Monaldo,1988*). Buoy data nearest in time to the TOPEX time are used in the comparisons.

In general, good agreement has been found between buoy and TOPEX wave heights for various regions. *Gower (1996)* found excellent agreement off the west coast of Canada and *Ebuchi and Kawamura (1994)* had favorable comparisons around Japan. Similar success was obtained using GEOSAT wave height and wind speed observations by *Dobson et al. (1987)* for selected buoys in the North Atlantic, North Pacific, and the Gulf of Mexico. We also find good agreement between buoy and TOPEX data in the Gulf of Mexico. Satellite wave height is compared with NDBC data from the buoy located near the crossover of TOPEX track 20 and 115 (Figure 4). The two data types are found to be in excellent agreement, within about a 0.16 m rms difference and a correlation of about 0.98. Wind speeds have also been found to be in excellent agreement, within 1.25 m/s rms and a correlation of about 0.85 using the algorithms of *Brown (1981)*, *Witter and Chelton (1991)*, and *Freilich and Challenor (1994)*.

The climatological means for significant wave heights and wind speeds derived from TOPEX are compared with the three NDBC buoys located in the central Gulf of Mexico near 26°N (shown in Figure 1) for the 1993–1995 period. TOPEX climatological seasonal means for wave heights and wind speeds at grid points along 26°N near these buoy locations, and buoy monthly means with standard deviations (smoothed over 3 months) are presented in Figure 5. TOPEX wave heights compare very well with the buoy measurements, particularly at higher wind speeds. Agreement is well within the buoy standard deviation of about 0.6 m for wave height. Wind speeds compared well for the higher wind speeds but not as well for the lower wind speeds, as anticipated. Standard deviation of buoy wind speeds averaged about 2.7 m/s.

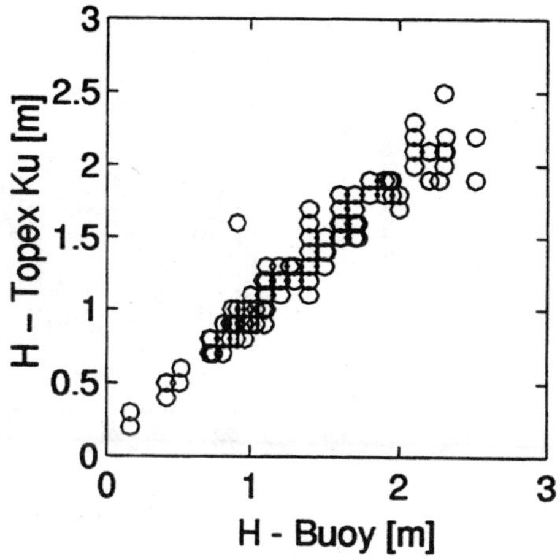

Figure 4. TOPEX significant wave heights compared with wave heights from the NDBC buoy in the northwestern Gulf of Mexico near 27N, within 30 minutes time of TOPEX passes. The spacing between the buoy location and TOPEX footprint is less than 10 km.

Twenty years (1956–1975) of hindcast wave data were output at 3-hr intervals for selected locations in the Gulf of Mexico by the U.S. Army Corps of Engineers (*Hubertz and Brooks, 1989*) using the Wave Information Study (WIS) wind-driven shallow-water wave model (*Corson, Resio, and Vincent, 1980; Resio, 1982*). The model has a one hour time resolution and a 0.25 degree grid. These model data, which compares favorably with in situ buoys, are valuable for analyzing statistics over long time periods and in comparisons with satellite measurements for cross validations. The model data provide broad area coverage like TOPEX but also provide the time resolution that TOPEX lacks. Providing that the model and TOPEX produce similar statistics, which preliminary comparisons indicate, the data sets can be very complementary. For example, three years (1993–1995) of data from TOPEX track 21 are compared with the 20 years of data at WIS station 9, less than 5 km in distance from the ground track (Figure 6). TOPEX wave heights are higher in all height ranges by about 3 to 8%, except for wave heights between 1.0 and 1.5m, where the WIS measurements are about 16% higher. The model hindcasts extend over a considerably longer time period (1976-95) than that covered by satellite or buoy measurements.

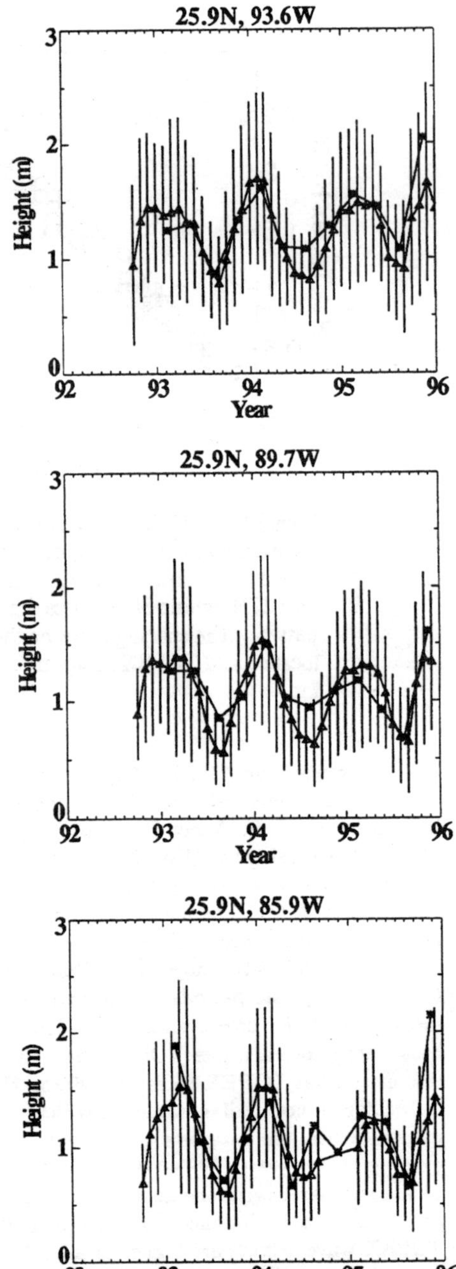

Figure 5a. Mean seasonal TOPEX wave heights (*) compared with buoy wave heights (Δ). The vertical bars indicate the standard deviation of wave heights calculated from hourly buoy measurements each month.

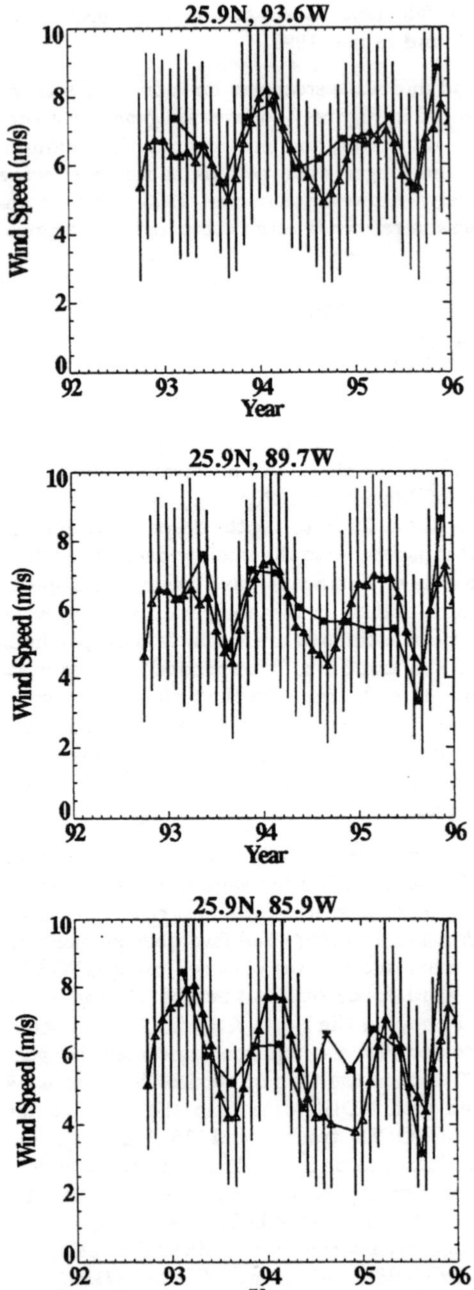

Figure 5b. Mean seasonal TOPEX wind speeds (*) compared with buoy wind speeds (Δ). The vertical bars indicate the standard deviation of wind speeds calculated from hourly buoy measurements each month.

DISCUSSION

This work demonstrates that broad area climatological maps can be produced using data from TOPEX. The resolution of these maps can be increased by using additional satellites such as the European remote sensing satellite ERS-1 (launched in July, 1991) and the GEOSAT Follow-on satellite (scheduled for launch in mid-year 1997). In addition, a long time series of global wind speed and wave height data are available from GEOSAT for 1985 to 1990. These satellite data are comparable in accuracy to the buoy data. Although time resolution is much poorer than the buoy data, horizontal coverage is far greater than ever possible with buoy data alone. The data are quite complementary and can be used in combined analyses.

Seasonal maps of wind speeds were also made for the 1990–1993 time period using meteorological stations along the periphery of the Gulf of Mexico by *Gutierrez de Velasco and Winant (1996)*. They find, as we do, that the smallest wind speeds occur off the coast of west Florida, Alabama, and Mississippi and increase towards the west. They find highest wind speeds near the Yucatan peninsula, where we also find high wind speeds. However, we find high wind speeds throughout the western Gulf of Mexico, where *Gutierrez de Velasco and Winant* lacked data coverage. In addition, they find as we find that the wind speeds can be described in terms of two seasons: fall and winter, and spring and summer. These seasons are also appropriate for H, T, and KA.

Encouraging results were also obtained from the WIS model. Validation of such a model is difficult but is possible using satellite data such as the TOPEX data set. Combined analyses of satellite data and model output can produce a Gulf of Mexico climatology which spans decades and perhaps more importantly, can produce real-time wave forecasts. More work is needed for the model validation.

The broad area climatology maps from satellite measurements are particularly valuable as they provide data where there are no buoy measurements and give the overall picture of the wind and wave climate. Annual and seasonal distributions are presented here but monthly and even smaller averaging periods are feasible when multiple satellite data are analyzed. Such climatological maps will be very valuable in planning operations such as: commercial drilling, spill dispersion prediction, underwater construction, amphibious landings, and mine sweeping.

CONCLUSIONS

Measurements of wave height and wind speed from TOPEX are accurate in the Gulf of Mexico. Sufficient data now exists for the development of wave and wind climatologies from satellite data. Additional satellites, some of which are already scheduled for launch, are needed to increase the temporal and spatial coverage. TOPEX wave and wind data are output every second which translates to about 7 km along track resolution. For near-shore realizations, 10 Hz wave and wind data, which are at about 700 m resolution, are needed. TOPEX currently outputs sea-surface heights at 10 Hz resolution but outputs wave heights at only 1 Hz resolution. However, 10 Hz wave heights are planned for the GEOSAT Follow-on mission. A combination of satellite and model data will ultimately produce the best wind and wave climatologies.

Acknowledgments

The TOPEX / POSEIDON data were obtained from the NASA Physical Oceanography Distributed Active Archive Center at the Jet Propulsion Laboratory, California Institute of Technology. Buoy data were obtained from the National Data Buoy Center, Stennis Space Center, MS. This work was supported by the Naval Research Laboratory's Ocean Executive Agent Office (Job Orders 73-7046-06 and 73-7075-06), the Office of Naval Research (Contract N0001496WX30445), and the National Data Buoy Center (Contract XA2310501) (NRL-SSC contribution NRL/PP/7332-97-0002).

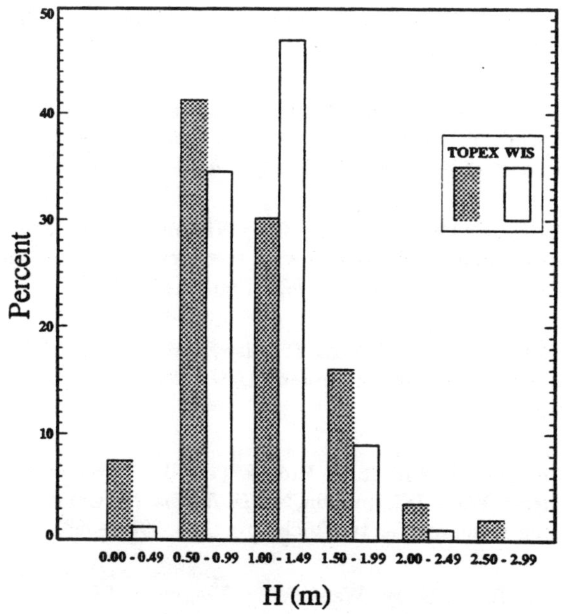

Figure 6. Percent occurrence of wave heights from TOPEX data along ground track 21 are compared with 20 years of data from WIS station 9, located about 5 km from the ground track.

REFERENCES

Brown, G.S., H.R. Stanley, and N.A. Roy, (1981). "The wind speed measurement capability of spaceborne radar altimeters," *IEEE J. Oceanic Eng.*, **OE-6**, pp 59–63.

Chelton, D.B., E.J. Walsh, and J.L. Macarthur, (1989). "Pulse compression and sea level tracking in satellite altimeters," *J. Atmos. and Oceanic Technol.*, **6**, pp 407–438.

Corson, W.D., D.T. Resio, and C.L. Vincent, (1980). "Wave Information Study of U.S. Coastlines; Surface Pressure Field Reconstruction for Wave Hindcasting Purposes, Report 1," *TR HL-80-11*, U.S. Army Engineer Waterways Experiment Station, Vicksburg, MS.

Dobson, E., F. Monaldo, J. Goldhirsh, and J. Wilkerson, (1987). "Validation of Geosat altimeter-derived wind speeds and significant wave heights using buoy data," *J. Geophys. Res.*, **92**, pp 10,719–10,731.

Ebuchi, N., and H. Kawamura, (1994). "Validation of wind speeds and significant wave heights observed by the TOPEX altimeter around Japan," *J. Oceanogr.*, **50**, pp 479–487.

Freilich, M.H., and P.G. Challenor, (1994). "A new approach for determining fully empirical altimeter wind speed model functions," *J. Geophys. Res.*, **99**, pp 25051–25062.

Fu, L.-L., E. J. Christensen, C. A. Yamarone Jr., M. Lefebvre, Y. Menard, M. Dorrer, and P. Escudier, (1994). "TOPEX / POSEIDON mission overview," *J. Geophys. Res.*, **99**, pp 24369–24381.

Gower, J.F.R., (1996). "Intercalibration of wave and wind data from TOPEX / POSEIDON and moored buoys off the west coast of Canada," *J. Geophys. Res.*, **101**, pp 3817–3829.

Gutierrez de Velasco, G. and C.D. Winant, (1996). "Seasonal patterns of wind stress and wind stress curl over the Gulf of Mexico," *J. Geophys. Res.*, *bf101*, pp 18,127–18,140.

Hasselmann, K., D.B. Ross, P. Miller, and W. Sell, (1976). "A parametric wave prediction model," *J. Phys. Oceanogr.*, **6**, pp 200–228.

Hubertz, J.M. and R.M. Brooks, (1989). "Gulf of Mexico Hindcast Wave Information," *WIS Report 19*, Coastal Engineering Research Center, Vicksburg, MS, 39181-0631.

Hwang, P. A., D. W. Wang, W. J. Teague, and G. A. Jacobs, (1997). "Effect of surface tilting on altimeter wind measurements," *J. Geophys. Res.*, (in review).

Hwang, P. A., W. J. Teague, G. A. Jacobs, and D.W.-C. Wang (1997). "Measurement of wind and wave parameters from satellite altimeters," , *Proceedings of the Fourth International Conference on Remote Sensing for Marine and Coastal Environments*, Orlando, FL, 17-19 March 1997, (in press).

Kinsman, B., (1965). *Wind Waves*, Prentice-Hall, Inc., 676 pp.

Monaldo, F., (1988). "Expected differences between buoy and radar altimeter estimates of wind speed and significant wave height and their implications on buoy-altimeter comparisons," *J. Geophys. Res.*, **93**, pp 2285–2302.

Resio, D.T., (1982). "The Estimation of Wind-Wave Generation in a Discrete Spectral Model," *WIS Report 5*, U.S. Army Engineer Waterways Experiment Station, Vicksburg, MS.

Witter, D.L., and D.B. Chelton, (1991). "A Geosat altimeter wind speed algorithm and a method for altimeter wind speed algorithm development," *J. Geophys. Res.*, **96**, pp 8853–8860.

Proceedings of the Seventh (1997) International Offshore and Polar Engineering Conference
Honolulu, USA, May 25-30, 1997

ISBN 1-880653-28-1 (Set); ISBN 1-880653-31-1 (Vol. III)

Effects of a Submerged Dike on the Statistical Properties of a Random Wave Field

Jaw-Guei Lin and John Z. Yim
National Taiwan Ocean University
Keelung, Taiwan, China

ABSTRACT

Effects of a submerged dike on the statistical properties of simulated random waves were studied experimentally. Deep water waves were generated mechanically by a computer controlled piston type wave generator according to the target JONSWAP spectrum. The variations of the surface elevations before and after the waves passing over the dike were measured. Possible changes of the spectral shapes and the probability distributions of wave heights and surface elevations were analyzed. Except for the possible energy loss due to the wave shoaling and breaking, the present results indicate that wave statistics will not be essentially (seriously) affected by the presence of the dike.

KEY WORDS: random waves, JONSWAP spectrum, wave height distribution

INTRODUCTION

Random wave fields are often treated as the combinations of denumerably many independent harmonic components, with their own frequencies and wavelengths, and thus travels freely on the water surface. Most of our present wave spectral models are based upon this assumption. Neglecting the wave-wave interaction, Phillips (1958), for example, derived his famous -5 power law by assuming that the only dominant factor for wave energy dissipation in deep water is through wave breaking. Later, Pierson & Moskowitz (1964) utilized this argument and celebrated the Pierson-Moskowitz (P-M) spectrum for fully developed seas. For developing wind sea, the modified P-M spectrum, i.e., the JONSWAP spectrum, due to Hasselmann et al. (1973) is now widely used. On the other hand, for the deterministic wave statistics, Longuet-Higgins (1952) derived the first theoretical base theory with an additional assumption of narrow-bandedness.

When a group of free waves propagates from deep to shallower waters, the longer waves will subject to the effects of shoaling, partial reflection, as well as eventual breaking; while the shorter waves will be relatively unaffected by the underlying topographical conditions. As a matter of fact, it was proposed that, in shallow water, waves spectral densities should have a form which deviate from those proposed for deep waters (Bouws et al., 1985, 1987; Young, 1988).

With the presence of an underwater obstacle, such as a submerged dike considered in this paper, waves should be more affected because of the changes of the water depth are more drastic than the pure topographical effects. A rather nature conjecture is therefore that the transmitted waves behind the dike will reform a new wave field which has rather different configuration than that before. However, it is quite surprising that even with some active wave breaking, the spectral forms of the wave field will not be seriously affected by the presence of a submerged trapezoidal bar (Battjes & Beji, 1992).

When wave breaking do not change the spectral forms of a wave field, it is interesting to find the influences on the statistical properties of the wave field. To study the possible effects of a submerged object on the statistical properties of wave fields, several experiments were conducted at the Laboratory of Ocean Engineering in the Department of Harbour and River Engineering, National Taiwan Ocean University. The empirical JONSWAP spectrum was used as target spectrum. This paper summarizes our present findings. The discussion is divided into three parts in this paper. The second part describes shortly the experimental configurations; the third part presents the results; and the discussions of the results is in the last section.

EXPERIMENTAL SETUP

The experiments were conducted in a large-dimensioned wave flume with a dimension of 100 x 3.0 x 3.0 (in meters). When facing the flume, a piston type wave generator is located at the right hand side of the tunnel, and is connected to a personal computer which is equipped with a software capable to generate three target spectra, namely Bretschneider-Mitsuyasu, Pierson-Moskowitz, and JONSWAP spectra. At the left end of the flume, an impermeable slope (s = 1:5) is installed for wave energy dissipation. Eight measuring stations, numbered from 1 to 8, were installed to measure the variations of the wave fields along the channel. These wave gauges are located respectively at 3.0, 12.0, 32.0, 42.0, 71.5, 73.4, 75.0 and 76.1 meters away from the wave paddle. The statistical properties of the measured results of the first four stations are reported in a accompanying paper (Yim & Lin, 1997). As can be seen from Fig. 1, the last four measuring stations are situated just in front of and behind the submerged dike, so that wave field properties from Stations 5 &6 can be compared with those from Stations 7 & 8.

Throughout the experiments, the water depth is kept at a constant depth of 1.5 meter. A problem associated with random wave simulation in wave flume is that, the measurements should be made long enough to achieve the statistical equilibrium on one side. Whereas the wave reflection and secondary reflections from the finite ends of the flume will inevitably affect the wave fields and lead to the misinterpretation of the results. After some observation, the recording time span of the surface elevations was set to 330 seconds (5.5 min.) with a sampling interval of 50 Hz. Whether this choice concerning sampling is justified or not will be considered elsewhere.

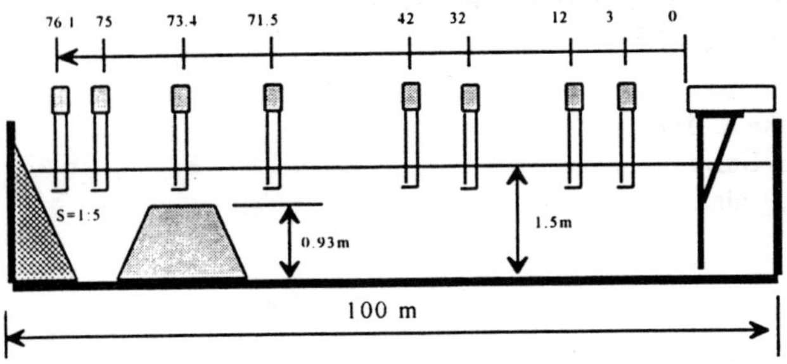

Fig. 1 Schematic representation of the experimental setup

To acquire the further insights as whether the fact that the spectral shapes are invariant to breaking or not, and the dependence of bandwidth or frequency on the spectral, the experimental conditions were varied by changing the parameters of the JONSWAP spectrum. As is well known, the typical JONSWAP spectrum has the form:

$$S(f) = \frac{\alpha g^2}{(2\pi)^4 f^5} \exp\left[-\frac{5}{4}\left(\frac{f}{f_p}\right)^{-4}\right]\gamma^{\beta} \qquad (1)$$

where f_p = the peak frequency; a = the so-called Phillips constant with a mean value of 0.0081; g = the peak enhancement factor with a mean value of 3.3; and

$$\beta = \exp\left[-\frac{(f - f_p)^2}{2\sigma^2 f_p^2}\right] \qquad (2)$$

with

$$\sigma = \begin{cases} \sigma_a = 0.07 & \text{for } f \le f_p \\ \sigma_b = 0.09 & \text{for } f > f_p \end{cases} \qquad (3)$$

s_a and s_b are, respectively, the deviations from the peak frequency to the left and to the right sides of the corresponding Pierson-Moskowitz spectrum. The parameters g, s_a and s_b can be called as shape parameters, whereas the parameter a and the peak frequency f_p are the so-called scale parameters (Evans & Kibblewhite, 1990). Various experimental conditions were achieved by changing the peak enhancement factor g or the peak frequency f_p. Altogether, a total of 66 experiments was conducted, and all analyses were carried out on an IBM 586 compatible personal computer. Table 1 summarizes these experimental conditions.

EXPERIMENTAL RESULTS

Fig. 2 shows four spectral density curves measured before and behind the dike. The experimental condition is a typical JONSWAP spectrum with a peak frequency f_p = 0.8 and g = 3.3. All four curves are quite similar, indicating that the presence of the dike has little, or even no effect on the spectral shapes of these four stations. The spectral densities of Station 6 are more pronounced than other stations around the peak frequency, showing a concentration of energy there. A a small hump exist near the second harmonic, i.e., $2f_p$ ~1.6 Hz, for the two stations, Stations 7 & 8, indicating that nonlinear waves were generated by the dike. Since no other random-like high frequency components can be seen even on logarithmic scaled plots (not shown here), it is believed that the small hump around $2f_p$ is not caused by wave breaking.

Table 1 Experimental conditions used

peak frequency (f_p) [Hz]	peak enhancement factor (g) [-]	gain* [-]	No. experiments
0.5	1.0	2.0	2
0.5	1.0	2.5	2
0.5	1.0	3.0	2
0.5	3.3	2.0	2
0.5	3.3	2.5	2
0.5	3.3	3.0	2
0.5	7.0	2.0	2
0.5	7.0	2.5	2
0.5	7.0	3.0	2
0.8	1.0	1.5	2
0.8	1.0	2.0	2
0.8	1.0	2.5	2
0.8	1.0	3.0	2
0.8	1.0	3.5	1
0.8	3.3	1.5	2
0.8	3.3	2.0	2
0.8	3.3	2.5	2
0.8	3.3	3.0	2
0.8	3.3	3.5	1
0.8	7.0	1.5	2
0.8	7.0	2.0	2
0.8	7.0	2.5	2
0.8	7.0	3.0	2
0.8	7.0	3.5	1
1.0	1.0	2.0	2
1.0	1.0	2.5	2
1.0	1.0	3.0	2
1.0	3.3	2.0	2
1.0	3.3	2.5	2
1.0	3.3	3.0	2
1.0	7.0	2.0	2
1.0	7.0	2.5	2
1.0	7.0	3.0	2

The wave case with more concentrated energy are shown in Fig. 3. The peak frequency f_p still remains the same, i.e., f_p = 0.8 Hz. However, the peak enhancement factor, g, is now raised up to a value of 7.0 to represent the active wind wave generation. For comparison, the spectral densities at Station 6 is also included. From the figure, there are a dip at approximately f = 0.4 Hz and a small hump at f = 0.2 Hz. The reasons for these are not clear at present. However, since these already appear at Station 5, the presence of the dike should not be the actual reason. Observing the fluctuations at the frequencies higher than the peak frequency, the components at the second harmonics were already aroused in front of the dike (Fig. 3, solid line), and did not show at Station 5(figure not shown here), one may have the conclusions that the waves were already nonlinear in front of the dike, and the nonlinearity was generated by the dike itself.

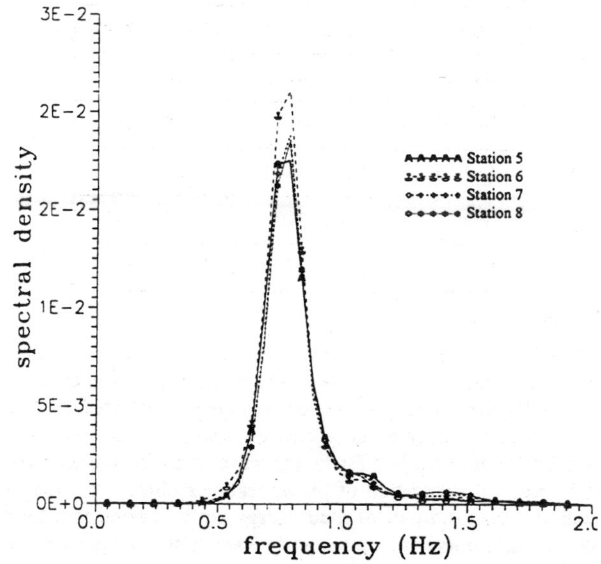

Fig. 2 Developments of spectral densities before and after passing a submerged dike. Experimental conditions used: f_p = 0.8 Hz, g = 3.3, gain (of the wave generator) = 1.5.

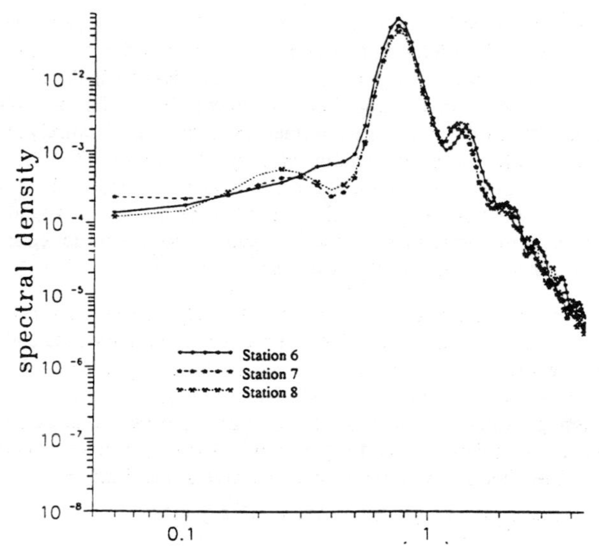

Fig. 3 Developement of spectral shapes before and after passing the dike. Experimenal conditions used: f_p = 0.8 Hz, γ = 7.0, gain of the wave generator = 3.5.

When wave spectral densities do not change appreciably, it is nature to ask whether other wave properties will be altered by the submerged dike. Fig. 4 shows the probability distributions of the wave heights at these four stations. The experimental conditions used are: f_p = 0.8 Hz, g = 1.0, and gain of the wave generator = 3.0. With the peak enhancement factor g = 1, the JONSWAP spectrum becomes P-M spectrum for fully developed sea. Theoretically, it represents a broad banded case, as compared with the typical JONSWAP spectrum. Three curves used here to model the wave height probability distributions are, respectively, normal (solid line), Rayleigh (broken line), and Weibull (dotted line) distribution curves. It seems that waves in front of (Figs. 4A & 4B) and right behind the dike (Fig. 4C) have distributions that can be described as Rayleigh distribution, whereas further away from the obstacle (Fig. 4D), waves are more adequately described with Weibull or even Gaussian distributions. In fact, from all the studied cases, all three models can be used to fit the wave height distributions rather satisfactorily, irrespective with or without the presence of a submerged obstacle. Table 2 summarizes the overall χ^2 goodness-of-fit test results.

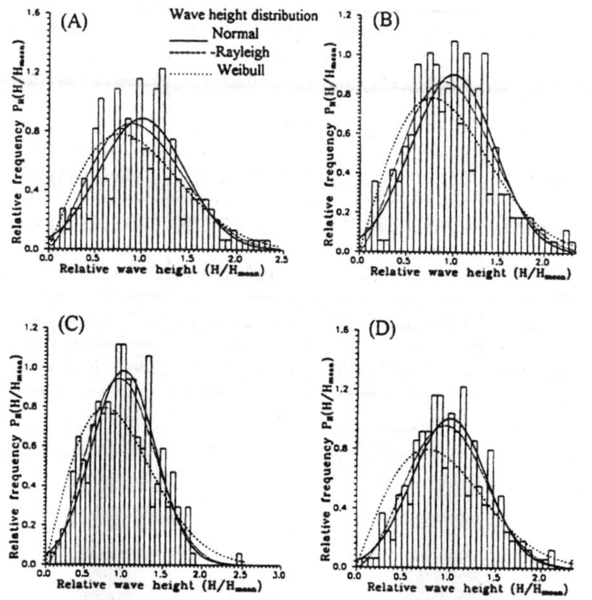

Fig. 4 Wave height distributions before and after propagating over the dike. Case studied: JONSWAP spectrum, f_p = 0.8 Hz, γ = 1.0, gain = 3.0.

Since the linear waves have symmetrical profiles, the water particles have therefore equal chance to stay just above and below the mean water surface. A nature consequence of the linear wave theory is that, the probability distributions of the surface elevations for the linear waves should be assume as Gaussian distribution. Waves with higher energies, on the other hand, have pronounced peaks and flatter troughs. An adequate model equation to describe the probability distribution of the surface elevations is the so-called Edgeworth's form of the type A Gram-Charlier series derived by Longuet-Higgins (1964). Huang et al. (1983) proposed a so-called Stokes model for the probability distribution of the surface elevations of finite amplitude waves. Fig. 5 shows the distributions of the surface elevation for the case with peak frequency f_p = 0.8 Hz, and peak enhancement factor, g = 7.0. From this figure, the curves from Gram-Charlier series expansions can be seen to fit better than the other two models. It is also interesting to note that the curves of the model proposed by Huang et al. (1983) in all four stations almost fall together with the curves of the normal distribution, making them indistinguishable. However, the deviations from linearity is not very serious for all the cases studied. Table 3 shows the mean values of the skewness and the kurtosis of the water surface. From this table, it can be said that, approximately one or two meters away from the submerged dike, i.e., Stations 5 & 8, the surface elevations resume their original states, and are approximately normally distributed.

Table 2 χ^2-goodness-of-fit test results for the models used for wave heights. Figures in the parenthesis denote standard deviation, figures in the square parenthsis are number of misfits.

model used / peak frequency [Hz]		Normal	Rayleigh	Weibull
0.5	Station 5	3.519 (1.311) [0]	3.177 (0.805) [0]	2.755 (0.869) [0]
	Station 6	3.697 (0.756) [0]	3.965 (0.486) [0]	3.352 (0.576) [0]
	Station 7	3.219 (0.522) [0]	2.739 (0.520) [0]	2.536 (0.480) [0]
	Station 8	2.919 (0.732) [0]	2.830 (0.806) [0]	2.492 (0.748) [0]
0.8	Station 5	7.003 (5.670) [1]	3.296 (0.605) [0]	4.375 (3.469) [0]
	Station 6	2.862 (0.848) [0]	3.296 (0.926) [0]	2.596 (0.779) [0]
	Station 7	3.937 (0.972) [0]	3.033 (0.846) [0]	2.831 (0.755) [0]
	Station 8	2.511 (0.714) [0]	2.448 (0.780) [0]	2.172 (0.747) [0]
1.0	Station 5	3.705 (2.008) [0]	3.016 (0.501) [0]	2.491 (0.815) [0]
	Station 6	3.501 (1.418) [0]	3.413 (0.862) [0]	2.574 (0.792) [0]
	Station 7	3.371 (1.476) [0]	4.377 (2.124) [0]	3.173 (1.342) [0]
	Station 8	3.185 (0.512) [0]	3.611 (0.924) [0]	2.942 (0.740) [0]

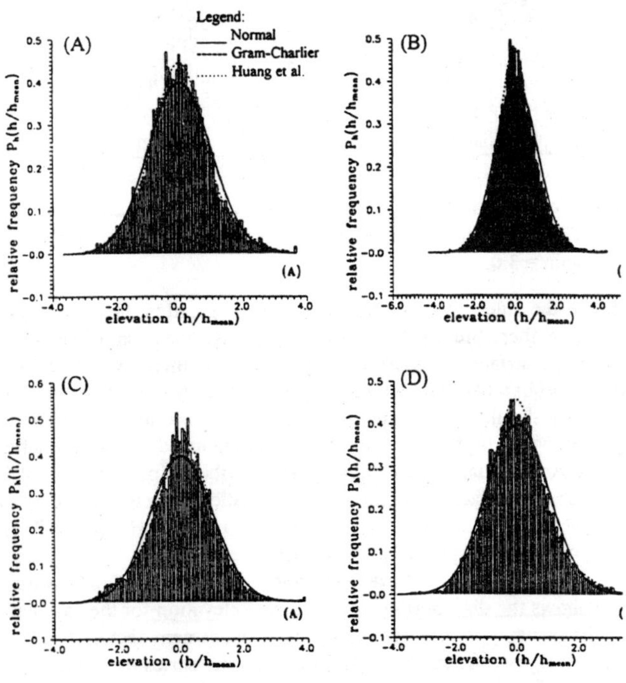

Fig. 5 Variations of surface elevations before and after propagating over the dike. Case studied: JONSWAP spectrum, f_p = 0.8 Hz, γ = 7.0, gain = 3.0.

Table 3 Skewness and kurtosis of surface elevations. Figures in the parenthesis denote standard deviation.

Peak frequency f_p [Hz]		skewness	kurtosis
0.5	Station 5	0.086 (0.059)	3.000 (0.143)
	Station 6	0.138 (0.029)	2.832 (0.031)
	Station 7	0.205 (0.076)	3.028 (0.070)
	Station 8	-0.009 (0.028)	2.841 (0.036)
0.8	Station 5	0.074 (0.080)	2.998 (0.094)
	Station 6	0.131 (0.026)	2.872 (0.017)
	Station 7	0.312 (0.228)	3.227 (0.229)
	Station 8	0.036 (0.032)	2.910 (0.046)
1.0	Station 5	0.121 (0.043)	3.089 (0.159)
	Station 6	0.147 (0.048)	2.988 (0.109)
	Station 7	-0.0503(0.055)	3.011 (0.112)
	Station 8	0.170 (0.062)	3.027 (0.160)

CONCLUSIONS

When waves of either synthesized or nature composition of linear components, traveling from deep water and confronting with the submerged obstacles, the shorter waves are relatively unaffected and the longer waves will certainly be subjected to changes. This is because more water bodies are involved for longer waves than for shorter ones. With the increase of dimension of the underwater objects, the longer waves tend to break, and part of their energy are dissipated in a form of turbulence. Such phenomenon naturally leads to the conclusion that the wave field characteristics will be altered considerably through the presence of the obstacle. Present analysis shows, however, that besides the lost of energy - probably due to the wave breaking, the wave spectra preserve their shapes even after passing through a submerged dike. Similar conclusions have also been reached by Battjes & Beji (1992).

Except for cases with peak frequency f_p = 0.5 Hz, which has a dominant wavelength of l = 6.24 m, all other cases considered in this paper belong to the deep water wave class. The bottom friction is therefore not a major cause to wave energy loss. The energy loss before and after passing through the submerged obstacle must be caused entirely by wave breaking. However, as demonstrated by Melville (1982, 1983), at least for uniform wave trains, wave breaking will leads to the so-called phenomenon of "frequency downshift". Whether this downshift may also occur for the simulated random waves or not is still not clear. At least, it is not observed for the cases shown here.

The wave height distributions of the mechanically simulated random waves can all be fitted to normal, Rayleigh and Weibull curves, and all have approximately equal chances to satisfy the χ^2 goodness-of-fit tests. The surface elevations are approximately Gaussian distribution. After passing through the obstacle, these characteristics are all essentially preserved as demonstrated in this paper. The question left is that what possible effects can be the submerged dike arouse? An intensive study in the near future is planed to answer this question.

REFERENCES

Battjes, JA and Beji, S (1992). "Breaking waves propagating over a shoal," *Proc. 23rd Int'l Conf. on Coastal Eng. ASCE.*

Bouws, E, Gnther, H, Rosenthal, W and Vincent, CL (1985). "Similarity of the wind wave spectrum in finite depth water. 1. Spectral form," *J. Geophy. Res.*, Vol. C90.

Bouws, E, Gnther, H, Rosenthal, W and Vincent, CL (1987). " Similarity of the wind wave spectrum in finite depth water. Part 2. Statistical relations between shape and growth stage parameters," *Dtsch. Hydrogr. Zeitschrift*, Vol. 40.

Evans, KC and Kibblewhite, AC (1990). "An examination of fetch-limited wave growth off the west coast of New Zealand by a comparison with the JONSWAP results," *J. Phys. Oceanogr.* Vol. 20.

Hasselmann, K, et al. (1973). "Measurements of wind-wave growth and swell decay during the Joint North Sea Wave Project (JONSWAP)," *Deut. Hydrogr. Zeitschrift*, Suppl. A. 8 No. 12.

Huang, NE, Long, SR, Tung, CC and Bliven, LF (1984). "The non-Gaussian joint probability density function of slope and elevation for a nonlinear gravity field," *J. Geophys. Res.*, Vol. C89.

Longuet-Higgins, MS (1952). "On the statistical distribution of the heights of sea waves," *J. Marine Res.*, Vol. 9.

Longuet-Higgins, MS (1964). "The effect of non-linearities on statistical distributions in the theory of sea waves," *J. Fluid Mech.*, Vol. 17.

Melville, WK (1982). "The instability and breaking of deep-water waves," J. Fluid Mech., Vol. 115.

Melville, WK, (1983). "Wave modulation and breakdown," *J. Fluid Mech.*, Vol. 128.

Phillips, OM (1958). "On some properties of the spectrum of wind-generated ocean waves," *J. Marine Res.*, Vol. 16.

Pierson, WJ, Jr. and Moskowitz, L (1964) "A proposed spectral form for fully developed wind seas based on the similarity theory of S. A. Kitaigorodskii," *J. Geophys. Res.*, Vol. 69.

Yim, JZ and Lin, JG (1997). "Effects of spectral widht on the statistical properties of simulated random wave fields," subjected to 7th ISOPE, Honolulu.

Young, IR (1988)"A shallow water spectral wave model," J. Geophy. Res. vol. C93.

Proceedings of the Seventh (1997) International Offshore and Polar Engineering Conference
Honolulu, USA, May 25-30, 1997

ISBN 1-880653-28-1 (Set); ISBN 1-880653-31-1 (Vol. III)

Solitons, Positons and Trains of Algebraically Decaying Waves

K.W. Chow and K. Tso
University of Hong Kong
Hong Kong

ABSTRACT

Recently the 'positons' have been introduced as new exact solutions for unidirectional wave propagation in nonlinear media. In this work positons are derived for nonlinear evolution equations for shallow water waves with three dimensional effects. In particular a train of algebraically decaying waves is obtained in the far field.

Keywords : solitons, positons, Kadomtsev - Petviashvili equation.

1. Introduction

The properties and propagation of long waves in fluids have been studied intensively for a long time. In particular the dynamics of solitary waves generated by ships moving in a confined channel or the open sea is important in both marine applications and military activities. Elegant theories have been developed to describe the unidirectional wave motion (Mei, 1986, Wu 1987). However, the understanding is not entirely satisfactory when three dimensional effects are taken into account. As an example the reflection of oblique solitary waves off a rigid wall has created some uncertainties and controversies even though the problem appears to be fairly simple (Miles, 1977).

A popular and plausible model for long waves in (2+1) (2 spatial and 1 temporal) dimensions is the celebrated Kadomtsev - Petviashvili (KP) equation (Aklyas, 1994), which is a generalization of the Korteweg - de Vries equation (KdV) in unidirectional wave motion.

There have been many elaborate studies regarding the emission of solitary waves by a travelling disturance in the literature. The purpose of the present work is to document a new class of solutions of KP and to examine if such solutions are applicable to the studies of ocean waves.

Recently the 'positons' have been introduced as new exact solutions for nonlinear evolution equations. Such solutions have been derived for several classical nonlinear evolution equations (NEEs) : KdV and the sine - Gordon equation. Positons in unidirectional wave propagation exhibit several novel features which differ fundamentally from classical solitons (Matveev, 1994) :

- Positons are weakly (usually algebraically, rather than exponentially) localized.

- The properties regarding the eigenvalue spectrum and the reflection coefficient are special.

- Two positons remain unchanged after mutual interactions. During a soliton - positon collision, the soliton remains unchanged but the positon is affected. Both the carrier wave and the envelope of the positon experience phase shifts.

These results are originally obtained by the applications of the Darboux transformation. Recent works have demonstrated that positons can also be derived by a special limiting procedure of a classical 2-soliton solution (Jaworski and Zagrodzinski, 1995).

The only drawback for the actual physical applications of positons is that they possess a singularity (or singularities) in the near field. Given that KP is derived under the assumptions of long waves and weak nonlinearity, one now regards a large amplitude as signs that these assumptions are being violated. Presumably new physics needs to be restored in the vicinity of the singularity in order to achieve a finite solution. Asymptotic matching must now be performed in an intermediate region. In quite a number of engineering applications only the far field conditions are required. Hence the use of 'positons' in the studies of ocean waves deserves further consideration. In fact point singularities are frequently used in classical hydrodynamics, e.g., in potential and low Reynolds number flows.

A remark regarding the limiting process is in order. One starts with a 2 - soliton solution but allows for (Jaworski and Zagrodzinski, 1995)

- complex conjugate wave numbers $(a+ib,\ a-ib)$,
- a suitably chosen complex phase, and
- a long wave limit $(a \to 0)$.

Positon in the KdV case is basically a point singularity. With three dimensional effects the structure of the singularities of KP is quite complex. The purpose of the present work is to initiate a study in this direction.

2. Analysis

The KP equation is

$$(u_t + 6uu_x + u_{xxx})_x = \pm 3u_{yy}, \tag{1}$$

where the positive (negative) sign corresponds to the positive (negative) dispersion regime. The $\pm$ is vital to the properties of the solutions since the KdV solitary wave is stable (unstable) to three dimensional perturbations in the negative (positive) dispersion regime. Most realistic ocean - wave situations fall under the regime of negative dispersion. No lump solution is known in that case, and this is consistent with the existing experimental data.

In this paper we shall focus exclusively on the negative dispersion regime :

$$(u_t + 6uu_x + u_{xxx})_x = -3u_{yy}. \tag{2}$$

The 2 - soliton solution is (Satsuma and Ablowitz, 1979 and the Appendix) :

$$u = 2(\log f)_{xx}, \tag{3}$$

$$f = 1 + \exp(\phi_1) + \exp(\phi_2) + M_0 \exp(\phi_1 + \phi_2), \tag{4}$$

$$\phi_n = \alpha_n x + \beta_n y - (\alpha_n^3 + \frac{3\beta_n^2}{\alpha_n})t,$$

$$M_0 = \frac{\alpha_1\alpha_2(\alpha_2-\alpha_1)^2 - \chi}{\alpha_1\alpha_2(\alpha_2+\alpha_1)^2 - \chi}, \tag{5}$$

$$\chi = \frac{(\alpha_1\beta_2 - \alpha_2\beta_1)^2}{\alpha_1\alpha_2}.$$

(A) A reference solution

We first utilize the technique of complex wavenumbers but shall not take the long wave limit yet. This will lead to exponentially localized solutions with no wave form in the far field. On choosing $\alpha_1 = \alpha_2^\star = a + ib$ and $\beta_1 = \beta_2^\star = p + iq$, one obtains

$$M_0 = \frac{(pb-aq)^2 - b^2(a^2+b^2)^2}{(pb-aq)^2 + a^2(a^2+b^2)^2}. \tag{6}$$

One special limit is $a \to 0$, $q \to 0$. We shall ignore scalar multiplicative factors of f in the intermediate calculations, since those quantities do not affect u. If $p^2 - b^4 > 0$, one has

$$f = \cosh(py) + r_1 \cos(bx - \Omega_1 t), \qquad u = 2(\log f)_{xx}, \tag{7}$$

$$r_1^2 = \frac{p^2}{p^2 - b^4} > 1, \qquad \Omega_1 = -b^3 - \frac{3p^2}{b}, \tag{8}$$

while for $p^2 - b^4 < 0$,

$$f = -\sinh(py) + r_2 \cos(bx - \Omega_2 t), \qquad u = 2(\log f)_{xx}, \tag{9}$$

$$r_2^2 = \frac{p^2}{b^4 - p^2}, \qquad \Omega_2 = \Omega_1. \tag{10}$$

(9) is illustrated in Figure 1. There is a sinusoidal curve along which the solution is singular. In addition there is a large displacement of the opposite sign at each local maxima and minima. No wave form is observed away from the singularity.

(B) Long wave limit - Positon

The more interesting limit is $p = ma, a \to 0$ with m being an order one constant. In this case

$$M_0 \sim -\frac{b^6}{a^2[(q-mb)^2 + b^4]} = -\frac{1}{a^2 r^2}. \tag{11}$$

One now argues that the phase factors $\exp(\phi_1^{(0)}) = ar\exp(-i\xi)$, $\exp(\phi_2^{(0)}) = ar\exp(i\xi)$ can be added to the appropriate terms in (4) without destroying the validity of the expansion. On taking the limit $a \to 0$ one obtains

$$f = \cos(bx + qy - \Omega_0 t - \xi) - \frac{b^3(x + my - \omega_0 t)}{\sqrt{(q-mb)^2 + b^4}}, \tag{12}$$

$$u = 2(\log f)_{xx},$$

$$\omega_0 = -3b^2 + \frac{3(2mqb - q^2)}{b^2}, \tag{13}$$

$$\Omega_0 = -b^3 + \frac{3q^2}{b}. \tag{14}$$

One recovers the 1-positon solution of KdV in earlier works by setting $m = q = 0$, $\xi = \pi/2$:

$$f = \sin(bx + b^3 t) - b(x + 3b^2 t), \qquad u = 2(\log f)_{xx}. \tag{15}$$

The phase factor ξ allows a slightly more general family of solutions by incorporating this free parameter.

(12) is illustrated in Figure 2. Trains of oblique waves propagate on either side of the singularity. The amplitude of the waves decays algebraically in the far field. A magnified picture of the wave train is provided in Figure 3.

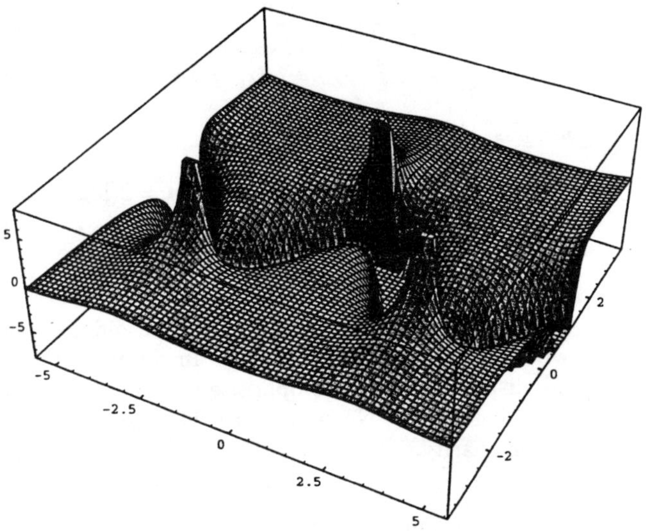

Figure 1 - $u = 2(\log f)_{xx}$ versus x and y with f given by (9), $b = 1$, $p = 0.5$, $t = 0$.

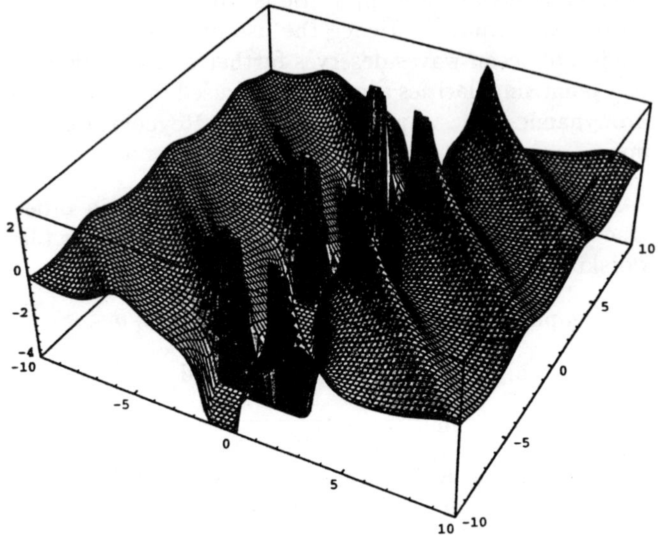

Figure 2 - $u = 2(\log f)_{xx}$ versus x and y with f given by (12), $b = 1$, $q = 1$, $m = 0$, $t = 0$.

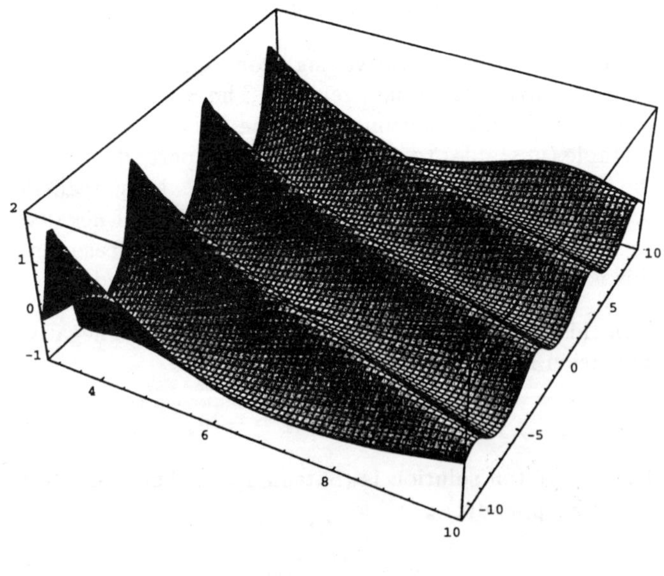

Figure 3 - Same as Figure (2), with focus on the wave field.

3. Discussions

We are not aware of any study on positons in (2+1) dimensions and the present work represents a preliminary step in this direction. We believe that positons enhance our capability to model the three dimensional wave field in the open ocean. Of particular relevance is the analytical formulation of trains of algebraically decaying waves in the far field.

4. Appendix - The bilinear transformation

The Hirota bilinear operator D is defined by :

$$D_x^m D_t^n g.f \tag{16}$$

$$= (\frac{\partial}{\partial x} - \frac{\partial}{\partial x'})^m (\frac{\partial}{\partial t} - \frac{\partial}{\partial t'})^n g(x,t) f(x',t')|_{x=x',t=t'}. \tag{17}$$

The bilinear form of KP (2) is

$$(D_x D_t + D_x^4 + 3D_y^2) f.f = 0, \qquad u = 2(\log f)_{xx}. \tag{18}$$

The bilinear operator is specially designed to handle exponential functions :

$$D_x^m \exp(\alpha x). \exp(\beta x) = (\alpha - \beta)^m \exp(\alpha x) \exp(\beta x). \tag{19}$$

Simple calculations now verify that (4, 5) solve (18).

Acknowledgements

Partial financial support is provided by the Committee of Research and Conference Grants, University of Hong Kong, under contract 335/064/0077.

References

Ablowitz M.J., Clarkson, P.A., (1991) *Solitons, nonlinear evolution equations and inverse scattering*, Cambridge University Press.

Akylas, T.R., (1994) Three dimensional long water - wave phenomena, *Annual Review of Fluid Mechanics*, volume 26, 191 - 210.

Beutler, R., (1993) Positon solutions of the sine - Gordon equation, *Journal of Mathematical Physics*, volume 34, 3098 - 3109.

Infeld, E., Rowlands, G., (1990) *Nonlinear waves, solitons, and chaos*, Cambridge University Press.

Jaworski, M., Zagrodzinski, J., (1995) Positon and positon - like solutions of the Korteweg - de Vries and sine - Gordon equations, *Chaos, Solitons and Fractals*, volume 5, 2229 - 2234.

Matveev, V.B., (1994) Asymptotics of the multi - positon - soliton τ function of the Korteweg - de Vries equation and the supertransparency, *Journal of Mathematical Physics*, volume 35, 2955 - 2970.

Mei, C.C., (1986) Radiation of solitons by slender bodies advancing in a shallow channel, *Journal of Fluid Mechanics*, volume 162, 53 - 67.

Miles, J. W., (1977) Resonantly interacting solitary waves, *Journal of Fluid Mechanics*, volume 79, 171 - 179.

Ohkuma, K., Wadati, M., (1983) The Kadomtsev - Petviashvili equation : the trace method and the soliton resonances, *Journal of the Physical Society of Japan*, volume 52, 749 - 760.

Satsuma, J., Ablowitz, M.J., (1979) Two dimensional lumps in nonlinear dispersive systems, *Journal of Mathematical Physics*, volume 20, 1496 - 1503.

Wu, T.Y., (1987) Generation of upstream advancing solitons by moving disturbance, *Journal of Fluid Mechanics*, volume 184, 75 - 99.

Proceedings of the Seventh (1997) International Offshore and Polar Engineering Conference
Honolulu, USA, May 25-30, 1997

ISBN 1-880653-28-1 (Set); ISBN 1-880653-31-1 (Vol. III)

Effect of Boundary on Ship Waves

Sing-Kwan Lee and Allen T. Chwang
University of Hong Kong
Hong Kong

ABSTRACT

A sink-source distribution method is combined with a finite difference method (FDM) in this paper for the simulation of ship navigation in restricted waterways. In this approach, only a single grid system is adopted to cover the whole fluid domain. Neither the grid regeneration employed in the conventional FDM nor the complex overlapping grid systems in the Chimera method is involved in our time-marching computation. The basic idea of this approach is theuse of singularities distributed inside a moving object with time dependent strength to fulfill the non-penetration condition on the body surface at each time step. Through this sink-source strength adjusting, the computation can be carried on with no need to change the original grid structure, except the vertical stretch due to the free surface movement. To examine the effectiveness of this method, cases of a Wigley hull ship moving on the open sea with zero and nonzero Froude numbers are first studied and results are compared with those obtained by the conventional FDM and by experiments. This method is then applied to cases of a ship advancing in a convergent channel to investigate the ship wave reflection and diffraction caused by the side boundaries. Finally, a mildly sloping channel case is used to simulate combined effects on ship wave refraction and reflection due to the bottom topography and sidewalls.

KEY WORDS: Ship Waves, boundary, sink-source distribution, finite difference method

INTRODUCTION

Wave agitation due to navigating ships inside restricted waterways is one of the interesting problems in hydrodynamics. The complicated interaction among ship waves, their reflection and diffraction from seawall boundaries produce wave disturbances quite differently from the unbounded sea situation. This problem is difficult to solve numerically. If a conventional FDM is used, numerical regridding is inevitable when ships change their locations inside the computational domain. As is well known, this time-consuming regridding process eventually causes the simulation economically impractical if calculation involves a large and complex region. To avoid the tedious grid regeneration, Benek *et al.* (1985) developed an overlapping grid technique called Chimera grid scheme which can be combined with other flow solvers to study the flow around moving bodies. Briefly, this scheme uses different grid system for every moving object and overlaps those independent grid systems on a background grid which covers the entire computational domain. To link the grids together, an interpolation procedure is performed to transfer information from one grid to another. Although the Chimera method has been employed for many different kinds of Computational Fluid Dynamics (CFD) applications, the application for problems with free surface waves is still under research.

Unlike the above-mentioned methods, a novel sink-source distribution method is used here for the simulation of ships navigating near an arbitrary-shaped boundary. In this approach, only a single grid system is adopted to cover the whole domain. Neither the grid regeneration employed in the conventional FDM nor the complex overlapping grid systems in the Chimera method is involved in our time-marching computation. The basic idea of this approach is the use of singularities distributed inside a moving object with time dependent strength to fulfill the non-penetration condition on the body surface at each time step. Through this sink-source strength adjusting, the computation can be carried on with no need to change the original grid structure, except the vertical stretch caused by the free surface movement. To examine the effectiveness of this method, cases of a Wigley hull ship moving on the open sea with zero and nonzero Froude numbers are first studied and results are compared with those obtained by the conventional FDM and by experiments. This method is then applied to cases of a convergent channel and mildly sloping channel to simulate the complex phenomenon due to the interaction between ship waves and boundaries.

METHODOLOGY

Formulation

Consider the problem of a ship advancing in a restricted waterway. A Cartesian coordinate system Oxyz is chosen such that z is directed vertically downward and x is along the ship sailing course with origin O located at the bow of the initial ship position. Plane Oxy is set on the clam water and plane Oxz as longitudinal ship centerplane as shown in Figure 1. Under the assumptions of incompressible and invicid fluid as well as irrotational motion, the velocity field can be defined in terms of a velocity potential Φ as,

$$\bar{V} = \nabla\Phi . \tag{1}$$

In our formulation, potential Φ is further decomposed into a regular part ϕ and a singular part ϕ_s where ϕ_s represents summation of sinks and sources with adjustable strengths,

$$\phi_s = \sum_{i=1}^{N} \frac{m_i}{|\bar{x} - \bar{x}_i|} , \tag{2}$$

where N is the number of isolated sources inside the body, $\bar{x}_i$ is the location of the ith source, and m_i is the strength of the ith source (negative m_i for sink).

It can be shown by direct substitution that the regular part ϕ satisfies the Laplace equation

$$\nabla^2\phi = 0 . \tag{3}$$

When surface tension is neglected on the free surface, the kinematic and dynamics boundary conditions on the exact free surface reduce to

$$\eta_t + (\phi_s + \phi)_x \eta_x + (\phi_s + \phi)_y \eta_y - (\phi_s + \phi)_z = 0 , \tag{4}$$

$$\phi_t + (\phi_s)_t + \frac{1}{2}|\nabla(\phi_s + \phi)|^2 - \frac{\eta}{Fr^2} = 0 , \tag{5}$$

where $Fr = U/\sqrt{gL}$ is the Froude number based on body length L, g the gravitational constant, η the free surface elevation, and subscript t denotes partial derivative with respect to t etc.. Here all the variables are nondimensional, with length scaled by ship length L, velocity by ship speed U, time by L/U, and potential function by UL.

On rigid boundaries such as ship hull and channel walls, the normal flux must vanish,

$$(\phi + \phi_s)_n = 0 , \tag{6}$$

where n represents the unit normal of a rigid boundary. It is assumed that the fluid is initially at rest, so the initial conditions are

$$\phi(x,y,z,0) = 0, \quad \phi_s(x,y,z,0) = 0, \quad \text{and} \quad \eta(x,y,0) = 0 . \tag{7}$$

Numerical Implementation

For enhancing the computational efficiency, a de-singularized scheme (Cao 1991) is applied to relieve the time-consuming surface source integration encountered in the traditional boundary integral method (Hess and Smith 1962). In this method, isolated sources are placed only inside a moving body instead of distributing source density on the body surface. Unlike the boundary integral formulation, no source is required on the free surface, the seawall and seabed boundaries in our approach. At each time step, the strengths of sources inside a moving body are adjusted according to equation (6) to fulfill the non-penetration condition. Once these sources are determined, potential ϕ on the free surface, the seawall and seabed boundaries is evaluated through equations (4) and (6). Usually, only few iterations are required for convergence. The overall procedure can be summarized as follows:

1. Move isolated sources inside the body to a new location and calculate the values required to maintain the non-penetration condition according to equation (6).
2. Use equations (4) and (6) to determine potential ϕ on the free surface, the seawall and seabed boundaries.
3. Update free surface elevation η by equation (5).
4. Solve Laplace equation (3) in the fluid domain by the FDM.
5. Calculate ϕ_n on the body surface and use equation (6) to recalculate the strengths of sources.
6. If the strengths of sources change beyond tolerance, go back to step 2, otherwise go to next step.
7. If the body arrives at its finial position, then stop computation, else go back to step 1.

To facilitate the numerical calculation, all equations are transformed into body-fitted coordinates, $t = \tau$, $\bar{x} = (\bar{\xi}, \tau)$ and solved in the transformed plane $(\bar{\xi}, \tau)$ using the finite-analytic method. For the sake of brevity, the details will not be mentioned here. Readers can refer to Lee (1995).

VALIDATION

In order to examine the effectiveness of our numerical method, cases of a Wigley ship hull moving in an open region with and without free surface effect are studied. Computational results are compared with the results by the FDM and with experiments to validate the accuracy of this new approach. For the convenience of comparison, all our computations are carried out for a Wigley ship with beam-length ratio 0.1 and draft-length ratio 0.0625.

Double-body Case

Prior to the complex free surface flow computation, a simple double-body case is first studied to estimate the accuracy of the de-singularized scheme. As is well known, although the de-singularized scheme provides an efficient approach in numerical computation, the manner of de-singularization, more precisely, how to select a proper de-singularization distance l_d, must be treated carefully. Usually, if the sources are too far from the boundary, the resulted linear system will be poorly conditioned and numerically unstable; if the sources are too close to the boundary, numerical computation of the increasingly singular kernel may significantly deteriorate the computational accuracy. In order to find out the proper de-singularization distance, numerical tests are performed. Two l_d s, namely 0.005 and 0.015, are chosen in our double-body computation. Here, 671 sources are used in all the computations. Results are shown in Figure 2. Apart from the de-singularized results, the FDM solution is also presented as a benchmark to evaluate the performance of the de-singularized method. As shown in Figure 2a, the equipotential lines at ship water plane calculated by using l_d equal to 0.005 (top) and 0.015 (bottom) are almost identical to each other. The de-singularized result with l_d equal to 0.005 is also compared with the FDM solution in Figure 2b and in good agreement with the FDM result. To have a further detailed comparison among the de-singularized results and the FDM solution, Figure 2c shows the potential value at domain centerline and ship waterline of those three numerical results. Except a slight decrease of the value at the bow and stern compared with the FDM solution, the de-singularized result is quite close to the FDM result. From our numerical tests we also find that, in the range of 0.005 to 0.015, the accuracy and convergence of solutions are not sensitive to the value of l_d.

Free Surface Flow

After the validation of the double-body cases, numerical calculations are performed for the free surface flows. Before applying the sink-source method, the FDM computation is first employed for the Fr = 0.289 case to obtain a reliable solution for accuracy check as in the double-body case. To ensure that the FDM solution is dependable, two sets of grid systems, i.e. the coarse grid (126×101×41) and the fine grid (251×115×41), are used to test the grid dependence of computations. At first, computations are done for linear waves. Figure 3 shows the linear wave patterns calculated by the coarse grid (bottom) and the fine grid (top). As seen in the figure, insufficient resolution from the coarse grid leads to more dissipative wave elevations than those from the fine grid. In Figure 4, the wave profiles along the ship side by coarse and fine grids are plotted and compared with the experimental result. Generally, the numerical results are in good agreement with the experimental ones except that the bow wave is underpredited especially for the coarse grid computation. Although the grid refinement can improve the result, as will be seen later, this bow wave underpredition is not mainly caused by insufficient grid resolution but by wave nonlinearity. To demonstrate the influence of the nonlinearity, fully nonlinear free surface conditions are applied at the exact water surface by moving the numerical grid with the wave movement. Figure 5 shows the nonlinear wave pattern (top) calculated by the fine grid. For convenience of comparison, linear solution is also plotted on the lower part of Figure 5. We note from Figure 5 that the nonlinearity does not change the wave pattern apparently in the far field. However, in the near field just along the ship side, the nonlinear wave profile shown in Figure 6 is obviously much closer to the experimental data than the linear wave profile, particularly for the first wave crest and trough at the ship bow.

To really validate the new sink-source distribution method, we now employ it to the free surface flow computation. Similar to the double-body case, the same amount of sources (671) are placed inside the body. The grid adopted here is rectangular with 151×81×25 points. In Figure 7, we compare the wave pattern obtained by the sink-source method with the pervious FDM fine grid result. Generally, the sink-source result is in good agreement with the FDM solution despite the gradient of the wave elevation of the sink-source result is a little bit more dissipitive than the FDM solution. This may be attributed to the coarseness of the computational grid used. To give a clear idea of the source strength variation due to the free surface effect, Figure 8 shows the first layer double-body and wavy sources under the waterplane. Because of the influence of free surface waves, wavy sources are oscillated around the values of double-body sources. For further examining the accuracy of the sink-source method, Figure 9 shows a comparison of wave elevations at three longitudinal cuts (y = 0.2 L, y = 0.3 L and y = 0.4 L). Generally, the sink-source results correspond well with the pervious FDM solution.

APPLICATIONS

Let us now consider the cases of ship moving in restricted waterways. Unlike the open sea situation, ship waves would be strongly affected by the sidewall and seabed boundaries. Reflection, diffraction and refraction of ship waves can be clearly observed in these simulations.

Convergent Channel

For a convergent channel shown in Figure 10, we use 300×11×11 grid points for our simulation. Here by utilizing the symmetric condition, only half channel domain needs to be calculated. In our computation, the Froude number Fr is set to be 0.3 and it is equivalent to water depth Froude number ($F_h = U / \sqrt{gh}$, where h is the depth of the channel) 0.671. Figure 11 presents wave patterns at three instance times when a ship passes three different locations in the channel. Here we can see that apparent wave diffraction occurs in the convergent part of the channel (Figure 11c). Also due to the wave reflection from the side wall, a sequence of symmetric diamonds behind the ship are presented.

Mildly Sloping Channel

The final computational case is a mildly sloping channel with 10 ship length long and 0.8 ship length wide. The ship is arranged to move from the deepest region with a depth of 0.2 ship length to the shallowest area with 0.1 ship length deep. Detailed geometry is shown in Figure 12. 501×21×11 grid points are used for this case. Froude number Fr is set to be 0.3 (the corresponding range of F_h is from 0.671 to 0.95). Wave patterns for a ship at three different locations in the channel are presented in Figure 13. Although the ship speed almost reaches the critical speed, shallow water solitons do not appear. This is mainly because the simulation channel is not long enough to let the ship continuously move for a long period. Hence the energy stored around the bow is not adequate to excite solitons. However wave refraction caused by the mildly sloping bottom topography can evidently be seen through the unsymmetric diamonds behind the ship. The reason for those unsymmetric diamonds is based on a well known truth that waves slow down in shallow waters. To clearly illustrate those unsymmetric diamonds, the top view of wave patterns for a mildly sloping channel and a constant depth convergent channel are plotted in Figure 14 for comparison.

CONCLUSIONS

A sink-source distribution method is combined with a finite difference method to simulate ships navigating in restricted waters. Unlike the common FDM computation, this method does not require the grid regeneration as ships change their positions inside the computational domain. Also no source is demanded to distribute on the seawall and seabed boundaries like the boundary integral method. In the computation, the strengths of sources are adjusted to accomplish the non-penetration condition on the ship hull. Detailed validation is done and confirms the accuracy of this method. Through numerical applications for a convergent channel and a mildly sloping channel, this sink-source distribution is demonstrated its capability for studying the boundary effect on ship waves.

ACKNOWLEDGMENTS

This research was jointly sponsored by the Hong Kong Research Grants Council under Grant No. HKU568/96E and by the University of Hong Kong under CRCG Grant No. 337/064/0043.

REFERENCES

Benek, J. A., Buning, P. G. and Steger, J. L. (1985). "A Chimera Grid Embedding Technique," AIAA Paper No. 85-1523.

Cao, Y. (1991). "Computation of Nonlinear Gravity Waves by a Desingularized Boundary Integral Method," Ph.D. Thesis, The University of Michigan.

Hess, J and Smith, A.M.O. (1962). "Calculation of Non-lifting Potential Flow about Arbitrary Three-dimensional Bodies," Douglas Aircraft Division Rep. E.S. 40622.

Lee, S. K. (1995). "A RANS/Laplace Coupling Method for Viscous Nonlinear Free Surface Flows," Ph.D. dissertation, Texas A&M University.

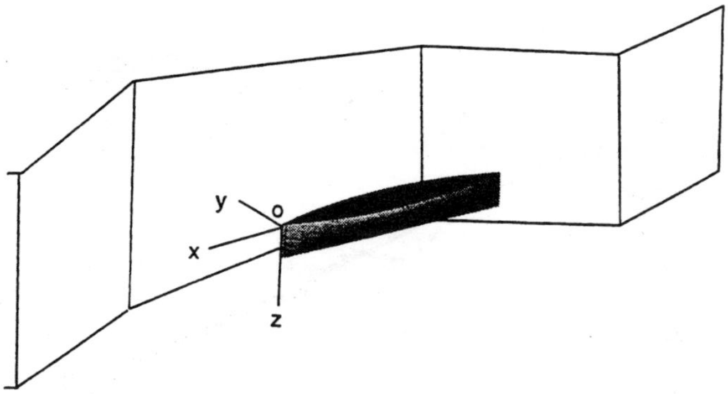

Fig. 1 Coordinate system

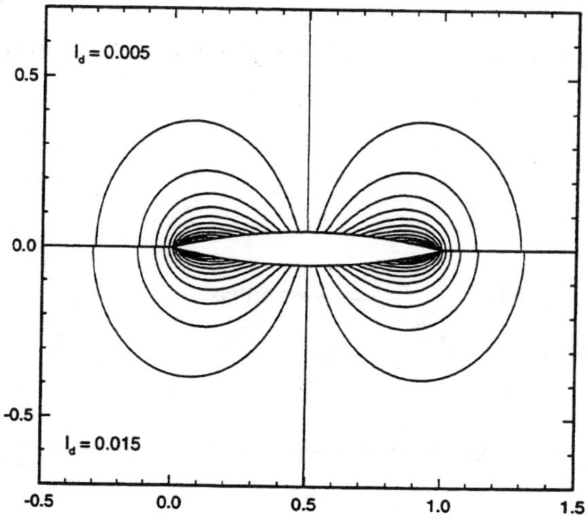

2 (a)

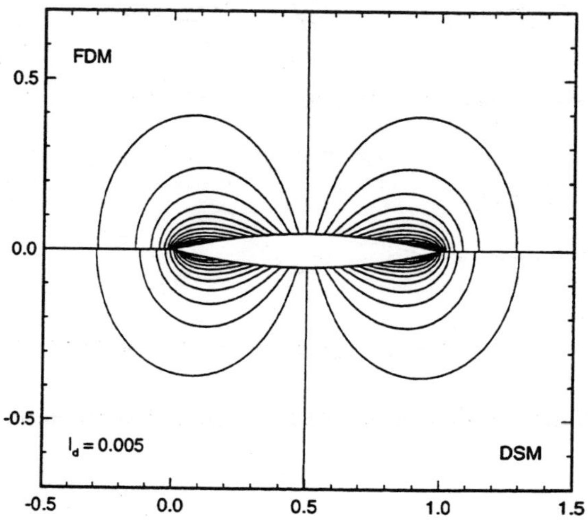

2 (b)

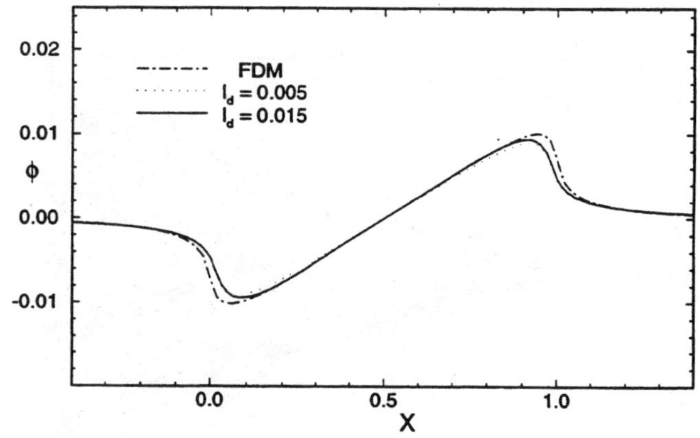

2 (c)

Fig. 2 (a) Equipotential lines on waterplane calculated by the de-singularized method (DSM)
(b) Equipotential lines on waterplane calculated by the FDM and the de-singularized method
(c) Potential value at domain centerline and ship waterline

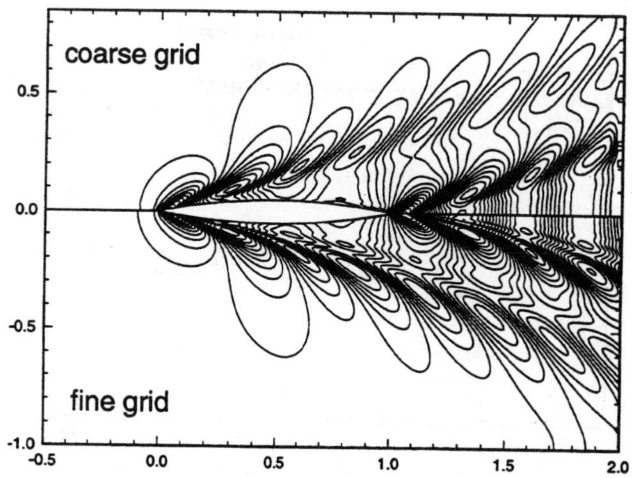

Fig. 3 Wave patterns by coarse and fine grids

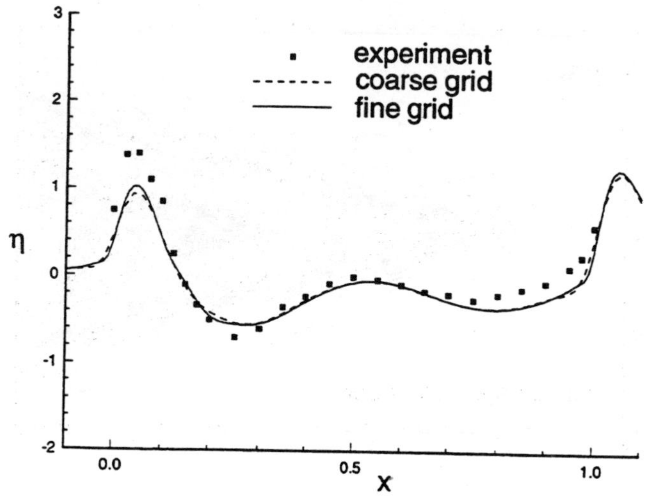

Fig. 4 Wave profiles along a ship hull side

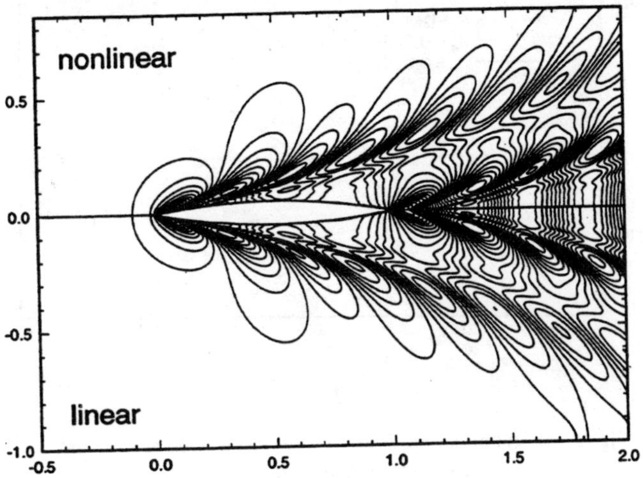

Fig. 5 Wave patterns by linear and nonlinear free surface conditions

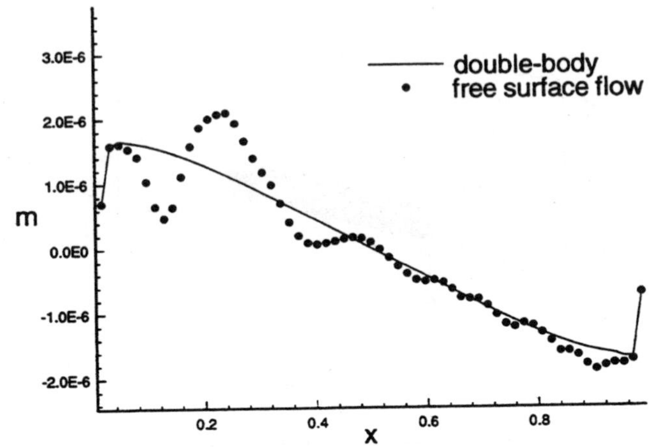

Fig. 8 Source strength of the first layer under the free surface

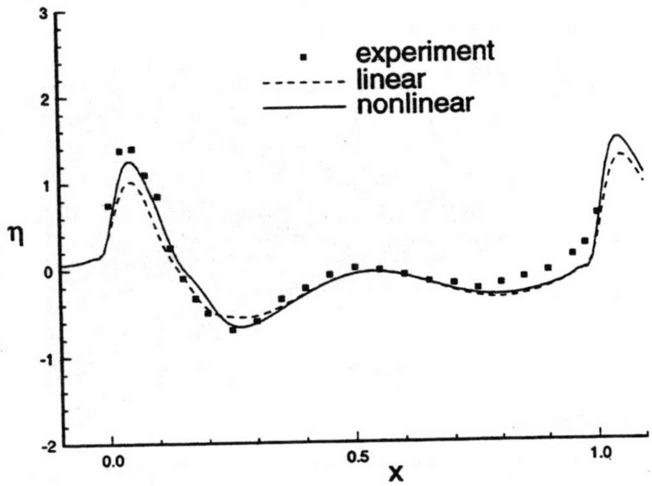

Fig. 6 Wave profiles along a ship hull side

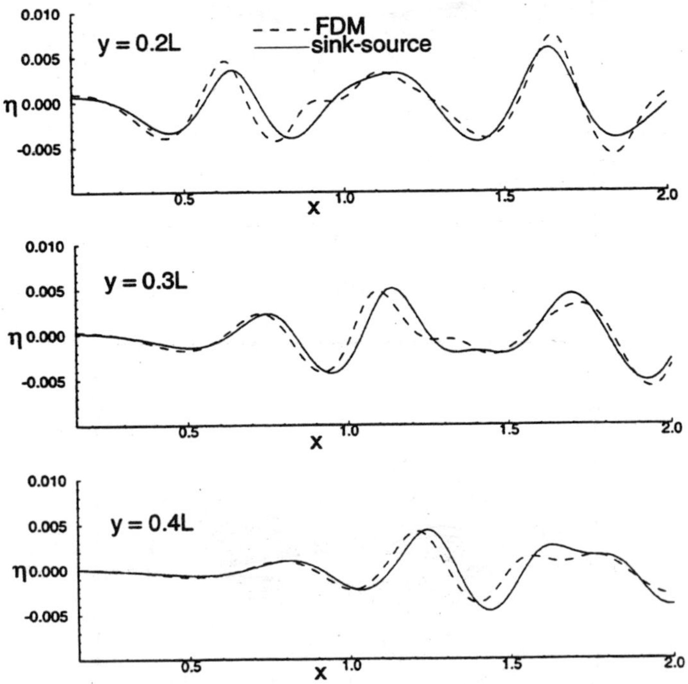

Fig. 9 Wave profiles computed by FDM and sink-source methods

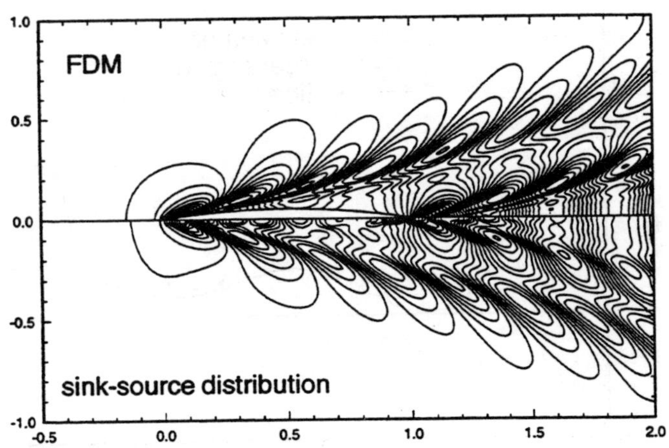

Fig. 7 Wave patterns computed by FDM and sink-source methods

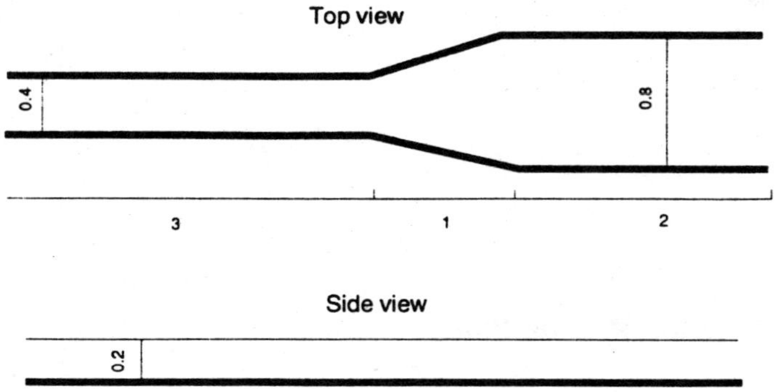

Fig. 10 Geometric configuration of a convergent channel

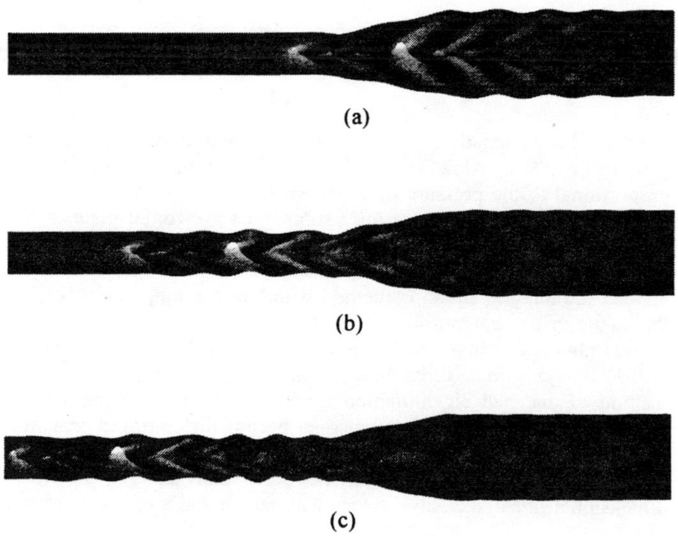

Fig. 11 Wave patterns in a convergent channel at three different instants

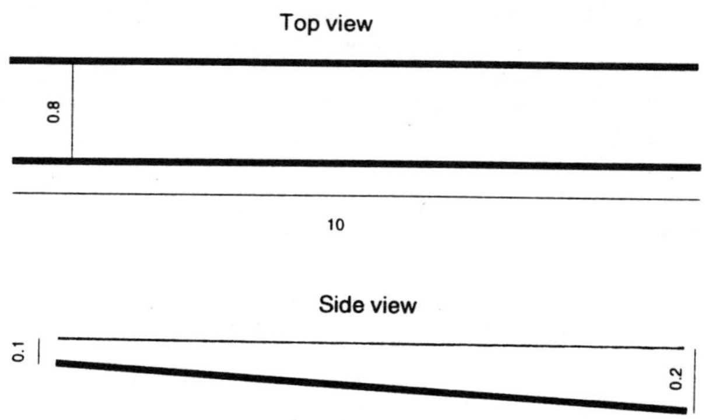

Fig. 12 Geometric configuration of a mildly sloping channel

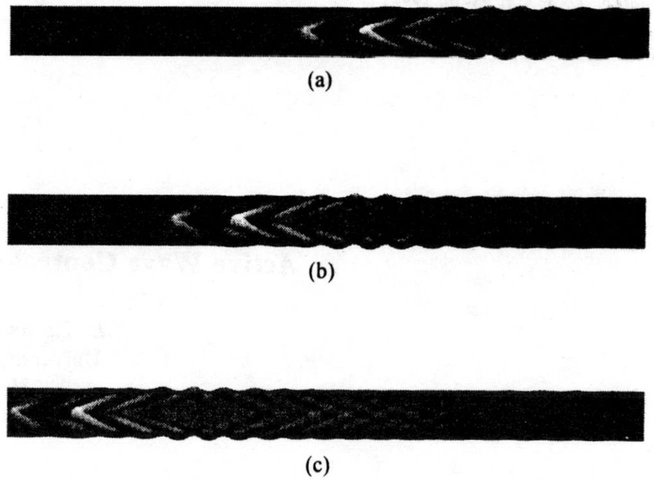

Fig. 13 wave patterns in a mildly sloping channel at three different instants

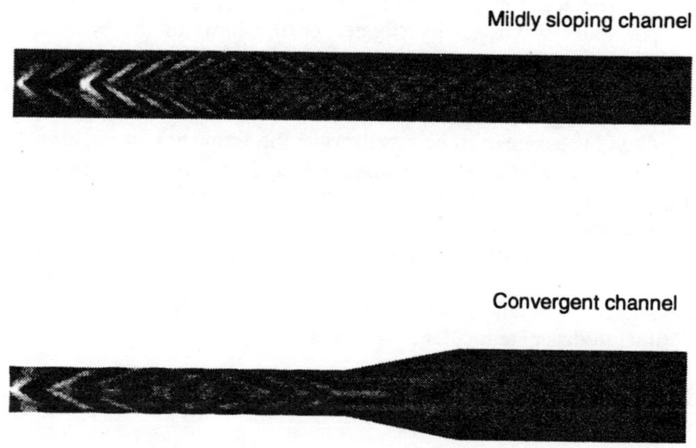

Fig. 14 Diamond region behind a ship

Proceedings of the Seventh (1997) International Offshore and Polar Engineering Conference
Honolulu, USA, May 25-30, 1997

ISBN 1-880653-28-1 (Set); ISBN 1-880653-31-1 (Vol. III)

Active Wave Control by a Pitching Porous Plate

T.L. Yip and Allen T. Chwang
University of Hong Kong
Hong Kong

ABSTRACT

The feasibility of using a pitching porous plate to actively control water waves is investigated on the basis of the linearized wave theory. The two-dimensional problem is formulated to deal with a fully submerged porous plate, pitching about its middle point and scattering an incident monochromatic wave. The pitching amplitude is assumed to be small, while the frequency of pitching is kept the same as that of the incident wave. The porous flow through the plate is governed by Darcy's law. The method of matched eigenfunction expansions is used to analyze the heights of reflected and transmitted waves. It is found that the heights of reflected and transmitted waves vary rapidly for small values of porous-effect parameter. A small value of porous-effect parameter is found to be optimal to dissipate incident wave energy.

KEY WORDS: porous plate, pitching plate, active controller, water waves, scattering

INTRODUCTION

here are many fixed barriers installed to reflect or diffract incident waves away from the protected area. The performance of a fixed barrier highly depends on its physical dimensions and the incident wave condition. Active control was first introduced by Milgram (1970) who found that the propagation of a monochromatic wave can be terminated by a hinged paddle absorber. With an emphasis on the usage of a pitching plate as a breakwater, Yip and Chwang (1996) suggested that a submerged pitching plate be used as an active wave-controller. The performance of a pitching plate was found by Yip and Chwang (1996) to be very sensitive to its submerged depth and dimension, although the phase and magnitude of pitching can modify the performance of the plate according to desired requirements. The height of radiation waves generated by pitching was found proportional to the pitching amplitude.

Since porous material has recently been utilized to dissipate wave energy, the porosity is introduced here to further improve the wave control. The first linearized analysis on a porous wavemaker was performed by Chwang (1983) who developed a theory for a porous vertical wavemaker making harmonic oscillations normal to its surface in a fluid of constant finite depth. In his two-dimensional analysis, he assumed that the flow in a porous wavemaker is governed by Darcy's law which states that the porous flow is linearly proportional to the pressure drop across the porous medium. After extensive investigations on a fully submerged horizontal plate as a breakwater by various researchers for 20 years, Yu and Chwang (1994) introduced a porous plate to dissipate the wave energy. They applied the integral equation method with Green's function to solve the two-dimensional problem of water waves above a submerged porous plate. Chwang and Wu (1994) extended Yu and Chwang's (1994) analysis to a three-dimensional problem and utilized the method of matched eigenfunction expansions to analyze the wave scattering by a horizontally submerged porous disk. In the present paper, the idea of active control is extended to a fully submerged porous plate and its characteristics are analyzed by a two-dimensional linearized wave theory with an emphasis on the porous effect.

THEORY

The two-dimensional problem is illustrated in Figure 1. In a Cartesian coordinate system, the origin is on the undisturbed free surface and the y-axis is pointing vertically upward. A porous plate is submerged horizontally in the fluid with its middle point below the origin. The plate thickness is negligible with respect to the water depth. The sea bottom is horizontal so that the water depth h is constant. The half-length and the submerged depth of the plate are denoted by a and h_1, respectively.

It is assumed that the fluid is inviscid and incompressible, and its motion irrotational. The velocity potential Φ exists,

$$\underline{U} = \nabla\Phi , \tag{1}$$

and satisfies the Laplace equation

$$\nabla^2\Phi = 0 . \tag{2}$$

For a monochromatic incident wave propagating in the x-direction, the incident velocity potential is

$$\Phi_0(x,y,t) = I_0 e^{ik_0(x+a)} \frac{\cosh k_0(h+y)}{\cosh k_0 h} e^{-i\omega t} , \tag{3}$$

where

$$I_0 = -i\frac{gH}{2\omega}, \tag{4}$$

in which g is the gravitational acceleration, H the wave height and ω the angular frequency of the incident wave, and k_0 is the wavenumber satisfies the dispersion relation

$$\omega^2 = gk_0 \tanh k_0 h. \tag{5}$$

Since the pitching is kept of the same frequency as the incident wave, the velocity potential is assumed to have the same time factor $\exp(-i\omega t)$,

$$\Phi(x,y,t) = \phi(x,y)e^{-i\omega t}, \tag{6}$$

where ϕ is the spatial velocity potential which satisfies the Laplace equation

$$\nabla^2\phi = 0. \tag{7}$$

The fluid domain is divided into four subregions: (1) the upstream subregion ($x<-a$, $-h<y<0$), (2) the interior subregion above the plate ($-a<x<a$, $-h_1<y<0$), (3) the interior subregion below the plate ($-a<x<a$, $-h<y<-h_1$), and (4) the downstream subregion ($a<x$, $-h<y<0$). The spatial velocity potential in subregion j is denoted by ϕ_j. Combining both kinematic and dynamic free surface boundary conditions, the linearized boundary condition on the free surface for ϕ_j's is

$$\frac{\partial\phi_j}{\partial y} = \frac{\omega^2}{g}\phi_j \qquad (j=1,2,4) \text{ at } y=0. \tag{8}$$

The vertical component of fluid velocity vanishes on the bottom so that the bottom boundary condition is

$$\frac{\partial\phi_j}{\partial y} = 0 \qquad (j=1,3,4) \text{ at } y=-h. \tag{9}$$

As shown in Figure 1, a pitching porous plate is positioned horizontally at $y=-h_1$. The plate pitches about $(0, -h_1)$, its middle point, with a maximum pitching angle α, which is kept very small so that the change of the mean water level due to pitching can be ignored. The displacement of pitching is described by

$$y = -h_1 + h_2 x e^{-i\omega t}, \tag{10}$$

where h_2 is the complex amplitude of pitching,

$$h_2 = \alpha e^{i\theta}, \tag{11}$$

in which α is constrained by

$$\alpha \le \sin^{-1}\frac{h_1}{a} \qquad \text{for } 0 \le \frac{h_1}{h} \le \frac{1}{2}, \tag{12}$$

$$\alpha \le \sin^{-1}\frac{h-h_1}{a} \qquad \text{for } \frac{1}{2} \le \frac{h_1}{h} \le 1, \tag{13}$$

and θ is the phase of pitching with respect to the incident wave. The vertical velocity of the plate v_0 along the length of plate is

$$v_0 = -i\omega h_2 x e^{-i\omega t} \qquad \text{for } -a<x<a \text{ at } y=-h_1. \tag{14}$$

Applying the Fourier series, we have

$$v_0 = -i2\omega h_2 a\left[\sum_{s=1}^{\infty}\frac{\sin q_s a}{(q_s a)^2}\sin q_s x\right]e^{-i\omega t} \qquad \text{for } -a<x<a \text{ at } y=-h_1, \tag{15}$$

where

$$q_s = \frac{(2s-1)\pi}{2a} \qquad (s=1,2,3,\ldots). \tag{16}$$

It is assumed that the porous plate is made of material with very fine pores. Thus, the normal velocity of the fluid passing through the porous plate obeys Darcy's law and is linearly proportional to the difference between the hydrodynamic pressure on the lower surface of the plate, $p_3(x,-h_1,t)$, and that on the upper surface of the plate, $p_2(x,-h_1,t)$ (see Chwang, 1983),

$$V(x,t) = \frac{b}{\mu}(p_3 - p_2) \qquad \text{for } -a<x<a \text{ at } y=-h_1, \tag{17}$$

where b is a material constant having the dimension of length, and μ is the dynamic viscosity of the fluid in porous flow. It should be noted that b is always greater than zero for a porous plate. b vanishes only for an impermeable plate. In the linear theory, the hydrodynamic pressure is related to the velocity potential by the Bernoulli Equation as

$$p = -\rho\frac{\partial\Phi}{\partial t} = i\rho\omega\phi e^{-i\omega t}, \tag{18}$$

where ρ is the constant density of the fluid. By the substitution of Eq. (18) into Eq. (17), we obtain

$$V = i\sigma(\phi_3 - \phi_2)e^{-i\omega t} \qquad \text{at } y=-h_1, \tag{19}$$

where

$$\sigma = \rho b\omega/\mu. \tag{20}$$

The boundary condition on the plate surface is

$$\frac{\partial\Phi}{\partial y} = V + v_0. \tag{21}$$

After substituting Eqs. (15) and (19) into Eq. (21), we have

$$\frac{\partial\phi_2}{\partial y} = \frac{\partial\phi_3}{\partial y} = i\sigma(\phi_3 - \phi_2) - i2\omega h_2 a\sum_{s=1}^{\infty}\frac{\sin q_s a}{(q_s a)^2}\sin q_s x \qquad \text{at } y=-h_1. \tag{22}$$

The general solutions of spatial velocity potentials satisfying all the above boundary conditions are

$$\phi_1 = [I_0 e^{ik_0(x+a)} + R_0 e^{-ik_0(x+a)}]\cosh k_0(h+y) + \sum_{m=1}^{\infty}R_m e^{k_m(x+a)}\frac{\cos k_m(h+y)}{\cos k_m h} \tag{23}$$

$$\phi_2 = \sum_{n=0}^{\infty}\left(A_n\frac{\cos p_n x}{\cos p_n a} + B_n\frac{\sin p_n x}{\sin p_n a}\right) \times \sinh p_n(h-h_1)[gp_n\cosh p_n y + \omega^2\sinh p_n y] - i2\omega h_2 a^2\sum_{s=1}^{\infty}\left[\frac{\sin q_s a\,\sin q_s x}{(q_s a)^3(1-i\sigma Q_s)}\right]\frac{gq_s\cosh q_s y + \omega^2\sinh q_s y}{\omega^2\cosh q_s h - gq_s\sinh q_s h_1}, \tag{24}$$

$$\phi_3 = \sum_{n=0}^{\infty}\left(A_n\frac{\cos p_n x}{\cos p_n a} + B_n\frac{\sin p_n x}{\sin p_n a}\right) \times [\omega^2\cosh p_n h_1 - gp_n \sin gp_n h_1]\cosh p_n(h+y) - i2\omega h_2 a^2\sum_{s=1}^{\infty}\left[\frac{\sin q_s a\sin q_s x}{(q_s a)^3(1-i\sigma Q_s)}\right]\frac{\cosh q_s(h+y)}{\sinh q_s(h-h_1)}, \tag{25}$$

$$\phi_4 = T_0 e^{ik_0(x-a)}\frac{\cosh k_0(h+y)}{\cosh k_0 h} + \sum_{m=1}^{\infty}T_m e^{-k_m(x-a)}\frac{\cos k_m(h+y)}{\cos k_m h}, \tag{26}$$

where the eigenvalues k_0 and k_m are the roots of the dispersion equation

$$\omega^2 = gk_0\tanh k_0 h = -gk_m\tan k_m h \qquad (m=1,2,3,\ldots), \tag{27}$$

p_n satisfies the porous dispersion equation

$$p_n \sinh[p_n(h-h_1)][\omega^2 \cosh p_n h_1 - g p_n \sinh p_n h_1]$$
$$= i\sigma[\omega^2 \cosh p_n h - g p_n \sinh p_n h]$$
$$(n=0,1,2,...), \quad (28)$$

and

$$Q_s = \frac{\omega^2 \cosh q_s h - g q_s \sinh q_s h}{q_s \sinh q_s(h-h_1)(\omega^2 \cosh q_s h_1 - g q_s \sinh q_s h_1)}. \quad (29)$$

In order to determine the unknown constant coefficients R_m, T_m, A_n, and B_n (m,n=0,1,2,...), the general solutions are matched at $x = -a$ and $x = a$ by the matching conditions

$$\phi_1 = \phi_j, \; \frac{\partial \phi_1}{\partial x} = \frac{\partial \phi_j}{\partial x}, \quad \text{at } x = -a, \quad (30)$$

$$\phi_4 = \phi_j, \; \frac{\partial \phi_4}{\partial x} = \frac{\partial \phi_j}{\partial x}, \quad \text{at } x = a, \quad (31)$$

where $j = 2$ for $-h_1 < y < 0$ and $j = 3$ for $-h < y < -h_1$. By straightforward integration over the water depth h, it can be shown that the set of eigenfunctions

$\{w_m\}=\{\cosh k_0(h+y), \cos k_1(h+y), \cos k_2(h+y),...\}$ (32)

is an orthogonal set and

$$\int_{-h}^{0} \phi_j w_m dy = \int_{-h_1}^{0} \phi_2 w_m dy + \int_{-h}^{-h_1} \phi_3 w_m dy, \quad (33)$$

$$\int_{-h}^{0} \frac{\partial \phi_j}{\partial x} w_m dy = \int_{-h_1}^{0} \frac{\partial \phi_2}{\partial x} w_m dy + \int_{-h}^{-h_1} \frac{\partial \phi_3}{\partial x} w_m dy, \quad (34)$$

where $m = 0,1,2,...$, $j = 1$ at $x = -a$, and $j = 4$ at $x = a$.

For numerical computations, we have to truncate the above infinite series into a set of linear algebraic equations. For M evanescent wave modes taken into consideration in Eq. (23) and Eq. (26), a set of (4M+4) linear equations will be formulated in terms of unknowns R_m, T_m, A_n, and B_n (m=0,1,2,...,M and n=0,1,2,...,N), and the total number of unknowns is (2M+2N+4). The number of evanescent wave modes M is reasonably selected according to the required accuracy set by Yu and Chwang (1993), and then N must be equal to M.

Results and Discussion

The dimensionless reflection coefficient K_r and transmission coefficient K_t are defined as

$$K_r = |R_0 / I_0| \text{ and } K_t = |T_0 / I_0|, \quad (35)$$

respectively. The energy coefficient K^2 is defined as

$$K^2=K_r^2+K_t^2 \quad (36)$$

to indicate the energy state of the whole fluid domain after wave transformation due to the plate. The energy state of incident waves is unit by definition. $K^2=1$ indicates that there is no energy gain or loss, while $K^2<1$ and $K^2>1$ indicate that there is some energy loss and some energy gain, respectively. The complex pitching-effect parameter P_g was defined by Yip and Chwang (1996) as

$$P_g = \omega h_2 a^2 / I_0, \quad (37)$$

which is a direct measure of the ratio of the pitching effect to the incident waves. In the presence of incident waves, a small value of $|P_g|$ indicates that the pitching effect is negligible while the pitching effect dominates for large values of $|P_g|$. Another important parameter for the present study is the porous-effect parameter G defined by Chwang (1983) and Chwang and Dong (1984) as

$$G = \sigma / k_0 = \rho b \omega / \mu k_0, \quad (38)$$

which is a direct measure of the porous effect to the incident waves. G=0 means that the plate is impermeable. On the other hand, as G approaches infinity, the plate is completely permeable to fluid; that is, there would be no plate at all.

Figures 2 and 3 show respectively the variations of K_r and K_t versus G at $2a/\lambda=0.4$, $h/\lambda=0.25$, $h_1/h=0.3$, and $\theta=0$, with $|P_g|=0.0$, 0.5, 1.0, 1.5 and 2.0. We note from Figures 2 and 3 that for fixed $|P_g|$, K_r and K_t are nearly unchanged for small values of G. As G increases, K_r decreases monotonically to zero while K_t sways around one. The slope of K_r is steeper for larger values of $|P_g|$. A large value of G has a weak effect on the plate in suppressing the vertical velocity component of fluid so that the wave transformation due to pitching and scattering becomes mild. As G tends to infinity, K_r and K_t tend to zero and one, respectively, which indicates that full transmission occurs. A very large value of the porous-effect parameter ($G\to\infty$) implies that the plate becomes transparent to the fluid and, consequently, there is no wave transformation any more.

As discussed by Yu and Chwang (1994), the loss of energy is affected by two factors, the porosity and the fluid discharge through the porous medium. The maximum dissipation of incident wave energy is found where the porosity is moderately large. This is why there is a rapid change of K^2 about G=1 as shown in Figure 4 which reflects the variation of K^2 versus G at $2a/\lambda=0.4$, $h/\lambda=0.25$, $h_1/h=0.3$, and $\theta=0$, with $|P_g|=0.0$, 0.5, 1.0, 1.5 and 2.0. Before this rapid change around unit G, K^2 decreases as G increases. The valleys of K^2 curves occur close to G=1, which agrees with Yu and Chwang's (1994) discovery on the maximum dissipation of incident waves. For G>1, K^2 approaches one, and there is no energy gain or loss in the fluid domain. Meanwhile, the pitching and scattering of the plate quickly lose their influences on the wave transformation.

Figures 5, 6 and 7 show the variations of K_r, K_t and K^2 versus θ at $2a/\lambda=0.4$, $h/\lambda=0.25$, $h_1/h=0.3$, and $|P_g|=1.0$, with G=0, 1, 2, 5 and 20. For an impermeable plate (G=0), the reflected and transmitted waves are found by Yip and Chwang (1996) to reach their maxima or minima, respectively, when the pitching is in phase or π out of phase with respect to incident waves. Here K_r and K_t show the shift of their maxima and minima with increasing G and, as a result, the peak of K^2 varies with different values of G. It implies that the porosity of the plate has adjusted the phase difference between radiation and scattered waves, and a higher porosity leads to a greater phase shift. To minimize the energy state of the whole system is no longer simply by setting the pitching and incident waves to be π out of phase. In other words, the choice of phase of pitching highly depends on the porous-effect parameter G.

Figures 8 and 9 illustrate the variations of K_r and K_t versus $|P_g|$ at $2a/\lambda=0.4$, $h/\lambda=0.25$, $h_1/h=0.3$, and $\theta=0$, with G=0, 1, 2, 5 and 20. We note from Figure 8 that K_r increases linearly with G, but the rate of its increase becomes moderate when G becomes large. On the other hand, the behaviour of K_t versus G is the resultant of two waves, namely, the transmitted wave due to the plate without pitching and the radiation wave generated by the pitching plate. With increasing G, the plate becomes more transparent so that the height of radiation waves decreases and that of transmitted wave approaches one. Thus, pitching and scattering effects disappear and K_t tends to one with only a minor change due to pitching.

Figures 10 and 11 demonstrate the variations of K_r and K_t versus $2a/\lambda$ at $h/\lambda=0.25$, $h_1/h=0.3$, $|P_g|=1$, and $\theta=0$, with G=0, 1, 2, 5 and 20. The case of G=0 agrees with the former results of Yip and Chwang (1996). This shows that the solutions developed in this paper reduce to those for an impermeable plate when the porous-effect parameter G vanishes. As $2a/\lambda$ tends to zero, K_r and K_t tend to zero and one respectively. It is because $2a/\lambda=0$ means the length of the plate is zero with respect to the wavelength, so that the plate is transparent to wave and wave transformation disappears. Like an impermeable plate, K_r and K_t fluctuate as $2a/\lambda$ increases. Although the variations of K_r and K_t diminish as G increases, they are still sensitive to the dimensionless plate length $2a/\lambda$.

Figures 12 and 13 show the variations of K_r and K_t versus h_1/h at $2a/\lambda=0.4$, $h/\lambda=0.25$, $|P_g|=1$, and $\theta=0$, with G=0, 1, 2, 5 and 20. We note from these two figures that K_r and K_t approach zero and one respectively as h_1/h ascends from zero to one. When the dimensionless submergence is set to one, the plate is located on to the sea bottom and the plate loses its influence on wave transformation. Against the variation of relative submergence, the peaks of K_r and K_t indicate the strong effects of pitching and scattering but the peaks disappear as the porosity increases. For a porous plate, the variations of K_r and K_t become moderate and the dimensionless submergence h_1/h has less importance. In comparison with other parameters, the relative water depth h/λ has a little effect on the variations of K_r and K_t except when h/λ is very small.

CONCLUSIONS

Active control is accomplished by adjusting the magnitude and phase of pitching so that a pitching plate generates desired radiation waves to modify the scattered waves around it. Porosity of the plate enhances the energy dissipation of incident waves. The effect of porosity also diminishes the pitching and scattering effects on the wave transformation. A small value of the porous-effect parameter is found to be optimal to dissipate incident wave energy. Further increase of the porous-effect parameter contributes little on controlling the wave transformation. Meanwhile, pitching has no effect on the phases of radiation and scattered waves, but porosity shifts the phases of radiation and scattered waves and makes the control more complicated. Careful control on pitching is definitely required to adapt the porous effect.

ACKNOWLEDGMENTS

This research was sponsored by the Hong Kong Research Grants Council under Grant No. HKU 294/93E.

REFERENCES

Chwang, A.T. (1983). "A porous-wavemaker theory," *J Fluid Mech*, Vol 132, pp 395-406.

Chwang, A.T. and Dong, Z.H. (1984). "Wave trapping due to a porous plate," *Proc of the 5th ONR symp naval hydrodynamics, Washington DC.*

Chwang, A.T. and Wu, J. (1994). "Wave scattering by submerged porous disk," *J Engrg Mech, ASCE*, Vol 120, pp 2575-2587.

Milgram, J.H. (1970). "Active water-wave absorbers," *J Fluid Mech*, Vol 42, pp 845-859.

Yip, T.L. and Chwang, A.T. (1996). "A pitching plate as an active water-wave controller," *Proc 6th Int Offshore and Polar Engrg Conf, Los Angeles, USA, ISOPE*, Vol 3, pp 408-413.

Yu, X. and Chwang, A.T. (1993). "Analysis of wave scattering by submerged circular disk," *J Engrg Mech, ASCE*, Vol 119, pp 1804-1817.

Yu, X. and Chwang, A.T. (1994). "Water waves above submerged porous plate," *J Engrg Mech, ASCE*, Vol 120, pp 1270-1282.

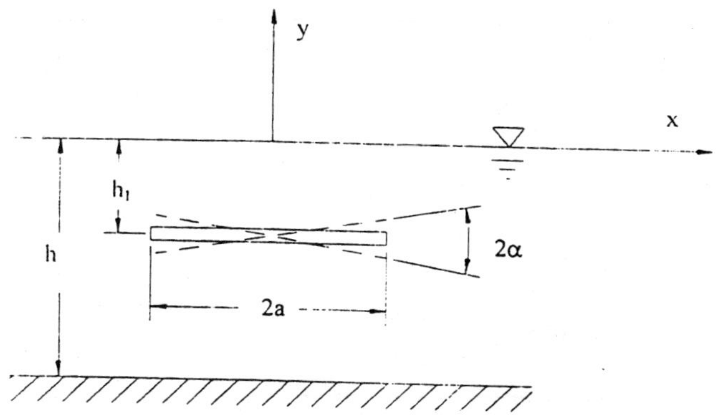

Fig. 1 Schematic diagram of a submerged pitching porous plate.

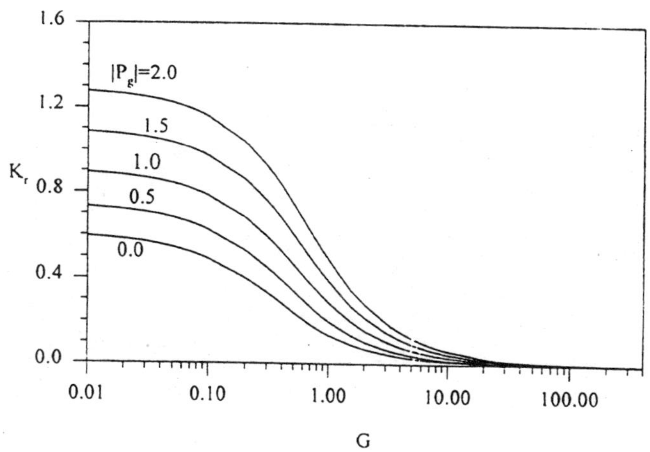

Fig. 2 Variation of K_r versus G at $2a/\lambda=0.4$, $h/\lambda=0.25$, $h_1/h=0.3$, and $\theta=0$.

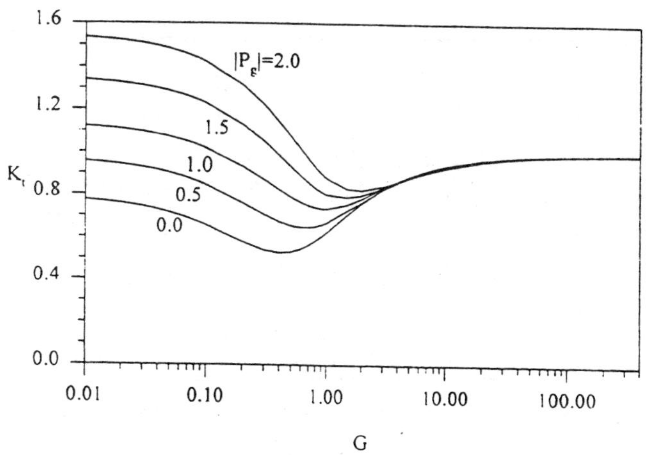

Fig. 3 Variation of K_t versus G at $2a/\lambda=0.4$, $h/\lambda=0.25$, $h_1/h=0.3$, and $\theta=0$.

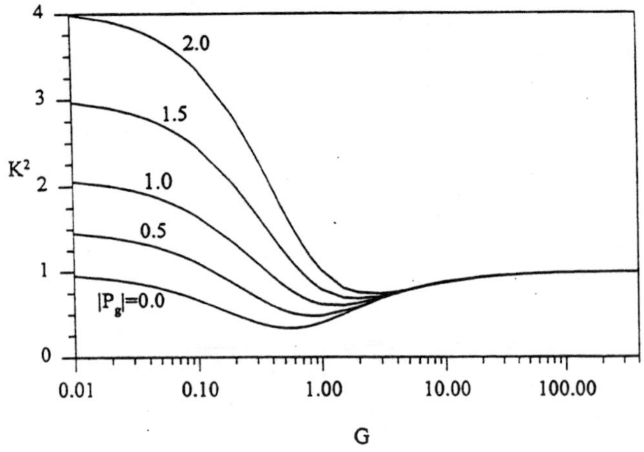

Fig. 4 Variation of K^2 versus G at $2a/\lambda=0.4$, $h/\lambda=0.25$, $h_1/h=0.3$, and $\theta=0$.

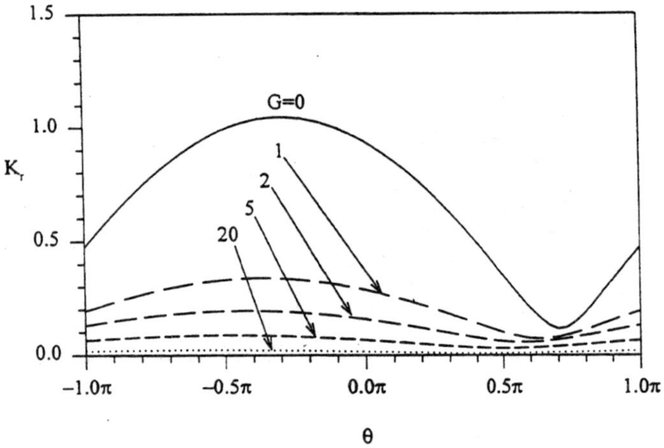

Fig. 5 Variation of K_r versus θ at $2a/\lambda=0.4$, $h/\lambda=0.25$, $h_1/h=0.3$, and $|P_g|=1$.

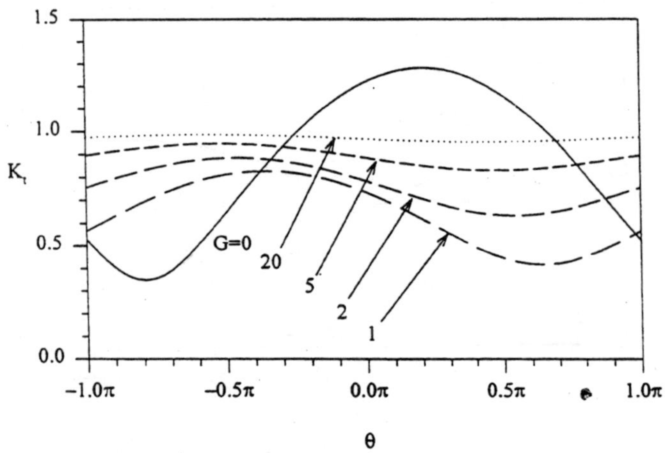

Fig. 6 Variation of K_t versus θ at $2a/\lambda=0.4$, $h/\lambda=0.25$, $h_1/h=0.3$, and $|P_g|=1$.

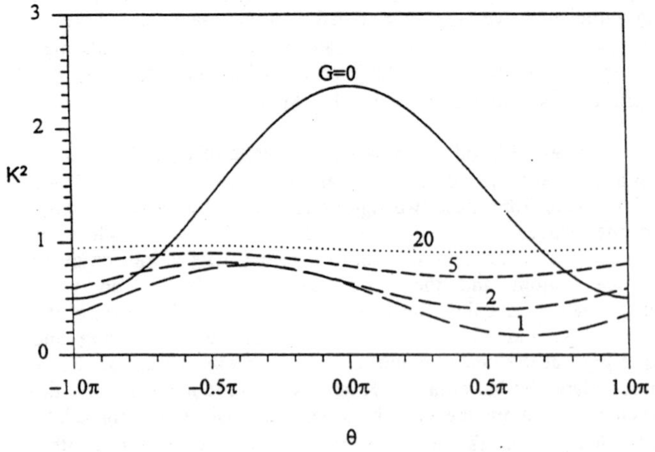

Fig. 7 Variation of K^2 versus θ at $2a/\lambda=0.4$, $h/\lambda=0.25$, $h_1/h=0.3$, and $|P_g|=1$.

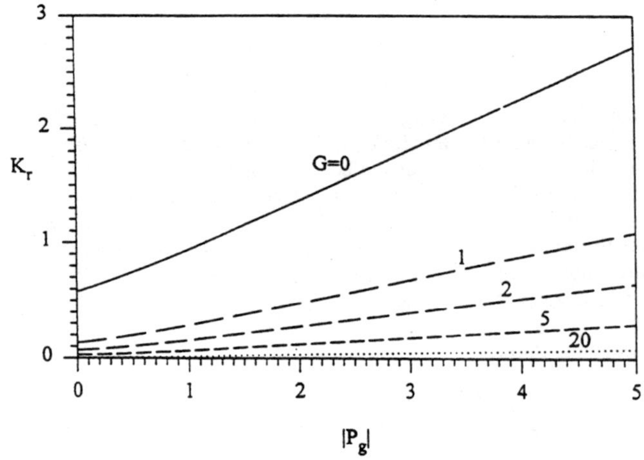

Fig. 8 Variation of K_r versus $|P_g|$ at $2a/\lambda=0.4$, $h/\lambda=0.25$, $h_1/h=0.3$, and $\theta=0$.

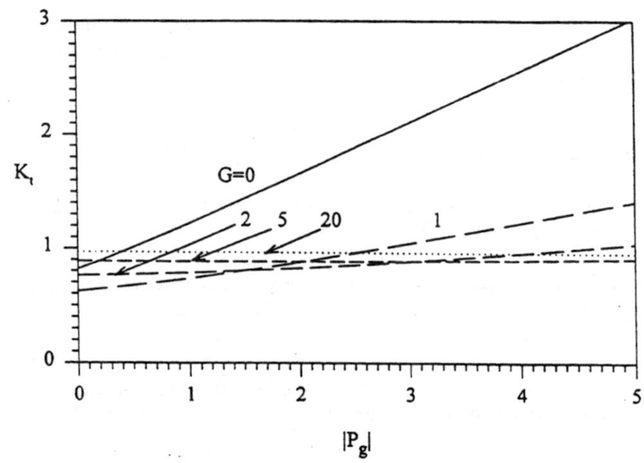

Fig. 9 Variation of K_t versus $|P_g|$ at $2a/\lambda=0.4$, $h/\lambda=0.25$, $h_1/h=0.3$, and $\theta=0$.

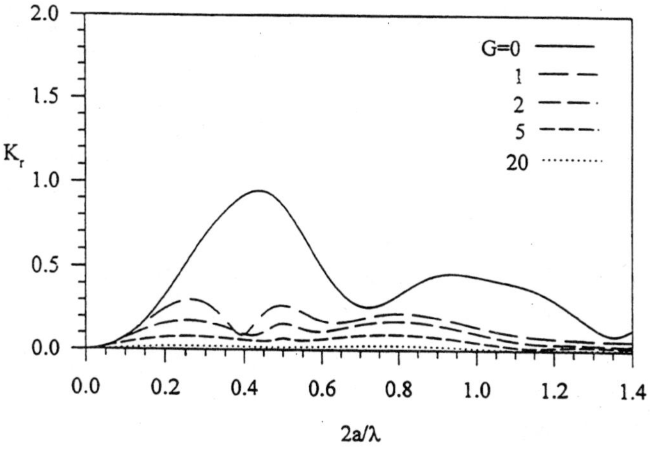

Fig.10 Variation of K_r versus $2a/\lambda$ at $h/\lambda=0.25$, $h_1/h=0.3$, $|P_g|=1$, and $\theta=0$.

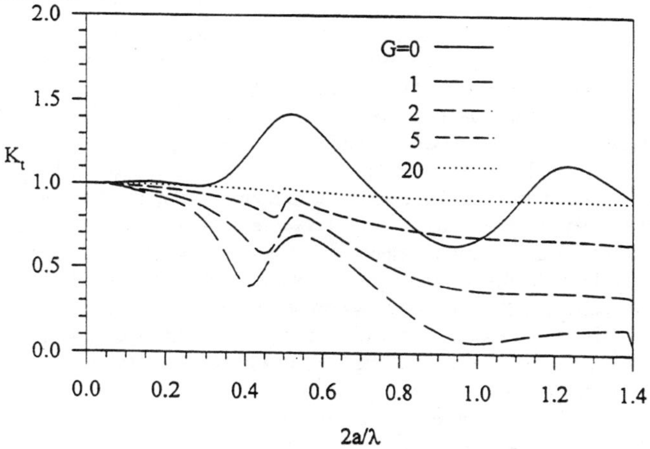

Fig. 11 Variation of K_t versus $2a/\lambda$ at $h/\lambda=0.25$, $h_1/h=0.3$, $|P_g|=1$, and $\theta=0$.

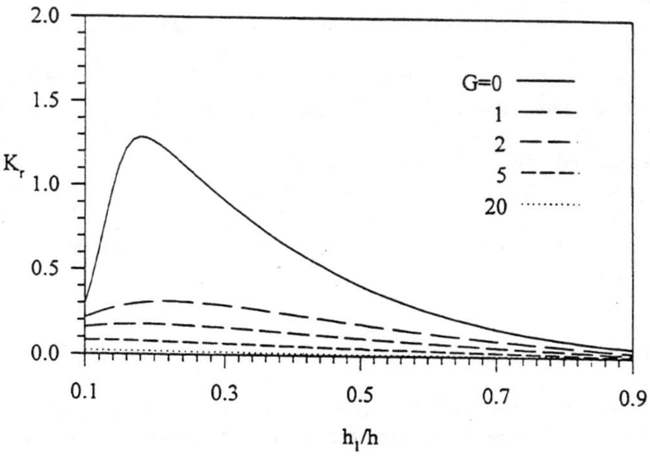

Fig. 12 Variation of K_r versus h_1/h at $2a/\lambda=0.4$, $h/\lambda=0.25$, $|P_g|=1$, and $\theta=0$.

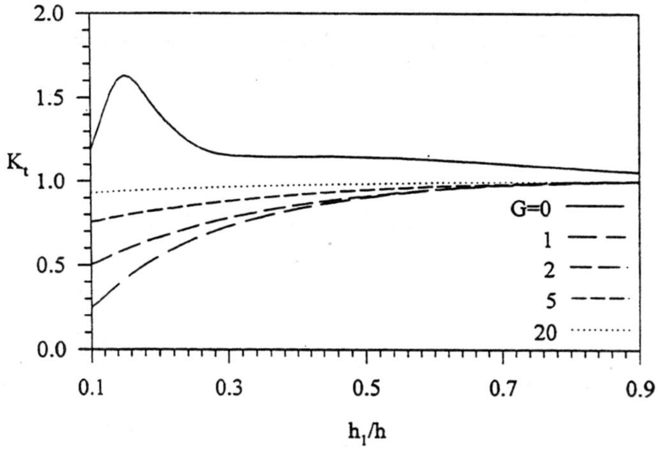

Fig. 13 Variation of K_t versus h_1/h at $2a/\lambda=0.4$, $h/\lambda=0.25$, $|P_g|=1$, and $\theta=0$.

Proceedings of the Seventh (1997) International Offshore and Polar Engineering Conference
Honolulu, USA, May 25-30, 1997

ISBN 1-880653-28-1 (Set); ISBN 1-880653-31-1 (Vol. III)

A Perturbation-DRBEM Model for Wave Refraction-Diffraction

S.S. Hsiao and J.H. Wu
National Taiwan Ocean University
Keelung, Taiwan, China

Y.F. Chiu
Institute of Harbour and Marine Technology
Taichung, Taiwan, China

ABSTRACT

A numerical model using perturbation technique and dual reciprocity boundary element method (DRBEM) is presented to investigate wave refraction-diffraction around an island or a offshore structure on a slowly varying topography. Using perturbation technique on the mild-slope equation, one can obtain an ordinary Helmholtz equation for constant depth and a corresponding inhomogeneous Helmholtz equation for the variable depth. The ordinary Helmholtz equation is easily solved by a regular BEM. Applying BEM to the inhomogeneous Helmholtz equation, the inhomogeneous part will result in a domain integral. This domain integral may be calculated by dividing the domain into a number of cells and by applying the DRBEM, and thus taking to the boundary. The solution of the problem in real depth is considered as a perturbation of the solution of the problem in constant depth. Therefore, once the solution of the case of flat bottom has been computed, the solution of variable depth can be easily obtained. The model is verified against the analytical solution for a circular island on a paraboloidal shoal by Homma(1950), and good agreements are obtained.

KEY WORDS: DRBEM, perturbation, wave refraction-diffraction.

INTRODUCTION

In the past few decades, various numerical and analytical methods have been proposed, to model the refraction and diffraction of water waves around ocean structures. Berkhoff(1972) proposed the well known mild-slope equation that simplified the three dimensional problems to a two dimensional one. Later Smith and Sprinks (1975) gave a formal derivation of the mild-slope equation. The mild-slope equation governs both refraction and diffraction. The proposed equation reduces to the well known "eikonal" equation. Bettess and Zienkiewicz (1976) simplified the equation to diffraction problems. Similarly, Mei(1978) also simplified the equation into diffraction problems. Based on "Galerkin-eigenfunction" method, Massel (1993) extended the mild-slope equation for cases where the mean bottom slope is not necessarily "mild".

As for numerical approach on mild-slope equation, Madsen and Larsen(1987) solve the equation based on finite difference method. Panchang and Pearce(1991) proposed a finite difference model based on iterative procedure designed for the mild-slope equation. Li and Anatasiou(1992) proposed a multigrid technique to other finite deffirence method. However, all the finite difference models have a common drawback, the boundary of a computational domain cannot be realistically modeled, as a results of the poor discretization of a boundary. Some recent developed techniques divided the domain into two regions. In the inner regions, finite elements are placed. In the outer region, different numerical treatment are applied. Chen and Mei(1974) first proposed an analytical eigenfunction expansion in the outer region. Followed by Bai and Yeung(1974), and Houston(1981), similar models are proposed. Tsay and Liu(1983) proposed a two-dimensional hybrid finite element method to study wave refraction and diffraction. Those model proposed are placed with a large number of finite elements in inner domain. Thus, the computer storage and time were formidable.

The BEM only requires a discretization on the boundary of a computational domain and it has been widely applied in engineering practice. Rangogni(1988) presented a model based on perturbation method with BEM. The model solved complicated analytical solutions on mild-slope equation. However, when application of the method to inner domain problems with varying topography, BEM seems incomplete, because integration in the inner regions become complicated. To overcome the difficulties associated with singular integrals, the dual reciprocity boundary element method(DRBEM) was first proposed by Nardini and Brebbia(1982). In this paper, A DRBEM model is proposed to overcome the deficiency in the model developed by Rangogni(1988).

GOVERNING EQUATION AND BOUNDARY CONDITIONS

Considering the structure placed on a slowly varying topography, the waves refraction-diffraction are induced by topography and ocean structure. The governing equation for the diffracted and refracted wave field is the mild-slope equaton, illustrated as followed:

$$\nabla \cdot (cc_g \nabla \phi) + \omega^2 \frac{c_g}{c} \phi = 0 \tag{1}$$

where, $\nabla = (\frac{\partial}{\partial x}, \frac{\partial}{\partial y})$, ϕ is the velocity potential, $c = \frac{\omega}{k}$ is the wave celerity, c_g is the group celerity ($c_g = nc$), $n = \frac{1}{2}(1 + \frac{2kh}{\sinh 2kh})$, k is the wave number and ω is the angular frequency, and satisfy the following dispersion relation:

$$\omega^2 = gk \tanh kh \tag{2}$$

As shown in Fig.1, the calculated domain in this paper is divided into two parts. An inner domain D_1 with a variable depth, and an semi-infinite outer domain D_0 having a constant water depth. In the outer domain D_0 where the water depth is constant ($h = h_0$). Thus, c, c_g and k are constants. The wave have no refraction and wave length does not changes. Thus, the governing equation can be rewritten as Helmholtz equation, and illustrated as followed:

$$\nabla^2 \phi + k^2 \phi = 0 \tag{3}$$

where, ϕ can be expressed as, the velocity potential function for incident wave (ϕ^i) and velocity potential for scattered wave(ϕ^s). Thus, $\phi = \phi^i + \phi^s$. As for the infinite far-field boundary S_∞, the velocity potential for scattered wave(ϕ^s) must satisfy far-field radiation conditions as follow:

$$\lim_{r \to \infty} \sqrt{r} (\frac{\partial}{\partial r} - ik) \phi^s = 0 \tag{4}$$

where, $r = \sqrt{x^2 + y^2}$ is the distance function, $i = \sqrt{-1}$ as the complex variable.

In the inner domain D_1, the wave field around the structure is governed by Eq.(1), the mild slope equation. Assumed the structure is solid and impermeable material, the surface of the structure S_1 must satisfy the following conditions:

$$\frac{\partial \phi_1}{\partial n} = 0 \quad , \quad on \ S_1 \tag{5}$$

where, n is the normal vector on the structure boundary S_1.

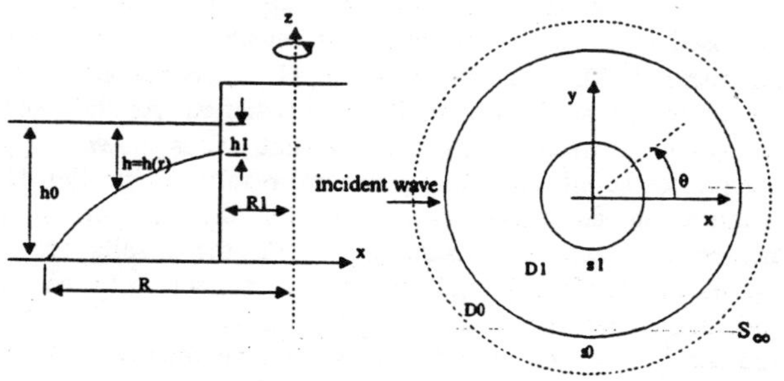

Fig.1 Definition Sketch of Computational Domain.

NUMERICAL METHODS

Perturbation Method

It is known that in the case of a flat bottom h=h_0, the refraction-diffraction problem becomes a pure diffraction one, governed by the Helmholtz equation, and can be solved by BEM with the desired accuracy. Therefore, assumed that h_1 is not too different from the depth h_0, the solution (ϕ_1) corresponding to the mild-slope topography is not too different from the solution corresponding to the constant depth.

It may be concluded that solution ϕ_2 for a fictitious depth h_2, where $h_1 \leqq h_2 \leqq h_0$, can be found linearly from ϕ_1 and ϕ_0.

$$\phi_2 = \phi_0 + \varepsilon(\phi_1 - \phi_0) \tag{6}$$

where, ε is the perturbation parameter, $0 \leqq \varepsilon \leqq 1$. The same goes for the phase and group velocities, which are expressed as follow:

$$c = c_0 + \varepsilon(c_1 - c_0) \tag{7}$$

$$c_g = c_{g0} + \varepsilon(c_{g1} - c_{g0}) \tag{8}$$

The Eq.(6), Eq.(7) and Eq.(8) are substitute into Eq.(1), and a polynomial equation in ε is obtained in which only zero'th and first order terms are retained. This equation must be true for all $\varepsilon \in [0,1]$, if and only if the coefficients of zore'th and first order of ε are both identically zero. After dividing by $c_0^2 c_{g0}$, the following equations are obtained:

$$\nabla^2 \phi_0 + k_0^2 \phi_0 = 0 \quad , k_0^2 = \frac{\omega^2}{c_0^2} \tag{9}$$

$$\nabla^2 \phi_1 + k_0^2 \phi_1 = 2k_0^2 \phi_0 (\frac{k_0}{k_1} - 1) - (\frac{\nabla c_{g1}}{c_{g0}} + \frac{\nabla c_1}{c_0}) \cdot \nabla \phi_0 \tag{10}$$

Eq.(9) is the ordinary Helmholtz equation for flat bottom problems and is similar to the governing equation in the constant depth (D_0) region. Eq.(10) is an inhomogeneous Helmholtz equation for the wave velocity potential ϕ_1 corresponding to the varying topography region(D_1). It is noted that the right-hand side of this equation is only a function of $h(x, y)$, and the constant depth potential ϕ_0. The solution ϕ_0 for Eq.(9) can be easily solved using BEM (Au and Brebbia, 1998). Once ϕ_0 has been computed, the potential ϕ_1 can be solved by Eq(10).

DRBEM

It is known that if BEM is applied to innomogeneous equation like Eq.(10), the inhomogeneous part will result in a domain integral. In this paper, DRBEM is applied to calculate this domain integral. First, the weighting function ϕ^* with constant depth for diffraction problems is considered, and is chosen as follow:

$$\phi^* = -\frac{i}{4} H_0^{(1)}(k_0 r) \tag{11}$$

where, r is the distance between a source point and a field point, $H_0^{(1)}(k_0 r)$ is the Hankel function of the first kind of zero order.

By applying the Green's identity, Eq.(10) is transform and expressed as:

$$C_{1i}\phi_{1i} - \int_s \phi_1 \frac{\partial \phi^*}{\partial n} ds + \int_s \phi^* \frac{\partial \phi_1}{\partial n} ds = \iint_{D_1} \phi^* b dA \tag{12}$$

where, $S = S_1 + S_0$, b is inhomogeneous part. Based on DRBEM theory(Partridge, Brebbia & Wrobel, 1992), an approximated assumption is made. Thus, let:

where, $S=S_1+S_0$, b is inhomogeneous part. Based on DRBEM theory(Partridge, Brebbia & Wrobel, 1992), an approximated assumption is made. Thus, let:

$$b=\sum_{k=1}^{N+L}\alpha_k f_k \tag{13}$$

where, N is the number of nodes on the boundary, L is the number of internal points, α_k are the coefficients to be determined. f_k are the expansion function, and can be expressed by the distance function r_k, from the source point to a specified point, and can be illustrated as:

$$f_k=1+r_k \tag{14}$$

If there exist a function $\hat{\phi}$ that satisfy (Partridge & Brebbia, 1990):

$$\nabla^2\hat{\phi}_k+k_0^2\hat{\phi}_k=f_k \tag{15}$$

Thus, Eq.(13), Eq.(14) and Eq.(15) substitute into Eq.(12), and can be rewritten as:

$$C_{1i}\phi_{1i}-\int_s\phi_1\frac{\partial\phi^*}{\partial n}ds+\int_s\phi^*\frac{\partial\phi_1}{\partial n}ds$$
$$=\sum_{k=1}^{N+L}\alpha_k[C_{2i}\hat{\phi}_{ik}-\int_s\hat{\phi}_k\frac{\partial\phi^*}{\partial n}ds+\int_s\phi^*\frac{\partial\hat{\phi}_k}{\partial n}ds] \tag{16}$$

According to the continuous properties of the fluids, and denoted on the assumed boundary S_0, the boundaries conditions are shown as follow:

$$\begin{cases}\phi_1=\phi^s+\phi^i\\ \frac{\partial\phi_1}{\partial n}=-(\frac{\partial\phi^s}{\partial n'}+\frac{\partial\phi^i}{\partial n'})\quad,\quad \vec{n}'=-\vec{n}\end{cases} \tag{17}$$

where, $\vec{n}'$ is the outward normal vector in the calculated domain D_0. Thus Eq.(16) can be rewritten as:

$$(C_{1i}+C_{0i})\phi_{1i}-\int_{s_1}(\phi_1\frac{\partial\phi^*}{\partial n}-\phi^*\frac{\partial\phi_1}{\partial n})ds-\int_{s_0}(\phi^i\frac{\partial\phi^*}{\partial n}-\phi^*\frac{\partial\phi^i}{\partial n})ds$$
$$=\sum_{k=1}^{N+L}\alpha_k[C_{2i}\hat{\phi}_{ik}-\int_s\hat{\phi}_k\frac{\partial\phi^*}{\partial n}ds+\int_s\phi^*\frac{\partial\hat{\phi}_k}{\partial n}ds] \tag{18}$$

Note that in the above equation, the point i is located in the calculated domain D_1, and boundaries S_1 and S_0. For the discretization of Eq.(18), constant elements on the boundary of domain D_1 are adopted. The boundary S_1 and S_0 are discretized into N_1 and N_0 constant elements, and let the length of the jth element be S_j. Eq.(18) is to be satisfied at all points on the boundary S_1 and S_0. Therefore, N, where $N=N_0+N_1$, linear algebraic equations are obtained, and is expressed in matrices as:

$$-[H_{ij}][\phi_{ij}]+[G_{ij}][q_{ij}]-[\bar{H}_{ij}][\phi_j^i]+[\bar{G}_{ij}][q_j^i]$$
$$=\sum_{k=1}^{N+L}\{-[H_{ij}][\hat{\phi}_{jk}]+[G_{ij}][\hat{q}_{jk}]\}[\alpha_k] \tag{19}$$

where, $H_{ij}=\int_{s_j}\frac{\partial}{\partial n}(-\frac{i}{4}H_0^{(1)}(k_0r))ds$,

$G_{ij}=\int_{s_j}(-\frac{i}{4}H_0^{(1)}(k_0r))ds$,

$q_{ij}=\frac{\partial}{\partial n}\phi_{ij}$, and $\alpha_k=[f_{ik}]^{-1}\cdot b$.

By solving the linear algebraic equations in Eq.(19), the wave potential in the boundary S_0 around the structure can be obtained. Using the wave potential on the boundaries, the wave potential in the domain D_1 at any point can be worked out.

COMPUTATIONAL RESULTS

Calculated Conditions

In this paper, based on perturbation method and DRBEM, the refraction and diffraction problems in wave are discussed. Homma's (1950) analytical examples for the interaction of waves with a circular island on a paraboloidal shoal (as shown in Fig.1, $h(R)=R_1(h/h_1)^2$), have been adopted as cases discussed in this paper. For the chosen computational examples, the topography is assumed to satisfy mild-slope equation. So far, Homma's (1950) analytical solution has been. the most popularly adopted case in a verification process (Berkhoff, 1972; Tsay and Liu, 1983; Rangogni, 1988). The varying water depth is one of the important factors influencing the wave transformation. Therefore, in this paper, three geometry(A, B and C) is considered. The wave period T=240sec, T=480sec, T=720sec and T=1440sec are considered as the conditions for incident waves. In all the cases, the radius(R_1) for the circular island is adopted as 10000m. The h_0 for three geometries is considered as 4000m. As for the water depth(h_1) around the island, geometry A is 3000m, geometry B is 1333.33m and geometry C is 1000m.

Model Verification and Calculations

To verify this numerical model, total of 4 incident periods (T=240sec, 480sec, 720sec, 1440sec) and 3 different values of the shoreline depth h_1 are calculated. The numerical results are compared with analytical solution (Homma, 1950) and other numerical results (Rangogni, 1988). The relative amplitudes (defined as the resulting wave amplitude divided by the incident wave amplitude in a given point) are compared with the analytical results on the boundary, in which are given in Fig.2 to Fig.4. It is demonstrated that the numerical results have a good agreement compared with analytical results. However, the results of the larger water depth ratio h_0/h_1=4.0, corresponding to the shorter wave period have some deficiency compared with analytical results.

Fig.5, shows the comparison between numerical results in the proposed model, with the analytical results and those obtained by Rangogni(1988). For geometry A(h_0/h_1=1.33) correspond to T=480sec, the compared results are similar and satisfied. As the water depth ratio(h_0/h_1) increase to 4.0, in the geometry C(as shown in Fig.6), the comparison results show that numerical results are smaller than those obtained in analytical results. As for the comparison between numerical results with Rangogni(1988), the results are in good agreement. Fig.7 and Fig.8 are the comparison between numerical results and analytical results along the sections of θ=0°,90°,180°, over the shoal, with geometry A and B corresponding to T=240sec. The numerical results are very close to the analytical ones.

As shown in the above results, the water depth variation parameter(h_0/h_1) is one of the factor that influences the accuracy of the model proposed. As the parameter is less than 3.0, the results in the model proposed provides satisfactory and accurate results. This is close to the conclusion obtained by Rangogni(1988). This occurs because that for solving the wave potential in the slowly varying topography region, perturbation method is adopted.

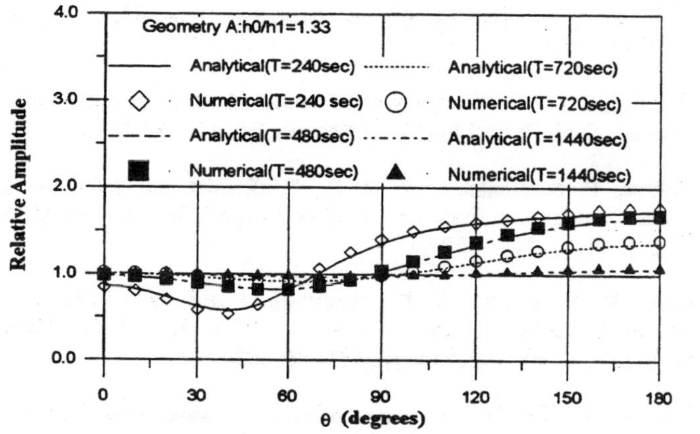

Fig.2 Relative amplitude along the boundary of the circular island (geometry A).

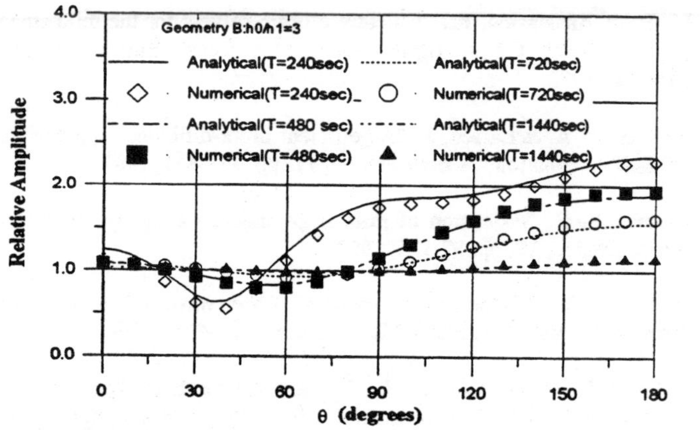

Fig.3 Relative amplitude along the boundary of the circular island (geometry B).

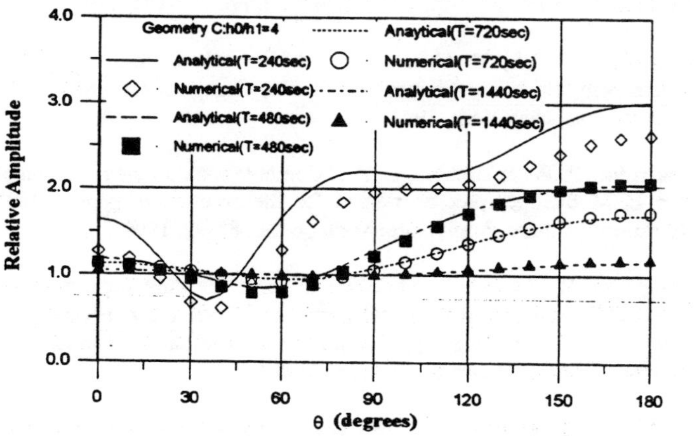

Fig.4 Relative amplitude along the boundary of the circular island (geometry C).

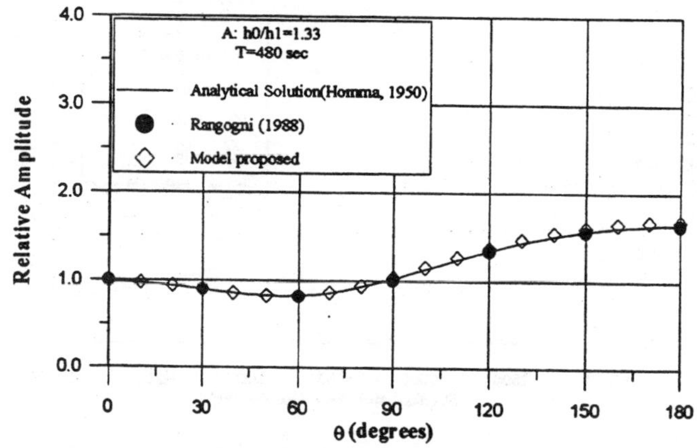

Fig.5 Comparison between analytical results(Homma,1950), Rangogni (1988) and the result of this model on relative amplitude along the boundary of the circular island (geometry A).

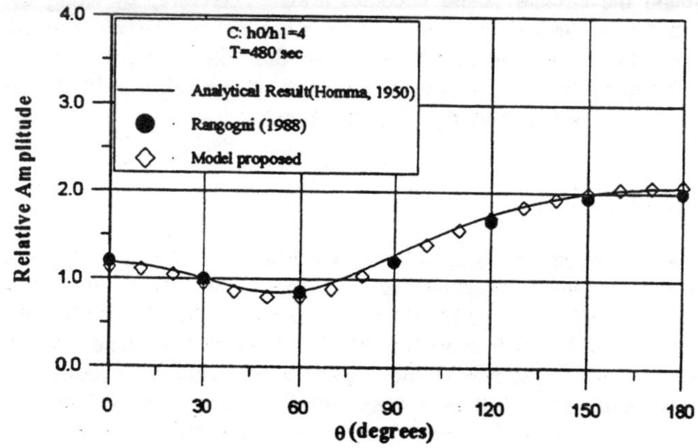

Fig.6 Comparison between analytical results(Homma,1950), Rangogni (1988) and the result of this model on relative amplitude along the boundary of the circular island (geometry C).

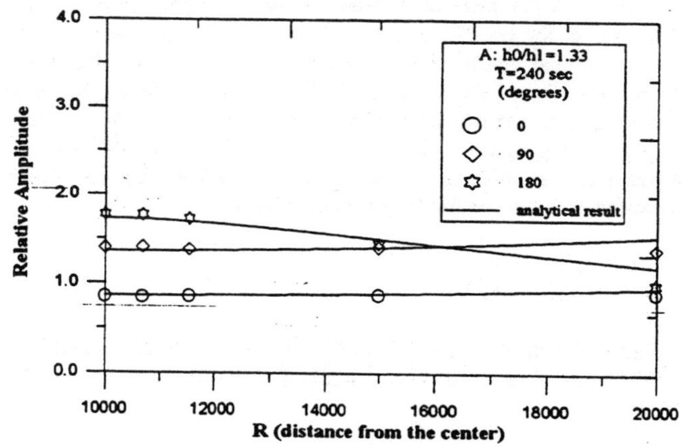

Fig.7 Comparison between analytical results(Homma,1950) and the result of this model on relative amplitude for sections (θ=0°,90°,180°) of the circular island (geometry A)

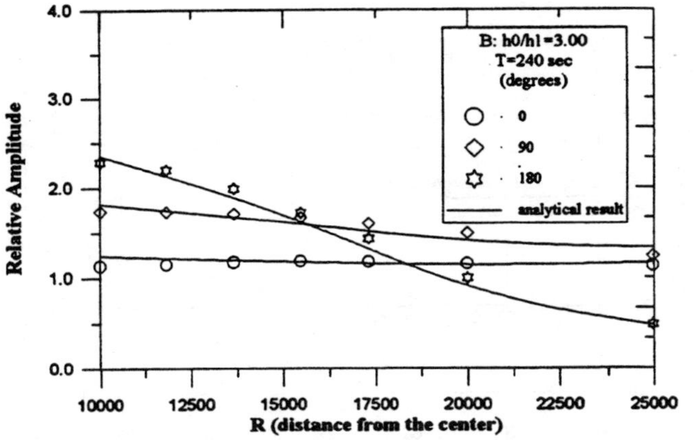

Fig.8 Comparison between analytical results(Homma,1950) and the result of this model on relative amplitude for sections (θ=0°,90°,180°) of the circular island (geometry B)

The ratio between h_0 and h_1 become larger, the shore line depth around the circular island becomes lower. Therefore, by using the perturbation method, the real conditions can not be fairly discussed. From the discussion above, it shows that the numerical model proposed seems to work satisfactory when the variation of the water depth is not very high ($h_0/h_1 \leq 3$ in this study).

CONCLUSIONS

A Perturbation-DRBEM(dual reciprocity boundary element method) model is proposed and developed for solving the wave refraction-diffraction problems. Comparisons are made between the proposed model with analytical results based on the same case by Homma(1950). The results show satisfy agreements. The results are also compared in good agreement with those obtained by Rangogni (1988). It shows that the validity of the model proposed.

It is also demonstrated that the numerical error relates to analytical solution increases with decreasing wave period and increasing depth ratio. For the depth ratio h_0/h_1 >3.0 and wave period T=240 sec, the numerical results and the analytical results show some disagreement. A maximun depth ratio of 3 seems to be the upper limit for the application of the present model. It should not a severe restriction in regard to similar coastal structures a depth ratio of 2 is rather high. Besides, the aim of this paper is to proposed a simple technique to reduce the system matrix which compared with direct BEM solutions of the governing equation. Therefore, the perturbation-DRBEM model proposed in this paper, is recommended to solve the wave refraction-diffraction problems within the indicated limits.

ACKNOWLEDGMENT

This research was supported by National Science Council of R.O.C., under Grant No. NSC-85-2611-E-019-003, and was greatly appreciated.

REFERENCES

Au, M. C. & Brebbia, C. A., "Diffraction of water waves for vertical cylinders using boundary elements" , Appl. Meth. Model., 7, pp. 106-114, 1983.

Bai, K. J. & Yeung, R., "Numerical solution of free-surface problems" , Proc.10th Symp.Naval Hydrodyn., Cambridge, MA, pp. 609-647, 1974.

Berkhoff, J. C. W., "Computation of combined refraction-diffraction" , 13th Int. Conf. Coastal Engng., Vancouver, pp. 471-490, 1972.

Bettess, P. & Zienkiewiccz, O. C., "Diffraction and refraction of surface waves using finite and infinite elements" , Intr. J. Num. Meth. Engng., 11, pp. 1271-1290, 1976.

Chen, H. S. & Mei, C. C., "Oscillations and wave forces in a man-made harbor in the open sea" , Proc. 10th Symp. Naval Hydrodyn., Cambridge, MA, pp. 573-596, 1974.

Homma, S., "On the behavior of seismic sea waves around circular island", Geophys. Mag., 1950, 21, pp.231-233.

Houston, J. R., "Combined refraction and diffraction of short waves using the finite element method" , Appl. Ocean Res., 3, pp. 163-170, 1981.

Li, B. & Anastasion, K., "Efficient elliptic solvers for the mild-slope equation using the multigrid technique" , Coast. Engng., 16, pp. 245-266, 1992.

Madsen, P. A. & Larsen, J., "An efficient finite-difference approach to mild-slpoe equation" , Coast. Engng., 11, pp. 329-351, 1987.

Massel, S. R. "Extension of mild-slope model for surface waves" , Coast. Engng., 19, pp. 97-126, 1993.

Mei, C. C., "Numerical method in water-wave diffraction and radiation" , Ann. Rev. Fluid. Mech., 10, pp. 393-416, 1978.

Nardini, D. & Brebbia, C. A., "A new approach to free vibration analysis using boundary elements" , Boundary Element Methods in Engineering, ed. C. A. Brebbia, Springer-Verlag, Berlin, New York, 1982.

Panchang, V. G. & Pearce, B. R., "Solution of the mild-slope wave problem by iteration" , Appl. Ocean Res., 3, pp. 187-199, 1991.

Partridge, P. W., Brebbia, C. A. & Wrobel, L. C., "The Dual Reciprocity Boundary Element Method" , Computational Mechanics Publications, New York, 1992.

Partridge, P. W. & Brebbia, C. A., "Computational implementation of the BEM dual reciprocity method for the solution of general field equation" , Comm. Appl. Num. Meth., 6, pp. 83-92, 1990.

Rangogni, R., "A simple procedure to solve the mild-slope equation using BEM and perturbation technique-part I" , Boundary Elements X, vol. 1, ed. C. A. Brebbia, Computational Mechanics Publications, Springer-Verlag, Berlin, pp. 331-343, 1988.

Smith, R. & Sprinks, T., "Scattering of surface waves by a conical island" , J. Fluid Mech., 72, pp. 373-384, 1975.

Tsay, T. K. & Liu, P. L. -F., "A finite element model for wave refraction and diffraction" , Appl. Ocean Res., 5, pp. 30-37, 1983.

Proceedings of the Seventh (1997) International Offshore and Polar Engineering Conference
Honolulu, USA, May 25-30, 1997

ISBN 1-880653-28-1 (Set); ISBN 1-880653-31-1 (Vol. III)

Nonlinear Shallow Water Waves Generated by Submerged Moving Slender Bodies: An Experimental Study

Michelle H. Teng
University of Hawaii at Manoa
Honolulu, HI, USA

Theodore Y. Wu
California Institute of Technology
Pasadena, CA, USA

ABSTRACT

In the present study, we examine the consistency and validity of the Boussinesq and KdV equations in describing nonlinear water waves generated by vertical slender bodies moving with near critical speed in a rectangular channel. Our study is focused on investigating the effect of disturbance length L on wave generation, and whether the two long wave models, which in theory require L to be much greater than water depth H, can actually be applied to cases where $L/H = O(1)$. Our numerical results based on the KdV and Boussinesq wave models show that, if L is sufficiently long, the dominant factor affecting wave amplitude and period will be the ratio of the maximum disturbance width (i.e., beam of a vertical strut) over the channel width, while L has little effect. This confirms Ertekin's (1984) and Mei's (1986) earlier results on the "blockage coefficient". When L is of the same order of H, we found that, as L decreases, it weakens the forcing strength significantly. Results from our towing tank experiments with Froude number ranging from 0.8 to 1.07 revealed that the long wave models give good predictions for resonantly forced long waves even when L is slightly shorter than water depth.

KEY WORDS: long waves, resonant forcing, Boussinesq and KdV models, hydrodynamics, solitary waves

INTRODUCTION

In 1982, Wu and Wu discovered from numerical simulations based on Wu's (1981) generalized Boussinesq model that upstream-advancing solitary waves (also called run-away solitons) can be generated periodically by a steadily moving pressure distribution moving with near critical speed on the free surface. Even though there had been earlier observations of run-away solitons in several experimental studies, Wu and Wu's study provided the first theoretical understanding of the phenomenon. Later, more theoretical analysis based on mass conservation and energy principles was given in Wu (1987). It is known that run-away solitons can be generated by different types of moving disturbances including surface pressure distribution (Wu and Wu 1982, 1987, Akylas 1984, Katsis and Akylas 1987), a bottom topography (Lee, Yates and Wu 1989, Teng and Wu 1990) and vertical slender bodies (Ertekin 1984, Ertekin, Webster and Wehausen 1984, 1986, Mei 1986, Ertekin and Qian 1990, Teng and Wu 1990). For resonantly forced long waves generated by moving disturbances in a rectangular channel that is not excessively wide, the run-away solitons advancing upstream of the disturbances have been observed to have uniform wave crest across the channel. In this case, 2-D models such as the KdV equation and the section-mean Boussinesq equations can be used to predict the wave amplitude and period. For waves generated in a wide channel or an unbounded region, the wave field will be three-dimensional, and 3-D models including the K-P equation (Katsis and Akylas 1987, Lee and Grimshaw 1990), the Green-Naghdi equations (Ertekin, Webster and Wehausen 1986) and the generalized Boussinesq equations (Wu and Wu 1982, 1987, Ertekin and Qian 1990) should be applied.

The present study is focused on studying forced long waves generated by moving vertical slender bodies in a uniform rectangular channel where the 2-D models, namely, the KdV equation and the section-mean Boussinesq equations, are applicable. In 1984, Ertekin carried out a series of experiments to study waves generated by moving ship models in a towing tank. His results showed that, the dominant factor that affects the wave amplitude and period is the disturbance blockage coefficient which is defined as the ratio of the strut beam over the channel width. These experimental results were later analyzed by Mei (1986) who derived a forced KdV equation to simulate Ertekin's experiment. Two specific cases with disturbance length about ten times the water depth were simulated, and the numerical results from Mei's KdV equation showed good agreement with Ertekin's experimental results for resonantly forced long waves generated in front of the ship models. (Shorter dispersive waves in the trailing region are three-dimensional and cannot be fully described by 2-D models.) Good agreement between Ertekin's (1984) experiments and the generalized Boussinesq equations were also found in Ertekin and Qian (1990). It is interesting to note that in both Mei's (1986) and Ertekin and Qian's (1990) studies, the detailed geometry of the slender body used in the numerical simulation and the experiments were actually not the same, but it was found that as long as the blockage coefficient was kept equal, the forced long waves predicted by the theory would agree very well with the experiment. In Teng and Wu (1990, 1992), previous study by Mei (1986) was extended to develop section-mean Boussinesq and KdV models for forced waves in arbitrarily shaped shallow water channels, and further experiments (Teng and Wu 1990) were carried out to verify the blockage-coefficient concept.

There are a few remaining questions that may need further studies. One is related to the validity criteria of the Boussinesq and KdV models in describing forced long waves by vertical struts whose length is small. Based on the perturbation theory used in developing the Boussinesq and KdV models, the moving disturbances need to be sufficiently long for the two models to be valid. However, exactly how long the disturbance should be can only be determined through experiments. Another question is the consistency between the Boussinesq and KdV equations in modeling resonantly forced long waves. Even though section-mean Boussinesq and KdV equations were developed in Teng and Wu (1990, 1992), detailed comparison between the two models was not pursued. Thus, the objective of the present study is to examine the effect of disturbance length on wave generation and the validity criteria of the Boussinesq and KdV models by carrying out both numerical and experimental studies. To focus on examining the effect of disturbance length, one type of disturbance, namely, submerged moving vertical strut, and one specific channel cross-sectional geometry, i.e., a rectangular channel, are studied here.

BOUSSINESQ AND KDV WAVE MODELS

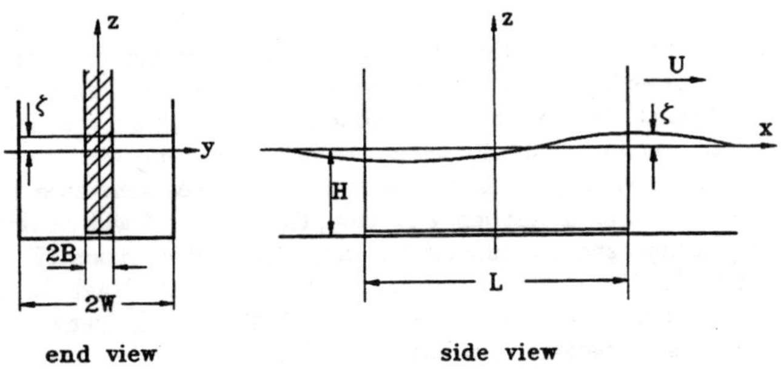

Figure 1. A sketch of a rectangular channel with a submerged moving slender body.

A sketch of the physical problem, namely, nonlinear long waves generated by a moving slender body in a uniform rectangular channel, is shown in figure 1. Assuming the channel is not excessively wide, we may integrate the continuity and momentum equations for inviscid and incompressible flow over the channel cross-sectional area to obtain the following section-mean equations for the dominant flow in the longitudinal direction:

$$A_t + (A\bar{u})_x = 0,$$
$$(A\bar{u})_t + (A\overline{uu})_x = -A\overline{p_x},$$

where $A(x,t)$ is the instantaneous cross-sectional area of the flow, u the longitudinal fluid velocity and p the pressure. The area A can be further written into $A = 2WH + 2W\tilde{\zeta} - 2BH$, where H is the unperturbed water depth, W the channel half-width and B the half-width of the disturbance. Here we have assumed that the strut extends all the way to the channel bed. In all expressions, the overlines denote quantities averaged over the cross-sectional area while the tildes indicate average over the channel width. All variables are nondimensionalized according to

$$x = x^*/\lambda, \quad (y,z) = (y^*,z^*)/H, \quad t = t^*c_0/\lambda, \quad u = u^*/c_0,$$
$$p = p^*/\rho c_0^2, \quad A = A^*/H^2$$

where $c_0 = \sqrt{gH}$ is the critical linear wave speed, ρ water density and λ the typical wavelength. The symbol $*$ here denotes dimensional quantities. For Boussinesq class of weakly nonlinear and weakly dispersive long waves, if we define $\alpha = a/H$ and $\epsilon = H/\lambda$ with a being the typical wave amplitude, we have $O(\alpha)$=$O(\epsilon^2)$ which implies that the nonlinear effect and the dispersive effect are of equal importance. Here small parameters $\alpha << 1$ and $\epsilon << 1$.

For shallow water long waves generated by weak disturbances moving along a rectangular channel that is not extremely wide, the dominant fluid motion is in the longitudinal direction while transversal and vertical flows are secondary, therefore, we can introduce the following perturbation expansion in terms of α:

$$u(x,y,z,t) = \alpha u_1(x,t) + \alpha^2 u_2(x,y,z,t) + ...$$
$$\zeta(x,y,t) = \alpha\zeta_1(x,t) + \alpha^2\zeta_2(x,y,t) + ...$$
$$\phi(x,y,z,t) = \alpha\phi_1(x,t) + \alpha^2\phi_2(x,y,z,t) + ...$$
$$B(x,t) = \alpha B_1(x,t) + \alpha^2 B_2(x,t) + ...$$

where ϕ is the velocity potential, scaled by $c_0\lambda$. Based on these expansions, we have $\overline{uu} = \bar{u}\bar{u} + O(\alpha^4)$, and applying this approximation along with the continuity equation, we can simplify the section-mean momentum equation to $\bar{u}_t + \bar{u}\bar{u}_x = -\overline{p_x}$, where the pressure term is solved by using the unsteady Bernoulli's equation

$$p + \phi_t + \frac{1}{2}[\phi_x^2 + \frac{1}{\alpha}(\phi_y^2 + \phi_z^2)] + z = 0$$

and the Laplace equation

$$\phi_{xx} + \frac{1}{\alpha}(\phi_{yy} + \phi_{zz}) = 0$$

to obtain

$$\overline{p_x} = \tilde{p}_{ax} + \tilde{\zeta}_x + \alpha^2(\tilde{\phi}_{2xt}|_{z=0} - \overline{\phi_{2xt}}) + O(\alpha^3), \qquad (1)$$

where p_a is the gage pressure at the free surface. For the specific problem studied here, ϕ_2 satisfies the following governing equation and boundary conditions:

$$\phi_{2yy} + \phi_{2zz} = -u_{1x}$$

$$\phi_{2z}|_{z=0} = \zeta_{1t}; \quad \phi_{2z}|_{z=-1} = 0; \quad \phi_{2y}|_{y=0} = B_{1t}; \quad \phi_{2y}|_{y=W} = 0,$$

where higher order terms have been neglected for consistency, and the dimensionless water depth H has been set to H=1. In addition, symmetry about the x-z plane is assumed without loss of generality. This mathematical set-up for ϕ_2 requires that the disturbance length L to be sufficiently long, as we assumed that the leading order of the derivative B_t (as well as B_x and B_{xtt} in later equations) is α, same as the order of B itself.

The solution for ϕ_2 for the present problem is obtained as

$$\phi_2(y,z;x,t) = f(x,t) - \frac{1}{2}b_{1t}(y-W)^2 - u_{1x}z + b_{1t}z - \frac{1}{2}u_{1x}z^2 + \frac{1}{2}b_{1t}z^2$$

where $f(x,t)$ is an undetermined constant of integration of order one and $b = B/W$. Substituting these results into (1) and assuming no pressure forcing at the free surface, i.e., $p_a = 0$, we have

$$\overline{p_x} = \tilde{\zeta}_x + \alpha(-\frac{1}{3}\bar{u}_{xxt} + \frac{1}{3}b_{xtt}).$$

Hence, the Boussinesq equations are given as

$$\zeta_t + [(1+\zeta-b)u]_x = b_t, \tag{2}$$

$$u_t + uu_x + \zeta_x - \frac{\alpha}{3}u_{xxt} = -\frac{\alpha}{3}b_{xtt}, \tag{3}$$

where the tildes and overlines are omitted for simplicity. This set of the Boussinesq equations can be used to describe waves traveling in opposite longitudinal directions simultaneously. For unidirectional waves under near-resonant forcing, the Boussinesq equations can be further simplified to an one-equation model, namely, the KdV model. Considering that waves vary slowly when observed from a coordinate that moves with the critical speed, we introduce a new coordinate system

$$s = x - t, \qquad \tau = \alpha t,$$

in which, the Boussinesq equations become

$$-\zeta_s + \alpha\zeta_\tau + [(1+\zeta-b)u]_s = -b_s + \alpha b_\tau$$

$$-u_s+\alpha u_\tau+uu_s+\zeta_s+\frac{\alpha}{3}u_{sss}-\frac{\alpha^2}{3}u_{ss\tau} = -\frac{\alpha}{3}(b_{sss}-2\alpha b_{ss\tau}+\alpha^2 b_{s\tau\tau}).$$

Further expanding all variables into a series of α and matching different terms at each order, we have, at $O(\alpha)$,

$$\zeta_1 = u_1, \quad \text{and} \quad b_1 = 0$$

which indicates that the KdV model is valid only for waves under weak forcing with $b = O(\alpha^2)$. At $O(\alpha^2)$, we obtain

$$2\zeta_{1\tau} + 3\zeta_1\zeta_{1s} + \frac{1}{3}\zeta_{1sss} = -b_{2s}$$

which can be transformed back into the original laboratory coordinate to yield the KdV equation

$$\zeta_t + \zeta_x + \frac{3}{2}\zeta\zeta_x + \frac{\alpha}{6}\zeta_{xxx} = -\frac{1}{2}b_x.$$

Rescaling x^* and t^* with H and H/c_0, we obtain the final Boussinesq model

$$\zeta_t + [(1+\zeta-b)u]_x = b_t, \tag{4}$$

$$u_t + uu_x + \zeta_x - \frac{1}{3}u_{xxt} = -\frac{1}{3}b_{xtt}, \tag{5}$$

and the KdV model

$$\zeta_t + \zeta_x + \frac{3}{2}\zeta\zeta_x + \frac{1}{6}\zeta_{xxx} = -\frac{1}{2}b_x, \tag{6}$$

where α has disappeared from the equations due to rescaling.

To investigate the effect of disturbance length L on wave generation based on these wave models, we carried out numerical computations to simulate run-away solitons generated by a slender body moving with critical speed in a rectangular channel. The forcing function (i.e., $b = B/W$) of the slender body examined here is given by

$$b(x,t) = \begin{cases} \frac{b_m}{2}[1-\cos(\frac{2\pi}{L}(x-Ft))], & \text{for } 0 \le x - Ft \le L; \\ 0, & \text{otherwise.} \end{cases} \tag{7}$$

where for resonant forcing, Froude number $F = 1$, and $b_m = B_{max}/W$ is the "blockage coefficient" discussed in Ertekin (1984), Ertekin, Webster and Wehausen (1984, 1986) and Mei (1986). Waves generated by the slender body with four different values of L, namely, $L = 20$, 10, 5 and 2, are investigated by solving the Boussinesq equations (4) and (5) and the KdV equation (6) numerically using two finite difference schemes. In all cases, F and b_m are kept the same with $F = 1$ and $b_m = 0.1$. The numerical results for waves generated by the disturbance having moved for a dimensionless time of $t = 200$ are presented in figure 2.

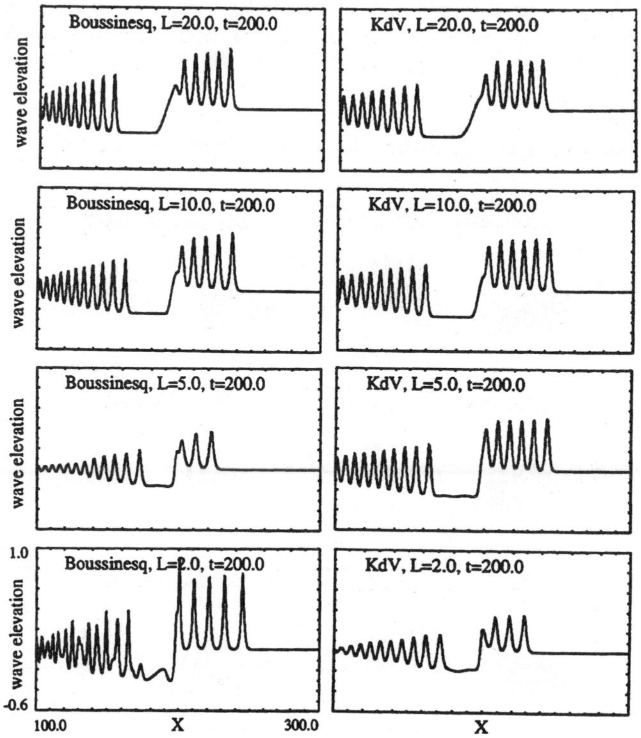

Figure 2. Numerical simulation of run-away solitons generated by moving slender bodies given by eqn.(7); left column: results based on the Boussinesq equations (4) and (5); right column: results based on the KdV equation (6). In all cases, $b_m = B_{max}/W = 0.1$ and Froude number $F = U/\sqrt{gH} = 1$.

In figure 2, numerical results based on the Boussinesq equations (4) and (5) are compared with that based on the KdV equation (6). We found that, when the disturbance length is sufficiently long, for example, $L = 20$ and 10, the two models give quite consistent results. However, as L is reduced to 5, the two wave models start to differ, especially in the case of $L = 2$ where the Boussinesq results do not seem to be physically plausible. Re-examining the Boussinesq equations, we found that the maximum magnitude of the higher order derivative term b_{xtt}, which equals to $4\pi^3 b_m F^2/L^3$ in the present case, increases significantly when L is reduced. When L becomes so short, this derivative term will become so large that its magnitude will no longer be of higher order as assumed in our perturbation expansion and matching, which causes the Boussinesq model to give wrong results. In fact, for Boussinesq class of shallow water long waves, $O(\zeta_t + u_x) = O(\alpha^2)$, hence, if we assume $O(b) = O(\alpha)$ for the Boussinesq model (eqns(2), (3)), then its continuity equation is not well matched and balanced. Based on this, it seems that for the particular problem investigated here, namely, nonlinear and dispersive waves generated by disturbances moving with near critical speed, it is appropriate to assume that the disturbance forcing strength to be of $O(\alpha^2)$ for both the Boussinesq and KdV models. Under this assumption, the modified and more consistent Boussinesq model becomes,

$$\zeta_t + [(1+\zeta)u]_x = b_t, \quad (8)$$

$$u_t + uu_x + \zeta_x - \frac{1}{3}u_{xxt} = 0. \quad (9)$$

(For related discussions on higher order terms in the Boussinesq equations for waves propagating in variable channels without forcing, please refer to Teng and Wu 1994.) Numerical simulations for the same previous cases with L =20, 10, 5 and 2 were performed again based on the modified Boussinesq equations (8) and (9), and the results are compared with the KdV results in figure 3.

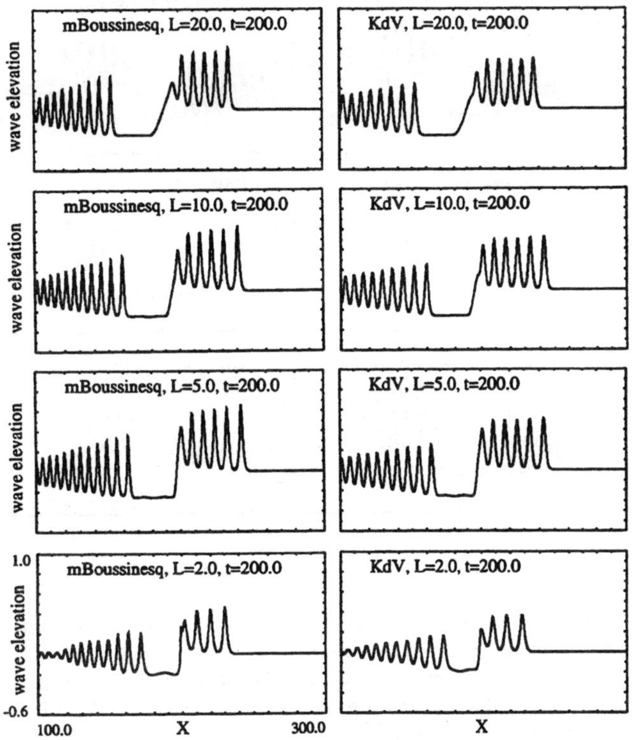

Figure 3. Numerical simulation of run-away solitons generated by moving slender bodies given by eqn.(7); left column: results based on the modified Boussinesq equations (8) and (9); right column: results based on the KdV equation (6). In all cases, $b_m = B_{max}/W = 0.1$ and Froude number $F = U/\sqrt{gH} = 1$.

From these results we can see that, the modified Boussinesq model and the KdV model give consistent results for all different values of L, with the wave amplitude predicted by the Boussinesq model being slightly larger than that by the KdV model. We also note that, if the disturbance length L is sufficiently long, it has little effect on wave amplitude and period as long as the "blockage coefficient" b_m is kept the same. This confirms Ertekin's experimental (1984) and Mei's KdV (1986) results. However, when L is further reduced, its effect on wave generation becomes more significant, as we can see from our numerical results for $L = 2$, where the forcing strength is seen to be greatly weakened compared with the cases where $L \geq 5$.

The question now is whether the theoretical and numerical results presented here reflect the physics in reality. To verify the theory, we need to carry out experiments and compare the experimental results with the theoretical predictions. In Mei's (1986) study, a similar KdV equation was derived, and the numerical results based on the KdV equation were compared with Ertekin's (1984) experimental results of resonantly forced long waves by moving ship models in a towing tank. The cases investigated in Mei's study were related to ship models whose length was about ten times the water depth, and the KdV results showed good agreement with the experiment. Hence, the focus of the present study is to examine the validity of the two wave models in describing forced waves generated by short slender bodies, for example, disturbances with $L/H = O(1)$, by carrying out experiments.

EXPERIMENTAL SET-UP AND RESULTS

The experiment was performed in a wave tank 7.5 m long, 0.75 m wide and 0.6 m deep. A narrower rectangular channel 10 cm wide was constructed inside the wave tank and two identical submerged vertical struts were towed by a carriage along the channel side walls to generate surface water waves. The detailed experimental set up is shown in figure 4. The curved surface of the strut is of the shape of a circular arc with B_{max}=0.66 cm and L=4.9 cm. In the experiment, three cases with different Froude numbers, namely, with $F = 0.801, 0.994$ and 1.071, were studied. In all these cases, the water depth was kept at 5 cm and the total distance the disturbance traveled is 6.4 m. In dimensionless form, the disturbance length $L = 0.98$. To account for the viscous effect in real flows, a mean boundary layer thickness $\delta = 0.5 \times 4.9\sqrt{\nu L/U}$ based on the Blassius solution for laminar boundary layer flow over a flat surface is added to B_{max}. Another 0.1 cm due to the uneveness of the channel side walls is also added to B_{max}. The effective blockage coefficient is calculated to be b_m = 0.166, 0.165 and 0.1646 for the cases with F=0.801, 0.994 and 1.071, respectively. In the experiment, the wave profile was measured by a resistance type wave gauge rigidly fastened to the carriage and moving with the struts with the same speed. The wave gauge was fixed at 12 cm in front of the disturbance and moved along the channel centerline. The data recorded by the wave gauge is a time sequence of wave elevation at the particular point that the gauge was located.

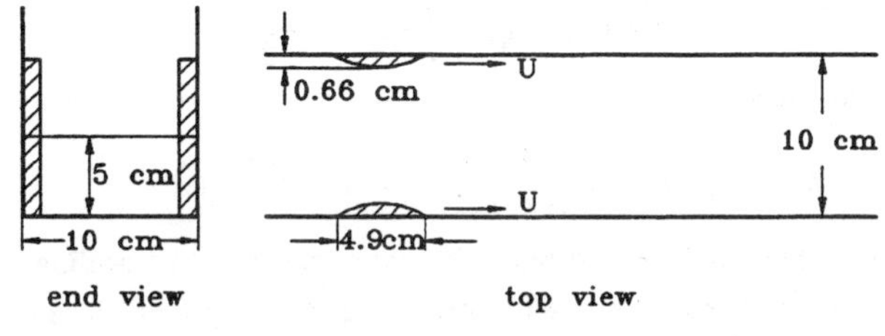

Figure 4. Experimental set-up to study nonlinear long waves generated by two moving slender bodies in a rectangular channel.

The measured wave profiles and the corresponding numerical simulations based on the KdV equation (6) are shown in figure 5. As we can see, even though the disturbance length L is as short as the water depth, the KdV model still gives good prediction for the actual waves generated in front of the disturbance, especially when F is close to one. Since we have shown that, the KdV equation and the (modified) Boussinesq equations are theoretically equivalent for resonantly forced long waves, we expect equally good agreement between the Boussinesq model and the experiments.

In our numerical simulations, we used the same values of the blockage coefficient and also the same detailed geometry (circular arc in this case) of the strut as in the experiment.

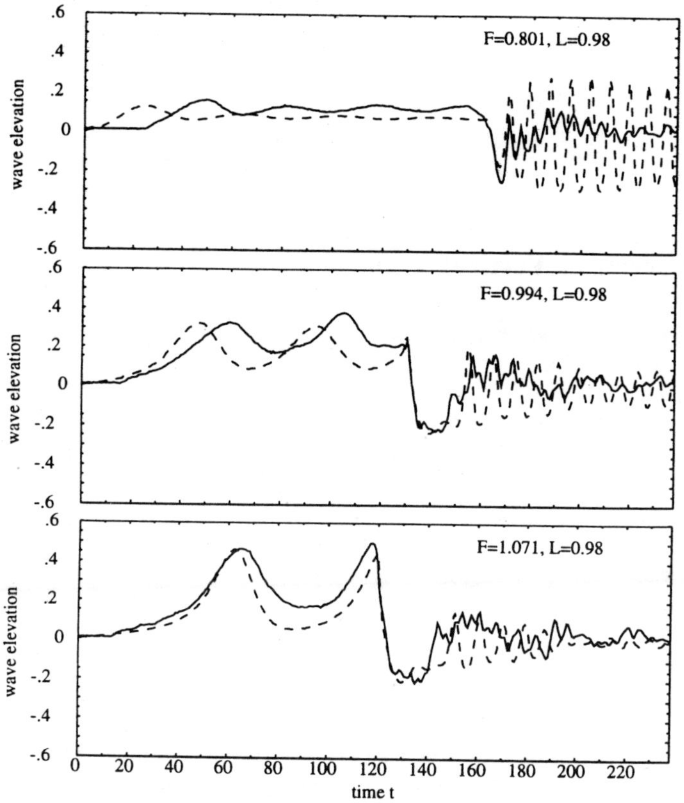

Figure 5. Comparison between measured (solid line) and theoretical (KdV, dashed line) wave profiles.

SUMMARY AND CONCLUSION

The original section-mean Boussinesq equations (2) and (3) as well as similar Boussinesq models developed in Teng and Wu 1990, 1992 require the disturbance to be sufficiently long and smooth for the wave models to be valid. This is because the higher order derivative terms such as b_{xtt} are very sensitive to the detailed geometry of a slender body as external forcing. For example, if the surface of a slender body has small ripples of short wavelength, the magnitude of b_{xtt} will be so large that it will become the dominant forcing term in the equations. However, existing experimental results show that, for resonantly forced long waves generated by a vertical strut, the dominant forcing factor is the ratio of the strut beam over the channel width, while the detailed geometry of the strut surface has little effect. Re-examining the order of different terms in the Boussinesq equations, we found that a more consistent Boussinesq model (modified Boussinesq equations (8) and (9)) can be reached by taking out the higher order terms. This modified Boussinesq model is shown to give consistent wave predictions as the well-matched KdV equation under different values of disturbance length.

To investigate the validity criteria of the KdV and the modified Boussinesq models as related to the disturbance length, experiments were performed to measure the nonlinear long waves generated by vertical struts moving with near critical speed in a rectangular channel. Our experiments show that, the KdV model gives good prediction for wave amplitude and period under resonant forcing even when the disturbance length is as short as the water depth. This indicates that the two long wave models have quite broad applicabilities, as far as the disturbance length is concerned. In addition, we note that nonlinear long waves can be generated by both long and short moving disturbances, therefore, in theoretical modeling of resonantly forced long waves, it may be more proper to use the typical wavelength rather than the length of the disturbance as a characteristic length.

ACKNOWLEDGEMENTS

The authors are grateful to the referees for their helpful comments which have been added in the present paper.

REFERENCES

Akylas, T.R. (1984) "On the excitation of long nonlinear water waves by a moving pressure distribution," *J Fluid Mech*, Vol.141, pp.455-466.

Ertekin, R.C. (1984) "Soliton generation by moving disturbances in shallow water: theory, computation and experiment," Ph.D. thesis, Department of Naval Arch., Offshore Engng, University of California, Berkeley.

Ertekin, R.C., Webster, W.C. and Wehausen, J.V. (1984) "Ship-generated solitons," *Proc. 15th Symposium on Naval Hydrodynamics,* Hamburg, pp.347-364.

Ertekin, R.C., Webster, W.C. and Wehausen, J.V. (1986) "Waves caused by a moving disturbance in a shallow channel of finite width," *J Fluid Mech,* Vol. 169, pp.275-292.

Ertekin, R.C. and Qian, Z. (1990) "Upstream solitons and wave resistance," *Engineering Science, Fluid Dynamics,* pp.29-43, World Scientific Publishing Co.

Katsis, C. and Akylas, T.R. (1987) "On the excitation of long nonlinear water waves by a moving pressure distribution. Part 2. Three-dimensional effects," *J Fluid Mech*, Vol.177, pp.49-65.

Lee, S.J., Yates, G.T. and Wu, T.Y. (1989) "Experiments and analyses of upstream-advancing solitary waves generated by moving disturbances," *J Fluid Mech,* Vol. 199, pp.569-593.

Lee, S.J. and Grimshaw, R.H.J. (1990) "Upstream-advancing waves generated by three-dimensional moving disturbances," *Phys. Fluids A*, Vol.2(2), pp.194-201.

Mei, C.C. (1986) "Radiation of solitons by slender bodies advancing in a shallow channel," *J Fluid Mech,* Vol. 162, pp.53-67.

Teng, M.H. and Wu, T.Y. (1990) "Generation and propagation of nonlinear water waves in a channel with variable cross section," *Engineering Science, Fluid Dynamics*, pp.87-108. World Scientific Publishing Co.

Teng, M.H. and Wu, T.Y. (1992) "Nonlinear water waves in channels of arbitrary shape," *J Fluid Mech,* Vol. 242, pp.211-233.

Teng, M.H. and Wu, T.Y. (1994) "Evolution of long water waves in variable channels," *J Fluid Mech,* Vol. 266, pp.303-317.

Wu, T.Y. (1981) "Long waves in ocean and coastal waters," *ASCE J Engng Mech*, Vol.107, pp.501-522.

Wu, D.M. and Wu, T.Y. (1982) "Three-dimensional nonlinear long waves due to moving surface pressure," *Proc 14th Symp on Naval Hydrodynamics,* pp.103-125. National Academy Press.

Wu, D.M. and Wu, T.Y. (1987) "Precursor solitons generated by three-dimensional disturbances moving in a channel," *Nonlinear Water Waves,* IUTAM Symposium, Tokyo, Japan (ed. K. Horikawa and H. Maruo), pp.69-75, Springer.

Wu, T.Y. (1987) "Generation of upstream advancing solitons by moving disturbances," *J Fluid Mech*, Vol.184, pp.75-99.

Proceedings of the Seventh (1997) International Offshore and Polar Engineering Conference
Honolulu, USA, May 25-30, 1997

ISBN 1-880653-28-1 (Set); ISBN 1-880653-31-1 (Vol. III)

Irregular Broken Wave Forces on Vertical Wall

Yu-cheng Li, Da-zhong Liu, Gui-ping Qi and Shiao-jun Su
Dalian University of Technology
Dalian, China

ABSTRACT

A method for the calculation of total horizontal wave forces by irregular broken waves on vertical wall was developed, which was based on physical model test, dimensionless analysis and assessment of real construction. The method for the determination of pressure distribution of horizontal forces on vertical wall as well as uplift forces on the bottom of the wall are also recommended. The agreement between calculated results by this method and real construction is quit good.

KEYWORDS: Irregular waves, broken waves, wave forces.

INTRODUCTION

Depends on wave parameters, the size of vertical wall and rubble mound foundation, in front of a vertical wall, there are three kinds of waves: namely standing waves, breaking waves and broken waves. The breaking waves occur when the height of rubble mound foundation is big enough, which may give larger impact force and may cause more problem on the safety and stability of vertical wall. The broken waves appear when the rubble mound foundation is small or underground and the depth before vertical wall is small, which will break usually at the points of antinodes before the wall. As well known, the investigation of wave forces by breaking or broken waves on vertical wall is still not enough. On the International Workshop on Wave Barriers in Deepwaters held in Japan 1994, the interaction of breaking waves with vertical walls has been discussed in detailed and more research are expected in the future. Based on regular wave test and analysis, Xou and Li(1965) developed a method for the calculation of broken wave forces on vertical walls, which has been practical used in some real breakwater design and construction in China and was adopted in the Chinese Harbor Engineering Design Specification (CHEDS) in 1974. So far the Goda's unified method (1973, 1974) has been widely recommended in the world. The advantage of Goda's method is that it can be used in various wave condition, the later model test in Japan proved that in case of low height of rubble mound foundation the calculated results by Goda's method coincide well with test data. The purpose of this study is to check the results by Goda's method and to develop a method for the calculation of irregular broken wave forces on vertical walls, which is based on irregular wave model test, dimensionless analysis and its assessment of real construction.

THE WAVE BEHAVIOR IN FRONT OF VERTICAL WALLS

When waves normally attack a vertical wall, in front of the wall the waves may be standing waves, broken waves or breaking waves. The broken wave is defined as wave break at the antinodes, so the distance between breaking point and vertical wall is equal to n(L/2), in which L is the wave length and n is an integer. The breaking wave is defined as the distance between breaking point and vertical wall is less than L/2. Based on regular wave test by Xou and Li(1965), in Chinese Harbor Engineering Design Specification the wave behavior in front of vertical wall is specified as follows : 1. For underground foundation or foundation with a small height above seabed ($\frac{d_1}{d} > \frac{2}{3}$, d_1 is the water depth above foundation, d is the water depth before the structure) : when $d \ge 2H$, it is standing wave ; when $d < 2H$ and bottom slope $i < 1/10$, it is broken wave. 2. For middle height foundation ($\frac{2}{3} \ge \frac{d_1}{d} > \frac{1}{3}$): when $d_1 \ge 1.8H$, it is standing wave;when $d_1 < 1.8H$,it is breaking wave. 3. For large height foundation ($\frac{d_1}{d} \le \frac{1}{3}$): when $d_1 \ge 1.5H$, it is standing wave; when $d_1 < 1.5H$, it is breaking wave.

The irregular wave test by Li and Liu(1992) indicated that in case of irregular waves, the waves become not so easy to break as for regular waves, which means that the critical water depth become shallower than the regular waves. Also in case of irregular waves, the wave height changes one by one in the wave train. So the critical water depth vary in some range. For irregular waves They recommended : 1. For underground foundation or foundation with a small height, the critical condition is d=(1.5~2.0H). 2. For middle height foundation, the critical condition is d=(1.1~1.8)H. 3. For large height foundation, the critical condition is d=(0.8~1.5)H. The final recommendation by the present authors will be given later in this paper.

CONDITION OF MODEL TEST

The model test was conducted in a wave-current flume 69m in

length, 2m in width, 1.8m in height in Dalian University of Technology. The test model was set with a distance of 40m in front of the wave paddle. The whole width of flume was divided into three sections uniformly, and the model was fixed in the side section. The wave reflection effect from the vertical wall model can be reduced significantly by such a way, and the error of incident significant wave height after setting the vertical wall is less than 5%. The water depth d before the vertical wall model is 40cm, the water depth d_1 above the foundation is 28 and 40cm respectively. The bottom slope i is 1/50 and 1/30 respectively. During the investigation, the total forces and wave pressure distribution have been measured in various test. The natural frequency of force gage is 12 to 18Hz and the natural frequency of pressure gage is 1KHz. The distribution of wave pressure gages is shown in Fig.1, on vertical face there are six gages and on the bottom there are four gages. The target wave spectrum during test are JONSWAP spectrum (with peak values γ =1.0 and 3.0) and Brethshneider-Mitsuyasu spectrum. The test range is as follows: $d_1/H_{1\%}$ = 1.41~2.10, d_1/d= 0.7~1.0, $H_{1\%}/\bar{L}$ = 1/10~1/20, in which $H_{1\%}$ means the wave height with 1% probability of exceedance and $\bar{L}$ is the mean wavelength.

STATISTICAL DISTRIBUTION OF WAVE FORCES

As well known, the wave height in deep water follows the Rayleigh distribution, and in shallow water it follows the Glufovski distribution [see Glufovski(1966)]. As concerning the statistical distribution of wave forces, no conclusion has been reported in literatures. In recent decades, the three parameters Weibull distribution has been recommended to describe the wave height statistical distribution. The probabilistic density function of Weibull distribution can be expressed as follows:

$$f(x)=\begin{cases}\alpha m(x-r)^{m-1}\exp[-\alpha(x-r)^m] & \text{when } x\ge r\\ 0 & \text{when } x<r\end{cases} \tag{1}$$

where m (m>0) is the shape parameter, α (α >0) is the scale parameter and r is the location parameter. Generally r equals to zero, so actually only two parameter m and α are needed to be determined.

When r = 0, the cumulative probability distribution is as follows:

$$F(x)=\exp[-\alpha x^m] \tag{2}$$

There is a theoretical relationship between α and m, which can be written by

$$\alpha=\Gamma^m(1+\frac{1}{m}) \tag{3}$$

where Γ is the Gamma function.

One important character of Weibull distribution is that the Weibull distribution (Equation 2) will automatically transfer to the Rayleigh distribution (Equation 4), when the shape parameter m equals two.

$$F(x)=\exp[-\frac{\pi}{4}x^2] \tag{4}$$

It is clear that for standing waves the wave force distribution is the same as the distribution of wave height, which means that the distribution of standing wave force is Rayleigh distribution. So Li and Liu(1994) tried to use the Weibull distribution for describing the force distribution of the both broken and breaking waves and the results are quite good. They (1994) found that for horizontal force there is a linear relationship between parameters of $\ln m$ and $\frac{d_1}{H_{1\%}}\tanh\frac{d_1}{d}$ as follows:

$$\ln m=0.383\left[\frac{d_1}{H_{1\%}}\tanh\frac{d_1}{d}\right]+0.183 \tag{5}$$

The correlation coefficient r of Eq.5 is 0.68 . They (1994) also found that for uplift forces there is a linear relationship between parameters $\ln m$, $d_1/H_{1\%}$ and d_1/d as follows:

$$\ln m=0.693-0.006\left\{\frac{d_1}{H}-\left[\left(\frac{d_1}{d}\right)^2-0.24\left(\frac{d_1}{d}\right)+0.55\right]\right\}^{-1} \tag{6}$$

The correlation coefficient r of Eq.6 is 0.74 .

When the calculated m value by Eq.5 or 6 is larger than two, the real value m used should be reduced to two. In Figure 2, some calculated wave force distribution are compared with test results, which means that the results are acceptable.

TOTAL HORIZONTAL FORCES BY BROKEN WAVES ON VERTICAL WALL

Usually the statistic method is used for the design of breakwater of vertical walls, because the weight of wall is very large and its natural frequency is quite high. In Chinese design code the one percent cumulative probability wave height is used as the design wave height. So the key point of our object is to develop a method of how to determine the one percent cumulative probability broken wave forces. Based on the main concept of Xou and Li(1965), the dimensionless total horizontal forces by irregular broken waves can be expressed as the following function

$$\frac{P_{1\%}}{\gamma H d_1}=f\left(\frac{d_1}{d},i,\frac{H}{L},\frac{H}{d_1}\right) \tag{7}$$

where $P_{1\%}$ is the one percent cumulative probability total horizontal forces by broken waves, in the later explanation the subscript 1% will be ignored.

During the derivation of Eq.7 , we will follow the following consideration, namely:

(1) When the relative height of rubble mound foundation increase and reach a value of d_1/d equals or less than 2/3, the wave behavior in front of vertical wall become breaking waves and the effect of bottom slope can be ignored. In its critical condition, the wave force by broken wave method should be coincide with the wave force by breaking wave method.

(2) When wave height is small, the wave behavior in front of vertical wall become standing waves. In its critical condition, the wave force by broken wave method should be coincide with the wave force by standing wave method.

(3) For standing waves the horizontal wave forces are affected by the value of wave steepness, but for breaking waves the wave forces are even not influenced by the value of wave steepness. So here we should consider two options: one is to consider the influence of wave steepness factor on broken wave forces, another is to ignore the wave steepness influence on broken wave forces.

Based on the above consideration, we found a factor of Kid to consider the joint influence of the bed slope i and the relative rubble mound foundation height d_1/d as follows:

$$Kid=1+3.2\left(\frac{d_1}{d}-\frac{2}{3}\right)i^{\left(\frac{d_1}{d}-\frac{2}{3}\right)} \tag{8}$$

From Eq.8 , it is clear that when $\frac{d_1}{d}=\frac{2}{3}$, Kid=1, which means that in this case the bottom slope effect can be ignored. For certain value of d_1/d, there is a maximum value of Kid as shown in Fig.3 .

As shown in Fig.4 , for irregular waves the influence of wave steepness parameter on total horizontal wave forces is not clear, but in regular wave condition it is a important factor on wave forces. As to consider more factors in the recommended method for the calculation of wave forces, it is better to put in the wave steepness factor. Then the total horizontal broken wave forces on vertical wall can be calculated as follows:

$$\frac{P}{\gamma H d_1} = \text{Kid}.K_{LH}\left(2.13\frac{H}{d_1} - 0.21\right) \quad (9)$$

where $K_{LH} = -0.00034\left(\frac{L}{H}\right)^2 + 0.023\left(\frac{L}{H}\right) + 0.746$ (10)

H is the wave height of one percent cumulative probability and L is the mean wave length. The results of Eq.9 are shown in Fig.5(line L2). In Fig. 5 line L1 is the best fitness line of the test data, but many test data are larger than the values given by line L1. So L1 may not give safe results, this is the reason we took line L2. The test data are usually below this line and line L2 is more safe for engineering application.

The comparison between author's recommended method, Chinese specification method and Goda's method are shown in Fig.6 . Usually the broken waves will occur only in the range of H/d_1<0.8~0.9,within this range we can find from Fig. 6 that discrepancy between the results by these three methods(Line 1, 2 and 3) are not so large.

WAVE PRESSURE DISTRIBUTION ON VERTICAL WALL

Based on model test data analysis, a linear wave pressure distribution can be recommended as shown in Fig.7 . The wave pressure on the toe of vertical wall p_b and the height of zero wave pressure point Z may be determined as follows:

$$p_b = 0.7p_s \quad \text{when } d/H \le 1.7$$
$$p_b = 0.55p_s \quad \text{when } d/H > 1.7 \quad (11)$$

and $Z = H_{1\%}$ (12)

Based on Eqs. 8~10, we can calculate both p_s and p_b. The measured p_b/p_s values are between 0.51~0.78, its total average value is 0.65 with a standard derivation σ of 0.0046 . The ratio of measured wave pressure on still water level p_{st} to the calculated wave pressure on still water level p_{sc} by total wave force are between 0.98~1.15, its mean value is 1.06 with a standard derivation σ of 0.003 . The ratio of calculated total wave force P_m from test wave pressure on still water p_s and Eq.11 and 12 to the measured wave force P are between 0.86~1.02, its mean value is 0.947 with a standard derivation σ of 0.0023 . The results means that the authors' recommended method may give quite reasonable results.

Some samples of measured wave pressure distribution on vertical wall and measured uplift pressure distribution on the bottom of vertical wall are shown in Fig.8 .

UPLIFT FORCES ON BOTTOM OF VERTICAL WALL BY BROKEN WAVES

In this study, a special test for the measurement of uplift forces on the bottom of vertical wall was conducted by measuring the uplift pressures. The configuration of pressure gages is shown in Fig.1 .

In the Chinese Harbor Engineering Design Specification (CHEDS), it is assumed that at the sea side of bottom the uplift pressure is the same value of horizontal wave pressure on the bottom of vertical wall, at the lee side of bottom the uplift pressure is equal to zero, and it is not a linear distribution between them. So the total uplift force P_u can be determined as follows:

$$P_u = \frac{1}{2}\mu p_b B \quad (13)$$

where B is the width of vertical wall and μ is a coefficient, which is usually less than one.

In Eq.13 , if we take P_u and p_b both by real measured values, The test value μ are between 0.7~1.16 . But we should use Eq.7 to determine the total horizontal force and to determine the calculated p_b by Eq.11 and 12, which is usually larger than the measured data. So in Eq.13, a suitable value should be chosen not directly from test. After analyzing the whole data, the recommended value of μ is 0.7, which is the same value as used already in CHEDS for regular waves.

CRITICAL CONDITION OF BROKEN WAVES AND STANDING WAVES

In section 2, the critical condition of broken waves, standing waves and breaking waves have been discussed. There is a range for the critical condition(d/H=1.5~2.0) from standing waves transferring to broken waves for irregular waves. According to the present recommended method for the calculation of broken waves, comparison has been made as listed in Table 1. As shown in this table, to take high critical condition(d/H=2.0) is better than to take low critical value(d/H=1.5). The reasons are as follows: firstly, at high critical condition, the difference of wave forces given by broken wave method and by standing wave method is smaller; secondly, the high critical condition is just the same value as given by regular wave method, so it would be more convenient for engineering application. The comparison between broken wave forces and breaking wave forces are given in Fig.9, at the critical condition, the wave forces at its critical condition are quite similar.

VERIFICATION BY EXISTED CONSTRUCTION

In Table 2, seven existed construction were selected, in which the wave behavior before breakwater is broken waves. The relative water depth d_1/d are between 0.67 to 1.0, relative wave height are between 1.3 to 2.62 and wave steepness are between 1/20.9 to 1/11.3 . In these samples the breakwater of first sample slided by wave attacking and the others are stable. It is shown by table 2 that authors' recommended method, Goda's method and CHEDS method both give good results and the results by these three method are quite close.

CONCLUSIONS

Based on the above analysis, the following conclusions may be reached:

1. The critical condition of irregular broken and standing waves in front of vertical wall is: when relative water depth before structure $d/H \ge 2.0$, there are standing waves; when $d/H < 2.0$, there are broken waves.

2. The total horizontal forces on vertical wall by broken waves can be calculated by authors' recommended method, as shown in Eqs. 8, 9 and 10.

3. The distribution of horizontal wave pressure can be determined by Eqs. 11, 12 and Fig 7.

4. The uplift force on bottom of vertical wall can be obtained by Eq. 13, in which the value of μ is 0.7 .

5. Verification of the authors' method by existed construction shows that the present study may give quite reasonable results. So this method may be recommended for engineering application.

ACKNOWLEDGMENT

This study is a part of the research project "The Investigation of Irregular Breaking and Broken Wave Forces on Vertical Wall", which was financially supported by both the National Natural Science Foundation of China and the Ministry of Communication Foundation for the Development of Design Specification.

REFERENCES

Glufovski, B. C.(1996), Investigation of Sea Wind Waves, Lenigrad 284c. (in Russian)

Goda,Y (1973), A New Method of Wave Pressure Calculation for the Design of Composite Breakwater, Report of Port and Harbor Research Institute, Vol. 12 No 3, pp31~70 (in Japanese) or Proc. 14th ICCE 1974, pp1702~1720.

Li, Y.C. (1990). Ch 1 Wave Behavior and its Forces against Vertical Wall, Wave Action on Maritime Structures, The Press of Dalian University of Technology, pp26~29.

Li, Y.C. and D.Z.Liu (1992). Study on Configuration of Irregular Wave in front of Vertical Wall, Port and Waterway Engineering , No 9, 1992, pp49~54. (in Chinese)

Li, Y.C. and D.Z.Liu (1994). The Statistic Characteristics of Irregular Breaking Wave Forces on Vertical Walls, Proc. of International Symposium on Wave Modeling, Vancouver, Canada 1994, pp 1424~1432.

Li, Y.C., D.Z.Liu, X.J.Su and G.P.Qi (1997), The Breaking Wave Forces by Irregular Waves on Vertical Wall, to be published in Jour. of Hydrodynamics, Ser. B 1997.

Tanimoto K. and S.Takahashi (1994), Japanese Experience on Composite Breakwaters, Proc. of Int. Workshop on Wave Barriers in Deepwaters, pp 1~24.

The Chinese Harbor Engineering Design Specification, The Ministry of Communication, P.R.China, 1987 edition, The Press of Peoples Communication, 1988.

Xou, M.T. and Y.C.Li etc.(1965), The Interaction between Waves and Vertical Wall, Jour. of Natural Science, Division of Civil Engineering, Vol. 1 No 2, pp 139~153.

Yu, Y.X. (1975), Broken Wave Forces, Jour. of DIT, 1975 No 2, pp 51~56.

Table 1 Dimensionless Horizontal Wave Forces $P/\gamma H d_1$

	Condition of calculation	L/H	Broken Wave Forces (authors' method)		Standing Wave Forces	Wave Forces by Goda's method	
			i=1/30	i=1/50		$H_{1/250}$	$H_{1\%}$
High critical	d=2.0H	10	1.34	1.32	1.27	1.17	1.04
		15	1.44	1.42	1.38	1.43	1.25
	d=12.5 d_1	20	1.52	1.50	1.44	1.59	1.38
	d=2.0H	10	1.08	1.04	1.13	0.98	0.93
		15	1.16	1.12	1.26	1.23	1.13
	d= d_1	20	1.23	1.18	1.32	1.38	1.26
Middle value	d=1.8H	10	1.52	1.50	1.36	1.32	1.14
		15	1.63	1.61	1.47	1.58	1.34
	d=1.25 d_1	20	1.72	1.70	1.84	1.73	1.46
	d=1.8H	10	1.23	1.18	1.21	1.09	1.10
		15	1.32	1.27	1.33	1.34	1.20
	d= d_1	20	1.39	1.34	1.69	1.48	1.32
Low critical	d=1.5H	10	1.62	1.60	1.41	1.41	1.19
		15	1.75	1.72	1.52	1.68	1.39
	d=1.25 d_1	20	1.84	1.81	1.90	1.82	1.50
	d=1.5H	10	1.32	1.27	1.26	1.16	1.04
		15	1.42	1.36	1.38	1.41	1.24
	d= d_1	20	1.49	1.44	1.74	1.54	1.35

Table 2 Verification by existed construction

No	Breakwater	d (m)	d_1 (m)	H (m)	L (m)	width of breakwater B (m)	Condition of stability	Safety factor of sliding K
1	Japanese 1	10.2	6.8	6.5	109	9.0	sliding	authors' 0.675 Goda's 0.818 CHEDS* 0.730
2	Japanese 2	10.0	7.2	5.4	92.3	8.5	nonsliding	authors' 1.024 Goda's 1.041 CHEDS 1.008
3	North breakwater in Basuo Port	9.5	8.5	4.9	71	9.0	nonsliding	authors' 1.492 Goda's 1.256 CHEDS 1.510
4	Malta Port	23.6	18.9	9.0	116	27.6	nonsliding	authors' 3.009 Goda's 2.701 CHEDS 2.654
5	Breakwater of Sajiao Power plant	6.8	5.9	3.5	54	8.0	nonsliding	authors' 1.488 Goda's 1.474 CHEDS 1.459
6	Breakwater in Qinhuangdao Port	7.4	7.4	4.8	54	9.3	nonsliding	authors' 1.378 Goda's 1.310 CHEDS 1.777
7	Western old Breakwater in Basuo Port	3.0	3.0	2.3	48	4.0	critical	authors' 1.052 Goda's 1.031 CHEDS 1.095

* in CHEDS the method was developed by Xou and Li, which was based on regular wave model test.

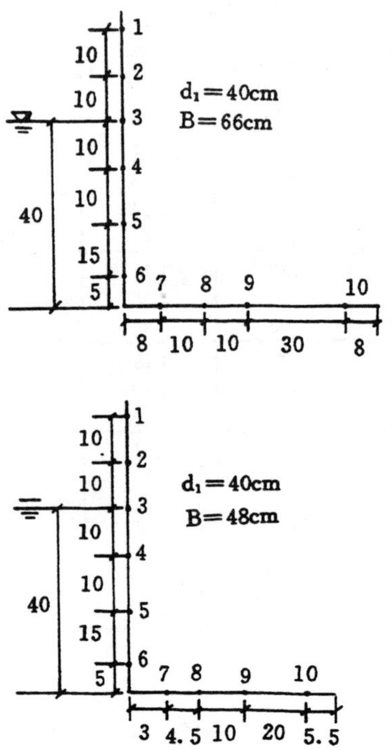

Fig.1 configuration of wave pressure gages

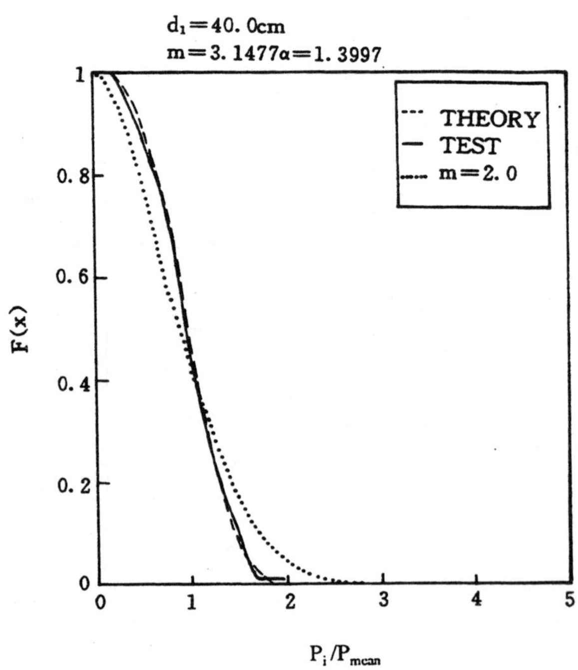

Fig. 2 distribution of wave force

Pi : wave force with a cumulative probability of i%

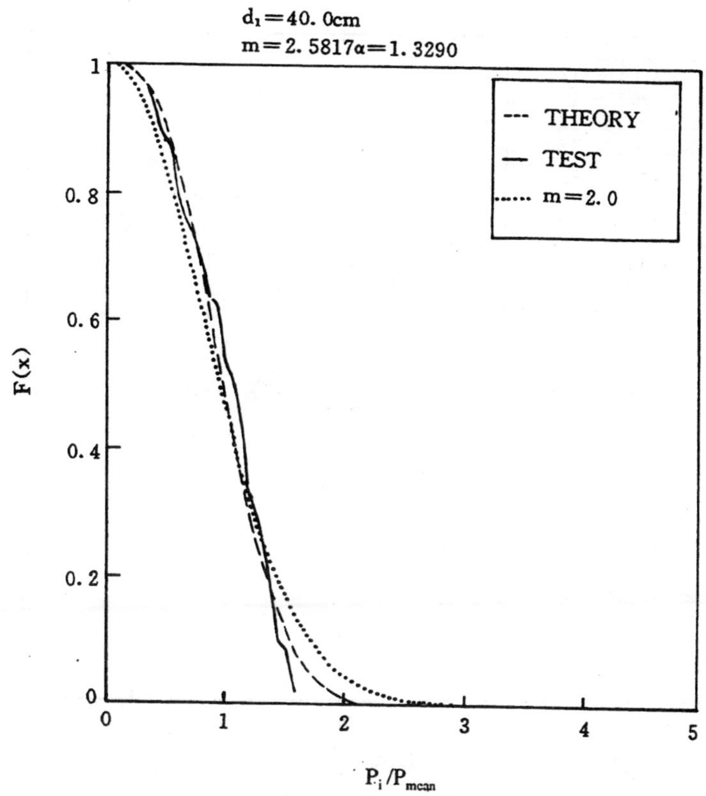

Fig. 2 (continue)

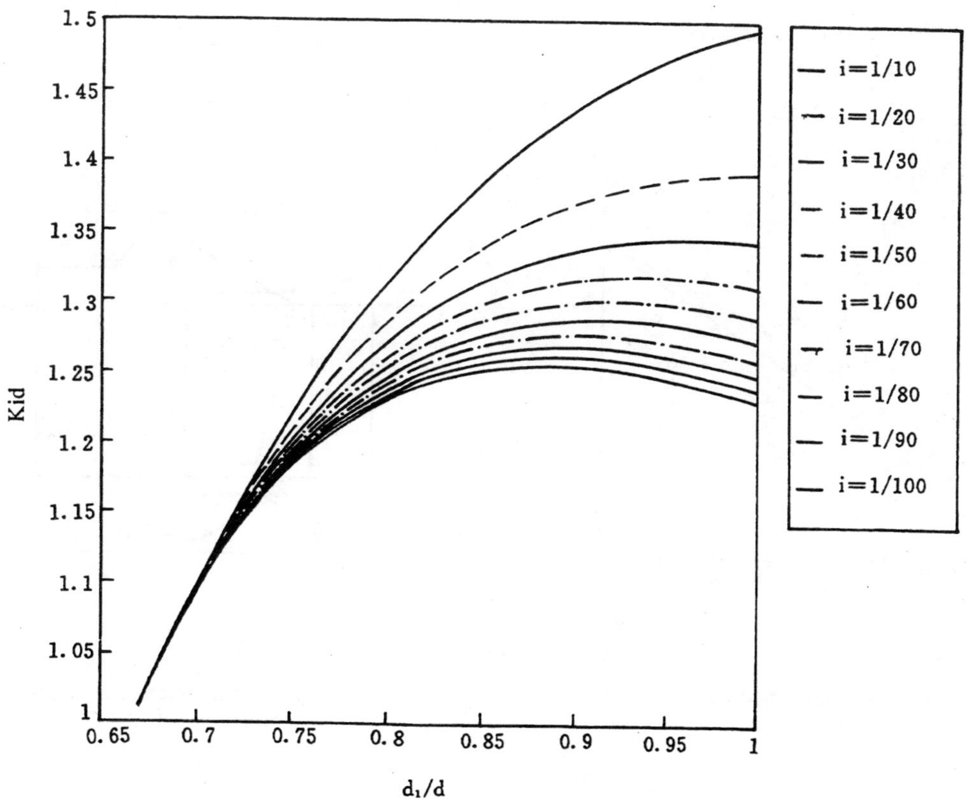

Fig.3 Kid related to d_l/d and i

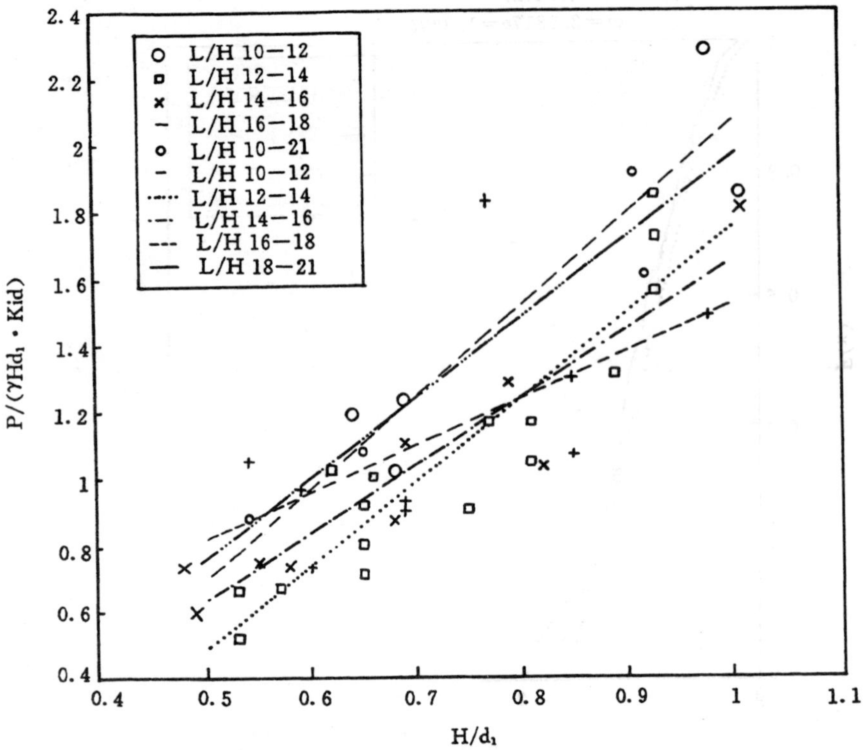

Fig.4 influence of wave steepness on wave forces

L1:Y=2.573*X-0.72 r=0.78

L2:Y=2.13*X-0.21 (used)

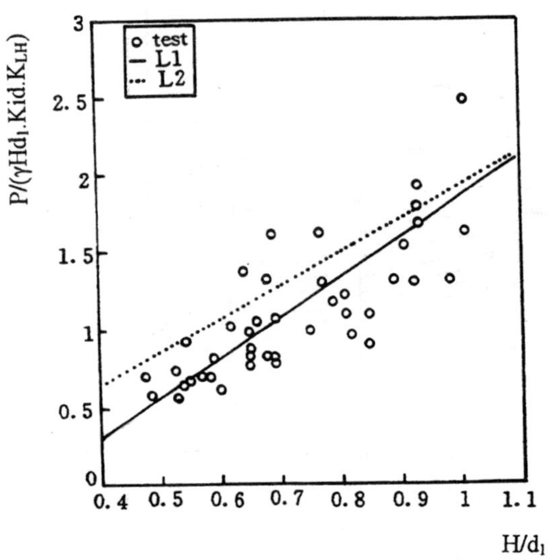

Fig.5 total horizontal forces

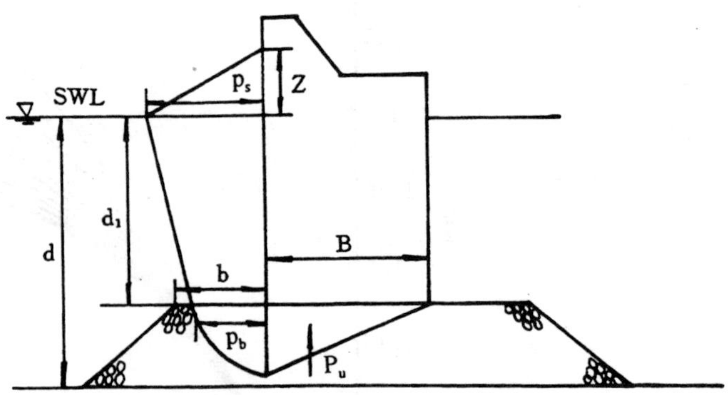

Fig.7 distribution of wave pressure

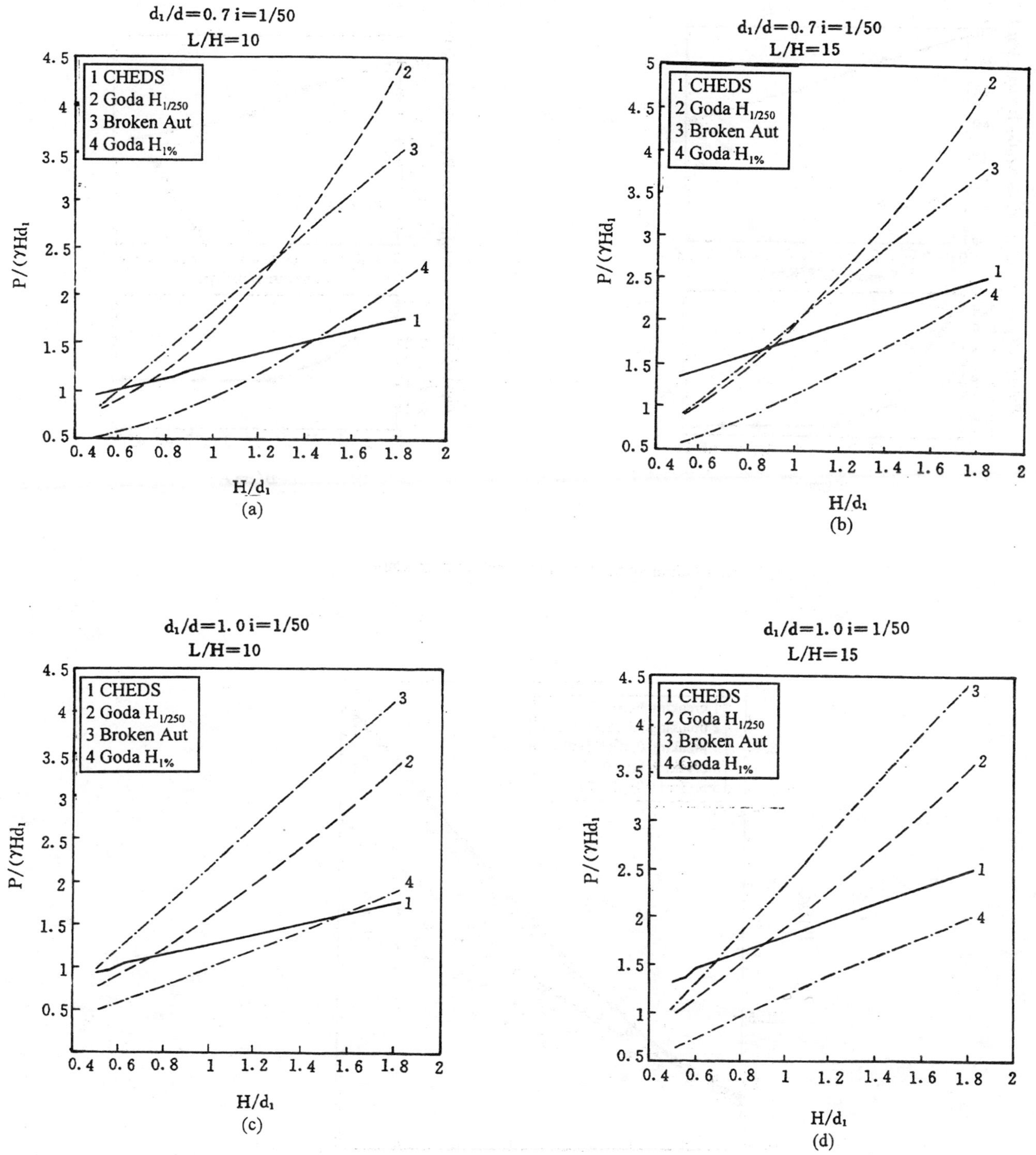

Fig. 6 comparison of various methods

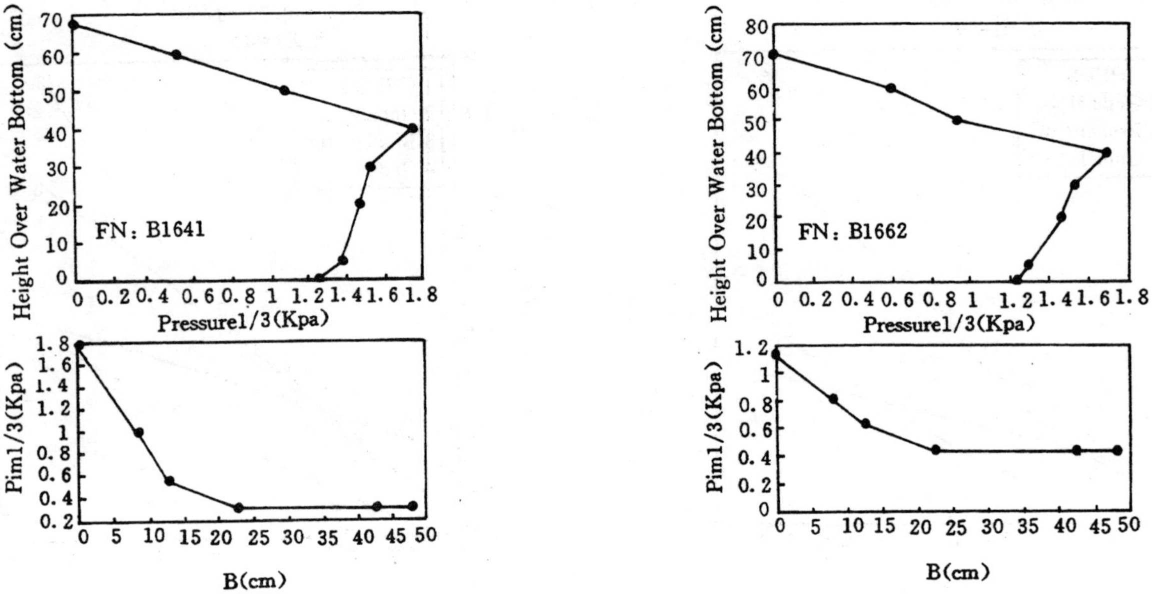

Fig.8 distribution of wave pressure and uplift pressure

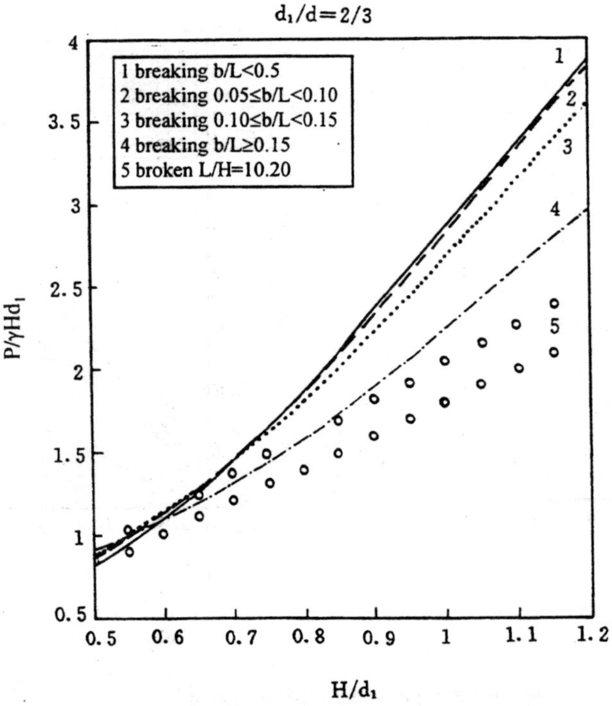

Fig.9 comparison of wave forces by breaking and broken waves

Proceedings of the Seventh (1997) International Offshore and Polar Engineering Conference
Honolulu, USA, May 25-30, 1997

ISBN 1-880653-28-1 (Set); ISBN 1-880653-31-1 (Vol. III)

Viscous Effects in Wave-Body Interaction

M. Landrini
INSEAN
Rome, Italy

M. Ranucci, C.M. Casciola and G. Graziani
University of Rome "La Sapienza"
Italy

ABSTRACT

The laminar flow about a circular cylinder beneath a Stokes waves train is numerically investigated.

An efficient grid free algorithm is developed by coupling an accurate boundary integral equation method for computing the velocity field with a viscous vortex method for solving the Navier Stokes equations near the body. The problem of generating the incident wave system is effectively circumvented by means of a perturbation formulation which assumes the Stokes wave solution as the base flow.

For $\beta \simeq 500$ and $K_c = \mathcal{O}(1)$ the systematic comparison with the available experimental values for the Fourier components of the loading is presented. An overall good agreement is observed, even for the added inertia coefficient which is known to be largely affected by viscous effects.
KEY WORDS: Nonlinear free surface flows, wave loading, wave induced vortex shedding, wave diffraction.

INTRODUCTION

The flow of an incompressible viscous fluid about a circular cylinder beneath a regular Stokes wave train is a typical problem of marine structures hydrodynamics. When the relevant wavelengths are comparable with the characteristic dimension of the body, the hydrodynamic loads are quantitatively dominated by the momentum exchange between the wave and the body. Consistently, it is widely accepted that the nonlinear potential flow theory effectively describes the force on the structure. In fact, the experimental analysis confirms that the vertical mean value of the hydrodynamic force and of both the second and third harmonics of the fluctuating components are well explained in terms of the inviscid wave diffraction in agreement with the predictions of weakly or fully nonlinear models (Ogilvie, 1963, Vada, 1987 and Liu et. al., 1992).

However, the potential theory fails to capture exhaustively the entire phenomenon. Actually, a significant reduction in the amplitude of the fundamental harmonic of the loading is emphasized by Chaplin (1984), with respect to purely inviscid models. This correction exhibits a marked dependance on the Keulegan–Carpenter number K_c. When no significant separation occurs, the observed $(K_c)^3$ behaviour has been ascribed to a Kutta–Joukowsky–like force originating by the streaming flow within the wave induced boundary layer. If both the wavelength and the submergence are much larger than the radius, the free surface effects can be neglected and the driving field is well approximated through an orbital flow. Concerning this problem, a considerable insight has been gained by Chaplin (1993) through the numerical solution of the full Navier–Stokes equations. For high Reynolds number and small wave amplitude, the major features of the steady state solution of the complete problem have been fully addressed in Yan and Riley (1996) by an elegant perturbation approach.

In order to analyze the phenomenon under more general conditions, i.e. when significant separation occurs, we have to consider the full Navier–Stokes equations for a free surface flow. More specifically, we discuss in detail different solutions for K_c ranging from 0.3 to 1.2 and for a fixed value of the frequency parameter $\beta = Re/K_c$, where Re is the Reynolds number. The value of β, 9120 in Chaplin's experiments, is presently 483 due to limits in the computational resources while our range of the Keulegan–Carpenter number is restricted by the occurrence of wave breaking at $K_c \simeq 1.2$, a value close to the experimental observations. The other relevant parameters, radius–to–wavelength ratio and submergences of the cylinder, are selected to match those considered in (Chaplin, 1984).

In these conditions, the wave dynamics is essentially nonlinear, owing to the small submergence of the cylinder, and it is tightly coupled to the vorticity field. On the other hand, the Reynolds number is large enough to confine the vorticity close to the body and away from the free–surface. We have then to consider the fully nonlinear behaviour of the free surface and flow separation near the body, but we may neglect the viscous effects at the interface.

The present approach couples, through a fractional step algorithm, a boundary integral equation method for the free surface (Casciola and Landrini, 1996) with a viscous vortex method for the vorticity (Graziani et al., 1995). A similar numerical procedure has been recently adopted also by Yeung et al. 1996. In particular, the wave pattern is described by following markers suitably distributed on the interface, where a dynamic equation for the tangential velocity component has to be solved. A particle method with a deterministic diffusion is used for the dynamics of the vorticity. The fluid velocity is expressed by an integral representation formula in terms of the vorticity in the field and of the velocity itself at the boundaries. This yields an integral equation which allows for enforcing both the normal velocity component at the body and the tangential component at the interface, which follows from the free surface dynamic condition. At the same time, the computed tangential velocity at the solid boundary allows for the generation of vorticity in order to satisfy the no-slip condition (Chorin, 1973). Since the number of discrete vortices in the field dramatically increases with the Reynolds number, the adoption of a fast summation technique for the rapid evaluation of the vorticity-induced component of the velocity is crucial for a reliable comparison with the experiments.

We introduce the mathematical model in the next section, where the free-surface evolution equations, the integral representation for velocity and the generation and diffusion of the vorticity are briefly recalled. The results are discussed by comparison with the available experimental and analytical data. Some conclusion are finally drawn in the last section.

MATHEMATICAL STATEMENT OF THE PROBLEM

We consider a circular cylinder $\mathcal{B}$ immersed beneath a regular wave train on the free surface $\mathcal{F}$. The water depth is assumed infinite, and the fluid domain $\mathcal{D}$ is unbounded in the horizontal direction. A coordinate system x–y attached to the body at rest is introduced, with the y–axis vertical upward. The motion of the viscous incompressible fluid within the domain $\mathcal{D}$ is described by the Navier–Stokes equations,

$$\nabla \cdot \boldsymbol{u} = 0 \quad , \qquad \frac{D\boldsymbol{u}}{Dt} = -\nabla p/\rho + \nu \nabla^2 \boldsymbol{u} + \boldsymbol{g} \tag{1}$$

where ρ is the fluid density, ν the kinematic viscosity, p the pressure an $\boldsymbol{g}$ the gravity acceleration. The standard boundary condition

$$\boldsymbol{\nu} \times \boldsymbol{u} = 0 \qquad \boldsymbol{\nu} \cdot \boldsymbol{u} = 0 \tag{2}$$

must be satisfied on the solid boundary $\partial\mathcal{B}$, and the kinematic and dynamic conditions

$$\boldsymbol{w} \cdot \boldsymbol{\nu} = \boldsymbol{u} \cdot \boldsymbol{\nu} \tag{3}$$

$$\boldsymbol{\tau} \cdot \boldsymbol{t} = 0 \qquad \boldsymbol{\nu} \cdot \boldsymbol{t} = p_a \tag{4}$$

hold at the free surface. In the above relations $\boldsymbol{\tau}$ and $\boldsymbol{\nu}$ are the unit vectors tangent and normal to $\partial\mathcal{D} = \mathcal{F} \cup \partial\mathcal{B}$ respectively, while $\boldsymbol{t}$ is the traction, p_a the atmospheric pressure and $\boldsymbol{w}$ the velocity of the free surface. We neglect the effects of the free surface boundary layer and assume the support of the body-generated vorticity to be confined away from the free interface. Consistently, the dynamic boundary condition on $\mathcal{F}$ simply requires the fluid pressure to be atmospheric, $p = p_a$. Under such hypothesis, for finite times and far from the body, the fluid motion reduces to an irrotational wave system. We consider the incoming Stokes wave field $\boldsymbol{u}_S = \nabla\phi_S$ as a basis flow and solve for the correction $\boldsymbol{v}$. Then, the fluid velocity is $\boldsymbol{u} = \boldsymbol{v} + \boldsymbol{u}_S$, with vorticity $\omega = \nabla \times \boldsymbol{u} \cdot \boldsymbol{e}_z = \nabla \times \boldsymbol{v} \cdot \boldsymbol{e}_z$. To be more specific, the initial conditions is $\boldsymbol{u} = \boldsymbol{u}_S$; soon after the cylinder appears and the correction field $\boldsymbol{v}$ arises in order to account for the diffraction of the incoming wave and for the body generated vorticity. Since the development of the actual transient is not relevant for an appropriate comparison with the experimental results in the steady state conditions, the adopted approach circumvents the problem of generating the incident wave field.

In the framework of a vorticity–velocity formulation, we consider a viscous vortex method coupled with the free surface dynamics. The algorithm, based on the operator splitting procedure sketched in figure 2, approximates the flow evolution by a sequence of diffusive and convective steps according to the Stokes and Euler equations, respectively. In particular, the convection step consists in solving the problem

$$\frac{D\omega}{Dt} = 0, \qquad \frac{D\boldsymbol{x}}{Dt} = \boldsymbol{u}, \qquad \boldsymbol{u} \cdot \boldsymbol{\nu} = 0 \quad \text{on } \partial\mathcal{B} \tag{5}$$

where the free surface evolves according to a purely inviscid dynamics, described in the following section. The diffusion of the vorticity ω is obtained through the solution of the following set of equations

$$\frac{\partial\omega}{\partial t} = \nu\nabla^2\omega, \qquad \boldsymbol{\nu} \cdot \nabla\omega = 0 \quad \text{on } \partial\mathcal{B}, \qquad \omega = \boldsymbol{\nu} \cdot \nabla\omega = 0 \quad \text{on } \mathcal{F} \tag{6}$$

which is to be completed with a suitable procedure of vorticity generation at the solid boundary (discussed in §). In order to describe the diffusion of the vorticity, a deterministic approach based on the related integral representation is adopted as discussed in Graziani et al. (1995) for flows without a free surface.

Free surface evolution equations

Let us now briefly describe the governing equations for the free surface motion. By considering a curvilinear coordinate ξ along the free surface, the motion of $\mathcal{F}$ is described by the kinematic equation

$$\frac{\partial \boldsymbol{x}(\xi,t)}{\partial t} = \boldsymbol{w}(\xi,t) \, , \tag{7}$$

where $\boldsymbol{x}(\xi,t)$ is the free surface profile and the free surface velocity $\boldsymbol{w}(\xi,t)$ must satisfy (3). In particular, when $\boldsymbol{w}$ is asumed to concide with the fluid velocity $\boldsymbol{u}$, ξ is a material parameter. Otherwise, we may adopt an Eulerian description of the free surface $y = \eta(x,t)$ and define $\boldsymbol{w} \equiv (0, \dot{\eta})$, with $\dot{\eta} = u_y - \eta_x u_x$. This choice, which is still consistent with (3), considers only vertical displacements of the free surface markers.

By neglecting the effect of viscosity at the free surface, the dynamic condition is expressed in the form

$$\frac{\partial u_\tau(\xi,t)}{\partial t} = ((\boldsymbol{w}-\boldsymbol{u})\cdot\nabla\boldsymbol{u})\cdot\boldsymbol{\tau} + u_\nu\boldsymbol{\nu}\cdot\frac{\partial \boldsymbol{\tau}(\xi,t)}{\partial t} + \boldsymbol{g}\cdot\boldsymbol{\tau} \, . \tag{8}$$

In the framework of the perturbation formulation presently adopted, after introducing the tangent vector $\boldsymbol{e}_\xi = \partial\boldsymbol{x}/\partial\xi$ and the covariant velocity component $v_\xi = \boldsymbol{v}\cdot\boldsymbol{e}_\xi$, the dynamic equation is recast in the form

$$\frac{\partial v_\xi(P,t)}{\partial t} = \partial_\xi \left\{ \boldsymbol{w}\cdot\boldsymbol{v} - \dot{\phi}_s - \frac{1}{2}u^2 - g\eta \right\} \tag{9}$$

where, as discussed above, the Stokes' potential $\phi_s(\boldsymbol{x},t)$ is a known function.

Kinematic problem

The knowledge of the fluid velocity is required in order to evaluate the inviscid evolution of the vorticity field and the motion of the free surface. In particular, the perturbation velocity field $\boldsymbol{v}$ in the fluid domain is expressed in terms of the Poincaré representation formula (Bassanini et al., 1991) as

$$\begin{aligned} \boldsymbol{v}(\boldsymbol{x}) \;=\; & -\nabla_x \cdot \int_{\partial\mathcal{D}} \boldsymbol{v}(\boldsymbol{y})\cdot\boldsymbol{\nu}(\boldsymbol{y}) G(\boldsymbol{y},\boldsymbol{x}) dS \\ & -\nabla_x \times \int_{\partial\mathcal{D}} \boldsymbol{v}(\boldsymbol{y})\times\boldsymbol{\nu}(\boldsymbol{y}) G(\boldsymbol{y},\boldsymbol{x}) dS \, . \\ & -\nabla_x \times \boldsymbol{k} \int_{\mathcal{D}_\omega} \omega(\boldsymbol{y}) G(\boldsymbol{y},\boldsymbol{x}) dA \end{aligned} \tag{10}$$

At each convective–step, a boundary integral equation is firstly solved to determine v_ν on $\mathcal{F}$ and v_τ on ∂B and then the velocity field in $\mathcal{D}_\omega$ is explicitely computed through (10).

The relevant integral equations, which are obtained by projecting (10) along the normal $\boldsymbol{\nu}$ to $\mathcal{F}$ or along the tangent $\boldsymbol{\tau}$ to ∂B, are approximated *via* a quadrature method based on the Euler-Mac Laurin summation formula (see *e.g.* Sidi and Israeli, 1988). By this method spectral accuracy is normally achieved for periodic densities, as in the case of a closed smooth body contour, and for compactly supported densities on the free surface, as they may be assumed for finite times. However, for a truncated computational domain, v_τ and v_ν as well as the perturbation with respect to the Stokes' wave height do not vanish at the edges and unphysical reflections may occur. This drawback is avoided by the introduction of artificial damping layers (Baker et al., 1989), which prevent the unwanted reflections by forcing the densities to vanish towards the edges of the computational domain. The accuracy of the computation is maintained under control by comparing different solutions for increasing domain lengths.

The direct evaluation of the velocity field through the discretized form of (10) usually requires a prohibitive amount of computational resources. Hence a fast summation algorithm based on the multipole expansion of the Poincaré kernels is adopted, as proposed by Greengard and Roklin (1987). and an operation count proportional to the number of particles is achieved. In order to further enhance the computational performances, an iterative solver is used instead of a direct matrix inversion.

Vorticity generation and diffusion

The sequence of advection and diffusion steps approximates the solution of the Navier–Stokes equations in free space. If the flow past a solid body is considered, the enforcement of the no–slip boundary condition is satisfied, following Chorin (1973), by inserting on the body contour, after each convective step, a vortex sheet with circulation density $\gamma = \dot{v}_\tau$ which is evaluated through (10). This vortex sheet is furtherly lumped in order to obtain the circulation of discrete vortices which are placed on the body contour. Finally, the newly generated vortices are diffused together with those already existing within the flow field. This step is accomplished by evaluating the effect of the diffusion on the nodes of a suitably defined mesh. By starting from the integral representation of the solution of the heat equation and by considering the point vortex approximation, the diffusion of the vorticity field after one time step $\Delta t = t - t_0$ is expressed as

$$\omega(\boldsymbol{x},t) = \sum_1^N \Gamma_j(t_0) F(\boldsymbol{x}-\boldsymbol{x}_j, \Delta t) \tag{11}$$

where the summation is extended to the N vortices at $\boldsymbol{x}_j$ with circulation Γ_j and

$$F(\boldsymbol{x}-\boldsymbol{x}_j, \Delta t) = \frac{1}{4\pi\nu\Delta t} \exp\left(-\frac{|\boldsymbol{x}-\boldsymbol{x}_j|^2}{4\nu\Delta t}\right) \tag{12}$$

is the fundamental solution of the heat equation.

DISCUSSION OF THE NUMERICAL RESULTS

In the following we analyze the interaction of a Stokes wave train with a submerged circular cylinder. In order to enlight the role of the viscosity in the phenomenon, the solutions of the Navier–Stokes equations are compared with those provided by an inviscid model as well as with the experimental results reported in Chaplin (1984).

In particular we focus the attention on Chaplin's Case–E ($kR = 0.206$, $h/R = 2.0$) since non–linear diffraction effects turn out to be more pronounced the shallower the submergence of the cylinder. In the present simulations the Keulegan–Carpenter number, $K_c = \pi ka/kRe^{-kh}$, ranges from 0.3 to 1.2. Actually, for larger values the wave breaks over the cylinder and a steady regime cannot be reached. Further, we assume infinite channel depth since the considered waves are 'nominally in deep–water'. Typically the free surface is described through 64 markers per wavelength and a timestep $1/160$ of the incoming wave period, T_S, is used.

The cylinder is placed at $x = 0$ and the wave train propagates from left to right. Consistently with the experiments (Chaplin, 1984) and with the inviscid results (Palm, 1992) no significant reflection occurs. A more complicated wave pattern (fig. 3) emerges on the lee side of the body, where the phase shift of the transmitted waves (Ogilvie, 1963), propagates with the group velocity of the incident wave train. Shorter waves ($\lambda_S/4, \lambda_S/9, ...$) gradually appear and propagate with a smaller velocity (Grue, 1992).

Figure 3 also shows the effect of the incident wave steepness $\epsilon = ka$ (*i.e.* the K_c number) and compares the viscous (solid lines) and inviscid (broken lines) solutions. Qualitatively, the wave diffraction dynamics appears to be unaffected by the vorticity generated at the body boundary. In more detail, while the viscous and inviscid solutions appear to be coincident for $K_c = 0.3$ and $K_c = 0.5$, an increasingly pronounced difference is detected for $K_c = 0.75$ and $K_c = 1.0$, the amplitude of the emitted shorter wavelengths resulting to be reduced in the viscous solution.

Consistently, the wave induced loading predicted by the viscous and by the inviscid model coincides for the smaller K_c while exhibiting significant discrepancies for the largest value. In particular, the time evolution of the two force components F_x, F_y is reported in figure 4 for $K_c = 1.0$. After an initial transient lasting about one period, where the two solutions are substantially superimposed, the viscous loads presents larger oscillation amplitudes. In both cases the apparent non–sinusoidal behaviour in time of the forces manifests the relevant presence of non–linear effects.

Since the differencies between the two solutions are related to the generation and to the evolution of the vorticity, let us examine in detail the dynamics of the rotational field.

Figures 5–7 show the vorticity distribution at different instants with the free surface profile reported at natural scale. The vorticity support is observed to remain well confined below the free surface. In particular, concerning figure 5 which displays the initial development of the vortical structure about the body, several features of the complex vorticity dynamics can be stressed. Two intense vorticity layers of opposite sign develop at the solid boundary driven by the clockwise rotating flow induced by the waves (first two plots). Afterwards, the diffusion spreads the vorticity farther from the cylinder. A thickened structure is generated and new vortex layers of opposite sign appear on the body (third plot). Finally, mainly forced by the external velocity field, the opposite signed layers interact, producing the well defined structure above the cylinder (last plot).

The evolution of this structure up to the second period is further analyzed in the next figure 6. More precisely, following a clockwise path around the cylinder, it becomes progressively detached from the body. In the meanwhile, the gross features of the vorticity field near the body described for fig. 5 are reproduced through the same dynamics, which eventually leads to the formation of a new structure (last plot in fig. 6) strikingly similar to that pointed out in the last plot of figure 5.

The periodicity of both the near wall vorticity and the free surface can be appreciated for $t/T_S = 10, 12, 14, 16$ in figure 7. For increasing time, the lee side vorticity feeds a large structure beneath the cylinder which evolves into a dipole–like shape and eventually detaches from the body. Due to its weak strength, it does not significantly affect the vorticity field near the body, which actually is periodically reproduced.

This behaviour can be quantitatively assessed by averaging over a period the circulation along circles C of increasing radius $\hat{r} = r/R$ concentric with the cylinder

$$\Gamma(nT_s, \hat{r}) = \frac{1}{T_s} \int_{nT_S}^{(n+1)T_S} \oint_{C(\hat{r})} \boldsymbol{u} \cdot d\boldsymbol{\ell} \, .$$

In particular, for $\hat{r} < 2$, Γ is independent on time, showing that a steady mean circulation has been estabilished about the cylinder. On the other hand, the *far* behaviour of the vorticity field clearly affects the distribution of the mean circulation for increasing distances.

In his analysis of the nonlinear behaviour of wave induced loads, Chaplin pointed out the key role of the mean circulation about the cylinder in explaining the marked decrease of the inertia coefficient with respect to the potential flow prediction. This difference clearly appears from the upper plot in figure 9, where the present inviscid solution (solid line) always exceeds Chaplin's experimental data. The correct behaviour is recovered by the viscous solution shown in the same plot (∘, •), thus enlighting the crucial role of the wave–induced boundary layer. It is worthwhile to discuss a viscous solution obtained by completely neglecting free surface effects (□, ■). In this case the fluid motion about a cylinder in an infinite domain is forced by the Stokes velocity field. As concerning the first harmonic of the hydrodynamic loads, even this simplified viscous solution reproduces the experimental data thus confirming the weak influence of the free surface on this force component.

From figure 9, which reports the vertical mean value, the second and the third harmonic of the loading, the other harmonics appear to be less dominated by viscous effects. The fully nonlinear inviscid results (solid lines) recover the second order theories (dashed lines in the second and third plot) for smaller K_c while introduce small corrections for the larger values. Interestingly, the inviscid and the viscous solutions (∘, •) are coincident for small wave amplitude while the viscosity appears to modify, although slightly, the inviscid results toward a better agreement with the experimental data for increasing K_c. Both the models slightly underpredict the $F_i^{(3)}$ harmonics, though still providing a reasonable agreement. Unlike the $F_i^{(1)}$ case, the viscous model without the free surface (□, ■) turns out to be completely inadequate in predicting these harmonics of the forces, the wave dynamics appearing now the key–mechanism which originates these components of the loading.

A special comment on the case $K_c = 1.2$ is finally due. In fact, for this wave amplitude a steady regime is not attained after 15 periods and the reported value is obtained as a mean over the last five periods. Unlike the lower K_c, the horizontal (filled symbols) and the vertical (open symbols) force components are different and the support of the vorticity (not shown) becomes closer to the free surface . More reliable results for the highest values of K_c could be obtained by modifying the dynamic free surface condition to account for viscous effects. In any case, from this wave amplitude on, the lagrangian description of the interface (in both the viscous and the inviscid models) indicates the evolution towards wave breaking in the form of a local steepening over the cylinder which terminates the computation.

CONCLUSION

The model proposed to solve the Navier–Stokes equation for the wave diffraction problem efficiently couples a boundary integral equation approach for the free surface with a grid–free vortex method with deterministic diffusion.

As the wave loading is concerned, a satisfactory agreement with Chaplin's experiments is obtained. In particular the decreasing behaviour of inertia coefficient for increasing K_c is recovered and the dominant role of viscosity is further stressed by considering a simplified viscous flow without free surface. The effects of the wave diffraction are more pronounced for the other force harmonics, which are nicely predicted by the inviscid model, though the viscosity seems to act toward a further improvement of the results.

Increasing differences in the wave amplitudes predicted by the viscous and the inviscid models are observed for large K_c. In these conditions the appearance of three–dimensional instabilities as well as the onset of turbulence, not considered in the present approach, are presumably relevant (Chaplin, 1993). Furthermore, possible interactions of the vorticity field with the free surface are expected, requiring a more complete treatment of the interface boundary condition presently under development.

REFERENCES

Chorin, A.J. (1973). "Numerical study of slightly viscous flow", *J. Fluid Mech.*, Vol. 57.

Graziani, G., Ranucci, M. and Piva, R. (1995). "From a Boundary Integral Formulation to a Vortex Method for Viscous Flows", *Comput. Mech.*, Vol. 15, No. 4.

Greengard, L. and Rokhlin, V. (1987). "A fast algorithm for particle simulations", *J. Comput. Phys.*, Vol. 73.

Baker, G. R., Meiron, D. I. and Orszag S. A. (1989) "Generalized vortex methods for free surface flow problems. II: Radiating waves", *J. Scient. Comp.*, Vol. 4.

Bassanini, P., Casciola, C. M., Lancia, M. R. and Piva, R. (1991). "A boundary integral formulation for the kinetic field in aerodynamics. Part I." *Europ. J. Mech.*, B/Fluids, Vol. 10, No. 6.

Casciola, C. M. and Landrini, M. (1996). "Nonlinear long waves generated by a moving pressure disturbance", *J. Fluid Mech.*, Vol. 325.

Sidi, A. and Israeli, M. (1988). "Quadrature methods for periodic singular and weakly singular Fredholm integral equations", *J. Scien. Comp.*, Vol. 3.

Chaplin., J.R. (1984). "Nonlinear forces on a horizontal cylinder beneath waves", *J. Fluid Mech.*, Vol. 147.

Chaplin., J.R. (1993) "Orbital flow around a circular cylinder. Part 2. Attached flow at larger amplitudes", *J. Fluid Mech.*, Vol. 246.

Cointe, R. (1989) "Nonlinear simulation of transient free surface flows", *5th International Conference Numerical Ship Hydrodynamics*. Hiroshima, Japan.

Grue, J. (1992) "Nonlinear water waves at a submerged obstacle or a bottom topography", *J. Fluid Mech.*, Vol. 244.

Liu, Y., Dommermuth, D.G., Yue, D.K.P. (1992) "A high–order spectral method for nonlinear wave–body interactions", *J. Fluid Mech.*, Vol. 245.

Ogilvie, T. F. (1963) "First and second order forces on a circular cylinder submerged under a free surface", *J. Fluid Mech.*, Vol. 16.

Palm, E. (1991) "Nonlinear wave reflection from a submerged circular cylinder under a free surface", *J. Fluid Mech.*, Vol. 233.

Ursell, F. (1950) "Surface waves in presence of a submerged circular cylinder, I e II", *Proc. Camb. Phil. Soc.*, Vol. 46.

Vada, T. (1987) "A numerical solution of the second–order wave–diffraction problem for a submerged cylinder of arbitrary shape", *J. Fluid Mech.*, Vol. 174.

Yan, B. and Riley, N. (1996) "Boundary–layer flow around a submerged circular cylinder induced by free–surface travelling waves", *J. Fluid Mech.*, Vol. 316.

Yeung, R.W., Cermelli, C. and Liao, S.W. (1996) "Vorticity fields due to rolling bodies in a free surface – Experiment and Theory", *21st Symposium on Naval Hydrodynamics*, Trondheim, Norway, 1996.

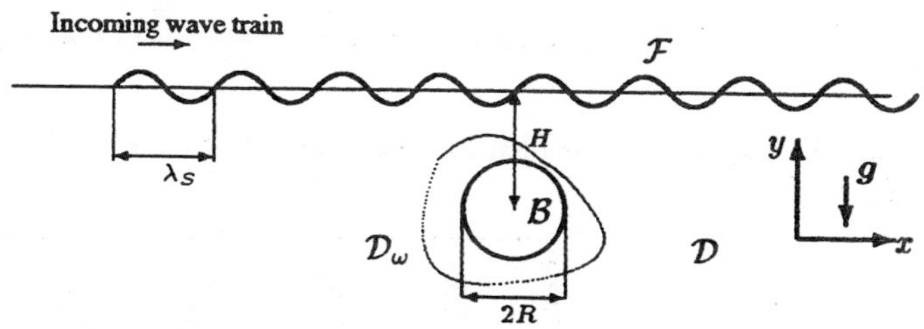

Figure 1: Geometry of the problem and nomenclature adopted

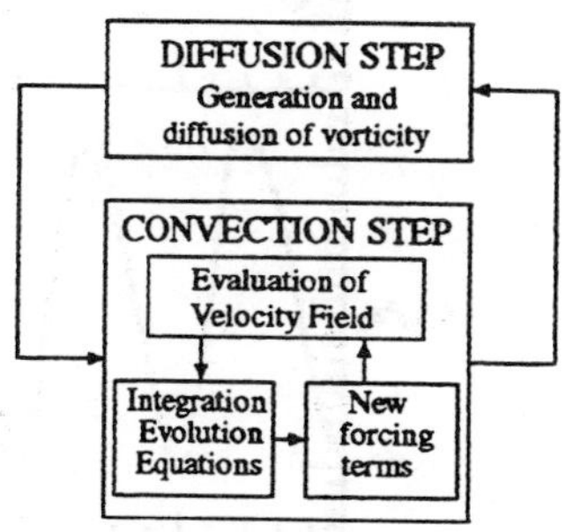

Figure 2: Flow chart of the operator splitting algorithm

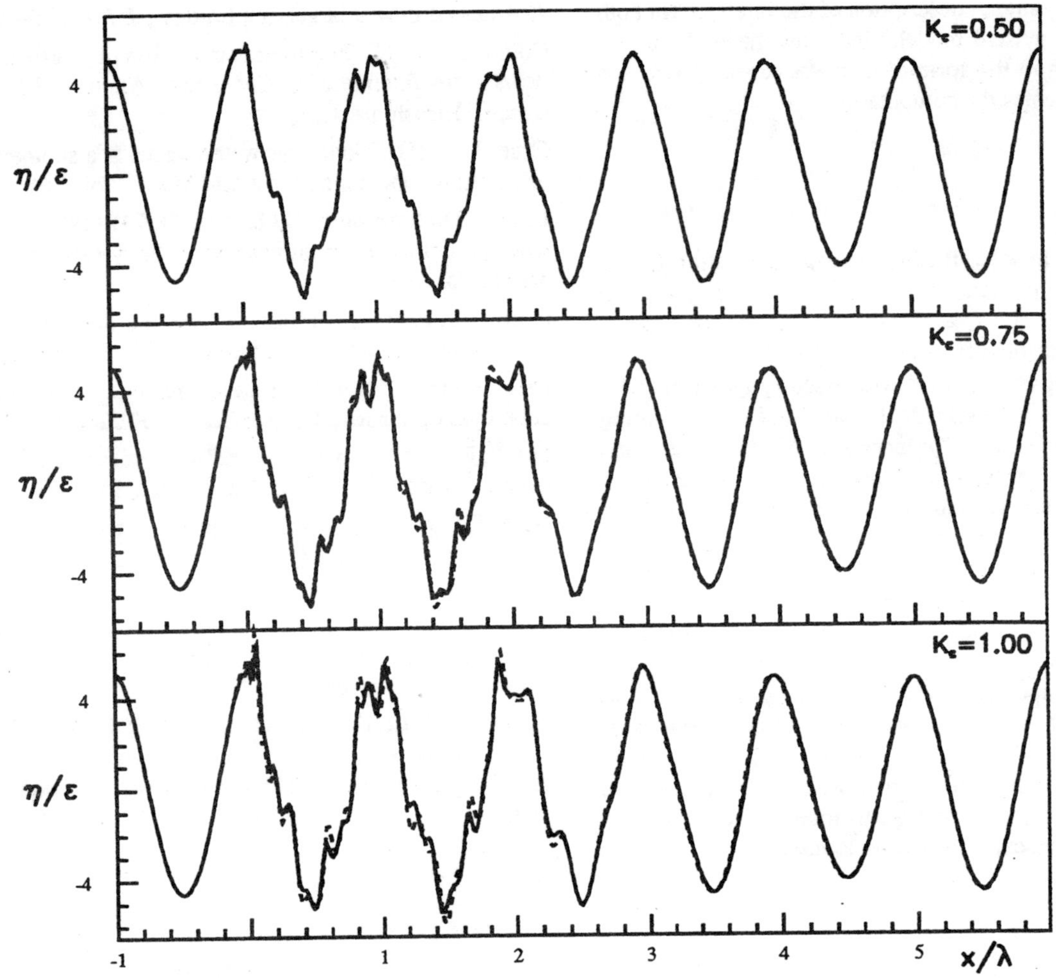

Figure 3: Free surface profiles for increasing K_c number. Comparison of viscous (solid lines) and inviscid (broken lines) solutions, $t/T_S = 10$.

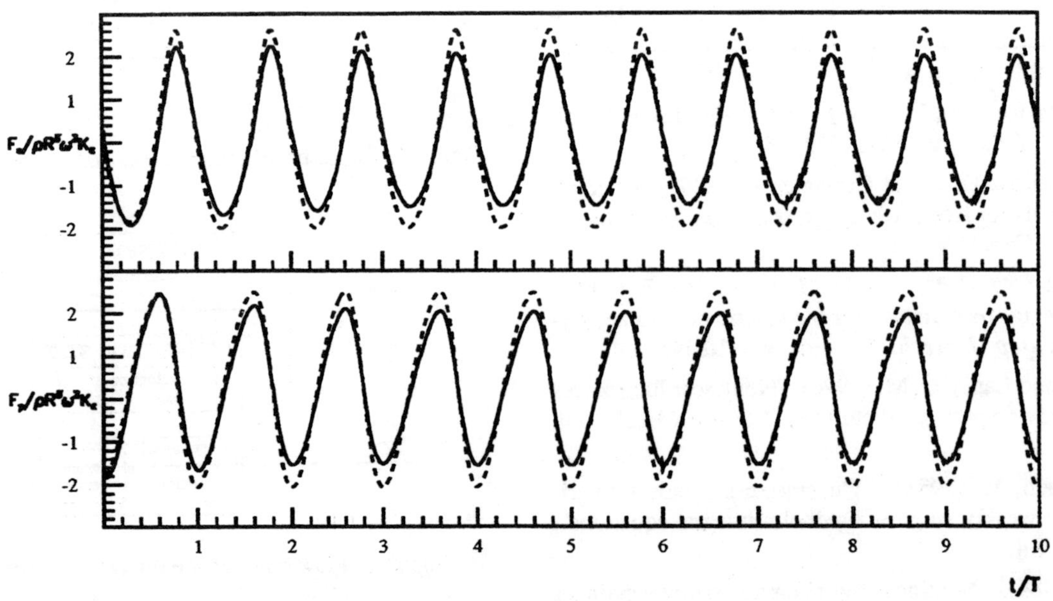

Figure 4: Evolution in time of horizontal (upper plot) and vertical (lower plot) force components ($K_c = 1.0$). Comparison of the viscous (solid lines) and inviscid (broken lines) solutions.

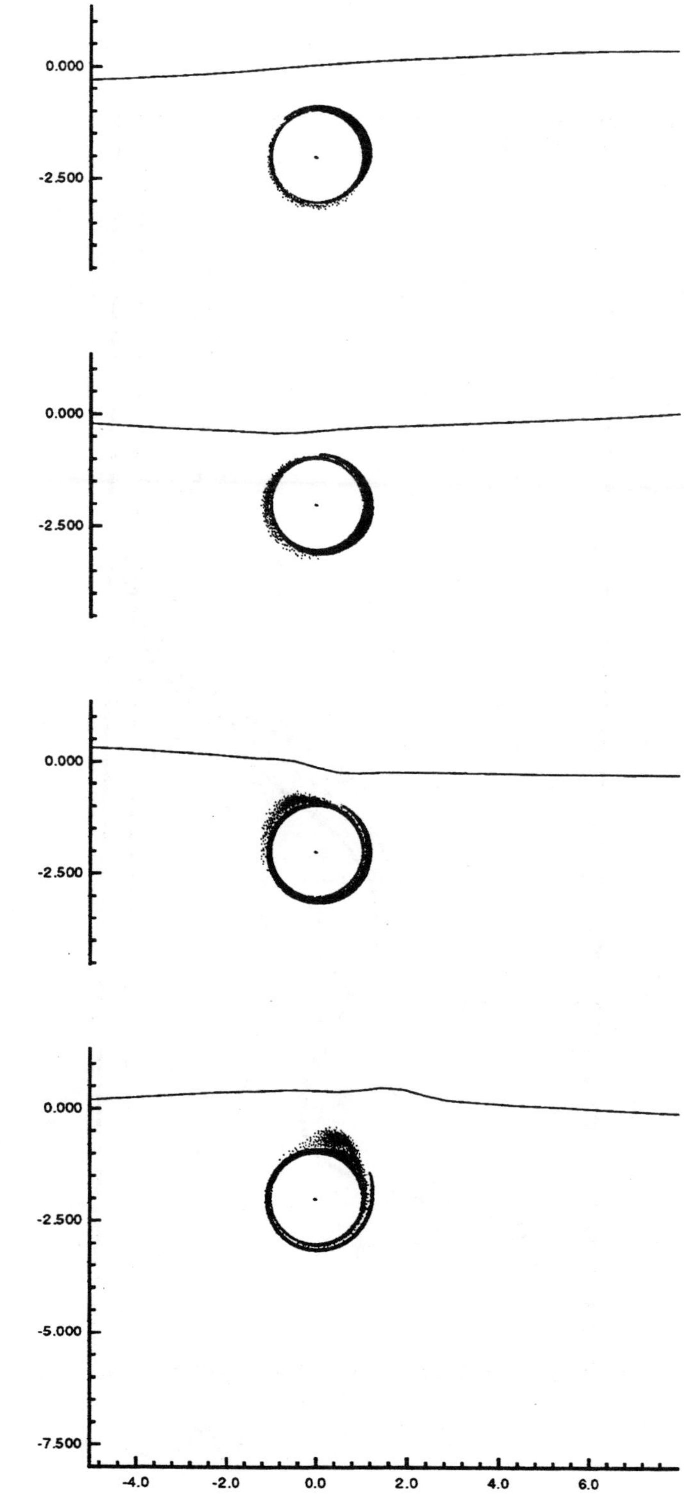

Figure 5: Vorticity contours and free surface profile for $K_c = 0.75$. From top down $t/T_S = 0.25, 0.5, 0.75, 1.0$. Red–yellow: positive (counterclockwise) vorticity. Blue–cyan: negative (clockwise) vorticity.

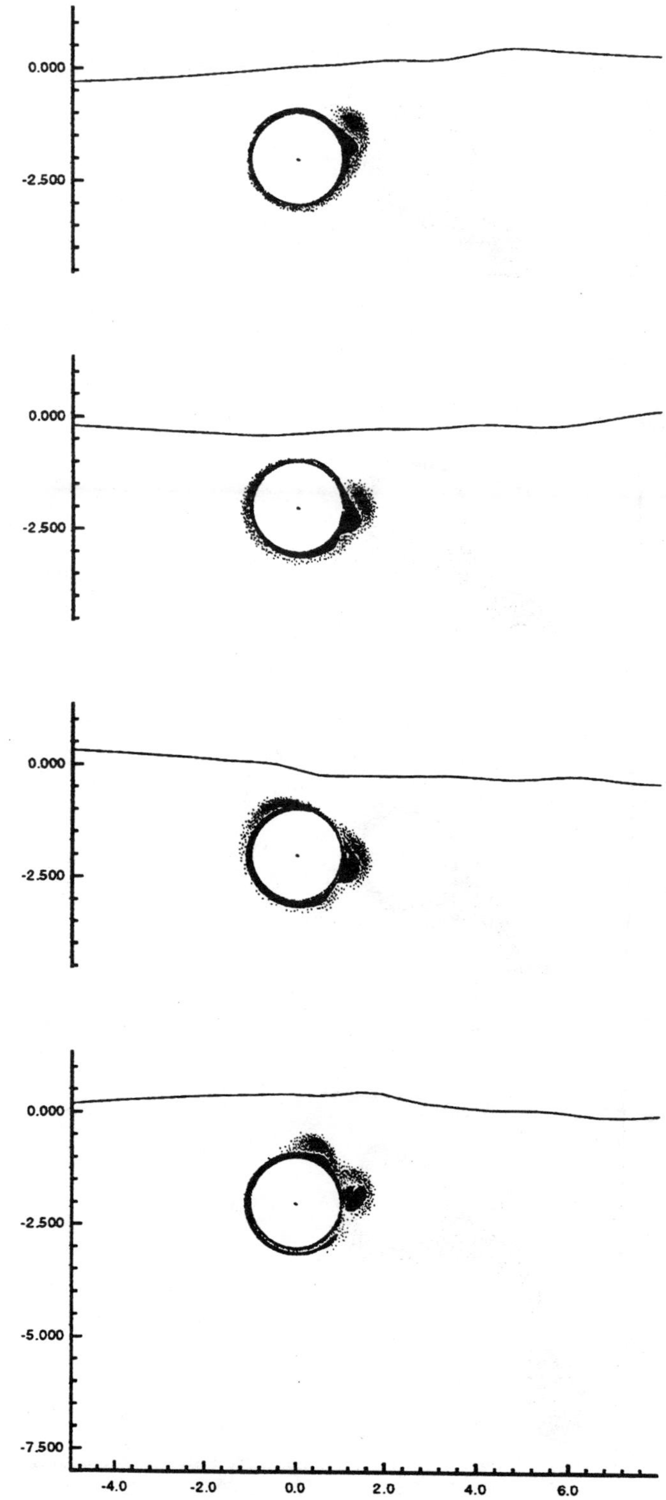

Figure 6: Vorticity contours and free surface profile for $K_c = 0.75$. From top down $t/T_S = 1.25, 1.5, 1.75, 2.0$. Red–yellow: positive (counterclockwise) vorticity. Blue–cyan: negative (clockwise) vorticity.

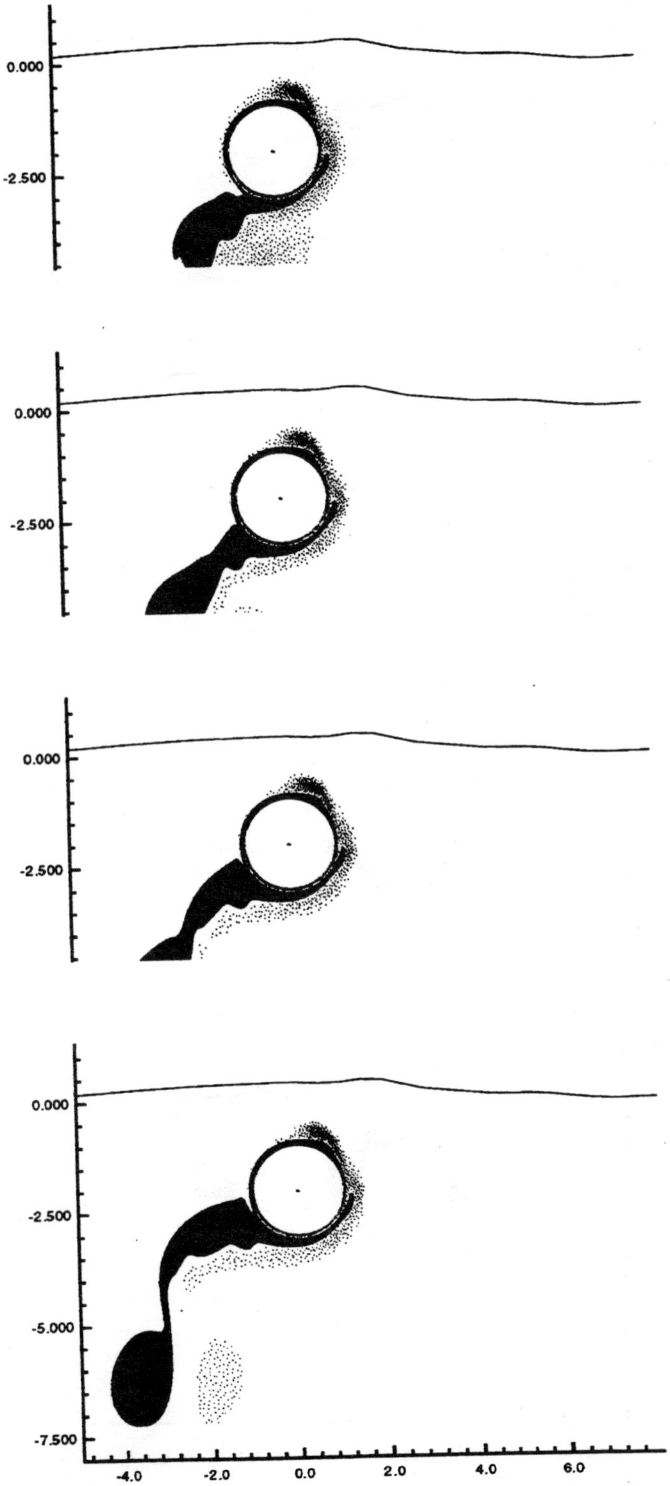

Figure 7: Vorticity contours and free surface profile for $K_c = 0.75$. From top down $t/T_S = 10., 12., 14., 16.0$. Red–yellow: positive (counterclockwise) vorticity. Blue–cyan: negative (clockwise) vorticity.

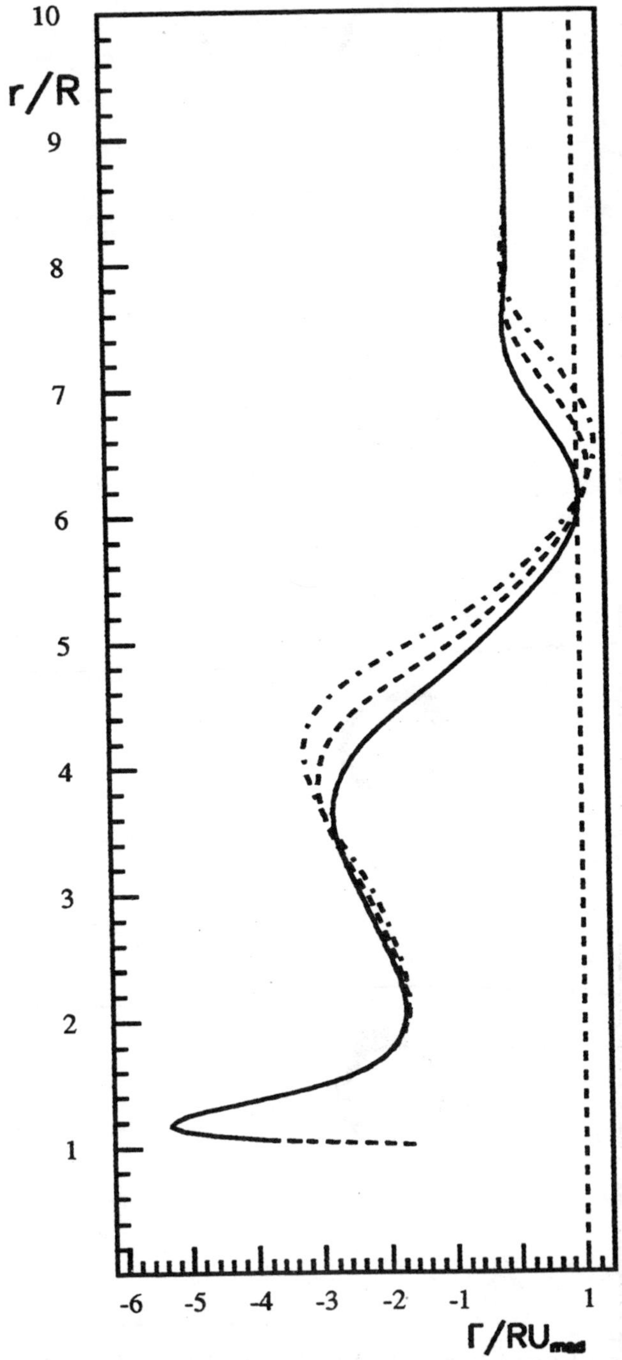

Figure 8: Total mean circulation about circles of increasing radius r. The three curves refer to $t/T_S = 18, 19, 20$ (solid, broken and dash–dot lines, respectively).

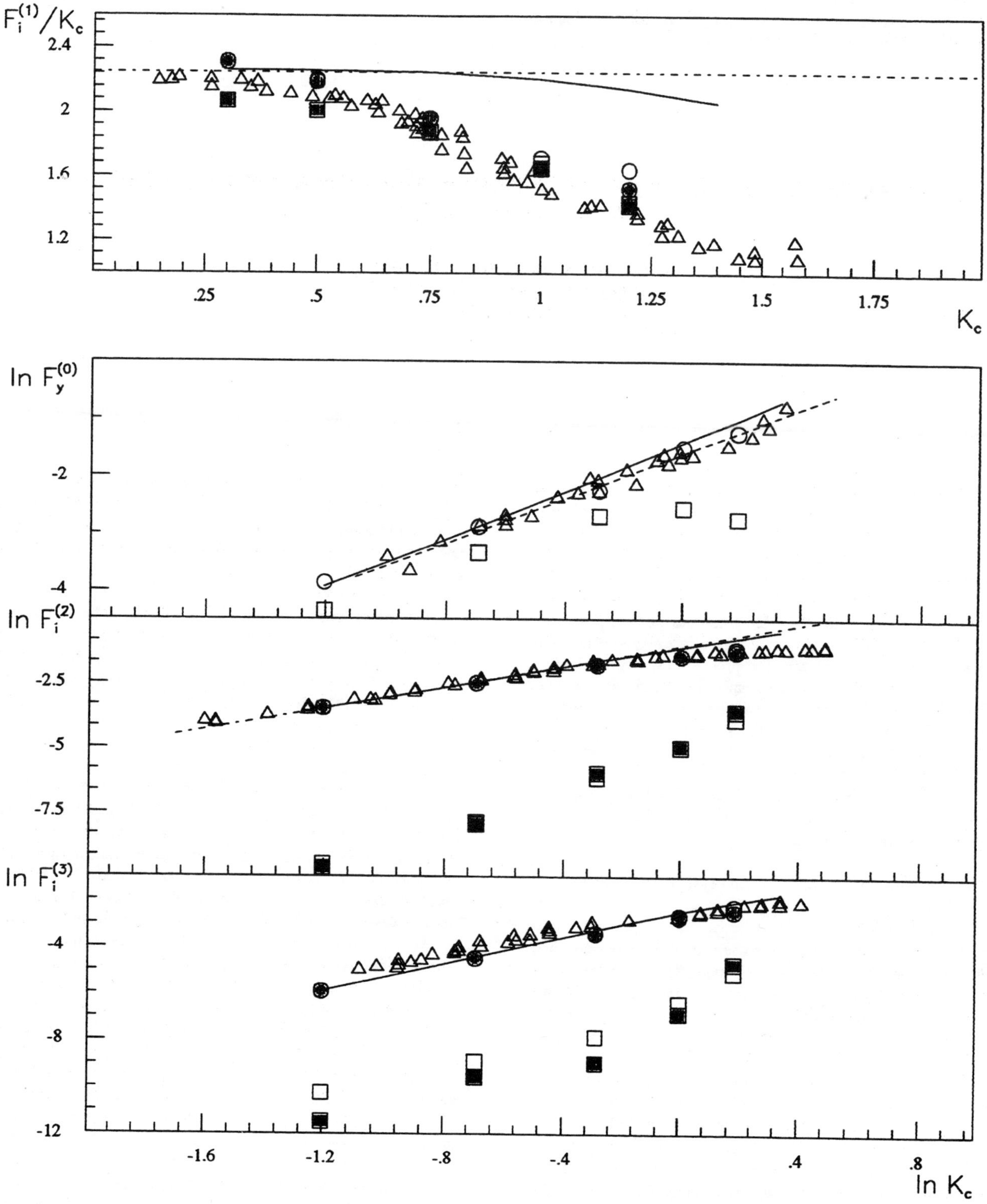

Figure 9: Amplitude of spectral components of the two hydrodynamics coefficients. Experimental data by Chaplin (1984), △. Present viscous solution, ○,●. Viscous forces acting on the cylinder in the case of wavy flow, □,■. Inviscid nonlinear solutions, —.

Proceedings of the Seventh (1997) International Offshore and Polar Engineering Conference
Honolulu, USA, May 25-30, 1997

ISBN 1-880653-28-1 (Set); ISBN 1-880653-31-1 (Vol. III)

Optimum Dynamic Response of a Floating Vessel Restrained by Cables: A Parametric Analysis

A. Carcaterra, M. Landrini and C. Lugni
Italian Ship Model Basin, INSEAN
Rome, Italy

ABSTRACT. The dynamics of a floating vessel restrained by cables is analysed and an optimum parametric analysis of the mooring lines is developed. A frequency domain approach to the linearized problem is adopted. Namely the hydrodynamic boundary value problem is recast in integral form by using the proper Green function and the relevant integral equations are numerically solved. A variational formulation for the cable dynamics is then proposed and its frequency response matrix is obtained in a closed analytical form by a Rayleigh-Ritz approach. The procedure is applied in the frame of an optimization technique and the sensitivity of the moored vessel response to some mooring line design parameters is investigated. Two basic advantages of the proposed procedure are enlightened. First the frequency approach avoids heavy numerical computations in the time domain. Moreover the impedance matrices formulation allows to update only the cable dynamic response during the iterative optimization. Numerical simulations are presented to show the effectiveness of the method.

Keywords: cable, floating body, mooring system, optimization.

1. INTRODUCTION

In this paper an optimum parametric analysis of the mooring lines of a floating vessel in waves is developed. This problem is of paramount importance at the design stage of moored ships, offshore structures and floating production storage and offloading systems.

The theoretical prediction of the dynamic response of such vessels to waves action requires the coupled solution of the hydrodynamic and structural problems. In principle, this is a non-linear partial differential problem to be solved (numerically) in time-domain. This complex approach is not suitable for dealing with optimum analysis where two basic obstacles are encountered. First, the non-deterministic behaviour of the wave exciting force compells to consider long time-simulation to obtain significant statistical information on the vessel response. Second, the search of an optimal configuration implies the analysis of several designs that makes the procedure unfeasible in practice. Therefore a simplified approach is proposed which relies on the linearization of the governing equations. A separate analysis of the vessel and the cables dynamics is then obtained by the impedance matrices formulation in the frequency domain. In this way a direct time simulation is avoided and a significant reduction of computational cost is obtained. Moreover, the hydrodynamic vessel impedance, not involved in the iterative optimization process, can be computed once and for all.

The fluid-body interactions are modelled within the framework of inviscid irrotational fluid dynamics while a submerged system of inextensible cables is considered. A variational formulation for the cable dynamics is proposed and a Rayleigh-Ritz solution is presented. A preliminary analysis of the catenary equation permits to find a suitable displacement field along the cable line. The Hamilton principle formulation leads to the Euler-Lagrange equations whose linearization permits to compute the frequency response matrix of the cable. In this way, for the mooring structure, a closed form analytical model is provided. A frequency domain approach to the linearized hydrodynamic problem is adopted. Namely the boundary value problem is recast in integral form by using the proper Green function and the relevant integral equations are solved via a zero-th order panel method. Once this is accomplished the linear behaviour of the freely floating vessel is fully described in terms of added mass, damping and restoring force coefficients, as well as by means of the exciting force.

In order to validate the effectiveness of the method, a cylindrical floating vessel constrained by two inextensible heavy cables is considered. A random exciting wave system is assumed in the optimization process. The optimal mooring cable configuration is selected acting on two simple design parameters: the position of the constraining cables with respect to the vessel and the initial cable tensions. The numerical results show an interesting behaviour of the cost function with respect to these quantities and optimal locations for the attachement points of the cables and their pre-tensions are determined.

Although the procedure developed is quite general, the example shown below is far from being of a practical value. In fact, only first order hydrodynamic forces are considered while it is well known how, in actual offshore structures, the mooring system is mainly forced by second order wave excited motions or by the presence of a current. Moreover, non potential flow forces are often of major importance, as those related to the vortex shedding from cables.

Finally, a significant exciting force can arise from the wind action. about the superstructures.

2. FORMULATION OF THE PROBLEM

The dynamic equations for the floating vessel can be written as follows:

$$\underline{\underline{M}}_V \ddot{\underline{X}}(t)=\sum_{i=1}^{N}\underline{F}_c^{(i)}\left[\underline{S}(t),\dot{\underline{S}}(t)\right]+\underline{F}_{hydr}\left[\underline{S}(\tau),\dot{\underline{S}}(\tau)\right] \quad 0\le\tau\le t$$

where $\underline{X}$, $\underline{S}=\{\underline{X},\dot{\underline{X}}\}$, $\underline{\underline{M}}_V$, $\underline{F}_c^{(i)}$, $\underline{F}_{hydr}$ are the vessel displacement vector, the state vector, the mass matrix, the *i-th* cable end tension and the hydrodinamic forces associated to the wave body interaction, respectively.

The restraining forces related to the cable actions depend only on the actual state of the system in a non-linear way, due to the nature of the dynamic governing equations of the catenary. On the contrary, due to the free surface, the hydrodynamic forces are non-linearly dependent on the whole state time-history (memory effects).

In the frame of an optimization technique the direct numerical simulation of the vessel motion in the time domain is not appropriate. In fact at least two relevant difficulties are encountered. First the solution of the non-linear differential equations always requires a high computational effort. Moreover the wave induced forces generally have a non-deterministic nature. Therefore the time evolution of the vessel motion is not significant, but rather its statistics (in terms, for example, of average value and standard deviation) should be considered. This would imply very time consuming simulations to obtain a significant statistics of the vessel response. The second problem is related to the optimum design of the mooring cables. In fact, although only the cables are modified, the exploration of different configurations require the analysis *ab initio* of the whole system (vessel and cables coupled).

Therefore to provide a model suitable for an optimization analysis, the following steps are followed:

- 1. linearization of the hydrodinamic and cable forces;
- 2. frequency domain representation of the motion;
- 3. independent frequency analysis of the system components: vessel and cables;
- 4. coupling of the sub-components in the frequency domain;
- 5. estimate of the vessel response statistics.

The first step leads to the following expressions of the forces:

$$\underline{F}_c^{(i)}=\underline{\underline{K}}_c^{(i)}\underline{X}(t)+\underline{\underline{C}}_c^{(i)}\dot{\underline{X}}(t)+\underline{\underline{M}}_c^{(i)}\ddot{\underline{X}}(t)$$

$$\underline{F}_{hydr}=\int_0^t \underline{\underline{K}}^{*}_{hydr}(t-\tau)\ \dot{\underline{X}}(t)\ d\tau\ +\ \underline{F}_w(t)$$

where mass, damping and stiffness matrices are introduced in the cable force expression using standard symbols, and the impulse-response matrix [5] is used to explicit the hydrodynamic force. In particular $\underline{F}_{hydr}$ is decomposed in the sum of two terms: the first, related to the radiation problem, takes into account the inertial and damping effects in the vessel motion; the second ($\underline{F}_w$) is associated with the diffraction problem and accounts for the forcing action of waves.

The linearization of the problems permits to transform the governing equation in the frequency domain (step 2.). Substituting the previous expression into the vessel equation of motion and using the Fourier transform, one obtains:

$$\underline{X}(\omega)=\left[-\omega^2\underline{\underline{M}}_V-\sum_{i=1}^{N}\underline{\underline{B}}_c^{(i)}(i\omega)-\underline{\underline{B}}_{hydr}(i\omega)\right]^{-1}\underline{F}_w(\omega)=\underline{\underline{H}}\ \underline{F}_w$$

where the impedence matrices have been introduced as:

$$\underline{\underline{B}}_c^{(i)}(i\omega)=\underline{\underline{K}}_c^{(i)}+i\omega\underline{\underline{C}}_c^{(i)}-\omega^2\underline{\underline{M}}_c^{(i)}$$

$$\underline{\underline{B}}_{hydr}(i\omega)=\underline{\underline{K}}_{hydr}+i\omega\underline{\underline{C}}_{hydr}(\omega)-\omega^2\underline{\underline{M}}_{hydr}(\omega)$$

The hydrodynamic damping and mass matrices are functions of the frequency as a consequence of the convolutory dependence of the hydrodynamic forces on the system state. The last two relationships show that the frequency response function of the whole system $\underline{\underline{H}}(i\omega)$ depends on the simple sum of the impedance matrices $\underline{\underline{B}}_c^{(i)}$, $\underline{\underline{B}}_{hydr}$, being their computations now completely independent. In this way the vessel and cables dynamic characteristics can be separately evaluated. Hence modifications involving the mooring system do not affect the hydrodynamic impedence that can be computed once during the whole iterative optimization procedure.

The frequency domain analysis of the vessel in waves, based on a classical integral formulation and solved via a zero-th order panel method, is described in section 3. The frequency response function of the cable, obtained by a Rayleigh-Ritz approach, is provided in section 4 (step 3.). The expression of the impedance cable matrix is there given in analytical form that, by avoiding heavy numerical computations, leads to a faster optimization analysis. The details of the vessel and cables coupling analysis to build the system transfer function $\underline{\underline{H}}(i\omega)$ are given in section 5 (step 4.).

The transfer function and the sea force spectrum allow to predict the vessel motion statistics. A cost function, based on a linear combination of the energy spectrum of each vessel degree of freedom is introduced as a measure of the system performances in a given sea-state (step 5., see section 6). Once some relevant design parameters are selected, an optimal solution is reached by minimizing the cost function. In particular the positions of the cable's attachment to the vessel and the cable's pre-tension are chosen as an example. The first parameter affects the coupling between the sub-components and not directely their frequency response functions. In this acception the parametric optimization involves the search of an optimum coupling between the cables and the restrained vessel.

3. VESSEL'S HYDRODYNAMIC FREQUENCY RESPONSE

The frequency domain approach to the linearized problem of a floating body is theoretically well assessed [1] and computationally robust (see [2] and the literature cited therein). Therefore the basic theory is just briefly recalled and a fiew details on the numerics are given.

The irrotational motion of an incompressible fluid about a floating body is considered. Here small amplitude oscillations of the vessel under the action of monochromatic plane waves of small steepness are considered. Hence a full linearization of the problem is permitted and, at the steady regime, a time harmonic dependence of the relevant variables can be assumed. As a result the complex spatial part $\varphi(P)$ of the velocity potential:

$$\Phi(P,t)=\Re(e^{i\omega t}\varphi(P))$$

is given in terms of the radiation potentials $\varphi_R(P)=i\omega\sum_{j=1}^{6}\xi_j\varphi_j$ and of a diffraction potential $\varphi_D=\varphi_0+\varphi_7$, being φ_0 the known incident wave potential and φ_7 the scattering potential. By introducing the standard Green function:

$$G(P,Q)=\frac{1}{r}+\frac{1}{\bar{r}}+\frac{2k}{\pi}\int_0^{\infty}\frac{e^{\upsilon(z+\zeta)}}{\upsilon-k}J_0(\upsilon R)\ \ d\upsilon$$

and applying Green's theorem, the relevant integral equations (j=1,..,6):

$$2\pi\varphi_j(P)+\int_{\partial\Omega_{c,e}}\varphi_j(Q)\frac{\partial G(P,Q)}{\partial n}\,dS=\int_{\partial\Omega_{c,e}}G(P,Q)\;n_j(Q)\;dS$$

$$2\pi\varphi_D(P)+\int_{\partial\Omega_{c,e}}\varphi_D(Q)\frac{\partial G(P,Q)}{\partial n}\,dS=4\pi\varphi_0(P)$$

for the radiation potentials and for the diffraction potential are obtained respectively. Once the integral equations are solved, added masses and damping coefficients follow from:

$$a_{jk}-\frac{i}{\omega}b_{jk}=\rho\int_{\partial\Omega_{c,e}}\varphi_j n_k dS$$

as well as the exciting force

$$X_k=i\omega\rho\int_{\partial\Omega_{c,e}}\varphi_D\frac{\partial\varphi_k}{\partial n}dS$$

corresponding to the given incident wave. The a_{jk} and b_{jk} coefficients are elements of the $\underline{\underline{M}}_{hydr}(\omega)$, $\underline{\underline{C}}_{hydr}(\omega)$ matrices respectively.

The numerical solution is obtained by a standard panel technique. The body surface is discretized by using flat quadrilateral and triangular panels. The $1/r+1/\hat{r}$ terms in the Green function are treated by Hess & Smith formulas [3], while Lines & Jespersen quadrature rules [4], with variable degree of accuracy, are adopted for the oscillatory part. The latter is computed following the algorithm described in [5].The code is tested by comparison with known accurate solutions and the convergence of the all results shown below are tested by doubling and doubling the number of panels. Finally the behaviour of the vessel, for a given sea spectrum, is readily obtained through superposition principle.

4. CABLE'S FREQUENCY RESPONSE: A RAYLEIGH-RITZ APPROACH

In this section a Rayleigh-Ritz method is applied to provide the frequency response of the in-plane motion of an inelastic cable and an approximated closed form solution is proposed.

One end of the cable is constrained to the sea bed by an anchor, while the second end is attached to the vessel. The co-ordinate reference system is shown in fig. 1, where the origin of the axes coincides with the point separating the suspended portion of the cable from the one lying on the sea bed.

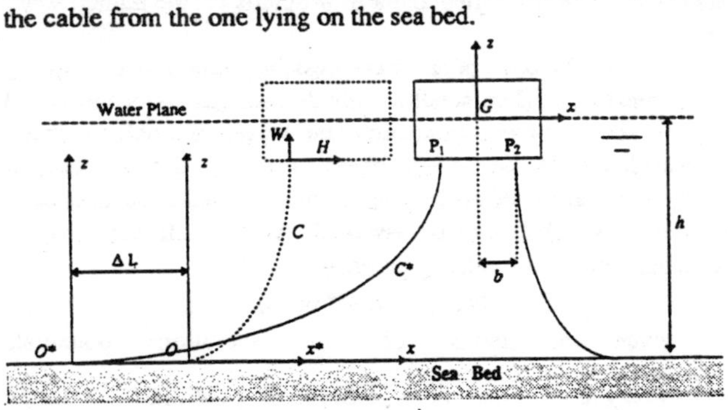

Figure 1

4.1 Catenary Static Relationships

Here some classical results of cable statics [6], useful to develop the next section, are resumed in short.

Let us consider the Hamilton principle for the cable in the static configuration C.

The potential energy of the system is:

$$U=\int_0^l \mu gz\,ds=\int_0^{x_L}\mu gz\sqrt{1+z'^2}\,dx$$

A constraint condition is associated to the inelasticity of the cable, i.e.:

$$V(z')=\int_0^{x_L}\sqrt{1+z'^2}\,dx-l=0$$

that imposes a constant cable length l.

The Hamilton principle leads to the isoperimetric variational problem:

$$\delta(U-\lambda V)=0\Rightarrow\delta\int_0^{x_L}(\mu gz-\lambda)\sqrt{1+z'^2}\,dx=0$$

where λ is a lagrangian multipler. The related Euler-Lagrange equation and its associated solution are:

$$\mu gz=\lambda+c_0\sqrt{1+z'^2}\Rightarrow z=\frac{c_0}{\mu g}\left[\cosh\left(\frac{\mu gx}{c_0}+c_1\right)+\frac{\lambda}{c_0}\right]$$

being λ, c_0, c_1 determined by imposing the isoperimetric condition and the boundary conditions:

$$z'(0)=0,\quad W=H\,z'(x_l),\quad z'(x_l)=\frac{\mu gl}{H}$$

where W, H and μ are the vertical and horizontal forces at the water plane of the cable and its mass per unit length respectively. Accordingly, the computed values for λ, c_0, c_1 and the corresponding cable shape are:

$$c_1=0,\;c_0=H,\;\lambda=-H,\Rightarrow z=\frac{H}{\mu g}\left[\cosh\left(\frac{\mu gx}{H}\right)-1\right]$$

The whole cable's minumum length l is determined by assuming that even in extreme sea conditions the anchor be not subjected to vertical load components [2,6]. This implies that in correspondence of the maximum estimate for the horizontal force H_M, the cable shape satisfies the condition z'(0)=0, i.e. W(0)=0. Simple mathematics leads to the expression:

$$l=h\left(\frac{2H_M}{\mu gh}+1\right)^{1/2}$$

If in operating conditions the horizontal force is H then the suspended cable length l_s and the length l_r of the portion at rest on the sea bed are respectively:

$$l_s=h\left(\frac{2H}{\mu gh}+1\right)^{1/2},\qquad l_r=l-l_s \qquad (1)$$

The cable shape is therefore:

$$z(x)=\frac{H}{\mu g}\left[\cosh\left(\frac{\mu gx}{H}\right)-1\right],\quad l_r\le s\le l,\;0\le x\le x_l$$

being s the abscissa along the cable.The correlation between H and the horizontal displacement u_w of the cable end at the water plane consequentely is:

$$u_w=l-l_s+\frac{H}{\mu g}\cosh^{-1}\left(1+\frac{\mu gh}{H}\right) \qquad (2)$$

4.2 Cable's Mode Shape Functions

Let us consider two different cable static configurations C and C* (see fig. 1), where C* is obtained from C by imposing a displacement vector $\underline{u}_w$ to the cable end at the water plane. Each of the cable's point is transferred from the position P(s) to P*(s), where the abscissa s is the same for both configurations due to the inelastic condition. The vector $\underline{u}$ expresses the displacement along the cable line in the horizontal and vertical directions respectively.

If s is the abscissa of the suspended portion of the cable and s=0 in correspondence of the origin of the axes, the inelastic condition gives:

$$l_r + s(x,H) = l_r^* + s(x^*,H^*) \qquad (3)$$

where

$$s(x,H) = \int_0^{x_1} \sqrt{1+z'^2}\,dx = \mu g \,\mathrm{senh}\left(\frac{\mu g x}{H}\right)$$

and the superscript * denotes the actual values of the quantities in the C* configuration. The abscissa x of the point P is modified in x* due to the horizontal displacement u. Moreover in C* the origin of the reference axes is displaced of the quantity l_r^* -l_r, because of the variation of the portion of the cable lying on the sea bed (see fig.1).

Therefore the following relationship holds:

$$x^* = x + u - (l_r^* - l_r) = x + u - \Delta l_r = x + \Delta x$$

The horizontal force changes also its value, hence Δ H=H * - H. From relationship (3) it follows:

$$s(x+\Delta x, H+\Delta H) - s(x,H) = -\Delta l_r$$

If small perturbations are accounted for, it is:

$$\frac{\partial s}{\partial x}dx + \frac{\partial s}{\partial H}dH = -dl_r,$$

$$\frac{\partial s}{\partial x}(du - dl_r) + \frac{\partial s}{\partial H}dH = -dl_r$$

and the law of distribution of the horizontal displacement is found:

$$du = \left[dl_r\left(\frac{\partial s}{\partial x} - 1\right) - dH\frac{\partial s}{\partial H}\right] \Big/ \left(\frac{\partial s}{\partial x}\right) \qquad (4)$$

For the vertical displacement v one has:

$$v = z(x^*,H^*) - z(x,H)$$

that, with the same assumption used for u , becomes:

$$dv = \frac{\partial z}{\partial x}(du - dl_r) + \frac{\partial z}{\partial H}dH \qquad (5)$$

Each displacement component depends on the differential quantities dH, dl_r that are related to the displacement $\underline{u}_w$ imposed at the cable end. Therefore the relationships hold:

$$dl_r = \left.\frac{\partial l_r}{\partial u_w}\right|_v du_w + \left.\frac{\partial l_r}{\partial v_w}\right|_u dv_w,$$

$$dH = \left.\frac{\partial H}{\partial u_w}\right|_v du_w + \left.\frac{\partial H}{\partial v_w}\right|_u dv_w \qquad (6)$$

The four derivatives in the previous formula must be evaluated to explicite, through the relationships (4) and (5), the dependence of du, dv on du_w, dv_w.

Using relationship (2), $\partial u_w / \partial H$ is easily computed as:

$$\left.\frac{\partial H}{\partial u_w}\right|_v = \left(\left.\frac{\partial u_w}{\partial H}\right|_v\right)^{-1} = \mu g\left[\cosh^{-1}\left(1+\frac{\mu g h}{H}\right) - \frac{2h}{l_s}\right]^{-1} = k_u$$

Then the total differential of the horizontal displacement is:

$$du_w = \left.\frac{\partial u_w}{\partial h}\right|_H dh + \left.\frac{\partial u_w}{\partial H}\right|_h dH = \left.\frac{\partial u_w}{\partial h}\right|_H dh + \frac{dH}{k_u}$$

If only a vertical displacement dv_w is imposed at the cable end, it is du_w=0 and therefore:

$$\left.\frac{\partial u_w}{\partial h}\right|_H \left.\frac{\partial h}{\partial v_w}\right|_u + \frac{1}{k_u}\left.\frac{\partial H}{\partial v_w}\right|_u = 0, \quad dh \equiv dv_w$$

$$\left.\frac{\partial H}{\partial v_w}\right|_u = -k_u \left.\frac{\partial u_w}{\partial h}\right|_H$$

Finally, using expression (2), one obtains:

$$\left.\frac{\partial H}{\partial v_w}\right|_u = -k_u\left\{\frac{H}{\mu g l_s} - \frac{l_s}{h} + \frac{1}{\mathrm{senh}\left[\cosh^{-1}(1+\mu g h/H)\right]}\right\}$$

Consider now the expression of dl_r provided by eq (1):

$$dl_r = \left.\frac{\partial l_r}{\partial h}\right|_H dh + \left.\frac{\partial l_r}{\partial H}\right|_h dH, \qquad dh \equiv dv_w \Rightarrow \left.\frac{\partial l_r}{\partial v_w}\right|_u =$$

$$\left(\frac{H}{\mu g l_s} - \frac{l_s}{h}\right) + \frac{h}{\mu g l_s}\left.\frac{\partial H}{\partial v_w}\right|_u$$

Finally it holds:

$$\left.\frac{\partial l_r}{\partial u_w}\right|_v = \left.\frac{\partial l_r}{\partial H}\right|_v \left.\frac{\partial H}{\partial u_w}\right|_v = \left.\frac{\partial l_r}{\partial H}\right|_h \left.\frac{\partial H}{\partial u_w}\right|_v =$$

$$= \left.\frac{\partial l_r}{\partial H}\right|_h k_u = -\frac{h}{\mu g l_s}k_u$$

Therefore the whole displacement cable field is determined and its expression can be represented by a convenient matrix form. Defining the two shape vectors $\underline{\varphi}_u$, $\underline{\varphi}_v$, short expressions replace eqs. (4), (5) :

$$du = \left[(s_x - 1)/s_x, -s_H/s_x\right]\begin{Bmatrix} dl_r \\ dH \end{Bmatrix} = \underline{\varphi}_u^T \underline{dc},$$

$$dv = z_x du + \left[-z_x, z_H\right]\begin{Bmatrix} dl_r \\ dH \end{Bmatrix} = \left[\underline{\varphi}_v + z_x\underline{\varphi}_u\right]^T \underline{dc}$$

$$\underline{du} = \begin{bmatrix} \underline{\varphi}_u \\ \underline{\varphi}_v + z_x\underline{\varphi}_u \end{bmatrix}^T \underline{dc} = \underline{\underline{\Phi}}\,\underline{dc}$$

where the subscripts x, H denote derivatives with respect to x and H respectively. The vector $\underline{dc}$ is related to the cable end displacement by the derivatives appearing in eq (6). With obvious meaning of the symbols one has:

$$\underline{dc} = \underline{\underline{D}}\,\underline{du}_w, \quad \Rightarrow \quad \underline{du} = \underline{\underline{\Phi}}\,\underline{\underline{D}}\,\underline{du}_w,$$

$$\underline{u}(x,t) = \left[\underline{\underline{\Phi}}(x)\underline{\underline{D}}\right]\underline{u}_w(t) = \begin{bmatrix} \underline{s}_u^T(x) \\ \underline{s}_v^T(x) \end{bmatrix}\underline{u}_w(t) = \underline{\underline{S}}(x)\,\underline{u}_w(t)$$

where the last transformation implies the introduction of a time dependent cable end motion. Therefore the cable field displacement is completely determined and all the terms appearing in the shape matrix are recovered in closed analytical form.

4.3 Cable's Frequency Response

Accordingly to the Hamilton principle the Rayleigh-Ritz method provides an approximated form for the governing equation of the cable motion. The kinetic and potential energies must be computed on the basis of the displacement field obtained in the previous section, so that their expression are:

$$2T = \underline{\dot{u}}_w^T \left[\int_0^{x_1} \mu \underline{\underline{S}}^T \underline{\underline{S}} \sqrt{1 + [z' + \underline{s}_u^T \underline{u}_w]^2}\,dx\right]\underline{\dot{u}}_w$$

$$U = \int_0^{x_1} \mu g\left[z + \underline{s}_v^T \underline{u}_w\right]\sqrt{1 + [z' + \underline{s}_v^T \underline{u}_w]^2}\,dx$$

The application of the Hamilton principle leads to the Euler-Lagrange equations. This is a system of two ordinary non-linear differential equations with respect to the time variable. A perturbation approach could be employed to study its solutions up to the wished order (examples of perturbation approach to non-linear cable dynamics by the Galerkin method are found in [7, 8]). To obtain the cable frequency response matrix the expansions must be limited to the second order for the kinetic and the potential energies. Therefore the kinetic energy is:

$$2T = \underline{\dot{u}}_w^T \left[\int_0^{x_1} \mu \underline{\underline{S}}^T \underline{\underline{S}} \sqrt{1+z'^2}\, dx\right] \underline{\dot{u}}_w = \underline{\dot{u}}_w^T \underline{\underline{M}}_c \underline{\dot{u}}_w$$

A simple and significant expression of the potential energy can be obtained by evaluating the work done by the cable end forces H and W [6]. Developing their expression in a Taylor series up to the first order (second order for the potential energy), one has:

$$2U = \underline{u}_w^T \begin{bmatrix} \left.\frac{\partial H}{\partial u_w}\right|_v & \left.\frac{\partial H}{\partial v_w}\right|_u \\ \left.\frac{\partial W}{\partial u_w}\right|_v & \left.\frac{\partial W}{\partial v_w}\right|_u \end{bmatrix} \underline{u}_w = \underline{u}_w^T \underline{\underline{K}}_c \underline{u}_w$$

where only the second row terms need to be explicited. The vertical force component is simply:

$$W = \mu g l_s = \mu g (l - l_r), \quad \Rightarrow \quad dW = -\mu g\, dl_r$$

hence:

$$\left.\frac{\partial W}{\partial u_w}\right|_v = -\mu g \left.\frac{\partial l_r}{\partial u_w}\right|_v, \qquad \left.\frac{\partial W}{\partial v_w}\right|_u = -\mu g \left.\frac{\partial l_r}{\partial v_w}\right|_u$$

where all the appearing derivatives have been considered in the previous section. The cable tranfer matrix is trivially determined by the expressions of $\underline{\underline{M}}_c$, $\underline{\underline{K}}_c$.

5. COUPLING IN THE FREQUENCY DOMAIN

Once the frequency response matrix of a single cable and of the floating vessel are determined, their coupling must be operated to provide the transfer matrix of the complete system.

Here a twin cable mooring line is considered, where the vessel center of mass G is the origin of the co-ordinate reference system as shown in fig.1.

The forcing wave produces a drift force F_D on the vessel inducing in the cables two different static tensions H_1 and H_2 depending also on the initial pre-tension H_0.

The system response is analysed around the static equilibrium configuration corresponding to the drift force application.

Two continuity conditions between the floating vessel and the mooring line must be imposed.

The dynamic equilibrium condition in the frequency domain reads:

$$-\omega^2 \begin{bmatrix} M & 0 & 0 \\ 0 & M & 0 \\ 0 & 0 & J \end{bmatrix} \begin{Bmatrix} X_G \\ Z_G \\ \Theta \end{Bmatrix} = -\omega^2 \underline{\underline{M}}_V \begin{Bmatrix} \underline{U}_G \\ \Theta \end{Bmatrix} =$$

$$= -\omega^2 \underline{\underline{M}}_V \underline{X} = \underline{F}_c^{(1)} + \underline{F}_c^{(2)} + \underline{F}_{rad} + \underline{F}_w$$

where M, J are the the vessel mass and the moment of inertia, X_G, Z_G the displacement components of the center of mass G (surge and heave motion respectively), Θ the vessel rotation (pitch motion), $\underline{F}_c^{(1)}, \underline{F}_c^{(2)}$ the cable end tensions and $\underline{F}_{rad}, \underline{F}_w$ the hydrodinamic forces associated to the radiation and the diffraction problems respectively. The transfer matrix $\underline{\underline{H}}(i\omega)$ of the moored floating body follows from the previous equation:

$$\underline{X} = \left[-\omega^2 \underline{\underline{M}}_V + \underline{\underline{B}}_c + \underline{\underline{B}}_{hydr}\right]^{-1} \underline{F}_w = \underline{\underline{H}}(i\omega) \underline{F}_w, \qquad (7)$$

$$\underline{\underline{B}}_c \underline{X} = \left(\underline{F}_c^{(1)} + \underline{F}_c^{(2)}\right), \qquad \underline{\underline{B}}_{hydr} \underline{X} = \underline{F}_{rad}$$

where $\underline{\underline{B}}_{hydr}$ is the complex impedance of the vessel, computed by solving the radiation problem through the described numerical procedure and $\underline{\underline{B}}_c$ is the impedance of the two coupled mooring cables.

The analysis developed in the previous section permits to provide the single cable impedance in its operating conditions. Basically the two different horizontal static forces H_1, H_2 must be estimated to calculate the corresponding mode shape functions. The problem is solved by computing the drift force F_D generated by the forcing wave system on the vessel and imposing the static equilibrium condition $F_D = H_1 - H_2$. Static considerations lead to: $H_2(u_w)=H_1(2u_{w0}-u_w)$, where u_{w0} is determined by eq. (2) posing $H=H_0$. Using the inverse of relationship (2) for each cable and $F_D = H_1 - H_2$, a non-linear system of equations in terms of H_1, H_2 is iteratively solved and $\underline{\underline{B}}_c^{(1)}, \underline{\underline{B}}_c^{(2)}$ definitely computed.

The second condition, expressing the kinematic continuity in the frequency domain, has the form:

$$\underline{U}_w^{(1)} = \underline{U}_G + \underline{\Theta} \times \underline{GP_1} = \underline{U}_G + \underline{\Theta} \times \begin{Bmatrix} a_1 \\ 0 \end{Bmatrix},$$

$$\underline{U}_w^{(2)} = \underline{U}_G + \underline{\Theta} \times \underline{GP_2} = \underline{U}_G + \underline{\Theta} \times \begin{Bmatrix} a_2 \\ 0 \end{Bmatrix}$$

where P_1, P_2 are the points in which the vessel is constrained to the cables (see fig.1). Using the following definitions:

$$\underline{M}_c = \underline{GP_1} \times \underline{F}_c^{(1)} + \underline{GP_2} \times \underline{F}_c^{(2)},$$

$$\underline{F}_c = \underline{F}_c^{(1)} + \underline{F}_c^{(2)} = \underline{\underline{B}}_c^{(1)} \underline{U}_w^{(1)} + \underline{\underline{B}}_c^{(2)} \underline{U}_w^{(2)}$$

and combining them with the second of eqs. (7), the partitioned impedence matrix of the coupled cables is determined:

$$\begin{Bmatrix} \underline{F}_c \\ |\underline{M}_c| \end{Bmatrix} = \underline{\underline{B}}_c \begin{Bmatrix} \underline{U}_G \\ \Theta \end{Bmatrix} =$$

$$= \begin{bmatrix} \underline{\underline{B}}_c^{(1)} + \underline{\underline{B}}_c^{(2)} & \underline{\underline{B}}_c^{(1)} \underline{a}_1 + \underline{\underline{B}}_c^{(2)} \underline{a}_2 \\ \underline{a}_1^T \underline{\underline{B}}_c^{(1)} + \underline{a}_2^T \underline{\underline{B}}_c^{(2)} & \underline{a}_1^T \underline{\underline{B}}_c^{(1)} \underline{a}_1 + \underline{a}_2^T \underline{\underline{B}}_c^{(2)} \underline{a}_2 \end{bmatrix} \begin{Bmatrix} \underline{U}_G \\ \Theta \end{Bmatrix}$$

Introducing this expression in the first of eqs. (7) a complete frequency domain model for the moored floating vessel is derived.

6. OPTIMUM PARAMETRIC ANALYSIS: STATISTICS OF THE RESPONSE AND COST FUNCTION

A simple optimization example is considered accounting for two design parameters. The first is the horizontal distance 2b between the two cable attachement points to the vessel, where the cables are assumed to be symmetric with respect to the z axis at rest. The second parameter is the pre-tension H_0 in the cable line.

The forcing wave system is generated on the semi-empirical basis of the Pearson-Moskovitz statistics that, for fully developed sea, states the following frequency spectrum of the wave elevation:

$$S_{PM}(\omega) = \frac{\alpha g^2}{\omega^5} \exp\left[-\beta (g / V \omega)^4\right]$$

where V is the wind velocity at a standard height on the free surface and α, β empirical parameters.

Once the linear diffraction problem is numerically solved, in terms of the forces F_{wj}, the power spectral density of each degree of freedom is computed in the form [9], [10]:

$$S_r(\omega) = S_{PM}(\omega) \left|\sum_j H_{rj}(i\omega) F_{wj}(\omega)\right|^2$$

The standard deviation related to each of the vessel degrees of freedom is therefore:

$$\sigma_r^2 = \int_{-\infty}^{+\infty} S_{PM}(\omega) \left| \sum_j H_{rj}(i\omega) F_{wj}(\omega) \right|^2 d\omega$$

The cost function can be assumed equal to a linear combination of the computed standard deviations:

$$f(H_0,b) = \alpha_x \sigma_x^2(H_0,b) + \alpha_z \sigma_z^2(H_0,b) + \alpha_\theta \sigma_\theta^2(H_0,b)$$

where the weighting (dimensional) coefficients are introduced to emphasize the wished oscillations reduction on selected degrees of freedom. This choice of the cost function is in some way related to the minimization of the kinetic energy of the floating body.

The optimal design parameters are dependent on the sea conditions, i.e. on the wind velocity. Two optimization examples are proposed in the following. The first analysis is performed in correspondence of a single assigned wind velocity (for example the maximum expected value) and the minimizing parameters are found. In the second test case two fixed wind velocities are considered (the maximum and the average values) and an optimal region design is recovered.

7. NUMERICAL SIMULATIONS

The numerical simulations are performed on a cilyndrical vessel with radius, depth, mass and giration beam equal to 5m, 5m, 378 tons, 0.5m, respectively, assuming a sea depth of 50 m ; the mooring cables have a weight per unit length of 1 kN / m.

In figs. 2, 3, 4 and 5 the relevant quantities for the moored vessel model, corresponding to a wind velocity of 15 m /s, a pre-tension of 120 kN and a horizontal distance of the cable attachment points from the center of mass equal to 1.5 m are shown.

In fig. 2 and 3 the added masses, the damping coefficients and the exciting forces are plotted versus frequency.

In the upper plots of fig. 4 the cable shape functions both for the higher and lower tensioned cables are shown. It can be noticed how the horizontal displacement field is more important for the latter, due to the high mean curvature value, than for the former.

In fig. 4 (lower plots) the elements of the frequency response matrices for the two cables are represented, where resonance and anti-resonance peaks are evident for each element of the transfer matrix.

In fig. 5 the frequency response of the vessel coupled with cables is shown. The diagonal and the out of diagonal terms are shown on the left and the right side respectively, and the related modal peaks clearly appear. From their observation it is apparent that the first mode, related to the surge motion, is less damped with respect to the heave and pitch modes, although the dominant damping, for the unrestrained body, should be due to the surge motion (see fig.2). Since in the analysed moored vessel the only introduced form of energy dissipation is due to the radiated waves, then it can be assessed that, due to the mooring line actions, most of the wave energy radiation is related to the pitch and the heave motion.

The previously discussed results are referred to a single mooring configuration, in which all the design parameters are assigned. In the frame of an optimization procedure several design solutions have to be explored. The weighting parameters in the cost function defined in section 6, are selected in order to emphasize the reduction of the pitch motion. Here two parameters are employed to reach the minimum of the cost function: the cable's attachment position and the pre-tension.

The initial pre-tension H_0 influences the values of the horizontal cable forces H_1 and H_2, arising under the drift force F_D as shown in section 5, and therefore the two cable impedance matrices singularly. On the contrary the effect of the attachment position on the vessel dynamics is involved only when coupling the cable responses, how appears from the expression of the coupled impedance matrix given in section 4, explicitely dependent on the vectors $\underline{a}_1$, $\underline{a}_2$. Since the vessel radius is 5 m, then the abscissa of the attachment point of each cable can vary in the range $0<b<5$m (see fig. 1).The pre-tension variation is instead investigated in the range $110<H_0<200$kN.

The sea condition, i.e. the wind velocity, affects the vessel dynamic response by two different mechanisms. The first is related to the frequency spectrum of the wave exciting force, whose peak location and average values are heavy dependent on the wind velocity. On the other hand the drift force F_D (also wind dependent) affects the initial cables configuration through horizontal force H appearing in the cable shape functions, and the impedance matrix of the mooring line indeed. Therefore we expect that an optimal design configuration, computed for an assigned wind velocity, is not still suitable in other operative sea conditions. Hence, in the attempt to satisfy different constraints on the vessel response, a compromise solution is generally required.

Here, considering the probability density function of the wind velocity, two characteristic values are selected: the first corresponding to the function peak, i.e. the more frequent wind velocity value or the average one, and the maximum expected value in correspondence to an assigned probability confidence limit. The optimization of the vessel response is then performed taking into account the minimum cost function in extreme and in average sea conditions, estimated in 15 m/s and 10 m/s respectively.

The first two plots in fig. 6 show the cost function surface versus pre-tension and attachment location for 10 m/s and 15 m/s, respectively. For wind velocity equal 10 m/s the best solution consists in anchoring the vessel in the middle (x=0 m), using the minimum considered pre-tension. On the co-ordinate plane the surface contour lines are represented.

For wind equal 15 m/s the cost function presents an interesting relative minimum, obtained varying the attachment position in the range $1.5 < x < 2.5$, corresponding to different pre-tension values; on the co-ordinate plane the contour lines are also shown. However in both cases the enhancement in the vessel performances is relevant: a reduction of about 70% and 100% for 15 m/s and 10 m/s respectively, passing from the worst to the best solution.

If the modification of the total frequency response function is observed during the optimization process, it can be seen that, approaching the optimal design parameters, the three modal peaks are displaced over the frequency axis. In the actual configuration the three natural frequencies, in order corresponding to surge, heave and pitch motions, belong to the frequency interval covered by the Pearson-Moskovitz distribution. The central heave modal peak appears not significantly influenced by the optimization process. On the contrary the surge peak tends to be displaced toward the origin of the frequency axis, while the pitch frequency is little increased. In other words the optimization process basically acts on the first and the third modes displacing them far from the peak of the Pearson-Moskovitz spectrum.

The analysis of the two cost function surfaces indicates that the found optimal parameters do not permit to satisfy both the minimal response requirements at the same time.

In the last plot the two contour line families, associated to the presented cases, are shown togheter. If both the design parametrs are modified following a contour line, than the cost function assumes the same costant value. For the two wind conditions a limit cost function value and a corresponding contour line are selected, namely the r and s curves (see the lowest plot in fig. 6). Thier intersection indicates the design point able to respect both the required limits. Moreover in this case the plane portion bounded by the two curves, allows to recover the optimal design region inside which the cost function is below the two wished limits.

8. CONCLUSIONS

An optimum parametric study based on a linearized frequency domain approach for moored vessels in waves is presented. Once the frequency response matrices of the freely floating body and of the mooring cables are separately computed, a suitable assemblement of the two frequency responses is operated by imposing the kinematic and the dynamic continuity conditions.The dynamic model for the cable is obtained by an Hamilton formulation and a Rayleigh-Ritz solution is employed to determine the cable frequency response in closed analytical form. The hydrodynamic vessel frequency response is obtained by numerically solving the integral equations for the radiation and diffraction potentials.

The formulation exhibits advantages in the frame of an optimal mooring system design analysis. In fact when the frequency response of the freely floating body is not interested in the optimization process, its response is definitely determined in a one shot computation. Then the obtained impedance matrix is coupled with that related to the cables. This last is computed at each optimitazion step but, due to its analytical form, a short computational time is needed and the procedure can so explore a wide range of design configurations.

ACKNOWLEDGEMENTS

This research activity was suppoted by Ministero dei Trasporti e della Navigazione through INSEAN Research Program 1994-96.

REFERENCES

[1] Mei, C. C., 'The Applied Dynamics of Ocean Surface Waves', Word Scientific, Singapore.

[2] O.M. Faltinsen, 'Sea Loads on Ships and Offshore Structures', Cambridge University Press,Ocean Technology Series, Cambridge, UK, 1990.

[3] Hess, J.L., Smith, A.M.O., 'Calculation of Potential Flow About Arbitrary Bodies', Progress in Aero. Sci., n.8(1966), 22-44.

[4] Lyness, J.N., Jespersen, D., 'Moderate Degree Symmetric Quadrature Rules for the Triangle', J. Inst. Maths. Appl., 15(1975),19-32.

[5] J.N. Newman, 'Marine Hydrodynamics', The MIT Press, 1977.

[6] H. Max Irvine, 'Cable Structures', The MIT Press Series in Structrural Mechanics, 1981.

[7] A. Luongo, G. Rega, F. Vestroni, 'Planar Non-Linear Free Vibrations of an Elastic Cable', International Journal of Non-Linear Mechanics, vol.19, 1984.

[8] F. Benedettini, G. Rega, R. Alaggio, 'Non-Linear Oscillations of a Four-Degree-of-Freedom Model of a Suspended Cable Under Multiple Internal Resonance Conditions', Journal of Sound and Vibration, vol.182(5), 1995.

[9] J.N. Newman, 'Algorithms for the free-surface Green function', Journal of Engineering Mathematics, vol. 19, 1985.

[10] L. Meirovitch, 'Elements of Vibration Analysis', Mc Graw Hill Book Company, Singapore, 1986.

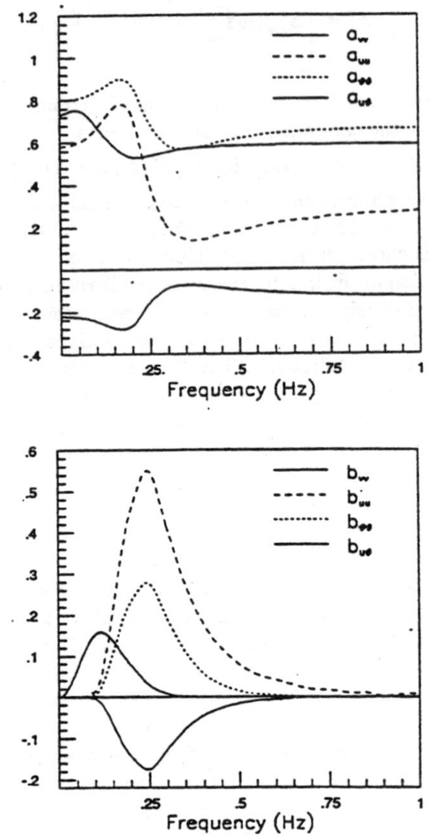

Figure 2: Non-dimensional added masses (upper plot) and damping (lower plot) coefficients

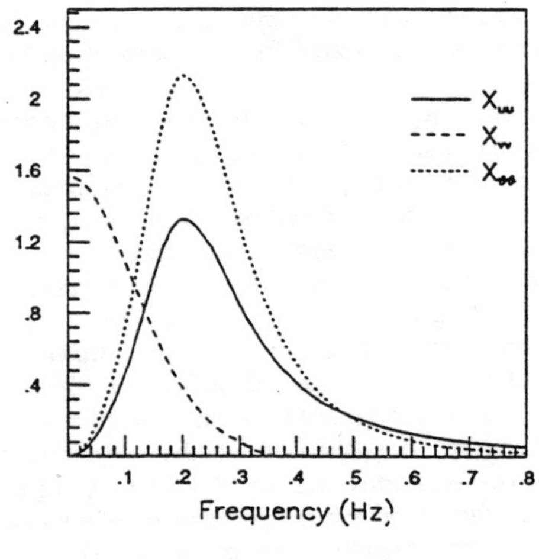

Figure 3: Non-dimensional exiting force

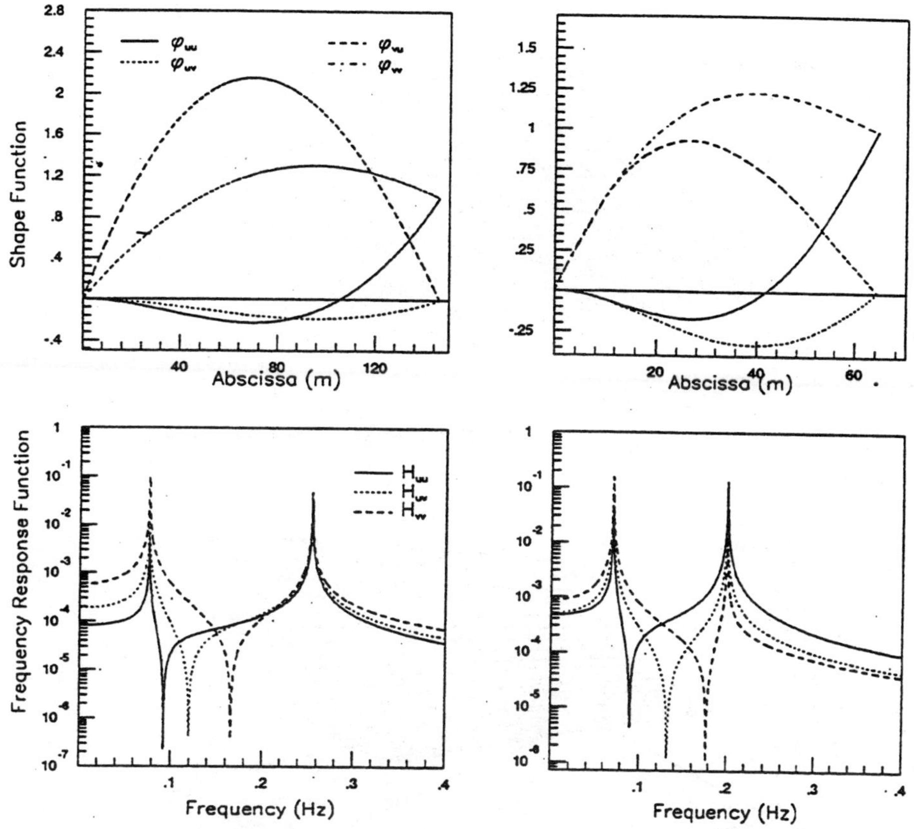

Figure 4: Shape functions (upper plots) and frequency response functions (lower plots) of the higher tensioned cable (left) and of the lower tensioned cable (right)

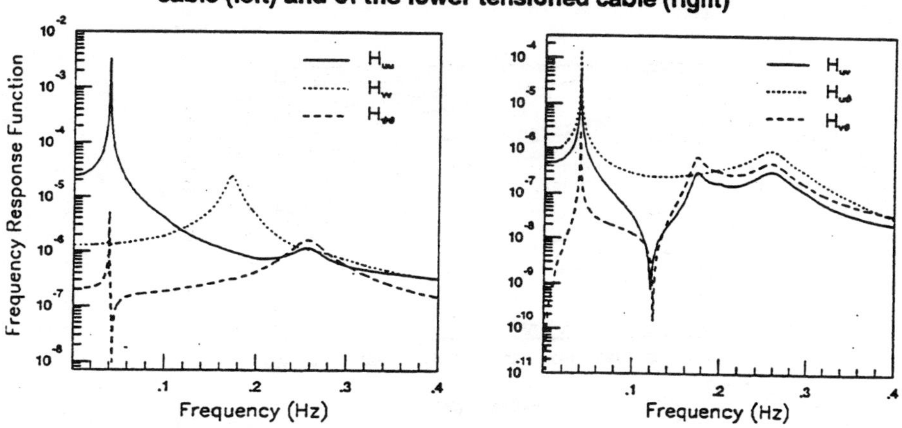

Figure 5: Frequency response functions of the moored vessel: on diagonal (left) and out of diagonal terms (right)

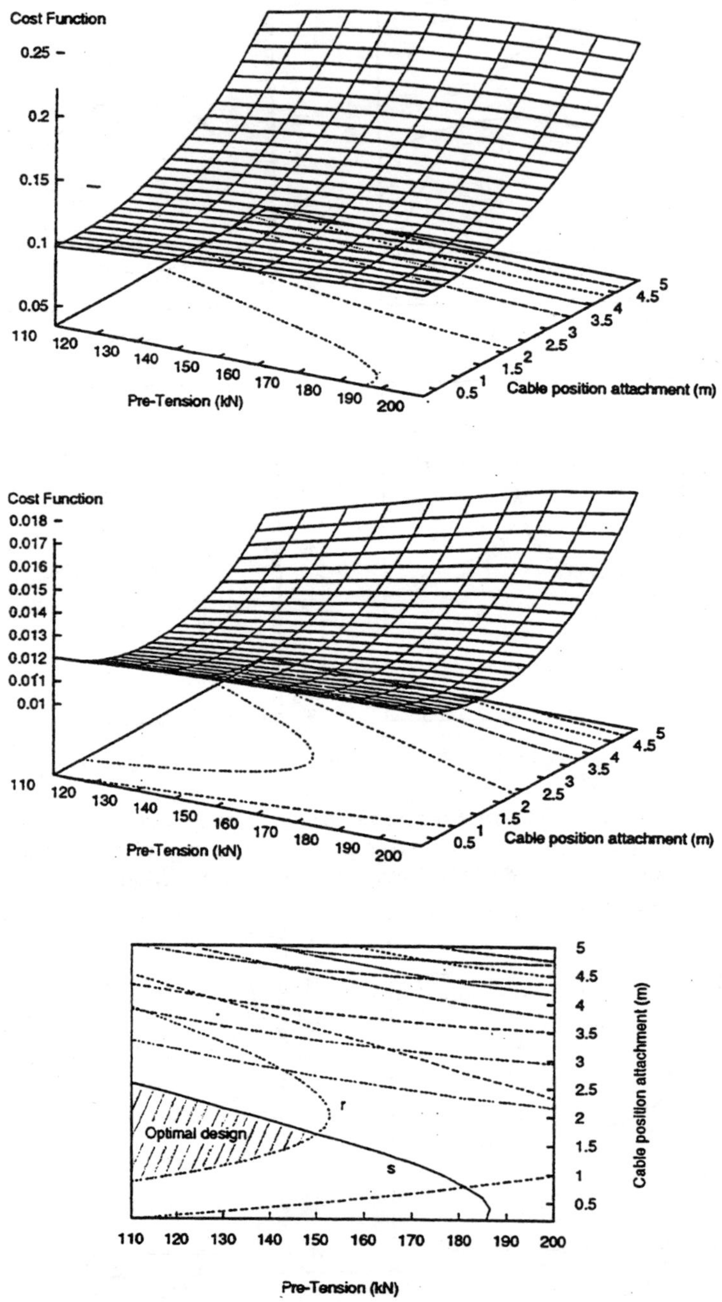

Figure 6 : Upper plots: cost function surfaces vs. cable pre-tension and attachment position for wind velocity *10m/s* (upper) and *15m/s* (lower). Lower plot: contour lines bounding the optimal design region

Proceedings of the Seventh (1997) International Offshore and Polar Engineering Conference
Honolulu, USA, May 25-30, 1997

ISBN 1-880653-28-1 (Set); ISBN 1-880653-31-1 (Vol. III)

Second-Order Approximate Solution of Nonlinear Wave Diffraction Due to a Cylinder in Directional Sea

Norimi Mizutani, **Nagoya University, Nagoya, Japan**
Takeshi Sanada, **Japan Port Consultants Ltd., Tokyo, Japan**
Akira Imai, **CTI Co. Ltd., Tokyo, Japan**
Koichiro Iwata, **Nagoya University, Nagoya, Japan**
Masumi Watanabe, **Chubu Electric Power Company, Nagoya, Japan**
Yoon-Koo Kang, **Port and Harbor Research Institute, Yokosuka, Japan**

ABSTRACT

The second-order approximated solution of the velocity potential is derived, without using the Green function, to express the diffracted waves from the isolated cylinder in the directional sea condition at a constant water depth. Laboratory experiments are carried out to verify the derived solution. In the experiments, multi-directional wave maker is employed to simulate the directional sea of four different kinds of directional spreading parameter. Moreover, two crossing waves, as the simplest multi-directional sea condition, are also generated. The wave forces in the principal direction which is normal to the wave maker and its transverse direction are measured with a three-component dynamometer.

First, calculated results of wave forces for the two crossing wave conditions are compared with experimental results to show that the present theory can predict the experimental results more accurately than the linear wave theory. Then, the numerical calculations of wave forces in directional sea condition are performed. The calculated results of the wave force in the principal direction are confirmed to be in good agreement with the experimental results.
It is also pointed out that the derivation of the present solution does not use the Green function but is given in analytical form; thus, the contributions of wave components are easily estimated.
KEY WORDS: nonlinear diffraction theory, directional sea, diffraction force, cylindrical structure

INTRODUCTION

Construction site of offshore structures is getting deeper and deeper where the waves are essentially multi-directional. Thus, the directionality of incoming waves should be properly estimated in the design of the structure. In the case of an isolated circular cylinder, the superposition of the linear solutions derived by MacCamy and Fuchs can be applied. In the real sea, however, the wave nonlinearity can play an important role not only on the wave forces acting but also on the wave field around it. The nonlinear effects on the wave diffraction for a case of an isolated cylinder in a regular wave train have been discussed by Kriebel (1990), Chau and Eatock Taylor (1992), Sanada et al. (1996), and importance of the nonlinearity is pointed out. The nonlinear solution of wave diffraction problem under the directional sea condition, however, has not yet been derived even in the case of an isolated cylindrical structure.

In this paper, the second-order solution of the velocity potential of nonlinear diffracted waves due to a surface piercing isolated cylinder with large diameter in the directional sea conditions was derived, and the validity of the present theorywas examined by comparing the calculated results with experimental ones .

THEORY

Governing Equation and Boundary Conditions

A surface piercing circular cylinder with radius, a, fixed on a sea floor at a constant depth, d, is considered (see Figure 1). It is convenient to introduce the cylindrical coordinate system (r, θ, z) with z-axis upward vertically with its origin on the still water level to describe the phenomenon. The assumption of irrotational motion of incompressible and inviscid fluid enable to introduce the velocity potential $\Phi(r, \theta, z; t)$ to describe the wave fields. The velocity

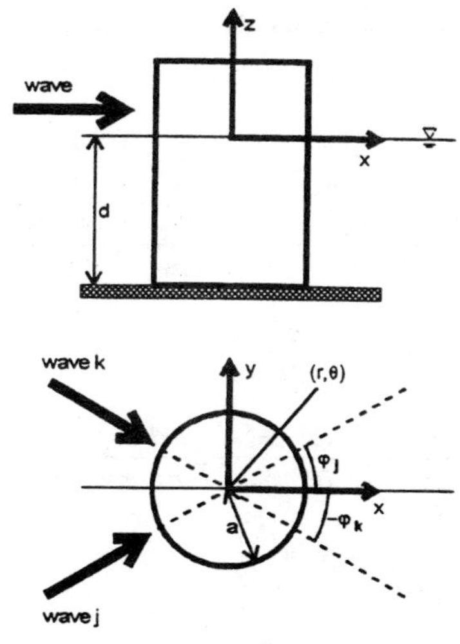

Figure 1 Definition sketch

potential $\Phi(r, \theta, z; t)$ is expanded into the perturbation series up to the second-order around the small parameter ε_j ($=k_{1j}H_{Ij}$) and ε_k ($=k_{1k}H_{Ik}$) where k_1 is the wave number in the first order solution, H_I is the incident wave height, and suffices j and k indicate the quantities corresponding to the wave-j and wave-k as shown in Figure 1.

$$\Phi(r,\theta,z;t) = \mathrm{Re}\left[\sum_{j=1}^{N}\varepsilon_j\phi_1^j(r,\theta,z)e^{-i(\sigma_j t-\delta_j)} + \sum_{J=1}^{N}\sum_{k=1}^{j}\varepsilon_j\varepsilon_k\Psi_{jk}\left\{\phi_{(+)}^{jk}(r,\theta,z)e^{-i\{(\sigma_j+\sigma_k)t-(\delta_j+\delta_k)\}} + \phi_{(-)}^{jk}(r,\theta,z)e^{-i\{(\sigma_j-\sigma_k)t-(\delta_j-\delta_k)\}}\right\}\right] \quad (1)$$

where,

$$\Psi_{jk} = \begin{cases}1 & (j \neq k)\\ 1/2 & (j = k)\end{cases}$$

and σ is the angular frequency and δ is the initial phase angle.

The solution of the first order component ϕ_1 can be expressed using the solution derived by MacCamy and Fuchs (1954) taking the wave direction into account. Thus, the second-order components, $\phi_{(+)}$ (component having a wave number of $k_1^j+k_1^k$) and $\phi_{(-)}$ (component having a wave number of $k_1^j-k_1^k$) will be derived. For the case of j=k, however, the term $\phi_{(-)}^{jk}$ which represents the velocity potential of the steady component will not be derived because it has no contribution on the free surface variation, the wave pressure and wave forces as far as the second-order theory is concerned.

The governing equations of these component of the velocity potential are given by the following Laplace equations.

$$\frac{\partial^2\phi_{(+)}^{jk}}{\partial r^2}+\frac{1}{r}\frac{\partial\phi_{(+)}^{jk}}{\partial r}+\frac{1}{r^2}\frac{\partial^2\phi_{(+)}^{jk}}{\partial\theta^2}+\frac{\partial^2\phi_{(+)}^{jk}}{\partial z^2}=0 \quad (2)$$

$$\frac{\partial^2\phi_{(-)}^{jk}}{\partial r^2}+\frac{1}{r}\frac{\partial\phi_{(-)}^{jk}}{\partial r}+\frac{1}{r^2}\frac{\partial^2\phi_{(-)}^{jk}}{\partial\theta^2}+\frac{\partial^2\phi_{(-)}^{jk}}{\partial z^2}=0 \quad (3)$$

The boundary conditions are

$$\frac{\partial\phi_{(+)}^{jk}}{\partial z}-\frac{(\sigma_j+\sigma_k)^2}{g}\phi_{(+)}^{jk} = \frac{i}{2g}\left\{\sigma_j\left(2\frac{\partial\phi_1^j}{\partial r}\frac{\partial\phi_1^k}{\partial r}+\frac{2}{r^2}\frac{\partial\phi_1^j}{\partial\theta}\frac{\partial\phi_1^k}{\partial\theta}+2\frac{\partial\phi_1^j}{\partial z}\frac{\partial\phi_1^k}{\partial z}-\phi_1^j\frac{\partial^2\phi_1^k}{\partial z^2}+\upsilon_k\phi_1^j\frac{\partial\phi_1^k}{\partial z}\right) + \sigma_k\left(2\frac{\partial\phi_1^j}{\partial r}\frac{\partial\phi_1^k}{\partial r}+\frac{2}{r^2}\frac{\partial\phi_1^j}{\partial\theta}\frac{\partial\phi_1^k}{\partial\theta}+2\frac{\partial\phi_1^j}{\partial z}\frac{\partial\phi_1^k}{\partial z}-\phi_1^k\frac{\partial^2\phi_1^j}{\partial z^2}+\upsilon_j\phi_1^k\frac{\partial\phi_1^j}{\partial z}\right)\right\} \quad ;z=0 \quad (4)$$

$$\frac{\partial\phi_{(-)}^{jk}}{\partial z}-\frac{(\sigma_j-\sigma_k)^2}{g}\phi_{(-)}^{jk} = \frac{i}{2g}\left\{\sigma_j\left(2\frac{\partial\phi_1^j}{\partial r}\frac{\partial\phi_1^{k*}}{\partial r}+\frac{2}{r^2}\frac{\partial\phi_1^j}{\partial\theta}\frac{\partial\phi_1^{k*}}{\partial\theta}+2\frac{\partial\phi_1^j}{\partial z}\frac{\partial\phi_1^{k*}}{\partial z}-\phi_1^j\frac{\partial^2\phi_1^{k*}}{\partial z^2}+\upsilon_k\phi_1^j\frac{\partial\phi_1^{k*}}{\partial z}\right) - \sigma_k\left(2\frac{\partial\phi_1^j}{\partial r}\frac{\partial\phi_1^{k*}}{\partial r}+\frac{2}{r^2}\frac{\partial\phi_1^j}{\partial\theta}\frac{\partial\phi_1^{k*}}{\partial\theta}+2\frac{\partial\phi_1^j}{\partial z}\frac{\partial\phi_1^{k*}}{\partial z}-\phi_1^{k*}\frac{\partial^2\phi_1^j}{\partial z^2}+\upsilon_j\phi_1^{k*}\frac{\partial\phi_1^j}{\partial z}\right)\right\} \quad ;z=0 \quad (5)$$

$$\frac{\partial\phi_{(+)}^{jk}}{\partial z}=0 \qquad \frac{\partial\phi_{(-)}^{jk}}{\partial z}=0 \qquad ;z=-d \quad (6),(7)$$

$$\frac{\partial\phi_{(+)}^{jk}}{\partial r}=0 \qquad \frac{\partial\phi_{(-)}^{jk}}{\partial r}=0 \qquad ;r=a \quad (8),(9)$$

where, $\nu(=\sigma^2/g)$ indicates the deep water wave number and suffix * implies complex conjugate.

Moreover, the solution should satisfy the radiation condition subjected at far field. Details of that are given in Appendix.

The components ϕ_1^j and ϕ_1^{j*} are given by the following equations using the MacCamy and Fuchs' theory.

$$\phi_1^j = -\frac{ig}{2\sigma_j k_1^j}\frac{\cosh k_1^j(z+d)}{\cosh k_1^j d}\sum_{m=-\infty}^{\infty} i^m A_m(k_1^j r)e^{im(\theta-\varphi_j)} \quad (10)$$

$$\phi_1^{j*} = \frac{ig}{2\sigma_j k_1^j}\frac{\cosh k_1^j(z+d)}{\cosh k_1^j d}\sum_{m=-\infty}^{\infty}(-i)^m B_m(k_1^j r)e^{-im(\theta-\varphi_j)} \quad (11)$$

where,

$$A_m(k_1^j r) = J_m(k_1^j r) - \alpha_m^j H_m^{(1)}(k_1^j r) \quad (12)$$

$$B_m(k_1^j r) = J_m(k_1^j r) - \beta_m^j H_m^{(2)}(k_1^j r) \quad (13)$$

and

$$\alpha_m^j = J_m'(k_1^j a)\big/H_m^{(1)\prime}(k_1^j a) \quad (14)$$

$$\beta_m^j = J_m'(k_1^j a)\big/H_m^{(2)\prime}(k_1^j a) \quad (15)$$

In the above equations, J_m is the m-th order Bessel function, $H_m^{(1)}$ and $H_m^{(2)}$ are the m-th order Hankel functions of the first and second kinds, respectively, and φ_j implies the propagation direction of wave-j. Substitution of Equations (10) and (11) into Equations (4) and (5) leads to the following equations.

$$\frac{\partial\phi_{(+)}^{jk}}{\partial z}-\frac{(\sigma_j+\sigma_k)^2}{g}\phi_{(+)}^{jk} = -\frac{ig}{8}\sum_{n=-\infty}^{\infty}i^n\left(\frac{1}{\sigma_k}g_{(+)n}^{jk}(r)+\frac{1}{\sigma_j}g_{(+)n}^{kj}(r)\right)e^{in\theta} \quad ;z=0 \quad (16)$$

$$\frac{\partial \phi^{jk}_{(-)}}{\partial z} - \frac{(\sigma_j - \sigma_k)^2}{g}\phi^{jk}_{(-)} = \frac{ig}{8}\sum_{n=-\infty}^{\infty} i^n\left(\frac{1}{\sigma_k}g^{jk}_{(-)n}(r) - \frac{1}{\sigma_j}g^{kj}_{(-)n}(r)\right)e^{in\theta} \quad ;z = 0 \tag{17}$$

where,

$$g^{jk}_{(+)n}(r) = g^{jkII}_{(+)n}(r) + g^{jkIS}_{(+)n}(r) + g^{jkSS}_{(+)n}(r) \tag{18}$$

$$g^{kj}_{(+)n}(r) = g^{kjII}_{(+)n}(r) + g^{kjIS}_{(+)n}(r) + g^{kjSS}_{(+)n}(r) \tag{19}$$

$$g^{jk}_{(-)n}(r) = g^{jkII}_{(-)n}(r) + g^{jkIS}_{(-)n}(r) + g^{jkSS}_{(-)n}(r) \tag{20}$$

$$g^{kj}_{(-)n}(r) = g^{kjII}_{(-)n}(r) + g^{kjIS}_{(-)n}(r) + g^{kjSS}_{(-)n}(r) \tag{21}$$

$$g^{jkII}_{(+)n}(r) = -\sum_{m=-\infty}^{\infty}\Big(\mu^j_{n-(m+1)}\mu^k_{m+1} - T_{jk}\mu^j_{n-m}\mu^k_m + \mu^j_{n-(m-1)}\mu^k_{m-1}\Big)J_{n-m}(k^j_1 r)J_m(k^k_1 r) \tag{22}$$

$$\begin{aligned} g^{jkIS}_{(+)n}(r) = &\sum_{m=-\infty}^{\infty}\Big(\mu^j_{n-(m+1)}\mu^k_{m+1}\alpha^k_{m+1} - T_{jk}\mu^j_{n-m}\mu^k_m\alpha^k_m + \mu^j_{n-(m-1)}\mu^k_{m-1}\alpha^k_{m-1}\Big)J_{n-m}(k^j_1 r)H^{(1)}_m(k^k_1 r) \\ &+ \sum_{m=-\infty}^{\infty}\Big(\mu^j_{n-(m+1)}\mu^k_{m+1}\alpha^j_{n-(m+1)} - T_{jk}\mu^j_{n-m}\mu^k_m\alpha^j_{n-m} + \mu^j_{n-(m-1)}\mu^k_{m-1}\alpha^j_{n-(m-1)}\Big)H^{(1)}_{n-m}(k^j_1 r)J_m(k^k_1 r) \end{aligned} \tag{23}$$

$$\begin{aligned} g^{jkSS}_{(+)n}(r) = -&\sum_{m=-\infty}^{\infty}\Big(\mu^j_{n-(m+1)}\mu^k_{m+1}\alpha^j_{n-(m+1)}\alpha^k_{m+1} - T_{jk}\mu^j_{n-m}\mu^k_m\alpha^j_{n-m}\alpha^k_m \\ &+ \mu^j_{n-(m-1)}\mu^k_{m-1}\alpha^j_{n-(m-1)}\alpha^k_{m-1}\Big)H^{(1)}_{n-m}(k^j_1 r)H^{(1)}_m(k^k_1 r) \end{aligned} \tag{24}$$

$$g^{jkII}_{(-)n}(r) = \sum_{m=-\infty}^{\infty}(-1)^m\Big\{\mu^j_{n-(m+1)}\mu^k_{m+1} + T_{jk}\mu^j_{n-m}\mu^k_m + \mu^j_{n-(m-1)}\mu^k_{m-1}\Big\}J_{n-m}(k^j_1 r)J_m(k^k_1 r) \tag{25}$$

$$\begin{aligned} g^{jkIS}_{(-)n}(r) = -&\sum_{m=-\infty}^{\infty}(-1)^m\Big\{\mu^j_{n-(m+1)}\mu^k_{m+1}\beta^k_{m+1} + T_{jk}\mu^j_{n-m}\mu^k_m\beta^k_m + \mu^j_{n-(m-1)}\mu^k_{m-1}\beta^k_{m-1}\Big\}J_{n-m}(k^j_1 r)H^{(2)}_m(k^k_1 r) \\ -&\sum_{m=-\infty}^{\infty}(-1)^m\Big\{\mu^j_{n-(m+1)}\mu^k_{m+1}\alpha^j_{n-(m+1)} + T_{jk}\mu^j_{n-m}\mu^k_m\alpha^j_{n-m} + \mu^j_{n-(m-1)}\mu^k_{m-1}\alpha^j_{n-(m-1)}\Big\}H^{(1)}_{n-m}(k^j_1 r)J_m(k^k_1 r) \end{aligned} \tag{26}$$

$$\begin{aligned} g^{jkSS}_{(-)n}(r) = &\sum_{m=-\infty}^{\infty}(-1)^m\Big\{\mu^j_{n-(m+1)}\mu^k_{m+1}\alpha^j_{n-(m+1)}\beta^k_{m+1} \\ &+ T_{jk}\mu^j_{n-m}\mu^k_m\alpha^j_{n-m}\beta^k_m + \mu^j_{n-(m-1)}\mu^k_{m-1}\alpha^j_{n-(m-1)}\beta^k_{m-1}\Big\} \\ &\cdot H^{(1)}_{n-m}(k^j_1 r)H^{(2)}_m(k^k_1 r) \end{aligned} \tag{27}$$

where, $\mu^j_m = e^{-im\varphi_j}$ and T_{jk} is defined by the following equation.

$$T_{jk} = 2\tanh k^j_1 d \tanh k^k_1 d - \frac{k^k_1}{k^j_1}\left(1 - \tanh^2 k^k_1 d\right) \tag{28}$$

Similarly, the expressions of $g_{(+)n}{}^{kj}$ and $g_{(-)n}{}^{kj}$ are obtained by Equations (22) through (27) with use of T_{kj} defined by Equation (29), instead of T_{jk}.

$$T_{kj} = 2\tanh k^k_1 d \tanh k^j_1 d - \frac{k^j_1}{k^k_1}\left(1 - \tanh^2 k^j_1 d\right) \tag{29}$$

Derivation of the Second-order Solution

The velocity potential of the second-order diffracted wave can be given as a sum of the bounded wave component (indicated by subsuffix L) and the free wave component (indicated by subsuffix F) and can be expressed by the following equations.

$$\phi^{jk}_{(+)}(r,\theta,z) = \phi^{jk}_{(+)L}(r,\theta,z) + \phi^{jk}_{(+)F}(r,\theta,z) \tag{30}$$

$$\phi^{jk}_{(-)}(r,\theta,z) = \phi^{jk}_{(-)L}(r,\theta,z) + \phi^{jk}_{(-)F}(r,\theta,z) \tag{31}$$

The solutions corresponding to the bounded wave component which satisfy the boundary conditions on the free surface and sea bed are given in the following equations.

$$\phi^{jk}_{(+)L}(r,\theta,z) = -\frac{ig}{8}\sum_{n=-\infty}^{\infty} i^n\left\{\frac{1}{\sigma_k}\varphi^{jk}_{(+)nL}(r,z) + \frac{1}{\sigma_j}\varphi^{kj}_{(+)nL}(r,z)\right\}e^{in\theta} \tag{32}$$

$$\phi^{jk}_{(-)L}(r,\theta,z) = \frac{ig}{8}\sum_{n=-\infty}^{\infty} i^n\left\{\frac{1}{\sigma_k}\varphi^{jk}_{(-)nL}(r,z) - \frac{1}{\sigma_j}\varphi^{kj}_{(-)nL}(r,z)\right\}e^{in\theta} \tag{33}$$

where,

$$\varphi^{jk}_{(+)nL}(r,z) = G^{jk}_{(+)}\frac{\cosh(k^j_1 + k^k_1)(z+d)}{\cosh(k^j_1 + k^k_1)d}g^{jk}_{(+)n}(r) \tag{34}$$

$$\varphi^{kj}_{(+)nL}(r,z) = G^{jk}_{(+)}\frac{\cosh(k^j_1 + k^k_1)(z+d)}{\cosh(k^j_1 + k^k_1)d}g^{kj}_{(+)n}(r) \tag{35}$$

$$\varphi^{jk}_{(-)nL}(r,z) = G^{jk}_{(-)}\frac{\cosh(k^j_1 - k^k_1)(z+d)}{\cosh(k^j_1 - k^k_1)d}g^{jk}_{(-)n}(r) \tag{36}$$

$$\varphi^{kj}_{(-)nL}(r,z) = G^{jk}_{(-)}\frac{\cosh(k^j_1 - k^k_1)(z+d)}{\cosh(k^j_1 - k^k_1)d}g^{kj}_{(-)n}(r) \tag{37}$$

$$G^{jk}_{(+)} = \frac{g}{g(k^j_1 + k^k_1)\tanh(k^j_1 + k^k_1)d - (\sigma_j + \sigma_k)^2} \tag{38}$$

$$G^{jk}_{(-)} = \frac{g}{g(k^j_1 - k^k_1)\tanh(k^j_1 - k^k_1)d - (\sigma_j - \sigma_k)^2} \tag{39}$$

The solutions above are approximated ones which cannot satisfy the governing equations (2) and (3) completely. Murakami et al. (1993) compared this type of solution with experimental results using regular waves to show good agreement between them. Thus, the solutions obtained here can be thought to give good approximation. After substituting Equations (32) and (33) into Equations (30) and (31), the coefficients in the free wave component are determined so that they satisfy the impermeable condition on the cylinder surface. Then the solution of the free wave components is obtained.

$$\phi^{jk}_{(+)F}(r,z) = -\frac{ig}{8}\sum_{n=-\infty}^{\infty} i^n \left\{ \frac{\cosh k^{jk}_{(+)}(z+d)}{\cosh k^{jk}_{(+)}d} E^{jk}_{(+)n,0} H^{(1)}_n (k^{jk}_{(+)}r) + \sum_{s=0}^{\infty} \frac{\cos\kappa^{jk}_{(+),s}(z+d)}{\cos\kappa^{jk}_{(+),s}d} E^{jk}_{(+)n,s} K_n(\kappa^{jk}_{(+),s}r) \right\} e^{in\theta} \tag{40}$$

$$\phi^{jk}_{(-)F}(r,z) = \frac{ig}{8}\sum_{n=-\infty}^{\infty} i^n \left\{ \frac{\cosh k^{jk}_{(-)}(z+d)}{\cosh k^{jk}_{(-)}d} E^{jk}_{(-)n,0} H^{(*)}_n (k^{jk}_{(-)}r) + \sum_{s=0}^{\infty} \frac{\cos\kappa^{jk}_{(-),s}(z+d)}{\cos\kappa^{jk}_{(-),s}d} E^{jk}_{(-)n,s} K_n(\kappa^{jk}_{(-),s}r) \right\} e^{in\theta} \tag{41}$$

in which K_n is the n-th order modified Bessel function of the second kind, $k_{(+)}^{jk}$, $\kappa_{(+),s}^{jk}$, $k_{(-)}^{jk}$ and $\kappa_{(-),s}^{jk}$ are the eigenvalues which are the positive roots of following dispersion relationships.

$$(\sigma_j+\sigma_k)^2 = gk^{jk}_{(+)}\tanh k^{jk}_{(+)}d = -g\kappa^{jk}_{(+),s}\tan\kappa^{jk}_{(+),s}d \tag{42}$$

$$(\sigma_j-\sigma_k)^2 = gk^{jk}_{(-)}\tanh k^{jk}_{(-)}d = -g\kappa^{jk}_{(-),s}\tan\kappa^{jk}_{(-),s}d \tag{43}$$

In Eq.(41), $H^{(*)}_n = H^{(1)}_n$ for $\sigma_j - \sigma_k > 0$ and $H^{(*)}_n = H^{(2)}_n$ for $\sigma_j - \sigma_k < 0$. The coefficients E in Equations (40) and (41) are determined by the impermeable conditions on the cylinder surface and given as in the following equations.

$$E^{jk}_{(+)n,0} = -\frac{\cosh k^{jk}_{(+)}d}{\cosh(k^j_1+k^k_1)d} \cdot Z^{(+)}_0 \left(\frac{1}{\sigma_k} g^{jk}_{(+)n}{}'(a) + \frac{1}{\sigma_j} g^{kj}_{(+)n}{}'(a) \right) \frac{G^{jk}_{(+)}}{k^{jk}_{(+)} H^{(1)}_n{}'(k^{jk}_{(+)}a)} \tag{44}$$

$$E^{jk}_{(+)n,s} = -\frac{\cos\kappa^{jk}_{(+),s}d}{\cosh(k^j_1+k^k_1)d} \cdot Z^{(+)}_s \left(\frac{1}{\sigma_k} g^{jk}_{(+)n}{}'(a) + \frac{1}{\sigma_j} g^{kj}_{(+)n}{}'(a) \right) \frac{G^{jk}_{(+)}}{\kappa^{jk}_{(+),s} K_n{}'(\kappa^{jk}_{(+),s}a)} \tag{45}$$

$$E^{jk}_{(-)n,0} = -\frac{\cosh k^{jk}_{(-)}d}{\cosh(k^j_1-k^k_1)d} \cdot Z^{(-)}_0 \left(\frac{1}{\sigma_k} g^{jk}_{(-)n}{}'(a) - \frac{1}{\sigma_j} g^{kj}_{(-)n}{}'(a) \right) \frac{G^{jk}_{(-)}}{k^{jk}_{(-)} H^{(*)}_n{}'(k^{jk}_{(-)}a)} \tag{46}$$

$$E^{jk}_{(-)n,1} = -\frac{\cos\kappa^{jk}_{(-),s}d}{\cosh(k^j_1-k^k_1)d} \cdot Z^{(-)}_s \left(\frac{1}{\sigma_k} g^{jk}_{(-)n}{}'(a) - \frac{1}{\sigma_j} g^{kj}_{(-)n}{}'(a) \right) \frac{G^{jk}_{(-)}}{\kappa^{jk}_{(-),s} K_n{}'(\kappa^{jk}_{(-),s}a)} \tag{47}$$

where

$$Z^{(+)}_0 = \frac{\int_{-d}^{0} \cosh(k^j_1+k^k_1)(z+d)\cosh k^{jk}_{(+)}(z+d)dz}{\int_{-d}^{0}\left\{\cosh k^{jk}_{(+)}(z+d)\right\}^2 dz} \tag{48}$$

$$Z^{(+)}_s = \frac{\int_{-d}^{0} \cosh(k^j_1+k^k_1)(z+d)\cos\kappa^{jk}_{(+),s}(z+d)dz}{\int_{-d}^{0}\left\{\cos\kappa^{jk}_{(+),s}(z+d)\right\}^2 dz} \tag{49}$$

$$Z^{(-)}_0 = \frac{\int_{-d}^{0} \cosh(k^j_1-k^k_1)(z+d)\cosh k^{jk}_{(-)}(z+d)dz}{\int_{-d}^{0}\left\{\cosh k^{jk}_{(-)}(z+d)\right\}^2 dz} \tag{50}$$

$$Z^{(-)}_s = \frac{\int_{-d}^{0} \cosh(k^j_1-k^k_1)(z+d)\cos\kappa^{jk}_{(-),s}(z+d)dz}{\int_{-d}^{0}\left\{\cos\kappa^{jk}_{(-),s}(z+d)\right\}^2 dz} \tag{51}$$

It is noted that the solution of $\phi_{(+)}^{jk}$ in case of j=k is confirmed to coincide with the analytical solution for a regular wave train derived by Sanada et al. (1996) and Appendix shows that the present solution satisfies the radiation condition.

LABORATORY EXPERIMENT

Laboratory experiments were conducted in the wave basin (30m in length, 23m in width, and 12m in depth) with a directional wave maker composed of 42 wave paddles at Electric Research Institute of Chubu Electric Power Company. The cylinder model with radius a=30cm was installed in the wave basin. The cylinder has enough height (100cm) and no wave overtopping occurs during the experiments. The wave forces in x and y directions were measured by means of a three-component dynamometer. The surface variations at six points around the cylinder (one point in front, two points at both sides, and three points at rear side) were also measured with electric capacitance-type wave gages.

Two kinds of experiments were conducted in this study. One is for two-crossing waves and the other is for multi-directional waves. Experimental conditions are listed in Tables 1 and 2. The target power spectrum was the Bretschneider-Mitsuyasu type spectrum and the Mitsuyasu-type directional spreading function was adopted. In generating the multi-directional random waves, the single summation method was employed.

RESULTS AND DISCUSSIONS

Two-Crossing Waves

Variations of dimensionless average wave forces, $\overline{C}\left(=2\overline{F}/\rho g\pi a^2\overline{H}\right)$, with the crossing angle under the condition that $k^j_1 a = 0.87$, $k^k_1 a = 0.42$ and d/a=2.33 are shown in Figures 2(a), (b) and (c). In these figures, curves indicated by LIN are results of the linear solution, ones indicated by PLS are the second-order component with a frequency of sum of frequencies of first order components; whereas, ones with MNS corresponds to the component with the subtract frequency, and curves indicated ALL are the total of the second-order solution. In the calculation, the lowest ten terms of the Fourier series are taken into account.

It is clear that the difference between the linear and second-order solutions increases with increasing wave steepness of component waves and the second-order solution is about 15% larger than the linear one in Figure 2(c). In the linear theory, the integration of wave pressure is performed from sea bottom to the still water level, while it is performed up to the free surface in the second-order. Thus, the wave pressure above the wave trough may play an important role on the second-order wave force. The experiment values tend to be a little larger than the second-order solution because the second-order cannot consider higher order nonlinearity as in the experimental results. It is clear, however, that the agreement of the second-order solution with the experiments is much better than the linear theory, and therefore the necessity of using nonlinear theory in the estimation of wave force is confirmed.

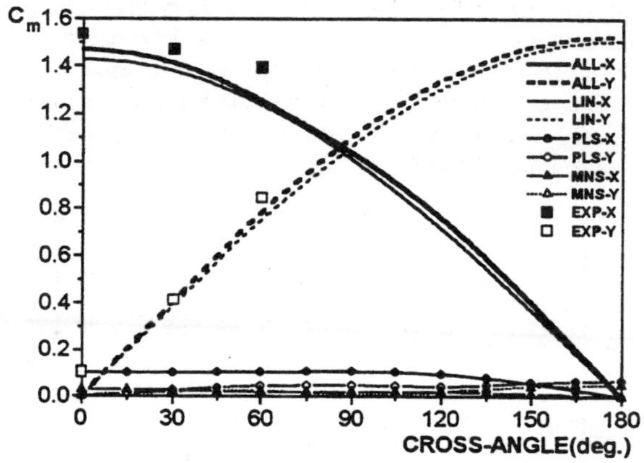

(a) $H_I^j / L_j = 0.02,\ H_I^k / L_k = 0.009$

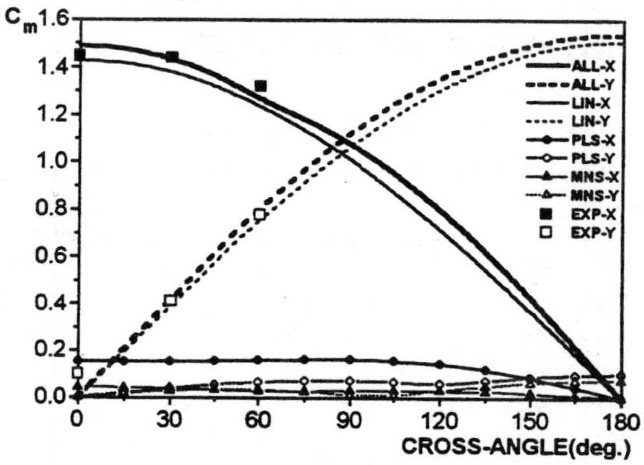

(b) $H_I^j / L_j = 0.03,\ H_I^k / L_k = 0.013$

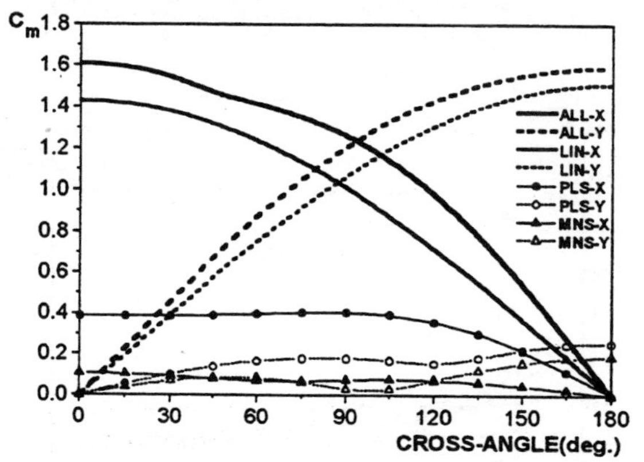

(c) $H_I^j / L_j = 0.07,\ H_I^k / L_k = 0.03$

Figure 2 Variations of dimensionless wave forces with crossing angle ($k_I^j a = 0.87,\ k_I^k a = 0.42$, d/a=2.33)

Dimensionless wave crest heights around the cylinder are shown in Figures 3(a) and (b), which correspond to the linear and second-order solutions, respectively. In the linear solution, the highest wave runup occurs on the x-axis which corresponds to the center of the first wave attack points of each wave (θ=7π/6 for wave-j and 5π/6 for wave-k). On the contrary, the location of highest wave runup approaches to θ=7π/6 where the first attack point of wave-j whose wave slope is larger than wave-k. This is due to the nonlinear effect in the free surface variation.

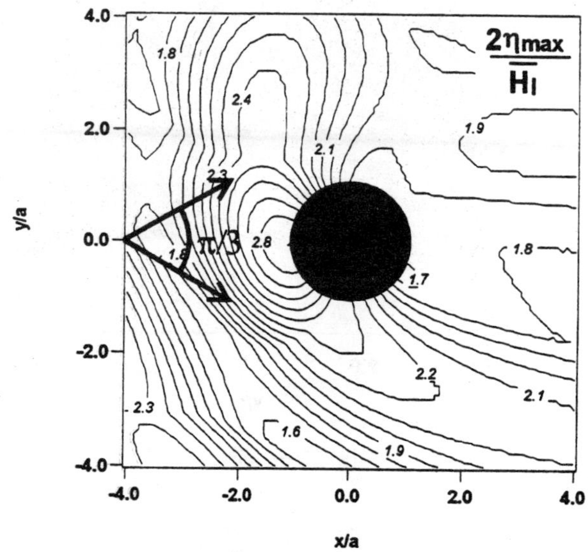

(a) Linear theory

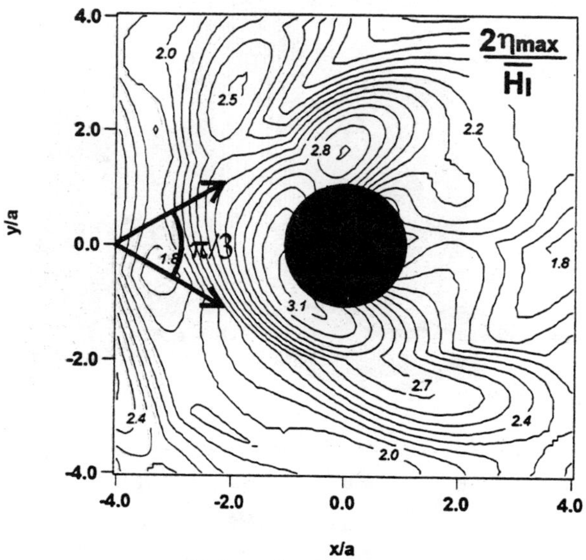

(b) Second-order theory

Figure 3 Dimensionless crest height distribution ($k_I^j a = 0.87,\ k_I^k a = 0.42,\ H_I^j / L_j = 0.03,\ H_I^k / L_k = 0.01,\ \varphi_j - \varphi_k = \pi / 3$)

The free surface profile around the cylinder at the phase that the wave force attains the maximum are shown in Figures 4(a) and (b). At this phase, the difference in the free surface between the linear and the second-order theories are clear at the rear face of the cylinder rather than the front face. That is, the free surface in the second-order solution is lower than the linear one at the rear face. The second-order free wave component may attribute this phenomena, which makes the difference in the water level between the front and rear faces of the cylinder. Thus, the wave force in the second-order theory is larger than that of the linear theory.

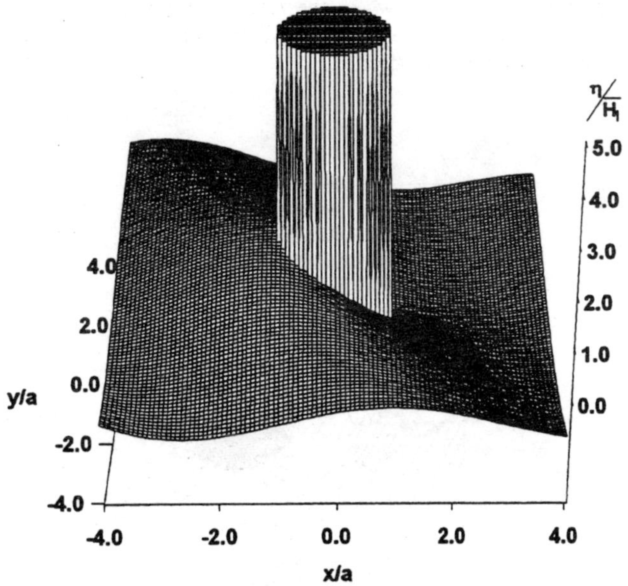

(a) Linear theory

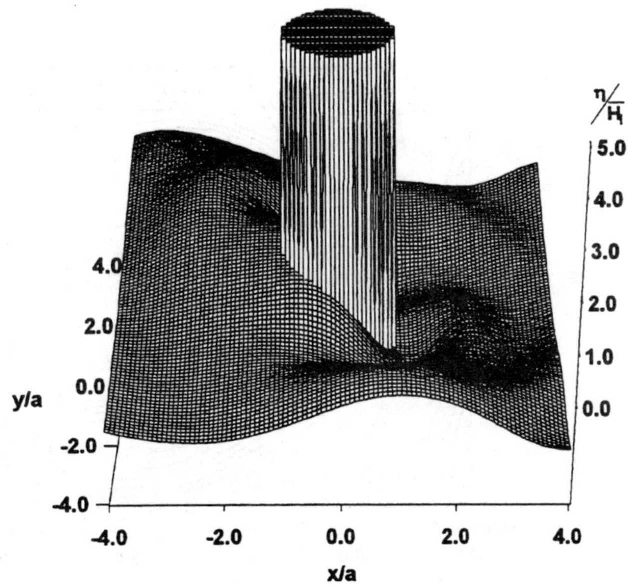

(b) Second-order theory

Figure 4 Dimensionless free surface profile

$(k_I^j a = 0.87,\ k_I^k a = 0.42,\ H_I^j / L_j = 0.06,\ H_I^k / L_k = 0.03,$

$\varphi_j - \varphi_k = \pi / 3)$

Multi-Directional Waves

In the calculation, it is hard to include all wave components because of the limit of computational capacity and long computational time. Thus, only nine waves are considered in the calculations. The wave heights of these component waves are determined by the power spectrum. First, the directional spreading functions are divided into nine sections. Then, the directional spreading function is integrated over each section. This integration gives the ratio of the wave energy of each component wave to the total energy. Then, the total energy is distributed using these ratio to each component wave. The wave height of each component wave is determined by the evaluated wave energy. The wave period of component wave is evaluated by following equation (Nagai, 1972).

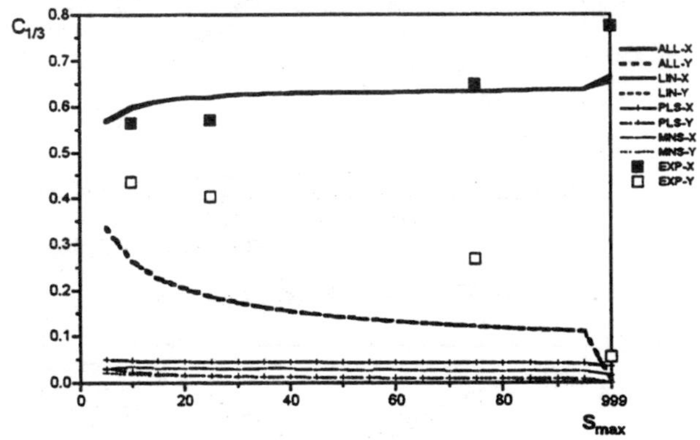

(a) $H_{1/3} / L_{1/3} = 0.04$

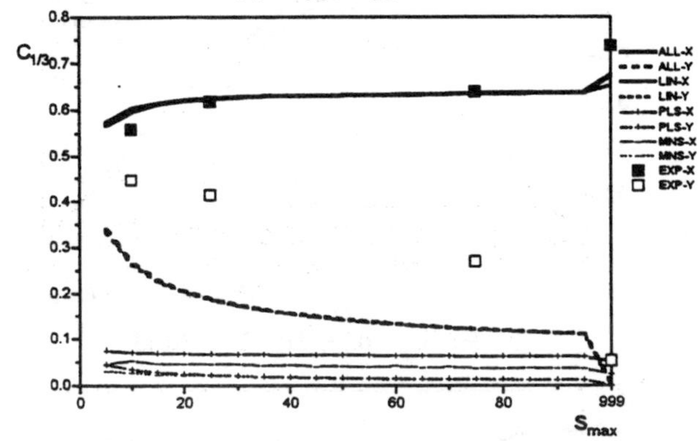

(b) $H_{1/3} / L_{1/3} = 0.06$

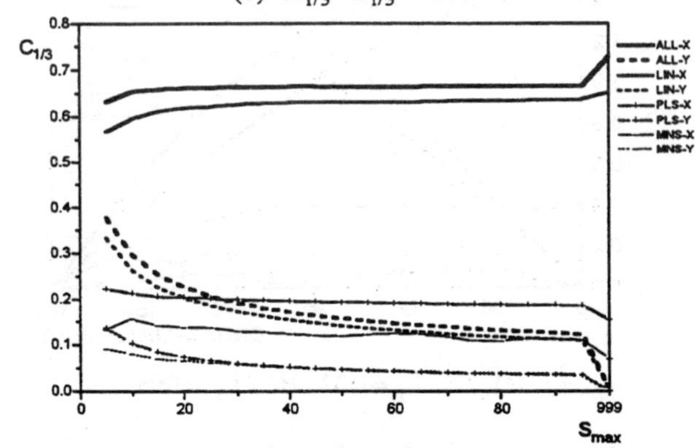

(c) $H_{1/3} / L_{1/3} = 0.08$

Figure 5 Variations of dimensionless wave forces with Smax

$$\frac{1}{T_j} = \frac{1}{0.9T_{1/3}}\left\{\frac{0.675}{\ln[2M/(2j-1)]}\right\}^{1/4}$$
$$= \frac{1.007}{T_{1/3}}\{\ln[2M/(2j-1)]\}^{-1/4} \qquad (52)$$

The initial phase of each component wave is given by random number generated in the range from 0 to 2π.

Figures 5(a), (b) and (c) show the variations of the dimensionless one-third highest wave forces, $C_{1/3}(=2F_{1/3}/\rho g\pi a^2 H_{1/3})$, with the direction spreading parameter, Smax. The values corresponding the principal direction (x) generally increase with Smax while their transverse components decrease. The computation results in the x direction agree well with the experimental results. Moreover, the second-order theory can give better approximation than the linear theory. This implies that the application of the nonlinear diffraction theory is important for more accurate estimation of the diffraction force.

The computed values corresponding to the y direction underestimate the experimental values. This may be attributed to the less components of the component waves in the multi-directional waves in the calculation. All the components of the oblique waves travel in positive direction along the x-axis, however, along the y-axis the waves travel in both positive and negative directions, and the effect of approximation of the wave field with finite wave components appear more clearly in the results of y direction. Thus, the approximation method of directional sea condition should be improved. Moreover, the target values of the direction spreading parameter and power spectrum are used in the calculation. The actual spreading parameter evaluated from experimental data are generally smaller than target ones. This is also a possible reason resulting in the poor agreement in the transverse wave force.

CONCLUDING REMARKS

The second-order analytical solution of the velocity potential of the diffracted waves from an isolated cylinder in the multi-directional sea has been derived. Through the investigation using two-crossing wave conditions, the validity of the present theory has been confirmed. In the computation of the wave force in the multi-directional sea condition, the present theory gives good estimation of the wave force in the principal direction even with the simple approximation method of the sea condition. In the estimation of the transverse wave force, the calculated results underestimate the experimental ones and approximation method of the sea condition should be improved.

ACKNOWLEGDEMENT

This research was conducted under the support of the Grant-in-Aids for Scientific Research from the Ministry of Education, Science and Culture of Japan, under Grant No.08555128 (Head, Koichiro IWATA).

REFERENCES

Chau FP and Eatock Taylor R (1992):"Second order wave diffraction by a vertical cylinder", *J. Fluid Mech.*, Vol.240, pp.571-599.

Kriebel DL (1990):"Nonlinear wave interaction with a vertical circular cylinder, Part I", *Ocean Engineering*, Vol.1, No.4, pp.345-377.

MacCamy RC and Fuchs RA (1954):"Wave forces on piles; A diffraction theory", *Beach Erosion Board, Tech. Memo.*, No.69, pp.1-17.

Murakami K, Yoshida A, and Irie I (1993):"Second-order wave interaction with vertical cylinder of arbitrary cross section", *Proc. of Coastal Eng., JSCE*, Vol.40, pp.846-850 (in Japanese).

Nagai K (1972):"Computation of refraction and diffraction of irregular sea", *Rep. Of Port and Harbor Res. Inst.*, Vol.11, No.2, pp.47-119 (in Japanese).

Sanada T, Iwata K, and Mizutani N (1996):"Second order exact theory of nonlinear wave diffraction due to a cylindrical structure", *J. Hydraulics, Coastal and Environmental Eng., JSCE*, No.551/II-37, pp.75-88 (in Japanese).

APPENDIX: Radiation Condition

The radiation conditions for the second order diffracted waves of free wave mode are given by Sommerfeld's radiation conditions as follows:

$$\lim_{r\to\infty}\sqrt{r}\left(\frac{\partial\phi^{jk}_{(+)F}}{\partial r} - ik^{jk}_{(+)}\phi^{jk}_{(+)F}\right) = 0 \qquad (A1)$$

$$\lim_{r\to\infty}\sqrt{r}\left(\frac{\partial\phi^{jk}_{(-)F}}{\partial r} \mp ik^{jk}_{(-)}\phi^{jk}_{(-)F}\right) = 0 \qquad (A2)$$

In Eq.(A2), the negative sign is adopted for $\sigma_j - \sigma_k > 0$ and the positive sign for $\sigma_j - \sigma_k < 0$. The derived velocity potentials given by Eqs.(40) and (41) satisfy these radiation conditions, which can be easily proved in a similar way to the first order solution and it is not given here.

Regarding the bounded wave mode solutions, they do not require the radiation condition and they are uniquely determined in terms of the first order solutions because of the non-homogeneous boundary conditions subjected on the free surface. However, it is important to investigate their behaviors at far field. At far field, the solutions given by Eqs.(23), (24), (26) and (27) can be approximated by introducing asymptotic forms of the cylindrical functions. Replacing complicated coefficients in front of cylindrical functions in each equation into one coefficient, these four equations can be rewritten as Eqs.(A3), (A4), (A5) and (A6).

$$g^{jkIS}_{(+)n}(r) = \sum_{m=-\infty}^{\infty} C^{jk}_{(+)m} J_{n-m}(k^j_1 r) H^{(1)}_m(k^k_1 r) + \sum_{m=-\infty}^{\infty} \overline{C}^{jk}_{(+)m} H^{(1)}_{n-m}(k^j_1 r) J_m(k^k_1 r) \qquad (A3)$$

$$g^{jkSS}_{(+)n}(r) = \sum_{m=-\infty}^{\infty} D^{jk}_{(+)m} H^{(1)}_{n-m}(k^j_1 r) H^{(1)}_m(k^k_1 r) \qquad (A4)$$

$$g^{jkIS}_{(-)n}(r)=\sum_{m=-\infty}^{\infty}C^{jk}_{(-)m}J_{n-m}\left(k^j_1 r\right)H^{(2)}_m\left(k^k_1 r\right)+\sum_{m=-\infty}^{\infty}\overline{C}^{jk}_{(-)m}H^{(1)}_{n-m}\left(k^j_1 r\right)J_m\left(k^k_1 r\right) \quad \text{(A5)}$$

$$g^{jkSS}_{(-)n}(r)=\sum_{m=-\infty}^{\infty}D^{jk}_{(-)m}H^{(1)}_{n-m}\left(k^j_1 r\right)H^{(2)}_m\left(k^k_1 r\right) \quad \text{(A6)}$$

We introduce the relationship Eq.(A7) among the Bessel function and the Hankel functions of the first and second kinds and asymptotic forms of the Hankel functions at $r\to\infty$ given by Eq.(A8).

$$J_n(x)=\frac{1}{2}\left(H^{(1)}_n(x)+H^{(2)}_n(x)\right) \quad \text{(A7)}$$

$$\left.\begin{aligned} H^{(1)}_n(x)&\cong\sqrt{\frac{2}{\pi x}}e^{i(x-\gamma_n)}\\ H^{(2)}_n(x)&\cong\sqrt{\frac{2}{\pi x}}e^{-i(x-\gamma_n)}\end{aligned}\right\} \quad \text{(A8)}$$

where $\gamma_n=(2n+1)\pi/4$.

Substitution of Eqs.(A7) and (A8) into Eqs.(A3), (A4), (A5) and (A6) yields the following equations.

$$\begin{aligned} g^{jkIS}_{(+)n}(r)\cong&\frac{1}{\pi\sqrt{k^j_1 k^k_1}}\sum_{m=-\infty}^{\infty}C^{jk}_{(+)m}\frac{1}{r}\left(e^{i\left\{\left(k^j_1+k^k_1\right)-\left(\gamma_{n-m}+\gamma_m\right)\right\}}+e^{-i\left\{\left(k^j_1-k^k_1\right)-\left(\gamma_{n-m}-\gamma_m\right)\right\}}\right)\\ &+\frac{1}{\pi\sqrt{k^j_1 k^k_1}}\sum_{m=-\infty}^{\infty}\overline{C}^{jk}_{(+)m}\frac{1}{r}\left(e^{i\left\{\left(k^j_1+k^k_1\right)-\left(\gamma_{n-m}+\gamma_m\right)\right\}}+e^{i\left\{\left(k^j_1-k^k_1\right)-\left(\gamma_{n-m}-\gamma_m\right)\right\}}\right)\end{aligned} \quad \text{(A9)}$$

$$g^{jkSS}_{(+)n}(r)\cong\frac{2}{\pi\sqrt{k^j_1 k^k_1}}\sum_{m=-\infty}^{\infty}D^{jk}_{(+)m}\frac{1}{r}e^{i\left\{\left(k^j_1+k^k_1\right)-\left(\gamma_{n-m}+\gamma_m\right)\right\}} \quad \text{(A10)}$$

$$\begin{aligned} g^{jkIS}_{(-)n}(r)\cong&\frac{1}{\pi\sqrt{k^j_1 k^k_1}}\sum_{m=-\infty}^{\infty}C^{jk}_{(-)m}\frac{1}{r}\left(e^{i\left\{\left(k^j_1-k^k_1\right)-\left(\gamma_{n-m}-\gamma_m\right)\right\}}+e^{-i\left\{\left(k^j_1+k^k_1\right)-\left(\gamma_{n-m}+\gamma_m\right)\right\}}\right)\\ &+\frac{1}{\pi\sqrt{k^j_1 k^k_1}}\sum_{m=-\infty}^{\infty}\overline{C}^{jk}_{(-)m}\frac{1}{r}\left(e^{i\left\{\left(k^j_1+k^k_1\right)-\left(\gamma_{n-m}+\gamma_m\right)\right\}}+e^{i\left\{\left(k^j_1-k^k_1\right)-\left(\gamma_{n-m}-\gamma_m\right)\right\}}\right)\end{aligned} \quad \text{(A11)}$$

$$g^{jkSS}_{(-)n}(r)\cong\frac{2}{\pi\sqrt{k^j_1 k^k_1}}\sum_{m=-\infty}^{\infty}D^{jk}_{(-)m}\frac{1}{r}e^{i\left\{\left(k^j_1-k^k_1\right)-\left(\gamma_{n-m}-\gamma_m\right)\right\}} \quad \text{(A12)}$$

Considering that the time dependent functions corresponding to $\phi^{jk}_{(+)L}$ and $\phi^{jk}_{(-)L}$ are given by $e^{-i(\sigma_j+\sigma_k)}$ and $e^{-i(\sigma_j-\sigma_k)}$, respectively, the first terms in the brackets in Eq.(A9), the first term in the first bracket and the second term in the second bracket in Eq.(A11), and terms given by Eqs.(A10) and (A12) are waves traveling with outgoing free waves. However, traveling directions of other terms, namely, the second terms in the bracket in Eq.(A9), the second term in the first bracket and the first term in the second bracket in Eq.(A11) depend on sign of (k_j-k_k) and $(\sigma_j-\sigma_k)$. This fact implies that the velocity potentials of the bounded wave components, $g^{jkIS}_{(+)n}(r)$ and $g^{jkIS}_{(-)n}(r)$, include the waves traveling with incoming free waves. These components arise from nonlinear interaction between the incident and diffracted waves of the first order and this interaction occurs at everywhere including at far field. This suggests that these waves are generated at far field and travel with incoming free waves. Considering this fact and these terms are necessary to satisfy the boundary conditions on the free surface, the solution derived in this study is reasonable.

Proceedings of the Seventh (1997) International Offshore and Polar Engineering Conference
Honolulu, USA, May 25-30, 1997

ISBN 1-880653-28-1 (Set); ISBN 1-880653-31-1 (Vol. III)

A Nonlinear Spectral Model for Directional Random Waves in Decreasing Depths

Laura Rebaudengo Landó, Renata Gentile and Giulio Scarsi
University of Genova
Genova, Italy

ABSTRACT

A wave propagation spectral model to second order of approximation is suggested, which provides the actual directional spectrum in decreasing depths, starting from the corresponding one in deep water. The model implements the procedure already proposed for shoaling, refraction and saturation processes by adding bottom friction; utilizes deep water directional spectra involving JONSWAP frequency spectra and frequency independent spreading functions; finally, it considers straight coastlines with slowly varying bathymetry, the extension to the usual three dimensional bottom not involving special difficulties. The bottom friction process is formalized through a scheme which gives the relevant directional spectrum for the effective flow regime (laminar, rough turbulent, smooth turbulent, transitional) in the bottom boundary layer, depending on the wave field and Nikuradse roughness parameter. The model was tested by comparing it with field data from Maa and the results are quite satisfactory.

KEY WORDS: Multidirectional random wave propagation; Bottom friction; Spectral models; Second order of approximation.

INTRODUCTION

The mathematical models available to quantify the modifications of multidirectional random waves propagating in decreasing depths up to the breaker zone are generally based on directional spectra and they differ from one another in the processes considered and in the order of approximation. Some of these spectral models are summarized well by Dalrymple (1992), also with reference to methods founded on the wave action.

Recently, Rebaudengo Landó et al. (1996) introduced a second order spectral model $[ADS]_2$ which takes into account the shoaling, refraction and saturation processes. The model needs the knowledge of the first order spectral model $[ADS]_1$ deduced by Scarsi (1995) on the basis of both the $[LW]_1$ model suggested by Le Méhauté & Wang (1982) for the shoaling and refraction processes and the $[TMAR]_1$ model proposed by Gentile et al. (1994) for the saturation process and it utilizes the procedure formulated by Laing (1986) for the specification of the second order spectra starting from the corresponding first order ones. The models $[ADS]_1$ and $[ADS]_2$ adopt deep water directional spectra constructed by mean JONSWAP frequency spectra and frequency independent spreading functions. Nevertheless, they can be specified with different values for the spectral parameters, especially as regards the enhancement and peak width parameters, and with frequency dependent spreading functions. In the following a brief description of the aforementioned models is given with reference to straight coastlines with slowly varying bathymetry, noting that the extension to a three dimensional bottom does not involve special difficulties, unless the bathymetry is so complex that diffraction becomes important.

In the present paper, the $[ADS]_1$ and $[ADS]_2$ models have been adapted to consider also energy dissipation due to bottom friction. This process is formalized by adjusting a procedure carried out by Rebaudengo Landó & Scarsi (1984) to evaluate the spectral bottom friction experienced by random waves. It takes into account the different flow regimes which may be present in the bottom boundary layer, that is laminar, smooth turbulent, rough turbulent, transitional ones. The adjustment, performed by assuming some relationships given by Hasselmann & Collins (1968), Hasselmann et al. (1973) and Collins (1972) as a starting point, allows the spectral energy dissipation to be obtained by a proper spreading of the global energy dissipation on the actual frequencies and directions in a way similar to that developed by Madsen et al. (1988), Graber & Madsen (1988). Thus, the $[ADS]_1$ and $[ADS]_2$ models are able to give the corresponding $[ADS^*]_1$ and $[ADS^*]_2$ models which consider the bottom friction process, jointly with the shoaling, refraction, saturation ones.

THE SPECTRAL WAVE PROPAGATION MODELS WITHOUT BOTTOM FRICTION

As mentioned in the Introduction, the spectral models relevant to the shoaling, refraction and saturation processes refer here to a bottom with straight and parallel contours and they are summarily described enough for practical use. The first order $[LW]_1$ model suggested by Le Méhauté & Wang (1982) provides the shoaling and refraction directional spectrum $[S_{SR}(f,\phi,\phi_m h)]_1$ on finite depth, the relevant spectrum $[S_o(f,\phi_o,\phi_{mo})]_1$ on deep water being considered in the usual form

$$[S_o(f,\phi_o,\phi_{mo})]_1 = [S_o(f)]_1 D_o(\phi_o,\phi_{mo}) \tag{1}$$

which involves the frequency spectrum $[S_o(f)]_1$ and the frequency independent directional spreading function $D_o(\phi_o,\phi_{mo})$. It turns out

$$[S_{SR}(f,\phi,\phi_m,h)]_1 = [S_o(f,\phi_o,\phi_{mo})]_1 \, K_{SH}^2(f,h) \cdot \cdot \, K_{RE}^2(f,\phi_o,h)\partial\phi_o/\partial\phi. \tag{2}$$

In the previous equations: the index o indicates deep water conditions; f is the frequency; h is the water depth; ϕ,ϕ_o and ϕ_m,ϕ_{mo} are the angles which indentify the wave component directions and the overall mean wave direction referred to the normal n to the bathymetrics, as shown in Figure 1; K_{SH} and K_{RE} are the shoaling and refraction coefficients. The angles ϕ,ϕ_o and ϕ_m,ϕ_{mo} are related to each other by the following inverse direction functions derived from Snell's law

$$\phi(f,\phi_o,h) = \sin^{-1}(\sin\phi_o \tanh kh) \tag{3}$$

$$\phi_o(f,\phi,h) = \sin^{-1}(\sin\phi/\tanh kh) \tag{4}$$

$$\phi_m(f,\phi_{mo},h) = \sin^{-1}(\sin\phi_{mo}\tanh k_p h) \tag{5}$$

$$\phi_{mo}(f,\phi_m,h) = \sin^{-1}(\sin\phi_m/\tanh k_p h) \tag{6}$$

where the index p indicates peak conditions and k is the wave number modulus obtained from the isotropic dispersion relationship

$$(2\pi f)^2 = gk\tanh kh \tag{7}$$

g being the gravity acceleration. The peak frequency is assumed unchanged in the wave propagation, that is $f_p = f_{po}$. The shoaling K_{SH} and refraction K_{RE} coefficients are provided by

$$K_{SH}(f,h) = \{2\cosh^2 kh/(2kh + \sinh 2kh]\}^{1/2} \tag{8}$$

$$K_{RE}(f,\phi_o,h) = \{\cos\phi_o/\cos[\sin^{-1}(\sin\phi_o\tanh kh)]\}^{1/2} \tag{9}$$

and the derivative $\partial\phi_o/\partial\phi$ is expressed by

$$\partial\phi_o/\partial\phi = \cos[\sin^{-1}(\sin\phi_o\tanh kh)]/\{\tanh^2 kh + - \sin^2[\sin^{-1}(\sin\phi_o\tanh kh)]\}^{1/2} \tag{10}$$

which is associated with the condition

$$\sin^2[\sin^{-1}(\sin\phi_o\tanh kh)]/\tanh^2 kh < 1 \tag{11}$$

that is automatically verified for the angles ϕ_o lying in the ranges

$$-\pi/2 < \phi_o < \pi/2 + \phi_{mo};\ \phi_{mo} < 0 \tag{12}$$

$$-\pi/2 + \phi_{mo} < \phi_o < \pi/2;\ \phi_{mo} > 0. \tag{13}$$

As Figure 1 clearly shows, Eqs. 12 and 13 ensure that only the wave components able to participate in the move towards the decreasing depths are correctly retained and, consequently, they supply the actual domain $\{\phi_o\}$ of ϕ_o.

The frequency spectrum which corresponds to the directional spectrum of Eq. 2 is expressed by

$$[S_{SR}(f,h)]_1 = \int_{\{\phi\}} [S_{SR}(f,\phi,\phi_m,h)]_1 d\phi \tag{14}$$

which leads to

$$[S_{SR}(f,h)]_1 = \int_{\{\phi_o\}} [S_o(f,\phi_o)]_1 K_{SH}^2(f,h) K_{RE}^2(f,\phi_o,h) d\phi_o \tag{15}$$

the $\{\phi\}$ domain being obtained from the $\{\phi_o\}$ one through Eq. 3. The directional spreading function $[D(f,\phi,\phi_m,h)]_1$, which allows the finite depth directional spectrum to be written in a form equivalent to that of Eq. 1, is given by

$$[D(f,\phi,\phi_m,h)]_1 = [S_{SR}(f,\phi,\phi_m,h)]_1/[S_{SR}(f,h)]_1 \tag{16}$$

and it becomes dependent on the frequency too.

Usually, the $[LW]_1$ model considers in deep water a frequency spectrum $[S_o(f)]_1$ of mean JONSWAP type (Hasselmann et al., 1973) and a directional spreading function $D_o(\phi_o,\phi_{mo})$ of cosine power type, that is

$$S_o(f) = \alpha_o g^2 (2\pi)^{-4} f^{-5} \exp\{-1.25(f/f_{po})^{-4}\} \cdot \cdot \exp\{\ln\gamma_o \exp[-0.5(f/f_{po} - 1)^2/\sigma_o^2]\} \tag{17}$$

$$D_o(\phi_o,\phi_{mo}) = \{\Gamma(s_o+1)/\sqrt{\pi}\Gamma(s_o+1/2)\}\cos(\phi_o - \phi_{mo})^{2s_o} \tag{18}$$

where

$$\alpha_o = (174/\tilde{U}_o)^{-2/3} \tag{19}$$

$$\gamma_o = 3.3 \tag{20}$$

$$\sigma_o = 0.07 \text{ for } f \le f_{po};\ \sigma_o = 0.09 \text{ for } f > f_p. \tag{21}$$

$$f_{po} = \tilde{U}_o g/U_o. \tag{22}$$

In the previous equations: s_o is the spreading index; α_o,γ_o and σ_o are, respectively, the equilibrium, enhancement and peak width dimensionless parameters; U_o is the wind speed at 10 m above the mean water level; $\tilde{U}_o$ is the dimensionless wind parameter related to the dimensionless fetch parameter $\tilde{x}_o$ by

$$\tilde{U}_o = 3.5\tilde{x}_o^{-1/3};\ \tilde{x}_o = gx_o/U_o^2. \tag{23}$$

noting that in the fully developed sea $\tilde{U}_o \le 0.13$ and $\tilde{x}_o \ge 19000$. Nevertheless, in this situation α_o and f_{po} must be evaluated by specifying Eqs. 19 and 22 with the values $\tilde{U}_o = 0.13$ and $\tilde{x}_o = 19000$.

The first order $[TMAR]_1$ model suggested by Gentile et al. (1994) provides the saturation frequency spectrum $[S_L(f,h)]_1$. It is obtained by a proper rederivation of the original $[TMA]_1$ model proposed by Bouws et al. (1985) by following the self-similarity shape hypothesis introduced by Kitaigoroskii et al. (1975) and using an extensive set of field data (TEXEL, MARSEN, ARSLOE).
The $[S_L(f,h)]_1$ spectrum is related in deep water to a frequency spectrum $[S_o(f)]_1$ of mean JONSWAP type and it is expressed by

$$\begin{aligned}[S_L(f,h)]_1 &= \alpha_* \tilde{U}_o^{2/3}\chi_p^{1/3} g^2 (2\pi)^{-4} f^{-5} \cdot \\ &\cdot \exp\{-1.25(f\sqrt{\chi}/f_{po}\sqrt{\chi_p})^{-4}\} \cdot \\ &\cdot \exp\{\ln\gamma_o \exp[-0.5(f\sqrt{\chi}/f_{po}\sqrt{\chi_p}-1)^2/\sigma_o^2]\} \cdot \\ &\cdot \{\chi^{-2}[1+2\sigma_h^2\chi/\sinh(2\sigma_h^2\chi)]^{-1}\} \end{aligned} \tag{24}$$

where

$$\alpha_* = 0.033 = \alpha_o/\tilde{U}_o^{2/3} \tag{25}$$

$$\sigma_h = 2\pi f(h/g)^{1/2} \tag{26}$$

$$\chi = 1/\tanh(\sigma_h^2\chi). \tag{27}$$

The first order $[ADS]_1$ model suggested by Scarsi (1995) provides the actual directional spectrum $[S_A(f,\phi,\phi_m,h)]_1$ from the $[LW]_1$ and $[TMAR]_1$ models. That spectrum, consequently, takes into account the shoaling, refraction and saturation processes and it is expressed by

$$[S_A(f,\phi,\phi_m,h)]_1 = [S_A(f,h)]_1[D(f,\phi,\phi_m,h)]_1 \tag{28}$$

where the directional spreading function $[D(f,\phi,\phi_m,h)]_1$ is given by Eq. 16 and the actual frequency spectrum $[S_A(f,h)]_1$ is defined by

$$[S_A(f,h)]_1 = [S_L(f,h)]_1;\ [S_{SR}(f,h)]_1 > [S_L(f,h)]_1 \tag{29}$$

and

$$[S_A(f,h)]_1 = [S_{SR}(f,h)]_1;\ [S_{SR}(f,h)]_1 \le [S_L(f,h)]_1 \tag{30}$$

because the saturation frequency spectrum represents a limit spectral form which cannot be exceeded.
The second order $[ADS]_2$ model suggested by Rebaudengo Landó et al. (1996) provides the actual directional spectrum $[S_A(f,\phi,\phi_m,h)]_2$ to second order of approximation obtained from the corresponding first order spectrum $[S_A(f,\phi,\phi_m,h)]_1$ by adopting and adjusting the procedure formulated by Laing (1986).
That spectrum is formally expressed by

$$[S_A(f,\phi,\phi_m,h)]_2 = [S_A(f,\phi,\phi_m,h)]^{(1)} + [S_A(f,\phi,\phi_m,h)]^{(2)} \tag{31}$$

where the contribution to first order is given by Eq. 28, that is

$$[S_A(f,\phi,\phi_m,h)]^{(1)} = [S_A(f,\phi,\phi_m,h)]_1, \tag{32}$$

and the contribution to second order is supplied by

$$\begin{aligned}[S_A(f,\phi,\phi_m,h)]^{(2)} &= \frac{1}{4}\int_{\{\phi\}}\int_o^f (B_s)^2[S_A(f',\phi',\phi_m,h)]_1 \cdot \\ &\cdot [S_A(f-f',\phi_s'',\phi_m,h)]_1 df' d\phi' + \\ &+ \frac{1}{2}\int_{\{\phi\}}\int_{\{f\}} (B_d)^2[S_A(f',\phi',\phi_m,h)]_1 \cdot \\ &\cdot [S_A(f+f',\phi_d'',\phi_m,h)]_1 df' d\phi'. \end{aligned} \tag{33}$$

The angles $\phi', \phi_s'', \phi_d''$ are shown in Figure 2 and identify the directions of $\boldsymbol{k}$, $\boldsymbol{k}-\boldsymbol{k}'$ and $\boldsymbol{k}+\boldsymbol{k}'$ respectively. The wave numbers vector $\boldsymbol{k}'$, which can be associated with each wave number vector $\boldsymbol{k}$, range in the $\{\boldsymbol{k}\}$ domain with the constraint that the angles ϕ_s'', ϕ_d'' fall in the $\{\phi\}$ domain defined for the first order spectrum.
The frequencies $f', f-f', f+f'$ are obtained from

$$(2\pi f')^2 = g|\boldsymbol{k}'|\tanh|\boldsymbol{k}'|h \tag{34}$$

$$(2\pi f - 2\pi f')^2 = g|\boldsymbol{k}-\boldsymbol{k}'|\tanh|\boldsymbol{k}-\boldsymbol{k}'|h \tag{35}$$

$$(2\pi f + 2\pi f')^2 = g|\boldsymbol{k}+\boldsymbol{k}'|\tanh|\boldsymbol{k}+\boldsymbol{k}'|h. \tag{36}$$

The quantities B_s and B_d are provided by

$$B_s = A_s - \boldsymbol{k}'\cdot(\boldsymbol{k}-\boldsymbol{k}')/(R'R_s'')^{1/2} + (R'+R_s'') + (R'R_s'')^{1/2} \tag{37}$$

$$B_d = A_d - \boldsymbol{k}'\cdot(\boldsymbol{k}+\boldsymbol{k}')/(R'R_d'')^{1/2} + (R'+R_d'') - (R'R_d'')^{1/2} \tag{38}$$

and the quantities A_s, A_d are given by

$$\begin{aligned} A_s = &\{[(|\boldsymbol{k}-\boldsymbol{k}'|^2 - R_s''^2)/(R_s'')^{1/2} + (|\boldsymbol{k}'|^2 - R'^2)/(R')^{1/2} \cdot \\ &/[(R')^{1/2} + (R_s'')^{1/2}] + 2[\boldsymbol{k}'\cdot(\boldsymbol{k}-\boldsymbol{k}') - R'R_s'']/(R'R_s'')^{1/2}\} \cdot \\ &/\{1 - |\boldsymbol{k}|\tanh|\boldsymbol{k}|h/[(R')^{1/2} + (R_s'')^{1/2}]^2\} \end{aligned} \tag{39}$$

$$\begin{aligned} A_d = &\{[(|\boldsymbol{k}+\boldsymbol{k}'|^2 - R_d''^2)/(R_d'')^{1/2} - (|\boldsymbol{k}'|^2 - R'^2)/(R')^{1/2}] \cdot \\ &/[(R_d'')^{1/2} - (R')^{1/2}] + 2[\boldsymbol{k}'\cdot(\boldsymbol{k}+\boldsymbol{k}') + R'R_d'']/(R'R_d'')^{1/2}\} \cdot \\ &/\{1 - |\boldsymbol{k}|\tanh|\boldsymbol{k}|h/[(R_d'')^{1/2} - (R')^{1/2}]^2\} \end{aligned} \tag{40}$$

where the quantities R', R_s'', R_d'' are deduced from

$$R' = (2\pi f')^2/g \tag{41}$$

$$R_s'' = (2\pi f - 2\pi f')^2/g \tag{42}$$

$$R_d'' = (2\pi f + 2\pi f')^2/g. \tag{43}$$

The second order frequency spectrum which corresponds to the directional spectrum of Eq. 31 is expressed by

$$[S_A(f,h)]_2 = \int_{\{\phi\}} [S_A(f,\phi,\phi_m,h)]_2 d\phi \tag{44}$$

and the second order spreading function $[D(f,\phi,\phi_m,h)]_2$ is obtained from

$$[S_A(f,\phi,\phi_m,h)]_2 = [S_A(f,h)]_2[D(f,\phi,\phi_m,h)]_2. \tag{45}$$

Two observations need to be made. A limit must be introduced for the lowest depth, as happens for the Stokian non linear waves, due to the fact that the second order model originates from a perturbation method. That limit may be specified starting from Eq. 31 referred to the zero-moments of the spectra, that is

$$[m_o]_2=[m_o]^{(1)}+[m_o]^{(2)} = \delta[m_o]'_1+\delta^2[m_o]'_2 \quad (46)$$

with δ being a proper perturbation parameter. The constraint is that the terms $[m_o]'_1$ and $[m_o]'_2$ keep the same order of magnitude. A check is needed to ensure that the sea states associated with the actual second order directional spectrum have stable random waves. The check is made by considering the heights H and periods T of each wave and by adopting the breaking condition formalized by Goda's relationship (1975) referred to a gentle slope

$$H_b = AgT^2\{1 - \exp[-0.75(2\pi/T)^2(h/g)]\} \quad (47)$$

where the coefficient $A = 0.15/2\pi$ is appropriate for random waves, according to Li and Dong (1993). However, only a small percentage of waves fall outside the range $H \leq H_b$, if the above mentioned limit is adopted.

THE WAVE PROPAGATION SPECTRAL MODELS WITH BOTTOM FRICTION

The model to first order of approximation suggested by Rebaudengo Landó & Scarsi (1980) to evaluate the first order mean energy dissipation $[P_{DI}]_1$ per unit area and per unit time within the bottom frictional layer of random waves, introduces the energetically equivalent regular wave EW associated with the frequency spectrum $[S(f,h)]_1$. It is substantially analogous to the one developed afterwards by Madsen et al. (1988). The model, which has been obtained by elaborating some relationships given by Hasselmann & Collins (1968), Collins (1972), Hasselmann et al. (1973) for rough turbulent flow regime, is able to take into account also the laminar, smooth turbulent and transitional flow regimes. It gives

$$[P_{DI}]_1=(1/2)\rho(\hat{U}_w)^3_{eq}(\hat{C}_w)_{eq}\cos(\vartheta_w)_{eq}G \quad (48)$$

ρ being the water density. In detail, $(\hat{U}_w)_{eq}$ is the amplitude of the bottom horizontal velocity $(U_w)_{eq}$ derived from the Airy theory involving the height H_{eq} and period T_{eq} of the EW, that is

$$(\hat{U}_w)_{eq}=(\pi H_{eq})/(T_{eq}\sinh k_{eq}h). \quad (49)$$

H_{eq} is expressed by the relationship

$$H_{eq}=\sqrt{8}\int_{\{f\}}\int_{\{\phi\}}[S(f,\phi,\phi_m,h)]_1 d\phi df \quad (50)$$

which ensures that the mean energy per unit area of the EW is equal to that of the random waves. T_{eq} is obtained through the implicit relationship

$$\begin{aligned}T_{eq} = {} & \{\tanh k_e h[1+ 2k_{eq}h/\sinh 2k_{eq}h]\cdot \\ & \cdot\int_{\{f\}}\int_{\{\phi\}}[S(f,\phi,\phi_m,h]_1 d\phi df\}^{-1}\cdot \\ & \cdot\int_{\{f\}}\int_{\{\phi\}}[S(f,\phi,\phi_m,h)]_1\cdot \\ & \cdot\tanh kh[1+2kh/\sinh 2kh](1/f)d\phi df \quad (51)\end{aligned}$$

which ensures that the mean energy flux per unit crest length of the EW is equal to that of the random waves, noting that the wave numbers k_{eq} and k are related to the frequencies $f_{eq}=1/T_{eq}$ and f through Eq. 7. $(\hat{C}_w)_{eq}$ and $(\vartheta_w)_{eq}$ are the amplitude and phase lead of the friction coefficient defined by Kajiura (1968) as

$$(C_w)_{eq}=(\hat{C}_w)_{eq}\exp[i(\vartheta_w)_{eq}] \quad (52)$$

i being the imaginary unit. They are evaluated by the relationships given in the Appendix for all the different flow regimes.
Finally, G is a function expressed by

$$G=\langle|\hat{\boldsymbol{u}}_w|\rangle/(\hat{U}_w)_{eq} \quad (53)$$

where $\langle|\hat{\boldsymbol{u}}_w|\rangle$ is a proper mean value averaged on the ensemble of the amplitudes $|\hat{\boldsymbol{u}}_w|$ of the bottom horizontal velocities $|\boldsymbol{u}_w|$ relevant to the wave components associated with $[S(f,\phi,\phi_m,h)]_1$, these amplitudes being given by

$$|\hat{\boldsymbol{u}}_w|=\{[(2\pi f)^2/\sinh^2 kh]\cdot 2[S(f,\phi,\phi_m,h)]_1 df d\phi\}^{1/2}. \quad (54)$$

Eq. 53 specified by Eq. 54 provides

$$\begin{aligned}G = {} & [1/(\hat{U}_w)_{eq}]\{\int_{\{f\}}\int_{\{\phi\}}[(2\pi f)^2/\sinh^2 kh]\cdot \\ & \cdot S(f,\phi,\phi_m,h)d\phi df\}^{1/2}. \quad (55)\end{aligned}$$

It is worth noting that all the integral forms have to be transferred, from an operative point of view, to equivalent summation forms.

The previous equations allow the first order directional spectrum $[S_D(f,\phi,\phi_m,h)]_1$, which takes into account bottom friction, to be defined. The knowledge of this spectrum is essential to deduce the first $[ADS^*]_1$ and second $[ADS^*]_2$ order models that give the corresponding directional $[S^*_A(f,\phi,\phi_m,h)]_1$, $[S^*_A(f,\phi,\phi_m,h)]_2$ and frequency $[S^*_A(f,h)]_1$, $[S^*_A(f,h)]_2$ actual spectra which consider the bottom energy dissipation, in addition to the shoaling, refraction and saturation processes.

A step by step procedure is described below to obtain the aforementioned $[ADS^*]_1$ and $[ADS^*]_2$ models noting that: (a) the total wave propagation extension λ, which goes from section 0 on deep water to section n on the finite depth under observation is divided into n steps of $\Delta\lambda$ length, the value of n being chosen so that

$$\Delta\lambda/L_{po} \leq 1/5 \quad (56)$$

where L_{po} is the peak length on deep water; (b) the bottom slope s_w is assumed to be gentle, typically

$$s_w \leq 0.005, \quad (57)$$

as happens for usual sea bottoms, offshore the breaker zone.

The suggested procedure develops in the following way, specified with reference to the generic $(j, j+1)$ step. The overall mean wave direction $(\phi_m)_j$, the first order directional $\{[S_A^*(f,\phi,\phi_m,h)]_1\}_j$ and frequency $\{[S_A^*(f,h)]_1\}_j$ actual spectra are known because they were obtained after using the procedure up to section j. The overall mean wave direction $(\phi_m)_{j+1}$ is derived from $(\phi_m)_j$ through Snell's law which leads to

$$(\phi_m)_{j+1}=\sin^{-1}\{\sin(\phi_m)_j\ (\tanh k_p h)_{j+1}/(\tanh k_p h)_j\ \}. \quad (58)$$

The first order directional spectrum $\{[S_{SR}(f,\phi,\phi_m,h)]_1\}_{j+1}$ that takes into account the shoaling and refraction processes is deduced from the actual directional spectrum $\{[S_A^*(f,\phi,\phi_m,h)]_1\}$, by a proper adjustment of Eq. 2. It is expressed by

$$\{[S_{SR}(f,\phi,\phi_m,h)]_1\}_{j+1}=\{[S_A^*(f,\phi,\phi_m,h)]_1\}_j[(K_{SH}^2)_{j+1}\cdot /(K_{SH}^2)_j][(K_{RE}^2)_{j+1}/(K_{RE}^2)_j][(\partial\phi)_j/(\partial\phi)_{j+1}]. \quad (59)$$

The corresponding frequency spectrum $\{[S_{SR}(f,h)]_1\}_{j+1}$ and directional spreading function $\{[D(f,\phi,\phi_m,h)]_1\}_{j+1}$ are provided by

$$\{[S_{SR}(f,h)]_1\}_{j+1}=\int_{\{\phi\}j+1}\{[S_{SR}(f,\phi,\phi_m,h)]_1\}_{j+1}d\phi \quad (60)$$

$$\{[D(f,\phi,\phi_m,h)]_1\}_{j+1}=\{[S_{SR}(f,\phi,\phi_m,h)]_1\}_{j+1}\cdot /\{[S_{SR}(f,h)]_1\}_{j+1} \quad (61)$$

according to Eqs. 14 and 16. In Eq. 59 the shoaling K_{SH} and refraction K_{RE} coefficients are evaluated by Eqs. 8 and 9 and the derivative $(\partial\phi)_j/(\partial\phi)_{j+1}$ is supplied by

$$(\partial\phi)_j/(\partial\phi)_{j+1}=\cos\{\sin^{-1}[(\sin\phi)_j\beta_{j+1,j}]\}/\{\beta_{j+1,j}^2+ -\sin^2[\sin^{-1}[(\sin\phi)_j\beta_{j+1,j}]]\} \quad (62)$$

where

$$\beta_{j+1,j}=(\tanh kh)_{j+1}/(\tanh kh)_j. \quad (63)$$

Eq. 62 must be associated with the condition

$$\sin^2\{\sin^{-1}[(\sin\phi)_j\beta_{j+1,j}]\}/\beta_{j+1,j}^2<1 \quad (64)$$

automatically verified if angles ϕ belong to the $\{\phi\}_j$ and $\{\phi\}_{j+1}$ domains that correspond to the $\{\phi_o\}$ domain given by Eqs. 12 and 13.

The first order bottom energy dissipation $\{[E_{DI}]_1\}_{j,j+1}$ per unit area is obtained from the mean energy flux loss due to bottom energy dissipation per unit time, evaluated by assuming the bottom depth as constant in the step and equal to that in section j, this assumption being consistent with Eqs. 56 and 57. Having in mind Eq. 48, it turns out

$$\{[E_{DI}]_1\}_{j,j+1}=\{[P_{DI}]_1\}_j\Delta\lambda/\{(c_g)_{eq}\}_j \quad (65)$$

where: $\{[P_{DI}]_1\}_j$ is the bottom energy dissipation per unit area and per unit time deduced from Eq. 48 after introducing the actual directional spectrum $\{[S_A^*(f,\phi,\phi_m,h)]_1\}_j$ in Eqs. 50 and 51; $\{(c_g)_{eq}\}_j$ is the wave group velocity of the EW given by

$$\{(c_g)_{eq}\}_j=\{[(gT_{eq}/4\pi)\tanh k_{eq}h][1+2k_{eq}h/\sinh 2k_{eq}h]\}_j. \quad (66)$$

The first order bottom friction directional spectrum $\{[S_D(f,\phi,\phi_m,h)]_1\}_{j,j+1}$ is deduced from a spread of $\{[E_{DI}]_1\}_{j,j+1}$ on all the frequencies and directions of the $\{f\}_j$ and $\{\phi\}_j$ domains, substantially in agreement with Graber & Madsen (1988). Thus, it turns out

$$\{[S_D(f,\phi,\phi_m,h)]_1\}_{j,j+1}=[1/\rho g]\{E_{DI}]_1\}_{j,j+1}\cdot \cdot\{[S_A^*(f,\phi,\phi_m,h)]_1\}_j\cdot\int_{\{f\}_j}\int_{\{\phi\}_j}\{S_A^*(f,\phi,\phi_m,h)]_1\}_j d\phi df \quad (67)$$

which correctly verifies the condition

$$\rho g\int_{\{f\}}\int_{\{\phi\}}\{[S_D(f,\phi,\phi_m,h)]_1\}_{j,j+1}d\phi df=\{[E_{DI}]_1\}_{j,j+1}. \quad (68)$$

The first order directional spectrum $\{[S_{SRD}(f,\phi,\phi_m,h)]_1\}_{j+1}$ that takes into account the shoaling, refraction and bottom friction processes, is deduced starting from the spectra defined by Eqs. 59 and 67. That spectrum is expressed by

$$\{[S_{SRD}(f,\phi,\phi_m,h)]_1\}_{j+1}=\{[S_{SR}(f,\phi,\phi_m,h)]\}_{j+1}+ -\{[S_D(f,\phi,\phi_m,h)]_1\}_{j,j+1} \quad (69)$$

and the corresponding first order frequency spectrum is provided by

$$\{[S_{SRD}(f,h)]_1\}_{j+1}=\int_{\{\phi\}_{j+1}}\{[S_{SRD}(f,\phi,\phi_m,h)]_1\}_{j+1}d\phi \quad (70)$$

The first order actual frequency spectrum $\{[S_A^*(f,h)]_1\}_{j+1}$ that considers all the processes is obtained by comparing the frequency spectrum given by Eq. 70 with the first order saturation frequency spectrum $[S_L(f,h)]_1$ expressed by Eq. 24. Thus, after adjusting Eqs. 29 and 30, it turns out

$$\{[S_A^*(f,h)]_1\}_{j+1}=\{[S_L(f,h)]_1\}_{j+1} \quad (71)$$

for

$$\{[S_{SRD}(f,h)]_1\}_{j+1}>\{[S_L(f,h)]\}_{j+1} \quad (72)$$

and

$$\{[S_A^*(f,h)]_1\}_{j+1}=\{[S_{SRD}(f,h)]_1\}_{j+1} \quad (73)$$

for

$$\{[S_{SRD}(f,h)]_1\}_{j+1}\leq\{[S_L(f,h)]_1\}_{j+1} \quad (74)$$

The corresponding first order directional spectrum $\{[S_A^*(f,\phi,\phi_m,h)]_1\}_{j+1}$ is deduced from $\{[S_A^*(f,h)]_1\}_{j+1}$ having in mind Eq. 28, that is

$$\{[S_A^*(f,\phi,\phi_m,h)]_1\}_{j+1}=\{[S_A^*(f,h)]_1\}_{j+1}\{D(f,\phi,\phi_m,h)]_1\}_{j+1} \quad (75)$$

where the directional spreading function is still provided by Eq. 61. Thus, the first order $[ADS^*]_1$ model, which takes into account the bottom friction jointly with the shoaling, refraction and saturation processes, is given by Eqs. 71, 73 and 75.

The $[ADS^*]_1$ model being obtained, the relevant second order $[ADS^*]_2$ model is formalized by the actual directional $\{[S^*_A(f,\phi,\phi_m,h)]_2\}_{j+1}$ and frequency $\{[S^*_A(f,h)]_2\}\}_{j+1}$ spectra to that order, noting that they are obtained from the ones to first order through Eqs. 31 to 45, after replacing $[S_A(f,\phi,\phi_m,h)]_1$ with $[S^*_A(f,\phi,\phi_m,h)]_1$. Of course, the limit for the lowest depth and control on the random wave stability are still examined with reference to Eqs. 46 and 47. Finally, it is worth noting that the $[ADS^*]_1$ and $[ADS^*]_2$ models can be referred to deep water frequency spectra $[S_o(f)]_1$ with different values for the enhancement and peak width parameters compared to those given by Eqs. 20 and 21.

COMPARISON WITH FIELD DATA

The suggested $[ADS^*]_2$ model was tested by considering the field data from Maa, 1993. The data are relevant to the "Hallowen" Northeaster of October 29-31, 1991, which generated a storm that pounded the U.S. East Coast. They were recorded at two wave stations located on the continental shelf off the Virginia Coast at water depths of $\sim$ *30* m and *10* m, the distance between the two stations, here named M and N, being *87* km. At Station M the wave energy reached its maximum on October 31, 03 hr and at Station N the maximum occurred *1* to *2* hrs later because of the time required for the random waves to travel between the two stations. Figure 3 (from Figure 2 by Maa) shows the observed frequency spectra $S_F = S_F(f,h)$ of the storm peak and Table 1 gives the values of the observed spectral maximum $(S_F)_{max}$ and significant wave height H_s. Validation of the model is carried out by taking into account the random wave propagation from Station M to Station N and by comparing, at the latter, the predicted data with the observed ones, a good reproduction of the wave conditions at Station M being ensured through the proper choice of deep water spectral parameter values.

The adopted procedure is developed in the following way.

Bottom description. The topography of the wave travelling region allows the bottom to be approximated by straight and parallel contours with an effective very gentle bottom slope, $s_w \sim 1.3 \cdot 10^{-4}$. The Nikuradse roughness parameter is assumed $\epsilon = 0.01$ m, as usual.

Deep water spectral parameters. The value $\phi_{mo} = -20$ degree referred to the normal to the bathymetrics is adopted for the overall mean wave direction associated with the directional spectrum. The values $f_{po} = 0.06$ Hz, $\gamma_o = 1.7$ and $\sigma_o = 0.1$ are assumed for the peak frequency, enhancement and peak width parameters of the frequency spectrum given by Eq. 17. The value $\tilde{U}_o = 0.13$ is adopted for the dimensionless wind parameter which appears in Eq. 19 supplying the equilibrium parameter α_o, that value being associated with a fully developed sea. The value s_o=*2* is introduced for the spreading index of the directional spreading function in Eq. 18. The frequency spectrum obtained exhibits the spectral maximum $(S_o)_{max}$ and spectral wave height H_{mo} given in Table 1.

Reproduced wave conditions at Station M. The preliminary use of the $[ADS^*]_2$ model from deep water to Station M provides the actual frequency spectrum $S^*_A = [S^*_A(f,h)]_2$ shown in Figure 3 and the values of the spectral maximum $(S^*_A)_{max}$ and spectral wave height H_{mo} given in Table 1. The value of the significant height H_s, put in the same table, is derived from that of H_{mo} through the relationship

$$H_s/H_{mo} = 0.88\exp\{0.023(\sigma_{hp}/2\pi)^{-0.87}\};\ \sigma_{hp} \leq 2 \quad (76)$$

suggested by Gentile (1996) after a re-examination of the one proposed by Thompson & Vincent (1985). The figure and the table show that the wave conditions reproduced at Station M are really close to the observed ones and thus they can be adopted as starting data for the validation of the model from Station M to Station N.

Model test. Predicted wave conditions at Station N. The use of $[ADS^*]_2$ model from Station M to Station N provides the frequency spectrum shown in Figure 3 and the values of the spectral maximum, spectral wave height, and thus significant wave height, given in Table 1. The local value of the overall mean wave direction is about *7* degrees.

	deep water		Station M			Station N		
	$(S_o)_{max}$ (m²/Hz)	H_{mo} (m)	$(S)_{max}$ (m²/Hz)	H_{mo} (m)	H_s (m)	$(S)_{max}$ (m²/Hz)	H_{mo} (m)	H_s (m)
obs.	-	-	83.1	-	8.05	14.5	-	3.90
pred.	327.3	12.48	83.4	8.01	8.30	13.5	3.20	3.80

Table 1 - Observed and predicted wave conditions.

The figure and the table show that the predicted wave conditions at Station N are in substantial agreement with the observed ones and this accordance makes the test quite satisfactory. In fact the spectral form is well reproduced, even if the predicted peak frequency is higher. The predicted values of the spectral maximum and significant wave height differ from the observed ones by 7% and 2.5%, respectively, noting that a variation $\sim \pm 50\%$ of the values adopted for the Nikuradse roughness parameter ϵ and overall mean wave direction ϕ_{mo} does not appreciably modify those values.

Finally, the following comments should be made. According to Maa (1993), the effect of bottom friction on random wave propagation can be disregarded if the distance to be travelled is not important. Figure 4, given as an example, confirms this circumstance, showing the modification that the predicted frequency spectrum would exhibit at Station N when the λ distance to be travelled decreases, the water depths at Station M and N being unchanged. A further decrease in λ does no longer modify the spectral form in a noticeable way. The same figure shows the presence of the short and long bound waves, which become important as the bottom friction effect decreases.

CONCLUDING REMARKS

The following main conclusions can be drawn.

1. A model is suggested which provides the directional and frequency spectra on decreasing depths to second order of approximation, with the corresponding spectra on deep water being known.
2. The model takes into account the shoaling, refraction, saturation and bottom friction processes. It refers to a bottom with straight and parallel contours, noting that the extension to a bottom with a three dimensional bathymetry does not involve special difficulties.
3. Bottom energy dissipation, towards which attention is specifically focused, is formalized through a model which considers the actual flow regime in the bottom frictional layer, depending on the random wave field and Nikuradse roughness parameter.
4. Bottom energy dissipation is important only with long distances to be travelled. In any case, surface energy dissipation due to the saturation process is prevalent.
5. A comparison of the model with field data from Maa gives very satisfactory results.

APPENDIX

FLOW REGIMES WITHIN THE BOTTOM LAYER

The specification of the flow regimes within the bottom frictional layer is carried out by means of Figure 5, suggested by Becchi et al. (1982), which shows the regions of the laminar (1), rough turbulent (3), smooth turbulent (5), transitional (2, 4, 6) flow regimes. The figure involves the Reynolds number RE, the Kajiura number M and the roughness parameter R, defined by

$$RE=(\hat{U}_w)_{eq}(\hat{A}_w)_{eq}/\nu \qquad (77)$$

$$M=(\hat{U}_w)_{eq}\epsilon/\nu \qquad (78)$$

$$R=(\hat{A}_w)_{eq}/\epsilon \qquad (79)$$

and referred to the EW. In the previous equations, ν is the kinematic viscosity, ϵ is the Nikuradse roughness parameter, $(\hat{A}_w)_{eq}$ is the orbital amplitude related to the bottom horizontal velocity amplitude $(\hat{U}_w)_{eq}$, given by Eq. 49 through the relationship

$$(\hat{A}_w)_{eq}=(\hat{U}_w)_{eq}T_{eq}/2\pi. \qquad (80)$$

AMPLITUDE AND PHASE LEAD OF THE FRICTION COEFFICIENT

The evaluation of both the amplitude $(\hat{C}_w)_{eq}$ and the phase lead $(\vartheta_w)_{eq}$ of the friction coefficient $(C_w)_{eq}$, Eq. 52, are made by the relationships suggested by Rebaudengo Landó & Scarsi (1984) on the basis of the Kajiura theory (1968), after taking into account the experimental data by Eagleson (1962), Ywagaki et al. (1967), Riedel et al. (1972), Kamphuis (1975), Jonsson & Carlsen (1976), Treolar & Abernethy (1978) for the laminar, rough turbulent and smooth turbulent flow regimes and introducing a first approach scheme for the transitional ones.

Those relationships, shown below, lead to results which are generally close to the analytical and experimental ones deduced afterwards.

Laminar flow regime (region 1 in Figure 5):

$$(\hat{C}_w)_{eq}=1/RE^{1/2} \qquad (81)$$

$$(\hat{\vartheta}_w)_{eq}=\pi/4=0.785 \qquad (82)$$

Rough turbulent flow regime (region 3 in Figure 5):

$$1/(\hat{C}_w)_{eq}^{1/2}=7.50\log\{(\hat{C}_w)_{eq}^{1/2}R/0.30\};\ R>2.67 \qquad (83)$$

$$(\hat{C}_w)_{eq}=0.094;\ R\leq 2.67 \qquad (84)$$

$$1/(\vartheta_w)_{eq}^{1/2}=0.30\log\{(\vartheta_w)_{eq}^{1/2}R/0.0045\};\ R>26 \qquad (85)$$

$$(\vartheta_w)_{eq}=\pi/4;\ R\leq 26 \qquad (86)$$

Smooth turbulent flow regime (region 5 in Figure 5):

$$1/(\hat{C}_w)_{eq}^{1/2}=6.70\log\{(\hat{C}_w)_{eq}RE/0.93\} \qquad (87)$$

$$1/(\vartheta_w)_{eq}^{1/2}=0.30\log\{(\vartheta_w)_{eq}RE/0.40\} \qquad (88)$$

Laminar-smooth turbulent transitional flow regime (region 6 in Figure 5):

$$\log(\hat{C}_w)_{eq}=2.425-1.616\log RE+0.127(\log RE)^2 \qquad (89)$$

$$\log(\vartheta_w)_{eq}=0.907-0.253\log RE \qquad (90)$$

Other transitional flow regime (regions 2 and 4, in Figure 5):

$$\log(\hat{C}_w)_{eq}=\log[(\hat{C}_w)_{eq}]_A+\{\log[(\hat{C}_w)_{eq}]_B-\log[(\hat{C}_w)_{eq}]_A\}W \qquad (91)$$

$$\log(\vartheta_w)_{eq}=\log[(\vartheta_w)_{eq}]_A+\{\log[(\vartheta_w)_{eq}]_B-\log[(\vartheta_w)_{eq}]_A\}W \qquad (92)$$

where

$$W=\{\log(RE)-\log(RE_A)\}/\{\log(RE_B)-\log(RE_A)\} \qquad (93)$$

noting that the quantities with the index A and B are obtained by using Table 2 in which the following equations are recalled too

$$RE_A=66R \qquad (94)$$

$$RE_A=49.15R^{1.06} \qquad (95)$$

$$RE_A=21.51R^{1.17} \qquad (96)$$

$$RE_B=1949.36R^{0.5} \qquad (97)$$

$$RE_B=218.36R^{1.17} \qquad (98)$$

Ranges of R	RE_A	RE_B	$\hat{C}_A$	$\hat{C}_B$	ϑ_A	ϑ_B
$R < 2.67$	(94)	(97)	(81)	(84)	(82)	(86)
$2.67 \leq R < 26$	(94)	(97)	(81)	(83)	(82)	(86)
$26 < R \leq 151$	(94)	(98)	(81)	(83)	(82)	(85)
$151 < R \leq 1162$	(95)	(98)	(89)	(83)	(90)	(85)
$R > 1162$	(96)	(98)	(87)	(83)	(88)	(85)

Table 2 - Reference equations for transitional flow regimes in regions 2 and 4. The simplified notations $\hat{C}=(\hat{C}_w)_{eq}$ and $\vartheta=(\vartheta_w)_{eq}$ are introduced.

REFERENCES

Becchi, I, Rebaudengo Landò, L, and Scarsi, G (1982). "Limits and critical roughness for laminar and turbulent flows in the boundary layer of progressive gravity waves," *Atti Accad Lig Sc e Lett*, Genova, Vol XXXVIII, pp 93-105.

Bouws, E, Gunter, H, Rosenthal, W and Vincent, CL (1985). "Similarity of wind wave spectrum in finite depth water. Part I: Spectral form," *J Geoph Res*, Vol 90, No C1, pp 975-986.

Collins, JI (1972). "Prediction of shallow water spectra," *J Geoph Res*, Vol 77, pp 2693-2707.

Dalrymple, RA (1992). "Wave propagation in shallow water," *Proc Short Course*, ICCE 92, pp 69-92.

Eagleson, PS (1962). "Laminar damping of oscillatory waves," *J Hydr Div*, ASCE, Vol 88, No HY3, pp 155-181.

Gentile, R (1996). "La propagazione delle onde random," *Atti 4° Congresso AIOM*, Padova, pp 71-80.

Gentile, R, Rebaudengo Landó, L, and Scarsi, G (1994). "Design wave on finite depth: a re-analysis of the TMA Model," *Proc 13th Int Conf on Offshore Mech and Arctic Eng*, Houston, Vol II, pp 49-56.

Goda, Y (1975). "Irregular wave deformation in the surf zone," *Coast Eng in Japan*, Vol 18, pp 13-26.

Graber, HC, and Madsen, OS (1988). "A finite-depth wind-wave model. Part I: Model description," *J Phys Ocean*, Vol 18, pp 1465-1483.

Hasselmann, K, and Collins, JI (1968). "Spectral dissipation of finite depth gravity waves due to bottom friction," *J Mar Res*, Vol 26 pp 1-12.

Hasselmann, K et al (1973). "Measurements of wind-wave growth and swell decay during the Joint North Sea Wave Project (JONSWAP)," *Deut Hydrogr Z*, A8, No 12, pp 9-95.

Jonsson, IG, and Carlsen, NA (1976). "Experimental and theoretical investigations in an oscillatory rough turbulent boundary layer," *Coastal Eng Lab and Hydr Lab*, Tech Univ Denmark, Basic Res Progr Rep No 12, pp 2-10.

Kajiura, K (1968). "A model of the bottom boundary layer in water waves," *Bull Earthq Res Inst*, Vol 46, pp 75-123.

Kamphuis, JW (1975). "Friction factor under oscillatory waves," *J Waterways, Harbors and Coast Eng Div*, ASCE, Vol 101, pp 135-144.

Kitaigorodskii, SA, Krasitskii, VP, and Zarlavskii, MM (1975). "On Phillips theory of equilibrium range in the spectra of wind-generated waves," *J Phys Ocean*, Vol 5, pp 410-420.

Laing, AK (1986). "Nonlinear properties of random gravity waves in the water of finite depth," *J Phys Ocean*, Vol 16, pp 2013-2030.

Le Méhauté, B, and Wang, JD (1982). "Wave spectrum changes on sloped beach," *J Waterway, Port, Coast and Ocean Eng*, ASCE, Vol 108, pp 33-47.

Li, Y, and Dong, G (1993). "Wave breaking phenomena of irregular wave combined with opposing current," *Proc 3rd Int Offshore and Polar Eng Conf*, Singapore, Vol 3, pp 64-70.

Maa, JPY (1993). " 'Halloween' wave transformation near Virginia Coast", *Proc Nat Conf Hydr Eng*, pp 1260-1265.

Madsen, OS, Poon, YK, and Graber, HC (1988). "Spectral wave attenuation by bottom friction: Theory," *Proc 21th ICCE*, ASCE, Vol 1, pp 492-504.

Rebaudengo Landó, L, and Scarsi, G (1980). "Sulla potenza dissipata dalle onde di gravitá progressive," *Atti Accad Lig Sc e Lett*, Genova, Vol XXXVII, pp 218-239.

Rebaudengo Landó, L, and Scarsi, G (1984). "La potenza dissipata dallo spettro di energia JONSWAP," *Ricerche sulle onde di mare*, Hydr Inst, Genova, Vol 1, pp 89-105.

Rebaudengo Landó, L, Gentile, R, and Scarsi, G (1996). "Multidirectional wave transformations on decreasing depths: a nonlinear model," *Proc 15th Int Conf on Offshore Mech and Arctic Eng*, Firenze, Vol II, pp 29-40.

Riedel, HP, Kamphuis, JW, and Brebner, A (1972). "Measurement of bed shear stresses under waves," *Proc 13th Conf*

Coast Eng, Vancouver, Vol 1, pp 587-603.

Scarsi, G (1995). "Caratteristiche meteomarine: trasformazioni sottocosta degli eventi di moto ondoso," *Giornate Italiane di Ingegneria Costiera,* AIPCN, Ravenna, pp 77-98.

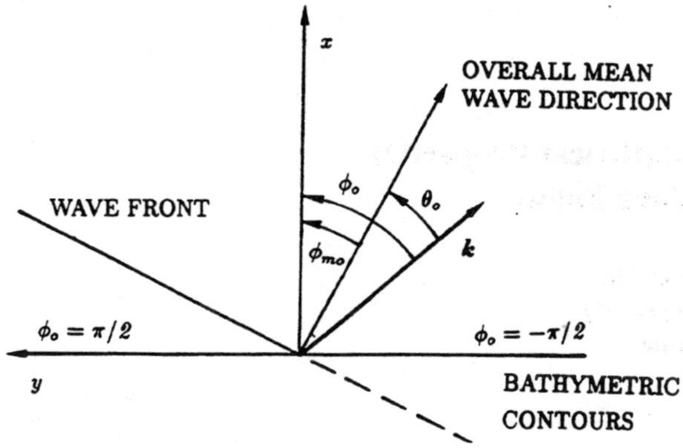

Figure 1 - Sketch in deep water.

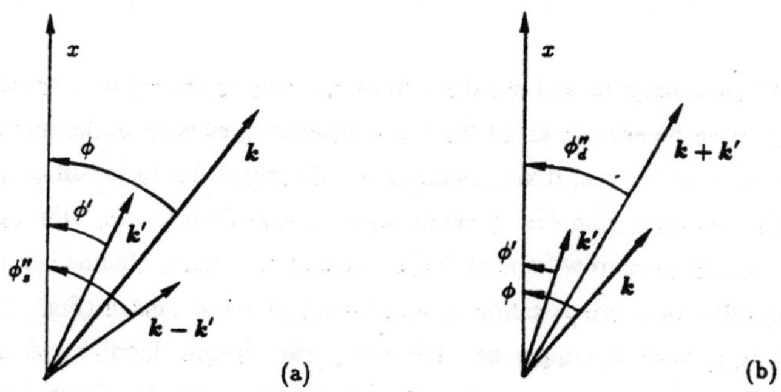

Figure 2 - Second order component k which arises from (a) the sum and (b) the difference of the two first order components k', $k \pm k'$.

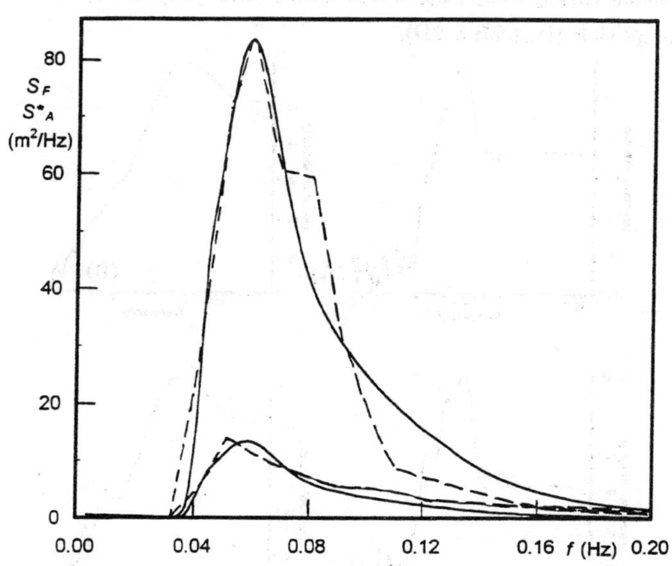

Figure 3 - Observed (- - -) and predicted (——) spectra at Stations M and N.

Thompson, EF, and Vincent, CL (1985). "Significant wave height for shallow water design," *J Waterway, Port, Coast and Ocean Eng,* ASCE, Vol 111, No 5, pp 828-842.

Treolar, PD, and Abernethy, CL (1978). "Determination of a bed friction factor for Botany Bay, Australia," *Coast Eng,* Vol 2, pp 1-20.

Ywagaki, Y, Tsuchya, Y, and Chen, H (1967). "On the mechanism of laminar damping of oscillatory waves due to bottom friction," *Bull Disas Prev Res Inst,* Kyoto Univ, No 116, pp 49-75.

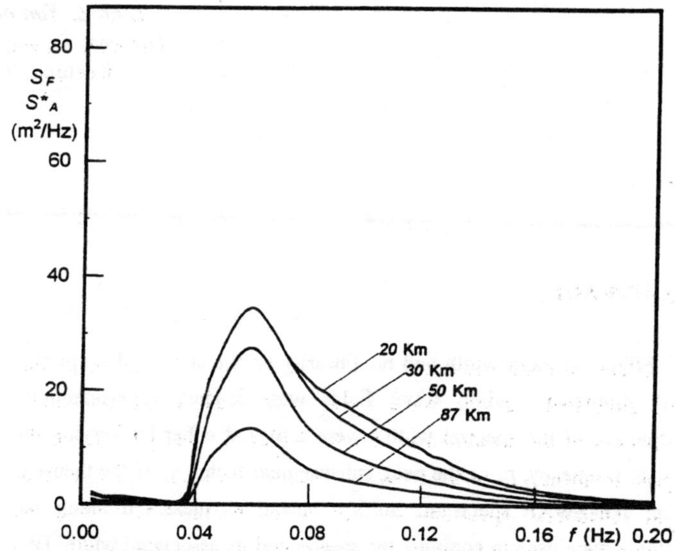

Figure 4 - Frequency spectra predicted at Station N with different distances to be travelled.

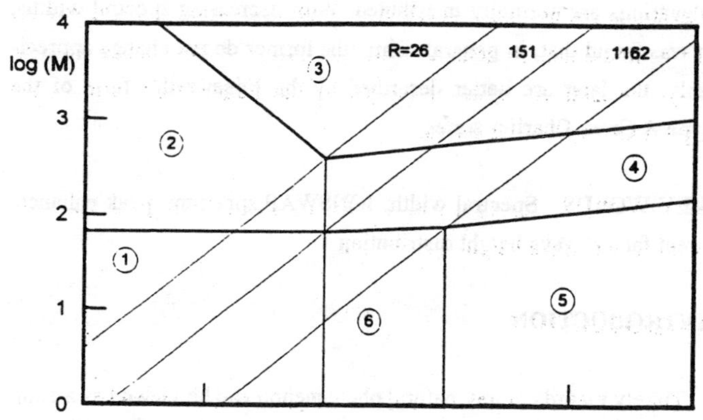

Figure 5 - Limits of the regions relevant to different flow regimes within the bottom frictional layer: (1) laminar; (2) rough turbulent; (5) smooth turbulent; (2, 4, 6) transitional regimes.

Proceedings of the Seventh (1997) International Offshore and Polar Engineering Conference
Honolulu, USA, May 25-30, 1997

ISBN 1-880653-28-1 (Set); ISBN 1-880653-31-1 (Vol. III)

Effects of Spectral Width on the Statistical Properties of Simulated Random Wave Fields

John Z. Yim and Jaw-Guei Lin
National Taiwan Ocean University
Keelung, Taiwan, China

ABSTRACT

Effects of band width and nonlinearity on the statistical properties of simulated random wave fields were studied experimentally. Changes of the spectral widths were achieved either by varying the peak frequency, f_p, or the peak enhancement factor, γ, of the theoretical JONSWAP spectrum. Surface elevations measured along the flume were used to compute the spectra and its associated width. Distributions of wave heights and surface elevations were computed and compared with theories. The results indicate that, wave heights in the flume can be fitted into the Rayleigh or Weibull models, and surface elevations are normally distributed. With decreasing spectral widths, it was found that, in general, while the former do not change appreciably, the later are better described by the Edgeworth's form of the type A Gram-Charlier series.

KEY WORDS: Spectral width, JONSWAP spectrum, peak enhancement factor, wave height distribution

INTRODUCTION

Timely records of any natural phenomenon can be viewed as a composition due to three influencing factors of, respectively, global, regional and random nature. The global influences are mostly due to large scale effects such as macroscopic climatological changes. Regional or local characteristics are caused by the specific geological or topographical conditions on site of measurement. Finally, occurrences of mostly unpredictable nature add in to the random part of measured results.

In an attempt to sort out these three influencing factors by comparing wave records obtained from two measuring stations in the northern part of Taiwan, it was found that, differences due to localities do exist. As seen from Fig. 1, while wave spectra from one location can be called as narrow-banded (Figs. 1A and 1C), those obtained from the other measuring station should be called broad-banded (Figs. 1B & 1D). Most strikingly are, however, wave height distributions on these two sites. It is seen from Fig. 2 that, wave heights for the site with narrow-banded spectra have distributions that are better described by the Gaussian (Figs. 2A & 2C), while wave heights of the station having broad band characteristics fit the Rayleigh model appropriately (Figs. 2B & 2D).

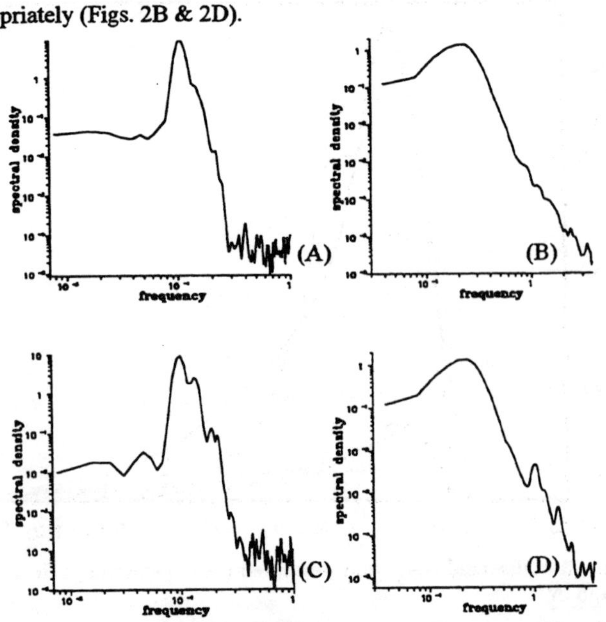

Fig. 1 Spectra measured on September 27 for KY (1A & 1C) and BTJ (1B & 1D). Time of measurement: Figs.A-D 12.00-3.00 am.

As is well known, Longuet-Higgins (1952) derived the Rayleigh model for narrow-banded wave fields through the assumption of Gaussianity of the surface fluctuations. Even though this model was shown to overpredict larger wave heights (Chakrabarti & Cooley, 1977), most researchers do find it rather satisfactory (Dattatri et al., 1979).

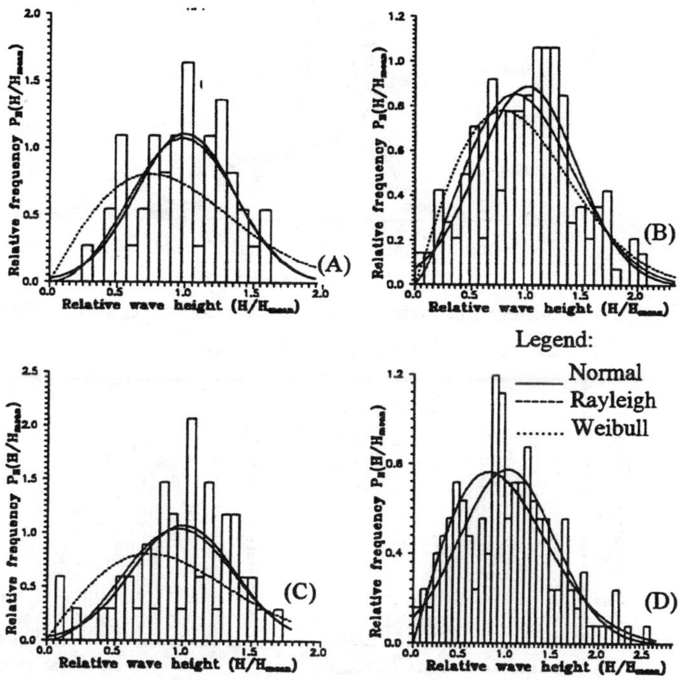

Fig. 2 Calculated wave height distributions for data in Fig. 1

A model that fits the larger wave tail better was first proposed by Forristall (1978). He used the two-parameter Weibull distribution to describe hurricane-generated waves. Later, this model was also used by Kimura (1981) for simulated random waves in a laboratory wave tank, by Mase (1989) to study possible correlation between wave grouping and wave height distribution, and by Sundar et al. (1993) for wave heights in a groin field.

Among these studies, the most interesting findings are those due to the works of Kimura (1981) and Sobey (1992). Kimura (1981) used a so-called two-dimensional Weibull to study joint distribution of wave heights and periods of mechanically generated random waves. He found that the shape parameter for the marginal distribution of wave heights has a value of 2. Sobey (1992), on the other hand, studied numerically simulated random wave fields and found that, wave heights "*differ little from Rayleigh distribution.*" Now, since Rayleigh distribution is a special case of Weibull distribution with the shape parameter equal to 2, the findings of these two researchers all point to the fact that, simulated random wave heights are Rayleigh distributed.

Random wave simulations can be carried out in a variety of ways. A review of the available techniques was presented earlier by Funke & Mansard (1980), and recently by Hughes (1993). Even though each proposed method has it own merits, as well as demerits, the underlying basic assumptions remain mostly unchanged. Among these assumptions, the most important two being linearity and narrow-bandedness.

According to Stokes wave theory, finite amplitude waves have a peaked crest and flat trough. Water particles thus stay longer beneath the mean water level than above. This changes the probability distribution of the surface elevations, as demonstrated first by Longuet-Higgins (1964), and later by Huang & Long (1980). Effects of spectral width on wave height distributions have been studied theoretically by Longuet-Higgins (1980), and experimentally by Larsen (1981). They all reach to the same conclusion that, spectral width has more profound influences than nonlinearity does.

Possible dependencies of wave height and surface elevation distributions on the broadness of the spectral shapes have also studied experimentally by Yim & Shiao (1996). Changes of spectral shapes were achieved through variations of the parameters of the JONSWAP spectrum. It was found that, irrespective of the spectral shapes, wave heights in the flume are to be approximated by Rayleigh distribution more satisfactorily than both the Gaussian and the Weibull distribution. In that paper it was also stated that, the experiments were considered to be rather unsuccessful in the sense that spectral shapes can only be marginally affected through changes of the values of the peak enhancement factor.

Acknowledging the fact that, previous experimental results could be misleading, new experiments were conducted carefully. The present paper resulted from these new experiments. In the following, we divide the remaining of this paper into three parts. In Section II a short description of the experimental apparatus will be given. Section III presents some of the experimental results, and section IV concludes this paper.

EXPERIMENTS

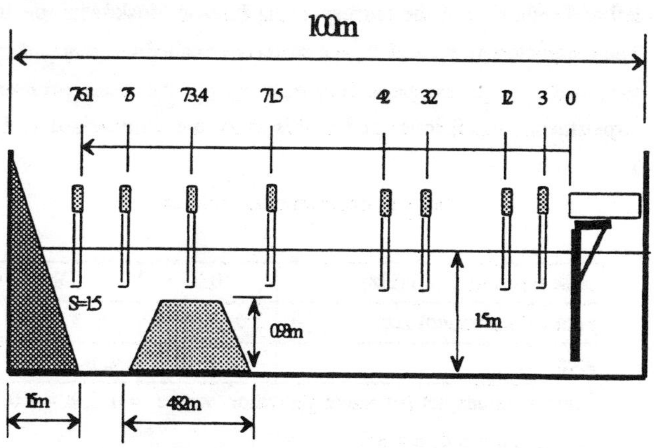

Fig. 3 Schematic representation of wave generating facilities

Experiments were conducted in a newly constructed wave flume in the Laboratory for Ocean Engineering, the Department of Harbour and River Engineering of National Taiwan Ocean University. The flume has a dimension of 100 × 3.0 × 3.0 (length × width × height, in meters). A more detailed description concerning the wave flume and data acquisition procedures can be found in Yim et al. (1996). The only difference being the locations of the measuring stations. Instead of six as in the previous experiments, eight wave gauges were used in this experiment. Locations of these stations are, respectively, 3.0, 12.0, 32.0, 42.0, 71.5, 73.4, 75.0 and 76.1 meters, away from the piston wave generator. However, only measured results of the first four stations will be used in this paper, the results of the last four measuring stations will be presented in an accompanying paper (Lin & Yim, 1997). Fig. 3 shows the experimental setups schematically.

As mentioned earlier, variations of spectral shapes were achieved by changing the parameters of the JONSWAP spectrum. The typical JONSWAP spectrum can be expressed as:

$$S(f) = \frac{\alpha g^2}{(2\pi)^4 f^5} \exp\left[-\frac{5}{4}\left(\frac{f}{f_0}\right)^{-4}\right] \gamma^a \tag{1}$$

where f_0 = the peak frequency; α = the so-called Phillips constant with a mean value of 0.0081; γ = the peak enhancement factor, having a mean value of 3.3; and the parameter a controls the width of the spectrum, with:

$$a = \exp\left[-\frac{(f - f_0)^2}{2\sigma^2 f_0^2}\right] \tag{2}$$

where

$$\sigma = \begin{cases} \sigma_a = 0.07 & \text{for } f \le f_0 \\ \sigma_b = 0.09 & \text{for } f > f_0 \end{cases} \tag{3}$$

σ_a and σ_b are deviations from the peak frequency to, respectively, the left and right side of the corresponding Pierson-Moskowitz spectrum. The parameters γ, σ_a, and σ_b are sometimes called the shape parameters, while α and the peak frequency, f_0, are the scale parameters. Experimental conditions used in this study are summarized in Table 1.

Table 1 Experimental conditions

peak frequency (f_p) [Hz]	0.5	0.8
peak enhancement factor (γ) [-]	1.0, 3.3, 7.0	1.0, 3.3, 7.0
gain* [-]	b, c, d,	a, b, c, d, e

*gain = values set for wave generator, with a = 1.5, b = 2.0, c = 2.5, d = 3.0, e = 3.5

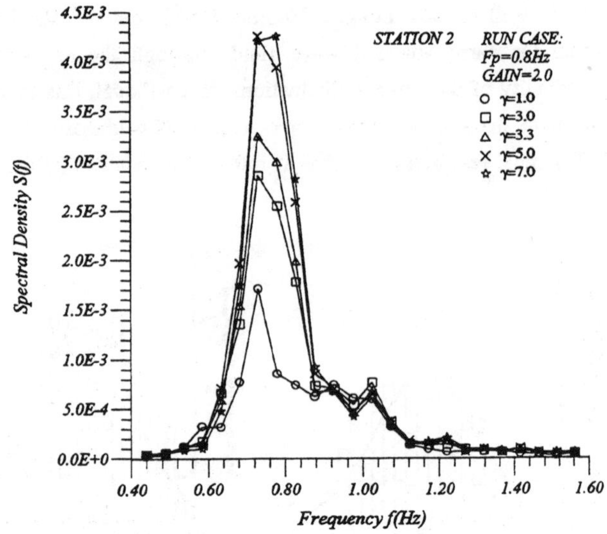

Fig. 4 Peak enhancement factor γ and its effects on spectral forms. Case studied: f_p = 0.8 Hz, γ = 1~7, gain = 2.0

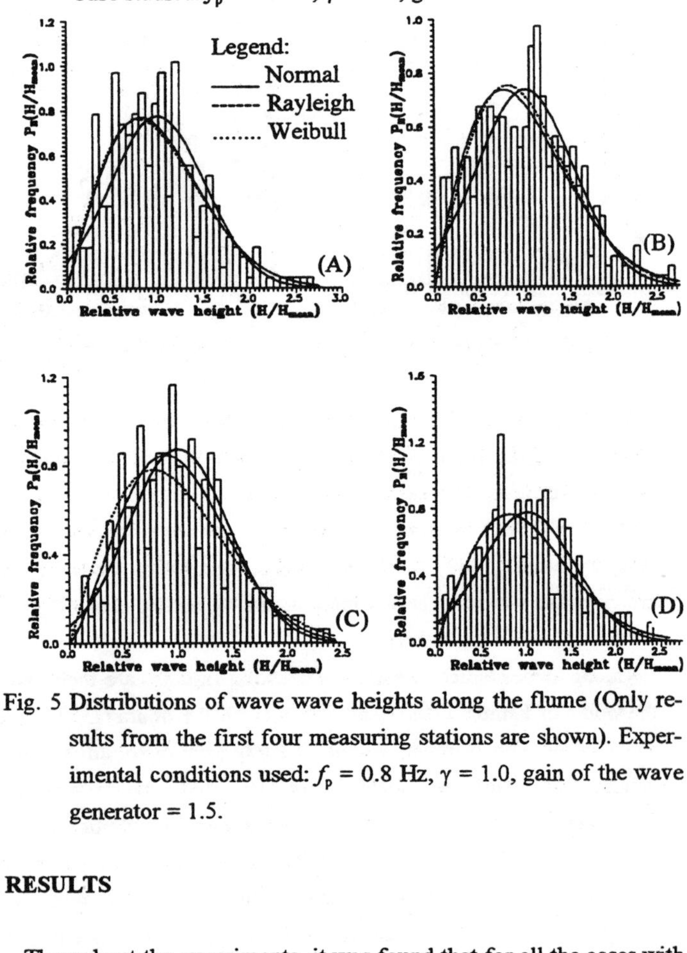

Fig. 5 Distributions of wave wave heights along the flume (Only results from the first four measuring stations are shown). Experimental conditions used: f_p = 0.8 Hz, γ = 1.0, gain of the wave generator = 1.5.

RESULTS

Throughout the experiments, it was found that for all the cases with lower peak frequency, f_p = 0.5, measured spectra were systematically broader than the target JONSWAP spectra, irrespective the values of the peak enhancement factor, γ, chosen. As for cases with higher peak frequencies, f_p = 0.8 and 1.0 Hz, the acquired spectra deviate from target spectrum for the case with $\gamma = 1$, i.e., Pierson-Moskowitz [P-M] spectrum, but are closer to the typical narrow-banded JONSWAP spectrum for cases with $\gamma > 1$. Fig. 4 shows the increments of spectral density in accordance with the values of pre-selected gains for the wave generator. Peak frequency used for this sets of experiments was 0.8 Hz, the data were all from Station 1.

As mentioned earlier, distributions of wave heights were fitted with three models, namely, the normal, the Rayleigh, and the Weibull distribution. Figure 5 shows the results of fitting wave heights of the first four measuring stations with these three models. The experimental conditions used for this case are: $f_p = 0.8$ Hz, $\gamma = 1.0$. It is often said that, since the two-parameter Weibull distribution has two adjustable parameters, it can be used to fit the experimental data more satisfactorily than the other two model used. Given this situation, and judging from the fact that the two curves representing the theoretical Rayleigh and Weibull almost coalesce with each other in Figs. 5A, 5B, 5D, it seems safe to say that, wave heights for this case are Rayleigh distributed.

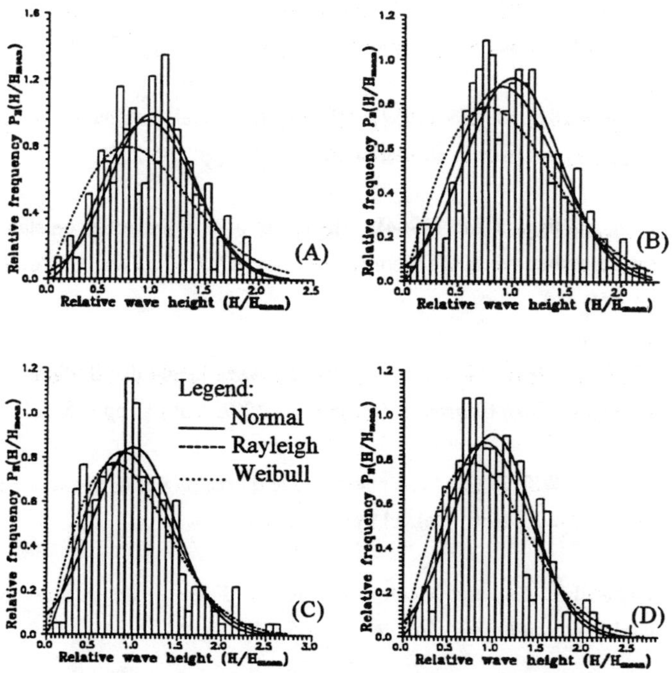

Fig. 6 Distributions of wave wave heights along the flume (Only results from the first four measuring stations are shown). Experimental conditions used: $f_p = 0.8$ Hz, $\gamma = 7.0$, gain of the wave generator = 3.5.

Figure 6 shows the probability distributions of wave heights with the peak frequency same as Fig. 5, but with a peak enhancement factor $\gamma = 7.0$. As can be seen from this figure, the curves representing Weibull model deviate from the Rayleigh distribution, indicating that wave heights in the flume are now more or less Gaussian. These results are rather peculiar, since spectra with more pronounced peaks represent concentration of wave energy, and are thus, narrow-banded. As mentioned earlier, the Rayleigh distribution of wave heights was derived by Longuet-Higgins based upon the assumption of narrow-bandedness. On the other hand, it is well known that finite amplitude waves are unstable to perturbations, which leads to energy concentration and eventually wave breaking (Melville, 1982; Su et al. 1982). It seems plausible to postulate that, other mechanism not considered in the original theory of Longuet-Higgins, such as wave breaking mentioned here, can cause these deviations. More studies are needed before any definite conclusion can be drawn.

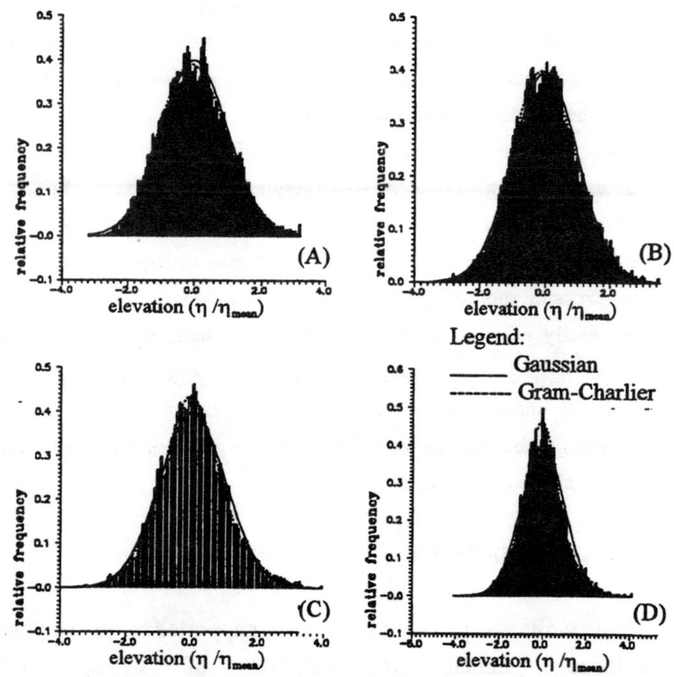

Fig. 7 Distributions of surface elevations along the flume (Only results from the first four measuring stations are shown). Experimental conditions same as Figure 6.

Figure 7 shows the ditributions of surface elevations along the wave flume for the experimental conditions same as Fig. 6. It can be seen that, the water surface at the first two stations are normally distributed (Figs. 7A & 7B), indicating that waves at these two stations are almost linear. This is also indicated by the fact that, the curves of the normal distribution and the Gram-Charlier series expansion are almost indistinguishable at these two stations. As the waves travelling down the flume, energies become concentrated, and waves are finite amplituded or nonlinear. The curves representing the Gram-Charlier series expansions gradually deviate from the Gaussian (Figs. 7C & 7D), showing the effect of nonlinearity are now more predominant, as demonstrated by Longuet-Higgins (1964).

DISCUSSION AND CONCLUSION

Effects of spectral shapes on the statistical properties of simulated random waves were studied experimentally. Changes of the spectral shapes were achieved through variations of the peak enhancement factor of the typical JONSWAP spectrum. Probability distributions of both wave heights and surface elevations are fitted to the theoretical models found in the literature. The present results indicate that, as wave energy concentrates wave heights will have a distribution that deviate from the theoretical Rayleigh. The present findings seem to contradict with the theoretical predictions, as waves with concentrated energies represent a narrow-banded spectra and thus should have a Rayleigh distribution for wave height. However, our present results shown in Figure 6 seem also to confirm our previous results of field data, one of which was shown in Fig. 2.

One plausible explanation of this seemingly contradictory results was that waves with concentrated energies tend to break on the water surface. Breaking of large amplitude waves generates turbulence, as well as small scale fluctuations on the water surface. In applying a zero-crossing method, these small scale fluctuations could be treated very easily as small amplitude waves, in a sense similar as the so-called "false waves" (Gimenez et al., 1994; Pires-Silva & Medina, 1994). It should be stressed that, more experimental evidences, whether in laboratory or from field, must be gathered before any definite conclusion can be drawn.

ACKNOWLEDGMENT

This work was partially financially supported by the National Science Council of Republic of China, NSC 83-0209-E-019-004. All the laboratory experiments were conducted by Mrs. T.-Z. Ding & D-S. Tsai. Their dedication and endeavor are deeply appreciated.

REFERENCES

Chakrabarti, SK, and Cooley, RP (1977). "Ocean wave statistics for 1961 North Atlantic storm," *J. Waterway Port Coastal & Ocean Div. ASCE* Vol 103 pp 433-448.

Dattatri, J, Raman, H, and Shankar, NJ (1979). "Height and period distribution for waves off Mangalore Harbour - West coast," *J. Geophys. Res.*, Vol. C84 pp. 3767-3772.

Forristall, GZ (1978). "On the statistical distribution of wave heights in a storm," *J. Geophys. Res.* Vol C83 pp 2353-2358.

Funke, ER, and Mansard, EP (1980). "On the synthesis of realistic sea states," *Proc. 17th Int'l Coastal Eng. Conf., ASCE*, pp. 2974-2991.

Gimenez, MH, Sanchez-Carratala, CR, and Medina, JR (1994) "Analysis of false waves in numerical sea simulations," *Ocean Engng.*, Vol. 21 pp. 751-764.

Huang, NE, and Long, SR (1980) "An experimental study of the surface elevation probability distribution and statistics of wind generated waves," *J. Fluid Mech.*, Vol. 101 pp. 179-200.

Hughes, SA (1993). "*Physical models and laboratory techniques in coastal engineering,*" World Scientific Pub. Co. Singapore, pp. 568.

Hudspeth, RT, and Chen, MC (1979). "Digital simulation of nonlinear random waves," *J. Waterway, Port, Coastal and Ocean Div., Proc. ASCE*, Vol. 105 pp. 67-85.

Kimura, A (1981). "Joint distribution of the wave heights and periods of random sea waves," *Coastal Eng. Japan*, Vol. 24 pp. 77-92.

Longuet-Higgins, MS (1952). "On the statistical distribution of the heights of sea waves," J. Marine Res., Vol. 9 pp. 245-266.

Longuet-Higgins, MS (1964). "The effect of non-linearities on statistical distributions in the theory of sea waves," *J. Fluid Mech.*, Vol. 17 pp. 459-480.

Mase, H (1989). "Groupiness factor and wave height distributions," *J. Waterway, Port, Coastal & Ocean Eng. ASCE* Vol 115 pp 105-121.

Melville, WK (1982). "The instability and breaking of deep-water waves," *J. Fluid Mech.*, Vol. 115 pp. 165-185.

Pires-Silva, AA, and Medina, JR (1994) "False waves in wave records," *Ocean Engng.*, Vol. 21 pp. 765-770.

Su, MY, Bergin, M, Marler, P, and Myrick, R (1982). "Experiments on nonlinear instabilities and evolution of deep gravity-wave trains," *J. Fluid Mech.*, Vol. 124 pp. 45-72.

Yim, JZ and Shiao, SS (1996). "Statistical properties of simulated random waves," *Proc. 6th Int'l Offshore & Polar Eng. Conf.*, Vol. III, Los Angles, USA, pp. 108-114.

Proceedings of the Seventh (1997) International Offshore and Polar Engineering Conference
Honolulu, USA, May 25-30, 1997

ISBN 1-880653-28-1 (Set); ISBN 1-880653-31-1 (Vol. III)

A Weibull-Stokes Model for the Distribution of Maximum Wave and Crest Heights

Raymond Nerzic
OPTIMER
Montpellier, France

Marc Prevosto
IFREMER
Brest, France

ABSTRACT

The present paper describes a model for the distribution of maximum wave and crest heights in a given sea state. The standard Rayleigh and Weibull distributions for wave and crest heights are modified using a third order Stokes expansion of the wave envelope. The maximum wave or crest height in a record is calculated using the asymptotic Gumbel distribution. It was validated by fitting the model distributions to North Sea measurement data and provided much better predictions of maximum crest and wave heights than standard models. To make up for the fact that directionality and spectral shape are not taken into account in the model, correction terms (specific to the studied site) are introduced and mainly one of them will give the optimized mean wave number to consider in the model. The proposed model is relatively easy to apply and could be an effective tool to determine extreme wave and crest heights for offshore structure design purposes.

KEYWORDS

Wave height, wave crest, distribution, extreme, nonlinearity.

INTRODUCTION

Standard models for the distribution of wave and crest heights are useful for computing maximum wave and crest heights, on the basis of significant wave heights, for the purposes of designing offshore structures. However, there can be differences of 10-20% in the results obtained with different models, which is unsuitable in terms of design criteria. These between-model discrepancies could be explained by the problem of accurately representing wave properties, especially strong and steep waves, in a given sea state. This is mainly due to nonlinearities in the wave kinematics which are not suitably taken into account in the models, nonlinearities which are controlled by directionality, spectral shape and wavelength-to-depth ratio.

In the present paper, maximum wave and crest heights computed by several models are compared, followed by a description of an analysis of maximum wave and crest heights measured at the Frigg offshore site in the North Sea. The analysis confirmed a marked effect of wave steepness on maximum crest heights, indicating that nonlinear effects occur in wave processes. But steepness did not have such an important effect on maximum crest to trough wave height.

This influence is generally not taken into account in wave and crest height distribution models, apart from those developed by Tayfun (1983) and Vinje (1989), as the nonlinear wave process is considered to be a distorted narrow- banded Gaussian process due to nonlinearity effects. A model was thus designed for the distribution of wave and crest heights, which combines a Rayleigh or Weibull distribution (describing wave and crest heights as a first-order Gaussian process) and third-order Stokes nonlinear wave expansion. Hereafter this is termed a Rayleigh-Stokes or Weibull-Stokes model.

Tests were performed by fitting distributions derived from this model, along with a maxima distribution model based on a Gumbel distribution, to measured data obtained at the Frigg site, in order to check its validity in representing maximum wave and crest heights.

This model can be applied in several different ways, depending on the wave height, period and steepness parameters considered in the Stokes expansion. Although the model seemed quite suitable for the case investigated, it has not yet been validated for other sea conditions. These points are discussed at the end of the paper.

STANDARD WAVE AND CREST HEIGHT DISTRIBUTION MODELS

The wave amplitude distribution model based on Rayleigh's distribution, is the standard model derived from the initial model proposed by Longuet-Higgins (1952). It is based on the assumption of a linear sea surface narrow-banded Gaussian process:

$$Q(x) = \mathrm{Prob}(X > x) = \exp\left(-\left(\frac{x}{\theta}\right)^2\right) \tag{1}$$

where:

Q is the exceedance probability (Q = 1 - F, where F is the probability distribution function)

X is the H/H_S ratio or the C/H_S ratio

H is the crest-to-trough height

C is the crest height

H_S is the significant wave height, $H_S = 4.\ m_0^{1/2}$

m_0 is the zero-order moment of the power spectrum, equal to the variance σ^2

θ is a scale parameter with a value of $(1/2)^{1/2}$, or 0.707 for wave heights, and of value $(1/8)^{1/2}$, or 0.354 for crests.

Empirical wave height distributions have also been fitted by a Rayleigh distribution for many applications (especially concerning North Sea and Gulf of Mexico situations), with θ parameters in the 0.63-0.70 range. Det Norske Veritas (1991) proposed a parameter $\theta = ((1 - c^2\eta^2)/2)^{1/2}$ for wave height distributions, where c is a constant (value $\approx$ 1) and η is a spectral width parameter (value $\approx$ 0.43), or θ = 0.638, and for the crest distribution a parameter $\theta = (1/8)^{1/2}$, i.e. the standard model.

After generalization of the Rayleigh distribution, use of the two-parameter Weibull distribution was also proposed:

$$Q(x) = \mathrm{Prob}(X > x) = \exp\left(-\left(\frac{x}{\theta}\right)^{\beta}\right) \quad (2)$$

where θ is the scale parameter and β is the shape parameter. For instance, the following data have been derived from the model of Forristall (1978) for wave height distributions based on measurements obtained in the Gulf of Mexico:

$$\theta = 0.681 \qquad \beta = 2.126 \quad (3)$$

and the ranges of data were derived from modelling fits reported by Krogstad (1985), for three North Sea sites and for significant wave height > 5 m:

$$\theta = 0.73 - 0.75 \qquad \beta = 2.37 - 2.50 \quad (4)$$

Semi-empirical models based on a Rayleigh distribution, that take the effects of certain nonlinearities into account, have also been put forward. Haring et *al.* (1976), for instance, proposed a wave height distribution model based on an analysis of wave measurements from the Gulf of Mexico, expressed as:

$$Q(x) = \mathrm{Prob}(X > x) = \exp(-2x^2(\alpha + \beta x)) \quad (5)$$

with the parameters $\alpha = 0.968$, $\beta = 0.176$. In addition, the equation of Jahns & Wheeler (1972) expresses wave crest distributions while taking some nonlinearities associated with depth effects into account:

$$Q(c) = \mathrm{Prob}(C > c) = \exp\left(-\frac{1}{2}\frac{c^2}{\sigma^2}\left(1 - \beta_1\frac{c}{d}\right)\left(\beta_2 - \frac{c}{d}\right)\right) \quad (6)$$

where d is the water depth and σ is the standard deviation of the free surface.

Based on an analysis of wave measurements from the Gulf of Mexico, Haring & Heideman (1978) proposed the parameters $\beta_1 = 4.37$ and $\beta_2 = 0.57$.

The above models have been applied, sometimes with slight differences in parameter choices, to determine maximum crest heights for offshore structure engineering design purposes.

However, the results obtained with these models differ substantially. This variability is highlighted in the following tables, which present the modes (most probable data) of the H_{max}/H_S and C_{max}/H_S ratios (H_{max} and C_{max} define maximum wave heights and crests) for 180 waves (number chosen to enable valid comparisons with the results of the analysis of the Frigg site measurements presented later) and for 1000 waves (corresponding to a duration of 1.5-3 h, depending on the mean period).

1 . Most probable H_{max}/H_s ratios

	Mode H_{max} / H_s	
Model	**N=180**	**N=1000**
Rayleigh $\theta = 0.707$	1.611	1.858
Rayleigh $\theta = 0.638$ (DNV)	1.456	1.679
Weibull $\theta = 0.681$ $\beta = 2.126$ (Forristall)	1.478	1.690
Weibull $\theta = 0.73$ $\beta = 2.37$ (Krogstad)	1.463	1.650
Weibull $\theta = 0.75$ $\beta = 2.50$ (Krogstad)	1.450	1.625
Haring	1.456	1.656

2 . Most probable C_{max}/H_s ratios

	Mode C_{max} / H_s	
Model	**N=180**	**N=1000**
Rayleigh $\theta = 0.354$	0.806	0.929
Haring ($H_S/d = 0.1$)	0.894	1.047
Haring ($H_S/d = 0.3$)	1.002	1.144

These results indicate that a Rayleigh distribution, as compared to the other models, tends to overpredict maximum wave heights and underpredict maximum crest heights. Computations obtained with the other models are quite close for maximum wave heights, and the Haring model highlights a marked effect of water depth on the C_{max}/H_S ratios.

Note also that, in some applications, maximum crest heights are often derived from maximum wave heights using nonlinear regular wave models.

MAXIMA DISTRIBUTION MODEL

If F(x) is the probability distribution function of a variable X (and f(x) the probability density function), then the maxima distribution in a sample of N variables, X_N, is expressed as:

$$G(y) = \mathrm{Prob}(X_N \leq y) = [F(y)]^N \quad (7)$$

$$g(y) = N[F(y)]^{N-1}f(y) \quad (8)$$

where G is the maxima distribution function and g is the probability density function.

Based on the extreme value theory, when N is high enough and the distributions resemble Weibull-distributions, the Gumbel distribution is the asymptotic law for maxima X_N:

$$G(y) = \mathrm{Prob}(X_N \leq y) = \exp\left(-\exp\left(-\frac{y - a_N}{b_N}\right)\right) \quad (9)$$

where a_N is the distribution mode and b_N is the scale parameter characterizing the variance The data are related to the parent distribution parameters as follows:

$$Q(a_N) = \frac{1}{N}$$
$$b_N = \frac{1}{Nf(a_N)} \quad (10)$$

where $Q(x) = 1 - F(x)$ is the exceedance probability

For a two-parameter Weibull distribution or a Rayleigh distribution (β=2), the expression is:

$$a_N = \theta \mathrm{Log}(N)^{1/\beta}$$
$$b_N = \frac{\theta}{\beta \mathrm{Log}(N)^{1-(1/\beta)}} = \frac{a_N}{\beta \mathrm{Log}(N)} \quad (11)$$

and for the other models, these parameters can be easily derived from the parent distributions.

ANALYSIS OF WAVE MEASUREMENTS FROM THE FRIGG SITE

The Gumbel distribution (9) was fitted to the wave crest and height maxima distributions measured by Elf at the Frigg offshore site located in the North Sea at 100 m water depth.

The wave data were obtained with a radar distancemeter (as described in Robin & Olagnon, 1991), which provided more accurate and reliable measurements of crest heights than could be obtained on the basis of buoy measurements. The analysis was performed on data collected over the 1980-1989 period, representing 18 500 measurement sequences of about 18 min each. The following parameters were considered in the analysis:

- Wave height $H_{1/3}$, with a maximum of 12.4 m over the study period
- Mean zero down-crossing period T_z
- Maximum wave height, with a maximum of 20.0 m
- Maximum crest height, with a maximum of 13.9 m
- Bandwidth $\varepsilon = \sqrt{1-\left(\frac{N_{DZC}}{N_{MAXL}}\right)}$ with a mean of about 0.7

where N_{DZC} is the number of zero down-crossings, and N_{MAXL} is the number of local maxima. These data were grouped by class intervals of $H_{1/3}$ and T_z, with bandwidths of 0.5 m in $H_{1/3}$ and 1 s in T_z (and sometimes wider when there were insufficient data available). The $H_{max}/H_{1/3}$ and $C_{max}/H_{1/3}$ ratio distributions in each class interval were then fit to Gumbel distributions using the nonlinear least squares method. These fittings demonstrated that the Gumbel distribution is suitable for predicting maxima distributions, in agreement with the theory put forward earlier, and as indicated in the examples presented in Figure 1.

Parameters a_N and b_N of the fitted distributions were closely dependent on the number of waves occurring during the measurement sequence, and therefore on the mean period T_z. For comparison of these parameters on the same basis, regardless of the real wave number N in the sequence, the distribution mode was also calculated for N = 180 waves, similar to the method described by Robin & Olagnon (1991), extrapolated from the fitted Gumbel distribution.

These results are presented in Figures 2 & 3, as a function of the steepness parameter s_z, defined as:

$$s_z = \frac{k_z}{2\pi}H_{1/3}, \tag{12}$$

where k_z is the mean wave number for a given sea state, calculated on the basis of the mean period T_z using the dispersion relation.

These results indicated the following trends:

- a slight increase in the $H_{max}/H_{1/3}$ mode with steepness and a relatively small scatter, with higher values than predicted by the fitted Rayleigh and Weibull models described earlier (this is undoubtedly due to the use of H_S instead of $H_{1/3}$, and H_S higher than $H_{1/3}$ in a nonlinear sea state)
- high scatter in the scale parameter, but with some overall stability
- a sharp increase in the $C_{max}/H_{1/3}$ mode with steepness and a relatively small scatter, with much higher values than predicted by the Rayleigh model ($\theta = 0.354$), and also higher than those computed by a Haring distribution ($H_s/d = 0.1$)
- clear dependency with steepness in the scale parameter, with high scatter.

The analysis highlighted the effect of steepness on maximum crest height distributions. This effect, due to nonlinearities in the wave kinematics, is taken into account in the Rayleigh-Stokes and Weibull-Stokes models described hereafter.

RAYLEIGH-STOKES MODEL

A model for the distribution of crest heights, using the regular-wave Stokes expansion to account for nonlinear effects, was proposed by Tayfun (1983) at second order and Vinje (1989) at third order. In this model, the sea surface process is considered to be a distortion of a first-order narrow-banded Gaussian process (with wave amplitudes modelled on a Rayleigh distribution), due to nonlinear effects which can be simulated with a second- or third-order Stokes expansion.

The first-order process is expressed in the following form:

$$k_m\eta_0(t) = \varepsilon(t)\cos(\theta(t)) \tag{13}$$

where:

$\eta_0(t)$ is the narrow-banded Gaussian process

k_m is a mean narrow-banded wave number

$\varepsilon(t)$ and $\theta(t)$ are, respectively, the envelope and phase of the process $[k_m \cdot \eta_0(t)]$, which are assumed to vary slowly

At infinite depth, the wave process can be expressed as follows, for a third-order Stokes expansion, validated on the assumption of a small $k_m \cdot \sigma_0$ parameter, where σ_0 is the root mean square of $\eta_0(t)$:

$$k_m\eta(t) = \varepsilon(t)\cos(\theta(t)) + b_2\varepsilon(t)^2\cos(2\theta(t)) + b_3\varepsilon(t)^3\cos(3\theta(t)) \tag{14}$$

where b_2 and b_3 are the parameters of the third-order Stokes expansion.

At infinite depth, $b_2 = 1/2$ and $b_3 = 3/8$. For the Frigg site situation, at 100 m depth, b_2 and b_3 were close to these levels. Model (17) could be reconsidered for shallower waters. If we consider as in (Vinje, 1989) the following relations between crest and wave heights (C and H) and wave amplitude (a) of the first order process:

$$H = 2a(1 + b_3(k_m a)^2) \tag{15}$$

$$C = a(1 + b_2(k_m a) + b_3(k_m a)^2) \tag{16}$$

The mode and scale parameters of the Gumbel law of the non-normalized maxima (Eq. 9) can be deduced (see Appendix 2) in a direct way, for maximum wave heights:

$$a_{HN} = 2a_N(1 + b_3(k_m a_N)^2)$$
$$b_{HN} = 2b_N(1 + 3b_3(k_m a_N)^2) \tag{17}$$

for maximum crest heights:

$$a_{CN} = a_N(1 + b_2(k_m a_N) + b_3(k_m a_N)^2)$$
$$b_{CN} = b_N(1 + 2b_2(k_m a_N) + 3b_3(k_m a_N)^2) \tag{18}$$

where a_{HN}, b_{HN}, a_{CN} and b_{CN} are, respectively, mode and scale parameters of a Gumbel distribution for wave and crest height maxima. a_N and b_N are parameters of a maxima distribution of $\eta_0(t)$, here considered as Rayleigh,

$$a_N = 4m_0^{1/2}(\mathrm{Log}(N)/8)^{1/2}$$
$$b_N = \frac{a_N}{2\mathrm{Log}(N)} \tag{19}$$

In the equations (17) and (18), the nonlinear steepness effects are represented by ka_N terms. These are clearly first-order effects for crests and solely second-order effects for wave heights, which is in agreement with the results of the analysis of measurements obtained at the Frigg site.

CORRECTED RAYLEIGH-STOKES MODEL

The third order Rayleigh-Stokes model which has been explained above is based on two fundamental assumptions. First, the model of the process of the free surface elevation is considered as an asymptotic narrow band model, secondly, the kinematics of the waves is unidirectional (in fact, the directional spreading complicates the second and third order terms) and we know that the non-linearity is strongly affected by this spreading (Longuet-Higgins, 1963).

To take into account, in a simple way, the error induced by the above hypotheses two coefficients of correction are introduced. A first one modifies the energy parameter m_0, and a second one corrects the mean wave number, or equivalently, the mean wave steepness.

So the new models for the mode and scale parameters of the wave and crest heights are the same as Eqs 17, 18 & 19, but with a_N replaced by $\alpha_a a_N$, and k_m replaced by $\alpha_k k_m$. Due to the complex effects of the spectral bandwidth and directional spreading on the law of the maximum heights, the coefficients α_a and α_k are not supposed to be the same for the mode parameter and the scale parameter.

The validity and the quality of this corrected model is based on the fact that on the site we consider that, for the severe sea states, there is a strong correlation between the energy (or H_S) the mean wave number (or T_z) the spectral bandwidth and the directional spreading. In this case, it is obvious that the correction coefficients α_a and $\alpha_k$are site dependent.

ANALYSIS OF THE FRIGG SITE WITH THE RAYLEIGH-STOKES MODEL

In a standard (non corrected) Rayleigh-Stokes model, there may be some problems in the choice of characteristic parameters of sea states to be taken into account when applying the above-described Rayleigh-Stokes model:

- the wave number, k_m, which can be calculated from different period parameters, i.e. mean period, peak period or significant period, but its value is dependent on the spectral shape.
- a first-order energy process, m_0, but the exact relation with the wave height $H_{1/3}$ is unknown.

These parameters were taken into account, in the present study, as follows:

- the wave number, k_m, was calculated from the period T_z, using the dispersion relation. It is called k_z in the sequel of the paper.
- the relation between the wave height $H_{1/3}$ and the first-order energy process, m_0, was represented using a relation derived from Stokes expansion:

$$H_{1/3} = 4m_0^{1/2}\left(1 + b_3\left(k\frac{H_{1/3}}{2}\right)^2\right) \qquad (20)$$

The results obtained with the Rayleigh and Rayleigh-Stokes model and their associated corrected version are presented in Figures 5-12. All the results in the sequel of the paper correspond to parameters calculated from the actual number of waves of the records. Two types of representation are used to compare the results of the models and the data. First, the $H_{max}/H_{1/3}$ mode and scale parameters (respectively $C_{max}/H_{1/3}$ mode and scale parameters) from the models and from the data are plotted in ascending T_z and $H_{1/3}$ values in Figures 5,7,9,11 (the way the T_z and $H_{1/3}$ values are examined is given in Figure 4.). Secondly, ratios between $H_{max}/H_{1/3}$ modes and scale parameters (respectively $C_{max}/H_{1/3}$ modes and scale parameters) derived from the on-site measurements and as computed by the models, and this as a function of the mean steepness (Fig. 6,8,10,12).

For the corrected Rayleigh (resp. Rayleigh-Stokes) models the coefficients of correction α_a (resp. α_a and α_k) are tuned so as to minimize the R.M.S. of the error between the law parameters given from the model and from the data. The coefficients of correction fitted to the data are given in Table 3. The bias and standard deviation of the error between the models and the data are given in Table 4.

These results lead to some main remarks.

Wave Height.

In Figure 5 to 8 we have only plotted the results corresponding to the Rayleigh and the fitted corrected Rayleigh laws. In fact, as we can verify in Tables 3 & 4 there is no significant difference between the fitted corrected Rayleigh and the fitted corrected Rayleigh-Stokes models although the slight increasing tendency with the high steepnesses for the mode is corrected with the introduction of the third order term in the Rayleigh-Stokes model. The fitting is good but produces different coefficients α_a and α_k for the mode and the scale parameters.

Wave Crest.

As it has been previously observed, the effect of non-linearities strongly needs to be taken into account to model the crest maxima. The use of a Rayleigh-Stokes model is here necessary. The fitted corrected Rayleigh model is unable to follow the variations of the Gumbel parameters with the mean steepness (see Fig. 9-12(top)).

In using the standard (non corrected) Rayleigh-Stokes model as above-described, (Fig. 9-12(middle)), the variation with steepness is corrected in a great part, for the mode and the scale parameters. But a slight bias remains.

The fitting of the corrected Rayleigh-Stokes model (Fig. 9-12(bottom)) eliminates this bias, but with different coefficient of correction for the mode parameter and the scale parameter. For the mode parameter there is no correction of the energy term (see Table 3), but a large correction of the steepness term (α_k=0.7). In fact, this correction corresponds to a correction on the mean wave number which leads to consider a value of mean period higher than T_z. and so closer to the T_p (and in fact equal to T_p for a Jonswap spectrum with γ=7).

3 . Fitted correction coefficients

	Mode $H_{max}/H_{1/3}$		Scale $H_{max}/H_{1/3}$		Mode $C_{max}/H_{1/3}$		Scale $C_{max}/H_{1/3}$	
Models	α_a	α_k	α_a	α_k	α_a	α_k	α_a	α_k
Rayleigh	0.96		0.81		1.12		1.12	
Rayleigh Stokes	0.96	0.66	0.79	0.82	1.03	0.70	0.89	1.12

4 . Bias and Standard deviation of the based-Rayleigh laws

	Mode $H_{max}/H_{1/3}$		Scale $H_{max}/H_{1/3}$		Mode $C_{max}/H_{1/3}$		Scale $C_{max}/H_{1/3}$	
Models	**Bias**	**St. Dev.**	**Bias**	**St. Dev.**	**Bias**	**St. Dev.**	**Bias**	**St. Dev.**
Rayleigh	.059	.016	.030	.013	-.089	.032	-.009	.012
Fitted Rayleigh	-.005	.017	.000	.012	.005	.032	.000	.012
Rayleigh Stokes	.076	.015	.040	.013	.002	.018	.012	.009
Fitted Rayleigh Stokes	.008	.015	.000	.012	.005	.015	.001	.009

CORRECTED WEIBULL-STOKES MODEL

Wave Height.

The fact that the correction coefficients are different for the mode and the scale parameters is a problem because it does not permit to associate to the height or crest a new parent distribution. In fact, as it has been already observed by several authors (Forristall, 1978), (Krogstad, 1985) among others, a Weibull law fits better to the distribution of height than a Rayleigh law. In an other part we verify easily than the mode and scale parameters of the law of the maxima associated with a Weibull law are given by two corrected Rayleigh laws (see the Appendix 1 at the end of this paper). In applying these relations to the results obtained previously we obtain:

$$\beta = 2*0.96/0.81 = 2.37$$
$$\theta_W = 0.766 \text{ with } T_z = 10s \qquad (21)$$
$$\theta_W = 0.775 \text{ with } T_z = 5s$$

So, we find with a different method a value of the shape parameter very close to the one found by Krogstad (1985)(see Eq. 4), but a scale parameter slightly higher.

The Weibull model ($\theta = 0.77$, $\beta = 2.37$) equivalent to the two corrected Rayleigh models as we have seen in Figures 5-8(bottom) gives a good prediction of the maxima on all the range of T_z, $H_{1/3}$ and steepness.

The use of the Stokes expansion permits to improve a little the model for high steepness. The fitting of the corrected Weibull-Stokes model gives

$$\theta_W = 0.77, \beta = 2.38, \alpha_a=1, \alpha_k=0.7. \quad (22)$$

The comparisons between model and data for these parameters are given in Figures 13 and 14. We observe from the values of bias and standard deviation given in Table 5 that the fitting is as good as the fitting of the two independent Rayleigh-Stokes models (Table 4). The relative standard deviation is about 1% for the mode and 10% for the scale with low bias.

Wave Crest.

Although it is often considered that the Rayleigh law fits well to the crest distribution, we followed the same methodology as adopted for the wave height. So, we tried to know if a corrected Weibull-Stokes model permitted to fit the correction coefficients in the same way for the mode and the scale parameters. The tuning of the different coefficients shows here again that the unavailability to fit a unique Rayleigh-Stokes model can be remove in using a Weibull-Stokes. The optimized coefficients are:

$$\theta_W = 0.78, \beta = 2.18, \alpha_a=1 \text{ and } \alpha_k=0.7 \quad (23)$$

The comparisons between model and data for these parameters are given in Figures 15 and 16 and show a very good agreement. We observe here again from the values of bias and standard deviation given in Table 5 that the fitting is as good as the fitting of the two independently corrected Rayleigh-Stokes models (Table 4). The relative standard deviation is about 1% for the mode and 10% for the scale with very low bias.

5 . Bias and Standard deviation of the based-Weibull laws

	Mode $H_{max}/H_{1/3}$		Scale $H_{max}/H_{1/3}$		Mode $C_{max}/H_{1/3}$		Scale $C_{max}/H_{1/3}$	
Model	**Bias**	**St. Dev.**	**Bias**	**St. Dev.**	**Bias**	**St. Dev.**	**Bias**	**St. Dev.**
Weibull Stokes	.001	.017	.002	.012	.000	.014	.000	.009

CONCLUSIONS

The measured data from the Frigg site show that the parameters of a fitted asymptotic Gumbel law of maxima is dependent of the steepness of the sea state. This dependency is observed very strong for the crest heights but slight for the wave heights. This effect of the non-linearity of the wave kinematics in infinite depth has not been taken into account in the models proposed in the past and explains some discrepancies between the different fitted models.

Based on a simple narrow-band model of the free surface elevation of a unidirectional sea state expanded to the third order, new formulations of the parameters of the asymptotic Gumbel law may be derived which depend of a mean wave number (or mean wave steepness).

The use of a so-called Rayleigh-Stokes model does not permit to adjust in a same way the mode and the scale parameters of the Gumbel law which have been estimated from the data. So, by extension, a Weibull-Stokes model is proposed. A coefficient of correction α_k is introduced on the mean wave number arbitrary taken as k_z (T_z transformed by the dispersion relation) in the initial Weibull-Stokes model. This coefficient of correction takes into account in a globalized way the characteristics of directional spreading and spectral shape specific to the site. This corrected Weibull-Stokes model fits very well the data base of the Frigg site in spite of a great dispersion of the scale parameter (the scatter of the scale parameter could be mainly related to prediction uncertainties for this parameter).

The proposed Corrected Weibull-Stokes (CWS) model furnishes a very simple unified asymptotic parametrized Gumbel law for the non-normalized maxima of the half wave height and the crest height:

$$G(x) = \text{Prob}(X_N \le x) = \exp\left(-\exp\left(-\frac{x - a_{XN}}{b_{XN}}\right)\right) \quad (24)$$

with the mode and scale parameters given by:

$$a_{XN} = a_N(1 + b_2(k_m a_N) + b_3(k_m a_N)^2)$$
$$b_{XN} = b_N(1 + 2b_2(k_m a_N) + 3b_3(k_m a_N)^2) \quad (25)$$

and,

$$a_N = \theta_W \frac{H_{1/3}}{2} \text{Log}(N)^{1/\beta}$$
$$b_N = \frac{\theta_W}{\beta \text{Log}(N)^{1-(1/\beta)}} \frac{H_{1/3}}{2} = \frac{a_N}{\beta \text{Log}(N)} \quad (26)$$

In infinite depth,

$$b_2 = 0 \text{ (half wave height)}, b_2 = 1/2 \text{ (crest height)}, b_3 = 3/8 \quad (27)$$

and for the Frigg site in the North Sea,

$$\theta_W = 0.78, \ k_m = 0.7k_z$$
$$\beta = 2.38 \text{ (half wave height)}, \ \beta = 2.18 \text{ (crest height)} \quad (28)$$

with k_z given by T_z transformed by the dispersion relation.

More complicated models could be developed which would take explicitly into account the directional spreading and the spectral shape, but this CWS model shows already a very good agreement with the measurements compared to other models proposed in the past. In analysing other sites in deep water, it will be interesting to determine which parameters are actually site-dependent (only α_k, or others).

The effect of water depth, which has not been considered here, due to the relatively great depth, could be introduced in an adapted narrow band model, and the same methodology used to validate the CWS model in shallow waters.

ACKNOWLEDGEMENTS

The present study was performed within the SCHOS (Surges and Wave Characteristics by Satellite Observation) project of the Clarom, co-funded by the French Fonds de Soutien aux Hydrocarbures. The participants and contributors were Ifremer, IMG/LEGI, Météomer, Optimer, STNMTE, Elf and EPSHOM. Their permission to publish is gratefully acknowledged here.

REFERENCES

Det Norske Veritas (1991). "Classification Notes n 30.5 - Environmental conditions and environmental loads", March 1991.

Forristall, G.Z. (1978). "On the statistical distribution of wave heights in a storm", *J. Geophys. Res.*, vol. 83, pp. 2353-2358.

Haring, R.E. & al (1976). "Extreme wave parameters based on continental shelf storm wave records", *Proc. of the 15th Coastal Eng. Conf.*, Honolulu, July 11-17.

Haring, R.E. & Heideman J.C. (1978). "Gulf of Mexico rare wave return periods", *Proc. OTC*, n°3230, Houston, May 1978.

Janhs, H.O. & Wheeler, J.D. (1972). "Long term wave probabilities based on hindcasting of severe storms", *Proc. OTC*, n°1590,

Houston, May 1972.

Krogstad, H.E. (1985). "Height and period distribution of extreme waves", *Applied Ocean Research*, vol. 7, no. 3, pp 158-165.

Longuet-Higgins, M.S. (1952). "On the statistical distribution of the heights of sea waves", *J. Marine Res.*, 11, pp. 245- 266.

Longuet-Higgins, M.S. (1963). "The effect of non-linearities on statistical distributions in the theory of sea waves", *J. Fluid Mech.*, vol. 17, pp. 459-480.

Robin, A., Olagnon, M. (1991). "Occurence of extreme waves with respect to significant wave height", *Proc. of the 10th Int. Conf. on Offshore Mechanics and Artic Engineering*, vol. 2, pp. 1-11.

Tayfun, M.A. (1983), "Nonlinear effects on the distribution of crest-to-trough wave heights", *Ocean Eng.*, vol. 10, no. 2, pp. 97-106.

Vinje T. (1989). "The statistical distribution of wave heights in a random seaway", *Applied Ocean Res.*, vol. 11, no. 3.

APPENDIX 1

We calculate hereunder the relations between the parameters of a Weibull law and the parameters of two Rayleigh laws corrected differently which give the same mode and scale parameters of the asymptotic Gumbel law of maxima.

We use the formulae given in Eq. 11 and we equal the mode and scale parameters calculated with the different models. It gives:

$$a_N = \alpha_a\theta_R \mathrm{Log}(N)^{1/2} = \theta_W \mathrm{Log}(N)^{1/\beta}$$
$$b_N = \alpha_b\frac{\theta_R}{2\mathrm{Log}(N)^{1/2}} = \frac{\theta_W}{\beta \mathrm{Log}(N)^{1-(1/\beta)}} \qquad (29)$$

with α_a and α_b the two coefficients of correction of the same Rayleigh law of parameter θ_R, θ_W and β the scale and shape parameter of the Weibull law, N the number of waves.

Eqs. 29 gives the following relations:

$$\beta = 2\frac{\alpha_a}{\alpha_b}$$
$$\theta_W = \theta_R\frac{\alpha_a}{\mathrm{Log}(N)^{\frac{1}{\beta}-\frac{1}{2}}} \qquad (30)$$

We can observe that the relation between the two scale parameters is dependent of the number of waves.

APPENDIX 2

Based on the extreme value theory, when N is high enough and the distribution of a random variable A resembles Weibull-distributions, the Gumbel distribution is the asymptotic law for maxima:

$$\mathrm{Prob}(Y \le y) = \exp\left(-\exp\left(-\frac{y-a_N}{b_N}\right)\right) \qquad (31)$$

where:

a_N is the distribution mode and b_N is the scale parameter which are related to the parent distribution parameters as follows:

$$Q_A(a_{AN}) = \frac{1}{N}$$
$$b_{AN} = \frac{1}{Nf_A(a_{AN})} \qquad (32)$$

where:

$Q(x) = 1 - F(x)$ is the exceedance probability

$f(x)$ the density function.

If we consider a new random variable T which is related to A by a strictly monotonic function Θ:

$$T = \Theta(A) \qquad (33)$$

The parameters of the corresponding Gumbel asymptotic law obey to the same relations as Eq. 32.

$$Q_T(a_{TN}) = \frac{1}{N}$$
$$b_{TN} = \frac{1}{Nf_T(a_{TN})} \qquad (34)$$

then, as the two probability laws are related by:

$$Q_T(t) = Q_A(\Theta^{-1}(t))$$
$$f_T(t) = \frac{f_A(\Theta^{-1}(t))}{\Theta'(\Theta^{-1}(t))} \qquad (35)$$

then, using Eq. 32 and 34,

$$Q_T(a_{TN}) = Q_A(\Theta^{-1}(a_{TN})) = Q_A(a_{AN}) \Rightarrow$$
$$\Theta^{-1}(a_{TN}) = a_{AN} \Rightarrow \qquad (36)$$
$$a_{TN} = \Theta(a_{AN})$$

and,

$$b_{TN} = \frac{1}{N}\frac{\Theta'(\Theta^{-1}(a_{TN}))}{f_A(\Theta^{-1}(a_{TN}))} = \frac{1}{N}\frac{\Theta'(a_{AN})}{f_A(a_{AN})} = b_{AN}\Theta'(a_{AN}) \qquad (37)$$

which give the final relations,

$$a_{TN} = \Theta(a_{AN})$$
$$b_{TN} = b_{AN}\Theta'(a_{AN}) \qquad (38)$$

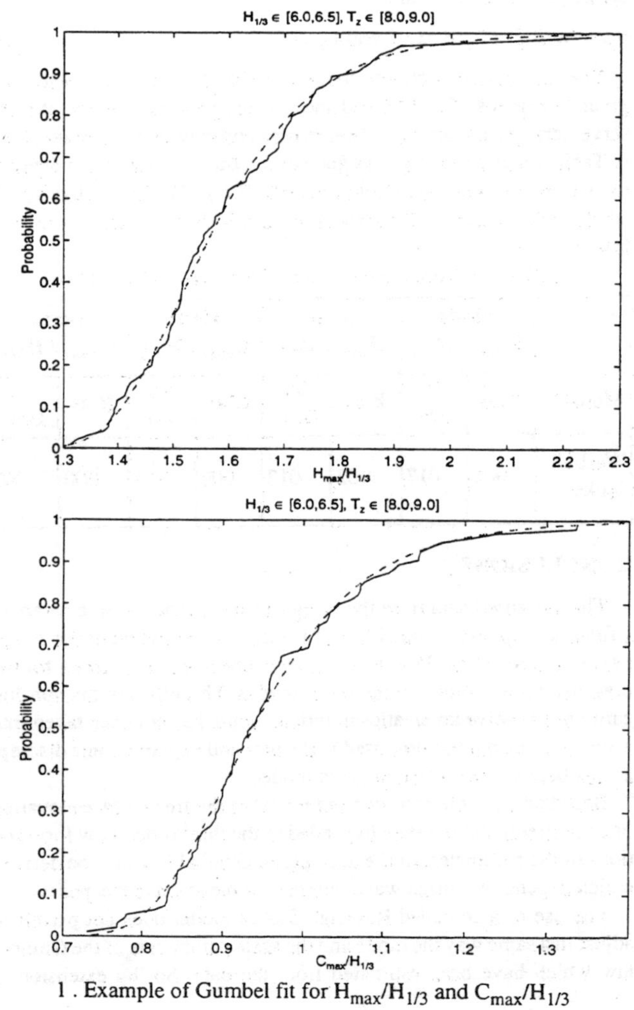

1 . Example of Gumbel fit for $H_{max}/H_{1/3}$ and $C_{max}/H_{1/3}$

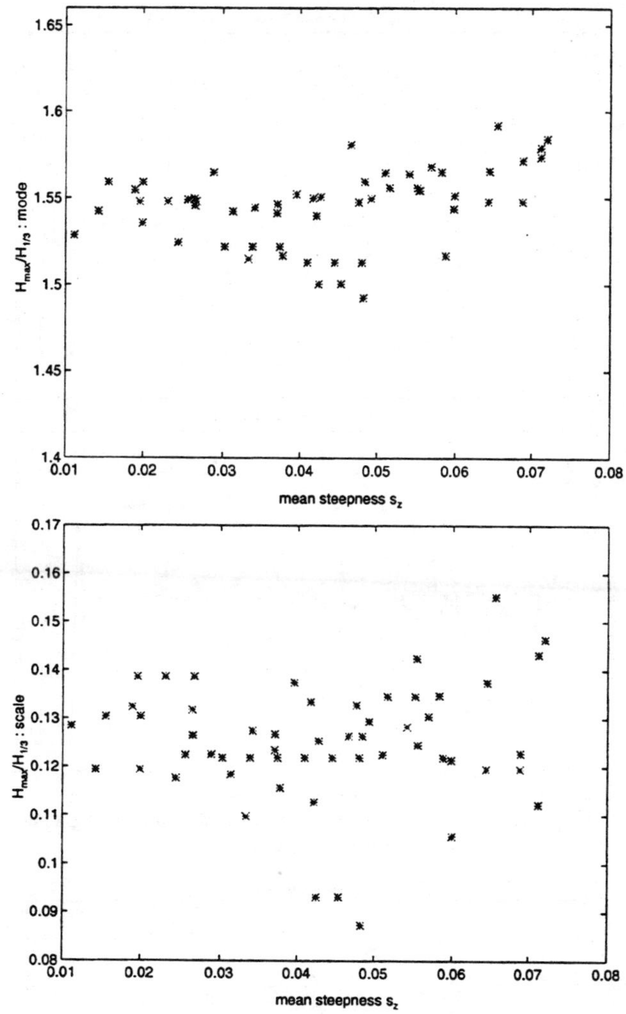

2 . $H_{max}/H_{1/3}$: Data mode (top) and scale (bottom) vs steepness

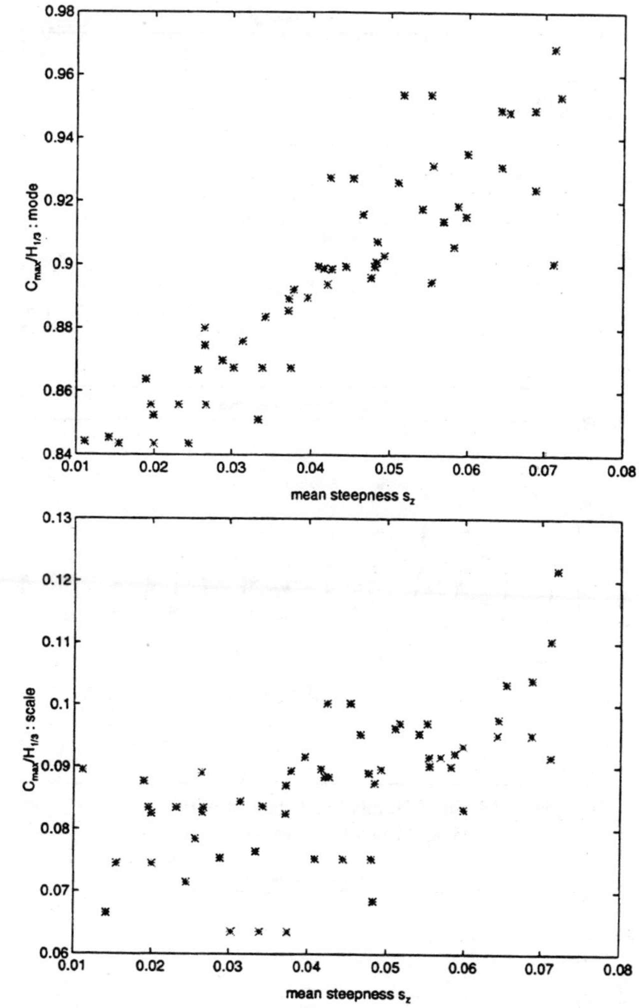

3 . $C_{max}/H_{1/3}$: Data mode (top) and scale (bottom) vs steepness

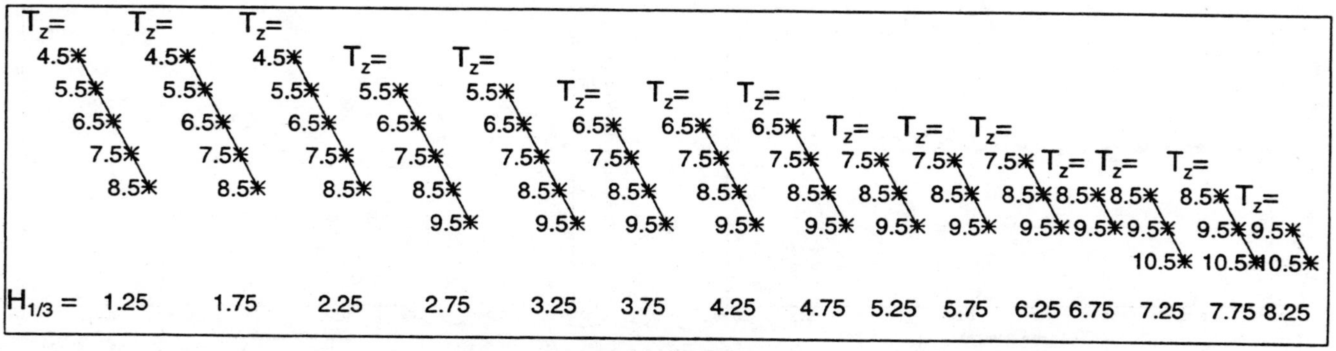

4 . Classes of T_z and $H_{1/3}$

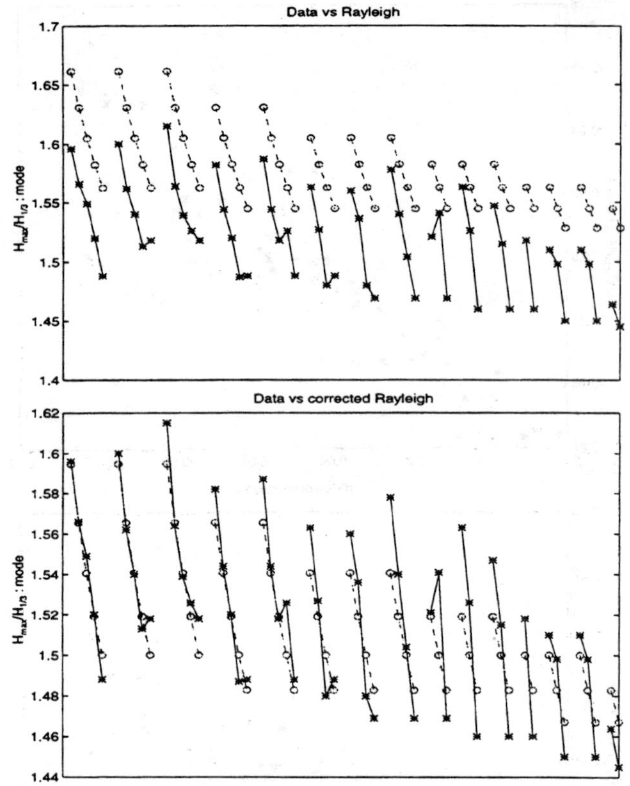

5 . $H_{max}/H_{1/3}$: Mode of Gumbel distribution with increasing T_z and $H_{1/3}$. Model (- o -) vs data (-*-)

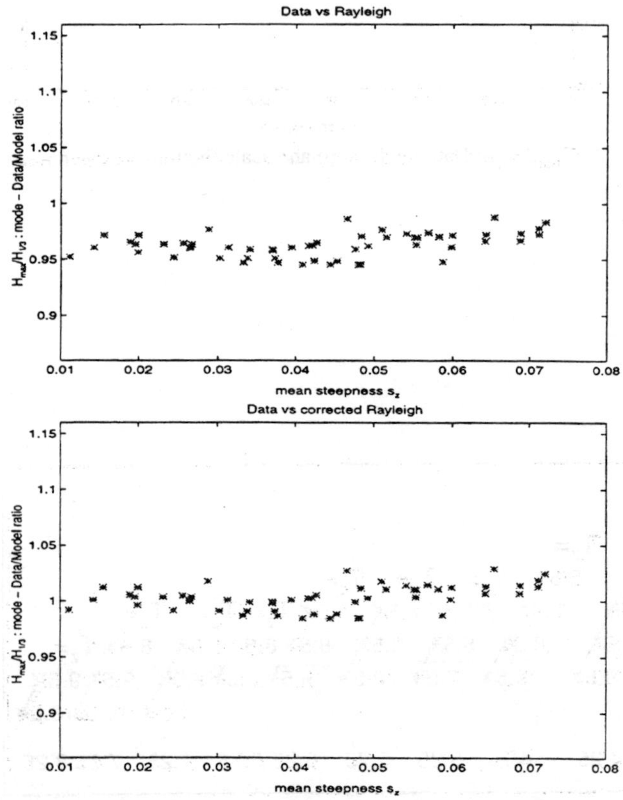

6 . $H_{max}/H_{1/3}$: Model-Data mode ratio of Gumbel distribution vs steepness

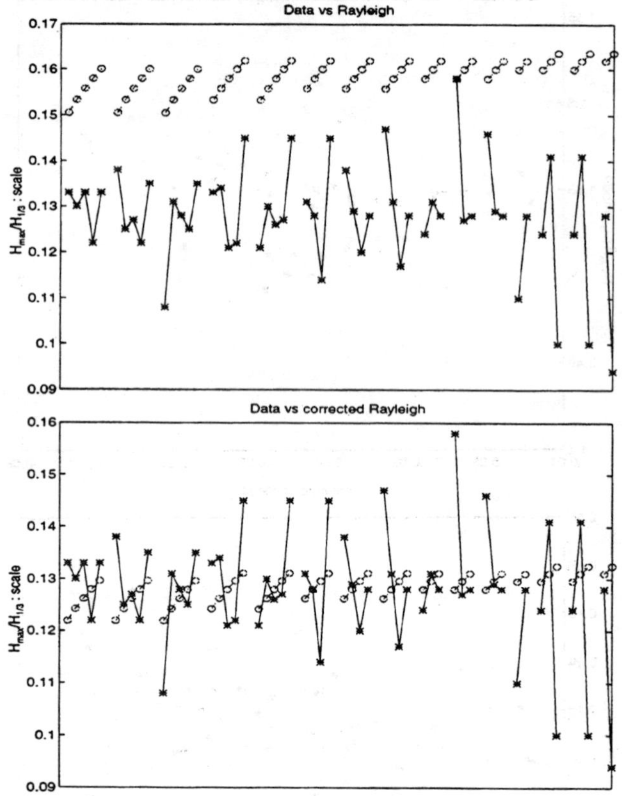

7 . $H_{max}/H_{1/3}$: Scale of Gumbel distribution with increasing T_z and $H_{1/3}$. Model (- o -) vs data (-*-)

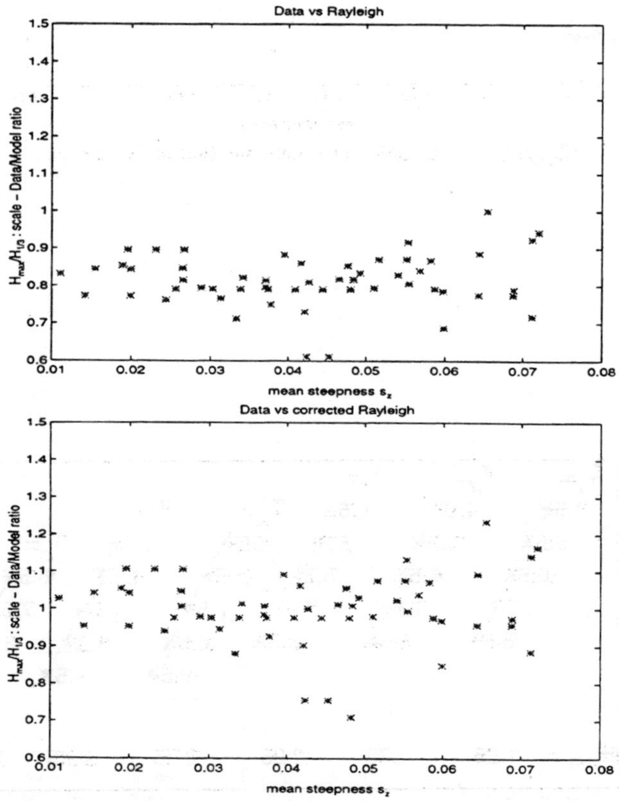

8 . $H_{max}/H_{1/3}$: Model-Data scale ratio of Gumbel distribution vs steepness

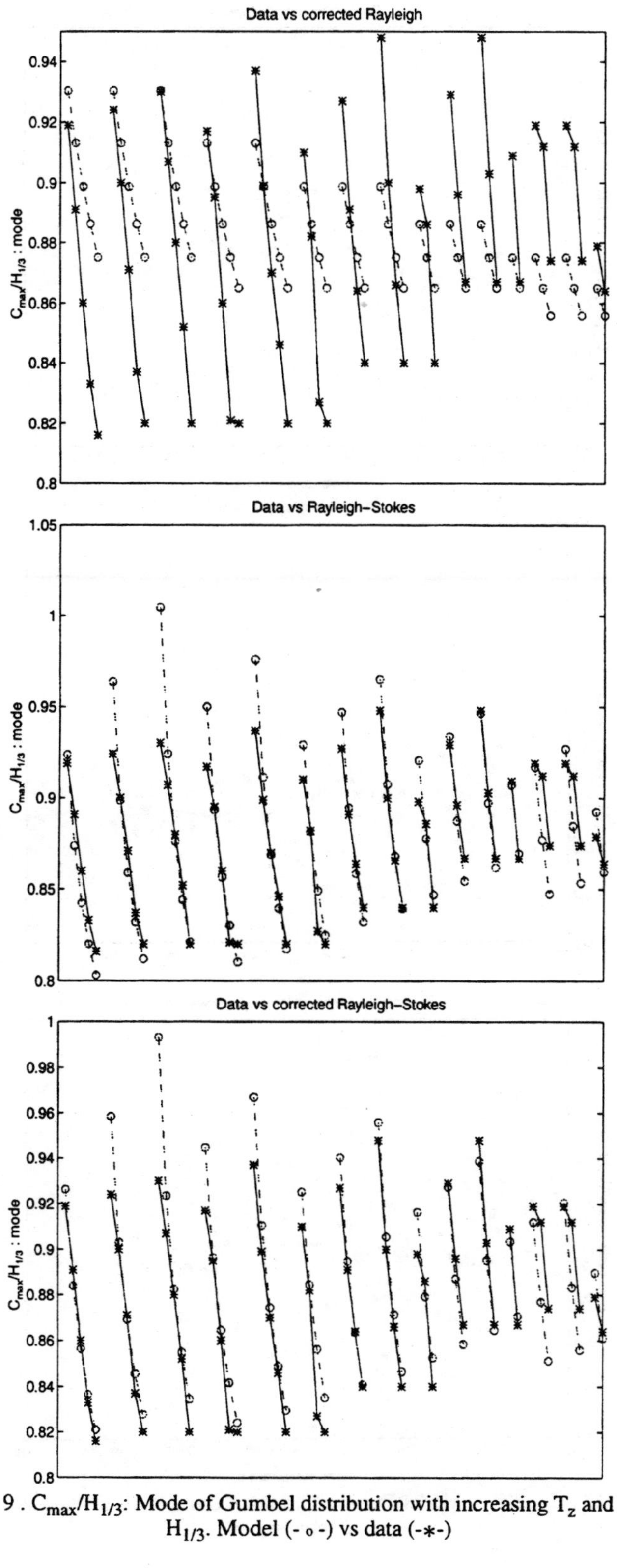

9 . $C_{max}/H_{1/3}$: Mode of Gumbel distribution with increasing T_z and $H_{1/3}$. Model (- o -) vs data (-∗-)

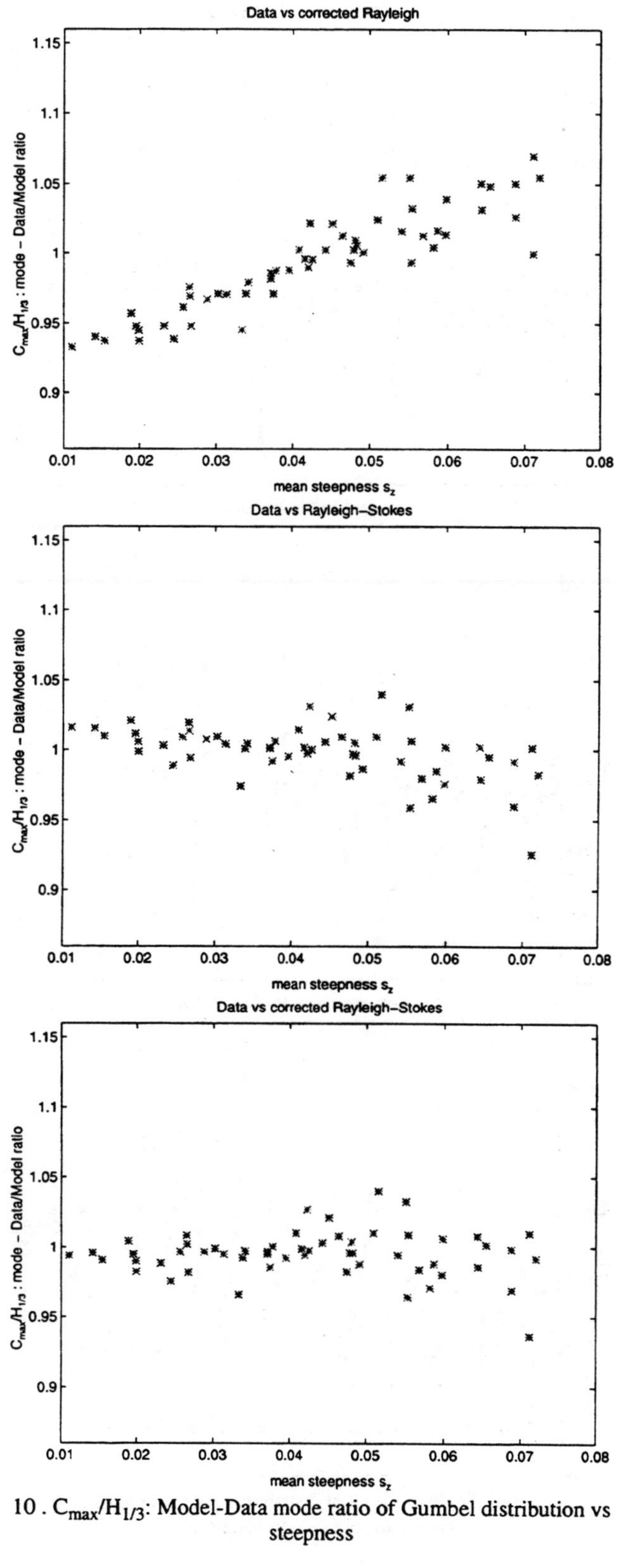

10 . $C_{max}/H_{1/3}$: Model-Data mode ratio of Gumbel distribution vs steepness

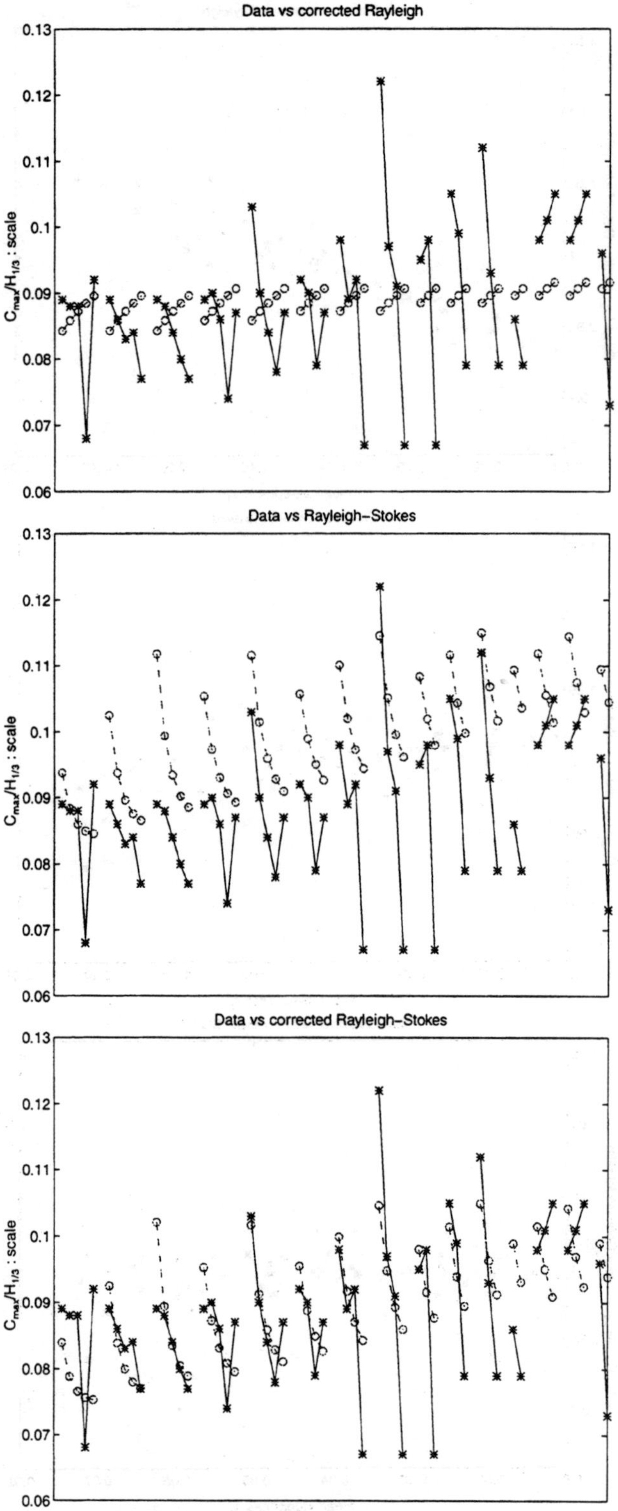

11 . $C_{max}/H_{1/3}$: Scale of Gumbel distribution with increasing T_z and $H_{1/3}$. Model (- ∘ -) vs data (-∗-)

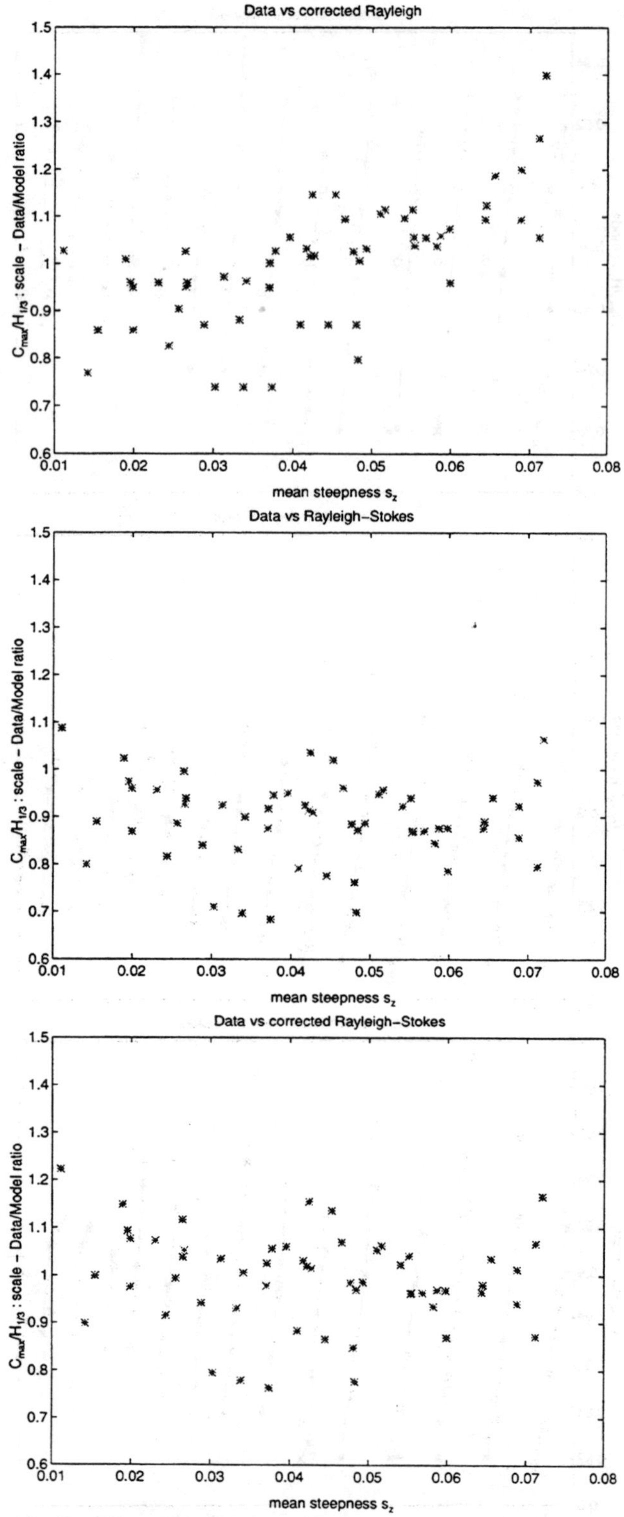

12 . $C_{max}/H_{1/3}$: Model-Data scale ratio of Gumbel distribution vs steepness

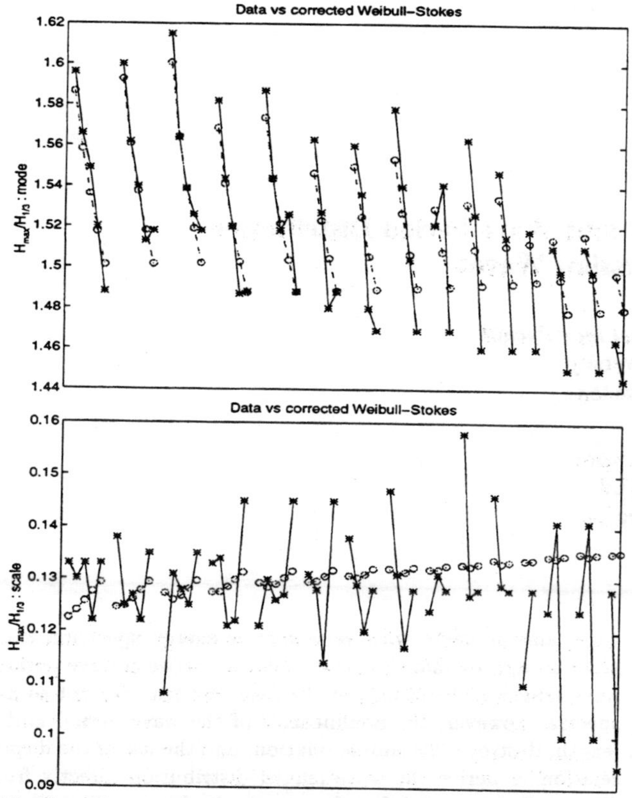

13 . $H_{max}/H_{1/3}$: Mode (top) and scale (bottom) of the Gumbel distribution with increasing T_z and $H_{1/3}$. Model (- o -) VS DATA (-*-)

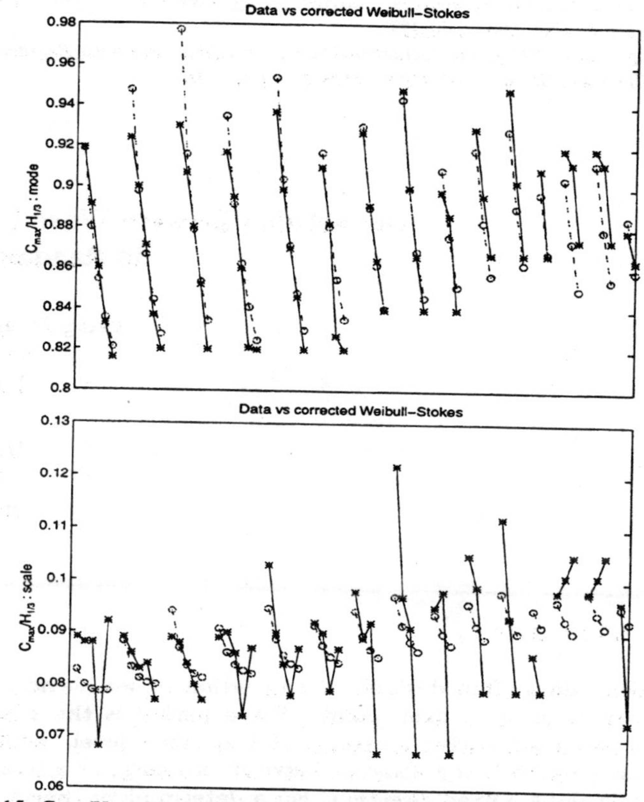

15 . $C_{max}/H_{1/3}$: Mode (top) and scale (bottom) of the Gumbel distribution with increasing T_z and $H_{1/3}$. Model (- o -) vs data (-*-)

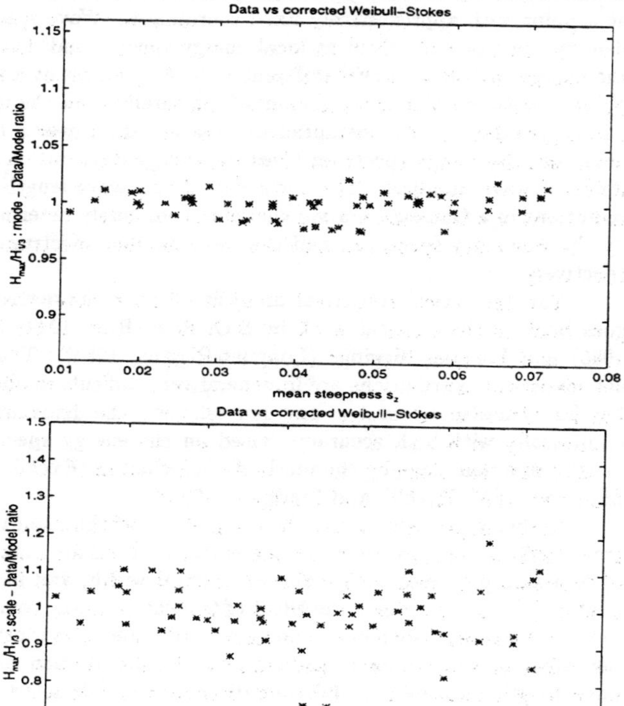

14 . $H_{max}/H_{1/3}$: Model-Data scale ratio of Gumbel distribution vs steepness (mode - top, scale - bottom)

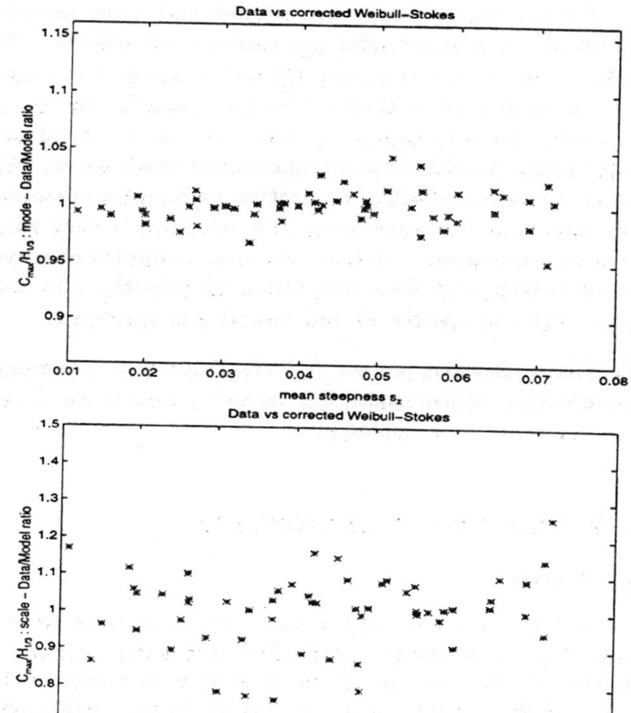

16 . $C_{max}/H_{1/3}$: Model-Data scale ratio of Gumbel distribution vs steepness (mode - top, scale - bottom)

Proceedings of the Seventh (1997) International Offshore and Polar Engineering Conference
Honolulu, USA, May 25-30, 1997

ISBN 1-880653-28-1 (Set); ISBN 1-880653-31-1 (Vol. III)

The Relation Between Wave Length and Wave Period Distributions in Random Gaussian Waves

Georg Lindgren and Igor Rychlik
Lund University
Lund, Sweden

Marc Prevosto
IFREMER
Brest, France

ABSTRACT

Wave data often describe the variation of sea elevation with time, at a fixed point. Wave period is the time between successive crossings of the mean level, while wave length is the distance between crossings at a fixed time along a fixed direction. For a deterministic, harmonic wave, the dispersion relation describes the relation between wave period and wave length.

For random waves, wave period and wave length, as well as wave amplitude, are random quantities. We study these distributions for Gaussian waves and show that the dispersion relation does not describe the relation except for very narrow spectra, and its uncritical use leads to considerable underestimation of short waves. We also consider the effect of truncation of high frequencies, and show that the wave length distribution is very sensitive to truncation. Further, we give examples of wave period/length and wave amplitude distributions in sea states with unidirectional and directional spectrum.

KEY WORDS: Gaussian process, JONSWAP spectrum, regression approximation, Slepian model, spectral width, crest width, directional spectrum, wave amplitude.

INSTRODUCTION AND NOTATION

Introduction

In a random sea, wave characteristics such as wave period and wave length are random quantities. Their distributions can be interpreted as long run empirical distributions when observing the sea at a single point over a long time interval, for wave period, and over a long distance at a single instant of time, for wave length. Thus, there is no immediate relation between the two distributions, but they can both be calculated from the energy spectrum of the waves. The dispersion relation relates wave length to wave period for each elementary wave that build up the total wave package, and in waves with very narrow energy spectrum this also gives an approximately correct relation between wave period and wave length distributions in the resulting sea. For not so narrow spectra, however, the nonlinearity of the wave period and wave length, destroys this simple relation, and the use of the dispersion relation to derive the wave length distribution directly from an observed wave period distribution can lead to considerable errors.

Most existing wave data statistics are based on measurements of the sea elevation at a fixed point on the sea, or at least at a point with approximately fixed coordinates. Wave spectral density functions are given as local energy spectra and describe the energy distribution over different wave frequencies at a fixed point. Altitude radar measurements from satellite, on the other hand, give data on the instantaneous sea elevation over a large area, and the energy spectrum gives the energy distribution over different wave numbers. The wave period and wave length distributions in a Gaussian sea are therefore completely determined by the frequency spectrum, and the wave number spectrum, respectively.

The theoretical statistical analysis of wave characteristics goes back to the classical work by S.O. Rice (Rice, 1944; Rice, 1945) and Longuet-Higgings (Longuet-Higgins, 1956). The exact statistical distributions are in general very difficult to obtain, but for Gaussian sea states, one can calculate the distributions numerically with high accuracy, based on the energy spectrum density function alone by the methods described in (Rychlik and Lindgren, 1993; Rychlik and Lindgren, 1995).

In this paper we shall use the methods of (Rychlik and Lindgren, 1993) to compare the two types of distributions for a number of Gaussian processes with different spectral width, and in particular we shall investigate the effect of truncation in the common Pierson-Moskowitz frequency spectrum. We shall also illustrate the effect of a directional spectrum on the distribution of the wave length measured in different directions in relation to the main wave direction.

The joint distribution of wave period/length and amplitude can also be found by the methods in (Rychlik and Lindgren, 1993). We give examples of such distributions for unidirectional and directional waves.

Basic notation

We consider first a one-dimensional wave profile along a certain direction, and denote the wave process by $\xi(t,x)$, where t is time and x is the distance from the origin in the chosen direction. We assume the mean level to be zero and define the following wave characteristic variables:

$$T = \text{half zero crossing wave period} \quad (1)$$
$$= \text{time from zero down- to upcrossing,}$$
$$T^{mM} = \text{min-Max wave period} \quad (2)$$
$$= \text{time from local wave minimum to maximum,}$$
$$L = \text{half zero crossing wave length} \quad (3)$$
$$= \text{distance from zero down- to upcrossing,}$$
$$L^{mM} = \text{min-Max wave length} \quad (4)$$
$$= \text{distance from local wave minimum to maximum,}$$
$$H^{mM} = \text{min-Max wave amplitude} \quad (5)$$
$$= \text{height difference between minimum and maximum.}$$

Here, the definition of H^{mM} applies both to the time and the space definitions.

The statistical distribution of these wave quantities shall be interpreted in the ergodic sense, e.g. the distribution of T is equal to the long run empirical distribution of the time between zero down- and upcrossings, under stationary conditions, while L is the long run empirical distribution along an infinite section on the surface.

SPECTRAL PROPERTIES

Unidirectional spectrum

The distribution of the wave quantities depends in a complicated way on the spectral properties of the wave. For Gaussian waves it is possible to calculate the distributions with arbitrary accuracy by means of dedicated computer software to be described below. The routines need as input the energy spectrum, either the *wave frequency spectrum* or the *wave number spectrum*. They are related to the correlation structure of the waves as follows.

For the time process $\{\xi(t,x_o), t \in \Re\}$ we have the wave frequency spectrum $R^{(\Omega)}(\omega)$ such that

$$r^{(\Omega)}(\tau) = Cov(\xi(t,x_o),\, \xi(t+\tau,x_o)) = \int_0^\infty \cos(\omega\tau) R^{(\Omega)}(\omega)\,d\omega. \quad (6)$$

For the space process $\{\xi(t_0,x),\, x \in \Re\}$, the wave number spectrum $R^{(K)}(\kappa)$, is defined by

$$r^{(K)}(\rho) = Cov(\xi(t_o,x),\xi(t_o,x+\rho)) = \int_0^\infty \cos(\kappa\rho) R^{(K)}(\kappa)\,d\kappa. \quad (7)$$

In linear wave theory with unidirectional waves the dispersion relation states that frequency ω and wave number κ are related through $\omega^2 = g\kappa \tanh(\kappa h)$, where g is earth acceleration and h is the water depth. For deep water ($h \approx \infty$) the relation simplifies to

$$\omega^2 = g\kappa, \quad (8)$$

and this relation shall be used in the sequel. It implies the following relation between the two spectral densities,

$$R^{(\Omega)}(\omega) = \frac{2\omega}{g} R^{(K)}(\omega^2/g). \quad (9)$$

In the examples, we shall use a JONSWAP frequency spectrum, with center frequency ω_m and peak enhancement factor $\gamma \geq 1$,

$$R^{(\Omega)}(\omega) = \frac{g^2\alpha}{\omega^5} \cdot e^{-\beta(\omega_m/\omega)^4} \cdot \gamma^{\exp(-(1-\omega/\omega_m)^2/\sigma_\omega^2)}, \quad (10)$$

and truncate it at ωmax, i.e. (10) holds for $0 \leq \omega \leq \omega\text{max}$. We use the standard form with $\sigma_\omega = 0.07$ on the low frequency side and 0.09 on the high frequency side; cf. (Gran, 1992).

As special case we shall also take $\gamma = 1$ and get the standard Pierson-Moskowitz frequency spectrum, with

$$R^{(\Omega)}(\omega) = \frac{g^2\alpha}{\omega^5} \cdot e^{-\beta(\omega_m/\omega)^4}, \quad \text{for } 0 \leq \omega \leq \omega\text{max}. \quad (11)$$

In the examples we always truncate the spectra at some high frequency, and restrict ω to $0 \leq \omega \leq \omega\text{max}$, where ωmax is a truncation limit. In the examples this limit is taken as a multiple of the center frequency ω_m. This truncation is necessary in order to get a finite average number of mean level crossings in the space process, and a non-zero mean wave length. As we shall see, the wave length distributions are heavily affected by the choice of the truncation limit. The higher the truncation limit, the more distorted is the wave length distribution.

Directional spectrum

Up til now we have only considered the wave profile along a specified direction and used the dispersion relation for unidirectional waves. If we consider a wave surface with directional spreading, the wave length distribution will depend of the observation direction in relation to the spectral directions. However, by projecting the directional spectrum wave number spectrum on to the observation direction we obtain the appropriate wave number spectrum for use in the further analysis.

As before, we consider waves along the $x-$ axis. With a directional frequency spectrum $G(\theta,\omega)$, with $\omega \in \Re$, $-\pi/2 \leq \theta \leq \pi/2$, the covariance between $\xi(0,0)$ and $\xi(\tau,\rho)$ is given by

$$Cov(\xi(0,0),\xi(\tau,\rho)) = \int_{-\infty}^{\infty}\int_{-\pi/2}^{\pi/2} \cos(\rho\kappa_\omega \cos\theta + \omega\tau)\, G(\theta,\omega)\, d\theta\, d\omega.$$

Setting $\tau = 0$ and substituting $\kappa_\omega = \frac{\omega^2}{g}\cos\theta$, with the inverse

$$\omega_\kappa(\theta) = \sqrt{\frac{g\kappa}{\cos\theta}},$$

we can simplify this to

$$Cov(\xi(0,0),\xi(0,\rho)) = \int_{-\infty}^{\infty}\int_{-\pi/2}^{\pi/2} \cos(\rho\frac{\omega^2}{g}\cos\theta)\, G(\theta,\omega)\, d\theta\, d\omega = \int_0^\infty \cos(\kappa\rho) R_{\text{dir}}^{(K)}(\kappa)\, d\kappa,$$

where

$$R_{\rm dir}^{(K)}(\kappa) \tag{12}$$
$$= \frac{1}{2}\int_{-\pi/2}^{\pi/2} \frac{g}{\omega_\kappa(\theta)} \{G(\theta,\omega_\kappa(\theta)) + G(\theta,-\omega_\kappa(\theta))\}\, d\theta,$$

for $\kappa > 0$. This is therefore the wave number spectrum along the direction $\theta = 0$. It should be compared with the expression for the frequency spectrum,

$$R^{(\Omega)}(\omega) = \int_{-\pi/2}^{\pi/2} \{G(\theta,\omega) + G(\theta,-\omega)\}\, d\theta, \tag{13}$$

for $\omega > 0$.

Spectra in other directions are obtained by rotating the directional spectrum before integration.

Moment relations

For Gaussian waves the average wave period and wave length are expressed in terms of the spectral moments in the measured direction,

$$M_k = \int x^k\, R(x)\, dx.$$

The explicit expressions are

$$T_z = E(T) = \text{mean zero crossing period}$$
$$= \pi\sqrt{M_0^{(\Omega)}/M_2^{(\Omega)}}, \tag{14}$$
$$T_{mM} = E(T^{mM}) = \text{mean min-Max period}$$
$$= \pi\sqrt{M_2^{(\Omega)}/M_4^{(\Omega)}}, \tag{15}$$
$$L_z = E(L) = \text{mean zero crossing wave length}$$
$$= \pi\sqrt{M_0^{(K)}/M_2^{(K)}}, \tag{16}$$
$$L_{mM} = E(L^{mM}) = \text{mean min-Max wave length}$$
$$= \pi\sqrt{M_2^{(K)}/M_4^{(K)}}. \tag{17}$$

The irregularity index α, defined as the mean number of wave crests per mean level downcrossing, is for a Gaussian wave given by $\alpha = \frac{\sqrt{M_4 M_0}}{M_2}$. The relation (9) implies that the irregularity measure for the space wave $\xi(0,x)$ contains higher spectral moments than that for the time wave $\xi(t,0)$. Specifically, for unidirectional waves, $M_k^{(K)} = g^{-k} M_{2k}^{(\Omega)}$, and thus, with $M_k = \int \omega^k R^{(\Omega)}(\omega)\, d\omega$,

$$\frac{\alpha^{(K)}}{\alpha^{(\Omega)}} = \left(\frac{M_8 M_2^2}{M_4^3}\right)^{1/2} \geq 1. \tag{18}$$

That the space wave always has a larger irregularity index than the time wave is due to the fact that individual short waves travel at a lower speed than long ones, and therefore can cause fewer mean level crossings in the time process than in the space process.

CROSSING DISTRIBUTIONS

Regression approximation

The statistical distribution of crossing distances in a random process, such as wave length, wave period, or wave amplitude, are in general very hard to express in closed analytic form. If the random process representing the wave is differentiable (with probability one), a wave trough or wave crest is equivalent to a zero crossing by the time or space derivative, respectively. The half wave period and wave length are the time difference or distance between a downcrossing and an upcrossing of the mean level, respectively.

The statistical distribution of these crossing distances is defined by means of the Kac-Slepian horizontal window conditioning, and is equal to the limit, as the observation interval tends to infinity, of the corresponding empirical distribution. For details about this interpretation, see e.g. (Leadbetter et al., 1983, Chap.10) and (Lindgren and Rychlik, 1991).

In general, the probability density function for the distribution of the time T from an arbitrary origin 0 to the first upcrossing zero of certain random processes $\xi(\tau)$ is given by Durbin's formula,

$$f_T(t) = f_{\xi(t)}(0) \cdot E(\xi'(t)^+ I(t) \mid \xi(t) = 0); \tag{19}$$

(Durbin, 1985; Rychlik, 1987; Lindgren and Rychlik, 1991). Here, $x^+ = \max(0,x)$ and $I(t)$ is the indicator function of the event

$$\{\xi(s) < 0 \text{ for all } s \in (0,t)\}. \tag{20}$$

Thus, the density of the first passage time T involves the expectation of the variable $\xi'(t)^+ I(t)$, which in turn depends on an infinite number of random variables. The expectation in (19) is therefore not readily computable.

For a Gaussian process, one can approximate the crossing distance distribution with high and controllable accuracy by replacing the infinite dimensional random variable $\xi'(t)^+ I(t)$ by a finite dimensional one. This is done by replacing $I(t)$ by the indicator $I_N(t)$ of the event

$$\{\xi(s) < 0 \text{ for } s = s_k,\ k = 1,\ldots,N\}, \tag{21}$$

where $s_1,\ldots,s_N$ are suitably chosen time points between 0 and t. The theoretical justification for this procedure, called a *regression approximation*, can be found in (Lindgren and Rychlik, 1991). The choice of $s_1,\ldots,s_N$ can be made in a recursive fashion, and by taking a moderately large N, one obtains high numerical accuracy. In most cases, $N = 5$ suffices.

A min-Max wave period of the random wave $\xi(t,x)$ is defined as the difference in time between an upcrossing zero for the time derivative $\frac{\partial \xi(t,x_0)}{\partial t}$ and the next downcrossing zero. Similarly, an min-Max wave length is the distance between an upcrossing and the following downcrossing zero of $\frac{\partial \xi(t_0,x)}{\partial x}$. The half wave period and wave length are similarly defined as the time difference and distance between zero down- and upcrossings for the process. Now, if $\xi(t,x)$ is a Gaussian process, the derivative is also Gaussian, and thus, crossings methods for Gaussian processes can be applied in both cases.

The regression approximation can also be used to calculate the distribution of wave amplitude H^{mM}, the height difference between a wave minimum and the following wave maximum; for details, see (Rychlik and Lindgren, 1993).

Numerical tools

To calculate zero crossing and other wave characteristic distributions one needs dedicated algorithms that can calculate finite dimensional integrals like $E(\xi'(t)^+ I_N(t))$ in (19). The authors have developed a package of such algorithms for use in crossings and extreme value theory. The technique is described in (Rychlik and Lindgren, 1993). Special adaptions of the algorithms for use in wave analysis and for calculations of wave distributions are collected in WAT, a toolbox of MATLAB algorithms, described in (Rychlik and Lindgren, 1995).[1]

With WAT one can calculate, with desired high accuracy, the theoretical density for the joint distribution of wave length/wave period and amplitude variables of different types, i.a. the time and height differences between a local minimum and the next maximum, zero upcrossing waves, etc.

The algorithms need as input the covariance function of the Gaussian process, and that can be of any form, analytical or as a tabulated function. The process can also be specified by its spectral density function, in which case the covariances are calculated by means of the Fast Fourier Transform.

It should be emphasized that the density functions calculated by WAT give numerically accurate values for the *exact* theoretical density functions if the process is Gaussian. To the authors knowledge, all other methods suggested in the literature, are based either on fitting a member of a standard class of statistical distributions to empirical wave data, on a degenerated narrow band approximation of the wave, or on replacement of the wave by a simple function. Such empirically based methods cannot describe the relation between wave period and wave length distributions. In particular, they cannot be used to predict wave length properties from wave period measurements taken at a single site.

EXAMPLES

We shall consider Gaussian waves with Pierson-Moskowitz, or more general JONSWAP spectrum. In the first example we consider processes with standard Pierson-Moskovitz (PM) wave frequency spectrum, and in particular study the effect of truncation of high frequencies. We also study a process where the wave number frequency is of PM-type. A JONSWAP spectrum, with peak enhancement factor $\gamma = 5$ will be used as an example of a more narrow spectrum.

In all examples, standard deviation is set to 1, and time and frequency normalized to make the average (half) wave period equal to π, which means that the wave frequency spectrum has its first two even spectral moments $M_0^{(\Omega)} = M_2^{(\Omega)} = 1$. We use the routine `wave_t` in the WAT toolbox (Rychlik and Lindgren, 1995) to calculate the density of the zero crossing period and wave length, and of the min-Max wave period and wave length. The routine `wave_th` is used to calculate the joint density of the min-Max wave period/length and the corresponding wave amplitude.

Example 1: PM frequency spectrum

We first let the wave frequency spectrum be of standard truncated Pierson-Moskovitz type (11) and truncate it at $c\omega_m$ with $c = 2, 3, 4$. Since $M_4^{(\Omega)} = \infty$ for the untruncated spectrum, this is needed in order to prevent a too irregular wave length pattern. The spectrum is normalized after truncation to have (half) mean wave period $T_z = \pi$. For $c = 3$ the irregularity index is $\alpha^{(K)} = 2.1144$ and $\alpha^{(\Omega)} = 1.2330$, respectively.

Figure 1 shows the probability density functions for the zero crossing period and wave length, for a Gaussian process with this truncated Pierson-Moskovitz time spectrum. As seen, there is not much similarity between the two densities, and this is true both for the zero crossing and for the min-Max definition. The wave period density is rather concentrated around the mean zero crossing period, while the distribution of the zero crossing wave length is much more spread out, and also has a characteristic oscillatory behaviour. This is caused by short waves of small amplitude, which may need one, two, or more wave lengths in order to complete an excursion below the zero level.

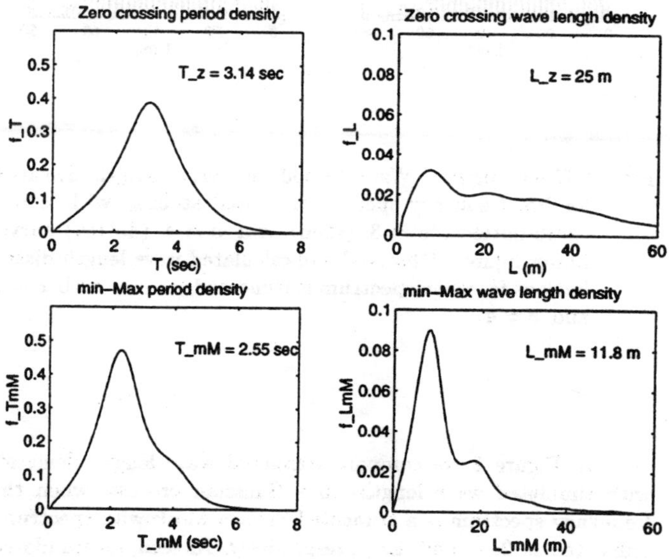

FIG. 1: Probability density functions for wave period and wave length for Gaussian waves with PM wave frequency spectrum, truncated at $3\omega_m$.

The effect of truncation on wave period and wave length density is shown in Figure 2. Truncation of the frequency spectrum at 2, 3, 4, times the dominant frequency ω_m has a drastic effect on the wave length distribution, but affects the wave period distribution only in that the distribution becomes slightly more concentrated. With $c = 4$ the wave length distribution has a dominating content of very short waves, and with $c = 2$ the main wave length is determined by the main frequency. We remind the reader that the spectra are always normalized to have $T_z = \pi$.

The sensitivity to truncation has been noticed in practise in the analysis of ocean waves, i.a. as a consequence of the choice of sampling interval. A short sampling interval will allow spectrum estimates at high frequencies, thereby increasing the irregularity index α, as well as it permits observation of really short waves and cycles.

[1] See also `http://www.maths.lth.se/matstat/staff/georg/watinfo.html`.

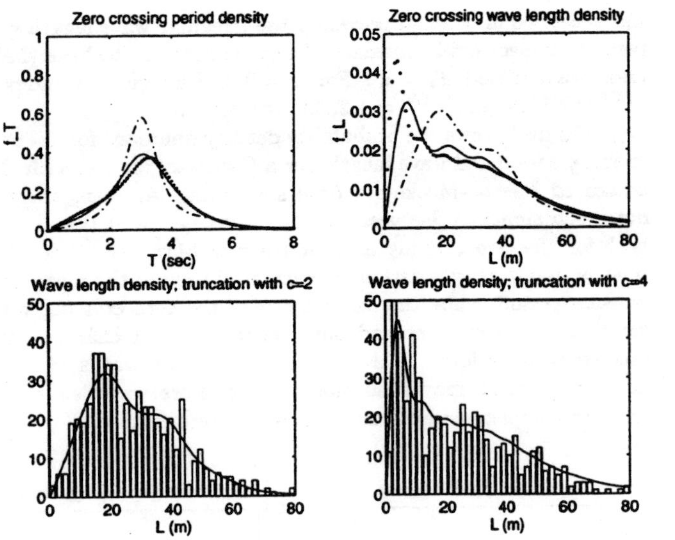

FIG. 2 Upper figure: Wave period and wave length densities for PM frequency spectra, truncated at $c\omega_m$ with $c = 2$ (dash-dotted), $c = 3$ (solid), and $c = 4$ (dotted) curve. Lower figures: Observed and calculated wave length distributions for when spectrum is truncated at $c\omega_m$ with $c = 2$ and $c = 4$.

In Figure 2 we compare computed wave length densities with simulated wave lengths in a Gaussian process, when the frequency spectrum is a truncated Pierson-Moskowitz spectrum, truncated at $2\omega_m$ and $4\omega_m$, respectively. As seen, for simulated Gaussian data, the calculated densities agree perfectly with the observed wave length distributions.

Example 2: PM wave number spectrum

The wave number spectrum has much power in high frequencies when the frequency spectrum is a Pierson-Moskowitz spectrum. As an example of a spectrum with less high frequencies we shall consider a wave number spectrum of Pierson-Moskowitz type,

$$R^{(K)}(\kappa) = \frac{g^2\alpha}{\kappa^5} \cdot e^{-\beta(\kappa_m/\kappa)^4},$$

again truncated at $3\kappa_m$ and normalized to make half mean wave period $T_z = \pi$. Then the irregularity index is $\alpha^{(K)} = 1.2330$ and $\alpha^{(\Omega)} = 1.0493$, respectively. In this case all wave length distributions will be identical to the wave period distributions in the previous example. Figure 3 shows the wave period and wave length densities in this very regular case.

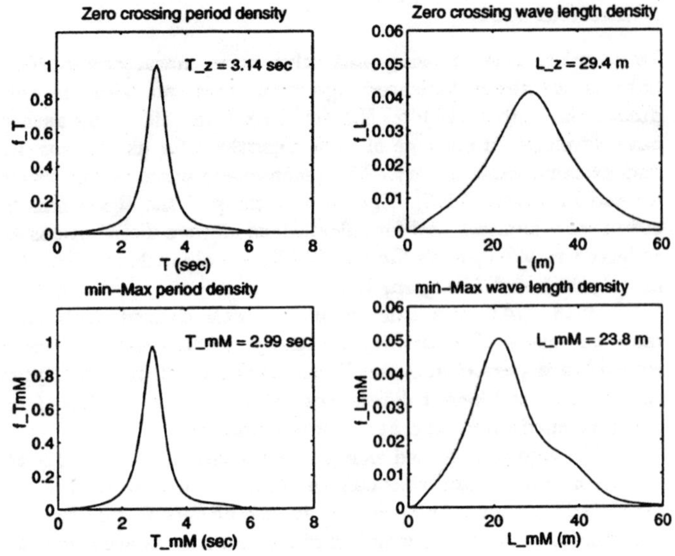

FIG. 3 Probability density functions for wave period and wave length for a Gaussian wave with truncated Pierson-Moskowitz wave number spectrum.

Example 3: JONSWAP frequency spectrum

A JONSWAP frequency spectrum has density

$$R^{(\Omega)}(\omega) = \frac{g^2\alpha}{\omega^5} \cdot e^{-\beta(\omega_m/\omega)^4} \cdot \gamma^{\exp(-(1-\omega/\omega_m)^2/\sigma_\omega^2)}. \tag{22}$$

We take $\beta = 1.25$ and let $\sigma_\omega = 0.07$ for $\omega \leq \omega_m$ and $= 0.09$ for $\omega > \omega_m$. We use a peak enhancement factor $\gamma = 5$ and truncate at $3\omega_m$. After truncation, we normalize the spectrum to make $T_z = \pi$.

The irregularity index is $\alpha^{(K)} = 2.4047$ and $\alpha^{(\Omega)} = 1.2192$, respectively. The zero crossing wave period and wave length are rather similar to those for the pure PM-spectrum, with a very spread out zero crossing wave length distribution. Again, we notice the oscillatory wave length distributions in Figure 4.

It should also be noticed that the wave length density is bimodal, with peaks around $L_{\text{short}} = 5.8$ m and $L_{\text{long}} = 36.7$ m, corresponding to wave numbers $\kappa_{\text{high}} = 0.55$ and $\kappa_{\text{low}} = 0.0856$, respectively. The normalized wave number spectrum has its peak at $\kappa_{\text{peak}} = 0.072$, corresponding to a true wave length $L = 43.6$ m, slightly longer than L_{long}.

The bivariate density for min-Max wave period/length and the corresponding amplitude (wave height) is also calculated for this spectrum. This is done by the routine wave_th in (Rychlik and Lindgren, 1995). The result is as level curves shown in Figure 5. Note that the oscillatory character in the wave length distribution in Figure 4 is also seen here in the wave length/amplitude density.

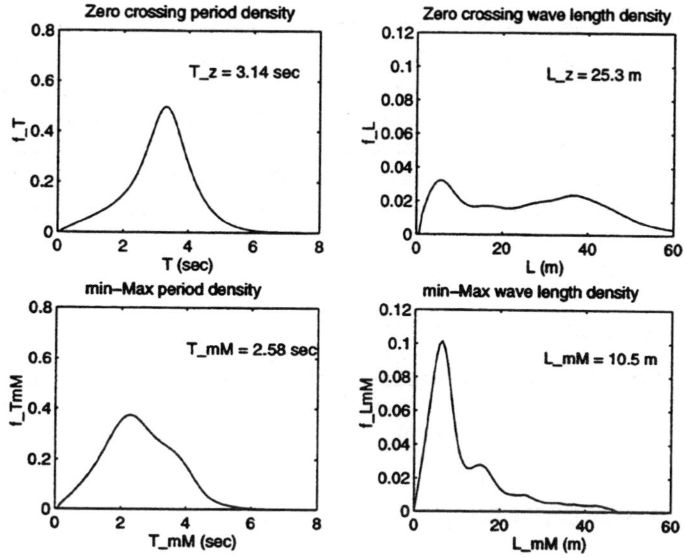

FIG. 4 Probability density functions for wave period and wave length for a Gaussian process with JONSWAP spectrum, truncated at $\omega = 3\,\omega_m$, $\gamma = 5$.

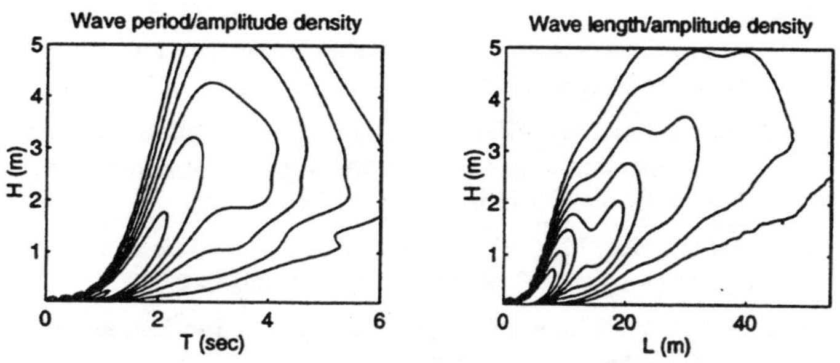

FIG. 5 Isolines for the joint density functions for min-Max wave period (left) and wave length (right) and amplitude for a Gaussian process with JONSWAP spectrum, truncated at $\omega = 3\,\omega_m$, $\gamma = 5$.

Example 4: Directional PM spectrum

The (half) wave length in a sea with unidirectional spectrum is defined as the distance between mean level crossings, measured in the direction of the waves. In a sea with directional spreading one can, with a slight abuse of terminology, define wave length in an arbitrary direction, relative to the main direction of the waves. Thus, a *directional wave length* in direction θ_0 is defined as the distances between a zero downcrossing and the next upcrossing in a section of the sea with direction θ_0 at a fixed moment.

By rotating the directional spectrum we can always assume $\theta_0 = 0$ and consider the sea level process $\{\xi(t_0, \rho), \rho \in \Re\}$. We can calculate its wave number spectrum according to eqn. (12), and hence calculate the distribution of the directional wave lengths for different directions. By taking the direction perpendicular to the main wave direction we get the statistical distribution of the *wave width.*

We use the standard Pierson-Moskowitz frequency spectrum (11) with $\omega > 0$ and a spreading function $D(\theta) = \frac{2}{\pi}\cos^2\theta$, $-\pi/2 \le \theta \le \pi/2$. Thus the directional spectrum is

$$G(\theta,\omega) = R^{(\Omega)}(\omega)\frac{2}{\pi}\cos^2\theta, \quad 0 \le \omega \le \omega_{\max}, \qquad (23)$$

which has main direction $\theta = 0$. We compute the wave number spectrum in the two directions $\theta = 0$, and $\theta = \pi/2$ and calculate the corresponding distribution of zero crossing distances. This will give us the distribution of the usual wave length, measured in the wave direction, and the wave width, measured along the wave crests.

The results are shown in Figure 6, together with the wave length distribution for an equivalent unidirectional wave, as in Exampel 1. The mean wave length and wave width are also shown; they are 34.4 m and 45.9 m, compared to 25 m, for a unidirectional sea with the same frequency spectrum.

The bivariate density for min-Max wave length and amplitudes across and along waves are illustrated in Figure 7.

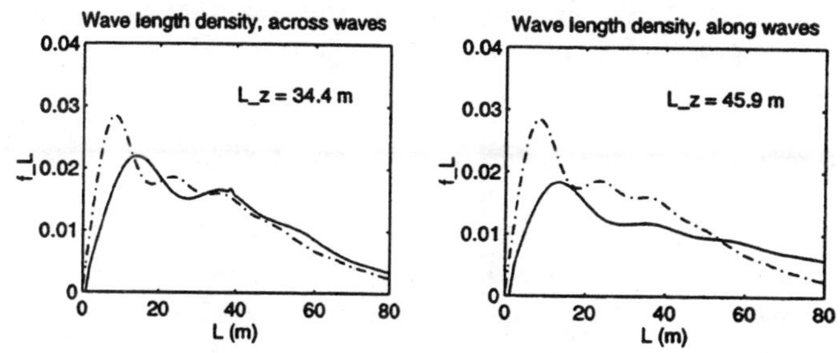

FIG. 6 Probability density functions for directional wave length (solid curves), for directional PM frequency spectrum (23). Left figure: across waves in direction $\theta = 0$; right figure: along waves in direction $\theta = \pi/2$. Dashed curve: wave length distribution for unidirectional waves.

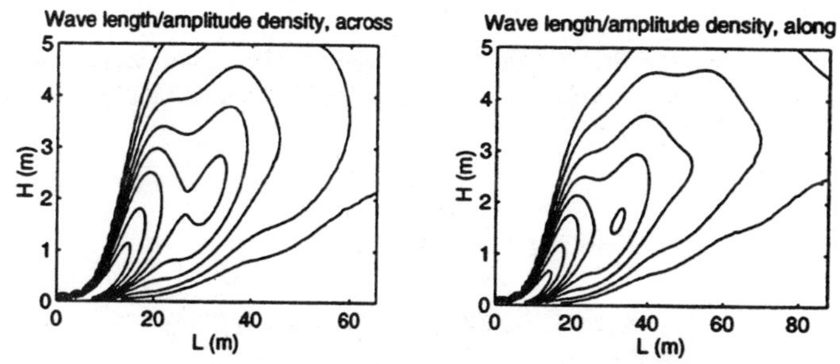

FIG. 7 Isolines for the joint density functions for min-Max wave length and amplitude for a Gaussian process with directional PM frequency spectrum. Left figure: across waves in direction $\theta = 0$; right figure: along waves in direction $\theta = \pi/2$.

AVERAGING AND THE DISPERSION RELATION

For a pure wave at infinite depth, relation (8) between (angular) frequency and wave number becomes $\kappa_0 = \omega_0^2/g$, and thus, half wave length $L = \pi/\kappa_0$ and half wave period $T = \pi/\omega_0$ are related through the equation

$$L = \frac{gT^2}{\pi}. \qquad (24)$$

One may ask, whether this relation is valid, or at least approximately valid in distribution, also for a random wave, i.e. if the distribution of L is (approximatively) equal to the distribution of $\widetilde{L} = \frac{gT^2}{\pi}$. We call $\widetilde{L}$ a *phantom wave length*, corresponding to the wave period T.

Now, if we denote by $f_T(t)$ the probability density of the true wave period, the distribution of $\widetilde{L} = gT^2/\pi$ has density

$$f_{\widetilde{L}}(x) = \frac{1}{2}\sqrt{\frac{\pi}{xg}}\, f_T\left(\sqrt{\frac{\pi x}{g}}\right). \tag{25}$$

We can therefore easily compare the two distributions, $f_L(x)$ and $f_{\widetilde{L}}(x)$. Figure 8 shows the result for Examples 1-3, the PM-frequency and wave number spectrum, and the JONSWAP spectrum. As seen, the true wave length distribution is quite different from the phantom distribution of gT^2/π as soon as the spectrum has a reasonable content of high frequencies. It is only for a Pierson-Moskowitz wave number spectrum that the phantom distribution of $\widetilde{L}$ has any similarity to that of L.

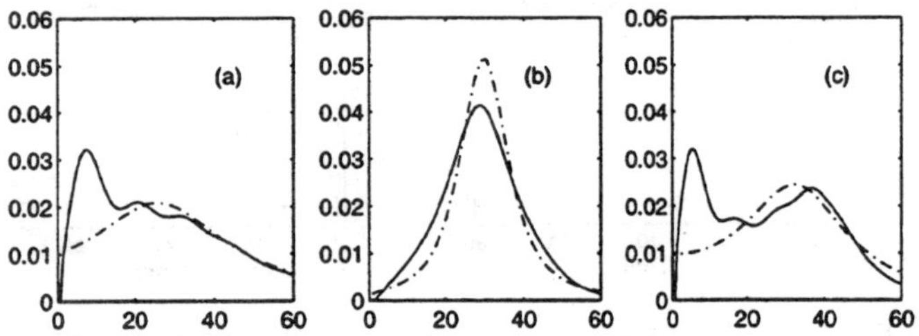

FIG. 8 Density of wave length L (solid line) and phantom wave length $\widetilde{L}$ (dash dotted line) for the three unidirectional spectra: (a) PM frequency spectrum, (b) PM wave number spectrum, (c) JONSWAP frequency spectrum with $\gamma = 5$.

For unidiriectional waves, we get the mean values of L and $\widetilde{L}$ from (16), with $\sigma_T^2 = V(T)$ and $E(T) = T_z = \pi\{M_0^{(\Omega)}/M_2^{(\Omega)}\}^{1/2}$,

$$E(L) = g\pi\sqrt{\frac{M_0^{(\Omega)}}{M_4^{(\Omega)}}},$$

$$E(\widetilde{L}) = \frac{g}{\pi}E(T^2) = \frac{g}{\pi}\left(V(T) + E(T)^2\right)$$

$$= g\pi\alpha^{(\Omega)}\sqrt{\frac{M_0^{(\Omega)}}{M_4^{(\Omega)}}} + \frac{g\sigma_T^2}{\pi}.$$

The relation between the two mean values involves the fourth spectral moment in the period spectrum, and the variance of T. Note that $\sigma_T^2 \to 0$ as $\alpha^{(\Omega)} \to 1$, and hence $E(\widetilde{L}) \to E(L)$ in that case. Otherwise, the exact form of the wave frequency spectrum for high frequencies is more important for the L-distribution than for the T-distribution. This must be kept in mind when theoretical wave frequency spectra are used to evaluate wave distributions, with or without truncation of high frequencies.

ACKNOWLEDGEMENT

Part of this work was supported by Office of Naval Research under Grant N00014-93-1-0841 and carried out when the first two authors visited the Center for Stochastic Processes, Chapel Hill.

References

Durbin, J. (1985). The first-passage density of continuous gaussian process to a general boundary. *J. Appl. Probab.*, 22:99–122.

Gran, S. (1992). *A course in Ocean Engineering*. Elsevier, Amsterdam.

Leadbetter, M., Lindgren, G., and Rootzén, H. (1983). *Extremes and related properties of random sequences and processes.* Springer-Verlag, New York.

Lindgren, G. and Rychlik, I. (1991). Slepian models and regression approximations in crossing and extreme value theory. *International Statistical Review*, 59:195–225.

Longuet-Higgins, M. (1956). The statistical analysis of a random, moving surface. *Phil. Trans. Ser. A*, 249:321–387.

Rice, S. (1944). The mathematical analysis of random noise. *Bell Syst. Techn. J.*, 23:282–332.

Rice, S. (1945). The mathematical analysis of random noise. *Bell Syst. Techn. J.*, 24:46–156.

Rychlik, I. (1987). Regression approximations of wavelength and amplitude distributions. *Advan. Appl. Prob.*, 19:396–430.

Rychlik, I. and Lindgren, G. (1993). Crossreg, a technique for first passage and wave density analysis. *Probability in Engng. Inform. Sciences*, 7:125–148.

Rychlik, I. and Lindgren, G. (1995). *WAVE analysis toolbox – a tutorial.* Dept of Mathematical Statistics, Lund University. ISSN 0281-1944.

Proceedings of the Seventh (1997) International Offshore and Polar Engineering Conference
Honolulu, USA, May 25-30, 1997

ISBN 1-880653-28-1 (Set); ISBN 1-880653-31-1 (Vol. III)

Spectral Modeling of Unidirectional Nonlinear Wave Propagation Over Arbitrary Depths

S. Beji
Istanbul Technical University
Istanbul, Turkey

K. Nadaoka
Tokyo Institute of Technology
Tokyo, Japan

Abstract

A weakly-nonlinear and dispersive wave equation recently developed by the authors (Beji and Nadaoka, 1997) is used for formulating a spectral wave model describing transformations of narrow-banded unidirectional waves traveling over variable bathymetry. The performance of the model is tested against the measured data for harmonic generation over constant depth as well as nonlinear random wave propagation over varying depth. The comparisons indicate good agreement with the measurements and establish the reliability of the model. In closing, a semi-empirical dissipation term is formulated for simulating the energy loss due to breaking waves.

Key Words

Nonlinear spectral model, Varying bathymetry, Nonlinear random waves, Cnoidal waves, Stokes waves, Breaking waves.

1. Introduction

Recent years have witnessed a constantly increasing interest towards the modeling of the nonlinear aspects of ocean waves. The trend originated from the need to explain definite observations which could not be accounted for by linear models. Wave skewness related sediment transport, influence of breaking on the surf-zone processes, and effects of harmonic generation on the characteristics of a wave field are the most striking examples of such phenomena (Doering and Bowen, 1986; Nadaoka *et al.*, 1989; Kojima *et al.*, 1990). For practical applications the nonlinear wave models are not yet in common use, however, there are evidences that when augmented with appropriate generation and dissipation mechanisms, these models may well be the prototypes of the commercial models to come.

In this work a spectral model is developed using the recently introduced wave equation of Beji and Nadaoka (1997). The derivation is based on a Fourier series representation of the surface elevation with spatially varying amplitudes and phases. The resulting evolution equations are numerically solved for various test cases to demonstrate the capabilities of the model. Also, a semi-emprical dissipation term is formulated to represent energy loss due to wave breaking.

2. Wave Model and Spectral Formulation

The present work uses a nonlinear unidirectional wave equation (Beji and Nadaoka, 1997) which may be viewed as a generalized form of the Korteweg and deVries (KdV) equation:

$$C_g\eta_t + \frac{1}{2}C_p(C_p + C_g)\eta_x - \frac{(C_p - C_g)}{k^2}\eta_{xxt}$$
$$-\frac{C_p(C_p - C_g)}{2k^2}\eta_{xxx} + \frac{1}{2}\left[C_p(C_g)_x + (C_p - C_g)(C_p)_x\right]\eta \quad (1)$$
$$+\frac{3}{4}g\left(3 - 2\frac{C_g}{C_p} - \frac{k^2C_p^4}{g^2}\right)\left(\eta^2\right)_x = 0,$$

where C_p, C_g, and k are respectively the phase and group velocities and the wave-number computed according to the linear theory dispersion relation for a dominant wave frequency ω and a given local depth h, the subscripts x and t indicate partial differentiation with respect to space and time, respectively.

Unlike the KdV equation, the applicable domain of equation (1) is not restricted to only shallow waters but covers the entire range of relative depths from very shallow to infinitely deep water. For relatively shallow water waves the equation renders the KdV equation as its special case and for deep water it admits the second-order Stokes waves as solution. When the incident wave frequency coincides with the prescribed dominant frequency of the model equation the linear shoaling characteristics of the incident wave are described exactly, in accordance with the energy flux concept.

Since the form of equation (1) resembles very closely to the KdV equation it is possible to adopt all the numerical methods applicable to the KdV equation. The most straightforward approach is to use finite-difference approximations, as reported by Beji and Nadaoka (1995, 1997). Here, the so-called spectral domain formulation is employed by representing the surface displacement as a Fourier series with spatially varying harmonic amplitudes:

$$\eta(x,t) = \sum_{n=-\infty}^{+\infty} A_n(x)e^{i[\omega_n t - \int k_n(x)dx]}, \quad (2)$$

where $A_n(x)$ is the spatially varying complex wave amplitude, ω_n the radian frequency which is equal to $n\Delta\omega$, $\Delta\omega$ being the frequency of resolution. $k_n(x)$ is the spatially varying wave-number determined according to the linear dispersion relation of equation (1) for the local depth $h(x)$ and the radian frequency ω_n:

$$C_g\omega_n - \frac{1}{2}C_p(C_p+C_g)k_n + \frac{(C_p-C_g)}{k^2}\omega_n k_n^2 - \frac{C_p(C_p-C_g)}{2k^2}k_n^3 = 0. \quad (3)$$

Substituting (2) into (1) and neglecting only the third-order spatial derivatives of $A_n(x)$ on the premise that the spatial variation of $A_n(x)$ is slow, result in the following second-order nonlinear differential equation that determines the spatial variation of each complex component:

$$i\alpha_2\frac{d^2A_n}{dx^2} + \alpha_1\frac{dA_n}{dx} + (i\alpha_0+\alpha_s)A_n = i\beta\sum_{m=-\infty}^{+\infty}(k_m+k_{n-m})A_mA_{n-m}\,e^{-i(k_m+k_{n-m}-k_n)x}, \quad (4)$$

where the free index n covers the range from $-\infty$ to $+\infty$. The coefficients are given as follows

$$\begin{aligned}
\alpha_0 &= C_g\omega_n - \frac{1}{2}C_p(C_p+C_g)k_n + \frac{(C_p-C_g)}{k^2}\omega_n k_n^2 - \frac{C_p(C_p-C_g)}{2k^2}k_n^3,\\
\alpha_1 &= \frac{1}{2}C_p(C_p+C_g) - 2\frac{(C_p-C_g)}{k^2}\omega_n k_n + \frac{3}{2}\frac{C_p(C_p-C_g)}{k^2}k_n^2,\\
\alpha_2 &= \left(\frac{3}{2}C_pk_n - \omega_n\right)\frac{(C_p-C_g)}{k^2},\\
\alpha_s &= \frac{1}{2}[C_p(C_g)_x + (C_p-C_g)(C_p)_x] + \frac{(C_p-C_g)}{k^2}\left(\frac{3}{2}C_pk_n-\omega_n\right)(k_n)_x,\\
\beta &= \frac{3}{4}g\left(3 - 2\frac{C_g}{C_p} - \frac{k^2C_p^4}{g^2}\right),
\end{aligned} \quad (5)$$

in which the spatial derivatives of k_n higher than the first have been neglected as in the wave model itself. $(k_n)_x$ appearing in α_s is computed from (3) by differentiating it with respect to x. Note that α_0 is the linear dispersion relation of the wave model and is identically zero in virtue of equation (3). However, if the wave numbers in (2) are selected as bound wave numbers then α_0 is not zero any more and it must be retained. In principle, the wavenumbers may be chosen either way, here they are selected as free wave numbers according to (3), which numerically proved to be a better choice except in the simulation of the Stokes second-order waves.

Following Bryant (1973) we multiply (4) by $-\frac{i}{\alpha_2}\exp\left(-i\frac{\alpha_1}{\alpha_2}x\right)$ so that it becomes

$$\frac{d}{dx}\left(\frac{dA_n}{dx}e^{-i\frac{\alpha_1}{\alpha_2}x}\right) = i\frac{\alpha_s}{\alpha_2}A_ne^{-i\frac{\alpha_1}{\alpha_2}x} + \beta\sum_{m=-\infty}^{+\infty}\frac{(k_m+k_{n-m})}{\alpha_2}A_mA_{n-m}\,e^{-i(k_m+k_{n-m}-k_n+\frac{\alpha_1}{\alpha_2})x}, \quad (6)$$

where α_0 has been set to zero. The spatial variations of the linear shoaling term α_sA_n and the nonlinear term A_mA_{n-m} are relatively slower and therefore these terms may be assumed locally constant, which in turn allows an integration of (6) with respect to x:

$$\frac{dA_n}{dx} = -\frac{\alpha_s}{\alpha_1}A_n + i\beta\sum_{m=-\infty}^{+\infty}\alpha_{nm}\,A_mA_{n-m}\,e^{-i(k_{n-m}+k_m-k_n)x},$$

$$\text{where} \quad \alpha_{nm} = \frac{k_{n-m}+k_m}{\alpha_1+\alpha_2(k_{n-m}+k_m-k_n)}. \quad (7)$$

It should be remarked that the linear shoaling characteristics of the generic equation are exactly preserved in the above formulation.

Equation (7) in its present form is not suitable for numerical treatment and should be manipulated further. The summations are first re-arranged to run in the positive range only. Then, a change of variable is introduced $A_n(x) = P_n(x)\exp\left(i\int k_n dx\right)$, which removes the sine and cosine functions as in Bryant (1973). Finally, $P_n(x)$ is set to $\frac{1}{2}[a_n(x) - ib_n(x)]$ so that the evolution equations for the real variables $a_n(x)$ and $b_n(x)$ are obtained

$$\frac{da_n}{dx} = -\frac{\alpha_s}{\alpha_1}a_n - k_nb_n + \beta\sum_{m=1}^{N-n}\alpha_{nm}^-(a_mb_{n+m} - a_{n+m}b_m) + \frac{1}{2}\beta\sum_{m=1}^{n-1}\alpha_{nm}^+(a_mb_{n-m} + a_{n-m}b_m), \quad (8)$$

$$\frac{db_n}{dx} = -\frac{\alpha_s}{\alpha_1}b_n + k_na_n - \beta\sum_{m=1}^{N-n}\alpha_{nm}^-(a_ma_{n+m} + b_mb_{n+m}) - \frac{1}{2}\beta\sum_{m=1}^{n-1}\alpha_{nm}^+(a_ma_{n-m} - b_mb_{n-m}), \quad (9)$$

where

$$\alpha_{nm}^+ = \frac{k_{n-m}+k_m}{\alpha_1+\alpha_2\delta_{nm}^+}, \qquad \alpha_{nm}^- = \frac{k_{n+m}-k_m}{\alpha_1+\alpha_2\delta_{nm}^-},$$

$$\delta_{nm}^+ = k_{n-m}+k_m-k_n, \qquad \delta_{nm}^- = k_{n+m}-k_m-k_n.$$

For N number of frequency components, the free index n runs from 1 to N, resulting in $2N$ number of nonlinearly coupled first-order differential equations for the unknown components $a_n(x)$ and $b_n(x)$. Once the $a_n(x)$s and $b_n(x)$s are obtained the free surface may be constructed simply as $\eta = \Sigma a_n\cos(\omega_nt) + b_n\sin(\omega_nt)$. Various numerical integration techniques (*e.g.*, Bulirsch-Stoer, Runge-Kutta, Adams-Bashford-Moulton) are available for the integration of (8) and (9). Here, the Runge-Kutta fourth-order formulation is preferred as it proved to be the fastest while being as reliable as the other more sophisticated integration methods.

3. Numerical Simulations

The evolution equations derived in the previous section are now used for the purposes of ascertaining their reliability and exploring the capabilities of the wave model. The first case is the simulation of harmonic generation in shallow water.

3.1. *Harmonic Generation in Shallow Water*

Chapalain *et al.* (1992) conducted a series of experiments concerning nonlinear shallow water waves undergoing harmonic generation over constant water depth. The experiments were done for four different cases named respectively as the trial A, C, D, and H. The experimental conditions and wave parameters are given in Table 1.

Trial	h(cm)	T(s)	$\epsilon = a_0/h$	$\mu = kh$	$U_r = \epsilon/\mu^2$
A	40	2.5	0.105	0.528	0.38
C	40	3.5	0.105	0.371	0.76
D	30	2.5	0.118	0.452	0.58
H	40	3.0	0.084	0.433	0.45

Table 1

All the experiments listed in Table 1 are numerically simulated using the evolution equations given by (8) and (9). Figure 1 shows the simulated and measured harmonic components for the trials A, C, D, and H, respectively. Overall, the numerical simulations agree well with the measurements and establish confidence in the spectral model. The beat lengths are slightly underestimated; this is probably due to the limitation of the wave model to relatively narrow-banded cases. Since the third and fourth harmonics fall outside the narrow-band range of the primary wave this restriction is somewhat violated.

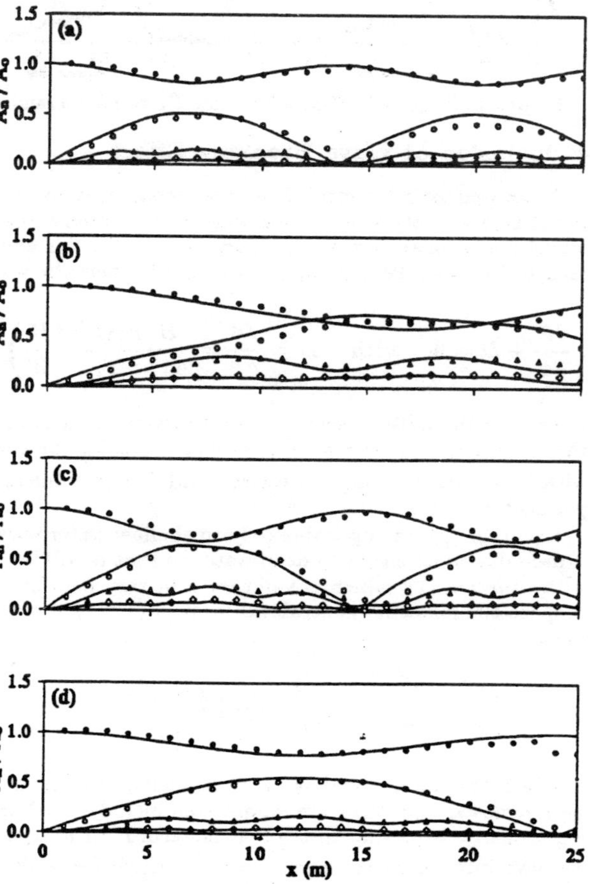

Figure 1: Experimental data of Chapalain et. al (1992) compared with the predictions of the spectral model for the trial A, C, D, and H. Scatter: experiment, solid line: computation.

3.2. *Nonlinear Wave Transformations over A Submerged Bar*

Nonlinear wave transformations over a submerged bar were investigated by Beji and Battjes (1993) in laboratory experiments that revealed the determining factors shaping the transformed wave spectrum. The waves were first observed to undergo harmonic generation due to nonlinear interactions in the shoaling region and then the generated bound harmonics were released behind the bar as the water depth increased again. For relatively long waves it was observed that an initially narrow-banded, sharply peaked spectrum was transformed to a broad-banded, double-peaked spectrum. On the other hand, the short wave evolutions were found to be not substantial. The details about the experimental setup and conditions can be found in Beji and Battjes (1993).

The experimental measurements for long and short sinusoidal waves and for random waves with initially JONSWAP type spectral shape are simulated using equations (8) and (9). Figures 2a and 2b show the comparisons with the experimental measurements respectively for long and short periodic waves at six different stations. The Fourier components of the recorded surface displacement at Station 1 (not shown) serves as the incoming boundary condition. In the computations five harmonic components $\omega_n = n\omega_0, n = 1, ..., 5$ were used with ω_0 corresponding to the primary wave frequency. As it is seen from the figures the model predictions agree very closely with the measurements both for the wave profiles and phases. The only exception is Station 7 of figure 2a, where obviously the narrow-bandedness of the wave model itself prevents a complete agreement.

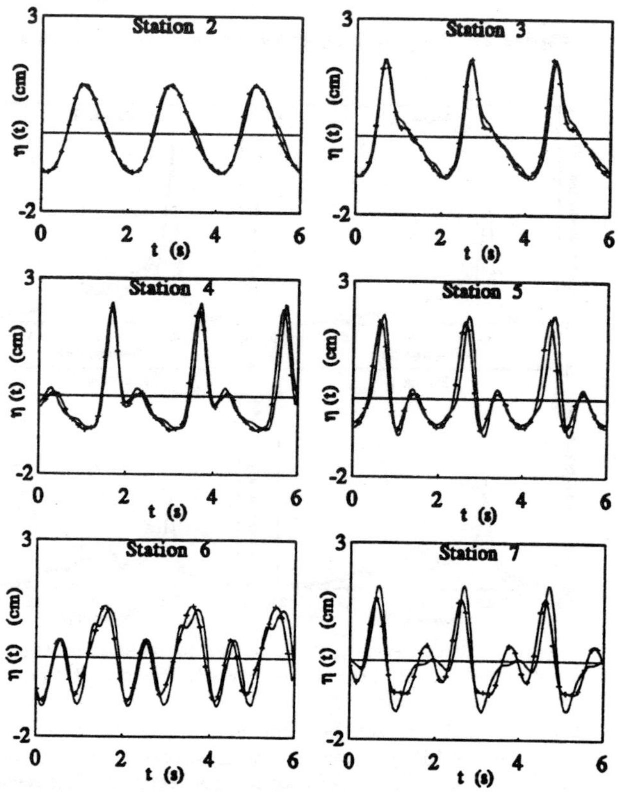

Figure 2a: Comparisons of the experimental measurements of initially sinusoidal wave propagation over a submerged bar (-) with the numerical simulations (+) for $T = 2$ s waves. Station 2: upslope 1:20, water depth 0.16 m, station 3 and 4: horizontal bottom, water depth 0.1 m, station 5: downslope 1:10, water depth 0.18 m, station 6: downslope 1:10, water depth 0.3 m, station 7: horizontal bottom, water depth: 0.4 m.

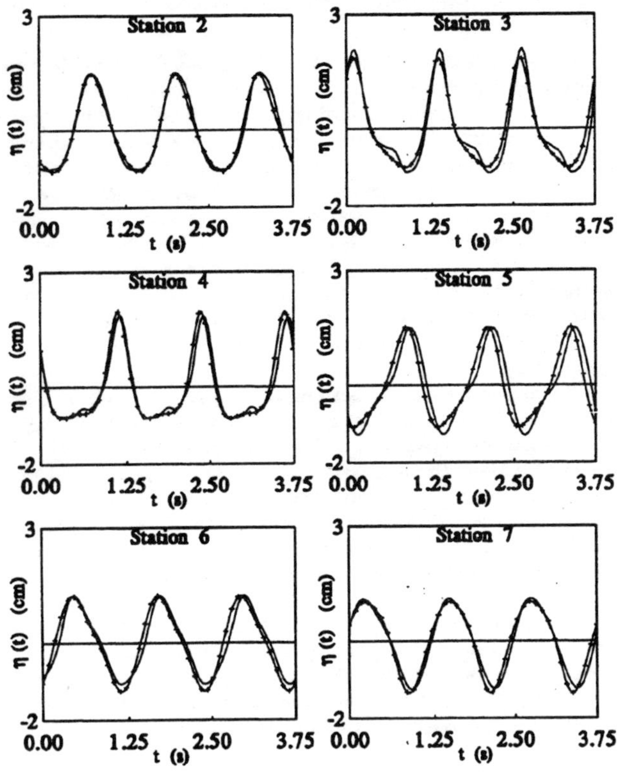

Figure 2b: Same as in figure 2a but for $T = 1.25$ s waves.

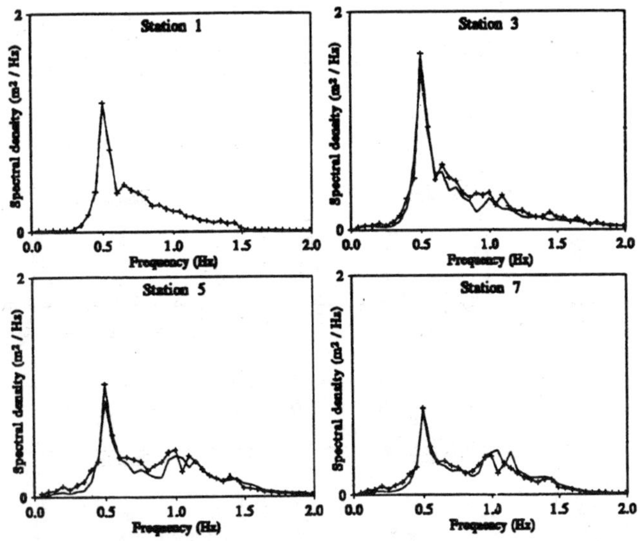

Figure 3: Comparisons of the experimental measurements (-) of the spectral evolutions over a submerged bar with the computations (+) for $T_p = 2$ s random waves.

Figures 3 and 4 depict the same comparisons for random waves having JONSWAP type incident wave spectrum with peak periods of $T_p = 2$ and $T_p = 1.25$ seconds, respectively. The records of the surface elevation at station 1 were divided into 40 segments of 512 data points and then each segment was Fourier transformed. Out of the 256 unique pairs the first 60 Fourier components, which covered a frequency range of 0.05-3.0 Hz, were found to be quite sufficient to represent the incident wave spectrum hence the spectral model was run for 40 different realizations with $N = 60$ using the measured Fourier components as the incoming boundary condition at station 1. The spectra shown in figures 3 and 4 were obtained after ensemble averaging all the realizations; no frequency averaging was done. Each spectrum has 80 degrees of freedom and 16 % normalized standard error. The overall agreement is good for both sets of measurements and the small discrepancies are attributed to the inherently narrow-banded nature of the wave model, which manifests itself as errors in the wavenumbers well outside the neigborhood of the primary wave wavenumber, which was taken to correspond to the peak period of the incident wave field.

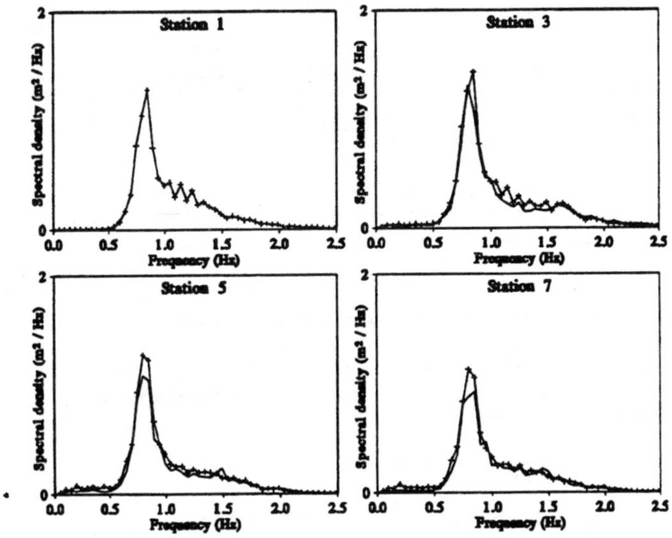

Figure 4: Same as in Figure 3 but for $T_p = 1.25$ s random waves.

4. Inclusion of Wave Breaking Effect

A semi-empirical approach is now considered to account for the dissipative role of wave breaking so as to render the spectral model operational even for breaking waves. To this end we first refer to Battjes'(1986) dissipation model for periodic waves:

$$\frac{\partial P}{\partial x} + D = 0 \quad \text{with} \quad D = \frac{D'}{cT} = \frac{B}{4\gamma^3}\frac{\rho g H^2}{T}\left(\frac{H}{h}\right)^4, \qquad (10)$$

where ρ is the water density, g the gravitational acceleration, H the wave height, T the wave period, $P = EC_g$ the energy flux with $C_g = (gh)^{1/2}$ (shallow water), and B and γ the calibration parameters.

Equation (10) is applicable only to shallow water waves and it is desirable to extend it to deep water so that it will be in accord with equation (1), which is valid for arbitrary depths. Thus, we propose the following generalized form of (10) as the dissipation model

$$\frac{H_x}{H} = -\frac{B}{2\pi\gamma^3}\frac{\omega}{C_g}\left(\frac{gH}{C_p^2}\right)^4. \qquad (11)$$

in which the linear shoaling term implicitly present in (10) has been removed since the wave model itself includes the linear shoaling properly. Note that instead of the term H/h in equation (10) we have introduced the term gH/C_p^2, which for shallow waters reduces to H/h. For deep water when h is very large H/h vanishes (so does the dissipation term D) and predicts no dissipation effect. On the other hand, for deep water the term gH/C_p^2 tends to kH and still predicts a non-zero dissipation due to breaking so long as the deep water wave steepness kH is appreciable. Obviously, this feature accords well with the expectations and agrees with Miches' breaking criteria (Beji, 1995).

If the wave is represented as $a(x)\cos\phi + b(x)\sin\phi$, ϕ being the phase angle, then, according to (11), the spatial rates of reduction of the amplitude components due to breaking are

$$\frac{a_x}{a} = -\frac{B}{2\pi\gamma^3}\frac{\omega}{C_g}\left(\frac{gH}{C_p^2}\right)^4, \quad \frac{b_x}{b} = -\frac{B}{2\pi\gamma^3}\frac{\omega}{C_g}\left(\frac{gH}{C_p^2}\right)^4 \qquad (12)$$

respectively for $a(x)$ and $b(x)$. Of course this formulation is true only for simple sinusoidal wave forms. The evolution equations contain not only a single component but a number of higher harmonics. It is therefore necessary to introduce some approximations to use equation (12) in (8). A plausible approach is to assume that each harmonic component is dissipated according to (12). This idea is also supported by the experimental measurements of Beji and Battjes (1993). Thus, for the wave breaking dissipation portions of (8) and (9) one can write

$$\frac{(a_n)_x}{a_n} = -\frac{B}{2\pi\gamma^3}\frac{\omega_n}{(C_g)_n}\left(\frac{gH}{C_p^2}\right)_n^4,$$
$$\frac{(b_n)_x}{b_n} = -\frac{B}{2\pi\gamma^3}\frac{\omega_n}{(C_g)_n}\left(\frac{gH}{C_p^2}\right)_n^4, \qquad (13)$$

in which the wave height for each component is computed from $H_n = 2\sqrt{(a_n^2 + b_n^2)}$. Thus, the final evolution equations are

$$\frac{da_n}{dx} = -\frac{\alpha_s}{\alpha_1}a_n - k_n b_n - \frac{B}{2\pi\gamma^3}\frac{\omega_n}{(C_g)_n}\left(\frac{2gH}{C_p^2}\right)_n^4 a_n + \beta\sum_{m=1}^{N-n}\alpha_{nm}^-(a_m b_{n+m} - a_{n+m}b_m) + \frac{1}{2}\beta\sum_{m=1}^{n-1}\alpha_{nm}^+(a_m b_{n-m} + a_{n-m}b_m), \qquad (14)$$

$$\frac{db_n}{dx} = -\frac{\alpha_s}{\alpha_1}b_n + k_n a_n - \frac{B}{2\pi\gamma^3}\frac{\omega_n}{(C_g)_n}\left(\frac{2gH}{C_p^2}\right)_n^4 b_n - \beta\sum_{m=1}^{N-n}\alpha_{nm}^-(a_m a_{n+m} + b_m b_{n+m}) - \frac{1}{2}\beta\sum_{m=1}^{n-1}\alpha_{nm}^+(a_m a_{n-m} - b_m b_{n-m}). \qquad (15)$$

For a simple demonstration we compare the results of the numerical solution of (14) and (15) with the exact solution of equation (10) (Battjes's (1986) original model which can be solved analytically for constant depth) for the experimental measurements of Horikawa and Kuo (1966) for breaking waves over a horizontal bottom. To account for the effects of modifications made it was necessary to select a slightly different value for the parameter B. Battjes (1986) uses $B = 2$, but since always $gH/C_p^2 > H/h$ and $\sqrt{gh} > C_g$, this value gives higher dissipation rate for the generalized model. After a few trials, $B = 1$ was found to be sufficiently good. Figure 5 shows the comparisons for $h = 0.1$ m depth and $H_b/h = 0.8$ for the linearized equations, that is, when $N = 1$ in (14). The agreement with the Battjes' (1986) original model is good and supports the validity of the approximations made in deriving (12). When compared with Horikawa and Kuo's (1966) data it is clear that the model predictions fall in the range given by these authors.

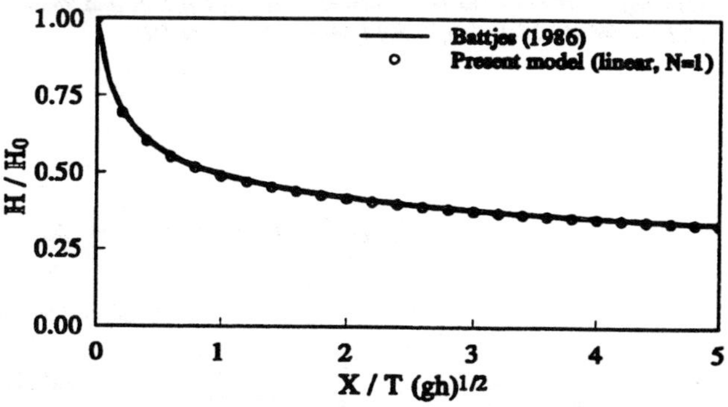

Figure 5: Wave height decay due to breaking over a horizontal bottom. The experimental conditions are from Horikawa and Kuo's (1966) measurements: $h = 0.1$ m and $H_b/h = 0.8$, H_b is the initial breaking waveheight.

5. Concluding Remarks

A spectral domain formulation of a recently developed nonlinear wave model is presented in the form of nonlinearly coupled first-order differential equations. The evolution equations describe the linear shoaling and second-order nonlinear transformations of each spectral component over arbitrary depths. Various tests are performed to check the performance of the proposed evolution equations against the measurements and the model is found to be sufficiently reliable for practical use. Furthermore, a semi-empirical formulation is suggested to account for the dissipative action of wave breaking. The proposed dissipation model works well for linear periodic waves; results for nonlinear random waves will be reported later.

References

Battjes, JA (1986). "Energy dissipation in breaking solitary and periodic waves," *Comm. on Hydraulic and Geotech. Eng.*, Internal Report, Delft Univ. of Tech., Report num. 86-5.

Beji, S (1995). "Note on a nonlinearity parameter of surface waves," *Coastal Eng.*, Vol 25, pp 81-85.

Beji, S and Battjes, JA (1993). "Experimental investigation of wave propagation over a bar," *Coastal Eng.*, Vol 19, pp 151-162.

Beji, S and Battjes, JA (1994). "Numerical simulation of nonlinear wave propagation over a bar," *Coastal Eng*, Vol 23, pp 1-16.

Beji, S and Nadaoka, K (1995). "Directional wave simulations by a time dependent nonlinear mild slope equation," *Proc. 2 nd Int Conf on Mediterranean Coastal Env*, Vol 3, pp 1929-1943.

Beji, S and Nadaoka, K (1997). "A time-dependent nonlinear mild-slope equation for water waves," *Proc. Roy. Soc. Lond. A*, Vol 453, pp 319-332.

Bryant, PJ (1973). "Periodic waves in shallow water," *J. Fluid Mech.*, Vol 59, pp 625-644.

Doering, JC and Bowen, AJ (1986). "Shoaling surface gravity waves: A bispectral analysis," *Proc 20 th Int Conf on Coastal Eng*, Vol 1, pp 150-162.

Kojima, H, Ijima, T and Yoshida, A (1990). "Decomposition and interception of long waves by a submerged horizontal plate," *Proc 22 th Int Conf on Coastal Eng*, Vol 2, pp 1228-1241.

Horikawa, K and Kuo, C-T (1966). "A study on wave transformation inside surf zone," *Proc. 10th Coastal Eng. Conf.*, Vol 1, pp 217-233.

Nadaoka, K, Hino, M and Koyano, Y (1989). "Structure of the turbulent flow field under breaking waves in the surf zone," *J Fluid Mech*, Vol 204, pp 359-387.

Proceedings of the Seventh (1997) International Offshore and Polar Engineering Conference
Honolulu, USA, May 25-30, 1997

ISBN 1-880653-28-1 (Set); ISBN 1-880653-31-1 (Vol. III)

The Measurement of Acceleration Fields Under Waves by DPIV

T.P. Dewhirst, M.L. Jakobsen and C.A. Greated
University of Edinburgh
Edinburgh, United Kingdom

Abstract

A multiple CCD camera system for full field velocity and acceleration measurements using Digital Particle Image Velocimetry (DPIV) analysis methods is presented. A sequence of three or four single or multiple exposure frames can be acquired within sub-milliseconds of each other to enable velocity and both Eulerian and Lagrangian acceleration fields to be measured. This work will present some preliminary results for the velocity and acceleration fields measured beneath a plunging breaker impinging on a flat vertical wall.

Keywords

Digital Particle Image Velocimetry, Forces, Accelerations

Introduction

Particle Image Velocimetry (PIV) is a well established experimental technique which enables highly-detailed, instantaneous full-field velocity maps of fluid flows to be constructed. This is achieved by seeding the flow with small neutrally buoyant particles and illuminating the region of interest with a pulsed sheet of light. PIV has been used in a number of applications including studies of water waves and flows around marine structures with great success and is now widely accepted as a standard measurement technique (Adrian, RJ, 1991).

In conventional PIV, multiple images of the seeding particles are captured on one frame of photographic film, or more recently high resolution CCD arrays. The resulting image is then analysed by auto-correlation either part optically, part digitally in the case of photographic film, or entirely digitally.

The limitations of this method are discussed in a previous paper by the authors (Dewhirst *et. al.*, 1996), with a system constructed from multiple CCD arrays being found to be the best solution to the problems of low dynamic range and directional ambiguity in the velocity measurement, and the low frame rate of conventional CCD cameras. This approach has already been undertaken by numerous people using both two array systems and multiple array systems to produce high-resolution, high-frame rate cameras (Raffel, M, 1995, Willert, C, 1995, Racca, RG, 1992), and the result is a system which can capture a sequence of four images of the flow under investigation with any desired time separations.

Now that four images can be acquired, the measurement of accelerations becomes a possibility, and this is the area being studied by the authors. The need to measure accelerations in flows can be clearly seen in the testing of, for example, offshore structures where wave loading plays a major part in the design of the structure, and it is with this in mind that one of the first uses of the acceleration measurement would be to derive a force measurement by applying, for example, Morison's equation to the structure in question.

As the acceleration depends on the accuracy of the velocity measurements, the source of errors in PIV measurements is investigated for fully digital acquisition and processing.

Force Measurements

Measurements of a force, applied by a fluid flow to an interface, for example a wall, can be implemented within both of the two common specifications of fluid dynamics; the Lagrangian and the Eulerian specification. In the Lagrangian specification a small sub-volume (e.g. a buoyant particle) can be tracked over a longer time interval to reveal local information about the velocity and total acceleration (which directly relates to the force) of the fluid. In the Eulerian specification the fluid flow is investigated at fixed points to reveal velocity and both temporal and spatial velocity gradients which can be used to calculate the total acceleration. In wave force calculations the convective acceleration term is often ignored, but this approach is not always acceptable as the total derivative of the velocity can be 10 or 15% higher than the temporal derivative alone (Sarpkay & Issacson, 1981). Using a multiple frame PIV system to acquire two or more velocity measurements the two concepts can be implemented.

The Lagrangian specification can be implemented by following a group of particles within an interrogation area over, for example, three frames and using cross correlation to get two velocity measurements of the group at two different times. If the first two frames are exposed with a short time interval to ensure optimum conditions for a velocity measurement, and the third frame is then exposed after a considerably longer time interval this will allow an eventual acceleration to change the velocity of the particle group. The third frame will then be correlated with the second frame to determine the second velocity by cross correlation.

The velocity varies in both time and space thus the difference can then be related as a first order approximation to the total acceleration of the fluid (the total derivative of the velocity in time):

$$\frac{d\underline{U}(\underline{x},t)}{dt} \simeq \frac{\Delta\underline{U}(\underline{x},t)}{\Delta t}$$

The Eulerian specification instead makes velocity measurements in each interrogation area at two different times. Two pairs of frames are acquired, where each of the frames in each pair is exposed with a short time interval to give the velocity measurements. The time interval between the two pairs of frames is then considerably longer to allow eventual accelerations to change the velocity in the interrogation areas. At each point the difference between the two velocity measurements will be related to the temporal derivative of the velocity, and the spatial difference between neighbouring velocity measurements will relate to the spatial derivative of the velocity. Finally these can be combined to give the total derivative of the velocity in time:

$$\frac{d\underline{U}(\underline{x},t)}{dt} = \frac{\partial\underline{U}(\underline{x},t)}{\partial t} + \frac{\partial\underline{U}(\underline{x},t)}{\partial\underline{x}} \cdot \underline{U}(\underline{x},t)$$

The limitations of the method of cross correlation PIV are discussed in the paper presented by Keane and Adrian (1992). All the optimisation criteria apply to the methods described and it also illustrates some basic problems/differences between the Eulerian and Lagrangian methods.

Generally the time intervals must be chosen carefully, especially between the two velocity measurements, which must be assumed to be acquired within a time interval where the acceleration within the fluid is almost constant.

The method based on the Lagrangian specification is limited by the need of the correlation routine to track/identify a particle group through all three frames, especially when accelerations are acting on the group. Any spatial gradients in the velocity field will tend to scatter the group and change its pattern. The correlation peak will therefore be reduced as a result and the detection of the correct peak might fail. This can be expressed by the following criteria, following the notation and theory developed by Adrian (1988) which relates the size of the correlation peak to the spatial velocity gradient within an interrogation area :

$$\mathrm{M}|\Delta\underline{U}(\underline{x},t)|_{d_I}\Delta t < d_\tau$$

Also the requirement for two very different time intervals between the three exposures means the the velocity measurements must be done within a large dynamic range (implying the use of a large interrogation area). This and the limited resolution of the CCD's are the main the reasons why analysis by cross correlation was employed.

Bias effects due to velocity gradients within relative large interrogation areas (as this method could imply) can be reduced by choosing a smaller interrogation area and moving it with the particle group (Keane, RD & Adrian, RJ, 1992), although as will be shown, this will lead to a larger systematic error in the velocity measurement.

System

The present system that has been constructed is illustrated in figure 1. This is a flat schematic of the system with the actual layout of the optics being 3-dimensional to allow for simpler construction and ease of image manipulation (figure 3). The cameras that are used in the construction of the device are Pulnix TM-9700's, which have a cell size of 11.6μm(H) x 13.3μm(V) in an array of 768(H) x 484(V) pixels. The CCD array itself is interline transfer and is progressive scan, with the output being selectable between interlaced and non-interlaced. The cameras are fully asyncronously resetable with a minimum full-frame integration time of 1/16000s (62.5μs). An area of continuing work is the accuracy avaliable with DPIV measurements. The fact that acceleration calculations require derivatives of velocities to be made implies that small errors in a velocity measurement can have a large effect on the acceleration if the difference between velocities is also small. It is with this in mind that an analysis of the DPIV technique was undertaken to see if there were any large sources of systematic error. It was found that using computer generated test data of randomly distributed gaussian particle images with $d_\tau/d_{pix} \approx 2$ (Prasad *et. al.*, 1992) there were three main sources of error, a systematic error possibly due to the effect of incomplete particle images in the extracted image sections, a sinusoidal error due to the

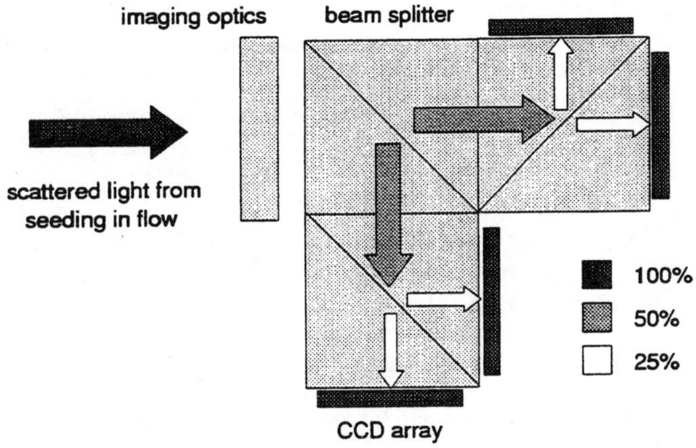

Figure 1: Optical layout of multiple CCD system

three point sub-pixel estimators (Prasad *et. al.* 1992) and a random error. These effects can be seen clearly in figure 2a) & b), with displacements from -10..10 pixels being measured at 0.1 pixel intervals. Each point on the graphs being the result of 1296 statistically independent measurements, and so a reliable mean and standard error on the mean can be found. The source of the errors can be seen to originate from two different sources, as the running average over 10 samples for both the parabolic peak fit and the gaussian peak fit are shown to be identical (figure 2b).

Although these effects appear very small, the results were obtained from idealised particle images, and so under real sampling conditions the effect could be much worse, as stated by Prasad *et. al.* (1992), and this is an area where more work is yet to be done, although this has been only qualitatively confirmed by the authors so far. Removal of incomplete particle images by a blob removal technique and also applying a kernel of varying types to the raw extracted image has as yet proved inconclusive in the removal of the systematic error.

Application

The application choosen initially is that of waves of different types impinging onto a vertical plain wall, with both sinusoidal and plunging breakers being currently investigated. Preliminary results from the plunging breakers are presented here.

The vector maps in figure 4 illustrate the two velocity fields and the Lagangian and the Eulerian acceleration field of the fluid flow formed by a plunging breaker. The wave has just experienced impact with a vertical wall and is beginning to withdraw from the wall at the time of acquisition. The PIV images are acquired next to the wall and just above the mean

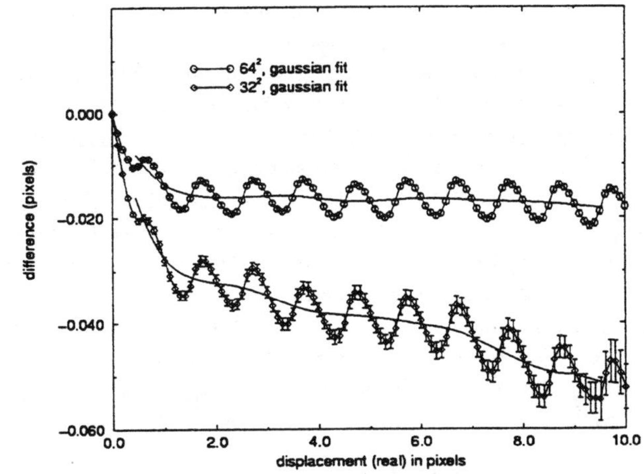

a) effect of interrogation size on error

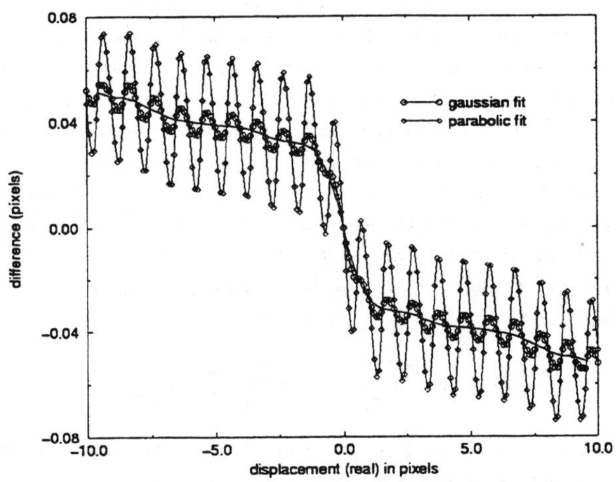

b) effect of different
3-point estimators

Figure 2: Systematic errors in DPIV measurements

Figure 3: Multiple CCD system

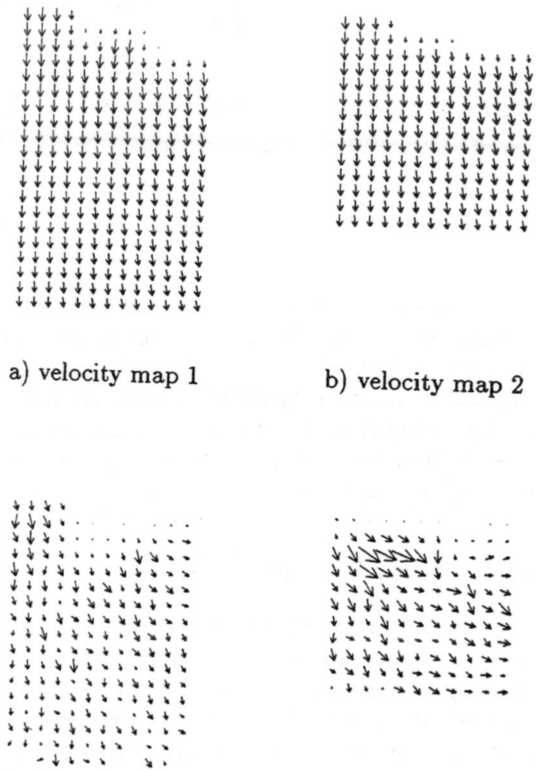

a) velocity map 1 b) velocity map 2

c) Lagrangian acceleration d) Eulerian acceleration

Figure 4: velocity and acceleration vector maps

water level. The physical area contained by the PIV image is of a size of 0.15m x 0.11m and relates to the main wavelength of the plunging breaker at the wall of approximately 1.5m. The time interval between the two velocity measurements (the time interval between frames 2 and 3, δt_{23}) is 35ms or 7 times the time interval between frames correlated for determination of velocity (time interval between frames 1 and 2, or frames 3 and 4).

The flow in the given window and phase is clearly moving and is accelerated downwards toward the mean water level, and similarly away from the wall.

The simplest way to present the acceleration of this fluid flow, and at the same time to gain a simple noise reduction, is spatial averaging of vectors of the entire PIV image. With a given size of the PIV image relative to the main wavelength of the wave, local details still appear clearly and in perfect correspondence with the observations based on the vector maps. Figure 5 illustrates the averaged accelations of both methods versus different time intervals, δt_{23}. Both the Lagrangian and the Eulerian method measure a significant negative vertical acceleration (downwards) and a positive horizontal acceleration (away from the wall). As the time interval, δt_{23}, increases, the measurements of the accelerations are averaged in time over a similarly increasing period. Thus according to the given phase of the first frame (fixed in time) the time-averaged acceleration should slightly decrease as the averaging period increases. This effect is also apparent in the measurements of the vertical acceleration of both mehods.

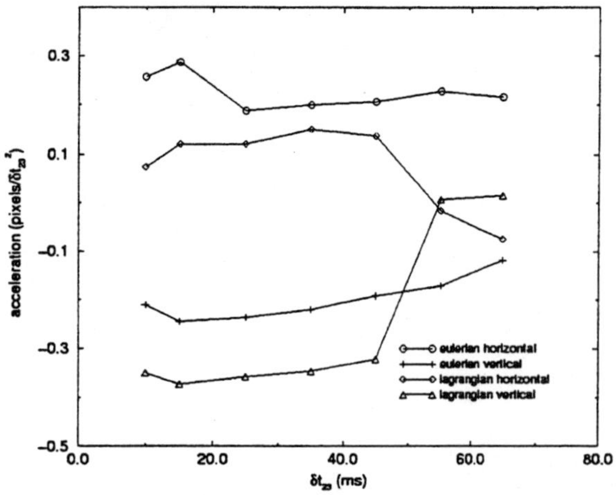

Figure 5: Measured acceleration with varying δt_{23}

The two methods indicate a certain displacement between their measurements. This is caused by the velocity bias mentioned above, and which under given conditions only affects the Lagrangian method. Further, the Lagrangian method has a cut-off when the time interval, δt_{23}, is extented beyond 40-50ms, as the cross-correlation between particle groups in frames 2 and 3 disappears.

Conclusion/Future Work

The immediate future work is to concentrate on the reduction or removal of systematic errors in the calculation of the acceleration to yield a more accurate measurement, which can then be used to calculate forces acting on boundaries within the fluid flow. As one of the requirements of PIV is that the accelerations present within interrogation regions should be negligible, the measurement of strongly accelerating flows with any accuracy becomes a problem. It is intended that further work will establish if deformation of the interrogation area will improve the quality of acceleration measurement, following the work of Hind & Christy (1996) and Huang *et. al.* (1993a & 1993b). This is a promising line of approach as increasing the interrogation area should reduce the effects of the systematic errors, as well as providing a more reliable velocity estimate due to the deformation of the interrogation area.

Bibliography

Adrian, RJ (1991) "Particle Imaging Techniques for Experimental Fluid Dynamics" *Ann. Rev. Fluid Mechanics*, pp261-304, Vol 23

Adrian, RJ (1988) "Statistical properties of particle image velocimetry measurements in turbulent flow" *Laser Anemometry in Fluid Mechanics*, LADOAN Institute Superior Technico, Lisbon, Portugal, pp115-129, Vol III

Dewhirst, TP, Jakobsen, ML, and Greated, CA (1996) "Multiple CCD camera DPIV for Force and Acceleration Measurements in High Speed Flows" *Proc. IMechE Optical Methods and Data Processing in Heat and Fluid Flows*, pp251-258

Hind, AK & Christy, JRE (1996) "A Method to Correct for Velocity Gradients in the Analysis of PIV Recordings" *Proc. IMechE Optical Methods and Data Processing in Heat and Fluid Flows*, pp201-210

Huang, HT, Fielder, HE & Wang, JJ (1993a) "Limitation and Improvement of PIV. Part I: Limitation of Conventional Techniques due to Deformation of Particle Image Patterns" *Experiments in Fluids*, 15:168-174

Huang, HT, Fielder, HE & Wang, JJ (1993b) "Limitation and Improvement of PIV. Part II: Particle Image Distortion, a Novel Technique" *Experiments in Fluids*, 15:263-273

Keane, RD & Adrian, RJ (1992) "Theory of cross-correlation analysis of PIV images" *Applied Scientific Reaserch*, pp191-215, Vol 49 No 3

Prasad, K, Adrian, RJ, Landreth, CC, Offutt,PW (1992) "Effect of Resolution on the Speed and Accuracy of Particle Image Velocimetry Interrogation" *Experiments in Fluids*, 13:105-116

Racca, RG, Stephenson, O & Clements, RM (1992) "High-speed video analysis system using multiple shuttered charge coupled device imagers and digital storage" *Optical Engineering*, pp1369-1374, Vol 31 No 6

Raffel, M, Kompenhans, J, Stasicki, B, Bretthauer, B & Meier, GEA (1995) "Velocity measurement of compressible air flows utilizing a high-speed video camera" *Experiments in Fluids*, pp204-215, Vol 18

Sarpkay & Isaacson (1981) *Mechanics of waves on offshore structures*, Van Nostrand Reinhold Company Inc

Willert, C, Stasicki, B, Raffel, M & Kompenhans, J (1995) "A digital video camera for application of particle image velocimetry in high-speed flows" *SPIE International Symposium on Optical Science, Engineering, and Instrumentation*, SPIE Proceedings Article No. 2546-19, 9-14 July

Proceedings of the Seventh (1997) International Offshore and Polar Engineering Conference
Honolulu, USA, May 25-30, 1997

ISBN 1-880653-28-1 (Set); ISBN 1-880653-31-1 (Vol. III)

A Free-Surface Pressure Distribution Solution for the Radiation and Diffraction Problems of Large-Diameter Piles

G.P. Miao, Y.X. You and Y.Z. Liu
Shanghai Jiao Tong University
Shanghai, China

ABSTRACT

The equivalence between the free-surface pressure distribution and the floating body motion forms a new kind of inverse problems. For the radiation and diffraction problems of large-diameter piles, corresponding free-surface pressure distribution solutions are obtained in the present paper. The study shows that a first kind Fredholm integral equation with a complex kernel can still be formed for the present inverse problem and the regularization method may be effectively used to obtain a stable unique solution. The effects of the parameters, such as water depth , vibration or incident frequencies etc., on the free-surface pressure distribution are also discussed in the paper in detail.

KEY WORDS: Large-diameter pile, pressure distribution, inverse problem, radiation, diffraction

INTRODUCTION

Free surface pressure distribution solutions for the radiation and diffraction problems of floating bodies denote such free surface pressure distributions by which the fluid response (outside the region occupied by the body) agrees completely with the one caused by the body with a given motion. The equivalence between the free-surface pressure distribution and the floating body motion forms a new kind of inverse problems.

The research on the equivalence problem is of importance in many aspects. In order to give an insight into the mechanism of body motion effects on the fluid response, the motion characteristics of fluid fields caused by free surface pressure distributions have been investigated by several researchers(Ertekin, Webster and Wehausen, 1986) since in some cases the body surface boundary condition may be troublesome in the fluid computation. However, the treatment method comes mainly from instinct. A series of important questions in the formulation of the corresponding inverse problems, for example, whether or not an equivalent free surface pressure distribution is uniquely existing, what is the explicit relationship between the free-surface pressure distribution and the body motion etc., haven't been systematically investigated yet. To answer these questions is of importance for the solution of the present equivalence inverse problems. Moreover, a potential application lies in that one may replace the motion of a real body by a corresponding free surface pressure distribution in some special cases。

A wave generation problem corresponding the equivalence between wave making by an oscillating flat plate and that by the pulsating air pressure in a two-dimensional wave tank was investigated by the present authors(Miao, You and Liu, 1996). In the linear potential regime, a first kind Fredholm integral equation with a complex kernel that determines the free-surface pressure distribution is obtained, the regularized solution and the effects of the regularization parameter on the regularized solution are further discussed in You and Miao(1996). The method is extended to the corresponding three-dimensional problem in the present paper, that is, we want to obtain corresponding free-surface pressure distribution solutions for the radiation or diffraction problems of large-diameter piles. The study shows that a first kind Fredholm integral equation with a complex kernel can still be formed for the present inverse problem and the regularization method may be effectively used to obtain a stable unique solution.

FORMULATION OF THE PROBLEM

There exists a vertical large-diameter pile of a radius a in waters of constant depth h with one end mounted on the sea bottom and the other piercing the free surface. The diffraction problem treats the disturbance of piles to the incident waves, while the radiation problem deals with the hydrodynamic response of an oscillating pile in still waters. A cylindrical coordinate system (r,θ,z) is defined with the origin on the undisturbed free surface and the oz axis pointing vertically upward, as shown in Fig.1.In the linear potential regime, the radiation and diffraction potentials of a large-diameter pile in the fluid field $r \ge a$ may be determined by the normal velocity $f(\theta,z)$ on the pile surface(MacCamy and Fuchs, 1954).

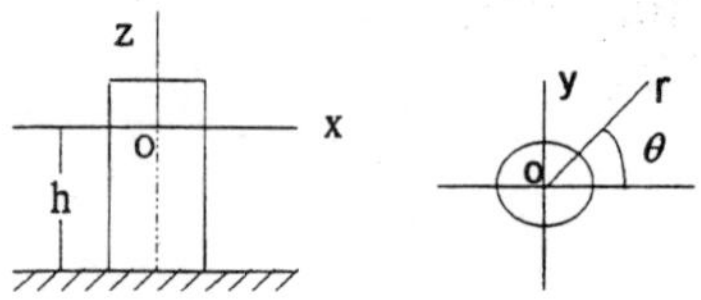

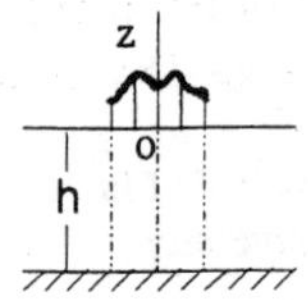

Fig.1 The cylindrical coordinate system Fig.2 Pressure Distribution

Now, we turn to consider such an inverse problem, i.e. seeking a pressure distribution $\mathrm{Re}\{\tilde{p}(r,\theta)e^{i\omega t}\}$ on the free surface $r \le a$ so that the velocity potential generated by the pressure distribution in the fluid field $r \ge a$ can imitate the radiation or diffraction potentials of the large-diameter pile, as shown in Fig.2, where Re{A} represents the real part of A, $\tilde{p}(r,\theta)$ and ω are known as spatial mode and frequency of the pressure distribution respectively.

Under the action of the pressure distribution, the velocity potential in the fluid field can be denoted as $\phi_I(r,\theta,z,t) = \mathrm{Re}\{\phi(r,\theta,z)e^{-i\omega t}\}$ where $\phi(r,\theta,z)$ is the spatial part of $\phi_I(r,\theta,z,t)$ should be governed by a three-dimensional Laplace equation ,the bottom condition, the radiation condition in the infinity, and the linear free surface condition $\frac{\partial\phi}{\partial z} - \frac{\omega^2}{g}\phi = \begin{cases} \frac{i\omega}{\tilde{\rho} g}\tilde{p}, & r \le a; \\ 0, & r \ge a \end{cases}$ on $z=0$. To determine the pressure distribution, we further ask that ϕ also satisfies the boundary condition on the mean position of the large-diameter pile as that in the radiation or diffraction problems of the pile, i.e.

$$\frac{\partial\phi}{\partial r}\Big|_{r=a} = f(\theta,z) \tag{1}$$

in which $\tilde{\rho}$ is the density of water, and g the acceleration of gravity.

For convenience sake, we denote $p(r,\theta) = \frac{i\omega}{\tilde{\rho} g}\tilde{p}(r,\theta)$, and hereafter refer p as the pressure distribution.

For the special case of a concentrated pressure at the acting point(ρ,α) on the free surface $\rho \le a$, we may write $p(r,\theta) = \frac{\delta(r-\rho)}{r}\delta(\theta-\alpha)$,the velocity potential at the field point (r,θ,z) may then be expressed as(Miao and Liu, 1989)

$$\phi\,(r,\theta,z) = \frac{1}{2\pi}\sum_{n=0}^{\infty}\varepsilon_n\phi_n(r,z)\cos n(\theta-\alpha) \tag{2}$$

in which $\varepsilon_0 = 1; \varepsilon_n = 2\ (n \ge 1)$,and

$$\phi_n(r,z) = \pi i k_0 q_0 \frac{\cosh k_0(z+h)}{\cosh k_0 h}\begin{Bmatrix} H_n^{(1)}(k_0 r)J_n(k_0\rho) \\ J_n(k_0 r)H_n^{(1)}(k_0\rho)\end{Bmatrix} + 2\sum_{j=1}^{\infty} k_j q_j \cos k_j(z+h)\begin{Bmatrix} K_n(k_j r)I_n(k_j\rho) \\ I_n(k_j r)K_n(k_j\rho)\end{Bmatrix}\begin{Bmatrix} r>\rho \\ r<\rho\end{Bmatrix} \tag{3}$$

where $q_0 \equiv \frac{1}{\tanh k_0 h + k_0 h \operatorname{sech}^2 k_0 h}$, $q_j \equiv \frac{1}{\sin k_j h + k_j h / \cos k_j h}$,

k_0 is the wave number determined by the dispersion relation $\frac{\omega^2}{g} = k_0 \tanh k_0 h$ and k_j are determined by $\frac{\omega^2}{g} = -k_j \tan k_j h$ 。

In the case of r>a, equation (3) may also be rewritten as

$$\phi_n(r,z) = \pi i k_0 q_0 \frac{\cosh k_0(z+h)}{\cosh k_0 h} H_n^{(1)}(K_0 r)J_0(k_0\rho) + 2\sum_{j=1}^{\infty} k_j q_j \cos k_j(z+h)K_n(k_j r)I_n(k_j\rho) \tag{4}$$

For an arbitrary pressure distribution $p(\rho,\alpha)$ $(\rho \le a)$, the velocity potential generated at the field point $(r,\theta,z)(r > a)$ may then be written as(Miao and Liu, 1989)

$$\phi(r,\theta,z) = \int_0^a\int_{-\pi}^{\pi}\left\{\frac{1}{2\pi}\sum_{n=0}^{\infty}\varepsilon_n \cos n(\theta-\alpha)\phi_n(r,z)\right\} p(\rho,\alpha)d\rho d\alpha \tag{5}$$

where $\phi_n(r,z)$ is determined by Eq.(4).

Let the function $P(\rho,\alpha)$ be even with respect to α and expressed in terms of the Fourier series, i.e.

$$P(\rho,\alpha) = \sum_{m=0}^{\infty} p_m(\rho)\cos m\alpha \tag{6}$$

the integration in Eq.(5) with respect to α becomes

$$I = \int_{-\pi}^{\pi}\cos n(\theta-\alpha)\cos m\alpha d\alpha$$
$$= \int_{-\pi}^{\pi}\left[\cos n\theta\cos n\alpha - \sin n\theta \sin n\alpha\right]\cos m\alpha d\alpha \tag{7}$$

it is easily obtained from the orthorgonality of trigonometric functions that

$$\phi(r,\theta,z) = \sum_{n=0}^{\infty}\cos n\theta\int_0^a \phi_n(r,z)p_n(\rho)d\rho \tag{8}$$

Substituting Eq.(4) in the above equation yields

$$\phi(r,\theta,z) = \sum_{n=0}^{\infty}\cos n\theta\int_0^a\left[\pi i k_0 q_0 \frac{\cosh k_0(z+h)}{\cosh k_0 h} H_n^{(1)}(k_0 r)J_n(k_0\rho) + 2\sum_{j=1}^{\infty} k_j q_j \cos k_j(z+h)K_n(k_j r)I_n(k_j\rho)\right]p_n(\rho)d\rho \tag{9}$$

We suppose that the function $f(\theta,z)$ is an even function of the argument θ, then it may also be expressed in terms of the Fourier series as

$$f(\theta,z) = \sum_{n=0}^{\infty} f_n(a,z)\cos n\theta \tag{10}$$

and the boundary condition (1) may be rewritten as

$$\frac{\partial\phi}{\partial r}\Big|_{r=a} = \sum_{n=0}^{\infty} f_n(a,z)\cos n\theta \tag{11}$$

According to Eq.(9) and Eq.(11), we have the following first kind Fredholm integral equations

$$\int_0^a\left[\pi i k_0^2 q_0 \frac{\cosh k_0(z+h)}{\cosh k_0 h} H_n^{(1)\prime}(k_0 a)J_n(k_0\rho) + 2\sum_{j=1}^{\infty} k_j^2 q_j \cos k_j(z+h)K_n{}'(k_j\rho)\right]P_n(\rho)d\rho = f(a,z) \tag{12}$$

for different n $(n = 1,2,...,\infty)$, in which, $-h \le z \le 0$.

For convenience sake, we denote it as

$$\int_0^a K_n(z,\rho)P_n(\rho)d\rho = f_n(a,z) \quad (n=1,2,\ ...)\ ,-h \le z \le 0 \qquad (13)$$

in which $K_n(z,\rho)$ denotes the expression in brackets in Eq.(12). $P_n(\rho)$ represents the solution of the inverse problem。

Remark 1. For the diffraction problem, the spatial mode of the incident wave potential propagating in the ox direction may be expressed as

$$\phi_{ic} = \frac{\overline{A}\,g}{i\omega}\frac{\cosh k_0(z+h)}{\cosh k_0 h}e^{ik_0 r\cos\theta}$$

$$= -\frac{i\overline{A}g}{\omega}\frac{\cosh k_0(z+h)}{\cosh k_0 h}\left[2\sum_{n=1}^{\infty} i^n J_n(k_0 r)\cos n\theta + J_0(k_0 r)\right] \qquad (14)$$

where $\overline{A}$ is the wave amplitude, ω is the wave frequency, then we have

$$f(\theta,z) = -\frac{\partial\phi_{ic}}{\partial r}\Big|_{r=a} =$$

$$\frac{i\overline{A}\,g\,k_0}{\omega}\frac{\cosh k_0(z+h)}{\cosh k_0 h}\left[2\sum_{n=1}^{\infty} i^n J_n'(k_0 a)\cos n\theta + J_0'(k_0 a)\right] \qquad (15)$$

and finally

$$f_n(a,z) = \frac{i\overline{A}\,g\,k_0}{\omega}\frac{\cosh k_0(z+h)}{\cosh k_0 h}\left[2i^n J_n'(k_0 a)\right]$$

$$f_0(a,z) = \frac{i\overline{A}\,g\,k_0}{\omega}\frac{\cosh k_0(z+h)}{\cosh k_0 h}J_0'(k_0 a) \qquad (16)$$

For a complete diffraction potential, all n terms need to be solved by truncating to a suitably large number N. However, only the case of $n = 1$ is needed to determined the diffraction forces on the pile due to its special configuration.

Remark 2. For the radiation problem, we assume that the vibration displacement of the pile is $x = X(z)e^{-i\omega t}$ with $X(z)$ being the amplitude of the vibration and the normal velocity at any point (r,θ,z) on the pile surface then becomes $v_n = -i\omega X(z)\cos\theta e^{-i\omega t}$. In this case, we need only to solve Eq.(13) for $n = 1$, that is

$$f(a,z) = -i\omega X(z) \qquad (17)$$

It is well known that the first kind Fredholm integral equation (13) is in general ill-posed ,that is, the solution of the equation (13) may not depend continuously on f of the right side(You and Miao, 1996). Tikhonov regularization method is adopted to restore the stability of the solution. To this end, we assume that $p_n \in H^1[0,a]$, $f_n \in L_2[-h,0]$ where $L^2[-h,0]$, $H^1[0,a]$ are a complex Hilbert space and a complex Sobolev space respectively. The Tikhonov functional is taken as $M^\alpha(p_n,f_n)$ = $\|Ap_n - f_n\|_{L^2}^2 + \alpha\Omega(f_n)$ in which $\Omega(p_n) = \int_0^a \{\left|\frac{dp_n}{ds}\right|^2 + |p_n(s)|^2\}ds$ is a stable functional, α(>0) is a suitably chosen constant known as a regularization parameter. The regularized solution p_n^α of equation (13) can then be obtained by minimizing the functional $M^\alpha(p_n,f_n)$, which satisfies following the boundary value problem(You and Miao, 1996)

$$\begin{cases}\int_0^a K_n^*(s,\rho)p_n^\alpha(\rho)dx - \alpha\{\frac{d^2p_n^\alpha(s)}{ds^2} - p_n^\alpha(s)\} = b_n(s) \\ p_n^\alpha(0) = 0, \quad p_n^\alpha(a) = 0\end{cases} \qquad (18)$$

where, $K_n^*(s,\rho) = \int_{-h}^0 K_n(z,\rho)\overline{K_n}(z,s)dz$, $b_n = \int_{-h}^0 \overline{K_n}(z,s)f_n(a,z)dz$.

The problem (18) may be solved by the usual difference method, i.e.

$$\sum_{i=1}^n hK^*(s_j,\rho_i)p_n^{\alpha,i} - \alpha\{\frac{p_n^{\alpha,j-1} - 2p_n^{\alpha,j} + p_n^{\alpha,j+1}}{h^2} - p_n^{\alpha,j}\} = b(s_j) \qquad (19)$$

in which, $h = \frac{a}{n+1}, x_j = s_j = h\times j(j = 1,\cdots m)$, and $p_n^{\alpha,0} = p_n^\alpha(0)$ $= 0$, $p_n^{\alpha,m+1} = p_n^\alpha(a) = 0$

The equation (19) may also be rewritten in a matrix form as

$$(A_m + \alpha\{{}^{C_m}/_{h^2} + I_m\}\tilde{p}_n^{m,\alpha} = B_m \qquad (20)$$

where $\tilde{p}_n^{m,\alpha} = (p_n^{\alpha,i})_m^T$ is the approximate solution of the problem(18), and I_m is a unite matrix of m order.

NUMERICAL RESULTS AND DISCUSSION

For the diffraction problem of large-diameter pile, we discuss the behavior of the free surface pressure distribution in detail. In the following computational examples, both the amplitude of the incident waves and the radius of the pile are taken as 1.0. The accuracy of $\tilde{p}_1^{m,\alpha}$ is examined by comparing the wave diffraction force on the pile obtained with the present free-surface pressure distribution method and that obtained from the corresponding direct problem, as shown in Table1 for the cause of $h = 10.0$, $\omega = 5.0$. The effects of the regularization parameter on the convergence and stability of the solution are shown in Fig.3. It is shown from these results that the regularization parameter plays an important role in controlling the convergence, stability of the solution, and the stable functional $\Omega(p_n)$ corresponds "a filter", which plays a role in "filtering" high-frequency oscillation parts of the solution. Hence, if α is too small then smoothness becomes unimportant, and the solution may be not convergent and stable. On the other hand, if α is too large then the solution has very good convergence and stability, but the accuracy of the solution will be very bad. A very crucial problem is how to find an suitable regularization parameter. An useful and effective method introduced in reference[3] is still suitable to our present problem. Secondly, The effects of the water depth and incident frequency on p_1^α are also discussed in detail. We show in Fig.4 that the effects of the frequency on p_1^α, where h is 10.0, and ω is taken as 3.6,4.0, 4.3 and 5.0 respectively, the corresponding regularized parameters are 1.9786×10^{-10}, 2.3157×10^{-10}, 3.5746×10^{-10}, 6.7537×10^{-10} respectively. In Fig 5, we show the effects of the water depths on p_1^α, where ω is 5.0, and h is taken as 8.0, 9.0,10.0 respectively, the corresponding regularized parameters are 7.9789×10^{-10}, 7.2539×10^{-10}, 6.7537×10^{-10} respectively. It is easily seen from these figures that the effects of the frequency on p_1^α are larger than that of the water depth on p_1^α. In addition, in order to investigate the behavior of the free surface pressure distribution, we calculate also the regularized solutions of p_0, p_2, and p_3, as shown in Fig. 6, where h and ω are 10.0 and 5.0 respectively, and the corresponding regularized parameters are 4.6876×10^{-9}, 9.7836×10^{-11} and 5.5683×10^{-11} respectively.

CONCLUSIONS

It is shown from the present results and analysis that the equivalence between the free-surface pressure distribution and the floating body motion taking a vertical pile as an example in the paper can be reached by seeking a free surface pressure distribution solution to the corresponding inverse problem.. A first kind Fedholm integral equation with a complex kernel can still be formed for the present inverse problem and the regularization method can be effectively used to obtain a stable unique solution.

ACKNOWLEDGEMENT

This project is supported by the National Nature Science Foundation of China.

REFERENCES

Ertekin,RC, Webster,WC, and Wehausen,JV(1986). " Waves Caused by a Moving Disturbance in a Shallow Channel of Finite Width," J Fluid Mech., Vol 169, pp275-191.

Kress, R(1989). "Linear Integral Equations," Springer-Verlag, World Publishing Corp.

MacCamy, RC and Fuchs, RA(1954). "Wave Forces on Piles: A Diffraction Theory," US Army Beach Erosion, Tech Memo, No69.

Miao, GP, and Liu, YZ(1989). "On Velocity Potentials due to Pulsating Pressure Distributions on the Free Surface of Infinite Depth Waves and the Radiation Conditions for Second Order Diffraction Problems," Shipbuilding of China, No107, pp1-13.

Miao, GP, You, YX, and Liu, YZ(1996). "An Inverse Problem of Water Generation in a Water Tank," J Hydrodynamics, SerB, No4, pp94-102.

You, YX and Miao, GP(1996). " On the Regularization Method of the First Kind Fredholm Integral Equation with a Complex Kernel and Its Application," J Appl Math and Mech, to appear.

1st-order diffraction		Real part	Imaginary part
Direct problem		0.2061808	0.04644297
Inverse	$\alpha = 6.7537 \times 10^{-7}$	0.2096683	0.04067206
problem	$\alpha = 6.7537 \times 10^{-10}$	0.2071179	0.04491685

Table 1 The effects of regularization parameter on the accuracy of the regularized solution, where the diffraction forces are nondimensionalized with the factor of $1/\tilde{\rho} g \bar{A} a^2$.

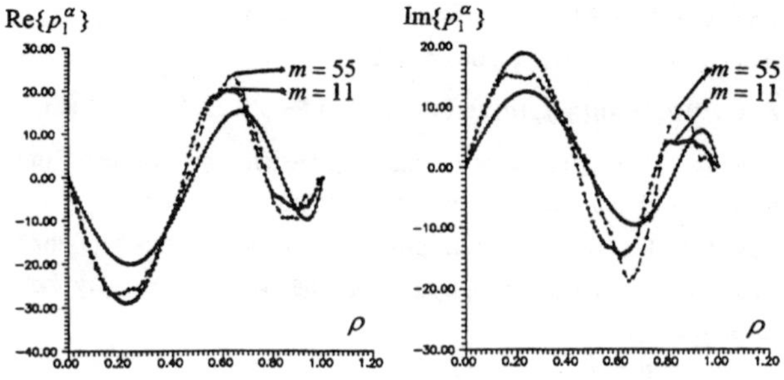

Fig.3 The effects of the regularization parameter on the stability and convergence of the regularized solution, where $h = 10.0$, $\omega = 5.0$, full lines: $\alpha = 6.7537 \times 10^{-10}$, dashed line $\alpha = 6.7537 \times 10^{-12}$.

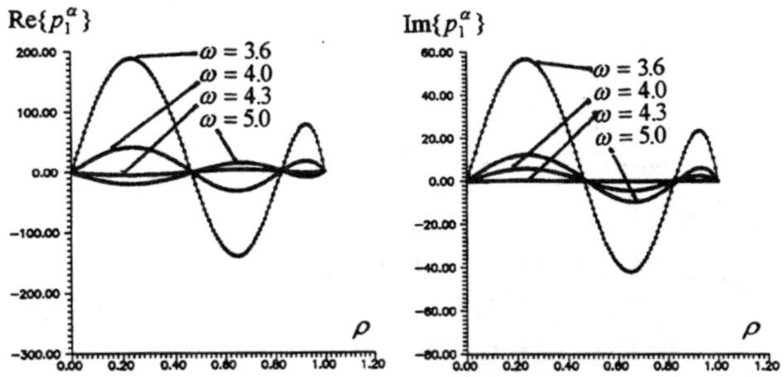

Fig.4 The effects of the pressure pulsating frequency on the regularization solution p_1^α.

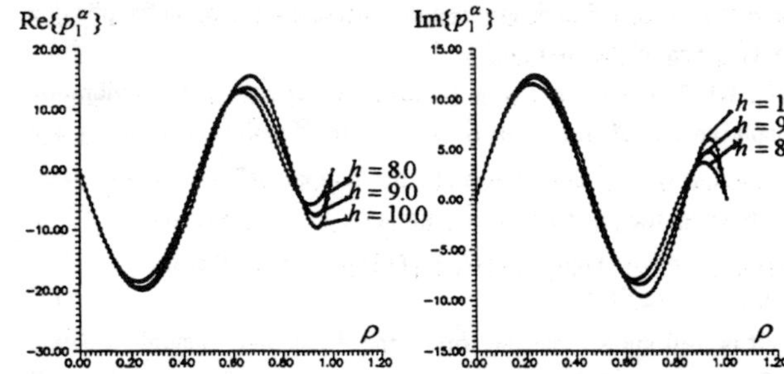

Fig.5 The effects of the water depth on the regularization solution. p_1^α.

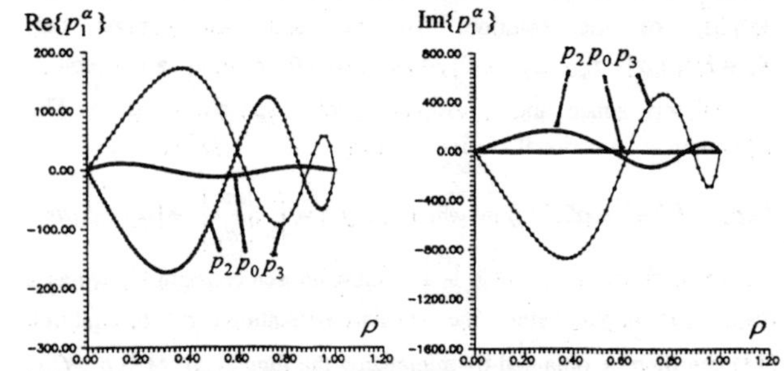

Fig.6 The regularization solutions of p_0, p_2 and p_3.

Proceedings of the Seventh (1997) International Offshore and Polar Engineering Conference
Honolulu, USA, May 25-30, 1997

ISBN 1-880653-28-1 (Set); ISBN 1-880653-31-1 (Vol. III)

Predictions of Morison Type Forces in Irregular, High Reynolds Number Waves

Geir Moe
Norwegian University of Science and Technology
Trondheim, Norway

Ove Tobias Gudmestad
STATOIl
Stavanger, Norway

ABSTRACT

This paper summarizes some key points about the Morison formula, its background and some experiences on its application. An attempt is made to categorize the different types of flows-pile combinations for which the Morison formula applies. For irregular, high Reynolds number flows it is attempted to identify the parameters that characterize a specific flow case, and how to define these parameters. Some new analyzes are presented.
The most important feature of this paper may be in the area of directional waves where an extremely simple method, the so-called 'principal flow approach', is shown to model the directionality of the waves so that the unidirectional results can be used to predict the actual wave force. This approach proved to be extremely accurate in the range tested, namely for Keulegan-Carpenter numbers up to 35.

INTRODUCTION

The Morison formula for prediction of the forces on a long circular cylinder was originally proposed by Morison et Al (1950) on an ad hoc basis by considering two limiting cases. The first case was one in which the fluid motion was small relative to body dimensions, so that the flow remained unseparated, and potential theory was valid, yielding a force equal to twice the displaced mass. (Provided the cylinder diameter is short compared to the spatial variation of the flow). In the other limiting case the fluid motion in each direction was very large, so that the flow could be considered as piecewise steady and the force was given empirically as a so-called form drag force. Denoting cylinder radius, water density and velocity and acceleration as R, ρ, u, and $\dot{u}$ respectively the «Morison equation» then predicted a force per unit length equal to the sum of these two terms, but adjusted them by two empirical coefficients C_m and C_d (so-called force coefficients), viz.

$$f = C_m \rho \pi R^2 \dot{u} + C_d \rho R u|u| \qquad (1)$$

Please note that the name «equation» was chosen for (1), suggesting mathematical rigor and thus encouraging unquestioning acceptance. Furthermore, the names of the coefficients, C_m and C_d, were unfortunately identical to the names of the corresponding coefficients in the two above mentioned limiting situations. In the seventies Sarpkaya established that these coefficients for a stationary cylinder in a unidirectional, harmonic, zero mean flow depended upon the Keulegan-Carpenter number KC, the Reynolds number Re and the average surface roughness ratio k/D in which k is the surface roughness diameter and $D=2R$ the cylinder diameter, see e.g. Sarpkaya et Iscaacson (1981). However, users and regulatory bodies were unclear on these matters. To exemplify, even the highly respected DNV rules (1982) prescribed the drag coefficient C_d as identical to the steady flow drag value (According to Section B2.1.5 and Figure B.1) Further even as late as in the 1989 DNV rules the limit for where to apply Morison's formula was set to the point where the ratio of wavelength L to diameter D exceeded 5. An example where this criterion is clearly incorrect is for a circular cylinder in long waves of low amplitude, or at a depth where the wave induced particle motion amplitude is small. Then $L/D>5$ could occur together with a very low Keulegan-Carpenter number. The latter means that there will be no separation, so that potential theory should be used, and not Morison's formula.

A very high number of experiments and some full scale measurements have been performed to determine the forces on cylinders in waves under conditions where the flow separates, so that Morison's formula may apply. It has turned out that the Morison's formula has been reasonably successful as a predictive tool, and no attempt to replace it or improve it e.g. by adding extra terms, has been successful, (Bendat and Piersol, 1996). It is however necessary to be discriminating concerning the flow conditions under which the Morison formula is used. In table 1 is shown a number of the most actual regular flow cases.

CLASSIFICATION OF FLOW SITUATIONS IN WHICH MORISON'S FORMULA APPLIES

A large number of experiments have dealt with a case in which the flow has been rectilinear and sinusoidal and perpendicular to the axis of a stationary test cylinder. This situation will be denoted 'the base case' herein, see Table 1, Case 1A. A current in-line with the oscillatory flow may also be added, Case 1B. In Table 1 is also shown other regular flow situations about a circular cylinder: Case 2 is as Case 1, but the flow direction is here at an inclined angle to the cylinder, Case 3 has a 'wave flow' in the same plane as the cylinder, i.e. a flow having circular or elliptical particle orbits, while in Case 4 a current in the same plane has been added to the flow components of Case 3, possibly at an inclined angle to the cylinder. Case 5 has a wave flow rotating around the cylinder axis, while in Case 6 a horizontal current and waves that propagate in directions perpendicular to the current acts on a vertical cylinder. Cases 7 and 8 covers various proximity effects, while Case 9 covers short cylinders giving rise to 3 D local flows. A literature survey of several of these cases is presented in Moe (1989).

The presented list is by no means exhaustive, and can easily be expanded. The orientation of the axes of the ellipse relative to the cylinder has thus been ignored in Case 3 and 4, and in Case 5 the cylinder can be given a skewed position relative to the wave front, and a current can also be added, while in Case 6 the cylinder could be inclined (in two directions). Finally the short cylinder in Case 9 could be positioned in all the above mentioned positions, relatively to the wave and current flows.

RELATIVE VELOCITY FORMULATION

It can be proven that a 2 D flow is uniquely determined by the difference between the undisturbed flow velocity u and the cylinder velocity $\dot{x}$, commonly called 'the relative velocity' $u_{rel} = u - \dot{x}$. If the relative velocity thus defined is identical for two flows, then the relative velocity field will be equal everywhere, and the only difference in the pressure fields will be a pressure gradient that accelerates the fluid particles in one system relatively to the other. Thus consider (A) as the first system a fixed cylinder in a velocity field $u=u(t)$ for which the force is given by (1), and (B) another system with a moving cylinder with velocity $\dot{x}$ in a flow with undisturbed velocity $\hat{u}$ so that the relative velocity is equal to u for the first system, $u_{rel} = \hat{u} - \dot{x} = u$. Then the force on the moving cylinder must be modified by the amount $\rho\pi R^2\ddot{x}$ which results in the well known expression for the Morison force for the moving case

$$f = \rho\pi R^2(C_m\dot{u}_{rel} + \ddot{x}) + C_d\rho R u_{rel}|u_{rel}| = \rho\pi R^2\{C_m\dot{u} - (C_m - 1)\ddot{x}\} + C_d\rho R u_{rel}|u_{rel}| \quad (2)$$

Here it should be stressed that the force coefficients C_d and C_m will be identical for the two flow situations, as long *as* $u=u_{rel}$. Up to now only harmonic or plane circular or elliptical orbits have been considered, possibly with an additional current. Equation (2) will often give rise to biharmonic relative velocities u_{reb} however, since the cylinder will often vibrate at a resonance period that is different from the wave period. Then the problem will have one more nondimensional parameter, e.g. a Keulegan-Carpenter number based on the cylinder velocity amplitude and frequency, KC_x, and the experiments must therefore be performed with biharmonic relative velocity. One series of such experiments has been performed for a case meant to represent a TLP in extreme waves, Moe & Al (1987), with a platform period to wave period ratio in the range 3-6, a moderate wave KC number, and a low platform motion KC_x number. It turned out that the damping of the TLP motion was not changed by the presence of the waves, so that it could best be modeled by disregarding the wave-cylinder velocity interaction, i.e. a so-called independent flow field model seemed most appropriate, viz.

$$f = \rho\pi R^2(C_m\dot{u}_{rel} + \ddot{x}) + \rho R(C_d u|u| - C_{dx}\dot{x}|\dot{x}|) \quad (3)$$

Please note that the high frequency velocity u in these tests was considerably larger than $\dot{x}$, so that the difference between formula (2) and (3) in this situation is significant. There is more literature to be found on biharmonic flows, however the authors are not aware of any such papers which interpret such data in the light of formula (3) above. This would be highly desirable, to provide an independent check of the findings, and also to determine its range of validity. Furthermore the API Recommended Practice RP2A recommends the use of the independent flow field model rather than formula (2) for cases where the structural motion is small.

A limiting situation to biharmonic flow is when the 'wave' period becomes infinitely long, i.e. the waves are replaced by a steady current, Case 1B. Some early experiments for this situation have been performed by Verley & Moe (1978) and has been widely reported on, notably in Sarpkaya & Isaacson (1981). Based on this Laya & Al. (1981) predicted that the relative velocity assumption for cylinders dynamically excited by waves will only be valid if the flow separation and vortex shedding processes both have time to adjust to the instantaneous flow velocity and cylinder motion, i.e. behaved 'quasisteady'. Then, they concluded, the wave KC number and KC_x for the cylinder motion must both exceed 10-15.

Attempts have been made by Iwagaki & Al (1983) to produce modified KC and Re numbers that would make it possible to use results from pure oscillatory flows, Case 1A, to predict the force coefficients for the present situation, Case 1B. In the parameter range tested, they were reasonably successful, but Sarpkaya & Al. (1984) used the same approach and two similar alternative KC number definitions, and concluded that in a wider parameter range no such equivalence could be found.

CROSS FLOW PRINCIPLE

Another concept that, if valid, will decrease the number of necessary test cases is the so-called 'cross flow principle'. It is intended for members that are long compared to their cross sectional dimensions, and states that the force depends only on the velocity component normal to the member axis, while the velocity component parallel to the member axis has no effects. The cross flow principle has been shown to work well in steady, subcritical flows by Hoerner (1964) and in steady, supercritical flows by Horton & Al (1981). In oscillatory flow situations it (usually) works well for the inertia term, but for the drag term it works only at high KC numbers, say above 20. (Garrison, 1985) If the cross flow principle is valid then Case 3 could be analyzed by using the horizontal component with the force coefficients from Case 1. However Stansby & Al (1983) and Sarpkaya (1984) both found a significant influence of the along-pile component for subcritical flows in the KC range 10 to 30. For Reynolds number in the range 100 000 to 500 000 Bearman (1988) found no such difference, but the data were collected from different sources at somewhat different Re numbers and there was an appreciable spread in the experimental results, so the conclusion was somewhat tentative. Overall it appears that the cross flow principle may be a good approximation for most cases. The exception is in the intermediate KC number range, but the variability even here is probably reduced if less 'pure' situations are encountered, i.e. situations more influenced by flow turbulence, surfaces roughness and irregularities in the waves.

ON THE ANALYSIS OF HIGH REYNOLDS NUMBER, IRREGULAR FLOW DATA

Most of the tests to determine Morison type force coefficients have been carried out on test cylinders in the order of 1 to 5 cm in diameter, and flow velocity amplitudes of 5-10 cm/s, resulting in Reynolds numbers of 500 to 5 000, i.e. firmly in the subcritical range. Under extreme wave conditions for jacket type members, the Reynolds number is typically 10^6-10^7, i.e. in the supercritical and transcritical regimes. Also the more modest waves that dominate in fatigue analyses will normally be at Re numbers higher than the critical one. For these situations, namely high Reynolds number irregular wave situations, only a moderate amount of data is available from which to determine the force coefficients. The available data can be analyzed in different ways, and classified according to different flow classes, and arranged according to a representative nondimensional numbers, but it may be worthwhile first to briefly consider the purpose of such analyses.

Strictly speaking it is not necessary that use of Morison's formula for a given particle kinematics history generates identical predicted and measured force histories, but rather that the character of both records are the same. The goal is to be able to prescribe force coefficients to be used with future seastates so that representative results are generated in the following areas:

1. The frequency content of the irregular wave force must be modeled correctly. Of special interest is to have the right magnitude of force components near the resonant frequencies of the structure, since resonant structural response may dominate in the total response, even if the loading around resonance is a small fraction of the total loading. Resonance for fixed structures will normally be in the high frequency end of the wave spectrum. The frequency decomposition of the force is also required in order to predict force attenuation with depth and correlation between forces that act on neighbouring members.

2. The extremes of the wave forces are of vital importance, since usually these govern the extreme response of the structure.

3. It is not sufficient that the average force is correct. The split into inertia and drag parts must also be correct, since the drag term attenuates faster with depth, but has a much wider probability density and therefore contributes more to the highest force peaks, than the inertia term.

If a perfect fit between predicted and measured force time series can be found, then all the above will of course be satisfied. Even then, it remains to find a rule for how to select force coefficients for a given seastate.

CLASSIFICATION AND NONDIMENSIONAL NUMBERS DETERMINING C_D AND C_M IN IRREGULAR FLOWS

Let us first of all consult the list of regular flow cases in Table 1, and attempt a generalization from flows with circular particle orbits to irregular ocean waves with waves propagating in a mean direction with no or moderate directional spread about this mean direction. Cases 1 and 2 may occur only under laboratory conditions, while Cases 3 and 4 represent important real situations where the flow and the cylinder are in the same plane, which are typical for offshore locations, while Case 5, rotating flow, will occur for horizontal members parallel to the wave front. Case 6, current and waves from different directions, will typically occur off a gently sloping beach where the waves are refracted so that they will approach the beach at almost an right angle, while the currents usually run parallel to the beach. In the following attention will concentrate on various versions of Cases 3.

The governing parameters for irregular wave Case 3 situations must most likely involve a generalization of the three parameters identified for regular flow Case 3 situations, namely cylinder roughness and generalized versions of the Reynolds number and the Keulegan Carpenter number. Irregular flows need more parameters to be described however, and the force coefficients must be expected to depend on the same additional parameters. Among such additional parameters are the spectrum shape and the directional spread.

A generalization of Reynolds number and Keulegan-Carpenter number has been suggested previously in Moe & Bostrøm (1989) and has been presented in Moe (1990). These are in addition to cylinder diameter D and kinematic viscosity ν based on the standard deviation of the water particle velocity σ_u and acceleration $\sigma_{\dot{u}}$ where u represents the particle velocity component in the mean flow direction and is perpendicular to the test cylinder, and σ_u and $\sigma_{\dot{u}}$ can be determined from the spectrum of particle velocity or surface elevation, if information on mean direction and directional spread is available. Then the characteristic velocity, and the characteristic time which represents the mean zero up-crossing period of the particle velocity process, are

$$u_c = \sqrt{2}\sigma_u \qquad t_c = 2\pi\sigma_u / \sigma_{\dot{u}} \tag{4}$$

Furthermore the generalized Reynolds and Keulegan-Carpenter numbers are

$$\mathrm{Re}' = \sqrt{2}\sigma_u D / \nu \qquad KC' = 2\sqrt{2}\sigma_u^2 / (\sigma_{\dot{u}} D) \tag{5}$$

This KC number is determined by comparison of the transport derivative term in the Navier-Stokes equation to the partial time derivative, and is expected to give information about 'degree of separation', notably on when separation will start. An alternative KC definition based on a comparison between the average contribution to the wave force from the drag and inertia terms, respectively, has been developed by Bishop (1984) (see also Moe (1990) for a brief presentation). This expression can not be compared directly to the above, since it involves the fourth moment of the particle velocity, but for normally (i.e. Gaussian) distributed particle velocities, it reduces to $\sqrt{2}\,KC'$. (In the following for convenience the apostrophe is omitted on KC and Re.) This means that these two KC numbers would order experiments in the same way, but would position the results differently by a factor of 1.4 relative to each other. Possibly none of these KC numbers will give a good fit to regular wave results, since different types of flow may exhibit different behaviour. Some other proposed procedures for determination of the Re and KC numbers are mentioned in Isaacson et Al (1991)

ON ESTIMATION TECHNIQUES FOR FORCE COEFFICIENTS FROM RANDOM WAVE DATA

There exists in the order of 10 commonly used methods for determination of the force coefficients from measured records of irregular particle velocities and the corresponding wave force data. In early work in this field regular wave and force segments were fitted to the irregular time histories on a wave by wave basis, but this method was found to be unreliable, Borgman (1969). Then various statistical approaches have been introduced. Some of these have been summarized in Moe & Overvik (1989) and Isaacson & Al. (1991). The latter authors carried out a simulation study based on irregular particle kinematics and a force record synthesized directly from the kinematics histories by Morison's formula with $C_d=0.8$ and $C_m=2.0$. Then the standard estimation methods were applied to the synthesized kinematics and force records, and it was concluded that least squares fitting of time series was the most reliable technique, while especially the method of moments, but also the least squares fit to the force spectra gave inaccurate results, and it was suggested that the deficiencies in these methods could explain much of the scatter in reported values for Morison force coefficients. In the simulations the C_m coefficient was quite accurately determined in the KC range 5 to 20 by the spectral least square fitting technique, while the estimated C_d value in this KC range was 10-20 % too large. In Moe & Overvik (1989) results obtained by this method on data from two large measuring programmes were reported. (OTS and Christchurch Bay respectively, see Moe & Overvik, 1989.) For all analyzed seastates the KC number was in the range 8-15. Indeed for the OTS data the C_m coefficient was found to be reasonably stable and in agreement with analogous regular wave runs, while the Christchurch Bay results deviated, however they came from data with a current roughly perpendicular to the waves, i.e. a situation classified as 'Case 6' in Table 1. The estimated drag coefficients in Moe & Overvik (1989) were quite large, however, typically 1.1-1.2 and a reduction by about 15% as indicated by the simulations study, would be more in line with expectations.

At these KC numbers the inertia term dominates and determination of the drag term will therefore necessarily be less accurate. The simulation study identifies a systematic error in the results from this procedure, however. Also the question on how robust the method is with respect to inaccuracies in the data may be raised. The spectra of force, particle velocity and particle acceleration describe important and stable features of an irregular wave force. However the force is split into drag and inertia terms solely on the basis of the shapes of the spectra, relying on the fact that the ratio between the spectra of the linearized drag term and the inertia term is a constant times the square of the frequency. In retrospect it is clear that this procedure is dangerous, since the higher order part of the drag term not only increases the total area of the spectrum, but more important, it alters the spectrum shape. Of special influence here is the third order term giving additional contributions at frequencies around three times the spectral peak frequency. Further, for narrow spectra this procedure will become gradually more sensitive to inaccuracies, and will thus be unreliable.

Najafian & Burrows (1994) made a thorough statistical analysis of the error, e, made when estimating the force coefficients in the Morison formula by means of the standard least square fitting of force coefficients based on the time histories of force and particle kinematics. They compared the variance of the measured force, $E[F_0^2]$, and the variance of the force determined by Morison's formula based on the measured particle kinematics and the fitted force coefficients, $E[F^2]$. Then they showed that

$$E[F_0{}^2] = E[F^2] + E[e^2] \qquad (6)$$

i.e. that the variance of the force based on mean square fitting of time series will always be smaller than the variance of the measured force. They also suggested a corrective approach by which the variances of the measured force and the estimate by particle kinematics and Morison formula were set equal to each other, and the correlation between measured and calculated force was maximized. The result turned out to be that C_d and C_m should both be increased by the same magnification factor. They concluded that this adjustment would eliminate some of the discrepancies between the reported results from different investigations. In their conclusion they also stated that the least square fitting of time series is not suitable for long records of wave data.

EXPERIMENTAL BASIS

Results from the following measurement programmes will be considered herein

1. Ocean Test Structure: OTS (Exxon)
2. Christchurch bay
3. Gothenburg Ocean Basin tests-SSPA (Shell)

Results from our reanalyses of data from the two former programmes have been presented earlier, see e.g. Moe & Overvik (1989). Data from the SSPA tests have generously been provided by Shell Development Company in Houston. A short description can be found in Rodenbusch & Källström (1986), and a more detailed exposition in Rodenbusch & Gutierres (1983). The data are from experiments with a 1 m wide test cylinder moving through a ocean test basin of 3 m depth at paths programmed such that the relative velocities would resemble certain registered particle velocities histories, and are summarized in Table 2. Please note that the flow in these experiments is not a wave flow, Case 3 in Table 1, since there is no vertical velocity component, so that the situation belongs under Case 1 in Table 1. Actually the cylinder is also short, near the bottom and has an end plate at the surface (Case 9, Case 7A and Case 7B). A discussion of the error sources associated with this experimental arrangement can be found in Sarpkaya (1990).

In Table 2 the following codes are used to describe the runs: Cases 1A and 1B represent two different 5 min. long records of hurricane Camilla, and 2A and 2B represent two different 5 min. runs from OTS. (The OTS records are not from the same storms as those analyzed by Bostrøm, 1987). These four base cases have been used to create the 14 cases presented in Table 2 by applying suitable multiplication factors, M_x for the in-line direction, and M_y for the transverse direction. The four base cases have $M_x = M_y = 1$ and are in Table 2 denoted $1A_2$, $1B_2$, $2A_2$, and $2B_2$. From Table 2 it can be seen that another four cases have been run with unaltered in-line velocity and zero transverse velocity (i.e. unidirectional flows), and there are three runs with halved transverse velocity. The last 3 rows in Table 2 represents unidirectional waves with halved in-line velocity. It can be seen from the Re numbers given in Table 2 that the tests cases are in the critical and supercritical range. They have been analyzed by least squares fitting of the timeseries by Shell. In our reanalysis the force coefficients were determined by the method of mean square fitting between the spectra of the measured force, and the force spectrum calculated from the particle kinematics, Moe & Bjørheim (1990). The resulting force coefficients are shown in Table 2. It is quite apparent that the inertia coefficients for the high KC number cases are significantly overestimated, as was to be expected on the basis of the previously cited paper by Isaacson & Al. (1991). Therefore also the so-called total force coefficient, C_f, has been determined and entered into the last column of Table 2. It is intended to determine the maximum force in a regular wave situation on the form $C_f \rho R u_0{}^2$ where u_0 is the particle velocity amplitude. C_f is determined by

$$C_f = C_d\left\{1+\left\{\frac{\pi^2 C_m}{2(KC)C_d}\right\}^2\right\} \quad (7)$$

Before the results are discussed, it will be explained how the directional spread is treated.

EFFECTS OF WAVE DIRECTIONALITY

The kinematics of real ocean waves are always three dimensional, i.e. the waves have directional spread. Typically they will have one dominant direction, and contributions from other directions that taper off to both sides. In rules and standards the contribution to the variances is often specified on the form $cos^{2n}\theta$ where n typically is in the range 2-8. In such waves slender structures will again experience separated flows, and an extension of the Morison formula appears an obvious choice to model the resulting forces. Most investigators have used the usual Morison formula (1), simply extending it by defining forces, velocities and accelerations as vector quantities. The analysis of the SSPA data was done by a somewhat different method called 'the velocity tracking method', see Rodenbusch & Källström (1986) and Rodenbusch & Gutierres (1983). The idea is that the direction of the velocity vector is determined at each time step, and used in the Morison formula together with the component of the acceleration in the same direction, to determine a force which will then be colinear with the instantaneous velocity vector. This will eliminate accelerations associated with directional changes of the velocity vector, reducing the flow to a pseudo-plane model. In the work by Bostrøm (1987) and in later work in Trondheim an even simpler approach was adopted, namely to use the Morison formula for a plane flow situation involving only the acceleration and velocity components along the principal flow direction which is determined as the principal axis of the second order moments of the velocities. (Later called 'The principal direction approach'.) The motivation behind this choice of model is that considerable time is needed before the direction of a wake adjusts to a new instantaneous flow direction. Therefore the transverse force components observed for small or moderately large KC numbers would probably be created by vortex shedding from the flow in the principal direction. These forces might be large, but would often be of a random nature, especially in irregular flow with its inherent variability.

The SSPA data offers a unique opportunity to assess the hypotheses about directional spread, since as already explained the same water particle history in the principal wave direction has been combined with different transverse particle velocity histories. In Figure 1 a representative particle trace is shown indicating that the average transverse particle motions are nearly half of the average in-line motions (Case $1A_2$). The principal direction approach predicts identical forces for all these situations, since, as explained, it in essence ignores the transverse flow component. From Table 2 it can be seen that the predicted drag and inertia coefficients vary somewhat between such cases, but not much more than what is seen between two realization of the same seastate (,i.e. cases such as $1A_0$ and $1a_0$). The total force coefficient C_f is practically identical within both the 1A family and the 1B family which have KC numbers of 34 and 31 respectively, while within the 2A and 2B families which have KC numbers of 15 and 20 respectively, deviations of up to 5% from the average value are seen. Rodenbusch & Gutierres (1983) have estimated a value for the C_f coefficient for case $1A_0$ and it happens to be identical to the one found herein! (C_f=0.80) For the two cases at lower KC numbers it is interesting to notice that the largest deviation is seen for two cases representing realizations of the same seastate, namely $2A_0$ and $2a_0$. Further the cases with the highest KC numbers have the smallest deviations between members in the same family. Both features contradict the hypothesis that the differences are caused by the use of the principal direction approach, and points more in the direction of statistical variability. The records are after all relatively short, only about 5 min. at prototype scale.

In Figure 2 the variance spectra of measured forces for the four cases in the 1A family are shown. It can be seen that these spectra are practically identical! Noting from Table 2 that the drag coefficient for these almost identical cases varies with ±5 % and the inertia coefficient with ±10 %, it appears that least squares fitting of spectra is extremely sensitive to inaccuracies in spectral shape, increasing the writers' skepticism about this method. The principal direction approach, however, appears sound within the KC range tested, namely for KC numbers up to 35.

Chaplin & Al(1995) present results for the effects on the wave force from wave directionality based on wave tank experiments at subcritical flow and for KC numbers up to 15. Their KC number was based on the fundamental period velocity u_1 and the wave period T. The results were compared to predictions based on a vector form of Morison's formula, which was found to overpredict the force in directional seas by a few percent, and it was stated that higher overpredictions were expected at higher KC number. Their finding does not necessarily contradict the conclusions drawn herein, since the in-line component of the drag term in a vector formulation will be larger than in the principal direction approach, viz. $\{C_d\, u\sqrt{u^2+v^2}\}$ compared to $\{C_d u|u|\}$. Here it is assumed that u and v are velocities along the x- and y-axis which are in a plane perpendicular to the cylinder axis. Some researchers might also include w, the along-cylinder velocity component, into the above square root. Furthermore different Re numbers and different definitions of KC makes a direct comparison of results unwarranted.

RESULTS IN TERMS OF THE FORCE COEFFICIENT

In Figure 3 the force coefficients C_f from our SSPA analyses and the Bostrøm (1987) results are plotted versus the Keulegan-Carpenter number. For comparison a curve representing the often suggested values C_d=0.7 and C_m=1.8 is included. It appears that a better fit would be obtained if the drag coefficient had been increased to e.g. 0.8. It is not expected, however, that the same C_d and C_m coefficients will apply over the full KC range, so it seems to be somewhat misleading to refine on this fit.

Furthermore it may be noted that the discussions in this paper can serve as the background and justification for the recommendation of hydrodynamic coefficients given in a recent paper by Gudmestad et Moe (1996).

CONCLUSION

This paper deals with wave loading on cylinders due to high Reynolds number, irregular wave flows.

1. An attempt has been made
   A: to classify the various flow situations for which the Morison formula may be used
   B: to determine the nondimensional numbers which determine the values for the force coefficients
   C: to list suitable definitions for the most important of the above mentioned nondimensional numbers

2. Data from a large scale experiment have been analyzed to determine C_d and C_m coefficients using least square fitting of spectra of measured forces, velocities and accelerations. The results were disappointing, and thus in agreement with simulation studies which found systematic errors in predictions based on this approach. A so-called total force coefficient was also calculated, showing consistency in the results when plotted versus the KC number.

3. These data were ideally suited for analysis of the effect of wave directionality, and it was found that in the actual parameter range the so-called 'Principal direction approach' appeared to give excellent results. This approach is based on the assumption that the flow solely in the principal direction determines the Morison force. It is argued that the wake will not adjust to transverse velocity components, provided the KC number, and the ratio between the transverse and principal direction velocities, are not too large. And indeed: for all the four cases where data were available the transverse force component had no measurable influence on the force in the dominant wave direction.

ACKNOWLEDGMENT

The SSPA data discussed in this report were generously made available to the project on which this paper is based by Shell Development Company, Houston, Texas, while the costs of the project were covered by STATOIL, Stavanger. The support of both organizations is gratefully acknowledged.

REFERENCES

American Petroleum Institute: API RP2A WSD: "Recommended practice for planning, designing and constructing fixed offshore platforms, Working stress design", 20th edition, API, USA, 1993.

Bearman, P.W.: "Wave Loading Experiments on Circular Cylinders at Large Scale", Proc. of the Int. Conf. on the Behaviour of Offshore Structures (BOSS), Trondheim, June 1988, pp. 471-487.

Bendat, J. S., and Piersol, A. G.: "Decomposition of wave forces into linear and non-linear components", J of Sound and vibr., Vol. 106(3), pp 391-408, 1986

Bishop, J.R.: "Wave Force Investigations at the Second Christchurch Bay Tower", National Maritime Institute Report No. P.117, OT-0-82100, 1984.

Borgman, L.E.: "Ocean wave simulation for engineering design", Journal of Waterways & Harbour Div., ASCE, Vol. 95, No. WW4, pp. 557-583, 1969.

Bostrøm, T.: "Hydrodynamic Force Coefficients in Random Wave Conditions", Dr.ing. Thesis, Section of Structural Engineering, Norwegian Institute of Technology, March 1987.

Chaplin, J.R., Subbiah, K., Irani, M.: "Loading on a vertical cylinder in multidirectional waves", Journal of Offshore Mechanics and Arctic Engineering, Vol. 117, August 1995, pp.151-158.

Garrison, C.J.: "A Review of Drag and Inertia Forces on Circular Cylinders", Offshore Technology Conference (OTC), Paper No. 3760, May 1980, pp. 205-218.

Garrison, C.J.: "Comments on Cross-flow Principle and Morison's Equation", Journal of Waterway, Port, Coastal and Ocean Engineering, Vol. 11, No. 6, Nov. 1985.

Gudmestad, O. T. and Moe, G.: "Hydrodynamic coefficients for calculation of hydrodynamic loads on offshore truss structures", Marine Structures, 9 (1996), pp. 745-758.

Hoerner,S.F.: "Fluid Dynamic Drag", Private publisher: Hoerner Fluid Dynamics, 1964.

Horton, D.J., Heidman, J.C. and Mallard, W.W.: "Wind Tunnel Tests of Inclined Circular Cylinders", Offshore Technology Conference, Paper No. OTC 4122, Vol. 4, 1981, pp. 67-76.

Iwagaki, Y., Asano, T. and Nagai, F.: "Hydrodynamic Forces on a Circular Cylinder placed in Wave-current Co-existing Fields", Memoirs of the Faculty of Engineering, Kyoto University, Japan 1983, XLV (1), pp. 11-23.

Isaacson, M., Baldvin, J. and Niwinski, C.: "Estimation of drag and inertia coefficients from random wave data", Journal of Offshore Mechanics and Arctic Engineering, Vol. 113, May 1991, pp.128-135.

Laya, E. J., Connor, J. J., and Sunder, S. S.: "Hydrodynamic forces on flexible offshore structures", J of Engineering Mech (ASCE), Vol. 110, No. 3, March 1984.

Moe, G.: "Morison type wave loading", In: Tørum, A., Gudmestad, O.T. (Ed): Water Wave Kinematics, NATO ASI Series, Series E: Applied Sciences, Vol. 178, 1990, Kluwer Academic Publishers, Dordrecht.

Moe, G., Demirbilek, Z., Yttervoll, P. O.: "Morison's formula: Relative velocity versus independent flow field formulation for a case representing fluid damping", Sixth Int. symposium on OMAE, Houston, Tx, 1987

Moe, G. and Overvik, T., "The Use of the Morison Equation - a Review of Field Measurements", E&P Forum on Wave and Current Kinematics and Loading, Report no.3.12/156, paper no. C5, at IFP, Rueil Malmaison, France, Oct. 1989

Moe, G. and Bostrøm, T., "Wave Force Coefficients in Irregular Waves: The Christchurchbay Data", Division of Structural Engineering, The Norwegian Institute of Technology, Trondheim, Norway, Report No R-6-88, 1989.

Moe, G. and Bjørheim, L. G.: "Wave Force Coefficients in Irregular Flows: The SSPA Smooth Cylinder Data", Department of Structural Engineering, Report No R-11-90, Nov. 1990

Morison, J. R., O'Brien, M. P., Johnson, J. W., and Schaaf, S. A. "The forces excerted by surface waves on piles", Petroleum Trans., AIME, Vol. 189, pp 149-157

Najafian, G. and Burrows, R.: "Critical assessment of the least squares method used in derivation of Morison's force coefficients", Journal of Offshore Mechanics and Arctic Engineering, Vol. 116, Feb. 1994, pp. 1-6.

Rodenbusch, G. and Gutierres, C.A.: "Forces on Cylinders in Two-Dimensional flows", Technical Progress Report, Vol. 1, BRC 13-83,

May 1983, Bellaire Research Center (Shell Development Company), Houston, Tx.

Rodenbusch, G. and Källström, C.: "Forces on a Large Cylinder in Random 2-Dimensional Flow", Proc. of the 18th Annual Offshore Technology Conference, Houston, May 1986.

Sarpkaya, T.: "Quasi-2D Forces on a Vertical Cylinder in Waves", Discussion to paper by Stansby & Al (1983), Journal of Waterway, Port, Coastal and Ocean Engineering, ASCE, Vol. 110, No. 1, Feb. 1984, pp. 120-123.

Sarpkaya, T.: "On the effect of roughness on cylinders", OMAE-90, Houston, Feb. 1990.

Sarpkaya, T., Bakmis, C. and Storm, M.A: "Hydrodynamic forces from combined wave and current flow on smooth and rough circular cylinders at high Reynolds number", Proc. of the Sixteenth Offshore Technology Conference, Houston, Texas, OTC 4830, May 1984, pp. 455-462.

Sarpkaya, T. and Isaacson, M.: "Mechanics of Wave Forces on Offshore Structures", Van Nostrand Reinhold, New York, 1981.

Stansby, P.K., Bullock, G.N. and Short, I.: "Quasi-2-D Forces on a Vertical Cylinder in Waves", Journal of Waterway, Port, Coastal and Ocean Engineering, ASCE, Vol. 109, No. 1, pp. 128-132, Feb. 1983.

Verley, R.L.P. and Moe, G.: "The Forces on a Cylinder Oscillating in a Current". River and Harbour Laboratory, The Norwegian Institute of Technology, Report No STF60 A 79061, 1979.

Table 1
Some flow situations for which Morison's formula may be used. The cylinder may be considered infinitely long, and the fluid of infinite extent, if not indicated otherwise. The lengthscale at which the flowfield changes is large compared to cylinder diameter

Case	Situation
1A) (Base case) Sinusoidal, rectilinear flow 1B) As case 1A, but a colinear, steady current is added	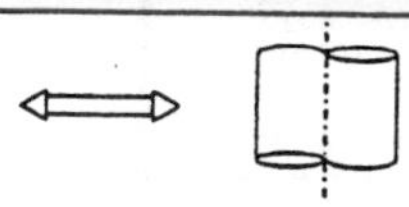
2A) Inclined cylinder Sinusoidal rectilinear flow 2B) Inclined cylinder. Flow field as for case 1B	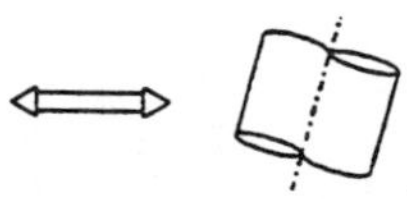
3) Cylinder in plane of particle velocity orbit (elliptical or circle)	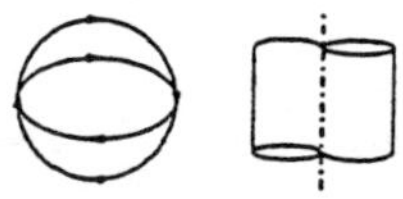
4) Waves, current and cylinder in one plane	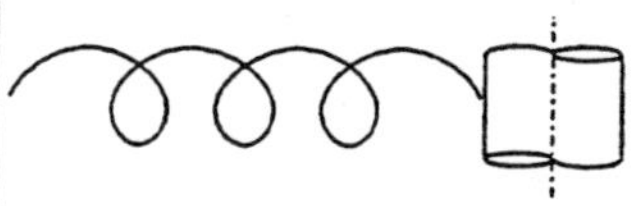
5) Rotating flow about cylinder axis (elliptical or circle)	
6) Current and waves in different directions	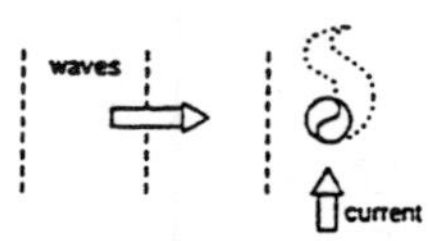
7) Surface or bottom proximity Vertical cylinders	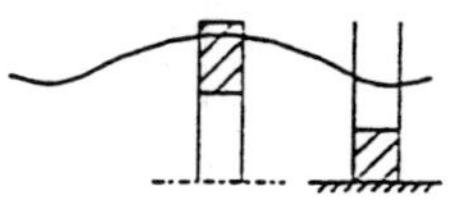
8) Along axis proximity to a wall, an object or the bottom or a free surface	
9) Short members	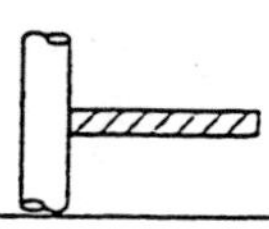

Table 2
SSPA test data and results of analysis
The results for all smooth random test cases
$M_x < M_y$: multipliers for the velocities in the x and y direction respectively

Case	M_x	M_y	KC	$Re \cdot 10^{-6}$	C_d	C_m	C_F
$1A_o$, $1a_o$	1.0	0.0	34	1.14	0.52/0.55	2.63/2.56	0.80/0.80
$(1A_1)$, $1a_1$	1.0	.5	35	1.15	- /0.60	- /2.42	0.79
$1A_2$, $1a_2$	1.0	1.0	34	1.16	0.61/0.56	2.20/2.46	0.78/0.79
$1B_o$	1.0	0.0	31	1.20	0.63	2.52	0.89
$1B_2$	1.0	1.0	31	1.20	0.63	2.59	0.90
$2A_o$, $2a_o$	1.0	0.0	15	0.58	0.61/0.54	1.93/2.07	1.27/1.40
$2A_1$, $(2a_1)$	1.0	.5	15	0.57	0.57/ -	2.02	1.34
$(2A_2)$, $2a_2$	1.0	1.0	14	0.55	- / 0.69	- / 1.92	1.35
$2B_o$	1.0	0.0	20	0.72	0.71	2.29	1.16
$2B_1$	1.0	.5	20	0.72	0.67	2.27	1.14
$2B_2$	1.0	1.0	20	0.72	0.77	2.01	1.09
1BH	0.5	0	16	0.60	0.58	2.01	1.24
2AH	0.5	0	7.6	0.29	0.47	2.02	2.62
2BH	0.5	0	10	0.36	0.63	2.02	1.99
2AS	1.0	1.0	37	0.58	0.44	2.72	0.74

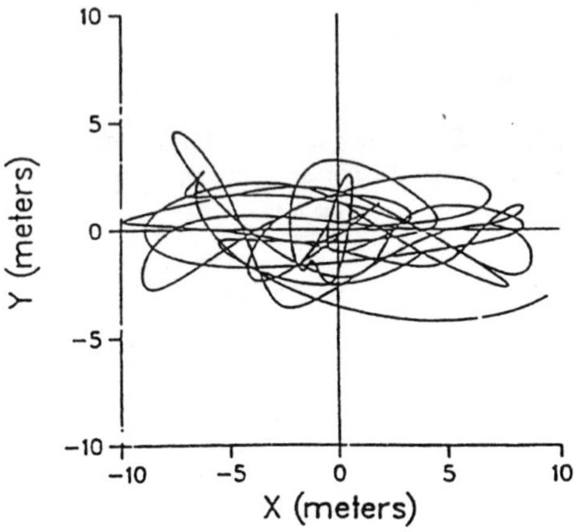

Fig 1 Example of oscillation path for one of the directional time series, namely $1A_2$. Rodenbusch and Gutierres (1983)

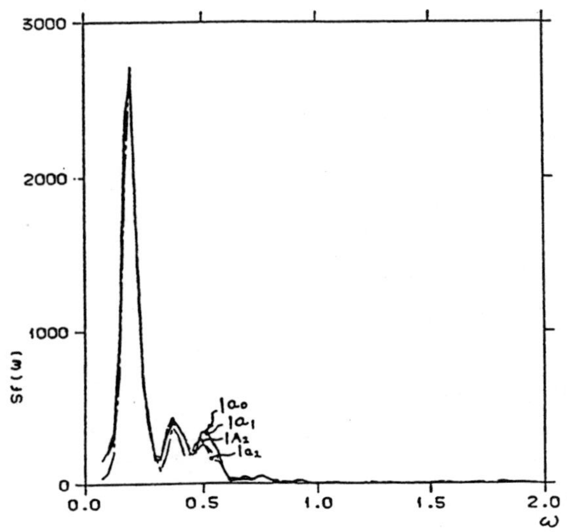

Fig 2 Spectra of measures in-line forces in 4 cases with the same in-line velocity component, but with varying transverse velocity component. Based on in-line velocity component only. The 4 cases are specified in Table 2.

Proceedings of the Seventh (1997) International Offshore and Polar Engineering Conference
Honolulu, USA, May 25-30, 1997

ISBN 1-880653-28-1 (Set); ISBN 1-880653-31-1 (Vol. III)

Estimation Method of Viscous Forces Acting on Floating Offshore Structures

Kunihiro Hoshino and Shunji Kato
Ship Research Institute, Ministry of Transport
Tokyo, Japan

Wataru Koterayama
Kyushu University
Kasuga, Japan

ABSTRACT

This paper deals with the method of estimating of the viscous forces acting on various kinds of offshore structures. Most of these are blunt-shaped element structures such as a column, a lower-hull and bracing. If we knew the basic data of hydrodynamic forces for such simple element structures, it would be possible to estimate the total hydrodynamic force using those data. In this paper we show how to combine basic data to estimate the total force considering the Reynolds number effect, three dimensional effect, and mutual interaction effect of each element. The developed method has been validated by comparing it with model experiments of three kinds of structures, TLP type, column footing type and semi-submersible type structure. Moreover, the applicability of this technique to the full-scale structure has been verified in a field experiment on a prototype offshore structure actually moored in the sea.

KEY WORDS: Floating offshore structure, Hydrodynamic force, Estimation method

1. INTRODUCTION

Most offshore structures consist of blunt-shaped elements of a column, a lower-hull and bracing. Knowledge of the viscous effect due to the vortex shedding from such elements is therefore essential to estimate the these hydrodynamic forces acting on the whole structure. Prediction of those hydrodynamic forces of an offshore structure is usually made by a tank experiment using a model of the whole structure. This require manufacturing a model for each target offshore structure. Moreover, it is difficult to incorporate the Reynolds number effect, for predicting drag coefficient of the full scale structure by this method.

Even though an offshore structure has a complicated shape, however, its fundamental structural elements are a combination simple shapes. Therefore, if the hydrodynamic force and the mutual interaction effects of each element structure are known, the total hydrodynamic force can be estimated without carrying out a model experiment. It is also possible to take into consideration the Reynolds number effect in the prediction of hydrodynamic force on a whole offshore structure using an existing study on the Reynolds-number effect on element structure.

The authors have conducted various fundamental research experiments on the hydrodynamic forces of circular cylinders with and without footing and of infinite and finite length. These are considered to be the element structure of an offshore structure.

This paper describes the estimation method of the drag force acting on an offshore structure, including the complete the Reynolds number effect, the three dimensional effect and the mutual interaction effect. To verify the validity of the technique, the estimated results were compared with the experimental results of models of three different kind of floating offshore structures. Applicability of the present method to an actual structure has further been verified by comparing the drag coefficient obtained from the estimation to that from the free decaying test of a full-scale offshore structure.

2. PROCEDURE FOR ESTIMATING THE DRAG COEFFICIENT

The estimation procedure developed in the present study of the drag coefficient for various kinds of offshore structures is as follows:

(1) First, the whole structure is divided into the element structures,.(for example: column, footing, pontoon and lower-hull, or etc.)

(2) The above elements are grouped according to whether cylinder are circular or rectangular.

(3) The 2-dimensional drag coefficient of each element is obtained using the existing experimental results including Reynolds number effect.

(4) The 3-dimensional effect is taken into consideration using the existing experimental results.

(5) The mutual interaction effect of each element is also taken into consideration using the existing experimental results.

(6) The total drag force is obtained by summing the drag forces acting on each element structure.

Flow chart of the estimation procedure for the various kinds of offshore structures is shown in Fig.1. The method is described in more detail in chapter 5, the because the procedure changes slightly with the composition style of the structure.

3. EXPERIMENTAL METHOD

<Tank experiment>

Forced oscillation tests were carried out using models of three types of offshore structures:

(a) TLP type, which consists of four columns and four horizontal square cylinders with rounded corners(Fig.2(a))

(b) Semi-submersible type, consisting of columns with a lower-hull (Fig.2(b))

(c) Column footing type, consisting of 12 columns with footing

(Fig.2(c))

Forced surging and heaving tests were carried out for the model(a), and only the forced surging test for model(b) and (c). For the TLP type(a) the heaving motion is important in relation to the ringing phenomenon, while for the semi-submersible(b) and column footing(c) types, the surging motion is most important because of the slow drift oscillation. These forced surging and heaving tests were carried out in still water with forced oscillation equipment. The hydrodynamic forces acting on the models were measured using a load-cell. Model(c) is a 1/25 scale model of a prototype offshore structure used in an at-sea experiment.

<At-sea experiment>

An outline of the at-sea experiment was reported by Ohmatu et al.(1989). The dimensions of the prototype offshore structure are indicated in parentheses in Fig.2(c). In the full-scale free surging test, the prototype offshore structure was pulled by the work vessel against the restraining force of the mooring lines using wire ropes and then suddenly released, and thus the free decay motion in the horizontal direction was made. The displacement was measured by an ultrasonic position meter installed in the seabed.

This free decaying test was carried out under very calm sea weather conditions.

4. DATA ANALYSIS

<Tank experiment>

The hydrodynamic forces acting on the model in a forced surging test are analyzed as follows. The damping term of the equation of motion of the offshore structure is expressed by the sum of the wave-making damping based on a potential flow theory and the viscous damping due to a fluid viscosity. However, a calculation by the 3-D panel method showed that the wave-making damping is very small in the present experiment range, and it can be neglected. Therefore, the following Morison's equation was used in the analysis of hydrodynamic forces acting on the models:

$$F_X = -(\rho\nabla C_A + M)\ddot{x} - \frac{1}{2}\rho A C_D \dot{x}|\dot{x}| \quad (1)$$

where ∇ is the volume of the model, M the mass of the model, A the projected area, C_A the added mass coefficient, and C_D the drag coefficient. Giving surging displacement $x = Xa\sin\omega t$, equation(1) can be linearized as follows,

$$F_X = (\rho\nabla C_A + M)X_a\omega^2\sin\omega t - \frac{4}{3\pi}\rho A C_D X_a^2\omega^2\cos\omega t \quad (2)$$

On the other hand, the measured in-line force in model experiments is expanded in a Fourier series as follows,

$$F_{X\,\mathrm{exp}} = A_0 + A_1\cos\omega t + B_1\sin\omega t + \cdots + A_n\cos n\omega t + B_n\sin n\omega t + \cdots \quad (3)$$

Comparing the first order term of frequency component ω of equations (2) and (3), we have C_D and C_A as;

$$\begin{cases} C_D = \dfrac{-A_1}{\dfrac{4}{3\pi}\rho A X_a^2\omega^2} \\ C_A = \dfrac{B_1}{\rho\nabla X_a\omega^2} - \dfrac{M}{\rho\nabla} \end{cases} \quad (4)$$

The hydrodynamic forces acting on the model in a forced heaving test have been analyzed as follows:

$$F_Z = -(M + \rho\nabla CA)\ddot{z} - \rho\omega\nabla B_{33}\dot{z} - \frac{1}{2}\rho A C_D \dot{z}|\dot{z}| - \rho g A_W z \quad (5)$$

where B_{33} is the wave-making damping coefficient for heaving motion, g the gravity acceleration, and A_W the water plane area of model. The same analysis is made, as in the forced surging test, and the related formula of C_A and C_D for the heaving motion is obtained as follows,

$$\begin{cases} C_D = \left(\dfrac{-A_1}{\omega Z_a} - B_{33}\right)\cdot\dfrac{3\pi}{4}\dfrac{1}{\rho A Z_a\omega} \\ C_A = \left(\dfrac{B_1}{Z_a\omega^2} + \dfrac{\rho g A_W}{\omega^2} - M\right)/\rho\nabla \end{cases} \quad (6)$$

where B_{33} is wave-making damping for heaving motion. B_{33} is calculated by 3-D panel method and it cannot be ignored in this case.

<At-sea experiment>

The time series fitting method was used to analyze the free decaying test in the at-sea experiment. The nonlinear optimization technique was applied for the analysis since the motion decays very rapidly. The model equation is assumed as equation (7):

$$\ddot{x} + \frac{N_1}{M}\dot{x} + \frac{N_2}{M}\dot{x}|\dot{x}| + \frac{A_1}{M}x + \frac{A_2}{M}x^2 + \frac{A_3}{M}x^3 = 0 \quad (7)$$

The coefficients N_1, N_2, A_1, A_2 and A_2 of the model equation are determined by minimizing the sum of the square of the instantaneous difference between the measured time series and the result of time series calculation of the model equation. The Powell method is used for minimization of the sum.

5. COMPARISON OF MEASURED AND ESTIMATED DRAG COEFFICIENT

The added mass coefficient C_A of all experimental models in the low Kc number region($1 \le Kc \le 13$) can be estimated in sufficient accuracy by the linear potential theory. This fact was confirmed by comparing the calculation results and experiments. Therefore, we discuss only the drag coefficient C_D in this section.

5.1 TLP TYPE OFFSHORE STORUCTURE

[C_D for surging motion]

The TLP type offshore structure shown in Fig.2(a) is divided into finite-length circular cylinders and horizontal square cylinders. The Reynolds number effects are first considered regarding the drag coefficient of a 2-D circular cylinder using existing experimental results (e.g. Sarpkaya, 1977). Then, the 3-D effect for a circular cylinder can be estimated using decrease factor(α) obtained from the existing experimental results of oscillating finite-length circular cylinders as shown in Fig.3 (Nakamura et al., 1991). The mutual interaction effect between columns is taken into consideration using the experimental results of a forced oscillating test of two circular cylinders by Ikeda et al.(Figs.4,5, Ikeda et al., 1987).

Secondarily, for the estimation of drag coefficient of a horizontal square cylinder with rounded corner, the experimental results by Kudo et al.(1980) and Tanaka et al.(1982) can be used. In Fig.6, the experimental results of Tanaka et al. were extrapolated to low KC number region using the result by Kudo et al. The horizontal cylinder is fixed between two columns as shown in Fig.2(a), and so is assumed to be 2-D cylinder because the two columns act as an end plate. The Reynolds number effect for a rectangular cylinder was disregarded because it is very small, as shown by the experimental results of Tanaka et al.

Finally, the drag forces of these four columns and four horizontal cylinders are summed to estimate total drag force.

Figure 7 shows estimated and measured drag coefficients C_D for surging motion versus Kc number. In this figure, the circles and triangles show the experimental results for the cases $Re=10^4$ and $Re=3.12\times10^4$, respectively. The solid lines indicate estimated drag coefficients which agree well with the measured one for two kinds of Reynolds number.

[C_D for Heaving motion]

The drag coefficient of the finite-length vertical circular cylinder, like the column for the heaving motion, is not known well. Therefore, a now forced heaving test of the finite-length vertical circular cylinder was performed. As an example of the experimental result, the frequency characteristic of the equivalently linearized damping coefficient is shown in Fig.8. Since no clear tendency could be found in the change of damping coefficient due to heaving amplitude, we can conclude that the drag force of the finite-length vertical circular cylinder for heaving can be disregarded because it was very small. Therefore, the whole drag coefficient can be estimated only from that of the horizontal square cylinder obtained from Fig. 6.

Figure 9 shows the drag coefficient C_D of the whole model for heaving motion versus Kc number. The horizontal axis in the figure is the Kc number defined using d of a column diameter. Circles in the figure mean measured C_D and the solid line means an estimated one. Although the drag force acting on columns is disregarded, the estimated and the measured values are in good agreement. Therefore, it is appropriate that the drag coefficient of the column of heaving can be disregarded. The great portion of the drag is believed to have been generated from the pontoon element of a TLP type offshore structure.

5.2 SEMI-SUBMERSIBLE TYPE OFFSHORE STRUCTURE

[C_D for Surging motion]

The elements of the semi-submersible type offshore structure are the columns and the lower-hull. The columns were a finite-length circular cylinders, and the lower-hull is a rectangular cylinder. The experimental results of the forced oscillating test on the rectangular cylinder by Kudo et al.(1980) were applied to estimate the drag coefficient of the lower-hull part. The experimental result of Kudo et al. is shown in Fig.10. Since that aspect ratio(d/B=0.6) was small, it was corrected by using the experimental result for a steady flow obtained by Nakaguchi et al.(Fig.11). Although change of the drag coefficient by aspect ratio(d/L) in the oscillating flow for a 3-dimensional circular cylinder is thought to be slightly less than the case of a steady flow, no big difference is expected. The experimental results of Kudo et al. were modified referring to the experimental result of Nakaguchi et al.(Fig.13). A comparison of the estimated C_D with the measured C_D is shown in Fig.12. The estimated C_D of the semi-submersible model agrees well with the measured C_D for the surging motion.

5.3 COLUMN FOOTING TYPE OFFSHORE STRUCTURE

[C_D for surging motion]

All elements of the column footing type offshore structure are circular cylinders. The column part is a 2-dimensional circular cylinder by the end plate effects of the footing, and the footing part is viewed as a 3-dimensional circular cylinder. Therefore, C_D of the column part is obtained from the experimental results of a 2-dimensional circular cylinder, and C_D of the footing part is estimated by correcting the drag coefficient of a 2-dimensional circular cylinder taking into consideration a 3-dimensional effect(see Fig.3). The mutual interaction effect is small for the column footing type offshore structure because the distance between columns is large, but the effect is considered in this process. The drag coefficient of the 2-dimensional circular cylinders is obtained from Sarpkaya's experimental results(Fig.13: 1986) at low Kc number. The value of β (=Re/Kc) in the present experiment is 1054 and is very close to Sarpkaya's experiment of β =1035. Comparison of the estimated C_D with the measured C_D is shown in Fig.14: they are in good agreement.

5.4 PROTOTYPE OFFSHORE STRUCTURE

An example of the fitting result of the measured value of the free surging motion is shown in Fig.15. Table 1 shows the damping coefficient obtained from the fitting. The damping coefficient N_2/M obtained from two experiments is 0.0759 and 0.0736. The drag coefficient C_D calculated from these values is 0.691 and 0.670, respectively.

Table 2 shows the estimation process of C_D of the prototype offshore structure. In this process the value of 3-D coefficient for the column is 1.0 because the footing is fixed to the end of the column. The Re number range is defined by the diameter of the column(d=2m) and the footing(d=4m), respectively. The drag coefficient of the 2-D circular cylinder used in the estimation is shown in Fig.16. Finally, comparison of the estimated with the measured C_D is shown in Table 2. The estimated C_D is slightly smaller, which is believed to be the effect of marine growth adhering to the column surface of the prototype structure.

6. CONCLUSIONS

A practical technique of estimating the drag coefficient for offshore structures was developed considering the Reynolds number effect, 3 dimensional effect and mutual interaction effect. The applicability of the developed method was validated by comparing the estimated drag coefficient with the results of a tank experiment, which was performed using offshore structure models of various forms. The applicability of the estimation method to the estimation of a full-scale structure was also verified using the value of the drag coefficient obtained from a free decaying test in the actual sea performed by the Ship Research Institute. The following conclusions were drowns:

(1) The drag coefficient for surging motion acting on various kinds of the offshore structures can be estimated with good accuracy by the present method.

(2) The drag coefficient for heaving motion of a TLP type offshore structure can be estimated by disregarding the drag force acting a column part. A practical and sufficiently accurate estimation can be made by taking into consideration the drag force of a rectangular cylinder part.

(3) The applicability of the present method to estimation of the drag forces acting on a full-scale offshore structure has been verified by the free decaying test of such a structure..

REFERENCES

Hoshino,K. et al. (1993), Hydrodynamic Forces Acting on Finite-length Circular Cylinder Oscillating in a Uniform Flow, *Proc 3rd Int Offshore and Polar Eng Conf* (ISOPE93),pp.242-253

Hoshino, K and Sato, H.(1994), Experimental Study on Hydrodynamic Forces Acting on an Oscillating Column with Circular Footing, *Proc 4th Int Offshore anf Polar Eng Conf* (ISOPE94), pp,173-181

Hoshino, K and Koterayama,W (1995), Hydorodynamic Forces Acting on an Oscillating Inclined Finite-length Circular Cylinder(in Japanese), Transactions of the West-Japan Society of Naval Architects, JAPAN, No.90, pp.103-113

Ikeda, Y , Horie, H and Tanaka, N (1987), Viscous Interference Effect Between Two Circular Cylinders in Harmonically Oscillating Flow(in Japanese), Journal of the Kansai Society of Naval Architects, JAPAN, pp.79-87

Kudo, K , Kinoshita, A and Nkato, M (1980) , Experimental Study on Hydrodynamic Forces Acting on the Oscillating Rectangular Cylinders(in Japanese), Journal of the Kansai Society of Naval Architects, JAPAN, pp83-90

Nakaguchi, H et al.(1968), An Experimental Study on Aerodynamic Drag of Rectangular Cylinder(in Japanese), Journal of the Japan Society for Aeronautical and Space Sciences, No.16, Vol.168, pp.1-6

Nakamura,M. ,Koterayama,W and Hoshino,K(1991) Three Dimensional Effect on the Hydrodynamic Forces Acting on an Oscillating Finite-length Circular Cylinder, *Proc. 1st Int Offshore and Polar Eng Conf* (ISOPE91), pp.143-148

Ohmatu, S et al. (1989). At-sea Experiment of Floating Offshore Plat form POSEIDON, *Proc 8th Offshore Mech and Arctic Eng Symp*

Sarpkaya, T (1977), In-Line and Transverse Forces on Cylinders in Oscillatory Flow at High Reynolds Number, Journal of Ship Re search, Vol.21, No.4, pp200- 216

Sarpkaya, T (1986), Force on a Circular Cylinder in Viscous Oscillatory Flow at low Keulegan-Capenter Numbers, *J Fluid Mech*, Vol.165, pp.61-71

Tanaka, N , Ikeda, Y and Nishino, K (1982), Experimental Study on Hydrodynamic Viscous Force Acting on Oscillating Bluff Body(in Japanese) , Proc 12th Ocean Engineering Symp ,The Society of Naval Architects of JAPAN, pp55-59

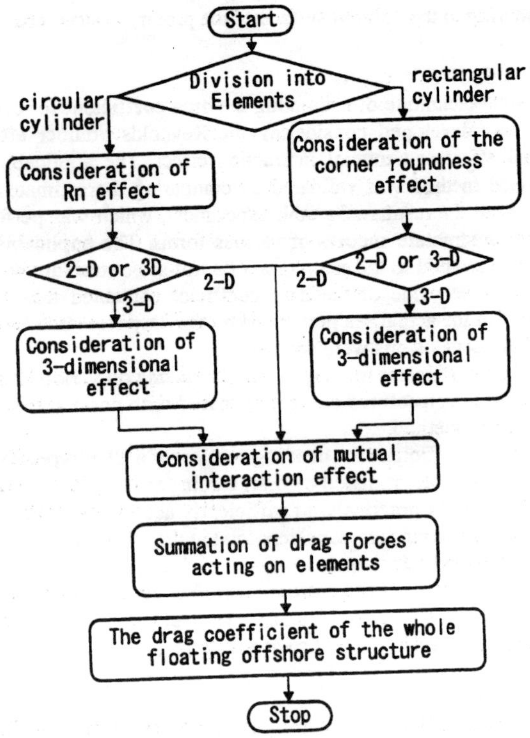

Fig.1 Flow chart of estimation of the drag coefficient

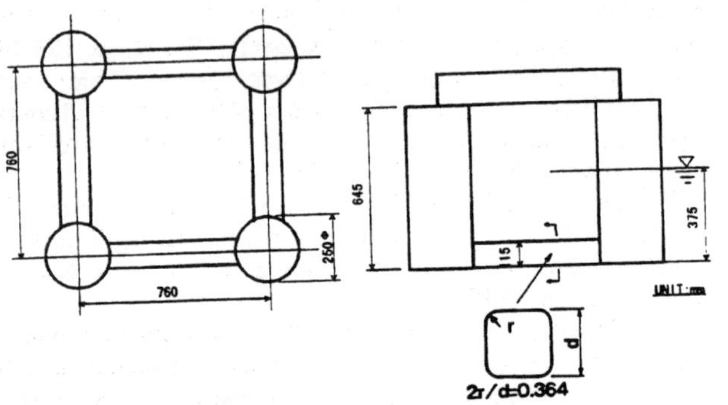

Fig.2(a) TLP type model

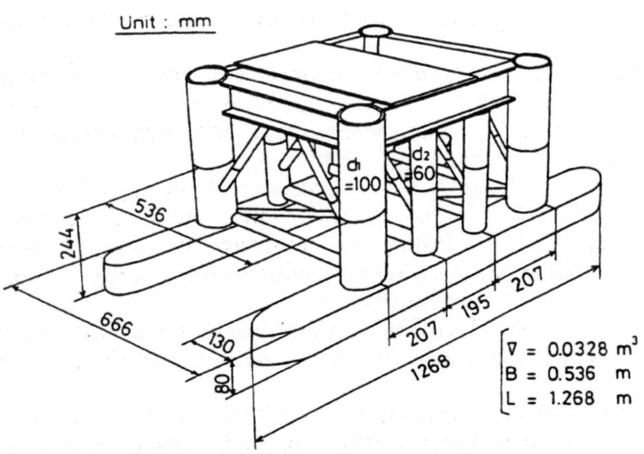

Fig.2(b) Semi-submersible type model

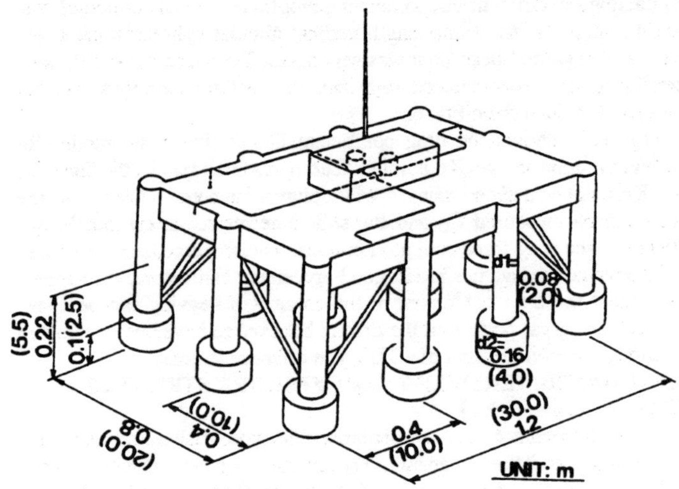

Fig.2(c) Column footing type model

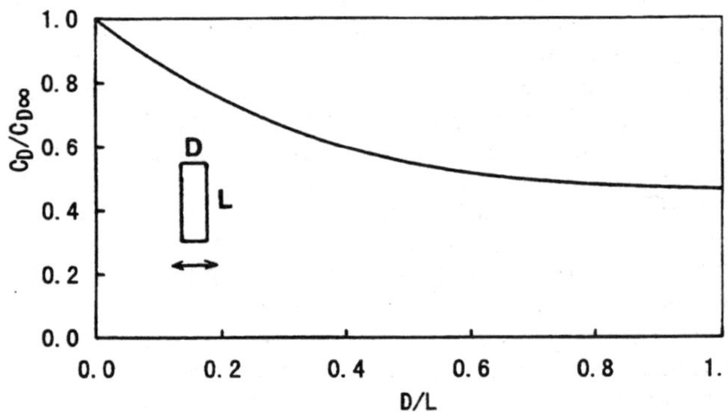

Fig.3 3-dimensional effect of finite-length circular cylinder in oscillating flow

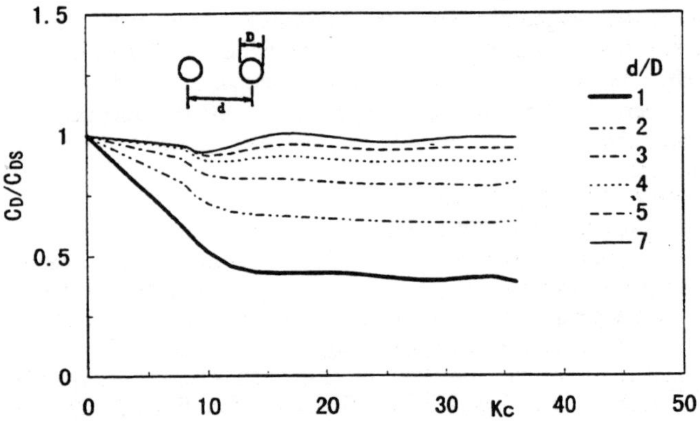

Fig.4 Interference of two circular cylinders in oscillating flow (Ikeda et al.,1987) (tandem arrangement)

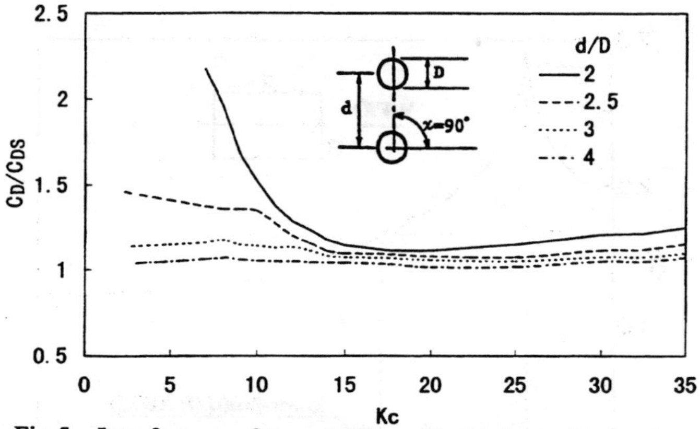

Fig.5 Interference of two circular cylinders in oscillating flow (Ikeda et al.,1987) (side by side arrangement)

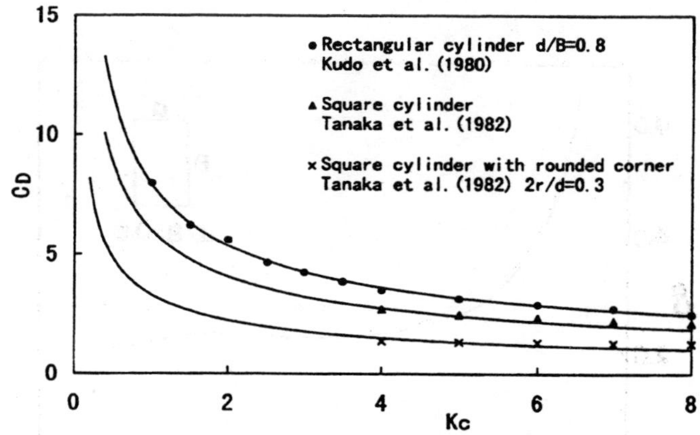

Fig.6 Drag coefficient of square cylinder in oscillating flow

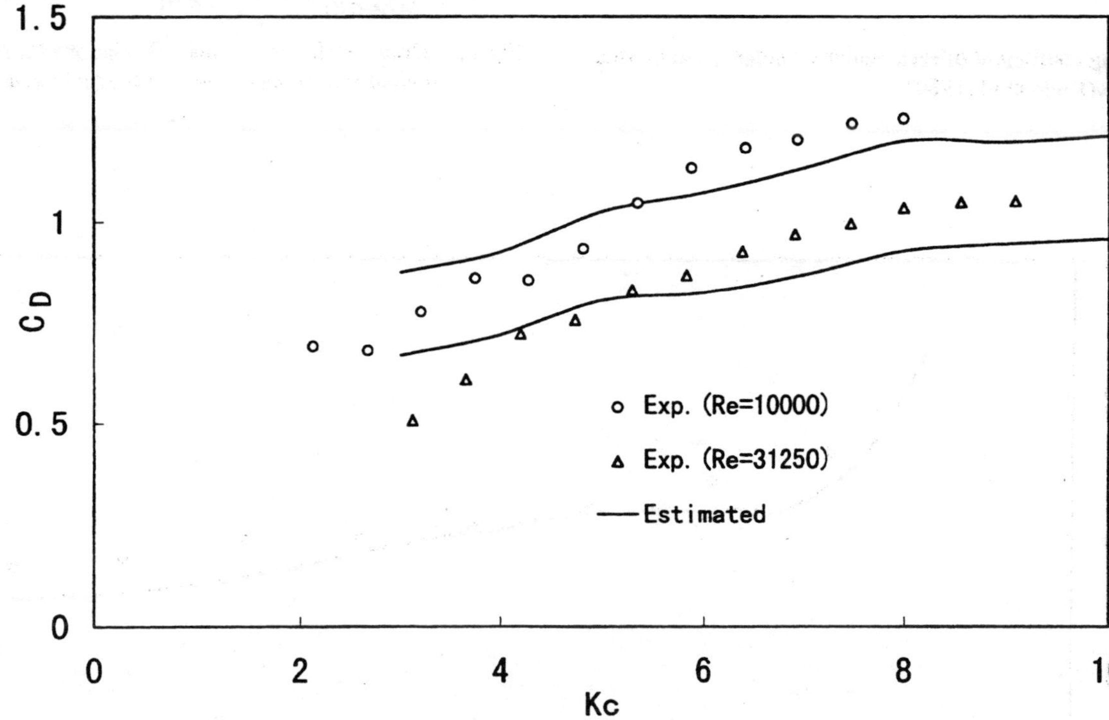

Fig.7 Comparison between measured and estimated drag coefficient for surge(TLP type)

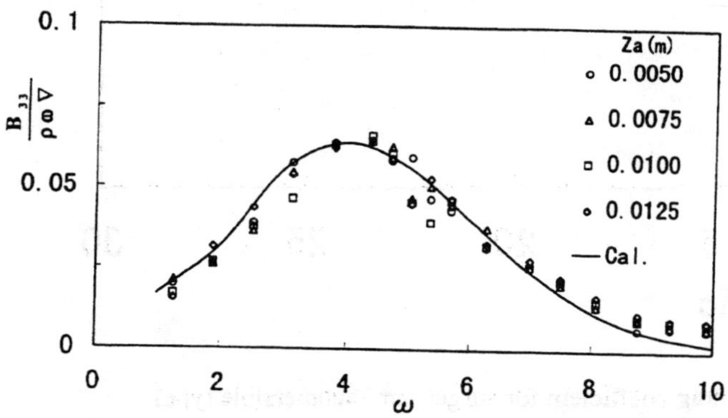

Fig.8 Equivalently linearized damping coefficient of finite-length circular cylinder for heave (L/D=1.5)

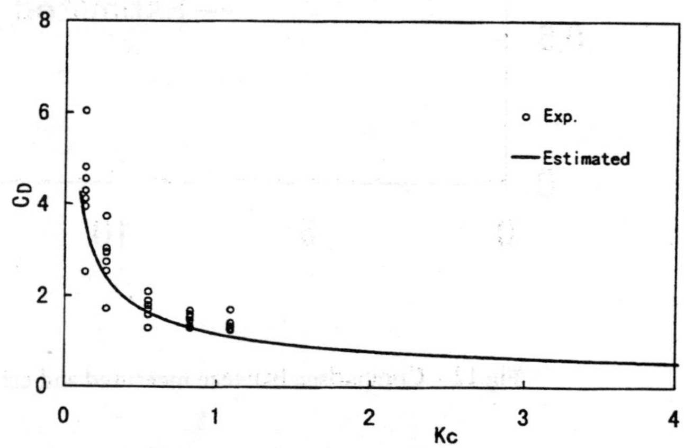

Fig.9 Comparison between measured and estimated drag coefficient for heave

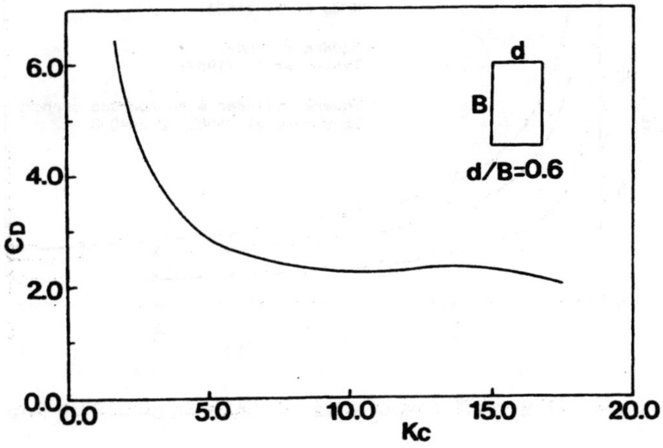

Fig.10 Drag coefficient of rectangular cylinder in oscillating flow(Kudo et al.,1980)

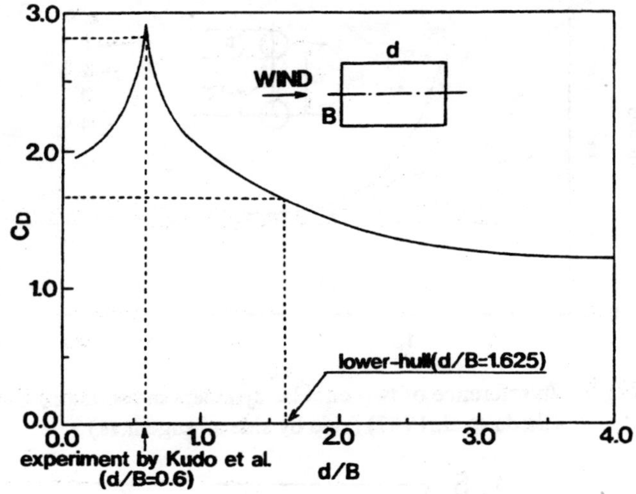

Fig.11 Drag coefficient versus d/B relations for rectangular cylinder in steady flow (Nakaguchi et al.,1968)

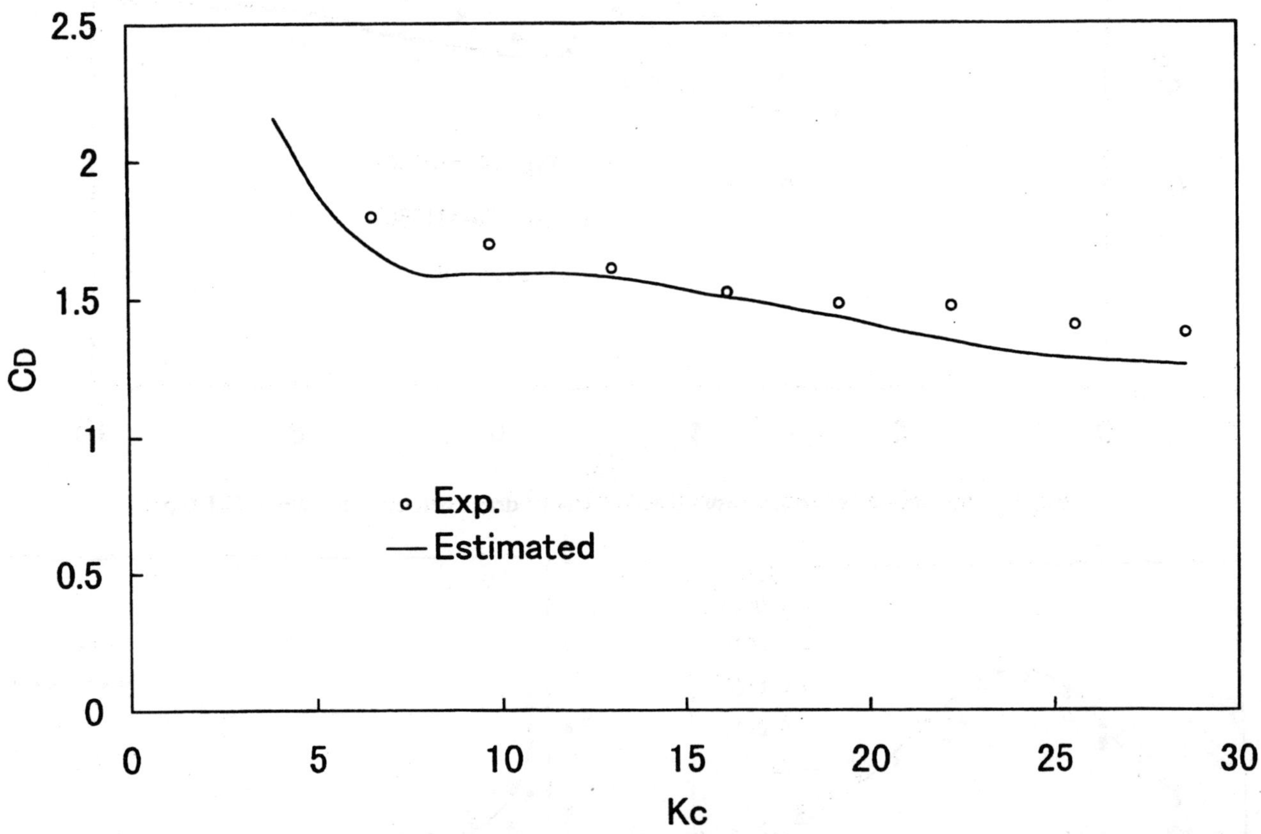

Fig.12 Comparison between measured and estimated drag coefficient for surge(Semi-submersible type)

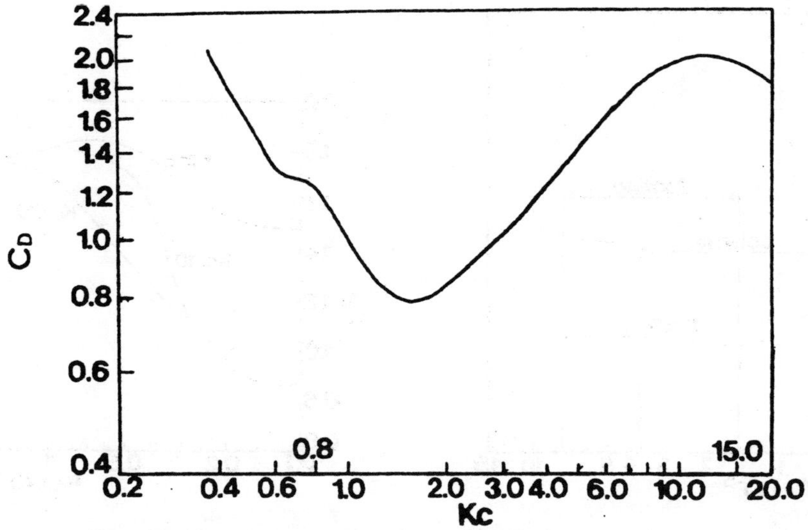

Fig.13 Drag coefficient versus Kc number for circular cylinder, β =1035 (Sarpkaya,1986)

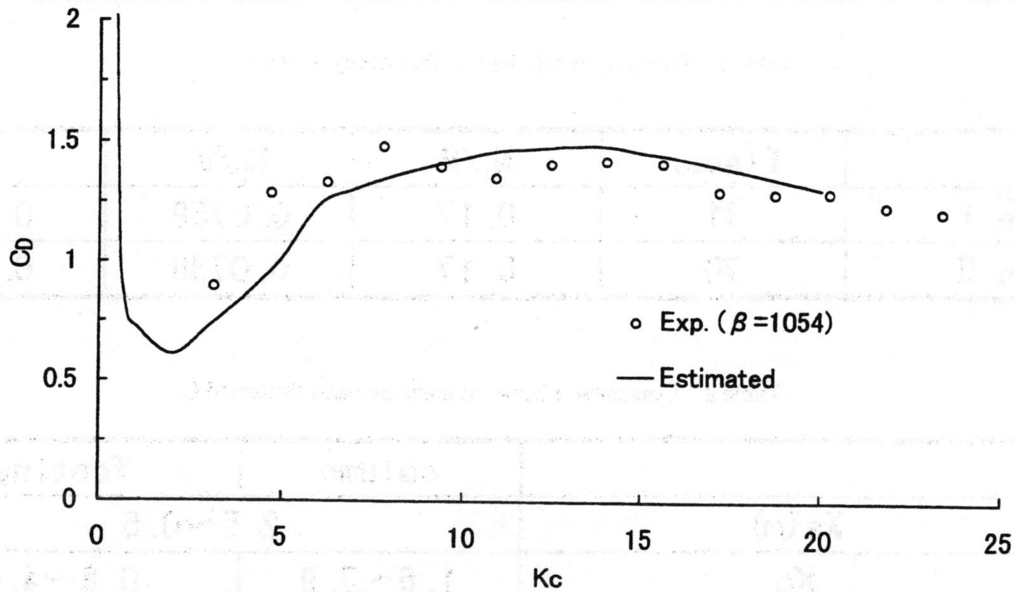

Fig.14 Comparison between measured and estimated drag coefficient for surge

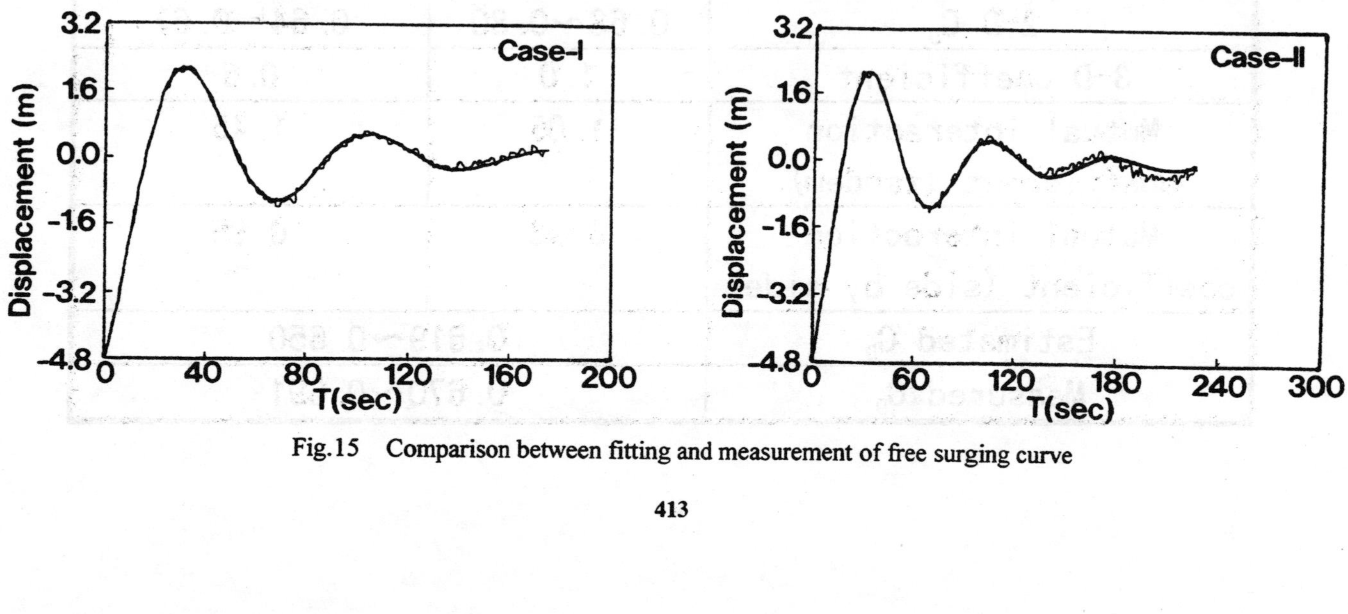

Fig.15 Comparison between fitting and measurement of free surging curve

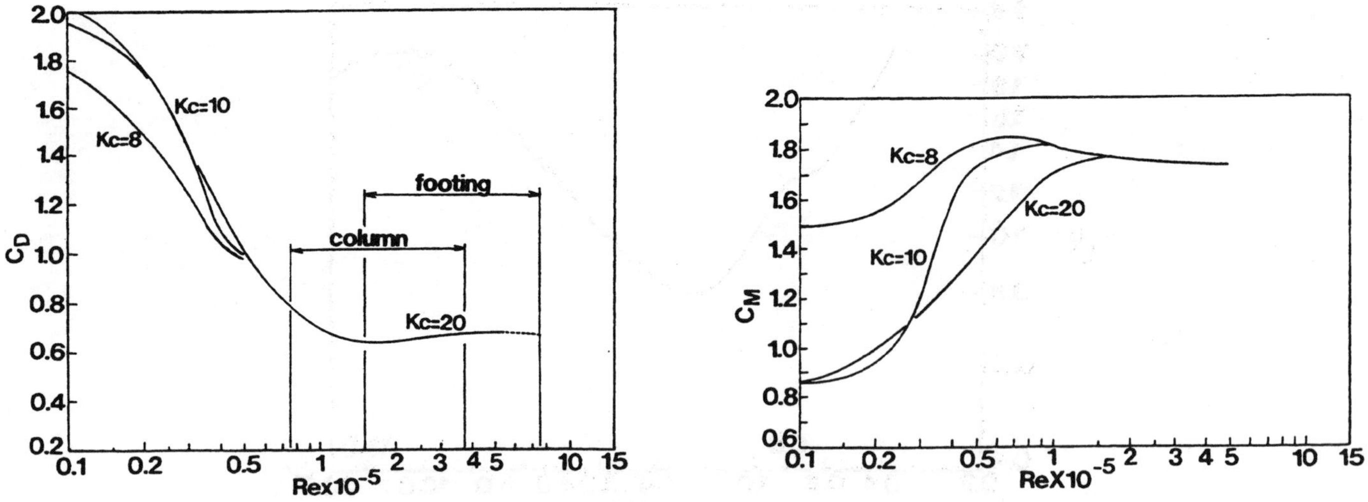

Fig.16 Drag coefficient versus Reynolds number for constant Kc number (Sarpkaya, 1977)

Table 1 Damping coefficient by free decaying tests

	T (sec)	N_1/M	N_2/M	C_D
Case I	71	0.17	0.0759	0.691
Case II	70	0.17	0.0736	0.670

Table 2 Comparison between measured and estimated C_D

	column	footing
Xa (m)	2.5~0.5	
Kc	1.6~7.9	0.8~4.0
Rex10^5	0.76~3.88	1.51~7.56
2-D C_D	0.68~0.80	0.64~0.67
3-D coefficient	1.0	0.5
Mutual interaction coefficient (tandem)	1.05	1.45
Mutual interaction coefficient (side by side)	0.98	0.95
Estimated C_D	0.619~0.650	
Measured C_D	0.670~0.691	

Proceedings of the Seventh (1997) International Offshore and Polar Engineering Conference
Honolulu, USA, May 25-30, 1997

ISBN 1-880653-28-1 (Set); ISBN 1-880653-31-1 (Vol. III)

Hydrodynamic Forces Acting on a Bottom Mounted Circular Cylinder

Ian A.R. Neill and Jon B. Hinwood
Monash University, Melbourne, Australia

Abstract:

Prediction of hydrodynamic forces acting on submarine pipelines lying on the seabed is essential for on-bottom stability calculations. Classical methodology is based on Morison's equation, for which non-dimensional inertial, drag and lift coefficients must be found from the vast array in the public domain or from relevant model tests. Generally the lift force is poorly predicted. More recently a *wake* model has been formulated, based on field data to which empirical constants were fitted, much of this has remained proprietary information and limited in application to pipelines with diameter greater than 400 mm.

An experimental programme using regular waves and currents with extensive flow visualization, has provided an insight into the physics of the problem. Traditional dimensionless force coefficients obtained using Stokes fifth order wave theory are compared to data from previous research and joint industry programmes. Analysis of flow visualization has produced correlations between the near field flow incorporating vortex motions and features in the force traces not predicted by conventional theory.

Key Words Hydrodynamic forces, Submarine pipelines, Flow visualization, Wave loading.

Nomenclature

C_{AW}	added mass coefficient of wake
$C_{M,D,L}$	inertial, drag, lift force coefficients
$C_{H,V}$	horizontal, vertical peak force coefficients
D	cylinder diameter
$F_{H,V}$	horizontal, vertical forces
H	wave height
Kc	Keulegan-Carpenter number
Re	Reynolds number
T	wave period
Uc	steady current
U_e	effective velocity
u_m	maximum horizontal flow velocity
u(t)	horizontal wave flow velocity at time t
u_0	horizontal velocity amplitude
W	wake velocity
α	current ratio
β	Re/Kc
ρ	fluid density
ω	wave frequency

1. Introduction

Many offshore oil and gas operations rely upon subsea pipelines to export their products to land. Extreme storm events have caused severe damage to many offshore pipelines around the world, due to gross vertical and horizontal displacements of the pipeline. Most recently hurricane Andrew damaged 454 subsea pipelines in the Gulf of Mexico in 1992, *Collins* (1995).

Prediction of the hydrodynamic forces acting on submarine pipelines, in contact with the seabed, is essential for accurate on-bottom stability analysis. Confident design necessitates accurate prediction of the hydrodynamic loading over a large range of flow parameters. Models used for calculating such forces are briefly presented.

Morison Force Model

Traditional hydrodynamic force prediction is based on the Morison (*Morison et al* (1950)) force formulation that treats the loading as a summation of a drag force and an inertial force. The common representation of the in-line force for a horizontal cylinder due to waves and current is given in Eq. 1, neglecting the effects of the vertical component of the wave velocity near the seabed. The lift force is given by classical aerodynamic formulation in Eq. 2;

$$F_H(t) = \frac{\pi}{4}\rho D^2 C_M \frac{du}{dt} + \frac{1}{2}\rho D C_D \left|Uc + u(t)\right| \left(Uc + u(t)\right) \qquad \text{Eq. (1)}$$

$$F_V(t) = \frac{1}{2}\rho D C_L \left(Uc + u(t)\right)^2 \qquad \text{Eq. (2)}$$

Selecting the appropriate force coefficients can become a contentious issue with large scatter in laboratory results and even more in those derived from field data. Apart from the difficulty in selecting the appropriate force coefficients, it is generally accepted that this classical representation poorly predicts the lift force.

Peak Force Model

The peak model uses only the flow velocity, neglecting inertial terms, to obtain peak coefficients of both the horizontal and vertical forces. These peak coefficients are obtained by non-dimensionalising the peak force in each wave cycle by the peak dynamic pressure times area.

$$C_{(H,V)} = \frac{F_{(H,V)\max}}{1/2\rho D u_m^2} \qquad \text{Eq. (3)}$$

Wake Model

The Morison force model, using constant force coefficients and undisturbed flow kinematics, does not accurately predict the time series of the hydrodynamic forces. In the presence of a pipe on or near a plane boundary a wake will develop downstream from the pipe. In oscillatory flow the wake redevelops in each half cycle whilst the wake from the previous half cycle is washed over the cylinder. This scenario becomes more complex for the case of combined wave - current flows and irregular waves.

Model tests of *Jacobsen et al* (1984) measured the local flow 0.16 D above a pipe resting upon a boundary, as well as the ambient velocity. Substituting these measured values of the local velocity into the Morison equation with constant force coefficients produced well predicted force traces. These coefficients were remarkably constant over a wide range of Kc and similar to those calculated in steady flow. Thus, the Morison force model is applicable to cylinders on a plane boundary, however, local flow kinematics must be used in the force prediction. *Jacobsen et al* (1984) attempted to model the wake by analogy to flow over a rearward facing step.

Lambrakos et al (1987) formulated the "wake model" as an attempt to improve the conventional model's ability to predict both the magnitude and temporal variation of the in-line and transverse forces. Field data, from the Pipeline Field Measuring Programme was used in generating this model which *Wolfram et al* (1987), *Verley et al* (1987) and *Chao et al* (1989) report as being validated by laboratory studies as part of the PIPESTAB project:

The wake model has a very similar form to that of the conventional model, except that the fluid velocity is replaced by an effective velocity that takes into account the wake re-encounter from the wake shed in the previous half cycle. The force coefficients are now time dependant. The wave force (per unit length) for a wave perpendicular to the cylinder's axis is given by:

$$F_H(t) = \frac{1}{2}\rho D C_D(t) U_e |U_e| + \frac{\pi}{4}\rho D^2 \left(C_M \frac{dU}{dt} - C_{AW}\frac{dW}{dt} \right) \qquad \text{Eq. (4)}$$

$$F_V(t) = \frac{1}{2}\rho D C_L(t) U_e^2 \qquad \text{Eq. (5)}$$

where U_e = effective velocity = U(t) + W(t)

W = wake velocity

C_{AW} = added mass coefficient for the wake velocity.

The effective flow velocity (U_e) is the ambient velocity (U) with the wake velocity (W) superimposed, thereby representing the flow that the pipe encounters. The theoretical model used to calculate the wake velocity is based on expressions by *Schlichting* (1968) for the velocity profile in the far wake behind a cylinder in steady flow and remote from any boundary. It assumes the boundary on which the cylinder lies can be represented by the plane of symmetry containing the common tangent to two cylinders in contact (much like the use of images in potential flow) and the wake flow is similar to that of a free pipe of twice the diameter. Since the presence of the boundary dramatically alters the wake behind a cylinder in free stream, this assumption is less realistic than the rearward facing step model of *Jacobsen et al* (1984).

The effective velocity, U_e, is larger and phase shifted relative to the ambient flow velocity. However, the predictive capacity of the lift force could not be significantly improved, so the lift and drag coefficients were taken as time-dependent. Such an approach adds increased complexity to this model, and ignores the findings of *Jacobsen et al* (1984) that constant force coefficients coupled with measured near pipe velocities provided an accurate time series prediction.

The wake model has claimed vast improvements to the Morison model by attempting to encompass the physics of the wake re-encounter into the model, but the result was a complex model that cannot be used readily as a tool and is limited to pipelines of diameter greater than 400 mm, *DnV* (1987).

Fourier Force Model

Due to the poor predictive capabilities of Morison type force formulations, particularly for the transverse component, a mathematical approach was adopted by *Fyfe et al* (1987) that analysed regular wave forces to obtain Fourier coefficients and phases, which were then used to predict irregular wave forces. A similar approach of Fourier analysis is described by *Bryndum et al* (1992).

For regular waves the forces are assumed periodic and therefore expressed as Fourier expansions,

$$F(t) = F_0 + \sum_{n=1}^{N} F_n \cos(n\omega t - \phi_n) \qquad \text{Eq. (7)}$$

Non-dimensional Fourier force coefficients were calculated by a direct Fourier transform and a least squares technique using experimental data normalized by $\frac{1}{2}\rho DU^2$ and using the following expressions:

$$C_H(t) = \frac{F_D(t)}{1/2\rho D U_m^2} = C_{Ho} + \sum_{n=1}^{9} C_{Hn} \cos(n\omega t - \phi_{Hn}) \qquad \text{Eq. (8)}$$

$$C_V(t) = \frac{F_L(t)}{1/2\rho D U_m^2} = C_{Vo} + \sum_{n=1}^{9} C_{Vn} \cos(n\omega t - \phi_{Vn}) \qquad \text{Eq. (9)}$$

It was claimed that this ninth order Fourier expansion gave a very accurate description of the hydrodynamic forces in regular waves. However, each description requires twenty Fourier coefficients and eighteen phases to uniquely define the hydrodynamics, although theoretically these numbers can be halved due to symmetry. *Fyfe et al* (1987) claimed only five Fourier coefficients adequately represented each force component, those being the mean and harmonics 1, 2, 3, 5 for the drag force, mean and harmonics 1, 2, 3, 4 for the lift force. These coefficients were presented as functions of either Kc or current ratio. *Bryndum et al* (1992) published Fourier coefficients and phases for both the in-line drag and lift forces as functions of Keulegan-Carpenter number, noting similar variations with this parameter as the traditional drag and lift coefficients in Morison's equation. It is the optimization of the phase that overcomes the failure of conventional force models.

In summary, Fourier decomposition is an elaborate curve fitting process in which there are too many disposable parameters to be practical.

Flow Visualization

An investigation of forces on a small cylinder in a sinusoidally oscillating flow and flow patterns around a sinusoidally oscillating cylinder and seabed was made by *Sumer et al* (1991). They oscillated a vertically mounted cylinder and plane boundary, that passed through the free surface, in otherwise still water. The flow was visualized with aluminium particles on the free surface, that the cylinder and seabed pierced. Such a testing method may give distorted patterns as the free surface will form gravity waves in response to the cylinder's motion. It was observed that a vortex pair was formed in each half cycle prior to the creation of the lee-wake vortex behind the cylinder. The passage of this lee-wake vortex over the pipe was associated with the peak in the lift force in each half period.

The present study is both complementary to and an extension of the work of *Sumer et al* (1991), as flow visualization was made in the same test as force measurements. The flow conditions were extended to include non-sinusoidal flows and combined wave and current flows.

2. Experimental Details

The experiments were conducted in the large wave-current flume at Monash University as shown in Figure 1. The flume comprises two working sections of different depths: 4 m and 2 m connected via a ramp. The shallower of these were used for the testing with a water depth of 1.55 m. Waves are generated by a bottom hinged paddle driven by a hydraulic ram and are passively absorbed by a slotted spending beach and baffles. Steady currents in the same direction as wave propagation are produced by recirculating water from behind the beach to the rear of the paddle.

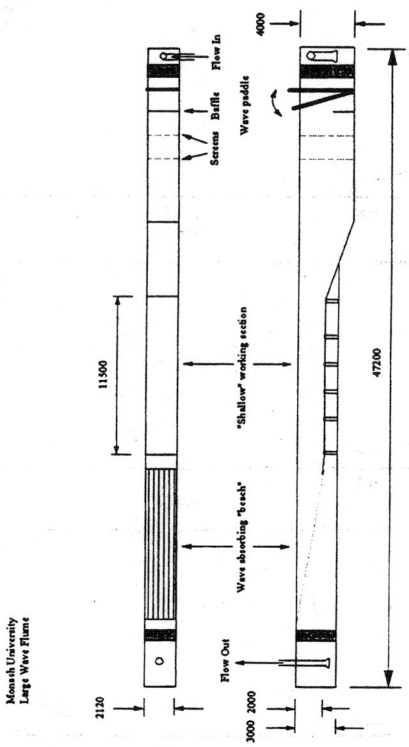

Figure 1 Large wave-current flume

Forces acting on a 1 m long section of 40 mm diameter aluminium tube, chosen so that the system's natural frequency was much larger than wave frequencies, were measured by two-component load cells, supporting the cylinder at either end. These load cells were housed in 300 mm lengths of tubing, the same diameter as the test pipe, to eliminate end effects and shelter the transducers from direct hydrodynamic loading themselves.

Water surface elevation above the pipe centre-line was measured using a capacitance type wave probe. The still water depth was 1.55 m for all tests.

Flow visualization was achieved by dye release through an array of nozzles in the vertical plane on the up-stream side of the cylinder and two nozzles piercing the test bed either side of the cylinder. The dye used was a mixture of fat-reduced milk and water, which maintained the contrast intensity of the dye while minimizing any density differential with the water in the flume. The visualized flow was highlighted with sheet lighting and recorded to video tape via an underwater miniature CCD camera. Superimposed on the flow visualization image were the force traces from the data acquisition monitor.

There were 69 experiments performed where the forces due to regular waves with and without the presence of a steady current were measured on a smooth cylinder; 25 were for the case of no current and 22 for each of the two current conditions. Flow visualization was performed for 27 of these tests. The range of parameters for the test programme are listed below;

T	1.7 - 3.3 [s]
H	< 650 [mm]
U_c	0.10, 0.18 [ms^{-1}]

3. Forces and Coefficients

Data from each test were ensemble averaged over the largest possible integer number of wave cycles in the data record, which ranged from 24 for the longest wave to 48 for the shortest wave. Representative time series of the water surface elevation and forces are given for a typical experiment in Figure 2 for the case of waves alone, and in Figures 3 and 4 for combined wave and current flows.

Peak forces occur prior to moments of maximum undisturbed velocity, under the wave crest and trough. The vertical force peaks occur slightly before extreme horizontal force values. Magnitudes of the two vertical force peaks in a wave cycle are not equal, for waves alone the larger peak generally occurs during the down-crossing half cycle, while for combined flows the larger peak occurs during the up-crossing half cycle. This phenomenon is summarized in Figures 6 and 7. The results of combined flow are in agreement with *Bryndum et al* (1992), in that the larger peaks occur in half periods where the combined velocity is its largest, whereas *Fyfe et al* (1987) observed the largest peaks in half cycles where the wave velocity opposes the current.

The phase difference between the peak forces and the peak ambient velocities is well described using the formulation of *Jacobsen et al* (1984) for the phase lag between zero up-crossing wave velocity and the wake induced peak velocity,

$$\frac{t_{w,max}}{T} = 1.2\left(\frac{0.8 + 0.064Kc}{Kc}\right) \quad \text{for } 6 < Kc < 82 \qquad \text{Eq. (10)}$$

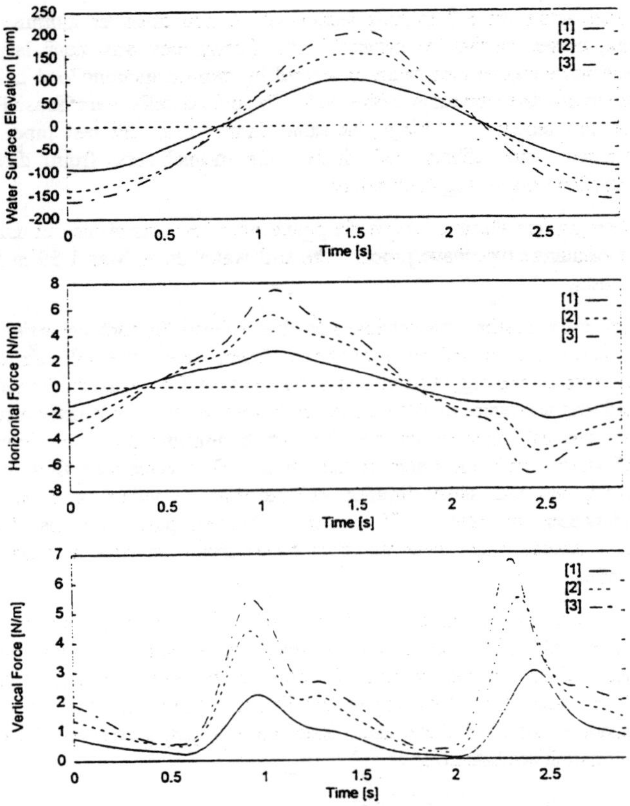

Figure 2 Time series of wave, horizontal and vertical forces. Uc=0, T=2.9 s: [1] Kc=13, [2] Kc=20, [3] Kc=25

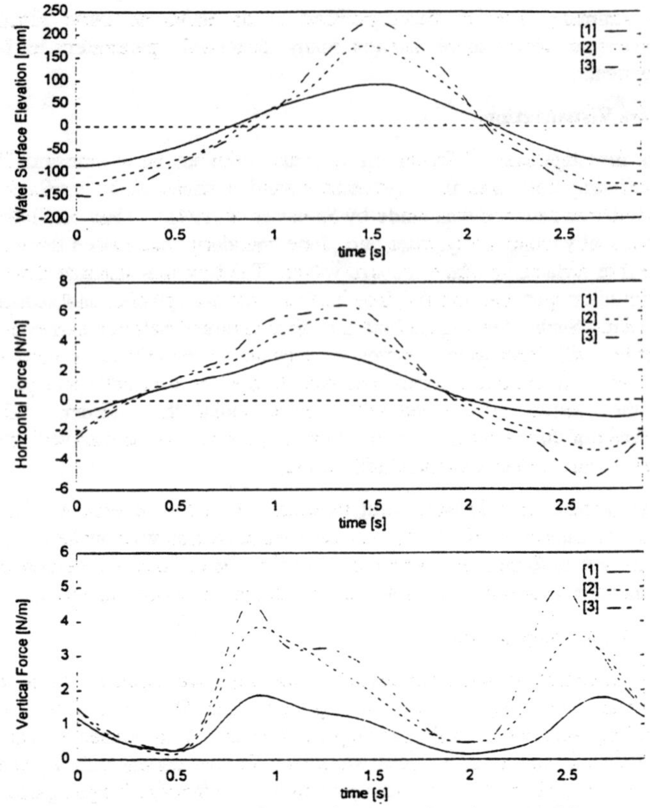

Figure 4 Time series of wave, horizontal and vertical forces. Uc=0.18 ms^{-1}, T=2.9 s: [1] Kc=26, [2] Kc=35, [3] Kc=41

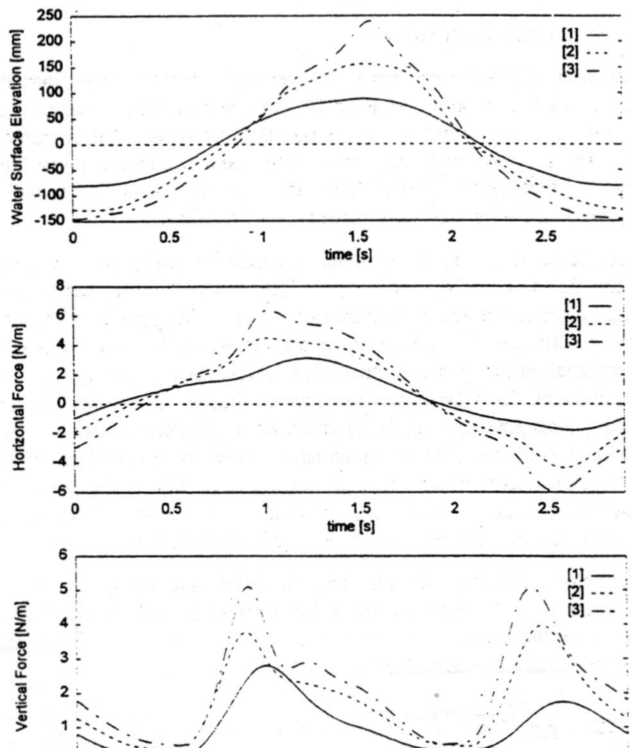

Figure 3 Time series of wave, horizontal and vertical forces. Uc=0.10 ms^{-1}, T=2.9 s: [1] Kc=19, [2] Kc=28, [3] Kc=35

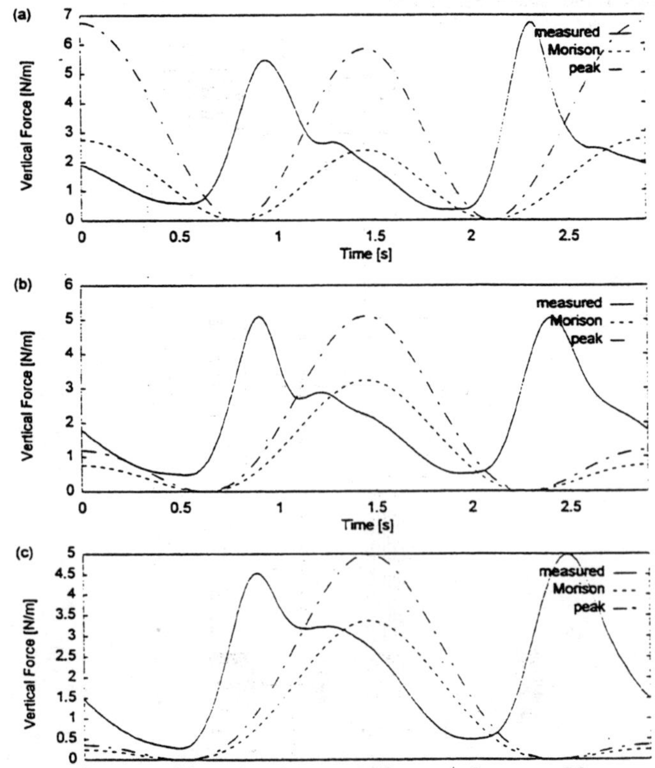

Figure 5 Measured and predicted vertical forces, T = 2.9 s (a) Uc=0 Kc=25, (b) Uc=0.10 ms^{-1} Kc=35, (c) Uc=0.18 ms^{-1} Kc=41

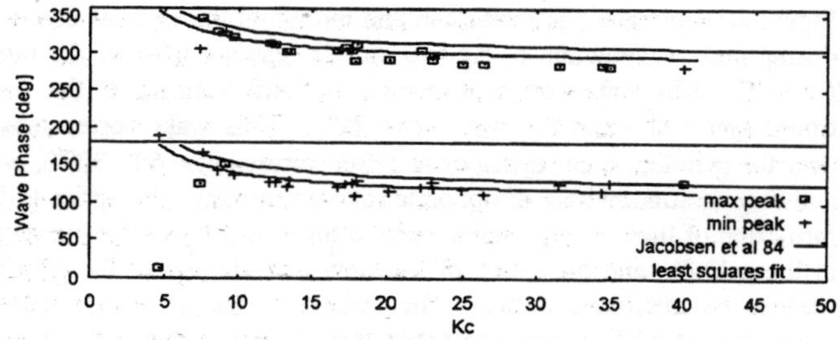

Figure 6 Position of vertical peak forces in wave cycle versus Kc, waves alone

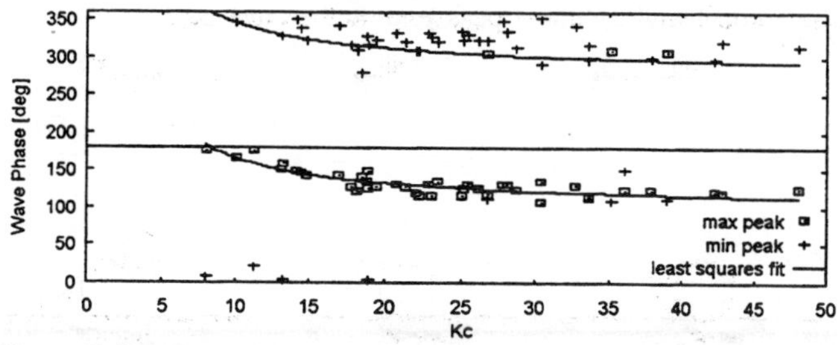

Figure 7 Position of vertical peak forces in wave cycle versus Kc, waves and current

A least squares regression of the data associated with the lower wave phase in Figure 5 gave coefficients of 0.75 and 0.038. The upper curve in this Figure 5 is the regression line translated up by 180^0, indicating that the two vertical force peaks are separated in time by half a wave cycle. Regression analysis of the combined flow in the same manner produced a gradient coefficient of twice the size of that obtained in pure wave flow, 1.5, and an offset coefficient of 0.021. This least squares fit is plotted in Figure 6 together with the same curve offset by 180^0, revealing that the vertical peak forces are not generally separated by half the wave period, as in pure wave flow. These smaller peaks occurring when the current and wave velocities are opposing, contain more variation in their phase position than other peaks due to the complex vortex motions existing at this stage in the combined flow (see Figures 11 and 12).

A comparison between the measured vertical force time history and those predicted by the Morison and peak force models is made in Figure 5 for three wave-current conditions. The Morison curves use a lift coefficient obtained from least squares fit of Eq. 2 to data, while the peak curves use vertical force coefficients obtained from Eq. 3. The coefficients used were obtained by fits to the experimental data and are therefore optimal. The poor fits are due solely to the form of these models, which assume a squared relationship to the undisturbed fluid velocity.

The suitability of the data to determine the Morison coefficients in the in-line direction is based on the method proposed by *Dean* (1976) and applied to a combined wave-current flow by *Chandler and Hinwood* (1981). The ratio, R, of change in C_M to change in C_D for a given change in error between the measured and predicted forces, is defined in Eq. (11), simplified here for conditions near the bed. For $R > 1$, a given change in error produces a larger change in C_M than in C_D, making the data more suitable for determining C_D than C_M; for $R < 1$, the opposite is true. Dean selected a ratio of 4 as the limit of acceptability, thus C_M are ill-conditioned for $R > 4$ and C_D for $R < 0.25$.

$$R = \frac{\Delta C_M}{\Delta C_D} = \frac{Kc_w}{\pi^2}\sqrt{2\alpha^4 + 6\alpha^2 + 0.75} \qquad \text{Eq. (11)}$$

where $Kc_w = u_0 T/D$ Kc based on wave velocity alone.

$\alpha = U_c/u_0$ current ratio.

Using Dean's suitability criteria, seven values of C_M were rejected, all of these tests were combined wave and current conditions with $Uc = 0.18\ ms^{-1}$.

The least squares fitted force coefficients, for waves alone, are plotted against Kc in Figure 8 together with coefficients from previous researchers. The kinematics required for the coefficient fitting were calculated from Stokes fifth order wave theory, which were very similar to those obtained by Fourier wave theory.

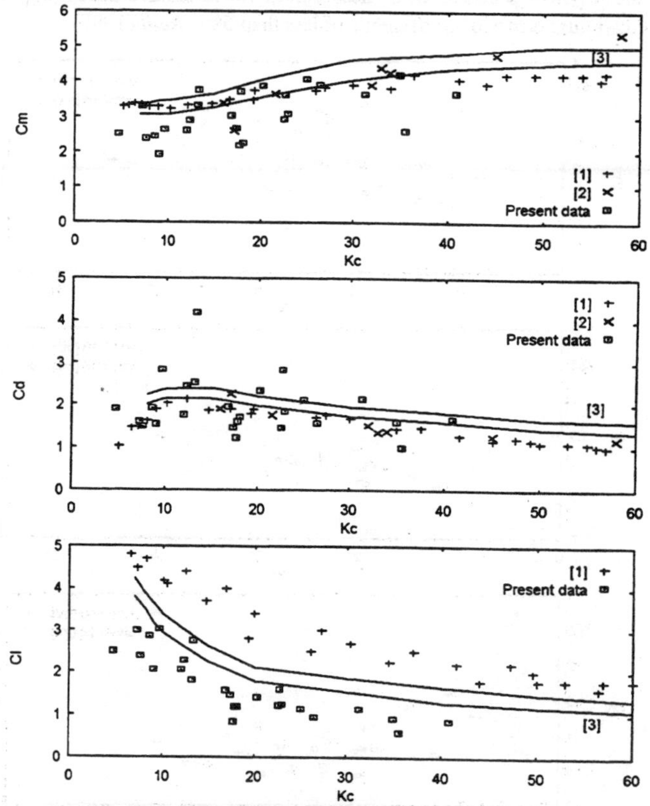

Figure 8 C_M, C_D and C_L versus Kc, Stokes fifth order
[1] *Sarpkaya and Rajabi* (1980) planar flow
[2] *Verley et al* (1987) planar flow, LST PIPESTAB (Marintek)
[3] *Bryndum et al* (1983) planar flow, (D.H.I.)

The force coefficients of *Sarpkaya and Rajabi* (1980), *Bryndum et al* (1983) and *Verley et al* (1987) are derived from sinusoidal planar flow experiments. *Sarpkaya and Rajabi* used a U-tube in which a pressure gradient exists as the fluid is forced into rectilinear harmonic motion, whereas the carriage technique of *Bryndum et al* and *Verley et al* fails to model this pressure gradient that would exist under a propagating wave.

For the present study the inertia coefficients are generally smaller than those obtained by the above authors, whilst the drag coefficients are scattered with few points below the curve of *Sarpkaya and Rajabi*. The lift coefficient is lower than the band of *Bryndum et al*. The lift coefficients obtained by *Sarpkaya and Rajabi* are based on an extreme force analysis, and are therefore naturally larger than those obtained by a least squares fit.

The consistently lower values of C_L could be attributable to a thick boundary layer that will reduce the velocity over the pipe, which would also lower C_D and C_M to a lesser extent.

Force coefficients from tests in combined regular wave and steady current are plotted against Kc in Figure 9. The characteristic velocity used in calculating Kc is the maximum flow velocity, $u_m = U_c + u_0$. The coefficients from the combined flow experiments are all reduced by increase in current. Comparisons to previous research are difficult due to inconsistent choices in characteristic velocities to define flow parameters and a dependence on current ratio as well as Kc.

The variation in all of the force coefficients is predominantly due to the physical process, with measurement uncertainties accounting for maximum errors in coefficients of less than 5%, *Neill* (1997).

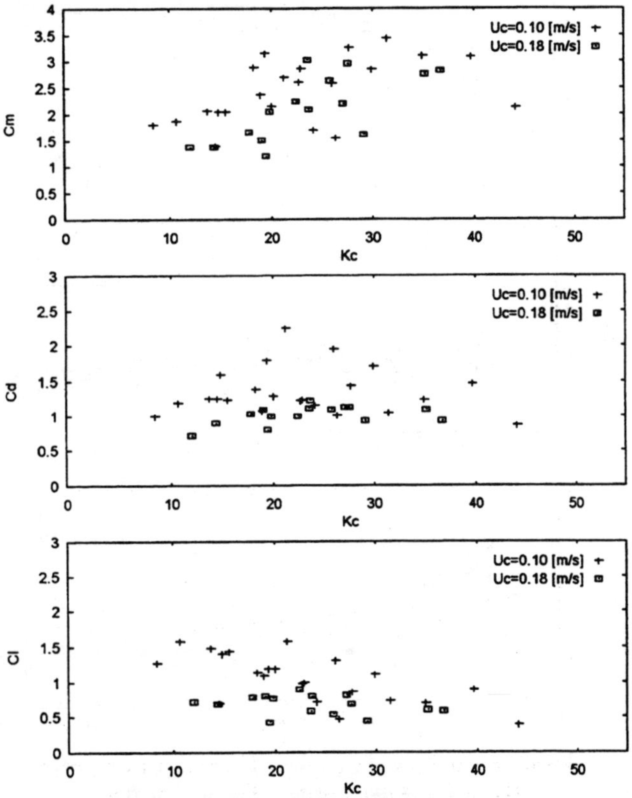

Figure 9 Morison force coefficients versus Kc, combined flow

4. Flow Visualization

Under pure wave action, with no mean current, two flow regimes could be identified. One pattern at low Kc (<10) evolving into the other at higher Kc (>10), with little change in the fundamental features other than timing and size for Kc up to 25. *Sumer et al* (1991) also identified two regimes, differentiating between them at Kc = 7. In the following figures vortex motions have been sketched from video records at six positions in the wave cycle, starting at the trough $\omega t = 0^0$. The wave is propagating from right to left, this also being the direction of the mean current when present. Arrows have been included above each frame to indicate the free stream velocity at that instant.

Figure 10 illustrates the formation and motion of the vortices observed during pure wave action of Kc = 6, and is representative of the range Kc < 10. The wake-vortex is present and still forming at the wave trough ($\omega t = 0^0$) and the crest ($\omega t=180^0$). This wake-vortex moves over the cylinder, then separates as a free vortex ($\omega t \approx 60^0, 300^0$), with a new separation bubble of opposite rotation forming. Just prior to the formation of the new separation bubble the flow follows the curvature of the cylinder and the speed of the flow over the top of the cylinder reaches its maximum value. The process of the separation bubble separating as a full vortex corresponds to a peak in the vertical force and occurs close to, but prior to the passage of the wave crest. From Figure 6 it is shown that the peak vertical forces occur approximately 40^0 prior to the wave crest or trough, with the major peak prior to the wave trough and the minor peak prior to the wave crest.

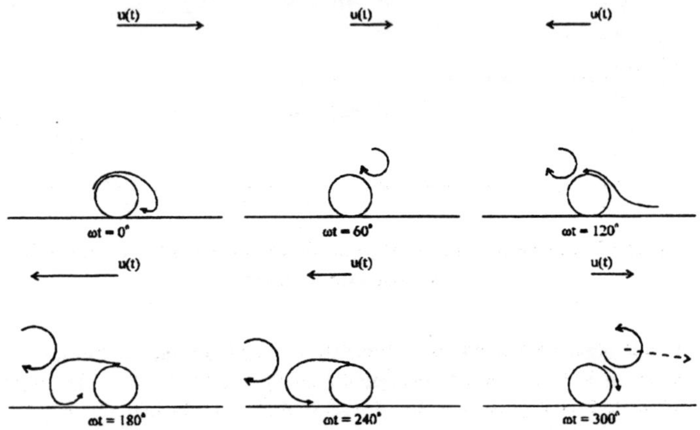

Figure 10 Vortex patterns in the range Kc < 10, $U_c = 0$. The case presented here Kc = 6, $U_c = 0$

A typical flow pattern for wave flow of Kc > 10 is shown in Figure 11. There is now obvious pairing of vortices, which move away from the cylinder under their self induced motion as well as being advected by the wave pressure gradient leaving an attached wake-vortex. The formation of the new vortex, that pairs with the old wake vortex, occurs just after flow reversal and produces a peak in the vertical force. The vertical force peaks now occur earlier than in the previous case and so frames of interest are $\omega t = 120^0$ and $\omega t = 300^0$ in Figure 11 which illustrate events immediately after a peak force event.

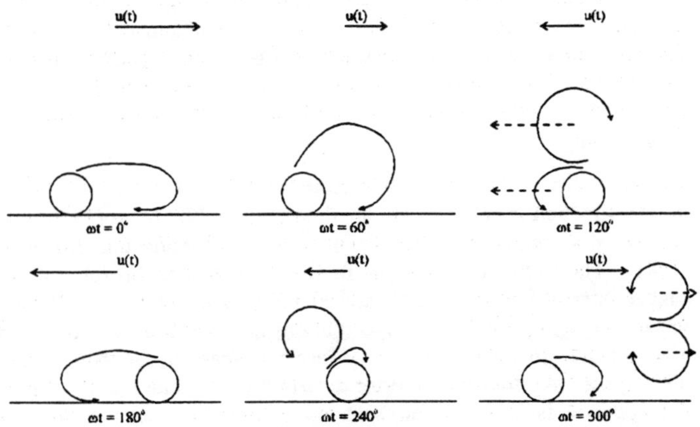

Figure 11 Vortex patterns in the range Kc > 10, $U_c = 0$. The case presented here Kc = 25, $U_c = 0$

For combined flow conditions two flow regimes were again identified, with the distinction between the two defined by a combined flow Kc of 25. Combining a current with the wave dramatically alters the flow pattern, as illustrated in Figure 12. For reasonably large current ratios (greater than 0.65) and low Kc there is more vertical displacement in the vortex motions. At the wave trough, a vortex pair, formed from the previous half cycle's wake vortex and a newly created vortex ($\omega t = 300^0$), moves upwards under self induced motion. There is no significant net pressure gradient during the passage of the wave trough as the current is opposing the upstream wave induced velocity, however once there is a significant downstream pressure gradient the vortex pair is advected in that direction. Whilst the vortex pair moves away from the cylinder a separation bubble is formed on the cylinder that evolves into an attached vortex producing the larger peak in the vertical force. This attached vortex grows until prior to the wave trough ($\omega t \approx 240^0$), when it starts to move back over the cylinder due to the pressure gradient of the wave field. Its high angular velocities causes separation and the formation of another vortex to which it pairs itself ($\omega t = 300^0$), this vortex formation produces the smaller of the vertical peak forces.

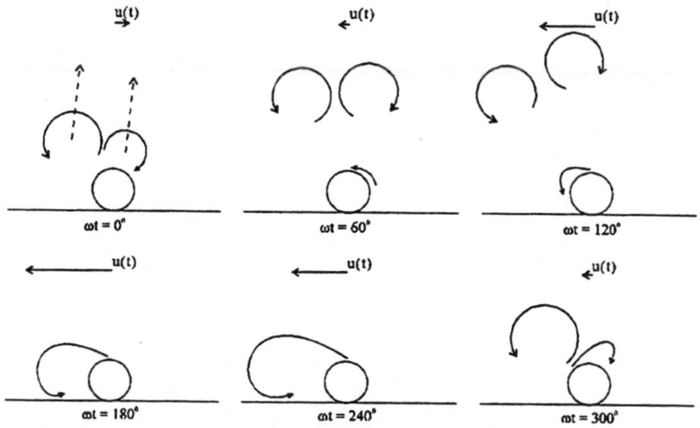

Figure 12 Vortex patterns in the range Kc < 20, U_c > 0. The case presented here Kc = 14, U_c = 0.1 ms^{-1}

For combined flow of larger Kc and hence lower current ratio, the vortex patterns are more symmetrical about the two half cycles. A representative pattern for Kc > 20 appears in Figure 12. The features in this case closely resemble those in Figure 10, the larger Kc regime in waves, but with significant phase changes. A vortex pair is formed by the wake vortex and the counter-rotating clockwise vortex which it forms as it reverses back towards the cylinder between $\omega t = 240^0$ and 300^0. This pair now pass over the cylinder, due to the wave field dominating the current, and progresses upstream. This passage occurs later in the cycle compared to the larger Kc wave flow due to the superposition of the constant current pressure gradient on the oscillating pressure gradient of the wave, given that flow separation occurs at some threshold (adverse pressure gradient). The secondary vertical force peak associated with the movement of this vortex pair away from the cylinder occurs closer to the wave trough than for the no current case, and the primary peak occurs earlier (compare Figures 10 and 12 for $\omega t = 60^0$). This results in a phase difference between the force peaks that is no longer 180^0, as noted in Figure 6.

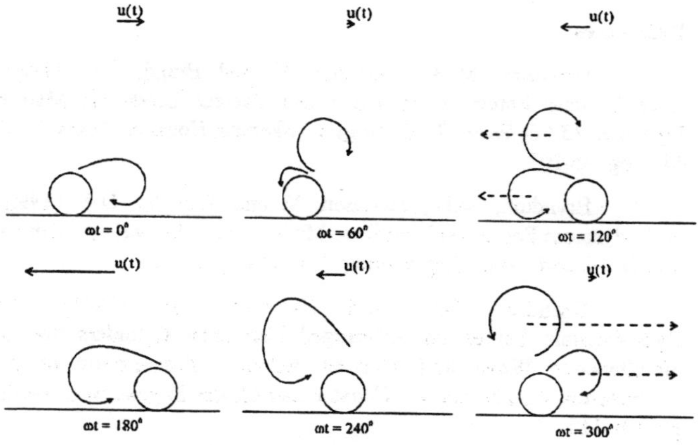

Figure 13 Vortex patterns in the range Kc > 20, U_c > 0. The case presented here Kc = 27, U_c = 0.1 ms^{-1}

5. Conclusions

Hydrodynamic force coefficients for a cylinder resting on a plane boundary have been determined from experiments in a wave-current flume. These tests investigated regular wave and combined regular wave and steady current loading, thereby modelling the true wave-current interaction, wave non-linearities and the two-dimensional structure of the flow field. Such a testing programme differs in technique from major studies in the past. In addition flow visualization was performed in a number of test conditions, reporting for the first time the flow patterns in combined wave-current flow.

The force coefficients have been presented against Kc, and for pure wave flow compared to three other studies revealing significantly smaller lift coefficients and more scatter in the in-line coefficients.

The predictive capacity of the Morison model, Eq. 1, is satisfactory for the in-line force, but fails badly for the lift force, Eq. 2, in estimating the force magnitude and phase. The peak force model should only be used if the phases of the forces are not required for horizontal or vertical forces. Hence both of these models and the Fourier decomposition and wake flow models have serious flaws and are not well suited to design of stable pipelines under a range of wave and current conditions. There is a need for a simple model that contains the true flow physics.

Typical flow patterns have been sketched for the two regimes identified under both wave and wave-current conditions. The cylinder's wake is composed of a single attached vortex which, as it is swept back over the cylinder, forms a new vortex. Depending on the flow conditions this new vortex becomes the new attached wake vortex, or forms a pair with the previous wake vortex and moves and never re-encounters the cylinder. Given this knowledge of the wake's structure a potential flow model has been developed to describe the near pipe flow field due to vortex interaction and will be reported separately.

Acknowledgment

The authors wish to thank Professor John Fenton for many valuable discussions and for use of his higher order wave theory programmes in this study.

References

Bryndum, M.B., Jacobsen, V. and Brand, L.P. (1983) Hydrodynamic Forces from Wave and Current Loads on Marine Pipelines *15th Offshore Technology Conference*, Houston, Texas, OTC 4454 pg. 95-102.

Bryndum, M.B., Jacobsen, V. and Tsahalis, D.T. (1992) Hydrodynamic Forces on Pipelines: Model Tests *Journal of Offshore Mechanics and Arctic Engineering* Vol. 114 pg. 231-241.

Chandler, B.D. and Hinwood, J.B. (1981) The Hydrodynamic Forces on Submerged Horizontal Cylinders due to Simultaneous Wave and Current Action *Proceedings of 5th Australasian Conference on Coastal and Ocean Engineering*, Perth, pg. 108-115.

Chao, J.C., Lambrakos, K.F. and Verley, R.L.P. (1989) Hydrodynamic Force Characteristics for Submarine Pipelines *21st Offshore Technology Conference*, Houston, Texas, OTC 6060 pg. 181-188.

Collins, J.I. (1995) Damage to Unburied Flowlines in the Gulf of Mexico during Hurricane Andrew *27th Offshore Technology Conference*, Houston, Texas, OTC 7859 pg. 503-512.

Dean, R.G. (1976) Methodology for Evaluating Suitability of Wave and Wave Force Data for Determining Drag and Inertia Coefficients *Behaviour of Offshore Structures*, Trondheim, Vol. 2 pg. 40-64.

DnV (1987) On-Bottom Stability Design of Submarine Pipelines *RP E305*.

Fyfe, A.J., Myrhaug, D. and Reed, K. (1987) Hydrodynamic Forces on Seabed Pipelines: Large-Scale Laboratory Experiments *19th Offshore Technology Conference*, Houston, Texas, OTC 5369 pg. 125-134.

Jacobsen, V., Bryndum, M.B. and Fredsøe, J. (1984) Determination of Flow Kinematics Close to Marine Pipelines and Their Use in Stability Calculations *16th Offshore Technology Conference*, Houston, Texas, OTC 4833 pg. 481-492.

Jacobsen, V., Bryndum, M.B. and Tsahalis, D.T. (1988) Prediction of Irregular Wave Forces on Submarine Pipelines *Proceedings of 7th International Conference on Offshore Mechanics and Arctic Engineering*, Houston, Texas, Vol. V pg. 23-32.

Lambrakos, K.F., Chao, J.C., Beckmann, H. and Brannon, H.R. (1987) Wake Model of Hydrodynamic Forces on Pipelines *Ocean Engineering* Vol. 14 No. 2 pg. 117-136.

Morison, J.R., O'Brien, M.P., Johnson, J.W. and Schaaf, S.A. (1950) The Force Exerted by Surface Waves on Piles *Petroleum Transactions, AIME* Vol. 189 pg. 149-154.

Neill, I.A.R. (1997) Hydrodynamic Loading of a Cylinder on a Plane Boundary Under Wave and Wave-Current Flows *PhD Thesis, Department of Mechanical Engineering, Monash University*.

Sarpkaya, T. and Rajabi, F. (1980) Hydrodynamic Drag on Bottom-Mounted Smooth and Rough Cylinders in Periodic Flow *12th Offshore Technology Conference*, Houston, Texas, OTC 3761 pg. 219-226.

Schlichting, H. (1968) Boundary Layer Theory 6th Edition *McGraw-Hill*, New York.

Sumer, B.M., Jensen, B.L. and Fredsøe, J. (1991) Effect of a Plane Boundary on Oscillatory Flow Around a Circular Cylinder *Journal of Fluid Mechanics* Vol. 225 pg. 271-300.

Wolfram, W.R., Getz, J.C. and Verley, R.L.P. (1987) PIPESTAB Project: Improved Design Basis for Submarine Pipeline Stability *19th Offshore Technology Conference*, Houston, Texas, OTC 5501 Vol. 3 pg. 153-158.

Proceedings of the Seventh (1997) International Offshore and Polar Engineering Conference
Honolulu, USA, May 25-30, 1997

ISBN 1-880653-28-1 (Set); ISBN 1-880653-31-1 (Vol. III)

A Simple Procedure for the Wave Impact Forces for Circular Structure Member in Breaking Wave Zone

Kyu Nam Cho and Jae Joon Yoon
Hong Ik University
Chochiwon, Korea

ABSTRACT

The characteristics of breaking wave forces are studied and the corresponding estimation procedures for them are proposed. For the characterization of the wave forces, three parts, i.e., drag force, inertia force, impact force are introduced and identified, respectively. Among drag force and inertia force and impact force, the impact force is mainly studied and a concise form of the force is proposed with the application calculation scheme.

Example calculations of impact force are presented by the proposed scheme for a circular structural member. For a typical sea condition and member size, the proposed procedures for the breaking wave forces calculation are well demonstrated. It is found that this procedure for wave impact forces calculation for cicular structure member is simple and easily applicable especially for the early design stage.

KEY WORDS: Impact force, surf zone, offshore structures

1. INTRODUCTION

The importance of the impact wave force on the vertical offshore circular structure member in the surf zone due to the breaking wave has been recognized recently. In this paper the characteristics on breaking wave forces and the corresponding estimation procedures for them are investigated.

For the characterization of the wave forces, three parts, drag force, inertia force, impact force are categorized and identified respectively. Among them the impact force is mainly studied and the concise form of the force is proposed with the application scheme for the design of offshore circular structure member. The resulting form proposed here for impact force is well coincided with former research results by other people.

Except the impact force, the so called Morison equation can be employed as a first estimate for the common offshore structure design. The drag force and inertia force are represented as conventionally for the profile except the breaking part. In the numerical example, for a typical sea condition and member size, the proposed procedure for the breaking wave forces calculation with linear wave theory is demonstrated.

The following sections describe theoretical backgrounds and a simple analysis procedures, numerical example.

2. BREAKING WAVE PROPERTIES

The concept of equivalent deepwater wave was introduced in order to relate the phenomena of wave breaking and other processes to the characteristics of the deepwater waves. The equivalent wave is a hypothetical one devised for the purpose of adjusting the heights of waves in reflection, diffraction and other transformations for the estimation of wave transformation and deformation when dealing with complex topographies.(C. A. Brebbia. S. Walker. Yoshimi Goda, 1985)

The height and period of the equivalent deepwater wave can be represented as below

$$H'_0 = K_d K_r (H_{1/3})_0 \qquad (1)$$
$$T_{1/3} = (T_{1/3})_0$$

where
H'_0 : equivalent deepwater height
$(H_{1/3})_0$: deepwater significant wave height
$(T_{1/3})_0$: significant wave period of deepwater waves
K_r : coefficient of random wave refraction
K_d : coefficient of random wave diffraction

The variation in wave height within the surf zone can be estimated by the following approximation.

$$H_{max} = \min\{(\beta_0 H'_0 + \beta_1 h), \beta_{max} H'_0, 1.8 K_s H'_0\} \quad \text{for } h/L_0 < 0.2 \qquad (2)$$
$$H_{max} = 1.8 K_s H'_0 \quad \text{for } h/L_0 > 0.2$$

where
K_s = shoaling coefficient

$\beta_0 = 0.052(H'_0/L_0)^{-0.38}\exp[20\tan^{1.5}\theta]$

$\beta_1 = 0.63\exp[3.8\tan\theta]$

$\beta_{max} = \max\{1.65, 0.53(H'_0/L_0)^{-0.29}\exp[2.4\tan\theta]\}$

In the transitional water depth, the relation among wave length, period, water depth can be written by using linear wave theory as shown below.

$$L = \frac{gT^2}{2\pi}\tanh\left(\frac{2\pi d}{L}\right) \quad (3)$$

3. BREAKING WAVE FORCE FOR CYLINDER MEMBER

The rigid vertical circular cylinder subjected to breaking wave receives the energy of incident waves. The two main identifiable wave forces on a vertical circular cylinder which may be a member of an offshore structure are the drag and inertia forces. They are both in line with the wave direction. For the cylinder in the surf zone the impact force due to the breaking wave should be considered in addition to drag and inertia forces.(Yoshimi Goda, 1985, K. N. Cho, 1992)

The wave forces can be written as below.

$$F_T = F_D + F_M + F_I \quad (4)$$

$$= \int_{-h}^{\eta}\frac{\rho}{2}C_D D u|u| dz + \int_{-h}^{\eta}\rho C_M\frac{\pi}{4}D^2\frac{du}{dt}dz + \int_{(1-\lambda)\eta_0}^{\eta_0} dF_I$$

for the model as shown in Fig. 1

where F_D : Drag force
F_M : Inertia force
F_I : Impact wave force due to breaking wave

The third part of equation (4) represents the impact wave force due to the wave breaking. The first and second parts of equation (4) can be easily calculated using well known Morison equation or other methods. The third one, however, is something to be seriously considered. F_I may be written as the impact slamming force, which is represented by following formula with the slamming coefficient, C_S.(O. Faltinsen, et al, B. L. Miller, 1977, P. Kaplan)

$$F_I = \frac{1}{2}\cdot\rho\cdot C_S\cdot D\cdot u^2 \quad (5)$$

Equation (5) with Cs equal to π is adopted as a simple representation of the breaking wave force on the vertical circular cylinder in this paper.

Maximum horizontal water particle velocity u can be found from Fig. 2.(W. M. Kim, 1983)

4. PROPOSED WAVE FORCE CALCULATION PROCEDURE

An empirical formula the forces on vertical cylinders in the presence of surface waves so called Morison equation can be used for the wave force calculation as shown in equation (4). They are first and second part. Approximate maximum wave force of drag and inertia forces as shown in equation (6) may be used alternatively for conservative design.(K. N. Cho, 1992, Yoshimi Goda, 1984)

$$F_{max} = (F_D)_{max} + \frac{(F_M)_{max}^2}{4(F_D)_{max}} \quad \text{for} \quad (2F_D)_{max} > (F_M)_{max} \quad (6)$$

$$F_{max} = (F_M)_{max} \quad \text{for} \quad (2F_D)_{max} < (F_M)_{max}$$

The values of $(F_D)_{max}$ and $(F_M)_{max}$ may be calculated as shown in equation (7).

$$(F_D)_{max} = \omega_0\cdot C_D\cdot D\cdot H^2\cdot K_D \quad (7)$$

$$(F_M)_{max} = \omega_0\cdot C_M\cdot D^2\cdot H\cdot K_M$$

where
ω_0 : specific weight of sea water
K_D : modified drag coefficient
K_M : modified inertia coefficient

K_D, K_M for vertical circular cylinder are found from Fig. 3.(W. M. Kim, 1983)

The procedure for calculation of the breaking wave forces for the circular cylinder member which is vertically situated in the surf zone can be summarized as below.

(1) Adjusting the heights of waves which may undergone refraction, diffraction and other transformations, resulting in equivalent wave heights values.

(2) Determining of the necessary parameters such as wave length, velocity of breaking wave, sea bottom slope, breaker height, rate of breaker entry, etc.

(3) Determining of drag and inertia force by using Morison equation or equation (7) for conservative design.

(4) Determining of impulsive breaker force by using equation (5)

(5) Determining of total breaking wave forces encountering into the vertical circular cylinder by combining drag and inertia force and impulsive breaker force.

(6) Designing of the cylinder section properties based on the input loads calculated in step (5)

5. BREAKING WAVE FORCES WITH LINEAR WAVE THEORY

For the sea condition as shown in Table 1, the proposed procedures for the breaking wave forces calculation with linear wave theory are carried out for demonstration.

· $H'_0 = K_d\cdot K_r\cdot H_0 = 0.9\times0.9\times14.3 = 11.6m$

$T_{1/3} = T_{(1/3)0} = 15\text{sec}$

$L_0 = 1.56\cdot T^2_{(1/3)} = 1.56\times15^2 = 351m$

$h/L_0 = 8/351 = 0.023 \quad < 0.2$

·

$(H_b)_{max} = Min\{(\beta_0\cdot H'_0 + \beta_1\cdot h), \beta_{max}\cdot H'_0, 1.8K_S\cdot H'_0\}$

$= Min\{8.85, 19.14, 20.88\} = 8.85\ m$

$\tan\theta = 1/20 = 0.05$

With linear wave theory breaking wave length, L_b can be calculated by equation (3)

$$L_b = \frac{gT^2}{2\pi}\tanh\left(\frac{2\pi h}{L_b}\right) \rightarrow L_b = 129.6m$$

Total breaker force for η_0 = 6.64, λ = 0.39

$$(F_I)_{max} = \frac{1}{2}\cdot\rho\cdot\pi\cdot D\cdot u^2\times\eta_0\times\lambda$$
$$= \frac{1}{2}\times1025\times3.14\times7\times15^2\times6.64\times0.39$$
$$= 6578\ KN = 671\ (Ton)$$

where u is calculated from Fig. 2.
Drag and inertia forces calculation

$$(F_D)_{max} = \omega_0\cdot C_D\cdot D\cdot(H_b)^2_{max}\cdot K_D$$
$$= 1.025\times1.0\times7\times8.85^2\times0.6=338\ (Ton)$$
$$(F_M)_{max} = \omega_0\cdot C_M\cdot D^2\cdot(H_b)_{max}\cdot K_M$$
$$= 1.025\times2.0\times7^2\times8.85\times0.14=124\ (Ton)$$

for $2(F_D)_{max} > (F_M)_{max}$,

$$(F_D+F_M)_{max} = (F_D)_{max} + \frac{(F_M)^2_{max}}{4(F_D)_{max}} = 349\ (Ton)$$

Finally total horizontal breaking wave forces for the vertical circular cylinder are

$$(F)_{max} = (F_D+F_M)_{max} + (F_I)_{max}$$
$$= 349+671 = 1020\ (Ton)$$

Fig. 4 shows the shear force and bending moment diagram for the numerical example.

6. DISCUSSIONS AND CONCLUSIONS

A simple analysis scheme for estimating impact forces for offshore circular cylinder member is developed. The theoretical backgrounds on which general analysis procedure is based are reviewed. Computational example are also demonstrated.

The results from the numerical analysis show that maximum impact force due to breaking wave is far more than the inertia and drag forces which are general design wave forces for offshore circular cylinder member design.

In the design of offshore circular structure member subjected to breaking wave forces, the first concise estimation procedure is needed. And a simpe scheme is proposed.

The proposed impact force estimation method can be easily applied and it is shown that the method is simple and reliable one especially for the early design stage.

REFERENCES

Brebbia, C. A, Walker,(1985) S, *Dynamic Analysis of offshore Structures*, Newnes-Butterworths.

Goda, Yoshimi(1985), *Random Seas and Design of Maritime Structures*, University of Tokyo Press.

Faltinsen, O. et al.,(1977) *Water Impact Loads and Dynamic Response of Horizontal Circular Cylinders in Offshore Structures*, OTC 2741.

Miller, B. L.(1977), *Wave Slamming Loads on Horizontal Circular Elements of Offshore Structure*, RINA.

Kaplan, P., Silbert, M. N.,(1976) *Impact Forces on Platform Horizontal Members in the Splash Zone*, OTC 2498

Cho. K. N.(1994), *Design and Analysis of Offshore Structures*, Iljungsa.

Yi, Y. S., Cho, K. N.(1988), *Impact Due to Slamming*, Hyundai Maritime Research Inst. Report.

Kim, D. Y., Cho, K. N.(1992), *Design of Observatory Tower*, Hyundai Maritime Research Inst. Report.

Goda. Yoshimi(1984), *Wave Forces on a Vertical Circular Cylinder : Experiments and a Proposed Method of Wave Force Computation*, Report of Port & Harbour Reseach Inst.

Society of Navel Architects of Korea(1993), *Introduction to Ship Design and Offshore Structures Design*, Dongmyunsa.

Kim, W. M.(1983), *Formulas for Hydraulics*, Kunwoosa.

Table 1. Design Condition Case

Significant wave height	14.3 m
Significant wave period	15 sec
Refraction Coefficient	0.9
Diffraction Coefficient	0.9
Design water depth	8 m
Sea bottom slope	1/20
Member Diameter	7 m
Shoaling Coefficient	1.0

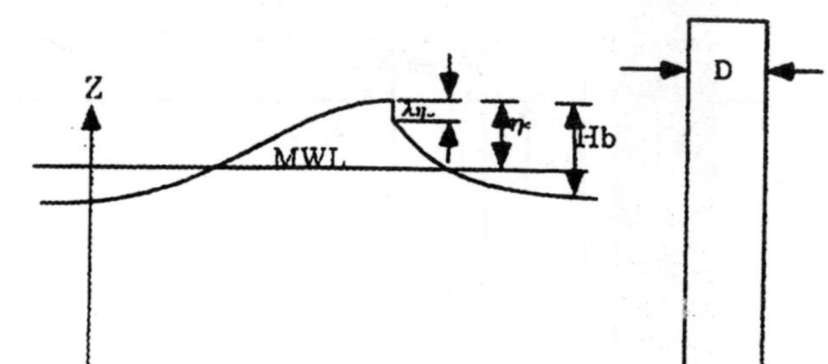

Fig. 1 Typical Breaking Wave Model

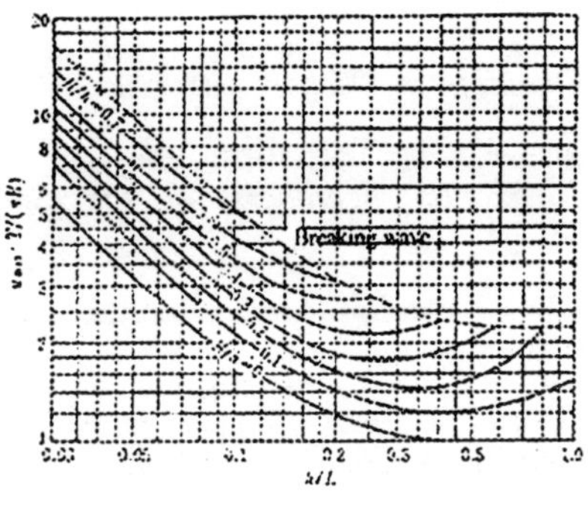

Fig. 2 Diagram for water particle horizontal velocity at crest

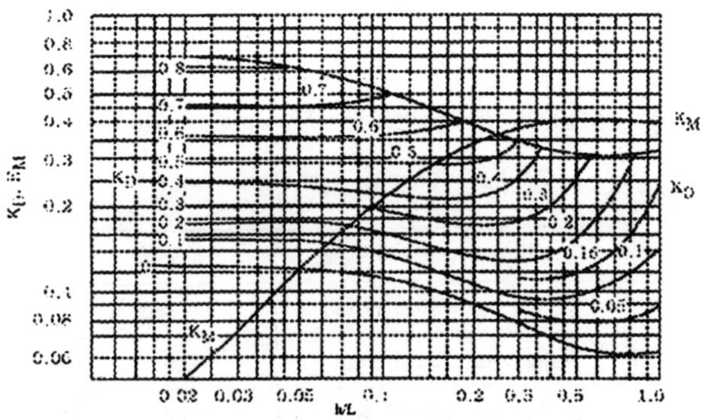

Fig. 3 Diagram of K_D, K_M values for wave force calculation on vertical cylinder

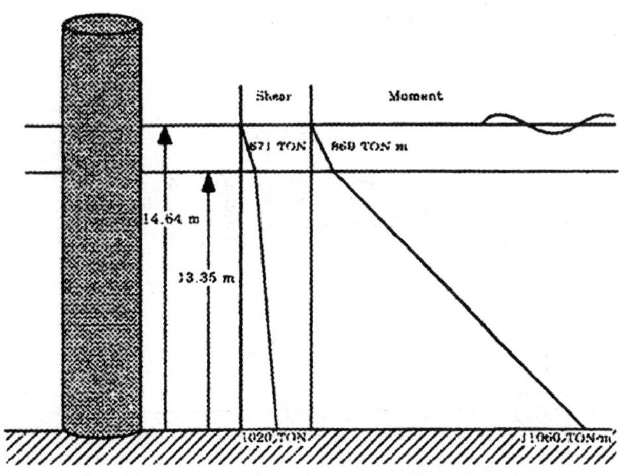

Fig. 4 Shear and Bending Moment Diagram for Circular Cylinder in Surf Zone

Proceedings of the Seventh (1997) International Offshore and Polar Engineering Conference
Honolulu, USA, May 25-30, 1997

ISBN 1-880653-28-1 (Set); ISBN 1-880653-31-1 (Vol. III)

The Force Acting on a Horizontal Cylinder Passing Through the Water Surface

M. Uchida
Kobe University of Mercantile Marine, Kobe, Japan
K. Hu
Register of Shipping of the People's Republic of China, Beijing, China
T. Nishijima
Kobe University of Mercantile Marine, Kobe, Japan

ABSTRACT

The load fluctuation acting on a propeller blade in the condition that the propeller blade exposes partially over the water surface had been investigated experimentally(Uchida and Naito, 1992). The experiments had made clear the interesting phenomenon that the negative torque acted on the blade at the moment when it was passing through the water surface from water to air. And the former report proposed the hypothesis that the negative torque was due to the force caused by the change oh the added mass of the blade along the rotational direction.

In order to confirm the above hypothesis, the forces acting on a horizontal cylinder passing through the water surface are experimentally investigated in this report. The experimental results are compared with the calculated results. The calculation is carried out according to the theory that the test cylinder is acted by a force system of 5 component, those are, the inertial force of real mass and added mass, the drag, the buoyancy and the force caused by the change of added mass. The last component acts in the same direction with the cylinder motion from water to air. It is a negative force called here the added fluctuation force. The calculated results are in approximate agreement with experimental results. The existence of the added fluctuation force component is confirmed.

KEYWORDS: Added Mass, Fluid Force, Oscillating Cylinder, Passing through Water Surface, Added Fluctuation Force, Negative Force

1. INTRODUCTION

It's not rare that a marine screw propeller is operated with its blade exposed partially over the water surface due to various reasons such as large diameter, sea state etc. It's well known that the load acting on a propeller in the above condition fluctuates to a great extent. The load fluctuation acting on a propeller shaft is composed of that acting on each blade of a propeller(Naito and Uchida, 1988). It is important to investigate the load fluctuation acting on not only the propeller shaft but also each blade of the propeller. One of the authors reported the experimental results with a great interest that the torque acting on the blade passing through the water surface from water to air is negative(Uchida and Naito, 1992). The reason was considered that the force caused by the change of the added mass of the blade along the rotational direction due to the difference of the fluid density.

In this report, the forces acting on a horizontal cylinder passing through the water surface are experimentally investigated in order to confirm the above hypothesis.

2. EXPERIMENTAL APPARATUS

Fig.1 illustrates a schematic diagram of the experimental apparatus. The experiments are carried out in the water tank with 2,000mm length, 2,050mm width and 870mm depth.

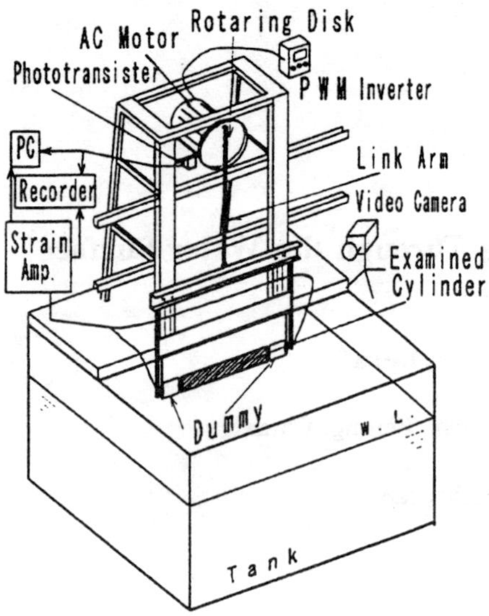

Fig. 1 Experimental apparatus.

Table 1 Examined Cylinders

Circular Cylinder		
Diameter	32	mm
Length	700	mm
Mass	755	g
Elliptic Cylinder		
Major Axis	128	mm
Minor Axis	32	mm
Length	700	mm
Mass	2230	g

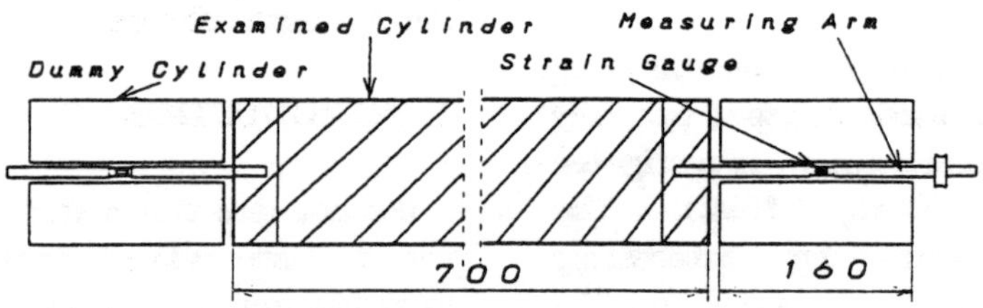

Fig. 2 Examined cylinder.

2.1 EXAMINED CYLINDER

Fig.2 illustrates a schematic diagram of the test cylinder. Two kinds of cylinder were tested. Those principal items are shown in Table 1. One is a circular cylinder with 32mm diameter, the other is a elliptic cylinder with 32mm minor–axis and 128mm major–axis. They have dummy cylinders of the same section on both ends of the examined cylinder. The examined cylinder is supported both sides by the measuring arms. The strain gauges are put on both measuring arms of the examined cylinder so as to measure forces of two components of the major–axis direction and the minor–axis direction. The dummy cylinders are attached to be free from the measuring arms. The test cylinder is given the motion by the oscillation generator.

2.2 OSCILLATION GENERATOR

The examined cylinder is oscillated vertically up and down by the oscillation generator consists of a rotating disk and link mechanism and is driven by AC motor controlled by a PWM inverter. The amplitude is 150mm and the frequency can be changed from 0.5Hz to 1.75Hz. In order to detect the angular position and the bottom position, two sets of a phototransistor and a white–black tape (1pc./rev. and 36pcs./rev.) are mounted on the stand frame and the rotating disk respectively.

3. CALCULATION METHOD

3.1 FUNDAMENTAL VIEW

The fundamental equation of motion can be expressed as follows.

$$F(t) = \frac{d}{dt}\{M(t)\cdot V(t)\}$$

$$= V(t)\cdot\frac{d}{dt}M(t) + M(t)\cdot\frac{d}{dt}V(t) \qquad (1)$$

$F(t)$: Force acting on a body

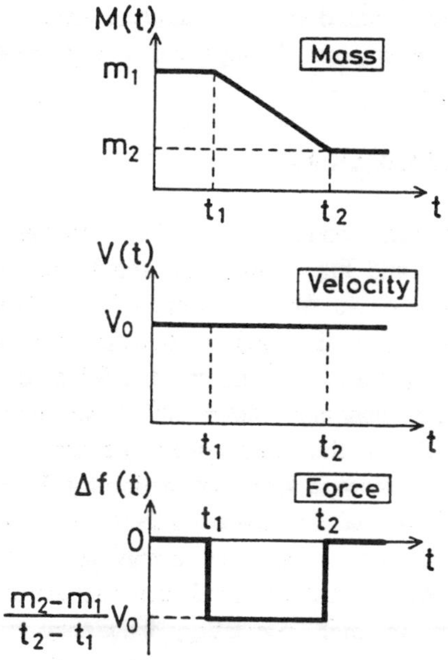

Fig. 3 Primitive model of transient period.

$M(t)$: Mass of a body
$V(t)$: Velocity of a body
t : Time

When the body moves at constant velocity V_0, and the change of the mass is assumed as follows,

$$M(t) = \begin{cases} m_1 & (t \leqq t_1) \\ \dfrac{m_2-m_1}{t_2-t_1}\cdot(t-t_1) + m_1 & (t_1 < t < t_2) \\ m_2 , & (t \geqq t_2) \end{cases} \quad (2)$$

m_1 : Mass before the change
m_2 : Mass after the change
t_1 : Start time of mass change
t_2 : End time of mass change

that is, the change rate of the mass with time is as follows,

$$\frac{d}{dt} M(t) = \begin{cases} 0 & (t \leqq t_1) \\ \dfrac{m_2-m_1}{t_2-t_1} & (t_1 < t < t_2) \\ 0 . & (t \geqq t_2) \end{cases} \quad (3)$$

Then, the external force must be changed as follows,

$$\varDelta f(t) = V_0\cdot \frac{d}{dt} M(t) = \begin{cases} 0 & (t \geqq t_2) \\ \dfrac{m_2-m_1}{t_2-t_1}\cdot V_0 & (t_1 < t < t_2) \\ 0 . & (t \leqq t_1) \end{cases} \quad (4)$$

If m_1 is larger than m_2 as shown in Fig.3, then the change of the external force will be negative, between t_1 and t_2 of time. Moreover, if the external force acting on a body before t_1 of time is smaller than an absolute value of $\varDelta f(t)$, the external force is negative between t_1 and t_2 of time.

Thus above, when a body decreases the mass in motion with a constant velocity, in order to keep the constant velocity, the negative force must act on the body at a certain condition. The negative means that the external force acts in the same direction with that of the motion.

3.2 ADDED FLUCTUATION FORCE

When a body is in motion with constant velocity in a homogeneous fluid, the external force mentioned above doesn't change. However, when a body is in motion passing through the water surface, not only the fluid dynamic force changes, but also the external force changes due to the change of added mass. This external force is called here the added fluctuation force.

3.3 TOTAL FORCE

The total force acting on the examined cylinder can be obtained by summing up the 5 component forces, that is, added fluctuation force F_{add}, inertial force of real mass and added mass, F_M, F_{Madd}, drag F_D, and buoyancy F_B. The drag coefficient of the elliptic cylinder was measured by the use of a circulation water channel.

$$F_{add} = V(t)\cdot \frac{d}{dt} m(t) \quad (5)$$

$$F_M = M\cdot A(t) \quad (6)$$

$$F_{Madd} = m(t)\cdot A(t) \quad (7)$$

$$F_D = \tfrac{1}{2}\cdot \rho(t)\cdot SP\cdot C\cdot V(t)^2\cdot C_D \quad (8)$$

$$F_B = \nabla\cdot(\rho(t)-\rho_w) \quad (9)$$

$$F_{Total} = F_{add} + F_M + F_{Madd} + F_D + F_B \quad (10)$$

$A(t)$: Acceleration of the test cylinder
C : Diameter of the circular cylinder or Minor−axis of the elliptic cylinder
C_D : Drag coefficient of the test cylinder
M : Mass of the test cylinder
$m(t)$: Added mass of the test cylinder
SP : Length of the test cylinder
$V(t)$: Motion velocity of the test cylinder

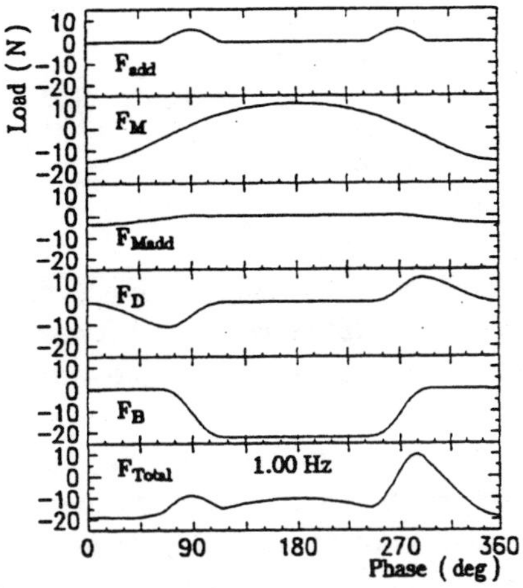

Fig. 4 An example of calculation.
(Elliptic cylinder, 1.00Hz, Passing through water surface)

$\rho(t)$: Fluid density
ρ_w : Water density
∇ : Volume of displacement of the test cylinder

When the test cylinder is passing through the water surface, the density of working fluid is treated simply as follows. The fluid density changes sinusoidally from one of water to one of air or from one of air to one of water.

An example of the calculated results for the elliptic cylinder passing through the water surface is shown in Fig.4, where it can be seen how each component and total force vary during one cycle of oscillating motion. The abscissa represents angular position counted from the bottom dead point. Consequently, the test cylinder passes through the water surface from water to air around 90° and from air to water around 270°. The upward forces acting on the cylinder are defined as the positive value. When the test cylinder passes the water surface from water to air, the added fluctuation force F_{add} acts same direction with the motion of the test cylinder. When it passes from air to water, F_{add} acts opposite direction with the motion of the test cylinder. Therefore the added fluctuation force F_{add} acts upward in the both cases.

4. EXPERIMENTAL RESULTS AND CALCULATION RESULTS

The experiments were carried out at 3 conditions, the first condition was that the cylinder oscillates in air, the second was in water, and the last was that the cylinder oscillates passing through the water surface. And, they were carried out in 6 cases of the oscillating frequency in the region of 0.5Hz~1.75Hz at 0.25Hz intervals. The results are shown in Fig's. 5~10 where the abscissa represents angular position counted from the bottom dead point. The upward forces acting on the cylinder are defined as the positive value.

4.1 OSCILLATION IN AIR

Fig.5 shows the results of the 32ϕ circular cylinder oscillated in air, and Fig.6 shows the results of the elliptic cylinder oscillated in air. When a cylinder is oscillated in air, the fluid dynamic force is small, so that the force acting on the cylinder is almost only the inertial force of the real mass. Then the maximum downward force acts on the examined cylinder at the bottom dead position (0°, 360°) and the maximum upward force acts at the top dead position (180°). It could be said that there is almost good agreement with the experiments and the calculations. The little discrepancy between experimental and calculated results, which appears in high oscillating frequency region, is considered to be due to the structural oscillation of the experimental apparatus.

4.2 OSCILLATION IN WATER

Fig.7 shows the results of the circular cylinder oscillating in water, and Fig.8 shows the results of the elliptic cylinder oscillating in water. In these cases, the fluid dynamic drag becomes very large compared to the case in air, and the inertial force of the added mass cannot be disregarded. The drag varies with the velocity. It reaches, therefore, the positive maximum at 270° and the negative maximum at 90°.

The drag of the circular cylinder is dominantly large among all components so that the total force becomes also absolutely maximum at about 90° and 270° of the angular position. In the case of elliptic cylinder, however, the inertia force of its mass is relatively large compared to drag force. That is the reason why the difference between the circular cylinder and the elliptic cylinder appeared in the figures, and which could be explained by the calculation reasonably.

4.3 MOTION PASSING THROUGH WATER SURFACE

The center of oscillating amplitude is adjusted just on the water surface. Consequently, the test cylinder passes through the water surface from water to air around 90° and from air to water around 270°. The gauge sensor is adjusted to indicate the zero value at the bottom dead position in the water, so that the buoyant force is measured to be negative in the air.

Fig.9 shows the results of the circular cylinder for the oscillation passing through the water surface, and Fig.10

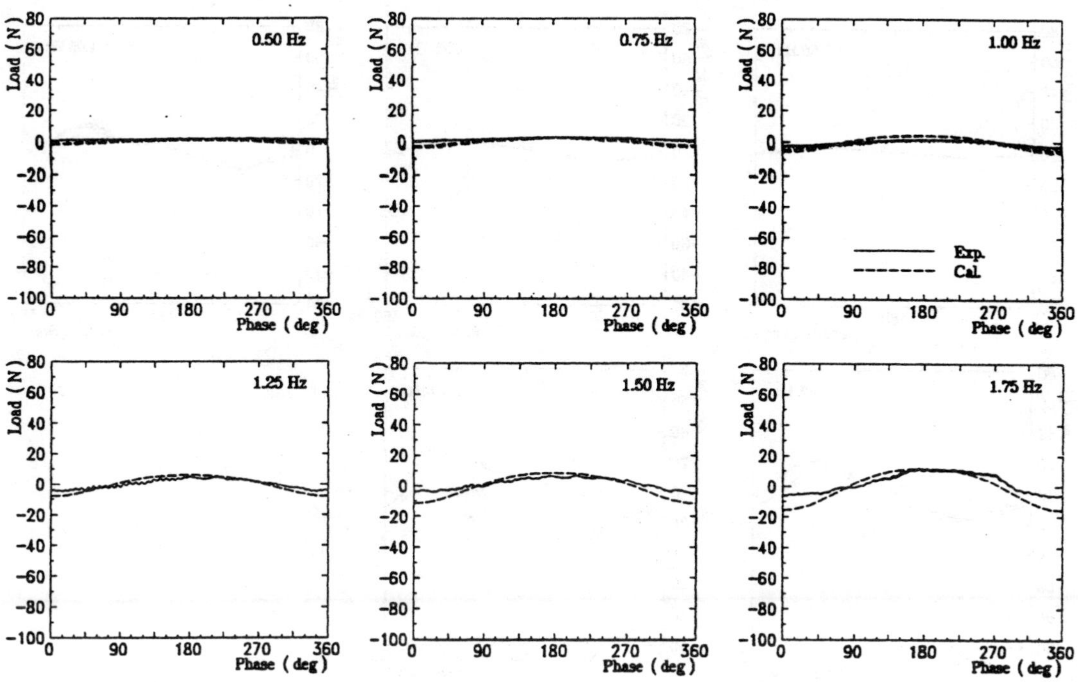

Fig. 5 Results of circular cylinder in air.

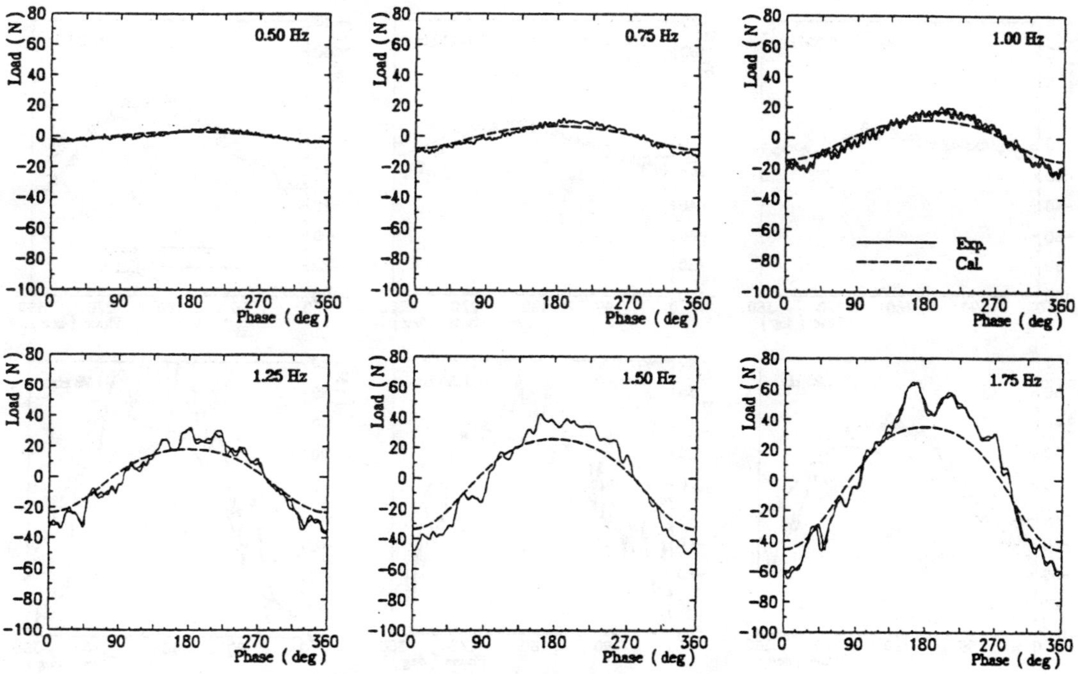

Fig. 6 Results of elliptic cylinder in air.

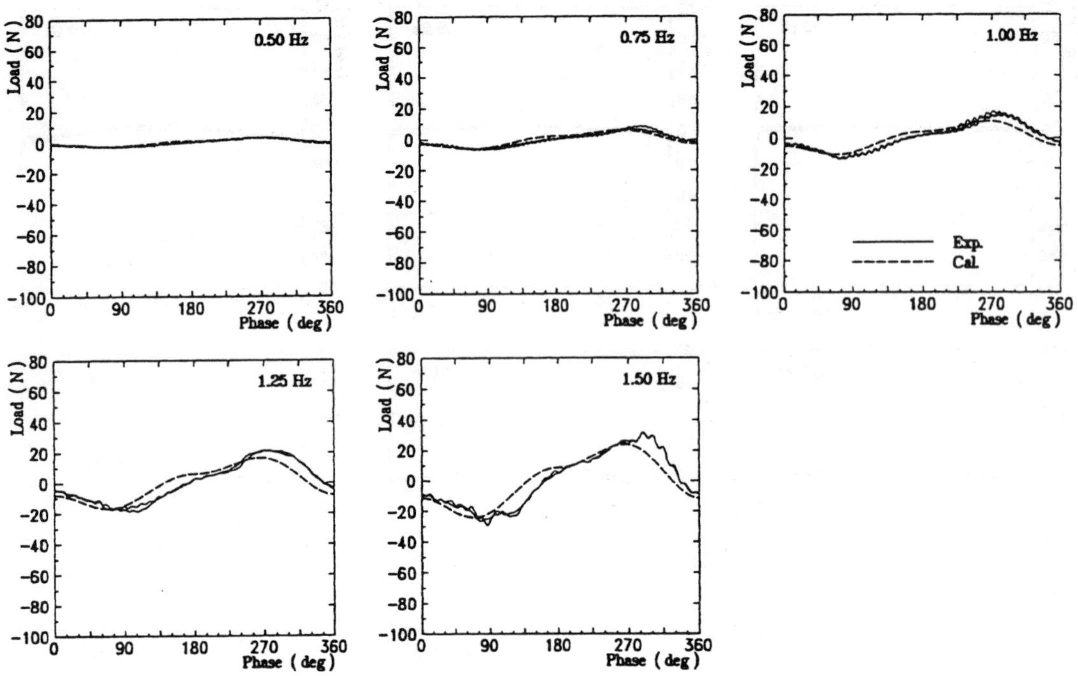

Fig. 7 Results of circular cylinder in water.

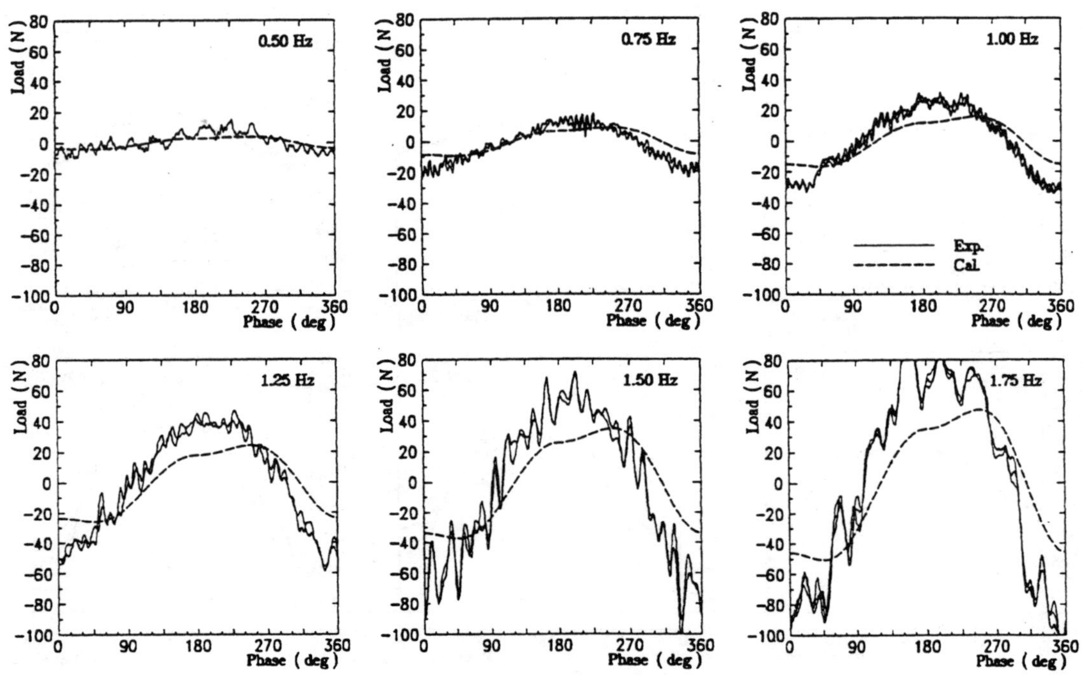

Fig. 8 Results of elliptic cylinder in water.

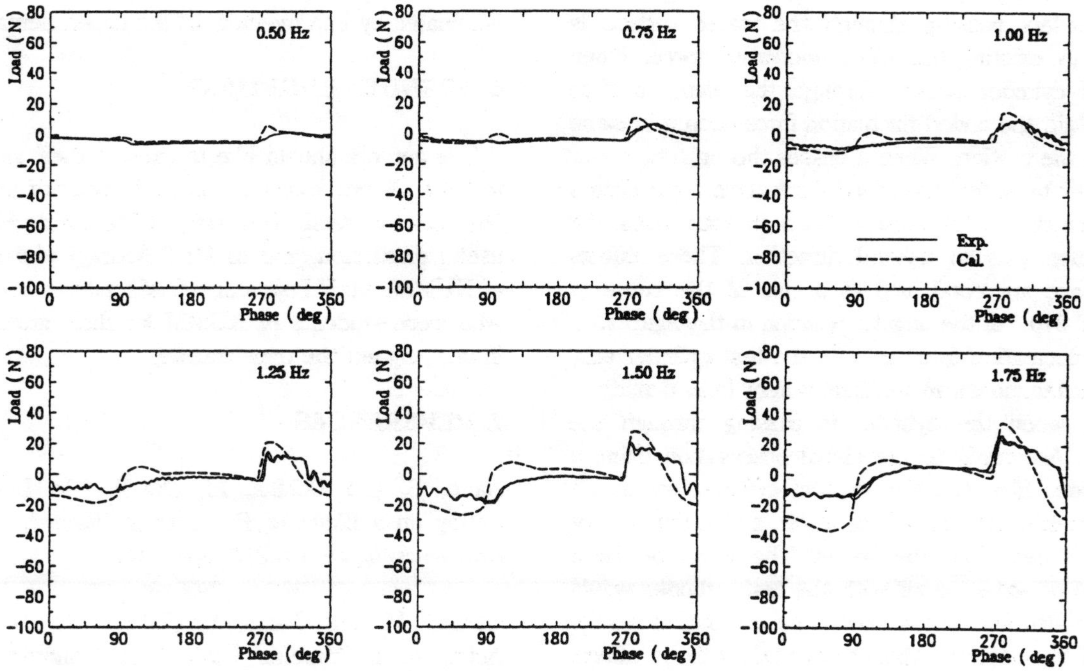

Fig. 9 Results of circular cylinder passing through water surface.

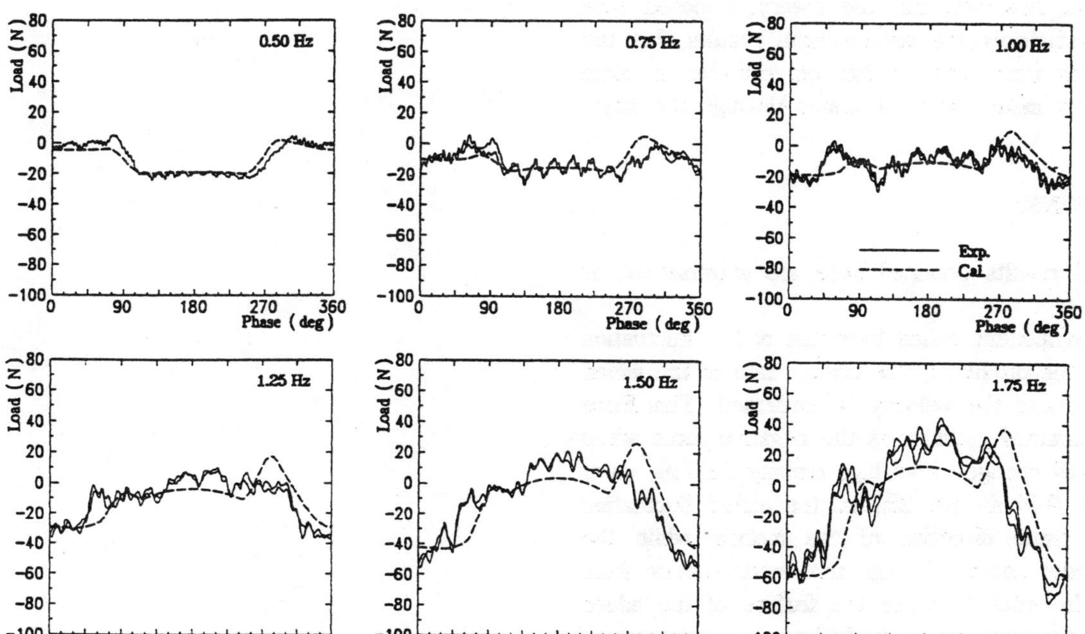

Fig. 10 Results of elliptic cylinder passing through water surface.

shows for the elliptic cylinder. The force acting on the examined cylinders passing through the water surface is characterized by existing the added fluctuation force. When the examined cylinder passes through the water surface from water to air, the added fluctuation force occurs in same direction with the motion. When it passes through the water surface from air to water, the added fluctuation force occurs in opposite direction to the motion. Then, in both cases, the added fluctuation acts in upward direction. These effects mentioned above are confirmed as a rise of the curve at around 90° and 270° of the angular position in the figures.

The added fluctuation force acts on the test cylinder only when the cylinder moves in the field where fluid density is changing, i.e. when the cylinder is passing through the water surface. According to the visual observation using a high speed video film, the circular test cylinder entrains a lot of water when it gets out of the water, and it draws a lot of air when it gets into the water. The mass of those entrained fluid of water or air with the test cylinder would affect on the added fluctuation force. The calculation results in the case of the circular cylinder have taken those effects into account. The transient time for changing the working fluid is made an estimate long on basis of the observation of the entrained fluid.

The difference between the experimental and calculated results is considered to be due to an adverse effect of the vibration of the test stand and due to that the wake flow induced by the test cylinder is not taken into consideration. It could be said, however, that the theory proposed here represents satisfactorily the experimental results that the added fluctuation force acts on the test cylinder in same direction with its motion when it passes through the water surface.

5. CONCLUSIONS

Experimental results obtained here are summarized as follows.

The force component called here the added fluctuation force which is the product of the change rate of the added mass with time and the velocity is confirmed. This force acts on the examined cylinder as the negative force when the working fluid changes from high density fluid as water to low density fluid as air, that is, the added fluctuation force acts in same direction of the motion while the examined cylinder passes through the water surface from water to air. In order to make the feature of the added fluctuation force more clear, it would be necessary to carry out further experiments using an elliptic cylinder with another major−minor axis ratio.

High frequency oscillations are observed in the results of the experiments. These are the natural frequency of the structure system of equipments. These have to be eliminated by improvement of the experimental apparatus.

6. ACKNOWLEDGEMENT

The authors should like to express their sincere gratitude to Prof. Sigeru Naito of Osaka University and Prof. Eiichi Nishikawa of Kobe University of Mercantile Marine for their useful discussion, and to Mr.D.Aburagi who is a technician of KUMM, Mr.Y.Fujinami, Mr.M.Ichikawa and Mr.M.Maeda who were students of KUMM for their ground cooperation in carrying out the experiments.

7. REFERENCES

Naito, S. and Uchida, M. (1988). "On Load Fluctuation Acting on a Blade of Propeller in Waves", *J. Kansai Soc. Nav. Arch. Japan*, No.210, pp61−68.

Uchida, M. and Naito, S. (1992). "On Load Fluctuation Acting on a Propeller Blade (On Condition that a Blade Rises Partially to Water Surface)", *J. Kansai Soc. Nav. Arch. Japan*, No.218, pp235−241.

Proceedings of the Seventh (1997) International Offshore and Polar Engineering Conference
Honolulu, USA, May 25-30, 1997

ISBN 1-880653-28-1 (Set); ISBN 1-880653-31-1 (Vol. III)

Stochastic Analysis of Combined Wave and Current Loading on a Vertical Cylinder from Large-Scale Laboratory Experiments

G. Najafian, R. Burrows and R.G. Tickell
University of Liverpool
Liverpool, United Kingdom

ABSTRACT

The data for this study comes from a series of large-scale tests conducted in the Delta flume of the Delft Hydraulics Laboratory in Netherlands (1993). In these tests two cylinders of diameters 210 and 500 mm have been exposed to both regular and random waves, with and without currents, where the currents have been simulated by towing the cylinder along the flume at a steady speed by means of a carriage. Measurements have been made with surface conditions corresponding to both smooth and rough conditions, with cylinders in both vertical and horizontal orientations. This paper is devoted to the stochastic (non-deterministic) analysis (as opposed to the deterministic wave-by-wave analysis) of the random wave load data on the two cylinders in the vertical orientation. Therefore, the results of this study are relevant to the non-deterministic analysis and design of offshore structures, which are inevitably exposed to random wave and current loading in the ocean environment.

KEY WORDS: Morison's equation, wave and current loading, stochastic analysis, roughness.

INTRODUCTION

Accurate estimation of wave loading on cylindrical members of offshore structures is a major design consideration. The most-widely accepted approach to the problem of predicting wave load on a submerged cylinder with a diameter (D) much smaller than the wavelength (L), $D/L < 1/5$, is due to Morison et al (1950). The drag and inertia coefficients of this equation cannot be predicted theoretically and hence must be determined from experimental data. Extensive small-scale experiments have been conducted to study the mechanism of wave loading on cylindrical members and to derive optimum coefficients for wave load prediction. However, it is well established that these non-dimensional coefficients depend on both the flow regime (represented by Reynolds number, Re) and the Keulegan-Carpenter number, Kc. For design applications, experiments must be carried out in the post-critical regime (high Reynolds numbers, say $Re > 10^5$ in wave-induced flow) with Kc values in the range $0 < Kc < 50$. The combination of these two conditions can be achieved only in large-scale experiments such as the series of the tests conducted at the Delta flume of the Delft Hydraulics Laboratory (DHL) since 1984. In these tests, smooth and rough cylinders (with a diameter of approximately equal to 0.50m) in both horizontal and vertical directions were exposed to both regular and irregular waves. The results of these tests can be found in Bearman et al (1985), Bearman (1988), Wolfram and Theophanatos (1989), Klopman and Kostense (1990), Chaplin (1991) and Chaplin et al (1992). In September and October 1993, a new series of tests were conducted in the flume (Bearman et al, 1994). The primary objectives of these tests were twofold: a) to study combined wave and current loading on a cylinder (where current was simulated by towing the cylinder along the flume at a steady speed by means of a carriage) and b) to examine wave loading on cylinders at higher Kc values, by introducing a test cylinder of smaller diameter than tested previously.

In these tests two cylinders of diameters 210 and 500 mm have been exposed to both regular and random waves, with and without currents. Measurements have been made with surface conditions corresponding to both smooth and rough conditions, with cylinders in both vertical and horizontal orientations. Some preliminary results from wave-by-wave analysis of irregular wave load data on both horizontal and vertical cylinders have been reported by Bearman et al (1994). This paper is devoted to the stochastic (non-deterministic) analysis (as opposed to the deterministic wave-by-wave analysis) of the random wave load data on the two cylinders in the vertical orientation. Therefore, the results of this study are relevant to the non-deterministic analysis and design of offshore structures, which are inevitably exposed to random wave and current loading in the ocean environment.

EXPERIMENTAL ARRANGEMENT

The Delta Flume. The experiments were conducted at the Delta flume of the Delft Hydraulics Laboratory (DHL) in Netherlands. The flume is 230m long, 5m wide and 7m deep over most of its length. Waves are generated by a hydraulically-driven plane waveboard and dissipate most of their energy at the other end of the flume on a sloping beach. In these experiments, the water depth was 5m. To simulate currents a towing carriage was constructed. The carriage is approximately 8m by 6m and runs on existing rails on the top of the flume walls. The carriage is driven by electric motors and can attain steady speeds of up to 1 m/sec in either direction. In these tests two cylinders were tested at the same time, one on the carriage and the other one at a fixed point in the flume, located at 164m from the paddle and 16m from the toe of the beach. The towing distance was approximately 110m, about half the length of the flume.

Test cylinders. Two test cylinders with nominal diameters of 500mm and 210mm were used in these experiments. The smaller cylinder had a total of five force measuring sleeves, each 105mm long, at levels 1 to 5 with corresponding elevations of z_1 = -3.5m, z_2 = -2.5m, z_3 = -1.5m, z_4 = -0.50m and z_5 = +0.5m, where z is the elevation with respect to mean water level (MWL) and is positive upwards. The two highest sleeves (Levels 4 and 5) are in the splash zone and hence would only be intermittently immersed. The large cylinder had two force sleeves, each 250mm long, at levels 2 and 3 with corresponding elevations of z_2 = -2.5m and z_3 = -1.5m (same elevations as Levels 2 and 3 for the small cylinder). All the force sleeves were capable of measuring both in-line (along the length of the flume, i.e. x direction) and transverse (across the channel, i.e. y direction) forces. The natural frequency of the force sleeves is about 10Hz and that of the entire system including the cylinder and the frame on which it is mounted is about 4Hz.

Roughness. The small cylinder was manufactured from stainless steel and had a turned surface finish which may be treated as hydraulically smooth. In-line with previous tests (Bearman et al, 1987), the large cylinder was completely covered with a layer of fine waterproof emery cloth, giving it an effective diameter of 505mm and a surface roughness height to diameter ratio of k/D=0.00012, which will be referred to as "smooth" in this paper.

To perform rough cylinder tests, the cylinders were covered with large roughness elements geometrically similar to those described by Wolfram and Theophanatos (1989). The roughness consisted of a pseudo-random arrangement of three different sizes of right square pyramids whose heights were the same as the dimensions of their corresponding bases, leading to an effective roughness of k/D=0.038 for each cylinder. The roughness elements were attached to the cylinders in sections, each of length 0.5D, consisting of two identical semi-circular shells.

Random waves. The random waves used in these tests were based on the JONSWAP spectrum. The significant wave height was 1.5m and the period was T_p=5.9 seconds, where T_p is the period associated with the peak of water surface elevation spectrum.

Cases. Different combinations of cylinder diameter, roughness condition and carriage velocity led to 15 different cases, as summarised in Table 1.

Instrumentation. Particle velocities were measured by means of electro-magnetic flow meters (EMF) or perforated ball velocity meters (PVM), of diameter 99mm, similar to those used at the Christchurch Bay Tower tests (Bishop, 1984). Both types of instruments recorded the components of velocity in the in-line (x) and vertical (z) directions. The original intention was to use PVMs for the small cylinder upper two force sleeves since the performance of EMF instruments is not satisfactory when they are not continuously immersed. However due to some technical problems (Mackwood and Bearman, 1995), the EMF signals were noisy and hence for some of the tests they were replaced by PVMs. The types of instruments used for measuring water particle velocities at different levels for all different cases are given in Table 1. Velocity meters were always positioned alongside each force sleeve so that there is no phase shift between measured velocities and forces.

Data acquisition. All channels of data were collected at 40Hz after analogue filtering at 20Hz.

Further source of information. For a more detailed description of the experimental arrangements includung those for the horizontal cylinders refer to Bearman et al (1994).

PRELIMINARY INVESTIGATION OF DATA AND DATA PREPARATION

Data quality checking. For all cases, the following have been calculated for in-line forces and water particle velocities:

a) The basic statistics of the signal, i.e. its mean, standard deviation, skewness and kurtosis have been calculated.

b) Their frequency spectra and probability distributions have been plotted.

c) Correlation coefficients and coherence functions between selective pairs of signals have been calculated.

Examination of the foregoing results revealed that:

1) A small number of signals were clipped, that is when the data were above a certain threshold, it remained constant at that level. The duration of each clipped section is short (less than one second) and hence it was decided to estimate the true value of the signal by fitting a cubic spline to the clipped section. The effect of fixing the clipped data in the foregoing manner on the calculated Morison's coefficients is negligible; however, it is an essential exercise for the comparison of the extremes of the observed and predicted forces. It is appreciated that the foregoing method for fixing the clipped data would inevitably lead to some inaccuracy in the extremes of the affected signals, but is better than leaving them untreated.

2) Some of the EMF velocity signals are noisy, resulting in unacceptably noisy acceleration signals, which in turn lead to a reduction in the inertia coefficient of Morison's equation because noise leads to an increase in the standard deviation of the acceleration signal. For example, for case 14, the level 1 inertia coefficient determined from (adjusted) time-domain Least Square Error (LSE) analysis with unfiltered data is 1.91 while that from filtered data (frequencies above 1 Hz removed from both in-line force and velocity) is 2.13, which shows an increase of about 11% as a result of filtering. The filtering was therefore carried out for all in-line forces and velocities. (The basic statistics of the filtered in-line forces are given in Table 2 and Morison's coefficients from (adjusted) LSE analysis of filtered forces and velocities are given in Table 3). It should be considered that the 1 Hz limit is well above the upper limit of important wave frequency range (about 0.4 Hz) and therefore frequencies above 1 Hz (in in-line velocities and forces) can safely be considered as noise.

Accuracy of the measured forces. The measured forces at Level 3 of the large cylinder are suspect (Mackwood et al, 1995). This is because in these experiments water could enter the cylinder through the gaps between the (Level 3) force sleeve and the cylinder, leading to a non-uniform pressure distribution around the inside of the force sleeve. As far as this study is concerned, the level 3 forces have also been analysed and the resultant Morison's coefficients are given in tabular form but they are not presented in figures and are excluded from the discussion and conclusions.

Derivation of velocity signals from perforated ball forces. The conversion of perforated ball forces into water particle kinematics is discussed by Bishop (1979) and Mackwood et al (1995). It was observed that it was difficult to choose drag and inertia coefficients (C_d and C_m) which result in velocity signals with correct phase shift, magnitude, spectral and probabilistic properties. In this study the coefficients have been chosen so that the phase and magnitude (standard deviation) of the velocity signals would be correct, because any inaccuracy in these two attributes of the signal would lead to unreliable drag and inertia coefficients from (time-domain) Least Square Error analysis. By implication, the frequency spectra and statistical properties of the derived velocity signals may be somewhat inaccurate, leading to less reliable values for Morison's coefficients when these coefficients are derived from spectral analysis or the method of moments, to be discussed later.

Residual current in the flume. Examination of the mean velocities reveals that there is a negative residual current (from beach toward the paddle) at the locations of the velocity meters. For example, for case 14, there is a measured velocity of -0.03, -0.05 and -0.07 m/sec at levels 1, 2 and 3, respectively. For cases 1 and 5 with a carriage speed of -1.00 and +1.00 m/sec, the measured mean velocities at level 1 are 1.05 and -1.11 m/sec, indicating a residual current of about -0.03 m/sec. The average residual current for all cases is about -0.03, -0.04 and -0.07 m/sec for levels 1, 2 and 3, respectively. Furthermore, it is observed that both measured mean velocities are higher (in magnitude) than the carriage speed of 1 m/sec. This is because the cylinder itself occupies part of the flume cross section, leading to a reduction in the effective area of the cross section, which in turn increases the mean velocity at the flume cross section.

THE EFFECT OF CURRENT ON SPECTRAL AND STATISTICAL PROPERTIES OF IN-LINE FORCES

In-line wave load prediction. According to Morison's equation (Morison et al, 1950), the in-line hydrodynamic force per unit length on a vertical cylinder of diameter D (D/wavelength < 1/5) is assumed to be the sum of a nonlinear drag component and a linear inertial component; that is,

$$F_p = F_d + F_i = \frac{C_d \rho D}{2}(\bar{u}+u)|\bar{u}+u| + \frac{C_m \rho \pi D^2}{4}\dot{u} \qquad (1)$$

where F_p is the predicted hydrodynamic load; F_d and F_i are the drag and inertial components of fluid loading; C_d and C_m are empirical drag and inertia coefficients; ρ is the fluid density; u and $\dot{u}$ are the undisturbed horizontal components of wave-induced water particle velocity and acceleration at the centre of the cylinder and $\bar{u}$ is the current velocity. $\dot{u}$ is assumed to be equal to the first local derivative of water particle velocity (Bearman et al, 1994).

The effect of current on mean forces. According to Morison's equation current will lead to a mean force in the direction of the current. This effect is particularly important for drag dominated cases in the presence of a strong current. For example, for case number 6 (large rough cylinder with a carriage speed of -1.0 m/sec) the level 2 mean force is nearly twice that of case 1 (smooth cylinder) (refer to Table 2). This is because roughness leads to higher Cd values and hence the drag loading increases.

The effect of current on frequency spectrum of force. Comparison of frequency spectra of level 2 forces for cases 6 and 8 (Figures 1 and 2, respectively) with carriage velocities of -1.00 and 1.00 m/sec reveals that due to Doppler effect, there is a significant shift toward higher frequencies when the carriage speed is negative (simulating a positive current) and that the opposite is true for negative currents.

The effect of current on the standard deviation of the inertial component of loading. Doppler shift has a significant effect on the magnitude of the inertial component of loading in that a shift toward higher frequencies in the velocity signal, will result in an acceleration signal with a higher variance. This is because the spectral densities of the acceleration signals are equal to ω^2 (ω: the angular frequency) times the spectral densities of the velocity signals and hence higher ω values lead to higher variance for the acceleration signals. Therefore, positive current leads to an increase in the magnitude of inertial component while a negative current has the opposite effect.

The effect of current on the standard deviation of the drag component of loading. Not only does current lead to a mean drag component, but also it increases the standard deviation of the drag component. The effect is independent of the direction of current and is particularly important in the case of rough cylinders which have a higher drag coefficient.

The effect of current on the standard deviation of force. The above arguments show that the effect of current on the standard deviation (s.d.) of force can be compound. Furthermore, as will be shown later, the current will lead to lower inertia coefficients, which makes the situation even more complex. However, according to the above arguments for drag-dominated cases, one would expect a significant increase in the standard deviation of the force. This expectation is verified by comparing the standard deviations of level 2 forces (Table 2) for cases 1, 3 and 5; 6, 7 and 8; 10, 11 and 12 and finally 13, 14 and 15.

The effect of current on skewness. Examination of skewness values in Table 2 shows that in agreement with the prediction of Morison's equation, the presence of current leads to an unsymmetrical force distribution, with the same sense of skewness as that of the current. As expected the skewness is very high for cases 13 and 15 in which the small rough cylinder (strongly drag dominated) was towed with a speed of 0.50 m/sec in the flume.

The effect of current on kurtosis. According to Morison's equation for Gaussian-distributed water particle kinematics, the non-linear drag component will lead to a force with a kurtosis greater than 3. Examination of the results reveals that with the exception of level 2 of case 3, this is always the case. For case 3, the kurtosis of measured velocity is 2.53, which is smaller than that of the force (kurtosis = 2.73). Therefore, even for this inertia-dominated case, the drag component has led to a force whose kurtosis is greater than that of the velocity signal. As a result of the current (e.g. cases 1 and 5), the drag component becomes more dominant, leading to higher kurtosis values.

The effect of current on the probability distribution of force. The probability distribution of force for inertia-dominated loading such as level 2 force of case 3 (the smooth large cylinder with no current) is nearly Gaussian, as shown in Figure 3. The probability distribution of force for the same cylinder in rough condition (case 7) is shown in Figure 4. It is observed that as a result of roughness, the drag component has become more prominent, leading to quite a significant departure from Gaussian distribution. It is also observed that the agreement between the observed and the theoretical 4-parameter Pierson-Holmes distribution (Pierson and Holmes, 1965) is quite good. (The 4 parameters of this theoretical distribution are obtained by equating its mean, standard deviation, skewness and kurtosis to those of the measured force). The probability distribution of the same rough cylinder with a carriage speed of 1 m/sec (case 6) is shown in Figure 5. Comparison of Figures 4 and 5 reveals that the effect of current on the probability distribution of force is paramount. While, in the absence of current, both tails of the distribution are underestimated by an equivalent (same variance) Gaussian distribution; in the presence of a strong current, the extreme value in the direction of current is significantly higher than that of the Gaussian distribution while the other extreme value is significantly lower. Examination of the probability distributions for all cases has revealed the following points:

1) From the two tails of the distribution, the one with the highest magnitude (the more important one) is always in good agreement with its theoretical 4-parameter Pierson-Holmes distribution.

2) In the presence of current, the tail with the lower magnitude (e.g. the negative tail in the presence of positive current) has a tendency to be larger (in magnitude) than its value from the theoretical distribution (Figure 6). The ratio between the magnitude of the observed extreme value to that predicted by the theoretical distribution is between 1.04 to 2.10. The highest ratios are for cases 15, 5, 13 and 8 with magnitude ratios of 2.10, 1.45, 1.43 and 1.37, respectively. This ratio for the remaining cases is close to one. A possible explanation for the underestimation of the low-magnitude tail could be due to asymmetries in the flow (including vortex-induced effects and residual currents) possibly justifying differential values of Morison's coefficients in parts of wave cycles. This matter is under further investigation.

For a detailed description of the effect of current on the drag and inertial components of loading on a cylinder exposed to a train of regular waves refer to Hedges (1983).

DERIVATION OF MORISON'S COEFFICIENTS

Least square error method (LSE). Morison's coefficients from this method are presented in Table 3. Since Morison's equation is not a precise model for wave loading on cylinders, there would always be an error term defined as the difference between the observed (F_o) and predicted (F_p) forces. The error term depends on the method of data analysis, and in the case of the LSE method, it satisfies the following relationship (Najafian and Burrows, 1994).

$$E[F_o^2] = E[F_d^2] + E[F_i^2] + E[e^2] = E[F_p^2] + E[e^2] \qquad (2)$$

Dividing both sides of the foregoing equation by $E[F_o^2]$ leads to

$$\frac{E[F_d^2]}{E[F_o^2]} + \frac{E[F_i^2]}{E[F_o^2]} + \frac{E[e^2]}{E[F_o^2]} = \frac{E[F_p^2]}{E[F_o^2]} + \frac{E[e^2]}{E[F_o^2]} = 1 \qquad (3)$$

or

$$a_d^2 + a_i^2 + a_e^2 = a_p^2 + a_e^2 = 1 \quad (4)$$

In the foregoing equation, a_d^2 and a_i^2 are measures of the contribution of the drag and inertial components of force to mean square (2nd statistical moment) of the observed force; a_p^2 is the ratio between the mean squares of the predicted and the observed forces and hence indicates what fraction of the 2nd moment of the observed force can be accounted for by Morison's equation. In contrast, a_e^2 (the error coefficient) indicates what fraction of the 2nd moment of the observed force cannot be accounted for by Morison's equation. Consequently, the 2nd moment of predicted force from the LSE method is always smaller than that of the observed force, and the underestimation is an increasing function of the error coefficient. It can therefore be concluded that the error coefficient is a measure of the reliability of Morison's equation in predicting the observed forces.

Adjusting the predicted forces. As previously mentioned, the 2nd moment of the force predicted from the LSE method is always smaller than that of the observed force. In order to resolve this problem, C_d and C_m values from LSE method must be multiplied by an adjustment factor (af) so that the mean squares of the observed and predicted forces would be equal. The adjustment factor is equal to the ratio between the root-mean-squares (rms) of the observed and predicted forces (Najafian and Burrows, 1994)

$$af = \frac{(F_o)_{rms}}{(F_p)_{rms}} = \frac{1}{\sqrt{1-a_e^2}} \quad (5)$$

Alternative methods. C_d and C_m values can also be determined by the spectral analysis and by the method of moments. In the spectral analysis, C_d and C_m values are determined so that the mean square error between the frequency spectra of the observed and predicted forces is minimised. In the method of moments, the 2nd and 4th statistical moments of the observed and predicted forces are equated. These methods (in particular, the method of moments) lead to some scatter in C_d and C_m values just because of sampling variability (Isaacson et al, 1991); in contrast the effect of sampling variability on C_d and C_m values from the LSE method is negligible. Therefore, in the absence of any significant phase shift between the measured forces and velocities, the LSE method is preferable. In this study, C_d and C_m values from both spectral analysis and the method of moments have been determined. It was observed that overall for drag-dominated cases the agreement between C_d values from different methods is good but that C_m values can vary significantly and that the reverse is true for inertia-dominated cases. For example for case 15 ($a_d^2 = 0.92$) C_d values from LSE, spectral analysis and the method of moments respectively are 1.29, 1.32 and 1.23 (good agreement) but the corresponding C_m values are 1.68, 1.05 and 2.53 (wide scatter). Corresponding C_d and C_m values for case 3 ($a_i^2 = 0.90$) are (0.73, 0.96, 1.25) and (1.82, 1.76, 1.66), respectively. Comparison of the observed and predicted forces has revealed that when there is a significant difference between Morison's coefficients from spectral analysis and the method of moments with those from LSE method, the predicted forces from these alternative methods are not in phase with the observed forces. It can therefore be concluded that C_d and C_m values from time-domain LSE analysis are more reliable (in agreement with Isaacson et al, 1991); therefore the forthcoming discussion will be based on the results from LSE analysis.

DISCUSSION OF THE RESULTS

Hydrodynamic parameters. Reynolds number, Keulegan-carpenter number and reduced velocity are respectively defined as

$$Re = \frac{\left(|\bar{u}| + \sqrt{2}\,\sigma_u\right) D}{\upsilon}$$

$$Kc = (\sqrt{2}\,\sigma_u T_z)/D \quad (6)$$

$$\bar{u}_r = (\bar{u}\,T_z)/D$$

$$T_z = (2\pi\sigma_u)/\sigma_{\dot{u}}$$

where σ stands for standard deviation and ν is the kinematic viscosity of water ($\nu = 1.15*10^{-6}$), T_z is the mean-crossing period of water particle velocity and other variables are as previously defined. The foregoing definitions are similar to those used by Bostrom et al (1986) except that T_p, the period associated with the peak of velocity spectrum has been replaced by T_z, and $\bar{u}$ has been added to account for the effect of current. Based on the foregoing equations, Reynolds numbers for the small and large cylinders are in the ranges $1.0*10^5$ - $2.1*10^5$ and $2.3*10^5$ - $7.3*10^5$, respectively. It can therefore be assumed that for both cylinders post-critical conditions have been achieved. The Kc values for the large and small cylinders are in the ranges 5-8 and 13-17, respectively. The reduced velocities for both cylinders vary between -17 to 12.

The effect of current and roughness on C_d values. C_d values for both cylinders are shown in Figure 7. It is observed that roughness has led to a substantial increase (about 2 times) in C_d values and that for rough cylinders current has led to a reduction (about 30%) in the values of this coefficient. The effect of current on smooth cylinders is less significant and C_d fluctuates about a mean value of about 0.70. Finally, Cd values for the small rough cylinder are somewhat higher that those for the large rough one.

The effect of current and roughness on C_m values. C_m values for both cylinders are shown in Figure 8. It is observed that the effect of roughness on this coefficient is much less significant than its effect on the drag coefficient. In the absence of current C_m is close to its theoretical value of 2 and is somewhat higher for rough cylinders. Furthermore, in all cases the current has led to a significant decline in C_m values. The decline is more substantial for the large rough cylinder.

The effect of current direction on a_i^2. The contributions of inertial loading to the mean square of the observed forces (equations 3 and 4) for different cases are presented in Table 3. It is observed that due to Doppler effect, a_i^2 for positive currents is always higher than its corresponding values for negative currents. For example, a_i^2 for case 2 is 0.61 while that for case 4 is only 0.41.

Comparison of the observed and predicted forces. The error coefficients from the LSE method (a_e^2) are presented in Table 3. The coefficient varies between 0.02 and 0.09, indicating that there is a good fit between the observed and predicted forces. In addition to a_e^2 the absolute and central correlation coefficients between the observed and predicted forces have also been calculated. These coefficients are respectively defined as

$$\rho^{(a)} = \frac{E[F_o \cdot F_p]}{(F_o)_{rms}(F_p)_{rms}}$$

$$\rho = \frac{E[(F_o - \bar{F}_o)(F_p - \bar{F}_p)]}{\sigma_{Fo}\sigma_{Fp}} \quad (7)$$

where overbar stands for the mean value. The difference between the absolute and central correlation coefficients is that the former is a measure of agreement for both the mean and the fluctuating parts of the observed and predicted forces, while in the latter the effect of means has been eliminated. From Table 3, ρ is always higher than 0.93, indicating that the observed and predicted forces are in phase. The absolute correlation coefficients are always very close to their corresponding ρ values, indicating that the mean of the observed and predicted forces are also in good agreement. The spectra of the observed and predicted forces are always in good agreement especially over the main wave frequency range (Figures 1 and 2). There are sometimes some disparities at the lower and higher frequency values, with the latter almost always being slightly overestimated.

The probability distributions of the predicted forces are generally in good agreement with their corresponding theoretical 4 parameter Pierson-Holmes distribution and with that of the observed forces (Figures 3, 4, 5 and 6). The low magnitude tails of the predicted forces are occasionally much smaller than their corresponding values for the observed forces. Possible reasons for this phenomenon have previously been highlighted.

CONCLUSIONS

Current has a very significant effect on both spectral (due to Doppler effect) and probabilistic properties of the force. While, in the absence of current, both tails of the distribution are underestimated by an equivalent (same variance) Gaussian distribution; in the presence of a strong current, the extreme value in the direction of current is significantly higher than that of the Gaussian distribution while the other extreme value is significantly lower. The results of this study are in reasonable agreement with those from Shell sponsored SSPA tests (Rodenbusch and Gutierrez, 1983) in that

- Roughness has led to a substantial increase in C_d values.
- C_d values for smooth cylinders are not significantly affected by current and that its value is about 0.70.
- Current leads to a significant reduction in C_d values for rough cylinders.
- Current leads to a significant reduction in C_m values for both smooth and rough cylinders.

The effect of partial roughness (case 9) on Morison's coefficients will be discussed in a forthcoming publication.

ACKNOWLEDGEMENTS

The European Community Large Installation Program, the UK Science and Engineering Research Council (through the Marine Technology Directorate Ltd), and the Offshore Safety Division of the UK Health and Safety Executive have jointly funded this project.

REFERENCES

Bearman, PW, Chaplin, JR, Graham, JMR, Kostense, JK, Hall, PF, and Klopman, G. (1985). "The Loading on a Cylinder in Post-Critical Flow beneath Periodic and Random Waves," *Proc BOSS Conf,* pp 213-225.

Bearman, PW, and Mackwood, PR (1987). "Wave Loading on Fixed Cylinders at High Reynolds Numbers," Final report on project FLE3, MTD Programme 'Fluid Loading in Offshore Engineering'.

Bearman, PW (1988). "Wave Loading Experiments on Circular Cylinders at Large Scale," *Proc BOSS Conf,* pp 471-487.

Bearman, PW, Chaplin, JR, Graham, JMR, Mackwood, P, Otta, A, Subbiah, K, and Wouters, J (1994). "Loading on a Cylinder at Large Scale in Irregular Waves with Currents," *Proc Int Symp Waves - Physical and Numerical Modelling,* Vancouver, Canada, pp 1133-1142.

Bishop, JR (1979). "Measurements of Wave Particle Motion at the Christchurch Bay Tower Using a Perforated Ball Instrument," NMI Report R55.

Bishop, JR (1984). "Wave Force Investigations at the Second Christchurch Bay Tower," NMI Report 177.

Bostrom, T, and Overik, T (1986). "Hydrodynamic Force Coefficients in Random Wave Conditions," *Proc 5th Int Offshore Mechanics and Arctic Engineering Symp,* Tokyo, Japan, pp 136-143.

Chaplin, JR (1991). "Loading on a Horizontal Cylinder in Irregular Waves," *Int J of Offshore and Polar Engineering,* Vol 1, pp 247-254.

Chaplin, JR, Greated, CA, Flintham, TP, and Skyner, DJ (1992). "Breaking Wave Forces on a Vertical Cylinder," *Dept of Energy,* Report OTH-90-324.

Hedges, TS (1983). "Effects of Currents on Wave Drag and Inertia Loads," *Proc Coastal Structures Conf,* Arlington, Virginia, USA, pp 461-473.

Isaacson, M, Subbiah, K and Baldwin, J (1991). "Force Coefficient Estimation from Random Wave Data," *Proc 1st Int Offshore and Polar Engineering Conf,* Edinburgh, UK, pp 149-157.

Klopman, G, and Kostense, JK (1990). " The Loading on a Vertical Cylinder in Random Waves at High Reynolds Numbers," IN *Water Wave Mechanics* (A Torum and OT Gudmestad, (eds)), Kluwer Academic, pp 679-699.

Mackwood, PR, Bearman, PW and Graham, JMR (1995). "Wave and Current Flows around Circular Cylinders at Large Scale," Final Report on EC Large Installation program, *Dept of Aeronautics, Imperial College,* London.

Morison, JR, O'Brien, MP, Johnson, JW, and Shaaf, SA (1950). "The Force Exerted by Surface Waves on Piles," *AIME, Petroleum Transactions,* Vol 189, pp 149-154.

Najafian, G, and Burrows, R (1994). "Critical Assessment of the Least Square Error Method Used in Derivation of Morison's Force Coefficients," *ASME, Journal of Offshore Mechanics and Arctic Engineering,* Vol 116, pp 1-6.

Pierson, WJ, and Holmes, P (1965). "Irregular Wave Forces on a Pile," *ASCE, Journal of the Waterways and Harbours Division,* Vol 91, No WW4, pp 1-10.

Rodenbusch, G, and Gutierrez, CA (1983). "Forces on Cylinders in Two-dimensional Flows" Technical Report BRC 13-83, Bellaire Research Center (Shell Development Company), Houston.

Wolfram, J, and Theophanatos, A (1989). "The Loading of Heavily-Roughened Cylinders in Waves and Linear Oscillatory Flow," *Proc 8th OMAE Conf,* pp 183-190.

Case No.	Cylinder size	Roughness condition	Carriage Speed ((m/secs))	Level 1 z = -3.5m	Level 2 z = -2.5m	Level 3 z = -1.5m	Level 4 z = -0.5m	Level 5 z = +0.5m	Effective cylinder diameter ((m))	Record length ((minutes))
01			-1.00	EMF*	PVM	PVM+	-	-	0.505	56
02			-0.50	EMF*	PVM	PVM+	-	-	0.505	48
03	Large	Smooth	0.00	EMF*	PVM	PVM+	-	-	0.505	33
04			+0.50	EMF*	PVM	PVM+	-	-	0.505	48
05			+1.00	EMF*	PVM	PVM+	-	-	0.505	55
06			-1.00	EMF*	PVM	PVM+	-	-	0.513	46
07	Large	Rough	Cylinder fixed	-	EMF	EMF+	-	-	0.513	174
08			+1.00	EMF*	PVM	PVM+	-	-	0.513	46
09	Large	Partial roughness	Cylinder fixed	-	EMF	EMF+	-	-	0.513	174
10			-0.50	EMF	EMF	EMF	PVM	PVM	0.210	68
11	Small	Smooth	Cylinder fixed	EMF	EMF	EMF	EMF	EMF	0.210	77
12			+0.50	EMF	EMF	EMF	PVM	PVM	0.210	68
13			-0.50	EMF	EMF	EMF	PVM	PVM	0.216	65
14	Small	Rough	Cylinder fixed	EMF	EMF	EMF	EMF	EMF	0.216	94
15			+0.50	EMF	EMF	EMF	PVM	PVM	0.216	65

Table 1: Definition of different cases together with types of instruments used for measuring water particle velocities. PVM, perforated ball velocity meter; EMF, electromagnetic flow meter; -, no velocity meter; *, no force measurement although there is a velocity meter; +, measured hydrodynamic force on the force sleeve is suspect.

Case	Level 1				Level 2				Level 3			
No.	Mean	s.d.	Skewness	Kurtosis	Mean	s.d.	Skewness	Kurtosis	Mean	s.d.	Skewness	Kurtosis
01	-	-	-	-	246	202	0.84	4.52	263	273	0.83	4.29
02	-	-	-	-	83.1	179	0.19	3.03	134	251	0.42	3.41
03	-	-	-	-	-3.0	179	0.02	2.73	32.2	239	0.24	3.18
04	-	-	-	-	-113	170	-0.30	3.34	-91.0	212	0.12	3.93
05	-	-	-	-	-240	218	-0.87	4.19	-263	217	-0.82	4.81
06	-	-	-	-	472	345	0.87	4.13	482	414	1.00	4.61
07	-	-	-	-	-17.0	237	-0.10	3.61	-22.3	262	-0.20	3.81
08	-	-	-	-	-502	340	-0.87	3.76	-513	369	-0.63	3.56
09	-	-	-	-	-18.8	224	-0.04	3.30	-8.7	195	-0.04	3.40
10	33.5	37.8	0.36	3.10	37.4	42.5	0.44	3.34	32.5	48.1	0.46	3.60
11	1.6	31.7	-0.19	3.40	-3.7	33.2	-0.13	3.50	-4.9	44.7	-0.12	3.78
12	-38.3	36.8	-0.43	3.25	-37.0	41.5	-0.38	3.48	-42.6	49.5	-0.38	4.04
13	52.7	65.2	1.15	4.95	57.7	74.0	1.20	5.25	57.4	85.0	1.18	5.49
14	-3.0	59.0	-0.49	5.67	-2.8	62.4	-0.48	5.66	-5.2	82.6	-0.34	5.35
15	-66.2	75.6	-1.35	5.81	-72.1	86.6	-1.28	5.64	-77.6	97.4	-1.02	4.92

Table 2: Basic statistics of (filtered) in-line forces (N/m). s.d., standard deviation; -, no hydrodynamic force measurement.

Case	Level 1								Level 2								Level 3							
No.	C_d	C_m	a_d^2	a_i^2	a_e^2	af	$\rho^{(a)}$	ρ	C_d	C_m	a_d^2	a_i^2	a_e^2	af	$\rho^{(a)}$	ρ	C_d	C_m	a_d^2	a_i^2	a_e^2	af	$\rho^{(a)}$	ρ
01	-	-	-	-	-	-	-	-	0.74	1.18	.78	.18	.04	1.02	.979	.954	0.83	1.27	.74	.21	.05	1.02	.977	.956
02	-	-	-	-	-	-	-	-	0.75	1.43	.33	.61	.06	1.03	.970	.965	1.16	1.55	.45	.48	.07	1.04	.965	.958
03	-	-	-	-	-	-	-	-	0.73	1.82	.07	.90	.03	1.01	.986	.986	0.93	2.00	.10	.80	.10	1.05	.948	.967
04	-	-	-	-	-	-	-	-	0.79	1.51	.52	.44	.05	1.02	.976	.970	0.67	1.62	.37	.54	.09	1.05	.954	.945
05	-	-	-	-	-	-	-	-	0.68	1.54	.82	.16	.02	1.01	.988	.974	0.64	1.39	.78	.16	.06	1.03	.991	.938
06	-	-	-	-	-	-	-	-	1.34	1.27	.90	.08	.02	1.01	.988	.971	1.38	1.32	.87	.10	.03	1.01	.987	.971
07	-	-	-	-	-	-	-	-	1.62	2.09	.26	.68	.06	1.03	.972	.971	1.53	1.88	.29	.65	.06	1.03	.968	.967
08	-	-	-	-	-	-	-	-	1.26	1.27	.95	.03	.02	1.01	.991	.977	1.18	1.40	.92	.06	.02	1.01	.989	.976
09	-	-	-	-	-	-	-	-	1.33	2.08	.19	.77	.04	1.02	.980	.980	0.77	1.56	.13	.82	.05	1.03	.974	.974
10	0.74	1.32	.73	.18	.09	1.05	.956	.935	0.77	1.46	.70	.21	.09	1.05	.957	.948	0.69	1.48	.64	.30	.06	1.03	.967	.961
11	0.62	1.89	.26	.70	.04	1.02	.981	.985	0.58	1.74	.28	.69	.03	1.02	.984	.985	0.65	1.89	.32	.65	.03	1.02	.984	.984
12	0.66	1.32	.79	.12	.09	1.05	.951	.931	0.62	1.42	.77	.15	.08	1.04	.960	.939	0.63	1.50	.76	.17	.07	1.04	.965	.950
13	1.30	1.49	.89	.09	.02	1.01	.991	.985	1.39	1.63	.88	.10	.02	1.01	.990	.985	1.38	1.64	.85	.12	.03	1.01	.988	.983
14	1.84	2.13	.65	.28	.07	1.04	.964	.966	1.68	1.96	.67	.26	.07	1.04	.964	.968	1.81	2.13	.68	.25	.07	1.04	.963	.967
15	1.28	1.54	.93	.05	.02	1.01	.991	.985	1.29	1.68	.92	.06	.02	1.01	.991	.985	1.24	1.73	.90	.07	.03	1.01	.987	.979

Table 3: C_d and C_m values from adjusted time-domain Least Square Error method. -, no hydrodynamic force measurement.

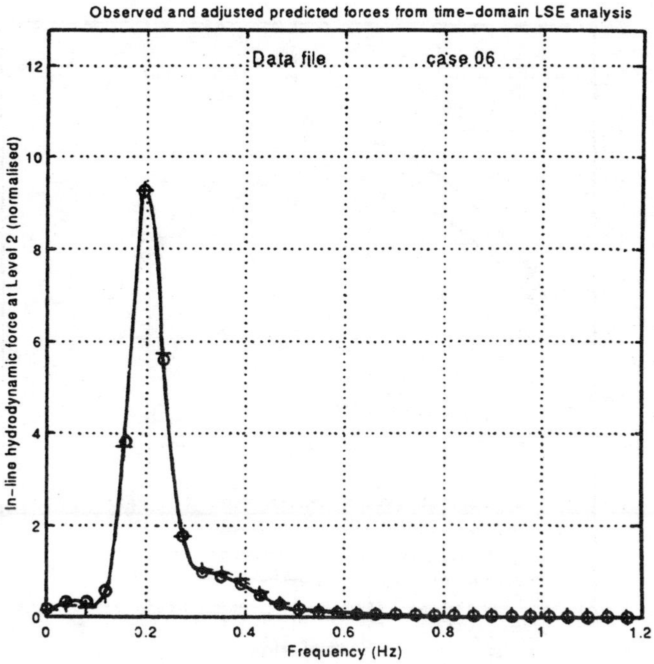

Figure 1. Comparison of the frequency spectra of the observed and predicted forces for a positive-current case. o, observed; +, predicted.

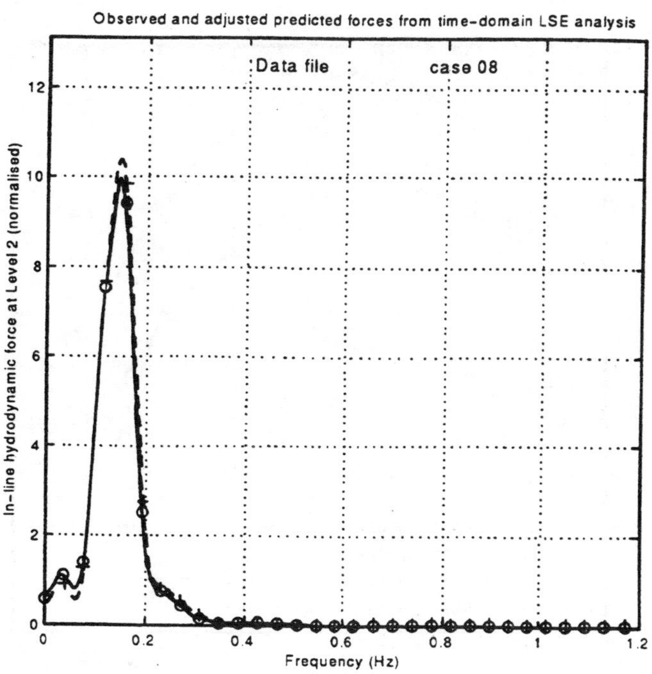

Figure 2. Comparison of the frequency spectra of the observed and predicted forces for a negative-current case. o, observed; +, predicted.

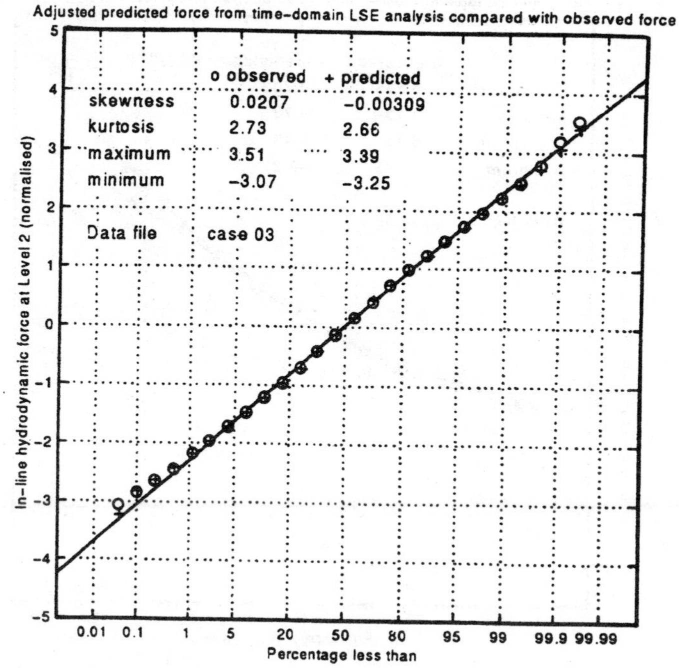

Figure 3. Probability distribution of an inertia-dominated force (large smooth cylinder with no current). o, observed; +, predicted; —, Gaussian.

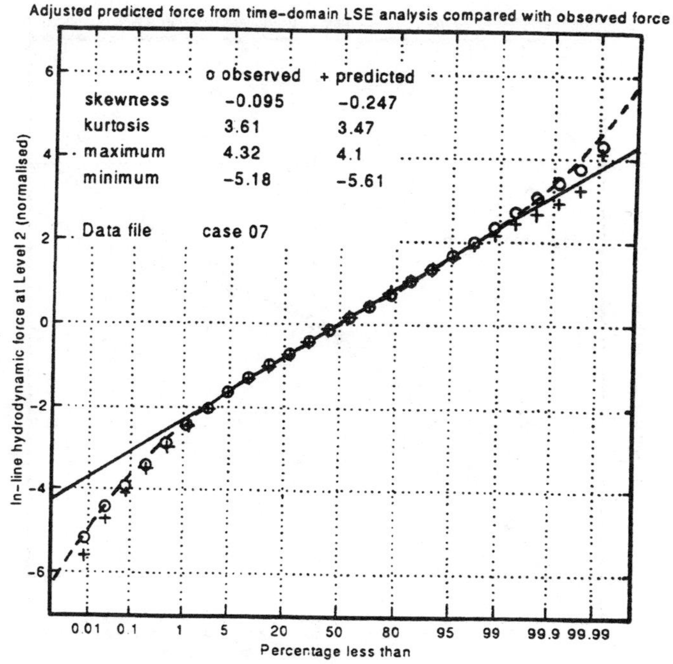

Figure 4. Probability distribution of force in the drag-inertia regime (large rough cylinder with no current). o, observed; +, predicted; —, Gaussian; ---, Pierson-Holmes.

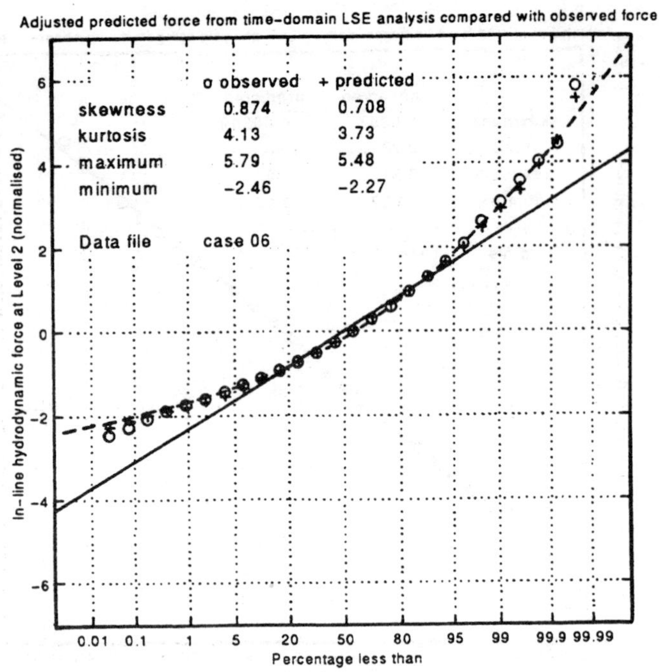

Figure 5. The effect of positive current on the force probability distribution (large rough cylinder). o, observed; +, predicted; —, Gaussian; ---, Pierson-Holmes.

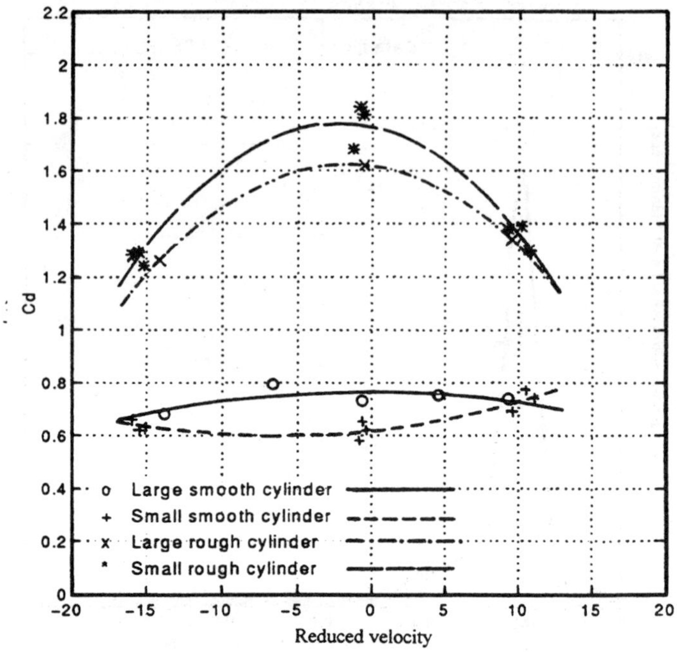

Figure 7. The effect of current on drag coefficient.

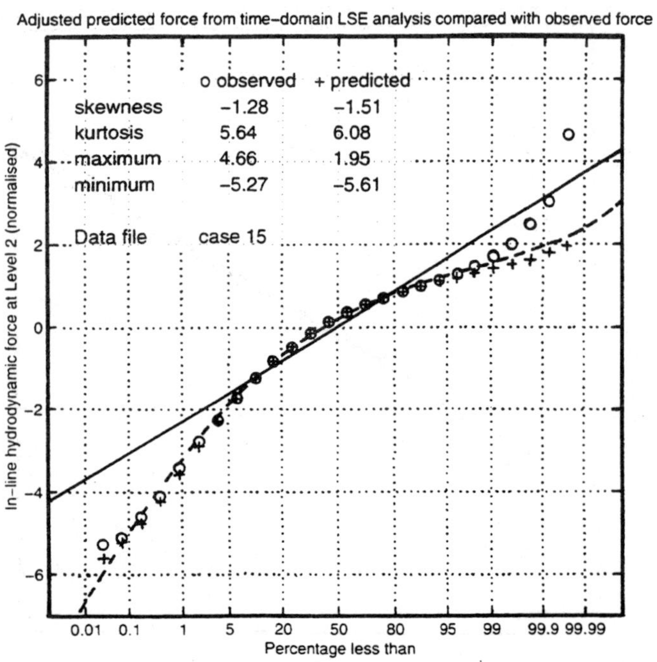

Figure 6. The effect of negative current on the force probability distribution (small rough cylinder). o, observed; +, predicted; —, Gaussian; ---, Pierson-Holmes.

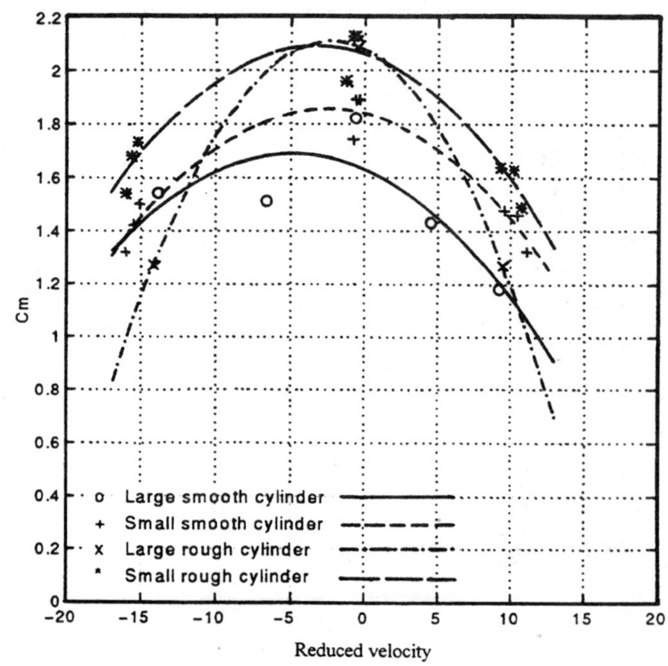

Figure 8. The effect of current on inertia coefficient.

Proceedings of the Seventh (1997) International Offshore and Polar Engineering Conference
Honolulu, USA, May 25-30, 1997

ISBN 1-880653-28-1 (Set); ISBN 1-880653-31-1 (Vol. III)

The Influence of Grouping on the Force Characteristics of Pairs of Vertical Surface-Piercing Cylinders

D. Smith and N. Haritos
University of Melbourne
Melbourne, Australia

ABSTRACT

This paper presents results of an investigation of the modifications to the in-line Morison force coefficients on individual vertical surface piercing cylinders within cylinder groups associated with group interference effects determined from an experimental study conducted in the University of Melbourne's Michell Laboratory wave flume. The current study investigates group interference effects at higher Keulegan-Carpenter numbers and in more detail within the previously identified critical close spacing region than previous work conducted within the department. This paper focuses on results associated with the current experimental program conducted on cylinder pairs oriented in-line and transverse to the wave direction in irregular wave conditions.

KEY WORDS: group interference effects, Morison forces, cylinders.

INTRODUCTION

Experimental programs investigating modifications to the Morison force coefficients due to group interference effects are necessitated by the lack of a theoretical solution to the determination of the individual forces on cylinders within cylinder groups in separated flow. Computation schemes whilst evolving rapidly are presently not able to model the complexities of separated flow on isolated cylinders over a wide Reynolds number range in "wavy" flows. Hence, the importance of small scale laboratory investigations which can provide valuable insights into group interference phenomena, although results are not directly applicable to full-scale offshore structures, such as in riser groups, due to the different Reynolds numbers regimes that small-scale and full-scale structures experience.

The current experimental study compared the variation in the in-line Morison force coefficients between the isolated cylinder condition and closely spaced cylinder groups. Modifications to the Morison force coefficients associated with group interference effects derived from tests in regular wave conditions have previously been reported in detail for the two cylinder case with cylinders oriented transverse to the wave direction (Haritos and Smith, (1994, 1995); Smith and Haritos (1995)) and in-line to the wave direction (Smith and Haritos 1996). This paper seeks to summarise succinctly this variation in regular waves and provide a comparison to a more comprehensive set of results obtained from irregular wave conditions.

FLOW INTERFERENCE EFFECTS IN CYLINDER PAIRS

Flow interference effects in cylinder pairs have previously been reported in steady flow conditions where the distinction between observed interaction effects for cylinder pairs aligned in-line, transverse and staggered to the flow direction has been made (Zdravkovich and Pridden, 1977; Zdravkovich, 1977). The little data available in oscillatory flow and even in wavy flows tentatively provides confirmation of similarly occurring effects in the different respective flow conditions (Sarpkaya and Isaacson, 1981).

For cylinders oriented in-line and transverse to the flow direction most available literature concerns steady flow conditions and these have been reviewed in detail by Zdravkovich (1977), Zdravkovich and Pridden (1977) and Sarpkaya and Isaacson (1981). Changes in flow mechanisms and flow patterns and their association with vortex shedding effects have been reported in these reviews particularly within the critical spacing ranges of S=3-3.8 and S<2, ($S=\ell/D$ where ℓ is the centre-to-centre cylinder spacing and D is the cylinder diameter) for in-line and side-by side groups respectively. The critical spacing regions, or regions of influence of interference effects, have been confirmed experimentally in wavy flows for in-line oriented cylinder pairs, S<≈3, (He, 1993; Haritos and He, 1992, 1993; Smith and Haritos, 1996) and transverse oriented cylinder pairs, S<≈2 (He and Haritos, 1991; Smith and Haritos, 1995).

EXPERIMENTATION

The experimental programme has been described previously in detail (Smith and Haritos, 1996) so only an overview is provided here for the sake of completeness. The study of interest was performed in the Michell Laboratory wave flume which is 65m in length from the Salter duck type wave generator, which optimally operates at a water

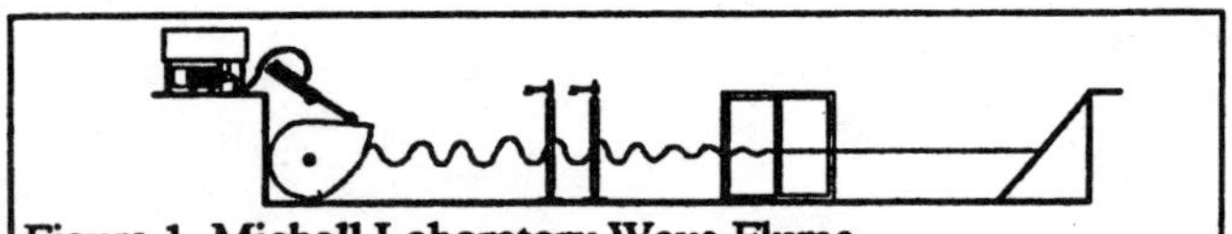

Figure 1 Michell Laboratory Wave Flume

depth of 1.0 m, to the low reflection beach. The wave flume is depicted schematically in Figure 1. Two sizes of cylinders, 16mm (C1 and C2) and 25mm (C3 and C4) diameter, were utilised in the experimental programme which investigated grouping effects over a range of orientations [$\theta = 0°, 45°, 90°$] and inter-cylinder spacings [S= 1.125, 1.25, 1.5, 1.75, 2.0, 2.5, 3.0 and ~17.0]. Wave generation and data sampling from instrumentation was facilitated by an XT-IBM compatible computer with digital-to-analog and analog-to-digital capabilities. Cylinder groups were subject to a range of both regular and irregular Pierson-Moskowitz waves (refer Smith and Haritos (1996) for more detail of actual hydrodynamic inputs utilised).

Bottom-pivoted cylinders fixed at their top end in the along and across wave directions by rigid rods connected to force transducers, themselves fixed to a stiff support frame, were utilised to capture total cylinder force measurements. Experimental measurements of water surface elevations adjacent to each individual cylinder were captured simultaneously with the force measurements using capacitance type wave probes.

METHODS OF ANALYSIS

Band pass filtering of experimental measurements for the regular wave tests were conducted as a standard procedure and was found to have only marginal influence on fitted force coefficients (Haritos and Smith, 1994). Frequency components below 0.3 Hz and above 6.0 Hz were removed from experimental measurements in this filtering. However, for the irregular wave records which had genuine frequency components below the lower cut-off frequency of 0.3 Hz (f_p = 0.25, 0.3, 0.35 and 0.4 Hz) no filtering was undertaken.

Experimental force measurements have been analysed on the basis of wave-by-wave fitting of the measured water surface elevation and by fitting of force models over a time series of prescribed length via water particle kinematics predictions based upon use of an appropriate wave theory.

WAVE-BY-WAVE FITTING

Fenton's stream function wave theory (Fenton, 1988) and Airy linear wave theory have both been utilised within this experimental study to "fit" the measured water surface elevation on a wave-by-wave basis and predict the water particle kinematics. Comparisons of measured and predicted water surface elevation profiles which elude to the goodness of water particle kinematics predictions have previously been reported (Haritos and Smith, 1994, 1995; Smith and Haritos, 1995, 1996) hence are not presented here. While not being a component of this study other studies conducted at the University of Melbourne (Daliri, 1996) which have included water particle velocity measurements have confirmed that Fenton's stream function theory provides better predictions than Airy wave theory.

TIME SERIES FITTING

Time series fitting has been applied to both regular and irregular wave conditions for the analysis of the in-line Morison force coefficients. Of the 50 seconds of useful data extracted from each regular and irregular wave test four over lapping time series were constructed [0-20, 10-30, 20-40, 30-50 sec] and analysed independently to determine the fitted force coefficients. Water particle kinematics for the time series analysis were predicted using linear superposition.

KC EVALUATION

In regular waves the Keulegan-Carpenter number is evaluated from linear Airy wave theory predictions of the peak velocity at the mean water level whereas in irregular waves KC has been determined by,

$$KC_{irr} = \frac{2\sqrt{2}\pi\sigma_\eta}{D\tanh(\kappa h)} \qquad (1)$$

where σ_η is the standard deviation of the time series extracted from the full record and κ is the wave number derived from the peak wave period.

IN-LINE CYLINDER FORCE PREDICTIONS

The in-line restraint force R(t) at the tip of the cylinder, according to the Morison force model becomes:

$$R(t) = \int(\alpha\dot{u} + \beta u|u|)\frac{z+h}{l}dz \qquad (2)$$

where $\alpha = \frac{\pi}{4}\rho C_m D^2$ and $\beta = \frac{1}{2}\rho C_d D$, l is the distance from the bottom of the tank to the tip restraint and h is the depth of water. C_m and C_d can be fitted using the method of least squares to measured portions of traces for R(t). These "portions" can extend over a single wave (wave-by-wave method) or over several waves or even the entire length of record.

CYLINDER IN-LINE MORISON FORCES

Regardless of whether the water particle kinematics were predicted by Airy or Fenton's theory it has been found that the fitted in-line force is virtually identical as reported previously (Haritos and Smith, 1994, 1995; Smith and Haritos, 1995, 1996). Variability between the water particle kinematics predictions by Fenton's and Airy theory has been found to be generally displayed in the force coefficients where coefficients fitted using Airy kinematics are generally smaller due to the relatively larger kinematics in comparison with Fenton's theory. The goodness of the fit, as exemplified by the correlation coefficient R^2, was typically found to be greater than 0.9 for the majority of regular wave-by-wave fits throughout this study. Generally, it was found that for lower amplitude input regular wave conditions the fitted and measured forces were very nearly sinusoidal and 90 degrees out of phase with the input water surface elevation as expected for low KC inertia dominant conditions. In contrast, the measured and predicted in-line cylinder forces for the higher amplitude regular waves were more irregular in form and were more closely in-phase with the water surface elevation.

Predicted cylinder forces for irregular waves utilising water particle kinematics predictions from the traditional linear superposition approach which have not been previously reported are presented in Figures 2 and 3 for sample irregular (f_p=0.4Hz) and regular waves (A=30mm, f=0.8Hz) respectively. While the predicted

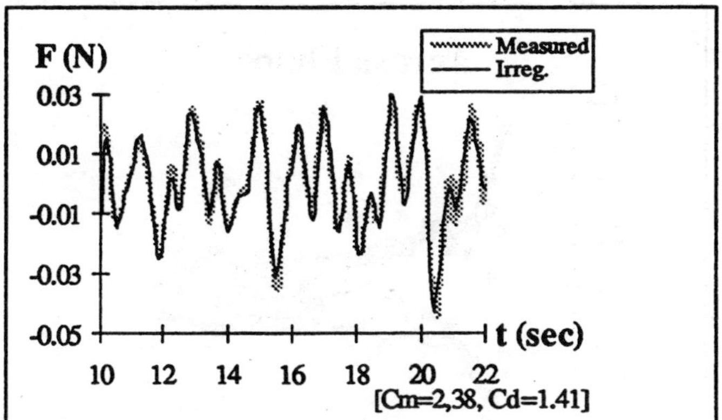

Figure 2 Measured and Predicted Forces
Time Series Method; 16mm Cylinder C2

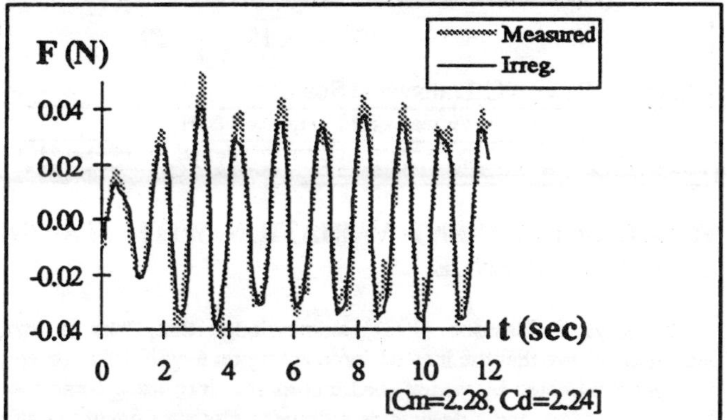

Figure 3 Measured and Predicted Forces
Time Series Method; 16mm Cylinder C1

forces do follow the measured forces reasonably well the actual peak in the in-line cylinder forces appears to be under predicted by the Morison equation based upon constant time-invariant force coefficients.

ISOLATED CYLINDER RESULTS

MORISON FORCE COEFFICIENTS

The Morison force coefficients obtained from the analysis of data associated with regular and irregular hydrodynamic excitation have been presented in Figures 4 and 5 respectively. The irregular wave

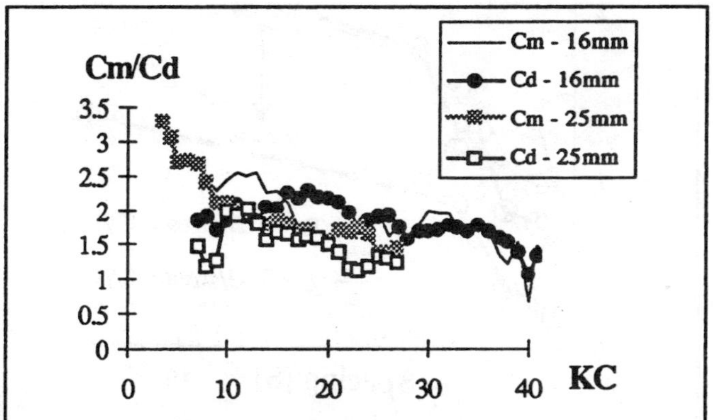

Figure 4 Morison Force Coefficients vs KC
Isolated Cylinder in Regular Waves

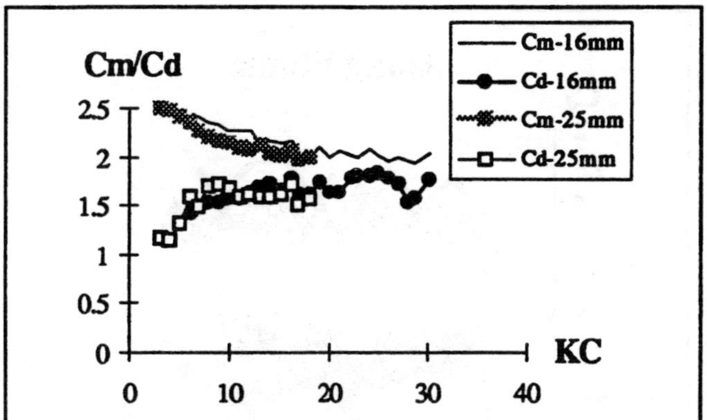

Figure 5 Morison Force Coefficients vs KC
Isolated Cylinder in Irregular Waves

drag and inertia force coefficient variation with KC is noticeably smoother than that obtained in regular wave conditions which may be a result of the greater number of samples taken over a much narrower range of β numbers for the irregular waves (fp=0.25-0.4 cf f=0.4-1.8 Hz). The inertia force coefficient from both cases approaches the potential value of 2 at higher KC while at lower KC the irregular wave coefficient does not increase as significantly as was found in regular waves. The drag force coefficient obtained from the irregular wave test data is very similar but at a decreased magnitude in comparison to C_d from regular wave tests.

CYLINDER PAIRS EXPERIMENTAL RESULTS

MORISON FORCE COEFFICIENT VARIATION WITH KC AND SPACING INFLUENCE

The C_m and C_d variation with KC for in-line aligned cylinder pairs in irregular wave conditions has been presented in Figures 6 and 7 respectively. Figure 6 indicates an obvious variation with spacing and a general modest decrease in C_m with increasing KC for tandem cylinders which was not evident previously in regular waves (He and Haritos, 1991; Haritos and He, 1992), particularly when results were collated by β number (β=180-200) (Smith and Haritos, 1996). The general form of the variation in C_d with KC given in Figure 7 is very

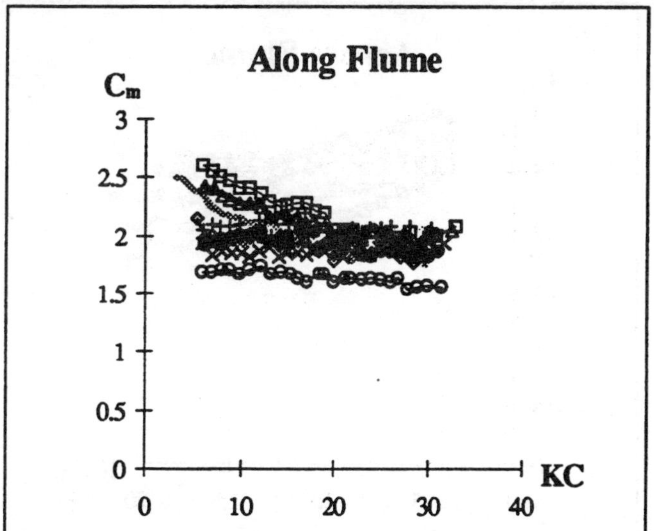

Figure 6 C_m vs KC In-line Spacing

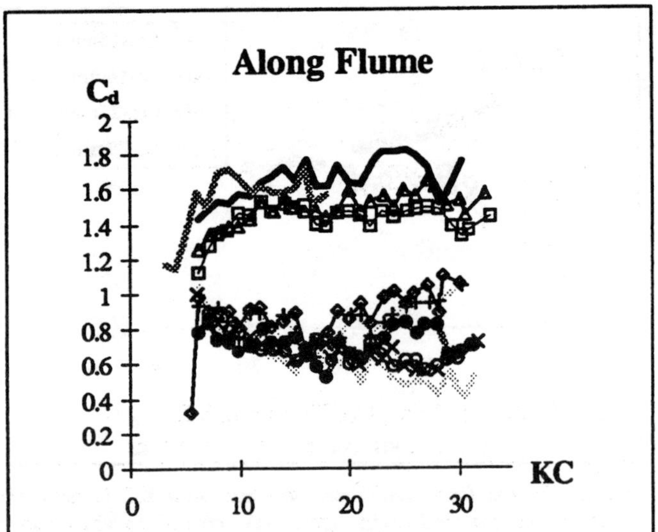

Figure 7 C_d vs KC In-line Spacing

Along Flume Legend [Figures 6 & 7]

S=17 C1/C2 · S=17 C3 · S=1.125 C1 · S=1.125 C2 · S=1.25 C1 · S=1.25 C2 · S=1.5 C1 · S=1.5 C2 · S=2.5 C1 · S=2.5 C2

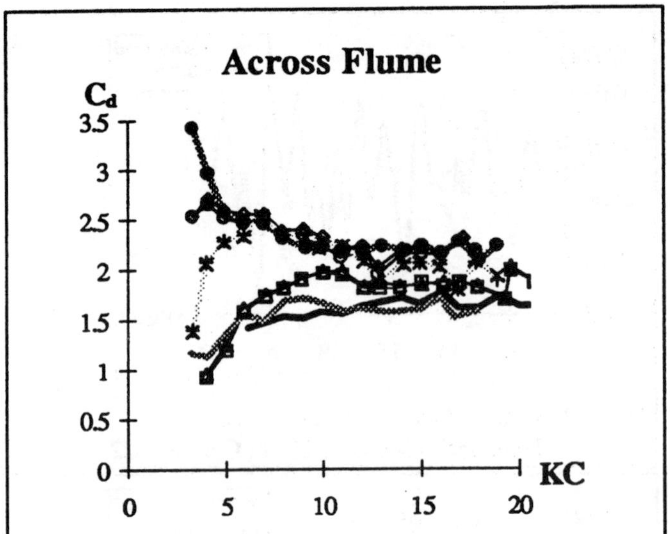

Figure 9 C_d vs KC Transverse Spacing

Across Flume Legend [Figures 8 & 9]

S=17 C1/C2 · S=17 C3 · S=1.125 C3 · S=1.125 C4 · S=1.25 C3 · S=1.25 C4 · S=1.5 C3 · S=1.5 C4 · S=2.5 C3 · S=2.5 C4

similar to results previously presented in regular waves (Smith and Haritos, 1996) although C_d here is not reduced as much at the closer cylinder spacings.

The inertia and drag force coefficient variations with KC over a range of inter-cylinder spacings for transverse aligned cylinders in irregular wave conditions have been presented in Figures 8 and 9 respectively. Little Cm variation with KC was found previously in regular wave conditions (He and Haritos, 1991) whereas in Figure 8 a genuine consistent decrease was found with increasing KC. A spacing influence is evident that resembles that found by Smith and Haritos (1995) in regular waves but is more distinct. While the Cd variation with KC in regular waves (Smith and Haritos, 1995) is markedly different to that in Figure 9 for irregular waves, the variation found by He and Haritos (1991) was similar in form for the larger spacings (S>1.5), although a significantly more obvious influence of spacing is evident in Figure 9.

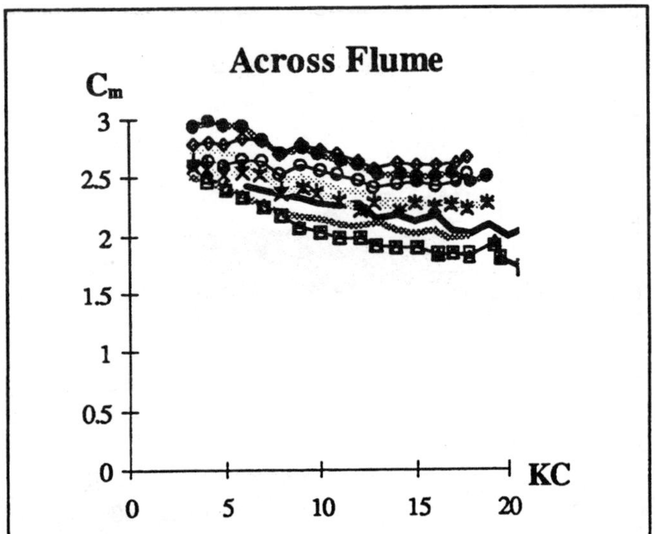

Figure 8 C_m vs KC Transverse Spacing

FORCE COEFFICIENT VARIATION WITH SPACING AND KC INFLUENCE

Sarpkaya & Isaacson (1981) acknowledge that there is every reason to believe that the inertial force acting on a cylinder in an array is as much affected by vortex shedding as the drag force component. In fact in tandem and side-by-side cylinder pairs significant variation in both coefficients was previously found in regular and irregular wave conditions (He and Haritos, 1991; Haritos and He, 1992, 1993; Haritos and Smith, 1994, 1995; Smith and Haritos, 1995, 1996) and was found here in irregular waves.

Tandem Cylinder Pairs C_d Variation with S

Previous results obtained by the authors (Smith and Haritos, 1996) for the drag force coefficient variation with inter-cylinder spacing in regular wave conditions, shown in Figure 10, revealed significant differences between the upstream and downstream

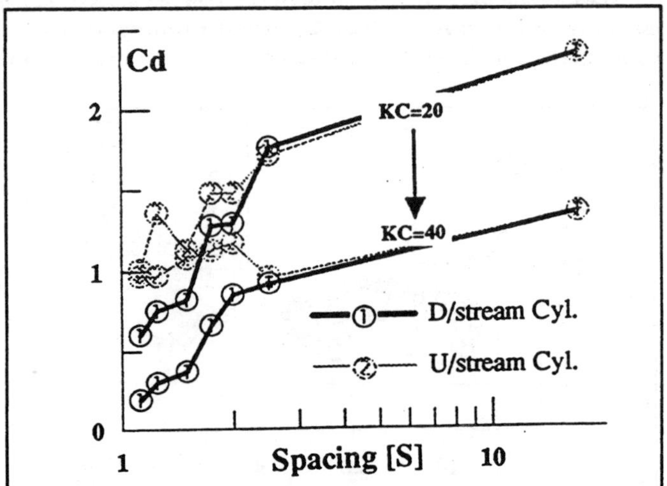

Figure 10 C_d vs S (Regular Wave Data)
TANDEM CYLINDERS

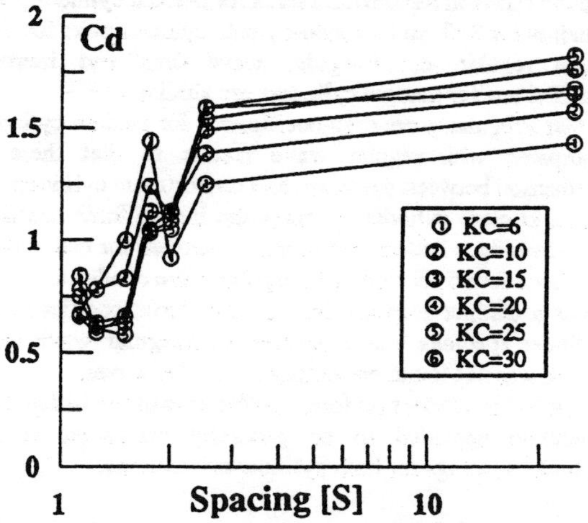

Figure 11 C_d vs S (Irregular Wave Data)
TANDEM CYLINDERS - Downstream Cylinder

cylinders which could be related to a representative plot of steady flow results presented by Sarpkaya and Isaacson (Figure 5.10, 1981). However, it was found that in irregular waves no significant differences occurred between the cylinders and that the upstream cylinder data was very similar to that presented in Figure 11 for the downstream cylinder. The upstream cylinder exhibited slightly less variation in C_d values with spacing, the major differences between the two cylinders occurring at the close cylinder spacings of S<1.75.

The C_d data in irregular waves is remarkably similar in form and magnitude to the upstream cylinder from regular waves (Smith and Haritos, 1996). Haritos and He (1993) similarly found that the regular and irregular wave C_d data was virtually identical when the greater variability associated with regular wave results is neglected. However, their C_d was generally larger than that in Figure 11 other than at their lowest spacing presented (S=1.5).

Tandem Cylinder Pairs C_m Variation with S

Comparison of the inertia force coefficient variation with inter-cylinder spacing in irregular and regular wave conditions from the current experimental program, in Figures 12 and 13 respectively,

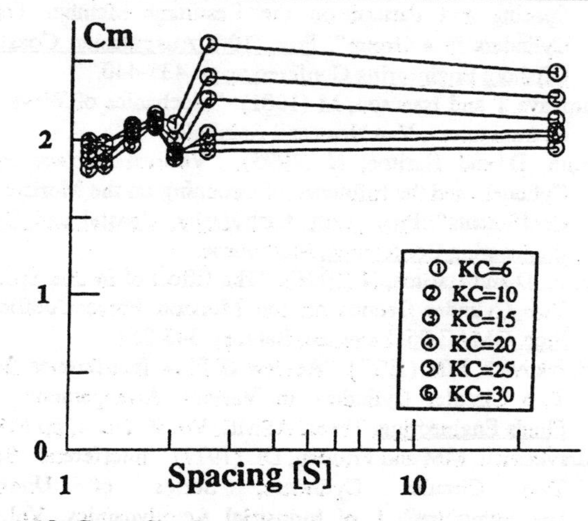

Figure 12 C_m vs S (Irregular Wave Data)
TANDEM CYLINDERS - Upstream Cylinder

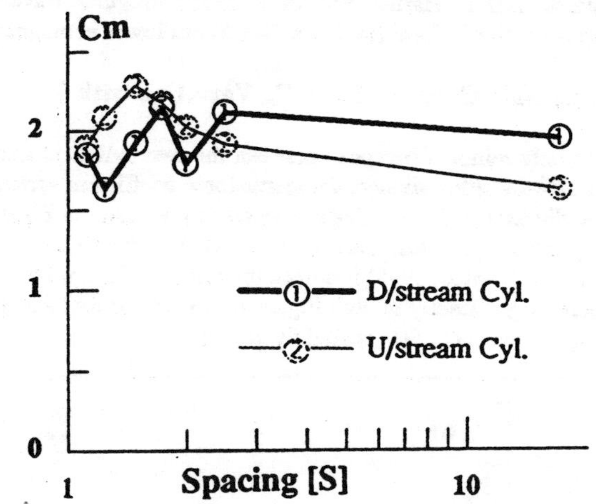

Figure 13 C_m vs S (Regular Wave Data)
TANDEM CYLINDERS

reveals that the C_m data is almost identical for the upstream cylinder. Haritos and He's (1993) C_m data was reasonably similar in form to that presented in Figure 12 but was of slightly greater magnitude. The downstream cylinder C_m data for irregular wave conditions was found to be virtually identical to that for regular wave tests associated with the current study which was presented at only KC=20 for clarity.

It is interesting that within the close cylinder spacing range, S<1.75, that any KC influence on C_m is minimal (even more so than that found by Smith and Haritos (1996) in regular waves) and the differences between the upstream and downstream cylinders also occur over this range.

Side-by-Side Cylinder Pairs C_d Variation with S

No distinguishable trend with inter-cylinder spacing was previously found for the drag force coefficient from data associated with the current experimental program (Haritos and Smith, 1995). A slight increase in C_d can be seen at close cylinder spacings, S<1.75, in the irregular wave data presented for cylinder C4 in Figure 14. Data was captured for both of the side-by-side cylinders simultaneously and C_d results were found to be virtually identical. The present data

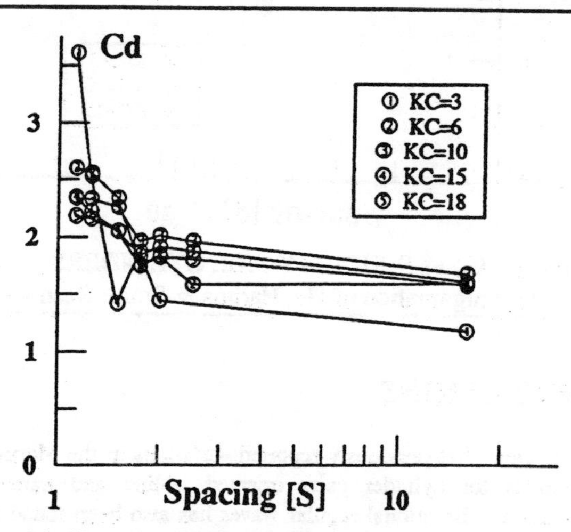

Figure 14 C_d vs S (Irregular Wave Data)
SIDE-BY-SIDE CYLINDERS - Cylinder C4

compares well to Haritos and He's (1993) irregular wave C_d data which is similar in form (except at S=1.5) but lower in magnitude.

Side-by-Side Cylinder Pairs C_m Variation with S

As only minor differences between the two cylinders were found in the side-by-side cylinders the inertia force coefficient variation with inter-cylinder spacing in irregular waves is presented in Figure 15 for one cylinder. An amalgamation of previously published C_m results (Smith and Haritos, 1995) has been presented in Figure 16 where Cm increases significantly in both Figure 15 and 16 for S<2 and generally the regular wave C_m data is slightly greater.

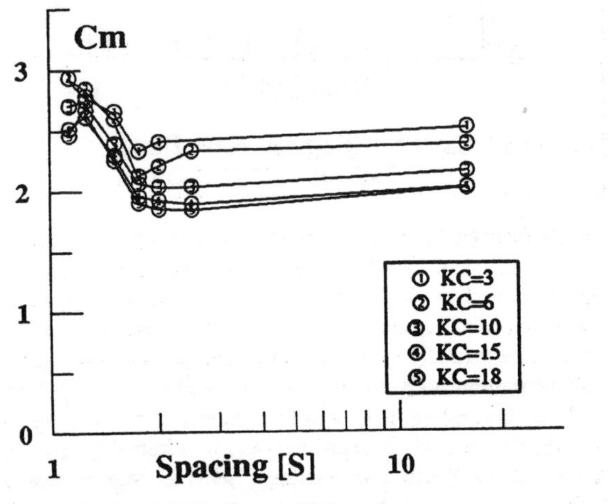

Figure 15 C_m vs S (Irregular Wave Data)
SIDE-BY-SIDE CYLINDERS - Cylinder C4

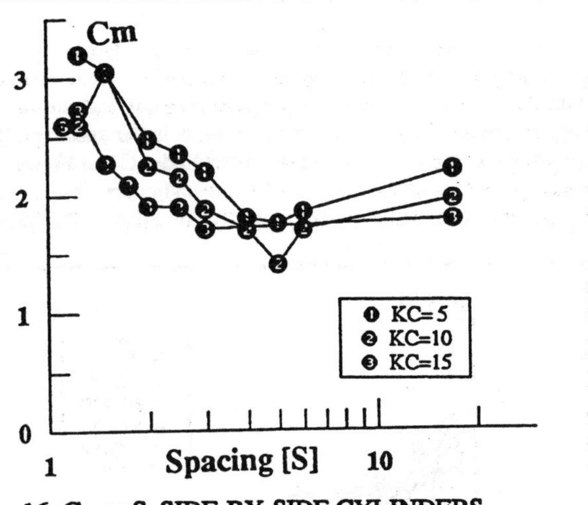

Figure 16 C_m vs S SIDE-BY-SIDE CYLINDERS
[Amalgamation of He, Haritos & Smith Results]

CONCLUSIONS

It is clear that previously reported variations in the Morison force coefficients for cylinder pairs oriented in-line and transverse to generated uni-directional regular waves has also been found to occur in a similar form in irregular wave conditions. Specifically,

i. that group interference effects as manifested in modifications to the Morison force coefficients for tandem cylinder pairs occur mainly for S<3, and for side-by-side cylinder pairs for S<2.

ii. that regular and irregular wave drag and inertia force coefficients for tandem cylinders are similar

iii. that irregular wave drag coefficients for tandem cylinders vary compared with regular wave results in that there is less distinction between upstream and downstream cylinders

iv. that at close cylinder spacings the inertia force coefficient for side-by-side cylinders was found to increase for both cylinders as has been found previously in regular wave conditions.

v. that a distinct increase in the drag force coefficient at close cylinder spacings was identified in irregular wave conditions which was not found previously in regular waves.

vi. that KC influences on force coefficients in the isolated cylinder condition appeared to be generally consistent at different cylinder spacings for both cylinder orientations.

REFERENCES

Daliri, R (1996). "Hydrodynamic Force Characteristics in the Splash Zone". Ph.D. Thesis (Under Examination), The University of Melbourne.

Fenton, JD (1988). "The Numerical Solution of Steady Water Wave Problems", Computers & Geoscience, Vol 14, No 3, 1988, pp 357-368.

Haritos, N and He, YQ (1992). "Group Interference Effects on Vertical Cylinders", Proc. ISOPE-92, Vol 3, San Francisco, pp 355-360.

Haritos, N and He, YQ (1993). "Wave Force and Flow Interference Effects on Vertical Cylinders", Proc. OMAE-93, Vol I, Glasgow, pp 73- 82.

Haritos, N and Smith, D (1994). "Hydrodynamic Force Characteristics of Closely Spaced Cylinder Groups", Proc. Australasian Structural Engineering Conference, Sydney, pp 361-366.

Haritos, N and Smith, D (1995). "The Effect of Spacing Transverse to the Wave Direction on the Morison Force Coefficients in Two Cylinder Groups", Proc. OMAE-95, Vol 1A, The Hague, pp 163-170.

He, YQ (1993). "Hydrodynamic Interference and Interactions of Multiple Circular Cylinders", Ph.D. Thesis, The University of Melbourne.

He, YQ and Haritos, N (1991). "The Influence of Relative Spacing and Pattern on the Resultant Member Force of Cylinders in a Group", Proc. 10th Australasian Coastal and Offshore Engineering Conference, pp 435-440.

Sarpkaya T and Isaacson, M (1981). "Mechanics of Wave Forces on Structures", Van Nostrand Reinhold.

Smith, D and Haritos, N (1995). "Vertical Surface Piercing Cylinders and the Influence of Grouping on the Morison Force Coefficients", Proc. 12th Australasian Coastal and Offshore Engineering Conference, Melbourne.

Smith, D and Haritos, N (1996). "The Effect of In-line Spacing of Two Cylinder Groups on the Morison Force Coefficients", Proc. OMAE-96, Florence, Italy, pp 345-352.

Zdravkovich, MM (1977). "Review of Flow Interference Between Two Circular Cylinders in Various Arrangements", J. of Fluids Engineering, Trans. ASME, Vol 99 No. 4, pp 618-633.

Zdravkovich, MM and Pridden, DL (1977). "Interference Between Two Circular Cylinders; Series of Unexpected Discontinuities", J. of Industrial Aerodynamics, Vol. 2, pp 255-270.

Proceedings of the Seventh (1997) International Offshore and Polar Engineering Conference
Honolulu, USA, May 25-30, 1997

ISBN 1-880653-28-1 (Set); ISBN 1-880653-31-1 (Vol. III)

Runup on Columns in Large Waves

A.J. Martin, W.J. Easson and T. Bruce
University of Edinburgh
Edinburgh, United Kingdom

Abstract

Runup is known to cause damage to deck plates in the vicinity of the columns of semi-submersibles. Previous studies have related the extent of the runup to the kinetic energy of the wave crest particle motion. However, these studies have been limited in the wave heights which were used.

In this study, wave runup on vertical cylinders is investigated with the use of video cameras. The waves are generated in a narrow wave flume and are representative of storm waves in the North Sea. Wave period and wave height are used as the independent variables and the extent of the runup is compared to the kinetic energy calculated from high order wave theory. It is shown that even high order wave theory is insufficient to calculate the extent of the runup, and that the measured values exceed the calculated values by about one third over a wide range of conditions.

Key words: Runup, large waves, columns, semi-submersibles.

1. Introduction

This paper describes an extension of the experiments reported by Bruce & Easson (1997), which dealt with the enhancement of waves due to structure blockage, such as might occur on the large columns of a semi-submersible or tension-leg platform. In that paper, the authors observed significant runup on the columns of a test model in large waves. The purpose of this study was to investigate whether the extent of runup could be easily predicted from the wave parameters.

The parameters for these experiments and those of Bruce & Easson were drawn from a particular range of short, steep waves which may have been responsible for deck damage on semi-submersibles. The spatial scale for these experiments is about 100:1.

2. Estimates of runup

Runup height is a function of the scattering parameter ka where k is the wavenumber and a is the cylinder radius. Linear scattering theory (e.g. Haney & Herbich, 1982) predicts that the runup ratio at the upwave side of the cylinder is given approximately by

$$\frac{r}{\eta_{\max}} = \left[1+4(ka)^2\right]^{1/2}$$

where r is the runup height (the elevation of the water on the cylinder above SWL) and $\eta_{\max}$ is the crest elevation. This approximation is valid for the limited range of ka studied here. Linear scattering is known to underpredict runup in large waves (Kriebel, 1993).

Hallermeier (1976) measured the extent of runup on a column in large waves, expressing the steepness as H/λ, where H is the wave height and λ is the wavelength. Hallermeier used solitary wave theory to predict the crest kinematics (the water depth was in the intermediate to shallow regime), and used the velocity head $u^2/2g$ to predict the height of the runup above the wave crest, where u is the maximum horizontal water particle velocity in the crest. He found that this calculation provided an upper limit to the extent of runup on a large column. His experiments covered a wide range of scattering regimes but with a maximum value of H/λ rather less than in these experiments.

Reference	ka	H/λ	s	d/λ
Hallermeier (1976)	0.023–0.5	0.006–0.06	0.001–0.01	0.05–0.13
Haney & Herbich (1982)	0.01–0.021	0.01–0.07	0.0007–0.011	0.08–0.7
Niedzwecki & Duggal (1990)	0.11–1.30	0.01–0.13	0.0016–0.02	0.28–3.32
This study	0.12–0.32	0.04–0.12	0.0064–0.02	0.26–0.69

Table 1: Comparative range of parameters

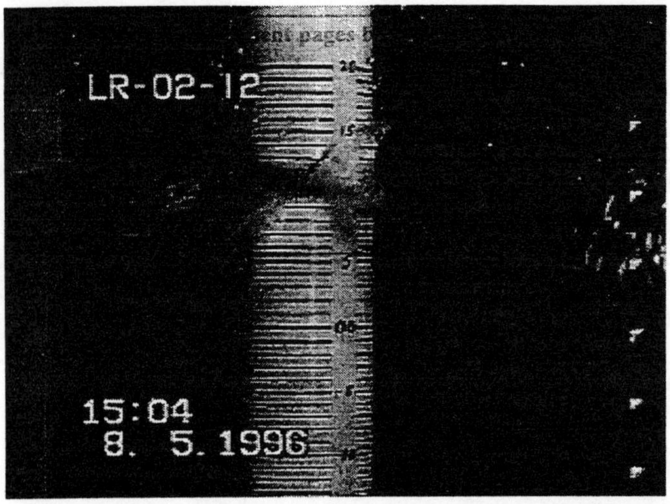

Figure 1: A typical runup event from the preliminary experiments (wave right to left)

Figure 2: A form of runup more typical of the longer waves

Haney & Herbich (1982) performed similar experiments on thin piles ($ka < 0.021$) and found that the velocity head predicted using Dean's streamfunction theory (fifth order) was a good predictor of runup height. Niedzwecki & Duggal (1990), with a wider range of steepnesses and in shorter waves, found that the velocity head calculated using linear theory underestimated the runup by a significant factor.

Kriebel (1993) presents an alternative prediction of runup, in which linear scattering theory is used to obtain runup estimates of each of the Fourier components. These values are combined by superposition.

A comparison of the parameters for these tests with those of Hallermeier, Haney & Herbich and Niedzwecki & Duggal is given in table 1. The steepness s used here is given by

$$s = \frac{H}{gT^2}$$

A limiting value of $s = 0.022$ is assumed, and experiments are performed at steepnesses of 50%, 70%, 80% and 90% of the maximum. The water depth d is such that a short wave regime applies.

3. Experimental set-up

The experiments were conducted in a wave channel of width 0.4m, quiescent depth 0.7m and length 20m. A cylinder of diameter 0.11m was mounted 13.5m from the flap-type absorbing wavemaker, on the centreline of the tank. Neglecting the effects of wall friction, this arrangement is equivalent to a series of identical cylinders placed along a line parallel to the incoming wave crests at a spacing of 0.4m. At the end of the tank, a foam wave absorber was mounted, and the absorbing wavemaker helped to reduce the effect of reflections.

Incoming wave heights were measured by three wire wave gauges. The wavemaker was programmed to generate sequences of monochromatic waves at the required steepness. Due to their large amplitude and reflections from the cylinder, the incoming waves did not remain stable enough to conduct long experiments, so the run time was limited to 32 seconds. Usually, the first wave to arrive at the cylinder had a steep front and was often close to breaking. Since the kinematics of a breaking wave would not be properly predicted by the theories being tested, these waves were removed from the record during analysis.

Five periods were used: 0.8s, 0.89s, 1.0s, 1.14s and 1.33s. The transfer function of the wavemaker limited the steepnesses which could be generated for the two longest periods.

4. Runup measurement and prediction

The measured runup height r_m was obtained by filming the cylinder through the glass side wall of the tank. Graduations at intervals of 10mm on the cylinder allowed runup height to be recorded. A timer on the video record allowed each incoming wave to be matched to the same wave in the wave gauge record.

The form of a typical runup event is shown in figure 1. The maximum extent of the runup generally occurs shortly after the crest passes the cylinder. This figure shows a wave of period 1.0s at about 90% of the limiting steepness.

Another runup event is shown in figure 2. This form of runup was more typical of the longer waves. Although the height of the runup is not great, the volume of water which is projected above the wave crest is much greater than in the previous case.

For each wave, runup heights were predicted using the velocity head:

$$r_p = \eta_{max} + \frac{u^2}{2g}$$

Maximum elevations and velocities were calculated both from linear theory, and also using a high order numerical wave theory. Rienecker & Fenton's (1981) Fourier approximation method was used. Sixteen Fourier components were used, which is the number recommended by Rienecker & Fenton for the wave parameters of interest.

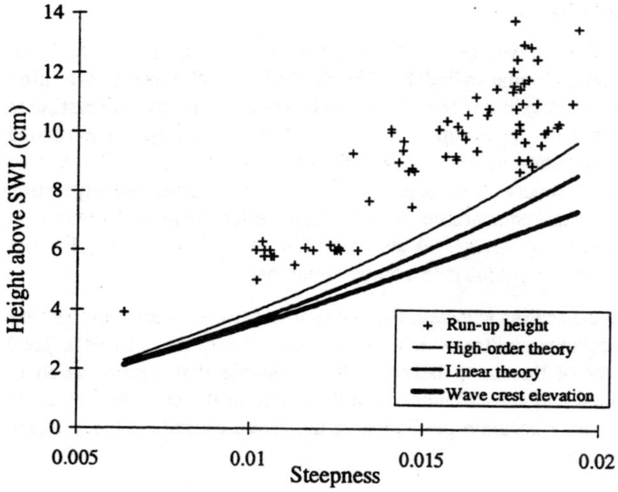

Figure 3: Runup measurements and head predictions for the 0.8s wave (crest elevation from high order theory)

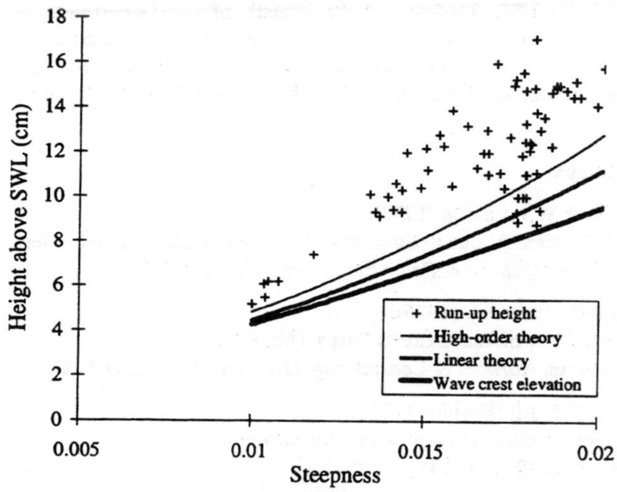

Figure 4: Runup measurements and head predictions for the 0.89s wave (crest elevation from high order theory

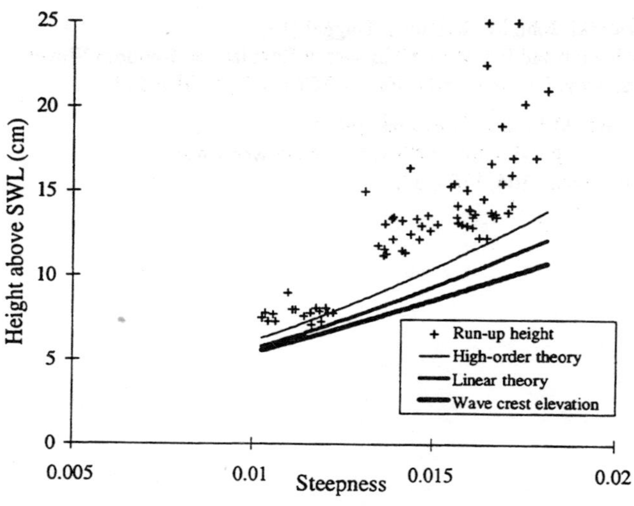

Figure 5: Runup measurements and head predictions for the 1.0s wave (crest elevation from high order theory

The readings from two wave gauges were used in the calculations, one 0.9m upwave of the cylinder and the other 2.5m upwave. For each gauge, zero-downcrossing wave heights were evaluated and these values were used to predict the runup height for each wave using both linear and high-order theory. Thus the ratio r_m/r_p was evaluated four times; for each of two gauges and for each of two wave theories. The results for each gauge were similar (the means differed by only a few percent), but the standard deviations for the gauge further from the cylinder were lower. This is probably because it was less contaminated by reflections. For this reason only the results for this gauge are presented here.

Initially, wave heights were evaluated for equivalent wave trains without the cylinder in place. However, since at the highest steepnesses the reflected wave from the cylinder appeared to be causing breaking in the incoming wave train near the wavemaker and thereby reducing the height of the waves reaching the cylinder, it was felt that the wave field with the cylinder in the tank should be used, explicitly matching each incoming wave with each runup event.

Figures 3–5 shows the results for particular wave periods. Four separate experimental runs at each period (each at different nominal steepnesses) are represented on this graph. Despite the scatter it is clearly seen that the runup height exceeds the predictions in all cases, and that runup is a function of wave steepness. This trend was seen for all periods tested and also in the predictions. In fact, the high-order prediction appears to give a *lower* bound to the runup height.

Figure 6 compares the runup ratio r_m/η_{max} to the velocity head predictions for the 0.8s period case. Linear scattering theory and Kriebel's superposition theory are also included. The linear scattering calculation, using an amplitude equal to half the wave height, fails even to predict the crest/trough asymmetry in the wave itself.

The simplest form of Kriebel's (1993) superposition method, which is used here, uses second order Stokes theory, in which the amplitude is equal to half the wave height. The runup is still underpredicted, but the method performs almost as well as the more complex high-order velocity head calculation. That the correspondence is poorer with increasing steepness reflects the fact that only two Fourier components were used, while the velocity head calculation used sixteen.

Figure 7 shows the ratios of measured runup to the runup predicted using high order theory for all of the cases. The scatter in these results reflects the randomness in the wave train, which was unavoidable at the steepnesses used. There is no clear trend in the ratio of measured to predicted runup with wave steepness.

Means and standard deviations for the ratios were calculated for each of the five wave periods. They are shown in figure 8. There is no clear relationship between the ratio and the wave period. It should be noted that the experiments with longer waves contained a narrower range of steepnesses than those with shorter waves, and so the results may not be directly comparable. For a period of 1.14s the highest steepness which could be produced by the wavemaker was 80% of the limiting value, while at a period of 1.33s the steepest wave was 50% of the limit.

The mean value from all of the waves of the ratio of measured runup to that predicted by linear theory was 1.48, with a standard deviation of 0.25. The mean ratio using the high-order theory was 1.33, with a standard deviation of 0.22. Thus whichever wave theory is used the runup height in steep, relatively short waves is underpredicted by a significant fraction.

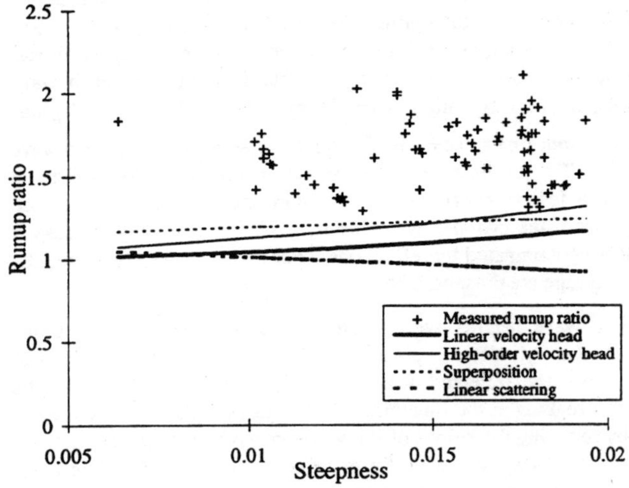

Figure 6: Runup ratio measurements for the 0.8s wave with predictions of the runup ratio by the various methods

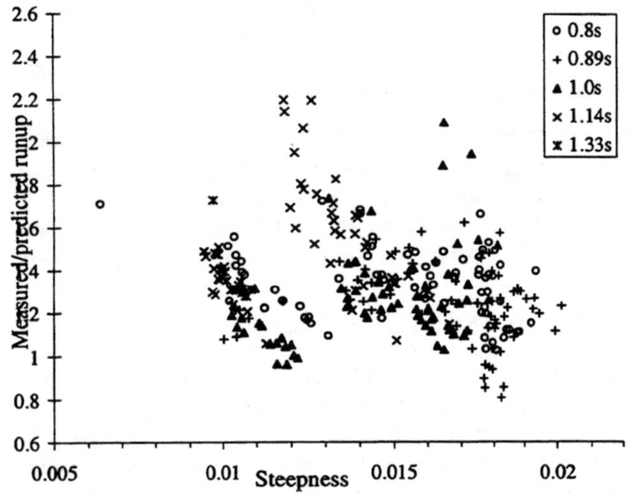

Figure 7: Ratio of measured runup to runup predicted by Rienecker & Fenton's wave theory

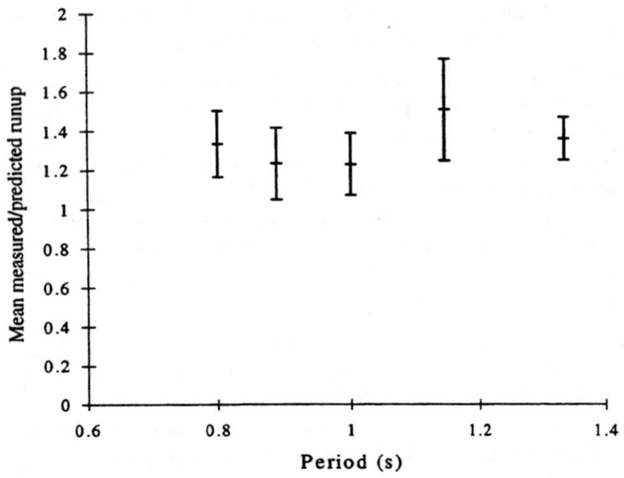

Figure 8: Mean runup ratios using high order theory (error bars show the standard deviation)

5. Conclusions

Steep, short waves generated in a laboratory tank cause significant runup on a circular cylinder. The method of adding crest stagnation head to the maximum crest elevation has been found to underestimate the extent of the runup. On average, measured runup exceeded linear prediction by 48% and a high order prediction by 33%. Linear scattering theory was poor, while a second order superposition method was comparable to the high order head calculation at moderate steepnesses. A superposition method with many Fourier coefficients might provide a better comparison.

Due to the scaling selected, the waves tested were steeper and shorter than previous studies which have found velocity head to be a good predictor of the runup height. It is possible that consideration of horizontal and vertical motions in the whole of the crest region might lead to more accurate prediction of the runup caused by these steep-fronted waves.

Acknowledgements

The authors gratefully acknowledge the financial support of the UK Health and Safety Executive. Helpful conversations with Dr Don Smith of the HSE, Professor Atilla Incecik of the Department of Naval Architecture, Newcastle University, and Mr Lewis Carlin are acknowledged. The high-order wave calculations were performed with the program RFWAVE by Dr Gert Klopman of Delft Hydraulics.

References

Bruce, T. & W.J. Easson (1997).
Wave Enhancement due to Structure Blockage in Semi-Submersibles.
Proc. 7th Int. Offshore & Polar Eng. Conf. (ISOPE 1997).

Hallermeier, Robert J. (1976).
Nonlinear Flow of Wave Crests Past a Thin Pile.
J. Waterways, Harbors & Coastal Eng. Div. (ASCE), **102**, 365–377.

Haney, J.P. & J.B. Herbich (1982).
Wave Flow Around Thin Piles and Pile Groups.
J. Hydr. Res., **20** (1) 1–14

Kriebel, David L. (1993).
Nonlinear Runup of Random Waves on a Large Cylinder.
Proc. Int. Conf. Offshore Mech. & Arctic Eng. (OMAE 1993), **1**, 49–55 (ASME)

Niedzwecki, John M. & Arun S. Duggal (1992)
Wave Runup and Forces on Cylinders in Regular and Random Waves.
J. Waterway, Port & Coastal Eng. (ASCE), **118** (6) 615–634

Rienecker, M.M. & J.D. Fenton (1981).
A Fourier approximation method for steady water waves.
J. Fluid Mech., **104**, 119–137.

Proceedings of the Seventh (1997) International Offshore and Polar Engineering Conference
Honolulu, USA, May 25-30, 1997

ISBN 1-880653-28-1 (Set); ISBN 1-880653-31-1 (Vol. III)

Wave Runup Effects on Large Diameter Caissons in Shallow Water

R. Armstrong and N. Haritos
University of Melbourne
Parkville, Australia

ABSTRACT

An extensive research program into the hydrodynamic effects of waves on large offshore caisson structures in shallow waters has been conducted. The study has used fine-wire capacitance surface elevation probes mounted on the cylinder together with video imaging to facilitate the recording of the maximum wave runup on two different diameter laboratory models. A newly designed force dynamometer has been used to measure the hydrodynamic induced base shear and overturning moments on these cylinders. The aim of the study was to compare the experimental results with estimated values using both linear diffraction theory and a recently developed numerical method of Isaacson & Cheung (1994), that is able to account for second order wave diffraction. This comparison was found to be generally "unfavourable" for the linear diffraction predictions. In the case of Isaacson and Cheung's predictions, the "match" with experimental results was considered to be quite satisfactory. However, the influence of the nonlinear distortions produced in the laboratory waves under high steepness, the "blockage" effect of the cylinder in the test section as well as the side wall reflections in the tank provided some discernible departures from these predictions, the extent of which was very much dependent upon the particular conditions under consideration.

KEYWORDS: hydrodynamics, wave runup, wave forces, caissons

INTRODUCTION

Wave runup may be defined as the maximum vertical elevation of the wave crest above the still water level at the water caisson boundary that occurs over one wave period on the upwave face of the caisson as shown in Figure 1. Wave runup is associated with increased horizontal loading of the structure. In more severe situations, uplift forces on the underside of decking and green water submergence of operational areas may occur as well.

Wave runup for linear waves is generally estimated using the linear diffraction theory of MacCamy and Fuchs (1954) and may be expressed as (Isaacson, 1978):

$$\eta_{r=a} = H\left|\sum_{m=0}^{\infty} \frac{\beta_m \cos(m\theta)}{2\pi\kappa a H_m^{(1)'}(\kappa a)}\right| \tag{1}$$

in which $\beta_m = 0$ for $m = 0$ and $\beta_m = 2i^m$ for $m > 1$, where η is the free surface elevation, κ the wave number, H the wave height and $H_m^{(1)}$ is the Hankel function of the first kind of order m; the prime denotes the derivative with respect to the argument. The radial measurement from the cylinder axis is r, the cylinder radius is a, and θ is the angle measured about the cylinder axis from the direction of wave propagation such that θ is zero at the rearmost point of the cylinder. Taking a limited number of terms in the power series, equation (1) may be simplified to:

$$\eta(\theta) = \frac{H}{2}\left[1 + 4(\kappa a)^2 \cos^2\theta\right]^{\frac{1}{2}} \tag{2}$$

This equation indicates that for linear theory runup is independent of water depth and wave steepness.

Kriebel (1990) has shown that for steep waves the runup may reach an elevation of more than twice the incident wave crest amplitude, and that the maximum nonlinear wave runup on a caisson can exceed that estimated by linear theory by more than 50 per cent. Hence, wave runup on large diameter caissons is influenced by the nonlinearity of the incident waves and as the steepness of the waves increases, the linear theory no longer satisfactorily predicts its level.

The total force on the cylinder up to the still water level and the overturning moment about the base in the direction of wave propagation is given by, Kriebel (1990) :

$$F(t) = \frac{2\rho g H}{\kappa^2} A(\kappa a)\tanh\kappa h \cos(\omega t - \alpha) \tag{3}$$

$$M(t)=\frac{2\rho g H}{\kappa^2}A(\kappa a) \times \left[\kappa h \tanh(\kappa h)+\operatorname{sech}(\kappa h)-1\right]\cos(\omega t-\alpha) \quad (4)$$

and: $A(\kappa a)=\left[J_1'^{\,2}(\kappa a)+Y_1'^{\,2}(\kappa a)\right]^{-\frac{1}{2}}$; $\alpha=\tan^{-1}\left(\frac{J_1'(\kappa a)}{Y_1'(\kappa a)}\right)$

The main purpose of the present test program was to provide an experimental investigation of the effects of wave runup from regular nonlinear wave inputs ranging from small amplitudes up to breaking wave conditions from which results could be compared against the predictions of the more recently available second order diffraction theory of Isaacson and Cheung (1994).

METHODS FOR PREDICTING WAVE RUNUP

Several methods are available of varying complexity and sophistication in the analysis of nonlinear interactions of waves with large offshore structures. Such methods may be categorized as either conforming to an analytical or a numerical approach.

Analytical Methods

Kim and Yue (1989) obtained a second order diffraction solution for the plane monochromatic wave case for vertically axisymmetric bodies using a boundary-integral formulation involving general order free-surface ring source Green functions, which appear to satisfy the second order boundary value problem for wave forces and free surface elevations for the finite water depth.

Kriebel (1992) presented a simple approximate theoretical method of predicting the maximum nonlinear wave runup on large circular cylinders. The method is based on linear diffraction theory and on the assumption that higher-order harmonic components of a nonlinear wave can be considered as if they were linear waves in order to estimate their individual runup contributions. Kriebel's simple formula for the maximum runup in highly nonlinear waves of any order is given by:

$$R=\overline{R}+H\left(\frac{R_1}{H}\right)+(2\eta_c-H)\left(\frac{R_2}{H}\right) \quad (5)$$

where η_c is the nonlinear crest amplitude obtained from any appropriate nonlinear wave theory; R_1/H is the dimensionless runup of the first harmonic found from linear diffraction theory based on the diffraction parameter κa; R_2/H is the dimensionless runup of the second harmonic and is approximately given by linear diffraction theory using a diffraction parameter $2\kappa a$.

R is the time averaged mean runup given by the expression:

$$\overline{R}=\frac{\kappa H^2}{4}\tanh(\kappa h) \times\left[\left(\frac{R_1}{H}\right)^2+4\left(\frac{R_2}{H}\right)^2\left(\frac{2\eta_c}{H}-1\right)^2\right] \quad (6)$$

To estimate η_c, Kriebel has suggested the use of Stokes fifth order or Stream Function Theories as appropriate.

Niedzwecki and Duggal (1992) provide an informative comparison of experimental studies on wave runup. They propose intuitively that the runup in a regular wave climate ought be able to be related to the conversion of the kinetic energy of the incident wave to potential energy associated with the increase in wave elevation and propose the following expression for wave runup on a cylinder:

$$\frac{R(0)}{H}=m\frac{u^2}{2g}+\frac{1}{2} \quad (7)$$

where $R(0)$ is the runup at the leading edge, $m=6.83$ and u is the maximum horizontal water particle velocity at the still water level based on linear wave theory.

Numerical Methods

The method of Isaacson and Cheung (1994) is based on an extended perturbation analysis that provides correction factors for the runup and maximum loads. The correction factors are applied to the prediction of linear diffraction theory and are independent of wave height. The runup and load maxima may be expressed directly in terms of a single correction factor:

$$\frac{R}{|\eta_1|}=1+\frac{H}{\lambda}\Delta_\eta \quad (8)$$

$$\frac{F\max}{|F_1|}=1+\frac{H}{\lambda}\Delta_f \quad (9)$$

$$\frac{M\max}{|M_1|}=1+\frac{H}{\lambda}\Delta_m \quad (10)$$

where Δ_η, Δ_f and Δ_m are dimensionless correction factors for runup, wave force and overturning moment respectively

The method assumes regular waves and is expected to produce accurate predictions when the Keulegan Carpenter number $K_c < 2$ and the Ursell number $U <$ O[0.1].

Another approach for fully nonlinear water wave problems has been presented by Beck and Scorpio (1995) and uses the Euler-Lagrangian time stepping procedure, utilising a fourth order Runga-Kutta method for the time stepping and simple isolated Rankine sources rather than the more complex panel distributions, which leads to computationally fast algorithms. The method has been applied to the diffraction of incident waves by a vertical cylinder with a diffraction parameter κa of 1.32 and relative depth h/a of 1.16, to predict wave runup and horizontal force on the cylinder. A comparison with predictions from other numerical methods (Finite Element Method and Boundary Integral Method) have shown that the desingularised method can be used successfully to solve fully nonlinear water wave problems.

EXPERIMENTAL PROGRAM

A parametric study of measured wave runup effects under regular nonlinear wave conditions on a laboratory model of a bottom-mounted vertical surface piercing circular caisson (see Figure 2) was recently undertaken. Although wave runup on the cylinder surface was of primary interest to this research program, an investigation of the base shear force and overturning moment on the caisson under a variety of wave conditions was also performed.

The caisson takes the form of a surface piercing bluff body of circular configuration whose horizontal dimension is a number of times greater than the water depth in which it is located. Two test cylinders were used in the program (D= 0.5 and 0.36 metres respectively) to permit experiments to be conducted for a range of conditions which can be described by the wave diffraction parameter κa up to a value of 2, a water depth to caisson radius ratio ranging from 0.5 to 0.7, a water depth to wave length length ratio ranging from 0.07 to 0.18, and for wave steepness values (H/L) up to 0.1 and into the breaking wave condition. (It is noted that there is a lack of published experimental data over this range of conditions).

The experimental results of the cylinder base shear force, overturning moment and surface runup from this extensive experimental program were then to be compared with the prediction of the numerical correction factor method of Isaacson and Cheung (1994).

Intially a pilot study was conducted on the 360mm diameter test cylinder in the range of regular wave conditions described by: Nominal wave amplitudes in 1.0m of water of 2.5, 5, 10 and 20 mm for wave frequencies of 0.6, 0.8 and 1.0 Hz leading to a d/L of 0.095, a D/L of 0.27 and H/L values ranging up to 0.035. The aim of the pilot test program was not only to validate the data acquisition procedure and the performance of the newly constructed dynamometer and cylinder surface elevation probes whilst at the same time providing results from a sub-set of wave conditions that could be used to compare against the predictions of Isaacson and Cheung (1994); but also to obtain information of the shoaling process of the waves on the raised floor of the experimental set up itself, and to obtain knowledge of the location of where wave tripping occurs and its distance from the change of grade of the floor, so that the cylinder could then be positioned to take full advantage of the maximum wave heights developed. Results from this study have been reported in Haritos & Armstrong (1996).

EXPERIMENTAL SETUP AND PROCEDURES

The experiments on wave runup effects on caissons located in shallow waters was performed as a similitude study in a two-dimensional wave flume 65m in length, 2.03m deep and 1.875m wide, with a water depth of 1.0m. Pairs of observation windows are positioned either side of the tank at three locations which provide excellent vantage points for viewing the generated wave profile and photographing experiments in progress. The wave generator is of the rigid "curved flap" Salter duck type which can be programmed to produce either regular or irregular wave profiles. Wave reflections are minimized by the use of an artificial beach sloped at approximately 45° at the far end. The beach has a satisfactory wave energy absorbing efficiency, with a reflection coefficient ranging between 3 to about 10 per cent, on its slope (McNamee, 1980).

A 16 channel A/D and single channel D/A data acquisition and control system operated by an IBM-compatible Personel Computer services the data capture during the performance of experimentation in the wave tank. The data for these experiments was initially sampled at 100Hz and then blocked averaged to obtain a lower effective sampling of 20Hz in an attempt to improve on signal quality. Recording was generally taken over a 100 second time period with wave generation taking place over a 50 second portion imbedded in the data capture so that the initial "build up" portion (waves reaching the cylinder after ~ 25 seconds) and the "tail end" portion containing the influence of any small wave reflections from the beach at the far end of the tank could also be observed.

The experimental program required the creation of shallow water waves and so a raised floor was constructed in the tank to simulate the desired wave conditions. The raised floor was required because the operating depth of the wave generator is in the range of 0.85m to 1.2m, with 1.0m being optimal. The cylinder set up was positioned adjacent to the centre pair of viewing windows. A diagrammatical representation of the false floor in the wave flume and the locality of the experimental set up is depicted in Fig. 2.

A simple economical dynamometer was designed and constructed to measure the wave forces on the laboratory model representing the vertical circular caisson. The six component dynamometer was instrumented to measure forces and moments in the inline, transverse and vertical directions. Fine wire surface mounted probes were also especially developed to record wave runup on the surface of the test cylinders for this test program. The probes in this experimentation were based upon the capacitance principle and were an adaptation of the design used for generally recording wave surface elevations in the tank. A similar concept of runup measurement has been used by Niedzwecki and Duggal (1992), but their probes were based on the resistance-type wave gauge. It was intended to instrument the test cylinder around one half of the circumference using a total of seven fine wire capacitance wave probes, attached to the sides and located such that the semicylinder was subtended into six equal arcs. The probes after being spot glued in their appropriate position on the outer wall face were then painted with a very thin coat of quick drying lacquer to prevent the ingress of moisture and to minimize any capillary rise in the gap between the cylinder wall face and the wire.

A major advantage in conducting a similitude study of this type to estimate the hydrodynamic responses of a large offshore structure is that account is automatically taken of the physical effects of the free and forced wave motions which must exist at the second order, which have been found difficult to incorporate into second order diffraction theories. Unfortunately, in "real" wave tanks, side wall reflections and the "blockage" effect of a model placed in a narrow tank add additional interaction effects normally not present in the prototype situation so results obtained from measurements in such a situation must be interpreted with these influences in mind.

Wave induced pressures under the model due to the gap influence would have an effect on the vertical force and hence on the overturning moment; thus an attempt was made to minimize its effect by positioning the cylinder's soffit some 3 to 5mm below the top of the false floor, and restricting the gap or annulus between the floor and the cylinder to approximately 1.5 mm.

DESCRIPTION OF MAJOR TESTS

The major series of tests were conducted on the 360mm and the 500mm diameter vertical circular cylinders located in a water depth of 125mm and then repeated in 160mm water depth. The tests on the 360mm diameter cylinder were in the range of conditions described by h/λ of 0.07 to 0.18, a D/λ of 0.18 to 0.44 and H/λ values ranging up to 0.11. The tests on the 500mm diameter cylinder were in the range of conditions described by h/λ of 0.07 to 0.18, a D/λ of 0.25 to 0.62 and H/λ values ranging up to 0.11. Wave frequencies used were: 0.6, 0.8, 0.9, 1.0 Ind 1.2 Hz.

In this series of tests, only two of the cylinder's surface wave probes did not capture data and these were positioned at 150° and 180° and so immediately after performing one set of experiments, the cylinder was rotated 180° and the tests repeated to provide surface elevation data for positions at 0° and 30° orientations. All six force components from the dynamometer were recorded. A typical "wetted surface profile" over half of the cylinder determined in this way is presented in Fig. 3.

Two "normal" wave surface elevation probes were located at the leading face of the cylinder, each to one side of it, for measuring the wave surface in the presence of the cylinder. The incident wave conditions were determined from these surface elevation probes for the condition when the cylinder was removed and the instrumentation recess for the test cylinder plugged for the same set of waves as used with the cylinder present.

During the tests the wave runup at the leading face of the cylinder was video recorded. The cylinder had been marked with graduations such that after running the tests the video recording could be viewed to establish the still water level, along with the highest wave runup for each set of test conditions and compared with the electronically captured data. On comparing this data, it was found that the electronically captured data did not always correlate well with the video recorded data, and so the maximum wave runup measurements at the leading face (0° position) obtained from the video recording considered to be the most direct and reliable of these measurements, were retained for presentation.

RESULTS FROM MAJOR STUDY

The results for the Major Study being presented here are those for the 500mm and 360mm diameter cylinder located in 160mm and 125mm water depths respectively. The results are for the base shear force, overturning moment and wave runup on the leading face of the cylinders and are presented as non-dimensional plots of experimental data versus those from various theoretical/ numerical methods. The values presented are applicable to *tank conditions* and any extrapolation may require consideration of blockage and side wall wave reflection effects not accounted for in the theoretical/numerical methods used in the computation.

Waves

Figure 4 depicts a 1.8 second portion of the wave profile for the 1.0Hz, 40mm deep water amplitude wave propagating in a water depth of 160mm, taken from the 50 second position in the wave record in the absence of the 500mm diameter cylinder and with the cylinder present. (The recorded surface time history was taken at the same position just forward of the leading face of the cylinder for all tests in both of these conditions). This figure clearly depicts that the presence of the cylinder causes the wave to distort, amplifying it's crest and attenuating it's trough slightly whilst introducing a minor phase shift.

Figure 5 depicts a 1.8 second portion of the wave profile for the 0.9Hz, 40mm deep water amplitude wave travelling in a water depth of 125mm, taken also from the 50 second position in the wave record in the absence of the 360mm diameter cylinder and with the cylinder present. Figure 6 depicts a similar trace for the 500mm diameter cylinder in 125mm water depth subjected to the same frequency and amplitude. Both figures provide good visual examples of the diffracted wave in front of the cylinder and the effect that the presence of the cylinder has in influencing the incident wave profile. The traces also clearly depict the nonlinear features of the incident waves where the wave crest amplitude generally exceeds half the wave height and exhibits horizontal asymmetry about a vertical line through the crest and vertical asymmetry about the still water level. The influence of the interaction of the side wall wave reflections with the incident and diffracted waves is a complex process which, although not easily visualized from these traces, can nonetheless be appreciated from them to some extent.

RESULTS FOR 500mm DIAMETER CYLINDER

Results for the 500mm diameter cylinder are presented in summary for comparison with the predictions from the theory of Isaacson and Cheung (1994) in Figures 7 to 9 for the 160mm water depth, in which the diffraction parameter, κa, is rounded at the first decimal place and where ascending values in sequence correspond to the five wave frequencies adopted for the tests of: 0.6, 0.8, 0.9, 1.0 and 1.2 Hz.. For the experiment the diffraction parameter κa was always greater than 0.6 and the Keulegan-Carpenter number was observed never to exceed 2, indicating that the flow regime is that of wave diffraction and that appreciable flow separation would therefore not occur. Hence, the cylinder would not likely experience oscillating lift forces acting transverse to the direction of wave motion due to vortex shedding.

Comparison of the measured to predicted (Isaacson and Cheung (1994)) longitudinal force and moment values in these figures indicate that, despite the limitations of the experimental setup in a "narrow" wave tank, generally good agreement is obtained except for waves with high steepness. The comparisons between the measured and predicted (Isaacson and Cheung (1994)) wave runup in Fig. 9 generally show much poorer agreement, which may be attributed to the interaction on the runup of the effects of blockage and side wall wave reflections. The blockage ratio (i.e., tank width to the model transverse dimension) for the setup is rather high at 1:3.75. (Chakrabarti (1994) suggests that the ratio be smaller than 1:5 to minimise these effects; see also Sarpkaya & Isaacson, (1981)).

RESULTS FOR THE 360mm DIAMETER CYLINDER

Results for the 360mm diameter cylinder are presented in

summary form enabling comparison of the experimental results with the theory of Isaacson and Cheung (1994) in Figures 10 to 12 for the 125mm water depth where once again the diffraction parameter κa is taken to the first decimal place for the same wave range as for the 500mm diameter cylinder.

Generally, the measured maximum forces compare reasonably well with the maximum forces predicted by Isaacson and Cheung's solution in Fig. 10. Poor agreement between theory and experiment is however observed for the $\kappa a = 0.6$ test condition, with the departure between theory and experiment commencing at low steepness and low Ursell number ($U_R \approx 0.3$) in this figure.

Isaacson and Cheung's moment solution is seen in Fig. 11 to generally underpredict, very slightly, the measured moment values at low steepness and slightly overpredicts at the higher steepnesses. The runup comparisons in Fig. 12 show that Isaacson and Cheung's theory generally slightly overpredicts runup in the 125mm water depth, with poor agreement occurring at high wave steepness.

Again, it should be noted that the experimental runup values will be influenced by blockage effects and by reflected waves emanating from the channel side walls, but since side wall reflections are orthogonal to the captured force and moment values in the inline direction, these reflections should not greatly influence these values. It is also interesting to have observed the similarity in the response curves for both force and moment of the 500mm diameter cylinder in the 160mm water depth with those of the 360mm diameter cylinder in the 125mm water depth, when subjected to waves of similar frequencies.

CONCLUDING REMARKS

Wave interaction with large diameter vertical cylinders in deep water has been extensively studied for its importance in offshore engineering, both experimentally and analytically, whereas the nonlinear diffraction of large cylinders in shallow water conditions has received relatively little attention in the literature. Consequently an experimental programme of research was conducted using two cylinders of different diameters (360mm and 500mm) and each subjected to the same subset of wave conditions in water depths of 125mm and 160mm in an attempt to provide experimental observations of nonlinear wave diffraction effects on these cylinders in shallow water conditions.

The major objective of the programme was to perform a similitude study in a narrow wave tank and to then compare the experimental data from the study with the analytical values obtained using linear diffraction theory and a recently published second order numerical approach by Isaacson and Cheung (1994). In this experimental test program, the hydrodynamic interaction effects on a large vertical cylinder, viz the horizontal force, moment and runup due to a range of wave conditons were measured using instrumentation developed for this purpose and the measurements then compared with the predicted values from linear diffraction theory and the numerical approach of Isaacson and Cheung (1994).

It was found in this programme that the linear theory consistently under estimated the quantities of interest, (which was to be expected), and that the predictions of Isaacson and Cheung (1994) based upon a second order diffraction theory generally provide reasonable lower bound estimates of the wave runup effects. Isaacson and Cheung's wave runup predictions were specifically found to give reasonable agreement with the experimental results for low wave steepness values (up to an Ursell number of approximately 0.3 and for Keulegan-Carpenter number less than 2), departing from (underpredicting) the experimental results for waves at high steepness ($H/\lambda > 0.05$). The experimental values of the base shear force and moment in particular are considered not to be significantly influenced by blockage and side wall wave reflection effects, whereas the wave runup, because it is enhanced by these effects would be less reliably estimated experimentally in a narrow tank, and this is reflected in the measurements made in these tests.

From this experimental program, it may be concluded that despite the limitations of "blockage" and the effects of side-wall reflections, model testing in wave flumes can provide confidence in the applicability/accuracy of numerical simulation techniques and simplified analytical methods derived therefrom for use in engineering design practice.

ACKNOWLEDGEMENTS

The authors wish to express their gratitude for the sponsorship by BHP Petroleum Pty Ltd of this experimental program which forms part of an Australian Research Council APA (Industry) award. All experimentation in association with this research was performed in the Michell laboratory wave tank at The University of Melbourne.

REFERENCES

Beck, R.F. and Scorpio, S.M. (1995) "A desingularised Boundary Integral method for fully nonlinear water wave problems", *Proc. 12th Austn. Fluid Mechs.* Conf., Univ. of Sydney, Dec., Vol 1, 255-258.

Chabrabarti, S.K. and Tam, W. A. (1995) "Gross and Local Wave Loads on a Large Vertical Cylinder - Theory and Experiment." *Offshore Technology Conf.* OTC 1818 .

Haritos, N. and Armstrong, R. (1996). "A Preliminary Model Study of Wave Runup on Caisson Type Structures", *Proc. OMAE96*, Florence, June 1996, Vol 1, pp 445-451.

Isaacson, M. and Cheung, K. (1994) "Correction Factors for Nonlinear Runup and Wave Forces on a Large Cylinder," Private communique with A/P N. Haritos.

Isaacson, M. (1978) "Wave Runup Around Large Circular Cylinder," *Proc. A.S.C.E. Jl. of Waterways, Port, Coastal and Ocean Division,* Vol. 104, No. WW1, Feb., pp.69-79.

Kim, M. H. and Yue, D.K.P. (1989) "The Complete Second-Order Diffraction Solution for an Axisymmetric Body. Part 1 Monochromatic Incident Waves," *Jl. Fluid Mechanics*, 166, pp 189-209.

Kriebel, D. (1990) "Nonlinear Wave Interaction with a Vertical Circular Cylinder, Part 1: Diffraction Theory," *Ocean Engineering*, Vol. 17, No. 4, pp345-377.

Kriebel, D. (1992) "Nonlinear Wave Runup on Large Circular Cylinders". *Civil Engineering in Oceans V : Proc. of International Conference,* A.S.C.E., College Station, Texas, Nov.

MacCamy, R. and Fuchs, R. (1950) "Wave Forces on Piles: A Diffraction Theory," Tech. Memo. No.69, Beach Erosion Board, Army Corps of Engineers, Washington , D.C.

McNamee, B.P. (1980) "A Study of Wave Forces on Cylinders in the Inertial Regime", Ph.D. Thesis, Dept. of Civil & Agricultural Eng., Univ. of Melbourne, Australia.

Niedzwecki, J. M. and Duggal, A. S. (1992) "Wave Runup and Forces on Cylinders in Regular and Random Waves", Proc. ASCE, *Jl. of Waterway, Port, Coastal, and Ocean Eng.*, Vol. 118, No. 6, Nov/Dec, pp 615-634.

Sarpkaya, T. and Isaacson, M. (1981) *Mechanics of Wave Forces on Structures*, Van Nostrand Reinhold.

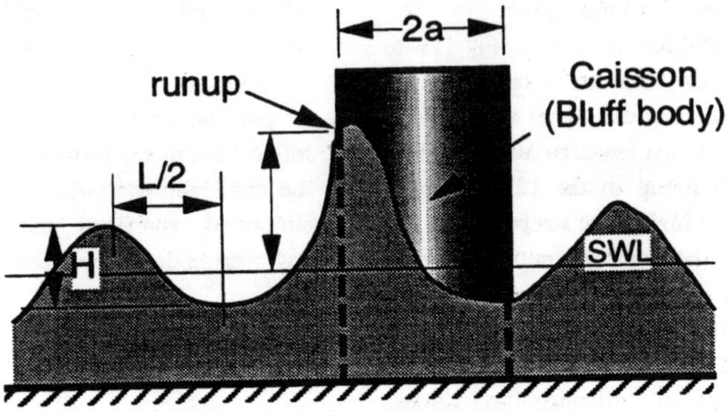

Figure 1: Representation of Wave Runup

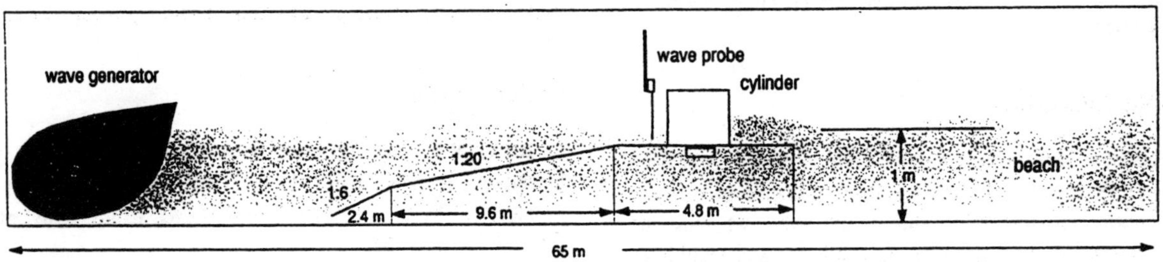

Figure 2: View of Experimental Setup

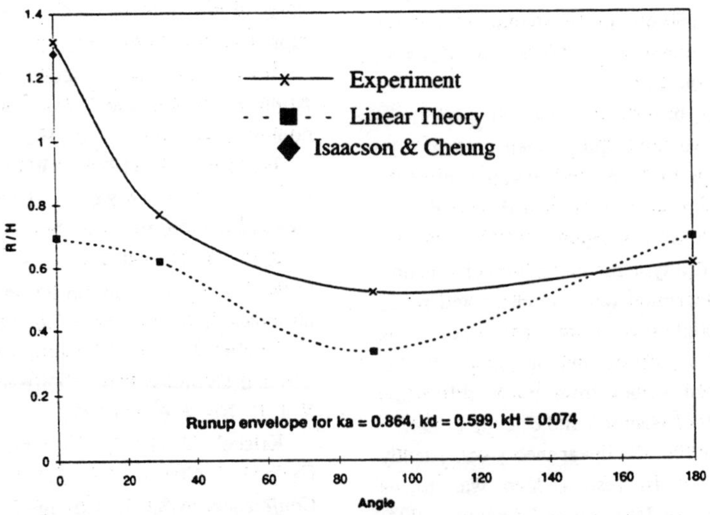

Figure 3: Wetted Profile of Cylinder

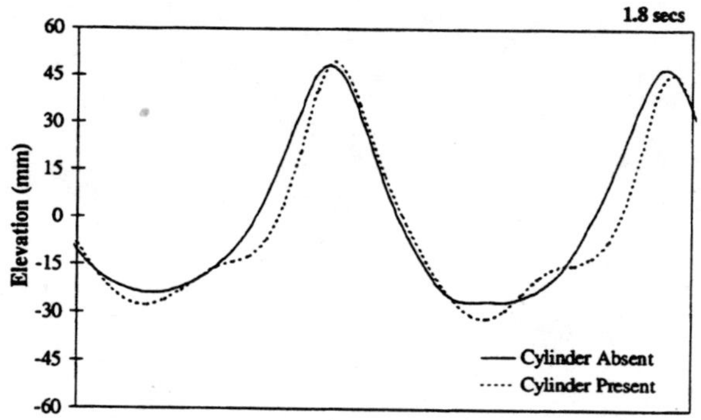

Figure 4: Sample Trace with/without Cylinder
(1.0 Hz wave, 160mm depth, 500mm cylinder)

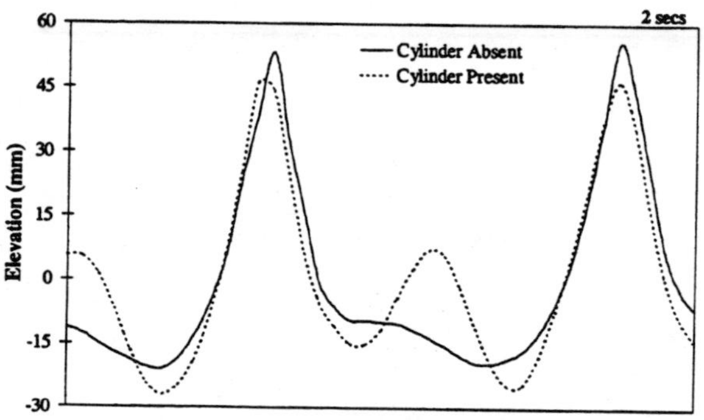

Figure 5: Sample Trace with/without Cylinder
(0.9 Hz wave, 125mm depth, 360mm cylinder)

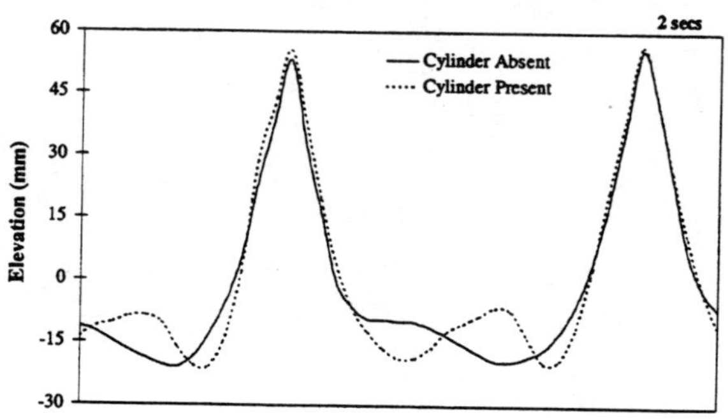

Figure 6: Sample Trace with/without Cylinder
(0.9Hz wave, 125mm depth, 500mm cylinder)

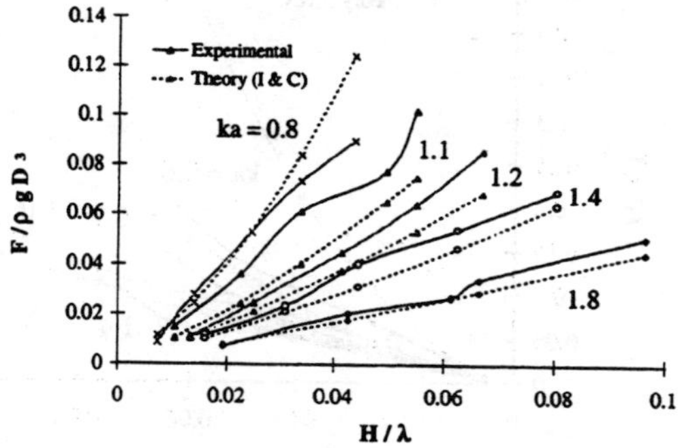

Figure 7: Comparison of measured and predicted
maximum longitudinal force
(500mm diameter cylinder in 160mm water depth)

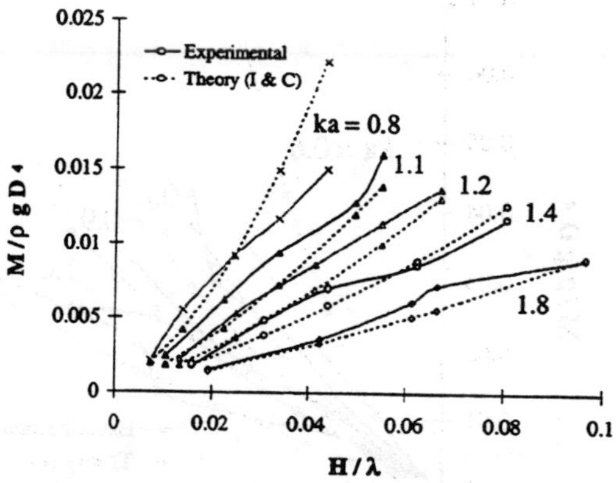

Figure 8: Comparison of measured and predicted maximum momen
(500mm diameter cylinder in 160mm water depth)

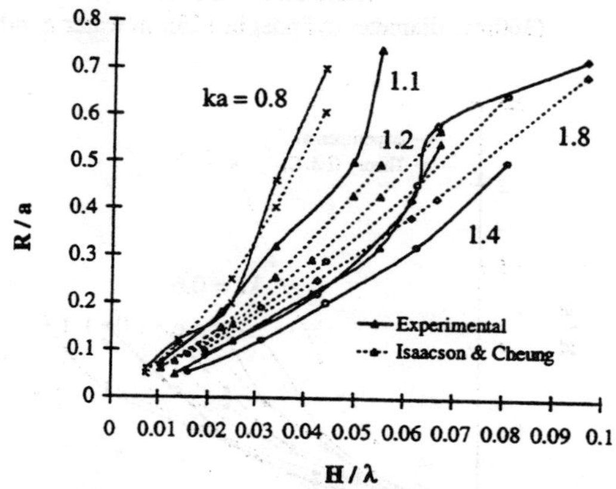

Figure 9: Comparison of measured and predicted maximum runup
(500mm diameter cylinder in 160mm water depth)

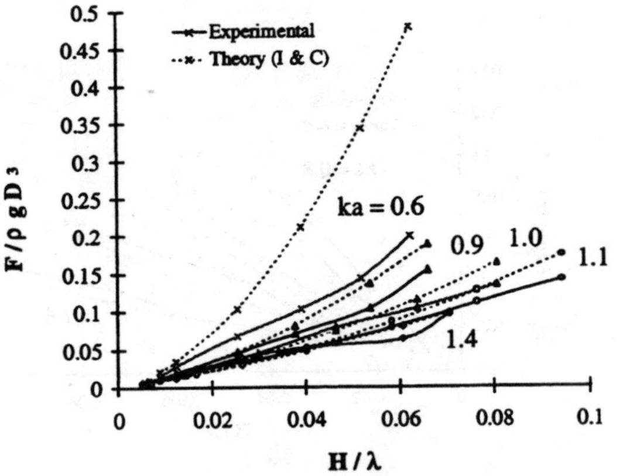

Figure 10: Comparisons of measured and predicted maximum longitudinal forces (360mm diameter cylinder in 125mm water depth)

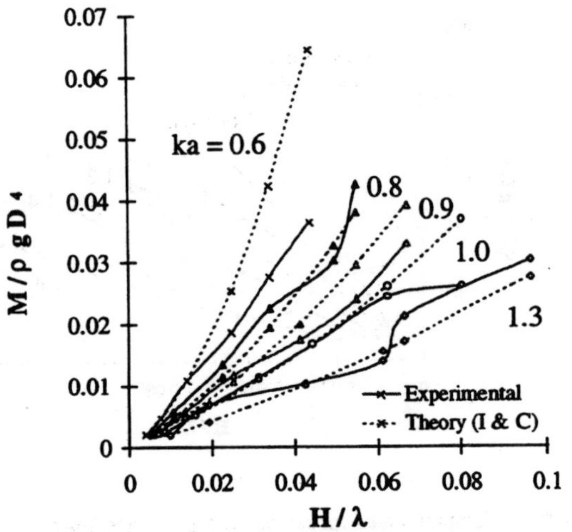

Figure 11: Comparisons of measured and predicted maximum moments (360mm diameter cylinder in 125mm water depth)

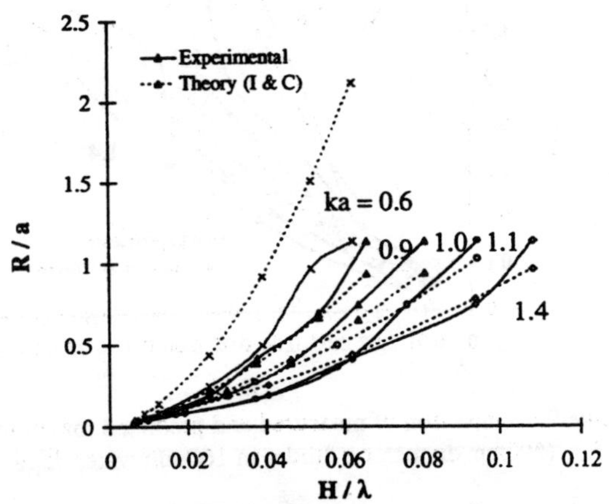

Figure 12: Comparisons of measured and predicted maximum runup (360mm diameter cylinder in 125mm water depth)

Proceedings of the Seventh (1997) International Offshore and Polar Engineering Conference
Honolulu, USA, May 25-30, 1997

ISBN 1-880653-28-1 (Set); ISBN 1-880653-31-1 (Vol. III)

Wave Enhancement Due to Structure Block in Semi-Submersibles

T. Bruce and W.J. Easson
University of Edinburgh
Edinburgh, United Kingdom

Abstract

Results of an experimental study of the effect of two bottom-mounted vertical cylinders on an incident monochromatic wave field are presented. The cylinder diameter:spacing ratios are representative of semi-submersible offshore drilling rigs. The purpose of this study was to investigate whether wave enhancement due to structure blockage caused a significant enhancement of wave elevations in the vicinity of such structures. Physical model tests have been performed to quantify wave elevation enhancement and wave runup on the cylinders exposed to monochromatic waves. It is shown that cylinder pairs can cause significant enhancement of the local wave field and that extreme amplitude enhancement - *jetting* - of up to a factor of 2.5 is possible.

Keywords: Wave enhancement, Wave-structure interaction, Runup, Semi-submersibles, Jetting

1. Introduction

Structure blockage is here defined as the influence on the waves of the presence of a structure. In a unidirectional flow, such as a river, the effect of blockage by a bridge pier would be to raise the elevation of water upstream and to increase the velocity at the constriction. However, neither of the effects is assumed or implied by the term blockage for the cases dealt with here. Indeed, blockage in waves produces a quite different result.

The particular application dealt with in this study is the blockage caused by two vertical cylinders, representative of the legs of a semi-submersible offshore drilling or production vessel.

The purpose of this study is to investigate whether wave enhancement due to structure blockage ought to be considered a significant effect in offshore design. It was decided that, in the first instance, the structure blockage effects could be represented by studying the effect on waves of a pair of bottom-mounted vertical cylinders. This is partially indicated by the work of Eatock-Taylor (1989), which shows only small differences in the diffraction effects due to the presence of the pontoons of the structure.

The objectives of the work were to look at the influence of leg diameter, leg spacing, wave height, wavelength and angle of wave incidence effect on wave elevations in the vicinity of the structures.

2. Experimental Set Up

The tests were performed in the three-dimensional wave basin at The University of Edinburgh The tank measures 24m x 10m, and is equipped with 80 independent wavemakers along one of the long sides, with wave absorbing material along the length of the other side.

Wave elevation records were recorded at eight locations near the model as shown in figure 1.

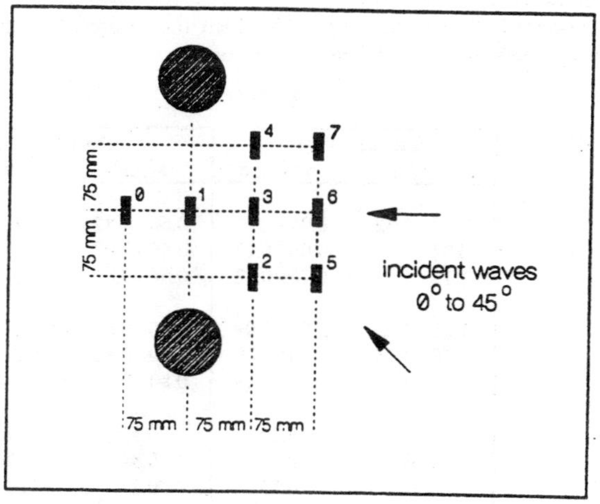

Figure 1: Wave gauge positions relative to test cylinders for 3-d tests.

The wave elevation at the eight wavegauges was recorded at 40Hz. All tests were video-taped for later analysis of wave runup and jetting phenomena (section 4). Tests were carried out for cylinders of diameter 51mm, 76mm and 102mm, arranged with centre-to-centre separations of 200mm, 300mm and 400mm. The sequence of tests carried out for each based upon a scale of 1:150. This test matrix covers a wide range of typical designs of semi-submersibles.

3. Wave Enhancement - Elevation Measurements

The wave generator was programmed to produce waves of three different steepnesses for a range of frequencies. Steepness is defined here in the standard temporal sense as

$$s = \frac{H}{gT^2}$$

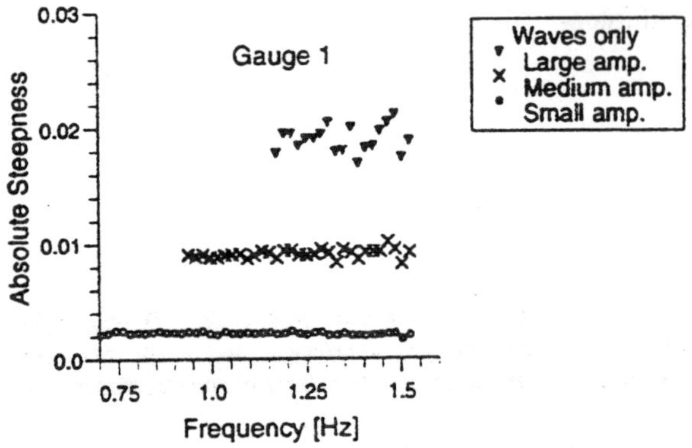

Figure 2: Absolute steepnesses of test waves without model in place (gauge 1).

To provide an indication of the calibration and accuracy of reproduction of the waves as measured in the tank, graphs of steepness have been produced for each wave gauge (figure 2). Due to the limits of the transfer function of the tank it is not possible to produce waves above a certain steepness at the lower frequencies. However, for a range of frequencies from 1.17Hz to 1.56Hz, waves of near maximum steepness were produced. The accuracy with which the desired heights were achieved decreases as the steepness increases.

As the purpose of these tests is to look at the possible incursion of water into the area of the superstructure of semi-submersibles, the actual elevation of the wave crest above the still water level (SWL) is of interest, rather than the trough to crest height. Figure 3 shows the maximum crest elevation between the legs as a fraction of wave height for the three steepnesses. It is a significant result that the relatively moderate waves of steepness 0.009 have a ratio of 0.55, whilst the waves of steepness 0.019 have a ratio of 0.6. This finding is discussed further by Easson (1997). In the plots of elevation enhancement which follow, the wave elevation is normalised by the elevation taken from this data set. To calculate the elevation of a wave above SWL from these plots the height should be multiplied by the appropriate asymmetry factor from figure 3, and then multiplied by the enhancement.

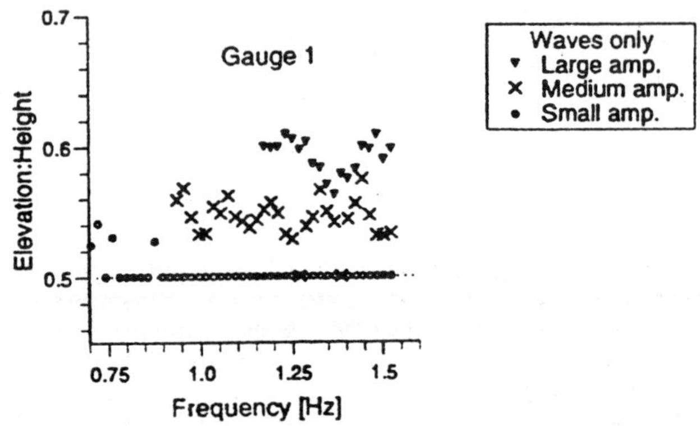

Figure 3: Maximum crest elevation as a fraction of wave height without model in place (gauge 1).

Figure 4 is presented to show the behaviour of the gauges in their relative positions for the small wave case. The enhancement is normalised by dividing the measured elevation by that measured without a model in position. The *central case*, a=300mm, D=76mm has been used here.

Frequency (Hz)		Period (s)		Amplitude (m)		Angle of
Model	Prototype	Model	Prototype	Model	Prototype	Incidence (^{0})
0.70 - 1.52	0.06 - 0.12	0.66 - 1.42	8.1 - 17.4	0.005 - 0.023	0.7 - 3.4	0
0.94 - 1.52	0.08 - 0.12	0.66 - 1.07	8.1 - 13.1	0.019 - 0.051	2.9 - 7.6	0
1.17 - 1.52	0.10 - 0.12	0.66 - 0.85	8.1 - 10.4	0.043 - 0.073	6.5 - 11.0	0
1.13 - 1.25	0.08 - 0.09	0.80 - 0.88	9.8 - 10.8	0.007 - 0.087	1.1 - 1.3	0, 22.5, 45
1.13 - 1.25	0.08 - 0.09	0.80 - 0.88	9.8 - 10.8	0.029 - 0.035	4.2 - 5.2	0, 22.5, 45
1.13 - 1.25	0.08 - 0.09	0.80 - 0.88	9.8 - 10.8	0.064 - 0.078	9.6 - 11.7	0, 22.5, 45
1.17	0.10	0.85	10.4	0.012 - 0.081	1.7 - 12.2	0, 22.5, 45
1.19	0.10	0.84	10.3	0.011 - 0.079	1.7 - 11.8	0, 22.5, 45
1.21	0.10	0.83	10.2	0.011 - 0.076	1.6 - 11.4	0, 22.5, 45
1.17	0.10	0.85	10.4	0.008, 0.032, 0.073	1.2, 4.9, 11.0	0 - 45
1.19	0.10	0.84	10.3	0.008, 0.031, 0.071	1.2, 4.7, 10.6	0 - 45
1.21	0.10	0.83	10.2	0.008, 0.030, 0.068	1.1, 4.6, 10.3	0 - 45

Table 1: Test parameters

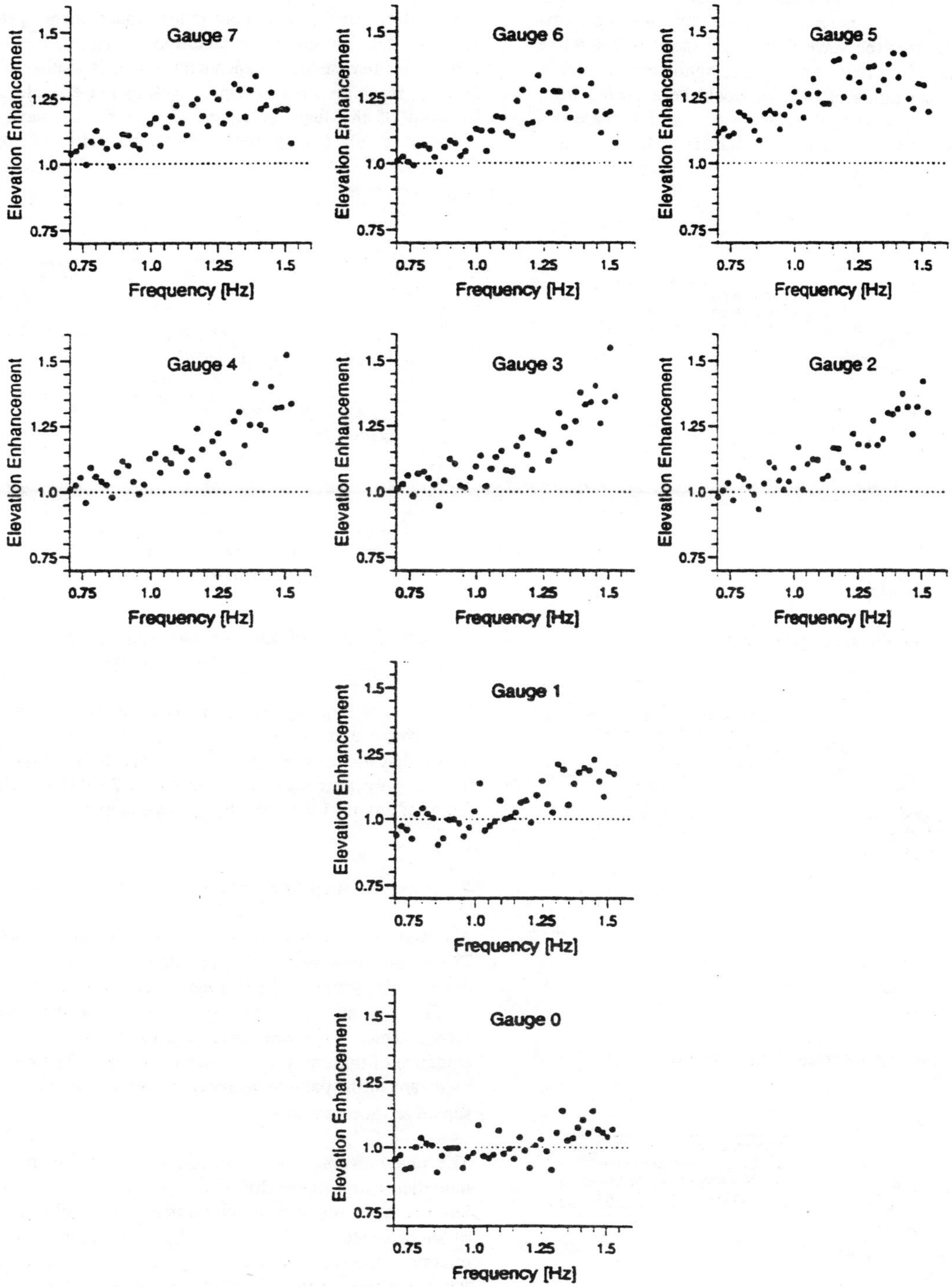

Figure 4: Elevation enhancement, small waves, central case

Gauge 1, between the legs, shows very little evidence of enhancement, and gauge 0, behind the legs, shows that the wave field is almost unchanged after passing the obstruction caused by the two legs. However, in front of the legs the wave field is enhanced by the wave scattering from the legs. The reflection is most significant for higher frequency waves, *ie* higher ratios of D/L, as would be expected from diffraction theory. At the gauges furthest in front of the legs there is a clear maximum in the enhancement at f~1.25Hz. The measured elevations for these small waves agreed with those given by linear diffraction theory.

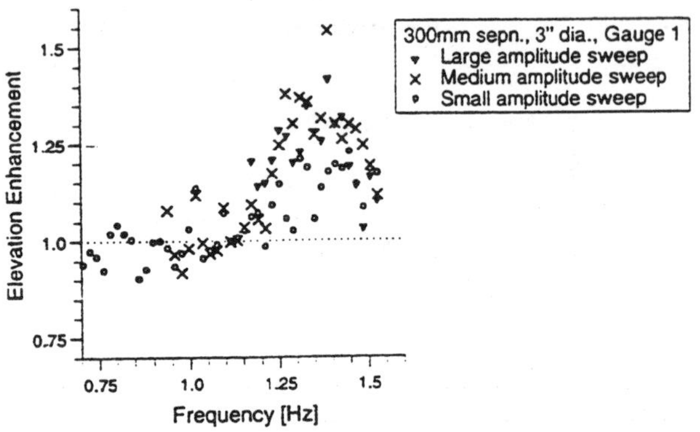

Figure 5: Elevation enhancement between legs.

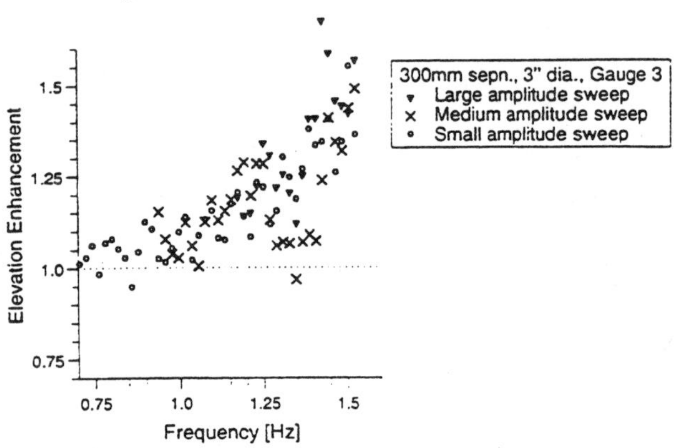

Figure 6: Elevation enhancement 75mm upwave of legs.

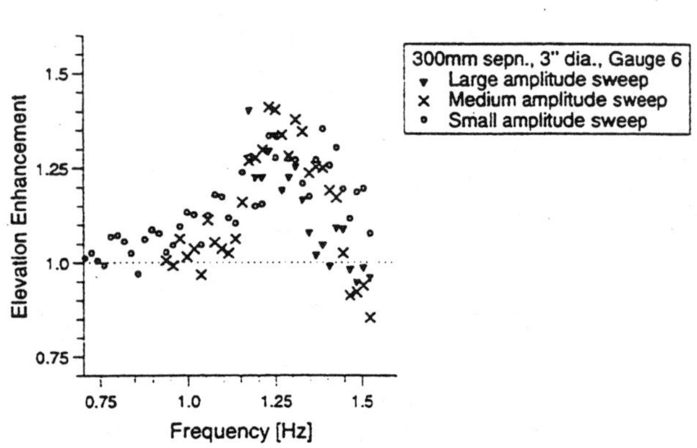

Figure 7: Elevation enhancement 150mm upwave of legs.

The effect of the legs on larger waves is plotted in figures 5 to 7, from elevation records of gauges along the bisector of the centre-line of the legs. Between the legs there is no enhancement up to frequencies of around 1.2Hz for any of the amplitudes. Above this frequency there is a marked increase in enhancement which is similar for all three wave heights, reaching a maximum of 30% increase in elevation at f~1.3Hz. In front of the legs, at gauges 3 and 6, the wave enhancement is apparent at all frequencies, and is generally, but not significantly, greater for the larger waves. Gauge 6 again shows a maximum in wave enhancement at f~1.25Hz.

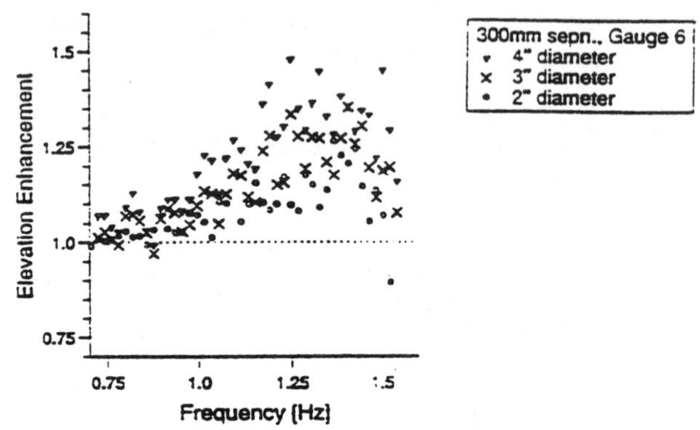

Figure 8: Effect of blockage ratio -elevation enhancement 150mm upwave of legs.

The effect of blockage ratio (the ratio of cylinder diameter to spacing) was investigated by varying the column diameter. Figure 8 clearly shows that the crest elevation is affected by the blockage ratio ranging from an enhancement factor of around 1.2 for the smallest cylinders to a factor of around 1.4 for the largest cylinders.

4. Wave Runup and Jetting

A subset of the tests reported in section 3 were recorded on video tape. The camera was mounted at the side of the tank, a few cm forward of the line of centres of the two cylinders. A zoom lens was used and the result is an almost perpendicular image of the waves attacking the model. These have been analysed for runup on the cylinder and for evidence of unusual effects, such as jetting. The records are calibrated by means of the painted scales on the model and the measurements from screen are accurate to ~10mm.

The phenomenon known as runup is well documented and may be modelled using linear diffraction (*eg* Sarpkaya & Isaacson, 1981), but has not been the subject of particularly detailed investigations in an offshore context. It occurs when, on reaching a vertical wall, the energy of the wave is directed upwards. Hallermeyer (1976) performed wet paper tests in the laboratory, and proposed that the runup be treated as a function of the kinetic energy of the crest water particle velocity. Haney & Herbich (1982) performed similar experiments, concluding that the runup was well predicted by the velocity head given by a Dean's 5th order model. Kriebel (1993) presents an alternative prediction of runup in which linear scattering theory is used to predict the runup due to each Fourier component. These values are then combined by superposition. In these tests the runup has been taken as the highest point of water on the leg.

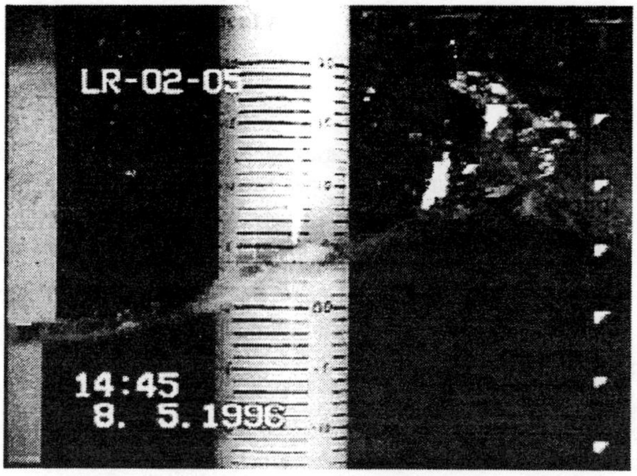

Figure 9: A jetting event.

The authors have found no independent record of 'jetting', but this may be the same phenomenon which is normally termed 'upwelling'. As this latter term is normally restricted in its use to the rise of water within the structure, the term jetting will continue to be used in this paper. Jetting was observed to occur at a point between and in front of the two model legs. It appears to be due to the interference of incident wave crests with scattered wave crests. Linear diffraction theory predicts the position of these wave enhancements but not their magnitude. Figure 9 shows a typical jetting event, taken from the video record. The jet appears as a finger of fluid rising above the wave with a small amount of forward or backward momentum. It will be demonstrated that jetting is an effect which occurs when the diffracted waves from the legs form a wave which exceeds the breaking limit. In the analysis of the video records, both the maximum elevation and the position with respect to the legs were recorded.

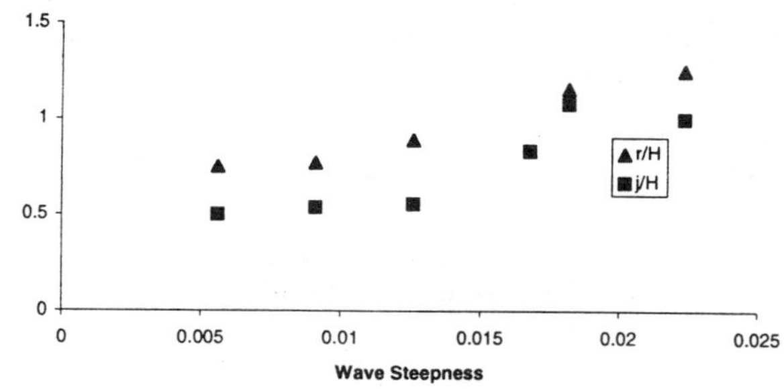

Figure 10: Jetting and runup vs steepness, central case, f=1.17Hz.

The graphs of runup and jetting are scaled with height. Thus a runup of 1.0 indicates that the water exceeded the SWL by a value equal to the height of that wave. Runup is observed for even the smallest waves. Jetting is only observed when the wave steepness is greater than half the limiting steepness (s>0.011). Figure 10 shows the runup and jetting as a function of amplitude for typical long waves of frequency 1.17Hz. In this example, the runup reaches values of 1.25H; whilst the jetting reaches values of 1.0H at a distance of around 0.65 cylinder spacings in front of the legs (190mm). The position of jetting seems to be a function of the wave steepness, moving towards the legs as the wave height increases, by around 10%.

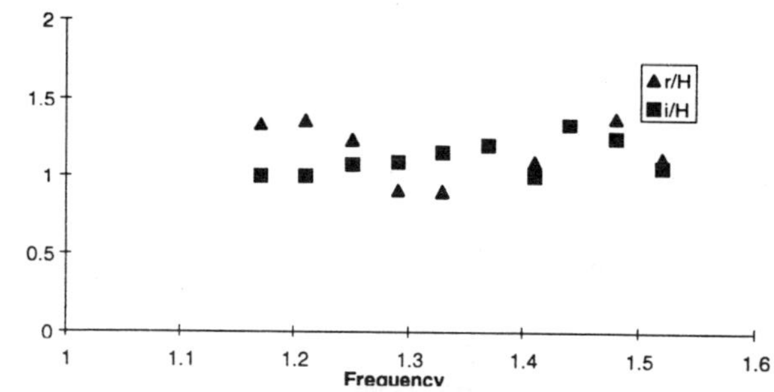

Figure 11: Jetting and runup vs. wave frequency, central case.

The variation of jetting and runup with frequency is plotted in figure 11. The waves have been chosen to have similar (large) steepnesses. The runup shows a wide range of values around a mean of 1.2H. The jetting has a very distinct trend, with a maximum of 1.35H at around 1.45Hz. Note, however, that these are relative enhancements, and the greatest absolute effect would occur for the longest (and therefore highest) waves. This variation of the jetting enhancement with period may be compared directly to the wave elevation plots, which show a maximum at about the same frequency. In these tests, the position of jetting away from the legs moved from around half a leg spacing out (T=0.85s) to around one third of a leg spacing at the shortest wavelength (T=0.66s).

Results from a subsequent, more detailed study of runup and jetting are presented in Martin *et al.* (1997) .

Effect of Angle of Wave Incidence

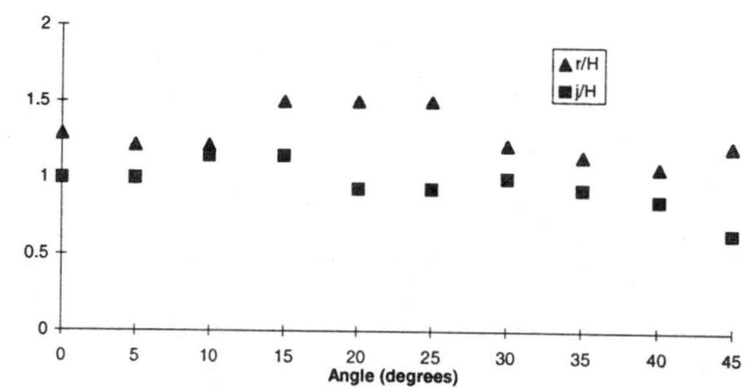

Figure 12: Jetting and runup vs angle of wave incidence, central case, f=1.17Hz.

Figure 12 shows the variation of runup and jetting with wave angle. A range of angles were used, up to 45^0. The runup is found to be independent of wave angle, as would be expected. However, the jetting falls of with angle, only remaining significant at angles of up to around 25^0.

Effect of Geometric Blockage Ratio

To illustrate the effects of blockage ratio, two examples are presented: varying the leg separation; and varying the leg diameter. When the leg separation is varied for a leg diameter of 102mm, the largest in the tests, the behaviour of the runup is unchanged. However, these legs are larger than those used in the previous tests (figure 10) and the runup is consistently higher. This suggests that runup is a function of leg diameter and the ratio of leg diameter to wavelength is probably the significant factor.

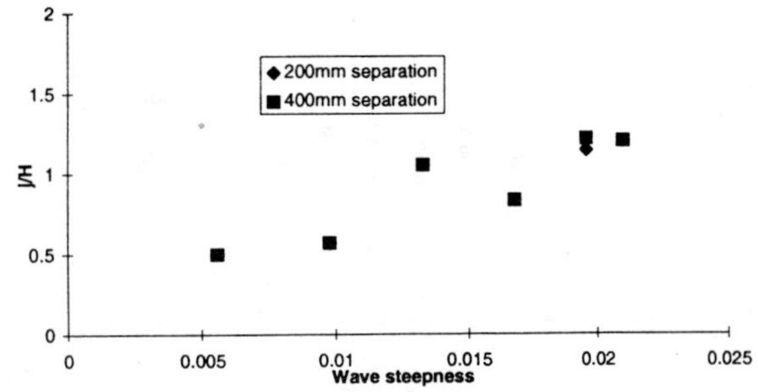

Figure 13: Jetting vs steepness for medium and high blockage - two spacings, leg diameter fixed.

For this fixed wavelength to diameter ratio, changing the separation does not seem to change the severity of the jetting to a noticeable extent, though the errors are large and the range of separations studied is small. The position of the jetting is the same for both cases, indicating that the effect is solely dependent on the wavelength, and relatively independent of the spacing for these ratios.

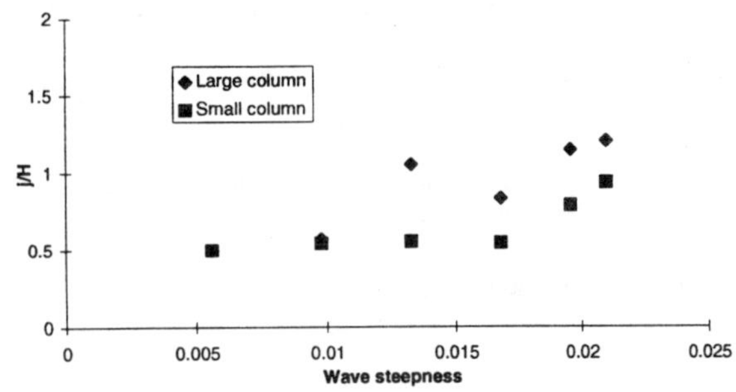

Figure 14: Jetting vs steepness for low and high blockage - two leg diameters, spacing fixed.

Varying the leg diameter for a fixed separation confirmed that the runup is greater for a larger leg diameter. The effect on jetting is even greater (figure 14). For a fixed separation to wavelength ratio, the jetting is 30-40% more severe for the high leg diameter. Therefore, the blockage ratio does play a part, in the sense that the amount of energy reflected from the legs determines the severity of the jetting.

5. Conclusions

- The horizontal asymmetry of the wave is shown to be around 0.65 for the steepest waves in these tests. For waves of steepness s=0.019 (75% of limiting steepness) the asymmetry is in excess of 0.6.
- Runup is observed in all cases and is independent of spacing. The extent of runup can be as great as 1.5 times the wave height above still water level. Runup is a function of wave steepness and a weak function of leg diameter.
- Jetting is a newly observed phenomenon, which the authors have not seen referred to elsewhere. It occurs due to non-linear diffraction effects.
- The magnitude of jetting may be up to 1.5 times the incoming wave height above SWL.
- Jetting occurs when the incoming waves steepness is about 50% of the limiting steepness.
- The magnitude of jetting varies with the diameter of the legs, but not significantly with the leg spacing.
- Jetting is a function of angle, but maintains high values to around 25^0.

Acknowledgements

The authors gratefully acknowledge the financial support of the UK Health and Safety Executive. The assistance of David Skyner, Peter Woodhead, Paul Nebel and Richard Yemm at various stages of the work is also gratefully acknowledged.

References

Easson, W.J. (1997). "Breaking Waves," *Proc. 7th Int. Offshore and Polar Eng. Conf. (ISOPE'97)*

Eatock-Taylor, R. & Sincock, P. (1989). "Wave Upwelling Effects in TLP and Semi-submersible Structures," *Ocean Eng.* 16, 3, pp281-306

Hallermeyer, R.J. (1976). "Non-linear Flow of Wave Crests Past a Thin Pile," *J.WPC&OE*, 102

Haney, J.P. & Herbich, J.B. (1982). "Wave Flow Around Thin Piles and Pile Groups," *J. Hydr. Res.*, 20, 1, pp1-14

Kriebel, D.J. (1993). "Nonlinear Runup of Random Waves on a Large Cylinder," *Proc. Int. Conf. Offshore Mech. & Arctic Eng. (OMAE 1993)*, (ASCE), 1, pp49-55

Martin, A.J., Easson, W.J. & Bruce, T. (1997). "Runup on Columns in Large Waves," *Proc. 7th Int. Offshore and Polar Eng. Conf. (ISOPE'97)*

Sarpkaya, T. & Isaacson, M. (1981). "Mechanics of Wave Forces on Offshore Structures," *Van Nostrand Reinhold*

Proceedings of the Seventh (1997) International Offshore and Polar Engineering Conference
Honolulu, USA, May 25-30, 1997

ISBN 1-880653-28-1 (Set); ISBN 1-880653-31-1 (Vol. III)

Hydrodynamic Analysis of Floating Bodies on Parallel Computers

B. Padmanabhan and R.C. Ertekin
University of Hawaii at Manoa
Honolulu, HI, USA

ABSTRACT

The solution of computationally large hydrodynamic problems using the Boundary Element Method (BEM) is limited by the storage and memory resources needed to assemble and solve the linear system of equations that result from a BEM formulation, and also by the CPU time. In this paper, we present a parallel implementation, which is based on the block-cyclic decomposition of system of matrices, to solve the classic 3-dimensional radiation and scattering problems of floating bodies under a BEM formulation. Particular attention is paid to the comparison of computational requirements in parallel and serial architecture machines.

KEYWORDS: Boundary Element Method, parallel computing, fluid-structure interaction.

INTRODUCTION

Boundary element formulations in the solution of wave-structure interaction problems offer a distinct advantage over volume based formulations in that the dimensionality of the problem is reduced by one. In 3-dimensional problems, for example, it is sufficient to discretize the bounding 2-dimensional surface. Moreover, in most wave-structure interaction problems, the goal is to find the pressure distribution on the structure surface, and hence the forces, which is obtained more economically with a boundary element formulation. The discretized integral equation results in a linear system of equations in the unknowns. From the implementation point of view, the major disadvantage is, therefore, the storage and computational costs associated with the method. The storage and computational costs in assembling the linear system grows rapidly with the size of the problem. Examples of large-scale problems include formulations that use the Rankine source as the fundamental solution. Such a formulation is necessary when an inhomogeneous free-surface condition or a non-flat bottom topography precludes the use of a Green function. Other examples include the hydroelastic response of very large floating structures, second-order hydrodynamic problems whose solutions lead to low- and high- frequency loads and responses, and transient wave propagation problems. Parallel computations offer the opportunity to solve these problems in a much more efficient manner. In this paper, we present a parallel implementation to solve the classic radiation and scattering problems of floating bodies.

The two major computationally intensive steps in a BEM formulation are the assembling of the linear system and the solution of the resulting linear system of equations. Parallel algorithms mitigate the computational limitations imposed by large problems by sub-dividing the problem into smaller sub-problems that can be handled by individual processors. The sub-division can be effected in two ways in a BEM formulation.

In the first method, the whole domain is divided into sub-domains with virtual internal boundaries. The boundary element formulation is applied to each sub-domain. The solution in each sub-domain is obtained after enforcing the assumed boundary conditions on the virtual boundaries. The procedure is continued iteratively until the boundary conditions at the internal boundaries are matched. The solution in each sub-domain is handled by one processor. Such a method is referred to as the domain decomposition method, see for example, Davies *et al.* (1996) and Kamiya *et al.* (1996).

The second method is based on partitioning the linear system matrix and assigning blocks of the matrix to different processors. From a BEM implementation point of view, an obvious partitioning of the matrix is to assign rows of the matrix to a given processor. The processor, thus, is responsible for assembling specific entries of the row. While the partitioning reflects the manipulation of rows in the LU factorization of a linear system, it is not optimal since some of the processors remain unutilized during the linear system solution phase of the problem. An alternate procedure is to use the 2-dimensional block cyclic partitioning of the matrix employed in the LU-decomposition of matrices on parallel computers as in Choi *et al.* (1992). Rather than assigning rows to processors, blocks of the matrix are assigned to processors. This method has recently been adopted by Flanery *et al.* (1996) in the BEM analysis of 2-D elastic wave-scattering problems.

In this paper, we extend the method of Flanery *et al.* (1996) to 3-D potential flow problems but modify their method by reducing processor communication during the assembly of the linear system of equations. In the following section, we give a brief overview of the boundary-value problem.

OVERVIEW OF THE BOUNDARY VALUE PROBLEM

Consider a body floating on the free-surface and in the presence of time-harmonic incoming waves in deep water. The origin of the right-handed coordinate system Oxyz is positioned at the still-water surface with the z-axis pointing up. Under the assumptions of linear potential theory, the total potential, Φ, of the flow is written as

$$\Phi = \mathrm{Re}[(\Phi_I + \Phi_B)e^{-i\omega t}] , \tag{1}$$

where Φ_I is the potential due to the incoming wave, and is given by

$$\Phi_I = \frac{-igA}{\omega} e^{kz} e^{i\bar{k}\cdot\bar{x}} , \tag{2}$$

where A is the amplitude of the incoming wave, ω the angular frequency, $\bar{k} = k(\cos\beta, \sin\beta, 0)$ the wave number vector, β the wave direction, g the acceleration due to gravity, ω and k are related by the dispersion relation, $\omega^2 = gk$, $\bar{x} = (x,y,z)$, and Φ_B is the disturbance potential due to the presence of the body, and is decomposed as

$$\Phi_B = \phi_7 + \sum_{j=1}^{6} \xi_j \phi_j . \tag{3}$$

In Eq. (3), ϕ_j is the velocity potential arising from the motion of the floating body, ξ_j the amplitude of motion in the six-degrees of freedom of the floating body, and ϕ_7 the diffraction potential.

The velocity potentials satisfy the following boundary-value problem in the domain Ω, bounded by the still-water surface, S_F, the body surface S_B and the control surface S_∞:

$$\nabla^2 \phi_j = 0 , \quad \bar{x} \in \Omega , \; j = 1,2,...7 , \tag{4}$$

$$-\omega^2 \phi_j + g\phi_{jz} = 0 , \; \bar{x} \in S_F \quad j = 1,2,...7 , \tag{5}$$

$$\frac{\partial \phi_j}{\partial n} = n_j , \; \bar{x} \in S_B , \quad j = 1,2,...,6 , \tag{6}$$

$$\frac{\partial \phi_7}{\partial n} = -\frac{\partial \Phi_I}{\partial n} , \; \bar{x} \in S_B , \tag{7}$$

$$\lim_{R \to \infty} \left\{ \sqrt{R} \left(\frac{\partial}{\partial R} - ik \right) \phi_j = 0 \right\} , \quad j = 1,2,...7 , \tag{8}$$

where $R = \sqrt{x^2 + y^2}$ is the radial distance in the horizontal plane, and $\bar{n}$ the normal vector out of the fluid defined by

$$n_j = \begin{cases} n_j & j = 1,2,3 \\ [(\bar{x} - \bar{x}_c) \times \bar{n}]_{j-3} & j = 4,5,6 \end{cases} \tag{9}$$

where $\bar{x}_c = (x_c, y_c, z_c)$ is the coordinate of the center of rotation of the body.

The boundary-integral formulation of the potentials ϕ_j is expressed as (see for example, Brebbia, 1984)

$$c(p)\psi(p) = \int_S \left[G(p,q)\frac{\partial \psi}{\partial n} - \psi \frac{\partial G(p,q)}{\partial n(q)} \right] dS , \tag{10}$$

where ψ is any one of the potentials ϕ_j, $G(p,q) = 1/|\bar{p} - \bar{q}|$ the Rankine source, p the field point (x,y,z), q the source point (x_o, y_o, z_o), $S = S_F U S_B U S_\infty$, and $c(p)$ is the included angle at p.

The surface S is discretized by quadratic isoparametric elements where the coordinates of points on the element and the potential and its derivatives on each element are expressed by

$$(\bar{x}, \psi, \psi_n) = \sum_{j=1}^{ND} N_j(s,t)(\bar{x}_j, \psi_j, \psi_{jn}) , \tag{11}$$

where $(s,t) \in [-1,1]$ are the element coordinates, $N_k(s,t)$ are the shape functions, ND is the number of nodes on the element. The subscript j refers to the nodal values of the variable. A mixture of 6-node triangular, 8-node quadrilateral and 9-node Lagrangian elements are used, the shape functions for which can be found , for example, in Brebbia *et al.* (1984). In discretizing the boundary, the control surface, S_∞, is a cylinder of finite radius and finite depth below the still-water surface. The radius of the cylinder is varied between five and six wavelengths and the cylinder is truncated at a depth of four wave lengths.

The discretized integral equation (10) is written in matrix form as

$$[H]\{\psi\} = [F]\{\psi_n\} , \tag{12}$$

where

$$H_{mn} = \sum_{e=1}^{EL} \sum_{j=1}^{ND} \delta_{nd} \hat{H}_{mj}^{(e)} + c_m \delta_{mn} , \tag{13}$$

$$F_{mn} = \sum_{e=1}^{M} \sum_{j=1}^{ND} \delta_{nd} \hat{F}_{mj}^{(e)} , \tag{14}$$

where $m,n = 1,2,...,N$, N is the total number of nodes, $d = (j,e)$ is a permutation that maps the local node number j of the e-th element to its global node number d, δ is the Kronecker-delta function and

$$\hat{F}_{mj}^{(e)} = \int_{\Gamma_e} G(p,q) N_j(s,t) \,|J(s,t)|\, ds dt , \tag{15}$$

$$\hat{H}_{mj}^{(e)} = \int_{\Gamma_e} \frac{\partial G(p,q)}{\partial n(q)} N_j(s,t) \,|J(s,t)|\, ds dt , \tag{16}$$

where Γ_e is the e-th element, $J(s,t)$ the Jacobian, and EL the total number of elements. When $m \neq n$, the integrals in Eq. (15) and (16) are integrated using an adaptive Gauss-Legendre quadrature. When the field point p, denoted by m is one of the nodes of an element, the degenerate triangular transformation (Lachat and Watson, 1976) is used to regularize the integrals. The terms, H_{mm}, are obtained from physical arguments by imposing a constant potential in Eq. (12) to obtain

$$H_{mm} = -\sum_{m \neq n} H_{mn} . \tag{17}$$

The Eq. (12), can be rearranged after application of the boundary conditions to obtain the linear system:

$$[A]\{X\} = \{b\} , \tag{18}$$

where $\{X\}$ is the vector of unknowns and $\{b\}$ the vector of right hand sides of the linear system. The next section explains the block cyclic decomposition.

THE BLOCK CYCLIC DECOMPOSITION

The block cyclic decomposition assumes a logical grid of processors $P \times Q$ so that the total number of processors is $N_p = PQ$. Each processor is identified by the grid coordinates (ip, jp). The elements of the matrix, A, and therefore all the nodes, are distributed cyclically on the grid, based on a block size, n_b. Thus, if $n_b = 3$, say, the first processor is assigned nodes 1, 2, 3; the second processor is assigned nodes 4, 5, 6, and so on, until all the nodes are distributed. Thus, an entry (m,n) of the global matrix has a local index (id,jd) in the processor (ip,jp).

The mapping of a given number of processors into a logical grid is achieved by calls to sub-programs in the BLACS (Basic Linear Algebra Communication Subprograms). The BLACS library provides utilities for communicating and manipulating common linear algebra objects like vectors and matrices (Dongara and van de Geijn, 1991). Once the mapping has been achieved, the geometry, element-node connectivity and boundary condition data are communicated to all the processors.

The block cyclic decomposition is achieved by having each processor evaluate the element integrals $\hat{H}_{mj}^{(e)}$ and $\hat{F}_{mj}^{(e)}$, which contribute to its portion of the matrix A and the right hand side vector $\mathbf{b}$. It follows that in calculating the terms H_{mn} and F_{mn}, the integrals $\hat{H}_{mj}^{(e)}$ and $\hat{F}_{mj}^{(e)}$, have to be calculated for all elements which share the node n. It is immediately seen that depending on the block size, n_b, and the processor grid, some of the element integrations will be repeated among processors in a processor grid column. For example, consider for the sake of simplicity a 9-node quadrilateral element, with nodes $\{1,2,3,8,9,10,5,6,7\}$ and a 6-node triangular element whose nodes are $\{5,7,11,6,12,4\}$. Let the nodes be distributed block cyclically with block size 2 over 4 processor columns, i.e., processor column one has nodes 1, 2, 9, 10; processor column two has nodes 3, 4, 11, 12; processor column three has nodes 5, 6; and processor column 4 has nodes 7, 8. In calculating the nodal contributions, therefore, the elemental integrations have to be performed four times over the quadrilateral element and three times over the triangular element, if communication between the processors is to be avoided.

The contribution to the diagonal terms in Eq. (17) are stored locally in each processor and then added across columns in parallel after assembly. The result is broadcast to all the processor columns of the same processor row, which then update the matrix A, if necessary. The right-hand side vector is also treated in a similar manner. In this case, however, the results are stored in first processor column of processor row. Parallel additions of vectors are facilitated by utility subprograms in BLACS and PBLAS (Choi *et al.*, 1994).

RESULTS

The parallel algorithm described was used to obtain the diffraction and radiation potentials to solve the problem of a sphere floating on the free-surface, with a draft equal to its radius, i.e., a hemisphere. The surface, $S_B US_F$ was subdivided by a grid of radial and circumferential curves. The surface, S_∞, which is cylindrical, is gridded by vertical and circumferential lines. Since, the grid is symmetric, the composite potential approach is used to reduce the size of the problem. Essentially, the composite potential approach decomposes the potential into even and odd parts taking advantage of the symmetry properties of the Green function. Detailed formulation of the composite potential approach can be found in Wu (1984) and Wu *et al.* (1993).

Four meshes with increasing number of nodes, 573, 753, 1217 and 2543, respectively, were used in the test cases. The wavelength used in each of the test cases was given by the non-dimensional wave number $ka = 1$, where a is the radius of the sphere. The nodes are numbered consecutively along the circumferential curves. The FORTRAN programs were run in double precision on the IBM-SP2 distributed memory parallel computer at the Maui High Performance Computing Center, Maui, Hawaii. The hydrodynamic force coefficients obtained by integrating the pressure on the body surface were verified with the available semi-analytical results in Hulme (1982) and numerical results in Yeung (1973). As expected, the accuracy of the results increased with a finer mesh and approached the semi-analytical results tabulated in Hulme (1982).

Time comparisons are made with a sequential program. Table 1 shows the CPU time in seconds for the assembling and solving phases for the serial program. As seen in Table 1, most of the CPU time is spent on the assembly of the matrices.

Table 1. CPU times for the serial program

N	Assembly	Solving	Total
573	862.3	4.3	866.6
753	1580	9.1	1589.1
1217	3844.2	34.6	3878.8
2543	12545.6	303.1	12848.7

The Tables 2-4 and 5-7 show the times when 8 processors are mapped into a 2×4 processor grid and a 4×2 processor grid, respectively, with $n_b = 1,4,8$. The efficacy of parallel programs are, however, gauged based on the speedup and efficiency. Speedup is defined as the ratio of the time taken by the sequential program to the time taken by the parallel program. Efficiency is defined as the ratio of the speedup to the number of processors. The speedup and efficiencies for the two processor grid configurations are compared in Figures 1-3. The results show that a 4×2 processor grid is substantially more efficient than the 2×4 processor grid for the block sizes considered. The results also show that block size of 4 achieves a better speedup, if not the same, for both configurations. Recall, that the number of redundant integrations performed in a block cyclic distribution of nodes depends on the processor grid, the block size and the numbering of nodes.

Table 2: CPU time for parallel program, 2×4 processor grid, $n_b = 1$.

N	Assembly	Solving	Total
573	440.1	2.6	442.7
753	807.7	6	813.7
1217	1964.2	15.5	1979.7
2573	6414.4	106.4	6520.8

Table 3: CPU time for parallel program, 2×4 processor grid, $n_b = 4$.

N	Assembly	Solving	Total
573	389.2	1.5	390.7
753	715.6	2.4	718.2
1217	1459.5	5.7	1465.2
2573	3246.3	68.3	3314.6

Table 4: CPU time for parallel program, 2×4 processor grid, $n_b = 8$.

N	Assembly	Solving	Total
573	396.3	1.3	397.6
753	719.6	2.2	721.8
1217	1608	7.1	1615.1
2573	4495.7	54.4	4550.1

Table 5: CPU time for parallel program, 4×2 processor grid, $n_b = 1$.

N	Assembly	Solving	Total
573	221.7	2.8	224.5
753	405.6	5.3	410.9
1217	982.2	16.5	998.7
2573	3209.6	116.2	3325.8

Table 6: CPU time for parallel program, 4×2 processor grid, $n_b = 4$.

N	Assembly	Solving	Total
573	226.9	2.7	224.5
753	407.7	3.2	410.9
1217	991.1	17.2	1008.3
2573	3199.9	71.2	3271.1

Table 7: CPU time for parallel program, 4×2 processor grid, $n_b = 8$.

N	Assembly	Solving	Total
573	227.5	1.7	229.2
753	410.7	2.7	413.4
1217	990.2	7.3	997.5
2573	2457.2	55.6	2512.8

The Tables 8-10 and Figures 4-6 show the results for 16, 25 and 64 processors, respectively. The processors are in a 4×4, 5×5 and 8×8 processor grid with block size, $n_b = 8$. While the expected increase in speedup is seen, the efficiency is actually lower. Higher speedups and efficiencies are, however, observed for the timing for solving phase of the calculations. The timings also reveal that using the same block sizes for the assembly and solving phases is not optimal. But using different block sizes for the two phases entails communication among the processors that will result in time overheads.

Table 8: CPU time for parallel program, 4×4 processor grid, $n_b = 8$.

N	Assembly	Solving	Total
573	204.3	1.3	205.6
753	364.5	1.4	365.9
1217	818.1	4.6	822.7
2573	2235.2	22.3	2257.5

Table 9: CPU time for parallel program, 5×5 processor grid, $n_b = 8$.

N	Assembly	Solving	Total
573	140.5	1.1	141.6
753	249.7	1.5	251.2
1217	623.9	3.8	627.7
2573	1704.6	17.7	1722.3

Table 10: CPU time for parallel program, 8×8 processor grid, $n_b = 8$.

N	Assembly	Solving	Total
573	57.9	2.3	60.2
753	108.7	2.7	111.4
1217	269.5	4.3	273.8
2573	808.5	15.3	823.8

CONCLUSIONS

The parallel computations presented in this paper are based on using the block-cyclic decomposition used in parallel linear system matrix solvers. The numbering of nodes in a 3-dimensional grid can, in general, be arbitrary. Since, the efficiency of the parallel algorithm depends on minimizing the number of redundant elemental integrations, the numbering of the nodes plays a crucial role in the efficiency of the algorithm. A more ordered numbering of nodes can, for example, be achieved using a renumbering algorithm as in Cuthill and McKee (1969) or reverse Cuthill-McKee algorithms (Gibbs *et al.*, 1976), used in finite element analysis. At this stage, it is not very clear what the optimum ordering of nodes, the block size and the processor grid are, to approach a speedup equal to the number of processors used. A variation on the parallel algorithm presented here is to store the computed integrals or the elements of the matrix on disk. In this manner, different block sizes may be used in the assembly and solving phases of the algorithm.

ACKNOWLEDGMENTS

This work is supported by the U.S. National Science Foundation, Grant No. BCS-8958346 received by the second author (RCE) through the Presidential Young Investigator award. Through the use of the Maui High Performance Computing Center, this research was sponsored in part by the Phillips Laboratory, Air Force Materiel Command, USAF, under cooperative agreement F29601-92-2-0001. The views and conclusions contained in this document are those of the authors and should not be interpreted as necessarily representing the official policies or endorsements, either expressed or implied, of Phillips Laboratory, the U.S. Government, The University of New Mexico, or the Maui High Performance Computing Center. SOEST Contribution No. 4375.

REFERENCES

Brebbia C. A., Telles, J. C. F., Wrobel, L. C. (1984). Boundary Element Techniques: Theory and Applications in Engineering. Springer-Verlag, New York, xiv+466 p.

Choi, J., Dongara, J. J., Pozo, R. and Walker, D. W. (1992). "ScaLAPACK: A Scalable Linear Algebra Library for Distributed Memory Concurrent Computers," Proc. Fourth Symposium on the Frontiers of Massively Parallel Computation, McLean, Virginia, USA, pp. 120-127.

Choi, J., Dongara, J. J., and Walker, D. W. (1994). "PB-BLAS: A Set of Parallel Block Basic Linear Algebra Subroutines," Proceedings of the Scalable High Performance Computing Conference," Knoxville, Tennessee, USA, pp. 534-541.

Cuthill, E. and McKee, J. (1969). " Reducing the Bandwidth of Sparse Symmetric Matrices," Proc. 24th National Conference Association for Computing Machinery, pp. 157-172.

Davies, A. J., Mushtaq, J. (1996). "The domain decomposition boundary element method on a network of transputers," *Boundary Element Technology XI*, 1996, pp. 397-406.

Dongara. J. J and Whaley, R. C. (1995). "A Users Guide to the BLACSv1.0," Technical Report UT CS-95-281, LAPACK Working Notes #94, University of Tennessee, Knoxville, USA, iv+55 pp.

Flanery, R. E., Gray, L. J., Drake, J. B. (1996). "Boundary Elements on Distributed Memory Architectures," *Boundary Element Communications*, Vol. 7, No. 2, pp. 57-60.

Gibbs, N. E., Poole, W. G., Stockmeyer, P. K. (1976). "An algorithm for reducing the bandwidth and profile of a sparse matrix, " SIAM Journal for Numerical Analysis, Vol. 13, No. 2, pp. 236-250.

Hulme, A. (1982). "The wave forces acting on a floating hemisphere undergoing forced periodic oscillations," *Journal of Fluid Mechanics*, Vol. 121, pp. 443-463.

Kamiya, N. Iwase, H. , Kita, E. (1996). "Boundary element cluster computing on distributed-parallel workstations," *Boundary Element Technology XI*, 1996, pp. 407-414.

Lachat, C. J. and Watson, J. O. (1976). "Effective numerical treatment of boundary integral equations: a formulation for three-dimensional elastostatics," *Intl. Journal of Numerical Methods in Engineering*, Vol. 10, pp. 991-1005.

Wu, Y. S. (1984). Hydroelasticity of floating bodies. Ph. D. Dissertation, Brunel University, U.K.

Wu. Y. S., Wang, D., Riggs, H. R., and Ertekin, R. C. (1993). "Composite singularity distribution method with application to hydroelasticity," *Marine Structures*, Vol. 6, pp. 143-163.

Yeung, R. W. (1973). "A singularity-distribution method for free-surface flow problems with an oscillating body," Report No. NA 73-6, College of Engineering, University of California, Berkeley, vi+124 pp.

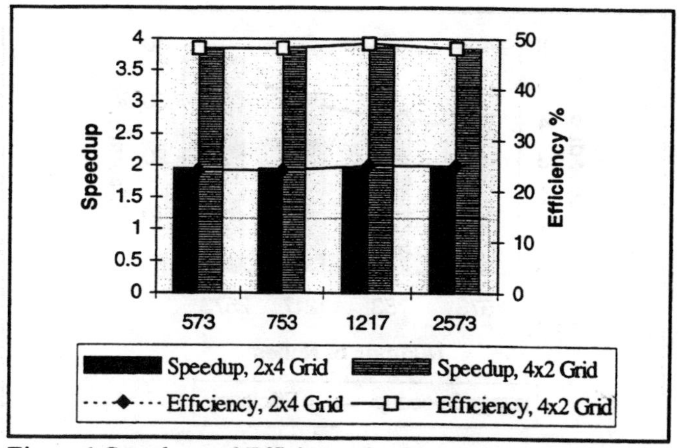

Figure 1 Speedup and Efficiency for 8 processors in 2×4 and 4×2 processor grids, block size 1.

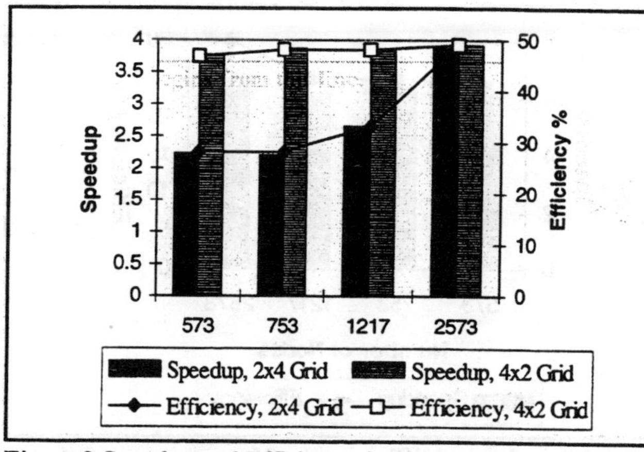

Figure 2 Speedup and Efficiency for 8 processors in 2×4 and 4×2 processor grids block size 4.

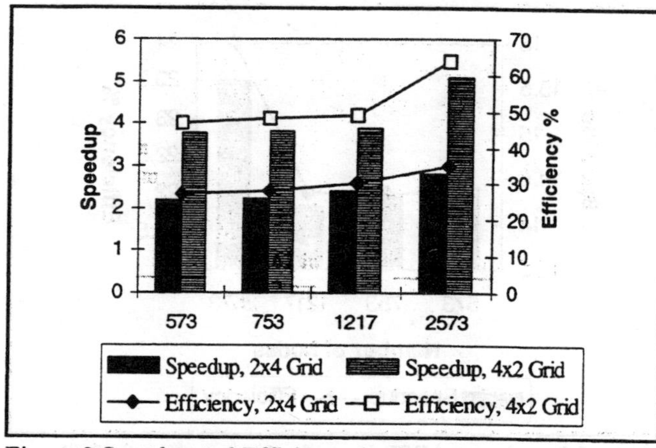

Figure 3 Speedup and Efficiency for 8 processors in 2×4 and 4×2 processor grids, block size 8.

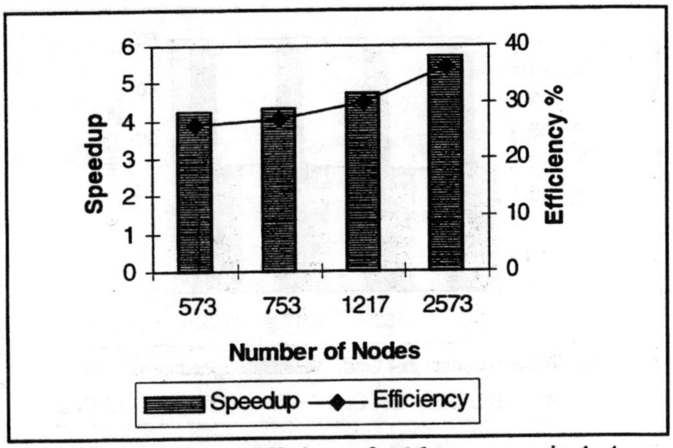

Figure 4 Speedup and Efficiency for 16 processors in 4×4 processor grid, block size 8.

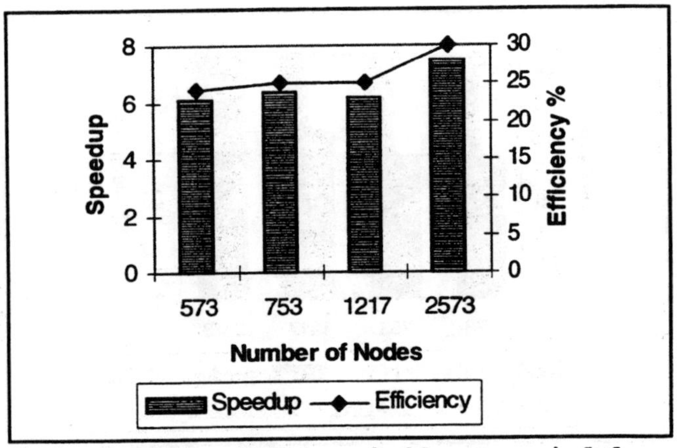

Figure 5 Speedup and Efficiency for 25 processors in 5×5 processor grid, block size 8.

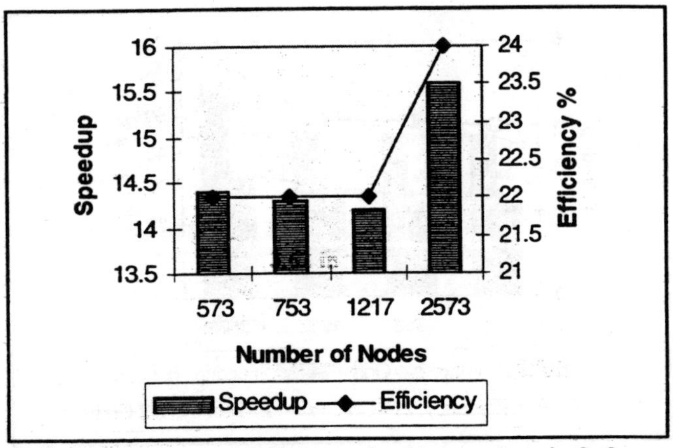

Figure 6 Speedup and Efficiency for 64 processors in 8×8 processor grid, block size 8.

Proceedings of the Seventh (1997) International Offshore and Polar Engineering Conference
Honolulu, USA, May 25-30, 1997

ISBN 1-880653-28-1 (Set); ISBN 1-880653-31-1 (Vol. III)

Drag Coefficients for Mooring Line Hydrodynamic Damping

G.J. Lyons, D.T. Brown and H.M. Lin
University College London
London, United Kingdom

ABSTRACT

For catenary moored systems low frequency excitation caused by random waves and wind loading results in resonant motion responses in the horizontal plane leading to high mooring line forces. Many of the loading mechanisms associated with this are well understood. However until relatively recently the fluid induced forces acting on the moorings were assumed to have little influence on the vessel dynamic response. It is now known that the catenary mooring system may under certain circumstances provide up to 80% of the total damping which may significantly reduce the peak line tensions.

Most of the previous experimental work performed to measure line damping has been associated with the relatively small geometric scale at which the model tests were performed (typically $^{1}/_{100}{}^{th}$). This and the fact that it is difficult to apply the results to non-specific field scenarios has led to severe difficulties in system design.

This paper presents results at significantly larger scale model testing work at University College London as part of a JIP study. This provides the added mass and drag contributions of sections of mooring line. The tests were carried out at large scale on sections of line to ensure that the Reynolds number flow regime was as close as possible to that encountered at full scale. A chain line segment was tested at various amplitudes and frequencies to represent a section of single catenary mooring line.

Because of the complex nature of the flow around oscillating lines it is essential to consider interaction between the low mooring frequency and wave frequency flows as it is known that simply superposing the two effects gives erroneous results. To this end a computer controlled biplanar carriage was designed and installed in the wave tank at UCL to simulate these motions with the line specimens. Sectional added mass and damping forces were derived from measurements of the loads on a range of segments. These can be used within a 'finite-element' formulation of mooring line behaviour to estimate the damping contributions for all lines allowing results to be applied to different field scenarios and weather conditions.

This paper describes the tests and provides some of the results of the drag coefficient measurements.

KEYWORDS: Mooring line, chain, hydrodynamic, damping, drag

INTRODUCTION

Throughout the offshore hydrocarbon producing areas of the world and in particular in the UK and Norwegian sectors of the North Sea, semisubmersibles and monohulls are increasingly being considered for floating production and storage.

Figure 1 indicates the primary external forcing mechanisms acting on a monohull vessel. The random loads result in the vessel undergoing small amplitude motions at wave frequencies and relatively large amplitude motions (particularly in surge) at low frequencies - often termed drift frequencies. The frequency of the slowly oscillating horizontal motions usually corresponds to the natural surge frequency of the mooring system. These phenomena are discussed further in Patel and Brown [1]. The damping associated with the resonant low frequency motions is relatively small and consequently high tensile loads are induced in the mooring lines. Accurate assessment of the individual damping levels is therefore critical to establishing peak mooring line loads.

The mooring lines themselves are made up of a number of components as shown in Figure 2. Lengths of chain are used near the vessel and sea-bed end points. These are relatively easy to terminate and have good wear properties. The suspended section of line is often of wire rope construction with the advantage of being typically only 20% of the cost and weight of chain for a similar breaking strength requirement. The mix of wire and chain depends on a number of parameters including strength requirement, water depth, and permitted riser offset.

There are usually considered to be four primary components contributing to damping of a moored vessel and its moorings. These are:

- current and viscous flow damping
- wind damping
- wave drift damping
- mooring line damping

There are a number of contributions to the overall damping from the mooring system. These are:

(i) Hydrodynamic drag damping - Depending on the water depth, line pretension and weight, azimuth angle, a relatively small surge motion can result in transverse motion over the centre section of the line that is many times larger than the surge amplitude itself [2] as indicated in Figure 1. The corresponding transverse drag force represents energy dissipation per oscillation cycle and thus can be used to quantify the line damping.

(ii) Vortex induced vibration - Vortex formation behind bluff bodies placed in a flow gives rise to unsteady forces at a frequency close to the Strouhal frequency. The forces cause line resonant response in a transverse direction to the flow and the vortex formation can become synchronised along the length resulting in the shedding frequency 'locking-in' to the line natural frequency [3]. This can result in a significant increase of in-line drag. It is generally considered that the effect is important for wire lines whereas for chains it is assumed negligible.

(iii) Line internal damping - Material damping caused by frictional forces between individual wires or chain links also contributes to the total damping. Work on this was undertaken by the authors at University College London on full scale mooring line concurrent with the work reported herein.

(iv) Damping caused by sea-bed interaction - Soil friction leads to reduced tension fluctuations in the grounded portion of line effectively increasing the line stiffness. Recent work [4] has shown that out of plane sea-bed friction and suction effects are negligible in deep water mooring situations whereas in-plane effects can significantly influence the peak tension values.

Although work has been carried out over a number of years investigating the dynamic response of floating systems to random forcing, it is only relatively recently that mooring line damping has been considered. Brown *et al* [5] provide a detailed review of previous research in this area.

The most significant drawbacks of previous work in the area of line hydrodynamic damping are associated with the relatively small scale at which model tests were performed and the fact that it is difficult to apply the results to field scenarios, i.e. water depths and chain sizes, other than those for which testing took place.

Consequently, the work carried out by the authors has achieved the following:

- Model testing at as large a geometric scale as feasible ensuring that the Reynolds number flow regime is as close as possible to that encountered at full scale.

- A systematic series of measurements on chain section so that the results can be applied to field scenarios with wide ranges of water depths, line sizes and spreads.

HYDRODYNAMIC DAMPING TESTS AT UCL

This work took place in the Department of Mechanical Engineering at University College London within the 'Uncertainties in Loads on Offshore Structures' (ULOS) Research Programme (JIP).

HYDRODYNAMIC TEST FACILITY

The existing wave tank was modified with the addition of a bi-planar horizontal motions carriage mounted above and within the tank, and further described in [6].

Design of oscillator systems

Two oscillators were needed to provide synchronised in-line and transverse movements between the wave and slow motion directions. The system is capable of creating complex motions, although the most complex used in this study have been combined low and high frequency motions in-line and transverse to the in-line (wave drift) motion.. The mooring segments are mounted on the oscillators as shown in Figure 4. Movements of the oscillators are controlled by computer which allows efficient set-up of motion amplitudes and periods as well as precise synchronisation of the bi-planar movements during the tests. This arrangement permits oscillations at any desired angle to the X direction, as well as in-line or transverse to it.

Mooring specimens

The objectives of the model tests were to estimate chain damping at as close as possible to full scale conditions; between full and $^1/_3$ scale was aimed for. Indeed it was considered that results for one chain size at a particular geometrical scale ratio may be applicable to other chain sizes at alternative scale ratios - subject to both flow regimes (Reynolds number and Keulegan Carpenter number) being similar.

Commonly used chain for mooring systems are Grade 3, Grade 4, ORQ or ORQ+20% (API RP 2F1, 1993). Floating production system usage requires Grade 4 or ORQ+20% Most chain manufacturers fabricate standard chain link, Figure 5. As this link makes up to 90% of the chain length, its geometry was used to make the chain samples.

While seeking the largest possible model scale, a suitable number of links was also required for each test as the tank available water depth was 1.5 m deep. By consideration of the tank dimensions, length of specimens, and available scale range nominal chain bar diameters of 25mm and 65mm were selected for the tests. To date only the 25mm bar chain has been tested.

Plastic materials with specific gravity of 1.1 - 1.3 were considered to make up the chain samples in order to achieve slightly negative buoyancy. This provided the correct form, but without a mass (inertia) penalty, which would otherwise require significantly more powerful motors. Plates 1 and 2 show a 25mm bar chain sample manufactured from Nylon 66. 9 links were used in the tests reported herein. Figure 6 indicates the model fabrication arrangement chosen. The CNC machine in the Department of Mechanical Engineering at UCL was used to manufacture the links of the chain.

In order to ensure that the chain did not significantly deform laterally or rotationally owing to fluid loading, a square section steel rod was passed through the chain to form a stiff stick of chain elements. This was appropriate since it was not realistic to apply any significant pre-tension to the chain model. In a mooring line in service this tension effectively suppresses any significant lateral motion over the specimen length in question. The steel rod was not considered to be of a dimension to significantly affect the hydrodynamic response.

Sensors and data acquisition system

Load cells fixed at both ends of the mooring segment measured the triaxial forces. These comprised Bofors proprietary load cells for the vertical component, on which strain gauged bars were vertically mounted to provide the transverse and in-line components. The mooring specimens were connected to these bars by universal joints so that no bending moment was imparted, only shear loads. The velocities of the carriage and oscillator for analysis purposes were tapped from their velocity controllers in analogue form. Accelerations were subsequently derived from these. Displacements were obtained from multi-turn potentiometers connected to the driving belt wheels. All acquired data were in analogue form. Data were collected using a proprietary package provided with the 12 bit ADC card used in the personal computer.

Large scale mooring specimen oscillating tests

A total of 320 hydrodynamic tests have been performed to date on the chain specimen. The results of other parts of the testing programme are reported in [7].

Each test was repeated in general five times in order to establish the statistical variability in results. In all cases the variability in drag coefficient with identical input conditions was small. For example Fig. 7 depicts derived time series data of chain position, velocity, and measured forces for combined LF and HF oscillations in the X direction. Each of the five graphs in Fig. 7 shows three curves indicating the mean value at each time step together with mean plus/minus one standard deviation. It is noted that these curves are virtually indistinguishable with the exception of the Y force values which are an order of magnitude smaller than the X or Z forces. Fig. 8 expands the X force data shown in Fig. 7 to amplify the point.

CALCULATION OF DRAG COEFFICIENTS

The calculation of drag coefficient for harmonically oscillating flow relative to a body is based on the appropriate term of the Morison equation. When there is bi-harmonic (i.e. two frequencies of oscillation) flow past a body the situation is not so simple. It is possible to define two drag (and inertia) coefficients, one appropriate to each of the two frequencies of oscillation. Furthermore Reynolds numbers and KC values may also be established for bi-harmonic flow based on the appropriate oscillation frequency and amplitude.

The drag coefficients presented herein are based on the chain bar diameter of 25mm, and the LF (drift frequency) definition. Velocities used to calculate the drag coefficient are based on the combined wave and drift oscillations. The appropriate drag and inertia values were obtained by multiplying the force record by $\sin \omega t$ or $\cos \omega t$, where ω is the frequency of drift oscillation, and integrating over an oscillation cycle. The Reynolds and Keulegan Carpenter values presented are howeverassociated with wave frequency oscillations so that the results can be compared on a common data base with data from other studies [7]. Furthermore this allows the influence on the drag coefficient of motions at wave frequencies to be more readily assessed.

TEST RESULTS

From statistical analysis of the various test series drag and inertia coefficients were derived. The mean value drag coefficient results are given in Figs 9, 10, and 11 plotted against HF oscillation frequency for a limited amount number of test conditions, ie

- LF/HF (single harmonics) Fig 9
- LF + HF (in-line) bi-frequency Fig 10
- LF + HF (transverse) bi-frequency Fig 11

Single frequency oscillations

Fig 9 shows the C_d results of single frequency fixed amplitude oscillations from both in-line and transverse motion tests.

They are plotted against angular frequency and grouped by KC number. For the lower KC values of 70 and 106 there is a clearly increasing trend with frequency. This is not so apparent with the other data. A seemingly anomalous result occurs at around 0.78 rad s^{-1} for KC values of 163 and 327. This may be owing to Reynolds number effects.

The drag coefficients consistently increase as the KC number is reduced for all three oscillation directions.

LF oscillation (0°) with superimposed HF oscillation (0°)

Fig 10 shows the results of LF drift motions with superimposed in-line HF wave motions. Two amplitudes of LF motion are considered (1.16m and 2.32m) with two amplitudes of HF motion (0.3m and 0.59m, ie double). By way of reference to the results for the case when no HF motion is applied, the results for LF amplitude at constant LF frequencies are plotted at 0 rad s^{-1} (HF frequency).

It had been anticipated that an increase in drag coefficient would result in each case which would be above the value for the state when no HF in-line motion was applied. For two cases (of lower HF amplitude of 0.3m) this is seen to be so. However, variable increase, and even reduction in drag coefficient occurs for the two cases of the HF higher amplitude of 0.59m.

The drag coefficients for these higher amplitudes are seen to be less than those for the lower HF amplitudes for any given HF (wave) frequency.

LF oscillation (0°) with superimposed HF oscillation (90°)

Fig 11 shows two cases of LF motion (amplitudes of 1.23m and 2,.7m) with superimposed transverse motion of constant amplitude of 0.29m, with varying frequency (HF). As anticipated the drag coefficients obtained are greater than for the cases when no HF motion is applied (data at 0 rad s^{-1}). The drag coefficients in the range covered increase with increasing frequency, but in a rather variable way.

The drag coefficients for equivalent amplitudes and frequencies are seen to be higher for the transverse HF superimposed motions than for the in-line superimposed HF motions.

CONCLUSIONS

- Drag coefficients have been calculated from measurement of fluid forces on vertically orientated sections of chain that have been tested of a range of drift frequencies (termed LF-low frequency) and amplitudes. Superimposed on these motions are wave oscillation frequencies (termed HF-high frequency) and amplitudes at orientations of 0°, 45° and 90° to the LF oscillation direction. LF and HF oscillations in isolation have also been tested. From a statistical assessment of results excellent repeatability has been obtained. Only the 0° and 90° results are presented herein.

- Evaluation of the drag and inertia coefficients obtained from the combined wave and drift (i.e. bi-harmonic) oscillation tests have highlighted the restrictions imposed by a Morison based drag formulation that was originally developed for uni-directional flow of constant amplitude and subsequently to harmonic flow. When bi-harmonic flow conditions exist it is necessary to define drag and inertia coefficients appropriate to each of the wave and drift oscillations. These coefficients for wave frequency motion with superimposed drift and drift frequency motion with superimposed waves must be used with care in design.

- Design procedures for prediction of dynamic line tension and line damping are usually based on either simplified frequency domain models or more sophisticated time domain representations that often utilise finite elements to represent sections of each mooring line.

- Appropriate results from this project to use within frequency domain models are the LF drag coefficient data with superimposed HF motions. Care must be taken to assess the likely motion amplitude of the line (both HF and LF) in order that a suitable C_d value is selected. The designer must furthermore allow for the HF oscillation direction as this influences the choice of drag coefficient.

- Time domain models calculate the instantaneous line shape based on a finite element formulation. Here it is theoretically possible to select from a database of drag coefficients that vary with inclination, and azimuth angle along the line. Furthermore variations in HF oscillation direction and HF and LF oscillation amplitude and frequency are possible. The work achieved in this project has provided a starting point for such a database of these drag coefficients. It expected that this work will continue to provide additional data.

- In the future it would be beneficial to extend the hydrodynamic work to consider wider ranges of orientations, motions, and mooring chain sizes. Wire sections should also be tested. This will allow the database of measurements to be substantially increased. Extensions to the work should also include for the effects of vortex induced vibrations which can act to increase the drag coefficient for wire substantially.

ACKNOWLEDGEMENTS

This work was sponsored by EPSRC through MTD Ltd and jointly funded with: Aker Engineering, Amoco (UK) Exploration Company, BP Exploration Operating Co. Ltd., Brown & Root Ltd., Exxon Production Research Company, Health and Safety Executive, Shell UK Exploration and Production, Den Norske Stats Oljeselskap a.s., Texaco Britain Ltd, EPSRC Grant Reference: GR/J23600

REFERENCES

1. Patel M H, and Brown D T, 1987, 'Analysis and design of catenary mooring systems', Advances in Underwater Technology, Ocean Science and Offshore Engineering, Vol 13: Economics of Floating Production Systems, SUT, London, Graham & Trotman,.

2. Huse E, and Matsumoto K, 1989, 'Mooring line damping owing to first and second order vessel motion', Proc OTC, Paper 6137.

3. Vandiver J K, 1988, 'Predicting the response characteristics of long flexible cylinders in ocean currents', Symposium on Ocean Structures Dynamics, Corvallis, Oregon.

4. Thomas D O, and Hearn G E, 1994, 'Deep water mooring line dynamics with emphasis on sea-bed interaction effects', Proc OTC, Paper 7488, Houston, USA.

5. Brown D T, Lyons G J and Lin H M, 1995, 'Advances in mooring line damping', Underwater Technology, Vol 21, Number 2, pp. 5-11.

6. Lin H M, Brown D T, and Lyons G J, 1996, 'Large scale testing for mooring line hydrodynamic damping', proc OMAE, Vol 1 part B, pp 79-86.

7. Brown D T, Lyons G J, Lin H M, 1997, ' Large scale testing for mooring line hydrodynamic damping contributions at combined wave and drift frequencies', BOSS 97, Delft.

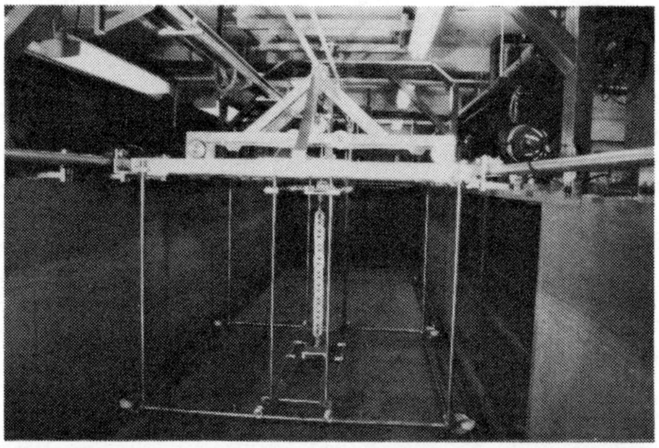

Plate 1 View from inside of tank of carriage with chain specimen

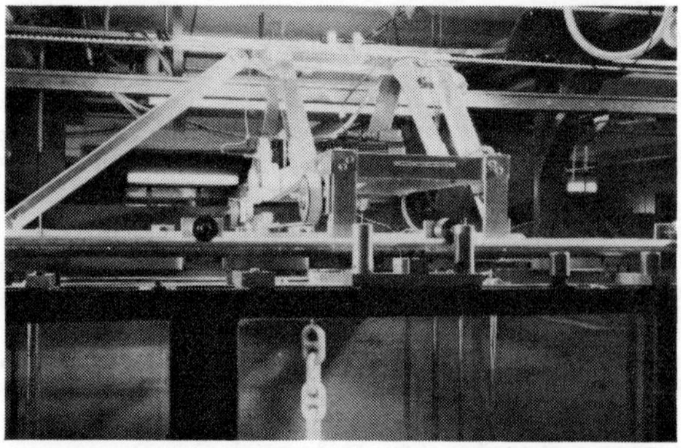

Plate 2 View of top of carriage showing transverse oscillator

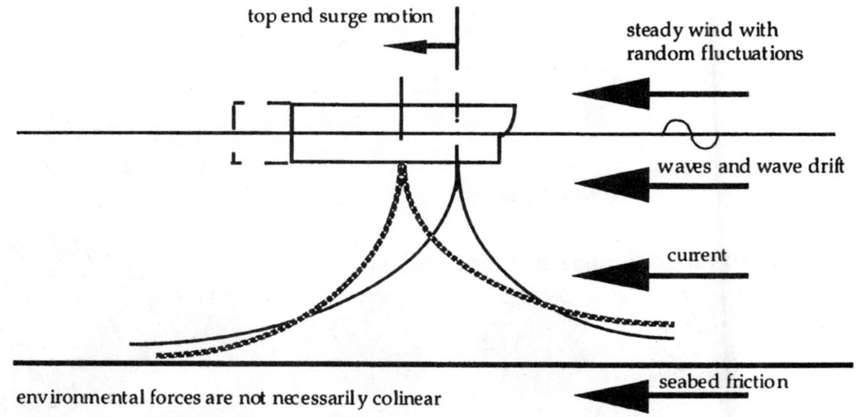

Fig 1. Environmental forces acting on a moored vessel and transverse motion of the catenary mooring lines

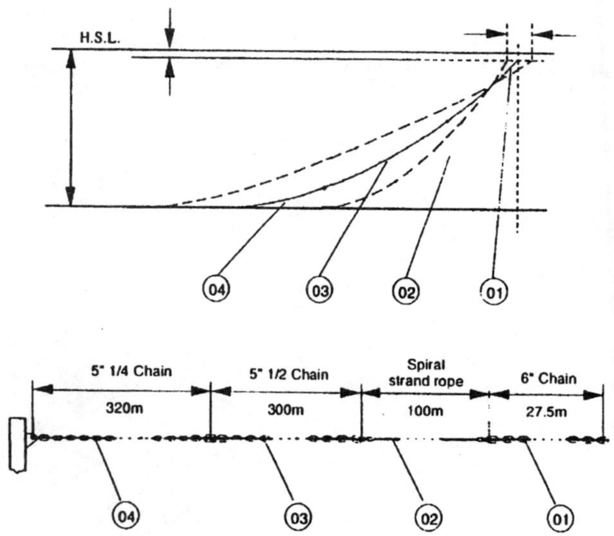

Fig 2. Typical mooring line components (shackles etc. not shown)

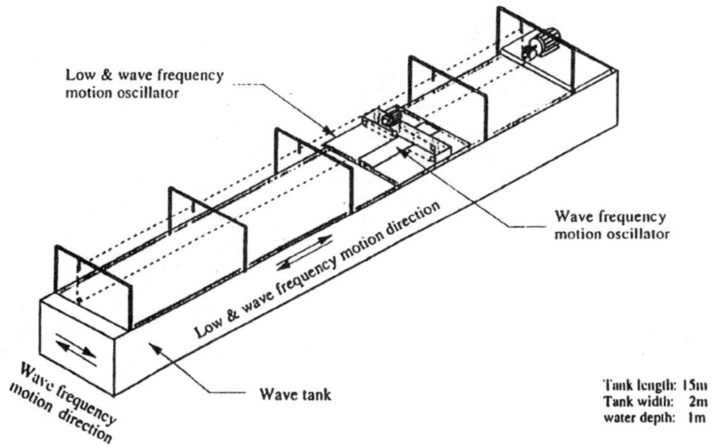

Fig 3. Longitudinal and transverse oscillators for mooring line specimens

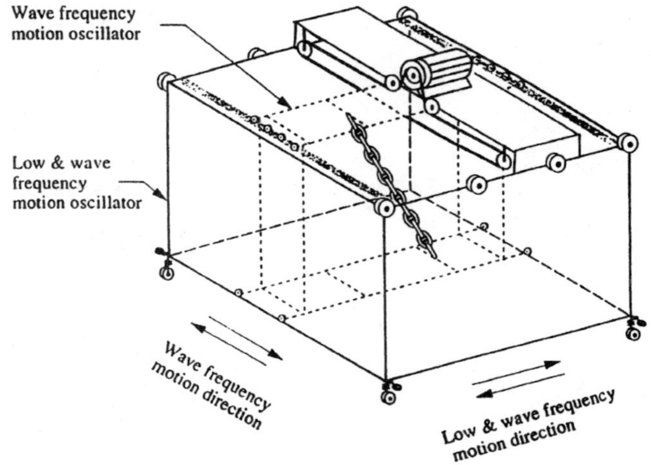

Fig 4. Transverse oscillator and mooring line specimen

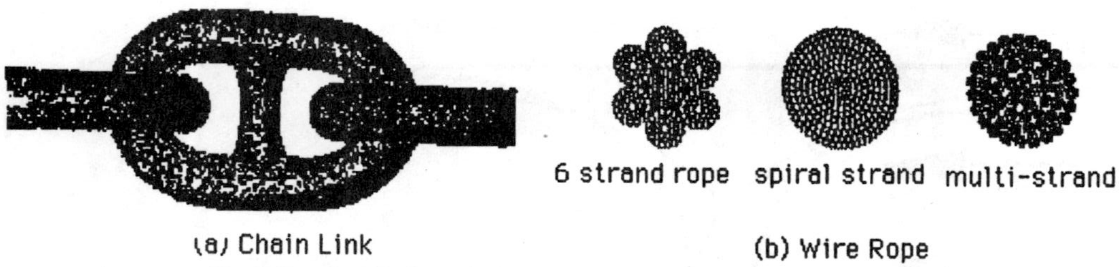

Fig 5. Standard chain and wire

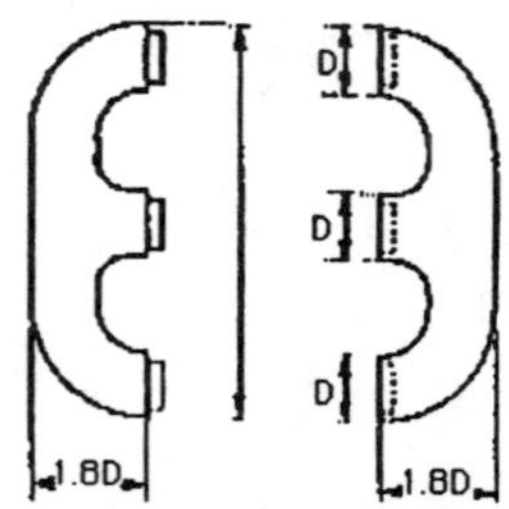

Fig 6. Fabrication scheme for model chain links

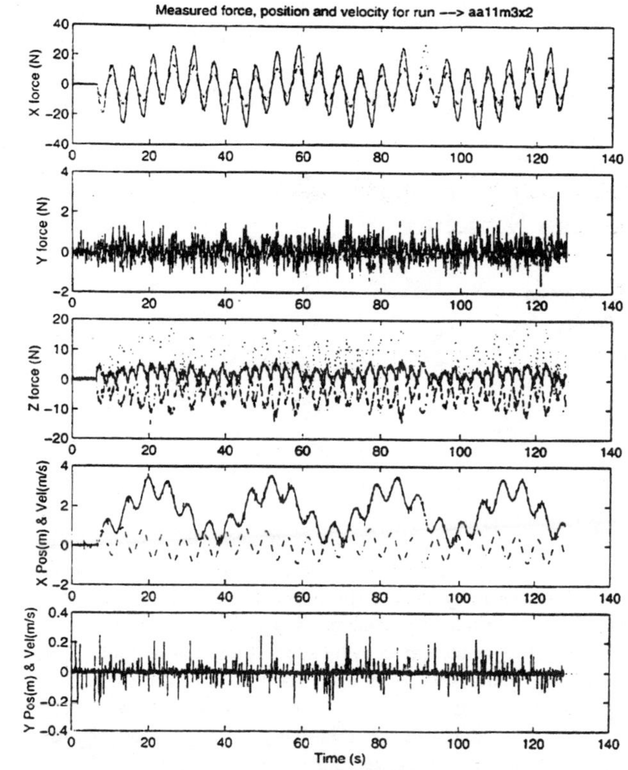

Fig 7.   Example time series

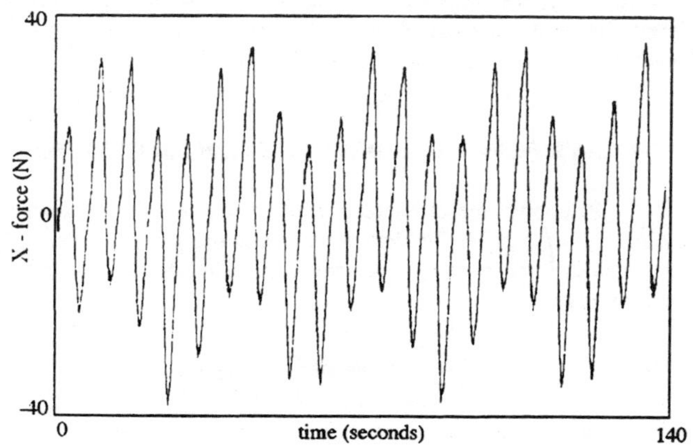

Fig 8.   Time series derived from 4 test runs for typical LF and HF combined motion in drift direction

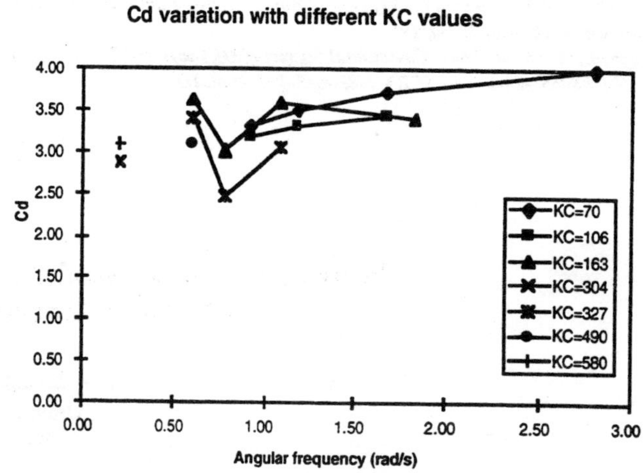

Fig 9.   Drag coefficients for single frequency and amplitude motions

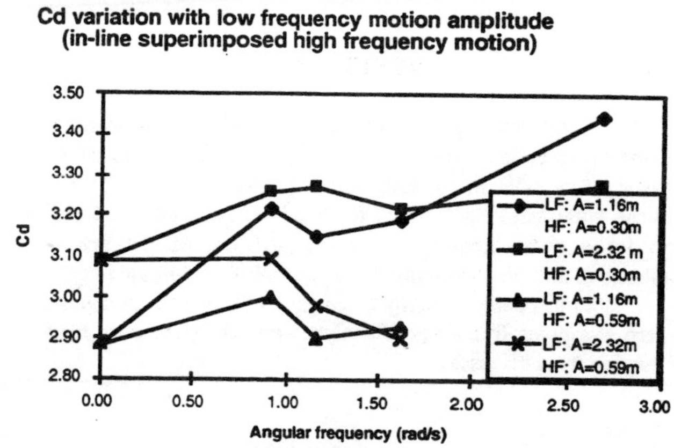

Fig 10.   Drag coefficients for combined in-line HF and LF motions

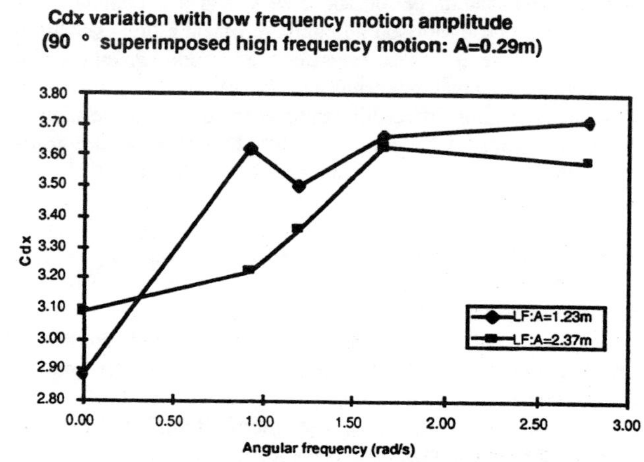

Fig 11.   Drag coefficients for combined transverse HF and in-line LF motions

Proceedings of the Seventh (1997) International Offshore and Polar Engineering Conference
Honolulu, USA, May 25-30, 1997

ISBN 1-880653-28-1 (Set); ISBN 1-880653-31-1 (Vol. III)

Heave Force on a Slender Submerged Body with Application to Form Optimization of Ocean Structure

A.N. Simos and J.A.P. Aranha
University of Sao Paulo
Sao Paulo, Brazil

ABSTRACT

In a recent work (Aranha & Pinto (1994)) an extension of the inertia term in Morison's Formula has been proposed for a submerged cylinder in all range of frequencies. With this formula one can study geometric forms for which the excitation forces are zero for a given frequency and the problem of the geometric synthesis of an ocean structure can eventually be addressed.

In this paper a family of geometry for which the excitation is zero at a given frequency is analyzed and the geometric features of this family is discussed.

1. INTRODUCTION

In a semi-submersible platform the natural frequencies are relatively small (natural period above 20 s) and it is common, in a first approach, to use Morison's formula to obtain a first estimate of the response. In general these estimates are in good agreement with experimental and numerical results, mainly due to the fact that in the low frequency regime diffraction can be ignored and the inertia term in Morison's formula takes properly care of the Froude-Krilov force. This point has been formally discussed by Chung (1975), for example.

The same approach cannot be used, however, to analyze the behavior of a tension leg platform (TLP), where the natural frequencies are high (natural periods below 4 s) and the diffraction effect is of importance at resonance. If one intends to proceed with an analytical tool that retains the simplicity and quality of Morison's equation, full diffraction effect must be incorporated in the expression of the exciting force.

The generalization of the inertia term of Morison's formula to the whole range of frequencies was derived in Aranha & Pinto (1994), the derivation being based in an old study done by Lamb in 1913 about the wave resistance of a submerged circular cylinder.

Although it may be certainly questionable to insist in simplified expressions in a time where numerical results and computer facilities are so well established, the fact is that exactly due to its simplicity these expressions can be used not solely as an *analysis* tool, where a geometry is given and the response is then determined, but they can also be used as a *synthesis* tool, where one looks to a geometric form that fits some desirable aspect in the response.

The objective of this paper is to exploit the derived expression just as a *synthesis* tool; being more specific, the intention here is to draw a geometry for which the exciting (heave) force is zero for a given frequency, that may be chosen equal to the resonance frequency of a TLP, for example, and to look how the geometry changes with the change of the tuning frequency. Although this exercise can be still far from a direct use by the oil industry, it can open some possibilities that may become of interest in certain context, mainly if one intends to use pontoons that have larger dimension and are closer to the free surface than those used nowadays.

2. HEAVE FORCE FOR DEEP SUBMERGED CYLINDER

Suppose the cross section of a relatively deep horizontal cylinder, with d being the distance between the section center O = (0,0) and the free surface; a harmonic wave induces at O a vertical velocity given by

$$w_0(t) = -i\omega A \cdot e^{-Kd} \cdot e^{-i\omega t}, \quad (1)$$

where ω is the wave frequency, $K = \omega^2/g$ the wavenumber and A is the wave amplitude; the inertia part of Morison's formula for the heave force is given by

$$f(t) = \rho S(C_m + 1) \cdot \frac{dw_0}{dt}, \quad (2)$$

with S being the cross section area and C_m the added mass coefficient. In this expression diffraction effect is almost completely ignored and the only geometric information needed, besides the area, is the added mass coefficient C_m.

In high frequency one should expect that the exciting force is influenced by the geometric details of the cross section and the intention, in the following, is to translate the geometric information into a suitable analytical form.

With this purpose in mind, and considering that the dimensions have been normalized by the length $a = S^{1/2}$, let W(z) be the complex potential related with a *vertical* motion of the cross section with *unit velocity* in an *unbounded* fluid, namely: in a fluid region without free surface. The function W(z) is analytic in the fluid region and can be expanded in the Laurent's series

$$W(z) = -\sum_{n=1}^{\infty}(-i)^n \frac{D_n}{z^n}, \quad (3)$$

with the coefficients $\{D_n;\ n = 1,2,...\}$ being *uniquely* associated with the cross section geometric form; these coefficients, in particular, are *real* if the section is symmetric with respect to the vertical axis that passes through the origin O, as it is assumed here.

The coefficients $\{D_n;\ n = 1,2,...\}$ contains then all geometric information needed and it seems plausible that the diffraction effect can be expressed in terms of them. In fact, consider the *function of form*

$$F(Ka) = \sum_{n=0}^{\infty}(-1)^n \frac{(Ka)^n}{n!}\frac{D_{n+1}}{D_1}, \quad (4)$$

uniquely defined for each geometry; as shown in Aranha & Pinto (1994), the heave force when the diffraction effect is considered can be written as

$$f(t) = \rho S(C_M + 1)\cdot F(Ka)\cdot \frac{dw_0}{dt}, \quad (5)$$

where (5) is essentially the inertia part of Morison's formula (2) corrected by the diffraction effect through the *function of form* F(Ka). From this point of view, this function translates in a suitable way all geometric information needed for the diffraction solution.

Notice that $F(Ka) \to 1$ when $Ka \to 0$, and so (5) recovers (2) in the low frequency limit, namely, when the diffraction effect is negligible. Also, $F(Ka) \equiv 1$ for a circle, and so (2) is correct for this geometry in the whole range of frequencies.

The purpose now is to explore the simplicity of (4) and (5) to construct a family of cross section that fit some desirable properties.

3. FAMILY OF TUNED CROSS SECTIONS

One considers here the one parameter family of *functions of form* defined by the expression

$$F(Ka) = 1 - \sin(\alpha Ka); \quad a = \sqrt{S}, \quad (6)$$

where the parameter α can be adjusted in such way that the corresponding *function of form* can be tuned to give zero exciting force at any desirable frequency. Expanding (6) in power series and using (4) the coefficients $\{D_n;\ n = 1,2,...\}$ can be determined and the series (3) can be summed; it turns out that:

$$W(z) = -\frac{i}{z} + \frac{1}{2}\left[\frac{1}{z-\alpha} - \frac{1}{z+\alpha}\right]; \quad (7)$$

Figure (1) displays how the geometry changes with the tuning parameter α, where for $\alpha = 0$ the geometry is a circle; with increasing α the circle is gradually distorted, maintaining the same area but changing the added mass coefficient; as a matter of fact, the added mass coefficient decreased steadily with α in the examples analyzed.

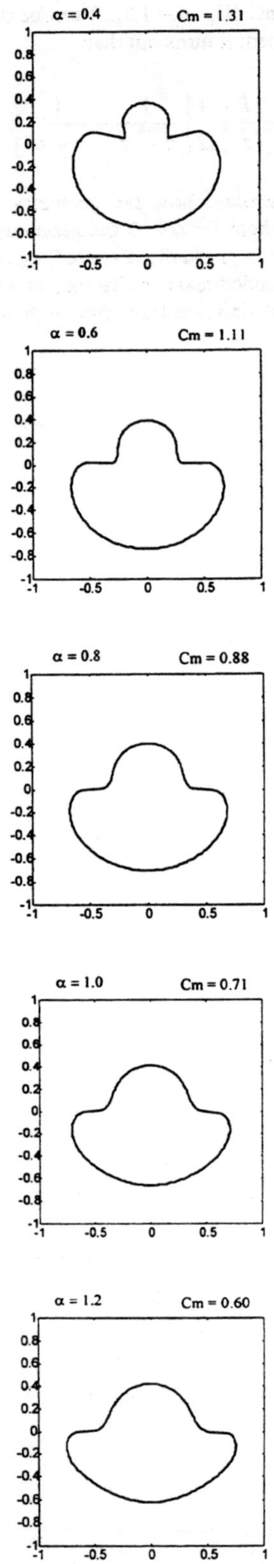

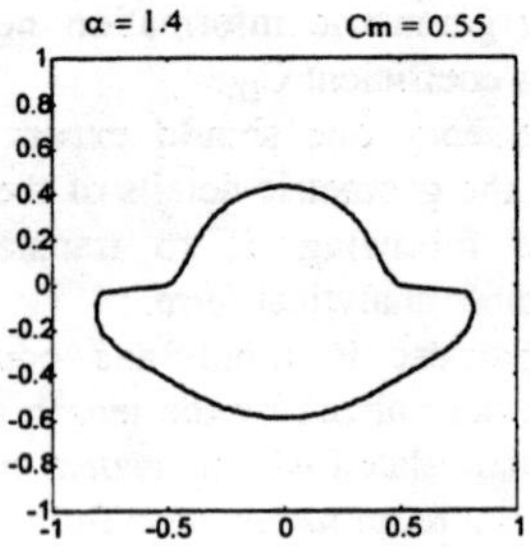

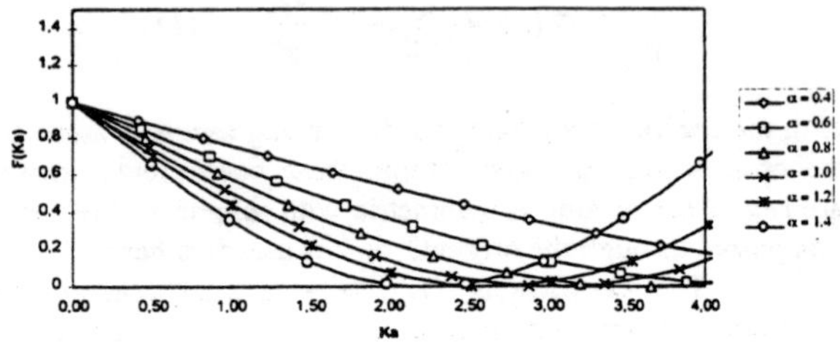

Fig. (1): *Normalized geometric sections ($a = s^{1/2}$; s: cross-section area) for the function of form:* $F(Ka) = 1 - sin(\alpha Ka)$ *(all geometries have the same displacement)*

For α = 1.4 (C_m = 0.55) the heave exciting force is essentially zero in the range of frequencies 2.0 ≤ Ka ≤ 2.50, corresponding to the period range 4.0 s ≤ T ≤ 4.5 s when a = 10 m. For this value of (a) one has also that the heave exciting force is 50% of the circle's value when T = 8.2 s, what shows the importance of the diffraction effect even for a relatively long wave.

4. CONCLUSION

In this paper it is presented some possible design applications of a recently developed formula for submerged bodies that extends, to the short wave regime, the inertia term of the classical Morison's formula. The basic idea is to use this expression as a *synthesis tool*, where the cross section pontoons geometry of a floating platform is draw starting with a chosen behavior of the heave response, related with the *function of form* F(Ka).

Although the final result should obviously be checked for a real structure, where the interference effects between pontoons and columns do play a role, and it should also be experimentally validated, it is believed that it opens a possibility that seems worth to be investigated by the oil industry.

REFERENCES

- Aranha,J.A.P. & Pinto, M.O. (1994): "Wave radiation by a deep submerged cylinder with application to ocean structure", Applied Ocean Research 15, 327-335;

- Chung, J.S. (1975): "A note on the two force equations for a floating platform", Journal of Hydronautics, Vol. 9, N. 4, pp.170-172;

- Lamb,H. (1945): *Hydrodynamics*, Dover Public. N.Y.

Proceedings of the Seventh (1997) International Offshore and Polar Engineering Conference
Honolulu, USA, May 25-30, 1997

ISBN 1-880653-28-1 (Set); ISBN 1-880653-31-1 (Vol. III)

Safety of Life at Sea: Lessons Learned from the Analysis of Casualties Involving Ferries

Giulio Russo Krauss
Università di Napoli "Federico II"
Napoli, Italy

Antonio Cardo
Università di Trieste
Trieste, Italy

ABSTRACT

The paper illustrates the data concerning casualties at sea where the ship was completely lost and special attention is paid to the loss of human lives as well as to the causes of these events. The characteristics of ferries are examined together with those elements that, as has also been demonstrated by the experience with serious accidents, can compromise their safety. Suggestions are put forward to prevent loading errors, to enhance escaping and abandoning the ship, to reduce the probability and minimize the danger of flooding, etc.

KEYWORDS: Safety, shipbuilding, ship design, regulations.

INTRODUCTION

A long term research on safety of life and the protection of environment in ship design has been undertaken in Italy with the financial support of Ministry of University, Scientific and Technological Research. This research involves the Universities of Genova (Mathematical modelling of propulsion system and its its behaviour in rough weather), Napoli (Analysis of casualties at sea, Suggestions for increased operational safety and regulations) and Trieste (Mathematical modelling of large amplitude ship motions and flooding process). The research is in progress (Francescutto, 1992, 1993, Mestro et al., 1992a, 1992b, 1993, Russo Krauss & Cardo, 1996). In this paper, some guidelines drawn from the analysis of casualties at sea are reported.

THE INTERNATIONAL DIMENSION OF MARITIME SAFETY

There are many ways in which a State can be directly involved in a casualty which has led to the loss of the ship.
The shipbuilding and the shipping sectors operate in a difficult market, characterized by a fierce international competition. Consequently, it is extremely hard, if not impossible, for a maritime State to unilaterally adopt safety measures concerning the construction and/or operation of the ships flying its flag. It is equally difficult to adopt measures aiming at limiting the navigability of its own waters and/or denying access to its ports to those ships which do not meet certain safety standards.
It is therefore clear that maritime safety, in terms of safety of the ship and/or safety of navigation, has an international dimension which concerns all States.
History tells us that progress in the field of maritime safety is above all the result of fatal casualties and that such a progress has been achieved only after the introduction of rules and regulations issued by supernational bodies and enforced by means of a control system deriving from international agreements.

ON SAFETY

When designing and building a ship the emphasis is usually placed on reaching given values for deadweight capacity, speed and power and on ensuring the rapidity of loading and unloading operations. Furthermore, the ship has to comply with the existing safety legislation. These objectives have to be attained at a cost which is convenient for the shipowner and remunerative for the shipyard. In the light of this general approach it is evident that a greater safety depends on the improvement of the legislation. Classification societies, insurance companies and other bodies generally accept this logic and their behaviour is certainly not devoid of negative consequences.
Designers should constantly consider safety a fundamental prerequisite and they should recommend and/or adopt all those solutions that - at zero or at very low cost - can be useful in order to reduce the risk or the number of casualties in the case of accidents.
In the light of the extremely high number of accidents which have resulted in losses of human lives or in serious cases of pollution, shipowners ought to understand that the legislation is bound to become stricter (also because of the pressure exerted by the public opinion which follow these events with growing attention) while costs are bound to become increasingly higher. There is (however) an important sector where certain measures could, at a limited cost, reduce the number of casualties at sea. Adequate planning and implementation of maintenance operations, for instance, are undoubtedly necessary and could even lead to economies. Equally, the participation of officers to training courses concerning the sailing of the ship (in centres equipped with simulators), fire prevention, etc., is certainly not expensive, though extremely useful.
Personnel on board should be fully aware of the need to be prepared for all emergences and this preparation is also to be achieved by means of the continuous and careful checking of all the necessary equipment (which is often not ready to use, being seldom utilized: an example is provided by

the equipment for the setting down of life boats), as well as the organization of and the participation in the appropriate drills.
Insurance companies should anyway favour those shipowners that adopt higher safety standards. They should also hire technical personnel entrusted with the task of directly controlling the general condition of the ship (especially when she is in port for routine or special maintenance) and all aspects related to safety (fire stations, the closing of watertight doors, the efficiency of life boat set down equipment, etc.).

SEA CASUALTIES WITH TOTAL LOSS OF THE SHIP

The sea casualties considered in this paper are those where merchant ships were involved (navy ships are therefore excluded) with a gross tonnage higher than 100 tons; these casualties entailed either the loss of the ship or a damage so serious that the ship had to be considered wrecked. The lack of data on ships with a gross tonnage lower than 100 tons, on the other hand, strongly penalizes fishing vessels which represent the vast majority of ships with a limited gross tonnage.
Traditionally, these casualties are classified according to the primary cause that has led to the total loss of the ship. The causes are subdivided as follows:

- fire: the accident is provoked by a fire or explosion; those ships that were lost because of a fire or explosion resulting from a collision, impact, etc., belong to one of the following groups:
- collision: the accident is provoked by the impact against another ship;
- impact: the accident is provoked by any impact (grounding, impact against cliffs, icebergs, wrecks, platforms, etc.) other than that against another ship;
- war: the accident is provoked by an air, sea or land attack aimed at the ship;
- unknown cause: the accident has an unknown cause since the ship has disappeared and no one has witnessed the event (survivors, bystanders);
- foundering: the sinking is provoked by a structural failure, by adverse sea and weather conditions, by a loss of stability or by any other reason other than those previously mentioned.

This paper does not consider those accidents due to a war. Therefore, the total number of ships lost is given by the sum of vessels belonging to the other five categories.
Fig. 1 shows the histogram of the total number of ships lost in the world from 1950 to 1993.
It is however more appropriate to consider the number of ships lost in relation with the number of operating ships (relative losses). Fig. 2 shows the histogram concerning the percentage of losses against the number of operating ships from 1950 to 1991.
As regards the losses with reference to the different types of ships and to the different causes of casualties, total losses have been taken into account for the 1983-1993 period.

	Foundering	Fire	Collision	Umknown	Impact	Total	%
Tankers	13	57	13	0	27	110	4.23
Bulk-Carriers	56	28	5	9	59	157	6.05
Cargo Ships	577	167	158	14	359	1275	49.11
Ferries	43	26	7	0	23	99	3.81
Containers	10	4	3	0	5	22	0.85
Fishing Vessels	329	126	75	11	146	687	26.46
Tugs	45	5	10	2	12	74	2.85
Other	71	21	30	0	50	172	6.63
Total	1144	435	301	36	680	2596	100
%	44.06	16.76	11.59	1.39	26.19	100	

Table. 1. Number of Ships Lost Between 1983 and 1993 According to Type and Cause

Table 1 shows the absolute values (number of ships lost) for certain types of ships (the bulk-carrier category also includes ore/oil carriers whereas the ferry group includes both roll on-roll off vessels and ferries for the transportation of railway trains).
In a previous paper (Russo Krauss & Cardo, 1996) a detailed analysis was made of the statistics concerning sea accidents (for the sake of brevity this analysis has not been included here). The analysis showed that:
- most of the accidents involving tankers, cargo ships and bulk-carriers which were lost due to impact and fire occurred with the ship in ballast, while foundering occurred with the ship fully loaded;
- most of the accidents involving ferries which were lost because of fire, impact and foundering, occurred when passengers and vehicles were on board;
- the highest number of collisions and impacts occurred in restricted waters, the highest number of fires occurred when the ship was at anchorage or in port, the highest number of foundering occurred on the open sea;
- the so called "flags of convenience" show a higher number of relative losses.
Table 2 indicates the number of human lives lost between 1983 and 1993 according to the various causes which have led to the total loss of the ship (besides the causes illustrated here, the total figures of column 7 also include the victims of accidents that have not led to the total loss of the ship). In Table 2 "Other Causes" indicate incidents which have not lead to the total loss of the ship (passengers falling over board or on the ship, fire confined to one compartment only, etc.). Figures in Table 2 are indeed very high: the average is 956 deaths a year, which means 2.6 deaths a day! Certain years were particularly tragic: in 1987, for instance, the average was 10.5 deaths per day.
These dramatic figures do not include the numerous victims of accidents involving ships with a gross tonnage lower than 100 tons, mostly fishing vessels.

	1 Foundering	2 Fire	3 Collision	4 Unknown	5 Impact	6 Total 1+2+3+4+5	7 Total 6+Other Causes	% With Ref. to 7
1983	570	53	25	0	15	663	671	6.38
1984	317	51	54	68	7	497	525	4.99
1985	440	94	14	32	12	592	619	5.89
1986	428	29	448	85	34	1024	1067	10.15
1987	523	29	3156	78	34	3820	3841	36.53
1988	561	83	63	29	23	759	763	7.26
1989	443	57	76	26	76	678	688	6.54
1990	149	40	72	128	0	389	389	3.70
1991	400	16	214	62	512	1204	1204	11.45
1992	119	22	63	24	7	246	257	2.44
1993	442	19	35	0	6	504	504	4.79
1983-1993	4392	493	4220	532	726	10363	10515	100
%	41.77	4.69	40.13	5.06	6.90	98.55	100	

Table. 2. Human Lives Lost Between 1983 and 1993 According to Causes.

A heavy toll of lives is taken by casualties involving the total loss of ferries among which some were flying the Italian flag. A few data on these ferries are reported hereafter:
* DONA FLORENTINA, fire followed by capsizing and foundering; may 1983; 56 deaths;
* ASSUNCION, foundering of the intact ship due to wrong loading and adverse weather and sea conditions; december 1985; 136 deaths;
* DONA JOSEFINA, foundering due to the flooding of the vehicle deck in the absence of a leak; april 1986; 196 deaths;
* HERALD OF FREE ENTERPRISE, capsizing due to the flooding of the vehicle deck in the absence of a leak; march 1987; 193 deaths;
* DONA PAZ, collision with the Victor and foundering; december 1987; 3157 deaths;
* ESPRESSO TRAPANI, capsizing and foundering of the intact ship due to wrong loading in ideal weather and sea conditions; april 1990; 13 deaths;
* MOBY PRINCE, collision with the Agip Abruzzo, fire of the whole ship; april 1991; 169 deaths;

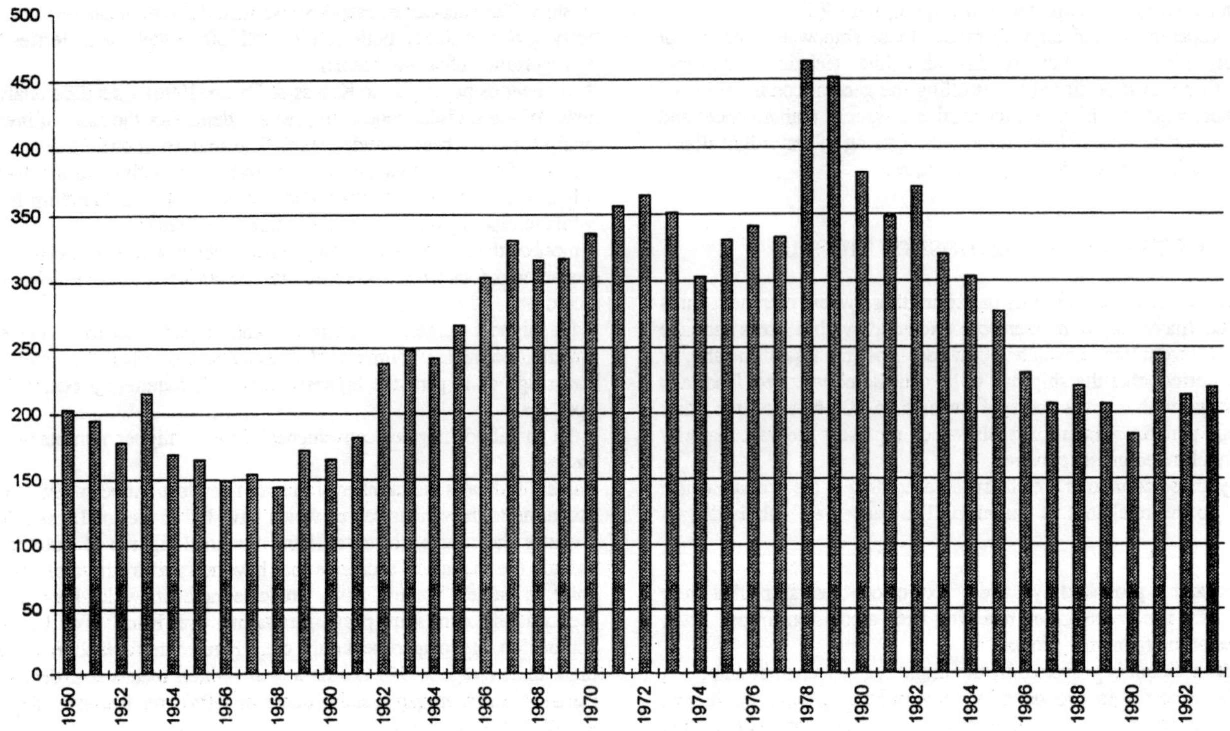

Fig. 1 Histogram of the total number of ships lost in the world from 1950 to 1993.

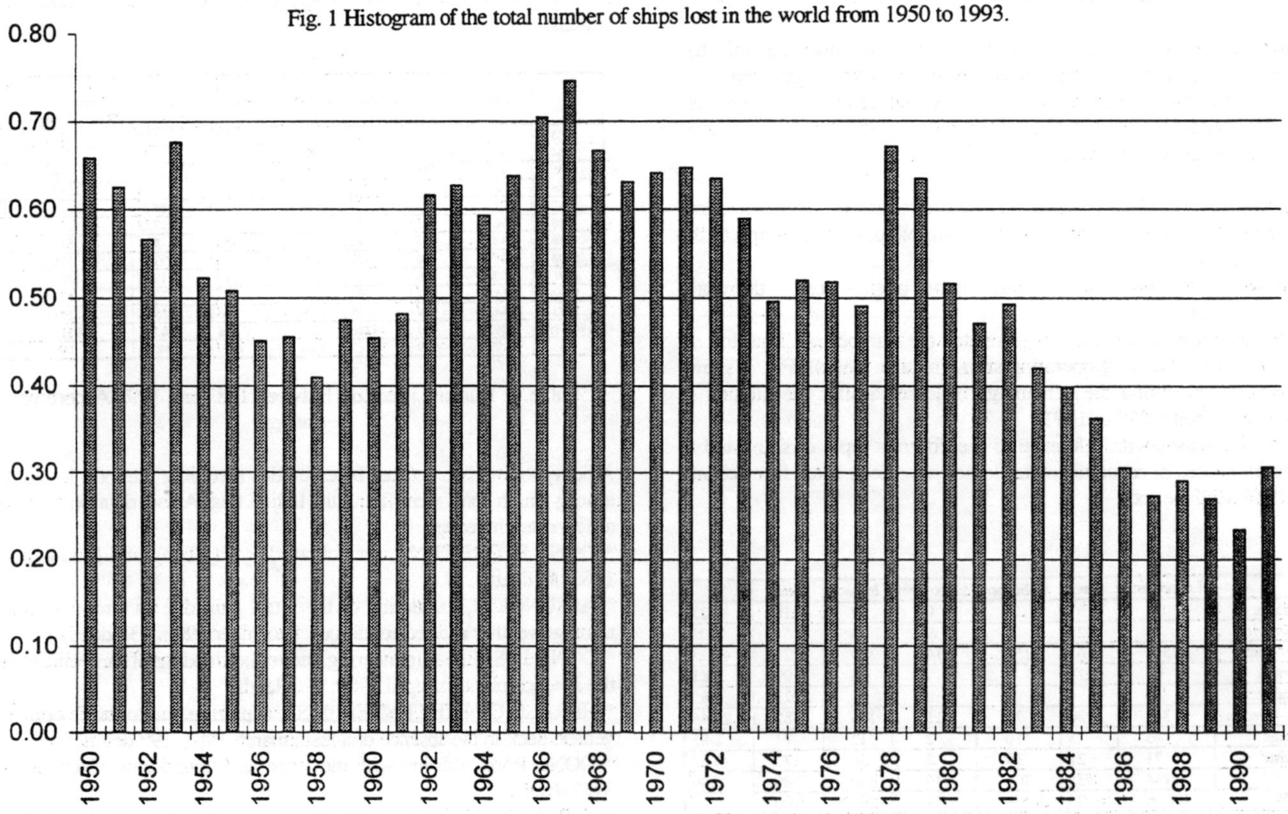

Fig. 2 Histogram of the percentage of losses against the number of operating ships from 1950 to 1991.

* SALEM EXPRESS, foundering due to impact; december 1991; 464 deaths;
* ESTONIA, in adverse weather and sea conditions, the ship heeled considerably and capsized and sank in a very short time; december 1994; out of 906 people on board 852 died;
* KALTIM MAS II, capsizing and foundering in adverse weather and sea conditions; april 1994; 48 deaths.

It can be estimated that, over a decade, not less than 50% of the deaths caused by the total loss of the ship can be attributed to casualties involving ferries!

CHARACTERISTICS OF FERRIES: ELEMENTS COMPROMISING THEIR SAFETY

Ferries are subdivided into two groups: vessels for the transportation of rolling-stock on fixed rails, where the rolling stock is loaded and unloaded on its own wheels; and vessels for the transportation of vehicles which are loaded and unloaded by means of its own wheels, and of goods arranged on pallets or in containers which are loaded and unloaded by means of wheeled vehicles. In both cases the ferry is also a ship used for the transportation of passengers [TP], railway passengers, drivers (articulated units, lorries, trailer trucks, etc.), car and motorcycle passengers, as well as passengers travelling on the vessel without any vehicle (with the exception of RO-RO ships used only for the transportation of vehicles carrying goods).

In certain cases, therefore, the RO-RO ship shows analogies with the container ship. The RO-RO is more flexible as it does not require massive port facilities or special shore-based equipment. On the other hand, the RO-RO has to cater for lorry drivers. For national cabotage and short-route sailing between neighbouring contries the favourite solution is and will continue to be the RO-RO ship vis-à-vis the container ship as the former can guarantee the connection with smaller ports and can better adjust to the spread organization of goods traffic. Moreover, since vehicles are accompanied by their owners or drivers, transport formalities are dealt with more rapidly, goods are handled with greater care, there is a lower risk of pilferage and a reduction in down time and costs. On an equivalent cargo basis, the RO-RO ship entails higher investments vis-à-vis a container ship. Indeed, the former requires doors, ramps and elevators and has to provide accomodation for the passengers on board with their vehicles. Futhermore, not all the space available for loading can be filled in due to the use of wheeled pallets, to the distance between the vehicles and to the lack of standardization of cargo units. The RO-RO ship has also the advantage of allowing for more rapid transshipment operations and streamlining transport from the port. The RO-RO ship also allows for the transportation of cargo whose dimensions do not necessarily comply with preset limits.

The main characteristic of RO-RO ships is that of having wide open areas so as to accomodate as many units as possible, to streamline their handling on board and to optimize their positioning with the objective to reach a higher value for the space utilization coefficient.

The stern opening - which is always present on these ships and whose door also serves as an approach ramp - must be particularly wide to allow for easy loading and discharge. All train ferries and some RO-RO ships operating on a short run on a frequent schedule also feature a bow door (twin shell doors, or more often, a bow visor) thereby dispensing with vehicles manoeuvring on board or coming aboard in reverse.

RO-RO ships operating on short routes (a few hours) and used for the transportation of passengers, cars and other vehicles only have a stern door, a single vehicle deck (bulkhead deck, main deck and loading deck) extending from stern to bow and for the whole width of the ship with the exception of the area at the far end of the stern where, in the side, two casings are located which accomodate certain services (ballasting and trim station, door control station, etc.) as well as stairs to higher public areas (covered and open areas, such as halls, bars, toilets, etc.).

RO-RO ships operating on average or long routes (between 10 and 36 hours) and used for the transportation of commercial vehicles only, usually have a single stern door and various vehicle decks. Vehicles are parked on the main deck (loading deck), in the hold below (access is ensured by an elevator) and on one or two of the upper decks (weather deck and, possibly, 'tween decks) where access is provided by ramps. Accomodations and essential services for the drivers (galley, canteen with bar, etc.) are located in the superstructure. Engines are located in the stern whereas the superstructure is found either aft or fore. On these ships the hold extends from the bow bulkhead of the engine room to the far end of the bow (70% of the ship length); the main deck and, possibly, the 'tween decks (generally used for cars and other low vehicles) extend through the whole lentgh and width of the ship. The space used for services (stairs, pipes/trunks) is either located at the centre of the ship and generally framed longitudinally (so as to reduce the beam clearance) or is subdivided into two blocks, one along the side shell and the other in the stern.

RO-RO ships used for the transportation of various types of vehicles as well as passengers are similar to the vessels already described. They only differ in that they do not have a vehicle deck/hold below the main deck, they have more (generally partial) decks used for cars (equipped with mobile ramps so as not to obstruct the vehicle deck below) as well as more decks (also with superstructures) for passengers (cabins, halls, restaurant, bar, game room, cinema, etc.). On these ships the space used for services (stairs, lifts, trunks/pipes) is located at the centre of the ship and is generally framed longitudinally. Other services (usually reserved to personnel and including the stability/ballast and trim stations, the door control station, etc.) are located in smaller blocks along the side and in the stern area.

The vulnerability of ferries with reference to the primary causes previously mentioned depends:

- as far as fire is concerned, on the presence of a considerable number of people (crew and passengers) and vehicles and on the type of cargo, often consisting in inflammable goods loaded on the vehicles;
- as far as collision is concerned, on the use of busy routes and ports;
- as far as impact is concerned, on sailing in restricted waters;
- as far as foundering is concerned, on adverse weather and sea conditions, on the loss of stability due to flooding or to the inappropriate positioning of the cargo on board and/or the omitted lashing of the vehicles, etc.

As regards the critical analysis of the circumstances under which the major accidents mentioned in this paper have occurred, the reader is referred to (Russo Krauss & Cardo, 1996). However, it has been considered useful to put forward the following suggestions ensuing from that past experience.

SUGGESTIONS TO REDUCE THE PROBABILITY OF FLOODING AND ITS NEGATIVE EFFECTS

The major risk for ferries is represented by flooding, regardless of its cause. Flooding generates an enormous free liquid surface and the consequent rapid reduction of stability which leads to capsizing and sinking in a very sort time. Some suggestions are presented here with the aim to reduce the proability of this event or to prolong the time elapsing between the accident and capsizing.

<u>Loading and discharge door</u>

The folding vertically hinged door (be it a stern or a bow door) has proved to be less reliable than the single section horizontally hinged door. The former should therefore be banned.

The scantling of the door should be based not on the maximum weight of the vehicle with the highest capacity, but on a much higher load. The door should also be designed by taking into account heavier operating conditions, even though realistic ones, like the case of a non continuous contact between the ramp and the quay.

The distance between the hinge pintles and between the blocking hooks should be reduced to avoid any distorsion when the door is exposed to the accelerations caused by the motions of the ship.
Regular compulsory ispections should be envisaged in order to verify the efficiency of the seals and of the blocking system; water jets could also be used to this purpose in order to ascertain that the door is perfectly closed and watertight.
Videocameras should be installed to enable officers on the bridge to control the closing of the door and the good functioning of the locking systems.
It should also be compulsory to write into the log-book that before the departure is has been ascertained that the door was perfectly closed.

Shape of the ship sides
The ideal solution to avoid the flooding of the main deck (loading deck) after the collision of the ferry with another ship is said to be the subdivision of the vehicle deck by means of longitudinal bulkheads at a distance from the side shell greater than B/5 (It has been ascertained that, in the case of a collision, the bow penetration was 20% greater than the ferry width).
This solution would be so demanding from the viewpoint of the ship operation that, as a result, there will be a dramatic decrease in the use of the vessel. Moreover, if this solution is not accompanied by a transversal compartmenting of the vessel, it would produce an eccentric flooding and the sudden inclination of the ship at considerable angles (It seems certain that a mere 7° inclination, if suddenly reached, can generate panic among all passengers).
Currently, all ferries have vertical sides (or almost vertical) through 75% of the maximum length of the vessel. The obligation could be envisaged to build sides that are externally sloped by at least 15° all along the length of the ship. The higher part of the sides should also be adequately designed. This would allow for a greater absorption of the impact energy with the following positive effects:
- the bow flare of the colliding ship would impact against the higher part of the side; in the case of a ship with a bulbous bow, the proposed solution would dramatically decrease the chance of the bulb producing a leak in the area between the bilge and the loading deck of the ferry thereby reducing the risk of flooding in the vehicle deck either below or near the waterline;
- under the same circumstances, the intact ferry would have a greater area stability diagram with a higher value for the capsizing angle; a greater stability would also be achieved in the case of leak with subsequent flooding, but it would take longer for the disastrous effects to take their toll and more people would have a chance to save their lives;
- the higher part of the side that is exposed to collisions could be adequately designed and box-like structures could be used which reduce the space available in areas not reserved for cars and therefore do not compromise the redditivity of the ship;
- on larger ships, the greater width of the higher areas could entail the disappearance of one superstructure deck, which, besides compensating for the higher costs of a ship with sloped sides, would lead to the lowering of the ship centre of gravity and to the consequent improvement of stability;
- the inclination of the side shell would allow for the setting down of life boats on the side opposite that of the collision (that is to say opposite the heeling side) even at large heeling angles (up to 15°);
- the greater width of the higher areas would make it possible to have a larger area on the embarkation deck available as muster station when the ship has to be abandoned.

Subdivision of the space used as vehicle deck
As an alternative to the above-mentioned solution, instead of introducing longitudinal bulkheads in the vehicle deck, one could suggest having large side shell volumes occupied by closed-cell material with a low specific gravity. In case of a leak, this would ensure the achievement of righting moments that can delay or avoid capsizing. In order to be effective this solution requires the sacrifice of a considerable area which would otherwise be used for the payload and a remarkable lengthening of handling times.

A further alternative is a transversal mobile subdivision of the vehicle deck. When not used, these bulkheads should take up a limited space and should be positioned in such a way that they do not excessively reduce the available space and do not hinder the loading and unloading of vehicles. Clearly, this solution entails that the ferry is to be loaded on a sector basis (from the bow to the stern) and that subdivision is to be carried out as each sector is loaded (the same holds true as far as discharging is concerned). This solution, therefore, must necessarily take into account the time required for the positioning and the stowing away of the mobile bulkheads.
The mobile bulkheads - they should be watertight and adequately high - could be arranged in the following ways:
- a couple of side-hung half-bulkheads (one per each side) made of one or more elements; when stowed away they would lie along the side shell; to transversally position the half-bulkheads they should be rotated by 90°, the left one in a counterclockwise direction and the right one in a clockwise direction; when closed, the two half-bulkheads should meet with a coaming located transversally on the deck (A similar solution has been developed by Mac Gregor);
- the bulkhead could consist in a top-hung door whose hinge pintles are located transversally on the deck; the bulkhead would rotate in a clockwise direction (if looked at from outside the right side).

As an alternative to mobile bulkheads, a fixed (structural) subdivision could be envisaged with openings wide enough to allow for the flow of vehicles from the stern to the bow. These openings should be closed with watertight doors as each compartment is filled in with vehicles and before the departure of the ship. Different types of doors could be used:
- a folding multi-panel door sliding orizontally along the deckhead of the vehicle deck and rotating on a polygonal shaft so as to be vertically positioned;
- a folding multi-panel door sliding longitudinally along the side and rotating on a polygonal shaft so as to be transversally positioned;
- a folding multi-panel door sliding longitudinally along the centerline plane of the ship or along any plane parallel to the centerline one and rotating on a polygonal shaft so as to be transversally positioned;
- a horizontally hinged top door; when open, the door is located below the structures of the vehicle deck;
- a door horizontally hinged to the vehicle deck; when open, the door lies on the deck of the vehicle area.
- a multi-panel door sliding transversally across the deck (the various elements of the door as well as the guide bars would not be located on the same transverse plane).

With these solutions it is of paramount importance to ensure that the doors provide a watertight closure. However, should this not be the case, flooding would occur very slowly in the compartments not directly affected by the accident, which would make it possible to intervene with the necessary means to pump water out of board and would allow all passengers to save their lives.

SUGGESTIONS TO AVOID LOADING ERRORS AND TO REDUCE THE RISKS RELATED TO CARGO

The overloading of commercial vehicles, the lack of data on their real weight and the difficulties in planning a rational loading are the primary cause of many casualties where human lives were lost. The designer can do nothing or very little to solve these problems, the legislator can do very much.

First of all, the obligation should be introduced to establish the fee for transportation on the basis of two elements: the length of the vehicle and its real weight. This obligation is the only way in which the shipowner

(and therefore the officers on board) will be interested in checking the real weight of the vehicle.
The docking of ferries should occur along quays located in the proximity of large areas where the loading of vehicles could be organized not only on the basis of their order of arrival but also of their weight.

Monitoring and positioning of cargo
Vehicles should arrive at the embarkation area long before the departure time of the ferry.
A boarding card (similar to that used on aircrafts) would have to be introduced for all passengers and vehicles. As regards the vehicles used for goods transportation, the boarding pass should also include the data concerning the vehicle itself, the owner, the driver, the type of goods carried and the total weight of the vehicle. The boarding pass should be issued by a specific office which would have to refer all vehicles or a given percentage of them to the weigh-bridge. The opening days and times of the weigh-bridge should be established so as to ensure an adequate service for all the ferries leaving from the port. If the weight that has been declared differs considerably from the weight actually measured, the boarding card should not be issued and an administrative action should be started against the owner and the driver of the vehicle. No boarding pass should be issued later than an hour before the departure time of the ship. In exceptional cases and upon written request of the captain, the departure time could be postponed thereby delaying the issuing of the last boarding pass. The vehicle can come aboard only if accompanied by the boarding card. This document should be issued in three copies: one is kept by the issuing office, one is collected by the embarkation officer and the third is kept by the driver of the vehicle. The office and the captain should keep the boarding cards for a given amount of time.

Prior to embarkation, the vehicles used for the transportation of goods (all or a few selected on a random sample basis) should be inspected in order to:
- identify possible stowaways,
- avoid the transportation of hazardous cargo (hazardous goods, explosives, weapons, etc.),
- check weather the cargo has been effectively positioned and lashed on the pallet,
- check the presence and the efficiency of the clamping hooks of the vehicle.

The ship should be equipped with an instrument to assess (even if not rigorously) the weight of the vehicle. Such an instrument would be used in exceptional cases or whenever the weigh-bridge is out of order.

Embarkation procedures and checks, the vehicle positioning and lashing, and every other operation directly related with safety should be entered in the log-book or in a specific register; the notes should clearly specify the name of the person directly responsible for every operation; he has to sign the notes concerning his responsibilities while the captain has to sign a declaration where he confirms that every operation has been subject to his supervision.

Parking areas where vehicles are not allowed (to ensure the positioning of mobile bulkheads and/or the closing of fixed bulkhead doors, watertight doors on a ramp connecting two decks, etc.) should have a different colour and show a no- parking sign. The decks that are meant for passengers only should show a sign indicating that no other cargo should be located there and inviting people to report any violation to the personnel on board and to the maritime authority at embarkation or disembarkation.

Videocameras should be installed on board so as to enable officers on the bridge to control what is going on in the parking areas (loading, closing of the doors, position of cargo on the vehicles, movement of the vehicles and presence of unauthorized persons in the parking area while the ship is sailing, etc.).

Clearly, it is on the road network of every civilized nation that the fight against the overloading of vehicles must be started. It is therefore indispensable for the police to avail themself of those tools that will make it possible to assess -even if not- the weight of a vehicle and ascertain that the maximum admissible load has not been exceeded. This would ensure a greater safety even on the road network, with a consequent marked reduction of mortality and a lower loss of goods. In the case of a flagrant violation, the vehicle has to be escorted to the nearest weigh-bridge/platform scale and, if necessary, banned from vehicular traffic.

Control of consumables and ballast
Every ferry should be equipped with systems capable to show the bridge the level of consumables (potable water, fuel oil, Diesel oil, lubricating oil) and ballast water for every tank so as to enable officers to have a more rapid and reliable knowledge of the weights involved. Maritime autorities should frequently check the good functioning of these systems. If a breakdown in the system is reported, the ship should remain in port. Manual probes should only be used to verify the good functioning of the automatic ones and to replace the latter if they break down at sea.

Control of draughts
To avoid uncertainties and inaccuracies and to ensure an immediate knowledge of relevant data automatic systems should be installed to measure draughts on both sides aft, fore and amidship.

Control of stability with reference to loading conditions
The ship departure should be authorized some time after completion of vehicle embarkation and after closing the embarkation door.
Given the need to ascertain that the ship has the requisite stability, it is necessary to carry out this control accurately, easily and rapidly. Once the data concerning payload, consumables and ballast are reliably known, it would be advisable to use a personal computer in order to perform the necessary calculations and elaborate the stability diagram necessary to verify whether international legislation has been complied with. This solution is perfectly feasible and not expensive. Moreover, its adoption would avoid any risk to persons and goods that may derive from superficiality and/or haste and/or incompetence and/or recklessness.
If the use of the computer remains optional and does not become mandatory, the time that has to elapse between the closing of the door and the departure of the ship should be established according to the availability of such a system for the control of stability. Furthermore, those ships that are not equipped with this system should comply with new compulsory criteria concerning the formulation of the "instructions for the Master" chapter which deals with stability. Nowadays, these "instructions" do not permit to carry out an easy and reliable verification of stability for conditions other than those originally studied by the shipyard technical department. The new criteria should at least envisage - besides normal cases - other possible loading conditions which do not meet safety or normative parameters. One should also consider, for instance, cases of vehicle overloading by 10%, 20% or 30% where cargo is not rationally positioned and the conditions for consumables and ballast are not the ideal ones (since nowadays these calculations are carried out with low cost computers in all shipyards, technical departments should be more attentive to safety issues and include these cases in the "istruction to the Captain" booklet even if their inclusion is not required by the law or by the shipowner). The behaviour of the ship at preset values of speed and rudder angles should also be studied and included in the booklet for all loading conditions (since on ferries the centre of gravity is high on the water plane, the tactical diameter is short-about 1.7 L-and the rudders have high lift, heeling during a turning can in some cases take up dangerous values). All cases dealt with in the booklet should be accompanied by notes in order to make officers aware of the dangers deriving from vehicle overloading, from an irrational positioning of vehicles on board as well as from a limited quantity of ballast. If possible, these notes should also indicate the appropriate measures to be taken in order to reduce risks.

SUGGESTIONS TO ENHANCE ESCAPING AND ABANDONING THE SHIP

The high number of human lives lost in accidents involving ferries is strictly related to the short time elapsing between the event and the capsizing of the vessel. Indeed, it is clear that the shorter the time available the smaller the number of people that can save their lives in any dangerous situation. In the case of ferries, however, there is not even the time to lower life boats. Moreover, if the event is immediately and clearly critical, panic will inevitably spread among the passengers (and not only among them) thereby making it practically impossible to control the situation and save one's life.

Many of the suggestions put forward in this paper certainly contribute to delay capsizing after the event and reduce the first negative effects: they are therefore effective in curbing panic, both in its forms and in the number of people stricken by it. This is why they can create the necessary conditions to ensure that the time available is used in the best possible way, asking for assistance, trying to save as many people as possible, trying to save the ship, etc.

In many instances of accidents involving ferries not a single rescue unit could be lowered (self-inflating raft or boat), only a few people could reach the boat deck and many could not even leave their own cabins. The attempt to abandon the ship was so chaotic that the number of drowned persons was very high even though the sea conditions were not particularly bad and the ship was near the coast.

Currently, the flooding of the parking deck produces a sudden and dramatic loss of stability with the consequent almost immediate heeling of the vessel at remarkable angles (even between 15° and 20°). At such angles it is impossible to lower the rescue units located on the side opposite the heeling. Furthermore, it would be extremely dangerous to lower the units located on the heeling side because this would produce the further heeling of the ship and the boats and rafts could be hit by the vehicles sliding off the deck or by their cargo.

A first suggestion is addressed to designers: the position of rescue units should not be too high on the vessel so as to avoid having to climb many stairs to reach them and to simplify their lowering into the water). Rescue units should also be located not too far away from public areas. The areas reserved for passengers should be ideally subdivided into zones and each of them should equipped with a sufficient number of rescue units. Obviously, adequate space should be provided near rescue units for people to gather before embarkation.

One should also bear in mind that, when the ship is heeled by more than 20°, it is extremely difficult to open a door and leave a cabin if that door is located on a longitudinal bulkhead. It is then practically impossible to open a door if the door opens outwards and if the cabin is located on the heeling side of the vessel. It is equally impossible to open a door if the door open inward and if the cabin is located on the side opposite the heeling side. It is therefore suggested that the cabin corridors should always be transverse, that all doors (not only those of the cabins) should be provided with a double opening system (outward and inward) and that a handle capable of taking the weight of a person trying to open the door should be fitted near the door itself on the side opposite that of the hinges.

A stair can become inaccessible when its slope - usually greater than that used in normal houses - adds to that of the heeling ship. Therefore, the connection between two decks should always be ensured by a double stair if the stair is exclusively or predominantly transversal. With this solution one of the stairs remains accessible even when the vessel is heeling. When a longitudinal stair ensures the connection between two decks, both sides of each step should be sloped by 30° so that, even at high heeling angles, the stair can be used along the side where the foothold is horizontal or slightly sloping. Obviously, to avoid any danger arising from the use of the stair in normal condition, the sloping sides should be painted in a bright colour so as to become clearly visible.

All public areas (including cabins) and, particularly, escape routes/escapes should be covered with flooring materials characterized by a high grip value so as to be accessible even at high heeling angles.

Handrails on both sides of the corridors and stairs should feature an adequate grip and be capable to take the weight of more people clutching at them in an attempt to walk when the ship is heavily heeling.

Outside promenade decks should be crossed by a sufficient number of transversal corridors to enable passengers to rapidly move to the side which is considered more suitable to abandon the heeling ship (if there is no time available to use life boats). In certain points deck railings should be constructed in such a way as to allow passengers to easily climb over them even at high heeling angles.

In case of emergency the windows of the cabins located on the higher decks should be opened or broken by means of a specific tool made available to passengers. They should be wide enough to allow passengers to escape through them. The same holds true for certain - cleary identifiable - windows located in public areas.

Emergency lights are indispensable and they should be equipped with automatically rechargeable batteries. These lights should be lit in all those areas and corridors that passengers have to go through in order to reach open decks. They should be especially installed in the areas where life boats are located. Low level lighting should also be provided.

Special attention must be paid to handicapped passengers for whom it would be more difficult to abandon the ship in case of accidents. They should be assigned special cabins, nearer to life boats and well known to personnel on board who should give these passengers special assistance in case of accident. Special cabins should also be envisaged for those people travelling with young children.

In public areas (halls, resaurants, etc.) furniture should be firmly fixed to the floor and placed in such a way as to ensure an escape route even at high heeling angles. Whenever possible, furniture should also be used as holding point for the passengers to clutch at when escaping.

Wall corners, window and door frames, buttresses, etc. should be covered with soft material to avoid people hurting themselves during the sudden heeling of the ship and the consequent attempt of the passengers to escape.

FURTHER SUGGESTIONS

A further suggestion does not concern ferries only, but other ships as well. As has been shown, the collision between a ferry and a ship with a bulbous bow causes an extremely dangerous situation on the ferry, which very rapidly leads to the capsizing and sinking of the vessel with all its passengers. Designers should study and adopt bulbous structures - obviously capable to take the stresses acting on them - that can easily deform under the action of the collision pressures so as to reduce the damage on the side of the collided ship. Besides using more deformable materials (for instance aluminium alloys, which however entail certain difficulties in connecting structures) a different orientation of bulb stiffeners could be studied and the preference could be given to a subvertical and subhorizontal arrangement vis-à-vis the horizontal and vertical one.

On ferries a deck officer should be exclusively or predominantly responsible for safety aspects after having attended a specific course. This officer should be entrusted with the task of supervising all control operations, the embarking and arrangement of vehicles on board, the functioning and perfect closing of the loading door and of watertight doors, the efficiency of life boats and of their set down devices, the presence in the correct position of individual rescue units, the efficiency of handrails and of emergency lights, the effective verification of stability before the ship departure, etc. This officer should also make aware and train the personnel on board for emergency (prevention and control). The officer should readily report in writing to the master as regards any aspect directly related to safety and every month he should write a report indicating what has to be done to preserve the efficiency of all safety elements and informing the captain of what he has done in the field of making aware and

training. Lastly, this officer should be the main advisor of the master as regards safety and during an emergency.
Every effort should be made - by the legislator in particular - to have ships equipped with a device similar to the "aircrafts' flight recorder". Such a device should receive and record all data related to safety (payload, consumables, ballast, closing of the loading door and of watertight doors, verification of stability, etc.) until the arrival in port. It should also record all radio communications with the ship. Such a device is now perfectly feasible and its cost would certainly not be particularly high, especially if it becomes mandatory to use it. Insurance companies should be very interested in the installation of a device similar to the "aircrafts' flight recorder" and they could therefore promote its development and installation.
Maritime traffic should be organized and controlled similarly to air traffic, though with the necessary semplifications deriving from the lack of the third dimension (altitude) and from the lower speed. The constant contact with one of the stations of the control network would make it possible:
- to know the exact weather and sea conditions the ship will encounter;
- to know the position of the other ships sailing in a wide area and their routes;
- to know about the traffic situation and the presence of ships in the gulfs, bays and ports that are being approached;
- to rapidly communicate emergency situations;
- to rapidly alert and send rescue units;
- etc.
All coast-states should organize a coastguard service similar to that already operating in the US, for example, and responsible for the control of maritime navigation, for search and rescue operations, for the control of fishing activity, for the prevention and fight against pollution, etc.

CONCLUSIONS

It has been recognized that, especially in the last few decades, the IMO contribution has been particularly valuable and, as a consequence, the international legislation on safety currently in force has become much more effective. Unfortunately, however, the number of serious casualties involving ferries is very high and the number of victims is impressive. The development of a specific legislation on ferries is therefore necessary.
After the serious accident in september 1994 of the ro-ro passenger ship "ESTONIA", flying the Estonian flag, which led to the death of 852 people, a working group was set up - with a 1,5 million dollar financial support by the Danish Maritime Authority, the Finnish Maritime Authority, the Norwegian Maritime Directorate, the Swedish Maritime Authority, the Danish Shipowners Association, the Finnish & Aland Shipowners Association, the Swedish Shipowners Association, the Marine Safety Agency (UK), the Bureau Veritas, the Germanischer Lloyd, the Det Norske Veritas and the IMO - with the task of improving the safety of ro-ro passenger ships.
Other governments, shipowners associations, classification societies and insurance companies should also launch similar initiatives to intensify efforts and find adequate solutions.
The compulsory structural subdivision of the vehicle deck would eccessively penalize the operational capability of ferriesand would entail higher investment costs. In due time this would lead to the disappearance of ferries from many routes among ports connected also by the road network and/or by railway, with the consequent increase in road traffic - especially as far as commercial vehicles are concerned - which already causes many serious accidents and the death of many people and pollutes the environment. Transportation fares among ports connected esclusively via sea routes would considerably increase since investment and management costs for any kind of transportation are always borne by the final users or, in the case of government subsidized lines, on the state and, therefore, on the public at large.
It is therefore necessary to take into account each of the numerous parameters that have an impact on safety so as to assess their individual contribution and cost and to implement a series of interventions aiming at improving safety with only a reasonable increase in investment and management costs. For this reason, not only must IMO rules and regulations be improved (it is clear, for instance, that the legislation on escapes has been especially developed for the case of fire while escaping can also occur when the ship is heeling) and become stricter, other bodies will also have to play a more active role: designers, training institutes at various levels, shipowners, insurance companies, classification societies, etc.
This has also been the objective of this paper, which has taken into account many aspects of the complex issue of safety on ferries. Topics have been considered that can look trivial, but that are often neglected even though they significantly contribute to endangering safety on these ships.
It must be stressed once again that the States are the only responsible for the protection of life at sea and for the safety of maritime transportation. This duty and committment cannot therefore confine itself to the mere ratification and adoption of IMO rules. Instead, it has to promote scientific research, improve crew vocational training, increase university curricula, improve the efficiency of port organization, increase surveillance and assistance at sea, etc. No State can ever delegate its own fundamental responsibilities, such as the protection of human life, to others [for instance, classification societies representing different interests (the State, insurance companies, shipowners, shipyards)]. In particular, the Italian State has to take up many responsibilities which in the past it has delegated to the Registro Italiano Navale and it has to do so by means of an adequate restructuring and improvement of the Ispettorato Tecnico della Marina Mercantile and of the Corpo delle Capitanierie di Porto. Our country would therefore be able to make up for the time lost vis-à-vis other more advanced countries which, despite the existence of prestigious classification societies (such as the Lloyd's Register in the United Kingdom) have developed a public structure able to autonomously control the aspects relating to the safety of life at sea.

ACKNOWLEDGEMENTS

This paper has been written within the framework of the research "Safety of life and protection of the environment in ship designing" financed by MURST.

REFERENCES

Russo Krauss, G., Cardo, A., 1996, "Sinistri Marittimi, Sicurezza e Salvaguardia della Vita Umana Relativamente alle Navi Traghetto", Studi Marittimi, Vol. 19, pp. 19-38.
Maestro, M., Marinò, A. and Russo Krauss, G., 1992a, "Overloading of Commercial Vehicles Offered for Sea Transport and Ro-Ro Ship's Safety", PRADS'92, 5th International Symposium on Practical Design of Ships and Mobile Units, Newcastle upon Tyne (UK), Proceedings v. 2, pp. 2.1337-2.1349.
Maestro, M., Marinò, A. and Russo Krauss, G., 1992, "In tema di sicurezza operativa della nave: gli accertamenti di stabilità in esercizio", Tecnica Italiana, Vol. 47, pp. 153-169.
Maestro, M., Marinò, A. e Russo Krauss, G., 1993, "Traffico di cabotaggio e navi ro-ro: considerazioni sulla sicurezza", Studi Marittimi, Vol. 16, pp. 13-23.
Francescutto, A., 1992, "Towards a Reliability Based Approach to the Hydrodynamic Aspects of Seagoing Vessels Safety", Proceedings 11th International Conference on Offshore Mechanics and Arctic Engineering - OMAE'92, Calgary, Vol. 2, pp. 169-173.
Francescutto, A.,1993, "Is it Really Impossible to Design Safe Ships?", Trans. Royal Institution of Naval Architects, Vol. 135, pp. 163-173.

Proceedings of the Seventh (1997) International Offshore and Polar Engineering Conference
Honolulu, USA, May 25-30, 1997

ISBN 1-880653-28-1 (Set); ISBN 1-880653-31-1 (Vol. III)

Survival Analysis of a Deep-Water Floating Offshore Platform About Its Critical Axis Including Coupling

Ravikiran S. Kota and Jeffrey M. Falzarano
University of New Orleans
New Orleans, LA, USA

Alexander Vakakis
University of Illinois
Urbana, IL, USA

ABSTRACT

Due to the movement of oil and gas exploration and production to deeper and deeper water depths, floating production systems are becoming ever more common. No longer are Column Stabilized Mobile Offshore Drilling Units (MODU) only being used for site-temporary exploratory drilling, but are increasingly being considered for semi-permanent production systems. As a result of this new purpose, analysis of their critical response when exposed to very extreme winds and waves must be an ever more important aspect of their design. Unfortunately, vessel stability criteria are inadequate to study extremely large amplitude motions leading to capsizing. With respect to ships, much work has been done using dynamical systems approaches to study capsizing, recently these same techniques have been extended and applied to study platform dynamics.

This paper describes the global nonlinear dynamics of a typical MODU about its critical, approximately quartering axis. It is well known that short and wide (i.e., small L/B ratio) twin hull vessels such as MODU's have minimum righting moment about an approximately quartering axis. In a previous paper, Falzarano and Kota (1996) showed that this axis was also a critical rotational motion axis. In order to understand this problem, a global transient dynamical systems analysis was undertaken which compared the vessel's response at all heading angles. In the present work, an attempt is made to answer the question of what is the effect of the out of plane coupling. To answer this question advanced de-coupling techniques known as nonlinear normal modes (Vakakis , et al, 1996) are utilized. An example application of these techniques applied to the capsized MODU *Ocean Ranger* is also included.

KEY WORDS: Nonlinear Dynamical Systems, Nonlinear Normal Modes, MODU, Dynamical Stability, Floating Platform Motion

BACKGROUND AND INTRODUCTION

This study is the result of a series of investigations which have analyzed the multiple degree of freedom nonlinear dynamics of marine vehicles using modern geometric methods over the last several years. These investigations have only been possible due to the developments outside the marine dynamics field in the applied mathematics and mechanics areas (Guckenheimer and Holmes, 1986). These previous investigations have considered steady state multiple degree of freedom ship motions at various headings (Falzarano, et al, 1991 and Taz Ul Mulk and Falzarano, 1994) and transient global nonlinear dynamics of single degree of freedom (Falzarano, et al, 1992) and transient global nonlinear dynamics of ship roll motion coupled with sway and yaw (Falzarano and Zhang, 1993 and Zhang and Falzarano, 1994). This work was subsequently extended to consider the nonlinear dynamics of floating offshore platforms (Falzarano, et al, 1995). Most recently that work focused on the nonlinear dynamics of floating offshore platforms about their critical axis (Falzarano and Kota, 1996). In that work it was realized that the coupling of the out-of-plane righting arm although small at small angles continuously increased and became quite large at the in-plane angle of vanishing stability. It was realized at that time that the powerful methods of nonlinear normal modes (NNM) being developed outside the marine field by Vakakis, et al (1996), and Shaw and Pierre (1993) might provide valuable insight into this important problem. This paper summaries the most significant results of this investigation; more complete details are described by Kota, (1997).

PROBLEM FORMULATION

Description of the Explicit Physical Model

The linearized six degree of freedom equations of motion of a vessel, in the vessel coordinate system (Chakrabarti, 1985), can be written in a vector form as

$$(M+A(\omega))\,\ddot{\underline{x}} + B(\omega)\,\dot{\underline{x}} + C\,\underline{x} + \underline{g}(\underline{x}) = \underline{F}(t) \qquad (1),$$

where the vector **x** represents the six degrees of motion, **M** is the 6x6 physical mass/inertia matrix, **A** and **B** are the hydrodynamic added mass and damping matrices, **C** is the linear part of the hydrostatic restoring force/moment, and the vector **g** is the nonlinear component of the restoring force/moment. Note that **A** and **B** are frequency dependent. The system is now modeled as a forced, nonlinear spring mass damper system which can be analyzed for response to any external input (initial conditions or forcing) using various techniques.

The focus of attention in this study is the dynamics of tilt about an arbitrary horizontal axis at an angle of orientation α with respect to the vessel's principal longitudinal axis. So, the roll and pitch equations are isolated from the general 6-DOF equation above, assuming that neither is coupled, to linear approximation, with any other degree of freedom statically or inertially. Thus, we have the following equations

$$\textit{Roll}: \quad [I_{44}+A_{44}(\omega)]\ddot{\varphi} + B_{44}(\omega)\dot{\varphi} + \Delta GZ(\varphi) = F_4\cos(\omega t+\gamma_4) \qquad (2a)$$

$$\textit{Pitch}: \quad [I_{55}+A_{55}(\omega)]\ddot{\theta} + B_{55}(\omega)\dot{\theta} + \Delta GZ(\theta) = F_5\cos(\omega t+\gamma_5) \qquad (2b).$$

In these equations φ and θ are the roll and pitch coordinates, while I_{ii} and Δ represent physical mass moment of inertia and displacement respectively. A_{ii} and B_{ii} are the frequency dependent hydrodynamic added moment of inertia and radiated wave damping respectively, $GZ(\cdot)$ is the nonlinear polynomial approximations to the static restoring arm. F_i is the amplitude and γ_i the phase of wave forcing at a frequency of ω.

Along the same lines, a single degree of freedom equation of motion for general tilt angle dynamics about an orientation angle α can be expressed as

$$\textit{Tilt}: \quad [I_\alpha+A_\alpha(\omega)]\ddot{\xi} + B_\alpha(\omega)\dot{\xi} + \Delta GZ(\xi) = F_\alpha\cos(\omega t+\gamma_\alpha) \qquad (3).$$

In this equation, ξ is the tilt coordinate. I_α, A_α and B_α are the physical mass moment of inertia, hydrodynamic added inertia and radiated wave damping respectively, evaluated about the tilt orientation axis, α using a simple coordinate transformation. F_α and γ_α are the amplitude and phase of the wave force(moment) again evaluated about the orientation axis.

When the coupling of out-of-plane rotation to in-plane rotation is considered, a 2-DOF analysis is essential. So, the two principal coordinates are - the in-plane rotation, ξ; and the out-of-plane rotation, ζ. This 2-DOF system is nonlinear and coupled dynamically and statically since the axis of orientation and axis of tilt need not necessarily be in the planes of symmetry of the vessel. Thus, the 2-DOF equations of motion in vector form are

$$(\mathbf{I}_\alpha+\mathbf{A}_\alpha(\omega))\ddot{\underline{z}} + \mathbf{B}_\alpha(\omega)\dot{\underline{z}} + \Delta\,\underline{GZ}(\underline{z}) = \underline{F}_\alpha\cos(\omega t+\underline{\gamma}_\alpha) \qquad (4).$$

In the equation above, the matrices , shown in bold, are full with non-zero cross-diagonal terms while the vector $\underline{GZ}(\underline{z})$ is nonlinear and coupled. The vector $\underline{z}$ is given by

$$\underline{z} = \begin{pmatrix}\xi\\ \zeta\end{pmatrix} \qquad (5).$$

This completes the problem formulation for the motions of a semi-submersible about an arbitrary tilt orientation axis.

THEORETICAL BACKGROUND

The methods of normal mode analysis are well established for linear MDOF discrete and continuous vibratory systems. However, the logical question to ask is what happens when the systems are nonlinear? Are the same techniques applicable and are the same or similar results realized? A powerful technique used to analyze normal modes in nonlinear systems is *nonlinear normal modes* (NNM). In the area of nonlinear normal modes, two competing analysis techniques have evolved, the first is more compact and mathematically elegant and easily extendable to continuous systems while the second is more heuristic yet possibly more generally applicable to a wider class of problems including damped and possibly forced systems.

In this section, the analytical techniques implemented to analyze the coupled dynamics of a MODU about its critical axis are described. These techniques are the theory of nonlinear normal modes which is the main emphasis of this paper. After an introductory overview, the two methodologies of analysis of multi-degree of freedom systems using NNMs are described.

Nonlinear Normal Mode Analysis

An efficient approach to understand the dynamics of multi-degree of freedom nonlinear systems is by the use of nonlinear normal modes. The idea of using nonlinear normal modes for such systems was first proposed by Rosenberg (Rosenberg,1966). Many researchers have extended his theory to seek normal modes of mainly discrete, homogenous, conservative nonlinear oscillators. Certain unique properties, like mode bifurcations and localizations, of NNMs were demonstrated (Vakakis, et al, 1996). Shaw and Pierre (1993) developed a fundamentally different definition and approach to obtain NNMs which is also applicable to non-conservative systems. The emphasis of this paper is on the application of the methodologies suggested by Shaw and Pierre (1993) and Vakakis, et al, (1996) to the problem of platform motions about their critical axis. Rosenberg's definition of nonlinear normal modes and his classification of two qualitatively distinct types of normal modes is presented followed by a brief description of the aforementioned methodologies. A more complete description is included in Kota (1997). Results are discussed in the next section.

The existence of normal modes and their useful application

in obtaining "natural" free vibrations and steady state forced vibrations is well-established for linear conservative systems with many degrees of freedom. This approach is also applicable to non-conservative systems for the special case of proportional damping (Caughey, 1960). Having known the normal modes of motion for an N-degree of freedom linear system, the transient response of the system can easily be expressed as a linear combination of the N normal modes using the modal analysis technique.

The central importance of normal modes in simplifying the analysis of coupled N-degree of freedom linear systems has motivated the search for a similar class of solutions for nonlinear systems. It has been shown (Rosenberg, 1966; Vakakis, et al, 1996) that although the principle of superposition does not hold for nonlinear systems, certain physical phenomena like resonance occur in the neighborhood of normal mode vibrations regardless of the system being linear or nonlinear. However, for nonlinear systems it is not possible to obtain normal modes in the same manner as for linear systems.

To obtain the normal modes of a linear system, expressed in vector form as

$$(\ddot{x}) + [A](x) = (0) \qquad (6)$$

the eigenvalues and the corresponding eigenvectors of the fundamental system matrix (A) are determined. These eigenvalues physically represent the squares of the natural frequencies of the system. Then, the eigenvector associated with each eigenvalue represents the normal mode of vibration of the system at that frequency. However, this procedure of obtaining eigenvalues and hence obtaining eigenvectors (normal modes) is only valid for linear systems. Demonstration of existence of eigenvalues has its roots in the theory of quadratic forms, and its application to vibration problems requires that the potential energy be a quadratic form (Rosenberg, 1966). For a general nonlinear system, the potential energy need not be a quadratic form. Hence, a new definition of normal modes is necessary which would not only be applicable to a general nonlinear system but is equally valid for linear systems.

According to Rosenberg, normal mode vibrations of an autonomous multi-degree of freedom system are the "vibrations-in-unison" of that system. This means that during normal mode vibration all masses of the system undergo synchronous periodic motions attaining their maxima, minima and equilibria at the same instant of time (Rosenberg, 1966). When the system is linear, the motion of every mass during a normal mode vibration is a simple harmonic function of time having the same frequency and the same zeroes as every other, i.e.

$$x_i(t)/x_j(t) = A_i\cos(\omega t)/A_j\cos(\omega t) \equiv c_{ij}, \; const \qquad (7)$$

It was shown that symmetric two degree of freedom systems (Vakakis and Rand, 1992) and homogenous n-degree of freedom systems (Rosenberg, 1966) are such that their normal mode vibrations satisfy (7) no matter how strongly nonlinear the systems or how large the displacements (Rosenberg, 1966). When the displacements of any two masses m_i and m_j during normal mode vibrations are such that

$$x_i(t)/x_j(t) \neq const \qquad (8),$$

then the normal mode vibration is defined to be *nonsimilar*.

One-Dimensional Modal Curve Approach

This approach is based on the assumption that the modal trajectory is a function of the displacement only and not the velocity of the single reference point. As a result, the trajectory is a one-dimensional curve. This approach is similar to Vakakis' (Vakakis, et al, 1996) Matched Asymptotic Expansion methodology for small oscillations of linearizable(nonlinearities with non-zero linear term) and nonlinearizable(nonlinearities with zero linear term) systems. In this work, however, it is chosen to use a power series expansion, in place of asymptotic series, for the modal curve based on the premise that the oscillations in vessel capsizing region are not necessarily small. Consider a general homogenous two degree of freedom system in the generalized coordinates x_1 and x_2, expressed as

$$\ddot{x}_1 + f_1(x_1,x_2) = 0 \qquad (9a)$$
$$\ddot{x}_2 + f_2(x_1,x_2) = 0 \qquad (9b)$$

x_1 where f_1 and f_2 are the nonlinear coupled stiffness functions. In the above, over-dots indicate time-derivatives. Assume a nonsimilar nonlinear normal mode of the form

$$x_2 = \hat{x}(x_1) \qquad (10)$$

where the function x is as yet unknown and arbitrary. Then, we have the following two relations

$$\dot{x}_2 = \hat{x}'\,\dot{x}_1 \qquad (11a)$$
$$\ddot{x}_2 = \hat{x}''\dot{x}_1^{\,2} + \hat{x}'\ddot{x}_1 \qquad (11b)$$

where primes imply derivatives with respect to x_1. Substituting (10)-(11a) into (9a) and (9b), we get

$$\ddot{x}_1 + f_1(x_1,\hat{x}(x_1)) = 0 \qquad (12a),$$
$$(\hat{x})'\ddot{x}_1 + (\hat{x})''\dot{x}_1^{\,2} + f_2(x_1,\hat{x}(x_1)) = 0 \qquad (12b).$$

Again, integrating (9a) with respect to x_1, we get

$$\dot{x}_1^{\,2} = -2\int_{s=X_1}^{x_1} f_1(s,\hat{x}(s))\,ds \qquad (13).$$

In the above equation, X_1 is the maximum amplitude attained by x_1. Using (12a) and (13) to eliminate the time derivatives of x_1 in (12b), we get the equation,

$$(\hat{x})''[-2\int_{s=X_1}^{x_1} f_1(s,\hat{x}(s))ds] + (\hat{x})'[-f_1(x_1,\hat{x}(x_1))] + f_2(x_1,\hat{x}(x_1)) = 0 \quad (14)$$

At this point, we can choose the form for the function x of x_1. It is convenient to choose x as a power series in x_1 with unknown coefficients. Then, by matching the powers of x_1 simultaneously starting from the lowest power, we obtain a set of algebraic equations in the unknown coefficients which can be easily solved. The function x would then be completely known and will represent the trajectory of the nonsimilar nonlinear normal mode in the configuration plane of (x_1,x_2). Moreover, using this modal approximation for x_2 in f_1, we can study the decoupled single degree of freedom system of (9). Note that the number of real solutions to the algebraic equations will indicate the number of real normal modes which should again be *at least* (modal bifurcations increase the number of modes (Vakakis, et al, 1996)) equal to the number of degrees of freedom of the system.

Invariant Manifold Approach

An alternative methodology to obtain the nonlinear normal modes was proposed by Shaw and Pierre (1993). In this approach, a normal mode motion is defined as one where all displacements and velocities are related to a single reference pair of displacement and velocity. This is in contrast to the earlier methodology wherein only the displacement of the reference point was the dependent variable. As a consequence of considering two dependent variables, the modal trajectory is no longer a curve but a manifold. A key feature of this method is its applicability to non-conservative systems also.

Recall for linear systems that the normal modes are mutually orthogonal. This leads to the *invariance property* of normal modes -i.e., vibrations started on any one of these modes will remain on that mode for all time with no participation whatsoever from any of the other modes. This property is also found to be valid for nonlinear systems. The *invariance property* of modal subspaces and the theory of invariant manifolds of dynamic systems form the basis for the definition of nonlinear normal modes as invariant manifolds in the phase space of the equations of motion governing the dynamical system. An invariant manifold, by definition, is a surface contained in the phase space of a dynamical system which has the property that orbits starting on the surface remain on the surface throughout the course of their dynamical evolution (Wiggins,1990).

Consider the N-DOF nonlinear autonomous system expressed as (Vakakis, et al, 1996),

$$\ddot{x}_i + f_i(\underline{x},\dot{\underline{x}}) = 0, \qquad i=1,2,....,n \qquad (15)$$

where $\underline{X}$ is a (nx1) vector of displacement and f a vector nonlinear function of displacement and velocity. The governing equations can be expressed in Cauchy standard first order form as

$$\dot{x}_i = y_i$$
$$\dot{y}_i = -f_i(x_i,y_i), \qquad i=1,2,......,n \qquad (16)$$

Now, we introduce, just as in the earlier method, the as yet unknown functions x and y which are functions relating the displacement and velocity at any point to a single reference pair, say, $(x_1,y_1) = (u,v)$ respectively, i.e.

$$x_i = \hat{x}_i(u,v)\ , \quad y_i = \hat{y}_i(u,v)\ , \qquad i=2,3,......,n \qquad (17)$$

Using these functional relationships in the equations of motion, we obtain

$$\dot{x}_i = \frac{\partial\hat{x}_i}{\partial u}\dot{u} + \frac{\partial\hat{x}_i}{\partial v}\dot{v}\ ,$$
$$\dot{y}_i = \frac{\partial\hat{y}_i}{\partial u}\dot{u} + \frac{\partial\hat{y}_i}{\partial v}\dot{v}\ , \qquad i=2,3,....,n \qquad (18)$$

Also, for the dynamics of the reference pair (u,v), we obtain

$$\dot{u} = v$$
$$\dot{v} = -f_1(u,v;\hat{x}_2,\hat{x}_3,...\hat{x}_n;\hat{y}_2,\hat{y}_3,...\hat{y}_n) \qquad (19)$$

Substituting (17)-(19) into (16) for $i = 2,3,...n$ and thus eliminating time from the equations, we obtain a pair of equations, in the unknown functions x(u,v) and y(u,v), for each mass point 'I'.

We assume a homogenous power series in u and v for x(u,v) and y(u,v) with unknown coefficients. By matching each power of u and v term by term starting from the lowest order, a system of simultaneous algebraic equations is obtained for each power. This can be solved for the unknown coefficients and the solution used to obtain the next higher order coefficients and so on. The functions x(u,v) and y(u,v) then represent the invariant modal manifolds for the given nonlinear system. Having obtained the solutions for the functions x(u,v) and y(u,v), (19) can be used to generate the modal dynamics on each mode. If there are no internal resonances or mode bifurcations, the number of real solutions coincides with the number of degrees of freedom of the system.

RESULTS

The geometric methods described in the earlier section are used to study the capsizing mechanism of a column-stabilized semi-submersible Mobile Offshore Drilling Unit (MODU)- *Ocean Ranger* (Royal Commission, 1982 and US Coast Guard, 1983). Using a single degree of freedom (SDOF) model, the existence of the critical rotation axis was established (Falzarano and Kota, 1996). In the present study, a 2-DOF analysis is performed, specifically about the critical axis, in order to investigate the importance of the coupling of the out-of-plane rotations to the in-plane rotations. To study the dynamics of this 2-DOF model, it is essential to decouple the two modes. This is achieved by using the NNM technique described in an earlier section implementing both - the modal curve approach and the invariant manifold approach. Then, the decoupled modes are analyzed as SDOF systems to study the dynamics.

The vessel is geometrically and inertially asymmetric about the critical axis. This suggests that a single degree of freedom analysis might not be adequate in determining the overall vessel stability. For every orientation, at every tilt angle, there are two righting moments, i.e., an in-plane moment and an out-of-plane moment. At small tilt angles, the magnitude of out-of-plane moment is comparatively small. However, from intermediate to large angles of tilt, its magnitude rapidly increases. The variation of the out-of-plane moment arm with tilt axis orientation is shown in fig. 1.

In the 2-DOF analysis presented here, the effects of damping and forcing have not been considered. The NNM technique is used to analyze nonlinear coupled oscillators, as discussed in the last section. Recall that the two methodologies - Modal Curve approach (Vakakis, et al 1996) and Invariant Manifold approach (Shaw and Pierre, 1993) used to obtain the normal modes of a coupled nonlinear oscillator, were discussed.

First, the dependence of the in-plane righting moment arm and the out-of-plane moment arm with the in-plane tilt angle and out-of-plane tilt angle were obtained. This was achieved by computing the moments against in-plane tilt angle at different fixed out-of-plane tilt angles. These out-of-plane tilt angles are vector sums of appropriate heel and trim. Euler angles were used for the transformations. Then, a homogenous cubic polynomial in two variables, i.e., the in-plane tilt and out-of-plane tilt angles, is used to obtain a surface fit approximation to model the stiffness in the equations of motion. A similar surface is obtained for the out-of-plane moment arm (fig. 2a and fig. 2b).

To implement the Modal Curve approach, eqn.(14) is used in which functions f_1 and f_2 are specified by the surfaces in fig. 2 (a) and (b) respectively times the inverse of the inertia matrix. In this application, x_1 and x_2 are the in-plane and out-of-plane motions respectively. The variable x was chosen to be a cubic polynomial function in x_1, with undetermined coefficients. By matching the powers of x_1, a set of algebraic equations in the unknown coefficients are obtained as described in the previous sections. As expected, two real solutions were obtained for the coefficient of the linear term suggesting two real modes. Fig. 3(a) and (b) represent the NNM's.

Using the modal approximation to x_2 from the equations of the above curves, the system can be decoupled in the in-plane mode from the out-of-plane mode. Then, the decoupled system can be solved like a SDOF system and the in-plane dynamics obtained. Fig. 4 shows the separatrices for each mode. These separatrices were obtained using the principle of conservation of energy for conservative systems.

Next, the invariant manifold approach is implemented to analyze the system. In this approach, the reference pair of coordinates are the in-plane displacement and velocity. As a result, the modal subspace is not a one-dimensional curve but a surface known as an invariant modal manifold by virtue of its invariance property described in the previous section.

As described in the last section, the second order system is expressed in Cauchy standard form, there-by, introducing velocity-dependence. Equations (18) and (19) form the basis of the formulation. In the present problem, the system possesses 2-DOF so that $n=2$ in eqns (18) and (19). The unknown functions x_1 and y_1 are chosen as bi-cubic polynomials in u and v consistent with the order of the polynomial approximations used for the restoring terms. Solving the system of algebraic equations yielded two real solution sets for the coefficients implying two real NNM's. Fig. (5) shows the two modes obtained.

Note that despite the system being undamped there is a velocity dependence in the functions representing the invariant modal manifolds. This dependence, however, only appears as even powers which can be eliminated using conservation of energy.

CONCLUSIONS

The results of this study emphasize the importance of using a dynamical systems approach to study the critical tilt dynamics of a marine vehicle. It was verified in previous studies that for hull forms similar to a semi-submersible, the effect of wave forcing is negligible (e.g., Falzarano, Vishnubholta and Kota, 1995). However, damping appears to be as important as for conventional hull forms, particularly in roll. Although, in this paper, viscous damping has not been explicitly calculated, the significant increase in vessel stability due to damping is clear from what was demonstrated in previous studies (Falzarano and Kota, 1996). This may justify the need to know viscous damping explicitly and that is the subject of a related study (Korpus and Falzarano, 1996).

The existence of a critical axis in proximity of the principal diagonal of the waterplane, was predicted and verified in previous work (Falzarano and Kota, 1996). Physically, this means that a tilting moment about the critical axis is likely to cause capsize more readily than moments about vessel longitudinal or transverse axes. This is so because the out-of-plane motion is unstable. This seems to have been the mechanism of capsize of the MODU *Ocean Ranger*. High waves and strong steady wind in approximately quartering direction caused a dangerous list of the vessel flooding one of its chain lockers (US Coast Guard, 1982).

For a more in-depth study of the motions about the critical axis, a 2-DOF analysis was performed. Preliminary stability analyses indicated strong static and dynamic coupling of the in-plane motions to the out-of-plane motions. Though in this study, the effects of damping and forcing have not been considered, the decoupling procedures seem to be readily applicable to at least damped systems (Shaw and Pierre, 1993). However, a major drawback of the procedure lies in the degree of computational complexity involved in solving the system of algebraic equations even for a simple cubic polynomial approximation for stiffness. It is well known that at least a fifth order polynomial is necessary to model the righting arm curve accurately (Falzarano, et al., 1992).

ACKNOWLEDGEMENTS

The authors would like to acknowledge the support of the National Science Foundation's Dynamic Systems and Control Program and its program manager Dr. Devendra Garg.

REFERENCES

American Bureau of Shipping(ABS), (1990), "Application of Dynamic Response Based Intact Stability Criteria for Column Stabilized Mobile Offshore Drilling Units", ABS, New York.

Caughey,T.K. (1960), "Classical Normal Modes in Damped Linear Dynamic Systems", *Journal of Applied Mechanics,* **27**.

Chakrabarti,S.K. (1987), *Hydrodynamics of Offshore Structures*, Springer-Verlag, New York.

Falzarano,J.M., Steindl, A., Troesch, A.W., and Troger,H. (1991), "Rolling Motion of a Ship Treated as a Bifurcation Problem", *Bifurcation, and Chaos: Analysis, Algorithims, Applications,* Birkhäuser, Basel.

Falzarano,J.M., Shaw,S.W., and Troesch,A.W. (1992), "Application of Global Methods for Analyzing Dynamical Systems to Ship Rolling and Capsizing", *International Journal of Bifurcation and Chaos in Applied Science and Engineering,* **2**,1.

Falzarano,J.M. and Troesch,A.W. (1990), "Application of Modern Geometric Methods for Dynamical Systems to the Problem of Vessel Capsizing with Water-On-Deck", *4th International Conference on the Stability of Ships and Ocean Vehicles,* Naples, Italy.

Falzarano,J.M. and Zhang,F. (1993), "Multiple Degree of Freedom Global Analysis of Transient Ship Rolling Motion", *Nonlinear Dynamics of Marine Vehicles: Modeling and Applications*, ASME Winter Annual Meeting, New Orleans, Louisiana.

Falzarano,J.M., Kota,R.S. and Vishnubhotla,S. (1995), "Survival Analysis of Deep-water Floating Offshore Platforms", *Brasil Offshore '95*, Rio de Janeiro, Brasil.

Falzarano,J.M. and Kota,R.S. (1996), "Survival Analysis of Deep-water Floating Offshore Platforms : Critical Axis", *Proceedings, ASME OMAE Conference*, Florence, Italy.

Guckenheimer,J. and Holmes,P. (1983), *Nonlinear Oscillations, Dynamical Systems and Bifurcations of Vector Fields*, Applied Mathematical Sciences **42**,Springer-Verlag, New York.

Korpus, RA and Falzarano, JM, (1996), "Prediction of Viscous Ship Roll Damping Using Unsteady Navier-Stokes," *Proceedings, ASME OMAE Conference,* Florence, Italy, to appear in: *J. of Offshore Mechanics and Arctic Engineering,*

Kota , RS, (1997),"Survival Analysis of Deep-Water Floating Offshore Platforms about Their Critical Axis" Master's Thesis, University of New Orleans, January.

Rosenberg,R.M. (1966), "On Nonlinear Vibrations of Systems with Many Degrees of Freedom", *Adv. in Applied Mechanics*, **9**.

Royal Commission (1982), "Report One :The Loss of the Semisubmersible Drill Rig *Ocean Ranger* and its Crew", *Royal Commission on the Ocean Ranger Disaster*, Canada.

Shaw,S.W. and Pierre,C. (1993), "Normal Modes for Nonlinear Vibratory Systems", *Journal of Sound and Vibration,* **164**.

Taz ul Mulk, M and Falzarano,J.M. (1994), "Complete Six Degrees of Freedom Nonlinear Ship Rolling Motion", *J. of Offshore Mechanics and Arctic Engineering*, **116**.

United States Coast Guard (1983) "*MODU Ocean Ranger Capsizing and Sinking in the Atlantic Ocean on 15 February 1982 with Multiple Loss of Life*", Marine Casualty Report.

Vakakis,A.F.,Manevitch,L.I., Mikhlin,Y.V., Pilipchuk,V.N. and Zevin,A.A. (1996), "*Normal Modes and Localization in Nonlinear Systems*", Wiley Series in Nonlinear Science, Wiley, New York.

Vakakis,A.F. and Rand,R.H. (1992), "Normal Modes and Global Dynamics of a Two Degree of Freedom Nonlinear System : I Low Energies, and II High Energies", *Int'l Journal of Nonlinear Mechanics*, **27**(5).

Wiggins,S. (1990), *Introduction to Applied Nonlinear Dynamical Systems and Chaos*, Texts in Applied Mathematics **2**, Springer-Verlag, New York..

Zhang,F and Falzarano,J.M. (1994), "Multiple Degree of Freedom Global Transient Ship Rolling Motion: Large Amplitude Forcing", ASME Winter Annual Meeting, *Symposium on Stochastic Dynamics and Reliability of Nonlinear Ocean Systems*, Chicago, Illinois.

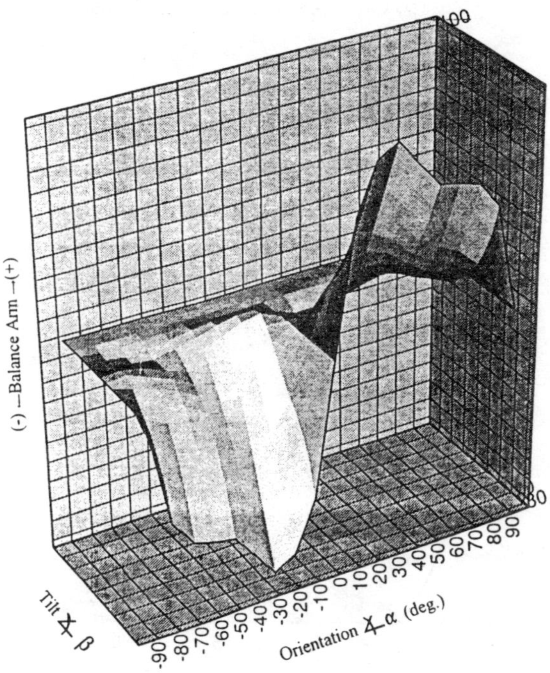

Fig. 1: Variation of Out-of-Plane Moment Arm with Tilt Axis Orientation

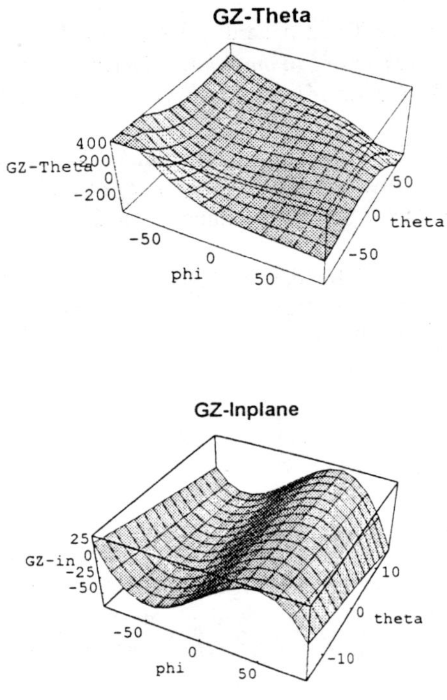

Fig. 2 : Surfaces of (a) In-Plane and
(b) Out-of-Plane, Restoring Arms

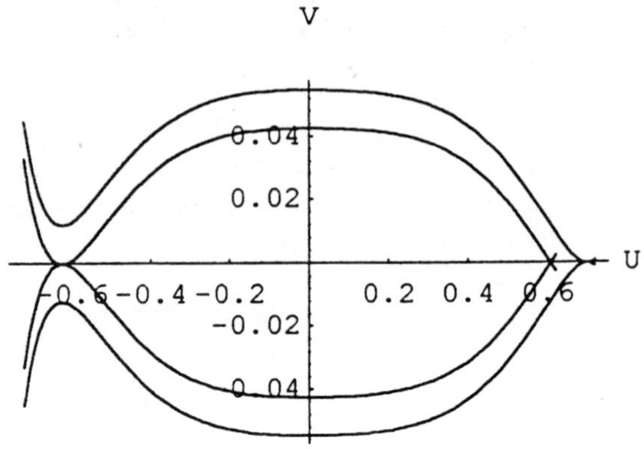

Fig. 4: In-Plane Dynamics for the decoupled system.

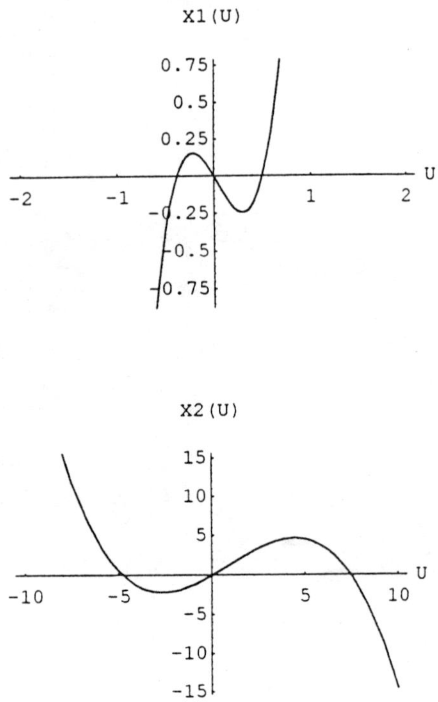

Fig. 3 : Nonlinear Normal Modes using Vakakis' approach

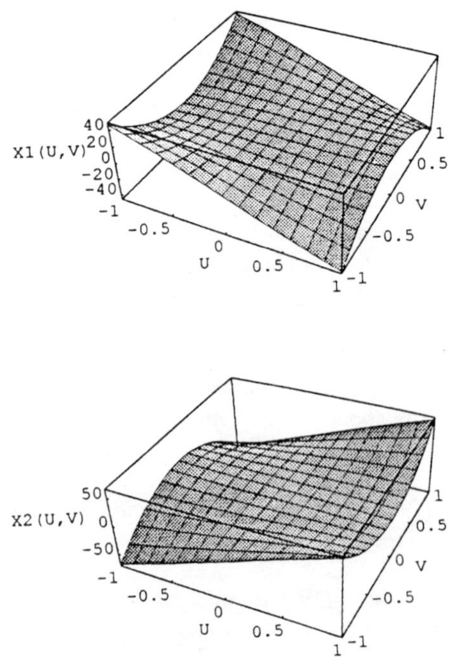

Fig. 5: Invariant Modal Manifolds

Proceedings of the Seventh (1997) International Offshore and Polar Engineering Conference
Honolulu, USA, May 25-30, 1997

ISBN 1-880653-28-1 (Set); ISBN 1-880653-31-1 (Vol. III)

On the Roll Radius of Gyration of Ro-Ro Passenger Ships

A. Papanikolaou, E. Boulougouris and D. Spanos
National Technical University of Athens
Athens, Greece

Abstract

The present paper is based on a study of the Ship Design Laboratory of NTUA concerning the specification of the Roll Radius of Gyration in the framework of the implementation of the so-called *Equivalent Model Test Method* following Res. 14 of the new SOLAS 95 regulations for Ro-Ro Passenger Ships. The objectives of the study were specifically to outline a methodology for the correct estimation of the roll radius of gyration of existing and new ships, to apply this methodology to a characteristic sample of existing Ro-Ro passenger ships, to review existing semi-empirical formulas for the determination of the roll radius of gyration of Ro-Ro passenger ships, as suggested by IMO, and to conclude, finally, on the practical implications of the correct estimation and reproduction in model scale of the roll radius of gyration to the validity of the Model Test results. The principal findings from the NTUA investigation were that existing, semiempirical formulas and approximations for the roll radius of gyration do not account for the special features of typical Ro-Ro passenger ship designs and can lead to significantly erroneous results, that the radius of gyration of modern Ro-Ro passenger ships is comparatively large, and that an underestimation of the actual roll radius of gyration and its modeling according to Res. 14, SOLAS 95, will eventually lead to an increased tendency to capsize the ship model in the specified seastate conditions.

Key Words: radius of gyration, Ro-Ro passenger ships, SOLAS95, equivalent model test method

1. Introduction
Background - Brief Review of the State of the Art

Recent IMO SOLAS Regulations for Ro-Ro passenger ships include the possibility of conducting model tests to prove the ability of a ship to survive in damage condition[1]. The ship model should copy the actual ship for both outer configuration and internal arrangement. It should also model properly the mass distribution of the actual ship, paying particular attention to scaling of radii of gyration in roll and pitch motions. The ship model should undergo survivability model tests to two different long crested irregular seaways defined by the following JONSWAP spectra, namely:

1. $T_p = 4\sqrt{H_s}$ with $\gamma = 3{,}3$ (1)
2. T_p is equal to the roll resonant period for the damaged ship without water on deck at the specified loading condition, but not higher than $6\sqrt{H_s}$ with $\gamma = 1.0$,

where H_S is the significant wave height (to be not exceeded with a probability of more than 10%), γ the peak enhancement factor and T_p the spectral peak period.

The difficulty in the execution of the above procedure is the determination of the roll resonant period for the damaged ship, without water on deck at the specified loading condition, when this value is not available from previous calculations. From the theoretical point of view the natural roll period of a ship can be obtained by the following formula :

$$T_R = 2\pi\sqrt{\frac{I_{xx}}{\Delta \cdot g \cdot GM}} \quad (2)$$

[1]Regarding the damage stability of existing and new Ro-Ro Passenger ships the SOLAS 95 conference adopted Resolution 14 stating that two or more Governments may conclude agreements modifying the requirements of Regulation II-2 / 8-1 in respect of every Ro-Ro Passenger Ship on regular scheduled voyages between designated ports in their territory and, if these safety requirements include specific stability requirements, they should not exceed those specified in the Annex to the Resolution (i.e. SOLAS 90 survivability standard with 50 cm of water accumulated on the vehicle deck, or, as an alternative, proof established by model tests carried out for an individual ship in accordance with a method developed by IMO, justifying that the ship will not capsize under the assumed damaged conditions. So far, 7 (seven) member countries of IMO have expressed their interest to conclude an agreement, as suggested above, namely United Kingdom, Ireland, Norway, Sweden, Finland, Denmark and Germany. Following the so-called Stockholm meeting in February 1996, the Marine Safety Agency (MSA) of United Kingdom released in May 1996 *Interim Guidance Notes for the Uniform Model Test Method*, a document intended to ensure uniformity in the methods employed in the construction and verification of the model as well as in the undertaking and analyses of the model tests.

where I_{xx} is total moment of inertia (mass and hydrodynamic) about the (longitudinal) roll axis, Δ the displacement, g the gravity acceleration and *GM* the transverse metacentric height. In the application of the above formula the main difficulty is the evaluation of the moment of inertia that consists of the mass moment of inertia and the added, hydrodynamic moment of inertia due to the acceleration of the fluid by the oscillating ship. The above formula can be found also written in the form:

$$T_R = 2\pi \frac{K}{\sqrt{(g \cdot GM)}} \quad (3)$$

where K = **c** B is the so-called roll radius of gyration, that might be expressed as percentage of the ship's beam. The coefficient **c** is generally varying between 0.38 and 0.44 for *surface ships*, but it might accept values as low as 0.32 and as high as 0.50 depending on the particular ship type *and her loading condition* (see, e.g., *Principles of Naval Architecture, 1989).* There seems to be no generally accepted state of the art on the estimation of the roll radius of gyration for Ro-Ro passenger ships. However, it can be expected, that modern Ro-Ro ship designs will tend to have **c** values close to above cited *upper limit* for surface ships, what should be supported by the following study.

Aims and Objectives of the Study

Following the so-called Stockholm meeting of the seven Northwest European countries and members of IMO, who have expressed their interest to conclude an agreement acc. to Res. 14 of SOLAS'95 in February 1996, the Marine Safety Agency (MSA) of United Kingdom released in May 1996 *Interim Guidance Notes for the Uniform Model Test Method,* a document intended to ensure uniformity in the methods employed in the construction and verification of the model as well as in the undertaking and analyses of the model tests. The text of this document is considered going beyond the corresponding text of originally approved IMO SOLAS'95 convention. Taking this into account, the relevant shipowners associations in Greece (Hellenic Chamber of Shipping, Greek Shipowners Association, Union of Greek Shipowners for Coastal Passenger Vessels) and the Hellenic Register of Shipping charged the Ship Design Laboratory of the National Technical University of Athens to consider the practical implications of the proposed *Interim Guidance Notes for the Uniform Model Test Method* on a sample of existing Greek Ro-Ro passenger ships and to critically review the proposed procedures.

Based on the outlined background, it was considered necessary to address herein the methodology for the Specification of the Roll Radius of Gyration for Ro-Ro Passenger ships. The study included the following tasks:

- *To collect the required technical data (structural plans, stability characteristics) for a sample of at least three (3) Greek Ro-Ro Passenger vessels in service on three (3) different routes.*
- *To develop a methodology for the estimation of the mass moment of inertia (and the mass roll radius of gyration) of existing ships and to apply the methodology to the selected sample of ships.*
- *To perform a seakeeping analysis of the selected vessels by suitable computer software in typical seaways of the service routes and to estimate the hydrodynamic, added moment of inertia and the related increase of the roll radius of gyration.*
- *To review the state of the art in the field and to outline the practical implications of the correct estimation of the roll radius of gyration on the validity of the equivalent model test method.*

2. Methodology

2.1 Theoretical estimation of the mass moment of inertia

The calculation of the mass moment of inertia of a ship is a complicate and tiresome task, that is usually omitted in the design process of a *merchant* ship. From the theoretical point of view the procedure, to be followed, is straightforward, considering the ship as a solid body and following the fundamental laws of basic mechanics. However, the accurate description of the ship components (namely their individual mass and mass moment of inertia, as well as their accurate position), contributing to the ship's mass (structural elements, machinery, outfitting and deadweight loads) is practically unknown and alternative ways must be considered to approach the problem in an approximate way. In the present project we tried to calculate the roll mass moment of inertia for a sample of Ro-Ro passenger ships in a more accurate way and to draw conclusions about the validity of existing semi-empirical formulas. For this purpose we selected a sample of 3 modern Ro-Ro passenger ships of the Greek fleet (see Table 1). The technical data, that were available to us, were the *TRIM & STABILITY BOOKLET* for all three of them, the *MIDSHIP SECTION* drawings and the *PROFILE INDEX & DECKS PLAN*. In addition, for two of the sample ships we used the *SHELL EXPANSION PLAN* and for one of them additional structural details of the sections. The limited extent of the available technical data made any attempt for more precise modeling of the structural elements of the studied ships impossible. However, the followed modeling procedure appears fully satisfactory for the purpose of the present study.

Ship No	1	2	3
Displacement [tons]	15500 (T=6.25m)	9833	11354
L.O.A. [m]	173.70	140.82	155.00
L.B.P. [m]	158.00	128.00	142.00
B [m]	24.00	22.40	22.80
T [m]	6.42 (max)	6.20	6.40
$D_{MAIN\ DK}$ [m]	8.80	8.01	8.00
$D_{UPPER\ DK}$ [m]	14.10	12.00	12.90
LS WEIGHT [tons]	10080	7091	7884
DEADWEIGHT [tons]	5420 (T=6.25m)	2742 (with 681W.B.)	3470 (with 915 W.B.)
SPEED	27.0 knots	19.0 knots	18.75 knots
No of Decks	9	8	7
No of Comp.	19	14	16
Year of Built	1995	1973	1972
Classification	ABS/HR	HR	HR

Table 1: General characteristics for the sample of selected Ro-Ro passenger ships

The total mass (displacement) of the ship was commonly divided into two main components, namely lightship weight and deadweight. For the deadweight, there is always a satisfactory subdivision, as given in the trim and stability booklet of the ship in question, while for the lightship weight, there is only the total weight and its center position available. Therefore several assumptions were required.

The lightship weight is commonly divided into three major components: the steel (structural) weight , the machinery weight and the outfit weight. According to design experience for Ro-Ro passenger ships, the structural weight is the largest component of the lightship. Thus, a good estimation of the ship's structural weight, including its distribution, is important for the calculation of the lightship's total moment of inertia. The methodology we used in this calculation was the "solid modeling" of a number of characteristic cross sections for each of the studied ships. The computer-aided procedure was based on the use of a well known software package, namely AUTOCAD v.12. A brief outline of the employed procedure can be found in Appendix A. It should be noted, that all calculations were performed with reference to a longitudinal axis through the center of gravity of the ship in question. Alternative ways of definition of the ship's rotational axis (e.g., through the ship's "rolling center" or the ship's center of floatation) lead to slightly different values for the mass moment of inertia and the related roll radius of gyration.

Each modeled section included both the longitudinal as well as the transverse structural members (see characteristic sketches in Figs 1 & 2). The length of the modeled longitudinal members was set equal to the ship's frame spacing. Consequently, a certain number of sectional modules were created for each ship. For the sample ship #1, for which the available data were most, the number of sections modeled where six (6) representing fairly well all structural elements along the ship. Using a composition of the above six sectional modules, an approximate 3D model of the whole ship was made. The resulting structural model was checked in terms of the resulting structural weight and mass centroid with semiempirical weight estimation procedures. After securing the validity of the generated structural model, the total mass radius of gyration of the ship #1 could be estimated. On the other side, for ships #2 and #3, the available data did not allow a detailed 3D modeling. Therefore the modeling was restricted to the midship section only. Assuming the remaining distribution of mass of ships # 2 and #3 similar to that for ship # 1, what is justified by common design features, the calculation of the total mass moment of inertia was completed by multiplication of the equivalent "midship radius of gyration", for each particular ship, by a reduction factor calculated through the 3D modeling developed for ship #1 . An interesting practical observation was made during the first stages of structural modeling of the ship's sections, namely the sectional value of the radius of gyration for the individual sections was practically reaching very quickly its final value, when taking into account the main structural elements (shell, decks, bulkheads, webs). Therefore, the details in the solid representation proved to be of small importance after considering the main structural components. This leads to the conclusion, that using the suggested modeling procedure we can estimate fairly accurately the roll radius of gyration, accounting for the ship's structure, and this will be not affected by inaccuracies in the total structural weight balance. Finally, the total moment of inertia has been obtained by multiplying the square of the resulting roll radius of gyration times the total mass of the ship's structural weight. The semiempirical method used for the estimation of the ship's steel weight is described by *S. A. Harvald & J. Juncher Jensen (1992)*. For the calculation of the Outfit Weight W_{OT} a simplified method was used, as originally proposed by *Careyette (Rawson-Tupper, 1979)*. Finally, the machinery weight could be estimated fairly accurately by using available data of the specific machinery manufacturers.

The vertical position of the center of gravity of each one of the above mentioned weight components was estimated based on empirical formulas as a fraction of the side depth to the upper deck. At this point it must be pointed out, that although the W_{OT} and W_M (except of the main engines) components were approached as groups of weights, leading to the estimation of the relevant weight magnitude and its centroid position, and assuming a distribution thereof over the ship's relevant spaces leading also to total mass moment of inertia, the mass moments of inertia of the individual components of the particular weight groups around their own centers of gravity was not included, due to the lack of data for the details of these weights. The same is also true for the individual weight components of the passengers and various other smaller loads included in the ship's deadweight.

Therefore the calculated mass moment of inertia obtained following the above described procedure is in fact **a lower limit of the true moment of inertia of the ship**. The results of calculation for each studied ship are presented in worksheets (see examples in Appendix A). They include also all relevant calculations made for the mass moments of inertia of the payload trucks and cars, according to a loading scenario.

Finally, we compared our results with equivalent results from existing semiempirical formulas of the international literature. The comparison leads to the conclusion, that the most promising formula appears to be that proposed originally by *Professor H. Kato*, because it includes all significant parameters typical to a ship design. However, the formula, originally developed in 1955, needs a modification as to the value of the so-called F parameter in order to reflect properly modern Ro-Ro passenger ship designs.

2.2 Theoretical estimation of the added (hydrodynamic) mass moment of inertia

The theoretical estimation of the hydrodynamic moment of inertia for an existing or new ship is a straightforward procedure, when the relevant computer algorithms are available. From the fundamentals of seakeeping hydrodynamics it is well known that the hydrodynamic roll moment of inertia A_{44} is a function of the ship's hull form, the frequency of oscillation ω, or alternatively of the period of oscillation T (is, for zero forward speed, equal to the wave period). The A_{44} moment depends also on the definition of the co-ordinate system and the longitudinal axis, over which it is defined. Three alternative longitudinal axes can be used for the calculation of A_{44}. These axes pass alternatively, firstly, through the center of roll, which lays approximately between the centers of gravity and buoyancy of the ship, secondly, through the center of gravity and thirdly, through the point defined by the intersection of the still waterplane and the vertical center plane. In the following, all calculations refer to an axis through ship's center of gravity.

Considering the roll motion of a specific vessel, we should compute first the function $A_{44}(T)$, which is a characteristic function for that ship, as described above. Generally, $A_{44}(T)$ proves to be for ships, in the range of periods of interest, a constant value. Assuming that the period range, over which this analysis takes place, includes the natural roll period of that ship, we can obtain a constant value for A_{44} which characterizes the ship in question. After calculation of the above value for the hydrodynamic (added) roll moment of inertia, assuming a certain value for the mass moment of inertia I_{44}, as suggested before, we can approximate the value of the roll resonant period of the ship, as requested for the purpose of this study (see, Figs. 3-5).

All computations were herein carried out by use of the computer program "NEWDRIFT v.6", running on a work station DEC-3000. This computer program has been originally developed for the seakeeping analysis of arbitrarily shaped three-dimensional bodies in long crested seaways *(see, A. Papanikolaou, 1985-1988)*. The computer algorithm is based on a 3D pulsating panel source method. The program evaluates all first and quasi second-order motions and loads, including drift deviations, forces and moments in regular waves, and is valid for arbitrary 3D bodies, finite or infinite water depth, arbitrary frequency of oscillation, with zero or nonzero forward speed, viscous effects included.

2.3 Experimental estimation of the roll radius of gyration for existing ships

Following the definition of the roll resonance period of a ship, namely Equ. (3), it is possible to determine, in general, for an existing ship the K value by the following pure experimental procedure:

1. Perform a hydrostatic *Inclining Experiment* and determine the GM value of the ship in the specific loading condition.
2. Perform a hydrodynamic *Induced Rolling ("Sallying")* experiment and measure the natural (resonance) period of roll. Taking into account the GM value, acc. to 1., the above equation can be solved for K to determine the roll radius of gyration.
3. Perform, as an alternative to 2., a *Roll Extinction* test and measure the period of the damped (free) roll motion that can be approximately set equal to the ship's natural period in roll. However, this procedure is less accurate and requires the knowledge of the roll damping and added mass moment coefficient, when the actual roll resonance period has to be determined.

The above procedure seems from the theoretical point of view straightforward, however for large Ro-Ro passenger ships practically impossible, due to the difficulty to excite the ship into a roll resonance situation, as required by the *Induced Rolling Experiment* or to conduct a *Roll Extinction Test* with sufficiently accurate results.

3. Overview of Results

3.1 Estimation of the mass moment of inertia for a sample of Ro-Ro passenger ships

The implementation of the above described approximate methodology for the evaluation of the mass moment of inertia of existing ships and the corresponding roll radius of gyration showed the following results for the selected sample of ships:

Ship No.	1	2	3
Radius of gyr. due to structure [% B]	43	42	42
Radius of gyr. due to lightship [% B]	42	37	40
Radius of gyr. due to total mass [% B]	39	35	38

Table 2: Roll Radius of Gyration due to Mass Moments of Inertia for a Sample of Three Ro-Ro Passenger Ships (*lower limit w/o hydrodynamic effects*)

The results confirm, to a great extent, the range of validity of existing semiempirical formulas. However, they also demonstrate the individual character of each studied ship. The resulting difference of about 12% between the dimensionless roll radius of gyration (as percentage of B) of the first two studied ships proves that more precise calculations have to be performed, before deciding on the value of the radius of gyration by use of simple formulas, e.g. setting it equal to 0.40 B. On the other hand, the methodology used herein gives a lower limit for the mass roll radius of gyration. Therefore, the roll radius of gyration of a modern Ro-Ro passenger ship will be eventually higher and, including the hydrodynamic (added) moment part, in several cases reaching or exceeding the value of 0.45 B.

3.2 Estimation of the hydrod. moment of inertia for a sample of Ro-Ro pass. ships

Following the results for the sample of the selected ships, it appears that the hydrodynamic effects **increase** the roll radius of gyration by a value of abt [1.5÷5.1] % B (ship beam). This percentage depends strongly on the particulars and the hull form of the specific vessel in question, as well as its mass distribution, that determines the percentage reference value of the mass moment of inertia. It should be pointed out that the related increase of the moment of inertia I_{44} due to the hydrodynamic term, as percentage of the mass moment of inertia, is much higher than the percentage increase of the radius of gyration, namely varying herein between 7 and 28 %. The percentage difference is due to the squaring procedure in obtaining the moment of inertia from the radius of gyration.

Ship No.	1	2	3
Radius of gyr. due to total mass [% B]	39	35	38
Overall Roll Radius of Gyration [% B]	44	38	40
Difference due to hydrod. term [% B]	5	3	2

Table 3: Total Roll Radius of Gyration for a Sample of Three Ro-Ro Passenger Ships (*lower limit including hydrodynamic effects*)

3.3 Review of Proposed Empirical Formulas by IMO members

A review of proposed semiempirical formulas by IMO members and by various authors from the international literature[2] suggests, that the results obtained thereafter can be very wrong. The formula suggested by IMO Resolution A 562 (14) gives herein relatively low values for the actual roll radius of gyration (including the hydrodynamic effects), whereas others, like H. Kato's formula, too high values. The formula suggested by the MSA of United Kingdom, namely to set the radius not higher than 0.4 B, is too simplistic and does not consider any particular features of the ship under investigation. After consultation with various other authorities, it was confirmed that there is a general lack of detailed information about the specification of the roll radius of gyration for Ro-Ro passenger ships (*see, Watanabe and Rathje, 1996*).

[2] see *Papanikolaou, A. et al (1996)* for details

4. Conclusions and Proposals

The principal findings from the NTUA investigation into the determination of the roll radius of gyration for three modern Ro-Ro passenger ferries, considering the approximation by IMO's formula after Resolution A 562(14) and the formula suggested by the Marine Safety Agency of United Kingdom (see, *Allan, 1996*), were the following:

1. The results for the sample of ships, investigated in the present study, suggest that:

.1 Proposed formulas and approximations by IMO and MSA of United Kingdom do not account for the special features of typical Ro-Ro passenger ship designs (large and high superstructures, multiple decks, high loading etc.) and can lead to significantly wrong results.

.2 In general, Ro-Ro passenger ships tend to have, due to the large longitudinal and vertical extent of superstructures and the number of decks above the main deck, comparatively large radius of gyration, namely in the order of 0.43 to 0.45 B (or even higher), compared to abt. 0.38 B for conventional merchant ships. This value includes the effect of the added (hydrodynamic) moment of inertia resulting from the fluid acceleration due to the oscillatory ship motion.

.3 The hydrodynamic (added) moment of inertia about the ship's roll axis[3] *increases the mass moment of inertia of the ship by about 10 to 25%, depending on the ship's hull form, the frequency of oscillation and the ship's mass distribution. The corresponding hydrodynamic (added) roll radius of gyration has been estimated*[4] *to be in the order of between 10-20% B. The increase of the overall roll radius of gyration due to the hydrodynamic term is, in terms of % B, in the order of between 2 and 5 % B, depending on the ship's hull form and the ship's mass distribution.*

.4 A review of existing semiempirical formulas for the roll radius of gyration suggests that the formula presented several years ago by H. Kato (1956), properly adjusted for modern Ro-Ro passenger ships, can be further developed to deliver a good approximation of the actual roll radius of gyration. The IMO A.562 formula delivers herein fairly good results, however when neglecting hydrodynamic effects. Applying the same formula to other ships led to less acceptable results.

.5 In view of SOLAS 95, Res. 14, and the specifications for the execution of the Equivalent Model Test Method, the effect of ***underestimating*** *the actual roll radius of gyration is the following:*

- *The roll period of the ship is becoming artificially smaller than in reality. This effect is enhanced due to neglect of damping and viscosity effects, that tend to increase, in general, the roll period of a ship calculated for the undamped roll motion.*
- *The ship, to be tested for a JONSWAP spectrum with a peak period T_P equal to the "ship's roll resonant period for the damaged ship without water on deck at the specified loading condition but not higher than 6 $(H_S)^{1/2}$"*[5]*, being reproduced in model scale with a smaller roll radius of gyration, will have, in general, an increased tendency to capsize, because either her estimated roll radius of gyration will appear to be less than 6 $(H_S)^{1/2}$, therefore the test will be executed directly at the ship's roll resonance period, or in case of T_P equal to 6 $(H_S)^{1/2}$ the roll resonance point of the ship will be still close to the testing point (slightly supercritical).*
- *The actual roll resonant period of the damaged ship will be even higher, when considering the effect of water on deck or the in the flooded compartments and the related decrease of GM.*

2. In view of the above, the following measures are recommended:

.1 The determination of the radius of gyration of the mass moment of inertia for Ro-Ro passenger ships should be established:

a. For existing ships:

aa. *experimentally*: by recalling data from the ship's stability experiment (determination of GM) and additionally a ship's *sallying experiment* or *roll extinction test* (determination of roll period).

ab. *numerically*: by an approximate method considering the radius of gyration of typical ship sections, approximating the ship's structure, including the ship's superstructure, outfitting and loading, and in addition, by including the hydrodynamic effects through a seakeeping calculation by use of a modern 3D panel method.

ac. The use of semi-empirical formulas, like that originally proposed by H. Kato and other (IMO and MSA-UK formula), should be considered with great caution, unless confirmed by a sufficient number of test cases for modern Ro-Ro passenger ship designs.

b. For new ships:

ba. *numerically*: by an exact method when using modern CAD systems for the ship's structural design and the ship's intact and damage stability. The hydrodynamic effects can be included through a seakeeping calculation by use of a proper software program (3D panel method).

3. In view of the real physics of the capsizing of damaged ships in waves, what is supported by recent results of the so-called NORDIC project *(see, Proc. RINA conference, 1996)*, suggesting that the capsizing process damaged Ro-Ro ships is a quasi-static phenomenon, it is recommended to consider the results of the 2nd test with a JONSWAP spectrum corresponding to the ship's roll resonant period with great caution, because the specified test conditions are very extreme and unlike in practice, from the exciting seaway point of view, while not sufficiently clear, from the reproduction of the actual radius of gyration in model scale point of view.

5. Acknowledgments

The Ship Design Laboratory of NTUA wishes to acknowledge the financial support to the present research effort by the Hellenic Register of Shipping.

The present study was effectively supported by technical information and administrative assistance provided to the research team of NTUA by Rear Admiral A. Kaloudis (Greek Shipowners Association for Passenger Ships), Messrs. Agapakis and Kyriakopoulos (Hellenic Register of Shipping), Messrs. I. Asmussen and H. Rathje (Germanischer Lloyd, Hamburg), Professor D. Vassalos (Univ. of

[3] Herein assumed that the reference longitudinal axis passes through the ship's center of gravity

[4] The calculation was performed for a sample of three ships by the 3D Panel code NEWDRIFT.

[5] Acc. to SOLAS/CONF.3/46 Annex, pp. 36-37.

Strathclyde) and Dr. I. Watanabe (Ship Research Institute, Japan). Their help is greatly acknowledged.

References

1. *Allan, T. (1996), "Comment on the Equivalent Model Test Method (Res. 14) - Determination of the Roll Radius of Gyration - Roll Period"*, MSA United Kingdom, Feb. 1996.

2. *Comstock, J. P., ed., (1967),* Principles of Naval Architecture, The SNAME, New York.

3. *Harvald, S. V., Juncher Jensen, J., (1992), "Steel Weight Estimation of Ships", Proc. PRADS'92 Conf., Vol. 2, Newcastle, pp. 1047-1059.*

4. *Ikeda, Y., Himeno, Y., Tanaka, N., (1978), "Components of Roll Damping of Ships at Forward Speed"*, Journ. Soc. Nav. Arch. Japan, Vol. 143.

5. *Kato, H. (1956), "On the approximate calculation of ship's rolling period"*, Journal of the Soc. Of Naval Arch. Of Japan, Vol. 89, 1956.

6. *Lewis, E. V., ed., (1989),* Principles of Naval Architecture, 2nd Rev., Vol. I & III, The SNAME, New York.

7. *Papanikolaou, A. (1988),* "NEWDRIFT: The six DOF three-dimensional diffraction theory program of NTUA-SDL for the calculation of motions and loads of arbitrarily shaped bodies in regular waves", NTUA-SDL Athens Report, Athens.

8. *Papanikolaou, A. (1985), "On Integral-Equation Methods for the Evaluation of Motions and Loads of Arbitrary Bodies in Waves", Journ. Ingenieur - Archiv (vol. 55), 17-29.*

9. *Papanikolaou, A. (1989), Ship Design - Methodology of Preliminary Design (Greek Textbook), SYMEON Publ., Athens.*

10. *Papanikolaou, A., Boulougouris, E., Spanos, D. (1996), "On the Specification of the Roll Radius of Gyration for Ro-Ro Passenger Ships in view of the Implementation of SOLAS 95 Res. 14 Equivalent Model Test Method", NTUA-SDL Report, June, Athens.*

11. *Rawson, K. J., Tupper, E. C. (1979),* Basic Ship Theory, *Vol. 2, LONGMAN Publ., London and New York.*

15. *Rathje, H. (1996), "Personal Communication", Germanischer Lloyd, Hamburg.*

16. *The Royal Inst. of Naval Architects (1996), Proc. Int. Conf. On the Safety of Passengers in Ro-Ro Vessels - Presenting the results of the Northwest European Research and Development Project at IMO, London, June 7.*

17. *Watanabe, I. (1996), "Personal Communication", Ship Research Institute, Tokyo.*

Appendix A - Structural Design Modeling Procedure

In the present chapter we describe briefly the computer-aided design procedure employed for solid modeling of the various ship sections and of ship components, in general. As outlined in the main text (under 3.1), the calculation of the mass moment of inertia of the midship sections for the sample ships was accomplished by use of the well known software package AutoCAD®, Advanced Modeling Extension (AME™). AME is both a solid and region modeling program. The region modeller allows the designer to create, edit and query closed 2D areas called *regions,* whereas the solid modeller supports the operations for enclosed three-dimensional volumes called *solids.* The particular version used in this project was the AME Release 2.1 (see, *Manuals Autodesk Ltd*).

Both modellers, namely the *region* and *solid* one, use a building-block approach to construct a solid or a region. The user may start with basic shapes, namely for solids the basic shapes are boxes, cylinders, tori (donuts) , spheres , wedges , and cones, whereas for regions the corresponding basic shapes are any AutoCAD entities, that enclose an area, such as Circles and Polygons or 2D Polylines . The user can combine (join) several solids to form a new composite region. He can also subtract solids from each other and find the intersection (say, the common volume) of neighboring solids. The same can be done with regions. AME permits the easy creation of regions from solids and vice versa. Using the appropriate commands the user can extrude (add height to) or revolve (sweep about an axis) any region or 2D AutoCAD object to create an new solid shape .

Finally, another interesting feature of AME is its ability to quantitatively analyze solids and regions. It can calculate the mass or area properties of solids or regions such as their center of gravity, mass, surface area, moments of inertia, products of inertia, radii of gyration, and more, as required in the framework of the present project.

Table 4: Ship Data used for Roll Motion Calculations by NEWDRIFT

	SHIP No 1	SHIP No 2	SHIP No 3	
L	158,00	128,00	142,00	m
B	24,00	22,40	22,80	m
T	6,42	6,20	6,40	m
∇	15240,1	9492,6	10912,8	m^3
trim	0,00	0,00	0,00	m
A_W	3410,9	2209,3	2607,4	m^2
Wetted Surface	4574,6	3064,4	3412,5	m^2
LCF	-13,660	-8,335	-7,095	m from MS
KB	3,630	3,602	3,814	m from BL
LCB	-5,22	-4,481	-5,930	m from MS
KG	10,790	9,231	9,874	m from BL
LCG	-5,22	-4,481	-5,930	m from MS
GM	2,597	1,504	1,851	m
GM_L	412,7	235,7	307,7	m
heading angle	90	90	90	deg
i_{xx} (mass only)	9,120÷10,56	8,512÷9,86	8,664÷10,03	m
i_{yy}	39,500	32,000	35,000	m
i_{zz}	39,500	32,000	35,000	m
$i_{46} = i_{45} = i_{56}$	0,000	0,000	0,000	m
Num. of Nodes	2×214	2×192	2×142	
Num. of panels	2×192	2×167	2×118	
Note: Origin of Co-ordinate System at Ship's Center of Gravity				

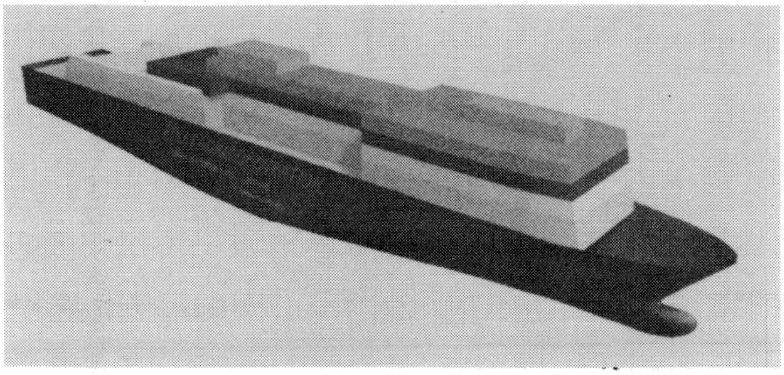

Fig. 1a 3D Solid Modeling of Ship # 1

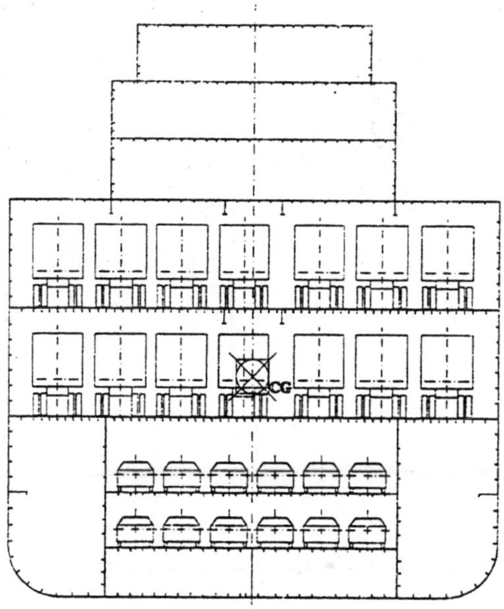

Fig. 2a Worst Loading Scenario for Ship # 1

Fig. 1b 3D Structural Modeling and Typical Loading of Midship Section Area for Ship # 1

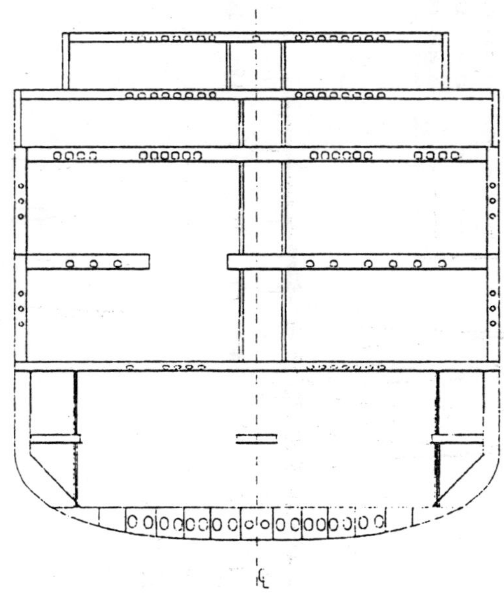

Fig. 2b Structural Modeling of Machinery Space for Ship # 1

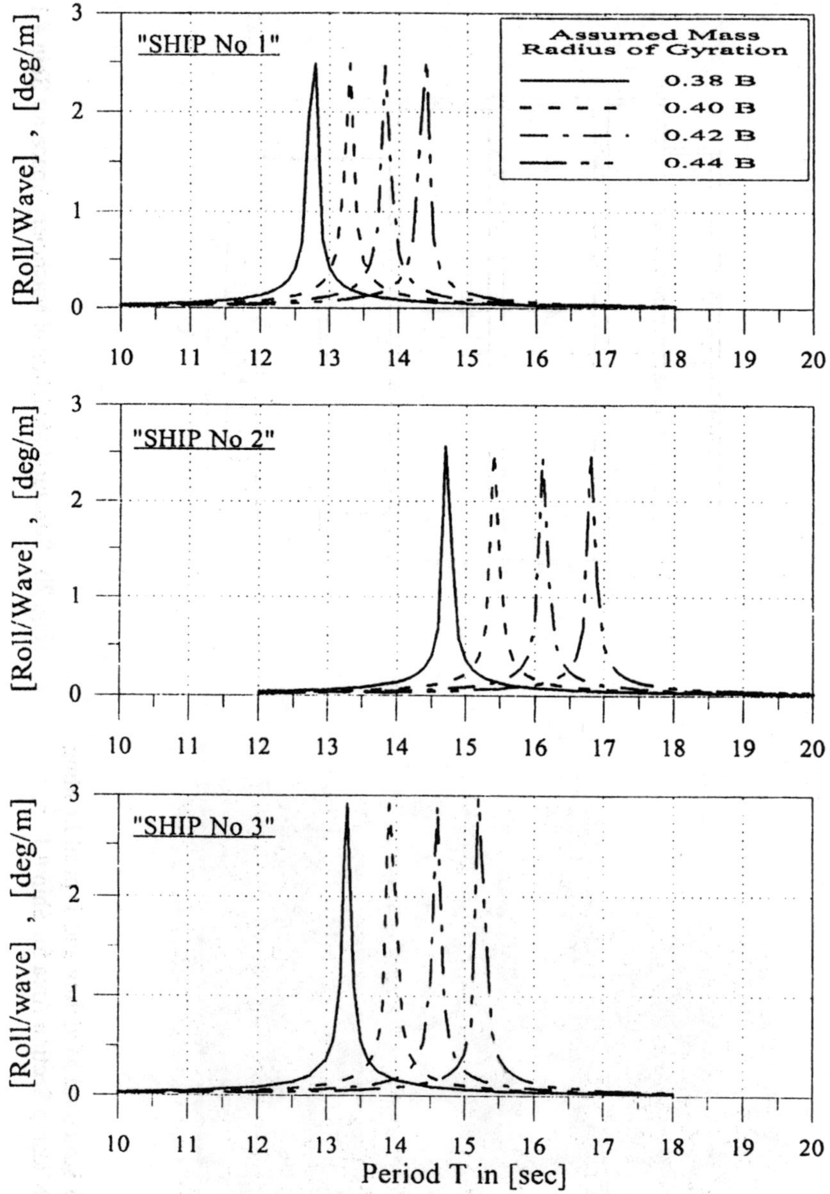

Fig. 3 Dimensionless Roll Motion Response Amplitude as a Function of Mass Radius of Gyration

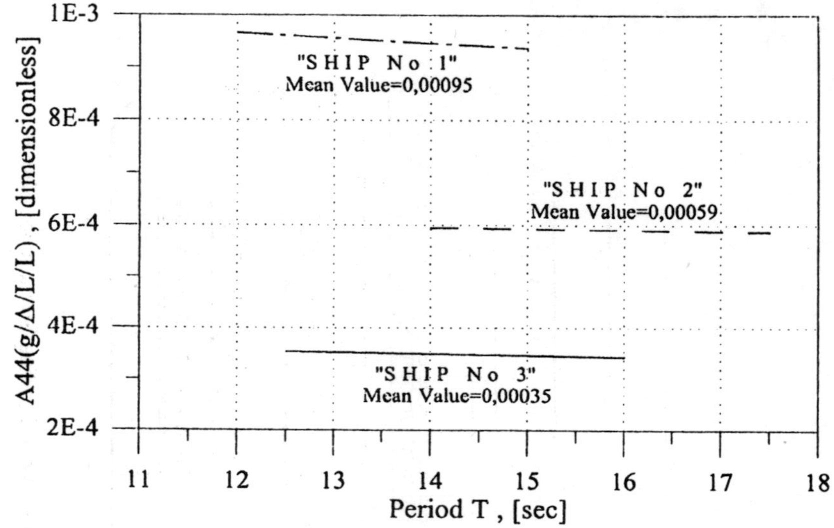

Fig. 4 Dimensionless Hydrodynamic Roll Moment of Inertia

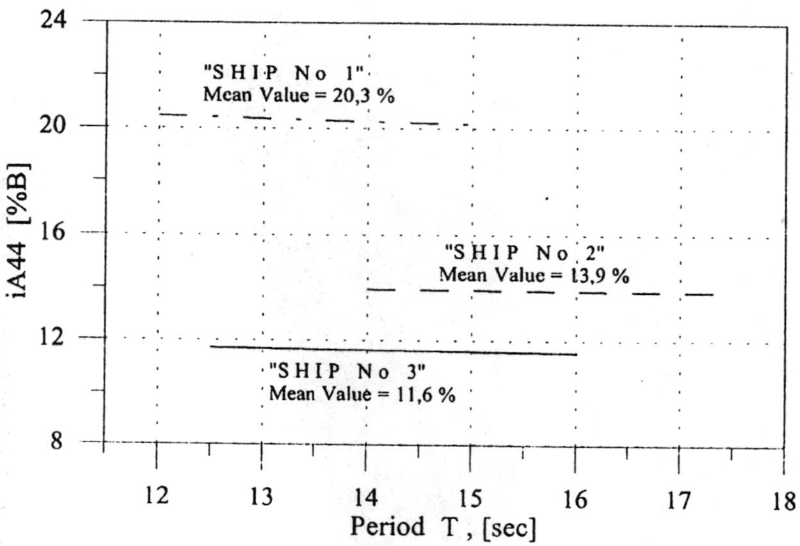

Fig. 5 Dimensionless Hydrodynamic Roll Radius of Gyration [%B]

Proceedings of the Seventh (1997) International Offshore and Polar Engineering Conference
Honolulu, USA, May 25-30, 1997

ISBN 1-880653-28-1 (Set); ISBN 1-880653-31-1 (Vol. III)

An Investigation on the Applicability of Simplified Mathematical Models to the Roll-Sloshing Problem

Alberto Francescutto and Giorgio Contento
University of Trieste
Trieste, Italy

ABSTRACT

In this paper, the possibility of obtaining an efficient description of the roll motion of a ship with liquids with free surface on board is discussed in detail. Available mathematical models with concentrated parameters are implemented and compared with scale model experiments. Critical points are evidenced by using a high efficiency Parameter Identification Technique.

Finally a model with good simulation capability is presented. The proposed mathematical model fits well the experimental data requiring the estimation of a reduced set of parameters. The method appears of great interest in the frame of current research on the damaged ship behaviour.

KEY WORDS: Sloshing, ship rolling, ship stability.

INTRODUCTION

Since the early papers of Van der Bosch and Vugts (1966) and Van der Bosch and Zwaan (1970) and even earlier with the work of Watts (1883,1885), the attempt to derive an appropriate mathematical model for the solution of the problem of the ship rolling with a partially filled tank on board has been attracting many researchers in the field of Naval Architecture. This is still an up-to-date task especially as regards the modelling of water inflow/outflow after damage. In Aeronautical Engineering the problem of the route stability of missiles and aircrafts with a free surface tank was also considered a problem of primary importance (Abramson, 1966). In the work of Van der Bosch et al. (1966) the main idea was to modify the traditional roll motion equation by simply adding the sloshing moment in the RHS of the equation as an additional exciting moment. The latter had to be evaluated experimentally from measurements on an isolated tank forced in a sinusoidal motion. By this approach the inherent complexity of developing a suitable mathematical modelling of the liquid motion in the tank could be avoided at the cost of large campaigns of experimental measurements. Later on Goodrich (1968) suggested a simple mathematical model constituted of two linear differential equations representing the roll angle and the slope relative to the ship of the free surface of the liquid (assumed flat) in the tank. There, the properties of the liquid in the tank were summarised by the natural frequency (derived from the linear wave theory), by a damping coefficient and by the classical static free surface correction factor. The excessive overlooking of the hydrodynamic aspects of the liquid motion in the adopted models pushed towards the searching of more realistic linear and nonlinear models (Faltinsen, 1974; Armenio et al., 1994) including CFD based approaches (Faltinsen, 1978; Mikelis et al., 1984; Pantazopoulos, 1992; Francescutto and Contento, 1994; Armenio et al., 1996a,b). More recently the occurrence of dramatic casualties at sea due to the flooding of the vehicle deck in RoRo ferries (Herald of Free Enterprice, Estonia) and the appearance of new typologies of tankers (double hull) and bulkcarriers affected by structural failures in the upper corners of the holds due to impulsive loads from cargo sloshing and by bulkhead failure due to flooding after damage respectively, has given a new impulse to the research both for a deeper understanding of the phenomena and for a rational approach to new rules. While the structural aspects are more related to a good prediction of liquid pressures and consequently to a detailed description of the fluid flow in the tank, the prediction of ship rolling amplitude for safety against capsize and sinking can be related to more simplified liquid models since only low order effects are primarily involved. In this perspective simplified mathematical models for the roll/sloshing coupled motions are welcome by Ship Safety Community.

In this paper a consistent method is developed and investigated. The equation for the sloshing motion is derived from the equivalent mechanical approach developed by Graham and Rodriguez (1952), whereas the rolling motion is considered to be described, for the time being, by a single nonlinear differential equation (Contento et al., 1996). The mathematical predictions are then compared with experimental steady roll amplitudes of a large scale model of a fishing vessel equipped with a flooded compartment with a constant amount of water and tested in a regular beam sea at zero forward speed. Four different liquid levels are analysed. A high efficiency Parameter Identification Technique is then developed to obtain the relevant parameters of the mathematical model (coefficients of the ODEs) and to discuss the model validity.

MATHEMATICAL MODELS PROPOSED IN THE LITERATURE

In the literature, several mathematical models for the roll/sloshing and stabilization problem have been presented and applied. Basically most models assume a quasi static behaviour of the liquid cargo (flat free surface) and/or the dynamics involved is that of a material particle located at the center of mass of the liquid. Then the quasi-static assumption is often "forgotten" and the analysis is conducted in a frequency range that includes the resonance of the tank with a consequent underestimate of the dynamic contributions. Sometimes the description of the liquid cargo dynamics is only contained in the natural frequency of the first surface mode derived from the wave theory.

The model proposed by Goodrich (1968) was derived from the theory of U-tube tank stabilizers for the extension to the design of free surface tanks. The model is written as follows

$$\begin{cases} \ddot{\varphi} + K\dot{\varphi} + \omega_\varphi{}^2\varphi - \mu\omega_\varphi{}^2\vartheta = \omega_\varphi{}^2\alpha\cos(\omega t) \\ \ddot{\vartheta} + k\dot{\vartheta} + \omega_t{}^2\vartheta = \omega_t{}^2\varphi - \dfrac{\omega_t{}^2 S}{g}\ddot{\varphi} \end{cases} \tag{1}$$

where φ and ϑ are the roll angle and the slope of the flat free surface of the liquid relative to the ship, K and k are the damping coefficients of the ship and of the tank respectively, ω_φ and ω_t are the natural frequencies of the ship (in the 'frozen' liquid condition) and of the tank respectively (from the linear wave theory), $\mu = \frac{1}{12}\frac{b^3 \ell}{\nabla \overline{GM}}$ is the static free surface correction factor, ℓ and B being the lenght and width of the tank, ∇ and $\overline{GM}$ the displaced volume and the metacentric height of the ship, S the distance between the mean liquid level and the rolling center. While the dynamics of the liquid in the isolated tank is implicitly contained in the natural frequency ω_t and the behaviour of the isolated ship is implicitly contained in ω_φ, the description of the coupled system remains ambiguous. It immediately appears that the term in $\ddot{\vartheta}$ is missing in the roll equation and the restoring capabilities of the ship are simply modified by the free surface static correction. The damping terms coefficients K and k have to be obtained from experimental tests.

Zdybek (1980) derived a similar mathematical model for U-tube stabilised ships. There, all the acceleration terms have been included and a different interpretation of the inertia coupling term in the tank equation has been given through solving the same problem (U-tube stabilizing tanks). In deriving this model, it is explicitly stated that the characteristics of the ship with frozen liquid must be used in the coupled system and the evaluation of the inertia coupling terms requires also the availability of results from experimental tests on the isolated tank. The analysis of this model has not been considered in the following due to the complexity of the tests needed to obtain these last terms (in addition to the damping coefficients).

Recently Ishida et al. (1996) presented a simplified model to study the behaviour of a ship in damaged condition, in particular for a RoRo with water on deck. The liquid motion is approximated by that of a material particle located at the center of mass of the liquid and the free surface of the liquid is assumed flat during the motion. The obtained system of equation in linearised form is given by eqns (2) where ϑ is the absolute slope of the free surface, b the breadth of the tank, h the liquid depth, M_F the mass of the fluid in the tank, a the distance of the bottom of the tank from the center of gravity G' of the ship including the 'frozen' liquid, Δ' and I_S' the displacement and the virtual moment of inertia of the ship about G'.

$$\begin{cases} \bullet \left[I_S' + M_F\left(a+\frac{h}{2}\right)\left(a+\frac{h}{2}+\frac{1}{12}\frac{b^2}{h}\right)\right]\ddot{\varphi} + \left[M_F\frac{1}{12}\frac{b^2}{h}\left(a+\frac{h}{2}\right)\right]\ddot{\vartheta} + \\ \quad \Delta'\left[\overline{G'M'} - \frac{M_F}{\rho\nabla'}\left(a+\frac{h}{2}+\frac{1}{12}\frac{b^2}{h}\right)\right]\varphi - M_F g\frac{1}{12}\frac{b^2}{h}\vartheta = A\sin(\omega t) \\ \bullet M_F\frac{1}{12}\frac{b^2}{h}\left(a+\frac{h}{2}+\frac{1}{12}\frac{b^2}{h}\right)\ddot{\varphi} + M_F\left(\frac{1}{12}\frac{b^2}{h}\right)^2\ddot{\vartheta} + \\ \quad M_F g\frac{1}{12}\frac{b^2}{h}\vartheta = 0 \end{cases} \tag{2}$$

The basic assumption on the mechanical equivalency of the liquid motion seen in the body fixed reference frame consists in the motion of a material point corresponding to the center of the liquid mass with flat free surface. After linarization, this reduces to the application of the metacentric method entailing the rotation of the point mass around the metacenter of the liquid volume.

In any case, the above assumptions lead to an incorrect estimate of the natural frequency of the isolated tank and consequently of the whole behaviour of the system. From eqn (2), the isolated tank equation is given by eqn (3)

$$M_F\left(\frac{1}{12}\frac{b^2}{h}\right)^2\ddot{\vartheta} + M_F g\frac{1}{12}\frac{b^2}{h}\vartheta = 0 \tag{3}$$

the natural frequency being therefore

$$\omega_t = \frac{1}{b}\sqrt{12gh} \tag{4}$$

whereas from linear wave theory it is well known that

$$\omega_t = \sqrt{\frac{\pi g}{b}\tanh\left(\frac{\pi h}{b}\right)} \tag{5}$$

which in very shallow liquid condition reduces to

$$\omega_t = \frac{\pi}{b}\sqrt{gh} \neq \frac{1}{b}\sqrt{12gh} \tag{6}$$

The Parameter Identification Technique briefly described in Appendix 2 has been applied to the above mathematical models to derive the best values of the 'unknown' coefficients, i.e. the damping terms and the effective wave slope coefficient which is usually neglected by previous authors. The simulation capability of the different models to reproduce the experimental values described in Appendix 1, is shown in Fig. 1. As one can see, the goodness of fit is generally poor in at least one of the two resonance frequencies. Moreover the identified parameters have an obscure interpretation since the failure of the methods in capturing the eigen-frequencies drives the best-fit procedure to balance the result with physically incorrect values of the dampings (often found to be negative) and of the effective wave slope (often found to be greater than unity).

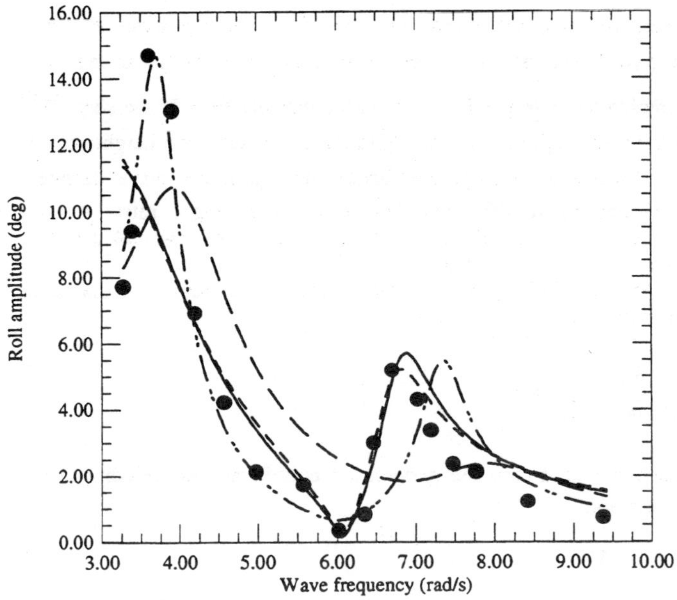

Figure 1. Simulation capability of the different models proposed in literature. The curves refer to stationary rolling. The used coefficients have been derived through a Parameter Identification Technique. [● exp. data; —— Goodrich (ship natural frequency ω_φ in the frozen liquid condition); - - - - Goodrich (ship natural frequency ω_φ without liquids); —·—·—Goodrich (best fit ship natural frequency ω_φ); — — — Ishida et al.]

PROPOSED MATHEMATICAL MODEL

From the analysis of the mathematical models above described, it results that the weak point is generally found at least in the description of the liquid motion in the tank. In particular the attempt to simplify the liquid motion reducing it to the motion of a material particle located in the center of gravity of the liquid and assuming a flat free surface leads to the isolated tank equation (3). The linear ordinary differential equation representing the motion of the free surface inviscid liquid in a rectangular tank at rest should be derived by integrating the potential U and the kinetic energy T of the elemental fluid mass over the whole volume of fluid. The results are briefly reported in eqns (7) for sake of compleness.

$$\begin{cases} \bullet\ T = \dfrac{1}{16}\dfrac{\pi\rho g^2 H^2}{\omega^2}\tanh\left(\dfrac{\pi h}{b}\right)\sin^2(\omega t) \\ \bullet\ U = \dfrac{1}{16}\rho g H^2 b \cos^2(\omega t) \end{cases} \tag{7}$$

H is the height of the standing wave in the tank. Consistently with the flat free surface approach, we assume $\vartheta = \dfrac{H}{b}$ and substitute it in T and U. Writing then the Lagrange equation, it follows

$$M_F \frac{1}{8}\frac{b^3}{\pi h \tanh\left(\dfrac{\pi h}{b}\right)}\ddot{\vartheta} + M_F g \frac{1}{8}\frac{b^2}{h}\vartheta = 0 \tag{8}$$

leading to the well known "isolated tank frequency" given by eqn (5).

It appears that eqn (8), i.e. the linear hydrodynamic method, and eqn (3), i.e. the quasi-static method, do not give the same potential and inertial contributions. This is believed to be one of the reasons why the fitting capability to experiments of the model given in eqn (2) is generally poor.

An attempt to improve the solution given by eqn (2) has been made by substituting the simple pendulum with a compound pendulum. Its inertia has to be determined through the best fit procedure. The results are reasonably good but the obtained values of the inertia of the pendulum have an obscure interpretation and cannot be simply assigned, therefore the model lacks of immediate application.

Still pursuing the hydrodynamic way, a further improvement of the simulation capability of a simplified model is here obtained by a novel approach. It is based on a study of Graham and Rodriguez (1952). They derived the equations governing the fluid flow in a 3D rectangular tank excited to roll, sway and yaw under the hypotesis of perfect fluid and linearized boundary conditions. The velocity potential and the related pressures exerted by the fluid on the walls of the tank were explicitly given as a function of the amplitude and frequency of the motion of the tank and of the tank dimensions. In the same paper some "equivalent" mechanical systems in term of input-output to the actual sloshing tank were described. These mechanical systems, in other words, give the same forces and moments exerted by the actual fluid sloshing when subject to the same forcing cause (motion of the tank). In Fig. 2 a schematic diagram of the equivalent mechanical system representing the liquid motion when the tank is subject to coupled sway/roll motion is reported. It is composed of a fixed mass M with moment of inertia I_y about the longitudinal axis y passing through $z = Z$ and an infinite set of undamped spring-point masses m_n allowed to move only in the direction parallel to the bottom of the tank. The spring constants are k_n and the point masses are located a distance z_n from the origin of the frame of reference (center of mass of the liquid at rest).

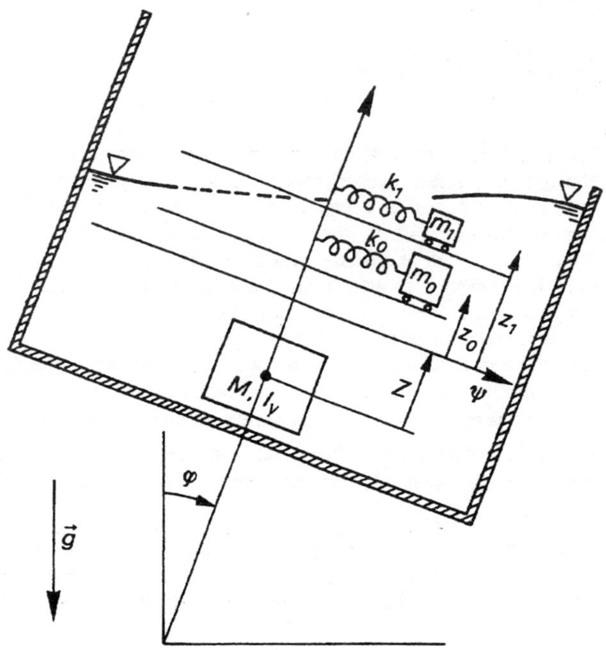

Figure 2. Equivalent mechanical system used to develop the mathematical model for the roll sloshing coupled motions as proposed in this paper.

According to the equivalence of the forces and moments required by the mechanical system and the fluid flow, the characteristics of the system are given by equations (9), where again h is the liquid depth, b the tank breadth, $r = \frac{h}{b}$ and M_F the mass of the fluid.

The parameter r plays an important role in determining the actual liquid motion (shallow/non shallow liquid condition). It must be stressed that despite its simplicity, the obtained equivalent mechanical system has well based hydrodynamic origins, i.e. its use is fully equivalent to the use of the linear solution of the hydrodynamic (inviscid) sloshing problem.

$$
\begin{cases}
\bullet \dfrac{m_n}{M_F} = \dfrac{8\tanh[(2n+1)\pi r]}{\pi^3(2n+1)^3 r} \\
\bullet \dfrac{z_n}{h} = \dfrac{1}{2} - \dfrac{\tanh\left[(2n+1)\frac{\pi}{2}r\right]}{(2n+1)\frac{\pi}{2}r} \\
\bullet M = M_F - \sum_{n=0}^{\infty} m_n \\
\bullet Z = -\sum_{n=0}^{\infty} \dfrac{m_n}{M} z_n \\
\bullet I_y = I_{F_y} - MZ^2 - \sum_{n=0}^{\infty} m_n z_n^2 \\
\bullet I_{F_y} = I_{S_y}\left\{1 - \dfrac{4}{1+r^2} + \dfrac{768}{\pi^5 r(1+r^2)} \sum_{n=0}^{\infty} \dfrac{\tanh\left[(2n+1)\frac{\pi}{2}r\right]}{(2n+1)^5}\right\} \\
\bullet I_{S_y} = \dfrac{1}{12} M_F\left[b^2 + h^2\right] \\
\bullet \dfrac{hk_n}{gM_F} = \dfrac{8\tanh^2[(2n+1)\pi r]}{\pi^2(2n+1)^2}
\end{cases}
\tag{9}
$$

This approach, in spite of the rather strong limitations connected with the use of a linear hydrodynamics to the description of quite large amplitude motions of a liquid with free surface, is undoubtly much more advanced than the classical approaches previously described implicitly or explicitly based on the static free surface approach. If for particular geometries these latter modes predict the actual motion quite well, they generally fail the prediction of the eigenfrequencies of the coupled roll/sloshing motions.

From the practical point of view, one important feature of the equivalent mechanical system is that the magnitude of the masses m_n for $n \geq 1$ is relevant only in very narrow range of the aspect ratio r close to r=0. In particular Graham and Rodriguez (1952) indicated that m_1 has a practical influence only for $r \leq 0.25$, being $m_1 \approx 0.125 m_0$ at $r \cong 0$.

As far as the ship rolling model is concerned, a nonlinear second order ODE is here considered allowing for a nonlinear contribution of the damping and of the restoring (Contento et al., 1996):

$$
I_S\ddot{\varphi} + D_{S_1}\dot{\varphi} + D_{S_3}\dot{\varphi}^3 + \Delta\overline{GZ}(\varphi) = \pi\alpha_0 s_w \Delta\overline{GM}\sin(\omega t) \tag{10}
$$

where I_S is the virtual moment of inertia of the ship about her center of gravity, D_{S_1} and D_{S_3} are the coefficients of the linear and cubic damping terms respectively, Δ is the displacement of the ship, $\overline{GZ}$ is the static righting arm and $\overline{GM}$ is the metacentric height, α_0 is the effective wave slope coefficient and s_w is the wave steepness corresponding to the wave height to wave lenght ratio. A brief discussion on the validity of the assumption of uncoupled roll is provided later in this section.

The coupled system is then obtained through the Lagrange equations:

$$
\frac{d}{dt}\left(\frac{\partial L}{\partial \dot{q}}\right) - \frac{\partial L}{\partial q} = 0 \tag{11}
$$

where q and $\dot{q}$ are the generalised coordinate and velocity of the system and

$$
L = T - U \tag{12}
$$

T and U being as usual the kinetic and potential energy of the whole system. It is assumed that the ship rotates about the center of rotation C_R. Here C_R is assumed to coincide with the center of mass of the whole "frozen" system.

With reference to Fig. 2 and to the previous positions, T and U are given by

$$
\begin{cases}
\bullet T = \frac{1}{2} I_S' \dot{\varphi}^2 + \frac{1}{2} I_y \dot{\varphi}^2 + \frac{1}{2} M\left(\overline{C_R G_F} + Z\right)^2 \dot{\varphi}^2 + \sum_{n=0}^{\infty} \frac{1}{2} m_n V_{m_n}^2 \\
\bullet U = \frac{1}{2} \Delta' \overline{G'M'} \varphi^2 + \sum_{n=0}^{\infty} \frac{1}{2} k_n \psi_n^2 - \sum_{n=0}^{\infty} m_n g \psi_n \sin(\varphi)
\end{cases}
\tag{13}
$$

where V_{m_n} is the velocity of the generic point mass m_n given by

$$
V_{m_n}^2 = \left[\left(\overline{C_R G_F} + z_n\right)\dot{\varphi} + \dot{\psi}_n\right]^2 \tag{14}
$$

G' and G_F being the center of mass of the ship including the 'frozen' liquid and the center of mass of the 'frozen' liquid respectively, ψ_n the displacement from the position at rest of the generic point mass m_n of the 'equivalent liquid', I_S' the virtual moment of inertia of the ship about C_R, M' the metacenter of the hull in the new loading condition corresponding to the displacement Δ'. By applying (11) and adding the damping and the forcing terms (non conservative forces), we obtain eqns (15).

In these equations the cross coupling damping terms have been neglected. The result is therefore a infinite set of equations, one for the ship rolling and the others for the point masses of the equivalent mechanical system. The application here presented refers to a relatively small value of $\overline{C_R G_F}$, therefore it results that one or at most two spring-masses are sufficient to describe the liquid motion as far as the liquid sloshing effect on rolling is concerned. The system of equations can therefore be conveniently reduced.

$$
\begin{cases}
\bullet \left[I_S' + I_y + M\left(\overline{C_R G_F} + Z\right)^2 + \sum_{n=0}^{\infty} m_n \left(\overline{C_R G_F} + z_n\right)^2 \right] \ddot{\varphi} + \\
\sum_{n=0}^{\infty} m_n \left(\overline{C_R G_F} + z_n\right) \ddot{\psi}_n + D'_{S_1} \dot{\varphi} + D'_{S_3} \dot{\varphi}^3 + \\
\Delta' \overline{G' Z(\varphi)} - \sum_{n=0}^{\infty} m_n g \psi_n \cos(\varphi) = \pi \alpha'_0 s_w \Delta' \overline{G' M'} \sin(\omega t) \\
\bullet \ m_n \left(\overline{C_R G_F} + z_n\right) \ddot{\varphi} + m_n \ddot{\psi}_n + D_{F_n} \dot{\psi}_n + \\
k_n \psi_n - m_n g \sin(\varphi) = 0 \ldots\ldots n = 0, \infty
\end{cases}
\tag{15}
$$

Finally it should be noted that for the isolated tank, the equation of the motion of the first spring-mass in the undamped condition becomes

$$m_0 \ddot{\psi}_0 + k_0 \psi_0 = 0 \tag{16}$$

and sustituting eqns (9) for n=0 one obtains

$$\ddot{\psi}_0 + \omega_0^2 \psi_0 = 0 \tag{17}$$

with ω_0^2 again given by eqn (5), i.e. the first natural frequency of the tank. Going on with $n \geq 1$, each spring-mass system corresponds to a specific higher order (n+1) oscillation mode.

The physical consistency of the proposed mathematical model rests among other on the validity of the approach based on the Lagrangian equations, i.e. mainly on the fact that the center of gravity G' selected as origin of the cartesian reference frame is effectively at rest during the system motion. This is a key point that often gives rise to hot discussions in the Naval Architecture community. The ship model in beam sea infact undergoes rolling motion and at least sway and heave motions: the origin of the reference frame is sometimes moved in the reduction point (point of minimum sway) for the coupled sway/roll motions (Hutchinson, 1991).

It is a generally accepted opinion that heave has negligible influence on coupled roll/sloshing in beam sea (completely different is the case of longitudinal sea), but sway has to be explicitly taken into account (Rakhmanin, 1994). In principle there is no further difficulty but the increased mathematical complexity of the model in including a third ODE accounting for swaying.

Actually, the influence of sway on the coupled roll/sloshing depends on the tank geometry and position on board of the ship and on the peculiarities of model testing. For reasons of practical convenience model tests are often conducted with partially restrained sway and yaw. The experiments reported in Appendix 1 were executed on a model free to sway but restrained to yaw. The fitting capability of the mathematical model was excellent in spite of non including a proper accounting of sway. In any case we agree on the fact that this subject needs further clarification, especially for geometries where the center of gravity of the water is much above floatation line.

We observe finally that the coordinates ψ_n do not correspond to the instantaneous slope of the liquid surface or to the height of the liquid of any point, i.e. the easy relationship between the slope of the free surface and the variable ψ of the static approach is here lost. The characteristics of the free surface are in any case obtainable by using the "equivalence" between the mechanical and the hydrodynamic sloshing model.

DISCUSSION AND CONCLUSIONS

The mathematical models given by eqn (10) for the uncoupled roll without liquids and by the linearized form of eqns (15) for the coupled roll/sloshing have been used to analyse the experimental data described in Appendix 1 and other available experiments (Armenio et al., 1996a,b). These data refer to the same ship model as described in Appendix 1 but in a different loading condition without liquids and with different tank position and geometry (rectangular). In these series of experiments the height of the bottom of the tank is $\overline{KK_c} = 0.197\text{m}$, i.e. higher than in the experiments presented in Table 2 and Table 3, so that a greater role of the liquid dynamics effects is expected. By the use of the Parameter Identification Technique briefly described in Appendix 2, the unknown coefficients (dampings and effective wave slope) of the equations have been derived. The obtained results are shown in Fig. 3 to 7 and the parameters derived are given in Table 1. In the loading condition 3, the lack of some experimental data in the second resonance peak has probably lead to an incorrect estimate of the tank damping which is believed to be responsible of the high roll computed value.

In the performed computations the added inertia term of the ship has been assumed to be that of the ship without liquids and therefore derived from her natural frequency (see Table 2). It can be immediately seen from the analysis of Fig. 3-7 that both the eigen-frequencies of the system are well captured within the investigated frequency range. It is worthwhile to stress that the eigenfrequencies of the tank/roll system are independent on the identified parameters (dampings and effective wave slope) being an intrinsic characteristic of the system itself. This result proves the good capability of the method. There is one more positive aspect. The mathematical model infact allows in principle the simulation of rectangular tanks only, whereas the compartment used in the experiments reported in Appendix 1 has a non negligible rounded corner between side walls and bottom (see Fig. 8). The vertical extension corresponds to the lowest liquid depth considered (10 cm) and the horizontal extension is 1/8 of the tank width. This rounded shape can in principle be responsible of a different behaviour of the liquid (mostly at low depths) facilitating the inception of liquid sprays. Despite this, the equivalent mechanical system seems to give the necessary main features for an accurate roll prediction even in this case.

As far as the quantitative aspect is concerned, the roll amplitude is reasonably predicted in the entire frequency range. The obtained linear roll damping coefficient has, from a practical point of view, the same value as in the case of the isolated (without liquids) forced roll. Moreover the tank damping coefficients lay within the range (some units percent of the critical damping) suggested by Faltinsen (1978) and also experimentally derived for the isolated tank case (Case and Parkinson, 1957) for a preliminary estimate.

Again, this is a very important result since in a preliminary stage of the evaluation of the roll behaviour of a ship with a flooded compartment or with a free surface liquid cargo, a rough prediction can be obtained by using such parameters.

The effective wave slope coefficient for the 'flooded' ship has been found to be generally lower than for the isolated rolling case. This is fully consistent with some remarks of Mikelis et al. (1984). They suggested that when the roll equation for a ship with free surface liquids is written in the form (10) with an additional contribution in the RHS given by the sloshing moment, the metacentric height used in the RHS (external wave excitation) must also be reduced by the use of the classical static free surface correction, being therefore lower than in the 'frozen' condition'.

The computations have been generally performed with one spring-mass only, i.e. n=0 in eqn (15). In Fig. 7 the case with n=1 and $D_{F_1} = 0$ is reported for sake of completeness. No substantial differences are observed apart from a slightly higher peak value at the first resonance. There, it is believed that the higher order simulation is closer to actual situation since from the experiments and from fully nonlinear numerical computations performed elsewhere (Francescutto and Contento, 1994) the occurence of a solitary wave superposed to the first mode standing wave has been generally observed. The parameters obtained in the case n=1 are basically the same as in the case n=0. They are reported in Table 1 in the last line. A moderate increase of the ship damping is observed probably due to the contribution to the rolling amplitude from the second oscillating mass.

Finally, it has to be observed that the use of a linearised version of some of the approaches does not affect the conclusions al least in the examined range of rolling amplitudes.

Table 1. - Identified parameters

	D_{S_1}	D_{F_0}	α_0
Model without liquids	2.02	-	0.682
Loading condition 1	1.950	3.530	0.464
Loading condition 2	1.782	4.964	0.477
Loading condition 3	2.101	1.998	0.490
Loading condition 4	1.738	5.940	0.499
Other data (n=0)	1.365	5.771	0.527
Other data (n=1)	1.744	5.534	0.539

ACKNOWLEDGEMENTS

This work has been performed under Contract code 1/LR343/95 between INSEAN, Roma and University of Trieste in the frame of INSEAN Research Plan 1994-96 financied by Ministry of Transport and Navigation with Law 8/8/95, n° 343.

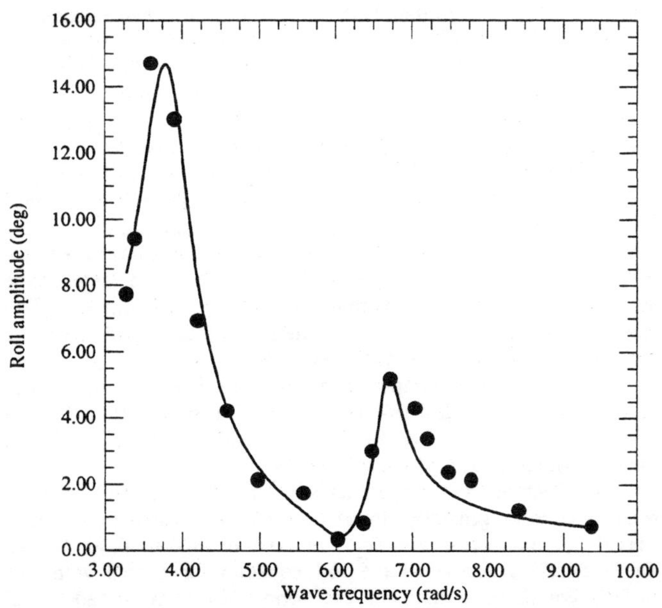

Figure 3. Simulation capability of the proposed model (eqn 15) for the loading condition 1 of Table 3. The values used for the coefficients are obtained by means of the parameter identification technique [● exp. data; — proposed model].

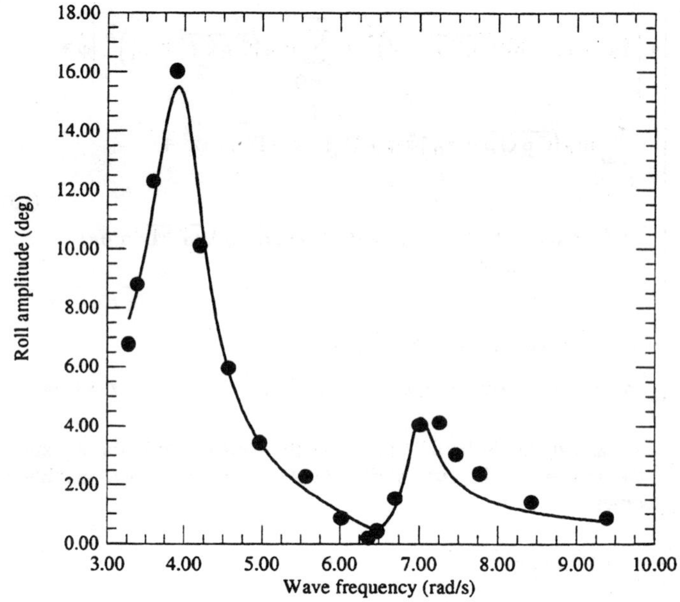

Figure 4. Simulation capability of the proposed model (eqn 15) for the loading condition 2 of Table 3. The values used for the coefficients are obtained by means of the parameter identification technique [● exp. data; — proposed model].

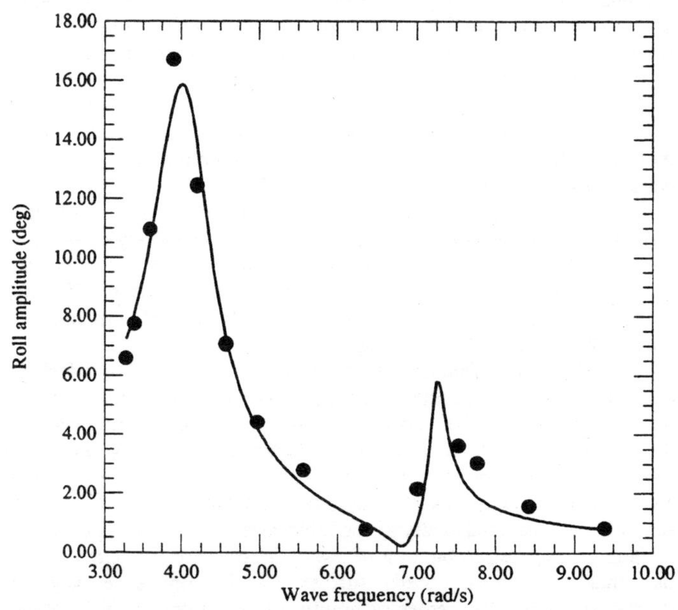

Figure 5. Simulation capability of the proposed model (eqn 15) for the loading condition 3 of Table 3. The values used for the coefficients are obtained by means of the parameter identification technique [● exp. data; — proposed model].

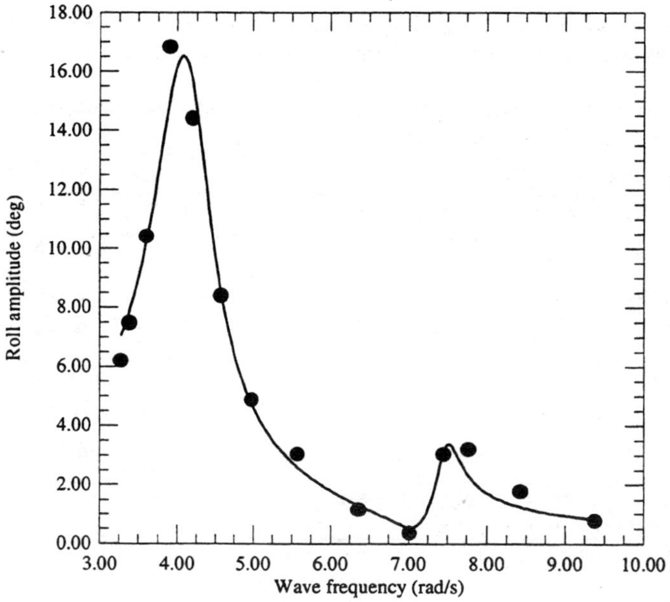

Figure 6. Simulation capability of the proposed model (eqn 15) for the loading condition 4 of Table 3. The values used for the coefficients are obtained by means of the parameter identification technique [● exp. data; — proposed model].

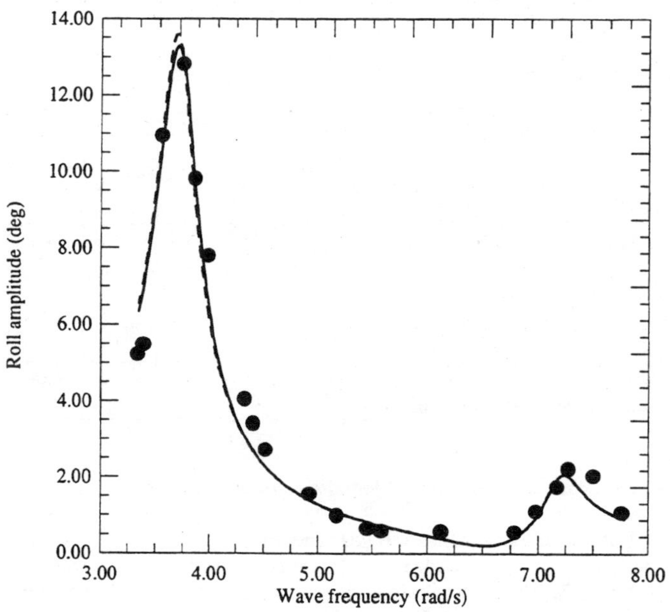

Figure 7. Simulation capability of the proposed model (eqn 15) for the experimental data given in [Armenio et al., 1996a,b]. The values used for the coefficients are obtained by means of the parameter identification technique [● exp. data; — proposed model (n=0); — proposed model (n=1)].

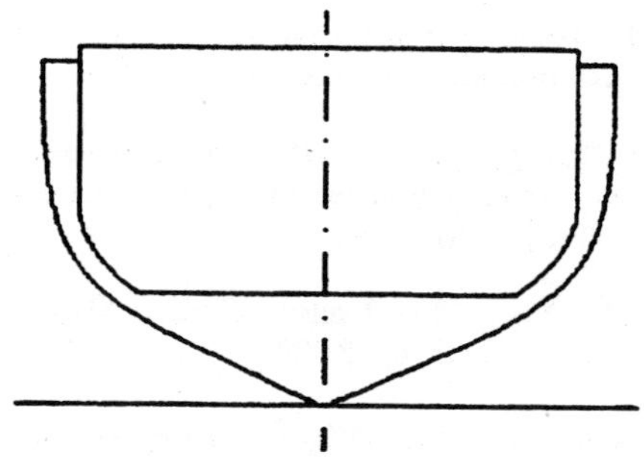

Figure 8. Schematic drawing of the ship transversal main section and tank arrangement used for the experiments reported in Francescutto and Contento (1994).

REFERENCES

Abramson, H.N., 1966, 'The Lateral Behaviour of Liquids in Moving Containers With Applications to Space Vehicle Technology', NASA Report SP-106.

Armenio, V., Cardo, A., La Rocca, M., Mele, P.,1994, 'Numerical Prediction of Roll Motion of a Ship with Liquids on Board in Regular Waves from Different Directions', Proceedings, 5th Int. Conference on Stability of Ships and Ocean Vehicles, STAB'94, Melbourne, Florida, USA, Vol. 2.

Armenio, V., La Rocca, M., Francescutto, A., 1996a, 'On the Roll Motion of a Ship with Partially Filled Unbaffled and Baffled Tanks: Part 1: Mathematical Modelling and Experimental Setup', Int. Journal of Offshore and Polar Engineering, Vol. 6, pp. 278-282.

Armenio, V., La Rocca, M., Francescutto, A., 1996b, 'On the Roll Motion of a Ship with Partially Filled Unbaffled and Baffled Tanks: Part 2: Numerical and Experimental Analysis', Int. Journal of Offshore and Polar Engineering, Vol. 6, pp. 283-290.

Case, K.M., Parkinson, W.C., 1957, 'Damping of Surface Waves in an Incompressible Liquid', Journal of Fluid Mechanics, Vol. 2, pp. 172-184.

Contento, G., Francescutto, A., Piciullo, M., 1996, 'On the Effectiveness of Constant Coefficients Roll Motion Equation', Ocean Engineering , Vol. 23, pp. 597-618.

Faltinsen, O., 1974, 'A Nonlinear Theory of Sloshing in Rectangular Tanks', Journal of Ship Research, Vol. 18, pp. 224-241.

Faltinsen , O., 1978, 'A Numerical Nonlinear Method of Sloshing in tanks with Two-Dimensional Flow', Journal of Ship Research, Vol. 22, pp. 193-202.

Francescutto, A., Contento, G., 1994, 'An Experimental Study of the Coupling between Roll Motion and Sloshing in a Compartment', Proceedings, 4th Int. Offshore and Polar Engineering Conference, Osaka, Japan, vol. 3, pp. 283-291.

Francescutto, A., Contento, G., Penna, R., 1994, 'Experimental Evidence of Strongly Nonlinear Effects in the Rolling Motion of a Destroyer in Beam Sea', Proceedings, 5th Int. Conference on Stability of Ships and Ocean Vehicles, STAB'94, Melbourne, Florida, USA, vol. 1.

Goodrich, G.J., 1969, 'Development and Design of Passive Roll Stabiliser', Transaction R.I.N.A., Vol. 111, pp. 81-95.

Graham, E.W., Rodriguez, A.M., 1952, 'The Characteristics of Fuel Motion Which Affect Airplane Dynamics', Journal of Fluid Mechanics, Vol. 19, pp. 381-3884.

Hutchinson, B.L., 1991, 'The Transverse Plane Motions of Ships', Marine Technology , vol. 28, pp. 55-72.

Ishida, S., Watanabe, I., Fujiwara, T., Murashige, S., Ogawa, Y., 1996, 'Study on Damage Stability with water on Deck of a Ro-Ro Passenger Ship in Waves', Journal of the Society of Naval Architects of Japan, Vol. 179, pp. 103-117.

Mikelis, N.E., Miller, J.K., Taylor, K.V., 1984, 'Sloshing in Partially Filled Liquid Tanks and its Effect on Ship Motions: Numerical Simulations and Experimental Verification', Transaction R.I.N.A., Vol. 126, pp. 267-280.

Rakhmanin, N., Zhivitsa, S., 'Prediction of Motion of Ships with Flooded Compartments in a Seaway', Proceedings, 5th Int. Conference on Stability of Ships and Ocean Vehicles, STAB'94, Melbourne, Florida, USA, vol. 2.

Pantazopoulos, M.S., 'Sloshing of Water on Deck of Small Vessels', Proceedings, 4th Int. Conference on Stability of Ships and Ocean Vehicles, STAB'90, Naples, Italy, vol. 1, pp. 58-65.

Van der Bosch, J.J., Vugts, J.H., 1966, 'Roll Damping by Free Surface Tanks', Report No. 83 S, Netherlands Ship Research Centre, Delft.

Van der Bosch, J.J., Zwaan, A.P., 1970, 'Roll Damping by Free Surface Tanks with Partially Raised Bottom', Report No. 280, Technische Hogeschool, Delft.

Watts, P., 1883, 'On a Method of Reducing of Ship Motions at Sea', Transactions I.N.A.,Vol. 24, pp. 165-190.

Watts, P., 1885, 'The Use of Water Chambers for Reducing the Rolling of Ships at Sea', Transactions I.N.A.,Vol. 26, pp. 30-49.

Zdybeck, T., 1980, 'The Use of Bench Test Results for calculating Roll Response of the Tank Stabilised Ship', Int. Shipbuilding Progress, Vol. 27, pp. 96-105.

APPENDIX 1. - Experimental results

A large campaign of experiments on a scale (1:12.5) model of an existing fishing vessel in regular beam sea equipped with a flooded compartment has been conducted in 1994 at the Hydrodynamic Laboratories of the Department of Naval Architecture of the University of Trieste. The experiments consist of four different series of mesurements of steady state roll in regular beam sea, each series with a different amount of liquid cargo ranging approximately from 18 % to 30% of the weight of the empty ship/model. The liquid level at four different stations in a transversal plane of the tank has been measured by capacitive transducers. The no-flooded condition has been tested too. Since the experimental set-up and the main geometrical and mechanical characteristics of the model and of the liquid cargo have been already described in a previous paper (Francescutto and Contento, 1994), only a brief summary is given in Table 2-3 and in Fig. 8. The incident wave steepness in the experiments was 1/50. The restoring curve GZ of the model in the five loading conditions exhibits a linear behaviour up to 30 degrees so the linear approximation has been regarded as satisfactory.

Table 2. - Ship model main dimensions and mechanical characteristics without liquids.

Displacement [N]	Δ	980.7
Lenght between perpendiculars [m]	L_{bp}	2.000
Lengh over all [m]	L_{oa}	2.456
Beam [m]	B	0.552
Draught mld [m]	T	0.214
Depth mld [m]	D	0.397
Trim [m]	δT	0.000
Vertical position of the center of gravity [m]	$\overline{KG}$	0.210
Long. position of the center of gravity (aft midsship section) [m]	LCG	0.006
Metacentric height [m]	$\overline{GM}$	0.077
Moment of inertia (long. axis) [kgm^2]	I_S	3.464
Natural frequency [rad/s]	ω_0	4.560
Length of the compartment [m]	l	0.400
Breadth of the compartment [m]	b	0.480
Long. position (aft mids) [m]	x_c	0.020
Vert. position of the bottom [m]	$\overline{KK_c}$	0.127

Table 3. - Loading conditions with liquids

Loading condition n°		1	2	3	4
Liquid depth [m]	h	0.10	0.12	0.14	0.16
Weight of the liquid [N]	W_F	175.2	212.	250.5	288.1
Vertical position of the center of gravity of the liquid [m]	$\overline{KG_F}$	0.179	0.189	0.200	0.210
Vertical position of the center of gravity of the ship with frozen liquid [m]	$\overline{KG'}$	0.205	0.206	0.208	0.210
Natural frequency of the ship in the frozen condition [rad/s]	$\omega_{\varphi,f}$	4.83	4.87	4.92	4.95
Natural frequency of the ship in the free surface cond [rad/s]	ω_φ	3.81	3.92	4.02	4.04

APPENDIX 2. - Parameter Identification Technique

It is rather evident that eqn (10) and eqn (15) allow a closed form solution only after linearization. In general numerical solutions are used and the obtained solutions depend in a nonlinear manner from the parameters (damping and restoring).

The present problem has been solved by means of a least square technique. The values of the parameters giving the minimum least square residual of the numerical solution obtained by integrating in the time domain (until the steady state is achieved) the differential equations and experimental data are obtained through a Levemberg-Marquard algorithm. This procedure is a high reliability one inasmuch as it avoids the use of approximate analytical solutions. For further details see Francescutto et al. (1994).

Proceedings of the Seventh (1997) International Offshore and Polar Engineering Conference
Honolulu, USA, May 25-30, 1997

ISBN 1-880653-28-1 (Set); ISBN 1-880653-31-1 (Vol. III)

Effect of Water Inside a Ship on Its Damage Stability

Roby Kambisseri, Tetsuya Hamano and Yoshiho Ikeda
Osaka Prefecture University
Osaka, Japan

ABSTRACT

This paper contains the study on the static and dynamic effects of flooded water on the stability of ships. We have done model experiments with different amounts of water in its compartments and with a damage hole to find the influence of the water inside the ship and a damage hole on the motion and damping of ships. These experiments show that the effect of flooded water on the roll motion of ships is to reduce the amplitudes. Damping is found to be more for the deck flooding than for the below-deck flooding. In some deck flooding cases, quasi-static heel, which is the mean roll angle, is observed and is different from the static heel.

KEY WORDS: Damage stability, flood water, model experiments, ship motion, survivability criteria, safety

INTRODUCTION

International rules for ensuring the safety of damaged ships are of two types - Deterministic (Sol92, 1992) and Probabilistic (A.265 & MSC.19(58)). Stability criteria in both the types consider the damage as loss of buoyancy, and the damage stability is assessed using the GZ (righting lever) curve derived from the undamaged portion of ships. The loss of RO-RO ships due to the flooding of their car-decks, which were above waterline, had shown that flooding into the upper decks can reduce the static stability of ships drastically, even leading to capsize. This adverse effect of the trapped water in compartments is to be included in the damage stability rules.

Damage stability rules prescribe not only stability criteria after a damage but also the required safety of damaged ships. Deterministic rules uses a factor-of-subdivision, which is used along with floodable length curves, to subdivide a ship. Probabilistic rules use a Required-subdivision-index, *R*. Attained-subdivision-index, *A* for the subdivision of a ship should not be less than *R*. Both these approaches have some drawbacks which need to be corrected. It is explained in the next section.

Safety of Ships

In the deterministic approach (*Chapter II-1, Part B, Reg.8 of* Sol92, 1992), the required safety is given by a factor-of-subdivision, which depends on length of ship, number of passengers, volumes of ship, machinery and passenger spaces. The ratio of compartment length to floodable length, for the mid-point of a compartment, should not be greater than the factor-of-subdivision. The concept behind the deterministic rule is explained by Welch (1915) and is to increase the damaged freeboard with the increase of safety requirement. The defect of the deterministic rule is that for a damage opening of a certain length, the probability of damage of bulkheads increases with increases of safety (Abell, 1920 & Comstock and Robertson, 1961).

In the probabilistic approach, the theory of which was proposed by Wendel (1961), the required safety is given by an Required-subdivision-index, *R*, which depends on length of ship, number of passengers (A.265 & MSC.19(58)). Probability of damage (p_i) of a space (one or more compartments) is multiplied with the probability of surviving (s_i) that damage. The sum of this product for all possible damages $\left(A=\Sigma p_i s_i\right)$ is the Attained-subdivision-index, *A*. The condition to satisfy the rule is that the Attained-subdivision-index should not be less than the Required-subdivision-index. Roby and Ikeda (1995) made a comparative study of the probabilistic rules and proposals by different countries for the improvement of the rules. The defect of the present probabilistic approach is that the criteria can be satisfied even when minor damages are present. It may be possible that a ship having a larger *A* value may be lost due to a smaller damage opening than that will make a ship with a smaller *A* value unsafe (Roby and Ikeda, 1996a).

Sen and Gerigk (1992) calls for the need for uniform distribution of survival capability along the length of a vessel and shows how it can be achieved, when applying the probabilistic damage stability rules, using a 'local' index of subdivision. Though the local index of subdivision will improve the overall safety of a transversely subdivided ship, the defects involved with the present probabilistic rules re-surfaces when there is additional subdivision using longitudinal bulkheads and decks, i.e. a region with a higher local index may not be safer than one with a lower index. Visualising and fixing reliable indices for all the possible cases will further complicate an already complicated procedure.

Since both the approaches tried so far did not succeed, it is necessary to adopt a new approach. A concept for the new approach was presented by the authors (Roby and Ikeda 1996a,b & Roby, Hamano and Ikeda 1996). According to it, safety level of a ship is to be determined by the value that could be lost by the loss of the ship. This safety level is achieved by making the ship survive all damages caused by a damage opening of a certain size, a bigger size is set for a vessel with a higher safety level.

Survivability Criteria

Damage stability criteria in the deterministic (Sol92, 1992) and probabilistic rules (A.265 & MSC.19(58)) use the after damage features of ships and are as given below.

In the deterministic rule, it is based on the righting lever (GZ) curve obtained after damage and equilization and should have certain minimum requirements for all damages, the length of the damage space being less than the floodable length.

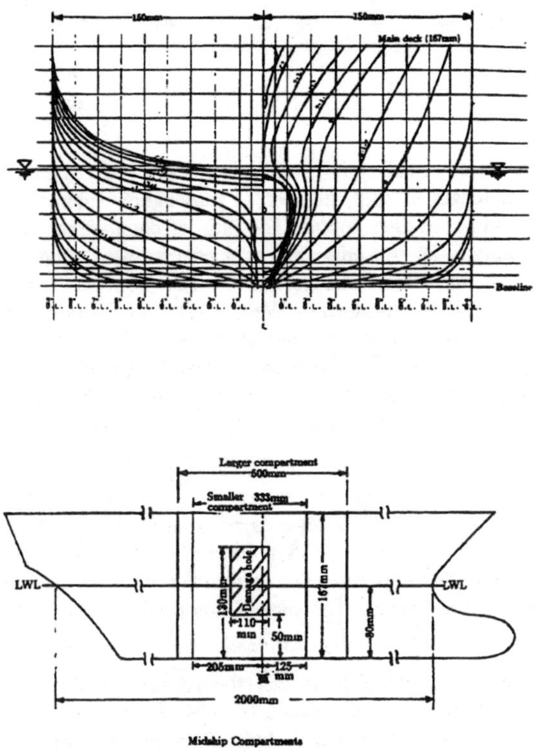

Figure 1 Body-plan and mid-ship compartments

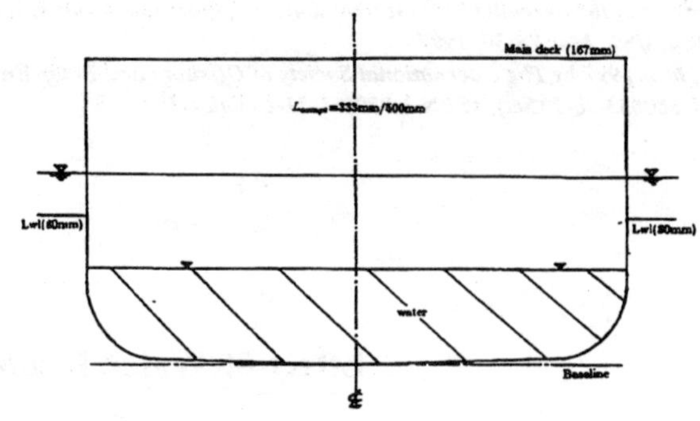

Figure 2 Section of below deck compartment

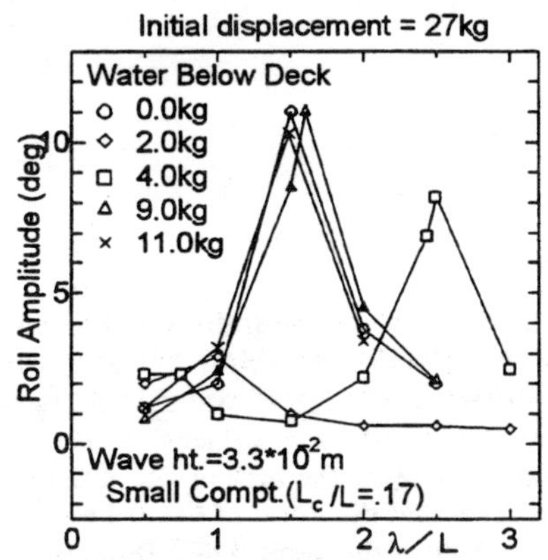

Figure 3 Roll motion amplitudes for below deck flooding

Table 1. Particulars of ship and model

	Ship	Model
L_{pp}	120.00 m	2.00 m
Breadth	18.00 m	0.30 m
Depth	10.00 m	0.167 m
Draft	4.80 m	0.08 m
$Disp._{vol}$	5900 m^3	0.027 m^3
C_b	0.569	0.569
KG	7.44m	0.124m
GM	1.95m	0.0325m

In the probabilistic rules for passenger ships (A.265), probability of survival is given by the following relation :

$$s_i = 4.9\left[\left(\frac{F_1}{B_2} - \frac{\tan\theta}{2}\right)(GM_R - MM_s)\right]^{\frac{1}{2}} \quad \text{but} \leq 1.0 \qquad (1)$$

where θ is the final angle of heel due to unsymmetrical flooding, F_1 is the effective upright damaged freeboard, B_2 is the breadth of the ship at the bulkhead deck, GM_R is the highest required intact metacentric height (in metres), and MM_s is the reduction in metacentric height as a result of flooding. Along with satisfying probabilistic criteria, a passenger ship has to have one compartment survivability and depending on the number of passengers a portion of length from the forward end need to have two compartment survivability.

In the probabilistic rules for cargo ships (MSC.19(58)), probability of survival is given by the following relation.

$$s_i = C\sqrt{0.5(GZ_{max})(range)} \quad \text{but} \leq 1.0 \qquad (2)$$

where C=1 if $\theta_e \leq 25°$, C=0 if $\theta_e > 30°$ and otherwise $C = \sqrt{\frac{30 - \theta_e}{5}}$; θ_e= final equilibrium angle of heel in degrees; *range* = range beyond θ_e,

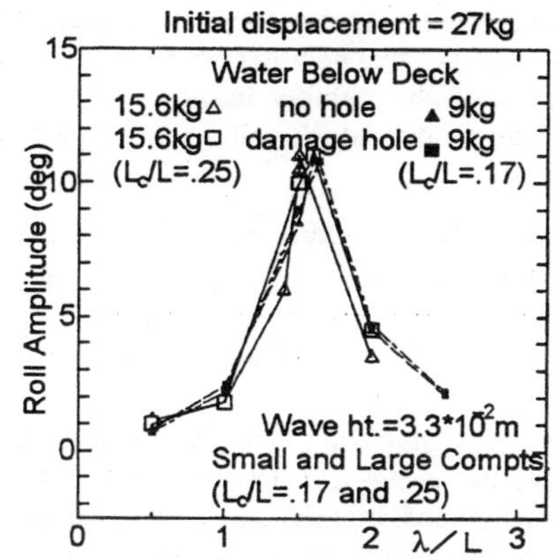

Figure 4 Roll motion amplitudes with and without damage hole

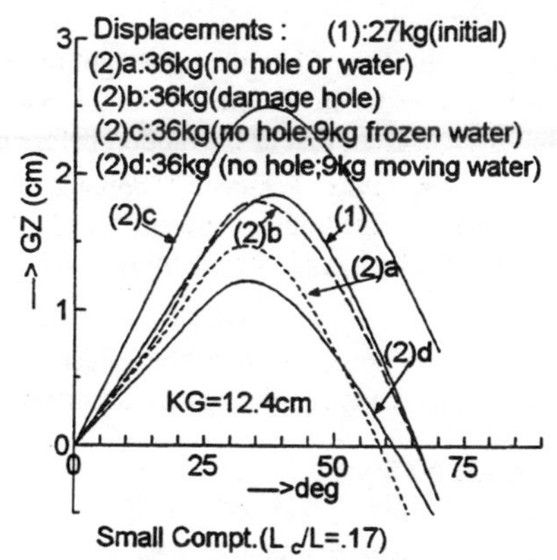

Figure 5 Righting lever curves for smaller compartment

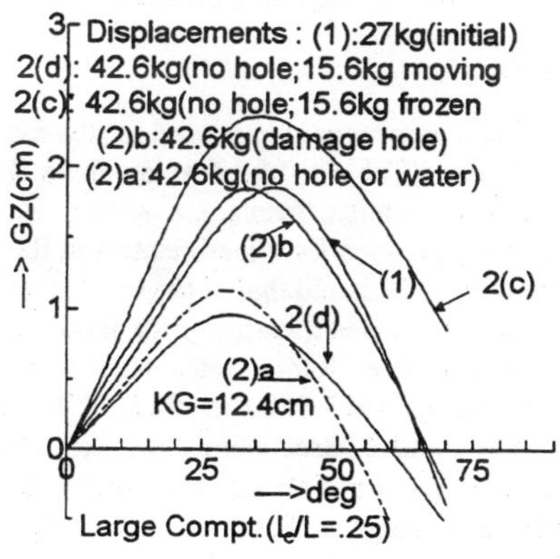

Figure 6 Righting lever curves for larger compartment

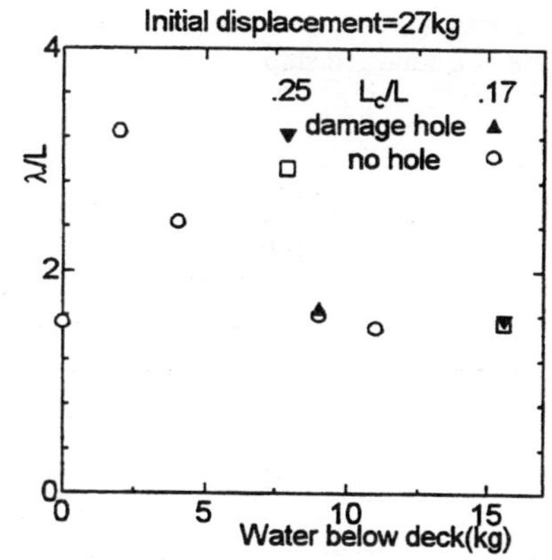

Figure 7 Natural period of ship with water in bottom compartment

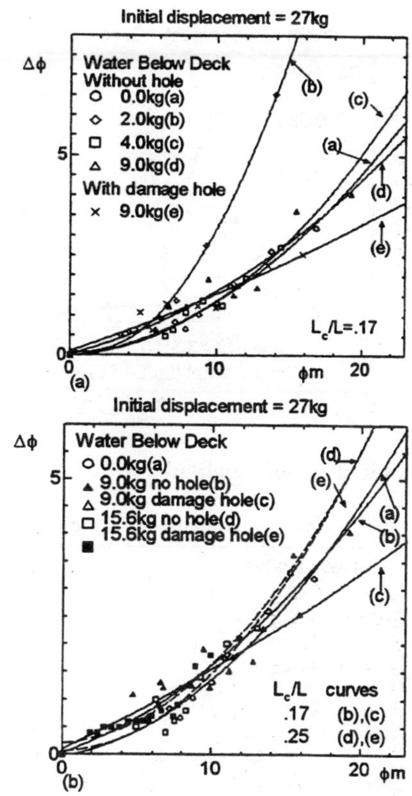

Figure 8 Roll extinction curve for below deck flooding : (a) without damage hole (b) with damage hole

but $\leq 20^{\circ}$ and GZ_{max}= max. GZ within the range, but ≤ 0.1m.

These rules does not take into account the loss of stability due to the presence of trapped water in upper compartments. Also, they does not specify the *sea state* that can be survived by a ship satisfying those damage stability criteriae.

A new criteria is required to predict the safety of damaged ships, in the sea states it has to operate. Since the criteria is for a damaged ship and flooding could be associated with a damage at sea, the criteria should consider the adverse effect of damage and flooding. Vassalos, Pawlowski and Turan (1996) gives a method to determine the survivable sea state of a damaged ship.

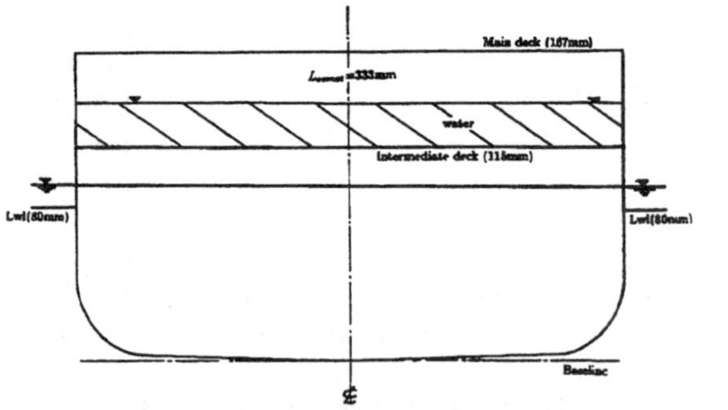

Figure 9 Section of compartment over intermediate deck

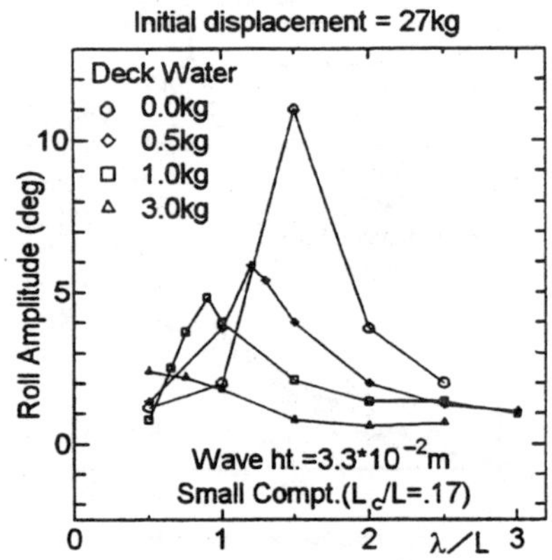

Figure 10 Roll motion amplitudes with water on deck

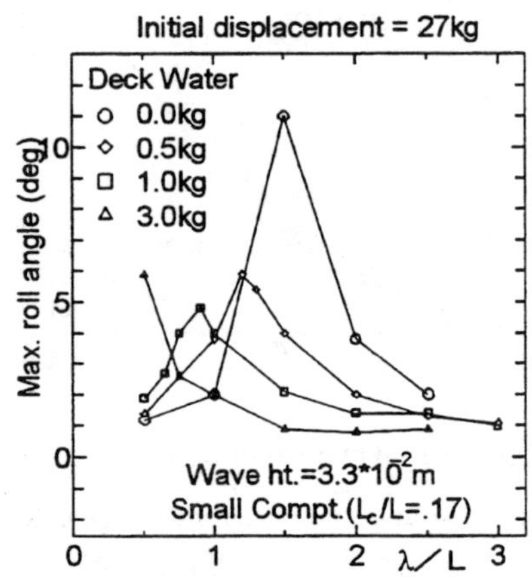

Figure 11 Maximum roll angles with water on deck

Effects of damage and flooding

Any stability criteria should ensure sufficient stability features for the ship to make it survive the environment in which it has to operate. Damage stability criteria should also incorporate features which will ensure sufficient survivability after damage. If a damage opening is in such a location on ship such that no water can enter the ship through it, then it will not change the stability features of the ship. We are concerned here of the damage which will result in water entering any region of the ship.

The following are the factors which will affect the floatability and stability of a ship after damage.

(a) Reduction in *reserve buoyancy* : A damage will reduce the reserve buoyancy of a ship resulting in the sinking of the ship or changes in it's draft, heel and trim.

(b) Change in *transverse stability* : Stability of ship in the transverse direction is changed by damage and flooding. Considering the flooded water as added weight, flooding shifts the curves of centers of buoyancy and gravity resulting in a different righting lever curve (Shimizu, Roby and Ikeda, 1996). Experiments and calculations have shown that heavy flooding of higher decks can cause sudden capsize.

(c) Change in *longitudinal stability* : Damage changes the longitudinal centers of gravity and buoyancy, which may result in heavy trim or even capsize by bow or stern.

(d) Change in *dynamic stability* : Change of the righting moment after damage and the motion of water inside the ship could make the motion of a damaged ship different from that of the motion before damage.

(e) Change in *wave exciting loads* : Presence of damage opening on the hull of the ship and the loads due to the motion of water inside and flow through the opening could change the exciting loads acting on a damaged ship.

Experiments, that we have done, to study the effect of damage and flooding on the motion and damping of ships are described below.

MOTION EXPERIMENTS

Loss of static stability and capsize due to damage and flooding of ships are well known. The effect of them on the dynamic stability, however, is not very clear. To study the effect of damage and flooding on the damping and motion of ships in waves, motion measurements were done with (a) no damage and water inside midship compartments and (b) damage opening on the midship compartments.

Experimental setup

A 1/60 scale model of a passenger ship is used for the experiments, the particulars of which are in Table 1. It's bodyplan and a sketch showing the position of midship compartments are shown in Fig. 1. Two midship compartments are used for experiments and the length of the smaller compartment is 0.333m and that of larger compartment is 0.5m. Model was kept beam to waves measuring the motions of heave and roll, keeping sway also free. Surge, pitch and yaw motions were prevented. Two wave heights of $3.3*10^{-2}$m and $6.67*10^{-2}$m with wave periods between 0.8sec and 1.98sec were used. Relationship between λ/L and wave period is given in Table 2. The initial draft of the model without any water or damage is 8cm, which corresponds to 4.80m (LWL) on the ship. The displacement of the model at this draft is about $27*10^{-3}m^3$ and of the ship is about $5900m^3$.

EXPERIMENTAL RESULTS AND DISCUSSION

Under-deck flooding without damage hole

Sketch of section of mid-ship compartments used for under deck flooding is shown in Fig. 2. Different stages of damage from intact to flooding and equalisation are modeled using the intact model at load draft and with 2kg, 4kg, 9kg and 11kg of water inside the smaller compartment. Roll motion amplitudes for them in regular beam seas are shown in Fig. 3. The following can be observed from this figure : a)Peak of the roll motion amplitudes occur at the resonant periods, which were obtained from free rolling tests and are plotted in Fig. 7, (b) Roll motion amplitudes with water inside the ship is less than the intact amplitudes and (c) Roll motion amplitudes are found to be the same as the intact when water inside is 9kg, which is the amount of water that will flood the compartment in case of a damage below waterline. We can conclude that the flooding of under-deck compartments will not increase the roll motion. Heave motion amplitudes for these cases were not much different from that for the initial draft.

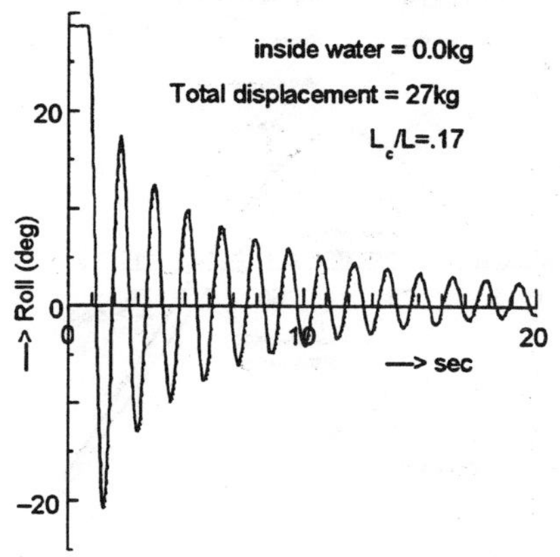

Figure 12 Roll decay curve without water on deck

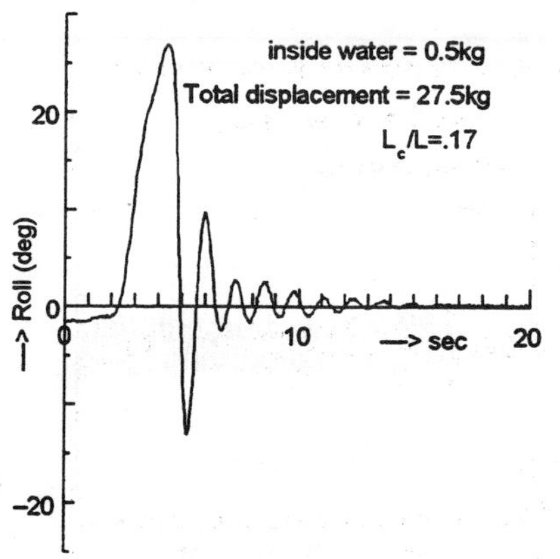

Figure 13 Roll decay curve for 0.5kg water on deck

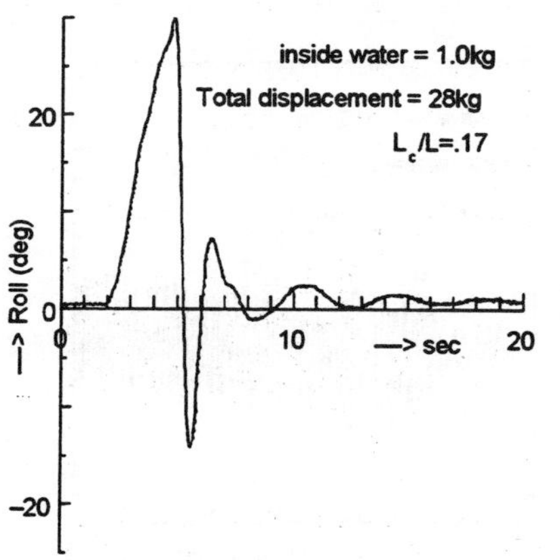

Figure 14 Roll decay curve for 1.0kg water on deck

Table 2. λ/L and Wave period (for model)

λ/L	0.5	1.0	1.5	2.0	2.5	3.0
Period (sec)	0.8	1.13	1.39	1.6	1.79	1.98

Under-deck flooding with damage hole

Motion measurements of model with and without damage hole are done for the smaller and larger compartments. If damaged, 9kg of water can flood the smaller compartment and 15.6kg of water can flood the larger compartment. Motions were measured for each compartment with this amount of water inside, both with and without damage hole. The roll motion amplitudes of damaged and undamaged compartments are plotted in Fig. 4 and are almost same. Heave motion amplitudes of these cases were same as that for initial draft. Righting levers (static) for the initial draft of the model, model with damage hole, and intact model with water inside for the smaller compartment are in Fig. 5 and for larger compartment are in Fig. 6. Here, when water is inside the intact model, righting levers are calculated (1) assuming water as frozen (or not changing shape) and (2) assuming water changing shape or moving when the ship is heeled, keeping water surface always parallel to the external calm water surface. Righting lever curves of intact and damaged cases are almost the same for both the compartments.

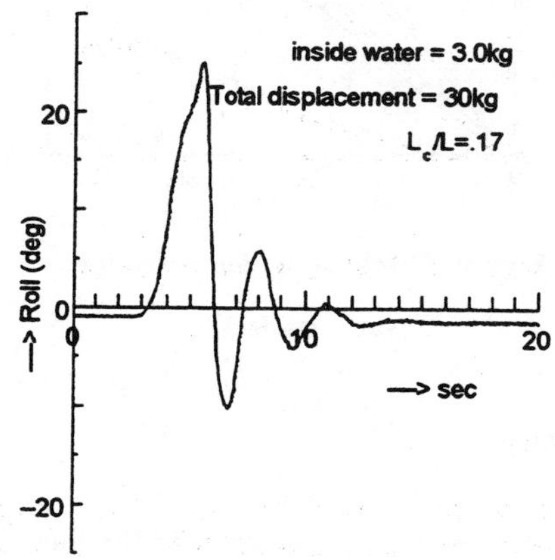

Figure 15 Roll decay curve for 3.0kg water on deck

Damping due to under-deck flooding

Free decay tests of the model are carried out in calm water. Roll extinction curves for under deck flooding is given in Fig. 8. For the roll range of the experiments, extinction curves of intact model, model with 9kg of water inside smaller compartment with and without damage are very close to each other. This and the similar righting lever curves explain similar motions in those cases. Similar situation can be found for 15.6kg of water, in the case of larger compartment. For 2kg of water inside the extinction curve is much steeper and the roll damping increases by inside water dramatically. These results explain that it's roll amplitudes are smaller, as shown in Fig. 3.

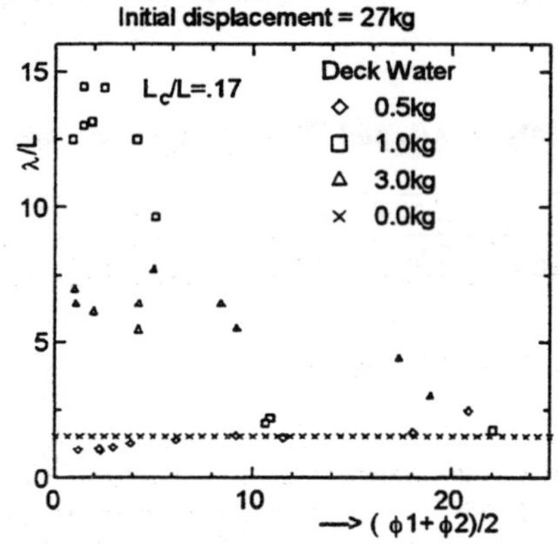

Figure 16 Mean roll and natural period for water on deck

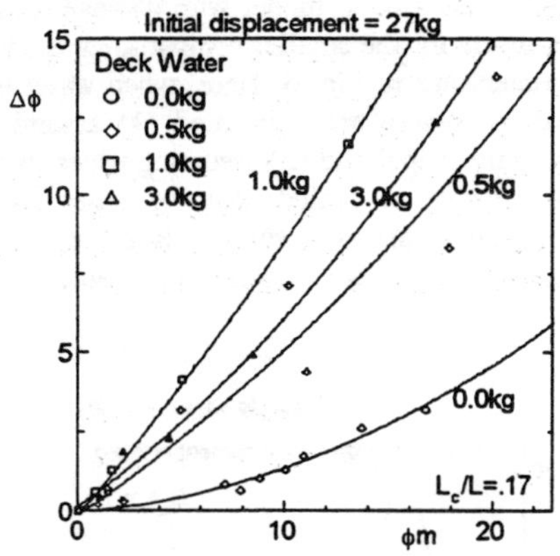

Figure 17 Roll extinction curve for water on deck

Over-deck flooding

When it comes to compartments above water level, the question that arises is that whether there is any probability for flooding that region in case of damage. Hutchinson (1995) shows that there is relation between the freeboard and wave height on the water accumulation on deck. Our studies also show that the flooding reduces with increase of freeboard (Roby, Hamano and Ikeda, 1996) and when the ship is running into waves, the flooding rate through a bow damage is very high to capsize the ship with in a few waves (Shimizu, Roby and Ikeda, 1996). Shin and Ohkusu (1986) found that (a) the deck wetness is higher for higher waves, (b) roll angles when deck wetness occurs is smaller for smaller wave periods and (c) rapid accumulation of shipping water on deck

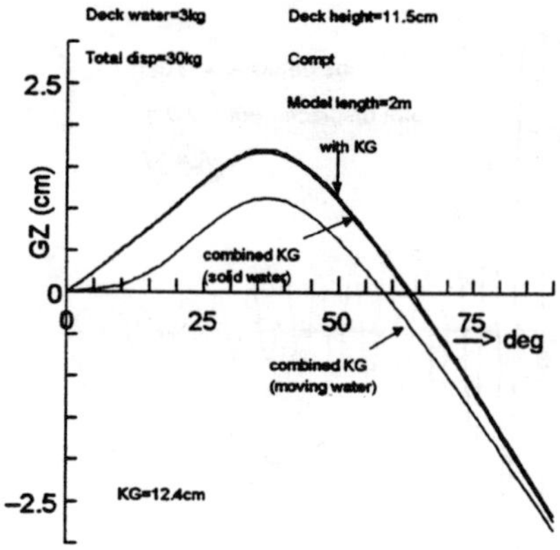

Figure 18 Righting lever curves for 3.0kg water on deck

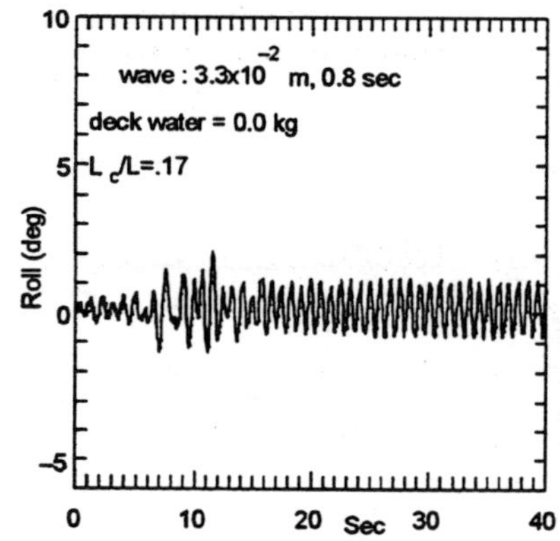

Figure 19 Roll motion without water on deck

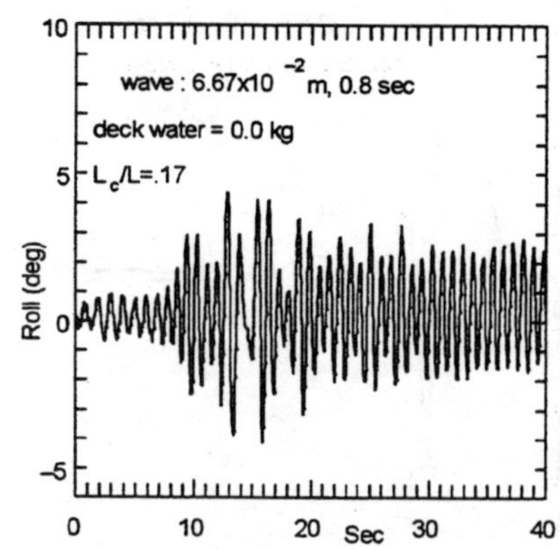

Figure 20 Roll motion without water on deck

caused by repeated deck wetting leads to capsize.

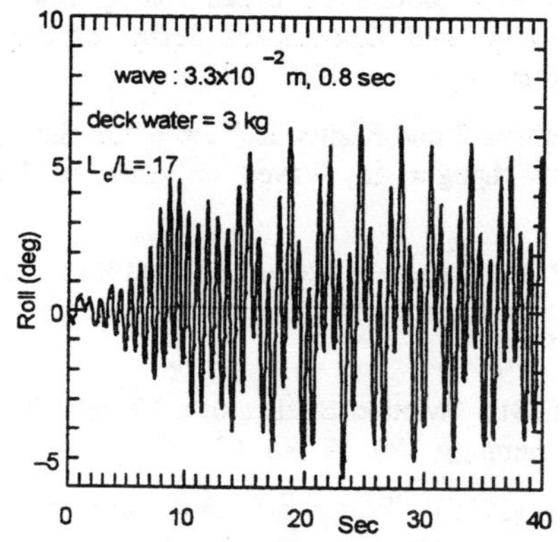

Figure 21 Roll motion with 3.0kg water on deck

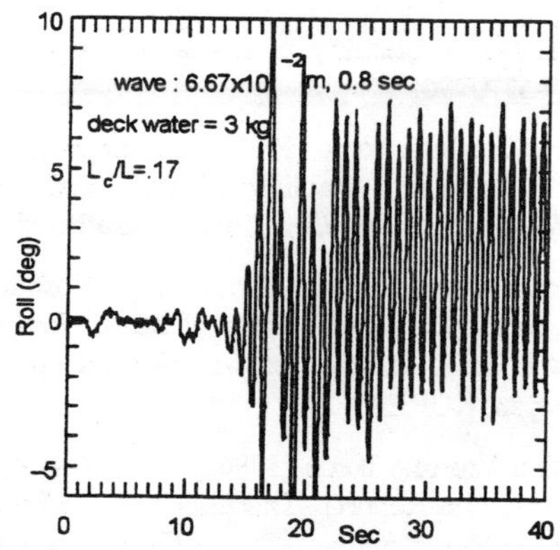

Figure 22 Roll motion with 3.0kg water on deck

Table 3. Deck water and heels (for model)

Wave period = 0.8 sec.
Wave-1 = 3.3*10^{-2}m wave.
Wave-2 = 6.67*10^{-2}m wave.

	Calm	Wave-1	Wave-2
Deck Water	Static heel	Quasi-static heel	
kg	degree	degree	degree
0.5	0.13	-.63	-0.12
1.0	0.30	-.12	+0.49
2.0	0.30	+.29	+1.20
3.0	0.30	+2.6 to-1.7	+2.25
4.0	0.37	+1.9 to-1.7	+1.8 to -1.1
4.6	0.05	-.3	+.69

A deck at 11.5cm above keel in the smaller compartment is made and the region between this intermediate deck and main deck is used for the water on deck study, sketch of which is given in Fig 9. The compartment length is only about 17% of total length and the amount of flooded water on deck is small.

Motion experiments were done for the initial draft of 8cm and with 0.5kg, 1kg and 3kg of water on deck. Roll motion amplitudes are plotted in Fig. 10, and maximum roll motion is in Fig. 11. Roll motion amplitudes and maximum roll angles for deck flooding are found to be much less than that for the initial draft. This shows that the effect of water on deck on the roll motion is damping. Water on deck studies by Dillingham (1981) also shows that the dynamic effect of the water can only be to damp the motion and not to excite it. The studies at Hokkaido University found that, sometimes, shallow water does not act as a roll damper, when the rolling frequency is higher than the second resonant frequency, of the inside water (Amagai, Ueno and Kimura, 1996). However, this is for small amount of water on deck, the roll motion generated by which may not be critical to ship safety. Heave motion amplitudes for intact model and with water on deck were the same.

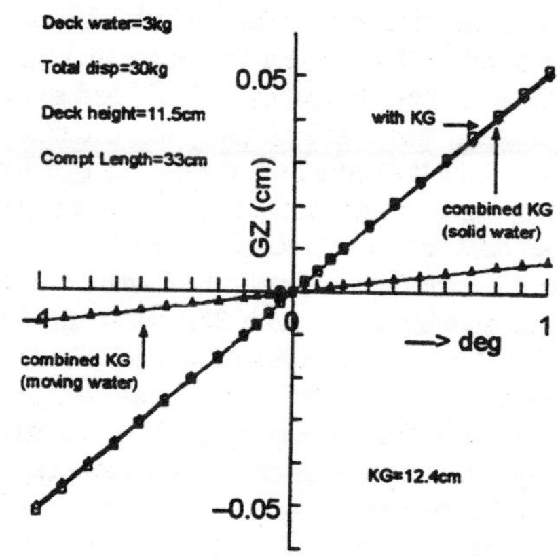

Figure 23 Righting lever curve for 3.0kg water on deck

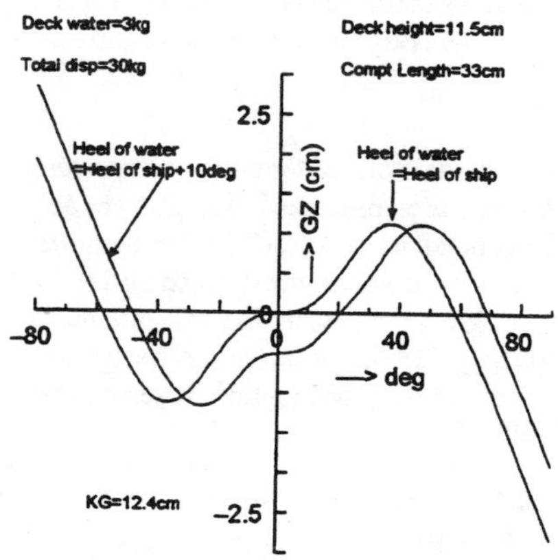

Figure 24 Shift of righting levers due to change in deck water surface angle

Damping due to over-deck flooding

Roll motion in roll decay tests with water over deck are quite different from that without any water. Roll decay traces of an intact model and those with 0.5kg, 1kg and 3 kg water on deck are given in Figs. 12 to 15. It is found that the natural period of roll changes with roll angles, which is plotted in Fig. 16. Fig. 17 is the roll extinction curve for deck flooding. Roll damping with deck water is much higher than that without water, which explains the reduction in the roll motion amplitudes in the deck flooding cases. Fig. 18 is the righting lever (static) curves for 3kg water on deck and is not too much different from the intact one. Sloshing of water on deck may change this curve in large roll amplitudes.

Quasi-static heel

In our experiments, static heel and quasi-static heel (in some cases) were observed for over-deck flooding. Static equilibrium heel angles and quasi static heel angles for different amounts of water on the intermediate deck are given in Table 3. Figs. 19, 20, 21 and 22 are the roll motion traces without and with 3kg water on deck for waves of period 0.8sec and heights $3.3*10^{-2}$m and $6.67*10^{-2}$m, respectively. In Fig. 21 the motion is irregular, the mean roll keeps on changing due to the shift of water from one side to the other, which takes about 2 to 3 rolls. Fig. 22 shows steady roll with mean roll angle (quasi-static heel) much higher than the static heel. Rakhmanin (1990) had shown that for a ship with cargo shifting, which has an un-symmetric static stability curve, the vessel rolls about the static heel angle and a quasi-static heel appears because of the un-symmetry of the righting lever curve which makes the roll angles different on both sides of static heel angle making the mean roll angle different from the static heel angle. Fig. 23 shows the calm water static stability curve with 3kg water on deck and is symmetric. So, it should not have a quasi-static heel. But the sloshing of water on deck during motion of ship in waves may change the center of gravity of deck water and thus the instantaneous static stability loss during dynamic motion making the instantaneous static righting lever curve different from the calm water one. Fig. 24 shows the righting lever curves, one for heel of water surface same as the ship's heel and the other for heel of water surface 10 degrees more than the ship's heel. The righting lever curve is not symmetrical in the latter case. In dynamic motion, the phase difference between the ship's heel and heel of water surface keep on changing, resulting in a different righting lever curve. Also, the harmonic changes of the restoring moment in waves, imposed over the static moment, leads to relevant fluctuations of the actual lever arm of stability (Bogdanov, Dimitrova and Kishev, 1990).

CONCLUSION

Motion studies of ship model with water inside compartments and with damaged compartments are presented in this paper. Following conclusions have been obtained. (1) Roll motion with water inside or with damage is found to be less than that for the initial draft. (2) Deck water is found to have greater effect on the roll motion damping than the below deck flooding. (3) The quasi-static heel during motion is observed in many cases of deck flooding and in some cases it keeps on shifting with each roll.

Authors also give an overview of their earlier studies on assessing the relative safety between ships.

REFERENCES

A.265. "Regulation on subdivision and stability of Passenger ships as an equivalent to Part B of Chapter II of the International Convention for the Safety Of Life at Sea, 1960," *IMO Resolution* A.265 (VIII), Adopted on 20 Nov. 1973.

Abell Westcott (1920):. "Safety of Life at Sea," *INA*, Vol. LXII.

Amagai Kiyoshi, Kimihiko Ueno and Nobuo Kimura (1996). "Characteristics of Roll Motion for Small Fishing Boats," *Second workshop on Stability and Operational Safety of Ships*, Osaka University, November.

Bogdanov P, Dimitrova S and Kishev R (1990). "On the influence of Restoring moment changes in Waves on Stability Estimations, " *STAB*'90.

Comstock John P and James B Robertson Jr.(1961). "Survival of Collision Damage Versus the 1960 Convention on Safety of Life at Sea," *SNAME*, Vol. 69.

Dillingham Jeff (1981). "Motion Studies of a Vessel with Water on Deck," *Marine Technology*, Vol. 18, No. 1.

Hutchinson Bruce L (1995) . "Water-on-Deck Accumulation Studies by the SNAME AdHoc RO-RO Safety Panel," *Contemporary Ideas on Ship Stability*, Univ. of Strathclyde.

MSC.19(58). "Adoption of Amendments to the International Convention for the Safety of Life at Sea, 1974," *IMO Resolution* MSC 19 (58), Adopted on 25 May 1990.

Rakhmanin NN (1990). " Stability of Damaged Ship during Ship Motions in Waves," *STAB*'90.

Roby Kambisseri and Yoshiho Ikeda (1995). "A Comparative Study of the Probabilistic Damage Stability Rules and Proposals, " *Contemporary Ideas on Ship Stability*, Univ. of Strathclyde.

Roby Kambisseri, Tetsuya Hamano and Yoshiho Ikeda (1996). "Studies on the Damage Stability of ships," *Proceedings of KOJAM '96 (Third KOREA-JAPAN joint workshop on ship & Marine hydrodynamics*, Taejon, KOREA), July.

Roby Kambisseri and Yoshiho Ikeda (1996a). "A New Approach to Damage Stability Rule (1st Report) - Discussion on the Present Rules and the Concept of the New Approach-," *J. of Kansai Society of Naval Architects, Japan*, No. 226.

Roby Kambisseri and Yoshiho Ikeda (1996b). "About Safety Assessment of Damaged Ships," *Second Workshop on Stability and Operational Safety of Ships*, Osaka, Japan, Nov.

Sen P and Gerigk MK (1992). "Some Aspects of a Knowledge-Based Expert System for Preliminary Ship Subdivision Design for Safety," Vol. 2, *PRADS*'92.

Shimizu Nobuyuki, Roby K and Yoshiho Ikeda (1996). "An Experimental Study on Flooding into the Car Deck of a RORO Ferry through Damaged Bow Door," *J. of Kansai Society of Naval Architects, Japan*, No. 225.

Shin C and Ohkusu M (1986). "The Effects of Deck Wetting on the Stability of Ships in Beam Seas," *STAB*'86.

Sol92 (1992). "*SOLAS* consolidated edition".

Vassalos D, Pawlowski M and Turan O (1996). "Criteria for Survival in Damaged Condition," *RINA International conf. on the Safety of Passengers in RO-RO Vessels*, 7 June.

Welch JJ (1915). " The Watertight Subdivision of Ships," *INA*, Vol. LVII.

Wendel K (1961). "The Evaluation of Subdivisions ; Translation of the paper **Die Bewertung von Unterteilungen** from the *Jahrbuch der Schiffbautechnischen Gesellschaft*".

Proceedings of the Seventh (1997) International Offshore and Polar Engineering Conference
Honolulu, USA, May 25-30, 1997

ISBN 1-880653-28-1 (Set); ISBN 1-880653-31-1 (Vol. III)

Characterisation of the Flooding Process of a Damaged Ro-Ro Vessel

D. Vassalos and L. Letizia
University of Strathclyde
Glasgow, United Kingdom

SUMMARY

Recent research at the University of Strathclyde culminated in the development of a numerical procedure for assessing the damage survivability of damaged Ro-Ro vessels and, using this as a basis, new survival criteria have been proposed and submitted to IMO for consideration by the international shipping community. This paper presents the results of a fundamental study aimed at enhancing the insight into one of the most dominant parameters affecting the survival of a Ro-Ro vessel, the water accumulation on the vehicle deck. The investigation represents an attempt to identify the most important contributing factors to the flooding process by performing a series of experiments using a scaled model of a typical Ro-Ro vessel. The matrix considered involves a range of ship design and environmental parameters in a number of simplified damage scenarios, building up to the more realistic damages in a way that allows for isolation of individual contributions to the water accumulation on the Ro-Ro deck. The results of the experiments, supplemented by and contrasted with results of numerical simulations, are presented and discussed, leading to recommendations for characterising the flooding process for general assessment of damage survivability.

INTRODUCTION

The limited understanding of the complex dynamic behaviour of a damaged vessel and the progression of flood water through the ship in a random sea state has, to date, resulted in approaches for assessing the damage survivability of ships that rely mainly on hydrostatic properties with potentially serious consequences concerning the loss of life and property whilst endangering the environment. The tragic accidents of the *Herald of Free Enterprise* and more recently of *Estonia* were the strongest indicators yet of the magnitude of the problem at hand, particularly when water enters the deck of ships with large undivided spaces, such as Ro-Ro vessels. The ship loss could be catastrophic as a result of rapid capsize, rendering evacuation of passengers and crew impractical, with disastrous (unacceptable) consequences. Concerted action to address the water-on-deck problem in the wake of these led to the proposal of new stability requirements, known as the *Stockholm Regional Agreement*, or more commonly as *SOLAS '90+50*, pertaining to compliance of existing Ro-Ro vessels with SOLAS '90 requirements whilst accounting for the presence of a maximum 0.5 m height of water on the vehicle deck. In view of the uncertainties in the current state of knowledge concerning the ability of a vessel to survive damage in a given sea state, an alternative route has been allowed which provides a non-prescriptive way of ensuring compliance and one hopes enhanced survivability, namely the "*Equivalence*" route, by performing model experiments in accordance with the requirements of the SOLAS regulation II-1/8. In response to these developments, the shipping industry, slowly but steadily, appears to be favouring the model experiments route, implicitly demonstrating mistrust towards deterministic regulations which, admittedly, lack solid foundations. An attractive alternative route to tackling the water-on-deck problem in a way that allows for a systematic identification of the most cost-effective and survivability-effective solutions has been introduced by the Ship Stability Research Centre (SSRC) at the University of Strathclyde, by making use of a mathematical/numerical model, developed and validated over the past nine years, describing the dynamic behaviour of a damaged ship in seaway whilst subjected to progressive flooding. This model was made the basis during the Joint North West European Project (JNWEP) for formulating and proposing rational survival criteria to deal with water on deck as part of the probabilistic procedure for assessing damage stability, [1]. A relevant paper was submitted to IMO and is currently being considered by the working group on harmonisation of probabilistic standards. In both the developed mathematical model and the ensuing criteria the process of water accumulation on the Ro-Ro deck as well as the actual amount of water are dominant features. In this respect, an acceptably accurate model of water ingress/egress is a prerequisite to undertaking any investigations on damage survivability.

Deriving from the above, this paper attempts to elucidate some of the basic characteristics of the flooding process considering a typical Ro-

Ro vessel, by presenting and discussing preliminary results from an extensive experimental programme aimed at enhancing understanding and insight of this complex phenomenon. Firstly, a brief introduction is of the state-of-the-art mathematical model presently undergoing validation at the SSRC.

MATHEMATICAL MODEL

To study effectively damage survivability one needs to put together a non-linear six-degrees-of-freedom seakeeping model (that allows the vessel to drift as well as changes in its mass, centre of mass and mean attitude relative to the mean waterplane with time - the same dependence of environmental excitation and hydrodynamic reaction forces on the changing underwater volume of the vessel must also be catered for), a water ingress model (that allows for multiple-compartment flooding in the presence of oscillatory flows in extreme wave conditions and at times of shear flows), a sloshing model (that allows for random inflow and outflow through multiple openings), together with their interaction. An attempt in this direction represents on-going efforts at SSRC, [2]. The use of such a "tool" in its complete form, however, for parametric investigations and routine design applications is not within reach at present. The model that has been the focus of research developments by the Strathclyde team during the recent past and is currently undergoing validation, allows for a vessel drifting with the centre of gravity updated instantaneously during progressive flooding. It is a non-linear, coupled six-degrees-of-freedom model comprising the following:

$$\{[M]+[M_w]+[A_\infty]\}\ddot{Q}+\{[\dot{M}_w]+[B_{Viscous}]\}\dot{Q}+\int_0^t[K(t-\tau)]\dot{Q}(\tau)d\tau =$$

$$=\vec{F}_{Wave}+\vec{F}_{Drift}+\vec{F}_{Wind}+\vec{F}_{Current}+\vec{F}_{Restoring}+\vec{F}_{Gravitational}-\vec{F}_{WOD}$$

(1)

with,

$[M]$: Generalised mass matrix.

$[M_w]$: Flood water moving independently of the vessel but with an instantaneous free surface parallel to the mean waterplane.

$[A_\infty]$: Generalised added mass matrix (asymptotic values)

$[\dot{M}_w]$: Rate of flood water matrix (acting as damping).

$[B_{Viscous}]$: Viscous damping matrix

$\int_0^t[K(t-\tau)]\dot{Q}(\tau)d\tau$: Convoluted radiation damping

$\vec{F}_i$: Various generalised force vectors comprising wave (1st and 2nd order), wind and current excitation as well as restoration and gravitational effects. All these are updated instantaneously as a function of the vessel attitude relative to the mean waterplane by using a database which spans the whole practical range of interest concerning heel, trim, sinkage, heading and frequency. The same applies to the hydrodynamic reaction forces. Excitation from shifting of cargo can also be considered.

$\vec{F}_{WOD}$: This force vector is now comprised of dynamic effects of flood water in contrast to its counterpart in the previous model which involves only gravitational effects.

The phase/amplitude difference between vessel roll and flood water motions will be determined by building a comprehensive database through a systematic series of model experiments using a sway-heave-roll bench test apparatus. In case the dynamic behaviour of the flood water is considerable and could prove to be dominating or heavily influencing the vessel behaviour, the dynamic system of vessel-flood water must be treated as two separate worlds interacting, using CFD techniques to describe flood water sloshing, as indicated above.

MODELLING THE WATER INGRESS

The correct modelling of the inflow/outflow of water through the damage opening is naturally of primary importance for a reliable simulation of the progress of events during the flooding process. Quantities like *time to capsize* depend in fact critically on the evolution of this phenomenon. On the basis of the assumption that a simplified technique can be acceptably accurate if supported by experimental evidence, the formulation attempted here does not diverge drastically from those employed by other researchers. However, it attempts to offer credibility to the adopted approach by providing a solid theoretical and experimental background. In this respect, the experimental investigation focuses mainly on the fundamental nature of the flooding process and attempts to elucidate the main contributing factors to its evolution. It is readily accepted that a characterisation of the flooding of damaged ship compartments in a random sea is a highly complicated problem, in need of strong assumptions, unless employing complex CFD techniques. Within the frame of the dynamic model presented here this route is clearly impractical, thus leading to the semi-empirical approach as the only viable alternative. In the approximate method adopted in the models described in the foregoing, water ingress is treated as an intermittent probabilistic event based on the calculation of the relative position between wave elevation and damage location. The mode of flow is affected largely by the hydrostatic pressure head and the area of the damage hole but this is influenced by dynamic effects, edge effect, shape of opening, wave direction and profile, water elevation on either side of the opening and damage location. The methodology proposed here leads to specific equations for the rate of flooding through the damage opening whilst offering flooding coefficients expressly measured experimentally for use with the derived equations. Considering, for example, damage below the bulkhead deck, a flooding scenario is depicted by the simplified picture shown in Figure 1, with the sea treated as a reservoir and the pressure distribution in the hold assumed hydrostatic. If Bernoulli's equation is applied at sections A and B, assuming that the total pressure head is maintained constant and the velocity is zero in the reservoir, the inflow velocity at point P can be calculated as follows:

$$h_{out}+\frac{P_{atm}}{\rho g}+0=h_{in}+\frac{P_{atm}}{\rho g}+\frac{v^2}{2g} \quad \rightarrow \quad v=\sqrt{2g(h_{out}-h_{in})} \qquad (2)$$

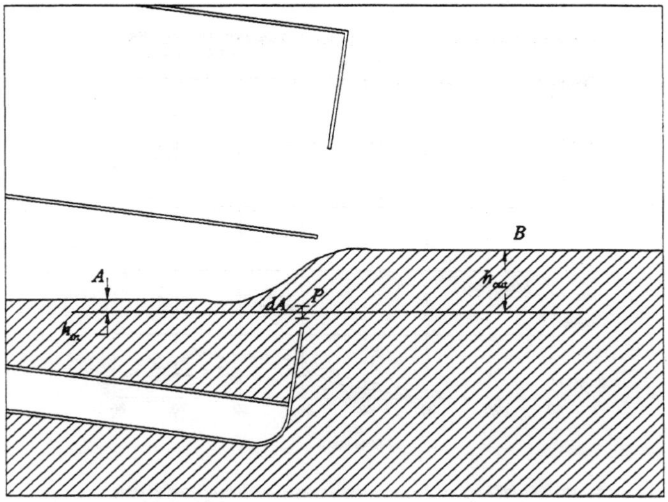

Figure 1: Water Ingress Main Parameters

and the flow rate through the horizontal layer around P:

$$dQ = K\sqrt{2g(h_{out} - h_{in})}dA \tag{3}$$

This expression reduces to the general form of those used for free-discharging orifices and notches when either h_{out} or h_{in} is negative, if the following limits are set:

$$\begin{cases} h_{in} = 0 \text{ if } h_{in} \le 0 \\ h_{out} = 0 \text{ if } h_{out} \le 0 \end{cases} \tag{4}$$

This takes care of those situations in which water is present only on one side of the damage. Of course, when h_{out} is less than h_{in}, the flow becomes negative, and water is expected to flow out of the compartment and into the sea. To accommodate for this the *pressure head* equation is put into the form:

$$dQ = K \quad sign(h_{out} - h_{in})\sqrt{2g|h_{out} - h_{in}|} \quad dA \tag{5}$$

with the same limits as above.

The rate of flooding and hence the time history of the flood water can be found by integrating dQ over the damage opening height, i.e.

$$Q(t) = K\int_{Damage} sign(h_{out} - h_{in})\sqrt{2g|h_{out} - h_{in}|}\, dA \tag{6}$$

Considering that $|h_{out} - h_{in}|$ represents the instantaneous down flooding distance which is relatively easy to compute, the problem of progressive flooding reduces to the evaluation of the coefficient K which can be done experimentally, i.e., K can be assumed to be an empirical function of relevant parameters to suite a more realistic case in which incident, diffracted and radiating waves as well as sloshing of the flood water are present. Preliminary experimental research, [2], demonstrated that the present formulation describes sufficiently well the trend of the rate of flooding with time as results in Figure 2 show.

Deriving from these results, it is worth noting also that the rate of flooding depends mainly on the relative water elevation as deduced from the argument presented in the foregoing and not the wave height itself. A second observation shows that the mean value of K appears to fall as soon as water is present on both sides of the damage opening. As a result of this, it would be to reasonable to adopt two coefficients intended for use according to the particular case encountered. The mean value for the flooding coefficient when pure inflow occurs (water only present external to the ship) is almost 1.4, whilst with water present on both sides it reduces to 0.8 when the damage opening faces the oncoming waves. If the damage opening is away from the oncoming waves, the corresponding values of the flooding coefficient appear to be approximately the same and close to 1.0.

EXPERIMENTAL PROGRAMME

Damage Scenarios and Test Conditions

To foster a better understanding of the water accumulation on Ro-Ro vehicle decks, a series of experiments has been planned and is taking place at the University of Strathclyde. In this paper the results of the first series of tests are presented, concerning mean asymptotic height of flood water on the vehicle deck h, as a function of wave and damage opening characteristics. The model used in these experiments is described in [3] and is a 1:42 scale model of a typical Ro-Ro vessel, the main particulars of which are shown in Table 1. The model accommodates two compartments open to the sea: the first extending from the double bottom to the bulkhead deck and from 52.6m to 65.9m from the aft perpendicular, giving a damaged freeboard of 1.08m and the second, - on the vehicle deck - extending from 22.25m to 97.25m from the aft perpendicular.

Table 1: Main Particulars of the Ro-Ro Vessel Used in the Experimental Investigation

L_{BP}	(length between perpendiculars)	=	131.0 m
B	(breadth)	=	26.0m
T	(design draught)	=	6.10m
D	(depth to uppermost continuous deck)	=	18.8m
D_{bd}	(depth to bulkhead deck)	=	7.8m
D_{db}	(depth to double bottom)	=	1.6m
Δ	(displacement)	=	12200 tonnes
C_b	(block coefficient)	=	0.582

The parameters investigated so far cover a range of regular and irregular waves as shown in Tables 2 and 3 and a number of damage scenarios as illustrated in Figure 3. These scenarios attempt to isolate the various contributions to the flooding process, as explained next:

- Scenario 1, describes damage of the vehicle deck only
- Scenario 2, describes damage both above and below the bulkhead deck but without penetration
- Scenario 3, is the same as scenario 2 but with penetration, allowing water to flow between the two compartments

All three scenarios were tested with a fixed and a heaving model to allow for an evaluation of the contribution of heave motion to the flooding process and of the interaction between the two processes. The damage opening in these scenarios was modelled according to a 100% SOLAS, i.e., damage length at the waterline equal to **(0.03L_s + 3.0)** metres or **11** metres, whichever is less, using a trapezoidal opening with sides at 15^0 to the vertical.

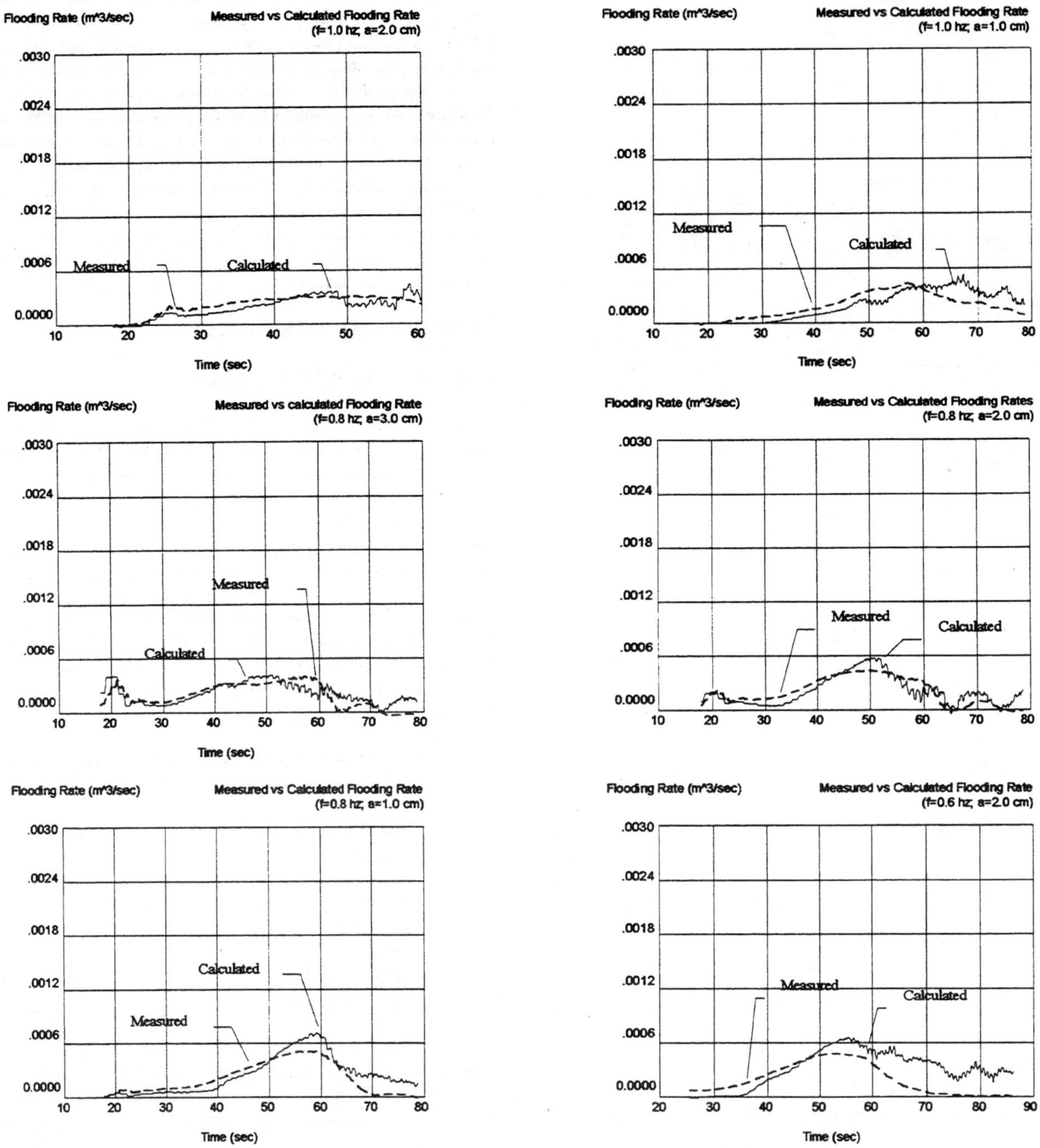

Figure 2: Experimental Flooding Curves

Scenarios 4 to 6, refer to different shapes of the damage opening, all satisfying SOLAS requirements, with scenario 6 in particular attempting to duplicate the side damage suffered by the European Gateway following collision with Speedlink Vanguard in 1982.

Table 2: Regular Wave Parameters

Wave Height *H* (m)	=	2.0, 3.0, 4.0, 5.0
Wave Period *T* (sec)	=	2.50, 1.25, 1.00, 0.83, 0.62

Table 3: Irregular Wave Parameters (JONSWAP Spectrum)

Significant Height H_s (m)	**Zero Crossing Period T_0 (sec)**
2.0	4.42
3.0	5.42
4.0	6.25
5.0	6.99

Roll motion, freeboard and KG are the parameters currently being investigated and the results will be published in the near future.

Presentation of Results and Discussion

The results derived from testing in regular waves are shown in Figures 4 (heaving model) and 5 (fixed model). With reference to the asymptotic height of water on the vehicle deck, the following observations can be made:

- *h* is clearly a function of both *T* and *H*. The variation of *h*, however, becomes more pronounced with the heaving model. This latter phenomenon is a direct result of the tuning/de-tuning between wave and heave motion. It would be reasonable, for example, to expect a reduction in flooding when the model moves in phase with the waves, whereas the reverse will be true when the model heaves in anti-phase with the waves. In addition, heave motion affects directly the flooding process both in terms of influencing the relative water elevation directly as well as causing water to be sucked into the damaged compartments as the model heaves downwards whilst giving rise to something of a "pumping" action, causing water to flood the vehicle deck through the triangular opening between the two compartments.

- The shape of the damage opening has a clear influence on *h*.

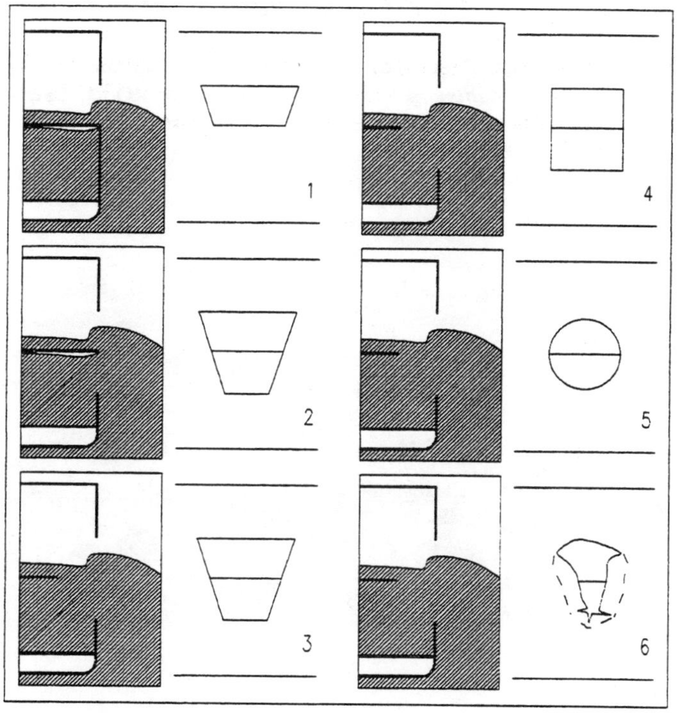

Figure 3: Damage Scenarios Considered in the Experimental Programme

It is worth noting that the water ingress and egress are fundamentally different flows. Water egress has more of a constant nature, depending mainly on *h*, whilst water ingress is intrinsically intermittent, especially for larger freeboards. As the flooding process evolves a steady (or stationary) state may be reached when the integral of the sum of inflow and outflow functions oscillates around zero, i.e. when, on the average, inflow and outflow neutralise each other. This leads to larger *h* values with increasing wave heights and decreasing wave periods, as a larger number of higher crests would reach the deck level. It has been observed though, that steep waves tend to diffract and break against the model side giving rise to highly distorted crests which carry considerably less water. This is the reason why there seems to be an optimal wave frequency for flooding even in the case of the fixed model.

As a direct consequence of the observations noted above, two important conclusions can be drawn. Firstly, the vessel motion influences considerably and directly the flooding process and conversely, flooding affects both the vessel motion and her attitude. It is essential, therefore, to take both phenomena into consideration when studying the evolution of either. Secondly, since flooding and ship motion are two compounded processes, testing in regular waves will be unsuitable as it can lead to deceptive results. This second conclusion can be better appreciated by considering that for long wave periods the vessel will be heaving in phase with the waves, hence the distance between sea water and vehicle deck will be nearly constant and flooding will be impaired. Furthermore, a comparison of the average height of water accumulated on deck between Figures 4, 5 (regular waves) and Figure 6 (irregular waves) clearly reveals that *h* is generally smaller in the presence of irregular waves, even though similarity in trends is evident. Finally, it may be observed that the amount of water accumulated in on deck in scenarios 1 and 2 is generally greater than the corresponding amount in scenario 3. This difference is attributed to the down flooding of water through the triangular penetration notch, from the vehicle deck to the lower compartment.

NUMERICAL SIMULATION RESULTS

Considering the findings described in the foregoing, a number of numerical tests were also undertaken for comparison purposes and to further enhance the insight into some of the interesting observations in the experimental results, considering damage scenarios 1, 2 and 3, for a free drifting vessel. Unlike the model used in the experiments, the ship employed in the numerical analysis is free to move and its behaviour is estimated in 3D in accordance with the model described above. The numerical model of a Ro-Ro vessel readily available was used for the purpose, the main particulars of which are given in Table 4. More details can be found in [4]. The damage scenario investigated considers a large compartment (comp. 1) and the car deck (comp. 2) open to the sea through a SOLAS trapezoidal damage opening amidships. The open deck extends longitudinally for 75.0m from -41.8m forward and transversely from side to side. The lower compartment occupies the space between double bottom and bulkhead deck for a length of 27.7m , from -19.0m forward. This arrangement yields a damaged freeboard of 0.7m.

Table 4: Main Particulars of the Ro-Ro Vessel Used in Numerical Investigation

L_{BP}	(length between perpendiculars)	=	126.1m
B	(breadth)	=	22.7m
T	(design draught)	=	5.625m
D	(depth to uppermost continuous deck)	=	12.6m
D_{bd}	(depth to bulkhead deck)	=	7.87m
D_{db}	(depth to double bottom)	=	1.2m
Δ	(displacement)	=	8807 tonnes
C_b	(block coefficient)	=	0.53
KG	(design KG)	=	10.10m
KM	(intact KM)	=	11.62m

In the numerical simulation, the damaged compartments are considered to be empty initially for damage scenarios 2 and 3 whilst in the case of scenario 1 some 4,000.0 tonnes of water is allowed in intact compartment 1 in order to achieve the same freeboard as in the other conditions. Once simulation begins, flooding of the damaged compartments is triggered twenty seconds later. This artifice allows for dynamic and numerical transients to die out before flooding starts, rendering it independent from the particular initial conditions chosen. All tests were performed with irregular waves using a JONSWAP spectrum, with constant values of Hs and KG, which were set to 2.45 m and 9.86 m, respectively. These define a point on the survivability boundary as determined by the model experiments [4]. This choice was aimed at amplifying the possible effects of the various parameters investigated.

In accordance with the experimental results, numerical simulations were also successful in demonstrating the importance of the triangular penetration notch between the two compartments in reducing the chance of water accumulation on the vehicle deck. In fact, in a fully dynamic case, flood water hardly ever manages to build up when this opening is accounted for and down flooding to the lower compartment is allowed. Unlike the experiments, however, the numerical simulations reveal a new aspect concerning the influence of different damage openings on the behaviour of a damaged vessel. It becomes evident from Figures 7 (scenario 1) and 8 (scenario 2) that the major difference between the output of these two damage conditions is that in scenario 1, the amount of water trapped in compartment 1 makes the vessel inherently more stable than when this space is open to the sea. This enhances the ability of the vessel to sustain a considerably larger amount of water on deck than the amount normally sufficient to lead to capsize. This observation is supported by the trends of the flood water realisation shown in Figure 8, which indicate how flood water in compartment 1 tends to flow out as the ship heels before capsizing. This, of course, is hardly surprising but it certainly helps further understanding on how important the interactions between vessel motion and water ingress and accumulation are. It is interesting to note, along the same lines, how well correlated the roll, heave and flood water realisations are in Figure 7 as indeed the only tangible accumulation of water on deck appearing in Figure 9 occurs in the presence of concurrent large heave and roll motions.

CONCLUDING REMARKS

The evidence presented in this paper offers important clues concerning specific contributions to the water accumulation on the Ro-Ro vehicle deck and the characterisation of the flooding process. It also demonstrates that there are good reasons to solicit due consideration of the shape and configuration of the damage openings in determining the amount of water accumulating on vehicle decks and discourage apparently innocuous simplifications often adopted to tackle the problem assessing the damage survivability of this type of vessels [5]. Considering, however, that modelling a damage opening in a way that reproduces reality exactly will not be possible and, in fact, not relevant, it will be particularly helpful for all concerned to appreciate that all that is necessary to progress further in this field is the definition of a generalised damage opening, based on acceptable statistical data. that is universally accepted and used in testing for damage survivability in both physical and numerical model tests. Thankfully, such convergence is becoming evident with the use of the opening described here as scenario 4.

It also very important to emphasise that the study of damage survivability involves two distinct but intrinsically interrelated and highly interacting processes, namely, ship motion and flooding. The non-stationarity in the vessel motion introduced by the water accumulation coupled with the intermittence of the flooding process itself and the severe non-linearities in the ensuing dynamic system, demand great care in dealing with the many issues of this complex problem.

REFERENCES

[1] Vassalos, D., Pawlowski, M. and Turan, O.: *"A Theoretical Investigation on the Capsizal Resistance of Passenger/Ro-Ro Vessels and Proposal of Survival Criteria"*, Final Report, Task 5, The Joint North West European R&D Project, March 1996.

[2] Letizia, L.: *"Damage Survivability of Passenger Ships in a Seaway"*, Ph.D. Thesis, Department of Ship and Marine Technology, University of Strathclyde, November 1996.

[3] DMI 88116: *"RO-RO Passenger Ferry Safety Studies Model Test for F10 - Final Report of Phase I"*, DMI Project Report to the UK Department of Transport, 1990.

[4] Dand, I.W.: *"Experiments with a Floodable Model of a RO-RO Passenger Ferry"*, MT Project Report to the Department of Transport, BMT Fluid Mechanics Ltd, 1990.

[5] The Glosten Associates, inc.: *"Water Accumulation on the Deck of a Stationary Ship"* - SNAME Ad Hoc RO-RO Safety Panel, Annex A, File No. 94209.01 - 8 February 1995.

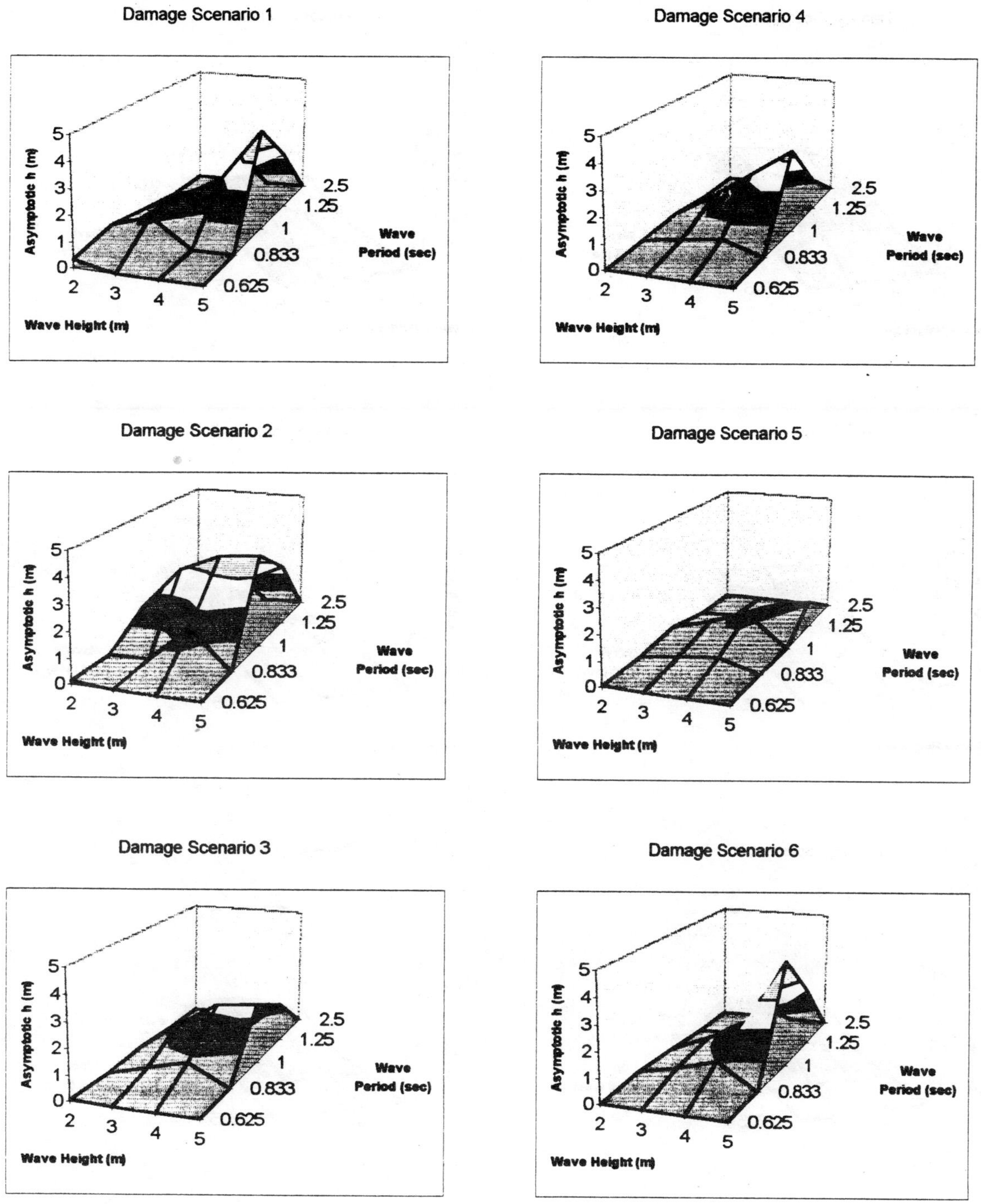

Figure 4: Variation of the Asymptotic Height of Flood Water on the Vehicle Deck with Regular Wave Parameters (Heaving Model)

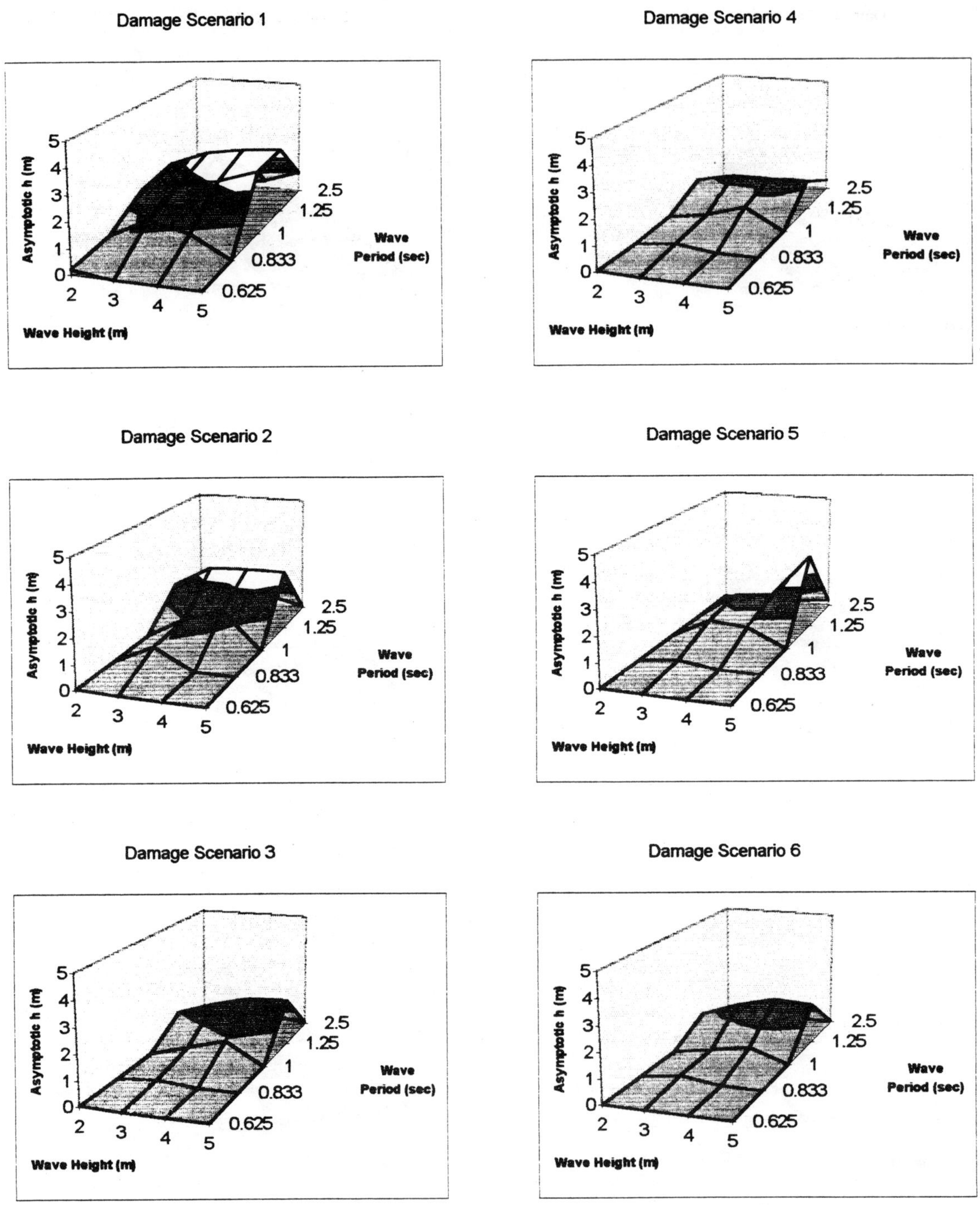

Figure 5: Variation of the Asymptotic Height of Flood Water on the Vehicle Deck with Regular Wave Parameters (Fixed Model)

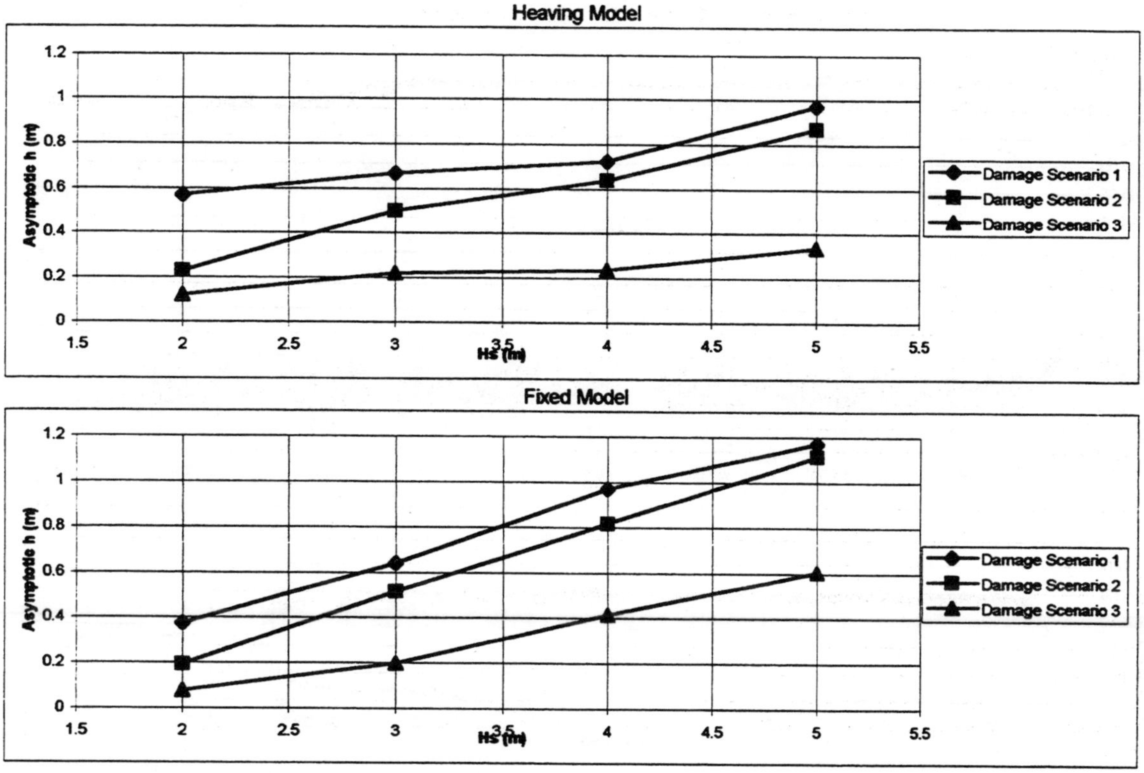

Figure 6: Variation of the Asymptotic Height of Flood Water on the Vehicle Deck with Significant Wave Height and Damage Condition

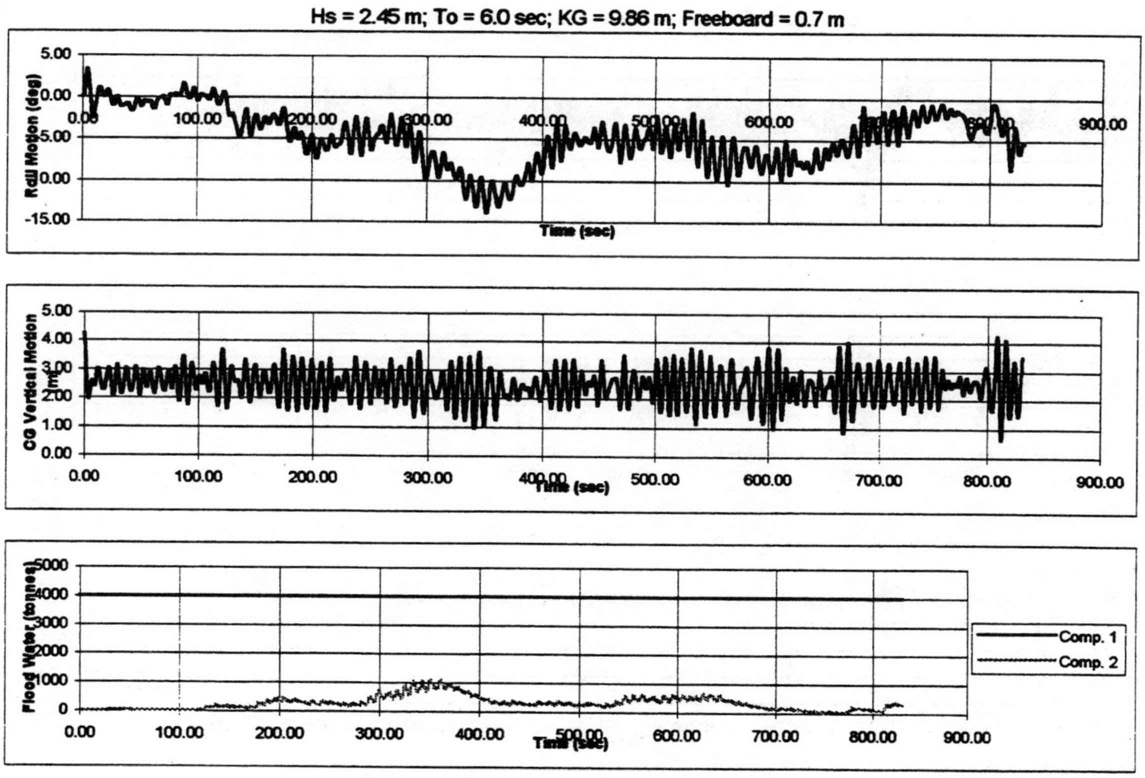

Figure 7: Simulation Results - Damage Scenario 1.

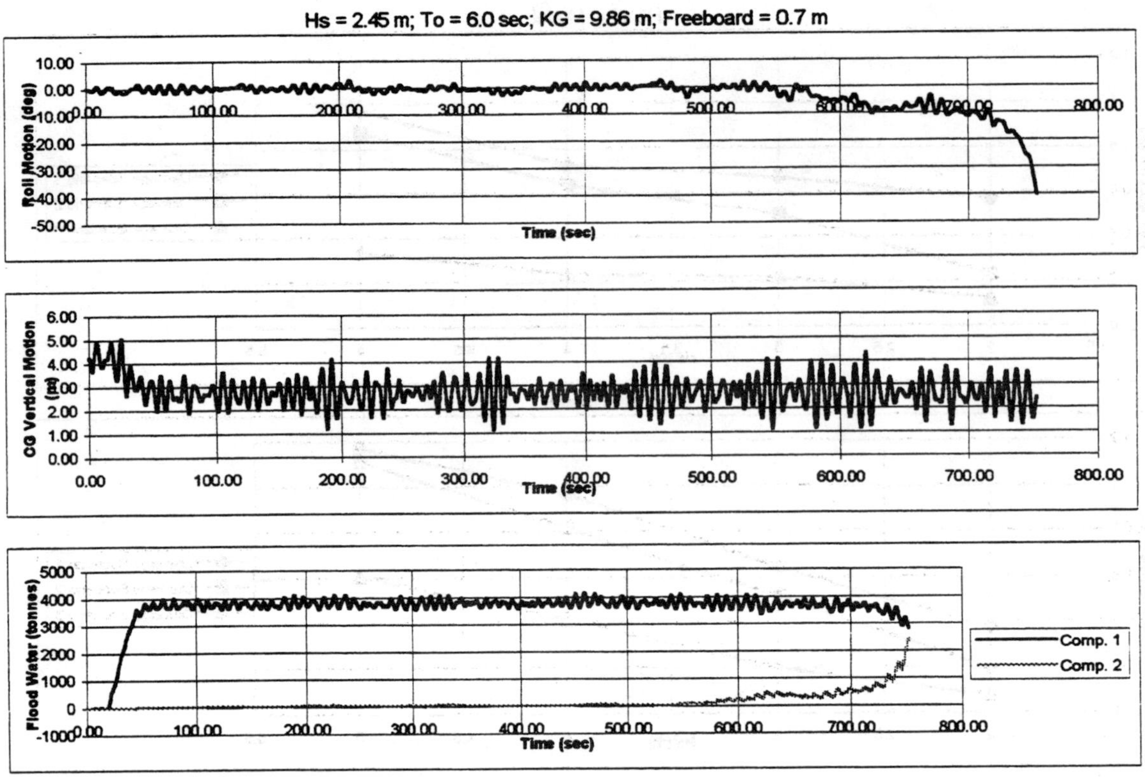

Figure 8: Simulation Results - Damage Scenario 2.

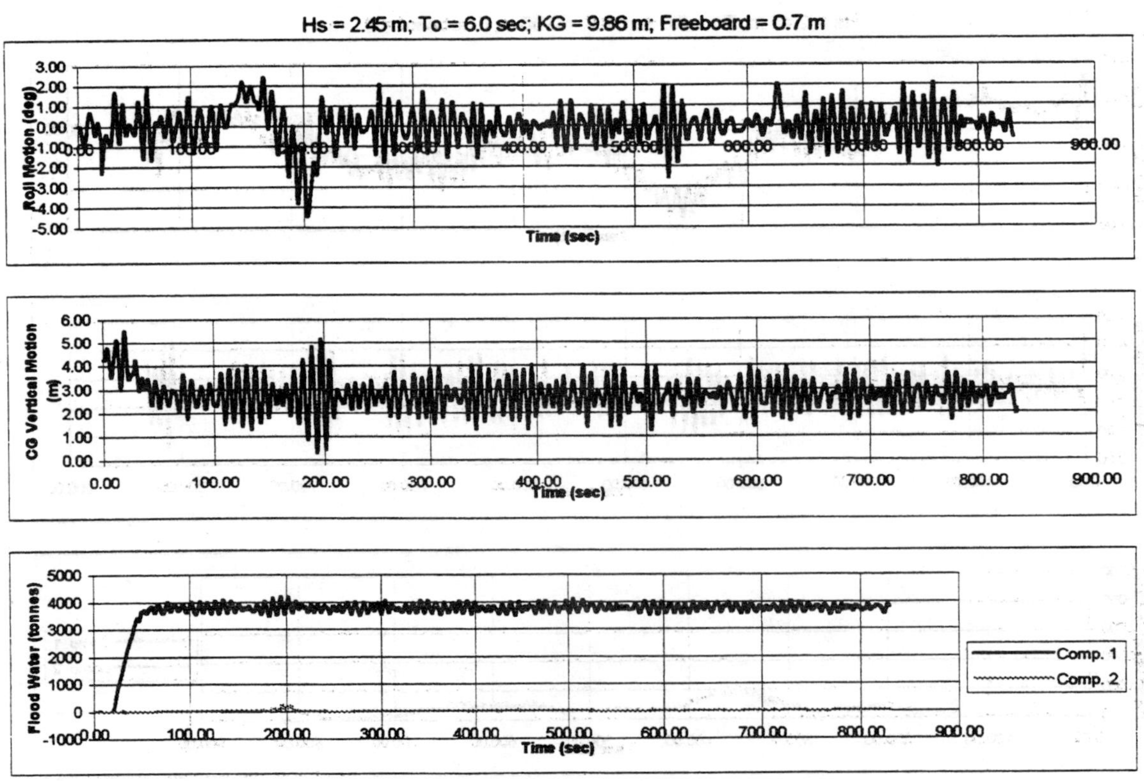

Figure 9: Simulation Results - Damage Scenario 3.

Proceedings of the Seventh (1997) International Offshore and Polar Engineering Conference
Honolulu, USA, May 25-30, 1997

ISBN 1-880653-28-1 (Set); ISBN 1-880653-31-1 (Vol. III)

Field Experiments and Numerical Prediction on Dynamics of a Light Floating Structure Moored in Deep Ocean

Wataru Koterayama and Hiroo Mizuoka
Kyushu University
Kasuga, Japan

Naoyuki Takatsu and Tetsuro Ikebuchi
Kawasaki Heavy Industries
Kobe, Japan

ABSTRACT

A floating structure for fish-gathering has been developed and moored in the deep sea south of Okinawa Island. The depth of the sea in this area is 1300m. The structure is moored with a composite mooring line of wire rope and chain and the displacement is about 94 tons, To establish a design for the small floating structure, a series of field experiments were carried out in which motions of the structure, current velocity and wave height were measured with motion sensors, a current meter and a wave rider buoy. Motions of the structure obtained from the field experiments were compared with numerical simulations. The dynamics of the mooring line were calculated using a lumped mass technique. Hydrodynamic coefficients used in the numerical simulations were obtained from theoretical calculation and model experiments in a wave tank. The design method for a small floating structure moored in deep sea were confirmed through the field experiments.

KEY WORDS Field experiment, Motions, Moored floating structure, Simulation technique

1 INTRODUCTION

Designing a floating offshore structure requires theoretical and experimental studies to determine the anticipated performance and safety. In the final stage of the design the performance under a working condition and the safety under the survival condition should be confirmed using numerical simulation programs developed on the basis of these studies.

Establishment of a numerical simulation method for analyzing the dynamics of a small floating structure for fishery or ocean observation is necessary; yet the difficulties confronted are much more serious than for a large oil rig: the incident wave height is relatively large in comparison with the scale of the floating structure, which leads to strong non-linear effects. The computer program developed for the numerical simulation of motions of a small floating structure must therefore be validated by experiments as was done by many researchers over the past twenty years for a large offshore structure.

In the case of deep sea mooring of a small structure another difficulty is that the mass of the mooring line is large compared to that of the floating structure, therefore, interference between the dynamics of this line and that of the structure itself must be considered; this interference is small enough to be neglected except for the phenomenon of so called slow drift oscillation for an oil rig with mass of more than ten thousand tons.

The results of field experiments on the motion of a structure moored in a 1300m deep sea area south of Okinawa island are reported here and the results will be compared with those of numerical simulation.

2 OBJECT STRUCTURE OF STUDY

Figure 1 shows a conceptual view of the structure and the mooring line; the structure was built to gather tuna and other fish. Its advantages for the fishing industry have already been confirmed by an investigation carried out by Okinawa Prefectural Fisheries Experiment Station. The total displacement including the mooring line is 94 tons, which includes the weight 30 tons of the suspended line in water.

The mooring line is composed of chain and wire rope sections, see Fig.1 for the layout and dimensions.The total length is 1807m including the part laid on the sea bottom. The diameter of the main deck at the water surface is 7m, and that of the column connected under the main deck is 1.9m A hexagonal footing is fixed at the lower end of the column at a depth of 7 meters. This is intended to increase the fish gathering capability and to serve as viscous damping for motions; it is composed of frame cylinders and nets as shown in Fig.1.

3 THEORETICAL MODEL

Motions of floating structure

The coordinate systems used to describe the motion of the

floating structure and mooring line are shown in Fig.2, where X and Z are the horizontal and vertical axis, θ the pitching angle, β_i the angle of the line between M_i and M_{i+1} to the X axis. The origin of the axis for the structure is fixed at its center of gravity and the cable is modeled based on the lumped mass method (Thresher and Nath, 1975); each mass in the figure is denoted by M_i. Motions of the structure and cable were assumed to be confined to two dimensions in this study.

As mentioned, the nonlinear effects on hydrodynamic forces are significant and the viscous force in particular has a great effect on the motion of the floating structure. The most popular means of describing the viscous force is based on the relative velocity concept (R.V.C) (Koterayama and Nakamura, 1988a). Under R.V.C the drag force F_D written in a modified form of the Morison equation is

$$F_D = \frac{1}{2}\rho C_D A_x(\dot{X} - u)|\dot{X} - u| \tag{1}$$

where ρ is water density, C_D the drag coefficient, A_x the projected area of a structure in the x direction, $\dot{X}$ the velocity of the structure, and u the x-component of orbital velocity of water particle in waves.

The R.V.C can be understood by intuition and is broadly used in a motion equation for a floating structure. A few papers (Moe and Verley, 1980; Koterayama and Nakamura, 1988b) pointed out the problems of this method when R.V.C is applied to a case in which the period of oscillation of a structure is different from that of the ambient flow field. Especially in the case of reduced velocity $V_r < 5.0$, where $V_r = \dot{X}_{max}T/D, \dot{X}$ is the maximum value of oscillating velocity of a structure, T the period of u which is the ambient flow field, and D the diameter of the structure, it is difficult to determine a single value of the drag coefficient for Equation (1) even in the R.V.C; this is because the value of the drag coefficient obtained from the experimental data of the structure's oscillating period differs from that of the ambient flow field period, therefore we get two different values of the drag coefficient from one experiment. This means the value of C_D is not constant during one cycle, which is longer one, of the flow field or oscillation of a structure.

The floating structure in this study has a deep draft, which makes it very difficult to apply eq.(1), because the flow field velocity u is the orbital velocity in waves and therefore varies with the depth from the water surface to the bottom of the structure. If we apply the R.V.C the equation has to include the integral form along the vertical axis from the water surface to the structure bottom, and the drag coefficient for each section has to be estimated. It is difficult to obtain these from experiments and the accuracy of the theoretical estimate can't be guaranteed.

The independent flow field concept (I.F.F.C) (Laya and Connor, 1984) is proposed to make up this lack of R.V.C. The drag force acting on a structure oscillating in an unsteady flow is described in I.F.F.C as follows:

$$F_D = \frac{1}{2}\rho C_{DX} A_x \dot{X}|\dot{X}| - \frac{1}{2}\rho C_{DWX} A_x u|u| \tag{2}$$

The physical meaning of the I.F.F.C is not easy to understand. Moe and Varley (1980) explained it as being based on the superposition of two independent flow fields, a far field which is unaffected by the structure motion and a near field resulting from that motion. The drag force is typically non-linear and therefore can't be superposed, but Laya and Connor (1984) showed that this concept can be applied to a practical problem with good accuracy in a small KC number region (KC< 10) and small reduced velocity region ($V_r < 5.0$). In present field experiments the maximum significant wave height $H_{\frac{1}{3}}$ through the experimental period was approximately 4m and the KC number (KC $= \pi H_{\frac{1}{3}}/D$, D the diameter of the main deck or the column of the floating structure) was smaller than 2 for the main deck and 6 for the column, and it is enough small to adopt the I.F.F.C..

The I.F.F.C also has an another advantage. Usually we carried out laboratory experiments, forced oscillation tests and wave force tests in a test tank to determine the hydrodynamic coefficients. We obtained C_{DX} from forced surging tests and C_{DWX} from wave force tests. If we want to apply the R.V.C to the present study, we have to carry out forced surging tests in waves to obtain the appropriate value of the drag coefficient for the various combinations of incident waves and motions of the floating structure; there would be numerous combinations of amplitude and frequency of forced oscillation and incident waves. For these reasons the motion of the structure was analyzed in this study on the basis of the I.F.F.C.. In case of large KC number (KC> 10 for the column), we have to consider the adoption of the R.V.C.

The motion equations based on the I.F.F.C are written as follows;

$$\left.\begin{aligned}
&\text{For the surging motion}\\
&\qquad (M + a_{XX})\ddot{X} + b_{XX}\dot{X} + \tfrac{1}{2}\rho A_x C_{DX}\dot{X}|\dot{X}|\\
&\qquad\quad - \tfrac{1}{2}\rho A_x C_{DWX} u|u| + a_{X\theta}\ddot{\theta} + b_{X\theta}\dot{\theta}\\
&\qquad\quad + \tfrac{1}{2}\rho A_x C_{DX\theta} l^2 \dot{\theta}|\dot{\theta}|\\
&\qquad\quad = F_{XW} + F_{XD} + T_X\\
&\text{For the heaving motion}\\
&\qquad (M + a_{ZZ})\ddot{Z} + b_{ZZ}\dot{Z}\\
&\qquad\quad + \tfrac{1}{2}\rho A_z C_{DZ}\dot{Z}|\dot{Z}|\\
&\qquad\quad - \tfrac{1}{2}\rho A_z C_{DWZ} w|w| + \rho g A_W Z\\
&\qquad\quad = F_{ZW} + F_{ZD} + T_Z\\
&\text{For the pitching motion}\\
&\qquad (J + a_{\theta\theta})\ddot{\theta} + b_{\theta\theta}\dot{\theta} + \tfrac{1}{2}\rho A_X l^3 C_{D\theta}\dot{\theta}|\dot{\theta}|\\
&\qquad\quad - \tfrac{1}{2}\rho A_\theta l C_{DW\theta}(u,w)|u,w| + a_{X\theta}\ddot{X} + b_{x\theta}\dot{X}\\
&\qquad\quad + \tfrac{1}{2}\rho A_x l C_{D\theta X}\dot{X}|\dot{X}| + W\cdot GM\cdot\theta\\
&\qquad\quad = F_{\theta W} + F_{\theta D} + T_\theta
\end{aligned}\right\} \tag{3}$$

in which

ρ	:	Water density
g	:	Gravitation
W	:	Displacement of floating structure
M	:	Mass of floating structure
A_W	:	Water plane area
J	:	Moment of inertia
GM	:	Metacentric height
a_{XX}, a_{ZZ}	:	Added mass for surging and heaving
$a_{\theta\theta}$	:	Added moment of inertia
$b_{XX}, b_{ZZ}, b_{\theta\theta}$	:	Wave damping coefficient
$a_{X\theta}, b_{X\theta}$	:	Coupling coefficient of inertia and damping force between surging and pitching
A_x, A_z	:	Projected area of floating structure for surging and heaving

$C_{DX}, C_{DX\theta}, C_{DZ}, C_{D\theta}$	:	Drag coefficient obtained from forced surging, heaving and pitching tests.
$C_{DWX}, C_{DWZ}, C_{DW\theta}$	:	Drag coefficients of x-, z- and θ--direction from wave force tests
u, w	:	Horizontal and vertical velocity of water particles in waves at the center of gravity of the floating structure
$F_{XW}, F_{ZW}, F_{\theta W}$	:	Linear wave force for surging, heaving and pitching
$F_{XD}, F_{ZD}, F_{\theta D}$	:	Drifting force for surging, heaving and pitching
T_X, T_Z	:	Mooring tension in x- and z-directions
T_θ	:	Moment caused by mooring line in θ-direction

The slow drifting forces don't affect motions so much in the case of one point mooring in the deep sea because the restoring force of the mooring system in X-direction varies from zero to a certain values and is usually too weak to induce the systematic slow drift oscillation against the large viscous damping forces acting on the mooring line and the floating structure.

The wave damping coefficients $b_{XX}, b_{ZZ}, b_{\theta\theta}, b_{x\theta}$ and linear external wave force $F_{XW}, F_{ZW}, F_{\theta W}$ for the structure are obtained from numerical calculation based on the 3-dimensional singularity distribution method (Sugiura, M., 1977). The added mass coefficient $a_{XX}, a_{ZZ}, a_{\theta\theta}, a_{X\theta}$ can be obtained from experiments and calculation, but this floating structure has a net and other small appendage and these makes difficult to calculate the added mass coefficient with good accuracy. Then these added mass coefficient were also determined from experiments.

Motions and tensions of mooring line

The tension T_X, T_Z and the motion of the mooring cable can be calculated using the lumped mass method (Thresher and Nath, 1975, Koterayama et al., 1992).

$$\begin{bmatrix} m_{11j} & m_{12j} \\ m_{21j} & m_{22j} \end{bmatrix} \times \begin{bmatrix} \ddot{X}_j \\ \ddot{Z}_j \end{bmatrix} = \begin{bmatrix} F_{Xj} \\ F_{Zj} \end{bmatrix} \tag{4}$$

in which,

$$\left.\begin{aligned} m_{11j} &= M_j + cos^2\bar{\beta}_j A_{tj} + (1 - cos^2\bar{\beta}_j) A_{nj} \\ m_{12j} &= m_{21j} = (A_{tj} - A_{nj}) sin\bar{\beta}_j cos\bar{\beta}_j \\ m_{22j} &= M_j + sin^2\bar{\beta}_j A_{tj} + cos^2\bar{\beta}_j A_{nj} \\ F_{Xj} &= T_j cos\beta_j - T_{j-1} cos\beta_{j-1} + f_{dxj} \\ F_{Zj} &= T_j sin\beta_j - T_{j-1} sin\beta_{j-1} + f_{dzj} - \delta_j \\ \bar{\beta} &= (\beta_j + \beta_{j-1})/2 \end{aligned}\right\} \tag{5}$$

A_{tj}, A_{nj}	:	Tangential and normal added mass of jth segment
F_{Xj}, F_{Zj}	:	Forces of jth segment
β_j	:	Angle of cable inclination
f_{dXj}, f_{dZj}	:	Drag force acting on j-th lumped mass

The drag and added mass coefficient of the mooring line are estimated from an experimental study (Ando and Yamanaka, 1973). The accuracy of the calculation using these values has been confirmed in laboratory experiments (Tasai et al., 1980) and field experiments (Koterayama et al., 1980)

The linear theory shows that hydrodynamic coefficients in the equations for motions of the floating structure in irregular waves depend on the frequency of incident waves, but it makes calculation procedure very complicated. In this study the wave frequency is fixed at the mean period of the incident irregular waves, in a small floating structure the wavelength is long enough incomparison with the scale of the structure to neglect the effect of wave frequency on the value of hydrodynamic coefficients, and therefore this approximation does not result in large error.

4 MODEL EXPERIMENTS

Model experiments to determine the added mass and drag coefficients were carried out in a wave tank 65m long, 5m wide and 7m deep. Forced surging, heaving, pitching tests and wave force tests were carried out using a 1/30 scale model to obtain the added mass coefficient $a_{XX}, a_{ZZ}, a_{\theta\theta}, a_{X\theta}$ and the coefficients $C_{DX}, C_{DZ}, C_{D\theta}, C_{DX\theta}, C_{DWX}, C_{DWZ}, C_{DW\theta}$, which are needed in the equation (1)~(3). To obtain these values the wave damping forces calculated by a linear theory were subtracted from the total damping forces obtained in experiments. Careful experimental arrangements are important to maintain high accuracy in the experiments because of the small scale of the model. The experimental methods were established earlier. (Koterayama and Nakamura, 1988b).

Forced surging tests

The added mass coefficients a_{xx} obtained from forced surging tests are shown in Fig.3, in which the theoretical calculation result is also indicated by a broken line. The theoretical results does not include the hexagonal footing and therefore it is smaller than the experimental results.

The drag coefficients C_{DX} for surging, which were obtained from surging tests, are shown in Fig.4. The horizontal axis is taken as KC number ($2\pi Xa/B$, X_a: amplitude of forced surging, B: diameter of the deck of the floating structure). These model experiments were carried out changing the period and amplitude of the forced oscillation. In the very small KC number region (KC< 1.0) the value of C_{DX} increases rapidly with the decrease of KC, while in the region of KC> 1.0 it is almost constant. In very small KC number region the boundary layer separation is not clear and the frictional component of the drag force acting on circular cylinders become main part of the drag force, and the drag coefficient is written in the following form (Tanaka et al. 1980),

$$C_D = 4\pi[\pi/(R_n \cdot KC)]^{\frac{1}{2}} \tag{6}$$

The equation indicates that in the very small KC number region the drag coefficient is in inverse proportion to $R_n^{\frac{1}{2}}$ and $KC^{\frac{1}{2}}$., and the results indicated in Figs.4 and 5 show this tendency and they are similar to that of a circular cylinder (Koterayama et al., 1994), because main elements of the structure are circular cylinders. This result is very convenient for use in numerical simulations. We can use the same drag coefficient value except in the very small KC number region.

As shown in eq.(3) the damping force caused by the surging motion has a comparatively large effect on the pitching motion because the pitching damping is small in comparison with the surging or heaving motion. The drag coefficient $C_{DX\theta}$ is also obtained from the forced surging tests as shown in Fig.5. In very small KC number region it decreases with the decrease of KC number and in the range of $KC > 1.0$ the value is steady. This tendency can be explained from the same reason as the case of C_{DX}. In the range of $KC < 0.5$ it is less than zero but this is not strange. It is just a problem of the action point of the drag forces in surging, when that point is higher than the center of gravity $C_{DX\theta}$ becomes positive and vice versa.

Figure 6 shows the result of C_{DZ} in the forced heaving test. The amplitude of this test wasn't large enough because the deck height of the model from the water surface was only 3cm. The tendency of C_{DZ} is slightly different from that of C_{DX}: the effect on the main deck in heaving is great and the wave damping forces acting on the deck are very large. The wave damping force was subtracted from the total drag forces when C_{DZ} was obtained from experiments, but the wave damping forces are larger than other modes and interference effects between the wave and viscous damping affect the value of the viscous damping force so that C_{DZ} is the function of not only KC number but also the wave frequency. The values of C_{DZ} denoted as the function of KC number are scattered by these effects.

Figure 7 shows the results $C_{D\theta}$ of forced pitching tests. It was difficult to change the amplitude over the wide range in these tests and the experiments were carried out on the basis of frequency. The results are therefore plotted on the horizontal axis of non-dimensional frequency ξ_B. Only the amplitudes of 2 and 4 degrees were selected for experiments, and from the results we found the value of $C_{D\theta}$ didn't change between them. The average non-dimensional frequency range in the field experiments is 1.1, which is the natural frequency of the pitching, and it is possible to determine a value of $C_{D\theta}$ from Fig.7.

$C_{D\theta X}$ can be also obtained from the forced pitching test, but this was difficult due to the mechanics of the forced pitching apparatus used in this study. The contribution of this term to the surging motion is not believed to be large because part of the damping force for the surging motion is much larger than this interference term. In the numerical simulation $C_{D\theta X}$ is neglected.

Wave force tests

The KC number range in wave force tests was limited to below 0.9, above 0.9 waves broke over the deck of the model and the hydrodynamic forces under such conditions are quite different from reality. In the real sea the KC number might be larger than the experimental values but in large waves a small floating structure heaves in the same phase of incident waves and the waves never break over the deck.

The velocity distribution induced by waves depends on the wave frequency and the drag coefficients become the function of the wave frequency and KC number. In the small KC number region the viscous damping forces obtained from wave force tests were very small in comparison with the wave damping forces. In the numerical simulation 0.53 is used for C_{DWX} and the value was determined from experimental results at KC= 0.9. Figure 4 shows that the value of C_{DX} is almost constant in the range of KC> 1.0 and this also seems to be true for C_{DWX}. The value of $C_{DX\theta}$ changed rapidly from minus to plus around KC= 1.0. It is difficult to estimate the value of $C_{DW\theta}$ from the experimental results in the range of KC< 0.9, and therefore we used the same value as that of $C_{DX\theta}$ 0.7 which is the value obtained from the forced oscillation tests in the range of KC> 1.0.

Figure 8 shows the results on C_{DWZ}, the tendency is similar to that of C_{DZ} except the range of KC< 0.45 but the value is larger. In this study the value of the fluid velocity used for non-dimensionalizing was determined to be that at the center of the gravitation. The projected area of the main deck was large and the fluid particle velocity was greatest at the water surface, so that, drag forces acting on the main deck became very large; this is the reason why C_{DWZ} is larger than C_{DZ}. In the numerical simulation the value of 4.4 obtained from Fig.8 at KC= 0.9 was used for KC> 0.9.

5 FIELD EXPERIMENT

Field experiments on measuring motions of the floating structure were carried out in August, 1995 southeast of Okinawa island as shown on the map in Fig.9. Pictures of the structure in the experiment are shown in Photo.1. This area is famous for many typhoons that attack every summer. During the experimental term there was one strong typhoon. Movements of the floating structure were measured during the storm with a motion sensor composed of three accelerometers and three piezoelectric oscillator gyroscopes. The latter measured angular velocity about its axis without any acceleration effect because this effect was canceled by the output of two oscillators; angle about the space fixed coordinate can then be obtained by integrating and transforming the angular velocity inside the motion sensor. The acceleration about the body fixed coordinate is transformed into a fixed space coordinate inside the motion sensor, and the displacement of the floating structure was obtained by integrating the acceleration. To avoid the effects of noise and error, a band pass filter ($1/15H_Z \sim 3H_Z$) was applied to the data. The moored floating structure oscillates at low frequency by the action of wind, current and non-linear wave forces and the low frequency component motions contained in the observed data might have been significant; regrettably, because of the lack of accuracy of the motion sensors we are unable to discuss this problem quantitatively but may be qualitatively. The incident wave height was measured with a wave rider buoy moored to the floating structure, and the current velocity at the depth of 4m was measured with a current meter fixed to the floating structure. In a laboratory experiment the phase difference between motions and waves is regarded as important for comparing experiments and theory. In this study the wave rider buoy measuring the amplitude and period of incident waves was moored to the floating structure with a synthetic rope and therefore it moved around within the 45m length of the rope. Maximum error in the phase of the incident wave is $2\pi \times 45/$ (wave length). Measurements were carried out every two hours, for 15min each time. The sampling time was $3H_z$. The electric power for the sensors, and PC for data recording and system control was supplied from a solar battery panel with chemical batteries for backup.

6 RESULTS AND DISCUSSION

The daily averages of significant values of incident wave height, surging, heaving and pitching motion are shown together with their mean periods in Fig.10. The horizontal axis indicates the day of the experiment. The correspondence of the heave amplitude and period to those of incident waves is very good. This verified the accuracy of the motion sensor system and its analyzing method. The amplitude and period of the surging are close to that of the wave when the wave height is large, although they are larger than the wave's when the incident wave height is small. These facts suggest the effects of the slow drift oscillation but as mentioned before this is not so systematic resonant phenomenon and in severe storm the slow drift oscillation is suppressed because of the large hydrodynamic damping force and the mooring tension increase. Regarding to pitching the amplitude increase with the increase of that of the incident wave except 17th and 18th. In these days the period of incident wave is about 5 sec, which is close to the resonance period of pitching, and the amplitude is large in comparison with that of the incident wave. On the whole the period of pitching is shorter than that of the incident wave and this suggests the resonance effect.

Figure 11 shows a time series of incident waves, surging, heaving and pitching motions which were measured on the 23rd of August when the strong typhoon hit. The maximum wave height was about 4.5m. The state of the sea became more severe toward midnight but the wave rider buoy was unable to send the signal for the time series to the station on the island because high waves blocked the way. In the figure the significant values of motions and the mean period obtained from spectrum analysis are shown. The significant amplitude of the heave is slightly smaller than that of the incident wave, this was caused by the effect of the constrain of the large mooring tension. Periods of the heave and surge are close to that of the wave. The period of the pitch is 5 sec and it is the natural frequency of the pitch.

The time series of motions obtained from the numerical simulations is shown in Fig.12. As mentioned in the section 5, spatial relation between the wave height sensor and structure are not so accurate in these field experiments as the usual laboratory experiments, and therefore it is not adequate to show results of experiments and simulations in same plot. The current velocity of 0.24 m/sec measured by the current meter was taken into consideration in the numerical simulation. On the whole, the simulation results show good agreement with the experimental ones. The significant amplitudes of the surge, heave and pitch obtained from the numerical simulation show fairly good agreements with the experimental results and all these of the simulation are smaller by 10% than experimental ones. The viscous damping force obtained from model experiments are too large for the real structure because of Reynolds effect. This will be researched in future works. The result of the numerical simulation on the mean period of the motions agreed fairly well to those of the field experiment.

In this field experiment the mooring line tension was not measured, and therefore only the numerical simulation can show the tension at the top of the mooring line. The average is roughly equal to the weight of the mooring line lifted by the floating structure. The amplitude of the variation even in a severe sea state is not large.

7 CONCLUSION

In this study field experiments on the dynamics of a small floating structure moored in the deep sea were carried out and the results were compared with those of numerical simulations using the viscous damping coefficient obtained from laboratory experiments. The main conclusions obtained were:

(1) The independent flow field concept is of good practical use to obtain the viscous damping forces acting on the floating structure in case of KC < 10.

(2) The slow drift surging motion was observed in the field experiment but it was suppressed in the severe sea state.

(3) Results of numerical simulations on the surging and heaving motion of the structure with viscous damping forces and hydrodynamic coefficients from linear theory showed sufficiently accurate agreement with field experiments.

(4) The amplitude of the variation of the mooring line tension is not as large as the average tension in deep sea mooring.

ACKNOWLEDGMENTS

Authors wish to thank Mr.Y.Saito for his valuable advice in this research project and Mr.T.Fujio for his help in the field experiments. They also express their sincere thanks to Mr.Y.Yosida for his contribution in the design of the measurement system. Ms.A.Yoshida helped us in preparing this manuscript.

This research was partly supported by a Grant-in-Aid for Scientific Research from the Ministry of Education and Science and Culture of Japan.

REFERENCES

Ando. A. and Yamanaka. K.,(1973). "Dynamics of Mooring Cables and Chains (Part 1: Hydrodynamic Characteristics of Chain)," *Abstract Note of the 22nd General Meeting of Ship Research Institute*, pp.110-113. (in Japanese)

Koterayama. W., Nagahama. T. and Ishii. H.,(1980). "Numerical Simulation of Buoy Dynamics by Lumped Mass Method," *Report of Research Institute for Applied Mechanics, Kyushu Univ.*, Vol.53, pp.19-36. (in Japanese)

Koterayama. W. and Nakamura.M(1988a)."Wave Forces on a Moving Cylinder," *Transaction of ASME, J. of Offshore Mechanics and Arctic Engineering*, Vol.110, No.4, pp.315-319

Koterayama,W. and Nakamura,M(1988b). "Hydrodynamic Forces Acting on a Vertical Circular Cylinder Oscillating with a Very Low Frequency in Waves," *Ocean Engineering*, Vol.15, No.3, pp.271-287

Koterayama,W., Mizuno.S., Marubayashi.K. and Ishibashi.M (1992). "Dynamic of Deep Sea Surface Buoy System for Ocean Mixed Layer Experiment," *La Mer*, Vol.30, No.3, pp.127-138

Koterayama W., Nakamura M. and Hu C., (1994) "Slow Drift Damping Force due to Viscous Force acting on Mooring Lines and Floating Structure," *Proc. of the BOSS'94*, Vol.2. pp.97-112.

Laya E. J. and Connor,J. J., (1984). "Hydrodynamic Forces on Flexible Offshore Structures," *Transaction of ASME, J. of Engineering Mechanics*, Vol.110, No.3, pp.433-447

Moe, G. and Verley,R.L.P.,(1980)"Hydrodynamic Damping of Offshore Structures in Waves and Currents," *Proc. of the Offshore Technology Conference*, Paper No.3798

Sugiura. M,(1977), "Some Calculation of Hydrodynamic Forces on Semi Submersible Bodies by Singularity Distribution Method," *J. of the Kansai Society of Naval Architects, Japan*, No.164, pp.33-42. (in Japanese)

Tanaka, N., Ikeda, Y., Himeno, Y. and Fukutomi, Y. (1980). "Experimental Study on Hydrodynamic Viscous Force Acting on Oscillating Bluff Body." *J. of the Kansai Society of Naval Architecture, Japan*, No.179, pp.35-44. (in Japanese)

Tasai. F., Suhara. T., Mitsuyasu. H., et al.,(1980). "Observations of the Kuroshio by Means of Mooring Buoy System," *Bulletin of Research Institute for Applied Mechanics, Kyushu Univ.*, Vol.52, pp.29-98. (in Japanese)

Thresher R. W. and Nath J. H., (1975) " Anchor-Last Deployment Simulation by Lumped Masses, " *J. of The Waterways Harbors And Coastal Engineering, ASCE*, Vol.1, No.WW4 Paper 11709, pp.419-433.

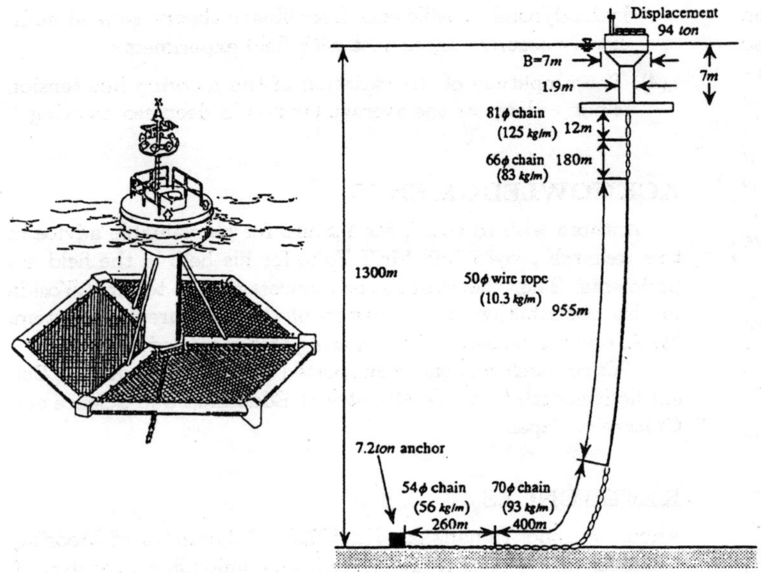

Fig.1 Conceptual views of the structure and mooring line.

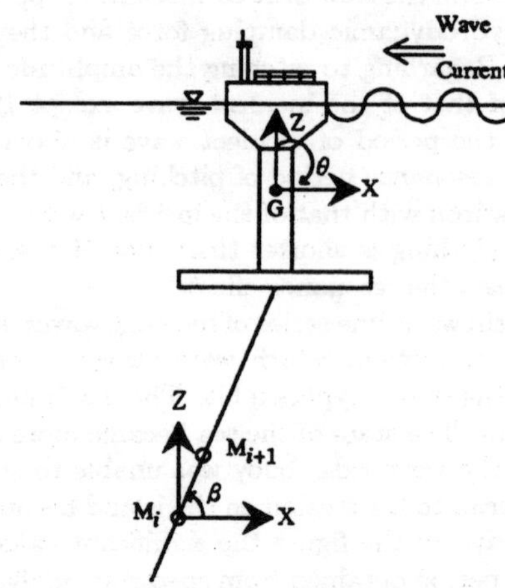

Fig.2 Coordinate systems for the floating structure and mooring line dynamics.

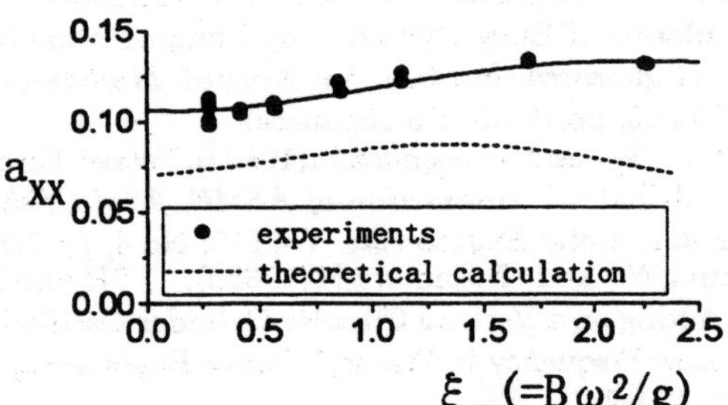

Fig.3 Comparison of added mass coefficient a_{xx} obtained from experiments and theoretical calculation

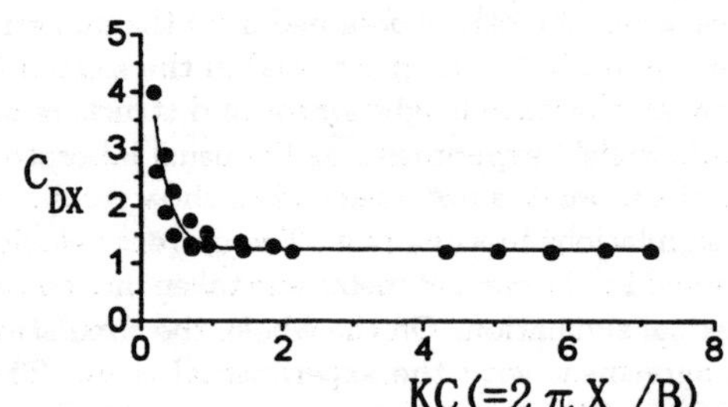

Fig.4 Drag coefficient C_{DX} obtained from forced surging tests.

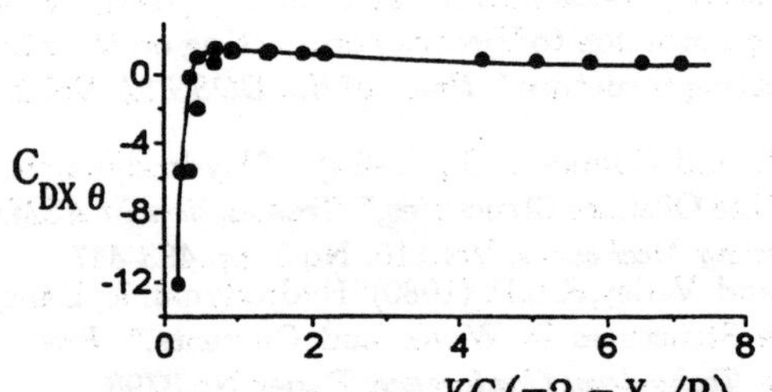

Fig.5 Drag coefficient $C_{DX\theta}$ obtained from forced surging tests.

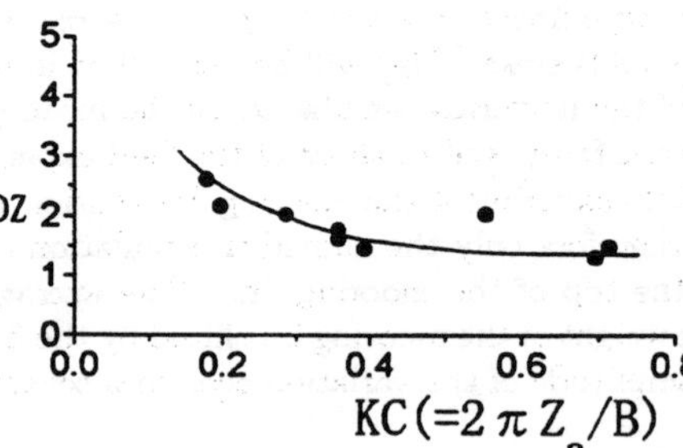

Fig.6 Drag coefficient C_{DZ} obtained from forced heaving tests.

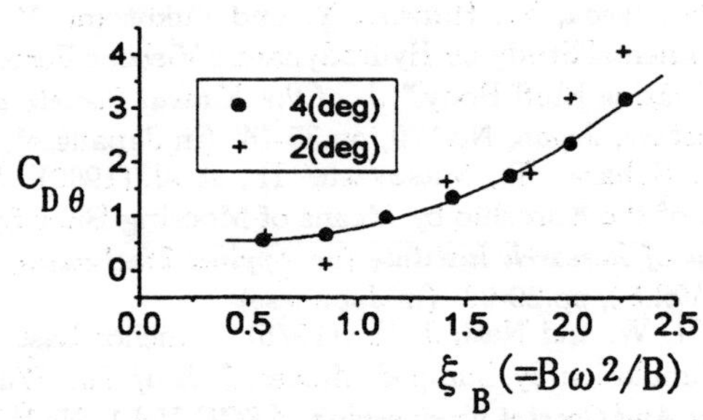

Fig.7 Drag coefficient $C_{D\theta}$ obtained from forced pitching tests.

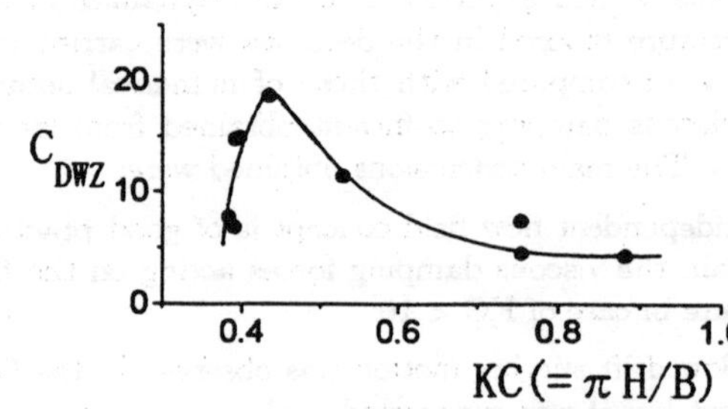

Fig.8 Drag coefficient C_{DWZ} obtained from wave force tests.

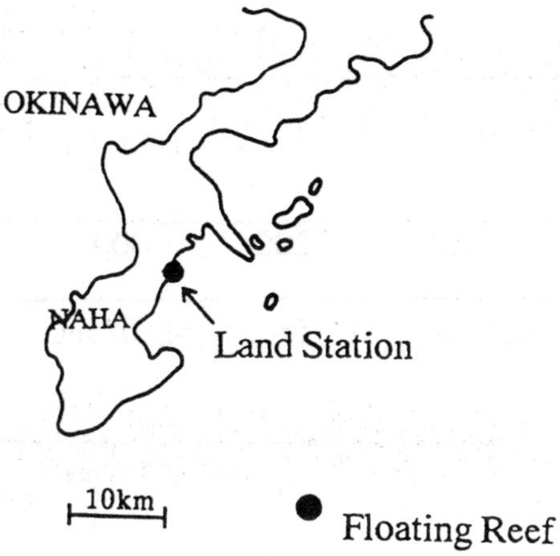

Fig.9 Area map of field experiments.

Photo.1 Pictures of the appearance of the floating structure.

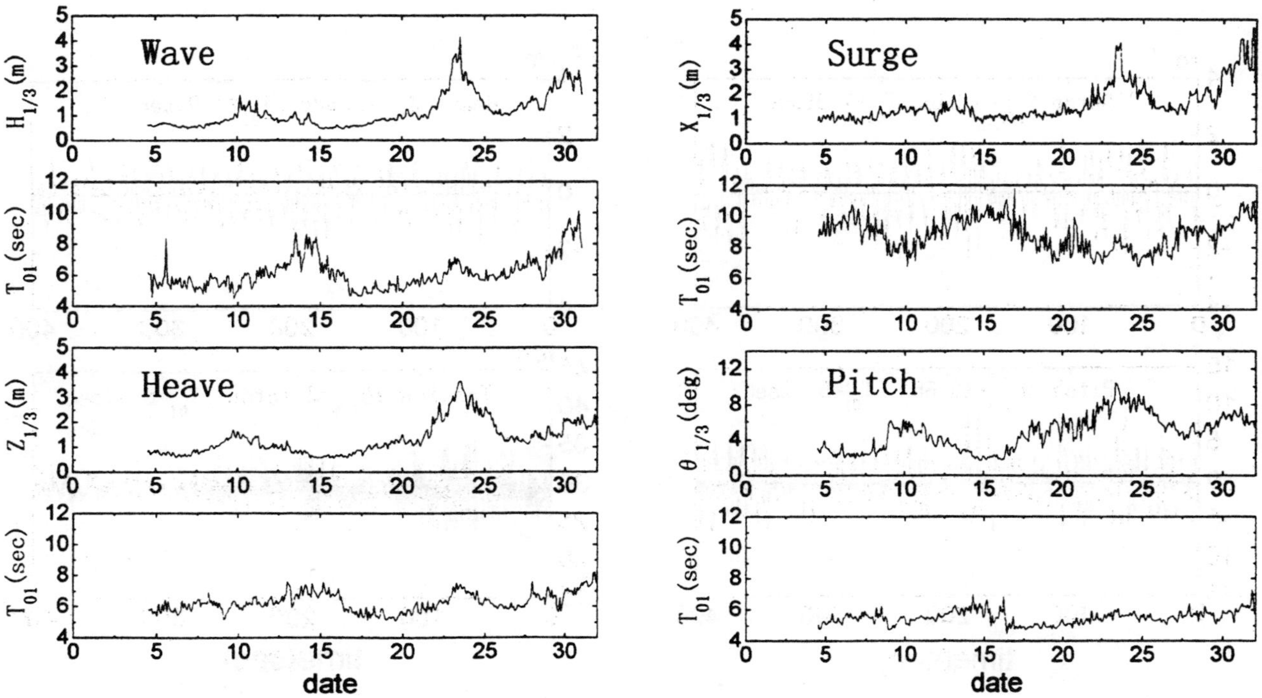

Fig.10 $\frac{1}{3}$ significant amplitudes of incident waves, surging, heaving and pitching motion and their mean periods.

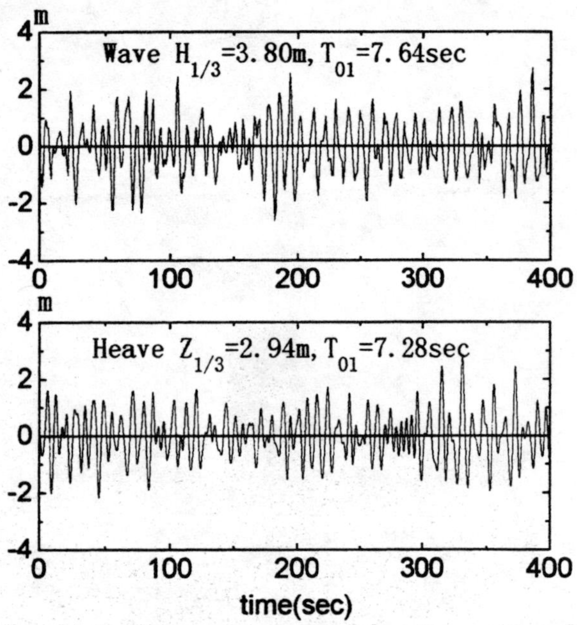

Fig.11 Time series of incident waves, surging, heaving and pitching obtained from the field experiment.

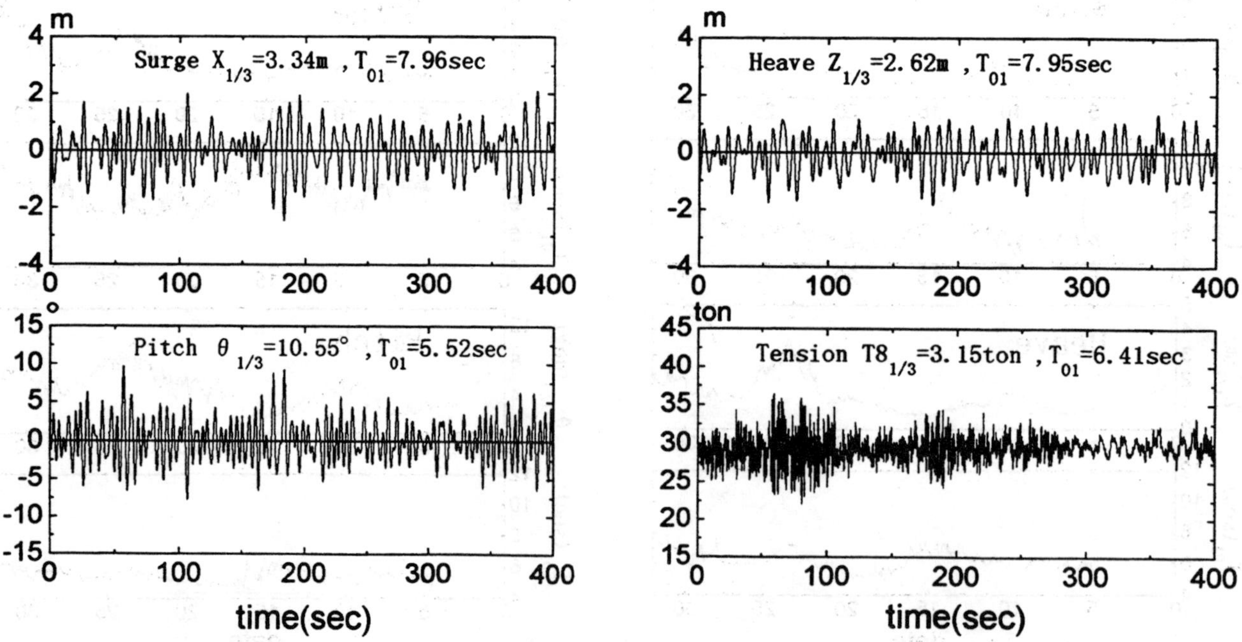

Fig.12 Time series of surging, heaving, pitching and mooring tension obtained from numerical simulation for the incident waves shown in Fig.11.

Proceedings of the Seventh (1997) International Offshore and Polar Engineering Conference
Honolulu, USA, May 25-30, 1997

ISBN 1-880653-28-1 (Set); ISBN 1-880653-31-1 (Vol. III)

Field Observation of Ship Motions Using Multi Camera Trackers

Katsuhiko Saito and Masayoshi Kubo
Kobe University of Mercantile Marine
Kobe, Japan

ABSTRACT

Although a few field observations of moored ship motions have been carried out, it is impossible to measure long period horizontal motions by using an accelerometer. In this study, an observation system of three dimensional motions of moored ship using multi camera trakers is described. The field measurement of a small craft motions is carried out and the results are compared with those by the other equipment. The applications of the system are also discussed in the paper.

KEY WORDS : Ship Motions, Field Observation, Multi Camera Trakers, Measurement Techniques

INTRODUCTION

Cargo handling at a wharf are often interrupted and sometimes mooring ropes are broken by large ship motions. Especially long-period ship motions in sway, surge and yaw which may result in interruption of cargo handling or breakage of mooring ropes (Sawaragi and Kubo, 1982). The wharf operation efficiency should be defined based on the allowable ship motions for cargo handling in terms of the type and size of a ship and a cargo handling equipment (Ueda and Shiraishi, 1988).

Some mathematical models (Saito and Spencer, 1994) considering non-linear long period motions have been developed and physical model tests (Bowers, 1992) have been carried out to simulate the motions in waves. These results should be compared with exact actual data (Zou and Bowers, 1993).

Only a few field observations of moored ship motions have been carried out because of several limitations, such as ability of instrument and operational cost. Moreover it was impossible to measure the motions of 6-degree of freedom at the same time including long period horizontal motions by the conventional measuring instrument using an accelerometer.

In this study, an observation method of three dimensional positions of the targets on a ship using the multi camera trackers is described as well as the 6-component motions of ship. This optical motion measurement system using two CCD video cameras is the most preferable in measurement of long period motions because we can measure the positions of the targets on a ship directly. The field measurements of small craft motions is carried out and the results of the high frequency motions are compared with those by the other equipment using gyroscopes, while the accuracy of the low frequency motions is confirmed by the indoor experiments.

Several applications of the above optical motion measurement system are also proposed in the paper.

THEORETICAL BACKGROUND

3D Position of Target

Figure 1 shows the definition sketch of a target and its 2D position on a picture, where (X,Y,Z) show the absolute coordinates, (x,y,z) show the relative coordinates of camera, point O(Xo,Yo,Zo) is the focus, (ω , ϕ , χ) are inclinations of camera around (X,Y,Z) axes, point P(X,Y,Z) is a 3D position of a target and point p(x',y') is its 2D position on a picture. Since there is point p on the line OP, Eq.(1) can be obtained.

$$x' = -c\,\frac{A11(X-Xo)+A12(Y-Yo)+A13(Z-Zo)}{A31(X-Xo)+A32(Y-Yo)+A33(Z-Zo)} + \Delta x$$

$$y' = -c\,\frac{A21(X-Xo)+A22(Y-Yo)+A23(Z-Zo)}{A31(X-Xo)+A32(Y-Yo)+A33(Z-Zo)} + \Delta y \qquad (1)$$

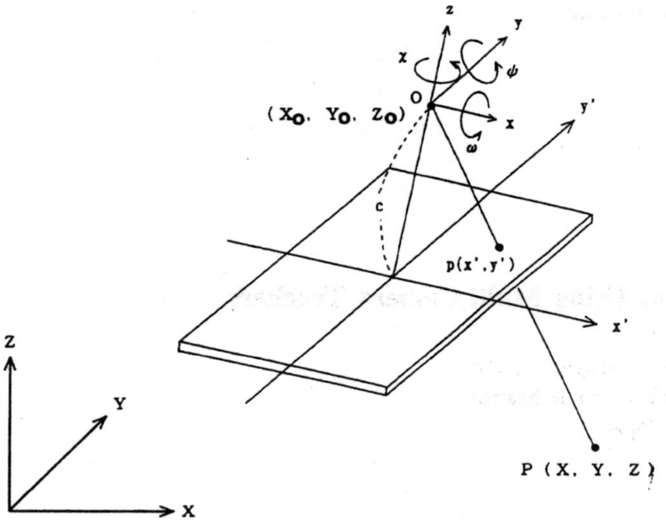

Fig.1 Definition sketch of a target and 2D image on a picture

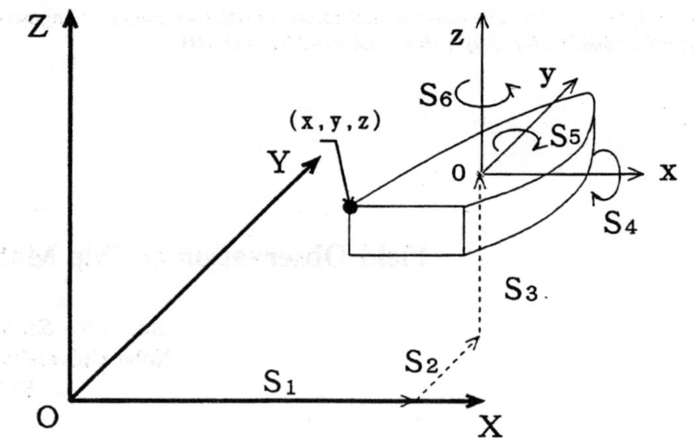

Fig.2 Definition sketch of a target on a ship

$A11=\cos\phi\cos\chi$, $A12=-\cos\phi\sin\chi$,
$A13=\sin\phi$,
$A21=\cos\omega\sin\chi+\sin\omega\sin\phi\cos\chi$,
$A22=\cos\omega\cos\chi-\sin\omega\sin\phi\sin\chi$,
$A23=-\sin\omega\cos\phi$,
$A31=\sin\omega\sin\chi-\cos\omega\sin\phi\cos\chi$,
$A32=\sin\omega\cos\chi+\cos\omega\sin\phi\sin\chi$,
$A33=\cos\omega\cos\phi$,
$\Delta x=x_o+x_1(k_1r^2+k_2r^4)$,
$\Delta y=y_o+y_1(k_1r^2+k_2r^4)$,
$r^2=x_1^2+y_1^2$, $x_1=x'-x_o$, $y_1=y'-y_o$

Where c is the focal distance, (x_o, y_o) are the aberrations of lens, (k_1, k_2) are lens distortion factors.

At first, two cameras are set to measure the 3D position of target. Before measurement, more than six known positions of calibrational targets are measured to know the eleven camera parameters, Xo, Yo, Zo, χ, ϕ, ω, x_o, y_o, c, k_1 and k_2 from Eq.(1). Since a target is taken a photograph by two camers at the same time, 3D position of target can be calculated by Eq.(1) where the eleven camera parameters are known after the calibration.

Ship Motions

Figure 2 shows a sketch of a target on a ship. The position of a target shows (X,Y,Z) in absolute coordinates and (x,y,z) in relative coordinates of the center of gravity of a ship. The relation between (X,Y,Z) and (x,y,z) are given by Eq.(2):

$$\begin{bmatrix} X \\ Y \\ Z \end{bmatrix} = \begin{bmatrix} S_1 \\ S_2 \\ S_3 \end{bmatrix} + \begin{bmatrix} B11 & B12 & B13 \\ B21 & B22 & B23 \\ B31 & B32 & B33 \end{bmatrix} \begin{bmatrix} x \\ y \\ z \end{bmatrix} \qquad (2)$$

$B11=\cos S_5\cos S_6$,
$B12=\sin S_4\sin S_5\cos S_6-\cos S_4\sin S_6$,
$B13=\cos S_4\sin S_5\cos S_6+\sin S_4\sin S_6$,
$B21=\cos S_5\sin S_6$,
$B22=\sin S_4\sin S_5\sin S_6+\cos S_4\cos S_6$,
$B23=\cos S_4\sin S_5\sin S_6-\sin S_4\cos S_6$,
$B31=-\sin S_5$, $B32=\sin S_4\cos S_5$
$B33=\cos S_4\cos S_5$

where S_1, S_2 and S_3 show the displacements of sway, surge and heave, S_4, S_5 and S_6 show the angles of pitch, roll and yaw, respectively. Then ship motions of six degree of freedom can be calculated by Eq.(2) after the measurement of 3D positions of three targets whose position (x,y,z) on a ship are already known.

MEASUREMENT SYSTEM

The measurement system using the multi camera trackers consists of a main unit, two CCD video cameras, a personal computer and a data processing software as shown in Fig.3. The target is recognized as a set of brighten or darken points contrasted with background image. The center of gravity of the set of brighten or darken points is calculated in the main unit. The targets are tracked automatically. In the PC, after 3D position of a target is calculated from the two 2D positions of a same target, the 6-component motions of ship are calculated from the three 3D positions of the targets on a ship.

INDOOR EXPERIMENT

The accuracy of the measurement is confirmed by the indoor experiments. Fig.4 shows the arrangement of the model. The 3-component motions, heave, pitch and roll, of an oscillation simulator are measured using the above system. The experiments are carried out in three types of the arrangement of the targets

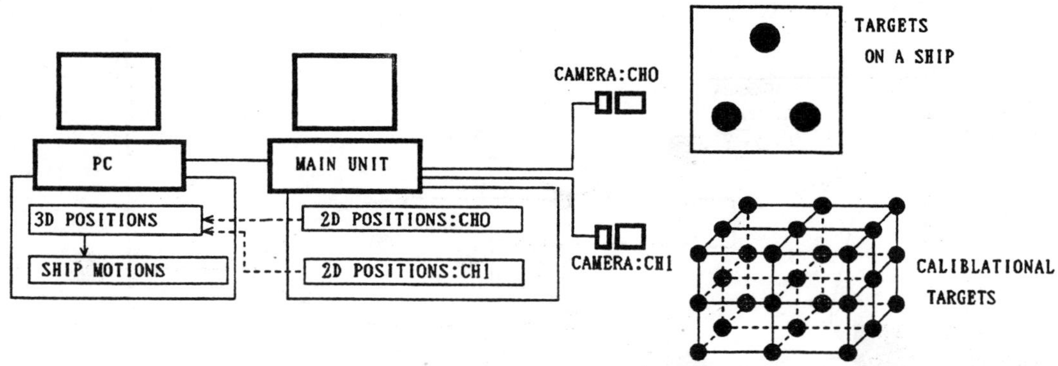

Fig.3 Multi camera trackers system

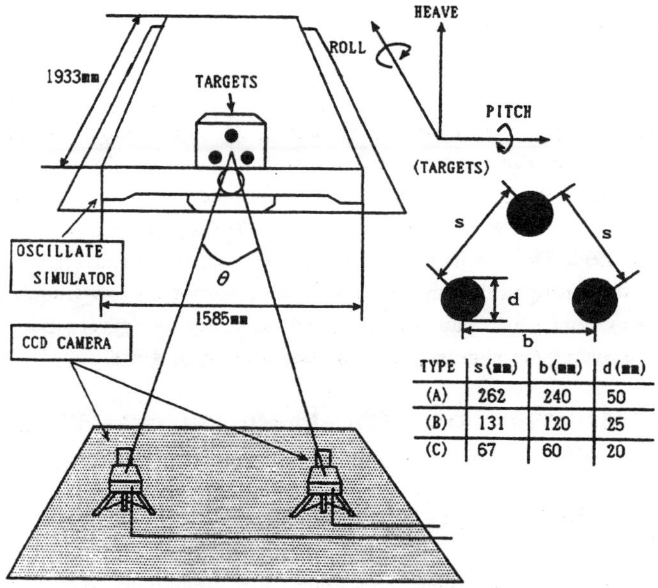

Fig.4 The arrangement of the indoor experiment

and three types of the arrangement of the cameras.

First of all, it is confirmed that the system can measure the low frequency motions (less than 0.05Hz) for which accelerometers are useless. We can also see from the results of the tests (Table 1) as follows. Since the accuracy of the 3D position of the target is good in the case that the angle θ between two cameras and targets is 90 degree, the ship motions can also be measured accurately in the case, while the accuracy is poor in the case of small angle. And the accuracy of the calculation results of the 6-component motions is good when the targets are arranged widely on a ship like (A)type in Fig.4, while the accuracy is not good for the arrangement of the targets in narrow area(type(C)).

FIELD MEASUREMENT

The field measurements of moored small craft (length:4.2m) movements are carried out in port of Kobe. The layout of the measurement is shown in Fig.5. The other equipment using gyroscopes also measures pitch and roll. We must use the black

Table 1 The results of the indoor experiment

θ	TARGETS	HEAVE	PITCH	ROLL	HEAVE & PITCH & ROLL
		68. 7mm	3. 85°	4. 68°	68. 7mm & 3. 85° & 4. 68°
		ERROR(%)	ERROR(%)	ERROR(%)	ERROR(%)
90	(A)	-1. 90	2. 35	2. 51	0 & -1. 56 & 4. 21
	(B)	0. 58	5. 74	0	-7. 10 & 1. 04 & 5. 35
	(C)	4. 09	-6. 98	4. 72	-14. 0 & -7. 31 & 3. 63
67. 5	(A)	1. 53	4. 95	1. 50	-0. 90 & -1. 57 & 2. 55
	(B)	-20. 2	4. 95	1. 29	-12. 9 & -0. 79 & 5. 10
	(C)	-7. 50	1. 03	5. 45	-17. 6 & -5. 67 & 5. 10
45	(A)	1. 46	1. 04	3. 00	-7. 04 & 0 & 2. 56
	(B)	1. 29	-7. 83	24. 0	-16. 6 & -12. 5 & 1. 71
	(C)	1. 36	-10. 7	6. 40	-18. 6 & -19. 2 & 4. 71

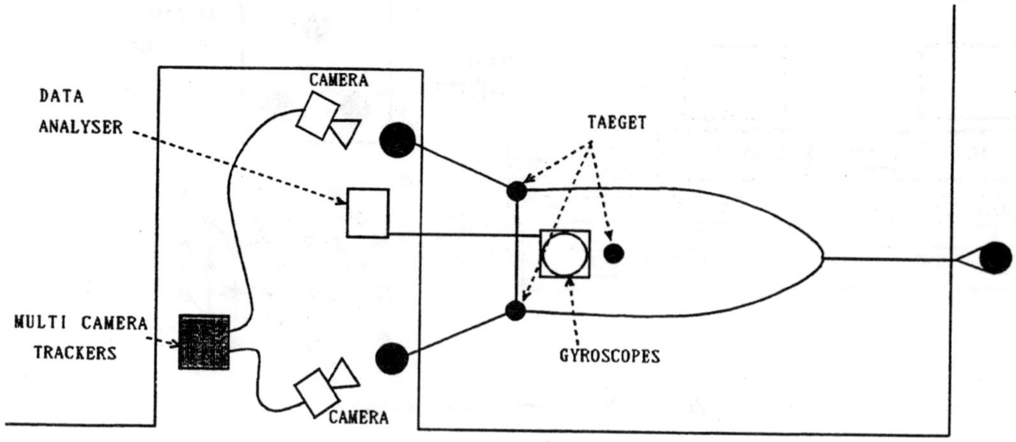

Fig.5 The arrangement of the field measurement

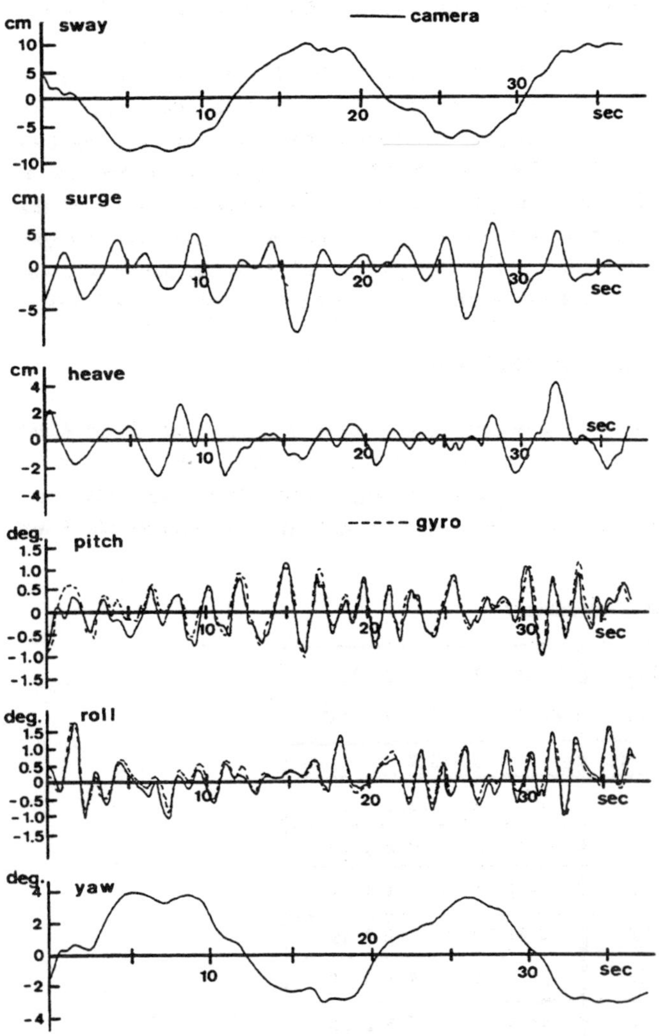

Fig.6 The results of the field measurement

circle targets with white background on the square board because the system can catch the moving targets automatically on actual sea under the influence of diffused reflections.

We found the low frequency (about 0.05Hz) sway and yaw in the field. The results of the measurement are shown in Fig.6. The system using multi camera trakers can measure the low frequency sway and yaw, and has as good accuracy as gyroscopes on measuring the high frequency motion of pitch and roll.

APPLICATION OF OPTICAL MOTION MEASURMENT SYSTEM

Allowable ship motions for cargo handling

Ship cargo handling operations are sometimes interrupted because of bad weather conditions such as rough waves or strong winds. The harbour tranquility index should be defined based on the allowable ship motions for cargo handling operations in terms of the type and size of ship and the cargo handling method. According to Ueda and Shiraishi(1988) the allowable ship motions for cargo handling are obtained provisionaly by the numerical simulations and then the provisional value are evaluated by cargo handling operators by means of the inquiry . Thus their revised allowable ship motions for cargo handling are not based on the actual field data of the ship motions.

Does cargo handling stop completely when the ship motions are over the allowable value ? Conversely is the cargo handling 100% possible at anytime while the ship motions are under the limitations ? A questionaire to the container crane operators in port of Kobe are carried out to consider the harbour calmness index again.

Figure 7 shows the ratio of limit judgement of cargo handling, where $X_{1/3}$, $Y_{1/3}$ and $Z_{1/3}$ are significant movement of sway, surge and heave, respectively, μ =0.5 means that 50% operators judge the limit of cargo handlings. Figure 8 shows the decreasing ratio of cargo handling efficiency, when μ '=0.5

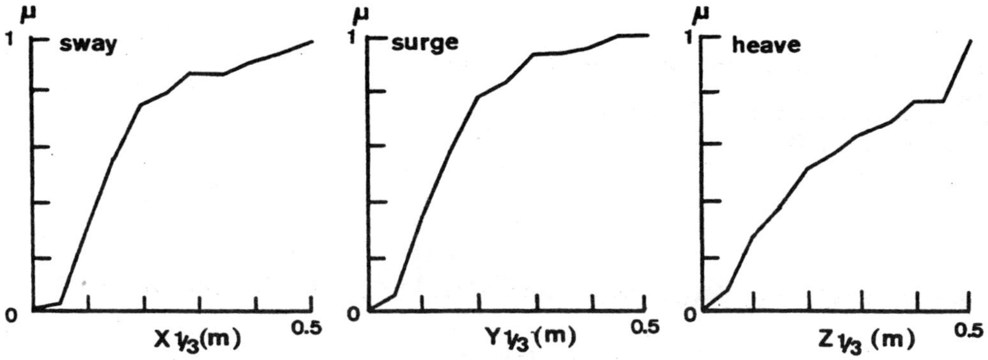

Fig.7 The ratio of limit judgement of cargo handling

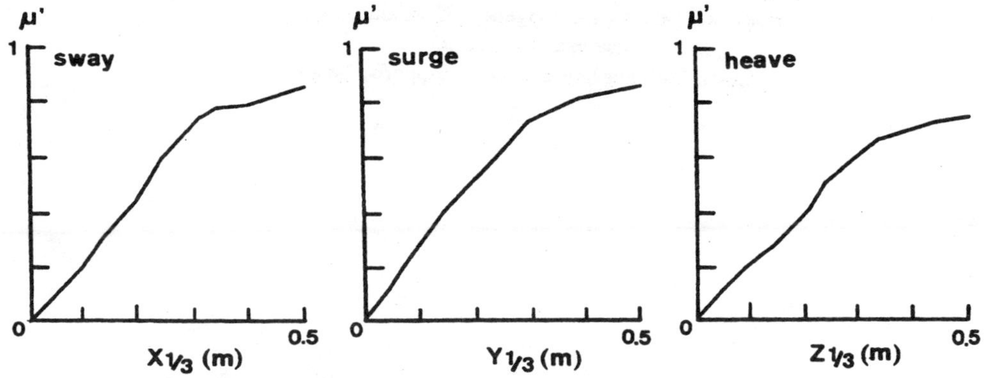

Fig.8 The decreasing ratio of cargo handling

means that the cargo handling efficiency is 50% compared with the best condition (μ '=0) of cargo handlings.

These "fuzzy allowable ship motions" should be confirmed and decided exactly based on many field observations and wharf operational records.

New navigational observation instrument of moored ship motions

There are many navigational instruments to help few navigators in a ship such as GPS, ARPA, electric chart, auto pilot and so on. But a navigational officer should do a final judgement of cargo handling interruptions without any instrumental support. So a new instrument for measuring ship motions in real time should be developed to keep the safe mooring and cargo handling in a harbour.

CONCLUSIONS

The conclusions in the measurements are summarized as follows:

(1) The targets colored black and white should be vivid to catch the moving targets automatically in the field observation.

(2) The system using multi camera trackers has as good accuracy as gyroscopes on measuring the high frequency motions.

(3) The system can also measure the low frequency motions accurately for which accelerometers are useless.

As the applications of the above optical motion measurement system, the following stages can be considered.

(1) A new instrument for measuring ship motions in real time should be developed to keep the safe mooring and cargo handling in a harbour.

(2) The allowable ship motions for mooring and cargo handling should be decided exactly based on many field observations and wharf operation records.

REFERENCES

Bowers E.C.(1992)."Physical Models of Harbours:the Proving with Field Data of a Model of Shoreham Harbour with a New Type of Wave-Maker Generating Short Crested Waves", H.R.Wallingford, U.K., Report SR335.

Saito K. and Spencer J.M.A.(1994)."A Mathematical Model of First- and Second-Order Ship Motions by 3D-BIEM with Perturbation Method", H.R.Wallingford, U.K., Report SR421.

Sawaragi T. and Kubo M.(1982)."Long-Period Motions of a Moored Ship Induced by Harbor Oscillations", Coastal Engineering in Japan, Vol.25.

Ueda S. and Shiraishi S.(1988)."The Allowable Ship Motions for Cargo Handling at Wharves", Report of the Port and Harbour Research Institute, Japan, Vol.27, No.4.

Zou Z. and Bowers E.C.(1993)."A Mathematical Model Simulation of a Ship Moored against a Quay in Shoreham Harbour with Comparison against Physical Model and Field Data, H.R.Wallingford, U.K., Report SR378.

Proceedings of the Seventh (1997) International Offshore and Polar Engineering Conference
Honolulu, USA, May 25-30, 1997

ISBN 1-880653-28-1 (Set); ISBN 1-880653-31-1 (Vol. III)

Observation of Long Period Wave and Ship Motion in Tomakomai-port

Tetsuya Hiraishi
Port and Harbour Research Institute, Yokosuka, Japan
Youichi Atsumi, Atsushi Kunita and Shin-ichiro Sekiguchi
Hokkaido Development Agency, Hokkaido, Japan
Tsutomu Kawaguchi
Nippon Data Service Co. Ltd., Sapporo, Japan

ABSTRUCT

Termination of cargo handling sometimes occurs in Tomakomai–port, located in the northern part of Japan. Observation of the profiles of long period waves and the motion of a large vessel moored at an berth is carried out in order to investigate the causes of termination. The obtained results demonstrate that the long period waves induce the large surging motion of the vessel by the resonance.

KEY WORDS : Field observation, Long period wave, Limitation of cargo handling, Surge motion

INTRODUCTION

Tomakomai–port is located in the northern part of Japan and is facing to the Pacific Ocean. The loading and unloading are sometimes terminated and the mooring ropes are broken by the large horizontal motions of vessels moored at the specified berth in the port. **Table–1** shows the numbers of broken mooring ropes at the berth with the depth of 14m. The wave date indicated in the table represents the significant wave heights and periods obtained at the offshore observation point. The table demonstrates that several accidents occur in the mooring ropes each year. The table shows the accidents which occur when the cargo vesels are berthed. In the other cases, the moored vessels are released before the rope break when the surge motion becomes very large.

The reports on similar accidents have been published(e.g. MacGee,1992). Some reserchers have suggested that the large surging motion which may cause the termination of cargo handling and breaking of mooring ropes are induced by the resonanse to the long period waves penetrating into harbors(e.g, Barthel et.al.,1994). We investigated the relation of the long period waves and motion of moored vessels in Tomakomai–port in field observation.

IMPLEMENTATION OF FIELD OBSERVATION

Figure 1 shows the topography of Tomakomai–port and the arrangement of observation facilities. The field observation was carried out from September 17th to 20th in 1994 when the cargo vessel **E.P.** was berthed at the quay with the depth of 14m. **ST.0** represents the offshore observation point where the water elevation

Table 1 Damage at –14m berth in Tomakomai–port

date	Ship name	deadweight	number of break
Jan. 28. 1985	SAPPORO-MARU	48. 884ton	1
Apr. 10	SAPPORO-MARU	48. 884	1
Sep. 2. 1986	NEPUTUNE CANOPUS	30. 703	3
Jun. 21. 1987	ASIAN THISTLE	37. 472	4
Nov. 19	EIRYU-MARU	30. 767	2
Nov. 26. 1988	LITO	28. 093	1
Nov. 13. 1989	SAPPORO-MARU	48. 884	4
Jan. 19. 1990	WORLD AGAMEMNON	19. 264	4
Nov. 20	DAPHNE OCEAN	38. 891	1
Jan. 26. 1991	MIYAGI-MARU	37. 582	1
Oct. 3	FAR EASTERN GRAIN	37. 159	1
Mar. 31. 1992	GOLD STAR	36. 264	2
Mar. 7	ASCEENSION	35. 191	2
Sep. 25	ANTOINS I ANGELICOUSSIS	36. 986	2
Oct. 26	SAPPORO-MARU	48. 884	2
Nov. 7	YOU XUAN	25. 891	1
Jan. 29. 1993	MOZU ARROW	23. 149	5
Jun. 4	ORIANA	36. 606	3
Aug. 29	PEGASUS	36. 764	1
Sep. 4	RUBIN ENERGY	36. 560	2
Oct. 9	SAPPORO-MARU	48. 884	4
Nov. 14	SAPPORO-MARU	48. 884	1
Mar. 10. 1994	GORTYS	38. 131	2
Sep. 18	ENERGY PIONEER	36. 493	4

profiles are measured for 20min every two hours by using an ultra-sonic wave gage. **ST.1**, **2** and **3** represent the observation station where the wave profiles in the port are measured at the same time when the offshore data are recorded.

Figure 2 shows the arrangements of mooring ropes for berthing. The heave and surge motions of moored vessel were photographed continuously in a video camera at the position **V.1**. The video images for 20min were taken into analysis every two hours.

We at first computed the power spectrum in FFT (Fast Fourier Transformation method) and the maximum and significant wave height and period in the zero-up crossing method analyzing the digital water elevation data measured at the points ST.0, ST.1, ST.2 and St.3. The power spectra and maximum and significant amplitudes of surge and heave motions were derived from analysis of the motion of marker on the vessel face taken in the video images. We define the representative long period wave height as

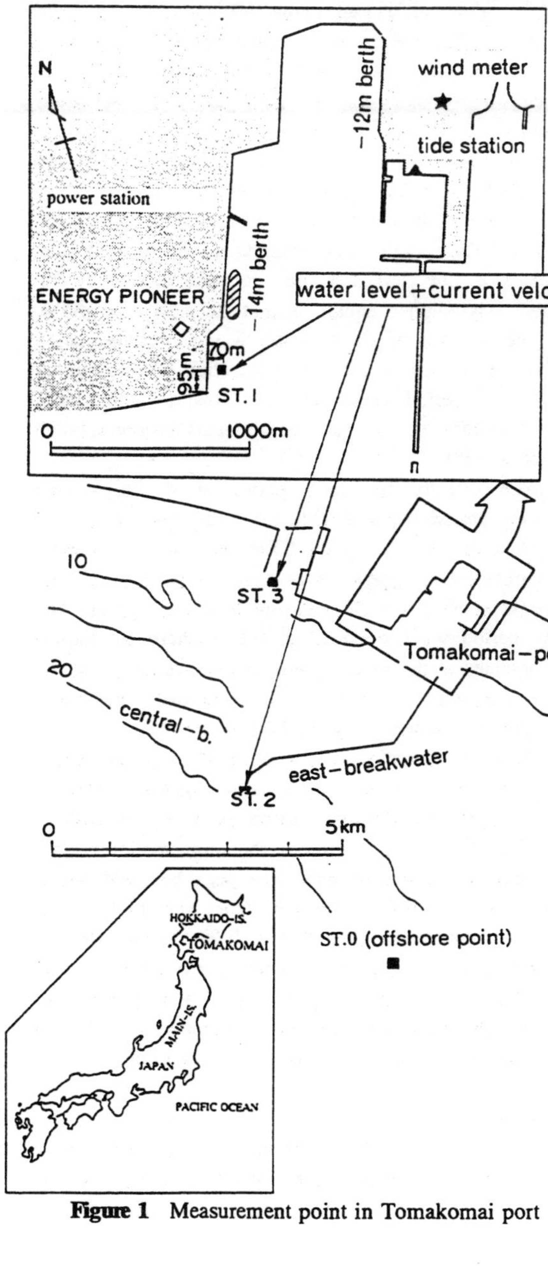

Figure 1 Measurement point in Tomakomai port

follows;

$$H_l = 4.0\sqrt{\int_{1/T2}^{1/T1} S(f)df} \tag{1}$$

where, Hl: representative long period wave height, $S(f)$: frequency spectrum, T_1: minimum period of long period wave component and T_2: maximum period of long period wave component. We employ 30s as T_1 and 600s as T_2 in Tomakomai-port.

OBSERVATION RESULT

(1) Long period wave height and motion amplitude

Figure 3 shows the variation of significant wave heights and periods obtained at **ST.0, ST.1** and **ST.2** during observation. At each graphs, black and white circles indicate the wave heights and periods, respectively. In the figures, the hatched parts represent the duration when the significant vessel motions are observed. At 8:30, September 18th, the motion had bocome so magnificiant that berthing might be dangerous. The vessel was released emergently on that time.

The wave heights at the offshore rapidly increase from the beginning of Sep.18. The largest significant wave height obtained at the offshore point **ST.0** is 4.6m at 14:00, Sep.18. The wave heights at **ST.2** are almost equal to those at **ST.0** as the point **ST.2** is located at the entrance of port. The wave heights at the observation point **ST.1** at the berth are much smaller than those obtained at the entrance and the offshore of the port. The significant wave heights at **ST.1** become only 50cm even when the vessel is released by the danger of rope breaking. According to the Japanese design standard, the acceptable limitation of significant wave height for the loading and unloading in a port is 50cm. Therefore, the wave height at the berth is not so large that the vessel should be released if the wave energy is only composed of the short period components.

Figure 4 shows the variation of vessel positions derived from the marker analysis in video camera. The upper figure shows the variation of maximum and significant surge amplitude obtained from 20min data. The lower figure is for the heave amplitudes. The significant heave amplitude is 0.2m at 0:00, Sep.18. It gradually increases and becomes 0.5m when the vessel is released. Ueda and Shiraishi(1988) propose the acceptable limitation of vessel displacement for loading and un-loading in berthing. They suggest the limitation for heave motion is 1m in the significant value. Therefore, the heave amplitude obtained in the field observation in

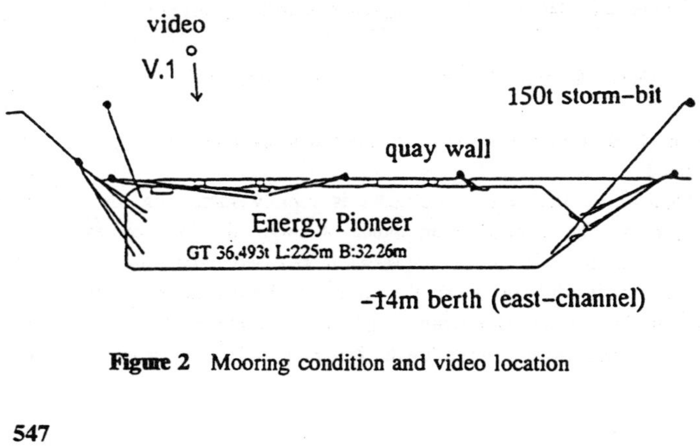

Figure 2 Mooring condition and video location

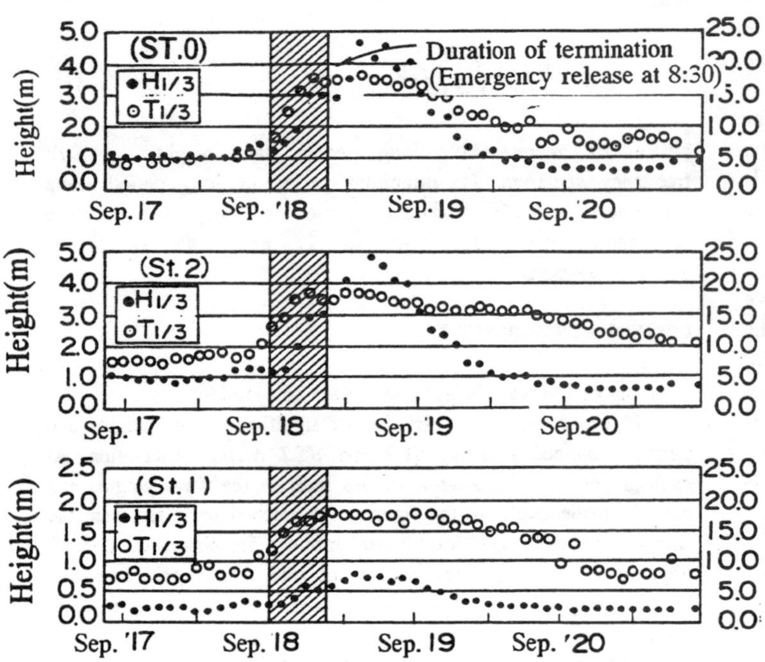

Figure 3 Variation of observed wave height and period

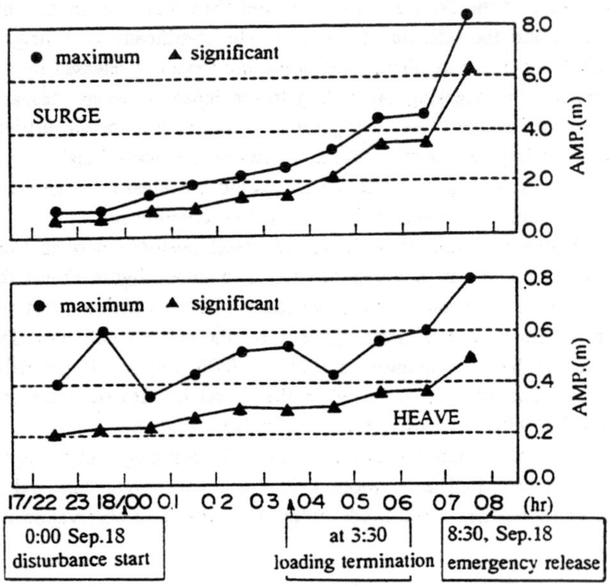

Figure 4 Variation of surge and heave amplitude

Tomakomai-port is not so large that the loading and mooring is terminated.

Meanwhile, the surge amplitudes is rapidly increasing after 0:00, Sep.18. The significant amplitude of surge motion becomes more than 6.0m at the time the vessel is released. According to Udeda and Shiraishi's proposal, the acceptable limitation for surge motion is 2m in the significant amplitude. Therefore, the surge amplitude is beyond the limitation for bernthing and the release of the vessel is reasonable. The main cause of termination and breaking of mooring ropes in Tomakomai-port is concluded to be the rapid increasing of surge displacements.

Figure 5 shows the analyzed frequency spectrum at **ST.0**, **ST.1** and **ST.0.** During observation, the vessel motion becomes remarkable from 0:00, Sep.18, and the loading is terminated at 3:30. The vessel is released at 8:30. The active wave condition is not obtained at the same time mentioned above because the wave profiles are analysed every two hours. The figure shows the spectrum obtained at the time near to those at loading termination and vessel releasing.

Figure 5(1) shows the observed spectrum at the offshore point **ST.0**. The spectrum at 0:00 has two peaks in the frequency range from 10^{-3} to 10^{-1} (1/s); one is the peak of wind wave components at 6×10^{-2} (1/s) and another is the peak of long period wave components at $f=10^{-2}$ (1/s). The density of the both peaks become higher as the time progresses. The spectrum variation at the harbor entrance **ST.2** is shown in Fig.5(2). The spectrum obtained at **ST.2** has two peaks at the same frequency to those in Fig.5(1). Fig.5(3) shows the spectrum at the berth **ST.1**. At **ST.1**, the density of the peaks of short period wave components is not so high. It's value changes slightly as the time progresses. The peak of long period components is not clear but the spectrum density at the frequency($f=10^{-2}$) is almost equal to that at **ST.0** and **ST.2**. Therefore, the energy level of short period waves becomes much smaller than those at the offshore by the protection of breakwaters but the energy of long period waves is equal to that at the offshore. The energy level of long period wave components becomes much larger at 4:00 and at 6:00.

To compare directly the long period wave height, we calculated the long period wave heights from the spectral density employing eq.(1) at the offshore point **ST.0** and the measurement point of inner harbor **ST.1**. **Figure 6** shows the comparison of long period wave height at **ST.0** and **ST.1** during field observation. The heights of long period wave obtained at the offshore are almost equal to those obtained at the inner harbor. It demonstrates that the long period waves generated in the offshore area easily penetrates into the port without the reduction by the breakwaters.

Figure 7 shows the variation of spectrum of surge and heave motions at 0:00, 4:00 and 6:00 on Sep.18, 1994. Fig.7(1) shows the variation of power spectrum of surge motion. At 0:00, the peak of spectrum exists at $f=10^{-2}$ (1/s). Therefore, the surge motion mainly consists of the long period components. The peak becomes higher as the time progresses. At 6:00, the peak becomes very high and its value is almost 50 times larger than that at 0:00. Meanwhile, the spectrum of heave motion in Fig.7(2) has low peaks corresponding to the short period wave's spectral peak. Also, the spectrum of heave motion have the mild peaks in low-frequency rigion but its strength is not so large compared with that of surge motion.

(2) Amplification factor

Figure 8 shows the ratio of the motion spectrum to the wave spectrum. The ratio is indicated as 'energy factor' in the figures. The energy factor A is almost 1.0 in the whole range of frequency for

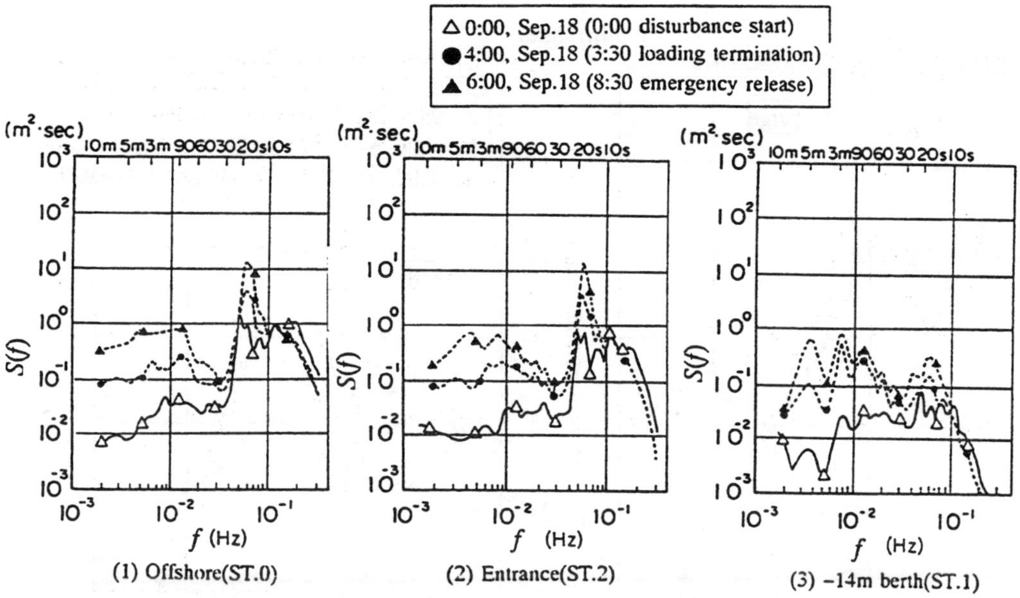

Figure 5 Wave frequency spectrum during vessel motion disturbance

heave motion shown in Fig.8(2). Therefore, the heave motion of vessel is corresponding to the water level elevation. In Fig.8(2), A becomes very large at the specified frequency range near to f=10^{-2}. Those frequencies are corresponding to those of long period waves. It demonstrates that the surge motion may be amplified by the long period wave components. The maximum amplification factor for surge motion is about 50 ~ 70. It means that the surge displacements become several times larger than the heights of long period waves.

The large energy factor A of surge motion indicates that the resonance of long period motion occurs. **Figure 9** shows the simple model of vessel and mooring hawsers. The moored vessel is enough short to wave length. It is, therefore, represented by a mass point M(kg). The vessel is usually moored with several elastic hawsers to the quay wall as shown in Fig.9. A hawser is approximately represented by a elastic spring line with the elasticity K(N/m). The natural period To is given by

$$To=2\pi\ (M/2K)^{1/2} \tag{2}$$

Assuming the mass is represented by DWT(Dead Weight Ton), a cargo(50,000DWT) moored in a single Manila hemp hawser(diameter 60mm, length 80m) has the natural period of 3.8min(4.4*10^{-3} Hz). For the cargo ship of 10,000DWT, the natural period becomes 1.6min (10^{-2} Hz). Therefore, the mooring system composed of a vessel and hawsers has the natural period near to the period of long period waves. The occurance of large surge motion in Tomakomai-port is expected to be induced by the resonance of the natural period of mooring system and the long period waves. As the energy of long period waves are not reduced by the breakwater in Tomakomai-port, the surge motion is induced even at the inner harbor protected by the breakwaters and groins. The prevension of penetration of long period waves into the inner harbor is important to prevent the cargo handling termination.

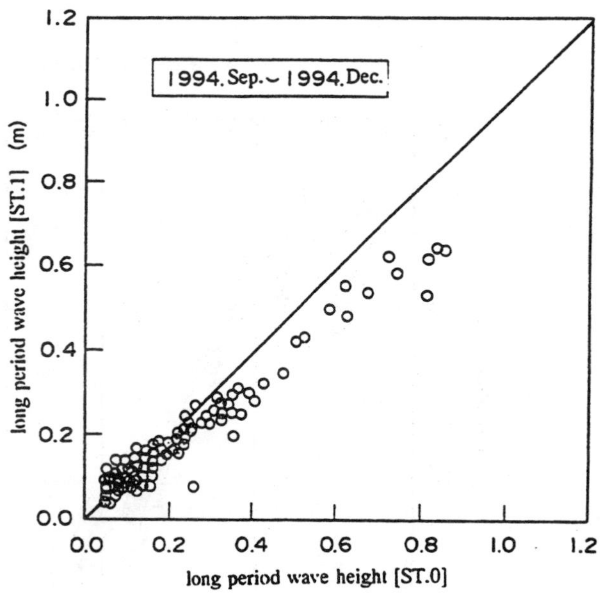

Figure 6 Cpmparison of long period wave height at offshore and ST.1

ALLOWABLE WAVE HEIGHT

For the safety operation, it is important to set the allowable long period wave height to cargo loading and un-loading. **Figure 10** shows the variation of long period wave heights and amplitudes of surge motion. During observation, the cargo loading is terminated at 3:30 and the emergency shift is carried out at 8:30. From 0:00 to 8:30, the long period wave height in front of the berth gradually increases. Meanwhile, the surge motion also increases as the wave

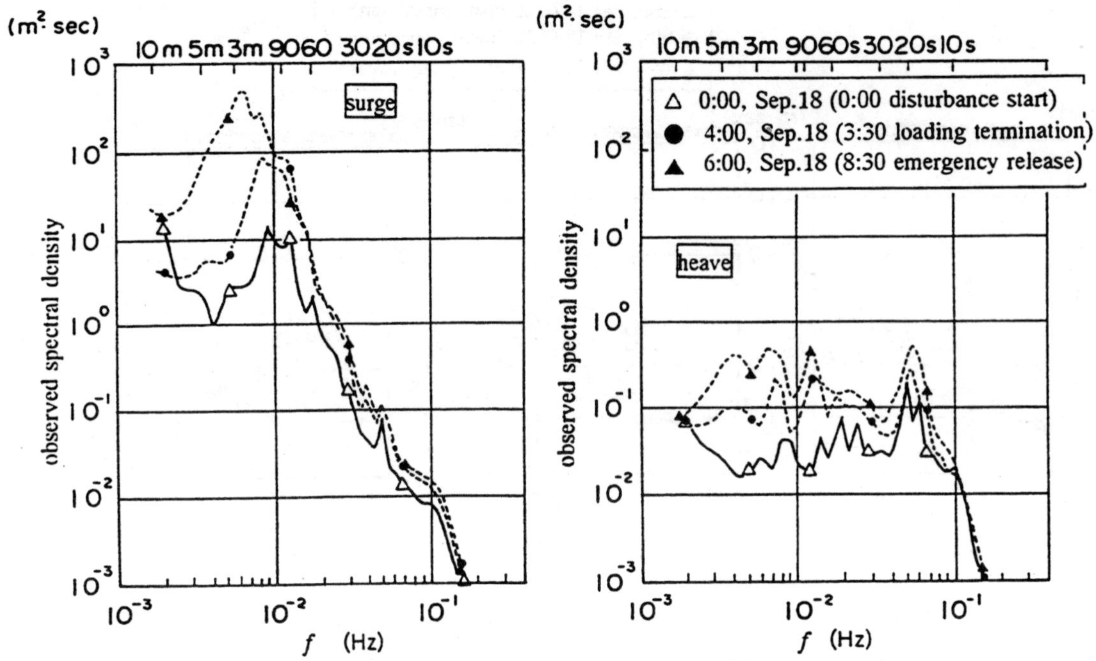

(1) Surge–motion

(2) Heave–motion

Figure 7 Frequency spectrum of ship motion

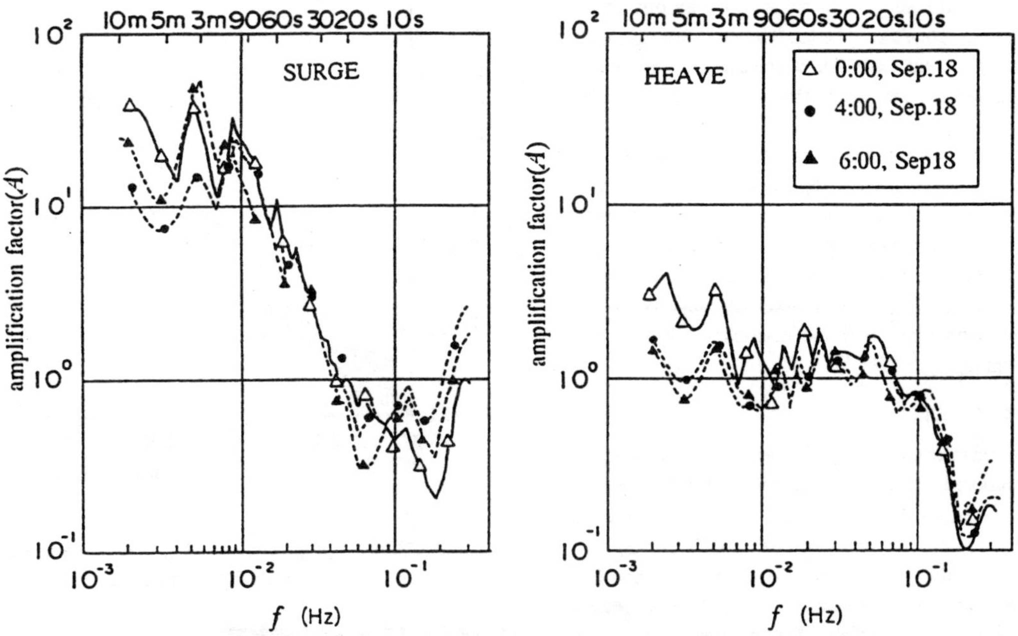

(1) Surge–motion

(2) Heave–motion

Figure 8 Variation of amplification factor

height increases. The long period wave height at **ST.1** is 15cm at the time when the cargo handling is terminated. The significant displacement of surge motion at the termination time is almost 2.0m. Ueda and Shiraishi(1988) shows the allowable level of surge displacement for short period wave. They proposes 1.0m in one–side of surge displacement. The allowble displacement in the both side becomes 2.0m. The observed limitation in Tomakomai–port is corresponding to the allowable level proposed by Ueda and Shiraishi. The allowable long period wave height in Tomakomai–port is, therefore, able to be set to 15 ~ 20cm.

Figure 11 shows the comparison of surge displacement and long period wave height at **ST.1**. The surge displacement increases linearly as the long period wave height increases. The amplitude becomes beyond 2.0m as the long period wave height becomes 15 ~ 20cm.

CONCLUSIONS

The following main conclusions are derived from the field observation in Tomakomai-port;

(1)Long period waves propagate to the inner harbor without any reduction by breakwaters.

(2)The surge motion is amplified by the resonance of the elastic mooring system and long period waves.

(3)The amplification factor of surge motion derived from spectral analysis becomes 50 ~ 70 at the maximum.

(4)The allowable long period wave height for loading is 15 ~ 20cm in Tomakomai-port.

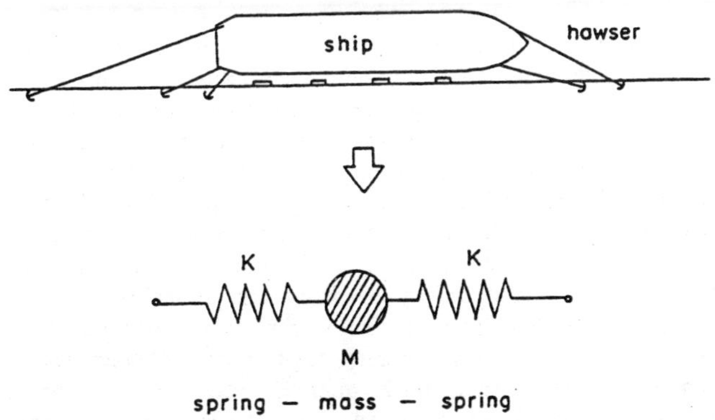

Figure 9 Model of vessel-mooring rope

REFERENCES

McGehee, D.D.(1991):Los Angeles and Long Beach harbors model enhancement program -measured response of moored ships to long-period waves-, *Technical Report CERC*-91-12.

Barthel, V., E.P.D.Mansard and D.D.MacDonald(1994): Penetration of long waves into a lagoon harbour and resulting ship motions, *Proc. Inter. Conf. on Hydro-Tech. Engi. for Port and Harbor Construction*, pp.333-351.

Ueda,S. and Shiraishi S.(1988):The allowable ship motions for cargo handling at wharves, *Rept. Port and Harbour Res. Inst.*, Vol.24, No.4, pp.3-61.

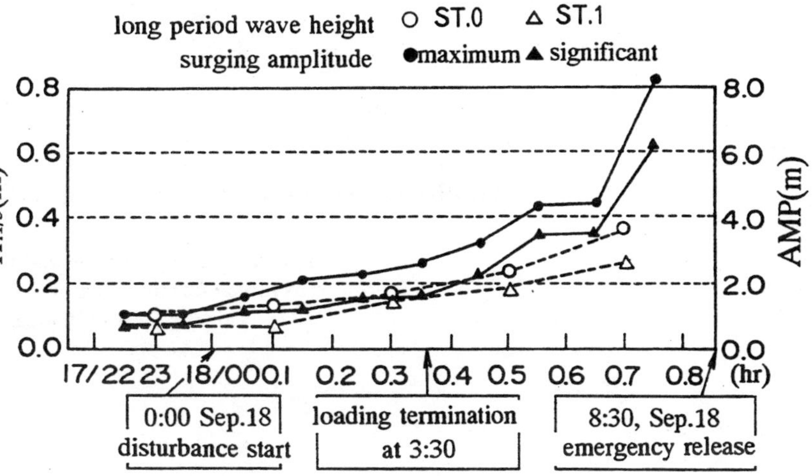

Figure 10 Variation of long period wave height and surge amplitude

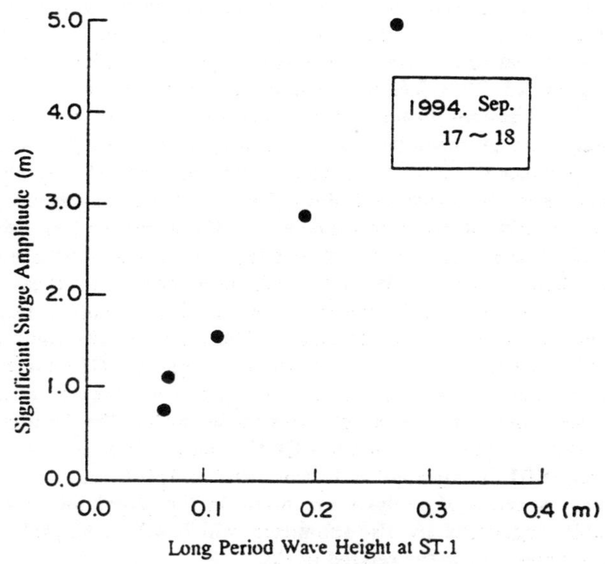

Figure 11 Comparison of long period wave height and surge amplitude osserved at **ST.1**

Proceedings of the Seventh (1997) International Offshore and Polar Engineering Conference
Honolulu, USA, May 25-30, 1997

ISBN 1-880653-28-1 (Set); ISBN 1-880653-31-1 (Vol. III)

Nonlinear Wave Loads on a Ship Advancing in Waves

Ming-Chung Fang and Cheng-Ming Liao
National Cheng Kung University
Tainan, Taiwan, China

ABSTRACT

Based on the strip theory, a method for predicting the nonlinear wave loads of a ship advancing in large oblique waves is proposed in the paper. With the previous well progress in the hydrodynamics, the hydrodynamic coefficients and forces can be well estimated. Therefore in the study, the authors consider the ship as a free-free beam and use the simple beam theory to estimate the corresponding sea loads along the ship hull, e.g. wave bending moment, shear force and torsion moment at any ship section. The numerical solutions for the linear analysis in frequency domain are firstly made and the accuracy has been verified by comparing with the existing experimental data and theory done by others. With the well developed linear theory , the analytical model of the nonlinear wave loads for a ship in large oblique waves are further developed. The time domain analyses including the instant grid generation technique (IGGT) for the corresponding sea loads are made. Besides, to avoid the numerical drift phenomena in the sway and yaw mode motion, the Artificial Restoring Force Technique(ARFT) and Digital Filter Technique(DFT) are applied to solve five coupled equations of motions. A series of analyses are made in the paper and some valuable suggestions are also submitted, which will be helpful for the researchers to do the related works.

The calculation procedures developed here may be considered as a useful analytical tool for predicting the ship motions and wave induced loads.

KEY WORDS: Nonlinear wave loads, Digital filter, Time domain

INTRODUCTION

Predicting the ship motions and the dynamic wave loads is an important basis for the seakeeping researches. Before the tool for dynamic analysis is well developed, the naval architect had been forced to use the ship's effective power performance in calm water and the ship's maximum bending moment in the static "0ne-over-twenty" wave as his main design criteria. Until now, the ship hydrodynamics has been well studied and the ship motion and dynamic wave loads are then widely applied for the ship design.

Since Korvin-Kroukovsky (1955) applied the strip theory to calculate the heave and pitch for a ship in head waves, the strip theory has been widely used to analyze the ship motion and wave loads. Jacobs (1958) firstly calculated the vertical shear force and the vertical bending moment by using the theory developed by Korvin-Kroukovsky (1955). Although the original strip theory (Korvin-Kroukovsky, 1955) was based on "physical intuition", it has been proved to be one of the most significant contributions in the field of seakeeping. Later, a modified strip theory was proposed by Gerritsma and Beukelman (1967), in which the close-fit methods are used to improve the computation of the sectional added mass and damping coefficients. Another new strip theories for heave and pitch motions were also derived, e.g. Soding(1969) and Tasai and Takaki(1969), to have identical forward-speed terms satisfying the Timman and Newman symmetry relationships. Salvesen, Tuck and Faltinsen(1970) used a new potential theory to solve the ship motion and sea load in regular waves and its validity was proved by comparing with the experimental data (Wahab, 1970) Using the same diffraction theory and different hydrodynamic force formulas, i.e. the Haskind-Newman relationship approach, Wahab and Vink (1975) also derived another method to calculate the sea loads and verified by previous experimental data(Wahab, 1970). Kim, Chou and Tein (1980) developed another strip method to estimate the ship motion, hydrodynamic pressure distribution and sea load by using different diffraction theory, i.e. diffraction force calculated directly by satisfying the diffraction boundary condition on a hull surface. Because of the mathematical model of Kim et. al.(1980) has been proved to be very useful in many respects of seakeeping, the authors used the same method and made some corrections on the original formulas for horizontal shear force, horizontal bending moment and torsional moment(Fang and Liao, 1996). The comparisons with the experimental data and other theory were also made and the agreements are generally good.

If the waves considered are large, the theories stated above can not be applied because of the small amplitude assumption. Recently there have been some authors made some deep studies on the nonlinear effect of the large wave on the ship motions and wave loads, e.g. Ohtsubo et. al.(1985) and Fujino and Yoon(1986). It would be more practical to use the large wave amplitude concept to predict the wave loads. Borresen and Tellsgard(1980) used the time domain simulation to analyze the nonlinear ship motion and wave loads in longitudinal waves, but they did not consider the case in oblique waves. In the present study, combing the analytical model predicting ship motions in oblique waves (Chen, 1994) and formulas for wave loads (Kim et. al., 1980), the authors develop an time simulation technique for predicting the nonlinear wave loads on a ship advancing in oblique waves. For saving the calculation time, an instant grid generation technique is applied. And an artificial restoring force technique and a digital filter technique are also included to avoid the numerical drifting phenomena in the paper. The general corresponding mathematical formulas are described as in the following sections.

EQUATIONS OF MOTIONS IN TIME DOMAIN

Considering the nonlinear effect of the large amplitude, the related hydrodynamic coefficients are all time-varying. Then, neglecting the effect of the surge mode, we can write five coupled equations of motions as below:

Heave mode(ζ):

$$\begin{aligned}&(M+M''_{\zeta\zeta}(t))\zeta+N_{\zeta\zeta}(t)\zeta+B_{\zeta\zeta}(t)\zeta+(M''_{\psi\zeta}(t))\psi+N_{\psi\zeta}(t)\psi+B_{\psi\zeta}(t)\psi\\&+(M''_{\eta\zeta}(t))\eta+N_{\eta\zeta}(t)\eta+B_{\eta\zeta}(t)\eta+(M''_{\chi\zeta}(t))\chi+N_{\chi\zeta}(t)\chi+B_{\chi\zeta}(t)\chi\\&+(M''_{\phi\zeta}(t))\phi+N_{\phi\zeta}(t)\phi+B_{\phi\zeta}(t)\phi=\mathrm{Re}\{F_\zeta(t)\}+R_\zeta(t)\end{aligned} \tag{1}$$

Pitch mode(ψ):

$$\begin{aligned}&(M''_{\zeta\psi}(t))\zeta+N_{\zeta\psi}(t)\zeta+B_{\zeta\psi}(t)\zeta+(I_{\psi\psi}+M''_{\psi\psi}(t))\psi+N_{\psi\psi}(t)\psi+B_{\psi\psi}(t)\psi\\&+(M''_{\eta\psi}(t))\eta+N_{\eta\psi}(t)\eta+B_{\eta\psi}(t)\eta+(M''_{\chi\psi}(t))\chi+N_{\chi\psi}(t)\chi+B_{\chi\psi}(t)\chi\\&+(M''_{\phi\psi}(t))\phi+N_{\phi\psi}(t)\phi+B_{\phi\psi}(t)\phi=\mathrm{Re}\{F_\psi(t)\}+R_\psi(t)\end{aligned} \tag{2}$$

Sway mode(η):

$$\begin{aligned}&(M''_{\zeta\eta}(t))\zeta+N_{\zeta\eta}(t)\zeta+B_{\zeta\eta}(t)\zeta+(M''_{\psi\eta}(t))\psi+N_{\psi\eta}(t)\psi+B_{\psi\eta}(t)\psi\\&+(M+M''_{\eta\eta}(t))\eta+N_{\eta\eta}(t)\eta+B_{\eta\eta}(t)\eta+(M''_{\chi\eta}(t))\chi+N_{\chi\eta}(t)\chi+B_{\chi\eta}(t)\chi\\&+(M''_{\phi\eta}(t))\phi+N_{\phi\eta}(t)\phi+B_{\phi\eta}(t)\phi=\mathrm{Re}\{F_\eta(t)\}\end{aligned} \tag{3}$$

Yaw mode(χ)

$$\begin{aligned}&(M''_{\zeta\chi}(t))\zeta+N_{\zeta\chi}(t)\zeta+B_{\zeta\chi}(t)\zeta+(M''_{\psi\chi}(t))\psi+N_{\psi\chi}(t)\psi+B_{\psi\chi}(t)\psi\\&+(M''_{\eta\chi}(t))\eta+N_{\eta\chi}(t)\eta+B_{\eta\chi}(t)\eta+(I_{\chi\chi}+M''_{\chi\chi}(t))\chi+N_{\chi\chi}(t)\chi+B_{\chi\chi}(t)\chi\\&+(M''_{\phi\chi}(t))\phi+N_{\phi\chi}(t)\phi+B_{\phi\chi}(t)\phi=\mathrm{Re}\{F_\chi(t)\}\end{aligned} \tag{4}$$

Roll mode(ϕ):

$$\begin{aligned}&(M''_{\zeta\phi}(t))\zeta+N_{\zeta\phi}(t)\zeta+B_{\zeta\phi}(t)\zeta+(M''_{\psi\phi}(t))\psi+N_{\psi\phi}(t)\psi+B_{\psi\phi}(t)\psi\\&+(M''_{\eta\phi}(t))\eta+N_{\eta\phi}(t)\eta+B_{\eta\phi}(t)\eta+(M''_{\chi\phi}(t))\chi+N_{\chi\phi}(t)\chi+B_{\chi\phi}(t)\chi\\&+(M+M''_{\phi\phi}(t))\phi+N_{\phi\phi}(t)\phi+B_{\phi\phi}(t)\phi=\mathrm{Re}\{F_\phi(t)\}\end{aligned} \tag{5}$$

where M''_{mk}, N_{mk} and B_{mk} represent the added mass, damping coefficient and restoring force, respectively. F_k is the exciting force. M is the ship mass. The subscripts m and k represent the mode of motion and direction of force, respectively, i.e. sway(η), heave(ζ), roll(ϕ), pitch(ψ) and yaw(χ).

The corresponding hydrodynamic coefficients are varied with time and the detail formulas can be referred to Chen(1994), in which the instant grid generation technique (IGGT) is used.

WAVE LOADS IN TIME DOMAIN

Generally the wave loads for a ship advancing in waves arouse from the following five components: (1) ship weight or ship inertia force (2) Froude-Krylov force due to incident wave (3) diffraction force (4) hydrodynamic inertia force, i.e. added mass, and damping force (5) restoring force.

Regarding the ship as a free-free beam, we can use the simple beam theory in material mechanics to calculate the corresponding sectional sea loads along the ship hull length. Because the ship is in motion and the dynamic equilibrium method, i.e. D'Alembert principle, can be used to calculate the corresponding sea loads at any ship section. The corresponding sea loads including vertical shear force, vertical bending moment, horizontal shear force, horizontal bending moment and torsional moment are shown in Figure 1.

The resultant sectional force and moment due to the vertical motion, horizontal motion and roll motion in large waves are shown in the following formulas:

Sectional Vertical Force:

$$\begin{aligned}K_\zeta(x,t)=K_{He}(x,t)&-\{(m_o(x,t)+M''_{HH}(x,t))\zeta+N''_{HH}(x,t)\zeta\}+R_\zeta(x,t)\\&-\{(-xm_o(x,t)-xM''_{HH}(x,t))\psi-xN''_{HH}(x,t)\psi\}\\&-\{M''_{SH}(x,t)\eta+N''_{SH}(x,t)\eta\}\\&-\{xM''_{SH}(x,t)\chi+xN''_{SH}(x,t)\chi\}\\&-\{(m_o(x,t)Z_{cx}(x,t)+M''_{RH}(x,t))\phi+N''_{RH}(x,t)\phi\}\end{aligned} \tag{6}$$

Sectional Horizontal Force:

$$\begin{aligned}K_\eta(x,t)=K_{Se}(x,t)&-\{M''_{HS}(x,t)\zeta+N''_{HS}(x,t)\zeta\}\\&-\{-xM''_{HS}(x,t)\psi-xN''_{HS}(x,t)\psi\}\\&-\{(m_o(x,t)+M''_{SS}(x,t))\eta+N''_{SS}(x,t)\eta\}\\&-\{(xm_o(x,t)+xM''_{SS}(x,t))\chi+xN''_{SS}(x,t)\chi\}\\&-\{(m_o(x,t)Z_{cy}(x,t)+M''_{RS}(x,t))\phi+N''_{RS}(x,t)\phi\}\end{aligned} \tag{7}$$

Sectional roll moment:

$$\begin{aligned}K_\phi(x,t)=K_{Re}(x,t)&-\{(m_o(x,t)Z_{cx}(x,t)+M''_{HR}(x,t))\zeta+N''_{HR}(x,t)\zeta\}\\&-\{(-xm_o(x,t)Z_{cx}(x,t)-xM''_{HR}(x,t))\psi-xN''_{HR}(x,t)\psi\}\\&-\{(m_o(x,t)Z_{cy}(x,t)+M''_{SR}(x,t))\eta+N''_{SR}(x,t)\eta\}\\&-\{(xm_o(x,t)Z_{cy}(x,t)+xM''_{SR}(x,t))\chi+xN''_{SR}(x,t)\chi\}\\&-\{(I_o(x,t)+M''_{RR}(x,t))\phi+N''_{RR}(x,t)\phi\}-R_\phi(x,t)\phi\end{aligned} \tag{8}$$

where

$m_o(x,t)$=instantaneous sectional mass

$Z_{cy}(x)$=the vertical distance between the instantaneous centroid and center of gravity, "+" represents the centroid is below center of gravity.

$Z_{cx}(x)$=the horizontal distance between the instantaneous centroid and center of gravity, "+" represents the centroid is to the right of center of gravity.

$xm_o(x,t)\psi$= instantaneous sectional vertical inertial force due to pitch

$xm_o(x,t)\chi$= instantaneous sectional horizontal inertial force due to yaw

$m_o(x,t)Z_{cx}(x,t)\phi$= instant sectional horizontal inertial force due to roll

$m_o(x,t)Z_{cy}(x,t)\phi$= instant sectional vertical inertial force due to roll

$m_o(x,t)Z_{cx}(x,t)(\zeta - x\psi)$=instant sectional horizontal inertial force due to roll

$m_o(x,t)Z_{cy}(x,t)(\eta + x\chi)$=instant sectional vertical inertial force due to roll

$I_o(x,t)\phi$= moment inertia force due to roll

$R_\zeta(x,t)$= instant restoring force per unit heave

$R_\phi(x,t)$= instant restoring moment per unit roll

$K_{He}(x,t)$=instant heave exciting force

$K_{Se}(x,t)$=instant sway exciting force

$K_{Re}(x,t)$=instant roll exciting moment

In order to evaluate the sea loads on an arbitrary section at $x=x_0$, fore or aft from the longitudinal center of gravity, we assume another variable ξ in $x=\xi+x_0$, where $x_0>0$ for a fore section. Then integrating the above three equations from x_0 to bow, we can obtain the wave loads at any section x_0 as below:

Vertical Shear Force:

$$F_v = \int_0^{L_2} K_\varsigma(\xi + x_o, t)\, d\xi \qquad (9)$$

Vertical Bending Moment:

$$M_v = -\int_0^{L_2} \xi K_\varsigma(\xi + x_o, t)\, d\xi \qquad (10)$$

Horizontal Shear Force:

$$F_L = \int_0^{L_2} K_\eta(\xi + x_o, t)\, d\xi \qquad (11)$$

Horizontal Bending Moment:

$$M_L = \int_0^{L_2} \xi K_\eta(\xi + x_o, t)\, d\xi \qquad (12)$$

Torsion Moment:

$$M_T = \int_0^{L_2} K_\phi(\xi + x_o, t)\, d\xi \qquad (13)$$

In the above integration, the underwater shape is different at any instant time, therefore every coefficient in the equation is the function of time or shape position. L_2 indicates the distance between the section concerned and fore end of the hull.

METHODS FOR AVOIDING THE NUMERICAL DRIFT

Some numerical drifting phenomena will occur on the sway and yaw modes while we solve the five coupled equations. To avoid this, two methods are used here:(1) Artificial Restoring Force Technique, ARFT. (2) Digital Filter Technique, DFT.

(1) Artificial Restoring Force Technique(ARFT):

The method is to add an artificial restoring force and moment in the equations of sway and yaw. The artificial force is related to the encounter frequency and the ship mass . The formula is

where ω is the encounter frequency. M is the ship mass for sway or moment of inertia for yaw. n is the coefficient for artificial restoring force.

(2) Digital Filter Technique (DFT):

Because the general solution for the frequency of causing the drift phenomena is zero and the required particular solution is equivalent to the encounter frequency, we only need to eliminate the frequencies which are lower than the required encounter frequency in the procedure of integration. Therefore the function of the filter used in the signal processing is applied in the present numerical calculation.

Generally the digital filter has two kinds: (1) FIR (Finite Impulse Response) filter (2) IIR (Infinite Impulse Response) filter. In order to make the phase constant and avoid the wave distortion, we adopt FIR filter. Besides, the higher order filter must be done to obtain sensitive truncation amplitude. According to FIR filter, the signal processing can be expressed as

$$\{p\}_j = \sum_{i=0}^{N} h(i).\{p\}_{j-N/2+1+i} \qquad (15)$$

where N is the number of filter, $h(i)$ is the coefficient of the filter, $\{p\}_j$ represents the output signal at jth step. The coefficient of the filter can be calculated by

$$h(i) = \frac{k(i)}{k(0) + \sum_{i=1}^{N/2} 2k(i)} \qquad if\ i = 1 \sim N/2$$

$$h(i) = h(N/2 - j)_{j=1\sim N/2} \qquad if\ i = N/2 + 1 \sim N \qquad (16)$$

$$i = 0 \Rightarrow k(i) = \frac{2f_c}{f_s}$$

$$i \neq 0 \Rightarrow k(i) = \frac{1}{i\pi}\sin(2i\pi\frac{f_c}{f_s}).w_B \qquad (17)$$

where f_c is the truncation frequency and f_s is the sampling frequency. $w_B(i)$ is Blackman window function and can be expressed as

$$w_B(i) = 0.42 + 0.5\cos(\frac{2\pi i}{N+1}) + 0.08\cos(\frac{4\pi i}{N+1}) \qquad (18)$$

Assume the value at jth step is known, then the value at next step $j+1$ can be calculated by the following procedures:

(1) Consider the displacement and velocity at $(j-N/2+1)$th step as the initial value and proceed the N steps time integration to obtain the displacement and velocity at each step.

(2) Apply the low pass filter to the calculated N+1 displacement and velocity, then subtract the value obtained from the $(N/2)$th step in (1) from the output value here.

(3) Substitute the value obtained in (2) into the equations of motions and loads to get the solution at step $j+1$.

(4) Proceed the time to the next step and repeat from procedure(1) again.

RESULTS AND DISCUSSION

The Series 60 ship with C_b=0.8 is selected as the numerical model for calculatingthe ship motion and wave loads at midship section. The validity of the present theory in linear application has been confirmed before (Fang and Liao, 1996). To simplify the present content, we only show the nonlinear analysis with large wave by time simulation here.

Figures 2-3 show the ship motions and wave loads in oblique wave without using the ARFT or DFT. We can see that the numerical drifting phenomena occur in the sway and yaw motions which lead to the unstable results in the wave loads. Therefore the techniques, ARFT or DFT, must be applied to avoid the drifting phenomena. Figure 4 shows the application of ARFT on the beam wave case. The results are indeed quite stable. However, the results for the horizontal mode vary with different artificial restoring coefficients, but there is almost no influence on the vertical modes. Because the artificial restoring force is an external force that will cause discontinuity while we draw the diagrams for shear force and bending moment, we therefore use DFT hereafter.

Before using the DFT, we must firstly decide what time interval and filter number should be selected. After a series of trial and test, we suggest that the time interval is selected as 1/60 of the wave period and the number of the filter is 30

Figures 5-6 show the time simulation for the ship motion and wave loads at μ=30° with Fr=0.15, λ/L =1.0 and H/λ=1/20. (μ is the wave direction, λ is the wave length, H is the wave height). From both figures, we can see the results are fairly stable. Generally the result for each time history is not symmetric with respect to the zero level which is the general phenomena in large amplitude analysis. In particular, the roll motion mode oscillates below zero level, which means the ship oscillates in an inclined status to the port side. Similarly the results for $\mu=60^\circ$ with $H/\lambda=1/30$ are shown in Figures 7-8 and are also stable without drifting phenomena. It proves that the DFT is really workable for the above two cases. There are some differences between two cases but not large. The oblique bow wave case, $\mu=120^\circ$, is shown in Figures 9-10. Again, the results are also quite stable because of the application of DFT. In Figure 9 the roll motion oscillates above the zero level which means the ship oscillates in an inclined status to the starboard side. The situation aroused from the wave comes from the port side while it is from the starboard for $\mu=60^\circ$ in Figure 7. Both confirmed with the physical phenomena. Comparing the sea loads results for these different wave directions, we find that the sea loads for bow wave are generally larger than those for stern wave. Besides, a special phenomena occurs in Figure 10 which shows two significant peaks in one wave period for the wave loads except the torsion moment. This interesting phenomena may be due to the phase difference and coupling effects of the motions.

SUMMARY

A calculated method for the nonlinear sea loads for a ship advancing in waves has been developed in the paper. The method described the procedure to yield results which, in general, can be referred by the naval architects in ship design. From the above analysis, some conclusions can be drawn as below:

(1) The ARFT method can retrieve the numerical drifting error in motion , but will cause an artificial external force on ship to make the sea load prediction impractical. Hence it may be suitable for motion prediction but not for sea loads.

(2) The DFT method not only can avoid the numerical drifting phenomena, but also make the sea load prediction more practical. However it must be careful to use the time interval and filter number to make the results stable. It is suggested to use 1/60 of wave period and N=30.

(3) A technique predicting the sea loads in large wave amplitude has been developed in the paper, however we still need the corresponding experimental data to further confirm its validity.

ACKNOWLEDGMENTS

The authors want to express their thanks to the National Science Council for the financial Support under the contract No. NSC-85-2611-E006-019. The authors also thank Mr. Lin, Han-Pi for his help to prepare the corresponding data.

REFERENCES

Borresen, R. and Tellsgard, F.(1980). "Time History Simulation of Vertical Motions and Loads on Ships in Regular head Waves of large Amplitude," Norwegian Maritime Research, No.2.

Chen, Shuan-Yu (1994). " Analysis for the nonlinear Motion and Dynamic Stability for a Ship advancing in Oblique Waves," Master Thesis in NCKU.

Fang, Ming-Chung and Liao, Cheng-Ming(1996). "Predictions of The Sea Loads for a Ship Advancing in Waves," Journal of SNAME, ROC. (To appear)

Fujino, M. and Yoon, B.S.(1986). "A Practical Method of Estimating ship Motions and Wave Loads in Large Amplitude Waves,"(3rd Report), International Shipbuilding Progress, Vol. 33, No.385, pp.159-172.

Gerritsma, J. and Beukelman, W.(1969). "Analysis of the Modified Strip Theory for the Calculation of Ship Motions and Wave Bending moments", international Shipbuilding Progress, Vol.14, No.156.

Jacobs, W. R.,(1958). "The Analytical Calculation of Ship Bending Moments in Regular Waves", Journal of Ship Research, Vol.2, No. 1.

Korvin-Kroukovsky, B. V.,(1955). "Investigation of Ship Motions in Regular Waves", Tran., SNAME, Vol.63.

Kim, C. H., Chou, F. S. and Tein, D.(1980). "Motions and Hydrodynamic Loads of a Ship Advancing in Oblique Waves", Tran., SNAME, Vol. 88, pp.225-256.

Ohtsubo, H., Kuroiwa, T. and Tamamoto, Y.(1985). "Structral Response of Ships among Rough Seas-Effects of Higher Frequency Vibration and Hydrodynamic Coefficients-",(1st Report), Journal of society of Naval Architects of Japan, Vol. 157, pp.391-400.

Soding, H.(1969). "Eine Modifikation der Streifenmethode," Schiffstechnik Bd.16,Heft 80.

Salvesen, N. Tuck, E. O. and Faltinsen, O.(1970) "Ship Motions and Sea Loads," Trans., SNAME Vol.78.

Tasai, F. and Takaki, M.(1969) "Theory and Calculation of Ship Responses in Regular Waves," (in Japanese), Symposium on Seaworthiness of Ships, Japan Society of Naval Architects.

Wahab, R.(1967). "Amidship Forces and Moments on a CB=0.80 Series 60 Model in Waves from Various Directions," Netherlands Ship Research Center TNO, Report 100S.

Wahab, R. and Vink, J. H.(1975). "Wave Induced Motions and Loads o Ships in Oblique Waves," International Shipbuilding Progress, Vol.22, No.249.

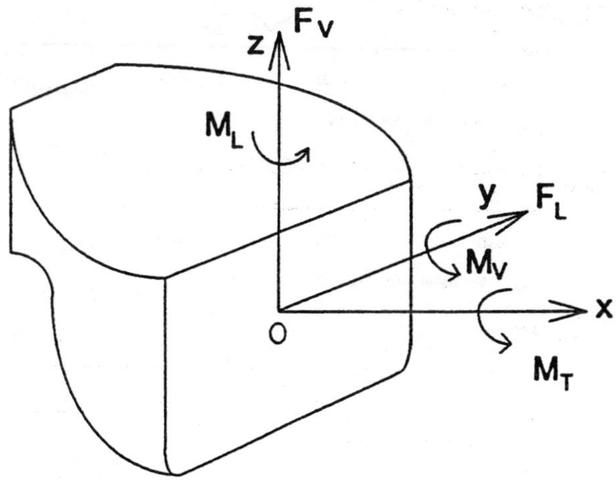

Figure 1. The diagram for the definition of sea loads

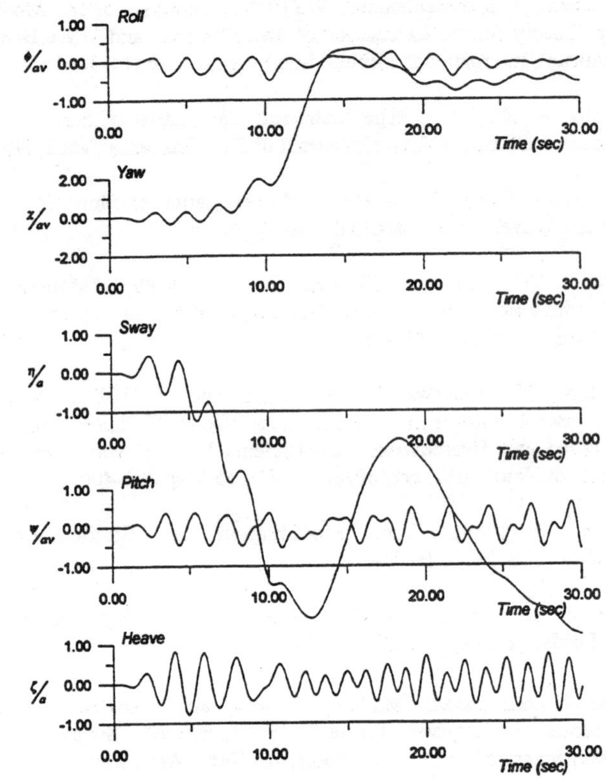

Figure 2. Time history of ship motion without using AFRT or DFT
(μ=60°, λ/L =1.0, H/λ=1/30, Fr=0.15)

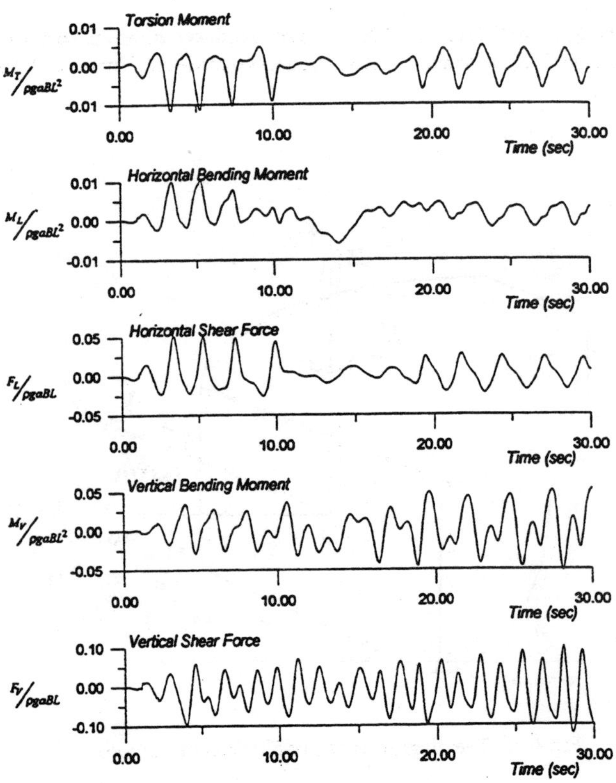

Figure 3. Time history of wave loads without using AFRT or DFT
(μ=60°, λ/L =1.0, H/λ=1/30, Fr=0.15)

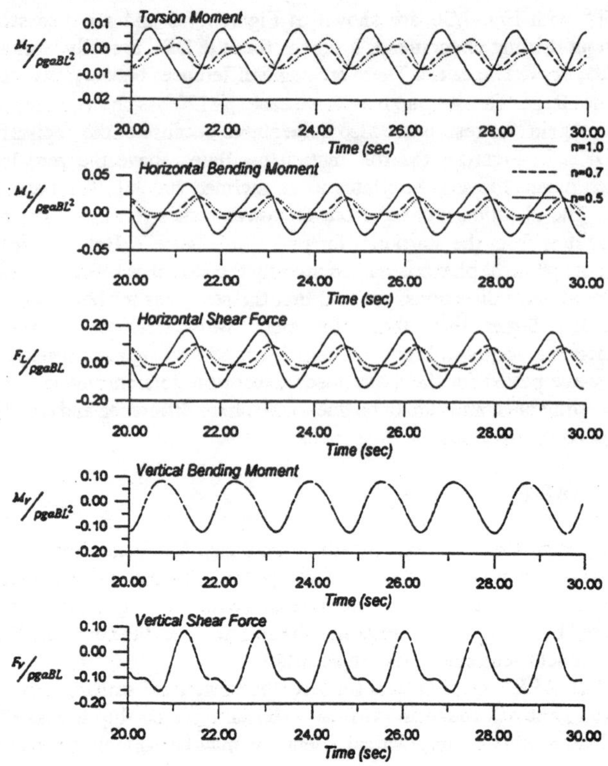

Figure 4. Time history of wave loads using AFRT with different artificial restoring coefficients
(μ=90°, λ/L =1.0, H/λ=1/30, Fr=0.0)

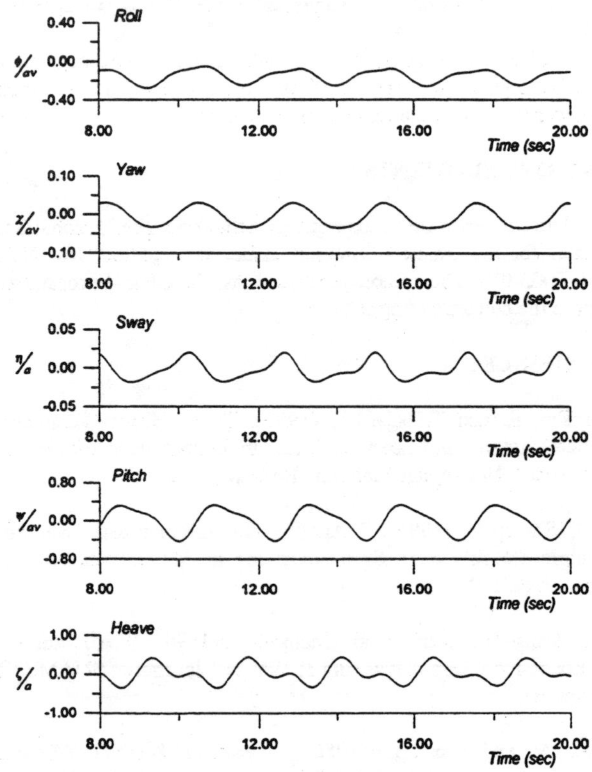

Figure 5. Time history of nondimensional ship motion
(μ=30°, λ/L =1.0, H/λ=1/20, Fr=0.15)

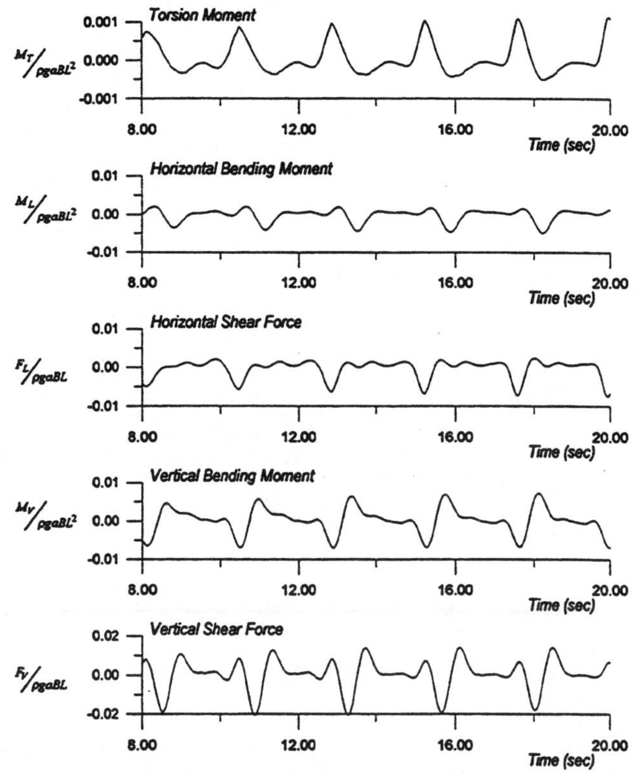

Figure 6. Time history of nondimensional wave loads
(μ=30°, λ/L =1.0, H/λ=1/20, Fr=0.15)

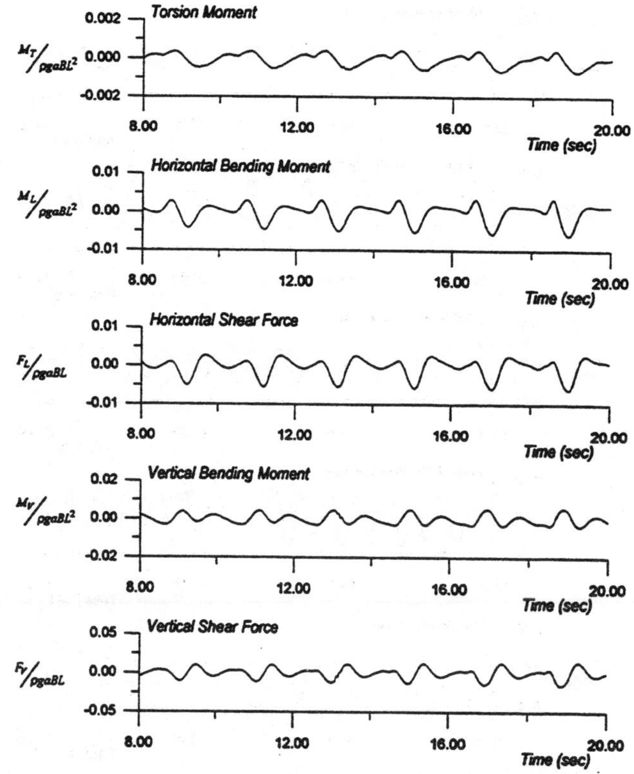

Figure 8. Time history of nondimensional wave loads
(μ=60°, λ/L =1.0, H/λ=1/30, Fr=0.15)

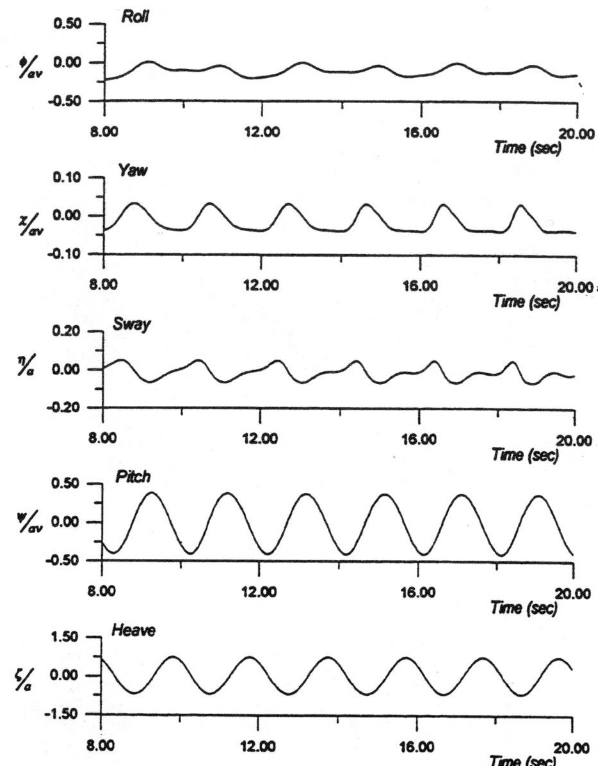

Figure 7. Time history of nondimensional ship motion
(μ=60°, λ/L =1.0, H/λ=1/30, Fr=0.15)

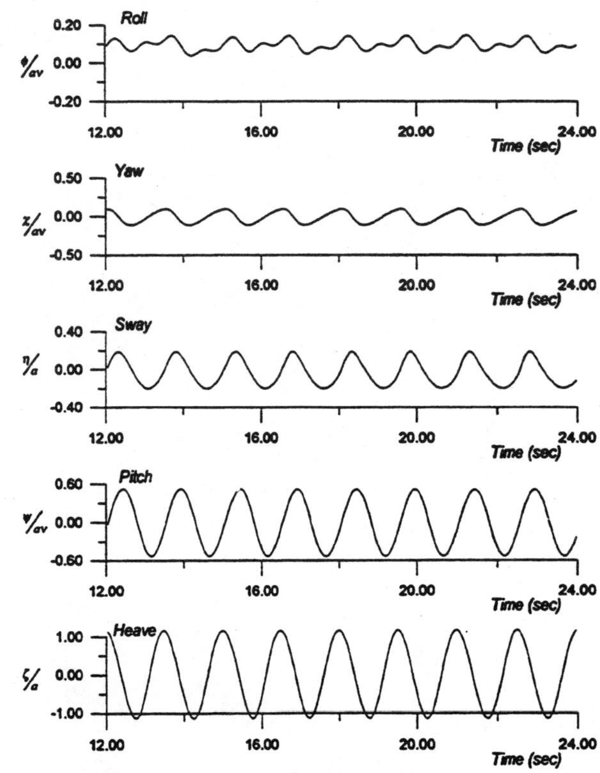

Figure 9. Time history of nondimensional ship motion
(μ=120°, λ/L =1.0, H/λ=1/30, Fr=0.05)

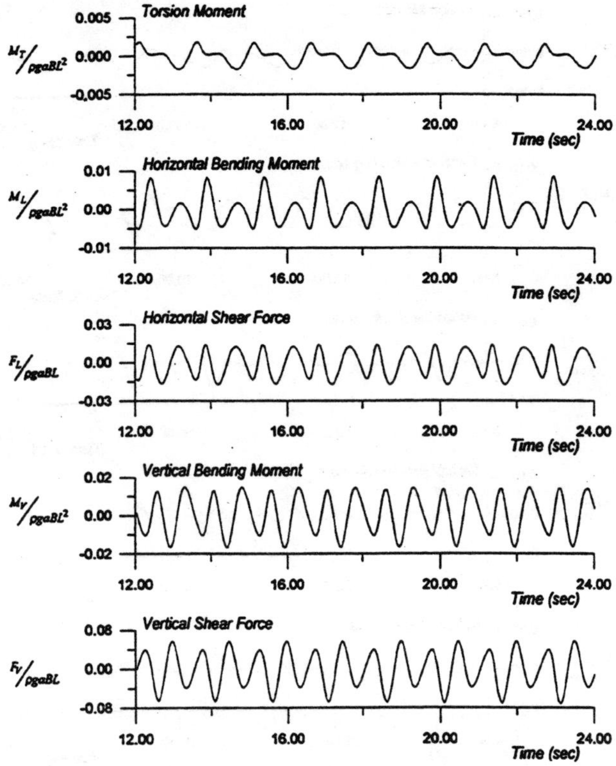

Figure 10. Time history of nondimensional wave loads (μ=120°, λ/L =1.0, H/λ=1/30, Fr=0.05)

Proceedings of the Seventh (1997) International Offshore and Polar Engineering Conference
Honolulu, USA, May 25-30, 1997

ISBN 1-880653-28-1 (Set); ISBN 1-880653-31-1 (Vol. III)

Time-Domain Prediction of Ship Motions with Experimental Verification

Dexiang Zhu and Mahmoud Katory
Hong Kong Polytechnical University
Hong Kong

ABSTRACT

The equations of motions of a ship traveling at a constant speed in regular oblique waves are developed and solved in the time-domain. The 3D time-domain Green function is adopted to formulate a unique expression for the scattering potential representing the perturbation caused by wave diffraction and ship motions when moving forward to the waves. In this way the dichotomy between radiation and diffraction adopted in the conventional seakeeping theories is discarded. Applications are made to predict the wave forces and ship motions at forward speed for a Wigley ship-hull in head waves and a catamaran ferry in oblique seas. Predictions are compared with published results for the Wigley ship-hull, and the agreement is good. For the catamaran, a model is built and tested in waves in a large towing tank and in a seakeeping basin in order to measure the six-degrees-of-freedom motions, forces and moments. Measurements are compared with predictions, and the agreement is generally good for the longitudinal motions, thus verifying the analysis. For the transverse motions, acceptable discrepancies exist mainly due to the exclusion of rudder forces and viscous damping in the analysis. The inclusion of such nonlinear factors in the time domain involves complex analysis and this problem must be left to a future research.

KEYWORDS: time-domain, ship motions, model testing.

INTRODUCTION

The majority of published numerical methods for the prediction of wave loading and ship motions are formulated in the frequency domain (Bruce, 1990). The pulsating-source distribution technique based on the frequency domain Green function (Wehausen and Laitone, 1960) is a favoured technique adopted to describe the flow around the hull, and this normally leads to good predictions at zero forward speed. In the presence of forward speed, however, it is difficult to satisfy the resulting boundary conditions by using the frequency domain Green function because the resulting formulation becomes too complex to solve (Inglis and Price, 1981, Newman, 1991).

An alternative approach is to formulate the ship motion problem in the time-domain using the time-domain Green function, which is not restricted to zero forward speed (Beck and Liapis, 1987, Pawlowski and Bass, 1991). This approach is adopted in the present work. The interaction between the forward moving oscillating ship-hull and the incident waves is formulated as an initial-value problem. The disturbance caused by ship motion is considered as a distribution of normal velocities on the wetted surface of the hull. A precise linear decomposition of the normal velocities is then obtained. In the same time, the integral equations for the impulsive and memory part of the modal scattering potential due to radiation and diffraction potentials and their interactions are also derived (Zhu, 1996). The modal potentials and hydrodynamic forces are determined iteratively with the ship motions by satisfying the ship-hull and a linearized free surface boundary conditions. Because diffraction and radiation problems are considered simultaneously, the basic dichotomy between radiation and diffraction adopted in the conventional seakeeping theories is now removed.

The time-domain solution, however, requires a time-stepping iteration. The hydrodynamic flow and ship motions are interacting, and the resulting expressions to solve for the velocity potential are lengthy and complex. A fast computer with large memory is essential. Careful programming is also required to avoid numerical instabilities. To overcome the many numerical and computational difficulties, the new breed high-speed large-memory multi-parallel-processors, known as the *super-computers*, are utilized.

COORDINATE SYSTEMS

The ship is assumed rigid and traveling with a constant mean forward speed $\bar{U}$ at an arbitrary heading in regular waves, and undergoing six-degrees-of-freedom oscillations. Referring to Figure 1, Three right-handed Cartesian coordinate systems are used. O_o-$x_o y_o z_o$ is a right-handed inertial coordinate system fixed in space, with $x_o y_o$-plane in the mean sea surface, z_o-axis points vertically upwards and x_o-axis in the direction of forward motion of the ship. $\hat{O}-\hat{x}\hat{y}\hat{z}$ is a ship-

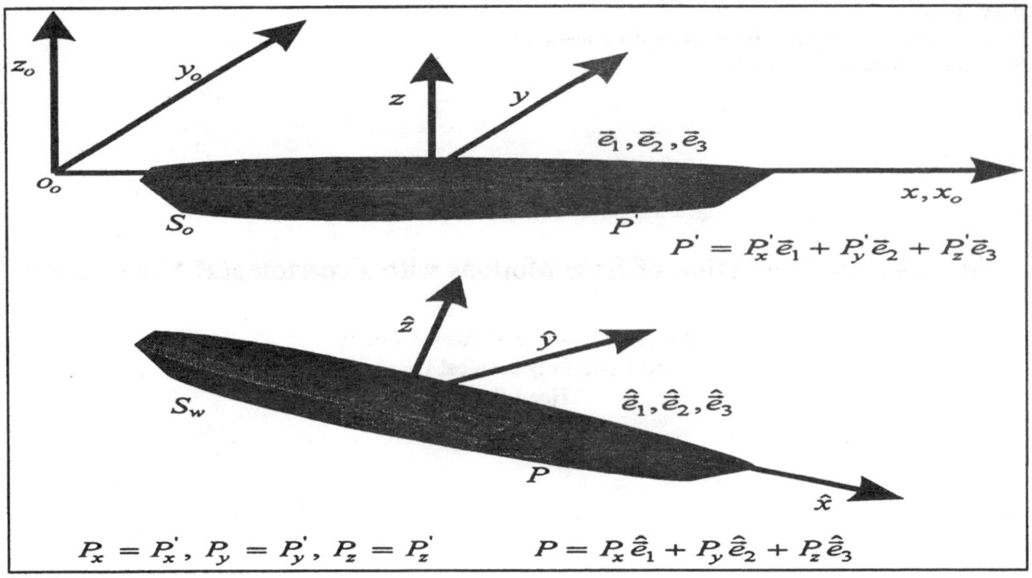

Figure 1: Coordinate systems

mounted coordinate system, $\hat{O}$ is at the centre of mass of the ship, $\hat{x}$-axis points towards the bow and $\hat{y}$-axis towards the port. *O-xyz* is a translating coordinate system moving with a velocity equal to the average velocity of the ship, such that when the ship is at rest the two systems, $\hat{O}$-$\hat{x}\hat{y}\hat{z}$ and *O-xyz*, coincide. The (x,y,z) axes always remain parallel to the (x_o, y_o, z_o) axes.

Initially at time = *0* the origins $\hat{O}$ and *O* coincide so that the coordinates of any point within the ship are:

$$x_o = x + Ut, \quad y_o = y \quad \text{and} \quad z_o = z \quad \text{at time} = t$$

Suppose that $\bar{e}_1, \bar{e}_2, \bar{e}_3$ and $\hat{\bar{e}}_1, \hat{\bar{e}}_2, \hat{\bar{e}}_3$ are the base unit vectors along the (x,y,z) and $(\hat{x}, \hat{y}, \hat{z})$ axes respectively; then the relation between these two systems of vectors can be expressed by tensor of rotation $\bar{\bar{R}}$. At a particular instant, let $\bar{X}_o$ be the radius vector of a point *P* on the ship surface with respect to the origin of space-fixed coordinate system O_o-$x_o y_o z_o$. If *P'* is the image point of *P* on the ship surface defined in the translating coordinate system *O-xyz*, and $\bar{X}$ is the radius vector of *P'* relative to the origin , then:

$$\bar{X}_o = \bar{x}_o + \bar{\bar{R}} \cdot \bar{X} \tag{1}$$

where $\bar{x}_o$ denotes the radius vector of the origin $\hat{O}$ of the ship-mounted system of axes relative to O_o, and $\bar{X}$ denotes the radius vector of *P'* relative to the origin of the translating system of reference. *P'* is the image point of *P* on the ship surface; that is, the coordinate of *P'* in the *Oxyz*-frame has precisely the same value as that of *P* in the $\hat{O}\hat{x}\hat{y}\hat{z}$-frame. Hence, the velocity of *P* is given by

$$\bar{V} = \frac{d\bar{X}_o}{dt} \tag{2}$$
$$= \bar{U} + \bar{V}_o + \bar{\omega} \times \bar{\bar{r}} \cdot \bar{X} + \bar{\omega} \times \bar{X}$$

with $\bar{V}_0$ signifying $\frac{d\bar{x}_o}{dt} - \bar{U}$, $\bar{\omega}$ denoting the angular velocity of the ship. $\bar{\bar{r}}$ is the tensor defined by:

$$\bar{\bar{R}} = \bar{\bar{I}} + \bar{\bar{r}} \tag{3}$$

where $\bar{\bar{I}}$ is the unit tensor.

FORMULATION OF THE SCATTERING POTENTIAL

The fluid is assumed incompressible and inviscid, and the motion irrotational without surface tension. Accordingly, there exists a velocity potential $\Phi(x_o, y_o, z_o)$ such that its gradient is equal to the velocity field $\bar{u}$ relative to the earth, i.e.:

$$\bar{u} = \nabla\Phi(x_o, y_o, z_o; t) \tag{4}$$

where ∇ denotes the gradient operator and Φ satisfies Laplace's equation.

Let the surface of the ship at all instants be given in the ship-mounted coordinate system $\hat{O} - \hat{x}\hat{y}\hat{z}$ by S_w: $\hat{F}(\hat{x}, \hat{y}, \hat{z}) = 0$. By transformation between translating and ship-mounted coordinate system, this function can also be written as: S_w: $F(x,y,z) = 0$. Here, the ship surface is denoted by S_w with reference to the *O-xyz* frame. Now, consider an imaginary surface S_{ow} given by $\hat{F}(x,y,z) = 0$. Named *reference configuration*, this surface is stationary with respect to the *O-xyx* frame and will coincide with the surface S_w when the ship is in its initial undisturbed position. The kinematics of the fluid flow is made dependent on the ship motion by the impermeability condition which must be fulfilled at points on the instantaneous wetted surface of the ship hull S_w which is:

$$(\nabla\Phi - \bar{V}) \cdot \bar{n} = 0 \tag{5}$$

where $\bar{n}$ is the unit normal vector to the surface S_w and directed towards the fluid field, $\bar{V}$ is the local velocity of the ship's wetted surface. This condition ensures that the fluid and ship velocities match at their interface. Using eqn. (2), eqn. (5) can be rewritten as:

$$(\nabla\Phi - \bar{U} - \bar{V}_o - \bar{\omega} \times \bar{X} - \bar{\omega} \times \bar{\bar{r}} \cdot \bar{X}) \cdot (\bar{N} + \bar{\bar{r}} \cdot \bar{N}) = 0 \tag{6}$$

where $\bar{N}$ denotes the unit normal vector in the *reference configuration* S_{ow} of the ship.

Assume that the ship is advancing in its reference configuration with a mean forward speed $\bar{U}$. The perturbation caused by the ship motion can be considered as a distribution of normal velocities V_N, relative to the translating coordinate system on the reference wetted surface of the hull S_o which is determined by the undisturbed waterplane $z = 0$. Accordingly, the impermeability condition on S_o becomes:

$$(\nabla\Phi - \bar{U}) \cdot \bar{N} - V_N = 0 \qquad (7)$$

In a linear theory, the velocity potential may be expressed in the form:

$$\Phi(x_o, y_o, z_z, t) = \Phi_u(x,y,z) + \Phi_w(x,y,z,t) + \Phi_s(x,y,z,t) \qquad (8)$$

where Φ_u is the potential induced in calm water by the forward motion of the ship in the absence of the normal velocity distribution V_N, Φ_w is the incident wave potential and Φ_s is the disturbance potential resulting from the appearance of $V_N \neq 0$. Following Pawlowski (1991), the so-called modal scattering potential used here includes the effects of radiation and diffraction and their interaction. The general normal velocity $\tilde{V}_N = V_N - \bar{N} \cdot \nabla\Phi_w$ is assumed to be expressed in terms of *modal amplitudes* $\beta(t)$ which depend on the ship motions, and a *shape function* $\Psi_N(\bar{X})$ which depends on the hull geometry, in the form:

$$\tilde{V}_N = \beta(t) \cdot \Psi_N(\bar{X}) \qquad (9)$$

The term *modal* is adopted to relate the disturbance potential Φ_s to the particular mode of excitation, as defined by the *shape function* Ψ_N. $\Phi_u(x,y,z)$ satisfies eqn. (7) with $V_N = 0$; therefore the impermeability condition for Φ_s on S_o becomes:

$$\bar{N} \cdot \nabla\Phi_s = \beta(t) \cdot \Psi_N(\bar{X}) \qquad (10)$$

The linear free surface boundary condition is assumed to be satisfied by Φ_s, hence:

$$\left[\left(\frac{\partial}{\partial t} - U\frac{\partial}{\partial x}\right)^2 + g\frac{\partial}{\partial z}\right]\Phi_s = 0 \qquad \text{on } z = 0 \qquad (11)$$

THE TIME DOMAIN EQUATIONS OF MOTION

The ship is considered as a rigid body and the origin of the coordinate system coincides with the centre of mass of the ship. The incident wave height is assumed to be small. The oscillations of the ship are considered small and accordingly the gyroscopic moments are neglected when the equations of ship motions are developed. To the first order approximation, the generalized equations of motion of the ship may be expressed in vector form as follows:

$$\mathbf{M} \cdot \dot{\bar{V}} = \bar{F} \qquad (12)$$

where

$\bar{F}$ is the generalized force vector,

$\mathbf{M}$ is the generalized inertia matrix,

$\bar{V}$ is the generalized velocity vector.

Let the generalized displacement vector of the centre of mass of the ship be represented by:

$$\bar{x}(t) = (x_1(t), x_2(t), x_3(t), \eta_1(t), \eta_2(t), \eta_3(t)) \qquad (13)$$

where $x_1(t)$, $x_2(t)$ and $x_3(t)$ are the surge, sway and heave displacements respectively; $\eta_1(t)$, $\eta_2(t)$ and $\eta_3(t)$ are the roll, pitch and yaw displacements respectively. The generalized normal velocity vector now becomes:

$$\bar{V}(t) = \frac{d\bar{x}(t)}{dt} \qquad (14)$$

In eqn. (12), the generalized force $\bar{F}$ comprises the hydrostatic restoring force $\bar{F}_s$ and the unsteady hydrodynamic pressure force $\bar{F}_k$. The hydrostatic restoring forces and moments $\bar{F}_s$ can be expressed as follows:

$$\bar{F}_s = \mathbf{c} \cdot \bar{x}(t) \qquad (15)$$

where **c** is the restoring force coefficient matrix.

The unsteady hydrodynamic forces are evaluated in terms of the velocity potential Φ_s by integrating the hydrodynamic pressure acting on the wetted hull surface, using the linearized Bernoulli's equation. The result is:

$$\bar{F}_k = -\rho \iint_{S_0} \frac{\partial \Phi_s}{\partial t} n_k ds + \rho U \iint_{S_0} \left(\frac{\partial \Phi_s}{\partial x}\right) n_k ds, \quad k = 1,2,\cdots,6 \qquad (16)$$

In deriving the above equation, the ship is assumed to be slender, and this obviates the need to evaluate the steady velocity potential (Newman, 1965). The notation n_k refers to the component of the generalized normal vector, defined by:

$$\begin{aligned}(n_1, n_2, n_3) &= \bar{n} \\ (n_4, n_5, n_6) &= (\bar{r} \times \bar{n})\end{aligned} \qquad (17)$$

$\bar{n}$ is a unit vector normal to the hull surface, $\bar{r}$ is the vector from a point on the hull surface to origin of coordinate system.

In discretized form, eqn. (12) can be rewritten as follows:

$$\frac{1}{\Delta t^2}\mathbf{M}\bar{x}(t_i) = \bar{F}_{fk} + \bar{F}_{sc} - \bar{F}_s + \frac{1}{\Delta t^2}\mathbf{M}(2\bar{x}(t_{i-1}) - \bar{x}(t_{i-2})) \qquad (18)$$

where

$\bar{F}_{fk}$ indicates the Froude-Krylov forces,

$\bar{F}_{sc}$ represents the scattering forces,

the subscript i means the i-th time step,

Δt is the time step size.

Integration of the pressure caused by the incident wave over the wetted hull surface gives the Froude-Krylov forces. The scattering forces can be similarly obtained by integrating the pressure distribution associated with the scattering potential. It is to be noted, however, that the scattering potential is dependent on the modal amplitudes $\beta(t)$ which, in turn, depend on the amplitudes of ship motions. The simulation commences by initially assuming that the scattering forces are zero. Substituting the expression of hydrostatic forces in eqn. (8), the solution for the ship motion $\bar{x}(t_i)$ may be written as follows:

$$\bar{x}(t_i) = (\mathbf{M} + (\Delta t)^2\mathbf{c})^{-1}\begin{Bmatrix}(\Delta t)^2(\bar{F}_{fk} + \bar{F}_{sc}) + \\ \mathbf{M}(2\bar{x}(t_{i-1}) - \bar{x}(t_{i-2}))\end{Bmatrix}, \quad i \geq 2 \qquad (19)$$

With assumed initial condition for $\bar{x}(0), \dot{\bar{x}}(0)$, the equation can be solved to yield a set of kinetic parameters $\bar{x}(\Delta t), \dot{\bar{x}}(\Delta t)$. The modal amplitudes may thus be determined enabling the scattering potentials and accordingly the scattering forces to be computed. With these newly computed scattering forces, plus the Froude-Krylov forces, the equation of motion is solved again, yielding a new set of kinetic parameters, and the process is repeated. Results are successively compared until the difference between the last two sets of values of modal amplitudes is smaller than a certain small value. After that, the next time step commences and so on until a wave cycle is completed. To economize in the calculations and reduce computing time, the starboard-port symmetry is taken into account and the expressions are simplified.

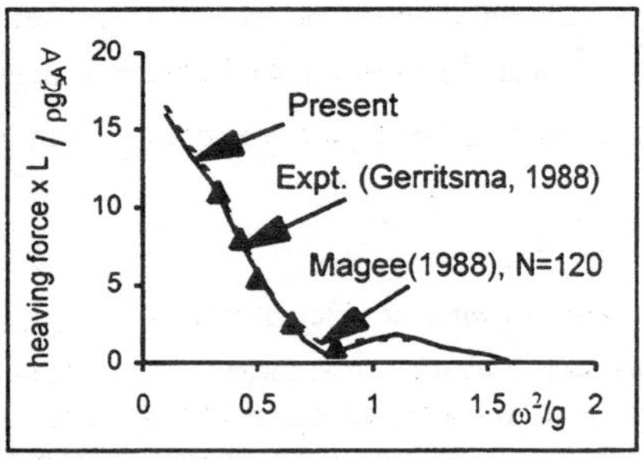

Fig. 2a: Wave exciting force in heave mode on a Wigley hull in 180° head waves, $F_r = 0.2$.

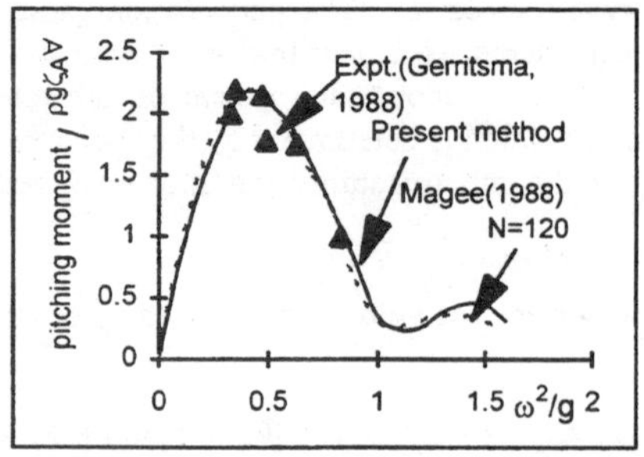

Fig. 2b: Wave exciting moment in pitching mode on a Wigley hull in 180° head waves, $F_r = 0.2$.

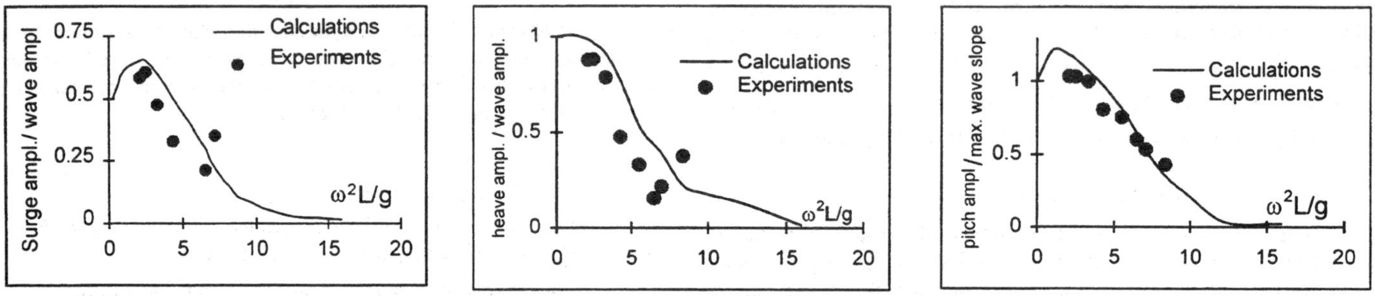

(a) (b) (c)

Fig.3(a-c): Surge, heave and pitch responses in 180° head waves for 119m catamaran, $F_r = 0.15$

VERIFICATION OF THE PREDICTION METHOD

Calculations are made for a Wigley ship-hull form, length-to-beam ratio $L/B = 10$ and a beam-to-draught ratio $B/T = 1.6$, moving in a regular head waves with a constant speed corresponding to a Froude number $F_r = 0.2$. The hull surface is represented by 160 panels, 20 lengthwise × 8 girth-wise. Details of the hull geometry are reported in the work of Zhu (1996). Results of the calculations for the heave wave exciting force and wave pitching moment are compared with experimental results carried out by Gerritsma (1988), and theoretical results by Magee and Beck (1988). The comparison is shown in Figs. 2a & 2b above. Good agreement between predictions and published results confirms the correctness of the numerical model.

Model experiments with a 2.83m model of a catamaran-ferry, length 119m displacement 6290 tonnes, were carried out in the seakeeping basin and the deep towing tank of the China Ship Scientific Research Centre (CSSRC), Wuxi, China. Geometrical details of the catamaran, and of the CSSRC testing facilities are given in the work of Zhu (1996). In the computer predictions, a series of time domain simulations were performed to determine the hydrodynamic forces, moments and motions for the catamaran as it travels with a constant speed in regular oblique waves of different headings. The waves were simulated using Airy wave theory, with a wave height similar to that adopted in the experiments. Predictions are compared with the model test results in Figs. 3(a-c) above and Figs. 4(a-c) below. Reasonably good agreement is achieved between predicted and measured longitudinal ship motions. Some discrepancy exist, however, between calculations and measurements for the transverse ship motions. The discrepancy is considered to be mainly due to the exclusion of the rudder forces and viscous effects in the analysis. All predicted results presented here are those obtained from the application of potential theory. The inclusion of viscous damping in the time domain involves complex analysis, and this problem must be left to a future research. Another source for the discrepancy is due to some differences which existed in the structural properties of the hull between the data used in the analysis and the physical model after it was fitted with the necessary equipment for the self propulsion test and the recording. Also, some difficulties were encountered while recording ship motions in the seakeeping basin. It was difficult to keep the self-propelled model on a straight course, in particular in the low frequency range. In the numerical simulations, the hull surface of the catamaran was discretized into 280 quadrilateral panels and in order to simplify the data processing the hull surface panels are generated directly from the ship hull table of offsets. This simplification in representing the hull surface may also be a source of discrepancy. The gyro-moments were neglected when the equations of motion were developed, and this may be another reason for the discrepancy, particularly in the cases involving oblique waves.

CONCLUSIONS

The equations of motions of a ship traveling at a constant speed in regular oblique waves are formulated and solved in the time-domain. The 3D-time domain Green function is adopted to formulate a unique expression for the scattering potential to represent the perturbation caused by the forward motion of the ship and her oscillations. In the formulation of the velocity potential to describes the hydrodynamic flow around the ship hull, the dichotomy between radiation and diffraction adopted in the conventional seakeeping theories is discarded. Because the velocity potential depends on the ship motions, an iterative solution is adopted. This requires somewhat extensive computations; the new breed of the *supercomputer* is utilized to solve the lengthy and complex analytical expressions.

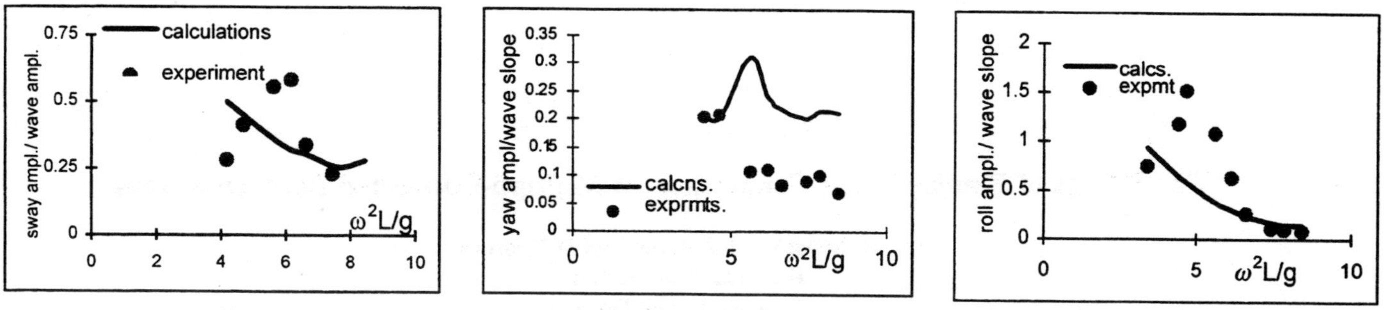

(a) (b) (c)

Fig. 4(a-c): sway, yaw and roll responses in 135°bow waves for 119m catamaran, $F_r = 0.15$

The newly developed 3D time-domain ship motion calculation method is applied to predict the wave forces and ship motions for a Wigley ship-hull in head waves and a catamaran ferry in oblique seas. Calculations are compared with published results of the Wigley ship-hull, and their agreement is good. For the catamaran, a model is built and tested in a large towing tank and in a seakeeping basin and the six-degrees-of-freedom motions, forces and moments are measured. Measurements are compared with predictions, and the agreement is generally good for the longitudinal motions, thus verifying the new method of calculations. For the transverse motions, some discrepancy exist mainly due to the exclusion of rudder forces and viscous damping in the analysis. The inclusion of such nonlinear factors in the time domain involves complex analysis and this problem must be left to a future research.

REFERENCES

Beck, R. F. and Liapis, S.: "Transient Motions of Floating Bodies at Forward Speed", Journal of Ship Research, Vol. 31, No. 3, pp. 164-176, 1987.

Bruce, L. H.: "Seakeeping Studies: A Status Report", Trans. SNAME, Vol. 98, 1990.

Gerritsma, J.: "Motions , Wave Loads and Added Resistance in Waves of Two Wigley Hull Forms", Report No. 804, Technical University of Delft, Netherlands, 1988.

Gradshteyn, I. S. and Razhik, I. M.: "Tables of Integrals, Series and Products", Academic Press, 1980.

Inglis, R. B. and Price, W. G.: "A Three Dimensional Ship Motion Theory —Comparison between Theoretical Predictions and Experimental Data of the Hydrodynamic Coefficients with Forward Speed", Trans. RINA, Vol. 124, 1981.

Magee, A. R. and Beck, R. F.: "Compendium of Ship Calculations Using Linear Time Domain Analysis", Report No. 310, Dept. Naval Arch. Marine Engineering, University of Michigan, Ann Arbor, Michigan, 1988.

Newman, J. N.: "The exciting forces on a moving body in waves", J. Ship Res., Vol. 9, p. 190, 1965.

Newman, J. N.: "The Quest for a Three-Dimensional Theory of Ship-Wave Interactions", Phil. Trans. R. Soc. Lond. A, 1991.

Pawlowski, J. S. and Bass, D. W.: "A Theoretical and Numerical Model of Ship Motion in Heavy Seas", SNAME Transaction, Vol. 99, pp. 319-352, 1991.

Wehausen, J. V. and Laitone, E. V.: "Surface waves", Handbuch der Physik, (Encyclopedia of Physics), ed. S. Fluegge, Vol. 9, Fluid Dynamics 3, Springerverlag, Berlin, 1960.

Zhu, D. X. and Katory, M.: "3D time domain numerical model for the prediction of ship motions in random seas", International Journal of Offshore and Polar Engineering, pp. 120-126, Vol. 5, No. 2, June 1995.

Zhu, D. X.: "3D Unified Time Domain Analysis for The Prediction of Ship Motions In Waves", PhD Thesis, The Hong Kong Polytechnic University, 1996.

Proceedings of the Seventh (1997) International Offshore and Polar Engineering Conference
Honolulu, USA, May 25-30, 1997

ISBN 1-880653-28-1 (Set); ISBN 1-880653-31-1 (Vol. III)

The Effects of Binding Hinge Connections in Multipli-Connected Pontoon Arrays

A. Nobakhti, G.J. Lyons and D.T. Brown
University College London
London, United Kingdom

ABSTRACT

Large arrays of interconnected pontoons have been developed in the past for structures such as bridges when connected in a line, and are also being considered for a range of projects such as floating harbours, and even airports when these arrays are extended to mat-like structures.

The benefits of these forms of structures are described.

Whereas a number of authors have performed experimental work to establish motions response and to develop numerical models for arrays with pure flexing, pure hinged, or otherwise rigid interconnections, little work has been carried out to date to examine the response of the array as a result of more realistic connections such as have already been introduced into service.

This paper presents the global structural response of pontoon arrays using such practical freely rotating inter-pontoon connections which bind beyond a limiting rotation. These are modelled theoretically for long arrays. The effects of variable loading are considered and comparisons with the results of model tests are made. Some interesting and unexpected observations arise which are accounted for in the theoretical solution developed. These relate to determining the locking and more particularly the unlocking criteria which are utilised within the global structural stiffness formulation.

It is shown that connections which initially lock as a result of pontoon section submergence on loading may be required to unlock as loading is increased (with possible subsequent locking/unlocking as loading is increased further or otherwise altered).

KEY WORDS: Pontoons, hydrostatic stiffness, hinges

INTRODUCTION

The aims of this work have been to consider the behaviour of a mat multipli-connected floating structure, with special hinge-rigid connections, owing to static and dynamic loading.

The principal differences between the work reported herein (which considers various static loading conditions) and that by others relate to the type of connections, and the arrangement of the structure. The more realistic hinge-rigid (locking hinge) connection is considered, rather than pure hinge, elastic, or rigid connections which have been otherwise used in analysis by other authors. The structure comprises a multipli-connected floating row, or mat (of interconnected rows).

The work reported herein relates principally to the development of the global stiffness matrix for hinge-rigid connections which may be used in static and dynamic loading, including that by waves.

OVERVIEW

From early times floating platforms as pontoon bridges have been used for crossing rivers. In the most primitive form these platforms mainly consisted of several pieces of wood joined by rope. The use of these systems has been developed, and increased in size such that they have even been considered for floating airports. Although construction of these airports requires much more elaborate and sophisticated engineering work, the ideas are similar to those used in building early floating bridges, namely, connecting several similar modules in an array consisting of one or several rows.

The connection between the constituent units perhaps plays an important role in determining the behaviour of multipli-connected systems. These may be catagorised as:

- **Hinge connections:** the translational displacement of two consecutive modules at the point of connection is the same, but the angle of rotation (pitch angle) can be different
- **Elastic connections:** the translational displacement of two adjacent modules at the point of connection is the same, but their angles of rotation may be different dependent on the elasticity of joints

- **Rigid connections:** translational displacement at the point of connection and angle of rotation of adjacent components are each the same for adjasent modules

- **Hinge-rigid connections:** the system behaves as a hinge connection, until the angle between two consecutive modules reaches a design threshold value, at which point the system acts rigidly. Depending on the design locking may be possible in either one or two directions of rotation.

APPLICATIONS OF MULTIPLI-CONNECTED FLOATING STRUCTURES

Wave Energy Devices

As a result of oil price rises in the mid 1970s, together with environmental concern over the use of fossil fuels serious consideration has been given to the exploitation of alternative energy resources. Wave energy has received much attention in countries with appropriate coastlines. Numerous studies have been carried out considering the utilisation of wave energy for number of prototype devices many of which use some kind of multipli-connected floating structures [1 to 3].

Wave Breakers

There is great demand for sheltered moorage of vessels including those for fishing and recreational purposes. Floating breakwaters have been employed at many new marina facilities to satisfy this demand .

Several new concepts have been studied and several more traditional approaches to floating breakwater construction found interesting applications. For example over ten floating breakwaters have been installed at different sites in the Pacific Northwest area [4-7].

Floating Marine Terminals

During the past three decades there has been increased interest in floating wharves to resolve port construction and operation problems [8]. The experiences gained in constructing floating wharves in Peru, Brazil, USA, Saudi Arabia, USSR, and many other countries have clearly demonstrated that under specific conditions (deep water, strong current, short periods suitable for construction, uncertain or poor soil conditions, erosion or sedimentation problems, high change in water level for example in the Caspian Sea, and construction in remote areas), floating marine terminals represent a very competitive solution compared to the fixed waterfront concept.

Other Uses

It is anticipated that there will be further growth in size of large industrial complexes, civil construction, and research facilities in the oceans owing to increasing demand for space resources utilisation. Some of these developments may require very large floating structures (VLFS) as compared with existing offshore structures. Floating industrial complexes consisting of, for example ocean thermal energy conversation (OTEC), power stations, fish processing, or manganese processing plants are likely to be very large. Other floating structures are also being considered such as airports, runways [13], entire floating cities [11], and oil storage.

Hinge-Rigid connections

Previous studies have considered connections between structures consisting of hinge, elastic, or rigid connections. Adee considered only hinge connections [4-7], whilst Takaki et al considered hinge, elastic, and rigid connections [11].

In the case of hinge-rigid connections the system behaves as a hinge connection until the angle between two consecutive modules reaches a design threshold value, at which point the system acts rigidly.,

The advantages of of the hinge-rigid connection are:

- less displacement in comparison with hinge connections

- less moment in the connections in comparison with rigid connections

Additionally, components of these forms of floating structures are often manufactured in the shipyard or factory, and are then transported to the assembly site. To ensure that components fit together in the assembly site, considering manufacturing tolerances, components must be manufactured in such way that joints have some looseness (allowance) to ensure that assembly is possible. These connections act in the hinge-rigid manner.

Theoretical Studies

Early work on floating breakwater design [4-7], considered the wave following impact. All these studies concentrated on measuring the transmission coefficient of the wave, which was found by dividing the transmitted wave height by the incident wave height. In the work it was assumed that both incident waves and the body motions were small, allowing the following simplifications to be made:

- The governing equations may be linearised and the breakwater treated as a linear system
- Linear superposition may be employed to simplify the computation of hydrodynamic effects

The work also separated the computation of the performance of a floating breakwater into three parts:

1) Formulate equation of motion
   (i) Calculate hydrostatic forces and moments
   (ii) Evaluate hydrodynamic coefficients in equations of motion
   (iii) Compute exciting forces on breakwater
   (iv) Solve for the motions and motion-generated waves
   (v) Compute static mooring-line response and calculate forces in the mooring lines
2) Solve for the waves diffracted by a rigidly restrained breakwater
3) Sum component to obtain total reflected and total transmitted waves

Tsinker [8] considered several kinds of multipli-connected floating structures including chain of pontoons with individual deck sections, and a double-cantilever system with suspended intermediate spans.

Piskorski, [9], investigated a floating pontoon bridge with limited rotational clearance between the pontoons. The work divided the pontoon bridge into two separated zones: the contact zone, characterised by the lack of voids between the pontoons, was considered as a rigid body; and open zones on either side of the contact zone. The work treated the open zone region as a super-element with linear elastic characteristics at its edge and considered the vibration in the open zone excited by the concentrated force at its edge. The work utilised Lagrange's formula and did not consider damping.

Xiling Che, et al [10], presented a method for predicting the loading and response of a very large floating structure (VLFS) in regular waves. They solved the problem of determining the motion of a system of integrated modules acted upon by waves. They applied a two-dimensional hydroelastic theory to a VLFS with multiple modules and extended it in the longitudinal direction. A modified strip theory, was employed to analyse the hydrodynamic coefficients. They did not consider the mooring system, but assumed that the system remained stationary on location. The system was modelled in two ways. In the first model, the longitudinal sequence of models was considered to form a single beam having varying shear and flexural rigidities. In the second model, the model was considered to be rigid, that is undeformable by the wave loading. The boundary-element method and Greens function methods were used to obtain hydrodynamic coefficients. They presented a numerical example with reference to an integrated system of semisubmersibles. As well as a simple model for engineering estimation.

Takaki and Tango [11] studied theoretically and experimentally the wave drift forces acting on a very large floating structure consisting of multiple barge type modules. They estimated the wave drift forces on the multiple floating structure with two module connectors, a rotationally hinged connector and a rotationally rigid connector. The effect of connector condition and the number of bodies on the wave drift forces was discussed in their paper. The three dimensional panel method was used to determine the hydrodynamic forces to take into account accurately the effect of hydrodynamic interactions among the modules, and the coupled equations of motions was solved directly. The number of modules was shown to affect the amplitude of motion in head sea conditions. Furthermore they discussed the effect of the connecting condition on the wave drifting forces.

EXPERIMENTAL WORK

Endo et al [12], conducted an experimental model analysis to estimate the modal properties such as natural frequencies, damping ratios, and mode shapes for a unit-linked large floating structure model. The fluid-structure coupled vibration tests were carried out using an impact hammer. A number of models, namely one, two, and eight unit linked models with and without anchors, were used in their study. Each unit consisted of a rectangular prism linked by some steel plates at a constant interval. A floating model was anchored by two spring-lines on the base edges of each unit.

The vertical acceleration responses were measured on the upper surfaces of the model by arranging acceleration sensors on each unit. The transfer functions defined by normalised input and output data were analysed by means of multi-channel modal analysis software. The model properties such as natural frequencies, damping ratios and mode shapes were estimated by carrying out curve fitting of the transfer functions obtained in the vibration tests.

Furthermore, the two-dimensional formulation of fluid-coupled free vibration analysis of the experimental model was presented by using a lumped mass model with the aid of the Boundary Element Method to take into account fluid-structure interaction. By numerical calculation, the coupled natural frequencies and the corresponding mode shapes were obtained. Finally, the results of experimental calculations were compared.

Basis of analytical approach

For the analysis of floating connected structures the finite element method may be used to obtain structural deformations, and equations of motion may be solved using normal coordinates.

The equation of motion can be written as:

$$[M_s]\{\ddot{u}\} + [C_s]\{\dot{u}\} + [K_s]\{u\} = \{F\} \quad (1)$$

where $[M_s]$, $[C_s]$ and $[K_s]$ are the structural mass, damping, and stiffness matrices, respectively, and {F} the vector of nodal forces resulting from the distributed fluid forces.

The generalised fluid forces may be decomposed into three components, these being the generalised restoring, wave exciting, and radiation forces, i.e.

$$\{F^*\} = \{F_H^*\} + \{F_D^*\} + \{F_R^*\} \quad (2)$$

Hence the equation of motion may otherwise be expressed as

$$[-\omega^2([M_s^*] + [M_f^*]) - i\omega([C_s^*] + [C_f^*]) + ([K_s^*] + [K_f^*])]\{p\} = \{F_D^*\} \quad (3)$$

where p is the normal coordinate, $[M_f^*]$ and $[C_f^*]$ are the generated added mass and damping coefficient matrices, and $\{F_H^*\}$ is written as $-[K_f^*]p$ where $[K_f^*]$ is the hydrostatic restoring matrix. It is the latter to which attention is now paid.

THEORETICAL DEVELOPMENT OF HYDROSTATIC STIFFNESS

The forces resulting from stiffness including hydrostatic components for the quasi-static response of a multipli-connected structure with several pontoon modules connected in a single row i may be written as:

$$[K_f(x)].\{x\} = \{F(x)\} \quad (4)$$

and the static equation for any individual pontoon (considered as wall-sided and with symmetry in the y plane) can be written as:

$$\begin{bmatrix} A & 0 \\ 0 & B \end{bmatrix} \cdot \begin{Bmatrix} z \\ \theta \end{Bmatrix} = \begin{Bmatrix} F \\ M \end{Bmatrix} \quad (5)$$

where $A = \rho g L b$ (6)

$B = \rho g L b d GM_l$ (7)

ρ	density of water
g	acceleration due to gravity
L	length of pontoon
L_1	distance between two pontoons
b	breadth of the pontoon
d	draft of pontoon in still water
z	vertical displacement
θ	pitch angle
GM_l	longitudinal metacentric height
F	force acting at centre of gravity of each pontoon
M	moment acting on centre of gravity of each pontoon

Figure 1shows the definition of coordinates used herein.

The static equation for two consecutive pontoons may be written as:

$$\begin{bmatrix} A & 0 & 0 & 0 \\ 0 & B & 0 & 0 \\ 0 & 0 & A & 0 \\ 0 & 0 & 0 & B \end{bmatrix} \cdot \begin{Bmatrix} z_1 \\ \theta_1 \\ z_2 \\ \theta_2 \end{Bmatrix} = \begin{Bmatrix} F_1 \\ M_1 \\ F_2 \\ M_2 \end{Bmatrix} \tag{8}$$

and the expanded equation for n pontoons will be as follows:

$$\begin{bmatrix} A & 0 & 0 & 0 & \cdots & \cdots & \cdots & 0 & 0 & 0 & 0 \\ 0 & B & 0 & 0 & \cdots & \cdots & \cdots & 0 & 0 & 0 & 0 \\ 0 & 0 & A & 0 & \cdots & \cdots & \cdots & 0 & 0 & 0 & 0 \\ 0 & 0 & 0 & B & \cdots & \cdots & \cdots & 0 & 0 & 0 & 0 \\ \cdots & \cdots & \cdots & \cdots & \cdots & \cdots & \cdots & \cdots & \cdots & \cdots & \cdots \\ \cdots & \cdots & \cdots & \cdots & \cdots & \cdots & \cdots & \cdots & \cdots & \cdots & \cdots \\ \cdots & \cdots & \cdots & \cdots & \cdots & \cdots & \cdots & \cdots & \cdots & \cdots & \cdots \\ 0 & 0 & 0 & 0 & \cdots & \cdots & \cdots & A & 0 & 0 & 0 \\ 0 & 0 & 0 & 0 & \cdots & \cdots & \cdots & 0 & B & 0 & 0 \\ 0 & 0 & 0 & 0 & \cdots & \cdots & \cdots & 0 & 0 & A & 0 \\ 0 & 0 & 0 & 0 & \cdots & \cdots & \cdots & 0 & 0 & 0 & B \end{bmatrix} \cdot \begin{Bmatrix} z_1 \\ \theta_1 \\ z_2 \\ \theta_2 \\ \cdots \\ \cdots \\ \cdots \\ z_{n-1} \\ \theta_{n-1} \\ z_n \\ \theta_n \end{Bmatrix} = \begin{Bmatrix} F_1 \\ M_1 \\ F_2 \\ M_2 \\ \cdots \\ \cdots \\ \cdots \\ F_{n-1} \\ M_{n-1} \\ F_n \\ M_n \end{Bmatrix} \tag{9}$$

According to Fig 1 the constraint relationship for hinged-connections for the above system is:

$$z_i + \left(\frac{L+L_1}{2}\right)\theta_i + \left(\frac{L+L_1}{2}\right)\theta_{i+1} = z_{i+1} \tag{10}$$

or

$$z_i = 0 - \left(\frac{L+L_1}{2}\right)\theta_i + z_{i+1} - \left(\frac{L+L_1}{2}\right)\theta_{i+1} \tag{11}$$

where: $j = i+1,\ \ k = j+1,\ \ \ell = k+1$

$$\alpha = 0;\ \beta = -\frac{L+L_1}{2};\ \gamma = 1;\ \psi = -\frac{L+L_1}{2}$$

To insert the constraint equation (10) into the system equation, this system must be first modified. The system K*X=F may be rewritten in the following way:

$$\sum_{s\neq i,j,k,\ell} K_{rs}x_s + K_{ri}x_i + K_{rj}x_j + K_{rk}x_k + K_{r\ell}x_\ell = p_r \tag{12}$$

where r is any row of the matrix K.

Introducing equation (11) in the place of z_i thus:

$$K_{rs}x_s + K_{ri}(\alpha + \beta x_{i+1} + \gamma x_{i+2} + \psi x_{i+3}) + K_{ri+1}x_{i+1} + K_{ri+2}x_{i+2} + K_{ri+3}x_{i+3} = p_r \tag{13}$$

$$K_{rs}x_s + (\beta K_{ri} + K_{ri+1})x_{i+1} + (\gamma K_{ri} + K_{ri+2})x_{i+2} + (\psi K_{ri} + K_{ri+3})x_{i+3} = p_r - \alpha K_{ri} = q_r \tag{14}$$

The above operation has removed the symmetry of the coefficient matrix K. This can be re-instated by multiplying the ith-row (14) by β, γ, ψ in turn and adding to the j, k, ℓth-rows respectively to obtain:

jth-row

$$\begin{aligned} &(\beta K_{js} + K_{js})x_s + (\beta^2 K_{ii} + \beta K_{ij} + \beta K_{ji} + K_{jj})x_j \\ &+(\beta\gamma K_{ii} + \beta K_{ik} + \gamma K_{ji} + K_{jk})x_k \\ &+(\beta\psi K_{ii} + \beta K_{i\ell} + \psi K_{ji} + K_{j\ell})x_\ell = \beta q_i + q_j \end{aligned} \tag{15}$$

kth-row

$$\begin{aligned} &(\gamma K_{ks} + K_{ks})x_s + (\gamma\beta K_{ii} + \gamma K_{ij} + \beta K_{ki} + K_{kj})x_j \\ &+(\gamma^2 K_{ii} + \gamma K_{ik} + \gamma K_{ki} + K_{kk})x_k \\ &+(\gamma\psi K_{ii} + \gamma K_{i\ell} + \psi K_{ki} + K_{k\ell})x_\ell = \gamma q_i + q_k \end{aligned} \tag{16}$$

ℓth-row

$$\begin{aligned} &(\psi K_{\ell s} + K_{\ell s})x_s + (\psi\beta K_{ii} + \psi K_{ij} + \beta K_{\ell i} + K_{\ell i})x_j \\ &+(\gamma\psi K_{ii} + \psi K_{ik} + \gamma K_{\ell i} + K_{\ell k})x_k \\ &+(\psi^2 K_{ii} + \psi K_{i\ell} + \psi K_{\ell i} + K_{\ell\ell})x_\ell = \psi q_i + q_\ell \end{aligned} \tag{17}$$

The matrix regains its symmetry. In the above matrix there is no i column, it is possible then to insert the constraint equation (11) in place of the ith-row. Thus:

$$-x + \beta x_j + \gamma x_k + \psi x_\ell = -\alpha$$

or

$$-x_i K_{ii} + \beta K_{ii}x_j + \gamma K_{ii}x_k + \psi K_{ii}x_\ell = -\alpha K_{ii} \tag{18}$$

Remembering that the stiffness matrix k must be symmetrical, we have to complete the i-column with $\beta K_{ii}x_i$ in the j-row:

$$0 = \beta K_{ii}x_i - \beta K_{ii}x_i$$

$$0 = \beta K_{ii}x_i - \beta K_{ii}(\alpha + \beta x_j + \gamma x_k + \psi x_\ell)$$

$$0 = \beta K_{ii}x_i - \beta^2 K_{ii}x_j - \beta\gamma K_{ii}x_k - \psi K_{ii}x_\ell - \alpha\beta K_{ii} \tag{19}$$

Finally the corrected rows have the following form:

jth-row

$$\begin{aligned} &(\beta K_{is} + K_{js})x_s + \beta K_{ii}x_i + (\beta K_{ij} + \beta K_{ji} + K_{jj})x_j \\ &+(\beta K_{ik} + \gamma K_{ji} + K_{jk})x_k \\ &+(\beta K_{i\ell} + \psi K_{ji} + K_{j\ell})x_\ell = \beta q_i + q_j + \alpha\beta K_{ii} \end{aligned} \tag{20}$$

kth-row

$$\begin{aligned} &(\gamma K_{is} + K_{ks})x_s + \gamma K_{ii}x_i + (\gamma K_{ij} + \beta K_{ki} + K_{kj})x_j \\ &+(\gamma K_{ik} + \gamma K_{ki} + K_{kk})x_k \\ &+(\gamma K_{i\ell} + \psi K_{ki} + K_{k\ell})x_\ell = \gamma q_i + q_k + \alpha\gamma K_{ii} \end{aligned} \tag{21}$$

ℓth-row

$$(\psi K_{is} + K_{\ell s})x_s + \psi K_{ii} x_i + (\psi K_{ij} + \beta K_{\ell i} + K_{\ell j})x_j$$
$$+(\psi K_{ik} + \gamma K_{\ell i} + K_{\ell k})x_k \qquad (22)$$
$$+(\psi K_{i\ell} + \psi K_{\ell i} + K_{\ell\ell})x_\ell = \psi q_i + q_\ell + \alpha\psi K_{ii}$$

Combining equations (20), (21), and (22) with equation (8) the system with two pontoons will be:

$$\begin{bmatrix} A & 0 & 0 & 0 \\ 0 & B & 0 & 0 \\ 0 & 0 & A & 0 \\ 0 & 0 & 0 & B \end{bmatrix} \cdot \begin{Bmatrix} z_1 \\ \theta_1 \\ z_2 \\ \theta_2 \end{Bmatrix} = \begin{Bmatrix} F_1 \\ M_1 \\ F_2 \\ M_2 \end{Bmatrix} \qquad (23)$$

From equations (18, 20, 21, and 22) we obtain the i, j, k, and ℓth-rows respectively:

$$-Az_1 + \beta A\theta_1 + \gamma Az_2 + \psi A\theta_2 = -\alpha A \qquad (24)$$

$$\beta Az_1 + B\theta_1 + 0 + 0 = M_1 + \beta F_1 + \alpha\beta A \qquad (25)$$

$$\gamma Az_1 + 0 + Az_2 + 0 = F_2 + \gamma F_1 + \alpha\gamma A \qquad (26)$$

$$\psi Az_1 + 0 + 0 + B\theta_2 = M_2 + \psi F_1 + \alpha\psi A \qquad (27)$$

Equations (24 to 27) can be written in the following form:

$$\begin{bmatrix} -A & \beta A & \gamma A & \psi A \\ \beta A & B & 0 & 0 \\ \gamma A & 0 & A & 0 \\ \psi A & 0 & 0 & B \end{bmatrix} \cdot \begin{Bmatrix} z_1 \\ \theta_1 \\ z_2 \\ \theta_2 \end{Bmatrix} = \begin{Bmatrix} -\alpha A \\ M_1 + \beta F_1 + \alpha\beta A \\ F_2 + \gamma F_1 + \alpha\gamma A \\ M_2 + \psi F_1 + \alpha\psi A \end{Bmatrix} \qquad (28)$$

So that the equation for a system with three pontoons will be:

$$\begin{bmatrix} -A & \beta A & \gamma A & \psi A & 0 & 0 \\ \beta A & B & 0 & 0 & 0 & 0 \\ \gamma A & 0 & A & 0 & 0 & 0 \\ \psi A & 0 & 0 & B & 0 & 0 \\ 0 & 0 & 0 & 0 & A & 0 \\ 0 & 0 & 0 & 0 & 0 & B \end{bmatrix} \cdot \begin{Bmatrix} z_1 \\ \theta_1 \\ z_2 \\ \theta_2 \\ z_3 \\ \theta_3 \end{Bmatrix} = \begin{Bmatrix} -\alpha A \\ M_1 + \beta F_1 + \alpha\beta A \\ F_2 + \gamma F_1 + \alpha\gamma A \\ M_2 + \psi F_1 + \alpha\psi A \\ F_3 \\ M_3 \end{Bmatrix} \qquad (29)$$

From equations (18, 20, 21, and 22) we obtain the i, j, k, and ℓth-rows respectively:

$$-Az_2 + BA\theta_2 + \gamma Az_3 + \psi A\theta_3 = -\alpha A \qquad (30)$$

$$(\gamma\beta A + \psi A)z_1 + 0 + \beta Az_2 + B\theta_2 + 0 + 0 =$$
$$(M_2 + \psi F_1 + \alpha\psi A) + (F_2 + \gamma F_1 + \alpha\gamma A) \qquad (31)$$

$$\gamma^2 Az_1 + 0 + \gamma Az_2 + 0 + Az_3 + 0 =$$
$$F_3 + \gamma(F_2 + \gamma F_1 + \alpha\gamma A) \qquad (32)$$

$$\gamma\psi Az_1 + 0 + \psi Az_2 + 0 + 0 + B\theta_3 =$$
$$M_3 + \psi(F_2 + \gamma F_1 + \alpha\gamma A) \qquad (33)$$

With consideration to the symmetry of the matrix (modifying sth row with sth column) and equation (30 to 33) the equation for a system with three pontoons can be written as following form:

$$\begin{bmatrix} -A & \beta A & 0 & \gamma\beta A + \psi A & \gamma^2 A & \gamma\psi A \\ \beta A & B & 0 & 0 & 0 & 0 \\ 0 & 0 & -A & \beta A & \gamma A & \psi A \\ \gamma\beta A + \psi A & 0 & \beta A & B & 0 & 0 \\ \gamma^2 A & 0 & \gamma A & 0 & A & 0 \\ \gamma\psi A & 0 & \psi A & 0 & 0 & B \end{bmatrix} \cdot \begin{Bmatrix} z_1 \\ \theta_1 \\ z_2 \\ \theta_2 \\ z_3 \\ \theta_3 \end{Bmatrix} =$$

$$\begin{Bmatrix} -\alpha A \\ M_1 + \beta F_1 + \alpha\beta A \\ -\alpha A \\ (M_2 + \psi F_1 + \alpha\psi A) + (F_2 + \gamma F_1 + \alpha\gamma A) \\ F_3 + \gamma(F_2 + \gamma F_1 + \alpha\gamma A) \\ M_3 + \psi(F_2 + \gamma F_1 + \alpha\gamma A) \end{Bmatrix} \qquad (34)$$

The above operation is applicable only for hinged-connections. If other constraints exist the above operations must be repeated for the new constraint with appropriate coefficients as follows:

For a rigid connection:

$$\theta_i = \theta_{i+1} \qquad (35)$$

and for a hinged-rigid connection:

$$-\varepsilon \geq \theta_{i+1} - \theta_i \geq \varepsilon \qquad (36)$$

where ε represents the angle for which the connection remains hinged.

This behaviour has been included by the authors in a suitable subroutine to enable inclusion of these forms of constraints, ie:

A hinge connection
B elastic connection
C rigid connection
D hinge-rigid connection in one direction
E hinge-rigid connection in two directions

ANALYTICAL RESULTS

These are presented at model scale to enable validation.

The analytical results of case D are shown in Fig 2. Figure 2(a) shows the general displacement of pontoons with different loads. The differences of two adjacent pitch angles are illustrated in Fig 2(b) in which it is clearly shown when this difference reaches the design threshold angle it remains constant (in one direction). The pitch angle and vertical displacement (heave) of any individual pontoon against load increment can be seen in Fig 2(c) and Fig 2(d) respectively.

The results of case E are shown in Fig 3. The explanation is similar to case D as shown in Fig 2, but in this case when the difference of two adjacent pitch angles reaches the design threshold angle this difference remains constant in both directions as shown in Fig 3(b). As a consequence a significantly different performance from the single direction locking case D is observed.

The analytical development includes for different connections, and it is generally shown to be good for hinge, elastic, rigid and hinge-rigid connections. However some difficulties arise in hinge-rigid connections. According to the method adopted, when the difference of pitch angle between two adjacent pontoons reaches the design threshold angle a change is made from hinge to rigid connection. The difficulty is that the current method is not always able to change rigid connections to hinge connections as appropriate (multiple states may exist). This is shown for example in Fig 3 when the joints 2 and 9 will be locked (of course in the opposite direction to joints 4, 5, 6 and 7) the joints 3 and 8 must change from rigid to hinge connections. However these joints (3 and 8) remain locked and as a result pontoons 1 and 9 rise and consequently since load and buoyancy must remain in equilibrium at all times greater displacement in pontoons 3, 4, 5, 6 and 7 results as is clearly shown in Fig 3(a and d).

EXPERIMENTAL WORK

Pontoon Model

The pontoon model used in this project was constructed in the workshop of the Department of Mechanical Engineering, University College London. The design was notionally 1:100 scale. The model was made of wooden hollow boxes (with lead ballast), of length 0.605 m, width 0.280 m, and height 0.200 m. Connections were designed to allow any pontoon rotations up to a specified value without contact to the adjacent pontoon.

Potentiometers - Data Logger

Potentiometers attached to ends of the model pontoons by a fine nylon line running over pulleys were used to measure the change in pitch and heave. The signals from these were connected to a PC with an Amplicon PC-126 analogue to digital conversion data acquisition card installed. The proprietary Waveview software was used to view the data stored before exporting for further analysis.

EXPERIMENTAL RESULTS

The experiments were divided into three test series. In each case the module connections were all of the hinge-rigid type, able to lock in two directions of rotation:

First test series

This consisted of five pontoons with specific free rotation between two adjacent pontoons in different load conditions and load increment as follows:

test	pontoon loaded	Figures
1A	1	4
1B	2	5
1C	2,4	6
1D	3	7

load applied in centre of gravity of pontoon indicated in 1kg increments
loaded pontoons numbered from end

Each above Figs 4 includes four parts a, b, c and d.

(a) displacement of joints in different load conditions
(b) difference of pitch angle of any two adjacent pontoons
(c) pitch displacement with increasing load of any individual pontoon
(d) heave displacement with increasing load of any individual pontoon

Figs 5 to 7show the only (a) and (b).

The results of test 1A where the load is applied on pontoon 1 are explained here as an example. Fig 4a shows the displacement of the joints along the length of the mutipli-connected structure. This and Figures 4(a) to 7(a) show that the displacement of the joint nearest to the loading point is always of greater magnitude than the displacement of the other joints.

Fig 4b shows the difference in rotations between adjacent pontoons and is a critical measure of the pitch measurement accuracy. Owing to manufacturing imperfections the expected results cannot be observed as clearly as the theoretical model predicts, but the pattern expected does exist and the difference in rotations increases with load increments near the loading point up to a certain threshold value beyond which there is no further increase in the difference in rotations. This threshold value is the maximum allowable free rotation at the joints.

Figs 4c and 4d show the pitch and heave displacement of each pontoon with increasing load. These figures show that the heave displacement of the loaded pontoon and the adjacent pontoon increase with increasing load. Heave and pitch displacements of the other pontoons do not necessarily follow the same pattern. The same conclusions can be drawn from Figs 5 to 7.

Second test series

In this case the value of θ was increased by increasing the looseness of the hinge-rigid connections compared with the first series. The case shown here (test 2, Fig 8) is for the same load condition and variation as test 1D, (Fig 7).

A significantly different behaviour is noted in displacement at the ends of the model Figs 7a and 8a, which is echoed in the pitch difference Figs 7b and 8b.

The differences in displacements between these two tests with similar load conditions and load increments are illustrated in Fig 9 showing that in this configuration there is greater displacement with greater looseness.

Third test series

This consisted of nine pontoons with some free rotation, which was the same as for the first test, and also with different load conditions as follows:

test	loaded pontoon(s) and loads	Figs
3A	2 (1kg increments)	
3B	3 (1kg increments)	
3C	4 (1kg increments)	
3D	5 (1kg increments)	10
3E	3, 7 (1kg increments)	
3F	3 (12 kg), 7 (1kg increments)	
3G	4, 8 (1kg increments)	
3H	4 (12 kg), 8 (1kg increments)	

load applied in centre of gravity of pontoon indicated in 1kg increments
loaded pontoons numbered from end

Only the results of a test case 3D where the load is applied on the central pontoon 5 are described herein, it is analogous to the 5 module case shown in case 1D (Fig 7), for which a similar response is seen. Fig 10a shows the displacement of the joints along the length of the mutipli-connected structure. This shows that the displacement of the joint nearest to the loading point has greater magnitude than the displacement of the other joints.

Fig 10b shows the difference in rotations between adjacent pontoons, the magnitude of which increase with increasing load increments up to a certain threshold value after which no further increase in the difference in rotations occurs.

Fig 10c shows the pitch displacement of each pontoon with increasing load. This figure shows that there is no pitch displacement in the loaded pontoon in this particular case (owing to symmetry) and pitch of the adjacent pontoon increases with increasing load.

Figs 10d shows the changing heave displacement of each pontoon with increasing load, and in particular that the heave displacement of the loaded pontoon and the adjacent pontoon increase with increasing load. Heave displacement of pontoons 2 and 8 are negligible and heave displacement of pontoons 1 and 9 rise up.

Comparison is made between the analytical model (Fig 3) and experiment (Fig 10a) for the case with the maximum load of 15kg at module 5 in Fig 11, and for variation in load at this module in Fig 12. Within the bounds of experimental accuracy good agreement is found.

CONCLUSIONS

The stiffness matrix formed has been shown to be adequate for all conditions for fully hinged, elastic, and rigid connections (shown herein for the linear array), and recently validated (but not reported herein) for the mat array. The methodology developed for the hinge-rigid connection has been shown to be adequate in most cases. However, an interesting condition arises in that different lock-unlock sequences take place for the upper and lower connections during loading variations. At present certain sequences are difficult to predict.

It is intended that this work be extended initially to quasi-static dynamic problems, and then to full dynamic wave loading for the mat form of array.

REFERENCES

1. 'Wave Energy- a design challenge', Ed Ronald Show, publ Ellis Horwood

2. 'Power Form Sea Waves', Ed B Count, publ Academic Press

3. 'Ocean Wave Energy Conversion', Ed Michael E. McCormick.

4. Adee, B H and Martin, W, 'Theoretical Analysis of Floating Breakwater Performance', 1974 Floating Breakwater conference papers, University of Rhode Island. Marine Technical Report Series Number 24, pp 21 - 39, 1974

5. Adee, B H, 'Analysis of Breakwater Performance', Proc symposium on modelling techniques, San Francisco, Calif, 3-5 September 1075a

6. Adee, B H., ' A Review of Developments and Problems in Using Floating Breakwaters', Offshore Technology Conference, Houston, Texas, May 1976

7. Adee, B H, 'Floating Breakwater Performance', Coastal Engineering, Chapter 159, Page 2777, 1976

8. 'Floating Ports Designed Construction Practices', Ed Gregory P. Tsinker

9. Piskorski R, ' Dynamics of an open zone of floating bridge', Polish Academy of Sciences, Gdansk Branch, Marine Technology Transactions, Vol. 2, pp 103-122, 1991

10. Xiling Che, Dayun Wang, Minglun Wang, and Yingfan Xu, 'Two Dimensional Hydroelastic Analysis of Very Large Floating Structures', Marine Technology, Vol. 29, No. 1, pp 13-24, Jan 1992

11. Takaki M, and Tango Y, 'Wave Drifting Forces on Multiple Connected Floating Structures', Hydroelasticity in Marine Technology, Ed Faltinsen O et al, p 403

12. Endo R, Kato T, Hamamoto T, Hayashi T, Tasaka, N, 'Experimental Modal Analysis on Unit-Linked Large Floating Structure Models', Proc of the Fifth International Offshore and Polar Engineering Conference, The Hague, The Netherlands, June 11-16, 1995

13. Wang S , Ertekin R C, van Stiphout A T F M, and Ferier P G P, 'Hydroelastic Response Analysis of a Box-link Floating Airport of Shallow Draft', Proc of the Fifth International Offshore and Polar Engineering Conference, The Hague, The Netherlands, June 11-16, 1995

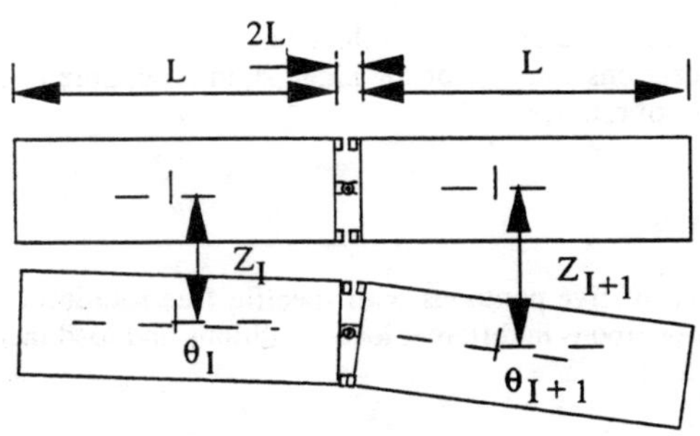

Fig 1 Hinge-rigid connections and local coordinates

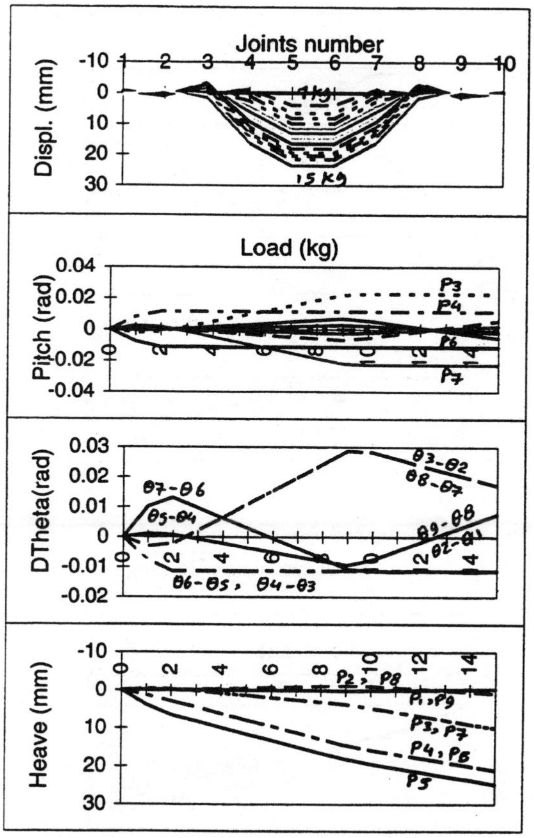

Fig 2 Theoretical response -- Load applied at pontoon (5) Hinged --rigid in one direction

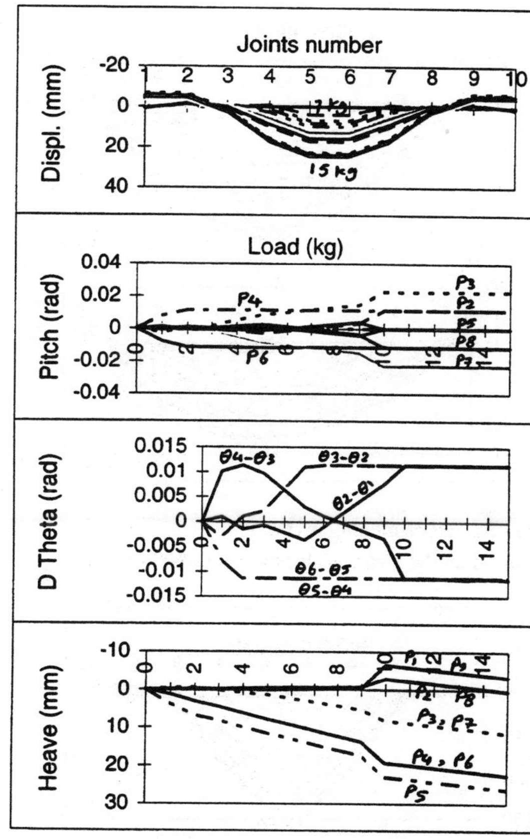

Fig 3 Theoretical response -- Load applied at pontoon (5) Hinged - -rigid in both direction

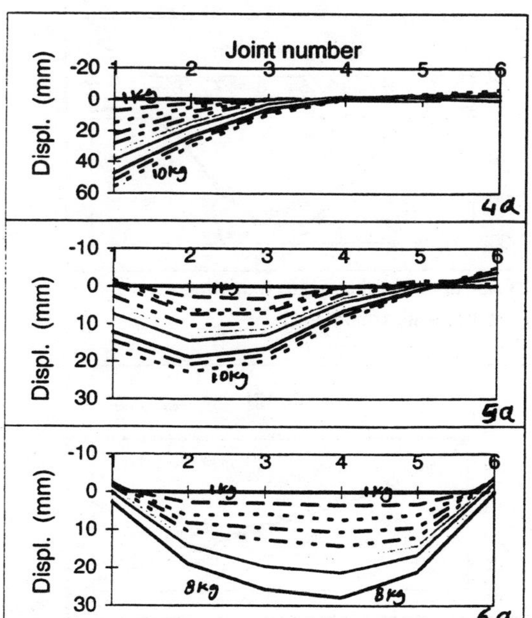

Fig 4a - 6a Experimental results (5 pontoon model) various locations of loading

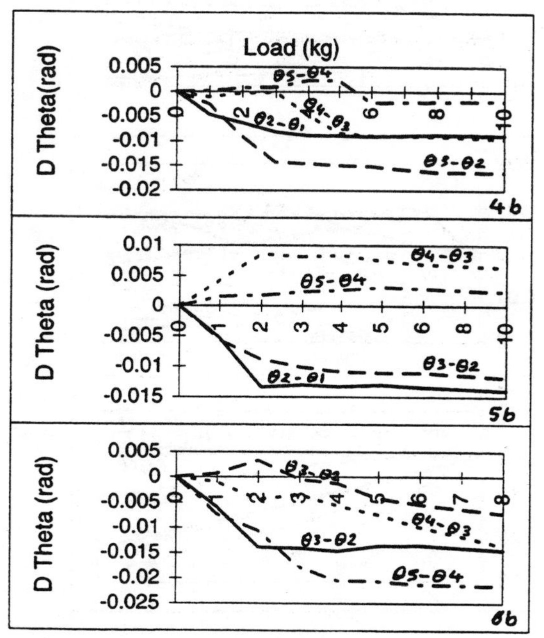

Fig 4b - 6b Experimental results (5 pontoon model) various locations of loading

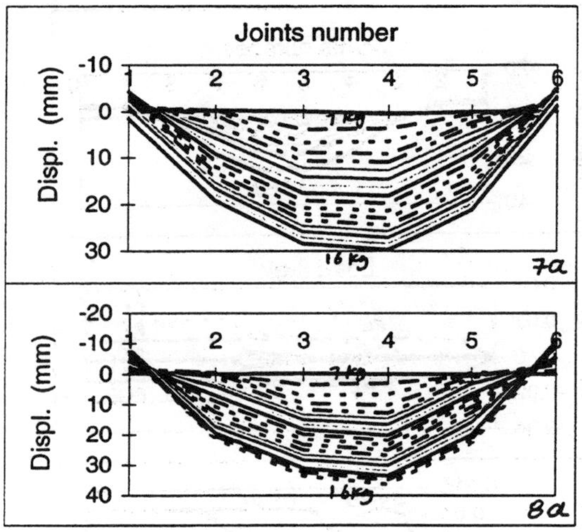

Fig 7a - 8a Experimental results
(5 pontoon model) various locations of loading

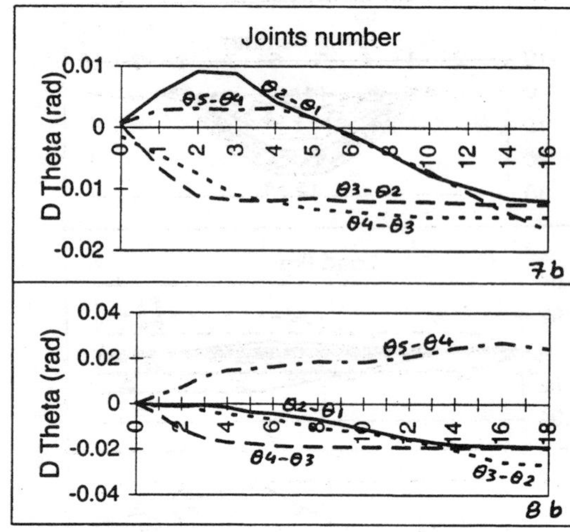

Fig 7b - 8b Experimental results
(5 pontoon model) various locations of loading

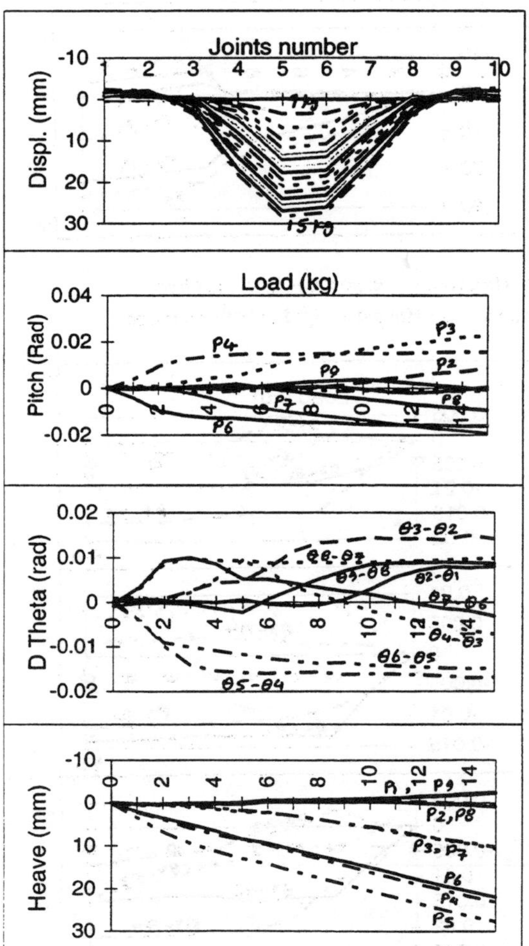

Fig 10 Experimental results 9 pontoon model
Load applied at central pontoon 5 - of Fig 3

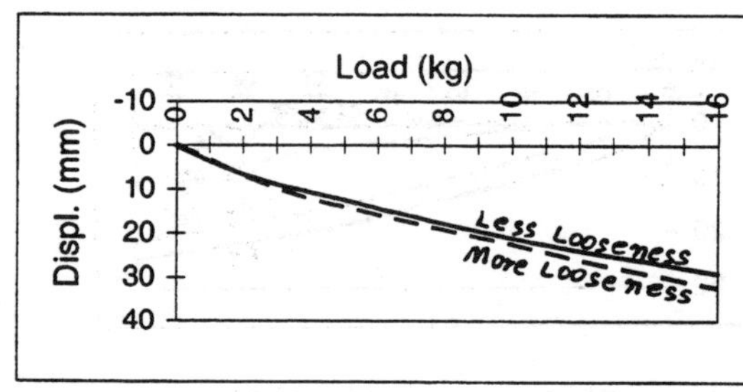

Fig 9 Effect of ε from figs 7a and 8a

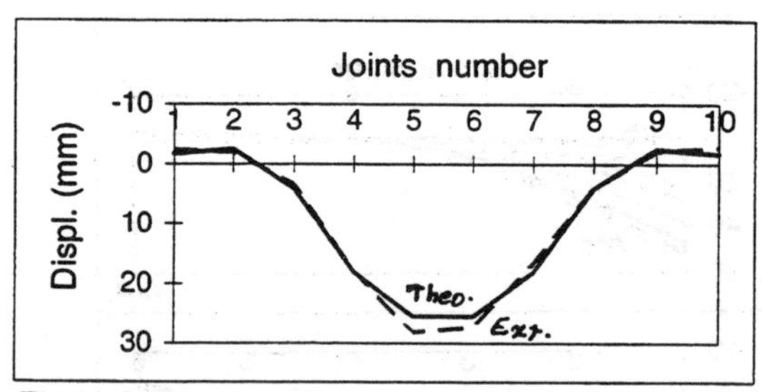

Fig 11 Comparison of theory and experiment
(Fig 3 and 10 for maximum load)

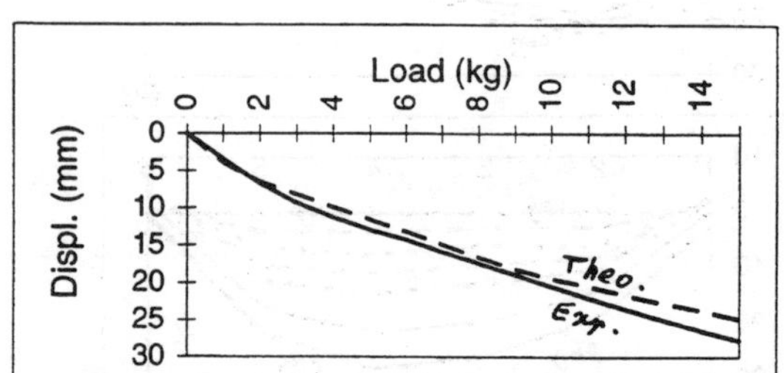

Fig 12 Comparison of theory and experiment
(Fig 3 and 10 for central pontoon)

Proceedings of the Seventh (1997) International Offshore and Polar Engineering Conference
Honolulu, USA, May 25-30, 1997

ISBN 1-880653-28-1 (Set); ISBN 1-880653-31-1 (Vol. III)

Dynamic Responses of a Spar-Buoy Boring Derrick in Waves

Koji Otsuka and Yoshiho Ikeda
Osaka Prefecture University, Sakai, Japan
Mikio Fukutomi
OYO Corporation, Tokyo, Japan
Hiroshi Aso
Zeni Lite Buoy Corporation, Okayama, Japan

ABSTRACT

Dynamic responses of a Spar-Buoy Boring Derrick (SBBD), which optionally has a stabilizer system, in regular waves are investigated. We measured hydrodynamic forces on an sinusoidally oscillating stabilizer model in still water, and responses of a SBBD model in regular waves. We also performed time-domain numerical simulations of the dynamic responses of a SBBD considering drag and inertia coefficients for each swing. Both experimental and numerical results for a 50 m-type SBBD demonstrate that the effect of the stabilizer on the pitch motion amplitude is not significant, because the setting point of the stabilizer is deep, and the length of the stay (lever of the damping moment) is relatively short comparing with the scale of the main body. However, the simulated results for a 16 m-type SBBD show that the pitch motion amplitude is reduced to 20 ~ 25% due to the effect of the stabilizer, since the length of the stay (lever of the damping moment) is relatively large comparing with the scale of the main body. The simulated results in the condition including a low-frequency and small-amplitude wave suggest that the effect of the hydrodynamic coefficients changing swing by swing is significant when the motion of the buoy includes a low-frequency component.

KEY WORDS: Spar buoy, dynamic response, viscous force, low-frequency motion

INTRODUCTION

Steel frame tower or guide tower platform has been commonly used for boring and solid investigation on site for less than 15 m depth. However, the use of these platforms involves some difficulties in installation and operation on site for higher than 15 ~ 20 m depth. For example, when the steel frame tower is installed, long anchor wires, which sometimes become an obstacle for passing ships, have to be set up to support the platform. Furthermore, a larger scale steel frame platform requires a larger scale floating crane and work ships for the installation.

A spar buoy platform, which is usually utilized for navigational marking, is relatively easy to install and operate in comparison with the steel frame and guide tower platforms, because it requires no anchor wires and is easily towed to the installation site. Fukutomi et al. (1995) proposed a new concept of a spar buoy platform, which is named "Spar-Buoy Boring Derrick (SBBD)", for boring and soil investigation on site for higher than 15 m depth (see Fig. 1). The SBBD mainly consists of stage, center pipe, buoyancy chamber, mooring chain, and sinker. The buoyancy chamber is located near the free surface, and provides a sufficient buoyancy to keep the system in a stable and upright position. They also proposed a new-type stabilizer system, which consists of stay and cone-shaped fin (see Fig.2), to reduce the steady inclination and the responses of SBBD in waves and current.

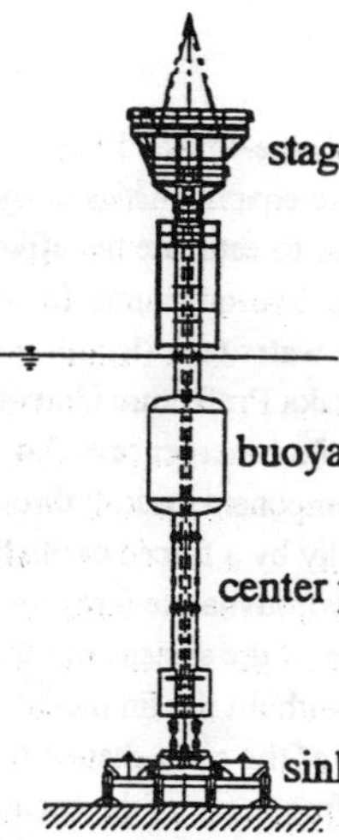

Fig. 1 Composition of basic-type SBBD

Prior to the development of the SBBD, we have to investigate the characteristics of hydrodynamic forces on the stabilizer and the dynamic responses of the SBBD in waves. Some recent researches on a spar platform (Mekha et al., 1996, Ran et al., 1996) have proposed time-domain simulation methods to estimate its motion in waves. Most of the simulation methods have used constant hydrodynamic coefficients relating to viscous forces.

However, the drag and added-mass coefficients depend on the Kuelegan-Carpenter number (Kc) and Reynolds number (Re) for the relative motion of the buoy and water particles. The accuracy of the hydrodynamic coefficients will become more important when the buoy motion includes low frequency component.

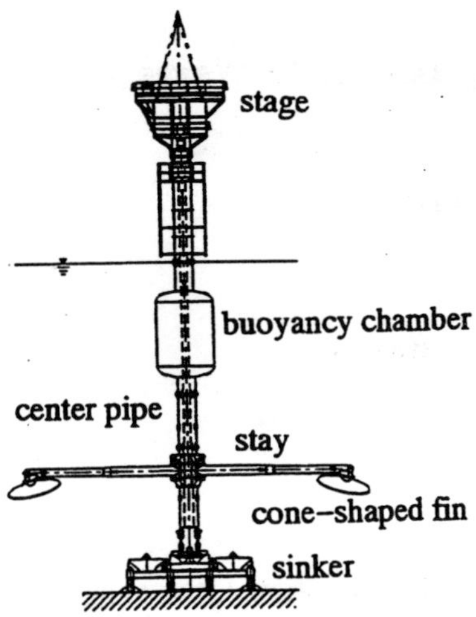

Fig. 2 Composition of SBBD with stabilizer

In this study, we investigated the characteristics of hydrodynamic forces on the stabilizer and the dynamic responses of a SBBD in regular waves. We measured hydrodynamic forces on an sinusoidally oscillating stabilizer model in still water, and responses of a SBBD model in regular waves. We also performed time-domain numerical simulations of the dynamic responses of the SBBD considering drag and inertia coefficients which are functions of Kc and Re numbers for the relative motion of the buoy and water particles.

EXPERIMENTS

Hydrodynamic Forces on Cone-Shaped Fin

It is necessary to clarify the characteristics of hydrodynamic coefficients of a cone-shaped fin, in order to estimate the effect of the stabilizer on the buoy motion. We measured hydrodynamic forces on a cone-shaped fin model (see Fig. 3) in a small water tank (length × width × depth = 4.0 m × 0.56 m × 1.5 m) in Osaka Prefecture University (see Fig. 4). The fin model is deeply submerged (submergence = 0.4 m) to avoid free-surface effect and connected to a 3-component loadcell through a thin strut. They are forced to oscillate sinusoidally by a forced oscillation mechanism mounted on the top of the tank. The hydrodynamic force on the fin model is obtained by subtracting the mass force of the system and the hydrodynamic force on the strut, which is measured without the fin model, from the total force. The diameter (D_F) and height (H_F) of the cone-shaped fin are 0.064 and 0.012 m, respectively. The circular frequency of the oscillation (ω) is 3.49 rad/sec and constant. The amplitude of the oscillation (Y_a) is varied from 0.01 to 0.20 m. The angle of attack of the fin (α) is set for 0, 30, 60 and 90 degrees.

The drag and added-mass coefficients (C_{DF} and C_{AF}) are obtained from the first order terms of Fourier expansion of the measured hydrodynamic force on the fin as follows.

$$C_{DF} = \frac{F_D}{\frac{4}{3\pi}\rho\frac{\pi}{4}D_F^2(Y_a\omega)^2} \qquad (1)$$

$$C_{AF} = \frac{F_A}{\rho\frac{\pi}{4}D_F^2 H_F Y_a\omega^2} \qquad (2)$$

where ρ denotes the density of water, F_D and F_A are the drag and added-mass force amplitudes, respectively.

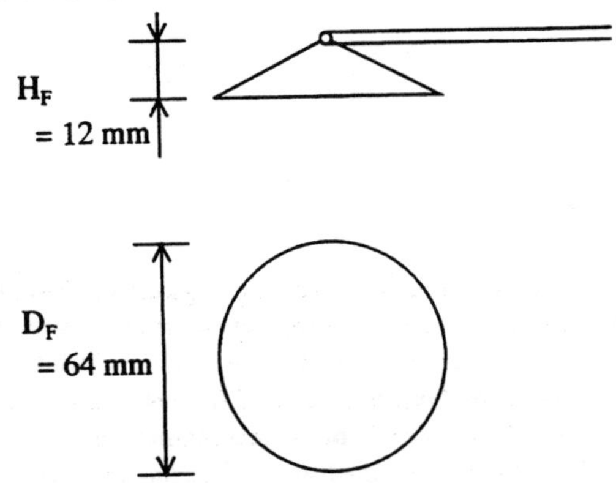

Fig. 3 Test model of cone-shaped fin

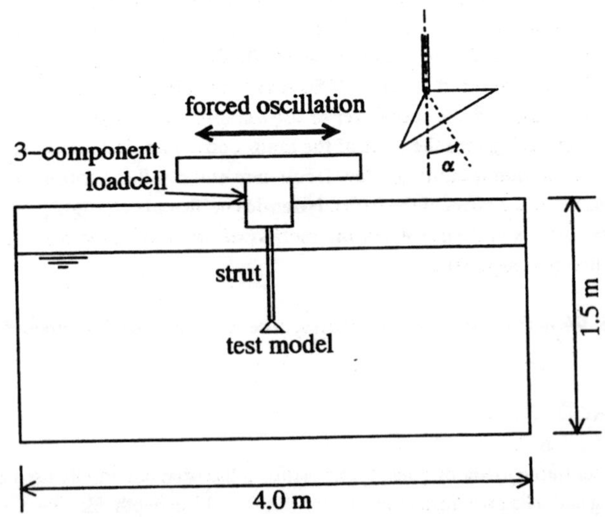

Fig. 4 Experimental arrangement for hydrodynamic force measurement of cone-shaped fin

Figures. 5 and 6 illustrate the obtained drag and added-mass coefficients, respectively. Both drag and added-mass coefficients increase with increasing the angle of attack. They also have some scattering associated with the oscillation amplitudes, even so they mainly depend on the angle of attack.

Then, the drag and added-mass coefficients are approximated by linear equations relating to the angle of attack, which are shown by the solid lines in Figs. 5 and 6. The approximate equations are input into a computer database, which is employed to calculate the hydrodynamic force on the fin when the buoy motion is simulated.

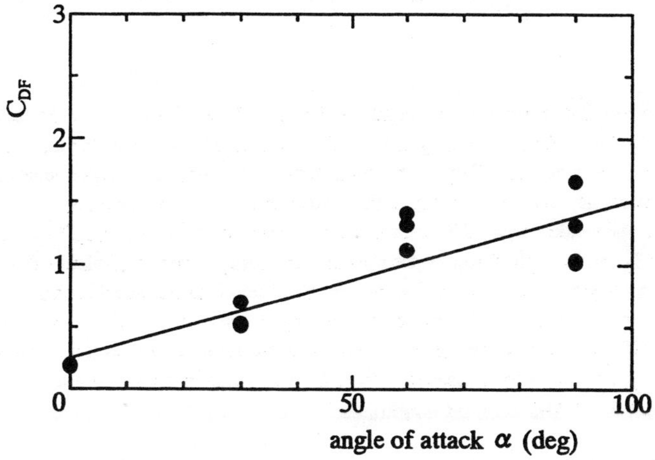

Fig. 5 Drag coefficient of cone-shaped fin

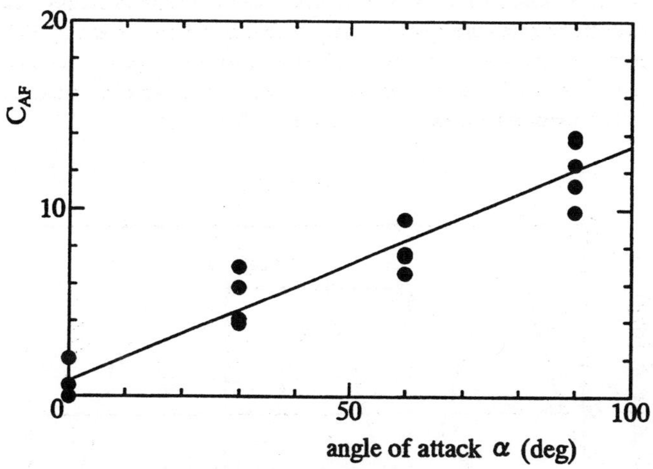

Fig.6 Added-mass coefficient of cone-shaped fin

Motion of SBBD in Regular Waves

The motion measurements of the SBBD in regular waves were carried out in a tank (length × width × depth = 60.2 m × 4.0 m × 5.4 m) in West Japan Fluid Engineering Laboratory Co., Ltd.. The tank is equipped with a wave maker which can generate regular and irregular waves, and the water depth can be adjusted to the experimental conditions. In the motion measurements, we used a 1/50 scaled test model of an actual SBBD for 50 m of water depth. The particulars of the 50 m-type SBBD (actual dimensions) are shown in Table 1. The test model is mounted on the middle of the tank. The motions of two markers in a steady state, which are put on the center pipe of the model, are recorded by a CCD camera through an observation window (see Fig.7). The pitch motion of the model can be obtained by analyzing the markers motion using an image processing technique (MVA-2000, EMTECH Co., Ltd.). In these experiments, the accuracy of the analysis is ± 0.1 degree.

Table 1 Particulars of 50 m-type SBBD

Main body	
length of main body (m)	55.6
sea bottom to mooring point (m)	2.5
diameter of center pipe (m)	1.016
length of buoyancy chamber (m)	4.884
diameter of buoyancy chamber (m)	3.218
mooring point to bottom of buoyancy chamber (m)	37.0
total mass of main body (kg)	5527
mooring point to center of gravity (m)	37.9
Stabilizer	
mooring point to connecting point (m)	7.7
length of stay (m)	12.0
diameter of stay (m)	0.45
height of fin (m)	0.58
diameter of fin (m)	3.211
attached angle of fin (deg)	20
total mass of stabilizer (kg)	498.5

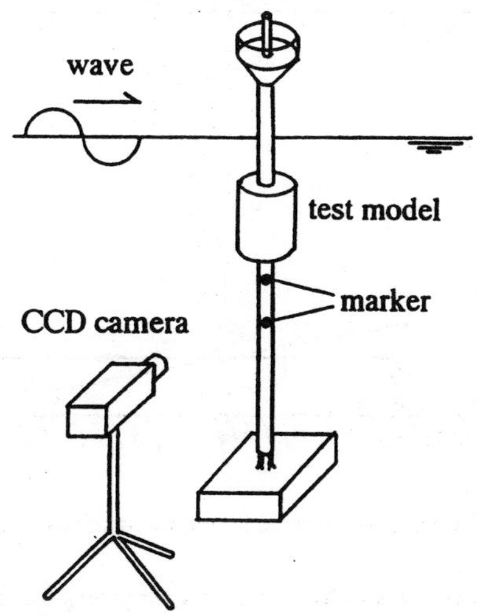

Fig.7 Experimental arrangement for motion measurement of SBBD

Table 2 Experimental conditions for motion measurement of SBBD

water depth h (m) (actual dimension)	stabilizer option	wave period T (sec) (actual time)	wave height H (m) (actual dimension)
1.0 (50)	—	0.566 (4.0)	0.02 (1.0)
		0.849 (6.0)	0.02 (1.0)
		0.990 (7.0)	0.06 (3.0)
		1.27 (9.0)	0.06 (3.0)
		1.41 (10.0)	0.12 (6.0)
		1.70 (12.0)	0.12 (6.0)
	○	0.566 (4.0)	0.02 (1.0)
		0.849 (6.0)	0.02 (1.0)
		0.990 (7.0)	0.06 (3.0)
		1.27 (9.0)	0.06 (3.0)
		1.41 (10.0)	0.12 (6.0)
		1.70 (12.0)	0.12 (6.0)

The experiments for the 50 m-type SBBD with and without stabilizer were carried out as shown in Table 2. The wave period (T) is varied from 0.566 to 1.70 sec, which correspond to 4.0~12.0 sec in actual time scale. The wave height (H) is varied from 0.02 to 0.12 m, which correspond to 1.0~6.0 m in actual dimension.

NUMERICAL SIMULATIONS

Motion Equation

Considering two-dimensional motion, we choose a coordinate system where the origin is at the mooring point of the buoy, and O-x_0 and O-z_0 coincide with the horizontal and vertical axes, respectively (see Fig. 8). The water depth and the height of the origin from sea bottom are represented by h and h_0, respectively. Waves come from left to right, and the pitch angle (θ) is defined by the angle between O-z_0 and O-z (the axis of the main body), where θ indicates a positive value in the clockwise case.

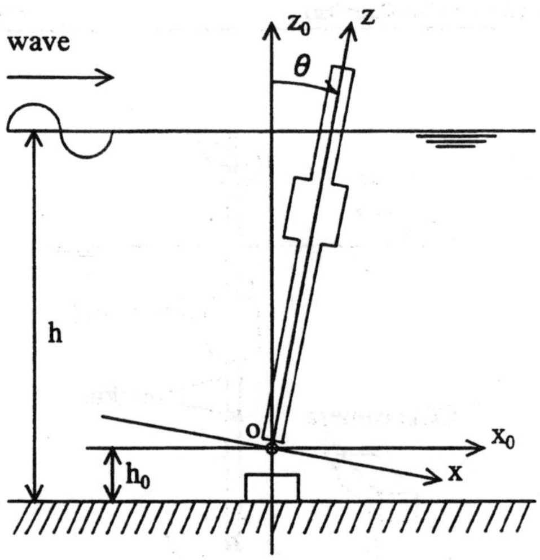

Fig. 8 Coordinate system for motion analysis of SBBD

The buoy is approximately fixed to a sinker by a hinge at the origin. Then, the motion of the buoy can be expressed by an motion equation of one degree of freedom as follows.

$$(I+I_S+i+i_S)\ddot{\theta} = M_R+M_D+M_{DS}+M_W+M_{WS} \tag{3}$$

where I, i, M_R, M_D and M_W denote the inertia moment, the added-mass, the restoring moment, the drag force relating to the relative incident flow, and the inertia force induced by waves, respectively, and the suffix S represents the component of the stabilizer.

The hydrodynamic forces on the main body, i, M_D and M_W, can be obtained by integrating the forces on a small sectional element of the main body along the axis as follows.

$$i = \int_0^l C_A(Kc(z),Re(z))\,\rho\frac{\pi}{4}D(z)^2z^2\,dz \tag{4}$$

$$M_D = \int_0^l C_D(Kc(z),Re(z))\,\frac{1}{2}\rho D(z)\,(v_x(z)-z\dot{\theta})\,|v_x(z)-z\dot{\theta}|\,z\,dz \tag{5}$$

$$M_W = \int_0^l [1+C_A(Kc(z),Re(z))]\,\rho\frac{\pi}{4}D(z)^2\dot{v}_x(z)\,z\,dz \tag{6}$$

where l denotes the wet length of the main body, C_D(Kc(z), Re(z)) and C_A(Kc(z), Re(z)) are drag and added-mass coefficients of the element (see the next section), D(z) is the diameter of the main body (center pipe or buoyancy chamber), and v_x is the water particle velocity in x-direction due to the incident wave. The hydrodynamic forces on the stabilizer, i_S, M_{DS} and M_{WS}, includes the forces on the stays and the cone-shaped fins, which can be calculated by using the relative incident flow velocities and accelerations and hydrodynamic coefficients of the fin shown in Figs. 5 and 6. Note that diffraction forces on the main body and the stabilizer are assumed to be neglected, since the diameters of them are much smaller than the wave length ($D/\lambda < 0.04$, λ: wave length).

Calculation Procedure

In the simulation program, two databases for hydrodynamic coefficients are loaded before calculating the motion of the buoy. One database files drag and added-mass coefficients of a circular cylinder in plane oscillatory flow, which are functions of Kc and Re numbers, shown by Sarpkaya et al. (1981). Another database keeps the drag and added-mass coefficients of the cone-shaped fin as a function of the angle of attack, which were obtained from hydrodynamic force measurements described before.

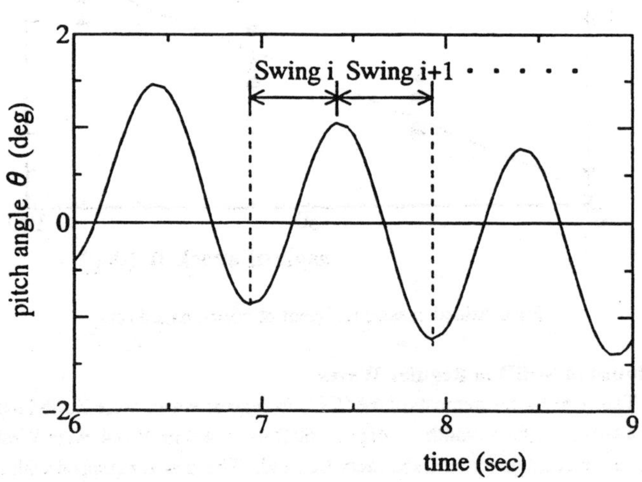

Fig. 9 Definition for each swing of SBBD motion

The motion of the buoy is calculated in time domain using a Runge-Kutta method. In this calculation, the hydrodynamic coefficients of sectional elements of the main body in Eqs. (4) ~ (6) are determined by the following procedure. A "swing" of the buoy is defined by a motion during a period ranging from a moment when $\theta = 0$ to the next moment when $\theta = 0$ as shown in Fig. 9. Firstly, the motion during a swing is simulated using constant C_D (= 1.17) and C_A (= 1.0) values, and the relative maximum velocity ($U_M(z)$) and double amplitude (Y(z)) of the sectional element in the considering swing are calculated. Kc(z) and Re(z) numbers of the sectional element at depth z are given by

$$Kc(z) = \pi Y(z)/D(z) \tag{7}$$

$$Re(z) = U_M(z)D(z)/\nu \tag{8}$$

where ν denotes the kinematic viscosity. Then, the drag and added-mass coefficients of the sectional element, $C_D(z)$ and $C_A(z)$, can be obtained from the database for the hydrodynamic coefficients of a circular cylinder described before. Using these coefficients, the motion of the swing is recalculated, and the Kc(z), Re(z), $C_D(z)$ and $C_A(z)$ are renewed. After the iterative calculation, the last calculated results are output, and the simulation of the motion for the next swing is started.

Simulation Condition

The simulations for a 50 m-type SBBD with and without stabilizer and a 16 m-type SBBD with and without stabilizer were carried out. The particulars of the 16 m-type SBBD and the simulation conditions are shown in Tables 3 and 4, respectively. For the 50 m-type SBBD, the simulation conditions are the same as those of the experiments except for a case including a low-frequency (T = 9.0 sec) and small-amplitude (H = 0.006 m) wave. For the 16 m-type SBBD, the wave period (T) is varied from 8.0 to 12.0 sec, and the wave height (H) is varied from 1.0 to 3.0 m.

Table 3 Particulars of 16 m-type SBBD

Main body	
length of main body (m)	55.6
sea bottom to mooring point (m)	2.5
diameter of center pipe (m)	1.016
length of buoyancy chamber (m)	4.884
diameter of buoyancy chamber (m)	3.218
mooring point to bottom of buoyancy chamber (m)	4.0
total mass of main body (kg)	2335
mooring point to center of gravity (m)	10.2
Stabilizer	
mooring point to connecting point (m)	3.0
length of stay (m)	12.0
diameter of stay (m)	0.45
height of fin (m)	0.58
diameter of fin (m)	3.211
attached angle of fin (deg)	20
total mass of stabilizer (kg)	498.5

RESULTS AND DISCUSSION

Figure 10 illustrates the time histories of measured pitch motions of the 50 m-type SBBD with and without stabilizer in the condition of T = 1.27 sec and H = 0.06 m. Note that the phase difference of the motion from the incident wave is not considered in this figure. In both cases, the wave-frequency motion amplitudes are almost the same. This means the effect of the stabilizer is not significant in this condition. The simulated results in the same condition is shown in Fig. 11. The simulated pitch motions have a low-frequency component, whose period coincides with the natural period of these models (T_n = 8.6 sec), since these records include the transient motions. The simulated wave-frequency motion amplitudes are in good agreement with the experimental results, and the effect of the stabilizer is small.

Table 4 Conditions for motion simulation of SBBD

water depth h (m) (actual dimension)	stabilizer option	wave period T (sec) (actual time)	wave height H (m) (actual dimension)
1.0 (50)	—	0.566 (4.0)	0.02 (1.0)
		0.849 (6.0)	0.02 (1.0)
		0.990 (7.0)	0.06 (3.0)
		1.27 (9.0)	0.06 (3.0)
		1.41 (10.0)	0.12 (6.0)
		1.70 (12.0)	0.12 (6.0)
	○	0.566 (4.0)	0.02 (1.0)
		0.849 (6.0)	0.02 (1.0)
		0.990 (7.0)	0.06 (3.0)
		1.27 (9.0)	0.06 (3.0)
		1.41 (10.0)	0.12 (6.0)
		1.70 (12.0)	0.12 (6.0)
16 (16)	—	8.0 (8.0)	1.0 (1.0)
		12.0 (12.0)	1.0 (1.0)
		12.0 (12.0)	3.0 (3.0)
	○	8.0 (8.0)	1.0 (1.0)
		12.0 (12.0)	1.0 (1.0)
		12.0 (12.0)	3.0 (3.0)
1.0 (50)	—	0.990 (7.0) + 9.0 (70.0)	0.06 (3.0) + 0.006 (0.30)

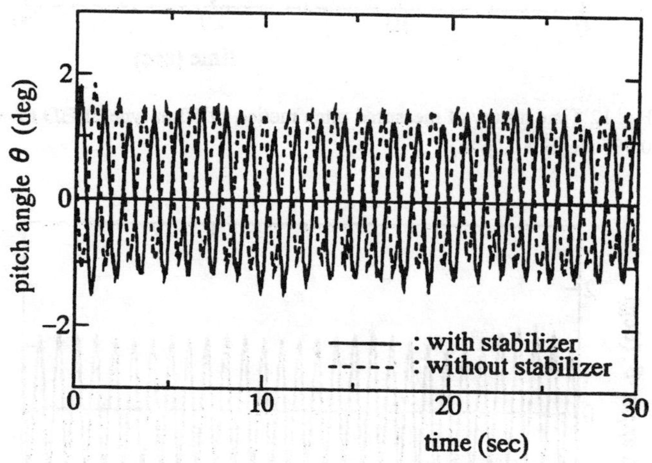

Fig. 10 Time history of measured pitch motion of 50 m-type SBBD (T = 1.27 sec, H = 0.06 m)

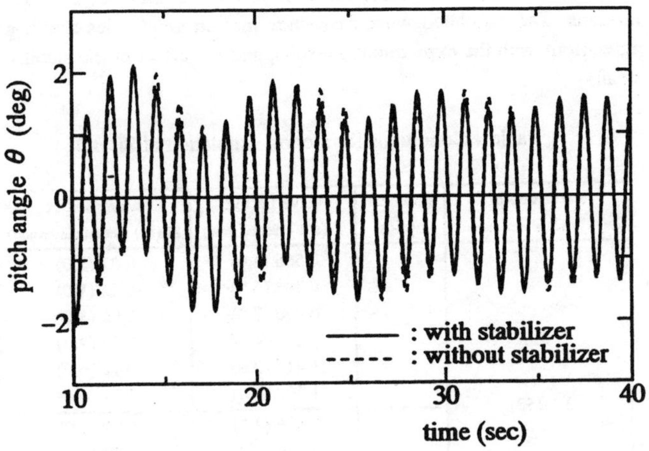

Fig. 11 Time history of simulated pitch motion of 50 m-type SBBD (T = 1.29 sec, H = 0.06 m)

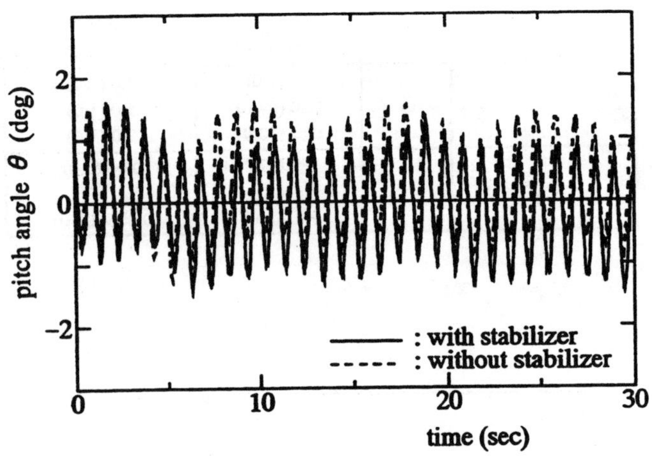

Fig. 12 Time history of measured pitch motion of 50 m-type SBBD (T = 0.990 sec, H = 0.06 m)

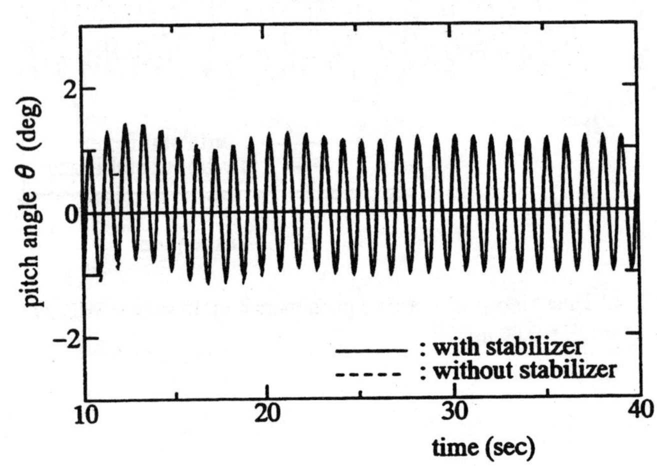

Fig. 13 Time history of simulated pitch motion of 50 m-type SBBD (T = 0.990 sec, H = 0.06 m)

Figure 12 shows the time histories of measured pitch motions of the 50 m-type SBBD with and without stabilizer in the condition of T = 0.990 sec and H = 0.06 m. These pitch motions have natural-frequency component even in the steady state. This kind of low-frequency motion is not observed in the other experimental data. The simulated results in the same condition, which is shown in Fig. 13, demonstrate that the natural-frequency motion only appear in the transient state. The low-frequency motion shown in Fig. 12 may be caused by a non-linear wave load, which is not considered in this simulation program. Another remarkable phenomena in Fig. 12 is a steady inclination of about 0.3 degrees in the case of the buoy without stabilizer. This may be also due to the non-linear effect of the incident waves.

Fig. 14 shows the measured and simulated pitch motion amplitudes $\bar{\theta}$ of the 50 m-type SBBD. The simulated results are generally in good agreement with the experimental results. The effect of the stabilizer on the pitch motion amplitude is not significant in the case of the 50 m-type SBBD, because the setting point of the stabilizer is deep, and the length of the stay (lever of the damping moment) is relatively short comparing with the scale of the main body.

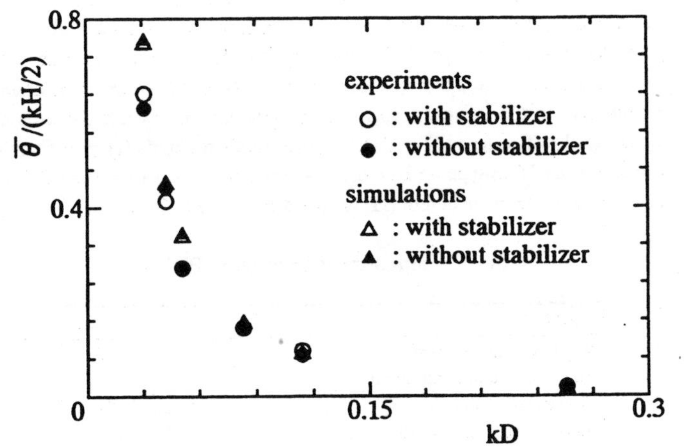

Fig. 14 Measured and simulated pitch motion amplitudes of 50 m-type SBBD

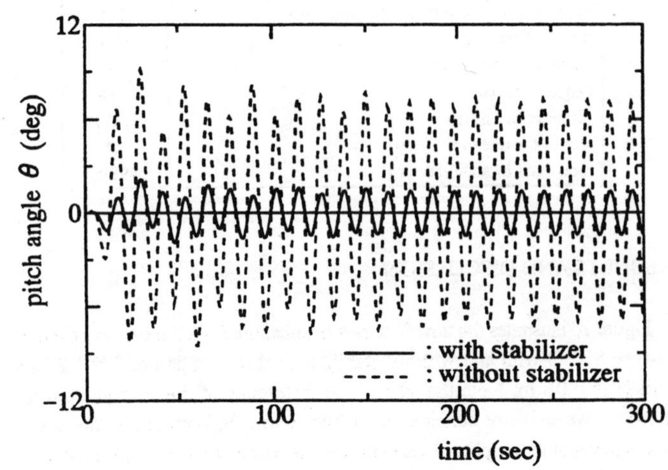

Fig. 15 Time history of simulated pitch motion of 16 m-type SBBD (T = 12.0 sec, H = 1.0 m)

Table 5 Simulated pitch motion amplitudes of 16 m-type SBBD

wave period T (sec)	wave height H (m)	stabilizer option	pitch motion amplitude $\bar{\theta}$ (deg)
8.0	1.0	—	3.20
		○	0.86
12.0	1.0	—	7.10
		○	1.45
12.0	3.0	—	21.0
		○	4.50

Figure 15 shows the time history of simulated pitch motion of the 16 m-type SBBD with and without stabilizer in the condition of T = 12.0 sec and H = 1.0 m. The pitch motion amplitude of the buoy with stabilizer is much smaller than that of the buoy without stabilizer. The similar tendency can be found in the other conditions shown in Table 5. The pitch motion amplitude is reduced to 20 ~ 25% due to the effect of the stabilizer. The effect of the stabilizer for the 16 m-type SBBD is much larger than that for the 50 m-type SBBD, because the length of the stay (lever of the damping moment) is relatively large comparing with the scale of the main body.

When we consider a low-frequency motion of a spar-buoy with a small diameter, it becomes more important to predict the low-frequency viscous damping force. Ikeda et al. (1990) pointed out that the low-frequency viscous damping force mainly depends on the difference of the drag coefficient between forward swing (the wave-frequency motion is in the same direction as the low-frequency motion) and backward swing (the wave-frequency motion is in the opposite direction to the low-frequency motion), and proposed an expression for low-frequency viscous damping coefficient C_{DL} as follows.

$$C_{DL} = \frac{1}{2}[C_D(Kc_F)(1+U_W/U_L)^2 - C_D(Kc_B)(1-U_W/U_L)^2] \qquad (9)$$

where Kc_F and Kc_B are the Kc numbers for the forward and backward swings, respectively, and U_W/U_L is the ratio of the maximum velocity of the wave-frequency motion to that of the low-frequency motion. The low-frequency viscous damping coefficient of a semisubmersible was analyzed by Otsuka et al. (1993) using this expression. In order to know the effect of the hydrodynamic coefficients changing swing by swing on the low-frequency motion of the buoy, we compared the simulated results using the present method with the results using constant C_D (= 1.17) and C_A (= 1.0) values in a condition including a low-frequency and small-amplitude wave (see Fig. 16). The wave-frequency motion amplitude is in good agreement with each other. However, the low-frequency motion amplitude obtained by using constant C_D and C_A values is larger than that using C_D and C_A values for each swing. This result suggests that the effect of the hydrodynamic coefficients changing swing by swing is significant when the buoy motion includes a low-frequency component.

CONCLUSIONS

In this paper, dynamic responses of a Spar-Buoy Boring Derrick (SBBD), which was proposed by Fukutomi et al. (1995), in regular waves have been experimentally and numerically investigated. The SBBD optionally has a stabilizer system, which consists of stays and cone-shaped fins.

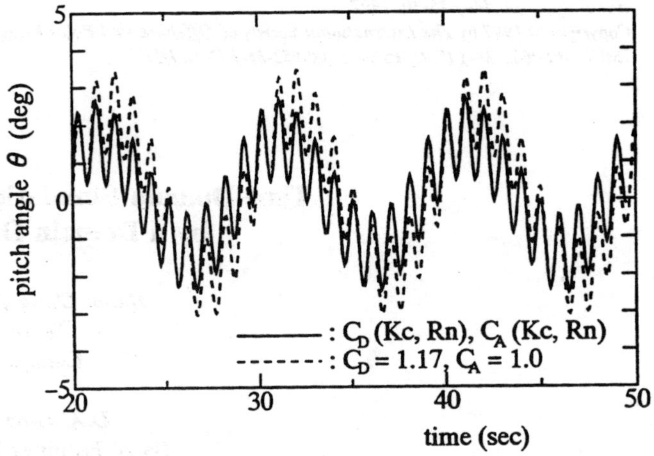

Fig. 16 Time history of simulated pitch motion of 50 m-type SBBD (T = 0.990 and 9.90 sec, H = 0.06 and 0.006 m)

Both experimental and numerical studies for a 50 m-type SBBD demonstrate that the effect of the stabilizer on the pitch motion amplitude is not significant, because the setting point of the stabilizer is deep, and the length of the stay (lever of the damping moment) is relatively short comparing with the scale of the main body. However, the simulated results for a 16 m-type SBBD show that the pitch motion amplitude is reduced to 20 ~ 25% due to the effect of the stabilizer, since the length of the stay (lever of the damping moment) is relatively large comparing with the scale of the main body.

The simulated results in the condition including a low-frequency and small-amplitude wave suggest that the effect of the hydrodynamic coefficients changing swing by swing is significant when the motion of the buoy includes a low-frequency component.

REFERENCES

Fukutomi, M, Nakazawa, T, Yokomine, M and Harada, J (1995). "Development of Spar-Buoy Boring Derrick ("SBBD")," *OYO Technical Report*, No 17, pp 55-86 (in Japanese).

Ikeda, Y and Fujiwara, A (1990). "Low-Frequency Viscous Damping Force Acting on a Cylinder and a Flat Plate," *Proc 9th Offshore Mech and Arctic Eng Conf*, Vol 1 part A, pp 189-197.

Mekha, BB, Weggel, DC, Johnson, CP and Roesset, JM (1996). "Effects of Second Order Diffraction Forces on the Global Response of Spars," *Proc 6th Int Offshore and Polar Eng Conf*, ISOPE, Vol 1, pp 273-280.

Otsuka, K, Kiku, T and Ikeda Y (1993). "Viscous Forces Acting on a Semisubmersible Oscillating with Low Frequency in Regular Waves," *Int J Offshore and Polar Eng*, IJOPE, Vol 3, No 1, pp 17-25.

Ran, Z and Kim, MH (1996). "Nonlinear Coupled Responses of a Tethered Spar Platform in Waves," *Proc 6th Int Offshore and Polar Eng Conf*, ISOPE, Vol 1, pp 281-288.

Sarpkaya, T and Isaacson, M (1981). *Mechanics of Wave Forces on Offshore Structures*, Van Nostrand Reinhold Ltd., New York, pp 98-99.

Proceedings of the Seventh (1997) International Offshore and Polar Engineering Conference
Honolulu, USA, May 25-30, 1997

ISBN 1-880653-28-1 (Set); ISBN 1-880653-31-1 (Vol. III)

Time-Domain Simulation of a Berthing DDG-51 Ship by a Domain Decomposition Approach

Hamn-Ching Chen and Miaomou Chen
Texas A&M University
College Station, TX, USA

D.A. Davis and Erick T. Huang
Naval Facilities Engineering Service Center
Port Hueneme, CA, USA

ABSTRACT

A Reynolds-Averaged Navier-Stokes (RANS) method has been employed in conjunction with a chimera domain decomposition technique for time-domain simulation of transient flow induced by a berthing DDG-51 ship undergoing translational and/or rotational motions. The method solves the mean flow and turbulence quantities using arbitrary combination of embedded, overlapped, or matched grids. The unsteady RANS equations were formulated in an earth-fixed reference frame and transformed into general curvilinear, moving coordinate systems. A chimera domain decomposition technique was employed to accommodate the relative motions between different grid blocks. Calculations have been performed for a DDG-51 guided missile destroyer in translational and rotational motions to demonstrate the capability of the chimera RANS approach for time-domain simulation of the ship and berthing structure interactions. The numerical solutions successfully captured many important features of the transient flow around berthing ships including the underkeel flow acceleration, separation around the bow and stern area, flow recirculation behind the ship, water cushion between the ship and harbor quaywall, and the complex interaction among bow, shoulder and stern wave systems.

KEYWORDS: Chimera Domain Decomposition, Translational Motions, Rotational Motions, Turbulent Flows, Free Surface Waves, Berthing Ships, Underkeel Clearance, Water Cushion.

INTRODUCTION

A large ship possesses enormous kinetic energy that could seriously damage the berthing structure as well as resulting in substantial economic and operational penalties. Fender systems are provided at a berth to absorb the kinetic energy of the berthing ship and to mitigate impact forces. The amount of the energy to be absorbed and the maximum impact force are the prevailing criteria for fender system design. Currently, the most commonly used fender system design methodology accounts for the influence of the ambient water with a simple constant coefficient (Lee et al., 1975; Plotkin, 1977; Keuning and Beukelman, 1979; Fontijn, 1980, 1988). In order to improve the fender system design, it is desirable to develop a reliable and robust method for the simulation of the structural and fluid interactions among the ship, fender system and the surrounding water. The model must be able to include the effects of the fluid viscosity, underkeel clearance and free surface variations for accurate prediction of the ship motion during typical berthing operations.

Recently, the authors (Chen and Chen, 1996a,b, Chen, Chen and Davis, 1996; Chen, Chen and Huang, 1996) have developed a chimera RANS method for time-domain simulation of complex free surface flow induced by 2D and 3D berthing ships undergoing simple lateral motions. The results obtained thus far clearly demonstrated the feasibility of the chimera RANS method for detailed resolution of the transient flow field induced by the ship and harbor interactions. In the present study, the method has been further generalized for time-domain simulation of a DDG-51 guided missile destroyer approaching a harbor quaywall in various combinations of translational and rotational motions.

CHIMERA RANS METHOD

In the present study, the Chimera RANS method of Chen, Chen and Davis (1995) and Chen, Chen and Huang (1996) has been further extended for time-domain simulation of transient flow induced by a berthing DDG-51 ship. In this approach, the transport equations for both momentum and turbulence quantities are solved using the finite-analytical approach of Chen, Patel, and Ju (1990). To solve for the pressure, the SIMPLER/PISO pressure-velocity coupling technique of Chen and Korpus (1993) is used. In the following, we shall briefly summarize the governing equations and numerical methods for the berthing simulations.

Consider the nondimensional Reynolds-Averaged Navier-Stokes equations for incompressible flow in general curvilinear coordinates $(\xi^i,t) = (\xi^1,\xi^2,\xi^3,t)$

$$U^i_{,i} = 0 \tag{1}$$

$$\frac{\partial U^i}{\partial t} + \left(U^j - W^j\right) U^i_{,j} + U^j W^i_{,j} + \left(\overline{u^i u^j}\right)_{,j} \qquad (2)$$
$$+ g^{ij} p_{,j} - \frac{1}{Re} g^{jk} U^i_{,jk} = 0$$

where U^i and u^i represent the mean and fluctuating velocity components. W^i is the grid velocity and g^{ij} is conjugate metric tensor. t is time, p is pressure, and $Re=U_oL/\nu$ is the Reynolds number based on a characteristic length L, a reference velocity U_o, and the kinematic viscosity ν. Equation (1) represents the continuity equation and equation (2) represents the mean momentum equation. The equations are written in tensor notation with the usual summation convention assumed. The subscripts, $_j$ and $_{jk}$, represent the covariant derivatives. In the present study, the two-layer turbulence model of Chen and Patel (1988) is employed to provide closure for the Reynolds stress tensor $\overline{u^i u^j}$.

The free surface boundary conditions for present berthing simulations consist of one kinematic and three dynamic conditions. The kinematic free surface boundary conditions can be expressed as

$$\frac{D\eta}{Dt} = \eta_t + \left(U^j - W^j\right) \eta_{,j} = 0 \qquad (3)$$

where η is the free surface wave elevation. For general viscous fluid flows, the dynamic conditions represent the continuity of stresses on the free surface. When the surface tension and free surface turbulent boundary layer are neglected, a set of simplified dynamic boundary conditions can be employed by simply extrapolating the velocities from the inner fluid domain and maintain constant total pressure on the free surface.

In the current chimera RANS method, the solution domain is first decomposed into a number of computational blocks. The body-fitted numerical grids for the ship and the harbor fluid domain are generated separately with the ship grid blocks completely embedded in the harbor grid. The ship grids are allowed to move with a prescribed velocity relative to the harbor grid in arbitrary combinations of translational and rotational motions. The PEGSUS program (Suhs and Tramel, 1991) is employed every time step to determine the interpolation information for linking grids. It outputs a complete set of information corresponding to the interpolation performed. The most important information is included in the IBLANK array, the interpolation and boundary lists, the interpolation stencils, and the maps showing boundary and interpolation node locations. The IBLANK array and interpolation stencils are incorporated into the present chimera RANS code to remove the hole points and update boundary conditions from the linking grids.

The overall solution procedure consists of an outer loop over time and an inner loop or sweep that iterates over the blocks of the grid. The discretization equations for pressure, velocity, and turbulence quantities form a system of tridiagonal matrices which was solved using an iterative ADI scheme. When a point is classified as a hole point (IBLANK=0) it must be removed from the solution process. To do this, the value at the hole point must remain unchanged over a time step to avoid affecting the solution. This is achieved by setting the diagonal elements to 1 and off-diagonal elements to zero in the tridiagonal matrix at the hole points. The points classified as hole boundaries also have an IBLANK value of 0, which removes them from the solution process. These values, however, must be updated by interpolated values from another grid. To ensure the proper conservation of mass and momentum between linking grid blocks, the grid-interface conservation technique of Hubbard and Chen (1994) and Chen and Hubbard (1996) has been employed to eliminate unphysical mass source resulting from the interpolation errors. More detailed description of the chimera RANS/free surface method and the associated data management system were given in Hubbard and Chen (1994, 1995) and Chen and Chen (1996a,b).

RESULTS AND DISCUSSION

In practical berthing maneuvers under tugboat assistance, the ship typically experience a combination of translational and rotational motions. In order to demonstrate the feasibility of the present chimera RANS method for realistic berthing maneuvers, calculations were performed for a DDG-51 guided missile destroyer in various combinations of lateral, oblique translational, and rotational motions at several different Froude (Fr) and Reynolds (Re) numbers. A complete summary of all the simulation results are given in Chen and Chen (1996a,b). For the sake of brevity, we shall present only the most important results for the Fr = 0.03 and Re = 10^6 case in combined translational/rotational motions. In particular, we shall consider a DDG-51 destroyer ship undergoing an oblique translational motion followed by a rotational motion, as shown in Figure 1.

For the full-scale DDG-51 destroyer considered here, the ship length at the designed waterline (DWL) is 466 ft, and the maximum ship draft below the bow bulb is 30.6 ft. In the present chimera domain decomposition approach, a 61 x 25 x 41 body-fitted numerical grid was constructed for accurate resolution of the turbulent boundary layer flow around a DDG-51 guided missile destroyer hull. This grid topology requires the specification of a branch cut along the front center plane (i.e., $y = 0$) where appropriate boundary conditions must be provided. In this study, the flow field at the front center plane is solved directly using a small 5 x 25 x 21 grid in the bow area which matches exactly with the first ship block. This ensures the complete consistency (exact matching) of the numerical solutions between the two ship grid blocks.

In present berthing simulations, a simple 61 x 151 x 21 rectangular grid block was used to represent the fully sheltered harbor. A fairly shallow water depth of 0.075 ship length (i.e., 34.95 ft) was chosen for all the test cases considered here. This corresponds to the minimum water depth of 34 feet and 11 inches encountered in practical berthing facilities. It should be noted that the minimum underkeel clearance for the present simulations is only 4.15 ft or about 0.89% of the ship length. This rather shallow water depth makes the construction of fully-connected grids very difficult and time-consuming even in the absence of ship motions. On the other hand, the chimera domain decomposition method enables us to accommodate arbitrary translational and rotational motions without tedious adaptive generation. Figure 2 shows partial views of the numerical grids for the combined translational/rotational motions at two different time steps. It is noted that part of the numerical grids (i.e., hole points) for the outer rectangular block have been removed from the computational domain using the PEGSUS program of Suhs and Tramel (1991).

For the berthing process considered here, the DDG-51 destroyer ship initially underwent a pure translational motion at an oblique angle of 10° towards the harbor quaywall from t = 0 to t = 0.8, followed by a simple rotational motion from t = 0.8 to t =1.03. This simulates a ship contacting a non-recoiling type fender. During the simple translation, the ship was accelerated from zero to U_o between $0 \leq t \leq 0.2$, maintaining a constant speed of U_o from t = 0.2 to 0.6, and then decelerated to zero from t = 0.6 to 0.8. The ship was then undergoing an angular acceleration from t = 0.8 to 0.85, followed by a constant angular velocity of 50° per unit time between t = 0.85 and 1.03. At the end of the simulation, the ship has rotated a total of 10° so that the ship keel plane is exactly parallel to the quaywall. For the present study, a total of 206 time steps were used with a constant time increment of Δt = 0.005 throughout the entire simulation. The computational domain and the corresponding ship positions at t = 0.25, 0.50, 0.80, and 1.03 are shown earlier in Figures 1(a) through 1(d), respectively.

Figures 3 through 5 show the computed longitudinal velocity (*U*), transverse velocity (*V*), and vertical velocity (*W*) contours under combined translational/rotational motions. The corresponding pressure (*p*) contours are also shown in Figure 6 to provide a complete description of the three-dimensional transient flows induced by the ship and quaywall interactions. Both the velocity and pressure solutions clearly indicated the presence of a very strong bow wave system which is caused by the large bow bulb and very narrow underkeel clearance. On the other hand, the stern wave system is rather weak due to the relatively shallow ship draft in the stern region. It is further noted that there is a distinct shoulder wave system in the middle section of the ship wake.

During the stage of oblique translational motions (i.e., $t \leq 0.8$), the transverse velocity contours are similar to those induced by the lateral motions shown in Chen and Chen (1996b). However, the longitudinal and vertical velocities were significantly modified due to the change of ship orientations. More specifically, the oblique ship motions produce a stronger recirculation region around the ship stern, while the recirculating flow in the bow region becomes weaker. For the lateral motion cases shown in Chen and Chen (1996b), the ambient fluid moves perpendicular to the ship center plane in a ship-fixed coordinate system. The recirculating flows around the bow and stern regions resulted purely from the three-dimensionality of the ship geometry. In the present oblique ship motions, however, the ambient velocity is no longer perpendicular to the ship center plane and the ambient flow from stern to bow (or vice versa) must be taken into account. For the ship orientation considered here, the oblique translational motion produces an ambient velocity of $U_O \sin 10°$ or $0.174\ U_O$ from the stern to bow, while the velocity component perpendicular to the ship center plane reduces slightly to $U_O \cos 10°$ or $0.985\ U_O$. This stern-to-bow flow has led to a significant increase of the flow recirculation in the stern region.

At t = 0.8, the ship came to a complete stop and changed to a simple rotational motion. It is seen from the numerical solutions at t = 0.93 and 1.03 that the ship rotation produces significant increases of both the velocity and pressure gradients around the bow area. This is clearly due to the rapid reduction of fluid gap between the ship bow and the harbor quaywall. Although there is no data available for direct comparison, the present results clearly demonstrated the capability of the present method for detailed resolutions of the underkeel flow acceleration, water cushion between the ship and harbor quaywall, and the complex wave patterns induced by arbitrary combinations of translational and rotational motions.

SUMMARY AND CONCLUSIONS

A chimera RANS method has been developed for time-domain simulation of transient flow induced by a berthing ship in a shallow harbor. The method solves the unsteady Reynolds-Averaged Navier-Stokes equations in conjunction with a chimera domain decomposition technique for detailed resolution of the hydrodynamic coupling among the ship, the harbor floor, and the harbor quaywall or other berthing structures. The use of chimera domain decomposition approach greatly facilitate the handling of relative motions between the ship and the berthing structures. Calculations have been performed for a DDG-51 guided missile destroyer in arbitrary combinations of translational and rotational motions. The numerical results successfully captured many important features of the transient flow around a berthing ship including the underkeel flow acceleration, separation in the ship wake, water cushion between the ship and the harbor quaywall, and the complex interaction among the bow, shoulder, and stern wave systems. With some further generalizations, the method can be readily employed for time-domain simulation of ship and structure interactions under the influence of ambient waves and currents.

ACKNOWLEDGMENT

This work was supported by the Office of Naval Research, Naval Exploratory Development Program in Facilities, under Contract No. N47408-95-C-0237. Most of the computations were performed on the Cray C-90 and YMP of Cray Research Inc. at Eagen, Minnesota under the sponsorship of Mr. Chris Hempel. The Texas A&M University Supercomputer Center also provided a large share of the computing time on its Cray J90 for code validations and preliminary calculations.

REFERENCES

Chen, H.C. and Chen, M. (1996a), "A Chimera RANS Method for Time-Domain Simulation of Berthing Ships in a Fully Sheltered Harbor," COE Report No. 349, Texas Engineering Experiment Station, Texas A&M University System, College Station, TX, 45 pages, May 1996.

Chen, H.C. and Chen, M. (1996b), "Time-Domain Simulation of a Berthing DDG-51 Ship in Translational and Rotational Motions," COE Report No. 351, Texas Engineering Experiment Station, Texas A&M University System, College Station, TX, 115 pages, September 1996.

Chen, H.C., Chen, M. and Huang, E.T. (1996), "Chimera RANS Simulations of Unsteady 3D Flows Induced by Ship and Structure Interactions," ***Flow Modeling and Turbulence VI***, edited by C.J. Chen, C. Shih, J. Lienau and R.J. Kung, pp. 373-380, Balkema, Rotterdam, 1996.

Chen, M, Chen, H.C. and Davis, D.A. (1996), "Numerical Simulation of Transient Flows Induced by a Berthing Ship," *Proceedings, Sixth International Offshore and Polar Engineering Conference*, Vol. IV, pp. 533-541, Los Angels, CA, May 26-31.

Chen, H.C. and Hubbard, B.J. (1996), "A Chimera RANS Method for Steady and Unsteady Incompressible Viscous Flows," *International Journal for Numerical Methods in Fluids*, submitted.

Chen, H.C. and Korpus, R. (1993), "A Multi-block Finite-Analytic Reynolds-Averaged Navier-Stokes Method for 3D Incompressible Flows," *Individual Papers in Fluids Engineering*, edited by F. M. White, ASME FED-Vol. 150, pp. 113-121, ASME Fluids Engineering Conference, Washington, DC, June 20-24.

Chen, H.C. and Patel, V.C. (1988), "Near-Wall Turbulence Models for Complex Flows Including Separation," *AIAA Journal*, Vol. 26, No. 6, pp. 641-648.

Chen, H.C., Patel, V.C. and Ju, S. (1990), "Solutions of Reynolds-Averaged Navier-Stokes Equations for Three-Dimensional Incompressible Flows," *Journal of Computational Physics*, Vol. 88, No. 2, pp. 305-336.

Fontijn, H.L. (1980), "The Berthing of a Ship to a Jetty," *Journal of the Waterway, Port, Coastal and Ocean Division*, ASCE, Vol. 106, No. WW2, pp. 239-259.

Fontijn, H.L. (1988), "On the Prediction of Fender Forces at Berthing Structures, Part I: Calculation Methods; Part II: Ship Berthing Related to Fender Structure," ***Advances in Berthing and Mooring of Ships and Offshore Structures***, Edited by E. Bratteland, pp. 62-166, Kluwer Academic Publishers.

Hubbard, B. J. and Chen, H.C. (1994), "A Chimera Scheme for Incompressible Viscous Flows with Applications to Submarine Hydrodynamics," AIAA Paper 94-2210, *25th AIAA Fluid Dynamics Conference*, Colorado Spring, CO, June 20-23.

Hubbard, B.J. and Chen, H.C. (1995), "Calculations of Unsteady Flows Around Bodies with Relative Motion Using a Chimera RANS Method," *Proceedings of the 10th ASCE Engineering Mechanics Conference*, Vol. II, pp. 782-785, University of Colorado at Boulder, Boulder, CO, May 21-24.

Keuning, J.A. and Beuhelman (1979), "Hydrodynamic Coefficients of Rectangular Barges in Shallow Water," *BOSS'79*, 2nd International Conference on Behaviour of Off-Shore Structures, Imperial College, London, England, *BHRA Fluid Engineering*, Vol. 2, Paper 55, pp. 105-124.

Lee, T.T., Nagai, S. and Oda, K. (1975), "On the Determination of Impact Forces, Mooring Forces and Motions of Supertankers at Marine Terminal," *7th Annual Offshore Technology Conference*, Houston, TX, Paper No. OTC 2211, pp. 671-678.

Plotkin, A. (1977), "Heave and Pitch Motions in Shallow Water Including the Effect of Forward Speed," *Journal of Fluid Mechanics*, Vol. 80, Part 3, pp. 433-441.

Suhs, N.E. and Tramel R.W. (1991), "PEGSUS 4.0 Users Manual," Arnold Engineering Development Center Report, AEDC-TR-91-8, Arnold Air Force Station, TN.

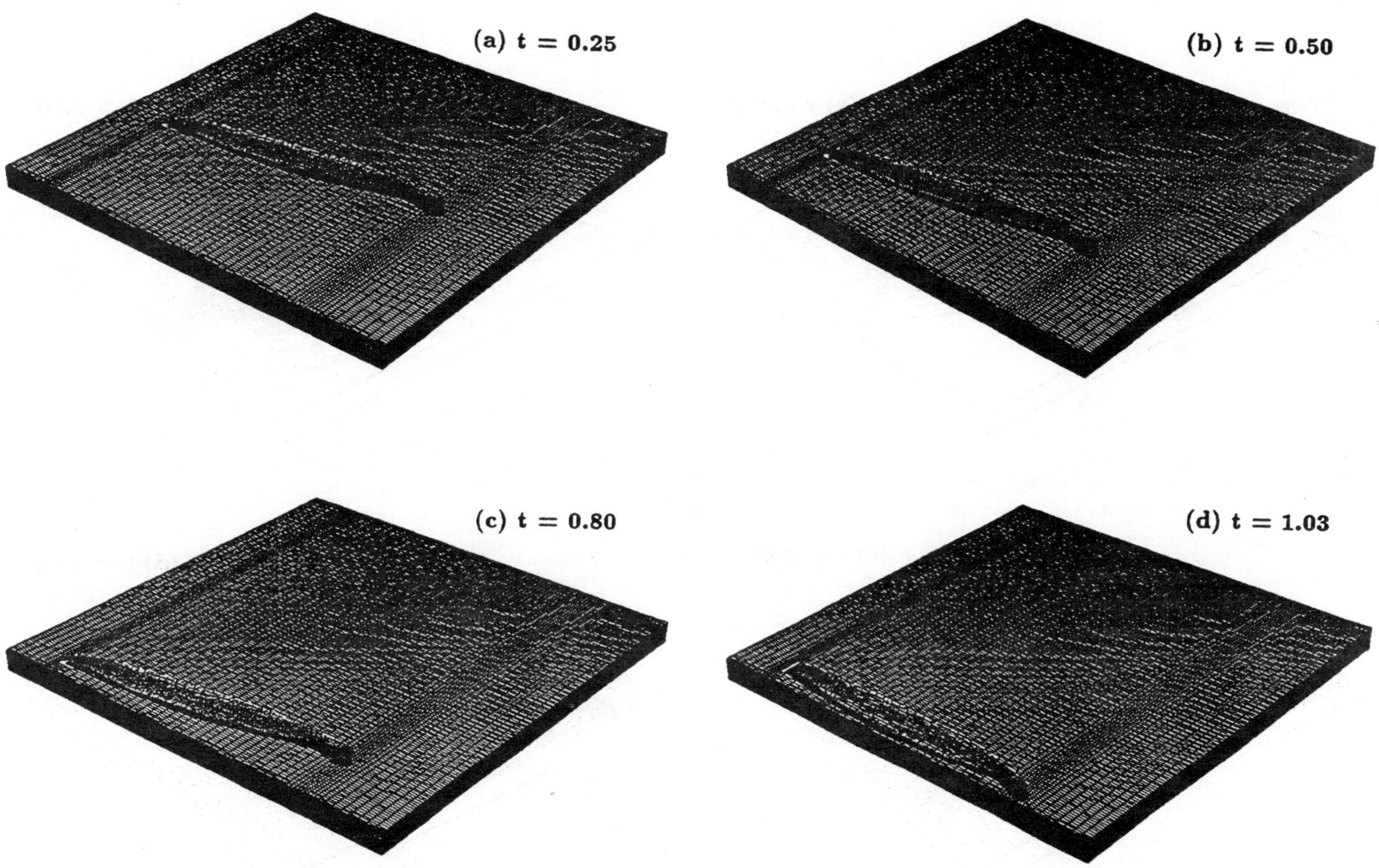

Fig. 1 DDG-51 ship in oblique translational and rotational motion.

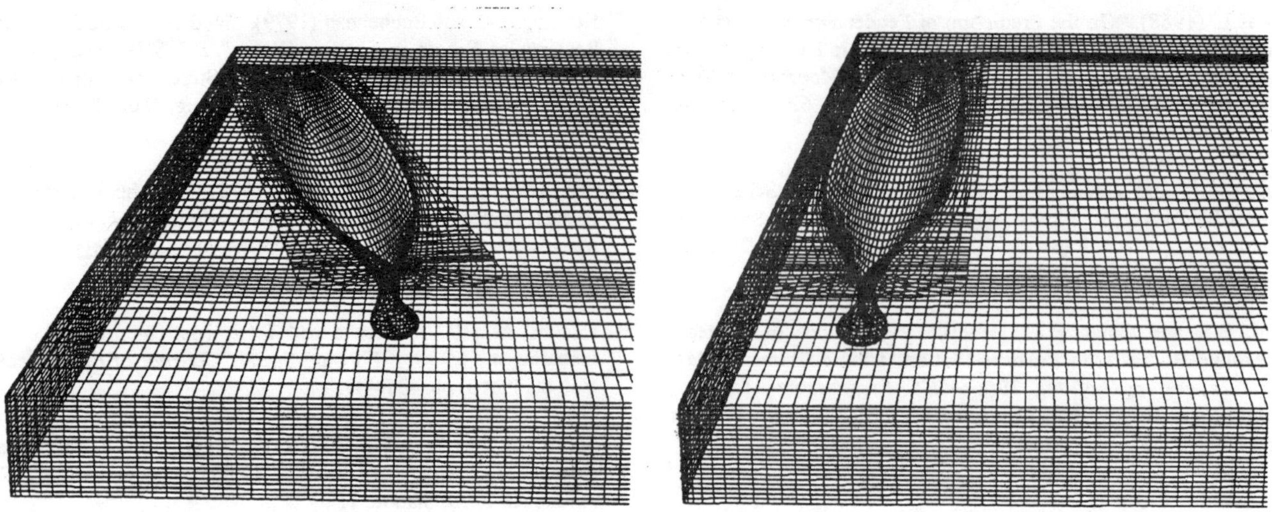

Fig. 2 Partial views of the chimera grid blocks for DDG-51 destroyer.

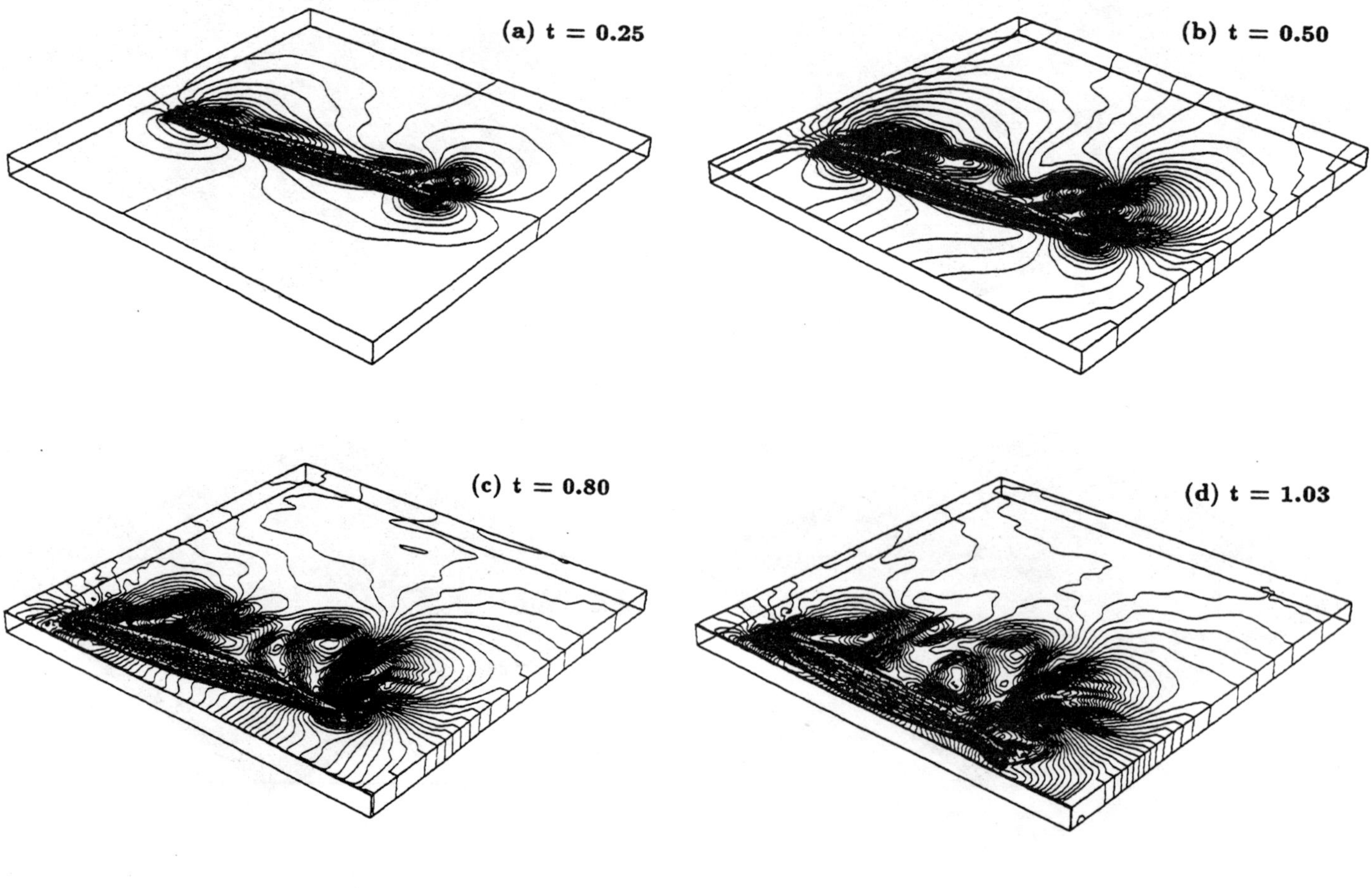

Fig. 3 Longitudinal velocity contours; Fr = 0.03; translational/rotational motions.

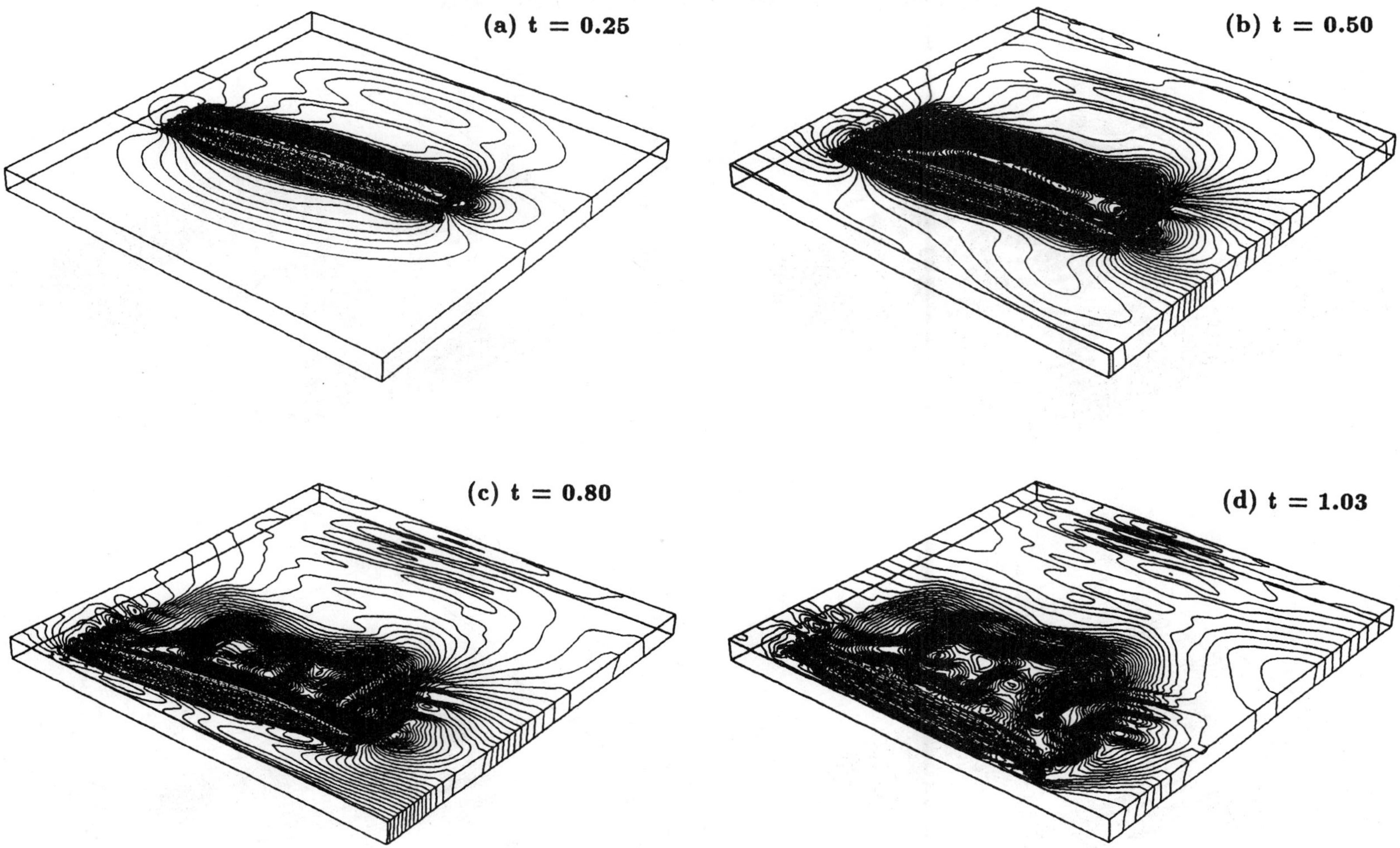

Fig. 4 Transverse velocity contours; Fr = 0.03; translational/ rotational motions.

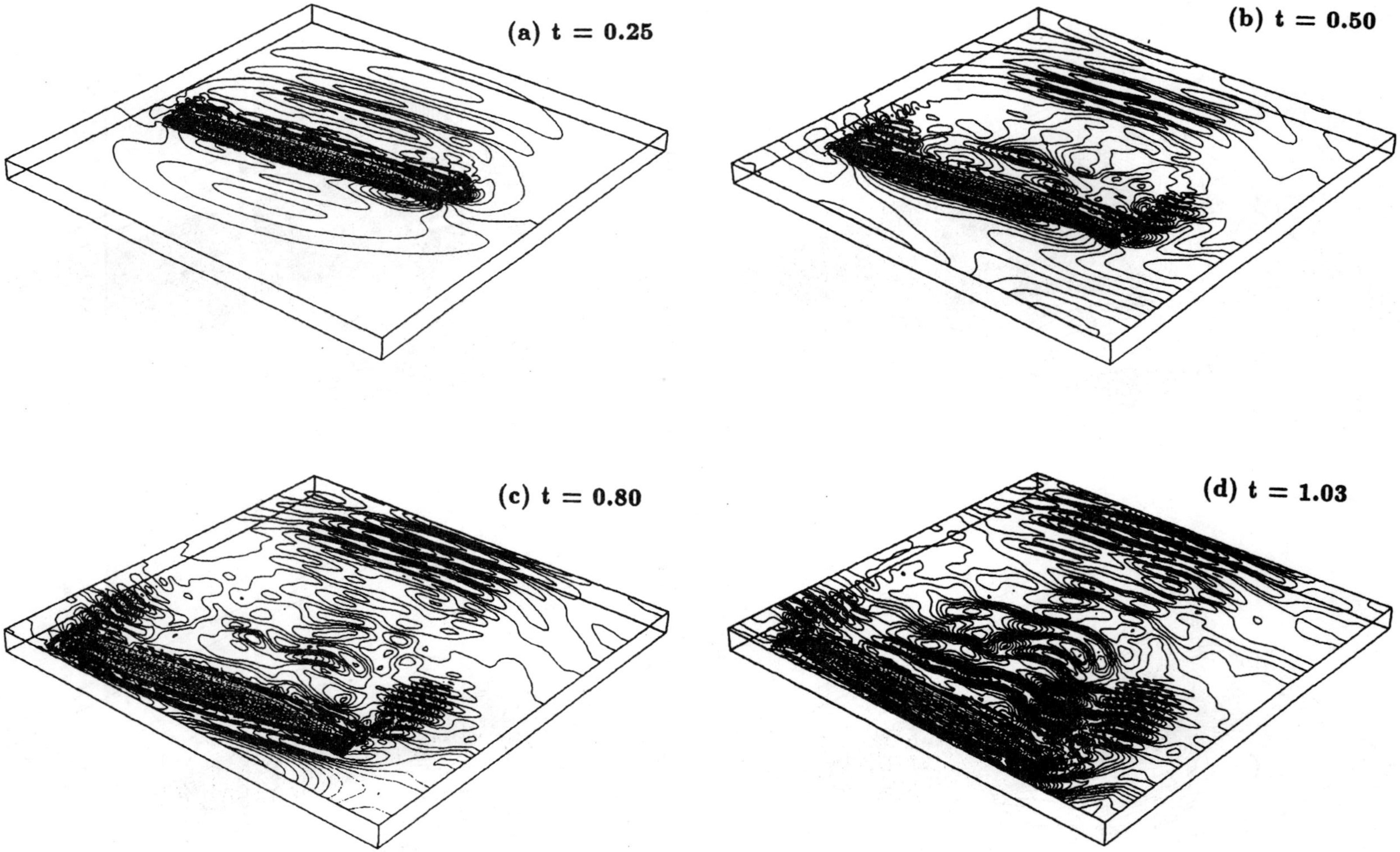

Fig. 5 Vertical velocity contours; Fr = 0.03; translational/rotational motions.

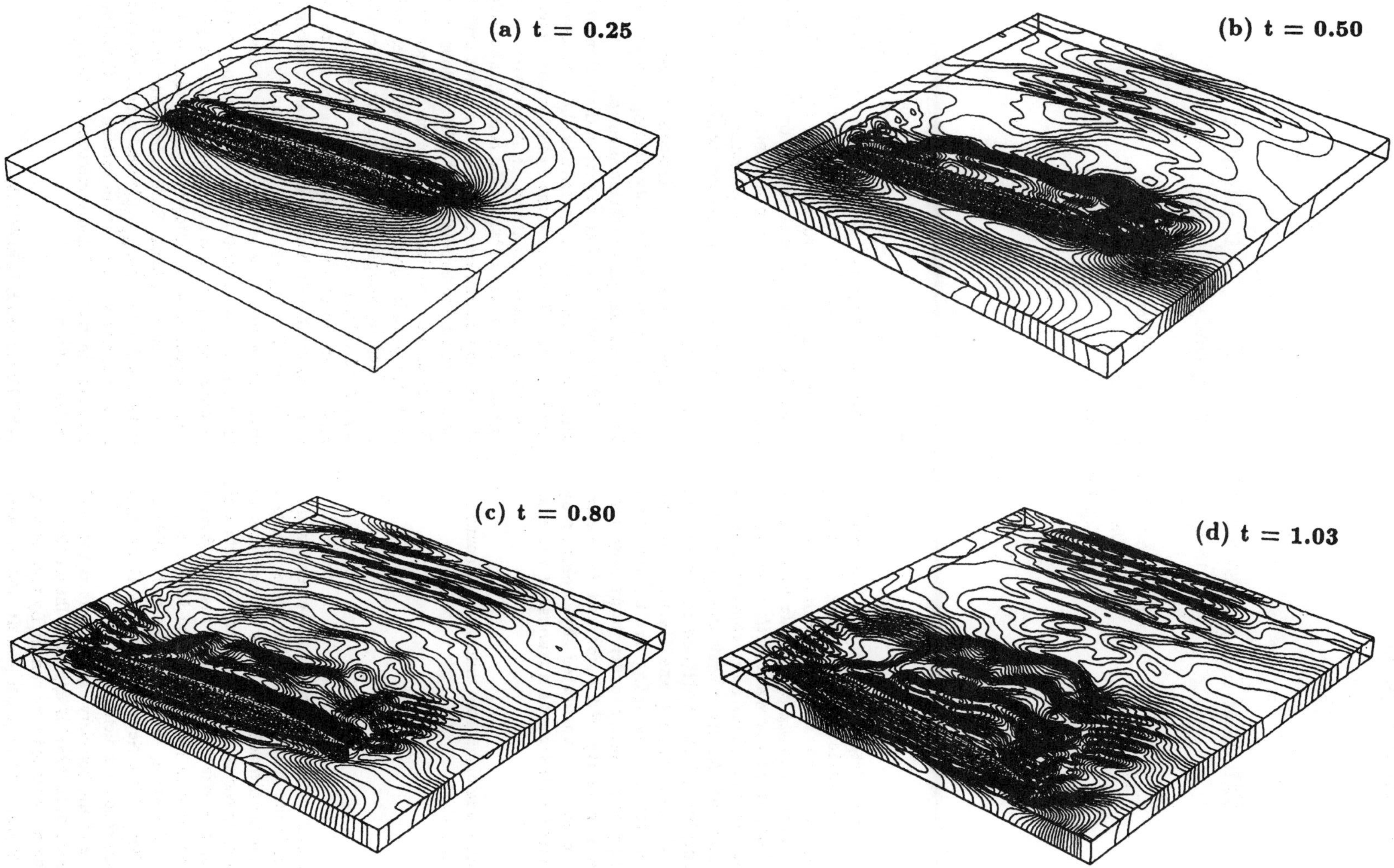

Fig. 6 Pressure contours; Fr = 0.03; translational/rotational motions.

Proceedings of the Seventh (1997) International Offshore and Polar Engineering Conference
Honolulu, USA, May 25-30, 1997

ISBN 1-880653-28-1 (Set); ISBN 1-880653-31-1 (Vol. III)

Effects of Slenderness and Flatness on Wave Drift Damping of a Moored Floating Body

Takeshi Kinoshita, Weiguang Bao and Rong Zhu
University of Tokyo
Tokyo, Japan

ABSTRACT

Floating marine structures moored in ambient waves may undergo slow drift motions with a large horizontal amplitude due to the resonance between the moored structures and slowly varying nonlinear wave excitation. The contribution of the wave drift damping is significant and sometimes dominant compared with the other components of the slow drift damping such as the viscous damping of the hull and the mooring damping. Effects of slenderness and flatness of the hull form on the wave drift damping are numerically studied by applying the higher order boundary element method to a semi-ellipsoid immersed in water. This numerical method is validated by comparing with a semi-analytical solution of a truncated vertical circular cylinder.

KEY WORDS: wave drift damping, slow drift motion, boundary element method, mooring

INTRODUCTION

There may be several methods, such as the eigen-function expansion method, the finite element method and the boundary element method, to solve the diffraction and radiation problems for marine structures floating in waves and slow current. The method of eigen-function expansion is effective to make it easier to have a physical insight to the problem (for example see Matsui et al , 1991, Emmerhoff and Sclavounos, 1992, Malenica et al , 1995 as well as Kinoshita and Bao, 1996). Nevertheless, it has a limitation to be applied to arbitrary shape of structures.

In the present work, a hybrid boundary element method(Yuen and Chau, 1987, Masuda and Nagai, 1988) is applied to the problem of floating structures in both waves and slow current. By this method, the fluid domain is divided into an inner domain and an outer domain by a virtual boundary of circular cylinder. In the inner domain, the boundary element method is applied, while in the outer domain an analytical solution, which satisfies the boundary conditions on the free surface and the sea bottom as well as the radiation condition, is adopted. On the surface in common, i.e. the virtual boundary, the continuation of the potential and its radial derivative is used to relate the solutions in these two domains. Therefore, it is necessary to discretize boundary elements only in a relatively small closed domain, which will save much computational effort.

The higher order boundary elements are used in the present work to solve the problem of a body floating in both waves and current. The wave drift damping is then computed.

Offshore structures often subject to the actions of both waves and current at the same time. Sometimes, they are towed to the working site in waves. To evaluate the hydrodynamic forces acting on the structures in these situations, it is necessary to investigate the interaction problem of waves and current, the results of which can also be used in the estimation of the wave drift damping of the floating bodies undergoing slow drift motions in irregular waves. Because of the complicated interaction between waves and current, the hydrodynamic forces caused by both waves and current cannot be obtained by simply summing up the forces due to waves and current individually. Especially for the very large bodies such as the offshore structures, the steady disturbance due to the bodies would greatly affect the result of the hydrodynamic forces.

In the following sections, the boundary value problem is discussed in section 2, where the formulae to evaluate the wave drift force and the wave drift damping are also presented. The eigen-function expansion used in the outer domain is given in section 3. Then, in section 4, the integral equation used to solve the boundary value problem is deduced. The discretization of the boundary surface is discussed in section 5 while some special treatments of integrals to raise accuracy are given in section 6. Calculated examples are presented in section 7 and some conclusion remarks are offered in the last section.

BOUNDARY VALUE PROBLEM

Coordinate System

As shown in Fig. 1, a right-handed rectangular coordinate system is adopted, the origin O of which is located on the still water surface. The fluid domain is designated by Ω which is surrounded by the body surface S_B, the free surface S_F , the sea bottom surface S_H and a surface S_∞ representing the boundary at infinitely far field. S_R is a virtual boundary surface which is located at a distance of r_0 from the origin O and is used to divide the fluid domain into an inner domain Ω_I and an outer domain Ω_R. C_B and C_R represent the water line of the surface S_B and surface S_R respectively, i.e. the intersection lines of these surfaces with the free surface. A cylindrical coordinate system $O-r\theta z$ is also used especially in the outer domain Ω_R including the virtual boundary surface S_R. The unit normal vector $\boldsymbol{n}$ on all the boundary surfaces is defined as positive when pointing out of the fluid domain.

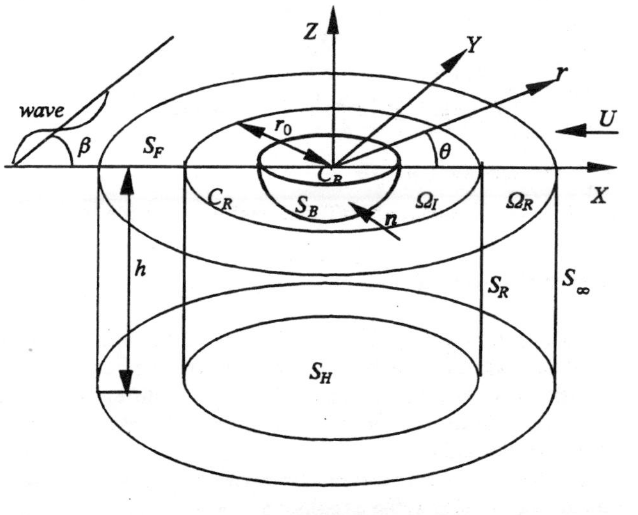

Fig.1 Coordinate systems.

A regular incident wave with an amplitude of ζ_0 is coming at an incident angle of β referring to the positive x-axis. The uniform current is flowing in the negative x-direction. The offshore structure is moored in water of depth h and is free to oscillate around its mean position at a frequency as a response to the incoming waves. The six modes of oscillation motion of the offshore structure, i.e. surge, sway, heave, roll, pitch and yaw, are defined as follows:

$$Z_j(t) = \mathrm{Re}\{\xi_j e^{-i\omega t}\} \quad (j=1\sim6) \tag{1}$$

$$\omega = \omega_0 - k_0 U\cos\beta \tag{2}$$

where

$Z_i(t)$: the displacement of the oscillation in mode j;
ξ_j : the complex amplitude of the corresponding oscillation;
ω : the encountering frequency;
ω_0 : the frequency of the incident waves;
k_0 : the wave number of the incident waves;
U : the velocity of the uniform current;
β : the incident wave angle referring to x-axis;
t : time.

Velocity Potential

It is assumed that the fluid is inviscid and incompressible and the flow is irrotational. There exists a velocity potential which can be expressed as:

$$\begin{aligned}\Phi_T(\boldsymbol{x},t) &= \Phi_s(\boldsymbol{x})+\Phi(\boldsymbol{x},t)\\ &= U\chi_s + \mathrm{Re}\{[\phi_D(\boldsymbol{x}) - i\omega\sum_{j=1}^{6}\xi_j\phi_j(\boldsymbol{x})]e^{-i\omega t}\}\\ &= U(\chi - x) + \mathrm{Re}\{[\phi_0(\boldsymbol{x})+\phi_7(\boldsymbol{x}) - i\omega\sum_{j=1}^{6}\xi_j\phi_j(\boldsymbol{x})]e^{-i\omega t}\}\end{aligned} \tag{3}$$

where

Φ_T : the total velocity potential;
Φ_T : the steady potential;
Φ : the unsteady potential;
χ_s : the steady potential due to the unit current velocity;
χ : the steady disturbance potential due to the unit current velocity;
ϕ_D : the diffraction potential;
ϕ_D : the incident wave potential;
ϕ_7 : the scattering potential;
ϕ_j : the radiation potential of unit oscillating velocity in mode j (j=1-6).

A small parameter τ_0 is introduced, which is defined as

$$\tau_0 = \omega_0 U/g \tag{4}$$

Then, the radiation potentials and the diffraction potential are expanded into power series of τ_0:

$$\phi_j = \phi_j^{(0)} + \tau_0\phi_j^{(1)} + O(\tau_0^2) \quad (j=1\sim6,\ D) \tag{5}$$

According to the power order of τ_0, we can obtain the well-known free surface conditions satisfied by potentials of different orders, i.e. $\phi_j^{(0)}$ and $\phi_j^{(1)}$ as follows:

[Zero-th order problem]

$$\frac{\partial\phi_j^{(0)}}{\partial z} - \nu_0\phi_j^{(0)} = 0 \quad \text{on} \quad z=0 \quad (j=1\sim6,\ D) \tag{6}$$

where

$$\nu_0 = \omega_0{}^2/g \tag{7}$$

[First order problem]

$$\frac{\partial\phi_j^{(1)}}{\partial z} - \nu_0\phi_j^{(1)} = f_j \quad \text{on} \quad z=0 \ (j=1\sim6,\ D) \tag{8}$$

where

$$f_j = -2k_0\cos\beta\phi_j^{(0)} + 2i\nabla(\chi-x)\cdot\nabla\phi_j^{(0)} - i\phi_j^{(0)}\frac{\partial^2\chi}{\partial z^2} \quad (j=1\sim6,\ D) \tag{9}$$

Similarly, the boundary conditions on the body surface are expressed as:

[Zero-th order problem]

$$\frac{\partial\phi_j^{(0)}}{\partial n} = \begin{cases} n_j & (j=1\sim6) \\ 0 & (j=D)\end{cases} \quad on \quad S_B \tag{10}$$

where

$$(n_1,n_2,n_3) = \boldsymbol{n} \tag{11}$$

$$(n_4,n_5,n_6) = \boldsymbol{x}\times\boldsymbol{n} \tag{12}$$

[First order problem]

$$\frac{\partial \phi_j^{(0)}}{\partial n} = \begin{cases} -\frac{m_j}{i\nu_0} & (j = 1 \sim 6) \\ 0 & (j = D) \end{cases} \quad on \quad S_B \tag{13}$$

where

$$(m_1, m_2, m_3) = -(n \cdot \nabla)\nabla(\chi - x) \tag{14}$$

$$(m_4, m_5, m_6) = -(n \cdot \nabla)[x \times \nabla(\chi - x)] \tag{15}$$

WAVE DRIFT FORCE AND WAVE DRIFT DAMPING

Once the velocity potentials are solved, the wave drift force can be evaluated either by so-called the near-field method of directly integrating the pressure over the wetted body surface or by the far-field method based on the momentum conservation. In the present work, the latter one is used.

According to Grue and Palm(1993), the wave drift force in x-direction can be obtained by

$$\overline{F_1} = \rho \int_{S_c} \overline{\left[\frac{1}{2} |\nabla \Phi_T|^2 n_1 - \frac{\partial \Phi_T}{\partial x}\frac{\partial \Phi_T}{\partial n} \right]} ds$$

$$- \frac{\rho}{2g} \int_{C_c} \overline{\left[\left(\frac{\partial \Phi_T}{\partial t}\right)^2 - U^2 \left(\frac{\partial \Phi_T}{\partial x}\right)^2 \right]} dl \tag{16}$$

where an over bar on the right hand side denotes time-average over a period is taken. This can also be expanded into the zero-th and the first order quantities according to the order in τ_0. For the zero-th order it is given by

$$\overline{F}_1^{(0)} = \frac{1}{4}\rho \int_{S_c} \left(\nabla\phi^{(0)} \cdot \nabla\phi^{(0)*} n_1 - \phi_x^{(0)}\phi_n^{(0)*} - \phi_n^{(0)}\phi_x^{(0)*} \right) ds$$

$$- \nu_0 \int_{C_c} \phi^0 \phi^{(0)*} dl \tag{17}$$

where the superscript * denotes the complex conjugate and the total zero-th potential is defined as

$$\phi^{(0)} = \phi_D^{(0)}(x) - i\omega_0 \sum_{j=1}^{6} \xi_j^{(0)} \phi_j^{(0)}(x) \tag{18}$$

The wave drift damping is as follows:

$$B_{11}(\omega_0) = -\frac{\omega_0}{g}\overline{F}_1^{(1)}$$

$$= -\frac{\rho\omega_0}{2g} \mathrm{Re}\left\{ \int_{S_c} \left(\phi_x^{(0)}\phi_n^{(1)*} + \phi_n^{(0)}\phi_x^{(1)*} - \nabla\phi^{(0)} \cdot \nabla\phi^{(1)*} n_1 \right) ds \right.$$

$$\left. + \int_{C_c} \left(\nu_0 \phi^{(0)}\phi^{(1)*} - k_0 \cos\beta \phi^{(0)}\phi^{(0)*} \right) n_1 dl \right\} \tag{19}$$

where the total first order potential is defined as

$$\phi^{(1)} = \phi_D^{(1)}(x) - i\omega_0 \sum_{j=1}^{6} \left[\xi_j^{(0)}\phi_j^{(1)}(x) + \xi_j^{(1)}\phi_j^{(0)}(x) \right.$$

$$\left. - k_0/\nu_0 \cos\beta \xi_j^{(0)} \phi_j^{(0)}(x) \right] \tag{20}$$

In (17) and (19) the surface integral is carried out over a control surface S_c at far field. The water line of S_c, i.e. the intersection of S_c with the free surface S_F, is designated as C_c, over which the line integral is pursued. It is proved by Emmerhoff and Sclavounos (1992) that these integrals have a property of conservation. This leads to the convenience that the control surface S_c can be placed at anywhere in the fluid domain as long as only the propagating wave terms in the potentials are kept. In the present work, these integrals are carried out over the virtual boundary S_R.

EIGEN-FUNCTION EXPANSION IN THE OUTER DOMAIN

Depending on the free surface conditions discussed in the previous section, potentials with different orders can be expanded into the following series in the outer domain.

The Steady Disturbance Potential

Since the steady potential satisfies a rigid wall boundary condition on the free surface, its solution in the outer domain can be written as:

$$\chi = \sum_{m=0}^{\infty}\sum_{n=0}^{\infty} R_{mn}^{(s)}(r) Z_m^{(s)}(z) \left(A_{mn}^{(s)} \cos n\theta + B_{mn}^{(s)} \sin n\theta \right) \tag{21}$$

where

$$R_{mn}^{(s)}(r) = \begin{cases} r_0^n / r^n & m = 0 \\ K_n\left(\frac{m\pi r}{h}\right) / K_n\left(\frac{m\pi r_0}{h}\right) & m > 0 \end{cases} \tag{22}$$

$$Z_m^{(s)}(z) = \cos m\pi(z + h)/h \tag{23}$$

and K_n is the second kind modified Bessel function of order n.

Unsteady Potentials

The unsteady potentials including the diffraction potential and radiation potentials are further divided into the zero-th order and the first order potentials. Since they satisfy different free surface conditions, the expansions for these potentials take different forms in the outer domain.

For the zero-th order potentials which satisfies a homogeneous free surface condition, the expansion is written as:

$$\phi_j^{(0)} = \sum_{m=0}^{\infty}\sum_{n=0}^{\infty} R_{mn}^{(0)}(r) Z_m^{(0)}(z) \left(A_{jmn}^{(0)} \cos n\theta + B_{jmn}^{(0)} \sin n\theta \right) \; (j = 1 - 7) \tag{24}$$

where

$$R_{mn}^{(0)}(r) = \begin{cases} H_n(k_0 r) / H_n(k_0 r_0) & m = 0 \\ K_n(k_m r) / K_n(k_m r_0) & m > 0 \end{cases} \tag{25}$$

$$Z_m^{(0)}(z) = \begin{cases} \dfrac{\cosh k_0(z + h)}{\cosh k_0 h} & m = 0 \\ \dfrac{\cos k_m(z + h)}{\cos k_m h} & m > 0 \end{cases} \tag{26}$$

and H_n is the first kind Hankel function of order n while k_m (m>0) is the positive solution of

$$\nu_0 = -k_m \tan k_m h \tag{27}$$

For the first order potentials, the free surface condition is inhomogeneous, additional terms should be added and the expansion takes the following form:

$$\phi_j^{(1)} = -2(i\frac{\partial}{\partial x} + k_0 \cos\beta)\frac{\partial \phi_j^{(0)}}{\partial \nu_0}$$

$$+ i\int_{S_F'} (2\nabla\chi \cdot \nabla\phi_j^{(0)} - \phi_j^{(0)}\frac{\partial^2 \chi}{\partial z^2}) G ds + \sum_{m=0}^{\infty}\sum_{n=0}^{\infty} R_{mn}^{(1)}(r) Z_m^{(1)}(z)$$

$$\left(A^{(1)}_{jmn}\cos n\theta + B^{(1)}_{jmn}\sin n\theta\right) \quad (j=1-7) \tag{28}$$

where S_F' denotes the part of the free surface in the outer domain. Due to the fast decay of the steady disturbance potential χ for the large distance, the contribution from the integral over S_F' in (3.2.5) is very small provided the radial distance r_0 of the virtual boundary S_R is sufficiently large. Therefore, this integral is ignored in the present work.

THE BOUNDARY INTEGRAL EQUATION

Green Function

To solve the boundary value problem in the inner domain, it is equivalent to solve the following integral equation:

$$C_P\phi_P = \int_S \left(\frac{\partial \phi_Q}{\partial n}G_{PQ} - \phi_Q\frac{\partial G_{PQ}}{\partial n}\right)ds_Q \tag{29}$$

where P is the field point which is located on the boundary surface S surrounding the inner domain Ω_I while Q is the source point on the same surface. Since the integral equation takes the same form for all the diffraction ($j=D$) and radiation potentials ($j=1$-6), instead of explicitly writing down the mode for each potential such as ϕ_{jP} or ϕ_{jQ}, a brief representation of ϕ_P or ϕ_Q will be used through out this section.

In (29), C_P is the shape coefficient at field point P, which has a value of $C_P = 2\pi$ for a smooth boundary surface. However, in general, the value of C_P can be determined by assuming a uniform potential, i.e.

$$\phi_P = \phi_Q = 1 \tag{30}$$

and consequently,

$$\frac{\partial \phi_Q}{\partial n} = 0 \tag{31}$$

This potential is certainly a solution of the Laplace equation so it is a solution of (29). Substituting (30) into (29) the shape coefficient C_P can be obtained in general by

$$C_P = -\int_S \frac{\partial G_{PQ}}{\partial n}ds \tag{32}$$

The Green function is designated by G_{PQ} in (29). In the present work, the fundamental solution for a pulsating source and its mirror imagine referring to the sea bottom S_H is used as the Green function, i.e.

$$G_{PQ} = \frac{1}{r_{PQ}} + \frac{1}{r_{PQ'}} \tag{33}$$

where

$$r_{PQ} = \sqrt{(x_P - x_Q)^2 + (y_P - y_Q)^2 + (z_P - z_Q)^2} \tag{34}$$

$$r_{PQ'} = \sqrt{(x_P - x_Q)^2 + (y_P - y_Q)^2 + (z_P + z_Q + 2h)^2} \tag{35}$$

This Green function has an advantage of simplicity although an integral over the free surface should be carried out. Anyway, the free surface integral is unavoidable when solving the first order problem even if more complicated Green function is used.

The Integral Equation

The boundary surface S surrounding the inner domain Ω_I consists of the body surface S_B, the part of the free surface S_F'' and the sea bottom S_H'' inside the virtual boundary and the artificial boundary S_R itself. Substituting the boundary conditions on these surfaces into (4.1.1) and using the expansion (3.1.1) for the value of potential and its derivative on the virtual boundary, the integral equation satisfied by the steady disturbance potential can be written as:

$$C_P\chi_P + \int_{S_F''}\chi_Q\frac{\partial G_{PQ}}{\partial n}ds + \int_{S_B}\chi_Q\frac{\partial G_{PQ}}{\partial n}ds$$
$$+ \sum_{m=0}^{\infty}\sum_{n=0}^{\infty}\int_{S_R}\left(\frac{\partial G_{PQ}}{\partial n} - \frac{\partial R^{(s)}_{mn}(r)}{\partial r}G_{PQ}\right)Z^{(s)}_m(z)$$
$$\left(A^{(s)}_{mn}\cos n\theta + B^{(s)}_{mn}\sin n\theta\right)ds = \iint_{S_B} n_1 G_{PQ}ds \tag{36}$$

Similarly, the integral equation for the zero-th and the first order unsteady potential can be obtained by substituting the corresponding boundary conditions and the expansion (24) or (28) into (29), i.e.

[Zero-th order problem]

$$C_P\phi^{(0)}_P + \int_{S_F}\phi^{(0)}_Q\left(\frac{\partial G_{PQ}}{\partial n} - \nu_0 G_{PQ}\right)ds + \int_{S_B}\phi^{(0)}_Q\frac{\partial G_{PQ}}{\partial n}ds$$
$$+ \sum_{m=0}^{\infty}\sum_{n=0}^{\infty}\int_{S_R}\left(\frac{\partial G_{PQ}}{\partial n} - \frac{\partial R_{mn}(r)}{\partial r}G_{PQ}\right)Z_m(z)$$
$$\left(A^{(0)}_{mn}\cos n\theta + B^{(0)}_{mn}\sin n\theta\right)ds = \int_{S_B} nG_{PQ}ds \tag{37}$$

where n is n_j which is defined in the boundary condition on the body surface, i.e. (11) and (12) for j = 1-6 respectively and it is understood that $n_7 = -\partial\phi_0/\partial n$.

[First order problem]

$$C_P\phi^{(1)}_P + \int_{S_F}\phi^{(1)}_Q\left(\frac{\partial G_{PQ}}{\partial n} - \nu_0 G_{PQ}\right)ds + \int_{S_B}\phi^{(1)}_Q\frac{\partial G_{PQ}}{\partial n}ds$$
$$+ \sum_{m=0}^{\infty}\sum_{n=0}^{\infty}\int_{S_R}\left(\frac{\partial G_{PQ}}{\partial n} - \frac{\partial R_{mn}(r)}{\partial r}G_{PQ}\right)Z_m(z)$$
$$\left(A^{(1)}_{mn}\cos n\theta + B^{(1)}_{mn}\sin n\theta\right)ds = -\int_{S_B}\frac{m}{i\nu_0}Gds$$
$$+\int_{S_F} f\,G_{PQ}ds - \int_{S_R}\Theta_1\frac{\partial G_{PQ}}{\partial n}ds + \int_{S_R}\Theta_2 G_{PQ}ds \tag{38}$$

where

$$\Theta_1 = -2\left(i\frac{\partial}{\partial x} + k_0\cos\beta\right)\frac{\partial\bar{\phi}^{(0)}_j}{\partial \nu_0} \tag{39}$$

$$\Theta_2 = \frac{\partial\Theta_1}{\partial n} \tag{40}$$

In (38), f and m are f_j (j =1-6, D) and m_j (j =1-6) defined in (9), (14) and (15) respectively and it is understood that m_D is null.

Comparing (37) with (38), the left hand side of both equations has the same form, which leads to the same coefficient matrix for either the zero-th order or the first order potential after discretization. This is a convenience in solving the problem.

DISCRETIZATION USING ISOPARAMETRIC ELEMENTS

In the present work, a set of second order isoparametric elements is used to discrete the integral equation given in the previous section. The following integral is taken as an example to discuss the discretization, i.e.

$$\int_S \phi_Q \frac{\partial G_{PQ}}{\partial n} ds \tag{41}$$

If the integral surface S is divided into n_e surface elements ΔS_l. (l = 1, 2, ..., n_e), the integral in (41) can be approximated as

$$\int_S \phi_Q \frac{\partial G_{PQ}}{\partial n} ds = \sum_{l=1}^{n_e} \int_{\Delta S_l} \phi_Q \frac{\partial G_{PQ}}{\partial n} ds \tag{42}$$

A local coordinate (ξ, η, ζ) is introduced to each surface element. The plane of (ξ, η) is perpendicular to the normal vector of the element. Considered in this local coordinate, the infinitesimal element ds can be expressed as

$$ds = \left|\frac{\partial x}{\partial \xi} \times \frac{\partial x}{\partial \eta}\right| d\xi d\eta = |D| d\xi d\eta \tag{43}$$

where

$$\frac{\partial x}{\partial \xi} = \left(\frac{\partial x}{\partial \xi}, \frac{\partial y}{\partial \xi}, \frac{\partial z}{\partial \xi}\right) \tag{44}$$

$$\frac{\partial x}{\partial \eta} = \left(\frac{\partial x}{\partial \eta}, \frac{\partial y}{\partial \eta}, \frac{\partial z}{\partial \eta}\right) \tag{45}$$

and the Jacobian $|D|$ can also be regarded as the absolute value of the normal vector. Therefore, the unit normal vector is given by

$$n = \left(\frac{\partial x}{\partial \xi} \times \frac{\partial x}{\partial \eta}\right) / |D| \tag{46}$$

Then, the integral over each element in (42) can be expressed as

$$\int_{\Delta S_l} \phi_Q \frac{\partial G_{PQ}}{\partial n} ds = \int_{\Delta S_l} \phi_Q \frac{\partial G_{PQ}}{\partial n} |D| d\xi d\eta \tag{47}$$

In each element, n_n nodes are selected. By means of basis function $N_k(\xi,\eta)$ (k = 1, 2, ..., n_n), the value of the potential at any point in the element ΔS_l can be expressed by the local coordinate (ξ, η), which is a linear combination of basis function N_k and the nodal value $\phi(k,l)$ at k-th nodes in l-th element, i.e.

$$\phi_Q(\xi,\eta) = \sum_{k=1}^{n_n} N_k(\xi,\eta)\phi(k,l) \tag{48}$$

Similarly, the shape of each element can also be expressed by the local coordinate by means of basis functions and the coordinate value of corresponding nodal points :

$$x_Q(\xi,\eta) = \sum_{k=1}^{n_n} N_k(\xi,\eta)x(k,l) \tag{49}$$

By substituting (48) into (47) the integral over each element is expressed as

$$\int_{\Delta S_l} \phi_Q \frac{\partial G_{PQ}}{\partial n} |D| d\xi d\eta$$

$$= \sum_{k=1}^{n_n} \phi(k,l) \int_{\Delta S_l} N_k(\xi,\eta) \frac{\partial G_{PQ}}{\partial n} |D| d\xi d\eta \tag{50}$$

In such a way, the unknown potential values at nodes are brought out of the integral. The remaining integrals only involve known quantities. Furthermore, the local coordinate (ξ, η) for each element is defined within the interval (-1, 1). The remaining integral can be numerically done by the Gauss integral identity.

The other integrals in the integral equations can be treated in a similar way. Finally, the integral equation is transferred to a group of algebraic equations for the unknown nodal values of potential.

SPECIAL TREATMENT

In the integral equation (38) for the first order unsteady potential derived in section 4, the integrand contains second order derivatives of steady potential, i.e. the m_j term in the body surface integral (see 14 and 15) and $\partial^2\chi/\partial z^2$ in f_j used in the free surface integral (see 9). In the numerical calculation, second order derivations would worsen the accuracy dramatically. They had better be avoided.

m_j Tterm On The Body Surface

By means of the Tuck theorem(Ogilvie and Tuck, 1969), we can prove the following integral identity:

$$-\int_{S_B} m_j G ds = \int_{S_B} \nabla(\chi - x)\cdot \nabla G n_j ds \tag{51}$$

Here, it is assumed that the body surface is orthogonal with the free surface at the intersection, i.e. at the water line and the boundary conditions for the steady potential have been used. Thus, the integral of the second order derivatives is replaced by the derivatives of first order only.

$\partial^2\chi/\partial z^2$ On The Free Surface

In a similar way, if the z-direction of the Tuck theorem(Newman, 1993) is applied to the free surface integral, the following identity can be obtained:

$$\int_{S_F} G\phi_j^{(0)}\chi_{zz} ds = \int_{S_F} \left(G\nabla_2\phi_j^{(0)} + \phi_j^{(0)}\nabla_2 G\right)\cdot\nabla_2\chi ds - \int_{C_R} G\phi_j^{(0)}\chi_n dl \tag{52}$$

where the subscript 2 denotes a two-dimensional gradient operator in the horizontal plane.

CALCULATED EXAMPLES

Vertical Circular Cylinder

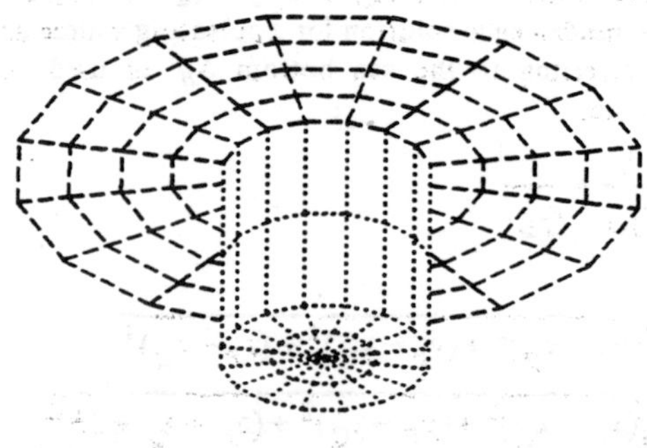

Fig. 2a Discretization of the body surface and the free surface for a circular cylinder.

A circular cylinder with radius a and draft $d=a$, which is floating in water of depth $h=2a$, is taken as an example. The cylinder is discretized into two facets in its draft and sixteen sections in its circumference. The radius of the virtual boundary cylinder is set as

$r_0 = 3a$. The free surface inside the virtual boundary is divided into four facets in its radial direction and sixteen sections in its circumference. The discretization of the cylinder as well as the free surface is shown in Fig. 2a. The incident wave angle β=0, i.e. opposite to the current direction. The calculated results are compared with the semi-analytical solutions(Kinoshita and Bao, 1996), which are expressed in an eigen-function expansion, to verify the present method.

Presented in Fig.3 is the wave exciting force in surge, heave and pitch direction respectively while the added mass and damping are shown in Fig. 4. It can be seen that the present results are in good agreement with the semi-analytical ones. The wave drift forces in x-direction are plotted in Fig. 5 while the wave drift damping is given in Fig. 6. Good agreement can be observed when compared with the semi-analytical results.

Ellipsoid

A series of ellipsoids with various ratios between breadth, draft and length are used to examine the effects of the slenderness and flatness of the body shape on the wave drift force as well as the wave drift damping. The length of these ellipsoids is set to $2a$ and various combinations of breadth and draft are listed in Table 1.

Breadth	$2a$	$2a$	$2a$	$2a$	$3/2a$	a	$1/2a$
draft	a	$3/4a$	$1/2a$	$1/4a$	$1/4a$	$1/4a$	$1/4a$

Table 1 Combination of breadth and draft for ellipsoids used in calculation.

The ellipsoid is fixed and the incident wave angle is again set to β=0. An example of discretization of the body surface and the free surface is shown in Fig.2b.

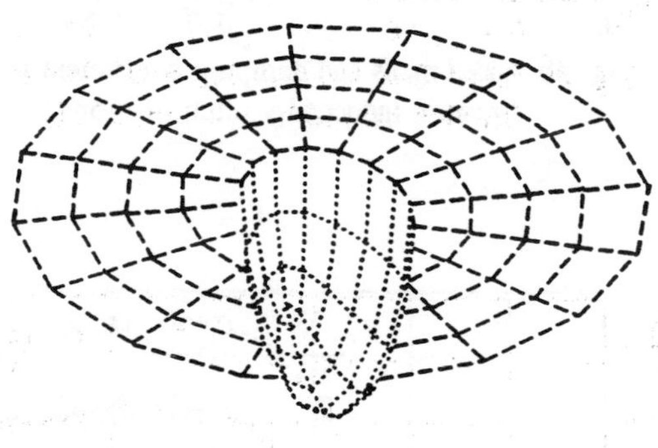

Fig. 2b Discretization of the body surface and the free surface for a ellipsoid.

Calculated wave drift force and wave drift damping are presented in Fig. 7 and Fig. 8 respectively. To show the effects of the slenderness and flatness more clearly, the ratio of the wave drift force to the wave drift damping is plotted in Fig.9. It can be observed that change in breadth or draft will affect this ratio in the same tendency for smaller wave numbers. For larger wave number, the breadth has much greater effect than the draft does. In the case of ellipsoids, when the wave number is greater than 1.1 the ratio between the wave drift force and the wave drift damping is dramatically increased as the breadth decreases. This implies a large amplitude of slow drift motion would occur.

CONCLUDING REMARKS

In the present work the problem of bodies floating in both waves and slow current is considered. The velocity potential is expanded into a perturbation series of small parameter τ_0. The problem is solved by a hybrid boundary element method. The fluid domain is divided into two regions by the virtual boundary. A higher-order boundary element method is applied in the inner region while the eigen-function expansion is adopted in the outer region. This method has an advantage of saving much computational effort by using fewer discretized panels especially when the free surface integral cannot be avoided.

In the comparison with the semi-analytical solutions for the case of a circular cylinder, good agreement is obtained although the number of used panels is only 192.

A series calculation for ellipsoids with various combination of breadth and draft is carried out to examine the effects of slenderness and draft of the body on the wave drift force as well as the wave drift damping. At small wave number, the breadth and draft have a similar effects on the change rates of the wave drift force and wave drift damping. However, The breadth affects the change rates dramatically at large wave number. This might cause a large excursion of slow drift motion.

REFERENCES

Emmerhoff, OJ and Sclavounos PD (1992). "The slow-drift motion of arrays of vertical cylinders." J Fluid Mech. Vol 242, pp 31-50.

Grue, J and Palm, E(1993). "The mean drift force and yaw moment on marine structures in waves and current." J Fluid Mech Vol 250, pp 121-145.

Kinoshita, T and Bao, W (1996). "Hydrodynamic forces acting on a circular cylinder oscillation in waves and a small current." J Mar.Sci Technol, Vol 1, No 3 pp 155-173.

Malenica, S, Clark, PJ and Molin, B (1995). "Wave and current forces on a vertical cylinder free to surge and sway." Appl Ocean Res. Vol 17, pp 79-90.

Masuda, K and Nagai, T (1988) "Study on nonlinear wave forces on plural vertical cylinder with arbitrary cross-sections (second report)." J Soc of Naval Archit of Japan. Vol 162, pp 222-229.

Matsui T, Lee SY, Sano K (1991). "Hydrodynamic forces on a vertical cylinder in current and Waves "(in Japanese). J Soc of Naval Archit of Japan. Vol 170, pp 277-287.

Newman, JN (1993)."Wave-drift damping of floating bodies in waves." J Fluid Mech. Vol 249, pp 241-259.

Ogilvie, TF and Tuck, EO (1969). "A rational strip theory of ship motions." Part 1 Rep 013 The Dept of Naval Archit and Marine Eng, The Univ. of Michigan, College of Eng.

Yuen, MMF and Chau, FP (1987). "A hybrid integral equation method for wave forces on three dimensional off-shore structures." J Offsh Mech and Arc Eng, Trans. ASME. Vol 109, No 3, pp 229-236.

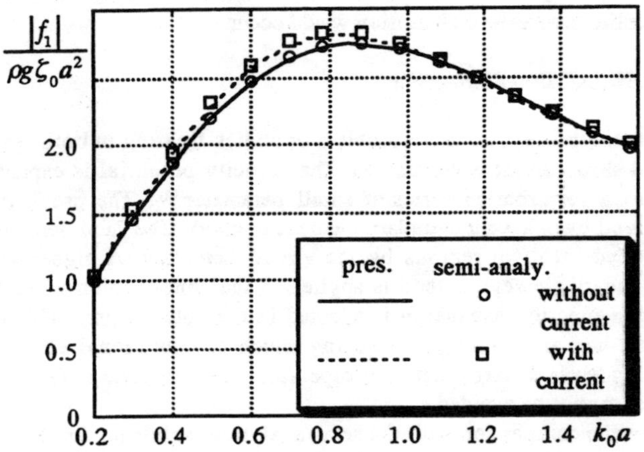

Fig. 3a Wave exciting forces in surge direction.

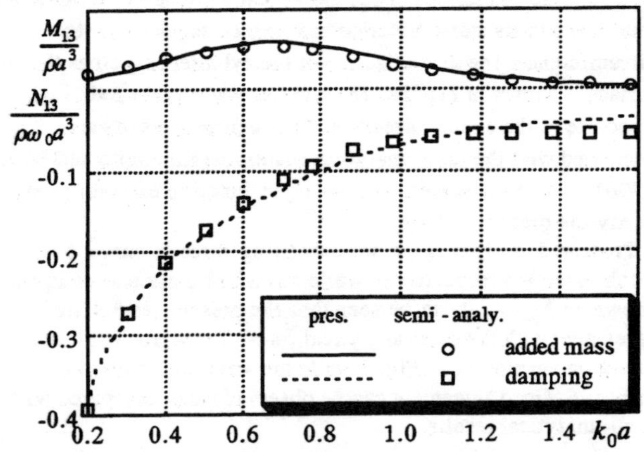

Fig. 4a Added mass and damping coefficient in surge direction induced by heave motion.

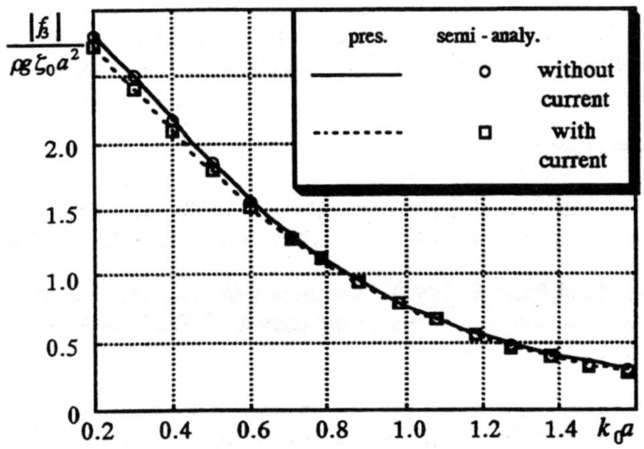

Fig. 3b Wave exciting forces in heave direction.

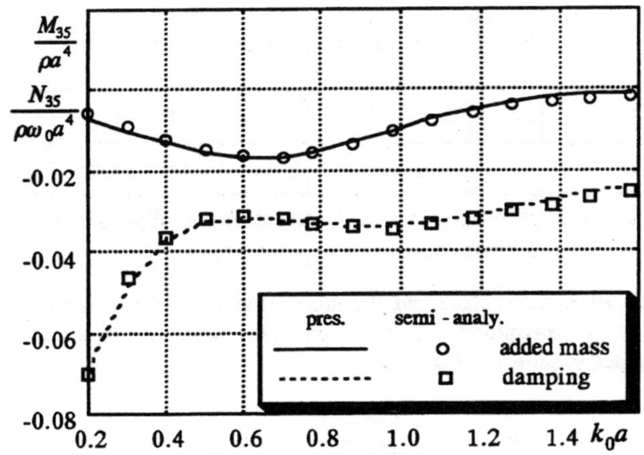

Fig. 4b Added mass and damping coefficient in heave direction induced by pitch motion.□

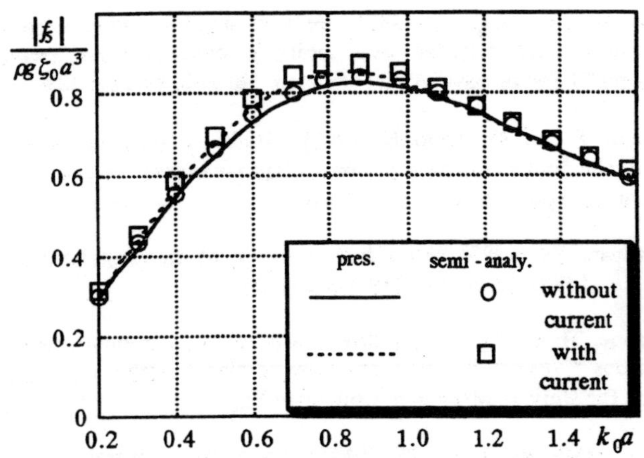

Fig. 3c Wave exciting pitch moments.

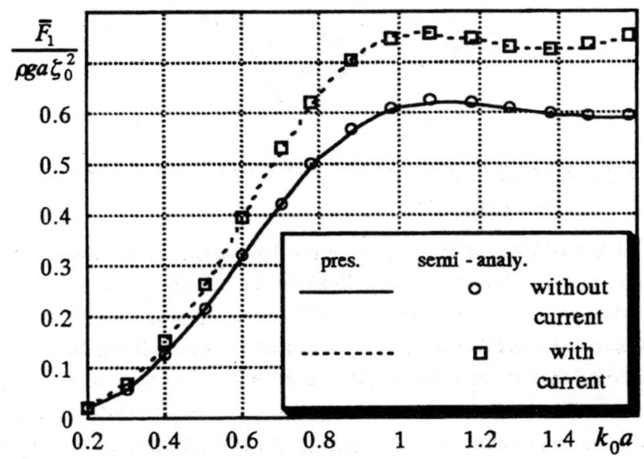

Fig. 5a Wave drift forces in surge direction for a fixed circular cylinder.

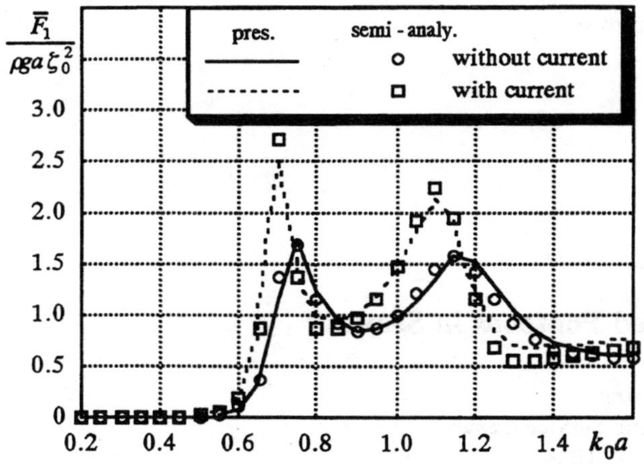

Fig.5b Wave drift forces in surge direction for an oscillating circular cylinder.

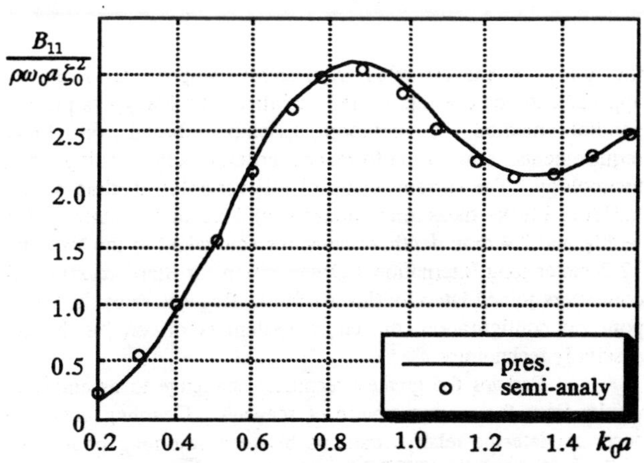

Fig. 6a Wave drift damping in surge direction for a fixed circular cylinder.

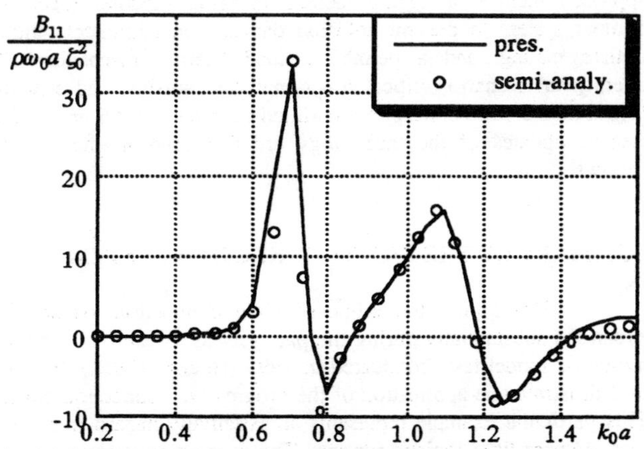

Fig. 6b Wave drift damping in surge direction for an oscillating circular cylinder.

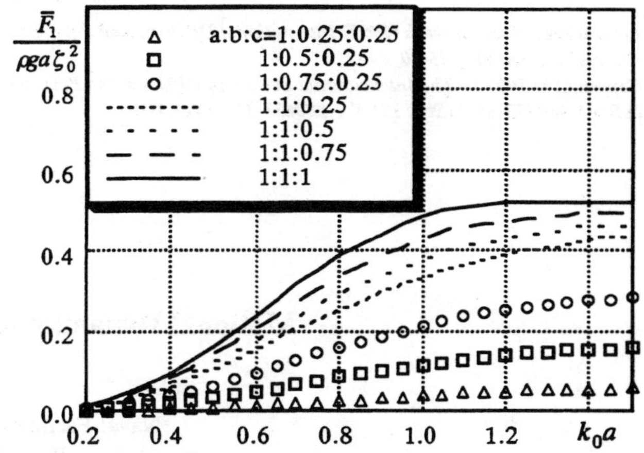

Fig. 7 Wave drift forces in surge direction for ellipsoids.

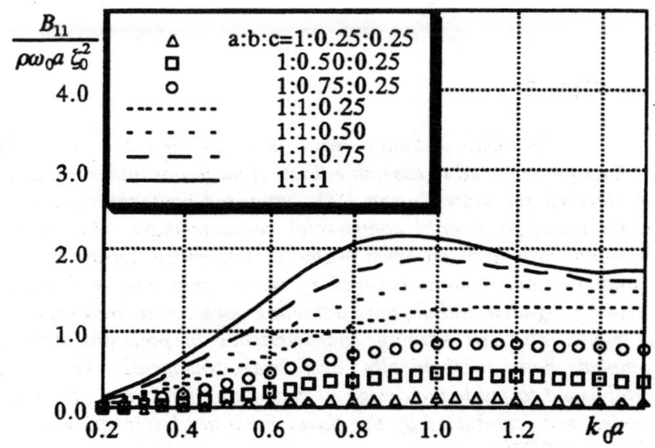

Fig. 8 Wave drift damping for ellipsoids.

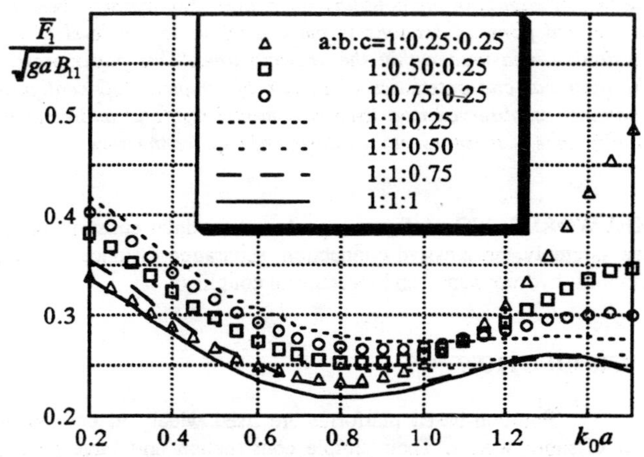

Fig. 9 Ratio between wave drift gorcd and wave drift damping for ellipsoids.

Proceedings of the Seventh (1997) International Offshore and Polar Engineering Conference
Honolulu, USA, May 25-30, 1997

ISBN 1-880653-28-1 (Set); ISBN 1-880653-31-1 (Vol. III)

Motional Dynamics of Coupled Pontoons in Seaways

Erick T. Huang
Naval Facilities Engineering Service Center
Port Hueneme, CA, USA

ABSTRACT

The United States Navy is developing the connecting technology required to assemble floating platforms under conditions of elevated sea states, using ISO-configured pontoons suitable for rapid transport aboard commercial containerships. Assembly of pontoon structures on open water is influenced substantially by rapid, random motions induced by waves, requiring a means for drawing together and aligning adjoining pontoons as the connection is made. The sea keeping characteristics of pontoons that are coupled both mechanically and hydrodynamically have been established to guide the design of connecting hardware. Hydraulic model tests conducted by the Naval Facilities Engineering Service Center (NFESC) indicate that a particular coupling apparatus may either reduce or augment wave-induced relative motion, depending on the layout and dynamic characteristics of the rigging system. Improper rigging could result in unacceptable excursions between sections - relative motions leading to collisions between pontoons or over tensioning of drawing lines - both problems that limit the success of engaging the coupling apparatus. The dynamic behavior of coupled pontoon sections is most sensitive to the level of pre-tensioning measured within the drawing lines. This paper presents the motional characteristics of a pair of coupled, ISO-configured pontoons as observed in a hydraulic model test, and addresses the significance of these tests to the design of rigid connectors..

KEYWORDS: ISO-configured, pontoons, modular construction, at-sea assembly, progressive connection, separation distance, dynamic motions, hydrodynamic and mechanical couplings.

INTRODUCTION

Pontoon-based platforms are used widely in both inland and offshore waters. Their simple construction and large payload capacity provide reliable working surfaces at a low cost. However, current Navy pontoons are not easily relocated by available sealift vessels due to their large size and blunt hull shape, and hence, are limited in their service watch circle. Modular construction techniques that allow on-site assembly from small components is one design approach to achieve the desired mobility. The Navy has proposed a general method of pontoon assembly based on logistical requirements, seaway performance criteria, and existing shipping technology. The current vision of this assembly method sees basic building blocks measuring (nominally) 12.2 m in length, 7.3 m in width, and 2.4 m in depth, dimensions equivalent to the size of three 12.2-meter long International Organization for Stardardization (ISO) containers joined laterally (Kane 1996). Regardless of the particular pontoon configuration, the entire system is driven by the on-site assembly technology.

Perhaps the greatest technical challenge to assembly at sea results from the random nature of seaways. Coupling devices must accommodate the relative motions between adjoining pontoons. The study completed by NFESC concludes that at-sea assembly can be accomplished by joining the pontoons progressively. Specifically, in the progressive connection, neighboring pontoons are drawn from an initial separation distance and aligned under increasing control so that relative motions are dampened gradually to amplitudes small enough to safely engage the connector components. Features required in this approach include: a rigging method to bring modules together, a bumper system to prevent collision damage, an alignment guide to initiate mating, and a permanent load bearing mechanism. The operational scenario, critical functional mechanisms, and structural layout of the basic concept are outlined in (Huang 1995 and 1996a). Recent updates of the technology are discussed in the text that follows.

SUMMARY OF THE RIGID CONNECTION SYSTEM

Developing the proposed rigid connection system has expanded substantially during the past year as a result of successful hydraulic model tests conducted in 1995 (Huang, 1996a). Figures 1 to 3 illustrate one application of the progressive connection concept. This particular example represents an essentially hazard free system that requires little logistic support. The connection system performs adequately in rough waters exceeding a sea state 3 condition, using two tugs with a 22.2 KN (5,000 lbs) bollard pull, at least one of which is also equipped with a winch of 66.7 KN (15,000 pound) capacity. One person on each pontoon, in addition to the tug crews, is required to connect rigging lines and trigger the final locking mechanism of the connection system. Figure 1 shows the general layout of the connection system, including four rigid connector assemblies (two at each end of the two mating pontoons), a pair of alignment pins, and marriage-bridle rigging that is attached to the end of each alignment pin. The other end of the bridle is attached to the winch on one of the tugs as shown in Figure 2. Figure 3 illustrates the structural details of the rigid connector assembly, consisting of a spring loaded stabbing pin, receptacle, roller racks, built-in crow bars, and restraining guillotines, which all fit into a steel frame. Actual layouts can vary to fit specific pontoon geometry and operational requirements. The inboard end of the stabbing pin bears against a spring support contained in a canister, which is mounted on rollers so that the pin and canister assembly can slide forward and backward in the ready and stowed modes, respectively. The pin and recoil canister at full extension are held at the inboard end. Further retraction of the stab pin as a result of pontoon collision passes the impact energy to the recoil spring. When a pin is properly engaged within its receptacle, a guillotine locking collar within a vertical guide channel drops from the deck to capture the recessed groove near the outboard end of the stabbing pin. The guillotine then bears against the guide channel preventing the pin from slipping out.

Figure 1. Connector System.

Figure 2. Baseline Concept.

Figure 3 Rigid connector assembly

Figure 3. Rigid Connector Assembly.

The alignment pin is simply a section of chain covered by an elastomeric sleeve. However, it performs the critical function of intermediate transition effectively, as experiments have shown. The combination of chain and sleeve is flexible enough to accommodate wild differential motions, yet sufficiently robust to withstand vigorous dynamic loads. Chains outperform rods or wire ropes as tendons because of strength and flexibility properties, and ability to absorb shock loads. The elastomeric sleeve prevents direct steel-to-steel abrasion and keeps the chain from tangling. The sleeve allows for substantial bending, yet is effective in restraining the differential translations between pontoons at short free lengths due to a relatively large cross section. Alignment pins are able to closely synchronize the random differential motions at the connection interface without suffering debilitating damage, and maintain the stabbing pins within close proximity to their respective receptacles long enough for connectors to engage.

The connection process begins the moment that pontoons are launched from the delivery ship. Tugs assume control of pontoons at ship side and tow them to an open site for connection. A pair of tugs and pontoons is aligned at a separation distance of roughly 12 m. With the spring-loaded stab pins extended, messenger lines carrying the split-leg bridles are passed between modules. With bridle lines connected to extended alignment pins, the tugs use opposing thrust to establish and maintain a reasonable level of pretension between bridle legs as shown in Figure 3, thus preventing snap loads. All the while, the winch draws the marriage bridle legs slowly until the alignment pins are led into the receivers. At close quarter, these pins assist in aligning the module ends and substantially reduce differential motions between the two. Further retraction of the bridle legs reduces the gap, and the stab pins are likely to bear against the face of the adjoining module if they do not find their respective matching receivers straight away. When entry is delayed, the spring-loaded pins function as bumpers to protect the modules, and small random excursions around the receivers continue until each stab pin finds its mark and glides in. Because this technique does not require the precise alignment of all connecting members simultaneously, it has a much better chance of working in the open seaways.

PERFORMANCE TEST

The performance of the recommended connection concept was demonstrated during a follow-on series of seakeeping tests at the Offshore Model Basin located at Escondido, California. These tests were conducted to: (1) observe the dynamic performance of the connection system, particularly the coordination among stabbing pins, alignment pins, and marriage bridles, as well as the fendering function of the stabbing pins, (2) quantify the motional dynamics of the pontoon couple and the associated coupling forces at various stages of the connection process, (3) investigate the influence of pretensioning on the motional dynamics, (4) quantify operational weather window, and (5) investigate manufacturing tolerance requirements. This paper addresses the first three areas.

Test Layouts

The test basin was about 90 m long, 15 m wide and 5 m deep. The test layout was similar to that used previously in 1:8 scale model tests (Huang 1995), with a few revisions to accommodate the larger scale models in the same basin. The U. S. Navy's high sea state pontoons as described by Kane (1996) were again selected as the test bed. Six 1:4 scale models of these pontoons were constructed from plywood and fiberglass. The model alignment pins were constructed roughly 50% as stiff as those used in 1995 experiment to check that characteristic on overall performance. Because the mobility and compatibility of moving parts in the rigid connector assembly are important to the performance of the entire connection system, these assemblies were built precisely to quarter scale using mild steel (A36) to evaluate the efficiency of mechanisms in use. An electric winch mounted on the leading pontoon (near the wave generator) served the function of the warping tug. The general layout of the test setup is shown in Figure 4. Mating pontoons were placed

Figure 4. General layout of the test setup.

under a mooring ring truss, near the center of the towing tank, at an initial separation of 1.5 m. Wire ropes payed from the electric winch were guided by a pulley through vacant receivers, passed over the water between sections, and attached to the ends of extended alignment pins on the trailing section. The trailing barge that had been intended to represent one of the supporting tugs used in previous tests was eliminated. Two spring-loaded mooring lines were hooked fore and aft to the outside ends of the adjoining pontoons as shown in Figure 4, in order to model the separation force that would be imposed by the sea anchors. The array was pretensioned to about 140, 350, 700, and 1040 N scale force (i.e. 9, 22, 45, and 67 KN pretension in full scale) with a free hanging dead weight attached to the end of the trailing mooring line strung through a series of pulleys. Each pontoon was attached to a motion sensing transducer (MST) that hung from the mooring ring truss.

Tests were conducted in calm water as well as in regular waves of scaled height 1.2 m (4 ft) with periods varying from 4 to 8 seconds, and also in irregular waves representative of sea states 3 and 4. The models were set up initially for tests in head seas. Additional tests were conducted at 15 degrees of head seas for selected cases. The connection process was modeled to resemble an actual operational scenario. Data were collected electronically and the tests were video taped.

Data Acquisition

Motions of individual pontoons were measured in six degrees of freedom using a separate MST for each mating structure. The distance between sections was measured by a rotary potentiometer connected by a string to provide a direct verification of the results observed by the MSTs. Tensions in bridle legs and pretension lines were measured by load cells. The instrumented stab pins measured shear, bending, and tensile loads. Unfortunately, the strain gauges did not function as designed and reported very little useable data. Wave conditions were recorded using two gauges located up stream along the center line of the model assembly and 5 m off the abeam with the mid-point between two pontoons. Time histories of pontoon motion, line tension, and incoming wave posture were recorded on a PC-based computer system. The motions measured by the MST were uncoupled with respect to the center of gravity (CG) of each model situation.

Data Analysis

The connection procedures, dynamic behavior of pontoons, and performance of the connector components are documented on a video tape (Huang 1996b). Figures 5 and 6 illustrate the raw data recorded from a typical test run. This particular test was conducted with two pontoons end connected in 5-second head seas of 1.2 meter waves with a 22.2 KN separation force (Figure 7). Figure 5 illustrates the time histories of line tension compared to separation distance. The connection is completed when the separation distance between the adjoining faces is reduced to nothing. As noted, the pretension remains essentially constant over the entire duration, while the tension loads recorded on the marriage bridles and mooring lines oscillate under wave actions. The high tensions in the bridle result from continued winching after the pontoons were pressing against one another in this test case, rather than from the occurrence of snap loads. Figure 6 verifies the active motion of pontoons in an elevated sea state. However, the primary concern to the connector design is the differential displacements at the connector interface rather than the behavior of individual pontoons. These relative motions may be derived from the motions of individual rigid bodies. The results of relative motions in surge, sway, heave, and pitch modes are illustrated in Figures 8, 9, 10 and 11. The joint distributions of relative kinematics that are directly associated to the coupling loads and critical to the structural layout of the connection system have been calculated. Figure 12 is a scatter diagram of velocity versus acceleration in relative surge at various stages of connection process. A completed set of these test results is documented by Huang (1997).

The charts on the next pages are presented in customary English units. Table 1 provides the conversion factors to obtain the same in SI system units.

Unit Conversions for	use in the following	charts
multiply	by	to obtain
kips	4.4482	kilonewtons
feet	.3048	meters

Table 1. Unit conversions table..

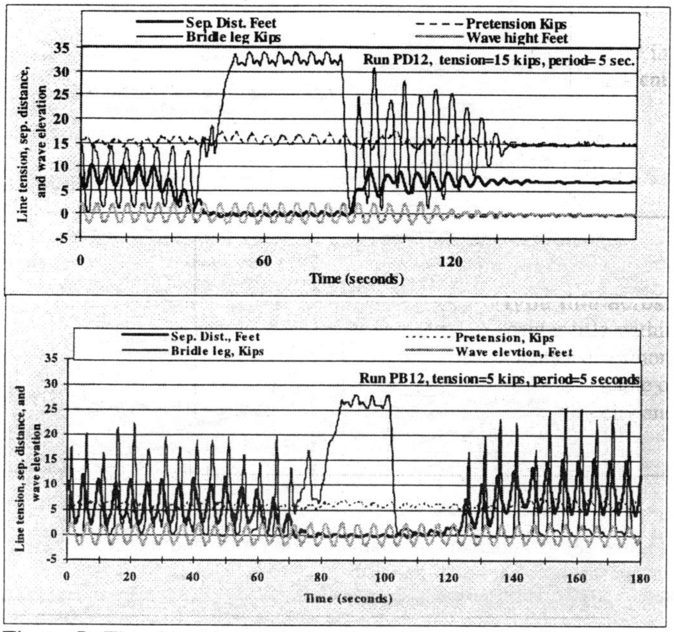

Figure 5. Time histories of line tension with separation distance.

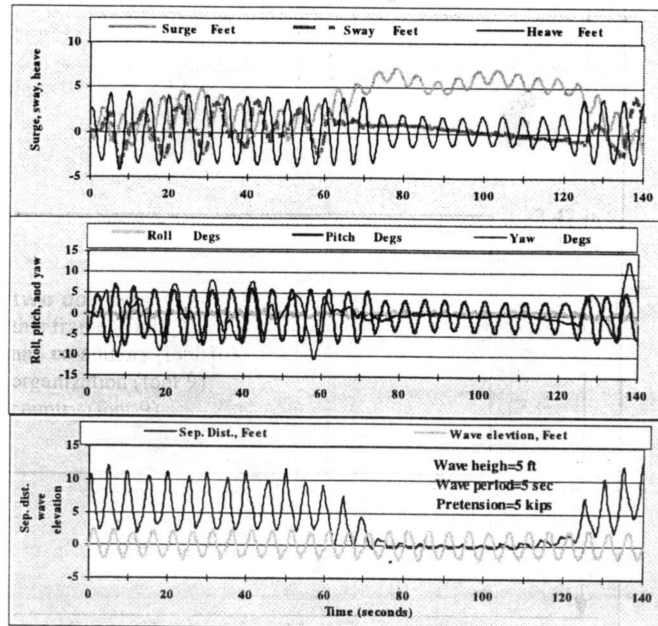

Figure 6. Time histories of a pontoon motion in an elevated sea state

Figure 8. Relative surge motion.

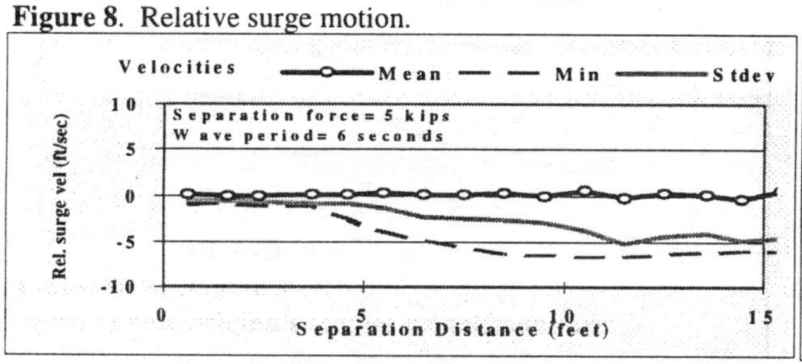

Figure 9. Relative heave motion.

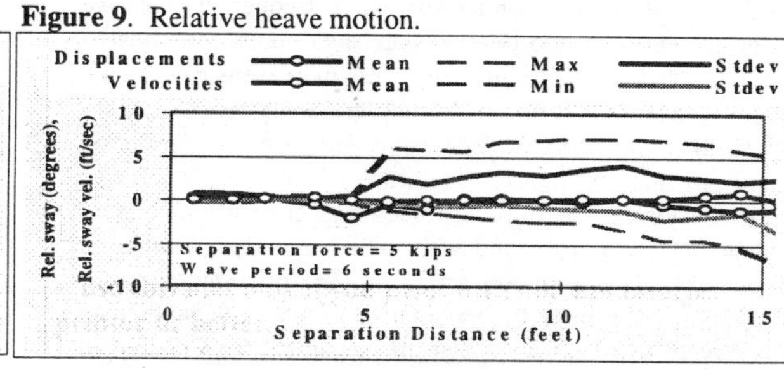

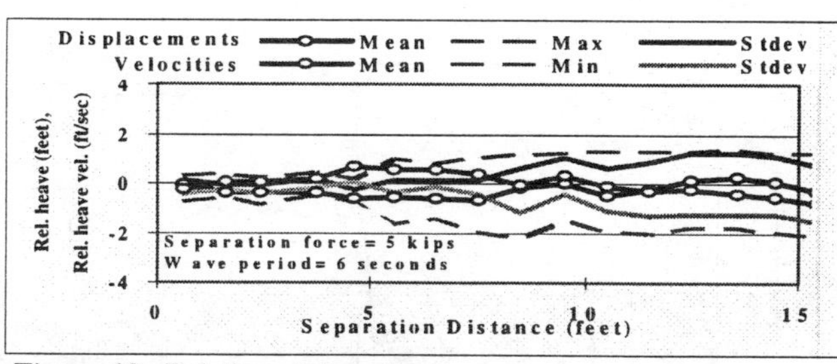

Figure 10. Relative pitch motion.

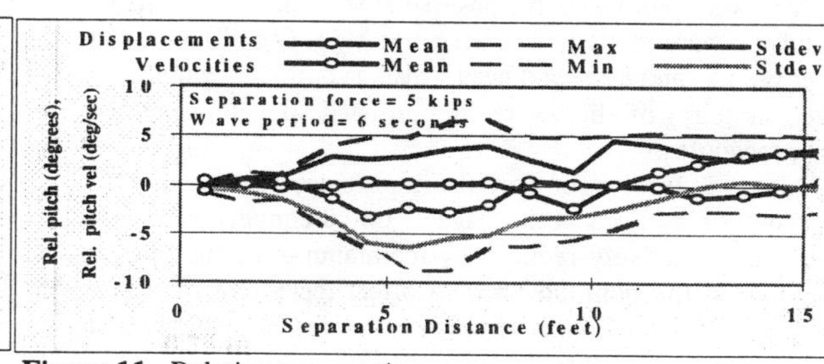

Figure 11. Relative yaw motion.

Figure 7. Two pontoons end connected in 5-second head seas.

DISCUSSION

The results of advanced model tests give support to the notion that pontoon modules configured for transport aboard containerships and assembly at sea are very responsive to prevailing sea state. This observation is perhaps the greatest technical challenge assembly on the water. Interface hardware used to complete the initial connection requires a combination of flexibility, strength, and shock absorbing capability to accommodate the large relative motions between pontoons. The progressive connection procedures proposed by the Navy offer ample opportunity for implementing these essential features that ease the transitional impact of drawing together and aligning pontoons for connection. Model tests confirmed that the most

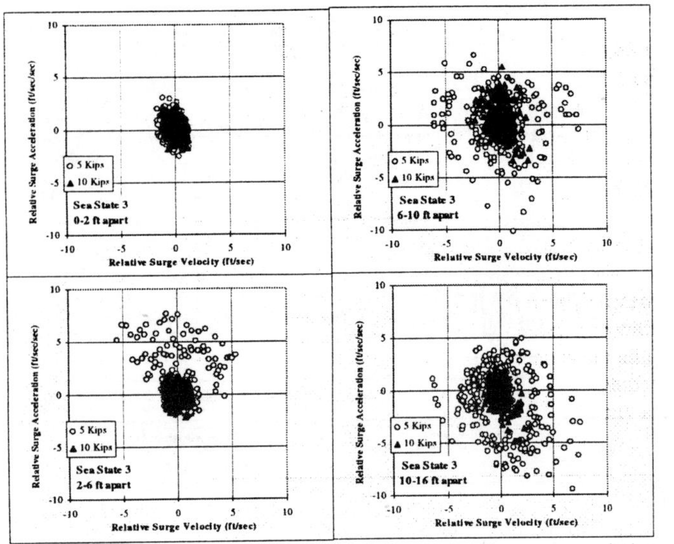

Figure 12. Scatter diagram of relative velocity versus relative acceleration.

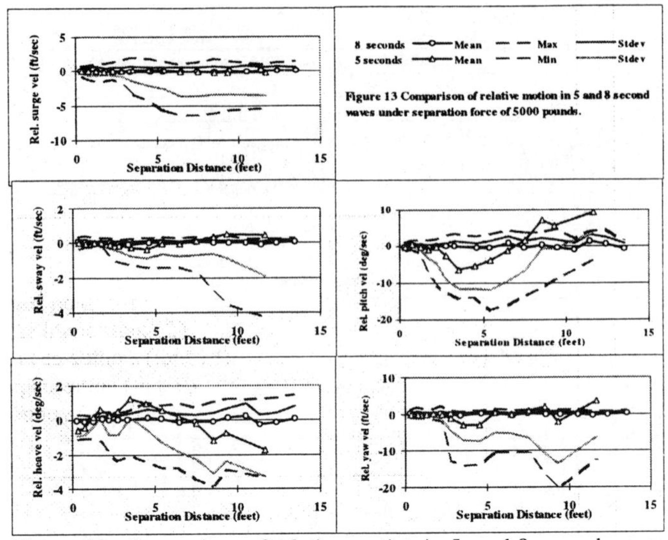

Figure 13. Comparison of relative motion in 5- and 8-second waves under separation force of 5,000 lbs.

Significant parameters influencing pontoon behavior during at-sea connections are wave frequency and the pretension force. Individual pontoons are as expected dependent on wave frequency. Relative motions at the interface of two pontoons are even more complicated because of phase differences which are even more sensitive to the wave frequency. Surge motion, in particular, is of most concern during the connection process. Excessive surge causes pontoon collisions and engages mechanical couplings of interface hardware. The subsequent asymmetric coupling forces further excite undesired lateral movements (sway, roll, and yaw) which could otherwise be minimized by taking the pontoon array into the oncoming waves. Figure 13 compares the pontoon motions in 5- and 8-second head waves, respectively, for pontoons that are separated by 1.5 m and witnessing 22.2 KN pretension. In this figure, the results of an 8-second test run were plotted on the positive side, while the results of the 5- second run were plotted on the negative side. The difference in behavior between 5- and 8-second head waves is dramatic enough to impact the strategy of hardware design and development of operational procedures.

This study further confirms the importance of pretension in controlling pontoon motions at the stage of initial connection. Surge motion could be effectively reduced by maintaining an appropriate separation force in the pontoon array to offset the forces of wave-induced surge. These separation forces may be realistically applied by sea anchors or tugs. Because wave-induced surge forces are highly frequency dependent, so is the required pretension to offset these forces. This relationship was clearly demonstrated in a series of tests with separation forces that varied between 9 and 90 KN. As noted in Figure 14, the significant surge that appears when pretension is 22 KN is reduced almost entirely when the pretension increases to 66 KN. At the same time, a substantial change in tension load imposed on the bridle legs as the pretension level is increased can be seen in Figure 15. At only 22 KN pretension, the bridle legs alternate between slack and taut, thus experiencing sharp shock loads. These shock loads, however, do not appear at 66 KN pretension, and the pontoons are drawn together without noticeable signs of inline oscillations. Although the tensions on the bridle legs are of comparable magnitudes to those occurring at 22 KN pretension, they appear almost static. The important implication is that pontoons may be drawn in at any time during the wave cycle without concern of the line snapping.

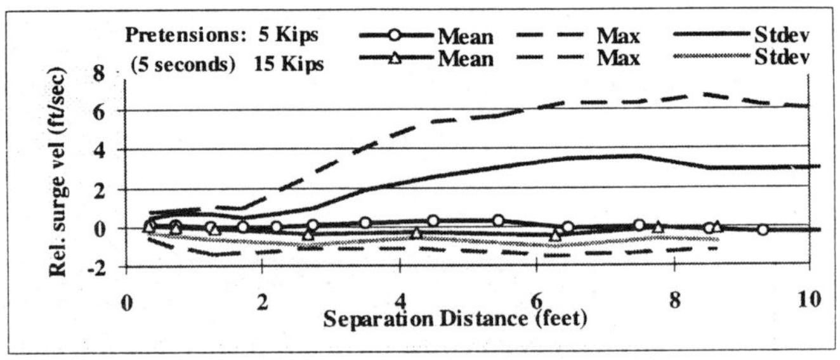

Figure 14 Comparison of relative surge velocities under separation forces of 5 and 15 thousad pounds, wave height = 4 feet, wave period = 5 seconds.

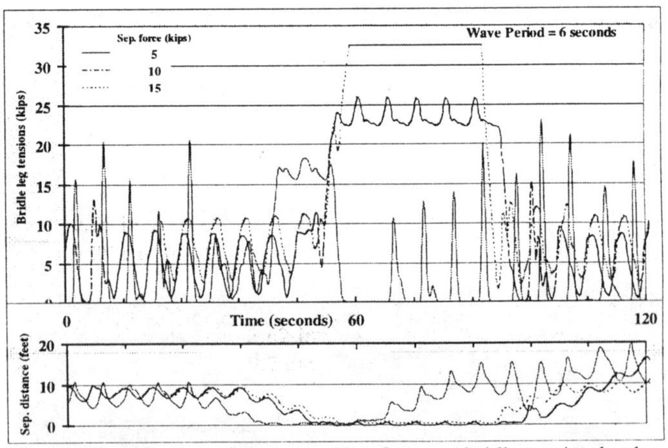

Figure 15. Influence of the separation force on bridle tension loads.

Figure 16 summarizes the pretension requirement, in which the horizontal axis indicates the wave frequency when the bridle legs begin to show signs of slacking at the pretension level indicated by the vertical axis. With the practical range of wind-generated seas, a higher pretension is required to maintain the connection process under control in seaways with shorter prevailing wave periods.

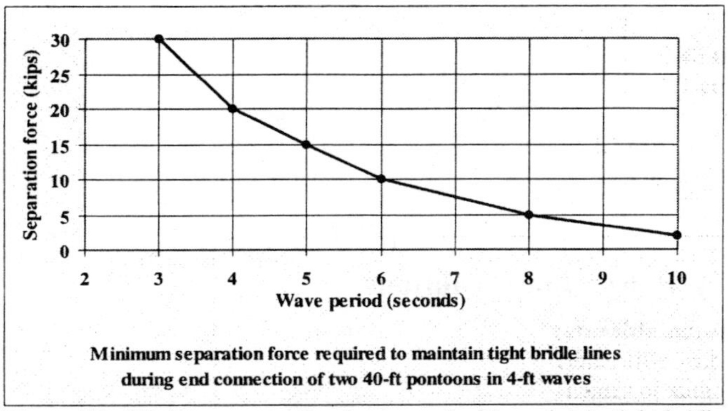

Figure 16. Minimum separation force required to maintain tight bridle lines during end connection of 12.2-m pontoons in 1.2-m waves.

CONCLUSIONS

Hydraulic model tests conducted by NFESC during the past year have demonstrated dramatically the feasibility of assembling large floating platforms on the open seas using prefabricated modules configured for rapid ocean transport aboard commercial containerships. At-sea assembly can be done even in rough water conditions using an approach in which connections between modules are completed progressively. The particular features required in such a progressive system include mechanisms for drawing and aligning modules, fenders to prevent collision damage at close quarters, and hardware strong enough to withstand wave induced loads. Experimental evidence suggests that a practical coupling device for at-sea assembly requires an appropriate combination of flexibility, strength, and shock load absorbing capability to survive the coupling forces resulting from large relative motion between modules. The effectiveness of the proposed at-sea assembly method was demonstrated using quarter scale models. Functional performance of all critically required mechanisms was confirmed for wave conditions exceeding sea state 3.

The greatest technical challenge to the design of coupling devices is the rapid, random motion between adjoining modules. The relative translational motions witnessed at the interface between two 12-m modules mating in sea state 3 conditions could exceed the amplitude of ambient waves. Effective control of relative motion, especially relative surge, is very important to the entire assembly procedure, as concluded from the test results. Assembly is easily and safely done by maintaining an adequate separation force within the pontoon array to offset surge forces generated by wave action. Although connections may be completed at lower magnitudes of separation force, this option comes at the expense of added weight and increased manufacturing costs as stronger, more durable components are needed. In addition, lower separation forces necessitate careful timing and, thus, greater experience on the part of the operating crew.

ACKNOWLEDGEMENT

This research and development effort is funded by the Office of Naval Research (ONR) as Task C-8: *Advanced Module Construction Technology* under the Replenishment Section (RM33U62) of the Navy Exploratory Development Technology Program LH2A - Facilities and Material.

REFERENCES

Kane, P., (1996). "Lots Development", *The Military Engineer*, Volume 88, Number 579, August-September, pp42-43.

Huang, E.T., (1995). "Conceptual Development of Open Sea Module Connection Techniques," Technical Memorandum TM-2067-AMP, Naval Facilities Engineering Service Center, Port Hueneme, California.

Huang, E.T., (1996a). "An Open Sea Modular Construction Method: Rigid Pontoon Connectors", *Sixth International Offshore and Polar Engineering Conference*, Los Angeles, CA.

Huang, E.T., (1996b). "Open Sea Module Connection System", Naval Facilities Engineering Service Center. In-house video presentation.

Huaang, E.T. (1997). "Open Sea Module Connection Techniques," Technical Memorandum (to be published), Naval Facilities Engineering Center, Port Hueneme, California..

Proceedings of the Seventh (1997) International Offshore and Polar Engineering Conference
Honolulu, USA, May 25-30, 1997

ISBN 1-880653-28-1 (Set); ISBN 1-880653-31-1 (Vol. III)

Interaction Between Twin Floating Power Plant Motions in Short-Crested Waves

Masaaki Ikeno and Ryoichi Kajima
Central Research Institute of Electric Power Industry
Abiko, Japan

ABSTRACT

A numerical model is developed by extending the equation of motion in the time domain with a retardation function of motion (Oortmerssen, 1976) to take into account directional spreading of irregular waves and interaction between plural body motions on the basis of the boundary element method. Hydraulic model tests of 1/100-scale twin barges are performed with a directional wave maker and the results compared with the calculated results. Simulation and experimental results show good agreement. The effects of the distance between the two barges on the motions and the difference between the motions caused by short- and long-crested waves are investigated. The effects of nonlinear mooring on surge and sway are also investigated.

KEY WORDS: floating power plant, plural body-interaction, short-crested waves, numerical simulation, hydraulic model tests

1. INTRODUCTION

The floating type power plant is one of the most attractive options in order to extend the available construction sites of nuclear or thermal power plants. In a typical concept of this type of power plant, each of twin plant barges is joined to a dolphin-link mooring system in a semi-closed basin surrounded by breakwaters and seawalls. In this design, there is concern that the motion of each plant increases more than that of a single plant due to interaction between the plants.

Oortmerssen (1979) investigated the interaction between two free floating body motions due to regular waves by the three-dimensional singular point distribution method and hydraulic model tests. Actual ocean waves, are however, irregular and multi-directional (short-crested), so the three-dimensional interaction among floating structures and nonlinear mooring force due to short-crested waves must be correctly estimated.

In this paper, first, a numerical model is developed by extending the equation of motion in time domain with a retardation function of motion (Oortmerssen, 1976) to take into account directional spreading of irregular waves and interaction between plural body motions on the basis of the boundary element method.
Secondly, hydraulic model tests of 1/100-scale twin plant barges are performed with a directional wave maker and the results compared with the calculated results. The characteristics of rubber fender models used for the mooring system are linear and nonlinear. Simulation and experimental results show good agreement. The effects of the distance between the two barges on the motion with six degrees of freedom are elucidated numerically, and the difference between twin plant motions caused by short- and long-crested waves is investigated. The nonlinear effects of the mooring system on surge and sway are also investigated.

2. NUMERICAL MODEL

2.1 Basic equations and boundary conditions

As shown in Fig. 1, the three-dimensional motions of each floating body consist of six degrees of freedom, so that the velocity potentials including the contributions from the motions of all the floating bodies of N_p can be expressed as follows:

$$\phi(x,y,z;t) = \{\phi_{in}(x,y,z) + \phi_{sc}(x,y,z) + \sum_{p=1}^{Np}\sum_{j=1}^{6} i\omega X_{jp} \exp(i\psi'_{jp})\, \phi_{jp}(x,y,z)\} \exp(i\omega t) \tag{1}$$

where t is time, ω the angular frequency, and ϕ_{in} and ϕ_{sc} are the potentials of incident and diffracted waves by N_p fixed bodies, respectively. ϕ_{jp} (j=1~6) is the radiation potential corresponding to the j-th mode of motion of the p-th floating body: surge, sway, heave, roll, pitch and yaw. X_{jp} and ψ'_{jp} represent the amplitude and phase in the j-th mode of motion of the p-th body due to monochromatic waves, respectively.

The fluid region being defined as shown in Fig. 2, the governing equation and the boundary condition on ϕ_{sc} are as follows:

$$
\begin{aligned}
&\Delta \phi_{sc} = 0 && (in\ \Omega)\\
&\frac{\partial \phi_{sc}}{\partial z} = \frac{\omega^2}{g}\phi_{sc} && (on\ S_F)\\
&\frac{\partial \phi_{sc}}{\partial z} = 0 && (on\ S_B)\\
&\frac{\partial \phi_{sc}}{\partial n} = -\frac{\partial \phi_{in}}{\partial n} && (on\ S_{Vp}\ ,\ p = 1\sim N_P)\\
&\lim_{R\to\infty} \sqrt{R}\left(\frac{\partial \phi_{sc}}{\partial R} - ik\phi_{sc}\right) = 0
\end{aligned}
\tag{2}
$$

where R is the horizontal distance between P inside Ω and Q on S_∞.

The governing equation and the boundary condition on ϕ_{jp} are as follows:

$$
\begin{aligned}
&\Delta \phi_{jp} = 0 && (in\ \Omega)\\
&\frac{\partial \phi_{jp}}{\partial z} = \frac{\omega^2}{g}\phi_{jp} && (on\ S_F)\\
&\frac{\partial \phi_{jp}}{\partial z} = 0 && (on\ S_B)\\
&\frac{\partial \phi_{jp}}{\partial n} = n_{jp} && (on\ S_{Vp})\\
&\frac{\partial \phi_{jp}}{\partial n} = 0 && (on\ S_{Vp'}\ ,\ p \neq p')\\
&\lim_{R\to\infty} \sqrt{R}\left(\frac{\partial \phi_{jp}}{\partial R} - ik\phi_{jp}\right) = 0\\
&&& (j = 1\sim 6\ ,\ p = 1\sim Np)
\end{aligned}
\tag{3}
$$

where n_{jp} is the normal component of unit velocity or unit angular velocity due to the j-th mode of motion of the p-th body and is represented as follows:

$$
\begin{pmatrix} n_{1p}\\ n_{2p}\\ n_{3p}\\ n_{4p}\\ n_{5p}\\ n_{6p} \end{pmatrix} = \begin{pmatrix} n_{xp}\\ n_{yp}\\ n_{zp}\\ (y-y_{op})\,n_{zp} - (z-z_{op})\,n_{yp}\\ (z-z_{op})\,n_{xp} - (x-x_{op})\,n_{zp}\\ (x-x_{op})\,n_{yp} - (y-y_{op})\,n_{xp} \end{pmatrix} \quad (p = 1\sim N_P)
\tag{4}
$$

By applying Green's formula and using the boundary condition of Eqs. (2) and (3), simultaneous linear equations of ϕ_{sc} and ϕ_{jp} on the submerged surface of N_p floating bodies are derived as follows:

$$
\phi_{sc}(P) = \lambda \sum_{p'=1}^{N} \iint_{S_{Vp'}} \{-\phi_{sc}(Q)\frac{\partial G(P,Q)}{\partial n} - G(P,Q)\frac{\partial \phi_{in}(Q)}{\partial n}\}\, dS_{Vp'}(Q)
\tag{5}
$$

$$
\begin{aligned}
\phi_{jp}(P) = \lambda[&\iint_{S_{Vp}} \{-\phi_{jp}(Q)\frac{\partial G(P,Q)}{\partial n} + G(P,Q)\, n_{jp}(Q)\}\, dS_{Vp}(Q)\\
&+ \sum_{\substack{p'=1\\(p'\neq p)}}^{Np} \iint_{S_{Vp'}} \{-\phi_{jp}(Q)\frac{\partial G(P,Q)}{\partial n}\}\, dS_{Vp'}(Q)\,]\\
&(j = 1\sim 6\ ,\ p = 1\sim N_p)
\end{aligned}
\tag{6}
$$

where $G(P,Q)$ is Green function (e.g., Oortmerssen, 1976), λ is a constant value of $1/4\pi$ inside Ω, the fluid region or $1/2\pi$ on S_{Vp}, the submerged surface of floating bodies. By solving Eqs. (5) and (6), ϕ_{sc} and ϕ_{jp} are obtained.

2.2 Wave-exciting force, added mass and wave damping coefficient

In the j-th mode, wave-exciting force on the p-th floating body is expressed as follows:

$$
F_{jp}\exp(i\psi_{jp}) = -i\rho\omega \iint_{S_{Vp}} n_{jp}(Q)\cdot\phi_{sc}(Q)\, dS_{Vp}(Q) \quad (j = 1\sim 6\ ,\ p = 1\sim N_P)
\tag{7}
$$

where F_{jp} and ψ_{jp} are the amplitude and the phase of the wave exciting force acting on the p-th floating body, respectively.

The added mass $a_{jpj'p'}$ and wave damping coefficient $b_{jpj'p'}$ in the j-th mode of the p-th floating body, caused by the motion in the j'-th mode of the p'-th floating body, is expressed as follows:

$$
a_{jpj'p'} - \frac{i}{\omega}b_{jpj'p'} = \rho \iint_{S_{Vp}} n_{jp}(Q)\cdot\phi_{j'p'}(Q)\, dS_{Vp}(Q) \quad (j,j' = 1\sim 6\ ;\ p,p' = 1\sim N_P)
\tag{8}
$$

2.3 Equations of motion of plural bodies subjected to short-crested waves

Defining the coordinates as shown in Fig. 1, the equation of motion is derived by extending that in the time domain with a retardation function of motion (Oortmerssen, 1976) to take into account directional spreading of irregular waves and interaction between plural body motions :

$$
\begin{aligned}
M_{jp}\ddot{x}_{jp}(t) + \sum_{p'=1}^{Np}\sum_{j'=1}^{6} \{&a_{jpj'p'}(\infty)\ddot{x}_{j'p'}(t)\\
&+ \int_0^t L_{jpj'p'}(t-\tau)\cdot\dot{x}_{j'p'}(\tau)d\tau + C_{jpj'p'}x_{j'p'}(t)\}\\
+ d_{jp}|\dot{x}_{jp}(t)|\dot{x}_{jp}(t) + R_{jp}(t) &= f_{jp}(t)\\
&(j = 1\sim 6\ ,\ p = 1\sim N_p)
\end{aligned}
\tag{9}
$$

where t and f denote the time and the wave frequency, respectively, x_{jp} represents the motion in the j-th mode of the p-th body in the time domain, dots denote time derivatives, M_{jp} is the mass or the inertia moment, $C_{jpj'p'}$ is the hydrostatic restoring force coefficient, d_{jp} is the viscous damping coefficient, R_{jp} is the nonlinear mooring force of each body, $a_{jpj'p'}(\infty)$ and $L_{jpj'p'}$ are the constant added mass and the retardation function of motion caused by interaction between the p-th and p'-th bodies, and respectively, and f_{jp} is the exciting force caused by short-crested waves.

$L_{jpj'p'}$ is calculated by using the wave damping coefficient in the frequency domain as follows:

$$L_{jpj'p'}(t) = 4\int_0^{\infty} b_{jpj'p'}(f)\cdot\cos 2\pi ftdf \qquad (10)$$

where $b_{jpj'p'}(f)$ is the wave damping coefficient in the frequency domain.

The wave exciting force due to multi-directional random waves is calculated by superposing the forces of all the component waves as follows:

$$f_{jp}(t) = \sum_f \sum_\theta F_{jp}(f,\theta)\cos\{2\pi ft + \varepsilon(f,\theta) + \psi_{jp}(f,\theta)\}$$
$$(j = 1\sim6,\ p = 1\sim N_p) \qquad (11)$$

where $F_{jp}(f,\theta)$ and $\psi_{jp}(f,\theta)$ are the amplitude and the phase of the wave exciting force, respectively, caused by incident component waves with frequency f and wave direction θ, and $\varepsilon(f,\theta)$ is the random initial phase corresponding to incident component waves with f and θ.

In this paper, the following Bretschneider-Mitsuyasu function is employed as the power spectrum $S(f)$ of irregular incident waves:

$$S(f) = 0.257H_s^2T_s(T_sf)^{-5}\exp\left(-1.03(T_sf)^{-4}\right) \qquad (12)$$

where H_s and T_s are the significant wave height and period, respectively.

The following Mitsuyasu function is employed as the directional spreading function $G(f,\theta)$ of short-crested waves (Mitsuyasu et al., 1975):

$$G(f,\theta) = G_o\cos^{2s}\left((\theta-\theta_a)/2\right) \qquad (13)$$

where θ_a represents the principal wave direction, and G_0 is the constant to satisfy the normalization. The parameter S is a function of the frequency f as follows (Goda and Suzuki, 1975; Takayama and Hiraishi, 1989):

$$S = \begin{cases} S_{max}(f/f_p)^5 & : f \le f_p \\ S_{max}(f/f_p)^{-2.5} & : f > f_p \end{cases} \qquad (14)$$

where f_p is the peak frequency of the power spectrum, and S_{max} is the maximum value of S.

Numerical integration of the equation of motion (9) at t=t by the Newmark-β method is performed and the plural floating body motions at t=t+Δt are solved.

3. HYDRAULIC MODEL TESTS

3.1 Experimental facilities

Experiments were carried out in the laboratory basin of the Central Research Institute of Electric Power Industry, Japan (Ikeno et al., 1992b; Hirakuchi et al., 1992; Ikeno et al., 1995). The basin shown in Fig. 3, was 45 m wide, 35 m long and 1.1 m deep, equipped with a 48-segmented wave maker (0.4 m wide per segment) capable of producing regular and uni- and multi-directional irregular waves. After that, the wave maker was extended and now has 72 segment paddles.

In short-crested wave experiments, input signals to individual segmented wave makers can be efficiently computed by the double and single summation model. In our experiments, input signals are usually computed by superposing component waves with 512 or 1024 frequencies and 90 directions in the double summation model, and 512 or 1024 component waves in the single summation model.

3.2 Experimental conditions

The model scale of the twin plant barges is 1/100 (horizontal size: 142 cm×142 cm, draft: 11.6 cm, water depth: 24 cm) as indicated in Table 1. Each plant model is box-shaped body made of acrylic plate and steel weights. As shown in Fig. 1, two sides of a barge model are connected to four units of a dolphin-link mooring system with two universal joints. Two types of resistance force characteristics of the actual rubber fender are investigated. One has linear relation between fender deformation and load. Two coil springs are used for the mooring system. The other has a plastic zone and hysteresis loss as shown in Fig. 11. Two nonlinear rubber fender models are used for the mooring system, and are made of silicone rubber.

Displacements of the twin plant models are measured simultaneously by twelve laser beam sensors. The measured displacements at six points of each plant model are converted into motions in six degrees of freedom at the center of gravity and displacements at the eight vertexes of the box-shaped body (Ikeno et al., 1992a; Ikeno, 1992). In the nonlinear rubber fender model tests, two compressive load sensors are installed in each mooring system and four mooring forces per plant are measured simultaneously as shown in Fig. 4.

The experimental waves are regular and irregular. In the irregular wave experiments, the significant wave height is 3 cm and the period 1-2s. The Bretschneider-Mitsuyasu frequency spectrum and Mitsuyasu directional spreading functions are employed as in Eqs.(12) and (13). To investigate wave directional spreading effects on the plant motion, the maximum values of the parameter S, S_{max}, in the Mitsuyasu directional spreading function are taken 10, 50 (short-crested) and infinity (long-crested).

A star array consisting of 4 capacitance type wave gauges are installed in the mooring area without the plant models, and the directional spectra of generated waves are checked. By the Bayesian Model (BDM; Hashimoto and Kobune, 1987) and the Maximum Likelihood Method (MLM; Capon, 1969), directional spectra are analyzed. The sampling time and data number per channel were 0.02 sec and 1,024 in the regular wave experiments, and 0.05 sec and 8,192 in the irregular waves.

4. INTERACTION BETWEEN TWIN BARGES IN FREQUENCY DOMAIN

4.1 Response function

We first discuss the results in the case of linear mooring force.

Figures 5 and 6 show amplitude ratio contours of the response functions of X_1 and X_5. They are computed with viscous damping by changing the frequency and the direction of incident waves (Ikeno et al., 1992b; Ikeno and Kajima, 1992c; 1993). The motions of twin barges (plant 1 and plant 2) are compared with those of a single barge. The gap width between plant 1 and plant 2 barges is 35 cm in the scale model. The plant barge dimensions are almost symmetrical with respect to the x-z and y-z planes. Comparing the figures of twin plants and a single plant, the responses of the twin plants near the resonant frequencies are 1.5~2.0 times larger than those of a single plant. Resonance phenomena caused by multi-reflection in the gap between the plants are found, which are affected by the ratio between the gap width and the wavelength. Thus, the response functions of twin plant motions in X_1 and X_5 modes are significantly different from those of single plant motion, which means that the interaction between twin plant motions must be considered.

On the other hand, for the response functions of X_2 and X_4, no significant differences among twin plants and a single plant are observed, because they are the motion components transverse to the x-axis of twin barges, and the effects of multi-reflection in the gap on X_2 and X_4 are relatively small.

4.2 Motion spectra

The expected value of the power spectrum of the motion $\bar{S}_{jp}(f)$ in the j-th mode of the p-th plant in short-crested waves is expressed as follows:

$$\bar{S}_{jp}(f) = \int_{-\pi}^{\pi} \hat{X}_{jp}^2(f,\theta)\, S(f)\, G(f,\theta) d\theta \quad (j=1\sim6\,;\, p=1\sim N_p) \tag{15}$$

where $\hat{X}_{jp}$ represents the amplitude ratio of the response function of the motions in the j-th mode of the p-th plant as shown in Figs. 5 and 6.

Next, the expected value of the cross spectrum between the motions in the j-th mode of the p-th plant and in the j'-th mode of the p'-th plant in short-crested waves is expressed as follows:

$$\bar{C}_o(f) - i\bar{Q}(f) = \int_{-\pi}^{\pi} \hat{X}_{jp}(f,\theta)\, \hat{X}_{j'p'}(f,\theta)\, S(f)\, G(f,\theta) \cdot \exp\left\{-i\left(\psi'_{jp}(f,\theta) - \psi'_{j'p'}(f,\theta)\right)\right\} d\theta \quad (j,j'=1\sim6\,;\, p,p'=1\sim N_p) \tag{16}$$

where $\bar{C}_o$ and $\bar{Q}$ are the expected values of the co- and quadrature-spectrum, respectively, and $\psi'_{jp}(f,\theta)$ represents the phase of the motion in the j-th mode of the p-th plant due to incident component waves with f and θ.

The expected value of the phase lag $\bar{\psi}'_{jpj'p'}(f)$ between the motions in the j-th mode of the p-th plant and in the j'-th mode of the p'-th plant in short-crested waves is expressed as follows :

$$\bar{\psi}'_{jpj'p'}(f) = \tan^{-1}(\bar{Q}(f)/\bar{C}_o(f)) \quad (j,j'=1\sim6\,;\, p,p'=1\sim N_p) \tag{17}$$

Figure 7 shows the computed and experimental power spectra of X_4 of plant 2. Incident wave conditions are H_s=3 cm, T_s=1 s, θ_a= 112.5° with S_{max}=10. In Fig. 7, the solid line shows the expected power spectrum by Eq. (15), the dotted line represents the result estimated by this numerical model, and the dash-dotted line shows the experimental result.

In calculating the estimated and experimental power spectra, Hanning's spectral window is applied forty times.

Figure 8 shows the computed and experimental cross spectra and phase lags of X_6 between plant 1 and plant 2 under the same incident wave conditions as in Fig.7, where Hanning's spectral window is also applied forty times.

According to Figs. 7 and 8, the estimated values of power and cross spectra show good agreements with the expected values. The experimental values also show good agreements with both expected and estimated spectra.

5. INTERACTION BETWEEN TWIN BARGES IN TIME DOMAIN

5.1 Retardation function and significant values of the motion

Figure 9 shows the comparison of the retardation functions of X_1 and X_5 between twin-plant and single-plant motions. The retardation functions are normalized by the mass and length of the barge. The figures show that retardation functions of the motion parallel to the x-axis for twin plants show larger values and last longer time than those of a single plant. This is caused by multi-reflection of radiated waves in the gap between two plants.

Figure 10 shows the significant values of the six-degrees-of-freedom motion in irregular waves with S_{max}=10 (short-crested) and infinity (long-crested) obtained by the zero-upcrossing method. Incident waves are H_s=3 cm, T_s=1.5 s and θ_a=90° . In the figure, the calculated motions of twin plants with gap length of 140 cm and 35 cm are compared with a single plant. The differences of twin plant motions by long- and short-crested waves can also be compared in the figure. In this case, the motions of plant 1 and plant 2 are identical, because incident waves are acting on both the plants symmetrically.

Comparing the twin and single plant motions in long-crested waves, the effects of interaction on plural plant motions are more significant in the parallel modes of motions to the x-axis, X_1 and X_5. As the separation distance becomes smaller, X_1 and X_5 are increasing. Next, comparing those in short-crested waves, effects of the interaction become smaller for X_1 and X_5 than those in long-crested waves. On the other hand, for X_2, X_3, X_4 and X_6, effects of the interaction in long- and short-crested waves are small.

5.2 Effects of nonlinear mooring force

Figure 11 shows the measured trajectories of the dynamic compressive load-deformation curves of the rubber fenders installed with one of two mooring systems corresponding to head-on waves. Incident wave conditions are H_s=3 cm, T_s=1 s and θ_a=90° . Figure 11 is the experimental results for long- and short-crested waves with S_{max}=10. The maximum values of the resistance force and the fender deformation in long-crested waves are a little greater than those in short-crested waves, because the wave force components in the direction of θ_a in long-crested waves become larger than those in short-crested waves (Ikeno et al., 1992b).

Figure 12 shows the maximum and significant values on the six-degrees-of-freedom motion and the four mooring forces in irregular waves obtained by the zero-upcrossing method. The incident waves are H_s=3 cm, T_s=1 s, and two incident angles θ_a of 90° and 135° are taken.

According to Fig. 11, when the compressive deformation of the rubber fender exceeds 0.6 cm, which corresponds to 20 % strain, the resistance force reaches a plastic zone and hysteresis loss occurs.

Thus, in Fig. 12, nonlinearity of the mooring force becomes significant when the double amplitude values of X_1, X_2 and the compressive loads of the fenders exceed 1.2 cm and 4 kgf, respectively, corresponding to the horizontal dash-dotted line. In the case of oblique wave incidence on the plant, the six-degrees-of-freedom motion and four mooring forces in short-crested waves are found to be greater than those in long-crested waves.

Figure 13 shows the experimental and computed results of sway motion X_2 as functions of incident wave height. Incident waves are regular with T=1 s, which is near the resonant frequency of surge and sway, and three incident angles of 90°, 112.5° and 135°. The hollow and solid symbols indicate the experimental and computed values, respectively. As incident wave height increases, the effects of nonlinear mooring on sway X_2 become more significant and X_2 increases more than that of the linear mooring. It is also seen that the computations can accurately reproduce the experimental results.

6. CONCLUSIONS AND REMARKS

In this paper, a numerical model was developed by extending the equation of motion in time domain with a retardation function of motion to take into account directional spreading of irregular waves and interaction between plural body motions using the boundary element method. Hydraulic model tests of 1/100-scale twin plant barges are performed with a directional wave maker to verify the computation results. The simulation and experimental results show good agreement. The effects of interaction on plural plant motions are more significant in the parallel modes of motions to the x-axis, X_1 and X_5. As the separation distance becomes smaller, the motions are increasing. Effects of the interaction in short-crested waves are smaller than those in long-crested waves. The effects of nonlinear mooring on surge and sway are more significant.

REFERENCES

Capon, J. (1969): High-Resolution Frequency - Wave Number Spectrum Analysis, Proc. Int. Electr. Electron Eng., Vol.57, pp.1408-1418.

Cummins, W.E. (1962): The Impulse Response Function and Ship Motions, Shiffstechnik, Bd.9, Heft 47, pp.101-109.

Dallinga, R.P., A.B. Aalbers and J.W.W. van der Vegt (1984): Design Aspects for Transport of Jack-Up Platforms on a Barge, Proc. of 18th Offshore Technology Conference, No.4733, Houston, U.S.A, pp.195-207.

Goda, Y. and Y. Suzuki (1975): Computation of Refraction and Diffraction of Sea Waves with Mitsuyasu's Directional Spectrum, Tech. Note Port and Harbour Res. Inst., No.230, 45p. (in Japanese)

Hashimoto, N. and K. Kobune (1987): Estimation of Directional Spectrum using the Bayesian Approach, and its Application to Field Data Analysis, Report of Port and Harbour Res. Inst., Vol.26, No.5, pp.57-100.

Hirakuchi, H., R. Kajima, T. Shimizu and M. Ikeno (1992): Characteristics of Absorbing Directional Wavemaker, Proc. 23rd Int. Conf. on Coastal Eng., ASCE, pp.281-294.

Ikeno, M., K. Marurama, H. Hirakuchi and R. Kajima (1992a): Characteristics of Plant Barge Motion in Short-Crested Waves, J. Hydraulic Coastal and Environmental Eng., 2-19, No.447, JSCE, pp.65-74. (in Japanese)

Ikeno, M., K. Marurama, H. Hirakuchi and R. Kajima (1992b): Simulation Model of Floating Power Plant Motion in Multi-Directional Irregular Waves, Proc. 2nd Int. Offshore and Polar Eng. Conf., Vol.3, ISOPE, pp.636-643.

Ikeno, M. and R. Kajima (1992c): Interaction between Twin Plant Barge Motions in Short-Crested Waves, Proc. Japanese Coastal Eng., Vol.39, JSCE, pp.806-810. (in Japanese)

Ikeno, M. (1992): Simulation Model of Floating Power Plant Motion in Short-Crested Waves, Abiko Res. Lab. Report of Central Res. Inst. of Electric Power Ind., No.U92023, 52p. (in Japanese)

Ikeno, M. and R. Kajima (1993): Simulation Model of Interaction between Two Power Plant Motions in Short-Crested Waves, Abiko Res. Lab. Report of Central Res. Inst. of Electric Power Ind., No.U92043, 77p. (in Japanese)

Ikeno, M., R. Kajima, M. Matsuyama and T. Sakakiyama (1995): Characteristics of Short-Crested Waves and Currents behind Offshore Man-Made Island type Power Plant, Proc. 5th Int. Offshore and Polar Eng. Conf., Vol.3, ISOPE, pp.168-175.

Isaacson, M. and S. Sinha (1986): Directional Wave Effects on Large Offshore Structures, J. Waterway, Port, Coastal and Ocean Engineering, ASCE, Vol.112, No.4, pp.482-497.

Mitsuyasu, H., F. Tasai, T. Subara, S. Mizuno, M. Okusu, T. Honda and K. Rikiishi (1975):Observation of the Directional Spectrum of Ocean Waves using a Clover-Leaf Buoy, J. Phys. Oceanography, Vol.5, No.4, pp.750-760.

Mynett, A.E., J. Bosma, J.A. Keuning and J. Gerritsma (1988): Laboratory Simulation of Ship Motions in Directional Seas, Proc. of BOSS'88, Int. Conf. on Behaviour of Offshore Structures, Trondheim, Norway, Vol.2, pp.755-772.

Pinkster, J.A. (1984): Numerical Modelling of Directional Seas, Symposium on Description and Modelling of Directional Seas, Technical University Denmark, No.C-1, pp.1-19.

Takayama, T. and T. Hiraishi (1989): Reproducibility of Directional Random Waves in Laboratory Wave Simulation, Report of the Port and Harbour Research Institute, Vol.28, No.4, pp.3-24.

Takezawa, S. and K. Kobayashi (1988): On the Motion Characteristics of Semi-Submersibles in Short-Crested Irregular Waves, Proc. of BOSS'88, Int. Conf. on Behaviour of Offshore Structures, Trondheim, Norway, Vol.2, pp.721-735.

van Oortmerssen, G. (1976): The Motions of a Moored Ship in Waves, N.S.M.B. Publication, No.510, 138p.

van Oortmerssen, G. (1979): Hydrodynamic Interaction between Two Structures, Floating in Waves, Proc. of BOSS' 79, Int. Conf. on Behaviour of Offshore Structures, Vol.1, pp.339-356.

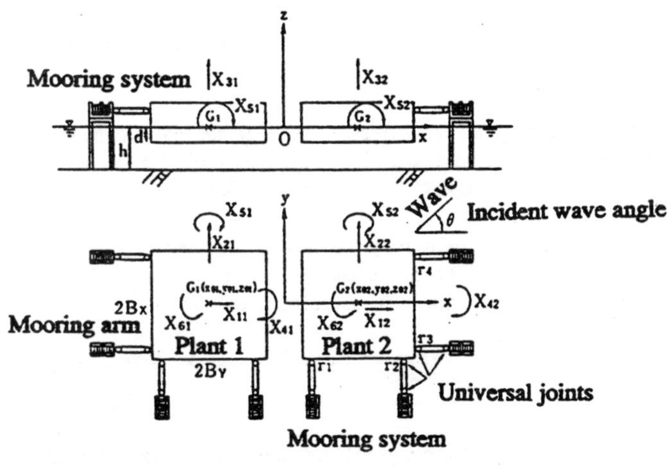

Fig. 1 Definition sketch for the six-degrees- of- freedom motion of plural plants

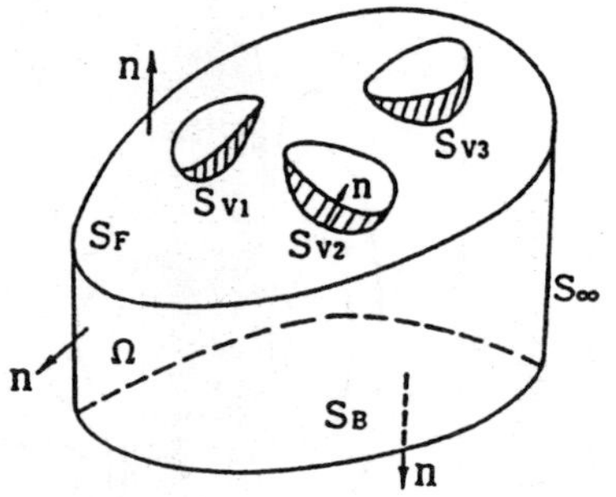

Fig. 2 The boundaries surrounding the fluid region

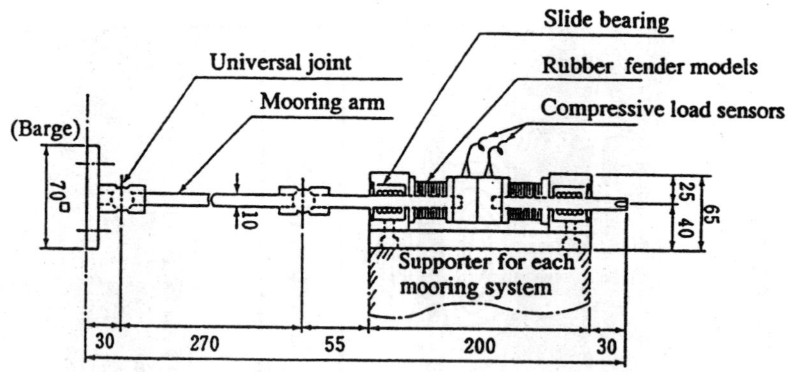

Fig. 4 Installation of rubber fender models and compressive load sensors in a mooring system

Table 1 Dimensions of each barge model

Dimensions per unit		Model	Prototype
Length× width ×draft		1.42×1.42 ×0.116m	142×142×11.6m
Mass		233.8 kg	2.338×10⁵ ton
Inertia moment	I_X	46.2 kg m²	4.62×10⁸ ton m²
	I_Y	47.3 kg m²	4.73×10⁸ ton m²
	I_Z	68.1 kg m²	6.81×10⁸ ton m²
Stiffness of coil springs per unit mooring system		3.882 kN/m	3.882×10⁴ kN/m
		1.274 kN/m	1.274×10⁴ kN/m
Stiffness of nonlinear rubber fender per unit mooring system		3.882 kN/m for 10% strain	3.882×10⁴ kN/m for 10% strain

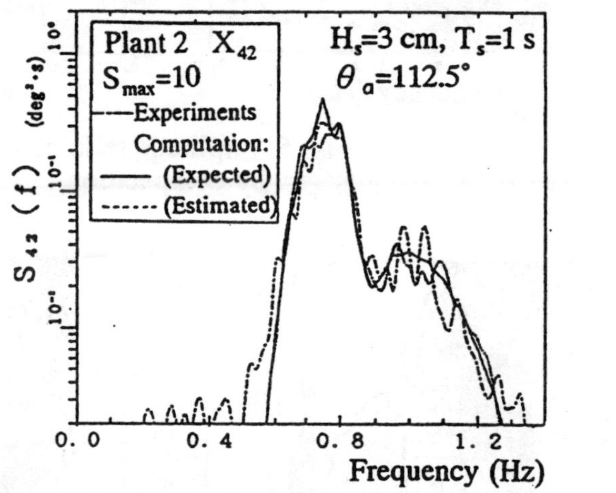

Fig. 7 Computed and experimental power spectra of X_{42}

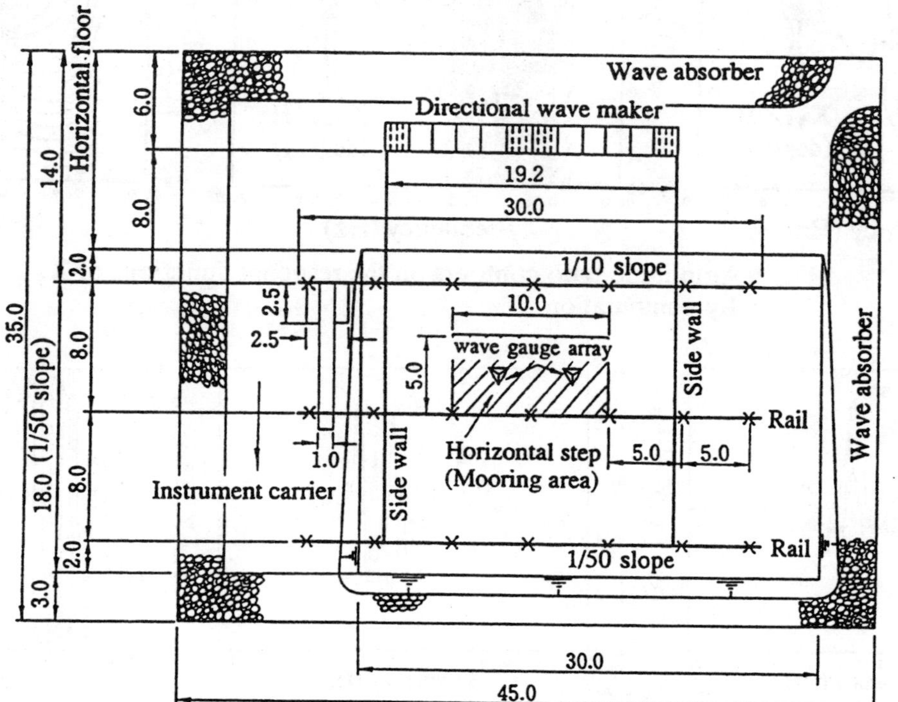

Fig. 3 Sketch of CRIEPI laboratory basin and experimental arrangement

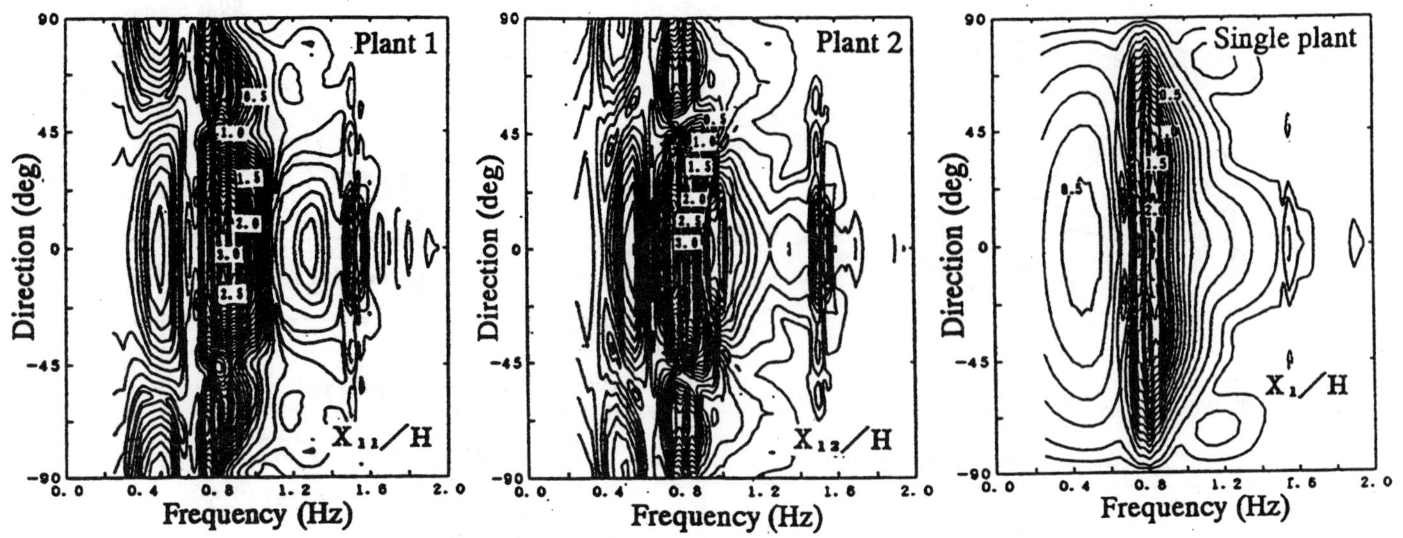

Fig. 5 Amplitude ratio contours on the response functions of X_1 by computation

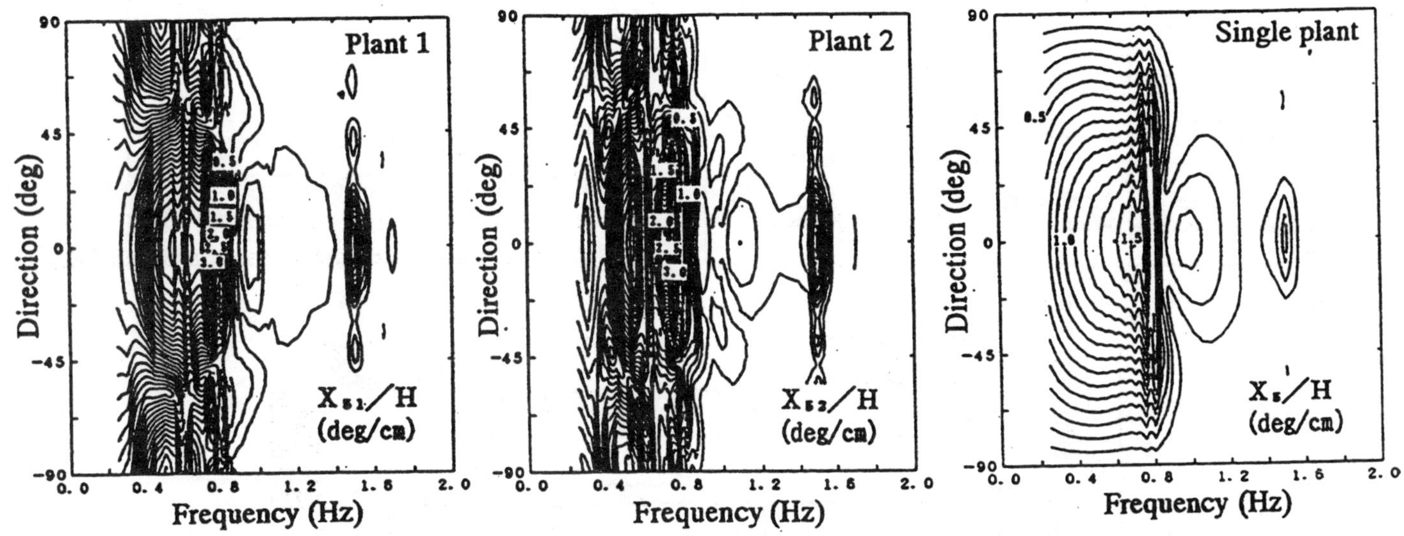

Fig. 6 Amplitude ratio contours on the response functions of X_5 by computation

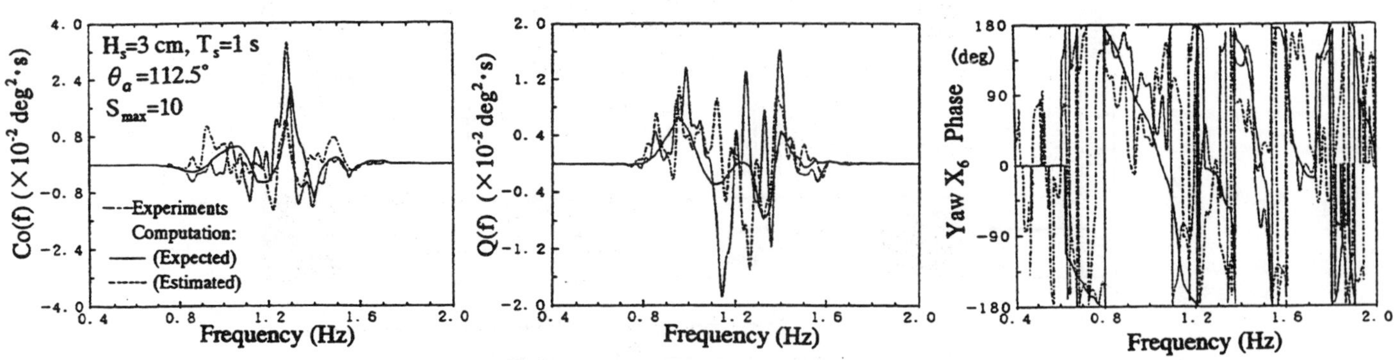

Fig. 8 Computed and experimental cross spectra and phase lags between X_{61} and X_{62}

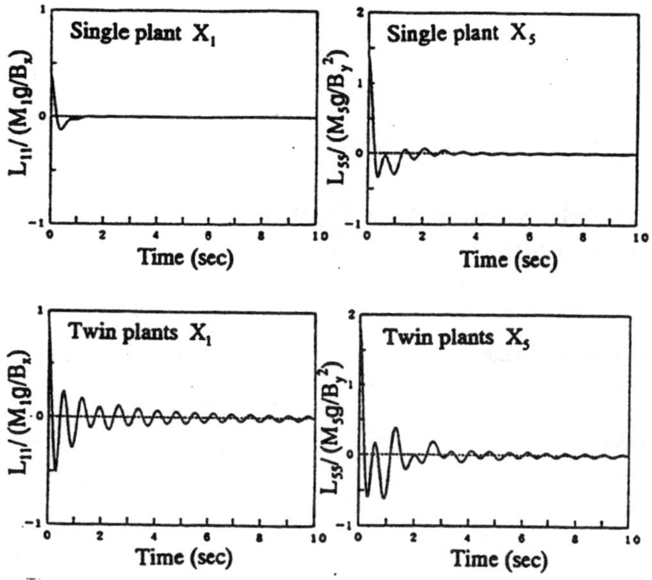

Fig. 9 Comparison among retardation functions of twin plants and a single plant

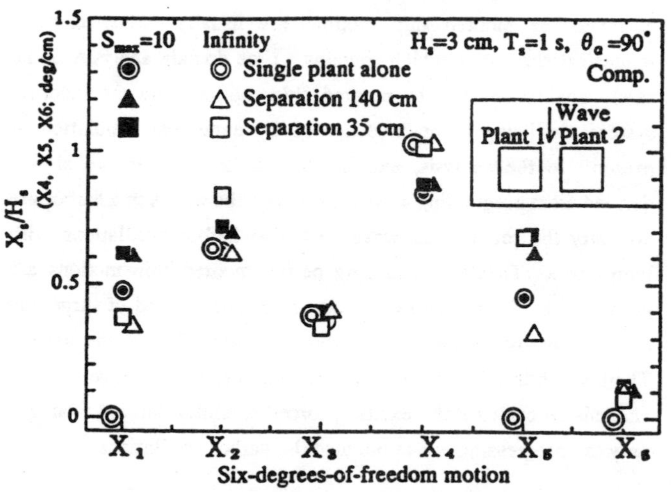

Fig. 10 Significant values on six-degrees-of-freedom motion in irregular waves by computation

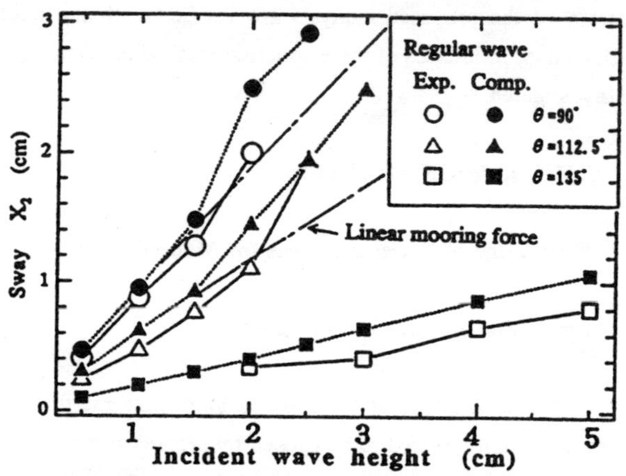

Fig. 13 Comparison between experimental and computed X_2 affected by nonlinear mooring force

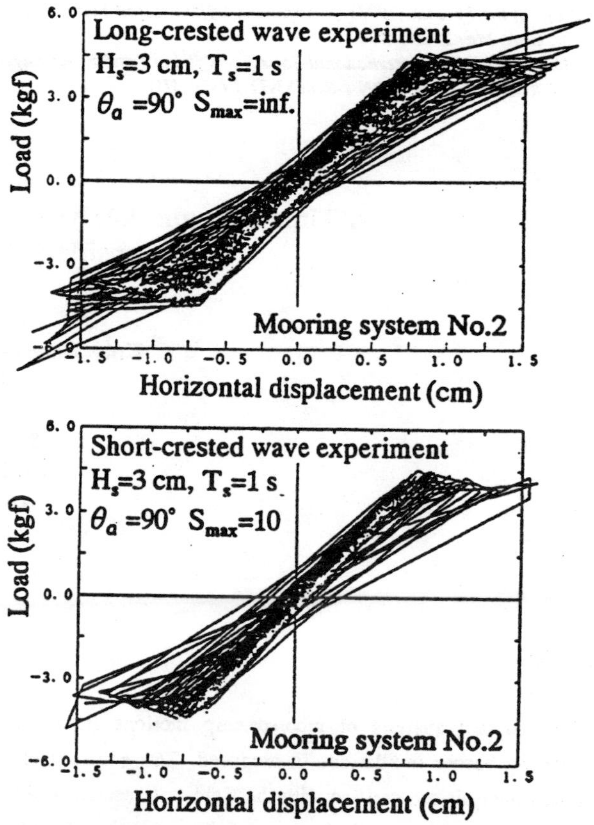

Fig. 11 Dynamic load-deformation curves of rubber fender models in irregular head-on waves

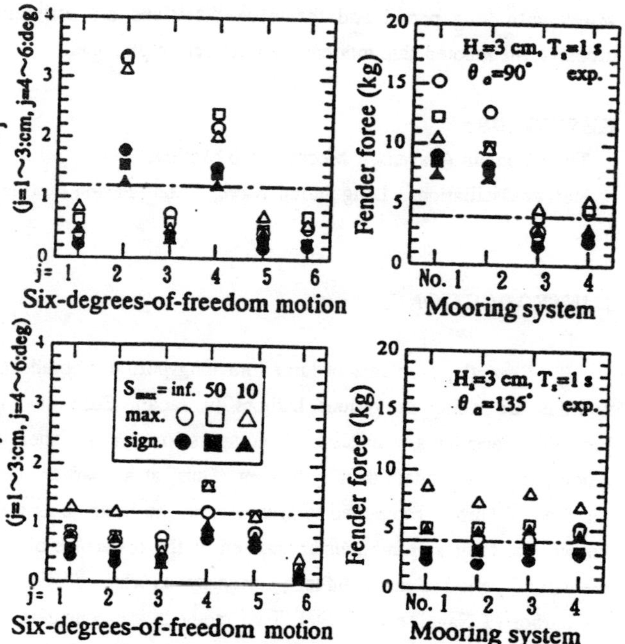

Fig. 12 Maximum and significant values of six-degrees-of-freedom motion and four mooring forces in irregular waves by experiments

Proceedings of the Seventh (1997) International Offshore and Polar Engineering Conference
Honolulu, USA, May 25-30, 1997

ISBN 1-880653-28-1 (Set); ISBN 1-880653-31-1 (Vol. III)

A Time Domain Analysis of Moored Ship Motions in a Harbor Considering Harbor Oscillations

Masayoshi Kubo
Kobe University of Mercantile Marine
Kobe, Japan

Shigeki Sakakibara
Yokohama Rubber Co.
Hiratsuka, Japan

ABSTRACT

A time domain analysis of moored ship motions in a harbor considering harbor oscillations is proposed. The accuracy of the proposed analysis are verified with the field test results. And it is necessary for the numerical simulation to improve the accuracy of the analysis of harbor oscillations. Even if the exciting forces of ship are evaluated by only progressing waves derived from the inside waves with long period and the wave directions are estimated properly, the moored ship motions are obtained in practice.

KEYWORDS :
Time Domain Analysis, Moored Ship Motion, Harbor Oscillation, Long Period Wave, Long Period Motion

1. INTRODUCTION

Recently, problems of large and long period moored ship motions are rising in various harbors in Japan. Such as rope breaking accidents, excessive compression and shearing deformations of fenders usually happen. Thus, at an earlier stage, the moored ship is compelled to escape from the berth. It is well-known that large ship motions are caused by the resonance of both the natural period of surge and long period waves which exist inside the harbor (T.Nagai, et al.,1994). Two methods are considered as countermeasure, the first one is to restrain the long period waves through the expansions of breakwaters, the other one is to control the natural period of surge through the improvement of the mooring system (S.Shiraishi, et al.,1996).

Under several external forces such as waves, winds and currents, the moored ship motions are usually estimated by a numerical simulation which contains a time domain analysis. Some safety evaluations for the moored ship motions and the mooring system are being to be investigated by numerical simulation at present. In the analysis, exciting forces for the ship are almost derived from progressing waves. However, the waves in a harbor are not only the progressing waves but also harbor oscillations with long period. The large and long period moored ship motions are generated by the resonance of both the natural period of surge and the long period waves (seich) which are amplified in the harbor. Then, when the ship is moored in the harbor, it is necessary for the analysis to evaluate the exciting forces carefully derived from not only the progressing waves but also the harbor oscillations.

In this paper, we propose a time domain analysis of moored ship motions in a harbor considering harbor oscillations. The accuracy of the proposed analysis are verified with the field test results. Also, influences of the evaluation method on exciting forces are investigated.

2. TIME DOMAIN ANALYSIS

2.1 Evaluation of Exciting Forces for Moored Ship

(1) Separation of inside harbor waves

In the analysis of the moored ship motions, the exciting forces are usually treated to be caused by progressing waves. Then, in estimating the exciting forces from inside waves near the ship, first the wave data is divided into some component waves by FFT.

Next, each unit wave exciting forces ($f_i^*=f_i/\zeta_i$) which are calculated by the frequency domain analysis are multiplied by the corresponding wave amplitudes (ζ_i). After that, those exciting forces (f_i) are superposed on each other. In this case, the unit wave exciting forces are derived from a diffraction potential for open sea, namely as a progressing wave. However, it is very difficult to assign the wave directions correctly. Because the directional dispersion of the inside waves is so large. So one wave direction as a predominant one is usually set by a visual observation and current data (Fig.-1).

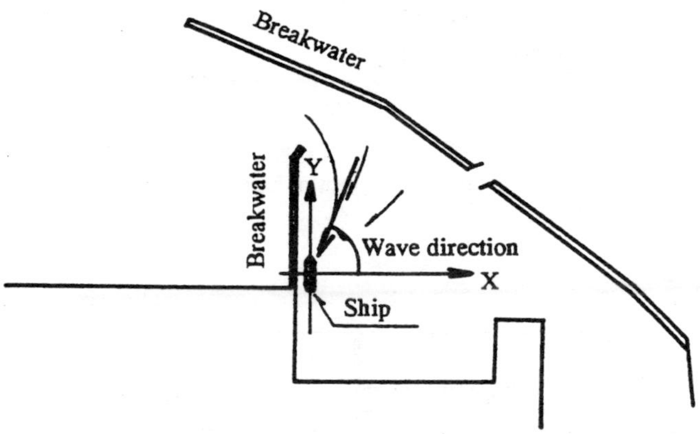

Fig.-1 Estimation of inside harbor wave direction

Essentially, waves inside harbor consist of progressing waves and harbor oscillations. The progressing waves enter from the harbor entrance directly. And the harbor oscillations are generated and amplified in the harbor. It is very important to consider the above effect for the evaluation of the exciting forces. Thus, the inside harbor waves should be separated into two components, progressing waves and harbor oscillations. It is well-known that periods of wind waves are about 10-15s. So, the waves of periods less than 20s are set to be progressing waves and the other longer waves are treated to be harbor oscillations. The wave direction of the progressing waves are established as same as that of ordinary analysis. The ordinary analysis for the exciting forces is treated all inside harbor waves as progressing waves.

(2) Calculation of exciting forces considering harbor oscillations

There are some kinds of analytical methods on harbor oscillations. However, they are only applied to calculate an amplification factor between inside and outside harbor waves (ζ_i/η_i, η_i: outside wave amplitudes, ζ_i: inside wave amplitudes). In case of the analytical method which calculate velocity potential functions, not only the amplification factor but also the velocity potentials and pressures inside harbor are obtained. Thus, the exciting forces of moored ship can be also derived from the calculation of the potentials at points of the submerged ship hull (T.Sawaragi, et al.,1981). In this case, the exciting forces are calculated using outside harbor waves directly. So, we can escape from the difficulty of estimating the wave directions. From the above, the total exciting forces (F) are evaluated by combining the progressing wave components and the harbor oscillations, as shown in Eq.(1).

$$F = \sum_{i}^{T\geq 20s} \eta_i (fh_i/\eta_i) \cos(\sigma_i t+\varepsilon_i) + \sum_{i}^{T<20s} \zeta_i (f_i/\zeta_i) \cos(\sigma_i t+\varepsilon_i)$$

$$= \sum_{i}^{T\geq 20s} \eta_i fh_i^* \cos(\sigma_i t+\varepsilon_i) + \sum_{i}^{T<20s} \zeta_i f_i^* \cos(\sigma_i t+\varepsilon_i) \qquad (1)$$

f_i^* : Unit wave exciting forces derived from progressing waves
fh_i^* : Unit wave exciting forces derived from harbor oscillations
η_i : Outside harbor wave amplitudes
ζ_i : Inside harbor wave amplitudes
σ_i : Angular frequency
t : Time
ε_i : Phase

In this paper, we use an analytical method for the harbor oscillation proposed by Lee (J.J.Lee,1971).

2.2 Numerical Simulation of Moored Ship Motions Considering Harbor Oscillations

As mentioned above, the proposed time domain analysis of moored ship motions considering harbor oscillations is carried out based on the given flow chart (Fig.-2).

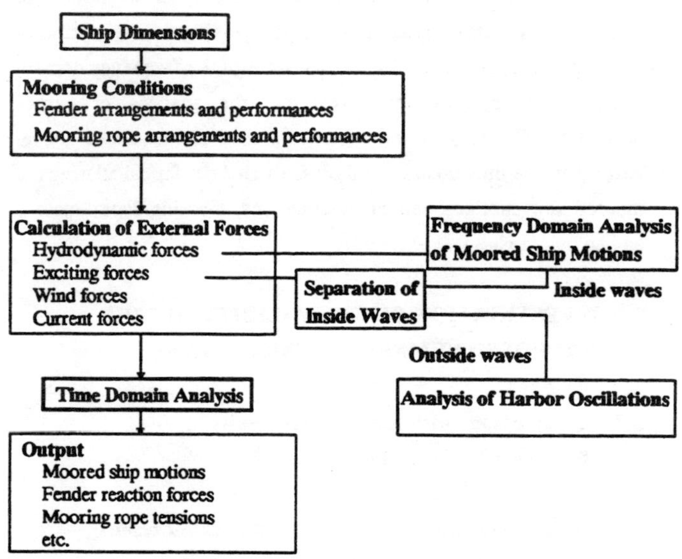

Fig.-2 (a) Flow chart of the time domain analysis of moored ship motions considering harbor oscillations

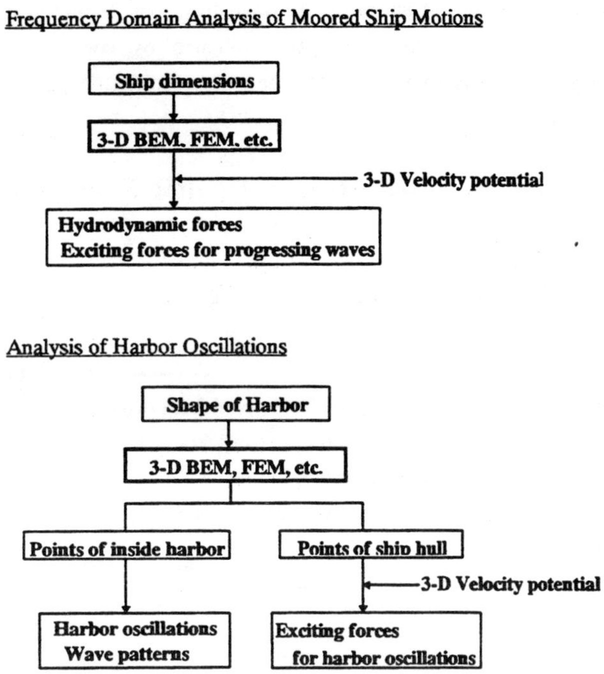

Fig.-2 (b) Flow chart of the time domain analysis of moored ship motions considering harbor oscillations

First, ship dimensions such as the ship length, breadth, draught and displacement are given and the mooring conditions which consist of mooring ropes and fenders are arranged. Then, the calculations of external forces such as hydrodynamic forces and exciting forces are carried out using the frequency domain analysis before the time domain analysis. The wind and current forces are included in the time domain analysis. Concerning the calculation of exciting forces, two analyses which consist of the frequency domain analysis (for progressing waves) and the analysis of harbor oscillations (for long period waves) are used as mentioned above. After that, the time domain analysis is carried out for calculating the moored ship motions, fender deformations, mooring rope tensions and so on (M.Kubo, et al.,1992).

3. INVESTIGATION ON THE VALIDITY OF THE PROPOSED NUMERICAL SIMULATIONS

3.1 Comparisons with the field test results for moored ship motions

(1) Harbor, ship mooring conditions and field test results

The harbor is a typical reclaimed one which is sheltered by breakwaters as shown in Fig.-3. The berth is also shown in Fig.-3.

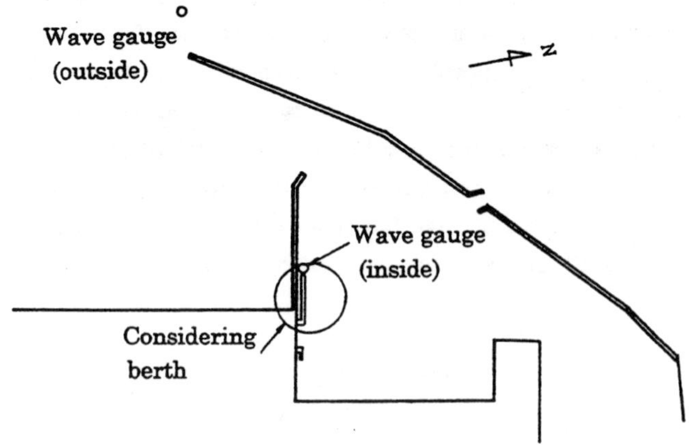

Fig.-3 Layout of harbor

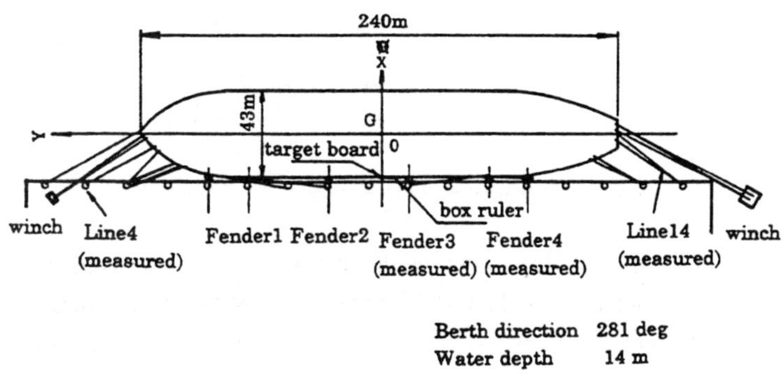

Fig.-4 Mooring condition and arrangement of sensors at the field test

The field test is carried out on January 29-February 4, 1995 for a bulk carrier (L=240m, B=43m, d=9.33m, W=75,590t). The observation items are the moored ship motions, mooring rope tensions, fender deformations, inside and outside harbor waves and winds. The mooring system consists of 14 lines of 60mm diameter PE ropes with a nylon rope and four lines for maneuvering and four pneumatic fenders 2000 $\phi \times$2000H-1.2k. The mooring condition and the arrangement of sensors at the field test are shown in Fig.-4.

The moored ship motions (sway, surge and heave) are measured by driving cameras against a target board (8.4m$\times$2m with lattices) and a box ruler installed on the ship hull. The mooring rope tension is measured by a steel bar with attached strain gauges which are inserted between the bit and the end of the mooring rope. The fender deformation is measured by a displacement meter and monitored by video cameras. Also, the inside and outside harbor waves and the winds are measured with wave gauges and a wind meter established near the berth, respectively.

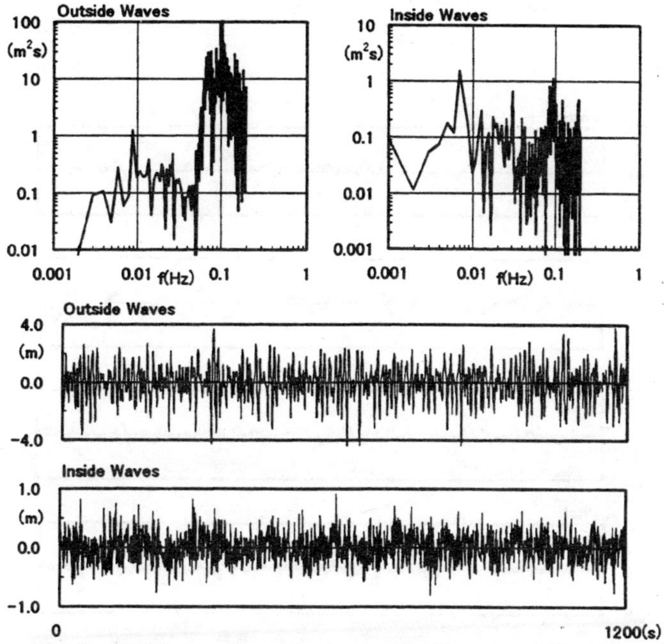

Fig.-5 The time series and spectra of inside and outside waves at the field test

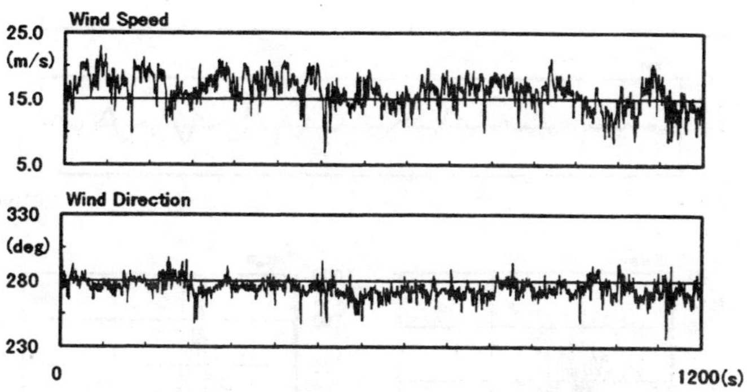

Fig.-6 The wind speed and direction at the field test

After berthing on January 29, the cargo handling is operated immediately. In accordance with a depression closing in the early morning on January 30, the weather grows worse and the operation is knocked off by the strong wind of more than U_{10}=15m/s, where U_{10} is the average speed. At that time when the moored ship motions attains almost maximum values, the observation is begun at 16:00 on January 31.

The time series and spectra of inside and outside waves from 16:00-16:20 are shown in Fig.-5. Also, Fig.-6 shows the speed and direction of the wind at that time. As shown in Fig.-5, it is recognized that the predominant periods of inside and outside waves are 146s and 11s, respectively. This is just a situation that the long period waves exist inside harbor as pointed out by KUBO (M.Kubo, et al.,1992). As mentioned above, the field test is carried out in a stormy weather, where wind speed is about U_{10}=18m/s and the waves are long period.

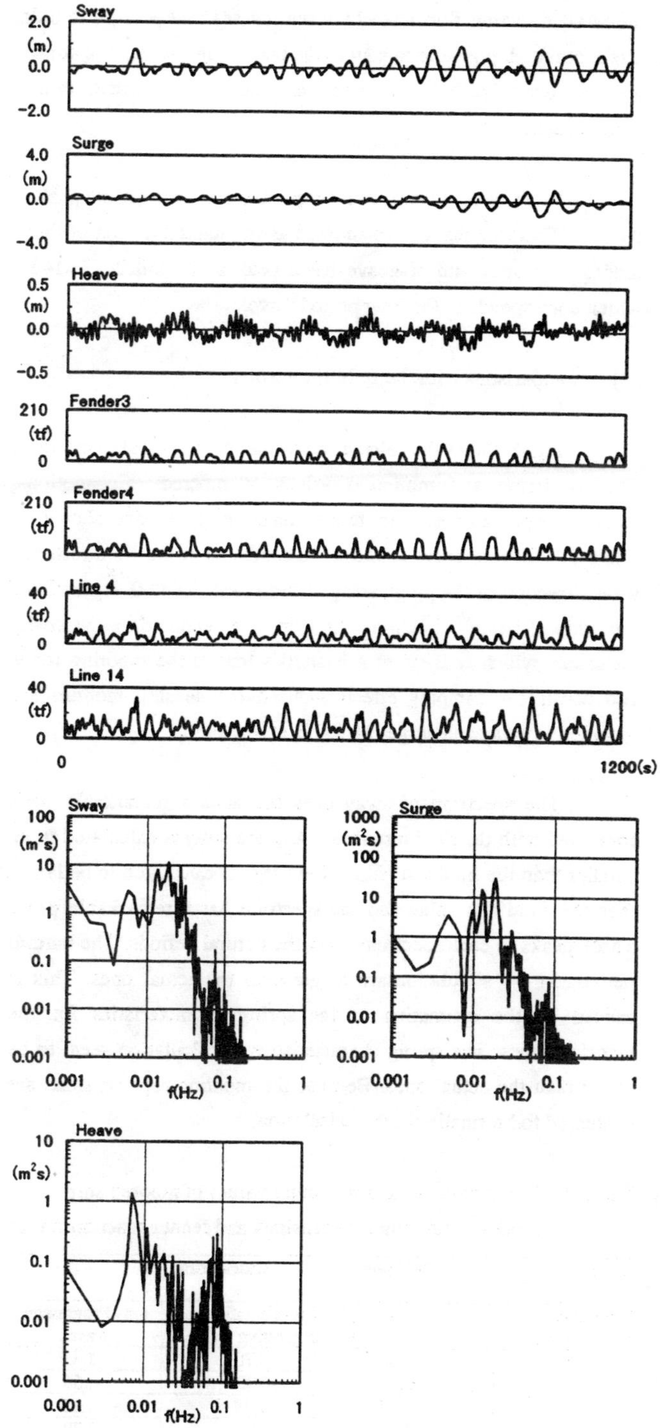

Fig.-7 Field test results of the moored ship motions, mooring rope tensions and fender reaction forces at the field test

Fig.-7 shows the field test results of the moored ship motions, mooring rope tensions and fender reaction forces calibrated by the displacement-reaction force characteristics. The surge amounts to 2m and the sway attains 1m offshore movement where the wind direction fluctuated between the sea side and berth side. The moored ship is able to stay at the berth safely under the stormy weather condition without rope accidents nor excessive fender deformations.

The spectra of sway and surge have almost the same peak at 0.02Hz. This frequency is considered as the natural period of those modes. The spectrum of heave has a peak at 0.0068Hz (T=146s) which corresponds to the long period wave.

(2) Comparisons with the field test results

Fig.-8 shows the calculation results of moored ship motions. The maximum and minimum values of moored ship motions, mooring rope tensions and fender reaction forces are shown in Table-1. With references to the wave condition at the field test, the wave direction of the progressing waves is set to be 60 degree from the ship's beam direction. And in the simulation, viscosity resistance which consists of a hysterisis loss of the mooring ropes and non-linear damping effects induced by the ship motions are included in sway and surge (M.Kubo, et al., 1980).

The spectrum of sway does not have a remarkable peak compared with the field test value. And the sway is calculated to be smaller than the field test value. The surge is calculated to be larger than the field test value and the spectrum has a remarkable peak. Those peaks in each spectrum show the natural periods. The natural periods in the simulation are longer than the actual ones. This is caused by the estimation of the spring characteristics for the mooring ropes. The spring characteristics in calculation seem to be softer than the actual ones. Because the mooring rope tensions are evaluated to be smaller in the calculation.

Table-1 The maximum and minimum values of moored ship motions, mooring rope tensions and fender reaction forces

Items		Field test	Calculation	
Evaluation of Exciting Forces		---	Harbor oscillations and progressing waves	only Progressing waves
Sway (m)	max.	1.00	0.70	1.43
	min.	-0.74	-0.33	-0.66
Surge (m)	max.	1.07	1.38	1.15
	min.	-1.32	-1.90	-2.29
Heave (m)	max.	0.26	0.13	0.19
	min.	-0.21	-0.13	-0.14
Fender 3 (tf)		80.0	44.8	78.2
Fender 4 (tf)		98.7	100.5	103.7
Line 4 (tf)		24.3	14.5	20.3
Line14 (tf)		38.6	11.9	21.6

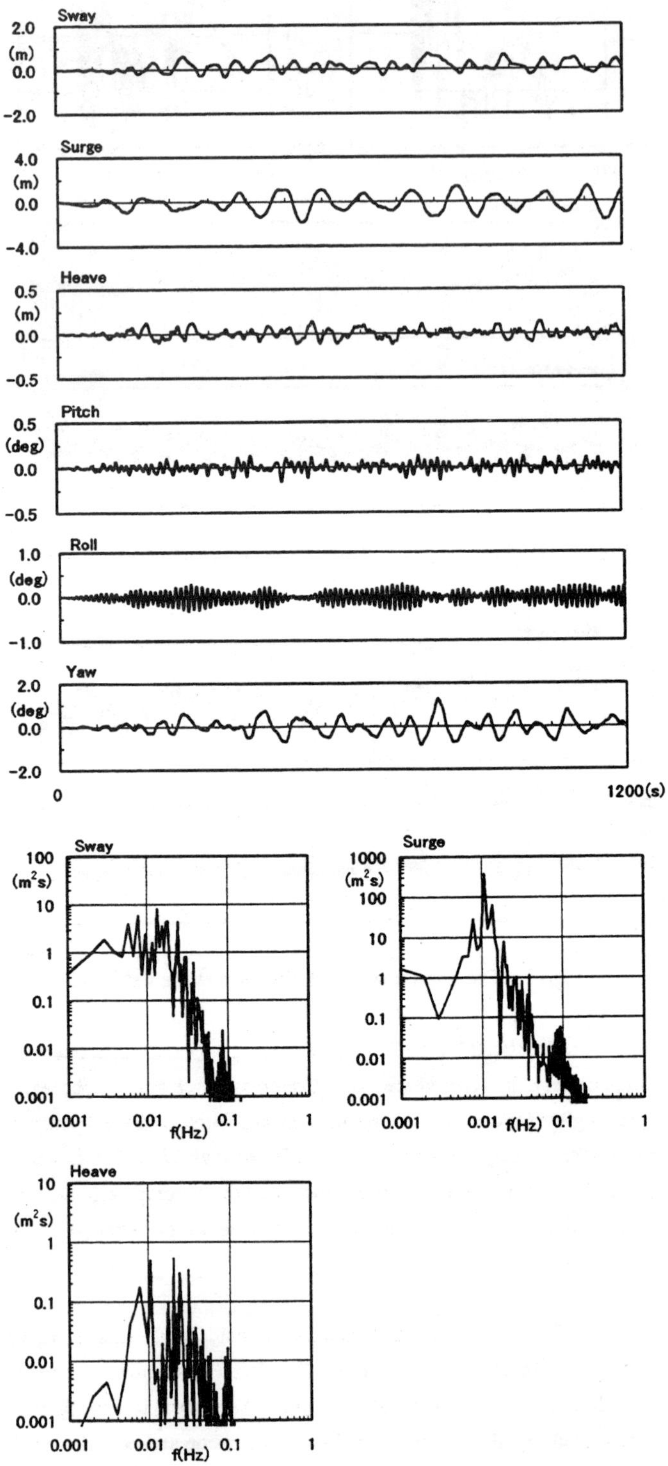

Fig.-8 Calculation results of the moored ship motions (Harbor oscillations + Progressing waves)

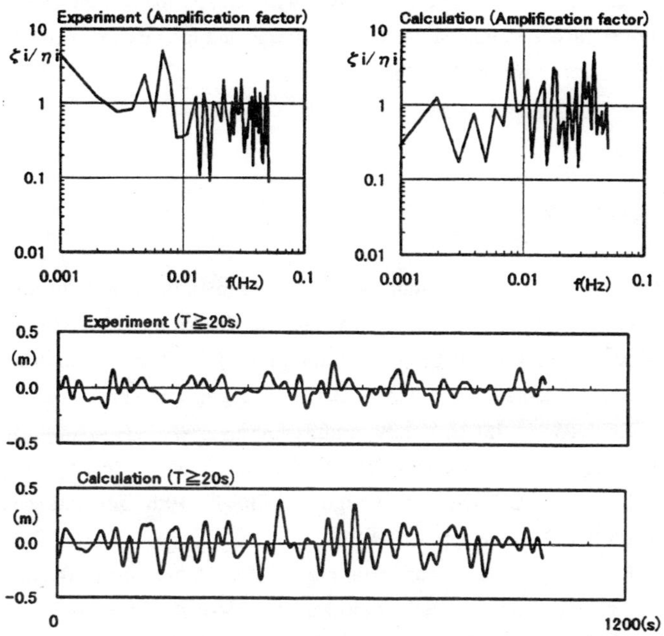

Fig.-9 Comparisons of amplification factors and time series of the inside waves

The heave does not have the larger and longer period motions as same as that of the field test (F=0.0068Hz). On the other hand, Fig.-9 shows comparisons of amplification factors of outside waves to inside ones (ζ_i/η_i) and time series of the inside waves from f=0-0.05Hz (T>20s). It is confirmed that the remarkable peak in amplification factor appears at 0.0068Hz in the experiment, however, the amplification factor of the calculation has some peaks, including the peak at 0.0068Hz. The amplification factors in Fig.-9 agree well with the spectra of heave in both experimental and calculation results. So, it can be considered that the difference in heave is attributed to the error in the analysis of harbor oscillations. Besides it is considered that the effect of vertical movements by roll appears at the field test about 0.1Hz in heave.

In the analysis of harbor oscillations used in this paper, the assumption that the water depth is constant is imposed and a loss of wave energy at the harbor entrance is ignored. It is necessary for the time domain analysis proposed in this paper to improve the accuracy of the analysis of harbor oscillations. However, it is confirmed that the numerical simulation of moored ship motions considering harbor oscillations can reproduce the long period ship motions in practice.

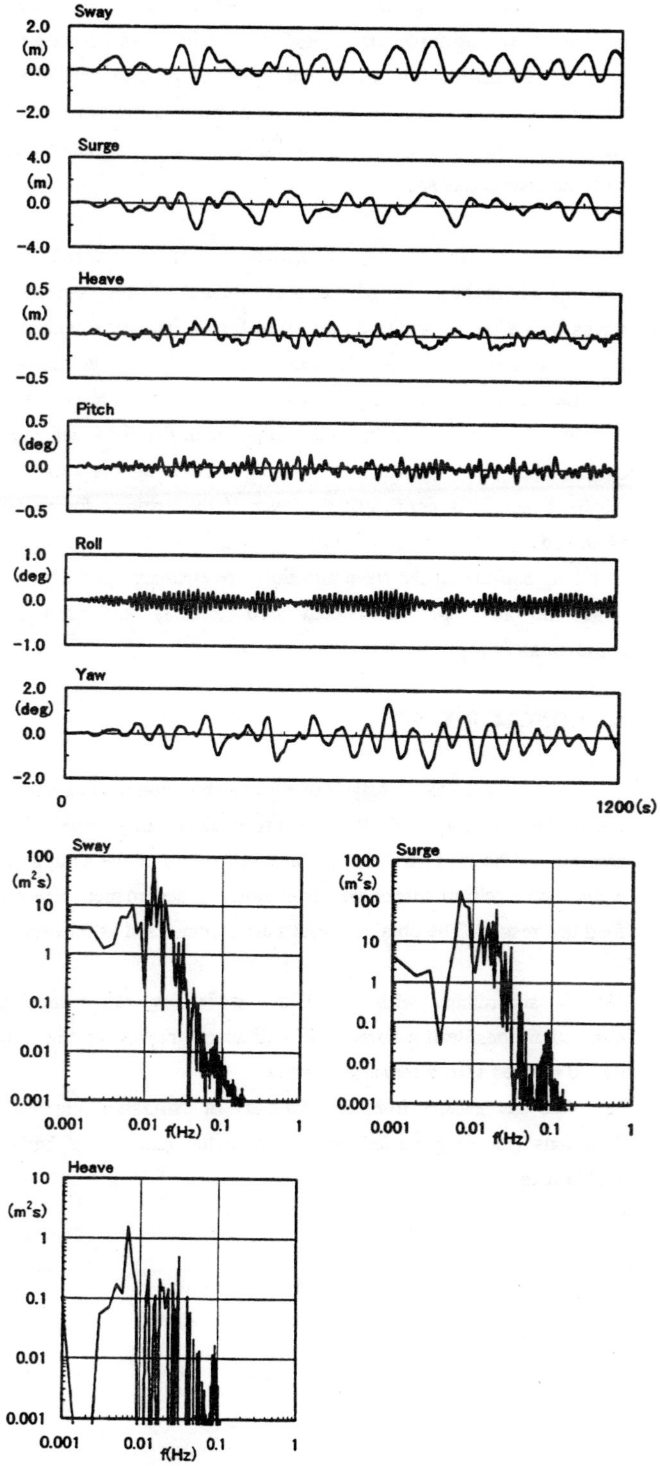

Fig.-10 Calculation results of the moored ship motions (only Progressing waves)

3.2 Influence of Evaluation Method for Exciting Forces on Moored Ship Motions

In order to investigate the influences of the evaluation method of exciting forces on moored ship motions, we compare the field test results with the calculation results of the case considering harbor oscillations and the case of only progressing waves (the ordinary analysis). The mooring conditions of both calculations are set to be completely same.

Fig.-10 shows the calculation results of moored ship motions, where the exciting forces are evaluated by only progressing waves of the inside waves as shown in Fig.-5. The maximum and minimum values of moored ship motions, etc. are given in Table-1. The spectra of sway and surge have a remarkable peak, respectively and those motions are estimated as larger than the field test ones. On the other hand, the calculation result in heave agrees well with the field test. It is confirmed that even if the exciting forces are evaluated by only progressing waves derived from the inside waves with long period and the wave directions are estimated properly, the large and long period moored ship motions are estimated approximately

4. CONCLUSIONS

A time domain analysis of moored ship motions in a harbor considering harbor oscillations is proposed in this paper. The accuracy of the proposed analysis and influences of the evaluation method on exciting forces are investigated in accordance with the field test results. The obtained results are summarized as follows.

(1) In accordance with the field test results, the numerical simulation considering harbor oscillations can reproduce the long period moored ship motions in practice.

(2) We can escape from the difficulty of estimating the wave directions (for long period waves) using the analysis of harbor oscillations.

(3) In order to improve the accuracy of the proposed time domain analysis, it is necessary to consider the improvements of the analysis of harbor oscillations.

(4) Even if the exciting forces are evaluated by only progressing waves derived from the inside waves with long period and the wave directions are estimated properly, the large and long period moored ship motions are estimated in practice.

REFEFENCES

J.J.Lee (1971)," Wave-Induced Oscillations in Harbors of Arbitrary Geometry", Journal of Fluid Mechanics, Vol.45, pp.375-394.

M.Kubo and V.Berthel (1992),"Experimental Study on Motions of Ships Moored at an Open Berth in a Lagoon", Proc.Coastal Engineering, JSCE, Vol.39, pp.831-835.

M.Kubo and S.Sakakibara (1992), "Influence on fender type of Operation Efficiency of Cargo Handling", 10th International Harbor Congress, pp.4.19-4.23.

M.Kubo, Y.Imaeda and M.Okumura (1980), "A Measuring Device for the Long Period Motions of Moored Ship and its Measurement", Review of Kobe University of Mercantile Marine, pp.289-298.

S.Shiraishi, M.Kubo, S.Ueda and S.Sakakibara (1996), " A Countermeasure by Mooring System for Moored Ship Motions under Long Period Waves",5th COSU'96, Buenos Aires, pp.207-216.

T.Nagai, N.Hashimoto, T.Asai, I.Tobiki, K.Ito, T.Toue, A.Kobayashi and T.Shibata (1994), "Relationship of a Moored Vessel in a Harbor and a Long Wave Caused by Wave Groups", Proc.24th ICCE, Kobe, Vol.1, pp.847-860.

T.Sawaragi and M.Kubo (1981), "Motions of a Moored Ship Induced by Harbor Oscillations", The Journal of Japan Institute of Navigation, Vol.65, pp.49-57.

Proceedings of the Seventh (1997) International Offshore and Polar Engineering Conference
Honolulu, USA, May 25-30, 1997

ISBN 1-880653-28-1 (Set); ISBN 1-880653-31-1 (Vol. III)

Hydrodynamic Response of a Bottom Mounted Flexible Membrane Structure in Gravity Waves

Amal C. Phadke and Kwok Fai Cheung
University of Hawaii at Manoa
Honolulu, HI, USA

Abstract

The present paper describes a numerical model for the interaction between a bottom-mounted fluid-filled membrane and surface gravity waves. The longitudinal dimension of the membrane is much larger than its sectional dimensions thus effectively yielding a two-dimensional problem. The membrane is elastic and is assumed to be very thin, massless and without any flexural rigidity. The static shape of the membrane is first computed by considering the equilibrium of the external loads with the membrane hoop stress. The linear differential equation governing membrane deformations due to dynamic loads is obtained from the classical membrane theory of shells and solved using the finite element method. Both the kinematic and dynamic boundary conditions are satisfied at the fluid-membrane interface. The motions of both the internal and external fluids are computed respectively by two boundary element models based on potential theory, which in turn are coupled with the finite element model. Through a numerical example, the proposed model is shown to be effective in simulating the dynamic behavior of the membrane in gravity waves.

Keywords

Hydrodynamics, fluid-filled membrane structures.

1 Introduction

Membrane and tension structures have been used extensively in structural and architectural design of buildings. Many analytical and numerical techniques are available to analyze these structures for a variety of engineering applications (Leonard [7]). These membrane structures change their geometry to accommodate external loads rather than increase stress levels, and therefore are efficient engineering structures in the severe environment. The dynamic behavior of a fluid-filled membrane can be easily tuned by adjusting the interior pressure. A number of feasibility studies have been carried out to investigate the applications of membrane structures in the ocean environment.

One of the earliest studies was carried out by Green [3] to investigate the dynamic behavior of pneumatic breakwaters. Fredriksen [2] also carried out an experimental study on the effects of fluid filled bags on wave attenuation. His study consisted of both a floating and a bottom-mounted pneumatic structure. Fok and Murabayashi [1] conducted full scale field experiments to study the feasibility of both bottom mounted and floating membrane structures for temporary storage of fresh water in the ocean. More recently, Ohyama *et al* [8] developed a numerical model to study the transmission and reflection characteristics of a water-filled bottom mounted membrane in two dimensions. They used two boundary element models to evaluate the external and internal fluid loads. The equation of motion was derived by considering the equilibrium of the external and internal pressure with the membrane hoop stress and inertia force. Zhao [11] applied a similar approach to study the dynamic response of a fluid-filled floating membrane.

The dynamic response of the membrane and internal fluid resembles that of internal waves with surface tension and involves a kinematic and dynamic boundary condition. Based on the membrane theory of thin elastic shells, the present study introduces a dynamic boundary condition at the fluid-membrane interface in addition to the kinematic condition. The membrane deformation is modeled more generally by the finite element method, which in turn is coupled with two boundary element models for the external and internal fluids respectively. To illustrate the present method, the problem is treated as linear and two-dimensional, and the dynamic response of a bottom-mounted fluid-filled membrane in gravity waves is considered.

2 Membrane Static Shape

The static shape of a sea-bed mounted membrane filled with a fluid of a specific density is initially unknown and must be calculated before the hydrodynamic analysis can be carried out. The static shape can be computed by considering the equilibrium of the forces acting on an element of the membrane. This problem is similar to calculating the shape of a liquid drop on a flat surface as discussed in Timoshenko and Woinowsky-Krieger [9]. This technique has been applied successfully to floating membrane structures (Hawthorne [5]).

Consider a bottom mounted fluid-filled membrane as shown in Fig. 1. The length of the structure is assumed to be large compared with its sectional dimensions, thus effectively yielding a two-dimensional problem. The membrane itself is considered massless for all practical purposes. A Cartesian coordinate system $X-Y$ and a cylindrical polar coordinate system $R-\theta$ are used in the derivation of the governing equations for the static shape. Since the membrane is symmetric about the y axis, computations need to be carried out only for one side of the membrane. In the absence of waves, the only external forces acting on the membrane are the hydrostatic pressures due to the internal and external fluids and a constant internal pressure in the internal fluid. A unique membrane shape may be computed by specifying the spread (x_0), height (y_0) and the internal pressure (P_0) of the membrane. The difference in pressure across the membrane is balanced by the hoop stress in the membrane:

$$\frac{N_0}{R} = (P_i - P_e) \tag{1}$$

where N_0 is the hoop stress in the membrane, R is the radius of curvature and $P_i - P_e$ is the difference between the internal and external pressure. In the absence of shear stress on the membrane, the hoop stress N_0 is constant throughout the membrane. Let s be distance measured along the membrane. Since $1/R = d\theta/ds$, eq. 1 can be rewritten as,

$$N_0\frac{d\theta}{ds} = (P_i - P_e) \tag{2}$$

Noting that $ds = dx/\cos\theta = dy/\sin\theta$ and upon substitution in eq. 2, the following two equations are obtained.

$$N_0 \sin\theta d\theta = (P_i - P_e)\,dy \tag{3}$$

$$N_0 \cos\theta d\theta = (P_i - P_e)\,dx \tag{4}$$

In the present case, the internal pressure P_i is composed of both the constant (P_0) and hydrostatic components, whereas the external pressure is only hydrostatic. The solution of Eq. 3 is given by

$$-N_0\cos\theta = P_o y - g(\rho-\rho_i)y_0 y + g(\rho-\rho_i)\frac{y^2}{2} + C \tag{5}$$

where ρ is the density of surrounding water, ρ_i is the density of the internal fluid and g is the acceleration due to gravity. Let $\mu = P_0 - g(\rho-\rho_i)y_0$. The constant of integration C and the hoop stress N_0 can be evaluated from the following boundary conditions.

$$y = y_0 \quad \text{at} \quad \theta = 0 \tag{6}$$

$$y = 0 \quad \text{at} \quad \theta = \theta_0 \tag{7}$$

where y_0 is the height of the membrane above the seabed and θ_0 is the angle at $x = x_0$. The constant of integration C and the hoop stress N_0 can be expressed in terms of θ_0 and y_0 as:

$$C = -N_0\cos\theta_0 \tag{8}$$

$$N_0 = -\frac{\mu y_o + g(\rho-\rho_i)y_0^2/2}{1-\cos\theta_0} \tag{9}$$

The solution to eq. (5) in terms of θ is

$$y = \frac{\mu}{(\rho-\rho_i)g}\left[-1+\sqrt{1-\frac{2N_0(\rho-\rho_i)g}{\mu^2}(\cos\theta-\cos\theta_0)}\right] \tag{10}$$

Once the y coordinate of the membrane is known, the x coordinate can be computed by integrating eq. (4):

$$x = N_0\int_0^\theta \frac{\cos\theta d\theta}{\mu + g(\rho-\rho_i)y} \tag{11}$$

Eq. (11) can not be solved immediately since θ_0 is unknown. However, x_0 or the spread of the membrane is known. The angle θ_0 can be evaluated iteratively by assuming some value of θ_0 and solving for N_0 and x. If the computed value of x is close to the specified spread of the membrane within a given tolerance, the iteration is stopped. Otherwise a new value of θ_0 is assumed and the process is repeated.

3 Hydrodynamic Analysis

The hydrodynamic analysis of a membrane structure involves the calculation of the hydrodynamic pressure acting on the membrane due to the incoming waves and subsequently the dynamic response of the membrane to waves. Since this problem is being treated as linear, the hydrodynamic load needs to be computed only on the undeformed membrane. The external and internal fluid problems are treated as two coupled boundary-value problems.

3.1 External Fluid Problem

With reference to Fig. 1, the water surrounding the membrane is assumed to be inviscid and incompressible and the flow irrotational. The sea bed is assumed to be horizontal and impermeable. These assumptions permit use of a velocity potential Φ satisfying the Laplace equation to describe the fluid motion.

Apart from the governing Laplace equation, the velocity potential Φ must also satisfy the free-surface, bottom, radiation and body boundary conditions. It is convenient to decompose the total potential Φ into the incident (Φ_W), diffraction (Φ_D) and radiation (Φ_R) potentials.

$$\Phi(x,z,t) = \Phi_W + \Phi_D + \Phi_R = \Re\left\{(\phi_W + \phi_D + \phi_R)e^{-i\sigma t}\right\} \tag{12}$$

where σ is the incident wave frequency. The kinematic boundary conditions on the membrane (S_b) are

$$\frac{\partial \Phi_D}{\partial n} = -\frac{\partial \Phi_W}{\partial n} \quad \text{on} \quad S_b \tag{13}$$

$$\frac{\partial \Phi_R}{\partial n} = \frac{\partial w}{\partial t} \quad \text{on} \quad S_b \tag{14}$$

Where n denotes distance in the normal direction. Green's second identity along with an infinite series form of the Green function (John [6]; Wehausen and Laitone [10]) is used to solve for the diffraction and radiation potentials. The solution of the diffraction problem for submerged structures is well established and need not be elaborated in great detail. Some details of the radiation problem are provided here. The Green function G and the radiation potential Φ_R are related by Green's second identity as given below

$$\Phi_R(x,z) = \int_S \left(\Phi_R \frac{\partial G}{\partial n} - G \frac{\partial \Phi_R}{\partial n} \right) dS \tag{15}$$

Since the Green function satisfies all the boundary conditions except the body boundary condition, after making use of eq. (14), the above equation reduces to the following in terms of the membrane velocity dw/dt.

$$\Phi_R(x,z) = \int_{S_b} \left(\Phi_R \frac{\partial G}{\partial n} - G \frac{\partial w}{\partial t} \right) dS_b \tag{16}$$

The diffraction and radiation problems are solved using the boundary element method. The fluid boundary along the membrane is discritized into a number of linear elements and the solution is obtained in terms of the membrane deformation. The choice of linear elements is purely a matter of convenience. The use of linear elements for the present problem assures that the boundary and the finite element nodes are precisely overlapped, thereby aiding in achieving an efficient coupling between the two models. The total external pressure (P_e) acting on the membrane is given by the Euler integral:

$$P_e = \rho \frac{\partial \Phi}{\partial t} + \rho g[(h-y) - w \cos\theta] \tag{17}$$

Application of the boundary element method to the diffraction and radiation problems results in the following matrix equation for the hydrodynamic force vector $\mathbf{F_{he}}$ corresponding to the term containing the velocity potential Φ in eq. (17):

$$\mathbf{F_{he}} = \mathbf{F} - \mathbf{M_a}\ddot{\mathbf{w}} - \mathbf{C}\dot{\mathbf{w}} \tag{18}$$

where $\mathbf{F}$ is the force vector due to the incident and diffracted waves, $\mathbf{M_a}$ and $\mathbf{C}$ are the added mass and damping matrices obtained from the the solution of the radiation problem.

3.2 Internal Fluid Problem

The internal fluid problem is treated very similarly to the external flow problem. With the assumption of potential theory, a velocity potential Φ_I satisfying the Laplace equation is used to define motion of the internal fluid. The velocity potential Φ_I satisfies the bottom boundary condition, and the kinematic boundary condition on the membrane surface is,

$$\frac{\partial \Phi_I}{\partial n} = \frac{\partial w}{\partial t} \quad \text{on} \quad S_b \tag{19}$$

As in the case of the external flow problem, the internal flow problem can be solved by making use of Green's second identity along with a simple Green function that satisfies the sea bed boundary condition. The Green function G_I is given by

$$G_I = \frac{1}{2\pi}\left[\ln\left(\frac{r}{h}\right) + \ln\left(\frac{r_1}{h}\right)\right] \tag{20}$$

where r is the distance between the source point and the field point, whereas r_1 is the distance between the image of the source point about the sea bed and the field point. The velocity potential Φ_I is expressed in terms of the unknown membrane velocity $\dot{w}$. The total internal pressure (P_i) acting on the membrane is given by the Euler integral:

$$P_i = P_0 + \rho g(h - y_0) + \rho_i g(y_0 - y) + \rho_i \frac{\partial \Phi_I}{\partial t} - \rho_i g w \cos\theta \tag{21}$$

In the boundary element model for the internal fluid, the hydrodynamic force vector $\mathbf{F_{hi}}$ corresponding to the term containing Φ_I in eq. (21) can be expressed in the following matrix form:

$$\mathbf{F_{hi}} = \mathbf{M}\ddot{\mathbf{w}} \tag{22}$$

where, $\mathbf{M}$ is the hydrodynamic mass matrix for the membrane.

4 Membrane Deformations

Owing to the assumption of small deformations, the hydrodynamic pressure needs to be computed on the undeformed shape of the membrane. The deformation in the membrane is computed by using the finite element method. The membrane deformation is assumed to be small, thus permitting the use of linearized equations. Following Han [4], the force equilibrium equations for a membrane are

$$\frac{1}{R}\frac{dN_\theta}{d\theta} = 0 \tag{23}$$

$$N_\theta\left[1 + \frac{1}{R}\frac{du}{d\theta} - \frac{1}{R}\frac{d^2w}{d\theta^2}\right] = R(P_i - P_e) \tag{24}$$

Where N_θ is the membrane hoop stress and R is the radius of curvature and u, w are the circumferential and radial displacements respectively. With the assumption of small deflection, the strain-displacement relationship (Timoshenko and Woinowsky-Krieger [9]) may be written as

$$\epsilon_\theta = \frac{1}{R}\left(\frac{du}{d\theta} + w\right) + \epsilon_0 \tag{25}$$

where, ϵ_o and ϵ_θ are respectively the initial and instantaneous circumferential strains in the membrane. In the present case, the circumferential strain and the membrane hoop stress are related by the following equation

$$\epsilon_\theta = K N_\theta \tag{26}$$

in which $K = (1-\nu^2)/E\delta$ where, ν, E and δ are the Poisson's ratio, elastic modulus and thickness of the membrane respectively.

After substituting for du/dt from eq. (25) into eq. (24), the following differential equation governing the deflection of the membrane is obtained:

$$\frac{1}{R}\frac{d^2w}{d\theta^2} + \frac{w}{R} = [1 + K(N_\theta - N_0)] - \frac{R}{N_\theta}(P_i - P_e) \tag{27}$$

Eq. (25) is used to calculate the final hoop stress in the membrane. The obvious boundary conditions are $u, w = 0$ at $\theta = \pm\theta_0$. Using these boundary conditions with eq. (25), it can be shown that,

$$N_\theta = N_0 + \frac{1}{Kl}\int_{-\theta_0}^{\theta_0} w d\theta = N_0 + \frac{I}{Kl} \tag{28}$$

Where l is the arc length of the membrane. Since,

$$P_0 - g(\rho - \rho_i)y_0 + g(\rho - \rho_i)y = N_0/R \tag{29}$$

using eqs. (28), (29), (17) and (21), eq. (27) becomes

$$\begin{aligned}\frac{1}{R}\frac{d^2w}{d\theta^2} + \frac{w}{R} &= \frac{1}{l}\left(\frac{1+\epsilon_0}{\epsilon_0}\right) I \\ &- \frac{R}{N_\theta}\left[\rho_i\frac{\partial\Phi_I}{\partial t} - \rho\frac{\partial\Phi}{\partial t} + g(\rho - \rho_i)w\cos\theta\right]\end{aligned} \tag{30}$$

The material properties of the membrane are included in the present analysis through the term containing the initial strain ϵ_0 in the above equation. The above equation can be solved using the finite element method. In the present analysis, the membrane is divided into a number of linear elements. The radial displacement w is assumed to vary linearly over an element as given below:

$$w(\theta) = \frac{\theta_2 - \theta}{\Delta\theta}w_1 + \frac{\theta - \theta_1}{\Delta\theta}w_2 = \sum_{j=1}^{2} f_j(\theta)w_j \tag{31}$$

where, w_j and f_j are respectively the nodal displacements and shape functions for an element. For each element, the residue produced by the governing differential equation due approximation of w as in Eq. (31) is minimized. This procedure transforms the governing differential equation into a set of linear algebraic equations of the following form:

$$(\mathbf{K_s} + \mathbf{K_h})\mathbf{w} = \mathbf{F_{he}} + \mathbf{F_{hi}} \tag{32}$$

where $\mathbf{K_s}$ is the structural stiffness matrix and $\mathbf{K_h}$ is the hydrodynamic stiffness matrix that arises from the terms containing w on the right hand side of eq. (30). After including the hydrodynamic loads due to the external and internal flows, matrix equation governing the motion of the membrane and fluids is given by

$$(\mathbf{M} + \mathbf{M_a})\ddot{\mathbf{w}} + \mathbf{C}\dot{\mathbf{w}} + (\mathbf{K_s} + \mathbf{K_h})\mathbf{w} = \mathbf{F} \tag{33}$$

Since the membrane is considered to be massless, it doesn't contribute to the the mass matrix. The mass matrix has contributions from both the external ($\mathbf{M_a}$) and internal ($\mathbf{M}$) fluids. Since w is periodic in time, the time dependence in the above equation can be removed, resulting in a set of complex simultaneous equations in frequency domain. These equations are solved using Gaussian elimination to obtain the membrane response.

5 Results and Discussion

To illustrate the present method, the dynamic response of a seabed-mounted membrane filled with fresh water is considered. The static shape of the membrane is evaluated prior to the dynamic analysis. Fig. 2 shows the static shape of a membrane computed using the theoretical development described in §2. The membrane spread and height are 5 m and 2 m respectively, and the internal pressure is 4200 N/m^2. The value of $E\delta$ is assumed to be 5300 N/m. Because of a small density difference between the internal and external fluids, the static shape is not exactly circular. However, it will be exactly circular if outer and inner fluid densities are the same.

Fig. 3 through 5 show the membrane response over a half wave cycle for three membrane breadth to incident wavelength ratios (B/L). At lower values of B/L, as in fig. 3, the response consists predominantly of the fundamental mode of vibration, where the midpoint of the membrane doesn't move appreciably. This is consistent with the numerical results of Ohyama *et al* [8]. When the breadth of the membrane is comparable to that of the incident wavelength (fig. 4), higher modes of vibration appear. For larger values of B/L it is possible to observe the formation of a standing wave over the surface of the membrane. This is quite similar to the interaction between surface and internal waves. However, in the present case, the hoop stress which is equivalent to the surface tension, is of more importance than the density difference. In fact, the dynamic boundary condition applied on the membrane surface is mathematically identical to the governing equation for membrane deformation obtained from the membrane theory of thin elastic shells. It is in this aspect that the present investigation differs from previous work where such equivalence is not recognized. Moreover, unlike the previous work, the present formulation can be extended into three dimensions without much difficulty. In order to assess the performance of the underwater membrane used as a flexible breakwater, it would be necessary to calculate transmission and reflection coefficients and it will be the topic of further investigation.

6 Conclusions

A linearized theory of fluid-filled membrane response in gravity waves is presented. It is recognized that the equation governing the membrane deformation is identical to a dynamic boundary condition satisfied on the membrane. A complete analysis is carried out in three stages. The static shape of the membrane is first estimated. The dynamic components of the internal and external pressures are then estimated. Finally, the membrane response to these external loads is computed. To illustrate the present method, the dynamic response of a seabed-mounted membrane filled with fresh water is considered. The results of the hydrodynamic analysis are presented in the form of a series of plots showing the deformed shape of the membrane over a half wave cycle for various values of B/L. The numerical results appear to compare well with results from Ohyama *et al* [8] and the method is shown to be effective in simulating the dynamic behavior of the membrane.

Acknowledgement

The material is based upon work supported by the National Science Foundation under Grant No. CTS-9408939.

References

[1] Y. Fok and E.T. Murabayashi. Stream-water storage in the ocean by using an impermeable membrane. Technical Report No. 152, Water Resources Research Center, University of Hawaii at Manoa, Honolulu, HI, May 1983.

[2] H.D. Fredriksen. Wave attenuation by fluid filled bags. *Journal of Waterways, Harbors and Coastal Engineering Division, ASCE*, 97(WW1):73–90, February 1971.

[3] J.L. Green. Pneumatic breakwaters to protect dredges. *Journal of Waterways, Harbors and Coastal Engineering Division, ASCE*, 87(WW2):67–87, May 1961.

[4] P.S. Han. *Interactive Non-linear Analysis of Wind-Loaded Pneumatic Membrane Structure*. PhD thesis, Department of Civil Engineering, University of British Columbia, May 1986.

[5] W.R. Hawthorne. The early development of the dracone flexible barge. In *Proceedings of the Institution of Mechanical Engineers*, volume 17, pages 52–81, London, 1960.

[6] F. John. On Motion of Floating Bodies, II - Simple Harmonic Motion. *Communications on Pure and Applied Mathematics*, 3:45–101, 1950.

[7] J.W. Leonard. *Tension Structures - Behavior and Analysis*. McGraw-Hill Book Company, New York, NY, 1988.

[8] T. Ohyama, M. Tanaka, T. Kiyokawa, T. Uda, and Y. Murai. Transmission and reflection characteristics of waves over a submerged flexible mound. *Coastal Engineering in Japan*, 32(1):53–68, 1989.

[9] S. Timoshenko and S. Woinowsky-Krieger. *Theory of Plates and Shells*. McGraw-Hill Inc., New York, NY, 1959.

[10] J. Wehausen and E. Laitone. Surface waves. In S. Flügge, editor, *Strömungsmechanik III*, volume IX of *Handbuch der Physik*, pages 446–778. Springer-Verlag, Berlin, Germany, 1960.

[11] R. Zhao. A complete linear theory for a two-dimensional floating and liquid-filled membrane structure in waves. *Journal of Fluids and Structures*, 9(8), November 1995.

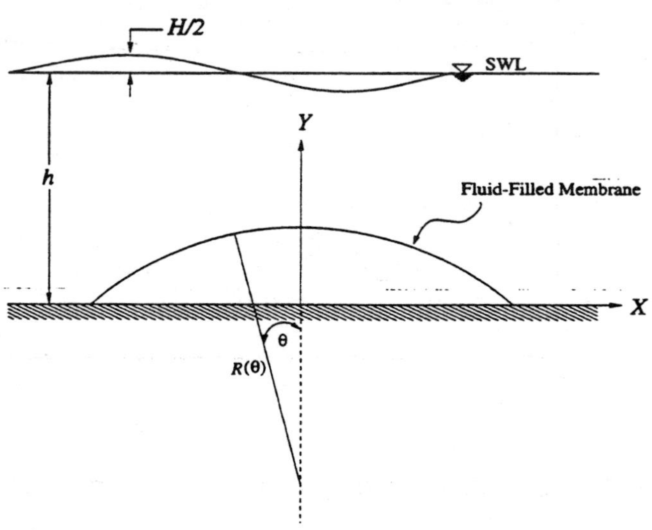

Fig. 1 Definition sketch of the problem

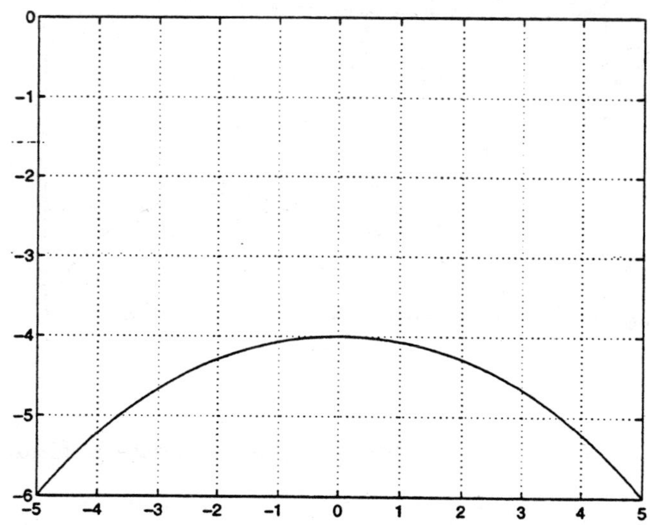

Fig. 2 Static shape of membrane ($x_0 = \pm 5$ m, $y_0 = 2$ m.)

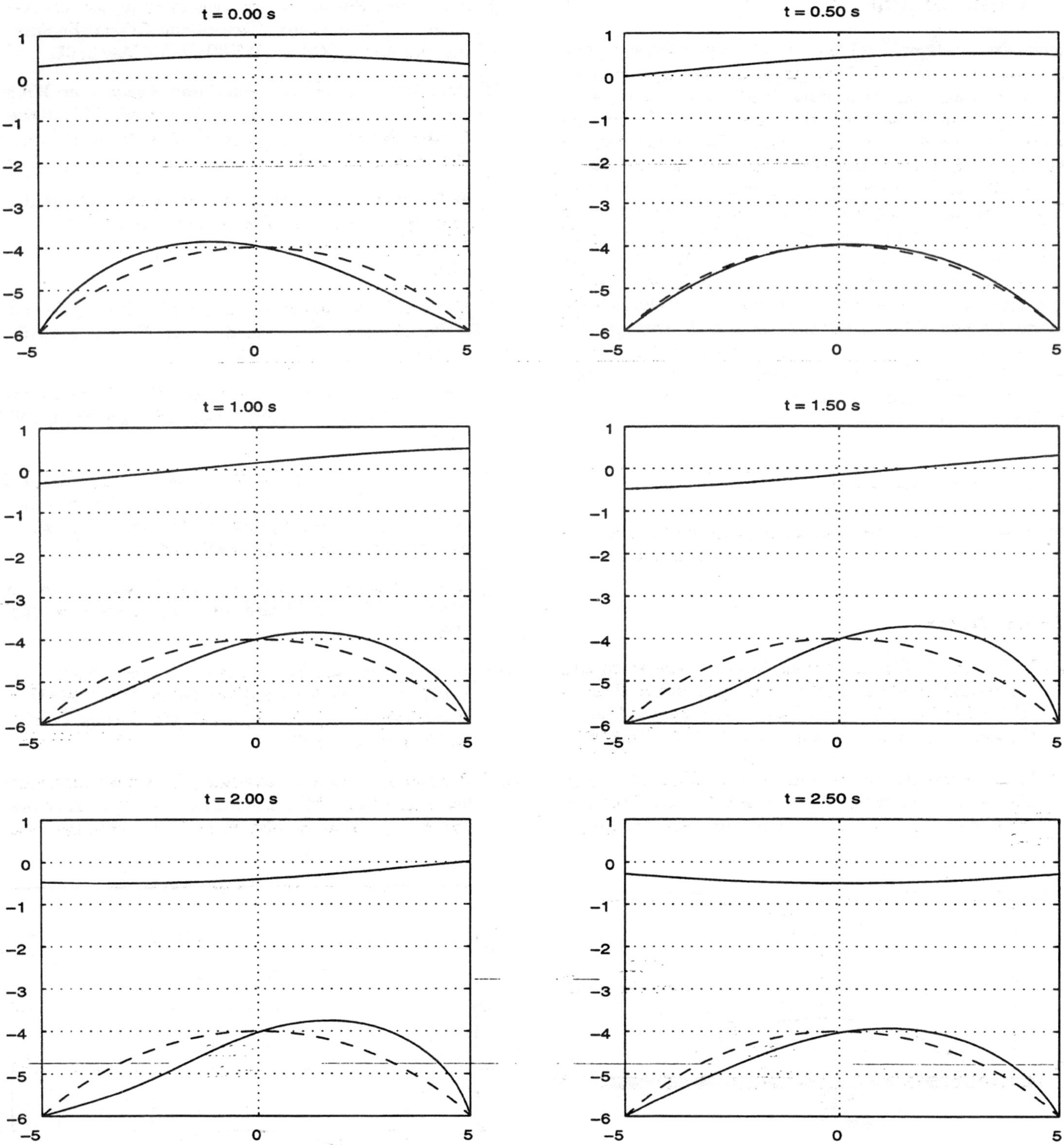

Fig. 3 Membrane response and incident wave profile over one half of the wave period ($B/L = 0.31$, $H = 1$ m, $T = 5$ s)

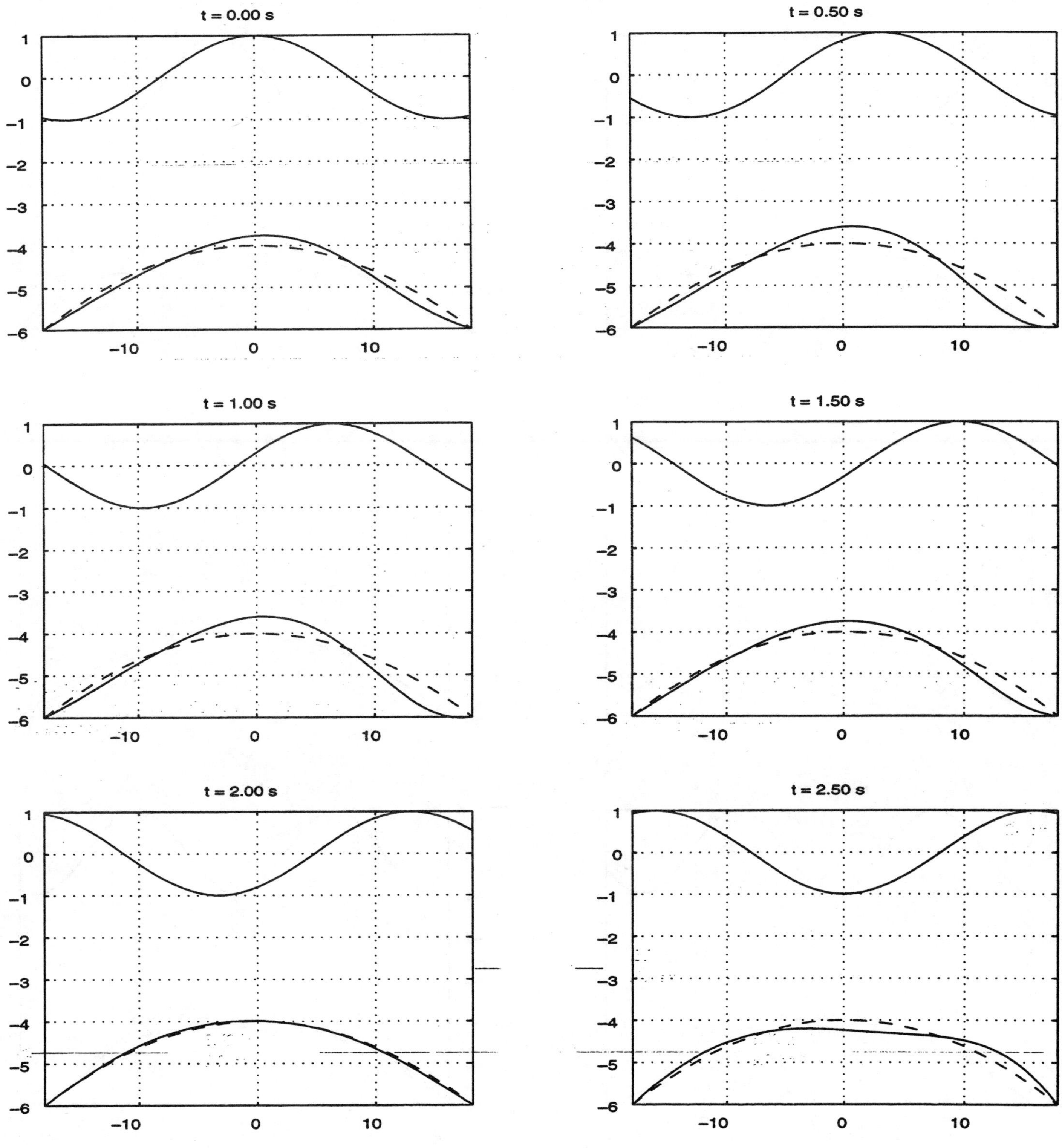

Fig. 4 Membrane response and incident wave profile over one half of the wave period ($B/L = 1.12$, $H = 2$ m, $T = 5$ s)

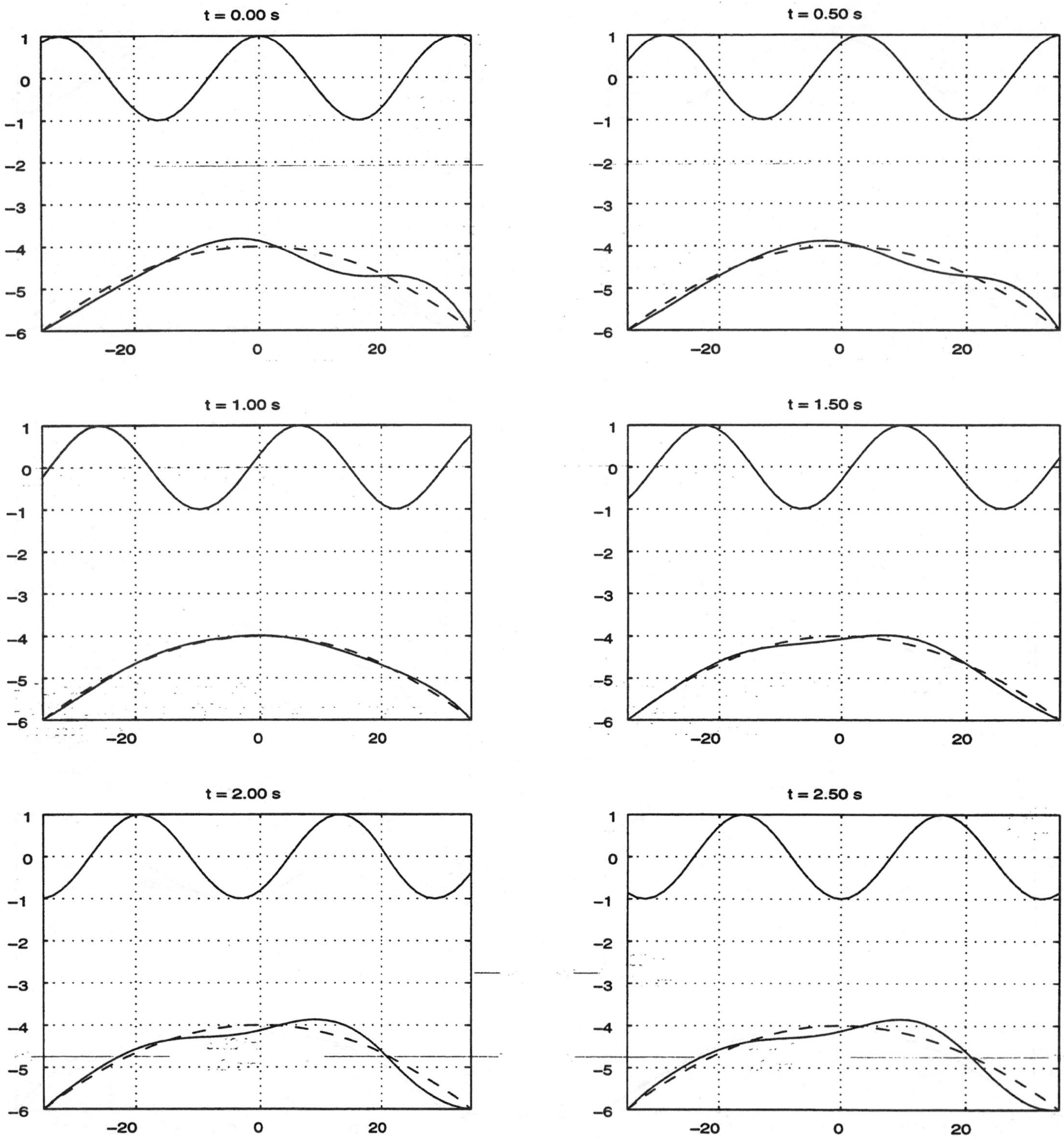

Fig. 5 Membrane response and incident wave profile over one half of the wave period ($B/L = 2.17$, $H = 2$ m, $T = 5$ s)

Proceedings of the Seventh (1997) International Offshore and Polar Engineering Conference
Honolulu, USA, May 25-30, 1997

ISBN 1-880653-28-1 (Set); ISBN 1-880653-31-1 (Vol. III)

Nonlinear Dynamic Motions of Discus Buoys

S.C.S. Yim
Oregon State University, Corvallis, OR, USA
K. Idris
Bandung Institute of Technology, Bandung, Indonesia
J.W. Leonard
University of Connecticut, Storrs, CT, USA

ABSTRACT

Kinetic and kinematic equations of a discus buoy are derived assuming large angles represented by Euler parameters. The three-dimensional coupled interaction of a floating discus and its mooring is studied. External loadings include hydrodynamic forces, tether tensions, wind loadings and system weight. Nonlinearities include large rotational and translational motions and non-conservative fluid loadings. At each instant in time, the two-point boundary-value mooring problem is solved by direct integration using a successive iterative algorithm to satisfy boundary conditions. Coupling between the buoy and the mooring is enforced by matching of the velocities at the attachment point. A predictor-corrector coupling algorithm is developed with multiple sizes of time steps used to provide stability for the separate mooring and buoy models. Numerical results are compared to experimental responses of the discus buoy subjected to regular waves.

INTRODUCTION

In the oceanic environment, mooring-buoy systems are used in various works as navigational aids, as ship and platform moorings, etc. Often, buoy systems float at the ocean surface and may be held to the bottom by mechanical devices that include mooring cables and anchors. Buoy systems come in various shapes, sizes, and mooring configurations, depending on their purpose and environmental constraints. The focus of this study is on discus buoys.

In addition to hydrodynamic loading from ocean waves and currents, wind, installation or deployment loads may be applied to the discus buoy system. Systems with small characteristic lengths are analyzed by methods that include fluid viscosity effects, while relatively large-diameter structures are analyzed by methods based on potential flow theory. The problem of simulating the response of buoy systems is complicated by the fact that the motion of a tethered buoy is affected by its mooring cables. The mooring cables also experience nonlinear hydrodynamic loads. Their behavior is highly nonlinear since their stiffness depends on the motions of the system. Since they are connected to the buoy, the dynamic cable stiffnesses are affected by the buoy motion.

Substantial efforts have been devoted to the development of a methodology to predict the responses of mooring-buoy systems. The dynamic response of moorings to hydrodynamic loading cannot be determined by closed-form analytical methods (Leonard and Nath, 1981). Direct integration method (Chiou and Leonard, 1991), have been utilized. Chiou and Leonard (1991) developed a direct integration algorithm for the static and dynamic analyses of three-dimensional nonlinear cable behavior. Only translational buoy motion was considered.

Models of wave-body interaction generally depends on the size of the body compared to the incident wave length. A floating body with small characteristic length may be analyzed by the Morison equation and for a large body case, analysis methods based on potential theory must be employed. The present work focuses on buoys which are slender compared to incident wave length. The Morison equation has been extended for dynamic analyses of ocean structures undergoing large displacement and have been widely used for analyses of oceanic structures assembled from slender cylindrical members (Liaw, *et al.*, 1989). The three-dimensional model of Chiou and Leonard (1991) was extended by Idris *et al.* (1997) to include small rotational motions of the buoy. Equations of motion of a rigid body undergoing large rotational motions can be derived by a Lagrangian formulation using Euler parameters. Application of these parameters in the ocean engineering field has been considered by Liaw *et al.* (1989).

The objective of this study is to develop and verify an algorithm to predict coupled dynamic motion of a moored discus buoy which may undergo large rotations and translations. This system is conceived as a tethered rigid buoy subjected to long crested waves and current with approach angle β. The characteristic length of the buoy in this study is considered small compared to the incident wave length. Three translational degrees of freedom and three rotational degrees of freedom of the buoy and their coupling are analyzed and the equations of motion of the buoy are formulated based on rigid body motions with large angular motion. A numerical algorithm is then formulated to allow coupling with the mooring algorithm of Chiou and Leonard (1991).

MOORING DYNAMICS

The mooring is modeled using an algorithm for a hydrodynamically loaded cable problem previously developed by Chiou and Leonard (1991). A fixed (global) coordinate system is defined with the origin at the still water line; S_0 is the mooring local coordinate along the scope; and ξ is an arbitrary material point on the mooring at a distance S_0 from one end of the mooring. The position of an arbitrary point on the mooring at time t is define as $x_i(S_0,t)$ and the tension components as $T_i(S_0,t)$, where I ranges from 1 to 3.

The governing equations were derived from the equilibrium of dynamic forces on an infinitesimal length at an arbitrary point ξ along the segment. Hydrodynamic loading per unit length includes buoyancy, added mass forces and drag forces. If strain ϵ is assumed to be small, then the governing equations can be written, in indicial notation, in terms of dependent variables of velocity as x_i and tension T_i (Idris *et al.*, 1997). The resulting set of equations constitutes a combined boundary-value and initial-value problem because of time evolution. The initial conditions for the problem are required to be specified in terms of S_0 and two sets of boundary conditions are also needed at each cable end. Approximation on the time derivatives can be made using an implicit Newmark-like formula. This approach converts the problem to a discrete two-point boundary-value problem at each instant in time.

The resulting nonlinear boundary-value problem can be converted to an iterative set of linear boundary-value problems using the Newton-Raphson method. The quasi-linearized two-point boundary-value problem then can be further decomposed into a set of initial-value problems so that trial solutions can be integrated numerically from one end to the other.

Solution of the boundary value problem is obtained by successive iteration. In each iteration, particular and homogenous mooring equations are integrated from the starting point to the other end of the mooring. Knowing the boundary conditions at the terminal end, one can evaluate the partial solutions at that end and use them to obtain appropriate coefficients α_j for the linear combination of partial solutions at the starting end and thus obtain new iterates to $\dot{x}_{Ti}$ and T_i. This approach was adopted by Idris, *et al.*, (1997); buoy response was approximated by a small angle model. In coupling that model to the mooring model convergence difficulties were encountered for rapid rotations of the buoy.

BUOY DYNAMICS WITH LARGE ANGULAR MOTIONS

A model of discus buoy dynamics based on large rotational motion is needed. The new formulation will use the same basic assumptions as previously except for that regarding the size of angular motion. Idris *et al.* *(1997)* used Euler angles, with small angular motion assumed, to describe rotational orientation of the buoy. When the possibility of large angular motions is admitted, the use of Euler angles has the disadvantage that singularities may occur at certain orientations of the body in space (Shabana, 1989). Moreover, Euler angles must be executed in a definite succession, one that may lead to a more difficult mathematical analysis.

The approach adopted here for the large angle formulation is to use Euler parameters, as recommended by Shabana (1989), to represent buoy angular orientation. Initially, Euler parameters may appear to be only a mathematical tool that does not have physical significance. However, their identities can be used to develop physical interpretations. In this section kinematics of the discus buoy using Euler parameters are presented and the equations of motion are derived in which all external moments will be expressed in Euler parameters.

A definition sketch of a buoy floating on the moving water surface and connected by a tether to the ocean bottom is depicted in Figure 1. Two cartesian coordinate systems are defined, a moving (local) coordinate system attached to the buoy, and a fixed (global) coordinate system with origin at the mean water line.

The origin of the buoy/local coordinate is located at the center of mass G a distance z_{cg} along the axis of symmetry from the origin of the fixed system at the initial time. Point T is the location of the tether connection point and $\underline{s}'_T$ is the local position vector to T from G. Point W is the location of application of loads from external sources, e.g. wind, and $\underline{s}'_W$ is the local position vector to W from G. Point B is the location of the center of buoyancy and is the point at which resultant hydrodynamic loads are assumed to act and $\underline{s}'_B$' is the local position vector to B from G. Note that the location of point B depends on the amount of buoy submergence.

Kinematics

Euler parameters represent the angular orientation of a local coordinate system z_i to the global coordinate system x_i. Based on Euler's theorem (Nikravesh *et al.* 1985), the transformation between two coordinate systems that have a coincident origin can be accomplished by a single rotation ø about a unique axis $\underline{u}$. Transformation of a particular point P on a rigid body can be written as

$$\underline{s} = \mathbf{A}\underline{s}' \tag{1}$$

where **A** is a 3x3 rotation matrix, $\underline{s}$ and $\underline{s}'$ represent position vectors of point P in the global and local coordinate systems, respectively. The rotation matrix can be presented in terms of Euler angles, direction cosines or Euler parameters. In this formulation Euler parameters are selected. They consist of four quantities

$$\underline{p} = [e_0, e_1, e_2, e_3]^T \tag{2}$$

The first three parameters can be defined as

$$\underline{e} = [e_1, e_2, e_3]^T = \underline{n}\ \sin(\frac{\phi}{2}) \tag{3}$$

where the unit vector $\underline{n}$ is a rotational axis and ø an angle of rotation about the axis. The fourth parameter is defined as

$$e_0 = \cos(\frac{\phi}{2}) \tag{4}$$

The Euler parameters are not all independent:

$$\underline{\Phi} = \underline{p}^T\underline{p} - 1 = 0 \tag{5}$$

The rotation matrix A can be expressed in term of Euler parameters as

$$\mathbf{A} = \begin{bmatrix} 2(e_0^2+e_1^2)-1 & 2(e_1e_2-e_0e_3) & 2(e_1e_3+e_0e_2) \\ 2(e_1e_2+e_0e_3) & 2(e_0^2+e_2^2)-1 & 2(e_2e_3-e_0e_1) \\ 2(e_1e_3-e_0e_2) & 2(e_2e_3+e_0e_1) & 2(e_0^2+e_3^2)-1 \end{bmatrix} \tag{6}$$

The motion of the buoy can be characterized by the translation vector of the point G, $\underline{r}^T = \{x_1, x_2, x_3\}^T$ and the rotational vector of the buoy around its center of gravity G, $\underline{\theta}^T = \{\theta_1, \theta_2, \theta_3\}^T$. As the buoy moves from an initial static equilibrium position to a position at time t, the position vector $\underline{r}_P$ of an arbitrary point P in the buoy can be written as

$$\underline{r}_P = \underline{r} + \mathbf{A}\underline{s}'_P \tag{7}$$

where $\underline{r}$ is the translational motion of point G and $\underline{s}_P'$ is the local position vector to P from G. Then, the acceleration and velocity at a general point P on the buoy can be obtained from time derivatives of Eq. (7):

$$\dot{\underline{r}}_P = \dot{\underline{r}} + \dot{\mathbf{A}}\underline{s}'_P$$
$$\ddot{\underline{r}}_P = \ddot{\underline{r}} + \ddot{\mathbf{A}}\underline{s}'_P \tag{8}$$

Using identities between Euler parameters and arbitrary vectors (Nikravesh *et al.*, 1985), one express the coupling components of velocity and acceleration as

$$\dot{\mathbf{A}}\underline{s}'_P = -2\mathbf{A}\tilde{\underline{s}}'_P\mathbf{L}\dot{\underline{p}} \tag{9}$$

$$\ddot{\mathbf{A}}\underline{s}'_P = -2\mathbf{A}\tilde{\underline{s}}'_P\mathbf{L}\ddot{\underline{p}} - \tilde{\omega}(2\mathbf{A}\tilde{\underline{s}}'_P\mathbf{L}\dot{\underline{p}}) \tag{10}$$

Now, the velocity and acceleration at point P can be written as

$$\dot{\underline{r}}_P = \dot{\underline{r}} - 2\mathbf{A}\tilde{\underline{s}}'_P\mathbf{L}\dot{\underline{p}} \tag{11}$$

$$\ddot{\underline{r}}_P = \ddot{\underline{r}} - 2\mathbf{A}\tilde{\underline{s}}'_P\mathbf{L}\ddot{\underline{p}} - \tilde{\omega}(2\mathbf{A}\tilde{\underline{s}}'_P\mathbf{L}\dot{\underline{p}}) \tag{12}$$

in which ω = skew symmetric angular velocity matrix = $2\mathbf{G}^T\mathbf{G}$, where matrices G and L are defined as

$$\mathbf{G} = \begin{bmatrix} -e_1 & e_0 & -e_3 & e_2 \\ -e_2 & e_3 & e_0 & -e_1 \\ -e_3 & -e_2 & e_1 & e_0 \end{bmatrix} \tag{13}$$

$$\mathbf{L} = \begin{bmatrix} -e_1 & e_0 & e_3 & -e_2 \\ -e_2 & -e_3 & e_0 & e_1 \\ -e_3 & e_2 & -e_1 & e_0 \end{bmatrix} \tag{14}$$

Kinetics

The equations of motion for an unconstrained rigid body in Lagrangian form (Shabana, 1989; Nikravesh *et al.*, 1985), can be written as

$$\frac{d}{dt}\left(\frac{\partial K}{\partial \dot{\underline{q}}}\right)^T - \left(\frac{\partial K}{\partial \underline{q}}\right)^T = \underline{g} \tag{15}$$

where K is the kinetic energy and $\underline{q}$, $\dot{\underline{q}}$ are the vector of generalized displacement and velocity of the body, $\underline{q} = \{\underline{r}\ \underline{p}\}^T$, and $\underline{g} = \{\underline{F}\ \underline{N}\}^T$ with $\underline{F} = \Sigma\,\underline{F}_p$ = sum of forces at all points P, and $\underline{N} = \Sigma\underline{N}_p$ = sum of moments of all $\underline{F}_p$ about G. The individual force vectors acting on the buoy are: buoy dry weight $\underline{W}_{DRY}$ at G, buoyancy $\underline{B}$ at B, hydrodynamic inertia force $\underline{F}_I$, drag force $\underline{F}_D$, Froude-Kryloff force $\underline{F}_{FK}$ (including added mass effects) at B, wind or other external force $\underline{F}_W$ at W, and tether tension $\underline{T}$ at T. The wind or other external force is given as a concentrated load which may vary with time. The buoy is assumed small compared to incident wave lengths. Thus, the added mass coefficient may be taken as frequency independent and radiation damping neglected. The external moment of forces $\underline{F}_p$ at point P is evaluated about the center of gravity G, and can be expressed as

$$\underline{N}_P = \underline{s}'_P \times \underline{F}_P \tag{16}$$

The resultant of hydrodynamic loads is assumed to be applied at the center of buoyancy B, the location of which depends on submergence.

Equation (15) is based on the assumption that the coordinates describing body orientation are independent. The Euler parameters are not independent and must satisfy Eq.(5). Therefore, the Lagrange multiplier technique is used to include this constraint (Nikravesh *et al.* 1985); the partial derivative of Eq.(5) with respect to $\underline{p}$ is

$$\frac{\partial \phi}{\partial \underline{p}} = 2\underline{p}^T \tag{17}$$

Now, the translational and rotational components of the equations of motion can be written as

$$\frac{d}{dt}\left(\frac{\partial K}{\partial \dot{\underline{r}}}\right)^T - \left(\frac{\partial K}{\partial \underline{r}}\right)^T = \underline{F} \tag{18}$$

$$\frac{d}{dt}\left(\frac{\partial K}{\partial \dot{\underline{p}}}\right)^T - \left(\frac{\partial K}{\partial \underline{p}}\right)^T + 2\underline{p}\sigma = \underline{N} \tag{19}$$

where σ is the Lagrange multiplier associated with the constraint of Eq.(5).

The kinetic energy of the body, K, can be written as

$$K = \frac{1}{2}\int \rho \dot{\underline{r}}_p^T \dot{\underline{r}}_p dV \tag{20}$$

Substituting Eq. (11) into Eq. (20), carrying out the matrix multiplication and utilizing the orthogonality principle, one can write the kinetic energy K as

$$= \frac{1}{2}\rho \left[\dot{\underline{r}}^T \ \dot{\underline{p}}^T\right] \int \begin{bmatrix} I & -2A\tilde{s}L \\ [-2A\tilde{s}L]^T & L^T\tilde{s}\tilde{s}L \end{bmatrix} d \tag{21}$$

Substituting Eq.(21) into Eqs.(18) and (19), one can write the equations of motion for a rigid body as

$$\begin{bmatrix} J_{rr} & J_{rp} & 0 \\ J_{pr} & J_{pp} & 2\underline{p} \end{bmatrix} \begin{Bmatrix} \ddot{\underline{r}} \\ \ddot{\underline{p}} \\ \sigma \end{Bmatrix} = \begin{Bmatrix} \underline{F} \\ \underline{N} + 8\dot{L}^T J_{pp} \dot{L}\dot{\underline{p}} \end{Bmatrix} \tag{22}$$

in which

$$J_{rr} = \int \rho I dV = \begin{bmatrix} m & 0 & 0 \\ 0 & m & 0 \\ 0 & 0 & m \end{bmatrix} \tag{23}$$

$$J_{rp} = -2A\int \rho \tilde{s} dV\ L = -2A \begin{bmatrix} 0 & mz_{cg} & 0 \\ mz_{cg} & 0 & 0 \\ 0 & 0 & 0 \end{bmatrix} L \tag{24}$$

$$J_{pp} = L^T \int \rho \tilde{s}\tilde{s} dV\ L = L^T \begin{bmatrix} i_{11} & 0 & 0 \\ 0 & i_{22} & 0 \\ 0 & 0 & i_{33} \end{bmatrix} L = L^T \tag{25}$$

$$J_{pr} = J^T_{rp} \tag{26}$$

where i_{kk} = principal moment of inertia for the buoy about the local axes; off diagonal moment of inertia matrix components are zero for an axi-symmetric body. Equation (22) has 7 equations with 8 unknowns. The eighth equation can be obtained from the second time derivative of the constraint condition Eq.(5) which can then be appended to obtain

$$\begin{bmatrix} J_{rr} & J_{rp} & 0 \\ J_{pr} & J_{pp} & 2\underline{p} \\ 0 & 2\underline{p}^T & 0 \end{bmatrix} \begin{Bmatrix} \ddot{\underline{r}} \\ \ddot{\underline{p}} \\ \sigma \end{Bmatrix} = \begin{Bmatrix} \underline{F} \\ \underline{N} + 8\dot{L}^T J_{pp} \dot{L}\dot{\underline{p}} \\ -2\dot{\underline{p}}^T\dot{\underline{p}} \end{Bmatrix} \tag{27}$$

The added inertia forces and moments generated by the wave loading can be added to Eq.(27). These force and moment components include acceleration and a quadratic velocity vector component so that Eq.(27) can be rearranged as

$$\begin{bmatrix} J_{rr}+M_{rr} & J_{rp}+M_{rp} & 0 \\ J_{pr}+M_{pr} & J_{pp}+M_{pp} & 2\underline{p} \\ 0 & 2\underline{p}^T & 0 \end{bmatrix} \begin{Bmatrix} \ddot{\underline{r}} \\ \ddot{\underline{p}} \\ \sigma \end{Bmatrix} = \begin{Bmatrix} \underline{F}_A \\ \underline{N}_A + 8\dot{L}^T J_{pp} \dot{L}\underline{p} \\ -2\dot{\underline{p}}^T\dot{\underline{p}} \end{Bmatrix} \tag{28}$$

Equation (28) is the equation of motion for a floating buoy, where $\underline{F}_A$ and $\underline{N}_A$ are the external forces and moments, respectively. The external force $\underline{F}_A$ and external moment $\underline{N}_A$ include added inertia force and added moment inertia components caused by the quadratic velocity vector, M_{rr} is the translational added inertia matrix, M_{rp} and M_{pr} are the coupled added mass components and M_{pp} is the rotational added inertia components.

COUPLING BUOY AND MOORING MODELS

Using Euler parameters, nonlinear buoy equations of motion were developed which cannot be solved in closed form. Therefore, a numerical procedure must be developed to obtain solutions in the time domain. These buoy equations of motion could be applied as boundary conditions for the mooring integration algorithm as performed by Idris *et al.* (1997). However, it will be difficult to iteratively linearize them and the same convergence difficulties encountered by Idris *et al.* (1997) as in the small angle solution algorithm are anticipated. Based on those difficulties and on the observed limitations of the algorithm of Idris *et al.* (1997), it was decided to integrate the buoy equations motion separately from the mooring algorithm. Then, the kinematic boundary conditions could be applied to one end of the mooring cable, and both mooring and buoy solutions could be coupled through the velocity of the buoy tether point: the velocity at the buoy tether point must equal the velocity at the end of the mooring. The selected temporal integration procedure for the buoy kinetics must be incorporated with the mooring solution algorithm and must facilitate coupling between the two algorithm.

After a single iteration to the solution at the end of a time interval the new solution may not satisfy equilibrium. The buoy may have a different draft and inclination angle. Its kinematics change and thus the velocity of the tether point also changes. In turn, the mooring tether must be recomputed with a new tether velocity as a boundary condition. Thus, an iterative algorithm is indicated. Therefore, the selected algorithm allows mooring/buoy kinematics and dynamics to be updated during each time step so that the buoy-mooring system is always in equilibrium.

In a predictor-corrector type of integration method, one first predicts a solution at the end of the time interval. The predicted solutions then can be used to update mooring kinematics and kinetics and also the hydrostatic properties of the buoy. The solution of the buoy equilibrium equation is then corrected using updated external forces and moments. The process continues until a convergent solution is obtained. This algorithm enables the mooring-buoy system to maintain equilibrium conditions at all instances. Two type of predictor-corrector methods were used. They are the modified single-step Euler method and the multi-step Adam-Moulton method. The Adam-Moulton method proved the most efficient and stable method. However, a single step Euler method is needed to start a multi-step integration method.

SOLUTION ALGORITHM

The system of equation of motion presented in Eq.(28) represents a set of eight nonlinear differential equation. To perform numerical integration, this set of second-order differential equations has to be transformed to a set of sixteen first-order equations which can be accomplished by considering Eq.(28) as a set of algebraic equation for the unknown acceleration vector $\{\ddot{\underline{r}}\ \ \ddot{\underline{p}}\ \ \sigma\}^T$ as

$$\begin{Bmatrix} \ddot{\underline{r}} \\ \ddot{\underline{p}} \\ \sigma \end{Bmatrix} = \begin{bmatrix} J_{rr}+M_{rr} & J_{rp}+M_{rp} & 0 \\ J_{pr}+M_{pr} & J_{pp}+M_{pp} & 2\underline{p} \\ 0 & 2\underline{p} & 0 \end{bmatrix}^{-1} \cdot \begin{Bmatrix} \underline{F}_A \\ \underline{N}_A + 8\dot{\underline{L}}^T J_{pp} \dot{\underline{L}} \underline{p} \\ -2\dot{\underline{p}}^T \dot{\underline{p}} \end{Bmatrix} \tag{29}$$

Then, the state vectors can be defined as

$$\underline{y} = \begin{Bmatrix} \underline{r} \\ \underline{p} \\ \dot{\underline{r}} \\ \dot{\underline{p}} \end{Bmatrix} \qquad \dot{\underline{y}} = \begin{Bmatrix} \dot{\underline{r}} \\ \dot{\underline{p}} \\ \ddot{\underline{r}} \\ \ddot{\underline{p}} \end{Bmatrix} = \underline{f}(\underline{y},t) \tag{30}$$

which gives a set of first-order differential equations. Now, solution for the next time step can be predicted and corrected by the Adam-Moulton method. During iteration, a relative error of the state vectors is calculated by

$$\text{error} = \sqrt{\sum\left[\frac{y_i - y_i^{*}}{y_i^{*}}\right]^2} \qquad i=1,2,...14 \tag{31}$$

where y_i^* is the preceding solutions. Iteration continues until the error is less then a specified tolerance. Solution of this differential equation can be used to define the state vector for the next interval. This process can be repeated until the end of the simulation time.

Multiple Time Steps

The present study has two numerical integration procedures. First, the buoy kinetics are integrated over one time step. Second, the cable kinematics and kinetics are spatially integrated over the mooring scope at the end of each time step. Both integration algorithms are combined to obtain an equilibrium solution at each time instant.

The buoy problem is integrated by a direct integration technique, in which very small integration time steps Δt give the best result. Large time step cause difficulties in obtaining a convergent solution and produce inaccurate solution in which the computed Euler parameters may violate the constraint equation, Eq.(5). On the other hand, the mooring algorithm formulation requires that Δt should not be so small that truncation errors (amplified by $1/\Delta t$) during numerical integration grow rapidly. A small Δt sometimes leads mooring computations to numerical instability (Chiou and Leonard, 1991).

This conflict on the size of integration steps Δt can be resolved by introducing two different integration steps: the mooring response may be computed with a relatively large integration step while buoy response may be computed with a smaller time step. The small time step can be defined as an integer division of the large one.

COMPARISON OF NUMERICAL AND EXPERIMENTAL RESULTS

Comparisons were made between numerical predictions and experimental results for experimental results for the cable-discus buoy system selected from regular wave tests conducted at the Oregon State University O.H. Hinsdale Wave Research Laboratory in April 1992.

For all simulations the mooring line is a 5/16 inch surgical rubber tube weighing 0.00072 lbs./ft. in water and having 2.41 ft. initial mooring length. The nonlinear tension-strain curve for the material is taken into account. The hydrodynamic coefficients for the mooring line were taken as constants, $C_{Dn} = 1.2$, $C_{Dt} = 0.03$, $C_A = 1.0$. The experimental data were recovered at 30 Hz.; thus a simulation time increment of 0.066 sec. was selected. A stream function wave theory was used to simulate the generated regular wave. An initial 30 sec. duration ramp of wave amplitude was used to start all simulations from rest. Each regular wave case was run for a total of 100 sec. The last 50 sec. of computational results were used for comparison to experimental results.

The comparison is for a 22 in. diameter, 6 in. deep and 21.4 lb. discus buoy commonly used by NDBC as a buoy model. The loading is a monochromatic wave train 2.5 ft. high with a 3.0 sec. period. A definition sketch for this problem is shown on Figure 2. The hydrodynamic coefficients for the discus buoy are given below. Because of uncertainties in the rotational and coupling added mass coefficients, numerical studies were conducted to determine the parametric sensitivity and optimize the rotational and coupling added mass terms.

Buoy Properties

Mass	0.644 slugs
Drag Coeff.	0.5
Added Mass Coeffs.	
in Heave & Surge	2.225 0.291
Added Mass Coeffs. (Optimal)	
in Pitch & Surge-Pitch	0.143 0.189

The comparison to experimental results of the predicted and measured response for the discus buoy are shown in Figure 3 (a-c) for both the current large angle model (labeled as KBLDYN94) and a prior small angle model (labeled as KBLDYNML). The simulated results of the large angle model (KBLDYN94) are closer to the experimental values than for the small angle model (KBLDYNML) for both translational and rotational motions.

CONCLUSIONS

A method to compute buoy motions coupled with hydrodynamically loaded mooring motions has been developed. The equations of motion of a floating buoy were derived allowing large angular motions within a formulation based on Euler parameters. All six degree of freedom for the buoy are considered. Coupling between translational and rotational degrees of freedom was included. The solution of the buoy was separated from that of the mooring line; coupling of the two problems was effected by iterative matching of the velocities at the tether connection point on the buoy.

The formulation provides a methodology to solve the coupled cable-buoy problem. The algorithm is based on a cable dynamic algorithm involving large displacements and nonlinear hydrodynamic forces which is integrated implicitly in time by a Newmark-like method. The nonlinear buoy equations of motion are integrated in time by a predictor-corrector method using a smaller time step than for the cable integrator; solutions are iteratively coupled at the end of each larger time step. The methodology was used to predict responses for discus buoys. Results were compared to experimental results. The buoy was excited by three different regular waves. The large angle model, as well as the small angle model, predicted translational motions for all regular wave cases which agreed well with experimental data. Rotational motion predictions when compared to experimental data were better predicted with the large angle model than by the small angle model.

ACKNOWLEDGMENTS

We gratefully acknowledge the support by the Naval Facilities Engineering Service Center, Contract Number N4708-94-C-7431, and the Office of Naval Research, Contract Number N00014-92-J-1221.

REFERENCES

Chiou, R., and Leonard, J.W., "Nonlinear Hydrodynamic Response of Curved, Singly-Connected Cables," Proceedings, 2nd International Conference on Computer Modelling in Ocean Engineering, Barcelona, Spain, Sept.30 - Oct.4, 1991.

Idris, K., Leonard, J.W., and Yim, S.C.S., "Coupled Dynamics of Tethered Buoy Systems," Ocean Engineering, Vol. 24, 1997, to appear.

Leonard, J.W., and Nath, J.H., "Comparison of Finite Element and Lumped Parameter Methods for Oceanic Cables," Engineering Structures, Vol. 3, 1981, pp. 153-167.

Liaw, C.Y., Shankar, N.J., and Chua, K.S., "Large Motion Analysis of Compliant Structures Using Euler Parameters," Ocean Engineering, Vol. 16, 1989, pp. 545-557.

Nikravesh, P.E., Wehange, R.A., and Kwon, O.K., "Euler Parameters in Computational Kinematics and Dynamics. Part 1," Transactions of the ASME, Vol. 107, 1985.

Shabana, A.A., Dynamics of Multibody Systems, John Wiley & Sons, Inc., New York, NY, 1989.

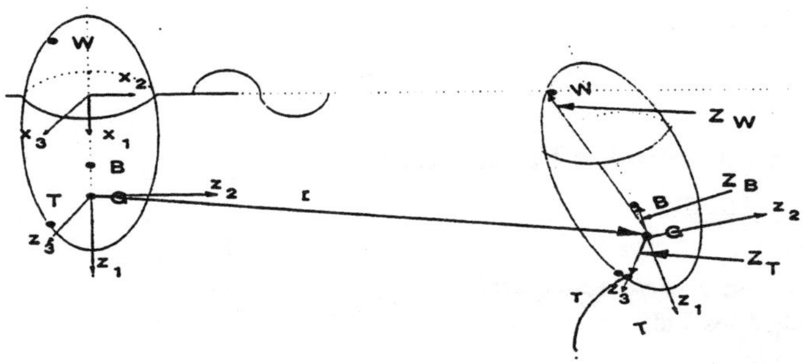

Figure 1. Definition Sketch of Buoy Vectors

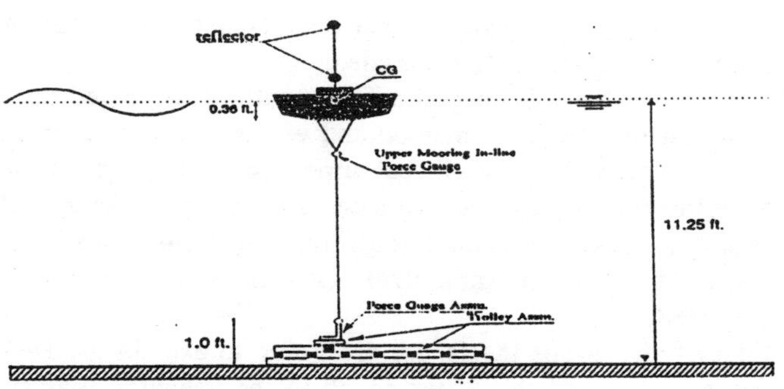

Figure 2. Configuration of Discus Buoy Test

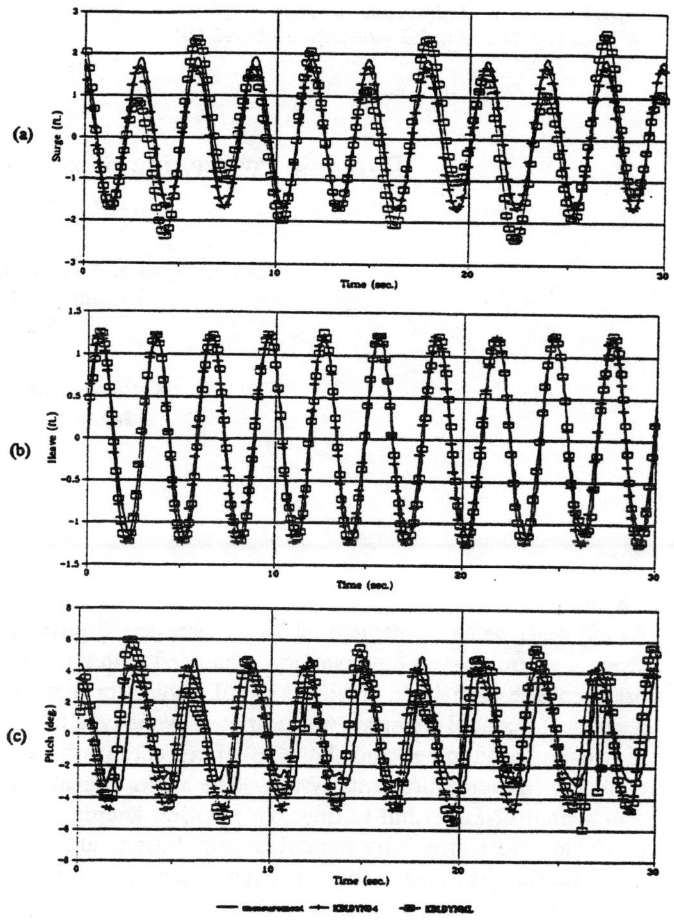

Figure 3. Surge, Heave and Pitch Displacements of Discus Buoy, H=2.5 ft., T=3.0 sec.

Proceedings of the Seventh (1997) International Offshore and Polar Engineering Conference
Honolulu, USA, May 25-30, 1997

ISBN 1-880653-28-1 (Set); ISBN 1-880653-31-1 (Vol. III)

Dynamic Positioning System Based on Nonlinear Programming for Offshore Platforms

Ikuo Yamamoto, Masami Matsuura and Youichi Yamaguchi
Mitsubishi Heavy Industries Ltd.
Nagasaki, Japan

Katsunori Shimazaki and Akio Tanabe
Japan National Oil Corporation
Chiba, Japan

Abstract
As the water depth of offshore oil fields increases, a position keeping control system on a platform equipped with thrusters has become essential. In previous control studies for position keeping, the control algorithm was restricted by the type of the platform, the number and the type of its actuators, and nonlinear characteristics of its system. The authors developed a new control algorithm to improve position keeping of a platform, and tested it by numerical simulations and tank experiments. They were the first experiments in the world conducted by reat time control of nonlinear programming.

Key words
Offshore platform, DPS, Real time control, Nonlinear programming, Simulation, Tank test

Introduction

The number of new oil bearing areas has been decreasing year by year. Therefore, the development of new oil resources at sea is moving its location gradually to deeper sea areas, that is, from existing continental shelves to areas where the continental shelf slopes down to deeper waters.

Further, from now on it can be expected that oil resource development in deeper sea areas will experience much more severe environmental conditions. In this situation, a floating production system is receiving more attention as a system suitable for marginal oil development (see Fig. 1 for constituent elements of a floating production platform system).

As small scale marine oil development takes place in deeper sea areas, position keeping has become more difficult, when using only a position keeping system with existing mooring methods. This is because of the technical and financial aspects. It is therefore necessary to apply a Dynamic Positioning System (hereinafter refered to as DPS) in conjunction with thrusters. In order to assess the performance of platform control with a DPS, that meets its specification requirements, DPS logic control development and behavior analysis simulation tools are needed.

The authors have developed a DPS control logic based on a nonlinear programming method and verified the efficiency of the control logic through real time control tests at a tank experimental facility. At the same time we have developed behavior analysis simulation tools, for a platform with a DPS, and established a technological assessment of that platform with DPS.

First of all, in this thesis, we are going to explain the DPS control logic and thereafter go on to the behavior analysis simulation tools, named DP-TOP (Dynamic Positioning system - Thruster force Optimal distribution Program) and DP-MAP (Dynamic Positioning system - Motion Analysis Program). An investigation procedure of the DPS with this tools will be carried out. Finally we are going to conclude with the results of an efficiency inspection of the DPS control logic by real time control tests, in a tank experiment.

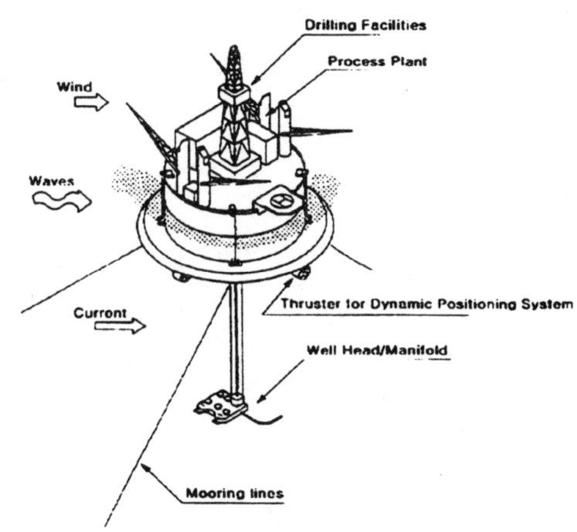

Fig.1 Example of a floating production system

Control algorithm

DPS logic constructs are based on a nonlinear programming method in order to widen the applicable use. In the past, in most of the cases (ex. Tamehiro, 1977; Akasaka, 1981), the control algorithm had been restricted by the type of platform, the number and type of actuators and the systems nonlinear characteristics.

However, this control logic has reduced those restrictions in order to widen logic applicability. Control logic consists of, as Fig. 2 shows, a thruster force distribution element, platform movement element, a command generation element and a thruster manipulation signal determination element (Yamamoto, 1991). The thruster force distribution element distributes total force and total moment to be established on each thruster. An algorithm of the thruster force distribution element is based on a nonlinear programming method, and this nonlinear programming method is carried out by the mixed penalty method and the modified quasi-Newton method, in order for it to be applied to marine platform motion, thruster restriction conditions, and mechanical cost functions using nonlinear characteristics. The algorithm is derived by the following steps.

Step 1:Input thrust value of the thrust initial state and thrust direction.

Step 2:Input mechanical conditions.

$$H_1 = Fxc - \sum_{i=1}^{n} Fxsi \qquad \text{Eq. 1}$$

$$H_2 = Fyc - \sum_{i=1}^{n} Fysi \qquad \text{Eq. 2}$$

$$H_3 = Nzc - \sum_{i=1}^{n} FxsiYsi - \sum_{i=1}^{n} FysiXsi \qquad \text{Eq. 3}$$

where,

Fxc,Fyc : Total force of the body axis fixed coordinate system x, y direction

Nzc : Total moment around the body axis fixed coordinate system z

Fxsi, Fysi : Body axis fixed coordinate system x,y direction component force of nth thruster

Xsi, Ysi : Body axis fixed coordinate system x,y direction fitting position of nth thruster

n : The number of thrusters.

Step 3:Input actuator control condition. Use thruster azimuth for the actuator and set the control condition as follows.

$$G_1 = F^2maxsi - (F^2xsi + F^2ysi) \qquad \text{Eq. 4}$$

$$G_2 = (F^2xsi + F^2ysi) - F^2minsi \qquad \text{Eq. 5}$$

where, Fmaxsi : Maximum capacity of nth thruster

Fminsi : Minimum capacity of nth thruster.

Step 4:Input cost function, which is the square sum of the following thrust and minimal energy.

$$f = \sum_{i=1}^{n} F^2si \qquad \text{Eq. 6}$$

Step 5:Convert cost function of Eq. 6 to total costfunction. Total cost function is set as follows, according to the limitations of the mechanical control condition and actuator control condition.

$$F = f + \psi(r, Hi) + \phi(t, Gj) \qquad \text{Eq. 7}$$

ψ of equation 7 should be as follows, according to the mechanical control condition.

$$\phi(r, Hi) = r\sum_{i=1}^{n} Hi \qquad \text{Eq. 8}$$

ψ should be as follows, according to the actuator control condition.

$$\phi(r, Hj) = t\sum_{j=1}^{n} \frac{1}{Gj} \qquad \text{Eq. 9}$$

r and t are parameters.

Step 6:Calculate the minimum value of total cost function, by the modified quasi-Newton method.

The optimum thrust distribution is found by Step 1 through Step. 6. In addition the optimum position will be also found even if the thruster position is variable. After determining the thrust distribution, the thruster positioning element command value with the thruster manipulation signal determination element is calculated. Consideration of the thrusters fixed characteristic should be included.

Total force and total moment to the thruster force distribution element is formulated from a command generation element. Based on a deviation signal from the platform position coordinates, position target value and azimuth target value are set, the proportional gain, the differential gain and the integration gain are multiplied, and the total force and moment for keeping a fixed position is calculated.

With this method, total force and total moment, which was calculated from the current and wind disturbance model,and a built in feedforward element as a prior control signal, can easily be bias added to the command generation element in order to input to thruster force distribution element. In the same manner, it is possible to expand various functions such as joystick manual control and unkown presumed disturbance controls.

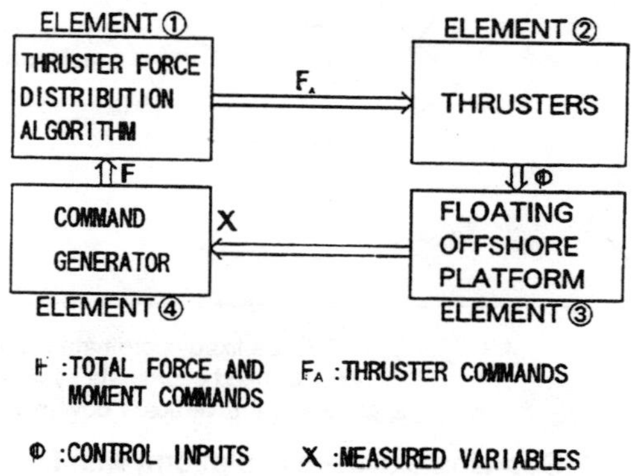

Fig.2 Position keeping control system

Behavioral analysis simulation tool

Behavioral analysis simulation tools, DP-TOP and DP-MAP, have been developed in order to assess the control ability of the platform with a DPS.

DP-TOP carries out static balance calculation of static external force functioned for platform by wind, wave and current, and that of the thruster thrust, and gets required minimum thruster capacity.

It is useful to examine the required thruster capacity in the process of initial plan with DPS. In addition, it is also possible to analyze the thrusters optimum position. Table 1 shows the calculation function of DP-TOP.

Table 1 Function of DP-TOP

Program	Calculation function	Average file input/output
DP-TOP	○Calculation of required thruster capacity by static balance calculation ○Optimization calculation of thruster position ○Data input/output by GUI	None

The program, drawn up using a personal computer and GUI (Graphical User Interface), was adapted for input/output, so that more users can use it easily. The entire program of DP-TOP consists of input/output components by GUI, and a calculation component, is formulated with FORTRAN language. The platform coordinate system is shown in Fig.3.

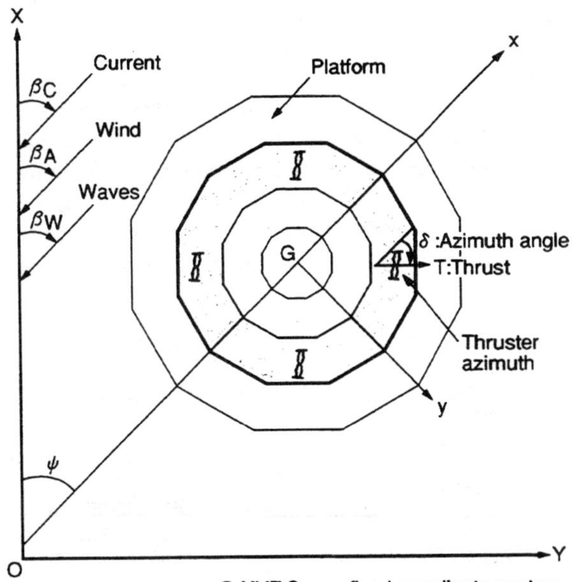

O-XYZ:Space fixed coordinate system
G-xyz:Body fixed coordinate system
Z and z axis are vertically downward

Fig.3 Coordinate system of platform with DPS

The basis of this programs, mathematical logic is to solve the thrusters thrust distibution problems with an established format, as a mathematical problem plan under several restricted conditions, in order to minimize assessment problems such as capability and cost.

In the GUI input/output component, dialogue manager is mainly used and a presentation manager is used for partial graphic image display. The purpose of the dialogue manager is to gather dialogue service based on the presentation managers functions, and present an interface which controls displays, handle variable, and application flow to window panels and messages.

The function of the DP-TOP and input/output data are shown in Table 2 and Table 3 respectively.

Table 2 Function of DP-TOP

Vessel type ● Semi - submersible ● Mono - hull ● Others
External force condition ● Steady external force Wave, Wind, and current
Thruster ● Azimuth
Type of analysis ● Analysis of required thruster capacity ● Anlysis of optimum thruster position

Table 3 Input/output data of DP-TOP

〈 Input data 〉 ● Displacement ● Representative length ● Representative wind pressure area
〈 Output data 〉 ● Necessary thruster capacity ● Thruster azimuth ● Evaluation function ● Optimum position coordinates

Fig.4 shows an example of a GUI screen. This is an analysis example based on the assumption that four thrusters of maximum capacity 80t, are installed on a mono-column type platform [Matsuura, 1995].

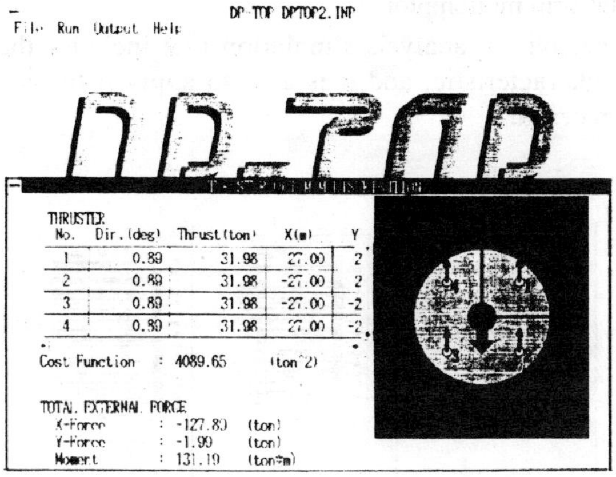

Fig.4 GUI Screen of DP-TOP

The DP-MAP is a dynamic analysis program utilized to simulate the motion of the platform with the DPS under compound fluctuating disturbances of wind, wave, and current. As Table 4 shows MOORFORC, which calls up mooring characteristic data, is included as a DP-MAP related program.

Table4 Function of DP-MAP and relevant program

Program	Calculation function	Average file input/output
DP-MAP	○Calculation based on historical data of platform with DPS ○Output plot of calculation results	⟨File output⟩ ○Plot data file ⟨File input⟩ ○Mooring characteristic table
MOORFORC	○Static mooring calculation ○Mooring characteristic table file call up	⟨File output⟩ ○Mooring characteristic table

DP-MAP was developed for a UNIX work station. This program consists of input data, data set component, fluid force calculation component, control component, movement analysis component, and output component. The program flow and the function of DP-MAP is shown in Fig.5.

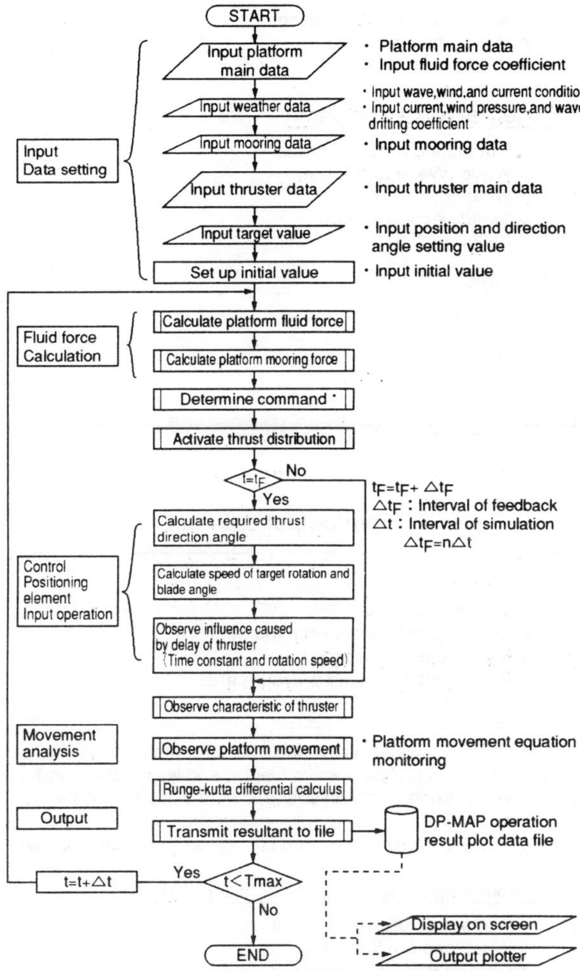

Fig.5 Flow chart of DP-MAP program

At the control section, based on input from the input component and data set component, you can carry out a command decision operation, thrust distribution operation and positioning element input operation, after adding the fluid and mooring force functions of the platform. At the movement analysis section, you can carry out a thruster characteristic operation and platform movement operation. Each time the program's loop operation cycles, time increment Δt is added and it moves to the output component, when t gets bigger than the simulation time set value T_{max}. At the output section, operation results can be observed in two ways, on a display screen and a plotter output, by transmitting the resultant data in the file.

Table 5 and Table 6 show the function of the DP-MAP and input/output data, respectively.

Table5 Function of DP-MAP

Vessel type

- Semi - submersible vessel
- Mono - hull vessel
- Others

External force condition

- Fluctuating external force
  Wave (Wave drift force)
  Wind
- Steady external force
  Current

Thruster

- Azimuth

Mooring

- Chain
- Wire
- Combination of chain and wire

Table6 Input/output data of DP-MAP

〈Input data〉

Platform main data

- Displacement and inertia moment
- Representative length
- Representative wind pressure area

Environmental condition data

- Unsteady wave (Wave direction change, average wave period, Significant wave height, and wave spectrum)
- Wind (Wind direction change, wind velocity, and wind spectrum)
- Current (Current direction change and current velocity)

Environmental external force coefficient data

- Wave drift force coefficient
- Wind load coefficient
- Current force coefficient

Mooring main data

- Mooring point coordinates

Thruster data

- Thruster position coordinates
- Maximum thruster capacity

Control gain data

- PID gain
- Calculation time interval
- Control time interval

Simulation data

〈 Output data 〉

Time history

- External force
- Movement (Roll,pitch,and yaw)
- Thruster azimuth angle
- Thrust
- Mooring force

Movement wake

Fig.6 shows an example of the display screen output of the DP-MAP. Fig.7 through Fig.9 show examples of the DPS simulation under wave, wind, and current disturbance in the case where four thrusters are loaded on the mono-column vessel. Fig.7 and Fig.8 show a time history and Fig.9 shows the platform motion plot.

This behavioral analysis simulation tool includes the fluid force characteristic, and it is able to apply it to the actual system design.

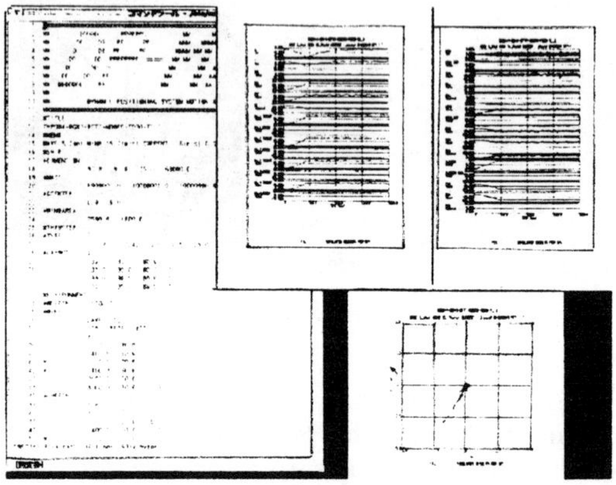

Fig.6 Display screen output of DP-MAP

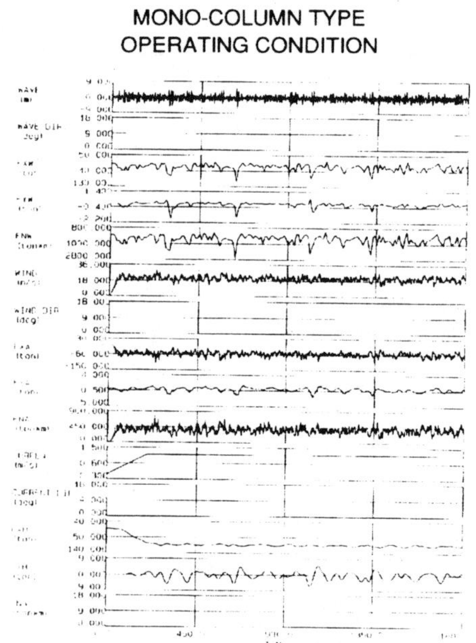

Fig.7 Simulated results for DPS
(Calculation result by DP-MAP
Mono-column type vessel configuration
4 thrusters)

MONO-COLUMN TYPE
OPERATING CONDITION

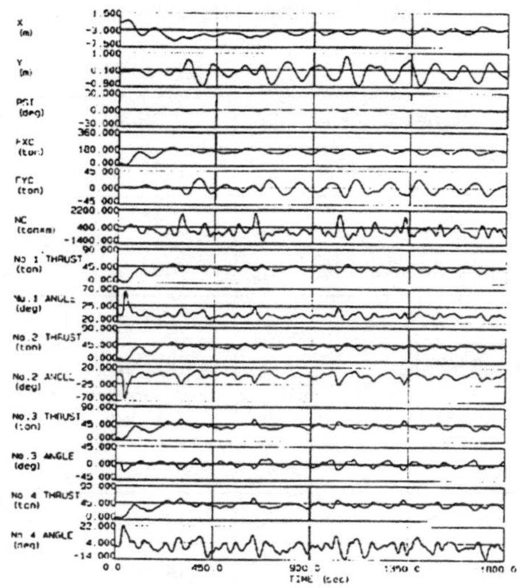

Fig.8 Simulated results for DPS
(Calculation result by DP-MAP
Mono-column type vessel configuration
4 thrusters)

MONO-COLUMN TYPE
OPERATING CONDITION

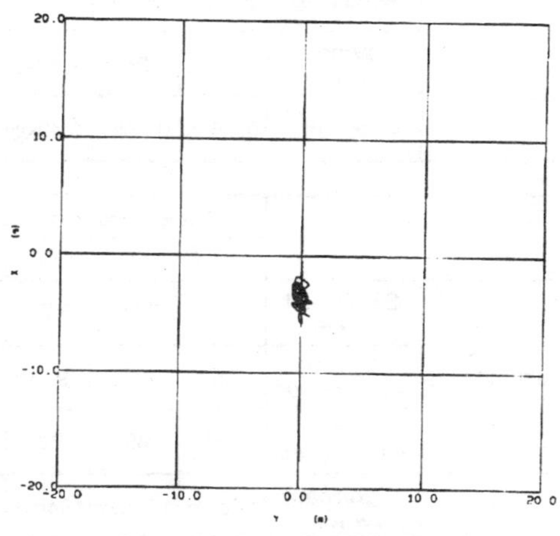

Fig.9 Simulation results for DPS
(Calculation result by DP-MAP
Mono-column type vessel configuration
4 thruster units)

Tank experiment

In order to examine the efficiency of the DPS control logic, we carried out a tank experiment in 1995. We incorporated DPS control logic into a personal computer (PC-9821XT) for control utilizing C programming language, and carried out the DPS test under wind, wave, and current disturbances using a scale model of 1/50 mono-column type platform with four azimuth thrusters. Fig. 10, Fig.11, photo 1, and photo 2 show the test overview diagram, the test equipment layout diagram, and the platform model, and test view respectively.

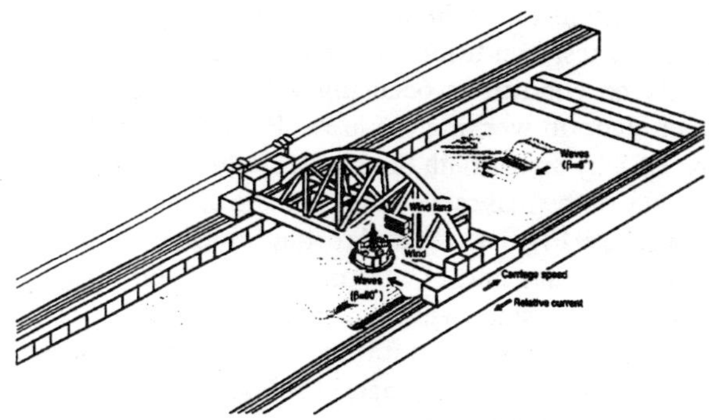

Fig.10 Model test overview diagram

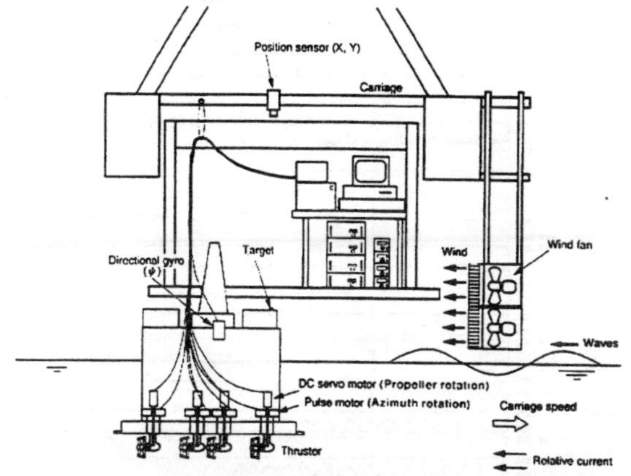

Fig.11 Model test equipment layout diagram

Photo 1 Mono-column type platform model

Photo 2 Test view

Table 7 details the main of platform model and Table 8 details the mode test conditions. Tank experiments were focused mainly on conditions occuring with high frequency at the ocean platform working sea area. For all the conditions, real time DPS by an azimuth thruster could be done within ±20 m accuracy of the real model. An example of results is are shown in Fig.12~Fig.14. We also implemented a test on the assumption that one of the thrusters suddenly broke down while a DPS system with four thrusters was being controlled. However, the DPS system had no problems, because the control logic functioned properly as Fig. 15 shows.

Beaufort scale 7 4 thrusters

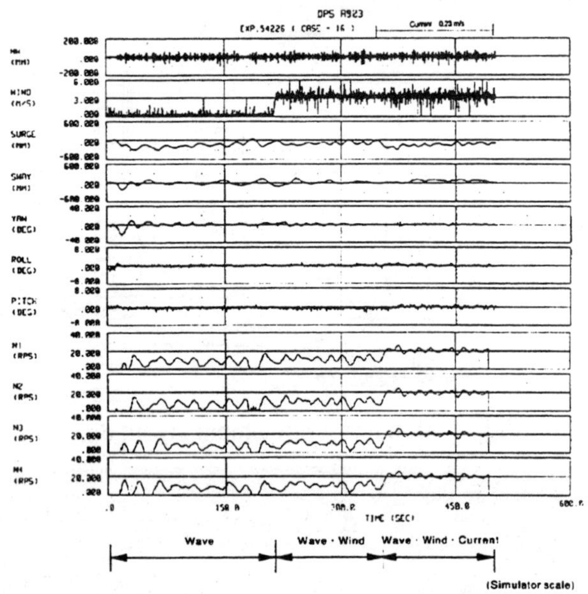

Fig.12 DPS test result

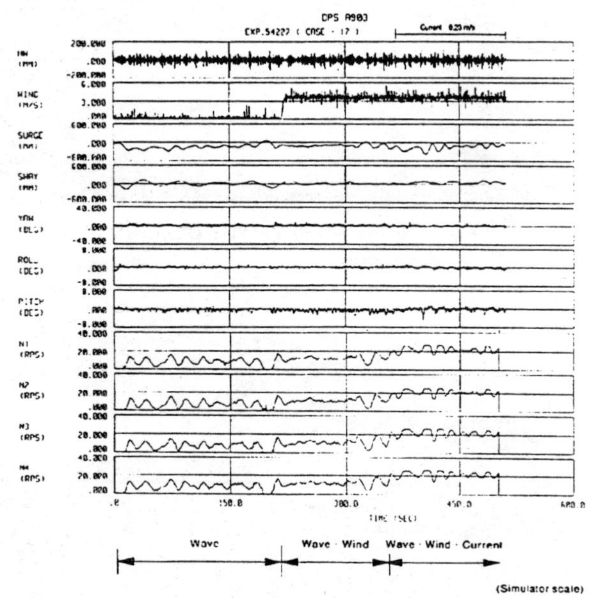

Beaufort scale 8 4 thrusters

Fig.13 DPS test result

Table7 Main components of platform model

Improved mono-column vessel configuration
Model reduction rate : 1/S=1/50

		Simulator	Actual model
Column external diameter	L_{CO}	1.36 m	68.0 m
Column internal diameter	L_{CI}	0.84 m	42.0 m
Well diameter	L_{HI}	0.42 m	21.0 m
Lower hull outer diameter	L_{HO}	1.64 m	82.0 m
Draft	d	500 mm	25.0 m
Lower hull height	D_H	100 mm	5.0 m
Displacement	△	530 kg	68,000 ton
Draft area	A_W	0.858 m^2	2145m^2
Metacenter height	GM	70 mm	3.5 m
Centroid height	KG	380 mm	19.0 m

Table8 Simulation test condition table

		Wind (Steady wind)			
		BFT 6	BFT 7	BFT 8	BFT 9
Actual model	Wind velocity	12.6m/6	15.7m/s	19.3m/s	22.6m/s
	Wind load	23 ton	36ton	54 ton	74 ton
Simulator	Wind velocity	1.8m/s	2.2m/s	2.7m/s	3.2m/s
	Wind load	0.18kg	0.29kg	0.43kg	0.58kg

		Wave (Unsteady wave)			
		BFT 6	BFT 7	BFT 8	BFT 9
Actual model	Significant wave	4.0m	5.2m	6.7m	8.5m
	Average wave period	7.5s	8.1s	9.0s	10.3s
	Average wave drift force	33t	49t	70t	79t
Simulator	Significant wave	80mm	104mm	134mm	170mm
	Average wave period	1.06s	1.15s	1.27s	1.46s
	Average wave drift force	0.26kg	0.38kg	0.55kg	0.62kg

BFT:Beaufort scale

		Current (Steady currert)		
		Cur.1	Cur.2	Cur.3
Actual model	Current velocity	0.8m/s	1.2m/s	1.6m/s
	Current force	50ton	113ton	201ton
Simulator	Current velocity	0.11m/s	0.17m/s	0.23m/s
	Current force	0.37kg	0.89kg	1.62kg

Beaufort scale 9 4 thrusters

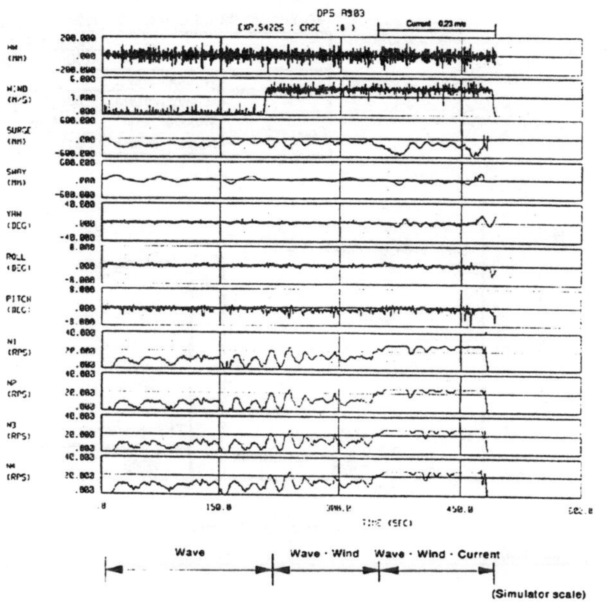

Fig.14 DPS test result

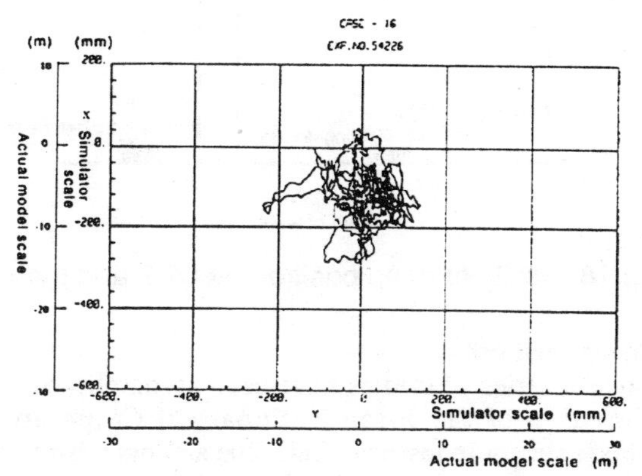

Beaufort scale 7 4 thrusters

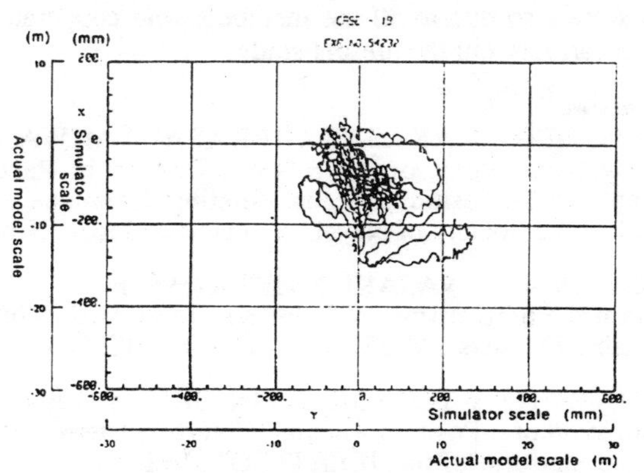

Beaufort scale 7 3 thrusters

Fig.15 DPS test result
(Influenced by the number of thrusters)

Discussion

The following items have been reached regarding the ability of the platform system to function with a DPS.

(1) We have developed the new DPS control logic based on a nonlinear programming method and examined its efficiency by enabling position keeping within ± 20 m of real model, with a real time control test tank experiment. In addition, we have developed control logic incorporating a separate feedforward system and improved the DPS accuracy of the tank experiment by more than 10%.

(2) Through the DP-TOP development, it is possible to easily presume the required thruster capacity. We have adopted the thruster thrust distribution method, which minimizes the evaluation functions of performance and cost by carrying out a static balance calculation of the static external force and thruster thrust with any number of thrusters. In addition, we have developed a method to optimize thruster position as Fig. 16 shows.

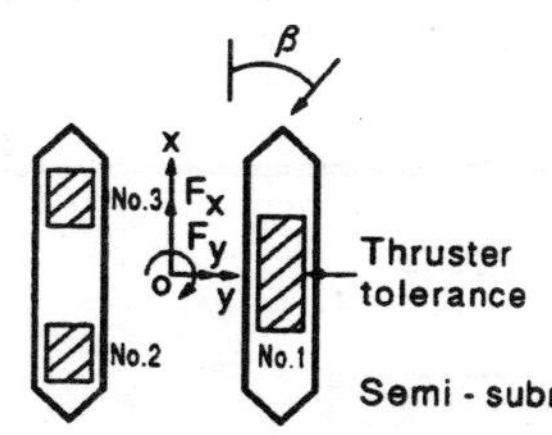

Unit tolerance

	x	y
No.1 thruster	0±15m	30±5m
No.2 thruster	-30±5m	-30±5m
No.3 thruster	30±5m	30±5m

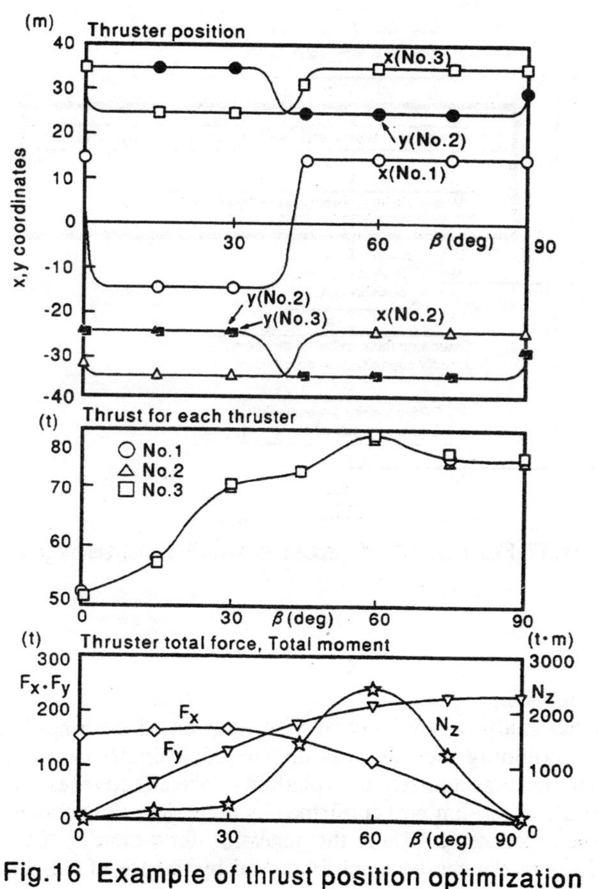

Fig.16 Example of thrust position optimization calculations

(3) As a result of the DP-MAP development, we have created a behavioral analysis method for a platform which keeps its position with a DPS system. It can consider mooring influence, and it also able to carry out platform behavioral analysis regardless of mooring and DPS examination. Fig. 17 shows DPS examination flow chart using DP-TOP and DP-MAP. The calculated result of the position keeping ability of a DPS system and mooring is as per Fig. 18.

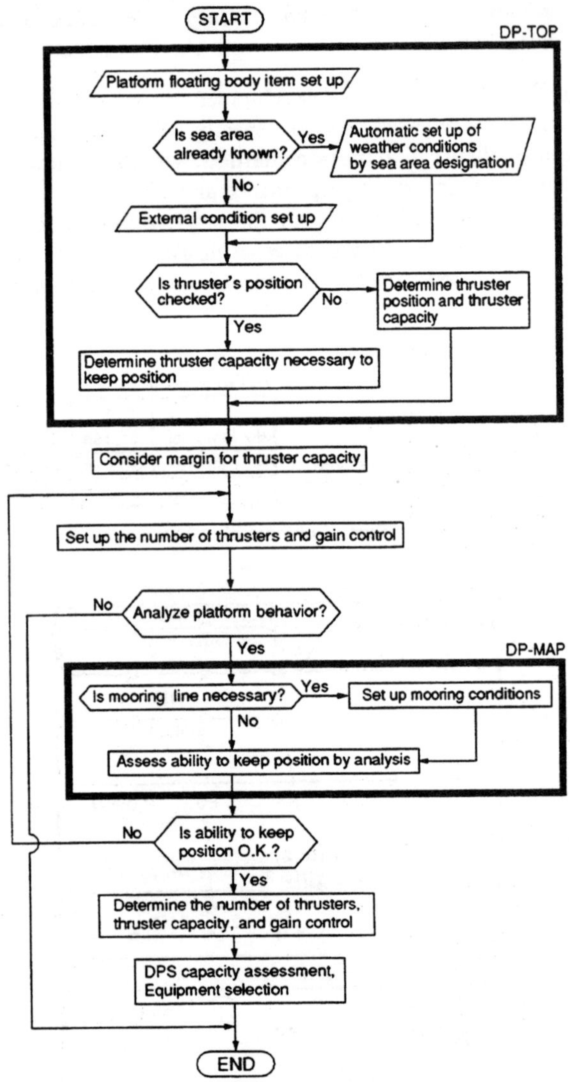

Fig.17 Flow chart of DPS procedure with DPS analysis program

Conclusion

In the study, a new control algorithm based on a nonlinear programming method was developed to improve position keeping characteristics of a platform. The effectiveness of the control algorithm was confirmed by numerical simulations and tank experiments. Then, the generality for designing the DPS of a platform can be much improved by the use of the control algorithm.

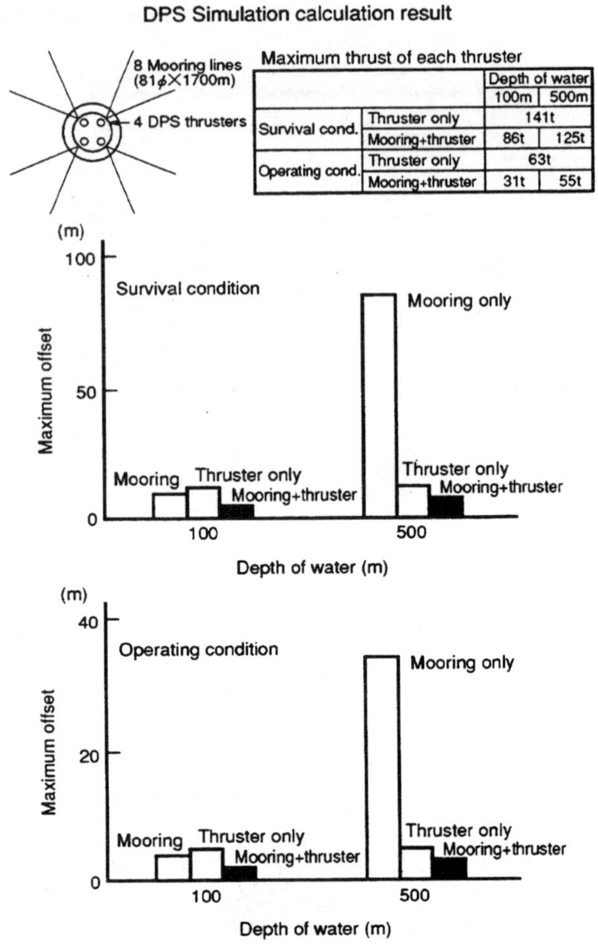

Fig.18 Ability to keep position by DPS and mooring

Acknowledgement

The present series of studies have been accomplished with the collaboration of the Japan National Oil Corporation and Mitsubishi Heavy Industries, Ltd. The authors wish to express their sincere gratitude to the Petroleum Engineering Laboratory of Technology Research Center of Japan National Oil Corporation for constant guidance and encouragement. Thanks are also due to all the members who cooperated with them in carrying out the present study.

References

M.TAMEHIRO N.AKASAKA H.KASAI E.MIWA (1977) "On Dynamic Positioning System Design in Particular Reference to the Positional Signal Filtering Technique", Journal of the Society of Naval Architects of Japan, November

N.AKASAKA A.IWASAKI E.MIWA (1981) "Dynamic Positioning of Floating Units with Rotatable Retractable Thrusters", MITSUBISHI JUKO GIHO, VOL.18, NO.2

I.YAMAMOTO Y.TERADA K.YOKOKURA (1991) "Application of a Position Keeping Control System to Floating Offshore Platform", Proc. IEEE IECON, Vol.3

M.MATSUURA Y.ISOZAKI Y.ISHIBASHI (1995) "Development of Mono-Column Type Hull Form with Passive Type Motion Damping Devices for Floating Production System", Proc. OMAE

Proceedings of the Seventh (1997) International Offshore and Polar Engineering Conference
Honolulu, USA, May 25-30, 1997

ISBN 1-880653-28-1 (Set); ISBN 1-880653-31-1 (Vol. III)

Control System Design and Model Experiments on Thruster Assisted Mooring System

Masahiko Nakamura
Kyushu University, Kasuga, Japan
Hiroyuki Kajiwara
Kyushu Institute of Technology, Iizuka, Japan
Wataru Koterayama and Tadahiro Hyakudome
Kyushu University, Kasuga, Japan

ABSTRACT

The paper is concerned with the dynamic positioning control of a moored floating platform model. Moorings are used to maintain the position of the platform against quasi-constant external forces induced by current and wind. The control purpose is to suppress the slow drift oscillation due to drifting force in irregular waves using thrusters so that they do not respond to the linear exciting force in the wave frequency range. Control of motions in the wave frequency region has to be ignored because the amplitude of those motions is smaller than that of the slow drift oscillation and enormous thruster power is needed to control them. The problem was formulated in the framework of H_∞ control, and was solved using the LMI (linear matrix inequalities) Control Toolbox on Matlab. Model experiments in oblique incident waves were carried out and some successful results are shown.

KEY WORDS: Thruster assisted mooring system, Floating platform, H_∞ control, Slow drift oscillation, Model experiments in oblique incident waves

NOMENCLATURE

A_{ij}	: Added mass
F_{xD}, F_{yD}, M_{xD}, M_{yD}, M_{zD}	: External force due to drag and wave damping
F_{xm}, F_{ym}, M_{xm}, M_{ym}, M_{zm}	: Force caused by mooring line
F_{xwave1}, F_{ywave1}, M_{xwave1}, M_{ywave1}, M_{zwave1}	: Linear wave exciting force
F_{xwave2}, F_{ywave2}, M_{xwave2}, M_{ywave2}, M_{zwave2}	: Drifting force in irregular waves
g	: Gravity acceleration
GM	: Metacentric height
GML	: Longitudinal metacentric height
I_{xx}, I_{yy}, I_{zz}	: Moment of inertia of platform
k	: Linear spring constant
K	: H_∞ controller transfer function
K_{act}	: H_∞ controller transfer function with actuator dynamics
L	: Length of platform
l_{th}	: Distance from origin o to thruster (see Fig.4)
l_x, l_y	: Position of point of application of drag force in steady flow
m	: Mass of platform
(o, x, y, z)	: Moving coordinate system
(O, X, Y, Z)	: Space-fixed coordinate system
p, q, r	: Angular velocity (see Fig.4)
u, v	: Translational velocity in x-, y-direction (see Fig.4)
u_a, v_a	: Relative velocity ($u_a = u - u_c$, $v_a = v - v_c$)
u_c, v_c	: Current velocity in x-, y-direction
$\tilde{X}$, $\tilde{Y}$	: Surge and sway of a level with the mooring point
$(0, 0, z_G)$	: Position of center of gravity
z_T	: Distance from origin o to thruster in z-direction
α	: Direction of current (see Fig.4)
θ_0	: Direction of waves (see Fig.4)
φ, θ, ψ	: Eulerian angles

1. INTRODUCTION

Floating offshore platforms are widely used in the exploration and development of ocean resources. They are required to maintain a given position under the external forces of ocean current, wind and waves. A mooring system or a dynamic positioning system using thrusters (Triantafyllou, 1979; Pinkster et al., 1986; Davison et al., 1987) is generally employed to maintain a position, however, the former system cannot completely prevent slow drift oscillation and the latter requires a great deal of energy because the thrusters must continually work against current and wind. A thruster assisted mooring system is needed which prevents the drift induced by these two quasi-constant external forces, and in which thrusters counteract the slow drift oscillation. Our concern here is the design of such a thruster controller. Slow drift oscillation appears as a result of resonance between the mooring system and the drifting force and the amplitude is very large, but it can be counteracted using thrusters because the exciting force is small. The PID controllers used to date tend to respond to the motion in the wave frequency range (see Fig.12). It should be noted that it is impossible to resist the wave frequency motion using thrusters because of the enormous amount of power needed and, in any event, this motion need not be controlled because of its small amplitude compared with the slow drift oscillation.

Our control aim is to suppress the slow drift oscillation using thruster activities that do not respond to wave frequency motion. We formulated the problem in the framework of H_∞ control (Doyle et al., 1992; Kajiwara et al., 1995; Nakamura et al., 1996) Which is convenient to enforce robustness to model uncertainty and express frequency-domain specifications such as bandwidth, low-frequency gain, etc, and solved the problem using the LMI Control Toolbox on Matlab (Gahinet et al., 1995).

Model experiments in oblique incident waves were carried out and successful results were obtained using the H_∞ controller.

2. MODEL AND EXPERIMENTAL METHOD

The configuration of a 1/100 scale floating platform model is shown in Fig.1. The length of the lower hull is 1 m, the maximum breadth is 0.64 m and displacement is 386 N. The model has three thrusters, two are on a diagonal line and one is on the x -axis (see Fig.4).

The experimental set-up is shown in Figs.2 and 3. Model experiments were carried out in oblique incident waves and surge, sway and yaw were controlled by the H_∞ controller. We knew the superiority of H_∞ control from experimental results in a beam sea (Nakamura et al., 1996).

A 4-point mooring by linear springs was employed. The No.1 spring is connected with the No.3 spring by wire through pulleys and the No. 2 spring is connected with the No.4 spring. When the wire shifts ΔL_1, mooring line 3 is rolled out and mooring line 1 is rolled in. When the wire shifts ΔL_2, mooring line 4 is rolled out and mooring line 2 is rolled in. If the adjustment of the mooring tension is made by the rolling in and out of the lines, the thruster's thrust to hold position against the current can be greatly reduced. The total system is fixed to a towing carriage that moves to generate the forward velocity, which simulates the current.

The motions (surge, sway and yaw) of the floating platform model were measured using a non-contact optical motion analyzer and roll and pitch were measured with a clinometer.

3. EQUATIONS OF MOTION OF FLOATING PLATFORM

The equations of motion were shown in detail by Nakamura et al. (1994 and 1995). An outline follows.

The moving coordinate system (o, x, y, z) coincides with the space-fixed coordinate system (O, X, Y, Z) at t=0 (Fig.4). Three thrusters are used to hold the platform in position. The relationships between space coordinate and velocity and that between Eulerian angle and angular velocity are

$$\begin{bmatrix} \dot{X} \\ \dot{Y} \end{bmatrix} = [\Omega]^{-1} \times \begin{bmatrix} u \\ v \end{bmatrix} \tag{1}$$

$$\begin{bmatrix} \dot{\varphi} \\ \dot{\theta} \\ \dot{\psi} \end{bmatrix} = \begin{bmatrix} 1 & \sin\varphi\tan\theta & \cos\varphi\tan\theta \\ 0 & \cos\varphi & -\sin\varphi \\ 0 & \sin\varphi\sec\theta & \cos\varphi\sec\theta \end{bmatrix} \times \begin{bmatrix} p \\ q \\ r \end{bmatrix} \tag{2}$$

in which

$$[\Omega]^{-1} = \begin{bmatrix} \cos\theta\cos\psi & \sin\varphi\sin\theta\cos\psi - \cos\varphi\sin\psi \\ \cos\theta\sin\psi & \sin\varphi\sin\theta\sin\psi + \cos\varphi\cos\psi \end{bmatrix} \tag{3}$$

The equations of motion of the floating platform are

$$\begin{bmatrix} m+A_{11} & 0 & 0 & mz_G+A_{15} & 0 \\ 0 & m+A_{22} & -mz_G & 0 & 0 \\ 0 & -mz_G+A_{42} & I_{xx}+A_{44} & 0 & 0 \\ mz_G+A_{51} & 0 & 0 & I_{yy}+A_{55} & 0 \\ 0 & 0 & 0 & 0 & I_{zz}+A_{66} \end{bmatrix} \times \begin{bmatrix} \dot{u} \\ \dot{v} \\ \dot{p} \\ \dot{q} \\ \dot{r} \end{bmatrix} = \begin{bmatrix} F_x \\ F_y \\ M_x \\ M_y \\ M_z \end{bmatrix} \tag{4}$$

$$\begin{aligned} F_x = &+(m+A_{22})rv_a - (mz_G - A_{24})pr + (m+A_{11})rv_c \\ &+F_{xD} + F_{xwave1} + F_{xwave2} + F_{xm} + F_{x\tau} \end{aligned} \tag{5}$$

$$\begin{aligned} F_y = &-(mz_G + A_{15})qr - (m+A_{22})ru_c - (m+A_{11})u_a r \\ &+F_{yD} + F_{ywave1} + F_{ywave2} + F_{ym} + F_{y\tau} \end{aligned} \tag{6}$$

$$\begin{aligned} M_x = &-\{(I_{zz}+A_{66}) - (I_{yy}+A_{55})\}qr + (mz_G - A_{42})ru_c \\ &+(A_{51}+mz_G)u_a r - mgGM\sin\varphi \\ &+M_{xD} + M_{xwave1} + M_{xwave2} + M_{xm} - z_T F_{y\tau} \end{aligned} \tag{7}$$

$$\begin{aligned} M_y = &+(mz_G - A_{42})rv_a - \{(I_{xx}+A_{44}) - (I_{zz}+A_{66})\}pr \\ &+(mz_G + A_{51})rv_c - mgGML\sin\theta \\ &+M_{yD} + M_{ywave1} + M_{ywave2} + M_{ym} + z_T F_{x\tau} \end{aligned} \tag{8}$$

$$\begin{aligned} M_z = &-\{(I_{yy}+A_{55}) - (I_{xx}+A_{44})\}pq - (A_{51}+A_{24})pu_a \\ &+(A_{15}+A_{42})qv_a - (A_{22}-A_{11})(uv - uv_c - u_c v) \\ &+M_{zD} + M_{zwave1} + M_{zwave2} + M_{zm} + M_{z\tau} \end{aligned} \tag{9}$$

Roll is not controlled variable, but the equation on roll is important to control yaw because yaw and roll are coupled. The same thing can be said of surge and pitch. Effects of heaving motion for another motions are small and heaving motion is not controlled, so that the equation on heave is omitted for simplicity. Hydrodynamic coefficients of the floating platform, linear wave exciting forces and drifting forces are calculated using the three-dimensional Singularity Distribution Method (Kyozuka and Inoue, 1987). Drifting forces in irregular waves are calculated using the method of Hsu and Blenkarn (1970).

Damping forces and Forces caused by mooring lines were shown in detail by Nakamura et al. (1994). The thruster's thrust F_{thi} determined by the input voltage to thruster V_{ini} can be written in a simple form (Nakamura et al., 1994),

$$F_{thi} = 0.3382 V_{ini} \left| V_{ini} \right| \quad (i = 1 \sim 3) \tag{10}$$

where subscript i indicates thruster number (see Fig.4). The time constant of the model thruster is almost zero and the thrust is generated at the same time the command is input, but an actual size thruster has a large time constant. Then, the model thrusters were operated with the time constant T_{thi}=0.7 sec in the experiments in consideration of the performance of a full size thruster (Leest and Bussemaker).

$$\frac{d}{dt} F_{thi} = -\frac{1}{T_{thi}} F_{thi} + \frac{1}{T_{thi}} f_{thi} \quad (i = 1 \sim 3) \tag{11}$$

f_{thi} is the input to the transfer function of the thruster dynamics and F_{thi} is the output (see Fig.7).

Forces in x-, y-direction and moment around the z-axis due to thrusters are represented as

$$\begin{cases} F_{x\tau} = F_{th3} \\ F_{y\tau} = F_{th1} + F_{th2} \\ M_{z\tau} = l_{th}(F_{th1} - F_{th2}) \end{cases} \tag{12}$$

In the experiments, the contoller K_{act} (see Fig.6) outputs $F_{x\tau}$, $F_{y\tau}$ and $M_{z\tau}$. Secondly F_{th1}, F_{th2} and F_{th3} are calculated by equation (10). Then the input voltage to the thruster V_{thi} is obtained by equation (12).

4. MATHEMATICAL MODEL OF MOORED FLOATING PLATFORM

$\dot{X}, \dot{Y}, \dot{\varphi}, \dot{\theta}, \dot{\psi}, \dot{u}, \dot{v}, \dot{p}, \dot{q}, \dot{r}$ are solved from Eqs. (1), (2) and(4).

$$\begin{bmatrix} \dot{X} \\ \dot{Y} \\ \dot{\varphi} \\ \dot{\theta} \\ \dot{\psi} \\ \dot{u} \\ \dot{v} \\ \dot{p} \\ \dot{q} \\ \dot{r} \end{bmatrix} = \begin{bmatrix} f_1(X,Y,\varphi,\theta,\psi,u,v,p,q,r,F_{x\tau},F_{y\tau},M_{z\tau},F_{xw},F_{yw},M_{zw}) \\ f_2(X,Y,\varphi,\theta,\psi,u,v,p,q,r,F_{x\tau},F_{y\tau},M_{z\tau},F_{xw},F_{yw},M_{zw}) \\ f_3(X,Y,\varphi,\theta,\psi,u,v,p,q,r,F_{x\tau},F_{y\tau},M_{z\tau},F_{xw},F_{yw},M_{zw}) \\ f_4(X,Y,\varphi,\theta,\psi,u,v,p,q,r,F_{x\tau},F_{y\tau},M_{z\tau},F_{xw},F_{yw},M_{zw}) \\ f_5(X,Y,\varphi,\theta,\psi,u,v,p,q,r,F_{x\tau},F_{y\tau},M_{z\tau},F_{xw},F_{yw},M_{zw}) \\ f_6(X,Y,\varphi,\theta,\psi,u,v,p,q,r,F_{x\tau},F_{y\tau},M_{z\tau},F_{xw},F_{yw},M_{zw}) \\ f_7(X,Y,\varphi,\theta,\psi,u,v,p,q,r,F_{x\tau},F_{y\tau},M_{z\tau},F_{xw},F_{yw},M_{zw}) \\ f_8(X,Y,\varphi,\theta,\psi,u,v,p,q,r,F_{x\tau},F_{y\tau},M_{z\tau},F_{xw},F_{yw},M_{zw}) \\ f_9(X,Y,\varphi,\theta,\psi,u,v,p,q,r,F_{x\tau},F_{y\tau},M_{z\tau},F_{xw},F_{yw},M_{zw}) \\ f_{10}(X,Y,\varphi,\theta,\psi,u,v,p,q,r,F_{x\tau},F_{y\tau},M_{z\tau},F_{xw},F_{yw},M_{zw}) \end{bmatrix} \tag{13}$$

Equation (13) is linearized around the equilibrium point in order to obtain a simplified linear mathematical model (Eq.14) for design of the controller.

$$\begin{cases} \dot{x} = Ax + Bu + Mw \\ y = Cx \\ z = Ex \end{cases} \tag{14}$$

$$\begin{cases} a_{ij} = \left. \dfrac{\partial f_i}{\partial x_j} \right|_{x=x^*, u=u^*, w=w^*}, \quad b_{ij} = \left. \dfrac{\partial f_i}{\partial u_j} \right|_{x=x^*, u=u^*, w=w^*} \\ m_{ij} = \left. \dfrac{\partial f_i}{\partial w_j} \right|_{x=x^*, u=u^*, w=w^*} \end{cases} \tag{15}$$

a_{ij} , b_{ij} and m_{ij} are components of matrix A , B , M. x is the state vector, u is the input vector, w is disturbance and y is measured output vector. x^* , u^* ,w^* are values of the state vector and input vector at the equilibrium point.

$$\begin{cases} x = \begin{bmatrix} X & Y & \varphi & \theta & \psi & u & v & p & q & r \end{bmatrix}^T \\ u = \begin{bmatrix} F_{x\tau} \\ F_{y\tau} \\ M_{z\tau} \end{bmatrix} \\ w = \begin{bmatrix} F_{xw} \\ F_{yw} \\ M_{zw} \end{bmatrix} = \begin{bmatrix} F_{xwavel} + F_{xwave2} \\ F_{ywavel} + F_{ywave2} \\ M_{zwavel} + M_{zwave2} \end{bmatrix} \\ y = \begin{bmatrix} \tilde{X} & \tilde{Y} & \psi \end{bmatrix}^T \end{cases} \tag{16}$$

This linear system is controllable and observable because of

rank[λ I-A B] = 10   ($\forall \lambda$(A))

rank[λ I-A^T C^T] = 10   ($\forall \lambda$(A))

Bode diagrams of the transfer functions of the mathematical model are shown in Fig.5. The peak of the bode diagrams ($F_{xw} \to \tilde{X}$, $F_{yw} \to \tilde{Y}$, $M_{zw} \to \psi$) shows the slow drift oscillation.

5. DESIGN OF H$_\infty$ CONTROL SYSTEM

From the discussion in section 1, the control purpose is to suppress the slow drift oscillation using thruster activities so that they do not respond to motion in the wave frequency range (linear wave exciting force). It is natural to formulate the problem from the viewpoint of shaping the closed-loop characteristics in the frequency domain, that is,

- Specification 1
  Reject the resonance characteristics (slow drift oscillation) at the natural period in the gain curves from F_{xw}, F_{yw}, M_{zw} to $\tilde{X}$, $\tilde{Y}$, ψ.
- Specification 2
  Make the gain curve F_{xw}, F_{yw}, M_{zw} to $F_{x\tau}$, $F_{y\tau}$, $M_{z\tau}$ as small as possible in the wave frequency 4~10 rad/sec.

The amplitude of the thruster's thrust must be in - 2 N <F_{th1}, F_{th2} and F_{th3}> 2 N in view of the limitation of the full-scale thruster performance.

Consider the interconnection structure as shown in Fig.6. Where the integral actions are included for perfect positioning, i.e., lower sensitivity, the derivatives of the thrust are need for mach more robust stability. The transfer function of plant G is calculated as

$$G = C(sI_{10} - A)^{-1}\begin{bmatrix} M & B \end{bmatrix} \tag{17}$$

The detail of the actuator dynamics is shown in Fig.7. The actuator dynamics is needed to operate the model thrusters as if they have a characteristic similar to a full size thruster.

W_1, W_2, W_3, W_4, W_5 and W_6 are weighting functions. These functions have to be chosen such that specifications 1 and 2 are satisfied to the greatest degree possible, which is difficult because of their tradeoff relation. W_1, W_3 and W_5 are selected so that the resonance of the floating structure (slow drift oscillation) is suppressed. W_2, W_4 and W_5 reflect the power spectrum of the linear wave exciting force and are selected so that the response of the thrusters to this force is minimal (see Fig.10). Our control problem is formulated as follows. Obtain a controller K to minimize H_∞ norm of transfer function from $F = \begin{bmatrix} F_{xw} & F_{yw} & M_{zw} \end{bmatrix}^T$ to $z = \begin{bmatrix} z_1 & z_2 & z_3 & z_4 & z_5 & z_6 \end{bmatrix}^T$ with a pole region constraint as shown in Fig.8, so that the following holds for some $\gamma > 0$.

$$\left\| T_{F \to z} \right\|_\infty < \gamma \tag{18}$$

A minimal value of γ is desirable. The guaranteed H_∞ performance level γ was 5.8. The pole region constraint as shown in Fig.8 is needed from the viewpoint of digital implementation. When the poles of the closed-loop system and the controller has a large value, the controller is not desirable because of the high gain, and the disk constraint with radius 35 is needed for the sampling period 0.01~(1/35)/3. The problem has muti-objective H_∞ control formulation and can't be solved by Riccati-equation based approach. We solved the problem using the LMI Control Toolbox on Matlab (Gahinet et al., 1995). The transfer function of the designed controller K_{act} is converted to the state-space equation and is transformed to a discrete-time system to carry out the model experiments. The input voltage to the thrusters is calculated by Eq.(10) after the thrust is calculated by Eq.(12). Frequency characteristics of the H_∞ controller are shown in Fig.10.

6. CONTROL OF MOORING LINES

When quasi-constant disturbances (current and wind) exist, the thrusters must generate thrust continuously to maintain position. Feedforward control of the mooring lines is used to avoid this:

(1) Calculate the average values of thrust ΔF_{th1}, ΔF_{th2} and ΔF_{th3} in some suitable period (for example, 4 or 6 times the period of the slow drift oscillation).

(2) Calculate the adjustment length of mooring lines ΔL_1, ΔL_2 so that the difference in tension of the lines in *x*-, *y*-direction becomes ΔF_{th3}, $\Delta F_{th1} + \Delta F_{th2}$.

$$\begin{cases} \Delta L_1 = \dfrac{\Delta F_{th1} + \Delta F_{th2} + \Delta F_{th3}}{4k\cos 45^\circ} \\ \Delta L_2 = \dfrac{\Delta F_{th1} + \Delta F_{th2} - \Delta F_{th3}}{4k\cos 45^\circ} \end{cases} \tag{19}$$

(3) Roll the lines in and out according to the calculated results.

If the tension of the mooring lines is adjusted once every 1 or 2 hours according to changes in the current (in a tidal current, the magnitude of the current changes over a period of 12 hours.), the energy to hold a position can be greatly reduced and the adjustment can be carried out easily by an automatic system or manually.

7. EXPERIMENTAL RESULTS

The irregular waves used in model experiments are shown in Fig.11. The average period is 1.0 sec and the significant wave height is 0.040 m. In actual size, these are 10 sec and 4 m, respectively. The direction of incident waves θ_0 (see Fig.4) is 30 deg.

The time series of motions of the floating platform and power spectra in case of no control is shown in Fig.12: at left is the experimental result and at right is the calculated result. The calculated results were obtained using nonlinear motion equations. From the top, surge, sway and yaw are shown. Their slow drift oscillation are clearly recognizable. The calculated response of surge and sway are in good agreement with the measured response. The calculated result of yaw is larger than the experimental; a more accurate model of yaw damping forces is required in the future.

Figure 13 shows the time series and power spectra of motions controlled by the PID controller and thruster's thrust. The gains of the PID controller were decided by ultimate sensitivity method and tuned. Frequency characteristics of the PID controller are shown in Fig.9. The height of the closed loop ($F_{xw} \Rightarrow \tilde{X}$, $F_{yw} \Rightarrow \tilde{Y}$ and $M_{zw} \Rightarrow \psi$) of the PID controller at the slow drift natural frequency is slightly lower than that of the H_∞ controller (Fig.10) at the same frequency. On the other hand the height of the closed loop ($F_{xw} \Rightarrow F_{x_\tau}$, $F_{yw} \Rightarrow F_{y_\tau}$ and $M_{zw} \Rightarrow M_{z_\tau}$) of the PID controller in the wave frequency band is higher than that of the H_∞ controller over the same band. Figure 13 is the experimental and calculated result. Although the amplitude of slow drift oscillation is small compared with the no control case (Fig.12), the thrust is found to vary greatly depending on wave frequency (ω = 4 ~ 8 rad/sec). Preventing motion in the wave frequency region is unrealistic because it requires a huge thrust of the thruster, and it is not necessary to control the wave frequency motion because its amplitude is small compared with the slow drift oscillation. The thruster response to the wave frequency motions is useless.

The results gained with the H_∞ controller are shown in Fig.14. The amplitude of the slow drift oscillation is almost zero as is the fluctuation of thrust in the wave frequency. The performance to reduce slow drift oscillation by the PID controller is 22% better than that of the H_∞ design. However the controlled slow drift oscillations are small enough and the control purpose is satisfied in both cases. The propeller horsepower is calculated on the assumption that the torque is in direct proportion to the thrust. The consumption of energy is calculated by integrating the propeller horsepower. In the case of H_∞ control, energy consumption was about 51 per cent of that with PID control. Therefore we conclude that the H_∞ controller is superior because of its lower energy requirements.

Figure 15 shows power spectra of the motion (surge, sway and yaw) under other irregular wave conditions. The top set of plots is the experimental results in irregular waves in which the average period is 0.76 sec. The bottom set of plots is the results in waves in which the average period is 1.42 sec. It is important to check the controller performance in various wave conditions because the position of the peak of the power spectrum of wave exciting force moves according to the average period of irregular waves. The amplitude of the slow drift oscillation is less than that without control. The figure thus shows that the H_∞ controller works effectively under various wave conditions.

Finally, current V_c was added to the irregular waves by moving the towing carriage at speed V_c (Fig.16). The direction of incident waves θ_0 and of current α (see Fig.4) is 30 deg respectively. From the top, incident waves, current velocity, surge, sway, yaw and thrusts of thruster 1, 2 and 3 are shown. The floating platform maintains its position despite the current and slow drift oscillation is suppressed, showing good performance by the controller. The thrusters are generating thrust continuously, however, which means they require a great deal of energy to maintain a position against the current. ΔF_{th1}, ΔF_{th2} and ΔF_{th3} are the average values of the thrust between t=360 sec and t=440 sec. Length of the mooring lines was adjusted after t=460 sec, and the average thrust became almost zero after t=550 sec. If the tension of the mooring lines is adjusted according to changes in the current, the energy to hold a position can be greatly reduced.

8. CONCLUSIONS

Model experiments on a thruster assisted mooring system controlled by H_∞ controller were carried out in oblique incident waves and the following conclusions drawn:

1. H_∞ control can prevent the slow drift oscillation of surge, sway and yaw in oblique incident waves.
2. H_∞ control can reduce the fluctuation of thrust in the wave frequency range and economize the energy needed to maintain a position compared with PID control.
3. The H_∞ controller works effectively under various wave conditions.
4. It was verified that the thruster assisted mooring system can greatly reduce the energy required to maintain a position if the tension of the mooring lines is adjusted in keeping with changes in current

Further study is needed to make a more accurate mathematical model and to test the performance of the controller under various wave and current directions. We also want to try to control position with steerable thrusters.

ACKNOWLEDGMENT

The authors are deeply indebted to Dr. Y. Kyozuka (Kyushu University) for his help in calculating the hydrodynamic coefficients of the floating platform.

REFERENCES

Davison, N. J. et al., (1987) : Application of an Alternative Concept in Dynamic Positioning to a Tanker Floating Production System, *OTC paper* No.5444

Doyle, J. et al., (1992) : Feedback Control Theory,*Macmillan*

Gahinet, P. et al., (1995) : LMI Control Toolbox, *The MathWorks*

Hsu, F. A. et al., (1970) : Analysis of Peak Mooring forces by Slow Vessel Drift Oscillation in Random Seas, *OTC paper* No.1159

Kajiwara, H. et al., (1995) : LMI-Based Design of Dynamic Positioning Control Laws for a Moored Floating Platform Model, *Proc. of the 14th SICE Kyushu Branch Annual Conference*, pp.181-184

Koterayama, W. et al., (1989) : Viscous Damping Forces for Slow Drift Oscillation of a Floating Structure Acting on the Hull and Mooring Lines, *Proc. of the Eighth International Conference on Offshore Mechanics and Arctic Engineering*, pp.497-503

Kyozuka, T. et al., (1987) : Second-order Wave Forces on a Semi-submersible, *Proc. of Third International Symposium on Practical Design of Ships and Mobile Units*, pp.181-184

Leest, H. V. et al., : The Performance and Characteristics of Thrusters, *Offshore South East Asia Conference Paper, No.40*, pp.65-82

Nakamura, M. et al., (1994) : Application of a Dynamic Positioning System to a Moored Floating Platform, *Proc. of the Fourth International Offshore and Polar Engineering Conference*, pp.190-197

Nakamura, M. et al., (1995) : Model Experiments on Thruster Assisted Mooring System, *Proc. of the Fifth International Offshore and Polar Engineering Conference*, pp.249-254

Nakamura, M. et al., (1996) : Model Experiments on Thruster Assisted Mooring System Controlled by H_∞ Controller, *Proc. of the Sixth International Offshore and Polar Engineering Conference*, pp.464-468

Pinkster, J. A. and Nienhuis, U., (1986) : Dynamic Positioning of Large Tankers at Sea, *OTC paper* No.5208

Triantafyllou, M. S., (1979) : The Design of a Dynamic Positioning System, *IEEE, CH1478-7*, pp.498-506

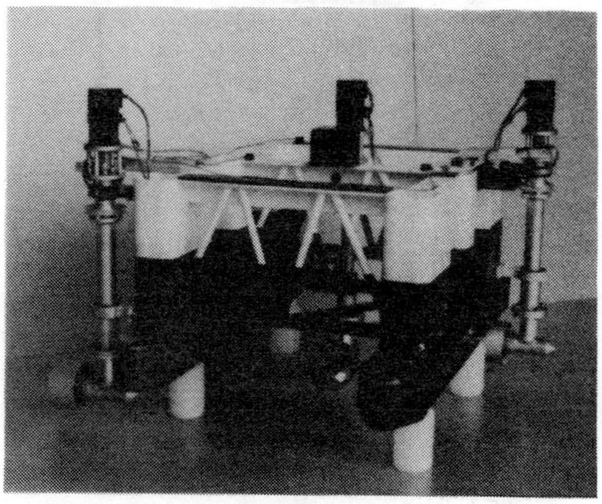

Length	1.00 m
Breadth	0.064 m
Displacement	386 N

Fig.1 Model of floating platform

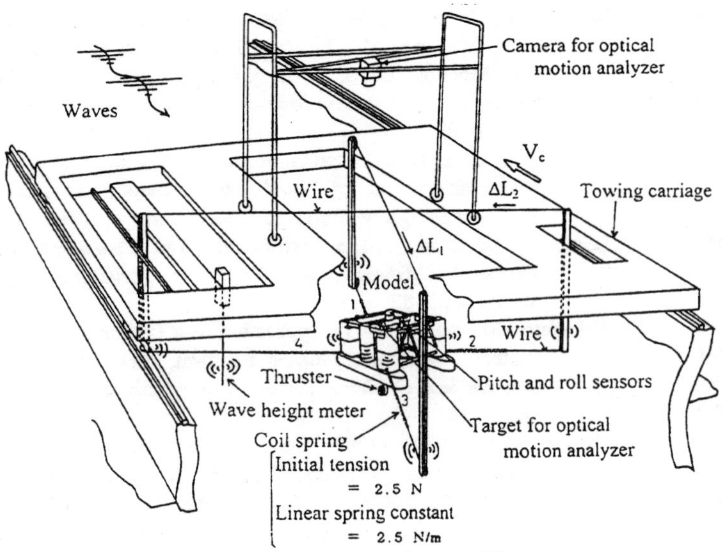

Fig.2 Experimental set-up (A)

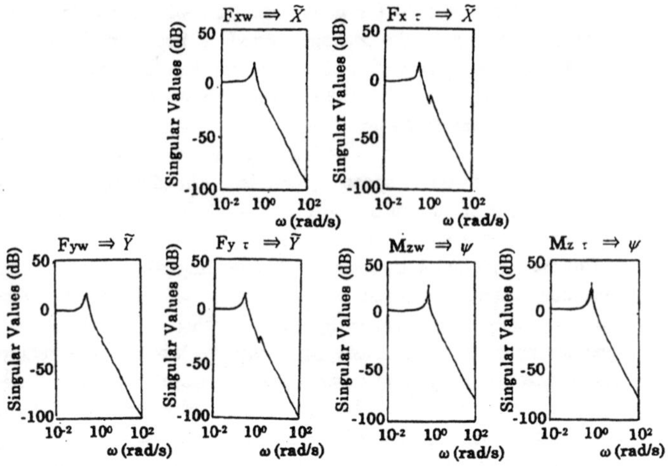

Fig.5 Gain curves of mathematical model

Fig.3 Experimental set-up (B)

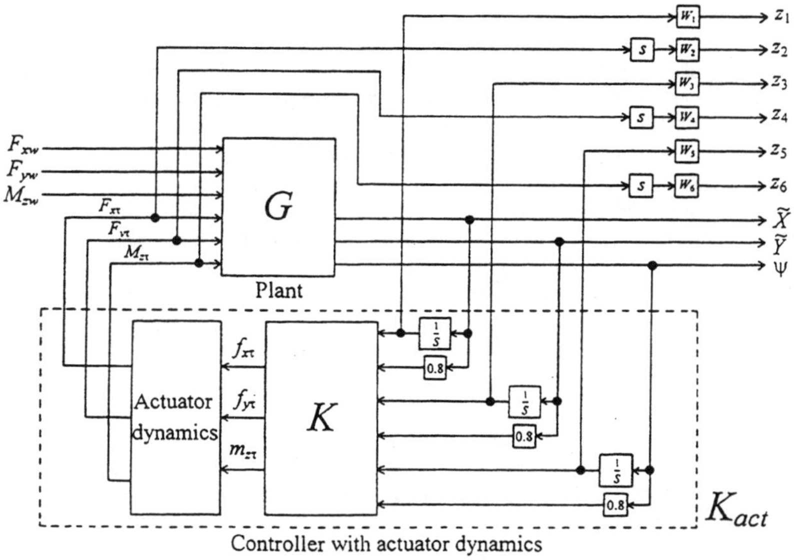

Fig.6 Interconnection structure of H$_\infty$ control system

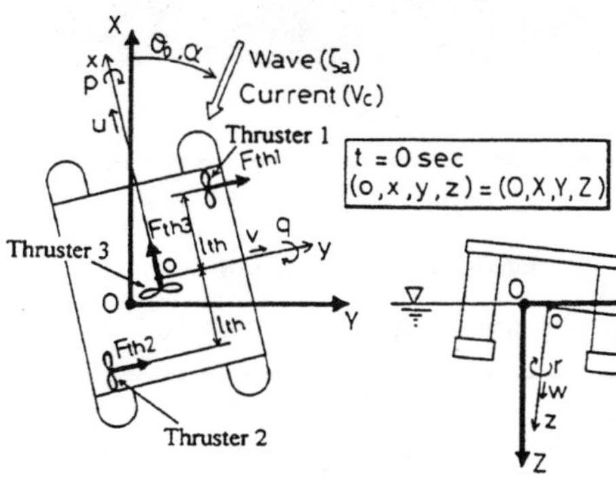

Fig.4 Coordinate system

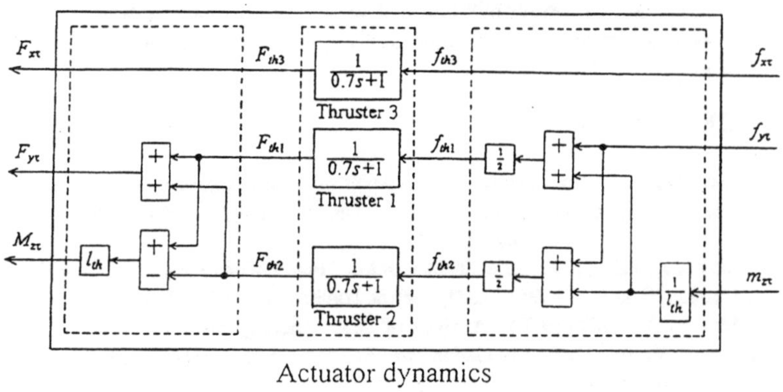

Fig.7 Actuator dynamics

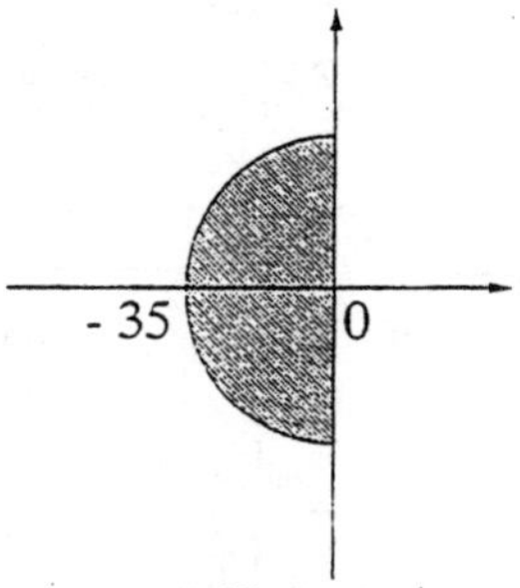

Fig.8 Specified pole region

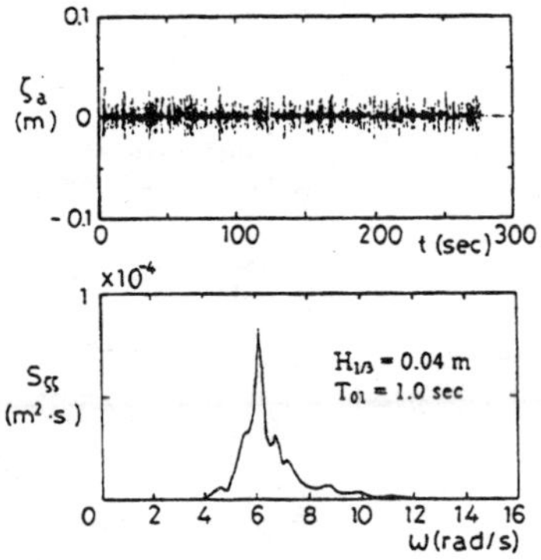

Fig.10 Incident waves

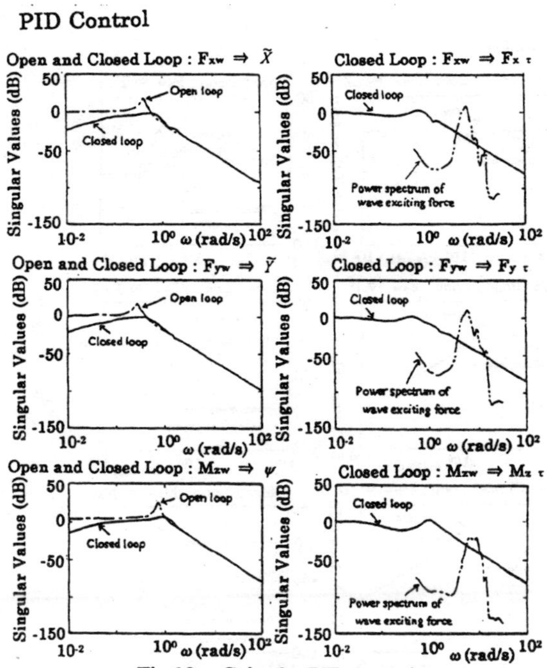

Fig.13 Gains by PID control

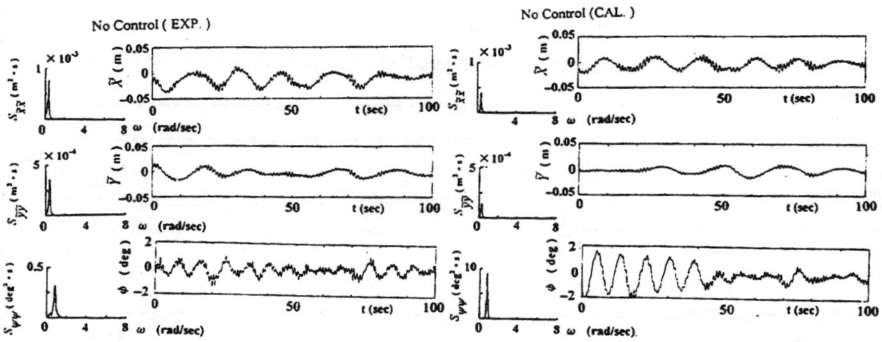

significant value of No control

	EXP.	CAL.
$\tilde{X}$ 1/3	0.049 m	0.031 m
T01	17.3 sec	16.4 sec
$\tilde{Y}$ 1/3	0.034 m	0.019 m
T01	20.2 sec	21.1 sec
φ 1/3	1.29 deg	4.93 deg
T01	8.21 sec	8.44 sec

Fig.11 Motions of floating platform (no control)

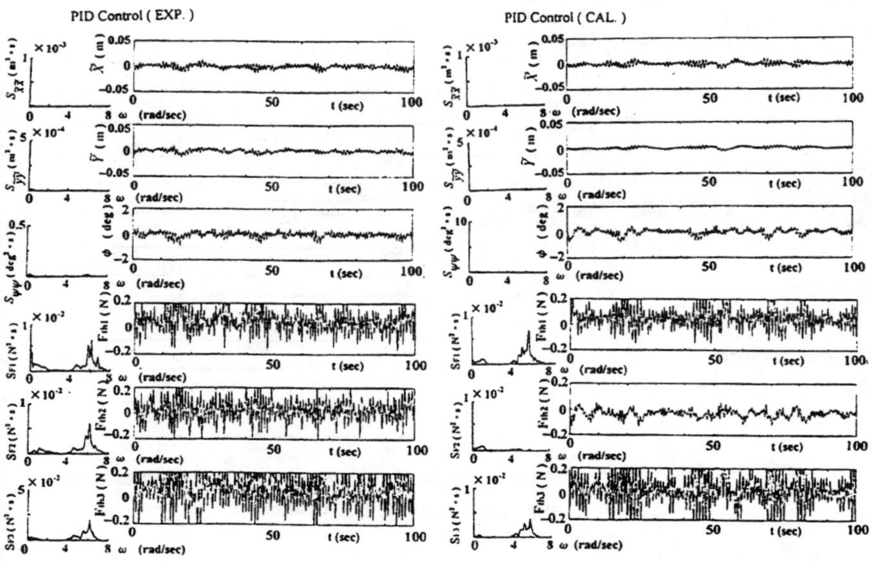

significant value of PID control

	EXP.	CAL.
$\tilde{X}$ 1/3	0.012 m	0.008 m
T01	13.8 sec	10.8 sec
$\tilde{Y}$ 1/3	0.012 m	0.006 m
T01	12.0 sec	13.9 sec
φ 1/3	0.58 deg	0.85 deg
T01	9.10 sec	8.61 sec
Fth1 1/3	0.33 N	0.31 N
T01	1.41 sec	1.26 sec
Fth2 1/3	0.31 N	0.13 N
T01	1.30 sec	2.81 sec
Fth3 1/3	0.53 N	0.54 N
T01	1.23 sec	1.20 sec

Fig.12 Motions of floating platform (PID control)

H_∞ Control

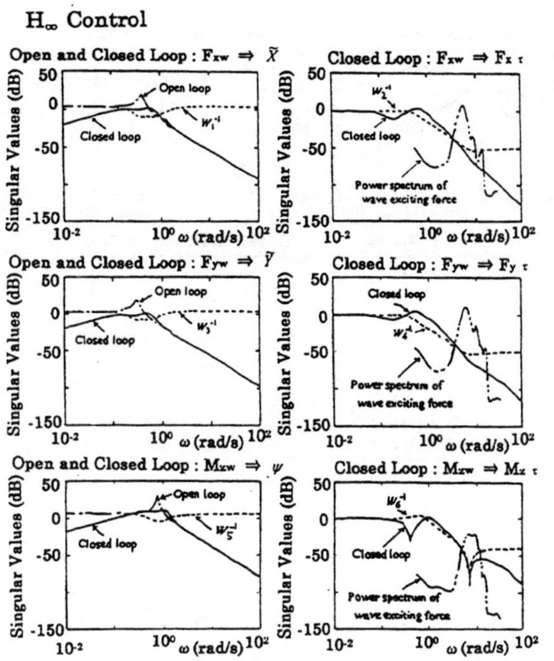

Fig.9 Gains by H_∞ control

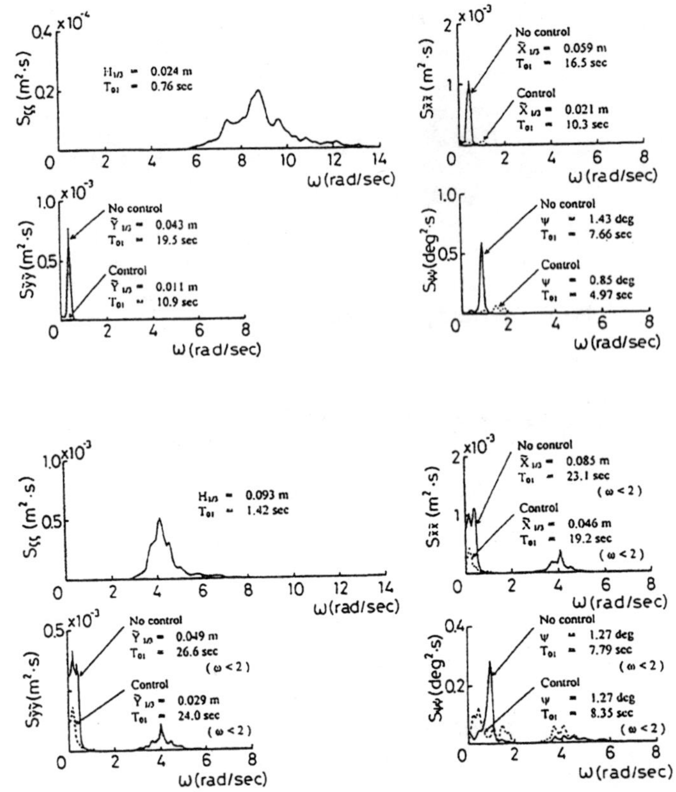

Fig.15 Motions of floating platform under various wave conditions

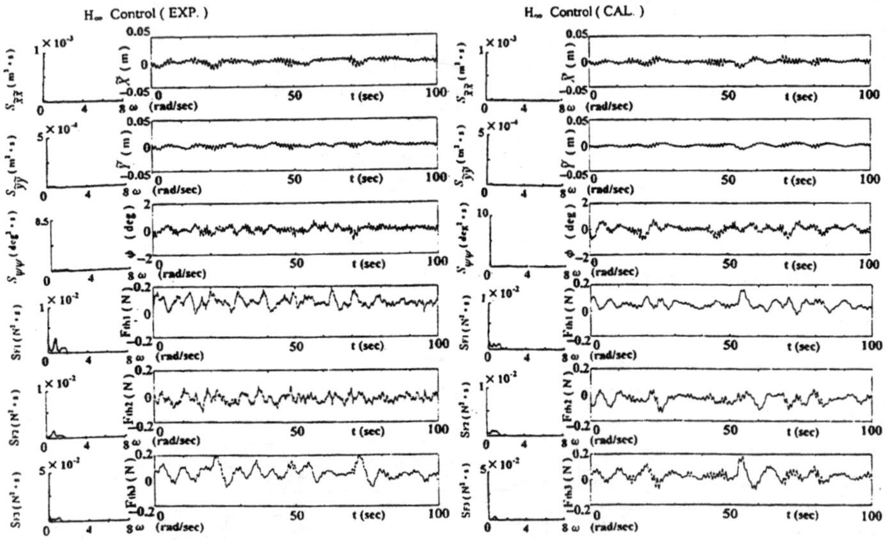

Fig.14 Motions of floating platform (H_∞ control)

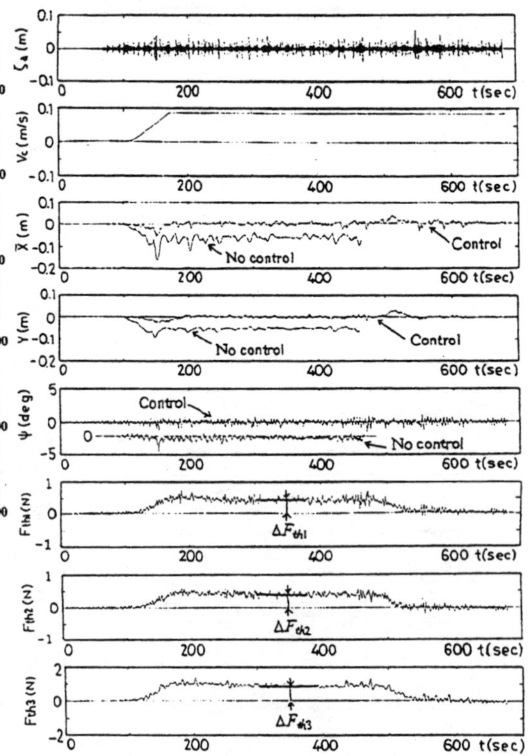

Fig. 16 Motions of floating platform and thrust of thrusters in waves and current (H_∞ control)

significant value of H_∞ control

	EXP.	CAL.
$\tilde{X}$ 1/3	0.019 m	0.011 m
T01	15.0 sec	11.0 sec
$\tilde{Y}$ 1/3	0.012 m	0.008 m
T01	11.7 sec	12.5 sec
ϕ 1/3	0.83 deg	1.18 deg
T01	6.35 sec	8.95 sec
Fth1 1/3	0.18 N	0.13 N
T01	6.40 sec	8.48 sec
Fth2 1/3	0.15 N	0.13 N
T01	6.59 sec	8.36 sec
Fth3 1/3	0.25 N	0.14 N
T01	8.93 sec	9.14 sec

Proceedings of the Seventh (1997) International Offshore and Polar Engineering Conference
Honolulu, USA, May 25-30, 1997

ISBN 1-880653-28-1 (Set); ISBN 1-880653-31-1 (Vol. III)

Estimation of Dynamic Towline Tension During Towing by the Probabilistic Method

Shoichi Hara
Ship Research Institute
Tokyo, Japan

ABSTRACT

The towline tension becomes exponentially large in case the distance between tug and tow becomes extremely large during towing. The dynamic towline tension has been composed of the cross-flow drag force and nonlinear restoring force. The nonlinear model using those force terms has been proposed in order to simulate the dynamic towline tension during towing. The effect of the inertia force term on the dynamic towline tension has also been evaluated. The probability density function, expected mean upcrossing frequency and expected largest value of the dynamic towline tension have been obtained. The theoretical calculation of the dynamic towline tension has been validated by the results of the numerical simulation.

KEY WORDS : Towing. Dynamic Towline Tension, Probability Density Function, Mean Upcrossing Frequency, Expected Maximum Value

INTRODUCTION

Various concepts on huge offshore structures for the space utilization in the ocean have recently been presented. Mega-Float Technology Association has been considering the 5000 m floating airport. As for the construction of this large scale offshore structures, it is hardly possible to manufacture all in the dockyard. The box shaped floating unit of the structure instead has to be constructed in the dockyard, towed to the site and installed by connecting each unit in the ocean. This floating unit can be 300 m long which is far large compared with the conventional towed floating bodies. Therefore, it is necessary to establish the safety evaluation method during towing. The mechanism of the towline force should be investigated in detail.

The towline tension during towing in waves has to be estimated concerning constant tension and fluctuating tension induced on the towline. The relative motion especially surge motion between tug and tow causes the dynamic towline tension due to cross-flow drag which is proportional with the square of velocity of the towline induced by the towline movement in water. This frequently has both the high frequency component of wave elevation and the low frequency component due to slow drift motion.

As for the researches on the probabilistic approach concerning the mooring force or wave loading on the marine structures, Ando et. al have presented the probability density function of the dynamic mooring line tension (Ando,1983) (Kato,1987). Borgman (1969) obtained the probability density function of the force due to current and waves on the structural elements of the offshore structures using the drag and inertia term. Vinje (1980) has presented the extreme distributions of the Morison-type of wave loading in current and waves and has also presented the asymptotic evaluation of mean upcrossing frequency in case of large wave loading. The expected largest value of loading function per unit time interval has been calculated by Naess(1983).

The towline tension becomes exponentially large in case the distance between tug and tow becomes extremely large during towing. The dynamic towline tension has been composed of the cross-flow drag force and nonlinear restoring force. The nonlinear model using those force terms has been proposed in order to simulate the dynamic towline tension during towing. The effect of the inertia force term on the dynamic towline tension has also been evaluated. The probability density function, expected mean upcrossing frequency and expected largest value of the dynamic towline tension have been obtained. The theoretical calculation of the dynamic towline tension has been validated by the results of the numerical simulation.

ESTIMATION OF DYNAMIC TOWLINE TENSION

The non-linear model has been proposed in order to obtain the probability density function of the dynamic towline tension during towing. The towed floating body through the wire rope by the towing carriage has been here assumed. The dynamic towline force f can be shown using the horizontal movement x in the following equation.

$$f(x) = c\dot{x}|\dot{x}| + \exp(ax) - 1 \quad (1)$$

where c and a are constants and $c, a > 0$ Suppose T_0 is the towline tension at the balanced point $x = 0$. The towline tension T can be shown by the horizontal movement x from it as $T = T_0 + f(x)$.

The dynamic towline tension from the balanced point (T_0)has only been considered here. Putting the following equations,

$$c\dot{x}|\dot{x}| = y_1(x) \tag{2}$$

$$\exp(ax) - 1 = y_2(x) \tag{3}$$

equation (1) can be replaced by the next equation.

$$f(x) = y_1(x) + y_2(x) \tag{4}$$

PROBABILITY DNSITY FUNCTION OF DYNAMIC TOWLINE TENSION

Probability Density Function $p(y_1)$ of Random Variable y_1

The probability density function $p(y_1)$ with respect to the random variable y_1 can be obtained from the equation (2).

(1) $\dot{x} > 0$,

$$y_1 \quad = \quad c\dot{x}^2 \; (c > 0) \tag{5}$$

Differentiating both sides of the equation(5), following equation can be obtained.

$$p(y_1) = \frac{1}{2\sqrt{cy_1}} p\left(\sqrt{\frac{y_1}{c}}\right), \; y_1 > 0 \tag{6}$$

(2)$\dot{x} < 0$,

$$y_1 \quad = \quad -c\dot{x}^2 \; (c > 0) \tag{7}$$

$$p(y_1) = \frac{1}{2\sqrt{-cy_1}} p\left(\sqrt{-\frac{y_1}{c}}\right), \; y_1 < 0 \tag{8}$$

Assuming $\dot{x}$ is a Gausian process, the probability density function $p(\dot{x})$ of random variable $\dot{x}$ is shown below.

$$p(\dot{x}) = \frac{1}{\sqrt{2\pi}\sigma_{\dot{x}}} \exp\left(-\frac{\dot{x}^2}{2\sigma_{\dot{x}}^2}\right) \tag{9}$$

Thus, from equations (6),(8) and (9), we divide into two cases such as $\dot{x} > 0$ and $\dot{x} < 0$. Considering those two cases, $p(y_1)$ can be shown as follows.

$$p(y_1) = \frac{1}{2\sqrt{2\pi}\sqrt{c\sigma_{\dot{x}}^2 |y_1|}} \exp\left(\frac{-|y_1|}{2c\sigma_{\dot{x}}^2}\right) \tag{10}$$

Probability Density Function $p(y_2)$ of Random Variable y_2

Differentiating both sides of the equation (3) to obtain the probability density function $p(y_2)$ of random variable y_2 and considering $y_2 + 1 > 0$, we have

$$x = \frac{\log(y_2 + 1)}{a} \tag{11}$$

If x follows a Gausian process, the probability density function $p(x)$ of random variable x becomes

$$p(x) = \frac{1}{\sqrt{2\pi}\sigma_x} \exp\left(-\frac{x^2}{2\sigma_x{}^2}\right) \tag{12}$$

$p(y_2)$ can be given from equations (11) and (12) as follows.

$$p(y_2) = \frac{1}{\sqrt{2\pi} a\sigma_x(1 + y_2)} \exp\left\{-\frac{\log^2(1 + y_2)}{2a^2\sigma_x{}^2}\right\} \tag{13}$$

Probability Density Function $p(\hat{f})$ of Random Variable $\hat{f}$

Nondimensionalizing the dynamic towline tension f, we put $\hat{f}$ as follows.

$$\hat{f} = \hat{y}_1 + \hat{y}_2 \tag{14}$$

where $\hat{f}, \hat{y}_1$ and $\hat{y}_2$ are given in the next equation.

$$\hat{f} = \frac{f}{a\sigma_x}, \hat{y}_1 = \frac{y_1}{a\sigma_x}, \hat{y}_2 = \frac{y_2}{a\sigma_x} \tag{15}$$

The probability density function $p(\hat{y}_1)$ of random variable $\hat{y}_1$ can be obtained from equation (10) and (15) as follows.

$$p(\hat{y}_1) = \frac{1}{2\sqrt{2\pi\alpha}\sqrt{|\hat{y}_1|}} \exp\left(-\frac{|\hat{y}_1|}{2\alpha}\right) \tag{16}$$

Where

$$\alpha = \frac{c\sigma_{\dot{x}}^2}{a\sigma_x} \tag{17}$$

The probability density function $p(\hat{y}_2)$ of random variable $\hat{y}_2$ can be obtained from the equations (13) and (15) as follows.

$$p(\hat{y}_2) = \frac{1}{\sqrt{2\pi}(a\sigma_x\hat{y}_2 + 1)} \exp\left\{-\frac{\log^2(a\sigma_x\hat{y}_2 + 1)}{2a^2\sigma_x{}^2}\right\} \tag{18}$$

If the random variables $\hat{y}_1$ and $\hat{y}_2$ are independent each other, the probability density function $p(\hat{f})$ of the summation of those is given from equation (16) and (18) as follows.

$$P(\hat{f}) = \int_{-\infty}^{\infty} \frac{1}{4\pi\sqrt{\alpha}\sqrt{|z|}\left\{a\sigma_x(\hat{f} - z) + 1\right\}} \times \exp\left[-\frac{|z|}{2\alpha} - \frac{\log^2\left\{a\sigma_x(\hat{f} - z) + 1\right\}}{2\alpha^2\sigma_x{}^2}\right] dz \tag{19}$$

where the range of integration is $-\infty < z < \infty$. The condition $a\sigma_x(\hat{f} - z) + 1 > 0$ should be satisfied.

(a)Case where the drag term becomes dominant

When the restoring force term a is small in the equation (1) and the standard deviation of the horizontal displacement x is also small, the assumption such as $-1 < a\sigma_x(\hat{f} - z) + 1 << 1$ can be valid. Then the following equation can be given.

$$\frac{\exp\left[-\log^2\left\{\frac{a\sigma_x(\hat{f}-z)+1}{2\alpha^2\sigma_x{}^2}\right\}\right]}{a\sigma_x(\hat{f} - z) + 1} \cong \exp\left\{-\frac{(\hat{f} - z)^2}{2}\right\}$$

Substituting this into the equation (19), we have

$$p(\hat{f}) \quad = \quad \int_{-\infty}^{\infty} \frac{1}{2\pi\sqrt{\alpha}\sqrt{|z|}} \exp\left(-\frac{\hat{f}^2}{2}\right) \cosh(\hat{f}z) \times \exp\left(-\frac{z^2}{2} - \frac{z}{2\alpha}\right) dz \tag{20}$$

Ando et.al.(1983) obtained the probability density function of the dynamic tension of mooring chain line including the inertia force term. They considered the correlation between the displacement and acceleration of the mooring line.

EXPECTED UPCROSSING FREQUENCY OF DYNAMIC TOWLINE TENSION

Expected maxima of the dynamic towline tension can be derived from the expected upcrossing frequency per unit time interval. The upcrossing frequency at first is shown in this section. The dynamic towline tension f assumes to be expressed in equation (1) using the horizontal displacement of one end of the towline x. $x(t)$ follows the normal random process having spectrum $S_x(\omega)$. The various random variables are defined as follows.

$$F(t) = \frac{f(t)}{a\sigma_x}, Y_1 = \frac{\dot{x}(t)}{\sigma_{\dot{x}}}, Y_2 = \frac{x(t)}{\sigma_x}, \alpha = \frac{c\sigma_{\dot{x}}^2}{a\sigma_x}, \beta = a\sigma_x \quad (21)$$

where,

$$\sigma_x^2 = \int_0^\infty S_x(\omega)d\omega \quad (22)$$

$$\sigma_{\dot{x}}^2 = \int_0^\infty \omega^2 S_x(\omega)d\omega \quad (23)$$

Substituting equation (21) into equation (1), ,$F(t)$ can be written as

$$F(t) = g(Y_1, Y_2) = \alpha Y_1|Y_1| + \frac{1}{\beta}\{\exp(\beta Y_2) - 1\} \quad (24)$$

Assuming that $Y_1(t)$ and $Y_2(t)$ follow the normal random process and both have a zero mean and a unit variance, the next equation can be given.

$$E[Y_1(t)] = 0,\ E[Y_2(t)] = 0$$
$$Var[Y_1(t)] = 1,\ Var[Y_2(t)] = 1$$

The expected value of upcrossing frequency $f_\xi{}^+$ where f exceeds a certain level ξ can be shown as follows according to Rice.

$$f_\xi{}^+ = \int_0^\infty P_{F\dot{F}}(\xi, \dot{f})\dot{f}d\dot{f} \quad (25)$$

where $P_{F\dot{F}}$ means the joint probability density function of two random variables $F(t)$ and $\dot{F}(t)$. From the marginal distribution principle, the following equation can be obtained.

$$P_{F\dot{F}}(f, \dot{f}) = \int_{-\infty}^{\infty} P_{Y_1 F\dot{F}}(y_1, f, \dot{f})\, dy_1 \quad (26)$$

Y_2 can be obtained from the equation (24).

$$Y_2 = \frac{1}{\beta}\log\{\beta(F - \alpha Y_1|Y_1|) + 1\} \quad (27)$$

The following equation can be obtained from the equation (27) by the variable transformation.

$$P_{Y_1 F\dot{F}}(y_1, f, \dot{f}) = P_{Y_1 Y_2 \dot{F}}(y_1, \log\frac{\beta(F - \alpha Y_1|Y_1|) + 1}{\beta}, \dot{f}) \quad (28)$$

Using the conditional probability density function $P_{\dot{F}|Y_1 Y_2}$ where y_1 and y_2 are given. $P_{Y_1 Y_2 F}$ can be written as follows.

$$P_{Y_1 Y_2 \dot{F}}(y_1, y_2, \dot{f}) = P_{\dot{F}|Y_1 Y_2}(\dot{f}|y_1, y_2)\, P_{Y_1 Y_2}(y_1, y_2) \quad (29)$$

The next equation can be obtained from the equation (24).

$$\dot{F}|_{Y_1=y_1, Y_2=y_2} = \frac{\partial g(Y_1, Y_2)}{\partial Y_1}\dot{Y}_1|_{Y_2=y_2} + \frac{\partial g(Y_1, Y_2)}{\partial Y_2}\dot{Y}_2|_{Y_1=y_1} \quad (30)$$

The next equation can be obtained from equations (24) and (30).

$$\dot{F}|_{Y_1=y_1, Y_2=y_2} = \exp(\beta y_2)\dot{Y}_2|_{Y_1=y_1} + 2\alpha|y_1|\dot{Y}_1|_{Y_2=y_2} \quad (31)$$

As $\dot{Y}_1$ and $\dot{Y}_2$ are the normal random variables, the conditional random variables such as $\dot{Y}_1|_{Y2} = y_2$ and $\dot{Y}_2|_{Y_1} = y_1$ are the normal random variables. From the equation (31), $\dot{F}|_{Y_1 Y_2}$ is also a normal random variable considering the equation (21).

$$\dot{Y}_2 = \frac{\dot{x}(t)}{\sigma_x}, \dot{Y}_1 = \frac{\ddot{x}(t)}{\sigma_{\dot{x}}}$$

Thus, the next relation can be obtained.

$$\dot{Y}_2 = \kappa Y_1 \quad (32)$$

where

$$\kappa = \sigma_{\dot{x}}/\sigma_x \quad (33)$$

Thus, we have the next equation from equation (32).

$$E\left[\dot{Y}_2|Y_1 = y_1\right] = E\left[\kappa Y_1|Y_1 = y_1\right] = \kappa y_1 \quad (34)$$

From the formula concerning the conditional random variable having a normal random process and equation (30), we have

$$E\left[\dot{Y}_1|Y_2 = y_2\right] = E\left[\frac{\ddot{Y}_2}{\kappa}|Y_2 = y_2\right] = \frac{1}{\kappa}\rho\frac{\sigma_{\ddot{Y}_2}}{\sigma_{Y_2}}y_2 \quad (35)$$

where

$$\rho = \frac{\sigma_{\ddot{Y}_2 Y_2}}{\sigma_{\ddot{Y}_2}\sigma_{Y_2}} \quad (36)$$

$$\sigma_{\ddot{Y}_2 Y_2} = E\left[\ddot{Y}_2 Y_2\right] = -\kappa^2 \quad (37)$$

Thus we have

$$\rho = -\frac{\kappa^2}{\sigma_{\ddot{Y}_2}\sigma_{Y_2}} = -\frac{\sigma_{\dot{x}}^2}{\sigma_x\sigma_{\ddot{x}}} \quad (38)$$

where

$$\sigma_{\ddot{x}}^2 = \int_0^\infty \omega^4 S_x(\omega)d\omega \quad (39)$$

The next equation can be obtained from equation (35) and (38).

$$E\left[\dot{Y}_1|Y_2 = y_2\right] = -\kappa y_2 \quad (40)$$

The mean $\hat{\mu}$ and variance $\hat{\sigma}^2$ of $\dot{F}|_{Y_1 Y_2}$ can be derived from equation (31).

$$\hat{\mu} = \kappa\exp(\beta y_2)y_1 - 2\alpha|y_1|y_2 \quad (41)$$

$$\hat{\sigma}^2 = 4\alpha^2\kappa^2 y_1{}^2\frac{1-\rho^2}{\rho^2} \quad (42)$$

Thus the conditional probability density function $P_{\dot{F}|Y_1 Y_2}$ of the conditional random variable$\dot{F}|_{Y_1 Y_2}$ can be obtained as the follows.

$$P_{\dot{F}|Y_1 Y_2}(\dot{f}|y_1, y_2) = \frac{1}{\sqrt{2\pi}\hat{\sigma}}\exp\left\{-\frac{1}{2}\frac{(\dot{f} - \hat{\mu})^2}{\hat{\sigma}^2}\right\} \quad (43)$$

$$E[Y_1] = E[Y_2] = 0$$

Assuming that Y_1 and Y_2 are independent normal random variables each other, the joint probability density function $P_{Y_1 Y_2}$ can be written as

$$P_{Y_1 Y_2}(y_1, y_2) = \frac{1}{2\pi}\exp\left\{-\frac{1}{2}(y_1^2 + y_2^2)\right\} \quad (44)$$

$f_\xi{}^+$ can be lead from the equations (25),(26),(28),(43) and (44).

$$f_\xi{}^+ = \frac{1}{2\pi}\int_{-\infty}^{\infty}\left[\frac{\tilde{\sigma}(y_1)}{\sqrt{2\pi}}\exp\left\{-\frac{1}{2}\left(\frac{\tilde{\mu}(y_1,y_2)}{\tilde{\sigma}(y_1)}\right)^2\right\}\right.$$
$$\left.+\tilde{\mu}(y_1,y_2)\Phi\left(\frac{\tilde{\mu}(y_1,y_2)}{\tilde{\sigma}(y_1)}\right)\right]\exp\left\{-\frac{1}{2}\left(y_1^2+y_2^2\right)\right\}dy_1 \quad (45)$$

Where

$$\Phi(x) = \frac{1}{\sqrt{2\pi}}\int_{-\infty}^{x}\exp\left(-\frac{t^2}{2}\right)dt \quad (46)$$

(a)Asymptotic value of $f_\xi{}^+$

A certain level ξ concerning the dynamic towline tension can be expressed from the equation (24) as follows.

$$\xi = \frac{1}{\beta}\{\exp(\beta y_2)-1\}+\alpha y_1|y_1|$$

And

$$y_2 = \frac{1}{\beta}\log\{\beta(\xi-\alpha y_1|y_1|)+1\} \quad (47)$$

Following Naess's (1983) procedure in order to obtain the asymptotic value of $f_\xi{}^+$, we put as follows.

$$\psi = \psi(y_1;\xi) = \frac{1}{2}(y_1^2+y_2^2)$$

$$= \frac{1}{2}\left[\frac{1}{\beta}\log\{\beta(\xi-\alpha y_1|y_1|)+1\}\right]^2+\frac{1}{2}y_1{}^2 \quad (48)$$

We have

$$\psi'(y_1;\xi) = y_1-\frac{1}{\beta}\frac{2\alpha|y_1|}{\{\beta(\xi-\alpha y_1|y_1|)+1\}}$$
$$\times\log\{\beta(\xi-\alpha y_1|y_1|)+1\} \quad (49)$$

Using the equations (41), (42) and (49), the following equation can be obtained.

$$\frac{\tilde{\mu}}{\tilde{\sigma}} = \psi'\frac{\beta(\xi-\alpha y_1|y_1|)+1}{2\alpha|y_1|}\cdot\frac{-\rho}{\sqrt{1-\rho^2}} \quad (50)$$

Putting as follows,

$$\frac{\beta(\xi-\alpha y_1|y_1|)+1}{2\alpha|y_1|}\cdot\frac{-\rho}{\sqrt{1-\rho^2}} = \gamma(y_1;\xi) \quad (51)$$

we have

$$f_\xi{}^+ = \frac{1}{2\pi}\int_{-\infty}^{\infty}\tilde{\sigma}(y_1)\left[\frac{1}{\sqrt{2\pi}}\exp\left\{-\frac{1}{2}(\psi'\gamma)^2\right\}\right.$$
$$\left.+\psi'\gamma\Phi(\psi'\gamma)\right]\exp(-\psi)dy_1 \quad (52)$$

If we put as follows in the equation (52),

$$\psi'\gamma = y_3(y_1;\xi) \quad (53)$$

we have

$$\Phi(y_3) = \frac{1}{\sqrt{2\pi}}\int_{-\infty}^{y_3}\exp\left(-\frac{t^2}{2}\right)dt \quad (54)$$

Differentiating the both sides of the equation (54) with respect to y_1, the following equation can be obtained from the equations (50) and (51).

$$\tilde{\sigma} = \frac{\kappa\exp(\beta y_2)}{\gamma}$$

From the equation (52) and above equation, we have

$$f_\xi{}^+ = \frac{\kappa\exp(\beta y_2)}{(2\pi)^{\frac{3}{2}}}\int_{-\infty}^{\infty}\frac{1+\gamma^2(y_1;\xi)\psi''(y_1;\xi)}{\gamma(y_1;\xi)}$$
$$\times\exp\left\{-\psi(y_1;\xi)-\frac{\gamma^2(y_1;\xi)\psi'^2(y_1;\xi)}{2}\right\}dy_1 \quad (55)$$

If we put as follows,

$$h(y_1;\xi) = \psi(y_1;\xi)+\frac{1}{2}\gamma^2\psi'^2(y_1;\xi) \quad (56)$$

equation (55) can be shown as follows.

$$f_\xi{}^+ = \frac{\kappa\exp(\beta y_2)}{(2\pi)^{\frac{3}{2}}}\int_{-\infty}^{\infty}\frac{1+\gamma^2(y_1;\xi)\psi''(y_1;\xi)}{\gamma(y_1;\xi)}$$
$$\times\exp\{-h(y_1;\xi)\}dy_1 \quad (57)$$

As ψ is not negative value, the contribution to the integral becomes large near the point where $h(y_1;\xi)$ becomes minimum. Therefore, having a series expansion of $h(y_1;\xi)$ near the point $y_1 = y_0$, we can put as

$$h(y_1;\xi)\cong h(y_0;\xi)+(y_1-y_0)^2h''(y_0;\xi)/2$$

Using *Laplace'smethod*, the integrated term of the equation (57) can be written as

$$\int_{-\infty}^{\infty}\frac{1+\gamma^2(y_1;\xi)\psi''(y_1;\xi)}{\gamma(y_1;\xi)}\exp\{-h(y_1;\xi)\}dy_1$$
$$= \frac{1+\gamma^2(y_0;\xi)\psi''(y_0;\xi)}{\gamma(y_0;\xi)}\exp\{-h(y_0;\xi)\}\sqrt{\frac{2\pi}{h''(y_0,\xi)}}$$

f_ξ^+ can be written from the equation (55) as follows.

$$f_\xi{}^+ = \frac{\kappa\exp(\beta y_2)}{2\pi}\frac{1+\gamma^2(y_0;\xi)\psi''(y_0;\xi)}{\gamma(y_0;\xi)\sqrt{h''(y_0;\xi)}}\exp\{-h(y_0;\xi)\} \quad (58)$$

Putting $\psi' = 0$, the next equation can be given from the equations (47) and (49).

$$\exp(\beta y_2) = 2\alpha y_2 \quad (59)$$

y_2 can be obtained from the equation (59). If we assume y_1 as y_0 taking minimum of $h(y_1;\xi)$, y_0 can be given as follows.

$$y_0 = \sqrt{\frac{1}{\alpha}\left(\xi-\frac{2\alpha y_2-1}{\beta}\right)} \quad (60)$$

$\psi(y_0;\xi)$ can be obtained from the equations (48) and (60).

$$\psi(y_0;\xi) = \frac{1}{2\beta^2}\log^2(2\alpha y_2-1)+\frac{1}{2\alpha}\left(\xi-\frac{2\alpha y_2-1}{\beta}\right) \quad (61)$$

If ξ is large enough, the expected upcrossing frequency per unit time interval can be obtained by putting $\psi(y_0;\xi) = h(y_0;\xi)$ as follows. y_2 can be obtained from the equation (59).

$$f_\xi{}^+ = \frac{\kappa \exp(\beta y_2)}{2\pi} \exp\{-\psi(y_0;\xi)\} = \frac{\kappa \exp(\beta y_2)}{2\pi}$$
$$\times \exp\left\{-\frac{1}{2\beta^2}\log^2(2\alpha y_2) - \frac{1}{2\alpha}\left(\xi - \frac{2\alpha y_2 - 1}{\beta}\right)\right\} \quad (62)$$

(b)Asymptotic solution in the drag force dominated case

If we assume $y_2 << 1/\beta$, the next equation can be obtained,

$$y_2 = \frac{1}{2\alpha - \beta} \quad (63)$$

Putting the equation (63) into the equation (60), we have

$$y_0 = \sqrt{\frac{1}{\alpha}(\xi - \frac{1}{2\alpha - \beta})} \quad (64)$$

Hence, the following equation can be given using the equation (61).

$$\psi(y_0;\xi) = \frac{1}{2\beta^2}\log^2\left(1 + \frac{\beta}{2\alpha - \beta}\right) + \frac{1}{2\alpha}\left(\xi - \frac{1}{2\alpha - \beta}\right) \quad (65)$$

Furthermore, if we assume $\frac{\beta}{\alpha} << 1$, we have

$$\psi(y_0;\xi) \cong \frac{\xi}{2\alpha} - \frac{1}{8\alpha^2} \quad (66)$$

When the drag force is dominant concerning the dynamic towline tension, the next equation can be valid.

$$f_\xi{}^+ = \frac{\kappa}{2\pi}\exp\left(-\frac{\xi}{2\alpha} + \frac{1}{8\alpha^2}\right) \quad (67)$$

(c) Asymptotic solution in the case considering drag force term without restoring force term

y_2 can be obtained by putting $\beta = 0$ in the equation (63). Then the next equation can be given substituting this into the equation (62).

$$f_\xi{}^+ = \frac{\kappa}{2\pi}\exp\left(-\frac{\xi}{2\alpha}\right) \quad (68)$$

EXPECTED LARGEST VALUE

The random variable can be defined as follows in order to obtain the probability distribution function of maximum value of $F(t)$ in the time interval T.

$$H = \alpha^{-1} max\{F(t); 0 \le t \le T\} \quad (69)$$

If it is assumed that the probability that the positive maxima exceeding a certain level η per unit time interval follows Poisson distribution, the probability having the expected number n can be given in the next equation.

$$p(n,T) = \frac{(\nu T)^n}{n!}\exp(-\nu T) \quad (70)$$

The probability that the random variable H does not exceed η is equivalent to the probability of maxima of $F(t)$ not exceeding a level $\alpha\eta$. Thus, the probability distribution function $P_H(\eta)$ can be expressed by this expected upcrossing frequency $f_{\alpha\eta}{}^+$.

$$\begin{aligned} P_H(\eta) &= Prob\{H \le \eta\} \\ &= Prob[max\{F(T); 0 \le t \le T\} \le \alpha\eta] \\ &= \exp\{-f_{\alpha\eta}{}^+ T\} \end{aligned} \quad (71)$$

The expected maxima of $F(t)$ in the time interval $0 \le t \le T$ can be given from the equation (69) as follows.

$$E[max\{F(t); 0 \le t \le T\}] = \alpha E[H] \quad (72)$$

Thus, the expected value of the random variable H is given by the next equation.

$$E[H] = \int_0^\infty \eta dP_H(\eta) \quad (73)$$

Putting $\xi = \alpha\eta$, we substitute the equation (62) into the equation (73) considering the equation (59). Hence,

$$\begin{aligned} P_H(\eta) &= \exp\left[-\frac{\kappa \exp(\beta y_2) T}{2\pi}\exp\left\{-\frac{y_2^2}{2}\right.\right. \\ &- \left.\left.\frac{1}{2}\left(\eta - \frac{\exp(\beta y_2) - 1}{\alpha\beta}\right)\right\}\right] \end{aligned} \quad (74)$$

Taking logarithm of both sides of the equation (74) two times, we have

$$\begin{aligned} \log[\{\log\{P_H(\eta))^{-1}\}] = \log\left[\frac{\kappa T}{2\pi}\exp\left\{\left(\beta y_2 - \frac{y_2{}^2}{2}\right)\right.\right. \\ - \left.\left.\frac{1}{2}\left(\eta - \frac{\exp(\beta y_2) - 1}{\alpha\beta}\right)\right\}\right] \end{aligned} \quad (75)$$

If we assume η which satisfies the next equation as η_0, P_H changes drastically near $\eta = \eta_0$ because the left hand side of the equation is double logarithm.

$$\begin{aligned} \log\left[\frac{\kappa T}{2\pi}\exp\left\{\left(\beta y_2 - \frac{y_2{}^2}{2}\right)\right.\right. \\ -\frac{1}{2}\left.\left.\left(\eta_0 - \frac{\exp(\beta y_2) - 1}{\alpha\beta}\right)\right\}\right] = 0 \end{aligned} \quad (76)$$

If we put

$$P_H(\eta) = \exp(-\zeta) \quad (77)$$

thus the next equation can be given from the equations (76) and (77).

$$\begin{aligned} \log\left[\frac{\kappa T}{2\pi}\exp\left\{\left(\beta y_2 - \frac{y_2{}^2}{2}\right)\right.\right. \\ -\frac{1}{2}\left.\left.\left(\eta - \frac{\exp(\beta y_2) - 1}{\alpha\beta}\right)\right\}\right] = \log\zeta \end{aligned} \quad (78)$$

Subtracting the equation (78) from the equation (75), we have

$$\eta = \eta_0 - 2\log\zeta \quad (79)$$

The next equation can be given from the equations (73) and (79).

$$E[H] = \eta_0 - 2\int_0^\infty \log\zeta\exp(-\zeta)d\zeta \quad (80)$$

Thus,

$$E[H] = \eta_0 + 2G \tag{81}$$

Where G stands for Euler constant. From the equation (75) we have

$$\eta_0 = 2\left\{\log\frac{\kappa T}{2\pi} + \left(\beta y_2 - \frac{y_2^2}{2}\right)\right\} + \frac{\exp(\beta y_2) - 1}{\alpha\beta} \tag{82}$$

The expected largest value can be obtained from the equations (72), (81) and (82).

$$E\left[max\left\{F(t); 0 \le t \le T\right\}\right] = \left[2\left\{\log\frac{\kappa T}{2\pi} + \left(\beta y_2 - \frac{{y_2}^2}{2}\right)\right\} + \frac{\exp(\beta y_2) - 1}{\alpha\beta} + 2G\right]\alpha \tag{83}$$

(a)Case for the drag force dominated

If we assume $\frac{\alpha}{\beta} << 1$, namely the drag force component of the dynamic towline tension is dominant, the next equation can be given.

$$E\left[max\left\{F(t); 0 \le t \le T\right\}\right] = 2\alpha\left\{\log\frac{\kappa T}{2\pi} + \frac{1}{8\alpha^2} + G\right\} \tag{84}$$

(b)Case considering only drag force

The next equation can be given using the equation (68).

$$E\left[max\left\{F(t); 0 \le t \le T\right\}\right] = 2\alpha\left\{\log\frac{\kappa T}{2\pi} + G\right\} \tag{85}$$

CAHRACTERISITICS OF PROBABILITY DENSITY FUNCTION OF DYNAMIC TOWLINE TENSION

Numerical Calculation of Probability Density Function

The method of calculating the equation (19) in order to obtain the probability density function $p(\hat{f})$ of the dynamic towline tension $\hat{f}$ is shown in this section. We divide the integrated function excluding the constant into two terms such as f_1 and f_2 where both are the function of z.

$$f_1(z) = \frac{1}{\sqrt{|z|}}\exp\left(-\frac{|z|}{2\alpha}\right) \tag{86}$$

$$f_2(z) = \frac{1}{a\sigma_x(\hat{f} - z) + 1}\exp\left[-\frac{\log^2\left\{a\sigma_x(\hat{f} - z) + 1\right\}}{2a^2\sigma_x^2}\right] \tag{87}$$

$f_1(z)$ has a singularity at $z = 0$. Concerning $f_2(z)$ the denominator is $a\sigma_x(\hat{f} - z) + 1 > 0$, then there exists no singularity in the range of $z < \hat{f} + \frac{1}{a\sigma_x}$. Thus we need to divide into cases according to the range of $\hat{f}$ from the view point that the range of integral includes the singularity namely $z = 0$ or not. We consider the following two cases using ϵ as an infinitesimal.

(1) $\hat{f} \ge -\frac{1}{a\sigma_x}$ ($z = 0$ is included in the integral range)

$$\int_{-\infty}^{\infty} dz = \int_{-\infty}^{-\epsilon} dz + \int_{-\epsilon}^{0} dz + \int_{0}^{\epsilon} dz + \int_{\epsilon}^{\hat{f}+\frac{1}{a\sigma_x}} dz \tag{88}$$

(2) $\hat{f} < -\frac{1}{a\sigma_x}$ ($z = 0$ is excluded in the integral range)

$$\int_{-\infty}^{\infty} dz = \int_{-\infty}^{\hat{f}+\frac{1}{a\sigma_x}} dz \tag{89}$$

Now we consider the behavior of the integrated term near the singularity.

(a)Integral of $f_1(z)$

The integrated term $f_1(z)$ in the equation (19) has the singularity and becomes infinity near $z = 0$. Thus the integral $f_1(z)$ in the range $0 < z < \epsilon$ is as follows.

$$\int_0^{\epsilon} f_1(z)dz = \int_0^{\epsilon}\frac{1}{\sqrt{z}}\exp\left(-\frac{z}{2\alpha}\right)dz \tag{90}$$

Letting $\sqrt{z} = t$ by the variable transformation, we have

$$\int_0^{\epsilon} f_1(z)dz = 2\int_0^{\sqrt{\epsilon}}\exp\left(-\frac{t^2}{2\alpha}\right)dt = 2\sqrt{2\pi\alpha}erf\left(\sqrt{\frac{\epsilon}{\alpha}}\right) \tag{91}$$

where $erf\left(\sqrt{\frac{\epsilon}{\alpha}}\right)$ stands for the error function. Therefore, there exists no inconvenience concerning the numerical calculation of the integral of the equation (88).

(b)Integral of $f_2(z)$

Concerning the integrated term in the equation (19), we consider the limit value of $f_2(z)$ in case the denominator approaches 0. Namely, we put

$$X = a\sigma_x(\hat{f} - z) + 1 \tag{92}$$

Assuming $X = 0$, we have

$$z = \hat{f} + \frac{1}{a\sigma_x} \tag{93}$$

We consider the behavior of the function when the denominator becomes 0. We focus on the following $g_3(X)$ excluding the constant terms in $f_2(z)$. Putting

$$g_3(X) = \frac{1}{X}\exp(-\log^2 X)\,(X > 0) \tag{94}$$

we further put $\log X = y$, then

$$X = \exp(y) \tag{95}$$

Thus, if X approaches 0, y approaches $-\infty$. Therefore, the limit value of $g_3(X)$ in case X approaches 0 can be obtained using y.

$$\begin{aligned}\lim_{y\to-\infty} g_3(y) &= \lim_{y\to-\infty}\frac{1}{\exp(y)}\exp\left\{-\log^2\exp(y)\right\} \\ &= \lim_{y\to-\infty}\exp(-y^2 - y) = 0 \end{aligned}\tag{96}$$

The integrated term $f_2(z)$ approaches 0 and never becomes singular when z value is obtained by the equation (93). Thus we have no numerical problem in the procedure of calculation.

Numerical Calculation Example of Probability Density Function of Dynamic Towline Tension

When we calculate the probability density function $p(\hat{f})$ with respect to the random variable $\hat{f}$ in the equation(19), the behavior of the integrated function $p(z)$ with respect to z with fixed value of $\hat{f}$ is shown in Fig.1. The figure shows the 5 different cases of $\hat{f}$ such as -5.0, -2.0, 0, 2.0 and 5.0 where $\alpha = 1.0$, $a\sigma_x = 2.0$. $p(z)$ shows a rapid change when z approaches $\hat{f} + \frac{1}{a\sigma_x}$ or 0. If $z = \hat{f} + \frac{1}{a\sigma_x}$, $p(z) = 0$. Fig.2 shows the probability density function $p(\hat{f})$ with 3 different values of $a\sigma_x$ such as 0.5, 1.0 and 2.0 in case of fixed value $\alpha = 1.0$. If $a\sigma_x$ becomes larger, $p(\hat{f})$ also becomes larger around $\hat{f} = 0$. It also becomes larger when $\hat{f}$ takes a large value. This tendency generally appears strongly with the degree of the non-linearity.

NMERICAL SIMULATION OF DYNAMIC TOWLINE TENSION

Probability Density Function of Dynamic Towline Tension

The calculation of the dynamic towline tension has been made on the condition that the irregular forced oscillation is given on one end of the towline with fixed another end. The towline has been assumed to be made of the wire rope (weight per length in water is 15.04 g/m). Its length is 26.396 m with 2 mm in diameter and the horizontal length between both ends is 26.32 m. The spectrum of Pierson-Moskowitz type has been used as a forced irregular oscillation. The wave data generated for 330 sec with the mean wave period of 3 to 6 sec at every 1 sec have been used as an irregular forced oscillation at the end. The constants of both a and c have to be obtained in order to calculate the dynamic towline tension f as given in equation (1). The constant a has been obtained from the static towline tension curve using an exponential curve approximation. The constant c has been obtained from the simplified method to estimate the dynamic towline tension. This method has been deduced from the simplified method of Shimada et al. (Shimada,1980) $C_{Dx\xi}$ (Hara,1996) has been resplaced by the constant c in the equation (1) considering the dynamic towline tension in the horizontal direction on the assumption of the taut condition of the towline.

Furthermore, the forced oscillation test of 3 different conditions from taut to slack towline has been done comparing with the theoretical calculation. The wire rope towline used in the experiment is 24.252 m long with 2 mm diameter. (weight in water per unit length : 14.17 g/m) The surge forced oscillation mode has been used as well as the heave mode. The results have been shown in the Appendix. Once the constant terms have been decided as for the horizontal displacement, the time series of the towline tension can be obtained. Fig.3 shows the time series of the towline tension obtained by the simplified method compared with the lumped mass method (Nakajima,1980). The forced irregular oscillation with the mean wave period 4 sec has been used. The results of the both methods agreed quite well.

The theoretical probability density function of the dynamic towline tension has been calculated using the standard deviation of the displacement and its velocity which have been obtained by the spectrum of the displacement with mean wave period 4 sec. Fig.4 shows the comparison between the theoretical calculation and numerical simulation results. The theoretical calculation agrees well with the numerical simulation results.

Fig.5 shows an example of the theoretical results of the probability density function of the dynamic towline tension considering the inertia term effect. If we consider only linear restoring force term besides the drag force term, the value in the range of large $\hat{f}$ becomes smaller than the case including only non-linear restoring force term. As for the case considering inertia force term and linear restoring force term, the probability density function becomes smaller in the range of large value of $\hat{f}$ compared with the case considering only linear restoring force term. The cancellation effect of the phase difference between inertia force term and restoring force term causes this result. The probability density function becomes generally smaller in the range of large $\hat{f}$ when it comes to taking account of the inertia force term. The theoretical model which includes the drag force term and the non-linear restoring force term shows the conservative estimation of the dynamic towline tension.

Calculation of Asymptotic Solution of Expected Upcrossing Frequency and Expected Largest Value

The comparison of both the expected upcrossing frequency and expected largest value between the theoretical model using both the non-linear and linear restoring force term has been made. As for the same case as previously mentioned, Fig.6 shows the numerical result of the mean upcrossing frequency concerning the dynamic towline tension using the equations (62) and (67). Fig.7 also shows the expected largest value using the equation of (83) and (84). There is a very small difference between them in the case of the small mean wave period of irregular forced oscillation. If the mean wave period becomes larger, the effect of the non-linear restoring force term will appear. This case can be regarded as the towing in well developed sea area of wind and waves where the non-linear restoring force term effect becomes larger. Hence, the investigation of the effect of the inertia force term has been shown as follows.

(a)Case considering inertia force term and drag force term

The dynamic towline tension f considering the inertia force term and drag force term can be shown in the next equation.

$$f = m\ddot{x} + c\dot{x}|\dot{x}| \tag{97}$$

where m stands for the constant proportional to the acceleration of the towline displacement. The expected mean upcrossing frequency can be obtained using $m_{x\xi}$ shown in the reference (Hara,1996).

$$f_\xi^+ = \frac{\kappa}{2\pi}\exp\left(-\frac{\xi}{2\alpha} + \frac{\alpha_1^2}{8\alpha^2}\right) \tag{98}$$

Where

$$\kappa_1 = \frac{\sigma_{\ddot{x}}}{\sigma_{\dot{x}}},\ \alpha_1 = \frac{m\sigma_{\ddot{x}}}{a\sigma_x} \tag{99}$$

The expected largest value can also be shown in the next equation.

$$E\left[max\left\{F(t): 0 \le t \le T\right\}\right] = 2\alpha\left\{\log\frac{\kappa_1 T}{2\pi} + \frac{\alpha_1^2}{8\alpha^2} + G\right\} \tag{100}$$

(b)Case considering inertia force term, drag force term and linear restoring force term

The dynamic towline tension f can be shown in the next equation.

$$f = m\ddot{x} + c\dot{x}|\dot{x}| + ax \tag{101}$$

If the stochastic process of the towline displacement is narrow-banded, the next relation between $\ddot{x}$ and x can be valid (Nakamura,1982).

$$\ddot{x} = \kappa_2 r x \tag{102}$$

Where

$$\kappa_2 = \frac{\sigma_{\ddot{x}}}{\sigma_x},\ r = \frac{\sigma_{x\ddot{x}}}{\sigma_x \sigma_{\ddot{x}}} \tag{103}$$

r is the correlation coefficient between $\ddot{x}$ and x. Thus, f can be shown using the equations (102) and (103).

$$f = c\dot{x}|\dot{x}| + (a - m\kappa^2)x \tag{104}$$

Where $\kappa = \frac{\sigma_{\dot{x}}}{\sigma_x}$. The equations of (98) and (100) correspond to the next equation.

$$f_\xi{}^+ = \frac{\kappa}{2\pi}\exp\left(-\frac{\xi}{2\alpha} + \frac{c_1^2}{8\alpha^2}\right) \tag{105}$$

$$E\left[max\left\{F(t); 0 \le t \le T\right\}\right] = 2\alpha\left\{\log\frac{\kappa T}{2\pi} + \frac{c_1^2}{8\alpha^2} + G\right\} \tag{106}$$

Where $c_1 = 1 - \frac{m}{a}\kappa^2$.

Fig.8 shows the comparison of the expected upcrossing frequency. Using the logarithmic scale as the vertical axis, all the equations (62), (67), (68), (98) and (105) show the straight lines with the same gradient of $-1/2\alpha$. Therefore, when the drag force term is dominant, the ratio of drag force term and restoring force term becomes larger and thus the expected upcrossing frequency also becomes larger. Fig.9 also shows the comparison of the expected largest value. Taking the logarithmic scale as the horizontal axis, the equations (83), 84), (100) and (106) show the straight lines with the same gradient of 2α. If the drag force term is dominant, the expected largest value thus becomes larger being proportional to α.

CONCLUSION

The following conclusions have been obtained from the estimation of the dynamic towline tension by the probabilistic method.

(1) The theoretical model for the dynamic towline tension considering the drag force term and non-linear restoring force term has been proposed. The equations for the probability density function, expected upcrossing frequency and expected largest value concerning the dynamic towline tension have been shown.

(2) The constants for the drag force term and non-linear restoring force term have been obtained in the theoretical model expressing the dynamic towline tension. Then its numerical simulation results have been compared with the results of lumped mass method. The theoretical calculation of the probability density function agrees well with the instantaneous distribution of the dynamic towline tension by the numerical simulation results.

(3) The effect of the drag force term, restoring force term and inertia force term of which the dynamic towline tension are consisted has been investigated. It has been confirmed that the proposed theoretical model shows a conservative estimate for the expected largest value concerning the dynamic towline tension.

The author expresses his sincere thanks for professor Shigeru Naito at university of Osaka who discussed the paper and gave many valuable suggestions. He would also like to appreciate senior researcher Kenji Yamakawa at Ship Research Institute for his help with the experiment and discussion of the paper. Professor Hyunkyoung Shin and Mr. Jonbae Kim at university of Ulsan helped with the experiment. The author thanks for their kind help.

REFERENCES

Ando, S, and Kato, S (1983). "On the Static and Dynamic Tensions on the Mooring Chains", Transactions of the West-Japan the Society of Naval Architects, No.66, pp.191-207 (in Japanese)

Kato, S (1987). "Kinoshita T., Hineno M.; "Probabilistic Treatment of the Non-linear Response on the Ocean Engineering," Bulletin of the Society of Naval Architects, No.699, pp.560-592 (in Japanese)

Borgman, LE(1969). "Ocean Wave Simulation for Engineering Design," Journal of Waterways and Harbors Division, ASCE, Vol.95, No.WW4, pp.557-583

Vinje, T(1980). "Statistical Distributions of Hydrodynamic Forces on Objects in Current and Waves," Norwegian Maritime Research, Vol.8 No.2,1980,pp.20-26

Naess, A(1983). "Prediction of Extremes of Morison-Type Loading - An Example of a General Method," Ocean Engineering, Vol.10, No.5, pp.313-324

Shimada, K, Kobayashi, M, and Hineno, M(1980). "Dynamic Effect on Tension of Mooring Lines," Transactions of the West-Japan the Society of Naval Architects, Vol.60, pp.113-127

Hara, S(1996). "Prediction of Dynamic Towline Tension During Towing by the Probabilistic Method," Journal of the Kansai Society of Naval Architects, Japan, No.226, pp.135-151 (in Japanese)

Nakajima, T(1980). "On the dynamic Analysis of the Various Type of Mooring Lines by the Lumped Mass Model," PhD thesis, University of Tokyo

Nakamura, S, Naito, S, and Koganeya, K(1982). "On Absorbing Wave Power in Irregular Waves," Journal of the Kansai Society of Naval Architects, Japan, No.185, pp.79-86 (in Japanese)

APPENDIX

COMPARISON OF SIMPLIFIED METHOD WITH EXPERIMENTAL RESULTS

The wire rope (weight per unit length in water :14.17 g/m) of which length is 24.252 m with 2 mm diameter has been used in the experiment. The towline tension at one oscillated end has been measured on the condition of being fixed at the other end. The experiment has been carried out in the towing tank of Ship Research Institute (length 50m×breadth 8m × depth 4.5 m). The surge and heave oscillating experiment have been done on the towline condition of 3 different degree of tautness. Namely, the horizontal distance between both ends are 23.0 m, 23.6 m and 24.05 m concerning surge oscillation mode and 24.201 m concerning heave. The oscillation amplitude is 10 mm, 20 mm and 30 mm and its frequency is between 0.1 Hz and 2.3 Hz.

Fig.A.1 shows one of the examples of the surge oscillation experimental results compared with the theoretical calculation using proposed simplified method and lumped mass method. This figure shows the case of most taut condition. The drag coefficients in the normal and tangential direction used in the calculation are 1.5 and 0.08 respectively. The added mass coefficients in each direction are 1.18 and 0.17. The 0 value in the figure shows the static towline tension and the maximum and minimum dynamic towline tension from the static tension have been shown as the experimental results. The maximum towline tension namely positive dynamic tension is mainly discussed in this section. Both theoretical calculation agree well with the experimental results concerning the positive dynamic tension. Fig.A.2 shows the comparison between the theoretical calculation and the experimental results concerning the heave oscillation experiment. As for the heave oscillation, the results of the simplified method agree well with the experimental results. The difference between 2 methods is small. It has been turned out that the proposed simplified method agrees well with the experimental results and there is no big difference compared with the lumped mass method which needs more computer time.

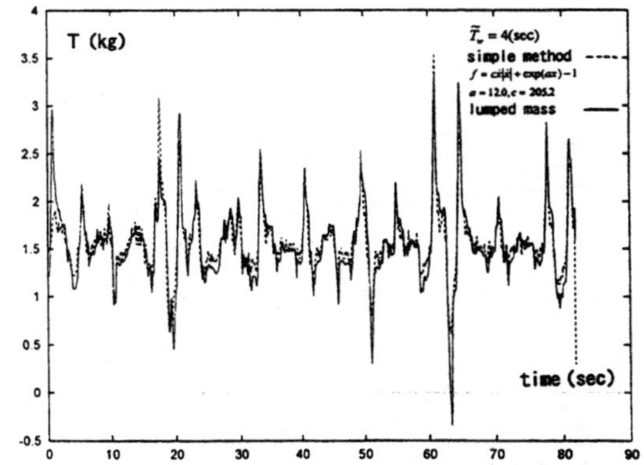

Fig. 3 Comparison of dynamic towline tension between simplified method and lumped mass method

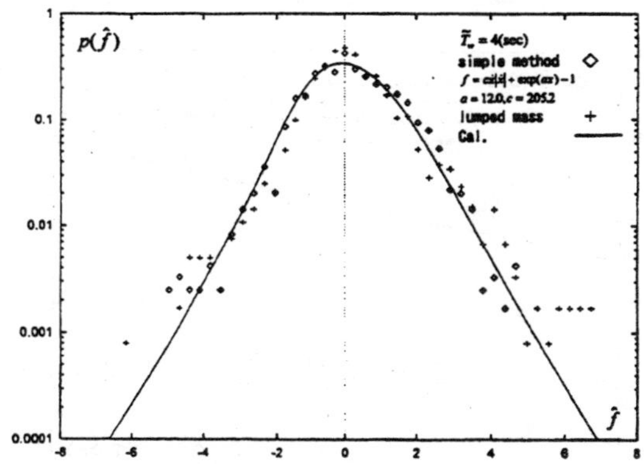

Fig. 4 Comparison of probability density function concerning dynamic towline tension between theoretical calculation and nu-

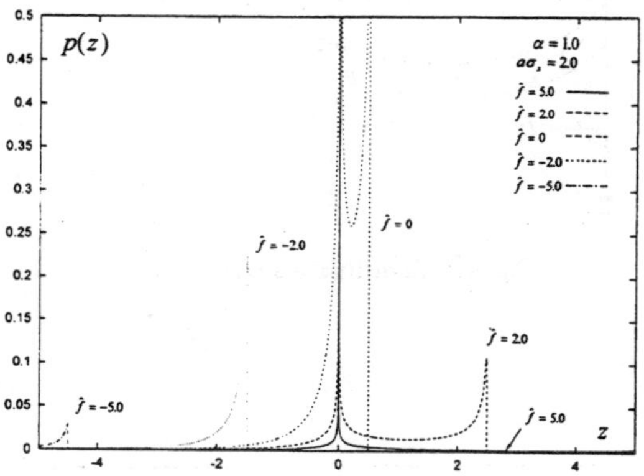

Fig. 1 Behavior of integrated function p(z)

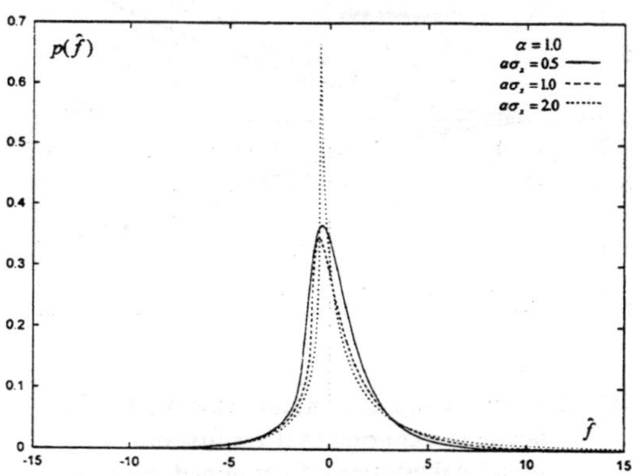

Fig. 2 Probability density function of dynamic towline tension

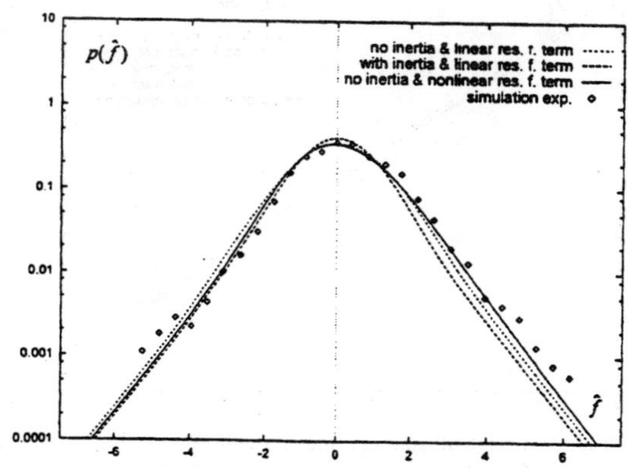

Fig. 5 Effect of inertia force on the dynamic towline tension

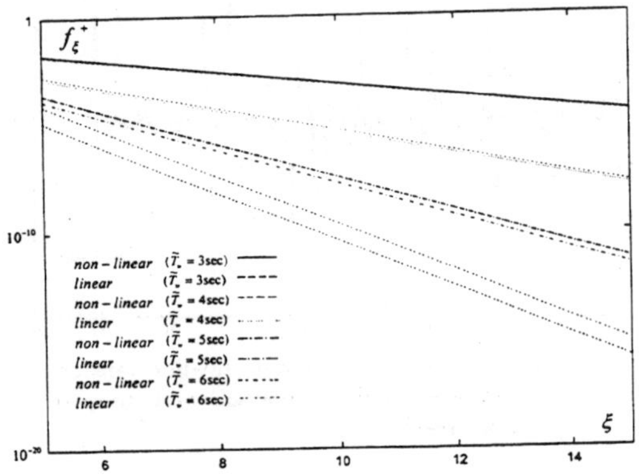

Fig. 6 Mean upcrossing frequency of dynamic towline tension

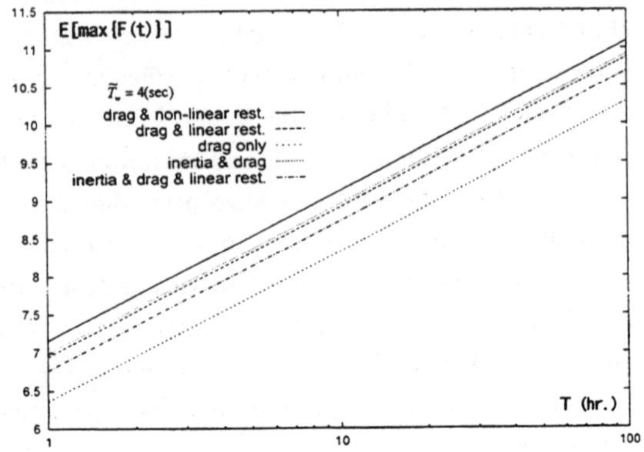

Fig. 9 Effect of inertia force on expected largest force

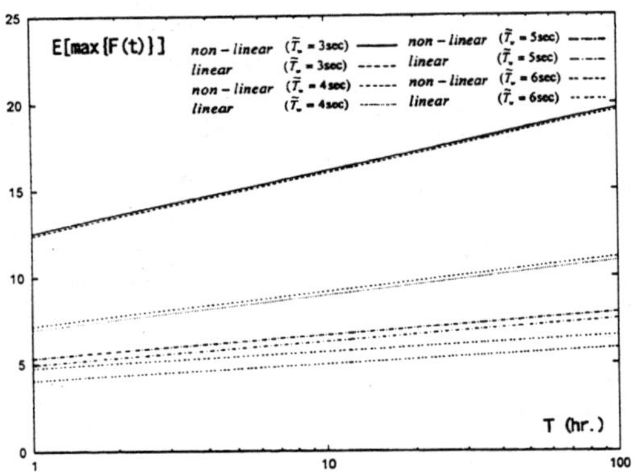

Fig. 7 Expected largest force of dynamic towline tension

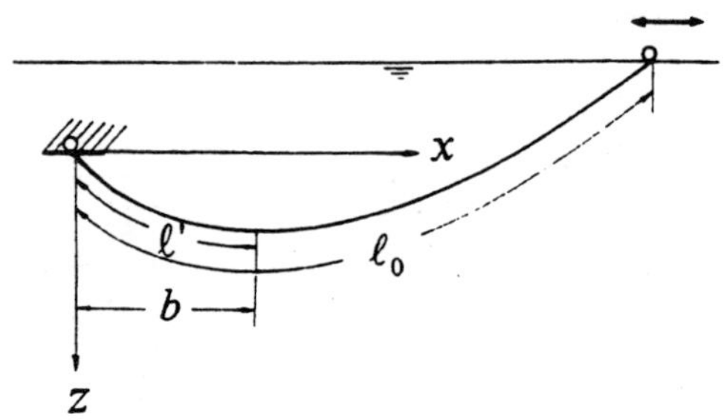

Fig.A.1 Coordinate system

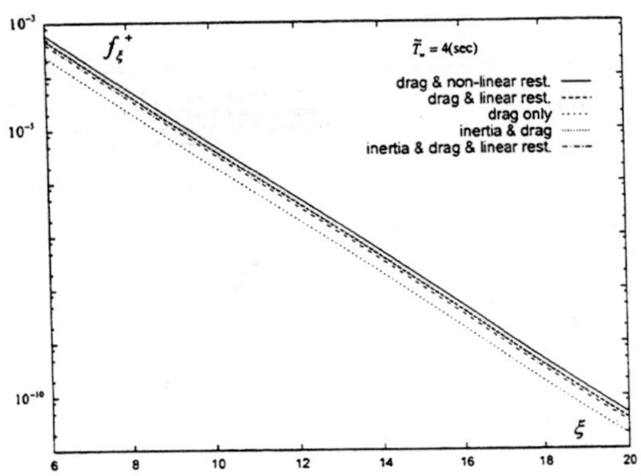

Fig. 8 Effect of inertia force on mean upcrossing frequency

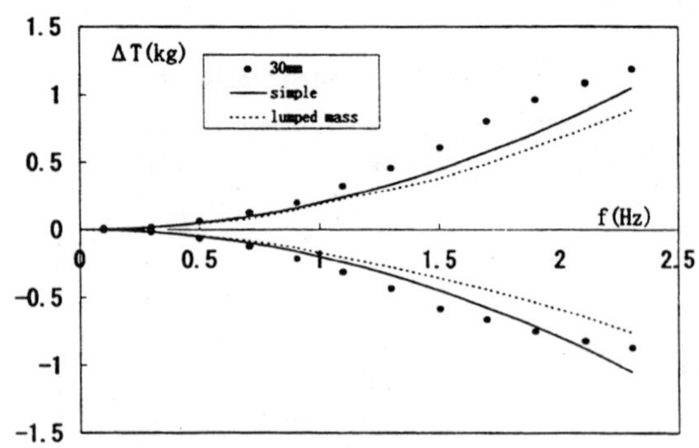

Fig.A.2 Comparison of dynamic towline tension between experimental results and theoretical calculation of simplified method and lumped mass method (surge forced oscillation x=24.201m)

Proceedings of the Seventh (1997) International Offshore and Polar Engineering Conference
Honolulu, USA, May 25-30, 1997

ISBN 1-880653-28-1 (Set); ISBN 1-880653-31-1 (Vol. III)

Noise-Induced Transitions in Moored Ocean Structural Responses

H. Lin and S.C.S. Yim
Oregon State University
Corvallis, OR, USA

ABSTRACT

Noise-induced transitions in nonlinear responses of a submerged moored structure are investigated from both global stability and probability points of view. The analysis procedure includes a stochastic Melnikov process to identify potential chaotic domains, and the Fokker-Planck equation to investigate and demonstrate response probabilistic characteristics. With the presence of weak perturbations, highly nonlinear phenomena including coexistence of attractors and chaos are found to exist. It is indicated that the presence of noise expands the potential chaotic domain in the parameter space and causes transitions between coexisting responses, which may not occur otherwise.

KEY WORDS: Moored ocean structure, noise, transition, Melnikov approach, Fokker-Planck equation, global stability, probability density function

INTRODUCTION

Complex nonlinear responses of moored ocean systems under deterministic settings has been elucidated analytically and numerically via period-doubling cascades and the (co)existence of harmonic, subharmonic and ultraharmonic responses (Gottlieb and Yim, 1992). Near resonances, coexisting (and competing) nonlinear moored structural responses have been found with different initial conditions, and the cascade of local bifurcations (and sudden explosions in some cases) often leads to chaotic responses.

An experimental investigation (Yim et al, 1993) examining nonlinear response behavior of a single-degree-of-freedom (SDOF) moored structural system identifies the existence of subharmonic and ultraharmonic as well as period-doubling bifurcations. Despite of good agreement between analytical predictions and numerical results, there are experimental observations which can not be explained using conventional deterministic analysis procedures. Figure 1 shows the time history of a sample experimental structural response subjected to "deterministic" monochromatic wave excitation. The response resides in a harmonic mode for a relatively long duration (about 120 seconds, Fig.1a) and then transitions to a subharmonic steady state (Figs.1b-c). The existence of unexpected transition indicates the presence of random noise caused by imperfect wave conditions (e.g., diffraction, reflection and re-reflection), which may lead the response trajectory to visit other coexisting attractors (Lin and Yim, 1995). Thus, this phenomenon and other transitions in responses are to be examined and interpreted in this study by taking into account the presence of inevitable perturbations in the wave excitation. In practice, the ocean field environment including wind, waves and current often contain a significant random component, which further necessitates stochastic analysis on randomly perturbed nonlinear responses.

In this paper, noisy moored structural responses are analyzed from a stochastic perspective to provide guidelines of interpreting the experimental observations and demonstrate pertinent probabilistic properties of response behavior. A generalized stochastic Melnikov process taking into account the presence of noise is developed. A mean-square criterion is derived accordingly to identify the noise effects on a possible chaotic domain. Random-noise effects on the nonlinear moored structural responses (regular and chaotic) and their interactive relationship are examined through transient and steady-state PDFs, which are obtained by solving the associated Fokker-Planck equation using path-integral solution (Markov process approach). The unexpected transition in response observed in the experiment is explained from a probabilistic perspective.

GOVERNING EQUATIONS

Assuming surge motion as the dominant system response and is uncoupled with other degrees of freedom, the multi-point moored structural system considered can be formulated as a single-degree-of-freedom (SDOF, in surge) submerged rigid body, hydrodynamically damped and excited nonlinear oscillator (Fig.2). The associated nondimensionalized governing equations are given by

$$\dot{x}_1 = x_2 \tag{1a}$$
$$\dot{x}_2 = -R(x_1) - \gamma x_2 + F_D(x_2) + F_I(\dot{x}_1, x_2) + \xi(t)$$

where x_1 and x_2 are nondimensionalized surge displacement and velocity (Gottlieb and Yim, 1992), respectively, and

$$R = \alpha\left[x_1 - \tau\left(\frac{l_1 + l_2}{l_1 l_2}x_1 - \beta\frac{l_1 - l_2}{l_1 l_2}\right)\right] \tag{1b}$$
$$l_{1,2} = [1 + (\beta \pm x_1)^2]^{1/2}$$

$$F_D = \mu\delta(u - x_2)\,|u - x_2| \tag{1c}$$

$$F_I = \mu\left[\frac{\partial u}{\partial t} + (u - x_2)\frac{\partial u}{\partial x_1}\right] \tag{1d}$$

with α, β, and τ denoting the parameters for restoring force R. δ and μ represent the parameters for Morison drag force; γ is the structural damping coefficient. The stochastic excitation component is denoted by a zero-mean, delta-correlated white noise, $\xi(t)$,

$$\langle \xi(t) \rangle = 0 \tag{1e}$$
$$\langle \xi(t)\xi(t') \rangle = \kappa\delta(t - t')$$

where κ is the noise intensity.

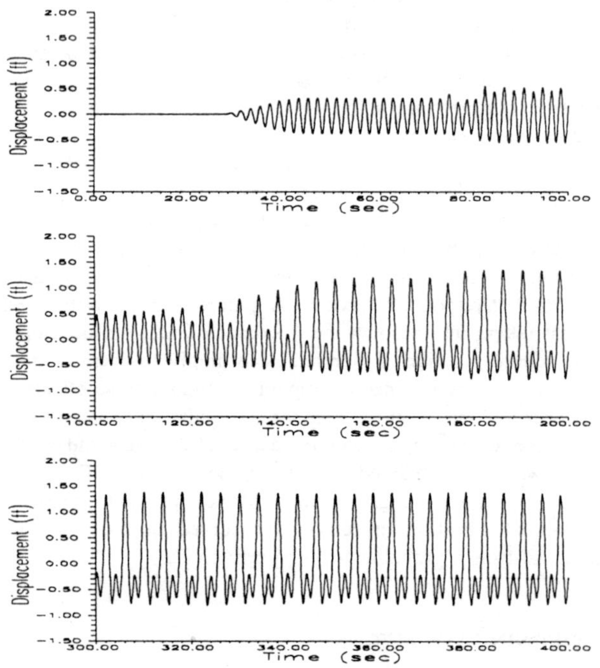

Fig. 1 Transition from harmonic to subharmonic in "deterministic" moored structural response (Test D2): a) transient harmonic, b) transition from harmonic to subharmonic, and c) steady-state subharmonic response

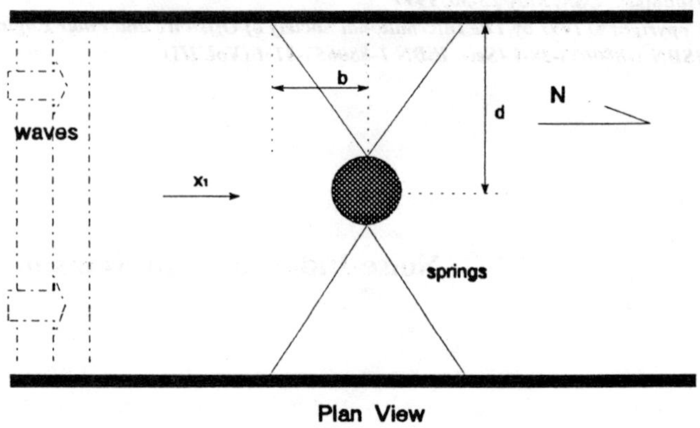

Fig. 2 Definition sketch of a submerged, hydrodynamically damped and excited nonlinear ocean structural system

METHODS OF ANALYSIS

Methods used in this study to analyze the noisy nonlinear response include stochastic Melnikov process and Markov approach. The stochastic Melnikov process is applied to identify possible chaotic domains in the parameter space. Transient and steady-state joint probability density functions (PDFs) resulted from the Markov approach are used to depict global information about the response behavior.

Melnikov Approach

Homoclinic connections corresponding to Eq. 1a have been identified to exist near the primary resonance in the associated averaged system applying nonlinear transformations and Bendixson criterion (Gottlieb and Yim, 1993). For small structural damping, the system near primary resonance contains homoclinic loops defined by the stable and unstable manifolds of the saddle. The effects of external noise on the averaged system are incorporated by using Rice noise representation and adopting the harmonics which are in tune with the dominant frequency. The pair of homoclinic orbits $(q_\pm{}^o(\gamma)) = (J_\pm(\theta), \Phi_\pm(\theta))$ is obtained by integrating the associated Hamiltonian system. $J_\pm(\theta)$ and Φ are functions of θ and θ_o, and parameters α_3^*, Ω^* and f are closely related to stiffness, drag coefficient, dominant frequency, and noise intensity. Taking into account the presence of random perturbations, the Melnikov function is given by (Lin and Yim, 1995)

$$M(\theta_o) = \int_{-\infty}^{\infty} F(q_\pm^0(\theta) \wedge G(q_\pm^0(\theta), \theta + \theta_o)d\theta = \int_{-\infty}^{\infty} (F_1 G_2 - F_2 G_1)\, d\theta \tag{2}$$

where F_1 and F_2 denote the components in a 2-D Hamiltonian vector field, and G_1 and G_2 represent the perturbations. When $M(\theta_o) = 0$ in Eq. 2, chaotic response may exist. The criterion based on the Melnikov function (Eq. 2) provides a necessary condition for the existence of chaotic response (Yim and Lin, 1992), and the noise effects should be represented in a mean-square sense

$$\gamma^{*}\left\langle\left(\int m_1^d d\theta\right)^2\right\rangle=\left\langle\left(\int m_2^d d\theta\right)^2\right\rangle+\sigma_{M_1^r}^2+\sigma_{M_2^r}^2+2\left\langle\!\left\langle\left(\int m_2^d d\theta\right)\left(A^{*}\int m_1^r d\theta\right)\right\rangle\!\right\rangle$$
$$+2\left\langle\!\left\langle\left(\int m_2^d d\theta\right)\left(A^{*}\int m_2^r d\theta\right)\right\rangle\!\right\rangle+2\left\langle\!\left\langle\left(\int m_1^r d\theta\right)\left(\int m_2^d d\theta\right)\right\rangle\!\right\rangle \tag{3}$$

where m_1^d and m_2^d/m_1^r and m_2^r are Melnikov components due to deterministic/random perturbations. Because m_1^r and m_2^r are uncorrelated, Eq. 3 can be simplified as

$$\gamma^{*}\left(\int m_1^d d\theta\right)^2 = \left(\int m_2^d d\theta\right)^2+\sigma_1^2+\sigma_2^2 \tag{4}$$

Where σ_1^2 and σ_2^2 are the variances associated to m_1^r and m_2^r, respectively. The positive terms, σ_1^2 and σ_2^2, indicate that the presence of noise lowers the threshold for chaos and enlarges the possible chaotic domain (mean-square sense) in the parameter space.

Fokker-Planck Equation and Path Integral Solution

The FPE corresponding to Eq. 1a is given by

$$\frac{\partial P(X,t)}{\partial t} = -\frac{\partial}{\partial x_1}\left[x_2 P(X,t)\right]-\frac{\partial}{\partial x_2}\Big[\left(-R(x_1)-\gamma x_2+F_D(x_2,t)+F_I(t)\right)$$
$$\times P(X,t)\Big]+\frac{\kappa}{2}\frac{\partial^2 P(X,t)}{\partial x_2^2} \tag{5}$$

with $X=[x_1\ x_2]^T$ and the short-time propagator $\Gamma(X',X,t;\ \tau)$ is given by

$$\Gamma(X',X,t;dt) = (2\pi dt)^{-\frac{1}{2}}\kappa^{-\frac{1}{2}}\exp\left[-R(x_1)-\gamma x_2+F_D+F_I-\frac{x_2'-x_2}{dt}\right]$$
$$\times\delta\left[x_2-\frac{x_1'-x_1}{dt}\right] \tag{6}$$

where X' and X represent the post-state and the prior-state, respectively. The PDF at the desired time can be obtained by applying the short-time propagator iteratively (Risken, 1984)

$$P(X,t) = \lim_{\substack{dt\to 0\\ N\to\infty}}\prod_{i=0}^{N-1}\int..\int \exp\Big[-dt\sum_{j=0}^{N-1}\Gamma(X_{j+1},X_j,t_j;dt)\,P(X_o,t_o)\Big]dX_i \tag{7}$$

NOISY NONLINEAR RESPONSES

Coexisting Attractors

Coexisting periodic responses in the moored system may be obtained near resonances with various initial conditions. Figure 3a shows two coexisting periodic responses (small- and large-amplitude) in the phase plane with initial conditions at (0,0) and (1,-2) respectively, where symbol "*" denotes the corresponding Poincaré points. Coexisting response attractors can be demonstrated in the probability domain via steady-state PDFs. Steady-state PDFs with various noise intensities are shown in Figs.3b-d.

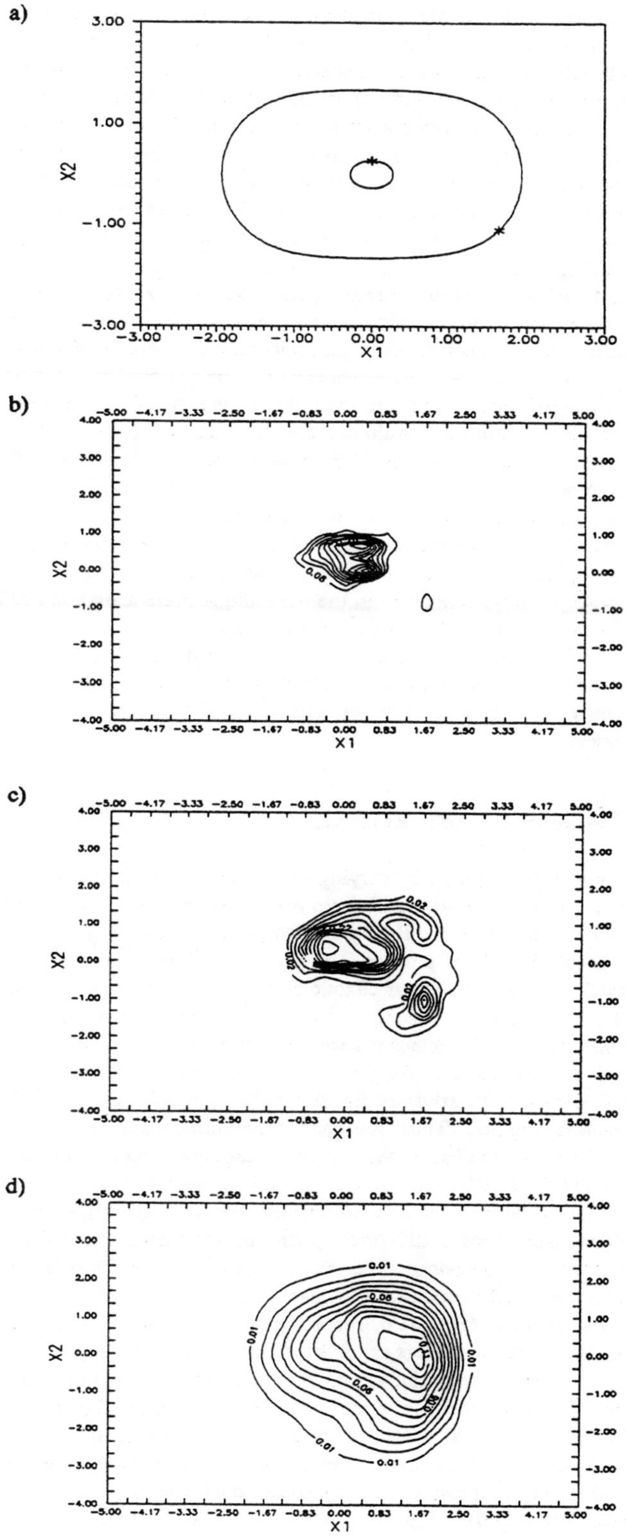

Fig. 3 Coexisting periodic responses: a) deterministic phase trajectories in phase plane; corresponding PDFs with increasing noise intensity: κ= b) 0.003, c) 0.007, and d) 0.4

The noise effects on the coexisting nonlinear responses are examined in Figs.3b-d. When the intensity of the noise present is low ($\kappa=0.003$), the PDF is concentrated at the domain of small-amplitude response (Fig.3b), indicating its stronger attracting strength relative to the coexisting large response attractor. Thus, in this case, the moored structural system oscillator mainly (most of the time) with small motions will have a small probability of excursion to larger amplitude motions (i.e. rarely visits the large amplitude domain). When the noise intensity is increased ($\kappa=0.007$), the domains of attraction of the coexisting responses begin to bridge as shown in Fig.3c. Characteristics of both attractors (small and large amplitudes) may exhibit in the moored structural response behavior. When the noise intensity is strong ($\kappa=0.4$), the attractors further emerge into a single attraction domain (Fig.3d), in which no obvious periodic response can be distinguished. The variation the boundary of the emerged attraction domain is stabilized, and the smooth curve and the stationary boundary of the PDF implies strong randomness in the system behavior.

Numerical results presented here demonstrate the effects of noise intensity on coexisting nonlinear moored structural responses, which may explain the transition in the experimental response (Fig.1). The presence of inevitable noise in the wave flume plays a bridging role between coexisting attractors as demonstrated (Fig.3b). The system was initially excited in a harmonic attraction domain. Due to the presence of very weak noise, the structural response transitions to the stronger subharmonic attractor and stays after a relatively long transient.

Noise-Induced Chaotic Response

It is well known that, by varying system parameters, a deterministic nonlinear response may shift from one attractor to another, even to chaos (Moon, 1987). It is indicated,based on criterion from the stochastic Melnikov process, the presence of random noise may expedite the occurrence of chaotic response in the parameter space. Effects on response transitions by these two parameters (excitation amplitude and noise intensity) are demonstrated and discussed here.

It is shown that with the presence of weak noise, the steady-state PDF starting with arbitrary initial conditions reflects the coexisting attractors (Fig.3c). Thus, precision of the initial conditions becomes less significant to observe the coexisting responses, and the associated PDF provides global information about the system behavior.

As an example, for the parameters specified in Fig.4 in the deterministic moored structural system, by increasing the excitation amplitude, the response shifts from a period-2 subharmonics, to a higher order subharmonics, to a chaotic attractor, and then to another chaotic attractor (Figs.4a-d, respectively). In a noisy environment, with the same parameters as its deterministic counterpart specified in Fig.4a, a similar transition among response attractors can be observed when the noise intensity is increased (Fig.5). Figure 5a (which is identical to Fig.4a) shows a deterministic period-2 subharmonic response ($\sigma^2 = 0.0$), and under the same system parameters, the moored structural response becomes chaotic with the presence of weak noise ($\sigma^2 = 0.01^2$, Fig.5b). When the noise intensity is further increased ($\sigma^2 = 0.02^2$), the response shifts to another chaotic attractor (Fig.5c). When the noise intensity is relatively strong, increasing noise intensity in the excitations increases the randomness in the response ($\sigma^2 = 0.07^2$, Fig.5d).

Note that the chaotic attractors observed via Poincaré map in both cases (deterministic (Figs.4b and c) and noisy (Figs.5b and c) are approximately of the same shape, indicating transitions induced by both excitation and noise parameter variations follow approximately the same route. Thus the noise intensity can also be considered as a controlling parameter between different response attractors in the parameter space.

CONCLUDING REMARKS

Noise-induced transitions among nonlinear response attractors of a hydrodynamically excited multi-point moored system is examined. The analytical and numerical results are summarized as follows:

(1) A criterion defined chaotic domain based on a generalized Melnikov function indicates that the presence of low intensity noise perturbations expedites possible occurrence of chaotic response.

(2) Probability density functions solved from the associated Fokker-Planck equation reflect the global information about system behavior and the relative strengths of coexisting attractors.

(3) Noise intensity is found to be a possible parameter governing response transitions.

(4) The "unexpected" response transition observed in experiment is interpreted in light of the analytical results, which further assesses that the presence and effects of noise may need to be taken into account for the analysis and design of engineering systems.

ACKNOWLEDGEMENTS

The authors gratefully acknowledge the financial support from the United States Office of Naval Research (Grant No. N00014-92-J-1221).

REFERENCES

Gottlieb, O., and Yim, S.C.S. (1992). "Nonlinear oscillations, bifurcations and chaos in a multi-point mooring system with a geometric nonlinearity." *Appl Ocean Res*, Vol 14, pp 241-257.

Gottlieb, O., and Yim, S.C.S. (1993). "Drag-induced instability and chaos in mooring systems." *J Ocean Eng*, Vol 29, pp 569-599.

Lin, H., and Yim, S.C.S. (1995). "Chaotic roll motion and capsizing of ships under periodic excitation with random noise". *Appl Ocean Res*, Vol 17, pp 185-204.

Moon, F.C. (1987). *Chaotic Vibrations: An Introduction for Applied Scientists and Engineers*. New York: John Wiley & Sons.

Risken, H. (1984). *The Fokker-Planck Equation: Methods of Solution and Applications*. Berlin Heidelberg: Springer-Verlag.

Yim, S.C.S., and Lin, H. (1992). "Probabilistic analysis of a chaotic dynamical system." *Applied Chaos*, 219-241. New York: John Wiley & Sons.

Yim, S.C.S., Myrum, M.A., Gottlieb, O., Lin, H. and Shih, I-Ming. (1993). *Summary and Preliminary Analysis of Nonlinear Oscillations in a Submerged Mooring System Experiment*, Report No. OE-93-03, Ocean Engineering Program, Oregon State University, 1993.

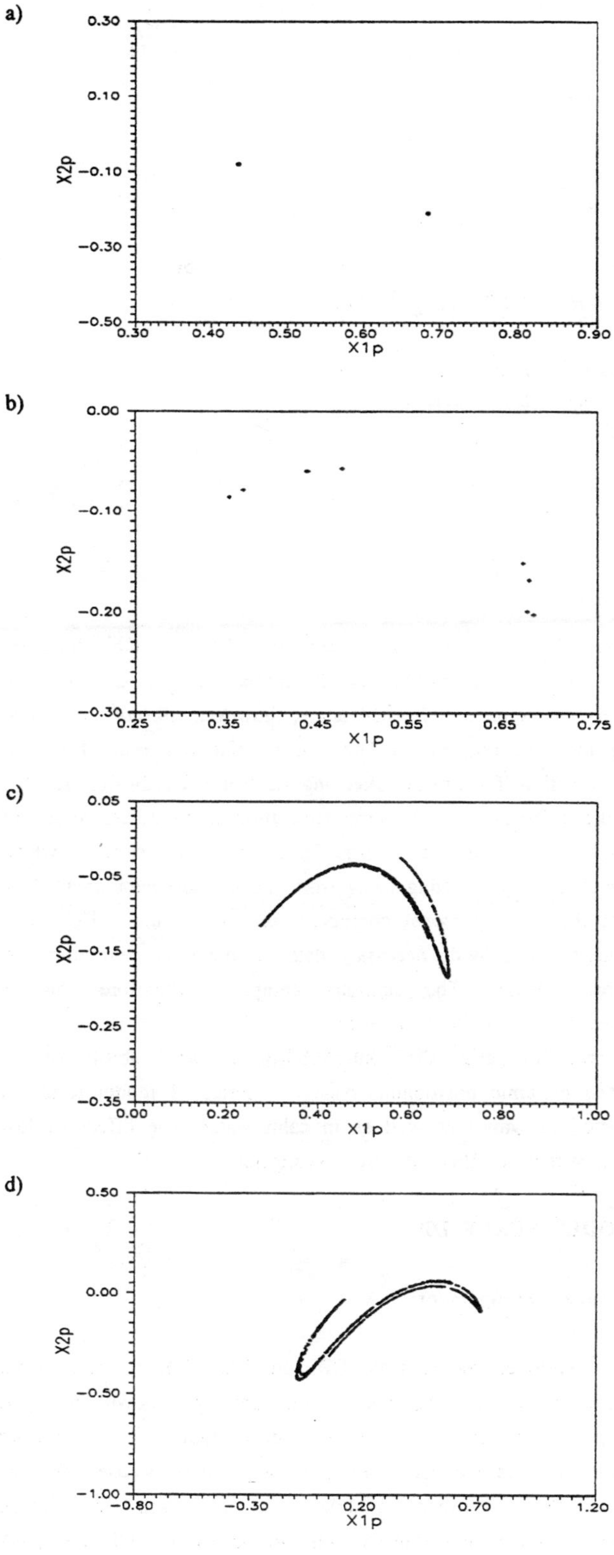

Fig. 4 Transition of system response with increasing excitation amplitude: A = a) 15, b) 16, c) 17 and d) 20

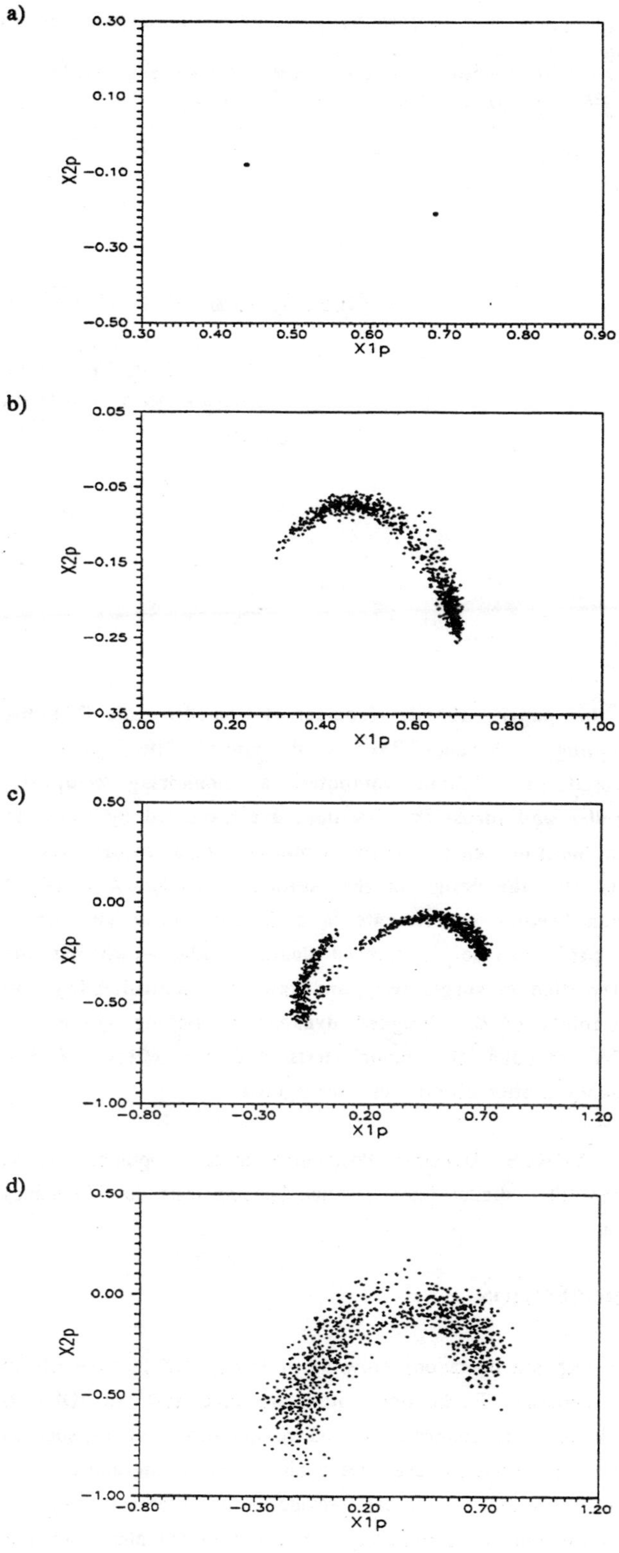

Fig. 5 Noise-induced transition of system response; σ^2 = a) 0.0^2, b) 0.01^2, c) 0.02^2 and d) 0.07^2

Proceedings of the Seventh (1997) International Offshore and Polar Engineering Conference
Honolulu, USA, May 25-30, 1997

ISBN 1-880653-28-1 (Set); ISBN 1-880653-31-1 (Vol. III)

Development of a Model Scale Dynamic Positioning System

Sa Y. Hong, Hyun J. Kim and C.M. Lee
Korea Research Institute of Ships and Ocean Engineering
Taejon, Korea

ABSTRACT

In this paper, the development of a model scale Dynamic Positioning System(DPS) is described. The system is composed of a main computer, a monitoring computer, controller and thrusters. Positions are measured by using an optical position sensor system. Optimal control theory(LQI) is adopted for the design of the position controller. A series of position keeping experiments in both calm water and waves were performed for a moored floating body, which has the 3-dof motion of surge, sway and yaw. The controllability and applicability of the designed dynamic positioning system are verified through the model tests and the effects of low frequency motion filters are investigated.

KEY WORDS: Dynamic Positioning System, optimal control, Kalman filter, digital filter, position keeping tests, low frequency motion

INTRODUCTION

As deep sea operations grow up, it is expected that the use of DPS increases. Furthermore, it is required that the DPS is capable of more sophisticated and complicated missions such as tracking operations for deep sea mining, cable laying and etc.

Model test of a DPS can provide useful information for not only developing a complicated prototype DPS but also evaluating a designed DPS. Nakamura and Koterayama et al.(1994, 1995) showed usefulness of simulation and model experiment of a thruster assisted mooring by using LQI control algorithm for sway motion.

In this study, a model scale dynamic positioning system is developed by using LQI control algorithm. Four rotatable thrusters are employed for providing the thrust necessary fir sophisticated missions in demand. Thruster angle, however, is not controlled during the test. (X, Y) position and heading angle are taken as object function for position keeping control. Hardwares for the tests are composed of 3 computers, thruster modules, a set of photo-electronic position sensors, and very soft springs which prevent the model drifting away from the measurement area. Only one of the three personal computers acts as a main CPU while the others just provide necessary data for the main computer and monitoring status. The auxiliary computers therefore can be replaced by independent modules..

In order to verify the controllability and applicability of the designed dynamic positioning system, a series of model tests are conducted in waves as well as in calm water. The effect of low frequency motion filters is also investigated.

A MODEL SCALE DPS

Hardware Arrangement

The developed model scale DPS consists of three main parts; measurement, monitoring, and control units as shown in Fig.1. Each part is designed as an independent module for easy further modification and extension. RS232C serial port is used for the communication between modules. In the case that serial communication is not effective due to its speed, A/D and D/A are used instead. 6-dof motion analog data is transferred to the control unit from the measurement unit.

The motion measurement system is devised by using a set of photo-electronic position sensors. Incident wave, tension of mooring springs, rpms and azimuthing angles from four thruster units as well as 6-dof platform motion, are converted to digital data by A/D and stored for analysis in the measurement unit. These signals are also used for the operation of monitoring system.

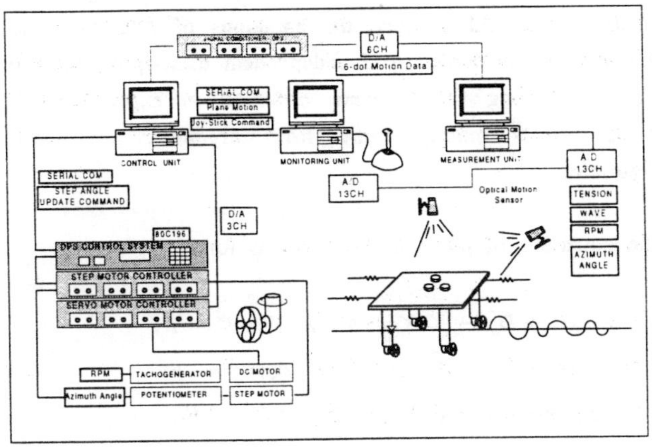

Fig. 1 Schematic Diagram of DP Model Test System

Status of the model can be monitored in real time through the monitoring system during the test and this enables an operator to respond appropriately to the variable situation. An example of the designed monitoring screen is given in Fig.2.

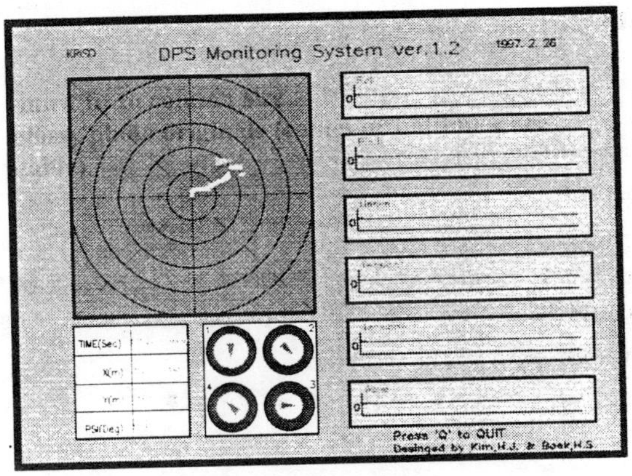

Fig.2 DPS Monitoring Screen

The control system digitizes motion data and filters them off high frequency components. Errors of X-Y position and heading are calculated by using these low frequency XY-plane motions. The magnitudes of thrust and moment required in each direction are determined from errors and position data. Then they are converted to the thrust command for thruster units by thrust allocation algorithm and the relations of rpm-voltage-thrusts. These rpm commands are sent to servo-motor drivers by D/A. In addition, it is possible to change the azimuthing angle of thruster unit by sending angle data to step motor controller based on 80C196 one-chip micom. Block diagram of the controllers is shown in Fig.3.

The function distribution to sub-modules reduces the burden of the main system, and subsequently the DPS can be split into independent modules. This makes it easy to change and upgrade some modules without changing others. Furthermore, modifications of the system for a more sophisticated mission such as route tracking in seakeeping test of ships can be achieved easily.

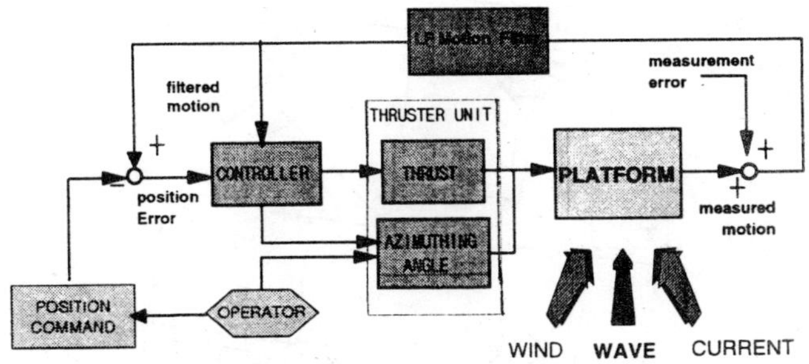

Fig.3 Block Diagram of DP Control System

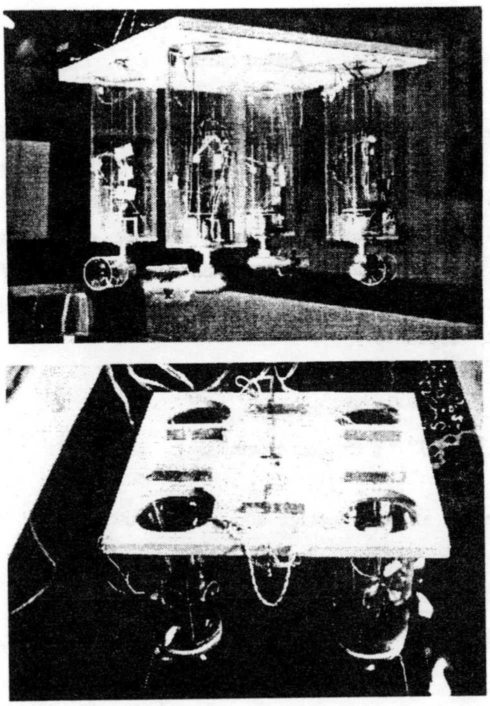

Fig.4 Photograph of Test Model

Thrust Allocation and the Characteristics of Thruster

The model has four ducted thrusters of which diameter is 14cm. Thruster unit consists of a DC servo motor for driving a propeller and a step motor to align thruster angle. Fig. 4 shows the photograph of the model. In the test, the active rotation of thruster unit resulted in so much disturbances in platform motions, that tests were performed with fixed azimuth angles. As shown in Fig.5, No.2 and No.4 thruster angles are set to produce moments required to keep heading angle, and no forces are induced in any direction. No.1 and No.3 thrusters which are perpendicular to each other, are used to generate required forces. The residual moments induced by No.1 and No.3 thrusters are compensated by No.2 and No.4 thrusters.

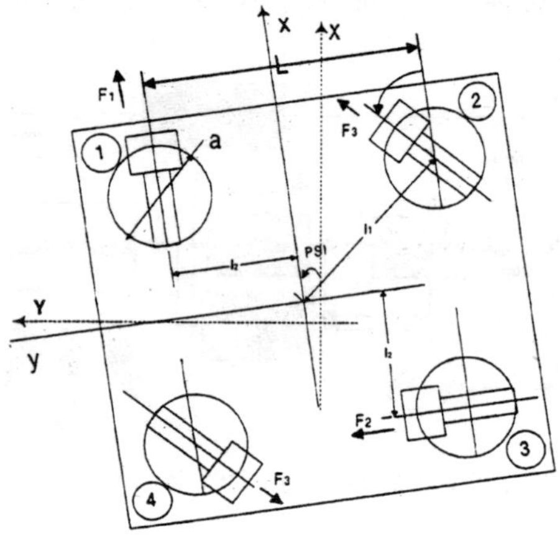

Fig.5 Coordinates System and Thruster Arrangement

Thruster performance tests were carried out in 2-D flume to find out the characteristics of the thruster. RPM and thrust are measured and fitted to equation. The measured thruster characteristics are shown in Fig.6. Because of asymmetry of thrust in clockwise and counter-clockwise rotation, the following equations are used in calculating control commands for both directions, respectively.

$V = 7.277 * T^{0.45}$ (clockwise)

$V = 4.630 * T^{0.55}$ (counter-clockwise)

where V is input voltage and T is thrust.

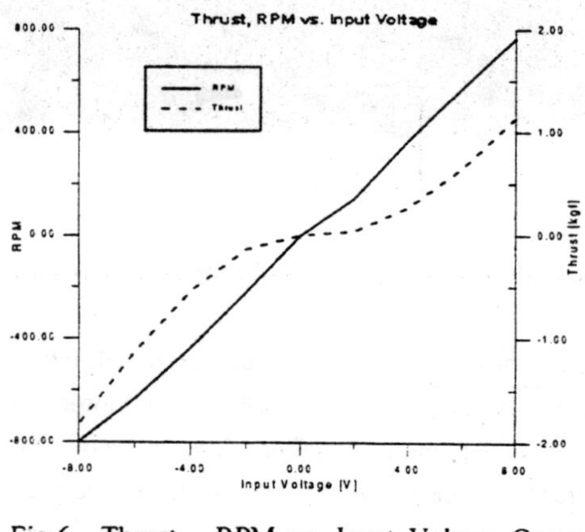

Fig.6 Thrust - RPM vs. Input Voltage Curve

THE DESIGN OF CONTROLLER AND FILTER

In Fig.5 , the cartesian coordinate system used in this paper is shown.

Since the model has 4 columns of which configuration is symmetric to x and y axes, the equations of motions in local coordinates are decoupled to 3 independent mass-spring equations. However, coupling effects were observed in experiments. The coupling effects mainly came from asymmetry of mooring configuration.

The equations of plane motions are as follows:

$$\begin{aligned}(M+a_{11})\dot{u}+D_u u+K_x X\cos\Psi+K_y Y\sin\Psi &= F_1\\ (M+a_{22})\dot{v}+D_v v-K_x X\sin\Psi+K_y Y\cos\Psi &= F_2\\ (I+a_{66})\dot{r}+D_r r+K_\psi \Psi &= -l_2F_1-l_2F_2+l_1F_3\ ,\end{aligned} \quad (1)$$

where M : mass, D : hydrodynamic damping,
a : added mass, K : spring constant,
F_1 : x-direction force in local coordinates,
F_2 : y-direction force in local coordinates,
F_3 : Moment, l_1, l_2 : moment arm

The velocities in global coordinates are transformed from local velocities.

$$\begin{aligned}\dot{X} &= u\cos(\Psi)-v\sin(\Psi)\\ \dot{Y} &= -u\sin(\Psi)+v\cos(\Psi)\\ \dot{\Psi} &= r\end{aligned} \quad (2)$$

The equations of nonlinear motions are linearized in order to design the controller with help of optimal control theory. The state vector is

$$\underline{x} = [u,\ X,\ v,\ Y,\ r,\ \Psi]^T.$$
$$\underline{u} = [\ F_1\ F_2\ F_3\ ]$$

The state equation is derived by linearizing nonlinear equations.

$$\begin{aligned}\underline{x} &= A\underline{x}+B\underline{u}\\ \underline{y} &= C\underline{x}\end{aligned} \quad ; \quad (3)$$

where

$$A=\begin{pmatrix} -\frac{D_u}{(M+a_{11})} & -\frac{K_X}{(M+a_{11})} & 0 & 0 & 0 & 0\\ 1 & 0 & 0 & 0 & 0 & 0\\ 0 & 0 & -\frac{D_r}{(M+a_{22})} & -\frac{K_y}{(M+a_{22})} & 0 & 0\\ 0 & 0 & 1 & 0 & 0 & 0\\ 0 & 0 & 0 & 0 & -\frac{D_\Psi}{(I+a_{66})} & -\frac{K_\Psi}{(I+a_{66})}\\ 0 & 0 & 0 & 0 & 1 & 0\end{pmatrix}$$

$$B=\begin{bmatrix} \frac{1}{(M+a_{11})} & 0 & 0 \\ 0 & 0 & 0 \\ 0 & \frac{1}{(M+a_{22})} & 0 \\ 0 & 0 & 0 \\ -\frac{l_2}{(I+a_{66})} & -\frac{l_2}{(I+a_{66})} & \frac{l_1}{(I+a_{66})} \end{bmatrix}$$

It is assumed that the position and heading angle are measurable. So, measurement matrix is

$$C=\begin{bmatrix} 0\,1\,0\,0\,0\,0 \\ 0\,0\,0\,1\,0\,0 \\ 0\,0\,0\,0\,0\,1 \end{bmatrix}.$$

To design the controller, Linear Quadratic Regulator controller with error integral(LQI) from optimal control theory is considered. The optimal controller minimizes the sum of state and control input energy. To reduce the steady state error, state variables to error dynamics are introduced as below:

$$\dot{\underline{e}} = \underline{y} - \underline{y}_c = C\underline{x} + D\underline{u} - \underline{y}_c\ , \tag{4}$$

where $\underline{y}_c$: the target position vector.

This linear system is reformulated as follows.

$$\begin{pmatrix} \dot{\underline{x}} \\ \dot{\underline{e}} \end{pmatrix} = \begin{pmatrix} A & 0 \\ C & 0 \end{pmatrix}\begin{pmatrix} \underline{x} \\ \underline{e} \end{pmatrix} + \begin{pmatrix} B \\ D \end{pmatrix}\underline{u} - \underline{y}_c \tag{5}$$

$$\underline{e} = [\,0\ I\,]\begin{pmatrix} \underline{x} \\ \underline{e} \end{pmatrix} = W\underline{z} \tag{6}$$

The cost function is given by

$$J = \frac{1}{2}\int_0^\infty \{\underline{e}^T Q\underline{e} + \underline{u}^T R\underline{u}\}dt = \frac{1}{2}\int_0^\infty \{\underline{z}^T W^T QW\underline{z} + \underline{u}^T R\underline{u}\}dt \tag{7}$$

and the control law is

$$\underline{u} = -K_x\underline{x} - K_e\underline{e}\ . \tag{8}$$

The motion of floating structures can be approximated by the sum of low-frequency large motion and wave-frequency small motion. The purpose of a DPS is to suppress the slow large drift motion. Therefore, it is not necessary for DPS to respond to the fast small motions and it is practically impossible to suppress the motion component by thrusters. Thus, the filtering of motions is important to increase control efficiency. In this study, digital filter with various orders and Kalman filter with a constant gain were applied and we investigated their performance of those filters in practical point of view.

The measured natural periods of surge, sway and yaw motions of the system are about 42, 68, 30 seconds, respectively. Wave spectrum used in experiments is ranged from 0.5 sec to 2.0 sec. Thus cut-off frequency of digital filter is designed to be 0.2 Hz. Because digital filter has inherent phase shift, it may deteriorate control performance when the order of the filter increases. In this study, filters with different order - 10, 20, 40th order - were tested, and compared with each other.

For the design of Kalman filter, wave frequency motions are approximated by the wave spectrum. To model the wave frequency motions, the wave spectrum is fitted by a linear system.

The wave spectrum is represented as the product of a transfer function and white noise spectrum as follow:

$$S_o(\omega) = |G(jw)|^2\, S_i(w), \tag{9}$$

where S_o : wave spectrum

S_i : white noise spectrum , $|S_i(jw)| = 1$

It is well known that the wave spectrum can be expressed by

$$G(s) = \frac{Ks^3}{s^4 + a_1 s^3 + a_2 s^2 + a_3 s + a_4}. \tag{10}$$

This system is represented by the state equation in time domain.

$$\begin{pmatrix} \dot{x}_1 \\ \dot{x}_2 \\ \dot{x}_3 \\ \dot{x}_4 \end{pmatrix} = \begin{pmatrix} 0 & 1 & 0 & 0 \\ 0 & 0 & 1 & 0 \\ 0 & 0 & 0 & 1 \\ -a_4 & -a_3 & -a_2 & -a_1 \end{pmatrix}\begin{pmatrix} x_1 \\ x_2 \\ x_3 \\ x_4 \end{pmatrix} + \begin{pmatrix} 0 \\ 0 \\ 0 \\ K \end{pmatrix} w$$

$$\underline{z} = [\,0\ 0\ 0\ 1\,]\,[\,x_1\ x_2\ x_3\ x_4]^T \tag{11}$$

The unknown coefficients a_1, a_2, a_3, a_4, K can be obtained by minimizing the difference between the wave spectrum and the fitting spectrum.

$$J = \left\{\sum_{k=1}^{N}\{S(w_i) - S_0(w_i)\}^2\right\}^{\frac{1}{2}} \tag{12}$$

Finally, a linear system representing wave frequency motions is

$$\dot{\underline{x}}_h = A_h\underline{x}_h + D_h\underline{w}_h\ ,\quad \underline{y}_h = C_h\underline{x}_h\ , \tag{13}$$

where $\underline{x}_h \in R^{12}$, $\underline{w}_h \in R^3$.

The linear system including the low frequency motion and the wave frequency motion is used for designing Kalman filter. The Kalman filter is the optimal estimator that minimizes error covariance under the assumption that statistics of noise processes is known. Kalman filter is given by

$$\dot{\hat{\underline{x}}} = A\hat{\underline{x}} + L(\,\underline{y} - C\hat{\underline{x}}\,)\ . \tag{14}$$

TEST RESULTS AND DISCUSSIONS

The platform model has 4 columns and each column is equipped with an independent thruster module which is composed of an azimuth unit and a thruster. Model scale are assumed to be 1/50. Main particulars of the model is summarized in Table 1.

Mooring line tension is considered as a steady external force similar to current and wind forces.

Carrying out a series of tests listed below, the performance of the developed system is investigated.

- ☐ Position keeping in calm water :
  - (X_c, Y_c, Ψ_c) = (0.1m, 0.1m, 15deg)
  - (0.2m, 0.1m, 15deg)
- ☐ Position keeping in waves :
  - (X_c, Y_c, Ψ_c) = (0.1m, -0.1m, 15deg)
  - Irregular Wave : ISSC Spectrum
  - ($H_{1/3}$ = 6 Cm, T_z = 0.88 sec)
  - w, w/o Low-frequency motion filter : digital filter and Kalman filter

Table 1 Principal Characteristics of Model

Items		Values
Symbol	Remarks	
a	Radius of Column	0.15 (m)
d	Draft	0.45 (m)
L	Distance between Columns	0.75 (m)
$\triangle$	Displacement	132.24 (Kg)
KG	Center of Gravity from Bottom	0.4725 (m)
K_{yy}	Radius of Gyration	0.495 (m)
k_{11}	Spring Constant	5.524(N/m)
k_{22}	Spring Constant	1.996(N/m)
k_{66}	Spring Constant	2.780(Nm/rad)

Fig.7 compares the result of position searching tests with that of simulation in calm water. The solid line denotes the measured values, while the dashed line the simulation. The abscissa means the time and the ordinate the position. Fig.7(a) represents the case of (X_c, Y_c, Ψ_c) = (0.1m, 0.1m, 15deg) and Fig.7(b) and (c) represent the case of (X_c, Y_c, Ψ_c) = (0.2m, 0.1m, 15deg). After transient behaviors, the platform maintains its position within allowable boundary of the target positions for both cases. As clearly seen in the figures, the designed control gain and thruster allocation works satisfactorily. The control gain is determined by LQI control theory by using linearized state equation. Hydrodynamic coefficients were obtained from free decay tests for each mode of plane motions.

It was found that some discrepancies between the results of experiment and simulation. In yaw mode, the difference is noticeable in transient region, the simulation result shows oscillation around the target angle and it decreases slowly. In sway mode, simulation results show oscillation around the target position.

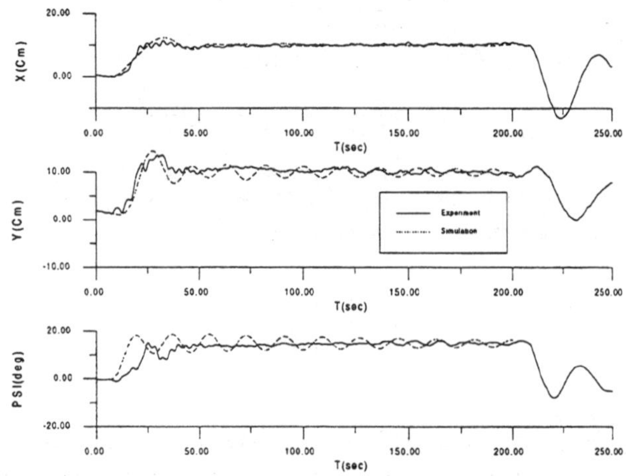

(a) Time Series for Target Position (X_c,Y_c,Ψ_c)=(10,10,15)

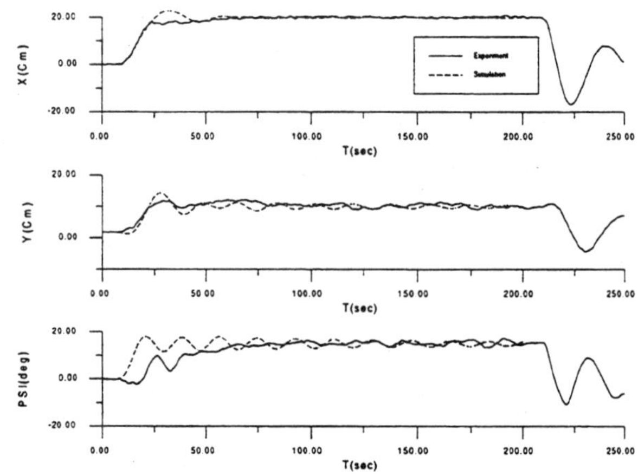

(b) Time Series for Target Position(X_c,Y_c,Ψ_c)=(20,10,15)

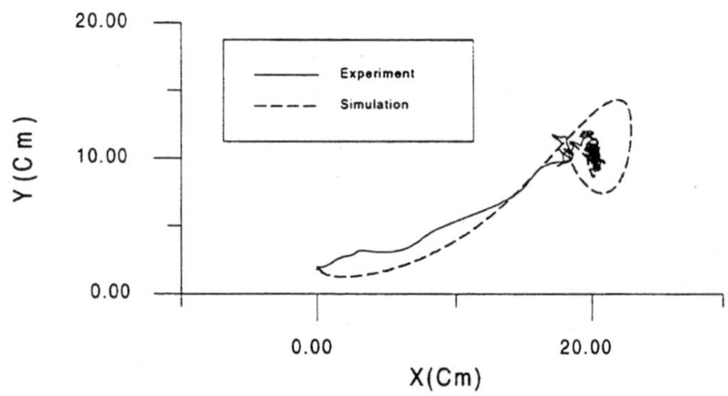

(c) Trajectory in X-Y Plane, Target (X_c,Y_c,Ψ_c)=(20,10,15)

Fig.7 Comparison of Simulation and Experiment in Position Keeping (Calm Water)

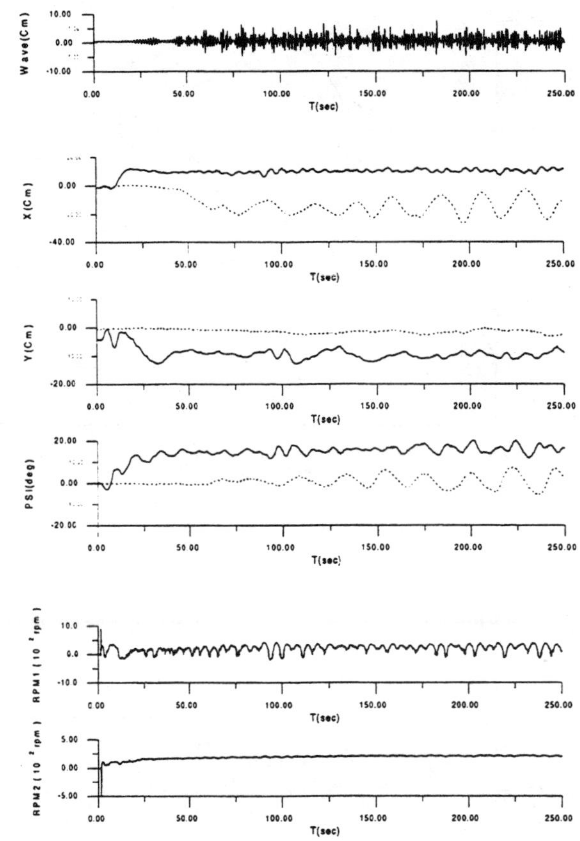

(a) Time Series (solid : w control, dash : w/o control)

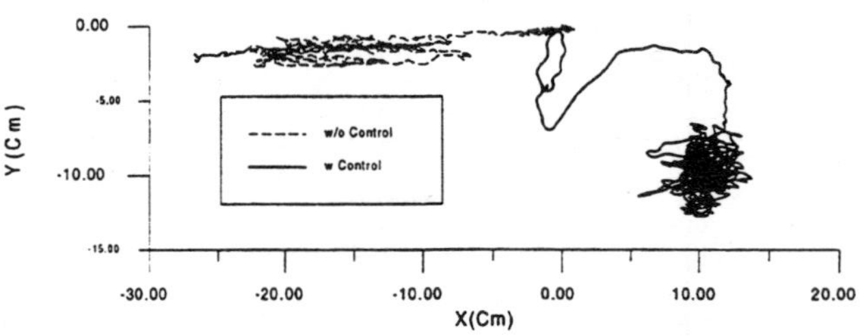

(b) Trajectory in X-Y Plane

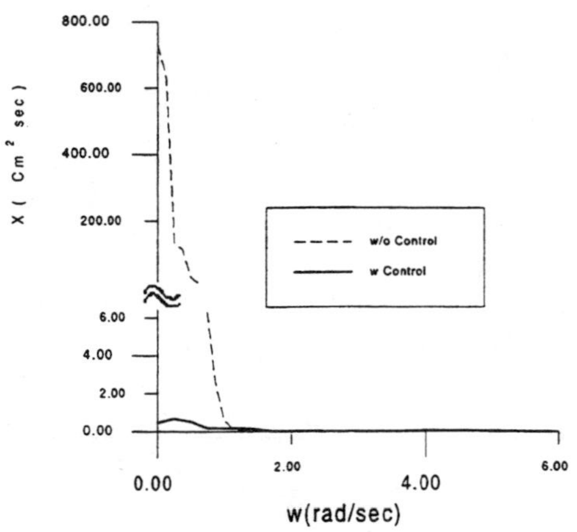

(c) Spectrum

Fig.8 Comparison of Surge Spectra in Irregular Waves

As the spring configuration is aligned in longitudinal direction, the spring force is dominant in surge mode. So, the estimation of spring constant in surge mode is more accurate than that of other modes. This may cause that spring and damping forces in sway and yaw are underestimated. Experimental results show much less overshoot in all modes. And the coupling effects in motion, which are not considered in the formulation, were occurred in experiments. In the transient region, yaw response is delayed due to asymmetry of mooring springs and the modelling errors in thrusts.

The control gains designed from simulation showed satisfactory performance and stability in calm water test, although some discrepancies between simulation and experiment exist. Also, a series of experiments in irregular waves were performed with the same control gains. Fig.8 represents the results of position keeping test in irregular waves. The target positions and heading angle were chosen as (X_c, Y_c, Ψ_c) = (0.1m, -0.1m, 15deg) and Kalman filter was used to remove wave frequency motion from total motion. The solid line denotes the controlled case while the dashed line the uncontrolled motion in head waves. In uncontrolled case, surge motion has about 10cm slow motion amplitude with 15cm offsets. But in position keeping case, the amplitude of surge motion is about 2cm and the mean offset is removed. In Fig.8(c), spectrum of surge motion shows significant difference between two cases. The force due to spring behaves in similar manner to wind or current force with heading angle. In this situation, the excursions in sway and yaw as well as surge are expected to be large. The controlled results show good performance not only in surge but also in sway and yaw. Fig.8(b) shows the trajectory in XY-plane. It can be seen that the controlled trajectory is confined to a circle within 5cm diameter in steady state.

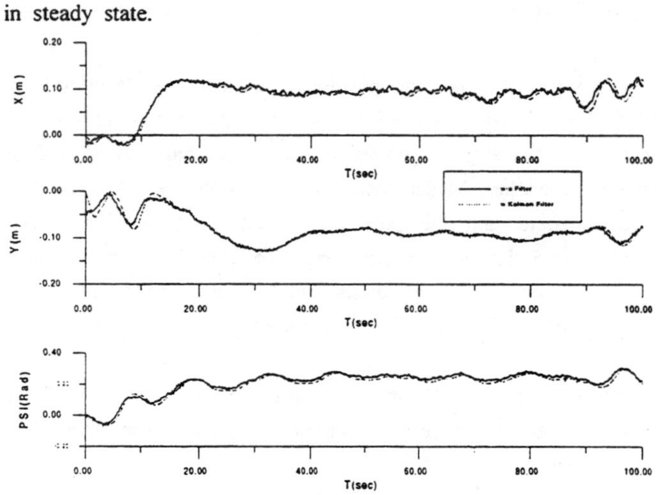

Fig.9 Time Series of Kalman Filtered and Unfiltered Plane Motion

Fig.9 compares the measured motion with the filtered one. The filtered motion is not converged to real one at initial stage, which causes large thrust rpm at intial time as shown in Fig.8(a). This resulted in longer transient behaviors.

When the motion filter is not used, the wave frequency motion due to wave exciting force results in the response of thrust with same frequency. Since this is expected to deteriorate thruster in practical sense, the filtering methods have been one of important fields in DP control system. Tests with various filters - i. e., digital filters with 10,20 and 40-th order and Kalman filter - are carried out and the results are shown in Fig.10.

In Fig.10, the wave spectrum , motion spectra and RPM spectra are given. Motion spectra do not show noticeable discrepancies except for the case that 40th order digital filter is applied to. While RPM spectra show distinguishable changes in wave frequency range and low frequency range. As the order of digital filters increases, the magnitude of high frequency components of RPM decreases. In RPM1, the unfiltered case shows the largest wave frequency components as expected. The wave frequency components in thrust are almost removed when the Kalman filter is used. When the order of digital filter is 20 or higher, it seems that the wave frequency components are removed sufficiently. But in the case of 40th digital filter, phase shift induces larger excursions in plane motion. This causes a dramatic increase of RPM1 in low frequency region.

The filters used in the test seem to have drawbacks for practical applications. Digital filters have inherent phase shift which grows as the order increases. The Kalman filter requires an exact modelling of the system and noise characteristics. Digital filter can be considered in practical application only if the proper choice of the order and cut-off frequency is made in accordance with the system and environments.

CONCLUSIONS

A model scale dynamic positioning system was developed by using four thrusters to control heading angle as well as position of floating structures in waves. A LQI algorithm was applied to design controller. Model tests in calm water for keeping the heading angle and the position showed good agreement with simulation. Satisfactory results were also obtained with the same system in waves. Among digital filters with various orders and Kalman filter, Kalman filter showed the best performance in irregular wave test.

Further experimental study is planned to include adaptive control algorithm for changeable system parameters and environments.

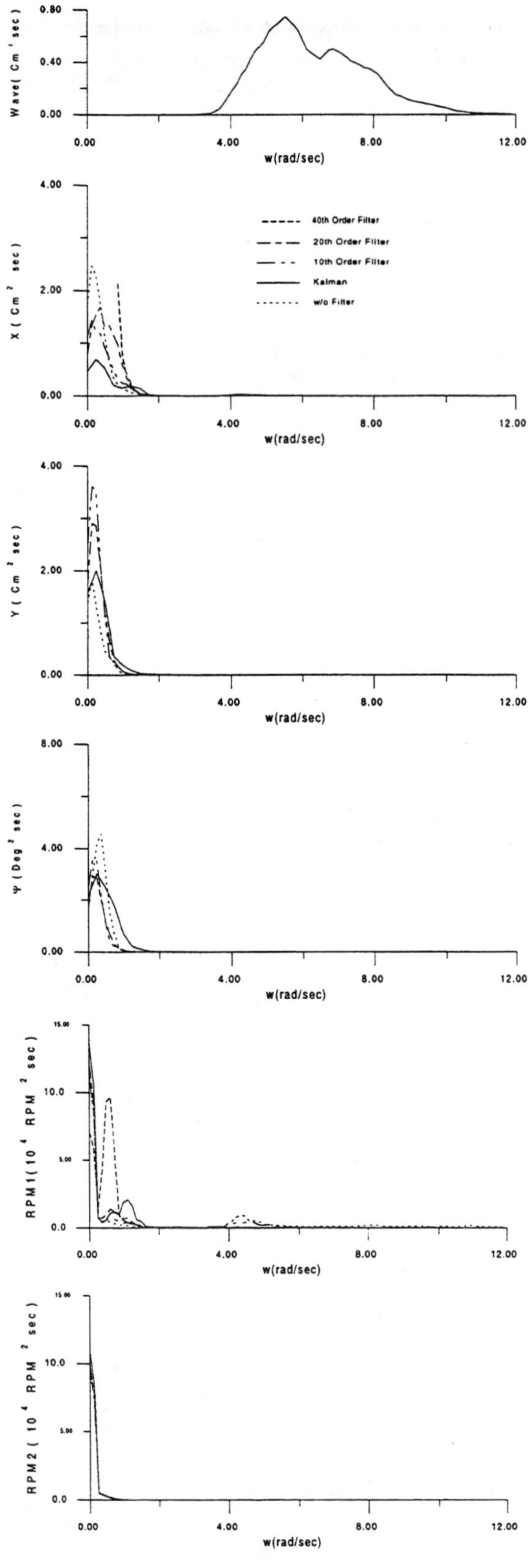

Fig.10 Comparison of Position Keeping Performance with Various Filters

ACKNOWLEDGEMENT

The authors appreciate MOST(Ministry of Science and Technology) for full support of this study. The authors also appreciate Mr. J.H. Kim for his cooperation during the model test.

REFERENCES

Fokatis, J. , Grimble, M. J, and Kouvaritakis, B. (1982). " A Comparison of Characteristics Locus and Optimal Designs for Dynamic Ship Positioning System. ", *IEEE Trans. Automat. Contr*, VOL. AC-27, NO.6.

Grimble, M. J. , Patton, R. J., Wise, D. A. (1980). " Use of Kalman Filtering Techniques in Dynamic Ship-Positioning Systems. ", *IEE Proceedings*, VOL. 127, pt.D,No.3 .

Kim, Hyun J. (1995). " The Design of Controller for DPS for Floating Structures with Optimal Control Theory", Master Thesis, SNU.

Morgan, M.J. (1978). *Dynamic Positioning of Offshore Vessels* , Tulsa, Oklahoma: Petroleum.

Nakamura,M., Koterayama, W., Kajiwara, H., Mitamura, T. (1994). " Application of a Dynamic Positioning System to a Moored Floating Platform", *Proceedings of 4th ISOPE,* Osaka, Japan.

Nakamura,M., Koterayama, W., Kajiwara, H. (1995). " Model Experiments of Thruster Assisted Mooring System", *Proceedings of 5th ISOPE,* Vol.2 pp249-254, Hague, Netherlands.

Saelid , S., Jenssen, N. A., and Balchen, J. G. (1983). "Design and Analysis of a Dynamic Positioning System Based on Kalman Filtering and Optimal Control ", *IEEE Trans. Automat. Contr*, VOL. AC-28, NO.3.

Hong, Sa Y., Kim, Hyun J., Lee, C. M. (1996)." Development of a Model Test Technique for DPS", *Proceedings of the 37th workshop on ocean engineering*, Korea.

Hong, Sa Y., et al. (1996). *Development of Dynamic Analysis Techniques for the Behaviors of Large Floating Structures III*, KIMM Report UCN012-1587 · D.

Proceedings of the Seventh (1997) International Offshore and Polar Engineering Conference
Honolulu, USA, May 25-30, 1997

ISBN 1-880653-28-1 (Set); ISBN 1-880653-31-1 (Vol. III)

Three-Dimensional Wave-Body Interactions in a Viscous Fluid: Heave Oscillation of a Submerged Vertical Cylinder

P. Ananthakrishnan
Florida Atlantic University
Boca Raton, FL, USA

ABSTRACT

Nonlinear axisymmetric viscous and inviscid solutions corresponding to heave oscillation of a submerged vertical cylinder are presented in this paper. The fully nonlinear free-surface flow problem is solved using finite-difference method based on curvilinear coordinates. A primitive-variables based fractional-step formulation is used for the analysis of viscous flow, while a mixed Eulerian-Lagrangian formulation for that of inviscid flow. New results obtained for a range of amplitudes and frequencies of body oscillation and mean depth of body submergence, demonstrate effects of free surface on the dynamics of body-generated vortices and on the vortical component of hydrodynamic force acting on the body. For example, a rapid radial transport of body-generated vortex rings, accompanied by breaking of a short wave by local instability, is observed in the case of shallow submergence and low frequency of oscillation. At intermediate frequency, interaction between vortical and wave motions results in the formation of a dipole-type pair of vortex rings. Comparison of viscous and inviscid solutions show that the vortex structures could significantly modify the near-field wave signature and also affect hydrodynamic force acting on the body. Present results of three-dimensional heave motion also reveal the occurrence of negative added mass at intermediate frequency when the submergence of the cylinder is small. Force calculations reaffirm that effect of viscosity on hydrodynamic forces is of significance at low and intermediate frequencies of body oscillation.

KEY WORDS: Wave-body interaction, vortical flow, free surface, wave mechanics, numerical method, hydrodynamics.

INTRODUCTION

Study of near-surface wave-body interactions is of significance in many ocean engineering and naval architecture applications. Prediction of motion response of ocean vehicles to surface waves, determination of wave loads on offshore structures, design of buoy systems for oceanographic measurements are examples of such applications. From a scientific viewpoint, the study is of importance for understanding vorticity generation, vortex dynamics, and flow turbulence in the presence of a free surface.

In engineering practice, wave-body interaction problem is often tackled expeditiously using the Morison-equation approach. Such an approach is of little value in problems involving, for example, slightly-submerged bodies. These problems are highly nonlinear and the interaction between wave and body-generated vortical motions could be a significant factor affecting the hydrodynamic loading on the body. A rigorous analysis of nonlinear wave-body interactions in viscous fluids is necessary for an accurate estimation of fluid loads on bodies near the free surface. Such an analysis is expected to eventually phase out the use of *ad-hoc* methods now prevalent in ocean engineering applications.

The specific purpose of the present work is to examine the importance of viscosity and free-surface nonlinearity in three dimensional wave-body interactions encountered in ocean-engineering applications. To achieve this objective, we have developed a robust and accurate finite-difference method based on curvilinear coordinates for the solution of the viscous free-surface flow problem. The present method is an extension of our earlier work developed for two-dimensional analysis (Ananthakrishnan, 1991; Yeung and Ananthakrishnan, 1992a). In order to quantify effects of viscosity, by comparison, we have also developed a finite-difference method to solve the counterpart inviscid-flow problem. In this paper, we describe these solution methods and present nonlinear axisymmetric viscous and inviscid solutions obtained for heave oscillation of a submerged vertical cylinder. These new results reveal some important effects of viscosity at low frequency of oscillation and thus, in a practical scenario, on the dynamics of compliant structures such as tension-leg platform which tend to have low natural frequency. Also, as observed in earlier two-dimensional studies (Hodges and Webster, 1986; Ananthakrishnan and Yeung, 1992), the flow is highly nonlinear even at very small amplitude of oscillation, when the body is slightly submerged. The presence of free surface affects evolution of body-generated vortices and thus the vortical-component of hydrodynamic force exerted on the body. These vortices also result in the generation of some peculiar and fascinating free-surface signatures in the near field, a finding which is of relevance in remote-

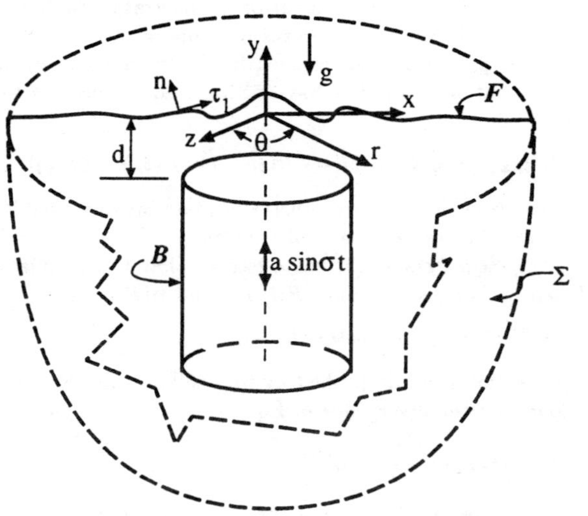

Fig. 1 Heave oscillation of a vertical cylinder below free surface.

sensing technology applied to deciphering the wave signature of submerged objects.

VISCOUS-FLOW FORMULATION

In this paper, the governing equations and all physical quantities are presented in a nondimensional form, with fundamental variables corresponding to density of fluid ρ^*, acceleration of gravity g^*, and radius of the cylinder R^*. We attach a * superscript, to denote quantities in the dimensional form. The problem under consideration pertains to heave oscillation of a circular vertical cylinder $\mathcal{B}$ under a free surface $\mathcal{F}$ as illustrated in Fig. 1. We denote the mean depth of the top surface of the cylinder as d ($\equiv d^*/R^*$), length of the cylinder as D ($\equiv D^*/R^*$), amplitude of body oscillation as a ($\equiv a^*/R^*$), and frequency of oscillation as σ ($\equiv \sigma^*\sqrt{R^*/g^*}$). Assuming this flow to be axisymmetric, we obtain the following Navier-Stokes equations written in cylindrical polar coordinate system $(r,\theta,y) \equiv (r^*/R^*, \theta, y^*/R^*)$:

$$\frac{\partial u}{\partial r} + \frac{u}{r} + \frac{\partial v}{\partial y} = 0 \tag{1}$$

$$\begin{aligned} \frac{\partial u}{\partial t} &+ u\frac{\partial u}{\partial r} - \frac{w^2}{r} + v\frac{\partial u}{\partial y} \\ &= -\frac{\partial P}{\partial r} + \frac{1}{Re}\left(\frac{\partial^2 u}{\partial r^2} + \frac{1}{r}\frac{\partial u}{\partial r} - \frac{u}{r^2} + \frac{\partial^2 u}{\partial y^2}\right), \end{aligned} \tag{2}$$

$$\begin{aligned} \frac{\partial v}{\partial t} &+ u\frac{\partial v}{\partial r} + v\frac{\partial v}{\partial y} \\ &= -\frac{\partial P}{\partial y} + \frac{1}{Re}\left(\frac{\partial^2 v}{\partial r^2} + \frac{1}{r}\frac{\partial v}{\partial r} + \frac{\partial^2 v}{\partial y^2}\right), \end{aligned} \tag{3}$$

$$\begin{aligned} \frac{\partial w}{\partial t} &+ u\frac{\partial w}{\partial r} + \frac{uw}{r} + v\frac{\partial w}{\partial y} \\ &= \frac{1}{Re}\left(\frac{\partial^2 w}{\partial r^2} + \frac{1}{r}\frac{\partial w}{\partial r} - \frac{w}{r^2}\frac{\partial^2 w}{\partial y^2}\right), \end{aligned} \tag{4}$$

where

$$P = p + y.$$

Here $(u, w, v) \equiv (u^*/\sqrt{R^*g^*},\ w^*/\sqrt{R^*g^*},\ v^*/\sqrt{R^*g^*})$ denote nondimensional radial, azimuthal, and vertical components of velocity, respectively, t ($\equiv t^*\sqrt{g^*/R^*}$) the nondimensional time, p ($\equiv p^*/\rho^*g^*R^*$) the gage pressure, and P the dynamic pressure with acceleration of gravity acting in the negative y direction. The Reynolds number Re is defined as

$$Re \equiv \frac{R^*\sqrt{R^*g^*}}{\nu^*}, \tag{5}$$

where ν^* is the coefficient of kinematic viscosity of the fluid. Because of the geometric and assumed flow axisymmetry, the flow variables u, w, v, p are functions of r, y, and t but not of θ.

Since, in the absence of atmospheric forcing and surface tension, the stress vector is zero on the free surface (Wehausen and Laitone, 1960), we have the following free-surface dynamic conditions:

$$\begin{aligned} -p &+ \frac{2}{Re}n_r^2\frac{\partial u}{\partial r} + \frac{2}{Re}n_y^2\frac{\partial v}{\partial y} \\ &+ \frac{2}{Re}n_r n_y\left(\frac{\partial u}{\partial y} + \frac{\partial v}{\partial r}\right) = 0, \end{aligned} \tag{6}$$

$$\begin{aligned} \frac{2}{Re}n_r\tau_{1r}\frac{\partial u}{\partial r} &+ \frac{2}{Re}n_y\tau_{1y}\frac{\partial v}{\partial y} \\ &+ \frac{1}{Re}(n_y\tau_{1r} + n_r\tau_{1y})\left(\frac{\partial u}{\partial y} + \frac{\partial v}{\partial r}\right) = 0, \end{aligned} \tag{7}$$

$$\frac{1}{Re}n_r\left(\frac{\partial w}{\partial r} - \frac{w}{r}\right) + \frac{1}{Re}n_y\frac{\partial w}{\partial y} = 0 \text{ on } \mathcal{F}, \tag{8}$$

where $\mathbf{n}$ and τ_1 denote unit-normal and tangential vectors on $\mathcal{F}$ in the meridional (θ = constant) plane. Since free surface is a material surface, its evolution is governed by the following kinematic condition

$$\frac{D\mathbf{X}}{Dt} = \mathbf{u}, \tag{9}$$

where D/Dt denotes the material time derivative and $\mathbf{X}$ the position vector of $\mathcal{F}$.

The body is forced to oscillate sinusoidally as

$$Y_{\mathcal{B}} = a\ \sin(\sigma t), \text{ for } t > 0. \tag{10}$$

Because of no slip and no flux, the fluid velocity on the body boundary is given by

$$u = 0;\ w = 0;\ v = a\sigma\ \cos(\sigma t) \text{ on } \mathcal{B}. \tag{11}$$

On the axis of symmetry ($r = 0$), we set

$$u = 0,\ w = 0, \text{ and } \frac{\partial v}{\partial y} = 0. \tag{12}$$

The body is set into motion from a quiescent state; *i.e.*, velocity and free-surface elevation are zero at initial time $t = 0$. On the Eulerian open boundary Σ, we assume the pressure to be static, *i.e.*,

$$P = 0 \text{ on } \Sigma \tag{13}$$

and determine velocity by a spatial extrapolation. Such an approximation seem to work well for sufficiently large duration of simulation, owing to radial decay of spreading waves by conservation of energy.

INVISCID-FLOW FORMULATION

Neglecting viscosity and assuming flow to be irrotational and axisymmetric, one can obtain the following governing equations for the velocity potential $\phi = \phi^*/R^*\sqrt{R^*g^*} \equiv \phi(r, y, t)$:

$$\nabla^2\phi = \frac{\partial^2}{\partial r^2}\phi + \frac{1}{r}\frac{\partial}{\partial r}\phi + \frac{\partial^2}{\partial y^2}\phi = 0, \tag{14}$$

which is nothing but the axisymmetric Laplace equation. The fluid pressure can be evaluated using the Euler's integral:

$$p = -y - \frac{\partial\phi}{\partial t} - \frac{1}{2}|\nabla\phi|^2. \tag{15}$$

Following Longuet-Higgins and Cokelet (1976), we consider free-surface conditions in the Lagrangian form as:

$$\frac{D\mathbf{X}}{Dt} = \mathbf{u}, \text{ where } \mathbf{X} \in \mathcal{F} \tag{16}$$

and

$$\frac{D\phi}{Dt} = \frac{1}{2}|\nabla\phi|^2 - Y \text{ on } \mathcal{F}, \tag{17}$$

where $Y \equiv Y(r, t)$ is the free-surface elevation. On the body, the no-flux condition yields the following Neumann condition for ϕ:

$$\begin{aligned} \frac{\partial\phi}{\partial y} &= a\sigma\cos(\sigma t), \text{ on top and bottom of } \mathcal{B} \\ \frac{\partial\phi}{\partial r} &= 0, \text{ on the side boundary of } \mathcal{B}. \end{aligned} \tag{18}$$

On the axis of symmetry in the fluid, we have

$$\frac{\partial\phi}{\partial r} = 0. \tag{19}$$

As done in the viscous-flow formulation, we assume pressure to be static on the open boundary. Approximately, this reduces to

$$\phi = 0 \text{ on } \Sigma. \tag{20}$$

The body is started from rest which corresponds to the following initial condition:

$$\phi = 0 \text{ and } Y = 0 \text{ on } \mathcal{F} \text{ at } t = 0. \tag{21}$$

SOLUTION OF THE VISCOUS-FLOW PROBLEM

The axisymmetric Navier-Stokes equations are solved in primitive variables using the *projection* fractional-step method (Chorin, 1968; Kim and Moin, 1985). In this method, the momentum equations without the pressure term are first solved to evaluate an intermediate auxiliary velocity field $\mathbf{u}^{aux}$. Symbolically this step can be written as

$$\mathbf{u}^{aux} = \mathbf{u}(t_n) + \delta t[\mathcal{D}(t_n) - \mathcal{C}(t_n)], \tag{22}$$

where t_n denotes the instant of discrete time, δt the time-step size which is determined by stability criteria of difference schemes, $\mathcal{D}(t_n)$ an explicit differencing of the diffusion term appropriate for large Reynolds-number flows, and $\mathcal{C}(t_n)$ a finite differencing of the convection term. In the present work we discretize the diffusion term by second-order central differencing. Following the numerical schemes suggested by Rai and Moin (1991), we have developed an elaborate second-order accurate upwind difference scheme for discretizing the convection term of the Navier-Stokes equations. Such schemes of higher-order accuracy enable accurate resolution of fine vortex structures developing in the flow field.

With the pressure term retained in the momentum equations, the discretization of the full Navier-Stokes equations can be written as

$$\mathbf{u}(t_{n+1}) = \mathbf{u}(t_n) + \delta t[-\nabla P(t_{n+1}) + \mathcal{D}(t_n) - \mathcal{C}(t_n)], \tag{23}$$

and, because of the pressure constraint in an incompressible fluid, $\mathbf{u}(t_{n+1})$ satisfies the equation of continuity: $\nabla \cdot \mathbf{u}(t_{n+1}) = 0$. By comparing Eqs. 22 and 23, one observes that the auxiliary field $\mathbf{u}^{aux}$ can be "projected" onto $P(t_{n+1})$ and $\mathbf{u}(t_{n+1})$ as:

$$\mathbf{u}^{aux} = \mathbf{u}(t_{n+1}) + \delta t\nabla P(t_{n+1}). \tag{24}$$

This projection is accomplished by first solving the Poisson equation, given by the divergence of Eq. 23

$$\nabla^2 P(t_{n+1}) = \frac{1}{\delta t}\nabla \cdot \mathbf{u}^{aux}, \tag{25}$$

for the pressure and then using the decomposition Eq. 24 for the divergence-free velocity $\mathbf{u}(t_{n+1})$.

The boundary conditions are implemented in the following manner. On the body boundary, by setting the normal component of $\mathbf{u}^{aux}$ to be the same as that of the known $\mathbf{u}(t_{n+1})$, one can obtain homogeneous Neumann condition for $P(t_{n+1})$:

$$\frac{\partial P(t_{n+1})}{\partial n} = 0, \text{ on } \mathcal{B}.$$

The above condition is justifiable for high Reynolds-number flow, as the one considered in this work. Using the normal stress condition (Eq. 6), we obtain a Dirichlet condition for pressure on the free surface:

$$P(t_{n+1}) \approx Y + \frac{2}{Re}n_r^2\frac{\partial u}{\partial r} + \frac{2}{Re}n_y^2\frac{\partial v}{\partial y} + \frac{2}{Re}n_r n_y\left(\frac{\partial u}{\partial y} + \frac{\partial v}{\partial r}\right),$$

with the right-hand side terms evaluated explicitly using the values corresponding to earlier time step. The approximation (Eq. 13) yields a homogeneous condition for $P(t_{n+1})$. The pressure Poisson equation (25) is solved subject to above Dirichlet conditions on $\mathcal{F}$ and Σ and Neumann condition on $\mathcal{B}$. We use the shear-stress conditions (Eqs. 7 and 8) to determine radial and azimuthal components of $\mathbf{u}(t_{n+1})$ on the free surface. Equation of continuity (1) is used to determine the vertical component of velocity on the free surface. By integrating the kinematic condition (9), we update the position of the free surface. On the open boundary, we evaluate $\mathbf{u}^*$ by second-order spatial extrapolation and $\mathbf{u}(t_{n+1})$ by completing the decomposition (Eq. 24).

SOLUTION OF THE INVISCID-FLOW PROBLEM

We implement the Eulerian-Lagrangian formulation of Longuet-Higgins and Cokelet (1976) in a finite-difference method to solve the present nonlinear, axisymmetric, free-surface flow problem. In this approach, we solve the Laplace equation (14) subject to homogeneous Dirichlet condition on Σ (Eq. 20), inhomogeneous Neumann condition on $\mathcal{B}$ (Eq. 18), homogeneous Neumann condition on the axis of symmetry (Eq. 19), and inhomogeneous Dirichlet condition on $\mathcal{F}$ (obtained by integrating the free-surface dynamic condition Eq. 17). The Laplace equation is discretized using second-order central differencing and solved directly using LU decomposition. The Lagrangian form of free-surface conditions, as given by Eqs. 16 and 17, are integrated using a two-step predictor-corrector method (Yeung, 1982):

Predictor:

$$\mathbf{X}^+ = \mathbf{X}(t_n) + \delta t \mathbf{u}(t_n)$$
$$\phi^+ = \phi(t_n) + \delta t[\frac{1}{2}|\nabla\phi|^2(t_n) - Y(t_n)],$$

Corrector:

$$\mathbf{X}(t_{n+1}) = \mathbf{X}(t_n) + \frac{\delta t}{2}[\mathbf{u}(t_n) + \mathbf{u}^+]$$
$$\phi(t_{n+1}) = \phi(t_n) + \frac{\delta t}{2}[\frac{1}{2}|\nabla\phi|^2(t_n) + \frac{1}{2}|\nabla\phi^+|^2 - Y(t_n) - Y^+],$$

where superscript + denotes the quantity evaluated at the predictor stage. The use of predictor-corrector algorithm requires solution of the Laplace equation twice at each time step.

BOUNDARY-FITTED COORDINATES

The governing flow equations defined in the physical space (r, y) are mapped to an uniform computational space by coordinate mapping, and thus the finite-difference methods discussed in earlier sections for viscous and inviscid solutions are implemented in the uniform mesh. The use of boundary-fitted coordinates facilitates accurate implementation of nonlinear boundary conditions and efficient discretization of the physical space for a given mesh size. For grid generation, we use a variational formulation and the concept of reference space, developed earlier for the analysis of two-dimensional nonlinear free-surface flow problems. Briefly, grid properties such as coordinate spacing and cell-area distribution are first specified in a reference space which is geometrically simpler than the actual physical space. The specified grid properties are replicated in the physical space using a variational formulation. For details and governing equations of this grid-generation procedure, please refer to Ananthakrishnan (1991) and Yeung and Ananthakrishnan (1992a).

RESULTS AND DISCUSSION

The convergence of the solution methods with respect to mesh size and time-step size was established by analyzing evolution of a small-smplitude standing wave and comparing the numerical solutions with analytical results, in a manner similar to that given in Ananthakrishnan (1991) and Yeung and Ananthakrishnan (1992b). Results for the wave-body interaction problem presented in this paper are obtained using a mesh size of (128 x 64) in the meridional (θ = constant) plane, with finer mesh spacings near the body and free surface where the flow gradient is large. The time-step size δt is so chosen as to satisfy the Courant condition based on both the celerity of radiating waves and the amplitude of body velocity. The aspect ratio of the cylinder is set to be two (*i.e.*, $D^*/R^* = 2$). In the viscous-flow calculations, the Reynolds number $Re = 10^3$. Inviscid and viscous solutions are obtained for a range of amplitude and period ($T = 2\pi/\sigma$) of oscillation and depth of submergence. The simulations are conducted for four periods of body oscillation in each case, barring the ones in which the occurrence of wave breaking forced the termination of the simulation. In the absence of wave breaking, steady state was attained within four periods of body oscillation. In the following paragraphs, we discuss the results presented in the form of instantaneous velocity fields, free-surface elevation, and time history of heave hydrodynamic force acting on the body.

Instantaneous Flow Fields.

Snapshots of velocity field in a meridional (θ = constant) plane are first presented. Fig. 2 shows the velocity-vector plots corresponding to a high ($T = 3$) frequency of oscillation and a large mean depth of submergence ($d = 1$) with amplitude of oscillation $a = 0.2$. One can observe the generation of vortex rings at the sharp edges both at the top and bottom of the body. The length of radiating waves being small, these vortex rings are not convected radially away from the body. Consequently, the fluid away from the vertical column about the body remains by and large irrotational. An instantaneous velocity field corresponding to intermediate period of oscillation ($T = 6$) with mean depth of submergence $d = 0.5$ and amplitude of oscillation $a = 0.1$ is shown in Fig. 3. Here the interaction between wake-vortex and radiating-wave motions results in the inception of a pair of counter-rotating vortex rings just outboard the top edge of the body. This interaction also results in the formation of a stagnation point beneath the free surface and above the vortex pair.

Instantaneous velocity fields for a low frequency of oscillation ($T = 10$) and a very small submergence ($d = 0.25$) with $a = 0.05$ are given in Figs. 4 and 5. The figures demonstrate the importance of convective effect of long wave motion, generated at low frequency of body oscillation, on the transport of vortices. The vortex rings generated at the top edge stretches rapidly in the radial direction under the free surface. Behind the lead vortex ring, a short wave is generated. The short wave continues to steepen and finally breaks at a later time because of local instability. Consequently a strong shear layer forms beneath the free surface. The effect of long waves is also felt on the bottom surface of the body. The phase-difference and interaction of outer-wave and body motions in this case are such that we do not notice inception of vortices near the bottom edge of the body unlike in the high-frequency shown in Fig. 2.

From Figs. 2 to 5, it is clear that the spatial evolution of vortex structures depends on the frequency of oscillation. One can show that the vortical component of hydrodynamic force is related to the moment of vorticity (see Lighthill, 1986). Thus it becomes apparent that the vortical force would depend on the frequency of oscillation. Hydrodynamic-force results presented in a later section confirm this observation.

Temporal Evolution of the Free Surface.

Next, lets examine the temporal evolution of the free surface for a range of governing parameters. Here, the free-surface elevation in a meridional (θ =constant) plane is presented in a characteristic $r - t$ plane, with successive curves displaced by δy for clarity. The wave evolution for $T = 6$, $d = 0.5$ and $a = 0.05$ for discretes times between 0 and $4T$ is given in Fig. 6. We observe a standing wave pattern right above the body as well as radially outgoing waves. The amplitude of the radiating waves decreases rapidly in the radial direction primarily as per conservation of wave energy in three dimensions and partly due to viscosity. The counterpart *inviscid-flow* result, plotted with a different scale for clarity, is shown on Fig. 7. This figure reveals generation of several higher-harmonic short waves in inviscid fluid due to free-surface nonlinearity. As per linear theory, wave amplitude in viscous fluid decays as $e^{-2\nu k^2 t}$ where ν is coefficient of viscosity and k the wave number. The wave-dampening effect of viscosity being dominant on short waves, we do not observe these short waves in the viscous fluid (Fig. 6).

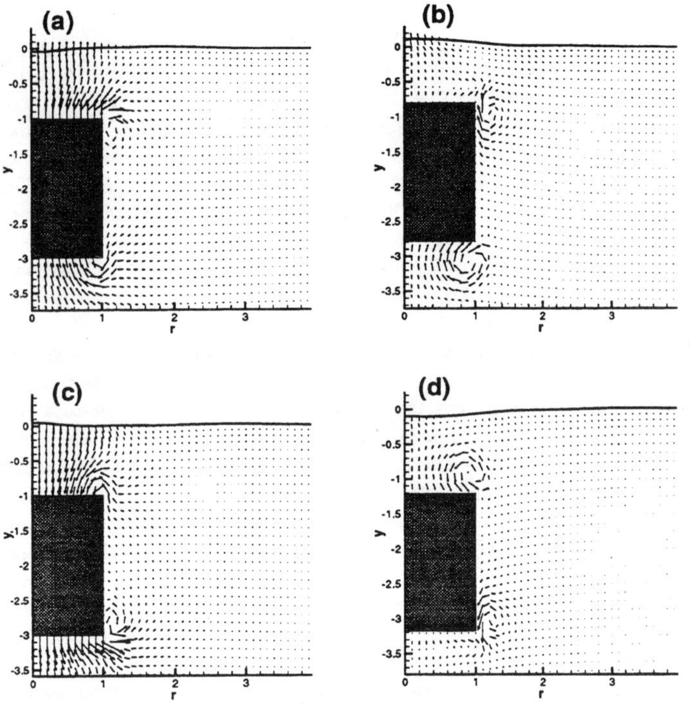

Fig. 2 Instantaneous velocity-vector fields corresponding to heave oscillation in viscous fluid for $T = 3$, $d = 1$, $a = 0.2$ at (a) $t/T = 2$; (b) $t/T = 2.25$; (c) $t/T = 2.5$; and (d) $t/T = 2.75$.

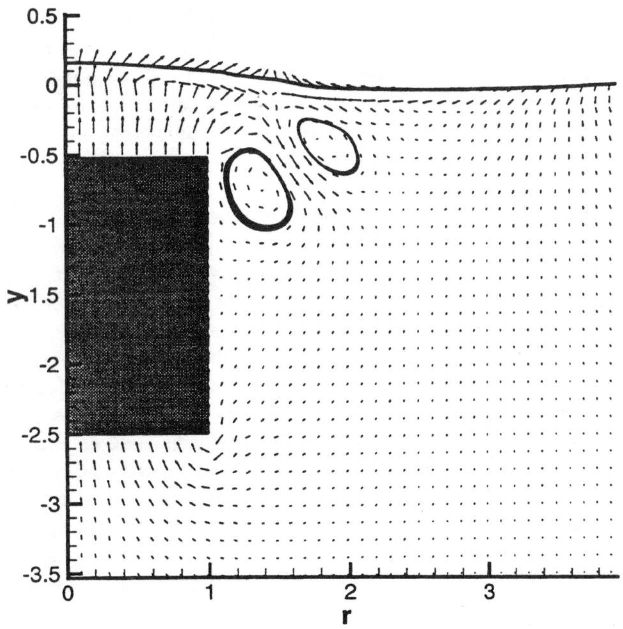

Fig. 3: Instantaneous velocity-vector fields corresponding to heave oscillation in viscous fluid for $T = 6$, $d = 0.5$, $a = 0.1$ at $t = 4T$.

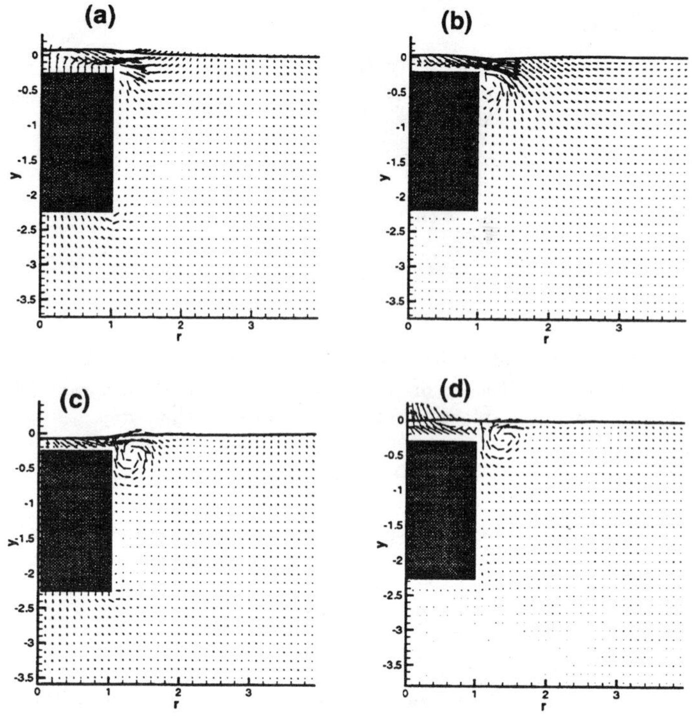

Fig. 4 Instantaneous velocity-vector fields corresponding to heave oscillation in viscous fluid for $T = 10$, $d = 0.25$, $a = 0.05$ at (a) $t/T = 1$; (b) $t/T = 1.25$; (c) $t/T = 1.5$; and (d) $t/T = 1.75$.

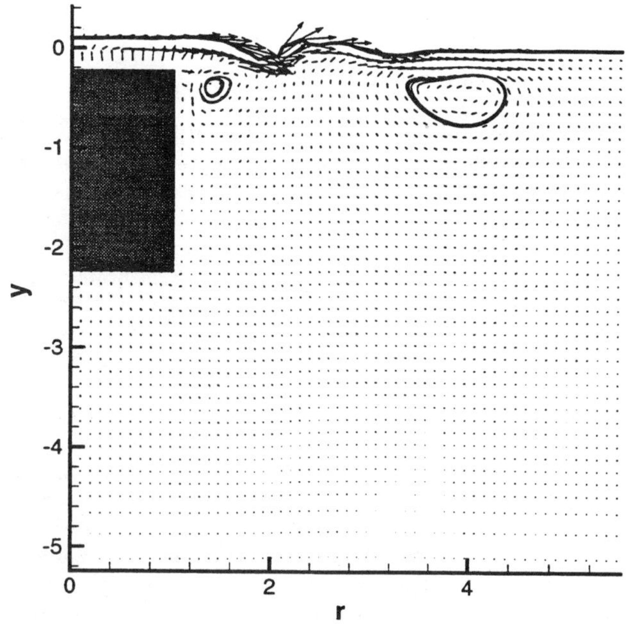

Fig. 5 Instantaneous velocity-vector fields corresponding to heave oscillation in viscous fluid for $T = 10$, $d = 0.25$, $a = 0.05$ at $t = 4T$.

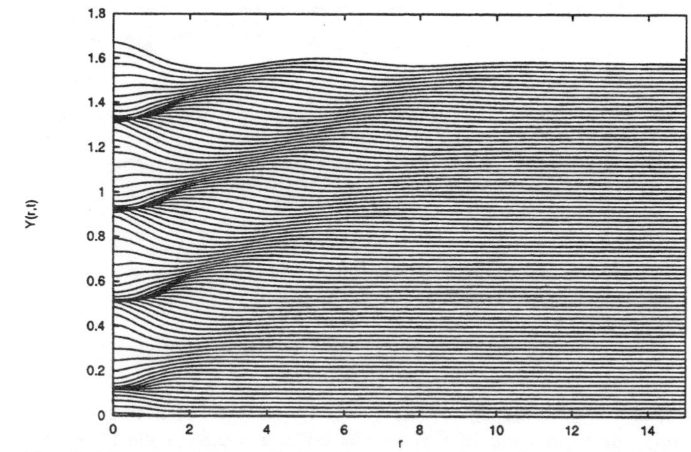

Fig. 6 Time evolution of the free-surface in a viscous fluid: $T = 6$, $d = 0.5$, $a = 0.05$, $t = (0, ..[0.05T], ...4T)$. Successive curves are displaced from the previous ones by $\delta y = 0.02$ for clarity.

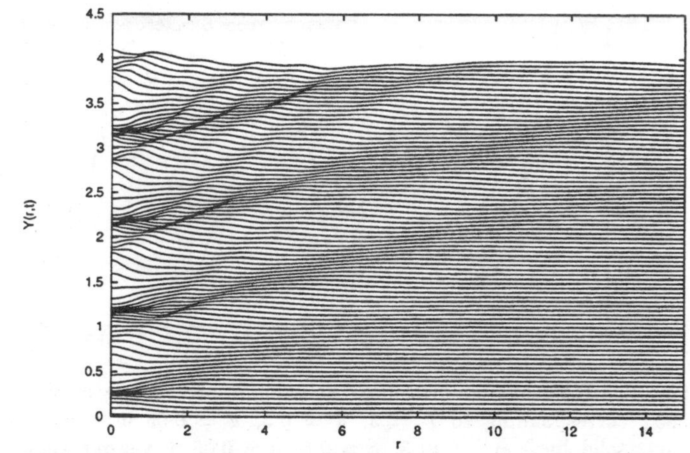

Fig. 7 Time evolution of the free-surface in inviscid fluid: $T = 6$, $d = 0.5$, $a = 0.05$, $t = (0, ..[0.05T], ...4T)$. Successive curves are displaced from the previous ones by $\delta y = 0.05$ for clarity.

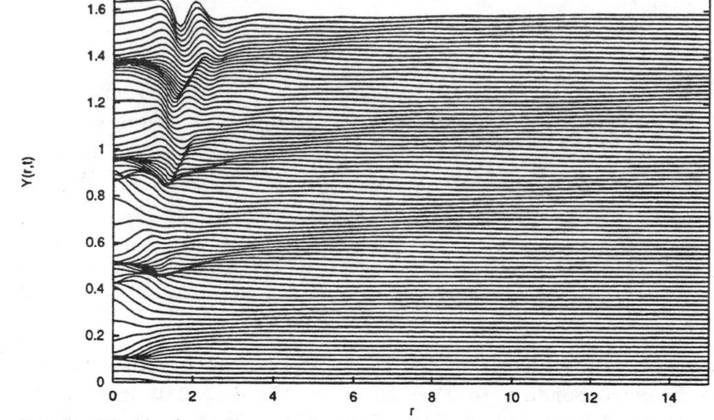

Fig. 8 Time evolution of the free-surface in a viscous fluid: $T = 10$, $d = 0.25$, $a = 0.05$, $t = (0, ..[0.05T], ...4T)$. Successive curves are displaced from the previous ones by $\delta y = 0.02$ for clarity.

The wave evolution corresponding to $T = 10$, $d = 0.25$ and $a = 0.05$ is shown in Fig. 8. At this shallow submergence of the body, the nonlinearity is so strong that we can observe short waves even in the viscous fluid. These short nonlinear waves are generated by the vortex ring expanding radially under the free surface (see Fig. 5). Interaction of between successive vortex rings and near-field free-surface shear layer results in the formation of a short wave outboard the edge of the cylinder. This short wave continues to steepen in time and eventually breaks because of local instability.

Time History of Hydrodynamic Force on the Cylinder.

The vertical component of the hydrodynamic force acting on the body is computed by integrating the dynamic pressure:

$$F_y = \int_{\mathcal{B}} P n_y d\mathcal{B}, \tag{26}$$

where $d\mathcal{B}$ denotes the differential body-surface area, P the dynamic pressure, and n_y the y component of the unit-normal vector inward the body. In some cases, because of the occurrence of wave breaking in the absence of dampening effect of viscosity, we were able to obtain long-time viscous solution but not the inviscid solution for the same amplitude of oscillation. In order to make a meaningful comparison between viscous and inviscid solutions, particularly in such cases, and to determine the importance of nonlinearity, we present our results as force per unit amplitude of oscillation (F_y/a). In the figures, a sine curve denoting body displacement is also plotted in order to depict the phase difference between the body motion and the heave hydrodynamic force. From the phase difference one can infer inertia and linear damping components of the heave hydrodynamic force.

The heave force in viscous fluid for three periods of oscillation $T = 3$, $T = 6$ and $T = 10$ all corresponding to $d = 0.5$ and $a = 0.05$ is shown in Fig. 9. This figure demonstrates the non-monotonicity of the phase difference between hydrodynamic force and body displacement with respect to frequency of oscillation. In particular, we notice the phase difference to be the largest at intermediate period of oscillation.

In the next three figures (Figs. 10, 11, and 12), we compare the hydrodynamic force for different mean depths of submergence. These figures correspond to periods of oscillation $T = 3$, 10, and 6, respectively. By studying these figures, we observe that vicinity of free surface affects both the amplitude and the phase of the hydrodynamic force. In particular, the difference of amplitude as well as phase of hydrodynamic force between that of $d = 0.5$ and of $d = 0.25$ is quite significant at the intermediate frequency of oscillation (Fig. 12). And interestingly, Fourier analysis of the time history of heave force reveals that the case $T = 6$, $d = 0.25$, $a = 0.05$ corresponds to a *negative* added mass! Occurrence of negative added masses has been reported in earlier studies involving slightly-submerged two-dimensional horizontal cylinders (Chung, 1977; Newman *et al.*, 1984; Hodges and Webster, 1986; Yeung and Wu, 1991; and Ananthakrishnan and Yeung, 1992). The new results presented in this paper show that negative added mass, which is due to formation of resonant standing waves above the body, would occur even in the three-dimensional vertical-cylinder case when the depth of body submergence is small.

Finally in Figs. 13 and 14, we compare the results of heave force in viscous and inviscid fluids, in order to determine the effect of viscosity in the radiation hydrodynamics problem. Fig. 13

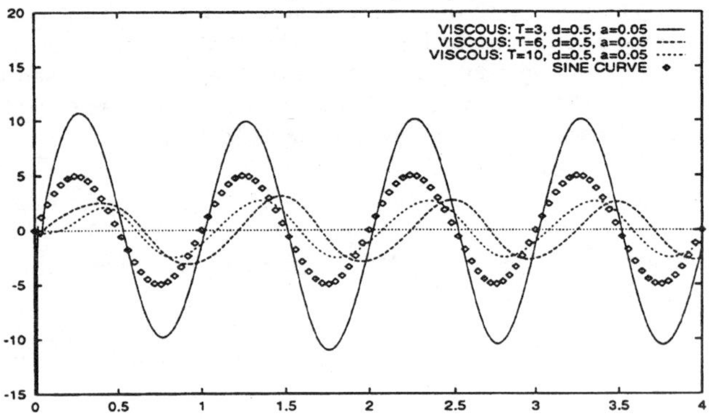

Fig. 9 Time history of the hydrodynamic force acting on the body corresponding to $T = 3$, $d = 0.5$, $a = 0.05$ (solid line), $T = 6$, $d = 0.5$, $a = 0.05$ (long-dashed line), and $T = 10$, $d = 0.5$, $a = 0.05$ (short-dashed line). Also plotted in symbols is a sine curve denoting body displacement.

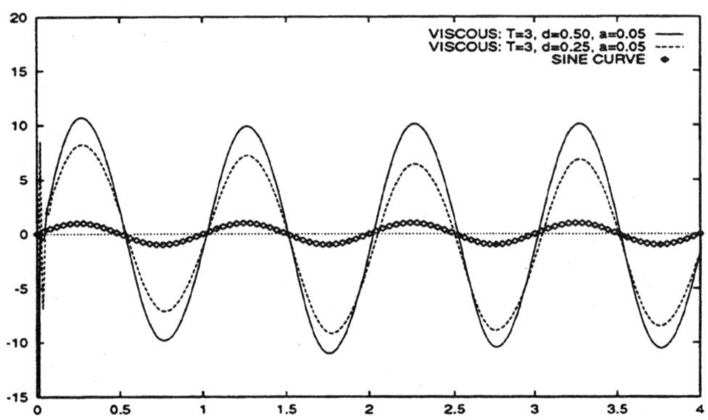

Fig. 10 Time history of the hydrodynamic force acting on the body corresponding to $T = 3$, $d = 0.5$, $a = 0.05$ (solid line) and $T = 3$, $d = 0.25$, $a = 0.05$ (dashed line). Sine curve plotted in symbols denotes body displacement.

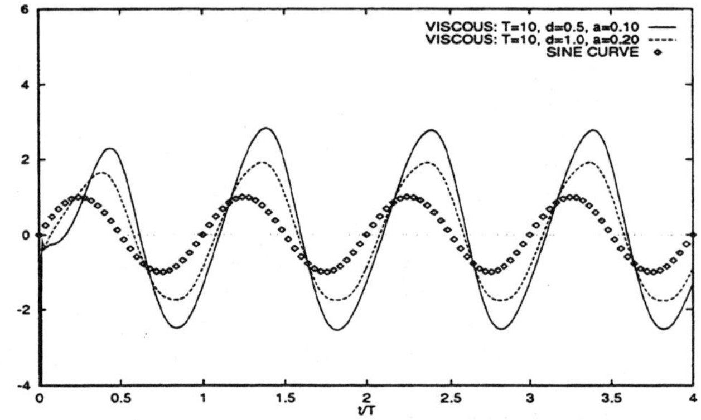

Fig. 11 Time history of the hydrodynamic force acting on the body corresponding to $T = 10$, $d = 0.5$, $a = 0.1$ (solid line) and $T = 10$, $d = 1.0$, $a = 0.2$ (dashed line). Also plotted in symbols is a sine curve denoting body displacement.

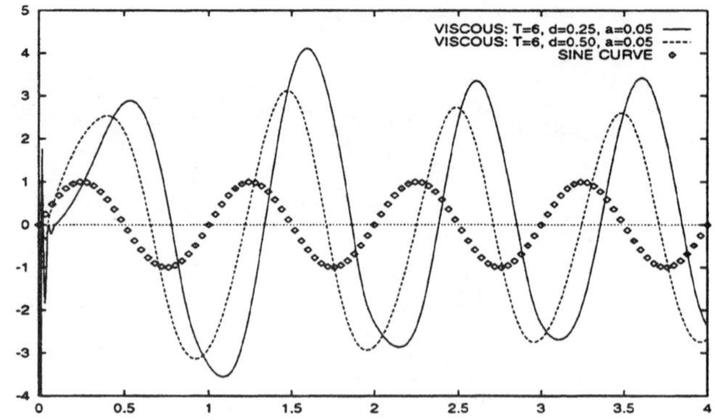

Fig. 12 Time history of the hydrodynamic force acting on the body corresponding to $T = 6$, $d = 0.25$, $a = 0.05$ (solid line) and $T = 6$, $d = 0.50$, $a = 0.05$ (dashed line). Also plotted in symbols is a sine curve denoting body displacement.

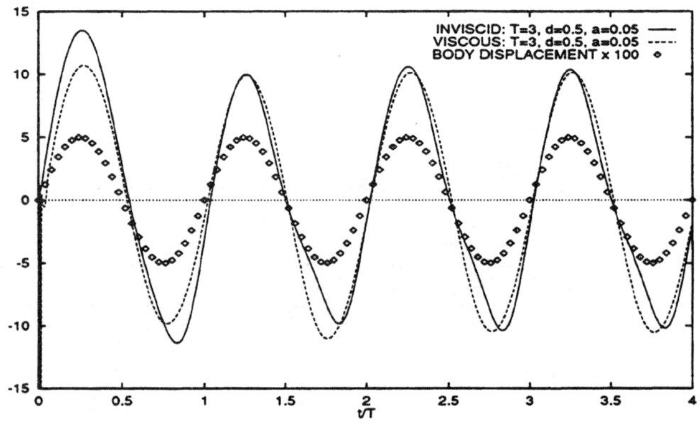

Fig. 13 Time history of the hydrodynamic force acting on the body corresponding to $T = 3$, $d = 0.5$, $a = 0.05$ in inviscid fluid (solid line) and $T = 3$, $d = 0.5$, $a = 0.05$ in viscous fluid (dashed line). Also plotted in symbols is a sine curve denoting body displacement.

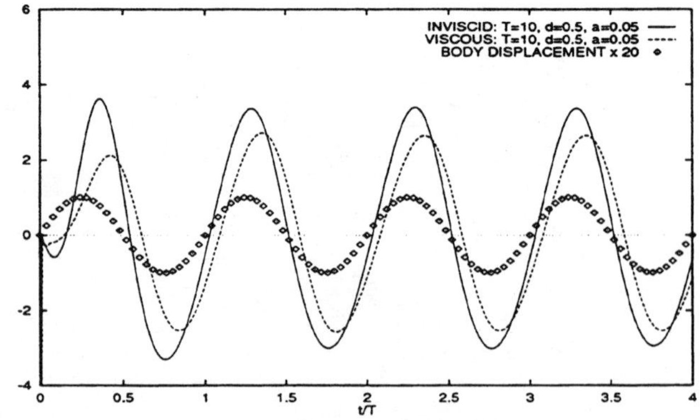

Fig. 14 Time history of the hydrodynamic force acting on the body corresponding to $T = 10$, $d = 0.5$, $a = 0.05$ in inviscid fluid (solid line) and $T = 10$, $d = 0.5$, $a = 0.05$ in viscous fluid (dashed line). Also plotted in symbols is a sine curve denoting body displacement.

corresponds to $T = 3$, $d = 0.5$, $a = 0.05$ while Fig. 14 to $T = 10$, $d = 0.5$, $a = 0.05$. We notice that effect of viscosity on the hydrodynamic (pressure) force is rather small at high frequency of body oscillation ($T = 3$), but quite significant at low frequency ($T = 10$). The earlier studies in this area have shown experimental results of hydrodynamic forces to deviate from that of potential-theory considerably at low frequencies. These and the present result seem to confirm that effect of viscosity on the hydrodynamic force, in particular the vortical force, is significant at low frequency of body oscillation.

CONCLUSION

In this work, we have developed a finite-difference solution method for the analysis of heave-oscillation of a submerged vertical cylinder in a viscous fluid, with a considerable effort spent in developing higher-order schemes capable of resolving fine vortical structures. Also developed for quantifying the effect of viscosity by comparison is an inviscid free-surface flow solver based on the mixed Eulerian-Lagrangian formulation. The results demonstrate the robustness and accuracy of the solution methods in capturing the vortical and wave structures and in predicting the hydrodynamic forces acting on the body.

The new results presented in this paper shed light to a better understanding of wave-body interactions in a viscous fluid. For example, it is found that the interaction between wave and body motions at low frequency of body oscillation could suppress the inception of vortices at the bottom edge of the body. The body-generated vortex rings are transported over a large radial distance by long-wave motions generated at low frequency. In particular, at low frequency of oscillation and small depth of submergence, we observe a series of vortex rings generated at the top edge of the body stretching underneath the surface. The vortex rings results in the generation of a steep wave just outboard the edge of the cylinder. Several higher-order waves, generated by nonlinearity, are observed in the inviscid fluid. These short radiating waves are dampened by viscosity. Force calculations show the significance of free-surface vicinity and of viscosity at low frequency of oscillation. The results also show occurrence of negative added mass, even in the case of a heaving vertical cylinder, at an intermediate frequency of oscillation when depth of submergence is small.

By abandoning the assumption of axisymmetry, the solution methods can be extended to consider other modes of rigid-body oscillations such as surge, sway, pitch, and roll. Results from laboratory experiments involving slightly-submerged vertical cylinders can be used to confirm the findings of the present study. These are our works in progress, results of which will be reported in the future.

ACKNOWLEDGEMENT

This work is supported primarily by the Department of Ocean Engineering at Florida Atlantic University. Support by Florida State University through allocation of super-computer resources is gratefully acknowledged.

REFERENCES

Ananthakrishnan, P (1991). *Surface waves generated by a translating two-dimensional body: effects of viscosity*, Ph.D. Thesis, Department of Naval Architecture and Offshore Engineering, University of California at Berkeley.

Ananthakrishnan, P, and Yeung, RW (1992). "Oscillation of a slightly-submerged cylinder in a viscous fluid," *Proc 7th Int Workshop on Water Waves and Floating Bodies*, Val de Reuil, France.

Chorin, AJ (1968). "Numerical solution of the Navier-Stokes equations," *Math of Computations*, Vol 22.

Chung, JS (1977). "Forces on submerged cylinders oscillating near a free surface," *J Hydronautics*, Vol 11, No 3.

Hodges, SB, and Webster, WC (1986). "Measurement of the forces on a slightly submerged cylinder," *Proc 21st American Towing Tank Conference*, Washington D.C.

Kim, J, and Moin, P (1985). "Application of a fractional-step method to incompressible Navier-Stokes equations," *J Comp Phys*, Vol 59.

Lighthill, J, (1986). "Fundamentals concerning wave loading on offshore structures," *J Fluid Mech*, Vol 173.

Longuet-Higgins, MS, and Cokelet, ED (1976). "The deformation of steep surface waves on water: I. A numerical method of computation," *Proc Roy Soc of London*, Vol A350.

Newman, JN, Sortland, B, and Vinje, T (1984). "Added mass and damping of rectangular bodies close to the free surface," *J Ship Research*, Vol 28, No 4.

Rai, MM, and Moin, P (1991). "Direct simulations of turbulent flow using finite-difference schemes," *J Comp Phys*, Vol 96.

Wehausen, JV, and Laitone, EV (1960). *Surface Waves, Handbuch der Physik*,ix, Springer-Verlag, Germany.

Yeung, RW (1982). "Numerical methods in free-surface flows," *Ann Rev Fluid Mech*, Vol 14.

Yeung, RW, and Wu, CF (1991). "Viscosity effects on the radiation hydrodynamics of horizontal cylinders," *J Offshore Mech and Arctic Eng*, Vol. 113.

Yeung, RW, and Ananthakrishnan, P (1992a). "Oscillation of a floating body in a viscous fluid," *J Eng Math*, Vol 26.

Yeung, RW, and Ananthakrishnan, P (1992b). "Vortical flows with and without a surface-piercing body," *Proc 19th Symp Naval Hydrodynamics*, Seoul, Korea.

Proceedings of the Seventh (1997) International Offshore and Polar Engineering Conference
Honolulu, USA, May 25-30, 1997

ISBN 1-880653-28-1 (Set); ISBN 1-880653-31-1 (Vol. III)

Vortex-Induced Vibration Tests of a Flexible Smooth Cylinder at Supercritical Reynolds Numbers

D.W. Allen and D.L. Henning
Shell E&P Technology Company
Houston, TX, USA

ABSTRACT

Tow tests have been conducted on flexible circular cylinders at supercritical Reynolds numbers in water. The tests were conducted at the Naval Surface Warfare Center's David Taylor Model Basin in Carderock, Maryland. Measurements were made of both drag and acceleration (due to vortex-induced vibration) of the cylinder.

An ABS pipe was used to achieve Reynolds numbers ranging from about 2×10^5 to 6×10^5, and a 5-9/16-in. diameter PVC pipe was used to achieve Reynolds numbers ranging from about 6×10^5 to 1.5×10^6. Tests were also conducted with aluminum inserts (strongbacks), made to fit just inside the test cylinders, in order to obtain stationary (rigid) cylinder drag measurements for comparison purposes. The test results for (relatively) smooth cylinders are presented in this paper.

KEY WORDS: vortex, vibration, riser, fatigue, Reynolds number, drag

INTRODUCTION

Vortex-induced vibration (VIV) of cylindrical structures is a fairly well-known phenomenon that can cause fatigue failure and/or excessive drag on a structure. In the ocean, many cylindrical structures are potentially subject to VIV from ocean currents. These structures include cables, risers, tendons, mooring lines and the hull of a spar-type structure.

One of the important parameters affecting VIV is known as the Reynolds number. The Reynolds number is defined as $Re = V*D/\nu$, where: V is the relative velocity of the flowing fluid that the cylinder experiences; D is the cylinder outside diameter, and ν is the kinematic viscosity of the fluid. Since the Reynolds number is proportional to both diameter and velocity, testing at Reynolds numbers that correspond to those of a prototype structure in which the velocity and diameter are often large (e.g., drilling risers with buoyancy and spar hulls in high currents), is quite difficult, especially in water where the associated forces are very large.

In the past, some researchers have performed high Reynolds number experiments by using air for the flowing fluid. Some of the more popular papers describing wind tunnel tests on stationary cylinders are those by Achenbach (1968), Jones et al. (1969), Roshko (1961), Schewe (1983), and Shih et al. (1992).

Reports of tests in water are more scarce. Rodenbusch and Gutierrez (1983) report on tests with a towed cylinder, not allowed to vibrate, that experienced Reynolds numbers over 2×10^6, however, suspected carriage vibrations prevented them from seeing a drop in drag coefficient for a smooth cylinder during the critical regime. (The critical Reynolds number regime is the flow regime in which the boundary layers become turbulent and is accompanied by a sharp reduction in the mean drag coefficient for a sufficiently smooth cylinder. This critical regime occurs at Reynolds numbers of about 1×10^5 - 1×10^5 and is dependent upon the cylinder's surface roughness and the free-stream turbulence of the flow. For purposes of this paper, Reynolds numbers above 1×10^5 are called critical, and those above 4×10^5 are called supercritical.) Some recent joint industry programs funded by oil companies have also produced drag measurements in water up to Reynolds numbers of 1.5×10^6, but the tests have been restricted to stationary cylinders or rigid (non-flexible) cylinders experiencing small forced oscillations (see also Jones et al. [1969] who forced a cylinder in air with small oscillations), and the information is not available in the public literature. To the authors' knowledge, while various field measurements have been made on cylinders such as marine risers experiencing currents (with all of the attendant inaccuracies of field measurements), controlled VIV measurements on flexible cylinders in water have been restricted to Reynolds numbers of about 350,000 (the latter are tests conducted by the author, but are unreported in the public literature) and are also quite scarce at these Reynolds numbers in air. This restriction is primarily due to the extremely large drag forces required to tow cylinders at high Reynolds numbers, thereby limiting the facilities that are capable of such tests (it is also possible to perform high Reynolds number tests in a wind tunnel or a large current tank such as the U.S. Navy's Large Cavitation Channel, but if the cylinder is allowed to vibrate, then there is a significant risk of propeller damage if the structural system should fail). Thus, controlled experimental information on VIV and the performance of suppression devices in water at supercritical Reynolds numbers is scarce. The objective of these experiments was to tow flexible and stationary (rigid) circular cylinders at speeds such that Reynolds numbers as high as possible could be achieved. A review of available tow and current tank facilities in the U.S. led to the decision to perform tests at the David Taylor Model Basin (DTMB), where Reynolds numbers as high as 2×10^6 could be achieved. Both smooth and rough cylinders were towed, and several vortex suppression devices were also tested (note that the "smooth" cylinders used in these tests have some surface roughness associated with them as reported later in the paper). The results for smooth cylinders are presented in this paper. Results for the rough cylinders (which had much higher surface roughnesses than the smooth cylinders) and for the cylinders equipped with vortex suppression devices may be presented or published at a later date.

TEST DESCRIPTION

Model Basin

All of the experiments were conducted in the David Taylor Model Basin, which is part of the Naval Surface Warfare Center in Carderock, Maryland. For these tests, the high-speed basin was used. This basin has dimensions of 21 ft wide by 2968 ft long (of tow area), including a pneumatic wavemaker at one end and a wave-absorbing beach at the other end (see Figures 1 and 2). For the first 1169 ft, the depth is 10 ft and then is ramped down for 112 ft, so that the depth is 16 ft for the final 1687 ft. Models are towed such that the 10-ft depth section is encountered first. DTMB's carriage #5 was used to tow the cylinders. This carriage has a velocity range of 0 to 85 ft/sec and has a maximum power of 4800 hp. The carriage is rectangular in shape and is approximately 70 ft long by 26 ft wide. Eight weight-bearing vertical drive wheels and 4 opposed pairs of horizontal driving guide wheels operate in tandem on each side of the carriage. An open rectangular test bay 31.5 ft long by 10 ft wide was used for mounting the removable towing bridge to which the struts, which DTMB called "parachute struts", were attached. The struts were mounted to a frame and the entire frame was locked to the carriage on a set of four pads.

Test Setup

Two model pipe sizes were used for the tests (Figure 3 is a drawing of the test setup). The first cylinder was a 3-in. 145 psi ABS pipe with a 3½-in. outside diameter (D) and a 0.247 wall thickness. The other cylinder was 5-in. Schedule 40 PVC pipe with an outside diameter of 5-9/16 (5.5625) in. and a 0.258 wall thickness. The 3½-in. ABS pipe had a modulus of elasticity of 220 ksi (provided by the pipe manufacturer and consistent with the authors' previous tests on similar ABS pipes), and the 5-9/16-in. PVC pipe had a modulus of elasticity of 457 ksi (determined from tensile and compressive tests).

The fundamental natural frequency in bending for a smooth pipe was computed to be (using a finite element program) 5.77 Hz for the 3½-in. pipe and 8.83 Hz for the 5-9/16-in. pipe. Pluck tests were later conducted on smooth 5-9/16-in. pipes, and the average measured natural frequency was 8.48 Hz. The difference between the computed and measured values, as well as the differences between the individual pluck tests, were both attributable to variations in the pipe samples.

The average surface roughness of each cylinder was determined using a confocal scanning microscope and measuring the average peak-to-trough height of the topographical image of five samples. The average roughness for the PVC cylinder was $k/D = 9.94 \times 10^{-5}$ (this is the average peak-to-trough height divided by the cylinder diameter), with the samples ranging from $k/D = 8.86 \times 10^{-5}$ to 1.09×10^{-4}. For the ABS cylinder, the average roughness was $k/D = 1.37 \times 10^{-4}$, with the samples ranging from $k/D = 1.21 \times 10^{-4}$ to 1.51×10^{-4}.

All pipes were cut and faced to a length of 78.5 in. The ends were slightly counter-bored to fit aluminum end caps (Figure 3). The end caps were 0.5-in. thick with a 0.25-in. recess cut on the outside of the cap. The 0.25-in. recess allowed the end caps to be centered in the pipe. With the addition of the end caps, the total test pipe length was 79 in. Pins that were 1.25-in. long were attached to the end plates using 1.25-in. - 12 jam nuts. The outer end of the 1.25-in. pins were smooth. The smooth ends were mounted in the trailing load cell rod eyes (the total distance between the rod eye centers was 83.75 in.). This allowed the test pipe ends to float in the rod eyes. The ends were designed for movement (flexing) of the test pipe center (both horizontally and vertically) of ±15 degrees.

Two 10,000-lb load cells were used to measure drag loads. One load cell was mounted on each end of the test cylinder. The load cells were pinned to the load cell mounting plate in the front. The load cell mounting plates were bolted to the DTMB parachute struts with 3 bolts (1 in.) on each plate. Holes were available in each parachute strut to run the load cell and accelerometer lead wires from the test cylinder up to the carriage instrumentation area. The 1.25-in. male rod end at the rear of the load cell was captured in a steel housing. This housing allowed the load cell to measure loads in the towing direction. The housing also protected the load cell against side loads generated by the in and out motion of the vibrating test cylinders (the motions of the ends during bending motions). The 1.25-in. pins and end caps were held to the plastic test cylinders with a 92-in. by 0.5-in. B-7 all-thread rod, which ran through the center of the test cylinders. An accelerometer holder was mounted in the center of the all-thread and made to fit snugly inside the test cylinder with leaf springs. Two holders were fabricated for the 5-9/16-in. test cylinders and two holders for the 3½- in. test cylinders. For each pipe size, one holder was fabricated for the configuration having a strongback insert (aluminum pipe) and one was fabricated for the configuration without a strongback. The accelerometer holder was a 1.5-in. thick piece of aluminum round stock machined to mount leaf springs on its periphery. An accelerometer mounting hole was also machined in the aluminum round stock. The accelerometer's x-axis was mounted to measure accelerations in-line with the current (the tow direction), and the y-axis was mounted to measure the crossflow accelerations. The accelerometer was positioned at the center of the pipe span in order to measure vibrations at the anti-node of first-mode bending. The all-thread rod was torqued to approximately 75 in.-lbs for the 3½-in. test cylinders and 150 in.-lbs for the 5-9/16-in. test cylinders. The nuts were locked in place by double-nutting. A thick washer was used under the all-thread rod nut on the port side. This prevented the 1.25-in. pin from pulling out of the male rod eye in the event of a pipe failure. An anti-rotation arm was mounted on the starboard side of the test assembly, outside of the struts, to restrict torsional motions of the test cylinders without affecting the drag force measurements.

A number of tests were conducted using a "strongback" mounted inside the test cylinder. The strongback consisted of a piece of 78-in. long aluminum tubing with an outside diameter just smaller than the test cylinder (thus there were two strongbacks: one for a 3½-in. test cylinder and one for a 5-9/16-in. test cylinder) and a wall thickness of approximately 0.375 in. The function of the strongback was to restrain the test cylinder from experiencing vortex-induced vibrations so that stationary cylinder drag could be measured. The presence of the strongback accomplished this by increasing the natural frequency in bending to a value that is higher than expected to be excited during the tests. The aluminum end caps for the test configurations using a strongback were 0.75 in. thick. A 0.25-in. recess was used to center the strongback and a 0.25-in. recess was used to center the pipe.

Test Procedure

For each test, the carriage was first backed up well into the shallow portion (10-ft depth) of the tank (the carriage was usually backed up at a speed of 3 knots). Once the turbulence had subsided to an acceptable level (this usually required at least 5 minutes to reduce the turbulence to less than about 3 percent), a carriage run was executed (after zeroing the load cells). Data were taken from just before the initial carriage acceleration until the carriage reached final rest or had begun reversing direction for the next run.

For most tests, the carriage was accelerated at 1 ft/sec^2. Once the first targeted test speed was reached, the carriage operator would signal with a horn on the carriage. The carriage operator would again signal when the deeper portion of the tank was reached. At that point, the DTMB test engineer would insure that a minimum of another 300-350 ft of tank was traversed before signaling the carriage operator to accelerate to the next speed. Once this speed was reached, the carriage operator would again signal with the horn, and the DTMB test engineer would insure that another 300-350 ft of tank was traversed before signaling the carriage operator to end the test by decelerating and stopping the carriage. The 300-350-ft length of the tank required for each test (note that about 500 ft was actually traversed for most tests) was determined by the need for at least about 50 cycles of transverse VIV motion to make accurate measurements for each test. For example, at 30 ft/sec, the cylinder might vibrate as low as 8 Hz, but 300 ft would insure 10 seconds of data and thus 80 cycles (8 Hz times 10 seconds) of data. Since the last part of the data for each test speed was selected for analysis (discussed below), this allowed for some VIV cycles to build to a steady state before the analyzed portion of the data was obtained. Note that there are lighted distance markers on the side of the tank every 30 ft; this allowed for easy observation by the DTMB test engineer.

At a speed of approximately 13.4 knots, the Froude number (with depth as the length scale) is unity. Degradation of the measurements was a concern for tests near this speed, however, examination of the data obtained in these experiments found the degradations (these degradations were estimated based upon the measurements of the surrounding data points and looking for changes at this tow speed, e.g., dips in the drag coefficients, etc.), if present at all, to be quite small for the transverse displacements and the drag force measurements (less than 3 percent), but substantial for the in-line displacement measurements (see discussion below in the section on Test Results). There was often significant spray and splashing near this speed.

Instrumentation and Data Acquisition

Tow speed measurements were made by feeding the carriage speed measurements (made by a magnetic pick-up on one of the drive wheels) directly into the data acquisition hardware. The excessive drag on the carriage sometimes caused, for the 5-9/16-in. cylinder tests, a slight reduction in the velocity that was preset by the carriage control system (due to some problems with the feedback loop). This was on the order of 1 to 1.5 percent. However, it should be noted that it is the actual tow velocity that is recorded with the results, since this speed reduction resulted in a reduction of the wheel r.p.m.

Turbulence was measured by a hot wire anemometer, with the results output to a strip chart recorder. These measurements were used to determine the time needed between tests for the turbulence to decay to an acceptable level (an intensity of about 3 percent or less).

The in-line drag forces were measured using custom load cells built by Houston Scientific Intl. (Model No. 3100-0010) with a range of 0 to 10,000 lb. Sixty feet of marine-type bronze armored cable was sealed to each cell with a 04-001 submersible seal. UNF mounting threads (1.25 - 12) were selected, which allowed the ends of the load cells to be fitted with matching male rod eyes (1.25 - 12). The load cells were each calibrated and checked for linearity prior to testing.

A calibration of the entire test assembly was conducted at Shell's Gasmer Road facility in Houston prior to shipping to DTMB. This assembly consisted of everything below the struts, including the end plates and a 5-9/16-in. PVC test cylinder. A 6-in. steel pipe was used to simulate the strut frame, and loads were applied to the PVC test cylinder by suspending custom-made steel canisters of sand from the vertically mounted test assembly (i.e., the assembly was rotated 90 degrees and loaded vertically instead of horizontally). A total of four canisters were used, and both smooth and rough pipes were loaded until yield (which occurred at around 3000-3200 lb). The load cells at the cylinder ends were compared with a load cell that measured the suspended weight of the assembly (this assembly load cell was zeroed with the unloaded assembly weight prior to the test so that it would measure the applied loads on the PVC pipe). The test results showed that the two cylinder load cells (one at each end) were within 1.5 percent of each other and that their combined readings were within 0.5 percent of the assembly load cell reading throughout the load range. Also, a pull test in air with the assembly mounted to the struts was later performed at DTMB with a load up to 1000 lb. and similar results were observed.

The water viscosity was determined by performing several temperature measurements during three days of the tests and using well-known values for the viscosity of freshwater at a given temperature. Since the measured value of the water temperature was 69.5 degrees Fahrenheit, the kinematic viscosity value used was 1.06×10^{-5} ft^2/sec. This value was also checked by taking a water sample and performing measurements using a viscometer at 69.5 degrees Fahrenheit.

Columbia HEVP-14 biaxial accelerometers were used to measure the accelerations. One accelerometer was mounted inside the accelerometer holder at the center of the test cylinder; a second accelerometer was located inside the bottom of the starboard strut; and a third accelerometer was located on the top of the strut frame on the bridge. Prior to the tests, the accelerometers were calibrated with a shaker assembly and were found to be accurate to within 1.7 percent over the range of 10 to 20 Hz.

The analog accelerometer voltage signals were first amplified using Columbia Model 9002 charge amplifiers. The amplified signals were then digitized by LABTECH NOTEBOOK data acquisition software, controlling a Data Translation DT2801 board, and stored on the hard disk drive of a Winbook laptop computer. The signals for the 5-9/16-in. cylinder tests were also stored in analog form by a TEAC XR510 tape recorder. The analog signals were low-pass filtered at 40 Hz (to eliminate tank noise, electrical noise, etc.) for the 3½-in. cylinder tests and at 50 Hz for the 5-9/16-in. cylinder tests, using Frequency Devices analog filters, before the data were acquired by LABTECH NOTEBOOK.

Data Analysis

Since two tests (i.e., two velocities) were usually performed for each carriage run, a stripping program was written for stripping out the best data for each test. This was accomplished by finding the range of data points with measured velocities that matched the test velocity and then using the last 2048 points. By using the last 2048 points as opposed to the first 2048 points, transients were best eliminated and, for the first test of each carriage run, the data from the 16-ft depth were used instead of any data that were taken during the 10-ft depth used for the approach.

The data analysis was performed using a custom data analysis package on an HP730 workstation, which performed the following steps for each test:

1) The raw accelerations were scaled according to the gain settings on the charge amplifiers and converted to the proper engineering units;
2) The accelerations were next Fast Fourier Transformed to produce frequency spectra of both the in-line and transverse directions;
3) Acceleration power spectra were computed from the frequency spectra for both the in-line and transverse directions.
4) The acceleration power spectra were filtered from a high-pass filter setting to a low-pass filter setting. The low-pass filter setting was 30 Hz for the 3½-in. test cylinder and 40 Hz for the 5-9/16-in. test cylinder. The high-pass filter setting was varied for each test in order to eliminate low frequency noise from the displacements resulting from double integration of the accelerations. This high-pass filter setting was chosen based upon a review of the power spectra and was an iterative process, but was usually about 3 Hz (high-pass filtering is required in order to obtain displacements since the frequency squared is in the denominator). A 10-point cosine taper was used in the filtering process.
5) The acceleration power spectra were twice integrated to compute rms displacements for that test.
6) The mean drag coefficients were computed from the mean drag forces (arithmetic average) that were measured by the in-line load cells.

Note that during testing, the above steps were performed to produce a preliminary determination of data validity using the Winbook computer.

The mean drag coefficients were computed from the formula $C_d = F_d/(0.5*\rho*V^2*D*L)$, where: C_d is the drag coefficient; F_d is the drag force; ρ is the density of the water in the tank (the specific weight of 62.44 lb/ft^3 divided by the gravitational constant); V is the tow velocity; D is the cylinder outside diameter; and L is the cylinder length of 79 in. (includes end plates).

TEST RESULTS

The following test sections present the test results. For simplicity, the results are plotted as a function of Reynolds number, however they could have been plotted against tow velocity or reduced velocity instead. Thus, it is important to note that there are parameters other than the Reynolds number that also influence the results.

For purposes of the following discussions, the subcritical Reynolds number range is defined as consisting of Reynolds numbers below 1×10^5; the critical Reynolds number range will be defined as approximately consisting of Reynolds numbers of 1×10^5 to 4×10^5, and the supercritical range defined as approximately 4×10^5 to 3.5×10^6 (the transcritical range consists of Reynolds numbers above 3.5×10^6).

Stationary Cylinder Results

Figure 4 is a plot of the drag coefficients, together with data from other researchers, for the tests with a strongback (aluminum pipe insert used to keep the pipe from vibrating). This figure shows a steep decrease in the drag coefficient through the critical Reynolds number range, known as the "drag crisis." The drag coefficient then continues to slightly decrease as the Reynolds number is increased through the supercritical range. Whereas the drag coefficients for the 3½-in. stationary cylinder were decreasing slightly up to a Reynolds number of just over 6×10^5 (the highest Reynolds numbers tested with this cylinder), the drag coefficients for the 5-9/16-in. cylinder show an increase for Reynolds numbers beginning at 8.8×10^5. Figure 4 indicates that the Reynolds number at which the drag coefficient for a stationary smooth cylinder reaches a minimum in, the supercritical range varies from about 3×10^5 to about 8×10^5. Thus, the data herein would appear to support a minimum in the upper end of this range. The differences between the present data and the data from most of the included resources in Figure 4 are most probably related to surface roughness, as is consistent in comparing the present data with the data from Shih et al (1992). However, other parameters such as free-stream turbulence and cylinder vibration can also have an effect (for these tests the turbulence intensity was less than about 3 percent and the cylinder in-line and transverse rms displacements were less than 0.0024D, as measured by the pipe accelerometer).

Vibrating Cylinder Results

Figure 5 is a plot of the results for the smooth ABS cylinder without a strongback (thus the cylinder is free to experience VIV). The substantial VIV resulted in a significant increase in the drag coefficients (about 50-80 percent). These results illustrate that significant VIV can, and does, occur at critical and supercritical Reynolds numbers and that the corresponding increase in drag coefficient is similar to that of subcritical Reynolds numbers.

Figure 6 is a plot of the results for the smooth PVC cylinder without a strongback, and shows both substantial vibration and significant increased drag due to the vibrations. As with many of the tests with this pipe diameter, the tests were limited by what the PVC pipe could structurally withstand, thus the velocities were not increased until the pipe "locked-out" (i.e., there is a velocity at which the first-mode VIV of the pipe will cease due to sufficient differences between the vortex shedding frequency and the natural frequency, but the pipe could not withstand the bending stresses associated with velocities this high).

Figure 6 shows one data point at which the in-line displacement was larger than that for surrounding data points. This test speed (24 ft/sec) consistently produced this phenomenon, and it should be noted that it occurs just above the speed at which the basin Froude number is unity. However, tests with a cylinder having a different diameter, or tests with a different facility are needed to determine whether this is a Froude number or a Reynolds number effect. Note that at this test speed (about 24 ft/sec), the reduced velocity, based upon the measured natural frequency of 8.48 Hz, is about 6.1. This is the approximate reduced velocity at which others have observed a phase change between the cylinder motions and the forces on the cylinder (e.g., see Wu [1989]). Thus, this phase change may also play a role in the large in-line motions at this test speed.

Figure 7 contains plots of the acceleration and displacement power spectra for the test with a Reynolds number of 1.39×10^6 (test speed of 31.7 ft/sec). The large in-line peak corresponding with the transverse (cross-flow) peak at about 8.5 Hz indicates that the cylinder is vibrating at an angle of about 0.6 degrees from the cross-flow direction. This is most likely due to the normal misalignment of the accelerometer and not due to actual cylinder vibration. The small magnitude of this misalignment is indicative of the high accuracy of these measurements.

SUMMARY AND CONCLUSIONS

Tests have been conducted on flexible and stationary cylinders in water at Reynolds numbers over 1.5×10^6. The results indicate that:

a) small levels of roughness can have a tremendous effect on the drag coefficient of supercritical Reynolds number flow past stationary and vibrating cylinders; and

b) significant VIV response is observed at supercritical Reynolds numbers and is accompanied by substantial increases in the mean drag coefficient.

These tests indicate that future test programs on cylinders experiencing flow at supercritical Reynolds numbers should carefully measure and report the values of surface roughness (in addition to other relevant parameters such as turbulence intensity), even for a cylinder thought to be "smooth." Careful studies of the effects of Reynolds number, surface roughness, and turbulence on the VIV response and drag of circular cylinders may allow more accurate determination of the response of a variety of cylindrical offshore structures exposed to high Reynolds number flows.

ACKNOWLEDGMENTS

A research project as demanding and ambitious as this one required tremendous support from a group of over 40 people who directly contributed to this project. The authors are especially grateful to Doug McMullen, who provided data analysis and testing support, Joe Haws, who provided data acquisition support, Dave McMillan, who provided drafting support, and Robert Patterson, who, with Shell E&P Technology Co., provided administrative and financial support. The authors are also especially appreciative of the staff at the Naval Surface Warfare Center - Towed Systems Branch who oversee the DTMB and without whom these tests would not have been possible.

REFERENCES

Achenbach, E. (1968). *Distribution of Local Pressure and Skin Friction Around a Circular Cylinder in Cross-Flow up to Re = 5x10⁶*, J. Fluid Mech., Vol 34, part 4, pp. 625-539.

Jones, G. W., Cincotta, J. J., and Walker, R. W. (1969). *Aerodynamic Forces on a Stationary and Oscillating Cylinder at High Reynolds Numbers*, NASA TR R-300, February.

Rodenbusch G. and Gutierrez, C. A. (1983). *Forces on Cylinders in Two-Dimensional Flows*, Technical Progress Report BRC 13-83, Shell Development Co., Houston, TX.

Roshko, A. (1961). *Experiments on the Flow Past a Circular Cylinder at Very High Reynolds Number*, J. Fluid Mech., Vol 10, pp. 345-356.

Schewe, G. (1983). *On the Force Fluctuations Acting on a Circular Cylinder in Crossflow From Subcritical up to Transcritical Reynolds Numbers*, J. Fluid Mech., Vol 133, pp. 265-285.

Shih, W.C.L., Wang, W.C.L., Coles, D., and Roshko, A. (1992). *Experiments on Flow Past Rough Circular Cylinders at Large Reynolds Numbers*, 2nd Intl. Colloquium on Bluff Body Aerodynamics and Applications, Melbourne Australia, December 7-10.

Wu, Z. J. (1989). *Current Induced Oscillations of a Flexible Cylinder*, Ph.D Thesis, Division of Structural Engineering, The Norwegian Institute of Technology, University of Trondheim, Norway.

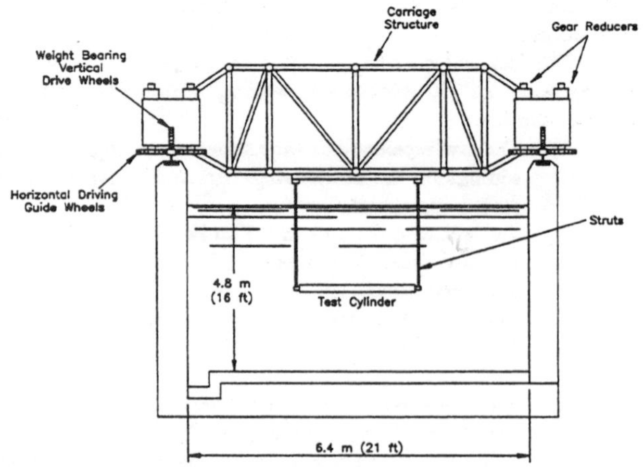

Fig. 1 Elevation View of Deep End of High Speed Basin and Carriage Number 5

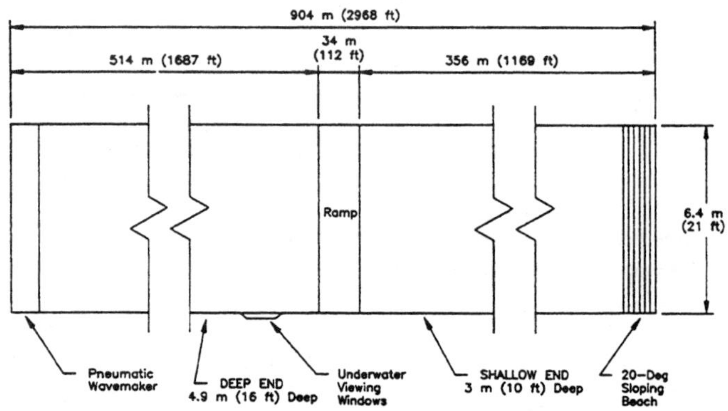

Fig. 2 Schematic Plan View of High Speed Basin

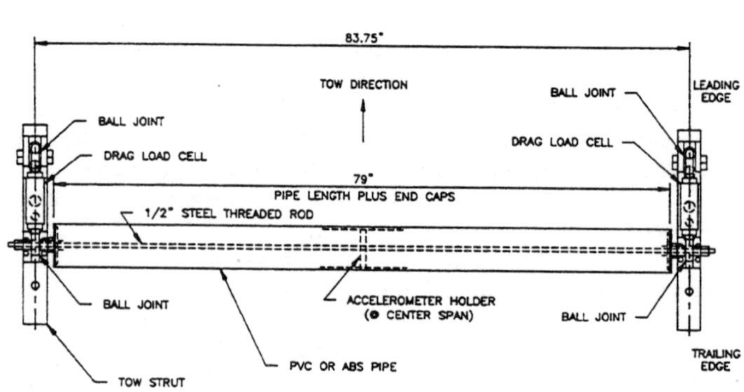

Fig. 3 Test Setup (Bottom View Looking Up)

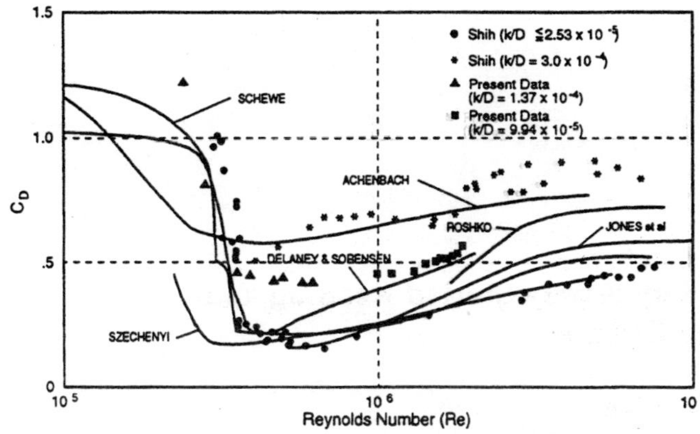

Fig. 4 Stationary Cylinder Drag Coefficients

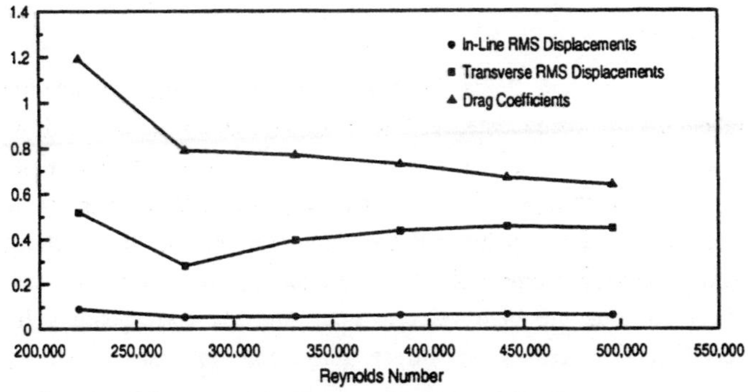

Fig. 5 Displacement and Drag Coefficient Results for 3 1/2 in. Diameter Smooth Cylinder Tests

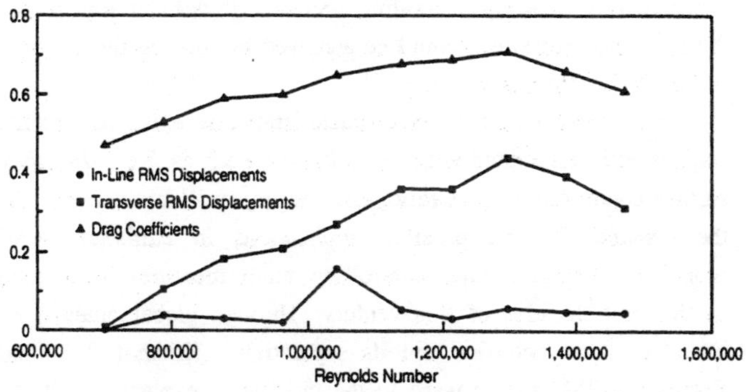

Fig. 6 Displacement and Drag Coefficient Results for 5 9/16 in. Diameter Smooth Cylinder Tests

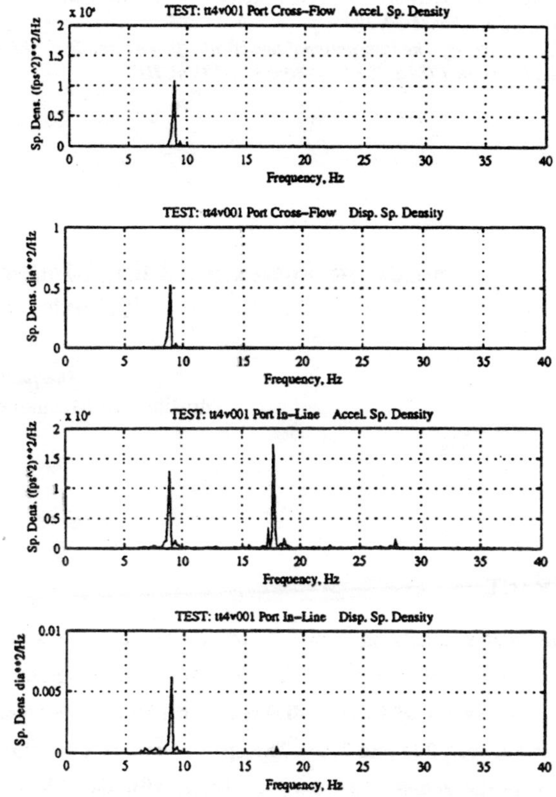

Fig. 7 Acceleration and Displacement Power Spectra

Proceedings of the Seventh (1997) International Offshore and Polar Engineering Conference
Honolulu, USA, May 25-30, 1997

ISBN 1-880653-28-1 (Set); ISBN 1-880653-31-1 (Vol. III)

Study on Vortex Shedding Modes from an Impulsively Started Rotating and Translating Circular Cylinder

Xuequan E and Caimao Luo
Institute of Mechanics, Chinese Academy of Sciences
Beijing, China

ABSTRACT

The vortex shedding mode from an impulsively started rotating and translating circular cylinder is investigated by compact scheme FDM (finite-difference method) for Reynolds number between 200 and 3000. The open boundary condition is utilized as the remote boundary condition with the 4th order accurate wall vortex generation formula proposed by us. The initial time step is increased from 10^{-8} to 10^{-2} till the time reaching the normal time step. The FDM scheme is 2nd accurate in space step and 1st accurate in time step. The streamline pattern at Re=1000, α=0.5 is a little different from previous calculated results, the present pattern indicates the double transposition phenomenon just as Coutanceau & Menard (1985b) pointed out. And at Re=3000, α=0.5, the lower vortex is found to be shedding prior to the upper primary vortex. The present vortex mode transition Re number is shifted from 20000 to at least 3000 compared with results calculated by Chang(1991) etc. Compared with the M. Coutanceau's flow visual work in France for Re less than 1000, the present results are good.

Key words: Wake, Vortex shedding mode, rotation and translation, FDM, saddle point.

1. INTRODUCTION

Because the flow around rotating and translating circular cylinder has two characteristics, limiting the vortex shedding and raising the lift force, so many scholars have been attracted to this problem investigation since the well known Magnus experiment in 1853. Asymptotic analysis, finite difference and flow visualization have been utilized for the problem study.

For inviscid fluids, Bachelor(1967)gave a clear explanation about the lift force existence. Due to the asymmetry of the two flows on each side of the cylinder, the speed on the lower side which moves downstream is higher than that on the upper side which moves upstream, leading to the asymmetrical pressure distributions produced by the Bernoulli effect and consequently to a transverse force towards the high velocity side. And the corresponding lift coefficient increases in direct proportion to the rotating-to-translating speed ratio α (=$\Omega a/U_\infty$, Ω being the angular velocity of rotating circular cylinder, a the radius of the cylinder, U_∞ the ambient flow velocity). Prandtl(1925) carried out the earliest visual experiment of flow past a rotating circular cylinder and argued that the maximum lift which could be achieved by rotating the cylinder in free stream flow is 4π.

For real fluids, the two characteristics also exist, but the lift coefficient is associated with many factors, such as, Re (=$2aU_\infty/\nu$), surface conditions, aspect ratio, end shapes, wall confinement. On the demand for the possible applications in ballistics, wind propulsion, boundary layer controlling, many measures were made in the first decades of this century. Through global analysis of boundary layer behavior and its separation , Krhan (1955) and Swanson (1961) put forward to the qualitative explanation of the experimental results. And many other scholars, such as Prandtl and Tietjens (1934), Ludwig(1964), Peller(1986), Chew (1987) and Tokumaru & Dimotakis(1993), tried to measure the lift force, pressure distribution around the cylinder. The results seemed to be particularly sensitive to the effects of aspect ratio and were rather scattered. M.Coutanceau & C. Menard (1985) investigated the early phase of the establishment of the flow past a circular cylinder started impulsively into rotation and translation through visualizing the flow patterns with solid tracers. Their experimental Re and α ranges are 200--1000 and 0--3.25 respectively. A lot of detailed photographs were present for the latter numerical contrasting.

Since M.Coutanceau & C. Menard (1985) presented their

clear near wake structure of the flow past a circular cylinder started impulsively into rotation and translation, this phenomena became the focus of the numerical study. Y.T.Chew, M. Cheng and S.C. Luo (1995) investigated the problem with Re=1000 and α=0--6.0, and argued that a critical α value of about 2 exists when a closed streamline begins to appear, below this critical value Karman vortex street occurring, above the value Karman vortex street stopping. However, Y.M.Chen, stopping. However, Y.M.Chen, Y.R. Ou & P.J. Pearlstein (1993) studied this problem with Re=200 and α=0.5--3.25, and indicated that rotation does not suppress vortex shedding for Re=200 and α=3.25. We noted that Y.T.Chew etc. computation domain is not as wide as Y.M.Chen etc., if the latter district larger, another vortex may be existed some where downstream. But the vortex shedding suppressing doesn't mean shedding one vortex only, to some degree the vortex shedding period becoming longer also means the same thing.

C.C. Chang & R.L. Chern (1991) investigated the flow past an impulsively started rotating and translating circular cylinder using hybrid vortex method (FDM + VIC (vortex in cell)) , proposed the three basic patterns of vortex shedding according to the stagnation points behavior associated with the primary upper and the primary lower vortices and gave the three basic regions of vortex shedding behind the cylinder (refer to Fig. 1), in which the lowest Re number for the R_∇ region is more than 10^4. And according to the communication with Prof. M. Coutanceau, the lowest Re number in their experiment is around 3000. Through direct difference method, we computed this problem again and obtained that the lowest Re number for the R_∇ region is near 2000.

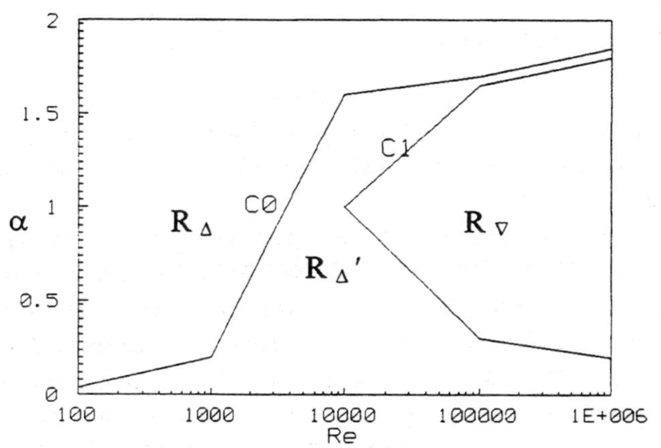

Fig.1 Three regions of vortex shedding behind the cylinder (from C.C.Chang & R.L. Chern (1991)).

2. Fundamental Equations

A non-rotating coordinate system translating with the cylinder is used during the present study. And the fluid is supposed to be incompressible, two dimensional. The fluid at infinity has a uniform velocity of magnitude of U_∞ in the x direction, and the cylinder rotates in the counterclockwise direction with angular velocity Ω, as shown in Fig. 2.

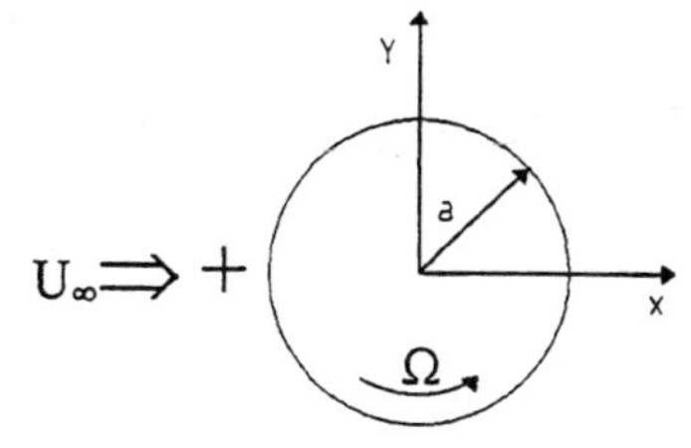

Fig.2 Definition sketch

Stream function/Vorticity formulation is adopted here. In two dimensions, the equations and boundary conditions are:

$$\frac{\partial\tilde{\omega}}{\partial\tilde{t}}-\frac{1}{\tilde{r}}\left[\frac{\partial}{\partial\tilde{r}}\left(\tilde{\omega}\frac{\partial\tilde{\psi}}{\partial\theta}\right)-\frac{\partial}{\partial\theta}\left(\tilde{\omega}\frac{\partial\tilde{\psi}}{\partial\tilde{r}}\right)\right]=\upsilon\nabla^2\tilde{\omega} \tag{1}$$

$$\tilde{\omega}=\nabla^2\tilde{\psi} \tag{2}$$

in which $\tilde{\omega},\tilde{\psi},\tilde{r},\tilde{\theta},\tilde{t}$ stand for dimensional vorticity, stream function, radial and orientation angular coordinates and time respectively.

$$\nabla^2=\frac{\partial^2}{\partial\tilde{r}^2}+\frac{1}{\tilde{r}}\frac{\partial}{\partial\tilde{r}}+\frac{1}{\tilde{r}^2}\frac{\partial^2}{\partial\theta^2}$$

$$\tilde{u}=-\frac{1}{\tilde{r}}\frac{\partial\tilde{\psi}}{\partial\theta},\tilde{v}=\frac{\partial\tilde{\psi}}{\partial\tilde{r}},\tilde{\omega}=\frac{1}{\tilde{r}}\left[\frac{\partial}{\partial\tilde{r}}(\tilde{v}\tilde{r})-\frac{\partial\tilde{u}}{\partial\theta}\right]$$

$$\tilde{r}=ae^{\pi\xi},\theta=\pi\eta,t=U_\infty\tilde{t}/a$$

Assuming cylinder radius a the characteristic length of r, a/U_∞ the time scale, and U_∞ the velocity scale, we obtained the following non-dimensional equations :

$$\frac{R}{2}\left[g(\xi,\eta)\frac{\partial\omega}{\partial t}+\frac{\partial}{\partial\eta}\left(\frac{\partial\psi}{\partial\xi}\omega\right)-\frac{\partial}{\partial\xi}\left(\frac{\partial\psi}{\partial\eta}\omega\right)\right]=\nabla^2\omega \tag{3}$$

$$\nabla^2\psi=g(\xi,\eta)\omega \tag{4}$$

$$\nabla^2=\frac{\partial^2}{\partial\xi^2}+\frac{\partial^2}{\partial\eta^2},g(\xi,\eta)=\pi^2e^{2\pi\xi}$$

in which, $R=\frac{2U_\infty a}{\upsilon},\psi=\frac{\tilde{\psi}}{U_\infty a},\omega=\frac{\tilde{\omega}a}{U_\infty},\alpha=\frac{\Omega a}{U_\infty}$

and the boundary conditions are:

non-slip condition over the cylinder surface:

$t\geq 0,\psi=0,\frac{\partial\psi}{\partial\xi}=\pi\alpha$, $\xi=0$

no restrict on ω, $\xi=0$

$\psi=2\sinh\pi\xi\sin\pi\eta$, $\xi\rightarrow\infty$

$$u=-\frac{1}{\pi e^{\pi\xi}}\frac{\partial\psi}{\partial\eta},\ v=\frac{1}{\pi e^{\pi\xi}}\frac{\partial\psi}{\partial\xi}$$

$$\omega=\frac{1}{\pi e^{\pi\xi}}\left[\frac{1}{e^{\pi\xi}}\frac{\partial}{\partial\xi}(ve^{\pi\xi})-\frac{\partial u}{\partial\eta}\right]$$

in which u, v are defined as:

$$u=\frac{\partial\psi}{\partial\xi},\ v=-\frac{\partial\psi}{\partial\eta}$$

Therefore, we can get:

$$\xi=0,\ \frac{\partial^2\psi}{\partial\xi^2}=g(\xi,\eta)\omega$$

3. NUMERICAL METHODS

The vorticity transport equations are obtained by ADI(alternative difference implicit) method.

Along η direction, we have

$$R\frac{g_{i,j}}{\Delta t}\omega_{i,j}^{n+1/2}-\frac{\omega_{i,j+1}^{n+1/2}-2\omega_{i,j}^{n+1/2}+\omega_{i,j-1}^{n+1/2}}{(\Delta\eta)^2}+\frac{R}{4}\frac{(u\omega)_{i,j+1}^{n+1/2}-(u\omega)_{i,j-1}^{n+1/2}}{\Delta\eta}=R\frac{g_{i,j}}{\Delta t}\omega_{i,j}^{n}+\frac{\omega_{i+1,j}^{n}-2\omega_{i,j}^{n}+\omega_{i-1,j}^{n}}{(\Delta\xi)^2}-\frac{R}{4}\frac{(v\omega)_{i+1,j}^{n}-(v\omega)_{i-1,j}^{n}}{\Delta\xi} \quad (5)$$

Boundary condition is:

when $\eta=0,\eta=1$, ω=0

Along the x direction, we have:

$$R\frac{g_{i,j}}{\Delta t}\omega_{i,j}^{n+1}-\frac{\omega_{i+1,j}^{n+1}-2\omega_{i,j}^{n+1}+\omega_{i-1,j}^{n+1}}{(\Delta\xi)^2}+\frac{R}{4}\frac{(v\omega)_{i+1,j}^{n+1}-(v\omega)_{i-1,j}^{n+1}}{\Delta\xi}=R\frac{g_{i,j}}{\Delta t}\omega_{i,j}^{n+1/2}+\frac{\omega_{i,j+1}^{n+1/2}-2\omega_{i,j}^{n+1/2}+\omega_{i,j-1}^{n+1/2}}{(\Delta\eta)^2}-\frac{R}{4}\frac{(u\omega)_{i,j+1}^{n+1/2}-(u\omega)_{i,j-1}^{n+1/2}}{\Delta\eta} \quad (6)$$

And the accuracy is $((Dx)^2,(Dh)^2,Dt)$. When $\xi=0$ and $\xi=\xi_\infty$, boundary conditions about w are as follows:

The 4th degree accuracy formula, when ξ=0:

$$\zeta(0,j)=2[4(\psi(2,j)-\pi\alpha\Delta\xi)-1.5(\psi(3,j)-2\pi\alpha\Delta\xi)+4(\psi(4,j)-3\pi\alpha\Delta\xi)/9-((\psi(5,j)-4\pi\alpha\Delta\xi)/16]/(\pi\Delta\xi)^2 \quad (7)$$

At infinity, we may choose closed boundary condition, ω=0 ($\xi=\xi_\infty$), or the open boundary condition (neglecting the viscosity diffusion effect and considering the vorticity transporting term), $\left.g(\xi,\eta)\frac{\partial\omega}{\partial t}+\frac{\partial}{\partial\eta}(\frac{\partial\psi}{\partial\xi}\omega)-\frac{\partial}{\partial\xi}(\frac{\partial\psi}{\partial\eta}\omega)\right|_{\xi=\xi_\infty}=0$, and similar to the Sommerfield radiation condition, this condition can be changed as:

$$g(\xi_\infty,\eta_0)\frac{\omega^{n+1}(\xi_\infty,\eta_0)-\omega^{n-1}(\xi_\infty,\eta_0)}{2\Delta t}+\frac{1}{2\Delta\eta}[(\frac{\partial\psi}{\partial\xi}\omega)^n(\xi_\infty,\eta_0+\Delta\eta)-(\frac{\partial\psi}{\partial\xi}\omega)^n(\xi_\infty,\eta_0-\Delta\eta)]$$

$$-\frac{1}{\Delta\xi}[(\frac{\partial\psi}{\partial\eta}\frac{\omega^{n+1}+\omega^{n-1}}{2})\Big|(\xi_\infty,\eta_0)-\frac{\partial\psi}{\partial\eta}\omega^n\Big|(\xi_\infty-\Delta\xi,\eta_0)]=0$$

The velocity field is solved by the Hermitian formula from the Poisson equation results. The u, v components are defined as:

$$u^{n+1}{}_{i,j}=\left(\frac{\partial\psi}{\partial\xi}\right)^{n+1}_{i,j},\ v^{n+1}{}_{i,j}=-\left(\frac{\partial\psi}{\partial\eta}\right)^{n+1}_{i,j}$$

Through Hermitian formula, from $\Psi^{n+1}{}_{i,j}$, we can obtain:

$$u^{n+1}{}_{i+1,j}+4\times u^{n+1}{}_{i,j}+u^{n+1}{}_{i-1,j}=\frac{3}{\Delta\xi}(\Psi^{n+1}{}_{i+1,j}-\Psi^{n+1}{}_{i-1,j}) \quad (8)$$

Boundary condition is:

when $\xi=0$, $u=\frac{\partial\psi}{\partial\xi}=\pi\alpha$

when $\xi=\xi_\infty$, $u=\frac{\partial\psi}{\partial\xi}=2\pi\cosh(\pi\xi_\infty)\sin(\pi\eta)$

And the same is for v component:

$$v^{n+1}{}_{i,j+1}+4\times v^{n+1}{}_{i,j}+v^{n+1}{}_{i,j-1}=-\frac{3}{\Delta\eta}(\Psi^{n+1}{}_{i,j+1}-\Psi^{n+1}{}_{i,j-1}) \quad (9)$$

And the Poisson equation is solved by point SOR method.

4. Results and Discussion

Present calculation is carried out for 200≤Re≤3000 and 0≤α≤3.25 on Pentium586 with main frequency 120MHz. The computation domain is 0≤ξ≤1.0, 0≤η≤2, or $a\le r\le e^{2\pi}a=23.14a$, 0≤θ≤2π.

We have carried out many computation cases, such as, Re=200, 300, 500, 550, 1000, 1700, 2000, 2500, 3000. And here we just present Re=200, 1000, 3000. For they are the typical cases and locate on the three regions R_Δ , $R_\Delta{}'$, R_∇ respectively. The streamline pattern at Re=1000, α=0.5 is a little different from previous calculated results, the present pattern indicates the double transposition phenomenon just as Coutanceau & Menard (1985b) pointed out. And at Re=3000, α=0.5, the lower vortex is found to be shedding prior to the upper primary vortex.

4.1 R_Δ mode, Re=200, α=0.5, 1.0

Fig. 3 shows series of the streamline patterns at τ=1,,4,8,9, 17, 18, 19. From τ=1 pattern, it indicates the existence of the upper right vortex and a bulge phenomenon of streamlines near the right lower side of the cylinder. τ=4 pattern shows the upper primary vortex (E_1) grows gradually in the upper wake and the bulge leads to the second lower vortex formation (E_2). And it should be noted that the saddle point (or closure point) of the upper vortex (S_1) doesn't exchange with S_2. And S_2 is near the cylinder surface before S_2 exchanges with S_3. The saddle point exchange phenomena is called transposition phenomena. τ=9 pattern shows this phenomena. τ=19 pattern repeats τ=9 very well, indicating the periodicity of the vortex shedding. Above all, the characteristics of the R_Δ are: the first upper vortex shedding first, only once saddle point exchange, and mode periodicity.

4.2 R_Δ' mode, Re=1000, α=0.5

τ=1 pattern shows the existence of the upper vortex E_1 and a bulge phenomenon which will result in the formation of the lower vortex E_2. τ=2 pattern confirms appearance of the lower vortex E_2, and the growth of vortex E_1. τ=3 pattern shows the transposition phenomenon between the saddle points S_1 and S_2, with vortex E_2 on its way to detach from the cylinder. τ=4 pattern indicates that vortex E_1 competes its lost saddle point S_1 and finally obtains it back from vortex E_2 at τ=5 pattern. For τ=4 pattern, the experiment result is also given for comparison . From now on, vortex E_1 wins back its lost saddle point position, and the vortex E_2 saddle point S_2 exchanges with that of the E_3' at τ=11. So this vortex shedding pattern is called double transposition mode. According to the Chien-Cheng Chang & R.L. Chern (1991), this phenomenon occurs at Re=3000, α=1, 1.5. But Coutanceau & Menard (1985b) obtain the same conclusion to the present paper. To examine the present model, the velocity distributions along x axis or y-axis at $\theta=0^0$, 90^0 are given in Fig. 5, it can be seen that the present results agree quite well with the experiment results.

4.3 The R_∇ mode, Re=3000, α=1.0

For Re=3000, the detailed evolving of vortex is shown in Fig.6. τ=1 pattern shows the existence of the upper vortex E_1 and a bulge phenomenon which will result in the formation of the lower vortex E_2. τ=2 pattern confirms appearance of the lower vortex E_2, and the growth of E_1. τ=3 pattern shows the transposition phenomenon between the saddle points of E_1 and E_2, and E_2 on its way to detach from the cylinder. And E_1 competes with E_2 for the saddle point, but E_2 never lose again, eventually leading to its shedding prior to the E_1 vortex. But the E_2 is rather weak with appending to the upper shedding vortex E_1. From τ=3 pattern, the E_1' is generated, and coalesces with E_1 at τ=6, resulting in stronger of the upper vortex E_1. And the complete shedding of E_1 occurs at τ=11. Therefore, this R_∇ mode has the following properties: (1) the lower vortex shedding first; (2) the upper primary vortex assimilating another upper vortex; (3) the lower vortex appending to the upper primary vortex.

5. Conclusion

It's successful to use the FDM to calculate the flow field of starting rotation and translation impulsively of circular cylinder. Through the careful consideration on boundary conditions, we obtain some different calculated results, and they seem more reasonable than previous calculated data. We give three typical cases, responding to the three shedding modes proposed by Chien-Cheng Chang & R.L. Chern (1991). But Re=1000, α=0.5 falls into the R_Δ' region, rather than the R_Δ region as C.C Chang & R.L.Chern, just as observed by Coutanceau and Benard (1985b). And compared with the minimum Re number being 20000 for R_∇ mode, the present model shows that the lower vortex shedding first occurs no later than Re=3000, α=1. And these results are just part of the whole research. We will present the whole results including all flow properties in the near future.

6. Acknowledgement

This study is supported by The State Key Laboratory of Ocean Engineering of Shanghai Jiaotong University and National Natural Science Fundation of China.

7. References

1. Batchelor, G.K. 1967 An Introduction to Fluid Dynamics. Cambridge University Press.

2. BADR,H.M., Coutanceau,M., Dennis, S.C.R. & Menard,C. 1990 Unsteady flow past a rotating circular cylinder at Reynolds number 10^3 and 10^4. J. Fluid Mech. 220, 459-484.

3. BADR,H.M. & Dennis, S.C.R. 1985 Time-dependent viscous flow past an impulsively started rotating and translating circular cylinder. J. Fluid Mech. 158, 447-488.

4. Chew, Y.T. 1987 Flow past a rotating cylinder. In Proc. Intl Conf. on Fluid Mechanics, Beijing, China.

5. Chew, Y.T., M.Cheng and S.C.Luo, 1995 A numerical study of flow past a rotating circular cylinder using a hybrid vortex scheme. J.F.M. 299, 35-71

6. Chien-Cheng Chang & Ruey-Ling Chern 1991 Vortex shedding from an impulsively started rotating and translating circular cylinder. J. Fluid Mech. 233, 265-298.

7. Coutanceau,M. & Menard,C. 1985 Influence of rotation on the near-wake development behind an impulsively started circular cylinder. J. Fluid Mech.158, 399-446.

8. Coutanceau,M. & Menard,C. 1985b Visualization of the eddy formation behind a rotating and translating circular cylinder. In Proc. Int. Symp. on Physical & Numerical Flow Visualization ASME-FED

Vol.22, pp9-19.

9. Krhan, E. 1955 The laminar boundary layer on a rotating cylinder in cross for. NAVORD Rep. 4022, Aerobal. Res. Rep. 288, U.S., NOL. Maryland.

10. Ludwig , G.R. 1964 An experimental investigation of laminar separation from a moving wall. AIAA Paper 64-6.

11 Peller, H. 1986 Thermofluiddynamic experiments with a heated and rotating circular cylinder in cross flow, Part2.1: Boundary layer profiles and location of separation points. Expts. Fluids 4, 223-231.

12. Prandtl, L. & Tietjens, O.G. 1934 Applied Hydro- and Aeromechanics (transl. J.P. Den Hartog 1957). Dover.

13. Swanson, W.M. 1961 The Magnus effect: a summary of investigation to date. Trans. ASMED: J. Basic Engng 461-470.

14. Tokumaru, P.T. & Dimotakis, P.E. 1993 The lift of a cylinder executing rotary motions in a uniform flow J.F.M 255, 1-10.

15. Yen-Ming Chen, Yuh-Roung Ou & Arne J. Pearlstein 1993 Development of the wake behind a circular cylinder impulsively started into rotatory and rectilinear motion. J. Fluid Mech. 253, 449-484.

16. Y.T.CHEW, M. CHENG & S.C. LUO 1995 A numerical study of flow past a rotating circular cylinder using a hybrid vortex scheme. J. Fluid Mech. 299, 35-71

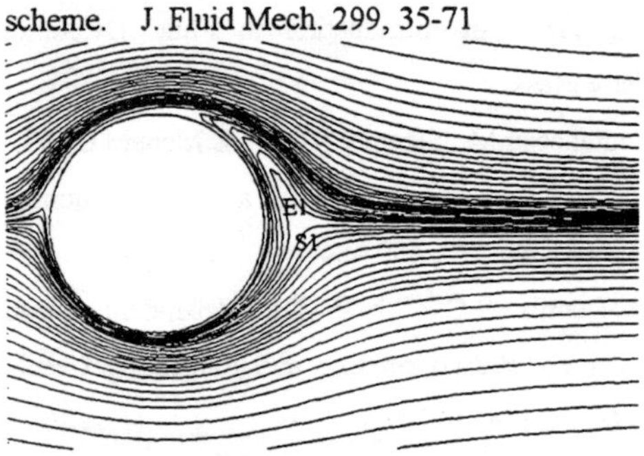

τ=1

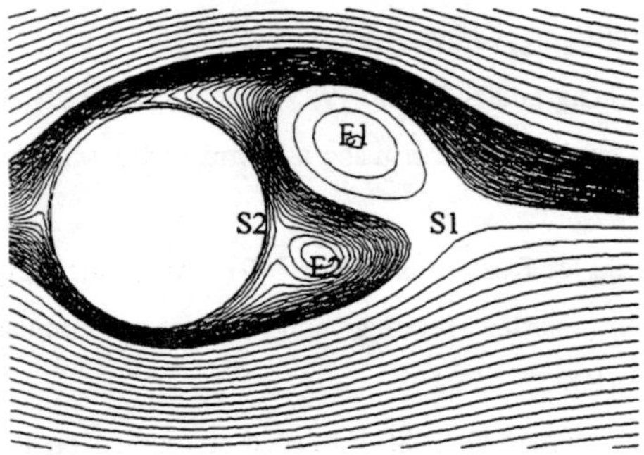

τ=4

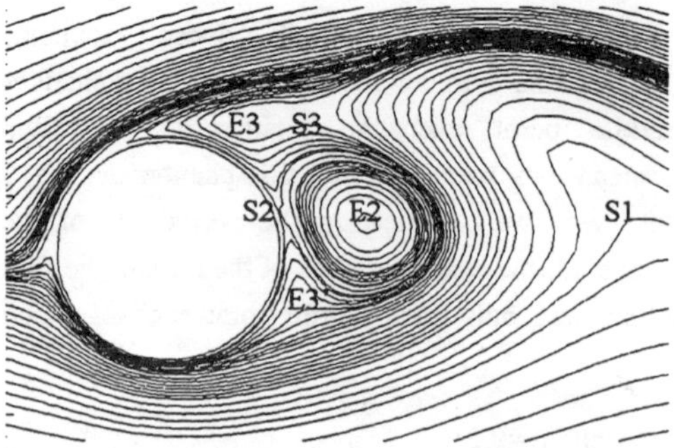

τ=8

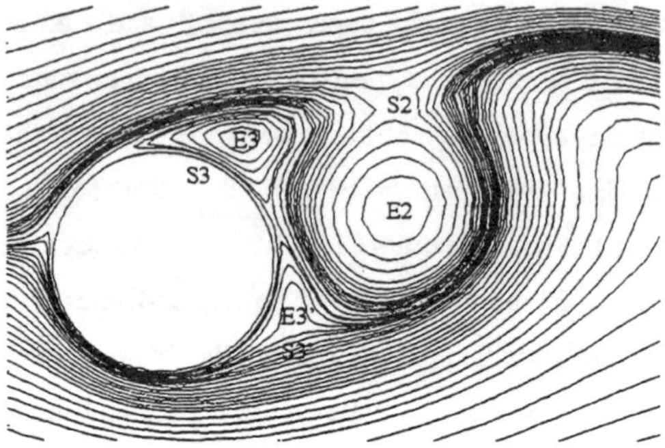

τ=9

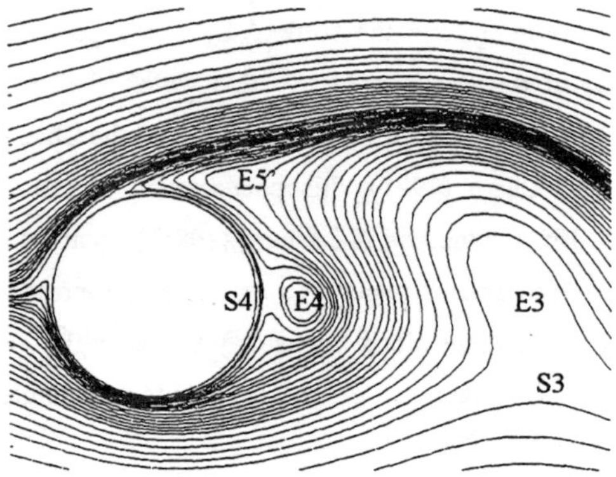

τ=17

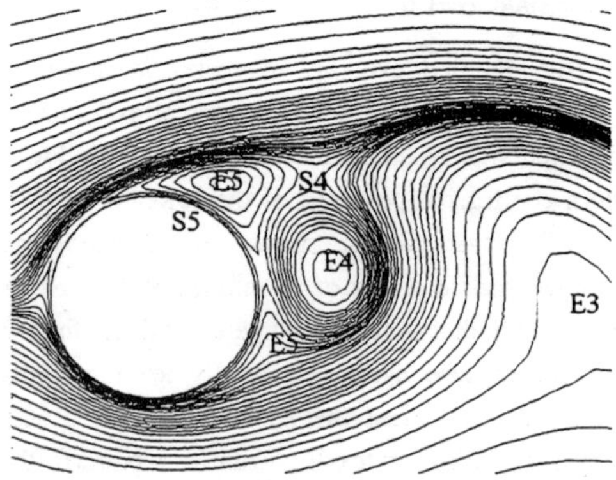

τ=18

For the graph caption , refer to the next page

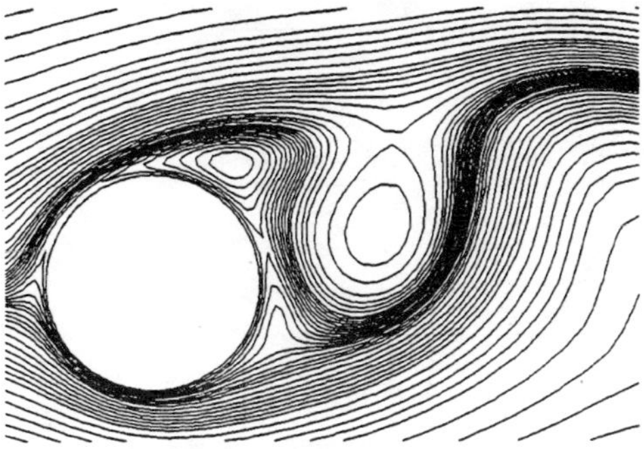

τ=19

Fig. 3 Re=200,α=0.5,Dt=0.02,Δξ=Δη=1/60, the streamline pattern for τ=1,4,8,9, 17, 18, 19, ψ=0, ±0.01, ±0.02, ±0.03, ±0.04, ±0.05, ±0.06, ±0.08, ±0.10, ±0.12, ±0.15, ±0.2, ±0.25, ±0.3, ±0.35, ±0.40, ±0.45, ±0.50, ±0.6, ±0.7, ±0.8, ±0.9, ±1.0,then the gap is 0.2.

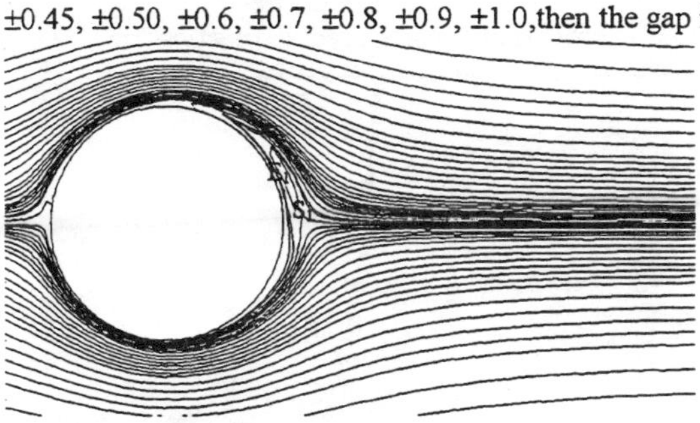

τ=1

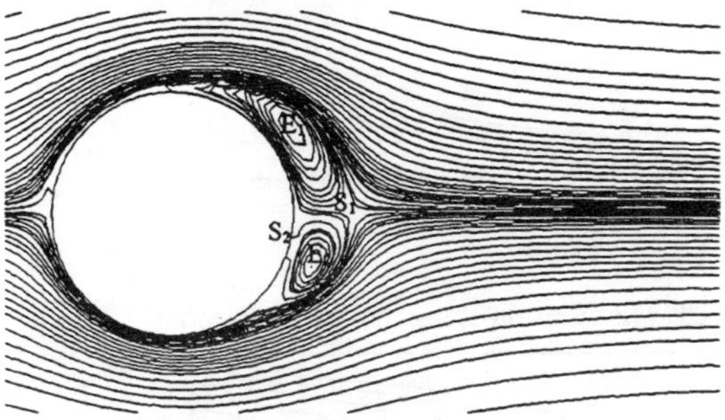

τ=2

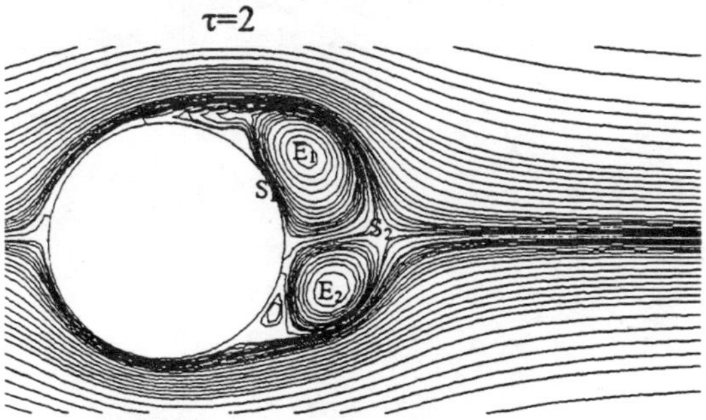

τ=3

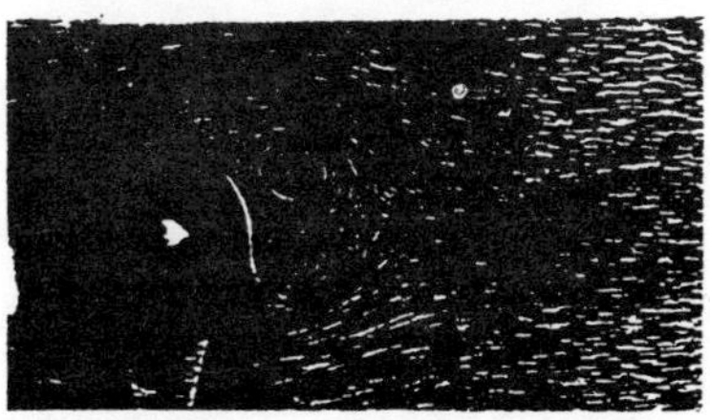

τ=4 experiment visual photograph by Coutanceau

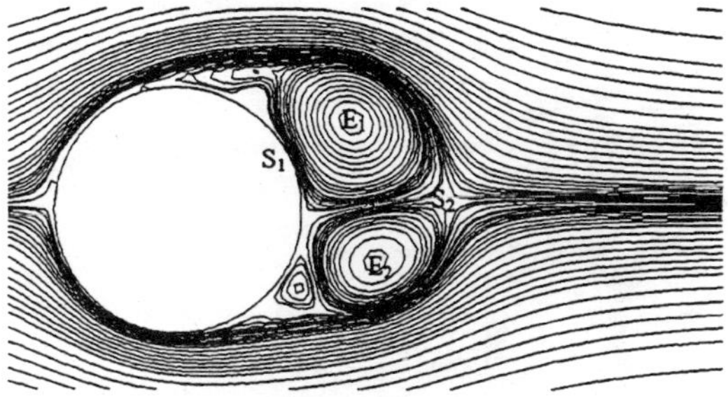

τ=4

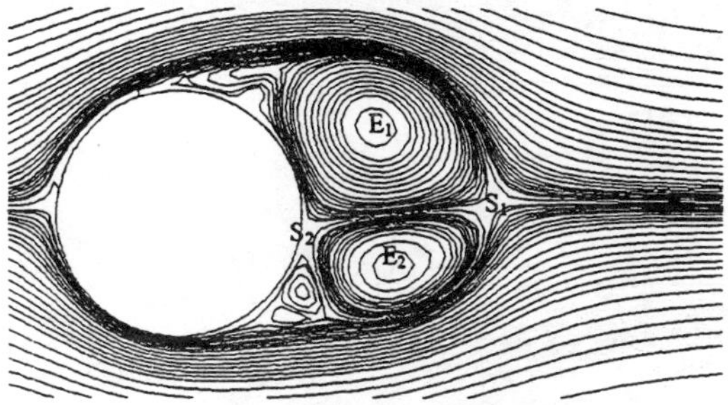

τ=5

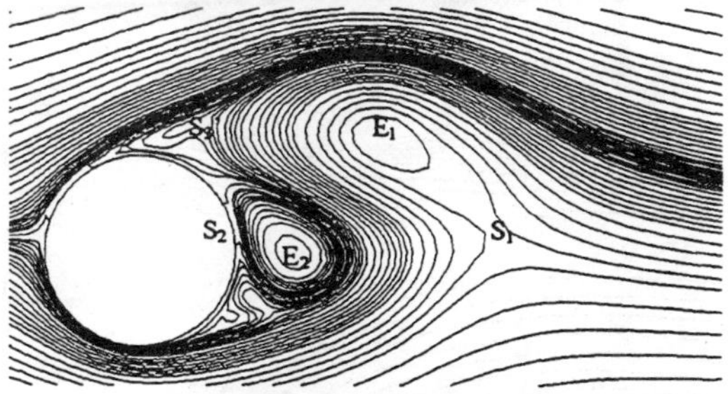

τ=10

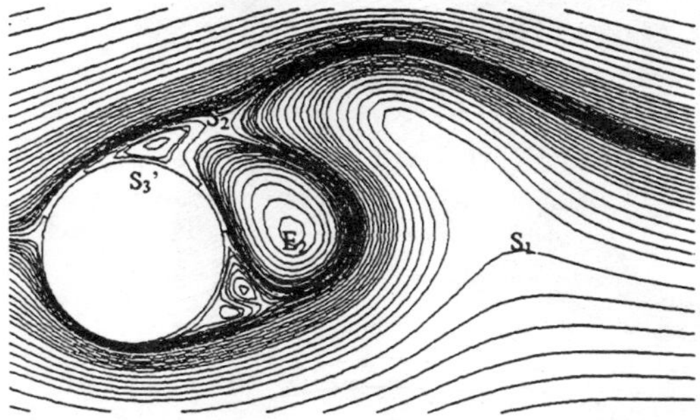

τ=11

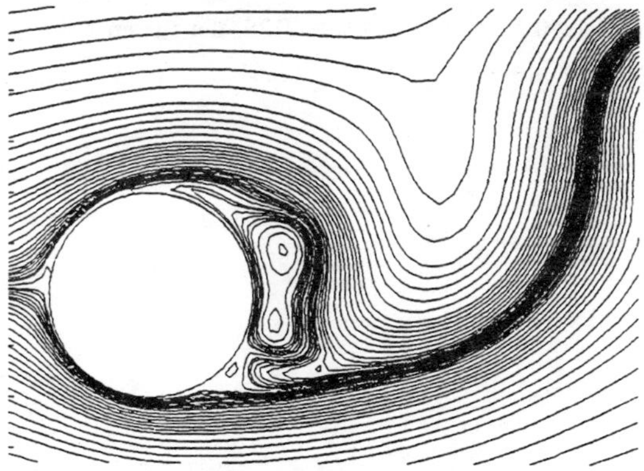

τ=14

For graph caption, refer to the next page

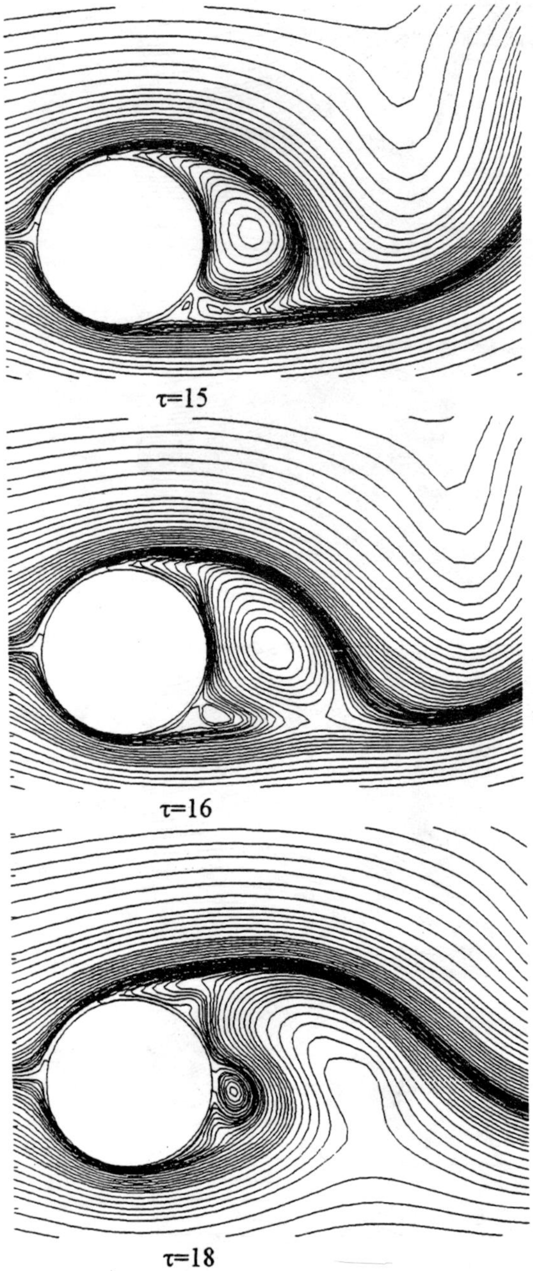

Fig. 4 Re=1000, α=0.5,Dt=0.01,Δξ=Δη=1/120, the stream function values are identical with the Fig. 3.

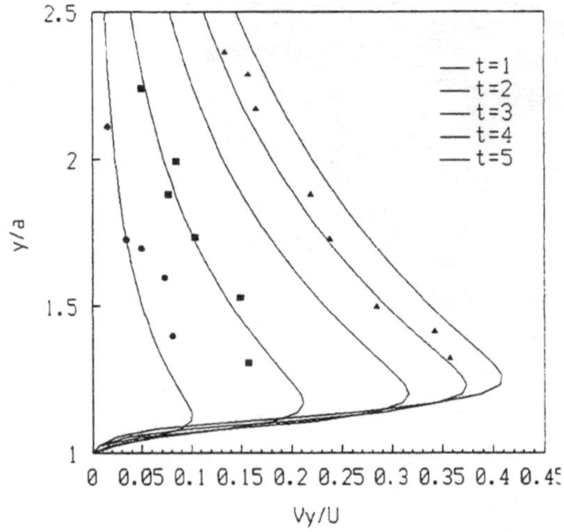

θ=90^0, V_y/U Distribution along the radial direction

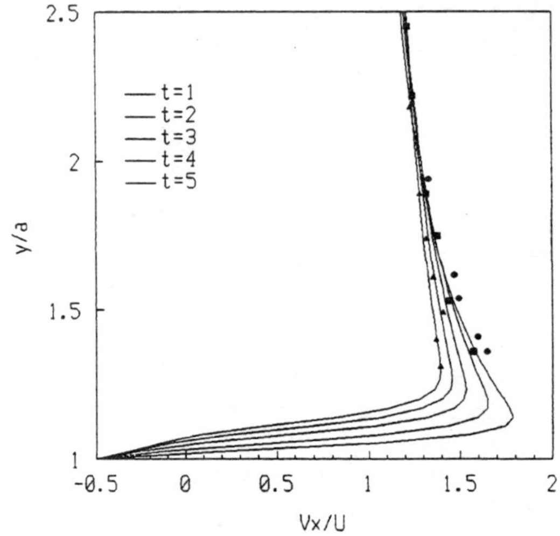

θ=90^0, V_x/U Distribution along the radial direction

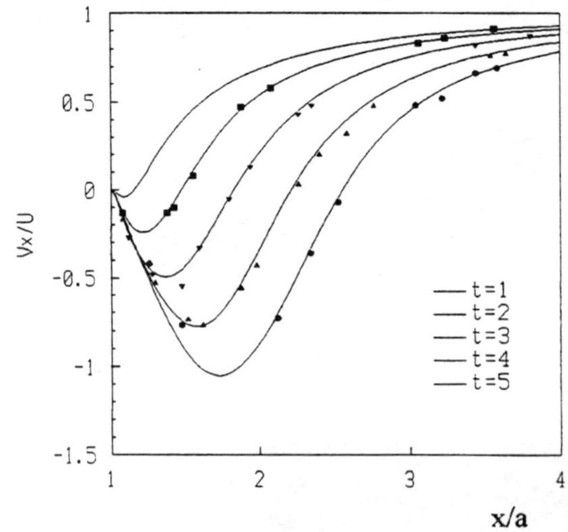

θ=0^0, V_x/U Distribution along the radial direction

Fig.5 Re=1000, α=0.5,Dt=0.01,Δξ=Δη=1/120. Comparison of velocity profiles for Re=1000, α=0.5

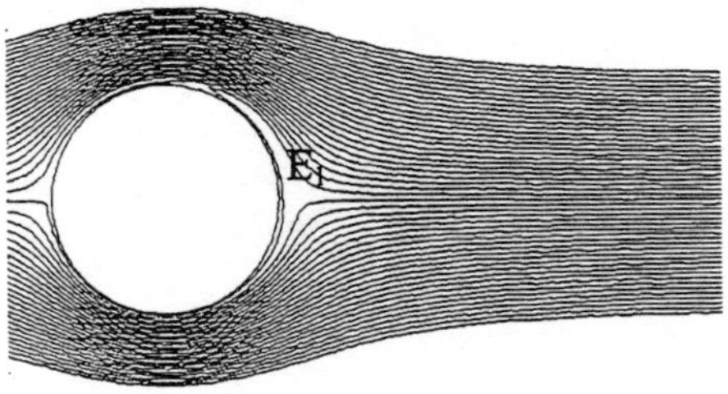

τ=1

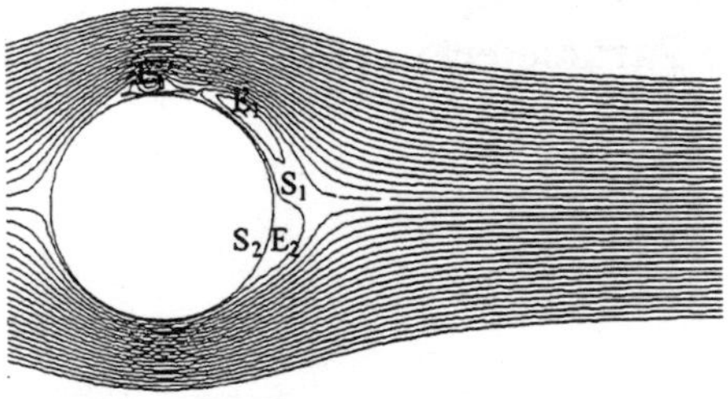

τ=2

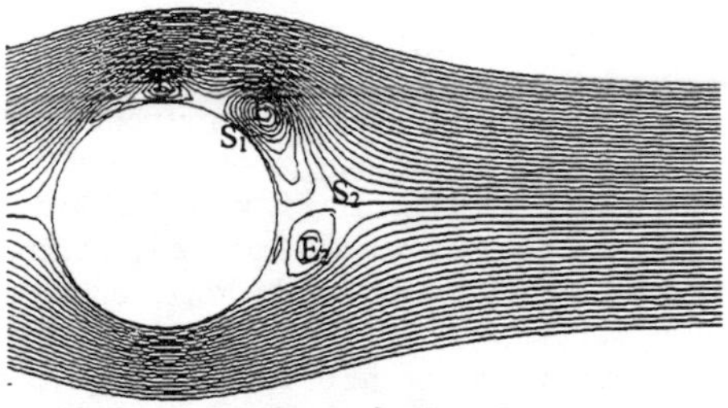

τ=3

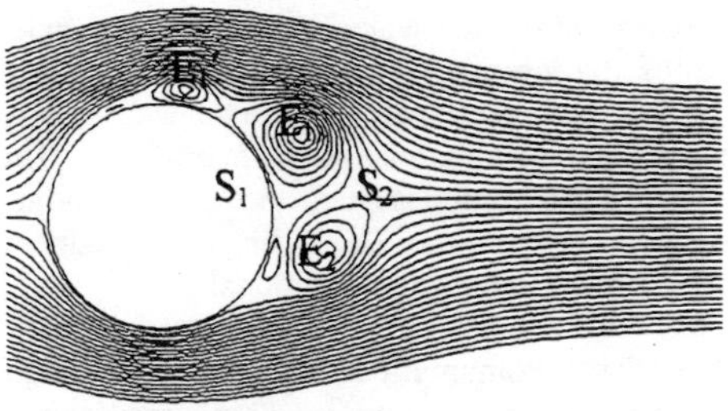

τ=4

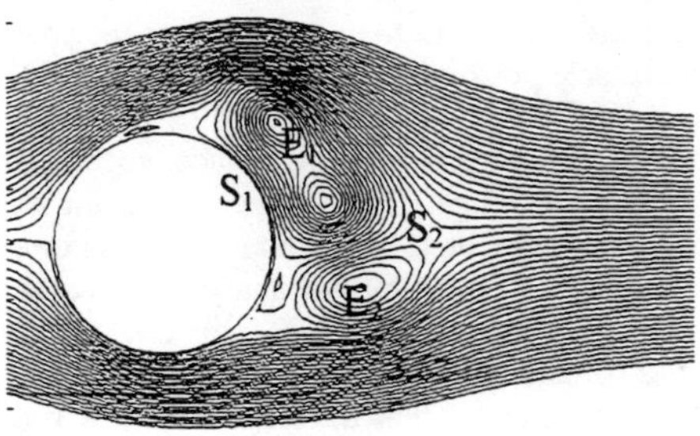

τ=6

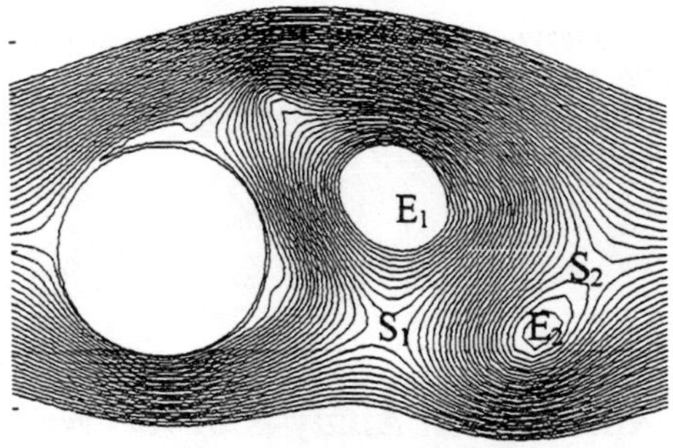

τ=11

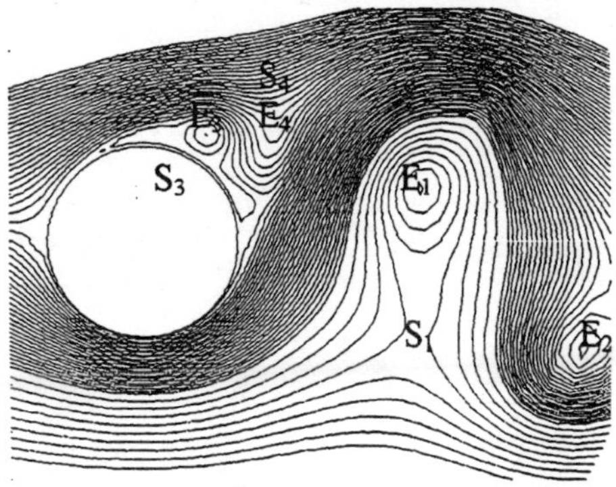

τ=13

Fig.6 Re=3000, α=1.0,Dt=0.01,$\Delta\xi$=1/180, $\Delta\eta$=1/90

Proceedings of the Seventh (1997) International Offshore and Polar Engineering Conference
Honolulu, USA, May 25-30, 1997

ISBN 1-880653-28-1 (Set); ISBN 1-880653-31-1 (Vol. III)

Experimental Analysis of the Wake Instabilities of an Oscillating Cylinder Near Synchronization: Influence of the Amplitude Parameter

O. Rodriguez
ONERA-Institut de Mécanique des Fluides de Lille
Lille, France

ABSTRACT

The large scale vortices in the wake of a rigid cylinder undergoing forced oscillations at amplitude values equal to 0.237, 0.475 and 1 are investigated throughout the synchronisation range by using a schlieren method. A precise description is given of the different regimes observed in the frequency range. The vortex shedding is locked to the cylinder oscillation so that the ratio between the two pulsations remains rational and keeps a constant value over the synchronisation range. The phase angle of the vortex shedding versus reduced velocity has been determined for every regime.It exhibits strong variations when the cylinder to vortex shedding frequencies ratio varies from 0.5 to 1.

KEY WORDS: Oscillating cylinders, resonant vibration, vortex shedding, visualization of flow, fluid-solid interactions.

1. INTRODUCTION

The hydrodynamic forces applied to a cable normal to a flow result from a fluid-structure coupling between the wake, which generates an oscillating crosswise force, and the cable, which oscillates due to this force. The phenomenon becomes critical when the wake frequency is near a cable vibration mode, causing an increase in the vibration amplitude and sudden changes of phase between the cable movement and the unsteady crosswise force. When the wake frequency locks onto the cable oscillation frequency, there is a large increase in drag.

Reviews of the research conducted in last years on the problem of synchronization of an oscillating obstacle with its wake were published by Parkinson (1974), Sarpkaya (1979), Bearman(1984) and Griffin and Hall (1991).

These works showed that large phase variations between the unsteady lift F_y and the cylinder position Y appear when a cylinder is subjected to forced oscillation near synchronization.These changes of phase mean that the instant of the oscillation cycle for which the vortex shedding generates maximum lift varies enormously with the reduced speed.

Contrary to the above properties, relatively well described, concerning the cylinder movement and applied forces, the near wake structure is still poorly known. Zdravkovich made a step in this direction by comparing flow visualizations from different authors concerning cylinders with forced or free oscillations (1982). He suggested that the changes of phase occurring in the unsteady lift near synchronization corresponded to changes of phase in the alternating vortex shedding process. The phase shift between the two shedding regimes characterizing synchronization can be considered equal to π. More recently, Ongoren and Rockwell produced wake visualizations of an oscillating cylinder using hydrogen bubbles. They correlated the signal transmitted by a hot film placed in the wake with the signal from a motion sensor located on the obstacle and supplying the instantaneous cylinder position (1988). The recorded phase shift was also equal to π.

Williamson and Roshko examined the wake of a moving cylinder pulled by trolley along an oscillating path and submerged in a tank of water at rest (1988). The optical method used to visualize the wake was to observe the path of suspended aluminum particles on the free surface. The results obtained showed that synchronization is characterized by shedding of a pair of vortices each oscillation half-cycle. Depending on the reduced frequency, the vortices shed are subjected to an additional pairing in the near wake or are subdivided, which results in the emergence of different regimes. The authors suggest that maximum lift coincides with the critical regime, defined as the regime for which the vorticity distribution is most concentrated in the near wake.

Feng (1968) showed that near synchronization, in addition to the phase variations between lift and obstacle position, there was a large hysteresis in the movement of a freely oscillating, weakly damped cylinder. The movements were higher in amplitude for an increasing reduced speed than for a decreasing speed over the same range. This nonlinear behavior was recently confirmed by Brika and Laneville (1993) who evidenced vortex shedding regimes similar to those described by Williamson and Roshko (1988).

An earlier analysis was aimed at determining the instantaneous structure of the alternating vortex street that develops behind a rigid cylinder heaving at a rate near synchronization and correlating this structure with the movement of the obstacle (Rodriguez (1996)). The synchronization range was scanned by varying the cylinder oscillation frequency in small steps while holding the reduced amplitude A/D constant at 0.237. The analysis method was based on the schlieren visualization technique used, with thermal marking of the alternating vortices (Rodriguez (1991)). This made it possible to describe the flow regimes in detail and determine the vortex shedding phase fairly accurately.

The present analysis was for the purpose of completing the above one by determining the influence of the amplitude parameter on appearance of the synchronization phenomenon and the flow structure characterizing it. The aim was to characterize the phenomenon for values of $A/D < 1$, corresponding to the maximum fluctuation recorded by Feng (1968) and Brika and Laneville (1993) during experiments on an oscillating free cylinder. The tests for this work were conducted for $A/D = 0.475$ and 1. Because of the influence of the lateral walls, higher values of A/D could not be achieved, although amplitudes equal to 1.5 or 2 diameters appear to have been observed in some cases (Bearman (1984)). This paper reviews the main results discussed by Rodriguez (1996) and compares the regimes observed for the three values of the reduced amplitude A/D considered.

2. EXPERIMENTAL SETUP

2.1. Installation

The tests were conducted in a water tank. The test section is vertical with a cross section of 140x104 mm^2. The model was a metal cylinder with a diameter $D = 1$ cm and a span equal to the width of the test section. The amplitude A of cylinder movement was adjustable and was determined by a movable cam (Fig.1). The far-field velocity was adjusted by an electromagnetic flowmeter with an accuracy of 0.5 percent. A position sensor integral with the drive system registered the cylinder movement. The signal transmitted was digitized and recorded. After processing, it gave the effective oscillation frequency N_C and the instantaneous position Y of the cylinder.The present tests were conducted for $N_C \leq 3.8$ Hz. The far-field velocity U was equal to 7 cm/s corresponding to a Reynolds number Re = 700. Under these conditions, $N_C D/U \leq 0.55$ and the reduced frequency variation interval was sufficiently large to include the synchronization phenomenon.

2.2. Visualization system

The cylinder wake was visualized by the schlieren technique. This technique is based on variations of the liquid refraction index with temperature. This property can be used to identify the individual turbulent structures of a wake. Its implementation is based on very low intensity thermal marking of the vortices starting from the separation lines (Rodriguez, 1991). In the case at hand, the cylinder was equipped over its entire span with two heating elements 1.5 mm in diameter, diametrically opposite one another, mounted parallel to the generatrices and flush with the cylinder surface near the separation points (Rodriguez (1996)).

The flow schlieren visualizations were recorded in video mode. The wake frequency N_V was determined by counting the number of vortices that crossed a fixed abscissa over a given time interval.

The vortex shedding phase was defined with respect to the cylinder position Y/A at the time of shedding. The vortices can be considered shed when they are no longer part of the recirculation region, i.e. when they are no longer supplied from the mixing regions resulting from separation. On this basis, the shedding time can be defined as the time when the contact between the vortex and the mixing region from which it arose is broken.

3. RESULTS

The results obtained show that vortex shedding tends to adapt to the obstacle movement by synchronizing with the cylinder oscillation frequency. The synchronization of the two oscillating systems occurs for frequency values such that:

$$N_C/N_V = p/q$$

where p and q are integers.

The graphs of N_C/N_V versus $S_C = N_C D/U$ show that for $p/q > 1/2$, synchronization persists when S_C varies over a given interval (Fig. 2). In particular, for $A/D = 0.237$, the existence of synchronization ranges corresponding to $N_C/N_V = 1/2$, 1 and 2 was observed. For $A/D = 0.475$, the synchronization ranges correspond to $N_C/N_V = 1/2$, 1, 2 and 3. However, for $A/D = 1$, only one synchronization range relative to $N_C/N_V = 1/2$ was observed for the frequency range imposed on the cylinder. High metastability was observed for $A/D = 0.475$. Subharmonic regimes were observed for $p/q = 1/5$, 1/4, 2/7, 1/3 and 2/5. It is likely that a small synchronization range also corresponds to each of these subharmonic regimes.

When the cylinder is stationary, the Strouhal number of the vortex street is $S_{VS} = 0.222$. Vibration of the cylinder for values of S_C that are small compared to S_{VS} causes modification of the wake due to crosswise movement of the vortices shed successively during the obstacle oscillation cycle. The flow regimes observed are characterized by the vortex shedding phase and frequency.

3.1. Subharmonic regimes corresponding to $N_C/N_V < 1/2$

The appearance and nature of the subharmonic regimes is practically independent of parameter A/D. For $A/D = 0.237$, the regime corresponding to S_C near 0.05 is a subharmonic regime characterized by shedding of the vortex from the lower separation point (called L-vortex) in $Y/A = 1$, 0 and -1 successively and by shedding of the vortex from the upper separation point (called U-vortex) in $Y/A = \pm 0.7$ approximately. Since the L-vortices and U-vortices are shed alternately, the result is a vortex system with a frequency equal to four times the cylinder oscillation frequency (Fig. 3a). The flow regime is that of synchronization at $S_C/S_V = 1/4$. The subharmonic regime observed for $S_C = 0.06$ and corresponding to $S_C/S_V = 2/7$ is characterized by the vortex shedding pattern illustrated in Fig.3b. Figures 3c and 3d show the vortex shedding patterns of regimes synchronized at $S_C/S_V = 1/3$ and $S_C/S_V = 2/5$. The subharmonic regimes observed for $A/D = 0.475$ are strictly the same as those recorded for $A/D = 0.237$. However, for $A/D = 1$, synchronization is observed at $N_C/N_V = 1/5$, a narrow synchronization range for $N_C/N_V = 1/3$ and the absence of synchronization at $N_C/N_V = 2/5$. In addition, for $A/D = 1$, the vortex shedding phase of certain subharmonic regimes is slightly different from the one recorded at lower values of A/D.

3.2. Subharmonic regimes corresponding to $N_C/N_V = 1/2$

If $A/D = 0.237$ and S_C approaches 0.10, the wake locks onto the synchronization corresponding to $S_C/S_V = 1/2$. The recirculation region becomes longer and vortex shedding occurs in a mode such that the U-vortices are shed in $Y/A = \pm 1$ and L-vortices in $Y/A = 0$ (Fig. 4). The vortex pattern in the near wake obeys a rule which is not consistent with the apparent symmetry of vortex shedding. In effect, considering an L-vortex shed on the axis, its position in the near wake depends on the direction of the current oscillation half-cycle since the angle between the mixing region from the separation point and the flow axis depends on the direction of movement of the cylinder (Fig. 4b). If the cylinder is moving in the increasing Y/A direction, the L-vortex shed faces outward. However, if the cylinder is moving in the decreasing Y/A direction, the L-vortex shed faces the flow axis. The same rule applies to the U-vortex. Alternate shedding of the L-vortices and U-vortices creates a wake with two vortices convected on either side of the axis, followed by two vortices convected along the axis. Since the vortices on the axis are somewhat confined, they diffuse less and remain smaller in size. The wake therefore includes two large contrarotating vortices followed by two smaller contrarotating vortices moving along the axis (Fig.4c).

If S_C increases more and approaches 0.13, the flow regime is still the same as for locking on the range of synchronization $S_C/S_V = 1/2$, but the vortex shedding process and wake structure are modified. In effect, it can be seen that in this case, each mixing region generates a vortex structure that is divided into two corotating vortices that follow each other and are convected into the far wake (Fig.5a). This may be due to the fact that the vorticity flux from the separation points is larger. The result is a (2L-2U) mode whereby two L-vortices are shed during one cylinder oscillation half-cycle followed by two U-vortices during the next half-cycle. Stretching of the mixing region and movements of the vortex shedding point over time means that each pair of two L- or two U-vortices includes a first vortex convected roughly on the axis and a second located on the outside part of the wake (Fig.5b). Since the external vortex is less confined than the other, it diffuses more easily and its size therefore increases somewhat more rapidly. Diffusion of the vortices places these corotating structures in contact and they finish by pairing in the far wake. Observing the phenomenon somewhat more closely, it can be seen that when a pair of two L-vortices is shed, the second vortex is shed at the same time as the first vortex of the subsequent pair of two U-vortices. The same thing occurs for shedding of the second vortex of the 2U pair. The regime considered can therefore be described slightly differently, although the two descriptions are complementary. It can thus be considered that the L-vortices and U-vortices are not shed in succession, but simultaneously, by consecutive pairs of (L-U) vortices, each consisting of one L-vortex and one U-vortex (Fig.5c). Since the recirculation region is relatively large, the length of each of the mixing region at the time of shedding is not the same for the two vortices. When the cylinder is moving in the positive Y direction, the mixing region relative to the L-vortex is larger than that relative to the U-vortex. The opposite is true when the cylinder is moving in the negative Y direction. Stretching of the mixing regions and the fact that these regions are roughly aligned in the relative flow direction cause a staggered distribution of the vortices. Each (L-U) vortex pair includes one vortex located on the external part of the wake and the other convected roughly along the flow axis. The shedding process gives rise to a sequence of vortices in which each external vortex is in contact with the internal vortex of the previous pair. These two structures are corotating and therefore aggregate in the far wake (Fig.5d). If the cylinder oscillation frequency increases and the vortex shedding frequency also increases, pairing occurs earlier since the convection velocity is relatively constant. After aggregation, the Strouhal number of the wake is divided by two and equal to S_C. If S_C continues to increase, pairing occurs closer and closer to the cylinder. The frequency of the vortex street is then equal to that of the cylinder, but shedding still occurs by pairs of (L-U) vortices. This regime corresponds to conditions on the boundary of synchronization range $N_C/N_V = 1/2$.

The structure of the range $N_C/N_V = 1/2$ is even more complex for $A/D = 0.475$. Four different flow regimes are observed in succession when S_C varies from 0.10 to 0.17. For $S_C = 0.10$, the L-vortices and U-vortices are shed at $Y = 0$ and $Y \approx \pm1$ respectively. Vortex shedding however exhibits a slight dissymmetry in $Y = +1$, where U-vortex is shed in two steps as two successive small corotating structures (Fig.6). The coalescence of these two structures downstream gives the far wake a more regular appearance with alternating contrarotating vortices. If S_C increases further, approaching 0.12, the flow regime is again locked on at $N_C/N_V = 1/2$, but the vortex shedding process and wake structure are different. In effect, it is observed that a U-vortex then an L-vortex are shed when Y/A decreases and is near 0.5 and -0.5 respectively. These two vortices are followed in $Y/A = 0$ by simultaneous shedding of an L-vortex and an U-vortex. This is qualified as (L-U) vortex pair shedding mode (Fig.7). In the present case, shedding by (L-U) pairs can occur in two configurations, according as the mixing region feeding the U-vortex extends farther downstream than the one relative to the L-vortex or vice versa. In the first case, the wake consists of a succession of contrarotating vortices. In the second case, it consists of two corotating vortices followed by two vortices rotating in the opposite direction. Both configurations were observed for $S_C = 0.121$, but the one corresponding to the second case is much more frequent. It should be noted that the phase shift separating the shedding of two corotating vortices causes a spatial distribution of the structures which makes it very difficult for them to coalesce in the far wake. This regime is therefore an intermediate regime changing from a wake with alternately shed contrarotating vortices to a wake with two successive corotating vortices. As will be seen below, only the last distribution is present for higher values of S_C in synchronization range $N_C/N_V = 1/2$.

In effect, when S_C approaches 0.14, a new change in the vortex shedding process is observed, although the regime remains synchronized at $N_C/N_V = 1/2$. If Y/A decreases, shedding of a U-vortex is observed at $Y = 0$ followed by an L-vortex at $Y = -0.5$. When Y/A increases, an L-vortex is shed at $Y = 0$ followed by a U-vortex at $Y = +0.5$ (Fig. 8). The vortices are again shed one by one, since the (L-U) pair shedding mode has disappeared, but the shedding mode is not alternating. The resulting wake exhibits two corotating vortices followed by two vortices rotating in the opposition direction. The phase shift separating shedding of the two corotating vortices causes a spatial distribution of the structures making it impossible for them to coalesce in the far wake visualized (Fig.8a). This distribution actually reflects the influence of the cylinder oscillation amplitude parameter and is the reason for the different regimes that appear at higher values of S_C.

When S_C increases further, the phase shift between shedding of an L-vortex and the following U-vortex gradually decreases. For $S_C > 0.16$, the regime observed is shedding of (L-U) vortex pairs at $Y = 0$ (Fig. 9). As was mentioned above, the vortex distribution in the wake does not allow coalescence of the corotating structures. The vortices even appear to aggregate in pairs, but this interaction concerns only structures with opposing vorticity. After aggregation, the Strouhal number is divided by two. Figure 9a illustrates this phenomenon clearly and shows that if vortex shedding takes place in a synchronization mode corresponding to $N_C/N_V = 1/2$, the far wake consists of turbulent structures whose vorticity is very complex and which corresponds to synchronization at $N_C/N_V = 1$. This regime is observed for $0.16 < S_C < 0.19$. In this range, shedding always occurs at $Y = 0$. The influence of amplitude parameter A/D appears very clearly here when these results are compared with those relative to $A/D = 0.237$. In effect for $A/D = 0.237$, vortex shedding by (L-U) vortex pairs is observed at $0.13 < S_C < 0.20$, but, considering the lower oscillation amplitude, the resulting wake is narrower and the vortices generated are closer to one another. In consequence, two successive corotating structures always ultimately aggregate (Rodriguez (1996)).

If $A/D = 1$ and $0.1 < S_C$, the regime locks onto synchronization at $N_C/N_V = 1/2$. The vortex shedding process then consists of shedding of two successive L-vortices during a cycle followed by two U-vortices during the next half cycle. Figure 10 shows the vortex shedding diagram and the arrangement of the turbulent structures for $S_C = 0.20$. It can be seen that the local flow direction when shedding occurs distributes the vortices according to a structure similar to that observed for $A/D = 0.475$ at the same regime (Fig. 9). It should be noted that for $A/D = 0.475$, shedding occurs by pairs of (L-U) vortices, which is not the case for $A/D = 1$.

3.3. Synchronised regime corresponding to $N_C/N_V = 1$

If $A/D = 0.237$ and S_C continues to increase above 0.2, the near wake tends to adapt to the cylinder oscillation conditions to allow convection downstream of the increasing vorticity flux. The aggregation mentioned above disappears as the vortices are now formed at the end of the recirculation region. There is therefore a transition from the shedding mode by (L-U) pairs to the conventional pattern whereby one L-vortex followed by one U-vortex are shed each half-cycle and we have $S_C = S_V$ (see Fig.2a). This regime continues over the range $0.20 \le S_C \le 0.26$ and corresponds to synchronization at $S_C/S_V = 1$.

For $A/D = 0.475$, the transition of synchronization range $N_C/N_V = 1/2$ to $N_C/N_V = 1$ is more gradual and occurs in presence of a metastable regime where the two synchronization modes alternate. This metastable regime continues for approximately $0.19 < S_C < 0.20$ (Fig. 11). When S_C is close to 0.19, the regime corresponding to $N_C/N_V = 1/2$ predominates, whereas when S_C is close to 0.20, the regime corresponding to $N_C/N_V = 1$

is more frequent. In all cases, vortex shedding takes place at Y = 0. If the region relative to the metastable regimes is within synchronization range $N_C/N_V = 1$, it corresponds approximately to $0.19 < S_C < 0.225$.

If, for A/D = 1, the cylinder oscillation frequency is increased up to the maximum value allowed by the apparatus, it is observed that the flow regime no longer varies. Synchronization remains in range $N_C/N_V = 1/2$. The wake structure differs very little from the diagram of Fig. 10 except that the mixing regions observed are more stretched and vortex individualization is therefore more difficult to establish (Fig.2c). Stretching of the mixing regions in the near wake has the effect of elongating the recirculation region and transferring further downstream the deformation of the large structures in a coordinate system related to the cylinder.

This phenomenon may result in the existence of lower pressure fluctuation amplitudes on the obstacle. This regime was observed at the maximum value allowed on the oscillating system, corresponding to $S_C = 0.368$. If it is confirmed that for A/D = 1, synchronization at $N_C/N_V = 1/2$ is accompanied by a decrease in the pressure fluctuations on the obstacle when S_C increases, the causes of autolimitation of the amplitude of a freely oscillating cable should probably be sought in this direction. It should be noted that this phenomenon was not observed by Williamson and Roshko (1988).

The vortex shedding process was analyzed in detail to determine how the phase varies when the wake changed from the synchronization range $N_C/N_V = 1/2$ to $N_C/N_V = 1$. As the cylinder movement was sinusoïdal, the phase is known only if the cylinder position and the direction of displacement are simultaneously known. So Fig.12 shows for A/D = 0.237 the instantaneous cylinder position Y/A and the direction of cylinder movement when the vortex is shed in the near wake.The analysis was made for a large number of vortices to minimize errors in evaluation of the shedding time and to smooth the dispersion related to small variations of the flow regimes. For a vortex shedding at a given cylinder position, different symbols are used according as the cylinder is moving in the negative or positive Y direction. It can be seen that the phase variation is continuous, very rapid and equal to π over the synchronization range at $N_C/N_V = 1$.

However, for A/D = 0.475, vortex shedding is subjected to a phase shift equal to $\pi/2$ when S_C varies from 0.20 to 0.22 on the range $N_C/N_V = 1$ (Fig. 13). The effect of the amplitude parameter on synchronization is therefore essential, since it strongly modifies the vortex shedding phase and therefore the phase of the loads applied.

Higher-order synchronizations relative to A/D = 0.475 are characterized, as for A/D = 0.237 (Rodriguez (1996)), by vortex shedding at Y = +1 or -1 followed by aggregation in the near wake. The existence of high metastability is observed.

4. CONCLUSION

At the outcome of this work, we can now make an initial analysis of the influence of the amplitude parameter on the synchronization phenomenon by comparing the results relative to A/D = 0.237, 0.475 and 1.

It is observed that the subharmonic regimes corresponding to $N_C/N_V = 1/2$ exhibit roughly the same structure for the three values of reduced amplitude. They are additionally observed for the same values of S_C.

In all cases, synchronization at $N_C/N_V = 1/2$ appears for $S_C = 0.10$. In this synchronization range, the vortex shedding modes and resulting wake structure differ for the three values of A/D. This result is essential for the part of the synchronization range corresponding to the largest oscillation frequencies. It is in effect there that the mechanisms allowing transition to the range $N_C/N_V = 1$ are established. For A/D = 0.237, shedding by (L-U) vortex pairs in the range $N_C/N_V = 1/2$ degenerates into synchronization at $N_C/N_V = 1$ by aggregation of two successive corotating vortices. For A/D = 0.475, shedding by (L-U) vortex pairs observed for $N_C/N_V = 1/2$ degenerates into synchronization at $N_C/N_V = 1$ by aggregation of two successive contrarotating vortices. In both cases, aggregation occurs in the far wake and moves towards the cylinder as S_C increases. These differences between regimes are accompanied by a difference in the phase of vortex shedding and therefore in that of the hydrodynamic forces applied. In effect, during the transition from the range $N_C/N_V = 1/2$ to $N_C/N_V = 1$, vortex shedding is shifted by π for A/D = 0.237 and by $\pi/2$ for A/D = 0.475. For A/D = 1, the differences are even more significant, since in this case synchronization does not appear to be able to be reached at $N_C/N_V = 1$.

If, as can reasonably be expected, the movement of a cable is directly related to the nature of the coupling on the above synchronization ranges, it is important to check whether other coupling modes exist for $0 < A/D < 1$. It is also important to determine when and how the transition from one mode to another occurs as A/D varies, since these are the phenomena controlling cable movement when the far-field speed increases or decreases.

The work will therefore be continued to achieve these objectives by performing tests for values of A/D that are sufficiently close to one another in order to be able to detect all the existing regimes.

Although the appearance of resonance is independent of the Reynolds number, the differences in flow regimes noted between this work and the literature make it necessary to examine the actual influence of the Reynolds number on the structure of these regimes.

ACKNOWLEDGEMENTS : This work was sponsored by the French Ministery for Defense, through the DRET G6.

REFERENCES

Parkinson, GV (1974). "Mathematical Models of Flow-induced Vibrations,". *Flow-Induced Structural Vibrations* (ed.Naudascher), 81-127.

Sarpkaya, T (1979). "Vortex-induced Vibrations: A Selective Review," *Trans ASME E: J Appl Mech*, Vol 26, 241-258.

Bearman, PW (1984). "Vortex Shedding from Oscillating Bluff Bodies," *Ann Rev Flui Mech*, Vol 6, 195-222.

Griffin, O.M, and Hall, MS (1991). "Review-Vortex Shedding Lock-on and Flow Control in Bluff Body Wakes," *Trans ASME: J Fluids Eng*, Vol 113, 531.

Zdravkovich, MM (1982). "Modification of Vortex Shedding in the Synchronisation Range," *Trans ASME I: J Fluids Eng*, Vol 104, 513-517.

Ongoren, A and Rockwell, D (1988). "Flow Structure from an Oscillating Cylinder-Part1-Mechanism of Phase Shift and Recovery in the Near Wake," *J Fluid Mech*, Vol 191, 197-223.

Feng, CC (1968). "Measurement of Vortex Induced Effects in Flow Past Stationnary and Oscillating Circular and D-Sections Cylinders," *M SC Thesis, University of British Columbia*, Vancouver, Canada.

Williamson, CHK and Roshko A (1988). " Vortex Formation in the Wake of an Oscillating Cylinder," *J Fluids Struct*, Vol 2, 355-381.

Brika,D and Laneville, A (1993). "Vortex- Induced Vibrations of a Long Flexible Circular Cylinder," *J Fluid Mech*, Vol 250, 481-508.

Rodriguez, O (1991). "Base Drag Reduction by Control of the Three-Dimentional Unsteady Vortical Structures," *Exp in Fluids*, Vol 11, 218-

226.

Rodriguez, O (1996)."Experimental Analysis of the Wake Instabilities of an Oscillating Cylinder near Synchronisation," *Sixth International Offshore and Polar Engineering Conference*, Los Angeles.

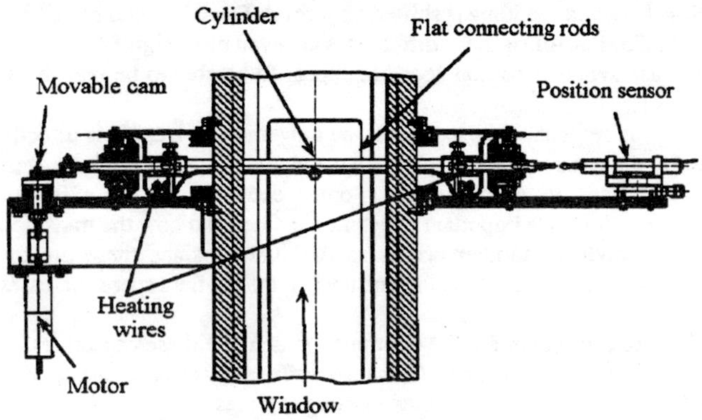

Fig. 1 Sketch of the test section and oscillating device (side view)

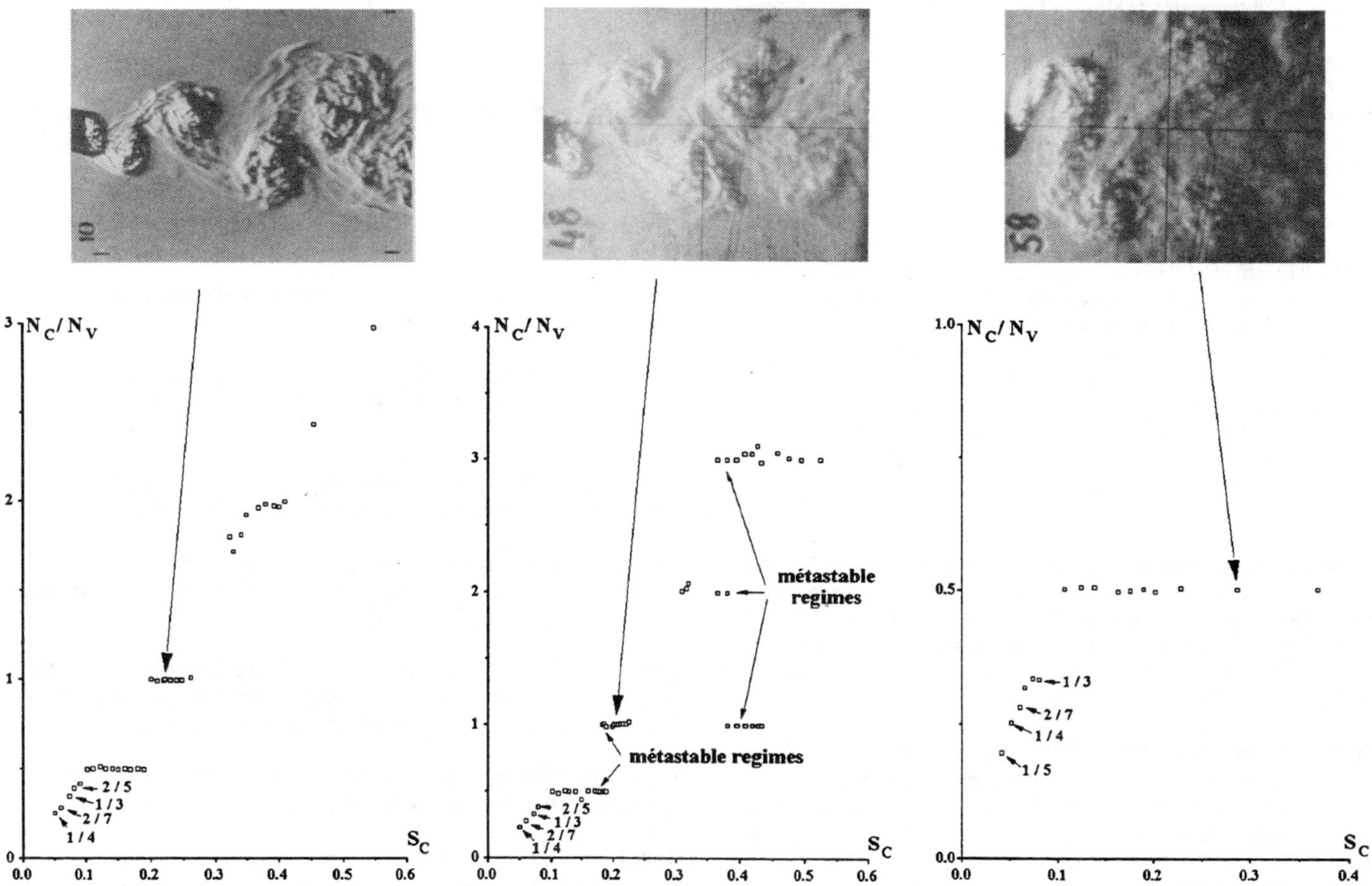

a) A / D = 0.237

b) A / D = 0.475

c) A / D = 1.

Fig. 2 Synchronized regimes versus the cylinder Strouhal number

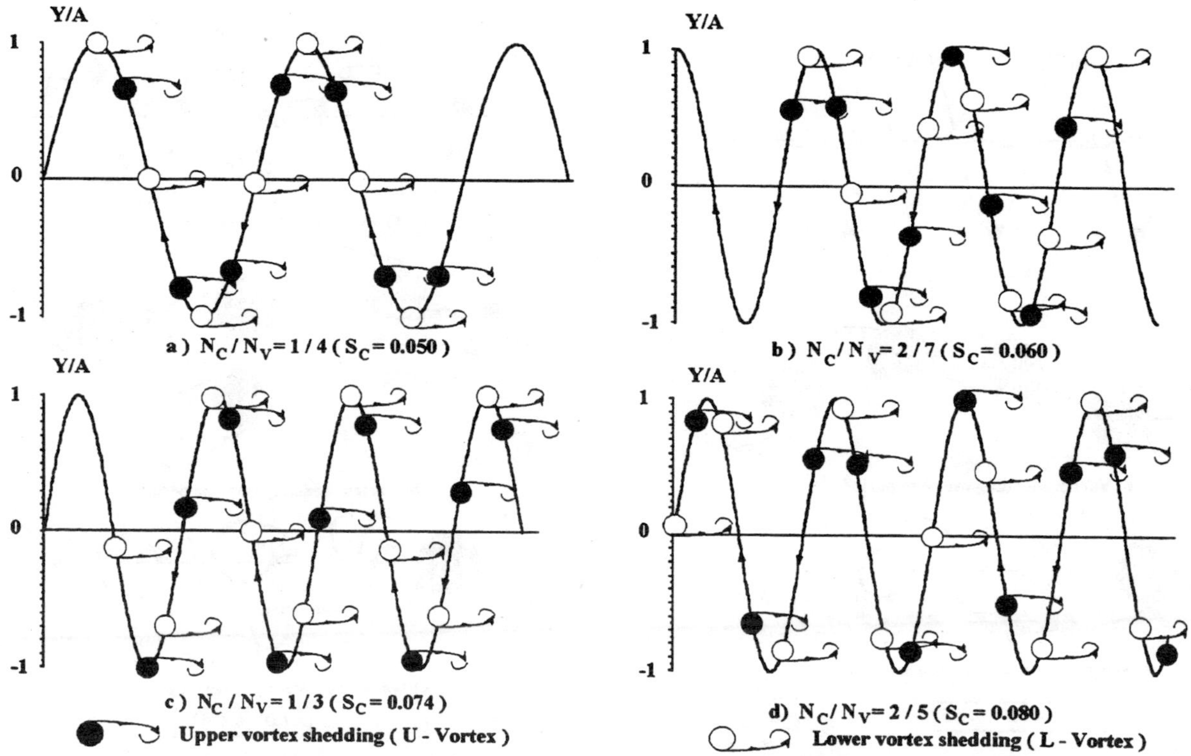

Fig. 3 Vortex shedding relative to subharmonic regimes

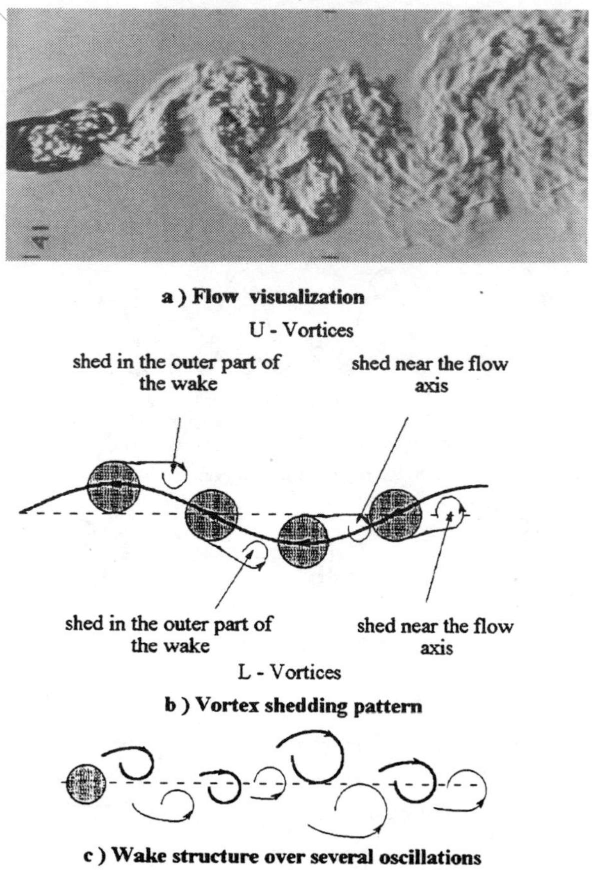

Fig. 4 Synchonization for $N_C / N_V = 1/2$ ($S_C = 0.109$)

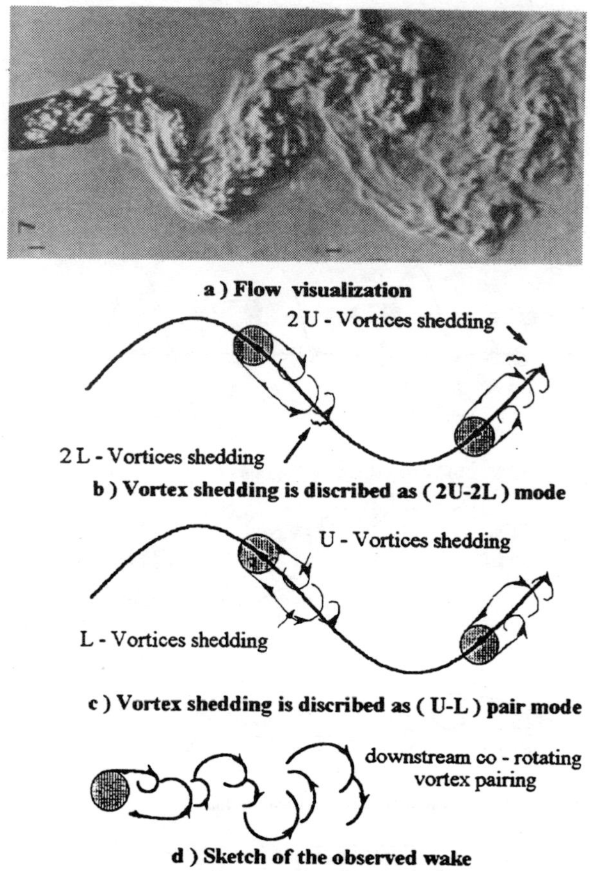

Fig. 5 Synchonization for $N_C / N_V = 1/2$ ($S_C = 0.141$)

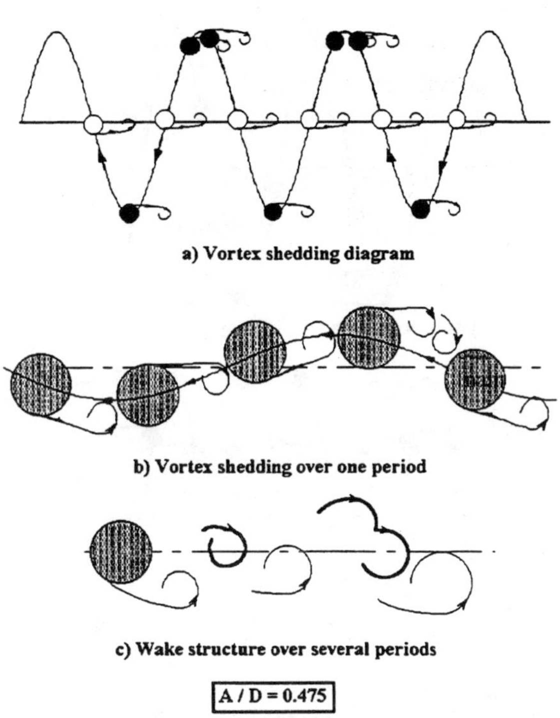

Fig. 6 Synchonization for $N_C / N_V = 1/2$ ($S_C = 0.101$)

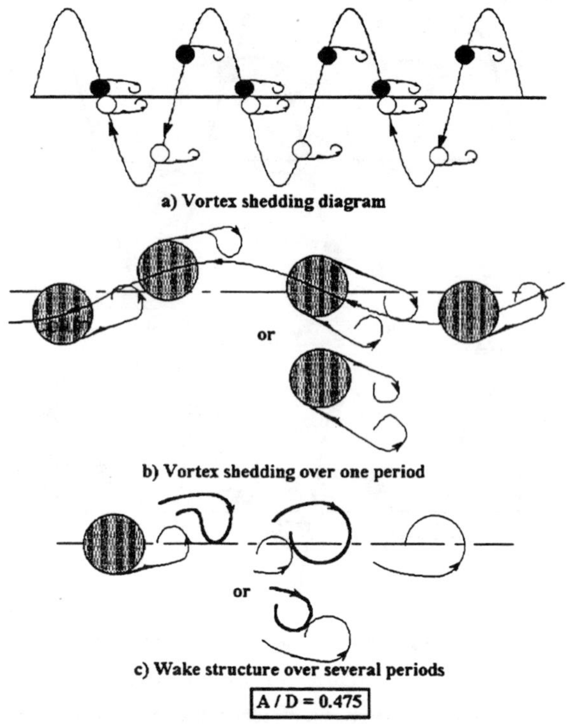

Fig. 7 Synchonization for $N_C / N_V = 1/2$ ($S_C = 0.121$)

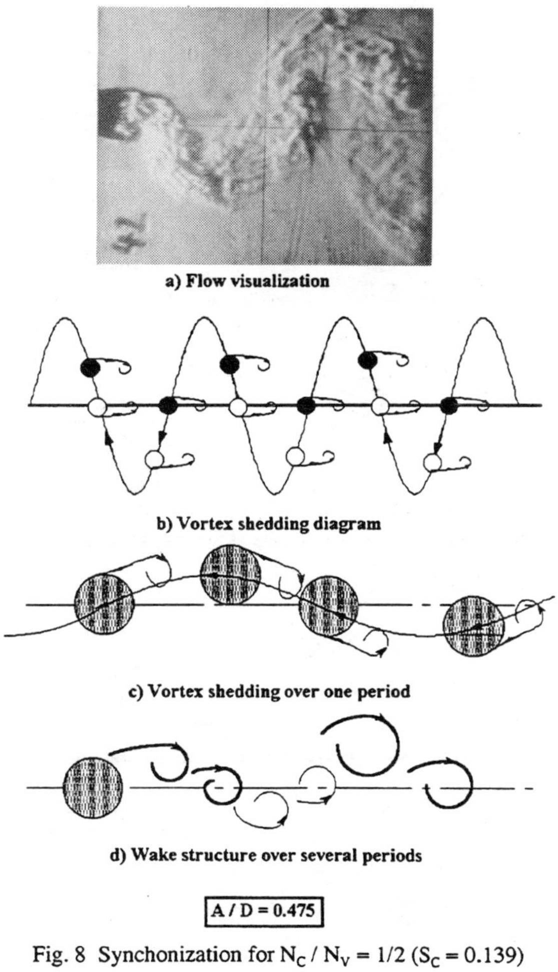

Fig. 8 Synchonization for $N_C / N_V = 1/2$ ($S_C = 0.139$)

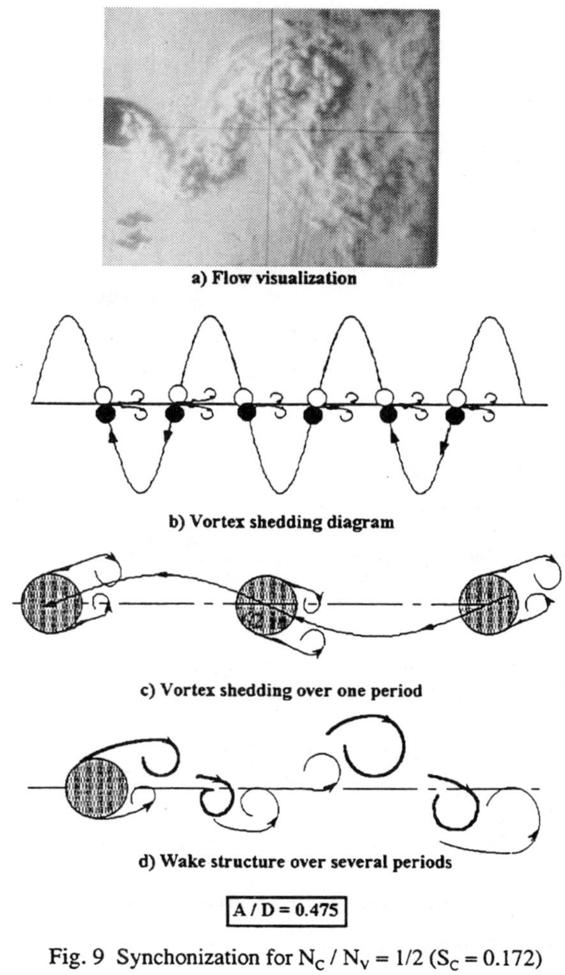

Fig. 9 Synchonization for $N_C / N_V = 1/2$ ($S_C = 0.172$)

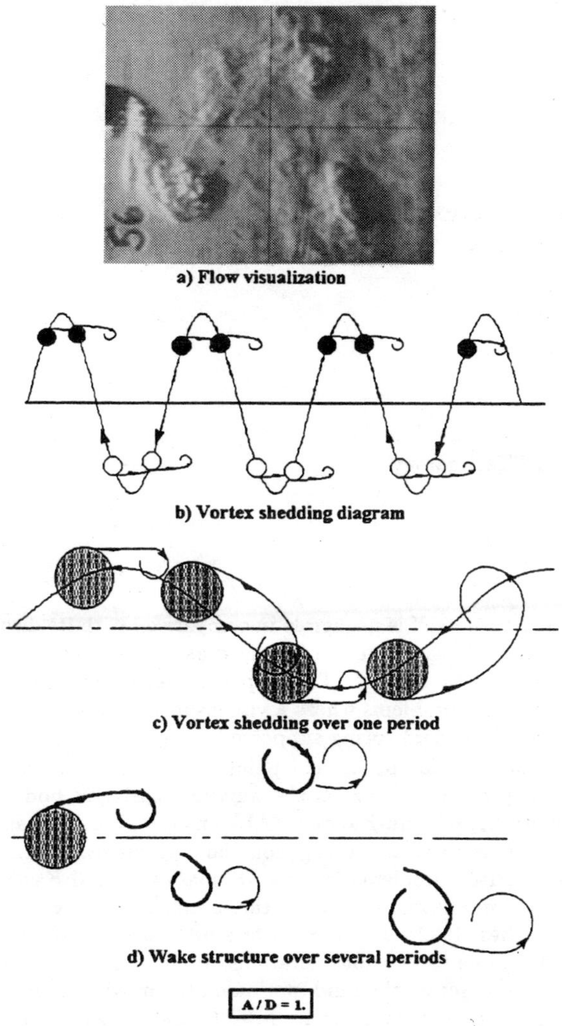

Fig. 10 Synchonization for $N_C / N_V = 1/2$ ($S_C = 0.202$)

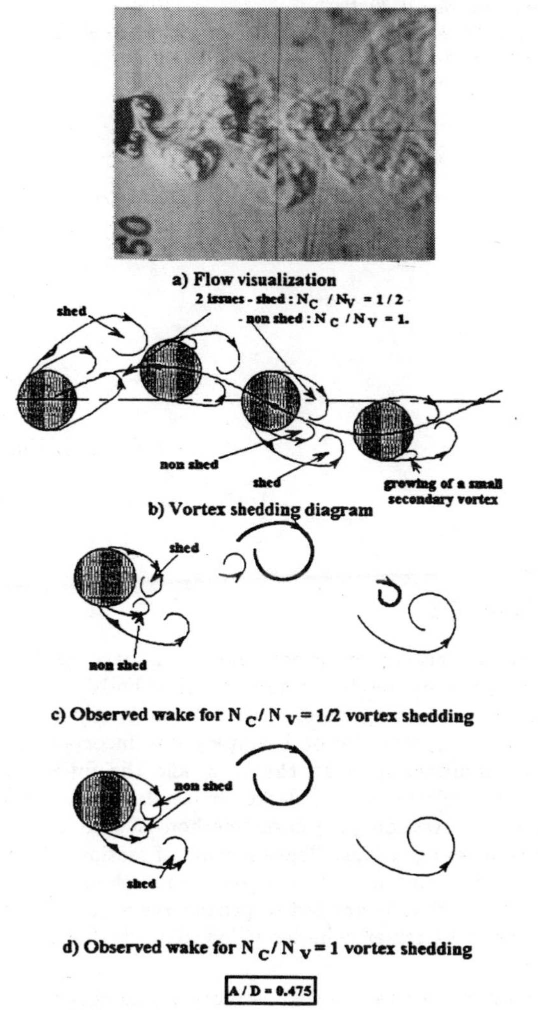

Fig. 11 Vortex shedding and metastable regimes for $0.19 \leq S_C \leq 0.2$

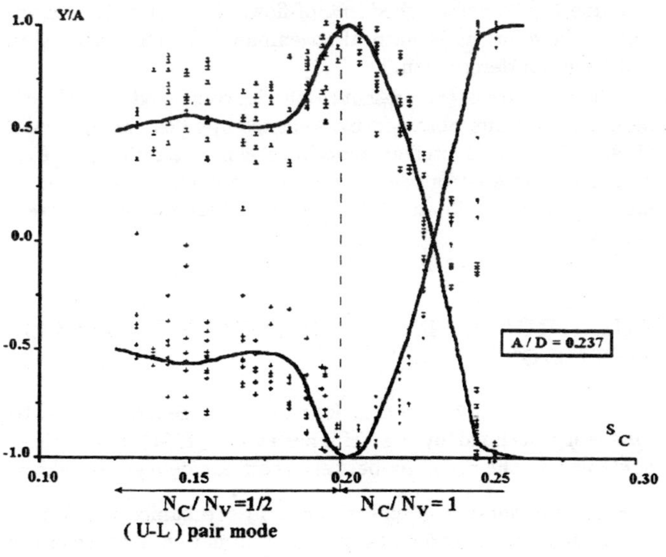

Fig. 12 Position of the cylinder at vortex shedding

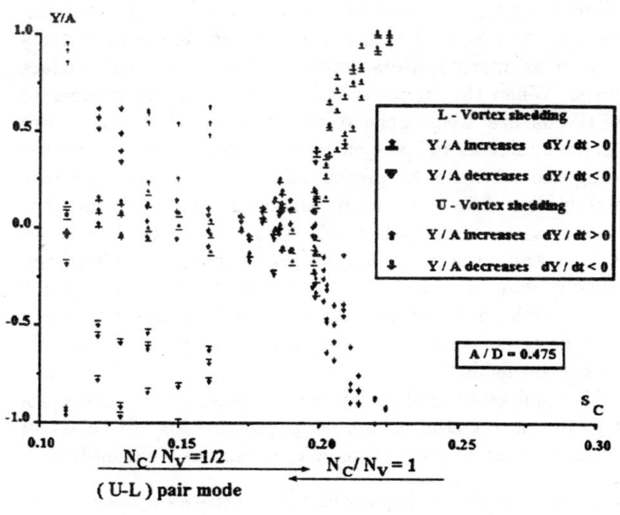

Fig. 13 Position of the cylinderat vortex shedding

Proceedings of the Seventh (1997) International Offshore and Polar Engineering Conference
Honolulu, USA, May 25-30, 1997

ISBN 1-880653-28-1 (Set); ISBN 1-880653-31-1 (Vol. III)

Numerical Prediction of the Response of a Vortex-Excited Cylinder at Low Reynolds Numbers

*Trond Stokka Meling**
Stavanger College
Stavanger, Norway

Joar Dalheim
Norwegian University of Science and Technology
Trondheim, Norway

ABSTRACT

An explicit computational method is developed in order to investigate the coupled motion of a rigid body-spring system and a viscous incompressible fluid. An arbitrary Lagrangian-Eulerian (ALE) formulation is employed to incorporate the interface conditions between the body and the fluid. The numerical formulation is a Galerkin finite element method, and a velocity correction projection method is used to solve the Navier-Stokes equations. The equation of motion of the cylinder is solved by a 4th-order Runge-Kutta method at every time step. The method is applied to predict the vortex-induced oscillations of a circular cylinder at low Reynolds numbers.

Keywords: vortex-excited oscillations, moving body, finite element, ALE

INTRODUCTION

Understanding vortex-induced vibrations is of great importance in the design of a variety of offshore engineering structures such as marine risers, mooring lines, and sub-surface pipelines. When the frequency of vortex shedding is close to one of the natural frequencies of the structure, the well-known lock-in phenomenon, or the synchronization between the vortex shedding and vibration frequency, can occur and the structure can exhibit amplified vibrations since energy is fed into the cylinder vibrations at small amplitudes of motion. For larger amplitudes and outside the lock-in region, energy is dissipated by fluid damping. The ability to predict the fluid forces on these structures with good accuracy and reliability is one of the key issues in design of such structures, especially when concerning deep-water oil fields.

It is required to deal with the viscous Navier-Stokes equations to develop a computational procedure for this class of fluid-structure interaction problems. When the deformation of the structure itself is negligible in comparison with its displacement, the structure is often idealized as a rigid body mounted on linear elastic springs. This paper presents a computational procedure for problems where a viscous incompressible fluid interacts with a rigid spring supported body.

From the computational point of view we have to treat a moving interface between the fluid and the rigid body. The arbitrary Lagrangian-Eulerian (ALE) method has been applied with success to such moving boundary problems, which arise in free surface problems (see e.g. Ramaswamy and Kawahara, 1987) and fluid-structure interaction problems (see e.g. Nomura and Hughes, 1992). In the present study, the ALE description is applied due to two considerations. First, the method is convenient to describe the fluid motion on the moving interface by the Lagrangian description in order to treat the compatibility conditions and the equilibrium between the fluid and the rigid body. Secondly, it is difficult to employ the Lagrangian description for the entire fluid domain because of severe mesh distortion due to the vortex shedding of flows through outer boundaries. Therefore it is natural to employ a mixed Lagrangian and Eulerian description.

The feature of the present method is compared with the experimental results obtained by Anagnostopoulos and Bearman (1992). They conducted an experimental investigation of the response characteristics of a vortex-excited cylinder at Reynolds numbers between 90 and 150, a regime where the vortex street is fully laminar.

THE ARBITRARY LAGRANGIAN-EULERIAN METHOD

The ALE description is founded on a general kinematical theory outlined by e.g. Hughes et al. (1981) where three domains and the mapping between them are defined as follows:

- *Spatial* domain is where the fluid mechanics problem is posed. It is regarded as generally in motion because of the moving interface to the adjoining structures.

*Visiting Researcher at Department of Aerospace Engineering and Engineering Mechanics, University of Texas, Austin, USA

- *Material* domain is to be thought of as the domain occupied at time $t = 0$ by the material particles which occupy the spatial domain at t (which is the time of interest). Hence, the material domain is also generally in motion.
- *Referential* domain is defined as fixed for all t, and its image at time t under prescribed mapping is the spatial domain.

Applied to the finite element analysis, the spatial domain is represented by a moving mesh and the referential domain by a reference state of the moving mesh. If the spatial domain is fixed throughout, the formulation can be regarded as an Eulerian description. Likewise, a Lagrangian description is obtained when the spatial domain always coincides with the material domain.

Ramaswamy and Kawahara (1987) outlined the following definition of the material time derivative of a physical quantity f in the ALE description:

$$\frac{\mathrm{D}f}{\mathrm{D}t} = \left.\frac{\partial f}{\partial t}\right|_{\chi_i} + (u_i - \hat{u}_i)\frac{\partial f}{\partial x_i} = \left.\frac{\partial f}{\partial t}\right|_{\chi_i} + c_i\frac{\partial f}{\partial x_i} \qquad (1)$$

where $\hat{u}_i$ is the mesh velocity in the ith direction and χ_i and x_i denote referential and spatial coordinates, respectively. $c_i = u_i - \hat{u}_i$ is the convective velocity in the ALE description, and when $\hat{u}_i = 0$ equation (1) is equivalent to the Euler co-ordinate description and to the Lagrangian co-ordinate description when $\hat{u}_i = u_i$.

BASIC EQUATIONS

The governing equations of the fluid

In Fig. 1 a general outline of the problem domain of interest is presented where $\Omega_G(t)$ is the domain of the moving rigid body, $\Omega_F(t)$ is the moving spatial domain upon which the fluid motion is described, and $\Gamma_I(t)$ is the moving interface between $\Omega_G(t)$ and $\Omega_F(t)$. When the rigid body $\Omega_G(t)$ moves, the interface $\Gamma_I(t)$ moves accordingly. $\Omega_F(t)$ is also surrounded by the boundaries Γ_u, on which the velocities are prescribed, and Γ_p, on which the tractions are prescribed. These boundaries are assumed to be fixed in space.

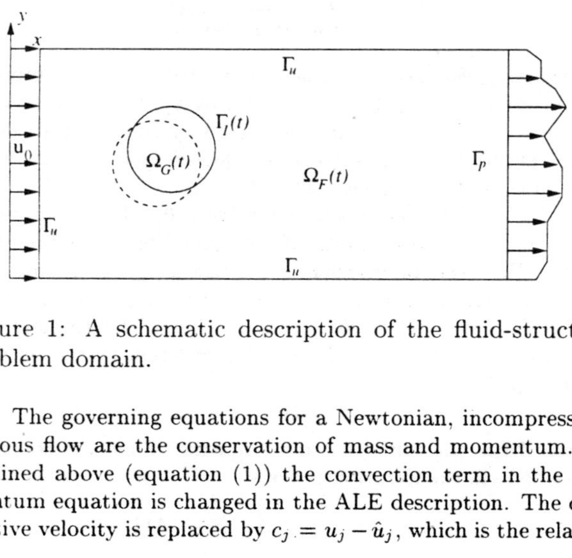

Figure 1: A schematic description of the fluid-structure problem domain.

The governing equations for a Newtonian, incompressible viscous flow are the conservation of mass and momentum. As outlined above (equation (1)) the convection term in the momentum equation is changed in the ALE description. The convective velocity is replaced by $c_j = u_j - \hat{u}_j$, which is the relative velocity of the fluid particle to the moving mesh. The ALE description of the non-dimensional Navier-Stokes equations may be written as follows

$$\frac{\partial u_i}{\partial x_i} = 0 \qquad \text{in } \Omega_F(t) \quad (2)$$

$$\left.\frac{\partial u_i}{\partial t}\right|_{\chi_i} + c_j\frac{\partial u_i}{\partial x_j} = -\frac{\partial p}{\partial x_i} + \frac{1}{Re}\frac{\partial^2 u_i}{\partial x_j \partial x_j} \qquad \text{in } \Omega_F(t) \quad (3)$$

where $Re = U_0 D/\nu$ denotes the Reynolds number. Here U_0 is the free-stream velocity, D the cylinder diameter and ν is the kinematic viscosity of the fluid.

The outer boundary conditions are given as

$$u = U_0,\ v = 0 \ \text{ on } \Gamma_u; \qquad p = P_0 \ \text{ on } \Gamma_p \qquad (4)$$

where U_0 and P_0 are prescribed constants. Since Γ_u and Γ_p is assumed to be fixed in space, the mesh velocity will be zero on these boundaries.

On the cylinder surface a no slip condition is applied, which implies that the fluid velocity must coincide with the cylinder velocity. As the mesh velocity equals the cylinder velocity on the moving interface $\Gamma_I(t)$, the following condition is applied

$$u_i = \hat{u}_i \qquad \text{on } \Gamma_I(t) \qquad (5)$$

The equation of motion for the cylinder

Considering motion in the transverse direction only and assuming rigid body motion, the non-dimensional equation of motion for the cylinder can be written as

$$\ddot{y} + \left(\frac{4\pi\zeta_s}{U_r}\right)\dot{y} + \left(\frac{4\pi}{U_r}\right)^2 y = \left(\frac{\rho D^2 l}{2m}\right) C_L \qquad (6)$$

where

$$U_r = \frac{U_0}{f_n D}, \quad (2\pi f_n)^2 = \frac{k}{m} \qquad (7)$$

Here ρ is the fluid density, l is the cylinder length immersed in the water, ζ_s is the structural damping ratio, k is the structural stiffness, m is the cylinder mass, f_n is the natural structural oscillation frequency, and y is the non-dimensional transverse cylinder displacement. The parameter U_r is known as the *reduced velocity*.

The lift coefficient C_L is calculated as the sum of the non-dimensional pressure and viscous shear forces around the cylinder surface

$$(C_L)_{press} = \frac{-\oint p \sin\theta \mathrm{d}\Gamma_I}{0.5\rho U_0^2 D}, \quad (C_L)_{shear} = \frac{\oint \tau_w \cos\theta \mathrm{d}\Gamma_I}{0.5\rho U_0^2 D} \qquad (8)$$

where $\theta = 0$ in the main flow direction and τ_w is the shear stress at the cylinder wall.

As mentioned in the section above, the no slip condition on the cylinder surface implies that

$$u_i = \dot{x}_i \qquad \text{on } \Gamma_I(t) \qquad (9)$$

NUMERICAL FORMULATION

A fractional step formulation is applied for the time discretization of the governing equations (2–3). An intermediate velocity is first computed by neglecting the pressure. The pressure field is then computed from a Poisson equation, and the velocity field is finally corrected by including the pressure effect. The intermediate velocity is computed by use of an explicit two-step Taylor-Galerkin formulation. The following is a resume of the algorithm described in Utnes and Ren (1995).

Fractional step algorithm for Navier-Stokes equations

For a time increment $\Delta t = t_{n+1} - t_n$, the algorithm is given as follows:

1. An intermediate velocity $\tilde{u}_i$ is computed at time t_{n+1} by integration of the reduced momentum equation omitting the pressure:

$$\tilde{u}_i^{n+1} = u_i^n - \int_{t_n}^{t_{n+1}} \left[c_j \frac{\partial u_i}{\partial x_j} - \frac{\partial^2 u_i}{\partial x_j \partial x_j} \right] dt \qquad (10)$$

2. The complete velocity u_i at t_{n+1} is computed by including the pressure field:

$$u_i^{n+1} = \tilde{u}_i^{n+1} - \int_{t_n}^{t_{n+1}} \frac{\partial p}{\partial x_i} dt \qquad (11)$$

3. The continuity equation must be satisfied by the final velocity field, i.e.:

$$\frac{\partial u_i^{n+1}}{\partial x_i} = 0 \qquad (12)$$

The first step in this algorithm is solved by use of the following predictor-corrector formulation:

$$\tilde{u}_i^{n+1/2} = u_i^n - \frac{\Delta t}{2} c_j^n \frac{\partial u_i^n}{\partial x_j} \qquad (13)$$

$$\tilde{u}_i^{n+1} = u_i^n - \Delta t \left[c_j^n \frac{\partial \tilde{u}_i^{n+1/2}}{\partial x_j} - \frac{\partial^2 u_i^n}{\partial x_j \partial x_j} \right] \qquad (14)$$

Here $\tilde{u}_i^{n+1/2}$ denotes the intermediate value predicted by convection of u_i^n a time increment $\Delta t/2$, and this value is then used for the convection in the correction step.

Steps 2 and 3 of the algorithm above, i.e., the computation of the complete velocity field satisfying the continuity equation, are combined through a derived pressure Poisson equation. By taking the divergence of the following discretized form of equation (11)

$$u_i^{n+1} = \tilde{u}_i^{n+1} - \Delta t \frac{\partial p^{n+1}}{\partial x_i} \qquad (15)$$

and then introducing the continuity equation (12), the resulting Poisson equation is written as

$$\frac{\partial^2 p^{n+1}}{\partial x_j \partial x_j} = \frac{1}{\Delta t} \frac{\partial \tilde{u}_j^{n+1}}{\partial x_j} \qquad (16)$$

An alternative form of the algorithm is to include the known pressure in the prediction of the intermediate velocity, and compute the pressure difference instead of the complete pressure. However, Ramaswamy and Kawahara (1987) and Ren and Utnes (1993) has observed that the removal of the pressure term in equation (10) is advantageous since no spurious pressure modes are obtained even if equal order interpolation is used for velocity and pressure.

Finite element formulation of the Navier-Stokes equations

A Galerkin finite element formulation is applied for the spatial discretization of the above algorithm. The velocity and pressure are both interpolated by the same shape functions:

$$u_i \approx \mathbf{N}\mathbf{U}_i, \qquad p \approx \mathbf{N}\mathbf{P} \qquad (17)$$

where $\mathbf{N}$ is the shape function vector, $\mathbf{U}_i$, $\mathbf{P}$ denote the nodal values of the velocity components and pressure, respectively. By introducing these expressions in a weighted residual formulation of the governing equations, the following set of matrix equations are obtained

$$\mathbf{M}^n \tilde{\mathbf{U}}_i^{n+1/2} = \mathbf{M}^n \mathbf{U}_i^n - \frac{\Delta t}{2} \mathbf{A}^n \mathbf{U}_i^n \qquad (18)$$

$$\mathbf{M}^n \tilde{\mathbf{U}}_i^{n+1} = \mathbf{M}^n \mathbf{U}_i^n - \Delta t \left[\mathbf{A}^n \tilde{\mathbf{U}}_i^{n+1/2} + \mathbf{L}^n \mathbf{U}_i^n \right] \qquad (19)$$

$$\mathbf{L}^n \mathbf{P}^{n+1} = -\frac{1}{\Delta t} \mathbf{D}_j^n \tilde{\mathbf{U}}_j^{n+1} \qquad (20)$$

$$\mathbf{M}^n \mathbf{U}_i^{n+1} = \mathbf{M}^n \tilde{\mathbf{U}}_i^{n+1} - \Delta t \mathbf{D}_i^n \mathbf{P}^{n+1} \qquad (21)$$

where the boundary integrals resulting from integration by parts of the diffusion term and the pressure Poisson equation are omitted due to the applied boundary conditions.

The global matrices associated with these equations are all time dependent due to the moving mesh, and are defined as follows

$$\mathbf{M}^n = \int_{\Omega^n} \mathbf{N}^T \mathbf{N} d\Omega \qquad (22)$$

$$\mathbf{A}^n = \sum_{e=1}^{n} (\bar{C}_j)_e \int_{\Omega_e^n} \mathbf{N}^T \frac{\partial \mathbf{N}}{\partial x_j} d\Omega \qquad (23)$$

$$\mathbf{D}_j^n = \int_{\Omega^n} \mathbf{N}^T \frac{\partial \mathbf{N}}{\partial x_j} d\Omega \qquad (24)$$

$$\mathbf{L}^n = \int_{\Omega^n} \frac{\partial \mathbf{N}^T}{\partial x_j} \frac{\partial \mathbf{N}}{\partial x_j} d\Omega \qquad (25)$$

where $(\bar{C}_j)_e$ is the element mean convective velocity component. The convection matrix is simplified by applying the "centroid convection velocity" method proposed by Gresho et al. (1984). This implies that the convective velocity, $c_j = u_j - \hat{u}_j$, is only evaluated at the element centroid.

Except for the pressure equation (20), the remaining system can be solved explicitly if a so-called mass lumping is applied. However, in transient problems a consistent mass matrix is desirable when solving the convection part of the system above (Lewis et al. 1995). This is achieved by employing the following iteration procedure: The equation system is of the form

$$\mathbf{M}\mathbf{U} = \mathbf{f} \qquad (26)$$

where $\mathbf{M}$ is the consistent mass matrix. Then the solution is obtained via the iteration process

$$\mathbf{U}_{k+1} = \mathbf{U}_k + \mathbf{M}_L^{-1}(\mathbf{f} - \mathbf{M}\mathbf{U}_k) \qquad (27)$$

where $\mathbf{M}_L$ is the lumped mass matrix, and k is the iteration number. This iteration converges very fast (approximately three iterations).

The pressure Poisson equation (20) is solved by a direct skyline profile solver and is by far the most CPU-expensive of the four equations listed above.

Coupling procedure

The non-dimensional fluid force obtained from equation (8) is assumed piecewise constant for every time step, and with only transverse rigid body motion, equation (6) is solved at the cylinder center using a 4th-order Runge-Kutta method as the time integration scheme. Hence, the fluid and structure problems are loosely coupled and the procedure for each global time step is outlined as follows:

- Obtain pressure and velocity fields from the velocity correction algorithm.
- Calculate the lift forces acting on the cylinder
- Compute the new cylinder displacement and velocity
- Update the grid velocity and node coordinates.

VORTEX-EXCITED OSCILLATIONS OF A CIRCULAR CYLINDER

In order to validate the model, comparison with the experimental results obtained by Anagnostopoulos and Bearman (1992) have been performed. They conducted a series of experiments on vortex-excited, cross-flow oscillations of a circular cylinder mounted elastically in a water channel. The Reynolds number ranged from 90 to 150, a regime where the wake is fully laminar. When the water velocity was increased while the cylinder was still oscillating, they captured the 'lock-in' phenomenon over the Reynolds number range 102-130.

We interpreted the experiments in terms of a finite element model employing the boundary conditions shown in Fig. 2. The cylinder was only allowed to move in the cross-flow direction. All the material properties are the same as those given by Anagnostopoulos and Bearman (1992) except that m, c and k are divided by the length of the submerged cylinder.

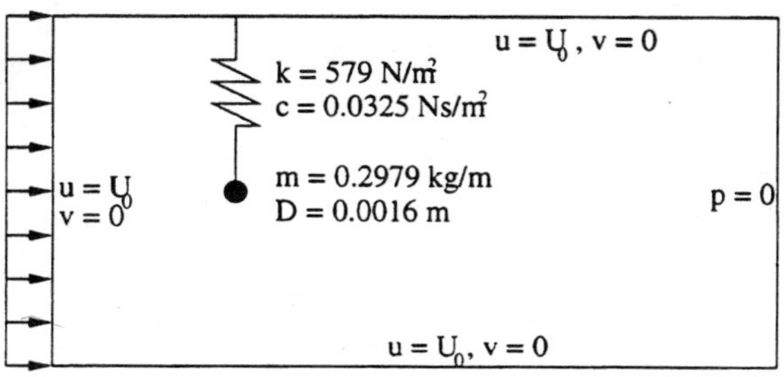

Figure 2: Boundary conditions and material properties.

We started the computations off at $Re = 90$ keeping the cylinder fixed until a full vortex-street was developed in the wake. Then we released the cylinder and conducted a series of computations for Reynolds numbers up to 145 by increasing the upstream velocity. Figures 4–6 shows the time histories of the non-dimensional amplitudes, drag and lift coefficients at $Re = 90$, $Re = 100$ and $Re = 135$.

At Reynolds numbers up to 97, which is below the the lock-in range of the experiment, the oscillation amplitudes are small and modulated because of the difference between the vortex-shedding frequency and the frequency of the cylinder vibrations. When the Reynolds number was increased to 98 and reduced velocity 5.3, synchronization between the cylinder oscillation frequency and the vortex-shedding frequency occurred at a frequency slightly below the natural frequency, as reported by Anagnostopoulos and Bearman (1992). Fig. 3 illustrates how the shedding frequency is shifted away from the natural shedding frequency to a value slightly below the cylinder vibration frequency at $Re = 100$.

Moreover, the amplitude increased significantly to 40% of the cylinder diameter. The vortex-shedding frequency was locked onto the cylinder oscillation frequency up to $Re = 130$, which is similar to the experiments. During the lock-in range the lock-in frequency also changed from a value slightly below the natural frequency of the cylinder to a value slightly above, a feature also notified in the experiments. For Reynolds numbers above 130 the shedding becomes unlocked and the vortices are shed at the Strouhal frequency ($St = f_s D/U_0$) and the cylinder oscillates at a frequency slightly above the natural.

It is well known that the drag increases with the amplitude of the cylinder oscillations and the results presented in Fig. 5 agree quite well with the results obtained by Sarpkaya (1978) for forced vibrations.

To exemplify the flow field during lock-in at $Re = 100$, contour lines of the pressure and vorticity and two velocity field snapshots are shown in Figures 7–9. The dark line is a stationary reference point corresponding to the initial location of the cylinder centroid. The transverse displacement of the freely oscillating cylinder is seen to heavily influence the surrounding velocity field as the wake region alternates between the upper and lower surfaces of the cylinder.

Fig. 10 summarizes the present computational results compared to the results obtained by Anagnostopoulos and Bearman (1992) and the numerical results obtained by Nomura (1994). Nomura (1994) applied the SUPG finite element method for the fluid problem where the coupled matrix equations are solved by a predictor-multicorrector method. The equation of motion for the cylinder is solved by the predictor-corrector version of the Newmark algorithm. Our experience is that the 4th-order Runge-Kutta method is more stable than the Newmark algorithm in this kind of coupled problems.

As shown in Fig. 10 the present method is capable of capturing the synchronization of the shedding and cylinder oscillation frequency, although the lock-in range occurs a little earlier than in the experiments due to the fact that our Strouhal frequencies are slightly larger than the ones obtained in the experiments. However, Anagnostopoulos and Bearman (1992) report that their Strouhal numbers are somewhat too low which they explain by having too small of an aspect ratio and by not using endplates.

In the experiments Anagnostopoulos and Bearman (1992) reached a maximum amplitude of 60% of the diameter, while the present method reached a maximum of 40%. The discrepancy might be explained by the loose coupling between the fluid and structure problems. The staggered solution procedure outlined above may cause small phase differences between the cylinder vibration and the oscillating lift force. Consequently, energy may be subtracted from the system instead of added in the model. To capture the highly non-linear interaction between the fluid and structure better, a stronger coupling between the fluid and structure systems than the one outlined here, is likely required.

The calculation of fluid forces and cylinder vibrations is also sensitive to the boundary conditions, size of the computational domain and the roughness of the finite element mesh. Moreover, three-dimensional effects are observed in the experiments, which we do not encounter. However, compared to Nomura (1994) the present method gives results closer to the experiments. Nomura (1994) reached a maximum amplitude of approximately 30% of the diameter and the lock-in range is somewhat narrower.

All the computations were run on an IBM RS6000 workstation. The CPU time for simulation on a grid consisting of 3876 bilinear elements was approximately $0.24s$ per time step requiring less than 4 MB RAM.

CONCLUDING REMARKS

We have described a computational method for prediction of vortex-excited oscillations of a $2D$ cylinder which is idealized as rigid body supported by linear elastic springs. The present method is able to capture the lock-in phenomenon, and the results obtained appear to be reasonable compared to experiments. However, to model the flow field around offshore structures like risers, the effect of turbulence must be taken into account. Therefore, a future task will be to include a turbulence model which accurately describes vortex-induced oscillations at high Reynolds numbers.

REFERENCES

Anagnostopoulos, P. and Bearman, P. W. (1992). "Response Characteristics of a Vortex-Excited Cylinder at Low Reynolds Numbers," *J Fluids and Structures*, Vol 6, pp 39-50.

Gresho, P. M., Chan, S. T., Lee, R. L. and Upson, C. D. (1984). "A Modified Finite Element Method for Solving the Time-Dependent, Incompressible Navier-Stokes Equations. Part 1: Theory," *Int J Num Meth Fluids*, Vol 4, pp 557-598.

Hughes, T. J. R., Liu, W. K. and Zimmermann, T. K. (1981). "Lagrangian-Eulerian Finite Element Formulation for Incompressible Viscous Flows," *Comp Meth App Mech and Eng*, Vol 29, pp 329-349.

Lewis, R. W., Ravindran, K. and Usmani, A. S. (1995). "Finite Element Solution of Incompressible Flows Using an Explicit Segregated Approach," *Archives Comp Meth Eng*, Vol 2, No 4, pp 69-93.

Nomura, T. and Hughes, T. J. R. (1992). "An Arbitrary Lagrangian-Eulerian Finite Element Method for Interaction of Fluid and a Rigid Body," *Comp Meth App Mech and Eng*, Vol 95, pp 115-138.

Nomura, T. (1994). "ALE Finite Element Computations of Fluid-Structure Interaction Problems," *Comp Meth in App Mech and Eng*, Vol 112, pp 291-308.

Ramaswamy, B. and Kawahara, M. (1987). "Arbitrary Lagrangian-Eulerian Finite Element Method for Unsteady, Convective, Incompressible Viscous Free Surface Flow," *Int J Num Meth Fluids*, Vol 7, pp 1053-1075.

Ren, G. and Utnes, T. (1993). "Comparison of Different Projection Methods for Solving the Time-Dependent Incompressible Navier-Stokes Equations via Implementation of a Finite Element Scheme," *Proc 8th Int Conf Num Meth Laminar and Turbulent Flow*, Swansea, pp 691-702.

Sarpkaya, T. (1978). "Fluid Forces on Oscillating Cylinders," *J Waterway Port Coastal Ocean Div ASCE*, Vol 104, pp 275-290.

Utnes, T. and Ren, G. (1995). "Finite Element Modeling of the Incompressible Navier-Stokes Equations," *Int J Comp Fluid Dyn*, Vol 4, pp 41-55

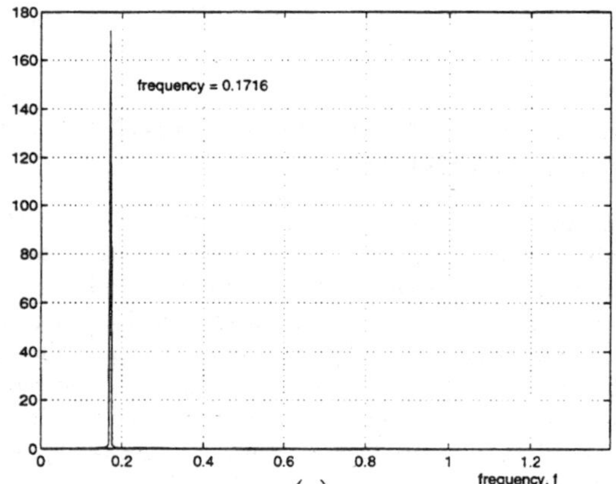

(a)

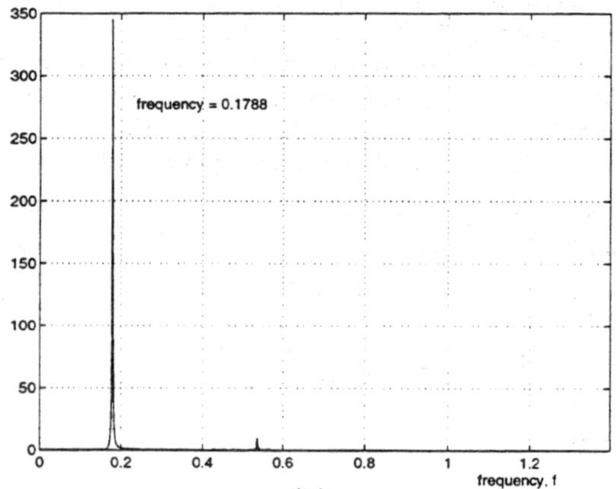

(b)

Figure 3: Frequency response of the transverse force history for an elastically mounted cylinder at $Re = 100$, $f_n = 0.1792$.
(a) Vortex shedding frequency when the cylinder is fixed.
(b) Vortex shedding frequency of the freely oscillating cylinder. The shedding frequency is synchronized with the natural structural frequency (f_n) indicating lock-in.

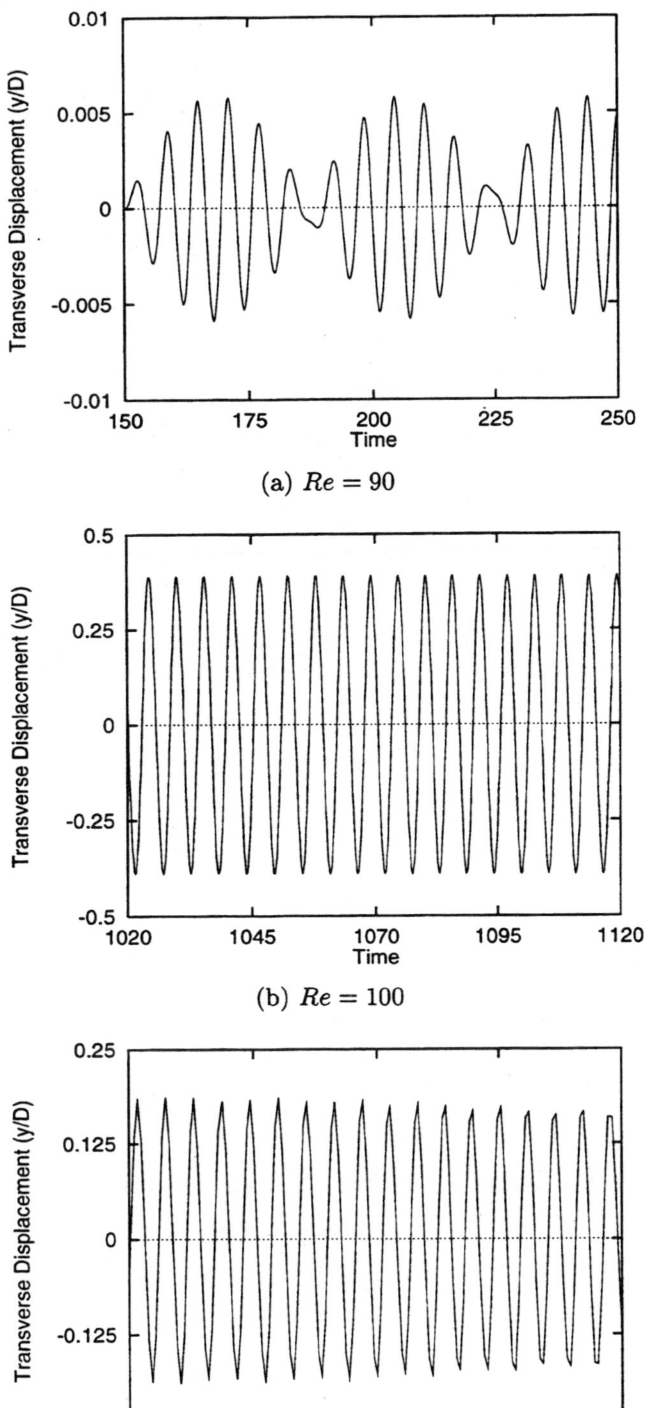

(a) $Re = 90$

(b) $Re = 100$

(c) $Re = 135$

Figure 4: Time histories of the computed non-dimensional amplitude at different Reynolds number.

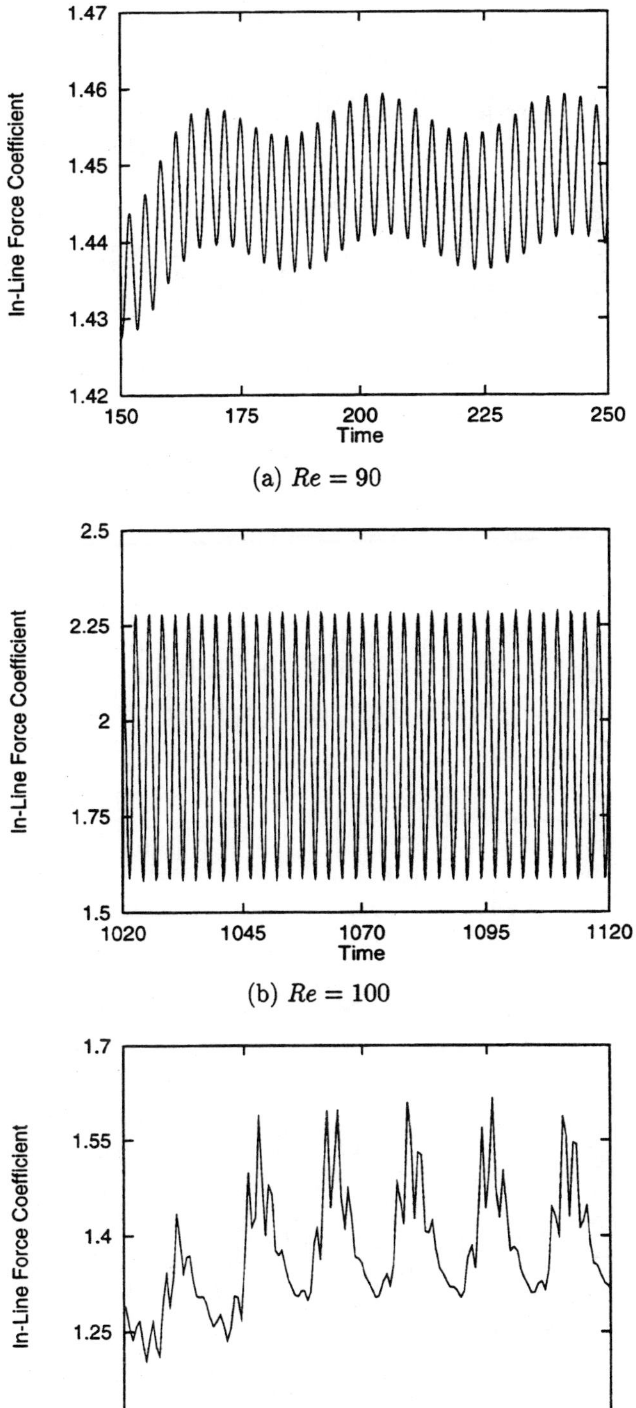

(a) $Re = 90$

(b) $Re = 100$

(c) $Re = 135$

Figure 5: Time histories of the computed non-dimensional drag force at different Reynolds number.

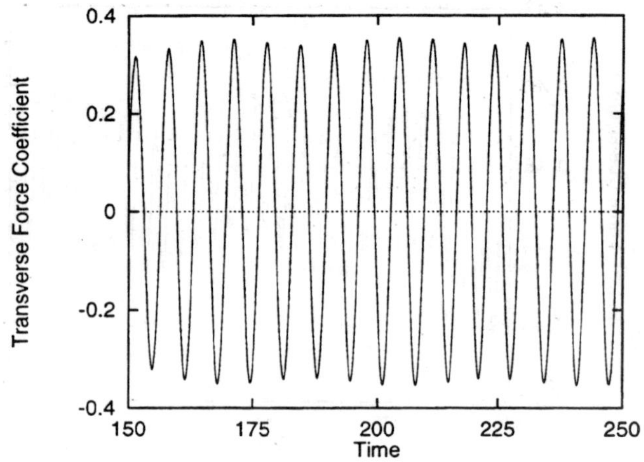

(a) $Re = 90$

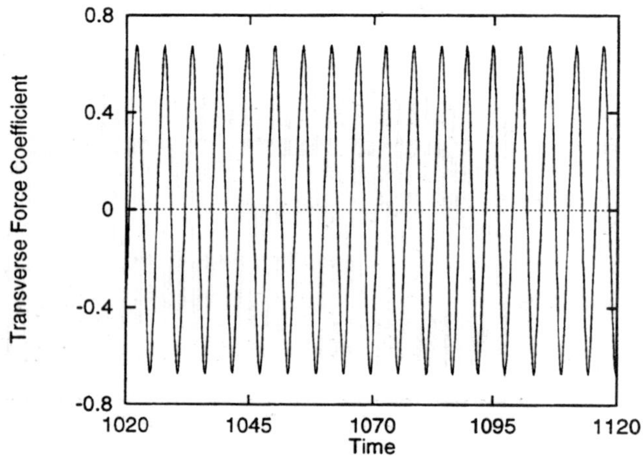

(b) $Re = 100$

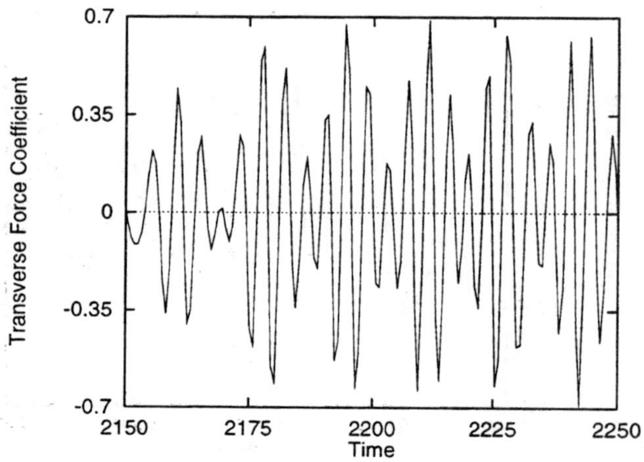

(c) $Re = 135$

Figure 6: Time histories of the computed non-dimensional lift force at different Reynolds number.

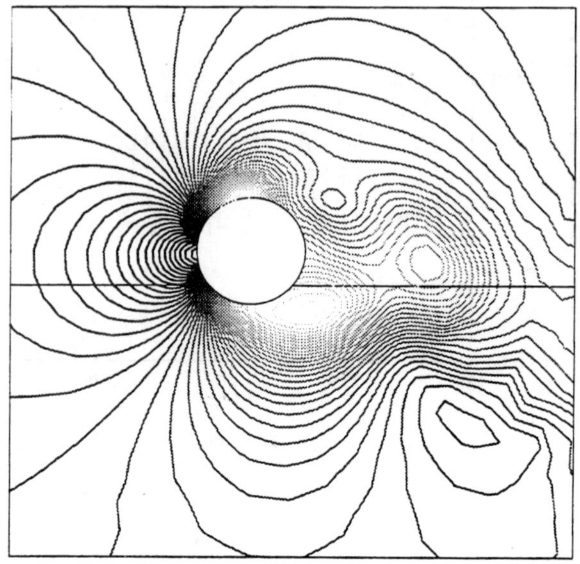

Figure 7: Contour lines of the pressure field around the cylinder at $Re = 100$. The dark line corresponds to $y = 0$ about which the cylinder oscillates.

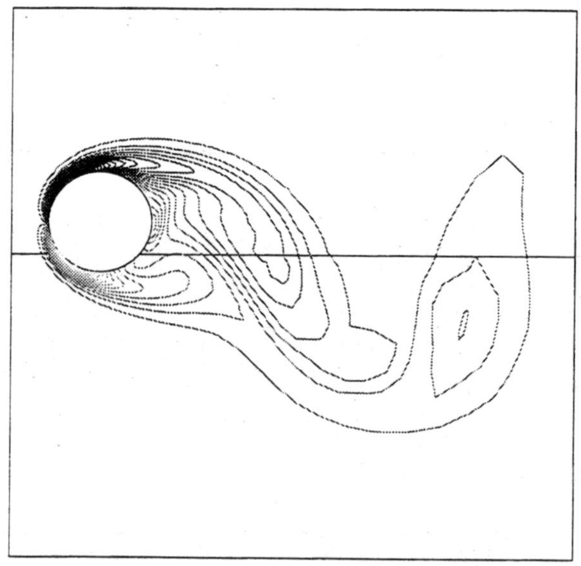

Figure 8: Contour lines of the vorticity, $\omega = \nabla \times \mathbf{v}$, around the cylinder at $Re = 100$. The dark line corresponds to $y = 0$ about which the cylinder oscillates.

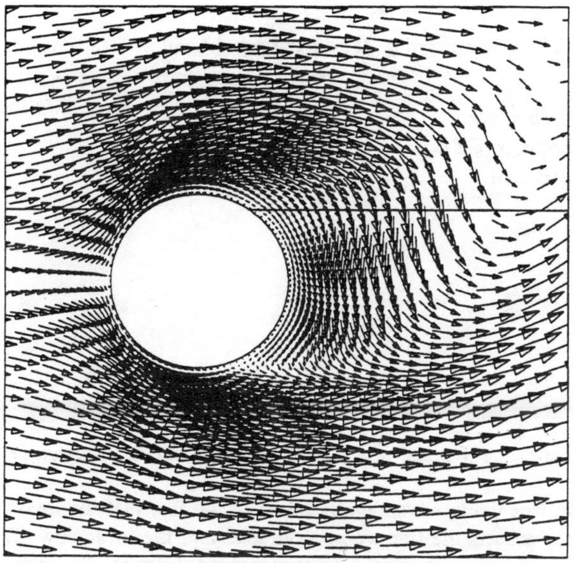

(a)

(b)

Figure 9: Velocity vector field around an elastically mounted cylinder oscillating freely in the transverse direction during lock-in, $Re = 100$, $f_n = 0.1792$. The dark line corresponds to $y = 0$ about which the cylinder oscillates.

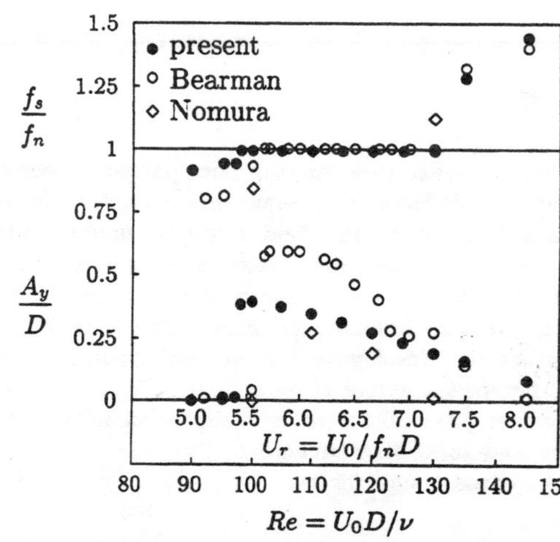

Figure 10: Comparison of computed oscillation amplitude and vortex shedding frequency with the experimental results obtained by Anagnostopoulos and Bearman(1992) and Nomura (1994).

Proceedings of the Seventh (1997) International Offshore and Polar Engineering Conference
Honolulu, USA, May 25-30, 1997

ISBN 1-880653-28-1 (Set); ISBN 1-880653-31-1 (Vol. III)

Experimental Nearfield Flow Maps of a Cylinder in a Reversing Flow

E.F. Medeiros and M.P. Brandão
Instituto de Aeronáutica e Espaço, CTA
San José dos Campos, Brazil

R.W. Miksad
University of Virginia
Charlottesville, VA, USA

ABSTRACT

The near-cylinder time-varying flow patterns associated with the transverse lift force are mapped in a reversing planar flow. A novel identification of flowfield changes, due to intermittent shedding and switching of vortices from one shoulder of the cylinder to the other, is accomplished using phase angle information of the fundamental harmonic of the lift. The flow maps reveal the development of vortex structures around the cylinder for several stages of an oscillation cycle. Analysis of the data provides physical insight into the relationship between the lift force and local flow structure. This may prove helpful in modeling hydrodynamic forces.

KEY WORDS: Flow Mapping, reversing flow, vortex shedding, lift on cylinder.

INTRODUCTION

An improved understanding of flow/structure interaction mechanisms is key to the efficient design of submerged bodies. Of interest (Bearman, 1984, Williamson, 1985, Ongoren and Rockwell, 1988) is the study of the asymmetric nearfield vortex-flow patterns which induce lift on a circular cylinder in a direction transverse to the flow oscillation line.

It is known (Obasaju, 1988) that a harmonically oscillating planar flow past a circular cylinder develops a wake of vortices that may intermittently switch between two dominant mirror-image patterns about the line of flow oscillation. That is, a 180° phase shift of the dominant vortices occurs as the transverse force alternates direction at same cycle-times in a flow oscillation.

In order to properly characterize the flowfields associated with the changing direction of the lift force, one must be able to sense (i.e. discriminate) between those flowfields associated with, say an upward lift force, and those associated with a downward (i.e. π-phase shifted) lift force. Then, a conditional ensemble average, to identify the characteristic flowfields at each stage, can be carried out over the respective records for upward and π-shifted downward lift force identified flowfields. An uncondicional (global) averaging operation would cancel out the image (i.e. switched vortex patterns) features of the flow.

In this work, advantage is taken of the fact that the total lift force signal reflects the transverse asymmetry of the local flow. Therefore, the lift force is used as a discriminator to identify the instantaneous nearfield flow patterns, including the predominant mirror-image *upwash* and *downwash* patterns due to switching. This flow-pattern discriminating technique is detailed in Medeiros (1995a). It allows the capture of the competing flow modes that occur intermittently during an experimental run of constant amplitude and constant frequency of flow oscillation f_o. The near-cylinder flow maps, presented here, result from the application of the referred technique to the reconstruction of the velocity field characteristic of each stage of the lift force cycle.

EXPERIMENTAL ARRANGEMENT

The measurements were carried out in an oscillating water tunnel shown in Fig. 1. The water tunnel is capable of generating data for the Keulegan-Carpenter ($KC=2\pi A/D$) number up to 25, i.e. over a flow range dominated by large-scale vortices. The peak Reynolds number (R_e), based on the undisturbed velocity amplitude (U) and the cylinder diameter D, can reach 30,000, i.e. the flow regime may be affected by a low-scale turbulence.

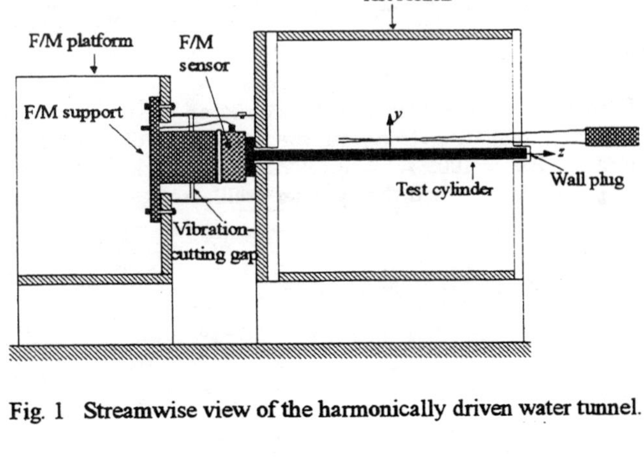

Fig. 1 Streamwise view of the harmonically driven water tunnel.

As the enclosed volume of water is harmonically driven past the cylinder with an amplitude A, the streamwise (u) and the cross-stream (v) fluid velocities, at a point in the flowfield, are measured simultaneously using a computer-positioned laser probe. The velocity probe was programmed to survey a grid of 1,788 measurement points closely distributed around the cylinder at its midispan plane as seen in Fig. 2. The inline and transverse loadings on the cylinder were measured using a high-sensitivity multi-axis load cell into which the cylinder was secured in the manner of a cantilever. For each measurement point, 150 cycles of fluid velocity, lift and inline forces were sampled.

The sampling of the cycles, according to whether they belong to the **D**- or **U**-mode of wake return, gave an average of 50 cycles per flow mode per measurement point. Data for every *KC* number were collected over 20 hours of tunnel operation at a sampling rate of 100Hz.

NEARFIELD FLOW MAPPING

After sampling the **D** and **U** flow modes from an experimental run, u- and v- fluid velocities are mode averaged according to the lift force signal indication of the mode of wake return. Velocity vector maps for selected instants of time (phases relative to the farflow velocity zero-crossing) are, then, constructed for showing the key features of the nearflow as shown in Fig. 3.

For improved visualization of the vortex structures around the stationary cylinder, a change of coordinate system, from the Eulerian to the Lagrangian framework, was essential. Thus, in the maps, one sees the cylinder 'moving' in still water. Accordingly, the driving flow force has been subtracted from the inline force component.

Quasi-symmetric flow regime ($KC = 10$)

As maps in Fig. 4 show, the flowfield around the submerged cylinder at $KC = 10$ depicts a *quasi-symmetric pattern* about the x-axis. At an early stage of the cylinder displacement from right to left, i.e. about 15°, the upstream (left side) flow is not quite aligned with the oscillation direction but is slightly at an angle due to the velocity field of the two decaying counter-rotating vortices of the previous half-cycle. This is clearly seen in the 345° map. As displacement is further increased, the top-half cylinder mass of water 'spirals' earlier than the bottom one causing, thus, an upward peak of lift to appear about 60° as indicated by the force trace placed below each map. When the velocity reaches a maximum at 90°, the streamlines are quite symmetric about the x-axis indicating nearly zero lift which is also in agreement with the indication of the force magnitude drawn at the center of the cylinder.

Over the deceleration range (90° to 180°), growth in both recirculation regions ('vortices') on the top and under the bottom halves of the cylinder is evident. At a phase of about 135°, the cylinder experiences maximum downward lift indicating that the lower recirculation region is stronger than the upper. As the end of the half-cycle, 180°, gets closer, the velocity is decreasing to zero, the deceleration is reaching a maximum, both counter-signed vortices appear shed (165° map) over the rear-half of the cylinder, and the flow induced inline and lift forces go to zero. The cylinder then reverses to move into its own wake from right to left. At phase 225° one notices the local fields of the shed vortices flowing past the cylinder over the same shoulder they previously grew and shed. Those vortices were so weak that they completely decayed during the early stages of flow reversal. The loading pattern will then repeat in like fashion to the previous half-cycle. The downwash mode at $KC = 10$ is not presented since it gave fewer cycles than the upwash one.

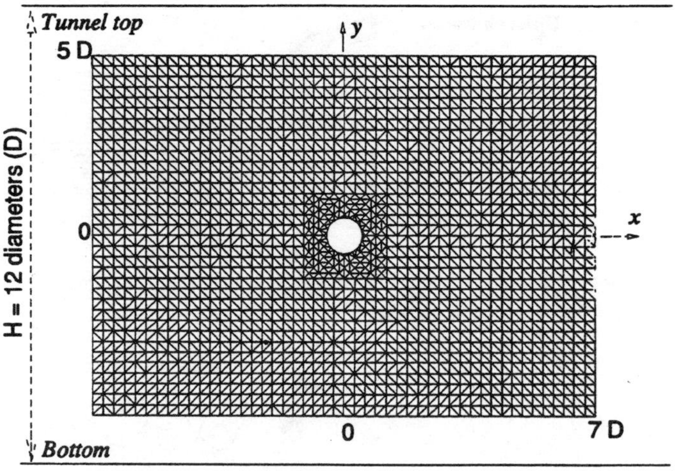

Fig. 2 Grid of measurement points. The triangulation offers a connected data structure for construction of streamline maps and contour plots.

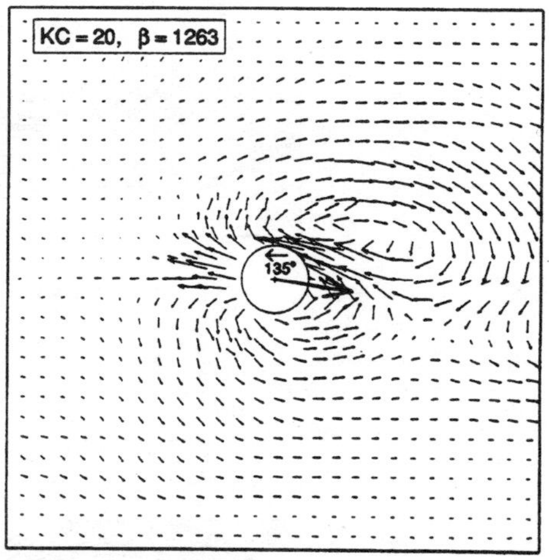

Fig. 3 Instantaneous flow map shown in the Eulerian frame of reference, i.e as viewed by an observer moving with the farflow.

The flow structures just described are not representative of a transverse-street wake, as suggested in Williamson's (1985) flow classification table. According to that table, $KC = 10$ fits right in the middle of a transverse regime characterized by the shedding of a single large vortex during each half-cycle. His outstanding flow pictures imply for *KC*s ranging from 7 to 13 a transverse street composed of vortices moving away from the nearfield flow region roughly perpendicular to the oscillation direction.

Flow visualizations conducted at the present water tunnel (Ventre, 1993) reveal not only the occurrence of a transverse-street regime, with switching back and forth between mirror-image modes about the x-axis, but also a second short-lasting symmetric regime similar to the one shown here. Thus, one may expect a significant sensitivity to the experimental apparatus on the flow structures around a cylinder in reversing flows.

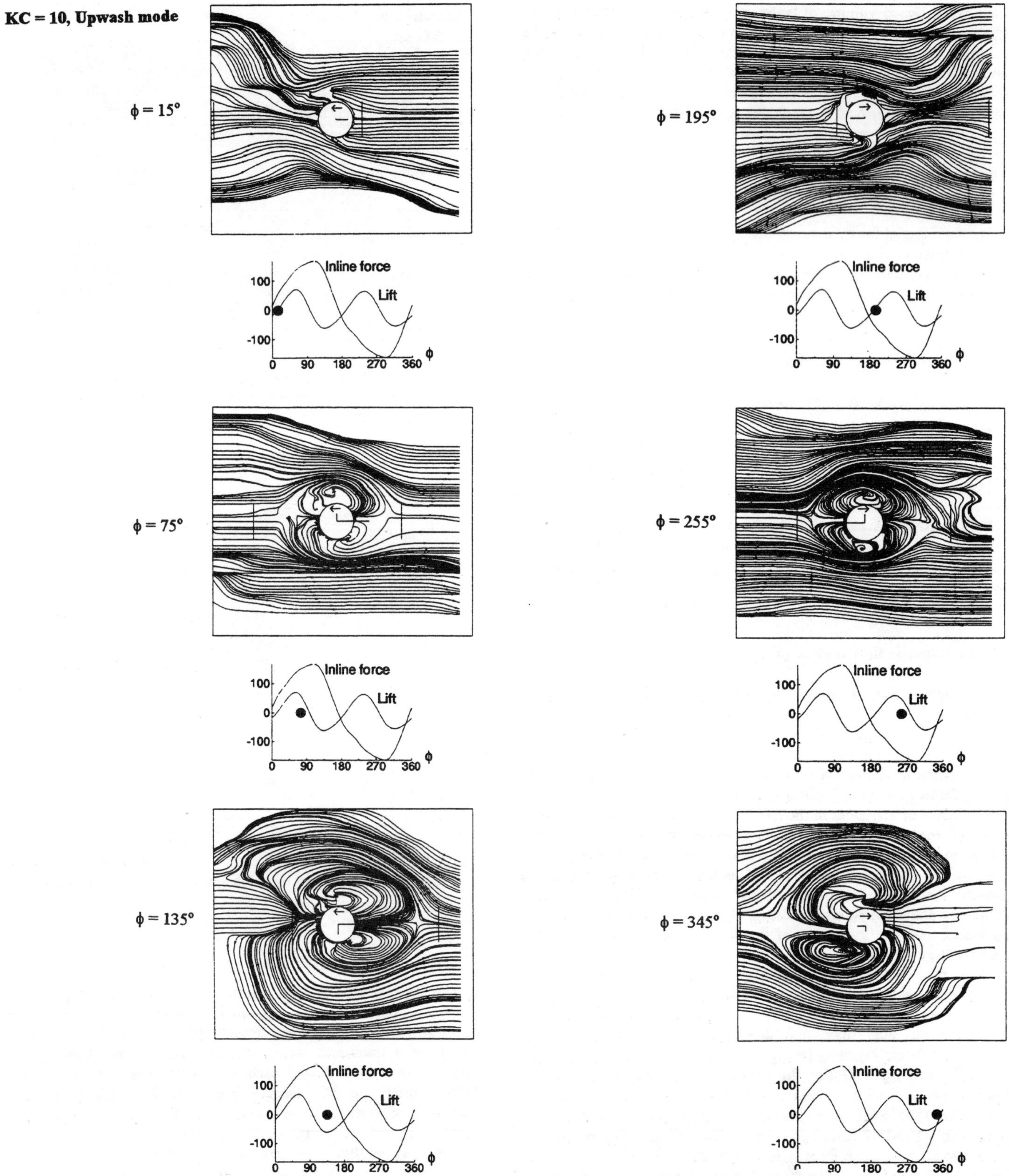

Fig. 4 Streamline maps for the upwash flow mode at $KC = 10$. ϕ is the cycle phase. Forces made dimensionless by $\rho D f_o^2/2$, where ρ is density of fluid. Segments with origin at the cylinder center show relative magnitude and direction of the loading.

Transverse flow regime ($KC = 15$)

The flow regime at $KC = 15$ consists of the shedding of one large vortex during the early stages of cylinder deceleration in each half-cycle. This large vortex convects out from either shoulder of the cylinder as represented in Fig. 5. At this flow amplitude the plots for the mode-averaged velocity fields clearly show a *transverse-street pattern*. The upwash and downwash flow modes are remarkably well resolved.

From the upwash mode plots, one sees a dominant flow activity occurring over the top-half of the cylinder. At 15° phase a street or 'water jet' contours the cylinder from the bottom-left to its top-right side. This jet results from the velocity field of the surviving counter-clockwise vortex that was shed during the previous half-cycle in the second quadrant of the (x,y) plane. The jet of fluid induces larger velocity over the top-half of the cylinder favoring, thus, an upper vortex shedding mode. The surviving recirculation region lasts over the entire acceleration range, from 0° to 90°, such that it pairs with the newly formed clockwise upper vortex. This pairing hastens the growth of the upper vortex which sheds early in the deceleration range, say about 135°. The pairing of the upper vortices appears so strong to cause (at about 165°) the lower unshed counter-clockwise vortex to move around on the rear-half of the cylinder inducing higher velocities over the cylinder. This latter trend will establish a returning-street of fluids at the early stages (195°) of the complementary half-cycle to upwash the cylinder just as in the previous half-cycle. The flow patterns will then repeat remarkably symmetric with respect to the axis when contrasting like-phases of the half-cycles.

Reduction of the lift data clearly shows (Medeiros, 1995b) significant steady transverse force in the range of KC from 10 to 16, i.e. in the transverse vortex-street regime. In particular at $KC = 15$ the steady lift magnitude amounts to 25% of the amplitude of the fluctuating zero-mean lift force.

The relationship between peaks of lift force and flow activity can be established by direct inspection of the velocity field plots and force waveforms. For instance, with the first lift force peak of upwash mode (U) occurring at a phase of about 70°, the flowfield shows the returning counter-clockwise 'vortex' (from the second quadrant) reaching the y-axis. Therefore, this lift peak is associated with the return of a previously shed counter-clockwise vortex over the cylinder's top. The second peak occurs at about 135° and it has negative sign. The main flow activity taking place about 135° corresponds to the shedding of the clockwise upper vortex. It is interesting to note the agreement between sign of lift and sense of flow circulation. It means that, the first lift peak has positive sign and the returning vortex has counter-clockwise rotation (positive); the second lift peak has negative sign and the shedding vortex has clockwise rotation (negative). This agreement reflects the Kutta-Joukowski lift-circulation theorem, where the signs of the cylinder velocity U match those of the x-axis.

The downwash flow mode, which occurred intermittently with the upwash mode in the same experimental run at $KC = 15$, is also shown in Fig. 5. Plots of the mode-averaged velocity fields for individual phases are quasi-image, relatively to the x-axis, of those from the upwash mode. It means that the nearfield flow activity is the same for both modal cycles.

Williamson (1985) describes the flow in the range $3 < KC < 15$ consisting of a single pair of vortices convecting away in a direction about 45 degrees to the flow oscillation. Here, this trend can be identified at phases 285° and 315° of upwash mode plots, for example. The second quadrant of the 45°-degree map is particularly helpful to reference this identification. Ventre's (1993) pictures also confirm the presently mapped flow features.

KC = 15, Upwash mode

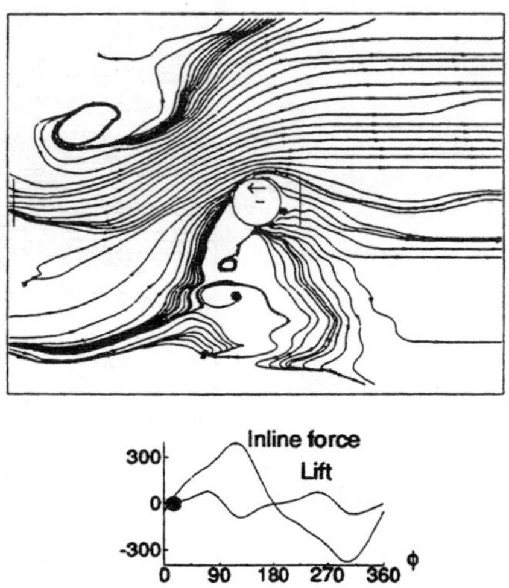

KC = 15, Downwash mode

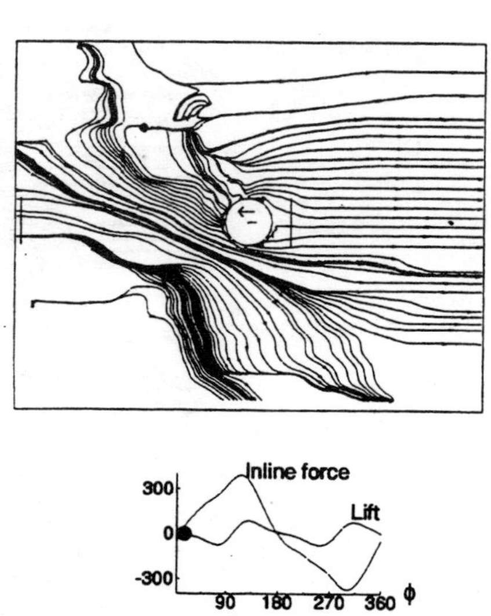

Fig. 5 Streamline maps for the upwash and downwash flow modes at $KC = 15$. Cylinder is 'moving' from right to left.

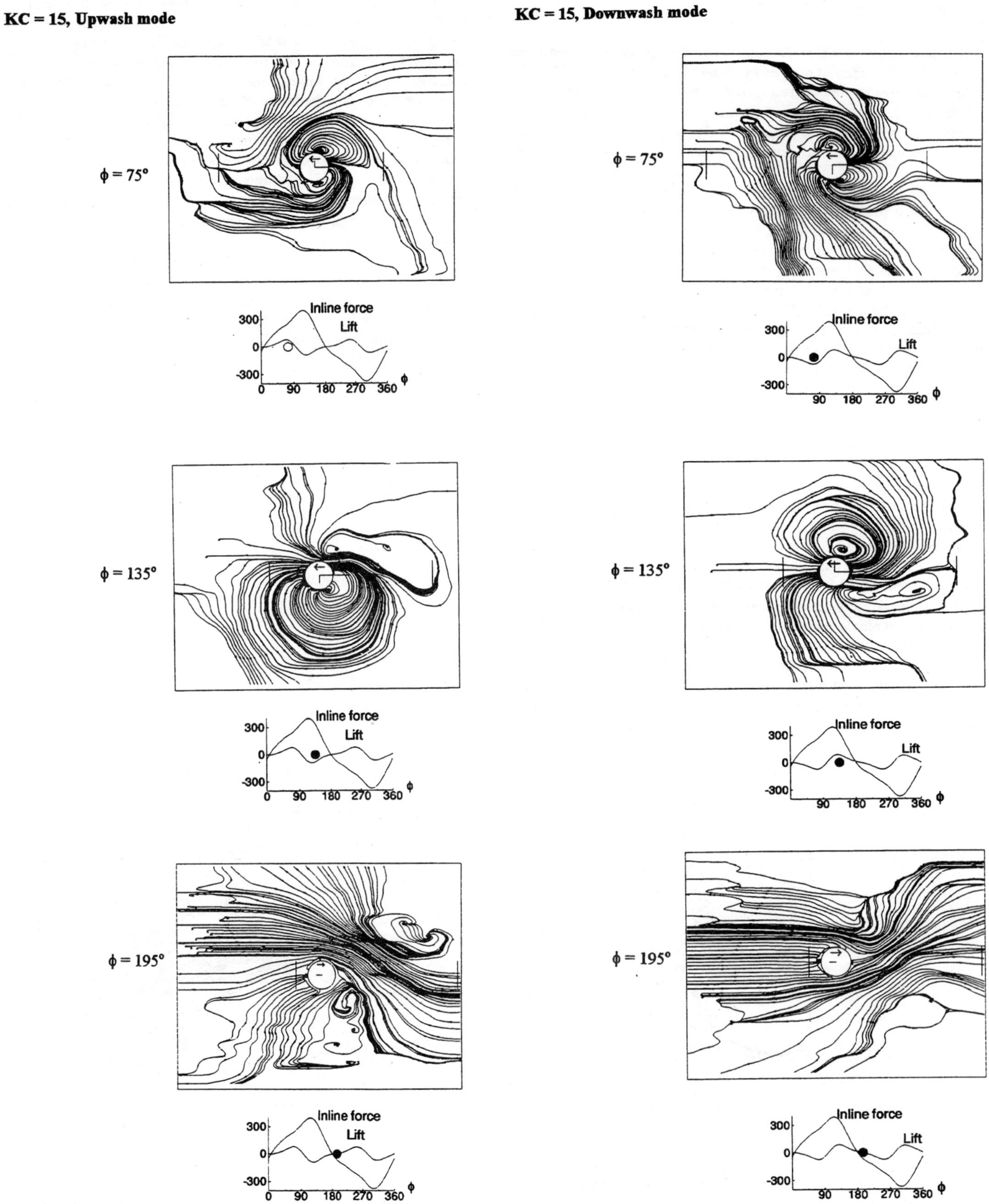

Fig. 5 continued KC = 15. Cylinder moves' from right to left, then, returns into its wake.

Double-pair vortex flow regime ($KC = 20$)

As the flow amplitude increases, the nearfield flow structure evolves from the transverse-street pattern to an *inline-street pattern.* The flow regime at $KC = 20$ consists of the formation of a single-pair of vortices in each half-cycle, therefore, making a 'double-pair wake' of vortices during a complete cycle. From each single-pair, one vortex sheds from alternate shoulders of the cylinder. Shed vortices convect and decay rather aligned with the oscillation direction. The bulk of the nearfield flow orbits around the cylinder during the oscillation cycle, i.e. the wake returns about alternate shoulders of the cylinder.

The flow mechanics in the upwash mode at $KC = 20$ can be described as follows:

At 15° phase, i.e., after having the cylinder 'moved' 1/12 of the half-cycle amplitude from right to left, the flowfield consists of a large counter-clockwise vortex which was shed during the previous half-cycle in the second quadrant of the (x,y) plane as seen on map 345°. The bulk of the flow upwashes the cylinder because of the induced velocity field of the surviving upper vortex. Two small recirculation regions behind the cylinder are visible at this early stage (45°). The returning vortex decays somewhere between 45° and 75°, and at this latter phase one notices the lower recirculation region outgrowing the upper. As the cylinder reaches halfway in the deceleration range, say about 165°, the lower larger vortex appears shed and the upper smaller vortex is just about to shed. The corresponding downwash map at phase 165° is also presented as a verification of the goodness of the conditional sampling technique as a flow-mode-discriminator approach. During the remaining stages of the deceleration range both vortices undergo decay – the upper vortex decaying faster than the lower. The cylinder then reverses and the returning wake downwashes it, as seen at 195°. In this second half-cycle the flow mechanics repeats and the flow patterns are symmetric about the x-axis when contrasting like-oscillation phases.

At $KC = 20$, the fundamental lift frequency is three times the frequency of water oscillation. This means that during a half-cycle the cylinder experiences three peaks of lift force. The flow mechanisms which cause these peaks can be identified by associating the force waveforms with the flow maps. For instance, the first upward lift peak of mode U occurs at about 55° phase and the maps imply that the primary flow activity is an upwash of the cylinder due to the return of the surviving upper vortex. When the second downward lift peak occurs (at about 100°), the lower vortex is outgrowing the upper. Therefore, the relevant flow mechanism causing this peak appears to be the reaching of maximum vortex outgrowth. The third lift peak is upward (at about 150°) and occurs simultaneously with the maximum growth of the upper vortex. In conclusion, the triplet lift peaks can be associated with a return-outgrowth-growth sequence of the upper-lower-upper vortices sequence. It is interesting to note that the shedding of the lower vortex is not included in this loading sequence. Probably, the core of the lower vortex (at the moment of the shedding) is far enough of the cylinder's transverse plane such that the shedding pulse causes only a mild lift force.

The downwash flow mode is nearly a perfect image-mode of the upwash one — both competing intermittently during a run with an average rate of intermittence of 6 cycles. Essential flow features mapped here, such as the development of the two large vortices per half-cycle and the alternation of cylinder's shoulders taken by the returning wakes, are reported by Williamson (1985) and Ventre (1993) as well.

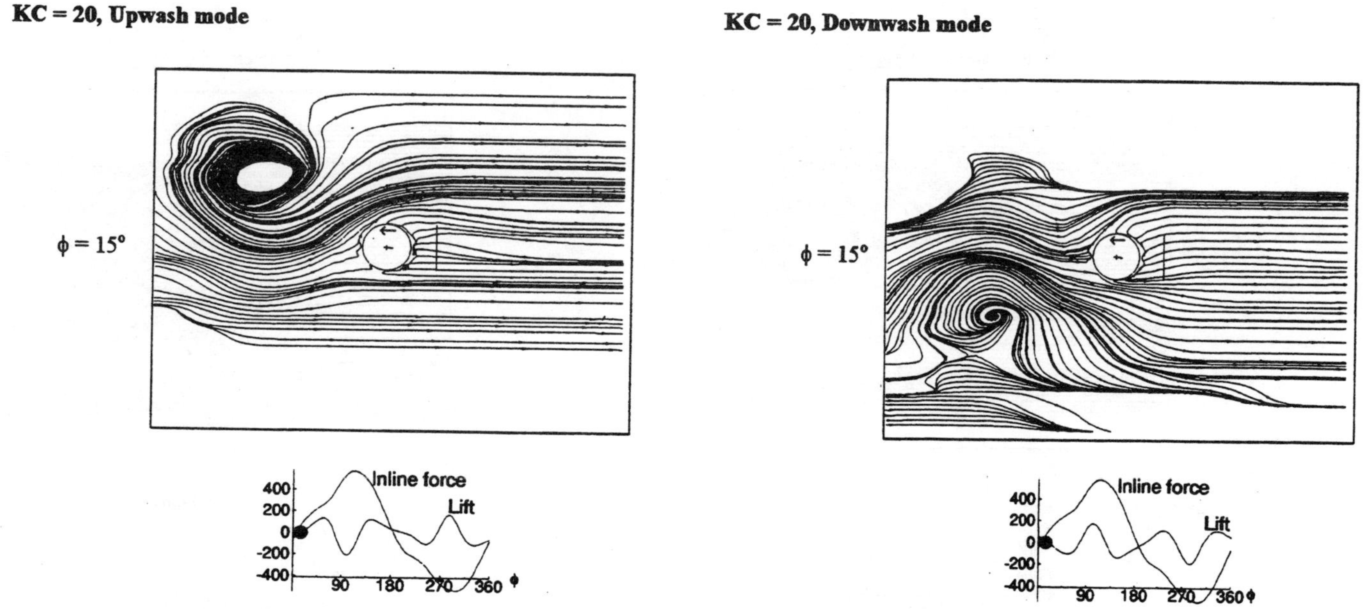

Fig. 6 Streamline maps for the upwash and downwash flow modes at $KC = 20$. Cylinder is 'moving' from right to left.

KC = 20, Upwash mode **KC = 20, Downwash mode**

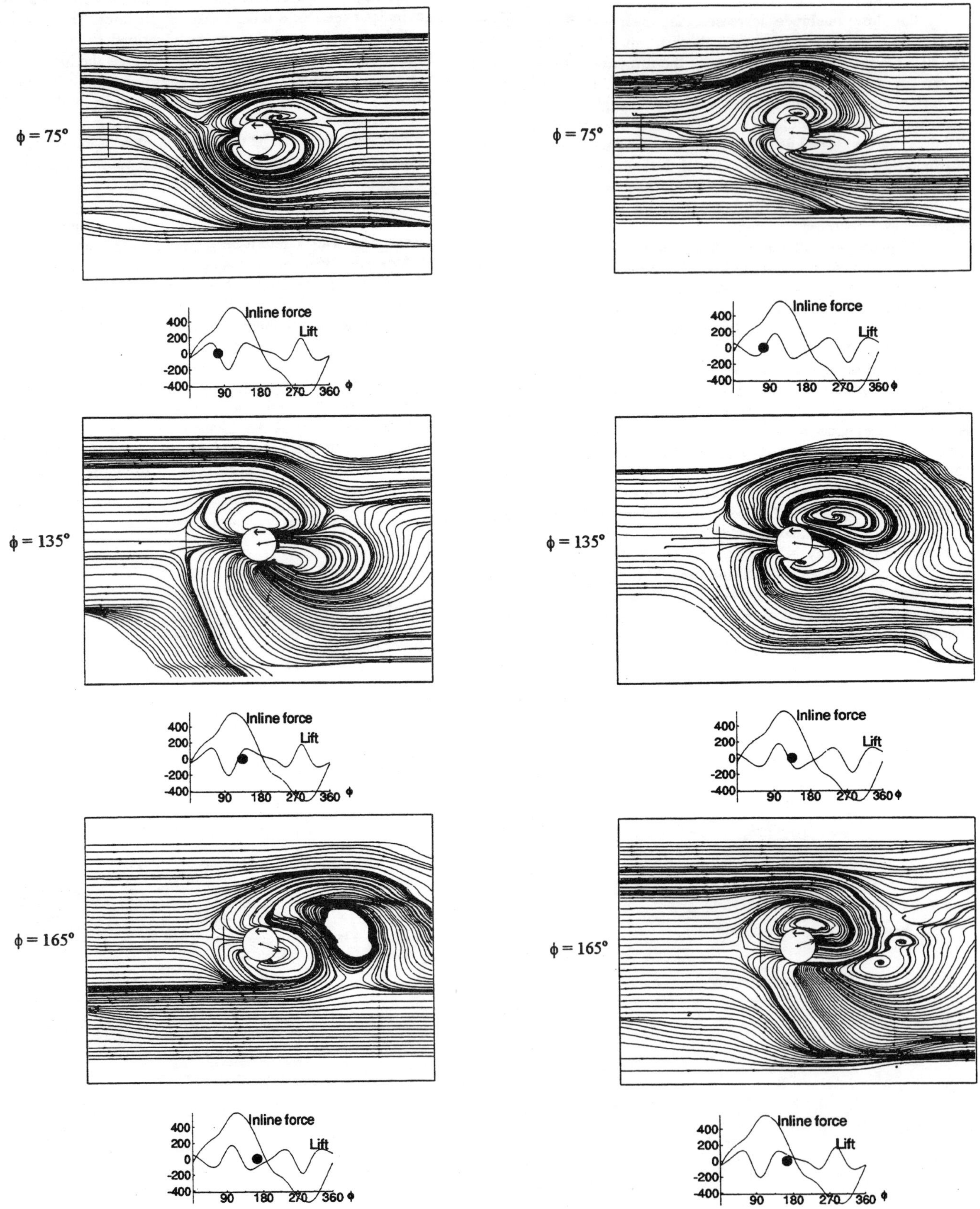

Fig. 6 continued KC = 20. Cylinder 'moves' from right to left, then, starts returning into its wake.

KC = 20, Upwash mode

$\phi = 195°$

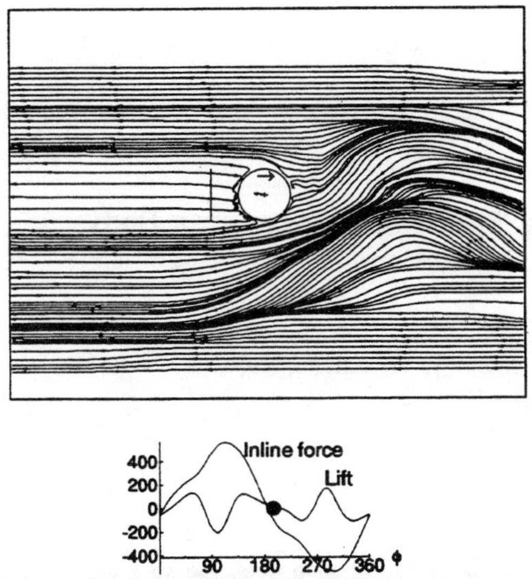

KC = 20, Downwash mode

$\phi = 195°$

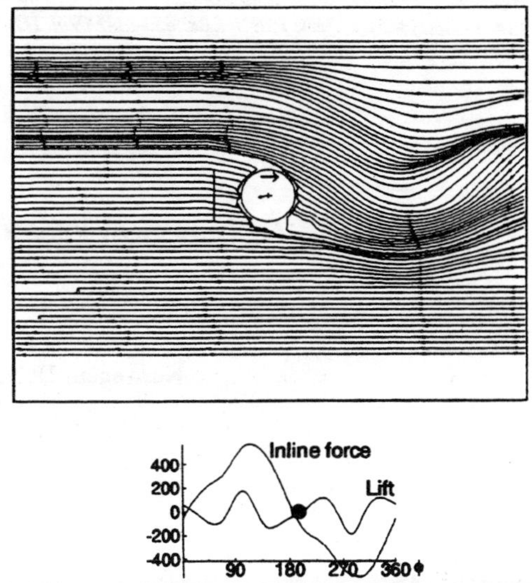

Fig. 6 continued Streamline maps for the upwash and downwash flow modes at *KC* = 20. Cylinder starts 'returning' into its wake.

CONCLUSION

Maps of the predominant flowfield structures around a circular cylinder in a reversing planar flow have been presented. An understanding of the lift force development has been acquired by visually correlating its waveform signatures with the instantaneous flow patterns of the mapped flowfield. The results suggest that insight as well as quantitative information on flow/structure interaction phenomena can be obtained by exploiting data from the hydrodynamic loading of a body.

ACKNOWLEDGMENTS

This material represents results of research supported in part by the U.S. National Science Foundation Engineering Research Center Program under grant No. CDR-8721512, the Texas Education Coordinating Board, Advanced Technology Project Grant No. 003658-413, and the Brazilian Air Force Department of Research and Development.

REFERENCES

Bearman, PW (1984). "Vortex Shedding from Oscillating Bluff Bodies", *Ann Rev Fluid Mech,* Vol 16, pp 195-222.

Medeiros, EF and Miksad, RW (1995a). "An Experimental Study of the Flow Structures and Hydrodynamic Forces of a Cylinder in a Reversing Flow," *Proc 9th Int Symp Offshore Eng* - BRASIL OFFSHORE'95, Rio de Janeiro, pp 73-87.

Medeiros, EF (1995b). "An Experimental Study of the Flow Structures and Hydrodynamic Forces of a Cylinder in a Reversing Flow," *Ph.D. Dissertation,* The University of Texas at Austin.

Obasaju, ED, Bearman, PW, and Graham, JMR (1988). "A Study of Forces, Circulation and Vortex Patterns around a Circular Cylinder in Oscillating Flow," *J Fluid Mech,* Vol 196, pp 467-494.

Ongoren, A, and Rockwell, D (1988). "Flow Structure from an Oscillating Cylinder Part 2. Mode Competition in the Near Wake," *J Fluid Mech,* Vol 191, pp 225-245.

Ventre, A-V (1993). "Flow Visualization of a Planar Oscillatory Flow around a Cylinder," *M.S. Thesis,* The University of Texas at Austin.

Williamson, CHK (1985). "Sinusoidal Flow Relative to Circular Cylinders," *J Fluid Mech,* Vol 155, pp 141-174.

Proceedings of the Seventh (1997) International Offshore and Polar Engineering Conference
Honolulu, USA, May 25-30, 1997

ISBN 1-880653-28-1 (Set); ISBN 1-880653-31-1 (Vol. III)

A Numerical Procedure for Predicting Transverse Forces on a Body Submitted to Lock-In Motion

Joar Dalheim
Norwegian University of Science and Technology, Trondheim, Norway
(presently at Det Norske Veritas)

ABSTRACT

An explicit computational procedure is developed to solve the coupled motion of a rigid body and a viscous incompressible fluid. In this study, the investigated body is a rigid 2D cylinder conducting forced sinusoidal motion in a viscous fluid flow. The arbitrary Lagrangian-Eulerian (ALE) formulation is employed in order to incorporate the interface conditions between the body and the fluid. The Galerkin finite element method is used for the spatial discretization of the fluid domain, and a velocity correction projection method is used to solve the Navier-Stokes equations. The fluid boundary motion is determined from the forced cylinder motion. The method is applied to predict the so-called transverse added mass and added damping of a circular cylinder under vortex shedding vibrations. The computed results are compared with experimental data reported in the literature, and the agreement appears to be quite good.

KEYWORDS: ALE, Finite Element Method, Forced Cylinder Oscillations, Transverse Forces.

1. INTRODUCTION

Extensive experimental studies have been performed on spring supported cylinders in steady, uniform flow, and it has been found that the drag force is quite regular, while the transverse force exhibits a large degree of variability. However, such experiments also show that the transverse motion is nearly regular (sinusoidal), and its amplitude is usually related to certain dimensionless parameters, such as reduced velocity, Reynolds number and some version of a stability parameter. Utilizing these dimensionless parameters one has been quite successful in fitting the experimental results onto a single curve (see e.g. Griffin [3] or Sarpkaya [10]). In this approach, the transverse force, which presumably is driving the motion is not paid much attention to. As a matter in fact, it is usually not reported, and in many cases probably not measured.

For more general applications it is of interest to know more about the exciting force. For example, some parts of a long flexible offshore riser will at a given vibration mode exhibit larger (and some parts smaller) amplitudes than a spring-supported segment at the same reduced velocity. It is then the total energy transmitted from the flow to the riser over its entire length which is of interest. This energy must balance the internally dissipated energy (with a possible adjustment for other contributors to the energy balance). A natural approach is then to perform forced vibration tests, and vary parameters such as reduced velocity and motion amplitude-to-diameter ratio, while measuring the lift force. Several experimental tests on forced vibrations of this kind have already been reported (see e.g. Sarpkaya [11] or Moe and Wu [8]).

In the present paper a numerical procedure for performing such forced vibrations tests is presented. The fluid domain is described by the finite element method, and the cylinder motion is given as prescribed boundary condition to the flow field. Computationally, one needs to treat the problems with a moving interface between the fluid and the rigid body. In the present study the ALE formulation is employed because of two considerations. First, it is convenient to describe the fluid motion on the moving interface by the Lagrangian description in order to treat the compatibility conditions and the equilibrium between the fluid and the rigid body. Secondly, it is apparently impossible to employ the Lagrangian description for the entire fluid domain because of severe mesh distortion due to vortex shedding and flow through outer boundaries. Therefore a mixed Lagrangian and Eulerian formulation seems a natural choice.

The remainder of this paper may be outlined as follows. In section 2, the ALE method is briefly reviewed. In section 3, the basic equations of motion of a fluid and solid are stated. In section 4, the discretized equations are presented. In section 5, the computational procedure including the force coefficient averaging is described. In section 6, some numerical results are presented and compared to reported experimental results, and finally some concluding remarks are presented in section 7.

2. THE ALE CONCEPT

The ALE description is based on a general kinematical theory described by e.g. Hughes [4] where three domains and the mapping between them are defined as follows:

Spatial domain is where the fluid mechanics is posed. It is regarded as generally in motion because of the moving interface to the adjoining structures.

Material domain is to be thought of as the domain occupied at time $t=0$ by the material particles which occupy the spatial domain at t (which is the time of interest). Hence, the material domain is also generally in motion.

Referential domain is defined as fixed for all *t*, and its image at time *t* under a prescribed mapping is the spatial domain.

In the context of finite element analysis, the spatial domain is represented by a moving mesh and the referential domain by a reference state of the moving mesh. If the spatial domain is fixed throughout, the formulation can be regarded as an Eulerian description. Likewise, a Lagrangian description is obtained when the spatial domain always coincides with the material domain. The important consequence of this concept is given as the following kinematical relation with respect to a physical property φ (see e.g. Donea [2]):

$$\dot{\varphi} = \varphi' + c_i \varphi_{,i} \quad (1)$$

where $\dot{\varphi}$ is the material time derivative, and φ' is the referential time derivative keeping coordinates in the referential domain constant and c_i is the convective velocity,

$$c_i = u_i - \hat{u}_i \quad (2)$$

where u_i represents the material velocity and $\hat{u}_i$ the mesh velocity. In the Eulerian description, the mesh velocity is zero, and in the Lagrangian description the convective velocity is zero because $\hat{u}_i = u_i$.

3. BASIC EQUATIONS

The ALE formulated Navier-Stokes equations

In Fig. 1 a general outline of the problem domain of interest is presented, where $\Omega_C(t)$ is the domain occupied by the moving cylinder. $\Omega_F(t)$ is the moving spatial domain upon which the fluid motion is described, and $\Gamma_I(t)$ is the interface between $\Omega_C(t)$ and $\Omega_F(t)$. When the rigid cylinder $\Omega_C(t)$ moves, the interface $\Gamma_I(t)$ moves accordingly.

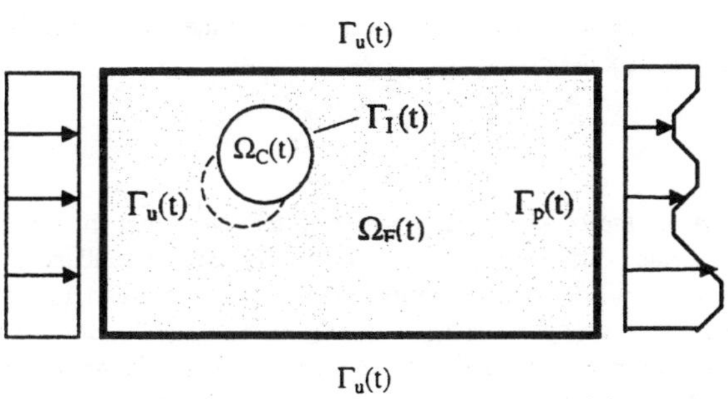

Fig. 1 General outline of the problem domain.

When the distribution of the mesh velocity $\hat{u}_i(\mathbf{x},t)$ is specified in some way on $\Omega_F(t)$ in accordance with the motion on $\Gamma_I(t)$, the following ALE description of the Navier-Stokes equations can be employed as the governing equations of the fluid domain:

$$\rho u_i' + \rho(u_j - \hat{u}_j)u_{i,j} = -p_{,i} + \mu u_{i,jj} + f_i \quad (3)$$

with the incompressibility constraint

$$u_{i,i} = 0 \quad \text{in } \Omega_F(t) \quad (4)$$

where ρ is the fluid density, f_i is the body force (e.g. gravity forces), p is the dynamic pressure and μ is the dynamic viscosity.

The outer boundary conditions are divided into two parts as shown in Fig. 1:

$$u_i = u_0 \quad \text{on } \Gamma_u \quad (5)$$

$$p = p_0 = \text{constant and traction-free on } \Gamma_p. \quad (6)$$

When Γ_u and Γ_p are fixed in space, the mesh velocity will be zero on these boundaries. Because of the compatibility requirement on the moving interface between the cylinder and the fluid, the velocity component on the moving interface $\Gamma_I(t)$ must coincide with the mesh velocity:

$$u_i^I = \hat{u}_i \quad \text{on } \Gamma_I(t). \quad (7)$$

The equation of motion of the cylinder

In the present problem the cylinder motion is fixed in the inline direction and forced to move in the transverse direction according to a prescribed function of time

$$y = A_y \sin(\omega t) \quad (8)$$

where A_y is the transverse motion amplitude and ω is the oscillation frequency. The coupling condition between the cylinder, mesh and the fluid is thus

$$u_1^I = \hat{u}_1^I = 0 \quad (9)$$

$$u_2^I = \hat{u}_2^I = A_y \sin(\omega t) \quad (10)$$

The fluid force acting on the cylinder

In the present computational procedure the fluid force acting on the cylinder is obtained from a numerical integration of the dynamic pressure and shear force on the cylinder interface.

$$F_{pres}(t) = -\oint_{\Gamma_I} p(\mathbf{x},t)\cdot \mathbf{n}\, d\Gamma \quad (11)$$

$$F_{shear}(t) = \oint_{\Gamma_I} \mu \frac{\partial u(\mathbf{x},t)}{\partial n} \cdot \mathbf{s}\, d\Gamma \quad (12)$$

where $F_{tot} = F_{pres} + F_{shear}$ and n and s is the outward normal and tangential unit vector respectively. Note that the unit vector is negative when pointing out of the fluid domain and into the cylinder.

4. SPATIAL DISCRETIZATION

The Navier-Stokes equations

A bilinear element formulation is used both in discretizing the velocity field and the pressure field. In mathematical terms this can be expressed as

$$\mathbf{u}(\mathbf{x},t) \approx \tilde{\mathbf{u}}(\mathbf{x},t) = \sum_{i=1}^{NPOIN} \mathbf{N}_i(\mathbf{x}) \cdot \mathbf{u}_i(t) = \mathbf{NU} \quad (13)$$

and similarly for the pressure, $p(\mathbf{x},t) \approx \mathbf{NP}$. The matrix **N** contains the interpolation polynomials for all elements, and **U** and **P** represent the nodal values of the velocity and the pressure, respectively. According to e.g. Lewis [7] equal order interpolation can be used without risking spurious pressure modes when employing the velocity correction method, which is the case in this work.

By employing the Galerkin weighted residual method combined with equation (13), the following discrete form of the Navier-Stokes equations (3) can be obtained (body forces omitted):

$$\mathbf{M}\frac{d\mathbf{U}}{dt} + \mathbf{AU} = -\frac{1}{\rho}\mathbf{DP} + (\mu/\rho)\cdot\mathbf{LU} \tag{14}$$

Here $\mathbf{M}$ is the mass matrix, $\mathbf{D}$ is the divergence matrix, $\mathbf{A}$ is the advection matrix and $\mathbf{L}$ is the weak form of the discrete Lagrangian. They are all time dependent because of the moving mesh. The matrix $\mathbf{A}$ contains information about the moving mesh, $\hat{u}_i$. Furthermore, Δt is the timestep and μ/ρ is the kinematic viscosity. The matrix $\mathbf{M}$ is defined as

$$\mathbf{M} = \int_\Omega \mathbf{N}^T \mathbf{N} \, d\Omega \tag{15}$$

and the other matrices are defined in a similar manner (see e.g. Utnes and Ren [14]).

The mesh velocity

The velocity field for the mesh, $\hat{u}_i$, is arbitrarily specified and restricted by the following boundary conditions:

$$\hat{u}_i = v_i \quad \text{on } \Gamma_I(t), \tag{16}$$

$$\hat{u}_i = 0 \quad \text{on } \Gamma_u \cup \Gamma_p \tag{17}$$

Hence it seems natural that $\hat{u}_i$ decays to zero with distance from the cylinder. In this study a geometric series is employed in order to make the mesh distortion minimal close to the cylinder interface. The movement of the mesh is restricted to an area around the flexible cylinder in order to make the matrix updating process quicker. Fig. 2 illustrates the moving mesh area. The fluid domain $\Omega_F(t)$ is partitioned into a fixed domain Ω_f and a deformable domain $\Omega_d(t)$:

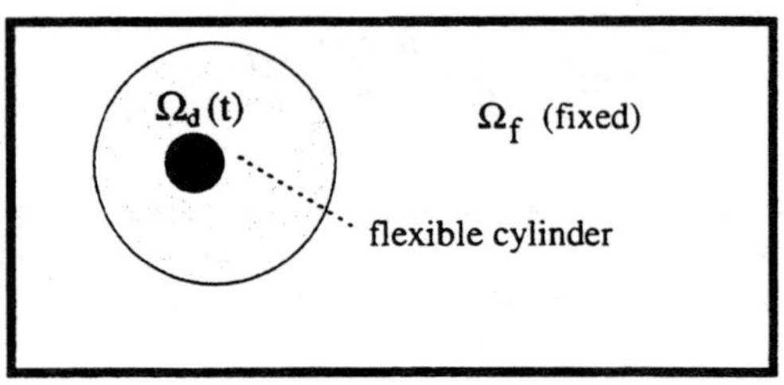

Fig. 2 Fluid domain divided into subdomains.

Node- and element positions within $\Omega_d(t)$ may vary with time according to the cylinder movement. The nodal displacements and velocities within $\Omega_d(t)$ are related to the cylinder motions by

$$\hat{x}_i^C(t) = \kappa^C(\mathbf{x})\cdot\delta_i(t) \tag{18}$$

$$\hat{u}_i^C(t) = \kappa^C(\mathbf{x})\cdot v_i(t) \tag{19}$$

Here, the superscript C indicates 'at node C', and $\kappa^C(\mathbf{x})$ is a real number $\in [0,1]$ at each node indicating how much this node is moving compared to the cylinder. As stated above, the value of κ is determined from a geometric series.

5. COMPUTATIONAL SOLUTION PROCEDURE

Solution of the flow field

The solution algorithm for the Navier-Stokes equations (14) is based on the formulation in Utnes and Ren [14]. The equation is solved by discretizing the time derivative using a forward Euler scheme and then employing a Chorin [1] projection method (Velocity Correction Method), where a predicted value of the velocity field is obtained by excluding the pressure values. Then the predicted velocity and the incompressibility condition is used to find an updated pressure value, which finally is employed to find the corrected value of the velocity. When calculating the predicted velocities, a two-step Taylor-Galerkin method is employed. In mathematical terms the computational procedure can be expressed as follows

1. Predict velocities:

$$\mathbf{M}^n\tilde{\mathbf{U}}^{n+\frac{1}{2}} = \mathbf{M}^n\mathbf{U}^n - \frac{\Delta t}{2}\left[\mathbf{A}^n\mathbf{U}^n\right] \tag{20a}$$

$$\mathbf{M}^n\tilde{\mathbf{U}}^{n+1} = \mathbf{M}^n\mathbf{U}^n - \Delta t\left[\mathbf{A}^n\tilde{\mathbf{U}}^{n+\frac{1}{2}} + \nu\mathbf{L}^n\mathbf{U}^n\right] \tag{20b}$$

2. Calculate new pressure values (Lewis [6]):

$$\mathbf{L}^n\mathbf{P}^{n+1} = \frac{\rho}{\Delta t}\mathbf{D}^n\tilde{\mathbf{U}}^{n+1} - \frac{\rho}{\Delta t}\mathbf{B}^n\mathbf{U}^{n+1} \tag{21}$$

3. Corrected velocities:

$$\mathbf{M}^n\mathbf{U}^{n+1} = \mathbf{M}^n\tilde{\mathbf{U}}^{n+1} - \frac{\rho}{\Delta t}\mathbf{D}^n\mathbf{P}^{n+1} \tag{22}$$

Here $\tilde{\mathbf{U}}^n$ represents predicted velocity at timestep n, and $\mathbf{B}$ is a boundary matrix defined as

$$\mathbf{B} = \oint_\Gamma \mathbf{N}^T(\bar{\mathbf{n}}\cdot\mathbf{N})\,d\Gamma$$

where $\bar{\mathbf{n}}$ is the outward unit normal vector. Note that in this study **BU** has a value different from zero only on the inflow boundary nodes. Except from the Poisson equation (21), this system can be solved explicitly if a so-called mass lumping is employed. However, in transient problems a consistent mass matrix is desirable when solving the advection part of the above system (Lewis [7]). This is achieved by employing the following iteration procedure:
The equation system is of the form

$$MU = f$$

where M is the consistent mass matrix. Then the solution is obtained from the iteration process

$$U_{i+1} = U_i + M_L^{-1}\left[f - MU_i\right]$$

where M_L is the lumped mass matrix, and i is the iteration number. This iteration converges very fast, usually three iterations are sufficient to yield accurate solutions.

The Poisson equation (21) is by far the most CPU-expensive of the four equations listed above to solve, and the solution algorithm for this equation is therefore crucial for the time cost of the whole calculation. The solution procedure stated above requires two initial conditions and two boundary conditions. All internal velocities in $\Omega_F(t)$ and outflow velocities on Γ_p are initially set to zero. The starting condition of the dynamic pressure is zero everywhere except on Γ_p , where a fixed value is prescribed. Then, for all timesteps Dirichlet boundary conditions for the velocities are explicitly imposed on Γ_u and Γ_I .

The interface boundary conditions are updated for each timestep according to (9) and (10).

Turbulence

The present numerical results are compared to experimental results which are obtained from relatively large Reynolds number from a computational point of view (Re > 10^4). In order to reduce the number of elements necessary for such high Reynolds numbers a two-equation turbulence model (k - ε) is employed. Details about the turbulence model can be found in Rodi [9]. The (k - ε) model is previously reported to give quite good results on 2D cylinders in fluid flows (see e.g. Utnes et al.[13] or Rodi[9]).

In the numerical tests (of section 6) the minimum (and inflow boundary) values of turbulent kinetic energy and dissipation was $k_{min} = 10^{-4}$ and $\varepsilon_{min} = 5\times10^{-5}$, and a smooth surface was assumed with a roughness parameter of $z_0 = 0.9\times10^{-8}$.

Time averaging of the lift coefficients

It is customary to approximate the lift forces as being harmonic and split them into two components, one in phase with the transverse acceleration and the other in phase with the transverse velocity (see e.g. Moe and Wu [8]). The force coefficients according to the cylinder motion (8) can then be written as

$$C_L = C_{my} \sin(\omega t) - C_{dy} \cos(\omega t) \tag{23}$$

where the force has been made dimensionless by dividing by $\rho DU^2/2$. The C_{my} term is often referred to as the transverse added mass (or inertia term) and represents additional mass to be accelerated when positive. C_{dy} represents the lift damping (also referred to as the drag coefficient [11]), and the minus sign indicates that a positive C_{dy} will add additional damping to the system.

Even though the cylinder motion is harmonic, the resulting fluid forces are not and the coefficients in (23) must therefore be time averaged in order to give useful information about the process (see e.g. Moe and Wu [8]). The time-averaging is obtained from a least square fitting procedure as described below:

In one numerical simulation, the value of C_L is stored at n different timesteps. Consequently, at the end of the simulation there exists n equations of the type

$$C_{L(i)} = C_{my} \sin(\omega t_i) - C_{dy} \cos(\omega t_i) \tag{24}$$

(24) can be written in on matrix form as

$$\mathbf{A} \times \mathbf{x} = \mathbf{b}$$

$$\begin{bmatrix} \sin(\omega t_1) & \cos(\omega t_1) \\ \sin(\omega t_2) & \cos(\omega t_2) \\ \vdots & \vdots \\ \sin(\omega t_n) & \cos(\omega t_n) \end{bmatrix} \begin{bmatrix} C_{my} \\ -C_{dy} \end{bmatrix} = \begin{bmatrix} C_{L(1)} \\ C_{L(2)} \\ \vdots \\ C_{L(n)} \end{bmatrix} \tag{25}$$

(25) is a system of n equations with only two unknowns, and is therefore overdetermined. However, if (25) is premultiplied with $\mathbf{A}^T$ the following 2×2 equation is obtained:

$$\underline{\mathbf{A}}\mathbf{x} = \underline{\mathbf{b}} \tag{26}$$

with solution

$$\mathbf{x} = \underline{\mathbf{A}}^{-1}\underline{\mathbf{b}} \tag{27}$$

where $\underline{\mathbf{A}} = \mathbf{A}^T\mathbf{A}$ and $\underline{\mathbf{b}} = \mathbf{A}^T\mathbf{b}$. It can be shown (see e.g. Strang [12]) that the above procedure is a least square approximation.

6. NUMERICAL RESULTS

The numerical results are obtained from a finite element grid of 3000 elements, and with boundary conditions as illustrated in Fig. 1. The oscillation frequency of the forced cylinder motion is set to ω = 1.26 rad/s, the cylinder diameter D = 1m and the kinematic viscosity $\nu = 10^{-4}$ m²/s. The fluid density used is ρ =1000 kg/m² and the timestep for all computations was in the order of $\Delta t \approx 0.005$. The oscillation amplitude is varied according to the amplitudes in the experimental tests (Sarpkaya [11]), Ay/D = 0.25, 0.5 and 0.75. The reduced velocity, given by ($U_r = 2\pi U_\infty/\omega D$), is varied by varying the inflow velocity $U_\infty \in$ [0.6 m/s - 1.4 m/s] giving a Reynolds number range of Re $\in$ [0.6 - 1.4]×10^4. This is corresponding to the Reynolds numbers for the water flows in Sarpkayas report, which are all in the subcritical region.

A true value of approximately 100 seconds was computed, which is equivalent to 20 periods of cylinder oscillation.

The time averaged numerical results are compared to time averaged experimental results in Fig. 3 - 5. The coefficients are plotted with respect to $U_r \times S_t$ rather than with respect to U_r alone ($S_t = f_sD/U_\infty$ is the Strouhal number, in which f_s is the vortex shedding frequency for a stationary cylinder). This is done to make the plot consistent with Sarpkaya[11], where the coefficients are plotted with respect to $U_r \times S_t$ in order to account for variations in Strouhal numbers and Reynolds numbers among the compared experimental results. The Strouhal number is numerically predicted from computing the flow around fixed cylinders at all investigated inflow velocities. The computed Strouhal number was found to be in the range $S_t \in$ [0.19 - 0.25].

In Fig. 3 is presented a comparison between experimental determined and computational determined lift coefficients for a forced oscillation amplitude of A_y/D = 0.25. Sarpkayas presentation of lift coefficients from various experimental tests are shown in the upper graphs, and the computed lift coefficients from the present study is shown in the lower graphs.

The agreement between the experimental and the computed results in Fig. 3 must be regarded as remarkable, considering the fact that the investigated phenomenon is the transverse vibration of a cylinder in a fluid flow and that the gradients of the force coefficients in the neighborhood of $U_r \times S_t$ = 1.0 is very large, but not discontinuous.

The agreement between the experimental and the computational peak values is striking, particularly for the drag coefficient (C_{dy}). However, the drop in the computed inertia coefficient (C_{my}) is not as distinct as the drop in the experimental inertia coefficients. This small discrepancy might be caused by one (or more) of the following factors:

Poor finite element mesh: The finite element mesh used for these computations is most certainly insufficient in terms of being able to describe the fluid flow exactly. This is primarily due to the relatively small number of elements, but also due to some skewness in the element boundaries. It is mainly the computational resources together

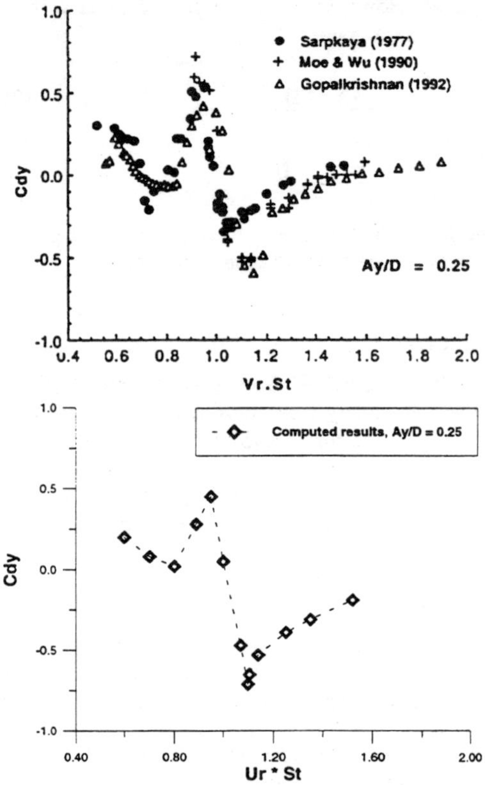

Fig. 3a Comparison of the drag coefficients from Sarpkaya[11] and the computed coefficients in the present investigation. Ay/D = 0.25.

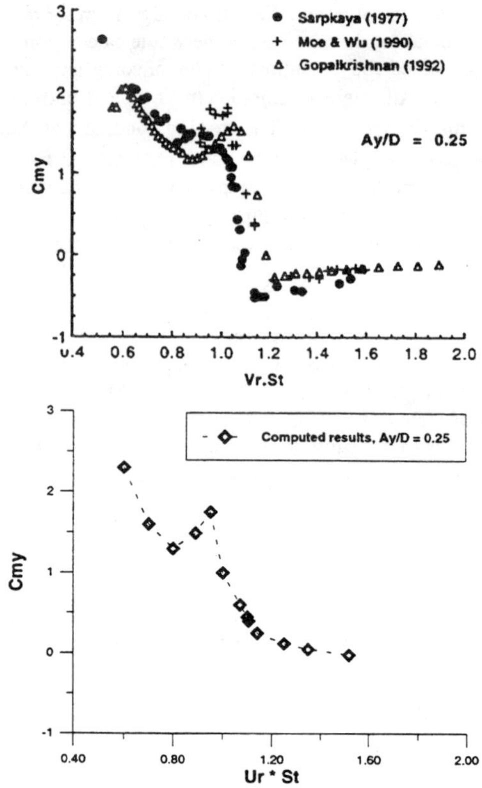

Fig. 3b Comparison of the inertia coefficients from Sarpkaya [11] and the computed coefficients in the present investigation. Ay/D = 0.25.

with available time that are dictating the rational number of elements to be used.

Insufficient accuracy of time-integration: The two-step Taylor-Galerkin procedure used in the present solution of the convection-diffusion equation is only of second order in time. A third or fourth order integration scheme might improve the computed results. An example of a third order scheme is found in Jiang and Kawahara [5], and a forth order scheme is presented in Ku et al. [6].

Turbulence model: The simple (k-ε)- model of turbulence used in this study might influence the vortex shedding in some undesirable way. Preliminary tests indicates that the computed results are rather sensitive to the choice of the minimum values of turbulent kinetic energy (k_{min}) and dissipation (ε_{min}).

REMARK 1 The three experimental results of Fig. 3 do also have some deviation from each other, and if the present computed results are compared to the results from Gopalkrishnan (1992) the computed and the experimental results are almost identical, especially for the drag coefficient.

A similar comparison between experimental and computational force coefficients is performed for a cylinder oscillation amplitude of $A_y/D = 0.5$, and the results are presented in Fig. 4 below.

From observing Fig. 4 it can be noted that the agreement between the experimental and the computational results is slightly reduced, compared to the previous figure. However, the accordance must still be regarded as very good in light of the complexity of the investigated phenomenon.

It is especially the peak values of the drag coefficient which differ in the two graphs. But, it must be considered that also among the different experimental test the coherence of the drag coefficients is reduced when the vibration amplitude is increased to $A_y/D = 0.5$. And if the computed results again are compared to the experimental results obtained by Gopalkrishnan (1992), then the peak values of the drag coefficients match each other quite well.

The value of $(U_r \times S_t)_{min}$ where the minimum drag occur, seems to vary within the different experimental results, but the computed minimum drag does not depart significantly from an average of the experimental minima in terms of the value of $(U_r \times S_t)_{min}$.

The same trends which are observed when the cylinder vibration amplitude increases from $A_y/D = 0.25$ to $A_y/D = 0.5$ are also seen when the amplitude increases to $A_y/D = 0.75$, only amplified.

From Fig. 6 it is clear that the agreement between experimental and computational results is not very strong for the large amplitude vibrations. However, the coherence among the experimental results is also highly reduced for large vibration amplitudes, especially for the drag coefficient (C_{dy}).

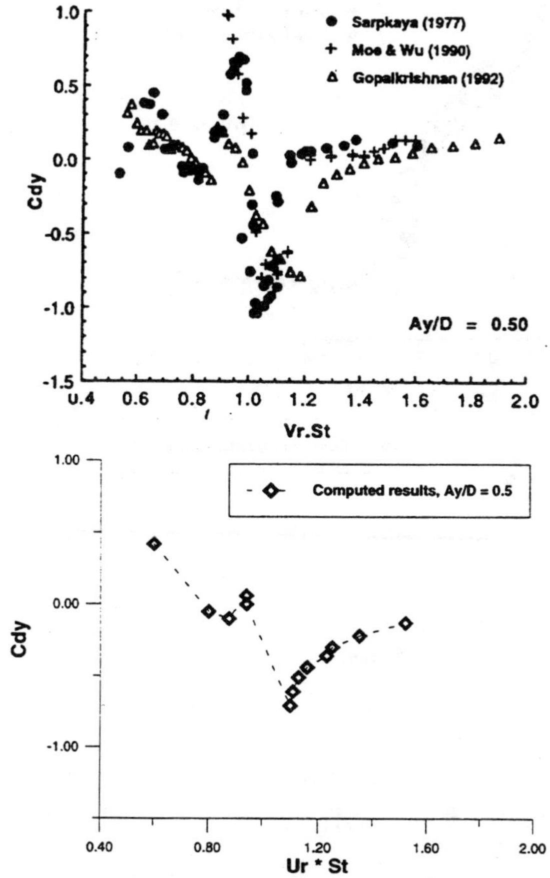

Fig. 4a Comparison of the drag coefficients from Sarpkaya[11] and the computed coefficients in the present investigation. Ay/D = 0.50.

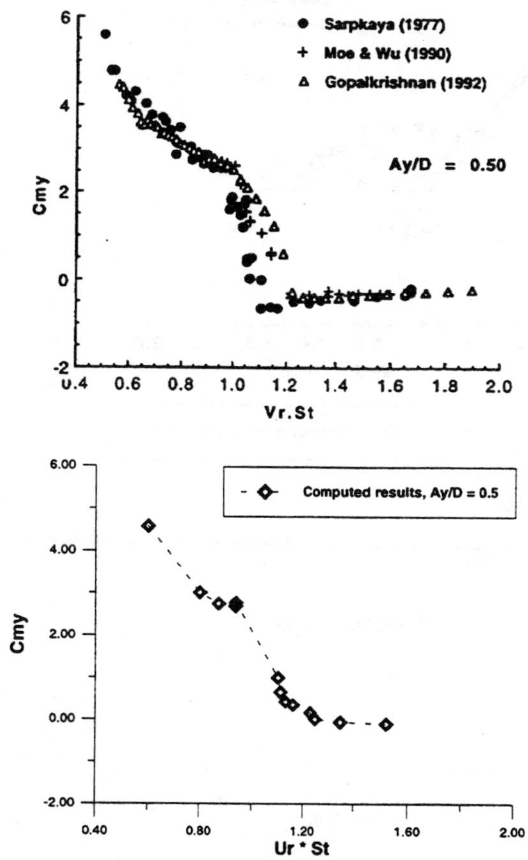

Fig. 4b Comparison of the inertia coefficients from Sarpkaya [11] and the computed coefficients in the present investigation. Ay/D = 0.50.

Again, it can be seen that the computed drag coefficient is in better agreement with Gopalkrishnans results than the other two experimental results. But one important difference is observed; the computed drag is always positive, while Gopalkrishnans experimental drag have some small negative values in a narrow ($U_r \times S_t$) range. In practice, this means that high amplitude vortex-induced vibrations can occur in a small velocity range (given small enough structural damping) for a suitable cylinder in Gopalkrishnans fluid, while the computational fluid in this case will not allow high amplitude vibrations of $A_y/D = 0.75$.

Another distinct difference between the experimental and the computed results is that the drop of the inertia (C_{my}) coefficient starts at lower ($U_r \times S_t$) values for the computed results than for the experimental results.

The reason for these discrepancies is believed to be connected to the factors presented earlier in this section. However, there is one additional factor that needs to be considered. The presented explicit numerical scheme is vulnerable to large deformations and accelerations of the elements. And the forced vibration tests of the present section are computed with the same timestep for all three amplitudes, which means that both the deformation and the acceleration of the elements are three times larger when $A_y/D = 0.75$ than what is the case when $A_y/D = 0.25$. This will in all likelihood affect the quality of the numerical prediction in an unfavorable way.

The remedy for the latter factor which is believed to cause inaccuracies is to increase the deformable mesh region around the cylinder (see Fig. 2), and decrease the timestep. This will result in a reduction in the skewness of the elements as well in the deformation per timestep.

The finite element mesh used in these numerical tests is shown at its maximum deformation configuration ($A_y/D = 0.75$) in Fig. 5. The skewness at the boundary between the fixed and the deformable elements is clearly displayed.

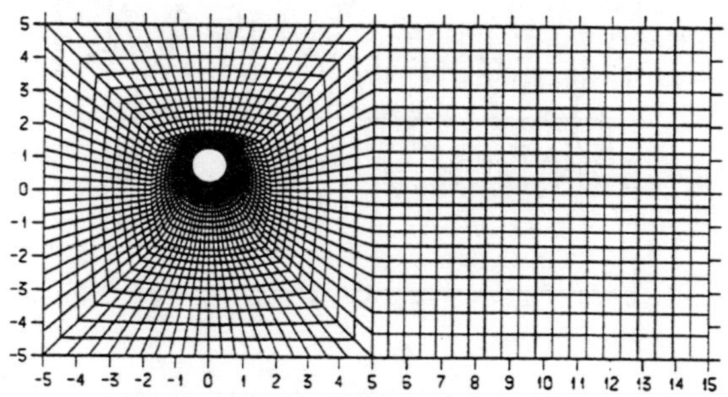

Fig. 5 Unfavorable deformation of finite element grid when Ay/D = 0.75.

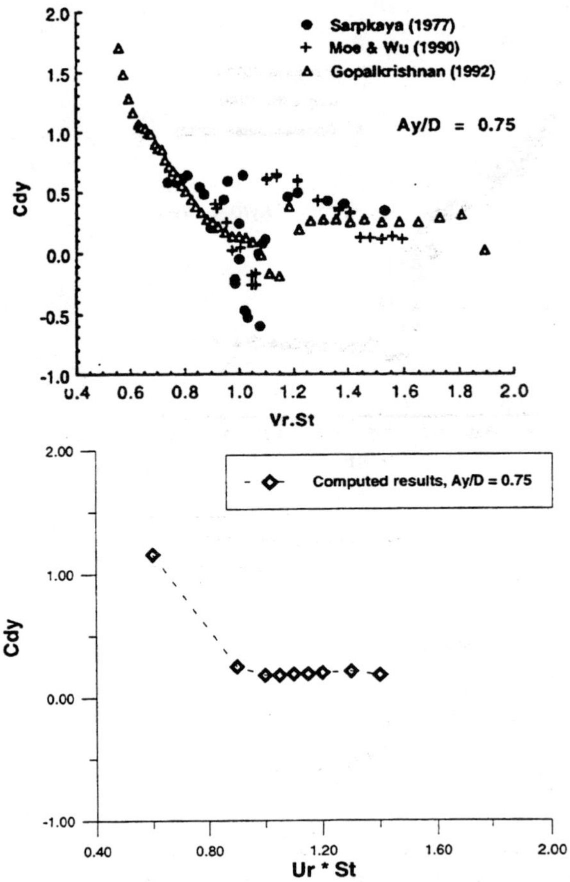

Fig. 6a Comparison of the drag coefficients from Sarpkaya[11] and the computed coefficients in the present investigation. Ay/D = 0.75.

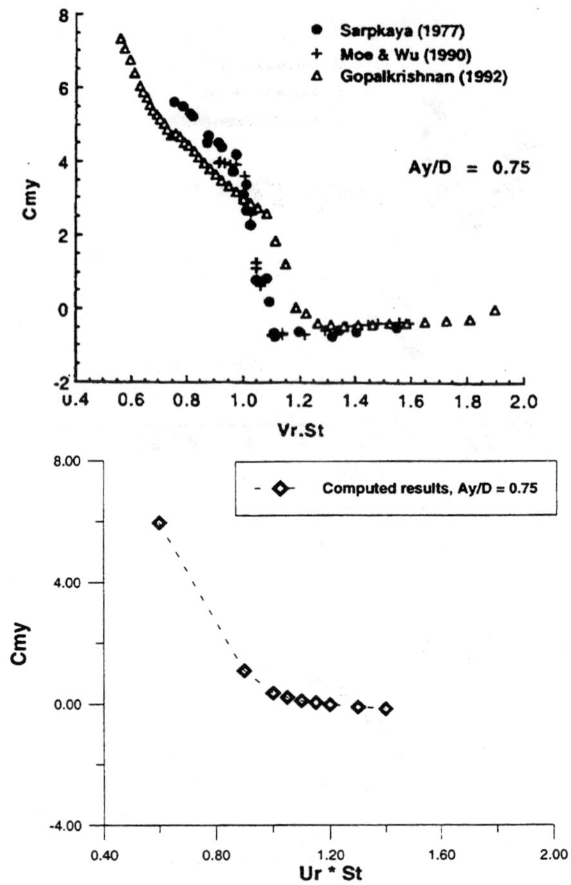

Fig. 6b Comparison of the inertia coefficients from Sarpkaya [11] and the computed coefficients in the present investigation. Ay/D = 0.75.

Irregular forces from harmonic motion

Another important phenomenon is investigated by the present numerical tests of forced vibration. Experimental tests show that the fluid force acting on the cylinder is highly irregular, even when the displacement is purely harmonic. This is clearly seen in Fig. 7 from Moe and Wu [8].

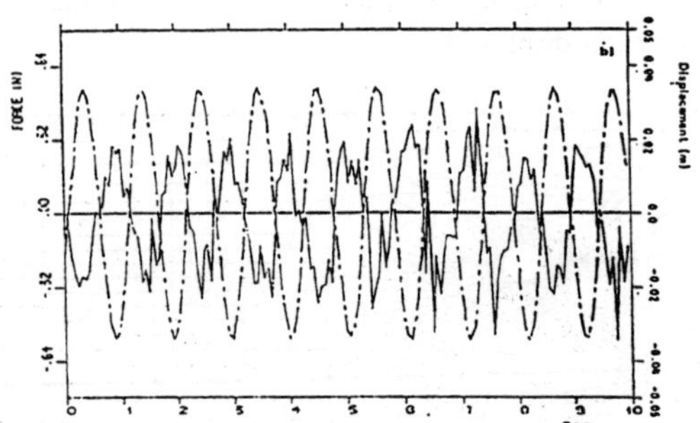

Fig. 7 Irregular transverse forces from forced harmonic oscillations. Inline fixed and Ur=5,51 (Moe and Wu [8]).

In order to find out whether the present numerical procedure would capture this phenomenon, several time series of force and displacement is computed with varying displacement amplitude, reduced velocity and Reynolds number. Some interesting results from these computations are presented in the Fig. 8 - 10.

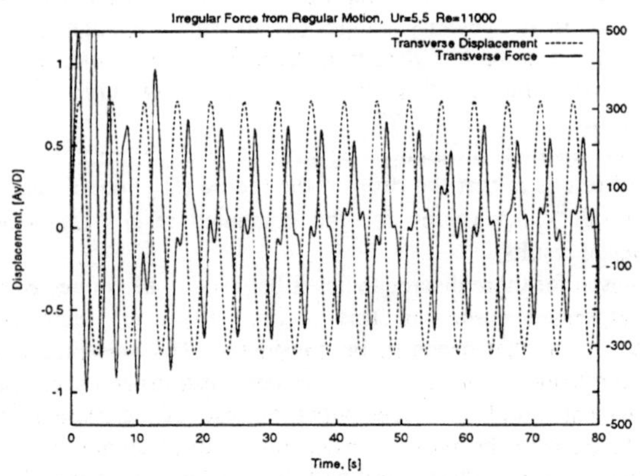

Fig. 8 Irregular force from regular motion. Ay/D=0.75, Ur=5,5 and Re=11000.

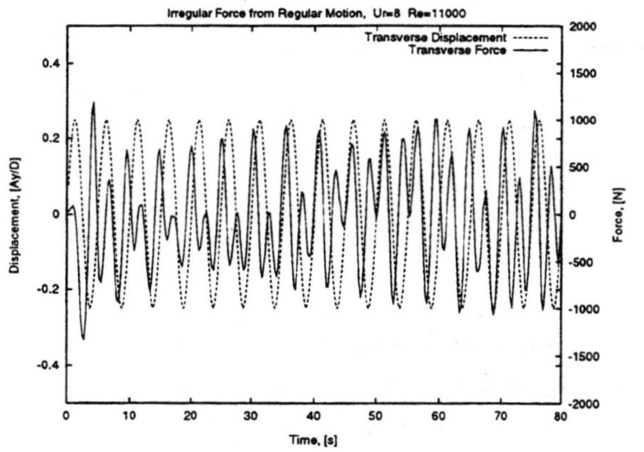

Fig. 9 Irregular force from regular motion.
Ay/D=0.25, Ur=8 and Re=11000.

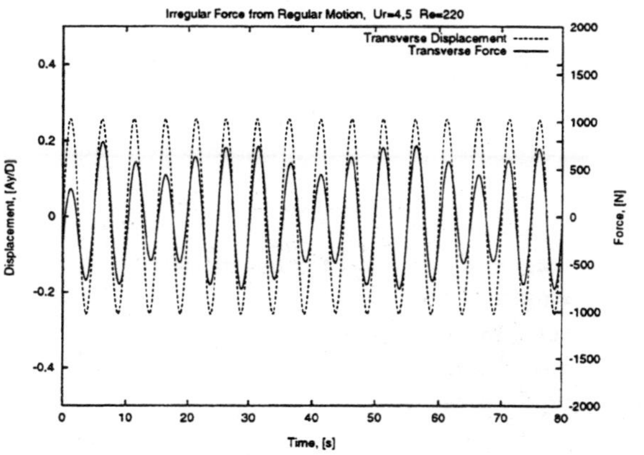

Fig. 10 Irregular force from regular motion.
Ay/D=0.25, Ur=4,5 and Re=220.

The figures above shows that the regularity of the force is extremely sensitive to Reynolds number, reduced velocity and vibration amplitude. However, the forces are, in agreement with the experimental results of Moe and Wu [8], generally highly irregular even though the displacements are not. Numerical tests from the present study indicates that the forces are more irregular for the high Reynolds numbers than for the low. In Fig. 10 the forces are nearly harmonic with an amplitude modulation, which was generally the case for the low U_r numbers.

7. CONCLUSIONS

A numerical procedure for predicting transverse forces on a body submitted to Lock-in motion is presented. An ALE finite element method with an incorporated Taylor-Galerkin discretization is used to solve for the fluid flow by using the Velocity Correction Method.

The numerical procedure is tested against reported experimental data from forced vibration tests, and the agreement between the experimental and the computational data indicates that the present computational procedure may be used with some confidence to predict the lock-in response of suitably idealized structures. Especially for small vibration amplitudes, the computed results are remarkably similar to the experimental results.

The computed fluid forces are, in agreement with reported experimental results, found to be highly irregular even though the displacements are perfectly regular.

The presented numerical procedure is composed of basic physical equations only, and is therefore well suited for adoption to other classes of fluid-structure interaction problems. The numerical computation permits determination of cylinder kinematics at any time (in the forced motion case this is trivial), as well as determination fluid forces on the cylinder at any time and in any position.

8. ACKNOWLEDGEMENT

The work was funded by a research scholarship from the Norwegian University of Science and Technology (NTNU). The author is gratefully acknowledging the advices from professor Geir Moe and Dr.Ing Torbjørn Utnes.

9. REFERENCES

[1] Chorin , A.J., *Numerical Solutions of the Navier- Stokes equations*, Mathematical Computation, 22, 745-762, 1968.

[2] Donea, J., Giuliani, S. and Halleux, J.P., '*An Arbitrary Lagrangian-Eulerian Finite Element Method for Transient Dynamic Fluid-Structure Interaction*', Computational Methods in Applied Mechanics and Engineering, Vol. 33 pp.689-723, 1982.

[3] Griffin, O.M, Skop, R.A. and Ramberg, S.E, '*The Resonant Vortex Excited Vibrations of Structures and Cable Systems*', Proceedings of the Offshore Technology Conference, Paper no. 2319, Houston, Texas, 1975.

[4] Hughes, T.J.R., Liu, W.K. and Zimmerman, T.K., '*Lagrangian-Eulerian finite element formulation for incompressible viscous flows*', Comput. Methods Appl. Mech. Engrg. no. 29, pp. 329-349, 1981.

[5] Jiang, C.B. and Kawahara, M., '*The Analysis of Unsteady Incompressible Flows by a Three-Step Finite Element Method*', International Journal for Numerical Methods in Fluids, Vol. 16, pp:793-811, 1993.

[6] Ku, H.C., Rosenberg, A. and Taylor, T.D., '*High- Order Time Integration Scheme and Preconditioned Residual Method for Solution of Incompressible Flow in Complex Geometries by the Pseudospectral Element Method*', in: Twelfth Int. Conf. On Num. Meth. in Fluid Dynamics, Oxford, England. pp:223- 227, 1990.

[7] Lewis, R.W., Ravindran, K. and Usmani, A.S., *Finite Element Solution of Incompressible Flows Using an Explicit segregated Approach*, Archives of Computational Methods in Engineering, Vol 2, 4, 69-93, 1995.

[8] Moe, G., Wu, Z.J., '*The Lift Force on a Cylinder Vibrating in a Current*', Journal of Offshore Mechanics and Arctic Engineering Volume 112, November 1990.

[9] Rodi, W., '*Turbulence models and their application in hydraulics*', State-of-the-art paper, IAHR, Delft, Holland, 1984.

[10] Sarpkaya, T., '*Vortex-Induced Oscillations. A Selective Review*', Journal of Applied Mechanics, Vol. 46, No.2, June 1979.

[11] Sarpkaya, T., '*Hydrodynamic Damping, Flow-Induced Oscillations, and Biharmonic Response*', Journal of Offshore Mechanics and Arctic Engineering, ASME, Vol. 117, pp:232-238, 1995.

[12] Strang, G., '*Linear Algebra and its Applications*', Third edition, Harcourt Brace Jovanovich, publishers San Diego, pp:156, 1986.

[13] Utnes, T., Eidsvik, K.J. and Moe, G., '*Dynamic separation over bluff bodies predicted based upon Navier-Stokes equations*', Hydraulic Research, 33,2, pp:219-242, 1995.

[14] Utnes, T., Ren, G., '*Finite Element Modelling of the Incompressible Navier-Stokes equations*', Computational Fluid Dynamics vol. 4, 1995.

Proceedings of the Seventh (1997) International Offshore and Polar Engineering Conference
Honolulu, USA, May 25-30, 1997

ISBN 1-880653-28-1 (Set); ISBN 1-880653-31-1 (Vol. III)

Control of Fluid-Structure Interaction Instabilities Using Circular Cylindrical Nutation Dampers

M.L. Seto and V.J. Modi
University of British Columbia
Vancouver, Canada

ABSTRACT

The paper studies damping of wind induced instabilities, using circular cylindrical nutation dampers, through a comprehensive test program having three distinct phases. To begin with, a parametric study of the damper, in conjunction with frequency response tests, is carried out using a specially designed and instrumented Scotch-Yoke type facility. It identifies important system variables contributing to significant energy dissipation. Results show that the optimum contributions of the system parameters can lead to an efficient damper, particularly if the operating conditions are conducive to wave-breaking. Next, visualization of the liquid sloshing modes, explaining the energy dissipation process at the fundamental level, is undertaken. It collaborates conclusions of the parametric study. Finally, wind tunnel tests with two- dimensional models substantiate, rather dramatically, the effectiveness of the nutation damper in arresting both vortex resonance and galloping type of instabilities. The damper continues to be effective even for the case when the structure is located in the wake of other structures, the situation frequently encountered in practice. A short video showing fluid dynamics of the damper and its effectiveness in arresting wind induced instabilites accompanies the presentation.

KEY WORDS: Nutation Damper, Vortex Resonance, Galloping, Vibration Suppression.

NOMENCLATURE

AR	aspect ratio of floating particle, l_p/d_p
c	structural damping of wind tunnel model
c_c	system critical damping
D	mean diameter of damper, $2R$
d	sectional diameter of toroidal damper, Fig. 1
d_p	particle diameter
F_s	sloshing force
H	width of the wind tunnel model
h	liquid height
L_m	length of the wind tunnel model
l_p	length of particle
M	total mass of the damper
M_a	added mass
M_l	liquid mass
t	time
U	nondimensional reduced wind speed, $V/\omega_n H$
V	freestream velocity
W	width of rectangular damper, Fig. 1
Y	model displacement, rms
ϵ_e	forced sinusoidal excitation amplitude, peak to peak
ζ_s	system damping factor, c/c_c
$\eta_{r,a}$	reduced aerodynamic damping
$\eta_{r,l}$	reduced liquid damping
ϕ	relative phase between F_s and ϵ_e
$\widehat{\omega}_e$	nondimensional excitation frequency, ω_n/ω_e
ω_e	forced excitation frequency, rad/s
ω_l	liquid natural frequency, rad/s
ω_n	natural frequency of the structure, rad/s
ρ_a	density of the air in the wind tunnel
ρ_p	floating particles seeding density, area covered as a percent of free surface

INTRODUCTION

The response of aerodynamically bluff bodies when exposed to a fluid stream has been a subject of considerable study for quite some time. The prevention of fluid-elastic vibrations of marine platforms, risers, towed sonars as well as ground based structures such as transmission lines, suspension bridges, tall buildings, etc. is of particular interest to engineers. Ever since the pioneering contribution by Strouhal, who correlated the periodicity of the vortex shedding with the diameter of a circular cylinder and the velocity of the fluid stream, there has been a continuous flow of important contributions resulting in a vast body of literature. This has been reviewed rather adequately by Cermak (1975) and Parkinson (1989).

A variety of devices have been developed which aim to arrest vibrations. They may be classified as:

(i) spoilers, which tend to change fluid dynamic characteristics of structures in such a way as to interfere with and weaken the exciting force;

(ii) dampers, which provide a mechanism for dissipation of energy or its transfer from one part of the system to the other.

The classical concept of helical strakes (Scruton and Walshe, 1957), shrouds and slats (Wong and Cox, 1980) belong to the first group; while the tuned mass dampers (Abdel-Rohman and Askar, 1996), Stockbridge dampers (Lawson, 1980), conventional oil dashpots and impact dampers (Reed, 1986) belong to the second category. Zdravkovich (1981, 1984) and Every et al. (1982) have presented in-depth reviews of contributions in this area.

NUTATION DAMPER

A nutation damper belongs to this category of energy dissipating devices (Figure 1). It can have a wide variety of geometries, from toroidal ring to rectangular or circular cross-section cylinders. Furthermore, it may carry internal devices in the form of baffles, screens, moving spheres, particle suspensions, etc. interacting with the sloshing motion. Such dampers are ideally suited for suppressing relatively low frequency motion. They have been in use for quite some time in controlling the librational motion of satellites (Anzai et al., 1981) where the period may range from around 90 minutes to 24 hours! The space environment dictates nutation dampers to be completely filled (no free surface). The dissipative wake of a sphere provides a mechanism for damping of the nutational motion. As the frequency of oscillations encountered in the wind-induced instability of bluff bodies, earthquake response of buildings, and wave-excited vibrations of offshore structures is relatively small, the authors thought it appropriate to explore applicability of the nutation dampers to this class of problems. The results show the nutation damper to be particularly attractive due to its efficiency, simplicity in design, ease of construction, and negligible maintenance as discussed by Tamura, Kousaka and Modi (1992). A recent paper by Modi, Welt and Seto (1995) as well as the thesis by Seto (1996) review the literature in the area of nutation damping of flow induced instabilities.

With this as background, the paper studies suppression of flow excited oscillations of structures using circular geometry nutation dampers. The comprehensive project represents synthesis of three distinct aspects:

(a) experiments aimed at damper performance as affected by system parameters to arrive at favorable combinations of design variables;

(b) visualization of the damper's free surface motion to provide physical appreciation of the dissipation process thus complementing the parametric study;

(c) wind tunnel experiments to assess effectiveness of the damper in suppressing flow induced instabilities.

Results show the circular cylindrical nutation damper to be quite effective in controlling the structural response during both vortex resonance and galloping instabilities.

PARAMETRIC STUDY OF DAMPER CHARACTERISTICS

Test Methodology

A circular cylindrical damper was tested at its liquid sloshing resonance frequency to assess the effect of system parameters such as liquid height, excitation amplitude and frequency, floating particle characteristics, etc. on the damping efficiency. To that end a Scotch-Yoke mechanism, connected to a horizontal platform was constructed. A strain gauge force balance is rigidly fixed to the platform, and the damper under test is mounted on the force balance (Figure 2). The balance consists of two galvanized steel plates supporting a horizontal stainless steel base which, in turn, carries the damper. Strain gauges, mounted differentially on both sides of each vertical plate, sense the sloshing force, F_s. The sensitivity of the force balance is 26 mV per gramme. The oscillating platform displacement was sensed using a Linear Voltage Differential Transducer (LVDT) having an accuracy of 3.40 V / cm.

Experimentally, the condition of sloshing resonance is established when the force F_s lags the displacement ϵ_e by 90°. A correlator (Hewlett Packard, 3721A) was used to measure the relative phase between the signals in real time.

The two signals, one from the force balance measuring the sloshing force F_s and the other from the platform displacement measuring the excitation ϵ_e, were recorded with no analog filtering. The quantities determined from these signals are: F_s, ϵ_e, ω_e (excitation frequency), and ϕ, the relative phase between the sloshing force and the platform displacement. These parameters identify the nondimensional sloshing force or added mass, M_a/M_l, and reduced damping $\eta_{r,l}$:

$$\frac{M_a}{M_l} = \left[\frac{F_s}{M_l \epsilon_e {\omega_e}^2} - 1\right] \cos\phi; \quad \eta_{r,l} = \frac{F_s}{2M_l \epsilon_e {\omega_e}^2} \sin\phi. \quad (1)$$

Generally, the parametric tests were performed by systematically changing variables associated with the liquid (height, Reynolds number) and excitation (frequency, amplitude).

Results and Discussion

As can be expected, through a planned variation of system parameters, the amount of information obtained is rather extensive. Here, for conciseness, only some typical results useful in establishing trends are presented.

Tests were conducted with a cylindrical geometry damper, systematically varying the liquid height and excitation frequency. The results are presented in Figure 3. It is apparent that the dissipation of energy improves at the lower end of the frequency range studied (0.4 Hz $< f_e <$ 0.55 Hz) as the liquid height diminishes. Note, the peaks are associated with resonance at the liquid fundamental frequency and higher harmonics. With an increase in the liquid height, the onset of the surface waves seems to be delayed and the dissipation level is reduced.

With some information concerning the damper performance in hand, the next logical step was to explore the possibility of improving the dissipation rate further. To that end, introduction of floating particles appeared attractive. The premise is that they may promote dissipation of energy through an increase in exposed area, higher particle drag and inelastic collisions among themselves as well as with the wall.

To assess the effect of floating particles, a systematic experimental program was formulated. It was immediately recognized that the number of variables involved can be rather large. To get some appreciation of the dissipation process, it was decided to focus attention on what appeared to be more important parameters. The investigation was limited to a set of particles described in Table 1.

Table 1 Floating particles used during the test-program.

Material	Diameter (in)	Aspect Ratio	Specific Gravity
surgical tube	0.066	1.0	1.1
poly-flo tube	0.25	0.5, 1.0, 2.0, 4.0	0.8
Perler beads	3/16	1.0	0.7-0.8
straw	2/16	1.0	0.5

The hollow cylindrical particles used in the study have diameter d_p and length l_p with particle aspect ratio l_p/d_p varied systematically as indicated. In general, for a given aspect ratio of particles, there was observed to be an optimal value of the seeding density leading to the peak reduced damping. For the optimum aspect ratio of 1, the peak $\eta_{r,l}$ of ≈ 2.2 represents an increase of around 30% (Figure 4). Note, a broad peak suggests relative insensitivity to the particle seeding density.

FLOW VISUALIZATION

In order to get better physical appreciation of the energy dissipation process through sloshing, a visual study of the liquid modes as affected by the damper geometry, liquid height, excitation frequency and floating particle parameters was carried out. The forced vibration facility described earlier was used to that end. During a test, a damper was mounted on the Scotch-Yoke mechanism (Figure 2) and the excitation frequency varied. The fundamental as well as subharmonics and higher harmonics were observed and recorded using still camera and S-VHS video. In particular, the video showed quite vividly evolution of the progressive waves, their collision amongst themselves and with the wall, disintegration and regeneration. The presence of floating particles and associated parameters clearly revealed their influence on the above process and thus helped understand the conditions leading to higher dissipation. The effectiveness of the dampers during structural dynamics tests in the wind tunnel, with vortex resonance and galloping, was also recorded.

In general, the flow visualization study confirmed the trends of the damper performance as predicted by the parametric analysis. The excitation frequency was gradually increased to observe the surface wave dynamics, which can be better appreciated on the video taken at the time. More important phases in the surface wave evolution around the fundamental sloshing resonance are schematically sketched (based on the video) in Figure 5.

At low frequencies, the standing wave mode is observed with the free surface pivoting about the line of nodes perpendicular to the direction of oscillation (90° and 270°) as shown in Figure 5(a). As the excitation frequency is increased, the free surface becomes flexible (i.e. nonplanar) and a propagating wave train appears as indicated in Figure 5(b). This changes to the colliding wave-type sloshing motion. Here the two circumferential waves follow the curved walls of the damper (in phase) and collide at the instant of the oscillation extrema (Figure 5c). At the sloshing resonance, the spectacular breaking of waves lead to a significant dissipation of energy as observed earlier in Figure 3. With a further increase in the frequency, the sloshing mode changes to stronger propagating waves with their ends attached to nodes mentioned before (Figure 5d). Next, the liquid acquires a swirl component which contributes little to energy dissipation (Figure 5e). Finally, with further deviation from the resonance, the free surface returns to the standing wave with rocking motion about the line of nodes (Figure 5f).

The effect of changes in h/R was also observed. Although, in general, the trends in the surface dynamics remained essentially the same, there were situations where certain phases became accentuated or weakened, and some cases got completely eliminated. This substantiated the variation in the damper performance with h/R as observed in Figure

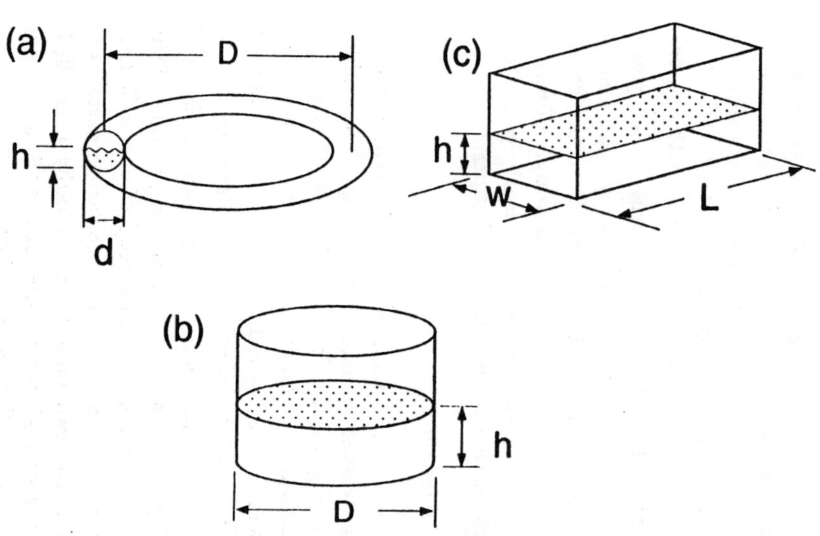

Fig. 1 Schematic diagrams showing several geometries of nutation dampers. A variety of internal devices can be used to further improve their performance. In this study, the focus is on the circular geometry.

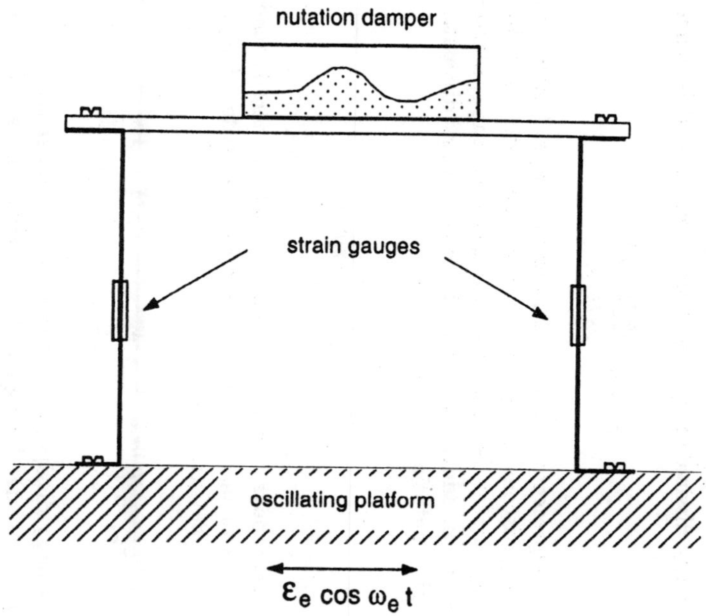

Fig. 2 The force balance used to measure the reduced liquid damping of nutation damper models under steady state excitation.

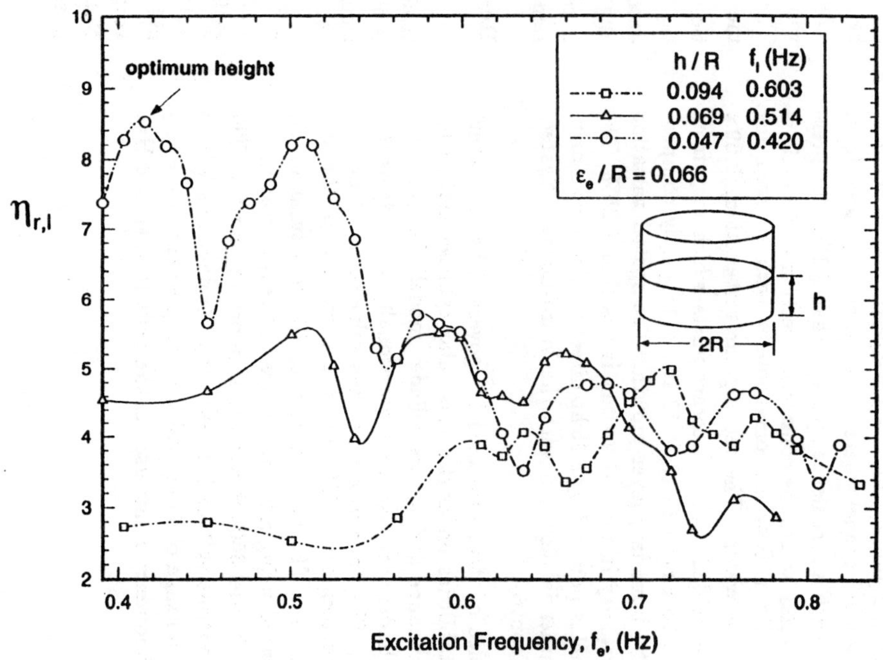

Fig. 3 Reduced liquid damping, $\eta_{r,l}$, for a circular nutation damper at different liquid heights. Note, a distinct improvement in dissipation as h/R decreases in the excitaiton range of $f_e < 0.55$ Hz.

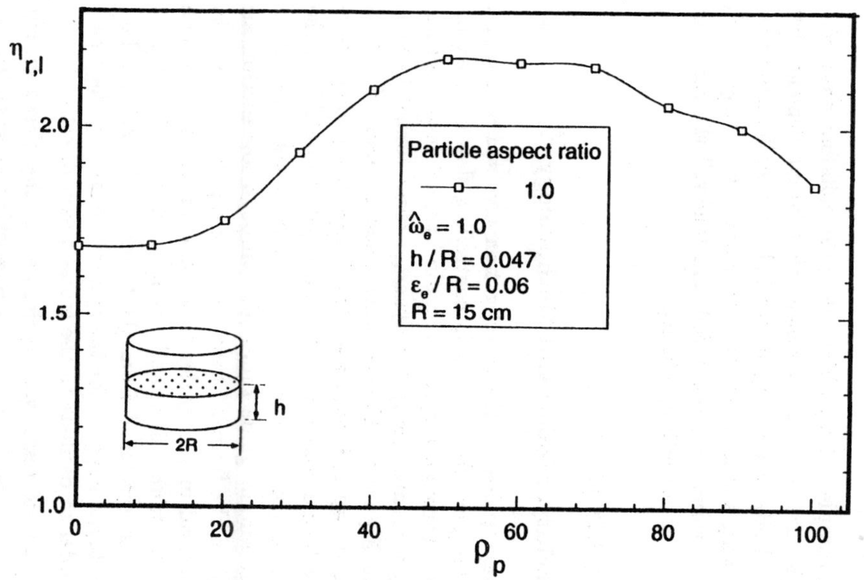

Fig. 4 Representative plot showing the effect of particle concentration on reduced damping for a circular damper. Note, the liquid height is held fixed at $h/R = 0.047$.

3. For example, with $h/R = 0.047$, the colliding wave condition (Figure 5c) was so violent that liquid elements left the main body, remaining suspended for a while, before returning with a splash. On the other hand, with larger h/R, the wave trains that preceeded and followed the resonant sloshing were essentially absent.

WIND TUNNEL TESTS

Test Facility, Models and Methodology

The next phase of the investigation involved assessment of the damper effectiveness in controlling flow-induced instabilities through a systematic wind tunnel test-program. The experimental setup for dynamic wind tunnel tests is shown in Figure 6. The structural model, mounted with a nutation damper, is spring loaded onto a heavy, rigid, free-standing frame constructed to fit around the outside of the wind tunnel (in the plane perpendicular to wind direction). Light aluminum guide tubes slide through low friction air bearings on the frame. These tubes are rigidly connected to the structural model by slender mounts protruding through slots in the tunnel floor and ceiling as shown. Four helical coil springs, stretched from the frame to the guide tubes, provide elastic support for the system permitting the structural model to oscillate in the cross-stream direction.

The dynamic tests were performed in the closed circuit laminar flow wind tunnel with a working cross-section of $0.69 \times 0.91 \times 2.44$m. This is a low-speed, low-turbulence, return type wind tunnel where air speed can be varied from 1.5-50 m/s with a turbulence level less than 0.1%. The pressure differential across the contraction section of 7:1 ratio is measured with a Betz micromanometer with an accuracy of 0.2 mm of water. The rectangular test-section, 0.69×0.91m, is provided with 45° corner fillets which vary from 13.25×13.25 cm to 12.10×12.10 cm to partly compensate for the boundary-layer growth. The spatial variation of mean velocity in the test-section is less than 0.25%. The wind tunnel is powered by a 15 hp direct current motor driving a commercial flow fan with the Ward-Leonard System for speed control.

The choice of structural model is governed by the nature of the instabilities under study; circular cylinder models for vortex resonance and square cylinder models for galloping. Square cylinders are also used, occasionally, to study vortex resonance. The nondimensional rms value of the structural model displacement, Y/H, serves as a measure of the system response. The displacement of the structural model was measured using a linear voltage differential transducer (Schaevits Model 3000 H.R.) with supporting circuitry. The resulting output voltage is linear with the deviation within 1.5% over a range of 6 cm, and it is independent of velocity over the frequency range used in the tests (less than 6 Hz).

The reduced aerodynamic damping, $\eta_{r,a}$, is a measure of the energy dissipated by a vibrating structure. It is defined as

$$\eta_{r,a} = \frac{4\pi(c/c_c)M}{\rho_a H^2 L_m}.$$

Here c represents the sum of inherent internal structural damping, viscous dissipation at the joints, etc. including the eddy current damping, which is also viscous in character. Of course, in addition, now we have the energy dissipation through a nutation damper,

$$\eta_{r,l} = \frac{F_s}{2M_l \epsilon_e \omega_e{}^2} \sin\phi.$$

From the above relations, it follows that the aerodynamic damping can be expressed in terms of the reduced liquid damping (Blevins, 1990) through

$$\eta_{r,l} = \left(\frac{\eta_{r,a}}{4\pi}\frac{1}{M/\rho_a L_m H^2} - \zeta_s\right)\frac{1}{M_l/M}.$$

Thus energy dissipation of the structure under test can be increased by introducing an eddy current damper. This can be used to advantage in avoiding overlap of the vortex resonance and galloping regions. The eddy current dissipation was provided by surrounding the oscillating aluminum guide tubes with steady magnetic fields. The system damping ratio, ζ_s, with no magnetic field applied was 0.0002 and could be increased to over 0.006 using the electromagnetic field. The level of damping was measured by observing, with no wind, the decay of free vibrations following a displacement. The effective mass of the moving system was determined by measuring the natural frequency of the model with an additional mass. A plot of additional mass versus $1/\omega_n{}^2$, where ω_n is the corresponding structural frequency, was used to determine the effective mass of a model (Bearman et al., 1987). As mentioned earlier, only a small sample of results is recorded here for brevity, however, it clearly points out remarkable performance of the nutation damper.

Results and Discussion

The effectiveness of a circular cylindrical damper in controlling both the vortex resonance and galloping instability is illustrated in Figure 7. Note, in absence of the nutation damper and with low inherent damping of $\eta_{r,a} = 2.90$, the galloping sets in at a low reduced wind speed of around $U = 0.8$. With the addition of eddy current damping ($\eta_{r,a} = 5.21$), the vortex resonance peak appears and the onset of galloping is delayed to $U \approx 3.6$. However, in presence of the nutation damper with a liquid mass ratio of only 1 % (liquid mass ratio = liquid mass / total oscillating mass), the vortex resonance and galloping are successfully controlled. Note, even at a very high wind speed of $U \approx 6$, it was not possible to excite the galloping instability.

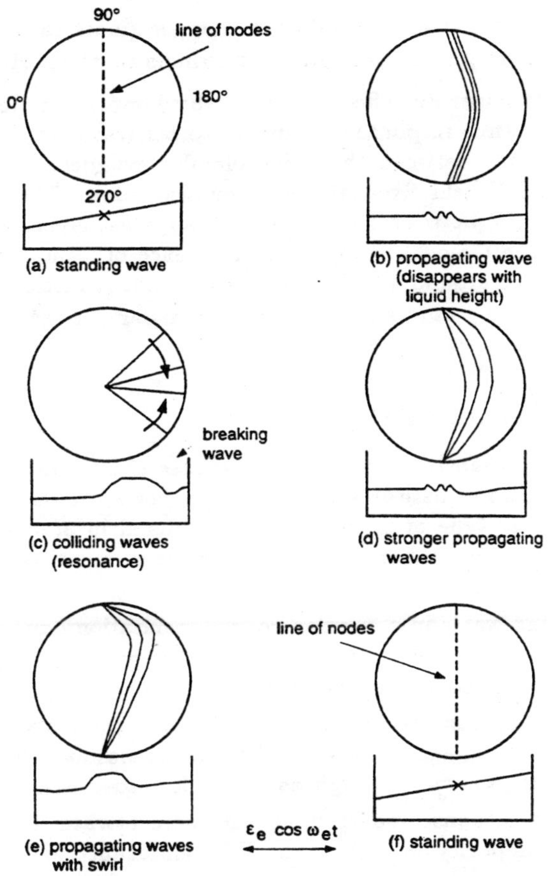

Fig. 5 Top and side views of the free surface modes in a circular damper as the forcing frequency, ω_e, increases from below to beyond the sloshing resonance. The damper is shown accelerating to the right: (a) standing wave; (b) propagating wave; (c) toroidal-type sloshing (resonance); (d) stronger propagating wave; (e) propagating waves with swirl; (f) standing wave.

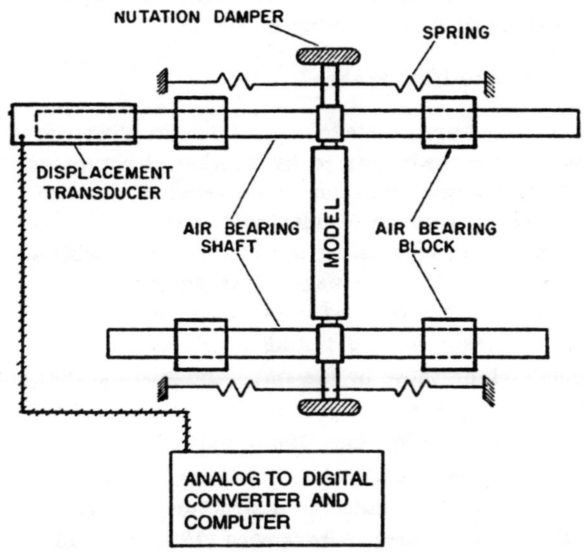

Fig. 6 Wind tunnel test setup to assess effectiveness of circular cylindrical nutation dampers in suppressing vortex resonance and galloping.

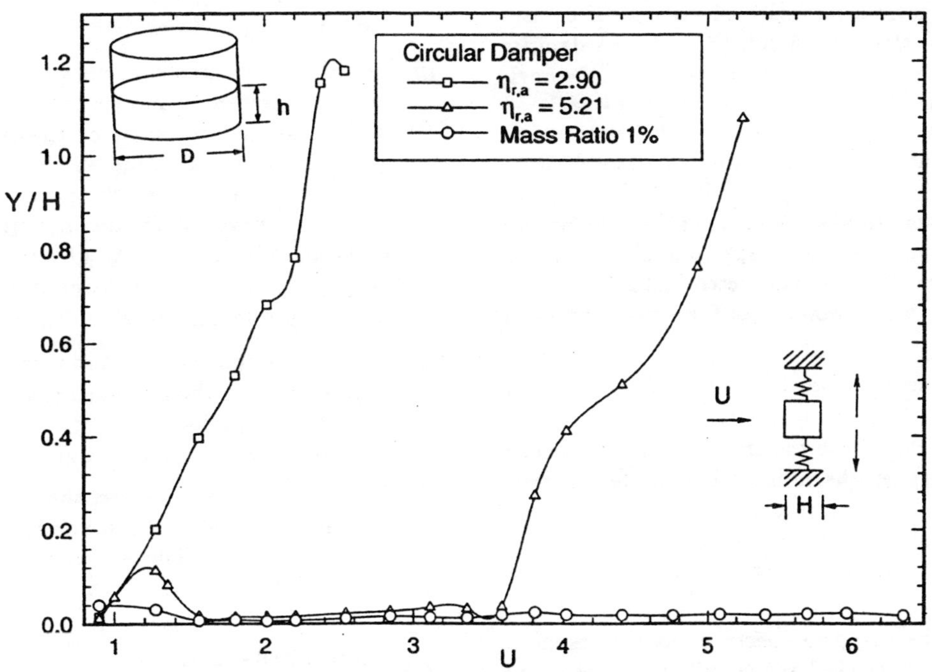

Fig. 7 The effectiveness of a circular nutation damper in arresting both vortex resonance and galloping instability of a square cylinder. -□- represents the system response with little damping, hence the square model gallops from the start. - Δ- presents the displacement data with system dissipation increased through introduction of eddy current dampers in order to separate vortex resonance from galloping. -o- shows the nutation damper suppressing both the vortex resonance and galloping over the entire range of the wind speed.

Figure 8 shows the effect of increasing the circular damper's liquid mass on the vortex resonance and galloping response of a square cylinder. Clearly, the structural oscillations decrease with an increase in the liquid mass. For the cases studied, the resonance peak was not very broad and it was easy to increase the wind speed past the peak. As expected, the effect of more liquid was primarily on the vortex resonance response, the galloping being essentially controlled at 1% of the mass ratio as seen earlier in Figure 7.

Structure Affected by Wake

It would be more appropriate to study dynamics of a structure located in the wake formed by another structure as it represents a realistic situation encountered in practice. Hence experiments were also conducted to assess effectiveness of the damper in suppressing wind induced instabilities of a structure located in the wake of an upstream structure. The problem has received considerable attention as indicated by Zdravkovich (1981) in his review paper. Later contributions include those by Blessman and Riera (1988), Su et al. (1990) and others. In general, devices such as helical strakes (Vickery and Watkins, 1962), slats (Wong, 1980) and shrouds (Walshe and Cowdrey, 1972) have proved to be of little value when mounted on a structure affected by the wake of other structures. Zdravkovich (1984) concludes that, in some cases, vibration suppression devices that affect the aerodynamics of the structure can have undesirable effects on the system dynamics. Thus, it appears that the flow interference can severely undermine the usefulness of vortex suppression devices normally effective with an isolated bluff body. On the other hand, the nutation damper being a dissipative device should be unaffected by the environmental conditions. This observation was substantiated by tests with in-line circular and square structural models (Figure 9). The nutation damper was mounted on a structural model located in the wake of a rigidly mounted identical structure upstream. In absence of a damper, the results showed a dramatic increase in oscillation amplitudes of the downstream structure. This is as expected and is attributed to higher excitation, which now arises from two sources of vortex shedding.

To begin with, the critical vortex resonance wind speed for a square (or a circular) cylinder was established in absence of a nutation damper. Now, with the critical wind speed held fixed, the spacing between the in-line cylinders was progressively reduced until the peak response of the downstream cylinder was observed. The critical distance corresponded to 6.5H for the square cylinders and 4.75H for the circular cylinders.

Figure 10 shows the response spectrum at the resonant wind speed, i.e. the vortex shedding from the upstream and downstream cylinders occurs at the natural frequency of the downstream cylinder. Note, a large amplitude of oscillations in absence of a nutation damper at resonance. However, with the addition of the nutation damper (mass ratio = 3.75%), the amplitude of oscillations reduces by about 90%.

Figure 11 illustrates effectiveness of the damper in reducing the dynamic response during the vortex resonance due to interference between the inline circular cylinders. The wind speed is held fixed at the resonance value. The effect of spacing between the cylinders is apparent giving the critical value for peak resonance, in absence of a nutation damper to be around 4.75H. Note a significant reduction in the response, by approximately 88%, for a damper mass of 4%.

CONCLUDING REMARKS

The comprehensive investigation assesses effectiveness of circular nutation dampers in controlling vortex resonance and galloping type of instabilities encountered in practice. Some of the more important results may be summarized as follows:

(i) Sloshing water in a circular geometry nutation damper provides an efficient mechanism for energy dissipation as represented by the reduced damping parameter $\eta_{r,l}$. It is a measure of dissipation with liquid compared to that with an equal mass of solid, i.e. no sloshing. The damper gave $\eta_{r,l}$ as high as 8.5! The broadband frequency response makes it all the more attractive besides simplicity of design, ease of maintenance and reduced cost. Nutation dampers have already been applied to prototype structures in Japan and Australia (Modi, Welt, and Seto, 1995).

(ii) Seeding of the free surface with particles permits more energetic collisions and leads to an increase in the bandwidth through merging of the fundamental and first sloshing harmonic. Generally lower aspect ratio, smaller diameter, and high specific gravity floating particles resulted in improved damping. Optimal values of aspect ratio and seeding density (1 and 50-70%, respectively) gave an increase in the reduced damping parameter by 30% with reference to the plain water case!

(iii) The flow visualization study provided better physical appreciation of the free surface dynamics leading to the efficient dissipation of energy. In general, breaking and collision of waves lead to higher damping.

(iv) Wind tunnel tests showed the damper to be quite successful in controling both vortex resonance and galloping forms of instabilities, even when a structure is located in the wake of another structure.

ACKNOWLEDGEMENT

The investigation reported here was supported by the Natural Sciences and Engineering Research Council of Canada, Grant No. A-2181.

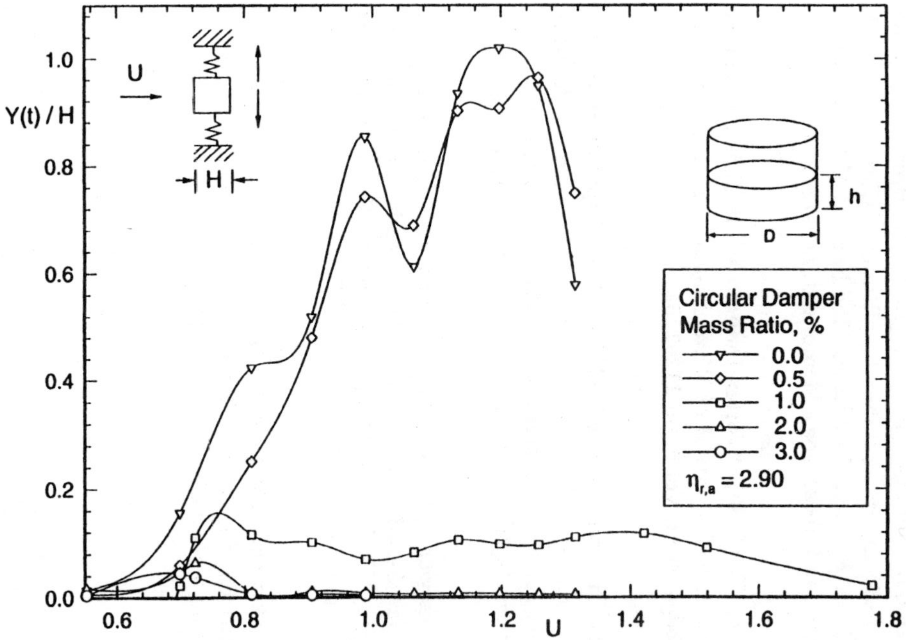

Fig. 8 Wind tunnel test results showing the effect of liquid mass ratio on the vibratory response of a square prism with a circular cylindrical damper.

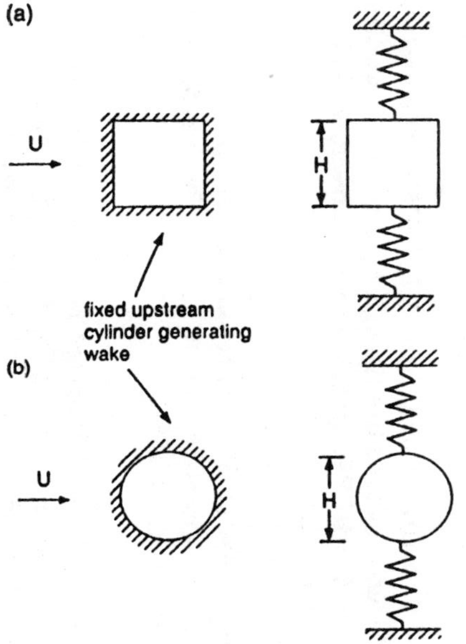

Fig. 9 Plan view of the interfering structural models arrangement: (a) inline square cylinders; (b) inline circular cylinders.

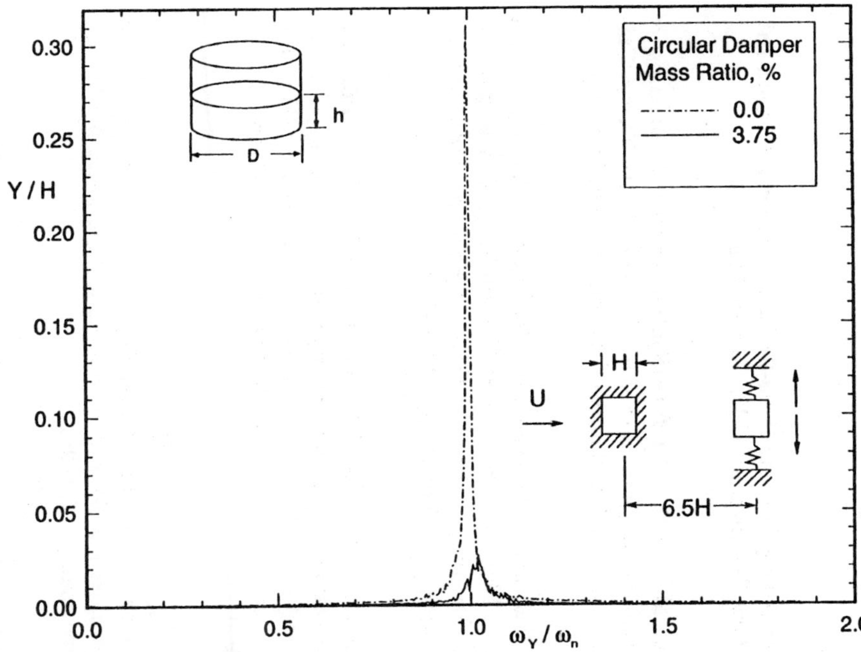

Fig. 10 Response of a square cylinder critically located in the wake of the other square cylinder at a wind speed corresponding to the resonance condition. Effectiveness of the nutation damper at resonance is also indicated.

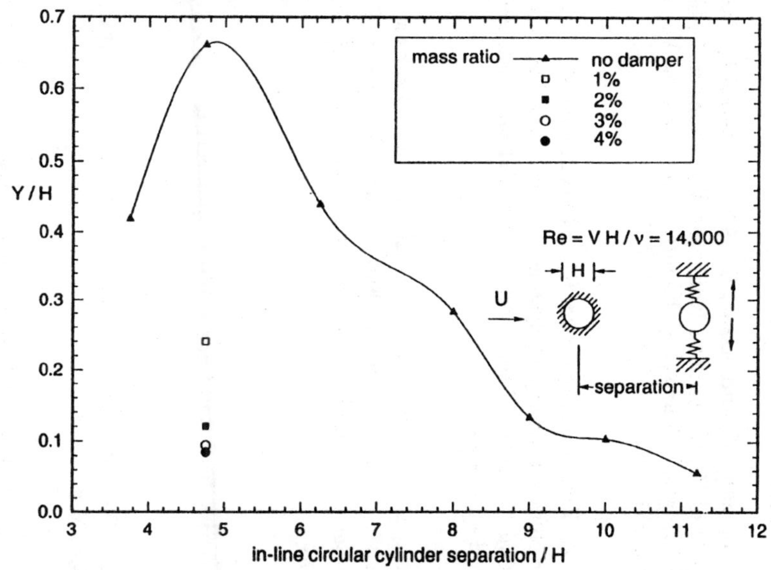

Fig. 11 Resonant response of the inline circular cylinder as affected by the spacing and nutation damper.

REFERENCES

Abdel-Rohman, M, and Askar, H (1996). "Control by Passive TMD of Wind-Induced Nonlinear Vibrations in Cable Stayed Bridges," *Journal of Vibration and Control*, Vol 2, No 2, pp 251–267.

Anzai, T, Ikeuchi, M, Igarashi, K, and Okanuma, T (1981). "Active Nutation Damping System of Engineering Test Satellite-IV," *22nd Congress of the International Astronautical Federation*, IAF 81-350.

Bearman, PW, Gartshore, IS, Maull, DJ, and Parkinson, GV (1987). "Experiments on Flow-Induced Vibration of a Square-Section Cylinder," *Journal of Fluids and Structures*, Vol 1, pp 19–34.

Blessman, J, and Riera, JD (1988). "Interaction Effects in Neighbouring Tall Buildings," *Wind Engineering*, Vol 1, pp 381–395.

Blevins, RD (1990). *Flow Induced Vibrations*, Van Nostrand Reinhold, N.Y.

Cermak, JE (1975). "Application of Fluid Mechanics to Wind Engineering," *Journal of Wind Engineering and Industrial Aerodynamics*, Vol 97, No 1, pp 9–38.

Every, MJ, King, R, and Weaver, DS (1982). "Vortex Excited Vibrations of Cylinders and Cables and their Suppression," *Ocean Engineering*, Vol 9, pp 135–157.

Lawson, TV (1980). *Wind Effects on Buildings*, Applied Science Publisher LTD, London, U.K.

Modi, VJ, Welt, F, and Seto, ML (1995). "Control of Wind-Induced Instabilities through Application of Nutation Dampers: A Brief Review," *Engineering Structures*, Vol 19, No 9, pp. 626–638.

Parkinson, GV (1989). "Phenomenon and Modelling of Flow-Induced Vibrations of Bluff Bodies," *Progress in Aerospace Sciences*, Vol 26, pp 169–224.

Reed, WH (1986). "Hanging-Chain Impact Dampers: A Simple Method for Damping Tall Flexible Structures," *Proceedings of the International Seminar on Wind Effects on Structures*, Toronto, Canada, pp 287–320.

Scruton, C, and Walshe, DEJ (1957). "A Means for Avoiding Wind-Excited Oscillations of Structures with Circular or Nearly Circular Cross-Section," *National Physical Laboratory*, Teddington, U.K., Report 335.

Seto, M (1996). *An Investigation on the Suppression of Flow-Induced Vibrations of Bluff Bodies*, Ph.D. Thesis, University of British Columbia, Vancouver, Canada, Chapter 1.

Su, TC, Lian, QX, and Lin, YK (1990). "Vibrations of a Pair of Elastically Supported Tall Building Models in a Uniform Stream," *Journal of Wind Engineering and Industrial Aerodynamics*, Vol 36, pp 1115–1124.

Tamura, Y, Kousaka, R, and Modi, VJ (1992). "Practical Application of Nutation Damper for Suppressing Wind Induced Vibrations of Airport Towers," *Journal of Wind Engineering and Industrial Aerodynamics*, Vol 41–44, pp 1919 – 1930.

Vickery, BJ, and Watkins, RD (1962). "Flow Induced Vibrations of Cylindrical Structures," *Proceedings of the 1st Australian Conference on Hydraulics and Fluid Mechanics*, Melbourne, pp 213–241.

Walshe, DE, and Cowdrey, CF (1972). "A Brief Study of the Effect of Shrouds on Buffet Amplitudes of Chimney Stacks," *National Physical Laboratory*, Teddington, U.K., Maritime Science and Technology Memorandum 2-72.

Wong, HY (1980). "Vortex-Induced Wake Buffeting and Its Suppression," *Journal of Wind Engineering and Industrial Aerodynamics*, Vol 6, pp 112–125.

Wong, HY, and Cox, RN (1980). "The Suppression of Vortex Induced Oscillations of Circular Cylinders by Aerodynamic Devices," *Proceedings of the 4th Colloquium on Industrial Aerodynamics*, Aachen, Germany, pp 185–204.

Zdravkovich, MM (1981). "Review and Classification of Various Aerodynamic and Hydrodynamic Means for Suppressing Vortex Shedding," *Journal of Wind Engineering and Industrial Aerodynamics*, Vol 7, pp 145–189.

Zdravkovich, MM (1984). "Reduction of Effectiveness of Means for Suppressing Wind-Induced Oscillations," *Engineering Structures*, Vol 6, pp 344–350.

Proceedings of the Seventh (1997) International Offshore and Polar Engineering Conference
Honolulu, USA, May 25-30, 1997

ISBN 1-880653-28-1 (Set); ISBN 1-880653-31-1 (Vol. III)

Wave Forces on Vertical Breakwaters by Wave Overtopping

Ching-Piao Tsai and Po-Hung Yeh
National Chung-Hsing University
Taichung, Taiwan, China

ABSTRACT

The wave force on vertical breakwater exerted by the standing wave with overtopping was studied experimentally in the present paper. In the experiment, vertical breakwaters with various heights under the action of a set of incident wave conditions were conducted in a two-dimensional wave channel. The relationship between height of breakwater and wave crest height in overtopping wave was firstly examined. Comparing with the situation of non-overtopping wave on the breakwater, the experimental results showed that wave forces due to wave overtopping were reducible 18% in maximum. Empirical formulas for estimating wave forces on breakwaters by wave overtopping were proposed in this study.

KEY WORDS: Wave force, vertical breakwater, standing wave, wave overtopping.

INTRODUCTION

Wave overtopping occurs when the crest of breakwater is lowered below the maximum water surface elevation. This phenomenon is allowed for the design of maritime structures in order to reduce the wave force on the wall and to save the construction cost. Thus the estimation of the quantity of wave overtopping has been become an essential part of the design of breakwaters. Nevertheless, most previous studies paid only attention to the volume rate of wave overtopping, rather than its influence on wave forces on breakwaters.

Wave overtopping is a complicated phenomenon arising from the interaction of incident waves and the structure. Studies of the feature of wave overtopping were examined by means of experiments, such as Saville and Caldwell (1953) for a vertical wall, Sibul (1955) for an inclined wall, and Iwagaki et al (1965) for effects of incident waves and mean water level on wave overtopping. The formula of wave overtopping rate given by Weggel (1976) was summarized in the Shore Protection Manual (1984) for various structure types. Endo and Miura (1983) obtained a relationship between wall height and wave crest height in wave overtopping, based on experimental results. Recently, Umeyama (1993) proposed a theoretical approach for estimating wave overtopping.

Wave forces on breakwaters are reducible from the effect of wave overtopping, comparing to those of non-overtopping waves on walls. When the wave overtopping is not severe, the Shore Protection Manual (1984) introduced that the wave pressure on the breakwater may be calculated by a truncated distribution result from the standing wave pressure profile. However, the error may become large when the amount of wave overtopping is significant. Because of the existence of the nonlinear nature of wave overtopping, it is difficult to obtain formulations by theoretical approaches. Thus a series of experiments were conducted in a two-dimensional wave flume for examining the wave forces by the wave overtopping. For simplicity in application, empirical formulas of wave forces are proposed in this investigation.

EXPERIMENTS

In the experiment, a set of incident wave conditions was designed to attack a vertical breakwater with variable crest heights for examining wave forces on walls by the wave overtopping. The experiment was conducted in a two-dimensional wave channel that was 100 m long, 2 m wide and 2 m high. The side walls at the observation sections of the channel were consisted of strengthen glass plates. A piston-type wave generator system, controlled by the D/A converter and the PC computer, was mounted at the other end of the channel. Though this wave generator system can generate irregular waves, only regular waves were generated in this investigation.

The sketch of the experimental set-up was shown in Fig. 1, in which a vertical breakwater with variable heights was installed at the end of the channel. In the experiment, the height (H_c) of breakwater above the still water level was varied from 10 cm up to the height that no wave overtopping occurred. Seven pressure transducers inserted in the wave-incident side wall of the breakwater were applied to measure the time histories of wave pressures. The transducers were located at +3 cm, 0 cm, -3 cm, -10 cm, -20 cm, -40 cm and -60 cm from the still water level, respectively. The depth of the water in front of the breakwater was kept 60 cm throughout the experiment. One capacitance type wave gauge set just in front of the wall was utilized to measure the wave profile and the crest of the wave overtopping, the other one immersed at the deeper water region was applied to check the incident wave condition.

The incident wave conditions and the heights of breakwater above

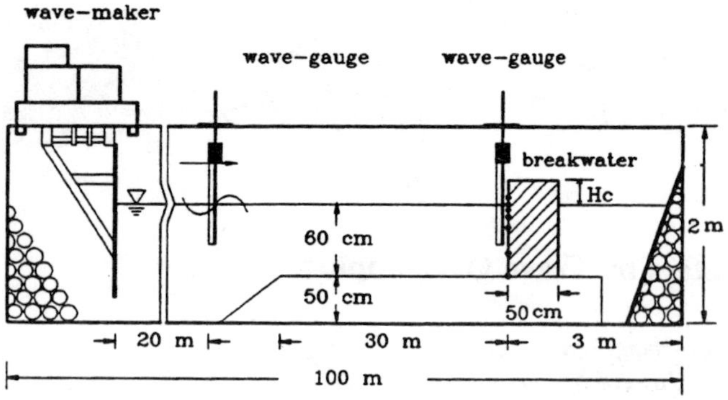

Fig. 1 The sketch of the experimental set-up. (no scale)

the still water level used in the experiment are listed in Table 1. These wave conditions were varied from 1.4 to 2.5 seconds periods corresponding with many different wave heights, from which the relative water depth (d/L, L was the wave length calculated from the linear wave theory) were ranged from 0.106 to 0.222 and the steepness of incident waves (H_i/L) were from 0.017 to 0.067. These incident wave conditions were generating non-breaking waves in front of breakwaters.

Table 1. Incident wave conditions for experiments

wave period T (sec)	relative depth d/L	wave height H_i (cm)	wave steepness H_i/L	height of breakwater from still water level H_c (cm)
1.4	0.222	10.53	0.0389	50, 10
		15.04	0.0556	50, 15, 10
1.6	0.183	10.23	0.0313	50, 10
		14.99	0.0458	50, 15, 10
		20.47	0.0626	50, 25, 20, 15, 10
1.8	0.157	10.17	0.0266	50, 10
		15.09	0.0395	50, 15, 10
		19.82	0.0519	50, 25, 20, 15, 10
		25.47	0.0666	50, 30, 25, 20, 15, 10
2.0	0.138	9.72	0.0223	50, 10
		14.74	0.0332	50, 15, 10
		19.83	0.0455	50, 25, 20, 15, 10
		25.29	0.0580	50, 30, 25, 20, 15, 10
2.2	0.123	10.18	0.0208	50, 10
		15.19	0.0311	50, 15, 10
		21.62	0.0422	50, 25, 20, 15, 10
		25.22	0.0516	50, 30, 25, 20, 15, 10
2.5	0.106	9.82	0.0173	50, 10
		15.09	0.0266	50, 15, 10
		20.10	0.0354	50, 25, 20, 15, 10
		25.052	0.0442	50, 30, 25, 20, 15, 10

EXPERIMENTAL RESULTS

Wave profile and crest height of wave overtopping

When the crest of a breakwater is much higher than the incident-wave crest, a nonlinear fully standing wave system is formed in front of the wall. While the crest of a breakwater is low compared to the incident-wave crest, wave overtopping occurs and the reflected wave from the wall interacting with the incident wave composes a more complicated wave system. Umeyama (1993) formulated the surface displacement of overtopping wave as a partial standing wave analyzed by the method of perturbation. Fig. 2 is the experimental result of the temporal water surface profiles measured just in front of the breakwater with different crest heights. It is seen that the typical surface profile of standing wave is depicted in the non-overtopping case (H_c = 50 cm), in which the crest of water surface is steeper and the trough is flatter. However, the asymmetric properties are found in the surface profiles due to wave overtopping. These asymmetric profiles are more significant in the lower height of breakwaters. It is also found from Fig. 2 that the surface crest of overtopping wave decreases when the height of breakwater is lowering.

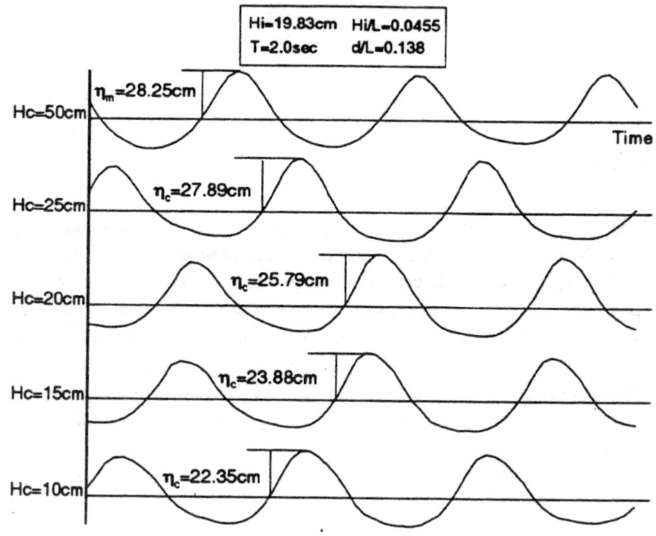

Fig. 2 Temporal water surface elevation for different crest heights of the breakwater.

Endo and Miura (1983) carried out an experimental work and obtained the relationship between the crest level of wave overtopping and the height of breakwater from experimental results, given by

$$\frac{\eta_c}{\eta_m} = 0.65 + 0.35\frac{H_c}{\eta_m} \tag{1}$$

in which η_c is the crest level of overtopping waves, η_m is the crest level of non-overtopping waves under the same incident wave conditions and H_c is the height of structure above the still water level. Fig. 3 illustrates the sketch of the definition of coordinate of the wave overtopping system.

The experimental result in this investigation shown in Fig. 4 is well in agreement with Endo and Miura (1983) for Eq. (1). An equation for the crest level η_m of non-overtopping wave related to the incident wave condition was also proposed by Endo and Miura (1983) based on their experiments in the form

$$\frac{\eta_m}{H_i} = 1.0 + (14.834 - 42.509\frac{d}{L})\frac{H_i}{L} \tag{2}$$

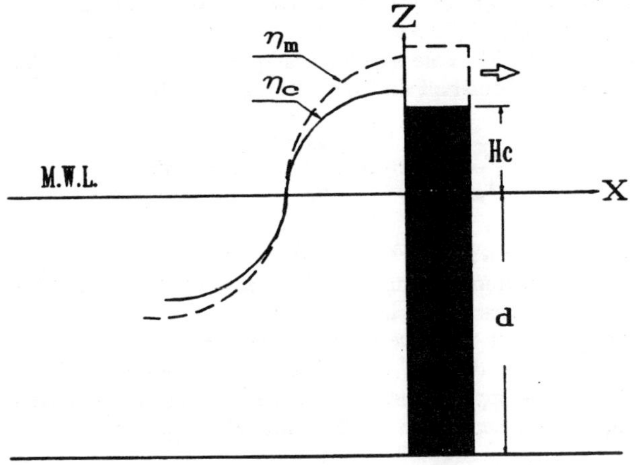

Fig. 3 The sketch of the coordinate system of wave overtopping.

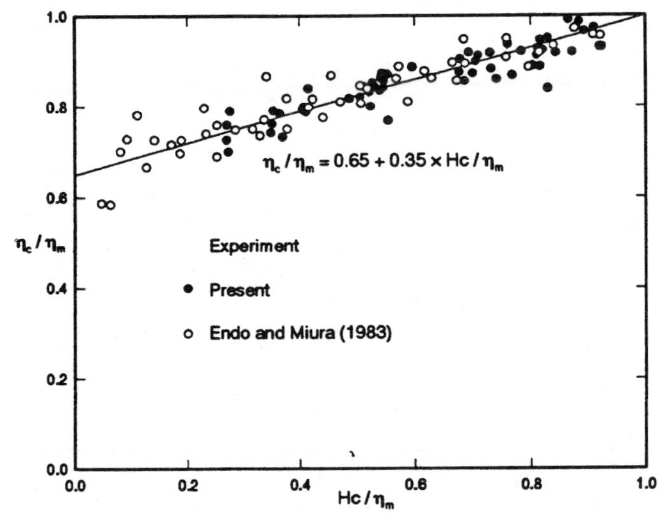

Fig. 4 Comparisons of experimental results for the crest level of overtopping wave.

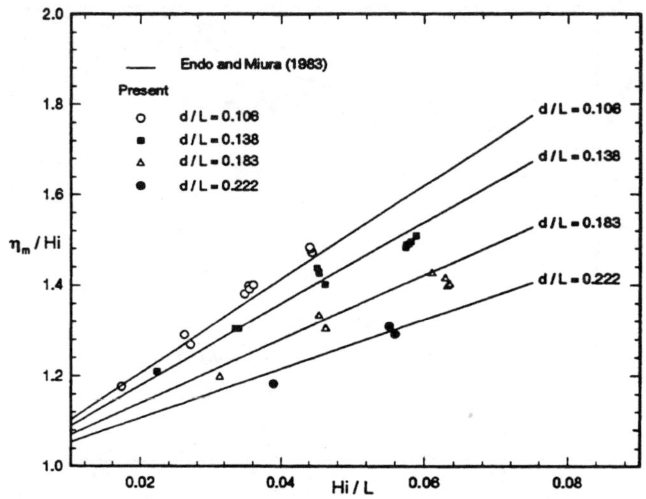

Fig. 5 Comparisons of experimental results for the crest level of non-overtopping wave.

where H_i is the incident wave height and L is the wave length of linear wave theory at water depth d. This experimental result shown in Fig. 5 also agrees well with the above equation. It is shown that the shallower water standing wave has higher crest level for the same incident wave steepness, that implies its nonlinearity is more obvious compared to that of deeper water. It is noted that the amplitude of standing waves equals to the incident wave height, if it was based on the linear wave theory. However, due to the nonlinearity of waves, the crest height of standing waves, η_m, is always higher than the incident wave height, H_i, that can be seen in Fig. 5.

Time-history of wave pressure

The time-histories of wave pressure on the wall at various locations

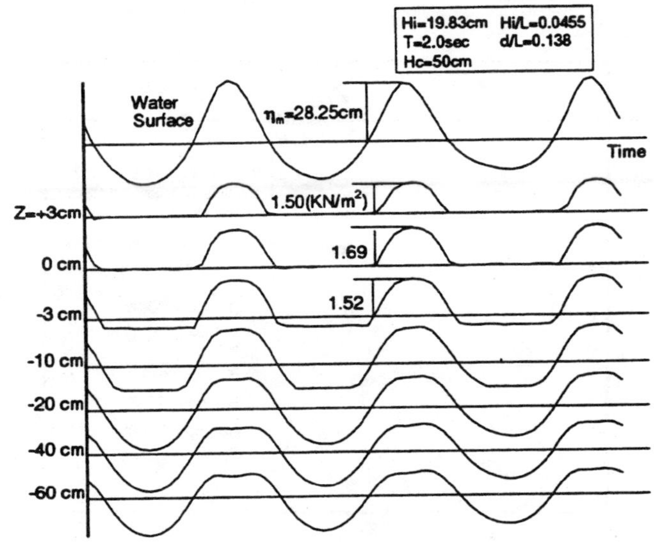

Fig. 6 Time-histories of non-overtopping wave pressures at various locations.

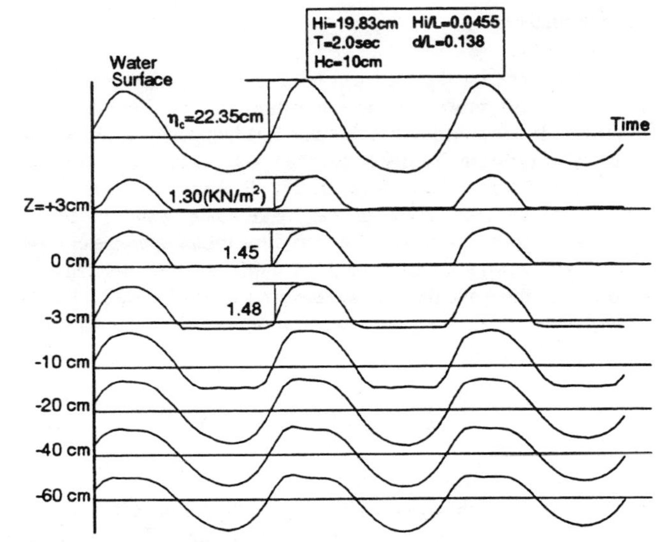

Fig. 7 Time-histories of overtopping wave pressures at various locations.

are depicted in Figs. 6 and 7 for non-overtopping and overtopping waves, respectively, in which the water surface elevation is also shown in figures. Fig. 6 shows the typical variation of standing wave pressure that slightly double humps appear in the time-history and where the maximum wave pressure occurs at the still water level. Owing to the asymmetric surface profile of overtopping wave, the asymmetric property also appears in the time-history of pressure, as shown in Fig. 7.

The vertical distribution profile of maximum wave pressure on the wall is depicted in Fig. 8, which shows obviously that the wave pressure was reduced due to wave overtopping by lowering the height of breakwater. The theoretical line shown in Fig. 8 is calculated by the Fourier solution of the full standing wave proposed by Tsai and Jeng (1994), from which the experimental result is well compared with the theory.

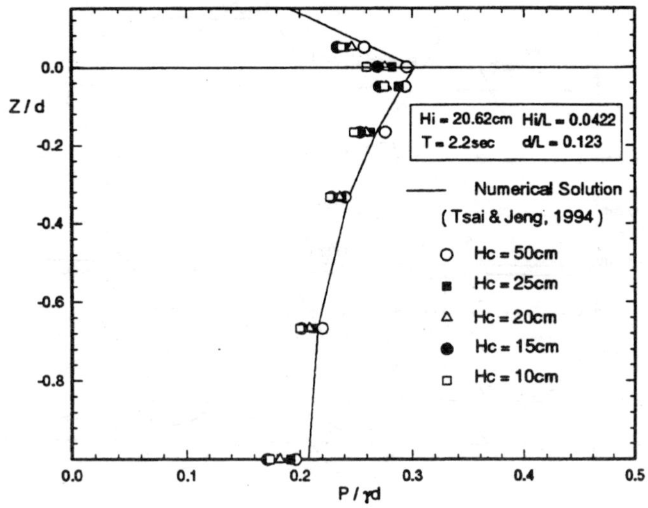

Fig. 8 Vertical distribution of maximum wave pressure on the wall.

Regression analysis

The wave force on the wall can be calculated by integrating the vertical distribution profile of wave pressure from the bottom to the water surface. It was stated above that a double hump may appear in the time-history of pressure. Thus the maximum wave force may not occur at the crest of water surface elevation. Fig. 9 shows maximum wave forces versus the crest level of the overtopping wave based on experimental results, in which the non-dimensional parameters are using as $F/\gamma d^2$ (γ is the specific weight of fluid) and η_c/d, respectively. It is shown that the wave force versus the crest level of wave is slightly dependent on the relative water depth. Nevertheless, the regression line can be simplified as the form

$$\frac{F}{\gamma d^2} = 0.02 + 0.58\frac{\eta_c}{d}, \quad C_c = 0.88 \tag{3}$$

in which C_c is the correlation coefficient that represents the deviation between the regression line and the experimental data. The wave force expressed in Eq. (3) is dependent only on η_c/d. It is conveniently to calculate the wave force when the incident wave condition and the height of the breakwater are specified, on the basis of Eqs. (1) and (2).

The reduction ratio of wave force due to wave overtopping can be defined as the ratio of the wave force of overtopping wave to the non-overtopping wave. Then the reduction ratio, F/F_m, related to the H_c/η_m is drawn in Fig. 10 and represented by the regression line as

$$\frac{F}{F_m} = 0.82 + 0.18\frac{H_c}{\eta_m}, \quad C_c = 0.85 \tag{4}$$

in which F_m is the wave force of non-overtopping wave, and H_c/η_m is defined as in equation (1). This result implies wave forces due to wave overtopping is reducible 18% in maximum.

It is noted that the wave force of non-overtopping wave can be calculated from the theories. If the Fourier solution theory (Tsai and Jeng, 1994) was applied, the force, $F_m/\gamma d^2$, versus the steepness of incident wave was depicted in Fig. 11 using d/L as a parameter.

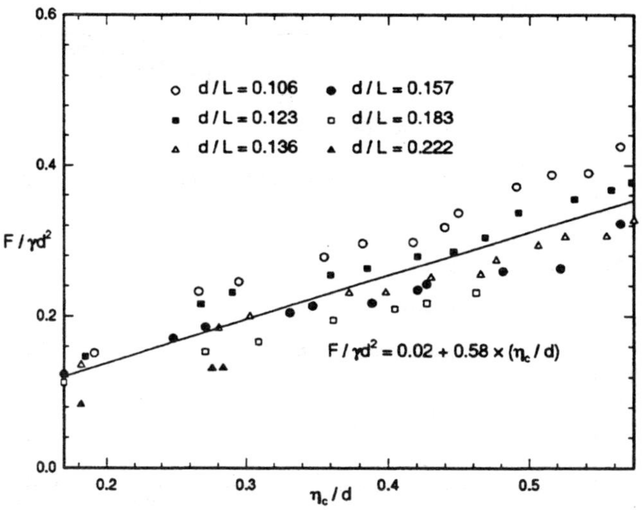

Fig. 9 Maximum wave force versus crest level of overtopping wave.

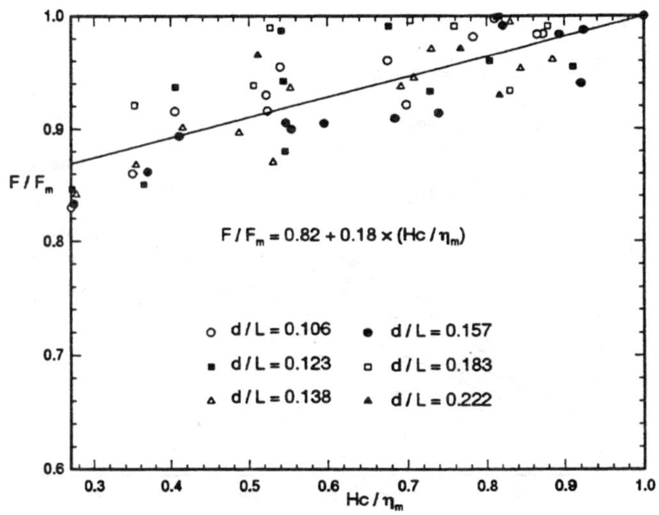

Fig. 10 Reduction ratio of wave force due to wave overtopping.

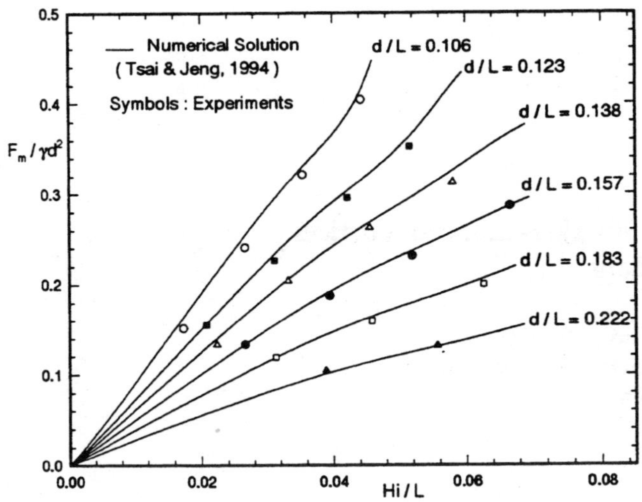

Fig. 11 Theoretical and experimental wave forces of non-overtopping wave.

CONCLUSIONS

This paper presented experimental results of wave forces on vertical breakwaters by standing waves with overtopping. From the results of the present study, the following conclusions are remarked.

The water surface profile as well as the time-history of wave pressure due to wave overtopping is similar to that of standing wave except for a slightly asymmetric property. The crest of water surface of overtopping wave is lower than that of wave non-overtopping when the same incident wave condition attacks on the breakwater. The wave force on breakwater is reducible due to wave overtopping. It is shown from experimental results that the reduction of wave force by wave overtopping can reach 18%, when compared to the non-overtopping wave. Empirical formulas for the wave force by wave overtopping has been suggested by the regression analysis.

ACKNOWLEDGEMENTS

This study was based on the work supported by the National Science Council, Republic of China, under grant No. NSC-85-2611-E-005-001.

REFERENCES

Endo, S, and Miura, A (1983). "Experimental Study on Sea Wave Movement Facing a Vertical Wall," Report of Res Inst of Industrial Tech, Nihon Univ, Japan, 20pp.

Iwagaki, Y, Shima, A, and Inoue, M (1965). "Effects of Wave Height and Sea Water Level on Wave Overtopping and Wave Run-up," *Coastal Eng in Japan*, Vol. 8, pp.141-151.

Saville, T and Caldwell, JM(1953). "Experimental Study of Wave Overtopping on Shore Structures," *Proc IAHR*, pp.261-269.

Sibul, OJ (1955). "Flow over Structure by Wave Action," *Trans AGU*, 36(1), pp.61-69.

Tsai, CP and Jeng, DS (1994). "Numerical Fourier Solution of Standing Waves in Finite Water Depth," *Applied Ocean Res*, 16(3), pp.193-195.

Umeyama, M (1993). "Wave Overtopping on Vertical Boundary and Water-Surface Displacement," *J WatWay, Port, Coastal & Ocean Eng*, 119(3), pp.243-260.

Shore Protection Manual (1984). Coastal Engineering Research Center, US Army Corps of Engineers, Washington, DC.

Weggel, JR (1976). "Wave overtopping equation," Proc 15th Coastal Eng Conf, ASCE, pp. 2737-2755.

Honolulu, USA, May 25-30, 1997
Copyright © 1997 by The International Society of Offshore and Polar Engineers
ISBN 1-880653-28-1 (Set); ISBN 1-880653-31-1 (Vol. III)

Coastal Dynamic Characteristics of Breaker-Induced Vortices in the Surf Zone

Daping Zhang and Shigenobu Tanaka
Public Works Research Inst., Ministry of Construction
Tsukuba, Japan

ABSTRACT

The characteristics of vortices in the surf zone were investigated by an approach in a large-scale wave-flume experiment using a three-video-camera system. Vortices associated with wave breaking can be largely classified into two: oblique and horizontal vortices. Conditions for seven types of vortices reaching bottom in the surf zone can be described by a combination of Galvin's breaker type index and Reynolds number of breaking waves. The spilling breakers provide a necessary condition for the occurrence of the multiple oblique vortices. Location and depth of the vortices reaching the bottom can be described quantitatively. The vortices reaching bottom, acted the bed material and lifted the sediment up into suspension; the suspended sediment is transported offshore by the mean offshore flow field, causing a net offshore sediment movement to form a bar. It is found that the breaker-induced vortices reaching the bottom is one of important factors for sediment transport in the nearshores.

KEY WORDS: Large-scale Experiment, Vortices, Breaker, Surf Zone, Sediment Transport

INTRODUCTION

The vortices produced by breaking waves were conducted by Miller (1976). The behavior of large-scale vortices, which were called the horizontal roller (i.e., a vortex with a horizontal axis and a circular cross-sectional shape), was examined concerning the mechanism of wave-height attenuation after breaking. This kind of vortex is called the horizontal vortex and studied by many researchers. Although Miller (1976) and many researchers also described the characteristics of horizontal vortex, intensive research on vortices induced by breaking waves has just begun only several years ago in some small-scale laboratories.

In the recent years, with the recent progress of measuring techniques of flow velocity, studies from a micro-scopic point of view have been conducted with the purpose of elucidating the nearshore hydrodynamics. Nadaoka et al. (1988, 1989) found the presence of "oblique vortex", which is a vortex with an oblique rotating axis, occurred under spilling breaker conditions, and explained that this oblique vortex could induce sediment suspension in the surf zone in the field. Zhang and Sunamura (1990) studied oblique vortex and four types of horizontal vortex in a small-scale laboratory. In generally, these studies, however, have been performed in the small-scale laboratories with a scale problem and no studies on conditions for formation of these vortices as well as characteristics of the vortices. For the scaling problems, this study discusses the characteristics of vortices in the surf zone through a large-scale wave-flume experiment, elucidates the conditions for formation of these vortices and characteristics of the vortices, and examines relationship between the vortex action and sediment suspension by direct inspects through a side-glass window.

BREAKER-INDUCED VORTICES REACHING BOTTOM

In order to examine the characteristics of vortices produced by breaking waves, a fixed-bed experiment was conducted at the Coastal Division, Public Works Research Institute, Ministry of Construction, Japan, using a 1/10 or 1/ 20 slope beach in a large-scale wave flume (150 m long, 2 m wide, and 5 m deep, Fig. 1). The period of laboratory waves ranged from 0.6 to 3.4 sec and the height of breaking waves from 10 to 96 cm. To examine the characteristics of vortices distributed in the whole surf zone, three video cameras were used (Fig. 1). The first camera was set up normal to the side of a glass window of the flume, the second camera was installed with an angle of 30 degrees behind the breaking point and the other was set up upper the sea level with an angle of 30 degrees behind the breaking point. The three-dimensional characteristics of vortices were examined on reproduced video pictures using both

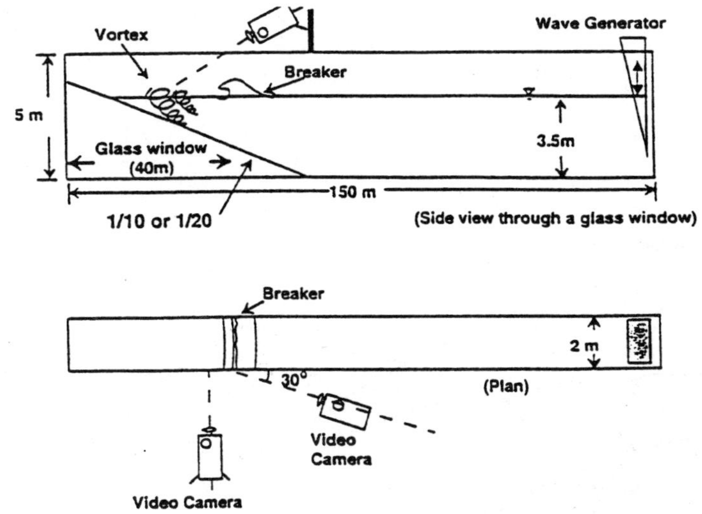

Fig. 1 Experimentsl equipment

breaker-induced air bubbles. Total experiment runs were 72.

Vortices formed just after wave breaking are classified into three types according to the direction of vortex axis: oblique vortex and two types of horizontal vortex, i.e., A-type and B-type horizontal vortices due to the classification way indicated by Zhang and Sunamura (1990) in their small-scale laboratory. The oblique vortex is like a tornado that has an obliquely stretched axis of rotation (Fig. 2a). The B-type horizontal vortex, which is shown by the lower diagram (Fig. 2c), looks a cultivator that has a horizontal axis of rotation. The A-type horizontal vortex, which is shown in the middle, is a hybrid between horizontal and oblique vortices (Fig. 2b). Namely, the horizontal vortex produces first in the upper part of water column and then it changes to the oblique vortex to reach bottom (Zhang and Sunamura, 1990). Comparing with oblique vortex, we found the A-type horizontal vortex's occurrence was seem almost likely to the B-type horizontal vortices, but it was stronger and the location from break point was shorter than that of the oblique vortex. Considering that the vortices reaching bottom are very important when we study sediment motion, we examined this kind of vortices.

The vortices reaching bottom are not only formed at the break point, but also in the whole surf zone depending on experimental conditions. Results obtained through the present experiment indicate that vortices reaching bottom are largely classified into seven: (1) the triple vortices (Fig. 3a), (2) three types of the double vortices (Fig. 3b), and (3) three types of the single vortex (Fig. 3c). The first vortex, i.e., the vortex formed just after wave breaking, is an oblique vortex, an A- type or a B-type horizontal vortex depending on experimental conditions, but the inshore vortex is always oblique vortices. It is unclear that why inshore vortex is always oblique vortex, which may be due to some difficult hydrodynamic characteristics between near break-point zone and in the inshore zone. In this experiment, it was also observed that there exist the vortices not reaching bottom; this phenomenon was found whenever the breaker height was extremely small.

For investigating systematically the occurrence condition for the seven types of characteristic vortex, two dimensionless parameters were introduced. One is the breaker-type index given by Galvin (1968), which is denoted as B_t in this study:

$$B_t = H_b/gT^2\tan\beta \qquad (1)$$

where H_b is the breaker height, T is the wave period, $\tan\beta$ is the bottom slope, and g is the acceleration due to gravity. The reason for the selection of this parameter is that the type of breaking waves is closely related to vortex types. The other parameter selected here is Reynolds number for breaking waves, which will be used as an approximate index to evaluate the viscous effect on fluid motion in the surf zone. The breaker height and the horizontal water-particle velocity at the crest of breaking waves were taken into account as a characteristic length and velocity, respectively. This particle velocity, U_b, can be assumed to be equal to wave velocity at the breaking point, i.e., $U_b \sim L_b/T$, where L_b is the length of breaking waves. Therefore, Reynolds number for breaking waves, Re, can be expressed as (Zhang and Sunamura, 1990):

$$Re = H_b L_b / \nu T \qquad (2)$$

where ν is the kinematic viscosity of fluid and was assumed to be 0.01 cm^2/sec in this study.

Conditions for the vortex occurrence are plotted using B_t and Re in Fig. 4. The three L-shaped solid lines are boundaries for the area of triple vortices, double vortices, and single vortex, respectively. Based on Galvin's (1968) criteria, breaker types were classified on reproduced video pictures in this study; the result is plotted along the y-axis in Fig. 4. It is indicated that the breaker type does not always coincide with the type of vortices, but spilling breakers provide a necessary condition for the occurrence of the triple oblique vortices. It is also indicated that the number of multiple vortices increases with increasing values of B_t and Re. The two dashed lines indicate the boundaries for oblique vortices, A-type and B-type horizontal vortex. Otherwise, for the steep (1/10) and medium (1/20) bottom slope conditions in this experiment, it is suggested that there would be more numbers of multiple vortices reaching bottom in a gentle slope beach in the field if we consider the effect of $\tan\beta$ in Eq. (2).

LOCATION, DEPTH AND WIDTH OF THE VORTICES REACHING BOTTOM

The locations and depths of the vortices reaching the bottom can be described respectively as: $l_{vn}/L_0=3.98\, n^{0.86}(H_b/gT^2\tan\beta)^{1.2}$ and $h_{vn}=0.69n^{-1.44}H_b$, where l_{vn} is the average horizontal distance from the break point to the location where an n-th vortex reaches the bottom, h_{vn} is the water depth of the vortex, n is the ordinal number (from the break point to the shore, n=1, 2•••), and L_0 is the deep-water wavelength.

ROLE OF VORTICES REACHING BOTTOM

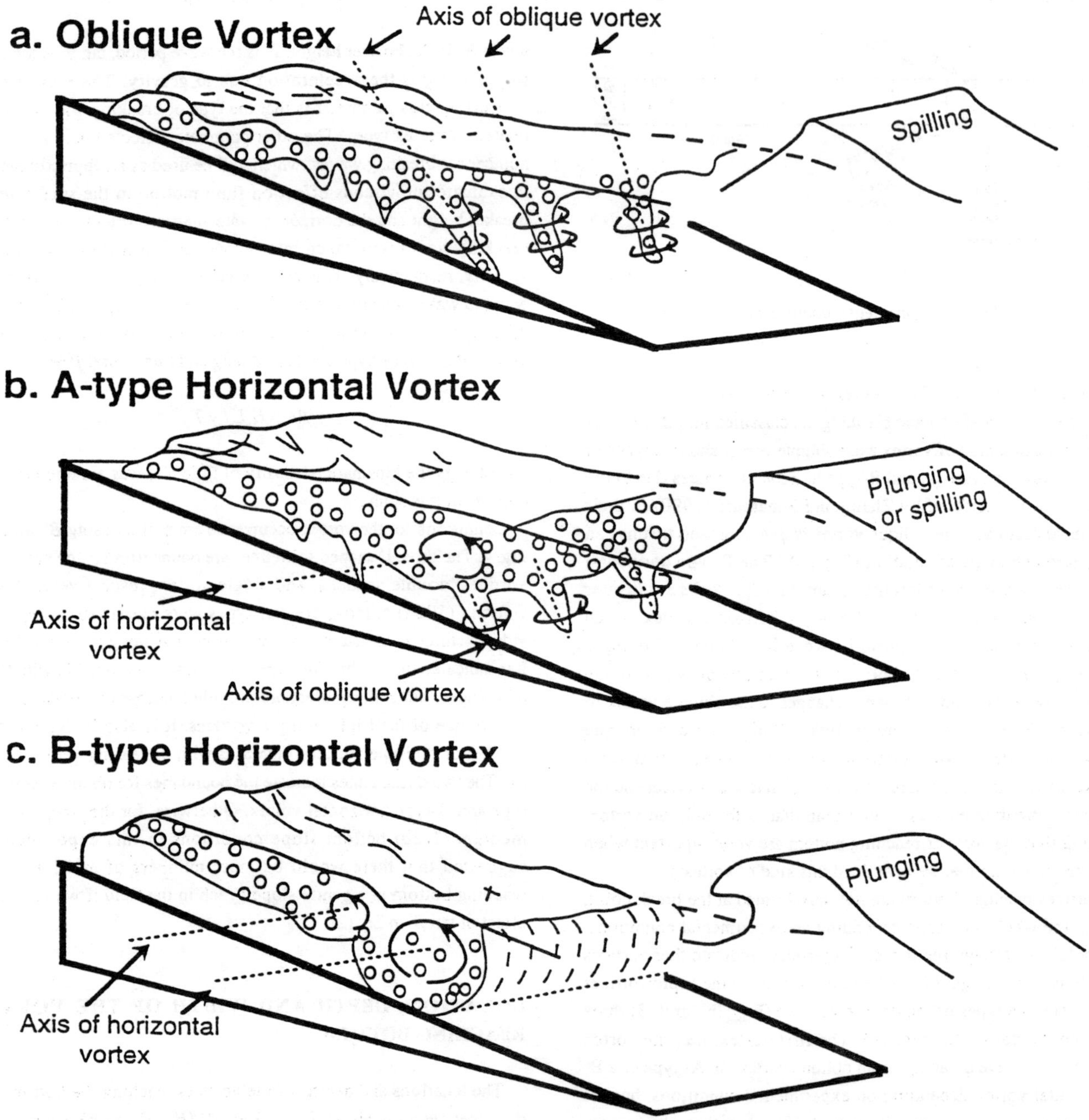

Fig. 2 Vortex types

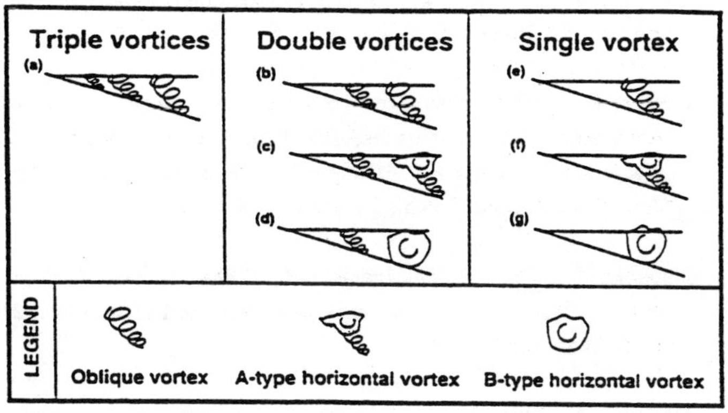

Fig. 3 Schematic diagram showing vortices reaching bottom

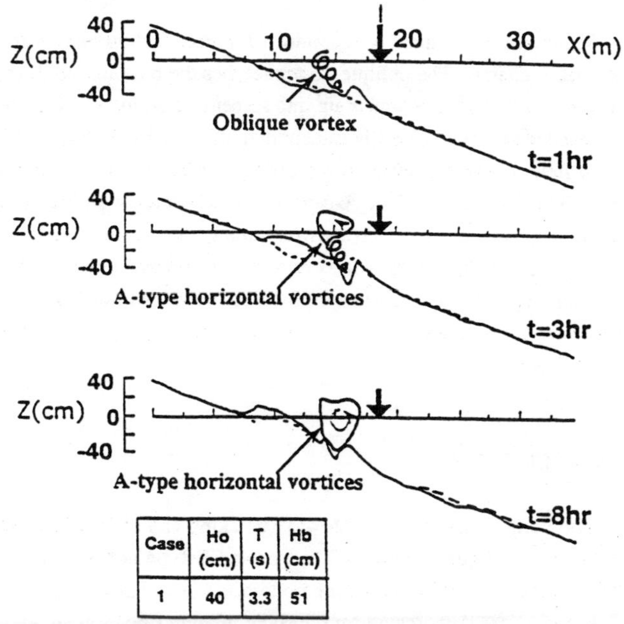

Fig. 5 An example of bar development and vortices reaching the bottom

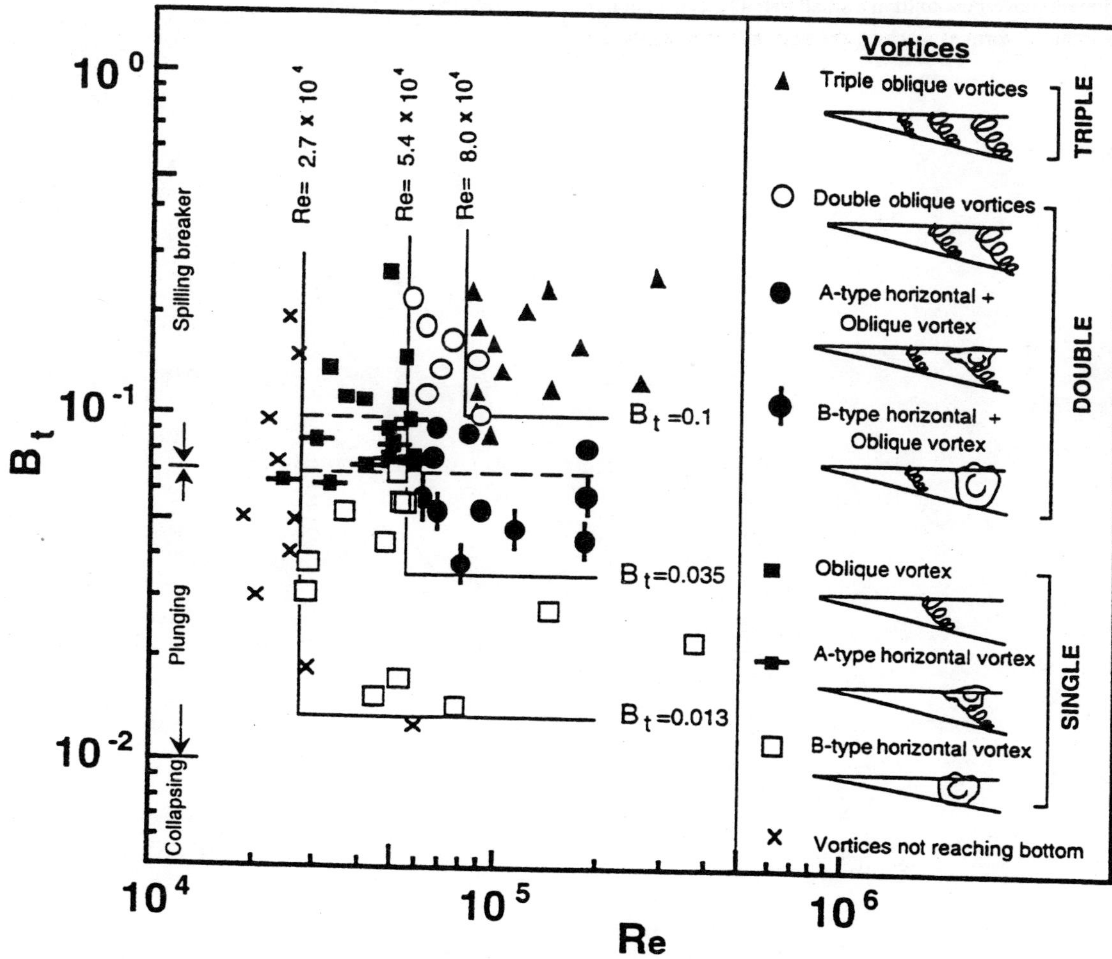

Fig. 4 Conditions for occurrence of vortices reaching bottom

movable-bed experiment was conducted. Figure 5 shows an example of the beach change. The oblique vortex reaches the bottom, whirls the bed material and lifts the sediment up into suspension acting as a tornado; the suspended sediment cloud is transported offshore by the mean offshore flow field, causing a net offshore sediment movement to form a small bar. While the action of the A-type horizontal vortex is similar to that of the oblique vortex, but it is somewhat stronger and forms a middle-sized bar. The B-type horizontal vortex digs the bed material like a cultivator to form a large bar. In this study, however, we could not find a way to evaluate the strength of the vortex quantitively, because it is too complicated.

CONCLUSIONS

The vortices associated with wave breaking can be largely classified into three: oblique vortex, and A-type and B-type horizontal vortices. Seven kinds of vortices reaching bottom in the surf zone of the present experiment are demarcated on the B_t-Re plane. Location and depth of the vortices reaching the bottom can be described quantitatively. The oblique vortex reaches bottom, whirls the bed material and lifts the sediment up into suspension acting as a tornado. The suspended sediment is transported offshore by the mean offshore flow field, which causes a net offshore sediment movement to form a small bar. The conclusion is that the breaker-induced vortices reaching the bottom is one important

REFERENCES

Galvin, C. J., Jr. (1968). "Breaker type classification on three laboratory beaches," J. Geophys. Res., Vol. 73, 3651-3659.

Miller, R. (1976). "Role of vortices in surf zone prediction: sedimentation and wave forces," *Beach and Nearshore Sediment (ed. by R.A. Davis, Jr. and R.L. Ethington) Soc. Economic Paleontologists Mineralogists*, Special Pub., Vol.24, 92-114.

Nadaoka, K., Ueno, S., and Igarashi, T. (1988). "Characteristics of bottom velocity and sediment suspension in the surf zone," *Proc. 21th ICCE*, 1646-1660

Nadaoka, K., Hino, M., and Koyano, Y. (1989). "Structure of the turbulent flow field under breaking waves in the surf zone," *J. Fluid Mech.*, Vol. 204, 359-389.

Zhang D. P. and Sunamura T. (1990). "Conditions for the occurrence of vortices induced by breaking waves," *Coastal Eng. Japan*, Vol. 33, 145-155.

Proceedings of the Seventh (1997) International Offshore and Polar Engineering Conference
Honolulu, USA, May 25-30, 1997

ISBN 1-880653-28-1 (Set); ISBN 1-880653-31-1 (Vol. III)

Experimental Study on Anchor Wire Type Automatic Installation System of Offshore Breakwaters

H. Yoneyama, S. Shiraishi and K. Nazato
Port and Harbour Research Institute
Yokosuka, Japan

H. Takahashi
Ministry of Transport
Yokohama, Japan

ABSTRACT

The authors have developed an automatic installation system for offshore breakwaters intended to save labor during caisson installation and to improve worker safety. By use of this system, a present winch operation which is directly hand-operated by several operators on a working craft can be automated, and working conditions on offshore constructions are expected to be improved. The automatic installation system includes programs that manage installation work using two associated units: a tugboat support system and an anchor wire system. Hydraulic model experiments on these two systems have been carried out, and the experimental results for the tugboat support system have already been reported. Accordingly, in this paper, the experimental results for the anchor wire system are primarily described and discussed in comparison with those for the tugboat support system. As a result, it has been validated that the anchor wire system can be used to install a caisson more stably and accurately than the tugboat support system.

KEY WORDS: Caisson, Offshore construction, Installation work, Winch, Anchor wire system, Tugboat support system, Automatization

INTRODUCTION

Recently, ports in Japan have been constructed at locations directly facing the Pacific Ocean. The installation of caissons for a breakwater in such open sea is frequently interrupted or suspended because of severe wave conditions. During this installation process, winches are hand-operated by several operators under sometimes dangerous conditions due to the operator's proximity to winches and taut wires. Therefore, it would be desirable to improve these working conditions in order to maintain the safety of workers on the caissons or near the winches. Moreover, new installation methods will be required to prepare for future decreases in the labor force as the present caisson installation process depends substantially on the intuition and experience of skilled workers.

On the basis of these needs, the authors developed an automatic installation system that enables the automatic construction of offshore breakwaters without the need for workers on a caisson and the deck of a working craft. This new system is intended to reduce the amount of labor required during caisson installation, and to improve the safety of workers. The system is composed of an Engineering Work Station (EWS, Toshiba AS4050GX) with A/D and D/A converters, winches operated by control signals from the EWS, and an observation system that includes a motion measurement system and an optical position measurement system used to measure caisson position and motion, etc. When a caisson is installed, the system uses the observation system to obtain data on caisson position and motion, wave height, and wire tension and displacement. This data is then used to predict the motions of the caisson through the use of the EWS on the working craft. The system then outputs control signals to each of the winches from the EWS to move the caisson to the optimum installation position. The automatic installation system repeats this procedure, and the caisson is finally installed at the target position near the end of the breakwater. The EWS includes programs enabling installation work to be performed using two associated units: a tugboat support system and an anchor wire system.

Hydraulic model experiments on the above two systems have been carried out, and the experimental results for the tugboat support system have already been reported (Shiraishi et al., 1994, 1995a, 1995b; Yoneyama et al., 1995, 1996a, 1996b). Accordingly, in the present paper, the experimental results for the anchor wire system are primarily described and discussed in comparison with those for the tugboat support system (Yoneyama et al., 1996c).

CURRENT CAISSON INSTALLATION

Photograph 1 shows the winch operation in the current installation method. Winch operators on the floating crane operate the winches by hand according to instructions and/or signs from the conductor on the caisson. As the operators are obliged to work near taut wires, their work is sometimes dangerous. If the automatic installation system is used for caisson installation, the present winch operation can be automated, and the safety of workers on

Photo 1 Winch operation under the current method

Photo 2 Working conditions of workers on the caisson

offshore constructions is expected to be improved.

Meanwhile, Photo 2 shows the working conditions of workers on the caisson during the current caisson installation. As seen in this picture, at present there are a large number of workers working on the caisson. This situation is very dangerous for the workers, as they are compelled to work on the moving caisson due to waves, and near taut wires. However, since a caisson installation method using the automatic installation system enables installation work to be performed with no workers on the caisson, labor-saving construction will be attained and workers will simultaneously be prevented from performing dangerous work.

NEW DEVELOPED AUTOMATIC INSTALLATION SYSTEM

The automatic installation system is capable of automatically installing a caisson at the target position according to the procedure described below:

(1) When a caisson is installed, the system obtains data on caisson position and motion, wave height, and wire tension and displacement through an observation system that includes a motion measurement system, an optical position measurement system, a wave gauge, etc.

(2) On the basis of the measured data, the system predicts the motions and position of the caisson through the use of an EWS mounted on a working craft.

(3) The system outputs control signals to each of the winches from the EWS to move the caisson to the optimum position for installation.

(4) The procedures described in (1) to (3) are repeated, and the caisson is finally installed at the target position in the vicinity of the end of the breakwater.

The control algorithm for the system has been developed by surveying and analyzing current installation methods for caissons in detail. In Japan, the caisson installation methods at locations directly facing open sea can be roughly classified into two types of procedures: the tugboat support system using tugboats and the anchor wire system using anchors.

Figures 1 (a) and (b) show the arrangement of wires, tugboats, and anchors in the tugboat support system and the anchor wire system, respectively. The EWS includes programs for these two elements of the automatic installation system. The system can execute caisson installation after either of them being selected. As shown in Figs. 1 (a) and (b), the configuration of the wires for the anchor wire system is more complicated than that for the tugboat support system. However, the former system is considered more effective in the installation of a caisson, as it is capable of installing a caisson more stably and accurately because of its configuration of wires stretched in four directions.

Figure 2 shows a block diagram of the system. This automatic installation system consists of the following five sections:

(a) Observation section

(b) Input/Output processing section

(c) Control section

(d) Display section

(e) Indication section

The automatic installation system can automatically execute caisson installation by operating the winches by use of a computer. However, in order to realize completely unmanned operations on the caisson during installation, an automatic wiring system and an automatic water-supply and drainage system are required. The former system is used to dispose and install wires on the caisson with no workers, and the latter system is used to automatically supply water to the caisson or drain water from it. These systems are linked with the automatic installation system during installation of the caisson. In this study, since those two systems have not yet been developed, only the automatic installation system has been validated by hydraulic model experiments.

HYDRAULIC MODEL EXPERIMENTS FOR THE ANCHOR WIRE SYSTEM

Model of the Anchor Wire Type Automatic Installation System

Hydraulic model experiments for the tugboat support system and the anchor wire system were carried out in a wave basin belonging to PHRI (Port and Harbour Research Institute). An irregular wave

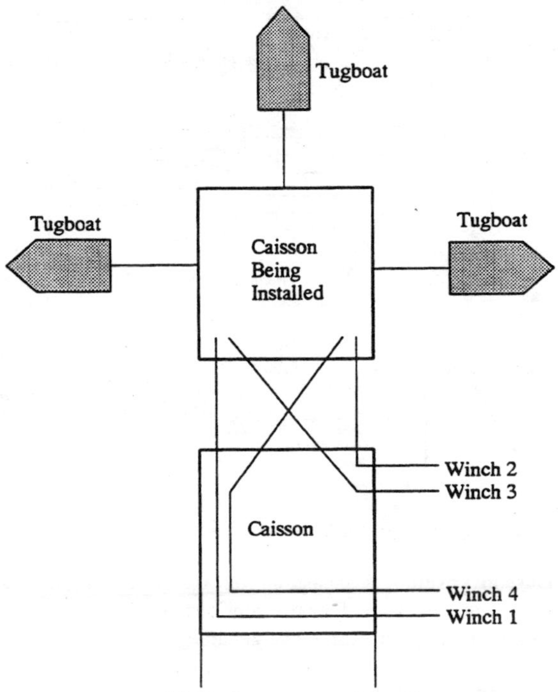

(a) Tugboat support system

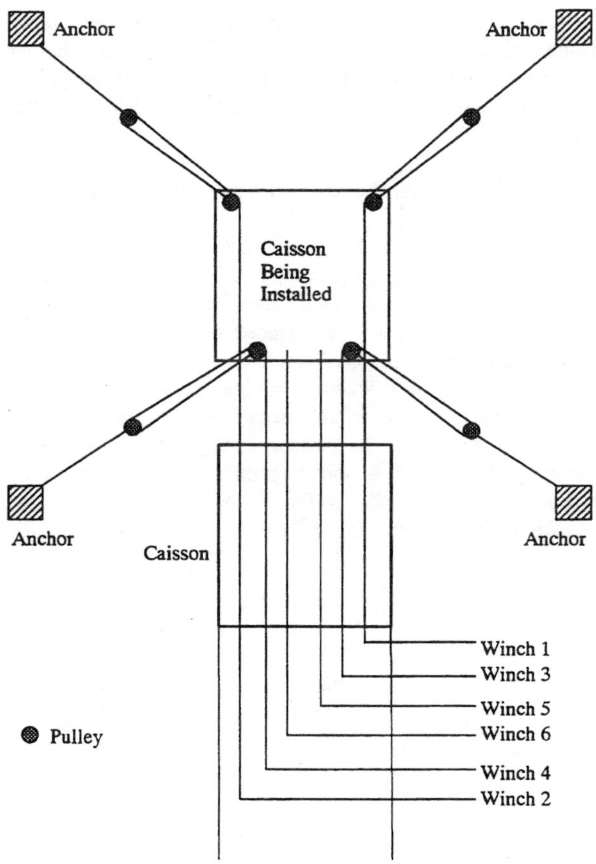

(b) Anchor wire system

Fig.1 Methods used during installation

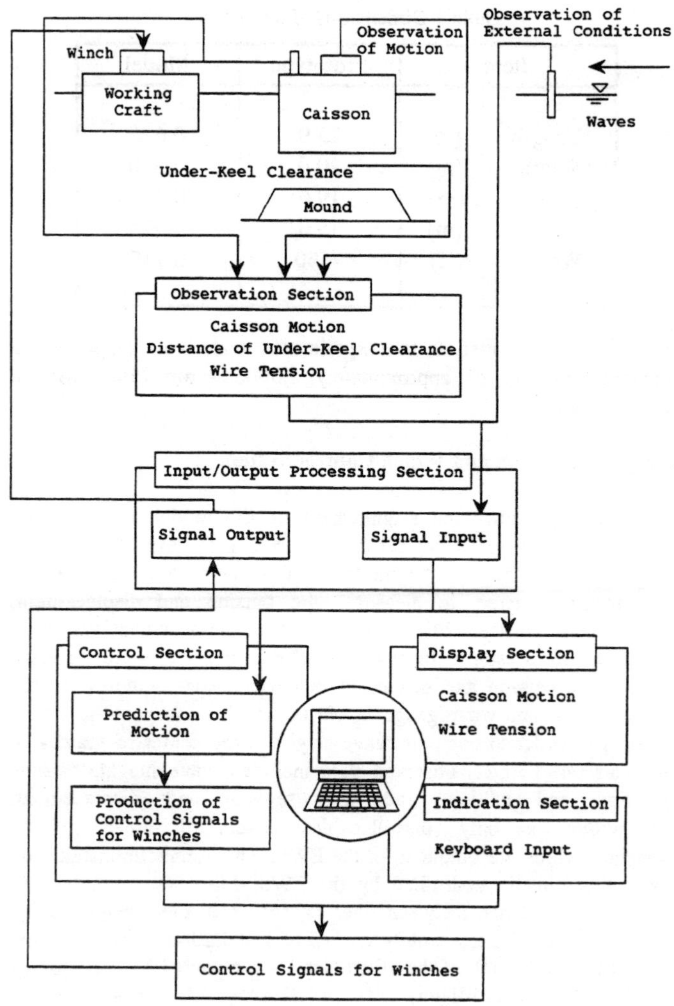

Fig.2 Block diagram of the automatic installation system

generator with two piston-type paddles is installed at one side of the basin, and wave absorbers are set up at the other three sides. The experimental model is a 1/40 scale model of a 7,500tf (7.35×10^4kN) caisson. Table 1 shows the dimensions of a model and prototype caisson. The model caisson is made of acrylic and has dimensions of 62.5cm x 50.0cm x 47.5cm in the model scale.

Photograph 3 shows the outward appearance of the model caisson and the model of the anchor wire type automatic installation system used in the hydraulic model experiments. The acrylic model caisson is afloat in front of the model breakwater. In the case of the anchor wire system, six winches fixed on the model breakwater are employed to wind or let out wire for controlling the motions and position of the caisson. In Photo 3, one of the four anchors can be seen to be installed at the bottom of the basin in front of the wave gauge. On the other hand, the space inside the floating caisson is divided into four compartments by acrylic boards. Water supply for caisson installation is performed for these cells by means of a hose connected to an acrylic water tank installed 1.1m above the water surface. The hose can be moved freely on a constant horizontal plane to change its position for supplying water according to the caisson motion and position. It is hand-operated by the person standing on

Table 1 Dimensions of a caisson

Item		Prototype	Model
Length	(m)	25.0	0.625
Width	(m)	20.0	0.500
Depth	(m)	19.0	0.475
Draft	(m)	15.0	0.375
Weight	(tf)	7500	0.117

the model breakwater. In the experiments, water was supplied at a constant velocity of approximately 5,000cm^3/min (50m^3/min in the prototype).

Data Observation and Winch Control System

Figure 3 shows the arrangement of the model and the data observation and winch control system used in the experiments. The six model winches on the model breakwater include tension and displacement meters to measure wire tension and displacement, respectively. On the other hand, the motion measurement system is fixed on the caisson to be installed. The four anchors are disposed around the caisson and connected to it by means of wires. In the experiments, one wave gauge is placed as close as possible to the caisson in order to compute wave height at the center of gravity of the caisson. Further, observed data includes wave height, caisson position and motion, and wire tension and displacement. Meanwhile, the only controlled data is control signals for the winches, which are output from the EWS. The automatic installation system is directly controlled by the EWS. Moreover, as shown in Fig.3, the distance between the caisson and the breakwater is 25.0cm (10.0m in the prototype) before installation work is begun. The stiffness of each of the wires Nos.1 - 6 modeled by coil springs is 100kgf/m (9.8 x 10^{2}N/m).

Figure 4 shows the components of caisson motion and the coordinate system in this automatic installation system. Note that the origin is defined at the center of one side of the breakwater, and that the system shows the absolute coordinates. Accordingly, since

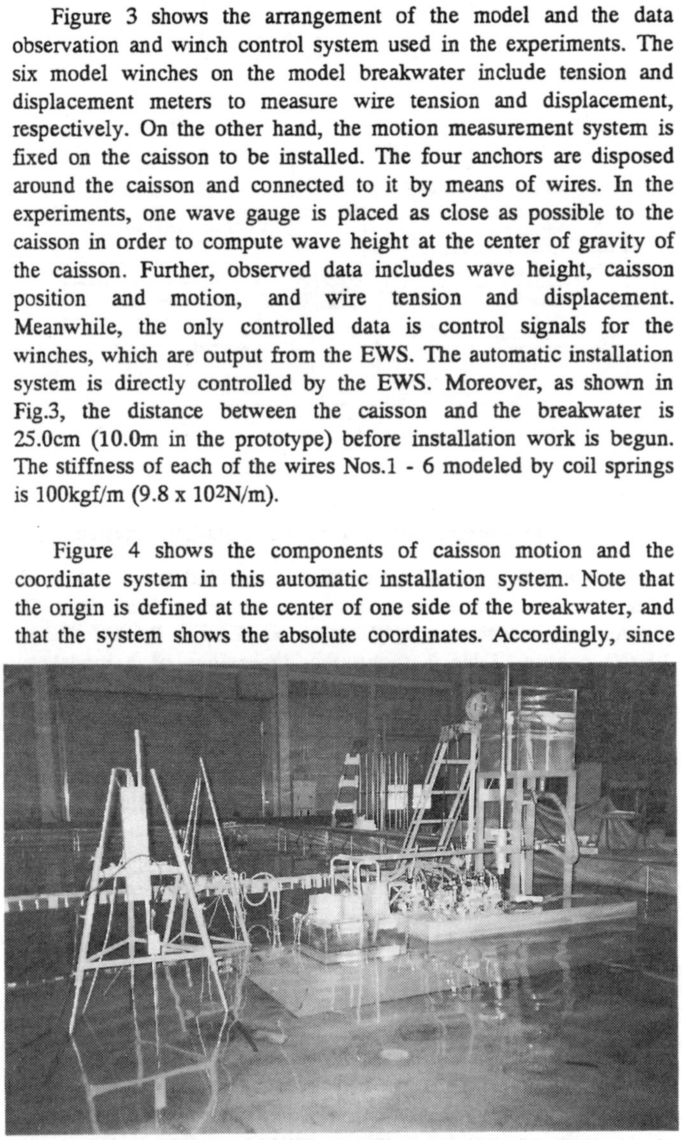

Photo 3 Outward appearance of the anchor wire type automatic installation system

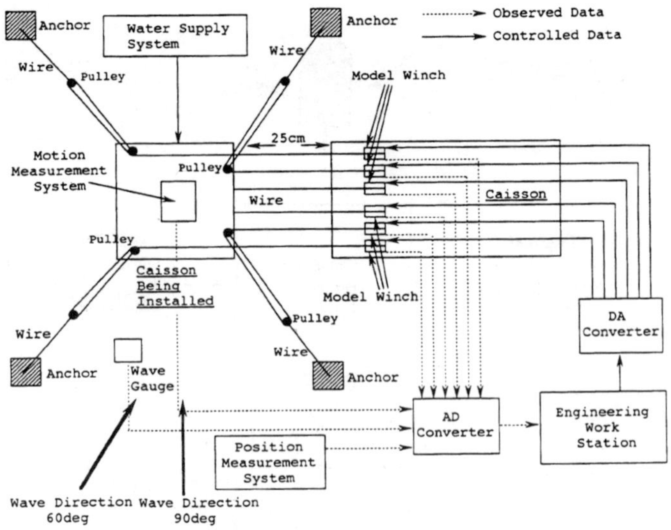

Fig.3 Arrangement of the data observation and winch control system

the caisson position is measured on the basis of this coordinate system, surge, sway, and heave motions of the caisson treated in this paper are different from those generally used.

Experimental Conditions

Experimental conditions such as significant wave heights and periods and wave directions are tabulated in Table 2. The hydraulic model tests were conducted twice for all cases in irregular waves. Further, the water depth in the experiments was 50cm (20.0m in the prototype).

View of Model Caisson Before and After Installation

Photographs 4 and 5 show the model caisson and winches before and after installation, respectively. Three units of the motion measurement system - a sensor unit, a computation unit, and a magnetic azimuth sensor - are placed on the caisson inside three boxes made of corrugated cardboard. These boxes are used to protect the units from dispersed water. A midget lamp is attached at the top of the left-side box; it is the mark used to measure the three-dimensional position of the caisson through the use of the position measurement system. At the beginning of the experiments, the caisson is located 25.0cm (10.0m in the prototype) away from the breakwater as shown in Photo 4. Finally, the caisson is

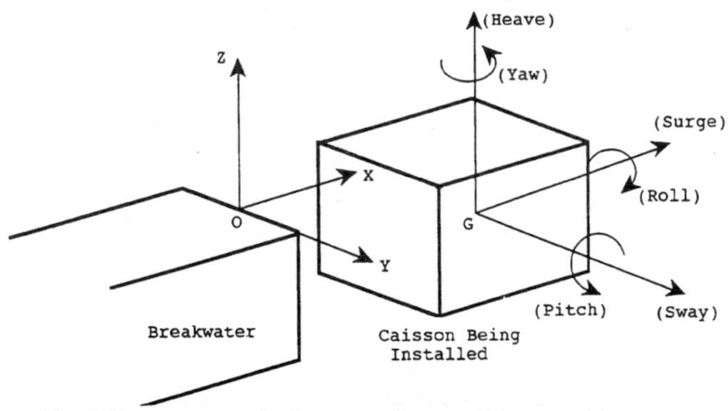

Fig.4 Components of caisson motion, and the coordinate system

Table 2 Experimental conditions

irregular wave

wave angle (°)	wave height (cm)	wave period (s)
90	1.25 (50.0)	0.95, 1.27, 1.58, 1.90 (6, 8, 10, 12)
	2.50 (100.0)	0.95, 1.27, 1.58, 1.90 (6, 8, 10, 12)
	3.75 (150.0)	0.95, 1.27 (6, 8)
60	1.25 (50.0)	0.95, 1.27, 1.58, 1.90 (6, 8, 10, 12)
	2.50 (100.0)	0.95, 1.27, 1.58, 1.90 (6, 8, 10, 12)
	3.75 (150.0)	0.95, 1.27 (6, 8)

※ This table shows model values. Prototype values are shown in parentheses below the model values.

automatically installed at the target position close to the end of the breakwater by use of the automatic installation system as shown in Photo 5. Installation errors were measured using a scale after caisson installation was completed. Here, the installation errors indicate discrepancies between the target position and the position of the installed caisson.

EXPERIMENTAL RESULTS

Trace of the Center of Gravity for the Caisson

Figures 5 (a) and (b) show a trace of the center of gravity of the caisson for a wave direction of 90deg in irregular waves. The significant wave periods in each figure are 0.95s (6.0s in the prototype) and 1.90s (12.0s in the prototype), respectively. Meanwhile, the significant wave height is 2.5cm (1.0m in the prototype). These figures are hard copies of the picture shown on the EWS display, which visually indicates the position and motions of the caisson during installation work in real time. The large picture on the left side of each figure shows a plane view (X-Y plane), while the small one on the right side shows an elevated view (X-Z plane). The inside of the two dotted lines shown over the breakwater in the large picture indicates the allowable range of the caisson motion. When the center of gravity of the caisson exceeds this range, winches control the caisson position and motion in the Y-direction, specifically sway motion, as shown in Fig.4. On the other hand, when the yaw motion of the caisson exceeds 2.0deg, winches control the rotation of the caisson on the horizontal plane. These beginning conditions for control can be determined optionally in the EWS.

From the comparison of Figs.5 (a) and (b), it is found that the sway and heave motions of the caisson slightly increase in long period waves ($T_{1/3}$=1.90s (12.0s in the prototype) than in short period waves ($T_{1/3}$=0.95s (6.0s in the prototype)). However, the difference between these two figures is not so clear, as the caisson motion is small in long period waves as well as in short period waves because the caisson is restrictively anchored to the sea bed by wires stretched in four directions. Further, heave motion gradually decreases as water is supplied to the caisson. The caisson is then finally installed at the target position in the vicinity of the breakwater as shown in Figs.5 (a) and (b). Although the experimental results for the tugboat support system (cf. Yoneyama et al., 1996a) are not shown in this paper as a result of space restrictions, here we compare them with the results for the anchor wire system. In the case of the tugboat support system, it was recognized that the caisson goes adrift owing to the effects of the wave drift force, especially in short period waves ($T_{1/3}$=0.95s (6.0s in the prototype)). On the other hand, for the anchor wire system, the drift amounts of the caisson are small compared with those of the tugboat support system. At the same time, it is found that the traces of the center of gravity for the caisson move nearly along the X-axis shown in Fig.4. Because, for the anchor wire system, the movement of the caisson is restricted by the wires stretched in four directions and the caisson installation is executed with more stability against waves. Consequently, it is concluded that the anchor wire system can be used to install the caisson more stably than the tugboat support system.

Time Histories of Caisson Motion

Figures 6 (a) and (b) show the time histories of caisson motion

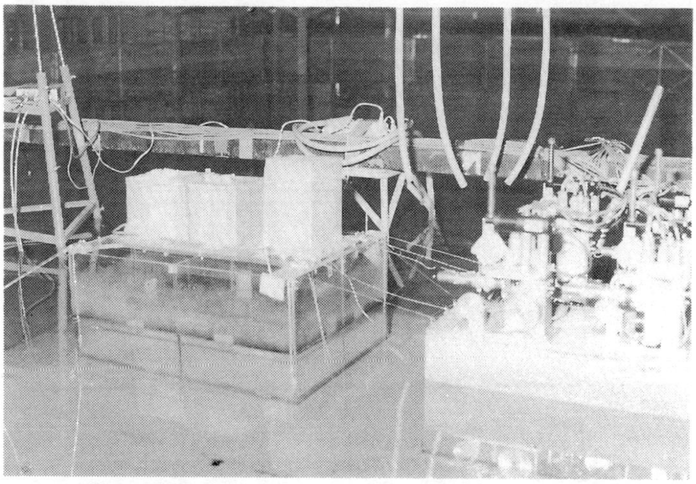

Photo 4 Model caisson and winches before installation

Photo 5 Model caisson and winches after installation

for a wave direction of 90deg in irregular waves. Each vertical axis denotes the motions of the caisson in the absolute coordinate system illustrated in Fig.4; each horizontal axis denotes the time from the beginning to the end of the experiments. Wave conditions correspond to those shown in Figs.5 (a) and (b).

The three figures from top to bottom in Figs.6 (a) and (b) show the surge, sway, and heave motions of the caisson, respectively. From these figures, it is clearly understood that the sway and heave motions are somewhat larger in long period waves ($T_{1/3}$=1.90s (12.0s in the prototype)) than in short period waves ($T_{1/3}$=0.95s (6.0s in the prototype)). Nevertheless, the sway and heave motions for the anchor wire system are shorter than those for the tugboat support system in both long and short period waves (cf. Yoneyama et al., 1996a). Further, heave motion decreases when water is added to the caisson, resulting in the achievement of stable and effective caisson installation. This tendency for the anchor wire system is similar to that for the tugboat support system (cf. Yoneyama et al., 1996a). At the end of the experiment, heave motion becomes almost constant, which is to say that distance between the mound and the bottom of the caisson nearly reaches zero. Meanwhile, sway motion does not indicate a constant value at that time. This means that since the caisson does not have enough weight to resist the wave force, incoming waves cause sway motion in it. In the experiment, the water-supply velocity is constant. However, in the actual

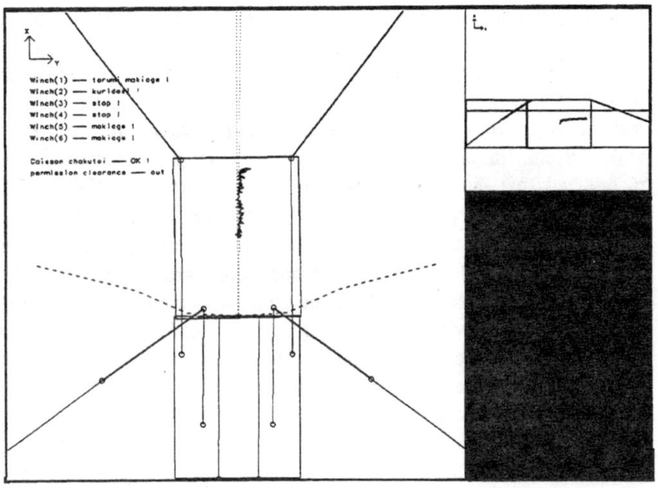

(a) $T_{1/3}$=0.95s(6.0s), $H_{1/3}$=2.5cm(1.0m), θ =90deg

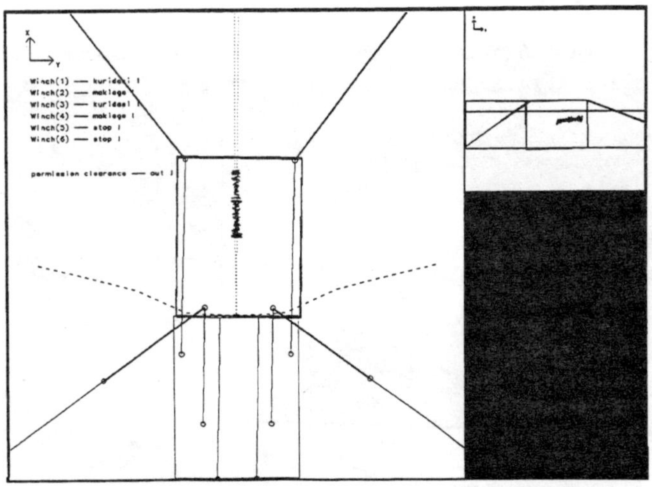

(b) $T_{1/3}$=1.90s(12.0s), $H_{1/3}$=2.5cm(1.0m), θ =90deg

Fig.5 Trace of the center of gravity for the caisson in irregular waves

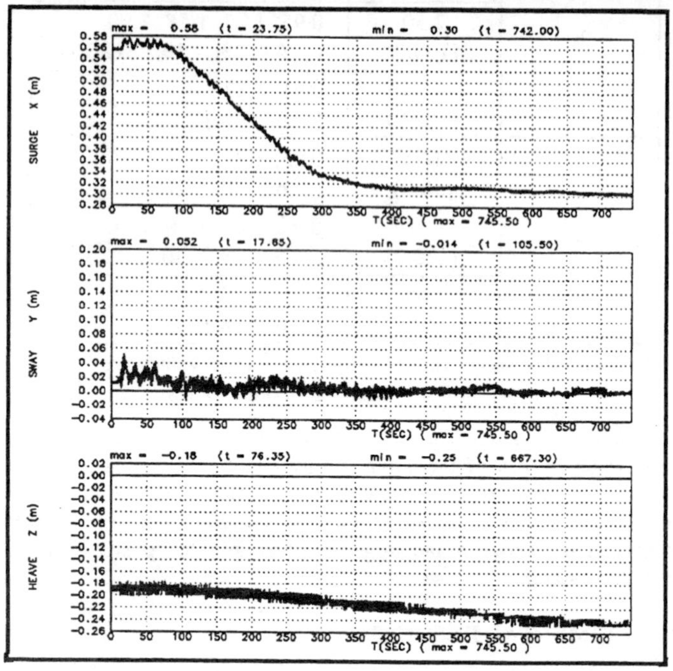

(a) $T_{1/3}$= 0.95s(6.0s), $H_{1/3}$=2.5cm(1.0m), θ =90deg

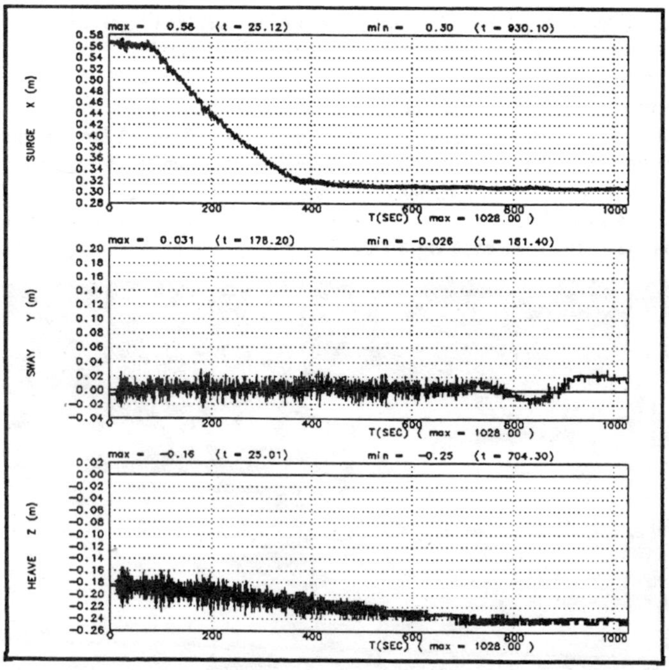

(b) $T_{1/3}$=1.90s(12.0s), $H_{1/3}$=2.5cm(1.0m), θ =90deg

Fig.6 Time histories of caisson motion in irregular waves (surge, sway, heave)

installation work, water is rapidly supplied just before the caisson is completely installed. If this situation can be reproduced in model experiments, the sway motion of the caisson prior to installation will probably decrease.

Maximum Values of Caisson Motion in Irregular Waves

Figures 7 (a) and (b) show the maximum values of sway motion in irregular waves with significant wave periods. Each figure contains the maximum sway motion for both the anchor wire system and the tugboat support system. The model experiments in irregular waves were conducted twice in all cases. Accordingly, the mean values of these two experimental results are plotted.

Figure 7 (a) represents the effects of wave heights on the sway motion of the caisson. It is clear that sway motion increases along with the wave height for the tugboat support system. On the other hand, sway motion for the anchor wire system also increases slightly with the wave height, except for a wave period of 1.27s (8.0s in the prototype) and a wave height of 3.75cm (1.5m in the prototype). In this case, measured sway data is considered unreliable, because in the other wave conditions sway motion for the anchor wire system indicates virtually the same value. Moreover, it is

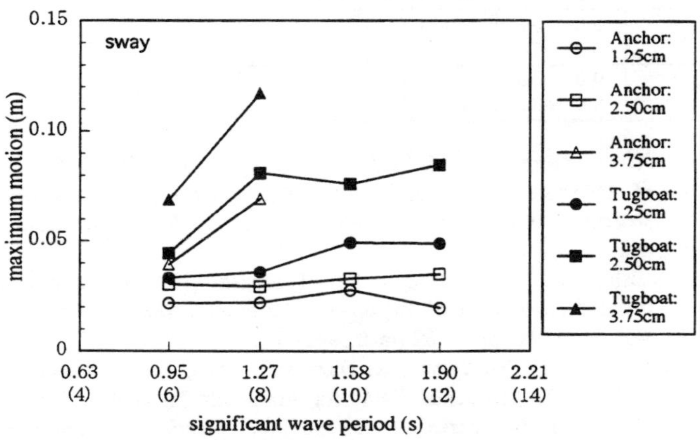

(a) Effect of wave height (θ =90deg)

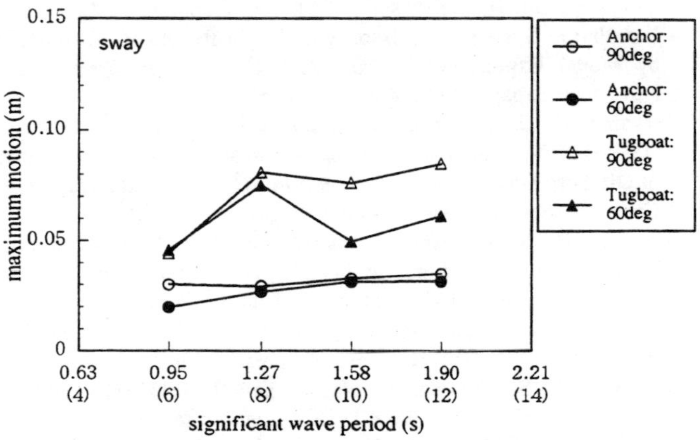

(b) Effect of wave direction ($H_{1/3}$=2.5cm(1.0m))

Fig.7 Maximum sway motion of the caisson in irregular waves

plainly recognized that in all cases, sway motion for the anchor wire system is remarkably smaller than for the tugboat support system. In other words, the anchor wire system can successfully reduce sway motion during caisson installation compared with the tugboat support system. The main reason for this is considered in the following way. For the anchor wire system, the confining forces by the four anchored wires can reduce the caisson motion. However, for the tugboat support system, since wire tension from three stands imitating tugboats is not enough large to control the caisson motion, caisson installation is not stable and the caisson motion becomes large compared with the anchor wire system.

Figure 7 (b) represents the effects of wave direction on the sway motion of the caisson. It is clear that in the anchor wire system, differences in sway motion with respect to wave direction do not appear for all wave conditions. Meanwhile, the sway motion of the tugboat support system is larger for a wave direction of 90deg than 60deg in long wave periods of 1.58s (10s in the prototype) and 1.90s (12.0s in the prototype). Moreover, it is apparent that sway motion for the anchor wire system is much smaller than for the tugboat support system. Accordingly, it is concluded that as for the anchor wire system, wave direction has no influence on the behavior of sway motion in the wave conditions adopted in this experiment, and that simultaneously the anchor wire system can be used to install the caisson more stably than the tugboat support system in the same way as shown in the previous figures.

Installation Errors

As previously mentioned, after the caisson was completely installed in the model experiments, installation errors were measured using a scale. Tables 3 and 4 show the installation errors obtained in irregular waves for the anchor wire system and the tugboat support system, respectively. According to the coordinate system illustrated in Fig.4, the installation errors in the normal line (X-direction) and perpendicular to the normal line (Y-direction) are represented as ΔX and ΔY, respectively. The experiments in irregular waves were carried out twice under identical wave conditions.

It is clear from Table 3 that there is no definite relationship between the installation errors and significant wave heights and periods for the anchor wire system. In other words, some installation errors are small even for large wave heights or long wave periods, when the errors would intuitively be thought to become large. In addition, the allowable range of installation error is approximately 0.2m to 0.5m for current caisson installation work. The range is equal to 0.5cm to 1.25cm in the model scale. Therefore, it is confirmed that by using the anchor wire system the caisson can be installed at the target position within the allowable range of installation error for all experimental wave conditions. On the other hand, for the tugboat support system as shown in Table 4, the installation errors are small in short wave periods or small wave heights. However, the errors tend to become large in long wave periods or large wave heights, and also frequently exceed the allowable limitation. It is therefore concluded from these results that the anchor wire system can be used to install the caisson with better accuracy than the tugboat support system.

Table 3 Installation errors for the anchor wire system in irregular waves

irregular wave

wave angle (deg)		90						60					
wave height (cm)		1.25 (50.0)		2.50 (100.0)		3.75 (150.0)		1.25 (50.0)		2.50 (100.0)		3.75 (150.0)	
installation error (cm)		ΔX	ΔY	ΔX	ΔY	ΔX	ΔY	ΔX	ΔY	ΔX	ΔY	ΔX	ΔY
wave period (s)	0.95	0.0	-0.2	0.0	-0.8	0.0	+0.1	0.0	0.0	0.0	-0.7	0.0	-0.5
	(6.0)	0.0	-0.3	0.0	-1.0	0.0	-0.8	0.0	+0.4	0.0	-0.7	0.0	-0.2
	1.27	0.0	-0.5	0.0	-0.8	0.0	-0.5	0.0	0.0	0.0	-0.9	+0.3	-1.0
	(8.0)	0.0	-0.9	0.0	-0.4	0.0	-0.2	0.0	+0.5	0.0	+0.1	0.0	-0.1
	1.58	0.0	-0.9	0.0	-0.7			0.0	-0.1	0.0	-0.3		
	(10.0)	0.0	-1.0	0.0	-0.6			0.0	+0.5	0.0	0.0		
	1.90	0.0	-0.6	0.0	-1.0			0.0	-0.6	0.0	-0.8		
	(12.0)	0.0	-0.8	0.0	-0.8			0.0	+0.4	0.0	-0.7		

Table 4 Installation errors for the tugboat support system in irregular waves

irregular wave

wave angle (deg)		90						60					
wave height (cm)		1.25 (50.0)		2.50 (100.0)		3.75 (150.0)		1.25 (50.0)		2.50 (100.0)		3.75 (150.0)	
installation error (cm)		ΔX	ΔY	ΔX	ΔY	ΔX	ΔY	ΔX	ΔY	ΔX	ΔY	ΔX	ΔY
wave period (s)	0.95	0.0	-0.5	0.0	0.0	0.0	+1.0	0.0	+1.1	0.0	0.0	+0.5	+1.6
	(6.0)	0.0	-0.4	0.0	-0.3	0.0	+1.5	0.0	+0.8	0.0	0.0	+0.4	+0.5
	1.27	0.0	+0.5	0.0	-0.5	-	-	0.0	+0.3	0.0	0.0	0.0	+1.2
	(8.0)	0.0	0.0	0.0	+2.4	0.0	+1.8	0.0	-0.5	0.0	+0.8	+0.3	+2.2
	1.58	0.0	+0.3	0.0	+0.3			0.0	-0.6	0.0	+0.4		
	(10.0)	0.0	+0.6	0.0	+2.0			0.0	0.0	+0.3	+0.6		
	1.90	0.0	+0.3	0.0	+1.0			0.0	0.0	+0.5	+0.3		
	(12.0)	0.0	+0.6	0.0	+0.6			0.0	+0.5	+0.4	-2.1		

CONCLUDING REMARKS

The authors have developed an automatic installation system for offshore breakwaters that enables the automatic construction of offshore breakwaters with no workers on a caisson and deck of a working craft. The system makes it possible to automate winch operation during caisson installation, saves labor in breakwater construction, and improves worker safety on caissons. The automatic installation system includes programs for two different types of installation methods: a tugboat support system and an anchor wire system. In this paper, the experimental results for the anchor wire system were primarily described, and were compared with those for the tugboat support system. As a result, it was experimentally confirmed that the anchor wire system can be used to install a caisson more stably and accurately than the tugboat support system. The authors are now considering conducting a full-scale model test using an actual caisson, to enable practical use of the system.

ACKNOWLEDGMENT

The authors would like to express their gratitude to the members of the committee of the Japan Dredging and Reclamation Engineering Association for their kind assistance and valuable suggestions concerning this research.

REFERENCES

Shiraishi, S., H. Yoneyama, H. Matsui, S. Ueda, and Y. Okayama (1994). "Experimental Study for Automatic Installation System of Offshore Structures", *Proc. Int. Workshop on Floating Structures in Coastal Zone*, Port and Harbour Res. Inst., pp.430-441.

Shiraishi, S., H. Yoneyama, and H. Takahashi (1995a). "Automatic Installation System of Breakwater Caissons", *The Doboku-seko*, Vol.36, No.10, pp.77-82 (in Japanese).

Shiraishi, S., H. Yoneyama, K. Nazato, S. Ueda, and H. Takahashi (1995b). "Experimental Study for Automatic Installation System of Offshore Structures", *Proc. 20th Joint Meeting UJNR Marine Facilities Panel*, pp.210-217.

Yoneyama, H., S. Shiraishi, K. Nazato, S. Ueda, H. Matsui, Y. Okayama, and H. Takahashi (1995). "Development of Automatic Installation System of Caisson with Tugboats and its Evaluation by Model Experiments", *Proc. 13th Ocean Engi. Symp.*, Soc. Nav. Arch. Japan, pp.459-466 (in Japanese).

Yoneyama, H., S. Shiraishi, K. Nazato, S. Ueda, and H. Takahashi (1996a). "Experimental Study on Automatic Installation System of Offshore Breakwaters", *Proc. 6th Int. Offshore and Polar Engi. Conf.*, Int. Soc. Offshore and Polar Engineers, pp.523-530.

Yoneyama, H., S. Shiraishi, K. Nazato, and H. Takahashi (1996b). "Development of Automatic Installation System of Breakwater Caisson (1st Report) -Evaluation of Tugboat Support System by Hydraulic Model Experiments-", *Report of Port and Harbour Res. Inst.*, Port and Harbour Res. Inst., (in press and in Japanese).

Yoneyama, H., S. Shiraishi, K. Nazato, and H. Takahashi (1996c). "Evaluation of Anchor Wire Type Automatic Installation System for Breakwater Caisson by Hydraulic Model Experiments", *Proc. TECHNO-OCEAN '96 Int. Symp.*, pp.287-292 (in Japanese).

Proceedings of the Seventh (1997) International Offshore and Polar Engineering Conference
Honolulu, USA, May 25-30, 1997

ISBN 1-880653-28-1 (Set); ISBN 1-880653-31-1 (Vol. III)

A Simple Device to Increase the Safety of Vertical Breakwaters

P. De Girolamo and A. Noli
University of Rome "La Sapienza", Rome, Italy
A. Togna
University of L'Aguila, L'Aguila, Italy
M. Pittori
Interprogetti s.r.l., Rome, Italy

ABSTRACT

It is generally accepted that the short-crestedness of wind-generated waves leads to a reduction in the total load exerted by the waves on long structures, compared to the case of long-crested, or unidirectional, waves. Nevertheless, the design of conventional vertical (caisson) deepwater breakwaters is usually carried out assuming long-crested waves because the length of each caisson (20 ÷ 40 m) is generally a small fraction of the wave crest length. In order to take into account the longitudinal wave pressure variability due to wave short-crestedness, a simple device ("spanner"), placed in the concrete crown of the caissons, can be used to connect adjacent caissons. In this paper an application of the spanners is presented and the advantages obtained with this solution are analyzed.

KEY WORDS: breakwaters, caisson, vertical, safety, short-crestedness, waves, spanner.

INTRODUCTION

In the last twenty years a growing attention has been devoted by researchers to the study of wind-generated waves. A particular effort has been given to the prediction of wave groups which are very important for the design of maritime structures because the highest waves of a sea state develop within them. An important characteristic of wind-waves groups is their three-dimensional shape and in particular their short-crestedness, or the finite length of the wave crest. This generally leads to a reduction of the total load exerted by the waves on structures, compared to the case of long-crested, or unidirectional, waves (Battjes, 1982; Boccotti, 1984, 1988; Boccotti et al., 1992; Franco et al., 1996). This reduction can be significant for long structures such as pipelines, storm surge barriers or floating structures. Therefore savings in their construction costs can be achieved by taking into account the finite length of the wave crest. Nevertheless, for the design of caisson breakwaters the incident waves are usually considered as long-crested, because the longitudinal dimension of the caissons is normally a small fraction of the wave crest length.

Today most vertical breakwaters are made by reinforced concrete caissons filled by sand or grouted rubble. The caissons are typically 20÷40 m long and are separated by joints with a thickness of 5÷10 cm, which are filled with plastic material after settlements. The role of the joints is to waterproof the connections between the caissons. The stability analysis of each caisson (e.g., against sliding and overturning) is usually carried out by assuming that the caissons are isolated (i.e., no load is transferred by the joints) and that the load acting on an upright section can be considered constant along the caisson.

In order to introduce in the caisson design the longitudinal wave pressure variation due to the short-crestedness of the highest waves, two different approaches can be followed.

The first one simply consists in building longer caissons. This solution has been adopted for example in Japan were 100 m long caissons have been built. This solution has two drawbacks. The length of the caissons is mostly determined by the maximum allowable size in the caissons manufacturing yard which makes this solution not a general one. Furthermore, in the case of soft soil conditions, differential settlements of the foundation give rise to increasing stress inside the structure as the caisson length increases.

The second solution consists in connecting the caissons only along the longitudinal direction, thereby leaving to each caisson the possibility of small vertical movements necessary to follow differential settlements of the foundation. The present work aims at presenting a simple device, hereafter called "spanner", which has been conceived for this purpose.

It should be mentioned that a number of vertical breakwaters have been subjected, without being damaged, to the action of high waves through the effect of the mentioned waterproofing joints which were provided with a resistance well beyond the supposed purposes. The peculiarity of the present proposal lies in determining the exact function of the spanners and in the fact that they are located in a part of the structure which is easily accessible for maintenance work.

Furthermore the proposed spanners can be easily installed also in existing structures in order to increase their safety.

In the following, section 2 describes the spanners and a general methodology which can be used in order to compute the safety factors of the caissons connected by the spanners. Section 3 presents the computation of the forces acting on the spanners. In section 4 some examples of application are described. The last section is devoted to discussion and concluding remarks.

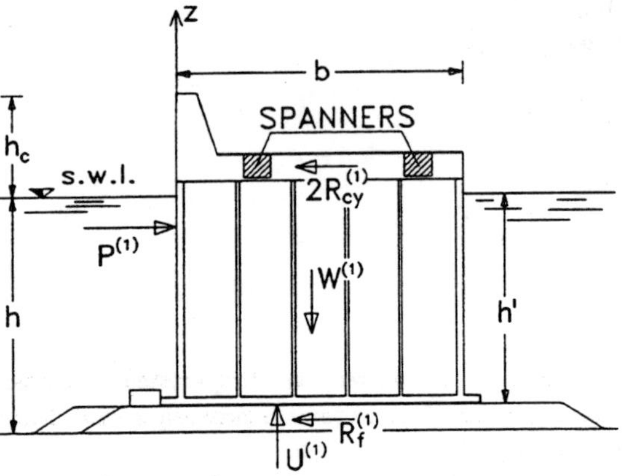

Fig 1 Loads acting on the most stressed caisson.

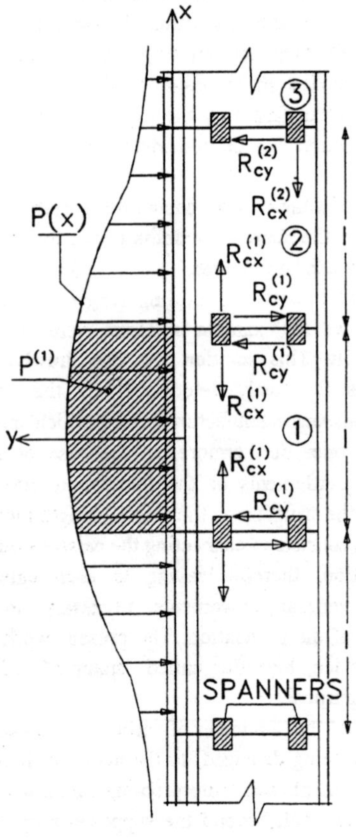

Fig. 2 Forces acting on the spanners.

THE SPANNERS

The spanners (see Figs. 1 and 2) must be of simple conception and, in case of damage, their replacement must be easy. Taking into consideration a caisson in sliding conditions, their main function is to transfer to the adjacent caissons the excess of horizontal wave load which is not balanced by the friction force of the sliding caisson. Nevertheless, the spanners may be subjected also to undesired stresses caused by differential settlement between adjacent caissons and by periodic oscillations of the caissons due to wave actions and spring effect of rubble mound and soil foundation. These undesired stresses, which are expected to be more relevant during the first years after the construction when the consolidation process of the foundation takes place, may cause the damage of the spanners reducing their future resistance. Therefore it is very important to locate the spanners in a part of the structure which is easily accessible to carry out maintenance work and the possibility of their replacement during the life time of the structure must be foreseen.

On the basis of these considerations, the spanners showed by Fig. 3 can be adopted. The spanners are made of reinforced concrete and are placed in seats of rectangular cross-section obtained in the concrete crown of the caissons as indicated by Figs. 1 and 2. A teflon sheet can be installed between the spanner and the seat in order to reduce the contact friction. The spanner reinforcement should be appropriately distributed in the concrete and, using for example crossed bars, the spanner should behave as a hinge also in presence of cracks in the concrete. The hinge function of the spanner, is pointed out by the presence of a section of minimum resistance as indicated in the scheme of Fig. 3.

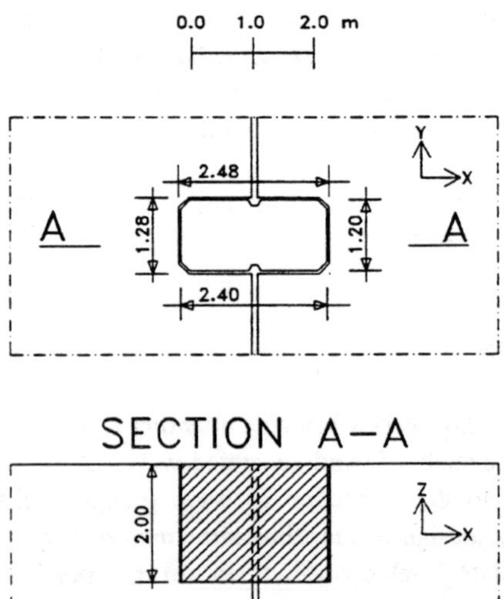

Fig. 3 Plan and cross-section of a spanner.

Fig. 2 shows the longitudinal variability of the wave load acting on a (deep) vertical breakwater when the waves are short-crested and the maximum expected wave height (H_{max}) takes place in the point

$x = y = 0$ (the vertical wall is placed along the line $y = 0$). The main function of the spanners develop when a caisson begins to slide under wave action. In this case the spanners of the most stressed caisson (e.g., caisson 1 of Fig. 2) transfer to the adjacent caissons only the excess of horizontal wave load which is not balanced by the friction force between the upright section and the rubble mound foundation. In turn, the spanners of the adjacent caissons may transfer the excess of horizontal force to the next caissons until a global equilibrium condition has been reached. The maximum number of caissons which can be mobilised by the spanners, "caisson array", depend on the spanner shearing strength.

Provided that the spanner shearing strength is known, the stability safety factors of the caissons array can be computed using the following methodology:

1) Computation of the maximum expected wave pressure acting on the vertical wall using existing formulae (e.g., Goda, 1985);
2) Computation of the longitudinal variability of the maximum expected wave pressure using the Boccotti (1988) or Battjes (1982) approaches;
3) Computation of the shear forces acting on the spanners;
4) Determination of the effective length of the caisson array;
5) Computation of the safety factors against sliding and overturning of the caisson array.

In the following, the Goda (1985) and the Boccotti (1988) approaches are used.

COMPUTATION OF WAVE PRESSURE AND FORCES

Provided that the highest wave in the design sea state (H_{max}) occurs in the point $x = y = 0$ and indicating with $H(x)$ the wave height variation along the longitudinal axis of the wall [$H(x)=H_{max}$ for $x = 0$], the wave induced pressure in the front of the vertical wall can be expressed by the relation:

$$p(x,z) = p[H(x),z] \quad (1)$$

where:

$$H(x) = H_{max}\, f(x) \quad (2)$$

Expression (1) can be evaluated using the Goda (1985) formulae, while the function $f(x)$, which ranges between one and zero, can be evaluated using the Boccotti (1988) theory (see Appendix A). For example, the function f has been represented in Fig. 4 versus the dimensionless abscissa x/L_p (L_p is the wave length characterized by the peak spectral period). The two curves n=5 and n=10 (n is the directional spreading parameter used in expression A.5 of Appendix A), have been obtained assuming a water depth h=19 m, a peak period T_p=10 s and a principal wave direction of incoming waves θ_0=0° (perpendicular to the vertical wall). Fig. 4 shows that, when the maximum wave height occurs in the point x/L_p=0, the length of the wave crest depends on the directional spreading parameter, but typically it can be assumed as $2 \div 2.5\ L_p$. This result has been verified experimentally by Boccotti et al. (1992).

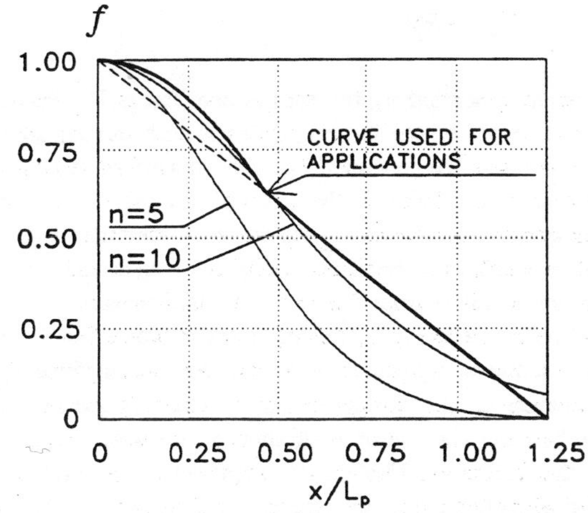

Fig. 4 Wave height variation in front of the verical wall in the case of waves impinging perpendicularly the wall.

With reference to Fig. 2 and assuming a sliding condition of one or more caissons, the computation of the forces transmitted by the spanners is straightforward. Assuming that the spanners are able to transmit only shear and compressive forces and that each pair of spanners equally bears the total shear force R_{cy}, the translational equilibrium condition along the y axis gives for the most stressed caisson (see caisson 1 of Fig. 2):

$$R_{cy}^{(1)} = \frac{1}{2}\left(P^{(1)} - R_f^{(1)}\right) \quad (3)$$

where:

$$P^{(1)} = \int_{-l/2}^{+l/2} \int_{-h'}^{\eta_{max}} p(x,z)\, dxdz \quad (4)$$

is the total horizontal wave load acting on caisson 1; η_{max} is the minimum value between h_c (see Fig. 1) and the maximum elevation to which the wave pressure is applied and

$$R_f^{(1)} = \left(W^{(1)} - U^{(1)}\right)\mu \quad (5)$$

is the total friction force acting on caisson 1. Furthermore, μ is the friction coefficient between the upright section and the rubble mound, $W^{(1)}$ is the weight of caisson 1 in still water and $U^{(1)}$ is the uplift force. If also caisson 2 (see Fig. 2) reaches sliding condition, the shearing force exchanged between caisson 2 and caisson 3 is:

$$R_{cy}^{(2)} = P^{(2)} + R_{cy}^{(1)} - R_f^{(2)} \quad (6)$$

where $P^{(2)}$ is the total horizontal wave load acting on caisson 2 and $R_f^{(2)}$ is the total friction force. Expression (6) can be generalised if the *i*-th caisson is in sliding condition:

$$R_{cy}^{(i)} = P^{(i)} + R_{cy}^{(i-1)} - R_f^{(i)} \qquad (7)$$

The moment generated by the compressive forces R_{cx} balances the moment generated by the shearing forces in each caisson. Neglecting the friction resistance generated between the upright section and the rubble mound foundation in the rotation of a caisson around the *z* axis, the compressive forces R_{cx} can be easily computed applying the translation equilibrium condition along the *x* axis and the rotation equilibrium condition around the *z* axis for each caisson.

With reference to Fig. 1, it is important to notice that the distance between the line of application of the spanner shearing force $R_{cy}^{(1)}$ and the foundation of the upright section is generally greater than the distance between the line of application of the wave horizontal load $P^{(1)}$ and the foundation. This leads to improve the equilibrium rotation condition along the *x* axis for caisson 1 and consequently to improve the distribution of the load on the foundation with respect to the case of the isolated caisson. In spite of this improvement, there is a worsening of the stability conditions for caisson 2 which contributes to redistribute the load along all the structure.

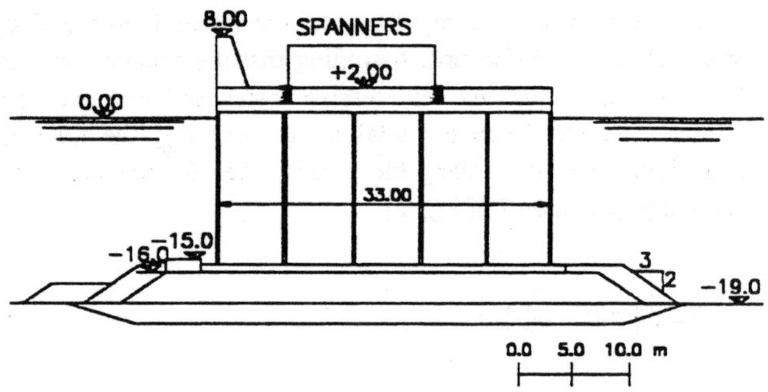

Fig. 5 Cross- section of the single caisson.

EXAMPLE OF APPLICATION

Fig. 5 shows the cross section of a caisson designed to resist (without spanners) to an incoming long-crested sea state characterized by a significant wave height H_s=6 *m* (H_{max}=1.8 H_s) and a peak period T_p=10 *s*. The caisson, hereafter called single caisson, has the following characteristics: length 27 *m*, width 33 *m*, dry density of the fill material γ_d=16.6 *kN*. The assumed safety factors against sliding and overturning are k_s=1.5 (sliding) and k_o=2.0 (overturning). Furthermore, the value μ = 0.6 has been considered for the friction coefficient. It is important to note that the assumed design parameters (e.g., safety factors) are not important for the description of the present design method.

Provided that the caissons are connected by spanners characterized by an ultimate shearing strength of 10000 *kN* (20000 *kN* for a couple of spanners), it is possible to determine the wave height increment which can be held up by the structure taking into consideration the short-crestedness of the incoming waves. In order to simplify the computation and with reference to Fig. 4, the function *f* has been approximated by the curve:

$$f = 1.-1.7458\left(\frac{x}{L_P}\right)^2 \quad \text{for} \quad \frac{|x|}{L_p} \leq 0.45824 \qquad (8)$$

$$f = 1.-0.8\frac{|x|}{L_p} \quad \text{for} \quad \frac{|x|}{L_p} > 0.45824 \qquad (9)$$

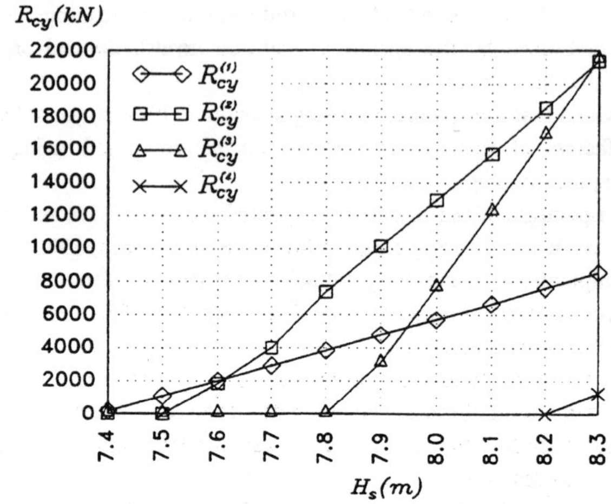

Fig. 6 Total shear force acting on each pair of spanners versus incoming H_s.

The determination of the caisson array length can be carried out using the diagram of Fig.6, where the total shearing force supported by each couple of spanners has been represented versus the incoming H_s. As shown in the figure, with an incoming $H_s \cong$ 8.25 *m*, the spanners connecting caissons 2 and 3 and caissons 3 and 4 ($R_{cy}^{(2)}$ and $R_{cy}^{(3)}$), reach their ultimate shearing strength (20000 *kN* for each couple) causing, with their failure, the interruption of the longitudinal continuity of the structure. When it happens, all the resistance of caisson 4 is mobilised because, starting from $H_s \cong$ 8.2 *m* some load is transferred from caisson 4 to caisson 5 by the spanners (see $R_{cy}^{(4)}$ in Fig. 6). Therefore it can be assumed that the maximum number of caisson which can be mobilised by the spanners is seven. Fig 7 shows the variation of the sliding safety factor k_s of the caisson array made by 7 caissons, versus the incoming H_s. In the same figure the variation of k_s for the single caisson is also represented. Looking at the figure, it can be seen that, for k_s= 1.5 the significant wave height increment obtained with the spanners is about 1.0 *m*. Almost the same result is given by Fig. 8 where the variation of the overturning safety factor k_o versus H_s is shown both for the caisson array and for the single caisson. It should be observed that by maintaining the same safety factors of the single caisson for the caisson array, the spanners never become operative during the life time of the structure (compare Figs. 7 and 8 with Fig. 6). Their function is just to guarantee a reserve of resistance of the structure.

Another application has been carried out in order to determine the volume reduction of the structure obtained with the application of the spanners, compared to the case of the single caisson. In this case, assuming the same design wave characteristics and safety factors used for the single caisson, the width *b* of the caisson has been determined.

The obtained result, b=25.7 m, shows that the application of the spanners give rise, in the present case, to a reduction of about 22% of the total volume of the structure.

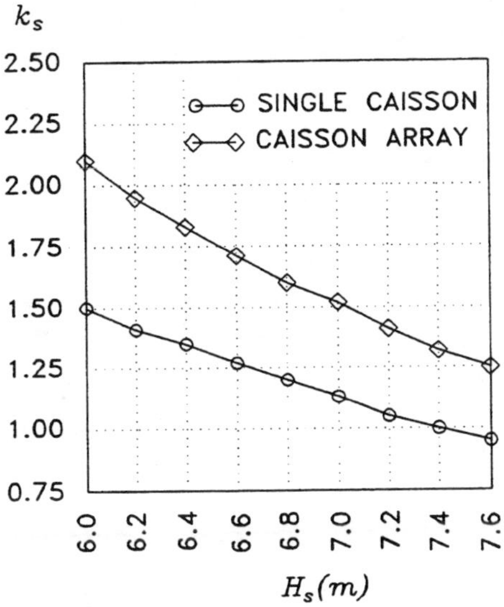

Fig. 7 Variation of the sliding safety factor versus H_s.

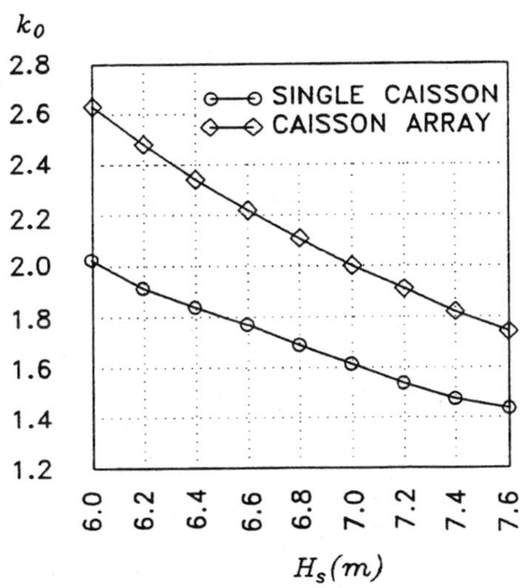

Fig. 8 Variation of the overturning safety factor versus H_s.

DISCUSSION AND CONCLUDING REMARKS

Caisson breakwaters are widely used in Japan and in Mediterranean Sea (mostly in Italy, see Franco, 1994), while harbour engineers of many countries are still rather doubtful about these structures, preferring traditional rubble mound breakwaters. Nevertheless, on the basis of the present knowledge it is possible to affirm that the uncertainties of rubble mound design criteria are still much more numerous and important than those affecting the vertical breakwaters design. On this matter it should be mentioned the reliability of Goda (1985) wave pressure formulae valid both inside and outside the surf zone. Furthermore, there is no information concerning total failures of caisson breakwaters also if in some cases horizontal caisson displacements have occurred owing to an incorrect evaluation of the design waves. Vertical breakwaters must be considered as a valid alternative to rubble mound structures because, respect to the latter, they are less costly in deep water, require less construction time and simpler maintenance and fundamentally they are safer if correctly designed. The present proposal aims to increase farther the safety of vertical breakwaters making use of simple devices and of recent results obtained in the hydraulic description of short-crested random sea waves.

The suggested device, besides allowing the reduction of the caisson width compared to the case of the single caisson, offers a number of advantages:

- it leaves to each caisson the possibility to follow, independently from the adjacent caissons, differential settlements of the foundation;
- it is located in a part of the structure which is easily accessible to carry out maintenance work. Therefore, it can be installed also in existing structures in order to increase, with limited costs, their safety;
- it improves the distribution of the loads on the foundation of the more stressed caisson and improves its overturning stability.

Finally it should be observed that the main function of the spanners is to guarantee a reserve of resistance of the structure. They may be designed by imposing that, under the design wave load, they never become operative during the life time of the structure.

ACKNOWLEDGEMENT

This work was founded by the Italian Ministry of University and the Scientific and Technological Research (M.U.R.S.T.).

REFERENCES

Boccotti, P (1984). "Sea waves and quasi determinism of rare events in random processes," *Proc Acc Naz Lincei, Rendiconti,* Vol 76, No 2, pp 119-127.

Boccotti, P (1988), "Refraction, reflection and diffraction of irregular gravity waves," *Excerpta of the Italian Contribution to the field of Hydraulic Engineering,* GNI, Padova (Italy): Libreria Progetto, Vol. 3, pp 47-89.

Boccotti, P, Barbaro, G and Mannino, L (1992). "An experiment at sea on mechanics of the wave groups," *23rd Int Conf on Coastal Engineering, ASCE,* Venice, Italy.

Battjes, JA (1982). "Effect of short-crestedness on wave loads on long structures," *Applied Ocean Research,* Vol 4, No 3, pp 165-172.

Franco, C, Van der Meer, JW and Franco L (1996). "Multidirectional wave loads on vertical breakwaters," *25th Conf. on Coastal*

Engineering, ASCE, Orlando, US.

Franco, L (1994). "Vertical breakwaters: the Italian experience," *Coastal Engineering*, Vol 22, pp 31-45.

Goda, Y (1985). "Random seas and design of maritime structures," *University of Tokyo Press*, Tokyo, Japan.

APPENDIX A

Assuming a sea state characterized by the principal wave direction θ_o and by the peak frequency $\omega_p = 2\,\pi / T_p$, the general theory of quasi-determinism of the highest waves within gravity irregular fields derived by Boccotti (1984) allows the computation of the function $f(x)$ in front of a vertical reflecting wall of infinite length (Boccotti, 1988) placed along the line $y=0$:

$$f(x) = \frac{\displaystyle\int_0^\infty \int_{\theta_o-\frac{\pi}{2}}^{\theta_o+\frac{\pi}{2}} \tilde{S}(\omega,\theta)\left\{\cos\left[\bar{k}_\omega \operatorname{sen}\theta \frac{x}{L_p}\right] - \cos\left[\bar{k}_\omega \operatorname{sen}\theta \frac{x}{L_p} - 2\pi\omega\left(-\frac{T^*}{T_p}\right)\right]\right\} d\theta\, d\omega}{\displaystyle\int_0^\infty \int_{\theta_o-\frac{\pi}{2}}^{\theta_o+\frac{\pi}{2}} \tilde{S}(\omega,\theta)\left[1-\cos\left(2\pi\omega\frac{T^*}{T_p}\right)\right] d\theta\, d\omega} \qquad (A.1)$$

where:

$$\bar{k}_\omega = \frac{2\pi\omega^2}{\tanh\left(\bar{k}_\omega \dfrac{h}{L_p}\right)} \qquad (A.2)$$

$$L_p = \frac{2\pi g}{\omega_p^2} \qquad (A.3)$$

and T^* is the abscissa of the absolute minimum of the function:

$$f(T) = \int_0^\infty \int_{\theta_o-\frac{\pi}{2}}^{\theta_o+\frac{\pi}{2}} \tilde{S}(\omega,\theta)\,\cos\left(2\pi\omega\frac{T}{T_p}\right) d\theta\, d\omega \qquad (A.4)$$

Assuming a JONSWAP frequency spectrum with $\cos^{2n}$ as directional spreading function, $\tilde{S}(\omega,\theta)$ becomes:

$$\tilde{S}(\omega,\theta) = \omega^{-5}\exp\left[-\frac{5}{4}\omega^{-4}\right]\exp\left\{\ln(\gamma)\exp\left[-\frac{(\omega-1)^2}{2\sigma^2}\right]\right\}\cos^{2n}(\theta-\theta_o) \qquad (A.5)$$

in which parameters γ and σ determine the spectrum shape (values assumed for the present work: $\gamma=3$ and $\sigma=0.08$). The function $f(x)$ assumes its maximum value, $f(x)=1$, for $x=0$ and tends to zero as x tends to $\pm\infty$. It should be noted that the function $f(x)$ is symmetric respect to the point $x=0$ only when the waves approach the wall perpendicularly ($\theta_o=0$). If $\theta_o \neq 0$ $f(x)$ is asymmetric. Furthermore it is observed that using the Boccotti theory it is possible to compute the function $f(x)$ also for vertical reflecting walls of finite length (e.g., see Boccotti, 1988).

Proceedings of the Seventh (1997) International Offshore and Polar Engineering Conference
Honolulu, USA, May 25-30, 1997

ISBN 1-880653-28-1 (Set); ISBN 1-880653-31-1 (Vol. III)

A Parametric Study on the Dynamic Response of Vertical Breakwaters Under Impulsive Wave Loads

E. Benassai, M. Calabrese and F. Silvestri
Università di Napoli Federico II
Naples, Italy

ABSTRACT

A dynamic analysis method, based on the equilibrium of a two-degree of freedom system, was used to simulate the behaviour of a vertical breakwater under breaking wave impact loads. Numerical analyses were carried out with reference to a model vertical breakwater, with various conditions of subsoil geometry, ground properties, shape and duration of impact wave loads. The results made it possible to assess the degree of importance of some hydrodynamic and geotechnical parameters on the dynamic response of the structure, pointing out the most apparent limits of simplified static approaches.

KEY WORDS: coastal structures, vertical breakwaters, dynamic analysis, breaking waves.

NOMENCLATURE

B	=	caisson width
c_x, c_φ	=	horizontal and rotational damping coefficients
D	=	thickness of the deformable subsoil
D_x, D_φ	=	equivalent damping ratios for horizontal and rotational motion
e	=	void ratio of the soil
F_h	=	horizontal wave thrust
F_u	=	uplift force
G	=	shear modulus of the soil
h_c	=	distance of point of application of the horizontal shear force from the base
k_x, k_φ	=	horizontal and rotational stiffness coefficients
l_G	=	distance of caisson gravity centre from the caisson base
M	=	moment at the caisson base
m	=	total mass of the system
m_c	=	caisson mass
m_g	=	geodynamic mass
m_h	=	hydrodynamic mass
p_a	=	atmospheric pressure
p'	=	mean effective stress in the soil
S	=	stiffness coefficient of the soil
$t_{d,h}$	=	total duration of the horizontal impact force
$t_{d,u}$	=	total duration of the uplift force
t_r	=	rise time of the impact force
T_n	=	natural period of free vibration
V	=	horizontal shear force
x, y	=	horizontal and vertical displacements of the caisson
Θ	=	polar moment of inertia of the caisson
ν	=	Poisson's ratio of the soil
φ	=	rotation of the caisson around the gravity centre

1. INTRODUCTION

Hydrodynamic wave loads on vertical breakwaters can affect the stability of the structure in conditions which lead to impulsive breaking wave pressures. The latter may be generated for particular combinations of wave conditions, bottom slope, berm and vertical wall profiles (Goda, 1985; Allsop et al., 1996; Calabrese et al., 1996).

Previous studies (Oumeraci, 1994; Takahashi et al., 1994; Oumeraci et al., 1995) on the various causes which contribute to the collapse of such monolithic structures have in fact confirmed that the impact loads of breaking waves not only yield localised effects, but can also trigger the failure of the breakwater-foundation system.

This gave rise to studies aimed at improving the reliability of the design procedures of vertical breakwaters, which is one of the main topics of the Research Project "PROVERBS" ('Probabilistic design tools for vertical breakwaters'), included in the European Programme "Marine Advanced Science and Technology" (MAST).

In presence of impulse loads, the motion of the monolithic structure (caisson) not only depends on the attributes of the dynamic excitation, but also on the geometry and inertia of the foundation and superstructure, as well as on the nature and deformability of the supporting ground. Therefore, a reliable prediction of the behaviour of the breakwater requires an analysis of the dynamic response of the caisson; this latter should rely on some geotechnical aspects which have been either unduly simplified or neglected in pseudo-static approaches. In most up-to-date dynamic analysis methods, the breakwater-foundation system is idealised assuming the caisson as a rigid block, the motion of which is expressed by two degrees of freedom in the vertical plane: the horizontal translation, x, and the rocking rotation, φ. The soil-caisson contact is usually supposed purely elastic (e.g. Goda, 1973, 1994; Benassai, 1975), visco-elastic (e.g. Oumeraci and Kortenhaus, 1994) or rigid-plastic (e.g. Oumeraci et al., 1995). The latter approach is suitable to analyse the triggering and the accumulation of permanent sliding displacements (i.e. the so called 'step-wise' failure mode); conversely, the others can predict the forced and free oscillatory motion around the steady-state equilibrium position.

In this paper a visco-elastic reaction model is assumed at the soil-caisson interface. The lumped parameters were assessed by the indications on the dynamic behaviour of a rigid foundation on elastic layer or half-space, given by Gazetas (1991). A parametric study was oriented to verify the influence of the thickness, stiffness and damping of the subsoil, as well as of the wave loading shape, on the oscillatory motion, namely the natural period and the amplitudes of displacement and shear force. The results made it possible to develop some critical considerations on the importance of the subsoil properties in assessing the dynamic response of the system.

2. FORMULATION OF THE MODEL

The analytical prediction of the dynamic response of the wave-breakwater-foundation system can be handled by a model in which the caisson is assumed as a rigid block, the motion of which in the vertical plane is expressed by three degrees of freedom: the horizontal (x) and vertical (y) translation and the rocking rotation (φ).

The vertical motion component y is usually ignored, while the rotation φ is often uncoupled, such as in the model proposed by Goda (1994); this makes it possible to reduce the system equilibrium to a single degree of freedom (SDOF) problem, with considerable analytical ease. In this study, however, sliding and rocking motions were assumed coupled, as in the approach followed by Oumeraci and Kortenhaus (1994).

In the model the dynamic coupling of the caisson with the fluid and ground masses was taken into account introducing two additional terms to the mass (m_c) of the caisson:

- the hydrodynamic mass (m_h), corresponding to the inertial water-caisson interaction;
- the geodynamic mass (m_g), corresponding to the inertial caisson-ground interaction.

The contact of the rigid mass ($m=m_c+m_h+m_g$) with the deformable subgrade (mound + subsoil) was expressed by lumped reactions, corresponding to a horizontal shear force, V, acting at a distance h_c from the base, and a moment, M, at the caisson base. They both result in an elastic and a viscous component, at least until the caisson starts to show plastic (frictional) sliding or non-contacting overturning motion with respect to the base.

Under the wave loads represented by given time-variable horizontal thrust $F_h(t)$ and uplift force $F_u(t)$, the dynamic equilibrium can be expressed in terms of the horizontal displacement $x(t)$ of the gravity centre G and the rotation $\varphi(t)$ around the same point (Fig. 1).

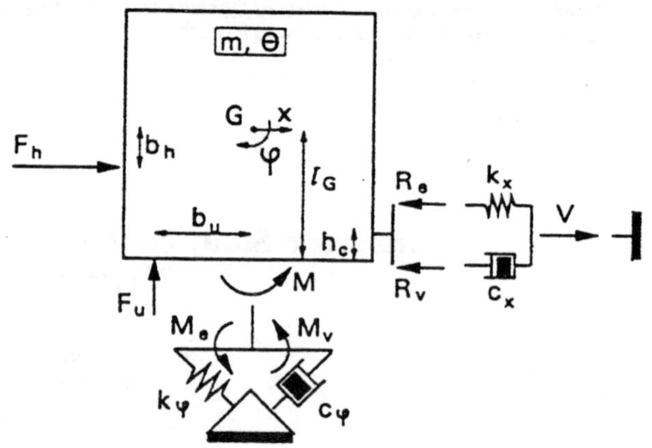

Fig. 1.- Sketch of the visco-elastic dynamic model

If Θ is the polar moment of inertia of the rigid mass around the gravity centre, k_x, k_φ the elastic stiffness and c_x, c_φ the viscous constants, the two equilibrium equations can be assembled into the system:

$$\begin{cases} m\ddot{x} + c_x\dot{x} - c_x(l_G-h_c)\dot{\varphi} + \\ \Theta\ddot{\varphi} - c_x(l_G-h_c)\dot{x} + \left[c_\varphi + c_x(l_G-h_c)^2\right]\dot{\varphi} - \end{cases}$$

$$\begin{cases} +k_x x - k_x(l_G-h_c)\varphi = F_h \\ -k_x(l_G-h_c)x + \left[k_\varphi + k_x(l_G-h_c)^2\right]\varphi = -F_h b_h + F_u b_u \end{cases} \tag{1}$$

which can be numerically integrated by a fourth-order Runge-Kutta algorithm. The output functions $x(t)$, $\varphi(t)$, and their derivatives can yield all the related quantities, such as the accelerations and the shear force $V(t)$, this latter given by the sum of elastic and viscous reactions:

$$V(t) = R_e(t) + R_v(t) = k_x\left[x-(l_G-h_c)\varphi\right] + c_x\left[\dot{x}-(l_G-h_c)\dot{\varphi}\right] = F_h(t) - m\ddot{x}(t) \tag{2}$$

3. INPUT PARAMETERS AND CALIBRATION

In the analysis discussed hereafter, the experimental and numerical results reported by Oumeraci and Kortenhaus (1994) were taken as a reference.+

In order to extend the Authors' sensitivity analyses, analytical expressions for the elastic subgrade stiffness k_x and k_φ were taken from the comprehensive work by Gazetas (1991) on the dynamic equilibrium of rigid block foundations on elastic continua.

The energy dissipation effects beneath the foundation, due to both radiation and internal damping (Dobry and Gazetas, 1985), were synthesized by the viscous constants c_x and c_φ , these latter, respecti-vely, expressed as $2D_x\sqrt{k_x m}$ and $2D_\varphi\sqrt{k_\varphi\Theta}$. According to Marinski and Oumeraci (1992), the values of the damping ratios D_x and D_φ can be assessed through pendulum tests on large-scale physical models. The model response was first calibrated on the same subgrade pattern pertaining to the physical model reported by Oumeraci and Kortenhaus (1994).

The indications by Gazetas (1991) do not provide any analytical expression for the stiffness of a rectangular block foundation on a deformable layer with thickness D; thus, the sliding and rocking stiffness constants k_x and k_φ were expressed by the formulae relevant to a circular base with radius R:

$$k_x = \frac{8GR}{2-\nu}\left(1+0.5\frac{R}{D}\right) \qquad k_\varphi = \frac{8GR^3}{3(1-\nu)}\left(1+0.17\frac{R}{D}\right) \quad (3)$$

in which G and ν are shear modulus and Poisson's ratio of the elastic medium. In eqn.(3) it was assumed $R = 9B/16$ (B is the caisson width), which yields coincidence between the above expressions and those relevant to a rectangular base when the subsoil is a halfspace ($D \rightarrow \infty$).

In the absence of direct experimental determination, the elastic parameters of the soil can be assessed on the basis of its nature, average confining stress p', and void ratio e (e.g. Hardin, 1978). A widespread correlation for the shear modulus G of sands is that originally proposed by Hardin and Richart (1963), confirmed by numerous Authors:

$$\frac{G}{p_a} = S\left(\frac{p'}{p_a}\right)^{0.5}\frac{(2.973-e)^2}{1+e} \quad (4)$$

where p_a is the atmospheric pressure.

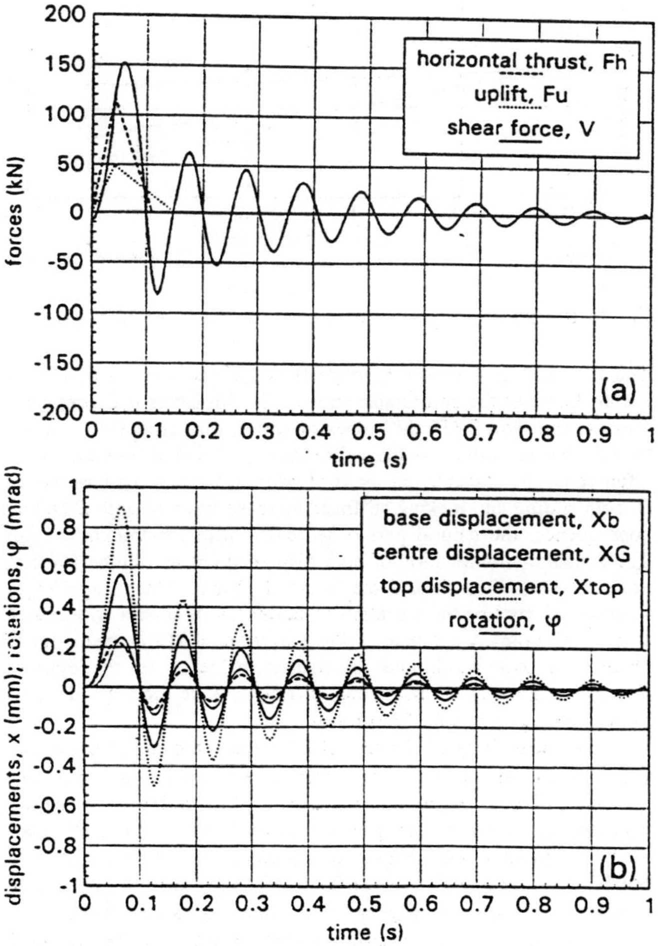

Fig. 2.- Forces (a) and motion (b) data from the analysis of the reference case.

For the sand layer in the reference model (D = 1.40 m), loaded by the caisson's weight, $p' \approx 45$ kPa. Assuming a stiffness coefficient S = 300 (average values for angular grained sands from literature) and $e \approx 1$, $G \approx 40$ MPa. Taking $\nu = 0.25$ (common value for a sand), k_x = 55 kN/m, k_φ = 103 kNm, which are very close to the values quoted by Marinski and Oumeraci (1992). From the pendulum test results reported in the same paper, the values of the damping ratios D_x and D_φ were both assumed equal to 7.5 %.

The horizontal thrust $F_h(t)$ and the uplift force $F_u(t)$ assumed as impact wave loads in the analysis, were idealised as triangular shape pulses with the same rise time (t_r = 0.04 s) but different duration ($t_{d,h}$ = 0.11 s, $t_{d,u}$ = 0.15 s), as shown in Fig. 2a. For plunging breakers with very small or no entrapped air, such a simplification is held not to affect the maximum displacement amplitudes and natural periods.

The output motion for the reference calibration case is represented in Fig. 2b in terms of time histories of displacements (of gravity centre, top and base of the caisson) and rotations. The time history of shear force $V(t)$ computed by eqn. (2) is plotted again in Fig. 2a. It can be noted that all the free vibration curves are in phase and show comparable natural periods (about 0.1 s). The appreciable variation in the displacement amplitudes along the caisson height confirms the importance of accounting for the rotational mode of oscillation in the predictions.

The results of this numerical analysis substantially agree with the test results reported by Oumeraci and Kortenhaus (1994), at least in terms of peak values for horizontal and rotational accelerations (about 1 m/s^2 and 0.6 rad/s^2, respectively).

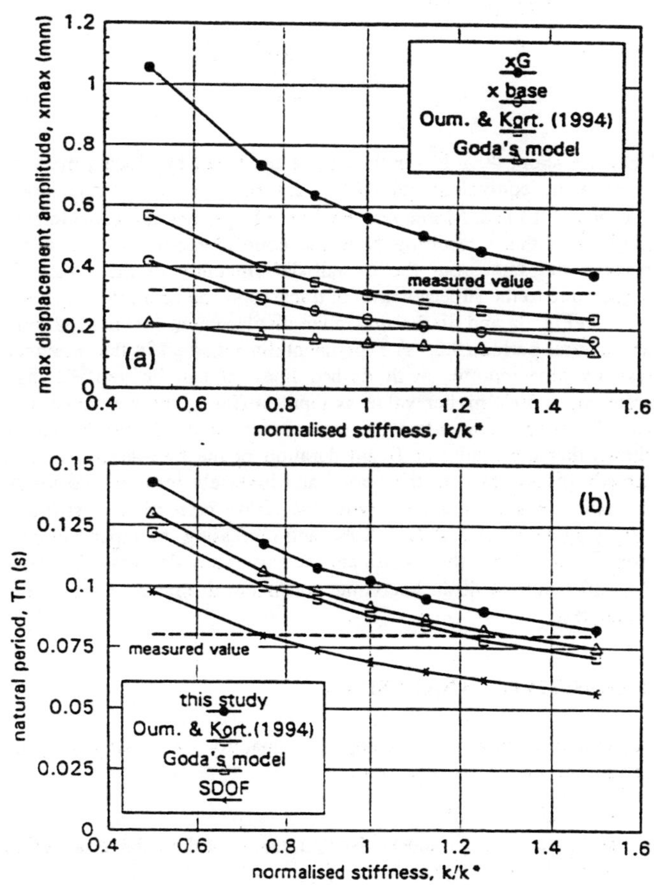

Fig. 3.- Variation of displacement (a) and period (b) with stiffness.

However, the computed natural periods and maximum displacement amplitudes resulted respectively higher for about 20% and 40% than the experimental ones. Also, the whole duration of the measured free vibration decay (about 0.6 s) results about 60% lower than the predicted one (1.0 s). Such differences may be explained, after verifying the influence of some input parameters on the results. At first, the sensitivity of the output to variations of sliding and rocking stiffness was explored in terms of displacement peak amplitude (x_{max}) and natural period (T_n).

This was accomplished by varying the values of G from 20 to 60 MPa, resulting in a ±50% variation of k_x and k_φ yielded by eqn. (3) around the values k^*, corresponding to G^*=40 MPa, assumed in the preliminary verification.

In Fig. 3a the calculated decrease of displacements of gravity centre (x_G) and base (x_{base}) with normalised stiffness (k / k^*) are compared to the data set obtained by Oumeraci and Kortenhaus (1994). Even if the numerical values do not agree, the trends shown appear quite similar; the shape of both curves is hyperbolic, corresponding to a linear increase of displacement with the subgrade deformability ($1 / k$). The chart also shows that the displacements computed by the coupled two-degree of freedom model are higher than those predictable with the method proposed of Goda (1994), in which the base sliding is uncoupled from the rocking motion, and the moving mass is twice that of the caisson, since it includes a geodynamic component of the same amount.

The influence of the stiffness variation on the natural period is shown in Fig. 3b: this time, the results of the coupled model are closer to those by Goda. The figure includes the values predictable for a S.D.O.F. system with mass m ($m=m_c+m_h+m_g$) and stiffness k_x by the simple expression:

$$T_n = 2\pi \sqrt{\frac{m}{k}} \tag{5}$$

They are about 40% lower than those predicted by Goda's method, because the equivalent mass ($2m_c$) assumed in the latter is larger than m ($\approx$1.13 m_c). Moreover, the S.D.O.F. periods are considerably lower than those resulting from the coupled model, in which the mass is the same, but the overall deformability of the system is higher, due to the superposition of translation and rotational modes.

In Fig. 3a and 3b the maximum displacement (about 0.32 mm) and natural period (0.08 s) experimentally recorded in the reference test are also reported with dashed lines; it can be verified that, assuming a 50% higher value as input stiffness, the computed x_{max} and T_n would coincide with the measurements. It can be easily shown that the total significant duration of the free-decay curve is directly proportional to the period and inversely proportional to the damping; as a consequence, with the above correction of stiffness (implying a decrease of T_n to 80%) and increasing the input damping from 7.5% to 10%, the significant decay duration decreases by about 7.5 / 10 * 80% ≈ 60%, becoming consistent to that detected in the model test.

4. SENSITIVITY ANALYSES

A parametric study was developed, to quantify the sensitivity of the predictions on the assessment of:

- layer thickness, D;
- shear stiffness, G, and Poisson's ratio, n, of the subsoil, assumed as an elastic halfspace;
- damping ratio of the soil, D_x and D_φ ;
- shape variations of the loading functions $F_h(t)$ and $F_u(t)$ at constant impulsive effort.

The model output will be hereafter synthesised in terms of:

- normalised natural period, $T_n / t_{d,h}$, where $t_{d,h}$ is the assumed duration of the horizontal impulse;
- normalised maximum base displacement, x_{max} / x_{stat}, representing the dynamic amplification of displacement with respect to $x_{stat}=F_{h,max} / k_x$, i.e. the displacement predictable by a pseudo-static approach referring to the single-degree-of-freedom system;
- normalised peak shear force, $V_{max}/F_{h,max}$.

Broadly speaking, the sensitivities of these two latter ratios are expected to be similar, because the shear force is proportional to the dynamic displacement, as shown by eqn. (2); if the system were purely elastic and with no rocking motion, the maximum shear force V_{max} would be given by $k_x x_{max}$, making the two ratios coincident.

The results of the analysis, of course, must be considered restricted to this case study, but the relative variations can be considered representative of an overall sensitivity of the numerical predictions to uncertainties in the assessment of the fundamental parameters.

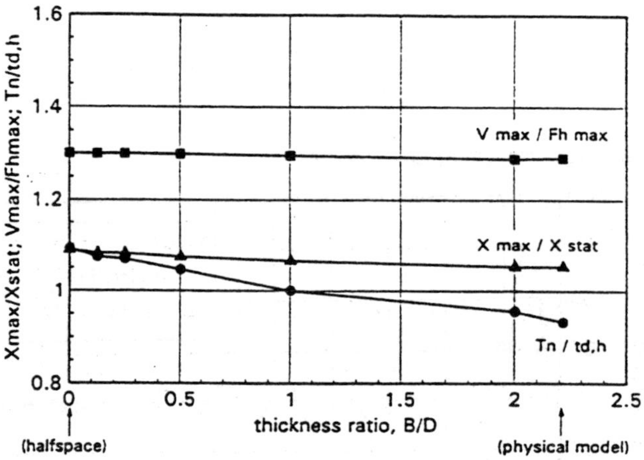

Fig. 4 - Results of the sensitivity analysis on the layer thickness.

The chart in Fig. 4 shows the dependence of the above variables on the thickness of the deformable subsoil, D; this latter is expressed in terms of the ratio B / D, the values of which span from 0 (halfspace) to 2.2, corresponding to the calibration problem discussed above. After eqns. (3), it can be easily verified that, for a given base width, B, both sliding and rocking stiffness increase linearly with B/D; as a consequence, the natural period decreases with the thickness ratio. The dynamic amplification of both displacement amplitude (ranging around 1.1) and the shear force (about 1.3) are, rather surprisingly, scarcely affected by the variation of thickness in such a wide range. It must be pointed out that, if the variation of B / D were due to changes in base width, the subsequent influence on the dynamic response might result considerably different, since the dependence of both k_x and k_φ on B is far from linear.

The sensitivity of the model to the assessment of the elastic parameters G and ν was tested taking the halfspace as a reference geometry, since it represents the most unfavourable deformability condition for a given homogeneous subsoil. As in the preliminary model verification, the stiffness G was assumed to vary for -50% to +50% times the reference value (G^* = 40 MPa); also, the sensitivity to Poisson's ratio ν was explored throughout the range 0.0÷0.5.

For both G and ν increasing, a stiffer response of the system is expected (see eqn. (3)): therefore, it is not surprising that, in both cases (Figs. 5 and 6), a reduction in natural period is observed.

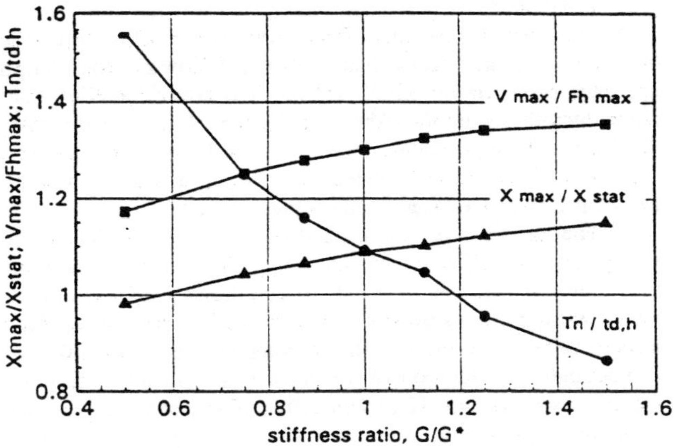

Fig. 5.- Results of the sensitivity analysis on the soil stiffness.

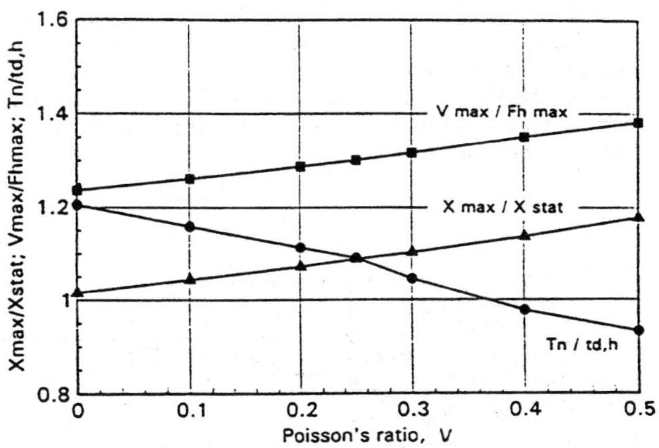

Fig. 6 - Results of the sensitivity analysis on the Poisson's ratio.

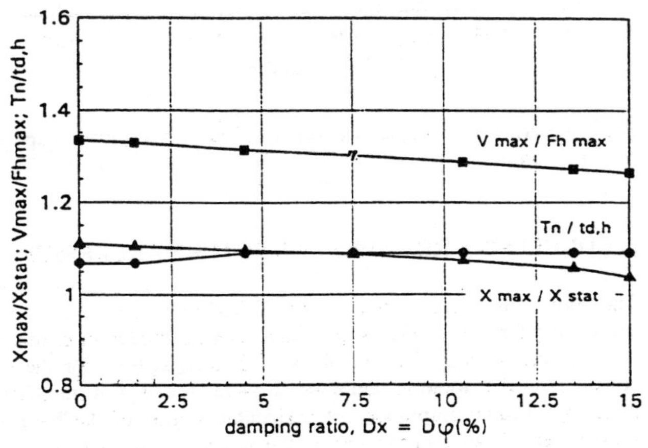

Fig. 7.- Results of the sensitivity analysis on the damping ratio.

On the other hand, the normalised displacement and shear force are seen to increase, showing dynamic effects becoming more significant simultaneously with the decrease of natural period T_n with respect to the load duration $t_{d,h}$.

The sensitivity of the dynamic response on the variability of the damping ratio of the subgrade is outlined in Fig. 7. In this parametric study, the values of the damping ratio were taken as equal for both sliding (D_x) and rocking (D_φ) modes, and varied from 0 to 15%. It must be observed that, even if this is a likely range for a physical model, it might not necessarily be so for a prototype. In fact, the total dissipated energy beneath a real caisson can be ascribed to the superposition of the "internal damping", due to viscous and frictional effects in the rubble mound and the subsoil, and the "radiation damping", due to the scattering of body and surface waves through the ground. The latter phenomena can in effect assume much more importance than what has been actually observed on the physical models by means of the well-known pendulum tests. For instance, when the thickness of the deformable subsoil increases, the radiation damping becomes more and more noticeable, as the waves are no longer reflected from the bedrock back to the foundation. Therefore, the increase of thickness of deformable subsoil should be associated to a proportional variation of the damped energy, and not limited to affect the subgrade deformability only, as was assumed in the parametric analysis discussed above.

Looking at Fig. 7, it appears that both the natural period (slightly increasing) and the amplification factors of base displacement and shear force (slightly decreasing) are scarcely affected by the assessment of damping. It must be verified, however, whether this still holds even for damping values that, for the above reasons, could result higher than those shown.

Another sensitivity analysis was carried out on the variations in shape of the loading functions $F_h(t)$ and $F_u(t)$, with no variation in the geometry and mechanical properties of the breakwater-subsoil system. Assuming a given impulsive effort (i.e. $\int_0^{t_d} F(t)dt$ = constant) and a t_d / t_r ratio fixed at 2.75 for the horzontal and at 3.75 for the uplift forces, the maximum amplitudes of both F_h and F_u reduce as the load duration becomes higher.

The variability of amplification ratios of base displacement and shear force versus normalised impact duration ($t_{d,h} / T_n$) is shown in Fig. 8. The variation of $t_{d,h} / T_n$ is directly proportional to the load duration, being the natural period constant (T_n = 0.12 s) throughout the whole parametric analysis.

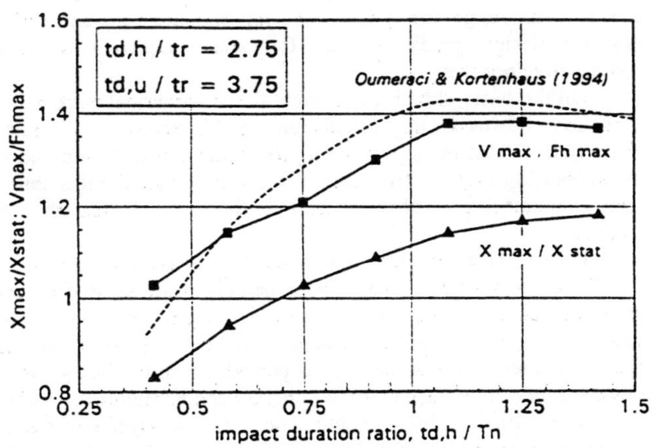

Fig. 8 - Results of the sensitivity analysis on the impact duration.

The natural period of vibration also resulted appreciably higher than those predictable by purely translational approaches, such as the simple single-degree-of-freedom system or Goda's model.

In addition, the coupled sliding-rocking equilibrium of the visco-elastic system was shown to yield non-coincident dynamic amplification factors for displacements and shear forces, the latter being about 20% higher.

The parametric studies showed a particular sensitivity of the model output to the assessment of the elastic parameters (G and ν), which can highly influence the response in terms of both amplification factors and natural period. This calls for more reliable geotechnical characterisation of the subsoil properties (e.g. stratigraphy, shear wave velocity, density) which need to be introduced into the dynamic model.

To fully characterise the model parameters, the total amount of dissipated energy (due to radiation and internal damping) into the supporting ground need to be quantified into comprehensive viscous coefficients. The variation of damping ratio was found of negligible influence on the natural period, and of little effect on the dynamic amplification of displacements and shear forces.

On the other hand, the total duration of the free decay of the caisson motion is inversely proportional to the damping ratio, the assessment of which can affect the vulnerability of a given breakwater to repeated wave loads.

In the sensitivity analyses reported above, the thickness of the subsoil was found to have little influence on the dynamic amplification factors of both displacement and shear forces, even if it affects their absolute amplitudes. Soil layering can also be responsible for significant modifications of the radiation damping, which should be proportional to the thickness of deformable subsoil underneath the caisson.

A further parametric study pointed out the outstanding importance of the impact load duration, which needs to be referred to the natural period in order to assess whether a dynamic analysis is needed to obtain safe predictions.

Broadly speaking, it appears that low duration/period ($t_{d,h}/T_n$) ratios (say, less than 0.5) tend to yield dynamic displacements and shear forces lower than those predictable statically and not affected by the load shape; therefore, in this range the static analysis is conservative and the dynamic design approach more economic.

On the contrary, for high $t_{d,h} / T_n$ ratios (more than 5), the dynamic amplification approaches to unity, and the static analysis becomes reliable.

In an intermediate range (such as $0.5 < t_{d,h} / T_n < 5$), the displacement amplification of the caisson centre can be significantly higher than unity, and the response affected by the impact load shape; in this range, the dynamic analysis should be the only reliable tool for design predictions, as the static approach is neither conservative nor trusty.

On the basis of both experimental and numerical work on such problems, however, the calculation model needs to be further improved, at least by implementing the behaviour of the mound into the supporting system. This objective can be pursued either through more sophisticated subsoil reaction models, or through FEM analyses.

Future developments of this model should also account for non-linear stress-strain behaviour of the soil, resulting in decreasing shear stiffness and increasing internal damping ratio with shear strain level. Soil non-linearity can justify some observed phenomena, such as the reduction of the natural period over time during the free vibration motion of the caisson (Oumeraci and Kortenhaus, 1994).

The final goal should be a two-degree-of-freedom comprehensive model, including visco-elastic reactions bounded by plastic sliding limits. In fact, the purely visco-elastic approach is unable to predict step-wise failure mechanisms due to progressive frictional sliding of the foundation.

As the plots clearly show, the dynamic amplification of both base displacement and shear force increase with comparable trends, featured by gradients that flatten out appreciably as $t_{d,h}/T_n$ becomes greater than unity.

The curve for the shear force factor is very close to that found by Oumeraci and Kortenhaus (1994), following the simplified approach of an elastic SDOF system; it is worth recalling that this latter hypothesis implies the coincidence of this factor with the displacement amplification, whereas in this study it was verified that, overall, the ratio between $V_{max} / F_{h,max}$ and x_{max} / x_{stat} ranges between 1.16 and 1.24, with an average of 1.20.

The last analysis was carried out to verify the dependence of the dynamic amplification of base and gravity centre displacement on the combined variation of the shape and duration of wave impact, at a constant impulse value. For any given load shape, corresponding to a fixed value of the ratio t_d / t_r (varying between 2 and 20) for both components, the duration of the wave loads on the same reference model caisson (T_n = 0.12 s) was increased to obtain impact duration ratios $t_{d,h} / T_n$ spanning from 0 to 2.

Fig. 9 reports the dynamic amplification factors of maximum displacements of base and gravity centre resulting from the analyses. It can be seen that the dynamic factors for the gravity centre are about twice those relevant to the caisson base, for which the amplification is never appreciably higher than unity.

The effect of the load shape is visible only for impact duration ratios higher than 0.25, and becomes negligible when t_d/t_r is higher than 10. When $t_{d,h} / T_n > 1$, the effect of the load shape on the dynamic amplification for the gravity centre becomes noticeable, and for high relative rise times (low t_d/t_r) the ratio x_{max}/x_{stat} is seen to decrease, approaching unity.

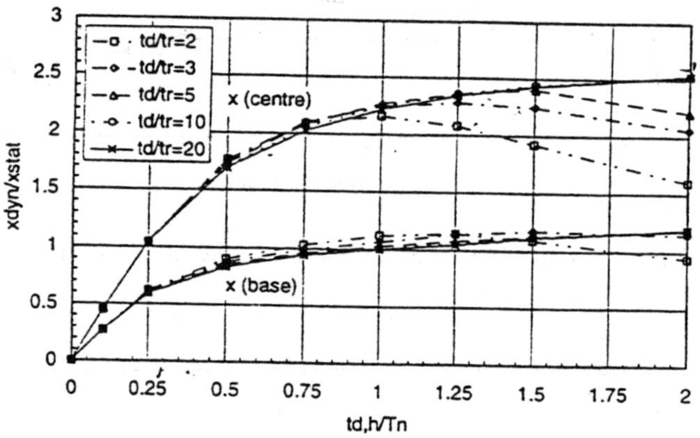

Fig. 9 - Variation of displacement amplification with load shape and duration.

5. CONCLUSIVE REMARKS AND FUTURE ADDRESSES

The above numerical results first of all highlight the importance of accounting for the coupled sliding and rocking modes in predicting the dynamic motion of the breakwater. In fact, at least for the case considered, the rotation amplitudes were found not negligible, implying unequal amplitudes of horizontal motion of the base and gravity centre. As a consequence, the maximum displacement can result significantly higher than that predictable through the model by Goda (1994), in which the dynamic equilibrium is uncoupled, since it is referred to the caisson base.

ACKNOWLEDGEMENTS

The work discussed here was in part supported under the PROVERBS project of the Ec's MAST III programme under contract MAS3-CT95-0041.

The Authors are gratefully indebted to Mr. G. Sorgenti degli Uberti for his essential contribution to the calibration of the model and the numerical analyses. The Authors are also grateful to the colleagues involved in the "PROVERBS" for the fruitful interaction developed, and intend to address this work as a further contribution to the scientific programme.

REFERENCES

ALLSOP N.W.H., VICINANZA D., CALABRESE M. & L. CENTURIONI (1996): "Breaking wave impact loads on vertical faces", *6th Int. Conf. ISOPE*, Los Angeles.

BENASSAI E. (1975): "The stability against sliding of breakwaters under action of breaking waves", *PIANC*, Vol.II n° 21.

CALABRESE M., VICINANZA D. & N.W.H. ALLSOP (1996): "Hydrodynamic loading on vertical faces - Influence of mound geometry", *Contribution to the 2nd MAST III Workshop*, Belfast.

DOBRY R. & G. GAZETAS (1985): "Dynamic stiffness and damping of foundations by simple methods", *Proc. "Vibration Problems in Geotechnical Engineering"*, ASCE, Detroit (Michigan).

GAZETAS G. (1991): "Foundation vibrations"; in *Foundation Engineering Handbook*, II edition, H.Y. Fang Ed., pp. 553-593. Van Nostrand Reinhold.

GODA Y. (1973): "A new method of wave pressures calculation for the design of composite breakwaters", *Rept.Port and Harbour Res. Inst.*, Vol 12, No.3, pp 3-29, Ministry of Transport.

GODA Y. (1985): "*Random seas and maritime structures*", University of Tokyo Press, Tokyo.

GODA Y. (1994): "Dynamic response of upright breakwaters to impulsive breaking wave forces", *Coastal Engineering, Special Issue on Vertical Breakwaters*, Vol 22, pp 135-158, Elsevier Science BV, Amsterdam.

HARDIN B.O. (1978): "The nature of stress-strain behaviour for soils - State of the Art", Proc. *"Geotechnical Engineering Division Specialty Conference on Earthquake Engineering and Soil Dynamics"*, ASCE, Pasadena (California).

HARDIN B.O. & F.E. RICHART (1963): "Elastic wave velocities in granular soils", *Journal of Soil Mechanics and Foundations Division*, ASCE, Vol. 89, No. SM1.

MARINSKI JG & OUMERACI H (1992): "Dynamic response of vertical structures to breaking wave forces - review of the CIS experience", *Proc. MAST G6-S Final Overall Workshop*, Lisbon.

OUMERACI H. (1994): "Review and analysis of vertical breakwater failures - lessons learned", *Coastal Engineering, Special Issue on Vertical Breakwaters*, Vol 22, pp 3-30, Elsevier Science BV, Amsterdam.

OUMERACI H. & KORTENHAUS A. (1994): "Analysis of the dynamic response of caisson breakwaters", *Special issue on vertical breakwaters. Coastal Engineering*, Vol 22, No 1-2, Elsevier Science, Amsterdam.

OUMERACI H., KORTENHAUS A. & KLAMMER P. (1995): "Displacement of caisson breakwaters induced by breaking wave impacts", *Proc. ICE Conference on Coastal Structures and Breakwaters*, London.

TAKAHASHI S. TANIMOTO K. & SHIMOSAKO K. (1994): "Dynamic response and sliding of breakwater caissons against impulsive breaking wave forces", *Proceedings of Workshop on Wave Barriers in Deep Waters*, Port and Harbour Research Institute, Yokosuka, Japan.

Proceedings of the Seventh (1997) International Offshore and Polar Engineering Conference
Honolulu, USA, May 25-30, 1997

ISBN 1-880653-28-1 (Set); ISBN 1-880653-31-1 (Vol. III)

Wave Transmission Past Slotted Barriers

Michael Isaacson, Sundarlingam Premasiri and Gang Yang
University of British Columbia
Vancouver, Canada

ABSTRACT

The present paper outlines the numerical calculation of wave interactions with a thin vertical slotted barrier extending from the water surface to some distance above the seabed, and describes laboratory tests undertaken to assess the numerical model. The numerical model is based on an eigenfunction expansion method and utilizes a boundary condition at the barrier surface which accounts for energy dissipation within the barrier. Comparisons with previous numerical studies for related situations show close agreement and a selection of results based on the method are presented for the transmission, reflection, and energy dissipation coefficients for a partially submerged porous barrier. In addition preliminary comparisons with experimental measurements are given and the choice of suitable friction and added mass coefficients to model the permeability of the breakwater is described.

KEYWORDS: breakwaters, coastal engineering, coastal structures, permeability, wave transmission, waves.

INTRODUCTION

Breakwaters are widely used to provide economical protection from waves in harbours and marinas. In certain situations, breakwaters in the form of thin, rigid, pile-supported vertical barriers which extend some distance down from the water surface have been used or considered. These have the advantages of allowing water circulation, fish passage and sediment transport beneath the breakwater, and may be relatively economical by providing protection closer to the water surface where wave action is most pronounced. Predictions of wave interactions with such structures have been obtained previously by a number of authors for the case of an impermeable barrier on the basis of linear wave diffraction theory (e.g. Liu and Abbaspour, 1982, Losada et al. 1992, and Abul-Azm, 1993).
In some instances, a permeable barrier, such as a slotted vertical barrier made from timber planks, may be preferred. For example, this may be selected in an effort to reduce unwanted wave reflections on the upwave side of the barrier. Thus, the prediction of wave interactions with a permeable or slotted thin vertical barrier is also of interest. Several authors have considered such predictions for permeable structures that extend from the free surface to the seabed. A primary feature of such interactions is that wave energy is absorbed within the structure. Most formulations for permeable structures assume the velocity through the porous medium is proportional to the pressure gradient, with a complex proportionality constant that accounts for possible phase differences between the velocity and pressure gradient (e.g. Sollitt and Cross, 1972; Sulisz, 1985; Yu, 1991). This description can be related to the physics of the flow within the structure on the basis of both frictional and inertial effects. Comparisons with experimental measurements of transmission and reflection coefficients for rubblemound breakwaters generally exhibit satisfactory agreement (e.g. Sollitt and Cross, 1972; Yu, 1995). Related studies include those by Kakuno (1983), Hagiwara (1984) and Bennet *et al.* (1993).

The purpose of the present paper is to summarize a recent extension of an eigenfunction method to the case of a thin permeable barrier which extends from the water surface to some distance above the seabed; and to compare predictions of the method with preliminary experimental results for a slotted breakwater.

EIGENFUNCTION EXPANSION METHOD

A normally incident, regular, small amplitude wave train of height H and angular frequency ω propagates in water of constant depth d past a thin permeable vertical barrier as shown in Fig. 1. The barrier extends downwards a distance h below the still water level. A Cartesian coordinate system (x, z) is defined with x measured in the direction of wave propagation and z measured upwards from the seabed. The fluid is assumed incompressible and inviscid, and the flow irrotational. The fluid motion can therefore be described by a velocity potential Φ which satisfies the Laplace equation within the fluid region. In addition, the wave height is assumed sufficiently small for linear wave theory to apply. Consequently, Φ is subject to the usual boundary conditions, linearized where appropriate, at the seabed, free surface and far field (see, for example, Sarpkaya and Isaacson, 1981), and may thus be expressed, using complex notation, in the form:

$$\Phi(x,z,t) = \mathrm{Re}[C\,\phi(x,z)\exp(-i\omega t)] \tag{1}$$

where

$$C = -\left(\frac{igH}{2\omega}\right)\frac{1}{\cosh(kd)} \quad (2)$$

Also, Re[ ] denotes the real part of the argument, $i = \sqrt{-1}$, t is time, k is the wave number, and g is the gravitational constant.

The two-dimensional potential ϕ in Eq. 1 is denoted ϕ_1 and ϕ_2 in regions 1 and 2 respectively, corresponding to the upwave and downwave sides of the barrier (see Fig. 1). Along x = 0, the pressure (and hence the velocity potential) and the horizontal velocity are equated along the matching boundary within the fluid, and are applied to a suitable boundary condition along the barrier surface. The conditions along the matching boundary are thus:

$$\left.\begin{aligned}\phi_1 &= \phi_2 \\ \frac{\partial\phi_1}{\partial x} &= \frac{\partial\phi_2}{\partial x}\end{aligned}\right\} \quad \text{for } 0 \le z \le a \quad (3)$$

where a = d - h (see Fig. 1).

The boundary condition along the permeable barrier may be developed on the basis of the formulation of Sollitt and Cross (1972), and as adopted by Yu (1995) for a thin vertical barrier extending to the seabed. This may be expressed as:

$$\frac{\partial\phi_1}{\partial x} = \frac{\partial\phi_2}{\partial x} = -iK(\phi_2 - \phi_1) \quad \text{for } a \le z \le d \quad (4)$$

where K = G/b, b is the barrier thickness and G is a permeability parameter which is generally complex. Equation 4 corresponds to the fluid velocity normal to the barrier being proportional to the pressure difference across the barrier, with a complex constant of proportionality so that the real part of G corresponds to the resistance of the barrier and the imaginary part of G corresponds to the phase differences between the velocity and the pressure due to inertial effects. Various authors have related the permeability parameter, G, to the physics of the flow within the barrier in different ways. In the present paper the method of Sollitt and Cross (1972) is followed and G is expressed by:

$$G = \frac{\varepsilon}{f - i\,s} \quad (5)$$

where ε is the porosity of the barrier (defined as the fraction of area occupied by the slots), f is a friction coefficient and s is an inertia coefficient given by

$$s = 1 + C_m\left(\frac{1-\varepsilon}{\varepsilon}\right) \quad (6)$$

In Eq. 6, C_m is an added mass coefficient. It is noted that although the barrier is taken to have zero thickness with respect to the wave diffraction problem under consideration, the barrier width is considered non-zero with respect to the flow within the barrier. The friction coefficient, f, comes from a linearization of the velocity squared term associated with the head loss across the barrier. In the original formulation of Sollitt and Cross (1972) f is calculated implicitly using the Lorentz principle of equivalent work so that the nonlinear effects of wave steepness are retained. This requires an iterative procedure, and in the present paper the formulation of Yu (1995) is followed such that f is treated simply as a constant which is assumed to be known.

Expressions for ϕ_1 and ϕ_2 which satisfy the seabed, free surface, and radiation conditions, as well as the above conditions along x = 0, may be developed in terms of a single set of coefficients A_m which are initially unknown:

$$\phi_1 = \phi_w - \sum_{m=0}^{\infty} A_m \cos(\mu_m z)\exp(\mu_m x) \qquad (x \le 0) \quad (7)$$

$$\phi_2 = \phi_w + \sum_{m=0}^{\infty} A_m \cos(\mu_m z)\exp(-\mu_m x) \qquad (x \ge 0) \quad (8)$$

Here ϕ_w represents the incident wave potential which is given as:

$$\phi_w = \cosh(kz)\exp(ikx) \quad (9)$$

Also, μ_m for $m \ge 1$ are the positive real roots of the following equation, taken in ascending order:

$$\mu_m d \tan(\mu_m d) = -\frac{\omega^2 d}{g} \qquad \text{for } m \ge 1 \quad (10)$$

μ_0 itself corresponds to the imaginary root of the above equation, such that $\mu_0 = -ik$, with the wave number k being given as the real root of the corresponding equation:

$$kd\tanh(kd) = \frac{\omega^2 d}{g} \quad (11)$$

Thus, Eqs. 7 and 8 each represent the incident wave train combined with a superposition of a propagating mode (m = 0) and a series of evanescent modes ($m \ge 1$) which decay with distance away from the barrier. They satisfy all the relevant boundary conditions, except that the conditions of pressure continuity along the matching boundary and the boundary condition at the barrier surface are still needed to determine the coefficients A_m. These conditions give rise to the following set of equations for A_m:

$$\sum_{m=0}^{\infty} A_m \cos(\mu_m z) = 0 \qquad \text{for } 0 \le z \le a \quad (12)$$

$$\sum_{m=0}^{\infty} A_m(\mu_m - 2iK)\cos(\mu_m z) = -\mu_0\cos(\mu_0 z) \qquad \text{for } a \le z \le d \quad (13)$$

Equations 12 and 13 are integrated in an appropriate way and then added to develop a suitable matrix equation for A_m. Thus, each equation is first multiplied by $\cos(\mu_n z)$, then integrated with respect to z over the appropriate domain of z (i.e. from z = 0 to a, or from z = a to d), and the resulting two equations are then added. This gives rise to a matrix equation for A_m:

$$\sum_{m=0}^{\infty} C_{mn} A_m = b_n \qquad \text{for } n = 0, 1, \ldots, \infty \quad (14)$$

where

$$C_{mn} = f_{mn}(0, a) + (\mu_m - 2iK)\, d\, f_{mn}(a, d) \quad (15)$$

$$b_n = -\mu_0\, d\, f_{0n}(a, d) \quad (16)$$

$$f_{mn}(\alpha,\beta) = \int_{\alpha}^{\beta} \cos(\mu_m z)\cos(\mu_n z)\,dz$$

$$= \begin{cases} \left[\dfrac{\sin(\mu_- z)}{\mu_-} + \dfrac{\sin(\mu_+ z)}{\mu_+}\right]_{\alpha}^{\beta} & \text{for } m \neq n \\ \dfrac{1}{4\mu_m}\left[2\mu_m z + \sin(\mu_m z)\right]_{\alpha}^{\beta} & \text{for } m = n \end{cases} \quad (17)$$

Also $\mu_- = \mu_m - \mu_n$, and $\mu_+ = \mu_m + \mu_n$.

In a numerical solution to the problem, Eq. 14 is truncated to a finite number of terms N, and thus becomes a complex matrix equation of rank N which can be solved for the first N unknown coefficients A_m. Once these have been calculated, the various quantities of engineering interest may readily be obtained. The method is described in more detail by Yang (1996). If required, it may readily be extended in the usual way to the cases of oblique waves and/or irregular waves (e.g. Losada et al., 1992, 1994).

The (real) transmission and reflection coefficients, denoted K_t and K_r respectively, are defined as the appropriate ratios of wave heights: $K_t = H_t/H$ and $K_r = H_r/H$, where H_t and H_r are the transmitted and reflected wave heights respectively. These are given in terms of A_n by:

$$K_t = |1 + A_0| \quad (18)$$

$$K_r = |A_0| \quad (19)$$

From considerations of energy conservation, these are related to the energy dissipation coefficient K_e:

$$K_r^2 + K_t^2 + K_e = 1 \quad (20)$$

where K_e is the proportion of the incident wave energy flux that is dissipated by the barrier.

Various limits to the above solution are known. As $G \to 0$, the solution should correspond to the case of an impermeable barrier. As $kd \to 0$ while $h/d < 1$ (long wave limit), the solution should correspond to the undisturbed incident wave field on either side of the barrier, so that $K_t \to 1$ and $K_r \to 0$. And as $kh \to \infty$ or as $h/d \to 1$ (full barrier extending to deep water or the seabed), the solution should correspond to that of Yu (1995).

DESCRIPTION OF EXPERIMENTS

Experiments were carried out in the wave flume of the Hydraulics Laboratory of the Department of Civil Engineering at the University of British Columbia. The flume is 20 m long and 0.62 m wide. Waves are generated by a single paddle wave actuator located at the upwave end. An artificial beach which is covered by a mat of synthetic hair is located at the downstream end of the flume in order to minimize wave reflection. The vertical barrier to be tested was placed in the test section 10 m from the wave generator. Figure 2 shows a general view of the experimental setup. The permeable wave barrier is constructed of vertical panels, each of width w = 2.0 cm, and thickness, b = 1.3 cm, such that the porosity of the barrier can be varied by changing the dimensions of the slots between the panel members.

Tests were carried out with a constant water depth of 0.45 m and included both regular and irregular waves. Only preliminary results with regular waves are described here. Additional results are reported by Premasiri (1997). Tests have been conducted for barrier dimensions corresponding to half and fully immersed barriers (h/d = 0.5 and 1.0 respectively), porosities of 0.05, 0.1, 0.2, 0.3, 0.4 and 0.5, and for various wave conditions corresponding to a constant wave steepness. In each test, the water surface elevation was measured upwave and downwave from the wave barrier using twin wire capacitance wave probes. Three probes were used on the upwave side of the barrier at distances of 1.0, 1.2 and 1.5 m from the barrier, and two probes were used on the downwave side at distances of 1.0 and 1.2 m from the barrier. These probe positions were selected by considering the evanescent wave modes from the barrier, the wave paddle location and the method of reflection analysis.

Wave generation and data acquisition were controlled by a VAX workstation, using the GEDAP (General Experimental control, Data acquisition and Analysis Package) software package developed at the Coastal Engineering Laboratory of the National Research Council, Canada. The wave probes were sampled at a rate of 20 Hz for a duration of 14 sec for each test run. The wave records obtained from the probes were used to estimate the reflection and transmission coefficients for each test using simultaneous measurements of water surface elevation upwave and downwave of the barrier based on a least squares method (e.g. Isaacson, 1991). The incident wave height was obtained from measurements without the barrier in place.

NUMERICAL RESULTS

The numerical method has been validated by comparison with previous predictions for the limiting cases of an impermeable barrier above the seabed (Losada et al., 1992; Abul-Azm, 1993), and a permeable barrier resting on the seabed (Yu, 1995). In all cases, very close agreement has been obtained over all conditions of interest, such that the various sets of results are not distinguishable on ordinary plots and hence are not shown here.

The present method has then been used to generate numerical results for a permeable barrier extending from the free surface to some distance above the seabed for various values of f, C_m, kh and h/d. A selection of such results is presented in Figs. 3 and 4, which show the transmission coefficient, the reflection coefficient, and the energy dissipation coefficient, each as a function of kh for the case h/d = 0.5 and $\varepsilon/b = 3.85\ m^{-1}$. Further results are reported in Yang (1996). The figures follow expected trends, and correspond to the various limits indicated earlier: the transmission coefficient decreases with increasing kh and the reflection coefficient follows the opposite trend, while the energy dissipation coefficient K_e is non-zero accounting for the energy loss across the barrier. Also shown is the influence of the friction coefficient and the added mass coefficient. Figure 3 presents results for various values of the friction coefficient for a fixed added mass coefficient of 0.0, and Fig. 4 presents results for various values of the added mass coefficient with a fixed friction coefficient of 2.0. An increasing friction coefficient is seen quite clearly to reduce the transmission coefficient and increase the reflection coefficient as expected, while somewhat less obviously the energy dissipation coefficient decreases with increasing friction coefficient at longer wavelengths relative to the barrier depth. Increasing the added mass coefficient is seen to reduce the transmission coefficient and increase the reflection coefficient, but now clearly decreases the energy dissipation coefficient. In particular, it appears that the contribution of the added mass to the energy dissipation coefficient is very important.

COMPARISON WITH EXPERIMENTS

In order to assess further the numerical model, numerical predictions have been compared with the results of the experiments. Figure 5 shows a preliminary comparison of the measured and predicted transmission, reflection and energy dissipation coefficients as functions of kh for the case $h/d = 0.5$ and $\varepsilon = 0.05$. The numerical model is able to adequately reproduce the most important features of the experimental results, including the energy dissipation through the slotted barrier. The most notable errors occur in the energy dissipation coefficient and appear to stem directly from differences between the measured and predicted reflection coefficients. These differences are associated in part with the difficulties in measuring the reflection coefficient accurately and are more pronounced for larger wave numbers. Similar differences have also been observed by Sulisz (1985) and others.

In order to apply the numerical model suitable values of the friction and added mass coefficients are needed. These are in general not known and in the present study have been estimated on the basis of a best fit between the measured and predicted values of the transmission, reflection and energy dissipation coefficients. This procedure gives a friction coefficient $f = 2.0$ and an added mass coefficient $C_m = 0.0$, which compares favourably with the values used by Yu (1995) for breakwaters extending down to the seabed.

CONCLUSIONS

The present paper summarizes a numerical model for predicting wave interactions with a thin slotted vertical barrier extending from the water surface to some distance above the seabed. The approach used is based on an eigenfunction expansion method and utilizes a permeable boundary condition at the barrier surface which accounts for energy dissipation within the barrier.

Comparisons with previous numerical studies for the limiting cases of an impermeable barrier and a permeable barrier extending down to the seabed indicate close agreement. A selection of results for a partly submerged barrier has been presented. These show the transmission, reflection, and energy dissipation coefficients as functions of the dimensionless wave frequency, kh. The energy dissipation within the slotted barrier is related to friction and added mass coefficients and the influence of these parameters is discussed.

Laboratory tests have been carried out to provide a preliminary assessment of the numerical model. A preliminary comparison of corresponding numerical predictions with experimental results is given. The agreement is generally satisfactory and indicates that the present numerical method is able to adequately account for the energy dissipation by the slotted breakwater provided that the relevant coefficients are suitably chosen. Preliminary estimates of the friction and added mass coefficients are given for a slotted barrier.

REFERENCES

Abul-Azm, R.G. (1993). "Water diffraction through submerged breakwaters". *J. Waterway, Port, Coastal and Ocean Engrg.*, ASCE, 119(6): 587-605.

Bennet, G.S., McIver, P. and Smallman, J.V. (1993). "A mathematical model of a slotted wavescreen breakwater". *Coastal Eng.*, 18: 231-249.

Hagiwara, K. (1984). "Analysis of upright structure for wave dissipation using integral equation". *Proc. 19th Coastal Engrg. Conf.*, ASCE, Houston, Vol. III, pp. 2810-2826.

Isaacson, M. (1991). "Measurement of regular wave reflection". *J. Waterway, Port, Coastal and Ocean Engrg.*, ASCE, 117(6): 553-569.

Kakuno, S. (1983). "Reflection and transmission of waves through vertical slit-type structures". *Proc. Coastal Structures '83*, ASCE, Arlington, pp. 939-952.

Liu, P. L-F., and Abbaspour, M. (1982). "Wave scattering by a rigid thin barrier". *J. Waterway, Port, Coastal and Ocean Engrg. Div.*, ASCE, 108(4): 479-491.

Losada, I.J., Losada, M.A., and Roldan, A.J. (1992). "Propagation of oblique incident waves past rigid vertical thin barriers". *Appl. Ocean Res.*, 14: 191-199.

Losada, I.J., Losada, M.A., and Losada, R. (1994). "Wave spectrum scattering by vertical thin barrier". *Appl. Ocean Res.*, 16: 123-128.

Premasiri, S. (1997). *Experimental study of wave interactions with slotted barriers.* M.A.Sc. thesis, Dept. of Civil Engineering, University of British Columbia, Vancouver, Canada.

Sarpkaya, T., and Isaacson, M. (1981). *Mechanics of wave forces on offshore structures.* Van Nostrand Reinhold, New York.

Sollitt, C.K., and Cross, R.H. (1972). "Wave transmission through permeable breakwaters". *Proc. 13th Coastal Engrg. Conf.*, ASCE, Vancouver, Vol. III, pp. 1827-1846.

Sulisz, W. (1985). "Wave reflection and transmission at permeable breakwaters of arbitrary cross section". *Coastal Eng.*, 9: 371-386.

Yang, G. (1996). *Numerical model of wave effects on permeable vertical barriers above the seabed.* M.A.Sc. thesis, Dept. of Civil Engineering, University of British Columbia, Vancouver, Canada.

Yu, X-P. (1995). "Diffraction of water waves by porous breakwaters". *J. Waterway, Port, Coast. and Ocean Engrg.*, ASCE, 121: 275-282.

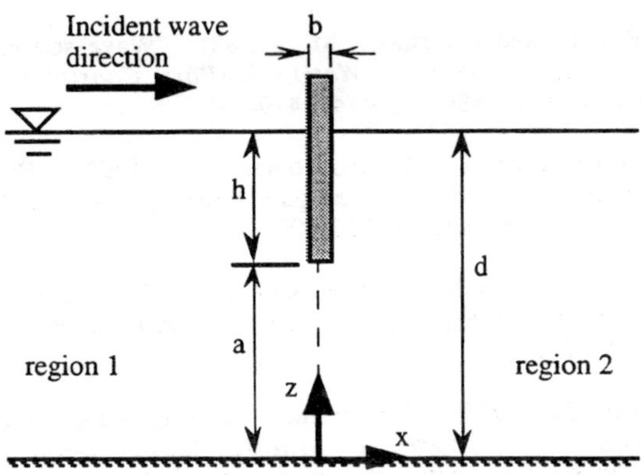

FIG. 1. Definition sketch.

FIG. 2. General view of the wave flume showing the barrier setup.

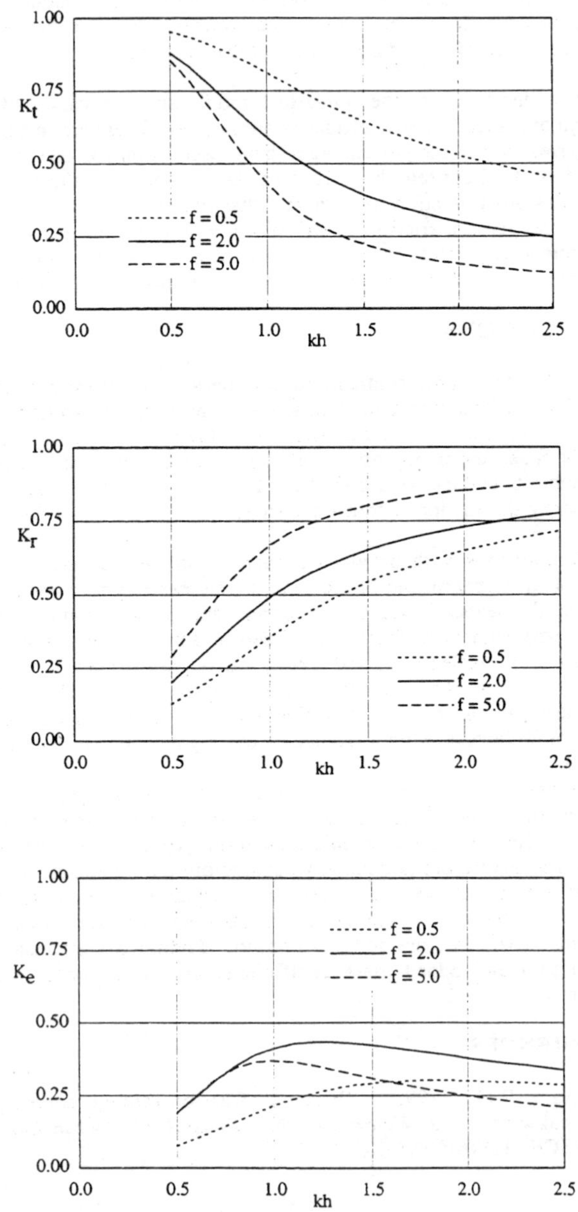

FIG. 3. Transmission, reflection and energy dissipation coefficients as functions of kh for h/d = 0.5, C_m = 2 and for various values of f.

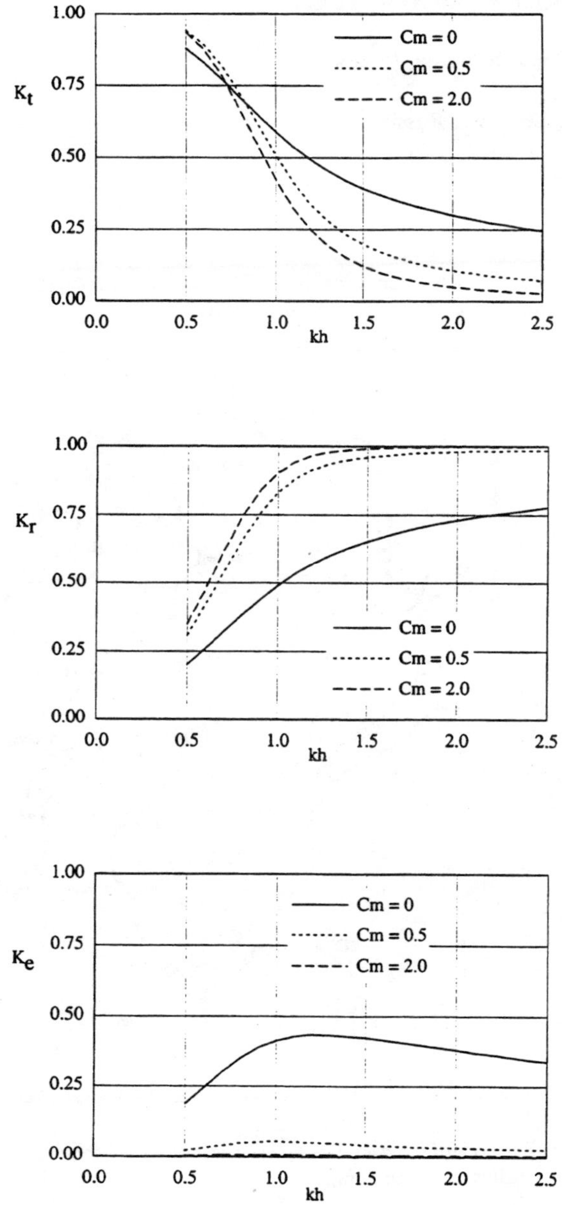

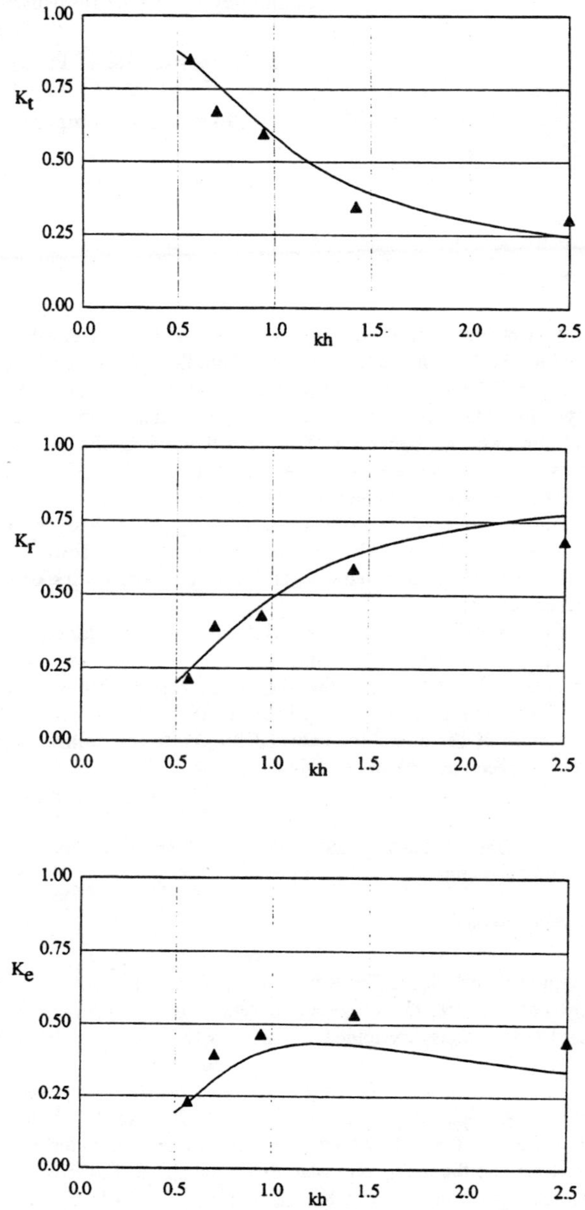

FIG. 4. Transmission, reflection and energy dissipation coefficients as functions of kh for h/d = 0.5, f = 2 and for various values of C_m.

FIG. 5. Comparison of measured and predicted transmission, reflection and energy dissipation coefficients as functions

Proceedings of the Seventh (1997) International Offshore and Polar Engineering Conference
Honolulu, USA, May 25-30, 1997

ISBN 1-880653-28-1 (Set); ISBN 1-880653-31-1 (Vol. III)

Hydrothermodynamic Model of the Ob and Tas Rivers Estuary

B.V. Arkhipov
Computing Center of the Russian Academy Sciences, Moscow, Russia
V.V. Solbakov
Moscow Institute of Physics and Technology, Moscow, Russia
A.S. Tsvetsinsky
Joint Stock Company "Eco-System", Moscow, Russia

ABSTRACT

A two-dimensional laterally averaged model was developed to describe the hydrodynamic and thermohaline processes in the Ob and Tas Rivers estuary. This model give the opportunity to simulate the tidal level and velocity oscillations, storm surges, thermohaline and ice processes. The base feature of the model is that it consists of two arms to take into account the complex three-dimensional geometry structure of the estuary.

Three types of characteristics may be described by the model: synoptic process characteristics, climatic seasonal and subannual variations, and seldom repetition characteristics of the considered values.

In this article the tidal oscillations were calculated and amplitude, and the phase distribution of the M_2 tidal wave along estuary were obtained. Also, the month averaged values of the salinity field were calculated. The basic feature of this field - the mean location of the boundary between the fresh and saline water is in good agreement with observation.

KEY WORDS: Estuary, tide, bay, mathematical modeling, thermohaline structure

INTRODUCTION

The region of the Ob and Tas Rivers estuary is characterized by very complex climatic, ice, hydrology and others nature conditions (Atlas Arktiki, 1985). This is determined primarily by the high latitude of the estuary (Fig. 1.).

The estuary of Ob and Tas Rivers is located on the south-western part of the Kara Sea (from 67°N to 63°N, from 70°E to 80°E). It is a long stretched estuary with two arms - the main one is an extension of the Ob River and the lateral arm is an extension of the Tas River. The Obskaja Bay is connected with Kara Sea in the north (A-A` section, Fig. 1). The length of it is approximately 800 km, the width changes from 30 km to 70 km, and the maximum depth is approximately 25m. So, the average relation of the width to length for Ob Estuary is about 0.05 and is quiet enough to neglect by the lateral variations of the hydrodynamic characteristics. The Tasovskaja Bay is connected to the Obskaja bay at the latitude 69°N. The length of Tas Estuary is 300km, it`s width is in the range of 10-45 km, and the maximum depth is approximately equal to 12m (Atlas Arktiki, 1985).

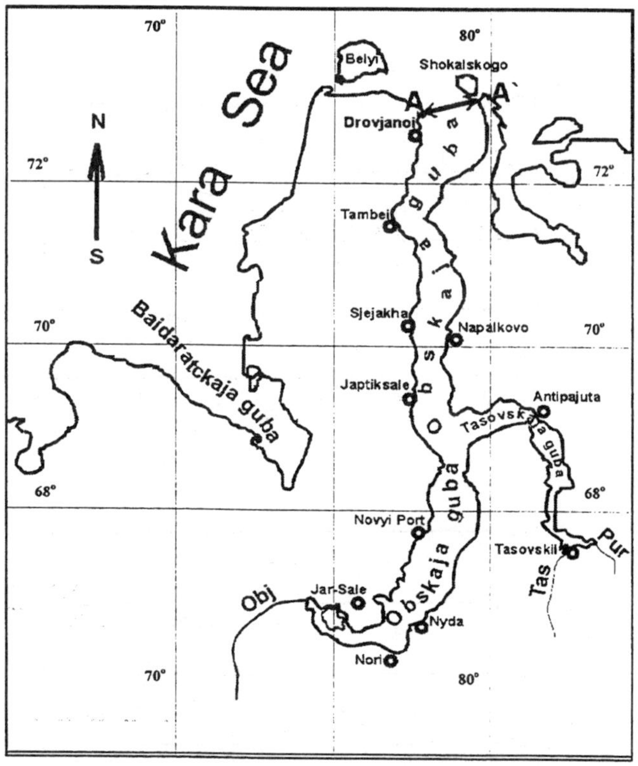

Fig. 1. Estuary of Ob and Tas rivers region. Cross-section A-A' is indicated on the map

The sources of fresh water discharge are the rivers Ob, Pur, Tas and so on. Theirs combined annually averaged discharge is about 530.5 km^3/year in the Obskaja Bay and about 60 km^3/year in Tasovskaja Bay (rivers Pur and Tas). The seasonal variability of the fresh water income is shown in Fig. 2 (Ivanov, 1980; Ivanova, 1984).

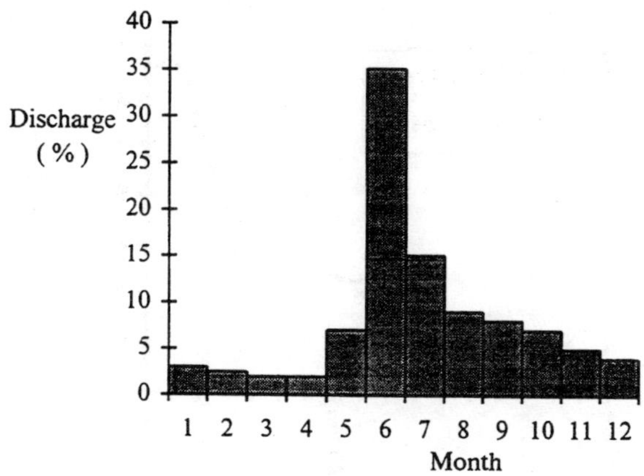

Fig. 2. Rivers discharge distribution

Atmosphere conditions are the following (Territory climate..., 1991). The monthly averaged temperature ranges from -28°C in January to +14°C in July. Alteration of the average monthly mean temperature at the north and at the south parts is shown in Fig. 3. The precipitation amount is in the range of 30 to 63 mm per month of rain in the summer and 3-37cm. per month of snow in the winter. The average wind velocity ranges from 4.5 to 7.8 m/s. Characteristic features of the bay region are high cloudiness (6-8 forces) and high relative humidity (75-90%).

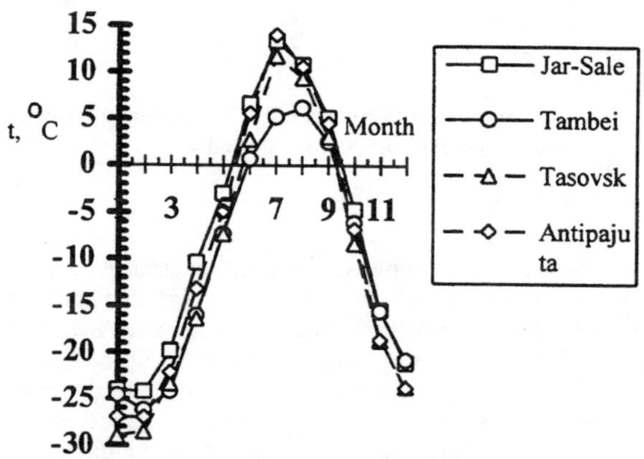

Fig. 3. Air temperature for the north and south part of the Obskaja and Tasovskaja Bays

Until now only few attempts were made to construct a numerical model of the of Ob and Tas Rivers estuary. In work of Molchanov (1976), a two-dimensional depth-averaged model was made and used to describe wind currents and ionic constituents spreading. In the article of Pavlov and Stanovoi (1983), wind currents are calculated on the basis of a simplified stationary three-dimensional model.

In this paper we present some results of a two-dimensional width-averaged numerical model of the Obskaja and Tasovskaja Bays. Because of the average relation of the width to length for the estuary is about 0.05 it is seemed reasonably to employ the laterally averaged model to simulate the main hydrodynamic and thermohaline processes in the estuary.

The purpose of the present work is to simulate tidal level and current oscillation and the salinity intrusion process into the estuary. To take into account the complex three-dimensional estuary shape, we constructed the model, consisting of two arms. The connection of the main channel (Ob Bay) with the lateral one (Tas Bay) is realised through the passing of the water level information in forward direction (from the main channel to the lateral one), and about flow rate, temperature and salinity in the backward direction (from the lateral channel to the main).

MODEL FORCING AND PARAMETERS

Because we examine the tide and salt intrusion processes, there are two types of external forcing: conditions on the open boundary and land-water discharge.

The Obskaja Bay is connected with the Kara Sea in the north part, and we specify the open boundary location at the A-A\` transact where the sea level elevation and salinity distribution are prescribed. The Tas Bay is connected to the Obskaja Bay in the point shown by the arrow in Fig. 4.

For the sea surface height on the open boundary, the semi-diurnal spring tide conditions were considered as M_2 wave with the mean range 120 cm and 12.36 h period.

The water salinity at the open boundary on the surface varies from 7-12‰ to 20-25‰ during the year. It is formed by the complex interaction of Kara Sea water with estuary water. According to data in Atlas Arktiki (1985), the salinity was taken to be appropriate to the August value - 10‰ on the surface, 20‰ on the bottom.

The fresh water discharge was appropriate to August and was equal to 23700 m^3/s and 2700 m^3/s for the Obskaja and Tasovskaja (rivers Pur and Tas) Bays accordingly.

Parameters of the numerical grid for the main channel and lateral one are shown in Fig. 4. Horizontal steps were equal to 10 km for both bays that correspond to 80 and 30 cells accordingly. The vertical step is shown in the Table 1. The maximum cell number at the vertical direction is 18 for the Ob bay, and 12 for the Tas. The time step is about 0.5 h and it is limited by the Kourant condition. The detailed model description is presented in the appendix.

Table 1. Cells thickness

Cell Number	Thickness, m
1	0.8
2-6	0.5
7-9	1
10-...	2

SIMULATION RESULTS

Calculations were made during three month with periodically forcing conditions (after each month), as far as the stationary state (with exactness of tidal period) was achieved.

Fig. 5 illustrates measured (Sailing directions, 1992) and calculated mean spring range along the estuary, a - for Ob Bay, b - for Tas Bay. At most locations a satisfactory coincidence was obtained. The better coincidence is in the Tas bay and in the southern part of Ob Bay. At the northern part of Ob Bay, the calculation results are somewhat worse. The possible reason of such deficiencies may be the abrupt crook of the main channel in this place, where a relatively wide bed becomes narrower.

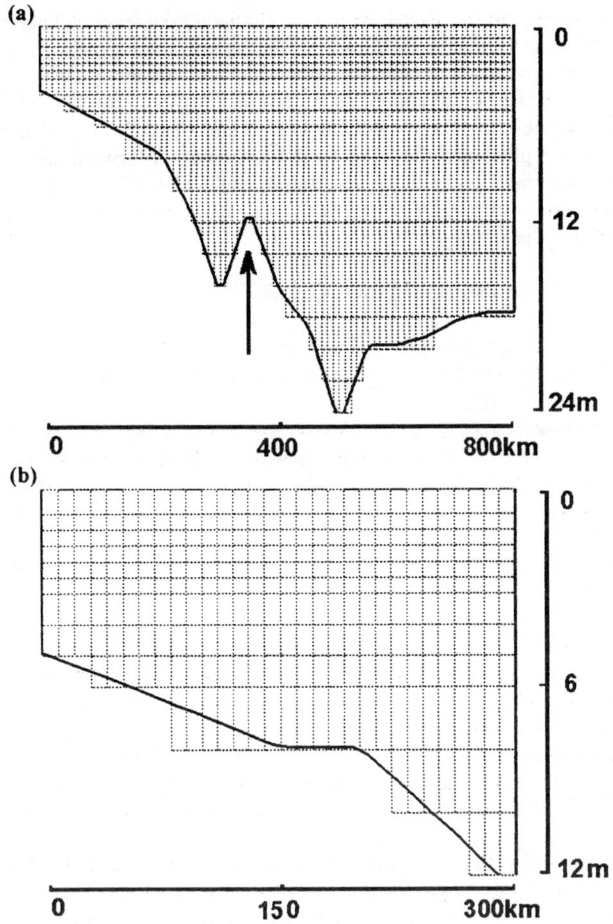

Fig. 4. Grid for numerical scheme, a - Obskaja bay, b - Tasovskaja Bay

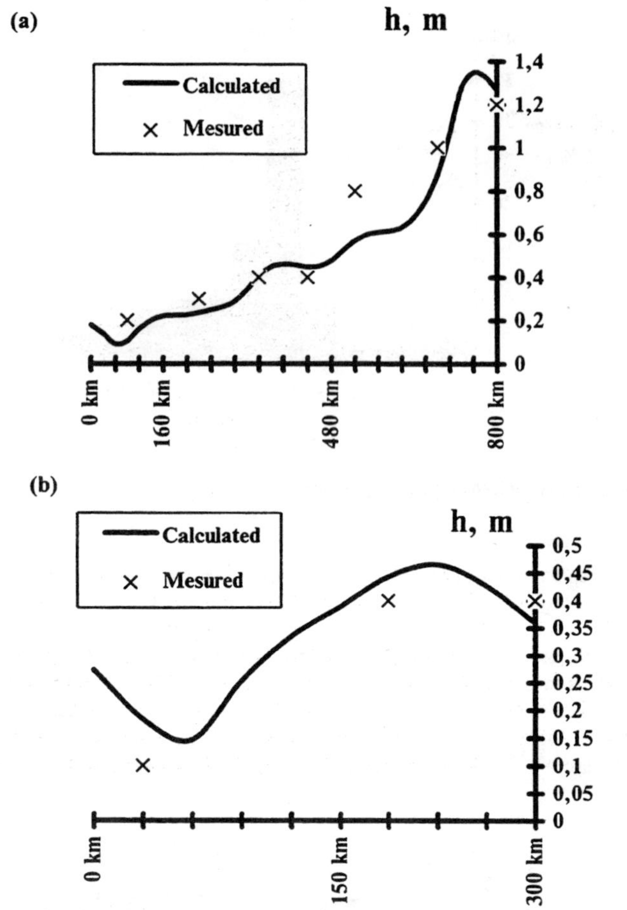

Fig. 5. Tidal range, a - Ob Estuary, b - Tas Estuary (solid line - calculated, cross marks -measured)

Fig. 6 illustrates the time lag of the high water moment (tide phase) relative to the open liquid boundary (in the north of the Ob Estuary). From it we can see that the wave runs along the bay during 23 hour approximately, corresponding to a phase velocity 9.7 m/s.

In the fig. 7 salinity field is presented. The main observed feature of the salinity field (Sailing directions, 1992) was the mean location of the boundary between the fresh and saline water. This boundary is determined by the isohaline 0.5 $^{o}/_{oo}$. From (Vasiljev, 1983; Sailing directions,1992) average location of the boundary is in Sjejaha - Napalkovo transaction. This transaction is shown in the Fig. 7 by the dashed line. The isohaline 1 $^{o}/_{oo}$ in the calculated field lies in the same location. The southern part of the Ob Bay and Tas Bay have a salinity less then 0.5 $^{o}/_{oo}$, so the water there is fresh.

Fig. 8 illustrates the tidal velocity and turbulent viscosity (cm^2/s) fields. It is evident that the maximum tide current velocity is about 0.6m/s. Observations indicate that the tidal velocity is in range 0.5-1m/s (Sailing directions, 1992). Fig.8b illustrates that the viscosity maximum equals to 10 cm^2/s and the minimum is about 0.001 cm^2/s.

CONCLUSIONS

The obtained results show satisfactory coincidence of measured and calculated data as far as we consider the tide amplitude and phase values. If we concern with the salinity field we see that the boundary between fresh and saline water location is in concord with the observation. At the present time, the comparison of the model results with observations was bounded by a lack of nature information in author's disposal. The future investigation will be directed to correct these deficiencies of the input data and some model decisions in order to get the better quantitative coincidence.

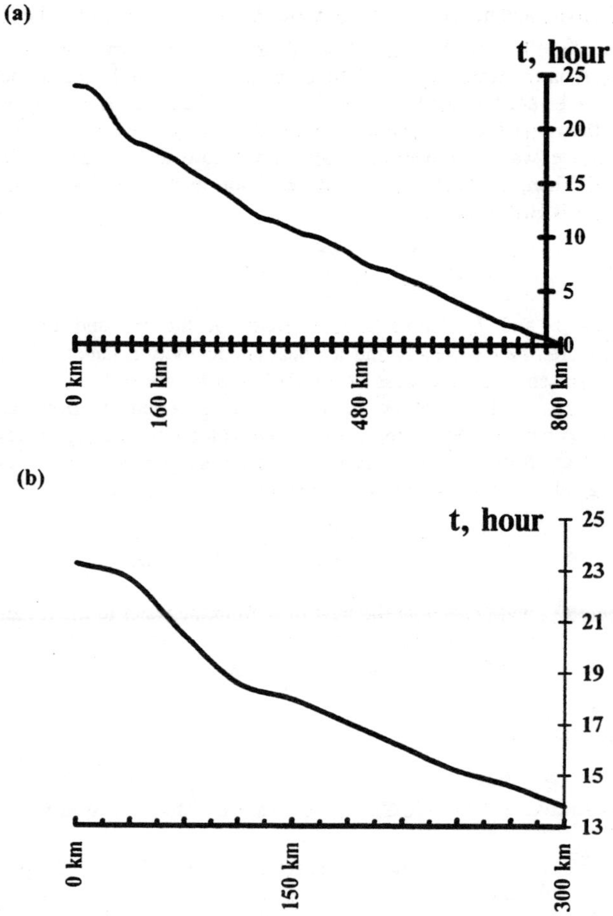

Fig. 6. The time lag of the high water moment relative to open boundary, a - Ob Estuary, b - Tas Estuary (the phase of M_2-wave)

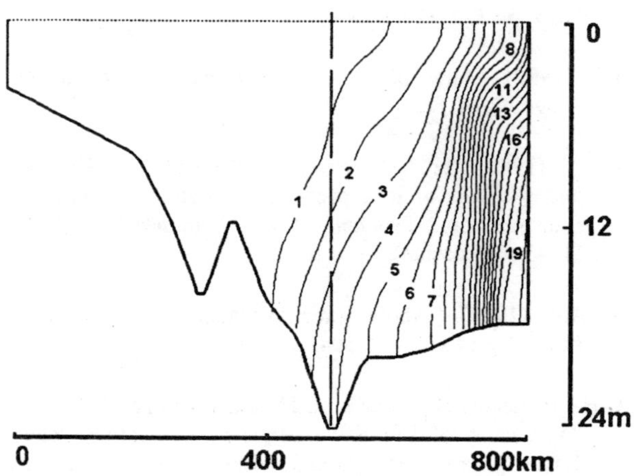

Fig. 7. Salinity ($^0/_{00}$) field appropriated to mean August condition

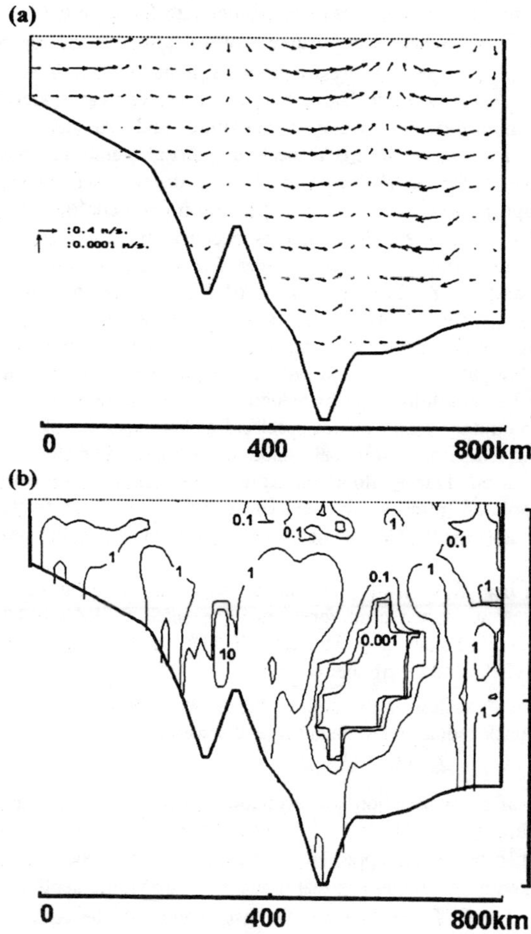

Fig. 8. Tide velocity (a) and viscosity (b) in the Ob Bay

APPENDIX. MODEL FORMULATION

Width-averaged equations system consists of the continuity equation (1), momentum equation (2) and salinity and heat equation (3) (Vasiljev, Kvon and Tchernishova, 1974; Blumberg, 1977; Lavelle, Cokelet and Cannon, 1991):

$$\frac{\partial}{\partial x}(B\cdot u)+\frac{\partial}{\partial z}(B\cdot w)=q(x,z) \qquad (1)$$

$$\frac{\partial}{\partial t}(B\cdot u)+\frac{\partial}{\partial x}\left(B\cdot u^2\right)+\frac{\partial}{\partial z}(B\cdot u\cdot w)=-g\cdot B\cdot\frac{\partial \zeta}{\partial x} - B\frac{\partial}{\partial x}\left(\frac{P_S}{\rho_0}\right)+\frac{\partial}{\partial z}\left(B\cdot A_z\frac{\partial u}{\partial z}\right)-\alpha\cdot u\cdot|u|\cdot\left|\frac{\partial B}{\partial z}\right| \qquad (2)$$

$$\frac{\partial}{\partial t}(B\cdot F)+\frac{\partial}{\partial x}(B\cdot u\cdot F)+\frac{\partial}{\partial z}(B\cdot w\cdot F)=\frac{\partial}{\partial z}\left(B\cdot K_z\frac{\partial F}{\partial z}\right)+r(x,z) \qquad (3)$$

In Eq. 2 the hydrostatic distribution of pressure is used:

$$P=\int_z^{\zeta}(\rho\cdot g)dz\approx\rho_0\cdot g\cdot\zeta+\int_z^{0}(\rho\cdot g)dz=\rho_0\cdot g\cdot\zeta+P_S\,,$$

and the density of water depends on its temperature and salinity (Gill, 1982): $\rho = \rho(T,S)$, where z is positive upward (origin at the mean tidal level), and x is positive west-northward along the bay axis. t denotes time, $u(x,z,t)$, $w(x,z,t)$ denote horizontal and vertical velocity components, appropriately, water temperature and salinity are designated by $F=T(x,z,t)$, $S(x,z,t)$ respectively, $B(x,z)$ stands for bay width, $\zeta(x,t)$ stands for sea level. $A_z(x,z,t)$, $K_z(x,z,t)$ denote viscosity and diffusivity, appropriately. The bottom and lateral friction coefficient is denoted by α=0.0025. g stands for gravity acceleration. ρ - reference density, q- water inflow, r - heat income due to the evaporation, precipitation and rivers. For the system of Eq. 1-3 the boundary conditions were specified. At the free surface they were kinematics condition, wind stress, salt (zero) and heat flows. At the bottom we used friction law, absence heat and salt flow, zero normal velocity, at the mouth we had conditions for the velocity, salinity and temperature. At the open boundary sea level was specified as M_2. Also during ebb the salinity and temperature were calculated, using "upwind" difference advection equation. During flood the salinity and temperature were linearly interpolated from its computed value at the end of ebb to the prescribed value (Lavelle, Cokelet and Cannon, 1991; Oey, Mellor and Hires, 1985)

The surface heat flux mainly determines temperature distribution in the bay. In our work we used formulates presented in Jirka et al. (1978), Ryan, Harleman and Stolzenbach (1974), Parkinson and Washington (1979) and Oberhuber (1990).

The net heat flux flow is the sum of the short radiation Q_{sn}, long radiation Q_a, back radiation Q_{br}, latent Q_l and sensible flow Q_c:

$$Q = Q_{sn} + Q_a - Q_{br} - Q_l - Q_c \ .$$

Absorption of short radiation was accounted as follows. If the heat flux of I_0 is falling to the surface of the water, then, the quality of the flow $\beta \cdot I_0$ is absorbed at the upper layer, and the rest is absorbed lower with the weakening coefficient η: that is the heat flux on the depth z is equal to the $(1-\beta)\cdot I_0 \cdot \exp(-\eta \cdot z)$. We use β and η to be equal to 0.45 and 0.6 m^{-1} appropriately (Jirka et al., 1978). The wind stress is defined by the equations (see for example Lavelle, Cokelet and Cannon, 1991):

$$\tau_w = \frac{\rho_a}{\rho_w}\cdot C_d \cdot V^2, \quad \tau_{wx} = \tau_w \cdot Cos(\theta_1 - \theta_2), \quad \tau_{wx} = \tau_w \cdot Sin(\theta_1 - \theta_2).$$

Here ρ_a denotes the air density, θ_1, θ_2 denote angles between the north direction and estuary axis or wind direction.

$$C_d = \begin{cases} 1.1 \cdot 10^{-3}, \ V < 6 \ m/\tilde{n} \\ (0.72 + 0.063 \cdot |V|) \cdot 10^{-3}, \ V > 6 \ m/s \end{cases}.$$

The vertical diffusivity and viscosity were described as follows.

$$A_0 = k_1^2 \cdot z^2 \cdot \left(1 - \frac{z}{H}\right)^2 \cdot \left|\frac{\partial U}{\partial z}\right| + k_2 \cdot z \cdot \tau \cdot \exp(-k_* \cdot z),$$

where $k_1 \approx 0.03$, $k_2 \approx 0.4$, $k_* \approx 0.1$, and z denotes the distance to the surface (Blumberg, 1977).

In stable-stratified water (Ri>0) the viscosity is decreased, while in unstable-stratified water (Ri<0) the viscosity is increased.

$$A_z = A_0 \cdot f(Ri), \quad f(Ri) = \begin{cases} \exp(-4.6 \cdot Ri), Ri < 0 \\ \exp(-1.1 \cdot Ri), Ri > 0 \end{cases},$$

$$Ri = -\frac{g}{\rho} \cdot \frac{\partial \rho}{\partial z} \cdot \left(\frac{\partial u}{\partial z}\right)^{-2}.$$

Diffusivity and heat conductivity were $K_z = k \cdot A_z$, $k = 0.14$. The ice model is briefly described as follows. A horizontal uniform plate of ice is placed on the water surface (Klimok et al., 1991). On the ice surface snow can be accumulated. On the snow surface heat flux is determined as for the water surface. The beginning of the ice is the moment when the surface water temperatures becomes below temperature of ice melting. Then, the heat conductivity equation in the ice layer, whose thickness is H(t), is solved:

$$c_i \cdot \rho_i \cdot \frac{\partial T_i}{\partial t} = k_i \cdot \frac{\partial^2 T_i}{\partial z^2}, z \in (-H(t), 0)$$

Where c_i and T_i denote heat capacity of the ice and the ice temperature, the ice density and ice heat conductivity is denoted by ρ_i and k_i. The snow of thickness of $h=h(t)$ lies at the ice surface, a linear distribution of temperature in the snow is assumed. Boundary conditions have the following form. If the melting or freezing process takes place, then the temperature on the ice surface is equal to the melting value and the phase transformation equation is used.

$$L_i \cdot \rho_i \cdot \frac{\partial H}{\partial t} = \left(Q_a - k_i \cdot \left.\frac{\partial T_i}{\partial z}\right|_{z=H} \right) - Q_{wi} - k_i \cdot \left.\frac{\partial T_i}{\partial z}\right|_{z=0}.$$

Where Q_{wi} and Q_a denote the heat flow from the water to the ice and heat flux from ice to air, the latent heat of ice production is denoted by L_i. This equation determines the ice thickness.

REFERENCES

"Atlas Arktiki" (1985). *GUGK SSSR*, Moskva, 204p (in Russian).

"Sailing directions of Ob and Enisei region" (1992). *GUNO*, 220p (in Russian).

"Territory climate of the oil and gas deposits on the Tasovskii and Jamal peninsulas" (1991). *Specializirovannii spravochnik.* Leningrad, Gidrometeoizdat, 222p (in Russian).

Blumberg A.F. (1977). "A two-dimensional Model of the Salinity Intrusion into Estuary," *J Hydrological Division*, Vol 103, No 3, pp 295-310.

Gill A.E. (1982). "Atmosphere - Ocean Dynamics", *Academic, San Diego*, Calif, 662 p.

Ivanov V.V. (1980). "Hydrology regime of the Western Siberia rivers and problem of it impact on the water resources territory redistribution," *Problemy Arktiki i Antarktiki*, Vol. 55, pp. 20-43 (in Russian).

Ivanova A.A. (1984). "About water exchange estimation in the Ob bay," *Tr. AANII*, Vol. 394, pp. 5-9.

Jirka G.H., Watanabe B., Octavio, C.H. and Harleman, D.R.F. (1978). "Mathematical Predictive Models for Cooling Ponds and Lakes. Part A: Model Development and Design Consideration," *Ralph M. Parson Lab. for Water Res. and Hydro*, Rep No 238, MIT, Cambridge, 109 p.

Klimok V.I., Bugrov A.N., Dunets T.V., and Bekisheva, Ch.B. (1991). "Numerical modeling of the thermal and ice regime in the high mount water reservoir," *Teoryja i Metodi*

Matematicheskogo Modelirovanija Zadach Okrugajuschej Sredy. Bishkek: "Ilim" (in Russian).

Lavelle J.W., Cokelet E.D. and Cannon G.A. (1991). "A model Study of Density Intrusions into and Circulation within a Deep, silled or Estuary," *J geophysical Res.* Vol 96, No C9, pp 16.769-16.810.

Molchanov V.N. (1976). "Numerical model of the water circulation on the mouth seashore with accounting thermal and ionic flows effects," *Tr. AANII*, Vol. 314 , pp. 36-43 (in Russian).

Oberhuber J.M. (1990). "Simulation of the Atlantic Circulation with a coupled sea ice-mixed layer - isopycnal general circulation model," *Max - Planck - Institut fuer Meteorogie*, Rep No 59.

Oey L.Y., Mellor G.L. and Hires R.I. (1985). "A three-dimensional simulation of the Hudson-Raritan estuary. Part 1," *J Physical Oceanography*, Vol 15, pp 1675-1692.

Parkinson C.L., and Washington W.M. (1979). "A Large-Scale Numerical Model of Sea Ice," *J Geophys Res*, Vol 84, pp 311 - 337.

Pavlov V.K. and Stanovoi V.V. (1983). "Climatic calculation of the flow-wind currents in the Ob bay," - *Tr. AANII*, Vol. 380, , pp. 49-54 (in Russian).

Ryan P.J., Harleman D.R.F. and Stolzenbach, K.D. (1974). "Surface heat loss from cooling ponds," *Water Resources Res*, Vol 10, No 5, pp 930 - 938.

Vasiljev A.N. (1976). "Interaction of the river and sea water in Ob mouth region," *Tr. AANII*, Vol. 14, pp.183-196 (in Russian).

Vasiljev O.F., Kvon V.I. and Tchernishova, R.T. (1974). "Temperature stratified current in long stretched water body," *Gidrotechnicheskoe stroiteljstvo*, No 4, pp 35-43 (in Russian).

Proceedings of the Seventh (1997) International Offshore and Polar Engineering Conference
Honolulu, USA, May 25-30, 1997

ISBN 1-880653-28-1 (Set); ISBN 1-880653-31-1 (Vol. III)

A Study of Anti-Collision Facilities of Quay Wall: A Two-Leg Fender

Su-Liek Shih and Hui-Ming Fang
National Taiwan Ocean University
Keelung, Taiwan, China

Te-Hsung Hsu
Hsu' Chemical Industrial Co. Ltd.
Keelung, Taiwan, China

ABSTRACT

In the processing of anchoring to the quay wall, a ship will swell to collies with the wall due to the outer coming effects, such as, wind, wave, current, water surface interference; and its own mass, approach speed, and approach angle. At that crash, a great damage may be caused between the ship and the wall, so that the anti-collision facilities of the quay wall is necessary. There were so many kinds of these facilities existed, they resisted the crash by its material advantage, such as, wool, rubber; or absorbed the crash energy by its structure advantage, such as, hollow structure, two-legs type, muti-legs type etc.. In the past experience, a rubber made two legs type fender is more efficient on energy dissipation during the crash, and that is the interested subject of this study.

Keeping the chemical quality of rubber production at the same grade, and changing the structure formation of two-legs type fender to make lots of model as testing bodies. To process compressive test of these shrinkage model, some observed results and their analysis are obtained. And that offer the basic knowledge of two-legs type fender to lead an advanced study of new formation of this type.

KEY WORDS: Anti-collisioin facilities, two-legs type fender, quay wall, response stress, absorbed energy, leg thickness, inclined angle of the leg, non-uniform thickness of the leg .

INTRODUCTION

An anti-collisioin facility equipped on the quay wall is used for the absorption of the energy and dissipation of the stress when the vessel is anchoring to the quay wall, that is to keep both the vessel and the quay wall away from damage due to the collision. Tang (1989), Pan(1970)proposed that the berth was used for small vessel, which length was less than 12m, needed no such facilities. But Lin(1981) had a comment that the berth used for large vessel should equip the anti-collisioin facilities on the quay wall, that he built a code of the construction as the design basis of the harbor engineering in Taiwan. According to the practical needs, many studies and reports on this topic was made, such as Baker(1969), Bijlsma(1970), Edward(1969), offer different ways on the absorption of the collision energy, and Chiang(1984)developed a numerical method with F.E.M. to calculate the variations of the stress during collisioin process. Dennis V. Padron et al.(1984) made a comparision of the rubber fender and wood fender, showed that rubber fender would be more advantageous.Wood fender was apt to be rotten and would be more expensive in maintenance than rubber made, so that the rubber fender would be the favorited developing ap. Based on the past studies, this research is mainly on the rubber fender, an advanced study of two-legs type fender was made. Some obtained results as, the effect of the leg thickness and its variations, and the effect of the inclined angle of the leg, that offer an optimal design of two-legs type rubber fender.

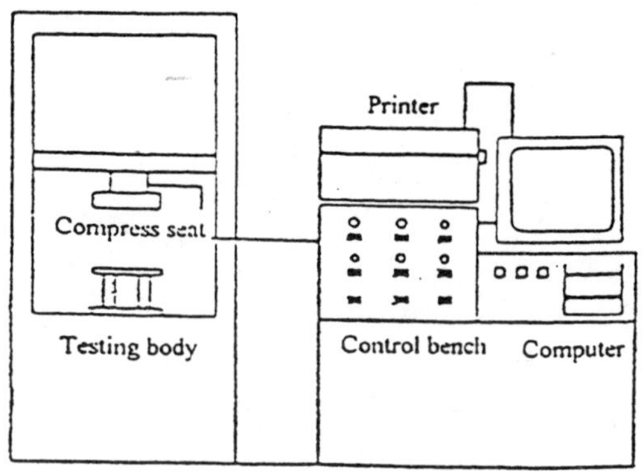

Fig 1 Brief sketch of testing bench

RESEARCH PROCESS

A rubber fender would be subject to damage during the collisioin process, and the resistance criteria of the rubber fender depend on the chemical composing and the structure function. The chemical composing is not our interesting, due to its limit effect in the improvement of the resistance ability, so that the unity of the chemical characteristics of the material is assumed. In this study, a research process is mainly on the structure function with laboratory investigations. Simulated models, made with different structure formation, are performing the compressive test; and the results of response stress, compress ratio, and the energy dissipation ability are obtained and discussed.

LABORATORY REQUIREMENTS

Labroatory requirements for compressive test involved A-D Converter, Detective Computer, and other relative soft ware. A brief sketch is depicted in Fig.1.

TESTING CONDITIONS

Kau(1975) proposed that compressive speed would be 5cm/min and the maximum compress ratio would be 50% for the testing of round type fender. There was a code built in Taiwan(1982), a comment of compressive speed range was 2cm/min to 8cm/min, and the maximum compress ratio was 50%. In that code, observations of 3 times continuous compress and depress were needed, the results of the 2nd and the last test should be match within 10% allowable limit, so that the average values of the two test results would be adapted for the study. Based on the code, the testing conditions were set up and the compressive speed is 5cm/min in this study.

TESTING BODIES CONDITIONS

Tesitng bodies involving different structure formations and their scales are depicted as, variation of the main body thickness in Fig.2a and Table 1-1, variation of the incline angle of legs in Fig.2b and Table 1-2, variation of non-uniform thickness of the leg in Fig.2c and Table 1-3.

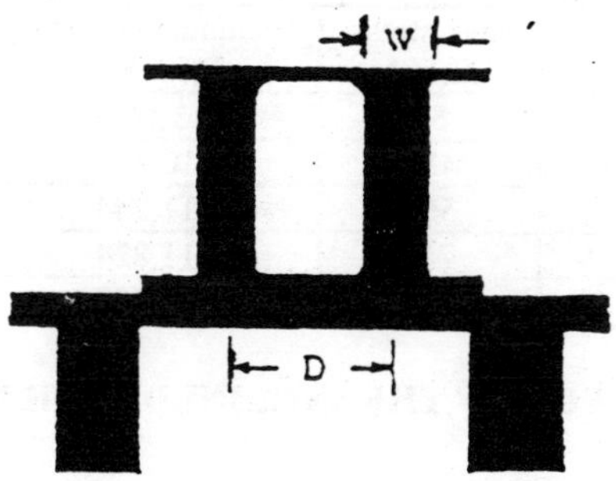

(a)

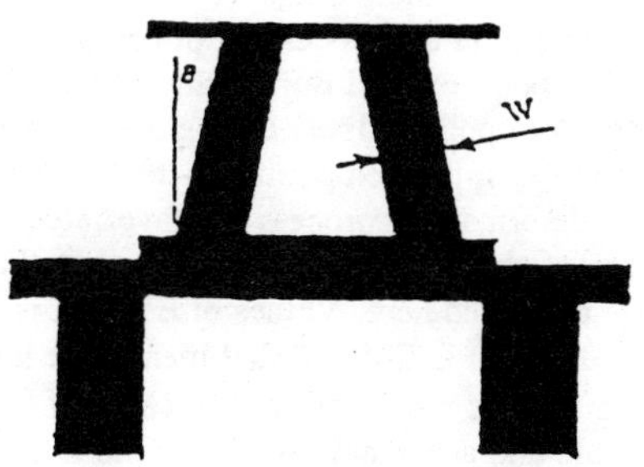

(b)

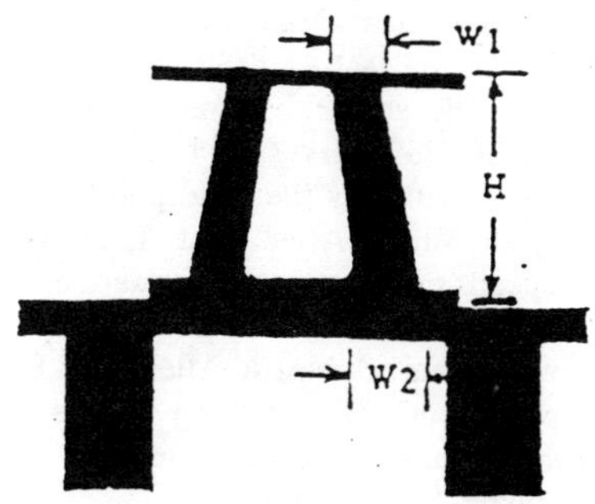

(c)

Fig 2 The definition sketch of testing body

Table 1-1

$\theta=0^0$	
W/H	0.14
	0.24
	0.26
	0.28
	0.3

Table 1-2

W/H	0.24	0.26	0.28
θ	5	5	5
	8	8	8
	10	10	10
	15	15	15

Table 1-3

$\theta=8^0$	
W_2/H	0.2
	0.22
	0.24
	0.26

W_1 : thickness of top end of non-uniform thickness
W_2 : thickness of bottom end of non-uniform thickness
W : uniform thickness
θ : inclined angel
H : height of leg
D : spacing between two legs(14cm)

TESTING PROCEDURE

Following the testing conditions, the laboratory test is performed that the compressive stress is directly normal to the fender surface. The testing procedure is to

(1). put the testing body on the right positioin and jacketed, no pre-stress is admitted.
(2). input the basic data of the testing body into the computer for self detect and auto adaptation of the observation data. The basic data are that the height, thickness and incline angle of the legs.
(3). set up the testing conditions, and process the test.
(4). repeat the testing for 3 times, and take the data according to the testing conditioins described.

ABSORBED ENERGY OF THE TESTING BODIES

Fig.3a is the stress-strain curve of observatioins, and the absorbed energy of the testing body will be the value of the area under the curve. The accumulated energy-compressive ratio curve expressed in Fig.3b. Both curves can be defined as,

(1). point A on the curve is the yield point, and the corresponding stress is yield stress. After point A, the curve stretches along mostly depends on the structure formation.

(2). point B on the curve is the cut point, and the corresponding stress equals to the value of point A. After point B, the curve stretches up abruptly, a critical condition of the fender resistance is limited that means the fender is invalid, and all the resistance energy transfer to the quay wall.

(3). the integrated area under the limeted curve of Fig.3a is the absorbed energy, and expressed in Fig.3b.

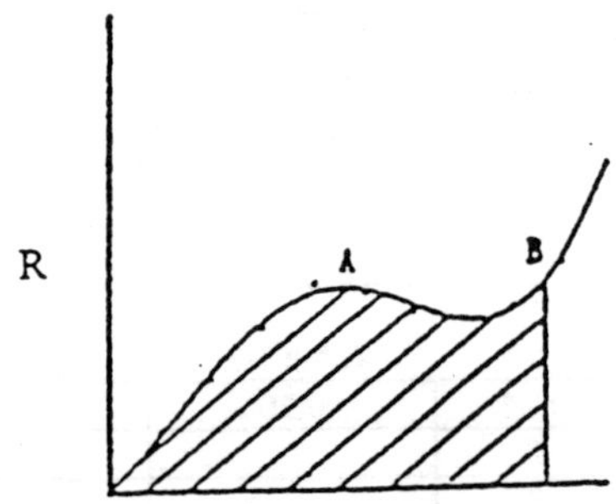

(a)

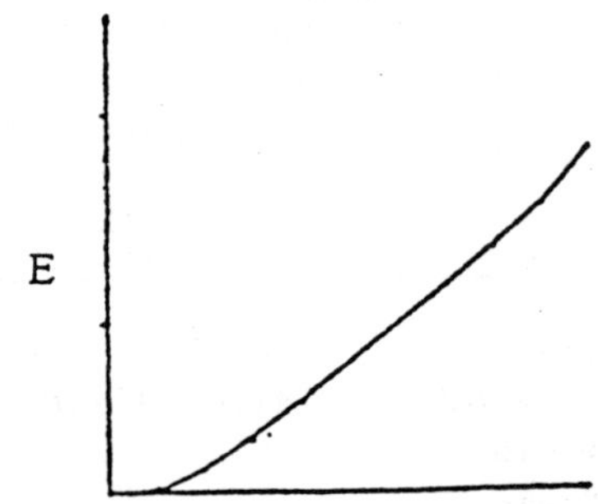

(b)

Fig 3 The definition skectch of response curve

OBSERVATIOINS AND DISCUSSIONS

Laboratory observations process in three times, involved the effect of the effect of the leg thickness and its variations, the effect of the inclined angle of the leg, the effect of non-uniform thickness of leg. The discussions followed as:

THE EFFECT OF THE LEG THICKNESS

Fig.4a and Fig.4b are selected results of the leg thickness (W/H=0.14,H=100mm), and some other observed data are listed in Table 2. A parameter E/R ,defined as the energy absorption of unit response force,is used to evaluate the effects. In Table 2, the maximum value 38.1 of E/R corresponding with dimensionless thickness W/H=0.14, but with a minimum value of E, that is not quite suit for practical use. Values of E at the range of thickness from W/H=0.24 to 0.28 are much more closer, and the values of R are all within the reasonable quantity requested, so that the recommendable range of the leg thickness are obtained.

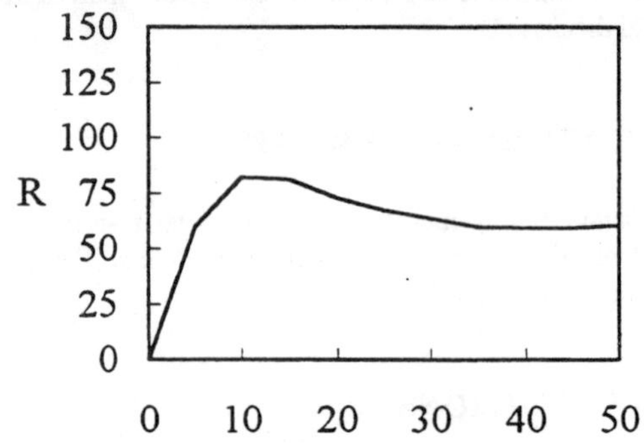

(a)

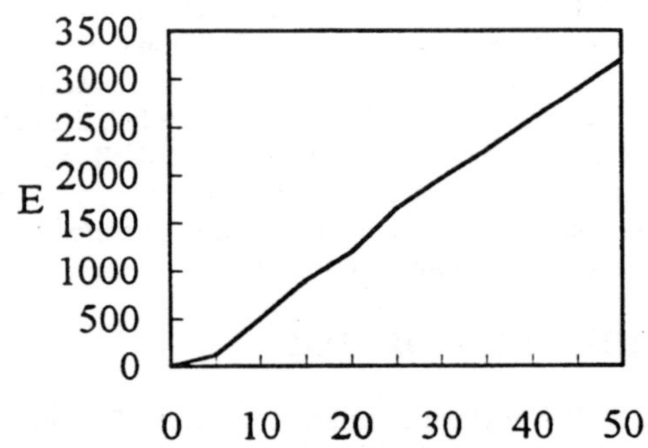

(b)

Fig 4 The response curve of stress & energy of W/H =0.14

Table 2

W/H	Max stress R(kg)	Accumulated Energy E(kg-mm)	E/R
0.14	83	3,165	38.1
0.24	450	10,426	23.1
0.26	560	11,244	20.1
0.28	740	11,824	15.9
0.3	900	10,802	12

THE EFFECT OF THE INCLINED ANGLE OF THE LEG

There are accepted thickness range(W/H=0.24 to 0.28) in the discussions before,and within that range,an advanced discussion involving different inclined angle ($\theta=5^0,8^0,10^0,15^0$) of the leg is proposed in Table 3. In the processing of compressive test, two types of the inflection point existed during the deformation process of the leg. One inflectioin point existed during the deformatioin precess of $\theta=5^0,8^0,10^0$,and two inflection points existed as $\theta=15^0$. One inflection point during deformation process is favoriated here due to the considerations of the material fatigue. so that the observations of $\theta=15^0$ are not discussed here. Values of E and R is reducing with the increasing of θ shown in Table 3,that means the larger incline angle makes the leg to bend more easily. The cases of $\theta=8^0$ indicate much more reasonable and acceptable values of the R and E, so the values of $\theta=8^0$ is selected and proposed.

Table 3a W/H=0.24

θ	Max stress R(kg)	Accumulated Energy E(kg-mm)	E/R
5	500	10,094	20.2
8	470	10,178	21.6
10	460	9,894	21.5
15	250	9,318	37.3

Table 3b W/H=0.26

θ	Max stress R(kg)	Accumulated Energy E(kg-mm)	E/R
5	560	10,094	20.2
10	470	9,894	21.5
15	220	9,318	37.3

Table 3c W/H=0.28

θ	Max stress R(kg)	Accumulated Energy E(kg-mm)	E/R
5	760	12,220	16.1
8	710	12,781	18
10	610	12,602	20.6
15	350	12,836	36.6

THE EFFECT OF THE NON-UNIFORM THICKNESS OF THE LEG

Observations of $\theta=8^0$ and $W_2/H=0.30$ at the bottom end involving different values of W_1/H at the top end are listed in Table 4. In reasonable range of R and E values, it shows that the formation of non-uniform thickness of the width $W_2/H=0.30$ at the bottom end and $W_1/H=0.26$ at the top end is the favorited one.

Table 4 $W_2/H=0.30$

W_1/H	Max stress R(kg)	Accumulated Energy E(kg-mm)	E/R
0.2	590	9,242	15.6
0.22	560	10,910	19.4
0.24	610	11,901	19.5
0.26	620	13,580	21.9

CONCLUSIONS

Some conclusions are obtained from observations and discussions:

1. The recommendable range of the thickness of the leg are W/H=0.24 to 0.28.
2. The values of accumulated energy E and maximum compressive stress R is reducing with the increasing of inclined angle θ, that means the larger θ makes the leg to bend more easily, and the optimal inclined angle of the leg is $\theta=8^0$.
3. The thickness of non-uniform leg with $W_2/H=0.30$ at the bottom end and $W_1/H=0.26$ at the top end is the favorited one.
4. The spacing between two legs is also the focus of this study ,but the observations shows that, in reasonable arrangement of structure formation ,no distinguish efect on the energy absorption.
5. The spacing between two units of the fender is surely an important factor in anti-collisioin process, but that belongs to observations of whole quay wall system functions, not like the interesting of this study just on the fender's sturcture formation only. In practical engineering job, the spacing between two units of the fender depends on the size of the approaching vessel and the type of the fender, and that almost depends on the information afforded by producer.

REFERENCE

Tang, L.W. and Shu, C. Y., (1989). "Harbor Engineering," *Published by the Association of Civil and Hydraulic Engineering,* Taiwan, R.O.C., in Chinese.

Lin, J. C., (1981). "The Construction Code of the Harbor in Taiwan,"*Published by the Taiwan Province Government, R.O.C.*, in Chinese.

Pan, J. C., (1980). "The Importance of the Fender," *The seasonal issue of China Port in R.O.C.*, vol.44, in Chinese.

Dennis V. Pardron and Edward H.Y. Han, (1984). "Fender System Problems in U.S. Ports," *ASCE.*

A. L. Baker, (1969). "New Type of Suspended Fender," *The Dock and Harbor Authority*, Sept., U.S.A..

T. J. Bijlsma, (1970). "Supper Fenders for Supper Tankers," *The Dock and Haubor Authority*, Apr., U.S.A..

Edward H. L. and Munck, (1969). "Modern Fendering System," *International Dredging and Port Construction* No.4, U.S.A..

Chiang, D. T., (1984). "A Study of P-△ Curve of the Rubber Fender," *master thesis,* Natioinal Taiwan Industrial Academy, R.O.C., in Chinese.

Fung, Y. S., (1982). "Design and Manufacture of the Rubber Fender,"*The seasonal issue of China Port in R.O.C.*,vol.26, in Chinese.

Kau, L. E., (1975). "Characteristics of the Fender," *The Proceeding of 22th Coastal Engineering speech in Japan*, in Japanese.

Proceedings of the Seventh (1997) International Offshore and Polar Engineering Conference
Honolulu, USA, May 25-30, 1997

ISBN 1-880653-28-1 (Set); ISBN 1-880653-31-1 (Vol. III)

Stability Assessment of Rubble Mound Breakwater and Armor Layer

Hajime Mase and Tomotsuka Takayama
Disaster Prevention Research Institute, Kyoto University
Uji, Japan

ABSTRACT

Stability of rubble mound breakwaters and armor layers is evaluated by using a neural network method. As parameters relating to the stability of rock and block slopes, following seven parameters are selected: the stability number; the damage level; the number of attacking waves; the surf similarity parameter; the permeability parameter; the dimensionless water depth in front of the structure; and the spectral shape parameter. The neural network method is confirmed to be useful for assessment of stability of the structures. In addition, sensitivity of some input parameters to predict the damage level is evaluated by utilizing the neural network method.

KEY WORDS: neural network, rubble mound breakwater, armor layer

INTRODUCTION

The stability of rocky slope under wave attack have been investigated by mainly hydraulic model tests, and empirical formulae have been proposed based on the test results. Van der Meer (1988) proposed a formula for the rocky slope stability based on a large number of irregular wave experiments. However, the formula is insufficient to represent the damage level. Kaku et al. (1991) and Smith et al. (1992) newly proposed an empirical formula to improve the degree of agreement between the measured and predicted damage levels. The formula, however, is not universal; which requires a change of the coefficients in the formula for different data source.

It is desired to make a universal formula or a methodology to arrange the experimental data having ambiguous mutual relationships between physical factors (Mase et al., 1995). This paper shows the efficiency of neural network method to assess the stability of rubble mound breakwaters and armor layers. A neural network modeled on the structure of brain is effective to deal with information having not clear relations between causes and effects. In addition to show the efficiency, sensitivity of some parameters governing the stability is evaluated by utilizing the neural network.

ANALYZED DATA

The data sets analyzed here are van der Meer's data (1988) and Kajima et al.'s data (1993). Van der Meer (1988) performed a large number of tests on the stability of rock slopes by changing the slope angle, the permeability of the breakwater, the stone size distribution, the mass density of stones, the wave spectral shape, the significant wave height and the period. We use the experimental data of small scale tests for static stability and large scale tests for scale effect.

Kajima et al. (1993) examined the deformation of armor layer of Tetrapods against waves over a design wave in a wave flume of 60 cm width and 34 m length. Random waves were generated in the experiments with the Wallops type frequency spectrum, and the spectral shapes of wind waves and narrow-band swells were employed. The significant wave period was changed, and the incident significant wave height also changed from once to twice the design wave height ($H_D = 7.3$ cm). The design wave height was determined from the Hudson formula for the armor units.

STABILITY FORMULAE

Van der Meer (1988) showed that dominant parameters concerning the stability of rubble mound breakwaters are: the significant wave height, H_s; the mean wave period obtained from moments of frequency spectrum, T_m; the relative density of stone, $\Delta = \rho_s / \rho_w - 1$; the nominal diameter of stone, D_{n50}; the permeability parameter of breakwater, P; the number of waves acting on the breakwater, N; and the slope angle of the breakwater, α. The damage level, S, was introduced as a parameter which indicates the degree of damage:

$$S = A / D_{n50}^2 \tag{1}$$

where A is the eroded area of the armor layer of the breakwater.

The stability formula proposed by van der Meer (1988) is as follows:

$$N_s = 6.2\, P^{0.18} (S/\sqrt{N})^{0.2} / \sqrt{\xi_m}\ ; \qquad \xi_m < \xi_c \tag{2}$$

$$N_s = 1.0\, P^{-0.13} (S/\sqrt{N})^{0.2} \sqrt{\cot\alpha}\ \xi_m^P\ ; \quad \xi_m \geq \xi_c \tag{3}$$

where $N_s = H_s / \Delta D_{n50}$ is the stability number and ξ_m is the surf similarity parameter (= $\tan\alpha / \sqrt{2\pi H_s\, g T_m^2}$).

The critical value, ξ_c, in Eqs.(2) and (3) is expressed by

$$\xi_c = (6.2\, P^{0.31} \sqrt{\tan\alpha})^{\frac{1}{P+0.5}} \tag{4}$$

The stability number N_s is calculated for given variables of ξ_m, S, N, P and $\cot\alpha$. The comparison of N_s between the measured value and predicted one by Eqs.(2) and (3) shows a good agreement.

Eqs.(2) and (3) can be rewritten as follows:

$$S = \sqrt{N}\left\{ N_s\sqrt{\xi_m} / (6.2\, P^{0.18}) \right\}^5 ; \qquad \xi_m < \xi_c \qquad (5)$$

$$S = \sqrt{N}\left\{ (N_s P^{0.13}) / (\xi_m^P \sqrt{\cot\alpha}) \right\}^5 ; \quad \xi_m \geq \xi_c \qquad (6)$$

The damage level S is calculated for given values of ξ_m, N_s, N, P and $\cot\alpha$. Agreement between the observations and predictions of S is not good (Mase et al., 1995).

Kajima et al. (1993) proposed a stability formula for the Tetrapod's armor layer based on the experimental results:

$$N_s = 8.5\,(S / N^{1/2})^{0.16}\,\xi_m \qquad (7)$$

or

$$S = (N_s\sqrt{\xi_m})^{6.25}\,N^{1/2} \qquad (8)$$

The agreement of N_s between the observed value and predicted one is good. However, the predicted S in the experimental cases of $H_s \cong 2H_D$ becomes larger than the observed S (Mase, 1995).

DATA PROCESSING BY NEURAL NETWORK

A neural network is an information processing system modeled on the structure of human brain, which is able to deal with fussy information whose interrelation is ambiguous (Nielsen, 1988; Nakano et al., 1989; Ichikawa, 1993).

An artificial system of human neurons is called a neuro, and a model of a neuron is called a unit. A model unit usually has many input signals and only one output signal, shown in Fig.1. The unit is connected by the passways corresponding to synaptic connections, and signals are transmitted in one direction. Input signals, x_i, from the incoming connections are summed up by being multiplied by the weights of the connections, w_i, and then the threshold value, γ_k, is subtracted from the summed signal. The modified signal is fed into the unit. The unit gives an output signal, y, through a nonlinear transfer function, $f(x)$. The output becomes an input to other units. A common transfer function is the sigmoid function. The sigmoid function has a characteristic that its differential is expressed by the sigmoid function.

In this study we adopt a three-layered network which consists of an input layer, a hidden layer and an output layer, shown in Fig.1. A set of data to the input layer is called an input pattern. The data are transmitted to a hidden layer, and transformed in the hidden units. The outputs from the hidden units are transmitted to the units of an output layer, and transformed. The outputs from the output units are the predicted values.

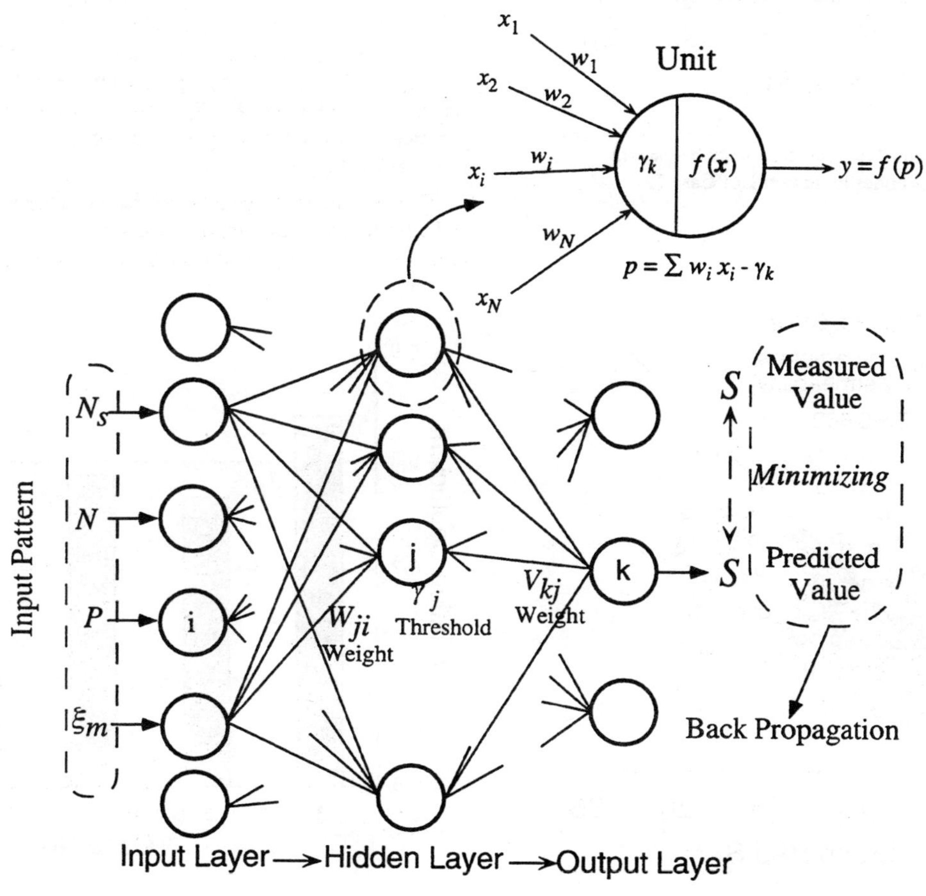

Fig.1 Unit and three-layered network.

The network is supplied not only with input patterns but also with teach signals (that is, desired output data or observed data).

The weights are modified step by step to minimize the mean square error between the predicted values and the teach signals. This procedure is called 'learning'. Achievements by learning are kept in the weights of the passways and in the thresholds of the units. A back propagation is one of the self-learning methods to give desired answers.

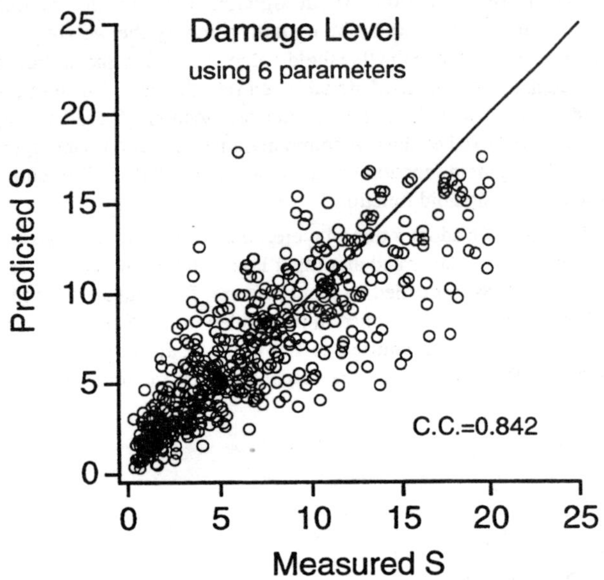

Fig.2 Comparison of measured damage levels and predicted ones by network of Case 1.

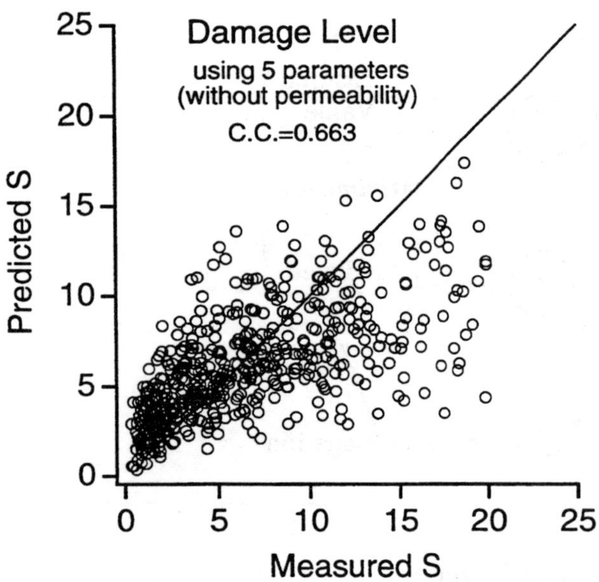

Fig.3 Comparison of measured damage levels and predicted ones by network of Case 4.

ASSESSMENT OF STABILITY OF ROCK AND BLOCK SLOPES

As a preliminary examination, we checked the characteristics of neural network by changing the numbers of hidden units and learning iterations with four input units of the permeability coefficient P, the number of waves N, the stability number N_s and the surf similarity parameter ξ_m when we predict the damage level S (Mase et al., 1995). We tried to construct a network by using 579 experimental data of van der Meer (1988). But it was impossible to construct a proper network. Therefore, we constructed the network by using the randomly selected 100 experimental data, and applied the network to 579 experimental data. However, the degree of agreement is not well improved compared with employing Kaku et al.'s formula (1991).

We added two more input parameters corresponding to the dimensionless water depth in front of the breakwater, h/H_s, and the spectral shape parameter. We constructed a network with 12 hidden units (double of input units) by using the randomly selected 100 experimental data of van der Meer (1988) and by 5000 learning iterations. Employing the network with six input units and 12 hidden units, we could predict the damage levels fairly well for van der Meer's data and Smith et al.'s data.

Here, we examine the sensitivity of some input parameters to predict the damage levels for the case of S<20 (static stability). Firstly we determine the three-layered networks with 12 hidden units by changing the input units:

(1) Case 1 employing six input parameters of P, N, N_s, ξ_m, h/H_s, and the spectral shape;
(2) Case 2 employing five parameters except h/H_s;
(3) Case 3 employing five parameters except h/H_s the spectral shape;
(4) Case 4 employing five parameters except P;
(5) Case 5 employing four parameters except and the spectral shape.

As input parameters, N, N_s, and ξ_m are always used, since these parameters are needed to represent the characteristics of breakwaters and waves at least.

Figure 2 shows the comparison between the measured damage levels and predictions by the neural network of Case 1, and Figure 3 shows the result by the network of Case 4; the correlation coefficients are 0.84 and

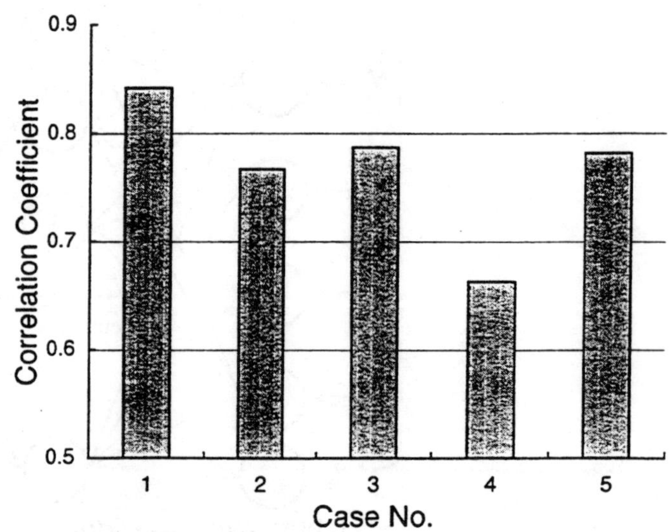

Fig.4 Correlation coefficients between measured and predicted damage levels.

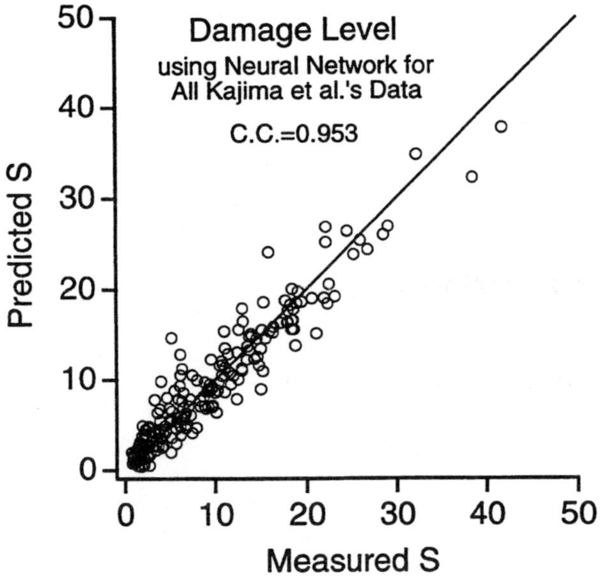

Fig.5 Comparison of measured and predicted damage levels for armor layer.

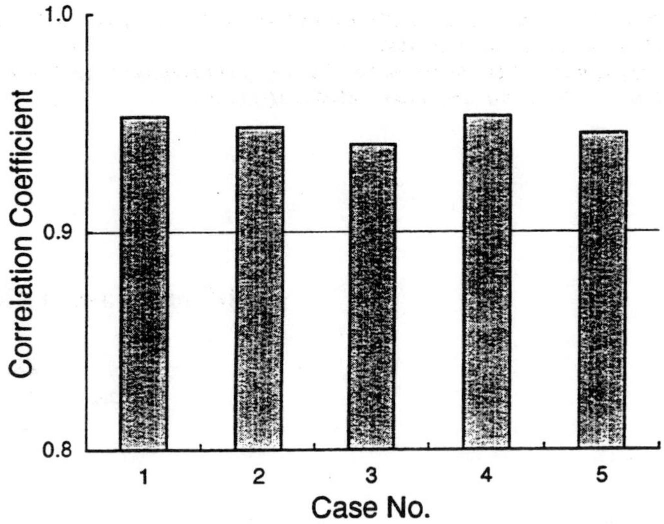

Fig.6 Correlation coefficients between measured and predicted damage levels for armor layer.

0.66, respectively. The correlation coefficients of Case 1 ~ Case 5 are summarized in Fig.4 from which it is seen that if the permeability coefficient is not employed as a input parameter, the correlation coefficient of Case 4 becomes smallest; which means that the permeability of rock slopes is significant. It is also realized from Fig.4 the parameter h/H_s is significant which represents the effect of wave height distribution indirectly.

We turn to Kajima et al.'s data. Predicted values of S by the neural network whose parameters were determined from van der Meer's data (1988) were compared with the observations of Kajima et al. (1993). An interesting point was that the predicted S becomes close to the observed S when the damage level S is large; which means that in the case of severe deformation, the stability of armor stones and Tetrapod becomes close. And it was considered that the neural network can grasp a pattern of deformation process (Mase, 1995).

Figure 5 shows the comparison of the observed S with the predicted S by the neural network with six input parameters of which weights and thresholds were determined from the Kajima et al.'s experimental data. The figure shows a good agreement between the observation and prediction. By changing the input parameters as Case 1 ~ Case 5, the correlation coefficients are obtained and arranged in Fig.6. Since the agreement between the measured and predicted S is originally very good, as seen in Fig.5, the difference between the correlation coefficients is little. Since the difference of sensitivity of input parameters may depend on the properties of structures (such as rocks or artificial blocks), detail investigations should be needed.

SUMMARY AND CONCLUSIONS

A neural network modeled on the structure of brain is said to be effective to deal with information having not clear relations between causes and effects. In this paper, we showed the efficiency of neural network method to assess the stability of rubble mound breakwaters and armor layers.

In addition to show the efficiency, sensitivity of some input parameters governing the damage level is evaluated by utilizing the neural network where the number of acting waves, the stability number, and the surf similarity parameter are always employed to represent at least the characteristics of breakwaters and waves. For the case of rubble mound breakwaters, the permeability coefficient and the dimensionless water depth were found to be significant. However, for the case of armor layer, the difference of sensitivity of input parameters was little.

ACKNOWLEDGMENT

This study was supported by the Grant-in-Aid for Developmental Scientific Research (No. 6555148), The Ministry of Education, Science, Sports and Culture, Japan.

REFERENCES

Ichikawa, H. (1993). "Layered neural network, " Kyoritu-Shuppan Company, Ltd., 184p. (in Japanese).

Kajima, R., Sakakiyama, T., Shimizu, T., Sekimoto, T., Kunisu, H. and Kyoya, O. (1993). "Estimation formula of deformation of wave-dissipating blocks under irregular waves, " Proc. Japanese Conf. Coastal Eng., Vol.40, pp.795-799 (in Japanese).

Kaku, S., Kobayashi, N. and Ryu, C. R. (1991). "Design formulas for hydraulic stability of rock slopes under irregular wave attack, " Proc. 38th Japanese Conf. on Coastal Eng., JSCE, 661-665 (in Japanese).

Mase, H., Sakamoto, M. and Sakai, T. (1995). "Neural network for stability analysis of rubble-mound breakwaters, " Jour. Waterway, Port, Coastal, and Ocean Eng., Vol.121, No.6, pp.294-299.

Mase, H. (1995). "Evaluation of artificial armour layer stability by neural network method," Proc. 26th Congress of IAHR, London, pp.341-346.

Nakano, K., Inuma, K., Neural Net Group and Kiritani, W. (1989). "Neuro computer, " Gijutu-Hyoronsha Company Ltd., 318p. (in Japanese).

Nielsen, R. H. (1988). "Neural computing: picking the human brain," IEEE Spectrum, 25(3), 36-41.

Smith, W. G., Kobayashi, N. and Kaku, S. (1992). "Profile changes of rock slopes by irregular waves, " Proc. 23rd Int. Conf. on Coastal Eng., ASCE, pp.1559-1572.

Van der Meer, J. W. (1988). "Rock slopes and gravel beaches under wave attack, " PhD thesis, Delf Univ. of Technology, Pub. No.396, Delft, The Netherlands, 152p.

Proceedings of the Seventh (1997) International Offshore and Polar Engineering Conference
Honolulu, USA, May 25-30, 1997

ISBN 1-880653-28-1 (Set); ISBN 1-880653-31-1 (Vol. III)

Oblique Wave Reflection from a Breakwater

Zhao-Chen Sun
Dalian University of Technology
Dalian, China

ABSTRACT

A new method is presented in this paper for separating oblique incident and reflected waves from coastal structures. The waves may be regular or irregular. The wave direction may be arbitrary. The technique is applicable to two wave gauges under a certain limitation. Two and three dimensional model tests have been done for vertical structures with and without air chambers. The results show that the method presented in this paper can be used to separate reflected waves from random incident waves, and can give accurate reflection coefficient.

INTRODUCTION

Ocean surface waves in coastal areas are usually the superposition of incident waves and reflected waves, especially in front of offshore or coastal structures. The incident waves are partly dissipated on or within the structure, partly transmitted through and/or over the structure, and partly reflected from the structure, depending on the type of the structure. But it is quite often that the reflection coefficients are unknown. It is necessary to separate reflected waves from incident waves for recorded wave data.

There are a few methods for separating the incident waves and reflected waves. In many studies, it is usually assumed that the incident waves are unidirectional, propagating perpendicular to the reflector. Thornton and Calhoun (1972), Morden et al (1976), and Goda and Suzuki (1976) used two wave gauges positioned on a line perpendicular to the reflector to decompose the wave field into incident and reflected wave contributions. The technique fails when the two wave gauge spacing is equal to half of the wave length. Mansard and Funke (1980) overcame this problem by using three wave gauges and least-square method, and this technique was extended by Zelt and Skjelbreia (1992) to linear arrays with an arbitrary number of wave gauges. To decompose the incident and reflected waves of a random, multidirectional sea from a coastal structure, a technique was proposed by Dickson et al (1995). The assumption of normally incident waves used in previous studies is relaxed. But angles of incidence is assumed to be small.

A new method is presented for separating the reflected waves from random, unidirectional incident waves. The technique is applicable to two wave gages arranged in arbitrary geometry deployed seaward of the breakwater. The assumption of normally incident waves in previous studies is relaxed. The incident angles can be arbitrary, nearly from -90° to +90°. When the incident waves are normal, the method developed by Goda is a special case of this technique. Two and three dimensional model tests have been done for vertical structures with and without air chambers. The results show that the method presented in this paper can be used to separate reflected waves from random incident waves, and can give accurate reflection coefficient.

MATHEMATICAL FORMULATION

In this section a new method is presented for calculating reflected waves and incident waves for recorded wave data in front of a reflector which may be a breakwater, seawall, sand bar, or steep foreshore. Two wave gauges are positioned seaward of the reflector in an arbitrary form, as shown in Fig. 1. The coordinates of the wave gauges are (x_1,y_1) and (x_2,y_2). For regular waves it is assumed that the incident waves can be expressed as

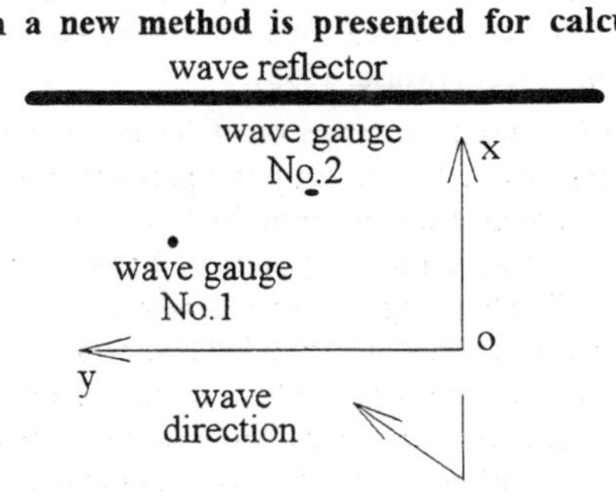

Fig.1 Definition Sketch

$$\eta_I = a_I \cos(kx\cos\theta + ky\sin\theta - \omega t + \phi_I) \quad (1)$$

If it is assumed that the incident wave angle is equal to the reflected wave angle, then the reflected waves can be expressed as

$$\eta_R = a_R \cos(-kx\cos\theta + ky\sin\theta - \omega t + \phi_R) \quad (2)$$

Where a_I and a_R are incident and reflected wave amplitude. ϕ_I and ϕ_R are the phase angles of incident and reflected waves. k is the wave number. θ is the incident wave angle between incident wave direction and x axis. ω is the wave angular frequency.

Superposition of incident waves and reflected waves gives total waves.

$$\eta = \eta_I + \eta_R = a_I \cos(\varphi_I - \omega t) + a_R \cos(\varphi_R - \omega t) \quad (3)$$

Where

$$\varphi_I = kx\cos\theta + ky\sin\theta + \phi_I,$$
$$\varphi_R = -kx\cos\theta + ky\sin\theta + \phi_R$$

At position 1, the total wave is

$$\begin{aligned}\eta_1 &= a_I \cos(\varphi_{I1} - \omega t) + a_R \cos(\varphi_{R1} - \omega t) \\ &= A_1 \cos\omega t + B_1 \sin\omega t\end{aligned} \quad (4)$$

At position 2, the total wave is

$$\begin{aligned}\eta_2 &= a_I \cos(\varphi_{I2} - \omega t) + a_R \cos(\varphi_{R2} - \omega t) \\ &= A_2 \cos\omega t + B_2 \sin\omega t\end{aligned} \quad (5)$$

Where

$$A_1 = a_I \cos\varphi_{I1} + a_R \cos\varphi_{R1} \quad (6)$$
$$B_1 = a_I \sin\varphi_{I1} + a_R \sin\varphi_{R1} \quad (7)$$
$$A_2 = a_I \cos\varphi_{I2} + a_R \cos\varphi_{R2} \quad (8)$$
$$B_2 = a_I \sin\varphi_{I2} + a_R \sin\varphi_{R2} \quad (9)$$

If the wave data at positions 1 and 2 are known, the coefficients A_1, B_1, A_2, B_2 in equations (4) and (5) can be determined by FFT analysis. Then solving equations (6) to (9), we can get the values of a_I, a_R, ϕ_I and ϕ_R. a_I and a_R can be determined by the following equations.

$$\begin{aligned}a_I = \frac{1}{2}\frac{1}{|\sin[(x_1 - x_2)k\cos\theta]|}[(B_1\cos D_1 - \\ A_1 \sin D_1 - B_2 \cos D_2 \\ +A_2 \sin D_2)^2 + (A_1 \cos D_1 + B_1 \sin D_1 \\ -A_2 \cos D_2 - B_2 \sin D_2)^2]^{1/2}\end{aligned} \quad (10)$$

$$\begin{aligned}a_R = \frac{1}{2}\frac{1}{|\sin[(x_1 - x_2)k\cos\theta]|}[(A_1\sin C_1 - \\ B_1 \cos C_1 - A_2 \sin C_2 \\ +B_2 \cos C_2)^2 + (A_1 \cos C_1 + B_1 \sin C_1 \\ -A_2 \cos C_2 - B_2 \sin C_2)^2]^{1/2}\end{aligned} \quad (11)$$

The reflection coefficient of the reflector can be determined through the following equation.

$$k_r = a_R / a_I \quad (12)$$

The assumption of regular waves is often violated in a natural coastal environment. In most cases, the recorded wave data is of the characteristics of random waves. It is necessary to get a technique to decompose a random incident and reflected sea state. Because the ocean surface elevation is a Gaussian process, its mean value has to be zero. If it is assumed that the incident waves are unidirectional random waves and, and are composed of wave components each with characteristics of regular wave, the incident and reflected waves for random seas can be expressed as

$$\eta_I = \sum_{m=1}^{M} a_{Im} \cos(k_m x\cos\theta + k_m y\sin\theta - \omega_m t + \phi_{Im}) \quad (13)$$

$$\eta_R = \sum_{m=1}^{M} a_{Rm} \cos(-k_m x\cos\theta + k_m y\sin\theta - \omega_m t + \phi_{Rm}) \quad (14)$$

Where a_{Im} and a_{Rm} are the amplitudes of wave components of incident and reflected waves, M is the total number of wave components, and k_m and ω_m are the wave number and angular frequency of the component wave and follow the following equation.

$$\omega_m^2 = k_m g \tanh(k_m d)$$

Where g is the gravitational acceleration, and d the water depth. At wave gauge position 1, the total wave can be expressed as

$$\begin{aligned}\eta_1 &= \sum_{m=1}^{M} a_{Im} \cos(\varphi_{I1m} - \omega_m t) + a_{Rm} \cos(\varphi_{R1m} - \omega_m t) \\ &= \sum_{m=1}^{M} A_{1m} \cos\omega_m t + B_{1m} \sin\omega_m t\end{aligned} \quad (15)$$

$$\eta_2 = \sum_{m=1}^{M} a_{Im}\cos(\varphi_{I2m} - \omega_m t) + a_{Rm}\cos(\varphi_{R2m} - \omega_m t)$$

$$= \sum_{m=1}^{M} A_{2m}\cos\omega_m t + B_{2m}\sin\omega_m t \quad (16)$$

Where

$$A_{1m} = a_{Im}\cos\varphi_{I1m} + a_{Rm}\cos\varphi_{R1m} \quad (17)$$

$$B_{1m} = a_{Im}\sin\varphi_{I1m} + a_{Rm}\sin\varphi_{R1m} \quad (18)$$

$$A_{2m} = a_{Im}\cos\varphi_{I2m} + a_{Rm}\cos\varphi_{R2m} \quad (19)$$

$$B_{2m} = a_{Im}\sin\varphi_{I2m} + a_{Rm}\sin\varphi_{R2m} \quad (20)$$

and

$$\varphi_{I1m} = k_m x_1\cos\theta + k_m y_1\sin\theta + \phi_{Im},$$

$$\varphi_{R1m} = -k_m x_1\cos\theta + k_m y_1\sin\theta + \phi_{Rm}$$

$$\varphi_{I2m} = k_m x_2\cos\theta + k_m y_2\sin\theta + \phi_{Im},$$

$$\varphi_{R2m} = -k_m x_2\cos\theta + k_m y_2\sin\theta + \phi_{2m}$$

If the water surface elevation at position 1 and 2 is known, the coefficients $A_{1m}, B_{1m}, A_{2m}, B_{2m}$ can be obtained by using Fourier analysis. Solving equations (16) to (19) we can get the wave amplitude of each component of both incident and reflected waves.

$$a_{Im} = \frac{1}{2}\frac{1}{|\sin[(x_1 - x_2)k_m\cos\theta]|}[(B_{1m}\cos D_{1m} - A_{1m}\sin D_{1m} - B_{2m}\cos D_{2m} + A_{2m}\sin D_{2m})^2 + (A_{1m}\cos D_{1m} + B_{1m}\sin D_{1m} - A_{2m}\cos D_{2m} - B_{2m}\sin D_{2m})^2]^{1/2} \quad (21)$$

$$a_{Rm} = \frac{1}{2}\frac{1}{|\sin[(x_1 - x_2)k_m\cos\theta|}[(A_{1m}\sin C_{1m} - B_{1m}\cos C_{1m} - A_{2m}\sin C_{2m} + B_{2m}\cos C_{2m})^2 + (A_{1m}\cos C_{1m} + B_{1m}\sin C_{1m} - A_{2m}\cos C_{2m} - B_{2m}\sin C_{2m})^2]^{1/2} \quad (22)$$

The observed waves are the summation of all the wave components and the observed wave energy is the summation of energy of each wave component. The amplitude of the observed incident and reflected waves can be obtained by the following equations.

$$A_I = \left(\sum_{m=1}^{M} a_{Im}^2\right)^{0.5} \quad (23)$$

$$A_R = \left(\sum_{m=1}^{M} a_{Rm}^2\right)^{0.5} \quad (24)$$

The reflection coefficient of the reflector is

$$k_r = A_R / A_I \quad (25)$$

It is noted that when the incident waves propagate normally to the reflector and the two wave gauges are positioned on a line perpendicular to the reflector, this technique is exactly as same as Goda's method.

It should be noted that the technique breaks down when $(x_1 - x_2)k_m\cos\theta = n\pi$ which means that the two wave gauges are on a line parallel with the reflector, or the incident wave angle is equal to $\pm 90^\circ$, which means that the incident waves propagate along the reflector, or the distance of the two wave gauges in x direction is equal to half of the wave length in x direction. In these instances, there is no difference between the wave data from the two wave gauges but only a phase shift of 180^0. The decomposition of incident and reflected waves can not be conducted.

The limitation being avoided, the technique can be used for spectral analysis. The spectral density curves for incident and reflected waves can be obtained as follows.

$$S_I(f_m) = \sum_{j=-p}^{p} H(j)\frac{0.5a_{I(m-j)}^2}{\Delta f} \quad (26)$$

$$S_R(f_m) = \sum_{j=-p}^{p} H(j)\frac{0.5a_{R(m-j)}^2}{\Delta f} \quad (27)$$

$$\sum_{j=-p}^{p} H(j) = 1 \quad (28)$$

where H(j) is a weighting function, Δf is the frequency interval, S_I and S_R are the spectral densities of incident and reflected waves.

PHYSICAL MODEL TESTS

To validate the technique, three dimensional model tests were carried out in a wave basin . The setup of 3D model tests is shown in figure 2.

Five wave gauges were used in the tests as shown in Fig. 2. The coordinates of the positions of the wave gauges are (-1.24, 0.49), (-0.7, 0.47), (-0.5, 0.0), (-0.97, -0.33) and (-1.42, -0.01) in meter. The water depth is 0.4m. The reflector is a vertical wall with a reflection coefficient close to 1. The wave conditions for regular waves are shown in Table 1.

For irregular waves, JONSWAP spectrum was used. The wave conditions are shown in Table 2.

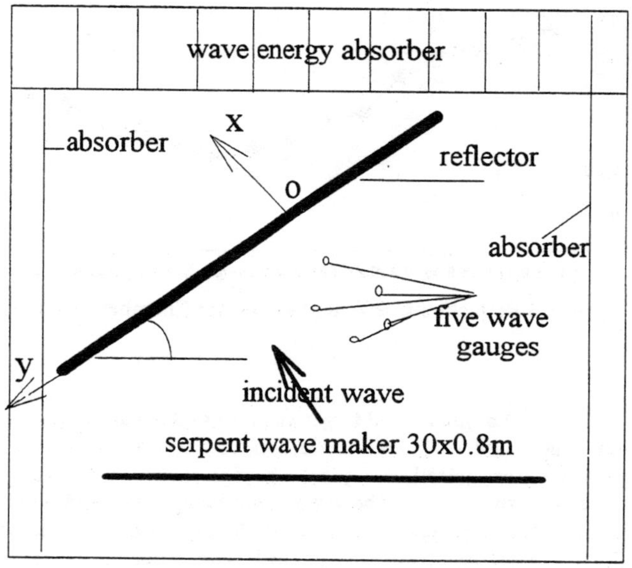

Figure 2 3D model test setup

Table 1 Regular wave conditions for 3D tests

wave height H (m)	wave period T (sec.)	wave direction θ	k_r
0.05	0.92	315^0	0.93
0.05	1.2	315^0	0.94
0.05	0.92	330^0	0.97
0.05	1.2	330^0	0.91

As the first step of the 3D tests, no reflector was installed. In this case, the reflection coefficient should be around zero. Only regular wave tests were done in this instance. The calculated reflection coefficient for the wave data is 0.08, which demonstrates that the technique is applicable with a reasonable accuracy.

When the reflector was installed, both regular and irregular waves were adopted. The reflection coefficients for regular waves are shown in the last column of Table 1. Because the reflector is a vertical wall, it is demonstrated that the results are reasonably good.

The reflection coefficients for irregular waves are shown in the last column of Table 2. It is shown that although the reflector is the same, the reflection coefficients for irregular waves are smaller than those of regular waves.

Table 2 Irregular wave conditions for 3D tests

significant wave height H (m)	mean wave period T (sec)	direction θ	γ	k_r
0.05	1.2	315	1	0.85
0.05	1.2	330	1	0.86
0.05	0.92	315	4	0.86
0.05	1.2	315	4	0.88
0.05	0.92	330	4	0.90
0.05	1.2	330	4	0.86

ENGINEERING APPLICATION

To maintain calmness in a harbor and for the berthing need, port authorities often require that the hydraulic structures of the harbor have reflectivity as low as possible. It is needed to present wave energy absorbing structures. It is also needed to determine the reflection coefficient for such structures.

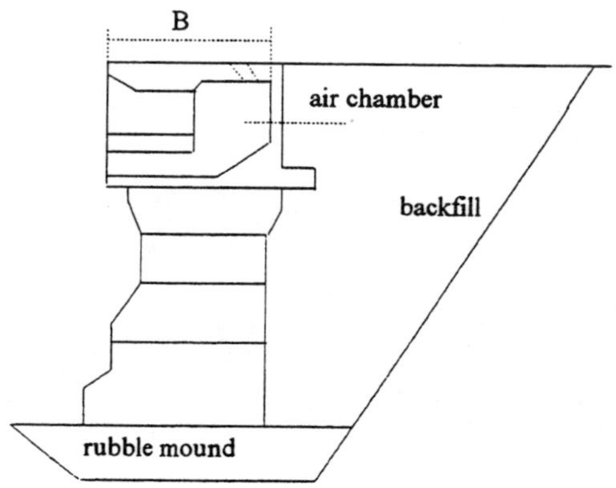

Figure 3 Wave Absorbing Wharf

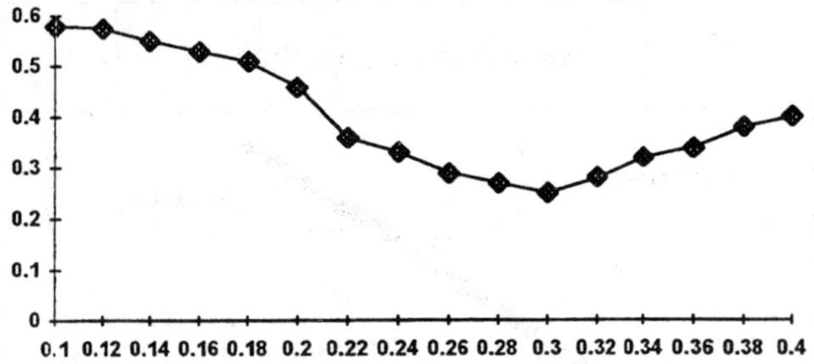

Figure 4 Reflection Coefficient kr via Air Chamber Size B/L_p

The technique is used to get an optimized geometry for a berthing wharf with least reflection coefficient. Five types of geometry were tested. A typical wharf is shown in Figure 3. It consists of five layers. The lower four layers are solid concrete blocks. The top layer is precast block with air chamber. Bretschneider spectrum was applied. The reflection coefficients vary for different size and form of the air chamber. Figure 4 shows the test results. In the figure, L_p is the wave length corresponding to the spectral peak frequency. It is noted that very low reflectivity can be achieved by adopting suitable size of the air chamber

DISCUSSION AND CONCLUSIONS

The main results of this study is the development of decomposition technique of unidirectional regular and irregular waves propagating obliquely to a coastal reflector, such as breakwater, seawall, sand bar, or steep foreshore. The technique is applicable to two wave gauges deployed seaward of the reflector. The assumption of normally incident waves in previous studies is relaxed. When the incident waves are normal and the two wave gauges are on a line perpendicular to the reflector, the method developed by Goda is a special case of this technique.

Model tests demonstrate that the technique can provide accurate estimates of the incident and reflected wave height, and reflectivity of the reflector, under the limitations of arrangement of the two wave gauges and incident wave direction.

REFERENCES

Dickson, W.S., Herbers, T.H.C. and Thornton, E.B. (1995), " Wave reflection from breakwater." J. Wtrwy., Harb., and Coast. Engrg. Div., ASCE, 121 (5), 262-268.

Goda, Y., and Suzuki, Y. (1976)," Estimation of incident and reflected waves in random wave experiments." Proc., 15th Int. Coast. Engrg. Conf., ASCE, 828-845.

Mansard, E.P.D., and Funke, E.R. (1980), " The measurement of incident and reflected spectra using a least square method." Proc., 17th Int. Coast. Engrg. Conf., ASCE, 154-172.

Morden, D.B., Richy, E.P., and Christensen, D.R. (1976), " Decomposition of co-existing random wave energy." Proc., 15th Int. Coast. Engrg. Conf., ASCE, 846-865.

Thornton, E.B., and Calhoun, R.J. (1972), " Spectral resolution of breakwater reflected waves." J. Wtrwy., Harb., and Coast. Engrg. Div., ASCE, 98 (4), 443-460.

Zelt, J.A., and Skjelbreia, J.E. (1992), " Estimating incident and reflected wave fields using an arbitrary number of wave gauges.", Proc., 23th Int. Coast. Engrg. Conf., ASCE, 777-789

Proceedings of the Seventh (1997) International Offshore and Polar Engineering Conference
Honolulu, USA, May 25-30, 1997

ISBN 1-880653-28-1 (Set); ISBN 1-880653-31-1 (Vol. III)

Nonlinear Wave Field Induced by Offshore Shoals and Vertical Structures

Eric C. Cruz and Toshio Aono
Toa Corporation
Yokohama, Japan

ABSTRACT

The horizontally two-dimensional transformations of waves due to a spherical shoal and vertical structures are simulated using Boussinesq equations with an extended dispersivity. A comparison of simulated wave heights is made with experiment data. Wave energy dissipation around structure discontinuities is modeled by a momentum exchange term. Results indicate that wave heights at a large distance from a diffracting rectangular island can be predicted well with a suitable choice of the structure damping coefficient. The model is applied to wave simulation around an existing offshore harbor and it is shown that radiated waves from the interior can be concurrently determined.

KEY WORDS: shoal, nonlinear wave, Boussinesq equations, diffraction, vertical structures, radiated waves

INTRODUCTION

The transformation of waves propagating in the offshore region is generally governed by refraction. Changes in wave direction and wave height are well predicted by linear refraction models. It is known, however, that refraction models based on the assumption of an infinitesimal wave become inadequate when the wave becomes increasingly steep or when the wave amplitude becomes significant relative to the water depth. In the presence of fixed structures, wave motion is affected by the abrupt decrease of the effective flow area. This condition must result in local amplification of wave energies and flow velocities adjacent to regions of discontinuities and further renders the infinitesimal wave assumption very limited. In this case, nonlinear wave models that are not based on this assumption can predict the wave evolution more closely.

Nonlinear models of the Boussinesq type have been employed in recent years in studies of wave transformation. The classical equations derived by Peregrine (1967) for long waves have been improved by later studies that extended their weak dispersivity so that they can be used up to the deep water limit (Madsen et al., 1991; Nwogu, 1993). This sufficient dispersivity in deep water and the leading-order nonlinearity in intermediate to shallow depths make the use of Boussinesq models desirable in wave studies than span a wider range of relative depths and wave amplitudes. The basic properties of a number of Boussinesq models have been investigated for wave propagation and wave shoaling (Madsen and Sorensen, 1992) and evolution of directional wave in shallow water (Nwogu, 1995).

An important class of problems where nonlinear dispersive wave models can be used is the diffraction of waves by vertical walls. On account of standing waves formed in front of reflective faces, wave nonlinearity becomes pronounced, while within the shadow zone, the low diffracted waves propagate essentially according to their phase celerities. It must follow that in the vicinity adjoining these regions, wave transformation is affected by both wave nonlinearity and dispersion. This is particularly important in situations where the diffractive wall is part of a system of structures where the region under investigation is neither close to nor far removed from them. It is then imperative to use a wave model that treats both wave nonlinearity and dispersion concurrently.

In this study, a set of Boussinesq-type equations is first used to simulate the transformation of regular waves around a spherical shoal to examine its ability to reproduce the transformed wave when caustics are present. It is then applied to the simulation of diffraction by vertical structures. Results of simulation are compared with experiment data or theory. These fundamental verifications are necessary in the final objective of simulating real sea waves on a seabed dotted with submerged features, such as shoals and reefs, as well as surface features such as breakwaters.

NONLINEAR WAVE MODEL

The model used in this study is a subset of the Boussinesq equations developed for porous beds(Cruz et al.1996):

$$\eta_t + \nabla \cdot [(h+\eta)\,\mathbf{u}] = 0 \tag{1}$$

$$
\mathbf{u}_t+\mathbf{u}\cdot\nabla\mathbf{u}+g\nabla\eta+\frac{h^2}{6}\nabla(\nabla\cdot\mathbf{u}_t)-\left(\frac{1}{2}+\gamma\right)h\nabla(h\nabla\cdot\mathbf{u}_t)
$$
$$
-\gamma gh\left[\nabla(h\nabla^2\eta)+\nabla(\nabla h\cdot\nabla\eta)\right]+\frac{f_w|\mathbf{u}|}{2(h+\eta)^2}u-f_D\nabla^2\mathrm{u}
$$
$$
-F_M+f\,\mathbf{u}=0 \qquad (2)
$$

where η denotes the water surface displacement from mean water level, **u** the depth-averaged horizontal velocity vector, the mean water depth, g the gravity acceleration, and x, y the horizontal coordinates and t time. Subscript t denotes differentiation with time while the gradient operator is $\nabla\equiv(\partial/\partial x,\partial/\partial y)$. γ denotes the celerity extension factor that extends the applicability of the model equation to deep water. $\gamma=1/15$ is an optimum value up to a relative depth $h/Lo=0.50$ with a maximum phase celerity error of three percent.

In shallow water, wave damping is modeled by a bottom friction term with coefficient f_w while wave breaking dissipation is represented by a diffusion type term with dissipation function f_D. F_M represents a flow impedance term to model the energy dissipation at structure discontinuities while f denotes the damping distribution function of the boundary damping region used to enforce the open boundary condition.

The model can be used to study nonlinear wave fields induced by bottom topographic features such as shoals, ridges and reefs as well as surface structures such as offshore platforms and breakwaters. It can also be used to quantify the nonlinear radiated waves induced by such features. This is significant when assessing the effects of structures on both shoreward and offshore tranquilities.

REFRACTION BY A SPHERICAL SHOAL

Eqs.(1) and (2) have been used to simulate the tranformation of regular waves around a spherical shoal. This problem is characterized by the combined occurrence of refraction on the shoal and diffraction on the sides and has been treated in the past by the use of linear wave models based on the mild slope equation (Berkhoff, 1972; Copeland, 1985). The combined effects of refraction and diffraction lead to the existence of caustics behind the shoal that renders the use of the linear refraction method invalid. A later method employed curvilinear coordinates to obtain close agreement between the directions of the coordinate system and the propagation of the wave (Isobe, 1986). This problem is here treated as a test case of the validity of the present nonlinear model to predict the wave heights in the presence of caustics.

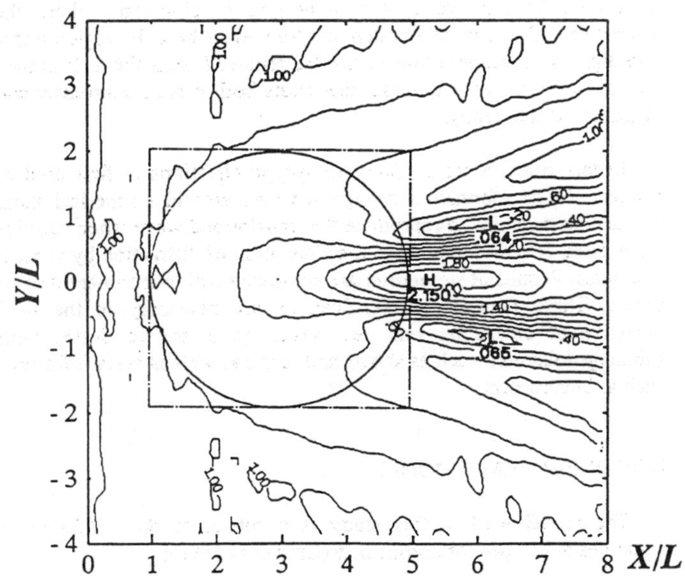

Fig. 1 Normalized wave height contours around offshore spherical shoal

Computations were carried out by attaching boundary damping regions at all sides to enforce the open boundary condition. Damping distribution $f(x,y)$ which is parabolic normal to the boundary and uniform parallel to it is the most efficient for deep water waves. The absorber width and maximum damping coefficient are obtained from the graphs in Cruz and Isobe (1994).

Fig.1 shows the computed contours of wave heights. Sectional distributions of wave heights are compared with the experiment data of Ito and Tanimoto (1972) in Fig.2. The refraction method indicates the caustic to occur immediately beyond the shoal where the resulting wave height is maximum. The model is seen to provide a close agreement with the data.

A small portion of the incident wave energy is scattered by the shoal seaward. Where the caustic would otherwise occur, a maximum wave height of $2.15H_{in}$, H_{in} being the incident wave height, is obtained. Simulation indicates the existence of extremal lows which are symmetrically disposed relative to the direction of wave propagation and beyond the caustic. These indicate that energy concentration and diffusion occur at the lee of the shoal resulting in the breakdown of the nondirectionality of the incident wave.

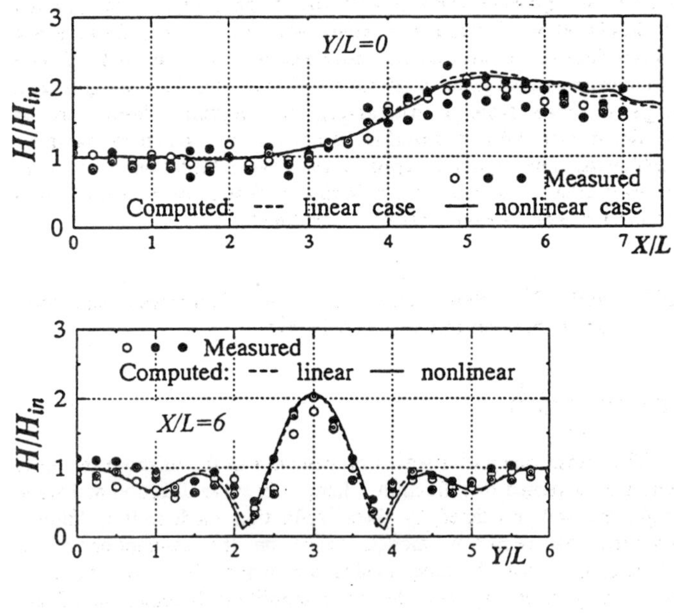

Fig. 2 Sectional distribution of wave heights of offshore shoal ($h/L_0=0.375$, $H_{in}/h=0.093$, $H/h_{min}=0.20$).

Wave evolution in the transverse direction is easily explained by a plot of the instantaneous surface displacement in Fig.3. The primary wave shoals on approach to the apex with a continually increasing bound second-order harmonic. On the increasing depth beyond the apex, this second-order harmonic starts to travel faster than the primary wave due to the larger depth on the sides. The primary and secondary crests become prominent in the flat region beyond the shoal as they travel at the same free wave celerity. The incident wave train becomes short-crested at the lee on account of these secondary crests that propagate obliquely due to the three-dimensional shape of the shoal. As seen in Fig.2 however, the effect of nonlinearity is small because of the rather low H/h at the apex and the fact that the region where this ratio is smallest is narrow.

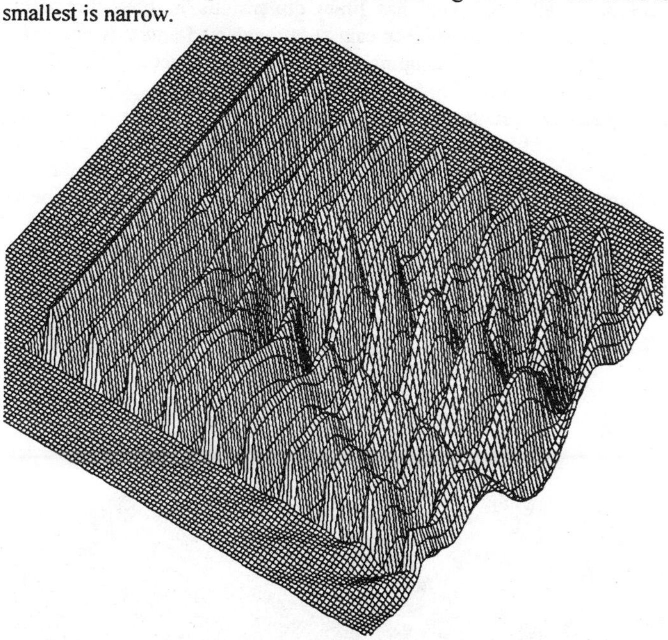

Fig. 3 Instantaneous water surface displacement around the shoal.

WAVE FIELD INDUCED BY VERTICAL STRUCTURES

Vertical structures lead to the formation of standing waves. The resulting large wave heights in front of them cause a steep surface gradient near flow discontinuities such as corners. Velocities in these areas undergo steep gradients in magnitude and direction. The structure damping term F_M. Eq.(2) represents the effect of this flow impedance and models the observed damping of nonbreaking waves behind offshore structures which could not be accounted for by bottom friction. F_M is formulated as a lateral mixing term

$$F_M = \nabla \cdot (\varepsilon \nabla \mathbf{u}) \tag{3}$$

where ε represents a flow-dependent momentum exchange coefficient which has the dimension of length^2/time. ε is a function of the size of eddies induced by flow impedance. Longuet-Higgins (1970) used a coefficient of the form

$$\varepsilon = N l \sqrt{g(h+\overline{\eta})} \tag{4}$$

where l is a typical length scale (taken as the surf zone width in nearshore wave studies), $\overline{\eta}$ the wave set-up and N is a number between 0 and 0.016.

In shallow water with an irregular topography, the large gradients of radiation stress cannot be balanced by changes in the mean water level alone and it is necessary to include bottom friction as a damping mechanism in wave models. This results in smaller currents that more closely match observed values. In deep water, however, this damping mechanism is insignificant and since radiation stresses are locally large due to reflection from and diffraction around a structure, damping must be induced by the structure through a form of momentum diffusion. In this study, the form of momentum diffusion coefficient in Eq.(4) is used with a length scale proportional to the effective depth at the structure discontinuity as follows:

$$\varepsilon = \alpha_M \sqrt{g}\left(h+\overline{\eta}\right)^{3/2} \tag{5}$$

where the coefficient α_M is null away from a structure discontinuity and has a maximum value at a structure corner. As a boundary condition, the normal velocity component is set to null at all structure points, including the corners.

Diffraction by a semi-infinite breakwater

Eqs.(1) and (2) have been used to model the classic example of wave diffraction by a semi-infinite breakwater. This problem has been analytically solved by Penny and Price (1952) in terms of Fresnel integrals. To be consistent with the assumptions of the theoretical solution, a minimum thickness of breakwater equal to the grid size is utilized. Because the theory also neglects the nonsingularity of the solution at the breakwater tip, it is necessary to use a small time step to avoid unstable velocities around the corners as the flow impedance F_M is set to zero.

Fig. 4a shows an instantaneous water surface displacement field, including the damping regions. The wave envelopes after 30 periods are used to obtain the diffraction coefficient K_d. This is plotted in Fig.4b for various locations together with the analytical solution. The general agreement with theory is clear. However, simulation gives higher K_d near the breakwater tips where the effect of wave nonlinearity is accentuated due to standing waves formed in front of the structure. The standing wave heights across a distant section from the tip are higher in the nonlinear model, especially at the locations of nodes and antinodes. In the shadow zone, the model completely agrees with theory since the wave nonlinearity is substantially reduced and wave propagation is governed by phase celerity alone.

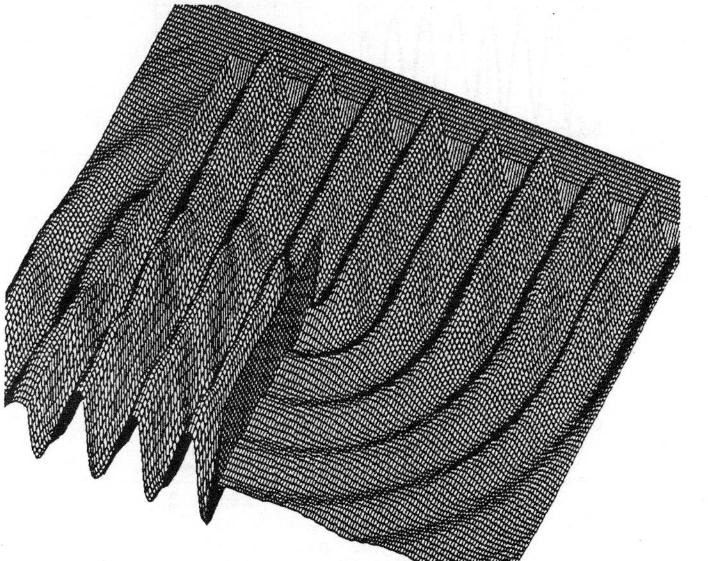

Fig. 4a An instantaneous water surface displacement around semi-infinite breakwater. *(H_{in}/h=0.10, h/L_o=0.23)*

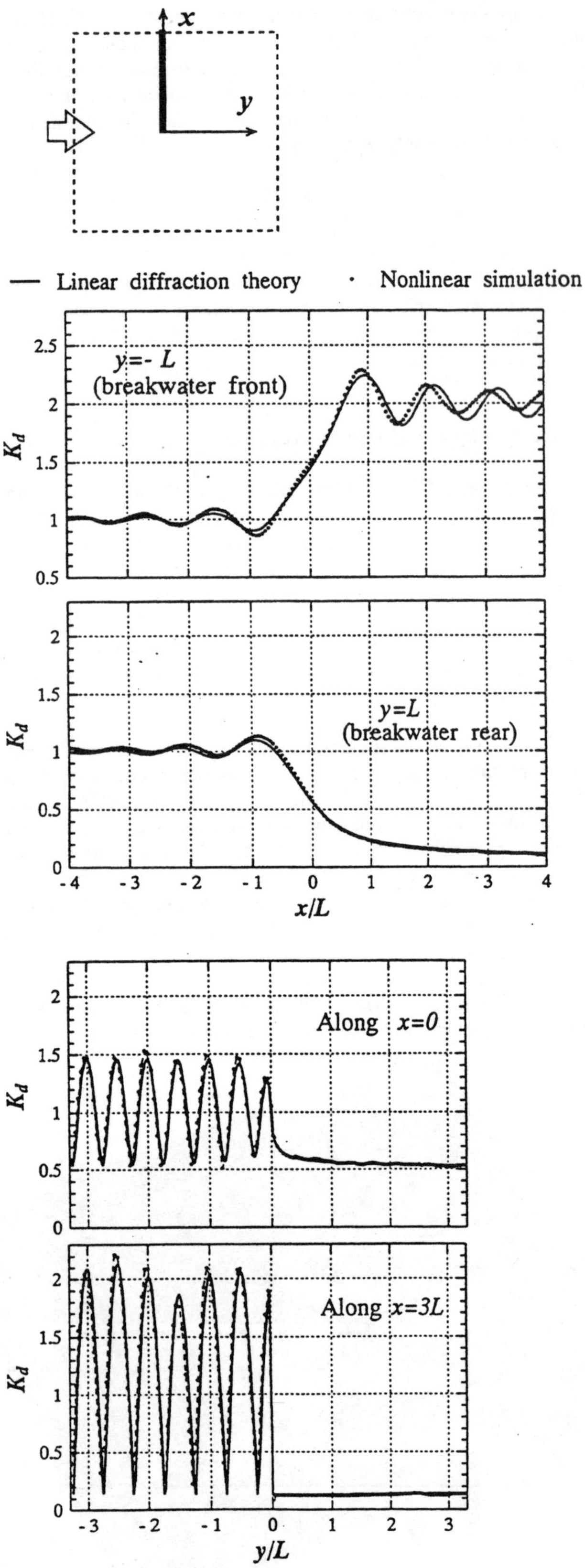

Fig. 4b Diffraction coefficients at selected sections of a semi-infinite breakwater *($H_{in}/h=0.10$, $h/L_0=0.23$)*

Diffraction by a rectangular island

The model has been used to determine the wave height at a distant transverse section of a rectangular island. The test location is located at a distance of more than eight wavelengths from the back of the structure. Unlike the semi-infinite breakwater which was assumed as a very thin wall, the island has finite dimensions in both directions. Consequently, flow impedance cannot be neglected and it is necessary to represent the actual damping induced by the structure through the term F_M in Eq.(2). Fig.5a shows an instantaneous water surface configuration around the island. and Fig 5b shows the resulting diffraction coefficients. The front is a reflection zone, the sides are progressive wave zones with scattered waves and the island lee is a diffraction zone. Fig. 6 shows the results for two simulations, one in which the structure damping is absent and another where it is nominally set. The simulated diffraction coefficients are compared with the wave height data of Harms (1980) which were obtained through careful measurements.

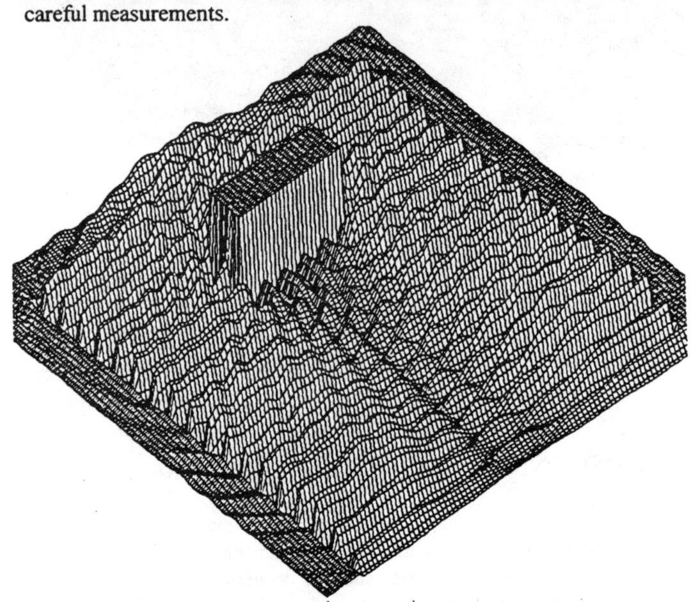

Fig 5a Instantaneous water surface field around a rectangular islanc
($H_{in}/h=0.12$, $h/L_0=0.373$)

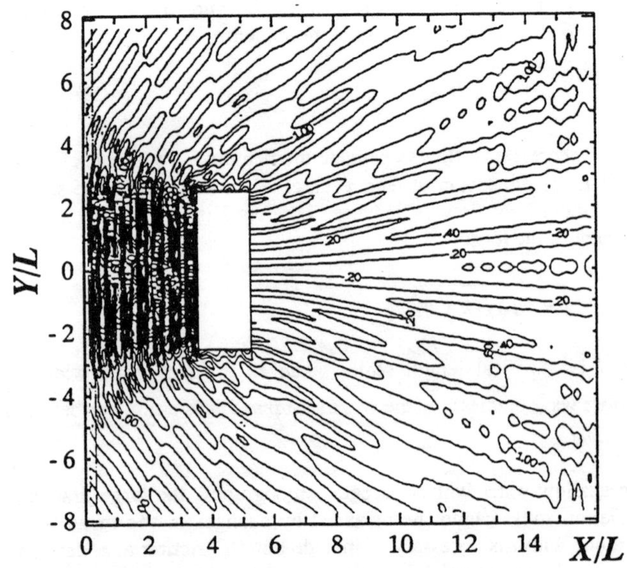

Fig. 5b Diffraction coefficients around rectangular island

Wave heights are predicted well by the simulation with α_M =0 except at the edges where the effect of structure damping is greatest. A second simulation with α_M =0.50 reduces the high velocities at the corners. Wave damping is significant at the edges because of the finite width (*1.74L*) of the island which makes the propagating wave less diffused by the ensuing diffraction at the back

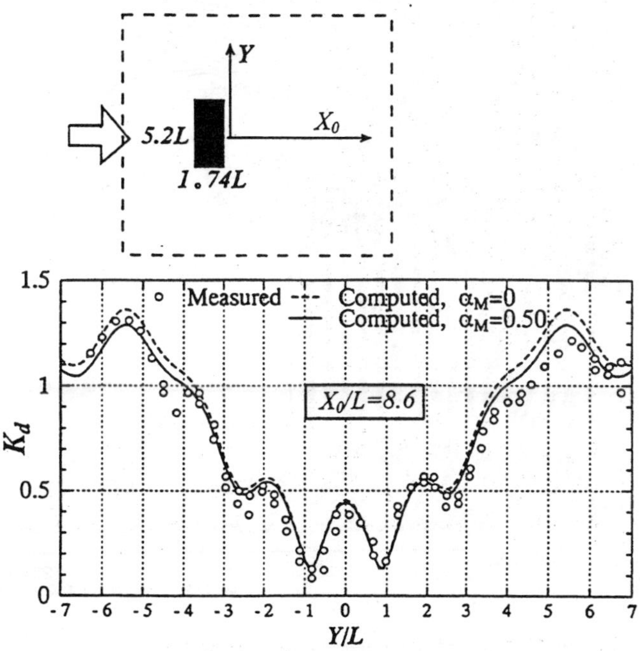

Fig. 6 Diffraction coefficients at a transverse section behind a rectangular island.

Wave field around an offshore breakwater

The model has been applied to the preliminary simulation of waves around an existing harbor in Japan. This harbor is used primarily to transport coal required for electricity generation. Because of the gaps in the breakwaters during construction, wave penetration was considerable in the shoremost region of the harbor. Structure dissipation was modeled with α_M =0.30 and bottom friction was included with f_w=*0.01*. Fig 7a shows the bottom configuration. The bed is 25 meters deep offshore and slopes at 0.0045 toward the shore with essentially parallel contours to the shoreline. The minimum depth onshore is about 10 meters. Due to this relatively deep water, the bottom friction term is negligible and no breaking zones appear anywhere so that f_D =*0* in Eq.(2).

Fig.8a shows an instantaneous wave field induced by a regular wave with the significant wave height and period. The wave heights obtained from the surface envelopes after 20 periods are shown in Fig. 8b for a nonlinear simulation. These show the extent to which the inner harbor is agitated by wave transformation. The long breakwater offshore causes substantial reflected waves while the small offshore vertical structure scatters the wave in all directions.

A linear simulation was carried out by neglecting the nonlinear terms in Eqs.(1), (2) and (3). This is the case of dispersive linear Boussinesq equations with linear structure damping. A surface configuration and the resulting wave heights are shown in Fig. 9. The concentration of wave energy at structure discontinuities for the nonlinear case is prominent particularly when the wall is normal to wave propagation or when it is in relatively shallow water. Wave nonlinearity also leads to higher radiated waves as well as those that shoal in shallow water.

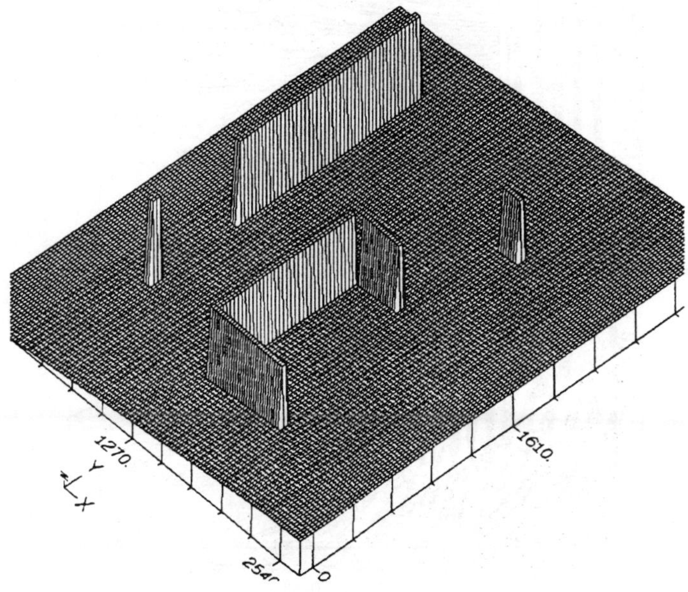

Fig. 7 Bottom configuration for simulation, Hitachinaka Harbor

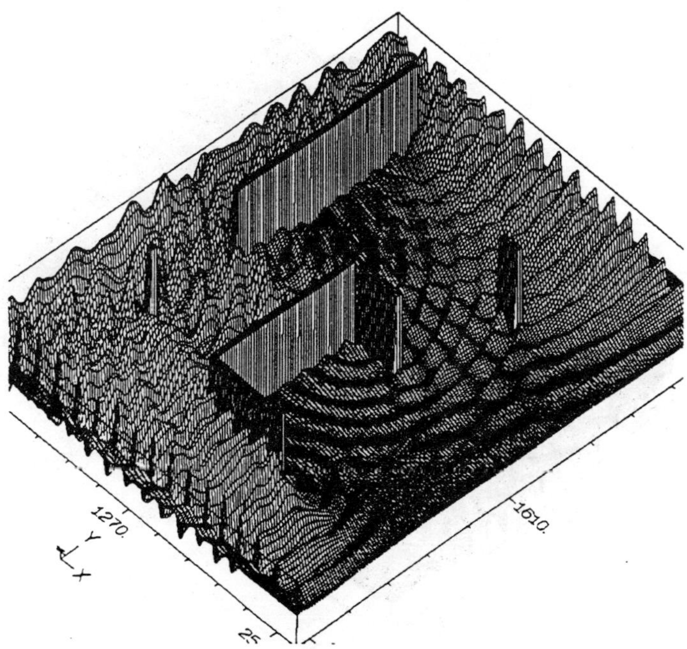

Fig. 8a Instantaneous water surface displacement, nonlinear simulation. (H_{in}=*3.28* meters, T_{in}=*12.30* sec)

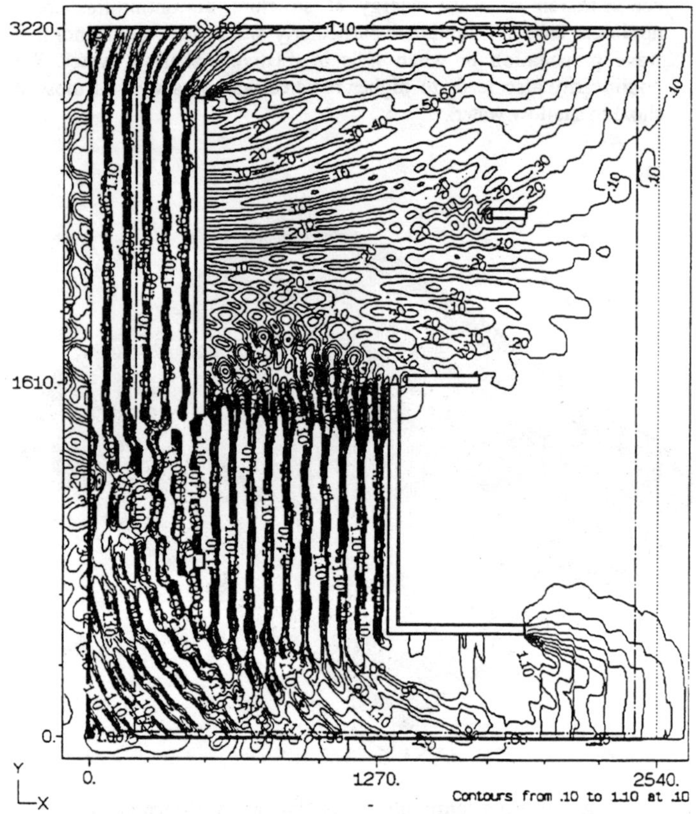

Fig 8b Normalized wave heights, nonlinear simulation.

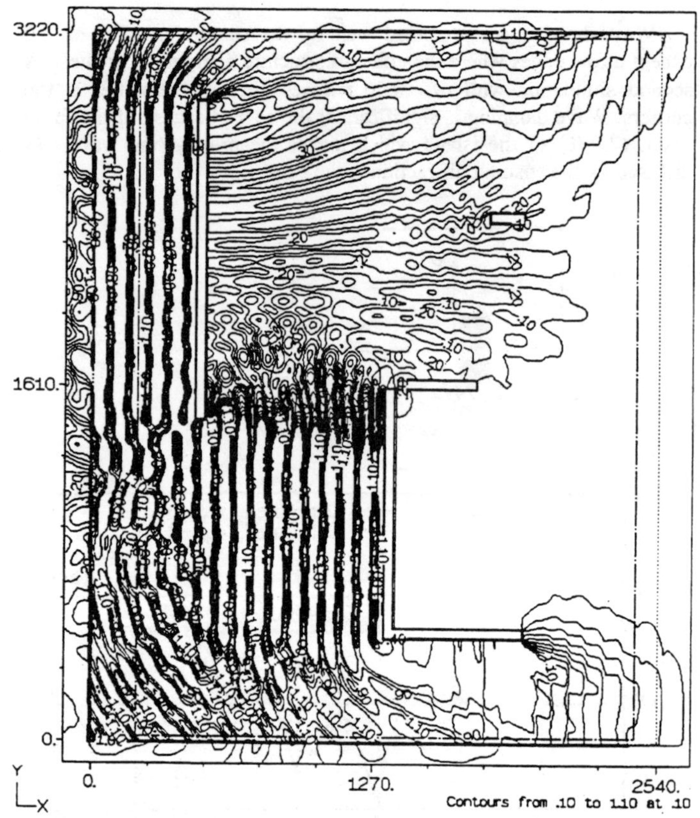

Fig. 9b Normalized wave heights, linear simulation.

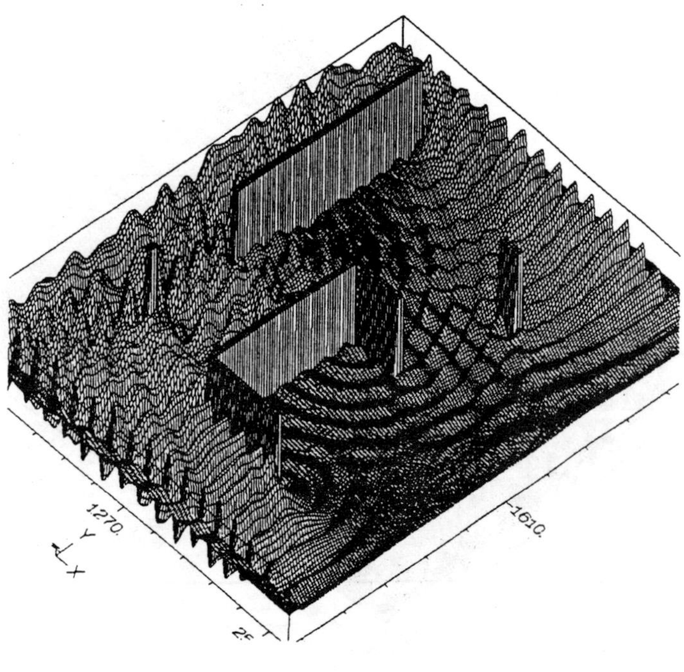

Fig. 9a Instantaneous water surface displacement, linear simulation ($H_{in}=3.28$ meters, $T_{in}=12.30$ sec)

An important property of the present model is its ability to determine the waves radiated out from the inner region and can be used to assess the effect of a system of structures on offshore tranquility. Fig. 10 shows the radiated waves from at the incidence region (*X=0*) and the outgoing wave along the side (*Y=0*). Vertical structures of the reflective type are seen to affect considerably their immediate vicinity

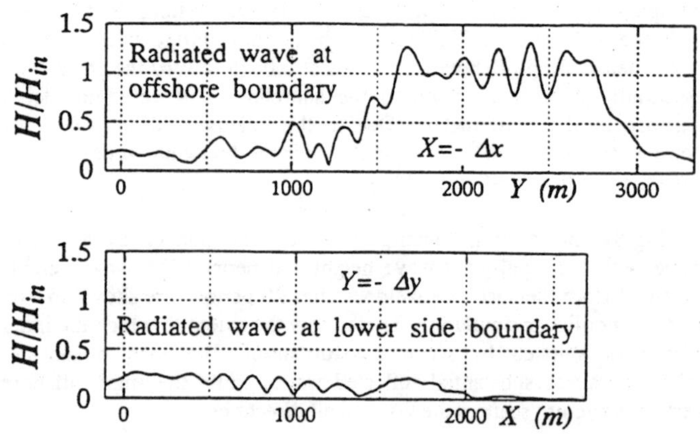

Fig. 10 Relative heights of radiated waves to offshore and side regions, nonlinear simulation.

CONCLUSIONS

An extended Boussinesq model is used to simulate the transformation of waves around an offshore shoal and vertical structures. It is found that wave evolution in the offshore region or in deep water is primarily dependent on frequency dispersion and that wave nonlinearity becomes significant in regions where bound waves are locally generated by depth constrictions or where waves steepen due to flow impedance. In the latter, the resulting steep velocity gradients result in physical wave damping at structure discontinuities and affect the wave field beyond them.

REFERENCES

Berkhoff, J.C.W. (1972). "Computations of Combined Refraction-Diffraction", *Proc 15th Int Conf Coastal Eng, ASCE*, pp 471-490.

Copeland, G. (1985). "*A Practical Alternative to the 'Mild Slope' Wave Equation*," Coastal Eng, Vol 9, pp 125-149.

Cruz, E.C. and Isobe, M. (1994). "Numerical Wave Absorbers for Short and Long Wave Modelling," *Proc Int Symp on Waves - Phys and Num Modelling*, Univ. British Columbia, Vol 2, pp 992-1001.

Cruz, E.C., Isobe, M. and Watanabe, A. (1996). "Boussinesq Equations for Wave Transformation on Porous Beds," (in press) *Coastal Eng*, CENG 00727.

Harms, V.W. (1980). "Diffraction of Water Waves by Isolated Structures," *J Waterw Port Coastal Ocean Div*, ASCE, Vol 106, No WW4, pp 494-497.

Isobe, M. (1986). "A Parabolic Refraction-Diffraction Equation in the Ray-front Coordinate System," *Proc 20th Int Conf Coastal Eng*, ASCE,

Ito, Y. and Tanimoto, K. (1972). "A Method of Numerical Analysis of Wave Propagation - Application to Wave Diffraction and Refraction," *Proc 13th Int Conf Coastal Eng*, ASCE, pp. 503-522.

Longuet-Higgins, M.S. (1970): "Longshore Currents Generated by Obliquely Incident Sea Waves," *J Geophys Res*, Vol 75, No 33, pp 6778-6801.

Madsen, P.A., Murray, R. and Sorensen, O.R. (1991). "A New Form of the Boussinesq Equations with Improved Linear Dispersion," *Coastal Eng*, Vol 15, pp 371-388.

Madsen, P.A. and Sorensen, O.R. (1992): "A New Form of the Boussinesq Equations with Improved Linear Dispersion Characteristics, Part 2. A Slowly-varying Bathymetry," *Coastal Eng*, Vol 18, pp 183-204.

Nwogu, O. (1993). "Alternative Form of Boussinesq Equations for Nearshore Wave Propagation," *J Waterw Port Coastal Ocean Eng*, ASCE, Vol 119, No 6, pp 618-638.

Nwogu, O. (1994). "Nonlinear Evolution of Directional Wave Spectra in Shallow Water," *Proc 24th Int Conf Coastal Eng*, ASCE, pp 467-481.

Peregrine, D.H. (1967). "Long Waves on a Beach," *J Fluid Mech*, Vol 27, Part 4, pp 815-827.

Penney, W.G. and Price, A.T. (1952). "The Diffraction Theory of Sea Waves and the Shelter Afforded by Breakwaters," *Philo Trans Royal Soc A*, Vol 244 (882), pp 236-253.

Proceedings of the Seventh (1997) International Offshore and Polar Engineering Conference
Honolulu, USA, May 25-30, 1997

ISBN 1-880653-28-1 (Set); ISBN 1-880653-31-1 (Vol. III)

Resonance Phenomenon Developed by Breakwater or Breakwater Fixed on Piles

C. Colmard and M. Bélorgey
Laboratoire de Mécanique des Fluides et Génie Civil
Le Havre, France

P. Roger
Société Bouygues Off Shore
Montigny-le-Bretonneux, France

Abstract

Many problems occur when the protected setting is in deep water. Many new systems were created like floating caissons, immersed flags, caissons fixed on piles, ARION systems, AGRIPEL absorbers... In this paper, we show that these breakwater types can present resonance phenomena developing when the protected setting is not totally absorbing. A simple calculation technique of these phenomena is described and applied to the « Fixed water wall » phenomenon developed by caissons fixed on piles.

Introduction

The system studied in the open sea (when the distance is great between the breakwater and the setting) behaves as a pass band filter (immersed flags for example) or a low pass filter.
In a range of periods the amplitude of the swells strongly decreases behind the breakwater. The comportment of this type of breakwater can be characterized by the swell transmission coefficient Kt : Kt=Ht/Hi. Ht is the swell amplitude transmitted behind the breakwater and Hi is the incident swell amplitude.
When we study the different papers relating to this type of structure, we observe that the breakwater is tested as if the port side was an infinite surface. For. the fact of using a perfect absorber beach (real or numerical) at the end of the wave flume, simulates the phenomenon, but is not representative of the reality.
For a good understanding of the phenomenon, we must consider the protected setting. This study must show the interaction between the protection systems and strongly reflecting settings like a steeply sloping sea bottom, (Monaco for example), or a large boat moored near the breakwater.

Phenomenon explanation :

To understand the phenomenon, we must divide the zone where the breakwater is built into three parts.

The port part is considered as an excited water body (oscillator).
The open seaside is the part which generates the energetic movement and transmits it to the water body in the port (excitation).
The part under the structure is a filter more or less active in accordance with the swell period (attenuation).

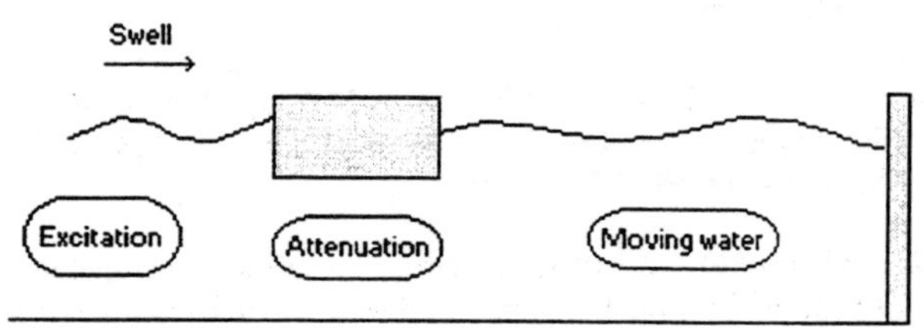

Fig. 1 : Modelization of the system « fixed water wall »

The water body in the port is actived by the opensea side excitation and partly attenuated by the structure. The incident attenuated swell goes toward the vertical wall and is reflected by it. Next, the swell goes back toward the structure. A part of the swell goes under the structure and the other part is reflected again by the immersed vertical surface of the breakwater.
The reflexions are manifold and constitute a swell system which moves the free surface. When the distance 2d (distance covered by a wave before it is reflected again by the structure) is a multiple of the wave length L, the swell reflected by the immersed part of the breakwater, is in phase with the incident excitation. It is in this configuration that the maximal amplitudes appear in the port.

Experimental study

One of the most outstanding system in this field is the system « fixed water wall ». This system is induced by caissons fixed on piles. Presently, it is developed by the Bouygues Offshore Company. This company owns an exclusive licence outside the Principality of Monaco. A theorical study was done by Bouchet and Manzone [1], to calculate the transmission and reflexion coefficient of the wave when the caisson is built in the opensea. In the same conditions, many studies realized by Bélorgey and Colmard [2], based on a numerical model and several Laser Velocimetry measure campaigns, show eddy phenomena developed by the system and the velocity field around the caisson.

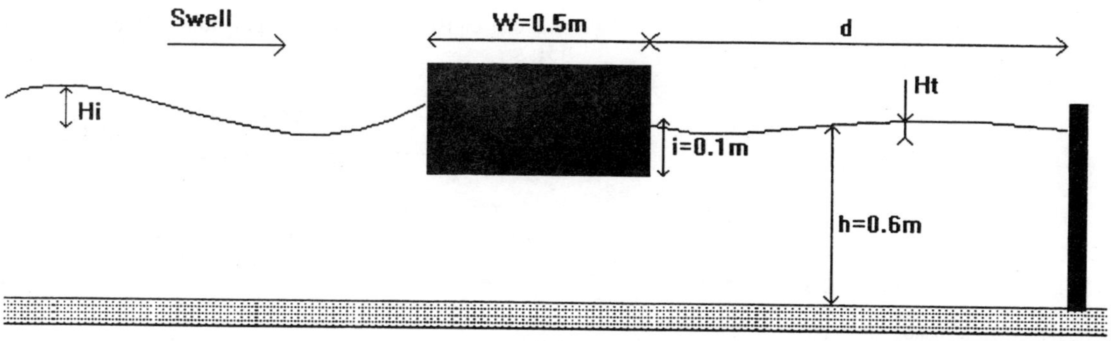

Fig 2. Experiment.

An article was also published concerning the numerical results of Sjöberg [3]. This program is based on the potential flow. The results of Bouchet are similar to ours and those of Götborg. An example is shown in Fig.3.

Experimental conditions

To observe the system « fixed water wall » in the resonance condition, we immersed 0,1 meter of a 0,5 meter side caisson in a water depth of 0.6 meter in the center of a wave flume (25 meters long and 0,8 meter wide).
The swell generated in the wave flume is monochromatic, and in order to limit the multieffects of reflexion between the structure and the wave maker, we put filters as wave aborbers between them (about 1 meter from the wave maker down stream).
The reflexion and transmission coefficients are obtained by the analysis of the time evolution of the free surface studied by three resistive probes.
At the wave flume end, a vertical wall is installed at a variable distance d of the caisson. The installation is described on Fig.2. We measured the opensea swell and the port swell for four swell periods and ten different positions of the vertical wall.

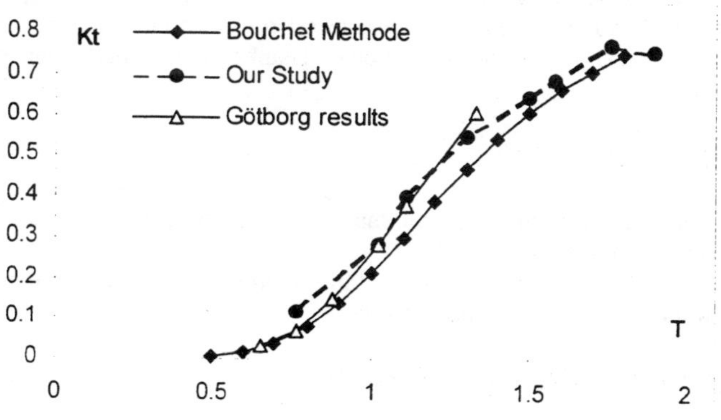

Fig. 3 Transmission coefficient evolution.

Experimental results

We see that the phenomenon « fixed water wall » when protecting the strongly reflecting setting includes resonance modes. When the system is resoning, the swell agitation in the port is more important than the opensea agitation. The higher amplitudes (peal - bonne height) measured in the port are as far as 3.2 more important than the swell heights measured in the opensea side (height induced by « loop and node » method [4]. The differences between maximum heights in the port divided by the incident height, give a coefficient Kpm. When the setting is totally absorbing (perfect absorber beach), we have Kpm = Kt. The results for different vertical wall positions in comparison with the caisson position are showed on the Fig. 4, 5, 6 and 7. We can observe that the resonance phenomena are developed when the double of the distance d is an integer multiple of wave length L.
The Fig.8 shows the Kpm evolution for the totally resonants conditions, according to swell periods.

Theoretical model

Presentation

It is possible to modelize simply the working of a caisson protecting a strongly reflecting setting. Using the free surface equation obtained from potential theory, we can obtain the maximal amplitudes measured in the port and we can compare them with the experimental results. To realize this calculation, we must know the magnitude and the phase of different swell amplitudes composing the flow. For that, we must determine the reflexion coefficient of the wall and of the caisson. The caisson reflexion coefficient is estimated by a swell energy balance-sheet.

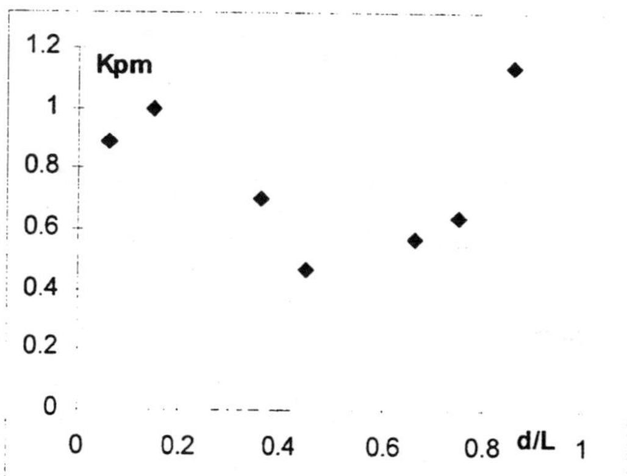

Fig. 4 Resonance phénomenon for T=1.15s.

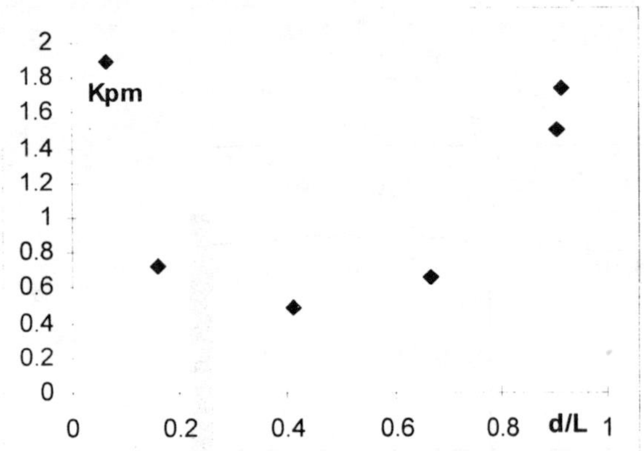

Fig. 5 Resonance phenomenon for T=1.3s.

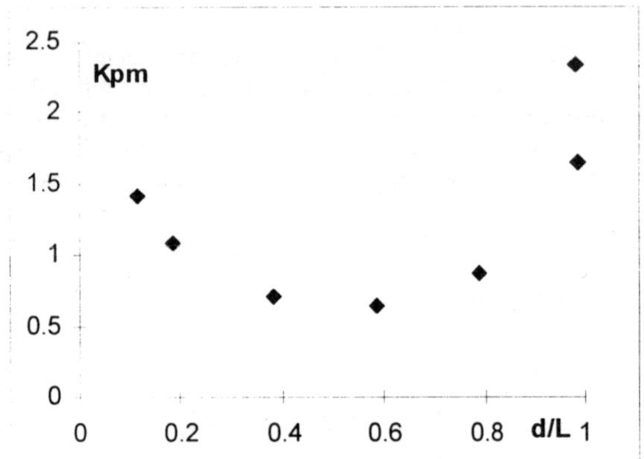

Fig. 6 Resonance phenomenon for T=1.5s.

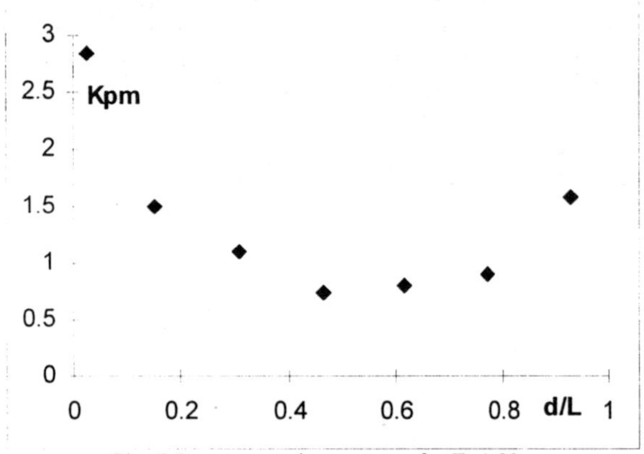

Fig. 7 Resonance phenomenon for T=1.82s.

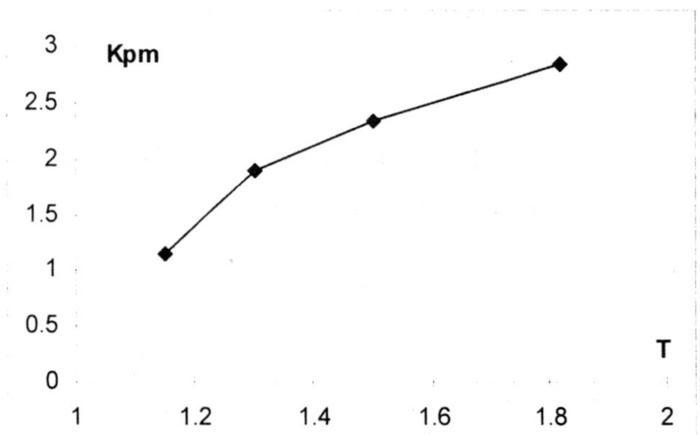

Fig. 8 Perfect resonance phenomenon.

Computation

We can consider the velocity potential in the port in the linear theory [5].

$$\phi(x,z,t) = \sum_{i=1}^{\infty} \frac{-a_{Ii}.g.\cosh k_i(h+z)}{\sigma_i.\cosh(k_i h)} \sin(k_i.x - \sigma_i.t + \varepsilon_{Ii})$$

$$(1) \qquad + \sum_{i=1}^{\infty} \frac{-a_{ri}.g.\cosh k_i(h+z)}{\sigma_i.\cosh(k_i h)} \sin(k_i.x + \sigma_i.t + \varepsilon_{ri})$$

Where aIi are incident swell amplitudes (i = 1) or swell amplitude reflected by the caisson wall (i > 1).
AIi are swell amplitudes reflected by the vertical wall.

$k_i = \frac{2.\pi}{L}$ $\quad \sigma_i = \frac{2.\pi}{T}$ $\quad \varepsilon_{ri}$ et ε_{Ii} are swell phase shift induced by the d distance.

Thanks to the potenticl expression, we determine the free surface expression :

$$\eta = \frac{1}{g}\left(\frac{\partial \phi}{\partial t}\right)_{z=0} \qquad \Rightarrow$$

$$(2) \quad \eta(x,t) = \sum_{i=1}^{\infty} \begin{bmatrix} a_{Ii}.\cos(k_i.x + \varepsilon_{Ii} - \sigma_i.t) \\ +a_{ri}.\cos(k_i.x + \varepsilon_{ri} + \sigma_i.t \end{bmatrix}$$

1) Computing aIi and ari amplitude :

We take the hypothesis that the swell is partially reflected by the vertical wall in order of Kpm. This coefficient is estimated equal to 0,33. This value is a mean value of experimental reflexion coefficients measured without caisson in the wave flume. Thanks to this coefficient, we can write :

(3) $a_{ri} = a_{Ii}.K_{rm}$

The swell reflected by a caisson again goes back towards the vertical wall in a proportion Krc.
This coefficient can be estimated in function of proportion of the energy of the swell in the immersed zone.

2) Computoring of energy percentage

The energy mean flow during the period T included in a fluid portion is written.

$$\Sigma = \frac{\rho}{T}\int_{-i}^{0}\int_{0}^{T} U.(\frac{p}{\rho} + gz)dz.dt$$

(4)

$$\Sigma = \frac{\rho.\pi.H^2.g}{T^2.sh(4.\pi.\frac{h}{L})}\int_{-i}^{0}\int_{0}^{T} ch^2(2.\pi.\frac{z+h}{L}).\cos^2(2.\pi.(\frac{t}{T}-\frac{x}{L})).dz.dt$$

$$\Sigma = \frac{C.\rho.g.H^2}{16.sh(4.\pi.\frac{h}{L})}\left(sh(4.\pi.\frac{h}{L}) - sh(4.\pi.(\frac{h-i}{L})) + \frac{4.\pi.i}{L} \right)$$

with C=L/T, i = immersion, h = depth, K = amplitude, L = wave length
From this expression, we deduct the energy percentage included in the immersed zone.

(5) $$E\% = \frac{sh(4.\pi.\frac{h}{L}) - sh(4.\pi.(\frac{h-i}{L})) + \frac{4.\pi.i}{L}}{sh(4.\pi.\frac{h}{L}) + \frac{4.\pi.h}{L}}$$

With this equation, we can define the caisson reflexion coefficient :

(6) $$Krc = Krm.\sqrt{\frac{sh(4.\pi.\frac{h}{L}) - sh(4.\pi.\frac{h-i}{L}) + \frac{4.\pi.i}{L}}{sh(4.\pi.\frac{h}{L}) + \frac{4.\pi.h}{L}}}$$

This parameters and E % are shown on the Fig.8 in accordance with the immersed height.

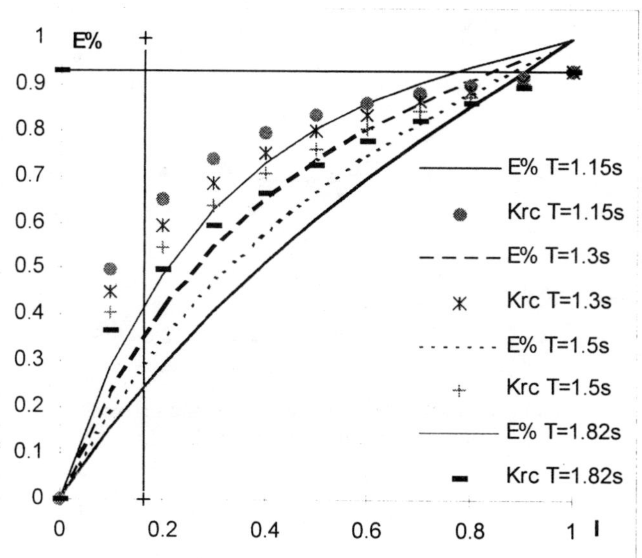

Fig. 9 Energy percentage included in the immersed zone

3) Phase shift expression

When the swell is reflected by the caisson wall, it is out of phase in comparison with the last incident swell. The phase shift has an expression as :

(7) $$\varepsilon = \frac{4.\pi.d}{L}$$

4) aIi and ari definitions

aI1 is the half excitation amplitude of the system. This amplitude is induced by the opensea swell, absorbed by the caisson. We write :

(8) $$a_{I1} = \frac{Hi}{2}.Kt$$

With Kt the coefficient calculated by Bouchet Method as

(9) $$Kt = \frac{1}{\sqrt{1+\left(\frac{\pi.W.h}{K.L.(h-i)}\right)^2}}$$

$$\text{avec } K = \frac{L}{2.\pi.(h-i)}\left(e^{-2.\pi.\frac{i}{L}} - e^{-2.\pi.\frac{h}{L}}\right)$$

The other amplitudes are determinated from the first :

(10) $a_{Ii} = a_{r(i-1)}.Krc$

(11) $a_{ri} = a_{Ii}.Krm.$

5) Balance-sheet

We can define the free surface equation in the port thanks to the equations 2, 3, 6, 8, 9 and 10.

(12) $$\eta(x,t) = \frac{Hi}{2}Kt.\sum_{n=0}^{m}\left[\begin{array}{l}\cos(kx-\sigma t+n.\varepsilon).Krm^n.Krc^n \\ +\cos(kx+\sigma t+n.\varepsilon).Krm^{n+1}.Krc^n\end{array}\right]$$

After 14 reflexions, the amplitudes are negligible ; in our computation we keep 16 reflexions.

Computation results

The Kpm calculated evolutions are compared to the experimental coefficients on the Fig. 10, 11, 12, 13.
The higher Kpm are shown on the Fig.14. We observe a good agreement between the experimental results and the theorical results.

Conclusion

The study of the phenomenon « fixed water wall » must take into account the protected setting. The reflecting character is very important.

We observe in the case where the setting is very reflecting, that the caisson fixed on piles is not an attenuation but is a wave amplifier.
The resonance phenomenon is better developed for big periods because the caisson reduces less the incident excitation than for small periods. Moreover, for small periods, the reflected swells are dissipated more quickly with water viscosity.
We have shown it is possible to simply modelize the phenomenon by using the wave lineary theories. We can optimize position of the caisson to protect the setting taking into account the usual swell of the area. On another hand, the analyse of the velocity field by the velocimetry laser around the breakwater allowed us to define an original constructive solution to avoid the resonance phenomena : the Bapbap system. The Bouygues Off Shore firm has taken out a patent for this system. The technical solution consists in modifiying the shape of the front and back walls in order to take advantage of the vertical component of the velocities to generate a continuous movement which remains near the structure without perturbing the mass of water in the port. The figure 15 shows the evolution of the kpm and kt coefficients versus the period of the wave for the simple Fixed Water Wall system and our constructive solution. These works are realized in collaboration with BOUYGUES OFF SHORE society.
This paper shows new research possibilities concerning resonance phenomena in other technologies like floating caissons, immersed flags, ARION systems, AGRIPEL absorbers...

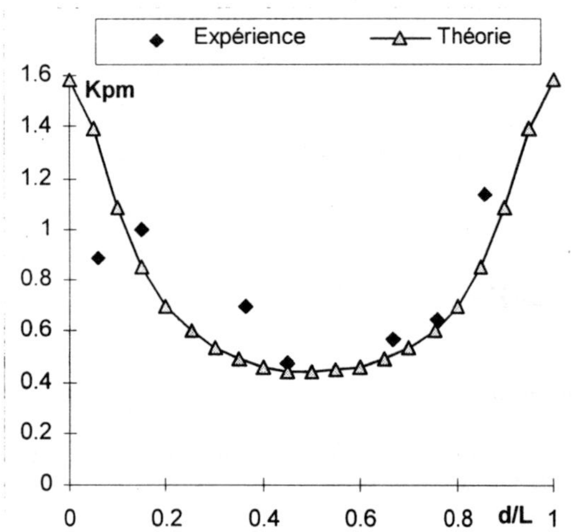

Fig. 10 Comparison between computation results and experimental results for T=1.15s.

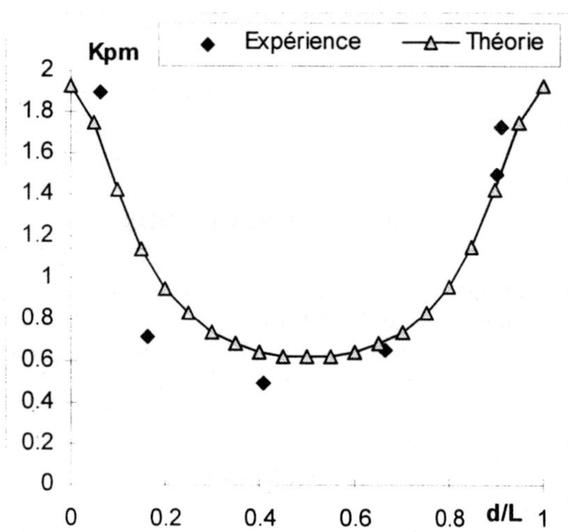

Fig. 11 Comparison between computation results and experimental results for T=1.3s.

Acknowledgement

We want to thank the firm BOUYGUES OFF SHORE for the financial support of this work.

References

[1].Bouchet R. Manzone J.M. , le mur d'eau. Dispositif nouveau de brise-houle en eau profonde, P.I.A.N.C. - A.I.P.C.N. - Bulletin 1986 - N°52 pp 60-77.

[2] Colmard C. Bélorgey M., Study of hydraulic functioning of break waters off « Fixed Water Wall » type, The proceedings of the sixth (1996) International OFFSHORE AND POLAR ENGINEERING CONFERENCE, Los Angeles, USA.

[3] Sjöberg A., Rectangular Ponton : Transferability of model test. Results to prototype conditions, 04/1988.

[4] Thèse de Monsieur Caminade D., interaction Houle-Structures. Etude de la transmission de la houle à travers une paroi perforée (type Jarlan) Université du Havre 1992 pp 41.

[5] Bonefille R., Cours d'hydraukique maritime 3e Edition, MASSON. pp 30.

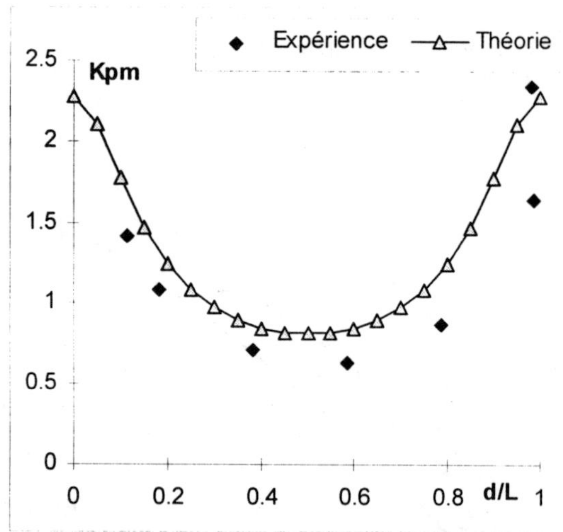

Fig. 12 Comparison between computation results and experimental results for T=1.5s.

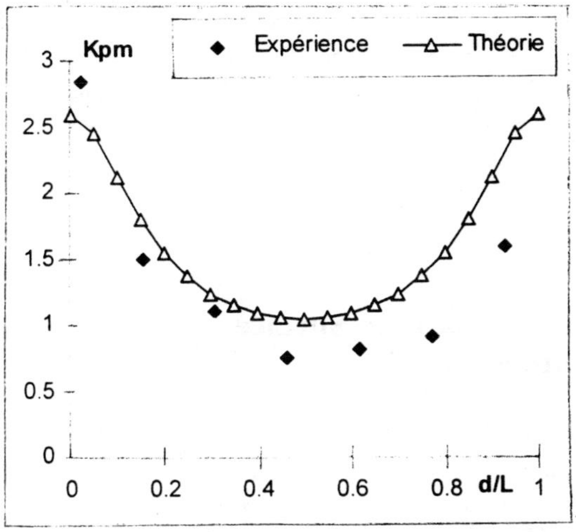

Fig. 13 Comparison between computation results and experimental results for T=1.82s.

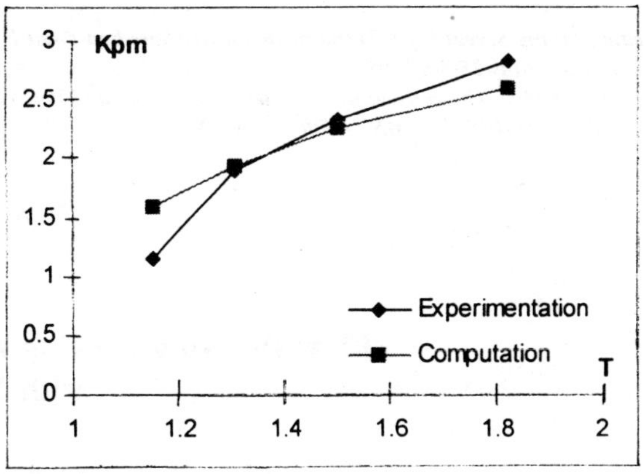

Fig. 14 Comparison between computation results and experimental results for perfect resonance.

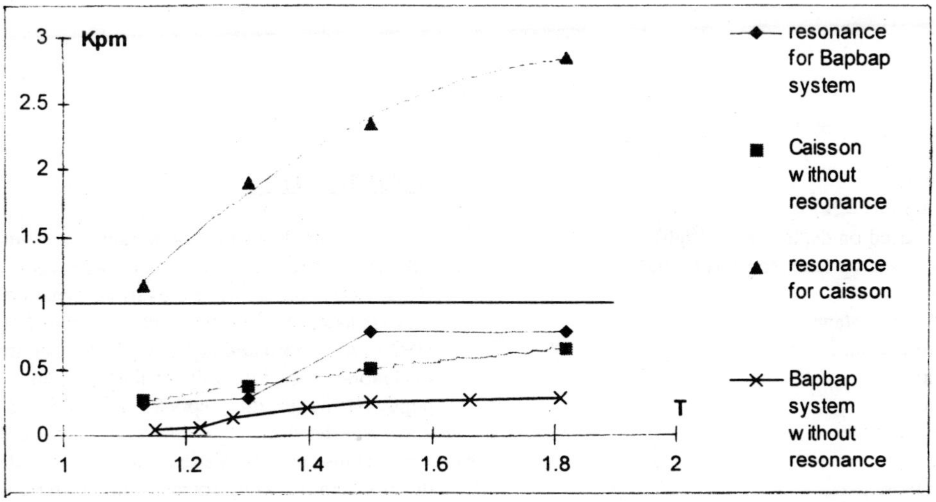

Fig. 15 Bapbap system performances

Proceedings of the Seventh (1997) International Offshore and Polar Engineering Conference
Honolulu, USA, May 25-30, 1997

ISBN 1-880653-28-1 (Set); ISBN 1-880653-31-1 (Vol. III)

Flow Due to a Moving Pressure Distribution on Free Surface with Finite-Depth Bottom

Noriaki Okita and Iskender Sahin
Western Michigan University
Kalamazoo, MI, USA

NOMENCLATURE

a	Half vehicle length
L	Total vehicle length (L=2a)
Fr_D	Froude number based on depth ($Fr_D^2=U^2/gh$)
Fr_L	Froude number based on total length ($Fr_L^2=U^2/gL$)
h	Depth of bottom
Iw	Integration in Fourier plane
P_{db}	Dynamic bottom pressure
P_o	Cushion pressure
R	Wave resistance
s	Fourier integral variable
t	Time
U	Speed of the vehicle
x	Horizontal coordinate
y	Vertical coordinate
β	Singularity location on s_{real}-axis
θ	Angle
η	Surface wave elevation
λ	Wave length
ρ	Density of Water
Φ, ϕ	Velocity potential

ABSTRACT

Surface elevation and dynamic bottom pressure profiles caused by a moving pressure distribution over the free-surface are obtained. A direct numerical integration approach for the linear, two-dimensional, and steady-state solution has been developed. The behavior of the surface elevation and bottom pressure profiles along with wave resistance for increasing Froude number or depth are presented. The agreement of the wave resistance calculations using the profiles obtained by the current method and the resistance expression given by Newman and Poole (1962) indicates that the current method can be used as a reliable tool for prediction as well as validation for other numerical approximation techniques.

INTRODUCTION

The flow due to a pressure distribution moving over the free-surface has been studied by many workers to obtain the flow parameters, such as wave elevation, dynamic bottom pressure, and wave resistance of air-cushioned-vehicles (ACV). Huang and Wong (1970), and Sahin and Hyman (1995) used the direct numerical evaluation of the integrals for three-dimensional problem with finite depth. However, they reported difficulties in evaluating the surface wave profiles due to the high frequency oscillation in the integrands at y=0. Haussling and VanEseltine (1974) obtained transient wave development by using spectro-finite difference method.

Newman, and Poole (1962) derived the wave resistance expressions from the linearized theory for rectangular and elliptic pressure distribution which is moving with the constant velocity over the canal of constant width and depth. Hogben (1966) and Everest and Hogben (1967) conducted several experimental study on the ACV of circular and rectangular plan forms, respectively. The comparison between experimental and theoretical wave resistance indicates that the linear theory prediction is reliable as far as the wave remains within the linear regime, or within the steepness limit.

In the present study the surface wave and dynamic bottom pressure profiles caused by a pressure distribution moving at constant speed over the free-surface with finite-depth bottom is studied. Ideal flow theory and linearized free-surface conditions are used, and the pressure distribution is considered to be constant over the ACV of length L. Direct numerical integration is used to evaluate the complex and real integrals of the analytical solution (Stoker, 1957). The nondimensional surface elevation and pressure profiles are examined in terms of the water depth (h) and Froude numbers based on depth (Fr_D) and length (Fr_L). Based on the surface elevation profiles, wave resistance calculations are calculated, and compared with the results of Newman and Poole (1962)

MATHEMATICAL FORMULATION

For a vehicle moving in -x direction with constant velocity U, the moving coordinate system, as shown in Fig. 1, is used. The flow domain is defined as a rectangle of $-\infty < x < +\infty$, and $-h < y < 0$, with the origin at the center of the ACV. The undisturbed free-surface lies at $y = 0$, and the sea bottom is at $y = -h$. By assuming ideal flow the following relationship exists:

$$\Phi(x,y,t) = Ux + \phi(x,y,t) \tag{1}$$

where Φ and ϕ are total and perturbation velocity potentials, respectively. By superposing the right-going uniform flow, the problem is made steady-state and the vehicle motion is brought to rest. The governing equation, boundary and initial conditions in the moving reference frame are given by Laplace's equation,

$$\nabla^2\phi(x,y,t) = 0 \qquad \text{for } -\infty < x < \infty, \quad -h < y < 0, \; t > 0 \tag{2}$$

The linearized dynamic and kinematics free-surface conditions,

$$\frac{P}{\rho} + g\eta + \phi_t + U\phi_x + \frac{U^2}{2} = 0 \qquad \text{at } y = 0 \tag{3}$$

$$\eta_t + U\eta_x - \phi_y = 0 \qquad \text{at } y = 0 \tag{4}$$

No flow condition across the bottom is:

$$\phi_y = 0 \qquad \text{at } y = -h \tag{5}$$

For the undisturbed state at t=0, one has

$$\phi(x,y,0) = \eta(x,0) = p(x,0) = \phi(x,y,0)_t = 0 \tag{6}$$

A general pressure distribution is defined as

$$P = P(x) \qquad \text{at } y = 0, \text{ for } t > 0 \tag{7}$$

The perturbation potential, ϕ, and its derivatives are bounded in $-\infty < x < \infty$, approaching 0 for $|x| \to \infty$.

Surface elevation (η) and dynamic bottom pressure (P_{db}) are given by:

$$\eta(x) = -\frac{1}{g}\left(\frac{P}{\rho} + \phi_t + U\phi_x\right) \qquad \text{at } y = 0 \tag{8}$$

and

$$P_{db} = -\rho\,\{\phi_t + U\phi_x\} \qquad \text{at } y = -h \tag{9}$$

The analytic solution to the above problem can be obtained by employing the Fourier transform technique. The expression for the perturbation potential, ϕ, which is a regular harmonic function in $-h < y < 0$, is (Stoker, 1957):

$$\phi(x,y,t) = \frac{1}{\sqrt{2\pi}} \int_{-\infty}^{\infty} A(s,t)\cdot\cosh\{s(y+h)\}\cdot e^{isx}\,ds \tag{10}$$

where

$$A(s,t) = \frac{isU\overline{P}}{\rho\cosh(sh)}\cdot\left\{\begin{aligned}&\frac{1}{s^2U^2 - gs\tanh(sh)}\\ &+\frac{1}{2\sqrt{gs\tanh(sh)}}\cdot\frac{e^{-it\left(sU+\sqrt{gs\tanh(sh)}\right)}}{sU+\sqrt{gs\tanh(sh)}}\\ &+\frac{1}{2\sqrt{gs\tanh(sh)}}\cdot\frac{e^{-it\left(sU-\sqrt{gs\tanh(sh)}\right)}}{sU-\sqrt{gs\tanh(sh)}}\end{aligned}\right\} \tag{11}$$

and $\overline{P}$ is the Fourier transform of P, in Eq. 7.

After taking the derivatives with respect to x, y, and t, and letting $t \to \infty$ for each partial component, all time-dependent terms vanish unless U^2/gh is unity. Also, the pressure distribution in the present case is constant for $|x| < a$, and is given by:

$$P(x) = \begin{cases} P_o & \text{for } |x| < a \\ 0 & \text{otherwise} \end{cases} \tag{12}$$

and the Fourier transform of P(x) is given by:

$$\overline{P}(s) = \frac{2P_o}{\sqrt{2\pi}}\frac{\sin(as)}{s} \tag{13}$$

The evaluation of the steady state surface elevation and pressure profiles requires only the steady-state part of ϕ_x. From Eq. 10, 11, and 13, it can be written as:

$$\phi_x(x,y) = \frac{-UP_o}{\rho\pi}\int_{path}\frac{\sin(a\cdot s)\cdot\cosh\{s\cdot(y+h)\}}{\left[U^2\cdot s - g\cdot\tanh(s\cdot h)\right]\cdot\cosh(s\cdot h)}\cdot e^{isx}ds \tag{14}$$

The expression above has three possible poles on real axis ($s_{real} = \pm\beta$, and $s_{real} = 0$); and it has infinitely many imaginary poles along imaginary axis. This fact makes it difficult to evaluate the integral by using the methods, such as calculus of residues with a closed contour. The path of integral in complex s-plane is shown in Fig. 2, as it extends from $-\infty$ to $+\infty$ on s_{real}-axis, and semi-circular path goes around the singularities. In evaluating the integral, the integration path needs to be modified, depending on the value of Fr_D.

For $Fr_D < 1.0$, the integrand has two simple poles on the real axis at $s_{real} = \pm\beta$, and $s_{real} = 0$ is not a pole, thus the path becomes the real axis. For $Fr_D > 1.0$, the path is simply the real axis because the denominator has no real roots. Also, from Rieman-Lebesgue theorem (Stoker, 1957), the integral vanishes for $|x| \to \infty$ for $Fr_D > 1$, thus the disturbance dies out both upstream and downstream; when $Fr_D < 1$, the integrand vanishes as $x \to -\infty$, thus the disturbance vanishes upstream.

It is possible to simplify the integration in Eq. 14 further. First, the integrand is divided into the real axis (Iw_R) and around semi-circle (Iw_C). For the straight path (ST), $s=s_{real}$, and Iw_R can be decomposed into real and imaginary parts as:

$$\mathrm{Re}\left(Iw_R\right)=PV\int_{-\infty\ ST}^{+\infty}\frac{\sin(a\cdot s)\cdot\cosh\{s\cdot(y+h)\}}{\left[U^2\cdot s-g\cdot\tanh(s\cdot h)\right]\cdot\cosh(s\cdot h)}\cdot\cos(s\cdot x)\ ds \tag{15}$$

$$\mathrm{Im}\left(Iw_R\right)=PV\int_{-\infty\ ST}^{+\infty}\frac{\sin(a\cdot s)\cdot\cosh\{s\cdot(y+h)\}}{\left[U^2\cdot s-g\cdot\tanh(s\cdot h)\right]\cdot\cosh(s\cdot h)}\cdot\sin(s\cdot x)\ ds \tag{16}$$

The integrands of $\mathrm{Re}(Iw_R)$ and $\mathrm{Im}(Iw_R)$ are even and odd functions, respectively, of both s and x. Thus, when the integral is carried out along a straight path, $\mathrm{Im}(Iw_R)$ cancels out, and $\mathrm{Re}(Iw_R)$ becomes:

$$\mathrm{Re}\left(Iw_R\right)=2\times PV\int_{0\ ST}^{+\infty}\frac{\sin(a\cdot s)\cdot\cosh\{s\cdot(y+h)\}}{\left[U^2\cdot s-g\cdot\tanh(s\cdot h)\right]\cdot\cosh(s\cdot h)}\cdot\cos(s\cdot x)\ ds \tag{17}$$

For the integration (Iw_c) along the semi-circular path (SC), s is changed to a polar form, $s=\pm\beta+\varepsilon e^{i\theta}$, where ε is small radius, and θ is the angle measured from + s_{real} axis, and substituting into Eq. 14, θ becomes the new integration variable for $[\pi<\theta<2\pi]$.

The integrand of Iw_C has a period 2π, and the real part of the integrand along the path on the positive side is a mirror image of the negative side about the middle of the integration range, $\theta=3\pi/2$. For the imaginary part, the positive side is mirror image of negative side about both $\theta=3\pi/2$ and θ-axis. Thus, as in the case of straight path integral, the imaginary part cancels, and the real part is simply twice as much as the $\mathrm{Re}(Iw_C)$ around one semi-circular path; 2) these integrands are different between positive and negative x, thus they contribute to asymmetry in the wave and pressure profiles between the upstream and downstream of the ACV. We find for $s=\beta+\varepsilon e^{i\theta}$

$$\mathrm{Re}\left(Iw_C\right)=2\times\mathrm{Re}\left[\int_{\pi\ SC}^{2\pi}\frac{\sin(a\cdot s)\cdot\cosh\{s\cdot(y+h)\}}{\left(U^2\cdot s-g\cdot\tanh(s\cdot h)\right)\cdot\cosh(s\cdot h)}\cdot i\varepsilon\cdot e^{i(sx+\theta)}\ d\theta\right] \tag{18}$$

where $s=\beta+\varepsilon e^{i\theta}$

Finally, ϕ_x for each (x,y) location is given by:

$$\phi_X(x,y)=-\frac{UP_O}{\rho\pi}\cdot Iw(x,y)=-\frac{UP_O}{\rho\pi}\cdot\left[\mathrm{Re}\left(Iw_R\right)+\mathrm{Re}\left(Iw_C\right)\right] \tag{19}$$

By using the surface elevation profile, the expressions for the wave resistance per unit width experienced by the craft can be obtained for comparison with the one given by Newman and Poole (1962).

Since the current case is steady-state, progressive wave with phase velocity U (with respect to the fixed coordinate system) exists in downstream of the disturbance, and upstream waves are diminished (Newman, 1977). For $Fr_D<1$, the wave resistance, R_1 (Lamb, 1945) can be evaluated by using amplitude A of the surface elevation in downstream and its wavelength, λ.

$$\frac{R_1}{w}=\frac{\rho g A^2}{4}\cdot\left(1-\frac{2kh}{\sinh(2kh)}\right) \tag{20}$$

where $k=\frac{2\pi}{\lambda}$

Another form of the wave resistance, R_2, can be defined as the horizontal component of the force acting under the craft. By assuming the surface profile is linear between the two consecutive data points, and by approximating the integral with the summation from leading to trailing edge of the disturbance, R_2 is defined as:

$$\frac{R_2}{w}=P_o\cdot\int_{LE}^{TE}dl\cdot n_x$$
$$=-P_o\sum_{i=1}^{N_{acv}}\eta_x\cdot\frac{\sqrt{\left(x_{i+1}-x_i\right)^2+\left(\eta_{i+1}-\eta_i\right)^2}}{\sqrt{\eta_{xi}^2+1}} \tag{21}$$

where η_i and η_{xi} are the surface elevation and its derivative with respect to x at the data point x_i, respectively.

The reference wave resistance for a two-dimensional, constant pressure distribution per unit width is given by (Newman and Poole, 1962):

$$\frac{R}{w}=\frac{4P_o}{\rho g}\cdot\frac{\sin^2\left(\frac{\beta L}{2}\right)}{1-\frac{2\cdot\beta h}{\sinh(2\cdot\beta h)}} \tag{22}$$

NUMERICAL EVALUATION

Direct numerical integration has been utilized in order to evaluate the integral of Eq. 19. Due to the even/odd properties of the integrand, the integration range in complex s-plane is from 0 to $+\infty$, of which three sub-intervals are used: Iw_1 (real axis, $0\le s_{real}\le\beta-\varepsilon$), Iw_C (semi-circular path), and Iw_2 (real axis, $\beta+\varepsilon\le s_{real}\le\infty$). Romberg integration of variable order with polynomial extrapolation and the trapezoidal rule (Press, et al, 1992) were used in evaluating the above real and complex integrals.

In evaluating Iw_C, instead of removing the singularity by changing the integration contour and using Cauchy's residue theorem, the integration along the semi-circular path was carried out by comparing the relative error between two iterations in terms of ε

(radius of semi-circular contour). For each (x,y) location, Iw_C is first evaluated, followed by the evaluation of Iw_1 and Iw_2. Since the integrand was of highly oscillatory nature, two types of subsectioning schemes were used in case the convergence could not be obtained. First subsection is done by roots of the integrand, and followed by equal space subsectioning.

Three different methods were tried to compute the Iw_2. First, several variable transformations were made for the computation of Iw_2 with no success. Secondly, the integration was carried out for each successive small interval (ds) until the returned value is smaller than either the accumulated error for Iw_2 or relative magnitude of the previous integration at this (x,y) location. However, this method had the difficulty of not properly determining the convergence since the amplitude of the integrand at y=0 does not decrease monotonically. Instead, Iw_2 was further subdivided by two: Iw_{2a} is evaluated for ($\beta+\varepsilon \le s_{real} \le BP$), and Iw_{2b} ($BP \le s_{real} \le \infty$) for y=0 was obtained by the following expression, and set equal to 0 for y=-h, provided that $\tanh(sh)\approx 1$,

$$Iw_{2b}(x,0) = -\frac{1}{2U^2}\cdot C\cdot\left[\begin{array}{l}\cos\left(D\cdot\frac{g}{U^2}\right)\cdot si\left\{D\cdot\left(BP-\frac{g}{U^2}\right)\right\}\\ +\sin\left(D\cdot\frac{g}{U^2}\right)\cdot ci\left\{D\cdot\left(BP-\frac{g}{U^2}\right)\right\}\end{array}\right]$$

$$\text{where}\quad \begin{cases} D = 2a & \text{for } x = \pm a \\ C = 1.0 \end{cases}, \text{ and } \begin{cases} D = |a-x| & \text{for } x \ne \pm a \\ C = \dfrac{A}{(a-x)} \end{cases} \tag{23}$$

The surface elevation and bottom pressure profiles can be calculated from the accurate evaluation of ϕ_x by substituting the numerical values in Eq. 8 and 9.
In order to evaluate R_2, η_x was evaluated by taking the first-order, forward finite-difference formula.

DISCUSSION OF THE RESULTS

All the results are nondimensionalized by following

$$(x,y,h,\lambda)' = (x,y,h,\lambda)/L\ ;\quad \eta' = \rho g\eta/P_o$$

$$P_{db}' = P_{db}/P_o\ ;\quad \frac{R'}{w} = \rho g R / P_o^2 w$$

where L is the ACV length, P_o is the cushion pressure, R is the wave resistance, and λ is the wave length of the profiles.

When the Froude number based on depth is greater than unity, the linear theory does not predict upstream or downstream waves (and the results show the same trend), thus the results for $Fr_D > 1$ are not considered here.

Common settings for all the cases include L=1.0 m, $P_o/\rho g$ =0.01 m, relative error criterion of 1.0×E-6 for integration, relative error criterion of 1.0×E-3 for ε iteration, BP=20 in Eq. 23, and no smoothing was used for the results. The computation with a FORTRAN program compiled with double-precision option was carried out in DEC Alpha workstations.

In Fig. 3a, wave resistances R_1 and R_2, which are calculated from the surface elevation profile under ACV and far downstream, respectively, are compared with the resistance curve, given by Eq. 22, as a function of Fr_D for h/L=0.9. Good agreement of R_1 and R_2 with the resistance curve ensures us that the surface profile in both region are reliable within the limit of linear theory. Also, the plot of steepness, $|2A|/\lambda$, for the downstream surface wave indicates that the waves are within the linear regime, or $|2A|/\lambda$ is below approximately 1/7 (Hogben, 1966).

In Fig 3b and 3c, the magnitudes of the first peak, downstream peak, and the wave length corresponding to the downstream peak (see Fig. 1) are plotted for surface elevation and bottom pressure profiles, respectively. It should be noted that the first peak may or may not occurs under the disturbance. For both surface elevation and bottom pressure, the downstream wave length is similar and increases as Fr_D increases; however, the curves for first and downstream peaks are different. The pressure peaks (Fig. 3c) are the increasing function of Fr_D, and decay in the amplitude between the first and downstream peaks can be seen up to $Fr_D \approx 0.84$, where $\lambda/L \approx 5.0$. As Fr_D increases, the peak curves for the surface (Fig. 3b) increase or decrease, and the critical points, such as maxima, minima, inflection points are found when the wave length is approximately integral multiple of the half length, a of the disturbance. Significant difference in wave elevation peaks is found in the neighborhood of the hollow. The difference near the primary hump region is due to the discontinuity at the trailing edge, and the peak under the craft is lower than the downstream peak.

The primary hump ($Fr_D \approx 0.6$, $\lambda/L \approx 2.0$) and the first hollow ($Fr_D \approx 0.42$, $\lambda/L \approx 1.0$) of the resistance curve (Fig. 3a) corresponds to the maxima and minima of the peak curves (Fig. 3b). Above primary hump Fr_D, the peak curve gradually decreases, and increases towards the absolute maximum. Because the profiles at the surface and the bottom are different by the multiplication factor in downstream, several aspects of these curves are explained by using the asymptotic approximation of ϕ_x for large x and $Fr_D < 1$ (Stoker, 1957) given by:

$$\phi_x(x,y) \approx \frac{4P_o}{\rho}\cdot\frac{\cosh\{\beta(y+h)\}}{\cosh(\beta h)}\cdot\frac{\beta U}{W'(\beta)}\sin(\beta a)\sin(\beta x) \tag{24}$$

where $\beta = 2\pi/\lambda$ is the non-zero root of the dispersion relationship, $W(s) = U^2 s^2 - gs\tanh(sh)$, and $W'(\beta)$ is the value of its derivative at β.

Hollows and humps are found when $L = n\lambda$ and $L = (n-0.5)\lambda$, corresponding to the minimum and maximum value of $\sin(\beta a)$. The plot of the dominant term $\beta U/W'(\beta)$ looks like "U" shape, and the curve remains nearly constant for Fr_D above the primary hump (n=1) up to certain Fr_D before increasing again, thus, the amplitude decreases gradually. The corresponding bottom pressure peak increases because βh decreases as Fr_D increase, and ϕ_x at the bottom (y=-h) is inversely proportional to $\cosh(\beta h)$.

In Fig. 4, the surface elevation and the bottom pressure profiles for h/L=0.9 and Fr_D = 0.354, 0.418, 0.612, and 0.758, are shown, corresponding to the second hump, the first hollow, the primary hump, and above the primary hump. At the leading and trailing edges of the disturbance ($x/L = \pm 0.5$), discontinuity presents due to the sudden change of the pressure from zero to P_o. Slight increase in the elevation is found in front of the leading edge. The surface elevation for Fr_D=0.4180 shows significant decay of the amplitude. On the other hand, the profiles for the humps show the elevation under the disturbance is lower than in downstream.

The same parameters as in Fig. 3 are plotted in Fig. 5 for Fr_L = 0.5477 against h/L. Except for the pressure peaks, the convergence of all the parameters towards non-zero values for higher h/L cases can be observed. The pressure peaks continue to decrease towards the non-disturbed (zero) state. In other words, the effect of the bottom becomes negligible approaching the infinite-depth case above h/L≈1.0, or h/λ≈1/2. The surface and bottom profiles for h/L= 0.4, 0.5, 1.0, and 2.0 are shown in Fig. 6.

CONCLUSION

A numerical integration routine was developed for the evaluation of surface elevation and bottom pressure with finite depth bottom, over which a constant pressure distribution moves with a uniform speed, U. The method utilizes variable-order Romberg's integration with the closed trapezoidal formula, in which the end points of the integration are is included. The approximation of the indefinite integral eliminated some of the solver variables, such as integration limit, termination criteria, and step size. Smooth profiles are obtained by the current method up to h/L≈10, where the noise starts to appear near disturbance and proceeds both upstream and downstream as h/L is increased.

From the absence of the upstream wave train and the comparison of the wave resistances, the current method successfully simulates the surface elevation in the entire domain, including under the disturbance. Since the current method is the numerical approximation of the analytical solution, the spacing of the physical grid does not affect the approximated value itself, thus this method would serve as a good validation tool for other numerical simulations. These simulations have flexibility in modeling complex pressure distribution, whereas this method requires the Fourier transform of the pressure distribution and appropriate path of the integral. This method can also be used to predict the steady-state surface elevation and bottom pressure profiles when an ACV is traveling in the canal of approximately equal width and fixed-depth.

REFERENCES

Doctors, L. J., and Sharma, S. D., (1972) "The Wave Resistance of an Air-Cushion Vehicle in Steady and Accelerated Motion," Journal of Ship Research, Vol. 16, No. 4, pp. 248-260.

Everest, J. T., and Hogben, B., (1967) "Research on Hovercraft over Calm Water," Transaction of the Royal Institution of Naval Architects, Vol. 109, pp. 311-326.

Haussling, H. J., and VanEseltine R. T., (1974) "A Combined Spectral Finite-Difference Method for Linear and Nonlinear Water Wave Problems," Naval Ship Research and Development Center Report No. 4580.

Hogben, B., (1966) "An Investigation of Hovercraft Wavemaking," The Journal of the Royal Aeronautical Society, Vol. 70, pp. 321-329.

Huang, T. T., and Wong, K. K., (1970) "Disturbance Induced by a Pressure Distribution Moving Over a Free-Surface," Journal of Ship Research, Vol. 14, No. 3, pp. 195-203.

Lamb, H., (1945) "Hydrodynamics," Dover Publications, NY.

Newman, J. N., (1977) " Marine Hydrodynamics," The MIT Press, MA.

Newman, J. N., and Poole, F. A. P., (1962) " The Wave Resistance of a Moving Pressure Distribution in a Canal," Schiffstechnik, Vol. 9, pp. 1-26.

Press, W. H., and Teukolsky, S. A., and Vetterling, W. T., and Flannery, B. P., (1992) "Numerical Recipes in FORTRAN: The Art of Scientific Computing," Cambridge University Press, NY.

Sahin, I. and Hyman, M., (1995) "Simulation of Three-Dimensional Finite-Depth Wave Phenomenon for Moving Pressure Distributions," Proceedings, 5th International Offshore and Polar Engineering Conference, The Hague, The Netherlands, June 1995.

Stoker, J. J., (1957) "Water Waves: The Mathematical Theory and Applications," Interscience, NY.

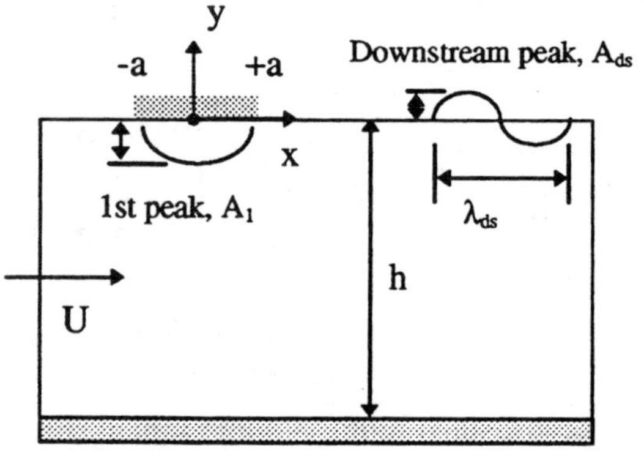

Fig. 1: Coordinate system (x,y)

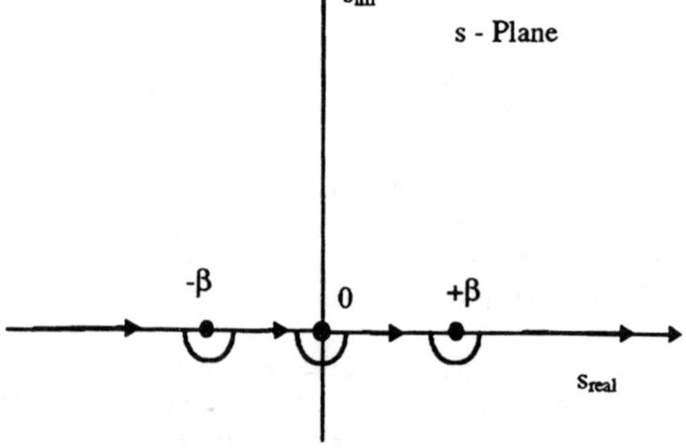

Fig. 2: Integration Path in s-plane

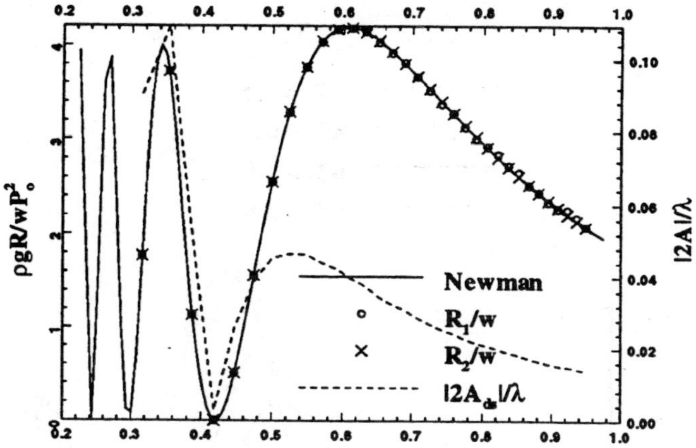

Fig. 3a: Wave resistance and Steepness versus Fr_D for h/L=0.9.

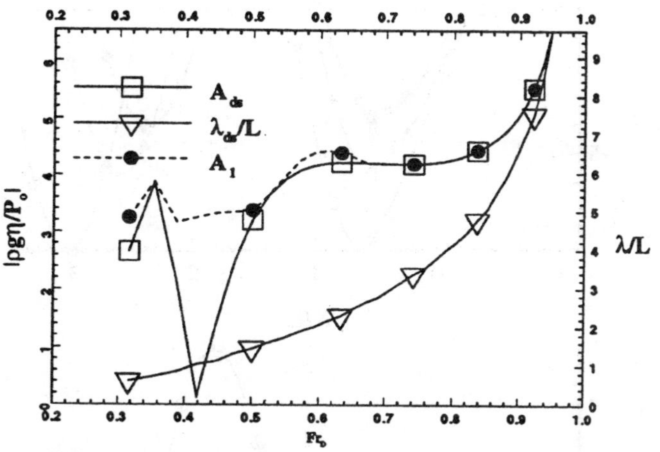

Fig. 3b: Peak values and wave length for surface elevation versus Fr_D for h/L=0.9.

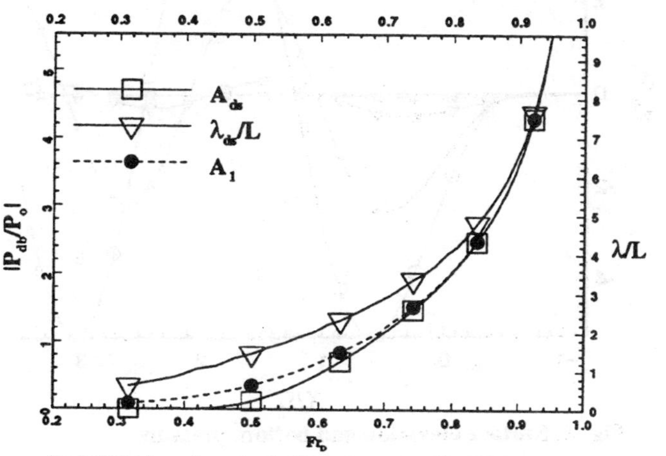

Fig. 3c: Peak values and wave length of bottom pressure versus Fr_D for h/L=0.9.

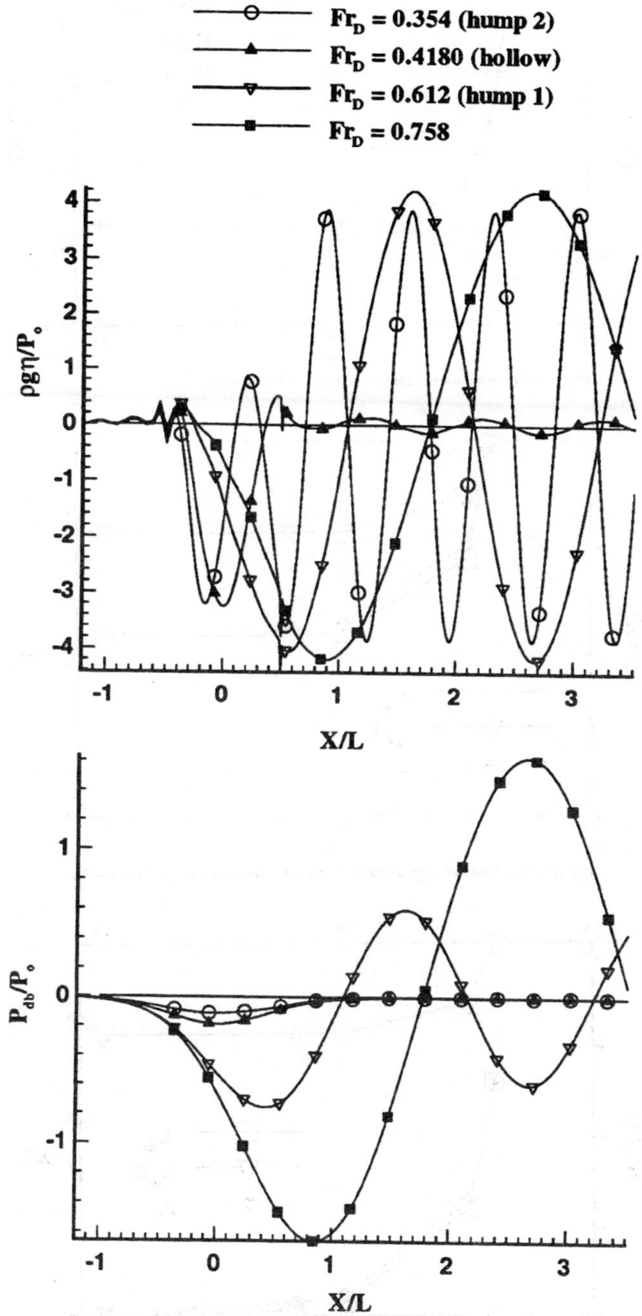

Fig. 4: Surface elevation and bottom pressure profiles for h/L=0.9.

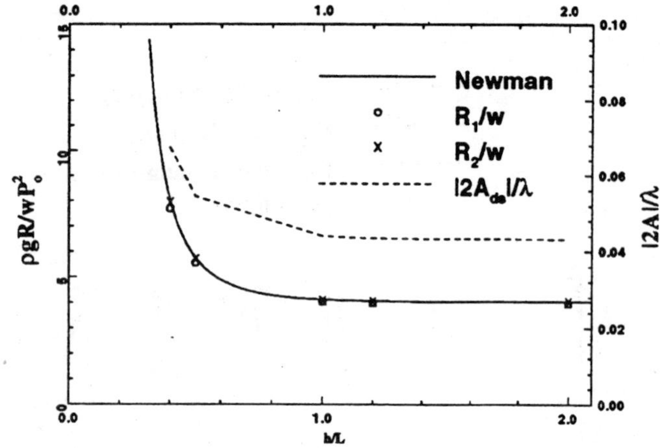

Fig. 5a: Wave resistance and Steepness versus h/L for Fr_L=0.5477.

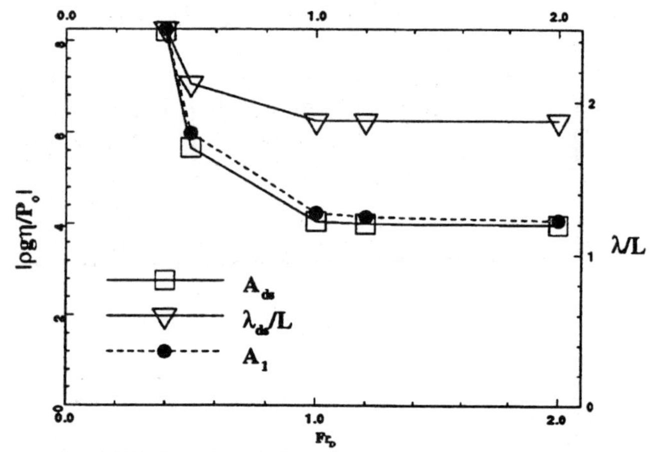

Fig. 5b: Peak values and wave length for surface elevation versus h/L for Fr_L = 0.5477.

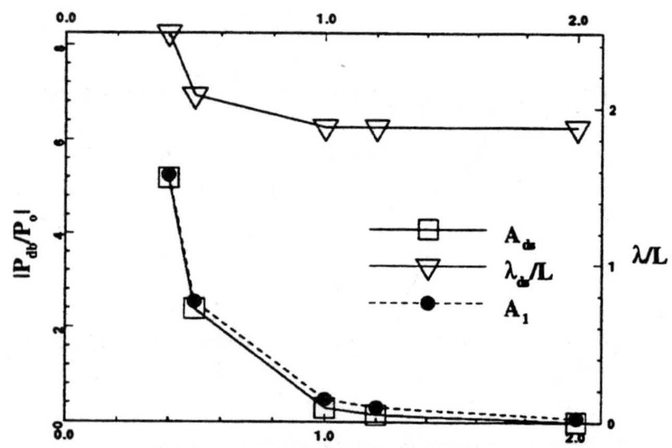

Fig. 5c: Peak values and wave length for bottom pressure versus h/L for Fr_L = 0.5477.

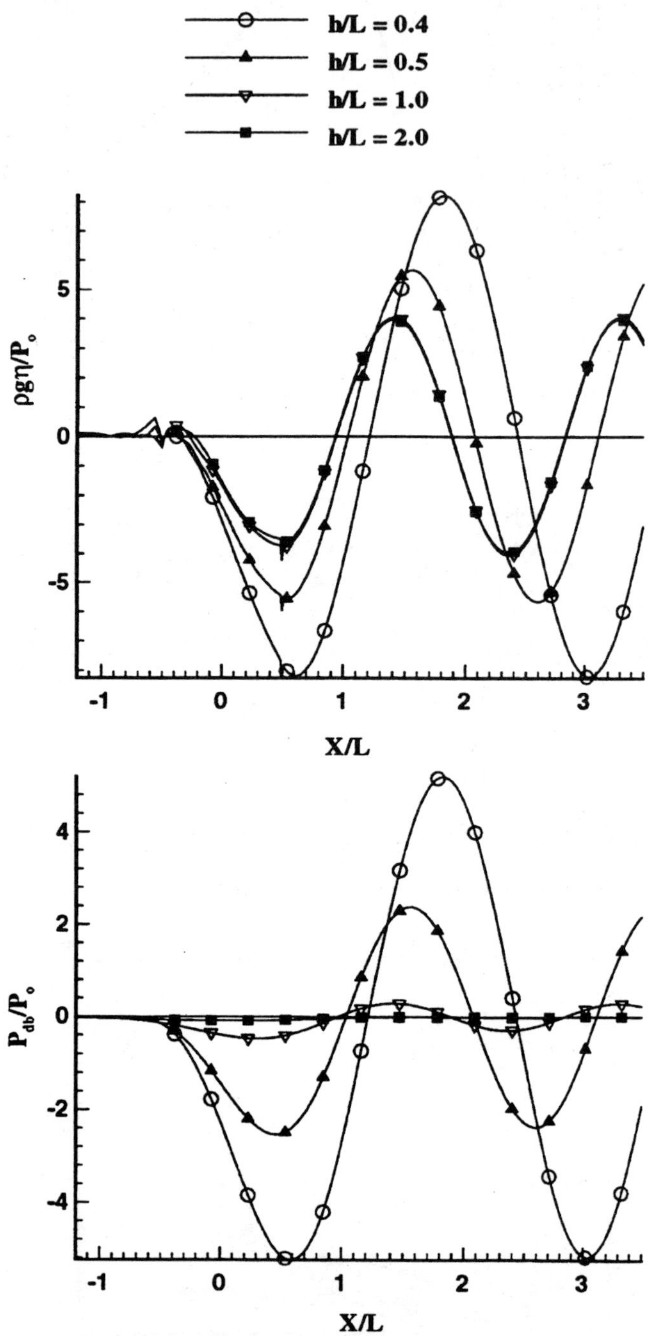

Fig. 6: Surface elevation and bottom pressure profiles for Fr_L=0.5477.

Proceedings of the Seventh (1997) International Offshore and Polar Engineering Conference
Honolulu, USA, May 25-30, 1997

ISBN 1-880653-28-1 (Set); ISBN 1-880653-31-1 (Vol. III)

Time Domain Analysis of Ship-Generated Waves in Harbour Using a Fast Hierarchical Method

Ding Li and Allen T. Chwang
University of Hong Kong
Hong Kong

ABSTRACT

In this paper, a new method for simulation of waves in a large domain is developed. In this method, the famous Barnes-Hut (BH) fast hierarchical algorithm is modified to evaluate the influence of all the singularities and an iterative solver of the Krylov-subspace type, called the Generalized Minimal Residual (GMRES) method, is used to solve the linear equation. The 5th-order Runge-Kutta-Merson method is used to integrate the governing equation in time. A cell-to-particle algorithm in BH is replaced by a cell-to-cell algorithm to improve the performance of the BH algorithm. A numerical code based on the new method, called FHWAVE, is developed. The accuracy and performance of FHWAVE are examined in comparison with a direct method. Numerical tests show that FHWAVE can reduce the CPU time and memory size significantly when the number of linear equations, N, becomes large. This method is applied to the time-domain simulation of three-dimensional waves generated by many moving bodies in a large domain. Some preliminary results on waves generated by three ships moving in Victoria Harbour are presented in this paper. The FHWAVE is a real 3D water wave model which opens up a new way for studying ship-generated waves, wind-generated waves and swells, including their propagation, reflection and interaction in harbours.

KEYWORDS: Ship-generated waves, Fast hierarchical algorithm, De-singularization approach, Boundary element method (BEM), Barnes-Hut (BH) algorithm

1. INTRODUCTION

The generation and propagation of ship waves have been studied extensively, but most of these studies have been focused on ship motions (Li and Zhou, 1986). Few works are about ship-wave propagation, reflection and interaction in harbour. In Hong Kong, the situation has been aggravated by the growing marine traffic and predominantly reflective seawalls along newly reclaimed land. Suitable computer software for simulation of ship-generated waves in harbor should be developed for subsequent investigation.

In this paper, ship-generated waves are analyzed by the time-stepping method in conjunction with a de-singularization approach. In the de-singularization approach, the singularities generating the flow field are outside the fluid domain. This allows the sigularity distribution to be replaced by isolated Rankine sources with the corresponding reduction in computational complexity (Beck et al., 1994 and Cao et al., 1991).

In numerical computations, an integral equation is usually discretized into a linear matrix equation. The boundary element method (BEM) computes the influence matrix at a cost of $O(N^2)$ operations, where N is the number of linear equations. The cost of solving the matrix equation is of $O(N^3)$ if a direct solver is used or $O(N^2)$ if an iterative solver is used. The computer memory seems to rise as $O(N^2)$ in order to store the influence matrix. This can become an obstacle to the practical use of a method when the domain is larger than nL^2, where $n \sim O(10)$ and L is the wavelength. The difficulty is due to the prohibitive computer requirements of the solution with small wavelength in comparison with domain size. For instance, if the wavelength of a ship wave is about L = 5m and the domain is roughly 100L*50L, then using five grid points per wavelength (which may be inadequate) would give 125,000 simultaneous, full-dominant equations, and the solution of such a system is formidable task using traditional methods.

All these facts strongly motivate us to find a better algorithm which costs less than $O(N^2)$. Fortunately, it turns out that there are clever divide-and-conquer algorithms which only take $O(N \log N)$ or even just $O(N)$ operations for this problem. A Fast Multipole Method (FMM) developed by Greengard and Rokhlin (1987) approximates the influence of far-field singularities with multipole expansions and costs much less than the direct computing of the effects of all individual singularities. Greengard's algorithm was extended by Scorpio et al. (1996) to include a de-singularized method with point singularities for the computation of nonlinear water waves. Cao and Maskew (1996) presented a mixed version of the boundary integral equation using both panel singularities and point singularities for the Boundary Value Problem (BVP) and a FMM algorithm to speed up the solution to the BVP.

In this paper, a new method for simulation of waves in a large domain is developed. In this method, the famous fast hierarchical algorithm developed by Barnes and Hut (BH) (1986) is modified to evaluate the influence of all the singularities and an iterative solver of the Krylov-subspace type, called the Generalized Minimal Residual (GMRES) method, is used to solve the linear equation. The 5th-order Runge-Kutta-Melson method is used to integrate the governing equation in time. The BH algorithm is not as accurate as the FMM, but faster than FMM. The BH algorithm is often used when 1% accuracy is desired, which is enough for most wave propagation problems in harbours. A particle-to-cell algorithm in BH is replaced by a cell-to-cell algorithm to improve the performance of the BH algorithm. A numerical code based on the new method, called FHWAVE, is developed. The accuracy and performance of FHWAVE are examined in comparison with a direct method and the results are presented. Finally, a preliminary result on waves generated by three ships moving in Victoria Harbour using FHWAVE is shown in the paper. In this preliminary case, 30,000 singularities are distributed over the free-surface and the ship is simplified by a sink-source pair.

2. MATHEMATICAL FORMULATION

An ideal, incompressible fluid is assumed and the surface tension is neglected in the present paper. The motion is started from rest so that the flow remains irrotational. This implies the existence of a velocity potential such that the fluid velocity is given by its gradient and the governing equation in the fluid domain is the Laplace equation. The problem to be solved is the one in which waves generated by bodies moving in water with a prescribed forward speed ($U_x(t)$, $U_y(t)$) starting from rest, as shown in Fig. 1.

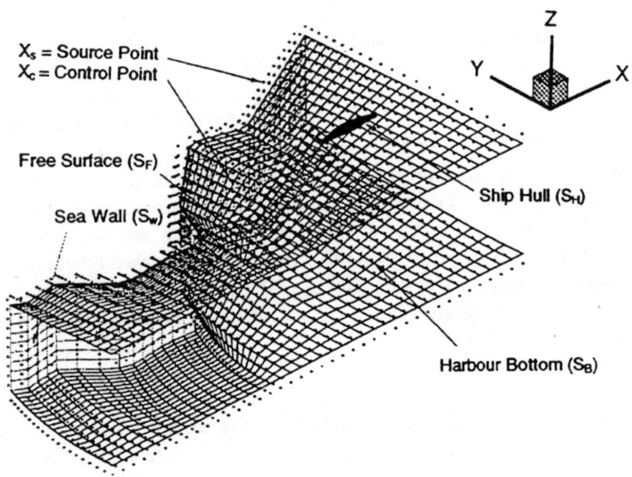

Fig. 1 Coordinate system and the problem domain

The problem is formulated in a fixed coordinate system. The z=0 plane coincides with the undisturbed water surface and z is positive upwards. The flow can then be described by a perturbation velocity potential, $\Phi(x,y,z,t)$, governed by the Laplace equation,

$$\nabla^2\Phi = 0 . \tag{1}$$

Boundary conditions must be applied on all surface surrounding the fluid domain: the free surfaces (S_F), the body surface (S_H), the seawall surface (S_W), the bottom (S_B) and the surrounding surface at infinity (S_∞). The non-penetration boundary condition on solid body S_H and bottom S_B is

$$\nabla\Phi \cdot \bar{n}_H = V_H \cdot \bar{n}_H \qquad \text{on } S_H , \tag{2}$$

$$\nabla\Phi \cdot \bar{n}_B = 0 \qquad \text{on } S_B , \tag{3}$$

where $\bar{n}_H$ and $\bar{n}_B$ are the unit outward normal of S_H and S_B respectively and V_H is the velocity of the body. The seawall reflection condition is

$$\nabla\Phi \cdot \bar{n}_W + \alpha\Phi = 0 \qquad \text{on } S_W , \tag{4}$$

where $\bar{n}_W$ is the unit outward normal of S_W and α is a complex reflection coefficient. The kinematic and dynamic conditions on the free surface S_F are

$$\frac{\partial\eta}{\partial t} = -\nabla\Phi \cdot \nabla\eta + \frac{\partial\Phi}{\partial z} \qquad \text{on } S_F , \tag{5}$$

$$\frac{\partial\Phi}{\partial t} = -g\eta - \frac{1}{2}\nabla\Phi \cdot \nabla\Phi \qquad \text{on } S_F , \tag{6}$$

where $\eta(x,y,t)$ is the free-surface elevation and g is the gravitational acceleration. Because we are solving an initial value problem with no incident waves, the fluid disturbance must vanish at infinity,

$$\nabla\Phi \to 0 \quad \text{as} \quad R \to \infty . \tag{7}$$

In addition, the initial values of the potential and free surface elevation are

$$\Phi = 0 \quad \text{and} \quad \eta = 0 \quad \text{for } t \le 0 . \tag{8}$$

In terms of a de-singularized source distribution, the velocity potential at any point in the fluid domain is given by

$$\Phi(\bar{x}) = \iint_\Omega \sigma(\bar{x}_s) G(\bar{x}, \bar{x}_s) d\Omega , \tag{9}$$

where $G(\bar{x}, \bar{x}_s) = \dfrac{1}{|\bar{x} - \bar{x}_s|}$ is the Rankine source Green function and Ω is the integration surface outside the fluid domain.

By applying the boundary conditions, the integral equations that must be solved to determine the unknown source strengths $\sigma(\bar{x}_s)$ are

$$\iint_\Omega \sigma(\bar{x}_s) G(\bar{x}_c, \bar{x}_s) d\Omega = \Phi_F(\bar{x}_c) , \quad \bar{x}_c \in \Gamma_d \tag{10}$$

and

$$\iint_{\Omega} \sigma(\vec{x}_s)\frac{\partial}{\partial n}G(\vec{x}_c, \vec{x}_s)d\Omega = \chi(\vec{x}_c), \quad \vec{x}_c \in \Gamma_n , \tag{11}$$

where $\vec{X}_s$ = a point on the integration surface, Ω,
$\vec{X}_c$ = a point on the real boundary,
Φ_F = the given potential value on the free surface at $\vec{X}_c$,
Γ_d = surface on which Φ_F is known,
χ = the given normal velocity on the solid boundaries at $\vec{X}_c$, and
Γ_n = surface on which χ is known.

3. THE MODIFIED BARNES-HUT ALGORITHM

Suppose we want to compute the velocity potential on a surface due to source distributions. We assume that

$$\Phi = \Phi_{far} + \Phi_{near} , \tag{12}$$

where Φ_{near} is a rapidly decaying potential in the near field and Φ_{far} is the far-field potential which costs expensively. However, when the ratio

$$\theta = \frac{r}{D} = \frac{\textit{distance of center of sources from control point}}{\textit{size of box containing sources}} \tag{13}$$

is very large, we can safely and accurately replace the sum over all sources in a box with one term at the center of the box (Fig. 2).

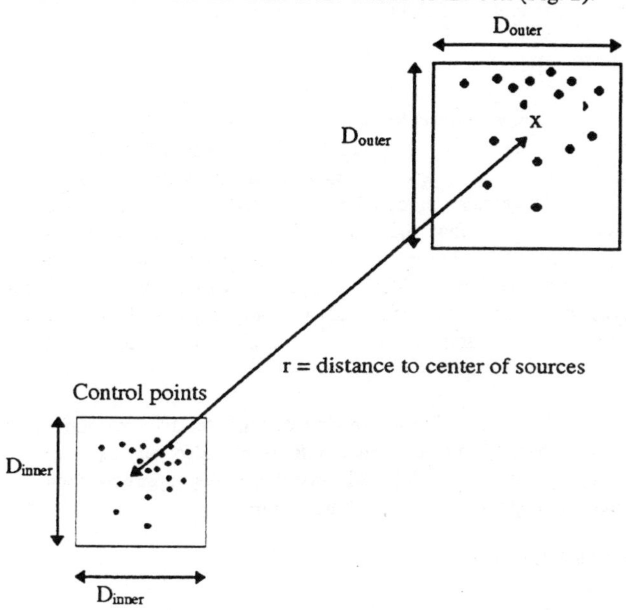

Fig. 2 An illustration of the relevant distances used in Φ_{far}

This idea can be applied for all control points recursively. A data structure is needed to subdivide space that makes this recursion easy. Here, a *quadtree* is used which begins with a square in the plane; this is the root of the *quadtree*. This large square can be divided into four smaller squares of half the perimeter and a quarter the area each; these are the four children of the root. Each child can in turn be divided into 4 *subsquares* to get its children, and so on.

The most interesting problem occurs when the sources are not uniformly distributed in their bounding box, so that many of the leaves of a complete *quadtree* would be empty. In this case, it makes no sense to store these empty parts of the *quadtree*. Instead, we continue to subdivide squares only when they contain more than 1 sources (or some small number of sources). This leads to the adaptive *quadtree* shown in Fig. 3 below. Note that there are exactly as many leaves as sources.

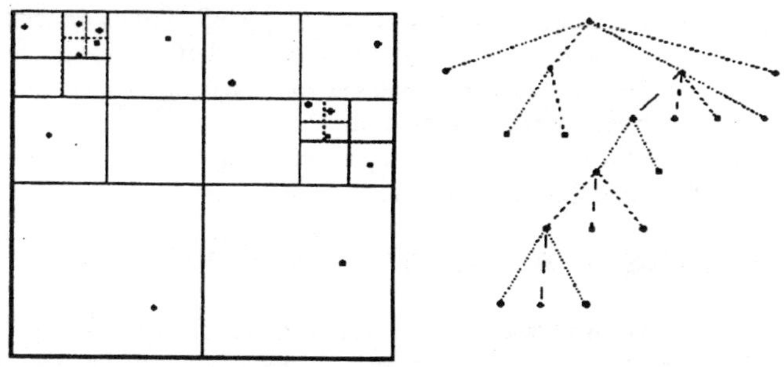

Fig. 3 Adaptive quadtree where no square contains more than 1 sources

Now, a fast hierarchical n-body algorithm, the modified BH algorithm, can be brought into our problem. In the BH algorithm, traversing the quadtree to compute the potential on each control point is named as cell-to-particles. We can modify Eq.(13) as follows (see Fig. 2):

$$\theta = \frac{r}{D_{outer} + D_{inner}} = \frac{\textit{distance of center of sources from center of control points}}{\textit{size of box containing sources + size of box containing control points}} \tag{14}$$

If θ is large enough, all the sources inside the outer box can be regarded as concentrated at the center of the outer box and all potentials at control points in the inner box can be evaluated at the center of the inner box due to the equivalent source at the center of the outer box. Hence, this algorithm is called the cell-to-cell algorithm. Briefly, the modified BH algorithm contains the following steps:

1) Build a Quadtree as described above.
2) For each subsquare in the quadtree, compute the center of geometry and total strength for all the sources it contains.
3) Traverse the tree to compute the potential at all control points. If the ratio θ is large enough, then we can use the cell-to-cell algorithm to compute the potentials at control points in the inner box due to all the sources in the outer box. We will compare θ to a user-supplied threshold θ_t to make this decision.

4. NUMERICAL SOLUTION PROCEDURE

A computer program, FHWAVE, is developed for computation of ship-generated waves in a large domain. FORTRAN90 is used in this code to improve the performance of the computation. The numerical solution procedure is as follows:

1) Distribute Rankine sources on a surface at a distance L_d from the free surface, as defined by Beck et al. (1994);
2) Setup initial conditions: $\Phi = 0$ and $\eta = 0$ for $t \le 0$;
3) Build an adaptive quadtree of sources and control points;
4) Use an 5th order Runge-Kutta-Merson algorithm for the time step;
5) Traverse the quadtree to calculate the potential and velocity on the free surface using the cell-to-cell algorithm;
6) Use an iterative method called GMRES to solve for the strengths of sources at the lowest level of the quadtree;
7) Re-calculate the strengths of sources along the quadtree from lower to higher levels;
8) Move the ship to a new position; and
9) Go to the next time step .

5. NUMERICAL TEST RESULTS AND DISCUSSION

We first test the numerical performance of FHWAVE for a simple problem where the wave is generated by the motion of a source-sink pair below the z=0 plane starting from rest. The linearized free surface condition is used. The problem has been non-dimensionalized by taking the depth of the disturbance, H, and the gravitational acceleration g to be unity. The distance between the source and sink is chosen to be 0.1. The pair moves at speed $U(t)=F_r(1-e^{-4t})$ along the positive x direction with the source and sink initially located at (0,0,-1) and (-0.1,0,-1) respectively, where the Froude number F_r is defined by $\frac{U_0}{\sqrt{gH}}$, U_0 being the steady velocity of the disturbance. We choose F_r to be unity in this example. The magnitude of the disturbance for both source and sink is given by $\sigma(t)=\sigma_0(1-e^{-4t})$, where $\sigma_0=0.75$. The time step is chosen to be 0.2 in the time-marching procedure. The free surface mesh grids are in the range of $0<x<20$ and $0<y<7.5$ (with a symmetry plane at y=0) and are divided into 40x15 elements (coarse case) or 60x20 elements (fine case). The free surface elevation is computed using both the direct method (the direct evaluation of the influence functions, e.g. threshold $\theta_t=\infty$) and FHWAVE. The numerical performance and accuracy of the different threshold θ_t (∞,8,6,4) are compared in coarse and fine cases. The computations are carried out on a DEC-ALPHA 600-5 workstation with 333MHZ CPU speed. It should be noted that for all results presented here, no filtering or smoothing is used.

Fig. 4 shows the CPU time and the norm of residual error, R_{err}, at different thresholds for coarse and fine cases after 100 time steps. The residual error is defined by $R_{err} = \sqrt{\frac{\Sigma(H_i-H_{0i})^2}{N}}$, where H_{0i} is the wave elevation on free surface calculated by the direct method and H_i by FHWAVE. It is clear that the CPU time of high threshold increases very rapidly with the number of sources N on the free surface. For N>2000, the direct method is too slow to tolerate. The new fast hierarchical algorithm produces a very significant reduction in CPU time and memory. For the low threshold, $\theta_t=4$, the accuracy of the results is also acceptable for some large domain application. This implies that for the FHWAVE code, we can choose different thresholds reasonably to march different requirement for CPU time and accuracy in practical applications. Fig. 5 shows the comparison of wave elevation contours between two methods for coarse and fine cases at the 90th time step.

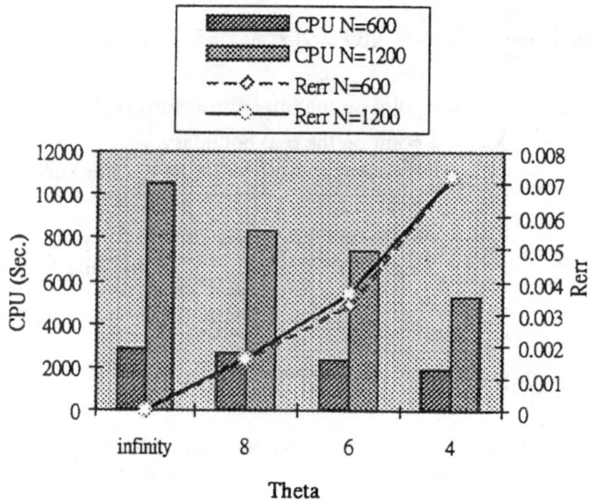

Fig. 4 CPU time and R_{err} comparison

Finally, the waves generated by three ships moving in Victoria Harbour using FHWAVE with threshold $\theta_t=4$ are simulated. In this case, 30,000 singularities are distributed over the free-surface near Tsim Sha Tsui in Kowloon and Central District on Hong Kong Island. The ships are simplified by three sink-source pairs with the same source strength and different velocities along three pre-designed paths. The computation only needs about 11MB of memory. Fig. 6 shows the 3D ship-generated waves in shade format. The animated results in time will be shown in the conference.

CONCLUSIONS

A new method for simulation of waves in a large domain is developed. The CPU time and memory requirement of the method depend on N(logN) and increase with threshold linearly. The accuracy of calculation can be controlled by the threshold. The numerical tests show that the computer program FHWAVE can reduce the CPU time and memory significantly when N becomes very large. The FHWAVE is a real 3D water wave model which opens up a new way for studying ship-generated waves, wind-generated waves and swells, including their propagation, reflection and interaction in harbours.

In our calculation, we find that the iterative method called GMRES for solving a large linear equation is also very expensive when N becomes huge. We will further develop a precondition method to improve the GMRES algorithm.

ACKNOWLEDGMENTS

This research was jointly sponsored by the Hong Kong Research Grants Council under Grant No. HKU 568/96E and by the University of Hong Kong under CRCG Grant No. 337/064/0043.

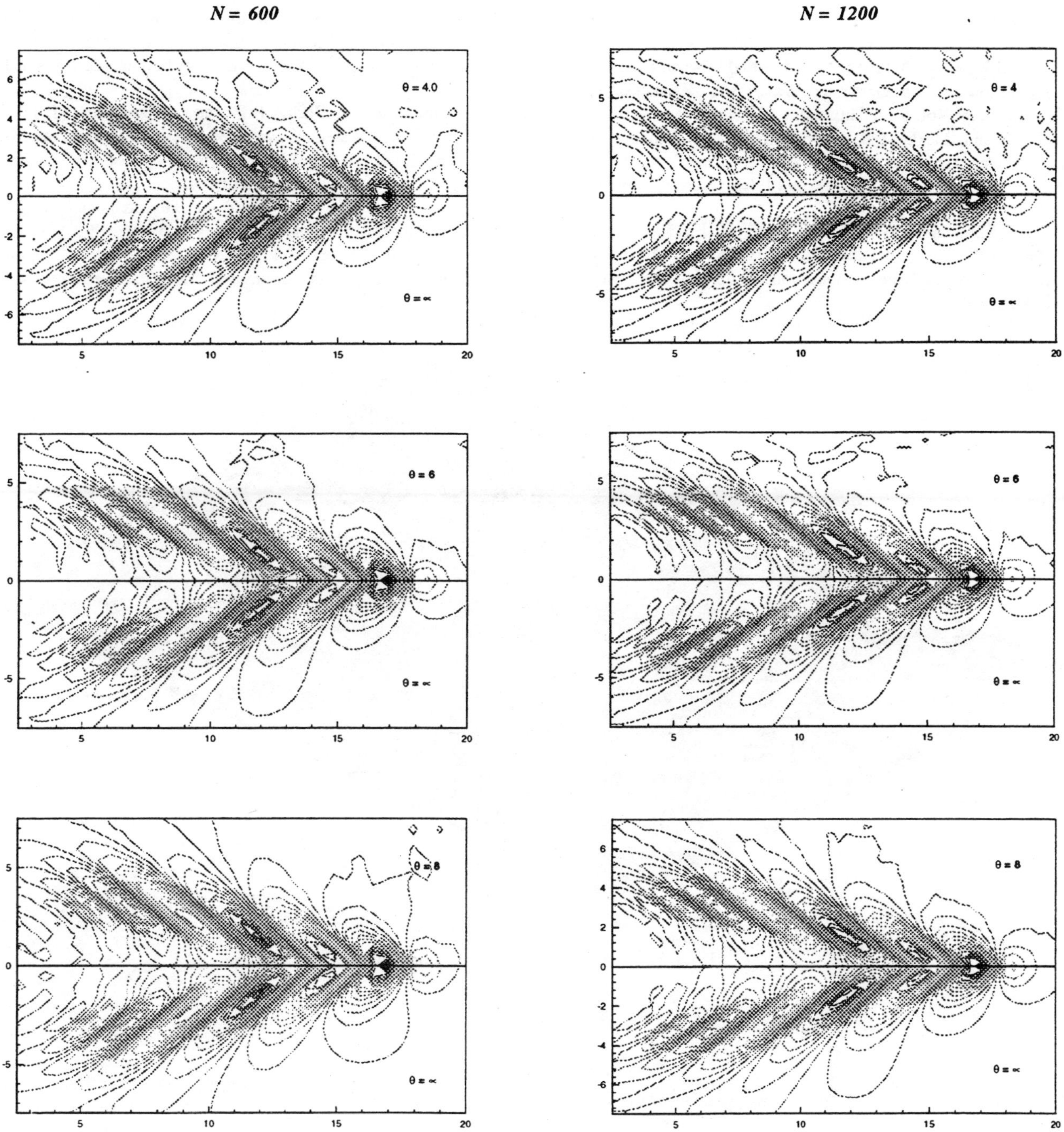

Fig. 5 Comparison of wave elevation contours

REFERENCES

1. Li, D and Zhou, L. D. (1986), "A Modified Dawson's Method for Computation of Ship Generated Waves and Resistance" China Ship Building, Vol. 5 (in Chinese).
2. Cao, Y.S. and Maskew, B. (1996), "Nonlinear Computations of Ship Waves Using a Mixed Boundary Integral Method with Multipole Acceleration," The Second International Conference on Hydrodynamics, Hong Kong.
3. Cao, Y. S., Schultz, W. W. and Beck, R. F. (1991), "Three-Dimensional Desingularized Boundary Integral Methods for Potential Problems," International Journal for Numerical Methods in Fluids, Vol. 12.
4. Beck, R.F., Cao, Y., Scorpio, S.M. and Schultz, W.W (1994), "Nonlinear Ship Motion Computations Using the Desingularized Method, " Proc. 20th Symp. Naval Hydrodynamics, Santa Barbara, CA, USA.
5. Scorpio, S.M., Beck, R.F. and Korsmeyer, F.T. (1996), "Nonlinear Water Wave Computations Using a Multipole Accelerated, Desingularized Method, " Proc. 21st Symp. Naval Hydrodynamics, Tronheim, Norway.
6. Barnes, J. and Hut, P. (1986), "A Hierarchical O(nlogn) Force Calculation Algorithm, " Nature, Vol. 324.
7. Greengard, L.F. and Rokhlin, V. (1987), "A Fast Algorithm for Particle Simulations, " Journal of Computational Physics, Vol. 73.

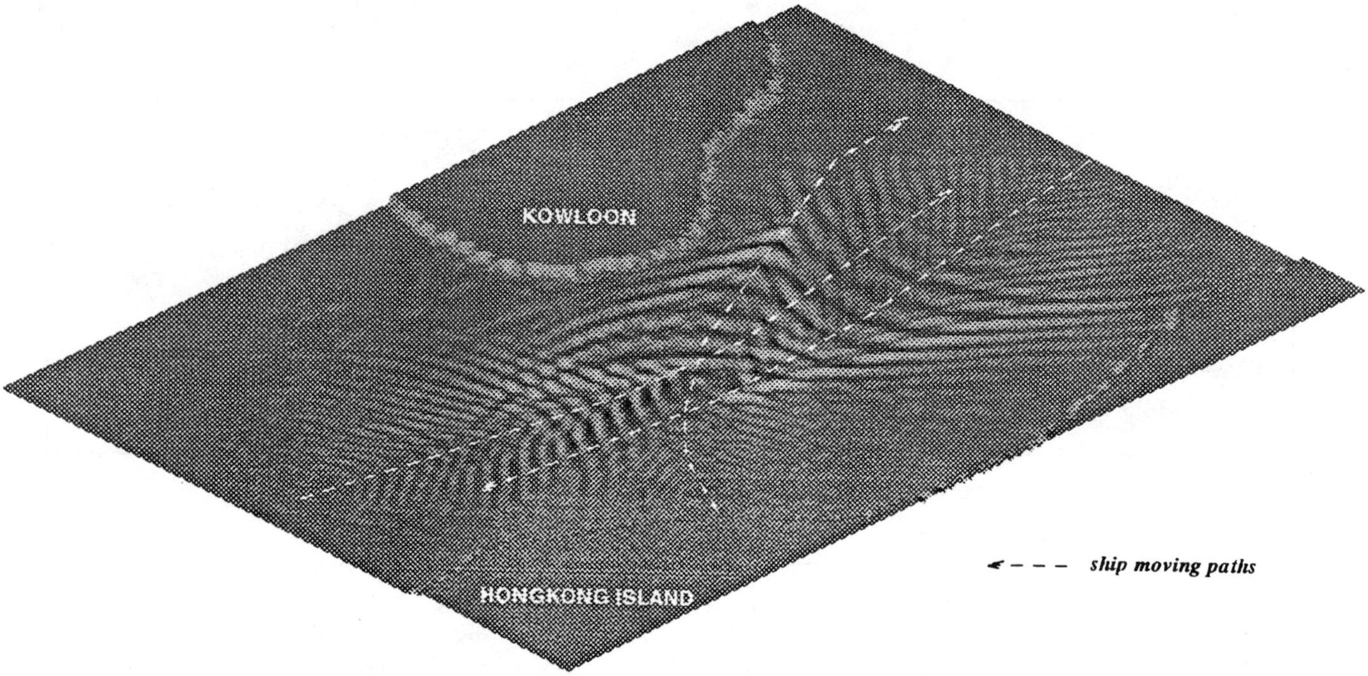

Fig. 6 Ship-generated waves in Victoria Harbour

Proceedings of the Seventh (1997) International Offshore and Polar Engineering Conference
Honolulu, USA, May 25-30, 1997

ISBN 1-880653-28-1 (Set); ISBN 1-880653-31-1 (Vol. III)

Numerical Simulation of Wave Run-Up and Breaking on a Sloping Beach*

Decheng Wan
Shanghai University, Shanghai, China
Guoping Miao
Shanghai Jiao Tong University, Shanghai, China
Shiqiang Dai
Shanghai University, Shanghai, China

ABSTRACT

Numerical simulations describing wave run-up on a sloping structure including the overturning, breaking and merging phenomena are presented. The motions are governed by the classical, two-dimensional Navier-Stokes equation for incompressible fluid. The numerical modeling of the flow is based on a piecewise linear version of the volume of fluid method(the VOF method). Computational results concerning the time evolution of the free surface and pressure distribution along water bed and slope boundary are given, showing how an initial solitary wave undergoes run-up, overturning, breaking and merging on the slope. It is found that most of the wave energy is lost after the wave breaking and merging.

KEY WORDS: VOF method, wave run-up, sloping beach

INTRODUCTION

Sloping structures are widely used as maritime structures for harbor and coastal engineering, which are constructed for various usage, such as breakwaters of ports, or river dikes in coastal zone or river side for various purposes(Li, 1990). The estimation of wave run-up and wave impact pressure is one of the main problems for the design of sloping structures. As waves propagate along the sloping structure, the free surface undergoes large deformation and the flow phenomena are quite complicated. Hence, for practical engineering application, enormous model tests have been conducted in the past(Saville, 1963). Numerical studies describing the evolution of wave run-up on a slope have been successfully made by numerous investigators using various methods based on the potential flow theory. Among the main contributions are the works of Longuet-Higgins(1976), Kim, Liu and Liggett(1983), Peregrine(1983), Tao(1984), and Zhou and Cheng(1987). But the process of the overturning and merging is less understood. The potential flow method seems to contain substantial shortcomings for the explanation of the wave motion during and after the stage of overturning. The stage of overturning is only the beginning of the nonlinear behaviors of breaking waves, and subsequent stages of the generation of wave impact load and vortices, which are of full nonlinear features, important for the safety of maritime structures. So it is necessary to further the study that is based on the solution of the Navier-Stokes equations and the computation of large free surface deformations.

So far, there are two special computational techniques for solving the problems of large free surface deformations, i.e., surface tracking and volume tracking (Floryan and Rasmussen, 1989). The volume tracking techniques have the potentiality for handling large surface deformations and surface folding and merging, which adopt a volumetric progress variable, such as marker particles in the marker and cell (MAC) technique(Harlow and Welch, 1966), and the cell volume fraction in the volume of fluid (VOF) technique(Hirt and Nichols, 1981), for Lagrangian transport of the interfaces. The MAC method involves Eulerian flow field calculations and Lagrangian liquid-particle movements. The velocity of a marker particle is determined by taking the average of the Eulerian velocities in its vicinity. One difficulty in using the MAC method is the possible creation of artificially high or low marker number densities in the cells, due to the irregularity of the flow field. However, the VOF technique provides a means of following fluid regions through an Eulerian mesh of stationary cells. The method needs a minimum of stored information, and because it follows regions rather than boundaries, it avoids all logic problems associated with intersecting surfaces. So the VOF method is much easier use and can be applied to handle surface folding and merging. In the present paper, the volume of fluid method is used to simulate the flow of wave run-up on a sloping structure. The Navier-Stokes equations in finite difference form are solved by time

* This paper is supported by Science and Technology Development Foundation of Shanghai Advanced School of China and National Natural Science Foundation of China.

marching scheme in an inflexible rectangular staggered mesh system. The governing equation of time dependence of the fractional volume of fluid F variable is discreticized by a type of donor-acceptor flux approximation. It is demonstrated that this method is applicable to the problem of a solitary wave run-up on a sloping structure that involves wave breaking and surface merging. The results of simulation show fairly good resemblance to the actual nonlinear wave motions including breaking of waves, surface overturning, surface merging, and occurrence of impact pressure.

MATHEMATICAL FORMULATION AND COMPUTATIONAL PROCEDURE

Governing Equations

Consider the two-dimensional motion of a incompressible viscous fluid and study the solitary wave evolution as it climbs up on a sloping structure(see figure 1). We choose a coordinate system where Ox coincides with the horizontal bottom and Oy is positive upward, and define (u, v) as the horizontal and vertical components of the velocity, t as the time, ρ as the density, ν as the kinematic viscosity of water, p as the pressure and g as the gravitational acceleration. Surface tension is neglected. The continuity equation and momentum equation are as followings:

$$\frac{\partial u}{\partial x}+\frac{\partial v}{\partial y}=0 \tag{1}$$

$$\frac{\partial u}{\partial t}+u\frac{\partial u}{\partial x}+v\frac{\partial u}{\partial y}=-\frac{1}{\rho}\frac{\partial p}{\partial x}+g_x+\nu\left[\frac{\partial^2 u}{\partial x^2}+\frac{\partial^2 u}{\partial y^2}\right] \tag{2}$$

$$\frac{\partial v}{\partial t}+u\frac{\partial v}{\partial x}+v\frac{\partial v}{\partial y}=-\frac{1}{\rho}\frac{\partial p}{\partial y}+g_y+\nu\left[\frac{\partial^2 v}{\partial x^2}+\frac{\partial^2 v}{\partial y^2}\right] \tag{3}$$

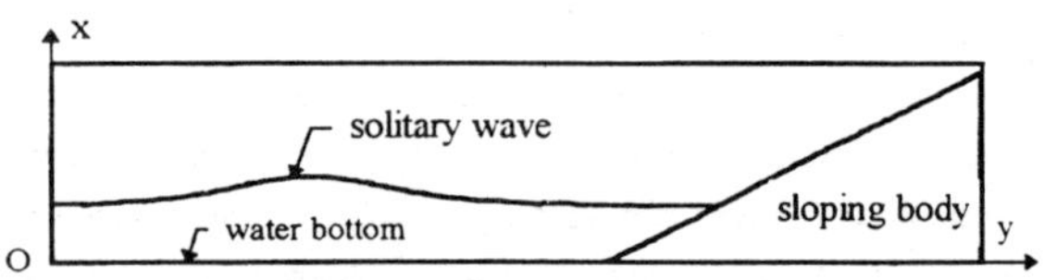

Fig 1 Definition sketch of physical plane

For overcoming the difficulty of solving strictly incompressible problem, it is available to allow limited compressibility effects, in which eq.(1) is replaced in computation with following equation:

$$\frac{1}{\rho c^2}\frac{\partial p}{\partial t}+\frac{\partial u}{\partial x}+\frac{\partial v}{\partial y}=0 \tag{4}$$

where c is the adiabatic speed of sound in the fluid. When c is enough large(eg. c is greater than 10^4), eq.(4) is close to eq.(1) through iteration procedure.

VOF Method

The basis of the VOF method is the fractional volume of fluid scheme for tracking free boundaries. In this technique, a function $\mathcal{F}(x,y,t)$ is defined whose value is unity at any point occupied by fluid and zero elsewhere. When averaged over the cell of a computational mesh, the average value of $\mathcal{F}$ in a cell is equal to F, which represents the fractional volume of the cell occupied by fluid. In particular, a unit value of F corresponds to a cell full of fluid, whereas a zero value indicates that the cell contains no fluid. Cells with F values between zero and one contain a free surface. The discontinuity in F propagates according to the equation:

$$\frac{\partial F}{\partial t}+u\frac{\partial F}{\partial x}+v\frac{\partial F}{\partial y}=0 \tag{5}$$

which states that F moves with fluid and the surface is a material line. Since F is a step function, standard finite difference approximations would lead to a smearing of the F function and interfaces would lose their definition. So a special flux approximation referred to as a Donor-Acceptor method(Johnson, 1970, Ramshaw and Trapp, 1976) is used to preserve the discontinuous nature of F. The essential idea is to use information about F downstream as well as upstream of a flux boundary to establish a crude interface shape, and then to use this shape in computing the flux. For the purpose of calculations, the free surface is approximated by a straight line through the cell. The location and slope of the line are determined by the average value of F in the cell (equal to the fractional volume) and by the gradient of F.. Numerical estimates of the derivatives of F are obtained by looking at fractional volumes of the neighboring cells. The VOF method depends on the ability to advect the volume fraction through the grid without smearing. This method allows us to follow interfaces beyond the point of reconnection, and is relatively simple and robust.

Computational Algorithm

The fundamental concept of the computational algorithm is similar to SOLA-VOF proposed by Nichols, Hirt and Hotchkiss (1980), Andersson and Benocci (1985). A staggered mesh system with rectangular cells of equal spacing is employed for the discretization, in which forward difference is used for time terms, centered difference in space for the viscous terms, and combined the donor cell and centered difference approximations for the convective terms. The detailed discretizational representation is well documented in Hirt and Nichols (1981), Wang (1994).

We can summarize the above procedure for one time step as follows: (a) explicit approximations to eq.(2) and (3) are used to give a first estimate of the new velocities; (b) the pressure is adjusted in an iterative manner in order that the continuity equation (4) can be satisfied; (c) F is updated through eq.(5) by the Donor-Acceptor flux approximation. Repetition of these steps will advance a solution through any desired time interval.

COMPUTATIONAL RESULTS

In this section we present and analyze the results of computations which cover the complete process of a solitary wave run-up, overturning, breaking, slip-down, merging and backward transmitting waves on a sloping structure. The computational mesh consists of 200 equally spaced cells in the horizational direction($\Delta x = 4cm$) and 80 equally spaced cells in the vertical direction($\Delta y = 2cm$). The slope of the structure is set to be 0.5. The water depth is D = 40 cm. The initial solitary wave amplitude is H = 20cm, which has its crest at $\mathbf{x_0}$ = **200 cm** and progresses from left to right. At T=0, the velocity field of fluid is obtained from the velocity potential. The principal phenomena of interest within each stage are described below.

Wave Run-up

Figure 2 illustrates the snapshots of instantaneous free surface profile, velocity field and pressure distribution for the stage of wave run-up on the slope. The displayed region is limited to the area of the wave run-up of interest. The vertical arrow lines below frame bottom represent the pressure distribution along the water bed and the slope wet boundary. As shown in Figure 2, during the wave run-up, the wave profile, particularly near the contact point where the free surface comes into contact with the slope boundary, becomes more and more straight and nearly skew parallel with the slope boundary. The water layer on the slope boundary also becomes thinner and the climbing velocity decreases with the process of time. The thin layer of water can be simulated because the VOF method follows regions rather than boundaries, mesh size just affects the numerical precise and does not affect track the fluid regions. The pressure distribution on the water and slope boundary is close similar to still water pressure except the tip area of wave front where there is small impact pressure (see T=1.103s cases). The free surface keeps stable during the whole wave run-up stage. The phenomenon of this stage is in agreement with the computational results reported by Zhou and Cheng(1987).

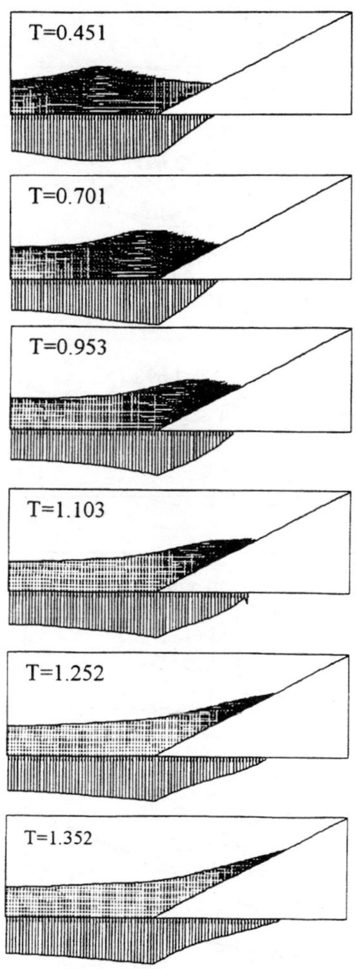

Fig. 2 The stage of wave run-up

Wave Breaking

Figure 3 shows a sequence of free surface profile, velocity field and pressure distribution for the stage of wave breaking. The maximum height of wave climbing is 103 cm(see the plot for T=1.401s) which is close to the result of 105cm given by Zhou and Cheng(1987). The wave breaks at the time where the tip front of the propagating wave steepens and becomes vertical(see T=1.500s), and then a jet of fluid is projected forward forming a characteristic overturning motion(see T=1.551s), which is spilling breaker(Duncan, Philomin, Behres and Kimmel, 1994). The vorticity and impact pressure are mainly located in the vicinity of the tip of the jet where the curvature of the free surface is important(see T=1.401s and T=1.500s). Under the action of gravity, the small jet plunges down into the surface below(see T=1.551s, T=1.601s and T=1.700s), and products second overturning and breaking(see T=1.753s), which is plunging breaker.

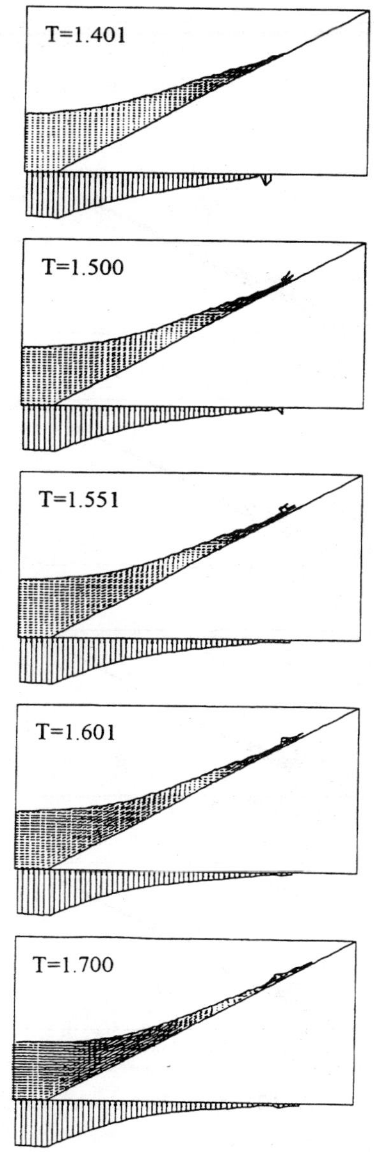

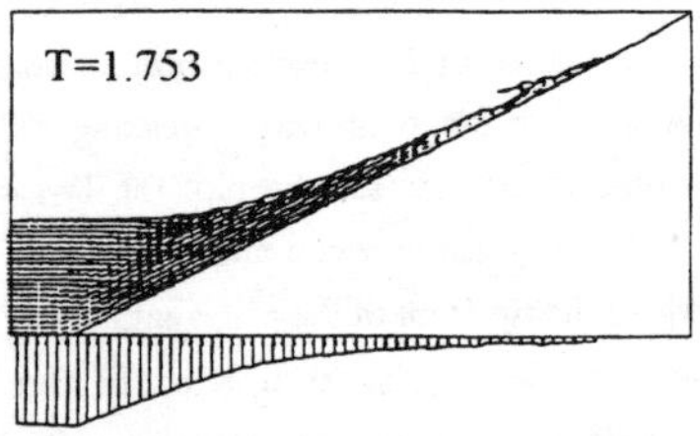

Fig 3 The stage of wave breaking

Wave Slip-down

Figure 4 is a sequence of free surface profiles, velocity fields and pressure distributions for the stage of wave slip down. Under the effect of gravity, the thin water layer slides down along the slope boundary, and gradually disappears on the slope with the process of time. In the later time of the slip-down, a water fovea are formed due to the momentum of the slipping fluid (see T=2.350s and T=2.450s). It is observed there appear many small ripples, and the irregular oscillation of pressure arises on the slope wet boundary. Some small negative pressure distributions come up near the area where the fluid just disappears and small vorticity begins to form.

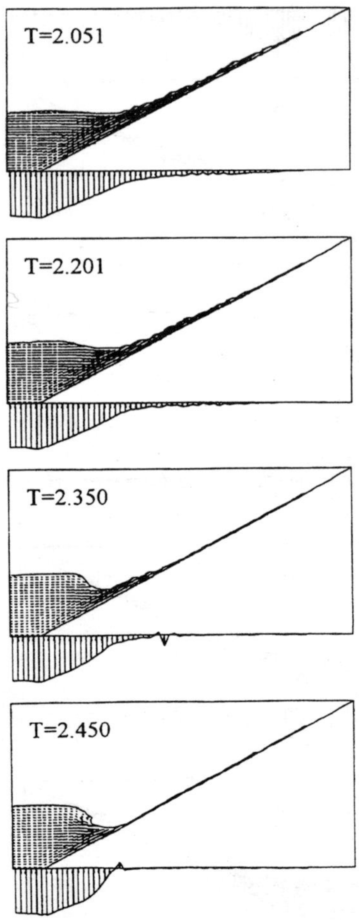

Fig 4 The stage of slip-down

Wave Merging

Figure 5 gives a close-up view of the active region of free surface profile, velocity field and pressure distribution at various instants for the stage of wave merging. Under the influence of the fluid of slip down, the fovea becomes large and nearly closed (see T=2.67s). The front of the water fovea forms a jet of liquid which is growing in size. Under the action of gravity, the jet falls and touches the surface below, entraining a volume of air(see T=2.683s), and then plunges down into the surface below and merges with the fluid below(see T=2.691s). The main effect of the impact of the jet is the generation of vortical motions and the formation of a small splash-up(see T=2.698s, T=2.704s and T= 2.709s). The change of pressure distribution is significant when the strong vorticity is generated due to the wave merging.

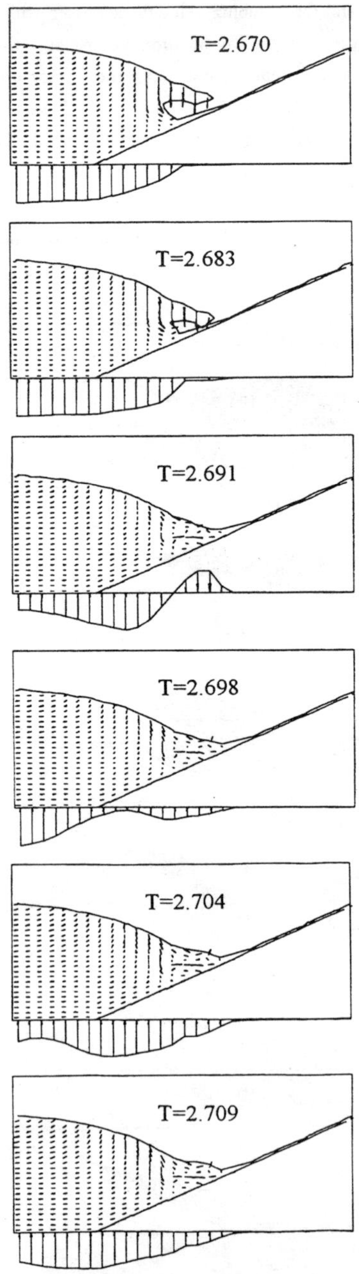

Fig 5 The stage of wave merging

Backward Transmitted Wave

Figure 6 illustrates a sequence of the interesting regions of free surface profiles, velocity fields and pressure distributions for the stage of backward transmitted wave. As the fluid is pushed forward by the merging bulk and the strong vorticity, a new small jet bulk is formed and begins to climb along the slope(see T=2.751s and T=2.950s). After the second small run-up, the bulk of fluid slides down again and progresses backward(see T=3.150s and T=3.450s). The free surface appears to be zigzag. The backward transmitted wave and the tail of tiny wave train can be observed in the cases of T=3.950s and T= 4.201s. These waves with small wave amplitudes and irregular wave profiles are completely different from the initial incident solitary wave. As shown in the cases of T=3.950s and T=4.201s, the wave loses most of its original energy due to the breaking and merging.

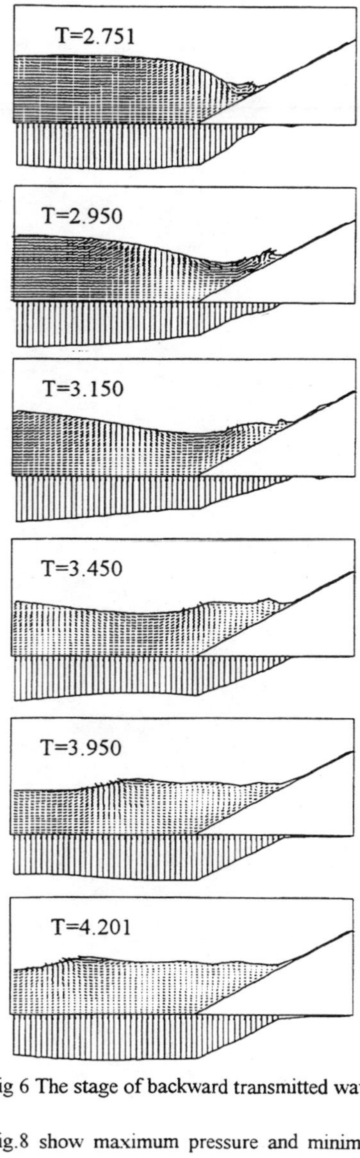

Fig 6 The stage of backward transmitted wave

Fig.7 and Fig.8 show maximum pressure and minimum pressure of flow fields at various time respectively. From these figures, we can seen the change of the scale of the pressure, the maximum impact pressure values and negative pressure amplitudes.

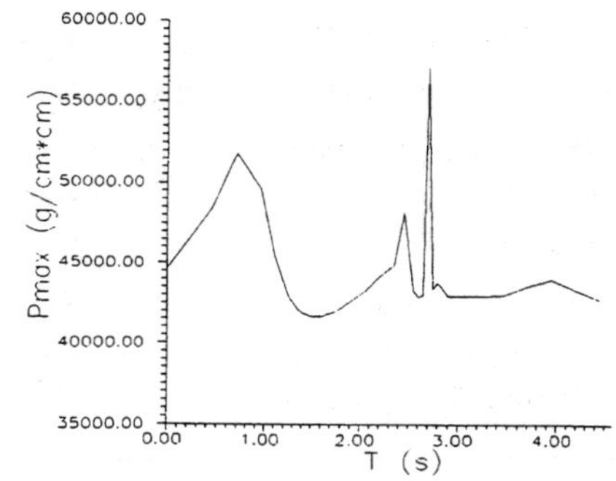

Fig.7 Maximum Pressure with Time

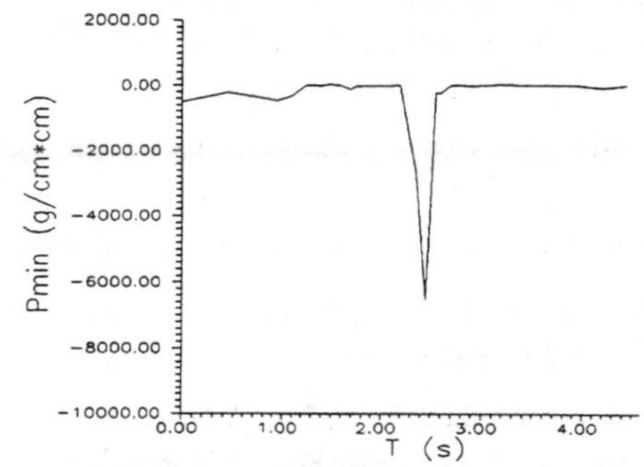

Fig.8 Minimum Pressure with Time

CONCLUSIONS

In this paper we have described that the simulation the fluid motion of a solitary wave propagating on a sloping structure by means of the VOF method. Computational results show good resemblance to the actual observations of wave run-up motions on a slope including overturning, breaking, merging, plunging process and air entertainment. The generation of vorticity by overturning and merging has also been shown. The change of pressure distribution and impact pressure due to merging and plunging has been found important during wave run-up process. Our results illustrate the effectiveness and potentiality of the VOF method to simulate the evolution of finite amplitude surface waves and to offer efficient tool for numerical design of maritime structures.

REFERENCES

Andersson, HI and Benocci, C (1985). "An Improved Pressure Iteration Technique for the SOLA Algorithm," in Numerical Methods in Laminar and Turbulent Flow, *4th Int Conf Swansea*, C Taylor et al. Eds., Pineridge, Swansea, UK.

Duncan, JH, Philomin, V, Behres, M and Kimmel, J (1994). "The Formation of Spilling Breaking Water Waves," *Phys Fluids*, Vol 6, pp 2558.

Floryan, JM and Rasmussen, H (1989). "Numerical Methods for Viscous Flows with Moving Boundaries," *Appl Mech Rev*, Vol 42, No 12, pp 323-341.

Harlow, FH and Welch, JE (1966). "Numerical Study of Large Amplitude Free Surface Motion," *Phys Fluids*, Vol 9, pp 842-851.

Hirt, CW and Nichols, BD (1981). "Volume of Fluid (VOF) Method for the Dynamics of Free Boundaries," *J Comput Phys*, Vol 39, pp 201-225.

Johnson, WE (1970). "Development and Application of Computer Programs Related to Hypervelocity Impact," Systems, Science and Software Report 3SR-353.

Kim, SK, Liu, PLF and Liggett, JA (1983). "Boundary Integral Equation Solutions for Solitary Wave Generation Propagation and Run-up," *Coastal Engineering*, Vol 7, pp 299-317.

Li, YC (1990). "Wave Action on Maritime Structures," The Press of Dalian University of Technology.

Longuet-Higgins, MS and Cokelet, ED (1976). "The Deformation of Steep Surface Waves on Water I. A Numerical Method of Computation," *Proc of the Royal Society*, Series A, Vol 350, No 1660, pp 1-26.

Nichols, BD, Hirt, CW and Hotchkiss, RS (1980). "SOLA-VOF: A Solution Algorithm for Transient Fluid Flow with Multiple Free boundaries," Los Alamos Scientific Laboratory Report LA-8355.

Peregrine, DH (1983). "Breaking Waves on Beaches," *Annu Rev Fluid Mech*, Vol 15, pp149-154.

Ramshaw, JD and Trapp, JA (1976). "A Numerical Technique for Low Speed Homogeneous Two Phase Flow with Sharp Interfaces," *J Comput Phys*, Vol 21, pp 438-453.

Saville, T (1963). "Wave Run-up on Compsite Slopes," *Proc. 6th ICCE*.

Tao, JH (1984). "Numerical Simulation of Wave Run-up and Wave Breaking on beaches," *J of Ocean*, Vol 6, No 5, pp692-700.(in Chinese)

Wang, YX (1994). "Numerical Wave Channel with Absorbing Wave Maker," *J of Hydrodynamics*, Ser A, Vol 9, No 2, pp 205-214.(in Chinese)

Zhou, GY and Cheng, YS (1987). "Numerical Simulation of Nonlinear Wave Run-up on Beaches," Technical Report of Department of Mechanics of Peking University.(in Chinese)

Proceedings of the Seventh (1997) International Offshore and Polar Engineering Conference
Honolulu, USA, May 25-30, 1997

ISBN 1-880653-28-1 (Set); ISBN 1-880653-31-1 (Vol. III)

Numerical Analysis of Dynamic Interaction Between Non-Linear Waves and Permeable Toe Over Sand Seabed in Front of a Seawall

Ayman M. Mostafa and Norimi Mizutani
Nagoya University
Nagoya, Japan

ABSTRACT

A combined BEM-FEM model has been developed to simulate the dynamic interaction between non-linear waves and an impermeable seawall with a submerged rubble toe over a sand seabed. A poro-elastic finite element model has been developed and used under the surface pressure computed by the BEM-FEM model. Modified Navier-Stokes equations are used to solve the flow inside the porous media and Biot's equations for the solid particles' displacement. The wave deformation and its induced pore-water flow are studied for different hydraulic properties and sizes of the toe. The effects of the seabed shear modulus and thickness on the solid particle displacement, pore-water pressure and dynamic stresses are investigated.

KEY WORDS: non-linear waves, sand seabed, permeable structures, seawall, wave-structure-foundation dynamic interaction, combined finite and boundary element method.

INTRODUCTION

The dynamic interaction between water waves and structures is one of the most important and complex coastal engineering problems. The existence of a permeable structure over a permeable seabed under the attack of non-linear water waves presents great difficulties for the solution of the problem. Therefore, simplified approximate solutions are common practice. These solutions may define either the wave field around the structure assuming it to be impermeable with a rigid foundation or may provide the wave-induced pore-water pressures and velocities inside the structure and foundation using the linear water wave theory.

The wave-induced pore-water pressure and solid particle displacements in a poro-elastic seabed were determined either analytically (Yamamoto, 1977; Madsen, 1978; Magda, 1994) or numerically using finite element method (Gatmiri, 1990) and, hence, the effective stresses could be found. In these studies, they aimed at testing the possibility of seabed liquefaction and/or shear failure under severe wave conditions. Others also reported that wave-induced liquefaction in the sand seabed is responsible for a collapse of the structure and they showed experimentally the settlement of concrete blocks by the pressure fluctuation (e.g. Maeno and Nago, 1988). Sakai et al. (1992) investigated the possibility of occurrence of the momentary seabed liquefaction at wave trough phase and its depth under the surf zone condition using the boundary layer approximation solution for Biot's equations developed by Mei and Foda (1981).

A mixed numerical model for the simulation of wave motion in rubble mound breakwaters was developed to check the dynamic stability of the seaward slope under severe wave attack using the modified Navier-Stokes equations (Hannoura and McCorquodale, 1985). To furnish the boundary pressure for this model, pressure transducers and a movie camera were used to record the dynamic pressure along the interface and to trace the outcrop point movement. For composite breakwaters, a finite element model based on Biot's equations (1941) was developed to compute the wave-induced solid's displacement and pore-water pressure in the seabed and rubble foundation (Mase et al., 1994). Linear standing water waves were used to provide the pressure along the boundary surfaces. In the latter approach, the wave and soil response were uncoupled and the effect of the rubble geometry on the wave was disregarded.

Ohyama and Nadaoka (1991) developed a numerical wave tank for the analysis of non-linear and irregular wave field using the Boundary Integral Equation Method (BIEM). They proved that a numerical wave absorption filter consisting of a sponge layer and Sommerfeld's radiation condition at the outer edge of the layer is applicable for a wide range of wave conditions. Moreover, Brorsen and Larsen (1987) introduced a non-reflective wave generator for use in the computational domain and proved its efficiency.

Few researchers coupled the wave and the structure in one computational scheme. The wave deformation and its induced pore-water pressures and velocities in case of wide crown submerged breakwaters were studied experimentally and numerically and the significance of non-linear dynamic interaction flow models was shown successfully (Mizutani et al., 1996b).

The foregoing flow model has been modified to study the coupled wave, structure and foundation non-linear dynamic interaction for a seawall with permeable toe resting over a porous seabed. This dynamic interaction is considered further to compute the wave-induced pore-water pressure, solid's displacement and dynamic stresses inside the porous media using Biot's equations. The Boundary Element Method (BEM) based on the time-domain solution of the boundary integral equation is utilized to solve the non-linear wave field. The Finite Element Method (FEM) is employed to solve the flow inside the porous media. The combined BEM-FEM model solves the wave field and porous media simultaneously at each time step in a time-marching scheme. A finite element poro-elastic model based on Biot's equations (1941) is developed to compute the solid particles' displacement and pore-water pressure inside the toe and seabed. The poro-elastic model needs the pressure on the surfaces which are subjected to the wave action. The pressure on these surfaces is computed by the BEM-FEM model at each time step.

The effects of hydraulic properties and size of the toe are tested to predict their corresponding impact on the flow inside the porous media and the wave deformation. Suitable toe dimensions are investigated for producing the minimum possible wave height at the seawall. Since the deformation of the porous media is a function of both the wave conditions and the properties of the porous media, the toe stability is studied for different seabed geotechnical properties and thickness.

GOVERNING EQUATIONS AND MODELS FORMULATION

This section describes the governing equations and the development of the BEM-FEM model for wave-structure-foundation non-linear dynamic interaction and the FEM poro-elastic model.

BEM Model for Non-Linear Waves

The governing equation for the velocity potential in the fluid domain is given by Poisson equation (Brorsen and Larsen, 1987).

$$\frac{\partial^2\phi}{\partial X^2}+\frac{\partial^2\phi}{\partial Z^2}=q^* \tag{1}$$

$$q^*=U^*(Z,t)\delta(X-X_s) \tag{2}$$

where ϕ is the velocity potential; X and Z are the horizontal and vertical coordinates, respectively; q* corresponds to the flux density at the wave-maker which equals U* at the wave-maker and is zero elsewhere; δ is the Dirac's delta function; and X_s is the X-position of the wave-maker.

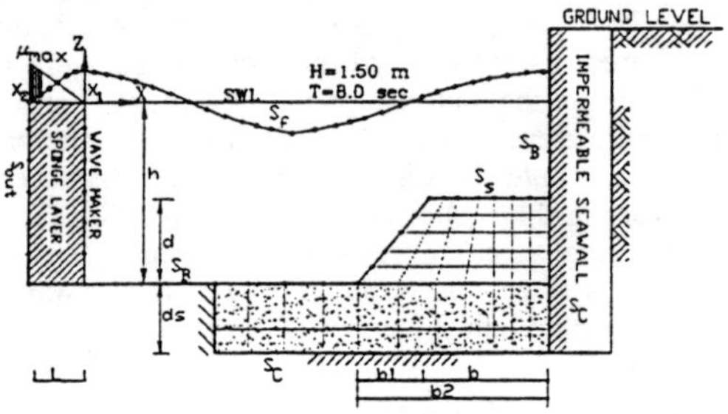

Fig.1 Nodes for the wave-toe-foundation dynamic interaction

Equation 1 is subjected to the following boundary conditions (Fig. 1):

$$\frac{\partial\phi}{\partial n}=n_z\frac{\partial\eta}{\partial t} \qquad \text{(on } S_f) \tag{3}$$

$$\frac{\partial\phi}{\partial n}=0 \qquad \text{(on } S_B) \tag{4}$$

$$\frac{\partial\phi}{\partial n}=V_n \qquad \text{(on } S_s) \tag{5}$$

$$\frac{\partial\phi}{\partial t}+\tfrac{1}{2}(\nabla\phi)^2+g\eta+\mu\phi-\int_{X_1}^{X}\frac{\partial\mu}{\partial X}\phi dX=0 \qquad \text{(on } S_f) \tag{6}$$

$$\frac{\partial\phi}{\partial X}=\frac{1}{\sqrt{gh}}\left(\frac{\partial\phi}{\partial t}+\mu\phi+\int_{X_2}^{X_1}\frac{\partial\mu}{\partial X}\phi dX\right) \qquad \text{(on } S_{out}) \tag{7}$$

$$\mu_{max}=(0.25-0.50)\sqrt{\frac{g}{h}} \tag{8}$$

where n_z is the directional cosine of the normal outward to the boundary with the Z axis; t is the time; V_n is the seepage velocity component normal to the boundary; h is the still water depth; η is the water surface elevation above SWL; and μ is a damping factor and is distributed linearly in the sponge layer. The coefficient in Eq.8 was determined such that minimum reflection would occur from the sponge layer. It was found that if the value of this coefficient ranges from 0.25 to 0.50 and the sponge layer width is not less than one wave length, the coefficient of reflection from the sponge layer will be less than 2% (Ohyama and Nadaoka, 1991). Equation 6 is reduced to the original dynamic boundary condition when $\mu=\partial\mu/\partial X=0$ and, therefore, represents the free surface condition for both the wave absorption filter and the flow regions. Using Green's theorem, the integral form of Eq. 1 can be written as follows (Brebbia and Walker, 1980):

$$\alpha(q)\phi(q)+\int_s\left(\phi\frac{\partial G_f}{\partial n}-\frac{\partial\phi}{\partial n}G_f\right)ds+\iint_\Omega q*G_f d\Omega=0 \tag{9}$$

where $\alpha(q)$ and $\phi(q)$ are the internal boundary angle and velocity potential at point (q), respectively. S represents the boundary and G_f is the Green's function defined as follows:

$$G_f=\ln(1/r) \tag{10}$$

where r is the distance between the point of interest (q) and other domain points. The relationship between U* and the horizontal velocity (U_o) at the wave-maker is given by Brorsen and Larsen (1987) as follows:

$$\int_{-h}^{\eta_s}U^*ds=2\int_{-h}^{\eta_o}U_o dz \tag{11}$$

where U_o, η_o and η_s are the horizontal water particle velocity and water surface elevation at the wave-maker computed by the linear wave theory and the actual water surface elevation above the wave-maker, respectively. The factor 2 in Eq. 11 has been used because of the bi-directional identical wave propagation from the wave-maker. However, a reflective boundary exists, therefore, a correction was made to account for the difference between η_s and η_o (Ohyama and Nadaoka, 1991).

$$U^*(j)=2U_o(j)\frac{(\eta_o+h)}{(\eta_s+h)} \tag{12}$$

The weighted residual method is applied to integrate the dynamic boundary condition on the free surface. Using linear elements and the weight factor ω, the resulting integral equation is:

$$\int\omega\left\{\frac{\partial\phi}{\partial t}+0.5\left(n_z^2\left[\frac{\partial\eta}{\partial t}\right]^2+\left[\frac{\partial\phi}{\partial s}\right]^2\right)+g\eta+\mu\phi-\int\frac{\partial\mu}{\partial x}\phi dx\right\}ds=0 \tag{13}$$

Linear interpolation functions are used to discretize any variable or its time derivative. Further details about this method of discretization are found in Ohyama and Nadaoka (1991). Using Taylor series expansion, the time derivative of any variable ψ at any time step (n) in the wave equations is expressed as a function of the time derivative in the previous time step (n-1) and the incremental value of the variable at this time step (n).

$$\left(\frac{\partial\psi}{\partial t}\right)^n=2\frac{\Delta\psi}{\Delta t}-\left(\frac{\partial\psi}{\partial t}\right)^{n-1} \tag{14}$$

Initial conditions

The water surface elevation above SWL, velocity potential and their derivatives are set to zero at the initial time step. Iterations for the location of the water surface are made at each time step in a time-marching scheme.

FEM Model for Flow in Porous Media

Based on Navier-Stokes equations, McCorquodale and Hannoura (1978) developed the flow equations in porous media and applied them to the case of rubble mound breakwaters. The resistance coefficients in the equations of motion were modified (Mizutani et al., 1996a) and used in this study.

$$\frac{\partial U}{\partial X}+\frac{\partial W}{\partial Z}=0 \tag{15}$$

$$A\frac{\partial U}{\partial t}+BU\frac{\partial U}{\partial X}+BW\frac{\partial U}{\partial Z}+C\frac{\partial P}{\partial X}+EU+FU\sqrt{U^2+W^2}=0 \tag{16}$$

$$A\frac{\partial W}{\partial t}+BU\frac{\partial W}{\partial X}+BW\frac{\partial W}{\partial Z}+C\frac{\partial P}{\partial Z}+EW+FW\sqrt{U^2+W^2}=0 \tag{17}$$

and

$$A=\frac{\left(1+\frac{1-m}{m}C_a\right)}{mg}, B=\frac{1}{m^2 g}, C=\frac{1}{\gamma}, E=\frac{6(1-m)\upsilon C_2}{gm^2D^2}$$

$$F=\frac{C_1(1-m)}{2gm^3D}, \quad P=p-\gamma Z$$

where U,W and P represent the horizontal and vertical seepage velocities and the pore-water pressure, respectively; m is the porosity; C_a is the added mass coefficient; g is the acceleration of gravity; υ is the kinematic viscosity of water; γ is the specific weight of water; D is the mean solid particle diameter, and C_1 and C_2 are the coefficients of drag for two dimensional flow. Values of C_a, C_1 and C_2 are estimated based on Mizutani et al. (1996a).

These equations are subjected to the following boundary conditions:

$$P=-\frac{\partial\phi}{\partial t}-\frac{1}{2}\left\{\left(\frac{V_n}{m}\right)^2+V_s^2\right\}-\gamma z \quad \text{(on } S_s\text{)} \quad (18)$$

$$V_n=W\cos\theta-U\sin\theta \quad \text{(on } S_s\text{)} \quad (19)$$

$$V_s=\frac{\partial\phi}{\partial S} \quad \text{(on } S_s\text{)} \quad (20)$$

$$V_n=0 \quad \text{(on } S_C\text{)} \quad (21)$$

where V_s is the water particle velocity in the tangential direction to the boundary; θ is the inclination of the toe slope on the X axis.

Along the interface between the toe and the foundation, the continuity of the pore-water pressure and the normal velocity component is considered. A finite element analysis based on the weighted residual technique is used to solve the governing equations in the porous media. Linear interpolation and weight functions (ϖ) and trapezoidal isoparameteric elements are selected to solve the problem. The governing equations reduce to:

$$\int\varpi\left(\frac{\partial U}{\partial X}+\frac{\partial W}{\partial Z}\right)dS_e=0 \quad (22)$$

$$\int\varpi\left(\begin{array}{c}A\frac{\partial U}{\partial t}+BU\frac{\partial U}{\partial X}+BW\frac{\partial U}{\partial Z}+C\frac{\partial P}{\partial X}+\\ EU+FU\sqrt{U^2+W^2}\end{array}\right)dS_e=0 \quad (23)$$

$$\int\varpi\left(\begin{array}{c}A\frac{\partial W}{\partial t}+BU\frac{\partial W}{\partial X}+BW\frac{\partial W}{\partial Z}+C\frac{\partial P}{\partial Z}+\\ EW+FW\sqrt{U^2+W^2}\end{array}\right)dS_e=0 \quad (24)$$

The time derivative of U and W at a time step (n) in the porous media equations is treated as in the BEM model (Eq. 14).

Initial conditions

All the variables (U,W and P) are set to zero at the initial time step.

Combined BEM-FEM Model

The aforementioned models are linked together in one model to compute the non-linear water surface elevations and the flow inside the toe and the foundation at each time step. The linkage between the BEM and the FEM models is made through the surface boundary conditions which include a mix from both models' variables. The time increment for this model has been selected to be less than 1/24 of the wave period (T). Time increments less than T/24 have been tested and found to be necessary with steep waves. The locations of the FEM nodal points along the outer boundary of the toe and foundation separating the water body from the porous media are the same as that of the BEM. The spacings are less than 1/20 of the wave length (L).

The FEM discretization has quadrilateral elements and the left hand side boundary of the foundation is closed using a smooth impermeable vertical surface that extends from the mud line to the rigid impervious bottom of the foundation (see Fig.1). This boundary closure reduces the number of computational nodes and limits them to the area of interest. It will be shown later in the subsequent sections that this boundary closure affects only its vicinity and does not affect the area of interest under or near the toe. The size of the coefficient matrix is large even after the closure of the left hand side foundation boundary. Therefore, a special matrix solution system has been developed and used to reduce the CPU run-time. In this matrix solution system, the coefficient matrix is rearranged and partitioned making use of the zero elements and the special properties of the partitioned matrices.

FEM Model for Poro-Elastic Deformations

The solid particle's displacement and the dynamic pressure are governed by the linear poro-elastic consolidation equations developed by Biot (1941). These equations are based on the following assumptions: (1) isotropy of the material; (2) linearity of stress-strain relations; (3) reversibility of stress-strain relations under final equilibrium conditions; (4) small strains; (5) water is incompressible, but it may contain air bubbles making their mixture a compressible one; and (6) water flows through the pores according to Darcy's law. The governing equations in this case may be written as follows:

$$G\nabla^2\zeta+\frac{G}{1-2\nu}\frac{\partial\varepsilon}{\partial X}=\frac{\partial p}{\partial X} \quad (25)$$

$$G\nabla^2\xi+\frac{G}{1-2\nu}\frac{\partial\varepsilon}{\partial Z}=\frac{\partial p}{\partial Z} \quad (26)$$

$$\frac{k}{\gamma}\nabla^2 p=\frac{m}{\beta}\frac{\partial p}{\partial t}+\frac{\partial\varepsilon}{\partial t} \quad (27)$$

where G is the shear modulus of the soil; k is the soil permeability; β is the apparent bulk modulus of water; ζ is the horizontal solid particle displacement; ξ is the vertical solid particle displacement; ν is the Poisson's ratio of the solids; and ε is the volume strain defined as:

$$\varepsilon=\frac{\partial\zeta}{\partial x}+\frac{\partial\xi}{\partial z} \quad (28)$$

The soil grains are about 30 times as incompressible as pure saturated water which itself is highly incompressible, but natural water in the seabed usually contains tiny gas bubbles due in part to organic processes. Therefore, in these equations the water and air are considered as one compressible single phase to avoid dealing with them separately. The compressibility of the mixture is given as follows (Verruijt, 1969):

$$\frac{1}{\beta}=\frac{1}{\beta_o}+\frac{1-S_r}{P_o} \quad (29)$$

where S_r is the degree of saturation; P_o is the absolute pore-water pressure; and β_o is the true bulk modulus of the water (2×10^9 N/m^2). The dynamic effective stresses in the solid skeleton are computed using Hooke's law as follows:

$$\sigma_{xx}=2G\left(\frac{\partial\zeta}{\partial X}+\frac{\nu}{1-2\nu}\varepsilon\right) \quad (30)$$

$$\sigma_{zz}=2G\left(\frac{\partial\xi}{\partial Z}+\frac{\nu}{1-2\nu}\varepsilon\right) \quad (31)$$

$$\tau_{xz}=G\left(\frac{\partial\zeta}{\partial Z}+\frac{\partial\xi}{\partial X}\right) \quad (32)$$

where σ_{xx} is the horizontal dynamic effective stress; σ_{zz} is the vertical dynamic effective stress; and τ_{xz} is the dynamic effective shear stress. These equations are subjected to the following boundary conditions along the interface between the water and the porous media (Eq.33), over the impermeable rough rigid horizontal bottom (Eq.34) and over the impermeable smooth rigid vertical walls (Eq.35):

p= wave pressure computed by the BEM-FEM model (33a)

$$\sigma_{nn}=\tau_{ns}=0 \quad (33b)$$

$$\zeta=\xi=\frac{\partial p}{\partial Z}=0 \quad (34)$$

$$\zeta=\tau_{xz}=\frac{\partial p}{\partial X}=0 \quad (35)$$

where σ_{nn} is the dynamic effective stress in the normal direction to the boundary and τ_{ns} is the dynamic shear stress in the tangential direction to the boundary.

A four node isoparameteric trapezoidal element is selected and linear interpolation and weight functions ϖ are utilized to solve the governing equations as follows:

$$\int_{S_e} \varpi\left(G\nabla^2\zeta + \frac{G}{1-2\nu}\frac{\partial\varepsilon}{\partial X} - \frac{\partial p}{\partial X}\right)ds_e = 0 \quad (36)$$

$$\int_{S_e} \varpi\left(G\nabla^2\xi + \frac{G}{1-2\nu}\frac{\partial\varepsilon}{\partial Z} - \frac{\partial p}{\partial Z}\right)ds_e = 0 \quad (37)$$

$$\int_{S_e} \varpi\left(\frac{k}{\gamma}\nabla^2 p - \frac{m}{\beta}\frac{\partial p}{\partial t} - \frac{\partial\varepsilon}{\partial t}\right)ds_e = 0 \quad (38)$$

The second order derivatives in the above equations are solved using Green-Gauss theorem as follows:

$$\iint_{S_e} \varpi\frac{\partial^2\psi}{\partial x^2}dxdz = \int_{\Gamma} \varpi\frac{\partial\psi}{\partial x}dz - \iint_{S_e}\frac{\partial\varpi}{\partial x}\frac{\partial\psi}{\partial x}dxdz \quad (39)$$

The line integral along the boundary cancels out for the interior domain points and appears only for the surface nodes. The integration along the outer boundary is automatically satisfied by the natural boundary conditions on that boundary. Furthermore, the time derivatives in Biot's equations are solved in the same way as in the BEM-FEM combined scheme (Eq.14). It is also crucial in this case to normalize the coefficient matrix due to the large discrepancies in the order of magnitude of the pressure and displacement. In a separate program subroutine, the dynamic effective stresses are computed inside the toe and foundation at each time step. The time increment, the location and the spacing of the nodes in the FEM mesh are the same as those in the BEM-FEM model.

Initial conditions

All the displacements and the pressure values are set to zero at the initial time step (*cold start*) in a time-marching scheme.

A CASE STUDY

The parameters used in this study are as follows: wave height is 1.50 m, wave period is 8.0 sec in a water depth of 9.0 m, normal wave incidence on an infinitely long smooth seawall and a finite thick sand seabed over an impermeable rough base. Different conditions for the toe and foundation are considered as listed in Table(1); bold values are those for the basic case study.

Table 1: Calculation conditions for the toe and foundation

Variable	Toe	Foundation
m	0.22,**0.28**,0.36	**0.24**
D (cm)	15,**30**,50	**0.10**
Crown length/wave length (b/L)	0.20,**0.35**,0.50	--
Slope length/wave length (b1/L)	0.0,**0.15**,0.30	--
Toe height/water depth (d/h)	0.33,**0.50**,0.66	--
Foundation depth/wave length (ds/L)	--	**0.11** to 0.60
Shear modulus GS & GF (10^8N/m^2)	**1.00**	0.50,**1.0**,5.0
(ν)	**0.33**	**0.33**
β or BULK (10^7 N/m^2)	**1.00**	**1.00**

RESULTS AND DISCUSSIONS

Before discussing the models results, it may be convenient to briefly explain the wave deformation mechanism in such structures. The incident wave is first obstructed by the permeable toe and, thus, a part of the wave is reflected back to the offshore side and the rest passes through and over it. Over the toe, the combined effects of wave shoaling and damping occur due to the change in the water depth and the dynamic interaction with the flow inside the toe. This leads to an increase in the wave height at the crown corner by shoaling and a decrease in the wave height as it propagates toward the seawall by wave damping. Thus, the transmitted wave over the toe depends on its hydraulic properties that control the amount of the reflected wave energy back to the sea and on the shoaling effect governed by the toe height. The wave damping along the toe crown will normally depend on its length and hydraulic properties. Moreover, in earlier investigations of the dynamic interaction between submerged breakwaters over impermeable seabed and non-linear waves, it was found numerically and verified experimentally that partial wave reflection occurs due to submerged permeable structures and damping occurs along their crown (Mizutani et al., 1996b). At the seawall, the wave over the toe is completely reflected forming a standing wave over the crown of the toe. This implies that the wave height at the face of the seawall depends on the hydraulic properties and size of the toe. Therefore, studying these effects helps to better understand the wave deformation mechanism and toe design.

In the subsequent sections, the term *magnitude* has been used to express the difference between the maximum and the minimum values of a variable during one wave period. The wave deformation in front of the seawall at different time steps for the basic case study is shown in Fig.2. It is obvious that the toe influences the wave profile along the wave tank, making it quite different from the linear standing waves, especially over the toe.

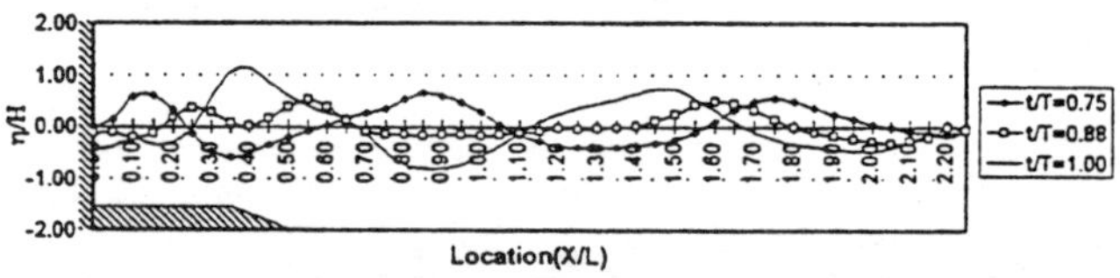

Fig.2 Water levels along the wave tank

To test the effect of the toe height (d), the BEM-FEM model has been used for three different toe heights ranging from 0.33 to 0.66 of the water depth (h) keeping all the other dimensions and properties the same as that in the basic case study. The corresponding variations in the water level (η_{rms}/H) along the wave tank during one wave period are presented in Fig.3. Before discussing the effect of the toe height, it is worth mentioning that the wave length in front of the seawall decreases as the toe height increases and, hence, the anti-node is getting closer to the seawall. It has been revealed that the minimum η_{rms}/H at the seawall occurs when the anti-node is at the crown corner (d/h=0.50) and increases if the anti-node is outside the crown (d/h=0.33), but it becomes maximum in case that the anti-node is inside the crown (d/h=0.66).

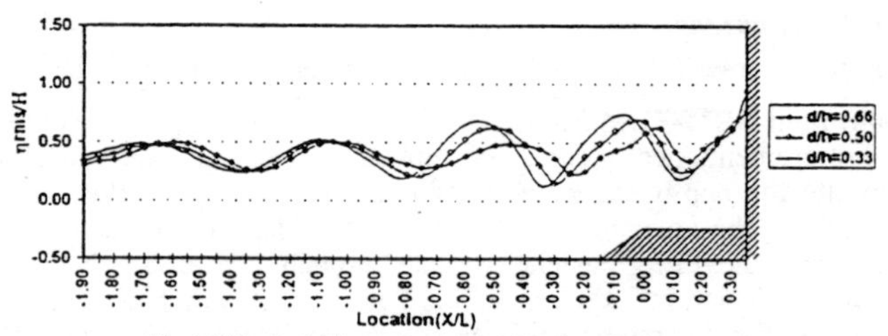

Fig.3 Effect of d/h on the water level variation

The wave height has been found to be the minimum of the three cases for d/h=0.33 and the maximum of them for d/h=0.66. This may be because of the increase in the wave height by the shoaling effect which is more pronounced than other effects. Furthermore, the effect of partial wave reflection is still pronounced and this can be realized since the wave height slightly increases when the toe height increases from 0.33 to 0.50 of the water depth and significantly increases as the toe height increases from 0.50 to 0.66 of the water depth. Therefore, it can be concluded that maximum wave reflection from the toe occurs if the crown corner coincides with the wave anti-node. Also, it has been found that increasing the toe height generally shifts the minimum water level up and may change the wave shape (Fig.4). It has been noticed that the water particle velocity on the crown of the toe increases as the height of the toe increases (Figs.5 and 6). Moreover, the rate of increase in the vertical velocity is higher than that of the horizontal velocity such that the magnitude of the maximum vertical velocity at the seawall becomes higher than the magnitude of the maximum horizontal velocity under the node for d/h=0.66.

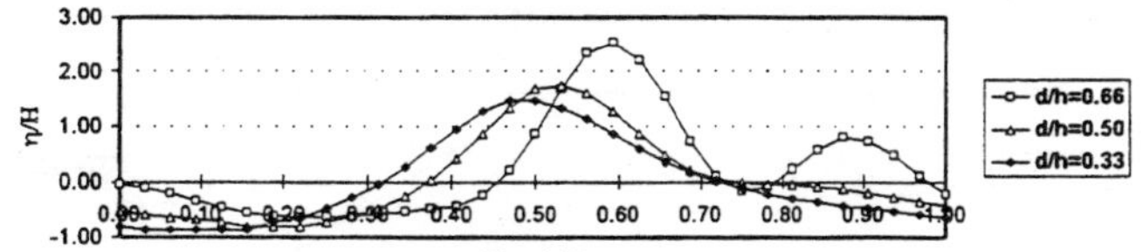

Fig.4 Effect of d/h on the water level at the seawall

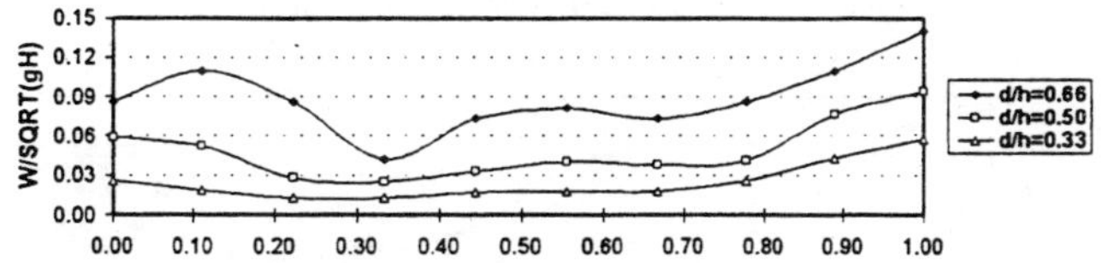

Fig.5 Effect of d/h on the crown vertical velocity magnitude

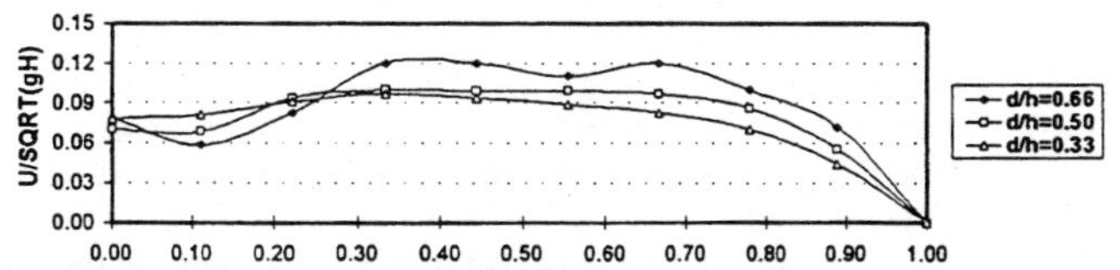

Fig.6 Effect of d/h on the crown horizontal velocity magnitude

The effect of the crown length is linked to the effect of its height since the toe height determines the location of the anti-nodes and the length of the crown determines the toe effect on the wave. In Fig.7, the anti-node is at a distance of 0.35 L from the seawall. The maximum wave reflection from the toe occurs for the case when the crown has a length of 0.35 L. In case of a crown length of 0.50 L, the wave height at the seawall is larger because the anti-node is inside the crown. In this case, less reflection is caused by the toe, but the wave damping along the toe is greater. The wave height becomes maximum if the crown length is 0.20 L because the anti-node is outside the toe crown and, hence, the reflection and damping effects of the toe are less than the previous two cases. The time variation of the water level in front of the seawall is generally affected by the crown length such that the highest water level experiences a phase delay as the crown length increases (Fig.7).

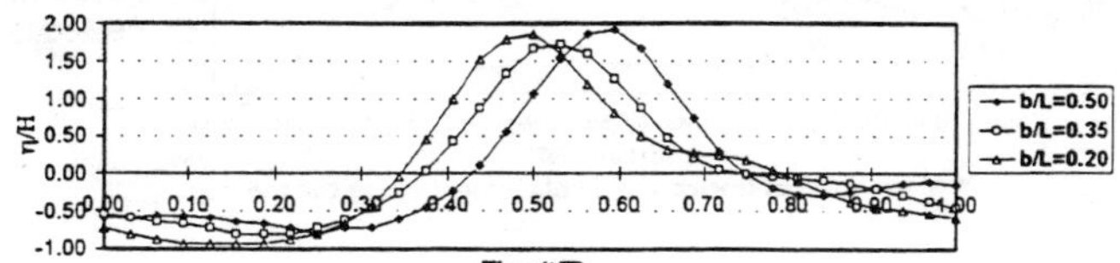

Fig. 7 Effect of b/L on the water levels at the seawall

The effect of the side slope of the toe on the wave deformation along the wave tank has also been investigated. It has been revealed that the side slope reduces the wave reflection and allows more wave energy to be transmitted to the seawall. The results presented in Fig.8 show that η_{rms}/H on the offshore side decreases and at the seawall increases as the side slope becomes flatter. The highest water level at the seawall increases with a phase delay as the toe slope becomes flatter, but the lowest water level is almost unchanged leading to an increase in the wave height (Fig.9).

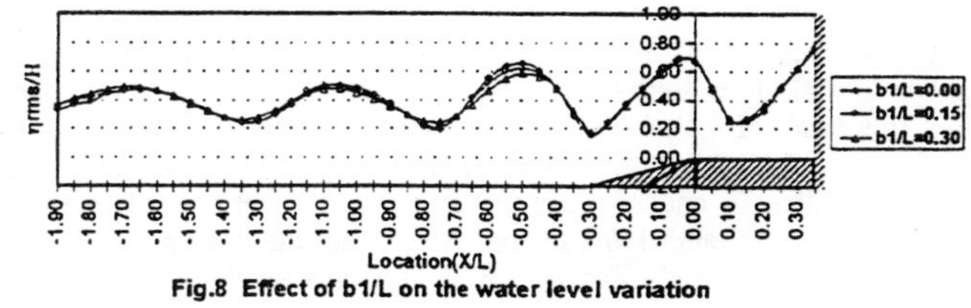

Fig.8 Effect of b1/L on the water level variation

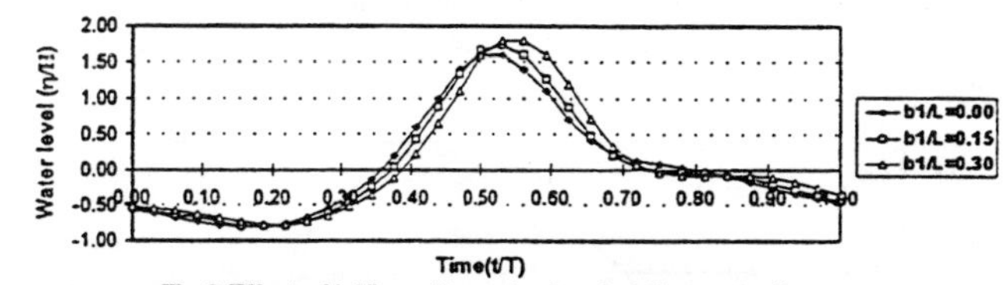

Fig.9 Effect of b1/L on the water level at the seawall

The effect of the toe porosity on the wave pressure and velocities on the crown has been investigated by changing the porosity from 0.22 to 0.36. It has been found that as the porosity increases, the wave that passes through the rubble increases and, hence, both the effects of shoaling over and reflection from the toe decrease. This leads to a reduction in the wave height at the seawall (Fig.10). As the porosity of the toe increases from 0.22 to 0.36, the water velocity inside the pores increases causing more wave energy dissipation (Figs.12 and 13) and reducing the wave heights and the pressure magnitude over the toe (Fig.11). The pore-water pressure and vertical velocity at the seawall are higher than those at the crown corner even if both are locations of wave anti-node. This may be accounted by the wave damping effect on the reflected wave back to the sea side.

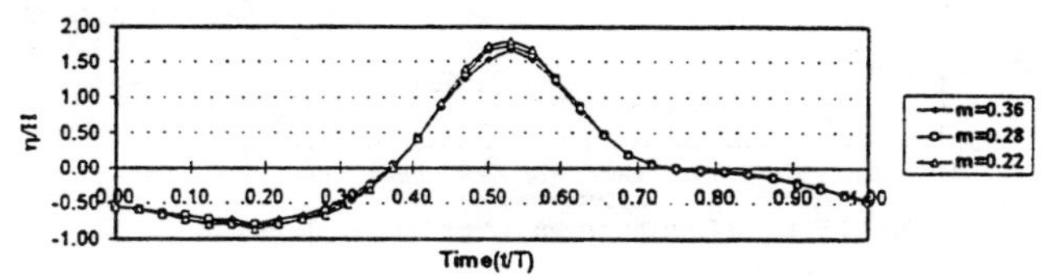

Fig.10 Porosity effect on the water level at the seawall

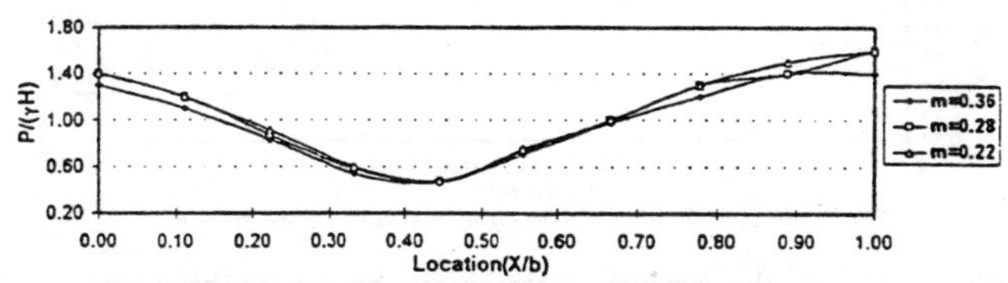

Fig.11 Porosity effect on the crown dynamic pressure magnitude

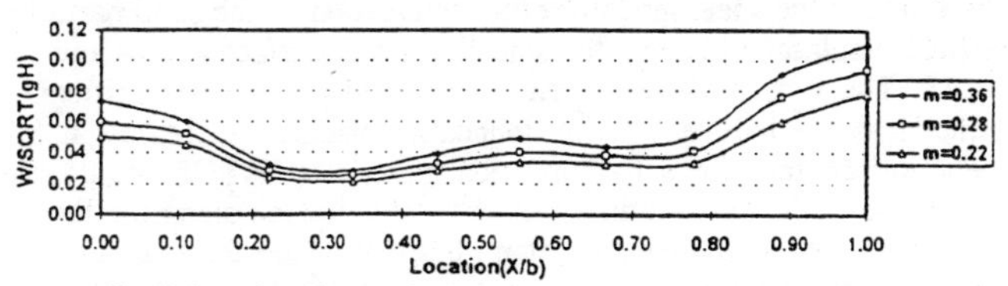

Fig.12 Porosity effect on the crown vertical velocity magnitude

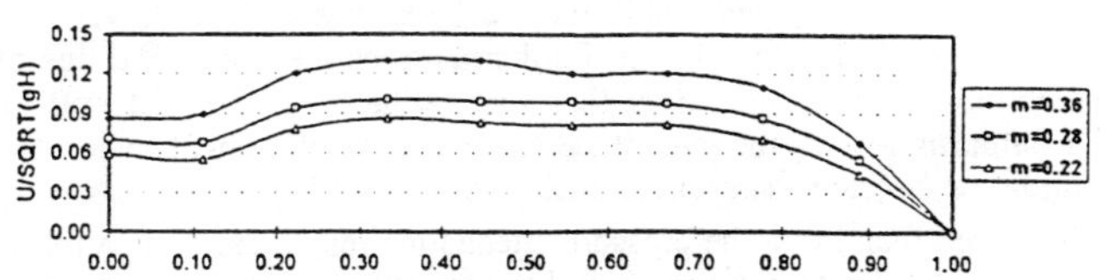

Fig.13 Porosity effect on the crown horizontal velocity magnitude

The effect of the toe material size on the pore-water pressure and velocity on the crown is studied. It has been noted that the rubble size has a similar effect to that of the porosity in the sense that small rubble has the same effect as the low porosity. The smaller the size of the rubble is, the higher the wave at the seawall and the pore-water pressure (Figs.14 and 15) and the slower the water velocities inside the toe (Figs.16 and 17) become. Moreover, the maximum vertical water particle velocity and the minimum horizontal velocity always occur at the seawall (anti-node), but the minimum vertical velocity and the maximum horizontal velocity occur almost at a distance of two-thirds the crown length from the seawall (i.e. 0.23 L) near the node.

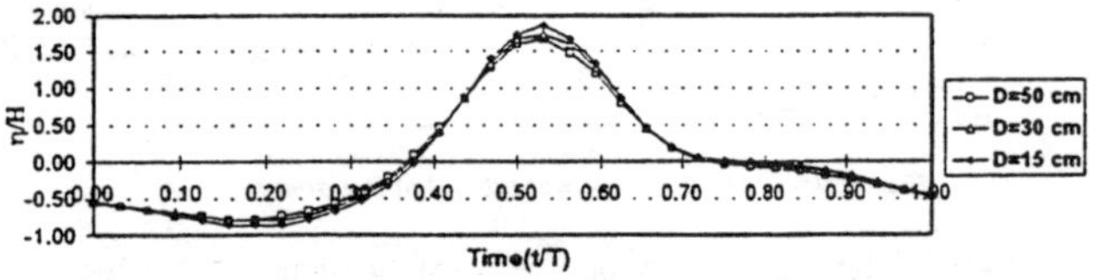

Fig.14 Rubble size effect on the water level at the seawall

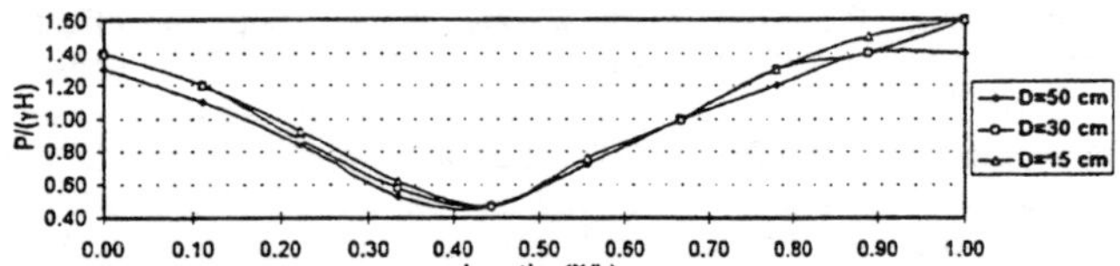

Fig.15 Effect of D on the crown dynamic pressure magnitude

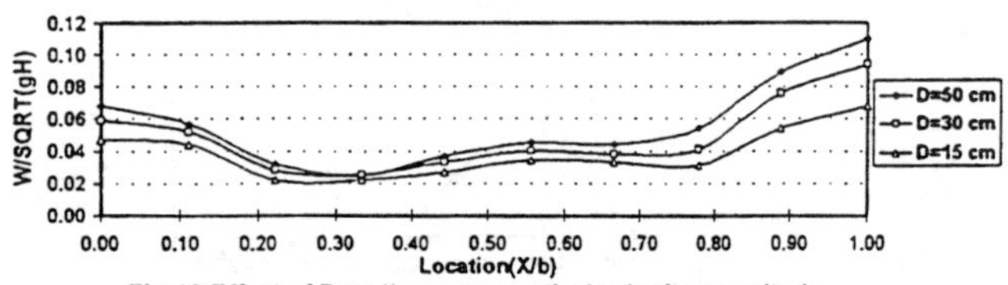

Fig.16 Effect of D on the crown vertical velocity magnitude

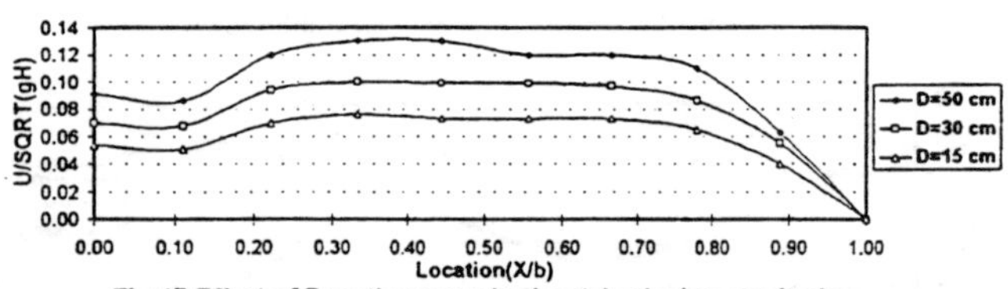

Fig.17 Effect of D on the crown horizontal velocity magnitude

The effect of the seabed geotechnical properties has also been investigated to test the dynamic stability of the toe and its foundation. The effect of the shear modulus of the foundation on the solid particles' vertical displacement along the smooth seawall surface is presented in Fig.18. The vertical displacement of the solid particle is significantly reduced as the hardness of the foundation material increases. It has been noticed that for soft foundation material (GF/BULK=5.0), the magnitude of the vertical displacement inside the toe along the face of the seawall remains almost constant, but that inside the foundation is damped non-linearly approaching the value for a rigid base. By increasing the foundation hardness (GF/BULK=10.0), the vertical displacement is damped inside the toe linearly and at a slower rate than that inside the foundation. Furthermore, as the shear modulus of the foundation becomes larger than that of the toe (GF/BULK=50.0), the damping rate in the magnitude of vertical displacement inside the foundation becomes constant and close to that inside the toe.

It has been found that the pressure attenuation inside the foundation is less for a harder foundation than that for softer one, but the pressure inside the toe is almost unchanged (Fig. 19). This may be accounted by the fact that the spatial distribution of the solid particle's displacement affects the pore-water pressure attenuation. The pore-water pressure inside the toe is not affected because the spatial distribution of the displacement is almost unaltered. The pore-water pressure inside the foundation is affected since the spatial distribution of the solids displacement inside the foundation is greatly affected by the shear modulus of its material. It can be concluded that generally the higher the shear modulus of the solids is, the less the solid particle's displacement (Figs.20 and 21) and the rate of pore-water pressure attenuation become (Fig.19) for GF/BULK>1.

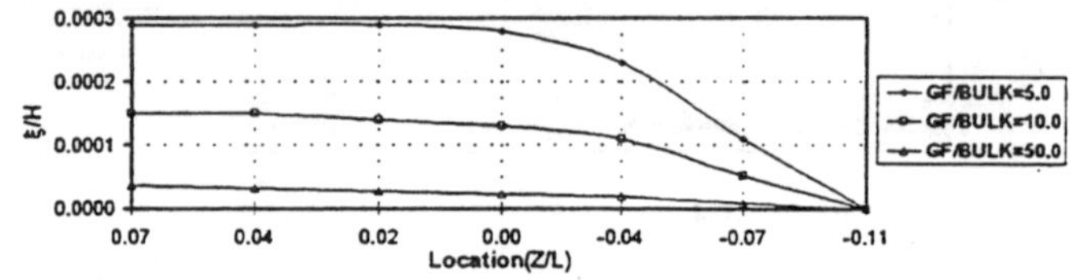

Fig.18 Magnitude of vertical displacement along the seawall face

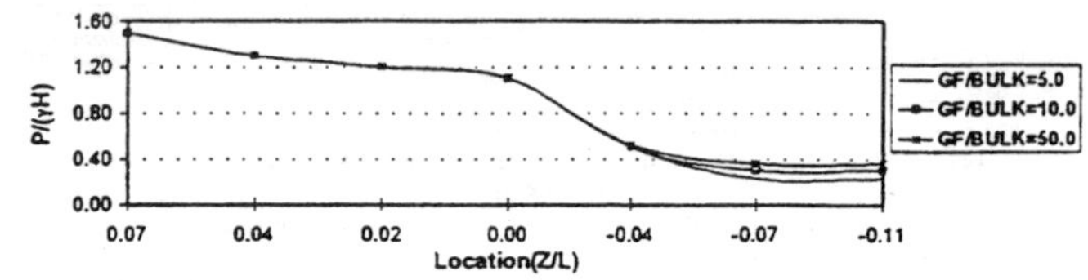

Fig.19 Magnitude of dynamic pressure along the seawall face

Gatmiri (1990) tested the effect of the seabed stiffness on the pore-water pressure attenuation inside the seabed using linear waves. He found that the pore-water pressure attenuates faster close to the mud line as the stiffness of the foundation decreases for a foundation thickness of 0.20 L. Of course, the existence of the porous toe over the mud line affects the pore-water pressure, but the comparison with the results in case of no toe helps to understand its effect. Therefore, the assumption that the solid particles remain fixed in their place affects the accuracy of computing the pore-water pressure inside the porous bed. If the value of GF/BULK is small, the error due to this assumption accumulates and becomes large.

It has been noted that the locations of the maximum vertical solid particle's displacement and the minimum horizontal solid's displacement along the toe base are always at the seawall. However, the locations of the maximum horizontal solid particle's displacement and the minimum vertical solid's displacement move under the standing wave node toward the seawall as the ratio between the shear modulus of the foundation and the apparent bulk modulus of water increases.

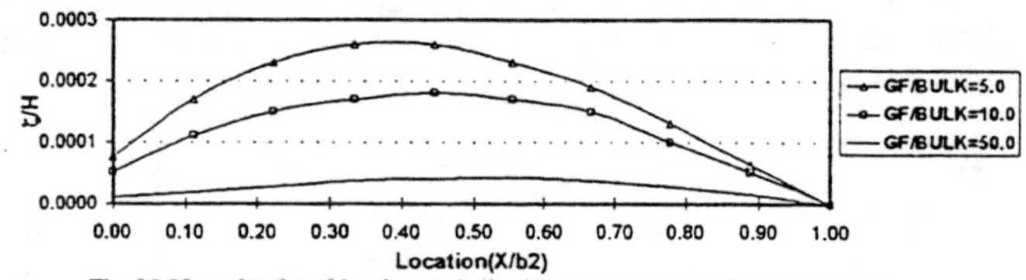

Fig.20 Magnitude of horizontal displacement along the toe base

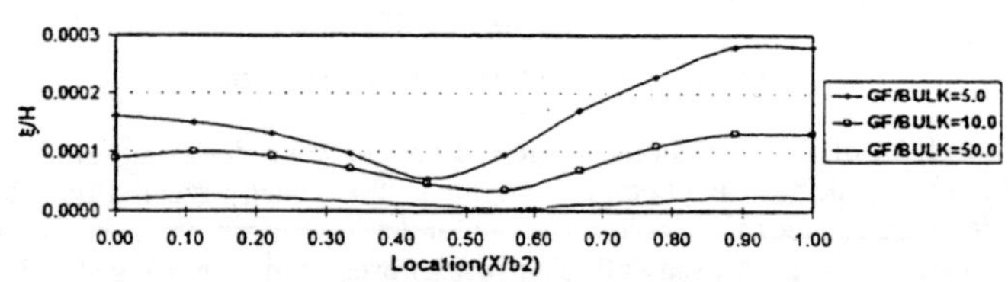

Fig.21 Magnitude of vertical displacement along the toe base

Since the solid particle's displacement inside the foundation is less as the soil becomes more rigid, the dynamic shear stresses along the base of the toe increase (Fig.22) and the variation in the dynamic vertical stresses decreases (Fig.23). The point of maximum shear stress along the toe base moves toward the seawall as the foundation becomes more rigid. It has also been found that the dynamic effective vertical stresses are not uniform along the toe base causing more settlement at its offshore end than near its middle section. The point of minimum dynamic effective vertical stress along the toe base moves from about the middle point (GF/BULK=5.0) toward the seawall as the shear modulus of the foundation increases and it coincides with it at GF/BULK=50.0. The maximum dynamic effective vertical stress along the toe base jumps from near its offshore slope (GF/BULK=50.0) to the seawall (GF/BULK=5.0) as the shear modulus decreases.

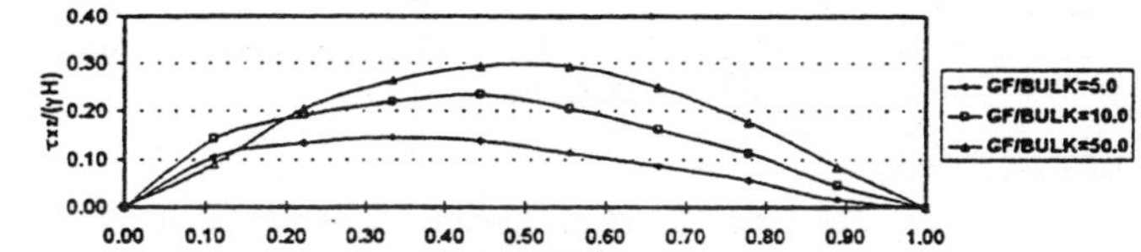

Fig.22 Magnitude of dynamic shear stresses along the toe base

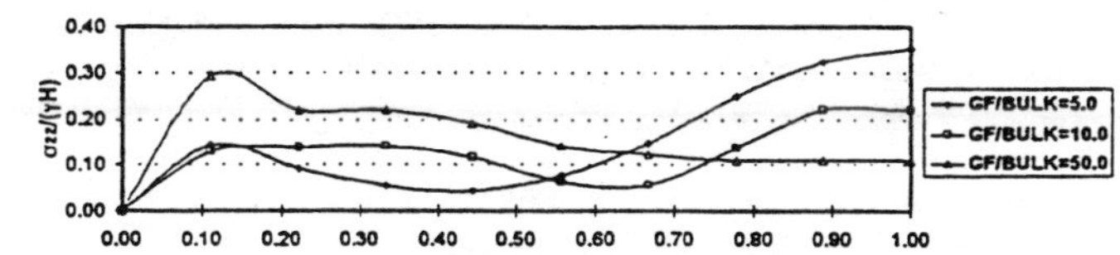

Fig. 23 Magnitude of dynamic vertical stresses along the toe base

The effect of shear modulus of the foundation on the solid's displacement along the crown of the toe has also been analyzed. It has been found that the points of the maximum horizontal displacement and the minimum vertical displacement are not fixed under the wave node; these move toward the seawall as the shear modulus of foundation increases (Figs.24 and 25). Thus, it can be concluded that the dynamic response of the toe depends greatly on the shear modulus of its foundation material.

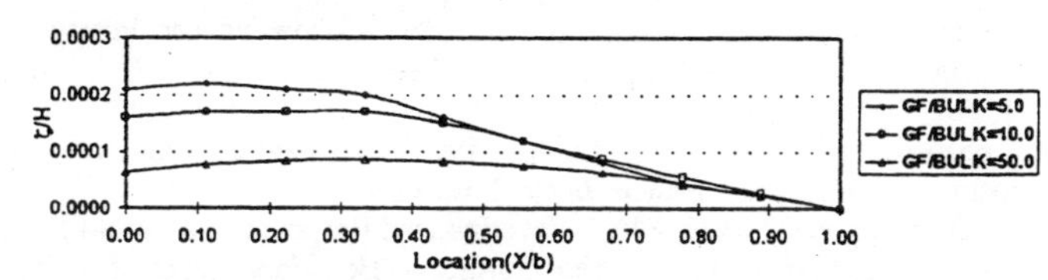

Fig.24 Magnitude of horizontal displacement along the toe crown

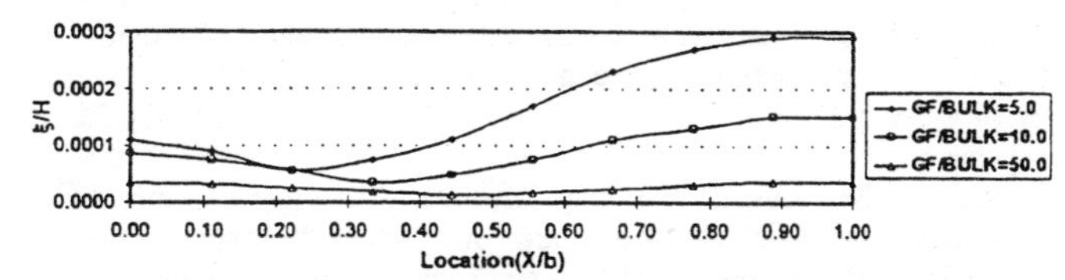

Fig. 25 Magnitude of vertical displacement along the toe crown

The effect of foundation thickness has also been investigated by changing it from 0.11 to 0.60 L. It has been noticed that the pore-water pressure attenuation curve inside the seabed depends on its thickness up to a certain limit and then it becomes independent of it (Fig.26). However, inside the toe, the pore-water pressure is almost independent of the seabed thickness. This implies that the effect of the seabed thickness on the pore-water pressure is negligible after certain value (app. 0.40 L in this case) and then the solution becomes similar to the case of infinitely thick sand seabed. Indeed, the foundation thickness affects the solid's displacement, as will be shown later, which in turn may affect the pore-water pressure. An increase in the pressure distribution occurs near the impermeable bottom along with a higher rate of pore-water pressure attenuation near the seabed level as the foundation thickness decreases. These results agree with those of Gatmiri (1990) except that his peak value of the pressure was shifted down due to the increase in the seabed thickness.

It has been found that the larger the foundation thickness is, the larger the vertical component of the solid particles' displacement along the seawall (Fig.27) and the less the horizontal displacements are along the base of the toe (Fig.28). The direction of the horizontal solid's displacement changes under the same wave phase as the foundation thickness exceeds 0.24 L and then the horizontal displacement increases as the foundation thickness increases. This increase starts from the area under the toe slope and its effect extends to the horizontal displacement along the toe base and the seabed in front of it. This occurs because the soil is not very hard and the foundation thickness is large and, therefore, the dynamic deformation of the structure may change due to the change in the foundation thickness.

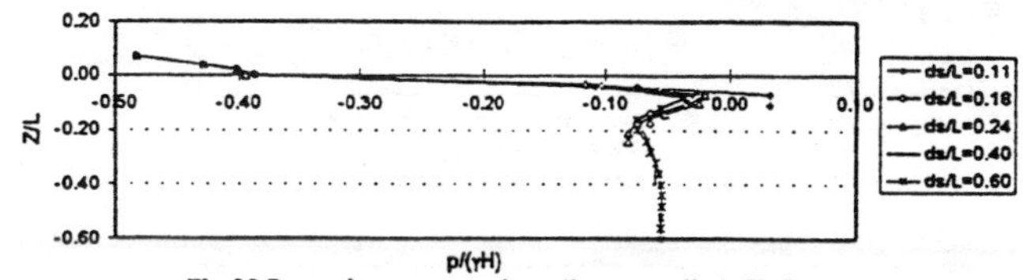

Fig.26 Dynamic pressure along the seawall at t/T=5.0

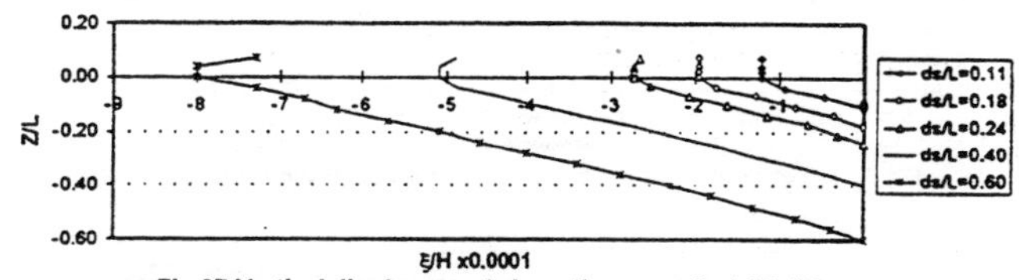

Fig.27 Vertical displacement along the seawall at t/T=5.0

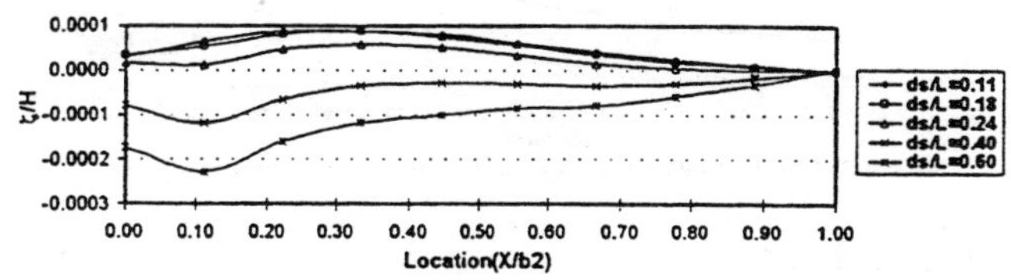

Fig. 28 Horizontal displacement along the toe base at t/T=5.0

These results differ from those obtained by Gatmiri (1990) who found that the horizontal displacement increases as the seabed thickness increases from zero to 0.20 L and then it decreases gradually with the increase in the seabed thickness. This proves that the response of the seabed will change if it is under and/or in front of a rubble toe. Therefore, the dynamic shear stresses decrease (Fig.29) and the dynamic effective vertical stresses increase along the toe base (Fig.30) as the foundation thickness increases. High values of horizontal solid displacement, dynamic shear and vertical stresses occur along the toe base under the slope as the seabed thickness increases. Therefore, care should be taken against slope failure over large thick sand seabed especially if its material is not very hard.

The effect of the foundation thickness on the solids' displacement on the toe crown has also been considered (Figs.31 and 32). It can be noticed that the horizontal solid's displacement on the crown decreases at the offshore side and slightly increases at the seawall side as the foundation thickness increases from 0.11 to 0.24 L even if the points of maximum vertical and horizontal solids' displacements almost remain in their places. The vertical solids' displacement at the crown corner is little affected by the seabed thickness till a value of 0.24 L, but it is significantly affected on the crown. If the foundation thickness increases from 0.24 to 0.60 L, the maximum horizontal solid's displacement occurs at the crown corner and its value is much less than the maximum vertical displacement on the crown. This also proves that the toe slope will be vulnerable to failure over large thick sand seabed even if the bed material is not soft. These results prove that the foundation thickness and shear modulus play a key role in the dynamic response of the toe and its foundation to water waves and, hence, are very important for the stability.

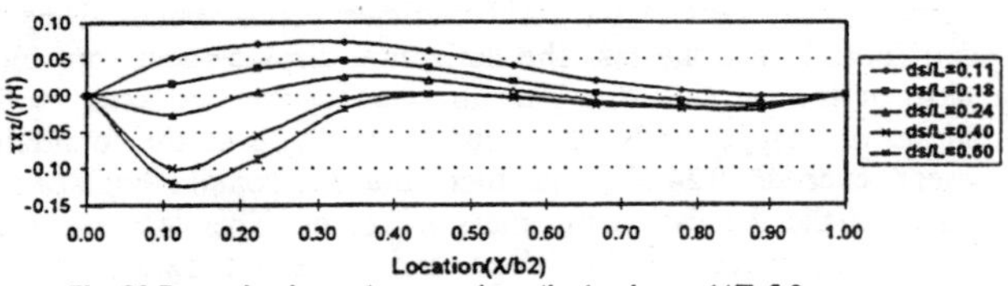

Fig. 29 Dynamic shear stresses along the toe base at t/T=5.0

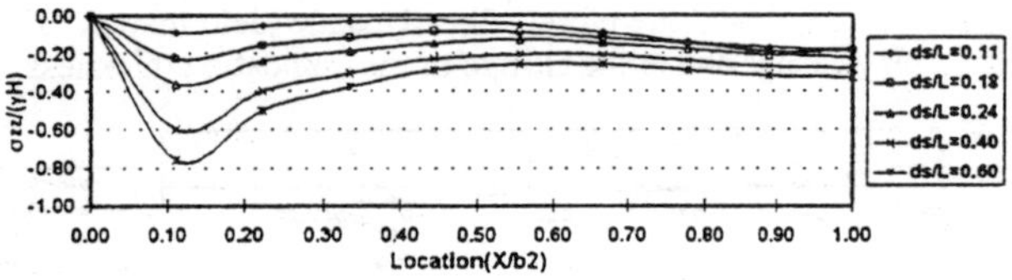

Fig.30 Dynamic vertical stresses along the toe base at t/T=5.0

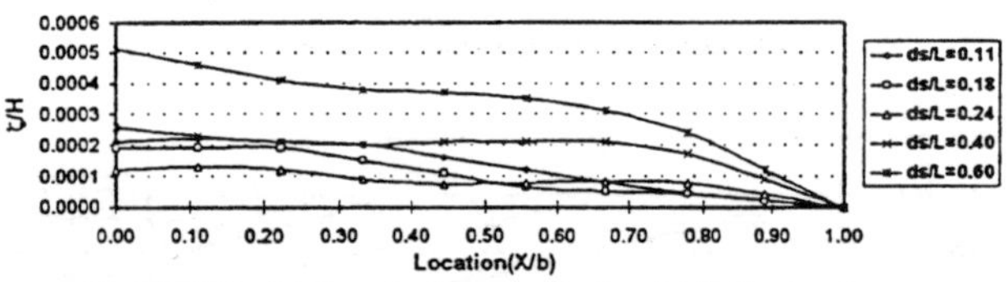

Fig.31 Effect of ds/L on the toe crown horizontal displacement

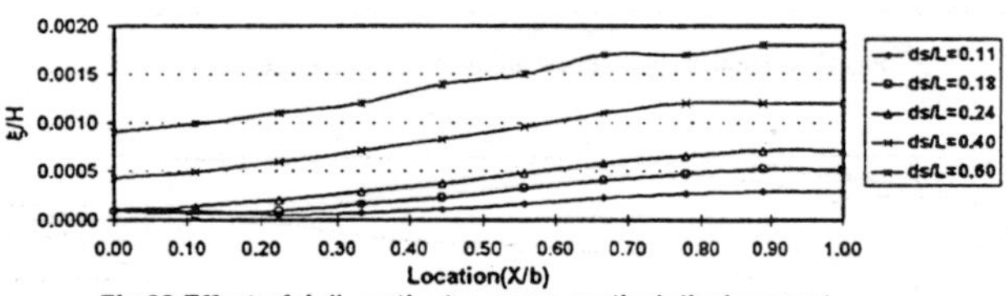

Fig.32 Effect of ds/L on the toe crown vertical displacement

CONCLUSIONS

- The standing wave height in front of a seawall depends on the toe properties and size. Improper selection of the toe size and material could increase the wave height at the seawall to more than 2.0 times of the incident wave height while an appropriately selected toe may reduce wave heights to less than 2.0 times of the incident wave height. Thus, the assumption of linear standing water waves in front of hydraulic structures equipped with a permeable toe leads to different results from those presented here especially when the toe size is comparable to the water depth.
- The side slope of the toe increases the wave height at the seawall and decreases the partial standing wave on the offshore side. For effective use of a permeable toe, its dimensions must be selected such that the crown corner is located at the first anti-node of the standing wave and the side slope is the steepest and yet safe for stability. The higher the porosity and the larger the size of the rubble stones are, the less the wave height in front of the seawall is.
- The locations of the maximum horizontal and the minimum vertical solid's displacement along the toe base and crown move toward the seawall as the shear modulus of the foundation increases. The pore-water pressure inside the foundation is significantly affected by the shear modulus of the foundation, but that inside the toe is almost not. Along the toe base, the dynamic effective vertical stresses distribution becomes uniform and the dynamic shear stresses increase as the shear modulus of the foundation increases.
- If the seabed thickness is less than 0.24 L, the dynamic effective vertical stresses and vertical solids' displacement increase, but the dynamic shear stresses and horizontal solids' displacement decrease as the seabed thickness increases. High values of the horizontal solids' displacement and dynamic shear and vertical stresses occur under the toe slope threatening its stability especially if the seabed material is not very hard and the seabed thickness is more than 0.24 L. This implies that short waves have more pronounced effect on the toe than long waves do. The pore-water pressure under the toe is almost independent of the seabed thickness if it is more than 0.40 L.
- The deformation mode and the stresses distribution in a porous toe and its sand seabed are significantly affected by the foundation thickness and shear modulus. A special care must be given to these parameters when constructing structures over deep sand seabed layer or even a small thickness of soft soils.

REFERENCES:

Biot, MA (1941). "General Theory of Three-Dimensional Consolidation," *J. Appl. Phys.*, Vol 12, pp 155-164.

Brebbia, CA and Walker, S (1980). "Boundary Element Techniques in Engineering," *Butterworths*, London.

Brorsen, M, and Larsen, J (1987). "Source Generation of Non-Linear Gravity Waves with the Boundary Integral Method," *Coastal Eng.*, Vol 11, pp 93-113.

Gatmiri, B (1990). "A Simplified Finite Element Analysis of Wave-Induced Effective Stresses and Pore Pressures in Permeable Sea Beds," *Geotechnique*, Vol 40, pp 15-30

Hannoura, AA, and McCorquodale, JA (1985). "Rubble Mounds: Numerical Modelling of Wave Motion," *J. WatWay, Port, Coastal Ocean Eng.*, ASCE, Vol 111, pp 800-816.

Madsen, OS (1978). "Wave-Induced Pore Pressures and Effective Stresses in a Porous Bed," *Geotechnique*, Vol 28, pp 377-393.

Maeno, S, and Nago, H (1988). "Settlement of a Concrete Block into a Sand Bed under Wave Pressure Variation," *In modelling Soil-Water-Structure Interactions, Kolkman et al. (ed.)*, pp 67-76, Balkema, The Netherlands.

Magda, W (1994). "Analytical Solution for the Wave-Induced Excess Pore-Pressure in a Finite-Thickness Seabed Layer," *24th International Conf. of Coastal Eng.*, ASCE, pp 3111-3125.

Mase, H, Sakai, T, and Sakamoto, M (1994). "Wave-Induced Porewater Pressures and Effective Stresses around Breakwater," *Ocean Eng.*, Vol 21, No. 4, pp 361-379.

McCorquodale, JA, and Hannoura, AA (1978). "Hydraulic Conductivity of Rockfill," *J. Hydraulic Research*, No 2, pp 123-137.

Mei., CC and Foda, MA (1981). "Wave-Induced Response in a Fluid-Filled Poro-Elastic Solid with a Free Surface-a Boundary Layer Theory," *Geophys. J.R. Astr. Soc.*, Vol 66, pp 597-631

Mizutani, N, Maeda, K, Mostafa, AM, and McDougal, WG (1996a). "Estimation of Resistance Coefficients and Numerical Analysis of Non-Linear Interaction between Wave and Permeable Submerged Breakwater," *Proc. of Coastal Eng.*, JSCE, Vol 43, pp131-135. (in Japanese)

Mizutani, N, McDougal, WG, and Mostafa, AM (1996b). "BEM-FEM Combined Analysis of Non-Linear Interaction between Wave and Submerged Breakwater," *25th International Conf. of Coastal Eng.*, ASCE, (in printing).

Ohyama, T, and Nadaoka, K (1991). "Development of a Numerical Wave Tank for Analysis of Non-Linear and Irregular Wave Field," *Fluid Dynamics Research*, Vol 8, pp 231-251.

Sakai, T, Hatanaka, K, and Mase, H (1992). "Wave-Induced Effective Stress in Seabed and Its Momentary Liquefaction," *J. WatWay, Port, Coastal Ocean Eng.*, ASCE, Vol 118, pp 202-206.

Verruijt, A (1969). "Elastic Storage of Aquifers," *In Flow through Porous Media, DeWiest, R.J.M. (ed.), Academic press*, London.

Yamamoto, T (1977). "Wave Induced Instability in Seabeds," *Proc. Coastal Sediments '77*, ASCE, pp 898-913.

Proceedings of the Seventh (1997) International Offshore and Polar Engineering Conference
Honolulu, USA, May 25-30, 1997

ISBN 1-880653-28-1 (Set); ISBN 1-880653-31-1 (Vol. III)

Distorted Flow Field and Induced Sand Drift Around Flexible, Artificial Seaweed

Makoto Ifuku
Ehime University, Matsuyama, Japan
Norihiko Sakata
Hirose Giken Co. Ltd., Osaka, Japan
Hidero Hayashi
Obayashi Corp., Tokyo, Japan

Abstract

Flow around artificial seaweed was computed by the Reynolds equation. Further, the effect of sway of artificial seaweed is considered in the basic equation. Sediment concentration is computed by the turbulent diffusion equation with a time-dependent turbulent diffusion coefficient which depends on the flow velocity. Sediment concentration at the reference level is evaluated by the author's model(1986), which is a modification of the Kalkanis' model(1965).

Near the artificial seaweed, flow is distorted shoreward at the boundary of vegetation. The organized vortex generated at the boundary of vegetation is closely associated with the transport of Reynolds stress. Moreover, the vortex produced near the boundary of vegetation carries its momentum by the sweep-like events down to the bottom. The agreement between computed results and experimental ones is satisfactory for the flow profile and the displacement of swaying. Temporal concentration of sediment at the point where the artificial seaweed is attached has four peaks, two of which are strong and the others weak, and the phase lag between the strong and weak peaks is approximately $\pi/2$.

The function of artificial seaweed is to facilitate shoreward sand movement near the boundary of vegetation and also to accelerate the accumulation of sand at both the fore and lee side of the attached point.

Key words : artificial seaweed, vortex, pick-up rate, sediment concentration

1. Introduction

So far such rigid structures as groins and detached breakwaters have contributed more or less to the erosion control of sandy beaches. After the advocacy of maintenance and servicing of waterfront environment including the landscape, a waves/sand drift control system that can replace the rigid structures has remained as a future possibility. With this background, a flexible structure using artificial seaweed has been proposed as a system that controls not merely the wave and sand drift but has an influence on the ecosystem.

The trial of waves/sand drift control system with artificial seaweed has its beginning in the 1960s. Roger (1968) attempted to look into the effect of wave/sand drift control with artificial seaweed by observation on-site. He reported that the effect of sand drift control could not be confirmed because long-term observation was impossible due to the fact that artificial seaweed or its substructure was destroyed by the waves during the observation. Price (1986) made a model with two-layer flow with different kinematic viscosity, the field of water particle velocity attached with artificial seaweed and also performed an experiment. He reported that the experimental result exhibited good agreement with the results of his model analysis. However, no hydraulic consideration has so far been made with the seaweed vegetation.

In Japan also the studies of wave/sand drift using artificial seaweed have extensively been carried out recently. Asano et al. (1988) studied the effect of wave control by the leaf-formed soft artificial seaweed. Such a soft type artificial seaweed is however problematic on the beach because it will very likely lie down on the bottom due to the adhesion of living things.

Oda et al. (1991) proposed a hard type artificial seaweed utilizing chemical fiber. They have experimentally attempted to study the wave/sand drift control mechanism, and obtained some important results toward the practical use of artificial seaweed.

Mase et al. (1989) studied the wave transformation by artificial seaweed attached on a sloping beach and obtained some basic materials for the practical use of the artificial seaweed. They pointed out that the practical use encounters many such problems as the selection of raw materials for the artificial seaweed, the attaching and vegetating process of the raw materials, and the prediction of waves and currents with the vegetation of the artificial seaweed.

2. Numerical analysis

2.1 Basic equations

(a) Flow

Considering the swaying of artificial seaweed, the momentum

equations of motion read

$$\frac{\partial u}{\partial t}+u\frac{\partial u}{\partial x}+w\frac{\partial u}{\partial z} = -\frac{1}{\rho}\frac{\partial p}{\partial x}+\frac{\partial \tau_{xx}}{\partial x}+\frac{\partial \tau_{zx}}{\partial z}+F_D \quad (1)$$

$$\frac{\partial w}{\partial t}+u\frac{\partial w}{\partial x}+w\frac{\partial w}{\partial z} = -g-\frac{1}{\rho}\frac{\partial p}{\partial z}+\frac{\partial \tau_{zx}}{\partial x}+\frac{\partial \tau_{zz}}{\partial z} \quad (2)$$

$$\frac{\partial u}{\partial x}+\frac{\partial w}{\partial z}=0 \quad (3)$$

where u and w are the fluid velocities in x and z directions respectively, ρ is the fluid density, p is the pressure.

The drag force, F_D in eq.(1) is given by

$$F_D=-\frac{1}{2}c_D\frac{1-\lambda_s}{l_s}u_r\sqrt{u_r{}^2+w_r{}^2} \quad (4)$$

where c_D is the drag coefficient, l_s is the attaching interval of artificial seaweed, λ_s is the opening rare[λ_s=(1-A_s)/$l_s h_s$; h_s :length of artificial seaweed, A_s:projected area of artificial seaweed], u_r and w_r are the relative velocities ($u_r= u-u_s$, $w_r= w-w_s$; u_s and w_s are the velocities of artificial seaweed in x and z directions respectively).

Moreover, τ_{xx},τ_{zx} and τ_{zz} are the Reynolds stresses and are as follows in the notation of Cartesian tensor

$$\left.\begin{array}{l}\tau_{ij}/\rho=(\nu+\nu_t)\left(\frac{\partial u_i}{\partial x_j}+\frac{\partial u_j}{\partial x_i}\right)\\ \nu_t=(c_s\Delta)^2\left[\left(\frac{\partial u_i}{\partial x_j}+\frac{\partial u_j}{\partial x_i}\right)\frac{\partial u_i}{\partial x_j}\right]^{1/2}\end{array}\right\} \quad (5)$$

where ν is the kinematic viscosity, ν_t is the eddy viscosity, c_s is an empirical constant, $\Delta=(\Delta x\Delta z)^{1/2}$($\Delta x$ and Δz are the horizontal and vertical grid intervals respectively).

The Poisson equation for the pressure reads

$$\frac{\partial^2 p}{\partial x^2}+\frac{\partial^2 p}{\partial z^2}=\rho\left[-2\left(\frac{\partial w}{\partial x}\frac{\partial u}{\partial z}-\frac{\partial u}{\partial x}\frac{\partial w}{\partial z}\right)+\frac{\partial A_x}{\partial x}+\frac{\partial A_z}{\partial z}\right] \quad (6)$$

with

$$\left.\begin{array}{l}A_x=\partial\tau_{xx}/\partial x+\partial\tau_{zx}/\partial z+F_D\\ A_z=-g+\partial\tau_{zx}/\partial x+\partial\tau_{zz}/\partial z\end{array}\right\} \quad (7)$$

(b) Sediment concentration

Sediment concentration in fluid is computed by the turbulent diffusion equation. The turbulent diffusion coefficients, K_{xx}, K_{zz}, are assumed to be

$$\left.\begin{array}{l}K_{xx}=a(a_L u^2+a_T w^2)/q\\ K_{zz}=a(a_T u^2+a_L w^2)/q\\ q=(u^2+w^2)^{1/2}\end{array}\right\} \quad (8)$$

where a is a constant of proportionality, a_T and a_L are the characteristic length($a_L = c_1\, a_T$; c_1 is an empirical constant).

Sediment concentration at the reference level, according to the author's numerical model(1986), is

$$C_0=2P\,\rho_s gdV/3/\int_0^{\delta_B}u\,dz \quad (9)$$

where ρ_s is the density of sediment ($\rho_s/\rho=2.65$), d is the diameter of sediment, V is the speed of sediment propagation, δ_B is the thickness of bed-layer(δ_B=2 d). In the present analysis, the speed of sediment propagation is set to be equal to the cross-sectional mean velocity of fluid in the bed-layer.

P is the probability that sediment is set in motion:

$$P=\frac{1}{\sqrt{2\pi}}\int_{B_*\Psi-1/\eta_0}^{\infty}\exp(-z^2/2)\,dz \quad (10)$$

where $B_*=4/(3c_L\eta_0)$ (c_L : lift coefficient, $1/\eta_0$ is an empirical constant), and $\Psi=(\rho_s-\rho)gd/\rho u_0{}^2$($u_0$ is the fluid velocity at the top of sand particle).

The fluid velocities, u_0, at the top of sand particle and in the bed-layer are interpolated by the spline method.

2.2 Numerical method

We will utilize the model of Asano et al. (1991) for the swaying of artificial seaweed caused by waves. The analytical region is 3 m length and a vegetated zone of 1 m length is set up in the center of the analytical region(Fig. 1). Moreover, the length h_s and the attaching interval l_s of artificial seaweed are 6 cm and 5 or 10 cm, respectively. The artificial seaweed is assumed to have a diameter of 1 mm, Young's modulus is 5x10^4 kgf/cm^2 and the opening rate λ_s is set to 0.2, 0.5 and 0.8.

The coefficient c_s in eq.(5) is set to be equal to 0.1. Einstein and El-Samni(1949) reported that the lift coefficient has a constant value 0.178 on the basis of measurement of the pressure difference. It was assumed here that the drag coefficient is 1.0, the lift coefficient is 0.2 and $1/\eta_0$ is 1.5 after Kalkanis(1965). The sediment diameter is 0.05 mm. The settling velocity is calculated by Rubey's formula(1933).

The horizontal grid interval Δx is 5 mm, the vertical distance from the bottom to the upper boundary is divided into 50 intervals. The intervals are determined so as to form a geometric progression.

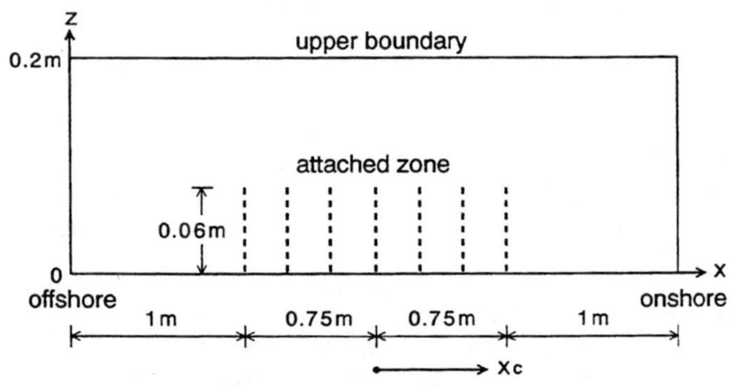

Fig. 1 Analytical region

2.3 Initial and boundary conditions

Oda et al. (1991) performed an experiment with regular waves, 4.4, 6.1 and 7.6 cm in wave height and 1.26, 1.58 and 1.90 s in wave period and studied the wave transformation and change of the bottom profile with the artificial seaweed. In this experiment, the damping rate of wave height is of the order of 5 % at both ends of the artificial seaweed vegetated zone. Considering that the damping rate of wave height is small enough in this experiment, it is assumed that the horizontal velocity at the upper boundary may be described by Airy wave theory. Since it is thought that the vertical distribution of velocity as measured manifests a good coincidence with the velocity calculated by Airy wave theory at a height of 10 cm or higher from the sea bottom, the height of the upper boundary from the bottom is set to be 2.5 times the length of the artificial seaweed. Then at the cross-shore boundaries the velocity near the bottom which is computed by one dimensional turbulent layer equation is given and the velocity calculated by Airy wave theory is given as that from the outer edge of the boundary layer to the upper boundary. The initial conditions are assumed to be without waves.

At the upper and cross-shore boundaries, the pressure is computed by Airy wave theory and at the bottom $\partial p/\partial z$ is set equal to zero. The concentration at the upper boundary is changed linearly and at the cross-shore boundaries the concentration which is computed by one dimensional diffusion equation is given.

3. Numerical results

3.1 Profiles of cross-shore velocity

Fig.2 shows the velocity distribution in the case where the cross-shore velocity at the upper boundary is given by Airy wave theory. In this figure, the marks ■, ○, ▲, ◆, and ● represent the experimental results in 0, $\pi/4$, $\pi/2$, $3\pi/4$ and π phases, respectively. The velocity is presumed to be positive onshore, and negative offshore. As for the phase of computed and experimental results, it is assumed to be 0 (zero) at a measuring point through which the wave crest passes. The measuring points are 2 cm from the offshore side of the artificial seaweed at the central position of the region where the artificial seaweed has been attached. The heights of measuring points are 2, 4, 6, 8 and 10 cm from the bottom.

In Fig.2 (a), it seems that the difference is great between the experimental and computed results in $\pi/4$ and $\pi/2$ phases at any height except for z = 2 cm. In any phase except $\pi/4$ and $\pi/2$, there may be a good correspondence between these both results. However, the measured data where the velocity amplitude decreases at the height of 4 cm from the bottom in $3\pi/4$ phase is not reproduced numerically. In Fig.2(b), the experimental and computed results almost coincide among them in the region above the upper edge of the artificial seaweed in 0 phase, but they do not manifest good agreement in the region beneath the upper edge of the artificial seaweed. As in Fig.2(a), the difference is relatively big between the experimental and computed results at any height in the $\pi/4$ and $\pi/2$ phases. Although the experimental result where the velocity amplitude reduces at the height of 4 cm from the bottom in $3\pi/4$ phase is not reproduced numerically, both results have an acceptable agreement at other heights. The reduction in velocity in π phase near the edge of the artificial seaweed is well reproduced.

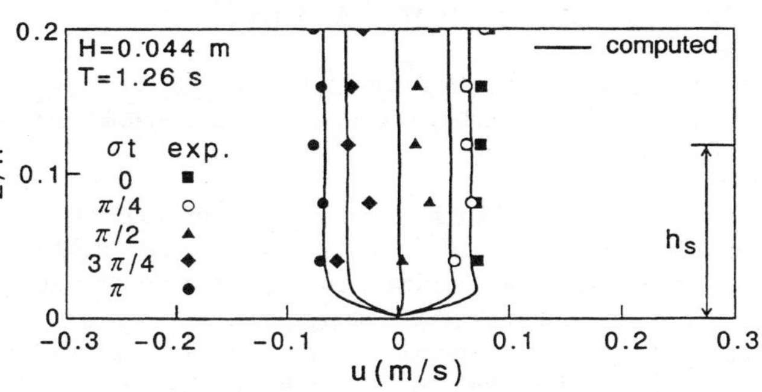

(a) H:4.4 cm, T:1.26 s, l_s:10 cm

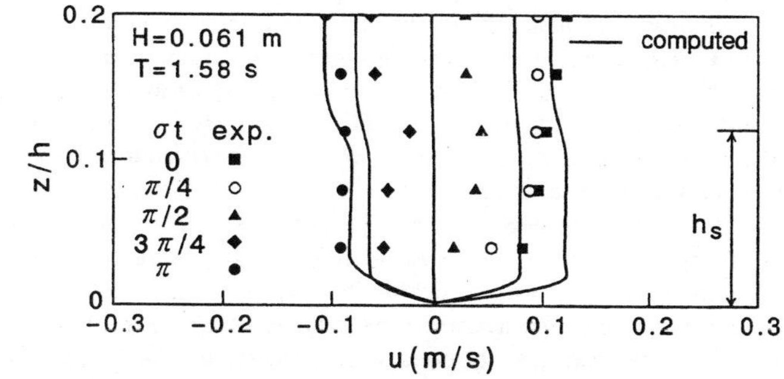

(b) H:6.1 cm, T:1.58 s, l_s:10 cm

Fig.2 Profiles of cross-shore velocity

3.2 Velocity vector

Fig.3 shows experimental and numerical results of the velocity vector at 2 cm from the shoreward side of the point attached with artificial seaweed. The velocity is assumed to be positive if the flow is onshore and negative if it is offshore. As a result of the computation, the longer axis of the ellipse connecting the tops of the velocity vectors at each time is tilted forward from the bottom to the edge of the artificial seaweed (referred to as "boundary of vegetation"). This inclination of the ellipse is similar to that of the experimental results. In the region above this boundary of vegetation, however, the longer axis of the ellipse as a result of the experiment tilts backward, while that of computed results tilts forward. Numerical results exhibit a rather large amplitude of the flow velocity in the vertical direction compared with experimental results.

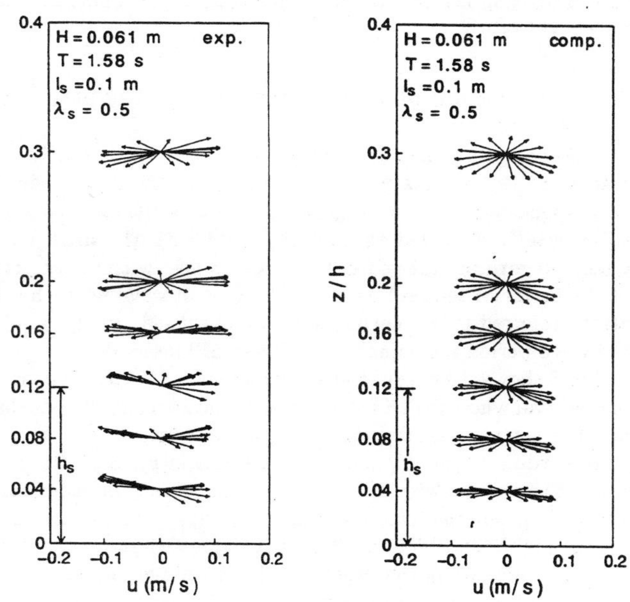

(a)experimental results (b)numerical results

Fig.3 Velocity vector

3.3 Time-averaged Reynolds stress and velocity

Fig.4 depicts the spatial distribution of Reynolds stress, $\overline{\tau}_{xz}/\rho$ and time-averaged velocity near the artificial seaweed. The position $x_c/l_s = 0$ in the figure is the center of the analytical region.

Though the cross-shore velocity near the boundary of vegetation is maximal at and just behind the vegetated point of artificial seaweed, it is maximal at a position by one layer higher at any other position. The vertical flow is produced at, just in front of, and just behind the point attached with artificial seaweed, while the vertical flow is downward at any other position. Note that clockwise circulation is generated on the shore side of the artificial seaweed. The distribution of the Reynolds stress is maximal near the boundary of vegetation near the artificial seaweed. It decreases as the flow goes toward the water surface and sea bottom.

Since both the time-averaged velocity and Reynolds stress become maximal near the boundary of vegetation, there must be large mixing and turbulence in this area. This manifests excellent agreement with the results of the study by Ikeda and Kanazawa (1994) who measured several factors in the flow inside and outside the submerging flexible vegetation.

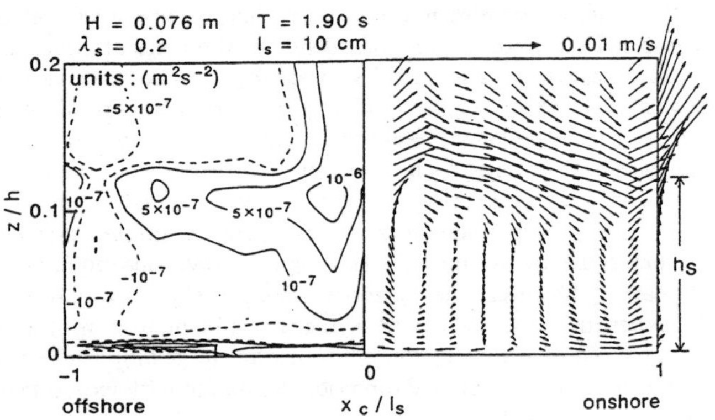

Fig.4 Spatial distribution of time-averaged Reynolds stress and velocity

3.4 Deformation of velocity profile near the artificial seaweed

Observing the temporal variation of velocity reveals the generation at the boundary of vegetation when the flow is reversed. It is suggested that the vegetation of the artificial seaweed has substantially distorted the velocity field with the artificial seaweed. Regarding, therefore, the velocity field without vegetation of the artificial seaweed as the mean flow of wave field, the deviations(referred to as "turbulence")are discussed. In this analysis, the wave period was equally divided in 512 intervals.

Fig.5 shows the case of the artificial seaweed with a 5 cm attaching interval where (a) and (b) represent the intensity of disturbance and the vertical distribution of Reynolds stress.

Regarding to the intensity of turbulence in Fig.5(a), its component in the cross-shore direction is maximal near the boundary of vegetation, but the intensity reduces gradually above this area as it goes higher and higher. Though it decreased gradually also below the boundary of the vegetation layer, it takes a value of about 80 % at the boundary of vegetation. The increase in the turbulence intensity near the boundary of vegetation is caused by the distortion of the velocity field due to the swaying of the artificial seaweed. The increase in the turbulence intensity near the bottom results in the reduction of velocity by the attaching of artificial seaweed. On the other hand, the vertical component of velocity increases gradually up to the vegetation layer boundary, while it remains almost constant in a zone higher than that. This may be because the turbulence intensity is uniform above the boundary of vegetation, while the turbulence energy is dissipated by the viscosity below the boundary of vegetation.

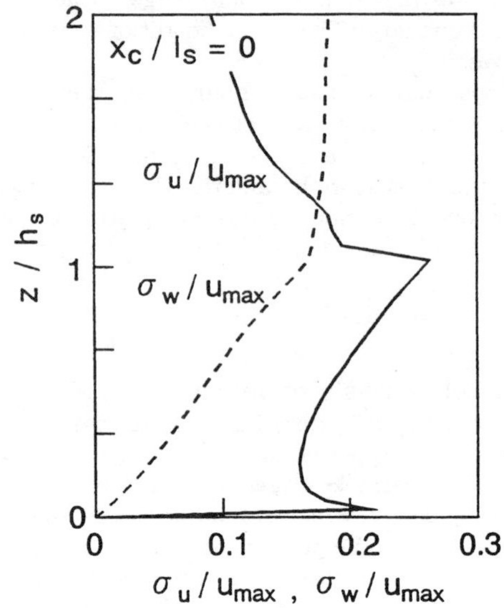

(a) intensity of turbulence

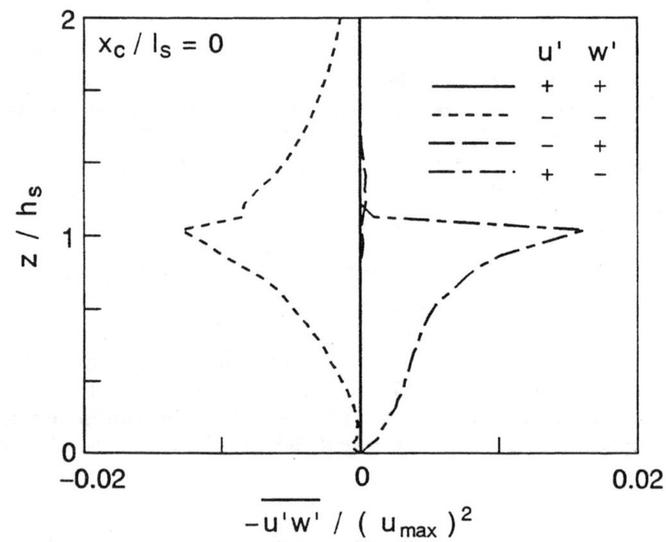

(b) Reynolds stress

Fig.5 Profiles of turbulence intensity and Reynolds stress

Fig.5(b) shows the fourth quadrant characteristics of transportation of momentum on the (u',w') plane. The perturbation with (u',w')=(+,-) increases gradually from the bottom to the boundary of vegetation and decreases sharply at the height 3 to 3.5 mm from the boundary of vegetation. The perturbation with(u',w')=(-,+) is about 10 % of the perturbation with (u',w')=(+,-) near the boundary of vegetation. The vortex produced near the boundary of vegetation carries its momentum by the sweep-like events down to the point near the bottom. As for the perturvation with (u',w')=(-,+), its momentum is shifted to the high frequency zone by the elementary wake action below the boundary of vegetation, thereby possibly causing the small transportation of momentum. The perturbation with (u',w')=(-,-) has a value of about 90 % of the perturbation with (u',w')=(+,-) near the boundary of vegetation and decreases gradually above this area.

3.5 Rotation and artificial seaweed parameter

Fig.6 shows the relationship between the rotation ω [$\omega = (\partial u/\partial z - \partial w/\partial x)/2$] near the boundary of vegetation at the point where the artificial seaweed is attached and the parameter relating to the attaching of the artificial seaweed S_p [S_p = (1-λ_s) h_s/l_s]. The blackened and open symbols in the figure represent 5 cm and 10 cm attaching intervals of the artificial seaweed respectively. In the case of 10 cm attaching intervals, the rotation increases with a reduction in the opening rate of the artificial seaweed at any frequency, thereby enlarging the distortion of flow near the boundary of vegetation. In the case of 5 cm attaching intervals of the artificial seaweed, the rotation increases as the opening rate of the artificial seaweed decreases at 0.79 Hz, thus presenting a tendency similar to that of 10 cm intervals. However, at 0.53 Hz and 0.63 Hz, the rotation leaps to its maximum at S_p=0.6. The gradient of rotation at 0.6 $\leq S_p \leq$ 0.96 is greater at 0.63 Hz.

It was confirmed that the upward flow is induced by the attaching of artificial seaweed and that there exists a relatively large flow shoreward near the boundary of vegetation. If the distortion of the flow field is related to the transportation mechanism of the sand particles, it seems that there must be an optimum amount of attaching of artificial seaweed to control sand drift.

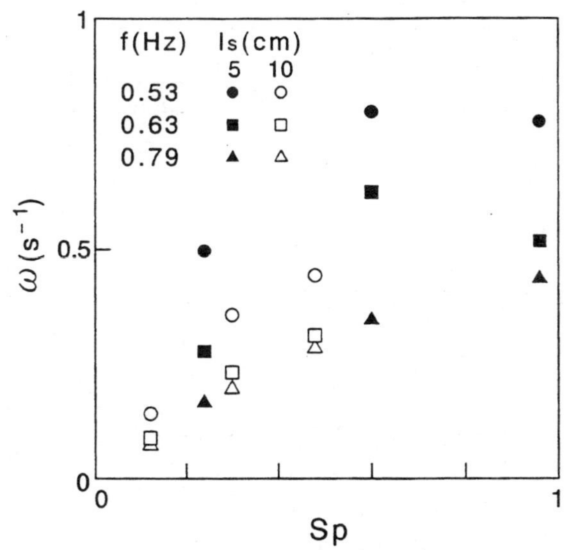

Fig.6 Rotation as a function of artificial seaweed parameter S_p

3.6 Displacement of artificial seaweed with swaying

Fig.7 illustrates the displacement of artificial seaweed. The blackened symbols and open symbols in the figure represent the analytical results by authors and those by Sakai et al. The experimental and analytical numerical by Sakai et al. have good agreement except in the case where the frequency is 0.5 Hz. As for the analytical results by the authors, the experimental and numerical results present good agreement if they are by wave at high frequency out of three waves. However, the numerical result at 0.53 Hz wave is greater than the experimental result for the displacement of artificial seaweed.

The possible reasons for this fact as follows: 1) The object of the experiment by Oda et al. is netting-formed seaweed. But in the present analysis, the seaweed was treated as a cylinder. Moreover, the links between the adjacent cylinders were not taken into account. 2) Only horizontal drag has been considered. 3) Evaluation method of drag coefficient in connection with 2).

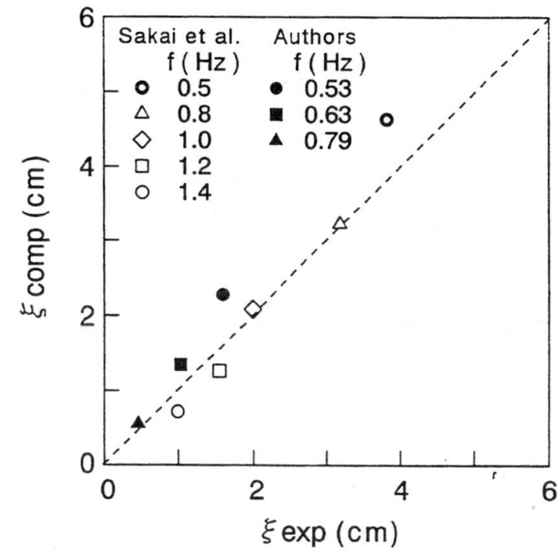

Fig.7 Displacement of artificial seaweed with swaying

3.7 Temporal variation of sediment concentration

Fig.8 depicts the temporal variation of sediment concentration at the point attached with artificial seaweed. The height from the bottom is 4 mm. The solid line, alternate long and short dashed line and broken line in the figure represent respectively the cases with 0.2, 0.5 and 0.8 opening rates. Let 0 (zero) be the phase when the wave crest passes through the center position $x_c/l_s = 0$ of the vegetated zone. Independent the opening rate of the artificial seaweed, the concentration becomes the greatest at a phase coming $\pi/25$ rad later than the phase where the wave crest passes through.

Though the concentration in bed-load layer presents its peak at a phase where the cross-shore velocity is maximum, the foregoing phase lag is due to advection and diffusion. The second peak where the concentration is about four-fifths of the first peak comes about π rad later than the first peak. A small bump-like peak has been produced about $\pi/2$ rad later than the second peak. Out of these two bump-like peaks, the concentration is higher at the peak produced about $\pi/2$ rad later than the second peak. The smaller the opening rate of artificial seaweed, the higher the concentration is. In the phase where such small bump-formed peaks arise, the cross-shore velocity is extremely small, thus contributing little to advection and diffusion. This means that the vertical velocity increases by reducing the opening rate; as a result, advection and diffusion effects in the vertical direction are enlarged.

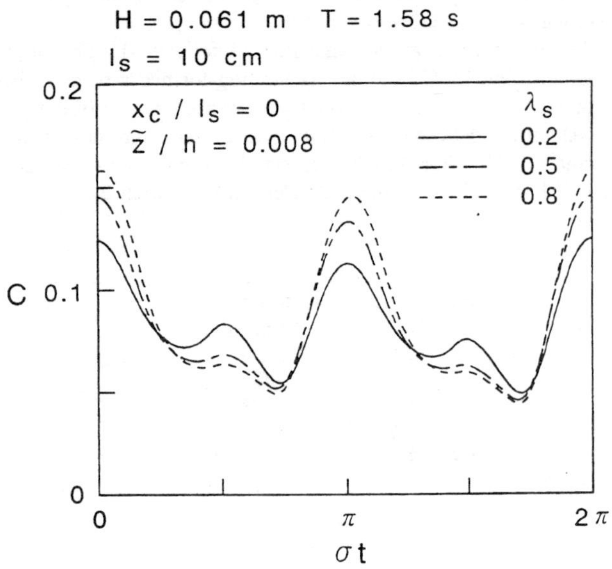

Fig.8 Temporal variation of sediment concentration near the bottom

3.8 Temporal variation of concentration with different attaching intervals

Fig.9 represents the sequential change of concentration due to the difference in the attaching intervals of artificial seaweed at the center of the vegetated zone at the height 4 mm from the bottom. The solid, alternate long and short dashed lines and broken line indicate 2, 5, and 10 cm of the attaching intervals, respectively. Let 0 (zero) be the phase where the wave crest passes though the center ($x_c/l_s = 0$) of the vegetated zone.

Fig.9(a) and (b) show the cases in which the opening rates are 0.2 and 0.5 respectively. In Fig.9(a), the first and second peaks are generated at $t/T = 0$ and 0.5 where the cross-shore velocity is maximum if the attaching intervals are 5 cm and 10 cm, respectively. Further, small bump-like peaks occur there where concentration is of the order of three-fourths of the first and second peaks.

In the case where the attaching interval is 2 cm, the first and second peaks are generated in phases near $t/T = 4.5$ and 0.95, respectively. The reasons for the phase where these first and second peaks are produced, shifts from the phase of $t/T = 0$ and 0.5, where the cross-shore velocity is maximum, is too far ahead of the velocity within the artificial seaweed. The concentration is greater at a 10 cm wide attaching interval, and smaller at 2 cm attaching interval. It is clear that in the case of a smaller opening rate (0.2) and narrower attaching interval, the velocity is low and the pick-up rate of sediment from the bottom is less.

In Fig.9(b), the concentration is the highest in the case where the attaching interval is at 10 cm wide as is in case of Fig.9(a). Though the concentration at 5 cm of attaching interval is of same order as with 10 cm, the concentration at 2 cm is much lower than them. Compared with Fig.9(a), the temporal variation in concentration is greater, the peak by the pick-up from the bottom near $t/T = 0$ and 0.5 is higher, and the small bump-like peaks near $t/T = 0.25$ and 0.75 by the vertical diffusion becomes even smaller.

This is because the opening rate that is greater than in the case of Fig.9(a) increases the cross-shore velocity and decreases the vertical velocity.

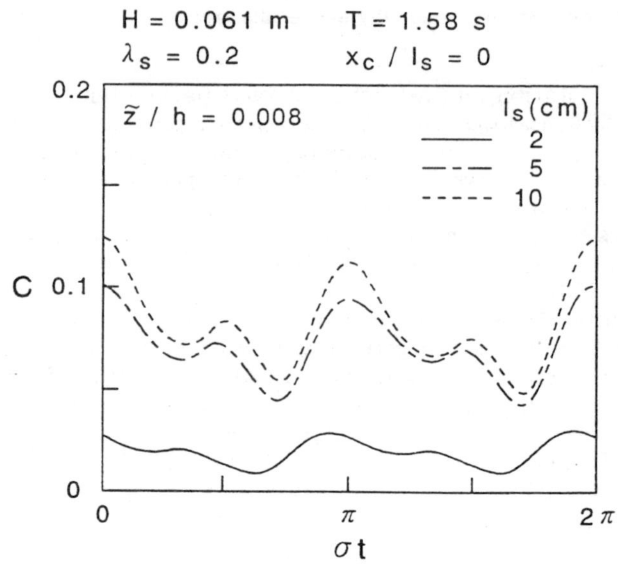

(a) opening rate : 0.2

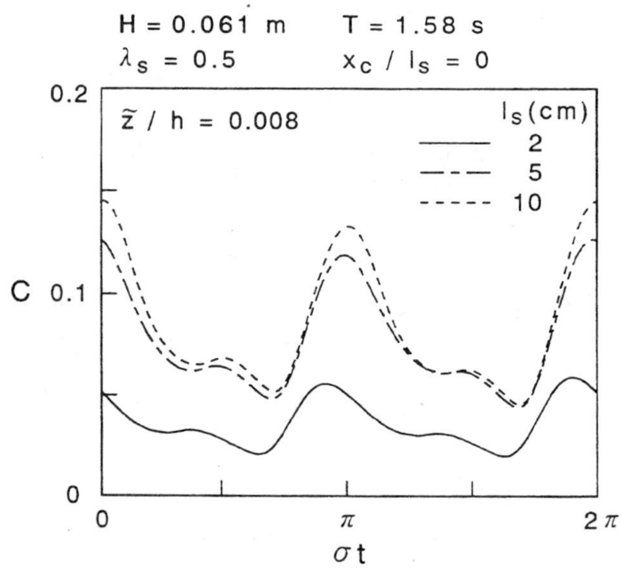

(b) opening rate : 0.5

Fig.9 Temporal variation of sediment concentration with different attaching interval

3.9 Profile of time-averaged sediment concentration

Fig.10 shows the vertical distribution of time-averaged concentration over a wave period at the attached point of artificial seaweed. The solid line and closely broken line in the figure represent the cases of 0.2 and 0.8 opening rates with 5 cm attaching interval of the artificial seaweed, and the roughly broken line and alternate long and short dashed line represent the cases of 0.2 and 0.8 opening rates with a 10 cm of attaching interval, respectively.

The time-averaged concentration reduces linearly upward from $z/h = 0.008$ in cases where the opening rate is high. In cases where the opening rate is 0.2, warping up to the neighborhood of $z/h = 0.116$ with 5 cm of the artificial seaweed attaching interval is sharper. The concentration at $z/h = 0.116$ is of the order of 10^{-2} for both 5 cm and 10 cm of attaching intervals. The gradient of concentration is the most acute among the four cases if the z/h is greater than 0.116, the attaching interval of artificial seaweed is 5 cm and the opening rate is 0.2.

The gradient of concentration above the boundary of vegetation becomes sharper in the case where the opening rare is 0.2 and the reduction in concentration is less acute than in cases with 0.8 of the opening rate. When this fact is taken into consideration, that is, the concentration reduces linearly in the case where the opening rate is 0.8, the foregoing relationship between the opening rate and vertical distribution of concentration may be due to the mechanism in which the vertical velocity increases near the point attached with artificial seaweed in which the opening rate is 0.2, and that the suspended sediment is taken into the vortex and then transported as the suspended sediment clouds up to near the boundary of vegetation by the vortex generated near the phase of $\pi/2$ and $3\pi/2$ where the cross-shore velocity reverses. The fact that the reduction in concentration above the area near the boundary of vegetation is smaller, for the narrower attaching interval(5 cm), implies that the influence of the vortex generated near the boundary of vegetation is greater.

The fact that the concentration is greater below the boundary of vegetation if the opening rate is as high as 0.8 is supposed to result from the increase of the pick-up rate of sediment from the bottom due to the increase of the cross-shore velocity, which is greater than that in the case where the opening rate is 0.2.

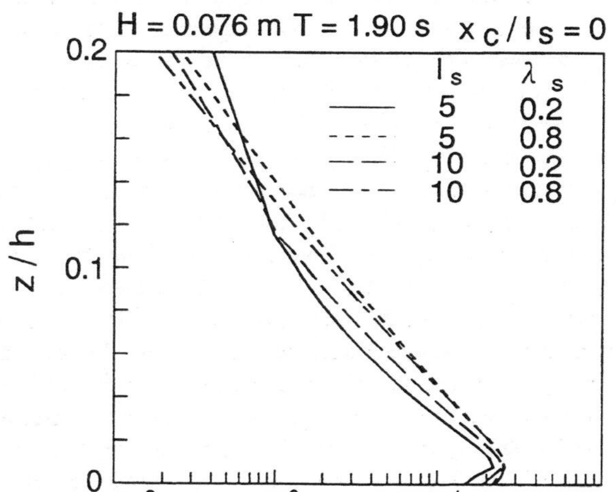

Fig.10 Profiles of time-averaged sediment concentration with different attaching intervals and opening rates of artificial seaweed

4. Conclusions

We examined the effect of the vegetation of the artificial seaweed on the flow and sand drift. The results obtained are summarized as follows:

(1) The experimental results and numerical results present good agreement in the cross-shore velocity distribution near the artificial seaweed if the period is relatively short, but they manifest some difference in the phase toward the shore if the period is longer.

(2) Both time-averaged cross-shore velocity and Reynolds stress become greatest near the vegetation layer boundary where there arises a large mixing and turbulence.

(3) The analytical result and numerical result almost coincide among them in the fluctuation displacement of the artificial seaweed in the case of low frequency waves except where the numerical result exceeds the experimental one.

(4) Four peaks (small and large) occur in the concentration near the bottom at the point attached with artificial seaweed. The influence of the vertical velocity due to the reduced opening rate of the artificial seaweed is greater on the two bump-like peaks.

(5) The gradient of the vertical distribution of time-averaged concentration changes at the boundary of vegetation. This is because the vertical velocity increases at the vegetated point of artificial seaweed and the vortex generated in the phase, where the cross-shore velocity reverses, traps the suspended sediment to transport it upward.

Acknowledgment: Part of this study was performed under the auspices of HOKD Marine Laboratory Workshop established for the development of a sand control system using artificial seaweed. We hereby express our deep thanks.

References

Asano T., S.Tsutsui and T.Sakai(1988): Wave attenuation due to seaweed layer, Proc 35th Japanese Conf. on Coastal Eng., pp.138-142.(in Japanese)

Einstein, H.A. and El-Samni El.S.A.(1949):Hydrodynamic forces on a rough wall, Rev. Mod. Phys., pp.520-524.

Ikeda S. and M.Kanazawa(1994): Flow with vegetation and three dimensional structure of the organized vortex generated at the boundary of vegetation, Proc. of Hydraulic Eng., JSCE, Vol.38, pp.345-350.(in Japanese)

Ifuku. M.(1986): Suspended sediment concentration in surf zone, Journal of Japan Society of Fluid Mechanics, Vol.5, No.4, pp.326-339.(in Japanese)

Ifuku M., M.Iwata, Y.Yonezawa and N.Sakata(1994):Flow and sand drift around artificial seaweed, Proc. 41st Japanese Conf. on Coastal Eng., pp.566-570.(in Japanese)

Kalkanis, P.(1965):Transportation of bed material due to wave action, U.S.Army, Coastal Res. Center, Tech. Memo 2, pp.1-114.

Oda K., K.Amano, K.Kobayashi and N.Sakata(1991): On the effect of stiff artificial seaweed in both reducing wave height and controlling transport of bed material, Proc. 38th Japanese Conf. on Coastal Eng., pp.331-335.(in Japanese)

Price, W.A., K.W.Tomlinson and J.N.Hunt(1968): The effect of artificial seaweed in promoting the build-up of beaches, Proc. 11th Conf. Coastal Eng., pp.570-578.

Roger, S.(1986): Artificial seaweed for shoreline erosion control, UNC Sea Grant Pub., UNC-SG-WP-8-6-4, 18p.

Rubey, W.W.(1933): Settling velocities of gravel, sand, and silt particles, Am.J.Sci.Ser.5, Vol.25, No.148, pp.325-338.

Sakai T., T.Asano and Y.Ueda(1989):Swaying of artificial seaweed under waves, Proc. of the 44th Annual Conf. of JSCE, 2, pp.692-693.(in Japanese)

Proceedings of the Seventh (1997) International Offshore and Polar Engineering Conference
Honolulu, USA, May 25-30, 1997

ISBN 1-880653-28-1 (Set); ISBN 1-880653-31-1 (Vol. III)

Modeling Horizontally Two Dimensional Wave-Current Coexistence Field Over Uneven Topography

Md. Hasanat Zaman and Hiroyoshi Togashi
Nagasaki University
Nagasaki, Japan

ABSTRACT

A horizontally two dimensional numerical model for the simulation of wave-current coexistence field has been formulated by means of vertical integration of the continuity equation and the equations of motion for gradually varying impermeable sea bottom. A semi-implicit finite difference technique has been employed for the solution. As an application, the established model is utilized to describe the highly interacted nonlinear wave-current field over a submerged mound of parabolic cross-section. The results show that the variation of the wave characteristics and the drop of the mean water level are very significant in the presence of current with uneven bottom topography.

KEY WORDS: Horizontally Two Dimensional Flow, Wave-Current Coexistence, Uneven Bottom, Finite Difference.

INTRODUCTION

Waves, currents and their combined effects are the predominant factors which directly influence the whole nearshore dynamic system, including sediment transport and the resulting beach topographic change, etc.. An uneven bottom topography, on the other hand, adds considerable effects to the phenomenon because it alters both the wave and current characteristics. Researches are thus necessary to discern the combined effects of waves, currents and topography variation. Hara *et al.* (1992) showed that when a solitary wave is distorted by some submerged structure, the wave itself tries to be stable by producing a reflected wave and a shelf turning to a secondary solitary wave and distributes the excess energy to them. Ohyama and Nadaoka (1992) described that during the passage of a regular wave over a submerged dike, higher harmonics are found to be generated which later are transformed into prominent free waves in the trailing side of the dike. Pudjaprasetya and Groesen (1996) studied the distortion of the solitary and cnoidal waves over varying bottom using modified quasi-homogeneous approximation method where the dynamic equation is obtained from the energy and mass conservation. Longuet Higgins and Stewart (1961 and 1962) are the pioneers who described the effects of the wave on current and vice versa based on the radiation stress concept which excludes the computation of wave induced current without successive iteration. Watanabe *et al.* (1986) proposed a numerical model considering linear time dependent mild slope equation where breaker-induced dissipation term is involved and the computation of the wave and current components are advanced through the evaluation of radiation stress. Kirby (1984) and Liu (1990) studied the effects of current on wave by applying the modified mild slope equation. In an early work of Prüser and Zielke (1990), it is showed that Boussinesq equations models give reasonable results when the wave length is ten times greater than that of the still water depth. They also added that if the corresponding wave length in still water is smaller than ten times but greater than six times of the still water depth then the existing Boussinesq equations models are applicable only for the case of following current. Many researches have been devoted to extend the Boussinesq equations for the reasonable evaluation of wave characteristics (McCowan, 1987, Madsen *et al.*, 1992, Sϕrensen *et al.*, 1994 and Hemming *et al.*, 1995), wave-current field (Sato and Kabiling, 1994 and Watanabe *et al.*, 1994), beach evolution, etc., but still these types of models are not yet widely applicable for the consideration of wave-current field in the deep sea.

So a direct method is necessary to describe the coexistence of wave-current field with minimum effort. In this regard our approach is to propose a more generally applicable numerical method which describes the coexistence of horizontally two dimensional wave-current field and where the traditional iterative concept is avoided to curtail the computational effort. It may be mentioned here that the present authors initially developed a one dimensional model for vertically two dimensional wave-current field (Zaman *et al.*, 1994, Togashi *et al.*, 1995 and, Zaman and Togashi, 1996a) which shows excellent agreement with the experimental results (Zaman and Togashi, 1996b).

GOVERNING EQUATIONS

By the conventional assumption of inviscid and incompressible fluid the wave motion can be described by the following continuity equation and equations of motion:

$$\frac{\partial u}{\partial x}+\frac{\partial v}{\partial y}+\frac{\partial w}{\partial z}=0 \qquad \text{Eq. 1}$$

$$\frac{\partial u}{\partial t}+\frac{\partial (uu)}{\partial x}+\frac{\partial (uv)}{\partial y}+\frac{\partial (uw)}{\partial z}+\frac{\partial}{\partial x}\left(\frac{p_d}{\rho}\right)=0 \qquad \text{Eq. 2}$$

$$\frac{\partial v}{\partial t}+\frac{\partial (uv)}{\partial x}+\frac{\partial (vv)}{\partial y}+\frac{\partial (vw)}{\partial z}+\frac{\partial}{\partial y}\left(\frac{p_d}{\rho}\right)=0 \qquad \text{Eq. 3}$$

$$\frac{\partial w}{\partial t}+\frac{\partial (uw)}{\partial x}+\frac{\partial (vw)}{\partial y}+\frac{\partial (ww)}{\partial z}+\frac{\partial}{\partial z}\left(\frac{p_d}{\rho}\right)=0 \qquad \text{Eq. 4}$$

where u, v and w are the velocity components in the horizontal x, y and, vertical z directions, respectively; $p_d = p+\rho gz$ is the dynamic pressure while p is the pressure, ρ the water density, g the gravitational acceleration and t is the time.

The kinematic boundary conditions on the free surface and at the bottom can be expressed as follows:

$$\frac{\partial \eta}{\partial t}+u^s\frac{\partial \eta}{\partial x}+v^s\frac{\partial \eta}{\partial y}-w^s=0 \qquad \text{Eq. 5}$$

$$u^b\frac{\partial h}{\partial x}+v^b\frac{\partial h}{\partial y}+w^b=0 \qquad \text{Eq. 6}$$

where η is surface elevation, h the still water depth and, s and b represent the values of the respective component at the free surface and at the bottom, respectively.

Based on the exact solution for linear waves propagating over a gradually varying steady current in constant water depth, the dynamic pressure for a wave-current coexistence field can generally be formulated in the following way:

$$p_d=\rho g\eta\frac{\cosh k(h+z)}{\cosh k(h+\eta)} \qquad \text{Eq. 7}$$

where the term $(h+\eta)$ in the denominator is the instantaneous water depth; $k(=\sqrt{k_x^2+k_y^2})$ is the resultant wave number in which $k_x(=k\cos\theta_w)$ and $k_y(=k\sin\theta_w)$ are the wave number in the x and y direction and θ_w is the incident wave angle.

The wave number k is related to the absolute angular frequency of the wave motion σ and the current velocities U and V in the x and y direction, respectively through the following dispersion relation:

$$\frac{\sigma^2}{g}\left(1-\frac{Uk_x}{\sigma}-\frac{Vk_y}{\sigma}\right)^2=k\tanh k(h+\eta) \qquad \text{Eq. 8}$$

Replacing Eq. 4 by the pressure assumption Eq. 7, the vertical integration of the Eqs. 1, 2 and 3 can be performed along with invoking the kinematic boundary conditions, Eqs. 5 and 6 to give the following governing equations:

$$\frac{\partial \eta}{\partial t}+\frac{\partial Q_x}{\partial x}+\frac{\partial Q_y}{\partial y}=0 \qquad \text{Eq. 9}$$

$$\frac{\partial Q_x}{\partial t}+\frac{\partial}{\partial x}\left(\frac{\beta Q_x^2}{h+\eta}\right)+\frac{\partial}{\partial y}\left(\frac{\beta Q_xQ_y}{h+\eta}\right)+\left(\eta g\chi_1+C^2\right)\frac{\partial \eta}{\partial x}+\eta g\chi_2\frac{\partial h}{\partial x}=0 \qquad \text{Eq. 10}$$

$$\frac{\partial Q_y}{\partial t}+\frac{\partial}{\partial y}\left(\frac{\beta Q_y^2}{h+\eta}\right)+\frac{\partial}{\partial x}\left(\frac{\beta Q_xQ_y}{h+\eta}\right)+\left(\eta g\mu_1+C^2\right)\frac{\partial \eta}{\partial y}+\eta g\mu_2\frac{\partial h}{\partial y}=0 \qquad \text{Eq. 11}$$

where $Q_x=\int_{-h}^{\eta}u\,dz$ and $Q_y=\int_{-h}^{\eta}v\,dz$; χ_1, χ_2 and, μ_1, μ_2 are the relative water depth dependent parameters in the x and y direction, respectively; β is the momentum correction factors and C is the local intrinsic wave celerity. χ_1, χ_2, μ_1, μ_2 and β can be expressed as follows:

$$\chi_1=\frac{E}{F+G}-1 \qquad \text{Eq. 12}$$

$$\chi_2=\frac{E}{F+G}-\frac{1}{\cosh k(h+\eta)} \qquad \text{Eq. 13}$$

$$\mu_1=\frac{E'}{F+G'}-1 \qquad \text{Eq. 14}$$

$$\mu_2=\frac{E'}{F+G'}-\frac{1}{\cosh k(h+\eta)} \qquad \text{Eq. 15}$$

$$\beta=\left[\frac{k(h+\eta)}{\tanh k(h+\eta)}\right]\cdot\frac{1}{2}\left[1+\frac{2k(h+\eta)}{\sinh 2k(h+\eta)}\right] \qquad \text{Eq. 16}$$

where

$$E=2\tanh k(h+\eta)+\frac{2\sigma U}{g}\cos\theta_w\left(1-\frac{k_xU}{\sigma}-\frac{k_yV}{\sigma}\right) \qquad \text{Eq. 17}$$

$$E'=2\tanh k(h+\eta)+\frac{2\sigma V}{g}\sin\theta_w\left(1-\frac{k_xU}{\sigma}-\frac{k_yV}{\sigma}\right) \qquad \text{Eq. 18}$$

$$F=k(h+\eta)+\frac{1}{2}\sinh 2k(h+\eta) \qquad \text{Eq. 19}$$

$$G=\frac{2\sigma U}{g}\cos\theta_w\left(1-\frac{k_xU}{\sigma}-\frac{k_yV}{\sigma}\right)\cosh^2 k(h+\eta) \qquad \text{Eq. 20}$$

$$G'=\frac{2\sigma V}{g}\sin\theta_w\left(1-\frac{k_xU}{\sigma}-\frac{k_yV}{\sigma}\right)\cosh^2 k(h+\eta) \qquad \text{Eq. 21}$$

Finally, Eqs. 9, 10 and 11 are obtained as the required governing equations for the simulation of a two dimensional wave-current coexistence field.

NUMERICAL METHOD

Eqs. 9, 10 and 11 have been discretized in a semi-implicit

technique (Dronkers, 1969 and Horikawa, 1988) where the computational points of discharges Q_x and Q_y and, surface elevation η (shown in the circle) are defined according to the mesh shown bellow in Fig. 1.

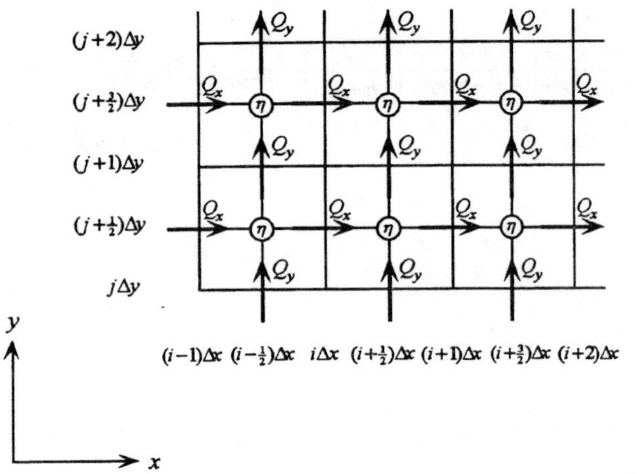

Fig. 1 Schematic view of the 2-D computational mesh

After discretization, Eqs. 9, 10 and 11 are transformed to the following finite difference equations:

$$\eta^{n+\frac{3}{2}}_{i-\frac{1}{2},j+\frac{1}{2}} = \eta^{n+\frac{1}{2}}_{i-\frac{1}{2},j+\frac{1}{2}} - \frac{\Delta t}{\Delta s}\left[Q^{n+1}_{xi,j+\frac{1}{2}} - Q^{n+1}_{xi-1,j+\frac{1}{2}}\right] - \frac{\Delta t}{\Delta s}\left[Q^{n+1}_{yi-\frac{1}{2},j+1} - Q^{n+1}_{yi-\frac{1}{2},j}\right] \qquad \text{Eq. 22}$$

$$A^{n+\frac{3}{2}}_{xi,j+\frac{1}{2}} Q^{n+2}_{xi-1,j+\frac{1}{2}} + B^{n+\frac{3}{2}}_{xi,j+\frac{1}{2}} Q^{n+2}_{xi,j+\frac{1}{2}} + C^{n+\frac{3}{2}}_{xi,j+\frac{1}{2}} Q^{n+2}_{xi+1,j+\frac{1}{2}} = D^{n+\frac{3}{2}}_{xi,j+\frac{1}{2}} \qquad \text{Eq. 23}$$

$$A^{n+\frac{3}{2}}_{yi-\frac{1}{2},j+1} Q^{n+2}_{yi-\frac{1}{2},j} + B^{n+\frac{3}{2}}_{yi-\frac{1}{2},j+1} Q^{n+2}_{yi-\frac{1}{2},j+1} + C^{n+\frac{3}{2}}_{yi-\frac{1}{2},j+1} Q^{n+2}_{yi-\frac{1}{2},j+2} = D^{n+\frac{3}{2}}_{yi-\frac{1}{2},j+1} \qquad \text{Eq. 24}$$

where

$$A^{n+\frac{3}{2}}_{xi,j+\frac{1}{2}} = -\frac{\Delta t}{4\Delta s}\left[\left(\frac{\beta}{h+\eta}\right)^{n+\frac{3}{2}}_{i-\frac{1}{2},j+\frac{1}{2}} \left(Q^{n+1}_{xi,j+\frac{1}{2}} + Q^{n+1}_{xi-1,j+\frac{1}{2}}\right)\right] \qquad \text{Eq. 25}$$

$$C^{n+\frac{3}{2}}_{xi,j+\frac{1}{2}} = \frac{\Delta t}{4\Delta s}\left[\left(\frac{\beta}{h+\eta}\right)^{n+\frac{3}{2}}_{i+\frac{1}{2},j+\frac{1}{2}} \left(Q^{n+1}_{xi+1,j+\frac{1}{2}} + Q^{n+1}_{xi,j+\frac{1}{2}}\right)\right] \qquad \text{Eq. 26}$$

$$B^{n+\frac{3}{2}}_{xi,j+\frac{1}{2}} = 1 + A^{n+\frac{3}{2}}_{xi,j+\frac{1}{2}} + C^{n+\frac{3}{2}}_{xi,j+\frac{1}{2}} \qquad \text{Eq. 27}$$

$$\begin{aligned}
D^{n+\frac{3}{2}}_{xi,j+\frac{1}{2}} = {} & Q^{n+1}_{xi,j+\frac{1}{2}} - \frac{\Delta t}{4\Delta s}\cdot\left(\frac{\beta}{h+\eta}\right)^{n+\frac{3}{2}}_{i-\frac{1}{2},j+\frac{3}{2}} \cdot\left(Q^{n+1}_{xi,j+\frac{3}{2}} + Q^{n+1}_{xi-1,j+\frac{3}{2}}\right) \\
& \cdot\left(Q^{n+1}_{yi-\frac{1}{2},j+2} + Q^{n+1}_{yi-\frac{1}{2},j+1}\right) + \frac{\Delta t}{4\Delta s}\cdot\left(\frac{\beta}{h+\eta}\right)^{n+\frac{3}{2}}_{i-\frac{1}{2},j+\frac{1}{2}} \\
& \cdot\left(Q^{n+1}_{xi,j+\frac{1}{2}} + Q^{n+1}_{xi-1,j+\frac{1}{2}}\right)\cdot\left(Q^{n+1}_{yi-\frac{1}{2},j+1} + Q^{n+1}_{xi-\frac{1}{2},j}\right) \\
& - \frac{\Delta t}{2\Delta s}\left[\left\{(\eta g\chi_1 + C^2)^{n+\frac{3}{2}}_{i+\frac{1}{2},j+\frac{1}{2}} + (\eta g\chi_1 + C^2)^{n+\frac{3}{2}}_{i-\frac{1}{2},j+\frac{1}{2}}\right\}\right] \\
& \cdot\left[\left\{\eta^{n+\frac{3}{2}}_{i+\frac{1}{2},j+\frac{1}{2}} - \eta^{n+\frac{3}{2}}_{i-\frac{1}{2},j+\frac{1}{2}}\right\}\right] \\
& - \frac{\Delta t}{2\Delta s}\left[\left\{(\eta g\chi_2)^{n+\frac{3}{2}}_{i+\frac{1}{2},j+\frac{1}{2}} + (\eta g\chi_2)^{n+\frac{3}{2}}_{i-\frac{1}{2},j+\frac{1}{2}}\right\}\right] \\
& \cdot\left[\left\{h^{n+\frac{3}{2}}_{i+\frac{1}{2},j+\frac{1}{2}} - h^{n+\frac{3}{2}}_{i-\frac{1}{2},j+\frac{1}{2}}\right\}\right]
\end{aligned} \qquad \text{Eq. 28}$$

$$A^{n+\frac{3}{2}}_{yi-\frac{1}{2},j+1} = -\frac{\Delta t}{4\Delta s}\left[\left(\frac{\beta}{h+\eta}\right)^{n+\frac{3}{2}}_{i-\frac{1}{2},j+\frac{1}{2}} \left(Q^{n+1}_{yi-\frac{1}{2},j+1} + Q^{n+1}_{yi-\frac{1}{2},j}\right)\right] \qquad \text{Eq. 29}$$

$$C^{n+\frac{3}{2}}_{yi-\frac{1}{2},j+1} = \frac{\Delta t}{4\Delta s}\left[\left(\frac{\beta}{h+\eta}\right)^{n+\frac{3}{2}}_{i-\frac{1}{2},j+\frac{3}{2}} \left(Q^{n+1}_{yi-\frac{1}{2},j+2} + Q^{n+1}_{yi-\frac{1}{2},j+1}\right)\right] \qquad \text{Eq. 30}$$

$$B^{n+\frac{3}{2}}_{yi-\frac{1}{2},j+1} = 1 + A^{n+\frac{3}{2}}_{yi-\frac{1}{2},j+1} + C^{n+\frac{3}{2}}_{yi-\frac{1}{2},j+1} \qquad \text{Eq. 31}$$

$$\begin{aligned}
D^{n+\frac{3}{2}}_{yi-\frac{1}{2},j+1} = {} & Q^{n+1}_{yi-\frac{1}{2},j+1} - \frac{\Delta t}{4\Delta s}\cdot\left(\frac{\beta}{h+\eta}\right)^{n+\frac{3}{2}}_{i+\frac{1}{2},j+\frac{1}{2}} \cdot\left(Q^{n+1}_{xi+1,j+\frac{1}{2}} + Q^{n+1}_{xi,j+\frac{1}{2}}\right) \\
& \cdot\left(Q^{n+1}_{yi+\frac{1}{2},j+1} + Q^{n+1}_{yi+\frac{1}{2},j}\right) + \frac{\Delta t}{4\Delta s}\cdot\left(\frac{\beta}{h+\eta}\right)^{n+\frac{3}{2}}_{i-\frac{1}{2},j+\frac{1}{2}} \\
& \cdot\left(Q^{n+1}_{xi,j+\frac{1}{2}} + Q^{n+1}_{xi-1,j+\frac{1}{2}}\right)\cdot\left(Q^{n+1}_{yi-\frac{1}{2},j+1} + Q^{n1+}_{yi-\frac{1}{2},j}\right) \\
& - \frac{\Delta t}{2\Delta s}\left[\left\{(\eta g\mu_1 + C^2)^{n+\frac{3}{2}}_{i-\frac{1}{2},j+\frac{3}{2}} + (\eta g\mu_1 + C^2)^{n+\frac{3}{2}}_{i-\frac{1}{2},j+\frac{1}{2}}\right\}\right] \\
& \cdot\left[\left\{\eta^{n+\frac{3}{2}}_{i-\frac{1}{2},j+\frac{3}{2}} - \eta^{n+\frac{3}{2}}_{i-\frac{1}{2},j+\frac{1}{2}}\right\}\right] \\
& - \frac{\Delta t}{2\Delta s}\left[\left\{(\eta g\mu_2)^{n+\frac{3}{2}}_{i-\frac{1}{2},j+\frac{3}{2}} + (\eta g\mu_2)^{n+\frac{3}{2}}_{i-\frac{1}{2},j+\frac{1}{2}}\right\}\right]
\end{aligned} \qquad \text{Eq. 32}$$

where Δs and Δt are the spatial and temporal increments, respectively.

The discharges will be evaluated at the integer time level and that of the surface fluctuation at the fractional time level. The nonlinear convective terms in the momentum equations have been linearized and properly expressed by the defined values of the unknowns. Numerical implementation of the scheme is direct and highly efficient as the discretized continuity equation, which is used for the computation of η is explicit. On the other hand, the momentum equation for each direction constitutes a tridiagonal coefficient matrix for every time step and can effectively be solved for the discharge (Q_x and Q_y) of the respective direction by the double-sweep algorithm. The initial condition for η is introduced at time level $(n+\frac{1}{2})\Delta t$ and, on the other hand, for Q_x and Q_y at level $(n+1)\Delta t$ to initiate the numerical computation. The computational tactics of this model is, η, at time level $(n+\frac{3}{2})\Delta t$ will be evaluated by the Eq. 22 with the known value of η at $(n+\frac{1}{2})\Delta t$ time level and, Q_x and Q_y at $(n+1)\Delta t$ time level and then, Eqs. 23 and 24, respectively, are employed to estimate Q_x and Q_y for $(n+2)\Delta t$ time level, using the calculated η at $(n+\frac{3}{2})\Delta t$ time level. The coupling of the above equations will continue up to the desired time level.

RESULTS AND DISCUSSIONS

As an application of this two dimensional numerical model, we consider a case where wave and current approach the mound with an angle θ_w and θ_c respectively. The computational domain which describes this physical problem is shown in Fig. 2. In the figure, h is the still water depth and U is the gradually varying steady current; l is the length and rh is the height of the mound; x and y are the horizontal coordinates, respectively.

Before going to the computation for wave-current-mound condition we applied the model for wave only (no current and mound), wave-current (mound is not in presence) and wave-mound (in the absence of current) conditions to study the effect of the individual factors.

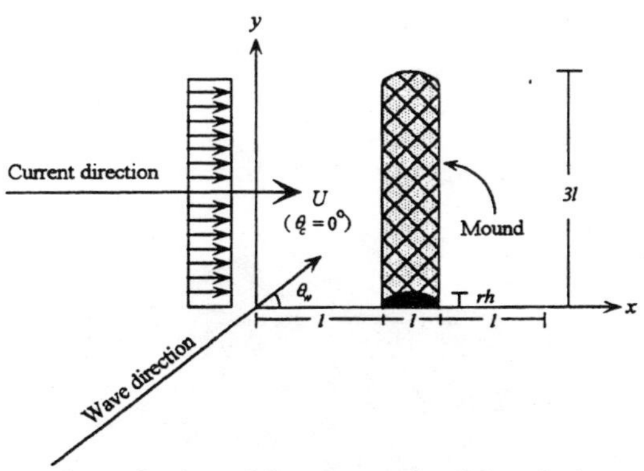

Fig. 2 Schematic view of the computational domain (not to the scale)

In these varieties of computations it is assumed that the still water depth h is 10m, the incident wave period T is 8s, incident wave height H_i is 0.6m, the current velocity U (normalized current parameter $U/C_i = 0.1$; where C_i is the wave celerity in the absence of current and mound) is 0.886m/s, the relative mound height r is 0.5 and length of the mound l is 80m and, the wave and current incident angles θ_w, θ_c are 30° and 0°. Here, the spatial increment is taken as $\Delta s = 2.0$m while the temporal increment is $\Delta t = 0.05$s. In the figures the scale factors for X and Y directions must be multiplied by 4.0 to express the distance in meter. Mound is set in between 20 and 40 of the scale factor of the figures in the X-direction. Figs. 3, 4 and 5 show the bird's view of the wave profiles for wave only, wave-mound and wave-current condition, respectively.

Fig. 3 shows that for wave only, the wave profile is uniform in length and in amplitude. Figs. 4 and 5, respectively, express that the wave profile is distorted due to presence of mound and presence of current elongates the wave length. It is also observed that the distortion to the wave profile imparted by the mound is recovered over the flat bottom with time. Tanaka (1990) described similar fact, where he mentioned that the wave itself readjusts the wave height along its crest to keep the crest-line straight when it is distorted by external source.

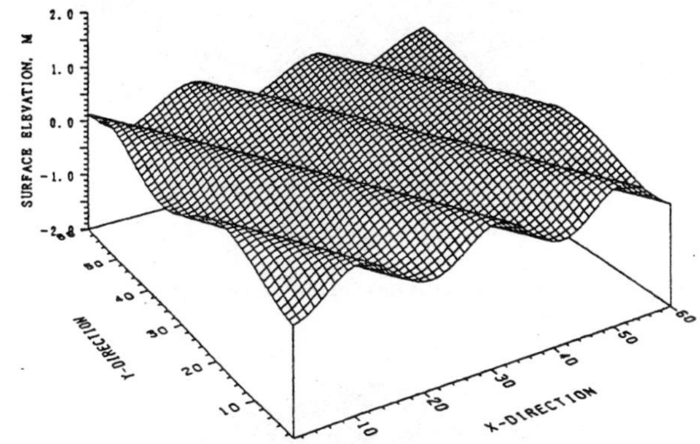

Fig. 3 Wave profile for wave only

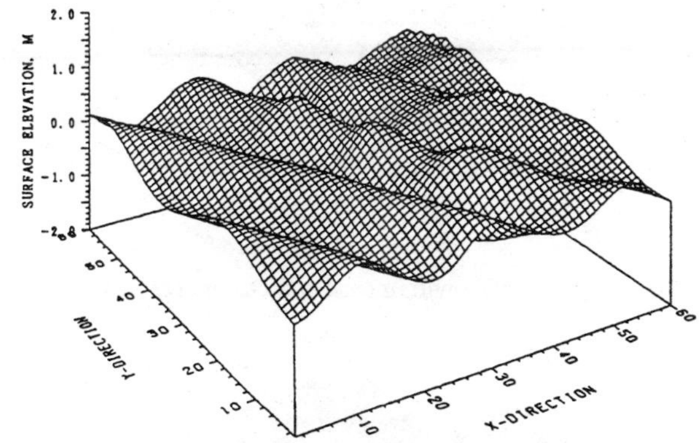

Fig. 4 Wave profile in the presence of mound

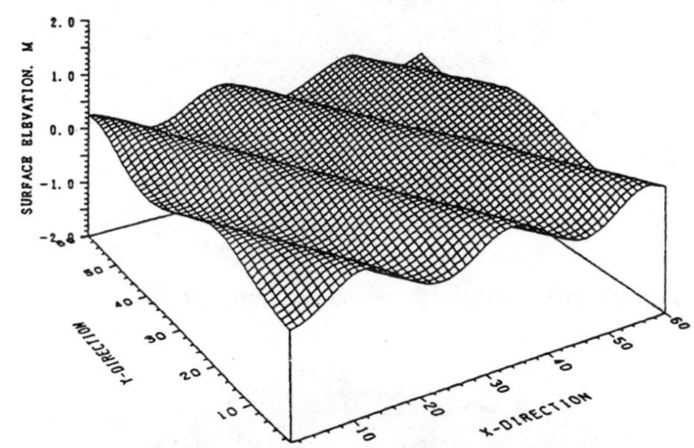

Fig. 5 Wave profile with current ($U/C_i = 0.1$)

Wave and Mound with Various Current Velocities

In this section the influences of various current velocities on the alike incident wave over the mound have been discussed. In the computation, it is assumed that h is 10m, T is 8s, H_i is 0.6m, θ_w is 30°, θ_c is 0°, r is 0.5 and l is 80m. The magnitude of the input current U (U/C_i = 0.06, 0.1 and 0.14) alters to 0.53m/s, 0.886m/s and 1.24m/s.

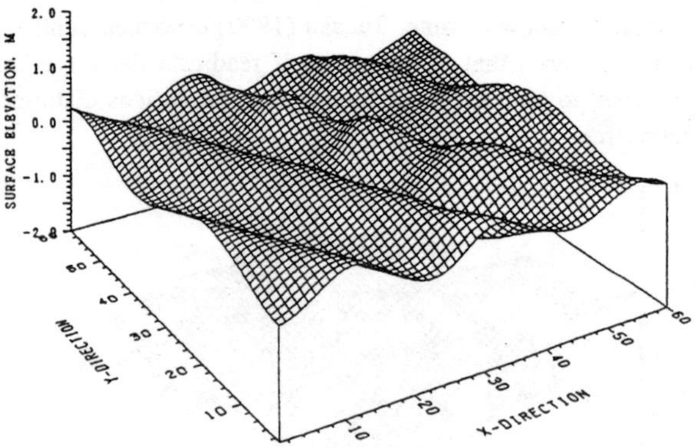

Fig. 6 Wave profile with mound and current (U/C_i = 0.06)

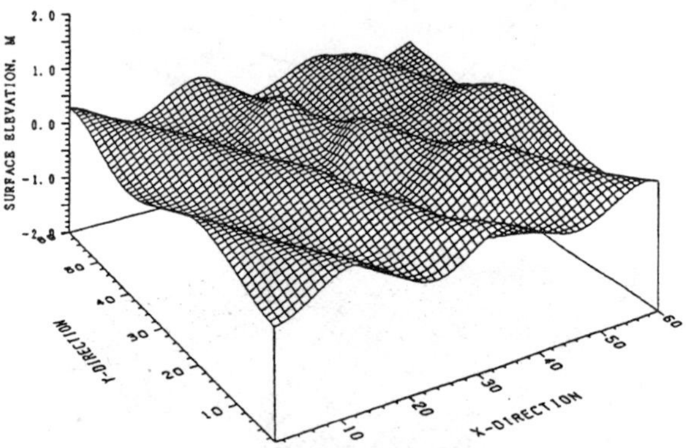

Fig. 7 Wave profile with mound and current (U/C_i = 0.1)

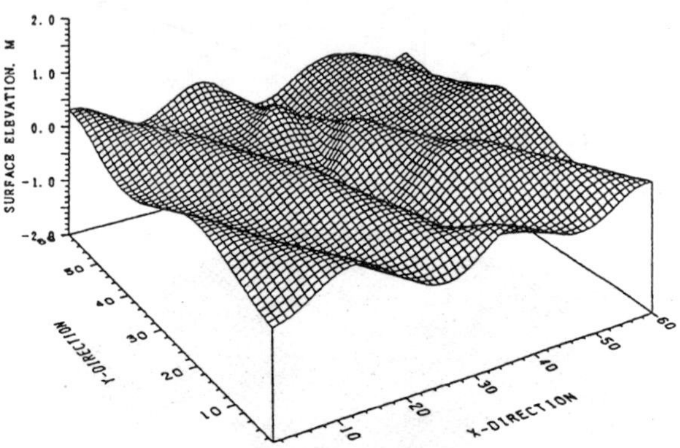

Fig. 8 Wave profile with mound and current (U/C_i = 0.14)

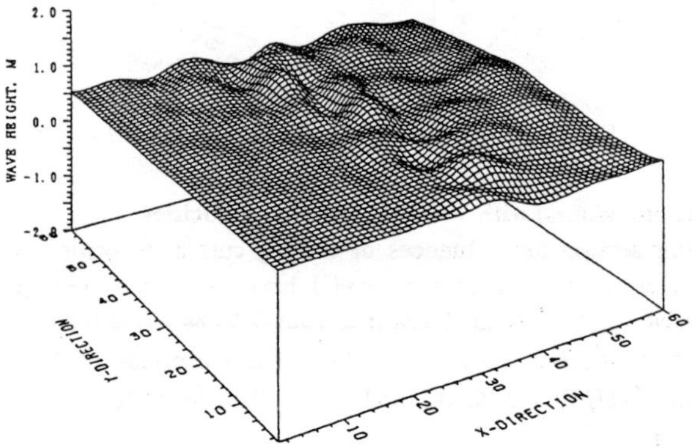

Fig. 9 Wave height distribution (U/C_i = 0.06)

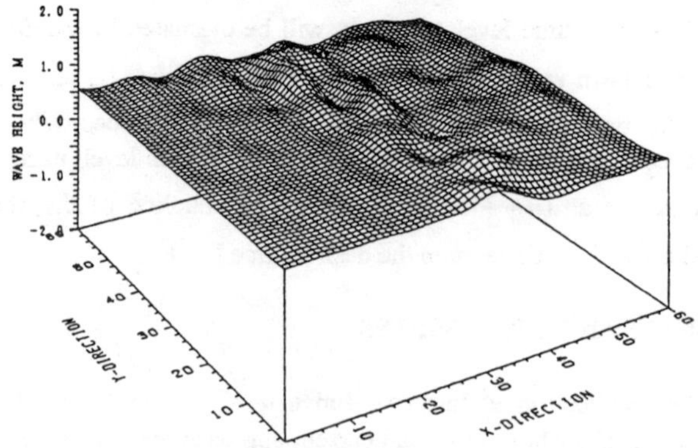

Fig. 10 Wave height distribution (U/C_i = 0.1)

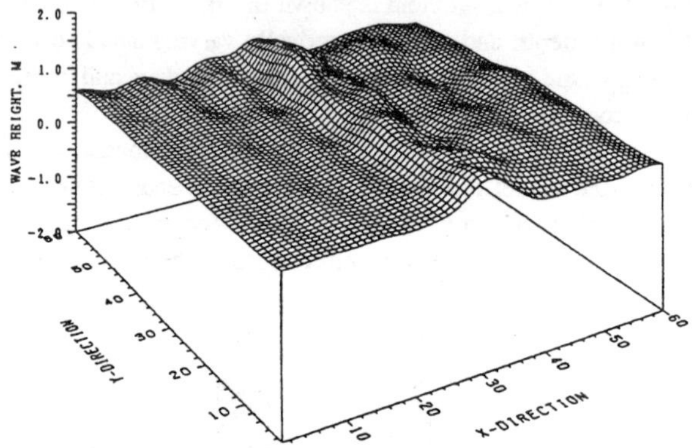

Fig. 11 Wave height distribution (U/C_i = 0.14)

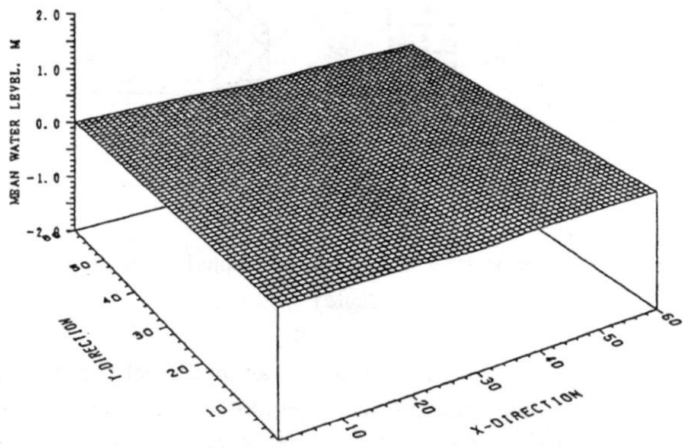

Fig. 12 Mean water level (U/C_i = 0.06)

Figs. 6, 7 and 8 display that the wave lengths increase along with distortion over the mound when the current velocity enlarges. Figs. 9, 10 and 11 indicate that the wave height over the mound is larger when the current velocity is stronger. On the other hand, Figs. 12, 13 and 14 exhibit that the drop of the mean water level ascent rapidly when the current velocity strengthens.

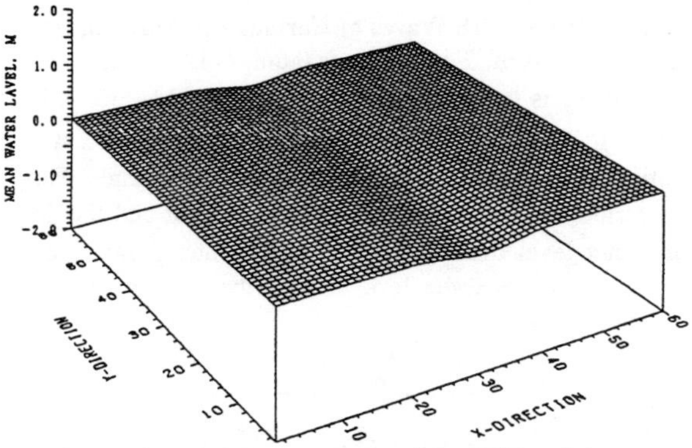

Fig. 13 Mean water level for U/C_i = 0.1

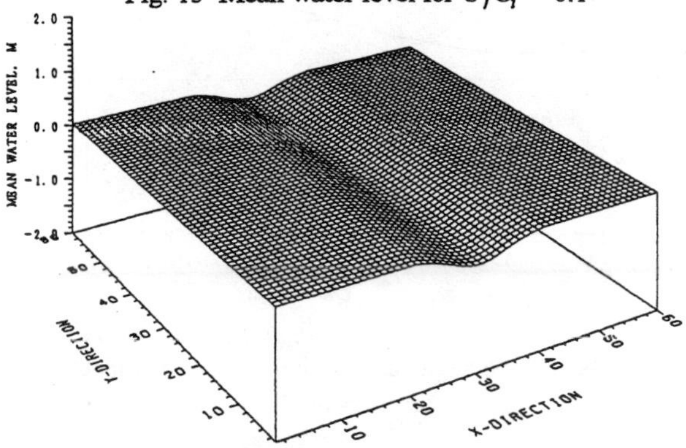

Fig. 14 Mean water level for U/C_i = 0.14

Wave and Current with Various Relative Mound Heights

In this section the effects of various relative mound heights on the fixed wave and current condition have been studied. In this event, it is assumed that h is 10m, T is 8s, H_i is 0.6m, U (U/C_i = 0.1) is 0.886m/s, θ_w is 30° and θ_c is 0° but r modifies to 0.4, 0.5 and 0.6.

Figs. 15, 16 and 17 represent that when incident wave and current conditions are fixed, the larger the relative mound height the greater the distortion in the wave profile over the mound.

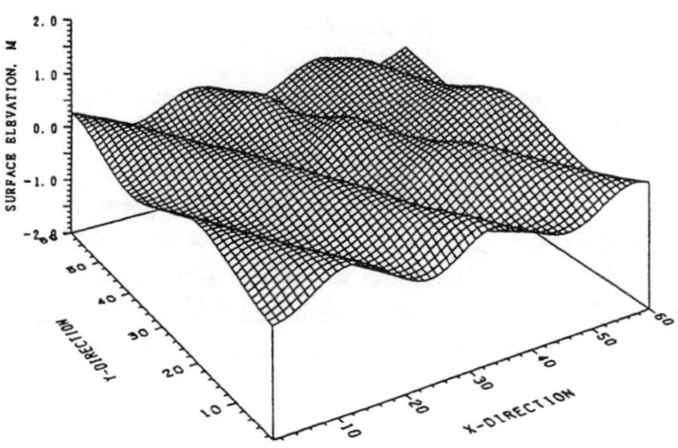

Fig. 15 Wave profile for r = 0.4

Figs. 18, 19 and 20 and, 21, 22 and 23 respectively describe that the increase of the relative mound height immensely contributes to the enlargement of the wave height and, drop of the mean water level over the mound.

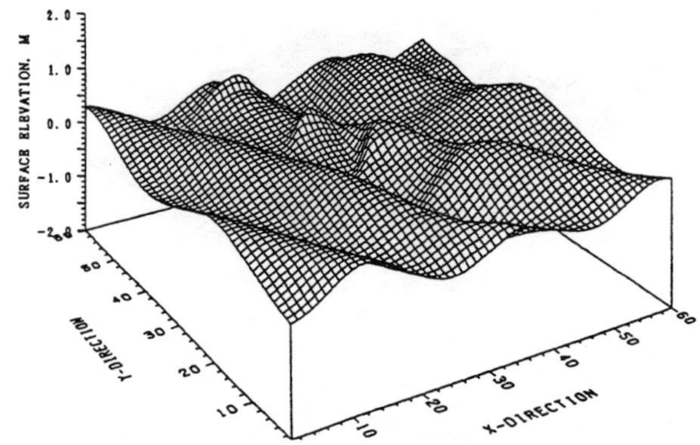

Fig. 16 Wave profile for r = 0.5

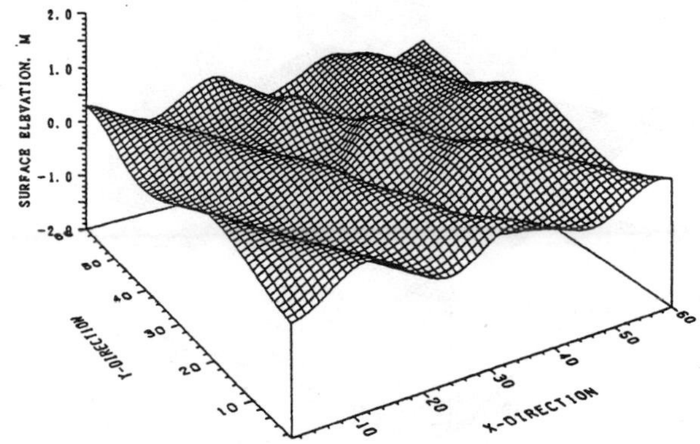

Fig. 17 Wave profile for r = 0.6

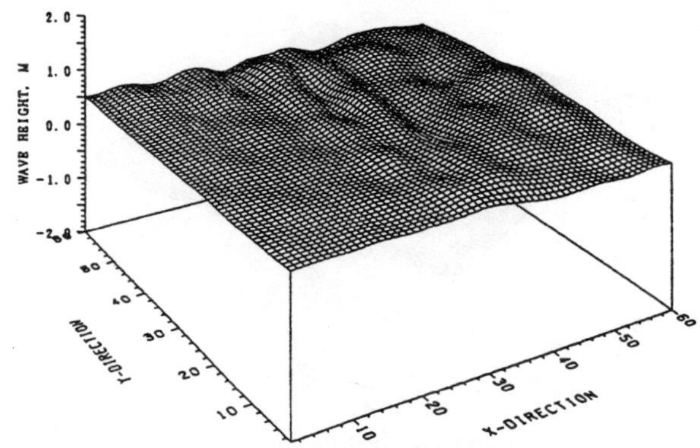

Fig. 18 Wave height distribution for r = 0.4

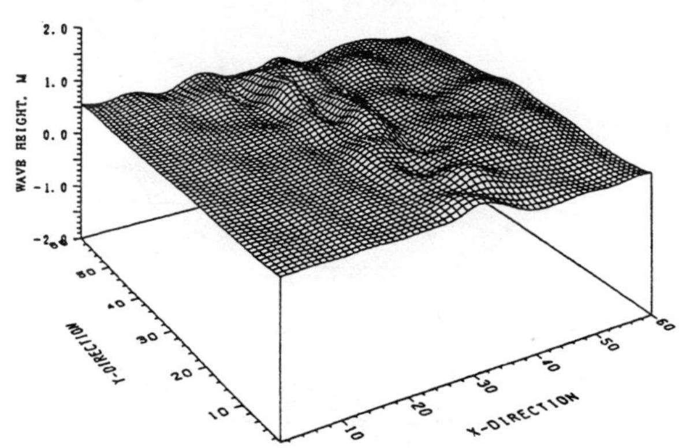

Fig. 19 Wave height distribution for r = 0.5

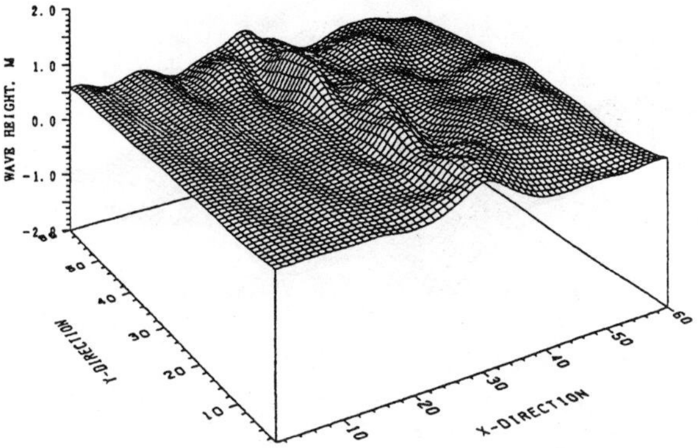

Fig. 20 Wave height distribution for $r = 0.6$

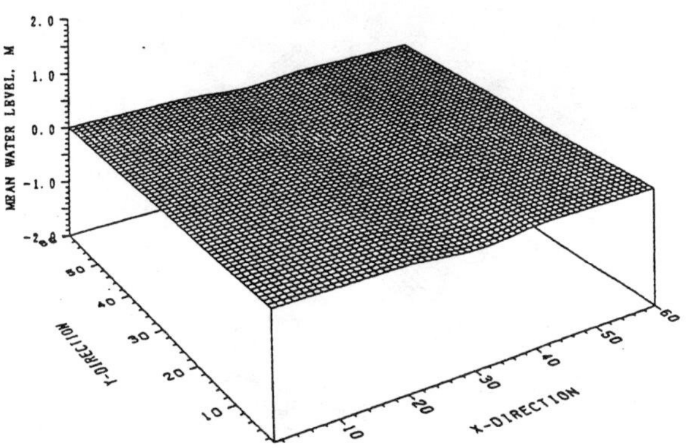

Fig. 21 Mean water level for $r = 0.4$

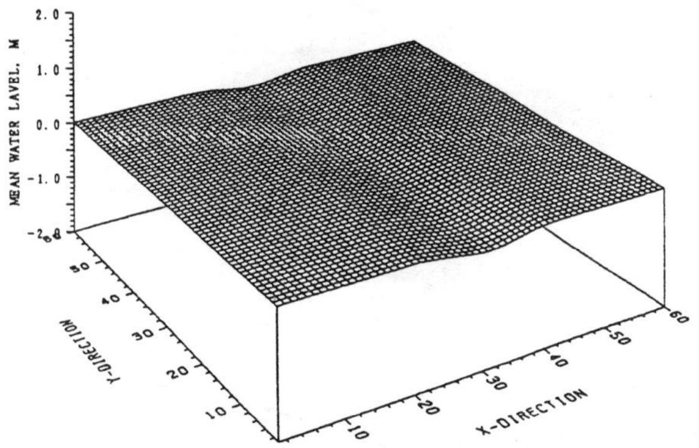

Fig. 22 Mean water level for $r = 0.5$

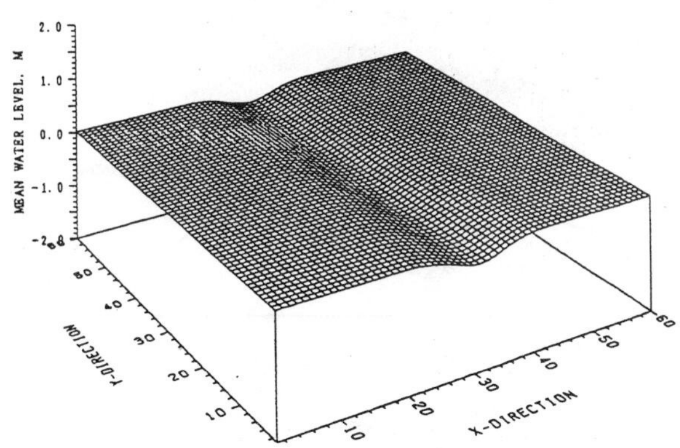

Fig. 23 Mean water level for $r = 0.6$

Current and Mound with Waves of Various Incident Angles

In this case, h is 10m, T is 8s, H_i is 0.6m, U is 0.886m/s, r is 0.5, l is 80m and θ_c is 0° but, θ_w alters from 0° to 15°, 30°, 45°, 60°, 75° and 90°. It is noticed that over the mound, the distortion in the wave profile enhances with the large incident wave angle. Figs. 24 and 25 describe the distribution of the wave height and the drop of the mean water level over the mound at its center point. The wave height H and the mean water level Δ have been normalized by the incident wave height H_i.

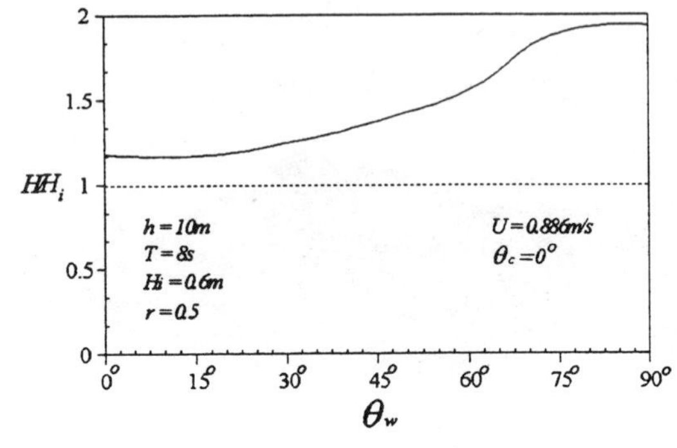

Fig. 24 Effects of wave angle on wave height (over the mound at its center point)

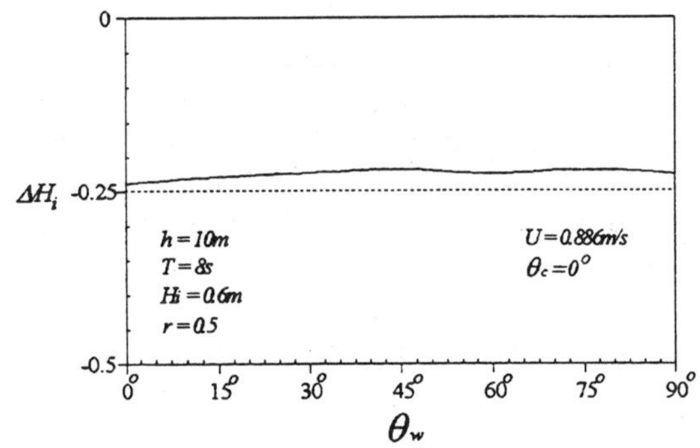

Fig. 25 Effects of wave angle on mean water level (over the mound at its center point)

Figs. 24 and 25 show that when the wave and current are in the same direction ($\theta_w = \theta_c = 0^\circ$), the increase of the wave height is the smallest and the drop of the mean water level is the largest over the mound. On the other hand, the wave height notably increases with the increase of the incident wave angle and the drop of the mean water level decreases for any inclination of the incident wave. It is beheld that the drop of the mean water level over the mound doesn't change significantly regardless of various incident wave angles.

CONCLUSIONS

The two dimensional model described here is fabricated through the vertical integration of the continuity equation and equations of motion. The discretization of the resulted equations are done following a semi-implicit technique which formulates a second order finite difference scheme and as an application, utilized to study the wave-current coexistence field over a submerged parabolic mound.

When the mound and current are not in presence, the wave profile is found to be uniform all over the domain. In the absence of mound it is noted that the wave length increases with current. When the current is not in presence, the wave profile is distorted or bent over the mound.

The average wave height and the drop of the mean water level over the mound increases with the increase of the current velocity.

It is discerned that the wave height and the drop of the mean water level grows rapidly over the mound when the relative mound height is large.

The variation of the incident wave angles (for a fixed current angle) also play significant role on the distribution of the wave height and mean water level over the mound.

REFERENCES

Dronkers, J. J. (1969). "Tidal Computations for Rivers, Coastal Areas and Seas," *J. Hydr. Div., Proc.* ASCE, Vol 95, No 1, pp 39-49.

Hara, M, T. Yasuda and Y. Sakakibara (1992). "Characteristics of a Solitary Wave Breaking Caused by a Submerged Obstacle," *Proc. 23rd Coast. Eng. Conf.*, pp 253-266.

Horikawa, K. (1988). "Nearshore Dynamics and Coastal Processes," University of Tokyo press, pp 522.

Kirby, J. T. (1984). "A Note on Linear Surface Wave-current Interaction over Slowly Varying Topography," *J. Geophys. Res.*, Vol 89, No C1, pp 745-747.

Liu, P. L. F. (1990). "Wave Transformation," *In The Sea* (ed. B. LeMehaute and D. M. Hanes), Wiley Interscience, pp 27-63.

Longuet-Higgins, M. S. and R. W. Stewart (1961). "The Changes in Amplitude of Short Gravity Waves on Steady Non-uniform Currents," *J. Fluid Mech.*, Vol 10, pp 529-549.

Longuet-Higgins, M. S. and R. W. Stewart (1962). "Radiation Stress and Mass Transport in Gravity Waves with Application to Surf Beats," *J. Fluid Mech.*, 13, pp 481-504.

Madsen, P. A. and O. R. Sϕrensen (1992). "A New Form of the Boussinesq Equations with Improved Linear Dispersion Characteristics. Part 2. A slowly Varying Bathymetry," *Coast. Eng., Elsevier Sci. Pub B. V.*, Vol 18, pp 183-204.

McCowan, A. D. (1987). "The Range of Application of Boussinesq Type Numerical Short Wave Models," *Proc. 22nd IAHR Congr., Laussane*, pp 378-384.

Ohyama, T. and K. Nadaoka (1992). "Modeling the Transformation of Nonlinear Waves Passing over a Submerged Dike," *Proc. 23rd Coast. Eng. Conf.*, pp 526-539.

Pudjaprasetya, S. R. and E. V. Groesen (1996). "Unidirectional Waves over Slowly Varying Bottom, Part II. Quasi-homogeneous Approximation of Distorting Waves," *Wave Motion, Elsevier*, Vol 23, No 1, pp 23-38.

Prüser, H. -H. and W. Zielke (1990). "Irregular Waves on a Current," *Proc. 22nd Coast. Eng. Conf.*, pp 1088-1101.

Sato, S and M. B. Kabiling (1994). "A Numerical Simulation of Beach Evolution Based on a Nonlinear Dispersive Wave-current Model," *Proc. 24th Coast. Eng. Conf.*, pp 2557-2570.

Sϕrensen, O. R., H. A. Schäffer, P. A. Madsen and R. Deigaard (1994). "Wave Breaking and Induced Nearshore Circulations," *Proc. 24th Coast. Eng. Conf.*, pp 2583-2594.

Tanaka, M. (1990). "Application of the Second-order Mode Coupling Equation to Coastal Engineering Problem," *Proc. 22nd Coast. Eng. Conf.*, pp 881-894.

Togashi, H., M. H. Zaman and X. Yu (1995). "Wave-current Interaction over a Submerged Mound of Trapezoidal Cross Section," Advances in Hydro-Science and-Engineering, II(Part B); *Proc. 2nd Int. Conf. on Hydro-Sci. and -Eng.*, CHES and IRTCES (ed.), Beijing, China, pp 1452-1459.

Watanabe, A., K. Murayama, T. Shimizu and T. Sakakiyama (1986). "Numerical Prediction Model of Three Dimensional Beach Deformation Around a Structure," *Coast. Eng. in Japan*, JSCE, Vol 29, pp 179-194.

Watanabe, A., K. Shiba and M. Isobe (1994). "A Numerical Model of Beach Change Due to Sheet-flow," *Proc. 24th Coast. Eng. Conf.*, pp 2785-2798.

Zaman, M. H., H. Togashi and X. Yu (1994). "Modeling Wave Current Interaction over an Uneven Sea Bottom," *Proc. Civil. Eng. in the Ocean*, JSCE, Vol 10, pp 25-30 (in Japanese).

Zaman, M. H. and H. Togashi (1996a). "Effects of Bottom Configurations on Wave-current Field," *Rep. the Faculty of Eng., Nagasaki University*, Vol 26, No 47, pp 317-321.

Zaman, M. H. and H. Togashi (1996b). "Experimental Study on Interaction among Waves, Currents and Bottom Topography," *Proc. Civil. Eng. in the Ocean*, JSCE, Vol 12, pp 49-54 (in Japanese).

Proceedings of the Seventh (1997) International Offshore and Polar Engineering Conference
Honolulu, USA, May 25-30, 1997

ISBN 1-880653-28-1 (Set); ISBN 1-880653-31-1 (Vol. III)

A Numerical Method for Solving Second Order Diffraction Problem by Using Green's Identity Formula

Keisuke Murakami, A. Yoshida and I. Irie
Kyushu University
Fukuoka, Japan

ABSTRACT

This paper presents a new approximate calculation method for three-dimensional nonlinear interactions between Stokes waves and submerged structures. In the mathematical formulations, Green's Identity formula is used to express the distributions of the velocity potential on a horizontal plane. A collocation method of matched eigenfunction expansion is also utilized to derive the linear relationships for the velocity potential and its normal derivative on the adjacent fluid regions. The numerical convergence of the second order solution is investigated and the validity of the present method is also investigated through wave tank experiments.

Key Words: Stokes wave, Green's Identity formula, collocation method of matched eigenfunction expansion

INTRODUCTION

Numerical calculation method based on the Haskind's relationship has been widely applied to the estimation of the second order wave forces acting on the offshore structures (e.g., Molin,1979; Rahman,1984; Etock Taylor,1987; Williams,1990). This method is efficient because the second order wave forces can be computed without solving the second order potential problem. However, it is impossible to apply this method to the computation of the second order wave fields around the structures and wave pressure distributions on them.

The analytical solution of the second order diffraction problem for a three dimensional structure have been studied (e.g., Kim,1989; Kriebel,1990; Mizutani,1996). These solutions can be applied only for the single circular cylinder or an axisymmetric body, and thus it is impossible to apply these solutions to the array of structures with arbitrary cross section.

In this paper, we propose the numerical method for the second order interaction between Stokes waves and submerged structures or floating ones with arbitrary cross section. The method is based on the Green's Identity Formula. The collocation method of matched eigenfunction expansion is used to simplify the mathematical formulation. The relationships between the truncation numbers of the series of eigenfunction expansion and the numbers of the collocation point are investigated to obtain the second order results with good accuracy. The validity of the present method is also investigated by comparing numerical results with experimental ones.

GENERAL FORMULATION

Fig–1 shows a coordinate system. x and y are the horizontal axes and z the vertical axis taken upward from the still water surface. An impermeable horizontal plate with a submerged depth of $qh(0 \leq q \leq 1)$ is placed in the uniform water depth h. The incident wave with the first order amplitude ζ_0 and frequency σ propagates with an incident angle θ measured from the positive x direction. The fluid region is divided into three subregions by the horizontal plate and imaginary boundary as shown in **Fig–1**. It is assumed that the fluid is both inviscid and incompressible, and its motion is irrotational. The velocity potential $\Phi(x, y, z, t)$ satisfies following Laplace equation in each subregion.

$$\frac{\partial^2 \Phi_i}{\partial x^2} + \frac{\partial^2 \Phi_i}{\partial y^2} + \frac{\partial^2 \Phi_i}{\partial z^2} = 0 \tag{1}$$

where the subscript $i(i = 1, 2, 3)$ means the number of the subregion.

The kinematic and dynamic free surface boundary conditions are written as follows:

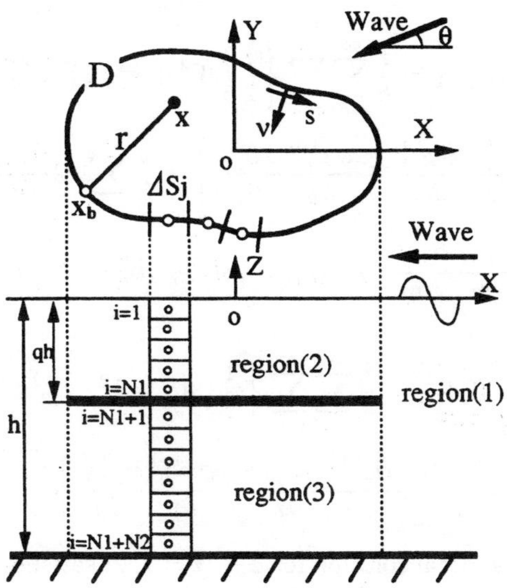

Fig.-1: Definition sketch

$$\frac{\partial \zeta_i}{\partial t} - \frac{\partial \Phi_i}{\partial z} + \frac{\partial \Phi_i}{\partial x}\frac{\partial \zeta_i}{\partial x} + \frac{\partial \Phi_i}{\partial y}\frac{\partial \zeta_i}{\partial y} = 0 \quad (2)$$

$$\frac{\partial \Phi_i}{\partial t} + g\zeta_i + \frac{1}{2}\left\{\left(\frac{\partial \Phi_i}{\partial x}\right)^2 + \left(\frac{\partial \Phi_i}{\partial y}\right)^2 + \left(\frac{\partial \Phi_i}{\partial z}\right)^2\right\} = Q_i \quad (3)$$

where g means a gravity acceleration. ζ_i and Q_i mean a water surface elevation and Bernoulli constant respectively.

The perturbation method is employed in order to obtain the first and second order boundary value problems. The velocity potential Φ_i, the water surface elevation ζ_i and the constant value Q_i are expressed in power series by introducing a parameter ϵ, where $\epsilon = k\zeta_0$ and k means a first order incident wave number.

$$\Phi_i(x,y,z,t) = \frac{g}{k\sigma}Re\left[\epsilon\phi_{1i}(x,y,z)\exp(-i\sigma t) + \epsilon^2\left\{\phi_{2i}^{(0)}(x,y,z) + \phi_{2i}^{(2)}(x,y,z)\exp(-i2\sigma t)\right\}\right] \quad (4)$$

$$\zeta_i(x,y,t) = \frac{1}{k}Re\left[\epsilon\eta_{1i}(x,y)\exp(-i\sigma t) + \epsilon^2\left\{\eta_{2i}^{(0)}(x,y) + \eta_{2i}^{(2)}(x,y)\exp(-i2\sigma t)\right\}\right] \quad (5)$$

$$Q_i(t) = \frac{g}{k}Re\left[\epsilon Q_{1i}(t) + \epsilon^2 Q_{2i}(t)\right] \quad (6)$$

where the exponent i means a pure imaginary number $\sqrt{-1}$. ϕ_{1i} means the first order potential function, and $\phi_{2i}^{(0)}$ and $\phi_{2i}^{(2)}$ mean the second order potential function. These functions satisfy the Laplace equation.

By expanding both Eqs.(2) and (3) in Taylor series about $z = 0$ and substituting Eqs.(4),(5) and (6) into them, the first and second order combined free surface boundary conditions are obtained as follows:

$$\frac{\partial \phi_{1i}}{\partial z} - \Gamma\phi_{1i} = 0 \quad (7)$$

$$\frac{\partial \phi_{2i}^{(2)}}{\partial z} - 4\Gamma\phi_{2i}^{(2)} = \frac{\eta_{1i}}{2k}\frac{\partial}{\partial z}\left\{\Gamma\phi_{1i} - \frac{\partial \phi_{1i}}{\partial z}\right\} + \frac{i}{k}\left\{\left(\frac{\partial \phi_{1i}}{\partial x}\right)^2 + \left(\frac{\partial \phi_{1i}}{\partial y}\right)^2 + \left(\frac{\partial \phi_{1i}}{\partial z}\right)^2\right\} \quad (8)$$

$$\frac{\partial \phi_{2i}^{(0)}}{\partial z} = \frac{\eta_{1i}}{2k}\frac{\partial}{\partial z}\left\{\Gamma\phi_{1i}^* - \frac{\partial \phi_{1i}^*}{\partial z}\right\} \quad (9)$$

where $\Gamma = \sigma^2/g$, and ϕ^* means the conjugate function.

FIRST ORDER BOUNDARY VALUE PROBLEM

The first order potential functions ϕ_{11} in subregion (1) and ϕ_{12} in subregion (2) satisfy the free surface boundary condition written in Eq.(7) and impermeable bottom boundary condition $\partial\phi_{1i}/\partial z = 0$ ($i = 1, 2$). The potential function ϕ_{13} satisfies the impermeable boundary conditions on the upper face and lower face of subregion (3). By using the eigenfunction expansion, the general solutions of $\phi_{1i}(i = 1, 2, 3)$ are given as follows (Ijima,1974):

$$\phi_{11}(x,y,z) = \varphi_{1I}(x,y)Z_1(z,k_0) + \sum_{n=0}^{\infty}\varphi_{11}^{(n)}(x,y)Z_1(z,k_n) \quad (10)$$

$$\phi_{12}(x,y,z) = \sum_{m=0}^{\infty}\varphi_{12}^{(m)}(x,y)Z_2(z,k_m^{'}) \quad (11)$$

$$\phi_{13}(x,y,z) = \sum_{s=0}^{\infty}\varphi_{13}^{(s)}(x,y)Z_3(z,s) \quad (12)$$

where φ_{1I} in Eq.(10) means an incident plain wave expressed as

$$\varphi_{1I}(x,y) = -i\exp\{-ik(x\cos\theta + y\sin\theta)\} \quad (13)$$

The eigenfunctions $Z_1(z,k_n)$, $Z_2(z,k_m^{'})$ and $Z_3(z,s)$ express the vertical distribution of the potential function and each function is expressed as follows:

$$\left.\begin{aligned} Z_1(z,k_n) &= \frac{\cos k_n(z+h)}{\cos k_n h} & (n = 0,1,2,\cdots) \\ Z_2(z,k_m^{'}) &= \frac{\cos k_m^{'}(z+qh)}{\cos k_m^{'} qh} & (m = 0,1,2,\cdots) \\ Z_3(z,s) &= \cos\frac{s\pi(z+qh)}{(1-q)h} & (s = 0,1,2,\cdots) \end{aligned}\right\} \quad (14)$$

k_n and $k_m^{'}$ are defined as the complex number when $n = 0$ and $m = 0$ ($k_0 = ik$ and $k_0^{'} = ik^{'}$), and also defined as the real number when $n \neq 0$ and $m \neq 0$. These values are given as a solution of Eq.(15).

$$\left.\begin{aligned} \frac{\sigma^2 h}{g} &= -k_n h \tan k_n h \qquad (n = 0, 1, \cdots) \\ \frac{\sigma^2 qh}{g} &= -k_m^{'} qh \tan k_m^{'} qh \quad (m = 0, 1, \cdots) \end{aligned}\right\} \qquad (15)$$

SECOND ORDER BOUNDARY VALUE PROBLEM

From Eqs.(1) and (4), the second order potential functions $\phi_{2i}^{(2)}$ and $\phi_{2i}^{(0)}$ $(i = 1, 2, 3)$ satisfy the Laplace equation in each subregion. $\phi_{2i}^{(2)}$ and $\phi_{2i}^{(0)}$ $(i = 1, 2)$ must satisfy the nonhomogeneous free surface boundary conditions as written in Eqs.(8) and (9), and the impermeable bottom boundary condition. On the other hand, $\phi_{23}^{(2)}$ and $\phi_{23}^{(0)}$ in the subregion (3) must satisfy the impermeable boundary conditions on the upper face and lower face of the subregion.

The time independent component $\phi_{2i}^{(0)}$ in the second order potential function does not contribute to the second order wave forces and water surface elevations. Since the main purpose of this study is to obtain the second order wave forces and wave fields around the structures, we only consider the time dependent component $\phi_{2i}^{(2)}$ in the following formulations and simplify the $\phi_{2i}^{(2)}$ as ϕ_{2i} .

The free surface boundary conditions in subregion (1) and (2) are nonhomogeneous because quadratic forcing terms appear in the right hand side of Eq.(8). The solution in each subregion may be obtained by separating the potential function ϕ_{2i} $(i = 1, 2)$ into the particular solution ϕ_{Li} and the general solution ϕ_{Fi} as follows.

$$\phi_{21}(x, y, z) = \phi_{2I}(x, y, z) + \phi_{F1}(x, y, z) + \phi_{L1}(x, y, z) \quad (16)$$

$$\phi_{22}(x, y, z) = \phi_{F2}(x, y, z) + \phi_{L2}(x, y, z) \quad (17)$$

ϕ_{Li} and ϕ_{Fi} satisfy the Laplace equation respectively. ϕ_{Li} satisfies the nonhomogeneous free surface boundary condition written as Eq.(8), and ϕ_{Fi} satisfies the homogeneous free surface boundary condition.

ϕ_{2I} in Eq.(16) means a potential function of the second order incident plain wave and it is expressed as

$$\left.\begin{aligned} \phi_{2I}(x, y, z) &= \varphi_{2I}(x, y) Z_1(z, i2k) \\ \varphi_{2I}(x, y) &= -\frac{i3}{8}\frac{\Gamma}{k}\frac{\cosh 2kh}{\sinh^4 kh} e^{-i2k(x\cos\theta + y\sin\theta)} \end{aligned}\right\} \quad (18)$$

By substituting Eqs.(10),(11),(16) and (17) into Eq.(8), the free surface boundary conditions for ϕ_{Fi} and ϕ_{Li} are obtained as follows:

$$\frac{\partial \phi_{Fi}}{\partial z} - 4\Gamma \phi_{Fi} = 0 \quad (i = 1, 2) \quad (19)$$

$$\frac{\partial \phi_{L1}}{\partial z} - 4\Gamma\phi_{L1} = \frac{i}{k}\sum_{p=0}^{\infty}\sum_{q=0}^{\infty}\left\{\frac{1}{2}\left(3\Gamma^2 + k_p^2\right)\varphi_{11}^{(p)}\varphi_{11}^{(q)} + \frac{\partial\varphi_{11}^{(p)}}{\partial x}\frac{\partial\varphi_{11}^{(q)}}{\partial x} + \frac{\partial\varphi_{11}^{(p)}}{\partial y}\frac{\partial\varphi_{11}^{(q)}}{\partial y}\right\} + \frac{i}{k}\sum_{p=0}^{\infty}\left[2\frac{\partial\varphi_{1I}}{\partial x}\frac{\partial\varphi_{11}^{(p)}}{\partial x} + 2\frac{\partial\varphi_{1I}}{\partial y}\frac{\partial\varphi_{11}^{(p)}}{\partial y} + \varphi_{1I}\varphi_{11}^{(p)}\left\{3\Gamma^2 + (k_p^2 - k^2)\right\}\right] \quad (20)$$

$$\frac{\partial \phi_{L2}}{\partial z} - 4\Gamma\phi_{L2} = \frac{i}{k}\sum_{p=0}^{\infty}\sum_{q=0}^{\infty}\left\{\frac{1}{2}\left(3\Gamma^2 + k_p^2\right)\varphi_{12}^{(p)}\varphi_{12}^{(q)} + \frac{\partial\varphi_{12}^{(p)}}{\partial x}\frac{\partial\varphi_{12}^{(q)}}{\partial x} + \frac{\partial\varphi_{12}^{(p)}}{\partial y}\frac{\partial\varphi_{12}^{(q)}}{\partial y}\right\} \quad (21)$$

The general solution for ϕ_{Fi} $(i = 1, 2)$ satisfies homogeneous free surface boundary condition, impermeable bottom boundary condition and Sommerfeld's radiation condition. The solution for ϕ_{Fi} can be expressed as follows in the same manner as the first order solution.

$$\phi_{F1}(x, y, z) = \sum_{n=0}^{\infty}\varphi_{21}^{(n)}(x, y) Z_1(z, k_n^{(2)}) \quad (22)$$

$$\phi_{F2}(x, y, z) = \sum_{m=0}^{\infty}\varphi_{22}^{(m)}(x, y) Z_2(z, k_m^{'(2)}) \quad (23)$$

$k_n^{(2)}$ and $k_m^{'(2)}$ are defined as the complex number when $n = 0$ and $m = 0$, and also defined as the real number when $n \neq 0$ and $m \neq 0$. These values are given as a solution of Eq.(24).

$$\left.\begin{aligned} \frac{(2\sigma)^2 h}{g} &= -k_n^{(2)} h \tan k_n^{(2)} h \qquad (n = 0, 1, \cdots) \\ \frac{(2\sigma)^2 qh}{g} &= -k_m^{'(2)} qh \tan k_m^{'(2)} qh \quad (m = 0, 1, \cdots) \end{aligned}\right\} \quad (24)$$

On the other hand, ϕ_{Li} $(i = 1, 2)$ must satisfy the nonhomogeneous free surface boundary condition, impermeable bottom boundary condition and Sommerfeld's radiation condition simultaneously. Though the particular solution of ϕ_{Li} is shown by Kim(1989), Kriebel(1990) and Mizutani(1996), these solutions cannot be applied to the submerged structures or floating ones directly. In this study, it is assumed that ϕ_{Li} can be approximately expressed as Eqs.(25) and (26) by the product of the eigenfunction and the function which expresses the nonhomogeneous term of the free surface boundary condition(Sabuncu, 1985).

$$\phi_{L1}(x, y, z) = \sum_{p=0}^{\infty}\sum_{q=0}^{\infty}\gamma_{pq} f_{pq}(x, y) Z_1(z, k_p + k_q) \quad (25)$$

$$\phi_{L2}(x,y,z) = \sum_{p=0}^{\infty}\sum_{q=0}^{\infty} \gamma'_{pq} f'_{pq}(x,y) Z_2(z, k'_p + k'_q) \quad (26)$$

where $\gamma_{pq}, \gamma'_{pq}$: f_{pq} and f'_{pq} are expressed as follows.

$$\gamma_{pq} = -\frac{i}{k}\frac{1}{(k_p+k_q)\tan(k_p+k_q)h + 4\Gamma} \quad (27)$$

$$\left.\begin{aligned}
f_{00}(x,y) &= \frac{1}{2}\left(3\Gamma^2 - k^2\right)\left(\varphi_{11}^{(0)} + 2\varphi_{1I}\right)\varphi_{11}^{(0)} \\
&+ \frac{\partial \varphi_{11}^{(0)}}{\partial x}\left(2\frac{\partial \varphi_{1I}}{\partial x} + \frac{\partial \varphi_{11}^{(0)}}{\partial x}\right) \\
&+ \frac{\partial \varphi_{11}^{(0)}}{\partial y}\left(2\frac{\partial \varphi_{1I}}{\partial y} + \frac{\partial \varphi_{11}^{(0)}}{\partial y}\right) \quad (p=0,\ q=0) \\
f_{p0}(x,y) &= f_{0p}(x,y) = \\
&\frac{1}{2}\left\{3\Gamma^2 + \frac{1}{2}\left(k_p^2 - k^2\right)\right\}\left(\varphi_{11}^{(0)} + \varphi_{1I}\right)\varphi_{11}^{(p)} \\
&+ \frac{\partial \varphi_{11}^{(p)}}{\partial x}\left(\frac{\partial \varphi_{1I}}{\partial x} + \frac{\partial \varphi_{11}^{(0)}}{\partial x}\right) \\
&+ \frac{\partial \varphi_{11}^{(p)}}{\partial y}\left(\frac{\partial \varphi_{1I}}{\partial y} + \frac{\partial \varphi_{11}^{(0)}}{\partial y}\right) \quad (p \neq 0) \\
f_{pq}(x,y) &= \frac{1}{2}\left(3\Gamma^2 + k_q^2\right)\varphi_{11}^{(p)}\varphi_{11}^{(q)} \\
&+ \frac{\partial \varphi_{11}^{(p)}}{\partial x}\frac{\partial \varphi_{11}^{(q)}}{\partial x} + \frac{\partial \varphi_{11}^{(p)}}{\partial y}\frac{\partial \varphi_{11}^{(q)}}{\partial y} \quad (p \neq 0,\ q \neq 0)
\end{aligned}\right\} \quad (28)$$

$$\gamma'_{pq} = -\frac{i}{k}\frac{1}{(k'_p+k'_q)\tan(k'_p+k'_q)qh + 4\Gamma} \quad (29)$$

$$\begin{aligned}
f'_{pq}(x,y) &= \frac{1}{2}\left(3\Gamma^2 + k'^2_q\right)\varphi_{12}^{(p)}\varphi_{12}^{(q)} \\
&+ \frac{\partial \varphi_{12}^{(p)}}{\partial x}\frac{\partial \varphi_{12}^{(q)}}{\partial x} + \frac{\partial \varphi_{12}^{(p)}}{\partial y}\frac{\partial \varphi_{12}^{(q)}}{\partial y}
\end{aligned} \quad (30)$$

The potential function ϕ_{23} in the subregion (3) is expressed as Eq.(31) in the same manner as Eq.(12).

$$\phi_{23}(x,y,z) = \sum_{s=0}^{\infty} \varphi_{23}^{(s)}(x,y) Z_3(z,s) \quad (31)$$

GREEN'S IDENTITY FORMULA

As shown in **Fig-1**, D is defined as a boundary of subregion (2), and ν as a unit vector normal to the boundary. $\mathbf{X}$ and $\mathbf{X_b}$ are the coordinates taken in the fluid region and on the boundary respectively. r is the distance between $\mathbf{X}$ and $\mathbf{X_b}$.

The first order potential functions φ_{1i} (i =1,2,3) and second order potential functions ϕ_{F1}, ϕ_{F2} and ϕ_{23} satisfy the Laplace equation, and the functions φ_{1i} and φ_{2i} (i =1,2,3) satisfy the following Helmholtz equations.

$$\nabla^2 \varphi_{\ell 1}^{(n)}(x,y) + \lambda_n^2 \varphi_{\ell 1}^{(n)}(x,y) = 0 \quad (32)$$

$$\nabla^2 \varphi_{\ell 2}^{(m)}(x,y) + \lambda'^2_m \varphi_{\ell 2}^{(m)}(x,y) = 0 \quad (33)$$

$$\nabla^2 \varphi_{\ell 3}^{(s)}(x,y) + \bar{s}^2 \varphi_{\ell 3}^{(s)}(x,y) = 0 \quad (34)$$

where ℓ ($\ell = 1,2$) means an order of ϵ and $\bar{s} = s\pi/(1-q)h$. λ_n and λ'_m equal to k_n and k'_m respectively in the first order, and also equal to $k_n^{(2)}$ and $k'^{(2)}_m$ in the second order.

The plain distribution of $\varphi_{\ell 1}^{(n)}, \varphi_{\ell 2}^{(m)}$ and $\varphi_{\ell 3}^{(s)}$ in each subregion can be expressed by means of the boundary integral equation(Green's Identity Formula) as follows by applying the Green's Theorem and Sommerfeld's radiation condition respectively.

$$\begin{aligned}
\varphi_{\ell 1}^{(n)}(X) = \alpha \int_D \Big\{ & \varphi_{\ell 1}^{(n)}(X_b)\frac{\partial}{\partial \nu}G(\lambda_n r) \\
& - G(\lambda_n r)\frac{\partial}{\partial \nu}\varphi_{\ell 1}^{(n)}(X_b)\Big\}\, ds
\end{aligned} \quad (35)$$

$$\begin{aligned}
\varphi_{\ell 2}^{(m)}(X) = \alpha \int_D \Big\{ & \varphi_{\ell 2}^{(m)}(X_b)\frac{\partial}{\partial \nu}G(\lambda'_m r) \\
& - G(\lambda'_m r)\frac{\partial}{\partial \nu}\varphi_{\ell 2}^{(m)}(X_b)\Big\}\, ds
\end{aligned} \quad (36)$$

$$\begin{aligned}
\varphi_{\ell 3}^{(s)}(X) = \alpha \int_D \Big\{ & \varphi_{\ell 3}^{(s)}(X_b)\frac{\partial}{\partial \nu}E(\bar{s}r) \\
& - E(\bar{s}r)\frac{\partial}{\partial \nu}\varphi_{\ell 3}^{(s)}(X_b)\Big\}\, ds
\end{aligned} \quad (37)$$

where $r = |\mathbf{X} - \mathbf{X_b}|$. $\alpha = -i/2$ when $\mathbf{X}$ is taken on the boundary D and $\alpha = -i/4$ when $\mathbf{X}$ is taken in each subregion.

In Eq.(35), $G(\lambda_n r)$ means a Hankel function of the first kind $H_0^{(1)}(\lambda_0 r)$ when $n = 0$, and a modified Bessel function of the second kind $K_0(k_n r)$ when $n \neq 0$. $G(\lambda'_m r)$ in Eq.(36) has a same meaning as $G(\lambda_n r)$. $E(\bar{s}r)$ means $\log \lambda_0 r$ when $s = 0$ and a Bessel function of the second kind $K_0(\bar{s}r)$ when $s \neq 0$.

COLLOCATION METHOD OF MATCHED EIGENFUNCTION EXPANSION

From the continuity condition of the velocity and pressure on the imaginary boundary between adjacent fluid regions, the potential functions must satisfy the following equations.

$$\phi_{\ell 1} = \begin{cases} \phi_{\ell 2} & (-qh < z < 0) \\ \phi_{\ell 3} & (-h < z < -qh) \end{cases} \quad (\ell = 1,2) \quad (38)$$

$$\frac{\partial\phi_{\ell 1}}{\partial\nu} = \begin{cases} \dfrac{\partial\phi_{\ell 2}}{\partial\nu} & (-qh < z < 0) \\ \dfrac{\partial\phi_{\ell 3}}{\partial\nu} & (-h < z < -qh) \end{cases} \quad (\ell = 1,2) \quad (39)$$

In the conventional formulation, the linear relationships for the potential function and its normal derivative on the imaginary boundaries are derived by integrating Eq.(38) and Eq.(39) about z with using the orthogonality of $Z_i(i=1,2,3)$. These integral procedures are very complicated even in the first order problem and it is not practical to apply this method to the second order problem. In this study, the collocation method of matched eigenfunction expansion is employed and the linear relationships for the potential function and its derivatives are derived (Yoshida, 1992).

As shown in **Fig-1**, the integral boundary D is divided into small line elements $\Delta S_j (j = 1 \sim N)$. The vertical strips on each ΔS_j are also divided into small surface elements of N_1 number on subregion (2) and N_2 number on subregion (3). Eqs.(38) and (39) are applied on the collocation point which is taken at the midpoint of each surface element. In accordance with the above procedures, the linear relationships for the second order potential function and its normal derivative are derived as follows:

$$\sum_{m=0}^{\infty} \varphi_{22}^{(m)}(i) Z_2(z_j, k_m^{'(2)}) - \sum_{n=0}^{\infty} \varphi_{21}^{(n)}(i) Z_1(z_j, k_n^{(2)}) = \varphi_{2I}(i) Z_1(z_j, 2ik) + \sum_{p=0}^{\infty}\sum_{q=0}^{\infty} \gamma_{pq} f_{pq}(i) Z_1(z_j, k_p + k_q) - \sum_{p=0}^{\infty}\sum_{q=0}^{\infty} \gamma'_{pq} f'_{pq}(i) Z_2(z_j, k'_p + k'_q)$$
$$(i = 1 \sim N) \quad (j = 1 \sim N_1) \quad (40)$$

$$\sum_{s=0}^{\infty} \varphi_{23}^{(s)}(i) Z_3(z_j, s) - \sum_{n=0}^{\infty} \varphi_{21}^{(n)}(i) Z_1(z_j, k_n^{(2)}) = \varphi_{2I}(i) Z_1(z_j, 2ik) + \sum_{p=0}^{\infty}\sum_{q=0}^{\infty} \gamma_{pq} f_{pq}(i) Z_1(z_j, k_p + k_q)$$
$$(i = 1 \sim N) \quad (j = N_1 + 1 \sim N_1 + N_2) \quad (41)$$

$$\sum_{m=0}^{\infty} \overline{\varphi_{22}^{(m)}}(i) Z_2(z_j, k_m^{'(2)}) - \sum_{n=0}^{\infty} \overline{\varphi_{21}^{(n)}}(i) Z_1(z_j, k_n^{(2)}) = \overline{\varphi_{2I}}(i) Z_1(z_j, 2ik) + \sum_{p=0}^{\infty}\sum_{q=0}^{\infty} \gamma_{pq} \overline{f_{pq}}(i) Z_1(z_j, k_p + k_q) - \sum_{p=0}^{\infty}\sum_{q=0}^{\infty} \gamma'_{pq} \overline{f'_{pq}}(i) Z_2(z_j, k'_p + k'_q)$$
$$(i = 1 \sim N) \quad (j = 1 \sim N_1) \quad (42)$$

$$\sum_{s=0}^{\infty} \overline{\varphi_{23}^{(s)}}(i) Z_3(z_j, s) - \sum_{n=0}^{\infty} \overline{\varphi_{21}^{(n)}}(i) Z_1(z_j, k_n^{(2)}) = \overline{\varphi_{2I}}(i) Z_1(z_j, 2ik) + \sum_{p=0}^{\infty}\sum_{q=0}^{\infty} \gamma_{pq} \overline{f_{pq}}(i) Z_1(z_j, k_p + k_q)$$
$$(i = 1 \sim N) \quad (j = N_1 + 1 \sim N_1 + N_2) \quad (43)$$

where $\overline{\varphi}(j)$ means a normal derivative of $\varphi(j)$.

On the other hand, the Green's Identity Formulas Eqs.(35),(36) and (37) are discretized as Eqs.(44),(45) and (46). In these equations, the coordinates $\mathbf{X}$ and $\mathbf{X_b}$ are taken on the line elements, and $\varphi(j)$ and $\overline{\varphi}(j)$ are assumed to be constant on them.

$$\sum_{j=1}^{N} \left(\overline{G_{ij}^{(n)}} - \delta_{ij}\right) \varphi_{\ell 1}^{(n)}(j) - \sum_{j=1}^{N} G_{ij}^{(n)} \overline{\varphi_{\ell 1}^{(n)}}(j) = 0 \quad (n = 0, 1, 2, \cdots n_\ell) \quad (44)$$

$$\sum_{j=1}^{N} \left(\overline{G_{ij}^{'(m)}} - \delta_{ij}\right) \varphi_{\ell 2}^{(m)}(j) - \sum_{j=1}^{N} G_{ij}^{'(m)} \overline{\varphi_{\ell 2}^{(m)}}(j) = 0 \quad (m = 0, 1, 2, \cdots m_\ell) \quad (45)$$

$$\sum_{j=1}^{N} \left(\overline{E_{ij}^{(s)}} - \delta_{ij}\right) \varphi_{\ell 3}^{(s)}(j) - \sum_{j=1}^{N} E_{ij}^{(s)} \overline{\varphi_{\ell 3}^{(s)}}(j) = 0 \quad (s = 0, 1, 2, \cdots s_\ell) \quad (46)$$

n_ℓ, m_ℓ and $s_\ell (\ell = 1, 2)$ mean truncation numbers of series of eigenfunction expansion. $G_{ij}^{(n)}$, $G_{ij}^{'(m)}$ and $E_{ij}^{(s)}$ mean the coefficients which are obtained by integrating $G(\lambda_n r)$, $G(\lambda'_m r)$ and $E(\overline{s}r)$ along each line element. $\overline{G_{ij}^{(n)}}$, $\overline{G_{ij}^{'(m)}}$ and $\overline{E_{ij}^{(s)}}$ also mean the coefficients which are obtained by integrating the normal derivative of $G(\lambda_n r)$, $G(\lambda'_m r)$ and $E(\overline{s}r)$ on ΔS_j.

Eqs.(40),(41),(42) and (43) give the linear relationships of the number of $2N\cdot(N_1+N_2)$, and Eqs.(44),(45) and (46) also give that of $N\cdot(n_\ell+m_\ell+s_\ell+3)$. The number of the unknown potentials on the line elements are $2N\cdot(n_\ell+m_\ell+s_\ell+3)$, so linear matrix equation can be obtained from Eqs.(40),(41),(42),(43),(44),(45) and (46) by letting the total number of the collocation points (N_1+N_2) equal to $(n_\ell+m_\ell+s_\ell+3)/2$.

NUMERICAL SOLUTIONS

The characteristics of numerical solutions were investigated for the submerged horizontal rectangular plate with $B/h = 2.0$ and $W/h = 2.0$, where B and W represent a length along the x axis and y axis respectively. The submerged depth is $0.5h$ and the incident wave of $kh = 1.0$ propagates from positive y direction. The length of the line elements ΔS_j were taken as $0.1h$. The numbers of the collocation points N_1 and N_2 were

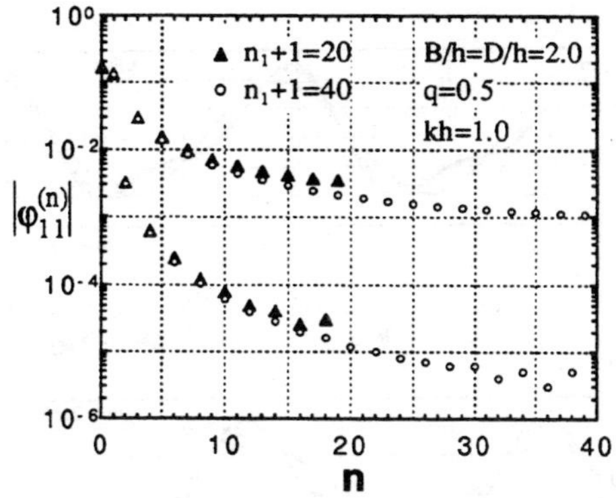

Fig.-2: First order numerical result on subregion(1)

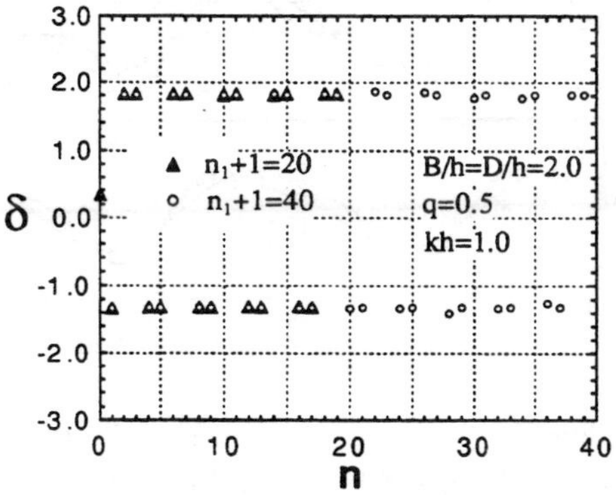

Fig.-3: Phase of the first order numerical result on subregion(1)

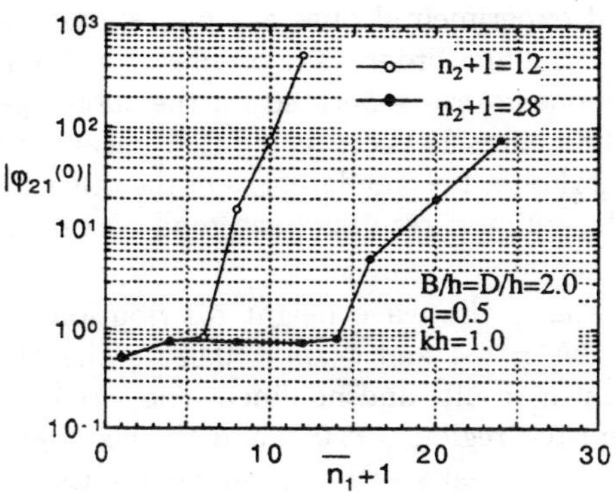

Fig.-4: Relationship between the truncation number $\overline{n}_1$ and the second order solution $\varphi_{21}^{(0)}$ for $n_2 = 11$ and 27 (n_2 and m_2 mean the truncation numbers of the series of the second order free waves)

taken as $N_1 + N_2 = n_\ell + 1$, $N_1 = m_\ell + 1$ and $N_2 = s_\ell + 1$ to obtain solutions with good accuracy (Yoshida,1992).

Fig.-2 shows the variation of the first order solution $\varphi_{11}^{(n)}$ on $(x, y) = (0, -W/2h)$ for $N_1+N_2 = 20$ and 40. The values of the series decrease monotonously with increase of n, and the values approach to a constant value on every term(n and $n+2$). The phase difference between n and $n+2$ is π as shown in **Fig.-3** and the convergence of the first order solution is confirmed.

Table.-1: Convergence of the second order solution with increase of the truncation numbers n_2 and m_2 in region(1) and (2) ($\Re[\phi]$ and $\Im[\phi]$ mean a real part and a imaginary part of the potential function ϕ)

n_2	$Re[\phi_{L1}]$	$Im[\phi_{L1}]$	$Re[\phi_{F1}]$	$Im[\phi_{F1}]$	$\lvert\phi_{L1}+\phi_{F1}\rvert$
7	-8.0162E+00	3.1627E+00	1.0138E+01	-2.7078E+00	2.1704
11	-2.1373E+01	5.9195E+00	2.3839E+01	-5.6206E+00	2.4832
15	-3.7051E+01	9.3978E+00	3.9266E+01	-9.0406E+00	2.2436
19	-5.7861E+01	1.3385E+01	6.0237E+01	-1.3098E+01	2.3931
23	-7.1219E+01	1.6152E+01	7.3452E+01	-1.5828E+01	2.2560
27	-7.7230E+01	1.7203E+01	7.9570E+01	-1.6922E+01	2.3527
31	-8.2273E+01	1.8287E+01	8.4509E+01	-1.7978E+01	2.2577
35	-9.5654E+01	2.0627E+01	9.7976E+01	-2.0351E+01	2.3387
39	-9.8758E+01	2.1310E+01	1.0100E+01	-2.1014E+01	2.2588
59	-1.2907E+02	2.6687E+01	1.3137E+02	-2.6416E+01	2.3195
m_2	$Re[\phi_{L2}]$	$Im[\phi_{L2}]$	$Re[\phi_{F2}]$	$Im[\phi_{F2}]$	$\lvert\phi_{L2}+\phi_{F2}\rvert$
3	-7.9165E-01	-2.1429E+00	3.5456E+00	2.8382E+00	2.8403
5	-6.5403E+00	-2.0728E+00	9.4537E+00	2.5883E+00	2.9586
7	-1.4682E+01	-1.5604E+00	1.7480E+01	2.1606E+00	2.8617
9	-2.6284E+01	-4.6795E-01	2.9120E+01	9.7720E-01	2.8819
11	-3.9200E+01	8.7850E-01	4.1998E+01	-3.1177E-01	2.8518
13	-5.1812E+01	2.4069E+00	5.4612E+01	-1.9013E+00	2.8544
15	-6.4163E+01	3.9124E+00	6.6956E+01	-3.3608E+00	2.8465
17	-7.0714E+01	4.8124E+00	7.3510E+01	-4.3104E+00	2.8410
19	-7.6677E+01	5.5721E+00	7.9466E+01	-5.0355E+00	2.8400
29	-1.0976E+02	1.0176E+01	1.1254E+02	-9.7759E+00	2.8315

In order to obtain the second order solution, the potential value of the phase locked waves expressed in Eqs.(25) and (26) must be computed by the first order solutions with the truncation numbers $n = \overline{n}_1$ and $m = \overline{m}_1$. In the two dimensional problem, Yoshida(1991) showed that the numbers of $\overline{n}_1$ and $\overline{m}_1$ influence the second order solution largely. **Fig.-4** shows the relationships between the truncation number $\overline{n}_1$ and the second order solution $\varphi_{21}^{(0)}$ for $n_2 = 11$ and 27, where n_2 and m_2 mean the truncation numbers of the series of the second order free waves in Eqs.(22),(23). In $n_2 + 1 = 12$, $\varphi_{21}^{(0)}$ diverges when $\overline{n}_1 + 1 > 6$, and $\varphi_{21}^{(0)}$ also diverges when $\overline{n}_1 + 1 > 14$ in $n_2 + 1 = 28$. These mean that the largest value of the eigenvalue $k_{n_2}^{(2)}$ included in ϕ_{Fi} must be greater than that of the eigenvalue $k_{\overline{n}_1}$ in ϕ_{Li}. In other words, n_2 and m_2 must be taken as $n_2 \geq 2\overline{n}_1 + 1$ and $m_2 \geq 2\overline{m}_1 + 1$ respectively.

Table-1 shows the convergence of the second order solution with increase of the truncation numbers n_2 and m_2 in subregion (1) and (2). Though the values of $|\phi_{Fi}|$ and $|\phi_{Li}|$ ($i = 1, 2$) increase with the increase of the truncation numbers n_2 and m_2, the sign of $\Re[\phi_{Fi}]$ and $\Im[\phi_{Fi}]$ are opposite to that of $Re[\phi_{Li}]$ and $\Im[\phi_{Li}]$ with the same values respectively, and thus the second order potential $|\phi_F + \phi_L|$ converges to a constant value.

Fig-5 shows the variation of the first and second order solutions corresponding to the number of collocation point

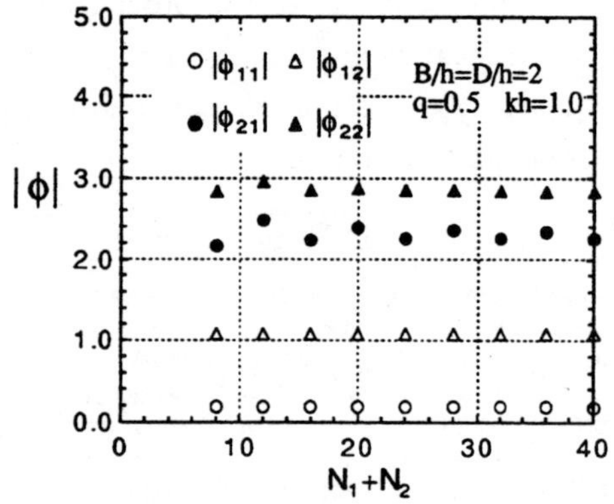

Fig.-5: Variation of the first and second order solutions corresponding to the number of collocation point $N_1 + N_2$

N_1+N_2. The potential functions are computed as follows:

$$\phi_{11} = \sum_{n=0}^{n_1} \varphi_{11}^{(n)} \quad , \quad \phi_{12} = \sum_{m=0}^{m_1} \varphi_{12}^{(m)} \tag{47}$$

$$\phi_{21} = \sum_{n=0}^{n_2} \varphi_{21}^{(n)} + \sum_{p=0}^{\overline{n_1}} \sum_{q=0}^{\overline{n_1}} \gamma_{pq} f_{pq} \tag{48}$$

$$\phi_{22} = \sum_{m=0}^{m_2} \varphi_{22}^{(m)} + \sum_{p=0}^{\overline{m_1}} \sum_{q=0}^{\overline{m_1}} \gamma_{pq}^{'} f_{pq}^{'} \tag{49}$$

Though the variations of the second order solutions are slightly larger than that of the first order ones, they converged to a constant value with increase of the collocation point. The collocation points are needed at least $N_1+N_2 = 10$ in the first order computation and $N_1+N_2 = 20$ in the second order one to obtain the solution with enough accuracy.

VERIFICATION OF NUMERICAL RESULTS BY EXPERIMENTS

The validity of the numerical method was investigated by comparing numerical results with experimental ones. The experiments were conducted with a wave tank of 18m in length, 20m in width and 40cm in depth. The horizontal plate with 80cm in length, 80cm in width and 12cm in submerged depth was place in the center of the wave tank. The incident waves with $kh = 1.0$, 1.4 and 1.6 propagated from positive y direction and wave profiles were measured along the y axis.

Fig-6 shows the computed wave profiles and experimental ones along the y axis. The incident wave height ζ_0/h is 0.036 and the relative wave length kh is 1.0. The thin solid line shows the computed linear results and the thick line also shows the computed nonlinear ones, and the broken line shows the experimental ones. The computed

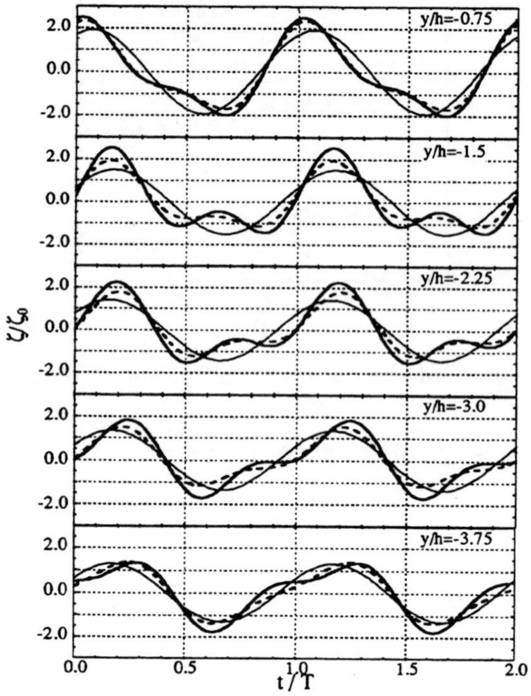

Fig.-6: Comparison of the computed wave profiles with the experimental ones $[kh = 1.0, \zeta_0/h = 0.036]$ ——— computed linear results, ━━━ computed nonlinear results, - - - - - experimental results

results and experimental ones are plotted with synchronizing both the first order wave profiles. The higher order wave components are generated on the submerged plate and the differences between computed linear results and experimental ones become larger. On the other hand, the computed nonlinear results show good agreements with experimental ones.

Fig.-7 shows the wave height distribution along the y axis for $kh = 1.4$ and $\zeta_0/h = 0.033$. The differences between linear results and nonlinear ones are observed in the transmitted region($y < 0$) , and the nonlinear results show good agreements with experimental ones.

The numerical method presented in this paper is based on the second order Stokes wave theory, and it is important to clarify the valid range for the incident wave amplitude. **Fig.-8** shows the computed wave profiles and experimental ones at the transmitted region $y/h = -1.875$ for $kh = 1.4$. The incident wave amplitude was varied from $\zeta_0/h = 0.033$ to 0.074. The Ursell number written in this figure is defined as $U_r = 2k^{'}\zeta_0(2\pi)^2/(qhk^{'})^3$.

Though the agreements between computed nonlinear results and experimental ones become worse with increase of the Ursell number, computed results show good agreements with experimental ones for $U_r = 20.6, 26.8$ and 30.5.

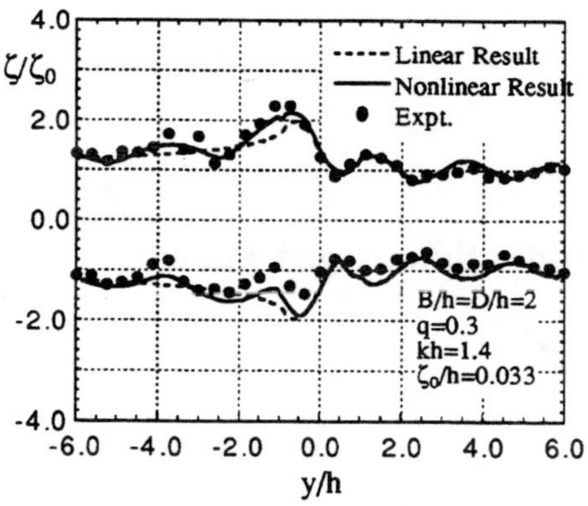

Fig.-7: Comparison of the computed wave envelope with experimental one[$kh = 1.4, \zeta_0/h = 0.033$]

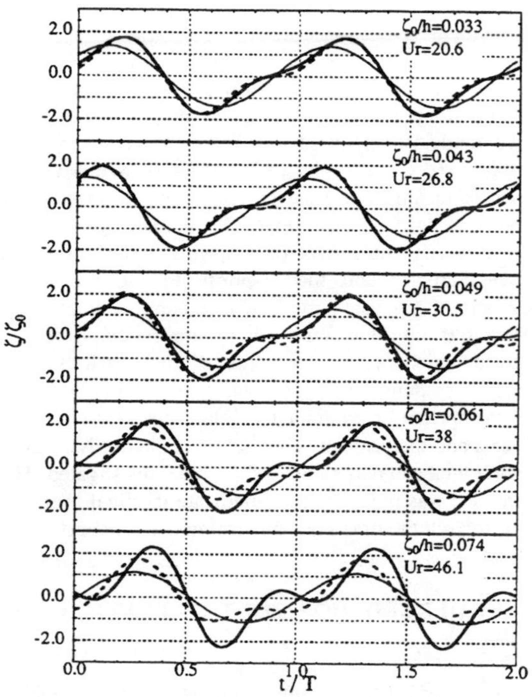

Fig.-8: Comparison of the computed wave profile with experimental one[$kh = 1.4, y/h = -1.875$]
——— computed linear results, ▬▬▬ computed nonlinear results, - - - - - experimental results

CONCLUSIONS

An approximate calculation method for the second order wave interactions with three-dimensional submerged structures was proposed. In the mathematical formulations, Green's Identity formula was used to express the distributions of the velocity potential on a horizontal plane. The collocation method of matched eigenfunction expansion was also utilized to derive the linear relationships for the velocity potential and its normal derivative on the adjacent fluid region. The relationships between the truncation numbers of the series of the eigenfunction expansion and the numbers of the collocation point were investigated to obtain the second order results with good accuracy. The validity of the present method was confirmed by comparing numerical results with experimental ones.

References

Eatock Taylor,R. and S.M.Hung(1987). "Second order diffraction forces on a vertical cylinder in regular waves," Applied Ocean Research, Vol.9, No.1.

Ijima,T., Y.Yumura, C.R.chou, A.Yoshida(1974). "Wave forces and scattering by submerged or semi-immersed cylinder of arbitrary shape," Proc. JSCE, Vol.228.

Kim,M.H., D.K.P.Yue(1989). "The complete second-order diffraction solution for an axisymmetric body, Part 1. Monochromatic incident waves," J.Fluid Mech., Vol.200.

Kriebel,D.L.(1990). "Nonlinear wave interaction with a vertical cylinder,Part I;Diffraction theory," Ocean Engineering, Vol.17, No.4.

Mizutani,N., T.Mada, S.Imai, K.Iwata, M.Watanabe (1996). "Second order approximate solution for the circular cylinder in Crossing Wave Fields," Proc. of Coastal Engineering, JSCE, Vol.43

Molin, B.(1979). "Second-order diffraction loads upon three-dimensional bodies," Applied Ocean Research, Vol.1, No.4.

Rahman,M.(1984). "Wave diffraction by large offshore structure: an exact second-order theory," Applied Ocean Research, Vol.6, No.2.

Sabuncu,T. and O.Goren(1985). "Second-order vertical and horizontal wave forces on a circular dock," Ocean Engineering, Vol.12, No.4.

Williams,A.N., A.G.Abul-Azm and S.A.Ghalayini(1990). "A comparison of complete and approximate solution for second-order diffraction loads on arrays of vertical circular cylinders," Ocean Engineering, Vol.17, No.5.

Yoshida,A., H.Kojima, R.Tsurumoto(1991). "Numerical nonlinear solution of wave-structure interactions with collocation method of matched eigenfunction expansions," Proc. JSCE, No.434/II-16.

Yoshida,A., K.Murakami, M.Ono(1992). "Three dimensional wave-structure interactions with collocation method of matched eigenfunction expansions," Proc. Coastal Engineering, JSCE, Vol.39.

Proceedings of the Seventh (1997) International Offshore and Polar Engineering Conference
Honolulu, USA, May 25-30, 1997

ISBN 1-880653-28-1 (Set); ISBN 1-880653-31-1 (Vol. III)

Small-Scale Beach Contour Modelling Behind a Submerged Plate

J. Lengricht and A. Schlenkhoff
University of Wuppertal
Germany

C. Fischer
Department of Inland Navigation and Waterways
Stralsund, Germany

ABSTRACT

Sediment transport in a small wave flume using light synthetic material was simulated. The constraints and uncertainties of scale effects inherent with small scale models are discussed and the model is applied towards the clarification of the influence of a horizontal submerged plate on beach evolution. The wave height reduction of the plate is commonly known, but pulsating flow underneath the plate, turbulent mixing and vortices generated by interaction of the plate and incoming waves have not been investigated sufficiently. These phenomena can affect the grain structure and increase the sediment transport in the beach zone. But the small-scale model tests indicate that wide regions of the beach zone are not loaded by waves at all and the locations where the incoming waves break out shift to the coastline. Consequently, a lower disintegration of the beach and lower transport rates of sediment in the shore zone can be expected.

KEY WORDS: Coastal hydrodynamics, experimental methods, small-scale model tests, permeable breakwaters, sediment transport.

INTRODUCTION

The intention of this model investigations was to study possible effects of beach erosion process induced by a submerged plate in front of the beach. Furthermore, the experiments carried out should gain the understanding of the hydrodynamic forces involved. Simulating a moveable bed evolution in a physical small scale model has to face a series of constraints, of which some are discussed in the following. The modelling material has to be composed such that it can react to the applied hydrodynamic forces. The scaling effects inherent in a physical model used for studying sedimentary problems are not as well understood as they are for fixed-bed models. Consequently, movable bed models results have to be carefully reviewed in the context of previous, similar models that have demonstrated success in reproducing prototype bed evolution. Examples of two-dimensional movable bed coastal models include studies of beach profile evolution, response to berm-breakwaters and bedform translation under unidirectional currents. Most of the studies have been carried out in large wave flumes or oscillating tanks. A compromise between the fixed-bed and the movable-bed model occurs when sediment tracer materials are used in fixed-bed models as a qualitative indicator of sediment erosion and deposition areas. Generic physical models as investigated here can be simplified or idealised in order to deduce the response for a variety of hydrodynamic forcing conditions. These data then can be used to develop or verify theoretical or empirical relationships between the flow characteristics and the response to the process. The major characteristic of a physical model depends on whether the process is a long or a short term one. With a long term model system changes can be detected which occur over extended time periods while with the latter one the response to the physical system with high intensity events such as storms can be examined. Within these experiments such short term events and the principle changes of the hydrodynamic process due to the horizontal submerged plate should be investigated. Therefore, the major interest was rather more to dedicate the final shape of the beach than figuring out the time scale governing the transport process.

THE SUBMERGED HORIZONTAL PLATE AS A BREAK-WATER

A submerged rigid horizontal plate used as a permeable breakwater (figure 1) has several advantages compared to conventional constructions. Its operation as a 'wave filter' allows advantageous solutions without blocking an ecologically desirable exchange of water between the open sea and the area to be protected.

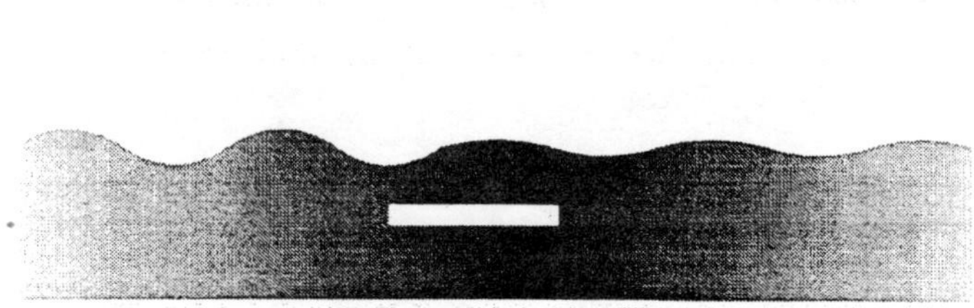

Figure 1 The submerged horizontal plate as a permeable breakwater

The effects of the submerged horizontal plate have been experimentally and numerically estimated by several researchers (Dick, 1968; Dattari et al., 1977; Guevel et al., 1985; Kaldenhoff and Fischer, 1988; Kojima et al, 1990;). As Graw (1993) reported the published results of transmission coefficients (c_t) in the wake of the plate differ distinctively from each other, because the physical understanding of the hydrodynamic condition creating a pulsating flow acting opposite to the wave propagation direction is only incomplete. The relation of transmission of initial wave energy, occurrence of a pulsating flow, phenomenon of vortex shedding, shear dissipation and turbulent mixing in the wake of the plate for different wave parameters and their combinations has not been clarified yet. Furthermore, no attention has been paid to possible effects on sediment transport, beach evolution or berm breakwaters stability. In figures 2 and 3 the wave attack against an unprotected beach and a beach sheltered with a submerged horizontal plate are presented. In figure 2 (unsheltered beach under wave attack) it can be seen that the initial wave with its origin height loads the shore intensely (see figure 2a) and soil is in suspension at a specific phase angle of the incoming wave just before the breaking process starts (see figure 2b).

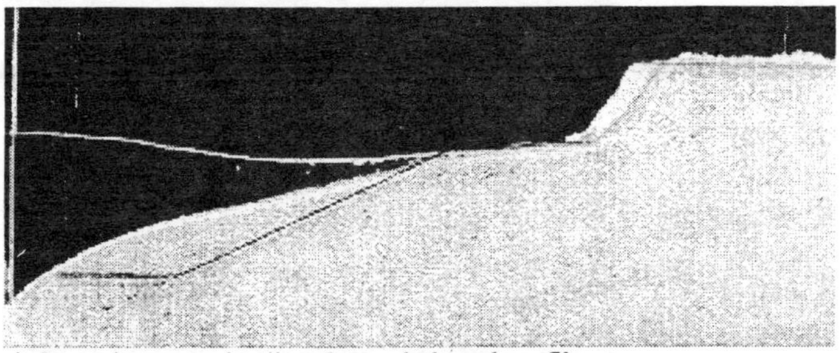

a) Incoming wave loading the unsheltered profile

b) Soil in suspension

Figure 2 Wave attack without beach protection

Figure 3 shows the wave propagation behind a submerged horizontal plate. It is illustrated that a plate interacting with the incoming wave influences the wave height behind the plate significantly. In this actual case an enormous wave height reduction is caused (see figure 3a) and the decomposition of the initial wave into waves of higher order which lead to notable changes in wave load can be clearly identified (see figure 3b).

Graw (1995) notified that the transformation of the initial wave into waves of higher order by the submerged horizontal plate has influence on the energy transmitted. As shown in figure 4 the experimentally determined c_t coefficients and frequency distributions for different wave periods in the wake of the plate imply that the wave energy transmitted to the area to be protected is reduced when the initial wave is broken up into waves of several orders. Nevertheless, it is pointed out by Graw (1995) that there is a considerable indication that the transformation of the initial wave into waves of higher order depends on the water depth.

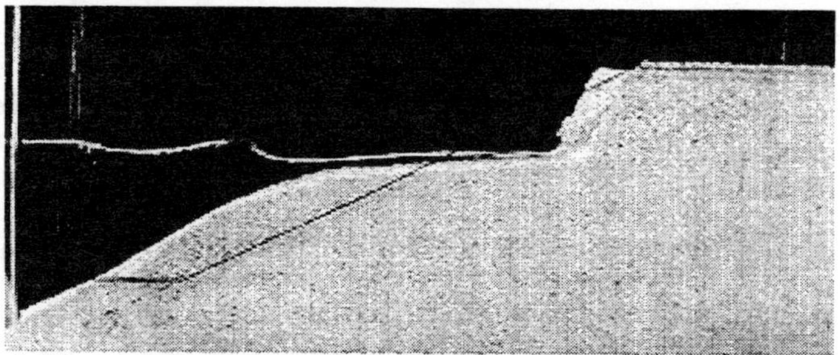

a) Wave height reduction behind the plate

b) Decomposition of the initial wave

Figure 3 Wave attack after wave height reduction due to interaction of the submerged horizontal plate and regular waves

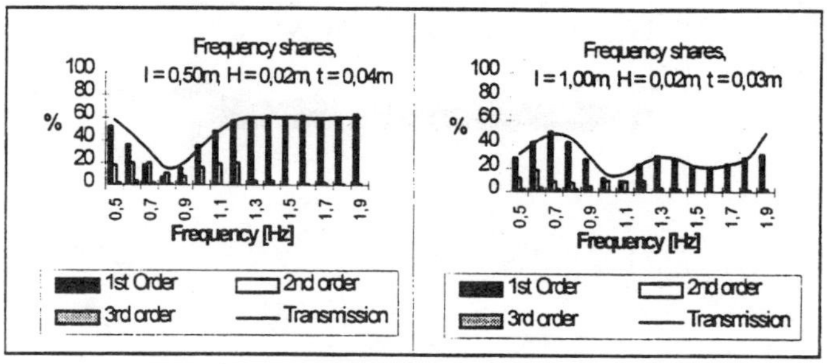

Figure 4 Frequency shares of higher harmonics and transmission coefficient c_t (Graw, 1995)

Figure 5 Turbulent mixing and vortex generation in the wake of the plate (wave propagation from the left to the right)

Van Gent (1993) recounted that the way the velocity moment is acting on the beach is of importance for the sediment transport process. Therefore, flow visualisation around the submerged horizontal plate was a first step to analyse the flow characteristics generated by the plate. Figure 5 shows that interacting with incoming waves vortices in the wake of the plate and turbulent mixing zones are created. It also gives an idea of induced vortex shedding, shear layers and turbulent mixing zones when propagating waves are interfered by the plate. Due to enhanced shear stresses it could be assumed that these phenomena influence the beach evolution as well.

Furthermore, in recent years finite element methods have been developed to serve as a useful method for calculating water wave scattering by submerged or semi-immersed bodies of arbitrary shape, e.g. the hybrid element method (HEM) which succeeded in reducing computing time and costs by using semi-analytical terms describing the problem. Numerical comparisons with some analytical solutions to particular bodies have proved the validity of this technique. The behaviour of the semi-submerged horizontal plate in sinusoidal waves, as an example of the application of the hybrid element method in connection with linear potential theory was demonstrated by Kaldenhoff and Fischer (1988). Their assumptions because of the linearity of the problem were irrotational flow, inviscid fluid, incompressible fluid, simple harmonic motion, small amplitudes, non-breaking waves, impermeable boundaries and a constant water depth.

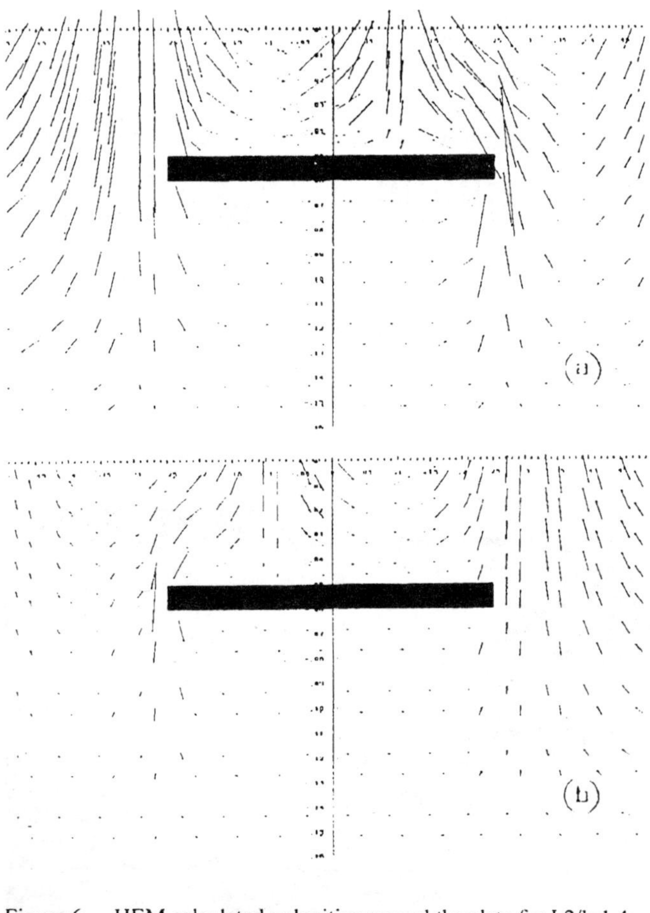

Figure 6 HEM calculated velocities around the plate for L2/l=1.4 sketch a) ωt=0.0 1 and sketch b)ωt=π/2

They calculated reflection and transmission coefficients, velocity vectors for distinct phase angles around the plate and kinetic energy distributions in the flow field. Figure 6 shows the calculated velocity filed around the semi-submerged horizontal plate for two different phase angles.

In figure 7 exemplary results for the distribution of the kinetic energy are shown. Kaldenhoff and Fischer reported that their numerical approach is a valuable tool for a pre-examination of the physical behavior around the semi-submerged horizontal plate in water of a finite depth, but no hint could be given so far whether the plate influences the beach contour development in the wake of the permeable breakwater or not.

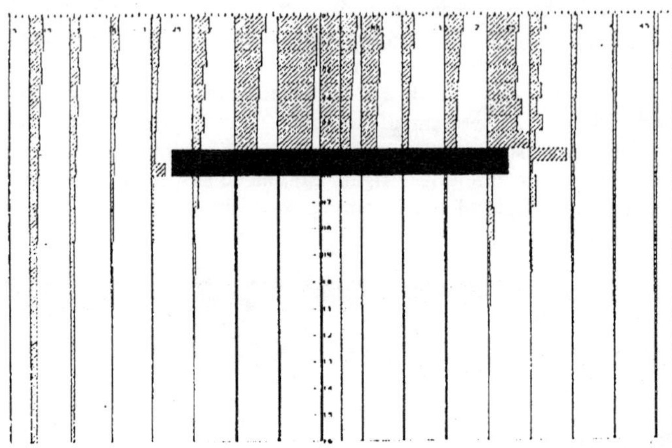

Figure 7 Kinetic energy distribution

Consequently, the intention of the investigations presented in this paper is to report about the simulation of sediment transport in the wake of a horizontal submerged plate and to give exemplary results of experiments in a small wave flume.

EXPERIMENTAL SET-UP AND TEST COURSE

The small-scale physical tests about excavation and sedimentation in the wake of the plate were carried out with regular waves at the University of Wuppertal in the Laboratory of the Institute of Geotechnics, Waste Management and Hydro-Sciences. The wave flume used is 24 m long, 0.3 m wide and 0.5 m high. The water depth was d = 0.25 m. Figure 5 gives a schematic view of the experimental set-up.

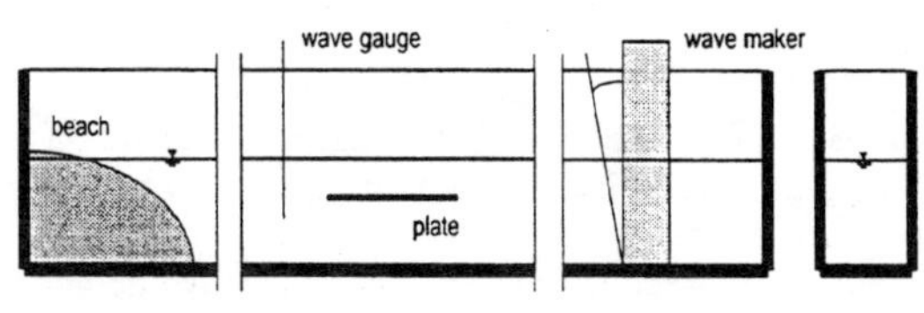

Figure 8 Experimental set-up in schematic view

The parameters used in the experiments were: plate length l = 0.40 m, submergence depth of t = 0.045 m, wave height H_i = 0.03 m and wave frequency f = 1.0 Hz.

The flow field in the wake of the plate was observed with a liquid crystal digital camera. The images taken as 640 x 480 pixel elements were stored and processed with an image processing station consisting of a host PC (equipped with Pentium processor running at 90 MHz and 16 MB RAM), a frame grabber, a RS-232C serial port, a SVGA colour monitor and a laser printer.

MODELLING OF SEDIMENT TRANSPORT

Hydrodynamic Similitude Requirements for Models

Scaling any of experimental results to nature or prototype size should include the following considerations. The prototype-to-model scale ratio of any parameter (X) in the prototype (p) to the value of the same parameter in the model (m) is represented in the following as:

$$K_X = X_p / X_m$$

Once one has chosen a set of scale ratios (mostly any arbitrary length scale (L) and the time scale (t) are selected) others can not be chosen independently. For instance, the scale ratio for velocity (v) can then be derived as :

$$K_v = K_L \;/ K_t$$

Completely similar models are models in which the values of all relevant dimensionless parameters in the prototype are maintained in the model. A prerequisite for a complete similarity is that the model is geometrically similar to the prototype (similarity in form). Other types of models are kinematically similar (similarity in motion) and dynamically similar (similarity in vectorial forces). For a more detailed discussion of principles of modelling techniques and dimensional analysis see Kaldenhoff (1992) and Kobus (1980).

The main forces involved in the process investigated here are inertial and gravitational forces and to some extent viscous forces, others are of inconsiderable importance. It is impossible to achieve a complete similitude where all force ratios are constant except for a prototype scale. However, experience indicates that almost any major problem can be simplified into the interplay of two major forces. This allows the similitude criteria to be developed theoretically. Inertial forces are always present in flow problems, so inertia needs to be balanced by one of the other forces. The most significant parameter is the *Froude* number (Fr) that expresses the relative influence of inertial and gravity forces, given by (ρ = fluid density, g = acceleration of gravity):

$$Fr = (\rho L^2 v^2 / \rho L^3 g)^{1/2} = (v / Lg)^{1/2};$$
$$\Rightarrow \text{scale ratios:} \quad K_v = (K_g K_L)^{1/2} \quad (1)$$

When viscous forces achieve a more dominate effect the ratio of inertial to viscous forces, i.e. the *Reynolds* number should be maintained (μ = dynamic coefficient of viscosity):

$$Re = (\rho L^2 v^2 / \mu v L) = (\rho L v / \mu)$$
$$\Rightarrow \text{scale ratios:} \quad K_v K_L K_\rho = K_\mu \quad (2)$$

Finally, it is worth to distinguish between the inertial forces caused by the two different types of acceleration, namely the convective and local acceleration. In terms of their physical units the ratio of the two forces is referred to as *Strouhal* number (Str):

$$Str = (\rho L^3)(v/t) / (v^2/L) = (L / vt)$$
$$\Rightarrow \text{scale ratios:} \quad K_L = (K_v K_t)^{1/2} \quad (3)$$

Maintaining the *Froude* model criterion this simply states that the length and velocity scale ratio determine a certain time scale. For example:

$$\textit{Froude}\text{ time scale:} \quad K_t = (K_L / K_g)^{1/2}$$
$$\text{velocity scale:} \quad K_v = (K_L K_g)^{1/2} \quad (4)$$

With a scale ratio of gravity of $K_g = 1$ this is reduced to:

$$\textit{Froude}\text{ time scale:} \quad K_t = (K_L)^{1/2}$$
$$\text{velocity scale:} \quad K_v = (K_L)^{1/2} \quad (5)$$

The time scale for *Reynolds* similarity can be derived to:

$$\textit{Reynolds}\text{ time scale:} \quad K_t = (K_L)^2 K_\rho / K_\mu$$
$$\text{velocity scale:} \quad K_v = K_\mu / K_L K_\rho \quad (6)$$

Using the same fluid properties it can be reduced to:

$$\textit{Reynolds}\text{ time scale:} \quad K_t = (K_L)^2$$
$$\text{velocity scale:} \quad K_v = 1 / K_L \quad (7)$$

From the above it is clear that both *Froude* and *Reynolds* similarity cannot be achieved at the same time. Applying *Froude* similarity criterion means offending against the viscous stress term. Consequently, the boundary layer in the model should be kept turbulent in order to minimise the effect of skin friction. But the turbulent dissipative processes will be maintained which govern the hydrodynamics.

Hydrodynamic Similitude Requirements for Movable-Beds

The hydrodynamic parameters involved in sediment transport can be given in terms of a complete set of dimensionless products. Kamphuis (1985) gave the hydrodynamic set as follows:

$$\Pi_H = f\left[\frac{H}{L};\frac{d}{L};\frac{x}{L};\frac{y}{L};\frac{z}{L};\frac{k}{L};\frac{t}{\sqrt{g/L}};\frac{\mu/\rho}{L\sqrt{gL}}\right] \quad (8)$$

The first five dimensionless products containing the wave height (H), the water depth (d) and the three coordinates (x, y, z) require that the model geometry and hydrodynamics be geometrically undistorted. The sixth dimensionless product will be the same when the boundary layer remains turbulent, i.e. when the roughness of the structure surface (k) is large enough. The seventh product requires that the model should be scaled according to *Froude* similarity criterion.

Finally, the eighth product is recognised as a flow *Reynolds* number. *Reynolds* similitude is usually not met in the model. However, neglecting viscosity is difficult to justify for sediment motion confined to boundary layer because model boundary layers can also be laminar smooth. Unfortunately, there is no practical alternative except trying to assure turbulent flow exists next to the movable bed.

Sediment Transport Similitude Requirements for Movable-Beds

Kamphuis (1985) used the sediment parameters as diameter (D), density (ρ_s) and shear stress at the bottom (τ_b) as well as the fluid parameters density (ρ) and a characteristically length (λ). The corresponding set of dimensionless products is given by:

$$\Pi_{So} = g\left[\frac{D v_*}{\mu/\rho};\frac{\rho v_*^2}{g(\rho_s-\rho)D};\frac{\rho_s}{\rho};\frac{\lambda}{D};\frac{\omega}{v_*}\right] \quad (9)$$

where (v_*) is the shear velocity and (f_W) the friction factor which can be estimated under oscillatory, turbulent flow in relation to the ratio of the roughness to the orbital amplitude at the bottom (k/a_b):

$$v_* = \sqrt{\tau_b / \rho}\;;\quad \tau_b = 0.5 f_W \rho v_b^2\;; f_W \approx 0.4(k/a_b)^{3/4} \quad (10)$$

Kamphuis notes that the above equations are the most appropriate for sediment transport that is driven by shear stress occurring within the bottom boundary layer. Kamphuis (1992) added the fifth parameter later which is the ratio of sediment fall speed to shear velocity (ω/v_*,). This parameter accounts for suspended transport occurring simultaneously with bed load transport.

The first dimensionless product in equation (9) is called grain size *Reynolds* number and the second is referred to as the densimetric *Froude* number. These two parameters may be recognised as the coordinates of the classical Shield's diagram for incipient motion in unidirectional flow. The third is the relative density of the sediment and the fourth is a relative length which can be scaled either by the wave height (H) or the water depth (d) for short wave or for long wave models respectively.

Dalrymple (1989) presented a similar dimensionless grouping of parameters for sediment transport, only he neglected the characteristic length and included the fall speed parameter or also called *Dean* number ($d_\omega = H / \omega T$) instead of (ω/v_*):

$$\Pi_S = g\left[\frac{D v_*}{\mu/\rho};\ \frac{\rho v_*^2}{g(\rho_S-\rho)D};\ \frac{\rho_S}{\rho};\ \frac{H}{\omega T}\right] \tag{11}$$

The complete similitude cannot be achieved in a model, however, any uncompleted set may be useful if appropriate.

Kamphuis (1991) formulated a general expression for sediment transport in the breaking zone as well where forces in addition to bottom shear stresses are present. He replaced the shear velocity (v_*) with a „basic velocity" given by a term including the breaking wave height (H_b) and replaced the characteristic length (λ) by the breaking wave height as well. This gave a breaking zone sediment transport relationship of:

$$\Pi_{Sb} = g\left[\frac{D\sqrt{gH_b}}{\mu/\rho};\ \frac{\rho g H_b}{g(\rho_S-\rho)D};\ \frac{\rho_S}{\rho};\ \frac{H_b}{D};\ \frac{\omega}{\sqrt{gH_b}}\right] \tag{12}$$

SELECTION OF SEDIMENT MATERIAL FOR MODELLING

Considering the effect of the horizontal submerged plate on the beach the transformation of the wave spectrum into higher frequencies shares might have the major impact to the sediment transport potential. The pulsating flow as well as the vorticities in the wake of the plate might have an impact but since the plate is situated well enough before the base of the beach slope only minor changes are expected.

However, simulating the translation process in the surf zone requires a material which is even able to react to small shear stresses. Since the intention was to study the general hydrodynamic behaviour of the flow behind the horizontal submerged plate the modelling material should react fast which also leads to a reduction of duration of the experiments. From this point of view the lightest material should be considered, but on the other hand the material must be heavy enough to withstand the direct forces due to wave breaking (turbulence dominated sediment transport). It was carried out that a minimum density of $\rho = 1.4$ g/cm³ is required in accordance with Noda (1972). The diameter of the material was chosen to D = 2 mm in order to reach a sufficient weight and therefore sediment dropping speed, which is required when both the suspended material transport and the bed load transport shall be investigated. The dropping speed was estimated to $\omega = 0.1$ m/s, which means that the time required for reaching the bottom after displacement by lifting is distinctively smaller than one wave period. It should be mentioned here that the material does not allow to reproduce any bedform evolution like dunes, bars or ripples as these features are simply washed out by the oscillating current. Furthermore, similar problems with lighter material and small scale models are reported in literature. The process of accretion and percolation as an important aspect of beach recovery can not be modelled as on the one hand the lighter material once moved on the beach is getting a larger relative weight and on the other hand the intergranular space is different and can be widen to a considerable extent. Finally, it should be mentioned that the material selected is hydrophobic and therefore it is able to remain on the surface when eroded from the beach face and can be transported until the next breaker let it sink. Once the equilibrium profile is reached this material can be washed back on the beach and the accretion process is balanced with the erosion one.

EXEMPLARY RESULTS

Compaction and Stability of the Unsheltered Beach Contour

Figure 9 shows the development of an initial beach contour with a slope angle of 1 : 2 under load of regular waves with an initial wave height of $H_i = 0.03$ m and a frequency of f = 1.0 Hz. The initial beach contour can be seen in figure 9a. In figure 9b the beach profile developed after a time interval of $\Delta t = 300$ s is shown and figure 9c represents the beach evolution after a time interval of $\Delta t = 450$ s.

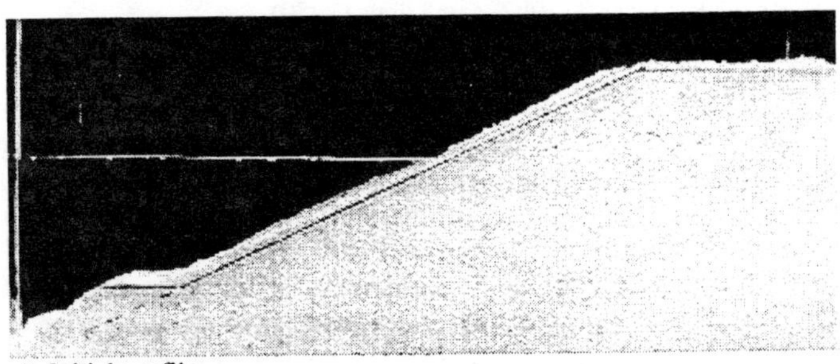

a) Initial profile

b) Beach contour after wave load duration $\Delta t = 300$ s

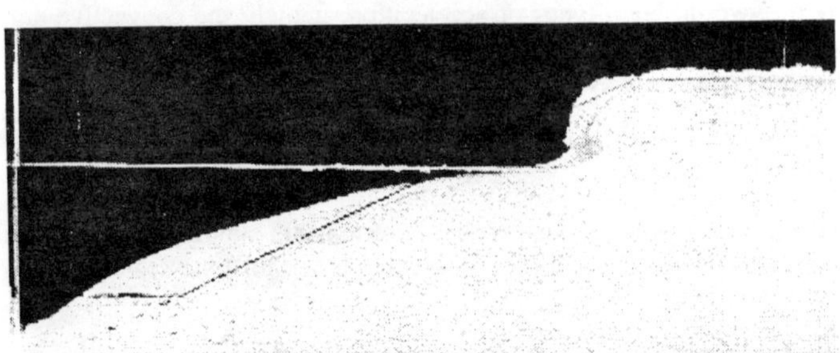

c) Beach profile after wave load duration $\Delta t = 450$ s

Figure 9 Beach contours without any breakwater sheltering the shore

While in the topmost part of the beach profile soil is excavated, this material is sedimented at the base of the slope. The beach contour is formed into a dynamically stable shape (see figure 9b) after a time interval of wave attack of about $\Delta t = 300$ s. There is no significant change of the profile visible after an additional wave load interval of $\Delta t = 150$ s (see figure 9c). Depending on the developing beach shape the breaking zone is situated at different locations and the profile is formed in such a manner that the resistance to the hydrodynamic load is minimised at all positions along the profile. Elements of the material eroded are replaced in positions where they are just stable. This means that at positions where waves break the local slope angle becomes small. In positions where the loading is less, e.g. before the breaking process starts, the angle can be larger.

The breaking process is mostly determined by the water depth and starts at a location where the water depth is approximately 0.04 m which is more or less about the breaking wave height itself. In upper positions along the slope, the loading is lower and the material has a relatively large resistance. This leads to both a parabolic beach profile beneath the still water level and to a very gentle slope in the breaking zone. This forms a S-shaped profile which is typical for the transport of coarse material. The profile development is so far in accordance with those observed for berm breakwaters by van Gent (1993).

Beach Development due to Interaction of Wave and Plate

For these test series the plate was fixed in front of the beach profile formed in the same way as for the experiments without any shore protection described above, i.e. with a beach slope of 1 : 2. The results are discussed for a submerged horizontal plate with a length of $l = 0.40$ m and submergence depth of $t = 0.045$ m, an initial wave height $H_i = 0.03$ m and a wave frequency $f = 1.0$ Hz.

In this particular frequency range the submerged plate acts as a wave filter. The wave is spilling over the plate and the surge forms a jet into the wake of the plate generating a large amount of turbulent motion. The transmission coefficient c_t decreases to less than 0.3 and a clear decomposition can be detected.

Compared to the case without protection a very different condition occurs (lower wave heights and higher frequencies) and the redeveloped waves tend to act on the beach like waves of higher order (figure 3). The width of the breaking zone is slightly increased, some waves can even reach the topmost beach zone. But due to the considerable height reduction, main parts of the beach become more stable. The beach profile development in the wake of the submerged horizontal plate is shown in figure 10.

Comparison Between Unsheltered Beach Development and Beach Development In The Wake Of The Submerged Horizontal Plate

In figure 11 profile development of the unsheltered beach profile is visible. The images of the initial beach profile and the profile after the wave load duration of $\Delta t = 450$ s are coupled by image processing. The processing algorithm averages greyscale values of each pixel of both the first (see figure 10a) and the second frame (see figure 10c). It can be seen that the amount of sedimentation exceeds the amount of excavation which can be explained by a significant loosening of the grain structure under wave load.

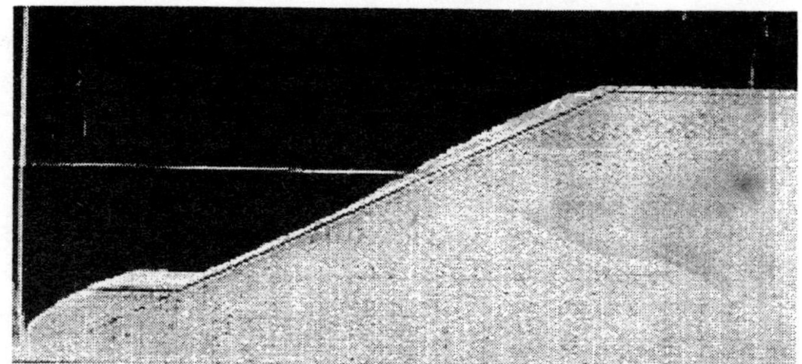

a) Initial profile

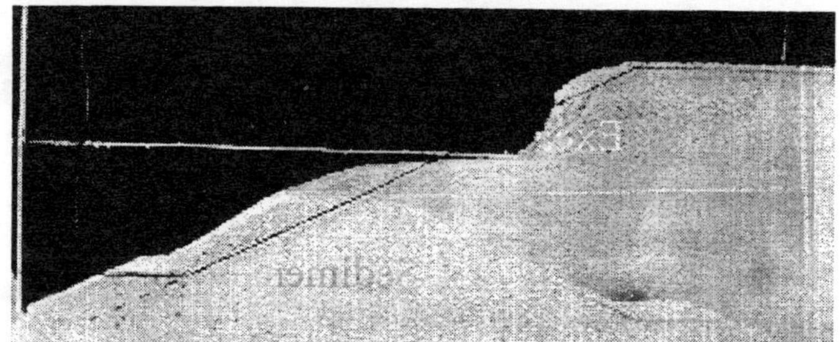

b) Beach contour after wave load duration $\Delta t = 300$ s

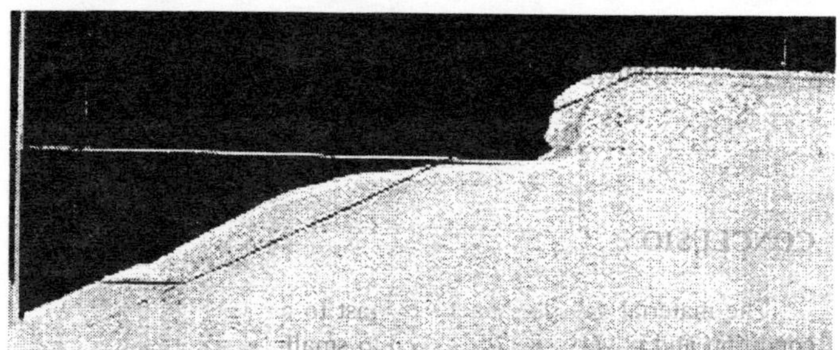

c) Beach profile after wave load duration $\Delta t = 450$ s

Figure 10 Beach development in the wake of a submerged horizontal plate

In figure 12 the beach evolution of the beach profile sheltered by the submerged horizontal plate is worked out by processing the images of the initial beach profile and the profile after a period of wave loading of $\Delta t = 450$ s. In this case the grain structure is not as loosened up as in the case of initial wave load on the unprotected beach which can be explained by the perceptible wave height reduction due to the interaction of the breakwater and the wave (see figure 2).

Compared to figure 11 the sediment transport shown for the protected beach contour is slightly smaller then for the material transport observed at the unsheltered shore. And in contrast to the unsheltered case the initial berm at the base of the beach is still visible which confirms that this area is not affected at all. Altogether, this indicates a positive effect of the submerged horizontal plate on disintegration of the surf zone and lower sediment transport rates in the shore zone, i.e. the profile stability. Consequently, this kind of breakwater might be useful to protect endangered beaches against shoreline erosion.

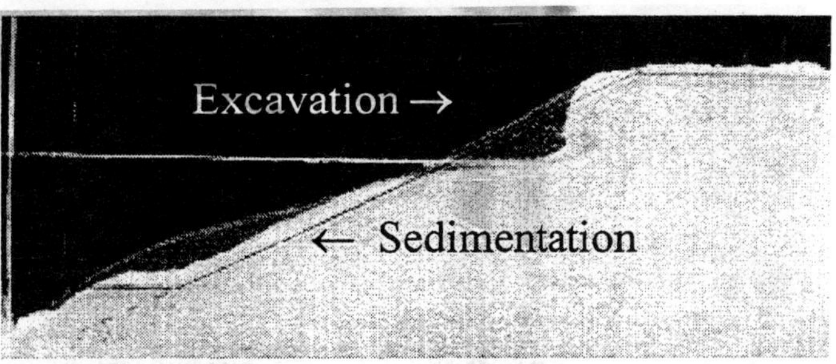

Figure 11 Evolution of the unsheltered beach profile

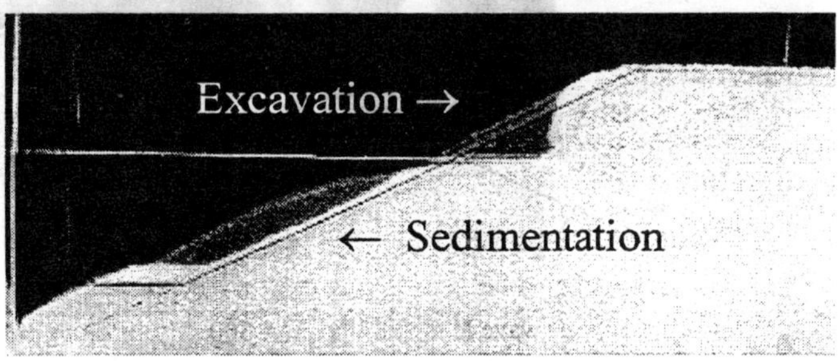

Figure 12 Excavation and sedimentation in the wake of the horizontal submerged plate

CONCLUSION

The material selected react very fast to any change of the loading condition and is very sensitive even to small shear stresses and forces. On the other hand the weight and the size is large enough in order to create a sufficient grade of compaction and resistance. Furthermore, the material is suitable to build up an equilibrium profile in a short duration. This makes the approach very attractive for generic experiments in small scale wave flumes. Due to unavoidable scale effects the model results should not be taken without care nor any sediment transport rates should be overpredicted. The approach is rather more a measuring technique for hydrodynamic forces than a measure for determining the transport rate itself. However, the model is more expressive than a simply fixed-bed model with sediment tracer.

The submerged horizontal plate in front of a beach has an obvious effect on the waves propagating into the beach zone. Beside the wave height reduction, the decomposition of waves into waves of higher order lead to most significant changes in the wave load and the development of the beach line. The locations where the incoming waves break out shift to the coastline. Even when some waves can reach the topmost part of the beach, wide regions of the whole beach zone are not loaded by waves at all and the profile is more stable than before. Consequently, a lower disintegration of the surf zone and lower sediment transport rates in the shore zone can be expected. Furthermore, it can be stated that the plate has a positive effect on the profile stability and might be useful for the protection of endangered beaches against shoreline erosion.

REFERENCES

Dalrymple, R. A. (1989). „Physical Modelling of Littoral Processes", in Recent Advances in Hydraulic Physical Modelling, R. Martins, (Ed.), Kluwer Academic Publ., Dordrecht, The Netherlands, pp 567-588.

Dattari, J., Jothi Shankar, J., and Raman, H. (1977). "Laboratory investigation of submerged platform breakwaters", Proceedings XVII IAHR Congress, Karlsruhe, Germany, 89-96.

Dick, T. M. (1968). "On solid and permeable submerged breakwaters", PhD Thesis, Queens University, Kingston, UK.

Gent, van M.R.A. (1993). „Berm Breakwaters -Hydrodynamics, Forces, and Profile Development-", Communications on Hydraulic and Geotechnical Engineering, Report No. 93-11. Faculty of Civil Engineering, Delft University of Technology, The Netherlands.

Graw, K.-U. (1993). "The submerged plate as a primary wavebreaker", Proceedings XXV IAHR Congress, Tokyo, Japan, Vol. C, 38-45.

Graw, K.-U. (1995). "Estimation of Ct-Coefficient for the submerged horizontal plate", Proceedings XXVI IAHR Congress, London, UK, Vol. 3, 245-250.

Guevel, P., Landel, E., Bouchet, R., and Manzone, J.M. (1985). "Le phénomène d'un mur oscillant et son application pour protéger un site còtier soumis à l'action de la houle", Bulletin de l'Association Technique Maritime et Aeronautique, Vol. 85, 229-245.

Hughes, S. A. (1993). „Physical Models and Laboratory Techniques in Coastal Engineering" in Advanced Series on Ocean Engineering - Vol. 7, Liu, P. L.-F. (Ed.), World Scientific Publishing, Singapore.

Kaldenhoff, H. and Fischer, C. (1988). „Transfer of Wave Energy around Structures - Illustrated for a Horizontal Plate" in Computational Methods in Flow Analysis, Niki, H. and Kawahara, M. (Ed.), Okayama University of Science, Japan, Vol. 2, 777-784.

Kaldenhoff, H. (1992). „Physical Modelling -Theory and Application", Short Course at OEC, IIT Madras, India.

Kamphuis, J. W. (1985). „On Understanding Scale Effects in Coastal Mobile Bed Models", in Physical Modelling in Coastal Engineering, R. A. Dalrymple, (Ed.) Balkema, The Netherlands, pp 141-162.

Kamphuis, J. W. (1991). „Physical Modelling" in Handbook of Coastal and Ocean Engineering, J. B. Herbich, (Ed.) Vol. 2, Gulf Pub-lishing Company, Houston, USA.

Kobus, H. (Ed.) (1980). „Hydraulic Modelling", German Association for Water Resources and Land Improvement, Bulletin 7, Paul Parey, Hamburg, Germany.

Kojima, H., Ijima, T. and Yoshida, A. (1990). "Decomposition and interception of long waves by a submerged horizontal plate", Proceedings 22nd ICCE, Delft, The Netherlands, 1228-1241.

Noda, E. K. (1972). „Equilibrium Beach Profiles Scale-Model Relationship", Journal of the waterways, Harbours and Coastal Division, ASCE, Vol. WW4, pp 511-528.

Proceedings of the Seventh (1997) International Offshore and Polar Engineering Conference
Honolulu, USA, May 25-30, 1997

ISBN 1-880653-28-1 (Set); ISBN 1-880653-31-1 (Vol. III)

Failure Mode Response Functions in Reliability-Based Design of Rubble-Mound Breakwaters

Aysen Ergin and Can Elmar Balas
Middle East Technical University
Ankara, Turkey

ABSTRACT

The reliability-based model applied in this paper, provides a valuable tool in the design of coastal structures which are characterized by large failure consequences and substantial capital expenditures. Therefore, the implementation of the model to the design of rubble-mound breakwaters has an utmost importance especially in countries such as Turkey, where the use of natural stone for breakwaters is very common due to its economy. The purpose of the present paper is to describe the model by implementing it to Mersin Yacht Harbor which will be constructed near the coastal city of Mersin located on the Mediterranean Sea, and compare the results of the reliability-based model which utilizes the failure functions of Hudson and Van der Meer, with the deterministic design. Mersin Yacht Harbor construction project is a public investment administrated by the Ministry of Transportation. This study is the first application of the reliability-based design for breakwaters in Turkey (Balas, 1996). In the reliability-based model, the hydraulic stability of the main armor layer which is the most important failure mode as dislocking of the armor units under the wave attack, is described to compare the effects of failure mode response functions derived from Hudson and Van der Meer equations.

KEY WORDS: Rubble mound breakwaters, rock armor stability, failure mode response functions, deterministic design, reliability-based design.

INTRODUCTION

The design of rubble mound breakwaters has been based on deterministic methods in which the resistance of the structure is defined in terms of the design load. In these methods, evaluation of structural safety depends upon the semi-empirical design formulae which give a relationship between resistance factors and wave characteristics. In such design formulae, the characteristic values of resistance and load parameters are used as recommended in common design practice, but their random variations are not explicitly considered in the evaluation of structural safety. However, the design of a coastal structure embodies a large degree of uncertainty and variability in the structural capacity and potential load intensities. Therefore, it is essential to evaluate the safety of rubble mound breakwaters by using probabilistic methods in which the design parameters are modelled as random variables.

Evaluation of structural safety depends upon the structural response characterized by various failure modes. A rubble mound breakwater is composed of several components, each of them having many failure modes associated with different failure mechanisms. The principal failure mechanisms for conventional type rubble mound breakwaters (illustrated in Figure 1) can be listed as:

I) Movement and breakage of the armor units on the front and rear face of the armor due to the wave action,
II) Breakage and sliding of the crown wall,
III) Erosion of toe and berm
IV) Geotechnical instabilities.

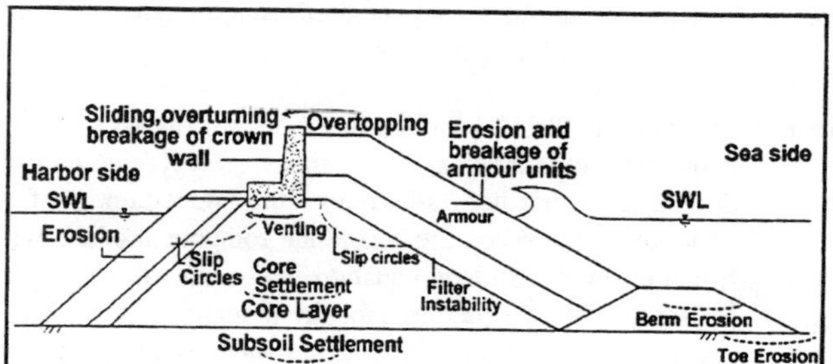

Figure 1. Principal failure mechanisms for conventional type rubble mound breakwaters.

In this paper, the most important failure mode, the hydraulic stability of the main armor layer, is discussed with a particular emphasis on different sources of uncertainties.

APPLICATION OF THE STRUCTURAL RELIABILITY MODEL

It is very difficult to evaluate the safety level of rubble mound breakwaters, due to sources of uncertainties inherent in the design intentions. Therefore, in the reliability-based design model utilized in this paper, the standard deviation which is a measure of scatter around the mean value, and the coefficient of variation which is a dimensionless ratio of standard deviation to mean, are used as measures of uncertainty.

The safety of the breakwater depends upon the combined affect of loads acting on the structure and the available strength. The limit state function specified for a particular failure mode, includes load and strength variables which are random in nature. These random variables are called the basic variables and can be designated by a vector as: $\vec{X}=(X_1,X_2,\ldots,X_n)$. The functional form of the basic variables consistent with the limit state is called the failure function: $g(\vec{X})=g(X_1,X_2,\ldots, X_n)$. Then, the safety of the structure can be ensured by specifying an acceptable value of exceedance probability of a damage level, which corresponds to a small probability of reaching a particular limit state, defined for a constant damage level as:

$$g(\bar{X}) = g(X_1, X_2, \ldots, X_n) = 0 \tag{1}$$

As a case study, the reliability-based design model is implemented to Mersin Yacht Harbor and the reliability of the main breakwater in the harbor is determined by using the limit state equations generated from the commonly used design equations, i.e. Hudson equation as used in the Shore Protection Manual (SPM, 1984) and Van der Meer (1988) equations, as presented in Table 1.

Name	Equations and Limit State Functions	
Hudson	$H_s/(\Delta D_{n50})=(K_D \cot\alpha)^{1/3}$ $g_1=Y_1\Delta D_{n50}(K_D \cot\alpha)^{1/3} - H_s$	(2) (3)
Van der Meer (plunging)	$H_s/(\Delta D_{n50})=6.2\ P^{0.18}\ \xi_m^{-1/2}\ (S/(N^{0.5}))^{1/5}$ $g_2=Y_2S^{0.2}P^{0.18}\cot\alpha^{0.5}\Delta D_{n50}-H_s(H_s/L_z)^{-0.25}N^{0.1}$	(4) (5)
Van der Meer (surging)	$H_s/(\Delta D_{n50})=1.0\ P^{-0.13}\ \xi_m^{P}\ (\cot\alpha)^{1/2}\ (S/(N^{0.5}))^{1/5}$ $g_3=Y_3S^{0.2}P^{-0.13}\cot\alpha^{(0.5-P)}\Delta D_{n50}-H_s(H_s/L_z)^{0.5P}N^{0.1}$	(6) (7)

Table 1. The limit state functions generated from the Hudson equation and Van der Meer equations.

The basic variables in Table 1 can be described as follows:

- g_i: Limit state functions; $g_i \leq 0$: failure, $g_i > 0$: no failure.
- Y_i: Stochastic variables signifying the uncertainty of the equations, lumped on the resistance component of the failure function. In the Hudson equation, the mean value of Y_1 is 1. In Van der Meer equations; for the plunging waves the mean value of Y_2 is 6.2, and for surging waves the mean value of Y_3 is 1.
- ξ_z: Dimensionless surf similarity parameter (Iribarren number): $\xi_z=(\tan\alpha\ L_z)/H_s$.
- H_s: Design significant wave height at the toe of the structure,
- α: Structure front face slope angle,
- L_z: Deep water wave length, $a_gT_z^2/2\pi$,
- a_g: Gravitational acceleration,
- T_z: Average wave period,
- D_{n50}: Nominal diameter of armor unit ($D_{n50} = (M_{50}/\rho_s)^{1/3}$;
- K_D: Dimensionless stability coefficient which signifies the degree of damage,
- Δ: Relative density of the armor unit, $\Delta= (\rho_s-\rho_w)/\rho_w$,
- ρ_s: Mass density of armor unit,
- ρ_w: Mass density of sea water,
- M_{50}: Fifty percent value of the mass distribution curve;
- P: Permeability coefficient of the structure;
- S: Damage level ($S = A_e/D_{n50}^2$) ;
- A_e: Erosion area in a cross section;
- N: Number of waves (storm duration).

In the design of main breakwater of Mersin Yacht Harbor, both deterministic and reliability-based models are applied to evaluate the nominal diameter D_{n50} by using both Hudson and Van der Meer equations. In these designs, two units of rough angular quarry stones are randomly placed on the armor layer with a slope of $\cot\alpha=2.5$. The unit weights of quarry stone and sea water are $\gamma_r= 2.7$ tons/m^3 and $\gamma_w=1.03$ tons/m^3, respectively.

The first part of the study constitutes the deterministic design, where the annual extreme significant design wave height in front of the main breakwater in Mersin (at 8 m. water depth), is determined from the Fisher-Tippett Type I distribution, which is the most accepted extreme value distribution for annual significant wave height data in practice, as follows (Ergin and Özhan, 1986):

$$P(H_s^L) = [\ \exp(-\exp(-\frac{H_s - 2.516}{0.6279}))\ ]^L \tag{8}$$

where,

- H_s^L : Maximum value of the significant wave height (meters) within a period of L years, at 8 m. water depth.
- $P(H_s^L)$: Non-exceedance cumulative probability of the extreme significant wave height within L years, at 8 m. water depth.

The significant wave heights predicted in front of the main breakwater for various return periods are presented in Table 2. For this project, the design significant wave height of the main breakwater is selected from Table 2, as 5.0 meters with a return period of 50 years.

Return Period (R_p) (years)	5	10	25	50	100
H_s (m.)	3.46	3.94	4.53	5.00	5.42
95% confidence interval	±0.04	±0.05	±0.08	±0.10	±0.11

Table 2. Significant wave height values predicted in front of the main breakwater for various return periods.

In the deterministic design, the current practice in Turkey is to use the "no-damage" criterion, i.e. (0-5)% damage level. For this condition, firstly, the nominal diameter D_{n50} is obtained from Hudson Equation (Equation 1), as $D_{n50}=1.44$ m. The stability coefficient in Hudson equation is given as $K_D=4$ for the trunk section subjected to non-breaking waves (SPM, 1984).

Secondly, the nominal diameter (D_{n50}) in the trunk section of the main breakwater is determined by using Van der Meer equations. The permeability coefficient of the breakwater is taken as P=0.4 (Van der Meer, 1988). For the Mersin region, the deep water wave steepness is determined from the wave climate study as $(H_s)_o/L_o = 0.042$ with a maximum standard deviation of 0.01 (Ergin and Özhan, 1986). Hence, the average wave period can be obtained by using this relation as

T_z=8.7 seconds. Then, the surf similarity parameter, which describes the form of wave breaking, is calculated as $\xi_z = 2.12$. The transition from plunging waves to surging waves on the structure is defined by using a critical surf similarity parameter ξ_c:

$$\xi_c = \left[6.2\ P^{0.31}\ (\tan\alpha)^{1/2}\right]^{\frac{1}{P+0.5}} \quad (9)$$

Hence, the critical surf similarity parameter is determined as ξ_c = 3.33. Since $\xi_z < \xi_c$, the form of wave breaking is plunging and Equation (4) can be used to describe the fast wave run-up after breaking that effects the stability of the breakwater. From the storm duration statistics performed for Mersin Yacht Harbor, the maximum average number of waves in storms can be taken as μ_{Nw}=2500 waves with a standard deviation of σ_{Nw} = 1720 waves (Özhan, et al., 1994). In Van der Meer equations, the definition of start of damage, i.e. S= 2, is the same as the "no-damage criterion" of Hudson, i.e. (0-5)% damage level. For this condition, the nominal rock diameter is obtained from Van der Meer Equation for plunging waves as D_{n50}=1.55 meters.

Second part of the study constitutes the reliability-based model, which takes into account the possible variations in the basic variables. In other words, the basic variables are characterized by their first and second moments of probability density functions, i.e. their mean and standard deviation in the model. Therefore, the basic variables defined in Table 1 are transformed into reduced variates by using their average values and standard deviations. The average values are the same as in the deterministic design and the following items with respect to mean biases and uncertainties are considered in the reliability model:

i) The design wave height is a time variant quantity which can be modelled as a stochastic variable. The uncertainties in the determination of design wave are inherent in wave hindcasting studies and the key parameter in the design is the significant wave height H_s. Therefore, the coefficient of variation of the significant wave height (δ_{Hs}), which is determined from wave hindcasting studies in deep water, varies between 15-20% (PIANC, 1992). In Turkey, the typical δ values for the deep water wave hindcasting studies carried out in the south east coast of the Mediterranean Sea, are also between 15 - 20% (Ergin and Özhan, 1986). These values may increase to 35% as refraction and shoaling, i.e. the typical shallow water effects, are taken into account in the nearshore (PIANC, 1992). In this study, δ_{Hs} is taken as 30 %.

The mean ($\mu_{Hs}{}^L$) and standard deviation ($\sigma_{Hs}{}^L$) values of the significant wave height due to statistical uncertainties can be determined by using the following equations derived for the Fisher-Tippet I distribution (Burcharth, 1992):

$$\mu_{H_s^L} = B - A\ \ln[-\ln(1-\frac{1}{L})] \quad (10)$$

The standard deviation of $H_s{}^L$, due to the limited sample size (n=24 years), can be determined from:

$$\sigma_{H_s^L} = \sqrt{\frac{A^2}{n}[1.109+0.514(-\ln[-\ln(1-\frac{1}{L})])+0.608(-\ln(-\ln(1-\frac{1}{L})))^2]} \quad (11)$$

where, A and B are the parameters of the Fisher-Tippet I distribution, and L is the lifetime of the structure. Hence, the coefficient of variation of the significant wave height due to statistical uncertainties is obtained as $\delta_{Hs}{}^L = \sigma_{Hs}{}^L/\mu_{Hs}{}^L$= 9.1 %. In these calculations, the parameters of the annual extreme Fisher-Tippett I distribution presented in Equation (8) are used and the lifetime of the structure is taken as L=50 years.

Finally, the total coefficient of variation (Ω_{Hs}) which includes the statistical uncertainties due to limited sample size and the uncertainties in the hindcasting model is evaluated as 32 % from the following equation:

$$\Omega_{H_s} = \sqrt{\delta_{H_s^L}{}^2 + \delta_{H_s}{}^2} \quad (12)$$

ii) The failure probabilities are evaluated from the available design formulae which are based upon various descriptions of failure mechanisms. As a result, the variability is often inherent in the derivation of these simple formulae, since they are fitted to experimental data points which are scattered in nature.

The reliability of the Hudson equation under random waves is examined by Van der Meer (1988), and it is found that, the stability coefficient has a variation coefficient of Ω_{KD}=18%. Hence, in this study, the reliability of the Hudson failure function is expressed by the normally distributed coefficient Y_1, which is the stochastic variable signifying the uncertainty of the equation as defined in Table 1. Then, the expected (mean) value of Y_1 can be taken as μ_{Y1}= 1, with a standard deviation of σ_{Y1}=0.18 and variation coefficient of Ω_{Y1}=18%. In the Van der Meer equation for plunging waves, the mean value of Y_2 is μ_{Y2}=6.2, with a standard deviation of σ_{Y2}=0.4 and variation coefficient of Ω_{Y2}=6.5% (Van der Meer, 1988). For surging waves the mean value of Y_3 is μ_{Y3}=1 by a standard deviation of σ_{Y3}=0.08 and a variation coefficient of Ω_{Y3}=8%. These variation coefficients reflect the random behavior of rock slopes, accuracy of damage measurements and curve fitting of test results. Reliabilities of these equations are expressed by normally distributed coefficients Y_2 and Y_3, respectively. These values are significantly lower than that for the Hudson equation. The variations in other parameters are discussed in Balas et al (1995); and in Balas and Hapoğlu (1995). The average and standard deviation values are listed in Table 3.

Basic Variables (X_i)	Distribution	Average (μ_{xi})	Standard deviation (σ_{xi})	Coefficient of variation (Ω) (%)
Y_1	Normal	1.0	0.18	18
$(D_{n50})_{Hudson}$	Normal	1.44 (m.)	0.07 (m.)	5
Y_3	Normal	6.2	0.4	6.5
$(D_{n50})_{V.D.Meer}$	Normal	1.55 (m.)	0.08 (m.)	5
H_s	Fisher-Tippett I	2.88 (m.)	0.81 (m.)	32
H_s/L_z	Normal	0.042	0.01	23
Δ	Normal	1.62	0.05	3
P	Normal	0.4	0.1	25
N	Normal	2500	1720	86
$\cot\alpha$	Normal	2.5	0.3	12

Table 3. Basic variables used in limit state equations of Hudson and Van der Meer.

Design by Hudson Failure Function

The first part of reliability-based design constitutes, determination of the size of armor units in the trunk section of main breakwater by using the design wave height parameters of H_s and $H_{1/10}$ in Hudson failure function. The section will be designed to survive under the extreme wave conditions without sustaining a damage that exceeds the predescribed damage level in its lifetime. The exceedance probability of that damage level is evaluated by using the inherent uncertainties in the related parameters of Hudson failure function, given in Table 3. The exceedance probabilities of several damage levels in the lifetime of the structure (L=50 years), are evaluated as a function of nominal diameter D_{n50} for both of the design wave heights H_s and $H_{1/10}$, as illustrated in Figures (2 and 5).

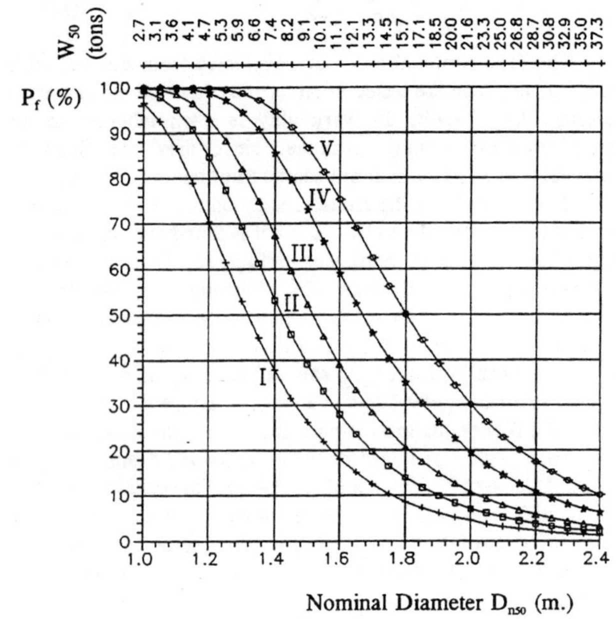

Figure 2 Design graph 1: Exceedance probabilities (P_f) as a function of nominal rock diameter (D_{n50}) in L=50 years lifetime of the structure, designed by using the wave height parameter H_s in Hudson failure function, for damage level (DL) alternatives of:
I: DL=(20-25)%, II: DL=(15-20)%, III: DL=(10-15)%, IV: DL=(5-10)%, V: DL=(0-5)%.

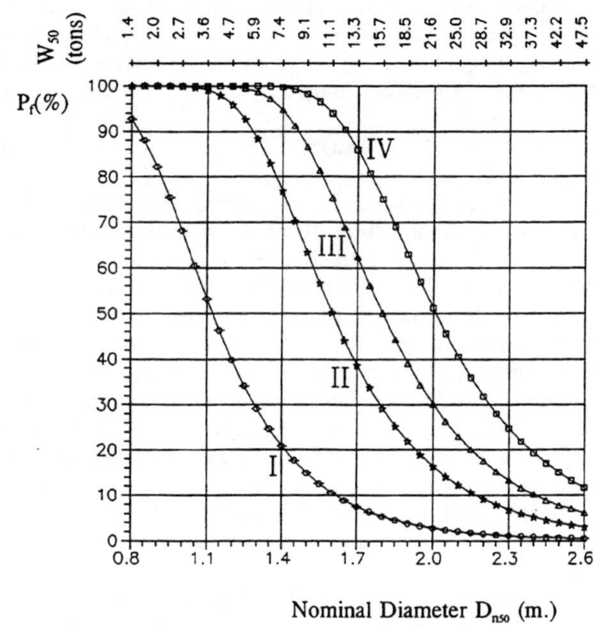

Figure 4 Design graph 3: Exceedance probabilities of the damage level DL=(0-5)%, as a function of nominal rock diameter D_{n50} obtained by using the wave height parameter H_s in Hudson failure function, for the following lifetime (L) alternatives of the structure:
I: L= 4 years, II: L=25 years, III: L=50 years, IV: L=100 years.

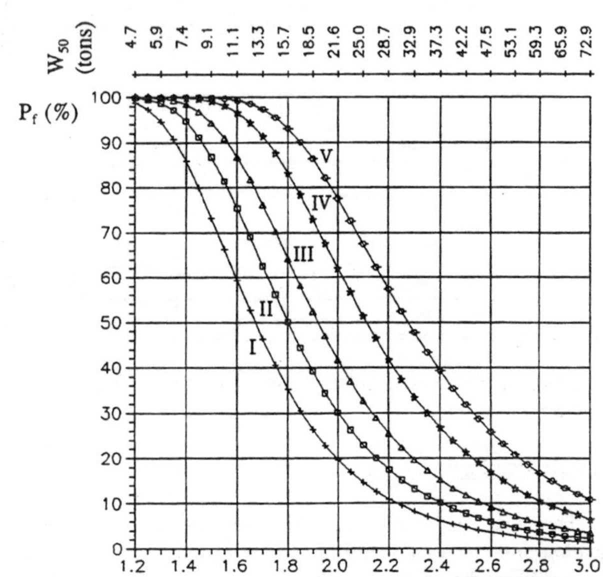

Figure 3 Design graph 2: Exceedance probabilities (P_f) as a function of nominal rock diameter (D_{n50}) in L=50 years lifetime of the structure, designed by using the wave height parameter $H_{1/10}$ in Hudson failure function, for damage level (DL) alternatives of:
I: DL=(20-25)%, II: DL=(15-20)%, III: DL=(10-15)%, IV: DL=(5-10)%, V: DL=(0-5)%.

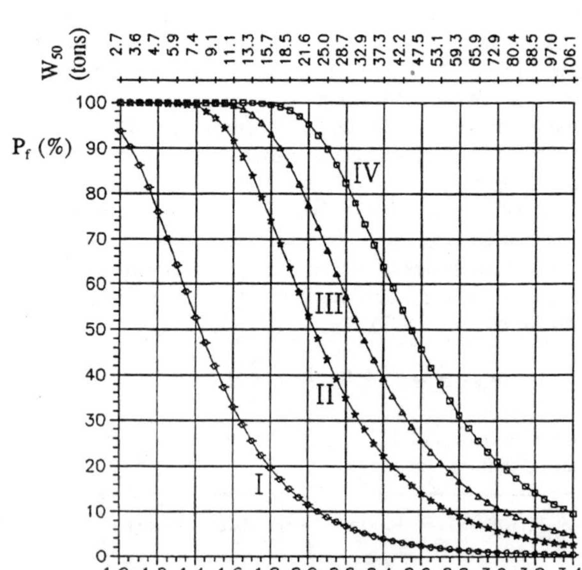

Figure 5 Design graph 4: Exceedance probabilities of the damage level DL=(0-5)%, as a function of nominal rock diameter D_{n50} obtained by using the wave height parameter $H_{1/10}$ in Hudson failure function, for the following lifetime (L) alternatives of the structure:
I: L= 4 years, II: L=25 years, III: L=50 years, IV: L=100 years.

For a constant lifetime of the structure ($L=50$ years), the following discussions can be carried out by considering Figures 2 and 3, where the wave height parameters of H_s and $H_{1/10}$ are used, respectively:

I) For a constant damage level (DL=constant), the nominal rock diameter (D_{n50}) increases as the exceedance probability (P_f) decreases.

II) For a constant design diameter (D_{n50}=constant), the damage level increases as the corresponding exceedance probability decreases. Hence, the exceedance probability also depends upon the damage level.

III) For a constant exceedance probability (P_f=constant), the damage level increases as the nominal rock diameter decreases. This leads to the planning and decision criteria in which the initial cost of the investment is decreased by selecting a smaller stone size, whereas the maintenance and repair costs of the breakwater is increased.

For another discussion, the damage level is taken as constant, e.g. DL=(0-5)%, and the exceedance probabilities of that damage level are evaluated for various lifetime alternatives of the breakwater as illustrated in Figure 4 and 5, where the wave height parameters of H^s and $H_{1/10}$ are used, respectively. For constant damage level DL=(0-5)%, it can be seen that;

I) For a constant rock diameter (D_{n50}=constant), as the lifetime of the structure (L) increases, the exceedance probability (P_f) also increases.

II) For a constant exceedance probability (P_f=constant), as the lifetime of the structure increases, the design rock diameter increases as well. Therefore, lifetime which depends upon the importance of the structure, is considered as a major decision parameter in design. Thus, selection of the design lifetime, which may lead to over-design or under-design, has to be made very carefully.

Figures (2-5) permit the rubble mound structure to be designed for several damage levels and failure consequences, and for design wave height and lifetime alternatives. Thus, an economical (cost/benefit) analyses can be carried out for an optimal design, where initial and maintenance costs of the structure can be optimized. In addition, the economic consequences of structural failure can be reflected in the model, by choosing a reliability level, which is defined as the non-exceedance probability of a damage condition during a specified reference period and under given environmental conditions. High reliability levels are selected, when the economic consequences of failure are serious, so that the public is not exposed to a risk, such as loss of property or life. In the reliability-based design, a moderate damage of (5-10)% can be allowed in the armor layer, that will be repaired economically in its lifetime. The acceptable exceedance probability of that damage level can be selected as 60% in 50 years lifetime, as suggested by Maritime Works Recommendations (PIANC, 1992). Hence, in the "reliability-based design where the wave height parameter H_s is employed in Hudson failure function", the nominal diameter of the quarry stone is determined as $D_{n50}=1.6$ m. (W_{50} = 11 tons) for a moderate damage level of DL=(5-10)% and an acceptable exceedance probability of $P_f=60\%$ in 50 years lifetime. The damage levels of (0-5)% and (10-15)% will be exceeded in the 50 years lifetime of the structure with a probability of 75.3% and 38.93%, respectively. Although these exceedance probabilities of damage levels (P_f) are high, they show reasonable agreements with the currently accepted exceedance probabilities (Burcharth, 1992).

In the "reliability-based design accomplished by using the wave height parameter $H_{1/10}$ in Hudson failure function", the nominal diameter of the quarry stone is determined as $D_{n50}=2.0$ m. (W_{50} = 21.6 tons) for a moderate damage level of (5-10)% and an acceptable exceedance probability of approximately 60% in 50 years lifetime. In the design, the damage levels of (0-5)% and (10-15)% will be exceeded in $L=50$ years, with a probability of 77.48% and 41.70%, respectively. The safety levels of the breakwater designed by the deterministic approach, can be investigated by taking the statistics of the basic variables (i.e. mean and uncertainty) presented in Table 3. Hence, in the structure designed by the "deterministic approach in which H_s wave height parameter is used in Hudson equation"; the "no-damage" level in the armor layer having a nominal diameter of $D_{n50}=1.44$ m. ($W_{50}=8.1$ tons), will be exceeded by a probability of $P_f=91.9\%$ in 50 years of lifetime, which may be considered too high to be accepted. In this structure, the damage levels of (5-10)% and (10-15)% will be exceeded with a probability of 80.7% and 61.0%, respectively.

In the "deterministic design where the $H_{1/10}$ wave height parameter is utilized in Hudson equation", the exceedance probability of the "no-damage" level in the armor layer which has a nominal diameter of $D_{n50}=1.81$ m. (W_{50} = 16.1 tons), can be obtained from Figure 3 as $P_f=93.1\%$ in 50 years of structural lifetime. In this structure, the damage levels of (5-10)% and (10-15)% will be exceeded in $L=50$ years, with a probability of 83.1% and 64.13%, respectively. The design parameters evaluated from the deterministic and reliability-based design approaches by using the Hudson failure function, are summarized in Table 4.

Design	Deterministic Design		Reliability-based Design	
H_d (m.)	$H_s=5.0$	$H_{1/10}=6.35$	$\mu_{Hs}=$	$\mu_{H1/10}=$
D_{n50} (m.)	1.44	1.81	1.6	2.0
W_{50} (tons)	8.1	16.0	11.0	21.6
DL (%)	0-5	0-5	5-10	5-10
P_f (%)	91.9	93.1	60	62.1
β_{HL}	1.65	1.81	2.1	2.07

Table 4. Design parameters evaluated from the deterministic and reliability-based design approaches by using the wave heights H_s and $H_{1/10}$ in Hudson equation.

By using Table 4, the deterministic and reliability-based design models can be compared as follows:

a) In the design of rubble mound structures, it is recommended to use the average of highest 10 percent of all waves $H_{1/10}$, as the design wave height in Hudson equation. On the other hand, the significant wave height H_s, has been generally employed in common practice. In reliability-based and deterministic models, the weight of the armor unit obtained by using the significant wave height is almost the half of the weight determined by utilizing $H_{1/10}$ in Hudson equation. As the design wave height increases, the uncertainties inherent in the wave height distribution are more reflected in the exceedance probabilities of the damage level. Consequently, the exceedance probability for $H_{1/10}$ is larger than the probability for H_s, as given in Table 4. Using the wave height parameters $H_{1/10}$ or H_s in Hudson equation still remains as a question, which should be studied in the light of safety levels determined by using prototype and model data. b) In deterministic design, a damage level of DL=(0-5)% is used as the common practice in Turkey. In reliability-based design, a damage level of DL=(5-10)% can be allowed in the breakwater by considering the maintenance possibilities in the armor layer; since the random variations associated with the resistance and load parameters, and the uncertainties inherent in the design are taken into account by the mean and variation coefficients of the design parameters. Thus, the reliability-based model is a more realistic model than the deterministic one.

c) The safety levels of breakwaters designed by the deterministic approach can be investigated by the reliability-based model. Hence, it is found that, the exceedance probability of the DL=(0-5)% damage level is in the range of (90-95)% in the 50 years lifetime of the structure. In reliability-based design, the exceedance probability of (5-10)% damage level is selected in the range of (60-70)%, depending on engineering judgement.

d) For rubble mound breakwaters, the current practice in the reliability-based design gives higher values of nominal rock diameter as compared with the deterministic one. This difference is quite expectable, since in the reliability-based design, the random variations associated especially with load parameters, and the uncertainties inherent in Hudson failure function are quite significant. The difference may increase as the damage level and its exceedance probability decreases.

This comparison of reliability and deterministic models showed that, the difference between these models is basically reflected in their design methodologies. The following steps are carried out in the deterministic design model:

I) The characteristic design wave height which is the dominating environmental load parameter, is determined by using an extreme probability distribution with the assigned return period.

II) For an assigned lifetime of the structure and for a selected return period, the encounter probability of damage (E_D) is evaluated. If the encounter probability of damage is considered as high, the return period is re-selected and the characteristic design wave height is calculated again.

III) For a selected damage level, the size of the armor layer is directly obtained from the design equation without considering the uncertainties associated in the design.

IV) Hydraulic model tests are carried out for the final design, since the uncertainties which may effect the preliminary design significantly, are not considered at all.

In the reliability-based model, the following steps are accomplished:

I) In the preliminary design, the characteristic design wave height is modelled by using an extreme probability distribution. The mean and variation coefficient of the distribution are utilized in the design.

II) A reliability level, i.e. the non-exceedance probability of a damage level is decided for the extreme environmental condition encountered in the lifetime of the structure.

III) In the reliability-based model which introduces a new design methodology for coastal structures, the most probable values of design parameters at the limit state condition are evaluated with their reliability level.

IV) For various design parameter alternatives, the size of the armor layer is obtained as a function of its reliability level as illustrated in Figures (2-5). Considering the optimum solution, the design parameters can be re-selected.

V) Hydraulic model tests are carried out for the final design, to examine the failure modes that are not covered by the design formula.

<u>Design by Van der Meer Failure Function</u>

In the reliability-based design of the main breakwater in Mersin Yacht Harbor, the significant design wave height H_s is utilized in Van der Meer failure function of plunging waves. The exceedance probabilities of various damage levels are evaluated by using the average and standard deviation values of basic variables listed in Table 5. These probabilities are obtained for a constant lifetime of the structure (L=50 years) and as a function of nominal diameter D_{n50}, as shown in Figure 6. This figure allows the design of the main breakwater, to be performed for several damage levels and failure consequences in its lifetime. For various lifetime alternatives of the breakwater, the exceedance probabilities can be examined for a constant damage level, as illustrated in Figure 7.

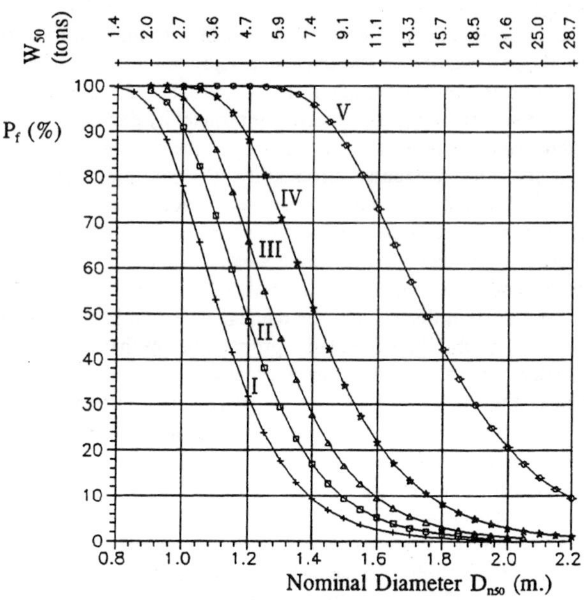

Figure 6 (Design graph 5): Exceedance probabilities as a function of nominal rock diameter in L=50 years lifetime of the structure, designed by using the wave height parameter H_s in Van der Meer failure function of plunging, for damage level alternatives of:
I: S=20, II: S=14, III: S=10, IV: S=6, V: S=2.

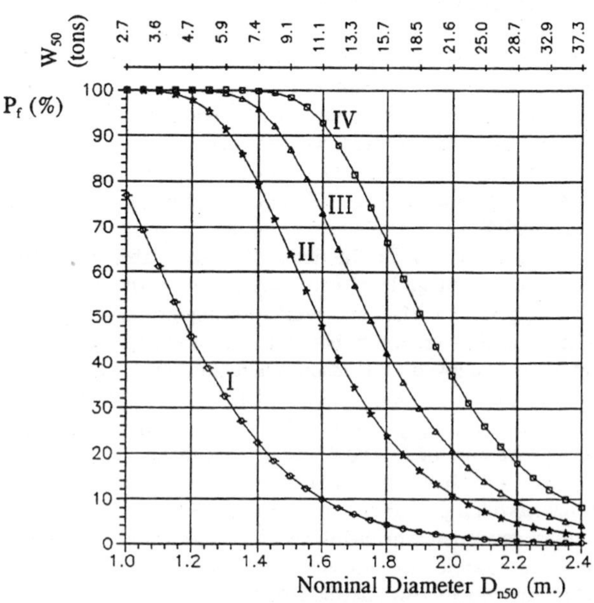

Figure 7 (Design graph 6): Exceedance probabilities of the damage level S=2, as a function of nominal rock diameter obtained by using the wave height parameter H_s in Van der Meer failure function of plunging, for the following lifetime (L) alternatives: I:L=4 years, II:L=25 years, III:L=50 years, IV:L=100 years.

The design curves displayed in Figures 6 and 7, allow the designer to conduct a cost/benefit analysis in order to select the optimal design of the structure and assess the economic consequences of failure. Similar discussions holds true for Figures 6 and 7; such as for a constant lifetime and damage level, as the rock diameter increases, exceedance probabilities decrease and on the contrast, for a constant rock diameter and damage level, as the lifetime of the structure increases, the exceedance probability also increases.

In the "reliability-based design performed by using wave height H_s in Van der Meer failure function of plunging waves", a moderate damage level of S=6, i.e. (5-10)% damage, can be allowed for the armor layer, that will be repaired economically in its lifetime. The acceptable exceedance probability of this damage level is also recommended as P_f=60%, in 50 years of structural lifetime (PIANC, 1992). Then, the nominal diameter and the weight of 50% size armor units in gradation, are obtained by using these commonly accepted design recommendations, as D_{n50}=1.35 m. and W_{50} = 6.6 tons, respectively. In the design, the damage levels of S=2 (0-5%) and S=10 (10-15%) will be exceeded in L=50 years with a probability of 98.06% and 35.48%, respectively.

In the "deterministic design performed by using the wave height parameter H_s in Van der Meer equation of plunging waves", the exceedance probability of S=2 in the armor layer which has a nominal diameter of D_{n50}=1.55 m., can be obtained from Figure 6, as P_f=80.52% in L=50 years of lifetime. In this structure, the damage levels of S=6 and S=10 will be exceeded in L=50 years with a probability of 27.42% and 12.57%, respectively. The design parameters evaluated from the deterministic and reliability-based design approaches by using Van der Meer failure function of plunging waves, are summarized in Table 5.

Design	Deterministic Design	Reliability-based Design
D_{n50} (m.)	1.55	1.35
W_{50} (tons)	10.1	6.60
DL (%)	0-5	5-10
P_f (%)	80.5	60.0

Table 5. Design parameters evaluated from the deterministic and reliability-based designs by using the wave height H_s in Van der Meer equation of plunging.

It can be seen from Table 5 that, the current practice in the reliability-based design in which a moderate damage level of S=6, i.e DL=(5-10)%, is accepted with an exceedance probability of 60%, gives higher values of nominal rock diameter as compared to the deterministic design accomplished for S=2, i.e DL=(0-5)% damage level, with an exceedance probability of 80% which may be considered too high to be accepted in L=50 years. This difference between deterministic and reliability-based design models is anticipated, since in the reliability-based design, the random variations and uncertainties in design parameters are considered.

COMPARISON OF FAILURE MODE RESPONSE FUNCTIONS

Exceedance probabilities of damage levels as a function of nominal diameter, obtained from the deterministic and reliability-based design models by using Hudson and Van der Meer equations are summarized in Table 6. Firstly, the deterministic model is considered for the comparison of the failure mode response functions used in the designs. It can be seen from Table 6 that, Hudson equation gives the nominal diameter of the armor layer as D_{n1}=1.44 m. and D_{n2}= 1.32 m. for damage levels of DL_1=(0-5)% and DL_2=(5-10)%, respectively. Hence, as the expected damage level increases, the nominal diameter of the armor layer decreases; whereas the exceedance probability of damage levels remains almost constant around 90%. Using Van der Meer equations in the deterministic model gives similar results, where the exceedance probability of damage levels remains almost constant around 80%. This difference in the exceedance probabilities, results from the difference in the uncertainty levels of the equations, which are signified by the stochastic variables Y_i, as given in Table 3. In the deterministic design, the comparison of the Hudson and Van der Meer equations shows that, the nominal diameter is effected by the damage level and Van der Meer equations give larger nominal rock diameter for the damage level of DL=(0-5)%. As the damage level increases, larger nominal diameters for the armor layer are obtained by using Hudson equation, as compared to ones obtained from Van der Meer equations.

Design Approach	Deterministic design				Reliability-based design			
Equation	Hudson		Van der Meer		Hudson		Van der Meer	
Damage level DL (%)	(0-5)	(5-10)	(0-5)	(5-10)	(0-5)	(5-10)	(0-5)	(5-10)
Exceedance Probability P_f (%)	91.91	96.1	80.5	80.4	60.0	60.1	60.3	61.09
Nominal Diameter D_{n50} (m.)	1.44	1.42	1.55	1.25	1.72	1.6	1.68	1.35

Table 6. Comparison of the design approaches

Secondly, the failure mode response functions of Hudson and Van der Meer are compared in the reliability-based design. In general, similar to the deterministic model, the rock diameter of the armor layer decreases as the damage level increases as it can be seen from Table 6. The computations for rock diameter using both Hudson and Van der Meer equations are carried out for the exceedance probability of P_f= 60%, which is the commonly used practical value in the reliability-based designs. It can be observed from Table 6 that, for the same damage levels, the Van der Meer equation gives smaller nominal diameter values as compared to Hudson equation.

The comparison of reliability and deterministic models shows that, the difference between these models is basically reflected in the exceedance probability concept, which reflects the difference in the design methodologies. In the reliability-based model, the design starts by setting the exceedance probability of damage level and is followed by calculating the corresponding nominal armor layer diameter. On the other hand, in deterministic model, the starting point of design is the determination of nominal armor layer diameter by using equations written for a given damage level, and the ending point is checking the corresponding exceedance probability of damage level.

For rubble mound breakwaters, the current practice in the reliability-based design, in which a moderate damage level can be accepted with an exceedance probability range of (60-70)%, gives higher values of nominal rock diameter as compared with the deterministic design performed for the same damage level by using both Hudson and Van der Meer equations, as shown in Table 6. This difference between deterministic and reliability-based design models is quite expectable, since in the reliability-based design, the random variations associated with the resistance and load parameters, and the uncertainties inherent in the design methodologies are taken into account. The difference increases as the damage level and its exceedance probability decreases.

CONCLUSIONS

Uncertainties inherent in the design of rubble mound breakwaters are relatively high as compared to classical construction operations. Assessment of these uncertainties, which is essential for the design of the structure, is possible in the reliability-based model. In this model, the exceedance probabilities of different damage levels are evaluated depending on the failure consequences in the lifetime of the structure. Thus, the reliability-based model permits the design to be performed for several damage levels and failure consequences, leading to an optimal design where initial and maintenance costs of the structure can be optimized.

In the reliability-based design, the combinations of (0-5)% damage level and low failure probabilities may lead to relatively large armor stone diameters in 50 years lifetime, since for the combinations, the effects of uncertainties become more significant. These combinations can be used for structures where the economic consequences of failure are serious. Thus, the economic consequences of structural failure can be reflected by choosing a reliability level, which is defined as the probability of survival of a structure during a specified reference period and under given environmental conditions. Low failure probabilities are selected, when the economic consequences of failure are serious, so that the public is not exposed to a risk, such as loss of property or life. Optimization of capital and maintenance costs is another factor to be considered in the design.

The reliability-based design is observed to give higher values of nominal rock diameter for the same damage level as compared to the deterministic design approach. This is due to the fact that, the reliability-based design incorporates the effects of random variations associated with the resistance and load parameters. It is also recommended to carry out hydraulic model tests for the final design of these structures, in order to examine the failure modes that are not covered by the design formula.

REFERENCES

Balas, CE (1996). "A Risk Management Model for Coastal Projects", *Unpublished Ph.D. Thesis* (In Progress), The Graduate School of Natural and Applied Sciences, Middle East Technical University, Ankara, Turkey.

Balas, CE, and Hapoğlu, H (1995). "Risk Management of Breakwaters: A Case Study of Finike Yacht Harbor", *Ocean Cities' 95 Conference*, Vol. 1 pp. 317-320, 19-23 November 1995, Monaco.

Balas, CE, Ergin, A and Yücemen, SM (1995). "Comparison of the Reliability-Based and Deterministic Design Procedures for Rubble Mound Breakwaters" *The Second International Conference on the Mediterranean Coastal Environment*, Vol 3, pp. 1767-1782, 24-27 October 1995, Tarragona, Spain.

Burcharth, HF (1992). "Reliability Evaluation of a Structure at Sea", *Design and Reliability of Coastal Structures*, Proceedings of the short course on design and reliability of coastal structures, Venice, Italy.

Ergin, A, and Özhan, E (1986). "Wave Hindcasting Studies and Determination of Design Wave Parameters for 15 Sea Regions", *Technical Report No: 35*, Coastal and Harbor Engineering Research Center, Civil Engineering Department, Middle East Technical University, Ankara, Turkey.

PIANC (1992). "Analyses of Rubble Mound Breakwaters", *Report of Working Group No: 12*, Permanent International Association of Navigation Congresses (PIANC), Brussels, Belgium.

SPM (1984). "Shore Protection Manual", *Coastal Engineering Research Center*, U.S. Army Waterways Experiment Station, U.S. Government Printing Office, Washington, D.C., U.S.A.

Van der Meer, JW (1988). "Rock Slopes and Gravel Beaches Under Wave Attack", *Delft Hydraulics Communication No: 396*, Delft, The Netherlands.

Özhan, E, Ergin, A, and Yalçıner, AC (1994). "Mersin Yacht Harbor Research Studies", *Technical Report No: 51*, Coastal and Harbor Engineering Research Center, Civil Engineering Department, Middle East Technical University, Ankara.

Proceedings of the Seventh (1997) International Offshore and Polar Engineering Conference
Honolulu, USA, May 25-30, 1997

ISBN 1-880653-28-1 (Set); ISBN 1-880653-31-1 (Vol. III)

Wave Blocking Efficiency of Flexible Membrane Breakwaters in Oblique Seas

S.T. Kee
Hyundai Institute of Construction Technology, Korea
I.H. Cho
Korea Research Institute of Ships and Ocean Engineering, Korea
M.H. Kim
Texas A&M University, College Station, TX, USA

ABSTRACT

The interaction of oblique incident waves with a tensioned, inextensible, vertical flexible membrane wave barrier hinged at the sea floor and a surface buoy is investigated in the context of two-dimensional linear wave-body interaction theory. First, analytic solutions were obtained for an idealized system without buoy. Second, a more practical system with the membrane tension provided by a buoy is investigated by numerical methods. For each case, both submerged and surface-piercing systems were considered. A two-domain boundary element method is developed based on a discrete membrane dynamic model and simple-source distribution over the entire fluid boundaries. Using the developed computer program, the performance of surface-piercing or submerged buoy/membrane wave barriers is tested with various waterdepths, membrane, buoy, and mooring characteristics and wave conditions including oblique wave headings. It is found that the efficiency of a submerged or surface-piercing buoy/membrane breakwater can be enhanced in oblique waves for certain design conditions. From our numerical examples, it can be concluded that the buoy/membrane wave barrier can function as a very effective breakwater even in oblique waves if it is properly designed.

INTRODUCTION

The vertical flexible membrane, which is light, inexpensive, reusable, and rapidly deployable, can be used as a portable and sacrificial breakwater. Since it can be easily removed, we expect minimum environmental impacts on various coastal processes. Using inflatable buoy, it can be air-dropped and self-erected. Its shape and mass can be easily controlled by filling with air or water (e.g. Ohyama et al., 1989; Broderick & Jenkins, 1993).

The performance of this kind of membrane breakwater has been studied by Thomson et al. (1992), Kim & Kee (1996), and Kee & Kim (1997). Another kind of flexible vertical breakwater using the elastic plate clamped at the sea floor was investigated by Lee & Chen (1990) and Williams et al. (1991,1992). The effectiveness of this kind of vertical breakwater can be improved by tuning structural responses (Sollitt et al, 1986; Abul Azm, 1994) or by adding structural porosity (Wang & Ren, 1993).

In Kim & Kee (1996), the interaction of regular waves with an idealized surface-piercing vertical flexible membrane hinged or elastically supported at the seabed and the mean free surface was considered. Both analytic and numerical solutions were developed and it was found that almost complete reflection was possible despite large vertically-sinusoidal membrane motions which tended to generate only exponentially decaying local (evanescent) waves in the lee side. Consequently, the efficiency was in general much higher than conventional floating breakwaters (e.g. Isaacson et al., 1994) or clamped-elastic-plate breakwaters.

In Kee & Kim (1997), more practical buoy-membrane systems were considered, for which a cylindrical buoy was attached to the top of membrane to provide sufficient tension needed for high performance. The diffraction/radiation of the buoy, dynamic coupling of buoy and membrane, and wave-membrane interaction were numerically solved. The system was highly efficient when the waves generated in the lee side by the buoy and membrane were mutually canceled. The radiated waves by a buoy tended to diminish the efficiency of the membrane-alone case unless the system was optimally designed. The numerical results were confirmed by a series of experiments conducted in the two-dimensional wave tank.

In this paper, the performance of a buoy/membrane wave barrier is investigated for oblique incident waves. The hydrodynamic interaction of oblique incident waves with the combination of rigid and flexible bodies was solved by the distribution of the simple sources (modified Bessel function of the second kind) that satisfy the Helmholz governing equation. The coupling of the buoy and membrane motions was taken into consideration through an appropriate boundary condition at the joint. The effects of taut, pretensioned mooring lines were also investigated.

The buoy that provides required tension on the membrane by buoyancy is assumed to be rigid and uniformly long in the longitudinal direction so that two-dimensional analysis can be applied. The membrane connecting the buoy and the seabed is assumed to be thin, inextensible, and free to move in the transverse direction while remaining uniform in the longitudinal direction. Under these assumptions, a two-dimensional hydro-elastic formulation was carried out in the context of linear wave-body interaction theory. Since boundary element method (BEM) was used, the shape of the buoy can be arbitrary but in this paper only a cylindrical section is investigated.

The interaction of oblique incident waves with a two-dimensional floating body without membrane has been investigated by many researchers (e.g. Bai, 1975; Rahman & Chen, 1993). In contrast to the previous studies for floating bodies only, a unique multi-domain boundary element method is developed in this paper to solve the interaction (both diffraction and radiation) of oblique incident waves with a thin vertical flexible membrane with a two-dimensional rigid buoy attached at its top. The numerical results are validated through energy relation and comparison with independently-developed analytic solutions. Using this computer program, the performance of various buoy/membrane systems in oblique incident waves is examined.

ANALYTIC SOLUTIONS

We consider the interaction of a vertical membrane wave barrier with monochromatic oblique incident waves. It is also assumed that wave and membrane motions are small so that linear theory is applicable. For analysis, the Cartesian coordinate system with the origin on the mean free surface and the y axis positive upward is used. Assuming ideal fluid and harmonic motion of frequency ω, the velocity potential can be written as $\Phi(x,y,z,t)=\mathrm{Re}[\phi(x,y)e^{ik_z z-i\omega t}]$, where $k_z = k_0 \sin\theta$ is the wave number component in the z direction and θ is the angle (wave heading) of incident waves with respect to the x axis. Then, the two-dimensional velocity potential of a monochromatic incident wave of amplitude A, wavenumber k_o, and wave heading θ $(0<\theta<\pi/2)$ is given by

$$\phi_0 = \frac{-igA}{\omega}\frac{\cosh k_o(y+h)}{\cosh k_o h}e^{ik_o\cos\theta x} \quad (1)$$

where $\omega^2 = k_o g \tanh k_o h$ with g and h being the gravitational acceleration and water depth, respectively.

The complex velocity potentials, ϕ_1 and ϕ_2, in two fluid domains 1 and 2 (see Fig.1), then satisfy the Helmholtz equation

$$\nabla^2\phi_l - (k_o^2\sin^2\theta)\phi_l = 0, (l=1,2) \quad (2)$$

as governing equation and the following linearized free-surface (Γ_F), bottom (Γ_b), and radiation conditions:

$$-\omega^2\phi_l + g\frac{\partial\phi_l}{\partial y} = 0 \quad (\text{on } \Gamma_F) \quad (3)$$

$$\frac{\partial\phi_l}{\partial n} = 0 \quad (\text{on } \Gamma_b) \quad (4)$$

$$\lim_{x\to\mp\infty}\left(\frac{\partial}{\partial x} \pm ik_o\cos\theta\right)\left(\begin{matrix}\phi_1-\phi_o \\ \phi_2\end{matrix}\right) = 0 \quad (\text{on } \Gamma_c) \quad (5)$$

where Γ_c is the vertical truncation boundary at far fields and $\mathbf{n}=(n_x,n_y)$ is the unit outward normal vector. For convenience, we set $A=1$ from now on.

To solve the above boundary value problem by the eigenfunction expansion method, the complex velocity potentials, ϕ_1 and ϕ_2, in two fluid domains 1 and 2 (see Fig.1) are written as

$$\phi_1 = -\frac{ig}{\omega}\{(e^{i\alpha_o x}+a_o e^{-i\alpha_o x})f_o(y)+\sum_{n=1}^{\infty}a_n e^{\alpha_n x}f_n(y)\} \quad (\text{in } \Omega_1) \quad (6)$$

$$\phi_2 = -\frac{ig}{\omega}\{b_o e^{i\alpha_o x}f_o(y)+\sum_{n=1}^{\infty}b_n e^{-\alpha_n x}f_n(y)\} \quad (\text{in } \Omega_2) \quad (7)$$

where

$$\alpha_n = \begin{cases} k_o\cos\theta & (for \quad n=0) \\ \sqrt{k_o^2\sin^2\theta + k_n^2} & (for \quad n=1,2,\dots) \end{cases} \quad (8)$$

and the eigen functions $f_n(y)$ are given by

$$f_n(y) = \begin{cases} \dfrac{\cosh k_o(y+h)}{\cosh k_o h} & (for \quad n=0) \\ \dfrac{\cos k_n(y+h)}{\cos k_n h} & (for \quad n=1,2,\dots) \end{cases} \quad (9)$$

The wavenumbers k_n are the solutions of the following equations

$$\begin{cases} k_o \tanh k_o h = \dfrac{\omega^2}{g} & (for \quad n=0) \\ k_n \tan k_n h = -\dfrac{\omega^2}{g} & (for \quad n=1,2,....) \end{cases} \tag{10}$$

where n=0 and n>1 correspond to propagating and evanescent (or local) modes, respectively.

The matrix equation for the unknowns a_n and b_n can be determined from the following linearized kinematic and dynamic boundary conditions on the membrane:

$$\frac{\partial\phi_1}{\partial x} = \frac{\partial\phi_2}{\partial x} = -i\omega\xi(y) \tag{11}$$

$$\frac{d^2\xi}{d^2 y} + \lambda^2\xi = \frac{i\omega\rho}{T}(\phi_2 - \phi_1) \tag{12}$$

in which $\lambda = \omega\sqrt{m/T}$ with T and m being the membrane tension and mass per unit length, respectively. In deriving the matrix equation for the unknowns a_n and b_n, the orthogonality of the eigen functions has also to be used. Here, it is assumed that the initial tension T is large so that the effects of dynamic tension can be neglected. The final form of the matrix equation and its derivation are detailed in Kee (1996).

When the system is submerged, an additional condition i.e. the continuity of hydrodynamic pressure and normal fluid velocity, has to be imposed along the vertical fictitious boundary Γ_f above the membrane:

$$\phi_1 = \phi_2 \ , \ \frac{\partial\phi_1}{\partial x} = \frac{\partial\phi_2}{\partial x} \quad \text{at } \Gamma_f \tag{13}$$

NUMERICAL METHODS

In practice, the membrane tension is given by the buoyancy of a large floater at the top. In this case, analytic solutions are in general not available. To solve this kind of problem, a boundary element method (BEM) was developed in Kee & Kim (1997) and, in this section, the numerical method is extended to the case of oblique incident waves. It is assumed, for simplicity, that the buoy is rigid and the heave motion of the buoy is negligible due to large initial tension.

The boundary value problem (2)-(5), (11) and (12) (and (13) if the system is submerged) has to be solved with additional kinematic and dynamic conditions on a rigid buoy. The kinematic boundary condition on the buoy is

$$\frac{\partial\phi_l}{\partial n} + i\omega\{\eta_1 n_x + \eta_3 n_\theta\} = 0 \qquad (\text{on } \Gamma_B) \tag{14}$$

where. The symbols η_1 and η_3 represent complex sway and roll responses. Since a buoy is connected to the membrane, coupled dynamic equations need to be solved. Unlike rigid body hydrodynamics, the body boundary condition on the membrane is not known in advance. Therefore, the membrane and buoy motions and velocity potentials need to be solved simultaneously. To solve the present boundary value problem, a two-domain boundary integral equation method using simple sources along the entire boundary is developed. Therefore, the computer program can be used for arbitrary bottom topography. Two auxiliary vertical boundaries (Γ_{c1}) and (Γ_{c2}) are located sufficiently far from the membrane such that the radiation condition (5) is valid.

The fundamental solution (Green function) of the Helmholtz equation (2) is

$$G = -\frac{1}{2\pi}K_o(k_z r) \tag{15}$$

where K_o is the modified zeroth-order Bessel function of the second kind and r is the distance from the source point (x', y') to the field point (x, y). Using recurrence formula, the normal derivative of G is given by

$$\frac{\partial G}{\partial n} = \frac{1}{2\pi}k_z K_1(k_z r)\frac{\partial r}{\partial n} \tag{16}$$

By applying Green's second identity in each of the regions to the unknown potentials ϕ_1, ϕ_2 and imposing relevant boundary conditions, the integral equations in each fluid domain can be written as

$$C\phi_l + \oint \frac{\partial G}{\partial n}\phi_l d\Gamma - \oint G\frac{\partial\phi_l}{\partial n} d\Gamma = 0 \tag{17}$$

where C= solid angle constant and the symbol $\oint$ represents the surface integral along the entire boundary surfaces. In (17), all the boundary conditions of ϕ_1 and ϕ_2 except for the dynamic boundary conditions of buoy and membrane can straightforwardly be implemented. To solve (17), each of half fluid domain is discretized by $N = N_F + N_c + N_b + N_m + N_B + N_F$ (if submerged) segments, there are $2N$ unknowns for ϕ_1 and ϕ_2, N_m unknowns for ξ, and two more unknowns η_1 and η_3. Therefore, $2N + N_m + 2$ linear simultaneous equation have to be solved. Since ξ and η are unknown, (17) cannot be solved for ϕ_1 and ϕ_2 independently. The equation (17) should be coupled with the equation of motion of the membrane and buoy. One way to solve this coupling problem is the use of an iteration method (Kee, 1996). In this paper, we solved the integral equations of the disturbance potentials and the equations of motion of the buoy and membrane simultaneously.

The discrete form of the membrane equation of motion for j-th element is given by

$$\rho i\omega(\phi_{1j} - \phi_{2j})l_j - T_j(\frac{\partial\xi}{\partial\zeta})_j + T_{j+1}(\frac{\partial\xi}{\partial\zeta})_{j+1} = -ml_j\omega^2\xi_j \quad (18)$$

$$(\frac{\partial\xi}{\partial\zeta})_j = (\xi_j - \xi_{j-1}) / \Delta\zeta_j \quad (19)$$

The symbol l_j is the length of the j-th segment, and $\Delta\zeta_j = \frac{l_j + l_{j+1}}{2}$.

The geometric boundary conditions at the seabed and the top connection point $(0,-R)$ are

$$\xi = 0 \text{ at } z = -h, \quad \xi = \eta_1 + R\eta_3 \text{ at } z = -R \quad (20)$$

Next, we consider the rigid-body motion of a buoy. As mentioned before, the heave response is neglected assuming large initial tension. The coupled equation of motion for sway and roll is then given by

$$M(-\omega^2)X = F_p - (K_{HS} + K_m)X - F_T + F_D \quad (21)$$

where $X = [\eta_1\ \eta_3]^T$, M=buoy mass matrix, K_{HS}=hydrostatic estoring coefficients, K_m=mooring stiffness coefficients, F_p=potential force, F_D=linearized drag force, and F_T=force at the connection point. The force F_T caused by membrane tension can be either restoring force or excitation. The detailed expressions of the variables used in the above equations are given in Kee (1996).

NUMERICAL RESULTS AND DISCUSSIONS

A two-domain boundary element program developed as described in the preceding section was used to demonstrate the performance of surface-piercing or submerged buoy/membrane wave barriers in oblique incident waves. The computational domain is defined as in Figure 1. The vertical truncation boundary is located 3-4 waterdepth away from the membrane to ensure that local wave effects are negligible. First, the numerical results were checked against the energy-conservation formula i.e. $R_f^2 + T_r^2 = 1$. The numerical prediction for the normal incidence case (θ=0) was also verified by a series of experiments (Kee & Kim, 1997).

For further verification, a surface-piercing or submerged vertical flexible membrane hinged at its top and on the sea floor are considered. For this kind of idealized system, analytic solutions can be obtained for arbitrary wave headings, as explained in the preceding section. In Figure 2a and 2b, the analytic solutions for a surface-piercing system are compared with the BEM solutions for two different wave headings and four different dimensionless membrane tensions $\tilde{T} = T / \rho g h^2$. For BEM solutions, N=240 was used. In both cases, the numerical solutions are in excellent agreement with analytic solutions. The curves generally migrate toward higher kh values as wave angle increases since the effective wave length in the x-direction becomes shorter with the increase of incidence angle.

In the following, the BEM code will be used to study more general cases i.e. surface-piercing or submerged buoy-membrane wave barriers with a buoy at the top and hinged on the sea floor. The top of membrane is attached to the bottom of a cylindrical buoy, as shown in Figure 1.

The performance of a buoy-membrane wave barrier should depend on various design parameters, such as membrane parameters, buoy parameters (e.g. radius=a, draft=d), mooring characteristics (e.g. stiffness=K), water depth h, and incident wave angles and frequencies. We first examine the performance of surface-piercing buoy-membrane breakwaters in oblique waves with varying various design parameters. In this study, the specific gravity of the buoy, the thickness to radius ratio t/a, and the dimensionless mass per area of membrane $m/\rho h$ are 1.4, 0.02, and 0.0025, respectively.

In Figure 3a and 3b, the reflection coefficients for $d/a = 1.5$, a / h=0.25, and T / Kh=0.1 (mooring-type 3) are plotted as function of dimensionless wave number kh and incident wave angle θ. We see generally good performance over a wide range of wave frequencies and angles. It is interesting to note that the performance in the long waves with large incident angles is poor. In Figure 4a and 4b, the performance of the same system with different mooring type (type 1) is shown. The efficiency of the dual mooring system (type 3) is higher in a wide range of wave frequencies and wave angles. The result shows the importance of mooring design for the performance of this kind of system in oblique waves.

Figure 5a and 5b show the performance of the same system considered in Figure 3 with reduced buoy draft $d/a = 0.5$. It can be seen that the efficiency for small kh values is greatly reduced with smaller draft (or membrane tension). Figure 6a and 6b show the performance of the same system considered in Figure 3 with softer mooring lines. Again, the performance for long waves is greatly reduced. The performance for short waves is also greatly reduced in the case of large incident angle. In Figure 7a and 7b, the reflection coefficients of the same system considered in Figure 3 are presented for smaller buoy size $a/h = 0.15$. Again, the efficiency in long waves is greatly diminished mainly due to reduced membrane tension. In general, sufficiently large a / h is needed to guarantee desirable performance for a wide range of incident wave frequencies and angles.

So far, we have investigated the efficiency of a surface-piercing buoy-membrane wave barrier with various design parameters. From now on, similar results will be presented for a submerged system, which has the same buoy and membrane characteristics as the surface-piercing system. First, the performance of a submerged system of mooring type 3 is shown in Figure 8a and 8b for the clearance $s/h = 0.125$ (s = the distance from the mean free surface to the top for the buoy), $a/h = 0.15$, and $T/Kh = 0.5$. It is interesting to see that the efficiency is greatly increased in larger wave incident angles. Figure 9a and 9b show the performance of the same submerged system considered in Figure 8 with increased buoy size $a/h = 0.2$. It is seen that the efficiency is greatly improved by slightly increasing buoy radius. In particular, the performance in long waves can be much better compared to the surface-piercing system of the same buoy size and mooring characteristics.

Finally, Figure 10a and 10b show the actual performance of a surface-piercing system (d/a=1.5, a/h=0.25, and type-3 T/Kh=0.7) in sea-state-3 irregular waves with θ=0. The experiment was conducted in the two-dimensional wave tank at Texas A&M University. The details are presented in Kim et al. (1996). We can clearly see that the proposed system can be a very effective wave barrier if it is properly designed.

CONCLUDING REMARKS

The interaction of oblique incident waves with a tensioned, inextensible, vertical flexible membrane hinged at the sea floor and attached to a rigid cylindrical buoy at its top, was solved in the context of two-dimensional linear wave-body interaction theory. Both submerged and surface-piercing buoy/membrane systems were considered. For a membrane without buoy, analytic solutions are obtained and used as a bench-mark result. To study more practical buoy-membrane systems, a two-domain boundary element program was developed based on a discrete-membrane dynamic model and simple-source distribution over the entire fluid boundaries.

Using the developed computer programs, the performance of surface-piercing or submerged systems in oblique waves was tested with varying buoy radius and draft, water depth, membrane length, mooring-line characteristics, and wave conditions. From these examples, it is shown that the flexible membrane can be a very effective wave barrier if properly designed. The efficiency can be even enhanced in oblique incident waves. In particular, sufficiently large membrane tension needs to be provided to guarantee high performance. It is also shown that the submerged system can perform much better than the surface piercing system of the same buoy size.

ACKNOWLEDGMENT

This research was sponsored by the Korea Research Institute of Ships & Ocean Engineering (KRISO) through a KRISO/TAMU cooperative research program. This work was also partly supported by the Offshore Technology Research Center through the National Science Foundation Engineering Research Centers Program, Grant Number CDR-8721512.

REFERENCES

Abul-Azm, A.G. (1994), "Wave diffraction by double flexible breakwaters" *Journal of Applied Ocean Research*, Vol.16, 87-99.

Bai, K.J. (1975) "Diffraction of oblique waves by an infinite cylinder" *J. of Fluid Mechanics*, Vol.68, 513-

Broderick, L.L. & Jenkins, C.H. (1993) "Experimental investigation of fluid-filled membrane breakwaters" *ASCE J. of Waterway, Port, Coastal & Ocean Engineering*, Vol 119, #6, 639-656.

Isaacson, M., Whiteside, N., Gardiner, R., & Hay, D. (1994), "Modelling of a circular section floating breakwater" *Intl. Symp. Waves - Physical & Numerical Modeling*, Vancouver.

Kee, S.T. (1996), "The interaction of vertical flexible membrane breakwaters with waves" *Ph.D. Thesis*, Texas A&M Univ.

Kee, S.T. & Kim, M.H. (1997) "Flexible membrane wave barrier. Part 2. Floating/submerged buoy-membrane system" *ASCE J. of Waterway, Port, Coastal & Ocean Engineering,* February, (In Press).

Kim, M.H. & Kee, S.T. (1996) "Flexible membrane wave barrier. Part 1. Analytic and numerical solutions" *ASCE J. of Waterway, Port, Coastal & Ocean Engineering,* Vol.122, No.1, 46-53.

Lee, J.F. & Chen, C.J. (1990) "Wave interaction with hinged flexible breakwater" *J. of Hydraulic Research*, Vol 28, 283-295.

Ohyama, T., Tanaka, M., Kiyokawa, T., Uda, T., & Murai, Y. (1989) "Transmission and reflection characteristics of waves over a submerged flexible mound" *Coastal Engineering in Japan*, Vol.32, No.1, 53-68

Rahman, M. & Chen, M. (1993), "Boundary element method for diffraction of oblique waves by an infinite cylinder" *J. of Engineering Analysis with Boundary Elements*, Vol.11, 17-24.

Sollitt, C.K., Lee, C.P., McDougal, W.G., & Perry, T.J. (1986) "Mechanically coupled buoyant flaps: theory and experiment" *Proc. 20th Coastal Engineering Conf.*, Vol.3, 2445-2459.

Thompson, G.O., Sollitt, C.K., McDougal, W.G. & Bender W.R. (1992) "Flexible membrane wave barrier" *ASCE Conf. Ocean V*, College Station, 129-148.

Wang, K.H. & Ren, X. (1993), "Water waves on a flexible and porous breakwater" *Journal of Engineering Mechanics*, Vol.119, 1025-1048.

Williams, A.N., Geiger, P.T., & McDougal, W.G. (1991), "Flexible floating breakwater" *ASCE J. of Waterway, Port, Coastal & Ocean Engineering*, Vol 117, #5, 429-450.

Williams, A.N., Geiger, P.T., & McDougal, W.G. (1992), "A submerged compliant breakwater" *J. of Offshore Mechanics & Arctic Engr.*, Vol 114, 83-90.

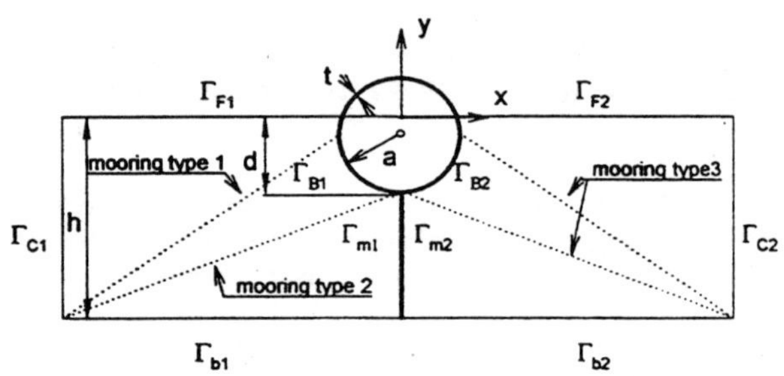

Figure 1. Coordinate system and integration domains for a surface piercing breakwater.

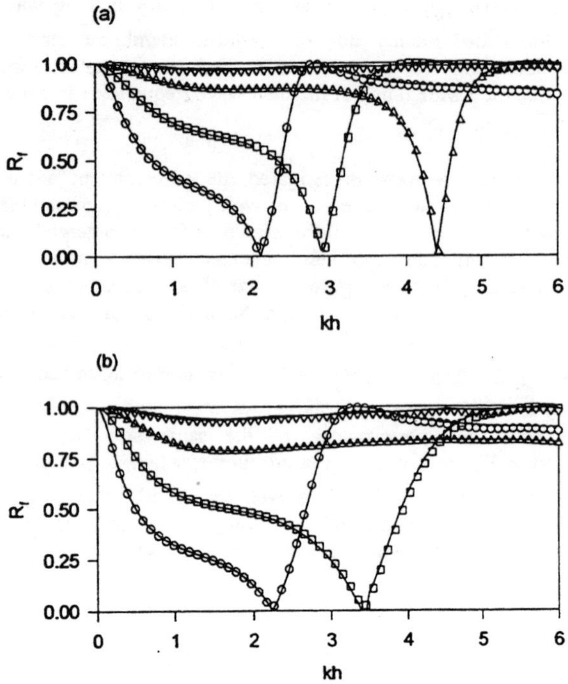

Figure 2. Reflection coefficients of a surface piercing hinged-hinged membrane breakwater as function of non-dimensional tension $\tilde{T}$ and wavenumber kh. The BEM results are for (a) $\theta = 15^{\circ}$, and (b) $\theta = 45^{\circ}$; $\tilde{T}$ = 0.064 (O), 0.127 (□), 0.255 (Δ), and 0.51 (∇); Solid lines are analytical solutions.

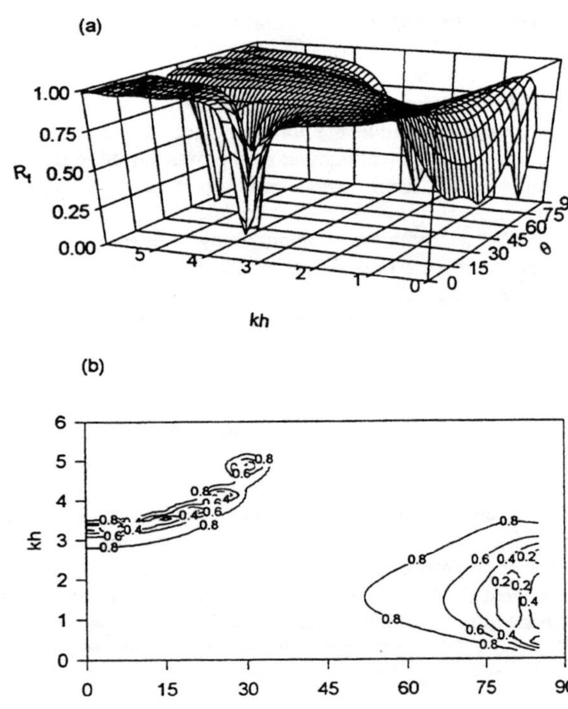

Figure 3. (a) Reflection coefficient of a surface-piercing system and (b) its contour as function of *kh* and θ for T/(Kh)=0.1 (mooring type 3), d/a=1.5, and a/h=0.25

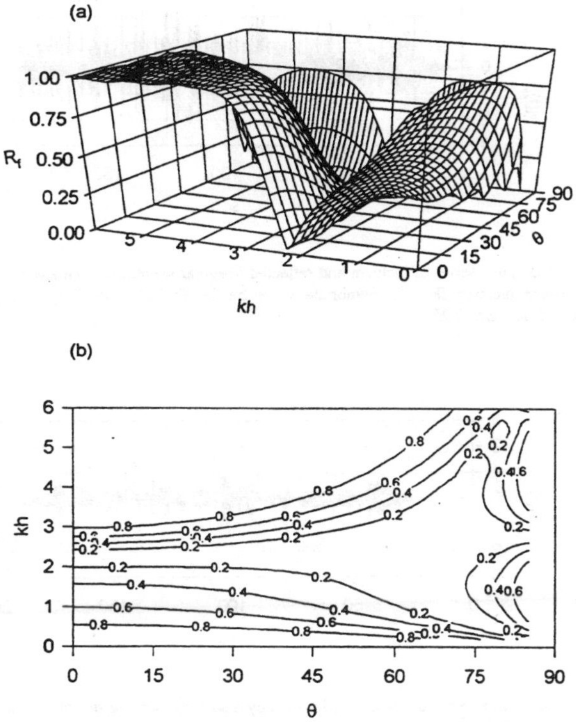

Figure 4. (a) Reflection coefficient of a surface-piercing system and (b) its contour line as function of *kh* and *θ* for T/(Kh)=0.1 (mooring type 1), d/a=1.5, and a/h=0.25

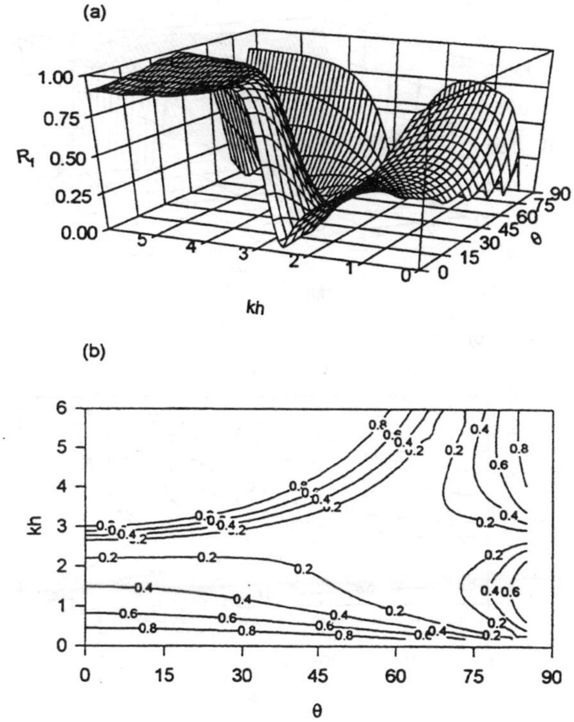

Figure 6. (a) Reflection coefficient of a surface-piercing system and (b) its contour line as function of *kh* and *θ* for T/(Kh)=1.0 (mooring type 3), d/a=1.5, and a/h=0.25

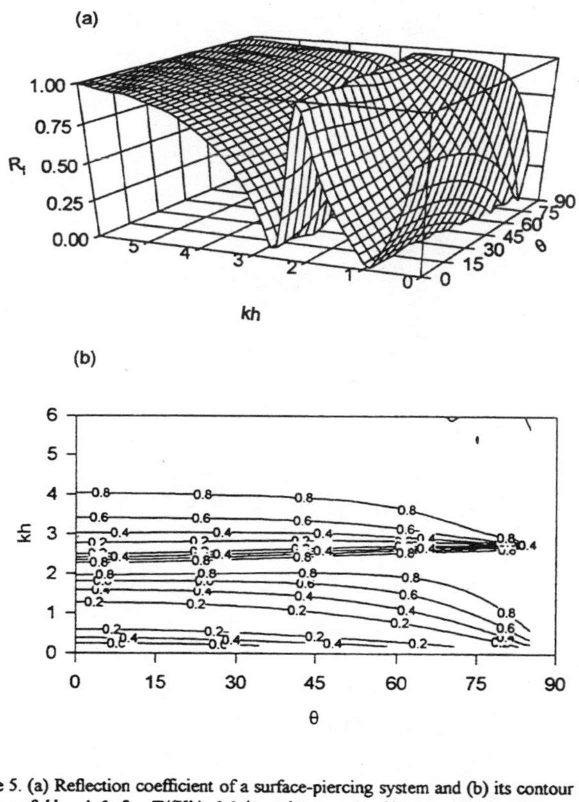

Figure 5. (a) Reflection coefficient of a surface-piercing system and (b) its contour line as function of *kh* and *θ* for T/(Kh)=0.1 (mooring type 3), d/a=0.5, and a/h=0.25

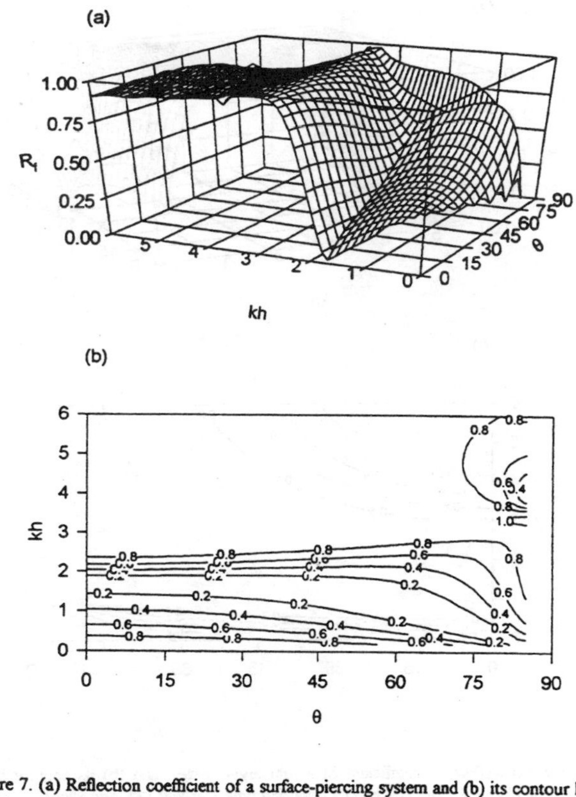

Figure 7. (a) Reflection coefficient of a surface-piercing system and (b) its contour line as function of *kh* and *θ* for T/(Kh)=0.1 (mooring type 3), d/a=1.5, and a/h=0.15

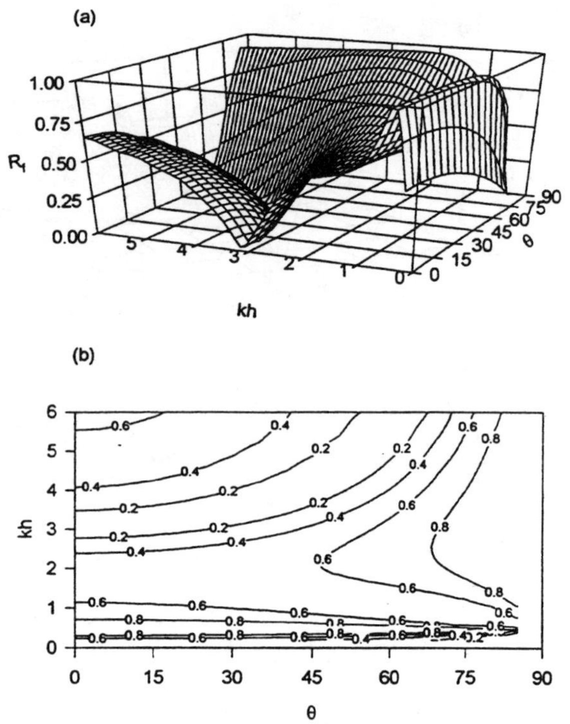

Figure 8. (a) Reflection coefficient of a submerged system and (b) its contour line as function of *kh* and θ for T/(Kh)=0.5 (mooring type 3), s/h=0.125, and a/h=0.15

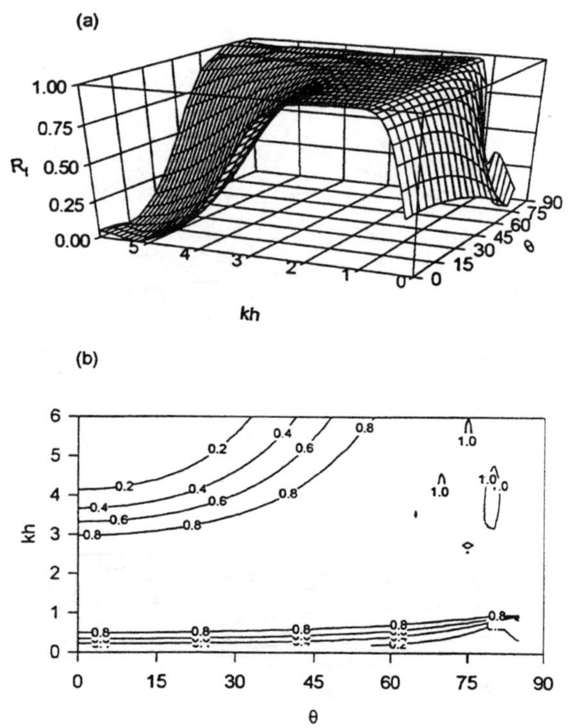

Figure 9. (a) Reflection coefficient of a submerged system and (b) its contour line as function of *kh* and θ for T/(Kh)=0.5 (mooring type 3), s/h=0.125, and a/h=0.2

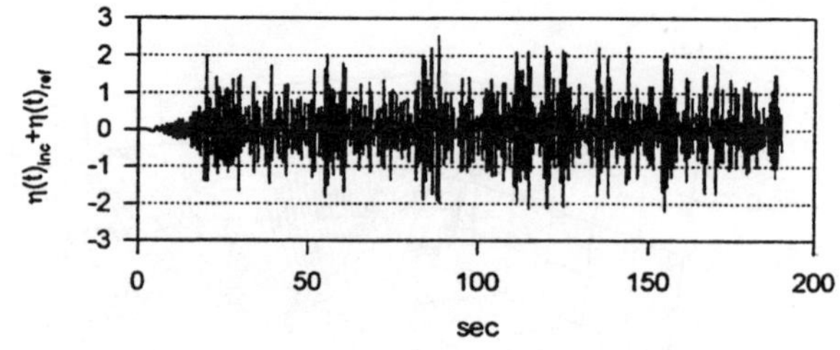

FIG. 10a. Time series of incident and reflected irregular waves between a wave paddle and a surface piercing flexible membrane wave barrier for T/(Kh)=0.7 (mooring type 3), d/a=1.5, and a/h=0.25

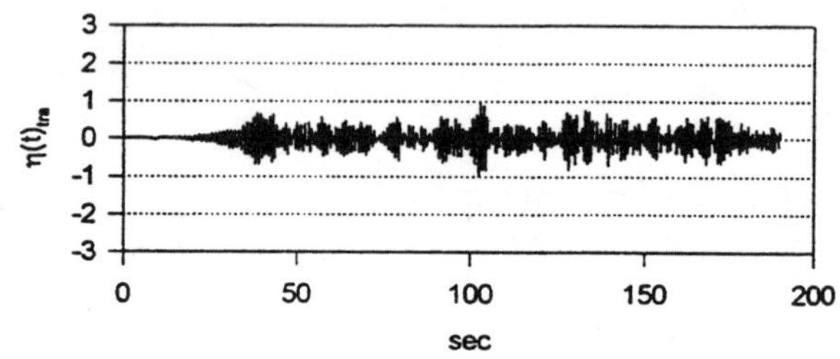

FIG. 10b. Time series of transmitted waves by a surface piercing flexible membrane wave barrier for T/(Kh)=0.7 (mooring type 3), d/a=1.5, and a/h=0.25

Proceedings of the Seventh (1997) International Offshore and Polar Engineering Conference
Honolulu, USA, May 25-30, 1997

ISBN 1-880653-28-1 (Set); ISBN 1-880653-31-1 (Vol. III)

Formation of River-Mouth Bar by Waves and Wave-Induced Current

Ichiro Deguchi and Su Kyeong Chun
Osaka University
Osaka, Japan

Abstract

In this study, characteristics of river mouth topography generated by waves and wave-induced current was first investigated through movable bed experiments. Then we examined applicability of two numerical procedures for predicting topography change around river mouth, that is, a modified one-line theory and a 3-D topography prediction model.

By comparing numerical and experimental results, it is found that modified one-line theory can apply for predicting bar formation of the small scale under the condition of a deep water depth at the river mouth and small wave incidence. On the other hand, change in water depth around river mouth calculated by the 3-D topography prediction model coincides well with the measured change in water depth provided that the interaction among topography change, wave transformation and change in wave-induced current is taken into account properly.

Key words

Coastal erosion, continuity of sediment transport, wave-induced current, river mouth bar, a modified one-line theory, a 3-D topography prediction model

1. Introduction

Coastal erosion has been a serious problem for recent years. It is said that the amount of the area lost from the nearshore region is more than 160 hectare / year in Japan (Tanaka et al., 1993). One of the most plausible agency for this erosion is the decrease in the volume of discharged sediment from the river due to the construction of various structures such as dams and reservoirs along rivers and improvement of river channel for preventing land from disaster such as flood and utilizing flesh water for agricultural, industrial use.

River mouth is interface of river and sea. Bed material is moving constantly, and it creates various types of river mouth topography. Blockage of river mouth and shoaling of fairway often take place by such sediment transport. It is necessary to comprehend the hydraulic function and role of the river mouth in sediment budget in coastal region around the river mouth.

There are many hydrodynamic forces with wide frequency range including fluid motion of the periods of few seconds (wind waves), few ten seconds(swell, and long waves in infragravity region), few minutes(long waves in infragravity region), a half day(tidal current) and so on. There is also river discharge that may fluctuate according to the rainfall. These fluid motions together with the density difference between fresh water(river discharge) and salt water(coastal region) make an extremely complicated flow field around river mouth.

Many studies have been carried out to comprehend flow field exactly and to solve the various problem around river mouth that come out in complicated natural system in the fields of coastal engineering, river engineering and sedimentalogy.

As for the topography around river mouth, Bruun (1978) proposed a gorge theory to analyze the minimum cross-section at the river mouth. Butakov(1971) investigated the generation process of mouth bar by discharged flow and their deformation process by waves. Özsoy(1977) conducted theoretical research about the topography change due to jet diffusion and settling of discharged sediment around tidal inlets. His results show that the pattern of sediment accretion around tidal inlets is decided by bottom friction, topography, settling velocity of sand and diffusion of discharged sediment. FitzGerald(1982, 1984) examined sand bar formation based on many field data of tidal inlets. His study qualitatively explained the process of sand bar formation on both coasts where tidal current and waves are predominant. However, it doesn't include the physical factor act on the system and deformation process of sand bar. Suga et al.(1987) also studied about formation of sand terrace in front of river mouth during a flood and its subsequent deformation to sand bar by wave action based on field data and laboratory experiments. Sawaragi and Deguchi(1984, 1988) have investigated the effect of waves and discharged flow on river mouth topography through experiments and proposed numerical model for predicting depositional pattern of discharged sediment in the coast from the river. Tanaka et al.(1993) and Ifuku et al.(1994) investigated the deformation process of the existing river mouth bar at specific rivers.

Although many enthusiastic investigations have been carried out, most of the above-mentioned studies were carried out under the limited conditions and it is generally impossible to determine the critical condition for the occurrence of river mouth blockage. Moreover, the exact prediction of mouth bar formation, enlargement process and deformation of the bar is far beyond of the routine work.

In this study, we investigated the dominating factor for the formation of river mouth bar and blockage of river mouth in different initial topography of the river mouth under different wave conditions through movable bed experiments. Then, applicability of two numerical procedures for predicting topography change around river mouth, i.e., shoreline prediction model based on a modified one-line theory and 3-D topography prediction model was examined.

2. Experiment of river mouth bar formation

2-1. Experimental apparatus, experimental cases and measured items

We carried out experiments to investigate formation process of river-mouth bar due to waves and wave-induced current in 3-D experimental basin of 17 m long and 10 m wide. The initial bottom topography is a simple shape as shown in Figure 1 to easily find out the relation of mouth bar formation and exerting forces. The bottom slope of the sea and the river bed are 1/10 and 1/200, respectively and the width of river channel B_r is 1.12 m.

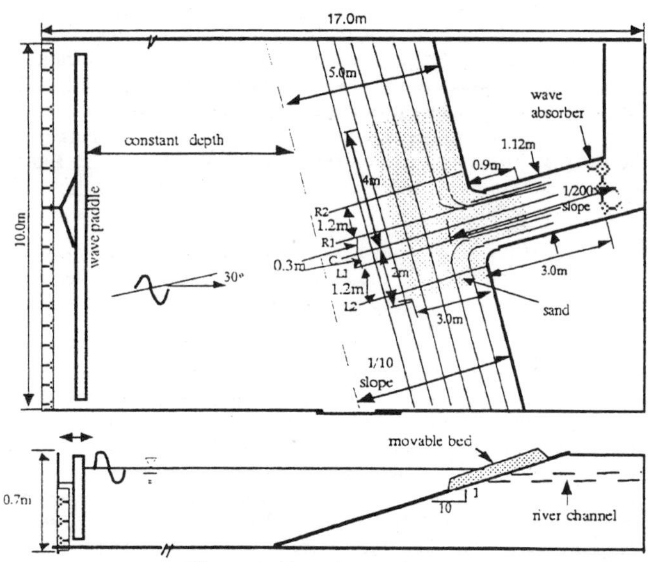

Figure 1 Experimental set-up

Total seven cases of experiments were carried out. Four cases of seven were conducted under the different wave conditions and another three cases were carried out on the different water depth at river mouth. These conditions are shown in Table 1. Where h_r is the depth at river mouth defined in Fig.1, H_o is the incident wave height at horizontal bottom, T is the wave period, L_o is the wave length at horizontal bottom, Q_y is the total longshore sediment transport rate and α is the Savage constant that will be mentioned later. P_l is the breaking wave energy flux in the longshore direction.

Incident wave angle was kept constant to be 30° at the horizontal bottom and water depth in offshore was varied in the region of 40 cm for h_r = 12 cm, 34 cm for h_r = 6 cm and 31 cm for h_r = 3 cm, respectively. The radius of the river mouth was 50 cm in all cases. To form movable bed, sand of median grain size(d_{50}) of 0.5 mm was laid uniformly out on the shadow region shown in Fig.1. The thickness of the sand layer was 6 cm.

Wave height, direction and the velocity of wave-induced currents were measured on the fixed bed and deformation of shoreline, change in bottom profile, longshore sediment transport rate were measured on the movable bed. Wave height was measured using capacitance type wave gage along measuring lines C, R1, L1, R2, L2 shown in Fig.1. Bottom topography was measured by a optical type bottom profiler along measuring line ML1 to ML6 shown in Fig.2 at 10 to 120 minutes after wave generation. Total longshore sediment transport rate was evaluated by measuring the volume of sediment transported downward from the shadow region shown in Fig.2.

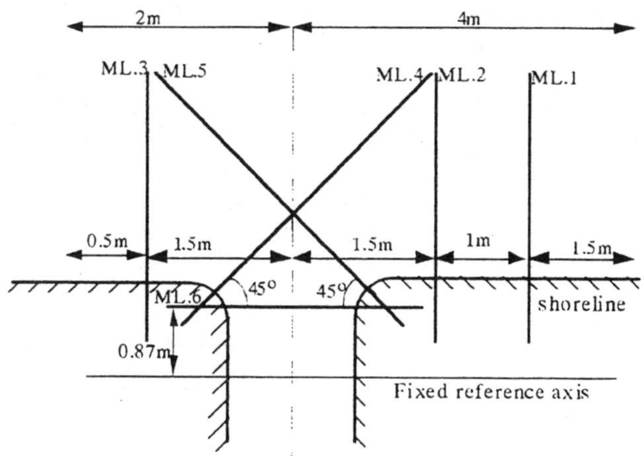

Figure 2 Measuring line in movable bed experiments

Table 1 Experimental conditions

CASE	h_r(cm)	H_0(cm)	T(sec)	H_0/L_0	H_0/L_0	h_r/H_0	H_0/d_{50}	Q_y (cm^3/1hr)	Q_y (cm^3/sec)	P_l (g · cm /sec^2)	α (cm^2/ g · sec)
W1	6.0	6.0	1.2	0.032	0.499	1.00	120	149,800	41.60	328.287	0.1267
W2	6.0	6.0	0.8	0.060	1.122	1.00	120	53,500	14.86	180.495	0.0823
W3	6.0	5.0	0.8	0.050	1.122	1.20	100	33,600 117,700 (3 hr)	9.33 10.90 (3hr)	154.943	0.0602 0.0703 (3hr)
W4	6.0	5.5	1.0	0.035	0.718	1.09	110	69,550	19.32	311.291	0.0621
W5	12.0	6.0	1.2	0.032	0.499	2.00	120	165,850	46.07	362.942	0.1269
W6	12.0	5.0	0.8	0.050	1.122	2.40	100	32,100	8.92	124.398	0.0717
W7	3.0	6.0	1.2	0.032	0.499	0.50	120	133,750	37.15	295.789	0.1256

The shifts of shoreline location are measured by measuring the distance from the fixed reference axis at 10, 30 and 60 minutes after wave generation.

2-2. Experimental results

Wave field, wave induced-current and longshore sediment transport

Except for the case W6, incident waves broke at offshore of the river mouth. In case W6, incident waves propagated into the river without breaking. Cross-shore distribution of wave height at the center line of all cases are shown in Fig.3. The ratio of wave height to water depth at wave breaking point in cases W1 ~ W5 and W7 in front of the river mouth was a little bit smaller than that on the coastal region. This means that the waves broke later or shallower than the other region on the coast.

In the up-ward of the river mouth, a relatively strong longshore current was developed in all cases. When the depth at the river mouth is shallow(case W7), the longshore current flowed downward smoothly even in the front of the river mouth. Even in the case where the depth at the river mouth was the same as the incident wave height(cases W1, W2) significant decrease in the longshore current took place. However, when the depth at the river mouth is deep relative to the incident wave height (cases W5, W6) decrease its velocity significantly and showed complex flow pattern in front of the river mouth.

$$Q_y = \alpha(EC_g)_b \sin\alpha_b \cos\alpha_b \quad (1)$$

where, E_b, C_{gb} and α_b is wave energy flux, group velocity and breaking angle at wave breaker point, respectively.

Characteristics of river mouth bar topography

Figure 4 shows the measured change of shoreline in the cases of W1, W3, W5, W7. When the water depth at the river mouth is the same as the incident wave height (case W1, Fig.(a)), Bar formed at the upstream of the river mouth and developed toward downstream side and it reached to the opposite side at 45 minutes after wave generation. At this time, river mouth was perfectly blocked. Although a small sand bar was also formed at the downstream side of the river channel, it did not develop and was observed upstream bar at 60 minutes after wave generation.

Until perfect blockage took place, shoreline in both upstream and downstream coasts retreated. After the river mouth was blocked, sediment was transported continuously in the longshore direction and retreat of shoreline stopped in the downstream coast.

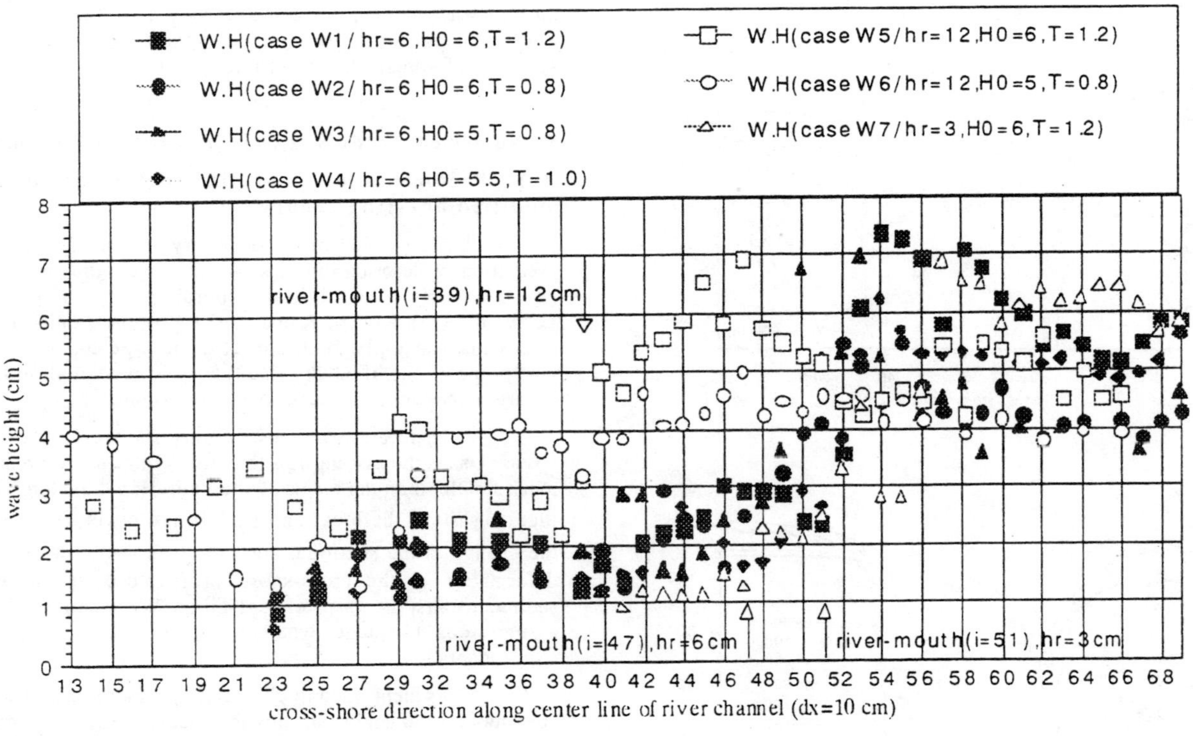

Figure 3 Cross-shore distribution of wave height along center line of river channel

Suspended sediment could be seen only in the region near the wave breaking point and a large part of sediment was transported as bed load. The bed load transported in the longshore direction generated the river mouth bar in front of the river mouth, their details illustrate in Fig.4.

The rate of total longshore sediment transport measured in the downward region is also shown in Table 1 together with the Savage coefficient α defined by the following equation:

In the case where the depth at the river mouth was about one-half of the incident wave height (case W7, Fig.(b)), bars were also generated at both sides of the river mouth and the bar at the upstream side developed in the longshore direction. However, it did not reach to the opposite side at 60 minutes after wave generation. Therefore, erosion of the coast in downstream side was still continuing at the time because sufficient amount of sediment from the upstream side did not supply to stop the erosion.

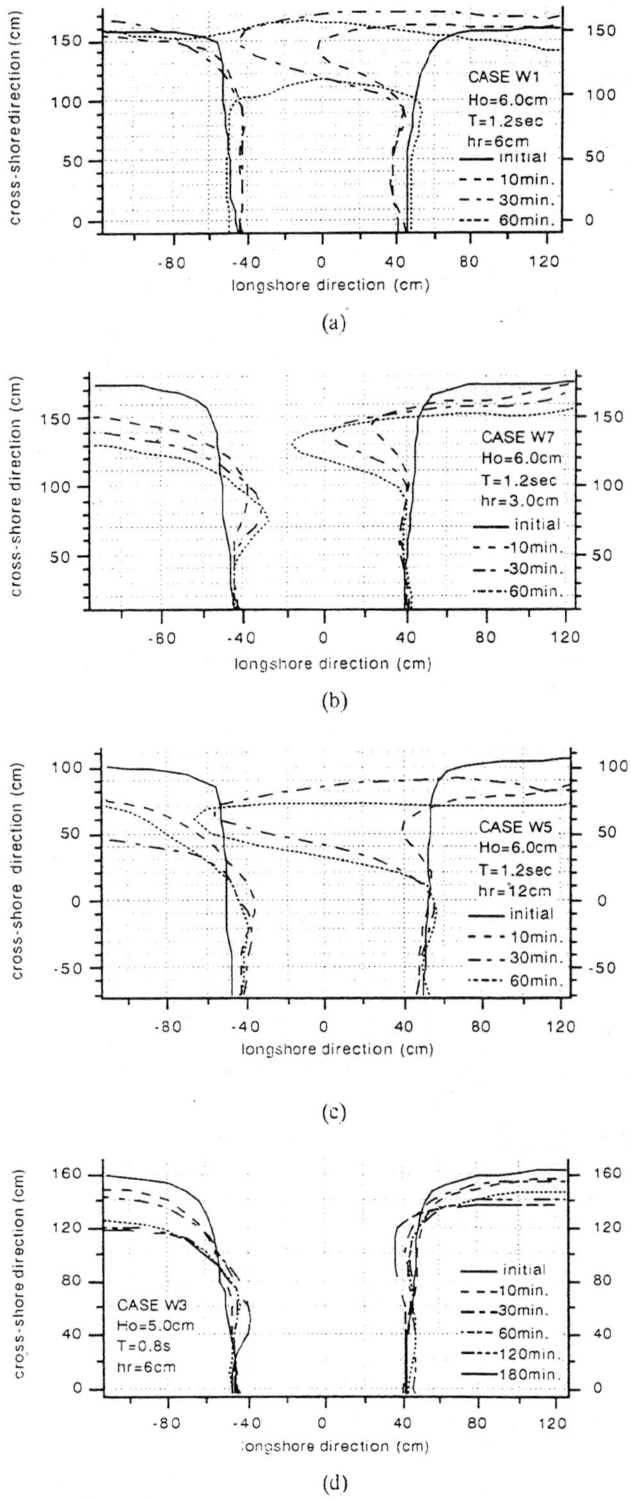

Figure 4 Measured change of shoreline; (a) case W1, (b) case W7, (c) case W5, (d) case W3

Figure 4 (c) shows the case where the depth at the river mouth was about twice of the incident wave height (case W5) and the sectional area of the river mouth was two times larger than that of the case W1 (Fig.4(a)). Although a large bar developed from the upstream side of the river mouth and the length of the bar was more than the initial width of the river mouth, there was a small gap at the river mouth because of the large erosion in the downstream coast.

Figure 4(d) shows the result of the case W3 where the incident wave height and period were a little smaller than the case W1 and with the same depth at the river mouth. In this case, total longshore sediment transport rate was a quarter less than that in the case W1 (see Table 1) and the scale of the bars at both sides of river mouth were small. The velocity of shoreline retreat at both coasts of upstream and downstream side seemed to stop at 2 hours after wave generation and even 3 hours after wave generation, the bar of the upstream side still remained small. In this case, sediment transported in the longshore direction deposited in front of the river mouth but it could not develop to pierce water surface.

The main results of experiments can be summarized as below;

First, bars are generated at both sides of the river mouth regardless of the incident wave conditions and depth at the river mouth. Bar of the upstream side is generally larger than that of the downstream side and its location is also deeper than that of the downstream side.

Second, the bar generated at the upstream side is strongly affected by the incident wave characteristics and depth at the river mouth.

Third, decrease in the sectional area of the river mouth is rapid in the case where the depth at the river mouth is shallow. It seems that a critical depth for the occurrence of the blockage exists under the same rate of longshore sediment transport.

3. Numerical models for river mouth bar formation

3-1. Prediction of shoreline deformation based on modified one-line theory

A so-called one-line theory is usually applied to the prediction of deformation of shoreline where the breaker angle is not so large and initial shoreline is almost straight. LeBlond(1975), Rea and Komar(1975) proposed a modified one-line theory that can apply to the shoreline of large curvature. Uda(1982) also proposed modified one-line theory to predict shoreline change of large curvature using curvilinear coordinate.

There are some restrictions for the application of the model. One is the assumption that the width of the breaker zone is much smaller than the radius of the shoreline curvature and other is that the shift in the location of shoreline δl is in proportional to the change in the sectional area δA. It is needless to say that the deformation of the cross-shore profile due to the cross-shore sediment transport must be negligible. We examined the first two assumptions to judge whether the model can apply to our experiments.

As mentioned in chapter 2, the radius of initial shoreline curvature was 50 cm in our movable bed experiments and the width of the breaker zone was a little bit larger than the radius. Figure 7 illustrates the relation between the shift in the location of the shoreline and the change in the sectional area along the measuring lines ML1 ~ ML5 in some cases. Although scattering of data is large, it seems that there exists correlation between two values. Although the first assumption is not satisfied, we applied Uda's method(1982) to the bar formation at the river mouth.

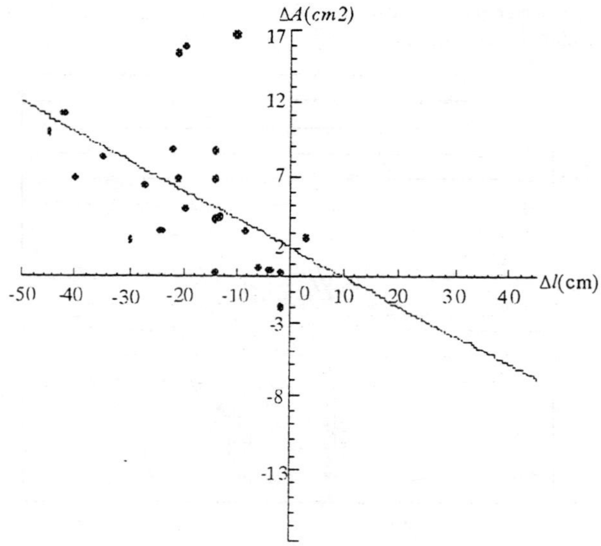

Figure 7 Relation between δl and δA

Fundamental equations of Uda's model are expressed in the coordinate system shown in Fig.8 as follows:

$$\frac{\partial q_{y,n}}{\partial y^*} + h\frac{\partial x_n^*}{\partial t} = 0 \qquad (2)$$

$$\beta = \tan^{-1}\left(\frac{\partial x}{\partial y}\right)_n \qquad (3)$$

$$q_n = \frac{1}{2}F_n \sin 2\alpha_{b,n} \qquad (4)$$

$$F_n = f\frac{W}{8}(H^2C_g)_{b,n} \qquad (5)$$

$$\alpha_{b,n} = \theta_{b,n} - \beta_{b,n} \qquad (6)$$

$$\Delta x_n = \Delta x_n^* \sin\left(\frac{\pi}{2} + \beta\right) \qquad (7)$$

$$\Delta y_n = \Delta x_n^* \cos\left(\frac{\pi}{2} + \beta\right) \qquad (8)$$

where, (x, y) is the cross-shore and longshore directions of Cartesian coordinate, (x^*, y^*) is the curvilinear coordinate that is perpendicular and tangential direction, h is the representative height of the topography change, β is the slope of shoreline, α_b and θ_b is the wave breaking angle in (x, y) coordinate and (x^*, y^*) coordinate, respectively, f is the constant proportional to the Savage factor and δx_n^* is the shift in the location of shoreline in the x^* direction.

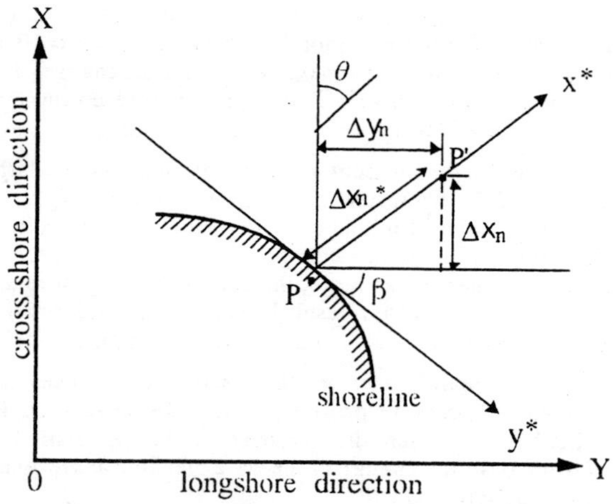

Figure 8 Coordinate system and definition of variables of Uda's model

To simulate deformation of the shoreline, we have to give the value of wave height, group velocity and wave direction at wave breaking point. These values are evaluated by solving the equation of energy conservation by finite difference method (for example, Sawaragi, 1995).

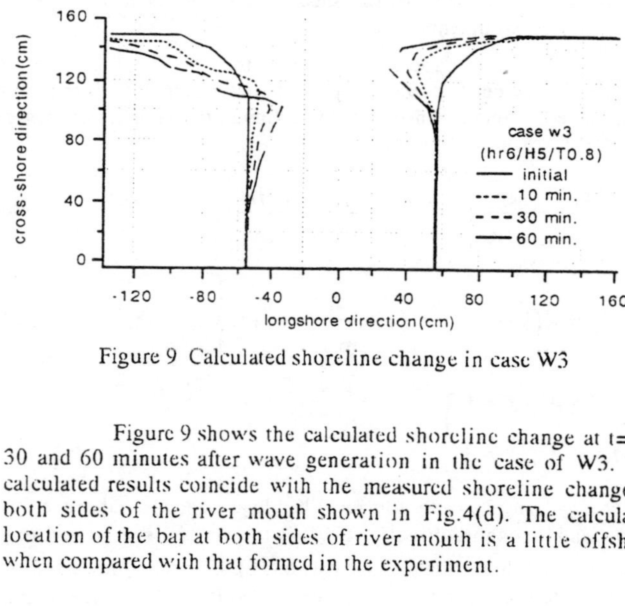

Figure 9 Calculated shoreline change in case W3

Figure 9 shows the calculated shoreline change at t=10, 30 and 60 minutes after wave generation in the case of W3. The calculated results coincide with the measured shoreline change at both sides of the river mouth shown in Fig.4(d). The calculated location of the bar at both sides of river mouth is a little offshore when compared with that formed in the experiment.

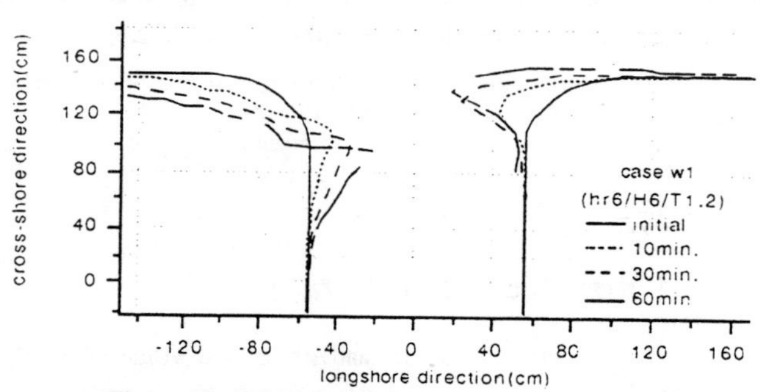

Figure 10 Calculated shoreline change in case W1

Figure 10 shows the calculated results of case W1 at t=10, 30 and 60 minutes. Shoreline deformation at t=10 minutes roughly coincides with the measured shoreline change. However, the bar of the upstream side did not grow toward downward as the experimental result did.

When the incident wave energy flux was not large as in the case of W3, the result of which is shown in Fig.9, a large deformation of shoreline did not take place and wave direction at breaking point also did not change significantly. While, when the incident wave energy was large, the amount of longshore sediment transport increased and as a result large shoreline deformation took place that altered the location of wave breaking point significantly.

The main reason for the discrepancy between measured and calculated shoreline profiles are that the change in incident wave field is not taken into account in the calculation. In the calculation, only the change of the wave breaking angle against the shoreline angle is evaluated.

3-2. Numerical model for the prediction of topography change

Sediment movement is usually in non-equilibrium state in the region where the spatial gradient of fluid motion is large. Sawaragi et al.(1984, 1988, 1995) developed a numerical model for predicting topographic change caused by such kind of non-equilibrium sediment transport around river mouth and navigation channel. In the model, an advection-diffusion equation is solved to evaluate non-equilibrium suspended sediment concentration, the effect of bottom slope on bed load transport is taken into account by assuming that the critical shear stress for bed load transport depends on the bottom slope.

In this paper, we applied the model developed by Sawaragi and Deguchi(1984) to predict the formation of river mouth bar. The equation of continuity for sediment transport is expressed by the following equation:

$$\frac{\partial h}{\partial t} = \frac{1}{1-\lambda}\left\{-Q_S + \left(\frac{\partial q_{bx}}{\partial x} + \frac{\partial q_{by}}{\partial y}\right)\right\} \tag{9}$$

$$Q_S = \left\{(1-r)c_0 w_f\left(1-\frac{u_*}{w_f}\right) + \bar{c}\,w_f\right\} \tag{10}$$

$$\left.\begin{array}{l} u_* > w_f ; r = 0 \\ u_* < w_f ; r = 1 \end{array}\right\} \tag{11}$$

where, q_{bx} and q_{by} are the local bed load transport rate in cross-shore and longshore directions, Q_S is the vertical sediment flux at the bottom, which is expressed by Eq.(10), λ is the void ration, c_0 is the reference concentration at the bottom, $\bar{c}$ is the depth-averaged suspended sediment concentration at steady state, w_f is the settling velocity of sediment, u_* is the shear velocity.

The local bed load transport rate is evaluated using Eq.(12) that is derived based on the Sleath's formula.

$$\left.\begin{array}{l} q_{bx} = 47\pi\sigma d_{50}^{\,2}(\psi-\psi_c)^{1.5}(U/\hat{u}_w) \\ q_{by} = 47\pi\sigma d_{50}^{\,2}(\psi-\psi_c)^{1.5}(V/\hat{u}_w) \end{array}\right\} \tag{12}$$

where, ψ, ψ_c, is Shields number and its critical value, s is the specific gravity of sand, σ is angle frequency, $\hat{u}_w$ is the amplitude of the water particle velocity due to waves just outside of the boundary layer, d_{50} is median grain size of sand.

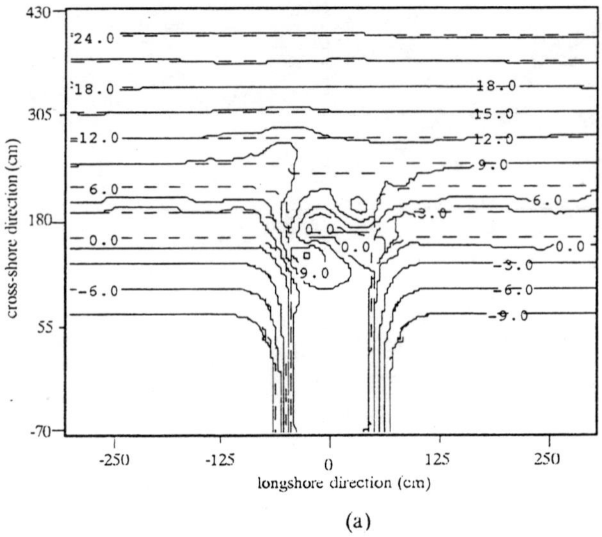

(a)

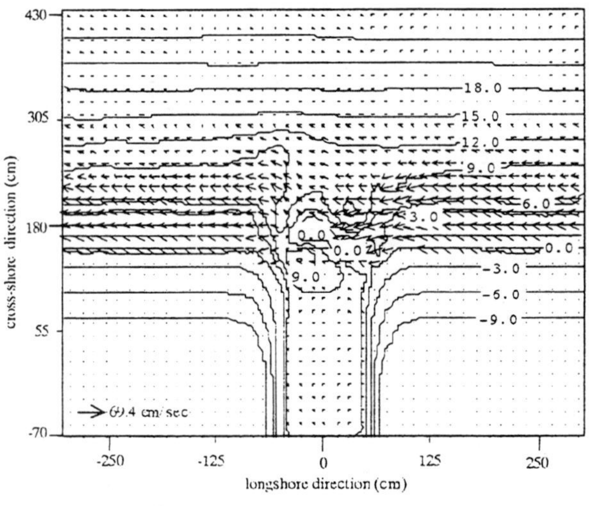

(b)

Figure 11 Calculated bottom topography (a) and wave-induced current (b), case W1

The depth-averaged sediment concentration is calculated by solving the following depth-averaged advection-diffusion equation:

$$\frac{\partial c}{\partial t} + U\frac{\partial c}{\partial x} + V\frac{\partial c}{\partial y} = \varepsilon_{sx}\frac{\partial^2 c}{\partial x^2} + \varepsilon_{sy}\frac{\partial^2 c}{\partial y^2} \tag{13}$$

where, ε_{sx}, ε_{sy} are the dispersion coefficients of suspended sediment in cross-shore and longshore directions, respectively. (U,V) are the time and depth-averaged wave-induced current velocity that are evaluated by solving a usual time and depth averaged equations of conservation of mass and momentum flux numerically. Wave field is obtained by solving the equation of conservation of wave energy flux to evaluate the radiation stress and the shear velocity at the bottom.

Figure 11(a) and 12(a) illustrate the calculated bottom topography change of the case W1 and W3 took place during 30 minutes. In both figures, initial bottom contours are shown by broken lines. Waves and wave-induced current were re-calculated on the bottom topography of 15 minutes after wave generation to obtain the topography change shown in these figures. Figure 11(b) and 12(b) show the calculated wave-induced current.

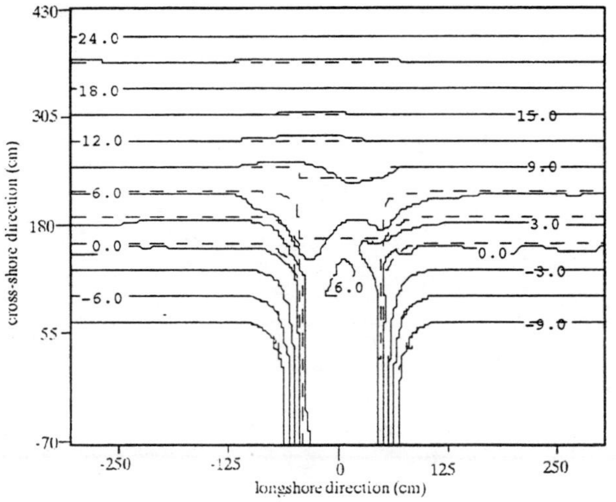

(a)

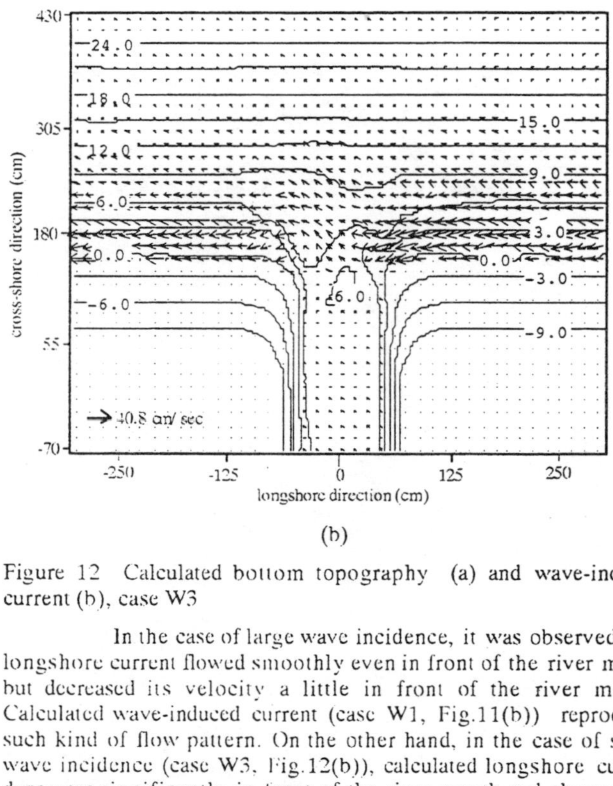

(b)

Figure 12 Calculated bottom topography (a) and wave-induced current (b), case W3

In the case of large wave incidence, it was observed that longshore current flowed smoothly even in front of the river mouth but decreased its velocity a little in front of the river mouth. Calculated wave-induced current (case W1, Fig.11(b)) reproduces such kind of flow pattern. On the other hand, in the case of small wave incidence (case W3, Fig.12(b)), calculated longshore current decreases significantly in front of the river mouth and changes its direction toward offshore. Such kind of flow pattern was also observed in the experiment.

Calculated topography change in both figures shows the deposition of sediment from the right hand side of the river mouth and erosion at both sides of the beach. The volume of the deposited sediment in the river mouth is larger in the case of large incident wave energy (case W1, Fig.11) than that in the case of small incident wave energy (case W3, Fig.12). In the case of small wave incidence, the dimension of calculated bar is a little smaller than that of measured bar as shown in Fig.4(d). While, in the case of large wave incidence, although the direction of deposition from the right hand side of the river mouth roughly coincides with the direction of the bar measured in the experiment as shown in Fig.4(a), the length of the calculated bar is shorter than that measured in the experiment.

When the bottom topography around the river mouth changes to some extent as shown in Fig. 11 and 12, wave and wave-induced current around the river mouth must also change significantly. To simulate perfect blockage of the river mouth, interaction between topographic change and wave and current fields have to be reproduced exactly as in the case of the modified one-line theory. However, in this study, we could not examine the adequate interval to interact topographic change and waves and current fields.

4. Conclusions

In this paper, we investigated the formation of river mouth bar under the conditions where longshore sediment transport by bed load predominated in the laboratory experiments. Then, the applicability of two numerical models, that is numerical model for predicting shoreline deformation based on modified one-line theory and numerical model of 3-dimensional topographic change, were examined by using these measured topographic change in the experiments.

The principal results of this experiment can be summarized as follows;

Bars are generated at both sides of the river mouth regardless of the incident wave conditions and depth at the river mouth. Bar of the upstream side is generally larger than that of the downstream side and its location is also deeper than that of the downstream sides.

Shape and dimension of the bar generated at the downstream side do not deeply depend on the wave condition and depth at the river mouth. While, bar generated at the upstream side is strongly affected by the incident wave characteristics and depth at the river mouth.

Decrease in the sectional area of the river mouth is rapid in the case where the depth at the river mouth is shallow. However, it seems that there is a critical depth for the occurrence of the blockage under the same rate of longshore sediment transport along the shoreline of the upstream side.

The two numerical models for predicting topographic change examined in this study can roughly reproduce topographic change around river mouth provided that the topographic change is not so large that can't affect incident wave field significantly. To exactly predict topographic change where a large deposition or erosion that affect incident wave field, we have to take into account the effect of interaction between topography, incident waves and wave-induced current. However, we could not examine the adequate interval to interact topographic change and waves and current fields in this study.

Acknowledgment

This study was carried out by as a part of the research sponsored by the Grant-in-Aid for Scientific Research of the Ministry of Science, Culture and education to DI (08650606).

References

Bruun, P.(1978), "*Stability of Tidal Inlets*", Elsevier, p.506

Butakov, A.N.(1971), "Study of Development and Deformation of Mouth Bar", *14th IAHR Conf.*, Paris

Deguchi, I. and Sawaragi, T.(1988),"Effects of Structure on Deposition of Discharged Sediment around River mouth", *Proc. 21st Int. Conf. on Coastal Eng.*, ASCE, pp. 1573~1587

FitzGerald, D.M.(1982),"Sediment bypassing at Mixed Energy Tidal Inlets", *Proc. 13th Conf. on Coastal Eng.*, pp.1094~1118

FitzGerald, D.M., Penland, S. and D.Nummedal(1984), "Sedimentation Processes along the East Friesian Island", *Proc.14th Conf. on Coastal Eng.*, pp.3051~3066

Ifuku, M., Yonezaw, Y., and Nakanishi, M.,"On the Profile Change of Sand Bar in the Hiji River", *Proc. of Japanese Coastal Eng.*, JSCE, Vol. 41, pp.491~495(in Japanese)

LeBlond, P.H.(1973),"On the Formation of Spiral Beaches", *Proc.13th Int. Conf. on Coastal Eng.*, pp.1331~1345

Ozsoy, E.(1977),"Suspended Sediment Transport near Tidal inlets",*Coastal Sediment'77*, ASCE, pp.914~926

Rea, C.C., and P.D.Komar(1975),"Computer Simulation Models of a Hooked Beach Shoreline Configuration", *J. of Sed. Pet.*, Vol. 45, No. 4, pp.866~872

Sawaragi, T., J. S. Lee and I.Deguchi(1984),"Study on Numerical Model for Predicting Wave-induced Current and Topographic Change around River Mouth", *Proc. of Japanese Coastal Eng.*, JSCE, Vol. 31, pp.411~415(in Japanese)

Sawaragi, T(1995),"*Coastal Engineering - Waves, Beaches, Wave-Structure Interactions*", Elsevier, p.479

Sleath, J.F.A(1978),"Measurement of bed load in oscillatory flow", *J. Waterway, Port, Coastal and Ocean Div., Proc. ASCE*, Vol. 104, No. WW4, pp. 291~307

Suga, K., Ishikawa, T., Nadaoka, K., and Tanaka, H.,"Formation of Sand Terrace in front of a River Mouth and its Decline", *Technical Note in Proc. of the Japan Society Civil Engineers*, No.381, pp. 227~230(in Japanese)

Tanaka, T., Shao. X., and Shuto., N.(1993),"Experimental and Theoretical study on Scouring of Sand bars by Overflowing Water", *Proc. of the Japan Society Civil Engineers*, No.479, pp.31~40(in Japanese)

Tanaka, S., Koarai, M., and Fukazawa, M.(1993),"Measurement on the Long-term Change of the Coastline in Japan by Using 1/50,000 Topographic Maps", *Proc. of Japanese Coastal Eng.*, JSCE, Vol. 40, pp. 416~420(in Japanese)

Uda, T.(1982),"An Application of a Shoreline Change Model to the Prediction of Change of a Spit in a River Mouth", *Proc. of the Japan Society Civil Engineers*, No.322, pp.77~87(in Japanese)

Proceedings of the Seventh (1997) International Offshore and Polar Engineering Conference
Honolulu, USA, May 25-30, 1997

ISBN 1-880653-28-1 (Set); ISBN 1-880653-31-1 (Vol. III)

Beach Response to Sea-Level Rise Around the Sea of Japan (East Sea)

Chang Bae Son, Hideaki Noda, Yuhei Matsubara and Masamitsu Kuroiwa
Tottori University
Tottori, Japan

Abstract

The present paper investigates long-term changes in mean sea-level as well as correlation among sea-level, atmospheric pressure and air temperature changes around the coast of the Sea of Japan (East Sea). Tidal data have been analyzed by moving average and the results showed that the sea-level is rising. Wave tank experiments demonstrated that both static retreat by water-level rise and additional retreat by wave action contributes to total retreat. Potential shoreline retreat was predicted in some places using models developed by Bruun, Dean and Mimura under various assumptions and results showed risk of serious loss of sandy beach.

Key words : sea-level rise, shoreline retreat, equilibrium beach profile, potential shore retreat

Introduction

In Japan and Korea, the population and industry tend to be concentrated in the coastal areas. As a result, significant rise in the sea-level could cause great damage, such as inundation and displacement of lowland and wetland, intensification of coastal storm flooding and coastal erosion. Recently, global warming and its environmental impact have become a concern worldwide. One of the important effects of global warming is a sea-level rise in the coastal zones. Many researchers have attempted to predict sea-level changes on global scale. The IPCC (Intergovernmental Panel on Climate Change) predicted a sea-level rise of between 19 *cm* and 95 *cm* (≒0.5 *cm/year*) by 2100 (IPCC, 1995). However the degree of sea-level rise depends strongly on regional characteristics, such as atmospheric pressure, sea-water temperature, ocean current, wind, topography and plate tectonics.(Murakami and Yamada, 1994). Therefore in order to accurately predict sea-level variation, it is important to accumulate data about local tendency of sea-level changes. In the present study, we investigated long-term sea-level, atmospheric pressure and air temperature changes along the coast of the Sea of Japan (East Sea). This region is associated with a relatively small tidal range as well as infrequent unusual climate changes.

When sea-level rises, the increase does not occur at a constant rate. This fluctuation also affects shore erosion. For example, storm surge transports large amounts of beach material to the deep offshore over the course of one or two days. In the present study, only problem concerning long-term sea-level rise and onshore-offshore transport were considered in order to find the basic relationship between sea-level rise and cross-shore profile changes. Wave tank experiments were carried out and the experimental results were compared with theoretical models developed by Bruun (1962), Dean (1991) and Mimura (1993). Using these models and analyzed long-term sea-level change results, shoreline retreat was predicted for 2 regions of Japan under various assumptions.

Long-term sea-level, atmospheric pressure and air temperature changes

Mean sea-level data from 46 tidal stations and atmospheric pressure and air temperature data from 38 meteorological stations in Japan and Korea were collected in order to analyze the long-term sea-level variation. Fig. 1 represents the location of tidal observation stations. Due to a lack of agreement between tidal and meteorological stations data were collected with considering of same latitude and proximity.

In the study areas, mean sea-level, atmospheric pressure and air temperature are all strongly dependent on the season. In general, the monthly mean sea-level is highest in summer and lowest in winter. The cause of this phenomenon may be related to seasonal changes of atmospheric pressure, sea-water temperature and other environmental change, such as currents and wind. Low atmospheric pressure and high sea-water temperature in summer result in high water-level, conversely high atmospheric pressure and low sea-water temperature in winter result in low water-level.

In order to analyze long-term changes, these seasonal variations must be eliminated. This can be accomplished by first determines the mean daily data based on hourly data collected at tidal and meteorological stations. Using the mean daily data, monthly mean data can be obtained. Finally, using the 12 months moving average, seasonal fluctuations can be eliminated. The resulting mean sea-level, atmospheric pressure and air temperature changes are presented in Table 1 with according to tidal stations and observation period. The symbols ▲ and ▽ in Table 1 indicate sea-level rise and fall, respectively.

Table 1 Annual mean sea-level, atmospheric pressure and air temperature changes

		No.	Obs.Station	Tidal Obs. Duration (year)			Annual mean sea-level change (cm/year)	Annual mean atmospheric pressue change (hPa/year)	Annual mean air temperature change(℃/year)	Annual mean sea-level change by air pressure and air temperature (cm/year)
				10<	10~20	21~32				
Japan	Hokkaido	1	Wakkanai			▲	0.312	0.032	0.042	○ 0.013 (24)
		2	Kutsugata	▽			-0.096			
		3	Ishikari	▲			2.172	0.181	0.767	○ 0.662 (6)
		4	Otaru	▲			0.685			
		5	Iwanai	▽			-0.167	0.076	0.338	× 0.296 (7)
		6	Esashi	▲			0.383	0.009	0.558	○ 0.605 (6)
		7	Matsumae	▲			0.011			
		8	Yoshioka	▲			0.067			
		9	Hakodate			▽	-0.028	0.02	0.056	× 0.042 (24)
		10	Ominato			▲	0.035	0.055	0.034	× -0.018 (26)
		11	Aomori			▲	2.010	0.028	0.056	○ 0.034 (23)
		12	Tappi	▲			0.045			
	Honshyu	13	Odomari			▽	-0.058	0.012	0.025	× 0.015 (25)
		14	Fukaura		▽		-0.335	0.04	0.052	× 0.018 (18)
		15	Iwasaki	▽			-1.049	0.414	0.464	× 0.096 (5)
		16	Toyama			▲	0.052	0.017	0.04	○ 0.028 (23)
		17	Maizuru-1			▲	0.220	0.033	0.038	○ 0.010 (26)
		18	Maizuru-2			▲	0.158	0.022	0.04	○ 0.022 (22)
		19	Saigo			▲	0.134	0.02	0.012	× -0.007 (30)
		20	Sakai			▽	-0.093	0.007	0.035	× 0.032 (24)
		21	Tonoura		▽		-0.046	-0.024	0.034	× 0.062 (17)
		22	Hamada	▲			0.738	-0.039	0.464	○ 0.549 (6)
		23	Shimonoseki			▽	-0.039	0.012	0.059	× 0.053 (28)
	Kyushyu	24	Izuhara			▲	0.172	0.021	0.043	○ 0.027 (26)
		25	Moji			▲	0.140	0.004	0.033	○ 0.032 (26)
		26	Hakata			▲	0.016	0.011	0.048	○ 0.042 (25)
		27	Sasebo			▲	0.016	0.014	0.038	○ 0.028 (26)
		28	Fukue			▽	-0.095	0.008	0.048	× 0.045 (24)
		29	Nagasaki			▲	0.096	0.012	0.049	○ 0.041 (23)
		30	Kuchinotsu			▽	-0.055	0.01	0.061	× 0.057 (23)
		31	Oura			▲	0.099	-0.003	0.047	○ 0.055 (22)
		32	Misumi			▲	0.178	0.005	0.031	○ 0.029 (30)
Korea	Donghae	33	Sokcho		▲		0.236	-0.012	0.008	○ 0.021 (21)
		34	Mukho	▲			0.230	-0.172	-0.001	○ 0.171 (9)
		35	Ullungdo		▽		-0.111	0.116	0.055	○ -0.057 (16)
		36	Pohang			▲	0.140	0.051	0.052	○ 0.006 (23)
		37	Ulsan			▽	-0.016	0.003	0.032	× 0.032 (30)
	Namhae	38	Pusan			▲	0.179	0.002	0.024	○ 0.024 (32)
		39	Gadukdo		▲		0.216	-0.004	0.029	○ 0.036 (9)
		40	Chungmu		▲		0.356	0.115	0.016	× -0.097 (18)
		41	Yosu			▲	0.118	0.022	0.024	○ 0.004 (29)
		42	Komundo		▲		0.533			
		43	Wando		▲		0.183	-0.011	0.066	○ 0.084 (12)
		44	Cheju			▲	0.439	0.004	0.017	○ 0.015 (31)
		45	Sogwipo		▲		0.947	-0.020	0.051	○ 0.076 (10)
		46	Chujado		▲		0.132			

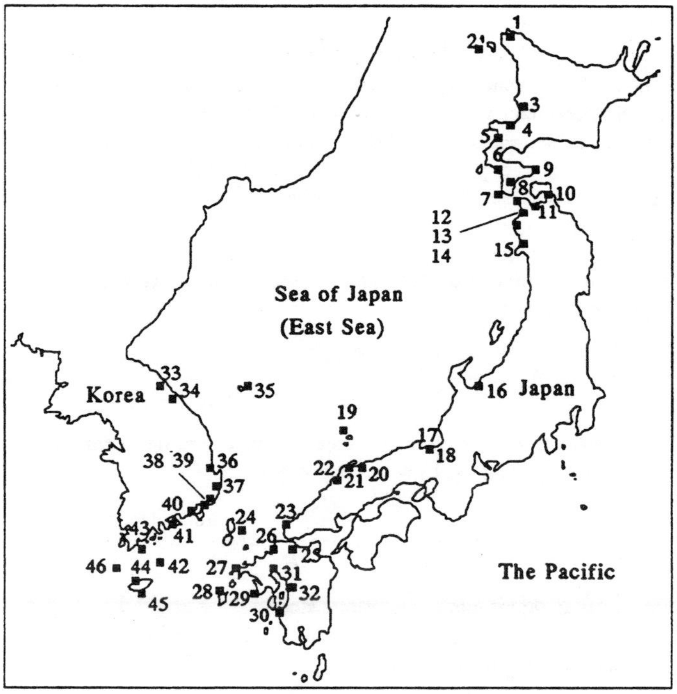

Fig. 1 Locations of tidal observation stations

Data collected over 20 years were used for long-term analysis. Results showed rise in 18 regions and a fall in 6 regions. The coast of the Sea of Japan (East Sea) has shown a tendency toward a slight rise during the past 3 decades. In particular, over the past 3 decades, air temperature has increased on average 0.036 ℃*/year* the overall study area. This value agrees with those reported by IPCC (1991), 0.3~0.6 ℃, over the past hundred years and provides evidence of global warming due to an increase in green house gases.

Sea-level and atmospheric pressure are inversely related such that sea-level rises 1 *cm* for each reduction in atmospheric pressure of 1 *hPa*. Sea-water temperature changes results density changes and ultimately sea-level variation. An increase in sea-water temperature of 1 ℃ results in approximately a 1.1 *cm* sea-level rise (Murakami and Yamada, 1994). In the present study, due to a lack of sea-water temperature data, maximum effect of air temperature was investigated under the assumption that sea-water temperature varies to the same degree as air-temperature. The right column of Table 1 lists the calculated sea-level variation due to the change in atmospheric pressure and air temperature using the above rate. Numbers in parentheses indicate the common period (*year*) for which sea-level, atmospheric pressure and air temperature data exist. The symbol ○ indicates that the calculated sea-level variation by atmospheric pressure and air temperature change showed the same tendency as estimated sea-level variation, and × indicates inverse case of ○. Based on the calculated results, contrary to the effects of seasonal variation, long-term sea-level variation and long-term atmospheric pressure and air temperature variations do not show a clear causality. Regions of contrary tendency in sea-level changes exist between tidal stations in close proximity and which may be related to the complex correlation among currents, sea-water circulation, topography, plate tectonics and air circulation.

Experiments

Experiments were carried out in a wave tank 24 *m* long, 0.6 *m* wide and 1.1 *m* deep with a plane initial beach slope of 1/10 on which sand of diameter 0.17 *mm* was placed. Regular waves were generated until an equilibrium profile was developed, followed by 5 *cm* increase in water-level and renewal of wave activity until a new equilibrium was achieved, such that the constant deep water wave steepness was 0.045 and period was 1 *sec*. Water depth increased 5 *cm* for each step from 40 to 55 *cm*. The resulting profile is presented in Fig. 2. The profiles maintained nearly the same equilibrium one and translated landward and upward during periods of rising water-level. The shoreline retreated by 40~50 % of the deep water wave length in the case of 5 *cm* water-level rise and it corresponded to 14.5 *cm* retreat in the case of 1 *cm* water-level rise. Fig. 3 shows the time variation of shoreline with dimensionless time T/t and dimensionless distance X_s/L_o in the case of constant deep water wave conditions. X_s represents the distance difference in shoreline between that before the sea-level rise and that after the sea-level rise. X_s/L_o at $T/t = 0$ means static retreat in the case of a water-level rise with no wave action. The additional retreat by wave action was revealed as a distance difference between shoreline position of the static retreat and that of the equilibrium profile achieved. Thus, based on the results of these tests, it has been found that both static retreat by water-level rise and additional retreat by wave action plays an important role in total retreat.

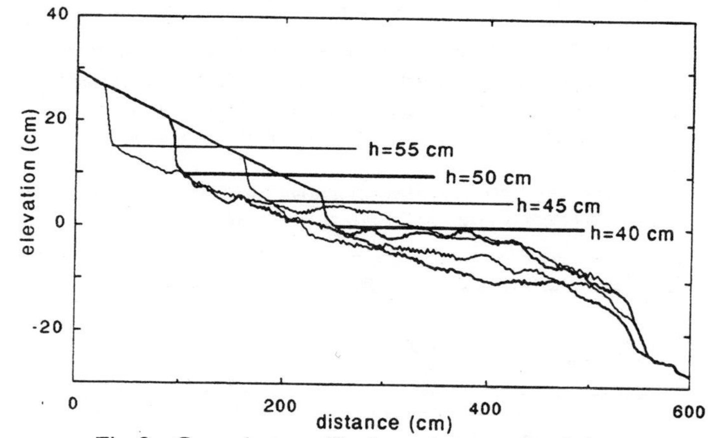

Fig. 2 Cross-shore profile changes to water-level rise

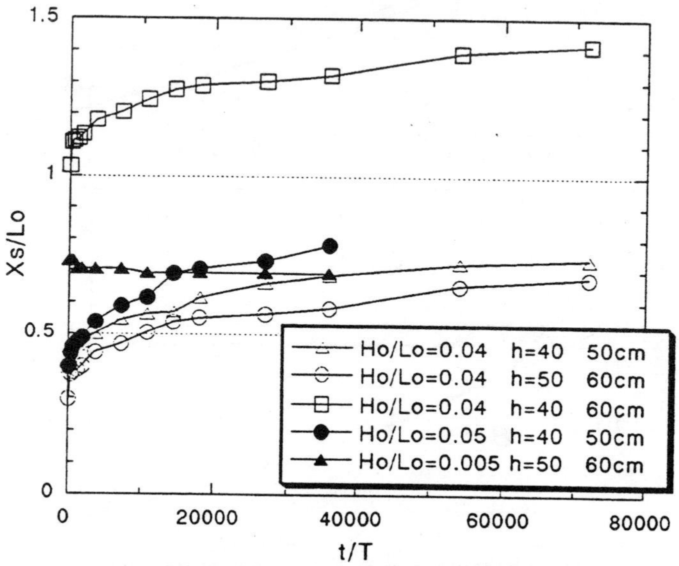

Fig. 3 Dimensionless time variations of shoreline

Prediction of shoreline retreat to sea-level rise

In the case of water-level rise, beach profile evolves new equilibrium toward the raised water-level and it includes the concept of equilibrium profile to sea-level rise such that beach erosion consists of additional retreat induced by waves as well as static retreat. The first well-known model that relates shoreline retreat to an increase in sea-level was proposed by Bruun (1962). Bruun's equation (1) was derived by assuming that the eroded volume in the upper beach is transported offshore such that the volume eroded is equal to the volume deposited. Dean and Mimura derived equation (3) and (4) with same geometrical concept by using the simplified equilibrium profile proposed by Dean (1977). Because these models are static (geometric), a sediment transport relationship is not required. However, these simple geometric models are more appropriate than the dynamic (process-based) models in the case of very long time scale problems such as long-term sea-level rise. Fig. 4 represents models of the beach profile response to sea-level rise proposed by Dean and Mimura. The thick line represents the initial beach profile and the thin line represents the beach profile following the rise in sea-level. Bruun proposed the following relationship between shoreline retreat and sea-level rise:

$$\Delta y = \frac{w_* s}{h_* + B} \tag{1}$$

where Δy is the shoreline recession, s is the increase in sea-level, w_* is the seaward limit of the active profile, h_* is the maximum depth of exchange of material between nearshore and offshore and B is the berm height. Bruun's model holds regardless of the shape of the beach profile.

As shown by Dean (1977), a simplified equilibrium profile for an open-coast beach is given by the power law curve as

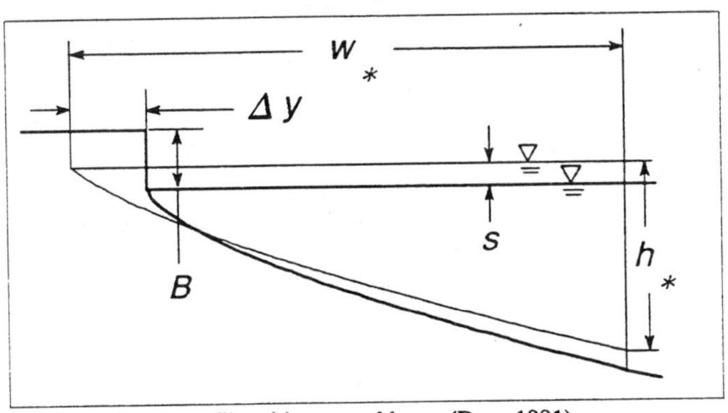

(a) Beach profile with squared berm (Dean,1991)

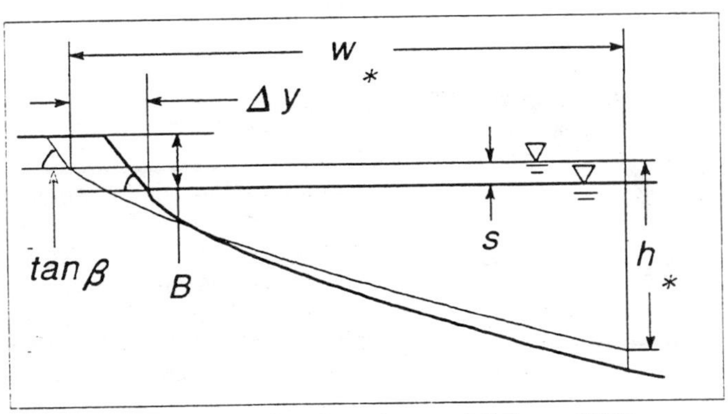

(b) Beach profile with foreshore slope tanß (Mimura,1993)

Fig. 4 Models of beach response to sea-level rise

$$h(y) = A(D)y^{2/3} \tag{2}$$

where h is the water depth at offshore distance y, A (dimension:$m^{1/3}$) is a scale parameter which depends primarily on sediment characteristics and D is sediment size. Scale parameter A governs the overall steepness of the profile. Although this simplified profile can not represent characteristic shapes of a beach profile, such as bars and troughs, it is applicable to long time scale beach response problem.

Using the same concept proposed by Bruun, Dean proposed the following relationship using the simplified beach profile given by equation (2). (Fig. 4. a)

$$-\Delta y + \frac{3h_* w_*}{5B}\left[1 - \frac{\Delta y}{w_*}\right]^{5/3} = \frac{3h_* w_*}{5B} - \frac{s w_*}{B} \tag{3}$$

Mimura (1993) suggested more realistic profile form with a foreshore slope (Fig. 4. b) which is given by:

$$-\Delta y + \frac{3h_* w_*}{5B}\left[1 - \frac{\Delta y}{w_*}\right]^{5/3} = \frac{3h_* w_*}{5B} - \frac{s w_*}{B} + \frac{0.5s^2 - sB}{B\tan\beta} \tag{4}$$

where $\tan\beta$ represents the foreshore slope.

Berm height B is calculated by the following relationship given by Takeda and Sunamura (1983).

$$B = 0.125 H_{bm}^{5/8} (g T_m^{2})^{3/5} \tag{5}$$

where H_{bm} is the mean breaking wave height and T_m is the mean wave period. Dean used breaking depth as a limit depth h_*, but in the present study we used a simple Birkemeier's relationship (1985) which defines the seaward limit of seasonal profile fluctuation, known as closure depth. Birkemeier evaluated Hallermeier's relationship (1978) using field measurements and proposed the following simplified equation.

$$h_* = 1.57(H_m + 5.6\sigma_H) \tag{6}$$

where H_m is the mean wave height and σ_H is the standard deviation

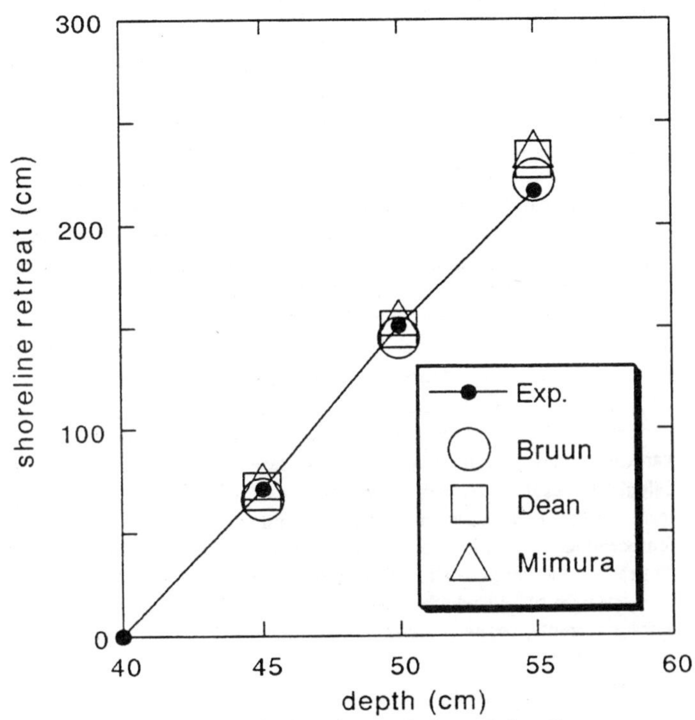

Fig. 5 Dimensionless time variations of shoreline

of wave height.

Foreshore slope tan β in equation (4) is calculated using Sumamura's relationship (1975).

$$\tan\beta = 0.45(D/H_o)^{0.5}(H_o/L_o)^{-0.3} \quad (7)$$

Shoreline retreat by experiments and calculated results by equation (1),(3) and (4) using experimental parameters is presented in Fig. 5 and show good agreement. Thus, these models are applicable to *2-D* beach response to sea-level rise problems that are not accompanied by longshore transport and varying time scale forcing.

Dimensionless form of equation (1) and (3) yield

$$\Delta y' = \frac{s'B'}{1+B'} \quad (8)$$

$$\Delta y' = s' - \frac{3}{5B'}\left[1-\left(1-\Delta y'\right)^{5/3}\right] \quad (9)$$

where $\Delta y' = \Delta y/w_*$, $s' = s/B$ and $B' = B/h_*$. Equations (8) and (9) are plotted in Fig. 6 (Dean, 1991). The relationship between H_o/L_o, s' and $\Delta y'$ is plotted in Fig. 7 in the case of T=6 *sec*, σ_H =7.5 and A=0.15 $m^{1/3}$. Letters "Bruun", "Dean" and "Mimura" in Fig. 7 represent calculated values by their relationships given by Equation (1), (3) and (4). From Fig. 6 and 7, the amounts of sea-level rise and slope of the beach profile appears to be more sensitive than wave conditions for the determination of shoreline recession due to sea-level rise.

Using the estimated sea-level rise data and time series of wave data (NOWPAS, 1994-95), we calculated potential shoreline recession for Maizuru-1 and Hamada in Japan for the next 5 decades. For the

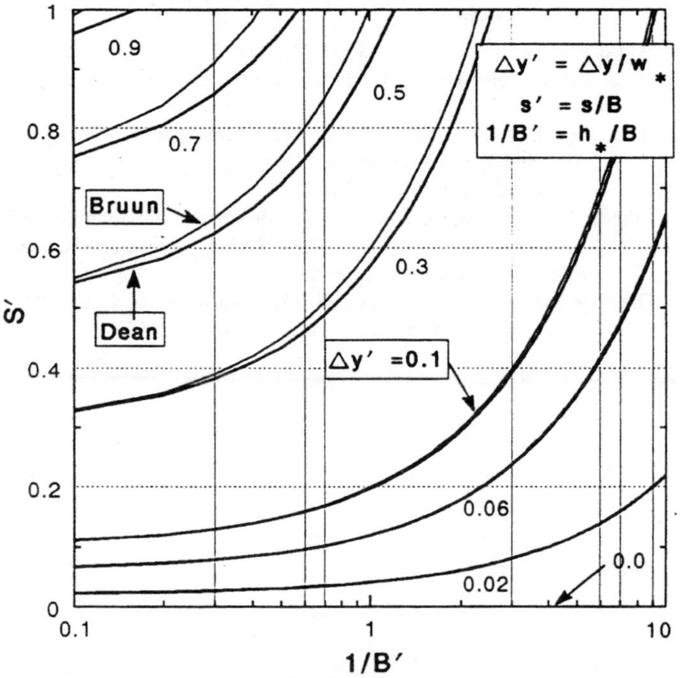

Fig. 6 Dimensionless shoreline change, Δy' vs dimensionless limiting depth, 1/B' and dimensionless sea-level rise, S'

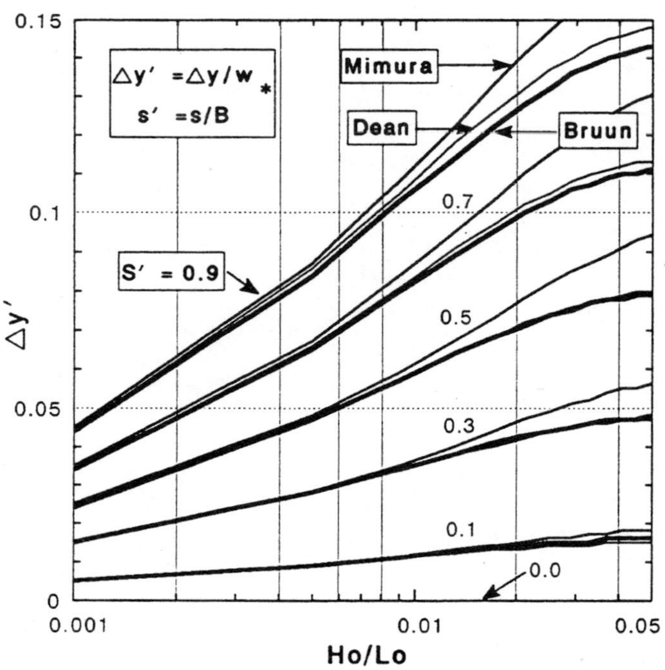

Fig. 7 Dimensionless shoreline change, Δy' vs dimensionless sea-level rise, S' and deep water wave steepness, H_o/L_o

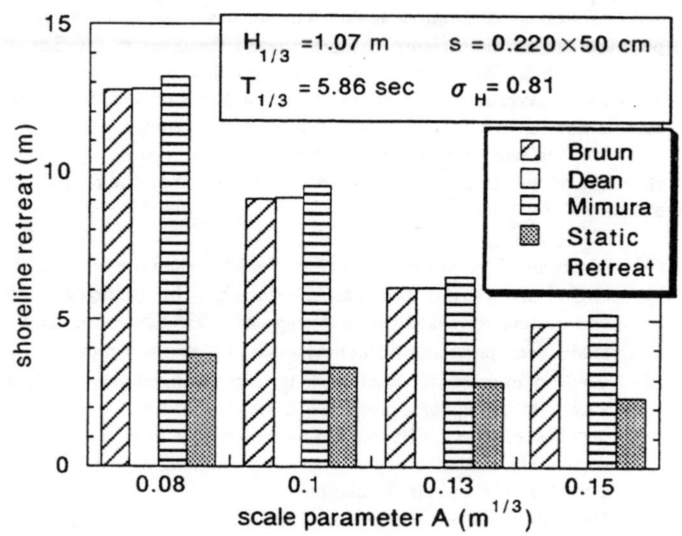

(a) Maizuru-1

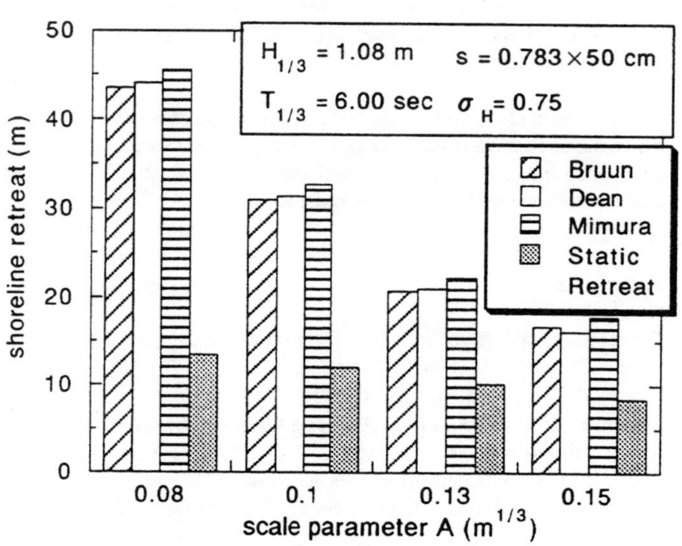

(b) Hamada

Fig. 8 Shoreline recession after 5 decades

calculation, we assumed that (1) there was no longshore sediment transport, (2) the rate of sea-level rise over 5 decades remains constant at the estimated rate same as estimated one, (3) the width of the sandy beach is infinite, (4) the wave conditions do not change significantly. Fig. 8 (a), (b) show the potential shoreline recession for the estimated rate of sea-level rise and some beach profile slopes (varying scale parameter A) in Mazuru-1 and Hamada, respectively. Static retreat in Fig.8 was calculated based on Fig. 4(b). Fig. 8 shows that shoreline retreat depends primarily on the degree of sea-level rise and steepness of the beach profile. Furthermore, based on the calculated results, a substantial portion of the sandy beach can be expected to disappeared in the case of using above assumptions. In Hamada, the shoreline recession is expected to be approximately 45 *m* based on a 0.783 *cm/year* sea-level rise and 0.08 $m^{1/3}$ scale parameter, A(sand size=0.2 *mm*, beach slope=0.029) after 5 decades.

Conclusions

The present paper investigated the long-term change in mean sea-level as well as correlation among long-term sea-level, atmospheric pressure and air temperature changes around the coast of the Sea of Japan (East Sea). Wave tank experiments were then carried out and the results were compared with calculated value based on the theoretical models developed by Bruun, Dean and Mimura. Using these models and the analyzed long-term sea-level changes results, potential shoreline retreats in 2 regions of Japan were predicted. The results are as follows.

(1) The coast of the Sea of Japan (East Sea) has shown a tendency to slight rise during the past 3 decades. In particular, air temperature increased by an average of 0.036 *℃/year* in all the study areas, providing further evidence of global warming .

(2) Contrary to seasonal variation, long-term sea-level variation and long-term atmospheric pressure and air temperature variations do not show clear causality and this may be related with complicated correlation among current, sea-water circulation, topography, plate tectonics and air circulation.

(3) Not only static retreat by water-level rise but also additional retreat by wave action plays an important role in total retreat.

(4) Based on the calculated result for Hamada (45*m* shoreline retreat for 0.783 *cm/year* sea-level rise and 0.08 $m^{1/3}A$), serious loss of sandy beach is predicted.

The present study investigated only the potential shoreline recession. Thus, based on these results, longshore transport and various time scale forcing, such as storm surge, tsunami and seasonal fluctuation that can cause shoreline recession must be considered.

References

Birkemeier, WA (1985). "Field Data on Seaward Limit of Profile Change." *J. Wtrwys Port Coast and Oc Eng Div*, ASCE, Vol 111, No 3. pp 598-602.

Bruun, P (1962). "Sea Level Rise as a Cause of Shore Erosion." *J. Wtrwys Harb Div*, ASCE, Vol 88, no 1, pp 117-130.

Dean, RG (1977). "Equilibrium Beach Profiles : U.S. Atlantic and Golf Coasts." Oc Eng Report, No 12, Univ of Delaware, Newark, De.

Dean, RG (1991). "Equilibrium Beach Profiles : Characteristics and Applications." *J Coastal Research*, Vol 7, No 1, pp 53-84.

Hallermeier, RJ (1978). "Uses for a Calculated Limit Depth to Beach Erosion." *Proc 16th Int Conf, Coastal Eng*, ASCE, Hamburg, pp1493-1512.

IPCC (1991). "Climate Change." Island Press, pp 272.

IPCC (1995)."Climate Change : The IPCC Second Assessment Report." Cambridge University Press

Mimura, N et al. (1993). "Impact Assessment of Sea-level Rise for Sandy Beach," *Proc 40th Japanese conf Coastal Eng*, pp 1046-1050. (in Japanese)

Murakami, K, and Yamada, K (1994). "Sea-level Change along the Japan Coast and its Causes," *Coastal Eng. in Japan*, Vol 37, No 2, pp 201-218.

Sunamura, T (1975). "Static Relationship among Beach Slope, Sand Size and Wave Properties." *Geographical Review of Japan*, Vol 48, No 7, pp 485-489.

Takeda, I and Suanmura, T (1983) "Profile Change of Sandy Beach for the Accretion Process," *Proc 30th Japanese conf Coastal Eng*, pp 254-258. (in Japanese)

The Ministry of Transport (1994-95). "Nationwide Ocean Wave Information Network for Ports and Harbours." CDIT.

Proceedings of the Seventh (1997) International Offshore and Polar Engineering Conference
Honolulu, USA, May 25-30, 1997

ISBN 1-880653-28-1 (Set); ISBN 1-880653-31-1 (Vol. III)

Effects of a Submerged Structure on the Tidal Current

S.Y. Boo
Korea Naval Academy
Chinhae, Korea

K. Hong
Korea Research Institute of Ships and Ocean Engineering
Taejon, Korea

ABSTRACT

A numerical prediction scheme of the tidal circulation due to a submerged structure was developed. The three-dimensional shallow water equation was solved using multi-level finite difference method. Time integration of the equations was done using three-time level explicit scheme. A staggered grid system was employed in the horizontal and vertical directions. The numerical model was tested and verified by solving wind-induced flow as well as tidal responses in a rectangular channel. Finally, circulation in a closed harbor was investigated, where a submerged flow control structure was placed. It was found that the vertical flow characteristics at flood tide are different from one at the ebb tide.

INTRODUCTION

Recently, researches on utilization of coastal zone have been being carried out. One of those works are to install a large floating structure, for instance, floating airport or complex plant (Inoue et al. 1994, Jung 1996). Even though those floating structure can provide various products or facilities, adverse effects on water quality near the site may be caused due to current change or slack water around the structure. Especially, in case a structure is placed at a partially closed area with relatively small entrance to reduce additional construction coast of floating breakwater, tidal-induced circulation or tidal flushing plays an important role for water quality. It is known that the water quality in basin depends on retention time of tide, flushing rate, velocity fields or mixing process.

The mathematical hydrodynamic model based on the three-dimensional shallow water equation can be treated various way. Finite difference method is one of the most popular scheme in solving the equations (Fusino&Tabeta 1993). Falconer (1890) and Jin&Krenenburg (1993) studied the circulation pattern in harbor in two- and three-dimensions, respectively. From those results, overall flushing increases if the length to breadth ration is made as close as unity, and velocities in a harbor depends on horizontal eddy viscosity. Finite element hydrodynamic model is also well implemented to the shallow water problem in which complex geometries can be efficiently represented with proper mesh distributions, for instance, Chen et al. (1991) and Kodama et al. (1996).

In this paper, a numerical scheme, therefore, was developed to predict tide-induced circulations in a harbor by implementing a finite difference method to the three-dimensional shallow water equation. The equation was solved using three-time level explicit scheme. A staggered grid system was employed in the horizontal and vertical directions. A multi-level scheme was used so that density variations were properly taken into account. Variable mesh system was utilized to effectively discretize the complex geometries as well.

The numerical model was verified by solving a wind-induced flow in a narrow channel as well as tidal responses in a rectangular channel. Asaeda et al (1994) pointed out strong convection can be produced by placing a V-shaped plated in a horizontal flow. Thus we placed a cross-shaped structure in a closed harbor and investigated effect on the circulation.

EQUATIONS OF MOTION

The origin of the Cartesian coordinate system (x, y, z) is placed at the undisturbed sea surface. The coordinates $x, y,$ and z are positive eastward, northward and upward, respectively. Basic equations of the multi-level model flow are based on the following continuity and conservation equations of momentum in the x and y direction,

$$\frac{\partial u}{\partial x}+\frac{\partial v}{\partial y}+\frac{\partial w}{\partial z}=0 \tag{1}$$

$$\frac{\partial u}{\partial t}+u\frac{\partial u}{\partial x}+v\frac{\partial u}{\partial y}+w\frac{\partial u}{\partial z}= -\frac{1}{\rho_o}\frac{\partial p}{\partial x}+fv+E_h(\frac{\partial^2 u}{\partial x^2}+\frac{\partial^2 u}{\partial y^2})+E_v\frac{\partial^2 u}{\partial y^2} \tag{2}$$

$$\frac{\partial v}{\partial t}+u\frac{\partial v}{\partial x}+v\frac{\partial v}{\partial y}+w\frac{\partial v}{\partial z}= -\frac{1}{\rho_o}\frac{\partial p}{\partial x}-fu+E_h(\frac{\partial^2 v}{\partial x^2}+\frac{\partial^2 v}{\partial y^2})+E_v\frac{\partial^2 v}{\partial z^2} \tag{3}$$

and the vertical momentum equation is simplified using assumption of a hydrostatic pressure distribution, which is valid when vertical acceleration is negligible compared to the vertical pressure gradient.

$$\frac{1}{\rho}\frac{\partial p}{\partial z}+g=0 \tag{4}$$

The physical variables in equation (1)-(4) are $u, v, w=$ velocity components in the x, y and z directions; $f=$ Coriolis parameter; $\rho_o=$ reference density of fluid; $\rho=$density of water; $E_h=$ horizontal eddy viscosity; $E_v=$ vertical eddy viscosity; $p=$ pressure; $g=$ gravity acceleration.

The boundary conditions on the free surface are represented as

$$\frac{\partial \zeta}{\partial t}+u\frac{\partial \zeta}{\partial x}+v\frac{\partial \zeta}{\partial y}-w=0 \tag{5}$$

$$E_v\frac{\partial u}{\partial z}=\frac{\tau_{xw}}{\rho_o},\quad E_v\frac{\partial v}{\partial z}=\frac{\tau_{yw}}{\rho_o} \tag{6}$$

where τ_{xw} and τ_{yw} are the wind shear stresses in x and y directions. The bottom boundary conditions are written as :

$$u\frac{\partial h}{\partial x}+v\frac{\partial h}{\partial y}+w=0 \tag{7}$$

$$E_v\frac{\partial u}{\partial z}=\frac{\tau_{xb}}{\rho_o},\quad E_v\frac{\partial v}{\partial z}=\frac{\tau_{yb}}{\rho_o} \tag{8}$$

where h is the water depth and τ_{xb} and τ_{yb} are the bottom shear stresses in x and y directions. The stresses on the free surface and bottom are formulated in the following way,

$$\begin{aligned}\tau_{xb} &= \gamma\rho_o u_b\sqrt{u_b^2+v_b^2}\\ \tau_{yb} &= \gamma\rho_o v_b\sqrt{u_b^2+v_b^2}\\ \tau_{xw} &= C_w\,\rho_a u_w\sqrt{u_w^2+v_w^2},\\ \tau_{xw} &= C_w\,\rho_a v_w\sqrt{u_w^2+v_w^2}\end{aligned} \tag{9}$$

where γ, u_b and v_b are the frictional coefficient and velocity components at the sea bottom. The symbol C_w denotes the drag coefficients at the surface, ρ_a the air density, u_w and v_w the velocities at the surfaces. The tidal elevation at the open boundary is given in the harmonic function and velocity gradient normal to the open boundary is assumed zero. If the open boundary is in the $y-z$ plane, the conditions are expressed as :

$$\zeta=\zeta_o\sin\frac{2\pi}{T_o}t$$

(10)

$$\frac{\partial u}{\partial x}=0,\quad \frac{\partial v}{\partial x}=0$$

where T_o and ζ_o are the period and amplitude of tide.

Nonslip condition is applied at the closed boundary surface so that the velocity parallel to the surface becomes zero and the normal velocity is also zero. The gradient of the elevation normal to the surface is taken as zero as well.

MULTI-LEVEL FINITE DIFFERENCE FORMULATION

The momentum equations of eqn. (2) and (3) are represented using three-time level explicit scheme in the following form :

$$\frac{u^{n+1}-u^{n-1}}{2\Delta t}+(u\frac{\partial u}{\partial x})^n+(v\frac{\partial u}{\partial y})^n+(w\frac{\partial u}{\partial z})^n= -(\frac{1}{\rho_o}\frac{\partial p_a}{\partial x})^n-(g\frac{\partial \zeta}{\partial x})^n-(\frac{g}{\rho_o}\frac{\partial}{\partial x}\int_z^{\zeta}\rho' dz)^n -fv^n+(E_h\frac{\partial^2 u}{\partial x^2}+E_h\frac{\partial^2 u}{\partial y^2})^{n-1}+(E_v\frac{\partial u}{\partial z^2})^{n-1} \tag{11}$$

$$\frac{v^{n+1}-v^{n-1}}{2\Delta t}+(u\frac{\partial v}{\partial x})^n+(v\frac{\partial v}{\partial y})^n+(w\frac{\partial v}{\partial z})^n= -(\frac{1}{\rho_o}\frac{\partial p_a}{\partial y})^n-(g\frac{\partial \zeta}{\partial y})^n-(\frac{g}{\rho_o}\frac{\partial}{\partial y}\int_z^{\zeta}\rho' dz)^n +fu^n+(E_h\frac{\partial^2 v}{\partial x^2}+E_h\frac{\partial^2 v}{\partial y^2})^{n-1}+(E_v\frac{\partial v}{\partial z^2})^{n-1} \tag{12}$$

where the vertical momentum equation (4) was integrated vertically and used to derive the above equations. A staggered grid system (see Fig. 1) is employed for the finite difference formulation of the above mentioned equations. In addition, a variable mesh is also used in the horizontal plane in order to effectively represent the complex boundary. In the vertical direction, multi-level scheme is utilized so that vertical variation of the density is considered. The nonlinear convective terms in equation (11) and (12) are computed using first order or second order upwinding difference. For instance, the convective term in the x-direction is derived using first-order upwinding scheme as :

$$u\frac{\partial u}{\partial x} = \frac{u_{ijk}+|u_{ijk}|}{2}\ \frac{u_{ijk}-u_{i-1jk}}{x_i-x_{i-1}} + \frac{u_{ijk}-|u_{ijk}|}{2}\ \frac{u_{i+1jk}-u_{ijk}}{x_{i+1}-x_i} \tag{13}$$

$$v\frac{\partial u}{\partial y} = \frac{\bar{v}_u+|\bar{v}_u|}{2}\ \frac{u_{ijk}-u_{ij-1k}}{y_j-y_{j-1}} + \frac{\bar{v}_u-|\bar{v}_u|}{2}\ \frac{u_{ij+1k}-u_{ijk}}{y_{j+1}-y_j} \tag{14}$$

$$w\frac{\partial u}{\partial z} = \frac{\bar{w}_u+|\bar{w}_u|}{2}\ \frac{u_{ijk}-u_{ijk+1}}{z_k-z_{k+1}} + \frac{\bar{w}_u-|\bar{w}_u|}{2}\ \frac{u_{ijk-1}-u_{ijk}}{z_{k-1}-z_k} \tag{15}$$

$$\bar{v}_u = \frac{1}{4}(v_{ijk}+v_{i+1jk}+v_{ij-1k}+v_{i+1j-1k})$$

$$\bar{w}_u = \frac{1}{4}(w_{ijk}+w_{i+1jk}+w_{ijk+1}+w_{i+1jk+1})$$

The equation (13) can also be written using the second-order scheme as :

$$u\frac{\partial u}{\partial x} = \frac{u_{ijk}+|u_{ijk}|}{2}\ \frac{\beta(\beta+2)u_{ijk}-(1+\beta)^2u_{i+1,jk}+u_{i-2jk}}{\beta(1+\beta)(x_i-x_{i-1})} + \frac{u_{ijk}-|u_{ijk}|}{2}\ \frac{-\alpha(\alpha+2)u_{ijk}+(1+\alpha)^2u_{i+1,jk}-u_{i+2jk}}{\alpha(1+\alpha)(x_{i+1}-x_i)}$$

$$\alpha=\frac{x_{i+2}-x_{i+1}}{x_{i+1}-x_i},\quad \beta=\frac{x_{i-1}-x_{i-2}}{x_i-x_{i-1}} \tag{16}$$

and other terms are derived by the similar manners (Boo 1996).

The stability of the explicit scheme is generally governed by the time increment and mesh sizes. The CFL condition for the external waves of the free surface gravity waves is written as :

$$\Delta t=\frac{\min(\Delta x, \Delta y)}{\sqrt{2gd}} \tag{17}$$

where d is the water depth. In case of the internal waves in the vertical direction, the stability is limited by the largest value of the vertical eddy viscosity and the smallest vertical mesh size due to vertical friction. The maximum time increment is computed using (Kowalik & Murty 1993) :

$$\Delta t=\frac{(\Delta z_m)^2}{2E_v} \tag{18}$$

where Δz_m is the minimum size of the vertical mesh.

VERIFICATION AND APPLICATION OF MODEL

The proposed numerical model needs to be verified in order to check accuracy in solving the three-dimensional shallow water equation. The validation process was carried out in the sequence from relatively simple problems to complex physical problems.

Wind-Induced Flow Simulation

The numerical verification of a steady wind-induced flow in a uniform rectangular channel was carried out. The momentum equation (2) can be simplified by considering wave elevation and vertical diffusion. The steady one dimensional equation is

$$g\frac{\partial \zeta}{\partial x}-E_v\frac{\partial^2 u}{\partial z^2}=0 \tag{19}$$

The equation (19) can be solved analytically if the boundary conditions are properly chosen. The conditions are that velocity at the bottom and vertically integrated velocity are zero and the stress at the surface is $\tau_{xa}=\rho E_v\frac{\partial u}{\partial z}$. The analytical solutions of the equation (19) can be written as (Cooper&Pierce 1977)

$$\frac{\partial \zeta}{\partial x}=\frac{3}{2}\frac{\tau_{xa}}{\rho g d} \tag{20}$$

$$u(z)=\frac{\tau_{xa}}{4\rho E_v}(d-z)(1-3\frac{z}{d}) \tag{21}$$

For this test, a rectangular channel with a length of $20km$, a width of $5km$ and a constant depth of $10km$ was used. The mesh size and other data for the present simulation are given as :

$$\Delta x=\Delta y=2km,\ \Delta z=0.5m,\ \tau_{xa}=0.1N/m^2$$
$$E_v=0.005m^2/s,\ \rho=1000kg/m^3$$

The numerical solutions of the elevation and vertical velocity were compared with the exact values in Fig. 2 and 3, respectively. Those values were taken on the section of the channel center. The numerical results are shown to be in good agreement with the theoretical solutions. In Fig. 4, velocity vectors along the channel were shown. It is clearly seen that the reverse flows against the surface flow and the circulating flow at the end of the channel exist.

One-Dimensional Tidal Flow Simulation

One-dimensional tidal response in the channel is also simulated in order to further investigate the numerical scheme. A square channel is chosen for this case since the analytical solution is available. In case of a constant depth, the analytic solutions based on the linearized shallow water equation was obtained in Lynch and Gray (1978). The detail formulations will be omitted here.

The east side of the channel is open while other sides are closed. The tidal input was appled to the east side and the condition is given as

$$\zeta=\zeta_o \sin\frac{2\pi}{T_o}$$

where amplitude was 2 meters and the period T_o was

12 hours. In this test case, the channel length and breadth are $15km$ and the water depth is $10km$. The mesh density and data used for the simulation are

$$\Delta x=\Delta y=1km,\ \Delta z=0.5m,\ \tau_{xb}=0.001N/m^2,$$
$$E_v=0.005m^2/s,\ E_h=100\,m^2/s$$

The simulation was started from the rest. Figure 5 shows the comparison of the computed elevation with the theoretical value after reaching a steady state. The present result agrees well with the theories. Computed velocity is also compared with the theoretical one in Fig. 6. The magnitude of the numerical value is shown to be bigger than one of the theoretical value. It may due to the present modeling in which three-dimensional effects and nonlinear terms were included.

Three-Dimensional Tidal Flow Simulation

As the third test case, three-dimensional tidal flow was simulated in the rectangular channel. The east side was open and a tidal forcing was imposed on the open boundary. The length and width of the channel are $20km$, and its depth is $10m$. Mesh sizes in the three directions and parameters are :

$$\Delta x=\Delta y=1km,\ \Delta z=1m,\ E_h=500N/m^2,$$
$$E_v=0.005m^2/s,\ \rho=1025kg/m^3$$

The tidal motion at the open boundary was given in the following form :

$$\zeta=(1+0.2\cos\frac{\pi y}{L_B})\ \sin\frac{2\pi}{T_o}$$

where L_B is the width of the open boundary, T_o the tidal period of 12 hours. Figure 7 and 8 are the velocity vector and contour of elevation at the time instant of 52 hours. Tidal induced circulations due to the elevation gradient were clearly presented around the mouth of the channel.

Tide-Induced Circulation in Closed Harbor

In the foregoing numerical test, it was shown that the present model is capable of treating a complex tidal response in a realistic seas. For the simplicity, the physical geometry chosen was again assumed to be square in shape with a constant depth, vertical side wall and asymmetric entrance. This assumption might be appropriate since a coastal area surrounded by islands or breakwaters would be the probable location for installing the barge-mounted plant so that the structure can be protected from the severe or stormy seas. However, in case the structure is placed in the closed region, a water quality problem may be caused due to waste water from the structure as well as variation of the tidal currents. It is, therefore, necessary to investigate the tidal flush in a closed area to provide environmental information before installing the structure.

The numerical test was carried out for a prototype harbor model with length and width of 400 meters and depth of 10 meters in which the harbor dimensions was chosen by taking into account the actual dimension of the floating structure installed. The principal dimension of a barge-mounted floating structure is presently decided to be about length × breadth=60 m× 40 m and its displacement will be about 12,000 tones (Jung 1996). East side of the harbor was assumed to be partially open and thus open boundary condition was applied to the boundary. A submerged flow control structure with cross-shape was also placed at the center of the harbor to generate vertical circulation. The schematics for the harbors and submerged structure at the bottom are drawn in Fig. 9. Dimensions of the bottom structure were chosen as length × breadth × depth= $240m\times20m\times2m$. A flow guide was placed at the entrance to cause asymmetric circulation. Length and width of the flow guide with vertical side was 80 meters and 20 meters, respectively and its depth was the same as the water depth. Parameters for this test problem are

$$\Delta x=\Delta y=20m,\ \Delta z=1m,\ \gamma=0.0026,$$
$$E_v=0.005m^2/s,\ E_h=100\,m^2/s,\ \zeta_o=1m$$

where ζ_o is tidal amplitude at the open boundary, and length of the harbor mouth was assumed to be 80 meters which was one fifth the harbor width.

Horizontal velocity vectors for two different depths of 8 and 9 meters at 6 hours were shown in Fig. 10 and 11. The vertical level of 8 meters are the same level as the top of the bottom structure. A possible adverse effects at 8 meters due to the bottom structure in the horizontal circulations are not presented while the circulations at the depth of 9 meters are reduced a lot by the structure. Additional effect on the vertical flow was examined in the followings. Figure 12 and 13 show the vertical velocity contours for the flood and ebb tide at the depth of 8 meters. Vertical circulations around the harbor entrance are shown to be much stronger in comparison with the values around the opposite area. Vertical circulation is improved due to the bottom structure. This could imply that waste materials or slits around the opposite area to the entrance might be accumulated. It can be stated that installation of a bottom structure is preferable for improvement of the vertical circulation. Further studies for various harbors or structure shape are required to obtain design information.

CONCLUSIONS

When a large complex plant is installed in a coastal area, it may cause water quality problem due to waste materials or current change. Therefore, it is necessary to develope a numerical prediction scheme and test a tidal circulation in advance of installation of the structure. In this paper, a prediction model for the tidal circulation in a harbor where a submerged structure was installed, was developed.

The three-dimensional shallow water equation was solved using finite difference method. Time integration of the equations was done using three-time level explicit scheme. A staggered grid system was employed in the horizontal and vertical directions. A multi-level scheme was used so that density variations were considered. Variable mesh in the horizontal plane was also utilized to effectively discretize the complex geometries.

The numerical model was firstly tested and verified by solving a wind-induced flow in a narrow channel as well as one-dimensional tidal response in a rectangular channel. The numerical results were in good agreement with the theories. Additionally, the numerical model was applied to analyze three-dimensional tidal response in a simplified bay and thus a basis in implementing the present scheme to circulation analysis in a complex coast area was established. Lastly, circulation in a closed harbor was investigated, where a submerged flow control structure was placed in order to cause vertical flow. The bottom structure was found to be preferable in improving the vertical circulation but horizontal circulations decreased. Further studies on effects of harbor shape or structure size are required to obtain information of water quality changes.

ACKNOWLEDGMENTS

This work has been supported by Korea Institute of Machinery & Materials and is a part of the first year research. This support is acknowledged.

REFERENCES

Asaeda, T., Son, P.H. & Armfield, S. (1994) "Vortex Convection Produced by V-Shaped Dihedral Obstruction", *J. Hydraulic Engineering*, Vol. 120, No. 11.

Boo, S.Y. (1996) "Analysis of Tidal Circulation Due to a Submerged Structure", 1st Year Report (in korean), KIMM, Daejon, Korea.

Chen, C.L. & Lee, K.K. (1991) "Great-Lakes River-Estuary Hydrodynamic Finite Element Method", *J. Hydraulic Engineering*, Vol. 117, No. 11.

Cooper, C.K. & Pierce, B.R. (1977) "A Three-dimensional Numerical Model to Calculate Currents in Coastal Waters Utilizing a Depth Varying Vertical Eddy Viscosity", Report No. 226, Ralph M. Parson Lab. for Water Resources and Hydrodynamics, MIT.

Falconer, R.A. (1990) "Numerical Modeling of Tidal Circulation in Harbors", *J. Waterway, Port, Coastal and Ocean Division*, Vol. 106, No. WW1.

Fujino, M & Tabeta, S. (1993) "Numerical Simulation of Tidal Currents by Means of Multi-Level Model", *JSNAJ*, No. 170.

Inoue, Y., Tabeta, S., Arai, M. & Nakazawa, K. (1994) "Feasibility Design of a Floating Airport and Investigation into its Effects on the Environment", - Effect of a Floating Airport on Ocean Current -, *JSNAJ*, Vol. 176.

Jin, X. & Kranenburg, C. (1992) "Quasi-3D Numerical Modeling of Shallow-Water Circulation", *J. Hydraulic Engineering*, Vol. 119, No. 4.

Jung, T.Y. (1996) "Development of Large Complex Plant for Utilization of Ocean Space", the First Year Report (in Korean), KIMM, Daejon, Korea.

Kodama, T., Wang, S. & Kawahara, M. (1996) "Model Verification on 3D Tidal Current Analysis in Tokyo Bay", *Intl. J. for Numerical Method in Fluid*, Vol. 22.

Kowalik, Z & Murty, T.S. (1993) "Numerical Modeling of Ocean Dynamics", Advanced Series on Ocean Engineering, Volume 5, World Scientific.

Lynch, D.R. & Gray, W.G. (1978) "Analytic Solutions for Computer Flow Model Testing", *J. Hydraulics Division, ASCE*.

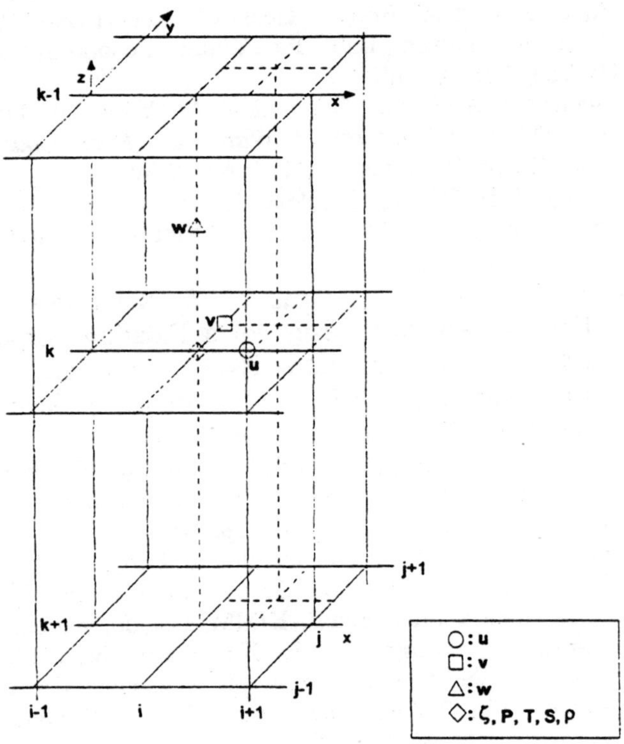

Fig. 1 Grid distributions in 3D spaces

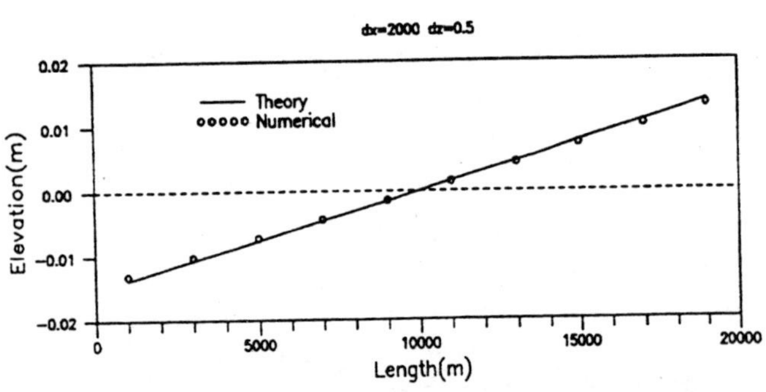

Fig. 2 Comparisons of wind-driven surface elevations with theory in the 1D channel

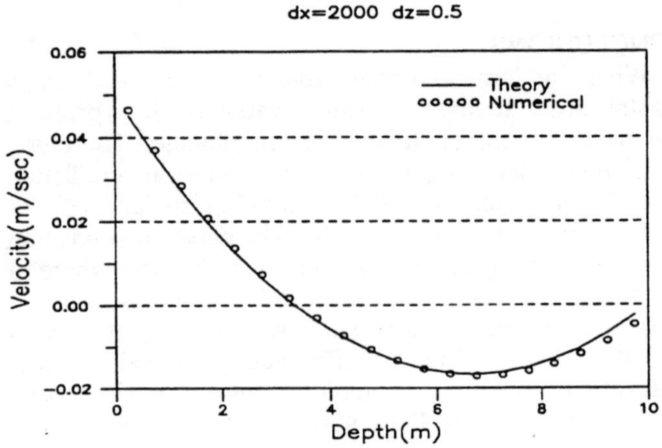

Fig. 3 Comparisons of wind-driven velocities with theory in the 1D channel

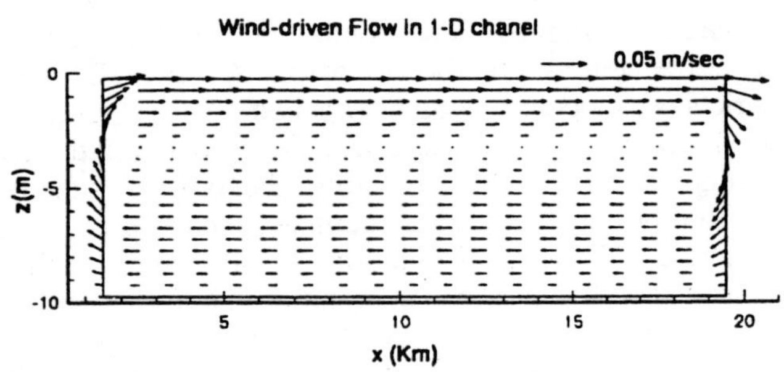

Fig. 4 Wind-Driven Velocity Vectors in 1-D Channel

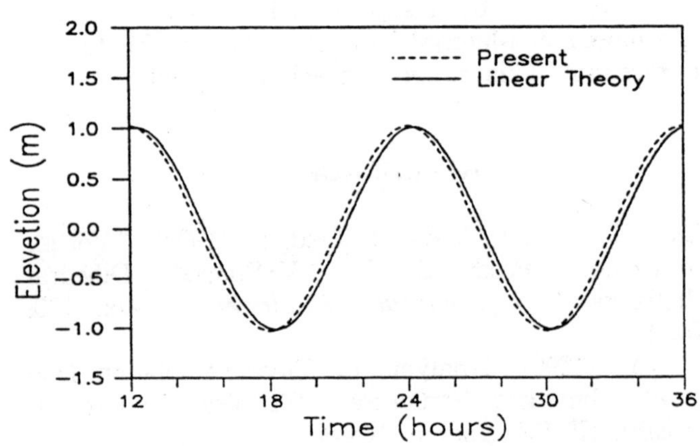

Fig. 5 Time histories of tidal elevation measured at the center of 1D channel

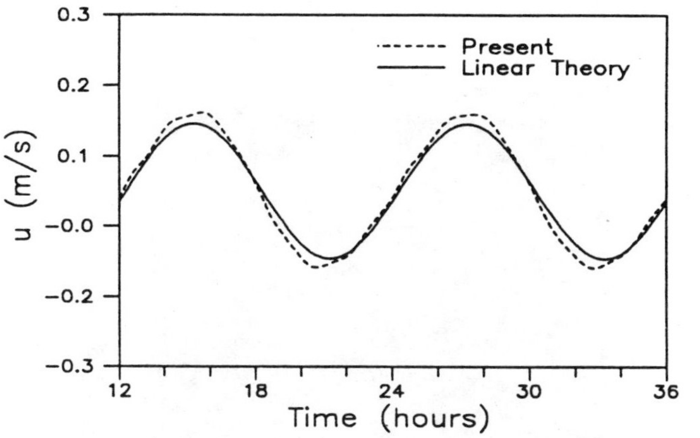

Fig. 6 Time histories of horizontal velocity of tide measured at the free surface in the 1D channel

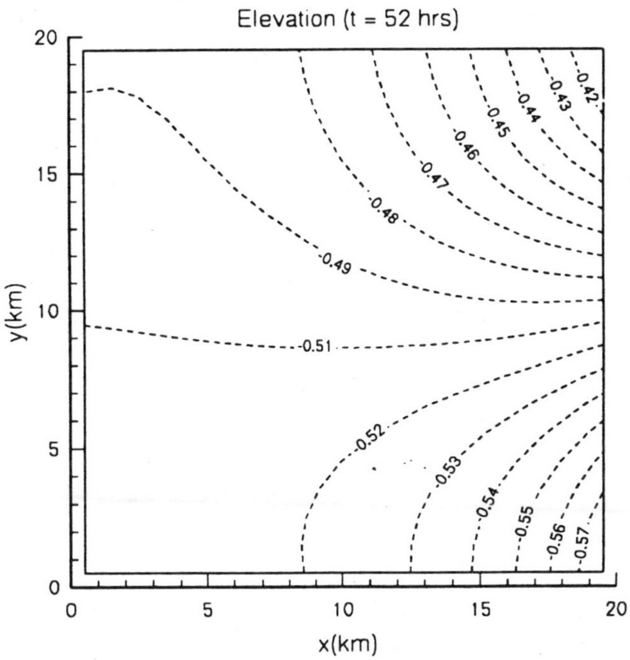

Fig. 8 Elevation contour for 3D tidal motion at t=52 hrs

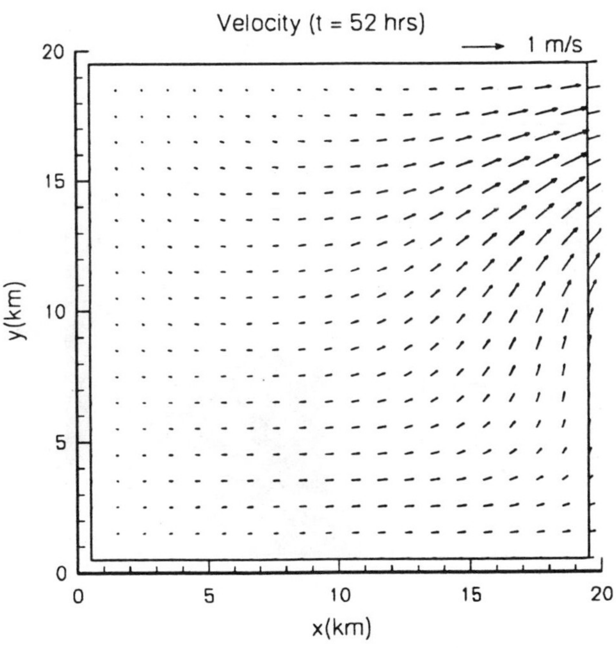

Fig. 7 Velocity vectors for 3D tidal motion at t=52 hrs

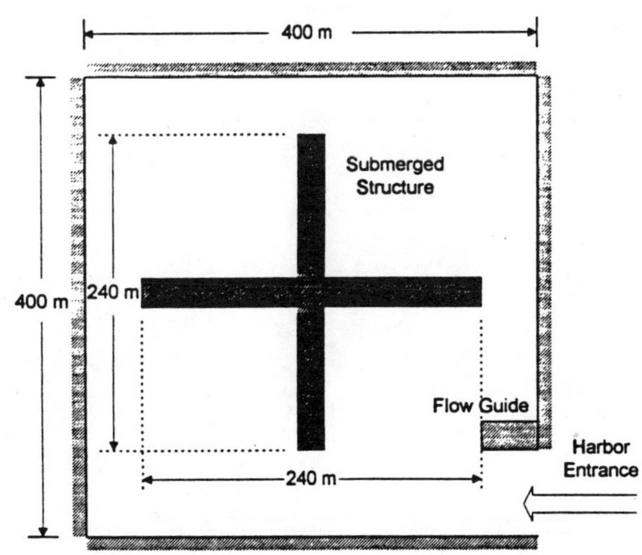

Fig. 9 Schematics of a closed harbor and submerged flow-control structure

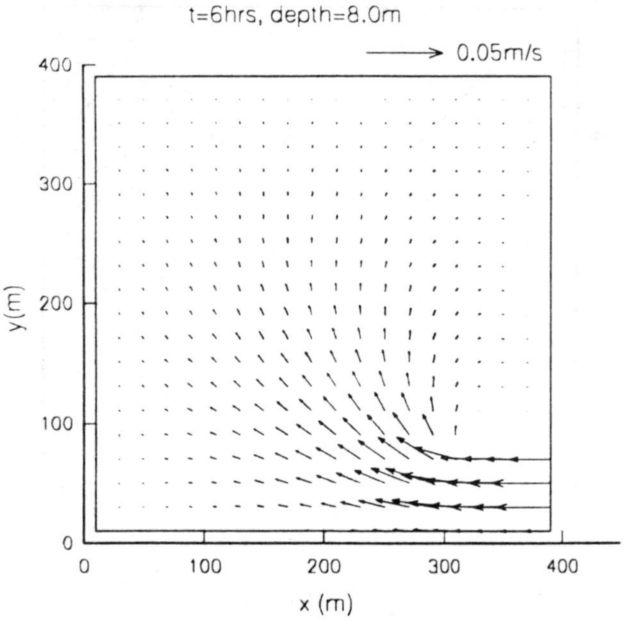

Fig. 10 Velocity fields at z=-8m in a closed harbor

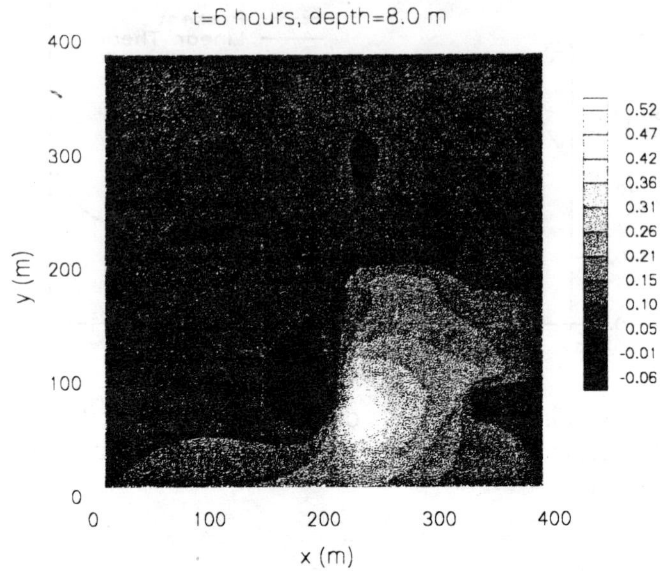

Fig. 12 Contour of vertical velocity fields at z=-8 m in a closed harbor at the flood tide

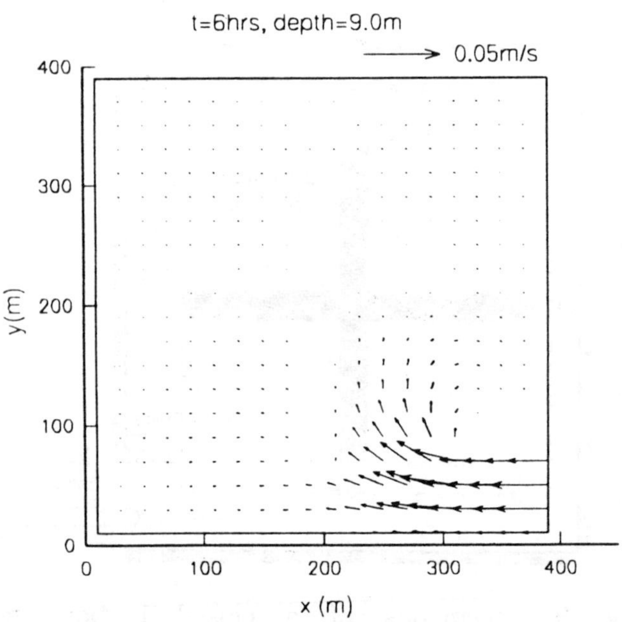

Fig. 11 Velocity fields at z=-9m in a closed harbor

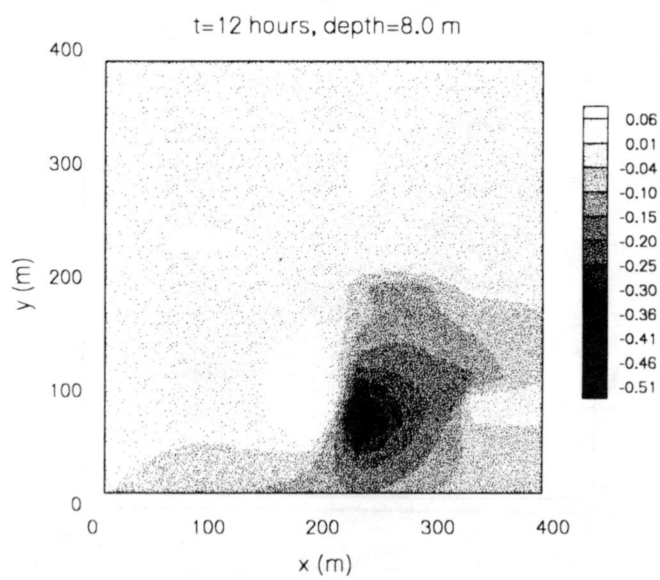

Fig. 13 Contour of vertical velocity fields at z=-8 m in a closed harbor at the ebb tide

Proceedings of the Seventh (1997) International Offshore and Polar Engineering Conference
Honolulu, USA, May 25-30, 1997

ISBN 1-880653-28-1 (Set); ISBN 1-880653-31-1 (Vol. III)

A Numerical Simulation of Vertical Convection Flow Induced by V-Shaped Plate

Chan-Kyu Yang, Keyyong Hong and Hark-Sun Choi
Korea Research Institute of Ships and Ocean Engineering
Taejon, Korea

ABSTRACT

In this paper, the flow around v-shaped plates placed in a uniform horizontal current is simulated by using a finite volume method and compared with the flow field visualized by a dye injection method. The flow properties are investigated particularly in viewpoints of the rising height of vortex head and the strength of the vertical convection, which varies with the configuration of the plate.

The flow patterns are examined by varying the dihedral angle(θ) and the inclination angle(ϕ) and the optimum configuration which induces the strongest vertical convection flow is determined.

Key Words : v-shaped plates, finite volume method, dye injection method, dihedral angle, inclination angle

1. INTRODUCTION

Recently, various structures such as floating plant and artificial harbour have been constructed in coastal region with a wide range of purposes. Such a facility disturbs the flow patterns around and may cause stagnant flow locally, which results in eutrophication and stratification of sea. It is therefore necessary to utilize an appropriate device to control the flow in favorable manner and to make the quality of seawater better.

Sea flow control device usually needs successive energy supply, which makes it necessary to design the system to use natural energy such as current and wave. Some kinds of a simple structure installed in sea may play an affirmative role by recirculating polluted seawater and refreshing it.

In this paper, a v-shaped plate is focused as a seawater recirculating system. Hunt et. al.(1980) revealed some flow phenomena of stratified fluid over three dimensional hill by a visualization technique. Takashi et. al.(1990) experimentally investigated the upwelling flow artificially induced by v-shaped plate, which mixes the stratified stagnant flow. And it can be applied to mitigate eutrophication and stratification of sea in the vicinity of the offshore structure.

A finite volume approach is adopted for numerical simulation, and the results are compared with the experimental results in viewpoint of rising height of vortex head with respect to the dihedral angle(θ) and the inclination angle(ϕ).

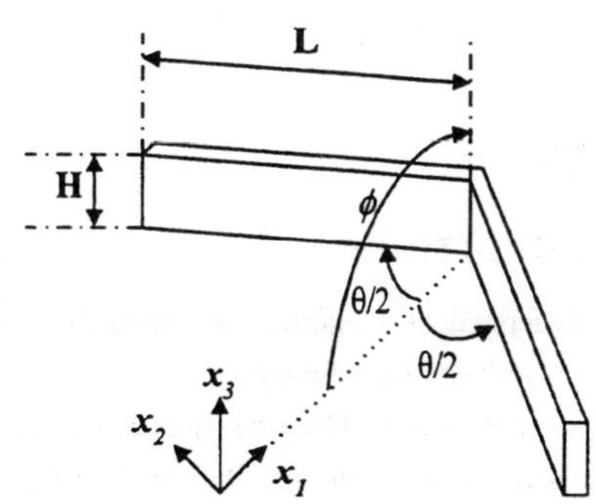

Fig.1 A Configuration of V-Shape Plate

2. MATHEMATICAL FORMULATION

2.1 GOVERNING EQUATIONS

Assuming steady state flow of viscous and incompressible fluid, the Navier-Stokes equation and the continuity equation in cartesian coordinate system(x_1, x_2, x_3) can be written in the following form.

$$\frac{\partial(\rho u_i)}{\partial x_i} = 0, \tag{1}$$

$$\frac{\partial(\rho u_i u_j)}{\partial x_j} = -\frac{\partial p}{\partial x_i} + \frac{\partial \tau_{ij}}{\partial x_j}, \tag{2}$$

where

$$\tau_{ij} = \mu\left(\frac{\partial u_i}{\partial x_j} + \frac{\partial u_j}{\partial x_i}\right),$$

μ = dynamic viscosity of fluid,

u_i = velocity in Cartesian coordinate system,

p = pressure,

ρ = density of the fluid.

Transformation of the governing equations in generalized coordinate system(ξ_1, ξ_2, ξ_3) gives

$$\frac{\partial}{\partial \xi_j}(\rho U_j) = 0 \tag{3}$$

$$\begin{aligned}
&\frac{\partial}{\partial \xi_1}\left(\rho U_1 u_i - \frac{g^{11}}{J}\mu\frac{\partial u_i}{\partial \xi_1}\right)\\
&+\frac{\partial}{\partial \xi_2}\left(\rho U_2 u_i - \frac{g^{22}}{J}\mu\frac{\partial u_i}{\partial \xi_2}\right)\\
&+\frac{\partial}{\partial \xi_3}\left(\rho U_3 u_i - \frac{g^{33}}{J}\mu\frac{\partial u_i}{\partial \xi_3}\right)\\
&= -\frac{1}{J}\frac{\partial \xi_1}{\partial x_i}\frac{\partial p}{\partial \xi_1} - \frac{1}{J}\frac{\partial \xi_2}{\partial x_i}\frac{\partial p}{\partial \xi_2} - \frac{1}{J}\frac{\partial \xi_3}{\partial x_i}\frac{\partial p}{\partial \xi_3}\\
&+\frac{\partial}{\partial \xi_1}\left(\mu\frac{g^{12}}{J}\frac{\partial u_i}{\partial \xi_2}\right) + \frac{\partial}{\partial \xi_1}\left(\mu\frac{g^{13}}{J}\frac{\partial u_i}{\partial \xi_3}\right)\\
&+\frac{\partial}{\partial \xi_2}\left(\mu\frac{g^{21}}{J}\frac{\partial u_i}{\partial \xi_1}\right) + \frac{\partial}{\partial \xi_2}\left(\mu\frac{g^{23}}{J}\frac{\partial u_i}{\partial \xi_3}\right)\\
&+\frac{\partial}{\partial \xi_3}\left(\mu\frac{g^{31}}{J}\frac{\partial u_i}{\partial \xi_1}\right) + \frac{\partial}{\partial \xi_3}\left(\mu\frac{g^{32}}{J}\frac{\partial u_i}{\partial \xi_2}\right),
\end{aligned} \tag{4}$$

where

$$U_i = \frac{u_j}{J}\frac{\partial \xi_i}{x_j}, \quad g^{ij} = \frac{\partial \xi_i}{\partial x_k}\frac{\partial \xi_j}{\partial x_k},$$

$$J = \frac{\partial(\xi_1, \xi_2, \xi_3)}{\partial(x_1, x_2, x_3)}.$$

2.2 BOUNDARY CONDITIONS

All the values of dependent variables are normally given at inlet boundaries and prescribed accordingly.

The profiles of variables may also be prescribed at outlet. However, so little is known about outlet that it is advisable in most of the cases (a) to place the outlet plane far downstream with no reverse flow so that the results in the region of interest is practically insensitive to the treatment at outlet and (b) to evaluate boundary values by extrapolation from upstream neighbours.

The condition of symmetry plane reduce to (a) no convective flux and (b) no diffusion flux normal to the boundary.

The velocity and the convective flux across the boundary are set to zero at the wall boundary.

3. NUMERICAL PROCEDURE

3.1 DISCRETIZATION BY FINITE VOLUME METHOD

The discretization is performed in the physical space following a finite volume approach, in which the computational domain is divided into contiguous quadrilateral control volume(Fig.2). A non-staggered grid is used where the scalar quantities and velocity components are all located at the geometric center of a control volume. The grid is arranged to make computer programming efficient and to minimize geometrical information because all the variables share the same control volume. However, such an arrangement is so defective in decoupling velocity and pressure fields that it may give rise to so-called "checker-board pattern". This drawback is remedied by the momentum interpolation technique suggested by Rhie and Chow(1983).

The continuity equation(eq.3) can be integrated over each control volume in the computational domain. For a typical point P (Fig.2), the integrated equation with $\Delta\xi = \Delta\eta = \Delta\zeta = 1$ can be expressed as

$$\begin{aligned}
&\left\{\frac{\rho}{J}\frac{\partial \xi_1}{\partial x_j}u_j\right\}_e - \left\{\frac{\rho}{J}\frac{\partial \xi_1}{\partial x_j}u_j\right\}_w\\
&+\left\{\frac{\rho}{J}\frac{\partial \xi_2}{\partial x_j}u_j\right\}_n - \left\{\frac{\rho}{J}\frac{\partial \xi_2}{\partial x_j}u_j\right\}_s\\
&+\left\{\frac{\rho}{J}\frac{\partial \xi_3}{\partial x_j}u_j\right\}_t - \left\{\frac{\rho}{J}\frac{\partial \xi_3}{\partial x_j}u_j\right\}_b = 0
\end{aligned} \tag{5}$$

The momentum equations(eq.4) are separated into three parts which are convection-diffusion, pressure and cross derivatives. The first term of convection-diffusion part is integrated as an example in the following equation, in which convection term consists of cell face velocities. Various schemes are used to estimate the cell face values in the convection term. In this paper, the cell face value is approximated by the Hybrid scheme which was developed by Spalding (1972).

$$\begin{aligned}
&\int_{\Omega(\xi_i)}\frac{\partial}{\partial \xi}\left(\rho U_1 u_i - \mu\frac{g^{11}}{J}\frac{\partial u_i}{\partial \xi_1}\right)d\Omega(\xi_i)\\
&= \left(\int \rho U_1\, d\xi_2 d\xi_3\right)_e (u_i)_e - \left(\int \rho U_1\, d\xi_2 d\xi_3\right)_w (u_i)_w\\
&- \left(\int \mu\frac{g^{11}}{J}\, d\xi_2 d\xi_3/\delta\xi_1\right)_e \{(u_i)_E - (u_i)_P\}\\
&+ \left(\int \mu\frac{g^{11}}{J}\, d\xi_2 d\xi_3/\delta\xi_1\right)_w \{(u_i)_P - (u_i)_W\}\\
&= (A_E + A_W)(u_i)_P - A_E(u_i)_E - A_W(u_i)_W
\end{aligned} \tag{6}$$

where

$\Omega(\xi_i)$ = volume of a control volume,

$A_E = D_e \max(0, 1-0.5|R|) + \max(-F_e, 0)$,

$A_W = D_w \max(0, 1-0.5|R|) + \max(F_w, 0)$,

$R = F/D$: cell Reynolds number,

$F = \left(\int \rho U_1\, d\xi_2 d\xi_3\right)$, $D = \left(\int \mu\frac{g^{11}}{J}\, d\xi_2 d\xi_3/\delta\xi_1\right)$.

Pressure terms are also integrated to give discretization form as

$$\int_{\Omega(\xi_i)} \frac{1}{J}\left\{-\frac{\partial\xi_1}{\partial x_i}\frac{\partial p}{\partial\xi_1}-\frac{\partial\xi_2}{\partial x_i}\frac{\partial p}{\partial\xi_2}-\frac{\partial\xi_3}{\partial x_i}\frac{\partial p}{\partial\xi_3}\right\}d\Omega(\xi_j)$$

$$\begin{aligned}&=-\left(\int\frac{1}{J}\frac{\partial\xi_1}{\partial x_i}d\xi_2 d\xi_3\right)_P(p_e-p_w)\\&\quad-\left(\int\frac{1}{J}\frac{\partial\xi_2}{\partial x_i}d\xi_3 d\xi_1\right)_P(p_n-p_s)\\&\quad-\left(\int\frac{1}{J}\frac{\partial\xi_3}{\partial x_i}d\xi_1 d\xi_2\right)_P(p_t-p_b)\end{aligned} \tag{7}$$

Cross derivative terms are discretized in the same manner and the remaining first derivative is expressed by central difference to be included in source term.

Finally, discretization form of momentum equation can be summarized and rearranged to

$$\begin{aligned}A_P(u_i)_P &= A_E(u_i)_E + A_W(u_i)_W + A_N(u_i)_N\\&\quad + A_S(u_i)_S + A_T(u_i)_T + A_B(u_i)_B + S_{u_i}\end{aligned} \tag{8}$$

where

$$S_{u_i} = S_U^{V_i} + C_1^i(p_w-p_e) + C_2^i(p_s-p_n) + C_3^i(p_b-p_t).$$

$$C_1^i = \left(\int\frac{1}{J}\frac{\partial\xi_1}{\partial x_i}d\xi_2 d\xi_3\right)_P$$

$$C_2^i = \left(\int\frac{1}{J}\frac{\partial\xi_2}{\partial x_i}d\xi_3 d\xi_1\right)_P$$

$$C_3^i = \left(\int\frac{1}{J}\frac{\partial\xi_3}{\partial x_i}d\xi_1 d\xi_2\right)_P$$

According to SIMPLEC algorithm, momentum equations are used to derive the velocity correction at cell center. Introducing under-relaxation factor(α_i), velocity corrections at center of a cell are given by the following equations.

$$\begin{aligned}(u_i)_P = &\left[D_1^i(p_w'-p_e') + D_2^i(p_s'-p_n') + D_3^i(p_b'-p_t')\right]\\&\alpha_v/\left(1-\alpha_v\textstyle\sum A_{NB}/A_P\right),\end{aligned} \tag{9}$$

where

$D_j^i = C_j^i/A_P$,

NB = geometrical center of neighbouring cell,

p' = correction pressure.

Based on the momentum interpolation, the corresponding velocity corrections at the cell-face is equal to

$$(u_i)_w = \alpha_v\overline{D_1^i}(p_W'-p_P')/\left(1-\alpha_v\overline{\textstyle\sum A_{NB}/A_P}\right) \tag{10}$$

The values with overbar indicate the averaging ones of west cell center and present cell center. By substituting eq.(10) into eq. (5), the final equation of pressure correction p' is obtained.

$$\begin{aligned}A_P p_P' &= A_E p_E' + A_W p_W' + A_N p_N'\\&\quad + A_S p_S' + A_T p_T' + A_B p_B'\end{aligned} \tag{11}$$

where

$$A_P = A_E + A_W + A_N + A_S + A_T + A_B$$

$$A_W = \left\{\alpha_v\frac{\rho}{J}\left(\frac{\partial\xi_1}{\partial x_1}\overline{D_1^1} + \frac{\partial\xi_1}{\partial x_2}\overline{D_1^2} + \frac{\partial\xi_1}{\partial\xi_3}\overline{D_1^3}\right)\right.\left./\left(1-\alpha_v\overline{\textstyle\sum A_{NB}/A_P}\right)\right\}_w$$

$$A_S = \left\{\alpha_v\frac{\rho}{J}\left(\frac{\partial\xi_2}{\partial x_1}\overline{D_2^1} + \frac{\partial\xi_2}{\partial x_2}\overline{D_2^2} + \frac{\partial\xi_2}{\partial\xi_3}\overline{D_2^3}\right)\right.\left./\left(1-\alpha_v\overline{\textstyle\sum A_{NB}/A_P}\right)\right\}_s$$

$$A_B = \left\{\alpha_v\frac{\rho}{J}\left(\frac{\partial\xi_3}{\partial x_1}\overline{D_3^1} + \frac{\partial\xi_3}{\partial x_2}\overline{D_3^2} + \frac{\partial\xi_3}{\partial\xi_3}\overline{D_3^3}\right)\right.\left./\left(1-\alpha_v\overline{\textstyle\sum A_{NB}/A_P}\right)\right\}_b$$

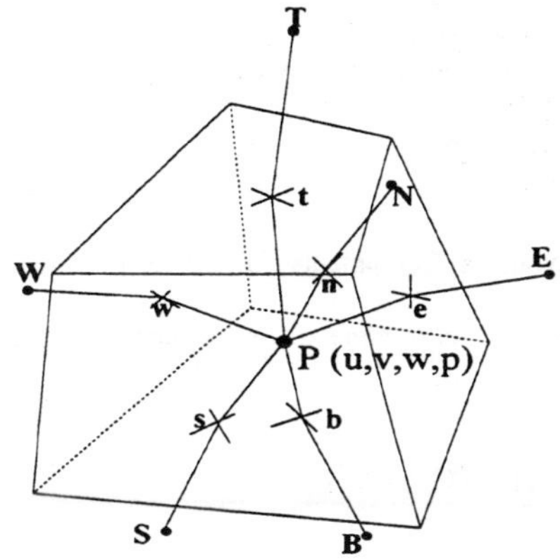

Fig.2 Configuration of a Cell

3.2 SOLUTION ALGORITHM

The strongly implicit procedure of Stone(1968) is used for solving the system of linear equations. An outlined steps of SIMPLEC algorithm(Jang et.al., 1986) are (a)At the first stage of calculation, all the dependent variables at each cell are assumed. (b)Momentum equation(eq.8) is solved with the assumed pressure to obtain velocity components. (c)Then pressure correction equation is solved to correct both velocities and pressure. (d)Returning back to the first step, the following steps are repeated until convergence is satisfied.

4. FLOW VISUALIZATION

For the comparison with the computed results, flow patterns around the V-shaped plate are investigated experimentally by a dye injection method which shows streak lines(JSME,1990 ;Wolfgang,1974). The experiment was performed in two dimensional flume whose length, width and depth are 25m, 60cm and 50cm, respectively. Two video cameras were set to record the unsteady motion of fluid and a camera to take a snapshot of streak lines(Fig.3).

The flow visualized phenomena are the generation and development of vortex from the side of structure and the shedding of it at the center-plane. So two colors of dye, CATIONIC RED RP and UNICION NAVY BLUE H-ER are simultaneously injected at upstream of side edge and vertex. The inner diameter of dye injection nozzle is 1mm. Series of horizontal plates clustered to bottom is located near inlet and used as a guide vane to make more uniform current. At

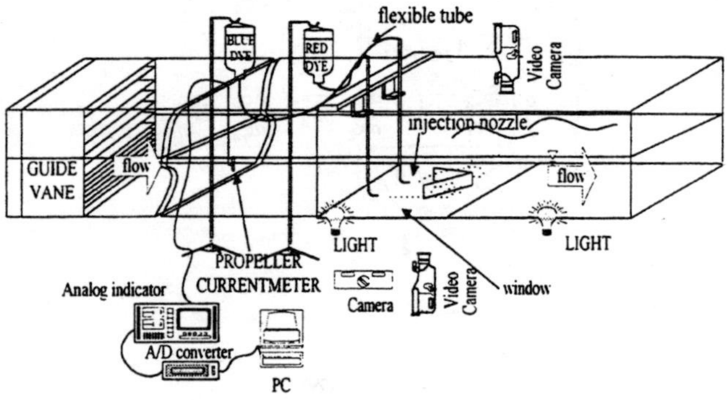

Fig.3 Layout of Equipments for the Experiment

upstream, a propeller currentmeter is placed to measure the current speed.

5. SIMULATION RESULT

5.1 EXAMPLES OF SIMULATION

In this study, the height(H) and the length(L) of structure are fixed. So the shape of structure is determined by dihedral angle (θ) and inclination angle (ϕ) as depicted in Fig.1. Computations are performed in two cases of inclination angle, 90° and 60°, at each case of which the dihedral angle is varied by 30°, 60°, 90°, 150° and 180°. Reynolds number defined

$$R_n = \frac{U_o H}{\nu}, \tag{12}$$

where

U_o = mean velocity uniformly imposed at inlet,

H = height of the plate (= 0.08 m),

ν = kinematic viscosity of the fluid($1.\times 10^{-6}$ m^2/sec).

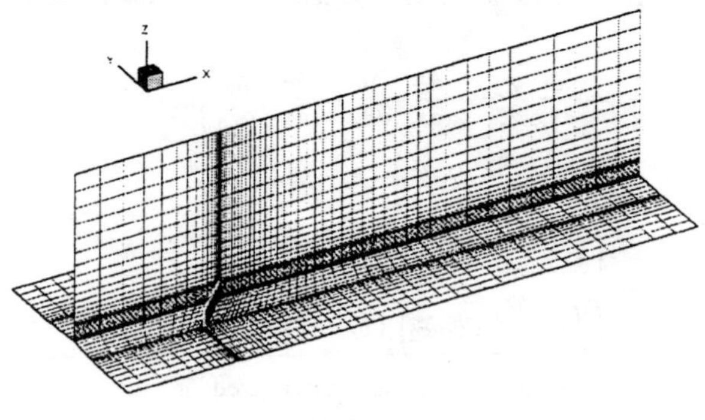

Fig.5 Mesh Configuration

is 8000. The variation of inclination angle and dihedral angle makes no change in dimensions of plate. However, the projected area is changed by the variation of angles. As an example, the projected area at ϕ=90° is larger than that at ϕ=60° at the same dihedral angle.

Computational domain and boundaries are depicted in Fig.4, where x, y and z denote x_1, x_2 and x_3 respectively. Inlet boundary is imposed on the plane at upstream and outlet boundary at downstream. The distance between inlet and structure is 3L and outlet and structure is 8L. The side planes are located at 2L far from the center-plane and treated as symmetry boundary.

Upper plane is simplified as symmetry boundary neglecting the surface wave and we observed no outstanding wave motion during the experiment. An example of grid system is depicted in Fig.5. Total grid number is 50×30×40 in the direction of streamwise, spanwise and depthwise, respectively.

Experiments are carried out at three Reynolds numbers near that of computational case (Table 2), among which the most similar condition is when Reynolds number is equal to 8000.

Table 1 Computational Cases

parameters	cases
Dihedral angle (θ)	30°,60°,90°,150°,180°
Inclination angle (ϕ)	60°,90°
Reynolds number (Rn)	8000

Table 2 Experimental Cases

parameters	cases
Dihedral angle (θ)	30°,60°,90°,150°,180°
Inclination angle (ϕ)	60°,90°
Reynolds number (Rn)	3760,5980 and 8000

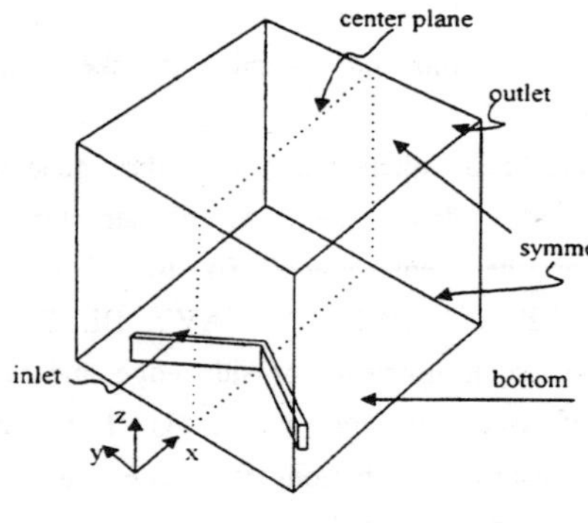

Fig.4 Computational Domain and Boundary

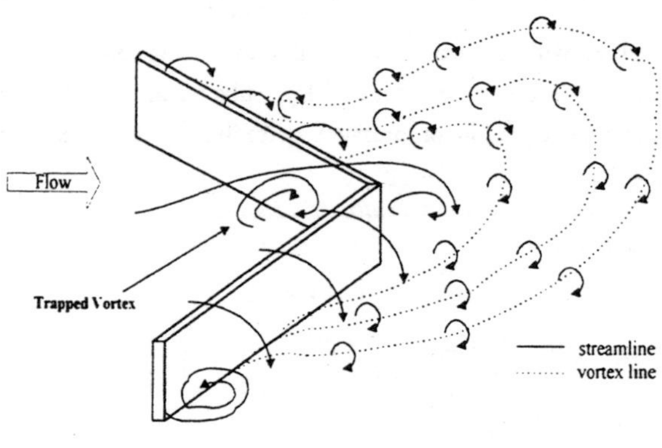

Fig.6 An Outline of Flow Pattern

The kinematic viscosity of fluid is $1.55\times10^{-6}(m^2/\sec)$ referred to temperature measuring 5°C.

5.2 RESULT AND DISCUSSION

With the help of both computational and experimental analyses of flow visualization, the flow pattern around the v-shaped plate is surveyed and outlined in Fig.6. Flow separates at the top of the plate and generates vortices. The vortices proceed along the plate and coalesce in the separation zone with ones from neighbouring regions and are shed downstream periodically(Fig.7). This horse-shoe shaped vortices stretch and rise due to self induced upward motion.

Water is mingled at the backside of the plate due to trapped vortices, which is seen from numerical simulation(Fig.11-Fig.13).

Some steady state streamline patterns which is numerically simulated are depicted in Fig.8 - Fig.10. A streamline at the center plane indicates the height of maximum vortex head and stream ribbon which is a series of streamlines illustrates the vortex generated from the side of structure. The starting point of each streamline has same height at upstream.

At θ=150°, the streamwise vorticity is insignificant with small effect on the upward motion. The decrement of dihedral angle makes stream ribbon twist, inducing strong vortices proceeding to center plane and helping the flow go upward. However, too small dihedral angle reduces upward motion due to the rapid decrement of the projected area on the cross section normal to flow.

Velocity distributions on the center plane confirm that the length of wake zone decreases and the upward motion increases at small dihedral angle. The exact difference of upward motion between at θ=90° and at θ=30° cannot be distinguished. So the spatially averaged induced velocity may be adopted as a reference value.

$$\bar{v}=\frac{1}{V}\int u_2\,dV \qquad \bar{w}=\frac{1}{V}\int u_3\,dV \tag{13}$$

where

$\bar{v}$ and $\bar{w}$: spatially averaged induced velocities,

u_2 and u_3 : spanwise and upward velocities,

V : total volume of computational domain,

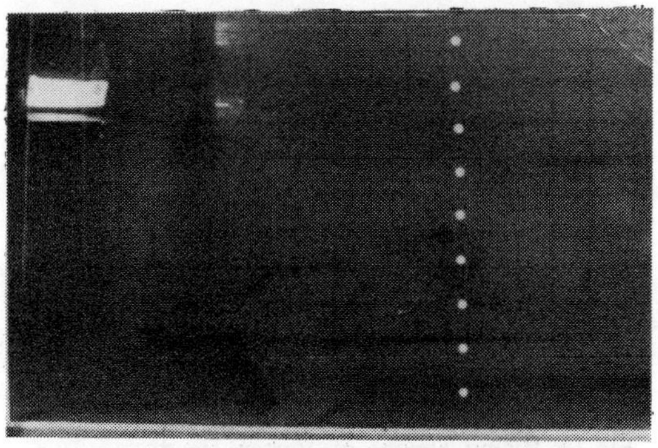

Fig.7 Periodical Vortex Shedding and Stretching

As a whole, upward velocity($\bar{w}$) appears to be higher than spanwise velocity($\bar{v}$) except for the case when dihedral angle is near 180°(Fig.14). The smaller the dihedral angle is, the larger the difference between spanwise and upward velocities in all inclination angles. Another tendency is that though the reduced inclination angle enhances the local upward motion of flow, the overall flow intensity is not improved since it causes the decrease of projected area.

The results show that v-shaped plate induces stronger vertical velocity with larger projected area and smaller dihedral angle. However, it is noticed that small dihedral angle brings about small projected area. Thus the small dihedral angle doesn't guarantee the improvement of vertically induced flow. The optimum dihedral angle for the strongest vertical convection flow as well as the strongest spanwise flow is found to be 90°, which is clearly revealed by the investigation of rising height of vortex head.

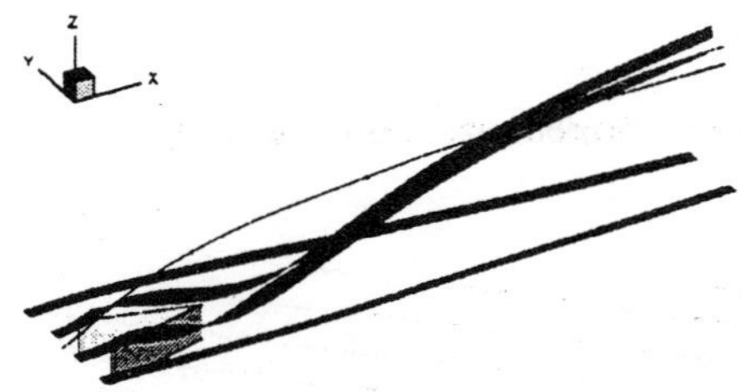

Fig.8 Stream line pattern (θ=30°, ϕ=90°)

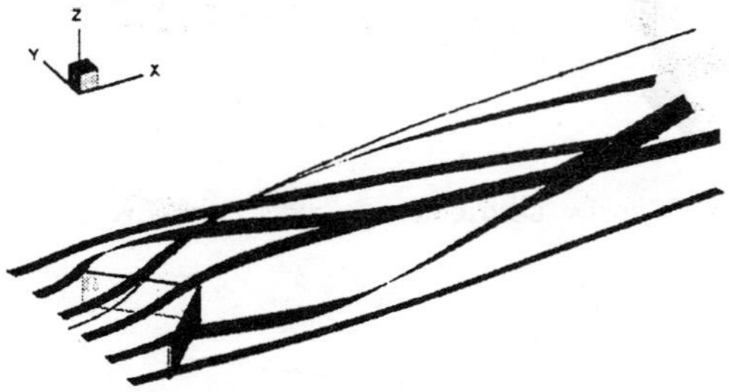

Fig.9 Stream line pattern (θ=90°, ϕ=90°)

Fig.10 Stream line pattern (θ=150°, ϕ=90°)

In Fig.15, the experimentally investigated results of rising height of vortex head are plotted with respect to dihedral angle. The rising height increases with the increment of Reynolds number and the maximum rising height is at θ=90° for all Reynolds numbers.

Computational results are compared with experimental results in Fig.16. Two results correspond qualitatively but show some discrepancies quantitatively.

6. CONCLUSION

The flow around v-shaped plates is simulated numerically and

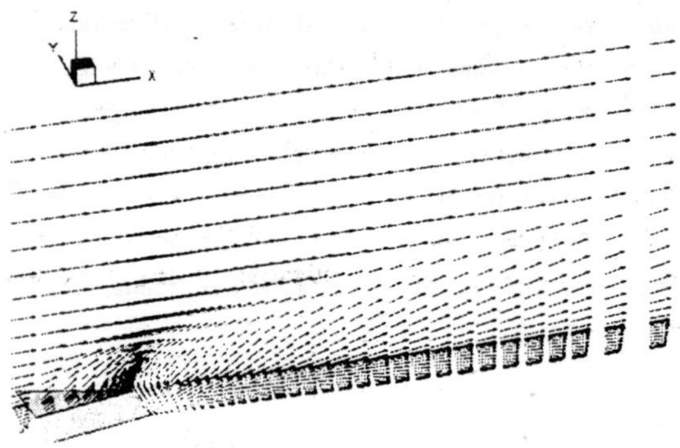

Fig.11 Velocity Distributions(θ=30°, ϕ=60°)

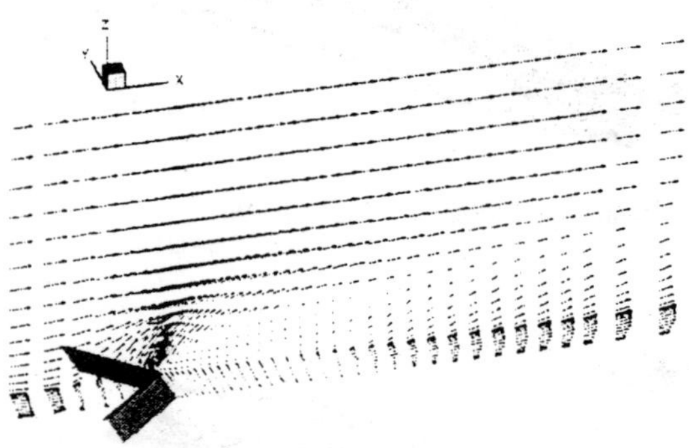

Fig.12 Velocity Distributions(θ=90°, ϕ=60°)

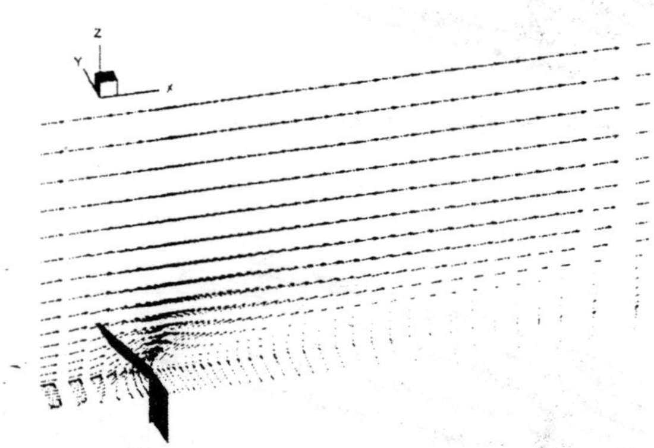

Fig.13 Velocity Distributions(θ=150°, ϕ=60°)

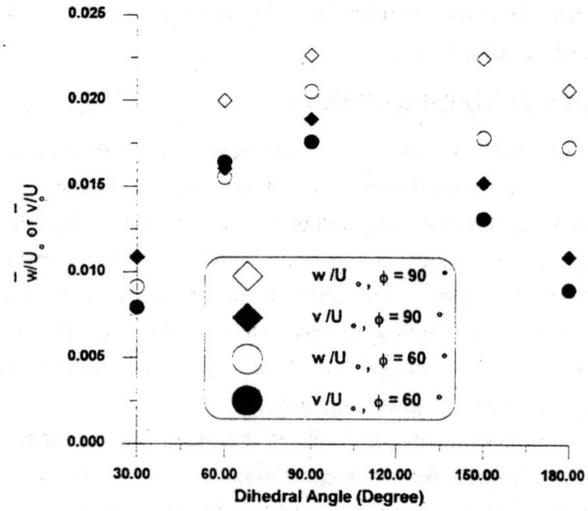

Fig.14 Velocities Induced by V-shaped plate

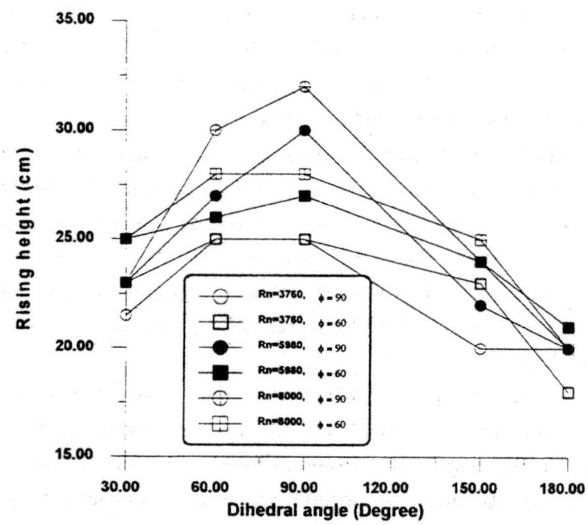

Fig.15 Rising Height of Vortex Head (Experiment)

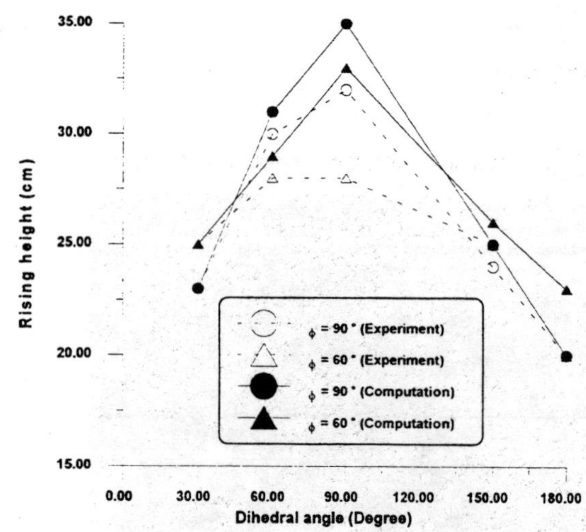

Fig.16 Rising Height of Vortex Head
(Experiment and Computation:Rn=8000)

visualized experimentally. Both results show a good agreement qualitatively but some discrepancies quantitatively. In review of results so far achieved, we could draw some remarkable conclusions as follows ;

(1) The vortices generated from the top portion of structure evolve along the plate and coalesce in the separation zone with ones from neighbouring regions to form a horse-shoe shaped vortices, which stretch and rise due to self induced upward motion.

(2) The smaller dihedral angle makes the greater difference between spanwise velocity and upward velocities.

(3) V-shaped plate produces stronger vertical velocity with larger projected area and smaller dihedral angle.

(4) Though the reduced inclination angle enhances the local upward motion of flow, the overall flow intensity is not improved since it causes the decrease of projected area.

(5) The optimum dihedral angle to induce the strongest vertical convection flow as well as spanwise flow is found to be 90°.

REFERENCES

J. C. R. Hunt and W. H. Snyder(1980)."Experiments on Stably and Neutrally Stratified Flow over a Model Three-Dimensional Hill" *J.Fluid Mech*, vol.96, part 4, pp.671-704.

D.S.Jang, R.Jetli and S.Acharya(1986)." Comparison of the PISO, SIMPLER and SIMPLEC Algorithms for the Treatment of Pressure-Velocity Coupling in Steady Flow Problems" *Numerical Heat Transfer*, Vol.10, pp.209-228.

Karki, K. C., and Patankar, S.V.(1988)."Calculation Procedure for Viscous Incompressible Flows in Complex Geometries" *Numerical Heat Transfer*, Vol.14, pp.295-307.

Patankar, S. V.(1980)."Numerical Heat Transfer and Fluid Flow" Hemisphere, Washington, D. C.

Rhie C.M. and Chow W.L.(1983),"A Numerical Study of the Turbulent Flow Fast an Isolated Airfoil with Trailing Edge Separation" *AIAA Journal*, Vol.21,pp.1525-1532.

Spalding D.B.(1972)."A Novel Finite Deference Formulation for Different Expressions Involving both First and Second Derivatives" *Int. J. Num.* Methods Eng., Vol.4, p.551.

Stone, H. L.(1968). Iterative Solution of Implicit Approximations of Multidimensional Partial Differential Equations" *SIAM J. Numer. Anal.*, Vol.5, pp.530-558.

Takashi Asaeda, Masanori Nakai, Nobuyuki Tamai and Kiyoshi Horikawa(1990)."Rising Current Production by a V-Shaped Plate" *Journal of JSCE*, vol 423, pp.83-90.

The Japan Society of Mechanical Engineers (1990). "Visualized Flow-Fluid Motion in Basic and Engineering Situations Revealed by Flow Visualization", Pergamon Press.

Wolfgang Merzkirch(1974)."Flow Visualization" Academic Press.

Proceedings of the Seventh (1997) International Offshore and Polar Engineering Conference
Honolulu, USA, May 25-30, 1997

ISBN 1-880653-28-1 (Set); ISBN 1-880653-31-1 (Vol. III)

Wave Interactions with an Infinite Array of 3-D Bodies

Takayuki Nakamura
Ehime University
Matsuyama, Japan

ABSTRACT

In order to clarify wave sheltering effects by an array of breakwaters, such as floating breakwaters and offshore breakwaters, a method of numerical analysis on the wave diffraction and radiation by an array of three-dimensional bodies is developed. The wave source distribution method is adopted in the analysis after having derived a Green's function in such a wave field. Using the numerical analysis, an effective arrangement of floating breakwaters for reducing incoming waves is extensively examined. Comparisons between the predicted results based on the three dimensional (3-D) analysis and a two-dimensional (2-D) analysis in a vertical plane are also given. For verifying the numerical results, laboratory experiments were carried out for an array of identical floating breakwaters with a rectangular shape. It was confirmed that the numerical analysis proposed here is very useful for predicting the complex wave field around the floating-breakwater array, including the directional characteristics of transmitted waves. It was also seen that the conventional 2-D analysis assuming an infinite length of the breakwater is insufficient for estimating the wave sheltring effect, especially under the condition in which a pitch length of the array is greater than a wave length.

Key Words: Arrayed floating breakwaters, Wave sheltering effect, Effective allocation, Green's function method, Dynamic response, Wave diffraction, Wave radiation

INTRODUCTION

Floating breakwaters are usually located as an array-structure. For instance, a series of floating pontoons in a linear array with some gap is very usual. In the existing practical designs, a necessary breadth of a floating breakwater in the wave propagation direction has been decided by considering a one-dimensional sheltering effect, say wave transmission coefficients. However, a method for determining a length of the breakwater in a transverse direction as well as a spacing between the breakwaters has not been so clear. In general, a transverse length of the breakwater has been determined only from the aspect of the structural mechanics. No or little attention has been paid on the aspect of wave sheltering effects. With respect to a spacing between the breakwaters, there is no rational way to estimate it in relation to the sheltering effect. Lack of understanding on plane wave interactions with an array of floating bodies may be a major reason for these uncertainties.

In the previous studies, it has become known that the wave pattern around a linear array of fixed offshore breakwaters is heavily dependent upon the ratio of a pitch length λ of the array to a wave length L of incident waves (e.g., Nakamura & Ono 1988; Dalrymple & Martin 1990; Nakamura 1994). For instance, in case of normally incident waves to the breakwater array, two-dimensional and short-crested wave patterns appear around the arrayed breakwaters when an incident wave length L is less than the pitch length λ of an array.

Furthermore, there are several studies in which wave interaction problems with multiple bodies have been dealt with. Simon (1982) studied the wave interaction problem with an array of axisymmetric bodies in heaving motion and used the so-called wide spacing approximation to derive the numerical solution. Other studies, but using different approaches and analytical models, are McIver & Evans (1984) and Kagemoto & Yue (1986). In these studies, however, major concerns are developing an efficient and accurate algorithm for the calculation of hydrodynamic forces on the multiple bodies. On the other hand, there are also several studies dealing with the similar problem in a mathematical sense to the wave interaction with an arrayed bodies. It is the sidewall effect of a narrow wave flume on a model body. Since the sidewall can be treated as a mirror wall hydrodynamically, the wave field about a model body placed in a narrow wave tank may be analogous to the one about an infinite array of model bodies. Such a sidewall effect has been studied by Yeung & Sphaier (1989), Thomas (1991), Linton & Evance (1992) and other researchers. However, their major concerns are again limited to the hydrodynamic forces on the body.

In this study, a numerical analysis on the wave diffraction and radiation by an infinite array of identical 3-D floating bodies is newly developed based on the wave source Green's function method. Here, the configuration of a floating body may be arbitrary. In the analysis, based on the similar concept to a directional wave spectrum, a method how to quantitatively assess the complicated wave field about a breakwater array is proposed. Using the numerical method proposed here, the effective arrangement of floating breakwaters for reducing incoming waves is examined extensively. Especially, changing a unit length of breakwaters as well as a spacing between the breakwaters, wave transmission characteristics through the breakwater array and dynamic responses of these breakwaters are clarified. Directional characteristics of the transmitted waves are also examined in detail to understand the complex wave field behind the breakwater array. In order to verify the numerical results, laboratory experiments were carried out for an array of rectangular floating bodies.

THEORETICAL ANALYSIS

Wave boundary-value problems

In the analysis, as seen in Fig.1, we assume that identical floating bodies with an arbitrary shape are periodically located in a line in the y-axis direction with a pitch length λ in a uniform water depth h. It is also assumed that the fluid is incompressible and inviscid. And the resultant wave motion is irrotational and can be described in a frame work of the small amplitude wave theory. In order to be able to deal with the much more general case, it is presumed that plane waves are incident to the arrayed bodies obliquely with an incident angle θ and the body configuration may be asymmetry about the longitudinal axis (x-axis). For the obliquely incident waves to the array, we can express the incident wave potential Φ_I as

$$\begin{aligned}\Phi_I(x,y,z,t) &= Real[-\frac{igH}{2\omega}\frac{\cosh\{k(h+z)\}}{\cosh(kh)}\exp\{i(k_x x+k_y y-\omega t)\}] \\ &= Real[-\frac{igH}{2\omega}\phi_I(x,y,z)\exp(-i\omega t)]\end{aligned} \tag{1}$$

where g = gravitational acceleration, H = incident wave height, k = progressive mode wave number, (k_x, k_y) = x and y components of a wave number vector, ω = angular frequency (= $2\pi/T$, T: wave period), h = water depth, and z = vertical coordinate measured from the still water level and positive for upward. And ϕ_I specifies the dimensionless complex amplitude of the incident wave potential.

Scattering problems

The scattered wave potential Φ_S by an array of floating bodies under the fixed condition at the equilibrium position can be assumed to have the following form

$$\Phi_S(x,y,z,t) = Real[-\frac{igH}{2\omega}\phi_S(x,y,z)\exp\{i(k_y y-\omega t)\}] \tag{2}$$

where ϕ_S = dimensionless complex amplitude of a scattered wave potential. As ϕ_S has a periodical nature in the y direction with a period λ, it can be expanded in a complex Fourier series

$$\phi_S(x,y,z) = \sum_{n=-\infty}^{\infty}\phi_{Sn}(x,z)\exp(iq_n y) \tag{3}$$

where ϕ_{Sn} = complex Fourier coefficient of the n-th harmonic and $q_n = 2n\pi/\lambda$.

The governing equation of ϕ_S can be derived by substituting Eq.(2) into the three-dimensional Laplace equation and is given by

$$\frac{\partial^2\phi_S}{\partial x^2}+\frac{\partial^2\phi_S}{\partial y^2}+2ik_y\frac{\partial\phi_S}{\partial y}-k_y^2\phi_S+\frac{\partial^2\phi_S}{\partial z^2}=0 \tag{4}$$

The boundary conditions for ϕ_S are i)kinematic and dynamic boundary conditions on the water surface, ii)impermeable condition on the sea bed, iii)radiation condition far from the body and iv)no flux condition on the submerged surface of the body. The final condition is specified as

$$\left.\frac{\partial\phi_I}{\partial n}+\frac{\partial\{\phi_S\exp(ik_y y)\}}{\partial n}\right|_{on\ S_B}=0 \tag{5}$$

where $\partial/\partial n$ indicates the normal derivative on the submerged surface and S_B = submerged surface of one of arrayed bodies located between $-\lambda/2<x<\lambda/2$.

Radiation problems

Assuming a small amplitude oscillation of the floating body in the different six modes, i.e., three translational modes (sway, heave and surge) and three rotational modes (pitch, yaw and roll) around axes passing the center of gravity of the body and parallel to the x, y and z axes, respectively, these can be given by

$$X_\alpha(t)=\hat{X}_\alpha\exp(-i\omega t)$$
$$(\alpha=1\sim6;\ 1=sway,\ 2=heave,\ 3=surge,\ 4=pitch,\ 5=yaw\ and\ 6=roll) \tag{6}$$

where $\hat{X}_\alpha$ = complex amplitude of each mode response. In order to unify the expression for both the translational and rotational motions,. rotational angles $\Theta_\alpha(\alpha=4\sim6)$ are redefined as the quantity having a length dimension by using the representative length of the body R_B

$$X_\alpha=\Theta_\alpha R_B \quad (\alpha=4\sim6) \tag{7}$$

It is noted that the absolute value of each mode amplitude of one floating body must be equal to that of other floating bodies in an array, but having a different phase lag in relation to the incident waves. Considering the phase difference due to spatial allocations of arrayed bodies, but which appears only for the case of obliquely incident waves, the radiation wave potential for each mode response can be written

$$\Phi_{R\alpha}(x,y,z,t)=-\frac{ig\hat{X}_\alpha}{\omega}\phi_{R\alpha}(x,y,z)\exp\{i(k_y y-\omega t)\} \quad (\alpha=1\sim6) \tag{8}$$

where $\phi_{R\alpha}$ = dimensionless complex amplitude of a radiation wave potential due to a unit amplitude of harmonic motion in each direction. Since $\phi_{R\alpha}$ is periodic in the y direction with period λ in the same manner as ϕ_S, it can be expanded in a complex Fourier series like Eq.(3). $\phi_{R\alpha}$ also satisfies the same governing equation, Eq.(4), and the same boundary conditions as ϕ_S except the kinematic boundary condition on the submerged surface.

Green's function

In this study, the wave source distribution method is used to calculate the scattered and radiated waves. In order to formulate the procedure, the Green's function for the wave field about an infinite array of 3-D bodies is necessary.

The Green's function G can be obtained by solving the same governing equation as ϕ_S and ϕ_{Rm} under the same boundary conditions with these, except the kinematic boundary condition on the submerged surface. Therefore, the governing equation of G is

$$\frac{\partial^2 G}{\partial x^2}+\frac{\partial^2 G}{\partial y^2}+2ik_y\frac{\partial G}{\partial y}-k_y^2 G+\frac{\partial^2 G}{\partial z^2}=\delta(x-\xi)\delta(y-\eta)\delta(z-\zeta) \tag{9}$$

where δ= Dirac delta function and (ξ,η,ζ)= x,y and z coordinates of a source point. Noting that G must be periodic in the y direction, it can be expanded into a complex Fourier series in the same manner as ϕ_S and $\phi_{R\alpha}$. The differential equation for the complex Fourier coefficient G_n of the expanded G is derived as follows

$$\frac{\partial^2 G_n}{\partial x^2}+\frac{\partial^2 G_n}{\partial z^2}-\{k_y+q_n\}^2 G_n=\frac{\exp(-iq_n\eta)}{\lambda}\delta(x-\xi)\delta(z-\zeta) \quad (n=-\infty\sim+\infty) \tag{10}$$

where the Fourier expansion of $\delta(y-\eta)$ has been adopted. Taking a Fourier transform of Eq.(10) with respect to x, the ordinary differential equation for $\tilde{G}_n$ that is a Fourier transform of G_n is obtained. After having solved this differential equation and applied the imposed boundary conditions, we can solve for $\tilde{G}_n$. By taking the inverse-Fourier transform of this solution and considering the Fourier expansion form of G_n like Eq.(3), the final solution for G is obtained

$$\begin{aligned} G(x,y,z;\xi,\eta,\zeta)=&-\frac{k^2-\sigma^2}{(k^2-\sigma^2)h+\sigma}\cosh\{k(h+z)\}\cosh\{k(h+\zeta)\} \\ &\cdot\left\{\sum_{n=N_-}^{N_+} P_n(x,y;\xi,\eta)+\sum_{n=N_--1}^{N_++1} Q_{0n}(x,y;\xi,\eta)\right\} \\ &+\sum_{m=1}^{\infty}\frac{k_m^2+\sigma^2}{(k_m^2+\sigma^2)h-\sigma}\cos\{k_m(h+z)\}\cos\{k_m(h+\zeta)\} \\ &\cdot\sum_{n=-\infty}^{\infty} Q_{mn}(x,y;\xi,\eta) \end{aligned} \tag{11}$$

where $\sigma=\omega^2/g$, and k_m= evanescent mode wave number satisfying the following equation

$$\omega^2=-gk_m\tan(k_m h) \qquad (m=1,2,3,\cdots) \tag{12}$$

Further, P_n, Q_{0n} and Q_{mn} are defined by

$$P_n=i\exp(i\mu_n|x-\xi|)\exp(iq_n|y-\eta|)/(\lambda\mu_n) \tag{13}$$

$$Q_{mn}=\exp(-\kappa_m|x-\xi|)\exp(iq_n|y-\eta|)/(\lambda\kappa_m) \tag{14}$$

in which μ_n, ν_n and κ_{mn} denote the characteristic wave numbers for different mode waves

$$\left.\begin{aligned} \mu_n&=\sqrt{k^2-(k_y+q_n)^2} \\ \nu_n&=\sqrt{(k_y+q_n)^2-k^2} \\ \kappa_{mn}&=\sqrt{k_m^2+(k_y+q_n)^2} \end{aligned}\right\} \tag{15}$$

N_+ and N_- in Eq.(11) are the upper and lower limit of integer n for which μ_n has a real root. In Eq.(11), the term proportional to P_n is the progressive mode component, and the terms including Q_{0n} and Q_{mn} are the evanescent mode components. Since the evanescent mode term damps rapidly with the distance from the source point, only the progressive mode term becomes important to describe the wave height distribution far from the body.

Regarding the progressive mode term in Eq (11), its wave number vector has a component μ_n in the x direction and (k_y+q_n) in the y direction, respectively. Therefore, we can see that there may appear several directional waves around the arrayed bodies, depending upon the values of N_+ and N_-. For instance, if N_-=1 and N_-=-1, then there are three directional waves, one is propagating in the same direction as that of incident waves (for n=0) and the other two (for n=1 and -1) are different directions. The latter two directional wave trains are newly generated by the interaction of incident waves with the periodically arrayed bodies. Generation mechanism of such directional wave trains can be explained intuitively by use of the Huygen's principle, as employed by Dalrymple & Martin (1990). Consequently, we can say that, even if long-crested wave trains are incident to the arrayed bodies, wave patterns around these may become short-crested.

We can estimate the generation of N-th higher mode of directional wave trains by examining whether or not there is a absolute value of n=N that gives a real root of μ_N. After some manipulations, this relation can be deduced as

$$\frac{\lambda}{L}\geq\frac{|N|}{1\mp\sin\theta} \qquad (N=\pm1,\pm2,\pm3,\cdots) \tag{16}$$

It is noted that the relation given by Eq.(16) is analogous to that for a periodical array of surface-piercing cylinders fixed on the sea bed (Nakamura 1994). In Eq.(16), the equality condition corresponds to the so-called wave resonance in the transverse direction.

Scattering and radiation potentials

In this study, the wave source distribution method (e.g., Mei 1990) was applied to calculate the scattering and radiation potentials about an infinite array of floating bodies. Applying the Green's function of Eq.(11) to the source distribution equation, the dimensionless complex amplitude ϕ_S and $\phi_{R\alpha}$ are given by

$$\phi_S(x,y,z)=\int_{S_b} f_S(\xi,\eta,\zeta)\,G(x,y,z;\xi,\eta,\zeta)\,dS(\xi,\eta,\zeta) \tag{17}$$

$$\phi_{R\alpha}(x,y,z)=\int_{S_b} f_{R\alpha}(\xi,\eta,\zeta)\,G(x,y,z;\xi,\eta,\zeta)\,dS(\xi,\eta,\zeta) \quad (\alpha=1\sim6) \tag{18}$$

where f_S and $f_{R\alpha}$= source strengths of the scattering and radiation potentials, respectively. These strengths are unknown at first and must be determined by use of the kinematic boundary conditions on the submerged surface of the body, e.g., Eq.(5) for f_S. After having determined the source strength, we can obtain ϕ_S and $\phi_{R\alpha}$ from Eqs.(17) and (18), respectively. The total velocity potential Φ_T about the arrayed floating bodies can be given by a linear summation of the scattering and radiation potentials in addition to the incident wave potential.

$$\Phi_T=-\frac{igH}{2\omega}\left\{\phi_I+\left(\phi_S+\sum_{\alpha=1}^{6}\hat{X}_\alpha\phi_{R\alpha}\right)\exp(ik_y y)\right\}\exp(-i\omega t) \tag{19}$$

where the dynamic displacement $\hat{X}_a$ can be determined by solving the equation of motion of the floating body with six degrees of freedom, which corresponds to the dynamic boundary condition on the oscillating submerged surface.

Parameters describing the wave field around an arrayed floating bodies

In order to understand the complex wave field around an arrayed floating bodies, an estimation method of the dominant wave propagation direction as well as the representative wave height is newly proposed based on the similar concept to the directional spectrum, i.e. the decomposition of wave energy according to wave propagation directions. Directional characteristics of the transmitted and reflected waves about the arrayed bodies are schematically shown in Fig.2, using the concept of the wave energy flux. The condition of $1<\lambda/L<2$ and normal incidence of waves to the array is assumed for simplicity. In the figure, Cg = group velocity, E_{T0} (or E_{R0}) = wave energy of a straightly transmitted (reflected) component, $E_{T\pm1}$ (or $E_{R\pm1}$) = wave energy of obliquely transmitted (reflected) components of $N_+=1$ and $N_-=-1$ mode in Eq.(16), respectively. For instance, the mode of $N_+=1$ corresponds to the wave train propagating obliquely in the left-hand side direction as seen in Fig.2.

In order to describe the complex wave field around the array, two parameters, K_T(or K_R) and T_n(or R_n) (n=1,2,3,···) are newly defined. The former parameter is used to describe the wave height characteristic on the transmission (or reflection) side of the array. The latter one is for describing directional characteristics of transmitted (reflected) waves. Definitions of these parameters are given by

$$K_T=\sqrt{\sum_{n=N_-}^{N_+} E_{Tn}/E_I}\ ,\quad K_R=\sqrt{\sum_{n=N_-}^{N_+} E_{Rn}/E_I} \tag{20}$$

$$T_j=E_{Tj}\Big/\sum_{n=N_-}^{N_+} E_{Tn}\ ,\ R_j=E_{Rj}\Big/\sum_{n=N_-}^{N_+} E_{Rn}\quad (j=0,\pm1,\pm2,\cdots) \tag{21}$$

where E_I = energy density of the incident wave per unit water surface, and N_+ and N_- are given by Eq.(16). The energy density of each component can be obtained analytically by considering the wave energy balance around the breakwater array. For instance, E_{Tn} is expressed by

$$\begin{aligned} E_{T0}&=[1+2Real\{\int\!\!\int_{S_s} f_\Sigma(\xi,\eta,\zeta)\ g_{p0}(\xi,\eta,\zeta)dS\}+|\int\!\!\int_{S_s} f_\Sigma(\xi,\eta,\zeta)\ g_{p0}(\xi,\eta,\zeta)dS|^2]\cdot E_I \\ E_{Tn}&=[|\int\!\!\int_{S_s} f_\Sigma(\xi,\eta,\zeta)\ g_{pn}(\xi,\eta,\zeta)dS|^2]\cdot E_I \qquad for\quad |n|\geq 1 \end{aligned} \tag{22}$$

In which, g_{pn} is the term concerning to the progressive mode of Green's function and is defined by

$$g_{pn}(\xi,\eta,\zeta)=\frac{1}{\lambda\mu}\frac{k^2+\sigma^2}{(k^2-\sigma^2)h+\sigma}\cosh(kh)\cosh\{k(h+\zeta)\}\exp[i(q_n\eta-k_n\xi)] \tag{23}$$

and f_Σ denotes the total source strength

$$f_\Sigma=f_S+\sum_{a=1}^{6}\hat{X}_a f_{Ra} \tag{24}$$

It is noted that, under the condition only the straightly transmitted wave component appears, the definitions of K_T and K_R are consistent with the conventional transmission and reflection coefficients, C_T and C_R, respectively.

EXPERIMENTAL SETUP

Model body

In the experiment, as seen in Fig.3, the model body of a rectangular cross section with additional vertical fins on the bottom side was used. In order to avoid complexities arising from the nonlinearity of catenary moorings and the drift motions due to wave forces, the model body was supported by a pair of flexible steel bars. Another advantage using the steel bar is that the stiffness of a supporting system can be easily adjusted to a necessary value by changing a diameter of the steel bar and its length.

Two different model bodies were used to change a pitch length λ of an array as shown in Table 1 and Fig.4. One has a transverse length of l_B=120cm (pitch length λ=150cm) and another one does l_B=40cm (pitch length λ=50cm). The spacing ratio r_s, defined by the ratio of an opening length between the neighboring two bodies to the pitch length λ, was kept constant at 0.2. In order to keep the stiffness of a supporting system at a constant value for per unit transverse length of a model body, the diameter of a steel bar was changed for these two different models. Corresponding values of dimensionless stiffness of one steel bar are specified in Table 1, where EI is the flexural rigidity of the bar. Other properties relating to the dynamics of the break-waters, i.e., a displaced weight, a mass moment of inertia and so on for per unit transverse length, were also set to the same values for these two models.

Wave basin and measurement system

Experiments were carried out in the three-dimensional wave basin at Ehime University. The basin is 15m length, 10m width and 1m depth. In the basin, 6 to 12 model bodies were periodically arrayed along the distance of 8m in the transverse direction, depending upon the pitch length of an array. Typical arrangement of model bodies for the pitch length of λ=150cm (l_B =120cm) is shown in Fig.5. Transmitted waves were measured at the location of 1.5 wave lengths apart from the arrayed bodies as seen in the figure. Incident waves were measured in a separated wave channel beside the main wave basin (see Fig.5).

Directional distribution of the transmitted wave energy and r.m.s. value of the transmitted wave height in the array direction were measured by use of linearly arrayed wave gages as shown in Fig.5. Considering the relation between the wave directional spectrum and the total wave energy density, we can easily derive the following relation

$$\sum_{n=N_-}^{N_+} E_{Tn}=\frac{1}{\lambda}\int_0^\lambda H_T^2\ dy\ \equiv\ (H_T)_{rms}^2 \tag{25}$$

where H_T = wave height on the transmission side of the array, $(H_T)_{rms}$= root mean squared value of H_T for the transverse pitch length λ. By using $(H_T)_{rms}$ of Eq.(25), we can rewrite the first equation of Eq.(20) as follows

$$K_T=(H_T)_{rms}/H \tag{26}$$

Using a similar technique to the method of measuring a directional spectrum, an index of directional distributions of transmitted waves T_n was obtained.

Wave conditions used in the experiment are as follows; a wave period T ranges from 0.65 sec. to 1.68sec. and a wave height H is fixed at about 4cm.

Water depth h is fixed at 40cm for all cases in the experiment.

RESULTS AND DISCUSSIONS

Wave transformations

Fig.6 is one of typical results on wave transmissions through an array of model breakwaters with a transverse length l_B=120cm (λ=150cm). In the figure, a well-known parameter L/B (B=breadth of a breakwater) is adopted as a second horizontal axis in addition to λ/L as a primary axis. Two distinctive computation results are shown with the experimental result ; one is given by the 3-D analysis described above and another is by the conventional 2-D analysis in a vertical plane (denoted by C_T in the figure). We can see that the dimensionless transmitted wave height K_T is greater than C_T for all range of λ/L. In particular, the difference between the two values becomes larger for the range $\lambda/L>1$ ($L/B<5$). This difference is very important for the practical design because the range of $L/B<5$ usually corresponds to the effective range of wave frequency of a floating breakwater. Such a difference is mainly caused by the effect of gap spacings among arrayed breakwaters. From the figure, it is also seen that the 3-D analysis is useful to predict the transmitted wave height, except some critical conditions at which the computed K_T shows a spike-like variation. These critical conditions coincide with $\lambda/L=1, 2$ in the figure, under which a wave resonance occurs in the transverse direction.

Fig.7 shows directional characteristics of the transmitted wave energy T_n(n=0, 1 and 2) through the breakwater array. Here, T_{+n} and T_{-n} are summed into T_n because of the symmetry of directional wave components. From this figure, when $\lambda/L>1$, obliquely transmitted wave components (T_1 and T_2) are predominant over a straightly transmitted one (T_0). This tendency can be confirmed by the computation result. The fact that K_T is greater than C_T as seen in Fig.6 may be directly caused by the predominance of obliquely transmitted wave components. From these facts, it can be easily estimated that a wave pattern behind the breakwater array becomes short-crested under the condition of $\lambda/L>1$. The discrepancy between the measured and computed results near the transverse resonance conditions, say $\lambda/L\approx 1$ and 2, may be caused by the small motions of asymmetric modes, like yawing and surging modes.

In Fig.8, the results of K_T and K_R for an array of model breakwaters with a transverse length l_B=40cm (λ=50cm) are shown. In this case, while the range of L/B is same as the case of a longer pitch length of Fig.6, the ratio of λ/L is less than 1 because of a shorter pitch length of the array. It is seen from the comparison between Fig.6 and Fig.8 that K_T decreases significantly for the range $L/B<5$ by using a shorter pitch length of the array, i.e., a shorter transverse length of a unit breakwater. It may be considered that the fact of no obliquely transmitted wave components results in the decrease of K_T.

By the examinations described above, it is concluded that a shorter pitch length of an arrayed breakwater than a wave length is recommended for the effective reduction of transmitted waves by an arrayed-breakwater system.

Dynamic response

Fig.9 shows one of typical results on the dynamic response of an array of model breakwaters to waves. In this case, a transverse length of a unit breakwater l_B is 120cm (λ=150cm). In the figure, the normalized horizontal displacement at the top the model breakwater $\hat{X}^{*}_{T}$ by an incident wave amplitude is plotted. For the sake of comparisons, two distinctive computation results by the 3-D analysis and conventional 2-D analysis in a vertical plane are also shown. From this figure, it can be seen that the difference between the prediction by the 3-D analysis and that by the 2-D analysis is little. Experimental results may be approximately predicted by the 2-D analysis in a vertical plane instead of the complex 3-D analysis.

CONCLUSIONS

Wave interactions with a periodically arrayed floating bodies have been examined from both the theoretical and experimental points of view. The parameter describing a complex wave field around an arrayed breakwater was derived theoretically. Furthermore, applicabilities of the conventional 2-D analysis in a vertical plane to the arrayed floating breakwater have been clarified.

It was confirmed that the numerical analysis proposed here is very useful for predicting the complex wave field behind the floating-breakwater array, including the directional characteristics of transmitted waves. In order to improve the effectiveness of the breakwater array, it is strongly recommended that a pitch length of the array should be less than a wave length of incoming waves.

It was also seen that the conventional 2-D analysis assuming an infinite length of the floating breakwater is insufficient for estimating sheltering effects of an array of floating breakwaters, especially under the condition in which a pitch length of the array is greater than a wave length. However, the 2-D analysis can be approximately used as an alternative method to the 3-D analysis to estimate the dynamic response of the floating breakwater array to waves without regard to the ratio of a pitch length of an array to a wave length.

REFERENCES

Dalrymple, R.A. and Martin, P.A.(1990). "Wave diffraction through offshore breakwaters. J. of the Waterway, Port, Coastal and Ocean Eng. Div., ASCE, Vol. 116, No.6, pp.727-741.

Kagemoto, H. and Yue, K.P.D. (1986). "Interactions among multiple three-dimensional bodies in waves: An exact algebraic m ethod , " J. Fluid Mech. Vol.166, pp.189-209.

Linton, C. M. and Evans, D. V. (1992). "The radiation and scattering of surface waves by a vertical circular cylinder in a channel," Phil. Trans. R. Soc. Lond. A, 338, pp.325-357.

McIver, P. and Evans, D. V. (1984). "Approximation of wave forces on cylinder arrays," Appl. Ocean Res., Vol.6, pp.101-107.

Mei, C.C. 1978. Numerical methods in water-waves diffraction and radiation. Annual Review of Fluid Mechanics, Vol.10, pp.393-416.

Nakamura, T. and Ono, T.(1988). " Analysis on wave diffraction about an infinite array of offshore breakwaters and wave resonance characteristics," Proc. Coast. Eng. in Japan, vol.35, pp.527-531.

Nakamura, T. and Ono, T. (1990). "Estimation method of two-dimensional wave diffraction around an array of offshore breakwaters including wave directional characteristics. Proc. Coast. Eng. in Japan, vol.37, pp.534-538.

Nakamura, T (1994). "Wave directional change by an array of breakwaters," Proc. Int. Symp. Waves-Physical and Numerical Modelling, Vol.II, pp.1069-1078.

Simon, M. J. (1982). "Multiple scattering in arrays of axisymmetric wave-energy devices. Part 1. A matrix method using a plane-wave approximation," J. Fluid Mech. Vol.120, pp.1-25.

Thomas, G. P. (1991). "The diffraction of water waves by a circular cylinder in a channel," Ocean Eng., Vol.18, pp.17-44.

Yeung, R. W. and Sphaier, S. H. (1989). "Wave interference effects on a truncated cylinder in a channel," J. Eng. Math., Vol. 23, pp.95-117.

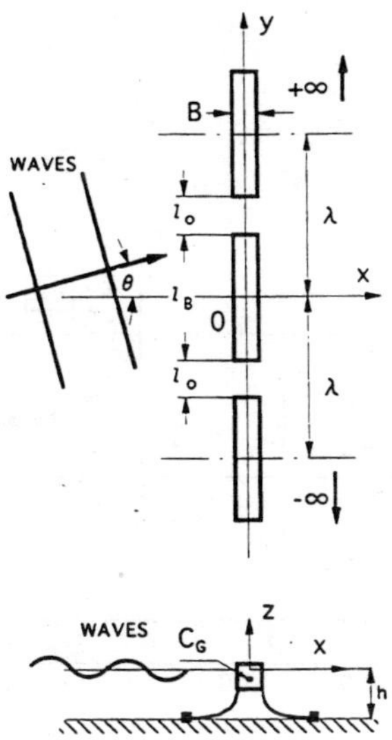

Fig.1 Analytical model and definition sketch.

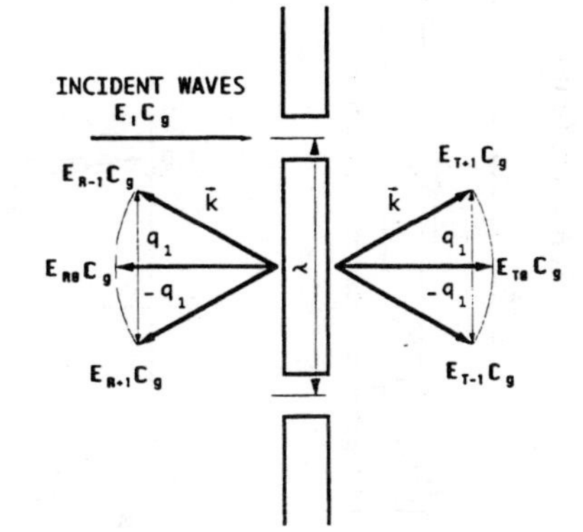

Fig.2 Directional characteristics of transmitted and reflected waves about an infinite array of floating breakwaters; in the case of 1<λ/L<2 and normal incidence of waves.

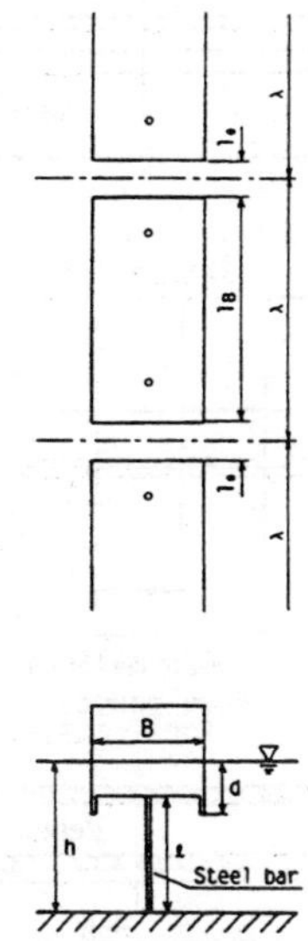

Fig.3 Model breakwaters and their supporting structures.

Table 1 Properties of model breakwaters.

Exp. case	I	II
Pitch length λ	150cm	50cm
Length of a breakwater l_B	120cm	40cm
Gap spacing l_0	30cm	10cm
Gap spacing ratio r_s	0.20	
Breadth of a breakwater B	30cm	
Draft of a breakwater d	14cm	
Weight of a unit breakwater W_B	21.08kgf	8.08kgf
Radius of giration	14.78cm	14.65cm
Diameter of a supporting bar	10mm	7.7mm
Number of supporting bars per unit breakwater	2	1
Length of a supporting bar l	28cm	
Dimensionless stiffness of a supporting bar ($W_B l^3/EI$)	0.3796	0.1890

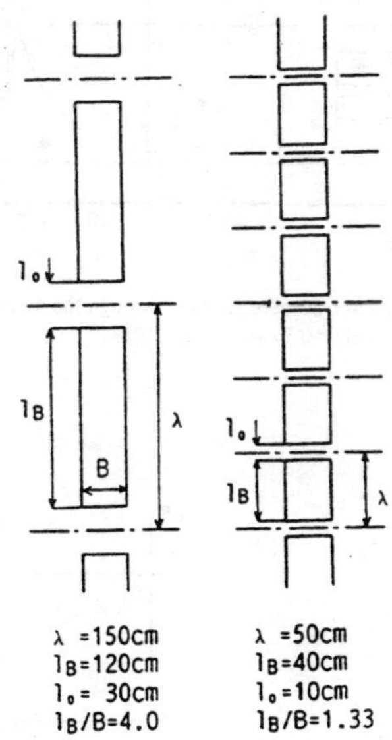

Fig.4 Plan view of the allocation of model breakwaters.

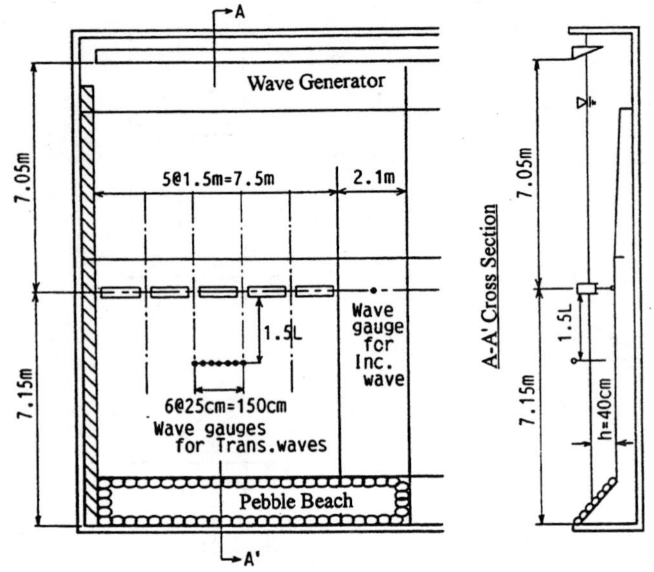

Fig.5 Experimental setup; in the case of λ=150cm (l_B=120cm and l_0=30cm).

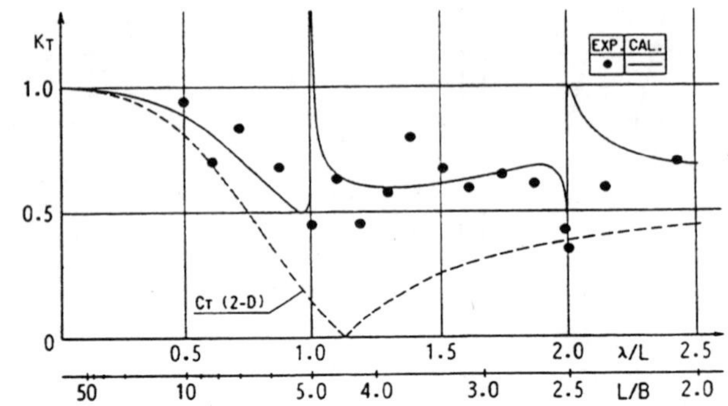

Fig.6 Dimensionless transmitted wave height K_T through the arrayed model breakwaters; in the case of l_B/B=4.0 and r_s=0.2.

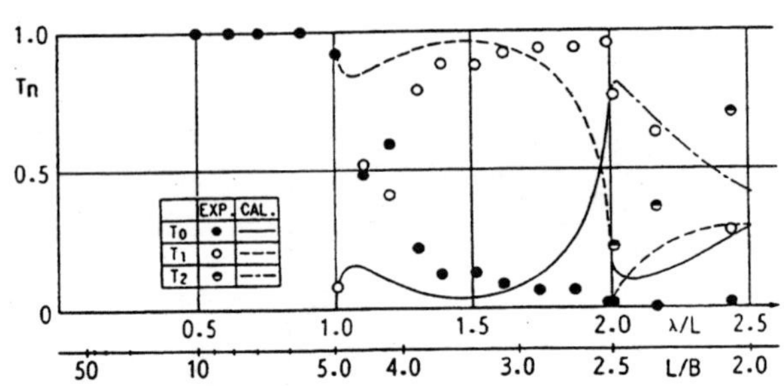

Fig.7 Directional distributions of transmitted waves through the arrayed model breakwaters; in the case of l_B/B=4.0 and r_s=0.2.

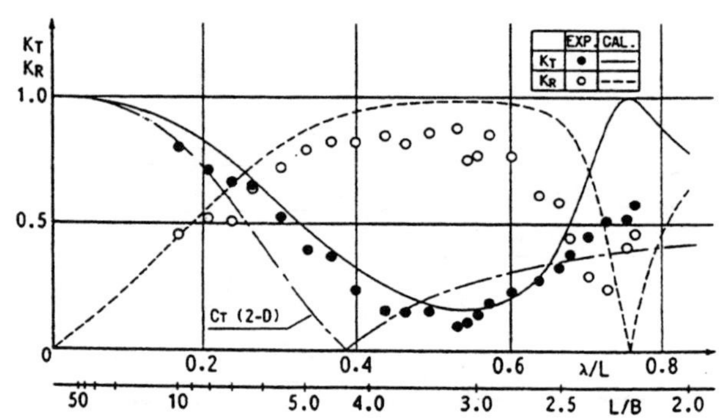

Fig.8 Dimensionless transmitted wave height K_T through the arrayed model breakwaters; in the case of l_B/B=1.33 and r_s=0.2.

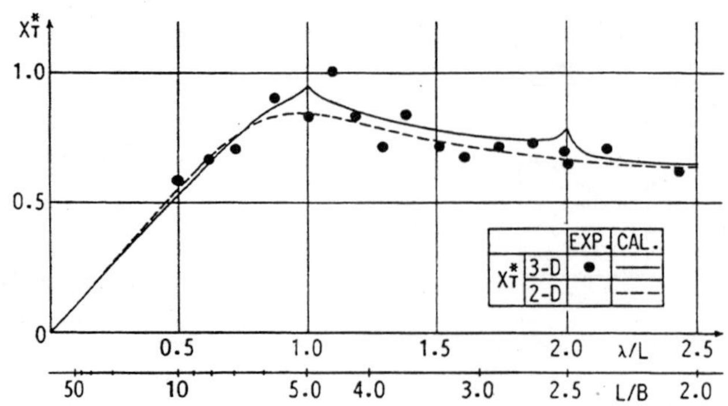

Fig.9 Sway mode responses of the arrayed model breakwaters; in the case of l_B/B=4.0 and r_s=0.2.

Proceedings of the Seventh (1997) International Offshore and Polar Engineering Conference
Honolulu, USA, May 25-30, 1997

ISBN 1-880653-28-1 (Set); ISBN 1-880653-31-1 (Vol. III)

In Situ Detection of the Settling Velocity of Suspended Particles by Means of Underwater Videomicroscope

Verena Baier and W. Bechteler
University of the Armed Forces
Neubiberg, Germany

Summary

The hydraulics of sediment transport is essentially influenced by the settling velocity and the size of the concerned solids. The measurement of these parameters has in the past mostly been performed for pretreated samples. Thus flocculation and aggregation processes which take place under real conditions are destroyed. In this paper a measuring technique is presented which allows the observation and quantification of velocity, size and shape of particles with a diameter of 5 - 500 µm direct in natural waters. The results have been compared with two different methods of a classical sedimentation analysis where a nearly perfect correlation has been found.

Key Words: ***velocity, fine material, in situ, particle measurement, shape***

1. Introduction

Hydrodynamic computations afford the knowledge of the settling velocity or the hydraulic diameter of the suspended solids. That parameter is of great significance for the forcast of erosion and deposition processes as well as the transport of solid bound pollutants (Maurer, 1994). Until ourdays the measurement of the settling velocity is mostly performed under laboratory conditions or computed by means of a charcteristic diameter which is as well evaluated for an often pretreated sample. Dying takes place and during measurement sometimes a dispergation is necessary, thus only the results for the single particles form the cumulative weight. In reality single particles mostly form aggregate particles. The flocculated (organic) or aggregated (minerals) material behave totally different than the single grains they consist of. The settling velocity may be less because of a lower density or even higher because of a greater volume. Especially for the fine material this will be important because they show higher tendency to flocculate. These processes are not satisfactorily understood today, because there are no techniques for the measuring of the settling velocity under real circumstances. For the forcasting of the transport and depostion behavior of flocs a sampling technique is not suitable at all. Thus the recently developed undewater videomicroscope (UWVM) represents a measuring unit to determine the suspended solids optically in natural waters up to a working depth of 100 m (Baier,1996).

2. Measuring technique

A CCD-camera has been equipped wih macro lenses to form the UWVM. It is waterproof and suits underwater insert (Fig. 1). The measuring volume spreads 1,2 x 0,9 mm.

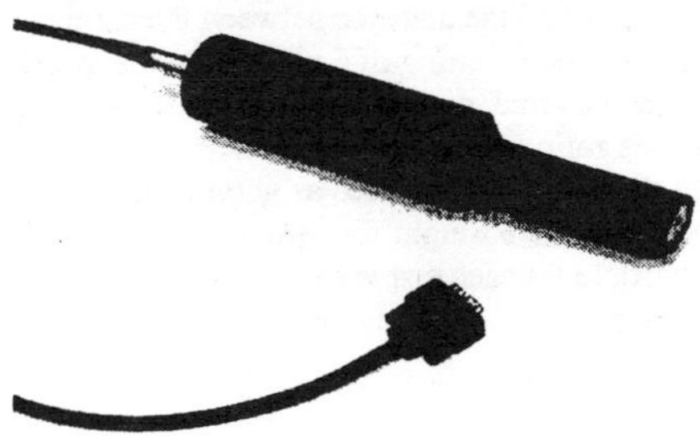

Fig. 1: underwater videomicroscope

focal depth	26.6 mm
measuring volume	0.914 x 1.219 mm
pixel size	1.79 x 2.38 µm
working distance	29.21 mm
field depth	0.20 mm

Tab. 1: optical data UWVM

For the illumination a pulsed Xenon flash is used. With a pulse frequency of up to 300 Hz the observed particles are frozen at several positions during their path between two light pulses. The image duration is normed with 20 ms thus the image frequency is 50 Hz. For the performed tests a light pulse frequency of 100 Hz has been applied because higher frequencies caused too much noise in the image. That means there are two positions of each frozen particle on each picture, if they have moved during flash pause. The distance of the centers of mass of each particle instances represent the coverd path during 10 ms thus a velocity is available.

The plexiglas cylinder serves for the destruction of turbulences, hence for the present the settling velocity under laminar flow conditions is available in laboratory tests.

Camera and flash are mounted on a sturdy aluminium plate in an angle of 45 degree. The camera is aligned in quadratur to the fall direction (Fig 2). To avoid turbulences and secondary movements in the measuring volume it is protected by a plexiglas cylinder. By means of LDA it has been ensured that there is no influence to the flow conditions by the instrument.

After transformation of the analoge video signals into digital pictures they may be stored and analyzed by digital image processing methods. The image processing strategy is partly automated and partly interactive. The program bv_veloc has been developed and implemented in Pascal on a personal computer. The image processing base routines are realized on a transputer (highly specified image processing unit) in the language OCCAM.

The first step is the selection of pictures with particle information. The algorithm is based on the detection of high gradients, that means steep transitions from low to high gray levels at the edges of the light particles on the dark background. For image enhancement a soft median filter is applied which only removes noise pixels but does not, as standard linear low pass smoothing operators, blur edges or even eliminate small particles. The result pictures are made available for the interactive analysis (Baier, 1996).

The investigator measures the smallest and longest diameter for each froozen matching particle pair and the distance between the centers of gravity. The input is a mouse guided setting of 6 points. On the graphic monitor a vector and the covered distance in µm is shown. The evaluated velocities, the axis ratio of the two marked diameters and the the mean value between them further indicated as geometric diameter are stored on a file. The cumulative weight for velocity and grain size for normally about 150 particle for each test is computed and shown in a xy-plot. If no solid density is known the cumulation concerning the total number is used.

3. Test performance and results

For the verification of this technique an investigation with three different materials has been performed. The tests are carried out in laboratory to allow calibration by materials with known properties, although the test circumstances are equal for field measurements.

Three materials with known density and visible differences in geometry have been used, the mean diameter in all samples was about 100 µm.

1.	glas beads	$\rho = 2{,}50\ g/cm^3$	spherical form
2.	aluminiumoxyd	$\rho = 2{,}00\ g/cm^3$	nearly spherical form, rough surface
3.	quarz sand	$\rho = 2{,}65\ g/cm^3$	ellipsoids, angular surface

All materials are sieved two times by different investigators. In Figure 7-9 only one line is shown for each material because there were no real differences.

The investigation of the settling velocity was at first performed by a classical settling tube. The supplied unit MACROGRANOMETER (Brezina, 1978) is based on a a two meter high plexiglas cylinder. A venetian blind as introduction system ensures the even release of the whole sample at test begin. At the end of the fall distance the weight is detected by a precision balance.

Figure 3 shows a photo of the unit MACRO-GRANOMETER. The results for the settling velocities may be taken from figure 4-6. This investigation as well has been performed two times for each material by different persons.

The settling velocities have been converted into a size distribution by means of the drag coefficient (eq. 1) proposed by Brezina (1979). The results are shown in Figure 7-9 with squares.

$$c_W = A\ Re^{-1} + B\ Re^{-0{,}5} + C \quad [-] \qquad \textbf{(Eq. 1)}$$

Fig. 3: Macrogranometer

The parameter A, B, and C depend upon the ratio of the main axes of the investigated grains normaly indicated as shape factor SF (Malaika and McNown, 1950).

The SediGraph from MICROMERITICS also based on the sedimentation principle. It is assumed that particles are dispersed in a fluid and settle in accordance with Stoke's Law. Thus by monitoring the rate at which particles settle and are removed from the monitored volume (decrease of concentration) a measurement of the cumulative size distribution can be obtained. The change of concentration is detected by collimated x rays.

The analysis program for the UWVM yield the settling velocity and the geometric diameter. During analyis a permanent manuel input of the material takes place. The video camera signals are stored on video tape to allow a second analysis of the same material by a second investigator. Figure 4-6 show the results of the velocity measured by UWVM in comparsion with that of the settling tests, the cumulative weight of the grain size may be taken from figure 7-9.

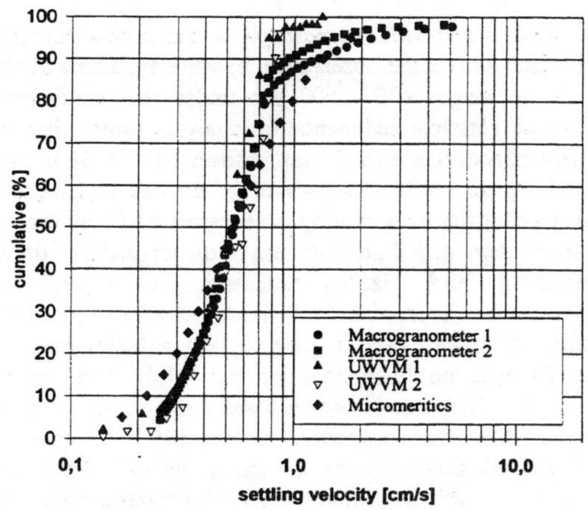

Fig. 4: velocity distribution of glas beads

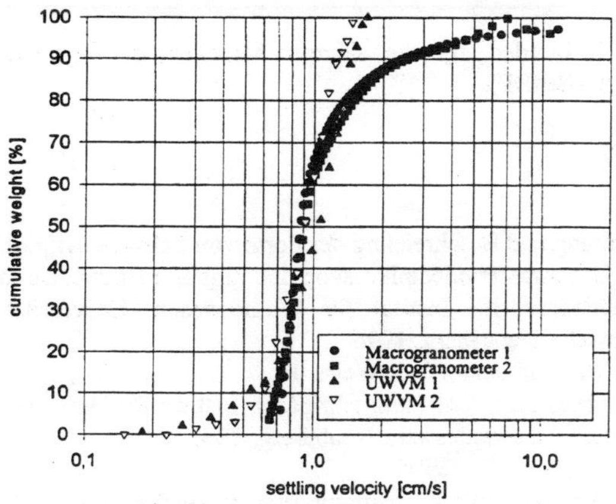

Fig. 5: velocity distribution of aluminium oxyd

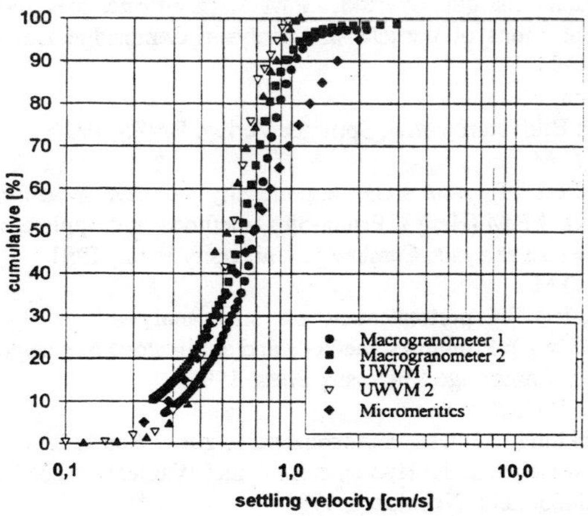

Fig. 6: velocity distribution of quarz sand

As shown in Figure 4-6 the results for the velocity measurement do not vary noticably in the range from 10 - 80 % for all materials. The reproduction by different investigators shows a variation less than 5 % for all samples which is an acceptable value. The deviation in the upper and lower parts of the measuring range are founded in the measuring technique of settling analysis. The diameter of about 100 µm lies near the lower limit of the measuring range of the MACROGRANOMETER. Thus the precision balance is adjusted very sensitive and that leads to the interpretation of even minor contamination of the water in the settling tube as data. That means the curve is more flat at upper and lower end of the data range. The results of aluminiumoxid from the SediGraph could not be used because the used fluids is a solvent for the material. Thus the x-rays are absorbed even after the last particles have settled.

The results for the sieve, the hydraulic and the geometric diameter show even more variation. The hydraulic diamater are cumputed with a shape factor SF' of 1.2 for all materials to show differences due to the form and the surface structure of the particles.

For glas beads (Fig. 7) as expected from theorie the deviation is low, only for small particles the cumulative weight differs. Maybe this is caused by a more rough surface of the fine portion.

For aluminiumoxyd (Fig. 8) a correlation between hydraulic and geometric diameter, but a smaller sieve diameter has been noticed. An explanation did not come up until now, more investigations are necessary to explain this phenomena. Maybe a rough surface does not slow down the settling velocity because of a higher drag coefficient as proposed by Brezina (1979).

As known from former investigations by other autors for the angular quarz ellipsoids a similar sieve and hydraulic diameter vary from the geomeric (mean) diameter determined by UWVM. For the measured axis ratio of 0,7 for that material this deviation of about 15 % is suspected for the hydraulic diameter but a little bit to high for the sieve diameter (McNown and Malaika, 1950, Mattews, 1991). Maybe this also is caused by the roughness of the surface.

Wether the deviation of the different diameters is of importance depends of the applications. The geometric diameter must not be the correct value in all circumstances, for example the sediment transport is more concerned with the hydraulic diameter. But it is even important to know about this measuring error when computing velocity or diameter by a the measured value.

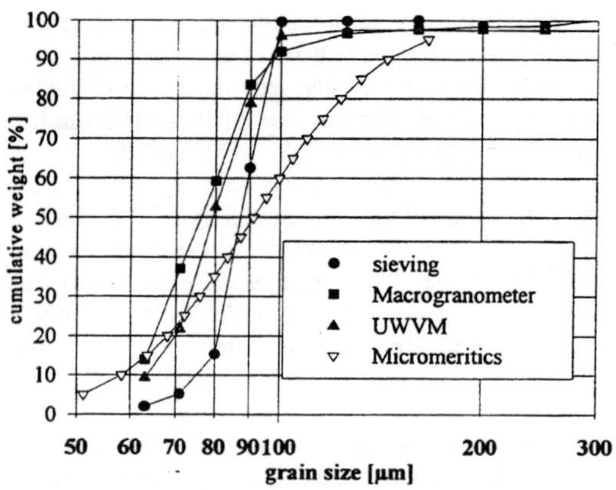

Fig. 7: grain size distribution glas beads

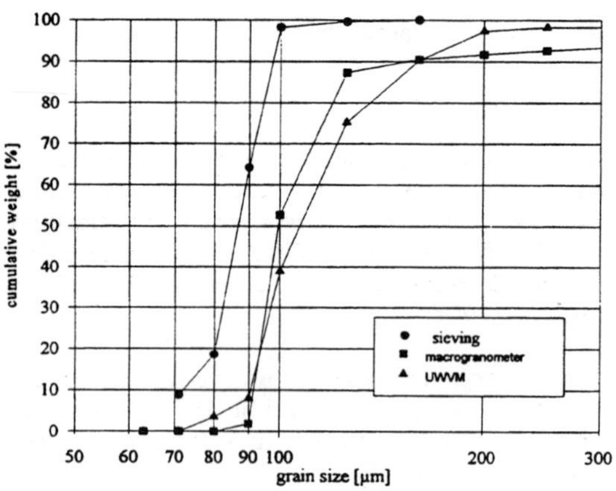

Fig. 8: grain size distribution aluminiumoxyd

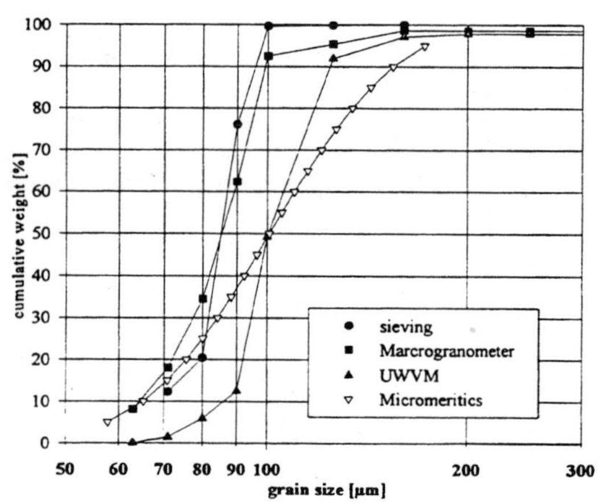

Fig. 9: grain size distribution quarzsand

4. Discussion

With the *underwater videomicroscope* a measuring technique is presented which allows the observation of velocity, size and shape of particles in the range of 5 - 500 µm under real conditions. The correlation with classical sedimentation analysis shows that velocity measurement is acceptable in the range form 10 - 90 % (cumulative weight) and reproductible. The advantage of this technique versus classical methods is that no sampling is necessary and that this method disperses with often needed pretreatment or dispergation of the sample in a measuring fluid. Natural processes like aggregation and flocculation may be observed and taken into consideration, that means the influence of those effects to the settling velocity may now be examined. Because no knowledge of the solid denity is needed, furthermore size, shape and surface texture are supplied, processes which are not understood completely ourdays may be investigated and possibly explained. Examples are the transportation of pollutants by aggregate particles and the sedimentation of activated sludge flocs in clearing plants. At the moment investigation of organic material in lakes take place.

5. Acknowledgement

The work is founded by the German Ministery of Research and Technology (BMBF).

References

Baier, V.:
Bestimmung und Beschreibung der Form von Schwebstoffpartikeln mittels Unterwasservideomikroskop und digitaler Bildverarbeitung, Dissertationsschrift, Institut für Wasserwesen, Universität der Bundeswehr, Neubiberg, 1996

Bechteler W., Vollmers H.-J., Weiss F.:
Feststofftransport, Vorlesungsskriptum, Institut für Wasserwesen, Universität der Bundeswehr Neubiberg, 1993

Brezina J.:
Particle size and settling rate distribution of sand-sized materials, 2. Europäisches Symposium für Partikelmeßtechnik, Nürnberg, 1978

Coakley J.P. Syvitski J.P.M.:
SediGraph technique, Syvitski, J.P.M.(Hrsg.): Princi-ples, methods and applicatons of particle size analysis, Cambridge University Press, 1991

Jähne Bernd:
Digitale Bildverarbeitung, Springer Verlag, Berlin, 1991

Matthews D.M.:
The effect of grain shape and density on size measurement, Syvitski, J.P.M.(Hrsg.): Princi-ples, methods and applicatons of particle size analysis, Cambridge University Press, 1991

Matthews D.M.:
The effect of pretreatement on size analysis, in: Syvitski, J.P.M.(Hrsg.): Principles, methods and applicatons of particle size analysis, Cambridge University Press, 1991

Maurer M., Bechteler, W.:
Entwicklung eines Gütekonzepts zur Beurteilung von Modellansätzen in der Hydromechanik und Wasserwirtschaft, DFG Abschlußbericht, Neubiberg, 1994

McNown J. S., Malaika J.:
Effects of particle shape on settling velocity at low Reynolds numbers, Transactions of the American Geophysical Union, Vol. 31, No. 1, 1950

Honolulu, USA, May 25-30, 1997

ISBN 1-880653-28-1 (Set); ISBN 1-880653-31-1 (Vol. III)

Measurement of Sand Beach Profiles in the Large Wave Flume

W. Dursthoff
University of Hannover
Hannover, Germany

O. Berend, R. Schmidt-Koppenhagen and W. Dursthoff
University of Hannover and Technical University of Braunschweig
Germany

ABSTRACT

In the Large Wave Flume in Hannover, wave-induced morphological alterations and scourings of sandy seabeds in shallow waters and at beaches of erodable material are investigated. Large-scale bed forms of about 250 m in length and 5 m in width have to be detected above water level as well as 5 m under the water level. To avoid usual deviations of electronic devices caused by sediment particles or temperature and gas content of the water a mechenical system was given priority. Nylon rolls mounted on a 7.6 m long sensor arm will be tracked over the seabed. From the measurements of x-position and the angularity of the sensor arm the bed profile can be calculated online by a computer. The acuracy of the system is in a range of ± 23 mm corresponding to ⊥ 0.4 % full scale caused by the compression of the sand.

1. PURPOSE OF INVESTIGATION

The Large Wave Flume is a notable facility for basic research in coastal engineering phenomena of the Coastal Research Center (Forschungszentrum Küste) of the University of Hannover and of the Technical University of Braunschweig. Water waves up to 2 m under quasi-prototype conditions can be simulated in the 300 m long, 7 m deep and 5 m wide flume. Besides various structures for coastal protection measures, sandy sea-beds in shallow waters and beaches of erodable material are investigated.

The quality and accuracy of the tests depends decisively upon the exact generation of waves in the channel and the consideration of disturbance variables (primarily reflections on the investigated structure). The Large Wave Flume possesses of an online absorption control system which is able to compensate for nearly all wave reflections directly at the wave paddle of the simulated beach profile and the natural frequency of the wave flume itself.

The equipment allows generating regular waves as well as natural motion of the sea over unlimited duration. This meet the requirements for the investigations of so called balanced sand profiles near the coastal line due to uniform wave attack and the change of the profile under new wave conditions. These tests are important for developing an optimized beach profile to withstand erosion processes during a storm surge or for example to measure wave forces onto partly covered pipeline sections in shallow water regions. The wave-induced morphological alterations and scourings have to be recorded immediately and in detail.

2. DEMANDS OF THE MEASURING DEVICE

Due to the large scale of the investigations carried out up to now, the sand elevation in the flume was manually detected by a plump bob. The hand lead was guided by a beam put on the handrails across the flume. Several times a day this procedure has to be done dot by dot over the complete distance where the channel was covered with sand. The system was double time consuming waiting for calm water conditions and then measuring step by step. To increase efficiency and accuracy it should be replaced by an automatic method handled by one person.

To observe the shifting of the sand in the flume a two-dimensional measuring system satisfying the following criteria is needed:

- high accuracy of repetition (better than ±10 mm),
- measuring the elevation of very large scale bed forms with a layer depth of the sand from 0 to 6 m in y-direction and up to 250 m in x-direction,
- measuring of sand profile below and above water level,
- software guided automatic collection of data as well as online analyzing and displaying the movements of the sand.

One of the significant result of the tests is to observe a so called balanced sand profile (see Fig. 1). A specific wave motion at a certain time duration will cause only minimal changes of the beach contour and sand movements. The accuracy between the different measurements sequences must be high so that changes in sedimentation can be attributed to wave action and not to measuring faults.

The sensor has to operate on the dry sand (detection through air) as well as under the water (detection trough sand and muddy water enriched with sediment particles). The device has to withstand influences through spray water, sand and changes in temperature. Furthermore the detection process should be able to start immediately in the end of the wave generation process to take advantage of the wave absorption time. Therefore the measuring system has to be reliable to avoid cost and time consuming repair works.

2.1 CRITERIA OF THE SYSTEM SELECTION

The following systems were considered in the selection process:

- acoustic based systems,
- optical systems,
- electromagnetic waves device (radar) and
- mechanical sensor system.

Acoustic systems analyze the time shift between sending an impulse and receiving the reflected answer. For once the sound velocity in the water depends of the temperature and the air and salt content of the water. Hence the supersonic transition time will be differ and the distance can not be calculated correctly. For the other the high sediment suspension and gas concentration scatters and distorts the returned sound of the sensor. Even with high quality supersonic devices only accuracy's worse than 1% of the measuring range can be achieved /1/.

Optical systems works in the same manner which would also not give usable results due to the blurring effect of the water. To archive optimal results using optical systems the wave channel would have to be drained and be refilled after the measurement. This procedure is time-consuming and by the flow of the water the sand profile to be measured would be destroyed.

A further alternative is the detection of the water depth with the help of electromagnetic waves (radar). However this procedure was eliminated due to extensive measurement techniques, due to financial reasons as well as not confirmed accuracy's of the results /2/.

Mechanical tools uses vertically driven sensorheads or a pulley rolling over the sand profile. The disadvantage of this procedure is that the original contour will be disformed due to the downward pressure of the measuring gauge. On the other hand there is the favorable effect that due to the direct contact onto the sand the described distortion factors of the other devices by temperature, suspension or gas content of the water do not have an influence of the recorded data. A pointing sensoring system realized using a motor driven linear telescopic device will bring a high degree of accuracy. But it is remarkable more time consuming to use than a continuously tracked pulley sensor system. Furthermore the guiding system of a telescopic device needs to be tightly sealed against water and sediment particles.

Finally a mechanical system was given priority. A pulley attached to a jib will be tracked over the sea bed (see Fig. 2). Such a measuring system can be used above and below the water level as well without any change of the control unit. Due to its uniqueness as a prototype the development of the required system was done in the above institution.

3. TECHNICAL DATA OF THE MEASURING DEVICE

Due to simplicity and after all the questions of realization the choice was made in favor of the mechanical solution. In this connection an about 7.60 m long sensor arm (jib) made from aluminum-pipe is used of which the upper end is pivoted on a test rig trolley (measuring platform) movable along the flume over a distance of about 250 m in the x-direction. Its lower end is attached to a three-divided pulley that rolls over the sand bedding. Linked momentumlessly to the above axis a precision shaft encoder is applied to determine the angular position of the sensor arm (resolution 0.001° ±0.0003°). The location of the measuring platform is recorded likewise by means of a standard shaft encoder (resolution 0.05° ±0.025° ≡ 0.26 mm deviation from intended driving path) which is mounted to the drive, see Fig. 3. The signals of the shaft encoder are evaluated by a particular PC-based countercard. A specific menu-guided program /3/ converts the data obtained by the shaft encoders to elevation profile, displays and stores the results.

3.1 STRUCTURE OF THE SENSOR ARM

Consisting of an aluminum-pipe framework the sensor arm measures 7.60 m in length, 1.30 m in width and 0.4 m in height while the thickness of the pipes is 50 mm in diameter. Compression joints connect the intersections of the structure which are additionally secured against shift. By favoring a framework arrangement a lightweight and yet stiff structure is obtained. Framework calculations were used to determine the static bowing under dead weight which causes a nonlinear systematic error in measurement due to a varying deflection of less than 6 mm over the entire range. The estimation of bending resistance was checked by a comparison of theoretically determined and experimentally gained natural frequency of the structure (f_{theo} = 7.54 Hz vs. f_{exp} = 7.78 Hz).

To prevent an unintended lift of the sensor arm during the measurement run caused by the turbulence of water around the structure a theoretical analysis was carried out. The maximum speed of the test rig trolley was determined for the case of equilibrium of hydrostatic and hydrodynamic uplift as well as dead weight. The measurement procedure can be started as soon as the current velocity has become less than 1 m/s; otherwise the forces exerted on the sensor arm by the current will cause an uplift force. This is the reason why the measuring system is normally used at the end of testing. Also under waves the scour around the sensor would strongly influence the results.

The shaft encoders are attached to the axes through specific bellows couplings to compensate misalignment and inaccuracies in displacement during transfer of the rotation. At the sensor arm an additional ball bearing of the coupling of the precision shaft encoder suppresses feedback of momentum and axial forces. This step was taken to protect the sensitive device during assembly and accidental misuse.

The sensor arm pulley consists of three side by side fastened nylon rolls with 300 mm in diameter and 50 mm in width each. The depth of penetration into the sand caused by the dead weight of the sensor arm was found to be in the range of 2 to 20 mm. The nylon rolls are sleeve mounted on a high-grade steel axis because this bearing application shows the least proneness to pollution and water.

3.2 CENTER OF PULLEY IMPACT

From the geometric parameters as there are diameter of rolls and radius of sensor arm as well as its variable angularity and x-position of the measuring platform along the flume the exact position of the pulley is calculated. The angular position of the sensor arm axis is detected by an incremental (relative) shaft encoder because as opposed to the absolute type it shows a far better resolution. According to their sense of rotation and angular position incremental shaft encoders generate two rectangular signals shifted in phase which are analyzed by a PC-based countercard. The number of counts represents the covered angel of rotation of the shaft encoders. It's possible to obtain the absolute angle of the sensor arm by evaluating the phase-shift of the two signals and handling reference signals. The measured angle α is equivalent to the actual angle of the wiper arm with the horizontal.

The x-position of the measuring platform is calibrated through reference labels along the wave flume (see Fig. 3). Its covered distance is calculated by the fixed diameter r_1 and the variable angle of rotation $\beta(x)$ of the wheels that is measured by the second standard shaft encoder:

$$x_{distance} = \beta \cdot r_1$$

From the angularity of the sensor arm and the position of the measuring platform the location of the sensor arm pulley is determined along the wave flume every 25 mm:

$$x = x_{label} + \beta \cdot r_1 + x_{offset} - \cos\alpha \cdot r_{arm}$$

$$y = y_{trolley} - \sin\alpha \cdot r_{arm} - r_2$$

3.3 ERROR HANDLING AND CORRECTION

The degree of the squeezing process of the rubber coated wheels exerting the drive of the measuring platform corresponds with the surrounding temperature and causes a systematic measurement error in the calculation of the x-position by temperature. At the end of each test this inaccuracy is subsequently compensated in the measuring software by an off-line calibration between the start and end position given by static labels mounted along the wave flume.

Moreover the actual dot of impact between pulley and sand profile deviates from the position calculates by the angularity of the shaft encoder in accordance to the elevation and slope of the sandy beach profile. Analyzing the obtained data train in the end of the measuring process allows to determine the exact beach profile as shown in Fig. 4. Depending of the pulley radius and the actual beach slope the corrected course can be calculated. The formula for a circle (shape of pulley) for a given position of the pulley center (x_m, y_m) is:

$$(x - x_m)^2 + (y - y_m)^2 = r_2^2$$

For the desired position to be corrected:

$$x = x_{position}$$

yields two possible intersection points by evaluating the quadratic equation:

$$y^2 + [-2 \cdot y_m] \cdot y + [y_m^2 - r_2^2 - (x_{position} - x_m)^2] = 0$$

$$y_{corrected} = \min\left(y_{position}\,;\, y_m - \sqrt{r_2^2 + (x_{position} - x_m)^2}\right)$$

If there is a solution the smaller one is to be compared with the measuring point. For the case the found solution is smaller than the y-value of the measuring point the correction is complete.

The PC-based measuring software is designed to handle all necessary procedures step by step in a user dialog. In the measuring course the computer gives a first online visualization of the measured profile onto the screen. After the correction processes the data will be stored and comparisons with the test series before can be plotted.

3.4 GEOMETRIC LIMITATION OF THE MEASURING DEVICE

The maximum slope in the direction of the sensor arm progress, which is to be allowed is equal to the tangent gradient starting at the point of the arc where the jib touches the sand bed with the radius of the length of the sensor arm. The maximum slope of the tangent is dependent on the angular position of the arm with respect to the horizontal level or it is dependent on the height (or depth) of the measuring point respectively (Fig. 5a).

When the sensor arm goes over a top of a dune, the maximum angle of the tangent to reversing point along the course before getting to the top must not become greater than the angle of the arm itself. Otherwise the measuring device would touch the top of the dune. The gradient of the tangent to reversing point is also dependent on the angular position of the sensor arm or on the height (or depth) of the measuring point respectively (Fig. 5b).

Concave troughs with a radius of curvature less than the radius of the pulley can, of course, not be resolved (Fig. 5c).

4 TESTING THE SENSOR DEVICE

To estimate the total error produced by the measuring apparatus a test drive over a pre-formed 183 m long sand profile in equilibrium was performed at a speed of 10 m/min. The test drive was repeated without any wave loads between the measurements. Initially the recorded data are compensated by scale distortion and calculation of center of impact. A comparison of the two profiles yields a measure for the accuracy of the system: within the entire range of the test drive (183.275 m) maximum differences amount to +22 and -23 mm respectively (vid. Fig. 6). Positive deviations (corresponding to material fill up) occur at a cavity in profile and negative ones (corresponding to material excavation) at a relative maximum in profile of the sandy beach. The mean difference between both profile sequences amounts to -7.1 mm and represents the measurement-related degree of compression of the sand. However, it has to be noted that the compression is the result of two measurement drives. Due to the wave load in between the measurement runs this influence on the profile sequence is largely made up for providing a common initial state. As under these conditions the sand is nearly homogeneously compressed by the sensor arm, a systematic error can be assumed. If the transport of sand induced by waves will be assessed, this error is cancelled.

In areas free of material compression (e. g. on the concrete sole, see left part of Fig. 6 between 60 m and 74 m) the margin of error lies in the range of ±2 mm. This value describes the accuracy of the measuring principle fairly well, yet depicting that in comparison to the partly greater local deviations on sand the deformation of the profile represents a weakness of a touching system. It can also be assumed that the different weight of the sensor arm being dry or submerged may have en influence on the deformation of the sand. However, as the load capacity of the sand under water is not as high as the capacity above the water level, both states are compensated. Dependent on particle size and compression of the sand, similar depths of penetration could be found out ranging between 2 and 23 mmm. The influence of the hydrostastic lift force on the value of static bowing of the sensor arm mentioned in para 3.1 can be neglected.

As opposed to acoustic and optic systems though, the obtained data are more trustworthy concerning their actual consistency. In relation to the gauging range of 6 m, an error of less than 0.4 % was found when the results of two tests were compared on the basis of a deviation of ± 23 mm measured.

5 SUMMARY

The measuring apparatus described above enables a quick (10 m/min) measurement of wave-induced sandmotions. The device proved to be very resistant against water, pollution and temperatur change in the course of the experiment. Expected influences on the sand profile by the mechanic touching system could be kept small because of the stable and lightweight construction. Applying a precision incremental shaft encoder with a force- and momentumless coupling that is true to the angle a high accuracy in repetition (max. deviation +22 /-23 mm, standard deviation $\sigma = 5.6$ mm, mean $\mu = -7.1$ mm) was achieved. However, for the range of ca. 6 m and the unfavorable surrounding conditions this could not be realized with ordinary methods so far. Due to the fine raster of 25 mm (smaller raster are possible) details in the morphologic changes become visible. Automated conduction of the measurement simplifies and speeds up the experiments considerably in the Large Wave Flume.

6 REFERENCE

/1/ Dette, H.H. Änderung der Vorstrandtopographie, Teil II.4 Tiefenmessung, Leichtweiß-Institut für Wasserbau, Abt. Hydromechanik und Küstenwasserbau, Technical University of Braunschweig, 1989

/2/ N.N. Übersicht über die derzeit verfügbaren Meßgeräte und -prinzipien und deren Genauigkeit, internal publication of the VEGA Grieshaber GmbH&Co. Company, Schiltach/Germany, 1992

/3/ Berend. O. Einrichtung zur Ermittlung von Strandprofilen im Großen Wellenkanal, large study thesis, Institut für Getriebetechnik im Maschinenbau, University of Hannover, 1993

Fig. 1 Sand profile in the Large Wave Flume, Coastal Research Center

Fig. 2 Measuring platform with mounted sensor arm in thearge Wave Flume

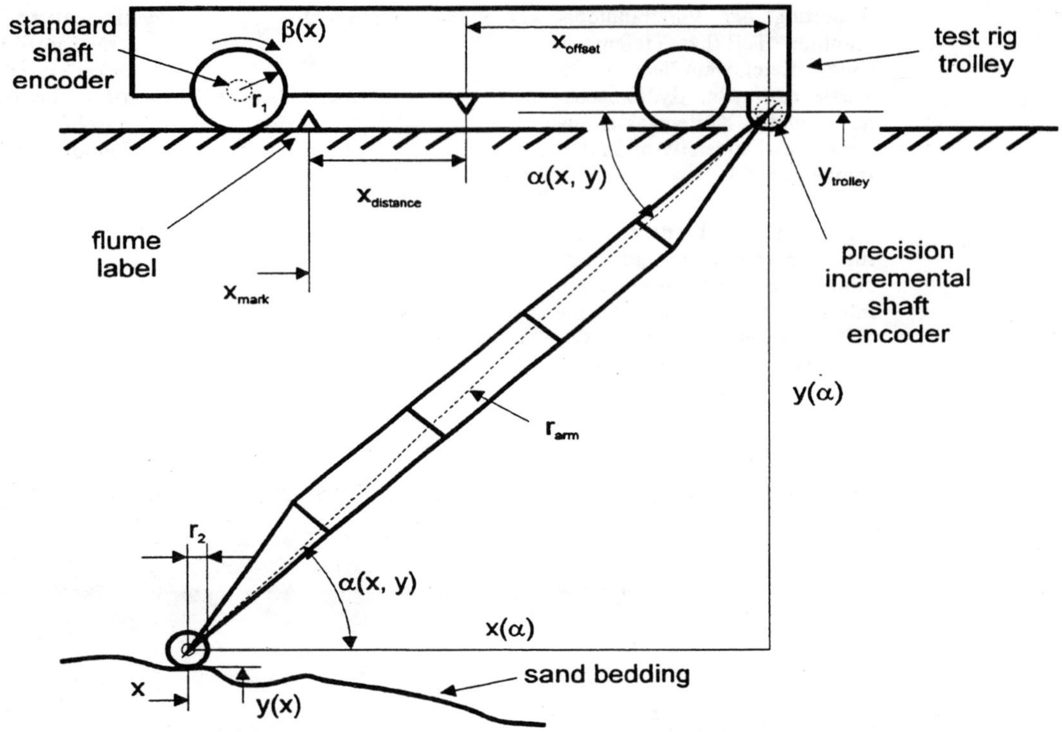

Fig. 3 Geometry of sensor arm and measuring platform with nomenclature

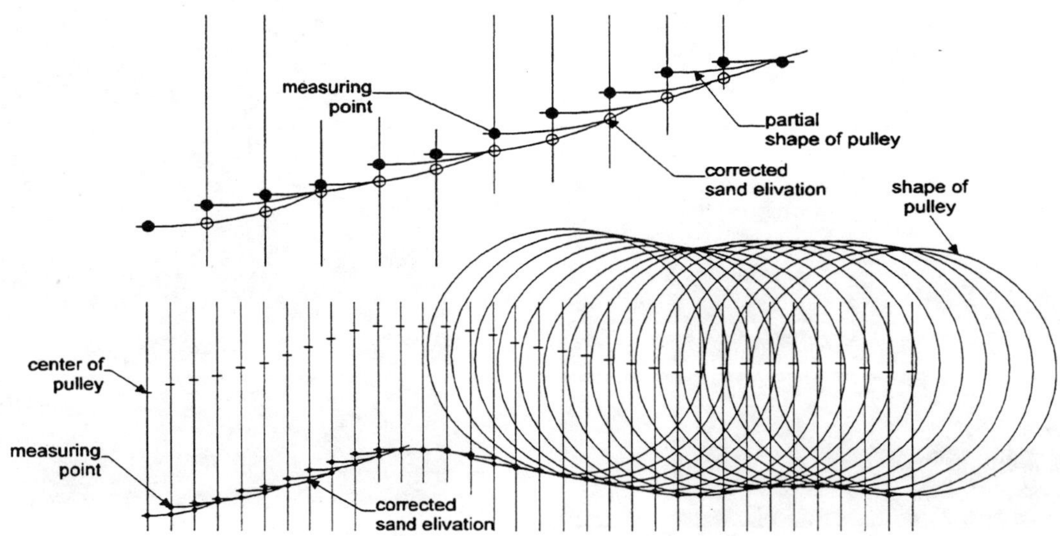

Fig. 4 Correction of measured profile sequence considering the pulley radius and the beach slope

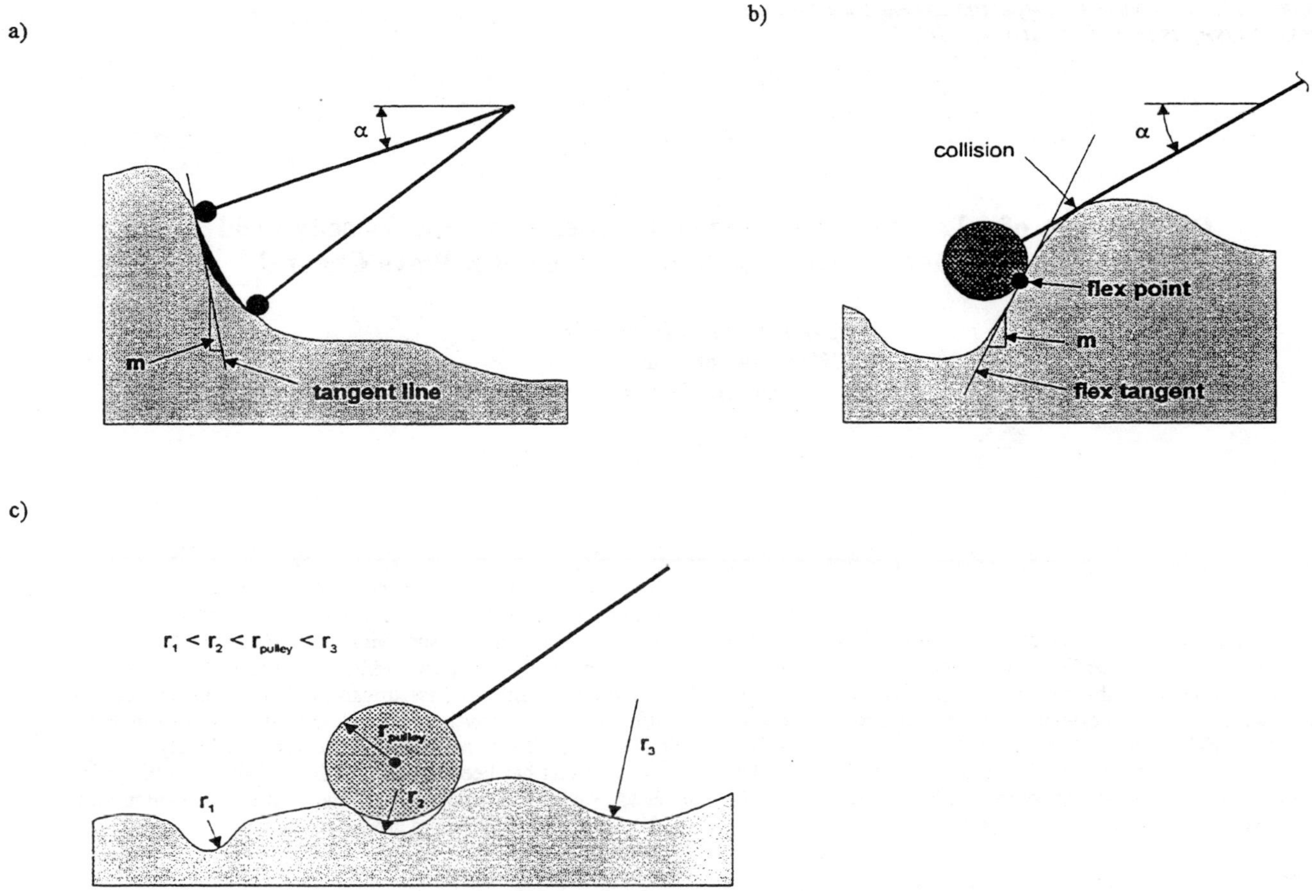

Fig. 5 Geometric limitations of the measuring device

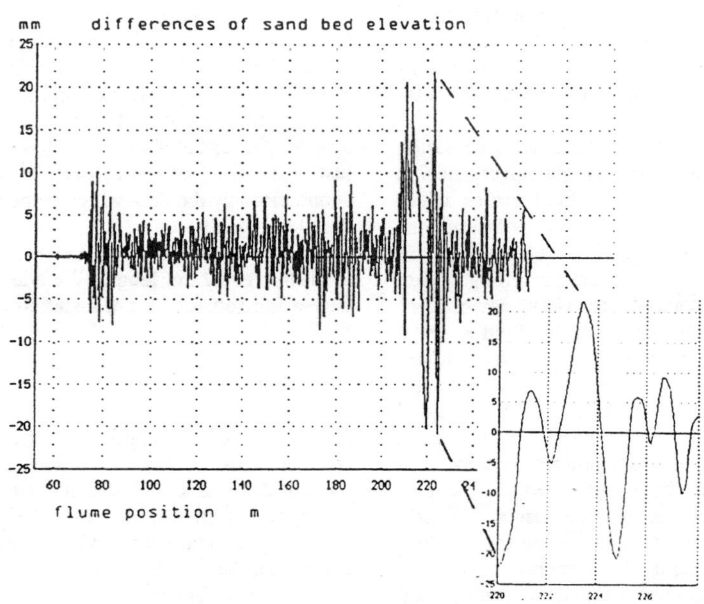

Fig. 6 Devition between two succeeding measurement runs

Proceedings of the Seventh (1997) International Offshore and Polar Engineering Conference
Honolulu, USA, May 25-30, 1997

ISBN 1-880653-28-1 (Set); ISBN 1-880653-31-1 (Vol. III)

Development of a Laser-Doppler-System for Measuring the Velocity Field Around a Cylinder Placed Close to the Bottom of a Wave Channel

W. Dursthoff and D. Mouazé
University of Hannover
Hannover, Germany

ABSTRACT

This paper describes the application of a fiber-optic LDA system to measure the velocity field, especially the boundary layer, around a circular cylinder placed near the bed of a large wave channel. The laser head is located inside the cylinder and the beams are focused in the flow near the surface of the cylinder through a transparent window. The main difficulties were to minimise the deformation of the interference pattern of the measuring control volume in order to obtain a sufficient signal in backscattering operation.

INTRODUCTION

The immersion of pipelines under seas represents an economical and safety method of the transport of fluids, gas or telecommunication cables. The wave-induced forces on submarine pipelines are distributed in several components. The gap between the pipeline and the sea bed induces a predominant lift force in the vertical direction. These loads on the structures are caused by the velocities features of the waves. The need to predict these force components reveals the importance to measure the velocity field around a cylinder in a wave channel.
The advantage of basic and applied research executed at large scales, as can be done in the *Large Wave Flume* in Hannover, lies in simulating hydrodynamic phenomena at quasi-natural scales which minimises scale effects. Detecting wave-induced flow velocities and turbulence intensities, for instance near a pipeline segment and his boundary-layer, plays an important role in physical modelling. Laser-Doppler-Anemometry (LDA) is an advanced and useful device (Ref. [2]). However, the use of a measuring technique of this kind in large-scale facilities involves some problems of technical and physical kind.
In order to make the use of the LDA technique in the context of current research, handling with the system has to be improved and adapted to the requirements of the large-scale facility and its operation. Preparatory work has therefore to be done in a smaller wave tank.
The employment of a two-component fiber-probe permits the non-intrusive measurement of small volumes in confined spaces. The fiber optics provide the possibility of significantly simplifying the handling and the traversing in back-scattering mode. The dimensional features of the probe, the small measuring control volume, as well as the coupling with an optic fiber to an Argon Ion laser, compensate to a certain degree the loss of scattered light intensity in „difficult" environment (light extinction of seeded water).
As an adequate solution to meeting the problem in large-scale hydraulic facility, it was decided to place the measuring system inside the cylinder, which means surveying out of the structure. This requires the assembling of a transparent optic system so that the beams and the back-scattering light can pass trough the hull of the cylinder without significant optical distortion. Correction of the chromatism and of the refraction of the beams made special measures necessary.
Specific interest has been attached to the movability of the cylinder in order to provide the measuring volume with every position necessary for detecting the velocity field around the structure.

EXPERIMENTATION SYSTEM

The cylinder in the wave flume

The measurements with laser Anemometry are performed in the wave flume of the Franzius Institut. A computer control system permits the control of the wave machinery. At the end of the channel, a slope (1:30) allows to absorb a part of the kinetic energy of the waves in order to reduce the wave reflection. The dimensions of the wave channel are in length L=120.0m, in width l=2.2m and 2.5m deep. The tests are accomplished at 40m of the wave generator. With the help of a wave gauge, the state of the oscillating surface can be determined just at the level of the cylinder. A second computer permits to save simultaneously the data which come from the resistive probe and from the LDV signals.
The cylinder consists of an empty piece of aluminium which diameter is 210mm and 2.2m length. We adopt a cylindrical coordinates system due to the rotation and translation of the measuring control volume. See Fig. 1;

The fiber optic LDA system

A LDV system is applied to the flow measurement in the wave channel. This system uses fiber cable conducting the laser beams without disturbances of the flow field in the measuring point. A fiber probe, adding fiber optics to a traditional LDV system, is suitable for flow measurement near a cylinder in wave channel because only small probe can be placed directly inside the structure, staying highly manoeuvrable. The schematic construction is shown in Fig. 2;
A beam, produced by a 5 Watt Argon Ion laser, is separated into two pairs of incident beams, green and blue, by a colour separator and laser beam splitters, which are included in an optical unit with 2 Bragg cells and an adaptator for coupling the traditional system to a new fiber

optic LDA probe. Four laser beams are transmitted to the probe by a single-mode polarisation preserving transmitting fibers. A fiber distribution unit (see Photo.1 & 2) serves as an interface between the transducer head, the transmitting fibres and the receiving fibres. These last one are terminated with a plug that fits the colour separator before the Avalanche-Photo-Diode for the reception of the scattered light.
A front lens, in stainless steel, is needed to complete the transducer. The optical elements, including the front lens, are achromatic. This means that the transducer can send with the same deviation angle both light wavelengths used, i.e. 488nm and 514nm.
The ratio of the focal length to the probe diameter results, in one hand, that the measuring volume is not too close to the probe for practical applications, and in the other hand, that the signal quality is sufficient in backscatter operation. The small measuring control volume allows measurements closer to the surface of the cylinder, and also increases the scattered light intensity. The aspect of the spatial resolution will be discussed in the next section.
The laser light is launched into optical fibers through fibre manipulators. In order to optimise the coupling with respect to the optical axis of the fibre, those manipulators adjust two important parameters of the laser beams:

-The beam angle
-The beam position

A screw on the fiber plug permits the beam to converge and to be focused on the fibre. A shutter provides an easy way of blocking the beam path to prevent the optic fibre to be burned when the beam is defocused. The fragile optical fibre is protected by a jacket in a steel spiral armed with Kevlar fibres and enclosed in silicone rubber. This protection makes it, tension, torsion and pressure resistant. The periodicity checks of the beams, and the observation of the interference patterns (with a short lens of 4mm focal length) reveals that the laser head is well adjust. With a total light intensity of 0.6W in the measuring control volume, the 51% lost of transmitted power in the fiber cable can be considered as normal after a length of 15 meters.

The transparent optic window

The measuring system is placed in the cylinder, so that the beams have to pass through a transparent window to reach the measuring point. In this case, an optical limitation occurs when the light crosses a number of interfaces of different refractive index. In the worse situation, the optical access becomes seriously compromised in presence of three different media: Air in the cylinder, Glass or Plexiglas of the transparent window, and finally, Water in the channel. The construction of the transparent optical material should minimise the disturbance of the incident and scattered light as they traverse the different middles. Other difficulties will create a lack of coincidence conditions:
-A small crossing angle and, also, a short focal length, allow to solve the problem of the surface curvature by the possibility to use a very small straight window.
-In back-scattering operation, secondary reflections from the optical material could be detected by the receiving optics in the probe head; Optical coatings are the most effective solution in producing 2% of rest reflection in our configuration.
-Due to the chromatic aberration, the different wavelengths of the beams will not submit the same deviation in a done middle. See Table 3 for the different refractive indexes:
The translation movement of the laser head excludes the use of aspherical lenses.
-The fluctuations of the temperature are too small to have an influence on the refraction.
-An excessive tilt will cause a lost of coincidence conditions in the measuring control volume. These three solutions could be envisaged:
In our configuration, it is impossible to fill the channel with a chemical liquid of higher refractive index than water, in order to create an index-matched flow system. In the first possibility, the cylinder is filled with water to reduce the number of media; But the electric motors for the traversing system must be isolated. In the two other cases, the window is composed of a prism of whole glass, or a prism of Plexiglas filled with water. At this point, an important condition is to observe a parallelism between the laser head and the first surface of the prism. Indeed, the same incidence angle of the four beams induces the same refraction angle over this surface. Then, the beams are crossing in the same point and the deformation of the interference pattern is minimised. The three possibilities are still tested by observing the fringe patterns in real conditions.
The inclination of the laser head induces a correction of the acquired velocity in the plane (Z, Y). Indeed, the tilted fringes will not exactly observe the component v. The third component (Y axis) of the velocity in the channel is in the order of 10% of the transversal velocity (Z axis). With a 30° inclination of the laser head, the calculation gives the correction, v=1.092va (See Fig.4)
The „inside measuring system" permits to avoid a reflected light from the surface, and then minimise the induced noise. This insures a detectable Doppler signal. This would not the case if the probe was outside the structure.
The tilted length of the measuring control volume determines the reachable spatial resolution. With an inclination of 30°, the spatial resolution is 0.11mm.

References

[1] BOADWAY, J.D., KARAHAN, E., 1981 Correction of Laser Doppler Anemometer readings for refraction at cylindrical interfaces, p.4. DISA Information, Measurement and Analysis.
[2] DURST, F., MELLING, A., WHITELAW, J.H., 1987 Principles and practice of Laser-Doppler-Anemometry. Braun G., Karlsruhe.
[3] DYBBS, A., EDWARDS, R.V., 1984 An index matched flow system for measurements of flow in complex geometries, p.171. Second International Symposium on Applications of Laser Anemometry to Fluid Mechanics.
[4] MOHR, B., ERTEL, H., 1986 Brechungsindexanpassung für Geschwindigkeitmessungen mittels Laser-Doppler-Anemometrie. Lehrstul für Strömungsmechanik. Universität Erlangen.
[5] NOURI, J.M., WHITELAW, J.H., YIANNESKIS, M., 1986 A refractive-index matching technique for solid/liquid flows, p.335. Third International Symposium on Applications of Laser Anemometry to Fluid Mechanics.
[6] PFORTE, H., 1989 Feinoptiker. VEB Verlag Technik Berlin.
[7] STIEGLMEIER, M., 1994 Halbleiter-Laser-Doppler-Anemometer unter Verwendung von Glasfasern und Integriert-Optischen Komponenten. Technischen Fakultät Erlangen.
[8] SUMMER, B.M., JENSEN, B.L., FREDSOE, J., 1986 Experimental investigation of turbulent oscillatory boundary layer flows, p.261. Third International Symposium on Applications of Laser Anemometry to Fluid Mechanics.
[9] UHLMANN, D.R., KREIDL, N.J., 1991 Optical properties of glass. The American Ceramic Society, Westerville.

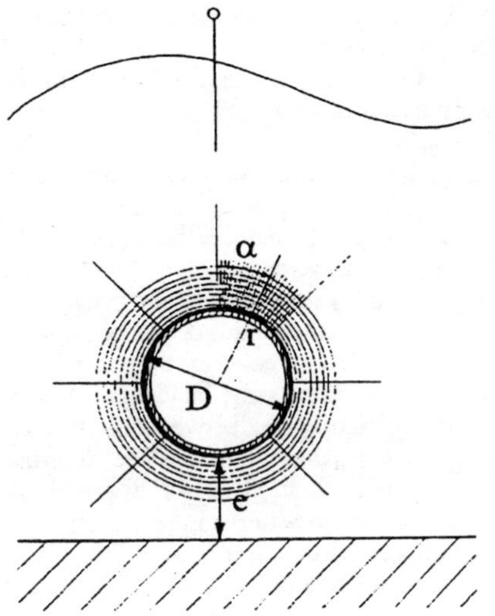

-D = 210 mm,
-e = 50 mm,
-e/D = 0.24,
-smooth cylinder

Fig.1 The coordinates with respect to the cylinder

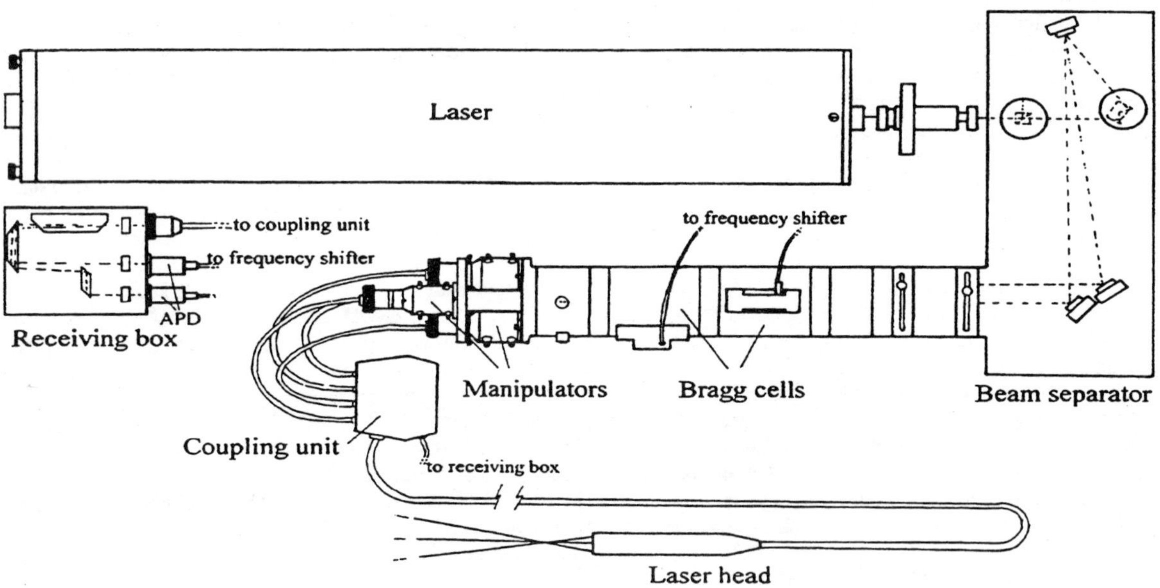

Fig.2 The schematic construction of the LDV system

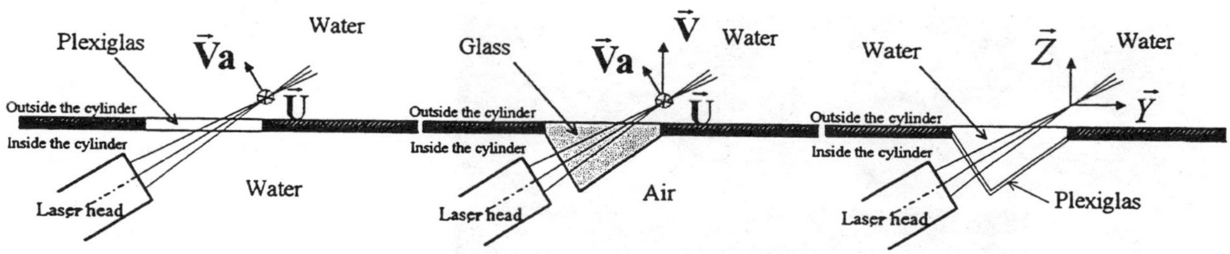

Fig.3 The transparent optical window

Wavelength nm	Focal length mm	Diameter mm	Length mm	Fringe spacing μm	Number of fringes
488	60	0.028	0.223	1.9671	14
514	60	0.029	0.234	2.074	14

Tab.1 Main characteristics of the measuring volume

27mm diameter	15mm beam spacing
205mm long, watertight	7.125° half-crossing angle
Ratio focal length/probe diameter=2.22	1.35mm beam diameter

Tab.2 Some specifications of the front optics

Wavelength nm	Refractive index of Water	Refractive index of Glass (Bk7)	Refractive index of Plexiglas (PMMA)
488	1.337	1.522	1.492
514	1.335	1.52	1.491

Tab.3 The refractive indexes of different media

1. Focussing ring
2. Beam angle plug
3. Beam direction plug
4. DANTEC Coupling adaptator

Photo.1 The coupling and fibre optic systems

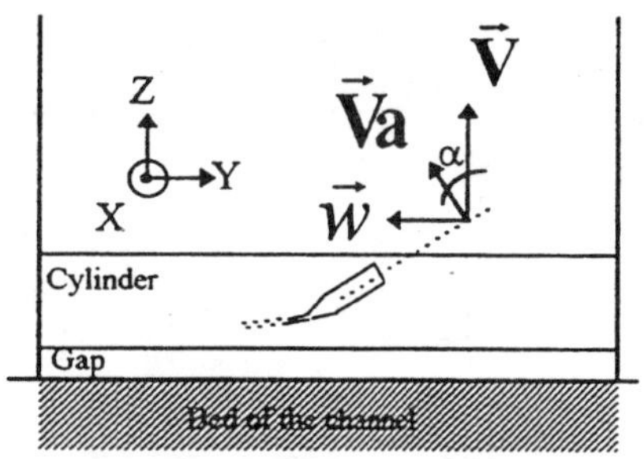

Fig.4 Correction of the tilted acquired velocity

$$v = {}^{va}\!/_{C}$$

$$C = \cos(\tan^{-1}({}^{w}\!/_{v}) - \alpha)\sqrt{1 + ({}^{w}\!/_{v})^2}$$

Proceedings of the Seventh (1997) International Offshore and Polar Engineering Conference
Honolulu, USA, May 25-30, 1997

ISBN 1-880653-28-1 (Set); ISBN 1-880653-31-1 (Vol. III)

Water Particle Velocity Field and Embedment of Artificial Fish Reef Under Wave Action

June-Q Kim
Toyo Construction Co., Tokyo, Japan
Koichiro Iwata, Norimi Mizutani and Atsushi Suzuki
Nagoya University, Nagoya, Japan
Teofilo Monge Ruffin, Jr.
United Nations Center for Regional Development, Nagoya, Japan

ABSTRACT

An experiment on cubical submerged structure like artificial fish reef is conducted to determine the flow field and mechanisms of local scouring and resultant embedment. Moreover, a numerical scheme for three-dimensional flow field computation including the vortex formations is developed. Based on the results, the pertinent characteristics of flow field are discussed in relation to the vortex formations. For movable bed, the vortex formation and the steady outward flows from the structure are very significant in local scouring which is directly responsible in the embedment of the structure. The relationships between the flow field due to variation of KC and Re numbers with the maximum values of scouring depth and embedment are discussed. Lastly, the flow patterns and vortex formations are well evaluated by the numerical computation; thus, the characteristics of the computed flow field explain well the generation mechanisms of local scouring.

KEY WORDS: artificial fish reef, Keulegan-Carpenter number, Reynolds number, vortex, local scouring, embedment, water particle velocity

INTRODUCTION

Artificial fish reefs are today a well established fishery enhancement technique that is being practiced around the world. For the past centuries, marine anglers have used artificial fish reefs to enhance their fishing efforts by placing rocks in bottom bed. Historical records show that the placement of rocks to enhance kelp production has been the common practice of Japanese fishermen since the late seventeenth century (Aomori Prefecture, 1980). This method can be very effective and is considered as the oldest and simplest enhancement technique. However, during storms and high waves, most of these rocks are moved or severely embedded; thus, many of the attached organisms are crushed and the sheltering of other living organisms in the zone of enhancement is greatly disturbed. Because of the poor stability and very low durability of these kinds of fishing reefs, it is more economical and efficient in the long run to build artificial reefs with specially designed structures and durable materials. Thus, a fundamental study on the significant characteristics of physical and geological aspects of artificial reef is deemed necessary. Presently, the lack of sufficient knowledge on how these reefs should be constructed for maximum benefit to both fisherman and environment of coastal areas prompted many researchers to improve the study on artificial reefs. The most common problems in shallow water are either the susceptibility to storm damage or the lack of suitable substrate and habitat. Consequently, research topics related to wave field, local scouring, and embedment of artificial fish reefs are very important. These topics are strongly related to stable design, mechanisms of fish gathering, and understanding of biological environment such as the distribution of plankton, fish eggs, and suspended materials.

There is no reliable investigation concerning the study of flow field around artificial fish reef in three-dimensional wavefield. For the case of local scouring and embedment problems in AFR in three-dimensional form, Sawaragi et al. (1984) conducted an experiment with sand layer on roughness floor in oscillatory flow tunnel and discussed the relationship between scoured area and void ratio of artificial fish reef. Previous researches were conducted mostly in two-dimensional form and mostly under oscillatory flow because of the complexities involved in the exact description of the characteristics and mechanisms of flow field, local scouring, and embedment of artificial fish reefs. Therefore, for in-depth study, this paper discusses experimentally and numerically the pertinent characteristics and interrelationship of water particle velocity field, scouring, and embedment of artificial fish reef under wave action in shallow-water depth in three-dimensional form.

NUMERICAL ANALYSIS

Governing equations

The governing equations consist of the continuity equation, Eq.(1), and Navier-Stokes equation for the incompressible fluid, Eqs.(2)~(4). The Cartesian coordinate system (x,y,z) is employed, where u, v, and w are the velocity components in respective directions of x, y, and z, in which z is upward positive with its origin being on the still water surface.

$$\frac{\partial u}{\partial x}+\frac{\partial v}{\partial y}+\frac{\partial w}{\partial z}=0 \tag{1}$$

$$\frac{\partial u}{\partial t}+\frac{\partial (u^2)}{\partial x}+\frac{\partial (vu)}{\partial y}+\frac{\partial (wu)}{\partial z}=-\frac{1}{\rho}\frac{\partial p}{\partial x}+\nu\left(\frac{\partial^2 u}{\partial x^2}+\frac{\partial^2 u}{\partial y^2}+\frac{\partial^2 u}{\partial z^2}\right) \tag{2}$$

$$\frac{\partial v}{\partial t}+\frac{\partial(uv)}{\partial x}+\frac{\partial(v^2)}{\partial y}+\frac{\partial(wv)}{\partial z}=-\frac{1}{\rho}\frac{\partial p}{\partial y}+\nu\left(\frac{\partial^2 v}{\partial x^2}+\frac{\partial^2 v}{\partial y^2}+\frac{\partial^2 v}{\partial z^2}\right) \quad (3)$$

$$\frac{\partial w}{\partial t}+\frac{\partial(uw)}{\partial x}+\frac{\partial(vw)}{\partial y}+\frac{\partial(w^2)}{\partial z}=-g-\frac{1}{\rho}\frac{\partial p}{\partial z}+\nu\left(\frac{\partial^2 w}{\partial x^2}+\frac{\partial^2 w}{\partial y^2}+\frac{\partial^2 w}{\partial z^2}\right) \quad (4)$$

where p is the pressure, g is the gravitational acceleration, ρ is the density of water, ν is the coefficient of kinematic viscosity of water, and t is the time.

Discretization

For the numerical computation of the governing equations, Eqs.(1)~(4), the finite difference method is employed. The staggered mesh is used in the discretization of governing equations and the computation domain. In the staggered mesh, the water particle velocities (u,v,w) are defined at the center of the cell surfaces and pressure p at the center of cell, as shown in Fig. 1. In the figure, Δx, Δy, and Δz are the dimensions of the cell in the directions of x, y, and z, respectively. Each cell in the computation domain is identified by subscripts i, j, and k. The viscosity terms of the governing equations are discretized with central difference, and the advective terms with donor difference. The donor difference is a mixture of upwind difference and central difference, and the ratio of their contributions to the computation of the advective term can be controlled by the weight α.

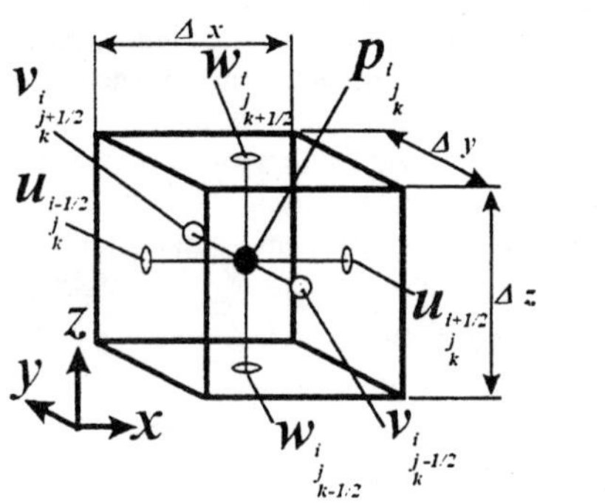

Fig. 1 Definition sketch of staggered mesh.

Boundary conditions

Submerged structures such as artificial fish reefs and seaweed feeding blocks are generally not so big as to change the free surface profile and the flow in the far field. Therefore, the computation domain can be reduced merely near the structure in the three-dimensional wave tank, as shown in Fig. 2. The free-slip conditions were given in the wall side and bottom boundaries. The non-slip conditions were subjected to the structure surfaces. The water particle velocities (u,v,w) and pressure (p) calculated with the Stokes third-order solution (two-dimensional, $v=0$) were given to the open boundaries, as shown by the shaded surface in Fig. 2.

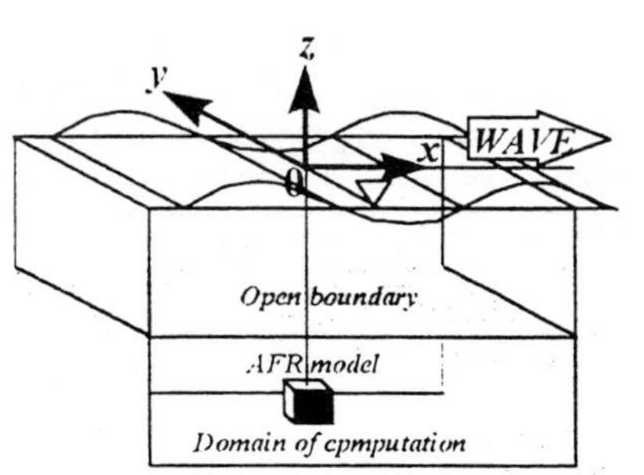

Fig. 2 Coordinates and computation domain.

Computational procedure

Introducing the explicit scheme, the velocities and pressure at advanced time step $(u,v,w,p)^{n+1}$ can be computed from the velocities and pressure at the present time step $(u,v,w,p)^n$ with the momentum equations, Eqs.(2), (3) and (4). The computed velocities $(u,v,w)^{n+1}$, however, may not satisfy the continuity equation for each cell in general. Therefore, $(u,v,w,p)^{n+1}$ must be adjusted to satisfy the continuity equation of each cell simultaneously. The SOLA scheme is used in this paper to iteratively and reasonably adjust $(u,v,w)^{n+1}$ to satisfy the continuity of each cell (Hirt et al., 1975). By repeating the computation of $(u,v,w)^{n+1}$ and through adjustment of $(u,v,w,p)^{n+1}$ with the SOLA scheme, the computation can be proceeded to the sufficient time for the solutions. The time step, Δt, for the computation is chosen to satisfy the Courant condition, Eq. (5), for every step of the computation. In Eq.(5), $|u|_{max}$, $|v|_{max}$ and $|w|_{max}$ are the respective maxima of u, v, and w at time step (n), and γ is the relaxation coefficient.

$$\Delta t<\gamma\cdot\min\left\{\frac{\Delta x}{|u|_{max}},\frac{\Delta y}{|v|_{max}},\frac{\Delta z}{|w|_{max}}\right\} \quad (5)$$

From the results of preliminary computations, the computation domain is determined to be 60cm, 40cm and 20cm in x, y, and z directions, respectively. A model of the cubic structure was installed at the center of computation domain, as given in Fig. 2. The conditions of numerical computation are listed in Table 1. The computation is performed under the conditions of the initial time step $\Delta t_i=0.01$s, the weight of donor difference $\alpha=0.5$, and the coefficient of relaxation for time step $\gamma=0.5$.

Table 1 Conditions of computation.

Still water depth (cm)	EXPT95, h_s=50.0; EXPT96, h_s=40.0
Wave period (s)	1.0, 1.4, 1.8
Wave height (cm) EXPT95	10.0, 13.0, 15.0, 18.0, 20.0
Wave height (cm) EXPT96	5.0, 7.5, 8.0, 9.5, 10.5, 12.0, 12.5, 15.5, 16.0, 18.0, 19.0, 22.0
$(\Delta x, \Delta y, \Delta z)$ (cm)	1.0, 1.0, 1.0

HYDRAULIC EXPERIMENTS

Laboratory investigation of water particle velocity field, local scouring, and embedment of artificial fish reef were performed in an indoor wave tank (25m in length, 0.7m in width, and 0.95 m in depth). In the wave tank, a 20cm high horizontal step with a leading slope of 1/15 was prepared, and sand pool (1.8m x 0.7m x 0.2m) was set at the middle part of the horizontal step. The sand pool was filled with sand of d_{50}= 0.038cm. Figure 3 gives the schematic diagram of the wave tank.

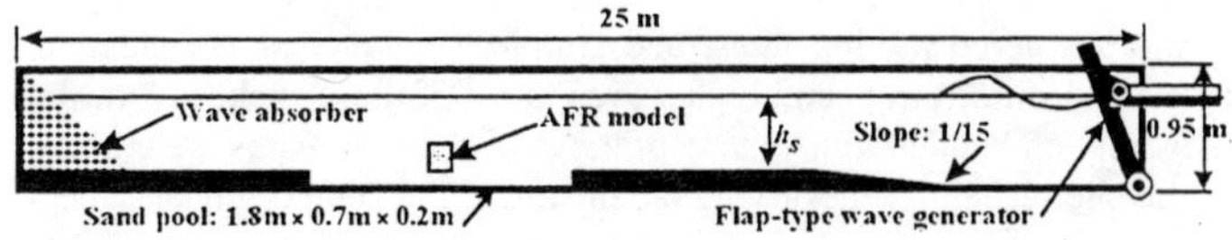

Fig. 3 Schematic diagram of wave tank.

Generally, actual artificial fish reefs were made of high density materials such as concrete and heavy steel. In this study, a cubical (8cm x 8cm x 8cm) model of artificial fish reef made of concrete was utilized. The prototype artificial fish reef are generally high permeable; however, high permeability makes the flow condition very complicated. Therefore, to investigate the fundamental mechanism of local scouring, an impermeable model was utilized. Two separate sets of experiment were conducted on both fixed and moving bed; the first set (EXPT95) was conducted with h_s = 50cm and the second set (EXPT96) was performed for h_s = 40cm, where h_s is the still water depth. Two methods of placement of cubical model were conducted in this study, as given in Fig. 4, and referred herein as CASE-I and CASE-II, respectively. In CASE-I, the wave propagation is perpendicular to the face of the cube model; whereas, in CASE-II, the face of the cubical model is rotated at an angle of 45° in the transverse direction referred from the direction of wave propagation. On movable bed experiment, regular waves were generated for one hour span, and the experimental conditions are given in Table 2. For each run, the maximum depth of local scour and displacement of the artificial fish reef were measured with a laser distancemeter.

The patterns of scour and deposition of sandy bottom were taken through 35mm-camera and video tape recorder. Also, the contact area of model with sandy bed after wave excitation was also measured. For fixed bottom experiment, the measurements of the water particle velocities and water surface profile were carried out at several points around the model after the steady state condition was attained. The electromagnetic velocimeter and electric capacitance wave gage were used for the velocity and water surface profile measurements, respectively. Also, the water particle movement around the structure was visualized by the hydrogen-bubble method and recorded in video camera.

Table 2 Experimental conditions for h_s=50cm and h_s=40cm.

T (s)	H (cm)						
1.0	5.0,	7.5,	10.5,	12.5,	15.5,	18.0,	21.0
1.4	5.0,	7.5,	9.5,	12.0,	15.5,	18.0,	22.0
1.8	5.0,	8.0,	9.5,	12.0,	16.0,	19.0,	22.0

WATER PARTICLE VELOCITY AROUND AFR

The numerical computations become stable and steady without any numerical vibration and divergence after three wave periods from wave generation; thus, this numerical scheme is applicable in the computation of velocity field around the bottom-seated structure. The numerical computation was verified to be applicable even in the case of low and high values of KC (=$U_{max}T/a$) and Re (=$U_{max}\ a/\nu$) numbers, where U_{max} is the maximum horizontal water particle velocity evaluated with the linear wave theory at z = -34.5cm in the absence of the structure, T is the wave period, and a (=8cm) is the representative length of the structure.

The vectors of the time-averaged velocity over one wave period around the structure is given in Fig. 5. It is confirmed from the results that relatively strong steady flows toward the wave propagation direction exist and the vortices formed in the onshore and offshore sides are asymmetrical. It is confirmed from the figures that the predominant outward flows are formed from the edges; moreover, the outward flows from the bottom are stronger than those in the upper layers. This phenomenon is merely responsible to the transport of the suspended sand away from the structure; the suspension of sand, however, is greatly attributed to the occurrence of vortices formed at the angled part of the structure. Figure 6 gives a typical example of velocity vectors of a fully developed vortex, where water particle velocity is at reversing instant, near the bottom at z=-39.5cm, the definition sketch of the phase angle is defined in Fig. 7.

Figure 8 shows the comparison of steady velocity fields of various KC and Re numbers. For small value of KC and Re numbers, the difference between the steady velocities near the bottom (z = -39.5cm) and at the upper layer (z=-34.5cm) is almost insignificant, as shown in Fig. 8(a). Figure 8(b), however, shows that a remarkable difference in flow pattern between the near bottom and at the upper layer for larger value of KC and Re numbers; and, it is confirmed that the steady velocity near the bottom is relatively stronger than those of the upper layers as the KC and Re number increases. In addition, as shown in the plan view of Fig. 5, very weak velocities are acting on the central part of the lateral faces of the structure. Similar condition, as given in Fig. 8, is also observed for the steady flows. The occurrence of very weak water particle velocity and inward flows result to local deposition. Consequently, it can be considered that scouring due to the water particle motion is not possible to occur at the central part of the four faces of the structure; however, deposition of suspended sand in this vicinity is more likely to occur.

Moreover, the time histories of computation and the third-order Stokes theory, without the structure, are compared and shown in Fig. 9. Based on the results, the time histories of the numerical computation and the Stokes theory are in good agreement, and this good agreement is recognized for all computational cases. Thus, it is verified that the numerical computation method reproduces well the water particle velocity field of Stokes third-order wave theory. The given characteristics of flow field are very important in the description of mechanisms of local scouring and embedment of artificial fish reef.

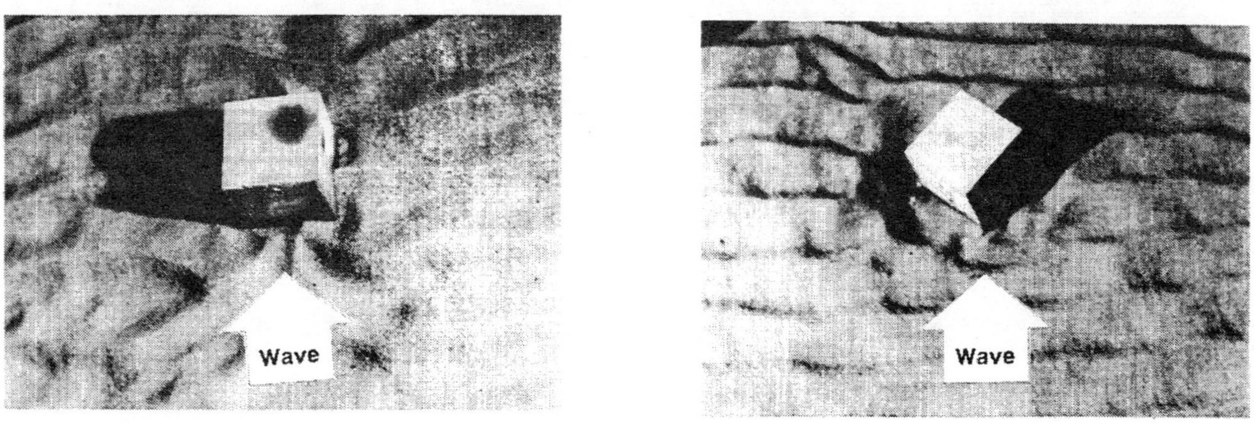

(a) CASE-I (b) CASE-I

Fig. 4 Methods of placement of the cubical model.

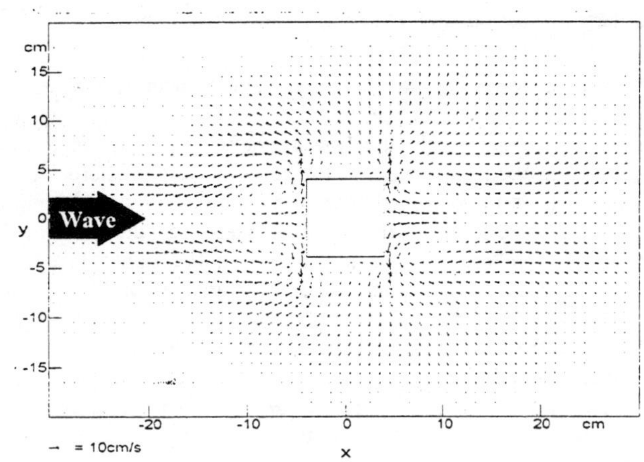

(i) z=-39.5cm

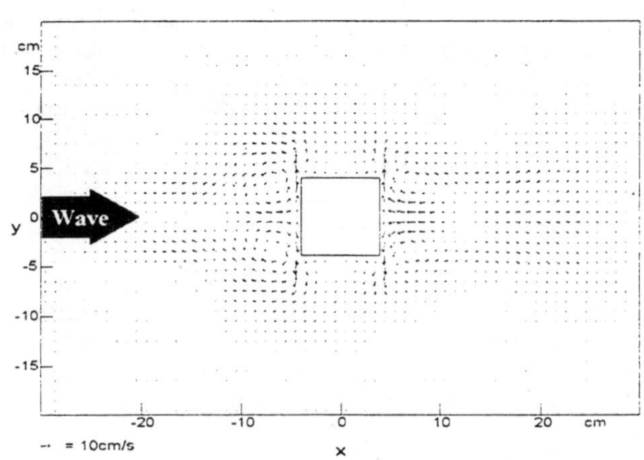

(ii) z=-38.5cm

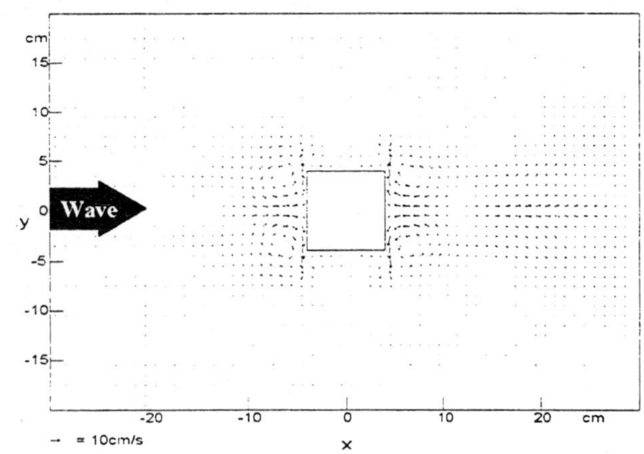

(iii) z=-37.5cm

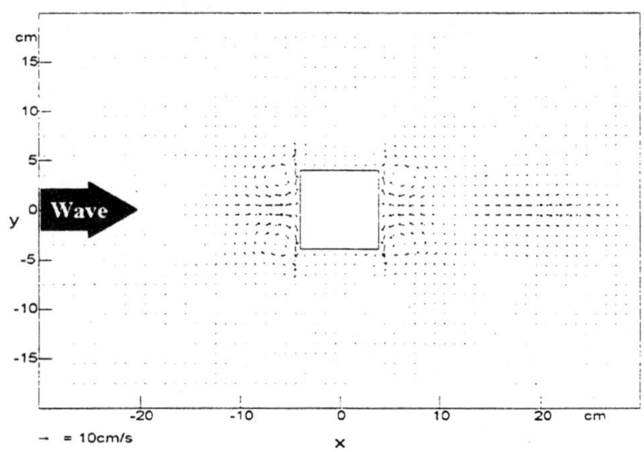

(iv) z=-36.5cm

(a) x-y plane

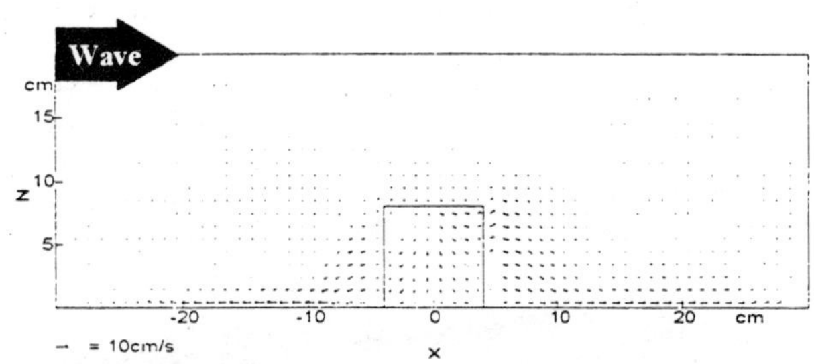

(i) y=-4.5cm

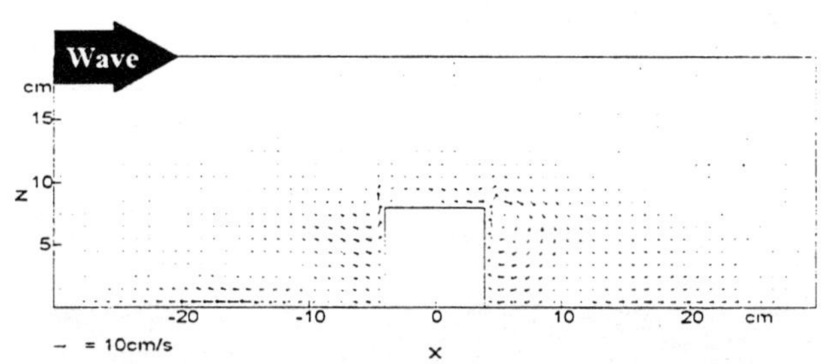

(ii) z=-2.5cm

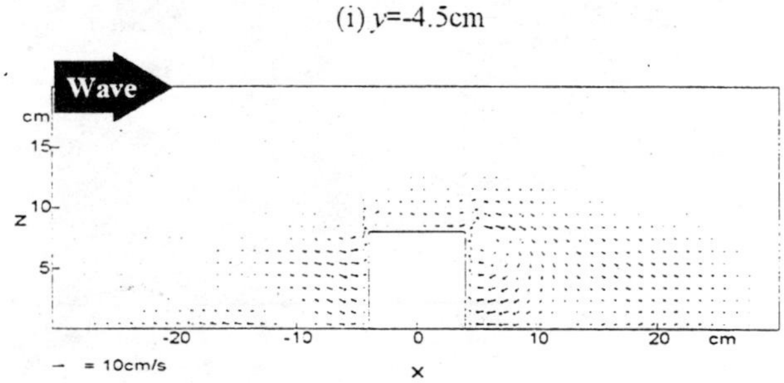

(iii) y=-1.5cm

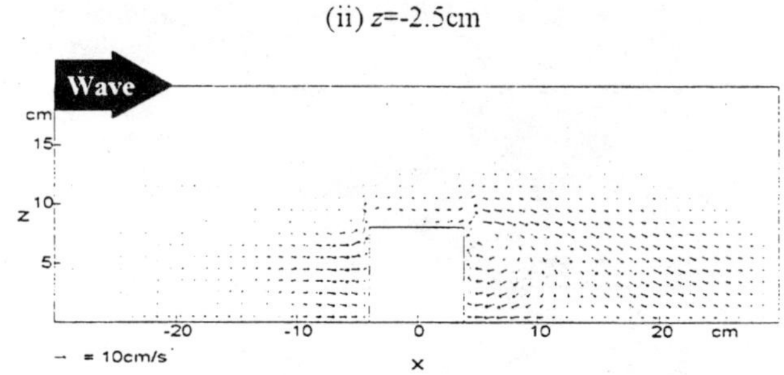

(iv) y=-0.5cm

(a) y-z plane

Fig. 5 Velocity vector of steady flow (T=1.4s, H=15.5cm, h_s=40cm)

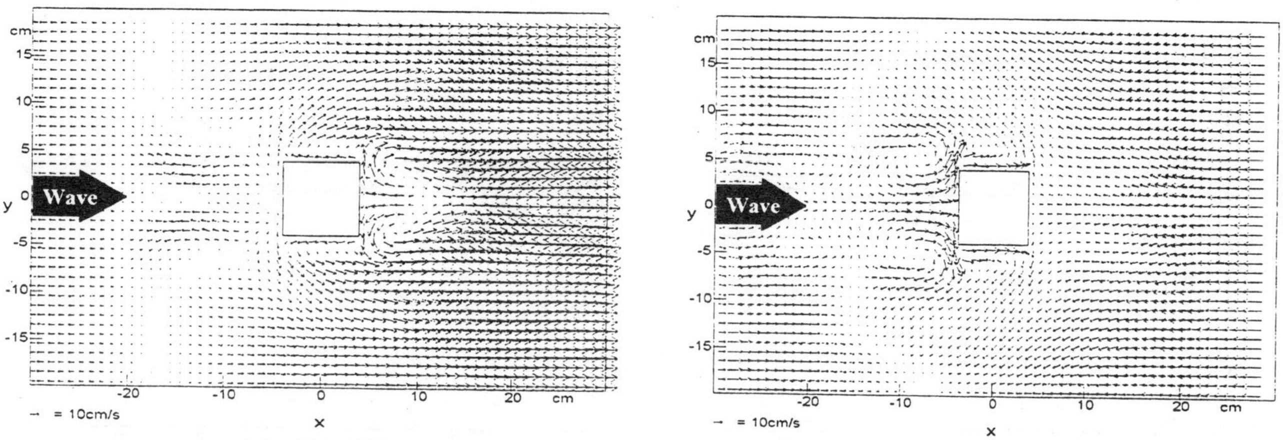

(a) Phase VI (b) Phase XII

Fig. 6 Velocity vectors near the bottom (T=1.4s, H=15.5cm, h_s=40cm)

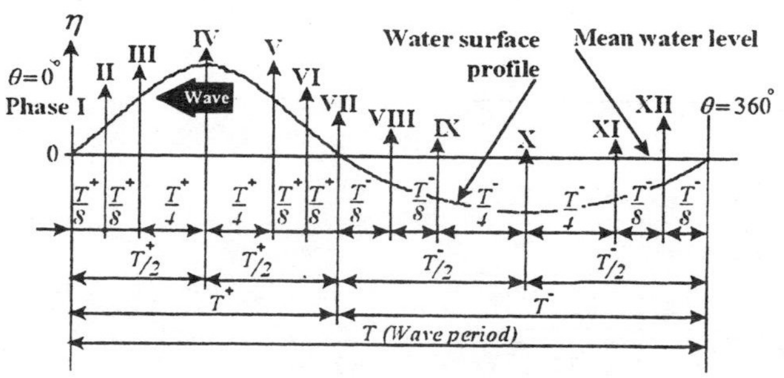

Fig. 7 Definition sketch of phase.

LOCAL SCOURING AND EMBEDMENT

The general patterns of local scouring are shown in Fig. 10, where the imaginary and solid lines indicate the initial and final positions of the structure, respectively. Scouring takes place at the angled part of the artificial fish reef model where vortex shedding occurs. It is clear that the scoured area are greater in the edge adjoining the lateral faces of the structure. Regardless of the value of h_s, the patterns of scouring varies significantly with the orientation of the model of artificial fish reef with the wave incidence. For a given wave condition, the scouring in CASE-II is larger than that of CASE-I, because the vertical edge adjoining the lateral surface of the model for CASE-II is directly fronting the wave incidence; as described in Section 4, predominant outward flows are easily formed in vertical edges of the structure (Kim et al., 1996). Based on these results, it can be concluded that the scouring around the artificial fish reef is generally influenced by the vortex formation and shedding around the artificial fish reef.

By flow visualization of the experiment and video analysis, the time variation of the maximum scour location for each corner of the structure is almost negligible and stays almost at the same location throughout the experimental run. The scour depth changes rapidly at the initial stage of the experiment; however, the rate of increment of scour depth decreases with time as it approaches the steady state condition. The relationship of KC number versus the maximum scour depth, h_s, normalized by characteristic length, a, is shown in Fig. 11. In general, the scour depth increases with increment of KC number; and, since the vortex formation is closely correlated with KC number, the depth of scour is considered to increase as the intensity of vortex becomes strong. The scour depth for CASE-II is relatively bigger than those of CASE-I.

The scour around the artificial fish reef makes it unstable and causes the embedment of the structure. It is observed in the experiment that the displacement of the structure becomes large as it moves toward the scoured area. Figure 12 shows the relationship between KC number and the dimensionless volume, V_e/V, where V_e is the embedded volume of the cubical structure and V is the volume of the structure. The embedded volume is defined as the volume of artificial fish reef beneath the initial bottom plane. Similar to the results of scour depth, the embedded volume increases with an increment of KC number. The orientation of the model structure is also significant in describing the relationship between the embedded volume and KC number, the results of CASE-II is relatively larger than those of CASE-I; this phenomenon is attributed to the stronger vortices acting on the edged-surface of CASE-II. Consequently, regardless of the value of h_s, tilting of the structure is generally observed for CASE-II; however, sliding movement of structure for CASE-I is observed for larger values of wave height.

Another physical factor to be considered in the stability of artificial fish reef is the contact area between the structure and the sandy bed. It was observed that the contact area of the base of the structure and sandy bed decreases with time. Thus, as the resistant force acting on the base decreases because of the reduced contact area, instability on the structure is most likely to occur. The gap between the base of the structure and the sandy bed is created by the enlargement of local scour at the initial stage; then, flow with strong vortex is further developed at the scoured part. This flow caused sand movements which produce clearance between corners of the artificial fish reef and the bottom. The reduced contact area is one of the main reasons of the tilting movement of the structure as typically described by CASE-II. Results show that the contact area of CASE-II after one-hour wave generation is always bigger than those of CASE-I (Kim et al., 1995).

Moreover, the results obtained in this study is comparable to the results of Kim et al. (1995) for different shapes of artificial fish reefs; thus, the fundamental knowledge and significant characteristics obtained from cubical artificial fish reef can be correlated to other shapes of artificial fish reefs.

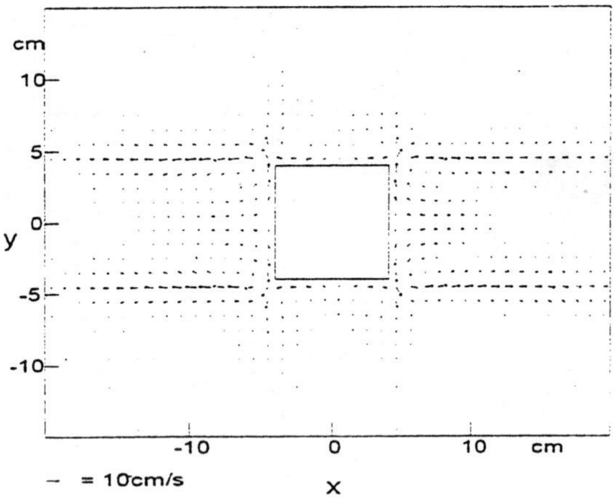

(i) z=-39.5cm

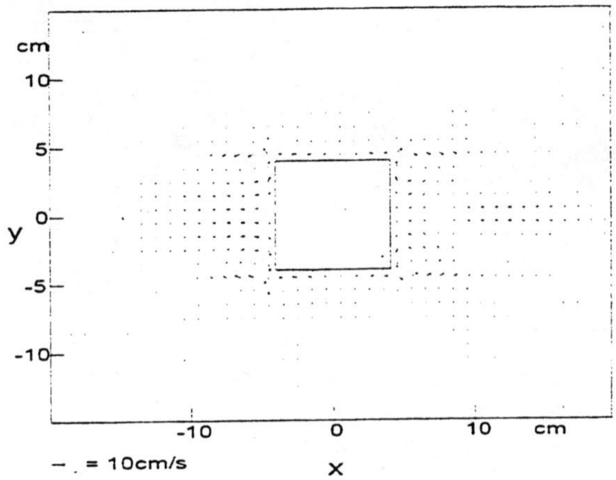

(ii) z=-34.5cm

(a) T=1.0s, H=10.5cm, KC=1.647, Re=10400

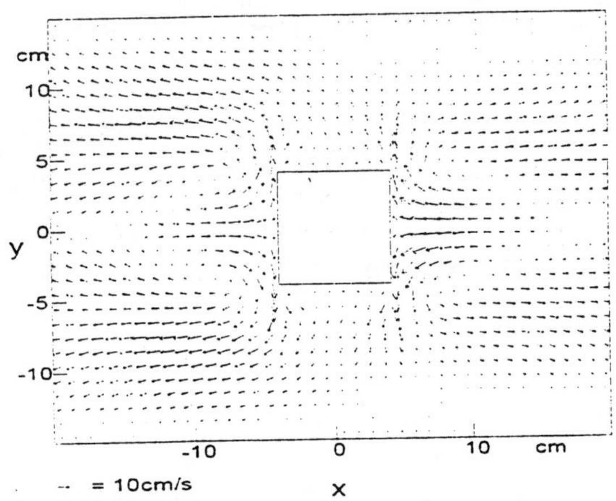

(i) z=-39.5cm

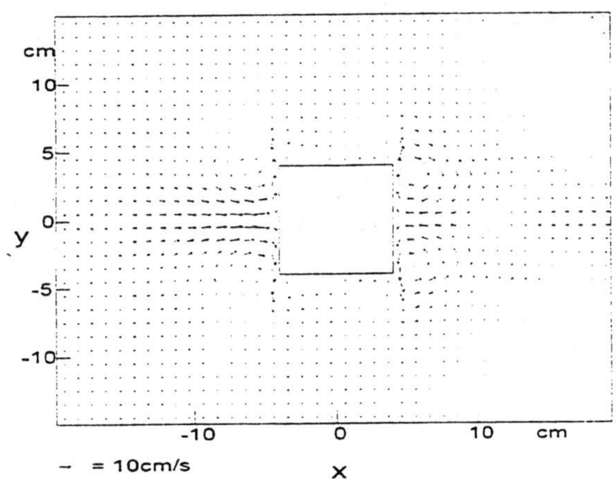

(ii) z=-34.5cm

(b) T=1.8s, H=16.0cm, KC=8.164, Re=28700

Fig. 8 Comparison of steady vectors and KC and Re numbers (h_s=40cm)

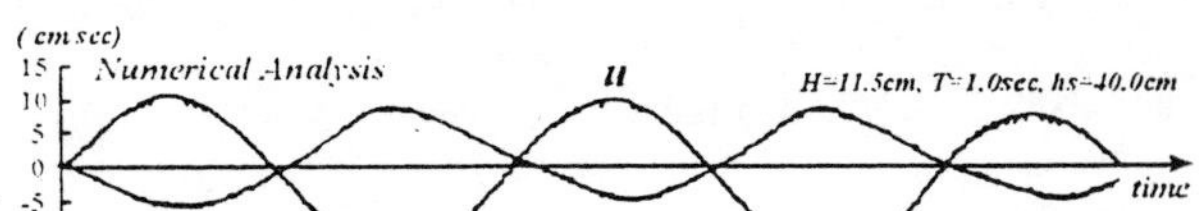

(a) Numerical computation

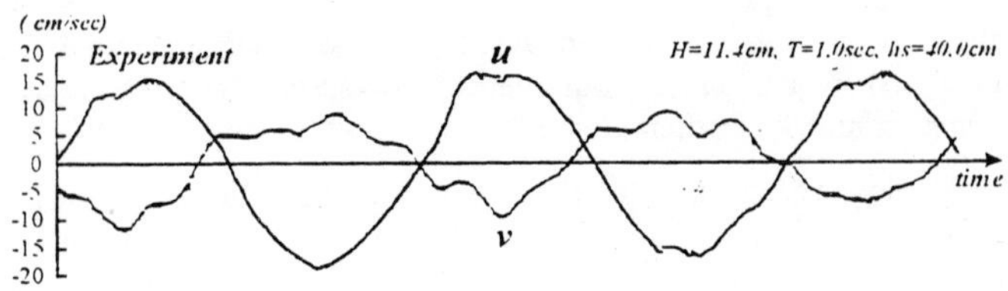

(b) Hydraulic experiment

Fig. 9 Comparison of water particle velocities near the cube; (x=-3.5cm, y=-6.5cm and z=-39.5)

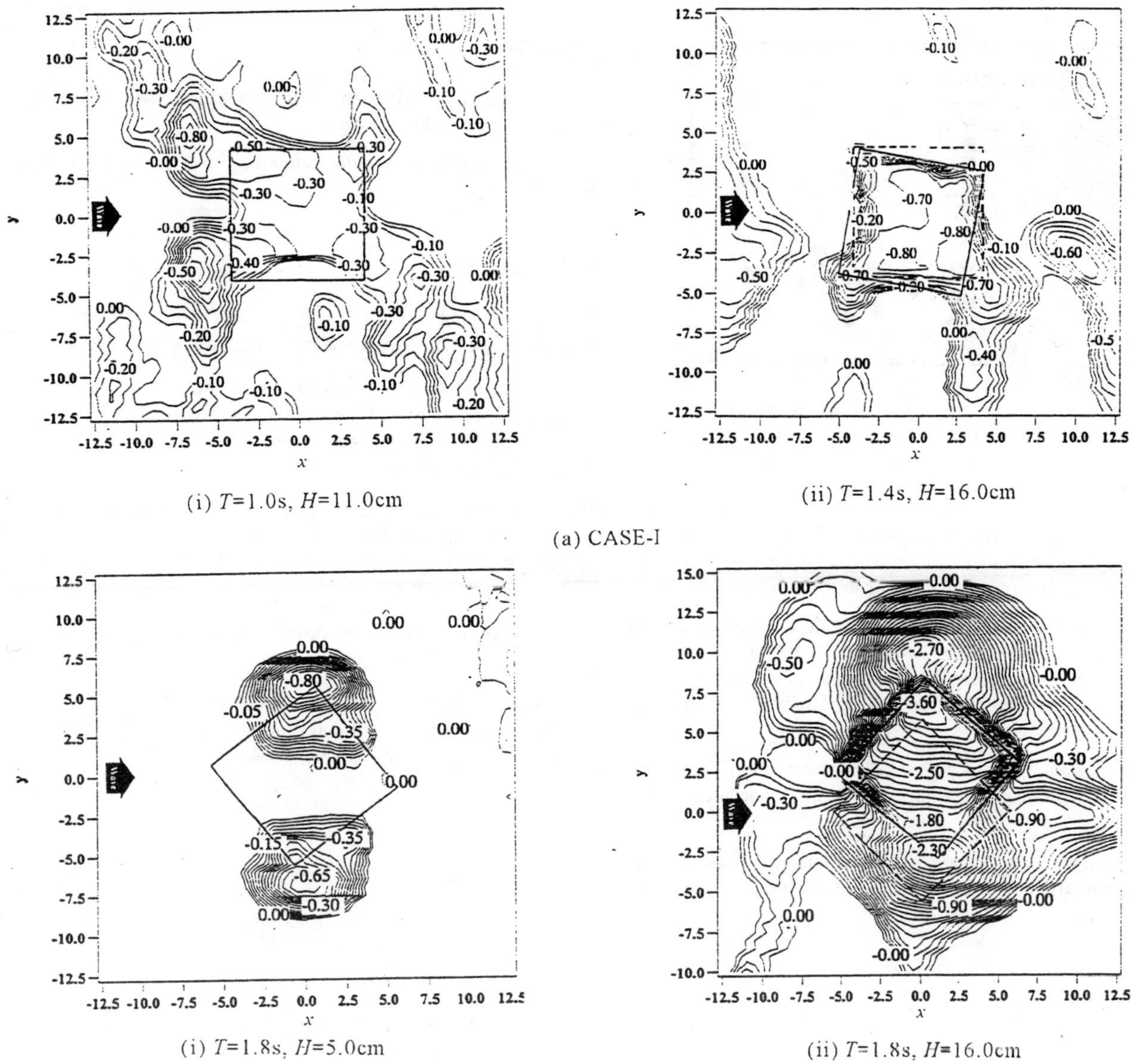

(i) T=1.0s, H=11.0cm

(ii) T=1.4s, H=16.0cm

(a) CASE-I

(i) T=1.8s, H=5.0cm

(ii) T=1.8s, H=16.0cm

(b) CASE-II

Fig. 10 Patterns of local scouring

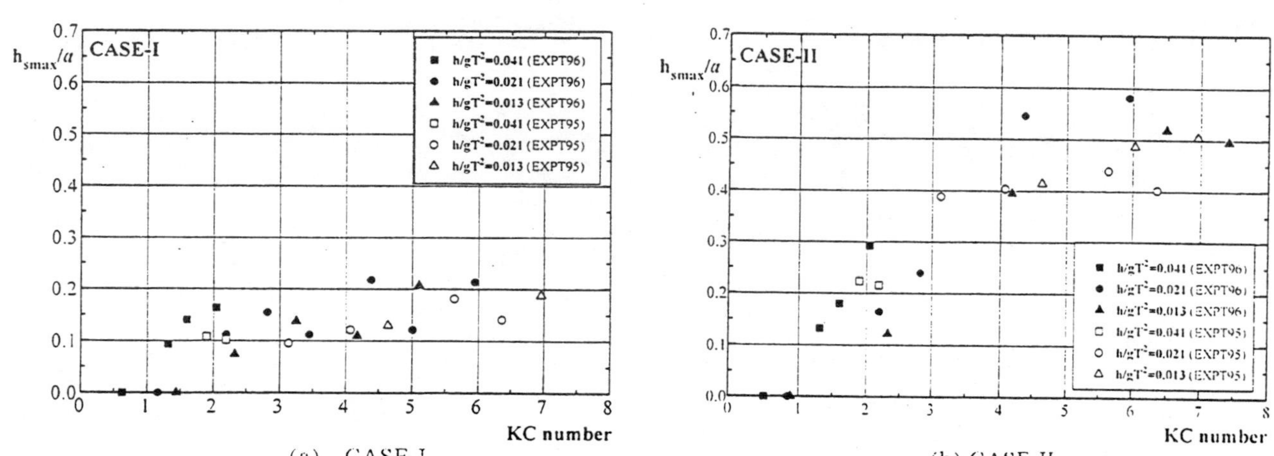

(a) CASE-I

(b) CASE-II

Fig. 11 Relationship of maximum scour, h_{smax}, and KC number.

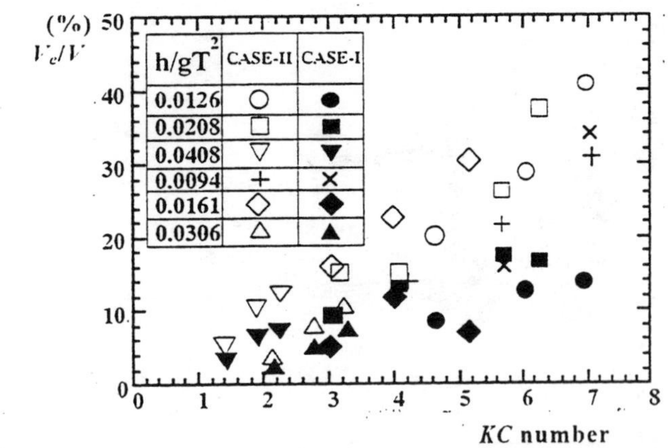

Fig. 12 Relationship of dimensionless volume and *KC*

CONCLUSIONS

This paper discusses the mechanism of local scouring and embedment of artificial fish reef under wave action for a shallow water depth. Significant results are summarized as follows:

(1) A numerical scheme for three-dimensional flow field computation including the vortex formations is developed. The flow patterns and vortex formations are well evaluated by the numerical computation; thus, the characteristics of the computed flow field explain well the generation mechanisms of local scouring.

(2) The vortex formation and the steady outward flows from the structure are very significant in local scouring and embedment of the structure.

(3) The fundamental knowledge and significant characteristics obtained for simple cubical structure is comparable to other shapes of artificial fish reefs.

ACKNOWLEDGEMENT

The great contribution of Mr. Natsuki JINNO in the experiment proper is deeply acknowledged.

This research is fully supported by the Kajima Foundation Research Grant.

REFERENCES

Hirt, C.W., Nichols, B.D., and Romero, N.C. (1975). "'SOLA' A Numerical Solution Algorithm for Transient Fluid Flows," *LA-5852*.

Kim, J.Q., Mizutani, N., Jinno, N., Furukawa, K., and Iwata, K. (1996). "Experimental Study on Characteristics of Local Scour and Embedment of Fish Reef by Wave Action," *Proc., Civil Eng. in the Ocean*, Vol. 11, pp. 243-247 (in Japanese).

Kim, J.Q., Mizutani, N., and Iwata, K. (1995). "Experimental Study on the Local Scour and Embedment of Fish Reef by Wave Action in Shallow Water Depth," *Proc., ECOSET'95, JIMSTF*, Vol. 1, pp. 168-173.

Publications of Aomori Prefecture Aquaculture Center (1980), Vol. 54(5), pp. 170-217 (in Japanese).

Sawaragi, T., Matsumi, Y., and Kitaguchi, M. (1984). "Local Scour Around Artificial Fish Reef and Countermeasure," *Proc., 31st Japanese Coastal Eng.*, pp. 345-349(in Japanese).

Proceedings of the Seventh (1997) International Offshore and Polar Engineering Conference
Honolulu, USA, May 25-30, 1997

ISBN 1-880653-28-1 (Set); ISBN 1-880653-31-1 (Vol. III)

Nonlinear Wave Effects on Wave Measurements at Different Depths

Igor Prislin
Deep Oil Technology, Inc.
Chula Vista, CA, USA

Jun Zhang
Texas A&M University
College Station, TX, USA

ABSTRACT

Point-measurements of wave induced dynamic pressures at certain depths from the mean water level are usually made to deduce surface wave elevation spectra and other wave properties. However, due to the exponential decay of wave induced pressure with the water depth, the ratios of signal to noise are low and the corresponding pressure head spectra have to be cut at certain period. Placing the instruments closer to the free surface, better signal resolution can be obtained, but nonlinear and directional wave effects become more pronounced. A new method for analyzing wave induced pressure is presented. Its efficiency is verified through model tests and comparisons with the linear and conventional wave theory. The results based on pressure measurements and the DHWM agree very well with predicted and corresponding measured wave elevation. It indicates that it is possible to improve interpretation of pressure measurements by deploying pressure transducers in relatively short distance from the free surface.

KEY WORDS: Directional waves, deterministic wave decomposition, short-crested waves, irregular nonlinear waves.

INTRODUCTION

Practices in wave measurements usually deploy pressure transducers in relatively large depths for reasons ranging from the protection of instruments to the validity of linear wave theory. However, due to the exponential decay of wave induced pressure with the water depth, the ratios of signal to noise are low. Thus, the corresponding pressure head spectra have to be cut at certain period as large as 5 to 7 second period. Hence, the information of waves of periods below 7 or 5 seconds is lost although important to the design of offshore and coastal structures. Deploying measuring instruments in depths relatively close to the mean water level can provide better signal to noise ratios of waves with relatively high frequencies but still crucial for engineering applications. However, the wave directionality and nonlinear wave-wave interactions become significant in the measurements. Linear wave theory may lead to inaccurate transformation of pressure head to elevation and nonlinear decomposition of measured pressure at the fixed point is necessary in the computation of surface elevation and other wave properties. The proposed study uses recently developed Directional Hybrid Wave Model (DHWM) [Prislin, 1996; Prislin *et al.*, *1996*] to decompose wave fields measured by pressure transducers and to predict wave elevation nearby the measurements. Very good agreement in the comparison of predicted and corresponding measured wave elevation indicates that it is possible to improve pressure measurements by deploying pressure transducers in relatively short distance from the free surface.

METHODOLOGY

The methodology for deterministic decomposition of irregular short crested surface waves is based on an iterative procedure of three steps for directional, phase, and nonlinear decomposition [Prislin *et al.*, 1996]. The novelty of the present method is in nonlinear wave modeling. The hybrid wave model incorporates the phase modulation approach and the conventional perturbation approach for wave-wave interactions [Zhang *et al.*, 1996]. The phase modulation approach for long-crested waves has been studied by Zhang *et al.* [1990, 1993, 1996]. The phase modulation solutions are extended by Hong [1993] and Prislin [1996] for short-crested waves.

In using the hybrid wave model, the free-wave components are grouped according to their frequency closeness and difference. The model accommodates conventional and phase modulated nonlinear wave methods. For the computation purpose, the wave frequency range of interest is divided into two low and two high frequency bands L_1, L_2 and S_1, S_2, respectively (see Figure 1). When two free-wave components are located at the same frequency band or two neighboring bands, their interactions are computed according to the conventional nonlinear wave theory. When the free-wave components belong to the short-wave bands S_1 or S_2, the short-wave components are modulated by the long-wave components from the bands L_1 and L_2. However, the condition for using the phase modulation method is that two interacting wave components are located at two bands that are separated by at least one other frequency band. For example, a wave component at S_2 is modulated by a wave component from L_1 and L_2, but a wave at S_1 is modulated only by a wave component from L_1.

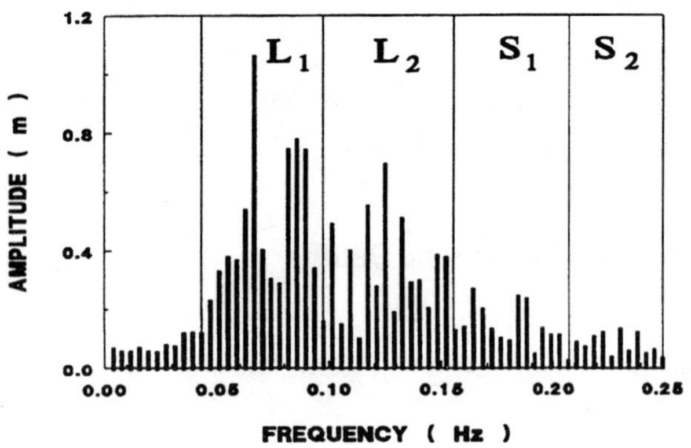

Figure 1 Frequency bands in the Hybrid Wave Model.
L_1, L_2 - long wave bands; S_1, S_2 - short-wave bands.

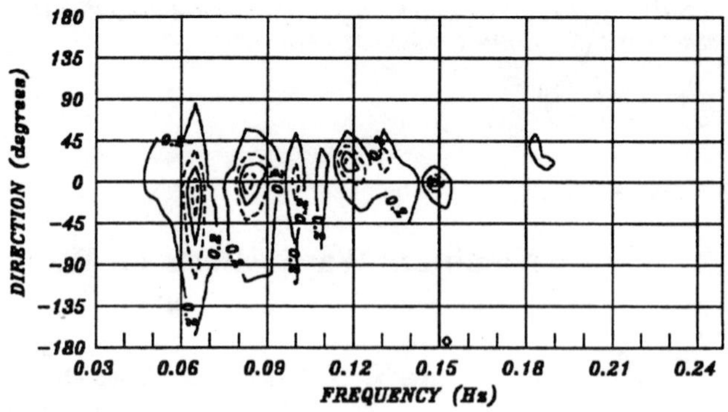

Figure 2 Directional wave spectra configuration.

In the iterative process, the conventional nonlinear interactions among free-wave components (from L_1 and L_2) are computed first and subtracted from the measurements. Then, the directions of propagation and initial phases of wave components are recomputed. Using improved results for the free-wave amplitudes, directions, and phases, the nonlinear effects due to the phase modulation of free-wave components in band S_1 are computed and subtracted from the previous modified measurements. Considering the newly modified measurements, the computation of directions and phases is repeated. Then, the conventional nonlinear wave interactions for wave components in L_2 and S_1 are computed and subtracted from the most recent modified measurements. Similarly, the iterative process is repeated for the wave components in S_2 frequency band. At the end of each iterative step, the maximum difference between two successive corresponding modifications is examined. Unless it is smaller than a prescribed tolerance, the iteration is repeated for the previous frequency band. A small difference between two modified values indicates that two successive nonlinear interactions between the same free-wave components are almost identical. Because the wave interactions are defined in a complex plane, the prescribed tolerance is applied to both real and imaginary parts of wave components. This guarantees that not only amplitudes, but also phases, are close to each other in two successive iterations.

EXPERIMENT

Wave Parameters. Laboratory generated waves are designed to simulate real ocean waves by combining waves of different frequencies and different directions. The simulated directional spectrum is shown in Figure 2. The directional spreading is relatively small in comparison to certain ocean waves, when wind waves and swell can propagate in quite different directions. The reason for limiting directional spreading in laboratory wave generation is to avoid strong reflections from the side walls of the basin. Although the reflections are always present, it seems that the contamination of the designed wave field by the reflected waves is not serious.

In addition to relatively narrow directional wave energy spreading, the wave reflection is avoided by two measures. First, the wave trains are run in a relatively short duration (120 seconds in the model scale) and a duration of time series of 51.2 seconds in the model scale is selected for the analyses. Second, data acquisition is delayed from the start of the wavemaker until the expected shortest waves with a wavelength around 0.7 m (model scale) reach the sensors located at about 17 m from the wavemaker. The delay is 30 to 40 seconds.

Array of Sensors Configuration. Two arrays of four wave gauges and three pressure sensors are used in the laboratory measurements. The pressure sensors are mounted directly on a square aluminum frame immersed about 1 meter below the free surface. One of the frame's diagonals is oriented perpendicular to the wavemaker (see Figures 3 and 4). The wave gauges are positioned right above the pressure sensors. The capital letters P and W, attached with numbers denote pressure sensors and wave gauges, respectively. The frame was made to be slender but rigid, spacious enough to carry all sensors, and simple for handling. The rigidity of the frame is necessary to prevent relative displacements of the sensors under wave loads. Such displacements could easily disturb the measured signal by influencing the relative phases that are crucial for determining wave field decomposition. The slenderness is important to avoid vortex shedding, which might influence the measured pressure signals. Aluminum, as an antimagnetic material, is selected to avoid electro-magnetic interference with the sensors. The frame is supported by the underwater towers laid at the bottom of the basin. The depth of the frame is adjustable. The square shape of the array configuration is chosen to make redundant measurements between pairs of sensors with a similar orientation. The redundancy is necessary for checking the accuracy of measured signals.

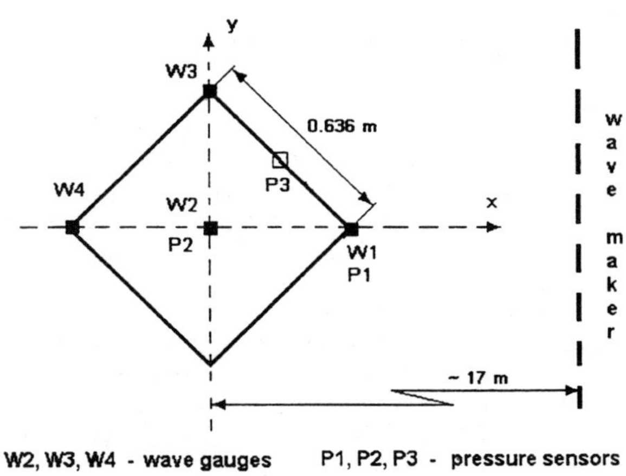

Figure 3 Array of sensors in the OTRC wave basin.

The OTRC Wave Model Basin. The OTRC wave basin is 45.7 m long, 30.5 m wide, and 5.8 m in water depth. At the center of the basin there is a deep pit with a length of 9.1 m, a width of 4.6 m, and an adjustable depth from 5.8 to 16.7 m. The pit was covered during our tests so that the depth of the water was constant over the basin. At the North side of the basin, there is a multi-directional 'snake-type' wave generator, which consists of 48 hinged flaps (wave boards) in a row, driven independently by servo-controlled linear hydraulic actuators. Each flap is 0.6 m wide and 9.84 m high. The wave maker is capable of generating multi-directional irregular waves, which is achieved through individually programmed drive signals for each wave board. The wave maker is capable of generating oblique waves of a maximum angle of approximately 60 degrees with respect to the longitudinal centerline of the basin. However, to avoid reflection of the oblique waves from the side walls of the basin, the directional wave angle was significantly reduced. The wave absorption in the basin was provided only at the opposite side of the wavemaker by progressive perforated metal panels (see Figure 4).

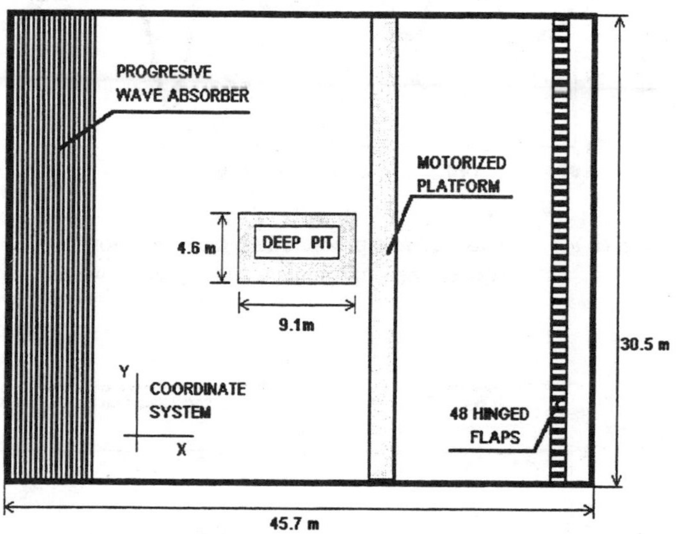

Figure 4 The OTRC multidirectional wave basin.

Wave Gauges. Capacitance-type wave gauges were used to measure surface elevations. The wave gauges were calibrated daily before the tests by the static immersion procedure. The calibrations showed that the curves were virtually linear within the calibration range with a standard deviation of 0.6 mm (0.1% of the calibration range).

Pressure Sensors. The SENSOTEC Model GW-100 Series Pressure Transducers of the range 0 to 1 psi were used. The accuracy of transducers expressed as standard deviation and specified by the manufacturer is 0.2% over the nominal pressure range with continuous resolution. The sensors were calibrated daily before the tests. The calibration was accomplished statically by immersing all sensors simultaneously to a specified depth of up to 0.55 m below the calm water level. Each reading was related to the hydrostatic pressure at a particular depth. Fresh water density, used for hydrostatic pressure computation, was determined based on the measured water temperature. The calibration results show that the curves were virtually linear in the range of interest. The average standard deviation was 0.7 mm in terms of the pressure head (0.2% of the calibration range).

RESULTS

The decomposition of a measured short-crested irregular wave fields are carried out using two different sets of measurements as input: the surface elevation at the wave gauges W1, W2, and W3, and dynamic pressure at sensors P1, P2, and P3, respectively. All results from now on are presented in a prototype scale. The length ratio of model to prototype is 1:25. The Froude number dynamic similarity is used for scaling time and pressure from the model to the prototype. The significant wave height, estimated from the frequency spectrum, is 7.3 m and nominal wave steepness, defined as the product of one half of the significant wave height and the wavenumber at the spectrum peak, is about 0.065. The duration of the analyzed records is 256 seconds and the sampling interval is 0.25 seconds. The single summation model, that is only one free-wave component per frequency is used for wave decomposition because the drive signals for the wave maker were created by the same principle.

Pressure Conversion. The time-series were filtered through a low-pass filter of cutoff frequency 0.25 Hz in order to avoid unrealistic augmentation of relatively small wave amplitudes at high frequencies when they are estimated from pressure measurements. In the case of low signal to noise ratio not only the measured pressure but also noise are exponentially amplified. Thus, conversion of the noise mixed with the pressure signal would show an exponential growth of wave amplitudes with the increase in frequency. Apparently, this exponential growth does not really represent wave amplitudes. In the pressure measurement of irregular waves, however, it is extremely difficult if not possible to remove the noise from the data because they are uncorrelated and the noise is randomly distributed over the whole wave-frequency range. To show the effect of noise on the pressure to surface elevation conversion, the average noise amplitude (defined here as the standard deviation of calibrated values) obtained from the pressure sensor calibration are converted to the wave surface amplitude and compared to the wave amplitude converted based on the measured pressure at two different depths (see Figure 5).

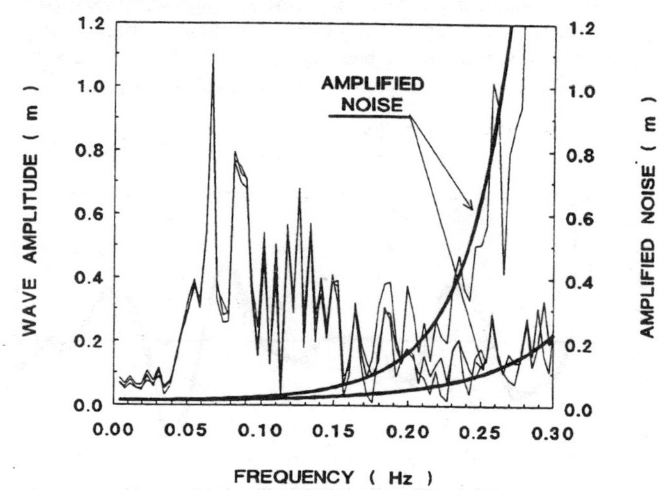

Figure 5 Wave amplitude spectra estimated from pressure measurements.

As a reference, the frequency spectra from surface elevation measurements are presented in Figure 6. It is interesting that there is a correlation between converted average noise and converted pressure for frequencies higher than 0.20 Hz. This demonstrates that the pressure signals above certain frequency are comparable or insignificant in comparison with the estimated average noise level. In addition, similar fluctuations of converted amplitudes based on two different pressure sensors at two different depths indicate that some wave effects due to the nonlinear wave-wave interactions on the pressure might be present at high frequency, the higher-order pressure components are, however, of at most the same or of a smaller order of magnitude than the recorded noise. If that was not the case, the curves of converted pressure and noise would differ notable. To conclude, if pressure measurements are used for decomposition of the ocean waves, a high pressure signal to noise ratio is necessary to get accurate results.

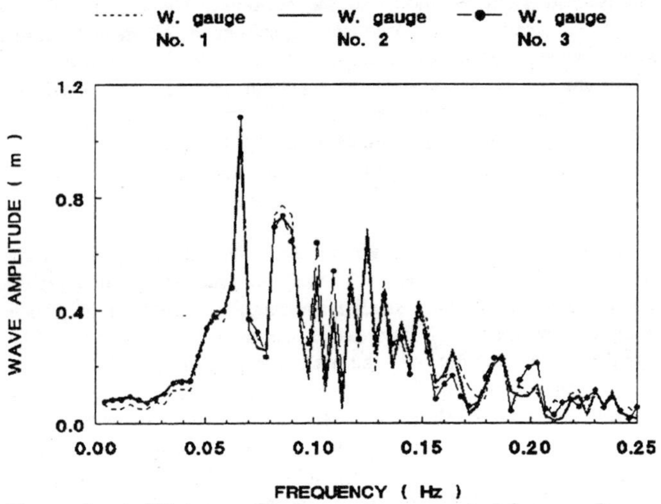

Figure 6 Wave amplitude spectra computed from surface elevation measurements.

Surface Elevation Predictions. The proposed methodology was tested for the prediction of surface elevations at location other than those of measurements used in the decomposition. Surface elevation at W4 were predicted, noticing that W4 were at the location different from the locations of the array of wave gauges W1 to W3 and pressure sensors P1 to P3 whose measurements were used in the decomposition (see Figure 3). The prediction of surface elevation W4 from surface elevation measurements W1 to W3 is presented in Figure 7. As expected, the agreement between the prediction and measurement is excellent.

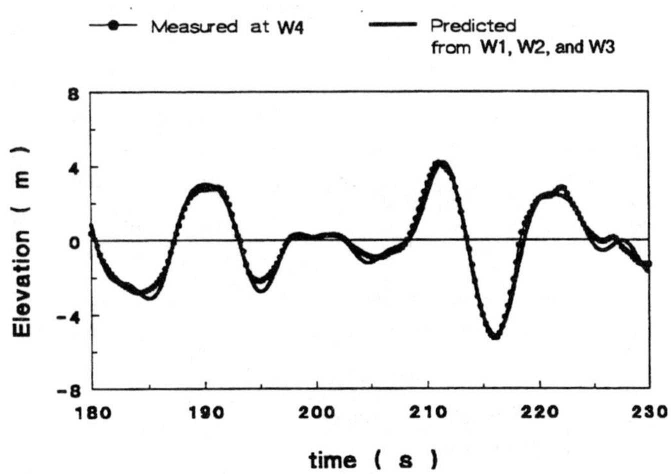

Figure 7 Measured and predicted surface elevation at W4 from W1, W2, and W3.

The interesting feature of the proposed methodology is the prediction of surface elevation from pressure measurements. Based on the decomposition of the wave field using the pressure measurements at P1, P2, and P3 to decompose the wave field the surface elevation at the location of wave gauge W4 were predicted accordingly to linear, conventional, and hybrid nonlinear wave models (see Figures 8,9,and 10, respectively). As expected, nonlinear effects are important in the decomposition of relatively steep, short-crested irregular waves, such as in the OTRC model tests. In addition to the pressure measurements, the velocity measurements in waves can be treated similarly.

The comparisons in Figures 8 to 10 illustrate the efficiency of the directional hybrid wave model (DHWM) versus linear and conventional second-order wave models. The prediction based on the directional hybrid wave model (DHWM) provides the best match with the measurement, especially at the trough of time between 210-220 seconds (see Figure 10).

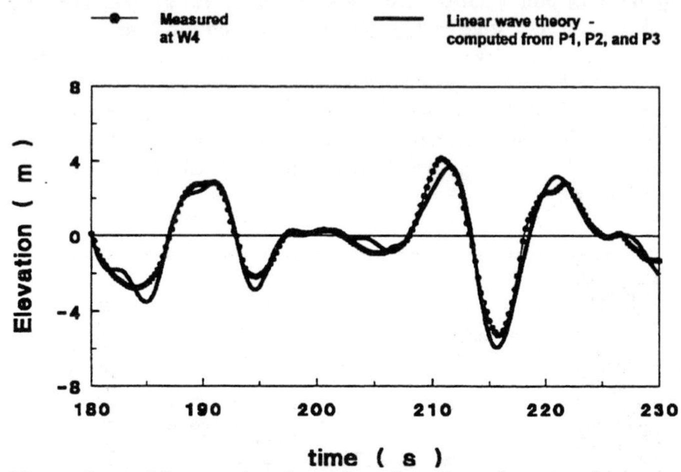

Figure 8 Measured and predicted surface elevation from pressure measurements using LINEAR wave theory.

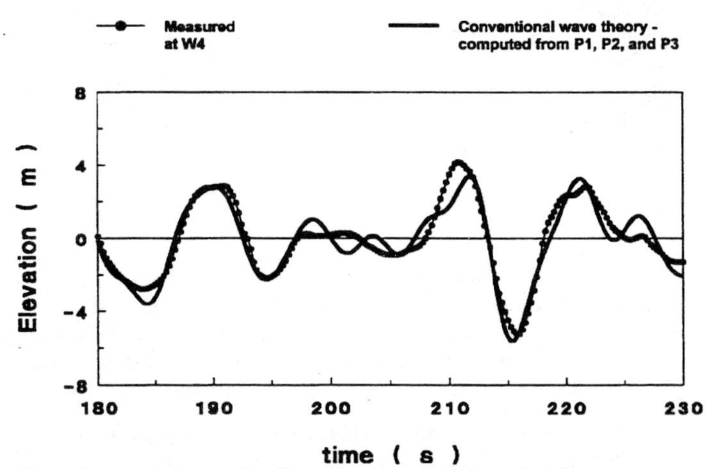

Figure 9 Measured and predicted surface elevation from pressure measurements using CONVENTIONAL wave theory

Prediction Improvement. The improvement in prediction due to the use of the modulation phase method is not accidental. Whereas the pressure sensors P1 and P2 were located at 8.25 meters below the calm water level, they were around 12 and 4 meters below the large wave crest and trough, respectively (in a duration between 210-220 seconds). Thus, the pressure induced by short-wave components riding on long-wave components was recorded differently under the crest or trough of steep long-wave components in comparison to the same short-wave components when it was riding on a calm water level. Linear wave theory totally ignores the interaction between long- and short-wave components and provides fairly accurate predictions. The conventional perturbation method includes nonlinear effects due to the wave-wave interaction up to second order and certainly gives much better predictions than linear wave theory at wave troughs but similarly less accurate results above the calm water level. The directional hybrid wave model (DHWM), [*Prislin*, 1996] however, provides the most accurate prediction. Differences between DHWM and conventional perturbation method are in favor of DHWM. That is because the conventional perturbation method misses certain higher order terms due to the phase modulation of short waves by long waves, which might be significant when the wavelength ratio of the short to long wave $\varepsilon_3 = L_2 / L_2$ becomes comparable to the long-wave steepness $\varepsilon_1 = k_1 L_1$ [*Zhang et al.*, 1993]. The phase modulated method may result in a fast convergent solution regardless of the ratio of long-wave steepness ε_1 to wavelength ratio ε_3 if $\varepsilon_3 \leq \frac{1}{2}$ [*Hong*, 1993]. The condition $\varepsilon_3 \leq \frac{1}{2}$ which is always satisfied because of the band division in our numerical scheme. In the OTRC test data the nominal wave steepness $\frac{1}{2} h_{1/3} k_o$ was 0.065, where $h_{1/3}$ is significant wave height of 7.3 m and k_o is the wave number at the spectrum peak of 0.0178 m^{-1}. In this case the convergence difficulties of the conventional method might be expected for the interactions of free-wave components of higher frequencies, say above 0.2 Hz (short wave band) with the free-wave components of the long-band L_1.

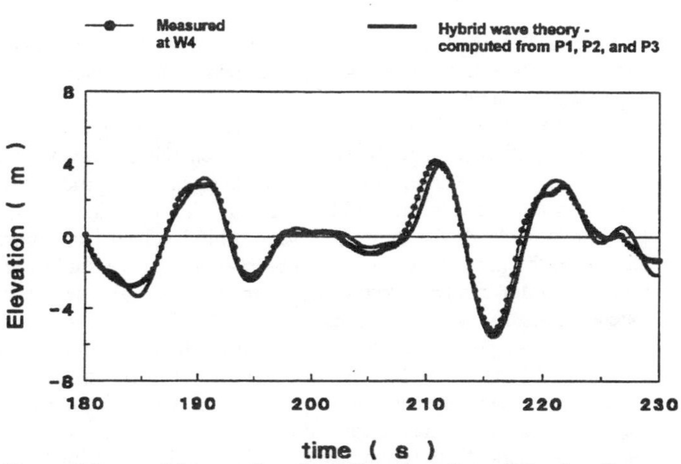

Figure 10 Measured and predicted surface elevation from pressure measurements using HDWM.

CONCLUSION

The DHWM gives better accuracy for wave decomposition than the linear and conventional up to second order wave models. Thus, it can be used not only for a routine decomposition of wave signals, but also to validate measurements other than those used in wave decomposition. The ability of DHWM to accurately decompose the nonlinear interactions among short waves riding on long waves, enables placing pressure sensors (and velocitymeters) closer to the free surface than usual, to better measure wave properties.

ACKNOWLEDGMENTS. This research was supported by the Offshore Technology Research Center, which is sponsored in part by National Science Foundation Engineering Research Centers Program Grant Number CDR-8721512 and the Joint Industry Program by Chevron, Exxon, Mobil, Shell, and Texaco. Without this support the research would not be possible.

REFERENCES

Forristall, G. Z., E. G. Ward, V. J. Cardone, and L. E. Borgmann, The directional spectra and kinematics of surface gravity waves in tropical storm Delia, *J. Phys. Ocean. 8*, 888-909, 1978.

Hong, K., Nonlinear interactions between directional dual component waves in deep water. A Dissertation, 143 pp., Texas A & M University, December 1993.

Longuet-Higgins, M. S., R. W. Stewart, Changes in the form of short gravity waves on long waves and tidal currents, *J. Fluid Mech., 8*, 565-583, 1960.

Longuet-Higgins, M. S., The propagation of short surface gravity waves on longer gravity waves, *J. Fluid Mech., 177*, 293-306, 1987.

Prislin, I., J. Zhang, and P. Johnson, Deterministic decomposition of Irregular Short-Crested Surface Gravity Waves, The Proceedings of the Sixth ISOPE Conference, 3, 1996.

Prislin, I., Nonlinear deterministic decomposition of short-crested ocean waves. A Dissertation, 162 pp., Texas A & M University, December 1996.

Zhang, J., L. Chen, M. Ye, and R. E. Randall, Hybrid wave model for unidirectional irregular waves, Part I, Theory and numerical scheme, *J. Applied Ocean Research (in press)*, 1996.

Zhang, J., K. Hong, and D. K. P. Yue, Effects of wavelength ratio on wave modeling, *J. Fluid Mech., 248*, 107-127, 1993.

Zhang, J. and W. K. Melville, Evolution of weakly nonlinear short-waves riding on long gravity waves. *J. Fluid Mech., 214*, 321-346, 1990.

Proceedings of the Seventh (1997) International Offshore and Polar Engineering Conference
Honolulu, USA, May 25-30, 1997

ISBN 1-880653-28-1 (Set); ISBN 1-880653-31-1 (Vol. III)

Experimental Study on the Estimation Methods of Wave Orbital Velocity

Leonardo Damiani and Michele Mossa
Polytechnic of Bari
Bari, Italy

ABSTRACT

The kinematics of a regular wave field has been investigated in order to determine the best mathematical representation. The paper describes the comparison between many theoretical models reported in literature and the experimental results obtained in a bidimensional channel at the laboratory of the Department of Water Engineering of the Polytechnic of Bari. First of all the classic wave theories have been discussed in order to evaluate the horizontal and vertical velocities. Then, the theoretical models (linear and non-linear), which link wave elevations and velocity components, have been tested by comparing the numerical results with the laboratory LDA measurements carried out in a channel with a sloped bottom. In the present paper the models proposed by Chakrabarti and by Koyama and Iwata have been applied, giving a relationship between time series of velocity components and wave elevations. Furthermore, a linear transfer function between the elevation amplitude spectra and the orbital velocity component ones has been investigated. The waves generated in the channel are regular (i.e. waves with a permanent form) non-linear waves which can be approximated with 2nd order Stokes in the deeper water sections of the channel, following the classic criteria.

KEY WORDS: Regular waves; Non-linear waves; Engineering approximations; Kinematics; Wave orbital velocity.

INTRODUCTION

A reliable description of the wave kinematics is very important to know the dynamic action of waves on structures and to predict coastal sediment transport processes. In order to obtain a correct understanding of the phenomena, many authors have recently proposed mathematical models and approximate methods able to describe the wave motion field both in deep and shallow water.

The problem has been well known since the last century, when the first classic theories were developed. Although these theories are still actual, they imply some hypotheses which are often not realistic and, thus, they need to be reviewed in the light of new numerical and experimental studies.

Relevant mathematical models can simulate the regular waves quite well within a range of validity in which the aforementioned hypotheses are quite acceptable. Regular waves can be reproduced in laboratory and are characterised by a constant form in time domain. Lately new methods for estimating wave orbital velocity have been studied in the light of experimental observations. These studies are preliminary for random waves which indeed can be studied by extending results on regular waves.

This paper presents a study of the transfer functions linking the local time history of elevations with the local velocity field. This procedure is useful because it is easier to measure in situ wave elevations (or to forecast them with statistical methods) than the velocity components. In recent years some methods which link wave elevations with velocity components have been studied, although the theoretical results are not yet supported enough by experimental results.

The merit of some of these methods is the simplification of the mathematical procedures. In fact, the increase of difficulties in a mathematical solution, to better the simulation of physical processes, does not always mean more reliable results. Moreover, some authors (*Dean, 1970*; *Graw, 1994*) agree that it is not always possible to evaluate all the characteristics of a wave field by using only one method with the same level of accuracy. On the contrary, it can be possible, for the same wave field, that a theory approximates a quantity better, and another one can be successfully used to interpret other quantities.

The aim of this paper is to investigate the wave velocity results obtained with some mathematical models and compare them with experimental measurements to determine when to use one model rather than another one. The models were used to interpret the same waves moving in a channel with a sloped bottom, where shoaling effects are evident.

The comparison of calculated velocities with the measured ones can be carried out both in time and frequency domains. In the present paper theories and measurements were compared in frequency domain, by defining a relative error.

THEORETICAL SOLUTIONS

Firstly, classic theories have been applied to obtain reference results for modern methods. To this purpose 2nd and 3rd order Stokes have been applied according to the typical classification criteria. Indeed almost all the investigated waves fall in the range of validity of 2nd and 3rd order Stokes theory.

Stokes waves have been studied by many researchers who gave different mathematical solutions. *Fenton (1985)* gave a decisive contribution to the understanding of fifth-order Stokes theory for steady waves. He stated that the waves travel at Stokes' first definition of wave speed. In Fenton's solution, some of the disadvantages of previous theories have been shown by assuming the expansion parameter ak in which a, having no physical significance other than that of being a length scale, is supposed to be equal to $H/2$. In such a way the expressions for the coefficients of higher order terms are functions of the dimensionless depth kh, so that the only unknown is the wave number k. The suggested procedure ensures the convergence of the series expansions for waves approaching the maximum steepness as observed by *Rebaudengo Landò and Scarsi (1995)*, who applied Fenton's assumptions to develop a non-linear model for directional random waves. As Fenton stated, the classic *Skjelbreira and Hendrikson (1960)* solution for higher order Stokes waves contains errors of the fifth order; moreover the theory suggested by Fenton has been shown to be quite accurate for waves shorter than 10 times the water depth.

The waves investigated in this study, as already said, can be generally approximated with the 3rd order Stokes theory. In this study a third order solution as reported in *Sawaragy (1995)* has been used while the equation in *CERC (1984)* has been employed to calculate the wave length.

The previous theories allow the prediction of both elevations and velocities, once depth, wave height and length (or period) are known; for practical purposes, it can be useful to evaluate the velocity field starting from elevation time history with the following equations:

$$u(t) = H_u\,(h,T,Z)\;\eta(t)$$
$$w(t) = H_w\,(h,T,Z)\;\eta(t+\Delta t) \qquad (1)$$

in which H_u and H_w are the transfer functions and the terms u and w are the horizontal and vertical orbital velocity components respectively, η is the wave elevation, Z is the distance from the bottom, h is the local depth, H and T are the wave height and period, k is the wave number and Δt represents the delay between the wave elevation values and vertical velocity ones, assumed to be equal to $T/4$ according to classic theories. It should be mentioned that the theoretical phase shifting among velocity components and elevations, assumed to be zero for the horizontal component and $\pi/2$ for the vertical one, should be reviewed in the light of the recent experiences (*Damiani and Mossa, 1996b*) showing different values of the aforementioned shifting. Sometimes it can be useful to replace the second equation (1) with a relationship between u and w by introducing the function $H_{u\text{-}w}$ which links the two velocity components. Thus the problem is the research of the better transfer functions linking the wave elevations with the velocity field.

The transfer functions can assume a linear form (*Koyama and Iwata, 1986*); in this way, if the elevations follow the small amplitude wave theory, the equations (1) become formally identical to Airy formulation. For non-linear waves Koyama and Iwata suggested a modified equation for the transfer function of the first equation (1), to evaluate horizontal velocity under the crests. Consequently, also the vertical velocity is modified by applying the function $H_{u\text{-}w}$.

The later two methods give a description of velocity field once the time history of wave elevations is known. It can be useful to decompose the wave surface profile into Fourier components.

Woltering and Daemrich (1995) applied the linear theory to each frequency component of a regular wave elevation spectrum obtaining the velocity component spectra. Basically they treated the spectrum like a superposition of small amplitude waves still resulting in a small amplitude wave. In this procedure, for each frequency component, a transfer function assuming the same equation proposed by Koyama and Iwata has been applied. The linear method of Koyama and Iwata gives the same results of the procedure of Woltering and Daemrich only if the considered waves are monochromatic, that is when the spectrum is chracterized by only one peak.

Woltering and Daemrich (1994) suggested also a Lagrangian approach to take into account the non-linearities in Stokes waves. They found that for wave groups the linear method overestimated the experimental velocities under the crests, while the non linear method gives more reasonable results.

Although Stokes' theory is not valid above the mean water level, it has often been used to find the velocity components between the trough and the crest. Measurements of wave kinematics indicate that the horizontal velocity is smaller at the crest and higher in the trough than predicted by Stokes higher order theories (*Gudmestad and Connor, 1986*). Many engineering approximate methods have been developed in order to go over these limits.

Gudmestad and Connor (1986) report some of these methods for non linear deep water waves, consisting in empirically modifed small amplitude wave theory equations (Wheeler and Chakrabarti methods) or in assuming a constant velocity potential above the still water level (Mo and Moan method). The first two methods predict higher trough velocities and lower crest velocities than the linear theory and the higher order Stokes theories. Gudmestad and Connor proposed also an approach to obtain second order correction terms to the engineering approximations, by following an expansion procedure. Particularly, in the present paper *Chakrabarti (1971)* method and its second order approximation (*Gudmestad and Connor, 1986*) have been applied to interpret the experimental velocity measurements in the whole wave field. The experiments of Gudmestad and Connor confirm the validity of the second order expansion of Chakrabarti method.

All the previous theories have been extended to random waves by *Vis (1980)*, *Guza and Thornton (1980)*, *Rebaudengo Landò and Scarsi (1995)* and other Authors. The former analysed the linkage between wave elevation amplitude spectra and velocity component ones. They found a good agreement with experimental results by using a linear transfer function not only for narrow band spectra, but also for spectra with a remarkable spreading in frequency or for clearly non linear waves.

Rebaudengo Landò and Scarsi (1995) developed a third order model able to describe both time history of the surface elevations and wave kinematics of unidirectional and multidirectional random waves obtaining a good agreement with the laboratory data.

The previous theories give reasonable results in the majority of waves of maritime interest. The open question is the definition of ranges in which the best method should be selected. A contribution in answering to this question was given by *Hattory (1986)* who suggested a method to recognise the best fitting theory by comparing the maxima and minima theoretical and experimental velocities. *Vis (1980)* assumed the error to be equal to the mean of the difference between the theoretical and measured spectral frequency components rated by the latter one.

The experiments hereafter presented have been used to analyse in frequency domain the validity of some of the previous theories for regular waves, and the criteria suggested by Vis have been used in evaluating the errors.

The velocity field was measured using a LDA system (*Damiani and Mossa, 1996a*). A 5 W Ar-Ion laser was used; in the transmitter, the incoming beam from the laser is divided into two pairs of laser beams of green and blue colors for measurements of two velocity components. At the same time a frequency shift is added to one beam of each beam pair to allow the measurements of the reversing flow.

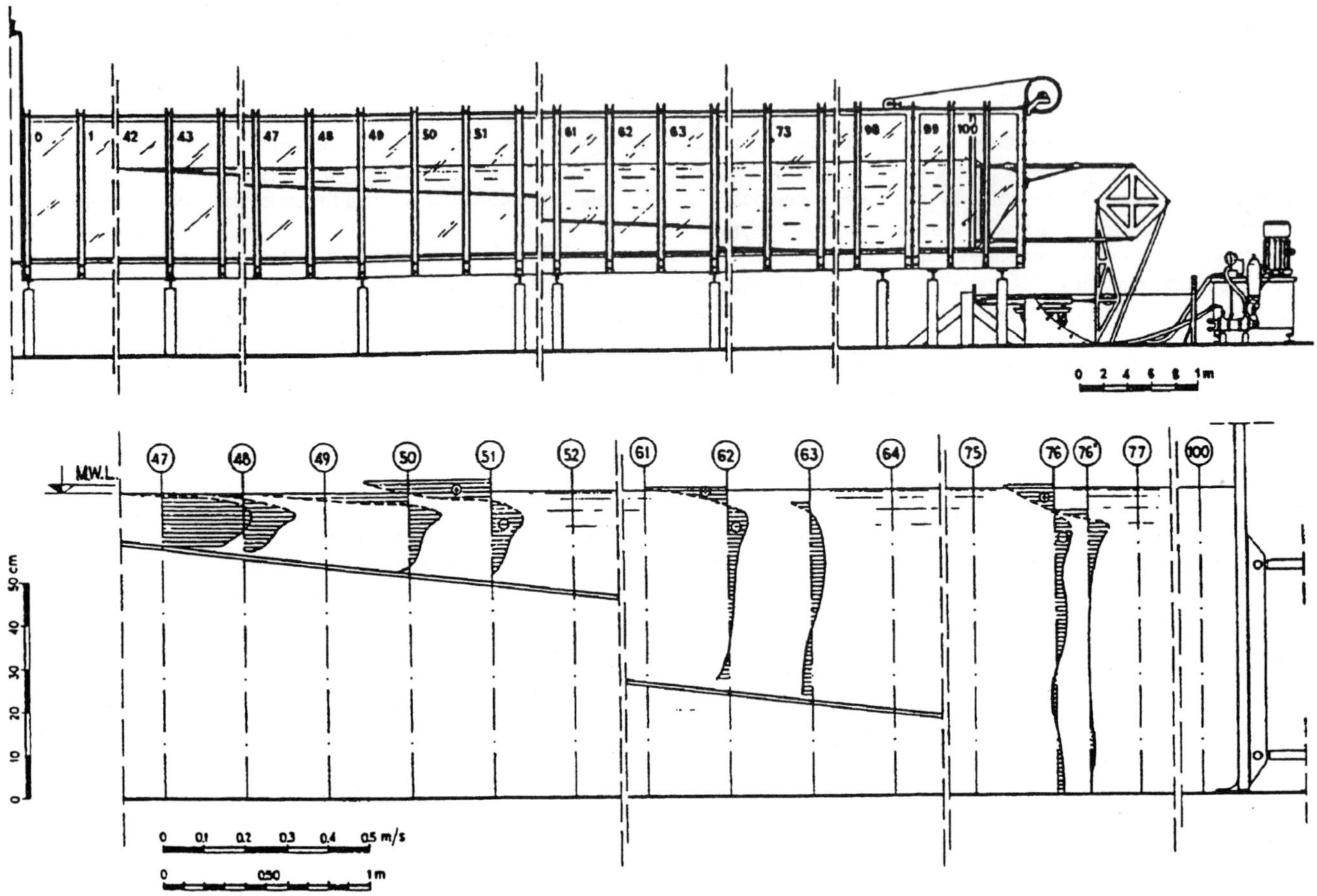

Fig. 1 Wave channel and horizontal mean velocity profiles.

EXPERIMENTAL SET-UP

The experiments were carried out in the wave flume at the laboratory of the Water Engineering Department of the Polytechnic of Bari (*figure 1*). The channel is about 45 m long and 1 m wide; its walls consist of crystal glass sheets 1.2 m high, supported by iron frames with a center to center distance of about 0.44 m, where wave measurement sections (numbered from the shoreline to the wave paddle) have been located. The channel has a 1/20 sloped wooden bottom from the shoreline to section 73 and a horizontal bottom from section 73 to the wave paddle; during the tests the mean water depth in the channel near the paddle was 0.7 m. The wave generation system consists of a flat paddle which receives a rotatory-translational motion, through a kinematic mechanism with an oleodynamic system driven by an electrical valve and controlled by a process computer.

A probe with optic fibre cable is connected to the transmitter using precision-adjustable manipulators. A Dantec LDA signal processor (58N40 FVA Enhanced), based on covariance techniques, was used. Wave elevations were measured by means of resistance probes simultaneously to velocity acquisition. This allows the phase shifting between elevations and velocity components to be evaluated.

In *table 1* the main characteristics of the tested waves are reported in the sections where measurements of both elevations and velocity components were made. L_A represents the wave length evaluated with Airy theory. The last two columns report the Ursell number and Goda non-linearities parameter respectively.

Figure 1 shows the experimental mean horizontal velocity for the first attack of *table 1*, together with the experimental channel. The plus sign (the left side of the diagrams with respect to the measurement section) indicates an onshore velocity direction, while the minus sign (the right side of the diagrams) shows an offshore mean velocity direction. It can be observed that in the sections nearest the shoreline a strong offshore directed current is present.

This is due to the breaking occurring around section 48, thus, the undertow current takes place and turbulence is the dominant action governing the process. This region will be disregarded in the discussion of results because of the poor consistency of the theoretical hypotheses.

In many cases the velocity components between trough and crest were measured. Unlike *Steve and Wind (1982)* the instrumentation used here permits to analyse these values.

SECTION	h [cm]	H [cm]	T [s]	L_A [m]	U	Π
ATTACK 1						
76'	70.00	9.97	1.60	3.43	3.42	0.0462
76	70.00	9.82	1.60	3.43	3.37	0.0455
63	48.80	8.73	1.60	3.05	6.99	0.0643
62	46.80	10.00	1.60	3.01	8.84	0.0783
51	22.60	11.36	1.60	2.24	49.38	0.2880
50	20.10	10.95	1.60	2.13	61.18	0.3419
48	16.01	13.85	1.60	1.92	124.42	0.6502
47	10.75	7.30	1.45	1.44	121.85	0.6064
ATTACK 2						
76'	70.00	15.60	1.50	3.12	4.43	0.0716
76	70.00	15.03	1.50	3.12	4.27	0.0690
63	46.60	15.50	1.50	2.76	11.67	0.1158
62	44.60	14.61	1.50	2.72	12.18	0.1159
60	40.20	14.19	1.50	2.62	14.99	0.1305
58	35.50	13.53	1.50	2.50	18.90	0.1498
56	31.70	15.26	1.50	2.39	27.36	0.2012
54	26.90	15.32	1.50	2.24	39.49	0.2638
52	22.20	15.32	1.50	2.07	60.00	0.3654
50	17.80	16.71	1.50	1.87	103.61	0.5820
48	13.80	8.38	1.43	1.59	80.61	0.4298

Table 1

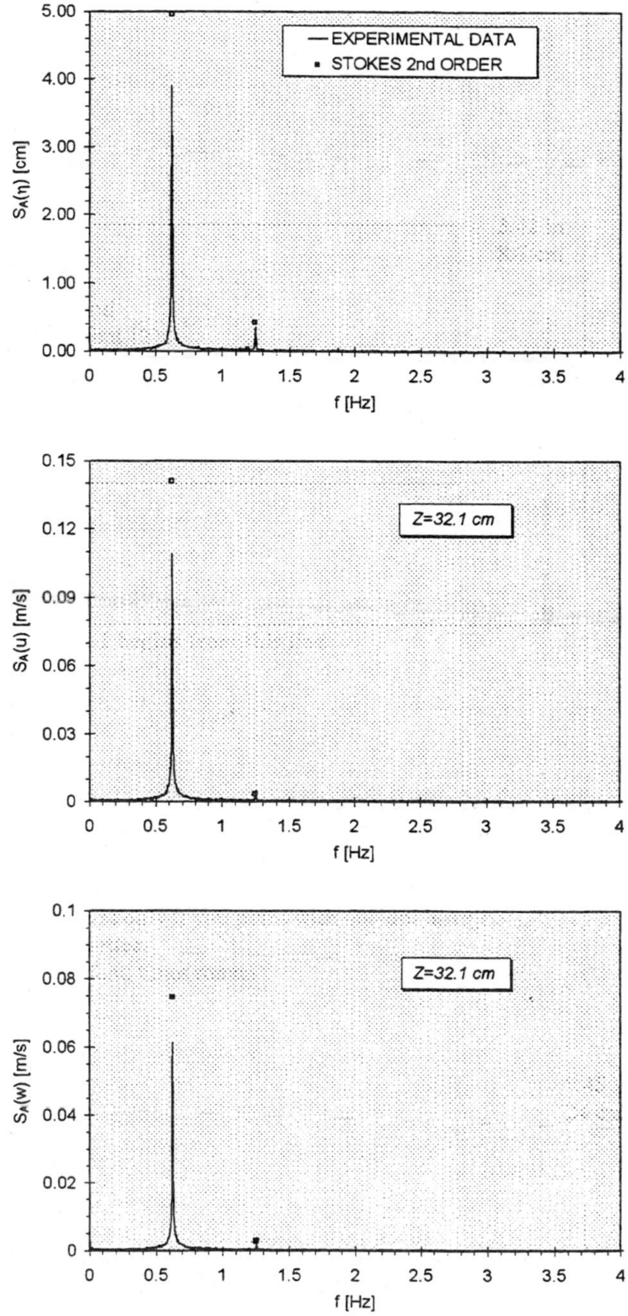

Fig. 2 Amplitude spectra of elevations, horizontal and vertical velocity components assessed in section 76.

RESULTS AND DISCUSSION

Figures 2 and *3* show the amplitude spectra of wave elevations, horizontal and vertical velocity components, obtained from measurements in a point of section 76 and 51. The velocities were gauged at a distance of *32* and *16* cm from the bottom respectively. It can be seen that in the offshore section the wave spectra present only a harmonic of fundamental frequency, with a typical 2nd order Stokes behaviour.

In *figure 2*, also the results of the 2nd order Stokes theory are shown. It has to be pointed out that the methods hereafter used to evaluate velocity components start from the knowledge of the real wave profile assessed in the channel. On the contrary, Stokes theory gives joint theoretical results for wave elevations and velocity components; this makes the comparison between Stokes results and those of all other methods here used difficult. For this reason in the following pages the results of 2nd and 3rd order Stokes calculations will not be presented, also because it has been verified that in all the examined situations they are worse than those obtained with the methods assuming the real elevations as an input.

When the wave goes up to the breaking region, other harmonics appear, clearly showing that the shoaling produces a deformation of the waves and the consequent increase of non-linearities.

Figure 3 shows that the vertical velocity component is more affected by non-linearities, indeed the rate between higher harmonic

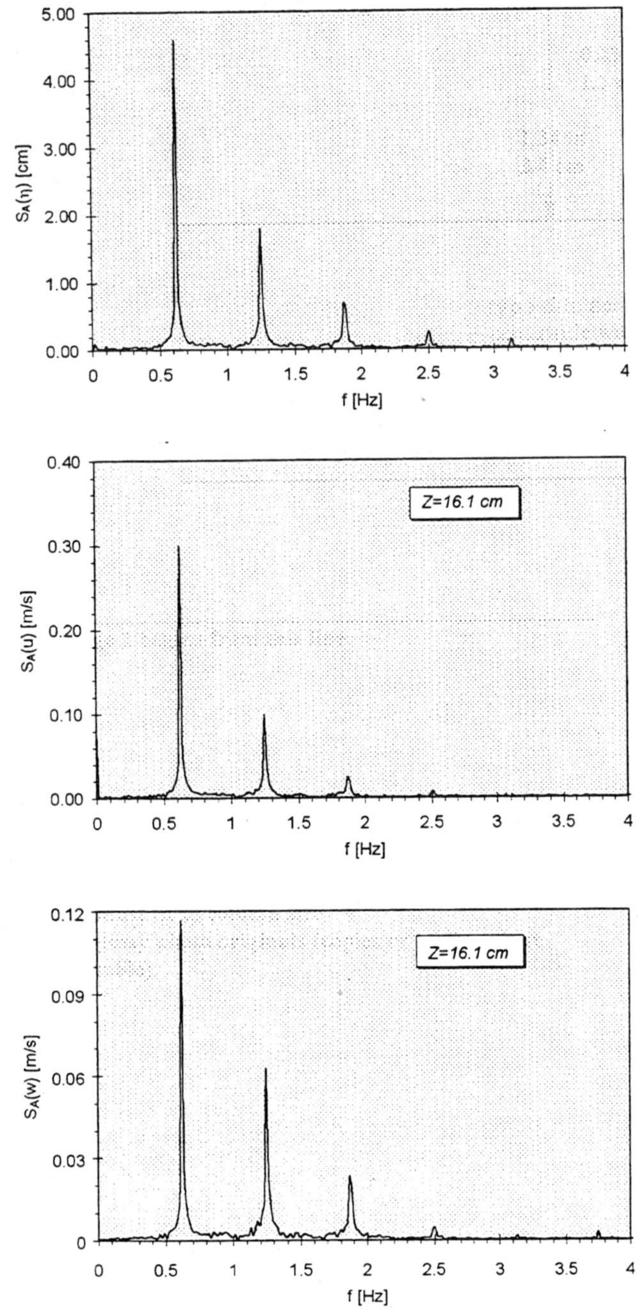

Fig. 3 Amplitude spectra of elevations, horizontal and vertical velocity components assessed in section 51.

and fundamental peak spectral density values is greater in vertical amplitude spectra than in the horizontal ones (*Damiani and Mossa, 1995a*).

The following analysis of velocity spectra has been limited to the first two peaks for homogeneity reasons, because the examined spectra in the deeper water sections have only two peaks as shown in *figure 2*.

Figures 4 and *5* show the first two peaks of previous velocity component spectra, together with the values coming from transfer function methods. For clarity reasons the two frequency peaks have been reported with different scales.

Chakrabarti's approach (*Chakrabarti, 1971*) and its 2nd order approximation (*Gudmestad and Connor, 1986*), respectively **CH I** and **CH II**, have been applied using time history elevation as an input. In the same way, the methods reported by *Koyama and Iwata (1986)* have been applied using equations (1) in which elevations are known. The transfer function has a linear form (**L**inear function in **T**ime **D**omain) or the modified form suggested by the Authors (**M**odified method in **T**ime **D**omain) and it assumes anyhow only one value for a fixed measurement point. Although the results are presented in spectral form, the previous methods propose a linkage between time series of elevations and velocity components, that is the water surface profile is not decomposed into Fourier components. Thus the adopted procedure consists of calculating the time history of velocity components firstly and then of assessing the relative

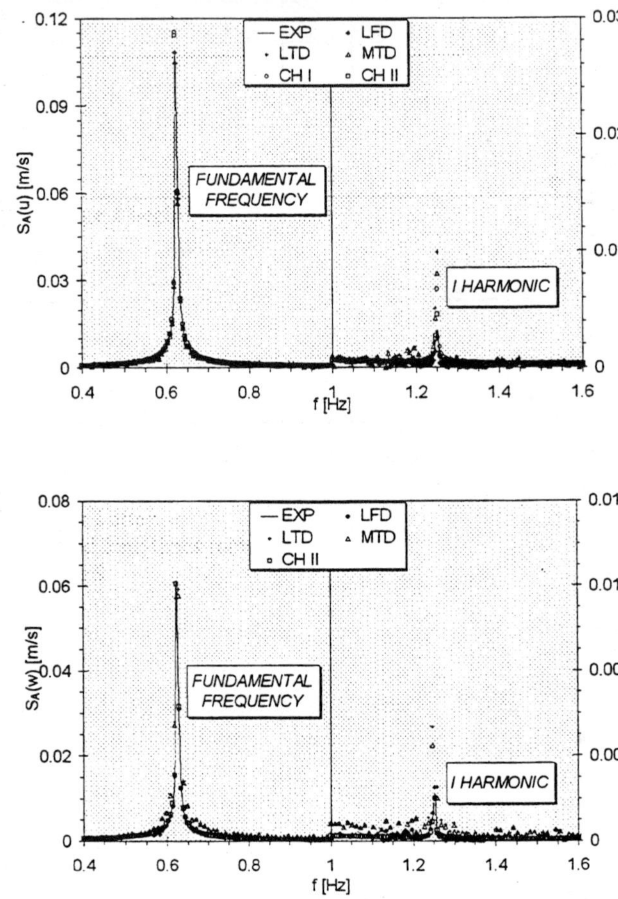

Fig. 4 Experimental and theoretical amplitude spectra of u and w components assessed in section 76 at Z=32.1 cm.

amplitude spectra by using FFT algorithm. In applying Koyama and Iwata methods, the value of Δt in equation (1) is the experimental time lag from a zero-upcrossing point to the next coming wave crest.

Finally, a linear transfer function method was applied to directly evaluate the velocity amplitude spectra in frequency domain, starting from wave elevation ones (**L**inear method in **F**requency **D**omain), as suggested by *Woltering and Daemrich* (*1995*). In this case the method was applied only to evaluate the fundamental peak and its harmonics by extending the computation to a narrow frequency band around each peak ($0.9\,f \div 1.1\,f$). In this case the transfer function assumes a different value for each Fourier component of the wave elevations and the phase shift between w and η is $\pi/2$.

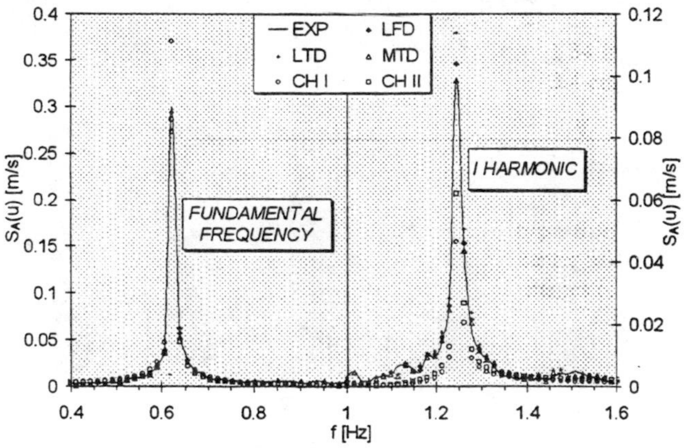

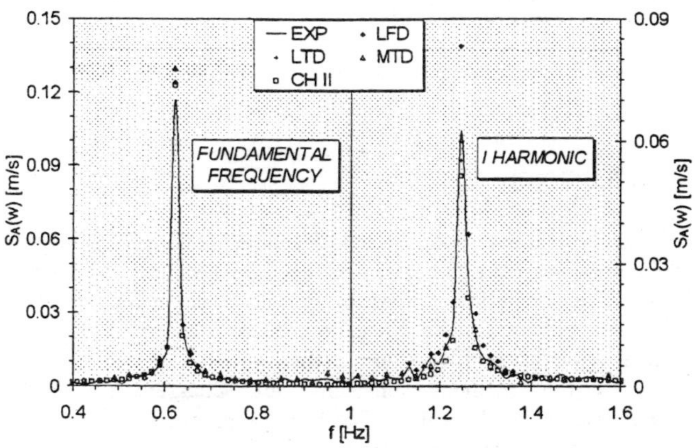

Fig. 5 *Experimental and theoretical amplitude spectra of u and w components assessed in section 51 at Z=16.1 cm.*

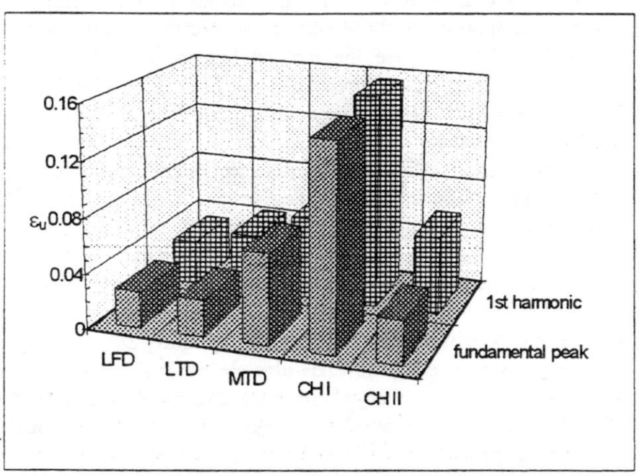

Fig. 6 *Mean relative errors of horizontal velocity at all measurement points.*

The figures confirm an acceptable approximation of almost all methods. The analysis was repeated for all measurement points to investigate the errors occurring when using different methods. For each frequency component the relative error can be evaluated as the difference between predicted and experimental values rated by experimental ones; in this way, the sign of the error indicates if the method overpredicts or underpredicts the experimental values. The average of previous absolute errors can be assumed as a quality index for the different methods. This procedure is the same one adopted by *Vis (1980)*; in this case only the first two frequency peaks were considered, whereas Vis calculated the mean error of all spectral frequency components for analysing random waves.

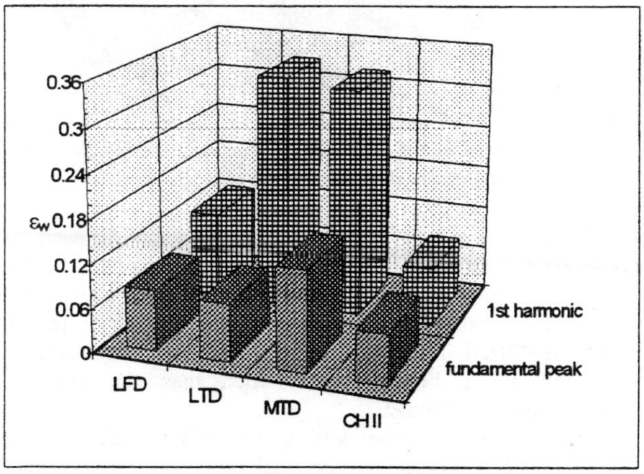

Fig. 7 *Mean relative errors of vertical velocity at all measurement points.*

Figures 6 and *7* report the errors on the first two frequency peaks obtained as the mean values of the absolute errors on all the spectra measured in the channel independently from the position of the measurement point (section and distance from the still water level).

First of all, it can be observed that all methods give a better approximation when predicting the horizontal velocity component u. Moreover, for both velocity components, the fundamental peak evaluation is generally more accurate than the 1st harmonic one. It has to be mentioned that the contribution of the fundamental peak is usually more important when determining the velocity components; this means that the influence of the error on the fundamental peak, even if smaller than the one on the 1st harmonic, could be more important. This justifies the choice made in the present paper to neglect the errors on the 2nd and following harmonics, usually too small with respect to the fundamental peak. On this subject it has to be underlined that *Vis (1980)*, analysing random waves, neglected all spectral density values smaller than 10% of peak one, considering them quite inaccurate.

In *figure 8*, in the same way with *Vis (1980)*, the previously discussed mean values between fundamental frequency peak and 1st harmonic errors are shown both for horizontal and vertical velocity components. It can be stated that the method which better approximates the experimental data is the linear transfer function applied in frequency domain concerning the horizontal velocity component, while for the vertical one Chakrabarti 2nd order is preferable; this confirms that not always the same method gives best results for all wave kinematics (*Dean, 1970*).

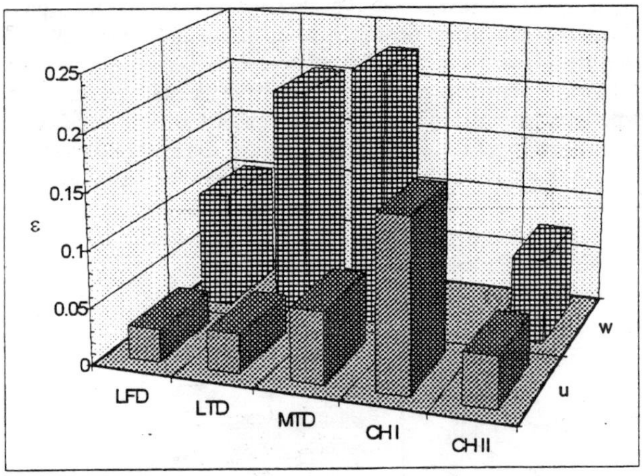

Fig. 8 *Mean errors of velocity components obtained by averaging the fundamental and 1st harmonic peak errors.*

In *figures 9* and *10*, the depth averaged errors for each method are shown against the sections where measurements have been made. Each bar represents the mean error in a section where many measurement points located at different distances from the bottom were investigated. The figure confirms that the *u* component can be assessed with a better approximation than the vertical one everywhere in the channel.

A tendency of all methods to worsen the horizontal velocity evaluation when approaching the breaking region can be noted. This consideration is evident for CH I which is not suitable in evaluating non-linearities. Chakrabarti's 2nd order approximation (CH II) was successfully introduced to refine the CH I.

About the LFD method it has to be noted that in spite of its simplicity and non rigorous assumption (it is not obvious that linear theory will adequately relate *u* and η for clearly non-linear waves), the results obtained are pretty well approximated.

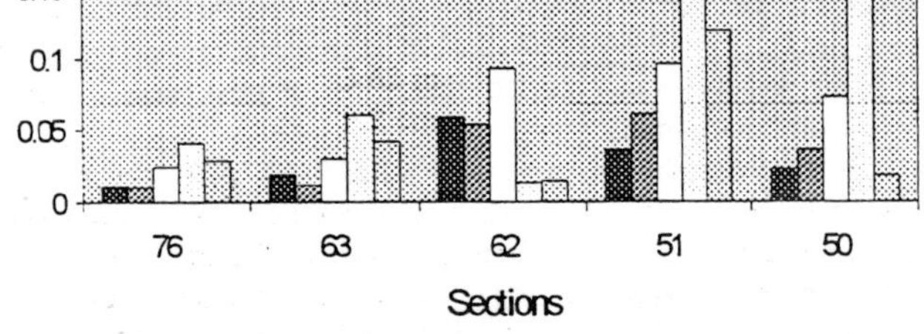

Fig. 9 *Mean errors in horizontal velocity predicting against the measurement sections.*

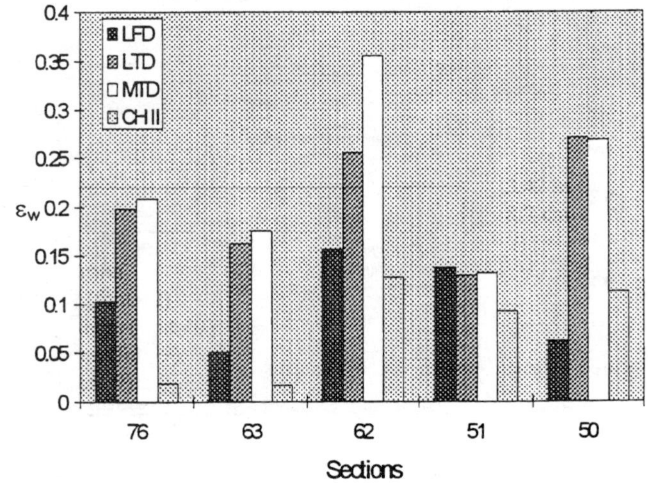

Fig. 10 *Mean errors in vertical velocity predicting against the measurement sections.*

CONCLUSIONS

In the present paper the velocity components of regular waves have been analysed by comparing theoretical values with measured ones. The theoretical values of velocity components have been obtained applying five approximate methods which predict the wave orbital velocities from wave elevations.

The five methods considered are Chakrabarti's approach (*Chakrabarti, 1971*) and its 2nd order approximation (*Gudmestad and Connor, 1986*), the two transfer function methods in time domain reported by *Koyama and Iwata* (*1986*) and a linear transfer function method in frequency domain.

The comparison between calculated and measured velocity amplitude spectra shows that all methods generally predict the motion field with an acceptable approximation. The CH I method is the only one that shows remarkable errors in the computing of horizontal component in those sections where the non-linearities are more visible. In this case the method has to be substituted with Chakrabarti's 2nd order approximation. It is also stated that all methods give a more reliable approximation of the fundamental peak than the first harmonic.

Moreover the predicted horizontal components are more accurate than the vertical ones. This is due to the fact that the vertical components are more strongly affected by non-linearities. The previous results have been depth averaged, since many measurement points have been analysed in each section from the bottom to the wave profile level. It has to be mentioned that no evident dependence has been found between the errors and the quantity *Z/h*.

The present investigation has also shown that a method giving the best approximation everywhere in the channel does not exist, and that it should be investigated the range of validity of each method. For practical proposals, LFD and CH II methods can be generally used better than the other ones.

Moreover it has to be observed that it is not always possible to obtain the best representation for both the velocity components by using the same method.

By averaging the obtained results, it can be said that for tested waves, the linear function method in frequency domain (LFD) can be generally used in predicting the horizontal velocity, with a mean error of about 4% and a maximum one smaller than 10%.

The most suitable method for predicting vertical velocity is Chakrabarti 2nd order approximation, with a mean error of about 6% and a maximum one smaller than 15%.

The same conclusion can be reached by analysing separately the fundamental peak and the first harmonic.

Acknowledgements. The authors are appreciative of the valuable suggestions and stimulating discussions with Prof. G. Scarsi from University of Genova and Prof. A. Petrillo of Polytechnic of Bari. Moreover the authors are grateful to the laboratory technicians for helping them in experimental procedures.

REFERENCES

Chakrabarti, SK (1971). Discussion on "Dynamics of single point mooring in deep water" *Journal of Waterway, Harbour and Coastal Engr. Division*, ASCE, 97 (WW3), pp 588-590.

CERC (1994). "Shore protection manual" *Department of the Army, US Army Corps of Engineers*, Washington, DC 20134.

Damiani, L, and Mossa, M (1996). "Indagine sul campo di moto di un'onda regolare" *Scritti in Onore di Mario Ippolito, Ass. Idrotecnica Italiana Sez. Campana - Dip. di Ing. Idraulica ed Ambientale Girolamo Ippolito*, Napoli, pp 227-238.

Damiani, L, and Mossa, M (1996). "Ricostruzione di un campo di moto ondoso" *Accademia Pugliese delle Scienze*, Bari - Italy.

Dean, RG (1970). "Relative validities of water wave theories" *J. Waterway and Harbors Div., Proc. of* ASCE, Vol 96, No WW1, pp 105-119.

Fenton, JD (1985). "A fifth-order Stokes theory for steady waves" *J. Waterway Port Coastal and Ocean Eng.*, ASCE, Vol 111, No 2, pp 216-234.

Graw, KU (1994). "Comparison of wave theories with velocity mesurement" *Proc. of Int. Symp. Waves-Physical and Numerical Modelling*, University of British Columbia, Vancouver, pp 561-569.

Gudmestad, OT, and Connor, JJ (1986). "Engineering approximation to nonlinear deepwater waves" *J. of Applied Ocean Research*, Vol 8, No 2, pp 76-88.

Guza, RT, and Thornton, EB (1980). "Local and shoaled comparisons of sea surface elevations, pressures, and velocities" *J. of Geophysical Research*, Vol 85, No C3, pp 1524-1530.

Hattori, M (1986). "Experimental study on the validity range of various wave theories" *Proc. of 20th Int. Conf. Coastal Engineering*, ASCE, Vol 1, pp 232-246.

Koyama, H, and Iwata, K (1986). "Estimation of particle velocities of shallow water waves by a modified transfer function method" *Proc. of 20th Int. Conf. on Coastal Engineering*, ASCE, Vol 1, pp 425-436.

Rebaudengo Landò, L, and Scarsi, G (1995). "Directional Random Wave Kinematics: Third Order Approximation" *Proc. of 5th Int. Offshore and Polar Eng Conference*, Vol III, pp 49-56.

Sawaragi, T (1995). "Coastal engineering-waves, beaches, wave-structure interactions" *Developments in Geotechnical Engineering*, 78, Ed. Elsevier.

Skjelbreia, L, and Hendrickson, J (1960). "Fifth order gravity waves theory" *Proc. of 7th Int. Coastal Engineering Conference*, ASCE, pp 184-196.

Steve, MJF, and Wind, HG (1982). "A study of radiation stress and set-up in the nearshore region" *Delft Hydraulics*, Publication No 267.

Ting, FCK, and Kirby, JT (1995). "Dynamics of surf-zone turbulence in a strong plunging breaker" *J. of Coastal Engineering*, Vol 24, pp 177-204.

Vis, FC (1980). "Orbital velocities in irregular waves" *Delft Hydraulics*, Publication No 231.

Woltering, S, and Daemrich, KF (1994). "Regular wave investigations of wave kinematics with lagrangian approach" *Proc. of International Symposium: Waves-Physical and Numerical Modelling*, Vancouver, pp 665-673 .

Woltering, S, and Daemrich KF (1995). "Experimental validity of wave theories and application to horizontal orbital velocities" *International Conference on Coastal and Port Engineering in Developing Countries*, Rio de Janeiro,Vol 3, pp 2411-2424.

Proceedings of the Seventh (1997) International Offshore and Polar Engineering Conference
Honolulu, USA, May 25-30, 1997

ISBN 1-880653-28-1 (Set); ISBN 1-880653-31-1 (Vol. III)

Numerical Model for Solving Boussinesq-Type Equations: Comparisons and Validation

Zhili Zou and Benhe Xui
Dalian University of Technology
Dalian, China

ABSTRACT

In order to solve the Boussinesq-type equations developed by the author (1996), two finite difference methods were used: a Predictor-Corrector method and a ADI method. Comparisons of the numerical results of the two methods were made for a variety of cases, including the wave diffraction on the gap of double breakwaters and the wave refraction on the topography with a deep channel. The comparisons with experiments results were also made to valid the two numerical schemes. The differences in computational efficiency and numerical results of the two methods were discussed.

KEY WORDS: Water wave, ADI, Boussinesq

1. INTRODUCTION

The Boussinesq equations are widely used in the computations of water waves in harbors and coastal area. These equations can take into account a variety of water wave phenomena such as wave refraction, diffraction and reflection with the nonlinearities being taken into account. The latter property can account for the wave energy transformations between different wave components in irregular waves and the asymmetry of the wave profiles and is the advantage of the Boussinesq equations over other methods for water wave motions.

But the classical Boussinesq equations (Peregrine (1967)) can only be applied in relatively shallow water and this limits the applications of the equations for the topography with depth changing greatly (for example, with a deep channel) and for the case of irregular waves. Recently many improvements have been made by developing a new type of Boussinesq equations with improved dispersion characteristics (Witting (1984), Madsen et al (1991) (1992), Nwogu (1993), Wei and Kirby (1995)). However, several inconsistencies existed in these improvements, including that the velocity distribution along depth is not improved, or that the nonlinearity of the equations is still limited to relatively shallow water. A form of higher order Boussinesq equations was developed by the author (1996) and the above weakpoints were removed in the work. This paper concerns with the numerical method used in solving the equations.

So far two numerical methods have been used widely in solving the Boussinesq equations: Predictor-Corrector method (P-C method) and ADI (Alternative Direction Implicit) method. But the comparison of the two method has not been made in the previous researches. This paper presents such a kind of researches. Smallman and Cooper (1989) used a Predictor-Corrector method to solve the Boussinesq equations and calculated the second order long waves, set-down, in wave groups. Madsen et al (1991,1992) developed a new form of Boussinesq equations and solved them by a kind of ADI method. The two difference schemes similar to those mentioned above were used in this paper to solve the Boussinesq equations and the comparisons of the numerical results of the two methods were made to investigate the accuracy of the two methods. Comparisons were also made with experimental results to valid the two methods.

2. EQUATIONS AND NUMERICAL SOLUTIONS

2.1 Equations

The author (1996) derived a following new form of Boussinesq type equations accurate to O($\mu^4, \varepsilon\mu^2, \varepsilon^2$), here $\mu = h/L$, $\varepsilon = H/h$, H, L and h are characteristic wave height, wave length and water depth.

$$\varsigma_t + \nabla \cdot [(h+\varsigma)\vec{u}] = 0 \tag{1}$$

$$\bar{u}_t + (\bar{u}\cdot\nabla)\bar{u} + g\nabla\varsigma + G = \frac{1}{2}d\nabla[\nabla\cdot(h\bar{u}_t)] - \frac{1}{6}d^2\nabla(\nabla\cdot\bar{u}_t)$$
$$+B_1 h^2\nabla[\nabla\cdot(\bar{u}_t + g\nabla\varsigma)] + B_2\nabla[\nabla\cdot(h^2\bar{u}_t + gh^2\nabla\varsigma)] \tag{2a}$$

$$G = \frac{1}{3}\nabla\left\{h^2\left[(\nabla\cdot\bar{u})^2 - \bar{u}\cdot\nabla^2\bar{u} - \frac{1}{10}\nabla^2(\bar{u}\cdot\bar{u})\right]\right\} \tag{2b}$$
$$+\frac{1}{2}\varsigma\nabla[\nabla\cdot(h\bar{u}_t)] - h\nabla\varsigma\nabla\cdot\bar{u}_t$$

here ς is the wave elevation, $\vec{u}$=(u, v) is the depth average horizontal velocity, h is still water depth and d is total water depth $d = h + \varsigma$. B_1 and B_2 are constants satisfying the relation $B_1 + B_2 = 1/15$. For constant water depth, B_1 and B_2 have exact values: B_1=1/15, B_2=0 and the resulting equations are exact; For varying water depth, when

B_1=2/59 and B_2=29/885, the equations possess the optimal shoaling characteristics (Zou (1996)). The above equations reduce to Madsen's equations (1992) when $B_1=B_2$=1/30 and the third order nonlinear terms and the second order derivatives of h are omitted, but the former was obtained by the consistent mathematical derivation while the latter was obtained by just adding the third order terms to the classical Boussinesq equations. The third order nonlinear terms of the equations (1)-(2) were omitted in the study of this paper and their contributions will be studied in other papers.

2.1 Predictor-Corrector Method

Staggered grids was used with depth-averaged velocity u and v at the time level n and wave elevation ς at time level n+1/2, as shown in Fig.1. The resulting equations with the third order nonlinear terms omitted were disretized in the following form of Predictor-Corrector method.

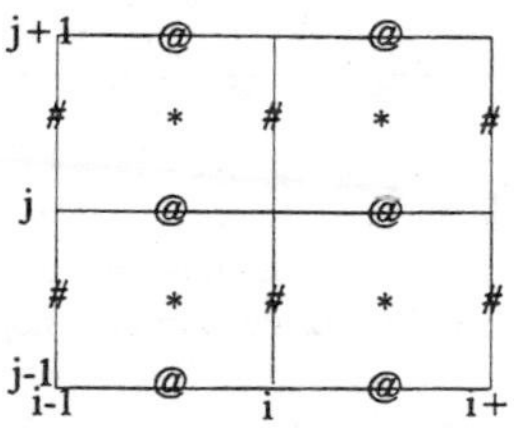

Fig.1 Staggered grid.

For P-C, # --u^n; @ --v^n; * --$\varsigma^{n+1/2}$.

For ADI, # -- P^n; @ -- $Q^{n+1/2}$; * -- ς^n.

$$\frac{\varsigma^{n+1/2}-\varsigma^{n-1/2}}{\Delta t}+\nabla\cdot(\vec{u}^n(h+\frac{\varsigma^{n+1/2}+\varsigma^{n-1/2}}{2}))=0 \quad (3)$$

$$\frac{\vec{u}^{n+1}-\vec{u}^n}{\Delta t}+(\vec{u}^{n+1/2}\cdot\nabla)\vec{u}^{n+1/2}+g\nabla\varsigma^{n+1/2}$$
$$=\frac{1}{2}\frac{h}{\Delta t}\left[\nabla(\nabla\cdot(h\vec{u}^{n+1}))-\nabla(\nabla\cdot(h\vec{u}^n))\right]\cdot$$
$$+(B_1-\frac{1}{6})\frac{h^2}{\Delta t}\left[\nabla(\nabla\cdot\vec{u}^{n+1})-\nabla(\nabla\cdot\vec{u}^n)\right]$$
$$+B_2\frac{1}{\Delta t}\left[\nabla(\nabla\cdot(h^2\vec{u}^{n+1}))-\nabla(\nabla\cdot(h^2\vec{u}^n))\right]$$
$$+B_1gh^2\nabla^2\nabla\varsigma^{n+1/2}+B_2g\nabla(\nabla\cdot(h^2\nabla\varsigma^{n+1/2})) \quad (4)$$

in which the partial derivatives with respect to space are discretized in central difference. The convection terms are nonlinear and how to devise their discretizations in the difference schemes would affect the presentations of the nonlinear property of the Boussinesq-type equations. Since the difference equations of P-C method can be nonlinear, the following nonlinear expressions were adopted for the convection terms.

$$(\vec{u}^{n+1/2}\cdot\nabla)\vec{u}^{n+1/2}=\frac{1}{4}((\vec{u}^{n+1}+\vec{u}^n)\cdot\nabla)(\vec{u}^{n+1}+\vec{u}^n) \quad (5)$$

In the numerical solution of the difference equations (3)-(4), the predicted value of u, v and ς were given by the linearized form of the equations. Then the predicted solutions were used as the initial value of the iteration solutions of equations (3)-(4). The final solution of the difference equations were obtained by the Seidel iteration.

2.2 ADI method

In order to apply ADI method, the equations (1)-(2) with the third order terms omitted are expressed in the following conservation form

$$\varsigma_t+P_x+Q_y=0 \quad (6)$$

$$P_t+(\frac{P^2}{d})_x+(\frac{PQ}{d})_y+gd\varsigma_x$$
$$=\frac{1}{2}h^2(P_{xxt}+Q_{xyt})+(B_1-\frac{1}{6})h^3((\frac{P}{d})_{xxt}+(\frac{Q}{d})_{xyt})$$
$$+B_2h((Ph)_{xxt}+(Qh)_{xyt})+B_1gh^3(\varsigma_{xxx}+\varsigma_{xyy})$$
$$+B_2gh((h^2\varsigma_x)_{xx}+(h^2\varsigma_y)_{xy}) \quad (7)$$

$$Q_t+(\frac{PQ}{d})_x+(\frac{Q^2}{d})_y+gd\varsigma_y$$
$$=\frac{1}{2}h^2(P_{xyt}+Q_{yyt})+(B_1-\frac{1}{6})h^3((\frac{P}{d})_{xyt}+(\frac{Q}{d})_{yyt})$$
$$+B_2h((Ph)_{xyt}+(Qh)_{yyt})+B_1gh^3(\varsigma_{xxy}+\varsigma_{yyy})$$
$$+B_2gh((h^2\varsigma_x)_{xy}+(h^2\varsigma_y)_{yy}) \quad (8)$$

where P and Q are the depth-integrated velocity components in the x- and y-direction respectively. The staggered grids shown in Fig.(1) were used in the discretization of the equations. The P and ς were at the n time level, Q was at the n+1/2 time level. With the above space- and time-staggered rectangular grids the equations (6)-(8) were discretized by using a time-centered implicit scheme. The ADI algorithm was invoked, and the resulting system of the difference equations was reduced to a three-diagonal system, which was solved by the efficient Double Sweep algorithm. The x-sweep equations read

$$\frac{\hat{\varsigma}^{n+1/2}-\varsigma^n}{\Delta t/2}+(\frac{5}{4}P_x^n-\frac{1}{4}P_x^{n-1})+(\frac{3}{4}Q_y^{n+1/2}+\frac{1}{4}Q_y^{n-1/2})=0 \quad (9)$$

$$\frac{\varsigma^{n+1/2}-\varsigma^n}{\Delta t/2}+(\frac{1}{4}P_x^{n+1}+\frac{3}{4}P_x^n)+(\frac{3}{4}Q_y^{n+1/2}+\frac{1}{4}Q_y^{n-1/2})=0 \quad (10)$$

$$\frac{P^{n+1}-P^n}{\Delta t}+\left(\frac{P^2}{d}\right)_x^{n+1/2}+\left(\frac{PQ}{d}\right)_y^{n+1/2}+g\hat{d}\varsigma_x^{n+1/2}$$
$$=\frac{1}{2}\frac{h^2}{\Delta t}\left[(P_{xx}^{n+1}-P_{xx}^n)+\frac{3}{2}(Q_{xy}^{n+1/2}-Q_{xy}^{n-1/2})-\frac{1}{2}(Q_{xy}^{n-1/2}-Q_{xy}^{n-3/2})\right]$$
$$+(B_1-\frac{1}{6})\frac{h^3}{\Delta t}\left[((P/h)_{xx}^{n+1}-(P/h)_{xx}^n))+\frac{3}{2}((Q/h)_{xy}^{n+1/2}-(Q/h)_{xy}^{n-1/2})\right.$$
$$\left.-\frac{1}{2}((Q/h)_{xy}^{n-1/2}-(Q/h)_{xy}^{n-3/2})\right]+B_2\frac{h}{\Delta t}\left[((Ph)_{xx}^{n+1}-(Ph)_{xx}^n)\right.$$
$$\left.+\frac{3}{2}((Qh)_{xy}^{n+1/2}-(Qh)_{xy}^{n-1/2})-\frac{1}{2}((Qh)_{xy}^{n-1/2}-(Qh)_{xy}^{n-3/2})\right]+$$
$$B_1gh^3(\hat{\varsigma}_{xxx}^{n+1/2}+\hat{\varsigma}_{xyy}^{n+1/2})+B_2gh((h^2\hat{\varsigma}_x^{n+1/2})_{xx}+(h^2\hat{\varsigma}_y^{n+1/2})_{xy})$$

(11)

in which $\varsigma^{n+1/2}$ and P^{n+1} are the unknown variable. $\hat{\varsigma}^{n+1/2}$ and $\hat{d}^{n+1/2}$ are the estimated values at the time level n+1/2. The y-sweep equations read

$$\frac{\tilde{\varsigma}^{n+1}-\varsigma^{n+1/2}}{\Delta t/2}+(\frac{3}{4}P_x^{n+1}+\frac{1}{4}P_x^n)+(\frac{5}{4}Q_y^{n+1/2}-\frac{1}{4}Q_y^{n-1/2})=0 \quad (12)$$

$$\frac{\varsigma^{n+1}-\varsigma^{n+1/2}}{\Delta t/2}+(\frac{3}{4}P_x^{n+1}+\frac{1}{4}P_x^n)+(\frac{1}{4}Q_y^{n+3/2}+\frac{3}{4}Q_y^{n+1/2})=0 \quad (13)$$

$$\frac{Q^{n+3/2}-Q^{n+1/2}}{\Delta t}+\left(\frac{Q^2}{d}\right)_y^{n+1}+\left(\frac{PQ}{d}\right)_x^{n+1}+g\tilde{d}\varsigma_y^{n+1}$$
$$=\frac{h^2}{2\Delta t}\left[(Q_{yy}^{n+3/2}-Q_{yy}^{n+1/2})+\frac{3}{2}(P_{xy}^{n+1}-P_{xy}^{n-1})-\frac{1}{2}(P_{xy}^n-P_{xy}^{n-1})\right]$$
$$+(B_1-\frac{1}{6})\frac{h^3}{\Delta t}\left[((Q/h)_{yy}^{n+3/2}-(Q/h)_{yy}^{n+1/2})+\frac{3}{2}((P/h)_{xy}^{n+1}-(P/h)_{xy}^n)\right.$$
$$\left.-\frac{1}{2}((P/h)_{xy}^n-(Q/h)_{xy}^{n-1})\right]+B_2\frac{h}{\Delta t}\left[((Qh)_{yy}^{n+3/2}-(Qh)_{yy}^{n+1/2})\right.$$

$$+\frac{3}{2}((Ph)^{n+1}_{xy}-(Ph)^{n-1}_{xy})-\frac{1}{2}((Ph)^{n}_{xy}-(Ph)^{n-1}_{xy})\Big]$$

$$+B_1gh^3(\tilde{\varsigma}^{n+1}_{yyy}+\tilde{\varsigma}^{n+1}_{xxy})+B_2gh((h^2\tilde{\varsigma}^{n+1}_x)_{xy}+(h^2\tilde{\varsigma}^{n+1}_y)_{yy}) \tag{14}$$

in which ς^{n+1} and $Q^{n+3/2}$ are unknown, $\tilde{\varsigma}^{n+1}$ and $\tilde{d}^{n+1}$ the estimated values at the time level n+1. The nonlinear convection terms should be linearized in order to apply the Double-Sweep method and this is the major difference between the ADI and P-C method discussed above, which will result in the difference in the nonlinear features of the numerical results of the two methods and will be discussed latter. So the following presentations of the convection terms were used in the

$$\left[\frac{P^2}{d}\right]^{n+1/2}_x=\left[\frac{P^nP^{n+1}}{d}\right]_x \tag{15}$$

$$\left[\frac{PQ}{d}\right]^{n+1/2}_y=\frac{1}{2\Delta y}\left[\frac{(P^{n+1}_{i+1,j+1}+P^{n}_{i,j+1})Q_{up}}{h_{i,j+1}+h_{i+1,j+1}+\hat{\varsigma}^{n+1/2}_{i,j+1}+\hat{\varsigma}^{n+1/2}_{i+1,j+1}}-\frac{(P^{n+1}_{i,j-1}+P^{n}_{i,j-1})Q_{down}}{h_{i,j-1}+h_{i+1,j-1}+\hat{\varsigma}^{n+1/2}_{i,j-1}+\hat{\varsigma}^{n+1/2}_{i+1,j-1}}\right] \tag{16a}$$

$$Q_{up}=(Q^{n+1/2}_{i+1,j+1}+Q^{n+1/2}_{i,j+1}+Q^{n+1/2}_{i+1,j}+Q^{n+1/2}_{i,j})/4 \tag{16b}$$

$$Q_{down}=(Q^{n+1/2}_{i+1,j-1}+Q^{n+1/2}_{i,j-1}+Q^{n+1/2}_{i+1,j-2}+Q^{n+1/2}_{i,j-2})/4 \tag{16c}$$

Similar presentations were adopted for the y-directions expressions. The solution process at each time level includes two steps: the first half time level is from n time level to n+1/2 time level and the values of $\varsigma^{n+1/2}$ and P^{n+1} are obtained by solving the equations (9)-(11); the second half time level is from n+1/2 time level to n+1 time level and the values of ς^{n+1} and $Q^{n+3/2}$ are obtained by solving equations (12)-(14).

The sponge boundary (Larsen and Dancy 1983) was applied for the open boundary conditions and found it effective.

3. COMPARISONS OF NUMERICAL RESULTS

In order to investigate the accuracy and efficiency of the above P-C method and ADI method, the numerical results of the two methods were compared for the following three cases. In the case (1) and (2) the input waves were cnoidal waves. The grid size was dx=dy=4m for both methods. The time step was 0.5s for the P-C method and 0.25s for the ADI method. In the case (3) the input waves were regular, unidirectional and directional waves. The grid size was dx=dy=0.167m and the time step was 0.05s for both the P-C method and theADI method.

3.1 Wave Refraction

Wave refraction on a plane slope at the incident angle of 30 degree was first computed. The slope is 1:100 and the water depth at input boundary was 10m. The wave height and period were 2.0m and 9.0s respectively. Fig.2 is the isolines of zero up- and down-crossing and the wave elevations along the slope. The results of P-C method and ADI method agreed well on the whole, but the wave elevations of the P-C method showed more stronger nonlinear feature than those of ADI method.

In order to investigate the differences in the nonlinear feature between the numerical results of the two methods further, the topography with a half of deep channel, which is 6m deep and begins 50 meters away from the input boundary, was considered. The water depth is 6m in the area outside the channel. The incident wave of 2m height and 6s period propagated along the channel direction. Fig.3 is the perspective plot of the wave elevations calculated by the ADI method and the wave elevations along three lines y=100m, 150m and 250m (y is perpendicular to the channel) by the two methods respectively. The perspective plot of the wave elevations calculated by the P-C method is not given in the figure

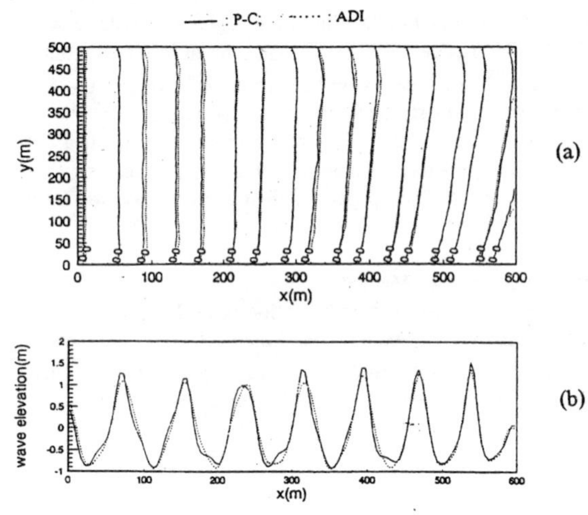

Fig.2 Cnoidal wave refraction on a slope (1:100)
(a) Isolines of zero up- and down-crossing;
(b) wave elevation along the slope.

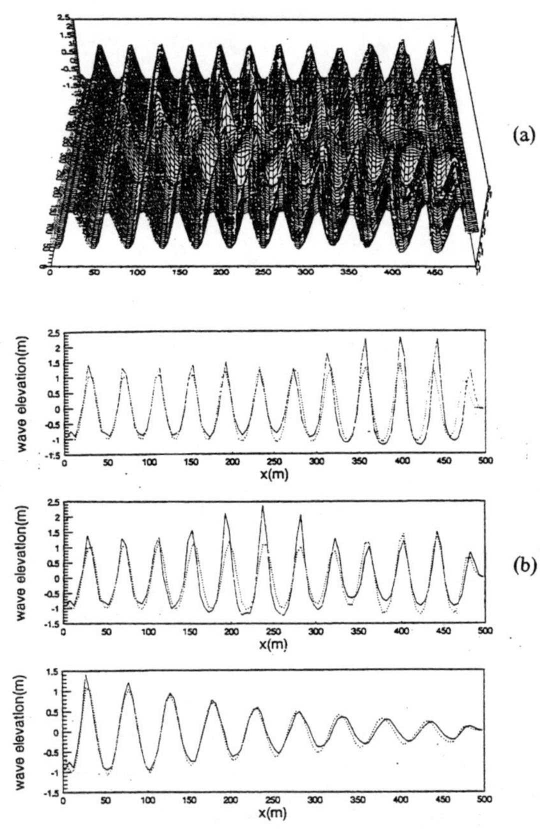

Fig.3 Cnoidal wave on a topography with a half channel .
(a) perspective plot of wave elevations;
(b) wave elevations at y=100m, 150m and 250 m
(from top to bottom) —— : P-C; ······ : ADI

and is similar to that of ADI method given in the figure, but the crests were sharper and higher as shown in detail in Fig.3b by the results of wave elevations. It is seen that the differences in the wave heights of the two methods are large at several position, P-C method gives greater heights than ADI method. They are apparently caused by the great change of the water depth at the beginning point of the channel because the higher heights of P-C's results at the section y=150m, which is closer to the channel, appear along the horizontal axis before the higher heights of P-C's results at the section y=100m, which is farther away from the channel than the section y=150. This demonstrates that the P-C method can take more nonlinearities into account than ADI method and this difference in nonlinearity showed itself strongly when the water depth changes greatly.

3.2 Wave Diffraction on a Rectangular Cylinder

Fig.4 is the results of wave diffraction on a rectangular cylinder, in which the perspective plot of wave elevations was given by the ADI method. The water depth was 6m. The cnoidal wave incident normally to the cylinder with 2m wave height and 6s wave period. Fig. 4b is the wave elevations at the sections y=100m, 150m and 250m respectively. P-C method gives a little greater wave heights than ADI method.

3.3 Wave Diffraction on the Gap of Double Breakwaters

The experimental results given by Wu et al (1995) were used to valid the numerical results and the numerical results of Mild Slope Equations used by the same authors were also presented in the figures. The breakwater gap is 4m and the water depth is 0.3m. Fig.5 and Fig.6 are the regular wave diffraction with 0.06m wave height and 1.41s wave period at the incident angle of 0^0 and 45^0 respectively. Fig. 7 is the unidirectional irregular wave diffraction with significant wave height 0.06m and period 0.93s at the incident angle of 0^0. Fig.8 is the directional wave diffraction with the same significant wave height , period and incident angle. The

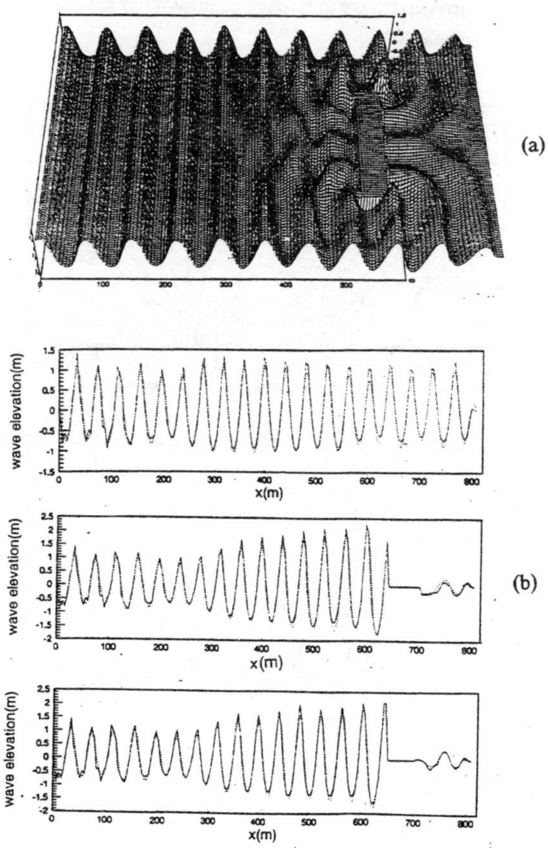

Fig.4 Cnoidal wave diffraction on a rectangular cylinder.
(a) perspective plot of wave elevations;
(b) wave elevations at y=100m, 150m and 250 m (from top to bottom)

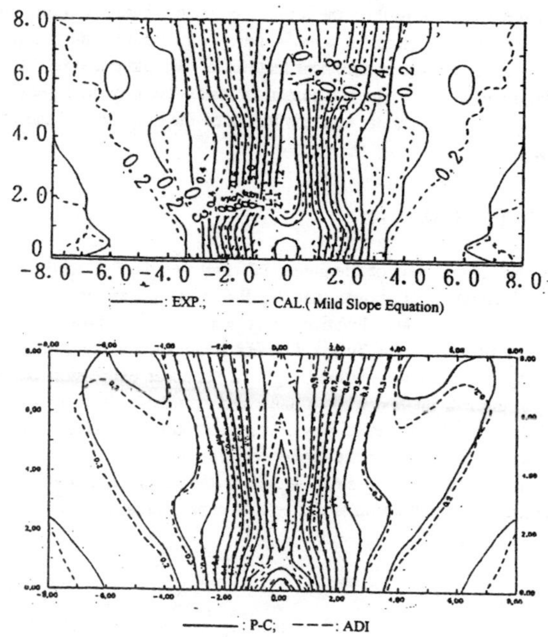

Fig.5 Isolines of relativewave height ratios of regular wave diffractionon breakwater gap. Incident angle 0°

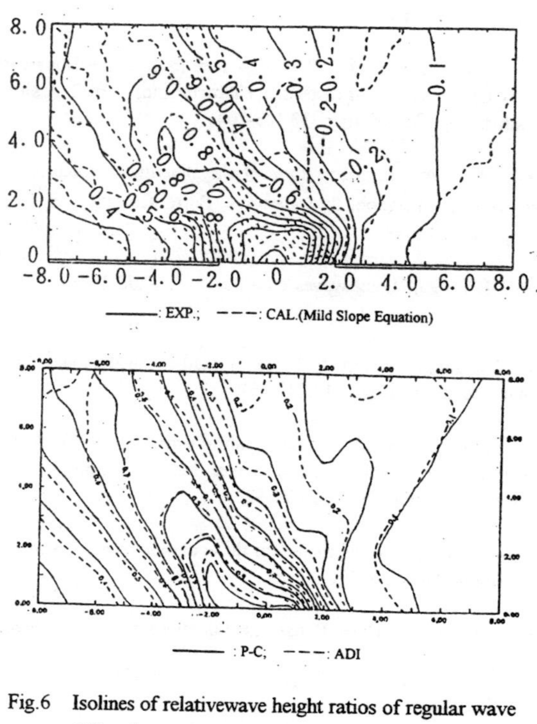

Fig.6 Isolines of relativewave height ratios of regular wave diffraction on breakwater gap. Incident angle 45°

wave spectrums were B-spectrum for both the unidirectional and directional irregular waves. The following directional distribution function for directional waves was used.

$$G(\omega,\theta)=\frac{2}{\pi}\cos^2\frac{\theta}{2}$$

It can be seen from these figures that the numerical results of P-C method and ADI method agree well on the whole, but some difference in quantity still exist and become large for some isolines in the figures and this may come from the difference in nonlinear features of two numerical results as discussed in the case (1). The two numerical results are in good agreement with the experimental results and this validated the accuracy of the two methods.

4. CONCLUSIONS

From the comparisons of the numerical results of the P-C method and the ADI Method, it is shown that the P-C method can present more nonlinear features of wave elevations in its numerical results than the ADI method and this is caused by the different treatment for the convection terms in the momentum equations, e.g., a nonlinear expression can be adopted in the P-C method while a linearized expression must be applied in the ADI method. In the efficiency of computations ADI method is much faster than P-C method, but the former requires smaller time step than P-C method. For the waves of small amplitude the numerical results of the two methods agree to the similar degree with experimental results.

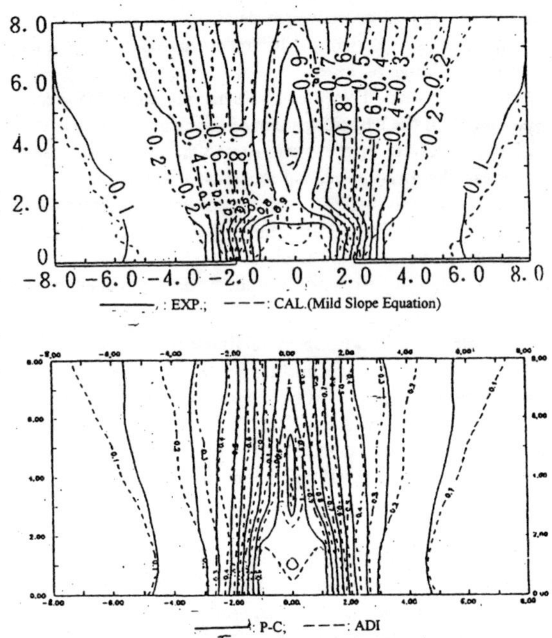

Fig.7 Isolines of relative wave height ratios of unidirectional irregular wave diffraction on breakwater gap. Incident angle 0°

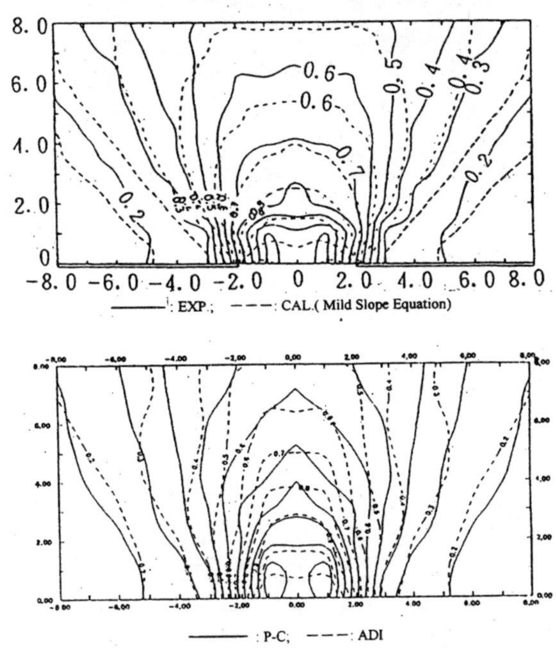

Fig.8 Isolines of relative wave height ratios of directional irregular wave diffraction on breakwater gap. Incident angle 0°

REFERENCES

Larsen J and Dancy H (1983). Open boundary in short wave simulations - a new approach. Coastal Eng., 7:285-297.

Madsen P A , Murray I R and Sorensen O R (1991). A new form of the Boussinesq equations with improved linear dispersion characteristics. Coastal Eng., 15:371-388.

Madsen P A and Sorensen O R (1992). A new form of the Boussinesq equations with improved linear dispersion characteristics. Part 2: A slowly varying bathymetry. Coastal Eng., 18:183-204.

Nwogu O (1993). An alternative form of the Boussinesq equations for nearshore wave propagation. J. waterway、Port、Coastal and Ocean Engineering, 119:618-638.

Peregrine D H (1996). Long waves on a beach. J. Fliud Mech. 27:815-827.

Smallman J V and Cooper A J (1989). A mathematical model for set down in harbors. Coastal Eng., 13:247-261.

Wei G, Kirby J T, Grill S T and Subramanya R (1995). A fully nonlinear Boussinesq model for surface waves. Part 1: Highly nonlinear unsteady waves. J. Fluid Mech. 294:71-92.

Witting J M (1984). A unified model for the evolution of nonlinear water waves. J Comput. Phys. 56: 203-236.

Zou Zhili (1996). Higher Order Boussinesq Equations, to be published.

Proceedings of the Seventh (1997) International Offshore and Polar Engineering Conference
Honolulu, USA, May 25-30, 1997

ISBN 1-880653-28-1 (Set); ISBN 1-880653-31-1 (Vol. III)

Simulation of Transient Bichromatic Second-Order Stokes Waves in a Two-Dimensional Wavetank

S. Zhang and A.N. Williams
University of Houston
Houston, TX, USA

ABSTRACT

The generation and propagation of second-order Stokes waves in a semi-infinite, narrow channel of uniform depth is simulated numerically. The wave motion is produced by the small-amplitude, bichromatic, oscillatory motion of a generic planar wavemaker. A novel second-order radiation boundary condition, previously developed for monochromatic waves, is extended herein to deal with a bichomatic wave system. Numerical results are presented for the second-order wave systems resulting from different wavemaker geometry and motion parameters. The suitability of the second-order radiation boundary condition is verified by monitoring the wave profile and through energy considerations on the computational domain. The accuracy of the computed results is checked by comparing the time-domain results with published frequency-domain solutions and laboratory data. The present approach provides an accurate and efficient technique to simulate the generation and propagation of nonlinear bichromatic Stokes waves in two dimensions.

INTRODUCTION

The proper interpretation of the experimental results obtained in a laboratory wave flume requires an appreciation of the limitations of the experimental setup and of the characteristics of the wave field produced, especially true if the waves are of finite-amplitude. In the present paper the nonlinear wave field produced in a laboratory flume consisting of a long, relatively narrow channel with a planar wavemaker at one end and an absorbing beach at the other is considered. Such facilities are common worldwide for the study of wave-structure interactions. Although the generated wave field is two-dimensional, this restriction is unimportant in many applications, and the use of a single planar wavemaker to produce the wave motion results in a high degree of control over the generated wave field.

In order to accurately predict the nonlinear wave field produced by planar wavemaker motions, several investigators have studied the problem using the Stokes expansion procedure in the frequency-domain. Hudspeth and Sulisz(1991) presented a complete second-order solution to the problem for both flap and piston-type wavemakers by an eigenfunction expansion approach. Monochromatic wavemaker motions were assumed and a set of complementary solutions was developed to satisfy the inhomogeneous boundary conditions on the wavemaker and free-surface. This solution has been extended to account for bichromatic wavemaker motions by Moubayed and Williams(1993) and, more recemtly, by Schaffer(1996). The frequency-domain solution yields the linear(first-order) and quadratic(second-order) transfer functions of wave elevation, from which the time history of the free-surface to second-order may be obtained.

An alternative approach to seeking a second-order solution in the frequency domain is to solve the wavemaker problem directly to this order in the time-domain. This technique has been successfully applied to analyze several wave-structure interaction problems in both two and three-dimensions by Isaacson and his co-workers(Isaacson and Cheung, 1990, 1991; Cheung and Isaacson, 1991; Isaacson and Ng, 1993; Ng and Isaacson, 1993) and was recently applied to the wavemaker problem by Zhang and Williams (1996). The analysis of Zhang and Williams(1996), however, was restricted to the study of monochromatic wavemaker motions. One feature of this latter analysis was the development and evaluation of a new second-order radiation boundary condition to be applied at the far-field boundary of the computational domain. The accuracy of this radiation boundary condition was verified by considering the forms of the computed second-order wave profiles and by monitoring the difference between the energy flux at the far-field boundary and the power input by the wavemaker. Relative errors of less than 2% were obtained in all test cases examined.

The present paper extends the application of the second-order radiation condition developed by Zhang and Williams (1996) to enable the generation and propagation of bichromatic second-order Stokes waves to be simulated. Potential flow theory is assumed and the initial-boundary value problems for the velocity potentials at both first- and second-order are solved by a high-order, boundary element approach. The bichromatic second-order potential is decomposed into free and forced wave components, and each of these consists of four separate terms corresonding to the double frequency, sum-frequency, and difference frequency of the two constituent wave components. The previously-developed second-order radiation boundary condition is modified to take into account the aymptotic properties of these bichromatic wave components. The frequency of the two constituent wave components. The previously-developed second-order radiation boundary condition is modified to take into account the aymptotic properties of these bichromatic wave components. The suitability of the second-order radiation boundary condition formulation and the correctness of its implementation are verified by monitoring the simulated wave profile, and through energy considerations on the computational domain. The accuracy of the computed results is also checked against published second-order frequency-domain solutions and laboratory data for bichromatic waves. It is concluded that the present approach provides and accurate and efficient technique to simulate the generation and propagation of bichromatic second-order Stokes waves in a laboratory wavetank and may provide a useful tool to analyze the results of experiments carried out in these type of facilities.

THEORETICAL DEVELOPMENT

The geometry of the problem is shown in Figure 1. A generic planar wavemaker is located at one end of a semi-infinite, two-dimensional channel. The still-water depth of the fluid in the channel is denoted by d. Cartesian coordinates are employed with $x = 0$ at the mean position of the wavemaker and $x > 0$ denoting the fluid domain, the z-axis is directed vertically upwards from an origin at the still-water level. The fluid is assumed to be inviscid and incompressible, and to undergo irrotational motion in response to the wavemaker oscillations. The fluid motion may therefore be described in terms of a velocity potential $\phi(x,z,t)$ such that the fluid velocity vector $\mathbf{q} = (u,w) = \nabla\phi$.

If the velocity potential $\phi(x,z,t)$ and the free-surface elevation $\eta(x,t)$ are assumed expressible in Stokes series, namely

$$\phi(x,z,t) = \phi^{(1)}(x,z,t) + \phi^{(2)}(x,z,t) + \dots \tag{1a}$$

$$\eta(x,t) = \eta^{(1)}(x,t) + \eta^{(2)}(x,t) + \dots \tag{1b}$$

then the full, nonlinear initial/boundary value problem may be replaced by a sequence of linear problems, one at each order of perturbation.

The wavemaker motion is assumed harmonic and bichromatic, therefore, the motion of the wavemaker can be expressed as

$$\zeta(z,t) = \sum_{j=1}^{2} \zeta_j(z,t) = \sum_{j=1}^{2} \zeta_{j0}(z)\sin(\omega_j t) \tag{2}$$

where $\zeta_{j0}(z)$ is the shape function for the wavemaker and the ω_j, j=1,2, are the radian frequencies of the wavemaker motions. The first-order velocity potential and free surface elevation are also decomposed into terms corresponding to their constituent frequency components, namely

$$\phi^{(1)}(x,z,t) = \sum_{j=1}^{2} \phi_j^{(1)}, \tag{3}$$

$$\eta(x,t) = \sum_{j=1}^{2} \eta_j^{(1)}, \tag{4}$$

then the initial/boundary problem at first-order is

$$\nabla^2\phi_j^{(1)} = 0 \quad \text{in the fluid domain } \Omega, \tag{5}$$

$$\frac{\partial\phi_j^{(1)}}{\partial z} = 0 \quad \text{on the channel bottom } \Gamma_B, \tag{6}$$

$$\frac{\partial\phi_j^{(1)}}{\partial z} - \frac{\partial\eta_j^{(1)}}{\partial t} = 0 \quad \text{on the quiescent free-surface } \Gamma_F, \tag{7}$$

$$\frac{\partial\phi_j^{(1)}}{\partial t} + g\,\eta_j^{(1)} = 0 \quad \text{on the quiescent free-surface } \Gamma_F, \tag{8}$$

$$\frac{\partial\phi_j^{(1)}}{\partial x} - \frac{\partial\zeta_j^{(1)}}{\partial t} = 0 \text{ on the mean wavemaker surface } \Gamma_w, \tag{9}$$

$$\left(\frac{\partial}{\partial t} + c_j\frac{\partial}{\partial x}\right)\phi_j^{(1)} = 0 \text{ on the radiation boundary } \Gamma_\infty, \tag{10}$$

for j = 1, 2, where $c_j = \omega_j/k_j$ is the celerity of the j-th wave component and the wavenumber of this component, k_j, is related to the frequency through the linear dispersion relation, $\omega_j^2 = gk_j\tanh(k_jd)$.

The boundary value problem at the second-order may be written as

$$\nabla^2\phi^{(2)} = 0 \quad \text{in } \Omega, \tag{11}$$

$$\frac{\partial\phi^{(2)}}{\partial z} = 0 \quad \text{on } \Gamma_B, \tag{12}$$

$$\frac{\partial\phi^{(2)}}{\partial z} - \frac{\partial\eta^{(2)}}{\partial t} = \frac{\partial\eta^{(1)}}{\partial x}\frac{\partial\phi^{(1)}}{\partial x} - \eta^{(1)}\frac{\partial^2\phi^{(1)}}{\partial z^2} \quad \text{on } \Gamma_F, \tag{13}$$

$$\frac{\partial\phi^{(2)}}{\partial t} + g\eta^{(2)} = -\eta^{(1)}\frac{\partial^2\phi^{(1)}}{\partial z\partial t} - \frac{1}{2}|\nabla\phi^{(1)}|^2 \quad \text{on } \Gamma_F, \tag{14}$$

$$\frac{\partial\phi^{(2)}}{\partial x} = -\frac{\partial\zeta_1}{\partial z}\frac{\partial\zeta_1}{\partial z} - \zeta_1\frac{\partial^2\phi^{(1)}}{\partial x^2} \quad \text{on } \Gamma_w, \tag{15}$$

together with a suitable radiation boundary condition on Γ_∞.

An inspection of the second-order wavemaker and free-surface boundary conditions reveals that the second-order wave system is made up of both free and forced waves, corresponding to solutions involving homogeneous or non-homogeneous boundary conditions on the free-surface, respectively. Associated with each of these wave systems are four frequency components: two double-frequency terms, at $2\omega_1$ and $2\omega_2$, and sum- and difference-frequency terms at $\omega_1 + \omega_2$ and $\omega_1 - \omega_2$. Therefore, the second-order potential and free-surface elevation will be expressed in the following forms,

$$\phi^{(2)}(x,z,t) = \psi_D^{(2)}(x,z,t) + \xi_D^{(2)}(x,z,t) + \psi_\pm^{(2)}(x,z,t) + \xi_\pm^{(2)}(x,z,t), \tag{16a}$$

$$\eta^{(2)}(x,t) = \Delta_D^{(2)}(x,t) + \Theta_D^{(2)}(x,t) + \Delta_\pm^{(2)}(x,t) + \Theta_\pm^{(2)}(x,t), \tag{16b}$$

where $\psi_D^{(2)}(x,z,t)$ and $\xi_D^{(2)}(x,z,t)$ are the potentials associated, respectively, with the forced and free waves at the double frequencies of the wave components, i.e., at frequencies $2\omega_1$ and $2\omega_2$; while $\psi_\pm^{(2)}(x,z,t)$ and $\xi_\pm^{(2)}(x,z,t)$ are the potentials associated, respectively, with the forced and free waves waves at the sum- and difference-frequencies. A similar decomposition holds for the second-order free-surface elevations (Eq. 16b).

The initial/boundary value problems for $\psi_D^{(2)}(x,z,t)$ and $\Delta_D^{(2)}(x,t)$, and $\xi_D^{(2)}(x,z,t)$ and $\Theta_D^{(2)}(x,t)$, are given by

$$\nabla^2(\psi_D^{(2)}, \xi_D^{(2)}) = 0 \quad \text{in } \Omega, \tag{17}$$

$$\frac{\partial}{\partial z}(\psi_D^{(2)}, \xi_D^{(2)}) = 0 \quad \text{on } \Gamma_B, \tag{18}$$

$$\frac{\partial \psi_D^{(2)}}{\partial z} - \frac{\partial \Delta_D^{(2)}}{\partial t} = \frac{\partial \eta_1^{(1)}}{\partial x}\frac{\partial \phi_1^{(1)}}{\partial x} - \eta_1^{(1)}\frac{\partial^2 \phi_1^{(1)}}{\partial z^2} + \frac{\partial \eta_2^{(1)}}{\partial x}\frac{\partial \phi_2^{(1)}}{\partial x} - \eta_2^{(1)}\frac{\partial^2 \phi_2^{(1)}}{\partial z^2} \quad \text{on } \Gamma_F, \tag{19a}$$

$$\frac{\partial \psi_D^{(2)}}{\partial t} + g\Delta_D^{(2)} = -\eta_1^{(1)}\frac{\partial^2 \phi_1^{(1)}}{\partial z \partial t} - \frac{1}{2}|\nabla \phi_1^{(1)}|^2 - \eta_2^{(1)}\frac{\partial^2 \phi_2^{(1)}}{\partial z \partial t} - \frac{1}{2}|\nabla \phi_2^{(1)}|^2 \quad \text{on } \Gamma_F, \tag{19b}$$

$$\frac{\partial \xi_D^{(2)}}{\partial z} - \frac{\partial \Theta_D^{(2)}}{\partial t} = 0 \quad \text{on } \Gamma_F, \tag{20a}$$

$$\frac{\partial \xi_D^{(2)}}{\partial t} + g\Theta_D^{(2)} = 0 \quad \text{on } \Gamma_F, \tag{20b}$$

$$\frac{\partial \psi_D^{(2)}}{\partial x} = 0 \quad \text{on } \Gamma_w, \tag{21a}$$

$$\frac{\partial \xi_D^{(2)}}{\partial x} = -\frac{\partial \zeta_1}{\partial z}\frac{\partial \zeta_1}{\partial z} - \zeta_1\frac{\partial^2 \phi_1^{(1)}}{\partial x^2} - \frac{\partial \zeta_2}{\partial z}\frac{\partial \zeta_2}{\partial z} - \zeta_2\frac{\partial^2 \phi_2^{(1)}}{\partial x^2} \quad \text{on } \Gamma_w, \tag{21b}$$

$$\left(\frac{\partial}{\partial t} + c_1\frac{\partial}{\partial x}\right)\left(\frac{\partial}{\partial t} + c_2\frac{\partial}{\partial x}\right)\psi_D^{(2)} = 0 \quad \text{on } \Gamma_\infty \tag{22a}$$

$$\left(\frac{\partial}{\partial t} + v_{11}\frac{\partial}{\partial x}\right)\left(\frac{\partial}{\partial t} + v_{22}\frac{\partial}{\partial x}\right)\xi_D^{(2)} = 0 \text{ on } \Gamma_\infty \tag{22b}$$

where v_{11} and v_{22} are the celerities of the second-order free-wave components at frequencies $2\omega_1$ and $2\omega_2$ and are defined by $v_{jj} = 2\omega_j/\kappa_j$ for j = 1, 2, where $4\omega_j^2 = g\kappa_j \tanh(\kappa_j d)$. It is noted that the celerities of the forced wave components at the double frequencies are the same as the those of the corresponding first-order components.

The initial/boundary value problems for $\psi_\pm^{(2)}(x,z,t)$ and $\Delta_\pm^{(2)}(x,t)$, and $\xi_\pm^{(2)}(x,z,t)$ and $\Theta_\pm^{(2)}(x,t)$ are given by

$$\nabla^2(\psi_\pm^{(2)}, \xi_\pm^{(2)}) = 0 \quad \text{in } \Omega, \tag{23}$$

$$\frac{\partial}{\partial z}(\psi_\pm^{(2)}, \xi_\pm^{(2)}) = 0 \quad \text{on } \Gamma_B, \tag{24}$$

$$\frac{\partial \psi_\pm^{(2)}}{\partial z} - \frac{\partial \Delta_\pm^{(2)}}{\partial t} = \frac{\partial \eta_1^{(1)}}{\partial x}\frac{\partial \phi_2^{(1)}}{\partial x} - \eta_1^{(1)}\frac{\partial^2 \phi_2^{(1)}}{\partial z^2} + \frac{\partial \eta_2^{(1)}}{\partial x}\frac{\partial \phi_1^{(1)}}{\partial x} - \eta_2^{(1)}\frac{\partial^2 \phi_1^{(1)}}{\partial z^2} \quad \text{on } \Gamma_F, \tag{25a}$$

$$\frac{\partial \psi_\pm^{(2)}}{\partial t} + g\Delta_\pm^{(2)} = -\eta_1^{(1)}\frac{\partial^2 \phi_2^{(1)}}{\partial z \partial t} - \eta_2^{(1)}\frac{\partial^2 \phi_1^{(1)}}{\partial z \partial t} - \left(\frac{\partial \phi_1^{(1)}}{\partial x}\frac{\partial \phi_2^{(1)}}{\partial x} + \frac{\partial \phi_1^{(1)}}{\partial z}\frac{\partial \eta_2^{(1)}}{\partial z}\right) \quad \text{on } \Gamma_F, \tag{25b}$$

$$\frac{\partial \xi_\pm^{(2)}}{\partial z} - \frac{\partial \Theta_\pm^{(2)}}{\partial t} = 0 \quad \text{on } \Gamma_F, \tag{26a}$$

$$\frac{\partial \xi_\pm^{(2)}}{\partial t} + g\Theta_\pm^{(2)} = 0 \quad \text{on } \Gamma_F, \tag{26b}$$

$$\frac{\partial \psi_\pm^{(2)}}{\partial x} = 0 \quad \text{on } \Gamma_w, \tag{27a}$$

$$\frac{\partial \xi_\pm^{(2)}}{\partial x} = -2\frac{\partial \zeta_1}{\partial z}\frac{\partial \zeta_2}{\partial z} - \zeta_1\frac{\partial^2 \phi_2^{(1)}}{\partial x^2} - \zeta_2\frac{\partial^2 \phi_1^{(1)}}{\partial x^2} \quad \text{on } \Gamma_w, \tag{27b}$$

$$\left(\frac{\partial}{\partial t} + c_+\frac{\partial}{\partial x}\right)\left(\frac{\partial}{\partial t} + c_-\frac{\partial}{\partial x}\right)\psi_\pm^{(2)} = 0 \quad \text{on } \Gamma_\infty \tag{28a}$$

$$\left(\frac{\partial}{\partial t} + v_+\frac{\partial}{\partial x}\right)\left(\frac{\partial}{\partial t} + v_-\frac{\partial}{\partial x}\right)\xi_\pm^{(2)} = 0 \quad \text{on } \Gamma_\infty \tag{28b}$$

where, in Eq. 28a, c_+ and c_- are the celerities of the sum- and difference frequency forced waves and are given by

$$c_+ = \frac{\omega_1+\omega_2}{k_1+k_2}, \tag{29a}$$

$$c_- = \left|\frac{\omega_1-\omega_2}{k_1-k_2}\right|. \tag{29b}$$

It should be noted that the wave numbers and frequencies associated with these waves do not satisfy linear dispersion relationships. In Eq. 28b, v_+ and v_- are celerities of the sum- and diffrence frequency free waves and are given by $v_+ = (\omega_1+\omega_2)/\kappa^+$ and $v_- = |\omega_1-\omega_2|/\kappa^-$ in which κ^+ and κ^- satisfy the following linear dispersion relationships, $(\omega_1+\omega_2)^2 = g\kappa^+ \tanh(\kappa^+ d)$, and $(\omega_1-\omega_2)^2 = g\kappa^- \tanh(\kappa^- d)$.

The velocity potentials at both first and second-order may be obtained through a boundary-element technique. Application of Green's second identity to the first-order potentials $\phi_j^{(1)}$, j = 1,2 and a singular solution of Laplace's equation in two dimensions, namely

$$G(\underline{x}; \underline{\zeta}) = \ln|\underline{x} - \underline{\zeta}|, \tag{30}$$

leads to the following integral equations

$$\gamma_o \phi_j^{(1)}(\underline{x}, t) = -\int_\Gamma \left\{ \frac{\partial \phi_j^{(1)}}{\partial n}(\underline{\zeta}, t)\, G(\underline{x}; \underline{\zeta}) - \phi_j^{(1)}(\underline{\zeta}, t)\frac{\partial G}{\partial n}(\underline{x}; \underline{\zeta}) \right\} d\Gamma, \tag{31}$$

for j = 1, 2, in which the contour Γ consists of Γ_B, Γ_F, Γ_W and Γ_∞. In equation (31), γ_o is the interior angle to Γ at $\underline{x}$. At a smooth boundary point γ_o takes the value π.

Applying Green's second identity separately to the fundamental solution of Laplace's equation, $G(\underline{x}; \underline{\zeta})$, and each of $\psi_D^{(2)}$, $\psi_\pm^{(2)}$, $\xi_D^{(2)}$ and $\xi_\pm^{(2)}$ in turn, yields four integral equations at second-order, namely

$$\gamma_o \Phi(\underline{x}) = - \int_\Gamma \left\{ \frac{\partial \Phi}{\partial n}(\zeta)\, G(\underline{x};\zeta) - \Phi(\zeta) \frac{\partial G}{\partial n}(\underline{x};\zeta) \right\} d\Gamma, \qquad (32)$$

where $\Phi(\underline{x}) = \left(\psi_D^{(2)}(\underline{x},t),\ \psi_\pm^{(2)}(\underline{x},t),\ \xi_D^{(2)}(\underline{x},t),\ \xi_\pm^{(2)}(\underline{x},t)\right)$. Note that in the present formulation the function $G(\underline{x};\zeta)$ does not contain a term corresponding to the reflection of the point ζ in the channel bottom. This additional term eliminates Γ_B from the contour of integration, but restricts the development to a horizontal bottom. The current formulation admits an arbitrary geometry for the channel bottom. Also, it should be noted that since the contour of integration is fixed, and the singular solution and its derivative are functions of geometry only, these quantities need be calculated only once during the simulation.

Special attention is given to corner points at the intersection of the wavemaker and free-surface and channel bottom, and intersection of the radiation boundary and free-surface and channel bottom. At these locations double nodes are used to provide two different normal directions and to allow the specification of different normal velocities on the two sides of the intersection point. Continuity of potential is imposed at these nodes (Grilli and Svendsen, 1990). On the wavemaker and channel bottom, the corner velocities are given explicitly through the wavemaker and bottom boundary conditions, on the free-surface and radiation boundaries, they may be calculated from neighboring points using appropriate one-sided finite-difference schemes.

NUMERICAL IMPLEMENTATION

The boundary contour Γ is now discretized into a number of small line elements with the velocity potential and its normal derivative being assumed to vary linearly within an element. Under this assumption, the line integrals involving the singular solution $G(\underline{x}, \zeta)$ and its normal derivative are carried out analytically. The integral equations, Eqs. 31 and 32, are thus replaced by a set of algebraic equations at first-order and four sets at second-order which involve the various velocity potentials and their normal derivatives at the nodal points on Γ. In order to solve the integral equations, Eqs. 31 and 32, it is required that either, $\phi^{(1)}$ or $\partial\phi^{(1)}/\partial n$, or, Φ or $\partial\Phi/\partial n$, be specified along any segment of the contour at each time step. In the present application, $\partial\phi^{(1)}/\partial n$ and $\partial\Phi/\partial n$ are specified on the wavemaker, radiation and bottom boundaries, while on the free-surface, $\phi^{(1)}$ and Φ are first obtained by numerical integration of the boundary condition, then $\partial\phi^{(1)}/\partial n$ and $\partial\Phi/\partial n$ are calculated from the corresponding integral equations.

The wavemaker boundary conditions at first and second-order are written in the form,

$$\frac{\partial \phi^{(1)}}{\partial n} \equiv - \frac{\partial \phi^{(1)}}{\partial x} = \omega\, \zeta_o \cos(\omega t) \qquad \text{on } \Gamma_w, \qquad (33)$$

$$\frac{\partial \xi_D^{(2)}}{\partial n} \equiv - \frac{\partial \xi_D^{(2)}}{\partial x} = \zeta_{10} \cos(\omega_1 t) \frac{\partial^2 \phi_1^{(1)}}{\partial x^2} - \frac{\partial \phi_1^{(1)}}{\partial z} \frac{\partial \zeta_{1o}}{\partial z} \sin(\omega_1 t) + \zeta_{20} \cos(\omega_2 t) \frac{\partial^2 \phi_2^{(1)}}{\partial x^2} - \frac{\partial \phi_2^{(1)}}{\partial z} \frac{\partial \zeta_{2o}}{\partial z} \sin(\omega_2 t) \text{ on } \Gamma_w, \quad (34)$$

$$\frac{\partial \xi_\pm^{(2)}}{\partial n} \equiv - \frac{\partial \xi_\pm^{(2)}}{\partial x} = \zeta_{10} \cos(\omega_1 t) \frac{\partial^2 \phi_2^{(1)}}{\partial x^2} - \frac{\partial \phi_1^{(1)}}{\partial z} \frac{\partial \zeta_{2o}}{\partial z} \sin(\omega_2 t) + \zeta_{20} \cos(\omega_2 t) \frac{\partial^2 \phi_1^{(1)}}{\partial x^2} - \frac{\partial \phi_2^{(1)}}{\partial z} \frac{\partial \zeta_{1o}}{\partial z} \sin(\omega_1 t) \text{ on } \Gamma_w, \quad (35)$$

where n is the vector normal to the domain boundary, pointing out of the fluid domain. It is noted that the wavemaker boundary condition for $\Psi_D^{(2)}$ and $\Psi_\pm^{(2)}$ are homogeneous. In Eqs. 34 and 35, vertical velocities are approximated using central-difference schemes while the second x-derivative terms are replaced by the corresponding z-derivative utilizing Laplace's equation, and then approximated by central differences. In the initial stages of the wavemaker motion, the stroke is modified by a so-called "ramp function." This ensures a smooth startup of the numerical process and so helps to avoid instabilities associated with large potential gradients. The simulation is therefore also relatively insensitive to the initial values assumed for the potentials and the values of the first and second potential derivatives computed in the very early stages of the simulation. On the free-surface, a five-point smoothing scheme is used to suppress the growth of high-frequency instabilities on that boundary.

The free-surface boundary conditions at first- and second-order are treated as follows: first, the velocity potentials $\phi^{(1)}$, $\psi_D^{(2)}$ and $\psi_\pm^{(2)}$ are obtained by integrating the dynamic free-surface boundary conditions, Eqs. 8, 19b and 25b, in time, and the normal derivatives $\partial\phi^{(1)}/\partial n$, $\partial\psi_D^{(2)}/\partial n$ and $\partial\psi_\pm^{(2)}/\partial n$, are then obtained from the kinematic free-surface boundary conditions, Eqs. 7, 19a and 25a. In the numerical implementation of the free-surface boundary conditions, central differences are used to evaluate x-derivatives, while backward differences are used for time derivatives. The first-order z-derivative terms in the second-order free-surface condition are obtained from the first-order boundary integral equation. Second derivatives with respect to z are replaced by corresponding x-derivatives utilizing Laplace's equation, and then approximated by a central-difference scheme.

The first-order radiation boundary condition, Eq. (10), is rearranged as

$$\frac{\partial \phi_j^{(1)}}{\partial n} \equiv \frac{\partial \phi_j^{(1)}}{\partial x} = - \frac{1}{c_j} \frac{\partial \phi_j^{(1)}}{\partial t} \qquad j = 1, 2 \qquad \text{on } \Gamma_\infty. \qquad (36)$$

The second-order radiation boundary conditions, Eq. 22 and 28, are each decomposed into two equations:

$$\frac{\partial \Phi}{\partial x} = - \frac{1}{u_1 + u_2} \frac{\partial \Phi}{\partial t} + \Lambda \qquad \text{on } \Gamma_\infty, \qquad (37a)$$

$$\frac{\partial \Lambda}{\partial t} = \frac{u_1 u_2}{u_1 + u_2} \frac{\partial^2 \Phi}{\partial z^2} \qquad \text{on } \Gamma_\infty. \qquad (37b)$$

where u_1 and u_2 are the two celerities in the boundary condition. In Eq. 37b the second derivative with respect to x has been replaced by the corresponding z-derivative by applying Laplace's equation. This rearrangement avoids the numerical computation of second derivatives.

Once the first- and second-order velocity potentials at each time step have been determined, the wave elevation to second-order may be obtained by combining the first- and second-order free-surface elevations given by the rearranged forms of Eqs. 8 and 14, namely

$$\eta^{(1)} = -\frac{1}{g}\frac{\partial\phi^{(1)}}{\partial t} \qquad \text{on } \Gamma_F, \qquad (38)$$

$$\eta^{(2)} = -\frac{1}{g}\frac{\partial\phi^{(2)}}{\partial t} + \frac{1}{g^2}\frac{\partial\phi^{(1)}}{\partial t}\frac{\partial^2\phi^{(1)}}{\partial z\partial t} - \frac{1}{2g}\left|\nabla\phi^{(1)}\right|^2 \text{ on } \Gamma_F. \quad (39)$$

VALIDATION OF NUMERICAL APPROACH

In Zhang and Williams (1996) the numerical aproach was validated for the case of monochromatic wavemaker motions. Therefore, the present discussion will concentrate on the bichromatic case only.

First the suitability of the radiation boundary conditions at first- and second-order for a channel of uniform depth are investigated. For simplicity, a full-draft piston wavemaker is assumed, that is, with reference to Figure 1, $S_b = S_o$, $h = 0$ and so $b = d$. The finite computational domain extends to $x = L$. Figure 2 presents the instantaneous first- plus second-order (i.e., total) dimensionless free-surface profiles $\eta(x,t)/S_o$ at various dimensionless times $\tau = t/T$, where T is the wavemaker oscillation period, for different locations of the radiation boundary, $x = L$. The dimensionless distance from the wavemaker is x/d. In each case locations corresponding to L/d = 10, 15 and 20 are chosen for $k_1d = 1$ and $k_2d = 1.25$, with $S_{1o}/d = S_{2o}/d = 0.15$. A comparison of the results from the figures indicates that the computed wave form to second-order is stable, and that no reflections from the finite radiation boundary are apparent.

The accuracy of the numerical approach is now investigated. Since the present theory is based on the Stokes perturbation expansion procedure, the boundary-initial value problem is actually solved over the fixed computational domain consisting of Γ_W, Γ_∞, Γ_F and Γ_B rather than the instantaneous fluid region. Therefore, for a given boundary element discretization and time-step, the accuracy of the numerical approach is assessed by calculating the difference between the net energy flux over the wavemaker, radiation and free-surface boundaries, and the rate of change of the kinetic energy of the fluid (see, for example, Stoker, 1957). It is noted that in the present application, the potential energy in the computational domain is constant. Therefore, the following quantity is computed:

$$\Delta F(t) = \rho \int_\Gamma \frac{\partial\phi}{\partial t}\frac{\partial\phi}{\partial n}\, d\Gamma - \frac{1}{2}\rho\frac{d}{dt}\int_\Gamma \frac{1}{2}\rho\,\phi\frac{\partial\phi}{\partial n}\, d\Gamma. \quad (40)$$

The quantity ΔF is computed only to $O(\varepsilon^3)$, where ε is of the order of the wave steepness, since terms of $O(\varepsilon^4)$ and above will involve the third-order solution for the velocity potential. Figure 3 presents the dimensionless time rate of change of kinetic energy, the dimensionless net outward energy flux and a relative error based on Eq. 40 for k_1d = 1, $k_2d = 1.5$, $S_{1o}/d = 0.2$, $S_{2o}/d = 0.1325$ and L/d = 15 and 20, respectively. All quantities in these figures have been nondimensionalized by the first-order, that is $O(\varepsilon^2)$, mean power for the wavemaker problem (Dean and Dalrymple, 1984). The instant the first-order waves reach the finite radiation boundary may be clearly seen in the figures. The dimensionless time rate of change of kinetic energy and the dimensionless net outward energy flux are equal and opposite and the relative errors at both first- and second-order are small compared to these quantities.

A comparison has also been made between the present approach and the second-order frequency domain solution of Moubayed and Williams (1994). The linear and quadratic transfer functions from the frequency-domain solution were used to construct the time-history of the free-surface elevation. The comparison of the free-surface elevations to second-order at x/d = 5 for $S_{1o}/d = S_{2o}/d$ = 0.2, $d/L_o^+ = 1$, $d/L_o^- = 0.005$, where $L_o^+ = 2\pi g/(\omega_1+\omega_2)^2$ and $L_o^- = 2\pi g/(\omega_1-\omega_2)^2$ is shown in Figure 3. The location of the radiation boundary is at L/d = 20. The comparison is shown over twenty-one sum-frequency wave periods, corresponding to a dimensionless time interval $3 \leq \tau^+ \leq 24$. The initial time period corresponding $0 \leq \tau^+ <$ 3 was not considered in order to minimize the influence of the wave generator ramp function. It can be seen from the figure that except for some discrepancies near $\tau^+ = 5$ (which are caused by the ramp function) the time-domain and frequency-domain solutions are in excellent agreement over the entire time period considered. Further comparison between the time and frequency-domain solutions, this time for $d/L_o^+ = 1$ and $d/L_o^- = 0.002$, are shown in Figure 5. Again, apart from the differences near $\tau^+ = 5$ excellent agreement is exhibited over the entire time period considered.

Finally, comparison has been made between the wave profiles computed using the present approach and the experimentally-measured free-surface elevations obtained by Schaffer(1996). The experiments were carried out in a wave flume 20 x 0.6 x 1.0 m. Two separate water depths were considered, 0.3 m and 0.6 m. The hydraulically driven wavemaker was operated in the piston mode. In the 0.3 m case, the bichromatic wave field was characterized by a pair of waves with a linear wave height of 0.03 m and frequencies of 0.495 Hz and 0.566 Hz. In the 0.6 m case, the bichromatic wave field was characterized by a pair of waves with a linear wave height of 0.12 m and frequencies of 0.60 Hz and 0.70 Hz. In the present case the required wavemaker stroke for the numerical model was obtained from the prescribed wave height using the linear wavemaker theory of Dean and Dalrymple(1984). Surface elevations are compared at one location in the tank, at x = 3.0 m over the tme interval from 60 to 120 seconds. These comparisons are presented in Figures 6 and 7. It can be seen that the agreement between the theoretical and experimental results is good, except when the surface elevations are small. However, it is noted that in Schaffer's experiments, a low pass filter was used to remove noise from his raw data, resulting in the loss of both high and low frequency components beyond the cut-offs. Also, it was noted by Schaffer that reflections of the difference-frequency components from the wave absorber at the far end of the tank resulted in a complex wave field in the tank and so the accuracy of the difference-frequency measurements is questionable. The slight differences between the experiments and the present model may be attributed to these facts. Therefore, it is concluded that the results of the present numerical model show good agreement with experimental results for bichromatic waves.

CONCLUSIONS

The generation and propagation of second-order Stokes waves in a semi-infinite, narrow channel of uniform depth has been simulated numerically. The wave motion was produced by the small-amplitude, bichromatic, oscillatory motion of a generic planar wavemaker. A novel second-order radiation boundary condition, previously developed for monochromatic waves, has been extended herein to deal with a bichomatic wave system. The suitability of the second-order radiation boundary condition has been verified by monitoring the wave profile and through energy considerations on the computational domain. The accuracy of the computed results has been further checked by comparing the time-domain results with published frequency-domain solutions and laboratory data. The present approach provides an accurate and efficient technique to simulate the generation and propagation of nonlinear bichromatic Stokes waves in two dimensions and represents a powerful tool for the analysis of experimental results in laboraotory facilities.

ACKNOWLEDGEMENT

This study was supported in part by the Texas Advanced Technology Program (Project No. 003562-111). This support is gratefully acknowledged.

REFERENCES

Cheung, K.F., and Isaacson, M., (1991). "On Two Alternative Field Solutions for Second-Order Wave Diffraction," Jl. Offshore Mechanics and Arctic Engineering, A.S.M.E., Vol. 113, No. 3, pp. 185-192.

Dean, R.G. and Dalrymple, R.A. (1984) Water Wave Mechanics for Scientists and Engineers. Prentice-Hall, Englewood Cliffs, New Jersey.

Grilli, S.T., and Svendsen, I.A., (1990). "Corner Problems and Global Accuracy in the Boundary Element Solution of Nonlinear Wave Flows ," Engineering Analyisis with Boundary Elements, Vol. 7, No. 4, pp. 178-195.

Hudspeth, R.T. and Sulisz, W. (1991). Stokes Drift in Two-Dimensional Wave Flumes. Jl. of Fluid Mechanics, **230**, pp. 209-229.

Isaacson, M. and Cheung, K.F., (1990). "Time-Domain Solution for Second-Order Wave Diffraction," Jl. Waterway, Port, Coastal, and Ocean Engineering, A.S.C.E., Vol. 116, No. 2, pp. 191-210.

Isaacson, M. and Cheung, K.F., (1991). "Second-Order Wave Diffraction Around Two-Dimensional Bodies by Time-Domain Method," Applied Ocean Research, Vol. 13, No. 4, pp. 175-186.

Isaacson, M. and Ng, J.Y.T. (1993). Time-Domain Second-Order Wave Radiation in Two Dimensions. Journal of Ship Research, Vol. 37, No. 1, pp. 25-33.

Moubayed, W.I. and Williams, A.N. (1994). Second-Order Bichromatic Waves Produced by a Generic Planar Wavemaker in a Two-Dimensional Wave Flume. Journal of Fluids and Structures, Vol. 8, pp. 73-92.

Ng, J. and Isaacson, M. (1993). Second-Order Wave Interaction with Two-Dimensional Floating Bodies by a Time-Domain Method. Applied Ocean Research, Vol. 15, pp. 95-105.

Schaffer, H.A. (1996) Second-Order Wavemaker Theory for Irregular Waves. Ocean Engineering, Vol. 23, pp. 47-88.

Stoker, J.J. (1957). Water Waves: The Mathematical Theory with Application. Wiley Interscience.

Zhang, S. and Williams, A.N. (1996). Time-Domain Simulation of the Generation and Propagation of Second-Order Stokes Waves in a Two-Dimensional Wave Flume. Part I: Monochromatic Wavemaker Notions. Journal of Fluids and Structures, Vol. 10, pp. 319-335.

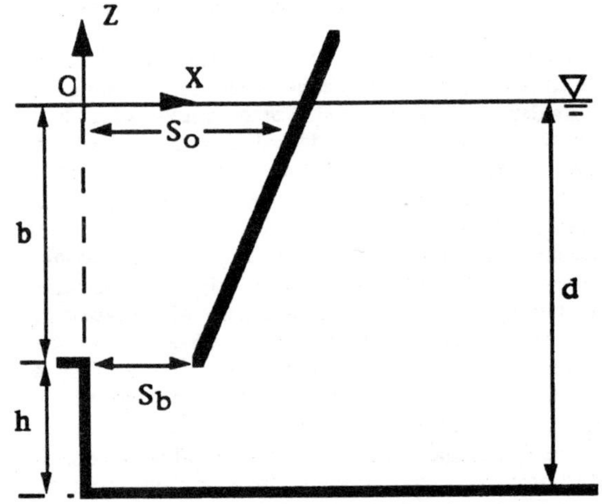

Figure 1: Definition sketch showing physical domain.

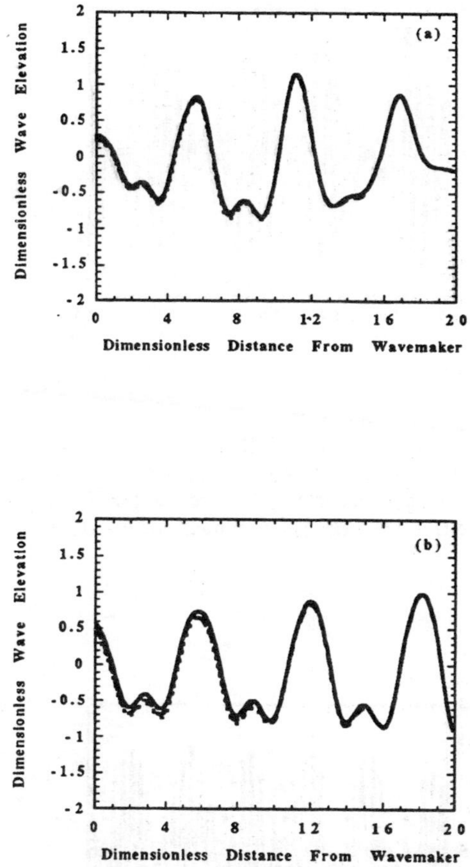

Figure 2: Dimensionless total (first- plus second-order) free-surface profile at dimensionless times $\tau = 5.22$ (a) and $\tau = 7.8$ (b) for different locations of the radiation boundary for $k_1d = 1.0$, $k_2d = 1.25$, $S_{10}/d = S_{20}/d = 0.15$. Notations: ---- L/d = 10; — — — L/d = 15; ——— L/d = 20.

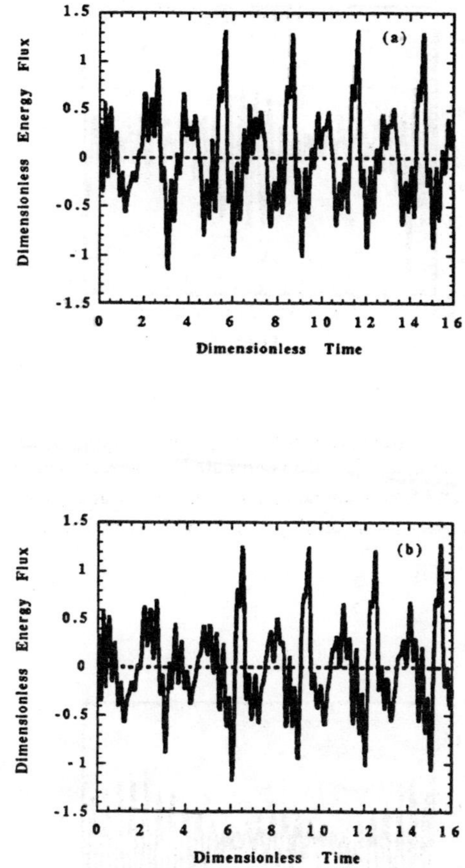

Figure 3: Variation of dimensionless net enery flux, dimensionless time rate of change of kinetic energy and relative error with time, for $k_1d = 1.0$, $k_2d = 1.5$, $S_{10}/d = 0.2$, $S_{20}/d = 0.1325$, and L/d = 15 (a) and L/d = 20 (b). Only $O(\varepsilon^3)$ terms, based on second-order solution, are shown. Notations: ——— time rate of change of energy; — — — net flux; ------ error.

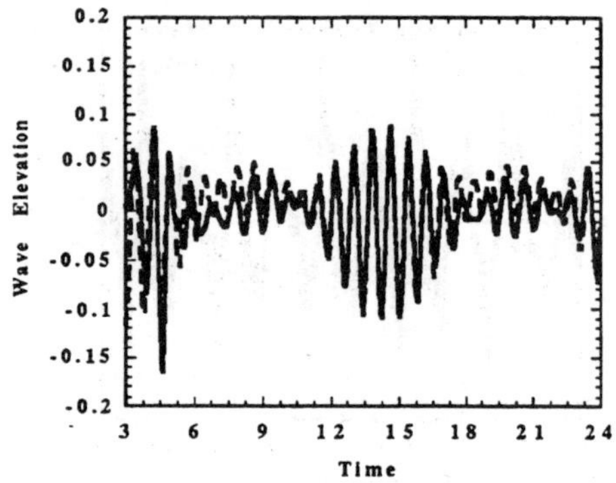

Figure 4: Comparison of total (first- plus second-order) wave elevation at x/L = 0.33 computed by present method (———) and frequency-domain solution of Moubayed and Williams (— — —) for $h/L_{o+} = 1.0$ and $h/L_{o-} = 0.005$.

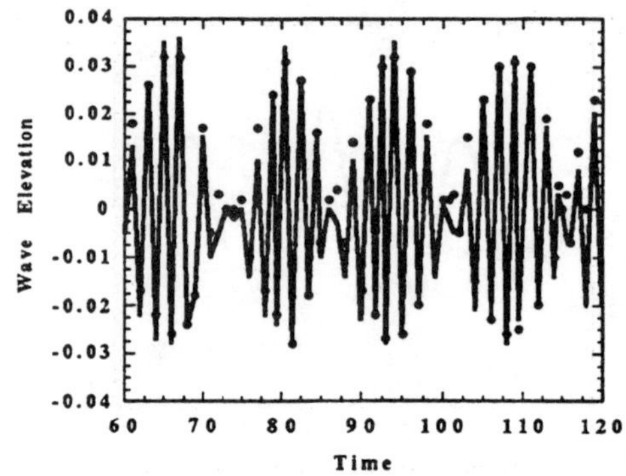

Figure 6: Comparison of total (first- plus second-order) wave elevation computed by present method (———) with experimental measurements of Schaffer (●) for $H_1 = H_2 =$ 0.03 m and $f_1 = 0.495$ Hz, $f_2 = 0.566$ Hz and d = 0.3 m.

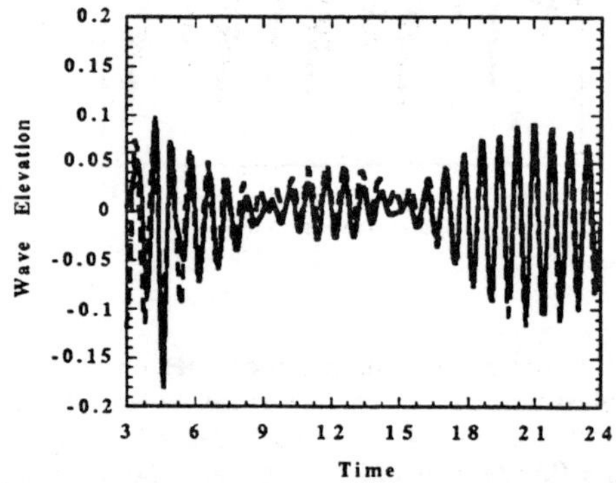

Figure 5: Comparison of total (first- plus second-order) wave elevation at x/L = 0.33 computed by present method (———) and frequency-domain solution of Moubayed and Williams (— — —) for $h/L_{o+} = 1.0$ and $h/L_{o-} = 0.002$.

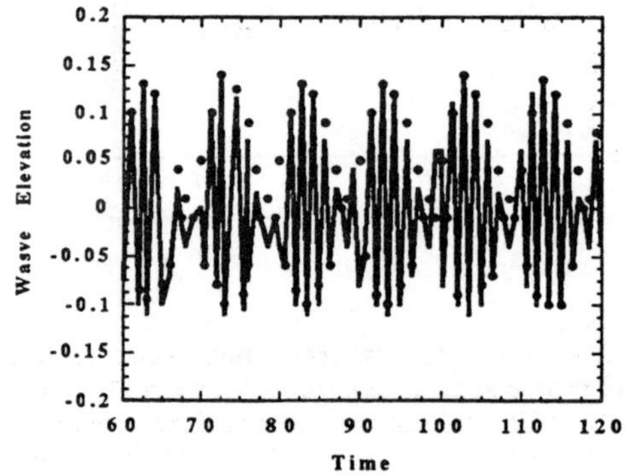

Figure 7: Comparison of total (first- plus second-order) wave elevation computed by present method (———) with experimental measurements of Schaffer (●) for $H_1 = H_2 =$ 0.12 m and $f_1 = 0.60$ Hz, $f_2 = 0.70$ Hz and d = 0.6 m.

Proceedings of the Seventh (1997) International Offshore and Polar Engineering Conference
Honolulu, USA, May 25-30, 1997

ISBN 1-880653-28-1 (Set); ISBN 1-880653-31-1 (Vol. III)

Vortex-Excited Vibration of a Vertical Circular Cylinder in Waves

Kenjirou Hayashi
National Defense Academy
Yokosuka, Japan

John R. Chaplin
City University
London, United Kingdom

Abstract

This study deals with some of the non-linear phenomena of the vortex-excited vibration of a circular cylinder in waves. Laboratory experiments have been performed to study the dynamic transverse response of a vertical circular cylinder in regular waves. The test cylinder was pivoted at its base and supported flexibly by spring at its top. The movement of the test cylinder in the direction of the inline force was restricted. The most remarkable result is the appearance of two types of peak in the amplitude of the vortex-excited vibration of the cylinder produced by perfect resonance coupled with waves and by vortex coupling. In the case of steady flow, perfect resonance appears in the range of locked-in, but in the case of wave, it appears only near to the ratio of wave frequency f_w to the natural frequency f_{nw} of the cylinder in still water $f_w/f_{nw}=1/2,1/3,1/4$,----(multi-appearance), elsewhere vortex coupling may occur for light damping in which the oscillation frequency is not simple a multiple of wave frequency.

Key Words : Vortex-excited vibration, Fluid Structure Interaction, Vertical Cylinder, Lift Force.

1. Introduction

The dynamic response of offshore structures to the wave forces is one of the most important factors in their design. There have been many instances reported where the failure of a structure is believed to have been caused by its response at frequencies of wave forces acting on it.

Wave forces acting on a vertical cylinder in waves are usually resolved into two components, inline force and a lift force. The in-line force acting in the direction of wave propagation is usually expressed by using Morison equation, and acts predominantly at a frequency equal to the wave frequency f_w. The lift force acting at right angles to the wave propagation is caused by vortex shedding and its predominant frequency is a multiple of that of in-line force. Therefore, the lift force give rise to a significant dynamic response in the cylinder.

A great number of experiments have been done in order to study the phenomenon of vortex-excited vibration of cylinder in steady flow. Summaries have been made by Blevins(1977), Sarpkaya(1979) and Bearman(1984). In this case, if the vortex shedding frequency approaches the natural frequency of a lightly damped cylinder, the vibration of cylinder becomes larger and this large vibration can drive the eddies to be shed with a frequency ranging between the natural frequency of the cylinder and the Strouhal frequency. This phenomenon is usually called " lock-in " between the frequency of vortex shedding and the frequency of the vibrating cylinder. Under "lock-in" conditions, large resonant vibration occurs and the lift forces are amplified both by the increase of vortex strength and by the improved correlation in the phase of vortex shedding along the cylinder axis.

A similar phenomenon may occur under certain conditions , if an elastically mounted vertical cylinder is placed in waves. However, this has not been sufficiently understood and relatively little work has been carried out into this interesting problem in planer oscillating flow and in waves. This is mainly due to the complexity of the phenomena, because the incident flow acting on a vertical cylinder in waves is oscillatory , varies with depth and possesses a vertical component. A comprehensive summary has been made by Sumer and Fredsoe(1994).

The work presents herein is an experimental investigation into the non-linear vortex-excited vibration of a vertical circular cylinder in regular waves. Emphasis is placed on the influence of the damping factor and Keulegan Carpenter number KC to the relationship between the transverse response and the ratio of wave frequency f_w to natural frequency f_{nw} of the cylinder.

2. Experiments

Laboratory experiments were carried out in the 18m long and 0.76m wide flume of the department of Civil

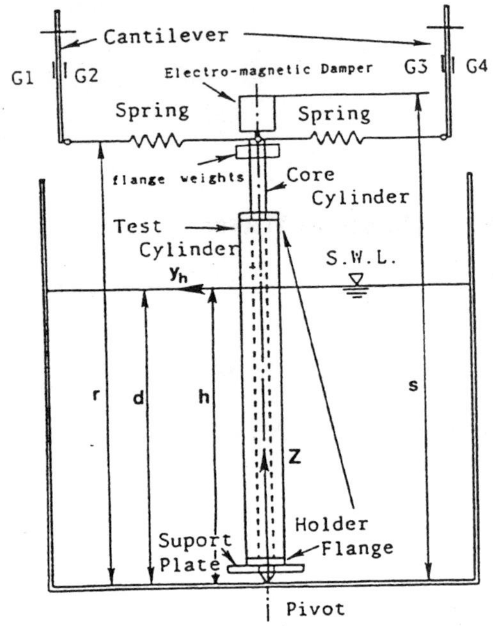

Figure 1 General arrangement of test cylinder

Engineering at the Liverpool University. The test cylinder was mounted in the flume at a section 7.35m from the paddle of the wave generator and 1.6m from the toe of the long beach with a slope of 1:6.4.

The general arrangement of the test cylinder is shown in Figure 1. A stainless steel hollow cylinder of 19.05mm outside diameter D, 0.536mm wall thickness and 985mm length, was used for the test cylinder. In order to give a high degree of rigidity to the test cylinder, both ends were connected to the aluminium core cylinder, of 10mm diameter 1090mm length, by holder flanges. The flange weights were attached to the core cylinder above the test cylinder to adjust the equivalent mass per unit length, m_e, of the test cylinder. The support plate(200x50x6mm) was attached to the holder flange at the bottom end of the test cylinder, and was pivoted at the level of the floor of the flume to prevent vibration in the in-line direction. The upper end of the core cylinder, above the water level, was mounted with springs only in the transverse (lift) direction. Each spring was connected to a strain-gauged steel cantilever by using a wire. Two strain gauges were fixed on each cantilever to measure its bending moment produced by the forces acting on the end of cantilever through the spring and wire. Strain gauges were connected into a wheatstone bridge circuit in the bridge conditioner to produce the output signal corresponding to the displacement of the test cylinder in the transverse direction.

In order to adjust the structural damping of the test cylinder, an electro-magnetic damper was used. The position of it is shown in Figure 1. It is composed of an electro-magnet, which is made of a coil and a soft iron, and conducting plate made of Aluminium,(Hayashi and Chaplin(1991)). It is easy to adjust the strength of the damping force of the test cylinder by changing the current used in the electro-magnet. An alternate current was used to prevent the generation of residual magnetism in the soft iron.

Free vibration tests of the test cylinder are performed in air and in still water. The damping factor, obtained from the logarithmic decrement, and the natural frequency of the test cylinder are measured by releasing its top end from an initial displacement and recording amplitude decay of the transient oscillation of the test cylinder. The equivalent mass per unit length of the test cylinder in water, m_e, is calculated from the measurements of natural frequency f_{nw} and stiffness of the test cylinder K in still water as follows,

$$m_e=\{K/(2\pi f_{nw})^2\}/\int_0^d(z/d).d.dz \tag{1}$$

where z is the distance from the bottom of the flume and d is the still water depth. Thus the value of m_e includes the mass of structure and the entrained fluid. The value of m_e is used in the calculation of the mass ratio, $m_e/\rho D^2$, and the combined mass ratio-damping parameter $K_s=2m_e(2\pi\zeta_{ta})/\rho D^2$, which is known as the Scruton number . It should be noted that ζ_{ta} is the damping factor of the test cylinder in air. ρ is the density of water.

The conditions for which measurements were made are shown in Table 1. In all cases the still water depth, d, was kept constant at 80cm and regular waves were used. The natural frequencies of the test cylinder both in air , f_{na}, and in still water, f_{nw}, and the damping factor both in air , ζ_{ta}, and in still water, ζ_{tw}, are the values which were obtained in the free vibration tests described above. The wave depth parameter , $kd=2\pi.d/L$, L=wave length, was calculated by using linear wave theory.

Table 1 Experimental Condition

Case	f_{na}	f_{nw}	CKC	f_w/f_{nw}	kd	ζ_{ta}	ζ_{tw}	Mass Ratio	Scruton number(Ks)
								$m_e/\rho D^2$	$2m_e(2\pi\zeta_{ta})/\rho D^2$
A1	1.52	1.46	$12^{+1.0}_{-2.1}$	0.237～1.07	0.67～7.98	0.001	0.004	15.7	0.20
A2	1.52	1.46	$20^{+4.7}_{-3}$	0.156～0.577	0.42～2.33	0.004	0.004	15.7	0.20
AB1	1.52	1.46	$12^{+2.1}_{-5.6}$	0.315～1.05	0.93～7.64	0.021	0.023	15.7	4.14
AB2	1.52	1.46	$20^{+2.8}_{-1}$	0.308～0.54	0.91～2.08	0.021	0.023	15.7	4.14

The relationships between the vortex-excited vibration of the test cylinder and the frequency ratio, f_w/f_{nw}, were measured in Case A1 and Case A2. The characteristic Keulegan Carpenter number, $CKC=U_{ms}.T/D$, where U_{ms} is the maximum horizontal water particle velocity at still water level, was fixed at about 12 in Case A1 and at about 20 in Case A2. In order to study the multi-appearance of the resonant frequency ratio, $f_w/f_{nw}=1, 1/2, 1/3$----, which was pointed out by Sawaragi, Nakamura and Miki(1977) and Isaacson and Maull(1981), the frequency ratio f_w/f_{nw} was changed widely from 0.237 to 1.07 in the Case A1 and from 0.166 to 0.577 in the Case A2 by changing the wave frequency f_w.

In order to study the influences of the damping factor on the vortex-excited vibration of the test cylinder, the same kind of measurements as for Case A1 and Case A2 were done for Case AB1 and Case AB2, but the damping factor in air , ζ_{ta}, was changed from 0.001 to 0.021 by using the electro-magnetic damper .

3. Results and Discussion

3.1 Damping factors

Figure 2 shows one of traces for decay of free oscillation of the test cylinder in air and in still water. Figure 3 shows the variation of the damping factors ζ_{tai} in air and ζ_{twi} in still water with the non dimensional amplitude Y_{hi}/D as the function of AC current I_e. Where Y_{hi} is the amplitude of i-th oscillation of the test cylinder displacement y_h at still water level. The values of ζ_{tai} and ζ_{twi} for Y_{hi}/D are defined as

$$\zeta_{tai} \text{ or } \zeta_{twi} = \{\ln(Y_{hi-2}/Y_{hi+2})\}/(2\pi \times 4) \qquad (2)$$

where Y_{hi-2} is amplitude of the (i-2)th period and Y_{hi+2} is amplitude of the (i+2)th period (see Figure 2). AC current I_e=0A was used in Case A1 and Case A2 . AC current I_e=4A was used in Case AB1 and Case AB2.

The value of ζ_{tai} at I_e=0A shows the structural damping factor which is mainly composed of damping factor of springs. It is nearly independent of the value of Y_{hi}/D. Slight increase of ζ_{tai} with increased amplitude Y_{hi}/D may be due to the non-linear characteristics of the structure damping of the test cylinder. The damping due to air can be expected to make negligible contribution in this case. Substituting ζ_{tai} at I_e=0A from ζ_{tai} at I_e=4A leaves the structural damping factor produced by the electro-magnetic damper.

The value of ζ_{twi} is independent of the amplitude effect only for low values of Y_{hi}/D and it becomes amplitude dependent at higher values of Y_{hi}/D. Substituting ζ_{tai} at I_e=0A from ζ_{twi} at I_e=0A shows the fluid damping factor ζ_{fi}. The constant value of ζ_{fi} at low value of Y_{hi}/D is explained theoretically by using the Stokes' solution introduced to explain the viscous effect of the cylinder at low Keulegan-Carpenter number KC (Hayashi and Chaplin, 1991). At large amplitude of Y_{hi}/D, the value of ζ_{fi} becomes amplitude effect and the drag coefficient obtained with damping factor corresponds well the drag coefficients of the circular cylinder in planner oscillatory flow at KC number above 2 (Hayashi and Chaplin, 1991). The values of ζ_{tai} and ζ_{twi} at Y_{hi}/D=0.1 are shown in Table 1 respectively.

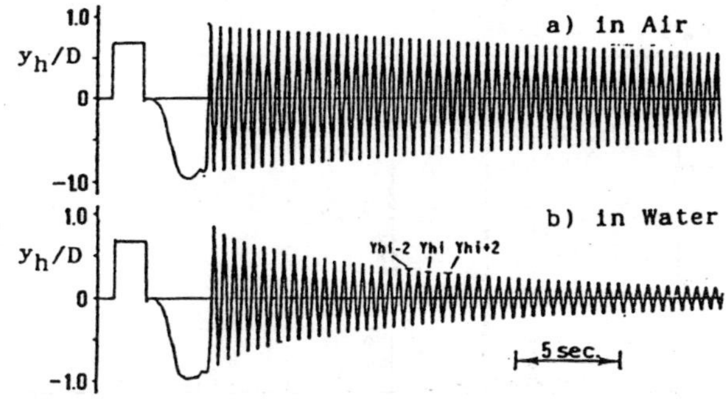

Figure 2 Decay of Free oscillation

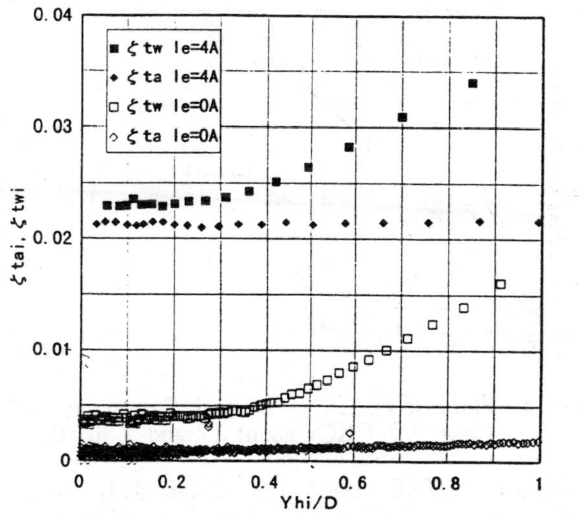

Figure 3 Variation of damping factors ζ_{tai} and ζ_{twi} with Y_{hi}/D

3.2 General description of the vortex-excited vibration of the test cylinder

The following statistical values were obtained for considering the irregularity of the vortex-excited vibration of the test cylinder in waves.

(1) The mean value of the half peak-to-peak amplitude of the vortex-excited vibration at mean water level (80cm above the bed). -------Y_{hm}

(2) The mean value of the frequency of the vortex-excited vibration . -------------f_{ym}

(3) The coefficient of vibration of the half peak-to-peak amplitude of the vortex-excited vibration defined as

C_{VY}=The standard deviation of the half peak-to-peak amplitude of vortex-excited vibration/($2 \cdot Y_{hm}$).

In order to show the general characteristics of the vortex-excited vibration of the test cylinder in waves, the relationship between the value of Y_{hm}/D, f_w/f_{nw} , C_{VY} and the frequency ratio f_w/f_{nw} for Case A1 and Case A2 are shown in Figure 4 and Figure 5.

The following main characteristics for each test cases are shown in these figures.

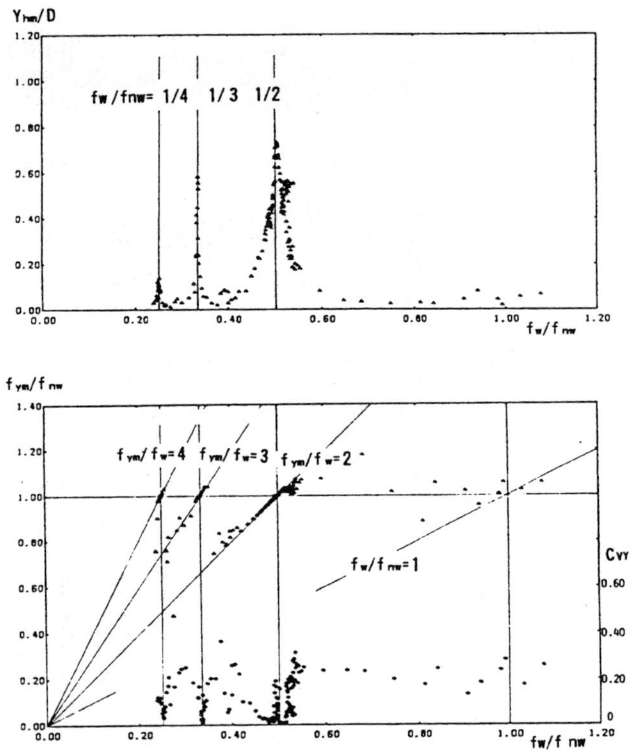

Figure 4 Plots of Y_{hm}/D, f_{ym}/f_{nw} and C_{VY} against f_w/f_{nw} for Case A1(CKC=about 12 and ζ_{ta}=0.001)

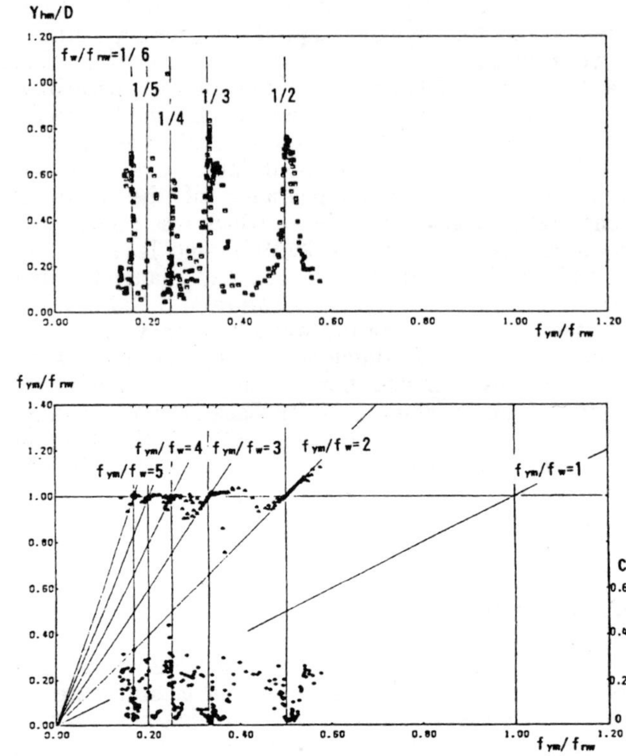

Figure 5 Plots of Y_{hm}/D, f_{ym}/f_{nw} and C_{VY} against f_w/f_{nw} Case A2 (CKC=about 20 and ζ_{ta}=0.001)

(A) Case A1 (CKC=about 12, ζ_{ta}=0.001, K_s=0.20)

The value of CKC was fixed at about 12 for this case. When the frequency ratio f_w/f_{nw} approaches around the value of f_w/f_{nw}=1/2, 1/3, 1/4, a) the value of Y_{hm}/D has a peak value, b)the value of f_{ym} is equal to f_{nw}, and c) the value of C_{VY} has very small value. when C_{VY} is less than 0.02, the time history of the vortex-excited vibration of the test cylinder y_h is very regular. It also shows that the amplitude of the vibration is not intermittent and is not modulate. We refer to these phenomena perfect resonant vibration.

The appearance of these peak values of Y_{hm}/D may be due to a) the resonance between the frequency of the harmonic component of the lift force and the natural frequency of the test cylinder in water f_{nw}, and b) the amplification of each harmonic component of the lift force by means of the increased vortex strength and the correlation in the phase of vortex shedding along the cylinder's axis. The maximum peak value of Y_{hm}/D occurs at f_w/f_{nw}=about 1/2 . This may be due to the domination of the second harmonic component of the lift forces coefficient $CL_c(2)$ at around CKC=12 observed for the vertical cylinder rigidly mounted in regular waves (Hayashi and Chaplin, 1989). The peak value of Y_{hm}/D at f_w/f_{nw}=about 1/4 is small compared with that at f_w/f_{nw}=about 1/2. This may be due to the lower value of the forth harmonic of the lift force coefficient $CL_c(4)$ and that the amplification of $CL_c(4)$ was suppressed by the existence of the large value of $CL_c(2)$.

B) Case A2 (CKC=about 20, ζ_{ta}=0.001, K_s=0.20)

The value of CKC was fixed at about 20 for this case. The perfect resonant vibration described above appear around f_w/f_{nw}=1/2, 1/3, 1/4, 1/5, 1/6. These phenomena may be also associated with the same factors as described above for CKC=about 12. In this case, The maximum peak value of Y_{hm}/D occurs at f_w/f_{nw}= about 1/3. However, the difference between maximum peak value and the other peak values is small, compared with that in Case A1. This may be due to the small difference between each harmonic component of the lift forces observed for the vertical cylinder rigidly mounted in regular waves (Hayashi and Chaplin, 1989).

As described above, the variation of the vortex-excited vibration of the test cylinder with f_w/f_{nw} depend on the value of CKC. The appearance of perfect resonant transverse vibration at f_w/f_{nw}=about 1/2 and 1/3, in the range of rms KC between 10 to 15, were reported by Sawaragi et al.(1977), Isaacson et al.(1981), and Bearman et al.(1987). However, the appearance of the resonant transverse vibration at f_w/f_{nw}=about 1/4, 1/5, 1/6 at CKC=about 20, have apparently not been previously reported. These multi-appearances of resonant transverse vibration should be noted as a significant feature of the resonance of a flexibly supported cylinder in waves.

The multi-appearances of the resonant transverse vibration of a circular flexibly-mounted cylinder in planer oscillatory flow, in the range of 7<KC number<50, have been reported by Hayashi et al.(1991 and 1995). Similar phenomena have been reported by McConnell and Park(1982) and Sumer and Fredsoe(1988), who oscillated a flexibly-mounted cylinder in still water in the range of 10<KC<100.

3.3 Detailed description of the vortex-excited vibration of the test cylinder

In order to study the details of the phenomena of the vortex-excited vibration of the test cylinder around $f_w/f_{nw}=1/2$ (for CKC=about 12) and 1/3 (for CKC=about 20), the relationship between f_w/f_{nw}, and Y_{hm}/D, f_{ym}/f_{nw} and C_{VY} for Case A1, Case A2, Case AB1, and Case AB2 are shown respectively in Figure 6 (a), (b), and (c) through Figure 9 (a), (b) and (c) respectively. The damping factor in air is ζ ta=0.001 for Case A1 and Case A 2. That for Case AB1 and AB2 is changed from 0.001 to 0.021 by using the elecro-magnetic damper.

In order to show the characteristics of the variation of the amplitude of the test cylinder with time t, several examples of the record of y_h on the U-V recorder for Case A1, for f_w/f_{nw} in the range of between 0.485 to 0.543, are shown with those of Case AB1 in Figure 10. In this figure, the results of Case A1 are classed with notation (a) and those of Case AB1 are classed with notation (b).

The following main characteristics for each test cases are shown in these figures.

(A) Case A1 (CKC=about 12, ζ_{ta}=0.001, K_s=0.20)

As the frequency ratio f_w/f_{nw} increased from 0.45 to 0.503, the value of Y_{hm}/D increases smoothly except for some small scatter around $f_w/f_{nw}=0.493$. It takes the peak value of 0.74 at $f_w/f_{nw}=0.503$. The frequency ratio f_{ym}/f_{nw} in this range of f_w/f_{nw} is also plotted from the relationship

$$f_{ym}/f_{nw} = 2(f_w/f_{nw}). \tag{3}$$

The value of C_{VY} has a small value in this range of f_w/f_{nw} except for the appearance of large value around $f_w/f_{nw}=0.493$. The small value of C_{VY} indicates the regularity of the time history of the vortex-excited vibration of the test cylinder as shown in Figure 10. The large value of C_{VY} shows the irregularity of the time history of the vortex-excited vibration. In this case, as shown at the runs 622w3(a) and 622w4(a) in Figure 10, the amplitude is intermittent and is modulated.

The peak value of Y_{hm}/D appears at $f_w/f_{nw}=0.503$ and not at $f_w/f_{nw}=0.500$ as might be expected. The time history of vibration of the test cylinder y_h is very regular as shown at the run 622w9(a) in Figure 10. The detail of this record is shown in Figure 11-(a) with the record of the displacement of the water surface elevation η. We can recognize that the frequency of vibration of the test cylinder f_{ym} is just two times of the wave frequency f_w.

The appearance of the peak value of Y_{hm}/D at about $f_w/f_{nw}=0.503$, not at $f_w/f_{nw}=0.500$, may be due to the increase of the natural frequency of the test cylinder in the vortex-excited condition from the natural frequency f_{nw} of the test cylinder in still water. This increase of the natural frequency may be caused by the variation of the added mass coefficient in the vortex-excited condition. It is worth noting that experimental results of Sarpkaya(1978) show the decrease of the added mass coefficient of a circular cylinder undergoing forced transverse oscillations in a uniform flow. The increases of the natural frequency of a vortex-excited circular cylinder vibrating in the transverse direction in planner oscillating flows have been shown in the experimental results obtained by Sumer and Fredsoe (1988) and Hayashi et al. (1995).

Returning to Figure 6, in the range of $0.503<f_w/f_{nw}<0.515$, the value of Y_{hm}/D decreases smoothly with increasing f_w/f_{nw}. The value of f_{ym}/f_{nw} still agrees with Eq. (3) indicating that the oscillation is coupled to the wave frequency f_w. Now, we call this phenomenon " wave coupling ". The value of C_{VY} is small because the amplitude of y_h is regular as shown at the runs 622w10(a) through 622w13(a) in Figure 10. In the range of $0.515<f_w/f_{nw}<0.519$, the value of Y_{hm}/D decreases with increasing f_w/f_{nw} and has a trough value of about 0.41 at f_w/f_{nw}=about 0.519. The value of f_{ym}/f_{nw} deviates downwards from the calculated value of Eq. (3) as shown in this figure. The value of C_{VY} becomes large because the amplitude of y_h is intermittent and modulates as shown at the run 622w14(a) and 622w15(a) in Figure 10 .

In the range of $0.519<f_w/f_{nw}<0.523$, the value of Y_{hm}/D increases rapidly from the trough value of $Y_{hm}/D=0.41$ and reaches around $Y_{hm}/D=0.55$. The corresponding values of f_{ym}/f_{nw} are around $f_{ym}/f_{nw}=1.018$. The value of C_{VY} is small, but it is not so small as the value of C_{VY} in the range of $0.49<f_w/f_{nw}<0.514$ because the amplitude of y_h modulates without intermittency in the amplitude of y_h as shown at 622w16(a) in Figure 10.

The detail of the variation of y_h and η with time for the data of run 622w16(a) are shown in Figure 11-(b). The variation of the phase angle between y_h and η can be found in this figure. This is due to the difference between the value of f_{ym} and $2f_w$. The appearance of modulation of the amplitude y_h may be due to the variation of the vortex-shedding strength which is caused by the variation of phase angle described above. Now, we call this phenomenon " vortex coupling " because Y_{hm}/D increases, and the value of y_{hm} is restricted to a constant value which is between the natural frequency of the test cylinder and the vortex-shedding frequency. The vortex-shedding frequency may be expressed by Eq. (3) in this figure.

At f_w/f_{nw} greater than about 0.523, some points are grouped around a constant value of Y_{hm}/D and the others decrease smoothly with increasing f_w/f_{nw}. When Y_{hm}/D is around 0.55, the value of f_{ym}/f_{nw} is around 1.018 and C_{VY} has a small value at about 0.05. When Y_{hm}/D decreases with increasing f_w/f_{nw}, the value of f_w/f_{nw} is scattered between 1.018 and the calculated value of Eq. (3). In this case, the value of C_{VY} is between 0.2 and 0.35 because the amplitude of y_h is intermittent and modulates as shown at the run 622w19(a) through 622w22(a) in Figure 10.

However, for the vortex-excited vibration of the test cylinder in the range of $0.31<f_w/f_{nw}<0.34$, this "vortex coupling" does not appear. The peak value of Y_{hm}/D appears around $f_w/f_{nw}=1/3$ and the value of f_{ym}/f_{nw} follows the calculated line from Eq. (4). The value of C_{VY} becomes less than 0.01 just around $f_w/f_{nw}=1/3$.

$$f_{ym}/f_{nw} = 3(f_w/f_{nw}). \tag{4}$$

(B) Case A2, (CKC=about 20, $_{ta}$=0.001, K_s=0.20)

In the range of $0.32<f_w/f_{nw}<0.336$, the value of Y_{hm}/D increases smoothly with increasing f_w/f_{nw}. The peak value of Y_{hm}/D appears at $f_w/f_{nw}=0.336$. In this range of f_w/f_{nw}, the value of f_{ym}/f_{nw} follows the calculated line from Eq. (4). The value of C_{VY} is less than 0.05. In the range of $0.336<f_w/f_{nw}<0.34$, Y_{hm}/D decreases rapidly with

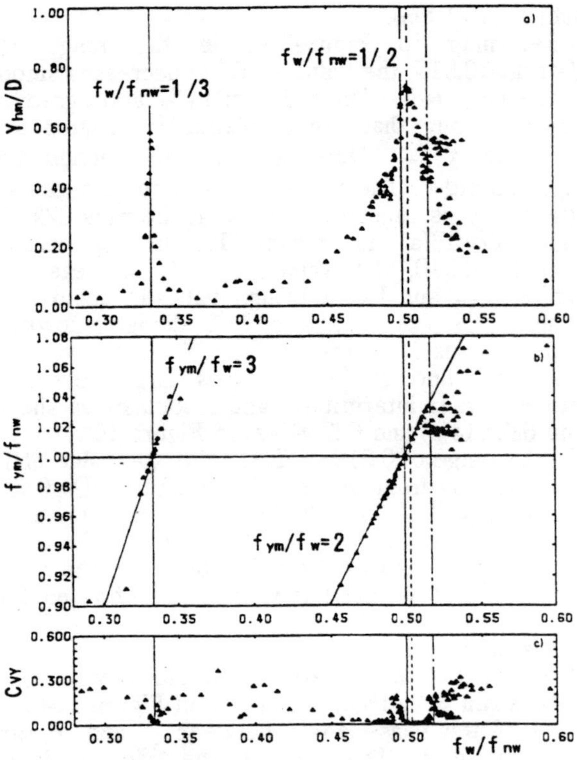

Figure 6 Detail plots of Y_{hm}/D, f_{ym}/f_{nw} and C_{VY} against f_w/f_{nw} for Case A1 (CKC=about 12 and ζ_{ta}=0.001)

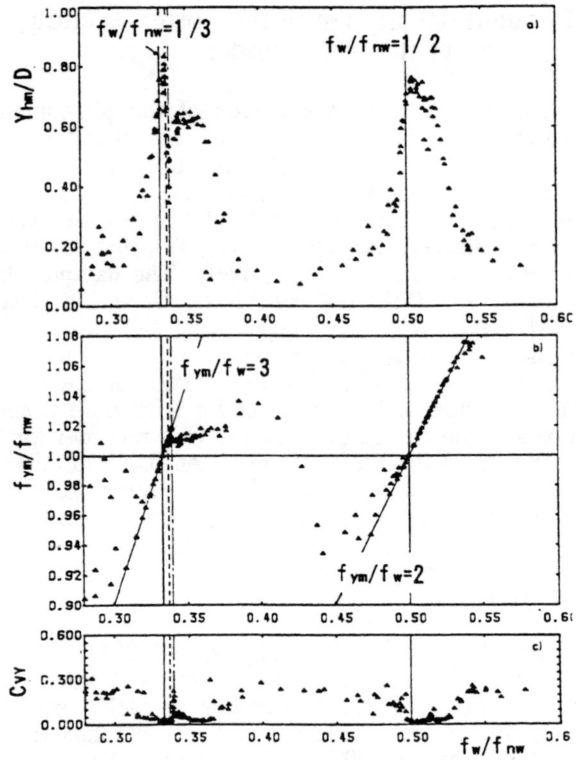

Figure 7 Detail plots of Y_{hm}/D, f_{ym}/f_{nw} and C_{VY} against f_w/f_{nw} for Case A2 (CKC=about 20 and ζ_{ta}=0.001)

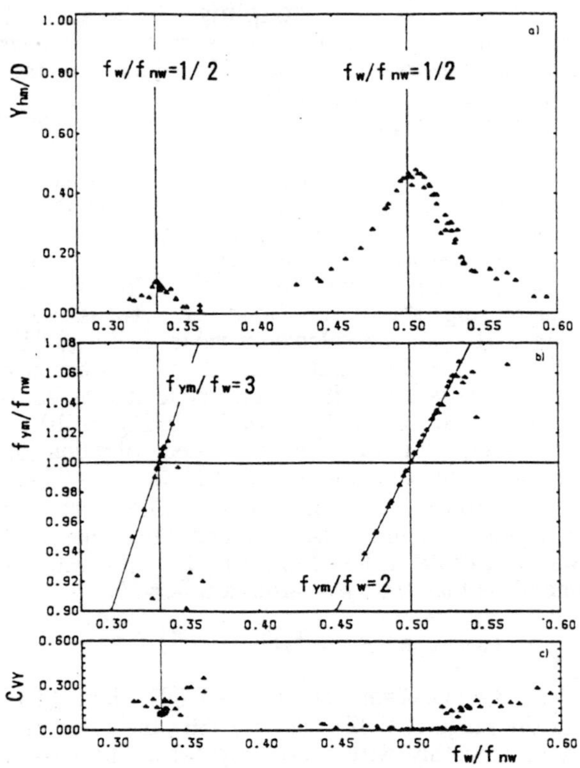

Figure 8 Detail plots of Y_{hm}/D, f_{ym}/f_{nw} and C_{VY} against f_w/f_{nw} for Case AB1 (CKC=about 12 and ζ_{ta}=0.001)

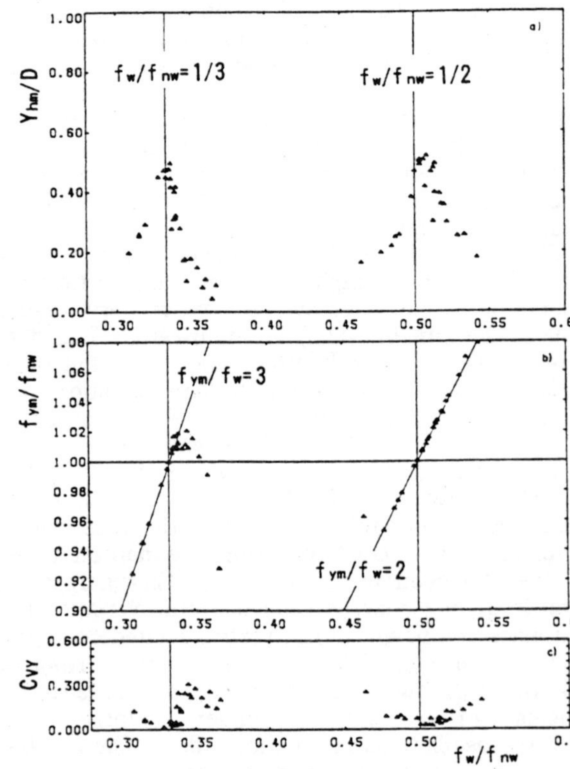

Figure 9 Detail plots of Y_{hm}/D, f_{ym}/f_{nw} and C_{VY} against f_w/f_{nw} for Case AB2 (CKC=about 20 and ζ_{ta}=0.021)

Figure 10 Variation of the amplitude of the test cylinder with time t in the range of f_w/f_{nw} between 0.485 to 0.543 [a) Case A1 (ζ_{ta}=0.001), b) Case AB1 (ζ_{tw}=0.021)]

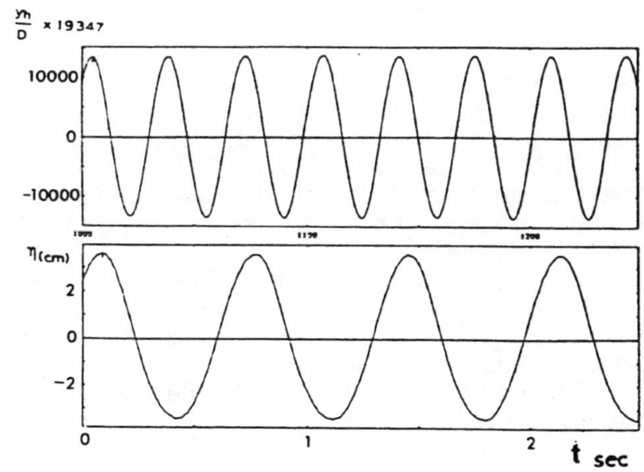

Figure 11-(a) Variation of y_h/D and water surface elevation η with time t for the data of Run No. 622w9(a) in Case A1, (f_w/f_{nw} =0.503)

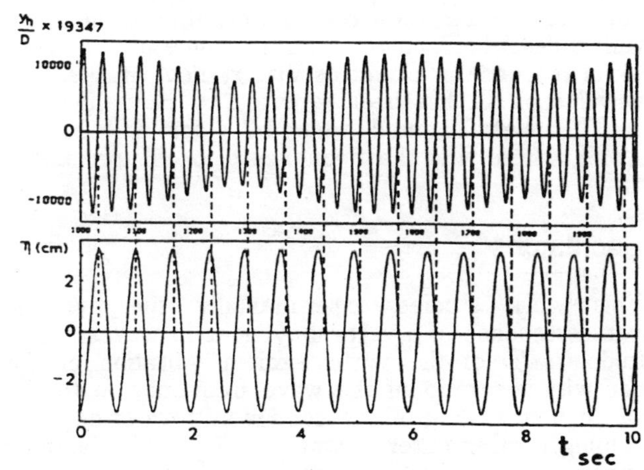

Figure 11-(b) Variation of y_h/D and water surface elevation η with time t for the data of Run No. 622w16(a) in Case A1, (f_w/f_{nw} =0.522).

increasing f_w/f_{nw} and it takes a trough value at f_w/f_{nw}=0.34. The value of f_{ym}/f_{nw} follows Eq. (4), but it deviates downward at around f_w/f_{nw}=0.34. The value of C_{VY} is less than 0.05, but it takes a large value around f_w/f_{nw}=0.34.

In the range of 0.34<f_w/f_{nw}<0.364 , Y_{hm}/D increases rapidly from the trough value and reaches approximately Y_{hm}/D=0.65. The value of f_{ym}/f_{nw} is plotted around f_{ym}/f_{nw}=1.01 and C_{VY} is less than 0.05. These results suggest that "vortex coupling" is occurring. In the range of 0.364<f_w/f_{nw}<0.38, the value of Y_{hm}/D decreases smoothly with increasing f_w/f_{nw}.

The vortex excited vibration described above is similar to that of Case A1 in the range of f_w/f_{nw} between

0.45 to 0.55. There is clear evidence of "vortex coupling" found in this case and the range of f_w/f_{nw} over which it occurs is wider than that of Case A1.

The appearance of the second peak value of Y_{hm}/D is $0.48<f_w/f_{nw}<0.55$. However, the deviation of f_{ym}/f_{nw} from the calculation line of Eq. (2) does not appear, suggesting that the oscillations are mainly "wave coupling " in this range.

(C) Case AB1 (CKC=about 12, ζ_{ta}=0.021, Ks=4.14), Case AB2 (CKC=about 20, ζ_{ta}=0.021, Ks=4.14)

In these cases, the damping factor in air ζ_{ta} is changed from 0.001 for Case A1 and Case A2 to 0.021. The appearance of "vortex coupling" can not be found for these two cases. Therefore, the dynamic response curves of Y_{hm}/D with f_w/f_{nw} are simpler compared with those of Case A1 and Case A2. The values of f_{ym}/f_w around $f_w/f_{nw}=1/2$ or 1/3 agree well with the calculated line from Eq.(3) or Eq.(4). Obviously the absence of "vortex coupling" for these two cases is due to the higher damping. The absence of lock-in for the dynamic response of a cylinder with high damping in steady flow has been reported by Unemara(1971).

The most remarkable result described above is the appearance of two types of peak of Y_{hm}/D, produced by the perfect resonance coupled with wave, "wave coupling", and by the vortex, "vortex coupling". The perfect resonance appears in the range of lock-in in the case of steady flow, but in waves, it appears only near to $f_w/f_{nw}=1/2$, 1/3, 1/4 ---------- as described above, elsewhere " vortex coupling " may occur as described above in the case of low damping.

The appearance of two types of peak of response in waves has been reported by Zedan et al.(1980). However, it is not clear whether their results were due to the same mechanism described above. The appearance of two types of peak of response in planer oscillatory flow, in the case of low value of K_s, has been reported by Hayashi et al. (1995) . It appears around $f_d/f_{nw}=1/2$ (for KC=10 and K_s=0.245), where f_d is the frequency of oscillatory flow .

4. Conclusions

This work studies experimentally the dynamic transverse response of a vertical cylinder in waves. The amplitude Y_{hm}/D of the vortex-excited vibration of the cylinder with the ratio of the wave frequency f_w to the natural frequency f_{nw} of the cylinder in still water depends on Keulegan Carpenter number KC and damping coefficient in air. The most remarkable result is the appearance of two types of peak in the amplitude Y_{hm}/D, which is produced by perfect resonance coupled with the waves, "wave coupling", and by vortex, "vortex coupling". In the case of steady flow, perfect resonance appears in the range of locked-in, but in wave, it appears only near to around $f_w/f_{nw}=1/2$, 1/3, 1/4,---(multi appearance), elsewhere "vortex coupling" may occur for light damping. The multi appearance of perfect resonance is more clear in case of higher values of KC and lower values of damping coefficient in air.

References

Brevins, R.D. (1977). " Flow-Induced Vibration, " Van Nostrand Reinhold.

Bearman, P.W. (1984). " Vortex Shedding from Oscillating Bluff Bodies ," Ann. Rev. Fluid Mechanics, Vol. 16.

Bearman, P.W. and Hall, P.F. (1987). " Dynamic Response of Circular Cylinder in Oscillatory Flow and Waves," Proc. of International Conference on Flow Induced Vibrations, Organized by BHRA, England.

Hayashi, K. and Chaplin, J. R. (1989). " Lift Forces on a Rigid Vertical Cylinder in Waves, "Proceedings of OMAE-89, Vol. 2 .

Hayashi, K. and Chaplin, J. R. (1991). " Damping of a Vertical Cylinder in Still Water," Proceedings of ISOPE-91, Vol. 3.

Hayashi, K., Ogihara, Y., Fujima, K. and Shigemura, T. (1991). " Vortex-excited Vibration of a Circular Cylinder in Planner Oscillatory Flow, " 10th Ocean Engineering Symposium, The Society of Naval Architects of Japan. (in Japanese)

Hayashi, K, Higaki F., Fujima K. and Shigemura T. (1995). "Forces acting on a Vortex-excited Vibrating Circular Cylinder in Planner Oscillatory Flow," Proceedings of Coastal Engineering, Japan Society of Civil Engineering, Vol. 42 (2). (in Japanese)

Isaacson, M. Q. and Maull, D. J. (1981). " Dynamic Response of Vertical Piles in Waves, " International Symposium on Hydrodynamics in Ocean Engineering, The Norwegian Institute of Technology.

McConnel, K.G. and Park, Y.S., (1982), " The Response and Lift-Force Analysis of an Elastically Mounted Cylinder Oscillating in Still Water, " Proceedings of Third International Conference on the Behavior of Off-Shore Structures, Vol. 2.

Sarpkaya, T. (1978). "Fluid Forces on Oscillating Cylinder, " Journal of the Water Ways, Port, Coastal and Ocean Division, ASCE. 104.

Sarpkaya, T. (1979). " Vortex-Induced Oscillation ", Journal of Applied Mechanics, Vol. 46, Jun.

Sawaragi, T., Nakamura,T. and Miki, H. (1977). " Dynamic Behaviour of a Circular Pile due to Eddy Shedding in Waves, " Coastal Engineering in Japan, Vol. 20.

Sumer, B. M. and Fredsoe, J. (1988). " Transverse Vibrations of an Elastically Mounted Cylinder Exposed to an Oscillating Flow, " Journal of Offshore Mechanics and Arctic Engineering, November , Vol. 110.

Sumer, B. M. and Fredsoe, J. (1994). " A Review on Vibrations of Marine Pipelines, " Proceedings of ISOPE-94, Vol. 2.

Unemara, S. (1971). " On the Vibration of Cylinders caused by Karman Vortex, " JSME, 14. 75.